Climate Change 2007
The Physical Science Basis

The Intergovernmental Panel on Climate Change (IPCC) was set up jointly by the World Meteorological Organization and the United Nations Environment Programme to provide an authoritative international statement of scientific understanding of climate change. The IPCC's periodic assessments of the causes, impacts and possible response strategies to climate change are the most comprehensive and up-to-date reports available on the subject, and form the standard reference for all concerned with climate change in academia, government and industry worldwide. Through three working groups, many hundreds of international experts assess climate change in this Fourth Assessment Report. The Report consists of three main volumes under the umbrella title Climate Change 2007, all available from Cambridge University Press:

Climate Change 2007 - The Physical Science Basis
Contribution of Working Group I to the Fourth Assessment Report of the IPCC
(ISBN 978 0521 88009-1 Hardback; 978 0521 70596-7 Paperback)

Climate Change 2007 - Impacts, Adaptation and Vulnerability
Contribution of Working Group II to the Fourth Assessment Report of the IPCC
(978 0521 88010-7 Hardback; 978 0521 70597-4 Paperback)

Climate Change 2007 - Mitigation of Climate Change
Contribution of Working Group III to the Fourth Assessment Report of the IPCC
(978 0521 88011-4 Hardback; 978 0521 70598-1 Paperback)

Climate Change 2007 - The Physical Science Basis is the most comprehensive and up-to-date scientific assessment of past, present and future climate change. The report provides:

• the most complete and quantitative assessment of how human activities are affecting the radiative energy balance in the atmosphere

• a more extensive assessment of changes observed throughout the climate system than ever before using the latest measurements covering the atmosphere, land surface, oceans, and snow, ice and frozen ground

• a detailed assessment of past climate change and its causes

• the first probabilistic assessment of climate model simulations and projections using detailed atmosphere-ocean coupled models from 18 modelling centres around the world

• a detailed assessment of climate change observations, modelling, and attribution for every continent

Simply put, this latest assessment of the IPCC will again form the standard scientific reference for all those concerned with climate change and its consequences, including students and researchers in environmental science, meteorology, climatology, biology, ecology and atmospheric chemistry, and policy makers in governments and industry worldwide.

From reviews of the Third Assessment Report – Climate Change 2001:

'The detail is truly amazing … invaluable works of reference … no reference or science library should be without a set [of the IPCC volumes] … unreservedly recommended to all readers.'
Journal of Meteorology

'This well-edited set of three volumes will surely be the standard reference for nearly all arguments related with global warming and climate change in the next years. It should not be missing in the libraries of atmospheric and climate research institutes and those administrative and political institutions which have to deal with global change and sustainable development.'
Meteorologische Zeitschrift

'… likely to remain a vital reference work until further research renders the details outdated by the time of the next survey … another significant step forward in the understanding of the likely impacts of climate change on a global scale.'
International Journal of Climatology

'The IPCC has conducted what is arguably the largest, most comprehensive and transparent study ever undertaken by mankind … The result is a work of substance and authority, which only the foolish would deride.'
Wind Engineering

'… the weight of evidence presented, the authority that IPCC commands and the breadth of view can hardly fail to impress and earn respect. Each of the volumes is essentially a remarkable work of reference, containing a plethora of information and copious bibliographies. There can be few natural scientists who will not want to have at least one of these volumes to hand on their bookshelves, at least until further research renders the details outdated by the time of the next survey.'
The Holocene

'The subject is explored in great depth and should prove valuable to policy makers, researchers, analysts, and students.'
American Meteorological Society

From reviews of the Second Assessment Report – Climate Change 1995:

' … essential reading for anyone interested in global environmental change, either past, present or future. … These volumes have a deservedly high reputation'
Geological Magazine

'… a tremendous achievement of coordinating the contributons of well over a thousand individuals to produce an authoritative, state-of-the-art review which will be of great value to decision-makers and the scientific community at large … an indispensable reference.'
International Journal of Climatology

'… a wealth of clear, well-organized information that is all in one place … there is much to applaud.'
Environment International

Climate Change 2007
The Physical Science Basis

Edited by

Susan Solomon
Co-Chair,
IPCC Working Group I

Dahe Qin
Co-Chair,
IPCC Working Group I

Martin Manning
Head, Technical Support Unit
IPCC Working Group I

Melinda Marquis   **Kristen Averyt**   **Melinda M.B. Tignor**   **Henry LeRoy Miller, Jr.**
Technical Support Unit, IPCC Working Group I

Zhenlin Chen
China Meteorological Administration

Contribution of Working Group I
to the Fourth Assessment Report of the
Intergovernmental Panel on Climate Change

Published for the Intergovernmental Panel on Climate Change

CAMBRIDGE
UNIVERSITY PRESS

CAMBRIDGE UNIVERSITY PRESS
Cambridge, New York, Melbourne, Madrid, Cape Town, Singapore, São Paolo, Delhi

Cambridge University Press
32 Avenue of the Americas, New York, NY 10013-2473, USA

www.cambridge.org
Information on this title: www.cambridge.org/9780521880091

First published 2007

Printed in Canada by Friesens

A catalog record for this publication is available from the British Library.

ISBN 978-0-521-88009-1 hardback
ISBN 978-0-521-70596-7 paperback

Please use the following reference to the whole report:

IPCC, 2007: *Climate Change 2007: The Physical Science Basis. Contribution of Working Group I to the Fourth Assessment Report of the Intergovernmental Panel on Climate Change* [Solomon, S., D. Qin, M. Manning, Z. Chen, M. Marquis, K.B. Averyt, M. Tignor and H.L. Miller (eds.)]. Cambridge University Press, Cambridge, United Kingdom and New York, NY, USA, 996 pp.

Cover photo:

The Blue Marble western and eastern hemispheres. These images integrate land, ocean, sea ice and clouds into a visual representation of the earth's climate system. They are based on space-borne earth observation data from NASA's MODIS (MODerate resolution Imaging Spectroradiometer) sensor aboard the TERRA and AQUA satellites. These images are part of the Blue Marble dataset which is freely available at http://bluemarble.nasa.gov. They are further documented in Stöckli, R., Vermote, E., Saleous, N., Simmon, R., and Herring, D. (2006). True color earth data set includes seasonal dynamics. EOS, 87(5):49, 55.

Foreword

Representing the first major global assessment of climate change science in six years, "Climate Change 2007 – The Physical Science Basis" has quickly captured the attention of both policymakers and the general public. The report confirms that our scientific understanding of the climate system and its sensitivity to greenhouse gas emissions is now richer and deeper than ever before. It also portrays a dynamic research sector that will provide ever greater insights into climate change over the coming years.

The rigor and credibility of this report owes much to the unique nature of the Intergovernmental Panel on Climate Change (IPCC). Established by the World Meteorological Organization and the United Nations Environment Programme in 1988, the IPCC is both an intergovernmental body and a network of the world's leading climate change scientists and experts.

The chapters forming the bulk of this report describe scientists' assessment of the state-of-knowledge in their respective fields. They were written by 152 coordinating lead authors and lead authors from over 30 countries and reviewed by over 600 experts. A large number of government reviewers also contributed review comments.

The Summary for Policymakers was approved by officials from 113 governments and represents their understanding – and their ownership – of the entire underlying report. It is this combination of expert and government review that constitutes the strength of the IPCC.

The IPCC does not conduct new research. Instead, its mandate is to make policy-relevant – as opposed to policy-prescriptive – assessments of the existing worldwide literature on the scientific, technical and socio-economic aspects of climate change. Its earlier assessment reports helped to inspire governments to adopt and implement the United Nations Framework Convention on Climate Change and the Kyoto Protocol. The current report will also be highly relevant as Governments consider their options for moving forward together to address the challenge of climate change.

Climate Change 2007 – the Physical Science Basis is the first volume of the IPCC's Fourth Assessment Report. The second volume considers climate change impacts, vulnerabilities and adaptation options, while the third volume assesses the opportunities for and the costs of mitigation. A fourth volume provides a synthesis of the IPCC's overall findings.

The Physical Science Basis was made possible by the commitment and voluntary labor of the world's leading climate scientists. We would like to express our gratitude to all the Coordinating Lead Authors, Lead Authors, Contributing Authors, Review Editors and Reviewers. We would also like to thank the staff of the Working Group I Technical Support Unit and the IPCC Secretariat for their dedication in coordinating the production of another successful IPCC report.

Many Governments have supported the participation of their resident scientists in the IPCC process and contributed to the IPCC Trust Fund, thus also assuring the participation of experts from developing countries and countries with economies in transition. The governments of Italy, China, New Zealand and Norway hosted drafting sessions, while the Government of France hosted the final plenary that approved and accepted the report. The Government of the United States of America funded the Working Group I Technical Support Unit.

Finally, we would like to thank Dr R.K. Pachauri, Chairman of the IPCC, for his sound direction and tireless and able guidance of the IPCC, and Dr. Susan Solomon and Prof. Dahe Qin, the Co-Chairs of Working Group I, for their skillful leadership of Working Group I through the production of this report.

M. Jarraud
Secretary General
World Meteorological Organization

A. Steiner
Executive Director
United Nations Environment Programme

Preface

This Working Group I contribution to the IPCC's Fourth Assessment Report (AR4) provides a comprehensive assessment of the physical science of climate change and continues to broaden the view of that science, following on from previous Working Group I assessments. The results presented here are based on the extensive scientific literature that has become available since completion of the IPCC's Third Assessment Report, together with expanded data sets, new analyses, and more sophisticated climate modelling capabilities.

This report has been prepared in accordance with rules and procedures established by the IPCC and used for previous assessment reports. The report outline was agreed at the 21st Session of the Panel in November 2003 and the lead authors were accepted at the 31st Session of the IPCC Bureau in April 2004. Drafts prepared by the authors were subject to two rounds of review and revision during which over 30,000 written comments were submitted by over 650 individual experts as well as by governments and international organizations. Review Editors for each chapter have ensured that all substantive government and expert review comments received appropriate consideration. The Summary for Policymakers was approved line-by-line and the underlying chapters were then accepted at the 10th Session of IPCC Working Group I from 29 January to 1 February 2007.

Scope of the Report

The Working Group I report focuses on those aspects of the current understanding of the physical science of climate change that are judged to be most relevant to policymakers. It does not attempt to review the evolution of scientific understanding or to cover all of climate science. Furthermore, this assessment is based on the relevant scientific literature available to the authors in mid-2006 and the reader should recognize that some topics covered here may be subject to further rapid development.

A feature of recent climate change research is the breadth of observations now available for different components of the climate system, including the atmosphere, oceans, and cryosphere. Additional observations and new analyses have broadened our understanding and enabled many uncertainties to be reduced. New information has also led to some new questions in areas such as unanticipated changes in ice sheets, their potential effect on sea level rise, and the implications of complex interactions between climate change and biogeochemistry.

In considering future projections of climate change, this report follows decisions made by the Panel during the AR4 scoping and approval process to use emission scenarios that have been previously assessed by the IPCC for consistency across the three Working Groups. However, the value of information from new climate models related to climate stabilization has also been recognized. In order to address both topics, climate modelling groups have conducted climate simulations that included idealized experiments in which atmospheric composition is held constant. Together with climate model ensemble simulations, including many model runs for the 20th and 21st centuries, this assessment has been able to consider far more simulations than any previous assessment of climate change.

The IPCC assessment of the effects of climate change and of options for responding to or avoiding such effects, are assessed by Working Groups II and III and so are not covered here. In particular, while this Working Group I report presents results for a range of emission scenarios consistent with previous reports, an updated assessment of the plausible range of future emissions can only be conducted by Working Group III.

The Structure of this Report

This Working Group I assessment includes, for the first time, an introductory chapter, Chapter 1, which covers the ways in which climate change science has progressed, including an overview of the methods used in climate change science, the role of climate models and evolution in the treatment of uncertainties.

Chapters 2 and 7 cover the changes in atmospheric constituents (both gases and aerosols) that affect the radiative energy balance in the atmosphere and determine the Earth's climate. Chapter 2 presents a perspective based on observed change in the atmosphere and covers the central concept of radiative forcing. Chapter 7 complements this by considering the interactions between the biogeochemical cycles that affect atmospheric constituents and climate change, including aerosol/cloud interactions.

Chapters 3, 4 and 5 cover the extensive range of observations now available for the atmosphere and surface, for snow, ice and frozen ground, and for the oceans respectively. While observed changes in these components of the climate system are closely inter-related through physical processes, the separate chapters allow a more focused assessment of available data and their uncertainties, including remote sensing data from satellites. Chapter 5 includes observed changes in sea level, recognizing the strong interconnections between these and ocean heat content.

Chapter 6 presents a palaeoclimatic perspective and assesses the evidence for past climate change and the extent to which that is explained by our present scientific understanding. It includes a new assessment of reconstructed temperatures for the past 1300 years.

Chapter 8 covers the ways in which physical processes are simulated in climate models and the evaluation of models against observed climate, including its average state and variability. Chapter 9 covers the closely related issue of the extent to which observed climate change can be attributed to different causes, both natural and anthropogenic.

Chapter 10 covers the use of climate models for projections of global climate including their uncertainties. It shows results for different levels of future greenhouse gases, providing a probabilistic assessment of a range of physical climate system responses and the time scales and inertia associated with such responses. Chapter 11 covers regional climate change projections consistent with the global projections. It includes an assessment of model reliability at regional levels and the factors that can significantly influence regional scale climate change.

The Summary for Policymakers (SPM) and Technical Summary (TS) of this report follow a parallel structure and each includes cross references to the chapter and section where the material being summarized can be found in the underlying report. In this way these summary components of the report provide a road-map to the content of the entire report and the reader is encouraged to use the SPM and TS in that way.

An innovation in this report is the inclusion of 19 Frequently Asked Questions, in which the authors provide scientific answers to a range of general questions in a form that will be useful for a broad range of teaching purposes. Finally the report is accompanied by about 250 pages of supplementary material that was reviewed along with the chapter drafts and is made available on CDRom and in web-based versions of the report to provide an additional level of detail, such as results for individual climate models.

Some key policy-relevant questions and issues addressed in this report and the relevant chapters

Question	Chapters
How has the science of climate change advanced since the IPCC began?	1
What is known about the natural and anthropogenic agents that contribute to climate change, and the underlying processes that are involved?	2, 6, 7
How has climate been observed to change during the period of instrumental measurements?	3, 4, 5
What is known of palaeoclimatic changes, before the instrumental era, over time scales of hundreds to millions of years, and the processes that caused them?	6, 9
How well do we understand human and natural contributions to recent climate change, and how well can we simulate changes in climate using models?	8, 9
How is climate projected to change in the future, globally and regionally?	10, 11
What is known about past and projected changes in sea level, including the role of changes in glaciers and ice sheets?	4, 5, 6, 10
Are extremes such as heavy precipitation, droughts, and heat waves changing and why, and how are they expected to change in the future?	3, 5, 9, 10, 11

Acknowledgments

This assessment has benefited greatly from the very high degree of co-operation that exists within the international climate science community and its coordination by the World Meteorological Organization World Climate Research Program (WCRP) and the International Geosphere Biosphere Program (IGBP). In particular we wish to acknowledge the enormous commitment by the individuals and agencies of 14 climate modelling groups from around the world, as well as the archiving and distribution of an unprecedented amount (over 30 Terabytes) of climate model output by the Program for Climate Model Diagnosis and Intercomparison (PCMDI). This has enabled a more detailed comparison among current climate models and a more comprehensive assessment of the potential nature of long term climate change than ever before.

We must emphasise that this report has been entirely dependent on the expertise, hard work, and commitment to excellence shown throughout by our Coordinating Lead Authors and Lead Authors with important help by many Contributing Authors. In addition we would like to express our sincere appreciation of the work carried out by the expert reviewers and acknowledge the value of the very large number of constructive comments received. Our Review Editors have similarly played a critical role in assisting the authors to deal with these comments.

The Working Group I Bureau, Kansri Boonpragob, Filippo Giorgi, Bubu Jallow, Jean Jouzel, Maria Martelo and David Wratt have played the role of an editorial board in assisting with the selection of authors and with guiding the initial outline of the report. They have provided constructive support to the Working Group Co-Chairs throughout for which we are very grateful.

Our sincere thanks go to the hosts and organizers of the four lead author meetings that were necessary for the preparation of the report and we gratefully acknowledge the support received from governments and agencies in Italy, China, New Zealand and Norway. The final Working Group I approval session was made possible by Mr Marc Gillet through the generosity of the government of France and the session was greatly facilitated by Francis Hayes, the WMO Conference Officer.

It is a pleasure to acknowledge the tireless work of the staff of the Working Group I Technical Support Unit, Melinda Marquis, Kristen Averyt, Melinda Tignor, Roy Miller, Tahl Kestin and Scott Longmore, who were ably assisted by Zhenlin Chen, Barbara Keppler, MaryAnn Pykkonen, Kyle Terran, Lelani Arris, and Marilyn Anderson. Graphics support and layout by Michael Shibao and Paula Megenhardt is gratefully appreciated. We thank Reto Stockli for kindly providing images of the Earth from space for the cover of this report. Assistance in helping the Co-Chairs to organize and edit the Frequently Asked Questions by David Wratt, David Fahey, and Susan Joy Hassol, is also appreciated. We should also like to thank Renate Christ, Secretary of the IPCC, and Secretariat staff Jian Liu, Rudie Bourgeois, Annie Courtin and Joelle Fernandez who provided logistical support for government liaison and travel of experts from developing countries and transitional economy countries.

Rajendra K. Pachauri	Susan Solomon	Dahe Qin	Martin Manning
IPCC Chairman	IPCC WGI Co-Chair	IPCC WGI Co-Chair	IPCC WGI TSU Head

Contents

A report of Working Group I of the Intergovernmental Panel on Climate Change

Summary for Policymakers

Drafting Authors:

Richard B. Alley, Terje Berntsen, Nathaniel L. Bindoff, Zhenlin Chen, Amnat Chidthaisong, Pierre Friedlingstein, Jonathan M. Gregory, Gabriele C. Hegerl, Martin Heimann, Bruce Hewitson, Brian J. Hoskins, Fortunat Joos, Jean Jouzel, Vladimir Kattsov, Ulrike Lohmann, Martin Manning, Taroh Matsuno, Mario Molina, Neville Nicholls, Jonathan Overpeck, Dahe Qin, Graciela Raga, Venkatachalam Ramaswamy, Jiawen Ren, Matilde Rusticucci, Susan Solomon, Richard Somerville, Thomas F. Stocker, Peter A. Stott, Ronald J. Stouffer, Penny Whetton, Richard A. Wood, David Wratt

Draft Contributing Authors:

J. Arblaster, G. Brasseur, J.H. Christensen, K.L. Denman, D.W. Fahey, P. Forster, E. Jansen, P.D. Jones, R. Knutti, H. Le Treut, P. Lemke, G. Meehl, P. Mote, D.A. Randall, D.A. Stone, K.E. Trenberth, J. Willebrand, F. Zwiers

This Summary for Policymakers should be cited as:

IPCC, 2007: Summary for Policymakers. In: *Climate Change 2007: The Physical Science Basis. Contribution of Working Group I to the Fourth Assessment Report of the Intergovernmental Panel on Climate Change* [Solomon, S., D. Qin, M. Manning, Z. Chen, M. Marquis, K.B. Averyt, M.Tignor and H.L. Miller (eds.)]. Cambridge University Press, Cambridge, United Kingdom and New York, NY, USA.

Introduction

The Working Group I contribution to the IPCC Fourth Assessment Report describes progress in understanding of the human and natural drivers of climate change,[1] observed climate change, climate processes and attribution, and estimates of projected future climate change. It builds upon past IPCC assessments and incorporates new findings from the past six years of research. Scientific progress since the Third Assessment Report (TAR) is based upon large amounts of new and more comprehensive data, more sophisticated analyses of data, improvements in understanding of processes and their simulation in models and more extensive exploration of uncertainty ranges.

The basis for substantive paragraphs in this Summary for Policymakers can be found in the chapter sections specified in curly brackets.

Human and Natural Drivers of Climate Change

Changes in the atmospheric abundance of greenhouse gases and aerosols, in solar radiation and in land surface properties alter the energy balance of the climate system. These changes are expressed in terms of radiative forcing,[2] which is used to compare how a range of human and natural factors drive warming or cooling influences on global climate. Since the TAR, new observations and related modelling of greenhouse gases, solar activity, land surface properties and some aspects of aerosols have led to improvements in the quantitative estimates of radiative forcing.

Global atmospheric concentrations of carbon dioxide, methane and nitrous oxide have increased markedly as a result of human activities since 1750 and now far exceed pre-industrial values determined from ice cores spanning many thousands of years (see Figure SPM.1). The global increases in carbon dioxide concentration are due primarily to fossil fuel use and land use change, while those of methane and nitrous oxide are primarily due to agriculture. {2.3, 6.4, 7.3}

- Carbon dioxide is the most important anthropogenic greenhouse gas (see Figure SPM.2). The global atmospheric concentration of carbon dioxide has increased from a pre-industrial value of about 280 ppm to 379 ppm[3] in 2005. The atmospheric concentration of carbon dioxide in 2005 exceeds by far the natural range over the last 650,000 years (180 to 300 ppm) as determined from ice cores. The annual carbon dioxide concentration growth rate was larger during the last 10 years (1995–2005 average: 1.9 ppm per year), than it has been since the beginning of continuous direct atmospheric measurements (1960–2005 average: 1.4 ppm per year) although there is year-to-year variability in growth rates. {2.3, 7.3}

- The primary source of the increased atmospheric concentration of carbon dioxide since the pre-industrial period results from fossil fuel use, with land-use change providing another significant but smaller contribution. Annual fossil carbon dioxide emissions[4] increased from an average of 6.4 [6.0 to 6.8][5] GtC (23.5 [22.0 to 25.0] $GtCO_2$) per year in the 1990s to 7.2 [6.9 to 7.5] GtC (26.4 [25.3 to 27.5] $GtCO_2$) per year in 2000–2005 (2004 and 2005 data are interim estimates). Carbon dioxide emissions associated with land-use change

---

[1] *Climate change* in IPCC usage refers to any change in climate over time, whether due to natural variability or as a result of human activity. This usage differs from that in the United Nations Framework Convention on Climate Change, where climate change refers to a change of climate that is attributed directly or indirectly to human activity that alters the composition of the global atmosphere and that is in addition to natural climate variability observed over comparable time periods.

[2] *Radiative forcing* is a measure of the influence that a factor has in altering the balance of incoming and outgoing energy in the Earth-atmosphere system and is an index of the importance of the factor as a potential climate change mechanism. Positive forcing tends to warm the surface while negative forcing tends to cool it. In this report, radiative forcing values are for 2005 relative to pre-industrial conditions defined at 1750 and are expressed in watts per square metre (W m^{-2}). See Glossary and Section 2.2 for further details.

[3] ppm (parts per million) or ppb (parts per billion, 1 billion = 1,000 million) is the ratio of the number of greenhouse gas molecules to the total number of molecules of dry air. For example, 300 ppm means 300 molecules of a greenhouse gas per million molecules of dry air.

[4] Fossil carbon dioxide emissions include those from the production, distribution and consumption of fossil fuels and as a by-product from cement production. An emission of 1 GtC corresponds to 3.67 $GtCO_2$.

[5] In general, uncertainty ranges for results given in this Summary for Policymakers are 90% uncertainty intervals unless stated otherwise, that is, there is an estimated 5% likelihood that the value could be above the range given in square brackets and 5% likelihood that the value could be below that range. Best estimates are given where available. Assessed uncertainty intervals are not always symmetric about the corresponding best estimate. Note that a number of uncertainty ranges in the Working Group I TAR corresponded to 2 standard deviations (95%), often using expert judgement.

CHANGES IN GREENHOUSE GASES FROM ICE CORE AND MODERN DATA

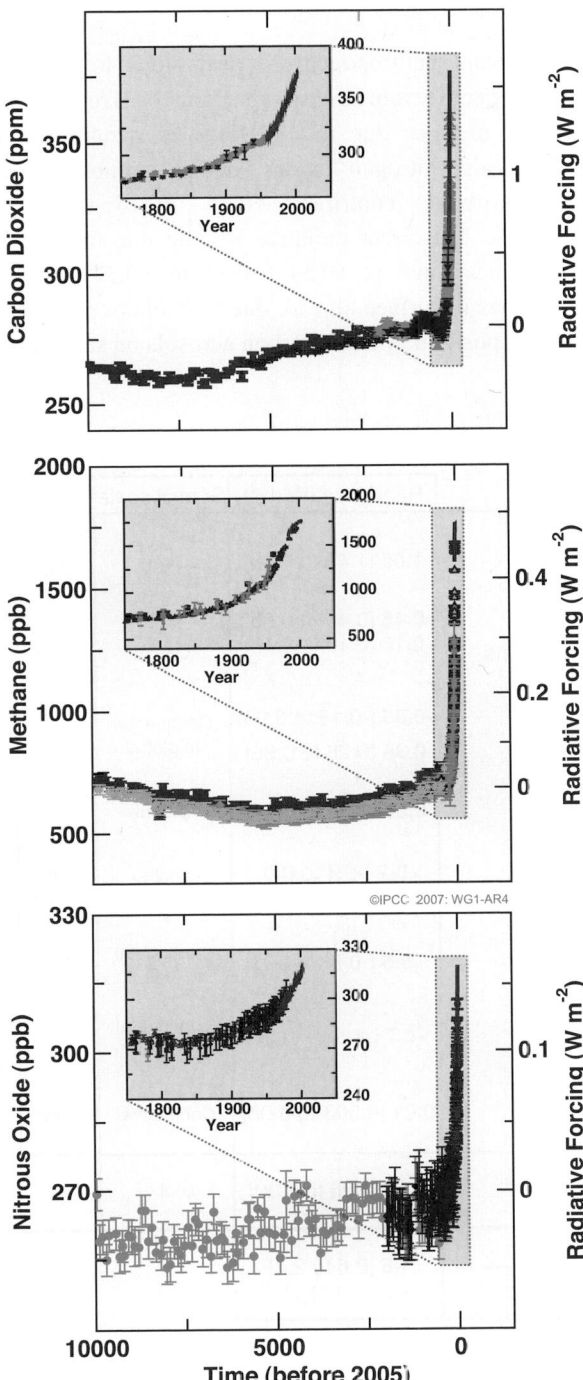

are estimated to be 1.6 [0.5 to 2.7] GtC (5.9 [1.8 to 9.9] $GtCO_2$) per year over the 1990s, although these estimates have a large uncertainty. {7.3}

- The global atmospheric concentration of methane has increased from a pre-industrial value of about 715 ppb to 1732 ppb in the early 1990s, and was 1774 ppb in 2005. The atmospheric concentration of methane in 2005 exceeds by far the natural range of the last 650,000 years (320 to 790 ppb) as determined from ice cores. Growth rates have declined since the early 1990s, consistent with total emissions (sum of anthropogenic and natural sources) being nearly constant during this period. It is *very likely*[6] that the observed increase in methane concentration is due to anthropogenic activities, predominantly agriculture and fossil fuel use, but relative contributions from different source types are not well determined. {2.3, 7.4}

- The global atmospheric nitrous oxide concentration increased from a pre-industrial value of about 270 ppb to 319 ppb in 2005. The growth rate has been approximately constant since 1980. More than a third of all nitrous oxide emissions are anthropogenic and are primarily due to agriculture. {2.3, 7.4}

The understanding of anthropogenic warming and cooling influences on climate has improved since the TAR, leading to *very high confidence*[7] that the global average net effect of human activities since 1750 has been one of warming, with a radiative forcing of +1.6 [+0.6 to +2.4] W m⁻² (see Figure SPM.2). {2.3., 6.5, 2.9}

- The combined radiative forcing due to increases in carbon dioxide, methane, and nitrous oxide is +2.30 [+2.07 to +2.53] W m⁻², and its rate of increase during the industrial era is *very likely* to have been unprecedented in more than 10,000 years (see Figures

Figure SPM.1. *Atmospheric concentrations of carbon dioxide, methane and nitrous oxide over the last 10,000 years (large panels) and since 1750 (inset panels). Measurements are shown from ice cores (symbols with different colours for different studies) and atmospheric samples (red lines). The corresponding radiative forcings are shown on the right hand axes of the large panels. {Figure 6.4}*

---

6 In this Summary for Policymakers, the following terms have been used to indicate the assessed likelihood, using expert judgement, of an outcome or a result: *Virtually certain* > 99% probability of occurrence, *Extremely likely* > 95%, *Very likely* > 90%, *Likely* > 66%, *More likely than not* > 50%, *Unlikely* < 33%, *Very unlikely* < 10%, *Extremely unlikely* < 5% (see Box TS.1 for more details).

7 In this Summary for Policymakers the following levels of confidence have been used to express expert judgements on the correctness of the underlying science: *very high confidence* represents at least a 9 out of 10 chance of being correct; *high confidence* represents about an 8 out of 10 chance of being correct (see Box TS.1).

3

SPM.1 and SPM.2). The carbon dioxide radiative forcing increased by 20% from 1995 to 2005, the largest change for any decade in at least the last 200 years. {2.3, 6.4}

- Anthropogenic contributions to aerosols (primarily sulphate, organic carbon, black carbon, nitrate and dust) together produce a cooling effect, with a total direct radiative forcing of −0.5 [−0.9 to −0.1] W m^{-2} and an indirect cloud albedo forcing of −0.7 [−1.8 to −0.3] W m^{-2}. These forcings are now better understood than at the time of the TAR due to improved *in situ*, satellite and ground-based measurements and more

comprehensive modelling, but remain the dominant uncertainty in radiative forcing. Aerosols also influence cloud lifetime and precipitation. {2.4, 2.9, 7.5}

- Significant anthropogenic contributions to radiative forcing come from several other sources. Tropospheric ozone changes due to emissions of ozone-forming chemicals (nitrogen oxides, carbon monoxide, and hydrocarbons) contribute +0.35 [+0.25 to +0.65] W m^{-2}. The direct radiative forcing due to changes in halocarbons[8] is +0.34 [+0.31 to +0.37] W m^{-2}. Changes in surface albedo, due to land cover changes and deposition of black carbon aerosols on snow, exert

RADIATIVE FORCING COMPONENTS

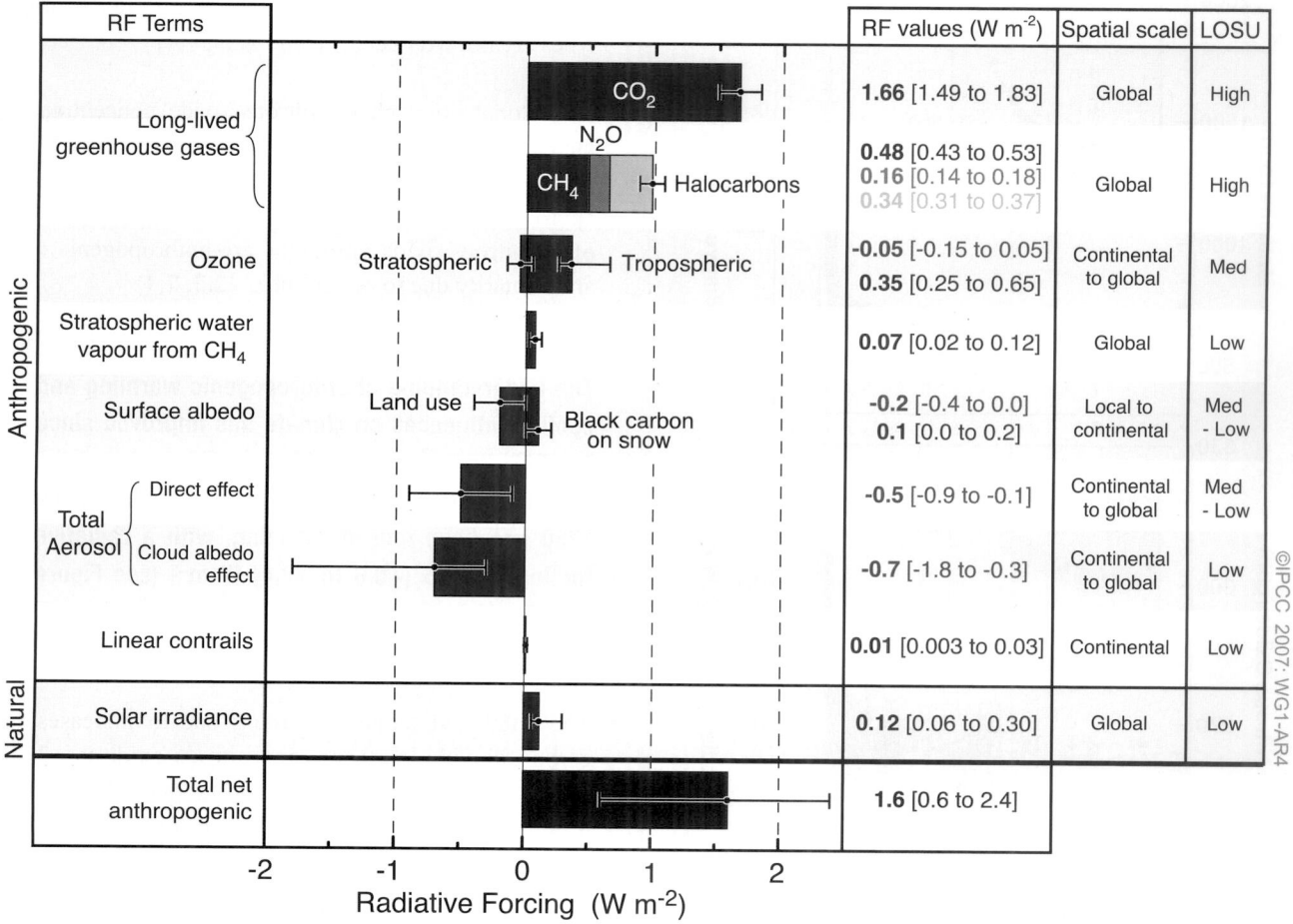

Figure SPM.2. *Global average radiative forcing (RF) estimates and ranges in 2005 for anthropogenic carbon dioxide (CO$_2$), methane (CH$_4$), nitrous oxide (N$_2$O) and other important agents and mechanisms, together with the typical geographical extent (spatial scale) of the forcing and the assessed level of scientific understanding (LOSU). The net anthropogenic radiative forcing and its range are also shown. These require summing asymmetric uncertainty estimates from the component terms, and cannot be obtained by simple addition. Additional forcing factors not included here are considered to have a very low LOSU. Volcanic aerosols contribute an additional natural forcing but are not included in this figure due to their episodic nature. The range for linear contrails does not include other possible effects of aviation on cloudiness. {2.9, Figure 2.20}*

---

[8] Halocarbon radiative forcing has been recently assessed in detail in *IPCC's Special Report on Safeguarding the Ozone Layer and the Global Climate System* (2005).

respective forcings of –0.2 [–0.4 to 0.0] and +0.1 [0.0 to +0.2] W m^{-2}. Additional terms smaller than ±0.1 W m^{-2} are shown in Figure SPM.2. {2.3, 2.5, 7.2}

- Changes in solar irradiance since 1750 are estimated to cause a radiative forcing of +0.12 [+0.06 to +0.30] W m^{-2}, which is less than half the estimate given in the TAR. {2.7}

Direct Observations of Recent Climate Change

Since the TAR, progress in understanding how climate is changing in space and in time has been gained through improvements and extensions of numerous datasets and data analyses, broader geographical coverage, better understanding of uncertainties, and a wider variety of measurements. Increasingly comprehensive observations are available for glaciers and snow cover since the 1960s, and for sea level and ice sheets since about the past decade. However, data coverage remains limited in some regions.

Warming of the climate system is unequivocal, as is now evident from observations of increases in global average air and ocean temperatures, widespread melting of snow and ice, and rising global average sea level (see Figure SPM.3). {3.2, 4.2, 5.5}

- Eleven of the last twelve years (1995–2006) rank among the 12 warmest years in the instrumental record of global surface temperature[9] (since 1850). The updated 100-year linear trend (1906 to 2005) of 0.74°C [0.56°C to 0.92°C] is therefore larger than the corresponding trend for 1901 to 2000 given in the TAR of 0.6°C [0.4°C to 0.8°C]. The linear warming trend over the last 50 years (0.13°C [0.10°C to 0.16°C] per decade) is nearly twice that for the last 100 years. The total temperature increase from 1850–1899 to 2001–2005 is 0.76°C [0.57°C to 0.95°C]. Urban heat island effects are real but local, and have a negligible influence (less than 0.006°C per decade over land and zero over the oceans) on these values. {3.2}

- New analyses of balloon-borne and satellite measurements of lower- and mid-tropospheric temperature show warming rates that are similar to those of the surface temperature record and are consistent within their respective uncertainties, largely reconciling a discrepancy noted in the TAR. {3.2, 3.4}

- The average atmospheric water vapour content has increased since at least the 1980s over land and ocean as well as in the upper troposphere. The increase is broadly consistent with the extra water vapour that warmer air can hold. {3.4}

- Observations since 1961 show that the average temperature of the global ocean has increased to depths of at least 3000 m and that the ocean has been absorbing more than 80% of the heat added to the climate system. Such warming causes seawater to expand, contributing to sea level rise (see Table SPM.1). {5.2, 5.5}

- Mountain glaciers and snow cover have declined on average in both hemispheres. Widespread decreases in glaciers and ice caps have contributed to sea level rise (ice caps do not include contributions from the Greenland and Antarctic Ice Sheets). (See Table SPM.1.) {4.6, 4.7, 4.8, 5.5}

- New data since the TAR now show that losses from the ice sheets of Greenland and Antarctica have *very likely* contributed to sea level rise over 1993 to 2003 (see Table SPM.1). Flow speed has increased for some Greenland and Antarctic outlet glaciers, which drain ice from the interior of the ice sheets. The corresponding increased ice sheet mass loss has often followed thinning, reduction or loss of ice shelves or loss of floating glacier tongues. Such dynamical ice loss is sufficient to explain most of the Antarctic net mass loss and approximately half of the Greenland net mass loss. The remainder of the ice loss from Greenland has occurred because losses due to melting have exceeded accumulation due to snowfall. {4.6, 4.8, 5.5}

- Global average sea level rose at an average rate of 1.8 [1.3 to 2.3] mm per year over 1961 to 2003. The rate was faster over 1993 to 2003: about 3.1 [2.4 to 3.8] mm per year. Whether the faster rate for 1993 to 2003 reflects decadal variability or an increase in the longer-term trend is unclear. There is *high confidence* that

---

[9] The average of near-surface air temperature over land and sea surface temperature.

CHANGES IN TEMPERATURE, SEA LEVEL AND NORTHERN HEMISPHERE SNOW COVER

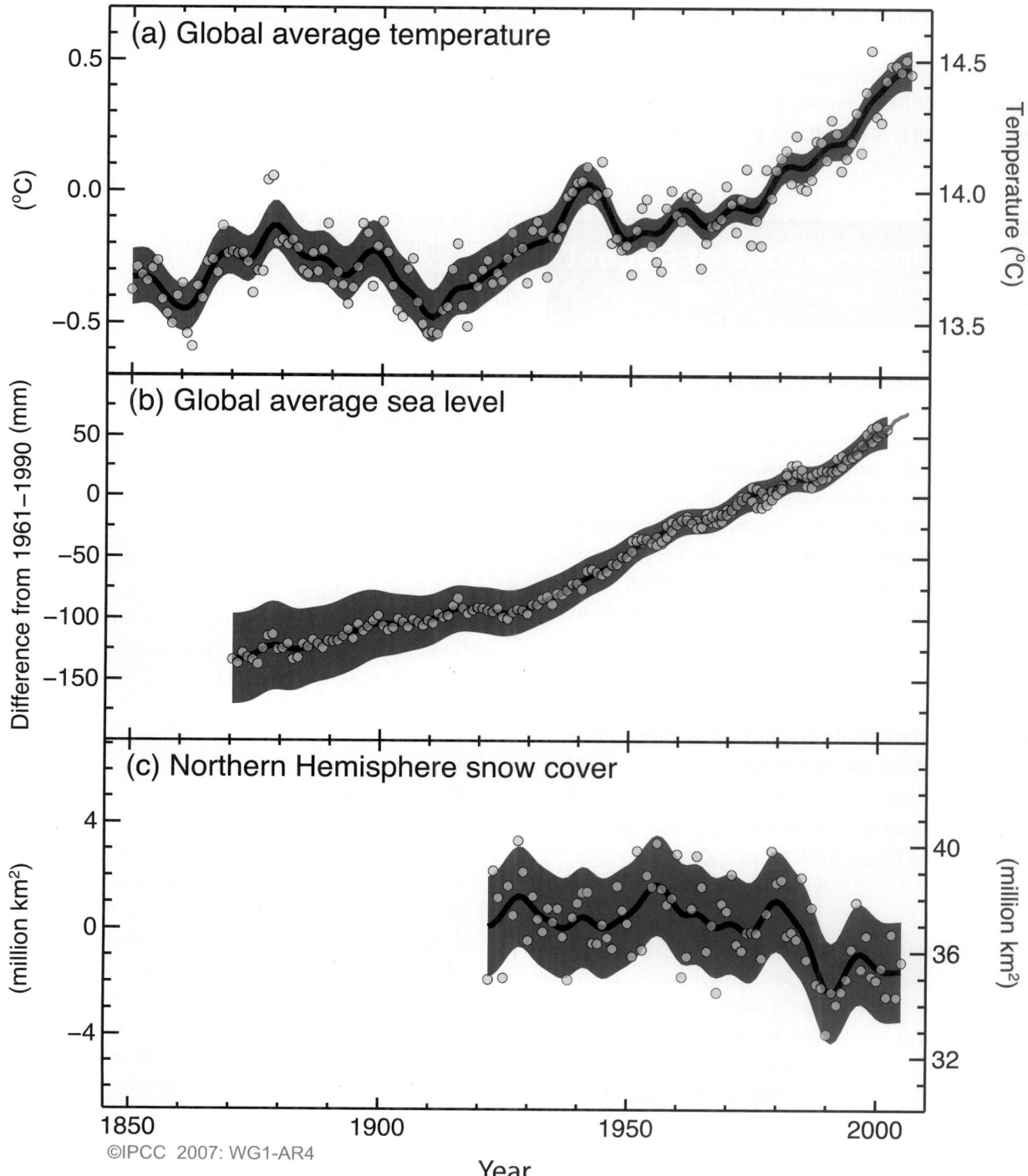

©IPCC 2007: WG1-AR4

Figure SPM.3. *Observed changes in (a) global average surface temperature, (b) global average sea level from tide gauge (blue) and satellite (red) data and (c) Northern Hemisphere snow cover for March-April. All changes are relative to corresponding averages for the period 1961–1990. Smoothed curves represent decadal average values while circles show yearly values. The shaded areas are the uncertainty intervals estimated from a comprehensive analysis of known uncertainties (a and b) and from the time series (c). {FAQ 3.1, Figure 1, Figure 4.2, Figure 5.13}*

the rate of observed sea level rise increased from the 19th to the 20th century. The total 20th-century rise is estimated to be 0.17 [0.12 to 0.22] m. {5.5}

• For 1993 to 2003, the sum of the climate contributions is consistent within uncertainties with the total sea level rise that is directly observed (see Table SPM.1). These estimates are based on improved satellite and *in situ* data now available. For the period 1961 to 2003, the sum of climate contributions is estimated to be smaller than the observed sea level rise. The TAR reported a similar discrepancy for 1910 to 1990. {5.5}

At continental, regional and ocean basin scales, numerous long-term changes in climate have been observed. These include changes in arctic temperatures and ice, widespread changes in precipitation amounts, ocean salinity, wind patterns and aspects of extreme weather including droughts, heavy precipitation, heat waves and the intensity of tropical cyclones.[10] {3.2, 3.3, 3.4, 3.5, 3.6, 5.2}

• Average arctic temperatures increased at almost twice the global average rate in the past 100 years. Arctic temperatures have high decadal variability, and a warm period was also observed from 1925 to 1945. {3.2}

• Satellite data since 1978 show that annual average arctic sea ice extent has shrunk by 2.7 [2.1 to 3.3]% per decade, with larger decreases in summer of 7.4 [5.0 to 9.8]% per decade. These values are consistent with those reported in the TAR. {4.4}

• Temperatures at the top of the permafrost layer have generally increased since the 1980s in the Arctic (by up to 3°C). The maximum area covered by seasonally frozen ground has decreased by about 7% in the Northern Hemisphere since 1900, with a decrease in spring of up to 15%. {4.7}

• Long-term trends from 1900 to 2005 have been observed in precipitation amount over many large regions.[11] Significantly increased precipitation has been observed in eastern parts of North and South America, northern Europe and northern and central Asia. Drying has been observed in the Sahel, the Mediterranean, southern Africa and parts of southern Asia. Precipitation is highly variable spatially and temporally, and data are limited in some regions. Long-term trends have not been observed for the other large regions assessed.[11] {3.3, 3.9}

• Changes in precipitation and evaporation over the oceans are suggested by freshening of mid- and high-latitude waters together with increased salinity in low-latitude waters. {5.2}

Table SPM.1. *Observed rate of sea level rise and estimated contributions from different sources. {5.5, Table 5.3}*

Source of sea level rise	Rate of sea level rise (mm per year)	
	1961–2003	1993–2003
Thermal expansion	0.42 ± 0.12	1.6 ± 0.5
Glaciers and ice caps	0.50 ± 0.18	0.77 ± 0.22
Greenland Ice Sheet	0.05 ± 0.12	0.21 ± 0.07
Antarctic Ice Sheet	0.14 ± 0.41	0.21 ± 0.35
Sum of individual climate contributions to sea level rise	1.1 ± 0.5	2.8 ± 0.7
Observed total sea level rise	1.8 ± 0.5[a]	3.1 ± 0.7[a]
Difference (Observed minus sum of estimated climate contributions)	0.7 ± 0.7	0.3 ± 1.0

Table note:
[a] Data prior to 1993 are from tide gauges and after 1993 are from satellite altimetry.

[10] Tropical cyclones include hurricanes and typhoons.

[11] The assessed regions are those considered in the regional projections chapter of the TAR and in Chapter 11 of this report.

- Mid-latitude westerly winds have strengthened in both hemispheres since the 1960s. {3.5}

- More intense and longer droughts have been observed over wider areas since the 1970s, particularly in the tropics and subtropics. Increased drying linked with higher temperatures and decreased precipitation has contributed to changes in drought. Changes in sea surface temperatures, wind patterns and decreased snowpack and snow cover have also been linked to droughts. {3.3}

- The frequency of heavy precipitation events has increased over most land areas, consistent with warming and observed increases of atmospheric water vapour. {3.8, 3.9}

- Widespread changes in extreme temperatures have been observed over the last 50 years. Cold days, cold nights and frost have become less frequent, while hot days, hot nights and heat waves have become more frequent (see Table SPM.2). {3.8}

Table SPM.2. *Recent trends, assessment of human influence on the trend and projections for extreme weather events for which there is an observed late-20th century trend. {Tables 3.7, 3.8, 9.4; Sections 3.8, 5.5, 9.7, 11.2–11.9}*

Phenomenon[a] and direction of trend	Likelihood that trend occurred in late 20th century (typically post 1960)	Likelihood of a human contribution to observed trend[b]	Likelihood of future trends based on projections for 21st century using SRES scenarios
Warmer and fewer cold days and nights over most land areas	*Very likely*[c]	*Likely*[d]	*Virtually certain*[d]
Warmer and more frequent hot days and nights over most land areas	*Very likely*[e]	*Likely (nights)*[d]	*Virtually certain*[d]
Warm spells/heat waves. Frequency increases over most land areas	*Likely*	*More likely than not*[f]	*Very likely*
Heavy precipitation events. Frequency (or proportion of total rainfall from heavy falls) increases over most areas	*Likely*	*More likely than not*[f]	*Very likely*
Area affected by droughts increases	*Likely* in many regions since 1970s	*More likely than not*	*Likely*
Intense tropical cyclone activity increases	*Likely* in some regions since 1970	*More likely than not*[f]	*Likely*
Increased incidence of extreme high sea level (excludes tsunamis)[g]	*Likely*	*More likely than not*[f,h]	*Likely*[i]

Table notes:

[a] See Table 3.7 for further details regarding definitions.

[b] See Table TS.4, Box TS.5 and Table 9.4.

[c] Decreased frequency of cold days and nights (coldest 10%).

[d] Warming of the most extreme days and nights each year.

[e] Increased frequency of hot days and nights (hottest 10%).

[f] Magnitude of anthropogenic contributions not assessed. Attribution for these phenomena based on expert judgement rather than formal attribution studies.

[g] Extreme high sea level depends on average sea level and on regional weather systems. It is defined here as the highest 1% of hourly values of observed sea level at a station for a given reference period.

[h] Changes in observed extreme high sea level closely follow the changes in average sea level. {5.5} It is *very likely* that anthropogenic activity contributed to a rise in average sea level. {9.5}

[i] In all scenarios, the projected global average sea level at 2100 is higher than in the reference period. {10.6} The effect of changes in regional weather systems on sea level extremes has not been assessed.

- There is observational evidence for an increase in intense tropical cyclone activity in the North Atlantic since about 1970, correlated with increases of tropical sea surface temperatures. There are also suggestions of increased intense tropical cyclone activity in some other regions where concerns over data quality are greater. Multi-decadal variability and the quality of the tropical cyclone records prior to routine satellite observations in about 1970 complicate the detection of long-term trends in tropical cyclone activity. There is no clear trend in the annual numbers of tropical cyclones. {3.8}

Some aspects of climate have not been observed to change. {3.2, 3.8, 4.4, 5.3}

- A decrease in diurnal temperature range (DTR) was reported in the TAR, but the data available then extended only from 1950 to 1993. Updated observations reveal that DTR has not changed from 1979 to 2004 as both day- and night-time temperature have risen at about the same rate. The trends are highly variable from one region to another. {3.2}

- Antarctic sea ice extent continues to show interannual variability and localised changes but no statistically significant average trends, consistent with the lack of warming reflected in atmospheric temperatures averaged across the region. {3.2, 4.4}

- There is insufficient evidence to determine whether trends exist in the meridional overturning circulation (MOC) of the global ocean or in small-scale phenomena such as tornadoes, hail, lightning and dust-storms. {3.8, 5.3}

A Palaeoclimatic Perspective

Palaeoclimatic studies use changes in climatically sensitive indicators to infer past changes in global climate on time scales ranging from decades to millions of years. Such proxy data (e.g., tree ring width) may be influenced by both local temperature and other factors such as precipitation, and are often representative of particular seasons rather than full years. Studies since the TAR draw increased confidence from additional data showing coherent behaviour across multiple indicators in different parts of the world. However, uncertainties generally increase with time into the past due to increasingly limited spatial coverage.

Palaeoclimatic information supports the interpretation that the warmth of the last half century is unusual in at least the previous 1,300 years. The last time the polar regions were significantly warmer than present for an extended period (about 125,000 years ago), reductions in polar ice volume led to 4 to 6 m of sea level rise. {6.4, 6.6}

- Average Northern Hemisphere temperatures during the second half of the 20th century were *very likely* higher than during any other 50-year period in the last 500 years and *likely* the highest in at least the past 1,300 years. Some recent studies indicate greater variability in Northern Hemisphere temperatures than suggested in the TAR, particularly finding that cooler periods existed in the 12th to 14th, 17th and 19th centuries. Warmer periods prior to the 20th century are within the uncertainty range given in the TAR. {6.6}

- Global average sea level in the last interglacial period (about 125,000 years ago) was *likely* 4 to 6 m higher than during the 20th century, mainly due to the retreat of polar ice. Ice core data indicate that average polar temperatures at that time were 3°C to 5°C higher than present, because of differences in the Earth's orbit. The Greenland Ice Sheet and other arctic ice fields *likely* contributed no more than 4 m of the observed sea level rise. There may also have been a contribution from Antarctica. {6.4}

Understanding and Attributing Climate Change

This assessment considers longer and improved records, an expanded range of observations and improvements in the simulation of many aspects of climate and its variability based on studies since the TAR. It also considers the results of new attribution studies that have evaluated whether observed changes are quantitatively consistent with the expected response to external forcings and inconsistent with alternative physically plausible explanations.

Most of the observed increase in global average temperatures since the mid-20th century is *very likely* due to the observed increase in anthropogenic greenhouse gas concentrations.[12] This is an advance since the TAR's conclusion that "most of the observed warming over the last 50 years is *likely* to have been due to the increase in greenhouse gas concentrations". Discernible human influences now extend to other aspects of climate, including ocean warming, continental-average temperatures, temperature extremes and wind patterns (see Figure SPM.4 and Table SPM.2). {9.4, 9.5}

- It is *likely* that increases in greenhouse gas concentrations alone would have caused more warming than observed because volcanic and anthropogenic aerosols have offset some warming that would otherwise have taken place. {2.9, 7.5, 9.4}

- The observed widespread warming of the atmosphere and ocean, together with ice mass loss, support the conclusion that it is *extremely unlikely* that global climate change of the past 50 years can be explained without external forcing, and *very likely* that it is not due to known natural causes alone. {4.8, 5.2, 9.4, 9.5, 9.7}

- Warming of the climate system has been detected in changes of surface and atmospheric temperatures in the upper several hundred metres of the ocean, and in contributions to sea level rise. Attribution studies have established anthropogenic contributions to all of these changes. The observed pattern of tropospheric warming and stratospheric cooling is *very likely* due to the combined influences of greenhouse gas increases and stratospheric ozone depletion. {3.2, 3.4, 9.4, 9.5}

- It is *likely* that there has been significant anthropogenic warming over the past 50 years averaged over each continent except Antarctica (see Figure SPM.4). The observed patterns of warming, including greater warming over land than over the ocean, and their changes over time, are only simulated by models that include anthropogenic forcing. The ability of coupled climate models to simulate the observed temperature evolution on each of six continents provides stronger evidence of human influence on climate than was available in the TAR. {3.2, 9.4}

- Difficulties remain in reliably simulating and attributing observed temperature changes at smaller scales. On these scales, natural climate variability is relatively larger, making it harder to distinguish changes expected due to external forcings. Uncertainties in local forcings and feedbacks also make it difficult to estimate the contribution of greenhouse gas increases to observed small-scale temperature changes. {8.3, 9.4}

- Anthropogenic forcing is *likely* to have contributed to changes in wind patterns,[13] affecting extra-tropical storm tracks and temperature patterns in both hemispheres. However, the observed changes in the Northern Hemisphere circulation are larger than simulated in response to 20th-century forcing change. {3.5, 3.6, 9.5, 10.3}

- Temperatures of the most extreme hot nights, cold nights and cold days are *likely* to have increased due to anthropogenic forcing. It is *more likely than not* that anthropogenic forcing has increased the risk of heat waves (see Table SPM.2). {9.4}

---

[12] Consideration of remaining uncertainty is based on current methodologies.

[13] In particular, the Southern and Northern Annular Modes and related changes in the North Atlantic Oscillation. {3.6, 9.5, Box TS.2}

GLOBAL AND CONTINENTAL TEMPERATURE CHANGE

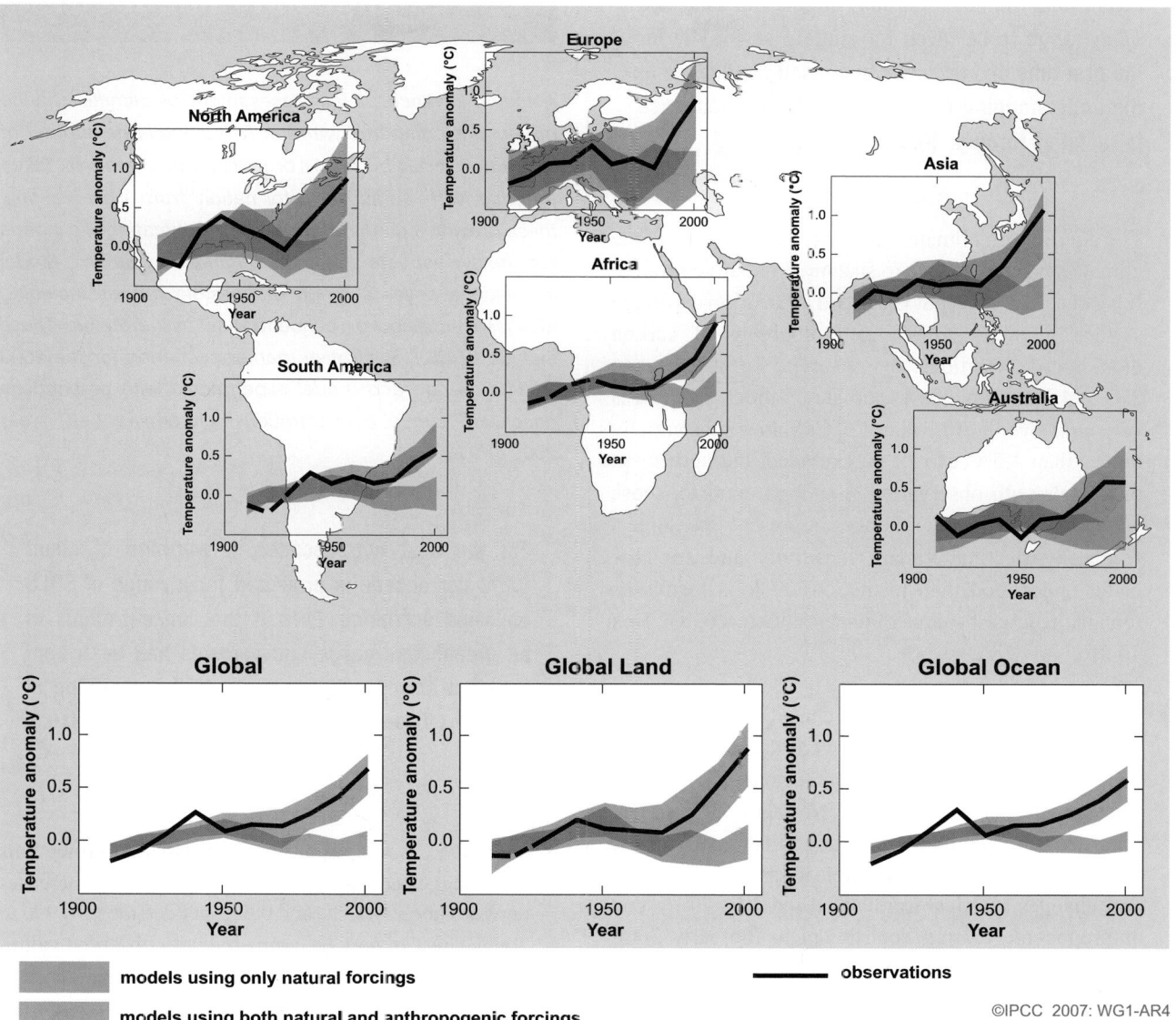

models using only natural forcings

models using both natural and anthropogenic forcings

observations

Figure SPM.4. *Comparison of observed continental- and global-scale changes in surface temperature with results simulated by climate models using natural and anthropogenic forcings. Decadal averages of observations are shown for the period 1906 to 2005 (black line) plotted against the centre of the decade and relative to the corresponding average for 1901–1950. Lines are dashed where spatial coverage is less than 50%. Blue shaded bands show the 5–95% range for 19 simulations from five climate models using only the natural forcings due to solar activity and volcanoes. Red shaded bands show the 5–95% range for 58 simulations from 14 climate models using both natural and anthropogenic forcings. {FAQ 9.2, Figure 1}*

Analysis of climate models together with constraints from observations enables an assessed *likely* range to be given for climate sensitivity for the first time and provides increased confidence in the understanding of the climate system response to radiative forcing. {6.6, 8.6, 9.6, Box 10.2}

- The equilibrium climate sensitivity is a measure of the climate system response to sustained radiative forcing. It is not a projection but is defined as the global average surface warming following a doubling of carbon dioxide concentrations. It is *likely* to be in the range 2°C to 4.5°C with a best estimate of about 3°C, and is *very unlikely* to be less than 1.5°C. Values substantially higher than 4.5°C cannot be excluded, but agreement of models with observations is not as good for those values. Water vapour changes represent the largest feedback affecting climate sensitivity and are now better understood than in the TAR. Cloud feedbacks remain the largest source of uncertainty. {8.6, 9.6, Box 10.2}

- It is *very unlikely* that climate changes of at least the seven centuries prior to 1950 were due to variability generated within the climate system alone. A significant fraction of the reconstructed Northern Hemisphere inter-decadal temperature variability over those centuries is *very likely* attributable to volcanic eruptions and changes in solar irradiance, and it is *likely* that anthropogenic forcing contributed to the early 20th-century warming evident in these records. {2.7, 2.8, 6.6, 9.3}

Projections of Future Changes in Climate

A major advance of this assessment of climate change projections compared with the TAR is the large number of simulations available from a broader range of models. Taken together with additional information from observations, these provide a quantitative basis for estimating likelihoods for many aspects of future climate change. Model simulations cover a range of possible futures including idealised emission or concentration assumptions. These include SRES[14] illustrative marker scenarios for the 2000 to 2100 period and model experiments with greenhouse gases and aerosol concentrations held constant after year 2000 or 2100.

For the next two decades, a warming of about 0.2°C per decade is projected for a range of SRES emission scenarios. Even if the concentrations of all greenhouse gases and aerosols had been kept constant at year 2000 levels, a further warming of about 0.1°C per decade would be expected. {10.3, 10.7}

- Since IPCC's first report in 1990, assessed projections have suggested global average temperature increases between about 0.15°C and 0.3°C per decade for 1990 to 2005. This can now be compared with observed values of about 0.2°C per decade, strengthening confidence in near-term projections. {1.2, 3.2}

- Model experiments show that even if all radiative forcing agents were held constant at year 2000 levels, a further warming trend would occur in the next two decades at a rate of about 0.1°C per decade, due mainly to the slow response of the oceans. About twice as much warming (0.2°C per decade) would be expected if emissions are within the range of the SRES scenarios. Best-estimate projections from models indicate that decadal average warming over each inhabited continent by 2030 is insensitive to the choice among SRES scenarios and is *very likely* to be at least twice as large as the corresponding model-estimated natural variability during the 20th century. {9.4, 10.3, 10.5, 11.2–11.7, Figure TS-29}

---

[14] SRES refers to the *IPCC Special Report on Emission Scenarios* (2000). The SRES scenario families and illustrative cases, which did not include additional climate initiatives, are summarised in a box at the end of this Summary for Policymakers. Approximate carbon dioxide equivalent concentrations corresponding to the computed radiative forcing due to anthropogenic greenhouse gases and aerosols in 2100 (see p. 823 of the TAR) for the SRES B1, A1T, B2, A1B, A2 and A1FI illustrative marker scenarios are about 600, 700, 800, 850, 1250 and 1,550 ppm respectively. Scenarios B1, A1B and A2 have been the focus of model intercomparison studies and many of those results are assessed in this report.

> Continued greenhouse gas emissions at or above current rates would cause further warming and induce many changes in the global climate system during the 21st century that would *very likely* be larger than those observed during the 20th century. {10.3}

- Advances in climate change modelling now enable best estimates and *likely* assessed uncertainty ranges to be given for projected warming for different emission scenarios. Results for different emission scenarios are provided explicitly in this report to avoid loss of this policy-relevant information. Projected global average surface warmings for the end of the 21st century (2090–2099) relative to 1980–1999 are shown in Table SPM.3. These illustrate the differences between lower and higher SRES emission scenarios, and the projected warming uncertainty associated with these scenarios. {10.5}

- Best estimates and *likely* ranges for global average surface air warming for six SRES emissions marker scenarios are given in this assessment and are shown in Table SPM.3. For example, the best estimate for the low scenario (B1) is 1.8°C (*likely* range is 1.1°C to 2.9°C), and the best estimate for the high scenario

(A1FI) is 4.0°C (*likely* range is 2.4°C to 6.4°C). Although these projections are broadly consistent with the span quoted in the TAR (1.4°C to 5.8°C), they are not directly comparable (see Figure SPM.5). The Fourth Assessment Report is more advanced as it provides best estimates and an assessed likelihood range for each of the marker scenarios. The new assessment of the *likely* ranges now relies on a larger number of climate models of increasing complexity and realism, as well as new information regarding the nature of feedbacks from the carbon cycle and constraints on climate response from observations. {10.5}

- Warming tends to reduce land and ocean uptake of atmospheric carbon dioxide, increasing the fraction of anthropogenic emissions that remains in the atmosphere. For the A2 scenario, for example, the climate-carbon cycle feedback increases the corresponding global average warming at 2100 by more than 1°C. Assessed upper ranges for temperature projections are larger than in the TAR (see Table SPM.3) mainly because the broader range of models now available suggests stronger climate-carbon cycle feedbacks. {7.3, 10.5}

- Model-based projections of global average sea level rise at the end of the 21st century (2090–2099) are shown in Table SPM.3. For each scenario, the midpoint of the range in Table SPM.3 is within 10% of the

Table SPM.3. *Projected global average surface warming and sea level rise at the end of the 21st century. {10.5, 10.6, Table 10.7}*

| Case | Temperature Change (°C at 2090-2099 relative to 1980-1999)[a] | | Sea Level Rise (m at 2090-2099 relative to 1980-1999) |
	Best estimate	*Likely* range	Model-based range excluding future rapid dynamical changes in ice flow
Constant Year 2000 concentrations[b]	0.6	0.3 – 0.9	NA
B1 scenario	1.8	1.1 – 2.9	0.18 – 0.38
A1T scenario	2.4	1.4 – 3.8	0.20 – 0.45
B2 scenario	2.4	1.4 – 3.8	0.20 – 0.43
A1B scenario	2.8	1.7 – 4.4	0.21 – 0.48
A2 scenario	3.4	2.0 – 5.4	0.23 – 0.51
A1FI scenario	4.0	2.4 – 6.4	0.26 – 0.59

Table notes:

[a] These estimates are assessed from a hierarchy of models that encompass a simple climate model, several Earth System Models of Intermediate Complexity and a large number of Atmosphere-Ocean General Circulation Models (AOGCMs).

[b] Year 2000 constant composition is derived from AOGCMs only.

MULTI-MODEL AVERAGES AND ASSESSED RANGES FOR SURFACE WARMING

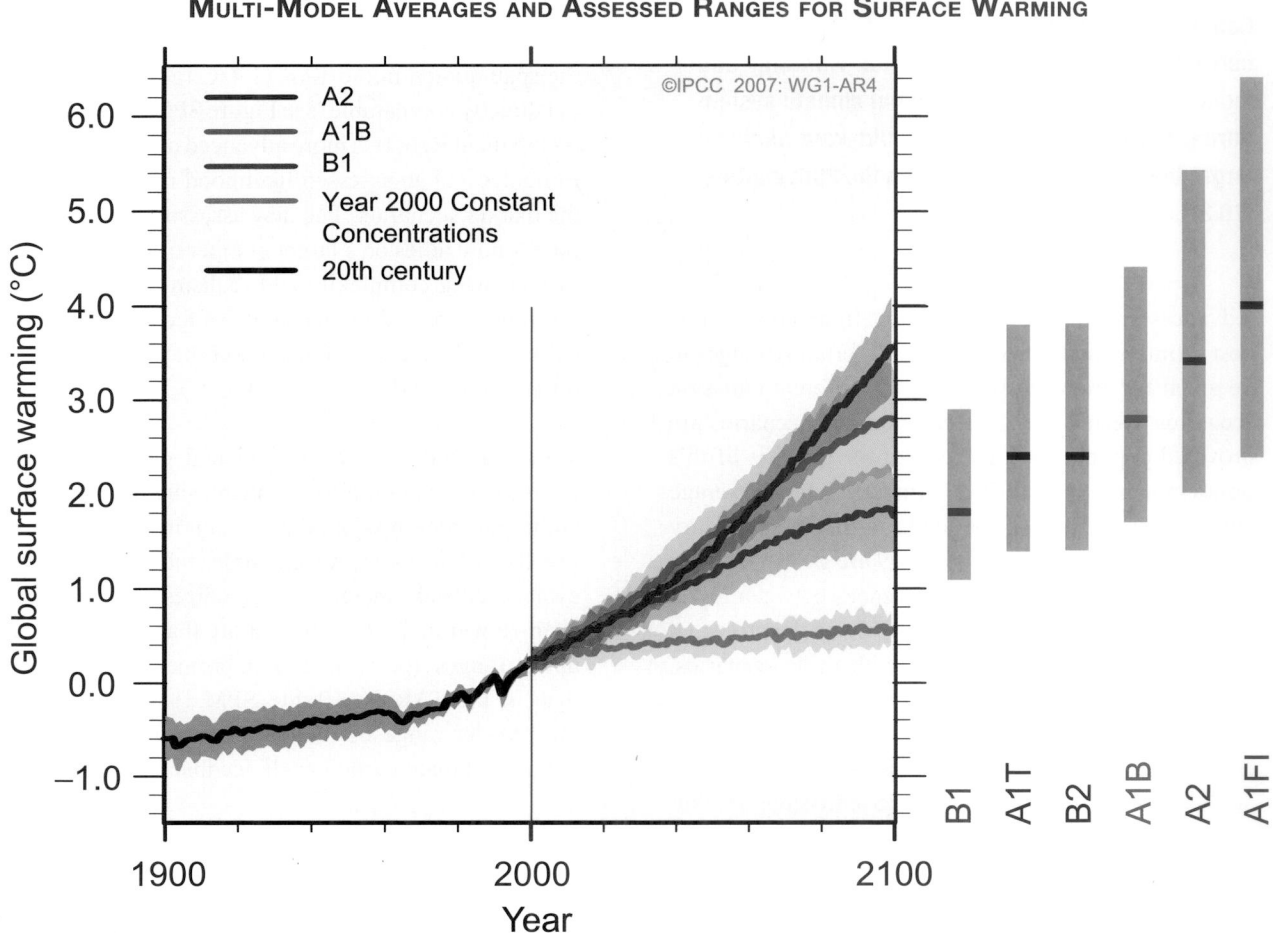

Figure SPM.5. *Solid lines are multi-model global averages of surface warming (relative to 1980–1999) for the scenarios A2, A1B and B1, shown as continuations of the 20th century simulations. Shading denotes the ±1 standard deviation range of individual model annual averages. The orange line is for the experiment where concentrations were held constant at year 2000 values. The grey bars at right indicate the best estimate (solid line within each bar) and the **likely** range assessed for the six SRES marker scenarios. The assessment of the best estimate and **likely** ranges in the grey bars includes the AOGCMs in the left part of the figure, as well as results from a hierarchy of independent models and observational constraints. {Figures 10.4 and 10.29}*

TAR model average for 2090–2099. The ranges are narrower than in the TAR mainly because of improved information about some uncertainties in the projected contributions.[15] {10.6}

- Models used to date do not include uncertainties in climate-carbon cycle feedback nor do they include the full effects of changes in ice sheet flow, because a basis in published literature is lacking. The projections include a contribution due to increased ice flow from Greenland and Antarctica at the rates observed for 1993 to 2003, but these flow rates could increase or decrease in the future. For example, if this contribution were to grow linearly with global average temperature change,

the upper ranges of sea level rise for SRES scenarios shown in Table SPM.3 would increase by 0.1 to 0.2 m. Larger values cannot be excluded, but understanding of these effects is too limited to assess their likelihood or provide a best estimate or an upper bound for sea level rise. {10.6}

- Increasing atmospheric carbon dioxide concentrations lead to increasing acidification of the ocean. Projections based on SRES scenarios give reductions in average global surface ocean pH[16] of between 0.14 and 0.35 units over the 21st century, adding to the present decrease of 0.1 units since pre-industrial times. {5.4, Box 7.3, 10.4}

---

[15] TAR poljcections were made for 2100, whereas projections in this report are for 2090–2099. The TAR would have had similar ranges to those in Table SPM.3 if it had treated the uncertainties in the same way.

[16] Decreases in pH correspond to increases in acidity of a solution. See Glossary for further details.

There is now higher confidence in projected patterns of warming and other regional-scale features, including changes in wind patterns, precipitation and some aspects of extremes and of ice. {8.2, 8.3, 8.4, 8.5, 9.4, 9.5, 10.3, 11.1}

- Projected warming in the 21st century shows scenario-independent geographical patterns similar to those observed over the past several decades. Warming is expected to be greatest over land and at most high northern latitudes, and least over the Southern Ocean and parts of the North Atlantic Ocean (see Figure SPM.6). {10.3}

- Snow cover is projected to contract. Widespread increases in thaw depth are projected over most permafrost regions. {10.3, 10.6}

- Sea ice is projected to shrink in both the Arctic and Antarctic under all SRES scenarios. In some projections, arctic late-summer sea ice disappears almost entirely by the latter part of the 21st century. {10.3}

- It is *very likely* that hot extremes, heat waves and heavy precipitation events will continue to become more frequent. {10.3}

- Based on a range of models, it is *likely* that future tropical cyclones (typhoons and hurricanes) will become more intense, with larger peak wind speeds and more heavy precipitation associated with ongoing increases of tropical sea surface temperatures. There is less confidence in projections of a global decrease in numbers of tropical cyclones. The apparent increase in the proportion of very intense storms since 1970 in some regions is much larger than simulated by current models for that period. {9.5, 10.3, 3.8}

PROJECTIONS OF SURFACE TEMPERATURES

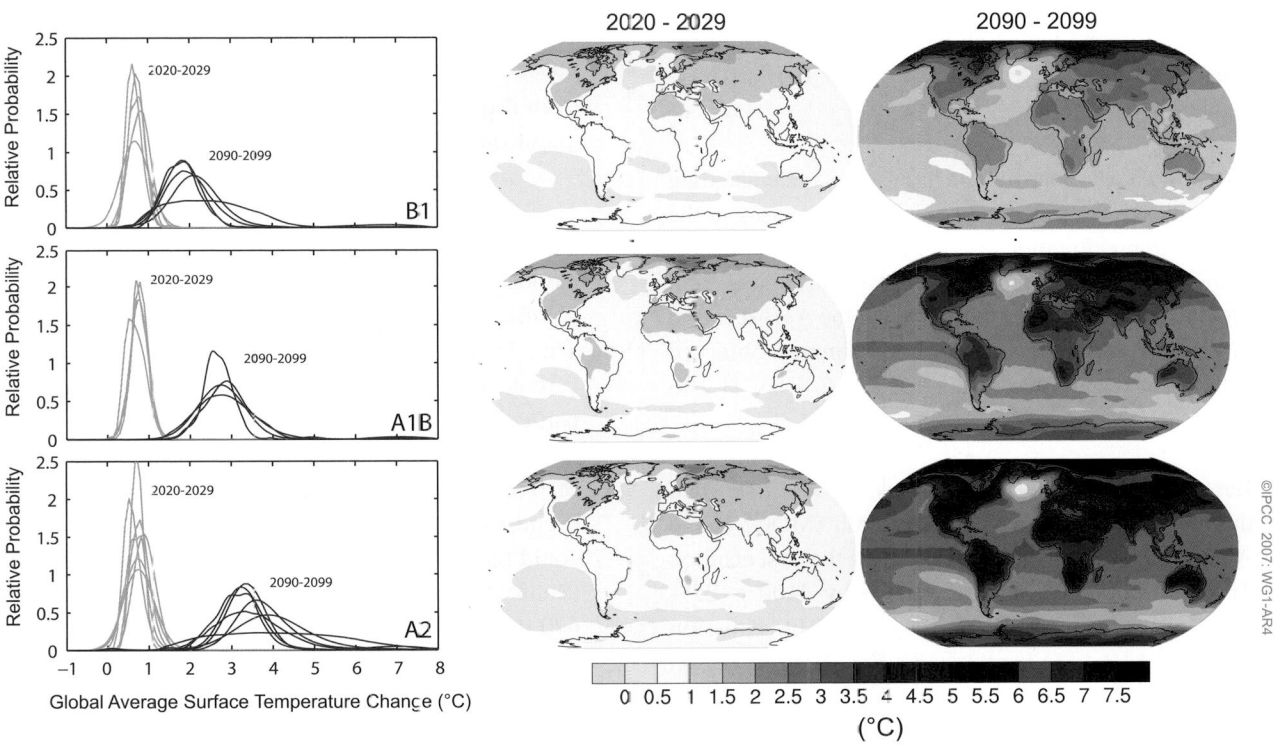

Figure SPM.6. *Projected surface temperature changes for the early and late 21st century relative to the period 1980–1999. The central and right panels show the AOGCM multi-model average projections for the B1 (top), A1B (middle) and A2 (bottom) SRES scenarios averaged over the decades 2020–2029 (centre) and 2090–2099 (right). The left panels show corresponding uncertainties as the relative probabilities of estimated global average warming from several different AOGCM and Earth System Model of Intermediate Complexity studies for the same periods. Some studies present results only for a subset of the SRES scenarios, or for various model versions. Therefore the difference in the number of curves shown in the left-hand panels is due only to differences in the availability of results. {Figures 10.8 and 10.28}*

PROJECTED PATTERNS OF PRECIPITATION CHANGES

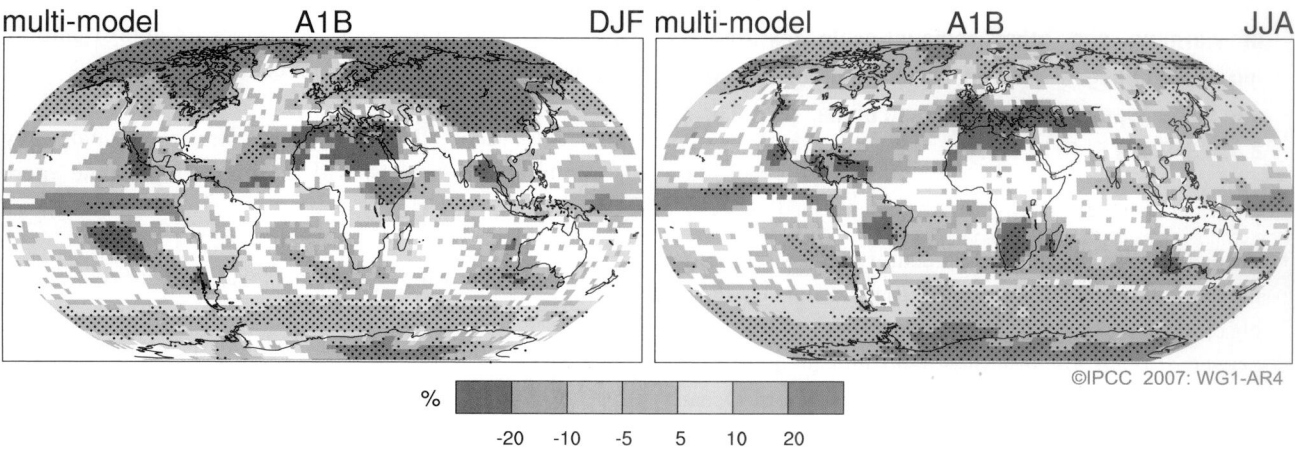

©IPCC 2007: WG1-AR4

Figure SPM.7. *Relative changes in precipitation (in percent) for the period 2090–2099, relative to 1980–1999. Values are multi-model averages based on the SRES A1B scenario for December to February (left) and June to August (right). White areas are where less than 66% of the models agree in the sign of the change and stippled areas are where more than 90% of the models agree in the sign of the change. {Figure 10.9}*

- Extratropical storm tracks are projected to move poleward, with consequent changes in wind, precipitation and temperature patterns, continuing the broad pattern of observed trends over the last half-century. {3.6, 10.3}

- Since the TAR, there is an improving understanding of projected patterns of precipitation. Increases in the amount of precipitation are *very likely* in high latitudes, while decreases are *likely* in most subtropical land regions (by as much as about 20% in the A1B scenario in 2100, see Figure SPM.7), continuing observed patterns in recent trends. {3.3, 8.3, 9.5, 10.3, 11.2 to 11.9}

- Based on current model simulations, it is *very likely* that the meridional overturning circulation (MOC) of the Atlantic Ocean will slow down during the 21st century. The multi-model average reduction by 2100 is 25% (range from zero to about 50%) for SRES emission scenario A1B. Temperatures in the Atlantic region are projected to increase despite such changes due to the much larger warming associated with projected increases in greenhouse gases. It is *very unlikely* that the MOC will undergo a large abrupt transition during the 21st century. Longer-term changes in the MOC cannot be assessed with confidence. {10.3, 10.7}

Anthropogenic warming and sea level rise would continue for centuries due to the time scales associated with climate processes and feedbacks, even if greenhouse gas concentrations were to be stabilised. {10.4, 10.5, 10.7}

- Climate-carbon cycle coupling is expected to add carbon dioxide to the atmosphere as the climate system warms, but the magnitude of this feedback is uncertain. This increases the uncertainty in the trajectory of carbon dioxide emissions required to achieve a particular stabilisation level of atmospheric carbon dioxide concentration. Based on current understanding of climate-carbon cycle feedback, model studies suggest that to stabilise at 450 ppm carbon dioxide could require that cumulative emissions over the 21st century be reduced from an average of approximately 670 [630 to 710] GtC (2460 [2310 to 2600] GtCO$_2$) to approximately 490 [375 to 600] GtC (1800 [1370 to 2200] GtCO$_2$). Similarly, to stabilise at 1000 ppm, this feedback could require that cumulative emissions be reduced from a model average of approximately 1415 [1340 to 1490] GtC (5190 [4910 to 5460] GtCO$_2$) to approximately 1100 [980 to 1250] GtC (4030 [3590 to 4580] GtCO$_2$). {7.3, 10.4}

- If radiative forcing were to be stabilised in 2100 at B1 or A1B levels[14] a further increase in global average temperature of about 0.5°C would still be expected, mostly by 2200. {10.7}

- If radiative forcing were to be stabilised in 2100 at A1B levels[14], thermal expansion alone would lead to 0.3 to 0.8 m of sea level rise by 2300 (relative to 1980–1999). Thermal expansion would continue for many centuries, due to the time required to transport heat into the deep ocean. {10.7}

- Contraction of the Greenland Ice Sheet is projected to continue to contribute to sea level rise after 2100. Current models suggest that ice mass losses increase with temperature more rapidly than gains due to precipitation and that the surface mass balance becomes negative at a global average warming (relative to pre-industrial values) in excess of 1.9°C to 4.6°C. If a negative surface mass balance were sustained for millennia, that would lead to virtually complete elimination of the Greenland Ice Sheet and a resulting contribution to sea level rise of about 7 m. The corresponding future temperatures in Greenland are comparable to those inferred for the last interglacial period 125,000 years ago, when palaeoclimatic information suggests reductions of polar land ice extent and 4 to 6 m of sea level rise. {6.4, 10.7}

- Dynamical processes related to ice flow not included in current models but suggested by recent observations could increase the vulnerability of the ice sheets to warming, increasing future sea level rise. Understanding of these processes is limited and there is no consensus on their magnitude. {4.6, 10.7}

- Current global model studies project that the Antarctic Ice Sheet will remain too cold for widespread surface melting and is expected to gain in mass due to increased snowfall. However, net loss of ice mass could occur if dynamical ice discharge dominates the ice sheet mass balance. {10.7}

- Both past and future anthropogenic carbon dioxide emissions will continue to contribute to warming and sea level rise for more than a millennium, due to the time scales required for removal of this gas from the atmosphere. {7.3, 10.3}

THE EMISSION SCENARIOS OF THE IPCC SPECIAL REPORT ON EMISSION SCENARIOS (SRES)[17]

A1. The A1 storyline and scenario family describes a future world of very rapid economic growth, global population that peaks in mid-century and declines thereafter, and the rapid introduction of new and more efficient technologies. Major underlying themes are convergence among regions, capacity building and increased cultural and social interactions, with a substantial reduction in regional differences in per capita income. The A1 scenario family develops into three groups that describe alternative directions of technological change in the energy system. The three A1 groups are distinguished by their technological emphasis: fossil-intensive (A1FI), non-fossil energy sources (A1T) or a balance across all sources (A1B) (where balanced is defined as not relying too heavily on one particular energy source, on the assumption that similar improvement rates apply to all energy supply and end use technologies).

A2. The A2 storyline and scenario family describes a very heterogeneous world. The underlying theme is self-reliance and preservation of local identities. Fertility patterns across regions converge very slowly, which results in continuously increasing population. Economic development is primarily regionally oriented and per capita economic growth and technological change more fragmented and slower than other storylines.

B1. The B1 storyline and scenario family describes a convergent world with the same global population, that peaks in mid-century and declines thereafter, as in the A1 storyline, but with rapid change in economic structures toward a service and information economy, with reductions in material intensity and the introduction of clean and resource-efficient technologies. The emphasis is on global solutions to economic, social and environmental sustainability, including improved equity, but without additional climate initiatives.

B2. The B2 storyline and scenario family describes a world in which the emphasis is on local solutions to economic, social and environmental sustainability. It is a world with continuously increasing global population, at a rate lower than A2, intermediate levels of economic development, and less rapid and more diverse technological change than in the B1 and A1 storylines. While the scenario is also oriented towards environmental protection and social equity, it focuses on local and regional levels.

An illustrative scenario was chosen for each of the six scenario groups A1B, A1FI, A1T, A2, B1 and B2. All should be considered equally sound.

The SRES scenarios do not include additional climate initiatives, which means that no scenarios are included that explicitly assume implementation of the United Nations Framework Convention on Climate Change or the emissions targets of the Kyoto Protocol.

---

[17] Emission scenarios are not assessed in this Working Group I Report of the IPCC. This box summarising the SRES scenarios is taken from the TAR and has been subject to prior line-by-line approval by the Panel.

A report accepted by Working Group I of the Intergovernmental Panel on Climate Change but not approved in detail

"Acceptance" of IPCC Reports at a Session of the Working Group or Panel signifies that the material has not been subject to line-by-line discussion and agreement, but nevertheless presents a comprehensive, objective and balanced view of the subject matter

Technical Summary

Coordinating Lead Authors:

Susan Solomon (USA), Dahe Qin (China), Martin Manning (USA, New Zealand)

Lead Authors:

Richard B. Alley (USA), Terje Berntsen (Norway), Nathaniel L. Bindoff (Australia), Zhenlin Chen (China), Amnat Chidthaisong (Thailand), Jonathan M. Gregory (UK), Gabriele C. Hegerl (USA, Germany), Martin Heimann (Germany, Switzerland), Bruce Hewitson (South Africa), Brian J. Hoskins (UK), Fortunat Joos (Switzerland), Jean Jouzel (France), Vladimir Kattsov (Russia), Ulrike Lohmann (Switzerland), Taroh Matsuno (Japan), Mario Molina (USA, Mexico), Neville Nicholls (Australia), Jonathan Overpeck (USA), Graciela Raga (Mexico, Argentina), Venkatachalam Ramaswamy (USA), Jiawen Ren (China), Matilde Rusticucci (Argentina), Richard Somerville (USA), Thomas F. Stocker (Switzerland), Ronald J. Stouffer (USA), Penny Whetton (Australia), Richard A. Wood (UK), David Wratt (New Zealand)

Contributing Authors:

J. Arblaster (USA, Australia), G. Brasseur (USA, Germany), J.H. Christensen (Denmark), K.L. Denman (Canada), D.W. Fahey (USA), P. Forster (UK), J. Haywood (UK), E. Jansen (Norway), P.D. Jones (UK), R. Knutti (Switzerland), H. Le Treut (France), P. Lemke (Germany), G. Meehl (USA), D. Randall (USA), D.A. Stone (UK, Canada), K.E. Trenberth (USA), J. Willebrand (Germany), F. Zwiers (Canada)

Review Editors:

Kansri Boonpragob (Thailand), Filippo Giorgi (Italy), Bubu Pateh Jallow (The Gambia)

This Technical Summary should be cited as:

Solomon, S., D. Qin, M. Manning, R.B. Alley, T. Berntsen, N.L. Bindoff, Z. Chen, A. Chidthaisong, J.M. Gregory, G.C. Hegerl, M. Heimann, B. Hewitson, B.J. Hoskins, F. Joos, J. Jouzel, V. Kattsov, U. Lohmann, T. Matsuno, M. Molina, N. Nicholls, J. Overpeck, G. Raga, V. Ramaswamy, J. Ren, M. Rusticucci, R. Somerville, T.F. Stocker, P. Whetton, R.A. Wood and D. Wratt, 2007: Technical Summary. In: *Climate Change 2007: The Physical Science Basis. Contribution of Working Group I to the Fourth Assessment Report of the Intergovernmental Panel on Climate Change* [Solomon, S., D. Qin, M. Manning, Z. Chen, M. Marquis, K.B. Averyt, M. Tignor and H.L. Miller (eds.)]. Cambridge University Press, Cambridge, United Kingdom and New York, NY, USA.

Table of Contents

TS.1 Introduction

In the six years since the IPCC's Third Assessment Report (TAR), significant progress has been made in understanding past and recent climate change and in projecting future changes. These advances have arisen from large amounts of new data, more sophisticated analyses of data, improvements in the understanding and simulation of physical processes in climate models and more extensive exploration of uncertainty ranges in model results. The increased confidence in climate science provided by these developments is evident in this Working Group I contribution to the IPCC's Fourth Assessment Report.

While this report provides new and important policy-relevant information on the scientific understanding of climate change, the complexity of the climate system and the multiple interactions that determine its behaviour impose limitations on our ability to understand fully the future course of Earth's global climate. There is still an incomplete physical understanding of many components of the climate system and their role in climate change. Key uncertainties include aspects of the roles played by clouds, the cryosphere, the oceans, land use and couplings between climate and biogeochemical cycles. The areas of science covered in this report continue to undergo rapid progress and it should be recognised that the present assessment reflects scientific understanding based on the peer-reviewed literature available in mid-2006.

The key findings of the IPCC Working Group I assessment are presented in the Summary for Policymakers. This Technical Summary provides a more detailed overview of the scientific basis for those findings and provides a road map to the chapters of the underlying report. It focuses on key findings, highlighting what is new since the TAR. The structure of the Technical Summary is as follows:

- Section 2: an overview of current scientific understanding of the natural and anthropogenic drivers of changes in climate;
- Section 3: an overview of observed changes in the climate system (including the atmosphere, oceans and cryosphere) and their relationships to physical processes;
- Section 4: an overview of explanations of observed climate changes based on climate models and physical

understanding, the extent to which climate change can be attributed to specific causes and a new evaluation of climate sensitivity to greenhouse gas increases;
- Section 5: an overview of projections for both near- and far-term climate changes including the time scales of responses to changes in forcing, and probabilistic information about future climate change; and
- Section 6: a summary of the most robust findings and the key uncertainties in current understanding of physical climate change science.

Each paragraph in the Technical Summary reporting substantive results is followed by a reference in curly brackets to the corresponding chapter section(s) of the underlying report where the detailed assessment of the scientific literature and additional information can be found.

TS.2 Changes in Human and Natural Drivers of Climate

The Earth's global mean climate is determined by incoming energy from the Sun and by the properties of the Earth and its atmosphere, namely the reflection, absorption and emission of energy within the atmosphere and at the surface. Although changes in received solar energy (e.g., caused by variations in the Earth's orbit around the Sun) inevitably affect the Earth's energy budget, the properties of the atmosphere and surface are also important and these may be affected by climate feedbacks. The importance of climate feedbacks is evident in the nature of past climate changes as recorded in ice cores up to 650,000 years old.

Changes have occurred in several aspects of the atmosphere and surface that alter the global energy budget of the Earth and can therefore cause the climate to change. Among these are increases in greenhouse gas concentrations that act primarily to increase the atmospheric absorption of outgoing radiation, and increases in aerosols (microscopic airborne particles or droplets) that act to reflect and absorb incoming solar radiation and change cloud radiative properties. Such changes cause a radiative forcing of the climate system.[1] Forcing agents can differ considerably from one another in terms of the magnitudes of forcing, as well as spatial and temporal features. Positive and negative radiative forcings contribute to increases and decreases, respectively, in

---

[1] 'Radiative forcing' is a measure of the influence a factor has in altering the balance of incoming and outgoing energy in the Earth-atmosphere system and is an index of the importance of the factor as a potential climate change mechanism. Positive forcing tends to warm the surface while negative forcing tends to cool it. In this report, radiative forcing values are for changes relative to a pre-industrial background at 1750, are expressed in Watts per square metre (W m^{-2}) and, unless otherwise noted, refer to a global and annual average value. See Glossary for further details.

Box TS.1: Treatment of Uncertainties in the Working Group I Assessment

The importance of consistent and transparent treatment of uncertainties is clearly recognised by the IPCC in preparing its assessments of climate change. The increasing attention given to formal treatments of uncertainty in previous assessments is addressed in Section 1.6. To promote consistency in the general treatment of uncertainty across all three Working Groups, authors of the Fourth Assessment Report have been asked to follow a brief set of guidance notes on determining and describing uncertainties in the context of an assessment.[2] This box summarises the way that Working Group I has applied those guidelines and covers some aspects of the treatment of uncertainty specific to material assessed here.

Uncertainties can be classified in several different ways according to their origin. Two primary types are 'value uncertainties' and 'structural uncertainties'. Value uncertainties arise from the incomplete determination of particular values or results, for example, when data are inaccurate or not fully representative of the phenomenon of interest. Structural uncertainties arise from an incomplete understanding of the processes that control particular values or results, for example, when the conceptual framework or model used for analysis does not include all the relevant processes or relationships. Value uncertainties are generally estimated using statistical techniques and expressed probabilistically. Structural uncertainties are generally described by giving the authors' collective judgment of their confidence in the correctness of a result. In both cases, estimating uncertainties is intrinsically about describing the limits to knowledge and for this reason involves expert judgment about the state of that knowledge. A different type of uncertainty arises in systems that are either chaotic or not fully deterministic in nature and this also limits our ability to project all aspects of climate change.

The scientific literature assessed here uses a variety of other generic ways of categorising uncertainties. Uncertainties associated with 'random errors' have the characteristic of decreasing as additional measurements are accumulated, whereas those associated with 'systematic errors' do not. In dealing with climate records, considerable attention has been given to the identification of systematic errors or unintended biases arising from data sampling issues and methods of analysing and combining data. Specialised statistical methods based on quantitative analysis have been developed for the detection and attribution of climate change and for producing probabilistic projections of future climate parameters. These are summarised in the relevant chapters.

The uncertainty guidance provided for the Fourth Assessment Report draws, for the first time, a careful distinction between levels of confidence in scientific understanding and the likelihoods of specific results. This allows authors to express high confidence that an event is extremely unlikely (e.g., rolling a dice twice and getting a six both times), as well as high confidence that an event is about as likely as not (e.g., a tossed coin coming up heads). Confidence and likelihood as used here are distinct concepts but are often linked in practice.

The standard terms used to define levels of confidence in this report are as given in the IPCC Uncertainty Guidance Note, namely:

Confidence Terminology	Degree of confidence in being correct
Very high confidence	At least 9 out of 10 chance
High confidence	About 8 out of 10 chance
Medium confidence	About 5 out of 10 chance
Low confidence	About 2 out of 10 chance
Very low confidence	Less than 1 out of 10 chance

Note that 'low confidence' and 'very low confidence' are only used for areas of major concern and where a risk-based perspective is justified.

Chapter 2 of this report uses a related term 'level of scientific understanding' when describing uncertainties in different contributions to radiative forcing. This terminology is used for consistency with the Third Assessment Report, and the basis on which the authors have determined particular levels of scientific understanding uses a combination of approaches consistent with the uncertainty guidance note as explained in detail in Section 2.9.2 and Table 2.11.

(continued)

---

[2] The IPCC Uncertainty Guidance Note is included in Supplementary Material for this report.

The standard terms used in this report to define the likelihood of an outcome or result where this can be estimated probabilistically are:

Likelihood Terminology	Likelihood of the occurrence/ outcome
Virtually certain	> 99% probability
Extremely likely	> 95% probability
Very likely	> 90% probability
Likely	> 66% probability
More likely than not	> 50% probability
About as likely as not	33 to 66% probability
Unlikely	< 33% probability
Very unlikely	< 10% probability
Extremely unlikely	< 5% probability
Exceptionally unlikely	< 1% probability

The terms 'extremely likely', 'extremely unlikely' and 'more likely than not' as defined above have been added to those given in the IPCC Uncertainty Guidance Note in order to provide a more specific assessment of aspects including attribution and radiative forcing.

Unless noted otherwise, values given in this report are assessed best estimates and their uncertainty ranges are 90% confidence intervals (i.e., there is an estimated 5% likelihood of the value being below the lower end of the range or above the upper end of the range). Note that in some cases the nature of the constraints on a value, or other information available, may indicate an asymmetric distribution of the uncertainty range around a best estimate. In such cases, the uncertainty range is given in square brackets following the best estimate.

global average surface temperature. This section updates the understanding of estimated anthropogenic and natural radiative forcings.

The overall response of global climate to radiative forcing is complex due to a number of positive and negative feedbacks that can have a strong influence on the climate system (see e.g., Sections 4.5 and 5.4). Although water vapour is a strong greenhouse gas, its concentration in the atmosphere changes in response to changes in surface climate and this must be treated as a feedback effect and not as a radiative forcing. This section also summarises changes in the surface energy budget and its links to the hydrological cycle. Insights into the effects of agents such as aerosols on precipitation are also noted.

TS.2.1 Greenhouse Gases

The dominant factor in the radiative forcing of climate in the industrial era is the increasing concentration of various greenhouse gases in the atmosphere. Several of the major greenhouse gases occur naturally but increases in their atmospheric concentrations over the last 250 years are due largely to human activities. Other greenhouse gases are entirely the result of human activities. The contribution of each greenhouse gas to radiative forcing

over a particular period of time is determined by the change in its concentration in the atmosphere over that period and the effectiveness of the gas in perturbing the radiative balance. Current atmospheric concentrations of the different greenhouse gases considered in this report vary by more than eight orders of magnitude (factor of 10^8), and their radiative efficiencies vary by more than four orders of magnitude (factor of 10^4), reflecting the enormous diversity in their properties and origins.

The current concentration of a greenhouse gas in the atmosphere is the net result of the history of its past emissions and removals from the atmosphere. The gases and aerosols considered here are emitted to the atmosphere by human activities or are formed from precursor species emitted to the atmosphere. These emissions are offset by chemical and physical removal processes. With the important exception of carbon dioxide (CO_2), it is generally the case that these processes remove a specific fraction of the amount of a gas in the atmosphere each year and the inverse of this removal rate gives the mean lifetime for that gas. In some cases, the removal rate may vary with gas concentration or other atmospheric properties (e.g., temperature or background chemical conditions).

Long-lived greenhouse gases (LLGHGs), for example, CO_2, methane (CH_4) and nitrous oxide (N_2O), are chemically stable and persist in the atmosphere over time scales of a decade to centuries or longer, so that their emission has a long-term influence on climate. Because these gases are long lived, they become well mixed throughout the atmosphere much faster than they are removed and their global concentrations can be accurately estimated from data at a few locations. Carbon dioxide does not have a specific lifetime because it is continuously cycled between the atmosphere, oceans and land biosphere and its net removal from the atmosphere involves a range of processes with different time scales.

Short-lived gases (e.g., sulphur dioxide and carbon monoxide) are chemically reactive and generally removed by natural oxidation processes in the atmosphere, by removal at the surface or by washout in precipitation; their concentrations are hence highly variable. Ozone is a significant greenhouse gas that is formed and destroyed by chemical reactions involving other species in the atmosphere. In the troposphere, the human influence on ozone occurs primarily through changes in precursor gases that lead to its formation, whereas in the stratosphere, the human influence has been primarily through changes in ozone removal rates caused by chlorofluorocarbons (CFCs) and other ozone-depleting substances.

TS.2.1.1   Changes in Atmospheric Carbon Dioxide, Methane and Nitrous Oxide

Current concentrations of atmospheric CO_2 and CH_4 far exceed pre-industrial values found in polar ice core records of atmospheric composition dating back 650,000 years. Multiple lines of evidence confirm that the post-industrial rise in these gases does not stem from natural mechanisms (see Figure TS.1 and Figure TS.2). {2.3, 6.3–6.5, FAQ 7.1}

The total radiative forcing of the Earth's climate due to increases in the concentrations of the LLGHGs CO_2, CH_4 and N_2O, and *very likely* the rate of increase in the total forcing due to these gases over the period since 1750, are unprecedented in more than 10,000 years (Figure TS.2). It is *very likely* that the sustained rate of increase in the combined radiative forcing from these greenhouse gases of about +1 W m^{-2} over the past four decades is at least six times faster than at any time during the two millennia before the Industrial Era, the period for which ice core data have the required temporal resolution. The radiative forcing due to these LLGHGs has the highest level of confidence of any forcing agent. {2.3, 6.4}

GLACIAL-INTERGLACIAL ICE CORE DATA

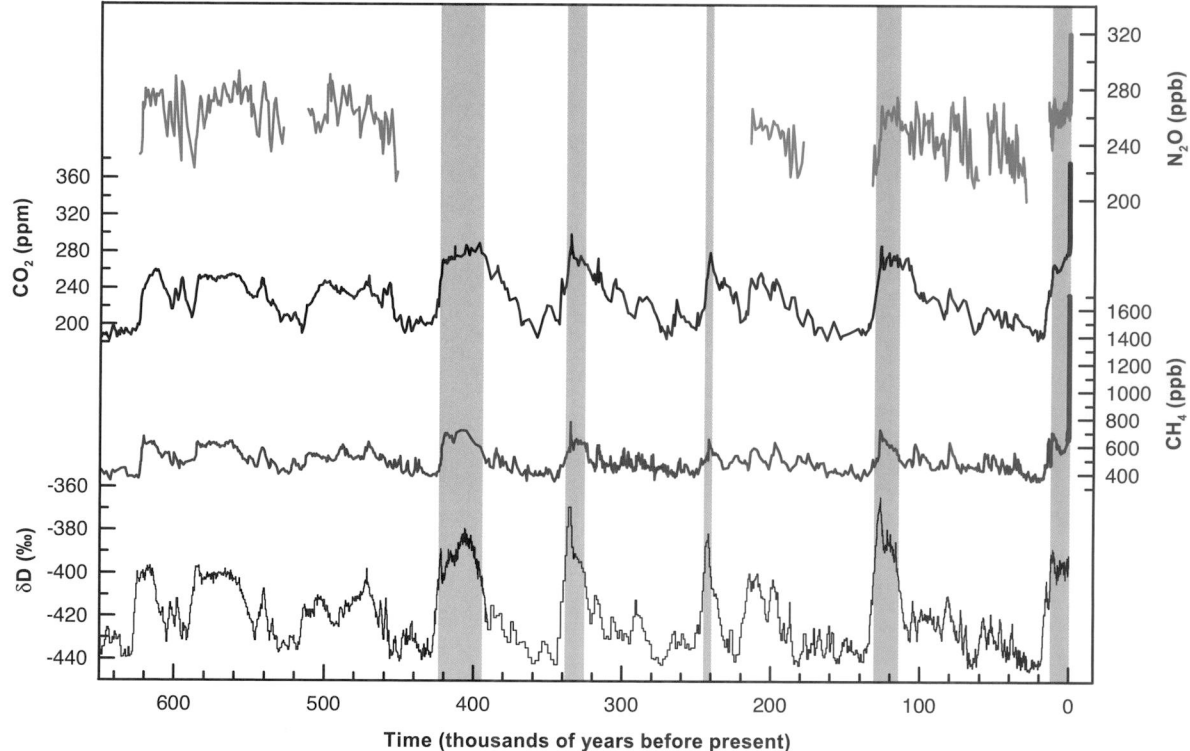

Figure TS.1. *Variations of deuterium (δD) in antarctic ice, which is a proxy for local temperature, and the atmospheric concentrations of the greenhouse gases carbon dioxide (CO_2), methane (CH_4), and nitrous oxide (N_2O) in air trapped within the ice cores and from recent atmospheric measurements. Data cover 650,000 years and the shaded bands indicate current and previous interglacial warm periods. {Adapted from Figure 6.3}*

CHANGES IN GREENHOUSE GASES FROM ICE CORE AND MODERN DATA

Figure TS.2. *The concentrations and radiative forcing by (a) carbon dioxide (CO₂), (b) methane (CH₄), (c) nitrous oxide (N₂O) and (d) the rate of change in their combined radiative forcing over the last 20 000 years reconstructed from antarctic and Greenland ice and firn data (symbols) and direct atmospheric measurements (panels a,b,c, red lines). The grey bars show the reconstructed ranges of natural variability for the past 650,000 years. The rate of change in radiative forcing (panel d, black line) has been computed from spline fits to the concentration data. The width of the age spread in the ice data varies from about 20 years for sites with a high accumulation of snow such as Law Dome, Antarctica, to about 200 years for low-accumulation sites such as Dome C, Antarctica. The arrow shows the peak in the rate of change in radiative forcing that would result if the anthropogenic signals of CO₂, CH₄, and N₂O had been smoothed corresponding to conditions at the low-accumulation Dome C site. The negative rate of change in forcing around 1600 shown in the higher-resolution inset in panel d results from a CO₂ decrease of about 10 ppm in the Law Dome record. {Figure 6.4}*

The concentration of atmospheric CO_2 has increased from a pre-industrial value of about 280 ppm to 379 ppm in 2005. Atmospheric CO_2 concentration increased by only 20 ppm over the 8000 years prior to industrialisation; multi-decadal to centennial-scale variations were less than 10 ppm and *likely* due mostly to natural processes. However, since 1750, the CO_2 concentration has risen by nearly 100 ppm. The annual CO_2 growth rate was larger during the last 10 years (1995–2005 average: 1.9 ppm yr⁻¹) than it has been since continuous direct atmospheric measurements began (1960–2005 average: 1.4 ppm yr⁻¹). {2.3, 6.4, 6.5}

Increases in atmospheric CO_2 since pre-industrial times are responsible for a radiative forcing of +1.66 ± 0.17 W m⁻²; a contribution which dominates all other radiative forcing agents considered in this report. For the decade from 1995 to 2005, the growth rate of CO_2

in the atmosphere led to a 20% increase in its radiative forcing. {2.3, 6.4, 6.5}

Emissions of CO_2 from fossil fuel use and from the effects of land use change on plant and soil carbon are the primary sources of increased atmospheric CO_2. Since 1750, it is estimated that about 2/3rds of anthropogenic CO_2 emissions have come from fossil fuel burning and about 1/3rd from land use change. About 45% of this CO_2 has remained in the atmosphere, while about 30% has been taken up by the oceans and the remainder has been taken up by the terrestrial biosphere. About half of a CO_2 pulse to the atmosphere is removed over a time scale of 30 years; a further 30% is removed within a few centuries; and the remaining 20% will typically stay in the atmosphere for many thousands of years. {7.3}

In recent decades, emissions of CO_2 have continued to increase (see Figure TS.3). Global annual fossil

CO_2 emissions[3] increased from an average of 6.4 ± 0.4 GtC yr^{-1} in the 1990s to 7.2 ± 0.3 GtC yr^{-1} in the period 2000 to 2005. Estimated CO_2 emissions associated with land use change, averaged over the 1990s, were[4] 0.5 to 2.7 GtC yr^{-1}, with a central estimate of 1.6 Gt yr^{-1}. Table TS.1 shows the estimated budgets of CO_2 in recent decades. {2.3, 6.4, 7.3, FAQ 7.1}

Since the 1980s, natural processes of CO_2 uptake by the terrestrial biosphere (i.e., the residual land sink in Table TS.1) and by the oceans have removed about 50% of anthropogenic emissions (i.e., fossil CO_2 emissions and land use change flux in Table TS.1). These removal processes are influenced by the atmospheric CO_2 concentration and by changes in climate. Uptake by the oceans and the terrestrial biosphere have been similar in magnitude but the terrestrial biosphere uptake is more variable and was higher in the 1990s than in the 1980s by about 1 GtC yr^{-1}. Observations demonstrate that dissolved CO_2 concentrations in the surface ocean (pCO$_2$) have been increasing nearly everywhere, roughly following the atmospheric CO_2 increase but with large regional and temporal variability. {5.4, 7.3}

Carbon uptake and storage in the terrestrial biosphere arise from the net difference between uptake due to vegetation growth, changes in reforestation and sequestration, and emissions due to heterotrophic respiration, harvest, deforestation, fire, damage by pollution and other disturbance factors affecting biomass and soils. Increases and decreases in fire frequency in different regions have affected net carbon uptake, and in boreal regions, emissions due to fires appear to have increased over recent decades. Estimates of net CO_2 surface fluxes from inverse studies using networks of atmospheric data demonstrate significant land uptake in the mid-latitudes of the Northern Hemisphere (NH) and near-zero land-atmosphere fluxes in the tropics, implying that tropical deforestation is approximately balanced by regrowth. {7.3}

Short-term (interannual) variations observed in the atmospheric CO_2 growth rate are primarily controlled by changes in the flux of CO_2 between the atmosphere and the terrestrial biosphere, with a smaller but significant fraction due to variability in ocean fluxes (see Figure TS.3). Variability in the terrestrial biosphere flux is driven by climatic fluctuations, which affect the uptake of CO_2 by plant growth and the return of CO_2 to the atmosphere by the decay of organic material through heterotrophic respiration and fires. El Niño-Southern Oscillation (ENSO) events are a major source of interannual variability in atmospheric CO_2 growth rate, due to their effects on fluxes through land and sea surface temperatures, precipitation and the incidence of fires. {7.3}

The direct effects of increasing atmospheric CO_2 on large-scale terrestrial carbon uptake cannot be quantified reliably at present. Plant growth can be stimulated by increased atmospheric CO_2 concentrations and by nutrient deposition (fertilization effects). However, most experiments and studies show that such responses appear to be relatively short lived and strongly coupled

Table TS.1. *Global carbon budget. By convention, positive values are CO_2 fluxes (GtC yr^{-1}) into the atmosphere and negative values represent uptake from the atmosphere (i.e., 'CO$_2$ sinks'). Fossil CO_2 emissions for 2004 and 2005 are based on interim estimates. Due to the limited number of available studies, for the net land-to-atmosphere flux and its components, uncertainty ranges are given as 65% confidence intervals and do not include interannual variability (see Section 7.3). NA indicates that data are not available.*

	1980s	1990s	2000–2005
Atmospheric increase	3.3 ± 0.1	3.2 ± 0.1	4.1 ± 0.1
Fossil carbon dioxide emissions	5.4 ± 0.3	6.4 ± 0.4	7.2 ± 0.3
Net ocean-to-atmosphere flux	-1.8 ± 0.8	-2.2 ± 0.4	-2.2 ± 0.5
Net land-to-atmosphere flux	-0.3 ± 0.9	-1.0 ± 0.6	-0.9 ± 0.6
Partitioned as follows			
Land use change flux	1.4 (0.4 to 2.3)	1.6 (0.5 to 2.7)	NA
Residual land sink	−1.7 (−3.4 to 0.2)	−2.6 (−4.3 to −0.9)	NA

---

3 Fossil CO_2 emissions include those from the production, distribution and consumption of fossil fuels and from cement production. Emission of 1 GtC corresponds to 3.67 GtCO$_2$.

4 As explained in Section 7.3, uncertainty ranges for land use change emissions, and hence for the full carbon cycle budget, can only be given as 65% confidence intervals.

CO₂ EMISSIONS AND INCREASES

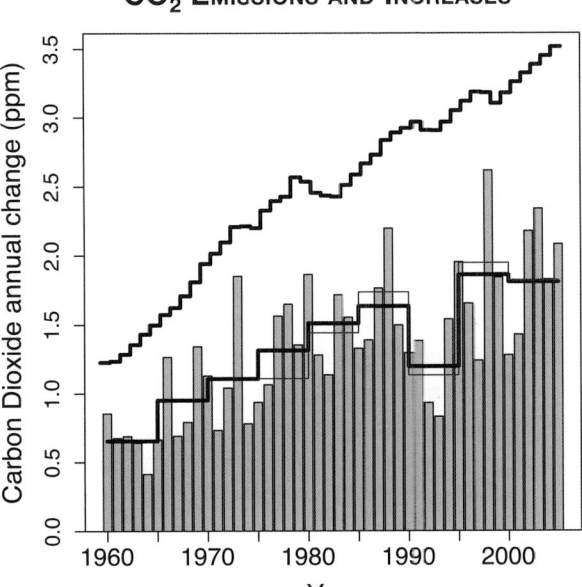

Figure TS.3. *Annual changes in global mean CO₂ concentration (grey bars) and their five-year means from two different measurement networks (red and lower black stepped lines). The five-year means smooth out short-term perturbations associated with strong ENSO events in 1972, 1982, 1987 and 1997. Uncertainties in the five-year means are indicated by the difference between the red and lower black lines and are of order 0.15 ppm. The upper stepped line shows the annual increases that would occur if all fossil fuel emissions stayed in the atmosphere and there were no other emissions. {Figure 7.4}*

to other effects such as availability of water and nutrients. Likewise, experiments and studies of the effects of climate (temperature and moisture) on heterotrophic respiration of litter and soils are equivocal. Note that the effect of climate change on carbon uptake is addressed separately in section TS.5.4. {7.3}

The CH₄ abundance in 2005 of about 1774 ppb is more than double its pre-industrial value. Atmospheric CH₄ concentrations varied slowly between 580 and 730 ppb over the last 10,000 years, but increased by about 1000 ppb in the last two centuries, representing the fastest changes in this gas over at least the last 80,000 years. In the late 1970s and early 1980s, CH₄ growth rates displayed maxima above 1% yr⁻¹, but since the early 1990s have decreased significantly and were close to zero for the six-year period from 1999 to 2005. Increases in CH₄ abundance occur when emissions exceed removals. The recent decline in growth rates implies that emissions now approximately match removals, which are due primarily to oxidation by the hydroxyl radical (OH). Since the TAR, new studies using two independent tracers (methyl chloroform and ¹⁴CO) suggest no significant long-term change in the global abundance of OH. Thus,

the slowdown in the atmospheric CH₄ growth rate since about 1993 is *likely* due to the atmosphere approaching an equilibrium during a period of near-constant total emissions. {2.3, 7.4, FAQ 7.1}

Increases in atmospheric CH₄ concentrations since pre-industrial times have contributed a radiative forcing of +0.48 ± 0.05 W m⁻². Among greenhouse gases, this forcing remains second only to that of CO₂ in magnitude. {2.3}

Current atmospheric CH₄ levels are due to continuing anthropogenic emissions of CH₄, which are greater than natural emissions. Total CH₄ emissions can be well determined from observed concentrations and independent estimates of removal rates. Emissions from individual sources of CH₄ are not as well quantified as the total emissions but are mostly biogenic and include emissions from wetlands, ruminant animals, rice agriculture and biomass burning, with smaller contributions from industrial sources including fossil fuel-related emissions. This knowledge of CH₄ sources, combined with the small natural range of CH₄ concentrations over the past 650,000 years (Figure TS.1) and their dramatic increase since 1750 (Figure TS.2), make it *very likely* that the observed long-term changes in CH₄ are due to anthropogenic activity. {2.3, 6.4, 7.4}

In addition to its slowdown over the last 15 years, the growth rate of atmospheric CH₄ has shown high interannual variability, which is not yet fully explained. The largest contributions to interannual variability during the 1996 to 2001 period appear to be variations in emissions from wetlands and biomass burning. Several studies indicate that wetland CH₄ emissions are highly sensitive to temperature and are also affected by hydrological changes. Available model estimates all indicate increases in wetland emissions due to future climate change but vary widely in the magnitude of such a positive feedback effect. {7.4}

The N₂O concentration in 2005 was 319 ppb, about 18% higher than its pre-industrial value. Nitrous oxide increased approximately linearly by about 0.8 ppb yr⁻¹ over the past few decades. Ice core data show that the atmospheric concentration of N₂O varied by less than about 10 ppb for 11,500 years before the onset of the industrial period. {2.3, 6.4, 6.5}

The increase in N₂O since the pre-industrial era now contributes a radiative forcing of +0.16 ± 0.02 W m⁻² and is due primarily to human activities, particularly agriculture and associated land use change. Current estimates are that about 40% of total N₂O emissions are anthropogenic but individual source estimates remain subject to significant uncertainties. {2.3, 7.4}

TS.2.1.3 Changes in Atmospheric Halocarbons, Stratospheric Ozone, Tropospheric Ozone and Other Gases

CFCs and hydrochlorofluorocarbons (HCFCs) are greenhouse gases that are purely anthropogenic in origin and used in a wide variety of applications. Emissions of these gases have decreased due to their phase-out under the Montreal Protocol, and the atmospheric concentrations of CFC-11 and CFC-113 are now decreasing due to natural removal processes. Observations in polar firn cores since the TAR have now extended the available time series information for some of these greenhouse gases. Ice core and *in situ* data confirm that industrial sources are the cause of observed atmospheric increases in CFCs and HCFCs. {2.3}

The Montreal Protocol gases contributed +0.32 ± 0.03 W m^{-2} to direct radiative forcing in 2005, with CFC-12 continuing to be the third most important long-lived radiative forcing agent. These gases as a group contribute about 12% of the total forcing due to LLGHGs. {2.3}

The concentrations of industrial fluorinated gases covered by the Kyoto Protocol (hydrofluorocarbons (HFCs), perfluorocarbons (PFCs), sulphur hexa-fluoride (SF$_6$)) are relatively small but are increasing rapidly. Their total radiative forcing in 2005 was +0.017 W m^{-2}. {2.3}

Tropospheric ozone is a short-lived greenhouse gas produced by chemical reactions of precursor species in the atmosphere and with large spatial and temporal variability. Improved measurements and modelling have advanced the understanding of chemical precursors that lead to the formation of tropospheric ozone, mainly carbon monoxide, nitrogen oxides (including sources and possible long-term trends in lightning) and formaldehyde. Overall, current models are successful in describing the principal features of the present global tropospheric ozone distribution on the basis of underlying processes. New satellite and *in situ* measurements provide important global constraints for these models; however, there is less confidence in their ability to reproduce the changes in ozone associated with large changes in emissions or climate, and in the simulation of observed long-term trends in ozone concentrations over the 20th century. {7.4}

Tropospheric ozone radiative forcing is estimated to be +0.35 [+0.25 to +0.65] W m^{-2} with a *medium* level of scientific understanding. The best estimate of this radiative forcing has not changed since the TAR. Observations show that trends in tropospheric ozone during the last few decades vary in sign and magnitude at many locations, but there are indications of significant upward trends at low latitudes. Model studies of the radiative forcing due to the increase in tropospheric ozone since pre-industrial times have increased in complexity and comprehensiveness compared with models used in the TAR. {2.3, 7.4}

Changes in tropospheric ozone are linked to air quality and climate change. A number of studies have shown that summer daytime ozone concentrations correlate strongly with temperature. This correlation appears to reflect contributions from temperature-dependent biogenic volatile organic carbon emissions, thermal decomposition of peroxyacetylnitrate, which acts as a reservoir for nitrogen oxides (NO$_x$), and association of high temperatures with regional stagnation. Anomalously hot and stagnant conditions during the summer of 1988 were responsible for the highest surface-level ozone year on record in the north-eastern USA. The summer heat wave in Europe in 2003 was also associated with exceptionally high local ozone at the surface. {Box 7.4}

The radiative forcing due to the destruction of stratospheric ozone is caused by the Montreal Protocol gases and is re-evaluated to be –0.05 ± 0.10 W m^{-2}, weaker than in the TAR, with a medium level of scientific understanding. The trend of greater and greater depletion of global stratospheric ozone observed during the 1980s and 1990s is no longer occurring; however, global stratospheric ozone is still about 4% below pre-1980 values and it is not yet clear whether ozone recovery has begun. In addition to the chemical destruction of ozone, dynamical changes may have contributed to NH mid-latitude ozone reduction. {2.3}

Direct emission of water vapour by human activities makes a negligible contribution to radiative forcing. However, as global mean temperatures increase, tropospheric water vapour concentrations increase and this represents a key feedback but not a forcing of climate change. Direct emission of water to the atmosphere by anthropogenic activities, mainly irrigation, is a possible forcing factor but corresponds to less than 1% of the natural sources of atmospheric water vapour. The direct injection of water vapour into the atmosphere from fossil fuel combustion is significantly lower than that from agricultural activity. {2.5}

Based on chemical transport model studies, the radiative forcing from increases in stratospheric water vapour due to oxidation of CH$_4$ is estimated to be +0.07 ± 0.05 W m^{-2}. The level of scientific understanding is low because the contribution of CH$_4$ to the corresponding vertical structure of the water vapour change near the tropopause is uncertain. Other potential human causes of stratospheric water vapour increases that could contribute to radiative forcing are poorly understood. {2.3}

TS.2.2 Aerosols

Direct aerosol radiative forcing is now considerably better quantified than previously and represents a major advance in understanding since the time of the TAR, when several components had a very low level of scientific understanding. A total direct aerosol radiative forcing combined across all aerosol types can now be given for the first time as -0.5 ± 0.4 W m^{-2}, with a medium-low level of scientific understanding. Atmospheric models have improved and many now represent all aerosol components of significance. Aerosols vary considerably in their properties that affect the extent to which they absorb and scatter radiation, and thus different types may have a net cooling or warming effect. Industrial aerosol consisting mainly of a mixture of sulphates, organic and black carbon, nitrates and industrial dust is clearly discernible over many continental regions of the NH. Improved *in situ*, satellite and surface-based measurements (see Figure TS.4) have enabled verification of global aerosol model simulations. These improvements allow quantification of the total direct aerosol radiative forcing for the first time, representing an important advance since the TAR. The direct radiative forcing for individual species remains less certain and is estimated from models to be -0.4 ± 0.2 W m^{-2} for sulphate, -0.05 ± 0.05 W m^{-2} for fossil fuel organic carbon, $+0.2 \pm 0.15$ W m^{-2} for fossil fuel black carbon, $+0.03 \pm 0.12$ W m^{-2} for biomass burning, -0.1 ± 0.1 W m^{-2} for nitrate and -0.1 ± 0.2 W m^{-2} for mineral dust. Two recent emission inventory studies support data from ice cores and suggest that global anthropogenic sulphate emissions decreased over the 1980 to 2000 period and that the geographic distribution of sulphate forcing has also changed. {2.4, 6.6}

Significant changes in the estimates of the direct radiative forcing due to biomass-burning, nitrate and mineral dust aerosols have occurred since the TAR. For biomass-burning aerosol, the estimated direct radiative forcing is now revised from being negative to near zero due to the estimate being strongly influenced by the occurrence of these aerosols over clouds. For the first time, the radiative forcing due to nitrate aerosol is given. For mineral dust, the range in the direct radiative forcing is reduced due to a reduction in the estimate of its anthropogenic fraction. {2.4}

TOTAL AEROSOL OPTICAL DEPTH

January to March 2001

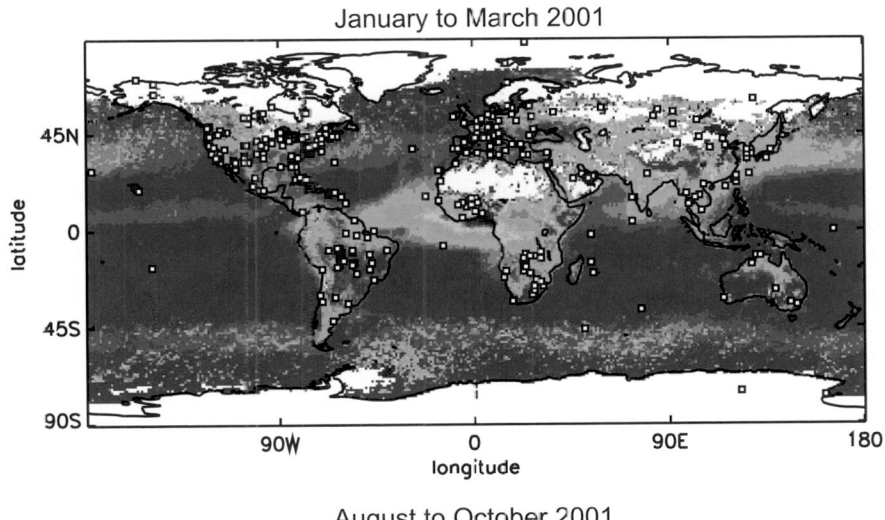

August to October 2001

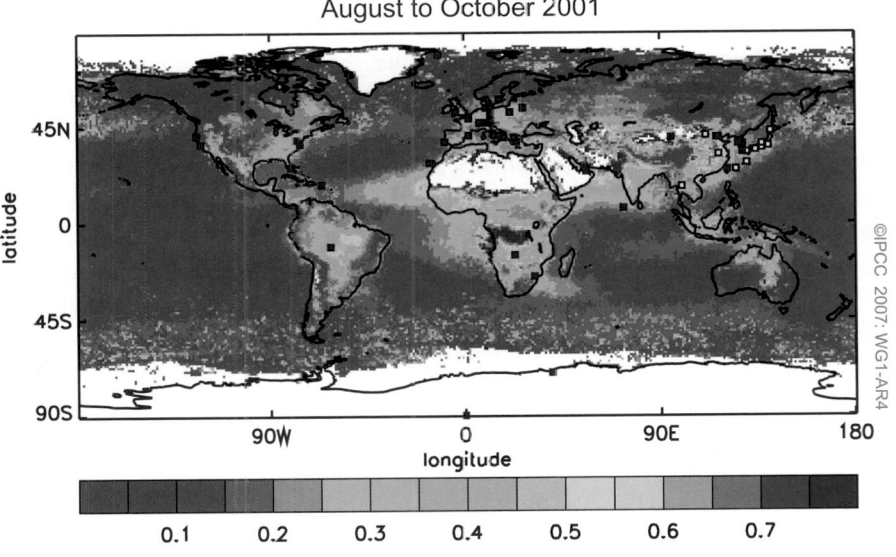

©IPCC 2007: WG1-AR4

Figure TS.4. *(Top) The total aerosol optical depth (due to natural plus anthropogenic aerosols) at a mid-visible wavelength determined by satellite measurements for January to March 2001 and (bottom) August to October 2001, illustrating seasonal changes in industrial and biomass-burning aerosols. Data are from satellite measurements, complemented by two different kinds of ground-based measurements at locations shown in the two panels (see Section 2.4.2 for details). {Figure 2.11}*

Anthropogenic aerosols effects on water clouds cause an indirect cloud albedo effect (referred to as the first indirect effect in the TAR), which has a best estimate for the first time of –0.7 [–0.3 to –1.8] W m^{-2}. The number of global model estimates of the albedo effect for liquid water clouds has increased substantially since the TAR, and the estimates have been evaluated in a more rigorous way. The estimate for this radiative forcing comes from multiple model studies incorporating more aerosol species and describing aerosol-cloud interaction processes in greater detail. Model studies including more aerosol species or constrained by satellite observations tend to yield a relatively weaker cloud albedo effect. Despite the advances and progress since the TAR and the reduction in the spread of the estimate of the forcing, there remain large uncertainties in both measurements and modelling of processes, leading to a low level of scientific understanding, which is an elevation from the very low rank in the TAR. {2.4, 7.5, 9.2}

Other effects of aerosol include a cloud lifetime effect, a semi-direct effect and aerosol-ice cloud interactions. These are considered to be part of the climate response rather than radiative forcings. {2.4, 7.5}

TS.2.3 Aviation Contrails and Cirrus, Land Use and Other Effects

Persistent linear contrails from global aviation contribute a small radiative forcing of +0.01 [+0.003 to +0.03] W m^{-2}, with a low level of scientific understanding. This best estimate is smaller than the estimate in the TAR. This difference results from new observations of contrail cover and reduced estimates of contrail optical depth. No best estimates are available for the net forcing from spreading contrails. Their effects on cirrus cloudiness and the global effect of aviation aerosol on background cloudiness remain unknown. {2.6}

Human-induced changes in land cover have increased the global surface albedo, leading to a radiative forcing of –0.2 ± 0.2 W m^{-2}, the same as in the TAR, with a medium-low level of scientific understanding. Black carbon aerosols deposited on snow reduce the surface albedo and are estimated to yield an associated radiative forcing of +0.1 ± 0.1 W m^{-2}, with a low level of scientific understanding. Since the TAR, a number of estimates of the forcing from land use changes have been made, using better techniques, exclusion of feedbacks in the evaluation and improved incorporation of large-scale observations. Uncertainties in the estimate include mapping and characterisation of present-day vegetation and historical state, parametrization of surface radiation processes and biases in models'

climate variables. The presence of soot particles in snow leads to a decrease in the albedo of snow and a positive forcing, and could affect snowmelt. Uncertainties are large regarding the manner in which soot is incorporated in snow and the resulting optical properties. {2.5}

The impacts of land use change on climate are expected to be locally significant in some regions, but are small at the global scale in comparison with greenhouse gas warming. Changes in the land surface (vegetation, soils, water) resulting from human activities can significantly affect local climate through shifts in radiation, cloudiness, surface roughness and surface temperatures. Changes in vegetation cover can also have a substantial effect on surface energy and water balance at the regional scale. These effects involve non-radiative processes (implying that they cannot be quantified by a radiative forcing) and have a very low level of scientific understanding. {2.5, 7.2, 9.3, Box 11.4}

The release of heat from anthropogenic energy production can be significant over urban areas but is not significant globally. {2.5}

TS.2.4 Radiative Forcing Due to Solar Activity and Volcanic Eruptions

Continuous monitoring of total solar irradiance now covers the last 28 years. The data show a well-established 11-year cycle in irradiance that varies by 0.08% from solar cycle minima to maxima, with no significant long-term trend. New data have more accurately quantified changes in solar spectral fluxes over a broad range of wavelengths in association with changing solar activity. Improved calibrations using high-quality overlapping measurements have also contributed to a better understanding. Current understanding of solar physics and the known sources of irradiance variability suggest comparable irradiance levels during the past two solar cycles, including at solar minima. The primary known cause of contemporary irradiance variability is the presence on the Sun's disk of sunspots (compact, dark features where radiation is locally depleted) and faculae (extended bright features where radiation is locally enhanced). {2.7}

The estimated direct radiative forcing due to changes in the solar output since 1750 is +0.12 [+0.06 to +0.3] W m^{-2}, which is less than half of the estimate given in the TAR, with a low level of scientific understanding. The reduced radiative forcing estimate comes from a re-evaluation of the long-term change in solar irradiance since 1610 (the Maunder Minimum) based upon: a new reconstruction using a model of solar magnetic flux variations that does not invoke geomagnetic, cosmogenic or stellar proxies; improved understanding of recent solar variations and their relationship to physical processes; and re-evaluation of the variations of Sun-like stars. While this leads to an elevation in the level of scientific understanding from very low in the TAR to low in this assessment, uncertainties remain large because of the lack of direct observations and incomplete understanding of solar variability mechanisms over long time scales. {2.7, 6.6}

Empirical associations have been reported between solar-modulated cosmic ray ionization of the atmosphere and global average low-level cloud cover but evidence for a systematic indirect solar effect remains ambiguous. It has been suggested that galactic cosmic rays with sufficient energy to reach the troposphere could alter the population of cloud condensation nuclei and hence microphysical cloud properties (droplet number and concentration), inducing changes in cloud processes analogous to the indirect cloud albedo effect of tropospheric aerosols and thus causing an indirect solar forcing of climate. Studies have probed various correlations with clouds in particular regions or using limited cloud types or limited time periods; however, the cosmic ray time series does not appear to correspond to global total cloud cover after 1991 or to global low-level cloud cover after 1994. Together with the lack of a proven physical mechanism and the plausibility of other causal factors affecting changes in cloud cover, this makes the association between galactic cosmic ray-induced changes in aerosol and cloud formation controversial. {2.7}

Explosive volcanic eruptions greatly increase the concentration of stratospheric sulphate aerosols. A single eruption can thereby cool global mean climate for a few years. Volcanic aerosols perturb both the stratosphere and surface/troposphere radiative energy budgets and climate in an episodic manner, and many past events are evident in ice core observations of sulphate as well as temperature records. There have been no explosive volcanic events since the 1991 Mt. Pinatubo eruption capable of injecting significant material to the stratosphere. However, the potential exists for volcanic eruptions much larger than the 1991 Mt. Pinatubo eruption, which could produce larger radiative forcing and longer-term cooling of the climate system. {2.7, 6.4, 6.6, 9.2}

TS.2.5 Net Global Radiative Forcing, Global Warming Potentials and Patterns of Forcing

The understanding of anthropogenic warming and cooling influences on climate has improved since the TAR, leading to *very high confidence* that the effect of human activities since 1750 has been a net positive forcing of +1.6 [+0.6 to +2.4] W m^{-2}. Improved understanding and better quantification of the forcing mechanisms since the TAR make it possible to derive a combined net anthropogenic radiative forcing for the first time. Combining the component values for each forcing agent and their uncertainties yields the probability distribution of the combined anthropogenic radiative forcing estimate shown in Figure TS.5; the most likely value is about an order of magnitude larger than the estimated radiative forcing from changes in solar irradiance. Since the range in the estimate is +0.6 to +2.4 W m^{-2}, there is very high confidence in the net positive radiative forcing of the climate system due to human activity. The LLGHGs together contribute +2.63 ± 0.26 W m^{-2}, which is the dominant radiative forcing term and has the highest level of scientific understanding. In contrast, the total direct aerosol, cloud albedo and surface albedo effects that contribute negative forcings are less well understood and have larger uncertainties. The range in the net estimate is increased by the negative forcing terms, which have larger uncertainties than the positive terms. The nature of the uncertainty in the estimated cloud albedo effect introduces a noticeable asymmetry in the distribution. Uncertainties in the distribution include structural aspects (e.g., representation of extremes in the component values, absence of any weighting of the radiative forcing mechanisms, possibility of unaccounted for but as yet unquantified radiative forcings) and statistical aspects (e.g., assumptions about the types of distributions describing component uncertainties). {2.7, 2.9}

The Global Warming Potential (GWP) is a useful metric for comparing the potential climate impact of the emissions of different LLGHGs (see Table TS.2). Global Warming Potentials compare the integrated radiative forcing over a specified period (e.g., 100 years) from a unit mass pulse emission and are a way of comparing the potential climate change associated with emissions of different greenhouse gases. There are well-documented shortcomings of the GWP concept, particularly in using it to assess the impact of short-lived species. {2.10}

For the magnitude and range of realistic forcings considered, evidence suggests an approximately linear relationship between global mean radiative forcing and global mean surface temperature response. The spatial patterns of radiative forcing vary between different forcing agents. However, the spatial signature of the climate response is not generally expected to match that of the forcing. Spatial patterns of climate response

GLOBAL MEAN RADIATIVE FORCINGS

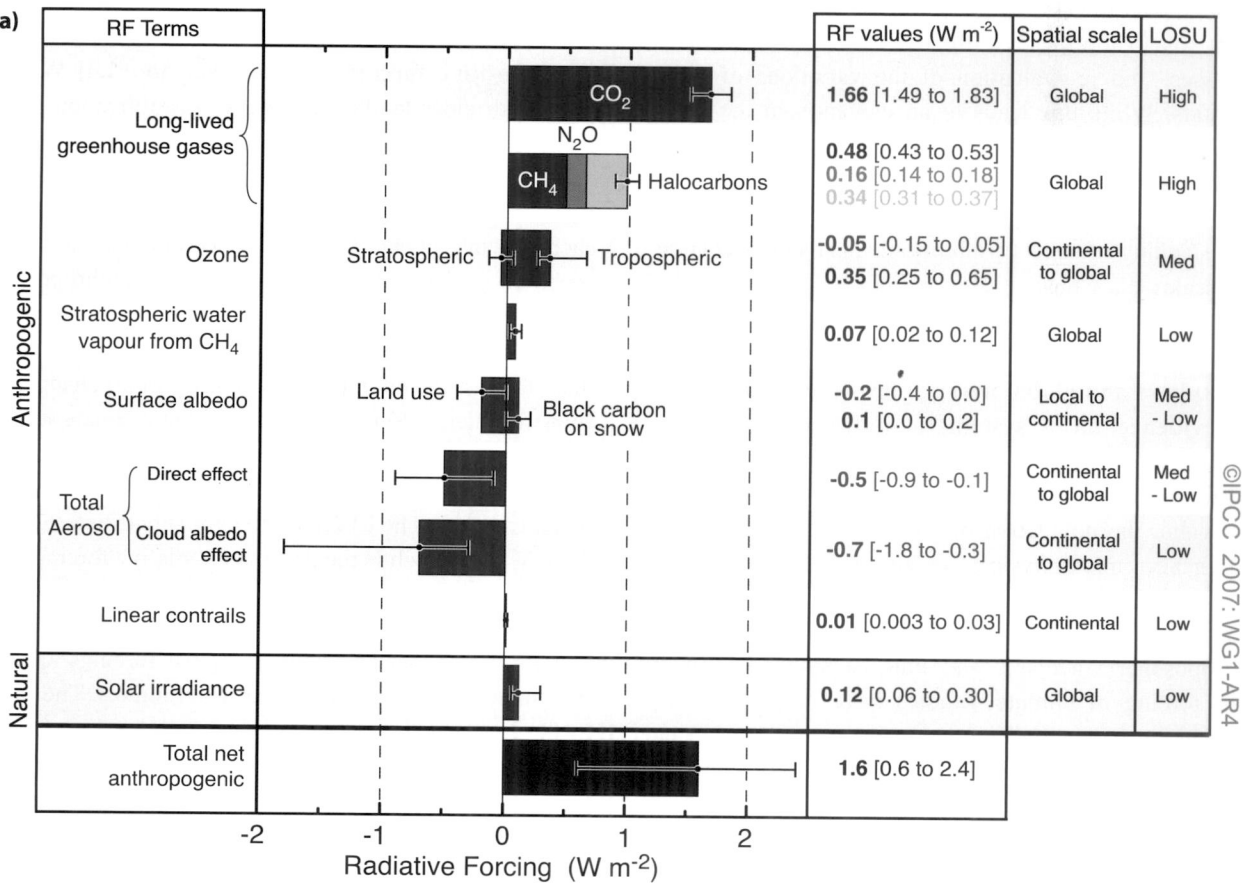

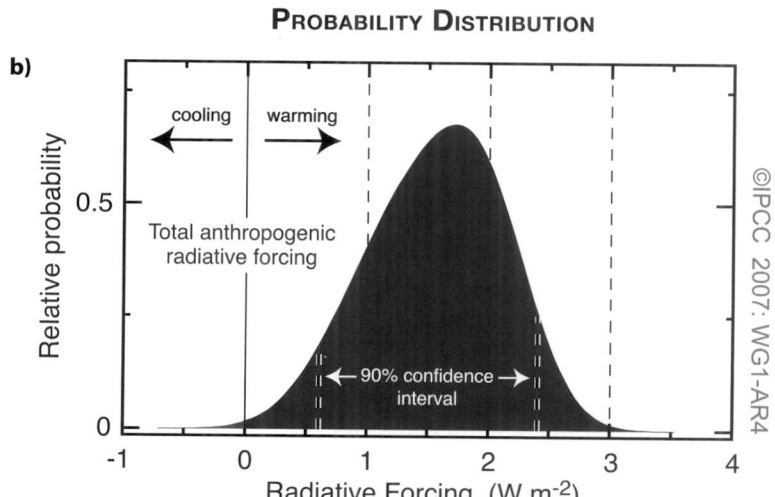

Figure TS.5. *(a) Global mean radiative forcings (RF) and their 90% confidence intervals in 2005 for various agents and mechanisms. Columns on the right-hand side specify best estimates and confidence intervals (RF values); typical geographical extent of the forcing (Spatial scale); and level of scientific understanding (LOSU) indicating the scientific confidence level as explained in Section 2.9. Errors for CH_4, N_2O and halocarbons have been combined. The net anthropogenic radiative forcing and its range are also shown. Best estimates and uncertainty ranges can not be obtained by direct addition of individual terms due to the asymmetric uncertainty ranges for some factors; the values given here were obtained from a Monte Carlo technique as discussed in Section 2.9. Additional forcing factors not included here are considered to have a very low LOSU. Volcanic aerosols contribute an additional form of natural forcing but are not included due to their episodic nature. The range for linear contrails does not include other possible effects of aviation on cloudiness. (b) Probability distribution of the global mean combined radiative forcing from all anthropogenic agents shown in (a). The distribution is calculated by combining the best estimates and uncertainties of each component. The spread in the distribution is increased significantly by the negative forcing terms, which have larger uncertainties than the positive terms. {2.9.1, 2.9.2; Figure 2.20}*

Table TS.2. *Lifetimes, radiative efficiencies and direct (except for CH_4) global warming potentials (GWP) relative to CO_2. {Table 2.14}*

Industrial Designation or Common Name (years)	Chemical Formula	Lifetime (years)	Radiative Efficiency (W m^{-2} ppb^{-1})	Global Warming Potential for Given Time Horizon			
				SAR[‡] (100-yr)	20-yr	100-yr	500-yr
Carbon dioxide	CO_2	See below[a]	[b]1.4x10^{-5}	1	1	1	1
Methane[c]	CH_4	12[e]	3.7x10^{-4}	21	72	25	7.6
Nitrous oxide	N_2O	114	3.03x10^{-3}	310	289	298	153
Substances controlled by the Montreal Protocol							
CFC-11	CCl_3F	45	0.25	3,800	6,730	4,750	1,620
CFC-12	CCl_2F_2	100	0.32	8,100	11,000	10,900	5,200
CFC-13	$CClF_3$	640	0.25		10,800	14,400	16,400
CFC-113	CCl_2FCClF_2	85	0.3	4,800	6,540	6,130	2,700
CFC-114	$CClF_2CClF_2$	300	0.31		8,040	10,000	8,730
CFC-115	$CClF_2CF_3$	1,700	0.18		5,310	7,370	9,990
Halon-1301	$CBrF_3$	65	0.32	5,400	8,480	7,140	2,760
Halon-1211	$CBrClF_2$	16	0.3		4,750	1,890	575
Halon-2402	$CBrF_2CBrF_2$	20	0.33		3,680	1,640	503
Carbon tetrachloride	CCl_4	26	0.13	1,400	2,700	1,400	435
Methyl bromide	CH_3Br	0.7	0.01		17	5	1
Methyl chloroform	CH_3CCl_3	5	0.06		506	146	45
HCFC-22	$CHClF_2$	12	0.2	1,500	5,160	1,810	549
HCFC-123	$CHCl_2CF_3$	1.3	0.14	90	273	77	24
HCFC-124	$CHClFCF_3$	5.8	0.22	470	2,070	609	185
HCFC-141b	CH_3CCl_2F	9.3	0.14		2,250	725	220
HCFC-142b	CH_3CClF_2	17.9	0.2	1,800	5,490	2,310	705
HCFC-225ca	$CHCl_2CF_2CF_3$	1.9	0.2		429	122	37
HCFC-225cb	$CHClFCF_2CClF_2$	5.8	0.32		2,030	595	181
Hydrofluorocarbons							
HFC-23	CHF_3	270	0.19	11,700	12,000	14,800	12,200
HFC-32	CH_2F_2	4.9	0.11	650	2,330	675	205
HFC-125	CHF_2CF_3	29	0.23	2,800	6,350	3,500	1,100
HFC-134a	CH_2FCF_3	14	0.16	1,300	3,830	1,430	435
HFC-143a	CH_3CF_3	52	0.13	3,800	5,890	4,470	1,590
HFC-152a	CH_3CHF_2	1.4	0.09	140	437	124	38
HFC-227ea	CF_3CHFCF_3	34.2	0.26	2,900	5,310	3,220	1,040
HFC-236fa	$CF_3CH_2CF_3$	240	0.28	6,300	8,100	9,810	7,660
HFC-245fa	$CHF_2CH_2CF_3$	7.6	0.28		3,380	1030	314
HFC-365mfc	$CH_3CF_2CH_2CF_3$	8.6	0.21		2,520	794	241
HFC-43-10mee	$CF_3CHFCHFCF_2CF_3$	15.9	0.4	1,300	4,140	1,640	500
Perfluorinated compounds							
Sulphur hexafluoride	SF_6	3,200	0.52	23,900	16,300	22,800	32,600
Nitrogen trifluoride	NF_3	740	0.21		12,300	17,200	20,700
PFC-14	CF_4	50,000	0.10	6,500	5,210	7,390	11,200
PFC-116	C_2F_6	10,000	0.26	9,200	8,630	12,200	18,200

Table TS.2 (continued)

Industrial Designation or Common Name (years)	Chemical Formula	Lifetime (years)	Radiative Efficiency (W m⁻² ppb⁻¹)	Global Warming Potential for Given Time Horizon			
				SAR[‡] (100-yr)	20-yr	100-yr	500-yr
Perfluorinated compounds *(continued)*							
PFC-218	C_3F_8	2,600	0.26	7,000	6,310	8,830	12,500
PFC-318	$c\text{-}C_4F_8$	3,200	0.32	8,700	7,310	10,300	14,700
PFC-3-1-10	C_4F_{10}	2,600	0.33	7,000	6,330	8,860	12,500
PFC-4-1-12	C_5F_{12}	4,100	0.41		6,510	9,160	13,300
PFC-5-1-14	C_6F_{14}	3,200	0.49	7,400	6,600	9,300	13,300
PFC-9-1-18	$C_{10}F_{18}$	>1,000[d]	0.56		>5,500	>7,500	>9,500
trifluoromethyl sulphur pentafluoride	SF_5CF_3	800	0.57		13,200	17,700	21,200
Fluorinated ethers							
HFE-125	CHF_2OCF_3	136	0.44		13,800	14,900	8,490
HFE-134	CHF_2OCHF_2	26	0.45		12,200	6,320	1,960
HFE-143a	CH_3OCF_3	4.3	0.27		2,630	756	230
HCFE-235da2	$CHF_2OCHClCF_3$	2.6	0.38		1,230	350	106
HFE-245cb2	$CH_3OCF_2CHF_2$	5.1	0.32		2,440	708	215
HFE-245fa2	$CHF_2OCH_2CF_3$	4.9	0.31		2,280	659	200
HFE-254cb2	$CH_3OCF_2CHF_2$	2.6	0.28		1,260	359	109
HFE-347mcc3	$CH_3OCF_2CF_2CF_3$	5.2	0.34		1,980	575	175
HFE-347pcf2	$CHF_2CF_2OCH_2CF_3$	7.1	0.25		1,900	580	175
HFE-356pcc3	$CH_3OCF_2CF_2CHF_2$	0.33	0.93		386	110	33
HFE-449sl (HFE-7100)	$C_4F_9OCH_3$	3.8	0.31		1,040	297	90
HFE-569sf2 (HFE-7200)	$C_4F_9OC_2H_5$	0.77	0.3		207	59	18
HFE-43-10pccc124 (H-Galden 1040x)	$CHF_2OCF_2OC_2F_4OCHF_2$	6.3	1.37		6,320	1,870	569
HFE-236ca12 (HG-10)	$CHF_2OCF_2OCHF_2$	12.1	0.66		8,000	2,800	860
HFE-338pcc13 (HG-01)	$CHF_2OCF_2CF_2OCHF_2$	6.2	0.87		5,100	1,500	460
Perfluoropolyethers							
PFPMIE	$CF_3OCF(CF_3)CF_2OCF_2OCF_3$	800	0.65		7,620	10,300	12,400
Hydrocarbons and other compounds – Direct Effects							
Dimethylether	CH_3OCH_3	0.015	0.02		1	1	<<1
Methylene chloride	CH_2Cl_2	0.38	0.03		31	8.7	2.7
Methyl chloride	CH_3Cl	1.0	0.01		45	13	4

Notes:

[‡] SAR refers to the IPCC Second Assessment Report (1995) used for reporting under the UNFCCC.

[a] The CO_2 response function used in this report is based on the revised version of the Bern Carbon cycle model used in Chapter 10 of this report (Bern2.5CC; Joos et al. 2001) using a background CO_2 concentration value of 378 ppm. The decay of a pulse of CO_2 with time t is given by

$$a_0 + \sum_{i=1}^{3} a_i \cdot e^{-t/\tau_i}$$ where $a_0 = 0.217$, $a_1 = 0.259$, $a_2 = 0.338$, $a_3 = 0.186$, $\tau_1 = 172.9$ years, $\tau_2 = 18.51$ years, and $\tau_3 = 1.186$ years, for t < 1,000 years.

[b] The radiative efficiency of CO_2 is calculated using the IPCC (1990) simplified expression as revised in the TAR, with an updated background concentration value of 378 ppm and a perturbation of +1 ppm (see Section 2.10.2).

[c] The perturbation lifetime for CH_4 is 12 years as in the TAR (see also Section 7.4). The GWP for CH_4 includes indirect effects from enhancements of ozone and stratospheric water vapour (see Section 2.10).

[d] The assumed lifetime of 1000 years is a lower limit.

are largely controlled by climate processes and feedbacks. For example, sea ice-albedo feedbacks tend to enhance the high-latitude response. Spatial patterns of response are also affected by differences in thermal inertia between land and sea areas. {2.8, 9.2}

The pattern of response to a radiative forcing can be altered substantially if its structure is favourable for affecting a particular aspect of the atmospheric structure or circulation. Modelling studies and data comparisons suggest that mid- to high-latitude circulation patterns are *likely* to be affected by some forcings such as volcanic eruptions, which have been linked to changes in the Northern Annular Mode (NAM) and North Atlantic Oscillation (NAO) (see Section 3.1 and Box TS.2). Simulations also suggest that absorbing aerosols, particularly black carbon, can reduce the solar radiation reaching the surface and can warm the atmosphere at regional scales, affecting the vertical temperature profile and the large-scale atmospheric circulation. {2.8, 7.5, 9.2}

The spatial patterns of radiative forcings for ozone, aerosol direct effects, aerosol-cloud interactions and land use have considerable uncertainties. This is in contrast to the relatively high confidence in the spatial pattern of radiative forcing for the LLGHGs. The net positive radiative forcing in the Southern Hemisphere (SH) *very likely* exceeds that in the NH because of smaller aerosol concentrations in the SH. {2.9}

TS 2.6 Surface Forcing and the Hydrologic Cycle

Observations and models indicate that changes in the radiative flux at the Earth's surface affect the surface heat and moisture budgets, thereby involving the hydrologic cycle. Recent studies indicate that some forcing agents can influence the hydrologic cycle differently than others through their interactions with clouds. In particular, changes in aerosols may have affected precipitation and other aspects of the hydrologic cycle more strongly than other anthropogenic forcing agents. Energy deposited at the surface directly affects evaporation and sensible heat transfer. The instantaneous radiative flux change at the surface (hereafter called 'surface forcing') is a useful diagnostic tool for understanding changes in the heat and moisture surface budgets and the accompanying climate change. However, unlike radiative forcing, it cannot be used to quantitatively compare the effects of different agents on the equilibrium global mean surface temperature change. Net radiative forcing and surface forcing have different equator-to-pole gradients in the NH, and are different between the NH and SH. {2.9, 7.2, 7.5, 9.5}

TS.3 Observations of Changes in Climate

This assessment evaluates changes in the Earth's climate system, considering not only the atmosphere, but also the ocean and the cryosphere, as well as phenomena such as atmospheric circulation changes, in order to increase understanding of trends, variability and processes of climate change at global and regional scales. Observational records employing direct methods are of variable length as described below, with global temperature estimates now beginning as early as 1850. Observations of extremes of weather and climate are discussed, and observed changes in extremes are described. The consistency of observed changes among different climate variables that allows an increasingly comprehensive picture to be drawn is also described. Finally, palaeoclimatic information that generally employs indirect proxies to infer information about climate change over longer time scales (up to millions of years) is also assessed.

TS.3.1 Atmospheric Changes: Instrumental Record

This assessment includes analysis of global and hemispheric means, changes over land and ocean and distributions of trends in latitude, longitude and altitude. Since the TAR, improvements in observations and their calibration, more detailed analysis of methods and extended time series allow more in-depth analyses of changes including atmospheric temperature, precipitation, humidity, wind and circulation. Extremes of climate are a key expression of climate variability, and this assessment includes new data that permit improved insights into the changes in many types of extreme events including heat waves, droughts, heavy precipitation and tropical cyclones (including hurricanes and typhoons). {3.2–3.4, 3.8}

Furthermore, advances have occurred since the TAR in understanding how a number of seasonal and long-term anomalies can be described by patterns of climate variability. These patterns arise from internal interactions and from the differential effects on the atmosphere of land and ocean, mountains and large changes in heating. Their response is often felt in regions far removed from their physical source through atmospheric teleconnections associated with large-scale waves in the atmosphere. Understanding temperature and precipitation anomalies associated with the dominant patterns of climate variability is essential to understanding many regional climate anomalies and why these may differ from those at the global scale. Changes in storm tracks, the jet streams, regions of preferred blocking anticyclones and changes in monsoons can also occur in conjunction with these preferred patterns of variability. {3.5–3.7}

TS.3.1.1 Global Average Temperatures

2005 and 1998 were the warmest two years in the instrumental global surface air temperature record since 1850. Surface temperatures in 1998 were enhanced by the major 1997–1998 El Niño but no such strong anomaly was present in 2005. Eleven of the last 12 years (1995 to 2006) – the exception being 1996 – rank among the 12 warmest years on record since 1850. {3.2}

The global average surface temperature has increased, especially since about 1950. The updated 100-year trend (1906–2005) of 0.74°C ± 0.18°C is larger than the 100-year warming trend at the time of the TAR (1901–2000) of 0.6°C ± 0.2°C due to additional warm years. The total temperature increase from 1850-1899 to 2001-2005 is 0.76°C ± 0.19°C. The rate of warming averaged over the last 50 years (0.13°C ± 0.03°C per decade) is nearly twice that for the last 100 years. Three different global estimates all show consistent warming trends. There is also consistency between the data sets in their separate land and ocean domains, and between sea surface temperature (SST) and nighttime marine air temperature (see Figure TS.6). {3.2}

Recent studies confirm that effects of urbanisation and land use change on the global temperature record are negligible (less than 0.006°C per decade over land and zero over the ocean) as far as hemispheric- and continental-scale averages are concerned. All observations are subject to data quality and consistency checks to correct for potential biases. The real but local effects of urban areas are accounted for in the land temperature data sets used. Urbanisation and land use effects are not relevant to the widespread oceanic warming that has been observed. Increasing evidence suggests that urban heat island effects also affect precipitation, cloud and diurnal temperature range (DTR). {3.2}

The global average DTR has stopped decreasing. A decrease in DTR of approximately 0.1°C per decade was reported in the TAR for the period 1950 to 1993. Updated observations reveal that DTR has not changed from 1979 to 2004 as both day- and night time temperature have risen at about the same rate. The trends are highly variable from one region to another. {3.2}

New analyses of radiosonde and satellite measurements of lower- and mid-tropospheric temperature show warming rates that are generally consistent with each other and with those in the surface temperature record within their respective uncertainties for the periods 1958 to 2005 and 1979 to 2005. This largely resolves a discrepancy noted in the TAR (see Figure TS.7). The radiosonde record is markedly less spatially complete than the surface record and increasing evidence suggests that a number of radiosonde data sets are unreliable, especially in the tropics. Disparities remain among different tropospheric temperature trends estimated from satellite Microwave Sounding Unit (MSU) and advanced MSU measurements since 1979, and all likely still contain residual errors. However, trend estimates have been substantially improved and data set differences reduced since the TAR, through adjustments for changing satellites, orbit decay and drift in local crossing time (diurnal cycle effects). It appears that the satellite tropospheric temperature record is broadly consistent with surface temperature trends provided that the stratospheric influence on MSU channel 2 is accounted for. The range across different data sets of global surface warming since 1979 is 0.16°C to 0.18°C per decade, compared to 0.12°C to 0.19°C per decade for MSU-derived estimates of tropospheric temperatures. It is likely that there is increased warming with altitude from the surface through much of the troposphere in the tropics, pronounced cooling in the stratosphere, and a trend towards a higher tropopause. {3.4}

Stratospheric temperature estimates from adjusted radiosondes, satellites and reanalyses are all in qualitative agreement, with a cooling of between 0.3°C and 0.6°C per decade since 1979 (see Figure TS.7). Longer radiosonde records (back to 1958) also indicate stratospheric cooling but are subject to substantial instrumental uncertainties. The rate of cooling increased after 1979 but has slowed in the last decade. It is *likely* that radiosonde records overestimate stratospheric cooling, owing to changes in sondes not yet taken into account. The trends are not monotonic, because of stratospheric warming episodes that follow major volcanic eruptions. {3.4}

GLOBAL TEMPERATURE TRENDS

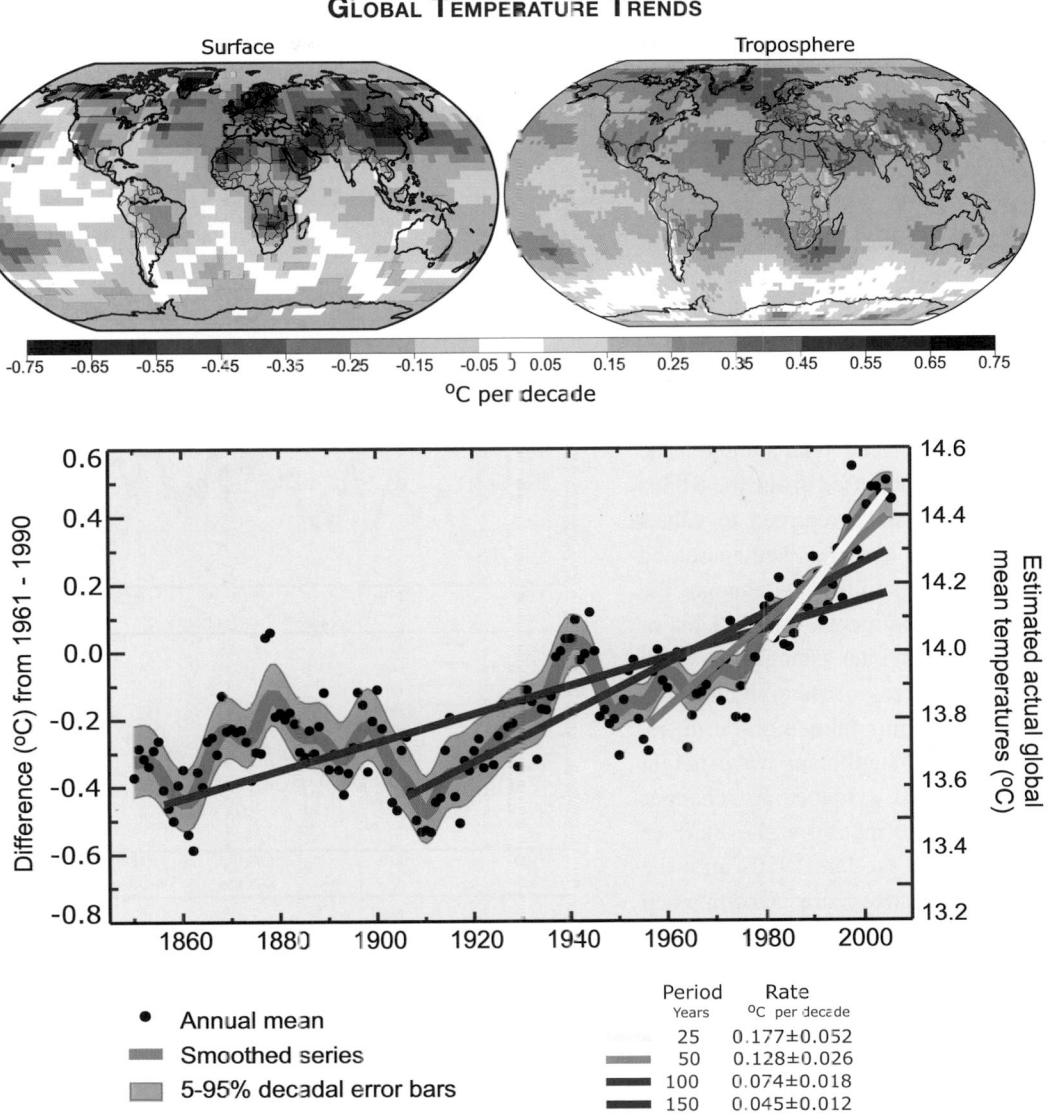

Figure TS.6. *(Top) Patterns of linear global temperature trends over the period 1979 to 2005 estimated at the surface (left), and for the troposphere from satellite records (right). Grey indicates areas with incomplete data. (Bottom) Annual global mean temperatures (black dots) with linear fits to the data. The left hand axis shows temperature anomalies relative to the 1961 to 1990 average and the right hand axis shows estimated actual temperatures, both in °C. Linear trends are shown for the last 25 (yellow), 50 (orange), 100 (purple) and 150 years (red). The smooth blue curve shows decadal variations (see Appendix 3.A), with the decadal 90% error range shown as a pale blue band about that line. The total temperature increase from the period 1850 to 1899 to the period 2001 to 2005 is 0.76°C ± 0.19°C. {FAQ 3.1, Figure 1.}*

TS.3.1.2 Spatial Distribution of Changes in Temperature, Circulation and Related Variables

Surface temperatures over land regions have warmed at a faster rate than over the oceans in both hemispheres. Longer records now available show significantly faster rates of warming over land than ocean in the past two decades (about 0.27°C vs. 0.13°C per decade). {3.2}

The warming in the last 30 years is widespread over the globe, and is greatest at higher northern latitudes. The greatest warming has occurred in the NH winter (DJF) and spring (MAM). Average arctic temperatures have been increasing at almost twice the rate of the rest of the world in the past 100 years. However, arctic temperatures are highly variable. A slightly longer arctic warm period, almost as warm as the present, was observed from 1925 to 1945, but its geographical distribution appears to have been different from the recent warming since its extent was not global. {3.2}

There is evidence for long-term changes in the large-scale atmospheric circulation, such as a poleward shift and strengthening of the westerly winds. Regional climate trends can be very different from the global average, reflecting changes in the circulations and interactions of the atmosphere and ocean and the other components of the climate system. Stronger mid-latitude westerly wind maxima have occurred in both hemispheres in most seasons from at least 1979 to the late 1990s, and poleward displacements of corresponding Atlantic and southern polar front jet streams have been documented. The westerlies in the NH increased from the 1960s to the 1990s but have since returned to values close to the long-term average. The increased strength of the westerlies in the NH changes the flow from oceans to continents, and is a major factor in the observed winter changes in storm tracks and related patterns of precipitation and temperature trends at mid- and high-latitudes. Analyses of wind and significant wave height support reanalysis-based evidence for changes in NH extratropical storms from the start of the reanalysis record in the late 1970s until the late 1990s. These changes are accompanied by a tendency towards stronger winter polar vortices throughout the troposphere and lower stratosphere. {3.2, 3.5}

Many regional climate changes can be described in terms of preferred patterns of climate variability and therefore as changes in the occurrence of indices that characterise the strength and phase of these patterns. The importance, over all time scales, of fluctuations in the westerlies and storm tracks in the North Atlantic has often been noted, and these fluctuations are described by the NAO (see Box TS.2 for an explanation of this and other preferred patterns). The characteristics of fluctuations in the zonally averaged westerlies in the two hemispheres have more recently been described by their respective 'annular modes', the Northern and Southern Annular Modes (NAM and SAM). The observed changes can be expressed as a shift of the circulation towards the structure associated with one sign of these preferred patterns. The increased mid-latitude westerlies in the North Atlantic can be largely viewed as reflecting either NAO or NAM changes; multi-decadal variability is also evident in the Atlantic, both in the atmosphere and the ocean. In the SH, changes in circulation related to an increase in the

OBSERVED AIR TEMPERATURES

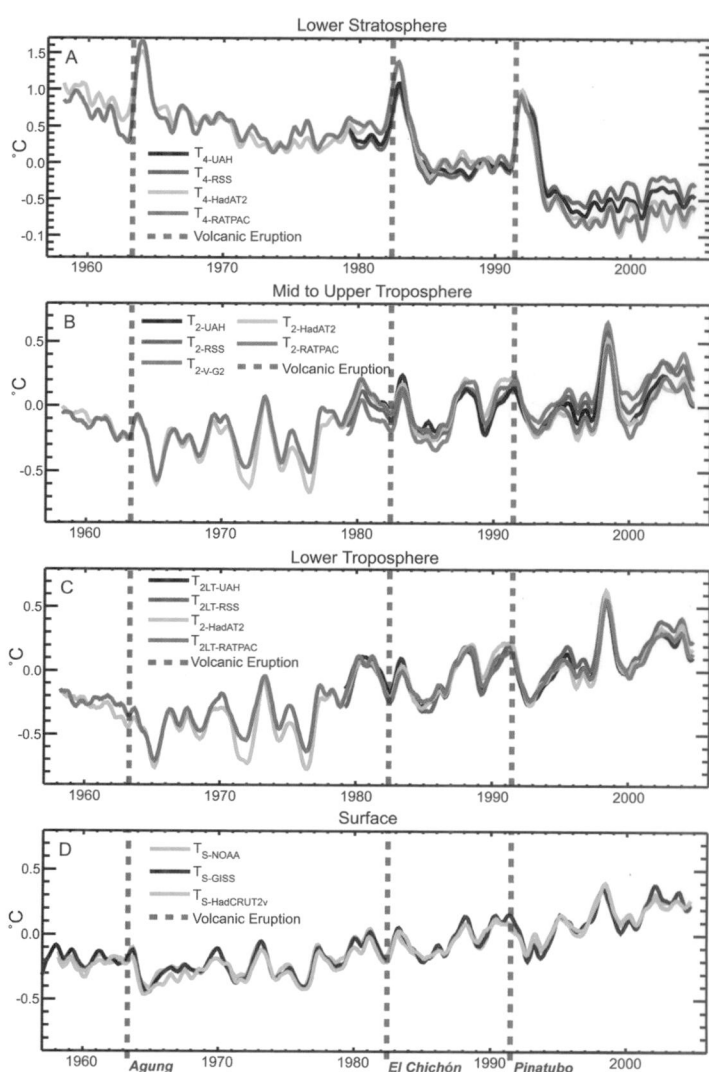

Figure TS.7. *Observed surface (D) and upper air temperatures for the lower troposphere (C), mid- to upper troposphere (B) and lower stratosphere (A), shown as monthly mean anomalies relative to the period 1979 to 1997 smoothed with a seven-month running mean filter. Dashed lines indicate the times of major volcanic eruptions. {Figure 3.17}*

SAM from the 1960s to the present are associated with strong warming over the Antarctic Peninsula and, to a lesser extent, cooling over parts of continental Antarctica. Changes have also been observed in ocean-atmosphere interactions in the Pacific. The ENSO is the dominant mode of global-scale variability on interannual time scales although there have been times when it is less apparent. The 1976–1977 climate shift, related to the phase change in the Pacific Decadal Oscillation (PDO) towards more El Niño events and changes in the evolution of ENSO, has affected many areas, including most tropical monsoons. For instance, over North America, ENSO and Pacific-

Box TS.2: Patterns (Modes) of Climate Variability

Analysis of atmospheric and climatic variability has shown that a significant component of it can be described in terms of fluctuations in the amplitude and sign of indices of a relatively small number of preferred patterns of variability. Some of the best known of these are:

- El Niño-Southern Oscillation (ENSO), a coupled fluctuation in the atmosphere and the equatorial Pacific Ocean, with preferred time scales of two to about seven years. ENSO is often measured by the difference in surface pressure anomalies between Tahiti and Darwin and the SSTs in the central and eastern equatorial Pacific. ENSO has global teleconnections.

- North Atlantic Oscillation (NAO), a measure of the strength of the Icelandic Low and the Azores High, and of the westerly winds between them, mainly in winter. The NAO has associated fluctuations in the storm track, temperature and precipitation from the North Atlantic into Eurasia (see Box TS.2, Figure 1).

- Northern Annular Mode (NAM), a winter fluctuation in the amplitude of a pattern characterised by low surface pressure in the Arctic and strong mid-latitude westerlies. The NAM has links with the northern polar vortex and hence the stratosphere.

- Southern Annular Mode (SAM), the fluctuation of a pattern with low antarctic surface pressure and strong mid-latitude westerlies, analogous to the NAM, but present year round.

- Pacific-North American (PNA) pattern, an atmospheric large-scale wave pattern featuring a sequence of tropospheric high- and low-pressure anomalies stretching from the subtropical west Pacific to the east coast of North America.

- Pacific Decadal Oscillation (PDO), a measure of the SSTs in the North Pacific that has a very strong correlation with the North Pacific Index (NPI) measure of the depth of the Aleutian Low. However, it has a signature throughout much of the Pacific.

POSITIVE PHASE OF NAO AND NAM

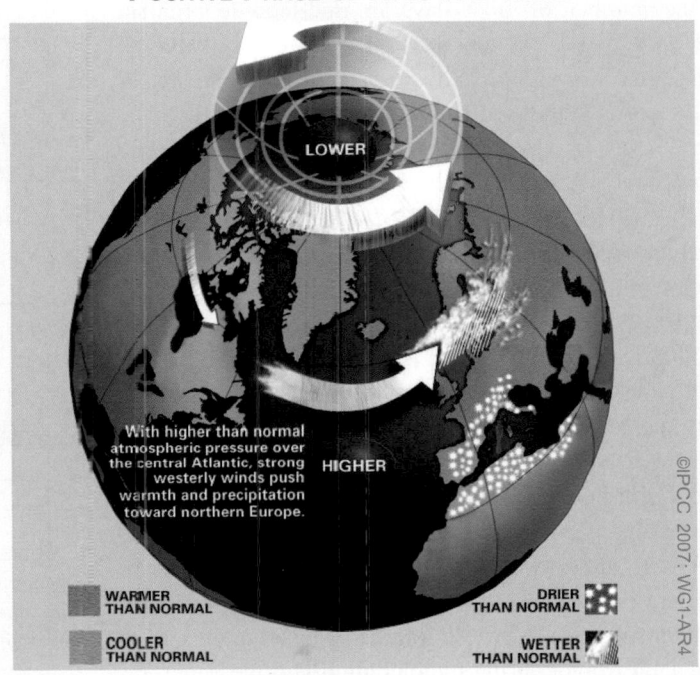

Box TS.2, Figure 1. *A schematic of the changes associated with the positive phase of the NAO and NAM. The changes in pressure and winds are shown, along with precipitation changes. Warm colours indicate areas that are warmer than normal and blue indicates areas that are cooler than normal.*

The extent to which all these preferred patterns of variability can be considered to be true modes of the climate system is a topic of active research. However, there is evidence that their existence can lead to larger-amplitude regional responses to forcing than would otherwise be expected. In particular, a number of the observed 20th-century climate changes can be viewed in terms of changes in these patterns. It is therefore important to test the ability of climate models to simulate them (see Section TS.4, Box TS.7) and to consider the extent to which observed changes related to these patterns are linked to internal variability or to anthropogenic climate change. {3.6, 8.4}

North American (PNA) teleconnection-related changes appear to have led to contrasting changes across the continent, as the western part has warmed more than the eastern part, while the latter has become cloudier and wetter. There is substantial low-frequency atmospheric variability in the Pacific sector over the 20th century, with extended periods of weakened (1900–1924; 1947–1976) as well as strengthened (1925–1946; 1977–2003) circulation. {3.2, 3.5, 3.6}

Changes in extremes of temperature are consistent with warming. Observations show widespread reductions in the number of frost days in mid-latitude regions, increases in the number of warm extremes (warmest 10% of days or nights) and a reduction in the number of daily cold extremes (coldest 10% of days or nights) (see Box TS.5). The most marked changes are for cold nights, which have declined over the 1951 to 2003 period for all regions where data are available (76% of the land). {3.8}

Heat waves have increased in duration beginning in the latter half of the 20th century. The record-breaking heat wave over western and central Europe in the summer of 2003 is an example of an exceptional recent extreme. That summer (JJA) was the warmest since comparable instrumental records began around 1780 (1.4°C above the previous warmest in 1807). Spring drying of the land surface over Europe was an important factor in the occurrence of the extreme 2003 temperatures. Evidence suggests that heat waves have also increased in frequency and duration in other locations. The very strong correlation between observed dryness and high temperatures over land in the tropics during summer highlights the important role moisture plays in moderating climate. {3.8, 3.9}

There is insufficient evidence to determine whether trends exist in such events as tornadoes, hail, lightning and dust storms which occur at small spatial scales. {3.8}

TS.3.1.3   Changes in the Water Cycle: Water Vapour, Clouds, Precipitation and Tropical Storms

Tropospheric water vapour is increasing (Figure TS.8). Surface specific humidity has generally increased since 1976 in close association with higher temperatures over both land and ocean. Total column water vapour has increased over the global oceans by 1.2 ± 0.3% per decade (95% confidence limits) from 1988 to 2004. The observed regional changes are consistent in pattern and amount with the changes in SST and the assumption of a near-constant relative humidity increase in water vapour mixing ratio. The additional atmospheric water vapour implies increased moisture availability for precipitation. {3.4}

ATMOSPHERIC WATER VAPOUR

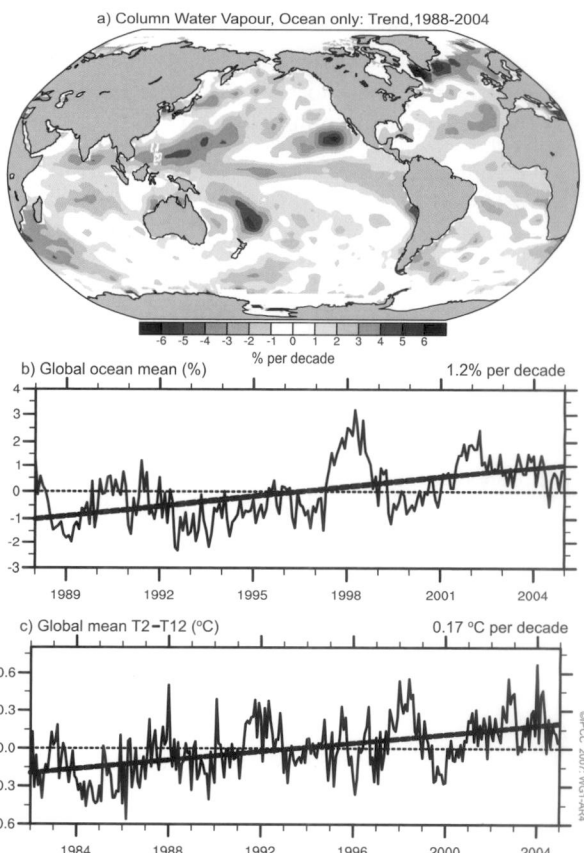

a) Column Water Vapour, Ocean only: Trend,1988-2004

-6 -5 -4 -3 -2 -1 0 1 2 3 4 5 6
% per decade

b) Global ocean mean (%)   1.2% per decade

c) Global mean T2-T12 (°C)   0.17 °C per decade

©IPCC 2007 WG1-AR4

Figure TS.8. *(a) Linear trends in precipitable water (total column water vapour) over the period 1988 to 2004 (% per decade) and (b) the monthly time series of anomalies, relative to the period shown, over the global ocean with linear trend. (c) The global mean (80°N to 80°S) radiative signature of upper-tropospheric moistening is given by monthly time series of combinations of satellite brightness temperature anomalies (°C), relative to the period 1982 to 2004, with the dashed line showing the linear trend of the key brightness temperature in °C per decade. {3.4, Figures 3.20 and 3.21}*

Upper-tropospheric water vapour is also increasing. Due to instrumental limitations, it is difficult to assess long-term changes in water vapour in the upper troposphere, where it is of radiative importance. However, the available data now show evidence for global increases in upper-tropospheric specific humidity over the past two decades (Figure TS.8). These observations are consistent with the observed increase in temperatures and represent an important advance since the TAR. {3.4}

Cloud changes are dominated by ENSO. Widespread (but not ubiquitous) decreases in continental DTR have coincided with increases in cloud amounts. Surface and satellite observations disagree on changes in total and low-level cloud changes over the ocean. However, radiation changes at the top of the atmosphere from the 1980s to 1990s (possibly related in part to the ENSO phenomenon) appear to be associated with reductions in tropical upper-level cloud cover, and are consistent with changes in the energy budget and in observed ocean heat content. {3.4}

'Global dimming' is not global in extent and it has not continued after 1990. Reported decreases in solar radiation at the Earth's surface from 1970 to 1990 have an urban bias. Further, there have been increases since about 1990. An increasing aerosol load due to human activities decreases regional air quality and the amount of solar radiation reaching the Earth's surface. In some areas, such as Eastern Europe, recent observations of a reversal in the sign of this effect link changes in solar radiation to concurrent air quality improvements. {3.4}

Long-term trends in precipitation amounts from 1900 to 2005 have been observed in many large regions (Figure TS.9). Significantly increased precipitation has been observed in the eastern parts of North and South America, northern Europe and northern and central Asia. Drying has been observed in the Sahel, the Mediterranean, southern Africa and parts of southern Asia. Precipitation is highly variable spatially and temporally, and robust long-term trends have not been established for other large regions.[5] {3.3}

Substantial increases in heavy precipitation events have been observed. It is *likely* that there have been increases in the number of heavy precipitation events (e.g., above the 95th percentile) in many land regions since about 1950, even in those regions where there has been a reduction in total precipitation amount. Increases have also been reported for rarer precipitation events (1 in 50 year return period), but only a few regions have sufficient data to assess such trends reliably (see Figure TS.10). {3.8}

There is observational evidence for an increase of intense tropical cyclone activity in the North Atlantic since about 1970, correlated with increases in tropical SSTs. There are also suggestions of increased intense tropical cyclone activity in some other regions where concerns over data quality are greater. Multi-decadal variability and the quality of the tropical cyclone records prior to routine satellite observations in about 1970 complicate the detection of long-term trends in tropical cyclone activity and there is no clear trend in the annual numbers of tropical cyclones. Estimates of the potential destructiveness of tropical cyclones suggest a substantial upward trend since the mid-1970s, with a trend towards longer lifetimes and greater intensity. Trends are also apparent in SST, a critical variable known to influence

GLOBAL MEAN PRECIPITATION

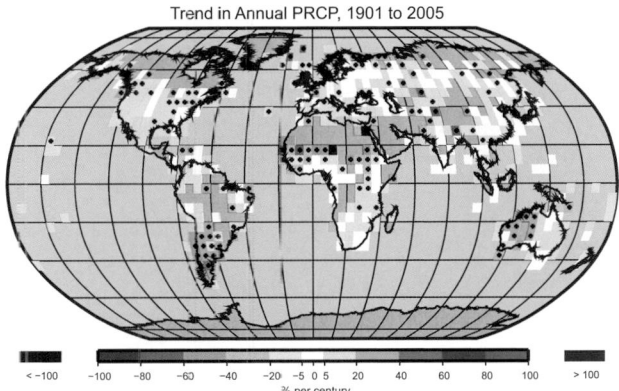

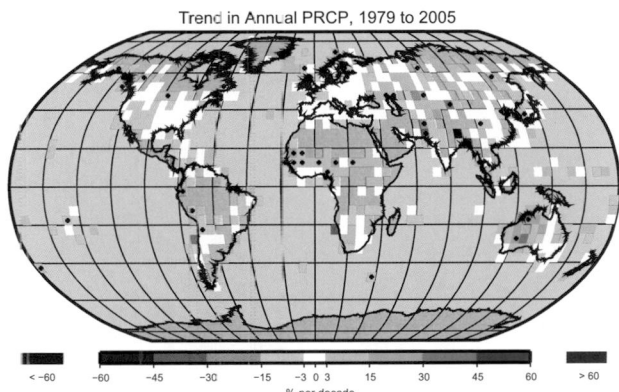

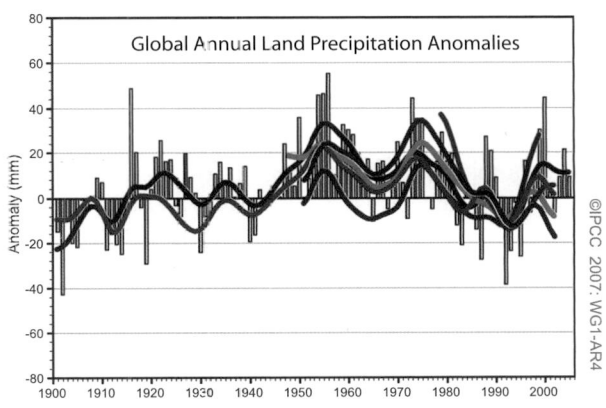

Figure TS.9. *(Top) Distribution of linear trends of annual land precipitation amounts over the period 1901 to 2005 (% per century) and (middle) 1979 to 2005 (% per decade). Areas in grey have insufficient data to produce reliable trends. The percentage is based on the 1961 to 1990 period. (Bottom) Time series of annual global land precipitation anomalies with respect to the 1961 to 1990 base period for 1900 to 2005. The smooth curves show decadal variations (see Appendix 3.A) for different data sets. {3.3, Figures 3.12 and 3.13}*

---

[5] The assessed regions are those considered in the regional projections chapter of the TAR and in Chapter 11 of this report.

ANNUAL PRECIPITATION TRENDS

Trend % per decade 1951 - 2003 contribution from very wet days

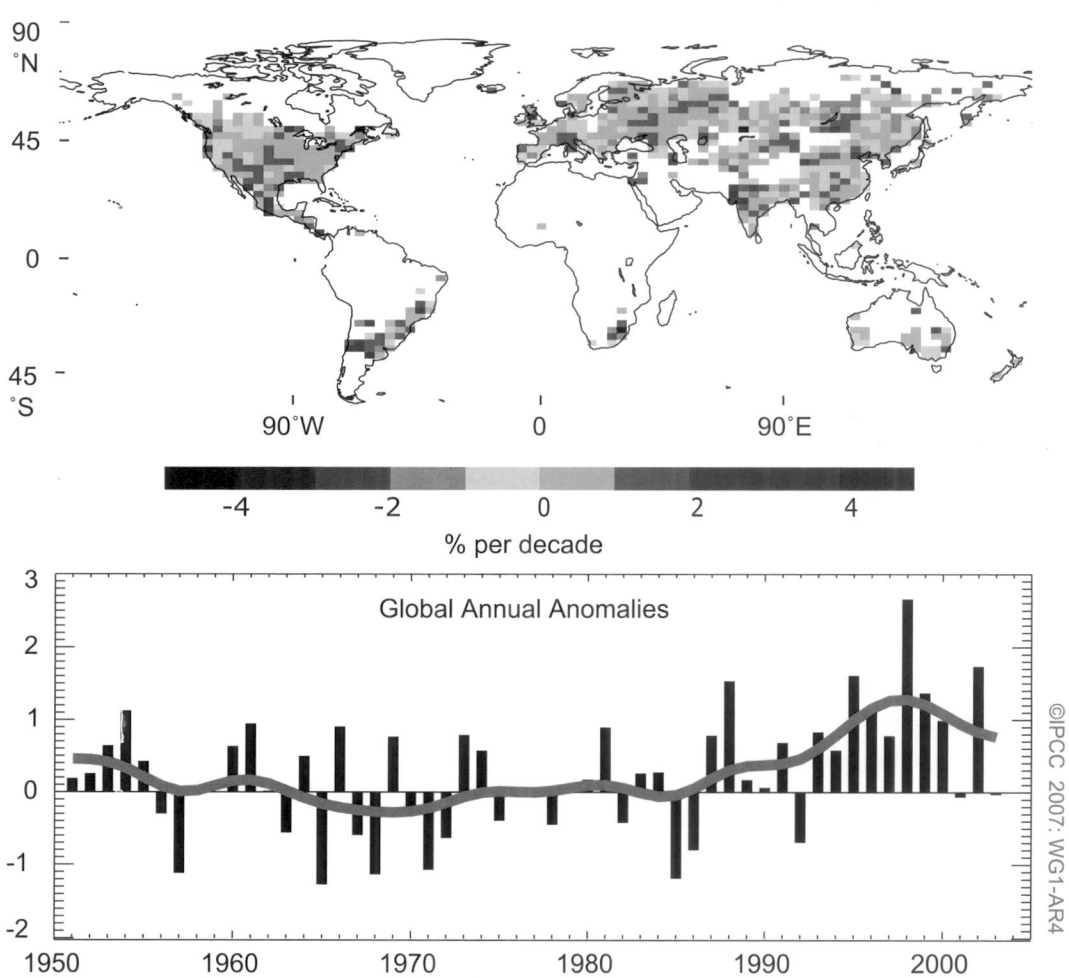

Figure TS.10. *(Top) Observed trends (% per decade) over the period 1951 to 2003 in the contribution to total annual precipitation from very wet days (i.e., corresponding to the 95th percentile and above). White land areas have insufficient data for trend determination. (Bottom) Anomalies (%) of the global (regions with data shown in top panel) annual time series of very wet days (with respect to 1961–1990) defined as the percentage change from the base period average (22.5%). The smooth orange curve shows decadal variations (see Appendix 3.A). {Figure 3.39}*

ANNUAL SEA-SURFACE TEMPERATURE ANOMALIES

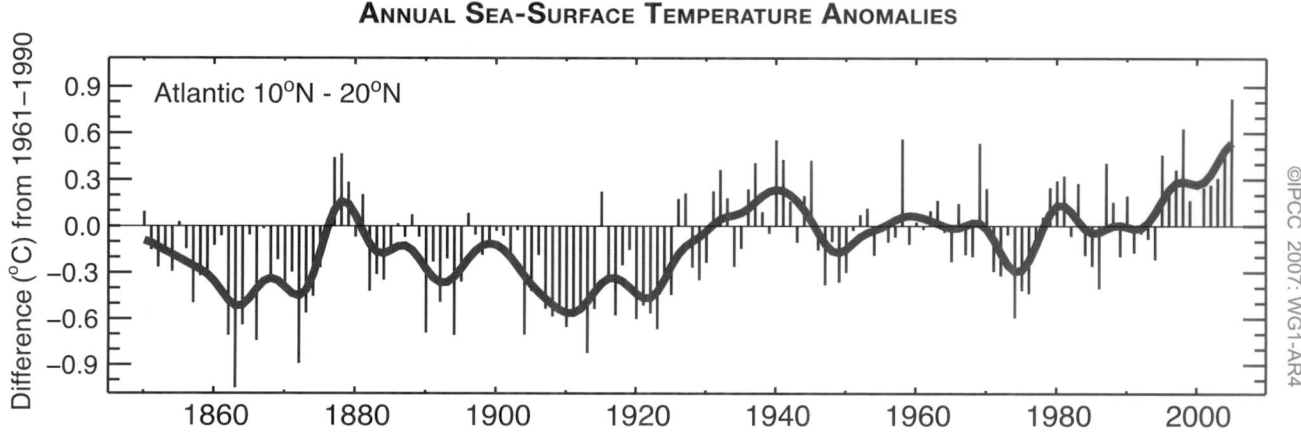

Figure TS.11. *Tropical Atlantic (10°N–20°N) sea surface temperature annual anomalies (°C) in the region of Atlantic hurricane formation, relative to the 1961 to 1990 mean. {Figure 3.33}*

tropical cyclone development (see Figure TS.11). Variations in the total numbers of tropical cyclones result from ENSO and decadal variability, which also lead to a redistribution of tropical storm numbers and tracks. The numbers of hurricanes in the North Atlantic have been above normal (based on 1981–2000) in nine of the years from 1995 to 2005. {3.8}

More intense and longer droughts have been observed over wider areas, particularly in the tropics and subtropics since the 1970s. While there are many different measures of drought, many studies use precipitation changes together with temperature.[6] Increased drying due to higher temperatures and decreased land precipitation have contributed to these changes. {3.3}

TS.3.2   Changes in the Cryosphere: Instrumental Record

Currently, ice permanently covers 10% of the land surface, with only a tiny fraction occurring outside Antarctica and Greenland. Ice also covers approximately 7% of the oceans in the annual mean. In midwinter, snow covers approximately 49% of the land surface in the NH. An important property of snow and ice is its high surface albedo. Because up to 90% of the incident solar radiation is reflected by snow and ice surfaces, while only about 10% is reflected by the open ocean or forested lands, changes in snow and ice cover are important feedback mechanisms in climate change. In addition, snow and ice are effective insulators. Seasonally frozen ground is more extensive than snow cover, and its presence is important for energy and moisture fluxes. Therefore, frozen surfaces play important roles in energy and climate processes. {4.1}

The cryosphere stores about 75% of the world's freshwater. At a regional scale, variations in mountain snowpack, glaciers and small ice caps play a crucial role in freshwater availability. Since the change from ice to liquid water occurs at specific temperatures, ice is a component of the climate system that could be subject to abrupt change following sufficient warming. Observations and analyses of changes in ice have expanded and improved since the TAR, including shrinkage of mountain glacier volume, decreases in snow cover, changes in permafrost and frozen ground, reductions in arctic sea ice extent, coastal thinning of the Greenland Ice Sheet exceeding inland thickening from increased snowfall, and reductions in seasonally frozen ground and river and lake ice cover.

These allow an improved understanding of how the cryosphere is changing, including its contributions to recent changes in sea level. The periods from 1961 to the present and from 1993 to the present are a focus of this report, due to the availability of directly measured glacier mass balance data and altimetry observations of the ice sheets, respectively. {4.1}

Snow cover has decreased in most regions, especially in spring. Northern Hemisphere snow cover observed by satellite over the 1966 to 2005 period decreased in every month except November and December, with a stepwise drop of 5% in the annual mean in the late 1980s (see Figure TS.12). In the SH, the few long records or proxies mostly show either decreases or no changes in the past 40 years or more. Northern Hemisphere April snow cover extent is strongly correlated with 40°N to 60°N April temperature, reflecting the feedback between snow and temperature. {4.2}

Decreases in snowpack have been documented in several regions worldwide based upon annual time series of mountain snow water equivalent and snow depth. Mountain snow can be sensitive to small changes in temperature, particularly in temperate climatic zones where the transition from rain to snow is generally closely associated with the altitude of the freezing level. Declines in mountain snowpack in western North America and in the Swiss Alps are largest at lower, warmer elevations. Mountain snow water equivalent has declined since 1950 at 75% of the stations monitored in western North America. Mountain snow depth has also declined in the Alps and in southeastern Australia. Direct observations of snow depth are too limited to determine changes in the Andes, but temperature measurements suggest that the altitude where snow occurs (above the snow line) has probably risen in mountainous regions of South America. {4.2}

Permafrost and seasonally frozen ground in most regions display large changes in recent decades. Changes in permafrost conditions can affect river runoff, water supply, carbon exchange and landscape stability, and can cause damage to infrastructure. Temperature increases at the top of the permafrost layer of up to 3°C since the 1980s have been reported. Permafrost warming has also been observed with variable magnitude in the Canadian Arctic, Siberia, the Tibetan Plateau and Europe. The permafrost base is thawing at a rate ranging from 0.04 m yr^{-1} in Alaska to 0.02 m yr^{-1} on the Tibetan Plateau. {4.7}

The maximum area covered by seasonally frozen ground decreased by about 7% in the NH over the

---

[6] Precipitation and temperature are combined in the Palmer Drought Severity Index (PDSI) considered in this report as one measure of drought. The PDSI does not include variables such as wind speed, solar radiation, cloudiness and water vapour but is a superior measure to precipitation alone.

Box TS.3: Ice Sheet Dynamics and Stability

Ice sheets are thick, broad masses of ice formed mainly from compaction of snow. They spread under their own weight, transferring mass towards their margins where it is lost primarily by runoff of surface melt water or by calving of icebergs into marginal seas or lakes. Ice sheets flow by deformation within the ice or melt water-lubricated sliding over materials beneath. Rapid basal motion requires that the basal temperature be raised to the melting point by heat from the Earth's interior, delivered by melt water transport, or from the 'friction' of ice motion. Sliding velocities under a given gravitational stress can differ by several orders of magnitude, depending on the presence or absence of deformable sediment, the roughness of the substrate and the supply and distribution of water. Basal conditions are generally poorly characterised, introducing important uncertainties to the understanding of ice sheet stability. {4.6}

Ice flow is often channelled into fast-moving ice streams (that flow between slower-moving ice walls) or outlet glaciers (with rock walls). Enhanced flow in ice streams arises either from higher gravitational stress linked to thicker ice in bedrock troughs, or from increased basal lubrication. {4.6}

Ice discharged across the coast often remains attached to the ice sheet to become a floating ice shelf. An ice shelf moves forward, spreading and thinning under its own weight, and fed by snowfall on its surface and ice input from the ice sheet. Friction at ice shelf sides and over local shoals slows the flow of the ice shelf and thus the discharge from the ice sheet. An ice shelf loses mass by calving icebergs from the front and by basal melting into the ocean cavity beneath. Studies suggest an ocean warming of 1°C could increase ice shelf basal melt by 10 m yr^{-1}, but inadequate knowledge of the largely inaccessible ice shelf cavities restricts the accuracy of such estimates. {4.6}

The palaeo-record of previous ice ages indicates that ice sheets shrink in response to warming and grow in response to cooling, and that shrinkage can be far faster than growth. The volumes of the Greenland and Antarctic Ice Sheets are equivalent to approximately 7 m and 57 m of sea level rise, respectively. Palaeoclimatic data indicate that substantial melting of one or both ice sheets has likely occurred in the past. However, ice core data show that neither ice sheet was completely removed during warm periods of at least the past million years. Ice sheets can respond to environmental forcing over very long time scales, implying that commitments to future changes may result from current warming. For example, a surface warming may take more than 10,000 years to penetrate to the bed and change temperatures there. Ice velocity over most of an ice sheet changes slowly in response to changes in the ice sheet shape or surface temperature, but large velocity changes may occur rapidly in ice streams and outlet glaciers in response to changing basal conditions, penetration of surface melt water to the bed or changes in the ice shelves into which they flow. {4.6, 6.4}

Models currently configured for long integrations remain most reliable in their treatment of surface accumulation and ablation, as for the TAR, but do not include full treatments of ice dynamics; thus, analyses of past changes or future projections using such models may underestimate ice flow contributions to sea level rise, but the magnitude of such an effect is unknown. {8.2}

---

latter half of the 20th century, with a decrease in spring of up to 15%. Its maximum depth has decreased by about 0.3 m in Eurasia since the mid-20th century. In addition, maximum seasonal thaw depth increased by about 0.2 m in the Russian Arctic from 1956 to 1990. {4.7}

On average, the general trend in NH river and lake ice over the past 150 years indicates that the freeze-up date has become later at an average rate of 5.8 ± 1.9 days per century, while the breakup date has occurred earlier, at a rate of 6.5 ± 1.4 days per century. However, considerable spatial variability has also been observed, with some regions showing trends of opposite sign. {4.3}

Annual average arctic sea ice extent has shrunk by about 2.7 ± 0.6% per decade since 1978 based upon satellite observations (see Figure TS.13). The decline in summer extent is larger than in winter extent, with the summer minimum declining at a rate of about 7.4 ± 2.4% per decade. Other data indicate that the summer decline began around 1970. Similar observations in the Antarctic

reveal larger interannual variability but no consistent trends during the period of satellite observations. In contrast to changes in continental ice such as ice sheets and glaciers, changes in sea ice do not directly contribute to sea level change (because this ice is already floating), but can contribute to salinity changes through input of freshwater. {4.4}

During the 20th century, glaciers and ice caps have experienced widespread mass losses and have contributed to sea level rise. Mass loss of glaciers and ice caps (excluding those around the ice sheets of Greenland and Antarctica) is estimated to be 0.50 ± 0.18 mm yr^{-1} in sea level equivalent (SLE) between 1961 and 2003, and 0.77 ± 0.22 mm yr^{-1} SLE between 1991 and 2003. The late 20th-century glacier wastage *likely* has been a response to post-1970 warming. {4.5}

Recent observations show evidence for rapid changes in ice flow in some regions, contributing to sea level rise and suggesting that the dynamics of ice

motion may be a key factor in future responses of ice shelves, coastal glaciers and ice sheets to climate change. Thinning or loss of ice shelves in some near-coastal regions of Greenland, the Antarctic Peninsula and West Antarctica has been associated with accelerated flow of nearby glaciers and ice streams, suggesting that ice shelves (including short ice shelves of kilometres or tens of kilometres in length) could play a larger role

CHANGES IN SNOW COVER

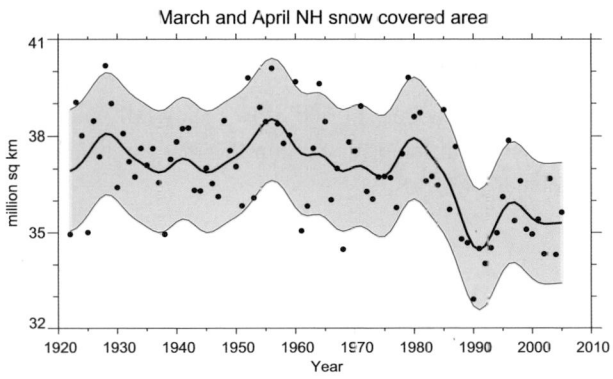

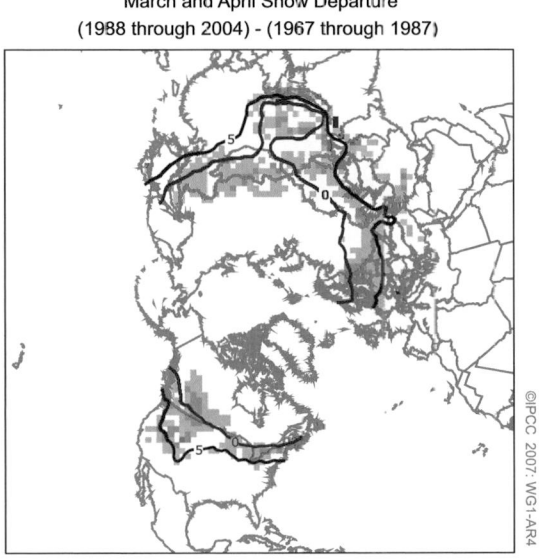

Figure TS.12. *(Top) Northern Hemisphere March-April snow-covered area from a station-derived snow cover index (prior to 1972) and from satellite data (during and after 1972). The smooth curve shows decadal variations (see Appendix 3.A) with the 5 to 95% data range shaded in yellow. (Bottom) Differences in the distribution of March-April snow cover between earlier (1967–1987) and later (1988–2004) portions of the satellite era (expressed in percent coverage). Tan colours show areas where snow cover has declined. Red curves show the 0°C and 5°C isotherms averaged for March-April 1967 to 2004, from the Climatic Research Unit (CRU) gridded land surface temperature version 2 (CRUTEM2v) data. The greatest decline generally tracks the 0°C and 5°C isotherms, reflecting the strong feedback between snow and temperature. {Figures 4.2, 4.3}*

in stabilising or restraining ice motion than previously thought. Both oceanic and atmospheric temperatures appear to contribute to the observed changes. Large summer warming in the Antarctic Peninsula region *very likely* played a role in the subsequent rapid breakup of the Larsen B Ice Shelf in 2002 by increasing summer melt water, which drained into crevasses and wedged them open. Models do not accurately capture all of the physical processes that appear to be involved in observed iceberg calving (as in the breakup of Larsen B). {4.6}

CHANGES IN SEA ICE EXTENT

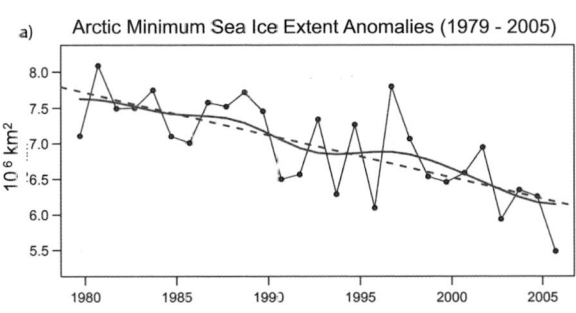

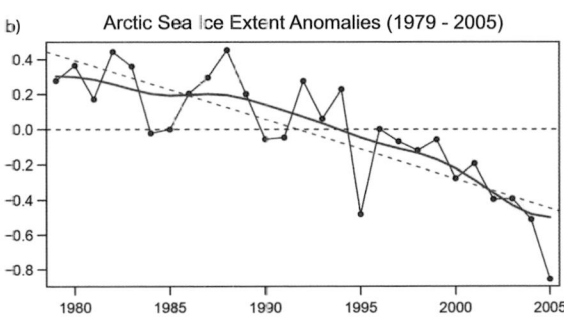

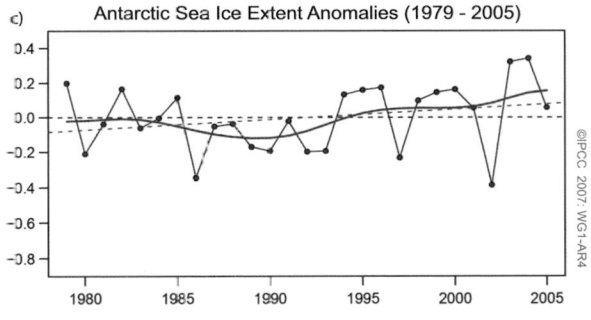

Figure TS.13. *(a) Arctic minimum sea ice extent; (b) arctic sea ice extent anomalies; and (c) antarctic sea ice extent anomalies all for the period 1979 to 2005. Symbols indicate annual values while the smooth blue curves show decadal variations (see Appendix 3.A). The dashed lines indicate the linear trends. (a) Results show a linear trend of −60 ± 20 x 10³ km² yr⁻¹, or approximately -7.4% per decade. (b) The linear trend is −33 ± 7.4 x 10³ km² yr⁻¹ (equivalent to approximately −2.7% per decade) and is significant at the 95% confidence level. (c) Antarctic results show a small positive trend of 5.6 ± 9.2 x 10³ km² yr⁻¹, which is not statistically significant. {Figures 4.8 and 4.9}*

The Greenland and Antarctic Ice Sheets taken together have *very likely* contributed to the sea level rise of the past decade. It is *very likely* that the Greenland Ice Sheet shrunk from 1993 to 2003, with thickening in central regions more than offset by increased melting in coastal regions. Whether the ice sheets have been growing or shrinking over time scales of longer than a decade is not well established from observations. Lack of agreement between techniques and the small number of estimates preclude assignment of best estimates or statistically rigorous error bounds for changes in ice sheet mass balances. However, acceleration of outlet glaciers drains ice from the interior and has been observed in both ice sheets (see Figure TS.14). Assessment of the data and techniques suggests a mass balance for the Greenland Ice Sheet of –50 to –100 Gt yr^{-1} (shrinkage contributing to raising global sea level by 0.14 to

0.28 mm yr^{-1}) during 1993 to 2003, with even larger losses in 2005. There are greater uncertainties for earlier time periods and for Antarctica. The estimated range in mass balance for the Greenland Ice Sheet over the period 1961 to 2003 is between growth of 25 Gt yr^{-1} and shrinkage by 60 Gt yr^{-1} (–0.07 to +0.17 mm yr^{-1} SLE). Assessment of all the data yields an estimate for the overall Antarctic Ice Sheet mass balance ranging from growth of 100 Gt yr^{-1} to shrinkage of 200 Gt yr^{-1} (–0.27 to +0.56 mm yr^{-1} SLE) from 1961 to 2003, and from +50 to –200 Gt yr^{-1} (–0.14 to +0.55 mm yr^{-1} SLE) from 1993 to 2003. The recent changes in ice flow are *likely* to be sufficient to explain much or all of the estimated antarctic mass imbalance, with recent changes in ice flow, snowfall and melt water runoff sufficient to explain the mass imbalance of Greenland. {4.6, 4.8}

RATES OF OBSERVED SURFACE ELEVATION CHANGE

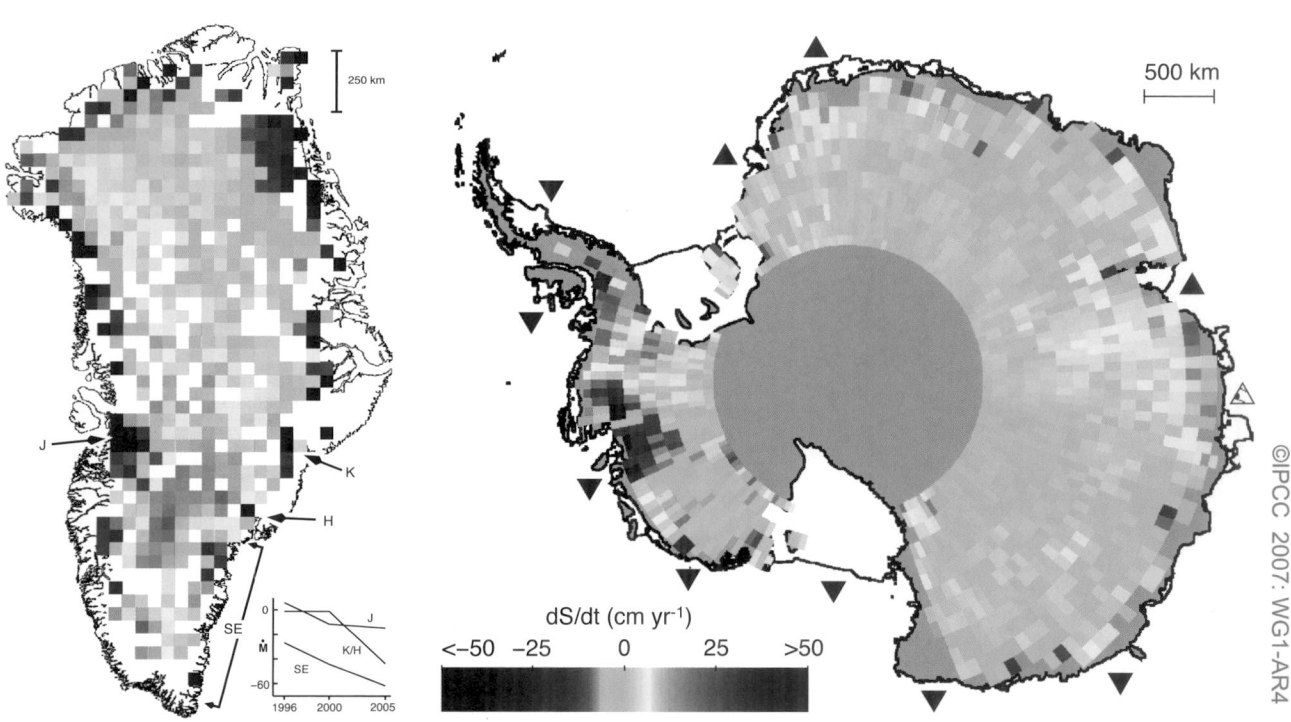

Figure TS.14. *Rates of observed recent surface elevation change for Greenland (left; 1989–2005) and Antarctica (right; 1992–2005). Red hues indicate a rising surface and blue hues a falling surface, which typically indicate an increase or loss in ice mass at a site, although changes over time in bedrock elevation and in near-surface density can be important. For Greenland, the rapidly thinning outlet glaciers Jakobshavn (J), Kangerdlugssuaq (K), Helheim (H) and areas along the southeast coast (SE) are shown, together with their estimated mass balance vs. time (with K and H combined, in Gt yr^{-1}, with negative values indicating loss of mass from the ice sheet to the ocean). For Antarctica, ice shelves estimated to be thickening or thinning by more than 30 cm yr^{-1} are shown by point-down purple triangles (thinning) and point-up red triangles (thickening) plotted just seaward of the relevant ice shelves. {Figures 4.17 and 4.19}*

TS.3.3 Changes in the Ocean: Instrumental Record

The ocean plays an important role in climate and climate change. The ocean is influenced by mass, energy and momentum exchanges with the atmosphere. Its heat capacity is about 1000 times larger than that of the atmosphere and the ocean's net heat uptake is therefore many times greater than that of the atmosphere (see Figure TS.15). Global observations of the heat taken up by the ocean can now be shown to be a definitive test of changes in the global energy budget. Changes in the amount of energy taken up by the upper layers of the ocean also play a crucial role for climate variations on seasonal to interannual time scales, such as El Niño. Changes in the transport of heat and SSTs have important effects upon many regional climates worldwide. Life in the sea is dependent on the biogeochemical status of the ocean and is affected by changes in its physical state and circulation. Changes in ocean biogeochemistry can also feed back into the climate system, for example, through changes in uptake or release of radiatively active gases such as CO_2. {5.1, 7.3}

ENERGY CONTENT IN THE CLIMATE SYSTEM

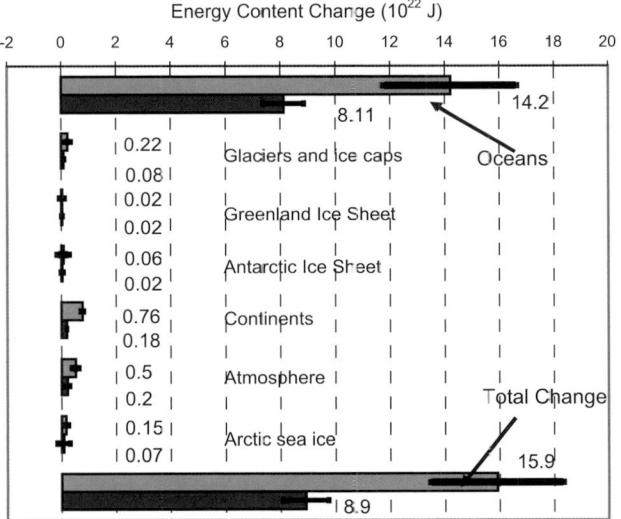

Figure TS.15. *Energy content changes in different components of the Earth system for two periods (1961–2003 and 1993–2003). Blue bars are for 1961 to 2003; burgundy bars are for 1993 to 2003. Positive energy content change means an increase in stored energy (i.e., heat content in oceans, latent heat from reduced ice or sea ice volumes, heat content in the continents excluding latent heat from permafrost changes, and latent and sensible heat and potential and kinetic energy in the atmosphere). All error estimates are 90% confidence intervals. No estimate of confidence is available for the continental heat gain. Some of the results have been scaled from published results for the two respective periods. {Figure 5.4}*

Global mean sea level variations are driven in part by changes in density, through thermal expansion or contraction of the ocean's volume. Local changes in sea level also have a density-related component due to temperature and salinity changes. In addition, exchange of water between oceans and other reservoirs (e.g., ice sheets, mountain glaciers, land water reservoirs and the atmosphere) can change the ocean's mass and hence contribute to changes in sea level. Sea level change is not geographically uniform because processes such as ocean circulation changes are not uniform across the globe (see Box TS.4). {5.5}

Oceanic variables can be useful for climate change detection, in particular temperature and salinity changes below the surface mixed layer where the variability is smaller and signal-to-noise ratio is higher. Observations analysed since the TAR have provided new evidence for changes in global ocean heat content and salinity, sea level, thermal expansion contributions to sea level rise, water mass evolution and biogeochemical cycles. {5.5}

TS.3.3.1 Changes in Ocean Heat Content and Circulation

The world ocean has warmed since 1955, accounting over this period for more than 80% of the changes in the energy content of the Earth's climate system. A total of 7.9 million vertical profiles of ocean temperature allows construction of improved global time series (see Figure TS.16). Analyses of the global oceanic heat budget have been replicated by several independent analysts and are robust to the method used. Data coverage limitations require averaging over decades for the deep ocean and observed decadal variability in the global heat content is not fully understood. However, inadequacies in the distribution of data (particularly coverage in the Southern Ocean and South Pacific) could contribute to the apparent decadal variations in heat content. During the period 1961 to 2003, the 0 to 3000 m ocean layer has taken up about 14.1 × 10²² J, equivalent to an average heating rate of 0.2 W m⁻² (per unit area of the Earth's surface). During 1993 to 2003, the corresponding rate of warming in the shallower 0 to 700 m ocean layer was higher, about 0.5 ± 0.18 W m⁻². Relative to 1961 to 2003, the period 1993 to 2003 had high rates of warming but in 2004 and 2005 there has been some cooling compared to 2003. {5.1–5.3}

Warming is widespread over the upper 700 m of the global ocean. The Atlantic has warmed south of 45°N. The warming is penetrating deeper in the Atlantic Ocean Basin than in the Pacific, Indian and Southern Oceans, due to the

GLOBAL OCEAN HEAT CONTENT (0 - 700 M)

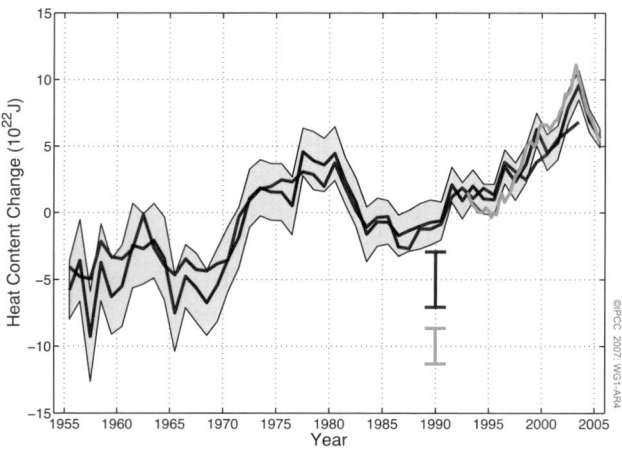

Figure TS.16. *Time series of global ocean heat content (10^{22} J) for the 0 to 700 m layer. The three coloured lines are independent analyses of the oceanographic data. The black and red curves denote the deviation from their 1961 to 1990 average and the shorter green curve denotes the deviation from the average of the black curve for the period 1993 to 2003. The 90% uncertainty range for the black curve is indicated by the grey shading and for the other two curves by the error bars. {Figure 5.1}*

deep overturning circulation cell that occurs in the North Atlantic. The SH deep overturning circulation shows little evidence of change based on available data. However, the upper layers of the Southern Ocean contribute strongly to the overall warming. At least two seas at subtropical latitudes (Mediterranean and Japan/East China Sea) are warming. While the global trend is one of warming, significant decadal variations have been observed in the global time series, and there are large regions where the oceans are cooling. Parts of the North Atlantic, North Pacific and equatorial Pacific have cooled over the last 50 years. The changes in the Pacific Ocean show ENSO-like spatial patterns linked in part to the PDO. {5.2, 5.3}

Parts of the Atlantic meridional overturning circulation exhibit considerable decadal variability, but data do not support a coherent trend in the overturning circulation. {5.3}

TS.3.3.2 Changes in Ocean Biogeochemistry and Salinity

The uptake of anthropogenic carbon since 1750 has led to the ocean becoming more acidic, with an average decrease in surface pH of 0.1 units.[7] Uptake of CO_2 by the ocean changes its chemical equilibrium. Dissolved

CO_2 forms a weak acid, so as dissolved CO_2 increases, pH decreases (i.e., the ocean becomes more acidic). The overall pH change is computed from estimates of anthropogenic carbon uptake and simple ocean models. Direct observations of pH at available stations for the last 20 years also show trends of decreasing pH, at a rate of about 0.02 pH units per decade. Decreasing ocean pH decreases the depth below which calcium carbonate dissolves and increases the volume of the ocean that is undersaturated with respect to the minerals aragonite (a meta-stable form of calcium carbonate) and calcite, which are used by marine organisms to build their shells. Decreasing surface ocean pH and rising surface temperatures also act to reduce the ocean buffer capacity for CO_2 and the rate at which the ocean can take up excess atmospheric CO_2. {5.4, 7.3}

The oxygen concentration of the ventilated thermocline (about 100 to 1000 m) decreased in most ocean basins between 1970 and 1995. These changes may reflect a reduced rate of ventilation linked to upper-level warming and/or changes in biological activity. {5.4}

There is now widespread evidence for changes in ocean salinity at gyre and basin scales in the past half century (see Figure TS.17) with the near-surface waters in the more evaporative regions increasing in salinity in almost all ocean basins. These changes in salinity imply changes in the hydrological cycle over the oceans. In the high-latitude regions in both hemispheres, the surface waters show an overall freshening consistent with these regions having greater precipitation, although higher runoff, ice melting, advection and changes in the meridional overturning circulation may also contribute. The subtropical latitudes in both hemispheres are characterised by an increase in salinity in the upper 500 m. The patterns are consistent with a change in the Earth's hydrological cycle, in particular with changes in precipitation and inferred larger water transport in the atmosphere from low latitudes to high latitudes and from the Atlantic to the Pacific. {5.2}

TS.3.3.3 Changes in Sea Level

Over the 1961 to 2003 period, the average rate of global mean sea level rise is estimated from tide gauge data to be 1.8 ± 0.5 mm yr^{-1} (see Figure TS.18). For the purpose of examining the sea level budget, best estimates and 5 to 95% confidence intervals are provided for all land ice contributions. The average

---

[7] Acidity is a measure of the concentration of H+ ions and is reported in pH units, where pH = –log(H+). A pH decrease of 1 unit means a 10-fold increase in the concentration of H+, or acidity.

LINEAR TRENDS OF ZONALLY AVERAGED SALINITY (1955 - 1998)

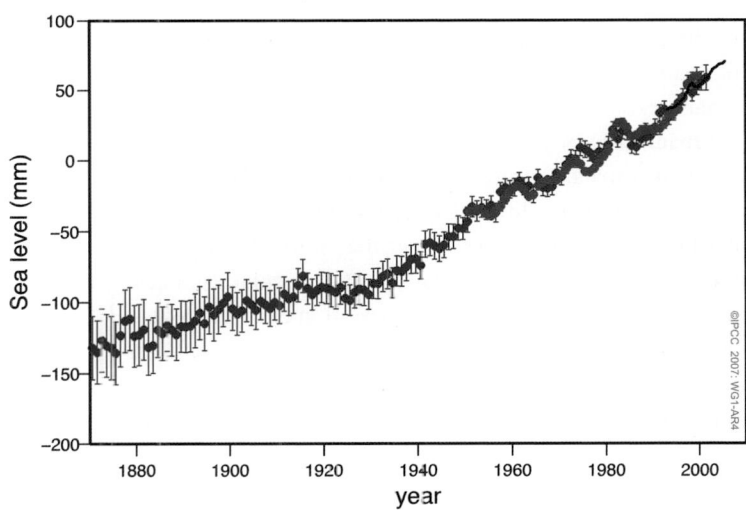

Figure TS.17. *Linear trends (1955–1998) of zonally averaged salinity (Practical Salinity Scale) for the World Ocean. The contour interval is 0.01 per decade and dashed contours are ±0.005 per decade. The dark, solid line is the zero contour. Red shading indicates values equal to or greater than 0.005 per decade and blue shading indicates values equal to or less than –0.005 per decade. {Figure 5.5}*

thermal expansion contribution to sea level rise for this period was 0.42 ± 0.12 mm yr^{-1}, with significant decadal variations, while the contribution from glaciers, ice caps and ice sheets is estimated to have been 0.7 ± 0.5 mm yr^{-1} (see Table TS.3). The sum of these estimated climate-related contributions for about the past four decades thus amounts to 1.1 ± 0.5 mm yr^{-1}, which is less than the best estimate from the tide gauge observations (similar to the discrepancy noted in the TAR). Therefore, the sea level budget for 1961 to 2003 has not been closed satisfactorily. {4.8, 5.5}

The global average rate of sea level rise measured by TOPEX/Poseidon satellite altimetry during 1993 to 2003 is 3.1 ± 0.7 mm yr^{-1}. This observed rate for the recent period is close to the estimated total of 2.8 ± 0.7 mm yr^{-1} for the climate-related contributions due to thermal expansion (1.6 ± 0.5 mm yr^{-1}) and changes in land ice (1.2 ± 0.4 mm yr^{-1}). Hence, the understanding of the budget has improved significantly for this recent period, with the climate contributions constituting the main factors in the sea level budget (which is closed to within known errors). Whether the faster rate for 1993 to 2003 compared to 1961 to 2003 reflects decadal variability or an increase in the longer-term trend is unclear. The

tide gauge record indicates that faster rates similar to that observed in 1993 to 2003 have occurred in other decades since 1950. {5.5, 9.5}

There is *high confidence* that the rate of sea level rise accelerated between the mid-19th and the mid-20th centuries based upon tide gauge and geological data. A recent reconstruction of sea level change back to 1870 using the best available tide records provides high confidence that the rate of sea level rise accelerated over the period 1870 to 2000. Geological observations indicate that during the previous 2000 years, sea level change was small, with average rates in the range 0.0 to 0.2 mm yr^{-1}. The use of proxy sea level data from archaeological sources is well established in the Mediterranean and indicates that oscillations in sea level from about AD 1 to AD 1900 did not exceed ±0.25 m. The available evidence indicates that the onset of modern sea level rise started between the mid-19th and mid-20th centuries. {5.5}

Precise satellite measurements since 1993 now provide unambiguous evidence of regional variability of sea level change. In some regions, rates of rise during this period are up to several times the global mean,

GLOBAL MEAN SEA LEVEL

Figure TS.18. *Annual averages of the global mean sea level based on reconstructed sea level fields since 1870 (red), tide gauge measurements since 1950 (blue) and satellite altimetry since 1992 (black). Units are in mm relative to the average for 1961 to 1990. Error bars are 90% confidence intervals. {Figure 5.13}*

Table TS.3. *Contributions to sea level rise based upon observations (left columns) compared to models used in this assessment (right columns; see Section 9.5 and Appendix 10.A for details). Values are presented for 1993 to 2003 and for the last four decades, including observed totals. {Adapted from Tables 5.3 and 9.2}*

| | Sea Level Rise (mm yr⁻¹) | | | |
| | 1961–2003 | | 1993–2003 | |
Sources of Sea Level Rise	Observed	Modelled	Observed	Modelled
Thermal expansion	0.42 ± 0.12	0.5 ± 0.2	1.6 ± 0.5	1.5 ± 0.7
Glaciers and ice caps	0.50 ± 0.18	0.5 ± 0.2	0.77 ± 0.22	0.7 ± 0.3
Greenland Ice Sheet	0.05 ± 0.12[a]		0.21 ± 0.07[a]	
Antarctic Ice Sheet	0.14 ± 0.41[a]		0.21 ± 0.35[a]	
Sum of individual climate contributions to sea level rise	1.1 ± 0.5	1.2 ± 0.5	2.8 ± 0.7	2.6 ± 0.8
Observed total sea level rise	1.8 ± 0.5 (tide gauges)		3.1 ± 0.7 (satellite altimeter)	
Difference (Observed total minus the sum of observed climate contributions)	0.7 ± 0.7		0.3 ± 1.0	

Notes:
[a] prescribed based upon observations (see Section 9.5)

while in other regions sea level is falling. The largest sea level rise since 1992 has taken place in the western Pacific and eastern Indian Oceans (see Figure TS.19). Nearly all of the Atlantic Ocean shows sea level rise during the past decade, while sea level in the eastern Pacific and western Indian Oceans has been falling. These temporal and spatial variations in regional sea level rise are influenced in part by patterns of coupled ocean-atmosphere variability, including ENSO and the NAO. The pattern of observed sea level change since 1992 is similar to the thermal expansion computed from ocean temperature changes, but different from the thermal expansion pattern of the last 50 years, indicating the importance of regional decadal variability. {5.5}

Observations suggest increases in extreme high water at a broad range of sites worldwide since 1975. Longer records are limited in space and under-sampled in time, so a global analysis over the entire 20th century is not feasible. In many locations, the secular changes in extremes were similar to those in mean sea level. At others, changes in atmospheric conditions such as storminess were more important in determining long-term trends. Interannual variability in high water extremes was positively correlated with regional mean sea level, as well as to indices of regional climate such as ENSO in the Pacific and NAO in the Atlantic. {5.5}

SEA LEVEL CHANGE PATTERNS

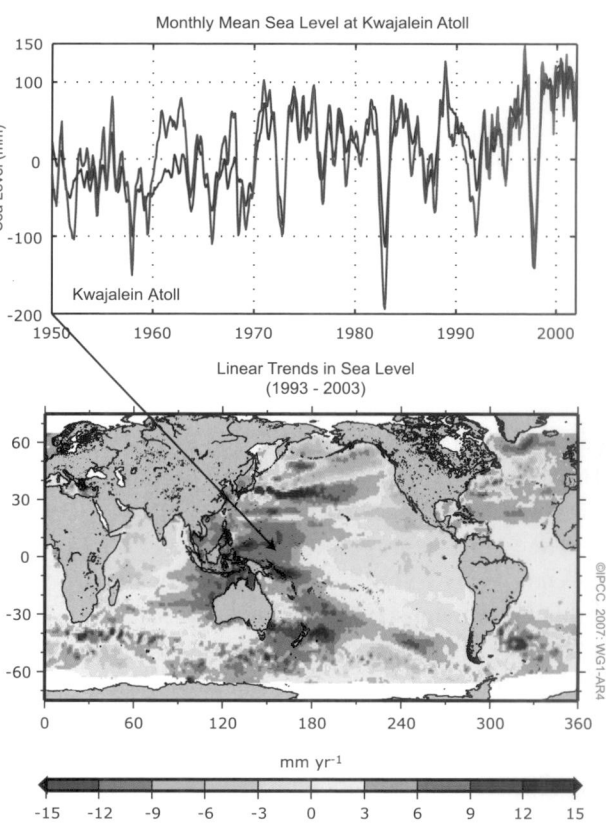

Figure TS.19. *(Top) Monthly mean sea level (mm) curve for 1950 to 2000 at Kwajalein (8°44'N, 167°44'E). The observed sea level (from tide gauge measurements) is in blue, the reconstructed sea level in red and the satellite altimetry record in green. Annual and semiannual signals have been removed from each time series and the tide gauge data have been smoothed. (Bottom) Geographic distribution of short-term linear trends in mean sea level for 1993 to 2003 (mm yr⁻¹) based on TOPEX/Poseidon satellite altimetry. {Figures 5.15 and 5.18}*

Box TS.4: Sea Level

The level of the sea at the shoreline is determined by many factors that operate over a great range of temporal scales: hours to days (tides and weather), years to millennia (climate), and longer. The land itself can rise and fall and such regional land movements need to be accounted for when using tide gauge measurements for evaluating the effect of oceanic climate change on coastal sea level. Coastal tide gauges indicate that global average sea level rose during the 20th century. Since the early 1990s, sea level has also been observed continuously by satellites with near-global coverage. Satellite and tide gauge data agree at a wide range of spatial scales and show that global average sea level has continued to rise during this period. Sea level changes show geographical variation because of several factors, including the distributions of changes in ocean temperature, salinity, winds and ocean circulation. Regional sea level is affected by climate variability on shorter time scales, for instance associated with El Niño and the NAO, leading to regional interannual variations which can be much greater or weaker than the global trend.

Based on ocean temperature observations, the thermal expansion of seawater as it warms has contributed substantially to sea level rise in recent decades. Climate models are consistent with the ocean observations and indicate that thermal expansion is expected to continue to contribute to sea level rise over the next 100 years. Since deep ocean temperatures change only slowly, thermal expansion would continue for many centuries even if atmospheric concentrations of greenhouse gases were stabilised.

Global average sea level also rises or falls when water is transferred from land to ocean or vice versa. Some human activities can contribute to sea level change, especially by the extraction of groundwater and construction of reservoirs. However, the major land store of freshwater is the water frozen in glaciers, ice caps and ice sheets. Sea level was more than 100 m lower during the glacial periods because of the ice sheets covering large parts of the NH continents. The present-day retreat of glaciers and ice caps is making a substantial contribution to sea level rise. This is expected to continue during the next 100 years. Their contribution should decrease in subsequent centuries as this store of freshwater diminishes.

The Greenland and Antarctic Ice Sheets contain much more ice and could make large contributions over many centuries. In recent years the Greenland Ice Sheet has experienced greater melting, which is projected to increase further. In a warmer climate, models suggest that the ice sheets could accumulate more snowfall, tending to lower sea level. However, in recent years any such tendency has probably been outweighed by accelerated ice flow and greater discharge observed in some marginal areas of the ice sheets. The processes of accelerated ice flow are not yet completely understood but could result in overall net sea level rise from ice sheets in the future.

The greatest climate- and weather-related impacts of sea level are due to extremes on time scales of days and hours, associated with tropical cyclones and mid-latitude storms. Low atmospheric pressure and high winds produce large local sea level excursions called 'storm surges', which are especially serious when they coincide with high tide. Changes in the frequency of occurrence of these extreme sea levels are affected both by changes in mean sea level and in the meteorological phenomena causing the extremes. {5.5}

TS.3.4   Consistency Among Observations

In this section, variability and trends within and across different climate variables including the atmosphere, cryosphere and oceans are examined for consistency based upon conceptual understanding of physical relationships between the variables. For example, increases in temperature will enhance the moisture-holding capacity of the atmosphere. Changes in temperature and/or precipitation should be consistent with those evident in glaciers. Consistency between independent observations using different techniques and variables provides a key test of understanding, and hence enhances confidence. {3.9}

Changes in the atmosphere, cryosphere and ocean show unequivocally that the world is warming. {3.2, 3.9, 4.2, 4.4–4.8, 5.2, 5.5}

Both land surface air temperatures and SSTs show warming. In both hemispheres, land regions have warmed at a faster rate than the oceans in the past few decades, consistent with the much greater thermal inertia of the oceans. {3.2}

The warming of the climate is consistent with observed increases in the number of daily warm extremes, reductions in the number of daily cold extremes and reductions in the number of frost days at mid-latitudes. {3.2, 3.8}

Surface air temperature trends since 1979 are now consistent with those at higher altitudes. It is *likely* that there is slightly greater warming in the troposphere than at the surface, and a higher tropopause, consistent with expectations from basic physical processes and observed increases in greenhouse gases together with depletion of stratospheric ozone. {3.4, 9.4}

Changes in temperature are broadly consistent with the observed nearly worldwide shrinkage of the cryosphere. There have been widespread reductions in mountain glacier mass and extent. Changes in climate consistent with warming are also indicated by decreases in snow cover, snow depth, arctic sea ice extent, permafrost thickness and temperature, the extent of seasonally frozen ground and the length of the freeze season of river and lake ice. {3.2, 3.9, 4.2–4.5, 4.7}

Observations of sea level rise since 1993 are consistent with observed changes in ocean heat content and the cryosphere. Sea level rose by 3.1 ± 0.7 mm yr^{-1} from 1993 to 2003, the period of availability of global altimetry measurements. During this time, a near balance was observed between observed total sea level rise and contributions from glacier, ice cap and ice sheet retreat together with increases in ocean heat content and associated ocean expansion. This balance gives increased

Table TS.4. *Recent trends, assessment of human influence on trends, and projections of extreme weather and climate events for which there is evidence of an observed late 20th-century trend. An asterisk in the column headed 'D' indicates that formal detection and attribution studies were used, along with expert judgement, to assess the likelihood of a discernible human influence. Where this is not available, assessments of likelihood of human influence are based on attribution results for changes in the mean of a variable or changes in physically related variables and/or on the qualitative similarity of observed and simulated changes, combined with expert judgement. {3.8, 5.5, 9.7, 11.2–11.9; Tables 3.7, 3.8, 9.4}*

Phenomenon[a] and direction of trend	Likelihood that trend occurred in late 20th century (typically post-1960)	Likelihood of a human contribution to observed trend	D	Likelihood of future trend based on projections for 21st century using SRES[b] scenarios
Warmer and fewer cold days and nights over most land areas	*Very likely[c]*	*Likely[e]*	*	*Virtually certain[e]*
Warmer and more frequent hot days and nights over most land areas	*Very likely[d]*	*Likely* (nights) [e]	*	*Virtually certain[e]*
Warm spells / heat waves: Frequency increases over most land areas	*Likely*	*More likely than not*		*Very likely*
Heavy precipitation events. Frequency (or proportion of total rainfall from heavy falls) increases over most areas	*Likely*	*More likely than not*		*Very likely*
Area affected by droughts increases	*Likely* in many regions since 1970s	*More likely than not*	*	*Likely*
Intense tropical cyclone activity increases	*Likely* in some regions since 1970	*More likely than not*		*Likely*
Increased incidence of extreme high sea level (excludes tsunamis)[f]	*Likely*	*More likely than not* [g]		*Likely* [h]

Notes:

[a] See Table 3.7 for further details regarding definitions.

[b] SRES refers to the IPCC Special Report on Emission Scenarios. The SRES scenario families and illustrative cases are summarised in a box at the end of the Summary for Policymakers.

[c] Decreased frequency of cold days and nights (coldest 10%)

[d] Increased frequency of hot days and nights (hottest 10%)

[e] Warming of the most extreme days/nights each year

[f] Extreme high sea level depends on average sea level and on regional weather systems. It is defined here as the highest 1% of hourly values of observed sea level at a station for a given reference period.

[g] Changes in observed extreme high sea level closely follow the changes in average sea level {5.5.2.6}. It is *very likely* that anthropogenic activity contributed to a rise in average sea level. {9.5.2}

[h] In all scenarios, the projected global average sea level at 2100 is higher than in the reference period {10.6}. The effect of changes in regional weather systems on sea level extremes has not been assessed.

confidence that the observed sea level rise is a strong indicator of warming. However, the sea level budget is not balanced for the longer period 1961 to 2003. {5.5, 3.9}

Observations are consistent with physical understanding regarding the expected linkage between water vapour and temperature, and with intensification of precipitation events in a warmer world. Column and upper-tropospheric water vapour have increased, providing important support for the hypothesis of simple physical models that specific humidity increases in a warming world and represents an important positive feedback to climate change. Consistent with rising amounts of water vapour in the atmosphere, there are widespread increases in the numbers of heavy precipitation events and increased likelihood of flooding events in many land regions, even those where there has been a reduction in total precipitation. Observations of changes in ocean salinity independently support the view

Box TS.5: Extreme Weather Events

People affected by an extreme weather event (e.g., the extremely hot summer in Europe in 2003, or the heavy rainfall in Mumbai, India in July 2005) often ask whether human influences on the climate are responsible for the event. A wide range of extreme weather events is expected in most regions even with an unchanging climate, so it is difficult to attribute any individual event to a change in the climate. In most regions, instrumental records of variability typically extend only over about 150 years, so there is limited information to characterise how extreme rare climatic events could be. Further, several factors usually need to combine to produce an extreme event, so linking a particular extreme event to a single, specific cause is problematic. In some cases, it may be possible to estimate the anthropogenic contribution to such changes in the probability of occurrence of extremes.

However, simple statistical reasoning indicates that substantial changes in the frequency of extreme events (and in the maximum feasible extreme, e.g., the maximum possible 24-hour rainfall at a specific location) can result from a relatively small shift of the distribution of a weather or climate variable.

Extremes are the infrequent events at the high and low end of the range of values of a particular variable. The probability of occurrence of values in this range is called a probability distribution function (PDF) that for some variables is shaped similarly to a 'Normal' or 'Gaussian' curve (the familiar 'bell' curve). Box TS.5, Figure 1 shows a schematic of a such a PDF and illustrates the effect a small shift (corresponding to a small change in the average or centre of the distribution) can have on the frequency of extremes at either end of the distribution. An increase in the frequency of one extreme (e.g., the number of hot days) will often be accompanied by a decline in the opposite extreme (in this case the number of cold days such as frosts). Changes in the variability or shape of the distribution can complicate this simple picture.

The IPCC Second Assessment Report noted that data and analyses of extremes related to climate change were sparse. By the time of the TAR, improved monitoring and data for changes in extremes was

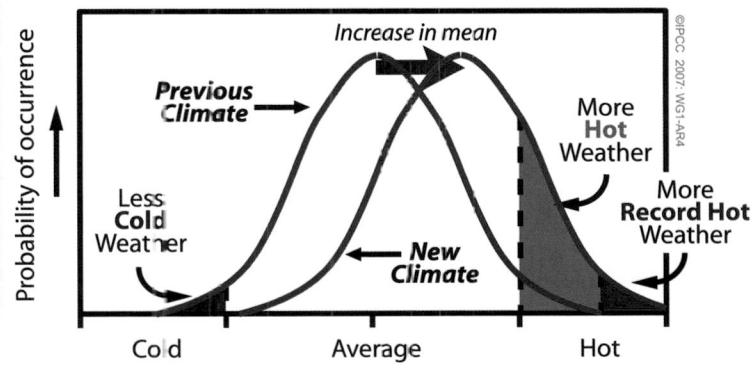

Box TS.5, Figure 1. Schematic showing the effect on extreme temperatures when the mean temperature increases, for a normal temperature distribution.

available, and climate models were being analysed to provide projections of extremes. Since the TAR, the observational basis of analyses of extremes has increased substantially, so that some extremes have now been examined over most land areas (e.g., daily temperature and rainfall extremes). More models have been used in the simulation and projection of extremes, and multiple integrations of models with different starting conditions (ensembles) now provide more robust information about PDFs and extremes. Since the TAR, some climate change detection and attribution studies focussed on changes in the global statistics of extremes have become available (Table TS.4). For some extremes (e.g., tropical cyclone intensity), there are still data concerns and/or inadequate models. Some assessments still rely on simple reasoning about how extremes might be expected to change with global warming (e.g., warming could be expected to lead to more heat waves). Others rely on qualitative similarity between observed and simulated changes. The assessed likelihood of anthropogenic contributions to trends is lower for variables where the assessment is based on indirect evidence.

that the Earth's hydrologic cycle has changed, in a manner consistent with observations showing greater precipitation and river runoff outside the tropics and subtropics, and increased transfer of freshwater from the ocean to the atmosphere at lower latitudes. {3.3, 3.4, 3.9, 5.2}

Although precipitation has increased in many areas of the globe, the area under drought has also increased. Drought duration and intensity has also increased. While regional droughts have occurred in the past, the widespread spatial extent of current droughts is broadly consistent with expected changes in the hydrologic cycle under warming. Water vapour increases with increasing global temperature, due to increased evaporation where surface moisture is available, and this tends to increase precipitation. However, increased continental temperatures are expected to lead to greater evaporation and drying, which is particularly important in dry regions where surface moisture is limited. Changes in snowpack, snow cover and in atmospheric circulation patterns and storm tracks can also reduce available seasonal moisture, and contribute to droughts. Changes in SSTs and associated changes in the atmospheric circulation and precipitation have contributed to changes in drought, particularly at low latitudes. The result is that drought has become more common, especially in the tropics and subtropics, since the 1970s. In Australia and Europe, direct links to global warming have been inferred through the extremes in high temperatures and heat waves accompanying recent droughts. {3.3, 3.8, 9.5}

TS.3.5 A Palaeoclimatic Perspective

Palaeoclimatic studies make use of measurements of past change derived from borehole temperatures, ocean sediment pore-water change and glacier extent changes, as well as proxy measurements involving the changes in chemical, physical and biological parameters that reflect past changes in the environment where the proxy grew or existed. Palaeoclimatic studies rely on multiple proxies so that results can be cross-verified and uncertainties better understood. It is now well accepted and verified that many biological organisms (e.g., trees, corals, plankton, animals) alter their growth and/or population dynamics in response to changing climate, and that these climate-induced changes are well recorded in past growth in living and dead (fossil) specimens or assemblages of organisms. Networks of tree ring width and tree ring density chronologies are used to infer past temperature changes based on calibration with temporally overlapping instrumental data. While these methods are heavily used, there are concerns regarding the distributions of available

measurements, how well these sample the globe, and such issues as the degree to which the methods have spatial and seasonal biases or apparent divergence in the relationship with recent climate change. {6.2}

It is *very likely* that average NH temperatures during the second half of the 20th century were warmer than any other 50-year period in the last 500 years and *likely* the warmest in at least the past 1300 years. The data supporting these conclusions are most extensive over summer extratropical land areas (particularly for the longer time period; see Figure TS.20). These conclusions are based upon proxy data such as the width and density of a tree ring, the isotopic composition of various elements in ice or the chemical composition of a growth band in corals, requiring analysis to derive temperature information and associated uncertainties. Among the key uncertainties are that temperature and precipitation are difficult to separate in some cases, or are representative of particular seasons rather than full years. There are now improved and expanded data since the TAR, including, for example, measurements at a larger number of sites, improved analysis of borehole temperature data and more extensive analyses of glaciers, corals and sediments. However, palaeoclimatic data are more limited than the instrumental record since 1850 in both space and time, so that statistical methods are employed to construct global averages, and these are subject to uncertainties as well. Current data are too limited to allow a similar evaluation of the SH temperatures prior to the period of instrumental data. {6.6, 6.7}

Some post-TAR studies indicate greater multi-centennial NH variability than was shown in the TAR, due to the particular proxies used and the specific statistical methods of processing and/or scaling them to represent past temperatures. The additional variability implies cooler conditions, predominantly during the 12th to 14th, the 17th and the 19th centuries; these are *likely* linked to natural forcings due to volcanic eruptions and/or solar activity. For example, reconstructions suggest decreased solar activity and increased volcanic activity in the 17th century as compared to current conditions. One reconstruction suggests slightly warmer conditions in the 11th century than those indicated in the TAR, but within the uncertainties quoted in the TAR. {6.6}

The ice core CO_2 record over the past millennium provides an additional constraint on natural climate variability. The amplitudes of the pre-industrial, decadal-scale NH temperature changes from the proxy-based reconstructions ($<1°C$) are broadly consistent with the ice core CO_2 record and understanding of the strength

NORTHERN HEMISPHERE TEMPERATURE RECONSTRUCTIONS

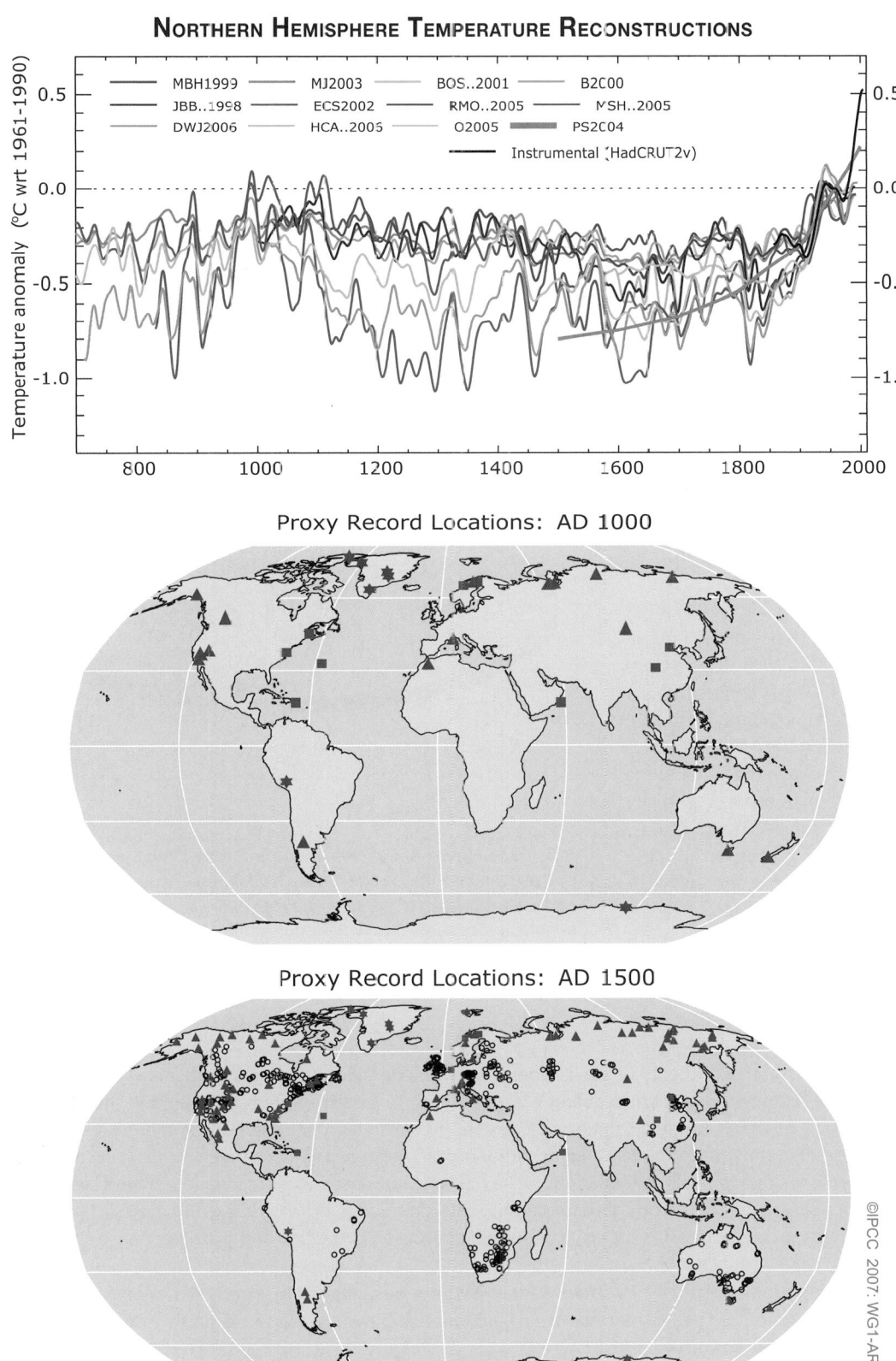

Figure TS.20. *(Top) Records of Northern Hemisphere temperature variation during the last 1300 years with 12 reconstructions using multiple climate proxy records shown in colour and instrumental records shown in black. (Middle and Bottom) Locations of temperature-sensitive proxy records with data back to AD 1000 and AD 1500 (tree rings: brown triangles; boreholes: black circles; ice core/ice boreholes: blue stars; other records including low-resolution records: purple squares) Data sources are given in Table 6.1, Figure 6.10 and are discussed in Chapter 6. {Figures 6.10 and 6.11}*

Box TS.6: Orbital Forcing

It is well known from astronomical calculations that periodic changes in characteristics of the Earth's orbit around the Sun control the seasonal and latitudinal distribution of incoming solar radiation at the top of the atmosphere (hereafter called 'insolation'). Past and future changes in insolation can be calculated over several millions of years with a high degree of confidence. {6.4}

Precession refers to changes in the time of the year when the Earth is closest to the Sun, with quasi-periodicities of about 19,000 and 23,000 years. As a result, changes in the position and duration of the seasons on the orbit strongly modulate the latitudinal and seasonal distribution of insolation. Seasonal changes in insolation are much larger than annual mean changes and can reach 60 W m^{-2} (Box TS.6, Figure 1).

The obliquity (tilt) of the Earth's axis varies between about 22° and 24.5° with two neighbouring quasi-periodicities of around 41,000 years. Changes in obliquity modulate seasonal contrasts as well as annual mean insolation changes with opposite effects at low vs. high latitudes (and therefore no effect on global average insolation) {6.4}.

The eccentricity of the Earth's orbit around the Sun has longer quasi-periodicities at 400,000 years and around 100,000 years. Changes in eccentricity alone have limited impacts on insolation, due to the resulting very small changes in the distance between the Sun and the Earth. However, changes in eccentricity interact with seasonal effects induced by obliquity and precession of the equinoxes. During periods of low eccentricity, such as about 400,000 years ago and during the next 100,000 years, seasonal insolation changes induced by precession are not as large as during periods of larger eccentricity (Box TS.6, Figure 1). {6.4}

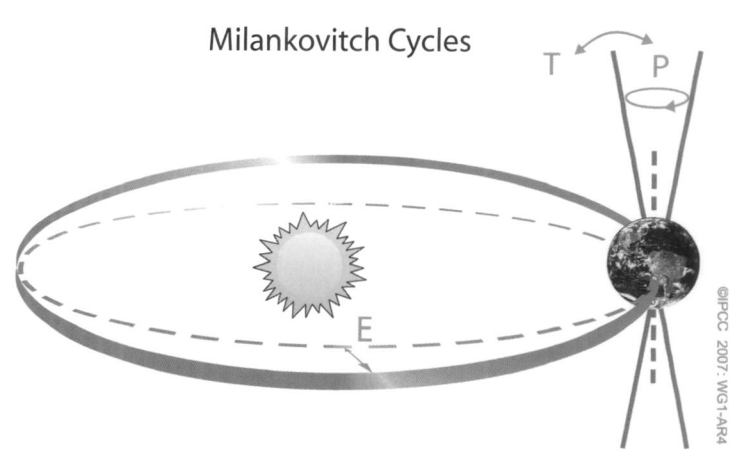

Box TS.6, Figure 1. *Schematic of the Earth's orbital changes (Milankovitch cycles) that drive the ice age cycles. 'T' denotes changes in the tilt (or obliquity) of the Earth's axis, 'E' denotes changes in the eccentricity of the orbit and 'P' denotes precession, that is, changes in the direction of the axis tilt at a given point of the orbit. {FAQ 6.1, Figure 1}*

The Milankovitch, or 'orbital' theory of the ice ages is now well developed. Ice ages are generally triggered by minima in high-latitude NH summer insolation, enabling winter snowfall to persist through the year and therefore accumulate to build NH glacial ice sheets. Similarly, times with especially intense high-latitude NH summer insolation, determined by orbital changes, are thought to trigger rapid deglaciations, associated climate change and sea level rise. These orbital forcings determine the pacing of climatic changes, while the large responses appear to be determined by strong feedback processes that amplify the orbital forcing. Over multi-millennial time scales, orbital forcing also exerts a major influence on key climate systems such as the Earth's major monsoons, global ocean circulation and the greenhouse gas content of the atmosphere. {6.4}

Available evidence indicates that the current warming will not be mitigated by a natural cooling trend towards glacial conditions. Understanding of the Earth's response to orbital forcing indicates that the Earth would not naturally enter another ice age for at least 30,000 years. {6.4, FAQ 6.1} of the carbon cycle-climate feedback. Atmospheric CO_2 and temperature in Antarctica co-varied over the past 650,000 years. Available data suggest that CO_2 acts as an amplifying feedback. {6.4, 6.6}

Changes in glaciers are evident in Holocene data, but these changes were caused by different processes than the late 20th-century retreat. Glaciers of several mountain regions in the NH retreated in response to orbitally forced regional warmth between 11,000 and 5000 years ago, and were smaller than at the end of the 20th century (or even absent) at times prior to 5000 years ago. The current near-global retreat of mountain glaciers cannot be due to the same causes, because decreased summer insolation during the past few thousand years in the NH should be favourable to the growth of glaciers. {6.5}

Palaeoclimatic data provide evidence for changes in many regional climates. The strength and frequency of ENSO events have varied in past climates. There is evidence that the strength of the Asian monsoon, and hence precipitation amount, can change abruptly. The palaeoclimatic records of northern and eastern Africa

and of North America indicate that droughts lasting decades to centuries are a recurrent feature of climate in these regions, so that recent droughts in North America and northern Africa are not unprecedented. Individual decadal-resolution palaeoclimatic data sets support the existence of regional quasi-periodic climate variability, but it is *unlikely* that these regional signals were coherent at the global scale. {6.5, 6.6}

Strong evidence from ocean sediment data and from modelling links abrupt climate changes during the last glacial period and glacial-interglacial transition to changes in the Atlantic Ocean circulation. Current understanding suggests that the ocean circulation can become unstable and change rapidly when critical thresholds are crossed. These events have affected temperature by up to 16°C in Greenland and have influenced tropical rainfall patterns. They were probably associated with a redistribution of heat between the NH and SH rather than with large changes in global mean temperature. Such events have not been observed during the past 8000 years. {6.4}

THE ARCTIC AND THE LAST INTERGLACIAL

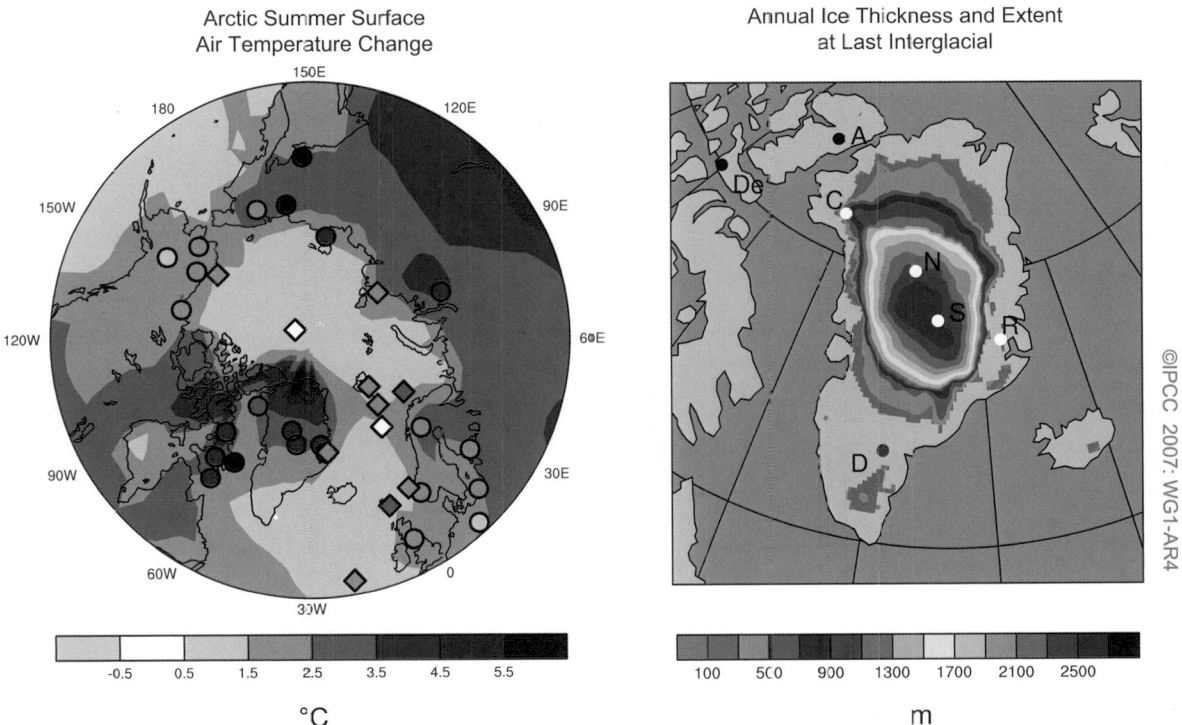

Figure TS.21. *Summer surface air temperature change relative to the present over the Arctic (left) and ice thickness and extent for Greenland and western arctic glaciers (right) for the last interglacial, approximately 125,000 years ago, from a multi-model and multi-proxy synthesis. (Left) A multi-model simulation of summer warming during the last interglacial is overlain by proxy estimates of maximum summer warming from terrestrial (circles) and marine (diamonds) sites. (Right) Extents and thicknesses of the Greenland Ice Sheet and western Canadian and Iceland glaciers at their minimum extent during the last interglacial, shown as a multi-model average from three ice models. Ice core observations indicate ice during the last interglacial at sites (white dots), Renland (R), North Greenland Ice Core Project (N), Summit (S, GRIP and GISP2) and possibly Camp Century (C), but no ice at sites (black dots): Devon (De) and Agassiz (A). Evidence for LIG ice at Dye-3 (D, grey dot) is equivocal. {Figure 6.6}*

Confidence in the understanding of past climate change and changes in orbital forcing is strengthened by the improved ability of current models to simulate past climate conditions. The Last Glacial Maximum (LGM; the last 'ice age' about 21,000 years ago) and the mid-Holocene (6000 years ago) were different from the current climate not because of random variability, but because of altered seasonal and global forcing linked to known differences in the Earth's orbit (see Box TS.6). Biogeochemical and biogeophysical feedbacks amplified the response to orbital forcings. Comparisons between simulated and reconstructed conditions in the LGM demonstrate that models capture the broad features of inferred changes in the temperature and precipitation patterns. For the mid-Holocene, coupled climate models are able to simulate mid-latitude warming and enhanced monsoons, with little change in global mean temperature (<0.4°C), consistent with our understanding of orbital forcing. {6.2, 6.4, 6.5, 9.3}

Global average sea level was likely between 4 and 6 m higher during the last interglacial period, about 125,000 years ago, than during the 20th century, mainly due to the retreat of polar ice (Figure TS.21). Ice core data suggest that the Greenland Summit region was ice-covered during this period, but reductions in the ice sheet extent are indicated in parts of southern Greenland. Ice core data also indicate that average polar temperatures at that time were 3°C to 5°C warmer than the 20th century because of differences in the Earth's orbit. The Greenland Ice Sheet and other arctic ice fields likely contributed no more than 4 m of the observed sea level rise, implying that there may also have been a contribution from Antarctica. {6.4}

TS.4 Understanding and Attributing Climate Change

Attribution evaluates whether observed changes are consistent with quantitative responses to different forcings obtained in well-tested models, and are not consistent with alternative physically plausible explanations. The first IPCC Assessment Report (FAR) contained little observational evidence of a detectable anthropogenic influence on climate. Six years later, the IPCC Second Assessment Report (SAR) concluded that the balance of evidence suggested a discernible human influence on the climate of the 20th century. The TAR concluded that 'most of the observed warming over the last 50 years

is likely to have been due to the increase in greenhouse gas concentrations'. Confidence in the assessment of the human contributions to recent climate change has increased considerably since the TAR, in part because of stronger signals obtained from longer records, and an expanded and improved range of observations allowing attribution of warming to be more fully addressed jointly with other changes in the climate system. Some apparent inconsistencies in the observational record (e.g., in the vertical profile of temperature changes) have been largely resolved. There have been improvements in the simulation of many aspects of present mean climate and its variability on seasonal to inter-decadal time scales, although uncertainties remain (see Box TS.7). Models now employ more detailed representations of processes related to aerosol and other forcings. Simulations of 20th-century climate change have used many more models and much more complete anthropogenic and natural forcings than were available for the TAR. Available multi-model ensembles increase confidence in attribution results by providing an improved representation of model uncertainty. An anthropogenic signal has now more clearly emerged in formal attribution studies of aspects of the climate system beyond global-scale atmospheric temperature, including changes in global ocean heat content, continental-scale temperature trends, temperature extremes, circulation and arctic sea ice extent. {9.1}

TS.4.1 Advances in Attribution of Changes in Global-Scale Temperature in the Instrumental Period: Atmosphere, Ocean and Ice

Anthropogenic warming of the climate system is widespread and can be detected in temperature observations taken at the surface, in the free atmosphere and in the oceans. {3.2, 3.4, 9.4}

Evidence of the effect of external influences, both anthropogenic and natural, on the climate system has continued to accumulate since the TAR. Model and data improvements, ensemble simulations and improved representations of aerosol and greenhouse gas forcing along with other influences lead to greater confidence that most current models reproduce large-scale forced variability of the atmosphere on decadal and inter-decadal time scales quite well. These advances confirm that past climate variations at large spatial scales have been strongly influenced by external forcings. However, uncertainties still exist in the magnitude and temporal evolution of estimated contributions from individual forcings other than well-mixed greenhouse gases, due, for

Box TS.7: Evaluation of Atmosphere-Ocean General Circulation Models

Atmosphere-ocean general circulation models (AOGCMs) are the primary tool used for understanding and attribution of past climate variations, and for future projections. Since there are no historical perturbations to radiative forcing that are fully analogous to the human-induced perturbations expected over the 21st century, confidence in the models must be built from a number of indirect methods, described below. In each of these areas there have been substantial advances since the TAR, increasing overall confidence in models. {8.1}

Enhanced scrutiny and analysis of model behaviour has been facilitated by internationally coordinated efforts to collect and disseminate output from model experiments performed under common conditions. This has encouraged a more comprehensive and open evaluation of models, encompassing a diversity of perspectives. {8.1}

Projections for different scales and different periods using global climate models. Climate models project the climate for several decades or longer into the future. Since the details of individual weather systems are not being tracked and forecast, the initial atmospheric conditions are much less important than for weather forecast models. For climate projections, the forcings are of much greater importance. These forcings include the amount of solar energy reaching the Earth, the amount of particulate matter from volcanic eruptions in the atmosphere, and the concentrations of anthropogenic gases and particles in the atmosphere. As the area of interest moves from global to regional to local, or the time scale of interest shortens, the amplitude of variability linked to weather increases relative to the signal of long-term climate change. This makes detection of the climate change signal more difficult at smaller scales. Conditions in the oceans are important as well, especially for interannual and decadal time scales. {FAQ 1.2, 9.4, 11.1}

Model formulation. The formulation of AOGCMs has developed through improved spatial resolution and improvements to numerical schemes and parametrizations (e.g., sea ice, atmospheric boundary layer, ocean mixing). More processes have been included in many models, including a number of key processes important for forcing (e.g., aerosols are now modelled interactively in many models). Most models now maintain a stable climate without use of flux adjustments, although some long-term trends remain in AOGCM control integrations, for example, due to slow processes in the ocean. {8.2, 8.3}

Simulation of present climate. As a result of improvements in model formulation, there have been improvements in the simulation of many aspects of present mean climate. Simulations of precipitation, sea level pressure and surface temperature have each improved overall, but deficiencies remain, notably in tropical precipitation. While significant deficiencies remain in the simulation of clouds (and corresponding feedbacks affecting climate sensitivity), some models have demonstrated improvements in the simulation of certain cloud regimes (notably marine stratocumulus). Simulation of extreme events (especially extreme temperature) has improved, but models generally simulate too little precipitation in the most extreme events. Simulation of extratropical cyclones has improved. Some models used for projections of tropical cyclone changes can simulate successfully the observed frequency and distribution of tropical cyclones. Improved simulations have been achieved for ocean water mass structure, the meridional overturning circulation and ocean heat transport. However most models show some biases in their simulation of the Southern Ocean, leading to some uncertainty in modelled ocean heat uptake when climate changes. {8.3, 8.5, 8.6}

Simulation of modes of climate variability. Models simulate dominant modes of extratropical climate variability that resemble the observed ones (NAM/SAM, PNA, PDO) but they still have problems in representing aspects of them. Some models can now simulate important aspects of ENSO, while simulation of the Madden-Julian Oscillation remains generally unsatisfactory. {8.4}

Simulation of past climate variations. Advances have been made in the simulation of past climate variations. Independently of any attribution of those changes, the ability of climate models to provide a physically self-consistent explanation of observed climate variations on various time scales builds confidence that the models are capturing many key processes for the evolution of 21st-century climate. Recent advances include success in modelling observed changes in a wider range of climate variables over the 20th century (e.g., continental-scale surface temperatures and extremes, sea ice extent, ocean heat content trends and land precipitation). There has also been progress in the ability to model many of the general features of past, very different climate states such as the mid-Holocene and the LGM using identical or related models to those used for studying current climate. Information on factors treated as boundary conditions in palaeoclimate calculations include the different states of ice sheets in those periods. The broad predictions of earlier climate models, of increasing global temperatures in response to increasing greenhouse gases, have been borne out by subsequent observations. This strengthens confidence in near-term climate projections and understanding of related climate change commitments. {6.4, 6.5, 8.1, 9.3–9.5}

(continued)

Weather and seasonal prediction using climate models. A few climate models have been tested for (and shown) capability in initial value prediction, on time scales from weather forecasting (a few days) to seasonal climate variations, when initialised with appropriate observations. While the predictive capability of models in this mode of operation does not necessarily imply that they will show the correct response to changes in climate forcing agents such as greenhouse gases, it does increase confidence that they are adequately representing some key processes and teleconnections in the climate system. {8.4}

Measures of model projection accuracy. The possibility of developing model capability measures ('metrics'), based on the above evaluation methods, that can be used to narrow uncertainty by providing quantitative constraints on model climate projections, has been explored for the first time using model ensembles. While these methods show promise, a proven set of measures has yet to be established. {8.1, 9.6, 10.5}

example, to uncertainties in model responses to forcing. Some potentially important forcings such as black carbon aerosols have not yet been considered in most formal detection and attribution studies. Uncertainties remain in estimates of natural internal climate variability. For example, there are discrepancies between estimates of ocean heat content variability from models and observations, although poor sampling of parts of the world ocean may explain this discrepancy. In addition, internal variability is difficult to estimate from available observational records since these are influenced by external forcing, and because records are not long enough in the case of instrumental data, or precise enough in the case of proxy reconstructions, to provide complete descriptions of variability on decadal and longer time scales (see Figure TS.22 and Box TS.7). {8.2–8.4, 8.6, 9.2–9.4}

It is *extremely unlikely* (<5%) that the global pattern of warming observed during the past half century can be explained without external forcing. These changes took place over a time period when non-anthropogenic forcing factors (i.e., the sum of solar and volcanic forcing) would be *likely* to have produced cooling, not warming (see Figure TS.23). Attribution studies show that it is *very likely* that these natural forcing factors alone cannot account for the observed warming (see Figure TS.23). There is also increased confidence that natural internal variability cannot account for the observed changes, due in part to improved studies demonstrating that the warming occurred in both oceans and atmosphere, together with observed ice mass losses. {2.9, 3.2, 5.2, 9.4, 9.5, 9.7}

It is *very likely* that anthropogenic greenhouse gas increases caused most of the observed increase in global average temperatures since the mid-20th century. Without the cooling effect of atmospheric aerosols, it is *likely* that greenhouse gases alone would have caused a greater global mean temperature rise than that observed during the last 50 years. A key factor in identifying the aerosol fingerprint, and therefore the amount of cooling counteracting greenhouse warming, is the temperature change through time (see Figure TS.23), as well as the hemispheric warming contrast. The conclusion that greenhouse gas forcing has been dominant takes into account observational and forcing uncertainties, and is robust to the use of different climate models, different methods for estimating the responses to external forcing and different analysis techniques. It also allows for possible amplification of the response to solar forcing. {2.9, 6.6, 9.1, 9.2, 9.4}

Widespread warming has been detected in ocean temperatures. Formal attribution studies now suggest that it is *likely* that anthropogenic forcing has contributed to the observed warming of the upper several hundred metres of the global ocean during the latter half of the 20th century. {5.2, 9.5}

Anthropogenic forcing has *likely* contributed to recent decreases in arctic sea ice extent. Changes in arctic sea ice are expected given the observed enhanced arctic warming. Attribution studies and improvements in the modelled representation of sea ice and ocean heat transport strengthen the confidence in this conclusion. {3.3, 4.4, 8.2, 8.3, 9.5}

It is *very likely* that the response to anthropogenic forcing contributed to sea level rise during the latter half of the 20th century, but decadal variability in sea level rise remains poorly understood. Modelled estimates of the contribution to sea level rise from thermal expansion are in good agreement with estimates based on observations during 1961 to 2003, although the budget for sea level rise over that interval is not closed. The observed increase in the rate of loss of mass from glaciers and ice caps is proportional to the global average temperature rise, as expected qualitatively from physical considerations (see Table TS.3). The greater rate of sea level rise in 1993 to 2003 than in 1961 to 2003 may be linked to increasing anthropogenic forcing, which has

GLOBAL AND CONTINENTAL TEMPERATURE CHANGE

Figure TS.22. *Comparison of observed continental- and global-scale changes in surface temperature with results simulated by climate models using natural and anthropogenic forcings. Decadal averages of observations are shown for the period 1906 to 2005 (black line) plotted against the centre of the decade and relative to the corresponding average for 1901 to 1950. Lines are dashed where spatial coverage is less than 50%. Blue shaded bands show the 5% to 95% range for 19 simulations from 5 climate models using only the natural forcings due to solar activity and volcanoes. Red shaded bands show the 5% to 95% range for 58 simulations from 14 climate models using both natural and anthropogenic forcings. Data sources and models used are described in Section 9.4, FAQ 9.2, Table 8.1 and the supplementary information for Chapter 9. {FAQ 9.2, Figure 1}*

likely contributed to the observed warming of the upper ocean and widespread glacier retreat. On the other hand, the tide gauge record of global mean sea level suggests that similarly large rates may have occurred in previous 10-year periods since 1950, implying that natural internal variability could also be a factor in the high rates for 1993 to 2003 period. Observed decadal variability in the tide gauge record is larger than can be explained by variability in observationally based estimates of thermal expansion

and land ice changes. Further, the observed decadal variability in thermal expansion is larger than simulated by models for the 20th century. Thus, the physical causes of the variability seen in the tide gauge record are uncertain. These unresolved issues relating to sea level change and its decadal variability during 1961 to 2003 make it unclear how much of the higher rate of sea level rise in 1993 to 2003 is due to natural internal variability and how much to anthropogenic climate change. {5.5, 9.5}

TS.4.2 Attribution of Spatial and Temporal Changes in Temperature

The observed pattern of tropospheric warming and stratospheric cooling is *very likely* due to the influence of anthropogenic forcing, particularly that due to greenhouse gas increases and stratospheric ozone depletion. New analyses since the TAR show that this pattern corresponds to an increase in the height of the tropopause that is *likely* due largely to greenhouse gas and stratospheric ozone changes. Significant uncertainty remains in the estimation of tropospheric temperature trends, particularly from the radiosonde record. {3.2, 3.4, 9.4}

It is *likely* that there has been a substantial anthropogenic contribution to surface temperature increases averaged over every continent except Antarctica since the middle of the 20th century. Antarctica has insufficient observational coverage to make an assessment. Anthropogenic warming has also been identified in some sub-continental land areas. The ability of coupled climate models to simulate the temperature evolution on each of six continents provides stronger evidence of human influence on the global climate than was available in the TAR. No coupled global climate model that has used natural forcing only has reproduced the observed global mean warming trend, or the continental mean warming trends in individual

GLOBAL MEAN SURFACE TEMPERATURE ANOMALIES

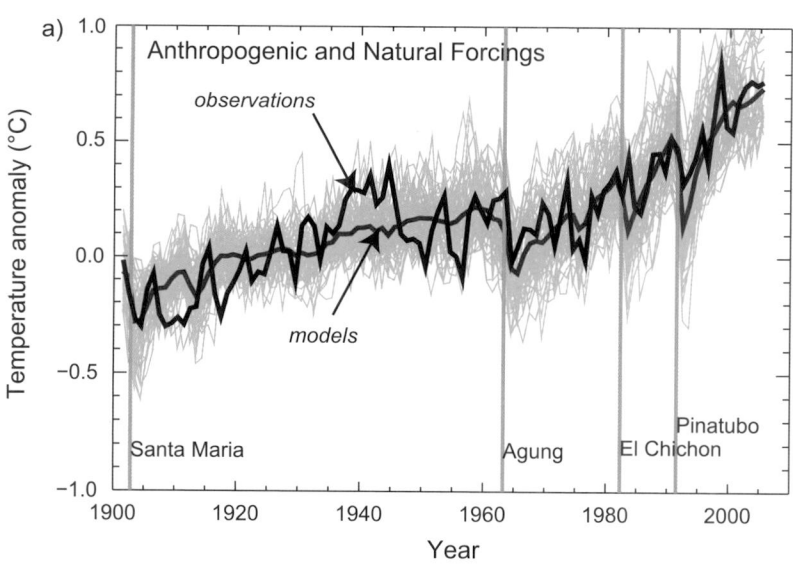

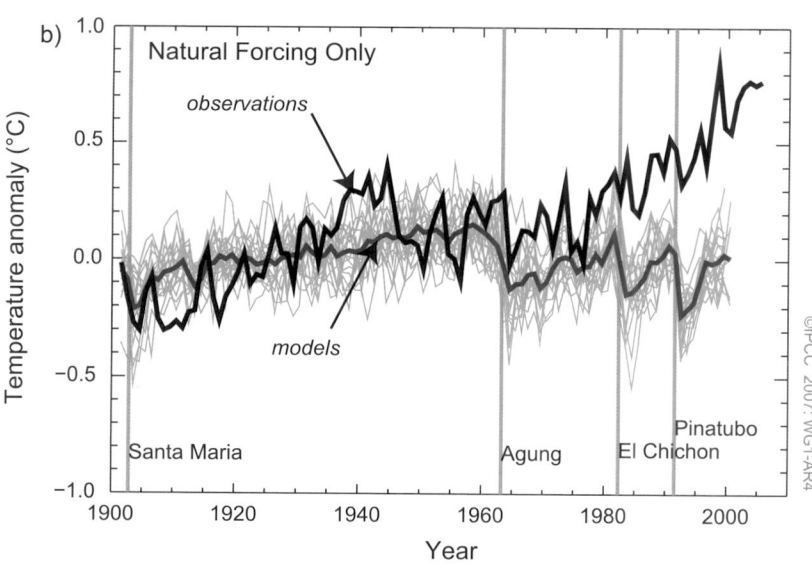

Figure TS.23. *(a) Global mean surface temperature anomalies relative to the period 1901 to 1950, as observed (black line) and as obtained from simulations with both anthropogenic and natural forcings. The thick red curve shows the multi-model ensemble mean and the thin yellow curves show the individual simulations. Vertical grey lines indicate the timing of major volcanic events. (b) As in (a), except that the simulated global mean temperature anomalies are for natural forcings only. The thick blue curve shows the multi-model ensemble mean and the thin lighter blue curves show individual simulations. Each simulation was sampled so that coverage corresponds to that of the observations. {Figure 9.5}*

©IPCC 2007: WG1-AR4

continents (except Antarctica) over the second half of the 20th century. {9.4}

Difficulties remain in attributing temperature changes at smaller than continental scales and over time scales of less than 50 years. Attribution results at these scales have, with limited exceptions, not been established. Averaging over smaller regions reduces the natural variability less than does averaging over large regions, making it more difficult to distinguish between changes expected from external forcing and variability. In addition, temperature changes associated with some modes of variability are poorly simulated by models in some regions and seasons. Furthermore, the small-scale

details of external forcing and the response simulated by models are less credible than large-scale features. {8.3, 9.4}

Surface temperature extremes have *likely* been affected by anthropogenic forcing. Many indicators of extremes, including the annual numbers and most extreme values of warm and cold days and nights, as well as numbers of frost days, show changes that are consistent with warming. Anthropogenic influence has been detected in some of these indices, and there is evidence that anthropogenic forcing may have substantially increased the risk of extremely warm summer conditions regionally, such as the 2003 European heat wave. {9.4}

DECEMBER - FEBRUARY SEA LEVEL PRESSURE TRENDS

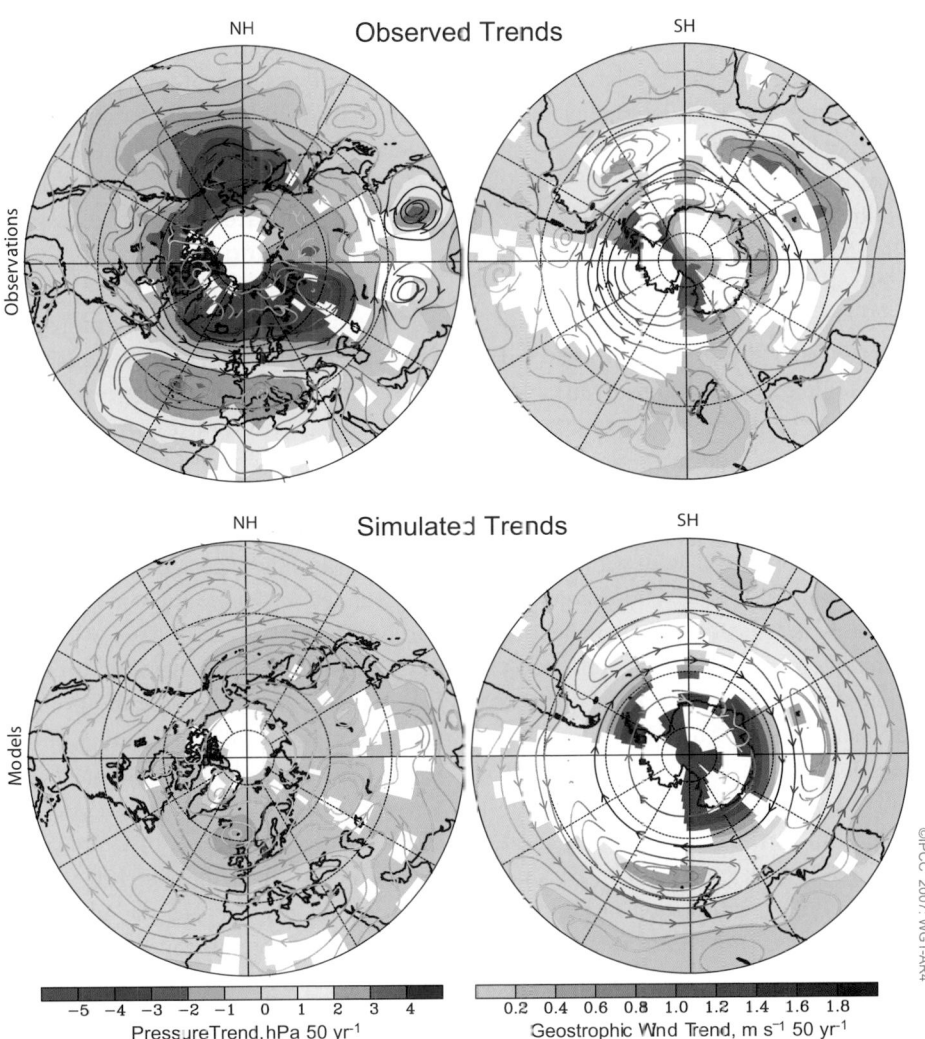

Figure TS.24. *December through February sea level pressure trends based on decadal means for the period 1955 to 2005. (Top) Trends estimated from an observational data set and displayed in regions where there is observational coverage. (Bottom) Mean trends simulated in response to natural and anthropogenic forcing changes in eight coupled models. The model-simulated trends are displayed only where observationally based trends are displayed. Streamlines, which are not masked, indicate the direction of the trends in the geostrophic wind derived from the trends in sea level pressure, and the shading of the streamlines indicates the magnitude of the change, with darker streamlines corresponding to larger changes in geostrophic wind. Data sources and models are described in Chapter 9 and its supplementary material, and Table 8.1 provides model details. {Figure 9.13}*

TS.4.3 Attribution of Changes in Circulation, Precipitation and Other Climate Variables

Trends in the Northern and Southern Annular Modes over recent decades, which correspond to sea level pressure reductions over the poles and related changes in atmospheric circulation, are *likely* **related in part to human activity (see Figure TS.24).** Models reproduce the sign of the NAM trend, but the simulated response is smaller than observed. Models including both greenhouse gas and stratospheric ozone changes simulate a realistic trend in the SAM, leading to a detectable human influence on global sea level pressure that is also consistent with the observed cooling trend in surface climate over parts of Antarctica. These changes in hemispheric circulation and their attribution to human activity imply that anthropogenic effects have *likely* contributed to changes in mid- and high-latitude patterns of circulation and temperature, as well as changes in winds and storm tracks. However, quantitative effects are uncertain because simulated responses to 20th century forcing change for the NH agree only qualitatively and not quantitatively with observations of these variables. {3.6, 9.5, 10.3}

There is some evidence of the impact of external influences on the hydrological cycle. The observed large-scale pattern of changes in land precipitation over the 20th century is qualitatively consistent with simulations, suggestive of a human influence. An observed global trend towards increases in drought in the second half of the 20th century has been reproduced with a model by taking anthropogenic and natural forcing into account. A number of studies have now demonstrated that changes in land use, due for example to overgrazing and conversion of woodland to agriculture, are *unlikely* to have been the primary cause of Sahelian and Australian droughts. Comparisons between observations and models suggest that changes in monsoons, storm intensities and Sahelian rainfall are related at least in part to changes in observed SSTs. Changes in global SSTs are expected to be affected by anthropogenic forcing, but an association of regional SST changes with forcing has not been established. Changes in rainfall depend not just upon SSTs but also upon changes in the spatial and temporal SST patterns and regional changes in atmospheric circulation, making attribution to human influences difficult. {3.3, 9.5, 10.3, 11.2}

TS.4.4 Palaeoclimate Studies of Attribution

It is *very likely* **that climate changes of at least the seven centuries prior to 1950 were not due to unforced variability alone.** Detection and attribution studies indicate that a substantial fraction of pre-industrial NH inter-decadal temperature variability contained in reconstructions for those centuries is *very likely* attributable to natural external forcing. Such forcing includes episodic cooling due to known volcanic eruptions, a number of which were larger than those of the 20th century (based on evidence such as ice cores), and long-term variations in solar irradiance, such as reduced radiation during the Maunder Minimum. Further, it is *likely* that anthropogenic forcing contributed to the early 20th-century warming evident in these records. Uncertainties are unlikely to lead to a spurious agreement between temperature reconstructions and forcing reconstructions as they are derived from independent proxies. Insufficient data are available to make a similar SH evaluation. {6.6, 9.3}

TS.4.5 Climate Response to Radiative Forcing

Specification of a *likely* **range and a most** *likely* **value for equilibrium climate sensitivity[8] in this report represents significant progress in quantifying the climate system response to radiative forcing since the TAR and an advance in challenges to understanding that have persisted for over 30 years.** A range for equilibrium climate sensitivity – the equilibrium global average warming expected if CO_2 concentrations were to be sustained at double their pre-industrial values (about 550 ppm) – was given in the TAR as between 1.5°C and 4.5°C. It has not been possible previously to provide a best estimate or to estimate the probability that climate sensitivity might fall outside that quoted range. Several approaches are used in this assessment to constrain climate sensitivity, including the use of AOGCMs, examination of the transient evolution of temperature (surface, upper air and ocean) over the last 150 years and examination of the rapid response of the global climate system to changes in the forcing caused by volcanic eruptions (see Figure TS.25). These are complemented by estimates based upon palaeoclimate studies such as reconstructions of the NH temperature record of the past millennium and the LGM. Large ensembles of climate model simulations have shown that the ability of models to simulate present climate has value in constraining climate sensitivity. {8.1, 8.6, 9.6, Box 10.2}

---

[8] See the Glossary for a detailed definition of climate sensitivity.

Analysis of models together with constraints from observations suggest that the equilibrium climate sensitivity is *likely* to be in the range 2°C to 4.5°C, with a best estimate value of about 3°C. It is *very unlikely* to be less than 1.5°C. Values substantially higher than 4.5°C cannot be excluded, but agreement with observations is not as good for those values. Probability density functions derived from different information and approaches generally tend to have a long tail towards high values exceeding 4.5°C. Analysis of climate and forcing evolution over previous centuries and model ensemble studies do not rule out climate sensitivity being as high as 6°C or more. One factor in this is the possibility of small net radiative forcing over the 20th century if aerosol indirect cooling effects were at the upper end of their uncertainty range, thus cancelling most of the positive forcing due to greenhouse gases. However, there is no well-established way of estimating a single probability distribution function from individual results taking account of the different assumptions in each study. The lack of strong constraints limiting high climate sensitivities prevents the specification of a 95th percentile bound or a very likely range for climate sensitivity. {Box 10.2}

There is now increased confidence in the understanding of key climate processes that are important to climate sensitivity due to improved analyses and comparisons of models to one another and to observations. Water vapour changes dominate the feedbacks affecting climate sensitivity and are now better understood. New observational and modelling evidence strongly favours a combined water vapour-lapse rate[9] feedback of around the strength found in General Circulation Models (GCMs), that is, approximately 1 W m^{-2} per degree global temperature increase, corresponding to about a 50% amplification of global mean warming. Such GCMs have demonstrated an ability to simulate seasonal to inter-decadal humidity variations in the upper troposphere over land and ocean, and have successfully simulated the observed surface temperature and humidity changes associated with volcanic eruptions. Cloud feedbacks (particularly from low clouds) remain the largest source of uncertainty. Cryospheric feedbacks such as changes in snow cover have been shown to contribute less to the spread in model estimates of climate sensitivity than cloud or water vapour feedbacks, but they

CUMULATIVE DISTRIBUTIONS OF CLIMATE SENSITIVITY

©IPCC 2007: WG1-AR4

Figure TS.25. *Cumulative distributions of climate sensitivity derived from observed 20th-century warming (red), model climatology (blue), proxy evidence (cyan) and from climate sensitivities of AOGCMs (green). Horizontal lines and arrows mark the boundaries of the likelihood estimates defined in the IPCC Fourth Assessment Uncertainty Guidance Note (see Box TS.1). {Box 10.2, Figures 1 and 2}*

---

[9] The rate at which air temperature decreases with altitude.

can be important for regional climate responses at mid- and high latitudes. A new model intercomparison suggests that differences in radiative transfer formulations also contribute to the range. {3.4, 8.6, 9.3, 9.4, 9.6, 10.2, Box 10.2}

Improved quantification of climate sensitivity allows estimation of best estimate equilibrium temperatures and ranges that could be expected if concentrations of CO_2 were to be stabilised at various levels based on global energy balance considerations (see Table TS.5). As in the estimate of climate sensitivity, a very likely upper bound cannot be established. Limitations to the concept of radiative forcing and climate sensitivity should be noted. Only a few AOGCMs have been run to equilibrium under elevated CO_2 concentrations, and some results show that climate feedbacks may change over long time scales, resulting in substantial deviations from estimates of warming based on equilibrium climate sensitivity inferred from mixed layer ocean models and past climate change. {10.7}

Agreement among models for projected transient climate change has also improved since the TAR. The range of transient climate responses (defined as the global average surface air temperature averaged over a 20-year period centred at the time of CO_2 doubling in a 1% yr^{-1} increase experiment) among models is smaller than the range in the equilibrium climate sensitivity. This parameter is now better constrained by multi-model ensembles and comparisons with observations; it is *very likely* to be greater than 1°C and *very unlikely* to be greater than 3°C. The transient climate response

Table TS.5. *Best estimate, likely ranges and very likely lower bounds of global mean equilibrium surface temperature increase (°C) over pre-industrial temperatures for different levels of CO_2-equivalent radiative forcing, as derived from the climate sensitivity.*

Equilibrium CO_2–eq (ppm)	Temperature Increase (°C)		
	Best Estimate	*Very Likely* Above	*Likely* in the Range
350	1.0	0.5	0.6–1.4
450	2.1	1.0	1.4–3.1
550	2.9	1.5	1.9–4.4
650	3.6	1.8	2.4–5.5
750	4.3	2.1	2.8–6.4
1000	5.5	2.8	3.7–8.3
1200	6.3	3.1	4.2–9.4

is related to sensitivity in a nonlinear way such that high sensitivities are not immediately manifested in the short-term response. Transient climate response is strongly affected by the rate of ocean heat uptake. Although the ocean models have improved, systematic model biases and limited ocean temperature data to evaluate transient ocean heat uptake affect the accuracy of current estimates. {8.3, 8.6, 9.4, 9.6, 10.5}

TS.5 Projections of Future Changes in Climate

Since the TAR, there have been many important advances in the science of climate change projections. An unprecedented effort has been initiated to make new model results available for prompt scrutiny by researchers outside of the modelling centres. A set of coordinated, standard experiments was performed by 14 AOGCM modelling groups from 10 countries using 23 models. The resulting multi-model database of outputs, analysed by hundreds of researchers worldwide, forms the basis for much of this assessment of model results. Many advances have come from the use of multi-member ensembles from single models (e.g., to test the sensitivity of response to initial conditions) and from multi-model ensembles. These two different types of ensembles allow more robust studies of the range of model results and more quantitative model evaluation against observations, and provide new information on simulated statistical variability. {8.1, 8.3, 9.4, 9.5, 10.1}

A number of methods for providing probabilistic climate change projections, both for global means and geographical depictions, have emerged since the TAR and are a focus of this report. These include methods based on results of AOGCM ensembles without formal application of observational constraints as well as methods based on detection algorithms and on large model ensembles that provide projections consistent with observations of climate change and their uncertainties. Some methods now explicitly account for key uncertainty sources such as climate feedbacks, ocean heat uptake, radiative forcing and the carbon cycle. Short-term projections are similarly constrained by observations of recent trends. Some studies have probed additional probabilistic issues, such as the likelihood of future changes in extremes such as heat waves that could occur due to human influences. Advances have also occurred since the TAR through broader ranges of studies of committed climate change and of carbon-climate feedbacks. {8.6, 9.6, 10.1, 10.3, 10.5}

These advances in the science of climate change modelling provide a probabilistic basis for distinguishing projections of climate change for different SRES marker scenarios. This is in contrast to the TAR where ranges for different marker scenarios could not be given in probabilistic terms. As a result, this assessment identifies and quantifies the difference in character between uncertainties that arise in climate modelling and those that arise from a lack of prior knowledge of decisions that will affect greenhouse gas emissions. A loss of policy-relevant information would result from combining probabilistic projections. For these reasons, projections for different emission scenarios are not combined in this report.

Model simulations used here consider the response of the physical climate system to a range of possible future conditions through use of idealised emissions or concentration assumptions. These include experiments with greenhouse gases and aerosols held constant at year 2000 levels, CO_2 doubling and quadrupling experiments, SRES marker scenarios for the 2000 to 2100 period, and experiments with greenhouse gases and aerosols held constant after 2100, providing new information on the physical aspects of long-term climate change and stabilisation. The SRES scenarios did not include climate initiatives. This Working Group I assessment does not evaluate the plausibility or likelihood of any specific emission scenario. {10.1, 10.3}

A new multi-model data set using Earth System Models of Intermediate Complexity (EMICs) complements AOGCM experiments to extend the time horizon for several more centuries in the future. This provides a more comprehensive range of model responses in this assessment as well as new information on climate change over long time scales when greenhouse gas and aerosol concentrations are held constant. Some AOGCMs and EMICs contain prognostic carbon cycle components, which permit estimation of the likely effects and associated uncertainties of carbon cycle feedbacks. {10.1}

Box TS.8: Hierarchy of Global Climate Models

Estimates of change in global mean temperature and sea level rise due to thermal expansion can be made using Simple Climate Models (SCMs) that represent the ocean-atmosphere system as a set of global or hemispheric boxes, and predict global surface temperature using an energy balance equation, a prescribed value of climate sensitivity and a basic representation of ocean heat uptake. Such models can also be coupled to simplified models of biogeochemical cycles and allow rapid estimation of the climate response to a wide range of emission scenarios. {8.8, 10.5}

Earth System Models of Intermediate Complexity (EMICs) include some dynamics of the atmospheric and oceanic circulations, or parametrizations thereof, and often include representations of biogeochemical cycles, but they commonly have reduced spatial resolution. These models can be used to investigate continental-scale climate change and long-term, large-scale effects of coupling between Earth system components using large ensembles of model runs or runs over many centuries. For both SCMs and EMICs it is computationally feasible to sample parameter spaces thoroughly, taking account of parameter uncertainties derived from tuning to more comprehensive climate models, matching observations and use of expert judgment. Thus, both types of model are well suited to the generation of probabilistic projections of future climate and allow a comparison of the 'response uncertainty' arising from uncertainty in climate model parameters with the 'scenario range' arising from the range of emission scenarios being considered. Earth System Models of Intermediate Complexity have been evaluated in greater depth than previously and intercomparison exercises have demonstrated that they are useful for studying questions involving long time scales or requiring large ensembles of simulations. {8.8, 10.5, 10.7}

The most comprehensive climate models are the AOGCMs. They include dynamical components describing atmospheric, oceanic and land surface processes, as well as sea ice and other components. Much progress has been made since the TAR (see Box TS.7), and there are over 20 models from different centres available for climate simulations. Although the large-scale dynamics of these models are comprehensive, parametrizations are still used to represent unresolved physical processes such as the formation of clouds and precipitation, ocean mixing due to wave processes and the formation of water masses, etc. Uncertainty in parametrizations is the primary reason why climate projections differ between different AOGCMs. While the resolution of AOGCMs is rapidly improving, it is often insufficient to capture the fine-scale structure of climatic variables in many regions. In such cases, the output from AOGCMs can be used to drive limited-area (or regional climate) models that combine the comprehensiveness of process representations comparable to AOGCMs with much higher spatial resolution. {8.2}

TS.5.1 Understanding Near-Term Climate Change

Knowledge of the climate system together with model simulations confirm that past changes in greenhouse gas concentrations will lead to a committed warming (see Box TS.9 for a definition) and future climate change. New model results for experiments in which concentrations of all forcing agents were held constant provide better estimates of the committed changes in atmospheric variables that would follow because of the long response time of the climate system, particularly the oceans. {10.3, 10.7}

Previous IPCC projections of future climate changes can now be compared to recent observations, increasing confidence in short-term projections and the underlying physical understanding of committed climate change over a few decades. Projections for 1990 to 2005 carried out for the FAR and the SAR suggested global mean temperature increases of about 0.3°C and 0.15°C per decade, respectively.[10] The difference between the two was due primarily to the inclusion of aerosol cooling effects in the SAR, whereas there was no quantitative basis for doing so in the FAR. Projections given in the TAR were similar to those of the SAR. These results are comparable to observed values of about 0.2°C per decade, as shown in Figure TS.26, providing broad confidence in such short-term projections. Some of this warming is the committed effect of changes in the concentrations of greenhouse gases prior to the times of those earlier assessments. {1.2, 3.2}

Committed climate change (see Box TS.9) due to atmospheric composition in the year 2000 corresponds to a warming trend of about 0.1°C per decade over the next two decades, in the absence of large changes in volcanic or solar forcing. About twice as much warming (0.2°C per decade) would be expected if emissions were to fall within the range of the SRES marker scenarios. This result is insensitive to the choice among the SRES marker scenarios, none of which considered climate initiatives. By 2050, the range of expected warming shows limited sensitivity to the choice among SRES scenarios (1.3°C to 1.7°C relative to 1980–1999) with about a quarter being due to the committed climate change if all radiative forcing agents were stabilised today. {10.3, 10.5, 10.7}

Sea level is expected to continue to rise over the next several decades. During 2000 to 2020 under the SRES A1B scenario in the ensemble of AOGCMs, the rate of thermal expansion is projected to be 1.3 ± 0.7 mm yr^{-1}, and is not significantly different under the A2 or B1 scenarios. These projected rates are within the uncertainty of the observed contribution of thermal expansion for 1993 to 2003 of 1.6 ± 0.6 mm yr^{-1}. The ratio of committed thermal expansion, caused by constant atmospheric composition at year 2000 values, to total thermal expansion (that is the ratio of expansion occurring after year 2000 to that occurring before and after) is larger than the corresponding ratio for global average surface temperature. {10.6, 10.7}

Box TS.9: Committed Climate Change

If the concentrations of greenhouse gases and aerosols were held fixed after a period of change, the climate system would continue to respond due to the thermal inertia of the oceans and ice sheets and their long time scales for adjustment. 'Committed warming' is defined here as the further change in global mean temperature after atmospheric composition, and hence radiative forcing, is held constant. Committed change also involves other aspects of the climate system, in particular sea level. Note that holding concentrations of radiatively active species constant would imply that ongoing emissions match natural removal rates, which for most species would be equivalent to a large reduction in emissions, although the corresponding model experiments are not intended to be considered as emission scenarios. {FAQ 10.3}

The troposphere adjusts to changes in its boundary conditions over time scales shorter than a month or so. The upper ocean responds over time scales of several years to decades, and the deep ocean and ice sheet response time scales are from centuries to millennia. When the radiative forcing changes, internal properties of the atmosphere tend to adjust quickly. However, because the atmosphere is strongly coupled to the oceanic mixed layer, which in turn is coupled to the deeper oceanic layer, it takes a very long time for the atmospheric variables to come to an equilibrium. During the long periods where the surface climate is changing very slowly, one can consider that the atmosphere is in a quasi-equilibrium state, and most energy is being absorbed by the ocean, so that ocean heat uptake is a key measure of climate change. {10.7}

---

[10] See IPCC First Assessment Report, Policymakers Summary, and Second Assessment Report, Technical Summary, Figure 18.

GLOBAL MEAN WARMING:
MODEL PROJECTIONS COMPARED WITH OBSERVATIONS

Figure TS.26. *Model projections of global mean warming compared to observed warming. Observed temperature anomalies, as in Figure TS.6, are shown as annual (black dots) and decadal average values (black line). Projected trends and their ranges from the IPCC First (FAR) and Second (SAR) Assessment Reports are shown as green and magenta solid lines and shaded areas, and the projected range from the TAR is shown by vertical blue bars. These projections were adjusted to start at the observed decadal average value in 1990. Multi-model mean projections from this report for the SRES B1, A1B and A2 scenarios, as in Figure TS.32, are shown for the period 2000 to 2025 as blue, green and red curves with uncertainty ranges indicated against the right-hand axis. The orange curve shows model projections of warming if greenhouse gas and aerosol concentrations were held constant from the year 2000 – that is, the committed warming. {Figures 1.1 and 10.4}*

TS.5.2   Large-Scale Projections for the 21st Century

This section covers advances in understanding global-scale climate projections and the processes that will influence their large-scale patterns in the 21st century. More specific discussion of regional-scale changes follows in TS.5.3.

Projected global average surface warming for the end of the 21st century (2090–2099) is scenario-dependent and the actual warming will be significantly affected by the actual emissions that occur. Warmings compared to 1980 to 1999 for six SRES scenarios[11] and for constant year 2000 concentrations, given as best estimates and corresponding *likely* ranges, are shown in Table TS.6. These results are based on AOGCMs, observational constraints and other methods to quantify the range of model response (see Figure TS.27). The combination of multiple lines of evidence allows likelihoods to be assigned to the resulting ranges, representing an important advance since the TAR. {10.5}

Assessed uncertainty ranges are larger than those given in the TAR because they consider a more complete range of models and climate-carbon cycle feedbacks. Warming tends to reduce land and ocean uptake of atmospheric CO_2, increasing the fraction of anthropogenic emissions that remains in the atmosphere. For the A2 scenario for example, the CO_2 feedback increases the corresponding global average warming in 2100 by more than 1°C. {7.3, 10.5}

---

[11] Approximate CO_2 equivalent concentrations corresponding to the computed radiative forcing due to anthropogenic greenhouse gases and aerosols in 2100 (see p. 823 of the TAR) for the SRES B1, A1T, B2, A1B, A2 and A1FI illustrative marker scenarios are about 600, 700, 800, 850, 1,250 and 1,550 ppm respectively. Constant emission at year 2000 levels would lead to a concentration for CO_2 alone of about 520 ppm by 2100.

Table TS.6. *Projected global average surface warming and sea level rise at the end of the 21st century. {10.5, 10.6, Table 10.7}*

| Case | Temperature Change (°C at 2090-2099 relative to 1980-1999) [a] | | Sea Level Rise (m at 2090-2099 relative to 1980-1999) |
	Best estimate	*Likely* range	Model-based range excluding future rapid dynamical changes in ice flow
Constant Year 2000 concentrations [b]	0.6	0.3 – 0.9	NA
B1 scenario	1.8	1.1 – 2.9	0.18 – 0.38
A1T scenario	2.4	1.4 – 3.8	0.20 – 0.45
B2 scenario	2.4	1.4 – 3.8	0.20 – 0.43
A1B scenario	2.8	1.7 – 4.4	0.21 – 0.48
A2 scenario	3.4	2.0 – 5.4	0.23 – 0.51
A1FI scenario	4.0	2.4 – 6.4	0.26 – 0.59

Notes:

[a] These estimates are assessed from a hierarchy of models that encompass a simple climate model, several Earth Models of Intermediate Complexity (EMICs), and a large number of Atmosphere-Ocean Global Circulation Models (AOGCMs).

[b] Year 2000 constant composition is derived from AOGCMs only.

Projected global-average sea level rise at the end of the 21st century (2090 to 2099), relative to 1980 to 1999 for the six SRES marker scenarios, given as 5% to 95% ranges based on the spread of model results, are shown in Table TS.6. Thermal expansion contributes 70 to 75% to the best estimate for each scenario. An improvement since the TAR is the use of AOGCMs to evaluate ocean heat uptake and thermal expansion. This has also reduced the projections as compared to the simple model used in the TAR. In all the SRES marker scenarios except B1, the average rate of sea level rise during the 21st century *very likely* exceeds the 1961–2003 average rate (1.8 ± 0.5 mm yr[-1]). For an average model, the scenario spread in sea level rise is only 0.02 m by the middle of the century, but by the end of the century it is 0.15 m. These ranges do not include uncertainties in carbon-cycle feedbacks or ice flow processes because a basis in published literature is lacking. {10.6, 10.7}

For each scenario, the midpoint of the range given here is within 10% of the TAR model average for 2090–2099, noting that the TAR projections were given for 2100, whereas projections in this report are for 2090–2099. The uncertainty in these projections is less than in the TAR for several reasons: uncertainty in land ice models is assumed independent of uncertainty in temperature and expansion projections; improved observations of recent mass loss from glaciers provide a better observational constraint; and the present report gives uncertainties as 5% to 95% ranges, equivalent to ±1.65 standard deviations, whereas the TAR gave

uncertainty ranges of ±2 standard deviations. The TAR would have had similar ranges for sea level projections to those in this report if it had treated the uncertainties in the same way. {10.6, 10.7}

Changes in the cryosphere will continue to affect sea level rise during the 21st century. Glaciers, ice caps and the Greenland Ice Sheet are projected to lose mass in the 21st century because increased melting will exceed increased snowfall. Current models suggest that the Antarctic Ice Sheet will remain too cold for widespread melting and may gain mass in future through increased snowfall, acting to reduce sea level rise. However, changes in ice dynamics could increase the contributions of both Greenland and Antarctica to 21st-century sea level rise. Recent observations of some Greenland outlet glaciers give strong evidence for enhanced flow when ice shelves are removed. The observations in west-central Greenland of seasonal variation in ice flow rate and of a correlation with summer temperature variation suggest that surface melt water may join a sub-glacially routed drainage system lubricating the ice flow. By both of these mechanisms, greater surface melting during the 21st century could cause acceleration of ice flow and discharge and increase the sea level contribution. In some parts of West Antarctica, large accelerations of ice flow have recently occurred, which may have been caused by thinning of ice shelves due to ocean warming. Although this has not been formally attributed to anthropogenic climate change due to greenhouse gases, it suggests that future warming could cause faster mass loss and greater

Projected Warming in 2090–2099

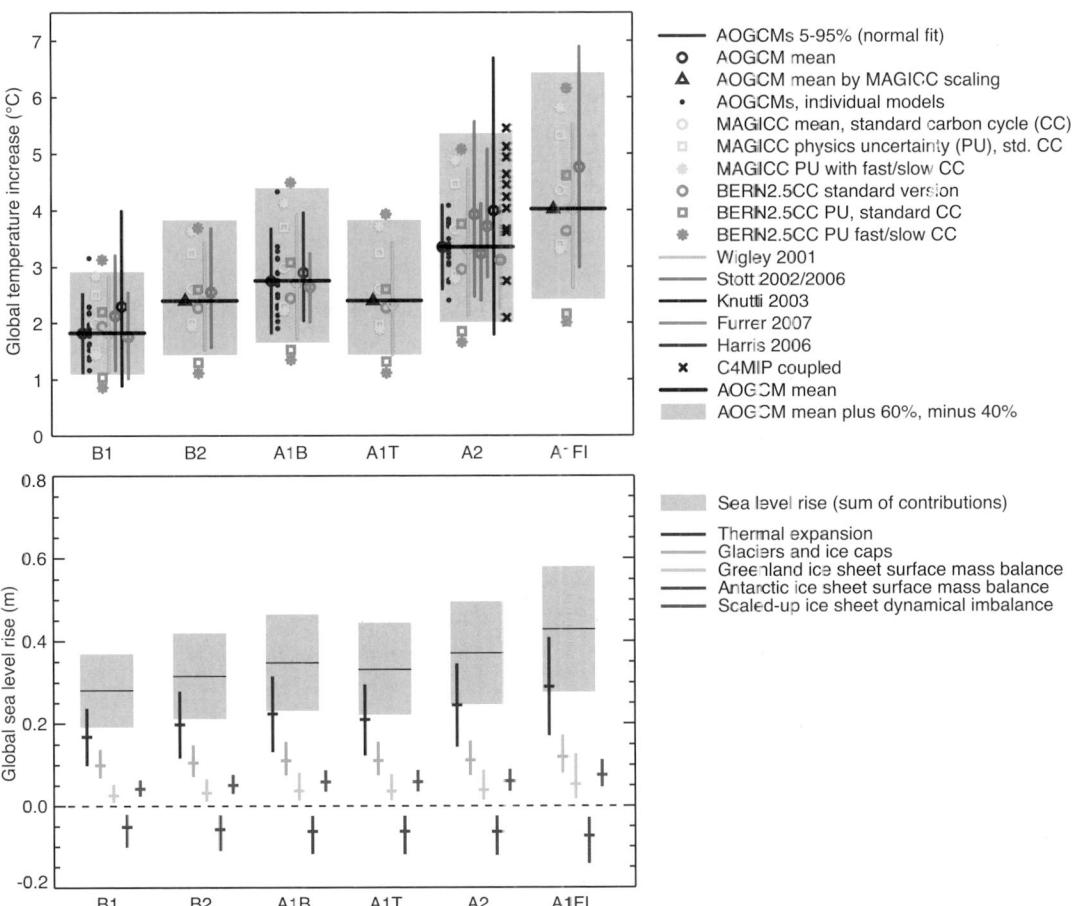

Figure TS.27. *(Top) Projected global mean temperature change in 2090 to 2099 relative to 1980 to 1999 for the six SRES marker scenarios based on results from different and independent models. The multi-model AOGCM mean and the range of the mean minus 40% to the mean plus 60% are shown as black horizontal solid lines and grey bars, respectively. Carbon cycle uncertainties are estimated for scenario A2 based on Coupled Carbon Cycle Climate Model Intercomparison Project (C4MIP) models (dark blue crosses), and for all marker scenarios using an EMIC (pale blue symbols). Other symbols represent individual studies (see Figure 10.29 for details of specific models). (Bottom) Projected global average sea level rise and its components in 2090 to 2099 (relative to 1980–1999) for the six SRES marker scenarios. The uncertainties denote 5 to 95% ranges, based on the spread of model results, and not including carbon cycle uncertainties. The contributions are derived by scaling AOGCM results and estimating land ice changes from temperature changes (see Appendix 10.A for details). Individual contributions are added to give the total sea level rise, which does not include the contribution shown for ice sheet dynamical imbalance, for which the current level of understanding prevents a best estimate from being given. {Figures 10.29 and 10.33}*

sea level rise. Quantitative projections of this effect cannot be made with confidence. If recently observed increases in ice discharge rates from the Greenland and Antarctic Ice Sheets were to increase linearly with global average temperature change, that would add 0.1 to 0.2 m to the upper bound of sea level rise. Understanding of these effects is too limited to assess their likelihood or to give a best estimate. {4.6, 10.6}

Many of the global and regional patterns of temperature and precipitation seen in the TAR projections remain in the new generation of models and across ensemble results (see Figure TS.28). Confidence in the robustness of these patterns is increased by the fact that they have remained largely unchanged while overall model simulations have improved (Box TS.7). This adds to confidence that these patterns reflect basic physical constraints on the climate system as it warms. {8.3–8.5, 10.3, 11.2–11.9}

The projected 21st-century temperature change is positive everywhere. It is greatest over land and at most high latitudes in the NH during winter, and increases going from the coasts into the continental interiors. In otherwise geographically similar areas, warming is typically larger in arid than in moist regions. {10.3, 11.2–11.9}

In contrast, warming is least over the southern oceans and parts of the North Atlantic Ocean. Temperatures are projected to increase, including over the North Atlantic and Europe, despite a projected slowdown of the meridional overturning circulation (MOC) in most models, due to the much larger influence of the increase in greenhouse gases. The projected pattern of zonal mean temperature change in the atmosphere displays a maximum warming in the upper tropical troposphere and cooling in the stratosphere. Further zonal mean warming in the ocean is expected to occur first near the surface and in the northern mid-latitudes, with the warming gradually reaching the ocean interior, most evident at high latitudes where vertical mixing is greatest. The projected pattern of change is very similar among the late-century cases irrespective of the scenario. Zonally averaged fields normalised by the mean warming are very similar for the scenarios examined (see Figure TS.28). {10.3}

It is *very likely* that the Atlantic MOC will slow down over the course of the 21st century. The multi-model average reduction by 2100 is 25% (range from zero to about 50%) for SRES emission scenario A1B. Temperatures in the Atlantic region are projected to increase despite such changes due to the much larger warming associated with projected increases of greenhouse gases. The projected reduction of the Atlantic MOC is due to the combined effects of an increase in high latitude temperatures and precipitation, which reduce the density of the surface waters in the North Atlantic. This could lead to a significant reduction in Labrador Sea Water formation. Very few AOGCM studies have included the impact of additional freshwater from melting of the Greenland Ice Sheet, but those that have do not suggest that this will lead to a complete MOC shutdown. Taken together, it is *very likely* that the MOC will reduce, but very *unlikely* that the MOC will undergo a large abrupt transition during the course of the 21st century. Longer-term changes in the MOC cannot be assessed with confidence. {8.7, 10.3}

PROJECTIONS OF SURFACE TEMPERATURES

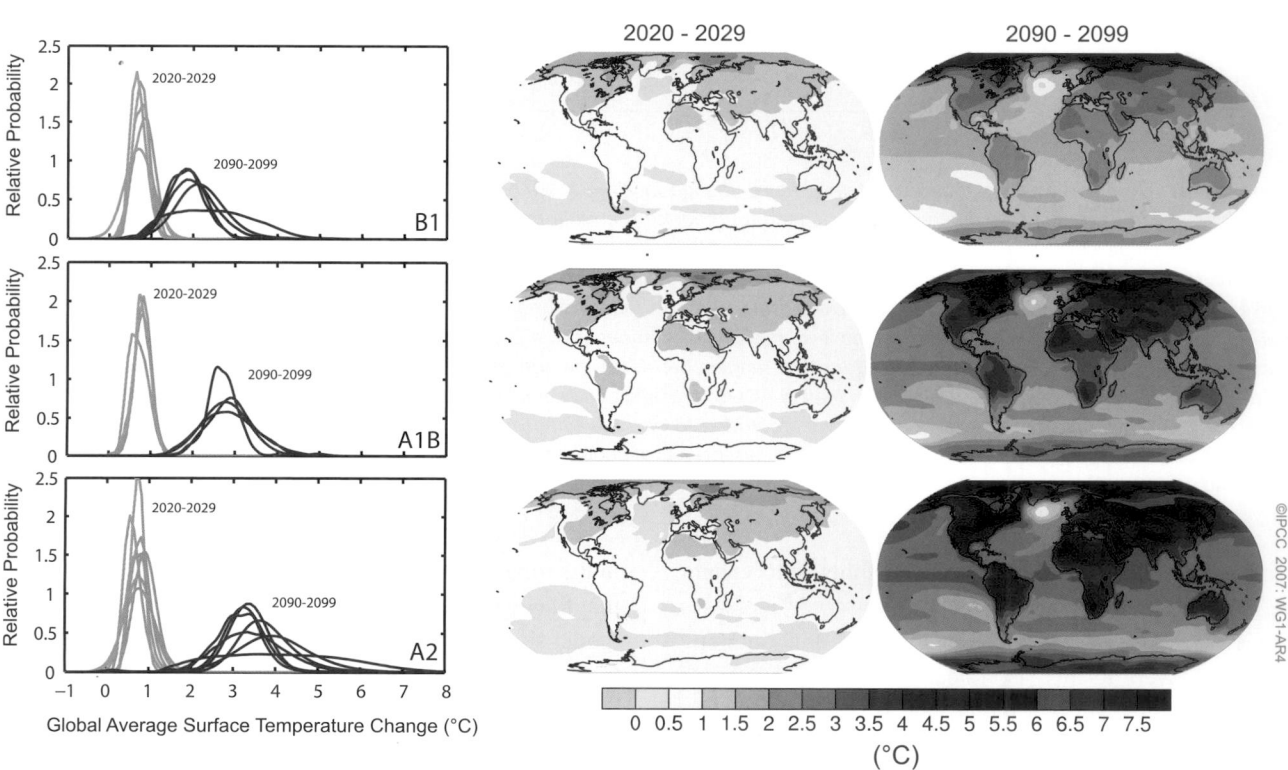

Figure TS.28. *Projected surface temperature changes for the early and late 21st century relative to the period 1980 to 1999. The central and right panels show the AOGCM multi-model average projections (°C) for the B1 (top), A1B (middle) and A2 (bottom) SRES scenarios averaged over the decades 2020 to 2029 (centre) and 2090 to 2099 (right). The left panel shows corresponding uncertainties as the relative probabilities of estimated global average warming from several different AOGCM and EMIC studies for the same periods. Some studies present results only for a subset of the SRES scenarios, or for various model versions. Therefore the difference in the number of curves, shown in the left-hand panels, is due only to differences in the availability of results. {Adapted from Figures 10.8 and 10.28}*

Models indicate that sea level rise during the 21st century will not be geographically uniform. Under scenario A1B for 2070 to 2099, AOGCMs give a median spatial standard deviation of 0.08 m, which is about 25% of the central estimate of the global average sea level rise. The geographic patterns of future sea level change arise mainly from changes in the distribution of heat and salinity in the ocean and consequent changes in ocean circulation. Projected patterns display more similarity across models than those analysed in the TAR. Common features are a smaller than average sea level rise in the Southern Ocean, larger than average sea level rise in the Arctic and a narrow band of pronounced sea level rise stretching across the southern Atlantic and Indian Oceans. {10.6}

Projections of changes in extremes such as the frequency of heat waves are better quantified than in the TAR, due to improved models and a better assessment of model spread based on multi-model ensembles. The TAR concluded that there was a risk of increased temperature extremes, with more extreme heat episodes in a future climate. This result has been confirmed and expanded in more recent studies. Future increases in temperature extremes are projected to follow increases in mean temperature over most of the world except where surface properties (e.g., snow cover or soil moisture) change. A multi-model analysis, based on simulations of 14 models for three scenarios, investigated changes in extreme seasonal (DJF and JJA) temperatures where 'extreme' is defined as lying above the 95th percentile of the simulated temperature distribution for the 20th century. By the end of the 21st century, the projected probability of extreme warm seasons rises above 90% in many tropical areas, and reaches around 40% elsewhere. Several recent studies have addressed possible future changes in heat waves, and found that, in a future climate, heat waves are expected to be more intense, longer lasting and more frequent. Based on an eight-member multi-model ensemble, heat waves are simulated to have been increasing for the latter part of the 20th century, and are projected to increase globally and over most regions. {8.5, 10.3}

For a future warmer climate, models project a 50 to 100% decline in the frequency of cold air outbreaks relative to the present in NH winters in most areas. Results from a nine-member multi-model ensemble show simulated decreases in frost days for the 20th century continuing into the 21st century globally and in most regions. Growing season length is related to frost days and is projected to increase in future climates. {10.3, FAQ 10.1}

Snow cover is projected to decrease. Widespread increases in thaw depth are projected to occur over most permafrost regions. {10.3}

Under several different scenarios (SRES A1B, A2 and B1), large parts of the Arctic Ocean are expected to no longer have year-round ice cover by the end of the 21st century. Arctic sea ice responds sensitively to warming. While projected changes in winter sea ice extent are moderate, late-summer sea ice is projected to disappear almost completely towards the end of the 21st century under the A2 scenario in some models. The reduction is accelerated by a number of positive feedbacks in the climate system. The ice-albedo feedback allows open water to receive more heat from the Sun during summer, the insulating effect of sea ice is reduced and the increase in ocean heat transport to the Arctic further reduces ice cover. Model simulations indicate that the late-summer sea ice cover decreases substantially and generally evolves over the same time scale as global warming. Antarctic sea ice extent is also projected to decrease in the 21st century. {8.6, 10.3, Box 10.1}

Sea level pressure is projected to increase over the subtropics and mid-latitudes, and decrease over high latitudes associated with an expansion of the Hadley Circulation and annular mode changes (NAM/NAO and SAM, see Box TS.2). A positive trend in the NAM/NAO as well as the SAM index is projected by many models. The magnitude of the projected increase is generally greater for the SAM, and there is considerable spread among the models. As a result of these changes, storm tracks are projected to move poleward, with consequent changes in wind, precipitation and temperature patterns outside the tropics, continuing the broad pattern of observed trends over the last half century. Some studies suggest fewer storms in mid-latitude regions. There are also indications of changes in extreme wave height associated with changing storm tracks and circulation. {3.6, 10.3}

In most models, the central and eastern equatorial Pacific SSTs warm more than those in the western equatorial Pacific, with a corresponding mean eastward shift in precipitation. ENSO interannual variability is projected to continue in all models, although changes differ from model to model. Large inter-model differences in projected changes in El Niño amplitude, and the inherent centennial time-scale variability of El Niño in the models, preclude a definitive projection of trends in ENSO variability. {10.3}

Recent studies with improved global models, ranging in resolution from about 100 to 20 km, suggest future changes in the number and intensity of future tropical cyclones (typhoons and hurricanes).

A synthesis of the model results to date indicates, for a warmer future climate, increased peak wind intensities and increased mean and peak precipitation intensities in future tropical cyclones, with the possibility of a decrease in the number of relatively weak hurricanes, and increased numbers of intense hurricanes. However, the total number of tropical cyclones globally is projected to decrease. The apparent observed increase in the proportion of very intense hurricanes since 1970 in some regions is in the same direction but much larger than predicted by theoretical models. {10.3, 8.5, 3.8}

Since the TAR, there is an improving understanding of projected patterns of precipitation. Increases in the amount of precipitation are *very likely* at high latitudes while decreases are *likely* in most subtropical land regions (by as much as about 20% in the A1B scenario in 2100). Poleward of 50°, mean precipitation is projected to increase due to the increase in water vapour in the atmosphere and the resulting increase in vapour transport from lower latitudes. Moving equatorward, there is a transition to mostly decreasing precipitation in the subtropics (20°–40° latitude). Due to increased water vapour transport out of the subtropics and a poleward expansion of the subtropical high-pressure systems, the drying tendency is especially pronounced at the higher-latitude margins of the subtropics (see Figure TS.30). {8.3, 10.3, 11.2–11.9}

Models suggest that changes in mean precipitation amount, even where robust, will rise above natural variability more slowly than the temperature signal. {10.3, 11.1}

Available research indicates a tendency for an increase in heavy daily rainfall events in many regions, including some in which the mean rainfall is projected to decrease. In the latter cases, the rainfall decrease is often attributable to a reduction in the number of rain days rather than the intensity of rain when it occurs. {11.2–11.9}

TS.5.3   Regional-Scale Projections

For each of the continental regions, the projected warming over 2000 to 2050 resulting from the SRES emissions scenarios is greater than the global average and greater than the observed warming over the past century. The warming projected for the next few decades of the 21st century, when averaged over the continents individually, would substantially exceed estimated 20th-century natural forced and unforced variability in all cases except Antarctica (Figure TS.29). Model best-estimate projections indicate that decadal average warming over each continent except Antarctica by 2030 is *very likely* to be at least twice as large as the corresponding model-estimated natural variability during the 20th century. The simulated warming over this period is not very sensitive to the choice of scenarios across the SRES set as is illustrated in Figure TS.32. Over longer time scales, the choice of scenario is more important, as shown in Figure TS.28. The projected warming in the SRES scenarios over 2000 to 2050 also exceeds estimates of natural variability when averaged over most sub-continental regions. {11.1}

Box TS.10. Regional Downscaling

Simulation of regional climates has improved in AOGCMs and, as a consequence, in nested regional climate models and in empirical downscaling techniques. Both dynamic and empirical downscaling methodologies show improving skill in simulating local features in present-day climates when the observed state of the atmosphere at scales resolved by current AOGCMs is used as input. The availability of downscaling and other regionally focused studies remains uneven geographically, causing unevenness in the assessments that can be provided, particularly for extreme weather events. Downscaling studies demonstrate that local precipitation changes can vary significantly from those expected from the large-scale hydrological response pattern, particularly in areas of complex topography. {11.10}

There remain a number of important sources of uncertainty limiting the ability to project regional climate change. While hydrological responses are relatively robust in certain core subpolar and subtropical regions, there is uncertainty in the precise location of these boundaries between increasing and decreasing precipitation. There are some important climate processes that have a significant effect on regional climate, but for which the climate change response is still poorly known. These include ENSO, the NAO, blocking, the thermohaline circulation and changes in tropical cyclone distribution. For those regions that have strong topographical controls on their climatic patterns, there is often insufficient climate change information at the fine spatial resolution of the topography. In some regions there has been only very limited research on extreme weather events. Further, the projected climate change signal becomes comparable to larger internal variability at smaller spatial and temporal scales, making it more difficult to utilise recent trends to evaluate model performance. {Box 11.1, 11.2–11.9}

CONTINENTAL SURFACE TEMPERATURE ANOMALIES:
OBSERVATIONS AND PROJECTIONS

legend:
- models using natural forcing only
- models using both anthropogenic and natural forcings
- projected changes (A1B scenario)
- range of anomalies with natural forcing only in 20th century simulations
- observations

©IPCC 2007: WG1-AR4

Figure TS.29. *Decadal mean continental surface temperature anomalies (°C) in observations and simulations for the period 1906 to 2005 and in projections for 2001 to 2050. Anomalies are calculated from the 1901 to 1950 average. The black lines represent the observations and the red and blue bands show simulated average temperature anomalies as in Figure TS.22 for the 20th century (i.e., red includes anthropogenic and natural forcings and blue includes only natural forcings). The yellow shading represents the 5th to 95th percentile range of projected changes according to the SRES A1B emissions scenario. The green bar denotes the 5th to 95th percentile range of decadal mean anomalies from the 20th-century simulations with only natural forcings (i.e., a measure of the natural decadal variability). For the observed part of these graphs, the decadal averages are centred on calendar decade boundaries (i.e., the last point is at 2000 for 1996 to 2005), whereas for the future period they are centred on calendar decade mid-points (i.e., the first point is at 2005 for 2001 to 2010). To construct the ranges, all simulations from the set of models involved were considered independent realisations of the possible evolution of the climate given the forcings applied. This involved 58 simulations from 14 models for the red curve, 19 simulations from 5 models (a subset of the 14) for the blue curve and green bar and 47 simulations from 18 models for the yellow curve. {FAQ 9.2.1, Figure 1 and Box 11.1, Figure 1}*

In the NH a robust pattern of increased subpolar and decreased subtropical precipitation dominates the projected precipitation pattern for the 21st century over North America and Europe, while subtropical drying is less evident over Asia (see Figure TS.30). Nearly all models project increased precipitation over most of northern North America and decreased precipitation over Central America, with much of the continental USA and northern Mexico in a more uncertain transition zone that moves north and south following the seasons. Decreased precipitation is confidently projected for southern Europe and Mediterranean Africa, with a transition to increased precipitation in northern Europe. In both continents, summer drying is extensive due both to the poleward movement of this transition zone in summer and to increased evaporation. Subpolar increases in precipitation are projected over much of northern Asia but with the subtropical drying spreading from the Mediterranean displaced by distinctive monsoonal signatures as one moves from central Asia eastward. {11.2–11.5}

SEASONAL MEAN PRECIPITATION RATES

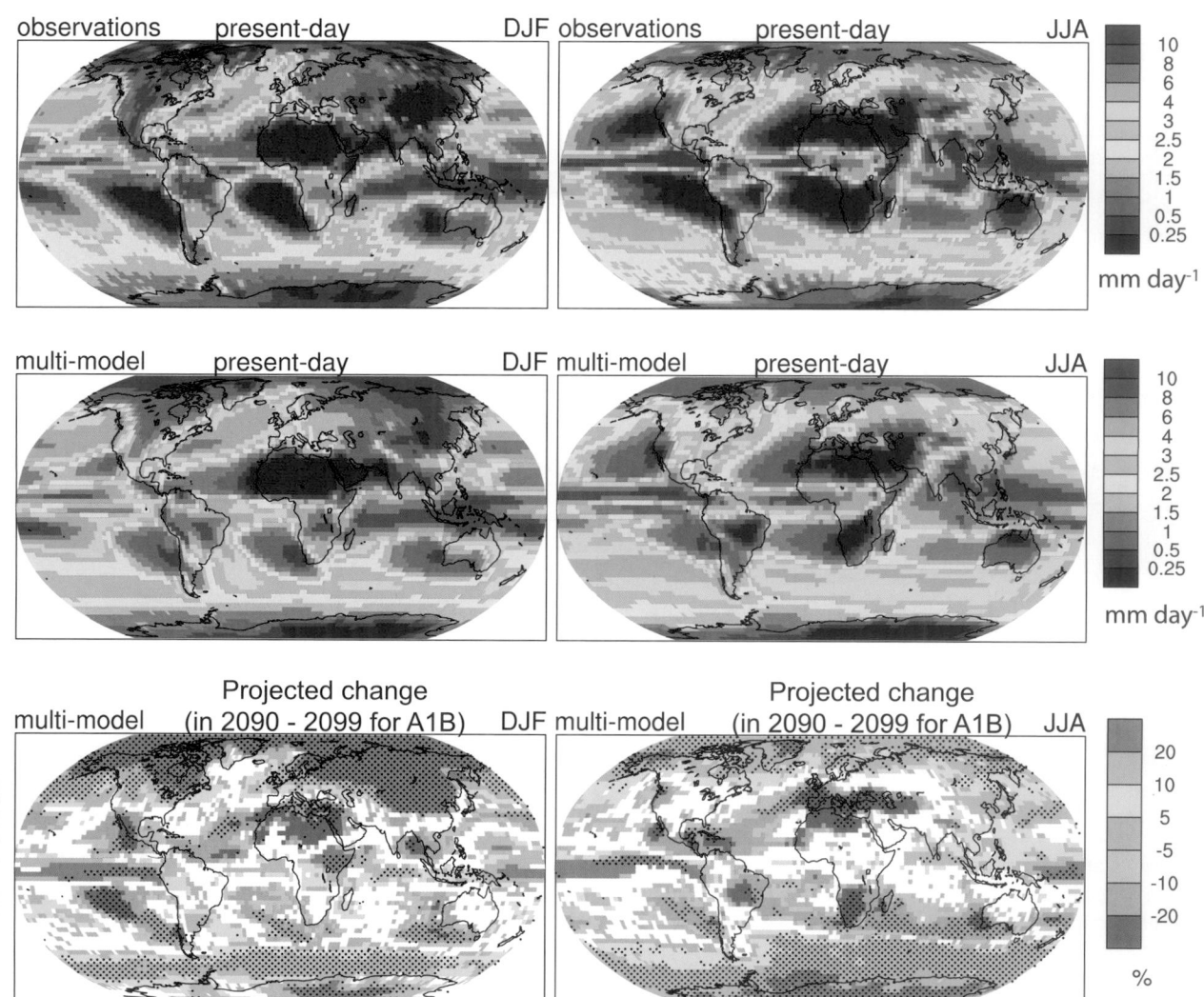

Figure TS.30. *Spatial patterns of observed (top row) and multi-model mean (middle row) seasonal mean precipitation rate (mm day⁻¹) for the period 1979 to 1993 and the multi-model mean for changes by the period 2090 to 2099 relative to 1980 to 1999 (% change) based on the SRES A1B scenario (bottom row). December to February means are in the left column, June to August means in the right column. In the bottom panel, changes are plotted only where more than 66% of the models agree on the sign of the change. The stippling indicates areas where more than 90% of the models agree on the sign of the change. {Based on same datasets as shown in Figures 8.5 and 10.9}*

In the SH, there are few land areas in the zone of projected subpolar moistening during the 21st century, with the subtropical drying more prominent (see Figure TS.30). The South Island of New Zealand and Tierra del Fuego fall within the subpolar precipitation increase zone, with southernmost Africa, the southern Andes in South America and southern Australia experiencing the drying tendency typical of the subtropics. {11.2, 11.6, 11.7}

Projections of precipitation over tropical land regions are more uncertain than those at higher latitudes, but, despite significant inadequacies in

modelling tropical convection and atmosphere-ocean interactions, and the added uncertainty associated with tropical cyclones, some robust features emerge in models. Rainfall in the summer monsoon season of South and Southeast Asia increases in most models, as does rainfall in East Africa. The sign of the precipitation response is considered less certain over both the Amazon and the African Sahel. These are regions in which there is added uncertainty due to potential vegetation-climate links, and there is less robustness across models even when vegetation feedbacks are not included. {8.3, 11.2, 11.4, 11.6}

TS.5.4 Coupling Between Climate Change and Changes in Biogeochemical Cycles

All models that treat the coupling of the carbon cycle to climate change indicate a positive feedback effect with warming acting to suppress land and ocean uptake of CO_2, leading to larger atmospheric CO_2 increases and greater climate change for a given emissions scenario, but the strength of this feedback effect varies markedly among models. Since the TAR, several new projections based on fully coupled carbon cycle-climate models have been performed and compared. For the SRES A2 scenario, and based on a range of model results, the projected increase in atmospheric CO_2 concentration over the 21st century is *likely* between 10 and 25% higher than projections without this feedback, adding more than 1°C to projected mean warming by 2100 for higher emission SRES scenarios. Correspondingly, the reduced CO_2 uptake caused by this effect reduces the CO_2 emissions that are consistent with a target stabilisation level. However, there are still significant uncertainties due, for example, to limitations in the understanding of the dynamics of land ecosystems and soils. {7.3, 10.4}

Increasing atmospheric CO_2 concentrations lead directly to increasing acidification of the surface ocean. Projections based on SRES scenarios give reductions in pH of between 0.14 and 0.35 units in the 21st century (depending on scenario), extending the present decrease of 0.1 units from pre-industrial times. Ocean acidification would lead to dissolution of shallow-water carbonate sediments. Southern Ocean surface waters are projected to exhibit undersaturation with regard to calcium carbonate ($CaCO_3$) for CO_2 concentrations higher than 600 ppm, a level exceeded during the second half of the 21st century in most of the SRES scenarios. Low-latitude regions and the deep ocean will be affected as well. These changes could affect marine organisms that form their exoskeletons out of $CaCO_3$, but the net effect on the biological cycling of carbon in the oceans is not well understood. {Box 7.3, 10.4}

Committed climate change due to past emissions varies considerably for different forcing agents because of differing lifetimes in the Earth's atmosphere (see Box TS.9). The committed climate change due to past emissions takes account of both (i) the time lags in the responses of the climate system to changes in radiative forcing; and (ii) the time scales over which different forcing agents persist in the atmosphere after their emission because of their differing lifetimes.

Typically the committed climate change due to past emissions includes an initial period of further increase in temperature, for the reasons discussed above, followed by a long-term decrease as radiative forcing decreases. Some greenhouse gases have relatively short atmospheric lifetimes (decades or less), such as CH_4 and carbon monoxide, while others such as N_2O have lifetimes of the order of a century, and some have lifetimes of millennia, such as SF_6 and PFCs. Atmospheric concentrations of CO_2 do not decay with a single well-defined lifetime if emissions are stopped. Removal of CO_2 emitted to the atmosphere occurs over multiple time scales, but some CO_2 will stay in the atmosphere for many thousands of years, so that emissions lead to a very long commitment to climate change. The slow long-term buffering of the ocean, including $CaCO_3$-sediment feedback, requires 30,000 to 35,000 years for atmospheric CO_2 concentrations to reach equilibrium. Using coupled carbon cycle components, EMICs show that the committed climate change due to past CO_2 emissions persists for more than 1000 years, so that even over these very long time scales, temperature and sea level do not return to pre-industrial values. An indication of the long time scales of committed climate change is obtained by prescribing anthropogenic CO_2 emissions following a path towards stabilisation at 750 ppm, but arbitrarily setting emissions to zero at year 2100. In this test case, it takes about 100 to 400 years in the different models for the atmospheric CO_2 concentration to drop from the maximum (ranges between 650 to 700 ppm) to below the level of two times the pre-industrial CO_2 concentration (about 560 ppm), owing to a continuous but slow transfer of carbon from the atmosphere and terrestrial reservoirs to the ocean (see Figure TS.31). {7.3, 10.7}

Future concentrations of many non-CO_2 greenhouse gases and their precursors are expected to be coupled to future climate change. Insufficient understanding of the causes of recent variations in the CH_4 growth rate suggests large uncertainties in future projections for this gas in particular. Emissions of CH_4 from wetlands are *likely* to increase in a warmer and wetter climate and to decrease in a warmer and drier climate. Observations also suggest increases in CH_4 released from northern peatlands that are experiencing permafrost melt, although the large-scale magnitude of this effect is not well quantified. Changes in temperature, humidity and clouds could also affect biogenic emissions of ozone precursors, such as volatile organic compounds. Climate change is also expected to affect tropospheric ozone through changes in chemistry and transport. Climate change could induce changes in OH through changes in humidity, and could alter stratospheric ozone concentrations and hence solar ultraviolet radiation in the troposphere. {7.4, 4.7}

CLIMATE CHANGE COMMITMENT

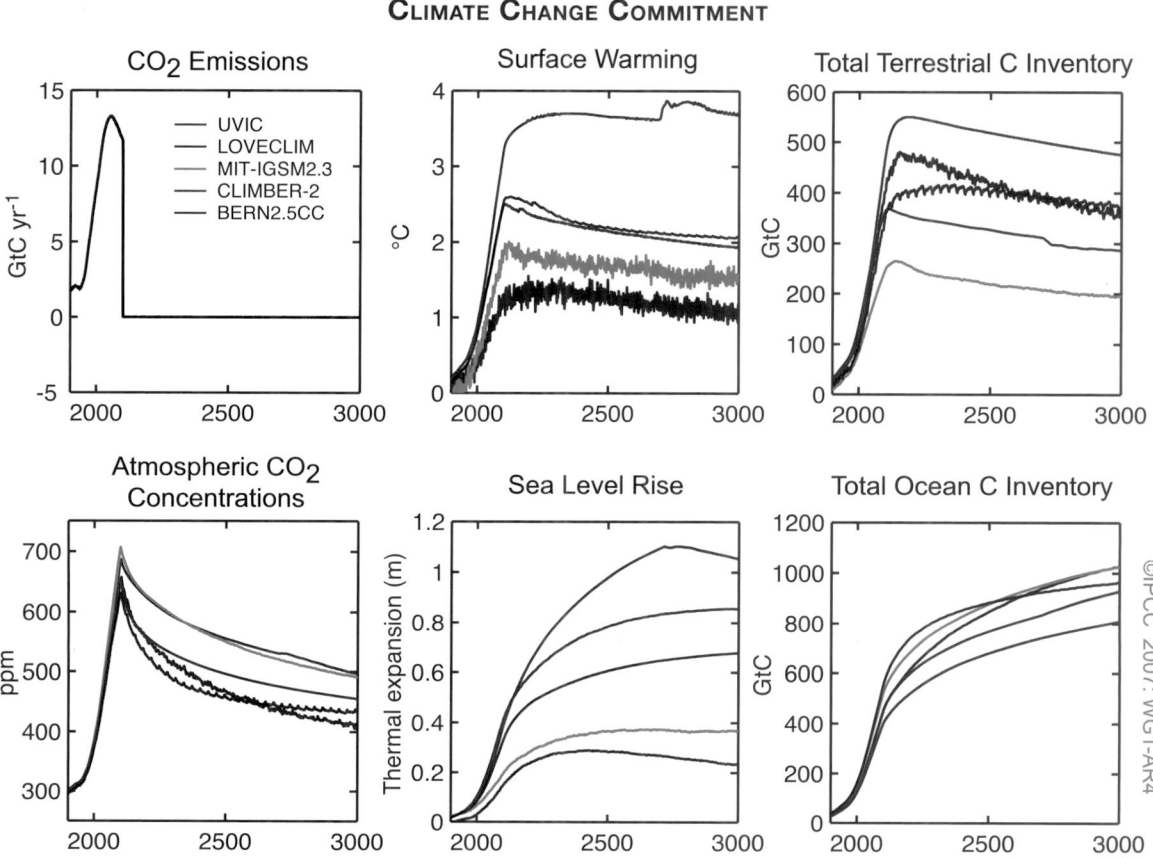

Figure TS.31. *Calculation of climate change commitment due to past emissions for five different EMICs and an idealised scenario where emissions follow a pathway leading to stabilisation of atmospheric CO$_2$ at 750 ppm, but before reaching this target, emissions are reduced to zero instantly at year 2100. (Left) CO$_2$ emissions and atmospheric CO$_2$ concentrations; (centre) surface warming and sea level rise due to thermal expansion; (right) change in total terrestrial and oceanic carbon inventory since the pre-industrial era. {Figure 10.35}*

Future emissions of many aerosols and their precursors are expected to be affected by climate change. Estimates of future changes in dust emissions under several climate and land use scenarios suggest that the effects of climate change are more important in controlling future dust emissions than changes in land use. Results from one study suggest that meteorology and climate have a greater influence on future Asian dust emissions and associated Asian dust storm occurrences than desertification. The biogenic emission of volatile organic compounds, a significant source of secondary organic aerosols, is known to be highly sensitive to (and increase with) temperature. However, aerosol yields decrease with temperature and the effects of changing precipitation and physiological adaptation are uncertain. Thus, change in biogenic secondary organic aerosol production in a warmer climate could be considerably lower than the response of biogenic volatile organic carbon emissions. Climate change may affect fluxes from the ocean of dimethyl sulphide (which is a precursor for

some sulphate aerosols) and sea salt aerosols, however, the effects on temperature and precipitation remain very uncertain. {7.5}

While the warming effect of CO$_2$ represents a commitment over many centuries, aerosols are removed from the atmosphere over time scales of only a few days, so that the negative radiative forcing due to aerosols could change rapidly in response to any changes in emissions of aerosols or aerosol precursors. Because sulphate aerosols are *very likely* exerting a substantial negative radiative forcing at present, future net forcing is very sensitive to changes in sulphate emissions. One study suggests that the hypothetical removal from the atmosphere of the entire current burden of anthropogenic sulphate aerosol particles would produce a rapid increase in global mean temperature of about 0.8°C within a decade or two. Changes in aerosols are also likely to influence precipitation. Thus, the effect of environmental strategies aimed at mitigating climate change requires consideration of changes in both greenhouse gas and aerosol emissions.

Changes in aerosol emissions may result from measures implemented to improve air quality which may therefore have consequences for climate change. {Box 7.4, 7.6, 10.7}

Climate change would modify a number of chemical and physical processes that control air quality and the net effects are *likely* to vary from one region to another. Climate change can affect air quality by modifying the rates at which pollutants are dispersed, the rate at which aerosols and soluble species are removed from the atmosphere, the general chemical environment for pollutant generation and the strength of emissions from the biosphere, fires and dust. Climate change is also expected to decrease the global ozone background. Overall, the net effect of climate change on air quality is highly uncertain. {Box 7.4}

TS.5.5 Implications of Climate Processes and their Time Scales for Long-Term Projections

The commitments to climate change after stabilisation of radiative forcing are expected to be about 0.5 to 0.6°C, mostly within the following century. The multi-model average when stabilising concentrations of greenhouse gases and aerosols at year 2000 values after a 20th-century climate simulation, and running an additional 100 years, is about 0.6°C of warming (relative to 1980–1999) at year 2100 (see Figure TS.32). If the B1 or A1B scenarios were to characterise 21st-century emissions followed by stabilisation at those levels, the additional warming after stabilisation is similar, about 0.5°C, mostly in the subsequent hundred years. {10.3, 10.7}

The magnitude of the positive feedback between climate change and the carbon cycle is uncertain. This leads to uncertainty in the trajectory of CO_2 emissions required to achieve a particular stabilization level of atmospheric CO_2 concentration. Based upon current understanding of climate-carbon cycle feedback, model studies suggest that, in order to stabilise CO_2 at 450 ppm, cumulative emissions in the 21st century could be reduced from a model average of approximately 670 [630 to 710] GtC to approximately 490 [375 to 600] GtC. Similarly, to stabilise CO_2 at 1000 ppm, the cumulative emissions could be reduced by this feedback from a model average of

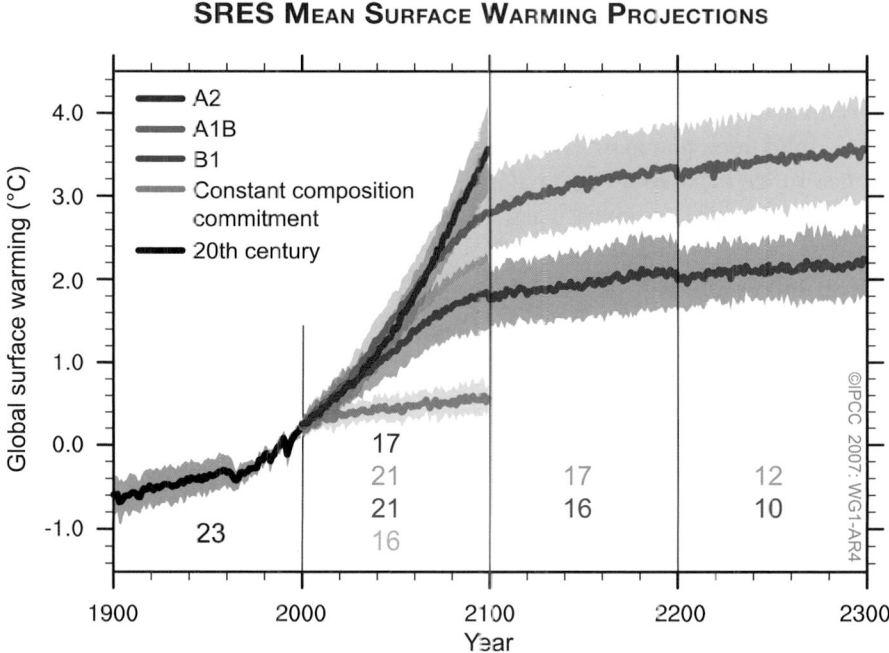

SRES Mean Surface Warming Projections

Figure TS.32. *Multi-model means of surface warming (compared to the 1980–1999 base period) for the SRES scenarios A2 (red), A1B (green) and B1 (blue), shown as continuations of the 20th-century simulation. The latter two scenarios are continued beyond the year 2100 with forcing kept constant (committed climate change as it is defined in Box TS.9). An additional experiment, in which the forcing is kept at the year 2000 level is also shown (orange). Linear trends from the corresponding control runs have been removed from these time series. Lines show the multi-model means, shading denotes the ±1 standard deviation range. Discontinuities between different periods have no physical meaning and are caused by the fact that the number of models that have run a given scenario is different for each period and scenario (numbers indicated in figure). For the same reason, uncertainty across scenarios should not be interpreted from this figure (see Section 10.5 for uncertainty estimates). {Figure 10.4}*

approximately 1415 [1340 to 1490] GtC to approximately 1100 [980 to 1250] GtC. {7.3, 10.4}

If radiative forcing were to be stabilised in 2100 at A1B concentrations, thermal expansion alone would lead to 0.3 to 0.8 m of sea level rise by 2300 (relative to 1980–1999) and would continue at decreasing rates for many centuries, due to slow processes that mix heat into the deep ocean. {10.7}

Contraction of the Greenland Ice Sheet is projected to continue to contribute to sea level rise after 2100. For stabilisation at A1B concentrations in 2100, a rate of 0.03 to 0.21 m per century due to thermal expansion is projected. If a global average warming of 1.9°C to 4.6°C relative to pre-industrial temperatures were maintained for millennia, the Greenland Ice Sheet would largely be eliminated except for remnant glaciers in the mountains. This would raise sea level by about 7 m and could be irreversible. These temperatures are comparable to those inferred for the last interglacial period 125,000 years ago, when palaeoclimatic information suggests reductions of polar ice extent and 4 to 6 m of sea level rise. {6.4, 10.7}

Dynamical processes not included in current models but suggested by recent observations could increase the vulnerability of the ice sheets to warming, increasing future sea level rise. Understanding of these processes is limited and there is no consensus on their likely magnitude. {4.6, 10.7}

Current global model studies project that the Antarctic Ice Sheet will remain too cold for widespread surface melting and will gain in mass due to increased snowfall. However, net loss of ice mass could occur if dynamical ice discharge dominates the ice sheet mass balance. {10.7}

While no models run for this assessment suggest an abrupt MOC shutdown during the 21st century, some models of reduced complexity suggest MOC shutdown as a possible long-term response to sufficiently strong warming. However, the likelihood of this occurring cannot be evaluated with confidence. The few available simulations with models of different complexity rather suggest a centennial-scale slowdown. Recovery of the MOC is *likely* if the radiative forcing is stabilised but would take several centuries. Systematic model comparison studies have helped establish some key processes that are responsible for variations between models in the response of the ocean to climate change (especially ocean heat uptake). {8.7, FAQ 10.2, 10.3}

TS.6 Robust Findings and Key Uncertainties

TS.6.1 Changes in Human and Natural Drivers of Climate

Robust Findings:

Current atmospheric concentrations of CO_2 and CH_4, and their associated positive radiative forcing, far exceed those determined from ice core measurements spanning the last 650,000 years. {6.4}

Fossil fuel use, agriculture and land use have been the dominant cause of increases in greenhouse gases over the last 250 years. {2.3, 7.3, 7.4}

Annual emissions of CO_2 from fossil fuel burning, cement production and gas flaring increased from a mean of 6.4 ± 0.4 GtC yr^{-1} in the 1990s to 7.2 ± 0.3 GtC yr^{-1} for 2000 to 2005. {7.3}

The sustained rate of increase in radiative forcing from CO_2, CH_4 and N_2O over the past 40 years is larger than at any time during at least the past 2000 years. {6.4}

Natural processes of CO_2 uptake by the oceans and terrestrial biosphere remove about 50 to 60% of anthropogenic emissions (i.e., fossil CO_2 emissions and land use change flux). Uptake by the oceans and the terrestrial biosphere are similar in magnitude over recent decades but that by the terrestrial biosphere is more variable. {7.3}

It is *virtually certain* that anthropogenic aerosols produce a net negative radiative forcing (cooling influence) with a greater magnitude in the NH than in the SH. {2.9, 9.2}

From new estimates of the combined anthropogenic forcing due to greenhouse gases, aerosols and land surface changes, it is *extremely likely* that human activities have exerted a substantial net warming influence on climate since 1750. {2.9}

Solar irradiance contributions to global average radiative forcing are considerably smaller than the contribution of increases in greenhouse gases over the industrial period. {2.5, 2.7}

Key Uncertainties:

The full range of processes leading to modification of cloud properties by aerosols is not well understood and the magnitudes of associated indirect radiative effects are poorly determined. {2.4, 7.5}

The causes of, and radiative forcing due to stratospheric water vapour changes are not well quantified. {2.3}

The geographical distribution and time evolution of the radiative forcing due to changes in aerosols during the 20th century are not well characterised. {2.4}

The causes of recent changes in the growth rate of atmospheric CH_4 are not well understood. {7.4}

The roles of different factors increasing tropospheric ozone concentrations since pre-industrial times are not well characterised. {2.3}

Land surface properties and land-atmosphere interactions that lead to radiative forcing are not well quantified. {2.5}

Knowledge of the contribution of past solar changes to radiative forcing on the time scale of centuries is not based upon direct measurements and is hence strongly dependent upon physical understanding. {2.7}

TS.6.2   Observations of Changes in Climate

TS.6.2.1   *Atmosphere and Surface*

Robust Findings:

Global mean surface temperatures continue to rise. Eleven of the last 12 years rank among the 12 warmest years on record since 1850. {3.2}

Rates of surface warming increased in the mid-1970s and the global land surface has been warming at about double the rate of ocean surface warming since then. {3.2}

Changes in surface temperature extremes are consistent with warming of the climate. {3.8}

Estimates of mid- and lower-tropospheric temperature trends have substantially improved. Lower-tropospheric temperatures have slightly greater warming rates than the surface from 1958 to 2005. {3.4}

Long-term trends from 1900 to 2005 have been observed in precipitation amount in many large regions. {3.3}

Increases have occurred in the number of heavy precipitation events. {3.8}

Droughts have become more common, especially in the tropics and subtropics, since the 1970s. {3.3}

Tropospheric water vapour has increased, at least since the 1980s. {3.4}

Key Uncertainties:

Radiosonde records are much less complete spatially than surface records and evidence suggests a number of radiosonde records are unreliable, especially in the tropics. It is likely that all records of tropospheric temperature trends still contain residual errors. {3.4}

While changes in large-scale atmospheric circulation are apparent, the quality of analyses is best only after 1979, making analysis of, and discrimination between, change and variability difficult. {3.5, 3.6}

Surface and satellite observations disagree on total and low-level cloud changes over the ocean. {3.4}

Multi-decadal changes in DTR are not well understood, in part because of limited observations of changes in cloudiness and aerosols. {3.2}

Difficulties in the measurement of precipitation remain an area of concern in quantifying trends in global and regional precipitation. {3.3}

Records of soil moisture and streamflow are often very short, and are available for only a few regions, which impedes complete analyses of changes in droughts. {3.3}

The availability of observational data restricts the types of extremes that can be analysed. The rarer the event, the more difficult it is to identify long-term changes because there are fewer cases available. {3.8}

Information on hurricane frequency and intensity is limited prior to the satellite era. There are questions about the interpretation of the satellite record. {3.8}

There is insufficient evidence to determine whether trends exist in tornadoes, hail, lightning and dust storms at small spatial scales. {3.8}

TS.6.2.2 Snow, Ice and Frozen Ground

Robust Findings:

The amount of ice on the Earth is decreasing. There has been widespread retreat of mountain glaciers since the end of the 19th century. The rate of mass loss from glaciers and the Greenland Ice Sheet is increasing. {4.5, 4.6}

The extent of NH snow cover has declined. Seasonal river and lake ice duration has decreased over the past 150 years. {4.2, 4.3}

Since 1978, annual mean arctic sea ice extent has been declining and summer minimum arctic ice extent has decreased. {4.4}

Ice thinning occurred in the Antarctic Peninsula and Amundsen shelf ice during the 1990s. Tributary glaciers have accelerated and complete breakup of the Larsen B Ice Shelf occurred in 2002. {4.6}

Temperature at the top of the permafrost layer has increased by up to 3°C since the 1980s in the Arctic. The maximum extent of seasonally frozen ground has decreased by about 7% in the NH since 1900, and its maximum depth has decreased by about 0.3 m in Eurasia since the mid-20th century. {4.7}

Key Uncertainties:

There is no global compilation of *in situ* snow data prior to 1960. Well-calibrated snow water equivalent data are not available for the satellite era. {4.2}

There are insufficient data to draw any conclusions about trends in the thickness of antarctic sea ice. {4.4}

Uncertainties in estimates of glacier mass loss arise from limited global inventory data, incomplete area-volume relationships and imbalance in geographic coverage. {4.5}

Mass balance estimates for ice shelves and ice sheets, especially for Antarctica, are limited by calibration and validation of changes detected by satellite altimetry and gravity measurements. {4.6}

Limited knowledge of basal processes and of ice shelf dynamics leads to large uncertainties in the understanding of ice flow processes and ice sheet stability. {4.6}

TS.6.2.3   Oceans and Sea Level

Robust Findings:

The global temperature (or heat content) of the oceans has increased since 1955. {5.2}

Large-scale regionally coherent trends in salinity have been observed over recent decades with freshening in subpolar regions and increased salinity in the shallower parts of the tropics and subtropics. These trends are consistent with changes in precipitation and inferred larger water transport in the atmosphere from low latitudes to high latitudes and from the Atlantic to the Pacific. {5.2}

Global average sea level rose during the 20th century. There is high confidence that the rate of sea level rise increased between the mid-19th and mid-20th centuries. During 1993 to 2003, sea level rose more rapidly than during 1961 to 2003. {5.5}

Thermal expansion of the ocean and loss of mass from glaciers and ice caps made substantial contributions to the observed sea level rise. {5.5}

The observed rate of sea level rise from 1993 to 2003 is consistent with the sum of observed contributions from thermal expansion and loss of land ice. {5.5}

The rate of sea level change over recent decades has not been geographically uniform. {5.5}

As a result of uptake of anthropogenic CO_2 since 1750, the acidity of the surface ocean has increased. {5.4, 7.3}

Key Uncertainties:

Limitations in ocean sampling imply that decadal variability in global heat content, salinity and sea level changes can only be evaluated with moderate confidence. {5.2, 5.5}

There is low confidence in observations of trends in the MOC. {Box 5.1}

Global average sea level rise from 1961 to 2003 appears to be larger than can be explained by thermal expansion and land ice melting. {5.5}

TS.6.2.4   Palaeoclimate

Robust Findings:

During the last interglacial, about 125,000 years ago, global sea level was *likely* 4 to 6 m higher than present, due primarily to retreat of polar ice. {6.4}

A number of past abrupt climate changes were *very likely* linked to changes in Atlantic Ocean circulation and affected the climate broadly across the NH. {6.4}

It is *very unlikely* that the Earth would naturally enter another ice age for at least 30,000 years. {6.4}

Biogeochemical and biogeophysical feedbacks have amplified climatic changes in the past. {6.4}

It is *very likely* that average NH temperatures during the second half of the 20th century were warmer than in any other 50-year period in the last 500 years and *likely* that this was also the warmest 50-year period in the past 1300 years. {6.6}

Palaeoclimate records indicate with high confidence that droughts lasting decades or longer were a recurrent feature of climate in several regions over the last 2000 years. {6.6}

Key Uncertainties:

Mechanisms of onset and evolution of past abrupt climate change and associated climate thresholds are not well understood. This limits confidence in the ability of climate models to simulate realistic abrupt change. {6.4}

The degree to which ice sheets retreated in the past, the rates of such change and the processes involved are not well known. {6.4}

Knowledge of climate variability over more than the last few hundred years in the SH and tropics is limited by the lack of palaeoclimatic records. {6.6}

Differing amplitudes and variability observed in available millennial-length NH temperature reconstructions, as well as the relation of these differences to choice of proxy data and statistical calibration methods, still need to be reconciled. {6.6}

The lack of extensive networks of proxy data for temperature in the last 20 years limits understanding of how such proxies respond to rapid global warming and of the influence of other environmental changes. {6.6}

TS.6.3   Understanding and Attributing Climate Change

Robust Findings:

Greenhouse gas forcing has *very likely* caused most of the observed global warming over the last 50 years. Greenhouse gas forcing alone during the past half century would *likely* have resulted in greater than the observed warming if there had not been an offsetting cooling effect from aerosol and other forcings. {9.4}

It is *extremely unlikely* (<5%) that the global pattern of warming during the past half century can be explained without external forcing, and *very unlikely* that it is due to known natural external causes alone. The warming occurred in both the ocean and the atmosphere and took place at a time when natural external forcing factors would *likely* have produced cooling. {9.4, 9.7}

It is *likely* that anthropogenic forcing has contributed to the general warming observed in the upper several hundred metres of the ocean during the latter half of the 20th century. Anthropogenic forcing, resulting in thermal expansion from ocean warming and glacier mass loss, has *very likely* contributed to sea level rise during the latter half of the 20th century. {9.5}

A substantial fraction of the reconstructed NH inter-decadal temperature variability of the past seven centuries is *very likely* attributable to natural external forcing (volcanic eruptions and solar variability). {9.3}

Key Uncertainties:

Confidence in attributing some climate change phenomena to anthropogenic influences is currently limited by uncertainties in radiative forcing, as well as uncertainties in feedbacks and in observations. {9.4, 9.5}

Attribution at scales smaller than continental and over time scales of less than 50 years is limited by larger climate variability on smaller scales, by uncertainties in the small-scale details of external forcing and the response simulated by models, as well as uncertainties in simulation of internal variability on small scales, including in relation to modes of variability. {9.4}

There is less confidence in understanding of forced changes in precipitation and surface pressure than there is of temperature. {9.5}

The range of attribution statements is limited by the absence of formal detection and attribution studies, or their very limited number, for some phenomena (e.g., some types of extreme events). {9.5}

Incomplete global data sets for extremes analysis and model uncertainties still restrict the regions and types of detection studies of extremes that can be performed. {9.4, 9.5}

Despite improved understanding, uncertainties in model-simulated internal climate variability limit some aspects of attribution studies. For example, there are apparent discrepancies between estimates of ocean heat content variability from models and observations. {5.2, 9.5}

Lack of studies quantifying the contributions of anthropogenic forcing to ocean heat content increase or glacier melting together with the open part of the sea level budget for 1961 to 2003 are among the uncertainties in quantifying the anthropogenic contribution to sea level rise. {9.5}

TS.6.4   Projections of Future Changes in Climate

TS.6.4.1   Model Evaluation

Robust Findings:

Climate models are based on well-established physical principles and have been demonstrated to reproduce observed features of recent climate and past climate changes. There is considerable confidence that AOGCMs provide credible quantitative estimates of future climate change, particularly at continental scales and above. Confidence in these estimates is higher for some climate variables (e.g., temperature) than for others (e.g., precipitation). {FAQ 8.1}

Confidence in models has increased due to:
- improvements in the simulation of many aspects of present climate, including important modes of climate variability and extreme hot and cold spells;
- improved model resolution, computational methods and parametrizations and inclusion of additional processes;
- more comprehensive diagnostic tests, including tests of model ability to forecast on time scales from days to a year when initialised with observed conditions; and
- enhanced scrutiny of models and expanded diagnostic analysis of model behaviour facilitated by internationally coordinated efforts to collect and disseminate output from model experiments performed under common conditions. {8.4}

Key Uncertainties:

A proven set of model metrics comparing simulations with observations, that might be used to narrow the range of plausible climate projections, has yet to be developed. {8.2}

Most models continue to have difficulty controlling climate drift, particularly in the deep ocean. This drift must be accounted for when assessing change in many oceanic variables. {8.2}

Models differ considerably in their estimates of the strength of different feedbacks in the climate system. {8.6}

Problems remain in the simulation of some modes of variability, notably the Madden-Julian Oscillation, recurrent atmospheric blocking and extreme precipitation. {8.4}

Systematic biases have been found in most models' simulations of the Southern Ocean that are linked to uncertainty in transient climate response. {8.3}

Climate models remain limited by the spatial resolution that can be achieved with present computer resources, by the need for more extensive ensemble runs and by the need to include some additional processes. {8.1–8.5}

TS.6.4.2 Equilibrium and Transient Climate Sensitivity

Robust Findings:

Equilibrium climate sensitivity is *likely* to be in the range 2°C to 4.5°C with a most likely value of about 3°C, based upon multiple observational and modelling constraints. It is *very unlikely* to be less than 1.5°C. {8.6, 9.6, Box 10.2}

The transient climate response is better constrained than the equilibrium climate sensitivity. It is *very likely* larger than 1°C and *very unlikely* greater than 3°C. {10.5}

There is a good understanding of the origin of differences in equilibrium climate sensitivity found in different models. Cloud feedbacks are the primary source of inter-model differences in equilibrium climate sensitivity, with low cloud being the largest contributor. {8.6}

New observational and modelling evidence strongly supports a combined water vapour-lapse rate feedback of a strength comparable to that found in AOGCMs. {8.6}

Key Uncertainties:

Large uncertainties remain about how clouds might respond to global climate change. {8.6}

TS.6.4.3 Global Projections

Robust Findings:

Even if concentrations of radiative forcing agents were to be stabilised, further committed warming and related climate changes would be expected to occur, largely because of time lags associated with processes in the oceans. {10.7}

Near-term warming projections are little affected by different scenario assumptions or different model sensitivities, and are consistent with that observed for the past few decades. The multi-model mean warming, averaged over 2011 to 2030 relative to 1980 to 1999 for all AOGCMs considered here, lies in a narrow range of 0.64°C to 0.69°C for the three different SRES emission scenarios B1, A1B and A2. {10.3}

Geographic patterns of projected warming show the greatest temperature increases at high northern latitudes and over land, with less warming over the southern oceans and North Atlantic. {10.3}

Changes in precipitation show robust large-scale patterns: precipitation generally increases in the tropical precipitation maxima, decreases in the subtropics and increases at high latitudes as a consequence of a general intensification of the global hydrological cycle. {10.3}

As the climate warms, snow cover and sea ice extent decrease; glaciers and ice caps lose mass and contribute to sea level rise. Sea ice extent decreases in the 21st century in both the Arctic and Antarctic. Snow cover reduction is accelerated in the Arctic by positive feedbacks and widespread increases in thaw depth occur over much of the permafrost regions. {10.3}

Based on current simulations, it is *very likely* that the Atlantic Ocean MOC will slow down by 2100. However, it is *very unlikely* that the MOC will undergo a large abrupt transition during the course of the 21st century. {10.3}

Heat waves become more frequent and longer lasting in a future warmer climate. Decreases in frost days are projected to occur almost everywhere in the mid- and high latitudes, with an increase in growing season length. There is a tendency for summer drying of the mid-continental areas during summer, indicating a greater risk of droughts in those regions. {10.3, FAQ 10.1}

Future warming would tend to reduce the capacity of the Earth system (land and ocean) to absorb anthropogenic CO_2. As a result, an increasingly large fraction of anthropogenic CO_2 would stay in the atmosphere under a warmer climate. This feedback requires reductions in the cumulative emissions consistent with stabilisation at a given atmospheric CO_2 level compared to the hypothetical case of no such feedback. The higher the stabilisation scenario, the larger the amount of climate change and the larger the required reductions. {7.3, 10.4}

Key Uncertainties:

The likelihood of a large abrupt change in the MOC beyond the end of the 21st century cannot yet be assessed reliably. For low and medium emission scenarios with atmospheric greenhouse gas concentrations stabilised beyond 2100, the MOC recovers from initial weakening within one to several centuries. A permanent reduction in the MOC cannot be excluded if the forcing is strong and long enough. {10.7}

The model projections for extremes of precipitation show larger ranges in amplitude and geographical locations than for temperature. {10.3, 11.1}

The response of some major modes of climate variability such as ENSO still differs from model to model, which may be associated with differences in the spatial and temporal representation of present-day conditions. {10.3}

The robustness of many model responses of tropical cyclones to climate change is still limited by the resolution of typical climate models. {10.3}

Changes in key processes that drive some global and regional climate changes are poorly known (e.g., ENSO, NAO, blocking, MOC, land surface feedbacks, tropical cyclone distribution). {11.2–11.9}

The magnitude of future carbon cycle feedbacks is still poorly determined. {7.3, 10.4}

TS.6.4.4   Sea Level

Robust Findings:

Sea level will continue to rise in the 21st century because of thermal expansion and loss of land ice. Sea level rise was not geographically uniform in the past and will not be in the future. {10.6}

Projected warming due to emission of greenhouse gases during the 21st century will continue to contribute to sea level rise for many centuries. {10.7}

Sea level rise due to thermal expansion and loss of mass from ice sheets would continue for centuries or millennia even if radiative forcing were to be stabilised. {10.7}

Key Uncertainties:

Models do not yet exist that address key processes that could contribute to large rapid dynamical changes in the Antarctic and Greenland Ice Sheets that could increase the discharge of ice into the ocean. {10.6}

The sensitivity of ice sheet surface mass balance (melting and precipitation) to global climate change is not well constrained by observations and has a large spread in models. There is consequently a large uncertainty in the magnitude of global warming that, if sustained, would lead to the elimination of the Greenland Ice Sheet. {10.7}

TS.6.4.5 Regional Projections

Robust Findings:

Temperatures averaged over all habitable continents and over many sub-continental land regions will *very likely* rise at greater than the global average rate in the next 50 years and by an amount substantially in excess of natural variability. {10.3, 11.2–11.9}

Precipitation is *likely* to increase in most subpolar and polar regions. The increase is considered especially robust, and *very likely* to occur, in annual precipitation in most of northern Europe, Canada, the northeast USA and the Arctic, and in winter precipitation in northern Asia and the Tibetan Plateau. {11.2–11.9}

Precipitation is *likely* to decrease in many subtropical regions, especially at the poleward margins of the subtropics. The decrease is considered especially robust, and *very likely* to occur, in annual precipitation in European and African regions bordering the Mediterranean and in winter rainfall in south-western Australia. {11.2–11.9}

Extremes of daily precipitation are *likely* to increase in many regions. The increase is considered as *very likely* in northern Europe, south Asia, East Asia, Australia and New Zealand – this list in part reflecting uneven geographic coverage in existing published research. {11.2–11.9}

Key Uncertainties:

In some regions there has been only very limited study of key aspects of regional climate change, particularly with regard to extreme events. {11.2–11.9}

Atmosphere-Ocean General Circulation Models show no consistency in simulated regional precipitation change in some key regions (e.g., northern South America, northern Australia and the Sahel). {10.3, 11.2–11.9}

In many regions where fine spatial scales in climate are generated by topography, there is insufficient information on how climate change will be expressed at these scales. {11.2–11.9}

1

Historical Overview of Climate Change Science

Coordinating Lead Authors:

Hervé Le Treut (France), Richard Somerville (USA)

Lead Authors:

Ulrich Cubasch (Germany), Yihui Ding (China), Cecilie Mauritzen (Norway), Abdalah Mokssit (Morocco), Thomas Peterson (USA), Michael Prather (USA)

Contributing Authors:

M. Allen (UK), I. Auer (Austria), J. Biercamp (Germany), C. Covey (USA), J.R. Fleming (USA), R. García-Herrera (Spain), P. Gleckler (USA), J. Haigh (UK), G.C. Hegerl (USA, Germany), K. Isaksen (Norway), J. Jones (Germany, UK), J. Luterbacher (Switzerland), M. MacCracken (USA), J.E. Penner (USA), C. Pfister (Switzerland), E. Roeckner (Germany), B. Santer (USA), F. Schott (Germany), F. Sirocko (Germany), A. Staniforth (UK), T.F. Stocker (Switzerland), R.J. Stouffer (USA), K.E. Taylor (USA), K.E. Trenberth (USA), A. Weisheimer (ECMWF, Germany), M. Widmann (Germany, UK)

Review Editors:

Alphonsus Baede (Netherlands), David Griggs (UK)

This chapter should be cited as:

Le Treut, H., R. Somerville, U. Cubasch, Y. Ding, C. Mauritzen, A. Mokssit, T. Peterson and M. Prather, 2007: Historical Overview of Climate Change. In: *Climate Change 2007: The Physical Science Basis. Contribution of Working Group I to the Fourth Assessment Report of the Intergovernmental Panel on Climate Change* [Solomon, S., D. Qin, M. Manning, Z. Chen, M. Marquis, K.B. Averyt, M. Tignor and H.L. Miller (eds.)]. Cambridge University Press, Cambridge, United Kingdom and New York, NY, USA.

Table of Contents

Executive Summary

Awareness and a partial understanding of most of the interactive processes in the Earth system that govern climate and climate change predate the IPCC, often by many decades. A deeper understanding and quantification of these processes and their incorporation in climate models have progressed rapidly since the IPCC First Assessment Report in 1990.

As climate science and the Earth's climate have continued to evolve over recent decades, increasing evidence of anthropogenic influences on climate change has been found. Correspondingly, the IPCC has made increasingly more definitive statements about human impacts on climate.

Debate has stimulated a wide variety of climate change research. The results of this research have refined but not significantly redirected the main scientific conclusions from the sequence of IPCC assessments.

1.1   Overview of the Chapter

To better understand the science assessed in this Fourth Assessment Report (AR4), it is helpful to review the long historical perspective that has led to the current state of climate change knowledge. This chapter starts by describing the fundamental nature of earth science. It then describes the history of climate change science using a wide-ranging subset of examples, and ends with a history of the IPCC.

The concept of this chapter is new. There is no counterpart in previous IPCC assessment reports for an introductory chapter providing historical context for the remainder of the report. Here, a restricted set of topics has been selected to illustrate key accomplishments and challenges in climate change science. The topics have been chosen for their significance to the IPCC task of assessing information relevant for understanding the risks of human-induced climate change, and also to illustrate the complex and uneven pace of scientific progress.

In this chapter, the time frame under consideration stops with the publication of the Third Assessment Report (TAR; IPCC, 2001a). Developments subsequent to the TAR are described in the other chapters of this report, and we refer to these chapters throughout this first chapter.

1.2   The Nature of Earth Science

Science may be stimulated by argument and debate, but it generally advances through formulating hypotheses clearly and testing them objectively. This testing is the key to science. In fact, one philosopher of science insisted that to be genuinely scientific, a statement must be susceptible to testing that could potentially show it to be false (Popper, 1934). In practice, contemporary scientists usually submit their research findings to the scrutiny of their peers, which includes disclosing the methods that they use, so their results can be checked through replication by other scientists. The insights and research results of individual scientists, even scientists of unquestioned genius, are thus confirmed or rejected in the peer-reviewed literature by the combined efforts of many other scientists. It is not the belief or opinion of the scientists that is important, but rather the results of this testing. Indeed, when Albert Einstein was informed of the publication of a book entitled *100 Authors Against Einstein*, he is said to have remarked, 'If I were wrong, then one would have been enough!' (Hawking, 1988); however, that one opposing scientist would have needed proof in the form of testable results.

Thus science is inherently self-correcting; incorrect or incomplete scientific concepts ultimately do not survive repeated testing against observations of nature. Scientific theories are ways of explaining phenomena and providing insights that can be evaluated by comparison with physical reality. Each successful prediction adds to the weight of evidence supporting the theory, and any unsuccessful prediction demonstrates that the underlying theory is imperfect and requires improvement or abandonment. Sometimes, only certain kinds of questions tend to be asked about a scientific phenomenon until contradictions build to a point where a sudden change of paradigm takes place (Kuhn, 1996). At that point, an entire field can be rapidly reconstructed under the new paradigm.

Despite occasional major paradigm shifts, the majority of scientific insights, even unexpected insights, tend to emerge incrementally as a result of repeated attempts to test hypotheses as thoroughly as possible. Therefore, because almost every new advance is based on the research and understanding that has gone before, science is cumulative, with useful features retained and non-useful features abandoned. Active research scientists, throughout their careers, typically spend large fractions of their working time studying in depth what other scientists have done. Superficial or amateurish acquaintance with the current state of a scientific research topic is an obstacle to a scientist's progress. Working scientists know that a day in the library can save a year in the laboratory. Even Sir Isaac Newton (1675) wrote that if he had 'seen further it is by standing on the shoulders of giants'. Intellectual honesty and professional ethics call for scientists to acknowledge the work of predecessors and colleagues.

The attributes of science briefly described here can be used in assessing competing assertions about climate change. Can the statement under consideration, in principle, be proven false? Has it been rigorously tested? Did it appear in the peer-reviewed literature? Did it build on the existing research record where appropriate? If the answer to any of these questions is no, then less credence should be given to the assertion until it is tested and independently verified. The IPCC assesses the scientific literature to create a report based on the best available science (Section 1.6). It must be acknowledged, however, that the IPCC also contributes to science by identifying the key uncertainties and by stimulating and coordinating targeted research to answer important climate change questions.

Frequently Asked Question 1.1
What Factors Determine Earth's Climate?

The climate system is a complex, interactive system consisting of the atmosphere, land surface, snow and ice, oceans and other bodies of water, and living things. The atmospheric component of the climate system most obviously characterises climate; climate is often defined as 'average weather'. Climate is usually described in terms of the mean and variability of temperature, precipitation and wind over a period of time, ranging from months to millions of years (the classical period is 30 years). The climate system evolves in time under the influence of its own internal dynamics and due to changes in external factors that affect climate (called 'forcings'). External forcings include natural phenomena such as volcanic eruptions and solar variations, as well as human-induced changes in atmospheric composition. Solar radiation powers the climate system. There are three fundamental ways to change the radiation balance of the Earth: 1) by changing the incoming solar radiation (e.g., by changes in Earth's orbit or in the Sun itself); 2) by changing the fraction of solar radiation that is reflected (called

'albedo'; e.g., by changes in cloud cover, atmospheric particles or vegetation); and 3) by altering the longwave radiation from Earth back towards space (e.g., by changing greenhouse gas concentrations). Climate, in turn, responds directly to such changes, as well as indirectly, through a variety of feedback mechanisms.

The amount of energy reaching the top of Earth's atmosphere each second on a surface area of one square metre facing the Sun during daytime is about 1,370 Watts, and the amount of energy per square metre per second averaged over the entire planet is one-quarter of this (see Figure 1). About 30% of the sunlight that reaches the top of the atmosphere is reflected back to space. Roughly two-thirds of this reflectivity is due to clouds and small particles in the atmosphere known as 'aerosols'. Light-coloured areas of Earth's surface – mainly snow, ice and deserts – reflect the remaining one-third of the sunlight. The most dramatic change in aerosol-produced reflectivity comes when major volcanic eruptions eject material very high into the atmosphere. Rain typically

(continued)

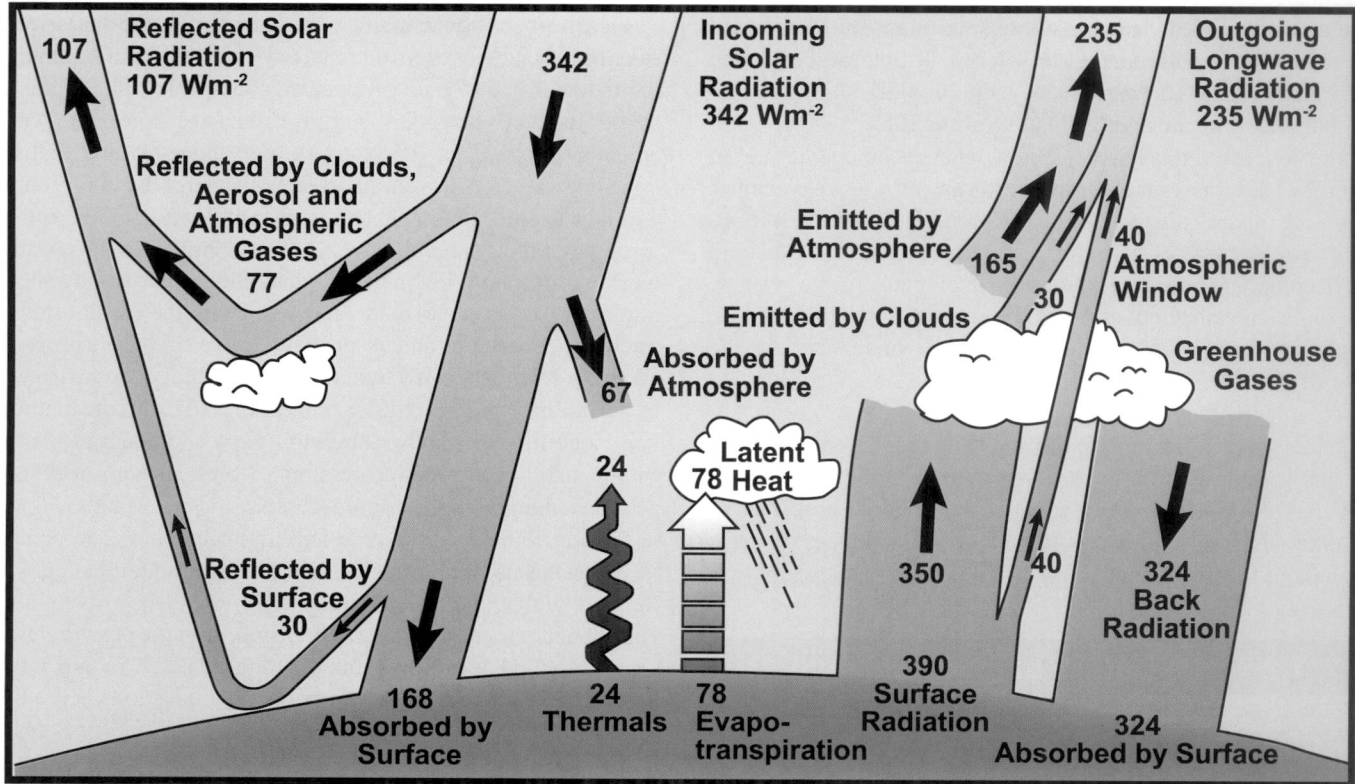

FAQ 1.1, Figure 1. *Estimate of the Earth's annual and global mean energy balance. Over the long term, the amount of incoming solar radiation absorbed by the Earth and atmosphere is balanced by the Earth and atmosphere releasing the same amount of outgoing longwave radiation. About half of the incoming solar radiation is absorbed by the Earth's surface. This energy is transferred to the atmosphere by warming the air in contact with the surface (thermals), by evapotranspiration and by longwave radiation that is absorbed by clouds and greenhouse gases. The atmosphere in turn radiates longwave energy back to Earth as well as out to space. Source: Kiehl and Trenberth (1997).*

clears aerosols out of the atmosphere in a week or two, but when material from a violent volcanic eruption is projected far above the highest cloud, these aerosols typically influence the climate for about a year or two before falling into the troposphere and being carried to the surface by precipitation. Major volcanic eruptions can thus cause a drop in mean global surface temperature of about half a degree celsius that can last for months or even years. Some man-made aerosols also significantly reflect sunlight.

The energy that is not reflected back to space is absorbed by the Earth's surface and atmosphere. This amount is approximately 240 Watts per square metre (W m^{-2}). To balance the incoming energy, the Earth itself must radiate, on average, the same amount of energy back to space. The Earth does this by emitting outgoing longwave radiation. Everything on Earth emits longwave radiation continuously. That is the heat energy one feels radiating out from a fire; the warmer an object, the more heat energy it radiates. To emit 240 W m^{-2}, a surface would have to have a temperature of around −19°C. This is much colder than the conditions that actually exist at the Earth's surface (the global mean surface temperature is about 14°C). Instead, the necessary −19°C is found at an altitude about 5 km above the surface.

The reason the Earth's surface is this warm is the presence of greenhouse gases, which act as a partial blanket for the longwave radiation coming from the surface. This blanketing is known as the natural greenhouse effect. The most important greenhouse gases are water vapour and carbon dioxide. The two most abundant constituents of the atmosphere – nitrogen and oxygen – have no such effect. Clouds, on the other hand, do exert a blanketing effect similar to that of the greenhouse gases; however, this effect is offset by their reflectivity, such that on average, clouds tend to have a cooling effect on climate (although locally one can feel the warming effect: cloudy nights tend to remain warmer than clear nights because the clouds radiate longwave energy back down to the surface). Human activities intensify the blanketing effect through the release of greenhouse gases. For instance, the amount of carbon dioxide in the atmosphere has increased by about 35% in the industrial era, and this increase is known to be due to human activities, primarily the combustion of fossil fuels and removal of forests. Thus, humankind has dramatically altered the chemical composition of the global atmosphere with substantial implications for climate.

Because the Earth is a sphere, more solar energy arrives for a given surface area in the tropics than at higher latitudes, where sunlight strikes the atmosphere at a lower angle. Energy is transported from the equatorial areas to higher latitudes via atmospheric and oceanic circulations, including storm systems. Energy is also required to evaporate water from the sea or land surface, and this energy, called latent heat, is released when water vapour condenses in clouds (see Figure 1). Atmospheric circulation is primarily driven by the release of this latent heat. Atmospheric circulation in turn drives much of the ocean circulation through the action of winds on the surface waters of the ocean, and through changes in the ocean's surface temperature and salinity through precipitation and evaporation.

Due to the rotation of the Earth, the atmospheric circulation patterns tend to be more east-west than north-south. Embedded in the mid-latitude westerly winds are large-scale weather systems that act to transport heat toward the poles. These weather systems are the familiar migrating low- and high-pressure systems and their associated cold and warm fronts. Because of land-ocean temperature contrasts and obstacles such as mountain ranges and ice sheets, the circulation system's planetary-scale atmospheric waves tend to be geographically anchored by continents and mountains although their amplitude can change with time. Because of the wave patterns, a particularly cold winter over North America may be associated with a particularly warm winter elsewhere in the hemisphere. Changes in various aspects of the climate system, such as the size of ice sheets, the type and distribution of vegetation or the temperature of the atmosphere or ocean will influence the large-scale circulation features of the atmosphere and oceans.

There are many feedback mechanisms in the climate system that can either amplify ('positive feedback') or diminish ('negative feedback') the effects of a change in climate forcing. For example, as rising concentrations of greenhouse gases warm Earth's climate, snow and ice begin to melt. This melting reveals darker land and water surfaces that were beneath the snow and ice, and these darker surfaces absorb more of the Sun's heat, causing more warming, which causes more melting, and so on, in a self-reinforcing cycle. This feedback loop, known as the 'ice-albedo feedback', amplifies the initial warming caused by rising levels of greenhouse gases. Detecting, understanding and accurately quantifying climate feedbacks have been the focus of a great deal of research by scientists unravelling the complexities of Earth's climate.

A characteristic of Earth sciences is that Earth scientists are unable to perform controlled experiments on the planet as a whole and then observe the results. In this sense, Earth science is similar to the disciplines of astronomy and cosmology that cannot conduct experiments on galaxies or the cosmos. This is an important consideration, because it is precisely such whole-Earth, system-scale experiments, incorporating the full complexity of interacting processes and feedbacks, that might ideally be required to fully verify or falsify climate change hypotheses (Schellnhuber et al., 2004). Nevertheless, countless empirical tests of numerous different hypotheses have built up a massive body of Earth science knowledge. This repeated testing has refined the understanding of numerous aspects of the climate system, from deep oceanic circulation to stratospheric chemistry. Sometimes a combination of observations and models can be used to test planetary-scale hypotheses. For example, the global cooling and drying of the atmosphere observed after the eruption of Mt. Pinatubo (Section 8.6) provided key tests of particular aspects of global climate models (Hansen et al., 1992).

Another example is provided by past IPCC projections of future climate change compared to current observations. Figure 1.1 reveals that the model projections of global average temperature from the First Assessment Report (FAR; IPCC, 1990) were higher than those from the Second Assessment Report (SAR; IPCC, 1996). Subsequent observations (Section 3.2) showed that the evolution of the actual climate system fell midway between the FAR and the SAR 'best estimate' projections and were within or near the upper range of projections from the TAR (IPCC, 2001a).

Not all theories or early results are verified by later analysis. In the mid-1970s, several articles about possible global cooling appeared in the popular press, primarily motivated by analyses indicating that Northern Hemisphere (NH) temperatures had decreased during the previous three decades (e.g., Gwynne, 1975). In the peer-reviewed literature, a paper by Bryson and Dittberner (1976) reported that increases in carbon dioxide (CO_2) should be associated with a decrease in global temperatures. When challenged by Woronko (1977), Bryson and Dittberner (1977) explained that the cooling projected by their model was due to aerosols (small particles in the atmosphere) produced by the same combustion that caused the increase in CO_2. However, because aerosols remain in the atmosphere only a short time compared to CO_2, the results were not applicable for long-term climate change projections. This example of a prediction of global cooling is a classic illustration of the self-correcting nature of Earth science. The scientists involved were reputable researchers who followed the accepted paradigm of publishing in scientific journals, submitting their methods and results to the scrutiny of their peers (although the peer-review did not catch this problem), and responding to legitimate criticism.

A recurring theme throughout this chapter is that climate science in recent decades has been characterised by the

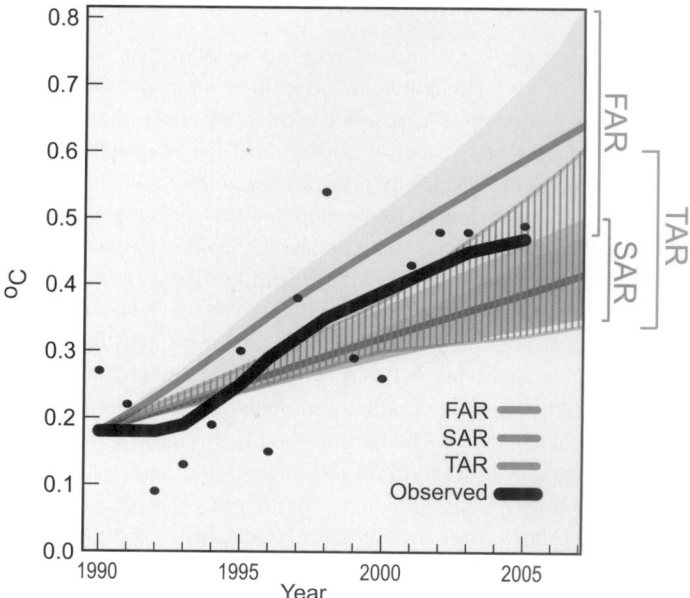

Figure 1.1. *Yearly global average surface temperature (Brohan et al., 2006), relative to the mean 1961 to 1990 values, and as projected in the FAR (IPCC, 1990), SAR (IPCC, 1996) and TAR (IPCC, 2001a). The 'best estimate' model projections from the FAR and SAR are in solid lines with their range of estimated projections shown by the shaded areas. The TAR did not have 'best estimate' model projections but rather a range of projections. Annual mean observations (Section 3.2) are depicted by black circles and the thick black line shows decadal variations obtained by smoothing the time series using a 13-point filter.*

increasing rate of advancement of research in the field and by the notable evolution of scientific methodology and tools, including the models and observations that support and enable the research. During the last four decades, the rate at which scientists have added to the body of knowledge of atmospheric and oceanic processes has accelerated dramatically. As scientists incrementally increase the totality of knowledge, they publish their results in peer-reviewed journals. Between 1965 and 1995, the number of articles published per year in atmospheric science journals tripled (Geerts, 1999). Focusing more narrowly, Stanhill (2001) found that the climate change science literature grew approximately exponentially with a doubling time of 11 years for the period 1951 to 1997. Furthermore, 95% of all the climate change science literature since 1834 was published after 1951. Because science is cumulative, this represents considerable growth in the knowledge of climate processes and in the complexity of climate research. An important example of this is the additional physics incorporated in climate models over the last several decades, as illustrated in Figure 1.2. As a result of the cumulative nature of science, climate science today is an interdisciplinary synthesis of countless tested and proven physical processes and principles painstakingly compiled and verified over several centuries of detailed laboratory measurements, observational experiments and theoretical analyses; and is now far more wide-ranging and physically comprehensive than was the case only a few decades ago.

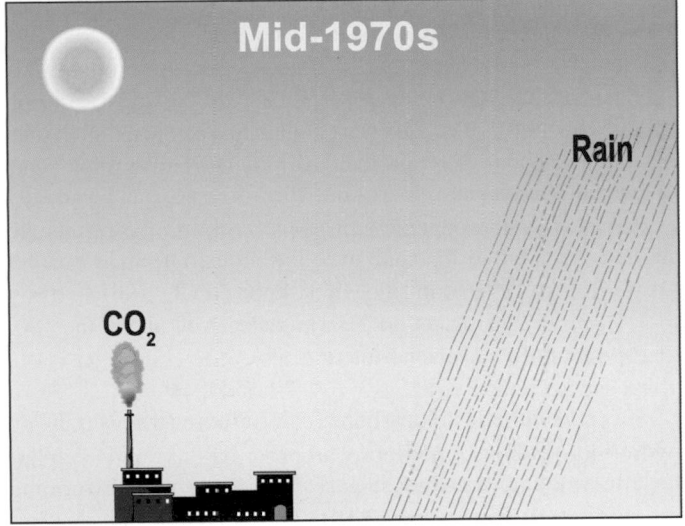
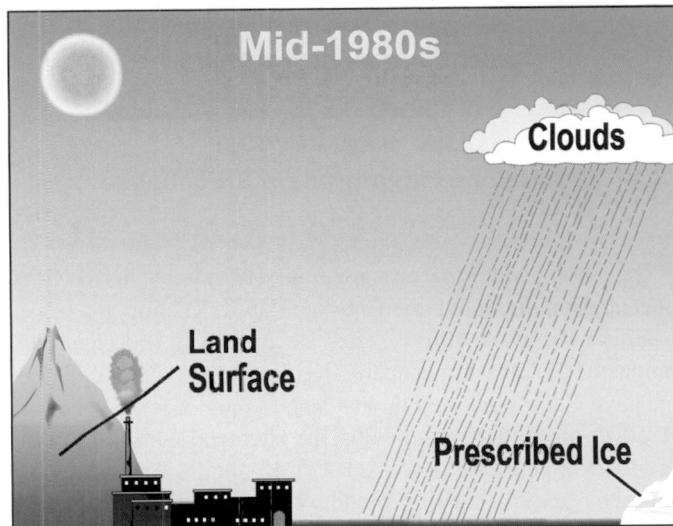
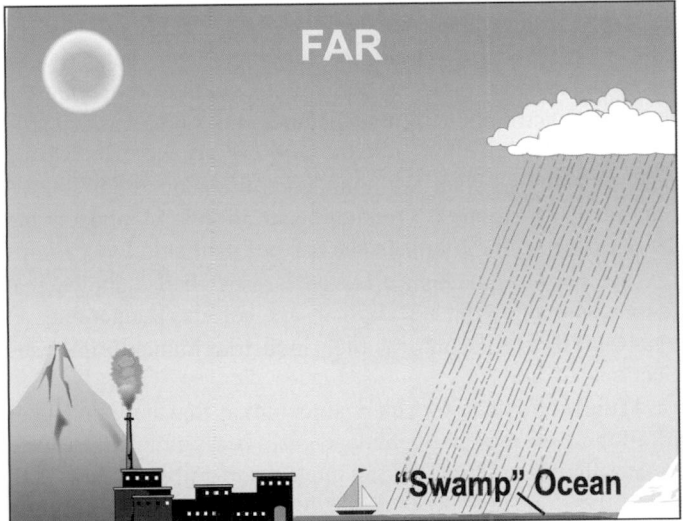
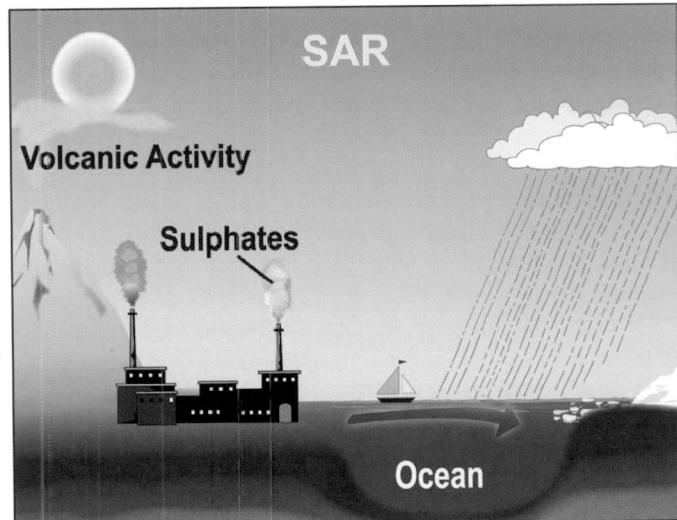
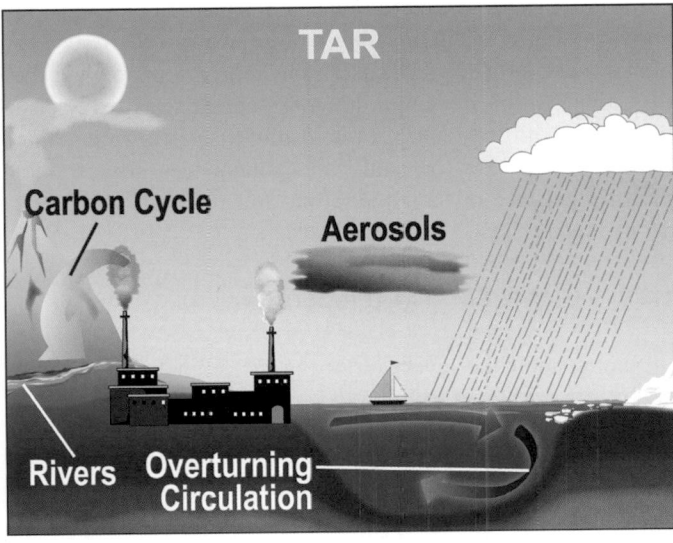
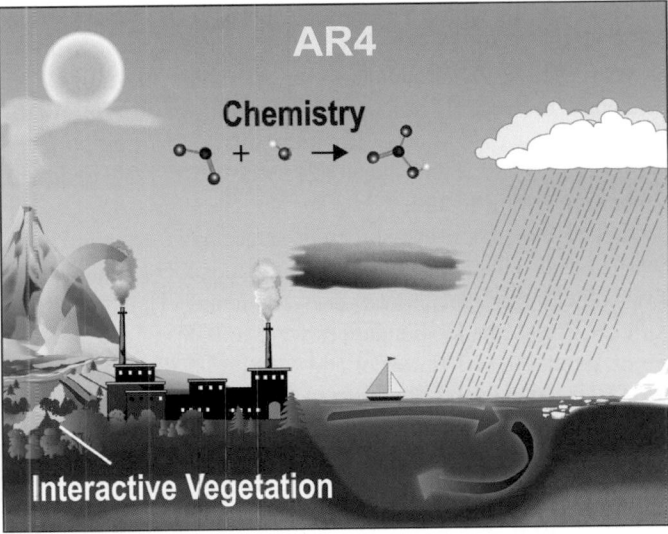

Figure 1.2. *The complexity of climate models has increased over the last few decades. The additional physics incorporated in the models are shown pictorially by the different features of the modelled world.*

1.3 Examples of Progress in Detecting and Attributing Recent Climate Change

1.3.1 The Human Fingerprint on Greenhouse Gases

The high-accuracy measurements of atmospheric CO_2 concentration, initiated by Charles David Keeling in 1958, constitute the master time series documenting the changing composition of the atmosphere (Keeling, 1961, 1998). These data have iconic status in climate change science as evidence of the effect of human activities on the chemical composition of the global atmosphere (see FAQ 7.1). Keeling's measurements on Mauna Loa in Hawaii provide a true measure of the global carbon cycle, an effectively continuous record of the burning of fossil fuel. They also maintain an accuracy and precision that allow scientists to separate fossil fuel emissions from those due to the natural annual cycle of the biosphere, demonstrating a long-term change in the seasonal exchange of CO_2 between the atmosphere, biosphere and ocean. Later observations of parallel trends in the atmospheric abundances of the $^{13}CO_2$ isotope (Francey and Farquhar, 1982) and molecular oxygen (O_2) (Keeling and Shertz, 1992; Bender et al., 1996) uniquely identified this rise in CO_2 with fossil fuel burning (Sections 2.3, 7.1 and 7.3).

To place the increase in CO_2 abundance since the late 1950s in perspective, and to compare the magnitude of the anthropogenic increase with natural cycles in the past, a longer-term record of CO_2 and other natural greenhouse gases is needed. These data came from analysis of the composition of air enclosed in bubbles in ice cores from Greenland and Antarctica. The initial measurements demonstrated that CO_2 abundances were significantly lower during the last ice age than over the last 10 kyr of the Holocene (Delmas et al., 1980; Berner et al., 1980; Neftel et al., 1982). From 10 kyr before present up to the year 1750, CO_2 abundances stayed within the range 280 ± 20 ppm (Indermühle et al., 1999). During the industrial era, CO_2 abundance rose roughly exponentially to 367 ppm in 1999 (Neftel et al., 1985; Etheridge et al., 1996; IPCC, 2001a) and to 379 ppm in 2005 (Section 2.3.1; see also Section 6.4).

Direct atmospheric measurements since 1970 (Steele et al., 1996) have also detected the increasing atmospheric abundances of two other major greenhouse gases, methane (CH_4) and nitrous oxide (N_2O). Methane abundances were initially increasing at a rate of about 1% yr^{-1} (Graedel and McRae, 1980; Fraser et al., 1981; Blake et al., 1982) but then slowed to an average increase of 0.4% yr^{-1} over the 1990s (Dlugokencky et al., 1998) with the possible stabilisation of CH_4 abundance (Section 2.3.2). The increase in N_2O abundance is smaller, about 0.25% yr^{-1}, and more difficult to detect (Weiss, 1981; Khalil and Rasmussen, 1988). To go back in time, measurements were made from firn air trapped in snowpack dating back over 200 years, and these data show an accelerating rise in both CH_4 and N_2O into the 20th century (Machida et al., 1995; Battle et al., 1996). When

ice core measurements extended the CH_4 abundance back 1 kyr, they showed a stable, relatively constant abundance of 700 ppb until the 19th century when a steady increase brought CH_4 abundances to 1,745 ppb in 1998 (IPCC, 2001a) and 1,774 ppb in 2005 (Section 2.3.2). This peak abundance is much higher than the range of 400 to 700 ppb seen over the last half-million years of glacial-interglacial cycles, and the increase can be readily explained by anthropogenic emissions. For N_2O the results are similar: the relative increase over the industrial era is smaller (15%), yet the 1998 abundance of 314 ppb (IPCC, 2001a), rising to 319 ppb in 2005 (Section 2.3.3), is also well above the 180-to-260 ppb range of glacial-interglacial cycles (Flückiger et al., 1999; see Sections 2.3, 6.2, 6.3, 6.4, 7.1 and 7.4)

Several synthetic halocarbons (chlorofluorocarbons (CFCs), hydrofluorocarbons, perfluorocarbons, halons and sulphur hexafluoride) are greenhouse gases with large global warming potentials (GWPs; Section 2.10). The chemical industry has been producing these gases and they have been leaking into the atmosphere since about 1930. Lovelock (1971) first measured CFC-11 ($CFCl_3$) in the atmosphere, noting that it could serve as an artificial tracer, with its north-south gradient reflecting the latitudinal distribution of anthropogenic emissions. Atmospheric abundances of all the synthetic halocarbons were increasing until the 1990s, when the abundance of halocarbons phased out under the Montreal Protocol began to fall (Montzka et al., 1999; Prinn et al., 2000). In the case of synthetic halocarbons (except perfluoromethane), ice core research has shown that these compounds did not exist in ancient air (Langenfelds et al., 1996) and thus confirms their industrial human origin (see Sections 2.3 and 7.1).

At the time of the TAR scientists could say that the abundances of all the well-mixed greenhouse gases during the 1990s were greater than at any time during the last half-million years (Petit et al, 1999), and this record now extends back nearly one million years (Section 6.3). Given this daunting picture of increasing greenhouse gas abundances in the atmosphere, it is noteworthy that, for simpler challenges but still on a hemispheric or even global scale, humans have shown the ability to undo what they have done. Sulphate pollution in Greenland was reversed in the 1980s with the control of acid rain in North America and Europe (IPCC, 2001b), and CFC abundances are declining globally because of their phase-out undertaken to protect the ozone layer.

1.3.2 Global Surface Temperature

Shortly after the invention of the thermometer in the early 1600s, efforts began to quantify and record the weather. The first meteorological network was formed in northern Italy in 1653 (Kington, 1988) and reports of temperature observations were published in the earliest scientific journals (e.g., Wallis and Beale, 1669). By the latter part of the 19th century, systematic observations of the weather were being made in almost all inhabited areas of the world. Formal international coordination of meteorological observations from ships commenced in 1853 (Quetelet, 1854).

Inspired by the paper *Suggestions on a Uniform System of Meteorological Observations* (Buys-Ballot, 1872), the International Meteorological Organization (IMO) was formed in 1873. Its successor, the World Meteorological Organization (WMO), still works to promote and exchange standardised meteorological observations. Yet even with uniform observations, there are still four major obstacles to turning instrumental observations into accurate global time series: (1) access to the data in usable form; (2) quality control to remove or edit erroneous data points; (3) homogeneity assessments and adjustments where necessary to ensure the fidelity of the data; and (4) area-averaging in the presence of substantial gaps.

Köppen (1873, 1880, 1881) was the first scientist to overcome most of these obstacles in his quest to study the effect of changes in sunspots (Section 2.7). Much of his data came from Dove (1852), but wherever possible he used data directly from the original source, because Dove often lacked information about the observing methods. Köppen considered examination of the annual mean temperature to be an adequate technique for quality control of far distant stations. Using data from more than 100 stations, Köppen averaged annual observations into several major latitude belts and then area-averaged these into a near-global time series shown in Figure 1.3.

Callendar (1938) produced the next global temperature time series expressly to investigate the influence of CO_2 on temperature (Section 2.3). Callendar examined about 200 station records. Only a small portion of them were deemed defective, based on quality concerns determined by comparing differences with neighbouring stations or on homogeneity concerns based on station changes documented in the recorded metadata. After further removing two arctic stations because he had no compensating stations from the antarctic region, he created a global average using data from 147 stations.

Most of Callendar's data came from World Weather Records (WWR; Clayton, 1927). Initiated by a resolution at the 1923 IMO Conference, WWR was a monumental international undertaking producing a 1,196-page volume of monthly temperature, precipitation and pressure data from hundreds of stations around the world, some with data starting in the early 1800s. In the early 1960s, J. Wolbach had these data digitised (National Climatic Data Center, 2002). The WWR project continues today under the auspices of the WMO with the digital publication of decadal updates to the climate records for thousands of stations worldwide (National Climatic Data Center, 2005).

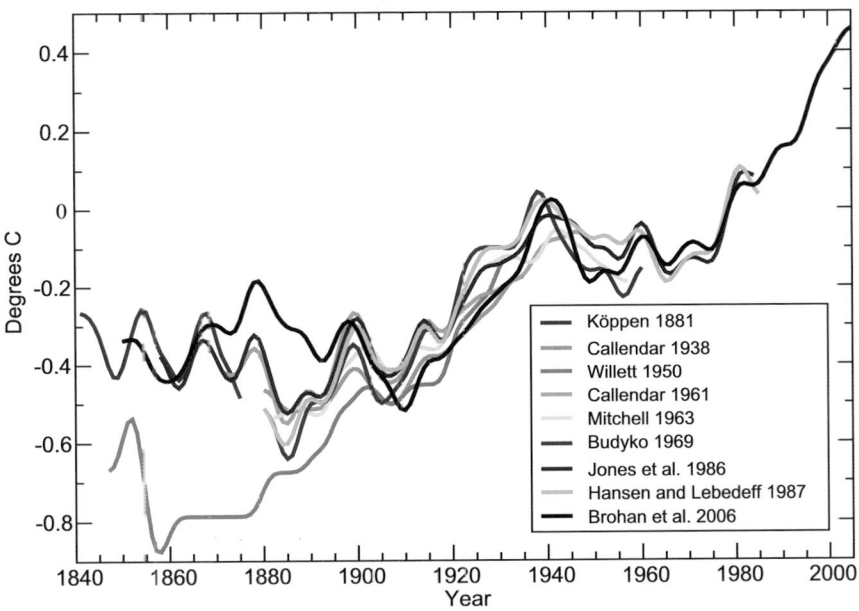

Figure 1.3. *Published records of surface temperature change over large regions. Köppen (1881) tropics and temperate latitudes using land air temperature. Callendar (1938) global using land stations. Willett (1950) global using land stations. Callendar (1961) 60°N to 60°S using land stations. Mitchell (1963) global using land stations. Budyko (1969) Northern Hemisphere using land stations and ship reports. Jones et al. (1986a,b) global using land stations. Hansen and Lebedeff (1987) global using land stations. Brohan et al. (2006) global using land air temperature and sea surface temperature data is the longest of the currently updated global temperature time series (Section 3.2). All time series were smoothed using a 13-point filter. The Brohan et al. (2006) time series are anomalies from the 1961 to 1990 mean (°C). Each of the other time series was originally presented as anomalies from the mean temperature of a specific and differing base period. To make them comparable, the other time series have been adjusted to have the mean of their last 30 years identical to that same period in the Brohan et al. (2006) anomaly time series.*

Willett (1950) also used WWR as the main source of data for 129 stations that he used to create a global temperature time series going back to 1845. While the resolution that initiated WWR called for the publication of long and homogeneous records, Willett took this mandate one step further by carefully selecting a subset of stations with as continuous and homogeneous a record as possible from the most recent update of WWR, which included data through 1940. To avoid over-weighting certain areas such as Europe, only one record, the best available, was included from each 10° latitude and longitude square. Station monthly data were averaged into five-year periods and then converted to anomalies with respect to the five-year period 1935 to 1939. Each station's anomaly was given equal weight to create the global time series.

Callendar in turn created a new near-global temperature time series in 1961 and cited Willett (1950) as a guide for some of his improvements. Callendar (1961) evaluated 600 stations with about three-quarters of them passing his quality checks. Unbeknownst to Callendar, a former student of Willett, Mitchell (1963), in work first presented in 1961, had created his own updated global temperature time series using slightly fewer than 200 stations and averaging the data into latitude bands. Landsberg and Mitchell (1961) compared Callendar's results with Mitchell's and stated that there was generally good agreement except in the data-sparse regions of the Southern Hemisphere.

Meanwhile, research in Russia was proceeding on a very different method to produce large-scale time series. Budyko (1969) used smoothed, hand-drawn maps of monthly temperature anomalies as a starting point. While restricted to analysis of the NH, this map-based approach not only allowed the inclusion of an increasing number of stations over time (e.g., 246 in 1881, 753 in 1913, 976 in 1940 and about 2,000 in 1960) but also the utilisation of data over the oceans (Robock, 1982).

Increasing the number of stations utilised has been a continuing theme over the last several decades with considerable effort being spent digitising historical station data as well as addressing the continuing problem of acquiring up-to-date data, as there can be a long lag between making an observation and the data getting into global data sets. During the 1970s and 1980s, several teams produced global temperature time series. Advances especially worth noting during this period include the extended spatial interpolation and station averaging technique of Hansen and Lebedeff (1987) and the Jones et al. (1986a,b) painstaking assessment of homogeneity and adjustments to account for discontinuities in the record of each of the thousands of stations in a global data set. Since then, global and national data sets have been rigorously adjusted for homogeneity using a variety of statistical and metadata-based approaches (Peterson et al., 1998).

One recurring homogeneity concern is potential urban heat island contamination in global temperature time series. This concern has been addressed in two ways. The first is by adjusting the temperature of urban stations to account for assessed urban heat island effects (e.g., Karl et al., 1988; Hansen et al., 2001). The second is by performing analyses that, like Callendar (1938), indicate that the bias induced by urban heat islands in the global temperature time series is either minor or non-existent (Jones et al., 1990; Peterson et al., 1999).

As the importance of ocean data became increasingly recognised, a major effort was initiated to seek out, digitise and quality-control historical archives of ocean data. This work has since grown into the International Comprehensive Ocean-Atmosphere Data Set (ICOADS; Worley et al., 2005), which has coordinated the acquisition, digitisation and synthesis of data ranging from transmissions by Japanese merchant ships to the logbooks of South African whaling boats. The amount of sea surface temperature (SST) and related data acquired continues to grow.

As fundamental as the basic data work of ICOADS was, there have been two other major advances in SST data. The first was adjusting the early observations to make them comparable to current observations (Section 3.2). Prior to 1940, the majority of SST observations were made from ships by hauling a bucket on deck filled with surface water and placing a thermometer in it. This ancient method eventually gave way to thermometers placed in engine cooling water inlets, which are typically located several metres below the ocean surface. Folland and Parker (1995) developed an adjustment model that accounted for heat loss from the buckets and that varied with bucket size and type, exposure to solar radiation, ambient wind speed and ship speed. They verified their results using time series of

night marine air temperature. This adjusted the early bucket observations upwards by a few tenths of a degree celsius.

Most of the ship observations are taken in narrow shipping lanes, so the second advance has been increasing global coverage in a variety of ways. Direct improvement of coverage has been achieved by the internationally coordinated placement of drifting and moored buoys. The buoys began to be numerous enough to make significant contributions to SST analyses in the mid-1980s (McPhaden et al., 1998) and have subsequently increased to more than 1,000 buoys transmitting data at any one time. Since 1982, satellite data, anchored to *in situ* observations, have contributed to near-global coverage (Reynolds and Smith, 1994). In addition, several different approaches have been used to interpolate and combine land and ocean observations into the current global temperature time series (Section 3.2). To place the current instrumental observations into a longer historical context requires the use of proxy data (Section 6.2).

Figure 1.3 depicts several historical 'global' temperature time series, together with the longest of the current global temperature time series, that of Brohan et al. (2006; Section 3.2). While the data and the analysis techniques have changed over time, all the time series show a high degree of consistency since 1900. The differences caused by using alternate data sources and interpolation techniques increase when the data are sparser. This phenomenon is especially illustrated by the pre-1880 values of Willett's (1950) time series. Willett noted that his data coverage remained fairly constant after 1885 but dropped off dramatically before that time to only 11 stations before 1850. The high degree of agreement between the time series resulting from these many different analyses increases the confidence that the changes they are indicating are real.

Despite the fact that many recent observations are automatic, the vast majority of data that go into global surface temperature calculations – over 400 million individual readings of thermometers at land stations and over 140 million individual *in situ* SST observations – have depended on the dedication of tens of thousands of individuals for well over a century. Climate science owes a great debt to the work of these individual weather observers as well as to international organisations such as the IMO, WMO and the Global Climate Observing System, which encourage the taking and sharing of high-quality meteorological observations. While modern researchers and their institutions put a great deal of time and effort into acquiring and adjusting the data to account for all known problems and biases, century-scale global temperature time series would not have been possible without the conscientious work of individuals and organisations worldwide dedicated to quantifying and documenting their local environment (Section 3.2).

1.3.3   Detection and Attribution

Using knowledge of past climates to qualify the nature of ongoing changes has become a concern of growing importance during the last decades, as reflected in the successive IPCC reports. While linked together at a technical level, detection and attribution have separate objectives. Detection of climate change is the process of demonstrating that climate has changed in some defined statistical sense, without providing a reason for that change. Attribution of causes of climate change is the process of establishing the most likely causes for the detected change with some defined level of confidence. Using traditional approaches, unequivocal attribution would require controlled experimentation with our climate system. However, with no spare Earth with which to experiment, attribution of anthropogenic climate change must be pursued by: (a) detecting that the climate has changed (as defined above); (b) demonstrating that the detected change is consistent with computer model simulations of the climate change 'signal' that is calculated to occur in response to anthropogenic forcing; and (c) demonstrating that the detected change is not consistent with alternative, physically plausible explanations of recent climate change that exclude important anthropogenic forcings.

Both detection and attribution rely on observational data and model output. In spite of the efforts described in Section 1.3.2, estimates of century-scale natural climate fluctuations remain difficult to obtain directly from observations due to the relatively short length of most observational records and a lack of understanding of the full range and effects of the various and ongoing external influences. Model simulations with no changes in external forcing (e.g., no increases in atmospheric CO_2 concentration) provide valuable information on the natural internal variability of the climate system on time scales of years to centuries. Attribution, on the other hand, requires output from model runs that incorporate historical estimates of changes in key anthropogenic and natural forcings, such as well-mixed greenhouse gases, volcanic aerosols and solar irradiance. These simulations can be performed with changes in a single forcing only (which helps to isolate the climate effect of that forcing), or with simultaneous changes in a whole suite of forcings.

In the early years of detection and attribution research, the focus was on a single time series – the estimated global-mean changes in the Earth's surface temperature. While it was not possible to detect anthropogenic warming in 1980, Madden and Ramanathan (1980) and Hansen et al. (1981) predicted it would be evident at least within the next two decades. A decade later, Wigley and Raper (1990) used a simple energy-balance climate model to show that the observed change in global-mean surface temperature from 1867 to 1982 could not be explained by natural internal variability. This finding was later confirmed using variability estimates from more complex coupled ocean-atmosphere general circulation models (e.g., Stouffer et al., 1994).

As the science of climate change progressed, detection and attribution research ventured into more sophisticated statistical analyses that examined complex patterns of climate change. Climate change patterns or 'fingerprints' were no longer limited to a single variable (temperature) or to the Earth's surface. More recent detection and attribution work has made use of precipitation and global pressure patterns, and analysis of vertical profiles of temperature change in the ocean and atmosphere. Studies with multiple variables make it easier to address attribution issues. While two different climate

forcings may yield similar changes in global mean temperature, it is highly unlikely that they will produce exactly the same 'fingerprint' (i.e., climate changes that are identical as a function of latitude, longitude, height, season and history over the 20th century).

Such model-predicted fingerprints of anthropogenic climate change are clearly statistically identifiable in observed data. The common conclusion of a wide range of fingerprint studies conducted over the past 15 years is that observed climate changes cannot be explained by natural factors alone (Santer et al., 1995, 1996a,b,c; Hegerl et al., 1996, 1997, 2000; Hasselmann, 1997; Barnett et al., 1999; Tett et al., 1999; Stott et al., 2000). A substantial anthropogenic influence is required in order to best explain the observed changes. The evidence from this body of work strengthens the scientific case for a discernible human influence on global climate.

1.4 Examples of Progress in Understanding Climate Processes

1.4.1 The Earth's Greenhouse Effect

The realisation that Earth's climate might be sensitive to the atmospheric concentrations of gases that create a greenhouse effect is more than a century old. Fleming (1998) and Weart (2003) provided an overview of the emerging science. In terms of the energy balance of the climate system, Edme Mariotte noted in 1681 that although the Sun's light and heat easily pass through glass and other transparent materials, heat from other sources (*chaleur de feu*) does not. The ability to generate an artificial warming of the Earth's surface was demonstrated in simple greenhouse experiments such as Horace Benedict de Saussure's experiments in the 1760s using a 'heliothermometer' (panes of glass covering a thermometer in a darkened box) to provide an early analogy to the greenhouse effect. It was a conceptual leap to recognise that the air itself could also trap thermal radiation. In 1824, Joseph Fourier, citing Saussure, argued 'the temperature [of the Earth] can be augmented by the interposition of the atmosphere, because heat in the state of light finds less resistance in penetrating the air, than in repassing into the air when converted into non-luminous heat'. In 1836, Pouillit followed up on Fourier's ideas and argued 'the atmospheric stratum…exercises a greater absorption upon the terrestrial than on the solar rays'. There was still no understanding of exactly what substance in the atmosphere was responsible for this absorption.

In 1859, John Tyndall (1861) identified through laboratory experiments the absorption of thermal radiation by complex molecules (as opposed to the primary bimolecular atmospheric constituents O_2 and molecular nitrogen). He noted that changes in the amount of any of the radiatively active constituents of the atmosphere such as water (H_2O) or CO_2 could have produced 'all the mutations of climate which the researches of geologists

Frequently Asked Question 1.2
What is the Relationship between Climate Change and Weather?

Climate is generally defined as average weather, and as such, climate change and weather are intertwined. Observations can show that there have been changes in weather, and it is the statistics of changes in weather over time that identify climate change. While weather and climate are closely related, there are important differences. A common confusion between weather and climate arises when scientists are asked how they can predict climate 50 years from now when they cannot predict the weather a few weeks from now. The chaotic nature of weather makes it unpredictable beyond a few days. Projecting changes in climate (i.e., long-term average weather) due to changes in atmospheric composition or other factors is a very different and much more manageable issue. As an analogy, while it is impossible to predict the age at which any particular man will die, we can say with high confidence that the average age of death for men in industrialised countries is about 75. Another common confusion of these issues is thinking

that a cold winter or a cooling spot on the globe is evidence against global warming. There are always extremes of hot and cold, although their frequency and intensity change as climate changes. But when weather is averaged over space and time, the fact that the globe is warming emerges clearly from the data.

Meteorologists put a great deal of effort into observing, understanding and predicting the day-to-day evolution of weather systems. Using physics-based concepts that govern how the atmosphere moves, warms, cools, rains, snows, and evaporates water, meteorologists are typically able to predict the weather successfully several days into the future. A major limiting factor to the predictability of weather beyond several days is a fundamental dynamical property of the atmosphere. In the 1960s, meteorologist Edward Lorenz discovered that very slight differences in initial conditions can produce very different forecast results.

(continued)

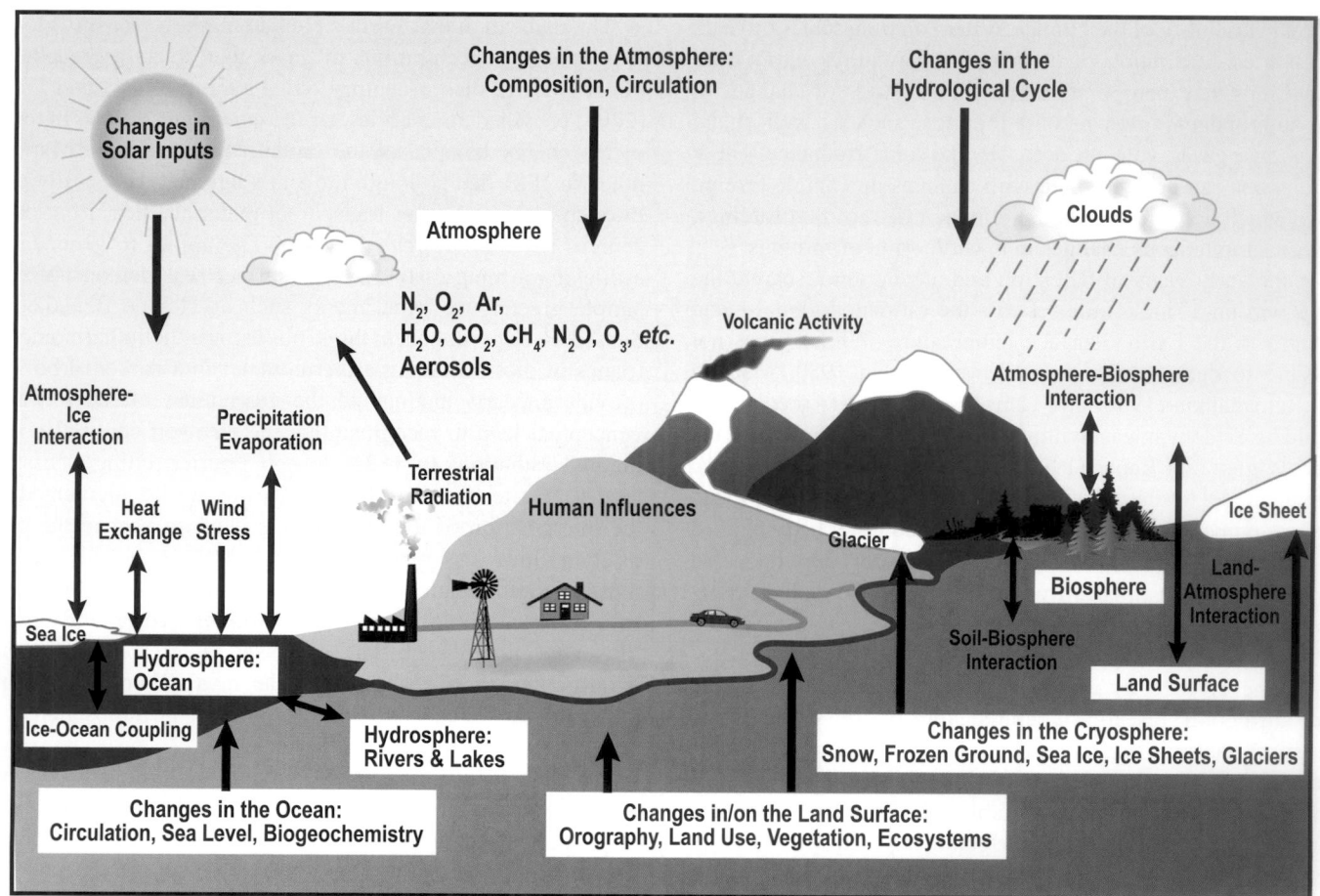

FAQ 1.2, Figure 1. *Schematic view of the components of the climate system, their processes and interactions.*

This is the so-called butterfly effect: a butterfly flapping its wings (or some other small phenomenon) in one place can, in principle, alter the subsequent weather pattern in a distant place. At the core of this effect is chaos theory, which deals with how small changes in certain variables can cause apparent randomness in complex systems.

Nevertheless, chaos theory does not imply a total lack of order. For example, slightly different conditions early in its history might alter the day a storm system would arrive or the exact path it would take, but the average temperature and precipitation (that is, climate) would still be about the same for that region and that period of time. Because a significant problem facing weather forecasting is knowing all the conditions at the start of the forecast period, it can be useful to think of climate as dealing with the background conditions for weather. More precisely, climate can be viewed as concerning the status of the entire Earth system, including the atmosphere, land, oceans, snow, ice and living things (see Figure 1) that serve as the global background conditions that determine weather patterns. An example of this would be an El Niño affecting the weather in coastal Peru. The El Niño sets limits on the probable evolution of weather patterns that random effects can produce. A La Niña would set different limits.

Another example is found in the familiar contrast between summer and winter. The march of the seasons is due to changes in the geographical patterns of energy absorbed and radiated away by the Earth system. Likewise, projections of future climate are shaped by fundamental changes in heat energy in the Earth system, in particular the increasing intensity of the greenhouse effect that traps heat near Earth's surface, determined by the amount of carbon dioxide and other greenhouse gases in the atmosphere. Projecting changes in climate due to changes in greenhouse gases 50 years from now is a very different and much more easily solved problem than forecasting weather patterns just weeks from now. To put it another way, long-term variations brought about by changes in the composition of the atmosphere are much more predictable than individual weather events. As an example, while we cannot predict the outcome of a single coin toss or roll of the dice, we can predict the statistical behaviour of a large number of such trials.

While many factors continue to influence climate, scientists have determined that human activities have become a dominant force, and are responsible for most of the warming observed over the past 50 years. Human-caused climate change has resulted primarily from changes in the amounts of greenhouse gases in the atmosphere, but also from changes in small particles (aerosols), as well as from changes in land use, for example. As climate changes, the probabilities of certain types of weather events are affected. For example, as Earth's average temperature has increased, some weather phenomena have become more frequent and intense (e.g., heat waves and heavy downpours), while others have become less frequent and intense (e.g., extreme cold events).

reveal'. In 1895, Svante Arrhenius (1896) followed with a climate prediction based on greenhouse gases, suggesting that a 40% increase or decrease in the atmospheric abundance of the trace gas CO_2 might trigger the glacial advances and retreats. One hundred years later, it would be found that CO_2 did indeed vary by this amount between glacial and interglacial periods. However, it now appears that the initial climatic change preceded the change in CO_2 but was enhanced by it (Section 6.4).

G. S. Callendar (1938) solved a set of equations linking greenhouse gases and climate change. He found that a doubling of atmospheric CO_2 concentration resulted in an increase in the mean global temperature of 2°C, with considerably more warming at the poles, and linked increasing fossil fuel combustion with a rise in CO_2 and its greenhouse effects: 'As man is now changing the composition of the atmosphere at a rate which must be very exceptional on the geological time scale, it is natural to seek for the probable effects of such a change. From the best laboratory observations it appears that the principal result of increasing atmospheric carbon dioxide... would be a gradual increase in the mean temperature of the colder regions of the Earth.' In 1947, Ahlmann reported a 1.3°C warming in the North Atlantic sector of the Arctic since the 19th century and mistakenly believed this climate variation could be explained entirely by greenhouse gas warming. Similar model predictions were echoed by Plass in 1956 (see Fleming, 1998): 'If at the end of this century, measurements show that the carbon dioxide content of the atmosphere has risen appreciably and at the same time the temperature has continued to rise throughout the world, it will be firmly established that carbon dioxide is an important factor in causing climatic change' (see Chapter 9).

In trying to understand the carbon cycle, and specifically how fossil fuel emissions would change atmospheric CO_2, the interdisciplinary field of carbon cycle science began. One of the first problems addressed was the atmosphere-ocean exchange of CO_2. Revelle and Suess (1957) explained why part of the emitted CO_2 was observed to accumulate in the atmosphere rather than being completely absorbed by the oceans. While CO_2 can be mixed rapidly into the upper layers of the ocean, the time to mix with the deep ocean is many centuries. By the time of the TAR, the interaction of climate change with the oceanic circulation and biogeochemistry was projected to reduce the fraction of anthropogenic CO_2 emissions taken up by the oceans in the future, leaving a greater fraction in the atmosphere (Sections 7.1, 7.3 and 10.4).

In the 1950s, the greenhouse gases of concern remained CO_2 and H_2O, the same two identified by Tyndall a century earlier. It was not until the 1970s that other greenhouse gases – CH_4, N_2O and CFCs – were widely recognised as important anthropogenic greenhouse gases (Ramanathan, 1975; Wang et al., 1976; Section 2.3). By the 1970s, the importance of aerosol-cloud effects in reflecting sunlight was known (Twomey, 1977), and atmospheric aerosols (suspended small particles) were being proposed as climate-forcing constituents. Charlson and others (summarised in Charlson et al., 1990) built a consensus that sulphate aerosols were, by themselves, cooling the Earth's surface by directly reflecting sunlight. Moreover, the increases in sulphate aerosols were anthropogenic and linked with the main source of CO_2, burning of fossil fuels (Section 2.4). Thus, the current picture of the atmospheric constituents driving climate change contains a much more diverse mix of greenhouse agents.

1.4.2 Past Climate Observations, Astronomical Theory and Abrupt Climate Changes

Throughout the 19th and 20th centuries, a wide range of geomorphology and palaeontology studies has provided new insight into the Earth's past climates, covering periods of hundreds of millions of years. The Palaeozoic Era, beginning 600 Ma, displayed evidence of both warmer and colder climatic conditions than the present; the Tertiary Period (65 to 2.6 Ma) was generally warmer; and the Quaternary Period (2.6 Ma to the present – the ice ages) showed oscillations between glacial and interglacial conditions. Louis Agassiz (1837) developed the hypothesis that Europe had experienced past glacial ages, and there has since been a growing awareness that long-term climate observations can advance the understanding of the physical mechanisms affecting climate change. The scientific study of one such mechanism – modifications in the geographical and temporal patterns of solar energy reaching the Earth's surface due to changes in the Earth's orbital parameters – has a long history. The pioneering contributions of Milankovitch (1941) to this astronomical theory of climate change are widely known, and the historical review of Imbrie and Imbrie (1979) calls attention to much earlier contributions, such as those of James Croll, originating in 1864.

The pace of palaeoclimatic research has accelerated over recent decades. Quantitative and well-dated records of climate fluctuations over the last 100 kyr have brought a more comprehensive view of how climate changes occur, as well as the means to test elements of the astronomical theory. By the 1950s, studies of deep-sea cores suggested that the ocean temperatures may have been different during glacial times (Emiliani, 1955). Ewing and Donn (1956) proposed that changes in ocean circulation actually could initiate an ice age. In the 1960s, the works of Emiliani (1969) and Shackleton (1967) showed the potential of isotopic measurements in deep-sea sediments to help explain Quaternary changes. In the 1970s, it became possible to analyse a deep-sea core time series of more than 700 kyr, thereby using the last reversal of the Earth's magnetic field to establish a dated chronology. This deep-sea observational record clearly showed the same periodicities found in the astronomical forcing, immediately providing strong support to Milankovitch's theory (Hays et al., 1976).

Ice cores provide key information about past climates, including surface temperatures and atmospheric chemical composition. The bubbles sealed in the ice are the only available samples of these past atmospheres. The first deep ice cores from Vostok in Antarctica (Barnola et al., 1987; Jouzel et al., 1987, 1993) provided additional evidence of the role of astronomical forcing. They also revealed a highly correlated evolution of temperature changes and atmospheric composition, which was subsequently confirmed over the past 400 kyr (Petit et al., 1999) and now extends to almost 1 Myr. This discovery drove research to understand the causal links between greenhouse gases and climate change. The same data that confirmed the astronomical theory also revealed its limits: a linear response of the climate system to astronomical forcing could not explain entirely the observed fluctuations of rapid ice-age terminations preceded by longer cycles of glaciations.

The importance of other sources of climate variability was heightened by the discovery of abrupt climate changes. In this context, 'abrupt' designates regional events of large amplitude, typically a few degrees celsius, which occurred within several decades – much shorter than the thousand-year time scales that characterise changes in astronomical forcing. Abrupt temperature changes were first revealed by the analysis of deep ice cores from Greenland (Dansgaard et al., 1984). Oeschger et al. (1984) recognised that the abrupt changes during the termination of the last ice age correlated with cooling in Gerzensee (Switzerland) and suggested that regime shifts in the Atlantic Ocean circulation were causing these widespread changes. The synthesis of palaeoclimatic observations by Broecker and Denton (1989) invigorated the community over the next decade. By the end of the 1990s, it became clear that the abrupt climate changes during the last ice age, particularly in the North Atlantic regions as found in the Greenland ice cores, were numerous (Dansgaard et al., 1993), indeed abrupt (Alley et al., 1993) and of large amplitude (Severinghaus and Brook, 1999). They are now referred to as Dansgaard-Oeschger events. A similar variability is seen in the North Atlantic Ocean, with north-south oscillations of the polar front (Bond et al., 1992) and associated changes in ocean temperature and salinity (Cortijo et al., 1999). With no obvious external forcing, these changes are thought to be manifestations of the internal variability of the climate system.

The importance of internal variability and processes was reinforced in the early 1990s with analysis of records with high temporal resolution. New ice cores (Greenland Ice Core Project, Johnsen et al., 1992; Greenland Ice Sheet Project 2, Grootes et al., 1993), new ocean cores from regions with high sedimentation rates, as well as lacustrine sediments and cave stalagmites produced additional evidence for unforced climate changes, and revealed a large number of abrupt changes in many regions throughout the last glacial cycle. Long sediment cores from the deep ocean were used to reconstruct the thermohaline circulation connecting deep and surface waters (Bond et al., 1992; Broecker, 1997) and to demonstrate the participation of the ocean in these abrupt climate changes during glacial periods.

By the end of the 1990s, palaeoclimate proxies for a range of climate observations had expanded greatly. The analysis of deep corals provided indicators for nutrient content and mass exchange from the surface to deep water (Adkins et al., 1998), showing abrupt variations characterised by synchronous changes in surface and deep-water properties (Shackleton et al., 2000). Precise measurements of the CH_4 abundances (a global quantity) in polar ice cores showed that they changed in concert with the Dansgaard-Oeschger events and thus allowed for synchronisation of the dating across ice cores (Blunier et al., 1998). The characteristics of the antarctic temperature variations and their relation to the Dansgaard-Oeschger events in Greenland were consistent with the simple concept of a bipolar seesaw caused by changes in the thermohaline circulation of the Atlantic Ocean (Stocker, 1998). This work underlined the role of the ocean in transmitting the signals of abrupt climate change.

Abrupt changes are often regional, for example, severe droughts lasting for many years have changed civilizations, and have occurred during the last 10 kyr of stable warm climate (deMenocal, 2001). This result has altered the notion of a stable climate during warm epochs, as previously suggested by the polar ice cores. The emerging picture of an unstable ocean-atmosphere system has opened the debate of whether human interference through greenhouse gases and aerosols could trigger such events (Broecker, 1997).

Palaeoclimate reconstructions cited in the FAR were based on various data, including pollen records, insect and animal remains, oxygen isotopes and other geological data from lake varves, loess, ocean sediments, ice cores and glacier termini. These records provided estimates of climate variability on time scales up to millions of years. A climate proxy is a local quantitative record (e.g., thickness and chemical properties of tree rings, pollen of different species) that is interpreted as a climate variable (e.g., temperature or rainfall) using a transfer function that is based on physical principles and recently observed correlations between the two records. The combination of instrumental and proxy data began in the 1960s with the investigation of the influence of climate on the proxy data, including tree rings (Fritts, 1962), corals (Weber and Woodhead, 1972; Dunbar and Wellington, 1981) and ice cores (Dansgaard et al., 1984; Jouzel et al., 1987). Phenological and historical data (e.g., blossoming dates, harvest dates, grain prices, ships' logs, newspapers, weather diaries, ancient manuscripts) are also a valuable source of climatic reconstruction for the period before instrumental records became available. Such documentary data also need calibration against instrumental data to extend and reconstruct the instrumental record (Lamb, 1969; Zhu, 1973; van den Dool, 1978; Brazdil, 1992; Pfister, 1992). With the development of multi-proxy reconstructions, the climate data were extended not only from local to global, but also from instrumental data to patterns of climate variability (Wanner et al., 1995; Mann et al., 1998; Luterbacher et al., 1999). Most of these reconstructions were at single sites and only loose efforts had been made to consolidate records. Mann et al. (1998) made a notable advance in the use of proxy data by

ensuring that the dating of different records lined up. Thus, the true spatial patterns of temperature variability and change could be derived, and estimates of NH average surface temperatures were obtained.

The Working Group I (WGI) WGI FAR noted that past climates could provide analogues. Fifteen years of research since that assessment has identified a range of variations and instabilities in the climate system that occurred during the last 2 Myr of glacial-interglacial cycles and in the super-warm period of 50 Ma. These past climates do not appear to be analogues of the immediate future, yet they do reveal a wide range of climate processes that need to be understood when projecting 21st-century climate change (see Chapter 6).

1.4.3 Solar Variability and the Total Solar Irradiance

Measurement of the absolute value of total solar irradiance (TSI) is difficult from the Earth's surface because of the need to correct for the influence of the atmosphere. Langley (1884) attempted to minimise the atmospheric effects by taking measurements from high on Mt. Whitney in California, and to estimate the correction for atmospheric effects by taking measurements at several times of day, for example, with the solar radiation having passed through different atmospheric pathlengths. Between 1902 and 1957, Charles Abbot and a number of other scientists around the globe made thousands of measurements of TSI from mountain sites. Values ranged from 1,322 to 1,465 W m^{-2}, which encompasses the current estimate of 1,365 W m^{-2}. Foukal et al. (1977) deduced from Abbot's daily observations that higher values of TSI were associated with more solar faculae (e.g., Abbot, 1910).

In 1978, the Nimbus-7 satellite was launched with a cavity radiometer and provided evidence of variations in TSI (Hickey et al., 1980). Additional observations were made with an active cavity radiometer on the Solar Maximum Mission, launched in 1980 (Willson et al., 1980). Both of these missions showed that the passage of sunspots and faculae across the Sun's disk influenced TSI. At the maximum of the 11-year solar activity cycle, the TSI is larger by about 0.1% than at the minimum. The observation that TSI is highest when sunspots are at their maximum is the opposite of Langley's (1876) hypothesis.

As early as 1910, Abbot believed that he had detected a downward trend in TSI that coincided with a general cooling of climate. The solar cycle variation in irradiance corresponds to an 11-year cycle in radiative forcing which varies by about 0.2 W m^{-2}. There is increasingly reliable evidence of its influence on atmospheric temperatures and circulations, particularly in the higher atmosphere (Reid, 1991; Brasseur, 1993; Balachandran and Rind, 1995; Haigh, 1996; Labitzke and van Loon, 1997; van Loon and Labitzke, 2000). Calculations with three-dimensional models (Wetherald and Manabe, 1975; Cubasch et al., 1997; Lean and Rind, 1998; Tett et al., 1999; Cubasch and Voss, 2000) suggest that the changes in solar radiation could cause surface temperature changes of the order of a few tenths of a degree celsius.

For the time before satellite measurements became available, the solar radiation variations can be inferred from cosmogenic isotopes (^{10}Be, ^{14}C) and from the sunspot number. Naked-eye observations of sunspots date back to ancient times, but it was only after the invention of the telescope in 1607 that it became possible to routinely monitor the number, size and position of these 'stains' on the surface of the Sun. Throughout the 17th and 18th centuries, numerous observers noted the variable concentrations and ephemeral nature of sunspots, but very few sightings were reported between 1672 and 1699 (for an overview see Hoyt et al., 1994). This period of low solar activity, now known as the Maunder Minimum, occurred during the climate period now commonly referred to as the Little Ice Age (Eddy, 1976). There is no exact agreement as to which dates mark the beginning and end of the Little Ice Age, but from about 1350 to about 1850 is one reasonable estimate.

During the latter part of the 18th century, Wilhelm Herschel (1801) noted the presence not only of sunspots but of bright patches, now referred to as faculae, and of granulations on the solar surface. He believed that when these indicators of activity were more numerous, solar emissions of light and heat were greater and could affect the weather on Earth. Heinrich Schwabe (1844) published his discovery of a '10-year cycle' in sunspot numbers. Samuel Langley (1876) compared the brightness of sunspots with that of the surrounding photosphere. He concluded that they would block the emission of radiation and estimated that at sunspot cycle maximum the Sun would be about 0.1% less bright than at the minimum of the cycle, and that the Earth would be 0.1°C to 0.3°C cooler.

These satellite data have been used in combination with the historically recorded sunspot number, records of cosmogenic isotopes, and the characteristics of other Sun-like stars to estimate the solar radiation over the last 1,000 years (Eddy, 1976; Hoyt and Schatten, 1993, 1997; Lean et al., 1995; Lean, 1997). These data sets indicated quasi-periodic changes in solar radiation of 0.24 to 0.30% on the centennial time scale. These values have recently been re-assessed (see, e.g., Chapter 2).

The TAR states that the changes in solar irradiance are not the major cause of the temperature changes in the second half of the 20th century unless those changes can induce unknown large feedbacks in the climate system. The effects of galactic cosmic rays on the atmosphere (via cloud nucleation) and those due to shifts in the solar spectrum towards the ultraviolet (UV) range, at times of high solar activity, are largely unknown. The latter may produce changes in tropospheric circulation via changes in static stability resulting from the interaction of the increased UV radiation with stratospheric ozone. More research to investigate the effects of solar behaviour on climate is needed before the magnitude of solar effects on climate can be stated with certainty.

1.4.4 Biogeochemistry and Radiative Forcing

The modern scientific understanding of the complex and interconnected roles of greenhouse gases and aerosols in climate change has undergone rapid evolution over the last two decades. While the concepts were recognised and outlined in the 1970s (see Sections 1.3.1 and 1.4.1), the publication of generally accepted quantitative results coincides with, and was driven in part by, the questions asked by the IPCC beginning in 1988. Thus, it is instructive to view the evolution of this topic as it has been treated in the successive IPCC reports.

The WGI FAR codified the key physical and biogeochemical processes in the Earth system that relate a changing climate to atmospheric composition, chemistry, the carbon cycle and natural ecosystems. The science of the time, as summarised in the FAR, made a clear case for anthropogenic interference with the climate system. In terms of greenhouse agents, the main conclusions from the WGI FAR Policymakers Summary are still valid today: (1) 'emissions resulting from human activities are substantially increasing the atmospheric concentrations of the greenhouse gases: CO_2, CH_4, CFCs, N_2O'; (2) 'some gases are potentially more effective (at greenhouse warming)'; (3) feedbacks between the carbon cycle, ecosystems and atmospheric greenhouse gases in a warmer world will affect CO_2 abundances; and (4) GWPs provide a metric for comparing the climatic impact of different greenhouse gases, one that integrates both the radiative influence and biogeochemical cycles. The climatic importance of tropospheric ozone, sulphate aerosols and atmospheric chemical feedbacks were proposed by scientists at the time and noted in the assessment. For example, early global chemical modelling results argued that global tropospheric ozone, a greenhouse gas, was controlled by emissions of the highly reactive gases nitrogen oxides (NO_x), carbon monoxide (CO) and non-methane hydrocarbons (NMHC, also known as volatile organic compounds, VOC). In terms of sulphate aerosols, both the direct radiative effects and the indirect effects on clouds were acknowledged, but the importance of carbonaceous aerosols from fossil fuel and biomass combustion was not recognised (Chapters 2, 7 and 10).

The concept of radiative forcing (RF) as the radiative imbalance (W m^{-2}) in the climate system at the top of the atmosphere caused by the addition of a greenhouse gas (or other change) was established at the time and summarised in Chapter 2 of the WGI FAR. Agents of RF included the direct greenhouse gases, solar radiation, aerosols and the Earth's surface albedo. What was new and only briefly mentioned was that 'many gases produce indirect effects on the global radiative forcing'. The innovative global modelling work of Derwent (1990) showed that emissions of the reactive but non-greenhouse gases – NO_x, CO and NMHCs – altered atmospheric chemistry and thus changed the abundance of other greenhouse gases. Indirect GWPs for NO_x, CO and VOCs were proposed. The projected chemical feedbacks were limited to short-lived increases in tropospheric ozone. By 1990, it was clear that the RF from tropospheric ozone had increased over the 20th century and stratospheric ozone had decreased since 1980 (e.g., Lacis et al., 1990), but the associated RFs were not evaluated in the assessments. Neither was the effect of anthropogenic sulphate aerosols, except to note in the FAR that 'it is conceivable that this radiative forcing has been of a comparable magnitude, but of opposite sign, to the greenhouse forcing earlier in the century'. Reflecting in general the community's concerns about this relatively new measure of climate forcing, RF bar charts appear only in the underlying FAR chapters, but not in the FAR Summary. Only the long-lived greenhouse gases are shown, although sulphate aerosols direct effect in the future is noted with a question mark (i.e., dependent on future emissions) (Chapters 2, 7 and 10).

The cases for more complex chemical and aerosol effects were becoming clear, but the scientific community was unable at the time to reach general agreement on the existence, scale and magnitude of these indirect effects. Nevertheless, these early discoveries drove the research agendas in the early 1990s. The widespread development and application of global chemistry-transport models had just begun with international workshops (Pyle et al., 1996; Jacob et al., 1997; Rasch, 2000). In the Supplementary Report (IPCC, 1992) to the FAR, the indirect chemical effects of CO, NO_x and VOC were reaffirmed, and the feedback effect of CH_4 on the tropospheric hydroxyl radical (OH) was noted, but the indirect RF values from the FAR were retracted and denoted in a table with '+', '0' or '−'. Aerosol-climate interactions still focused on sulphates, and the assessment of their direct RF for the NH (i.e., a cooling) was now somewhat quantitative as compared to the FAR. Stratospheric ozone depletion was noted as causing a significant and negative RF, but not quantified. Ecosystems research at this time was identifying the responses to climate change and CO_2 increases, as well as altered CH_4 and N_2O fluxes from natural systems; however, in terms of a community assessment it remained qualitative.

By 1994, with work on SAR progressing, the Special Report on Radiative Forcing (IPCC, 1995) reported significant breakthroughs in a set of chapters limited to assessment of the carbon cycle, atmospheric chemistry, aerosols and RF. The carbon budget for the 1980s was analysed not only from bottom-up emissions estimates, but also from a top-down approach including carbon isotopes. A first carbon cycle assessment was performed through an international model and analysis workshop examining terrestrial and oceanic uptake to better quantify the relationship between CO_2 emissions and the resulting increase in atmospheric abundance. Similarly, expanded analyses of the global budgets of trace gases and aerosols from both natural and anthropogenic sources highlighted the rapid expansion of biogeochemical research. The first RF bar chart appears, comparing all the major components of RF change from the pre-industrial period to the present. Anthropogenic soot aerosol, with a positive RF, was not in the 1995 Special Report but was added to the SAR. In terms of atmospheric chemistry, the first open-invitation modelling study for the IPCC recruited 21 atmospheric chemistry models to participate in a controlled study of photochemistry and chemical feedbacks. These studies (e.g., Olson et al., 1997) demonstrated a robust consensus about some indirect effects, such as the CH_4 impact on atmospheric chemistry, but great uncertainty about others, such as the prediction of tropospheric ozone changes. The model studies plus the theory of chemical feedbacks in the CH_4-CO-OH system (Prather, 1994) firmly established that the atmospheric lifetime of a perturbation

(and hence climate impact and GWP) of CH_4 emissions was about 50% greater than reported in the FAR. There was still no consensus on quantifying the past or future changes in tropospheric ozone or OH (the primary sink for CH_4) (Chapters 2, 7 and 10).

In the early 1990s, research on aerosols as climate forcing agents expanded. Based on new research, the range of climate-relevant aerosols was extended for the first time beyond sulphates to include nitrates, organics, soot, mineral dust and sea salt. Quantitative estimates of sulphate aerosol indirect effects on cloud properties and hence RF were sufficiently well established to be included in assessments, and carbonaceous aerosols from biomass burning were recognised as being comparable in importance to sulphate (Penner et al., 1992). Ranges are given in the special report (IPCC, 1995) for direct sulphate RF (−0.25 to −0.9 W m^{-2}) and biomass-burning aerosols (−0.05 to −0.6 W m^{-2}). The aerosol indirect RF was estimated to be about equal to the direct RF, but with larger uncertainty. The injection of stratospheric aerosols from the eruption of Mt. Pinatubo was noted as the first modern test of a known radiative forcing, and indeed one climate model accurately predicted the temperature response (Hansen et al., 1992). In the one-year interval between the special report and the SAR, the scientific understanding of aerosols grew. The direct anthropogenic aerosol forcing (from sulphate, fossil-fuel soot and biomass-burning aerosols) was reduced to −0.5 W m^{-2}. The RF bar chart was now broken into aerosol components (sulphate, fossil-fuel soot and biomass burning aerosols) with a separate range for indirect effects (Chapters 2 and 7; Sections 8.2 and 9.2).

Throughout the 1990s, there were concerted research programs in the USA and EU to evaluate the global environmental impacts of aviation. Several national assessments culminated in the IPCC Special Report on Aviation and the Global Atmosphere (IPCC, 1999), which assessed the impacts on climate and global air quality. An open invitation for atmospheric model participation resulted in community participation and a consensus on many of the environmental impacts of aviation (e.g., the increase in tropospheric ozone and decrease in CH_4 due to NO_x emissions were quantified). The direct RF of sulphate and of soot aerosols was likewise quantified along with that of contrails, but the impact on cirrus clouds that are sometimes generated downwind of contrails was not. The assessment re-affirmed that RF was a first-order metric for the global mean surface temperature response, but noted that it was inadequate for regional climate change, especially in view of the largely regional forcing from aerosols and tropospheric ozone (Sections 2.6, 2.8 and 10.2).

By the end of the 1990s, research on atmospheric composition and climate forcing had made many important advances. The TAR was able to provide a more quantitative evaluation in some areas. For example, a large, open-invitation modelling workshop was held for both aerosols (11 global models) and tropospheric ozone-OH chemistry (14 global models). This workshop brought together as collaborating authors most of the international scientific community involved in developing and testing global models of atmospheric composition. In terms of atmospheric chemistry, a strong consensus was reached for the first time that science could predict the changes in tropospheric ozone in response to scenarios for CH_4 and the indirect greenhouse gases (CO, NO_x, VOC) and that a quantitative GWP for CO could be reported. Further, combining these models with observational analysis, an estimate of the change in tropospheric ozone since the pre-industrial era – with uncertainties – was reported. The aerosol workshop made similar advances in evaluating the impact of different aerosol types. There were many different representations of uncertainty (e.g., a range in models versus an expert judgment) in the TAR, and the consensus RF bar chart did not generate a total RF or uncertainties for use in the subsequent IPCC Synthesis Report (IPCC, 2001b) (Chapters 2 and 7; Section 9.2).

1.4.5   Cryospheric Topics

The cryosphere, which includes the ice sheets of Greenland and Antarctica, continental (including tropical) glaciers, snow, sea ice, river and lake ice, permafrost and seasonally frozen ground, is an important component of the climate system. The cryosphere derives its importance to the climate system from a variety of effects, including its high reflectivity (albedo) for solar radiation, its low thermal conductivity, its large thermal inertia, its potential for affecting ocean circulation (through exchange of freshwater and heat) and atmospheric circulation (through topographic changes), its large potential for affecting sea level (through growth and melt of land ice), and its potential for affecting greenhouse gases (through changes in permafrost) (Chapter 4).

Studies of the cryospheric albedo feedback have a long history. The albedo is the fraction of solar energy reflected back to space. Over snow and ice, the albedo (about 0.7 to 0.9) is large compared to that over the oceans (<0.1). In a warming climate, it is anticipated that the cryosphere would shrink, the Earth's overall albedo would decrease and more solar energy would be absorbed to warm the Earth still further. This powerful feedback loop was recognised in the 19th century by Croll (1890) and was first introduced in climate models by Budyko (1969) and Sellers (1969). But although the principle of the albedo feedback is simple, a quantitative understanding of the effect is still far from complete. For instance, it is not clear whether this mechanism is the main reason for the high-latitude amplification of the warming signal.

The potential cryospheric impact on ocean circulation and sea level are of particular importance. There may be 'large-scale discontinuities' (IPCC, 2001a) resulting from both the shutdown of the large-scale meridional circulation of the world oceans (see Section 1.4.6) and the disintegration of large continental ice sheets. Mercer (1968, 1978) proposed that atmospheric warming could cause the ice shelves of western Antarctica to disintegrate and that as a consequence the entire West Antarctic Ice Sheet (10% of the antarctic ice volume) would lose its land connection and come afloat, causing a sea level rise of about five metres.

The importance of permafrost-climate feedbacks came to be realised widely only in the 1990s, starting with the works of

Kvenvolden (1988, 1993), MacDonald (1990) and Harriss et al. (1993). As permafrost thaws due to a warmer climate, CO_2 and CH_4 trapped in permafrost are released to the atmosphere. Since CO_2 and CH_4 are greenhouse gases, atmospheric temperature is likely to increase in turn, resulting in a feedback loop with more permafrost thawing. The permafrost and seasonally thawed soil layers at high latitudes contain a significant amount (about one-quarter) of the global total amount of soil carbon. Because global warming signals are amplified in high-latitude regions, the potential for permafrost thawing and consequent greenhouse gas releases is thus large.

In situ monitoring of the cryosphere has a long tradition. For instance, it is important for fisheries and agriculture. Seagoing communities have documented sea ice extent for centuries. Records of thaw and freeze dates for lake and river ice start with Lake Suwa in Japan in 1444, and extensive records of snowfall in China were made during the Qing Dynasty (1644–1912). Records of glacial length go back to the mid-1500s. Internationally coordinated, long-term glacier observations started in 1894 with the establishment of the International Glacier Commission in Zurich, Switzerland. The longest time series of a glacial mass balance was started in 1946 at the Storglaciären in northern Sweden, followed by Storbreen in Norway (begun in 1949). Today a global network of mass balance monitoring for some 60 glaciers is coordinated through the World Glacier Monitoring Service. Systematic measurements of permafrost (thermal state and active layer) began in earnest around 1950 and were coordinated under the Global Terrestrial Network for Permafrost.

The main climate variables of the cryosphere (extent, albedo, topography and mass) are in principle observable from space, given proper calibration and validation through *in situ* observing efforts. Indeed, satellite data are required in order to have full global coverage. The polar-orbiting Nimbus 5 satellite, launched in 1972, yielded the earliest all-weather, all-season imagery of global sea ice, using microwave instruments (Parkinson et al., 1987), and enabled a major advance in the scientific understanding of the dynamics of the cryosphere. Launched in 1978, the Television Infrared Observation Satellite (TIROS-N) yielded the first monitoring from space of snow on land surfaces (Dozier et al., 1981). The number of cryospheric elements now routinely monitored from space is growing, and current satellites are now addressing one of the more challenging elements, variability of ice volume.

Climate modelling results have pointed to high-latitude regions as areas of particular importance and ecological vulnerability to global climate change. It might seem logical to expect that the cryosphere overall would shrink in a warming climate or expand in a cooling climate. However, potential changes in precipitation, for instance due to an altered hydrological cycle, may counter this effect both regionally and globally. By the time of the TAR, several climate models incorporated physically based treatments of ice dynamics, although the land ice processes were only rudimentary. Improving representation of the cryosphere in climate models is still an area of intense research and continuing progress (Chapter 8).

1.4.6 Ocean and Coupled Ocean-Atmosphere Dynamics

Developments in the understanding of the oceanic and atmospheric circulations, as well as their interactions, constitute a striking example of the continuous interplay among theory, observations and, more recently, model simulations. The atmosphere and ocean surface circulations were observed and analysed globally as early as the 16th and 17th centuries, in close association with the development of worldwide trade based on sailing. These efforts led to a number of important conceptual and theoretical works. For example, Edmund Halley first published a description of the tropical atmospheric cells in 1686, and George Hadley proposed a theory linking the existence of the trade winds with those cells in 1735. These early studies helped to forge concepts that are still useful in analysing and understanding both the atmospheric general circulation itself and model simulations (Lorenz, 1967; Holton, 1992).

A comprehensive description of these circulations was delayed by the lack of necessary observations in the higher atmosphere or deeper ocean. The balloon record of Gay-Lussac, who reached an altitude of 7,016 m in 1804, remained unbroken for more than 50 years. The stratosphere was independently discovered near the turn of the 20th century by Aßmann (1902) and Teisserenc de Bort (1902), and the first manned balloon flight into the stratosphere was made in 1901 (Berson and Süring, 1901). Even though it was recognised over 200 years ago (Rumford, 1800; see also Warren, 1981) that the oceans' cold subsurface waters must originate at high latitudes, it was not appreciated until the 20th century that the strength of the deep circulation might vary over time, or that the ocean's Meridional Overturning Circulation (MOC; often loosely referred to as the 'thermohaline circulation', see the Glossary for more information) may be very important for Earth's climate.

By the 1950s, studies of deep-sea cores suggested that the deep ocean temperatures had varied in the distant past. Technology also evolved to enable measurements that could confirm that the deep ocean is not only not static, but in fact quite dynamic (Swallow and Stommel's 1960 subsurface float experiment Aries, referred to by Crease, 1962). By the late 1970s, current meters could monitor deep currents for substantial amounts of time, and the first ocean observing satellite (SeaSat) revealed that significant information about subsurface ocean variability is imprinted on the sea surface. At the same time, the first estimates of the strength of the meridional transport of heat and mass were made (Oort and Vonder Haar, 1976; Wunsch, 1978), using a combination of models and data. Since then the technological developments have accelerated, but monitoring the MOC directly remains a substantial challenge (see Chapter 5), and routine observations of the subsurface ocean remain scarce compared to that of the atmosphere.

In parallel with the technological developments yielding new insights through observations, theoretical and numerical explorations of multiple (stable or unstable) equilibria began. Chamberlain (1906) suggested that deep ocean currents could reverse in direction, and might affect climate. The idea did not gain momentum until fifty years later, when Stommel (1961) presented a mechanism, based on the opposing effects that temperature and salinity have on density, by which ocean circulation can fluctuate between states. Numerical climate models incorporating models of the ocean circulation were developed during this period, including the pioneering work of Bryan (1969) and Manabe and Bryan (1969). The idea that the ocean circulation could change radically, and might perhaps even feel the attraction of different equilibrium states, gained further support through the simulations of coupled climate models (Bryan and Spelman, 1985; Bryan, 1986; Manabe and Stouffer, 1988). Model simulations using a hierarchy of models showed that the ocean circulation system appeared to be particularly vulnerable to changes in the freshwater balance, either by direct addition of freshwater or by changes in the hydrological cycle. A strong case emerged for the hypothesis that rapid changes in the Atlantic meridional circulation were responsible for the abrupt Dansgaard-Oeschger climate change events.

Although scientists now better appreciate the strength and variability of the global-scale ocean circulation, its roles in climate are still hotly debated. Is it a passive recipient of atmospheric forcing and so merely a diagnostic consequence of climate change, or is it an active contributor? Observational evidence for the latter proposition was presented by Sutton and Allen (1997), who noticed SST anomalies propagating along the Gulf Stream/North Atlantic Current system for years, and therefore implicated internal oceanic time scales. Is a radical change in the MOC likely in the near future? Brewer et al. (1983) and Lazier (1995) showed that the water masses of the North Atlantic were indeed changing (some becoming significantly fresher) in the modern observational record, a phenomenon that at least raises the possibility that ocean conditions may be approaching the point where the circulation might shift into Stommel's other stable regime. Recent developments in the ocean's various roles in climate can be found in Chapters 5, 6, 9 and 10.

Studying the interactions between atmosphere and ocean circulations was also facilitated through continuous interactions between observations, theories and simulations, as is dramatically illustrated by the century-long history of the advances in understanding the El Niño-Southern Oscillation (ENSO) phenomenon. This coupled air-sea phenomenon originates in the Pacific but affects climate globally, and has raised concern since at least the 19th century. Sir Gilbert Walker (1928) describes how H. H. Hildebrandsson (1897) noted large-scale relationships between interannual trends in pressure data from a worldwide network of 68 weather stations, and how Lockyer and Lockyer (1902) confirmed Hildebrandsson's discovery of an apparent 'seesaw' in pressure between South America and the Indonesian region. Walker named this seesaw pattern the 'Southern Oscillation' and related it to occurrences of drought and heavy rains in India, Australia, Indonesia and Africa. He also proposed that there must be a certain level of predictive skill in that system.

El Niño is the name given to the rather unusual oceanic conditions involving anomalously warm waters occurring in the eastern tropical Pacific off the coast of Peru every few years. The 1957–1958 International Geophysical Year coincided with a large El Niño, allowing a remarkable set of observations of the phenomenon. A decade later, a mechanism was presented that connected Walker's observations to El Niño (Bjerknes, 1969). This mechanism involved the interaction, through the SST field, between the east-west atmospheric circulation of which Walker's Southern Oscillation was an indicator (Bjerknes appropriately referred to this as the 'Walker Circulation') and variability in the pool of equatorial warm water of the Pacific Ocean. Observations made in the 1970s (e.g., Wyrtki, 1975) showed that prior to ENSO warm phases, the sea level in the western Pacific often rises significantly. By the mid-1980s, after an unusually disruptive El Niño struck in 1982 and 1983, an observing system (the Tropical Ocean Global Atmosphere (TOGA) array; see McPhaden et al., 1998) had been put in place to monitor ENSO. The resulting data confirmed the idea that the phenomenon was inherently one involving coupled atmosphere-ocean interactions and yielded much-needed detailed observational insights. By 1986, the first experimental ENSO forecasts were made (Cane et al., 1986; Zebiak and Cane, 1987).

The mechanisms and predictive skill of ENSO are still under discussion. In particular, it is not clear how ENSO changes with, and perhaps interacts with, a changing climate. The TAR states '...increasing evidence suggests the ENSO plays a fundamental role in global climate and its interannual variability, and increased credibility in both regional and global climate projections will be gained once realistic ENSOs and their changes are simulated'.

Just as the phenomenon of El Niño has been familiar to the people of tropical South America for centuries, a spatial pattern affecting climate variability in the North Atlantic has similarly been known by the people of Northern Europe for a long time. The Danish missionary Hans Egede made the following well-known diary entry in the mid-18th century: 'In Greenland, all winters are severe, yet they are not alike. The Danes have noticed that when the winter in Denmark is severe, as we perceive it, the winter in Greenland in its manner was mild, and conversely' (van Loon and Rogers, 1978).

Teisserenc de Bort, Hann, Exner, Defant and Walker all contributed to the discovery of the underlying dynamic structure. Walker, in his studies in the Indian Ocean, actually studied global maps of sea level pressure correlations, and named not only the Southern Oscillation, but also a Northern Oscillation, which he subsequently divided into a North Pacific and a North Atlantic Oscillation (Walker, 1924). However, it was Exner (1913, 1924) who made the first correlation maps showing the spatial structure in the NH, where the North Atlantic Oscillation (NAO) pattern stands out clearly as a north-south oscillation in atmospheric mass with centres of action near Iceland and Portugal.

The NAO significantly affects weather and climate, ecosystems and human activities of the North Atlantic sector. But what is the underlying mechanism? The recognition that the NAO is associated with variability and latitudinal shifts in the westerly flow of the jet stream originates with the works of

Willett, Namias, Lorenz, Rossby and others in the 1930s, 1940s and 1950s (reviewed by Stephenson et al., 2003). Because atmospheric planetary waves are hemispheric in nature, changes in one region are often connected with changes in other regions, a phenomenon dubbed 'teleconnection' (Wallace and Gutzler, 1981).

The NAO may be partly described as a high-frequency stochastic process internal to the atmosphere. This understanding is evidenced by numerous atmosphere-only model simulations. It is also considered an expression of one of Earth's 'annular modes' (See Chapter 3). It is, however, the low-frequency variability of this phenomenon (Hurrell, 1995) that fuels continued investigations by climate scientists. The long time scales are the indication of potential predictive skill in the NAO. The mechanisms responsible for the correspondingly long 'memory' are still debated, although they are likely to have a local or remote oceanic origin. Bjerknes (1964) recognised the connection between the NAO index (which he referred to as the 'zonal index') and sea surface conditions. He speculated that ocean heat advection could play a role on longer time scales. The circulation of the Atlantic Ocean is radically different from that of the Indian and Pacific Oceans, in that the MOC is strongest in the Atlantic with warm water flowing northwards, even south of the equator, and cold water returning at depth. It would therefore not be surprising if the oceanic contributions to the NAO and to the Southern Oscillation were different.

Earth's climate is characterised by many modes of variability, involving both the atmosphere and ocean, and also the cryosphere and biosphere. Understanding the physical processes involved in producing low-frequency variability is crucial for improving scientists' ability to accurately predict climate change and for allowing the separation of anthropogenic and natural variability, thereby improving the ability to detect and attribute anthropogenic climate change. One central question for climate scientists, addressed in particular in Chapter 9, is to determine how human activities influence the dynamic nature of Earth's climate, and to identify what would have happened without any human influence at all.

1.5 Examples of Progress in Modelling the Climate

1.5.1 Model Evolution and Model Hierarchies

Climate scenarios rely upon the use of numerical models. The continuous evolution of these models over recent decades has been enabled by a considerable increase in computational capacity, with supercomputer speeds increasing by roughly a factor of a million in the three decades from the 1970s to the present. This computational progress has permitted a corresponding increase in model complexity (by including more and more components and processes, as depicted in Figure 1.2), in the length of the simulations, and in spatial resolution, as shown in Figure 1.4. The models used to evaluate future climate changes have therefore evolved over time. Most of the pioneering work on CO_2-induced climate change was based on atmospheric general circulation models coupled to simple 'slab' ocean models (i.e., models omitting ocean dynamics), from the early work of Manabe and Wetherald (1975) to the review of Schlesinger and Mitchell (1987). At the same time the physical content of the models has become more comprehensive (see in Section 1.5.2 the example of clouds). Similarly, most of the results presented in the FAR were from atmospheric models, rather than from models of the coupled climate system, and were used to analyse changes in the equilibrium climate resulting from a doubling of the atmospheric CO_2 concentration. Current climate projections can investigate time-dependent scenarios of climate evolution and can make use of much more complex coupled ocean-atmosphere models, sometimes even including interactive chemical or biochemical components.

A parallel evolution toward increased complexity and resolution has occurred in the domain of numerical weather prediction, and has resulted in a large and verifiable improvement in operational weather forecast quality. This example alone shows that present models are more realistic than were those of a decade ago. There is also, however, a continuing awareness that models do not provide a perfect simulation of reality, because resolving all important spatial or time scales remains far beyond current capabilities, and also because the behaviour of such a complex nonlinear system may in general be chaotic.

It has been known since the work of Lorenz (1963) that even simple models may display intricate behaviour because of their nonlinearities. The inherent nonlinear behaviour of the climate system appears in climate simulations at all time scales (Ghil, 1989). In fact, the study of nonlinear dynamical systems has become important for a wide range of scientific disciplines, and the corresponding mathematical developments are essential to interdisciplinary studies. Simple models of ocean-atmosphere interactions, climate-biosphere interactions or climate-economy interactions may exhibit a similar behaviour, characterised by partial unpredictability, bifurcations and transition to chaos.

In addition, many of the key processes that control climate sensitivity or abrupt climate changes (e.g., clouds, vegetation, oceanic convection) depend on very small spatial scales. They cannot be represented in full detail in the context of global models, and scientific understanding of them is still notably incomplete. Consequently, there is a continuing need to assist in the use and interpretation of complex models through models that are either conceptually simpler, or limited to a number of processes or to a specific region, therefore enabling a deeper understanding of the processes at work or a more relevant comparison with observations. With the development of computer capacities, simpler models have not disappeared; on the contrary, a stronger emphasis has been given to the concept of a 'hierarchy of models' as the only way to provide a linkage between theoretical understanding and the complexity of realistic models (Held, 2005).

The list of these 'simpler' models is very long. Simplicity may lie in the reduced number of equations (e.g., a single

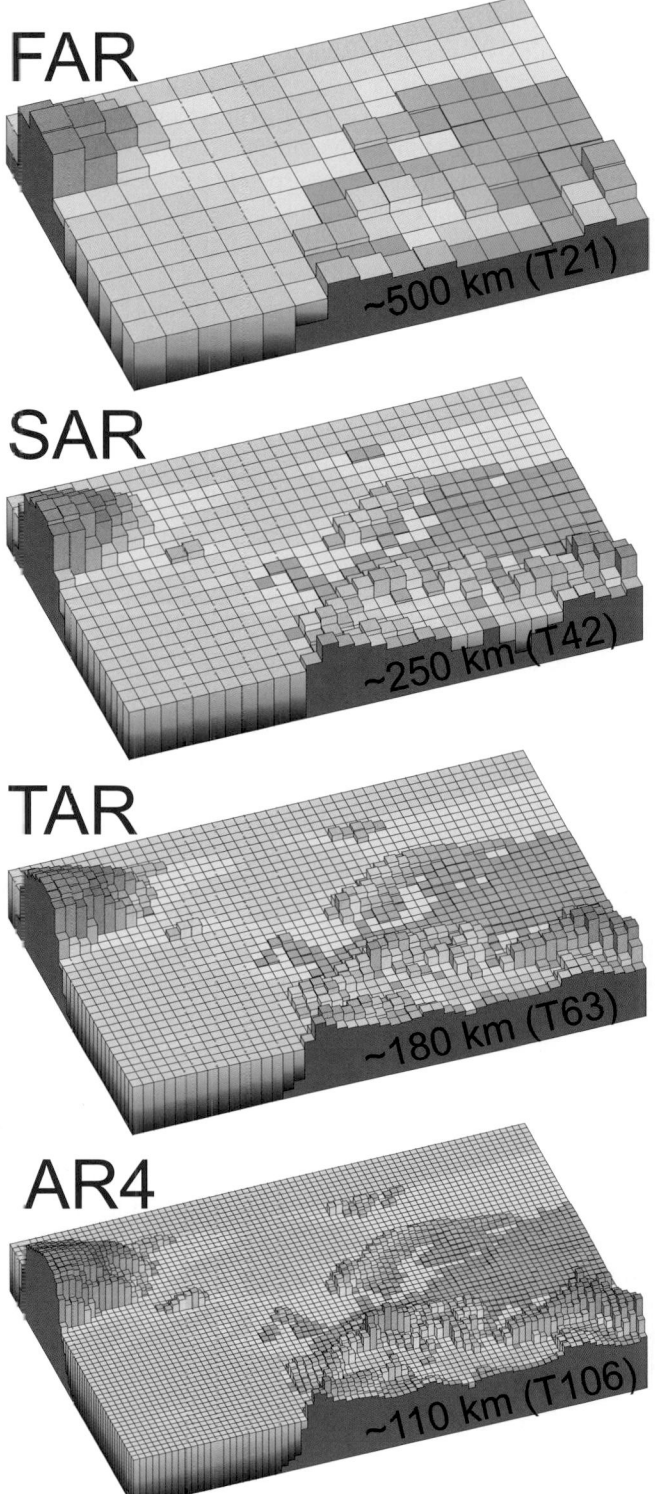

Figure 1.4. *Geographic resolution characteristic of the generations of climate models used in the IPCC Assessment Reports: FAR (IPCC, 1990), SAR (IPCC, 1996), TAR (IPCC, 2001a), and AR4 (2007). The figures above show how successive generations of these global models increasingly resolved northern Europe. These illustrations are representative of the most detailed horizontal resolution used for short-term climate simulations. The century-long simulations cited in IPCC Assessment Reports after the FAR were typically run with the previous generation's resolution. Vertical resolution in both atmosphere and ocean models is not shown, but it has increased comparably with the horizontal resolution, beginning typically with a single-layer slab ocean and ten atmospheric layers in the FAR and progressing to about thirty levels in both atmosphere and ocean.*

equation for the global surface temperature); in the reduced dimensionality of the problem (one-dimension vertical, one-dimension latitudinal, two-dimension); or in the restriction to a few processes (e.g., a mid-latitude quasi-geostrophic atmosphere with or without the inclusion of moist processes). The notion of model hierarchy is also linked to the idea of scale: global circulation models are complemented by regional models that exhibit a higher resolution over a given area, or process oriented models, such as cloud resolving models or large eddy simulations. Earth Models of Intermediate Complexity are used to investigate long time scales, such as those corresponding to glacial to interglacial oscillations (Berger et al., 1998). This distinction between models according to scale is evolving quickly, driven by the increase in computer capacities. For example, global models explicitly resolving the dynamics of convective clouds may soon become computationally feasible.

Many important scientific debates in recent years have had their origin in the use of conceptually simple models. The study of idealised atmospheric representations of the tropical climate, for example by Pierrehumbert (1995) who introduced a separate representation of the areas with ascending and subsiding circulation in the tropics, has significantly improved the understanding of the feedbacks that control climate. Simple linearized models of the atmospheric circulation have been used to investigate potential new feedback effects. Ocean box models have played an important role in improving the understanding of the possible slowing down of the Atlantic thermohaline circulation (Birchfield et al., 1990), as emphasized in the TAR. Simple models have also played a central role in the interpretation of IPCC scenarios: the investigation of climate scenarios presented in the SAR or the TAR has been extended to larger ensembles of cases using idealised models.

1.5.2 Model Clouds and Climate Sensitivity

The modelling of cloud processes and feedbacks provides a striking example of the irregular pace of progress in climate science. Representation of clouds may constitute the area in which atmospheric models have been modified most continuously to take into account increasingly complex physical processes. At the time of the TAR clouds remained a major source of uncertainty in the simulation of climate changes (as they still are at present: e.g., Sections 2.4, 2.6, 3.4.3, 7.5, 8.2, 8.4.11, 8.6.2.2, 8.6.3.2, 9.2.1.2, 9.4.1.8, 10.2.1.2, 10.3.2.2, 10.5.4.3, 11.8.1.3, 11.8.2.2).

In the early 1980s, most models were still using prescribed cloud amounts, as functions of location and altitude, and prescribed cloud radiative properties, to compute atmospheric radiation. The cloud amounts were very often derived from the zonally averaged climatology of London (1957). Succeeding generations of models have used relative humidity or other simple predictors to diagnose cloudiness (Slingo, 1987), thus providing a foundation of increased realism for the models, but at the same time possibly causing inconsistencies in the representation of the multiple roles of clouds as bodies interacting with radiation, generating precipitation and

influencing small-scale convective or turbulent circulations. Following the pioneering studies of Sundqvist (1978), an explicit representation of clouds was progressively introduced into climate models, beginning in the late 1980s. Models first used simplified representations of cloud microphysics, following, for example, Kessler (1969), but more recent generations of models generally incorporate a much more comprehensive and detailed representation of clouds, based on consistent physical principles. Comparisons of model results with observational data presented in the TAR have shown that, based on zonal averages, the representation of clouds in most climate models was also more realistic in 2000 than had been the case only a few years before.

In spite of this undeniable progress, the amplitude and even the sign of cloud feedbacks was noted in the TAR as highly uncertain, and this uncertainty was cited as one of the key factors explaining the spread in model simulations of future climate for a given emission scenario. This cannot be regarded as a surprise: that the sensitivity of the Earth's climate to changing atmospheric greenhouse gas concentrations must depend strongly on cloud feedbacks can be illustrated on the simplest theoretical grounds, using data that have been available for a long time. Satellite measurements have indeed provided meaningful estimates of Earth's radiation budget since the early 1970s (Vonder Haar and Suomi, 1971). Clouds, which cover about 60% of the Earth's surface, are responsible for up to two-thirds of the planetary albedo, which is about 30%. An albedo decrease of only 1%, bringing the Earth's albedo from 30% to 29%, would cause an increase in the black-body radiative equilibrium temperature of about 1°C, a highly significant value, roughly equivalent to the direct radiative effect of a doubling of the atmospheric CO_2 concentration. Simultaneously, clouds make an important contribution to the planetary greenhouse effect. In addition, changes in cloud cover constitute only one of the many parameters that affect cloud radiative interactions: cloud optical thickness, cloud height and cloud microphysical properties can also be modified by atmospheric temperature changes, which adds to the complexity of feedbacks, as evidenced, for example, through satellite observations analysed by Tselioudis and Rossow (1994).

The importance of simulated cloud feedbacks was revealed by the analysis of model results (Manabe and Wetherald, 1975; Hansen et al, 1984), and the first extensive model intercomparisons (Cess et al., 1989) also showed a substantial model dependency. The strong effect of cloud processes on climate model sensitivities to greenhouse gases was emphasized further through a now-classic set of General Circulation Model (GCM) experiments, carried out by Senior and Mitchell (1993). They produced global average surface temperature changes (due to doubled atmospheric CO_2 concentration) ranging from 1.9°C to 5.4°C, simply by altering the way that cloud radiative properties were treated in the model. It is somewhat unsettling that the results of a complex climate model can be so drastically altered by substituting one reasonable cloud parametrization for another, thereby approximately replicating the overall inter-model range of sensitivities. Other GCM groups have also

Frequently Asked Question 1.3
What is the Greenhouse Effect?

The Sun powers Earth's climate, radiating energy at very short wavelengths, predominately in the visible or near-visible (e.g., ultraviolet) part of the spectrum. Roughly one-third of the solar energy that reaches the top of Earth's atmosphere is reflected directly back to space. The remaining two-thirds is absorbed by the surface and, to a lesser extent, by the atmosphere. To balance the absorbed incoming energy, the Earth must, on average, radiate the same amount of energy back to space. Because the Earth is much colder than the Sun, it radiates at much longer wavelengths, primarily in the infrared part of the spectrum (see Figure 1). Much of this thermal radiation emitted by the land and ocean is absorbed by the atmosphere, including clouds, and reradiated back to Earth. This is called the greenhouse effect. The glass walls in a greenhouse reduce airflow and increase the temperature of the air inside. Analogously, but through a different physical process, the Earth's greenhouse effect warms the surface of the planet. Without the natural greenhouse effect, the average temperature at Earth's surface would be below the freezing point of water. Thus,

Earth's natural greenhouse effect makes life as we know it possible. However, human activities, primarily the burning of fossil fuels and clearing of forests, have greatly intensified the natural greenhouse effect, causing global warming.

The two most abundant gases in the atmosphere, nitrogen (comprising 78% of the dry atmosphere) and oxygen (comprising 21%), exert almost no greenhouse effect. Instead, the greenhouse effect comes from molecules that are more complex and much less common. Water vapour is the most important greenhouse gas, and carbon dioxide (CO_2) is the second-most important one. Methane, nitrous oxide, ozone and several other gases present in the atmosphere in small amounts also contribute to the greenhouse effect. In the humid equatorial regions, where there is so much water vapour in the air that the greenhouse effect is very large, adding a small additional amount of CO_2 or water vapour has only a small direct impact on downward infrared radiation. However, in the cold, dry polar regions, the effect of a small increase in CO_2 or

(continued)

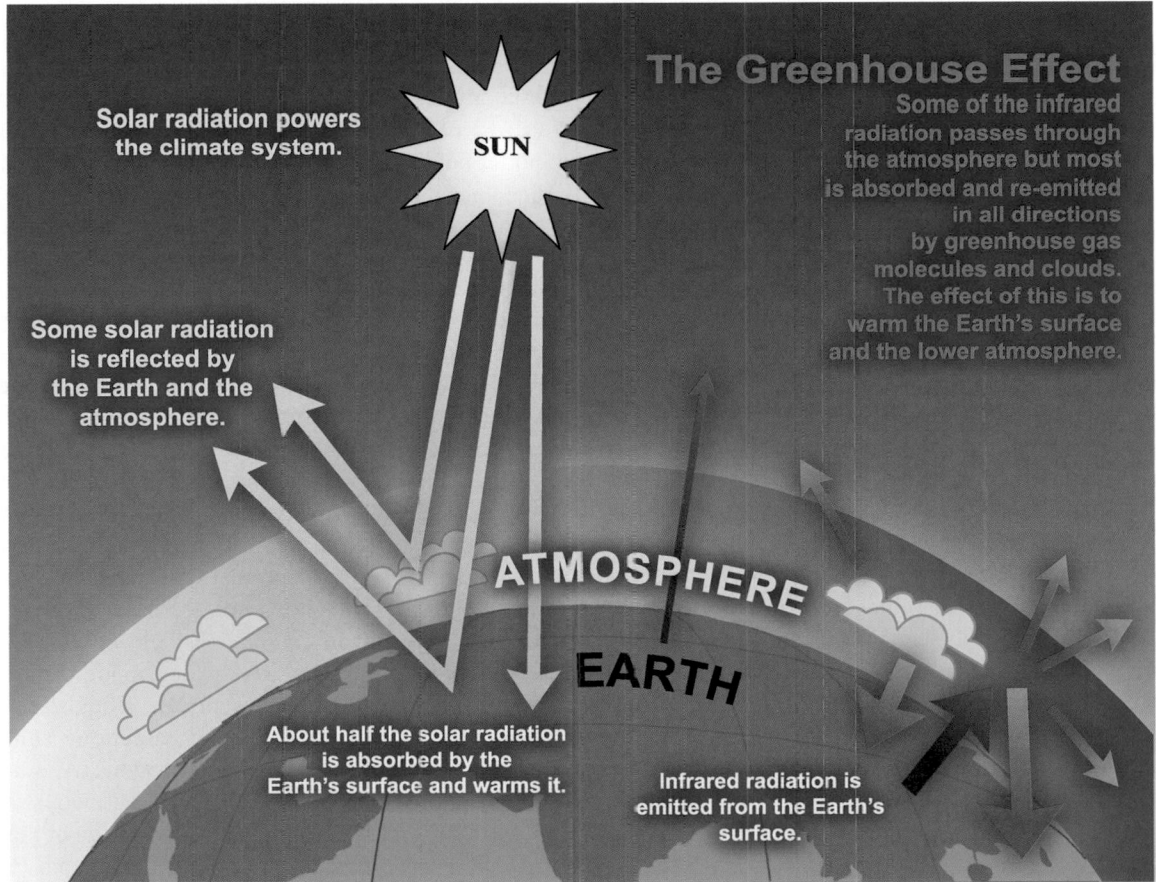

FAQ 1.3, Figure 1. *An idealised model of the natural greenhouse effect. See text for explanation.*

water vapour is much greater. The same is true for the cold, dry upper atmosphere where a small increase in water vapour has a greater influence on the greenhouse effect than the same change in water vapour would have near the surface.

Several components of the climate system, notably the oceans and living things, affect atmospheric concentrations of greenhouse gases. A prime example of this is plants taking CO_2 out of the atmosphere and converting it (and water) into carbohydrates via photosynthesis. In the industrial era, human activities have added greenhouse gases to the atmosphere, primarily through the burning of fossil fuels and clearing of forests.

Adding more of a greenhouse gas, such as CO_2, to the atmosphere intensifies the greenhouse effect, thus warming Earth's climate. The amount of warming depends on various feedback mechanisms. For example, as the atmosphere warms due to rising levels of greenhouse gases, its concentration of water vapour increases, further intensifying the greenhouse effect. This in turn causes more warming, which causes an additional increase in water vapour, in a self-reinforcing cycle. This water vapour feedback may be strong enough to approximately double the increase in the greenhouse effect due to the added CO_2 alone.

Additional important feedback mechanisms involve clouds. Clouds are effective at absorbing infrared radiation and therefore exert a large greenhouse effect, thus warming the Earth. Clouds are also effective at reflecting away incoming solar radiation, thus cooling the Earth. A change in almost any aspect of clouds, such as their type, location, water content, cloud altitude, particle size and shape, or lifetimes, affects the degree to which clouds warm or cool the Earth. Some changes amplify warming while others diminish it. Much research is in progress to better understand how clouds change in response to climate warming, and how these changes affect climate through various feedback mechanisms.

---

consistently obtained widely varying results by trying other techniques of incorporating cloud microphysical processes and their radiative interactions (e.g., Roeckner et al., 1987; Le Treut and Li, 1991), which differed from the approach of Senior and Mitchell (1993) through the treatment of partial cloudiness or mixed-phase properties. The model intercomparisons presented in the TAR showed no clear resolution of this unsatisfactory situation.

The scientific community realised long ago that using adequate data to constrain models was the only way to solve this problem. Using climate changes in the distant past to constrain the amplitude of cloud feedback has definite limitations (Ramstein et al., 1998). The study of cloud changes at decadal, interannual or seasonal time scales therefore remains a necessary path to constrain models. A long history of cloud observations now runs parallel to that of model development. Operational ground-based measurements, carried out for the purpose of weather prediction, constitute a valuable source of information that has been gathered and analysed by Warren et al. (1986, 1988). The International Satellite Cloud Climatology Project (ISCCP; Rossow and Schiffer, 1991) has developed an analysis of cloud cover and cloud properties using the measurements of operational meteorological satellites over a period of more than two decades. These data have been complemented by other satellite remote sensing data sets, such as those associated with the Nimbus-7 Temperature Humidity Infrared Radiometer (THIR) instrument (Stowe et al., 1988), with high-resolution spectrometers such as the High Resolution Infrared Radiation Sounder (HIRS) (Susskind et al., 1987), and with microwave absorption, as used by the Special Sensor Microwave/Imager (SSM/I). Chapter 8 provides an update of this ongoing observational effort.

A parallel effort has been carried out to develop a wider range of ground-based measurements, not only to provide an adequate reference for satellite observations, but also to make possible a detailed and empirically based analysis of the entire range of space and time scales involved in cloud processes. The longest-lasting and most comprehensive effort has been the Atmospheric Radiation Measurement (ARM) Program in the USA, which has established elaborately instrumented observational sites to monitor the full complexity of cloud systems on a long-term basis (Ackerman and Stokes, 2003). Shorter field campaigns dedicated to the observation of specific phenomena have also been established, such as the TOGA Coupled Ocean-Atmosphere Response Experiment (COARE) for convective systems (Webster and Lukas, 1992), or the Atlantic Stratocumulus Transition Experiment (ASTEX) for stratocumulus (Albrecht et al., 1995).

Observational data have clearly helped the development of models. The ISCCP data have greatly aided the development of cloud representations in climate models since the mid-1980s (e.g., Le Treut and Li, 1988; Del Genio et al., 1996). However, existing data have not yet brought about any reduction in the existing range of simulated cloud feedbacks. More recently, new theoretical tools have been developed to aid in validating parametrizations in a mode that emphasizes the role of cloud processes participating in climatic feedbacks. One such approach has been to focus on comprehensively observed episodes of cloudiness for which the large-scale forcing is observationally known, using single-column models (Randall et al., 1996; Somerville, 2000) and higher-resolution cloud-resolving models to evaluate GCM parametrizations. Another approach is to make use of the more global and continuous satellite data, on a statistical basis, through an investigation of the correlation between climate forcing and cloud parameters (Bony et al., 1997), in such a way as to provide a test of feedbacks between different climate variables. Chapter 8 assesses recent progress in this area.

1.5.3   Coupled Models: Evolution, Use, Assessment

The first National Academy of Sciences of the USA report on global warming (Charney et al., 1979), on the basis of two models simulating the impact of doubled atmospheric CO_2 concentrations, spoke of a range of global mean equilibrium surface temperature increase of between 1.5°C and 4.5°C, a range that has remained part of conventional wisdom at least as recently as the TAR. These climate projections, as well as those treated later in the comparison of three models by Schlesinger and Mitchell (1987) and most of those presented in the FAR, were the results of atmospheric models coupled with simple 'slab' ocean models (i.e., models omitting all changes in ocean dynamics).

The first attempts at coupling atmospheric and oceanic models were carried out during the late 1960s and early 1970s (Manabe and Bryan, 1969; Bryan et al., 1975; Manabe et al., 1975). Replacing 'slab' ocean models by fully coupled ocean-atmosphere models may arguably have constituted one of the most significant leaps forward in climate modelling during the last 20 years (Trenberth, 1993), although both the atmospheric and oceanic components themselves have undergone highly significant improvements. This advance has led to significant modifications in the patterns of simulated climate change, particularly in oceanic regions. It has also opened up the possibility of exploring transient climate scenarios, and it constitutes a step toward the development of comprehensive 'Earth-system models' that include explicit representations of chemical and biogeochemical cycles.

Throughout their short history, coupled models have faced difficulties that have considerably impeded their development, including: (i) the initial state of the ocean is not precisely known; (ii) a surface flux imbalance (in either energy, momentum or fresh water) much smaller than the observational accuracy is enough to cause a drifting of coupled GCM simulations into unrealistic states; and (iii) there is no direct stabilising feedback that can compensate for any errors in the simulated salinity. The strong emphasis placed on the realism of the simulated base state provided a rationale for introducing 'flux adjustments' or 'flux corrections' (Manabe and Stouffer, 1988; Sausen et al., 1988) in early simulations. These were essentially empirical corrections that could not be justified on physical principles, and that consisted of arbitrary additions of surface fluxes of heat and salinity in order to prevent the drift of the simulated climate away from a realistic state. The National Center for Atmospheric Research model may have been the first to realise non-flux-corrected coupled simulations systematically, and it was able to achieve simulations of climate change into the 21st century, in spite of a persistent drift that still affected many of its early simulations. Both the FAR and the SAR pointed out the apparent need for flux adjustments as a problematic feature of climate modelling (Cubasch et al., 1990; Gates et al., 1996).

By the time of the TAR, however, the situation had evolved, and about half the coupled GCMs assessed in the TAR did not employ flux adjustments. That report noted that 'some non-flux-adjusted models are now able to maintain stable climatologies of comparable quality to flux-adjusted models' (McAvaney et al., 2001). Since that time, evolution away from flux correction (or flux adjustment) has continued at some modelling centres, although a number of state-of-the-art models continue to rely on it. The design of the coupled model simulations is also strongly linked with the methods chosen for model initialisation. In flux-adjusted models, the initial ocean state is necessarily the result of preliminary and typically thousand-year-long simulations to bring the ocean model into equilibrium. Non-flux-adjusted models often employ a simpler procedure based on ocean observations, such as those compiled by Levitus et al. (1994), although some spin-up phase is even then necessary. One argument brought forward is that non-adjusted models made use of *ad hoc* tuning of radiative parameters (i.e., an implicit flux adjustment).

This considerable advance in model design has not diminished the existence of a range of model results. This is not a surprise, however, because it is known that climate predictions are intrinsically affected by uncertainty (Lorenz, 1963). Two distinct kinds of prediction problems were defined by Lorenz (1975). The first kind was defined as the prediction of the actual properties of the climate system in response to a given initial state. Predictions of the first kind are initial-value problems and, because of the nonlinearity and instability of the governing equations, such systems are not predictable indefinitely into the future. Predictions of the second kind deal with the determination of the response of the climate system to changes in the external forcings. These predictions are not concerned directly with the chronological evolution of the climate state, but rather with the long-term average of the statistical properties of climate. Originally, it was thought that predictions of the second kind do not at all depend on initial conditions. Instead, they are intended to determine how the statistical properties of the climate system (e.g., the average annual global mean temperature, or the expected number of winter storms or hurricanes, or the average monsoon rainfall) change as some external forcing parameter, for example CO_2 content, is altered. Estimates of future climate scenarios as a function of the concentration of atmospheric greenhouse gases are typical examples of predictions of the second kind. However, ensemble simulations show that the projections tend to form clusters around a number of attractors as a function of their initial state (see Chapter 10).

Uncertainties in climate predictions (of the second kind) arise mainly from model uncertainties and errors. To assess and disentangle these effects, the scientific community has organised a series of systematic comparisons of the different existing models, and it has worked to achieve an increase in the number and range of simulations being carried out in order to more fully explore the factors affecting the accuracy of the simulations.

An early example of systematic comparison of models is provided by Cess et al. (1989), who compared results of documented differences among model simulations in their representation of cloud feedback to show how the consequent effects on atmospheric radiation resulted in different model response to doubling of the CO_2 concentration. A number of ambitious and comprehensive 'model intercomparison projects' (MIPs) were set up in the 1990s under the auspices of the World Climate Research Programme to undertake controlled conditions for model evaluation. One of the first was the Atmospheric Model Intercomparison Project (AMIP), which studied atmospheric GCMs. The development of coupled models induced the development of the Coupled Model Intercomparison Project (CMIP), which studied coupled ocean-atmosphere GCMs and their response to idealised forcings, such as a 1% yearly increase in the atmospheric CO_2 concentration. It proved important in carrying out the various MIPs to standardise the model forcing parameters and the model output so that file formats, variable names, units, etc., are easily recognised by data users. The fact that the model results were stored separately and independently of the modelling centres, and that the analysis of the model output was performed mainly by research groups independent of the modellers, has added confidence in the results. Summary diagnostic products such as the Taylor (2001) diagram were developed for MIPs.

The establishment of the AMIP and CMIP projects opened a new era for climate modelling, setting standards of quality control, providing organisational continuity and ensuring that results are generally reproducible. Results from AMIP have provided a number of insights into climate model behaviour (Gates et al., 1999) and quantified improved agreement between simulated and observed atmospheric properties as new versions of models are developed. In general, results of the MIPs suggest that the most problematic areas of coupled model simulations involve cloud-radiation processes, the cryosphere, the deep ocean and ocean-atmosphere interactions.

Comparing different models is not sufficient, however. Using multiple simulations from a single model (the so-called Monte Carlo, or ensemble, approach) has proved a necessary and complementary approach to assess the stochastic nature of the climate system. The first ensemble climate change simulations with global GCMs used a set of different initial and boundary conditions (Cubasch et al., 1994; Barnett, 1995). Computational constraints limited early ensembles to a relatively small number of samples (fewer than 10). These ensemble simulations clearly indicated that even with a single model a large spread in the climate projections can be obtained.

Intercomparison of existing models and ensemble model studies (i.e., those involving many integrations of the same model) are still undergoing rapid development. Running ensembles was essentially impossible until recent advances in computer power occurred, as these systematic comprehensive climate model studies are exceptionally demanding on computer resources. Their progress has marked the evolution from the FAR to the TAR, and is likely to continue in the years to come.

1.6 The IPCC Assessments of Climate Change and Uncertainties

The WMO and the United Nations Environment Programme (UNEP) established the IPCC in 1988 with the assigned role of assessing the scientific, technical and socioeconomic information relevant for understanding the risk of human-induced climate change. The original 1988 mandate for the IPCC was extensive: '(a) Identification of uncertainties and gaps in our present knowledge with regard to climate changes and its potential impacts, and preparation of a plan of action over the short-term in filling these gaps; (b) Identification of information needed to evaluate policy implications of climate change and response strategies; (c) Review of current and planned national/international policies related to the greenhouse gas issue; (d) Scientific and environmental assessments of all aspects of the greenhouse gas issue and the transfer of these assessments and other relevant information to governments and intergovernmental organisations to be taken into account in their policies on social and economic development and environmental programs.' The IPCC is open to all members of UNEP and WMO. It does not directly support new research or monitor climate-related data. However, the IPCC process of synthesis and assessment has often inspired scientific research leading to new findings.

The IPCC has three Working Groups and a Task Force. Working Group I (WGI) assesses the scientific aspects of the climate system and climate change, while Working Groups II (WGII) and III (WGIII) assess the vulnerability and adaptation of socioeconomic and natural systems to climate change, and the mitigation options for limiting greenhouse gas emissions, respectively. The Task Force is responsible for the IPCC National Greenhouse Gas Inventories Programme. This brief history focuses on WGI and how it has described uncertainty in the quantities presented (See Box 1.1).

A main activity of the IPCC is to provide on a regular basis an assessment of the state of knowledge on climate change, and this volume is the fourth such Assessment Report of WGI. The IPCC also prepares Special Reports and Technical Papers on topics for which independent scientific information and advice is deemed necessary, and it supports the United Nations Framework Convention on Climate Change (UNFCCC) through its work on methodologies for National Greenhouse Gas Inventories. The FAR played an important role in the discussions of the Intergovernmental Negotiating Committee for the UNFCCC. The UNFCCC was adopted in 1992 and entered into force in 1994. It provides the overall policy framework and legal basis for addressing the climate change issue.

The WGI FAR was completed under the leadership of Bert Bolin (IPCC Chair) and John Houghton (WGI Chair) in a plenary at Windsor, UK in May 1990. In a mere 365 pages with eight colour plates, it made a persuasive, but not quantitative, case for anthropogenic interference with the climate system. Most conclusions from the FAR were non-quantitative and remain valid today (see also Section 1.4.4). For example, in terms of the greenhouse gases, 'emissions resulting from human activities are substantially increasing the atmospheric concentrations of the greenhouse gases: CO_2, CH_4, CFCs, N_2O' (see Chapters 2 and 3; Section 7.1). On the other hand, the FAR did not foresee the phase-out of CFCs, missed the importance of biomass-burning aerosols and dust to climate and stated that unequivocal detection of the enhanced greenhouse effect was more than a decade away. The latter two areas highlight the advance of climate science and in particular the merging of models and observations in the new field of detection and attribution (see Section 9.1).

The Policymakers Summary of the WGI FAR gave a broad overview of climate change science and its Executive Summary separated key findings into areas of varying levels of confidence ranging from 'certainty' to providing an expert 'judgment'. Much of the summary is not quantitative (e.g., the radiative forcing bar charts do not appear in the summary). Similarly, scientific uncertainty is hardly mentioned; when ranges are given, as in the projected temperature increases of 0.2°C to 0.5°C per decade, no probability or likelihood is assigned to explain the range (see Chapter 10). In discussion of the climate sensitivity to doubled atmospheric CO_2 concentration, the combined subjective and objective criteria are explained: the range of model results was 1.9°C to 5.2°C; most were close to 4.0°C; but the newer model results were lower; and hence the best estimate was 2.5°C with a range of 1.5°C to 4.5°C. The likelihood of the value being within this range was not defined. However, the importance of identifying those areas where climate scientists had high confidence was recognised in the Policymakers Summary.

The Supplementary Report (IPCC, 1992) re-evaluated the RF values of the FAR and included the new IPCC scenarios for future emissions, designated IS92a–f. It also included updated chapters on climate observations and modelling (see Chapters 3, 4, 5, 6 and 8). The treatment of scientific uncertainty remained as in the FAR. For example, the calculated increase in global mean surface temperature since the 19th century was given as 0.45°C ± 0.15°C, with no quantitative likelihood for this range (see Section 3.2).

The SAR, under Bert Bolin (IPCC Chair) and John Houghton and Gylvan Meira Filho (WGI Co-chairs), was planned with and coupled to a preliminary Special Report (IPCC, 1995) that contained intensive chapters on the carbon cycle, atmospheric chemistry, aerosols and radiative forcing. The WGI SAR culminated in the government plenary in Madrid in November 1995. The most cited finding from that plenary, on attribution of climate change, has been consistently reaffirmed by subsequent research: 'The balance of evidence suggests a discernible human influence on global climate' (see Chapter 9). The SAR provided key input to the negotiations that led to the adoption in 1997 of the Kyoto Protocol to the UNFCCC.

Uncertainty in the WGI SAR was defined in a number of ways. The carbon cycle budgets used symmetric plus/minus ranges explicitly defined as 90% confidence intervals, whereas the RF bar chart reported a 'mid-range' bar along with a plus/minus range that was estimated largely on the spread of published values. The likelihood, or confidence interval, of the spread of published results was not given. These uncertainties were additionally modified by a declaration that the confidence of the RF being within the specified range was indicated by a stated confidence level that ranged from 'high' (greenhouse gases) to 'very low' (aerosols). Due to the difficulty in approving such a long draft in plenary, the Summary for Policy Makers (SPM) became a short document with no figures and few numbers. The use of scientific uncertainty in the SPM was thus limited and similar to the FAR: a range in the mean surface temperature increase since 1900 was given as 0.3°C to 0.6°C with no explanation as to likelihood of this range. While the underlying report showed projected future warming for a range of different climate models, the Technical Summary focused on a central estimate.

The IPCC Special Report on Aviation and the Global Atmosphere (IPCC, 1999) was a major interim assessment involving both WGI and WGIII and the Scientific Assessment Panel to the Montreal Protocol on Substances that Deplete the Ozone Layer. It assessed the impacts of civil aviation in terms of climate change and global air quality as well as looking at the effect of technology options for the future fleet. It was the first complete assessment of an industrial sub-sector. The summary related aviation's role relative to all human influence on the climate system: 'The best estimate of the radiative forcing in 1992 by aircraft is 0.05 W m^{-2} or about 3.5% of the total radiative forcing by all anthropogenic activities.' The authors took a uniform approach to assigning and propagating uncertainty in these RF values based on mixed objective and subjective criteria. In addition to a best value, a two-thirds likelihood (67% confidence) interval is given. This interval is similar to a one-sigma (i.e., one standard deviation) normal error distribution, but it was explicitly noted that the probability distribution outside this interval was not evaluated and might not have a normal distribution. A bar chart with 'whiskers' (two-thirds likelihood range) showing the components and total (without cirrus effects) RF for aviation in 1992 appeared in the SPM (see Sections 2.6 and 10.2).

The TAR, under Robert Watson (IPCC Chair) and John Houghton and Ding YiHui (WGI Co-chairs), was approved at the government plenary in Shanghai in January 2001. The predominant summary statements from the TAR WGI strengthened the SAR's attribution statement: 'An increasing body of observations gives a collective picture of a warming world and other changes in the climate system', and 'There is new and stronger evidence that most of the warming observed over the last 50 years is attributable to human activities.' The TAR Synthesis Report (IPCC, 2001b) combined the assessment reports from the three Working Groups. By combining data on global (WGI) and regional (WGII) climate change, the Synthesis Report was able to strengthen the conclusion regarding human influence: 'The Earth's climate system has demonstrably changed on both global and regional scales since the pre-industrial era, with some of these changes attributable to human activities' (see Chapter 9).

Box 1.1: Treatment of Uncertainties in the Working Group I Assessment

The importance of consistent and transparent treatment of uncertainties is clearly recognised by the IPCC in preparing its assessments of climate change. The increasing attention given to formal treatments of uncertainty in previous assessments is addressed in Section 1.6. To promote consistency in the general treatment of uncertainty across all three Working Groups, authors of the Fourth Assessment Report have been asked to follow a brief set of guidance notes on determining and describing uncertainties in the context of an assessment .[1] This box summarises the way that Working Group I has applied those guidelines and covers some aspects of the treatment of uncertainty specific to material assessed here.

Uncertainties can be classified in several different ways according to their origin. Two primary types are 'value uncertainties' and 'structural uncertainties'. Value uncertainties arise from the incomplete determination of particular values or results, for example, when data are inaccurate or not fully representative of the phenomenon of interest. Structural uncertainties arise from an incomplete understanding of the processes that control particular values or results, for example, when the conceptual framework or model used for analysis does not include all the relevant processes or relationships. Value uncertainties are generally estimated using statistical techniques and expressed probabilistically. Structural uncertainties are generally described by giving the authors' collective judgment of their confidence in the correctness of a result. In both cases, estimating uncertainties is intrinsically about describing the limits to knowledge and for this reason involves expert judgment about the state of that knowledge. A different type of uncertainty arises in systems that are either chaotic or not fully deterministic in nature and this also limits our ability to project all aspects of climate change.

The scientific literature assessed here uses a variety of other generic ways of categorising uncertainties. Uncertainties associated with 'random errors' have the characteristic of decreasing as additional measurements are accumulated, whereas those associated with 'systematic errors' do not. In dealing with climate records, considerable attention has been given to the identification of systematic errors or unintended biases arising from data sampling issues and methods of analysing and combining data. Specialised statistical methods based on quantitative analysis have been developed for the detection and attribution of climate change and for producing probabilistic projections of future climate parameters. These are summarised in the relevant chapters.

The uncertainty guidance provided for the Fourth Assessment Report draws, for the first time, a careful distinction between levels of confidence in scientific understanding and the likelihoods of specific results. This allows authors to express high confidence that an event is extremely unlikely (e.g., rolling a dice twice and getting a six both times), as well as high confidence that an event is about as likely as not (e.g., a tossed coin coming up heads). Confidence and likelihood as used here are distinct concepts but are often linked in practice.

The standard terms used to define levels of confidence in this report are as given in the IPCC Uncertainty Guidance Note, namely:

Confidence Terminology	Degree of confidence in being correct
Very high confidence	At least 9 out of 10 chance
High confidence	About 8 out of 10 chance
Medium confidence	About 5 out of 10 chance
Low confidence	About 2 out of 10 chance
Very low confidence	Less than 1 out of 10 chance

Note that 'low confidence' and 'very low confidence' are only used for areas of major concern and where a risk-based perspective is justified.

Chapter 2 of this report uses a related term 'level of scientific understanding' when describing uncertainties in different contributions to radiative forcing. This terminology is used for consistency with the Third Assessment Report, and the basis on which the authors have determined particular levels of scientific understanding uses a combination of approaches consistent with the uncertainty guidance note as explained in detail in Section 2.9.2 and Table 2.11.

---

[1] See Supplementary Material for this report

The standard terms used in this report to define the likelihood of an outcome or result where this can be estimated probabilistically are:

Likelihood Terminology	Likelihood of the occurrence/ outcome
Virtually certain	> 99% probability
Extremely likely	> 95% probability
Very likely	> 90% probability
Likely	> 66% probability
More likely than not	> 50% probability
About as likely as not	33 to 66% probability
Unlikely	< 33% probability
Very unlikely	< 10% probability
Extremely unlikely	< 5% probability
Exceptionally unlikely	< 1% probability

The terms 'extremely likely', 'extremely unlikely' and 'more likely than not' as defined above have been added to those given in the IPCC Uncertainty Guidance Note in order to provide a more specific assessment of aspects including attribution and radiative forcing.

Unless noted otherwise, values given in this report are assessed best estimates and their uncertainty ranges are 90% confidence intervals (i.e., there is an estimated 5% likelihood of the value being below the lower end of the range or above the upper end of the range). Note that in some cases the nature of the constraints on a value, or other information available, may indicate an asymmetric distribution of the uncertainty range around a best estimate.

In an effort to promote consistency, a guidance paper on uncertainty (Moss and Schneider, 2000) was distributed to all Working Group authors during the drafting of the TAR. The WGI TAR made some effort at consistency, noting in the SPM that when ranges were given they generally denoted 95% confidence intervals, although the carbon budget uncertainties were specified as ±1 standard deviation (68% likelihood). The range of 1.5°C to 4.5°C for climate sensitivity to atmospheric CO_2 doubling was reiterated but with no confidence assigned; however, it was clear that the level of scientific understanding had increased since that same range was first given in the Charney et al. (1979) report. The RF bar chart noted that the RF components could not be summed (except for the long-lived greenhouse gases) and that the 'whiskers' on the RF bars each meant something different (e.g., some were the range of models, some were uncertainties). Another failure in dealing with uncertainty was the projection of 21st-century warming: it was reported as a range covering (i) six Special Report on Emission Scenarios (SRES) emissions scenarios and (ii) nine atmosphere-ocean climate models using two grey envelopes without estimates of likelihood levels. The full range (i.e., scenario plus climate model range) of 1.4°C to 5.8°C is a much-cited finding of the WGI TAR but the lack of discussion of associated likelihood in the report makes the interpretation and useful application of this result difficult.

1.7 Summary

As this chapter shows, the history of the centuries-long effort to document and understand climate change is often complex, marked by successes and failures, and has followed a very uneven pace. Testing scientific findings and openly discussing the test results have been the key to the remarkable progress that is now accelerating in all domains, in spite of inherent limitations to predictive capacity. Climate change science is now contributing to the foundation of a new interdisciplinary approach to understanding our environment. Consequently, much published research and many notable scientific advances have occurred since the TAR, including advances in the understanding and treatment of uncertainty. Key aspects of recent climate change research are assessed in Chapters 2 through 11 of this report.

References

Abbot, C.G., 1910: The solar constant of radiation. *Smithsonian Institution Annual Report*, p. 319.

Ackerman, T., and G. Stokes, 2003: The Atmospheric Radiation Measurement Program. *Phys. Today*, **56**, 38–44.

Adkins, J.F., et al., 1998: Deep-sea coral evidence for rapid change in ventilation of the deep North Atlantic 15,400 years ago. *Science*, **280**, 725–728.

Agassiz, L., 1837: Discours d'ouverture sur l'ancienne extension des glaciers. *Société Helvétique des Sciences Naturelles*, Neufchâtel.

Albrecht, B.A., et al., 1995: The Atlantic Stratocumulus Transition Experiment – ASTEX. *Bull. Am. Meteorol. Soc.*, **76**, 889–904.

Alley, R.B., et al., 1993: Abrupt increase in Greenland snow accumulation at the end of the Younger Dryas event. *Nature*, **362**, 527–529.

Arrhenius, S., 1896: On the influence of carbonic acid in the air upon the temperature on the ground, *Philos. Mag.*, **41**, 237–276.

Aßmann, R., 1902: Über die Existenz eines wärmeren Luftstromes in der Höhe von 10 bis 15 km. *Sitzungsbericht der Königlich-Preußischen Akademie der Wissenschaften zu Berlin, Sitzung der physikalisch-mathematischen Klasse vom 1. Mai 1902*, **XXIV**, 1–10.

Balachandran, N.K., and D. Rind, 1995: Modeling the effects of UV-variability and the QBO on the troposphere-stratosphere system. Part I: The middle atmosphere. *J. Clim.*, **8**, 2058–2079.

Barnett, T.P., 1995: Monte Carlo climate forecasting. *J. Clim.*, **8**, 1005–1022.

Barnett, T.P., et al., 1999: Detection and attribution of recent climate change: A status report. *Bull. Am. Meteorol. Soc.*, **80**, 2631–2660.

Barnola, J.-M., D. Raynaud, Y.S. Korotkevich, and C. Lorius, 1987: Vostok ice core provides 160,000-year record of atmospheric CO_2. *Nature*, **329**, 408–414.

Battle, M., et al., 1996: Atmospheric gas concentrations over the past century measured in air from firn at South Pole. *Nature*, **383**, 231–235.

Bender, M., et al., 1996: Variability in the O-2/N-2 ratio of southern hemisphere air, 1991-1994: Implications for the carbon cycle. *Global Biogeochem. Cycles*, **1**, 9–21.

Berger, A., M.F. Loutre, and H. Gallée, 1998: Sensitivity of the LLN climate model to the astronomical and CO_2 forcings over the last 200 kyr. *Clim. Dyn.*, **14**, 615–629.

Berner, W., H. Oeschger, and B. Stauffer, 1980: Information on the CO_2 cycle from ice core studies. *Radiocarbon*, **22**, 227–235.

Berson, A., and R. Süring, 1901: Ein ballonaufstieg bis 10 500m. *Illustrierte Aeronautische Mitteilungen*, Heft 4, 117–119.

Birchfield, G.E., H. Wang, and M. Wyant, 1990: A bimodal climate response controlled by water vapor transport in a coupled ocean-atmosphere box model. *Paleoceanography*, **5**, 383–395.

Bjerknes, J., 1964: Atlantic air-sea interaction. *Adv. Geophys.*, **10**, 1–82.

Bjerknes, J., 1969: Atmospheric teleconnections from the equatorial Pacific. *Mon. Weather Rev.*, **97**, 163–172.

Blake, D.R., et al., 1982: Global increase in atmospheric methane concentrations between 1978 and 1980. *Geophys. Res. Lett.*, **9**, 477–480.

Blunier, T., et al., 1998: Asynchrony of Antarctic and Greenland climate change during the last glacial period. *Nature*, **394**, 739–743.

Bond, G., et al., 1992: Evidence for massive discharges of icebergs into the glacial Northern Atlantic. *Nature*, **360**, 245–249.

Bony, S., K.-M. Lau, and Y.C. Sud, 1997: Sea surface temperature and large-scale circulation influences on tropical greenhouse effect and cloud radiative forcing. *J. Clim.*, **10**, 2055–2077.

Brasseur, G., 1993: The response of the middle atmosphere to long term and short term solar variability: A two dimensional model. *J. Geophys. Res.*, **28**, 23079–23090.

Brazdil, R., 1992: Reconstructions of past climate from historical sources in the Czech lands. In: *Climatic Variations and Forcing Mechanisms of the Last 2000 Years* [Jones, P.D., R.S. Bradley, and J. Jouzel (eds.)]. Springer Verlag, Berlin and Heidelberg, 649 pp.

Brewer, P.G., et al., 1983: A climatic freshening of the deep North Atlantic (north of 50° N) over the past 20 years. *Science*, **222**, 1237–1239.

Broecker, W.S., 1997: Thermohaline circulation, the Achilles heel of our climate system: will man-made CO_2 upset the current balance? *Science*, **278**, 1582–1588.

Broecker, W.S., and G.H. Denton, 1989: The role of ocean-atmosphere reorganizations in glacial cycles. *Geochim. Cosmochim. Acta*, **53**, 2465–2501.

Brohan P., et al., 2006: Uncertainty estimates in regional and global observed temperature changes: A new data set from 1850. *J. Geophys. Res.*, **111**, D12106, doi:10.1029/2005JD006548.

Bryan, F., 1986: High-latitude salinity effects and interhemispheric thermohaline circulations. *Nature*, **323**, 301–304.

Bryan, K., 1969: A numerical method for the study of the circulation of the world ocean. *J. Comput. Phys.*, **4**, 347–376.

Bryan, K., and M.J. Spelman, 1985: The ocean's response to a CO_2-induced warming. *J. Geophys. Res.*, **90**, 679–688.

Bryan, K., S. Manabe, and R.C. Pacanowski, 1975: A global ocean-atmosphere climate model. Part II. The oceanic circulation. *J. Phys. Oceanogr.*, **5**, 30–46.

Bryson, R.A., and G.J. Dittberner, 1976: A non-equilibrium model of hemispheric mean surface temperature. *J. Atmos. Sci.*, **33**, 2094–2106.

Bryson, R.A., and G.J. Dittberner, 1977: Reply. *J. Atmos. Sci.*, **34**, 1821–1824.

Budyko, M.I., 1969: The effect of solar radiation variations on the climate of the Earth. *Tellus*, **21**, 611–619.

Buys Ballot, C.H.D., 1872: *Suggestions on a Uniform System of Meteorological Observations*. Publication No. 37, Royal Netherlands Meteorological Institute, Utrecht, 56 pp.

Callendar, G.S., 1938: The artificial production of carbon dioxide and its influence on temperature. *Q. J. R. Meteorol. Soc.*, **64**, 223–237.

Callendar, G.S., 1961: Temperature fluctuations and trends over the Earth. *Q. J. R. Meteorol. Soc.*, **87**, 1–12.

Cane, M.A., S.C. Dolan, and S.E. Zebiak, 1986: Experimental forecasts of the El Niño. *Nature*, **321**, 827–832.

Cess, R.D., et al., 1989: Interpretation of cloud-climate feedback as produced by 14 atmospheric general circulation models. *Science*, **245**, 513–516.

Chamberlain, T.C., 1906: On a possible reversal of deep-sea circulation and its influence on geologic climates. *J. Geol.*, **14**, 371–372.

Charlson, R.J., J. Langner, and H. Rodhe, 1990: Sulfur, aerosol, and climate. *Nature*, **22**, 348.

Charney, J.G., et al., 1979: *Carbon Dioxide and Climate: A Scientific Assessment*. National Academy of Sciences, Washington, DC, 22 pp.

Clayton, H.H., 1927: *World Weather Records*. Smithsonian Miscellaneous Collection, Volume 79, Washington, DC, 1196 pp.

Cortijo, E., et al., 1999: Changes in meridional temperature and salinity gradients in the North Atlantic Ocean (30 degrees-72 degrees N) during the last interglacial period. *Paleoceanography*, **14**, 23–33.

Crease, J., 1962: Velocity measurements in the deep water of the western North Atlantic. *J. Geophys. Res.*, **67**, 3173–3176.

Croll, J., 1890: *Climate and Time in Their Geological Relations: A Theory of Secular Changes of the Earth's Climate*, 2nd ed. Appleton, New York, 577 pp.

Cubasch, U., and R. Voss, 2000: The influence of total solar irradiance on climate. *Space Sci. Rev.*, **94**, 185–198.

Cubasch, U., et al., 1990: Processes and modelling. In: *Climate Change: The IPCC Scientific Assessment* [Houghton, J.T., G.J. Jenkins, and J.J. Ephraums (eds.)]. Cambridge University Press, Cambridge, United Kingdom and New York, NY, USA, pp. 69–91.

Cubasch, U., et al., 1994: Monte Carlo climate change forecasts with a global coupled ocean-atmosphere model. *Clim. Dyn.*, **10**, 1–19.

Cubasch, U., et al., 1997: Simulation with an O-AGCM of the influence of variations of the solar constant on the global climate. *Clim. Dyn.*, **13**, 757–767.

Dansgaard, W., et al., 1984: North Atlantic climatic oscillations revealed by deep Greenland ice cores. In: *Climate Processes and Climate Sensitivity* [Hansen, J.E., and T. Takahashi (eds.)]. American Geophysical Union, Washington, DC, pp. 288–298.

Dansgaard, W., et al., 1993: Evidence for general instability of past climate from a 250-kyr ice-core record. *Nature*, **364**, 218–220.

Del Genio, A.D., M.-S. Yao, W. Kovari, and K.K.-W. Lo, 1996: A prognostic cloud water parameterization for global climate models. *J. Clim.*, **9**, 270–304, doi:10.1175/1520-0442.

Delmas, R.J., J.M. Ascencio, and M. Legrand, 1980: Polar ice evidence that atmospheric CO_2 20,000 yr BP was 50% of present. *Nature*, **284**, 155–157.

deMenocal, P.B., 2001: Cultural responses during the late Holocene. *Science*, **292**, 667–673.

Derwent, R., 1990: *Trace Gases and Their Relative Contribution to the Greenhouse Effect*. Report AERE- R13716, Atomic Energy Research Establishment, Harwell, Oxon, UK. 95 pp.

Dlugokencky, E.J., K.A. Masarie, P.M. Lang, and P.P. Tans, 1998: Continuing decline in the growth rate of the atmospheric methane burden. *Nature*, **393**, 447–450.

Dove, H.W., 1852: Über die geographische Verbreitung gleichartiger Witterungserscheinungen (Über die nichtperiodischen Änderungen der Temperaturverteilung auf der Oberfläche der Erde). *Abh. Akad. Wiss. Berlin*, V Teil, **42**, 3–4.

Dozier, J., S.R. Schneider, and D.F. McGinnis Jr., 1981: Effect of grain size and snowpack water equivalence on visible and near-infrared satellite observations of snow. *Water Resour. Res.*, **17**, 1213–1221.

Dunbar, R.B., and G.M. Wellington, 1981: Stable isotopes in a branching coral monitor seasonal temperature variation. *Nature*, **298**, 453–455.

Eddy, J.A., 1976: The Maunder Minimum. *Science*, **192**, 1189–1202

Emiliani, C., 1955: Pleistocene temperatures. *J. Geol.*, **63**, 538–578.

Emiliani, C., 1969: Interglacials, high sea levels and the control of Greenland ice by the precession of the equinoxes. *Science*, **166**, 1503–1504.

Etheridge, D.M., et al., 1996: Natural and anthropogenic changes in atmospheric CO_2 over the last 1000 years from air in Antarctic ice and firn. *J. Geophys. Res.*, **101**, 4115–4128.

Ewing, M., and W.L. Donn, 1956: A theory of ice ages. *Science*, **123**, 1061–1065.

Exner, F.M., 1913: Übermonatliche Witterungsanomalien auf der nördlichen Erdhälfte im Winter. *Sitzungsberichte d. Kaiserl, Akad. der Wissenschaften*, **122**, 1165–1241.

Exner, F.M., 1924: Monatliche Luftdruck- und Temperaturanomalien auf der Erde. *Sitzungsberichte d. Kaiserl, Akad. der Wissenschaften*, **133**, 307–408.

Fleming, J.R., 1998: *Historical Perspectives on Climate Change*. Oxford University Press, New York, 208pp.

Flückiger, J., et al., 1999: Variations in atmospheric N_2O concentration during abrupt climatic changes. *Science*, **285**, 227–230.

Folland, C.K., and D.E. Parker, 1995: Correction of instrumental biases in historical sea surface temperature data. *Q. J. R. Meteorol. Soc.*, **121**, 319–367.

Foukal, P.V., P.E. Mack, and J.E. Vernazza, 1977: The effect of sunspots and faculae on the solar constant. *Astrophys. J.*, **215**, 952–959.

Francey, R.J., and G.D. Farquhar, 1982: An explanation of C-13/C-12 variations in tree rings. *Nature*, **297**, 28–31.

Fraser, P.J., M.A.K. Khalil, R.A. Rasmussen, and A.J. Crawford, 1981: Trends of atmospheric methane in the southern hemisphere. *Geophys. Res. Lett.*, **8**, 1063–1066.

Fritts, H.C., 1962: An approach to dendroclimatology: screening by means of multiple regression techniques. *J. Geophys. Res.*, **67**, 1413–1420.

Gates, W.L., et al., 1996: Climate models – evaluation. In: *Climate 1995: The Science of Climate Change* [Houghton, J.T., et al. (eds.)]. Cambridge University Press, Cambridge, United Kingdom and New York, NY, USA, pp. 229–284.

Gates, W.L., et al., 1999: An overview of the results of the Atmospheric Model Intercomparison Project (AMIP I). *Bull. Am. Meteorol. Soc.*, **80**, 29–55.

Geerts, B., 1999: Trends in atmospheric science journals. *Bull. Am. Meteorol. Soc.*, **80**, 639–652.

Grootes, P.M., et al., 1993: Comparison of oxygen isotope records from the GISP2 and GRIP Greenland ice cores. *Nature*, **366**, 552–554.

Ghil, M., 1989: Deceptively-simple models of climatic change. In: *Climate and Geo-Sciences* [Berger, A., J.-C. Duplessy, and S.H. Schneider (eds.)]. D. Reidel, Dordrecht, Netherlands and Hingham, MA, pp. 211–240.

Graedel, T.E., and J.E. McRae, 1980: On the possible increase of atmospheric methane and carbon monoxide concentrations during the last decade. *Geophys. Res. Lett.*, **7**, 977–979.

Gwynne, P., 1975: The cooling world. *Newsweek*, April 28, 64.

Haigh, J., 1996: The impact of solar variability on climate. *Science*, **272**, 981–985.

Hansen, J., and S. Lebedeff, 1987: Global trends of measured surface air temperature. *J. Geophys. Res.*, **92**, 13345–13372.

Hansen, J., A. Lacis, R. Ruedy, and M. Sato, 1992: Potential climate impact of Mount-Pinatubo eruption. *Geophys. Res. Lett.*, **19**, 215–218.

Hansen, J., et al., 1981: Climate impact of increasing atmospheric carbon dioxide. *Science*, **213**, 957–966.

Hansen, J., et al., 1984: Climate sensitivity: Analysis of feedback mechanisms. In: *Climate Processes and Climate Sensitivity* [Hansen, J.E., and T. Takahashi (eds.)]. Geophysical Monograph 29, American Geophysical Union, Washington, DC, pp. 130–163.

Hansen, J., et al., 2001: A closer look at United States and global surface temperature change. *J. Geophys. Res.*, **106**, 23947–23963.

Harriss, R., K. Bartlett, S. Frolking, and P. Crill, 1993: Methane emissions from northern high-latitude wetlands. In: *Biogeochemistry of Global Change* [Oremland, R.S. (ed.)]. Chapman & Hall, New York, pp. 449–486.

Hasselmann, K., 1997: Multi-pattern fingerprint method for detection and attribution of climate change. *Clim. Dyn.*, **13**, 601–612.

Hawking, S., 1988: *A Brief History of Time*. Bantam Press, New York, 224 pp.

Hays, J.D., J. Imbrie, and N.J. Shackleton, 1976: Variations in the Earth's orbit: Pace-maker of the ice ages. *Science*, **194**, 1121–1132.

Hegerl, G.C., et al., 1996: Detecting greenhouse-gas-induced climate change with an optimal fingerprint method. *J. Clim.*, **9**, 2281–2306.

Hegerl, G.C., et al., 1997: Multi-fingerprint detection and attribution of greenhouse-gas and aerosol-forced climate change. *Clim. Dyn.*, **13**, 613–634.

Hegerl, G.C., et al., 2000: Optimal detection and attribution of climate change: Sensitivity of results to climate model differences. *Clim. Dyn.*, **16**, 737–754

Held, I.M., 2005: The gap between simulation and understanding in climate modelling. *Bull. Am. Meteorol. Soc.*, **86**, 1609–1614.

Herschel, W., 1801: Observations tending to investigate the nature of the sun, in order to find the causes or symptoms of its variable emission of light and heat. *Philos. Trans. R. Soc. London*, **91**, 265-318.

Hickey, J.R., et al., 1980: Initial solar irradiance determinations from Nimbus 7 cavity radiometer measurements. *Science*, **208**, 281–283.

Hildebrandsson, H.H., 1897: Quelques recherches sur les centres d'action de l'atmosphère. *Svenska Vet. Akad. Handlingar*, 36 pp.

Holton, J.R., 1992: *An Introduction to Dynamic Meteorology*, 3rd ed. Volume 48 of International Geophysics Series. Academic Press, San Diego, 511 pp.

Hoyt, D.V., and K.H. Schatten, 1993: A discussion of plausible solar irradiance variations 1700-1992. *J. Geophys. Res.*, **98**, 18895–18906

Hoyt, D.V., and K.H. Schatten, 1997: *The Role of the Sun in Climate Change*. Oxford University Press, Oxford, 279 pp.

Hoyt, D.V., K.H. Schatten, and E. Nesmes-Ribes, 1994: The hundredth year of Rudolf Wolf's death: Do we have the correct reconstruction of solar activity? *Geophys. Res. Lett.*, **21**, 2067–2070.

Hurrell, J.W., 1995: Decadal trends in the North Atlantic Oscillation: Regional temperatures and precipitation. *Science*, **269**, 676–679.

Imbrie, J., and K.P. Imbrie, 1979: *Ice Ages: Solving the Mystery*. Harvard University Press, Cambridge, 224pp.

Indermühle, A., et al., 1999: Holocene carbon-cycle dynamics based on CO_2 trapped in ice at Taylor Dome, Antarctica. *Nature*, **398**, 121–126.

IPCC, 1990: *Climate Change: The IPCC Scientific Assessment* [Houghton, J.T., G.J. Jenkins, and J.J. Ephraums (eds.)]. Cambridge University Press, Cambridge, United Kingdom and New York, NY, USA, 365 pp.

IPCC, 1992: *Climate Change 1992: The Supplementary Report to the IPCC Scientific Assessment* [Houghton, J.T., B.A. Callander, and S.K. Varney (eds.)]. Cambridge University Press, Cambridge, United Kingdom and New York, NY, USA, 200 pp.

IPCC, 1995: *Climate Change 1994: Radiative Forcing of Climate Change and an Evaluation of the IPCC IS92 Emission Scenarios* [Houghton, J.T., et al. (eds.)]. Cambridge University Press, Cambridge, United Kingdom and New York, NY, USA, 339 pp.

IPCC, 1996: *Climate Change 1995: The Science of Climate Change* [Houghton, J.T., et al. (eds.)]. Cambridge University Press, Cambridge, United Kingdom and New York, NY, USA, 572 pp.

IPCC, 1999: *Special Report on Aviation and the Global Atmosphere* [Penner, J.E., et al. (eds.)]. Cambridge University Press, Cambridge, United Kingdom and New York, NY, USA, 373 pp.

IPCC, 2001a: *Climate Change 2001: The Scientific Basis. Contribution of Working Group I to the Third Assessment Report of the Intergovernmental Panel on Climate Change* [Houghton, J.T., et al. (eds.)]. Cambridge University Press, Cambridge, United Kingdom and New York, NY, USA, 881 pp.

IPCC, 2001b: *Climate Change 2001: Synthesis Report. A contribution of Working Groups I, II, and III to the Third Assessment Report of the Intergovernmental Panel on Climate Change* [Watson, R.T., et al. (eds.)]. Cambridge University Press, Cambridge, United Kingdom and New York, NY, USA, 398 pp.

Jacob, D.J., et al., 1997: Evaluation and intercomparison of global atmospheric transport models using ^{222}Rn and other short-lived tracers. *J. Geophys. Res.*, **102**, 5953–5970.

Johnsen, S.J., et al., 1992: Irregular glacial interstadials recorded in a new Greenland ice core. *Nature*, **359**, 311–313.

Jones, P.D., S.C.B. Raper, and T.M.L. Wigley, 1986a: Southern Hemisphere surface air temperature variations: 1851-1984. *J. Appl. Meteorol.*, **25**, 1213–1230.

Jones, P.D., et al., 1986b: Northern Hemisphere surface air temperature variations: 1851-1984. *J. Clim. Appl. Meteorol.*, **25**, 161–179.

Jones, P.D., et al., 1990: Assessment of urbanization effects in time series of surface air temperature over land. *Nature*, **347**, 169–172.

Jouzel, J., et al., 1987: Vostok ice core: a continuous isotope temperature record over the last climatic cycle (160,000 years). *Nature*, **329**, 402–408.

Jouzel, J., et al., 1993: Extending the Vostok ice-core record of palaeoclimate to the penultimate glacial period. *Nature*, **364**, 407–412.

Karl, T.R., H.F. Diaz, and G. Kukla, 1988: Urbanization: Its detection and effect in the United States climate record. *J. Clim.*, **1**, 1099–1123.

Keeling, C.D., 1961: The concentration and isotopic abundances of carbon dioxide in rural and marine air. *Geochim. Cosmochim. Acta*, **24**, 277–298.

Keeling, C.D., 1998: Rewards and penalties of monitoring the Earth. *Annu. Rev. Energy Environ.*, **23**, 25–82.

Keeling, R.F., and S.R. Shertz, 1992: Seasonal and interannual variations in atmospheric oxygen and implications for the global carbon-cycle. *Nature*, **358**, 723–727.

Kessler, E., 1969: *On the Distribution and Continuity of Water Substance in Atmospheric Circulation*. Meteorological Monograph Series, Vol. 10, No. 32, American Meteorological Society, Boston, MA, 84 pp.

Khalil, M.A.K., and R.A. Rasmussen, 1988: Nitrous oxide: Trends and global mass balance over the last 3000 years. *Ann. Glaciol.*, **10**, 73–79.

Kiehl, J., and K. Trenberth, 1997: Earth's annual global mean energy budget. *Bull. Am. Meteorol. Soc.*, **78**, 197–206.

Kington, J., 1988: *The Weather of the 1780s over Europe*. Cambridge University Press, Cambridge, UK, 164 pp.

Köppen, W., 1873: Über mehrjährige Perioden der Witterung, insbesondere über die 11-jährige Periode der Temperatur. *Zeitschrift der Österreichischen Gesellschaft für Meteorologie*, **Bd VIII**, 241–248 and 257–267.

Köppen, W., 1880: Kleinere Mittheilungen (Conferenz des permanenten internationalen Meteorologen-Comitè's). *Zeitschrift der Österreichischen Gesellschaft für Meteorologie*, **Bd XV**, 278–283.

Köppen, W., 1881: Über mehrjährige Perioden der Witterung – III. Mehrjährige Änderungen der Temperatur 1841 bis 1875 in den Tropen der nördlichen und südlichen gemässigten Zone, an den Jahresmitteln. untersucht. *Zeitschrift der Österreichischen Gesellschaft für Meteorologie*, **Bd XVI**, 141–150.

Kuhn, T.S., 1996: *The Structure of Scientific Revolutions,* 3rd edition. University of Chicago Press, Chicago, 226 pp.

Kvenvolden, K.A., 1988: Methane hydrate – a major reservoir of carbon in the shallow geosphere? *Chem. Geol.*, **71**, 41–51.

Kvenvolden, K.A., 1993: Gas hydrates – geological perspective and global change. *Rev. Geophys.*, **31**, 173–187.

Labitzke, K., and H. van Loon, 1997: The signal of the 11-year sunspot cycle in the upper troposphere-lower stratosphere. *Space Sci. Rev.*, **80**, 393–410.

Lacis, A.A., D.J. Wuebbles, and J.A. Logan, 1990: Radiative forcing of climate by changes in the vertical distribution of ozone. *J. Geophys. Res.*, **95**, 9971–9981.

Lamb, H.H., 1969: The new look of climatology. *Nature*, **223**, 1209–1215.

Landsberg, H.E., and J.M. Mitchell Jr., 1961: Temperature fluctuations and trends over the Earth. *Q. J. R. Meteorol. Soc.*, **87**, 435–436.

Langenfelds, R.L., et al., 1996: The Cape Grim Air Archive: The first seventeen years. In: *Baseline Atmospheric Program Australia, 1994-95* [Francey, R.J., A.L. Dick, and N. Derek (eds.)]. Bureau of Meteorology and CSIRO Division of Atmospheric Research, Melbourne, Australia, pp. 53–70.

Langley, S.P., 1876: Measurement of the direct effect of sun-spots on terrestrial climates. *Mon. Not. R. Astron. Soc.*, **37**, 5–11.

Langley, S.P., 1884: *Researches on the Solar Heat and its Absorption by the Earth's Atmosphere. A Report of the Mount Whitney Expedition.* Signal Service Professional Paper 15, Washington, DC.

Lazier, J.R.N., 1995: The salinity decrease in the Labrador Sea over the past thirty years. In: *Natural Climate Variability on Decade-to-Century Time Scales* [Martinson, D.G., et al. (eds.)]. National Academy Press, Washington, DC, pp. 295–302.

Le Treut, H., and Z.-X. Li, 1988: Using meteosat data to validate a prognostic cloud generation scheme. *Atmos. Res.*, **21**, 273–292.

Le Treut, H., and Z.-X. Li, 1991: Sensitivity of an atmospheric general circulation model to prescribed SST changes: feedback effects associated with the simulation of cloud optical properties. *Clim. Dyn.*, **5**, 175–187.

Lean, J., 1997: The sun's variable radiation and its relevance to Earth. *Annu. Rev. Astron. Astrophys.*, **35**, 33–67.

Lean, J., and D. Rind, 1998: Climate forcing by changing solar radiation. *J. Clim.*, **11**, 3069–3093.

Lean, J., J. Beer, and R. Bradley, 1995: Reconstruction of solar irradiance since 1610: Implications for climate change. *Geophys. Res. Lett.*, **22**, 3195–3198.

Levitus, S., J. Antonov, and T. Boyer, 1994: Interannual variability of temperature at a depth of 125 m in the North Atlantic Ocean. *Science*, **266**, 96–99.

Lockyer, N., and W.J.S. Lockyer, 1902: On some phenomena which suggest a short period of solar and meteorological changes. *Proc. R. Soc. London*, **70**, 500–504.

London, J., 1957: *A Study of Atmospheric Heat Balance*. Final Report, Contract AF 19(122)-165, AFCRC-TR57-287, College of Engineering, New York University, New York, NY. 99 pp.

Lorenz, E.N., 1963: Deterministic nonperiodic flow. *J. Atmos. Sci.*, **20**, 130–141.

Lorenz, E.N., 1967: *On the Nature and Theory of the General Circulation of the Atmosphere*. Publication No. 218, World Meteorological Association, Geneva, 161 pp.

Lorenz, E.N., 1975: The physical bases of climate and climate modelling. In: *Climate Predictability*. GARP Publication Series 16, World Meteorological Association, Geneva, pp. 132–136.

Lovelock, J.E., 1971: Atmospheric fluorine compounds as indicators of air movements. *Nature*, **230**, 379–381.

Luterbacher, J., et al., 1999: Reconstruction of monthly NAO and EU indices back to AD 1675. *Geophys. Res. Lett.*, **26**, 2745–2748.

MacDonald, G.J., 1990: Role of methane clathrates in past and future climates. *Clim. Change*, **16**, 247–281.

Machida, T., et al., 1995: Increase in the atmospheric nitrous oxide concentration during the last 250 years. *Geophys. Res. Lett.*, **22**, 2921–2924.

Madden, R.A., and V. Ramanathan, 1980: Detecting climate change due to increasing carbon dioxide. *Science*, **209**, 763–768.

Manabe, S., and K. Bryan, 1969: Climate calculations with a combined ocean-atmosphere model. *J. Atmos. Sci.*, **26**, 786–789.

Manabe, S., and R.T. Wetherald, 1975: The effects of doubling the CO_2 concentration on the climate of a general circulation model. *J. Atmos. Sci.*, **32**, 3–15.

Manabe, S., and R.J. Stouffer, 1988: Two stable equilibria of a coupled ocean-atmosphere model. *J. Clim.*, **1**, 841–866.

Manabe, S., K. Bryan, and M.J. Spelman, 1975: A global ocean-atmosphere climate model. Part I. The atmospheric circulation. *J. Phys. Oceanogr.*, **5**, 3–29.

Mann, M.E., R.S. Bradley, and M.K. Hughes, 1998: Global-scale temperature patterns and climate forcing over the past six centuries. *Nature*, **392**, 779–787.

McAvaney, B.J., et al., 2001: Model evaluation. In: *Climate Change 2001: The Scientific Basis* [Houghton, J.T., et al. (eds.)]. Cambridge University Press, Cambridge, United Kingdom and New York, NY, USA, pp. 471–521.

McPhaden, M.J., et al., 1998: The Tropical Ocean – Global Atmosphere (TOGA) observing system: a decade of progress. *J. Geophys. Res.*, **103**(C7), 14169–14240.

Mercer, J.H., 1968: Antarctic ice and Sangamon sea level. *Int. Assoc. Sci. Hydrol. Symp.*, **79**, 217–225.

Mercer, J.H., 1978: West Antarctic ice sheet and CO_2 greenhouse effect: a threat of disaster. *Nature*, **271**, 321–325.

Milankovitch, M., 1941: *Kanon der Erdbestrahlungen und seine Anwendung auf das Eiszeitenproblem*. Belgrade. English translation by Pantic, N., 1998: *Canon of Insolation and the Ice Age Problem*. Alven Global, 636 pp.

Mitchell, J.M. Jr., 1963: On the world-wide pattern of secular temperature change. In: *Changes of Climate. Proceedings of the Rome Symposium Organized by UNESCO and the World Meteorological Organization, 1961*. Arid Zone Research Series No. 20, UNESCO, Paris, pp. 161–181.

Montzka, S. A., et al., 1999: Present and future trends in the atmospheric burden of ozone-depleting halogens. *Nature*, **398**, 690–694.

Moss, R., and S. Schneider, 2000: Uncertainties. In: *Guidance Papers on the Cross Cutting Issues of the Third Assessment Report of the IPCC* [Pachauri, R., T. Taniguchi, and K. Tanaka (eds.)]. Intergovernmental Panel on Climate Change, Geneva, pp. 33-51.

National Climatic Data Center, 2002: *Data Documentation for Data Set 9645 World Weather Records – NCAR Surface (World Monthly Surface Station Climatology)*. U.S. Department of Commerce, NOAA, National Climatic Data Center, Asheville, NC, 17 pp.

National Climatic Data Center, 2005: *World Meteorological Organization, World Weather Records, 1991-2000, Volumes I-VI*. U.S. Department of Commerce, NOAA, National Climatic Data Center, Asheville, NC, CD-ROM format.

Neftel, A., E. Moor, H. Oeschger, and B. Stauffer, 1985: Evidence from polar ice cores for the increase in atmospheric CO_2 in the past 2 centuries. *Nature*, **315**, 45–47.

Neftel, A., et al., 1982: Ice core sample measurements give atmospheric CO_2 content during the Past 40,000 Yr. *Nature*, **295**, 220–223.

Newton, I., 1675: Letter to Robert Hooke, February 5, 1675. In: Andrews, R., 1993: *The Columbia Dictionary of Quotations*. Columbia University Press, New York, 1090 pp.

Oeschger, H., et al., 1984: Late glacial climate history from ice cores. In: *Climate Processes and Climate Sensitivity* [Hansen, J.E., and T. Takahashi (eds.)]. American Geophysical Union, Washington, DC, pp. 299–306.

Olson, J., et al., 1997: Results from the Intergovernmental Panel on Climatic Change Photochemical Model Intercomparison (PhotoComp). *J. Geophys. Res.*, **102**(D5), 5979–5991.

Oort, A.H., and T.H. Vonder Haar, 1976: On the observed annual cycle in the ocean-atmosphere heat balance over the Northern Hemisphere. *J. Phys. Oceanogr.*, **6**, 781–800.

Parkinson, C.L., et al., 1987: *Arctic Sea Ice, 1973-1976: Satellite Passive-Microwave Observations*. NASA SP-489, National Aeronautics and Space Administration, Washington, DC, 296 pp.

Penner, J., R. Dickinson, and C. O'Neill, 1992: Effects of aerosol from biomass burning on the global radiation budget. *Science*, **256**, 1432–1434.

Peterson, T.C., et al., 1998: Homogeneity adjustments of in situ atmospheric climate data: A review. *Int. J. Climatol.*, **18**, 1493–1517.

Peterson, T.C., et al., 1999: Global rural temperature trends. *Geophys. Res. Lett.*, **26**, 329–332.

Petit, J.R., et al., 1999: Climate and atmospheric history of the past 420,000 years from the Vostok ice core, Antarctica. *Nature*, **399**, 429–436.

Pfister, C., 1992: Monthly temperature and precipitation in central Europe 1525-1979: quantifying documentary evidence on weather and its effects. In: *Climatic Variations and Forcing Mechanisms of the Last 2000 Years*. [Jones, P.D., R.S. Bradley, and J. Jouzel (eds.)]. Springer Verlag, Berlin and Heidelberg, 649 pp.

Pierrehumbert, R.T., 1995: Thermostats, radiator fins, and the local runaway greenhouse. *J. Atmos. Sci.*, **52**, 1784–1806.

Popper, K.R., 1934: *The Logic of Scientific Discovery*. English edition: Routledge, London (1992), 544 pp.

Prather, M., 1994: Lifetimes and eigenstates in atmospheric chemistry. *Geophys. Res. Lett.*, **21**, 801–804.

Prinn, R.G., et al., 2000: A history of chemically and radiatively important gases in air deduced from ALE/GAGE/AGAGE. *J. Geophys. Res.*, **105**, 17751–17792.

Pyle, J., et al., 1996: *Global Tracer Transport Models: Report of a Scientific Symposium, Bermuda, 10-13 Dec. 1990*. WCRP CAS/JSC Report No. 24 (World Meteorological Organization, TD-No.770, Geneva, Switzerland), 186 pp.

Quetelet, A., 1854: Rapport de la Conférence, tenue à Bruxelles, sur l'invitation du gouvernement des Etats-Unis d'Amérique, à l'effet de s'entendre sur un système uniform d'observations météorologiques à la mer. *Annuaire de l'Observatoire Royal de Belgique*, **21**, 155–167.

Ramanathan, V., 1975: Greenhouse effect due to chlorofluorocarbons: Climatic implications. *Science*, **190**, 50–52.

Ramstein, G., et al., 1998: Cloud processes associated with past and future climate changes, *Clim. Dyn.*, **14**, 233–247.

Randall, D.A., K.-M. Xu, R.C.J. Somerville, and S. Iacobellis, 1996: Single-column models and cloud ensemble models as links between observations and climate models. *J. Clim.*, **9**, 1683–1697.

Rasch, P.J., 2000: A comparison of scavenging and deposition processes in global models: results from the WCRP Cambridge Workshop of 1995. *Tellus*, **52B**, 1025–1056.

Reid, G.C., 1991: Solar irradiance variations and the global sea surface temperature record. *J. Geophys. Res.*, **96**, 2835–2844.

Revelle, R., and H.E. Suess, 1957: Carbon dioxide exchange between atmosphere and ocean and the question of an increase of atmospheric CO_2 during the past decades. *Tellus*, **9**, 18–27.

Reynolds, R.W., and T.M. Smith, 1994: Improved global sea surface temperature analyses using optimum interpolation. *J. Clim.*, **7**, 929–948.

Robock, A., 1982: The Russian surface temperature data set. *J. Appl. Meteorol.*, **21**, 1781–1785.

Roeckner, E., U. Schlese, J. Biercamp, and P. Loewe, 1987: Cloud optical depth feedbacks and climate modelling. *Nature*, **329**, 138–140.

Rossow, W.B., and R.A. Schiffer, 1991: ISCCP cloud data products. *Bull. Am. Meteorol. Soc.*, **72**, 2–20.

Rumford, B., Count, 1800: Essay VII. The propagation of heat in fluids. In: *Essays, Political, Economical, and Philosophical, A New Edition.* T. Cadell, Jr., and W. Davies. London, pp. 197–386. Also in: *Collected Works of Count Rumford, 1, The Nature of Heat* [Brown, S.C. (ed.)]. Harvard University Press, Cambridge, MA (1968), pp. 117–285.

Santer, B.D., J.S. Boyle, and D.E. Parker, 1996a: Human effect on global climate? Reply. *Nature*, **384**, 524.

Santer, B.D., T.M.L. Wigley, T.P. Barnett, and E. Anyamba, 1996b: Detection of climate change, and attribution of causes. In: *Climate Change 1995: The Science of Climate Change* [Houghton, J.T., et al. (eds.)]. Cambridge University Press, Cambridge, United Kingdom and New York, NY, USA, pp. 407–443.

Santer, B.D., et al., 1995: Towards the detection and attribution of an anthropogenic effect on climate. *Clim. Dyn.*, **12**, 77–100.

Santer, B.D., et al., 1996c: A search for human influences on the thermal structure of the atmosphere. *Nature*, **382**, 39–46.

Sausen, R., K. Barthel, and K. Hasselman, 1988: Coupled ocean-atmosphere models with flux correction. *Clim. Dyn.*, **2**, 145–163.

Schellnhuber, H.J., et al. (eds.), 2004: *Earth System Analysis for Sustainability*. MIT Press, Cambridge, MA, 352 pp.

Schlesinger, M.E., and J.F.B. Mitchell, 1987: Climate model simulations of the equilibrium climatic response to increased carbon-dioxide. *Rev. Geophys.*, **25**, 760–798.

Schwabe, S.H., 1844: Sonnen-Beobachtungen im Jahre 1843. *Astronomische Nachrichten*, **21**, 233.

Sellers, W.D., 1969: A climate model based on the energy balance of the Earth-atmosphere system. *J. Appl. Meteorol.*, **8**, 392–400.

Senior, C.A., and J.F.B. Mitchell, 1993: Carbon dioxide and climate: the impact of cloud parameterization. *J. Clim.*, **6**, 393–418.

Severinghaus, J.P., and E.J. Brook, 1999: Abrupt climate change at the end of the last glacial period inferred from trapped air in polar ice. *Science*, **286**, 930–934.

Shackleton, N., 1967: Oxygen isotope analyses and Pleistocene temperatures reassessed. *Nature*, **215**, 15–17.

Shackleton, N.J., M.A. Hall, and E. Vincent, 2000: Phase relationships between millennial-scale events 64,000–24,000 years ago. *Paleoceanography*, **15**, 565–569.

Slingo, J., 1987: The development and verification of a cloud prediction scheme for the ECMWF model. *Q. J. R. Meteorol. Soc.*, **113**, 899–927.

Somerville, R.C.J., 2000: Using single-column models to improve cloud-radiation parameterizations. In: *General Circulation Model Development: Past, Present and Future* [Randall, D.A. (ed.)]. Academic Press, San Diego and London, pp. 641–657.

Stanhill, G., 2001: The growth of climate change science: A scientometric study. *Clim. Change*, **48**, 515–524.

Steele, L.P., et al., 1996: Atmospheric methane, carbon dioxide, carbon monoxide, hydrogen, and nitrous oxide from Cape Grim air samples analysed by gas chromatography. In: *Baseline Atmospheric Program Australia, 1994-95* [Francey, R.J., A.L. Dick, and N. Derek (eds.)]. Bureau of Meteorology and CSIRO Division of Atmospheric Research, Melbourne, Australia, pp. 107–110.

Stephenson, D.B., H. Wanner, S. Brönnimann, and J. Luterbacher, 2003: The history of scientific research on the North Atlantic Oscillation. In: *The North Atlantic Oscillation: Climatic Significance and Environmental Impact* [Hurrell, J.W., et al. (eds.)]. Geophysical Monograph 134, American Geophysical Union, Washington, DC, doi:10.1029/134GM02.

Stocker, T.F., 1998: The seesaw effect. *Science*, **282**, 61–62.

Stommel, H., 1961: Thermohaline convection with two stable regimes of flow. *Tellus*, **13**, 224–230.

Stott, P.A., et al., 2000: External control of 20th century temperature by natural and anthropogenic forcings. *Science*, **290**, 2133–2137.

Stouffer, R.J., S. Manabe, and K.Y. Vinnikov, 1994: Model assessment of the role of natural variability in recent global warming. *Nature*, **367**, 634–636.

Stowe, L., et al., 1988: Nimbus-7 global cloud climatology. Part I: Algorithms and validation. *J. Clim.*, **1**, 445–470.

Sundqvist, H., 1978: A parametrization scheme for non-convective condensation including prediction of cloud water content. *Q. J. R. Meteorol. Soc.*, **104**, 677–690.

Susskind, J., D. Reuter, and M.T. Chahine, 1987: Clouds fields retrieved from HIRS/MSU data. *J. Geophys. Res.*, **92**, 4035–4050.

Sutton, R., and M. Allen, 1997: Decadal predictability of North Atlantic sea surface temperature and climate. *Nature*, **388**, 563–567.

Taylor, K.E., 2001: Summarizing multiple aspects of model performance in a single diagram. *J. Geophys. Res.*, **106**, 7183–7192.

Teisserenc de Bort, L.P., 1902: Variations de la température de l'air libre dans la zona comprise entre 8km et 13km d'altitude. *Comptes Rendus de l'Acad. Sci. Paris*, **134**, 987–989.

Tett, S.F.B., et al., 1999: Causes of twentieth century temperature change. *Nature*, **399**, 569–572.

Trenberth, K. (ed.), 1993: *Climate System Modeling*. Cambridge University Press, Cambridge, UK, 818 pp.

Tselioudis, G., and W.B. Rossow, 1994: Global, multiyear variations of optical thickness with temperature in low and cirrus clouds. *Geophys. Res. Lett.*, **21**, 2211–2214, doi:10.1029/94GL02004.

Twomey, S., 1977: Influence of pollution on shortwave albedo of clouds. *J. Atmos. Sci.*, **34**, 1149–1152.

Tyndall, J., 1861: On the absorption and radiation of heat by gases and vapours, and on the physical connection, *Philos. Mag.*, **22**, 277–302.

Van den Dool, H.M., H.J. Krijnen, and C.J.E. Schuurmans, 1978: Average winter temperatures at de Bilt (the Netherlands): 1634-1977. *Clim. Change*, **1**, 319–330.

Van Loon, H., and J. C. Rogers, 1978: The seesaw in winter temperatures between Greenland and northern Europe. Part 1: General descriptions. *Mon. Weather Rev.*, **106**, 296–310.

Van Loon, H., and K. Labitzke, 2000: The influence of the 11-year solar cycle on the stratosphere below 30 km: A review. *Space Sci. Rev.*, **94**, 259–278.

Vonder Haar, T.H., and V.E. Suomi, 1971: Measurements of the Earth's radiation budget from satellites during a five-year period. Part 1: Extended time and space means. *J. Atmos. Sci.*, **28**, 305–314.

Walker, G.T., 1924: Correlation in seasonal variation of weather. IX *Mem. Ind. Met. Dept.*, **25**, 275–332.

Walker, G.T., 1928: World weather: III. *Mem. Roy. Meteorol. Soc.*, **2**, 97–106.

Wallace, J.M., and D.S. Gutzler, 1981: Teleconnections in the geopotential height field during the Northern Hemisphere winter. *Mon. Weather Rev.*, **109**, 784–812.

Wallis, I., and I. Beale, 1669: Some observations concerning the baroscope and thermoscope, made and communicated by Doctor I. Wallis at Oxford, and Dr. I Beale at Yeovil in Somerset, deliver'd here according to the several dates, when they were imparted. Dr. Beale in those letters of his dated Decemb.18. Decemb. 29. 1669. and Januar. 3. 1670. *Philosophical Transactions (1665-1678)*, **4**, 1113–1120.

Wang, W. C., et al., 1976: Greenhouse effects due to man-made perturbations of trace gases. *Science*, **194**, 685–690.

Wanner, H., et al., 1995. Wintertime European circulation patterns during the late maunder minimum cooling period (1675-1704). *Theor. Appl. Climatol.*, **51**, 167–175.

Warren, B.A., 1981: Deep circulation of the World Ocean. In: *Evolution of Physical Oceanography* [Warren, B.A., and C. Wunsch (eds.)]. MIT Press, Cambridge, MA, pp. 6–41.

Warren, S.G., et al., 1986: *Global Distribution of Total Cloud Cover and Cloud Type Amounts Over Land.* DOE/ER/60085-H1, NCAR/TN-273 + STR, National Center for Atmospheric Research, Boulder, CO.

Warren, S.G., et al., 1988: *Global Distribution of Total Cloud Cover and Cloud Type Amounts Over the Ocean.* DOE/ER-0406, NCAR/FN-317 + STR, National Center for Atmospheric Research, Boulder, CO.

Weart, S., 2003: *The Discovery of Global Warming.* Harvard University Press, Cambridge, MA, 240 pp.

Weber, J.N., and P.M.J. Woodhead, 1972: Temperature dependence of oxygen-18 concentration in reef coral carbonates. *J. Geophys. Res.*, **77**, 463–473.

Webster, P.J., and R. Lukas, 1992: TOGA-COARE: The coupled ocean-atmosphere response experiment. *Bull. Am. Meteorol. Soc.*, **73**, 1377–1416.

Weiss, R.F., 1981: The temporal and spatial distribution of tropospheric nitrous oxide. *J. Geophys. Res.*, **86**, 7185–7195.

Wetherald, R.T., and S. Manabe, 1975: The effects of changing solar constant on the climate of a general circulation model. *J. Atmos. Sci.*, **32**, 2044–2059.

Wigley, T.M.L., and S.C.B. Raper, 1990: Natural variability of the climate system and detection of the greenhouse effect. *Nature*, **344**, 324–327.

Willett, H.C., 1950: Temperature trends of the past century. In: *Centenary Proceedings of the Royal Meteorological Society.* Royal Meteorological Society, London. pp. 195–206.

Willson, R.C., C.H. Duncan, and J. Geist, 1980: Direct measurements of solar luminosity variation. *Science*, **207**, 177–179.

Worley, S.J., et al., 2005: ICOADS release 2.1 data and products. *Int. J. Climatol.*, **25**, 823–842.

Woronko, S.F., 1977: Comments on "a non-equilibrium model of hemispheric mean surface temperature". *J. Atmos. Sci.*, **34**, 1820–1821.

Wunsch, C., 1978: The North Atlantic general circulation west of 50°W determined by inverse methods. *Rev. Geophys. Space Phys.*, **16**, 583–620.

Wyrtki, K., 1975: El Niño – the dynamic response of the equatorial Pacific Ocean to atmospheric forcing. *J. Phys. Oceanogr.*, **5**, 572–584.

Zebiak, S.E., and M.A. Cane, 1987: A model El Nino-Southern Oscillation. *Mon. Weather Rev.*, **115**, 2262–2278.

Zhu, K., 1973: A preliminary study on the climate changes since the last 5000 years in China. *Science in China*, **2**, 168–189.

2

Changes in Atmospheric Constituents and in Radiative Forcing

Coordinating Lead Authors:

Piers Forster (UK), Venkatachalam Ramaswamy (USA)

Lead Authors:

Paulo Artaxo (Brazil), Terje Berntsen (Norway), Richard Betts (UK), David W. Fahey (USA), James Haywood (UK), Judith Lean (USA), David C. Lowe (New Zealand), Gunnar Myhre (Norway), John Nganga (Kenya), Ronald Prinn (USA, New Zealand), Graciela Raga (Mexico, Argentina), Michael Schulz (France, Germany), Robert Van Dorland (Netherlands)

Contributing Authors:

G. Bodeker (New Zealand), O. Boucher (UK, France), W.D. Collins (USA), T.J. Conway (USA), E. Dlugokencky (USA), J.W. Elkins (USA), D. Etheridge (Australia), P. Foukal (USA), P. Fraser (Australia), M. Geller (USA), F. Joos (Switzerland), C.D. Keeling (USA), R. Keeling (USA), S. Kinne (Germany), K. Lassey (New Zealand), U. Lohmann (Switzerland), A.C. Manning (UK, New Zealand), S. Montzka (USA), D. Oram (UK), K. O'Shaughnessy (New Zealand), S. Piper (USA), G.-K. Plattner (Switzerland), M. Ponater (Germany), N. Ramankutty (USA, India), G. Reid (USA), D. Rind (USA), K. Rosenlof (USA), R. Sausen (Germany), D. Schwarzkopf (USA), S.K. Solanki (Germany, Switzerland), G. Stenchikov (USA), N. Stuber (UK, Germany), T. Takemura (Japan), C. Textor (France, Germany), R. Wang (USA), R. Weiss (USA), T. Whorf (USA)

Review Editors:

Teruyuki Nakajima (Japan), Veerabhadran Ramanathan (USA)

This chapter should be cited as:

Forster, P., V. Ramaswamy, P. Artaxo, T. Berntsen, R. Betts, D.W. Fahey, J. Haywood, J. Lean, D.C. Lowe, G. Myhre, J. Nganga, R. Prinn, G. Raga, M. Schulz and R. Van Dorland, 2007: Changes in Atmospheric Constituents and in Radiative Forcing. In: Climate Change 2007: The Physical Science Basis. Contribution of Working Group I to the Fourth Assessment Report of the Intergovernmental Panel on Climate Change [Solomon, S., D. Qin, M. Manning, Z. Chen, M. Marquis, K.B. Averyt, M.Tignor and H.L. Miller (eds.)]. Cambridge University Press, Cambridge, United Kingdom and New York, NY, USA.

Table of Contents

Executive Summary

Radiative forcing (RF)[1] is a concept used for quantitative comparisons of the strength of different human and natural agents in causing climate change. Climate model studies since the Working Group I Third Assessment Report (TAR; IPCC, 2001) give *medium* confidence that the equilibrium global mean temperature response to a given RF is approximately the same (to within 25%) for most drivers of climate change.

For the first time, the combined RF for all anthropogenic agents is derived. Estimates are also made for the first time of the separate RF components associated with the emissions of each agent.

The combined anthropogenic RF is estimated to be +1.6 [–1.0, +0.8][2] W m^{-2}, indicating that, since 1750, it is *extremely likely*[3] that humans have exerted a substantial warming influence on climate. This RF estimate is *likely* to be at least five times greater than that due to solar irradiance changes. For the period 1950 to 2005, it is *exceptionally unlikely* that the combined natural RF (solar irradiance plus volcanic aerosol) has had a warming influence comparable to that of the combined anthropogenic RF.

Increasing concentrations of the long-lived greenhouse gases (carbon dioxide (CO_2), methane (CH_4), nitrous oxide (N_2O), halocarbons and sulphur hexafluoride (SF_6); hereinafter LLGHGs) have led to a combined RF of +2.63 [±0.26] W m^{-2}. Their RF has a *high* level of scientific understanding.[4] The 9% increase in this RF since the TAR is the result of concentration changes since 1998.

— The global mean concentration of CO_2 in 2005 was 379 ppm, leading to an RF of +1.66 [±0.17] W m^{-2}. Past emissions of fossil fuels and cement production have *likely* contributed about three-quarters of the current RF, with the remainder caused by land use changes. For the 1995 to 2005 decade, the growth rate of CO_2 in the atmosphere was 1.9 ppm yr^{-1} and the CO_2 RF increased by 20%: this is the largest change observed or inferred for any decade in at least the last 200 years. From 1999 to 2005, global emissions from fossil fuel and cement production increased at a rate of roughly 3% yr^{-1}.

— The global mean concentration of CH_4 in 2005 was 1,774 ppb, contributing an RF of +0.48 [±0.05] W m^{-2}. Over the past two decades, CH_4 growth rates in the atmosphere have generally decreased. The cause of this is not well understood. However, this decrease and the negligible long-term change in its main sink (the hydroxyl radical OH) imply that total CH_4 emissions are not increasing.

— The Montreal Protocol gases (chlorofluorocarbons (CFCs), hydrochlorofluorocarbons (HCFCs), and chlorocarbons) as a group contributed +0.32 [±0.03] W m^{-2} to the RF in 2005. Their RF peaked in 2003 and is now beginning to decline.

— Nitrous oxide continues to rise approximately linearly (0.26% yr^{-1}) and reached a concentration of 319 ppb in 2005, contributing an RF of +0.16 [±0.02] W m^{-2}. Recent studies reinforce the large role of emissions from tropical regions in influencing the observed spatial concentration gradients.

— Concentrations of many of the fluorine-containing Kyoto Protocol gases (hydrofluorocarbons (HFCs), perfluorocarbons, SF_6) have increased by large factors (between 4.3 and 1.3) between 1998 and 2005. Their total RF in 2005 was +0.017 [±0.002] W m^{-2} and is rapidly increasing by roughly 10% yr^{-1}.

— The reactive gas, OH, is a key chemical species that influences the lifetimes and thus RF values of CH_4, HFCs, HCFCs and ozone; it also plays an important role in the formation of sulphate, nitrate and some organic aerosol species. Estimates of the global average OH concentration have shown no detectable net change between 1979 and 2004.

Based on newer and better chemical transport models than were available for the TAR, the RF from increases in tropospheric ozone is estimated to be +0.35 [–0.1, +0.3] W m^{-2}, with a *medium* level of scientific understanding. There are indications of significant upward trends at low latitudes.

The trend of greater and greater depletion of global stratospheric ozone observed during the 1980s and 1990s is no longer occurring; however, it is not yet clear whether these recent changes are indicative of ozone recovery. The RF is largely due to the destruction of stratospheric ozone by the Montreal Protocol gases and it is re-evaluated to be –0.05 [±0.10] W m^{-2}, with a *medium* level of scientific understanding.

Based on chemical transport model studies, the RF from the increase in stratospheric water vapour due to oxidation of CH_4 is estimated to be +0.07 [± 0.05] W m^{-2}, with a *low* level of scientific understanding. Other potential human causes of water vapour increase that could contribute an RF are poorly understood.

The total direct aerosol RF as derived from models and observations is estimated to be –0.5 [±0.4] W m^{-2}, with a

---

[1] The RF represents the stratospherically adjusted radiative flux change evaluated at the tropopause, as defined in the TAR. Positive RFs lead to a global mean surface warming and negative RFs to a global mean surface cooling. Radiative forcing, however, is not designed as an indicator of the detailed aspects of climate response. Unless otherwise mentioned, RF here refers to global mean RF. Radiative forcings are calculated in various ways depending on the agent: from changes in emissions and/or changes in concentrations, and from observations and other knowledge of climate change drivers. In this report, the RF value for each agent is reported as the difference in RF, unless otherwise mentioned, between the present day (approximately 2005) and the beginning of the industrial era (approximately 1750), and is given in units of W m^{-2}.

[2] 90% confidence ranges are given in square brackets. Where the 90% confidence range is asymmetric about a best estimate, it is given in the form A [–X, +Y] where the lower limit of the range is (A – X) and the upper limit is (A + Y).

[3] The use of 'extremely likely' is an example of the calibrated language used in this document, it represents a 95% confidence level or higher; 'likely' (66%) is another example (See Box TS.1).

[4] Estimates of RF are accompanied by both an uncertainty range (value uncertainty) and a level of scientific understanding (structural uncertainty). The value uncertainties represent the 5 to 95% (90%) confidence range, and are based on available published studies; the level of scientific understanding is a subjective measure of structural uncertainty and represents how well understood the underlying processes are. Climate change agents with a *high* level of scientific understanding are expected to have an RF that falls within their respective uncertainty ranges (See Section 2.9.1 and Box TS.1 for more information).

medium-low level of scientific understanding. The RF due to the cloud albedo effect (also referred to as first indirect or Twomey effect), in the context of liquid water clouds, is estimated to be –0.7 [–1.1, +0.4] W m^{-2}, with a *low* level of scientific understanding.

— Atmospheric models have improved and many now represent all aerosol components of significance. Improved *in situ*, satellite and surface-based measurements have enabled verification of global aerosol models. The best estimate and uncertainty range of the total direct aerosol RF are based on a combination of modelling studies and observations.

— The direct RF of the individual aerosol species is less certain than the total direct aerosol RF. The estimates are: sulphate, –0.4 [±0.2] W m^{-2}; fossil fuel organic carbon, –0.05 [±0.05] W m^{-2}; fossil fuel black carbon, +0.2 [±0.15] W m^{-2}; biomass burning, +0.03 [±0.12] W m^{-2}; nitrate, –0.1 [±0.1] W m^{-2}; and mineral dust, –0.1 [±0.2] W m^{-2}. For biomass burning, the estimate is strongly influenced by aerosol overlying clouds. For the first time best estimates are given for nitrate and mineral dust aerosols.

— Incorporation of more aerosol species and improved treatment of aerosol-cloud interactions allow a best estimate of the cloud albedo effect. However, the uncertainty remains large. Model studies including more aerosol species or constrained by satellite observations tend to yield a relatively weaker RF. Other aspects of aerosol-cloud interactions (e.g., cloud lifetime, semi-direct effect) are not considered to be an RF (see Chapter 7).

Land cover changes, largely due to net deforestation, have increased the surface albedo giving an RF of –0.2 [±0.2] W m^{-2}, with a *medium-low* level of scientific understanding. Black carbon aerosol deposited on snow has reduced the surface albedo, producing an associated RF of +0.1 [±0.1] W m^{-2}, with a *low* level of scientific understanding. Other surface property changes can affect climate through processes that cannot be quantified by RF; these have a *very low* level of scientific understanding.

Persistent linear contrails from aviation contribute an RF of +0.01 [–0.007, +0.02] W m^{-2}, with a *low* level of scientific understanding; the best estimate is smaller than in the TAR. No best estimates are available for the net forcing from spreading contrails and their effects on cirrus cloudiness.

The direct RF due to increases in solar irradiance since 1750 is estimated to be +0.12 [–0.06, +0.18] W m^{-2}, with a *low* level of scientific understanding. This RF is less than half of the TAR estimate.

— The smaller RF is due to a re-evaluation of the long-term change in solar irradiance, namely a smaller increase from the Maunder Minimum to the present. However, uncertainties in the RF remain large. The total solar irradiance, monitored from space for the last three decades, reveals a well-established cycle of 0.08% (cycle minimum to maximum) with no significant trend at cycle minima.

— Changes (order of a few percent) in globally averaged column ozone forced by the solar ultraviolet irradiance 11-year cycle are now better understood, but ozone profile changes are less certain. Empirical associations between solar-modulated cosmic ray ionization of the atmosphere and globally averaged low-level cloud cover remain ambiguous.

The global stratospheric aerosol concentrations in 2005 were at their lowest values since satellite measurements began in about 1980. This can be attributed to the absence of significant explosive volcanic eruptions since Mt. Pinatubo in 1991. Aerosols from such episodic volcanic events exert a transitory negative RF; however, there is limited knowledge of the RF associated with eruptions prior to Mt. Pinatubo.

The spatial patterns of RFs for non-LLGHGs (ozone, aerosol direct and cloud albedo effects, and land use changes) have considerable uncertainties, in contrast to the relatively high confidence in that of the LLGHGs. The Southern Hemisphere net positive RF *very likely* exceeds that in Northern Hemisphere because of smaller aerosol contributions in the Southern Hemisphere. The RF spatial pattern is not indicative of the pattern of climate response.

The total global mean surface forcing[5] is *very likely* negative. By reducing the shortwave radiative flux at the surface, increases in stratospheric and tropospheric aerosols are principally responsible for the negative surface forcing. This is in contrast to LLGHG increases, which are the principal contributors to the total positive anthropogenic RF.

---

[5] Surface forcing is the instantaneous radiative flux change at the surface; it is a useful diagnostic tool for understanding changes in the heat and moisture surface budgets. However, unlike RF, it cannot be used for quantitative comparisons of the effects of different agents on the equilibrium global mean surface temperature change.

2.1   Introduction and Scope

This chapter updates information taken from Chapters 3 to 6 of the IPCC Working Group I Third Assessment Report (TAR; IPCC, 2001). It concerns itself with trends in forcing agents and their precursors since 1750, and estimates their contribution to the radiative forcing (RF) of the climate system. Discussion of the understanding of atmospheric composition changes is limited to explaining the trends in forcing agents and their precursors. Areas where significant developments have occurred since the TAR are highlighted. The chapter draws on various assessments since the TAR, in particular the 2002 World Meteorological Organization (WMO)-United Nations Environment Programme (UNEP) Scientific Assessment of Ozone Depletion (WMO, 2003) and the IPCC-Technology and Economic Assessment Panel (TEAP) special report on Safeguarding the Ozone Layer and the Global Climate System (IPCC/TEAP, 2005).

The chapter assesses anthropogenic greenhouse gas changes, aerosol changes and their impact on clouds, aviation-induced contrails and cirrus changes, surface albedo changes and natural solar and volcanic mechanisms. The chapter reassesses the 'radiative forcing' concept (Sections 2.2 and 2.8), presents spatial and temporal patterns of RF, and examines the radiative energy budget changes at the surface.

For the long-lived greenhouse gases (carbon dioxide (CO_2), methane (CH_4), nitrous oxide (N_2O), chlorofluorocarbons (CFCs), hydrochlorofluorocarbons (HCFCs), hydrofluorocarbons (HFCs), perfluorocarbons (PFCs) and sulphur hexafluoride (SF_6), hereinafter collectively referred to as the LLGHGs; Section 2.3), the chapter makes use of new global measurement capabilities and combines long-term measurements from various networks to update trends through 2005. Compared to other RF agents, these trends are considerably better quantified; because of this, the chapter does not devote as much space to them as previous assessments (although the processes involved and the related budgets are further discussed in Sections 7.3 and 7.4). Nevertheless, LLGHGs remain the largest and most important driver of climate change, and evaluation of their trends is one of the fundamental tasks of both this chapter and this assessment.

The chapter considers only 'forward calculation' methods of estimating RF. These rely on observations and/or modelling of the relevant forcing agent. Since the TAR, several studies have attempted to constrain aspects of RF using 'inverse calculation' methods. In particular, attempts have been made to constrain the aerosol RF using knowledge of the temporal and/or spatial evolution of several aspects of climate. These include temperatures over the last 100 years, other RFs, climate response and ocean heat uptake. These methods depend on an understanding of – and sufficiently small uncertainties in – other aspects of climate change and are consequently discussed in the detection and attribution chapter (see Section 9.2).

Other discussions of atmospheric composition changes and their associated feedbacks are presented in Chapter 7. Radiative forcing and atmospheric composition changes before 1750 are discussed in Chapter 6. Future RF scenarios that were presented in Ramaswamy et al. (2001) are not updated in this report; however, they are briefly discussed in Chapter 10.

2.2   Concept of Radiative Forcing

The definition of RF from the TAR and earlier IPCC assessment reports is retained. Ramaswamy et al. (2001) define it as 'the change in net (down minus up) irradiance (solar plus longwave; in W m^{-2}) at the tropopause after allowing for stratospheric temperatures to readjust to radiative equilibrium, but with surface and tropospheric temperatures and state held fixed at the unperturbed values'. Radiative forcing is used to assess and compare the anthropogenic and natural drivers of climate change. The concept arose from early studies of the climate response to changes in solar insolation and CO_2, using simple radiative-convective models. However, it has proven to be particularly applicable for the assessment of the climate impact of LLGHGs (Ramaswamy et al., 2001). Radiative forcing can be related through a linear relationship to the global mean equilibrium temperature change at the surface (ΔT_s): $\Delta T_s = \lambda RF$, where λ is the climate sensitivity parameter (e.g., Ramaswamy et al., 2001). This equation, developed from these early climate studies, represents a linear view of global mean climate change between two equilibrium climate states. Radiative forcing is a simple measure for both quantifying and ranking the many different influences on climate change; it provides a limited measure of climate change as it does not attempt to represent the overall climate response. However, as climate sensitivity and other aspects of the climate response to external forcings remain inadequately quantified, it has the advantage of being more readily calculable and comparable than estimates of the climate response. Figure 2.1 shows how the RF concept fits within a general understanding of climate change comprised of 'forcing' and 'response'. This chapter also uses the term 'surface forcing' to refer to the instantaneous perturbation of the surface radiative balance by a forcing agent. Surface forcing has quite different properties than RF and should not be used to compare forcing agents (see Section 2.8.1). Nevertheless, it is a useful diagnostic, particularly for aerosols (see Sections 2.4 and 2.9).

Since the TAR a number of studies have investigated the relationship between RF and climate response, assessing the limitations of the RF concept; related to this there has been considerable debate whether some climate change drivers are better considered as a 'forcing' or a 'response' (Hansen et al., 2005; Jacob et al., 2005; Section 2.8). Emissions of forcing agents, such as LLGHGs, aerosols and aerosol precursors, ozone precursors and ozone-depleting substances, are the more fundamental drivers of climate change and these emissions can be used in state-of-the-art climate models to interactively evolve forcing agent fields along with their associated climate change. In such models, some 'response' is necessary to evaluate the

RF. This 'response' is most significant for aerosol-related cloud changes, where the tropospheric state needs to change significantly in order to create a radiative perturbation of the climate system (Jacob et al., 2005).

Over the palaeoclimate time scales that are discussed in Chapter 6, long-term changes in forcing agents arise due to so-called 'boundary condition' changes to the Earth's climate system (such as changes in orbital parameters, ice sheets and continents). For the purposes of this chapter, these 'boundary conditions' are assumed to be invariant and forcing agent changes are considered to be external to the climate system. The natural RFs considered are solar changes and volcanoes; the other RF agents are all attributed to humans. For the LLGHGs it is appropriate to assume that forcing agent concentrations have not been significantly altered by biogeochemical responses (see Sections 7.3 and 7.4), and RF is typically calculated in off-line radiative transfer schemes, using observed changes in concentration (i.e., humans are considered solely responsible for their increase). For the other climate change drivers, RF is often estimated using general circulation model (GCM) data employing a variety of methodologies (Ramaswamy et al., 2001; Stuber et al., 2001b; Tett et al., 2002; Shine et al., 2003; Hansen et al., 2005; Section 2.8.3). Often, alternative RF calculation methodologies that do not directly follow the TAR definition of a stratospheric-adjusted RF are used; the most important ones are illustrated in Figure 2.2. For most aerosol constituents (see Section 2.4), stratospheric adjustment has little effect on the RF, and the instantaneous RF at either the top of the atmosphere or the tropopause can be substituted. For the climate change drivers discussed in Sections 7.5 and 2.5, that are not initially radiative in nature, an RF-like quantity can be evaluated by

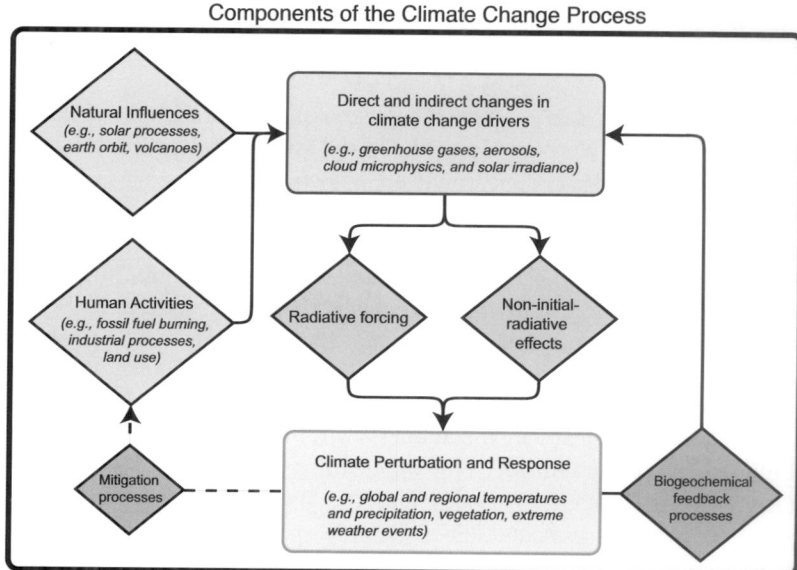

Figure 2.1. *Diagram illustrating how RF is linked to other aspects of climate change assessed by the IPCC. Human activities and natural processes cause direct and indirect changes in climate change drivers. In general, these changes result in specific RF changes, either positive or negative, and cause some non-initial radiative effects, such as changes in evaporation. Radiative forcing and non-initial radiative effects lead to climate perturbations and responses as discussed in Chapters 6, 7 and 8. Attribution of climate change to natural and anthropogenic factors is discussed in Chapter 9. The coupling among biogeochemical processes leads to feedbacks from climate change to its drivers (Chapter 7). An example of this is the change in wetland emissions of CH₄ that may occur in a warmer climate. The potential approaches to mitigating climate change by altering human activities (dashed lines) are topics addressed by IPCC's Working Group III.*

allowing the tropospheric state to change: this is the zero-surface-temperature-change RF in Figure 2.2 (see Shine et al., 2003; Hansen et al., 2005; Section 2.8.3). Other water vapour and cloud changes are considered climate feedbacks and are evaluated in Section 8.6.

Climate change agents that require changes in the tropospheric state (temperature and/or water vapour amounts) prior to causing a radiative perturbation are aerosol-cloud lifetime effects, aerosol semi-direct effects and some surface

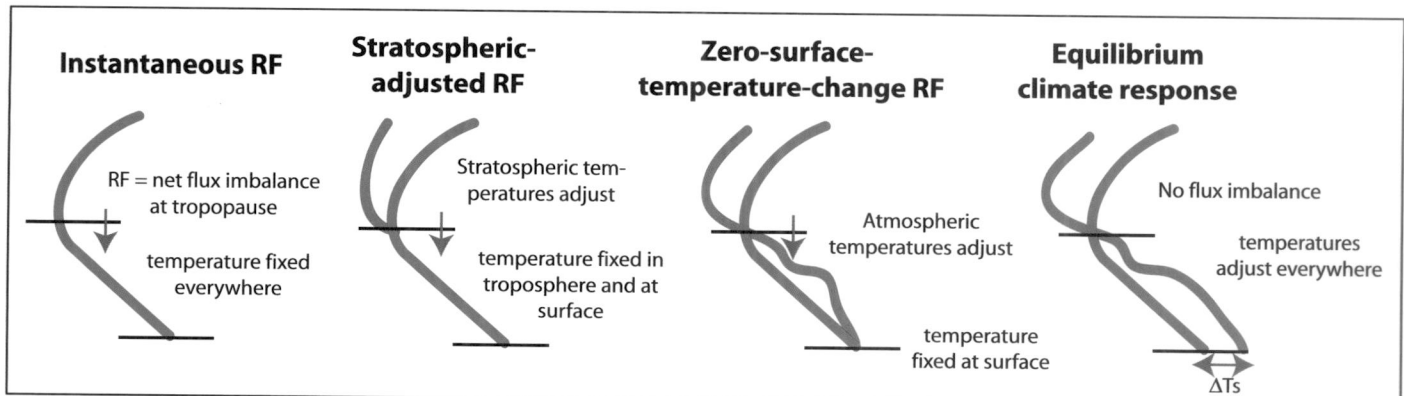

Figure 2.2. *Schematic comparing RF calculation methodologies. Radiative forcing, defined as the net flux imbalance at the tropopause, is shown by an arrow. The horizontal lines represent the surface (lower line) and tropopause (upper line). The unperturbed temperature profile is shown as the blue line and the perturbed temperature profile as the orange line. From left to right: Instantaneous RF: atmospheric temperatures are fixed everywhere; stratospheric-adjusted RF: allows stratospheric temperatures to adjust; zero-surface-temperature-change RF: allows atmospheric temperatures to adjust everywhere with surface temperatures fixed; and equilibrium climate response: allows the atmospheric and surface temperatures to adjust to reach equilibrium (no tropopause flux imbalance), giving a surface temperature change (ΔT_s).*

Frequently Asked Question 2.1
How do Human Activities Contribute to Climate Change and How do They Compare with Natural Influences?

Human activities contribute to climate change by causing changes in Earth's atmosphere in the amounts of greenhouse gases, aerosols (small particles), and cloudiness. The largest known contribution comes from the burning of fossil fuels, which releases carbon dioxide gas to the atmosphere. Greenhouse gases and aerosols affect climate by altering incoming solar radiation and outgoing infrared (thermal) radiation that are part of Earth's energy balance. Changing the atmospheric abundance or properties of these gases and particles can lead to a warming or cooling of the climate system. Since the start of the industrial era (about 1750), the overall effect of human activities on climate has been a warming influence. The human impact on climate during this era greatly exceeds that due to known changes in natural processes, such as solar changes and volcanic eruptions.

Greenhouse Gases

Human activities result in emissions of four principal greenhouse gases: carbon dioxide (CO_2), methane (CH_4), nitrous oxide (N_2O) and the halocarbons (a group of gases containing fluorine, chlorine and bromine). These gases accumulate in the atmosphere, causing concentrations to increase with time. Significant increases in all of these gases have occurred in the industrial era (see Figure 1). All of these increases are attributable to human activities.

- Carbon dioxide has increased from fossil fuel use in transportation, building heating and cooling and the manufacture of cement and other goods. Deforestation releases CO_2 and reduces its uptake by plants. Carbon dioxide is also released in natural processes such as the decay of plant matter.

- Methane has increased as a result of human activities related to agriculture, natural gas distribution and landfills. Methane is also released from natural processes that occur, for example, in wetlands. Methane concentrations are not currently increasing in the atmosphere because growth rates decreased over the last two decades.

- Nitrous oxide is also emitted by human activities such as fertilizer use and fossil fuel burning. Natural processes in soils and the oceans also release N_2O.

- Halocarbon gas concentrations have increased primarily due to human activities. Natural processes are also a small source. Principal halocarbons include the chlorofluorocarbons (e.g., CFC-11 and CFC-12), which were used extensively as refrigeration agents and in other industrial processes before their presence in the atmosphere was found to cause stratospheric ozone depletion. The abundance of chlorofluorocarbon gases is decreasing as a result of international regulations designed to protect the ozone layer.

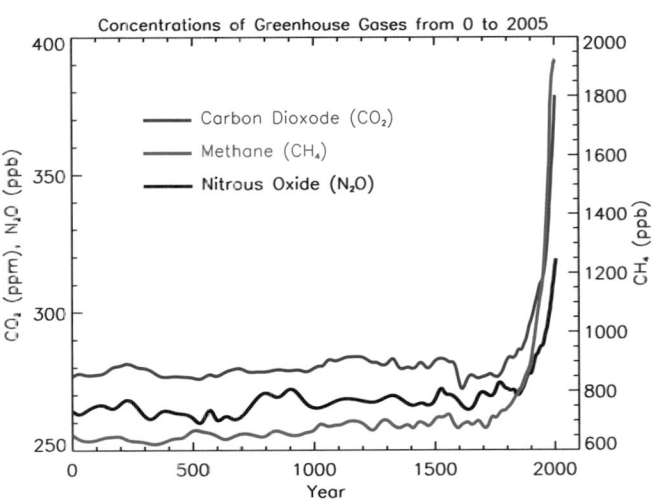

FAQ 2.1, Figure 1. *Atmospheric concentrations of important long-lived greenhouse gases over the last 2,000 years. Increases since about 1750 are attributed to human activities in the industrial era. Concentration units are parts per million (ppm) or parts per billion (ppb), indicating the number of molecules of the greenhouse gas per million or billion air molecules, respectively, in an atmospheric sample. (Data combined and simplified from Chapters 6 and 2 of this report.)*

- Ozone is a greenhouse gas that is continually produced and destroyed in the atmosphere by chemical reactions. In the troposphere, human activities have increased ozone through the release of gases such as carbon monoxide, hydrocarbons and nitrogen oxide, which chemically react to produce ozone. As mentioned above, halocarbons released by human activities destroy ozone in the stratosphere and have caused the ozone hole over Antarctica.

- Water vapour is the most abundant and important greenhouse gas in the atmosphere. However, human activities have only a small direct influence on the amount of atmospheric water vapour. Indirectly, humans have the potential to affect water vapour substantially by changing climate. For example, a warmer atmosphere contains more water vapour. Human activities also influence water vapour through CH_4 emissions, because CH_4 undergoes chemical destruction in the stratosphere, producing a small amount of water vapour.

- Aerosols are small particles present in the atmosphere with widely varying size, concentration and chemical composition. Some aerosols are emitted directly into the atmosphere while others are formed from emitted compounds. Aerosols contain both naturally occurring compounds and those emitted as a result of human activities. Fossil fuel and biomass burning have increased aerosols containing sulphur compounds, organic compounds and black carbon (soot). Human activities such as

(continued)

surface mining and industrial processes have increased dust in the atmosphere. Natural aerosols include mineral dust released from the surface, sea salt aerosols, biogenic emissions from the land and oceans and sulphate and dust aerosols produced by volcanic eruptions.

Radiative Forcing of Factors Affected by Human Activities

The contributions to radiative forcing from some of the factors influenced by human activities are shown in Figure 2. The values reflect the total forcing relative to the start of the industrial era (about 1750). The forcings for all greenhouse gas increases, which are the best understood of those due to human activities, are positive because each gas absorbs outgoing infrared radiation in the atmosphere. Among the greenhouse gases, CO_2 increases have caused the largest forcing over this period. Tropospheric ozone increases have also contributed to warming, while stratospheric ozone decreases have contributed to cooling.

Aerosol particles influence radiative forcing directly through reflection and absorption of solar and infrared radiation in the atmosphere. Some aerosols cause a positive forcing while others cause a negative forcing. The direct radiative forcing summed over all aerosol types is negative. Aerosols also cause a negative radiative forcing indirectly through the changes they cause in cloud properties.

Human activities since the industrial era have altered the nature of land cover over the globe, principally through changes in

(continued)

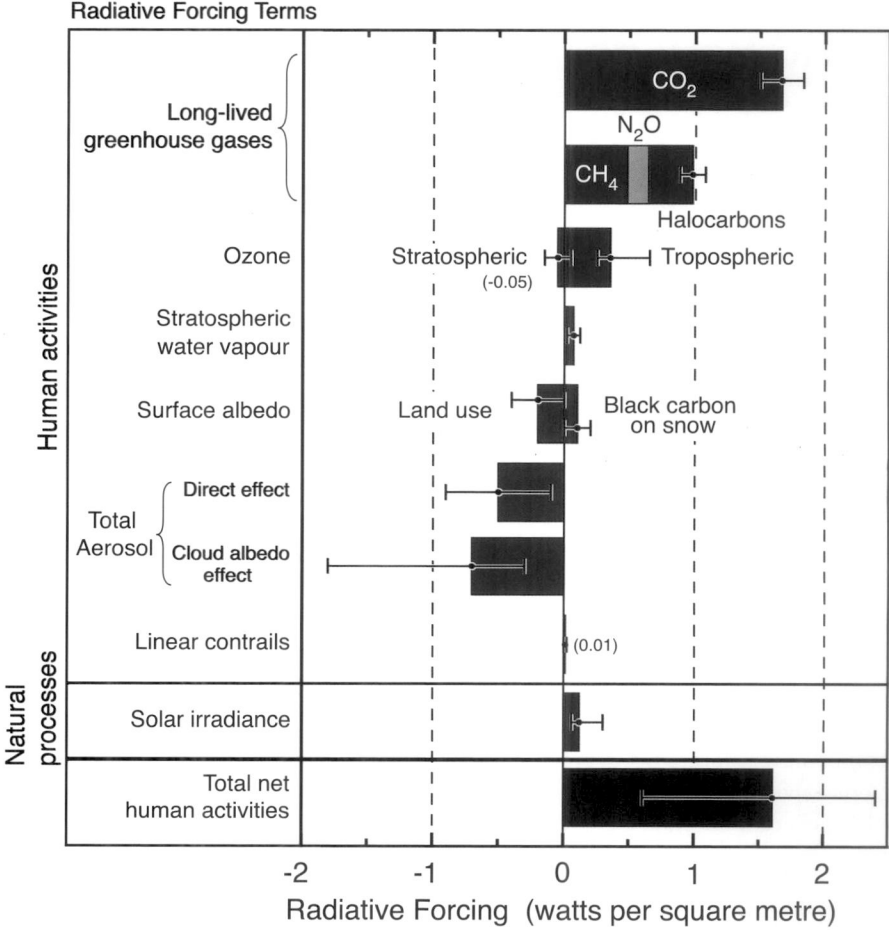

Radiative forcing of climate between 1750 and 2005

FAQ 2.1, Figure 2. *Summary of the principal components of the radiative forcing of climate change. All these radiative forcings result from one or more factors that affect climate and are associated with human activities or natural processes as discussed in the text. The values represent the forcings in 2005 relative to the start of the industrial era (about 1750). Human activities cause significant changes in long-lived gases, ozone, water vapour, surface albedo, aerosols and contrails. The only increase in natural forcing of any significance between 1750 and 2005 occurred in solar irradiance. Positive forcings lead to warming of climate and negative forcings lead to a cooling. The thin black line attached to each coloured bar represents the range of uncertainty for the respective value. (Figure adapted from Figure 2.20 of this report.)*

FAQ 2.1, Box 1: What is Radiative Forcing?

What is radiative forcing? The influence of a factor that can cause climate change, such as a greenhouse gas, is often evaluated in terms of its radiative forcing. Radiative forcing is a measure of how the energy balance of the Earth-atmosphere system is influenced when factors that affect climate are altered. The word radiative arises because these factors change the balance between incoming solar radiation and outgoing infrared radiation within the Earth's atmosphere. This radiative balance controls the Earth's surface temperature. The term forcing is used to indicate that Earth's radiative balance is being pushed away from its normal state.

Radiative forcing is usually quantified as the 'rate of energy change per unit area of the globe as measured at the top of the atmosphere', and is expressed in units of 'Watts per square metre' (see Figure 2). When radiative forcing from a factor or group of factors is evaluated as positive, the energy of the Earth-atmosphere system will ultimately increase, leading to a warming of the system. In contrast, for a negative radiative forcing, the energy will ultimately decrease, leading to a cooling of the system. Important challenges for climate scientists are to identify all the factors that affect climate and the mechanisms by which they exert a forcing, to quantify the radiative forcing of each factor and to evaluate the total radiative forcing from the group of factors.

croplands, pastures and forests. They have also modified the reflective properties of ice and snow. Overall, it is likely that more solar radiation is now being reflected from Earth's surface as a result of human activities. This change results in a negative forcing.

Aircraft produce persistent linear trails of condensation ('contrails') in regions that have suitably low temperatures and high humidity. Contrails are a form of cirrus cloud that reflect solar radiation and absorb infrared radiation. Linear contrails from global aircraft operations have increased Earth's cloudiness and are estimated to cause a small positive radiative forcing.

Radiative Forcing from Natural Changes

Natural forcings arise due to solar changes and explosive volcanic eruptions. Solar output has increased gradually in the industrial era, causing a small positive radiative forcing (see Figure 2). This is in addition to the cyclic changes in solar radiation that follow an 11-year cycle. Solar energy directly heats the climate system and can also affect the atmospheric abundance of some greenhouse gases, such as stratospheric ozone. Explosive volcanic eruptions can create a short-lived (2 to 3 years) negative forcing through the temporary increases that occur in sulphate aerosol in the stratosphere. The stratosphere is currently free of volcanic aerosol, since the last major eruption was in 1991 (Mt. Pinatubo).

The differences in radiative forcing estimates between the present day and the start of the industrial era for solar irradiance changes and volcanoes are both very small compared to the differences in radiative forcing estimated to have resulted from human activities. As a result, in today's atmosphere, the radiative forcing from human activities is much more important for current and future climate change than the estimated radiative forcing from changes in natural processes.

change effects. They need to be accounted for when evaluating the overall effect of humans on climate and their radiative effects as discussed in Sections 7.2 and 7.5. However, in both this chapter and the Fourth Assessment Report they are not considered to be RFs, although the RF definition could be altered to accommodate them. Reasons for this are twofold and concern the need to be simple and pragmatic. Firstly, many GCMs have some representation of these effects inherent in their climate response and evaluation of variation in climate sensitivity between mechanisms already accounts for them (see 'efficacy', Section 2.8.5). Secondly, the evaluation of these tropospheric state changes rely on some of the most uncertain aspects of a climate model's response (e.g., the hydrologic cycle); their radiative effects are very climate-model dependent and such a dependence is what the RF concept was designed to avoid. In practice these effects can also be excluded on practical grounds – they are simply too uncertain to be adequately quantified (see Sections 7.5, 2.4.5 and 2.5.6).

The RF relationship to transient climate change is not straightforward. To evaluate the overall climate response associated with a forcing agent, its temporal evolution and its spatial and vertical structure need to be taken into account. Further, RF alone cannot be used to assess the potential climate change associated with emissions, as it does not take into account the different atmospheric lifetimes of the forcing agents. Global Warming Potentials (GWPs) are one way to assess these emissions. They compare the integrated RF over a specified period (e.g., 100 years) from a unit mass pulse emission relative to CO_2 (see Section 2.10).

2.3 Chemically and Radiatively Important Gases

2.3.1 Atmospheric Carbon Dioxide

This section discusses the instrumental measurements of CO_2, documenting recent changes in atmospheric mixing ratios needed for the RF calculations presented later in the section. In addition, it provides data for the pre-industrial levels of CO_2 required as the accepted reference level for the RF calculations. For dates before about 1950 indirect measurements are relied upon. For these periods, levels of atmospheric CO_2 are usually determined from analyses of air bubbles trapped in polar ice cores. These time periods are primarily considered in Chapter 6.

A wide range of direct and indirect measurements confirm that the atmospheric mixing ratio of CO_2 has increased globally by about 100 ppm (36%) over the last 250 years, from a range of 275 to 285 ppm in the pre-industrial era (AD 1000–1750) to 379 ppm in 2005 (see FAQ 2.1, Figure 1). During this period, the absolute growth rate of CO_2 in the atmosphere increased substantially: the first 50 ppm increase above the pre-industrial value was reached in the 1970s after more than 200 years, whereas the second 50 ppm was achieved in about 30 years. In the 10 years from 1995 to 2005 atmospheric CO_2 increased by about 19 ppm; the highest average growth rate recorded for any decade since direct atmospheric CO_2 measurements began in the 1950s. The average rate of increase in CO_2 determined by these direct instrumental measurements over the period 1960 to 2005 is 1.4 ppm yr^{-1}.

High-precision measurements of atmospheric CO_2 are essential to the understanding of the carbon cycle budgets discussed in Section 7.3. The first *in situ* continuous measurements of atmospheric CO_2 made by a high-precision non-dispersive infrared gas analyser were implemented by C.D. Keeling from the Scripps Institution of Oceanography (SIO) (see Section 1.3). These began in 1958 at Mauna Loa, Hawaii, located at 19°N (Keeling et al., 1995). The data documented for the first time that not only was CO_2 increasing in the atmosphere, but also that it was modulated by cycles caused by seasonal changes in photosynthesis in the terrestrial biosphere. These measurements were followed by continuous *in situ* analysis programmes at other sites in both hemispheres (Conway et al., 1994; Nakazawa et al., 1997; Langenfelds et al., 2002). In Figure 2.3, atmospheric CO_2 mixing ratio data at Mauna Loa in the Northern Hemisphere (NH) are shown with contemporaneous measurements at Baring Head, New Zealand in the Southern Hemisphere (SH; Manning et al., 1997; Keeling and Whorf, 2005). These two stations provide the longest continuous records of atmospheric CO_2 in the NH and SH, respectively. Remote sites such as Mauna Loa, Baring Head, Cape Grim (Tasmania) and the South Pole were chosen because air sampled at such locations shows little short-term variation caused by local sources and sinks of CO_2 and provided the first data from which the global increase of atmospheric CO_2 was documented. Because CO_2 is a LLGHG and well mixed in the atmosphere, measurements made at such sites provide an integrated picture of large parts of the Earth including continents and city point sources. Note that this also applies to the other LLGHGs reported in Section 2.3.

In the 1980s and 1990s, it was recognised that greater coverage of CO_2 measurements over continental areas was required to provide the basis for estimating sources and sinks of atmospheric CO_2 over land as well as ocean regions. Because continuous CO_2 analysers are relatively expensive to maintain and require meticulous on-site calibration, these records are now widely supplemented by air sample flask programmes, where air is collected in glass and metal containers at a large number of continental and marine sites. After collection, the filled flasks are sent to central well-calibrated laboratories for analysis. The most extensive network of international air sampling sites is operated by the National Oceanic and Atmospheric Administration's Global Monitoring Division (NOAA/GMD; formerly NOAA/Climate Monitoring and Diagnostics Laboratory (CMDL)) in the USA. This organisation collates measurements of atmospheric CO_2 from six continuous analyser locations as well as weekly flask air samples from a global network of almost 50 surface sites. Many international laboratories make atmospheric CO_2 observations and worldwide databases of their measurements are maintained by the Carbon Dioxide Information Analysis Center (CDIAC) and by the World Data Centre for Greenhouse Gases (WDCGG) in the WMO Global Atmosphere Watch (GAW) programme.[6]

The increases in global atmospheric CO_2 since the industrial revolution are mainly due to CO_2 emissions from the combustion of fossil fuels, gas flaring and cement production. Other sources include emissions due to land use changes such as deforestation (Houghton, 2003) and biomass burning (Andreae and Merlet, 2001; van der Werf, 2004). After entering the atmosphere, CO_2 exchanges rapidly with the short-lived components of the terrestrial biosphere and surface ocean, and is then redistributed on time scales of hundreds of years among all active carbon reservoirs including the long-lived terrestrial biosphere and

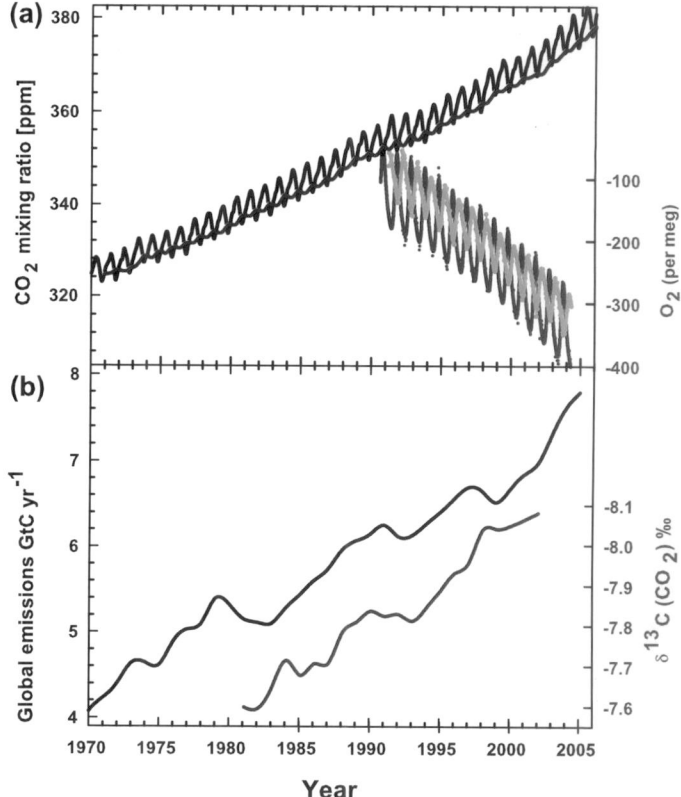

Figure 2.3. *Recent CO_2 concentrations and emissions. (a) CO_2 concentrations (monthly averages) measured by continuous analysers over the period 1970 to 2005 from Mauna Loa, Hawaii (19°N, black; Keeling and Whorf, 2005) and Baring Head, New Zealand (41°S, blue; following techniques by Manning et al., 1997). Due to the larger amount of terrestrial biosphere in the NH, seasonal cycles in CO_2 are larger there than in the SH. In the lower right of the panel, atmospheric oxygen (O_2) measurements from flask samples are shown from Alert, Canada (82°N, pink) and Cape Grim, Australia (41°S, cyan) (Manning and Keeling, 2006). The O_2 concentration is measured as 'per meg' deviations in the O_2/N_2 ratio from an arbitrary reference, analogous to the 'per mil' unit typically used in stable isotope work, but where the ratio is multiplied by 10^6 instead of 10^3 because much smaller changes are measured. (b) Annual global CO_2 emissions from fossil fuel burning and cement manufacture in GtC yr^{-1} (black) through 2005, using data from the CDIAC website (Marland et al, 2006) to 2003. Emissions data for 2004 and 2005 are extrapolated from CDIAC using data from the BP Statistical Review of World Energy (BP, 2006). Land use emissions are not shown; these are estimated to be between 0.5 and 2.7 GtC yr^{-1} for the 1990s (Table 7.2). Annual averages of the $^{13}C/^{12}C$ ratio measured in atmospheric CO_2 at Mauna Loa from 1981 to 2002 (red) are also shown (Keeling et al, 2005). The isotope data are expressed as $\delta^{13}C(CO_2)$ ‰ (per mil) deviation from a calibration standard. Note that this scale is inverted to improve clarity.*

---

6   CDIAC, http://cdiac.esd.ornl.gov/; WDCGG, http://gaw.kishou.go.jp/wdcgg.html.

deep ocean. The processes governing the movement of carbon between the active carbon reservoirs, climate carbon cycle feedbacks and their importance in determining the levels of CO_2 remaining in the atmosphere, are presented in Section 7.3, where carbon cycle budgets are discussed.

The increase in CO_2 mixing ratios continues to yield the largest sustained RF of any forcing agent. The RF of CO_2 is a function of the change in CO_2 in the atmosphere over the time period under consideration. Hence, a key question is 'How is the CO_2 released from fossil fuel combustion, cement production and land cover change distributed amongst the atmosphere, oceans and terrestrial biosphere?'. This partitioning has been investigated using a variety of techniques. Among the most powerful of these are measurements of the carbon isotopes in CO_2 as well as high-precision measurements of atmospheric oxygen (O_2) content. The carbon contained in CO_2 has two naturally occurring stable isotopes denoted ^{12}C and ^{13}C. The first of these, ^{12}C, is the most abundant isotope at about 99%, followed by ^{13}C at about 1%. Emissions of CO_2 from coal, gas and oil combustion and land clearing have $^{13}C/^{12}C$ isotopic ratios that are less than those in atmospheric CO_2, and each carries a signature related to its source. Thus, as shown in Prentice et al. (2001), when CO_2 from fossil fuel combustion enters the atmosphere, the $^{13}C/^{12}C$ isotopic ratio in atmospheric CO_2 decreases at a predictable rate consistent with emissions of CO_2 from fossil origin. Note that changes in the $^{13}C/^{12}C$ ratio of atmospheric CO_2 are also caused by other sources and sinks, but the changing isotopic signal due to CO_2 from fossil fuel combustion can be resolved from the other components (Francey et al., 1995). These changes can easily be measured using modern isotope ratio mass spectrometry, which has the capability of measuring $^{13}C/^{12}C$ in atmospheric CO_2 to better than 1 part in 10^5 (Ferretti et al., 2000). Data presented in Figure 2.3 for the $^{13}C/^{12}C$ ratio of atmospheric CO_2 at Mauna Loa show a decreasing ratio, consistent with trends in both fossil fuel CO_2 emissions and atmospheric CO_2 mixing ratios (Andres et al., 2000; Keeling et al., 2005).

Atmospheric O_2 measurements provide a powerful and independent method of determining the partitioning of CO_2 between the oceans and land (Keeling et al., 1996). Atmospheric O_2 and CO_2 changes are inversely coupled during plant respiration and photosynthesis. In addition, during the process of combustion O_2 is removed from the atmosphere, producing a signal that decreases as atmospheric CO_2 increases on a molar basis (Figure 2.3). Measuring changes in atmospheric O_2 is technically challenging because of the difficulty of resolving changes at the part-per-million level in a background mixing ratio of roughly 209,000 ppm. These difficulties were first overcome by Keeling and Shertz (1992), who used an interferometric technique to show that it is possible to track both seasonal cycles and the decline of O_2 in the atmosphere at the part-per-million level (Figure 2.3). Recent work by Manning and Keeling (2006) indicates that atmospheric O_2 is decreasing at a faster rate than CO_2 is increasing, which demonstrates the importance of the oceanic carbon sink. Measurements of

both the $^{13}C/^{12}C$ ratio in atmospheric CO_2 and atmospheric O_2 levels are valuable tools used to determine the distribution of fossil-fuel derived CO_2 among the active carbon reservoirs, as discussed in Section 7.3. In Figure 2.3, recent measurements in both hemispheres are shown to emphasize the strong linkages between atmospheric CO_2 increases, O_2 decreases, fossil fuel consumption and the $^{13}C/^{12}C$ ratio of atmospheric CO_2.

From 1990 to 1999, a period reported in Prentice et al. (2001), the emission rate due to fossil fuel burning and cement production increased irregularly from 6.1 to 6.5 GtC yr^{-1} or about 0.7% yr^{-1}. From 1999 to 2005 however, the emission rate rose systematically from 6.5 to 7.8 GtC yr^{-1} (BP, 2006; Marland et al., 2006) or about 3.0% yr^{-1}, representing a period of higher emissions and growth in emissions than those considered in the TAR (see Figure 2.3). Carbon dioxide emissions due to global annual fossil fuel combustion and cement manufacture combined have increased by 70% over the last 30 years (Marland et al., 2006). The relationship between increases in atmospheric CO_2 mixing ratios and emissions has been tracked using a scaling factor known as the apparent 'airborne fraction', defined as the ratio of the annual increase in atmospheric CO_2 to the CO_2 emissions from annual fossil fuel and cement manufacture combined (Keeling et al., 1995). On decadal scales, this fraction has averaged about 60% since the 1950s. Assuming emissions of 7 GtC yr^{-1} and an airborne fraction remaining at about 60%, Hansen and Sato (2004) predicted that the underlying long-term global atmospheric CO_2 growth rate will be about 1.9 ppm yr^{-1}, a value consistent with observations over the 1995 to 2005 decade.

Carbon dioxide emissions due to land use changes during the 1990s are estimated as 0.5 to 2.7 GtC yr^{-1} (Section 7.3, Table 7.2), contributing 6% to 39% of the CO_2 growth rate (Brovkin et al., 2004). Prentice et al. (2001) cited an inventory-based estimate that land use change resulted in net emissions of 121 GtC between 1850 and 1990, after Houghton (1999, 2000). The estimate for this period was revised upwards to 134 GtC by Houghton (2003), mostly due to an increase in estimated emissions prior to 1960. Houghton (2003) also extended the inventory emissions estimate to 2000, giving cumulative emissions of 156 GtC since 1850. In carbon cycle simulations by Brovkin et al. (2004) and Matthews et al. (2004), land use change emissions contributed 12 to 35 ppm of the total CO_2 rise from 1850 to 2000 (Section 2.5.3, Table 2.8). Historical changes in land cover are discussed in Section 2.5.2, and the CO_2 budget over the 1980s and 1990s is discussed further in Section 7.3.

In 2005, the global mean average CO_2 mixing ratio for the SIO network of 9 sites was 378.75 ± 0.13 ppm and for the NOAA/GMD network of 40 sites was 378.76 ± 0.05 ppm, yielding a global average of almost 379 ppm. For both networks, only sites in the remote marine boundary layer are used and high-altitude locations are not included. For example, the Mauna Loa site is excluded due to an 'altitude effect' of about 0.5 ppm. In addition, the 2005 values are still pending final reference gas calibrations used to measure the samples.

New measurements of CO_2 from Antarctic ice and firn (MacFarling Meure et al., 2006) update and extend those from Etheridge et al. (1996) to AD 0. The CO_2 mixing ratio in 1750 was 277 ± 1.2 ppm.[7] This record shows variations between 272 and 284 ppm before 1800 and also that CO_2 mixing ratios dropped by 5 to 10 ppm between 1600 and 1800 (see Section 6.3). The RF calculations usually take 1750 as the pre-industrial index (e.g., the TAR and this report). Therefore, using 1750 may slightly overestimate the RF, as the changes in the mixing ratios of CO_2, CH_4 and N_2O after the end of this naturally cooler period may not be solely attributable to anthropogenic emissions. Using 1860 as an alternative start date for the RF calculations would reduce the LLGHG RF by roughly 10%. For the RF calculation, the data from Law Dome ice cap in the Antarctic are used because they show the highest age resolution (approximately 10 years) of any ice core record in existence. In addition, the high-precision data from the cores are connected to direct observational records of atmospheric CO_2 from Cape Grim, Tasmania.

The simple formulae for RF of the LLGHG quoted in Ramaswamy et al. (2001) are still valid. These formulae are based on global RF calculations where clouds, stratospheric adjustment and solar absorption are included, and give an RF of +3.7 W m^{-2} for a doubling in the CO_2 mixing ratio. (The formula used for the CO_2 RF calculation in this chapter is the IPCC (1990) expression as revised in the TAR. Note that for CO_2, RF increases logarithmically with mixing ratio.) Collins et al. (2006) performed a comparison of five detailed line-by-line models and 20 GCM radiation schemes. The spread of line-by-line model results were consistent with the ±10% uncertainty estimate for the LLGHG RFs adopted in Ramaswamy et al. (2001) and a similar ±10% for the 90% confidence interval is adopted here. However, it is also important to note that these relatively small uncertainties are not always achievable when incorporating the LLGHG forcings into GCMs. For example, both Collins et al. (2006) and Forster and Taylor (2006) found that GCM radiation schemes could have inaccuracies of around 20% in their total LLGHG RF (see also Sections 2.3.2 and 10.2).

Using the global average value of 379 ppm for atmospheric CO_2 in 2005 gives an RF of 1.66 ± 0.17 W m^{-2}; a contribution that dominates that of all other forcing agents considered in this chapter. This is an increase of 13 to 14% over the value reported for 1998 in Ramaswamy et al. (2001). This change is solely due to increases in atmospheric CO_2 and is also much larger than the RF changes due to other agents. In the decade 1995 to 2005, the RF due to CO_2 increased by about 0.28 W m^{-2} (20%), an increase greater than that calculated for any decade since at least 1800 (see Section 6.6 and FAQ 2.1, Figure 1).

Table 2.1 summarises the present-day mixing ratios and RF for the LLGHGs, and indicates changes since 1998. The RF from CO_2 and that from the other LLGHGs have a high level of scientific understanding (Section 2.9, Table 2.11). Note that

the uncertainty in RF is almost entirely due to radiative transfer assumptions and not mixing ratio estimates, therefore trends in RF can be more accurately determined than the absolute RF. From Section 2.5.3, Table 2.8, the contribution from land use change to the present CO_2 RF is likely to be about 0.4 W m^{-2} (since 1850). This implies that fossil fuel and cement production have likely contributed about three-quarters of the current RF.

2.3.2 Atmospheric Methane

This section describes the current global measurement programmes for atmospheric CH_4, which provide the data required for the understanding of its budget and for the calculation of its RF. In addition, this section provides data for the pre-industrial levels of CH_4 required as the accepted reference level for these calculations. Detailed analyses of CH_4 budgets and its biogeochemistry are presented in Section 7.4.

Methane has the second-largest RF of the LLGHGs after CO_2 (Ramaswamy et al., 2001). Over the last 650 kyr, ice core records indicate that the abundance of CH_4 in the Earth's atmosphere has varied from lows of about 400 ppb during glacial periods to highs of about 700 ppb during interglacials (Spahni et al., 2005) with a single measurement from the Vostok core reaching about 770 ppb (see Figure 6.3).

In 2005, the global average abundance of CH_4 measured at the network of 40 surface air flask sampling sites operated by NOAA/GMD in both hemispheres was 1,774.62 ± 1.22 ppb.[8] This is the most geographically extensive network of sites operated by any laboratory and it is important to note that the calibration scale it uses has changed since the TAR (Dlugokencky et al., 2005). The new scale (known as NOAA04) increases all previously reported CH_4 mixing ratios from NOAA/GMD by about 1%, bringing them into much closer agreement with the Advanced Global Atmospheric Gases Experiment (AGAGE) network. This scale will be used by laboratories participating in the WMO's GAW programme as a 'common reference'. Atmospheric CH_4 is also monitored at five sites in the NH and SH by the AGAGE network. This group uses automated systems to make 36 CH_4 measurements per day at each site, and the mean for 2005 was 1,774.03 ± 1.68 ppb with calibration and methods described by Cunnold et al. (2002). For the NOAA/GMD network, the 90% confidence interval is calculated with a Monte Carlo technique, which only accounts for the uncertainty due to the distribution of sampling sites. For both networks, only sites in the remote marine boundary layer are used and continental sites are not included. Global databases of atmospheric CH_4 measurements for these and other CH_4 measurement programmes (e.g., Japanese, European and Australian) are maintained by the CDIAC and by the WDCGG in the GAW programme.

Present atmospheric levels of CH_4 are unprecedented in at least the last 650 kyr (Spahni et al., 2005). Direct atmospheric

---

[7] For consistency with the TAR, the pre-industrial value of 278 ppm is retained in the CO_2 RF calculation.

[8] The 90% confidence range quoted is from the normal standard deviation error for trace gas measurements assuming a normal distribution (i.e., multiplying by a factor of 1.645).

Table 2.1. *Present-day concentrations and RF for the measured LLGHGs. The changes since 1998 (the time of the TAR estimates) are also shown.*

Species[a]	Concentrations[b] and their changes[c]		Radiative Forcing[d]	
	2005	Change since 1998	2005 (W m^{-2})	Change since 1998 (%)
CO_2	379 ± 0.65 ppm	+13 ppm	1.66	+13
CH_4	1,774 ± 1.8 ppb	+11 ppb	0.48	-
N_2O	319 ± 0.12 ppb	+5 ppb	0.16	+11
	ppt	ppt		
CFC-11	251 ± 0.36	−13	0.063	−5
CFC-12	538 ± 0.18	+4	0.17	+1
CFC-113	79 ± 0.064	−4	0.024	−5
HCFC-22	169 ± 1.0	+38	0.033	+29
HCFC-141b	18 ± 0.068	+9	0.0025	+93
HCFC-142b	15 ± 0.13	+6	0.0031	+57
CH_3CCl_3	19 ± 0.47	−47	0.0011	−72
CCl_4	93 ± 0.17	−7	0.012	−7
HFC-125	3.7 ± 0.10[e]	+2.6[f]	0.0009	+234
HFC-134a	35 ± 0.73	+27	0.0055	+349
HFC-152a	3.9 ± 0.11[e]	+2.4[f]	0.0004	+151
HFC-23	18 ± 0.12[g,h]	+4	0.0033	+29
SF_6	5.6 ± 0.038[i]	+1.5	0.0029	+36
CF_4 (PFC-14)	74 ± 1.6[j]	-	0.0034	-
C_2F_6 (PFC-116)	2.9 ± 0.025[g,h]	+0.5	0.0008	+22
CFCs Total[k]			0.268	−1
HCFCs Total			0.039	+33
Montreal Gases			0.320	−1
Other Kyoto Gases (HFCs + PFCs + SF$_6$)			0.017	+69
Halocarbons			0.337	+1
Total LLGHGs			2.63	+9

Notes:

[a] See Table 2.14 for common names of gases and the radiative efficiencies used to calculate RF.

[b] Mixing ratio errors are 90% confidence ranges of combined 2005 data, including intra-annual standard deviation, measurement and global averaging uncertainty. Standard deviations were multiplied by 1.645 to obtain estimates of the 90% confidence range; this assumes normal distributions. Data for CO_2 are combined measurements from the NOAA Earth System Research Laboratory (ESRL) and SIO networks (see Section 2.3.1); CH_4 measurements are combined data from the ESRL and Advanced Global Atmospheric Gases Experiment (AGAGE) networks (see Section 2.3.2); halocarbon measurements are the average of ESRL and AGAGE networks. University of East Anglia (UEA) and Pennsylvania State University (PSU) measurements were also used (see Section 2.3.3).

[c] Pre-industrial values are zero except for CO_2 (278 ppm), CH_4 (715 ppb; 700 ppb was used in the TAR), N_2O (270 ppb) and CF_4 (40 ppt).

[d] 90% confidence ranges for RF are not shown but are approximately 10%. This confidence range is almost entirely due to radiative transfer assumptions, therefore trends remain valid when quoted to higher accuracies. Higher precision data are used for totals and affect rounding of the values. Percent changes are calculated relative to 1998.

[e] Data available from AGAGE network only.

[f] Data for 1998 not available; values from 1999 are used.

[g] Data from UEA only.

[h] Data from 2003 are used due to lack of available data for 2004 and 2005.

[i] Data from ESRL only.

[j] 1997 data from PSU (Khalil et al., 2003, not updated) are used.

[k] CFC total includes a small 0.009 W m^{-2} RF from CFC-13, CFC-114, CFC-115 and the halons, as measurements of these were not updated.

measurements of the gas made at a wide variety of sites in both hemispheres over the last 25 years show that, although the abundance of CH_4 has increased by about 30% during that time, its growth rate has decreased substantially from highs of greater than 1% yr^{-1} in the late 1 70s and early 1980s (Blake and Rowland, 1988) to lows of close to zero towards the end of the 1990s (Dlugokencky et al., 1998; Simpson et al., 2002). The slowdown in the growth rate began in the 1980s, decreasing from 14 ppb yr^{-1} (about 1% yr^{-1}) in 1984 to close to zero during 1999 to 2005, for the network of surface sites maintained by NOAA/GMD (Dlugokencky et al., 2003). Measurements by Lowe et al. (2004) for sites in the SH and Cunnold et al. (2002) for the network of GAGE/AGAGE sites show similar features. A key feature of the global growth rate of CH_4 is its current interannual variability, with growth rates ranging from a high of 14 ppb yr^{-1} in 1998 to less than zero in 2001, 2004 and 2005. (Figure 2.4)

The reasons for the decrease in the atmospheric CH_4 growth rate and the implications for future changes in its atmospheric burden are not understood (Prather et al., 2001) but are clearly related to changes in the imbalance between CH_4 sources and sinks. Most CH_4 is removed from the atmosphere by reaction with the hydroxyl free radical (OH), which is produced photochemically in the atmosphere. The role of OH in controlling atmospheric CH_4 levels is discussed in Section 2.3.5. Other minor sinks include reaction with free chlorine (Platt et al., 2004; Allan et al., 2005), destruction in the stratosphere and soil sinks (Born et al., 1990).

The total global CH_4 source is relatively well known but the strength of each source component and their trends are not. As detailed in Section 7.4, the sources are mostly biogenic and include wetlands, rice agriculture, biomass burning and ruminant animals. Methane is also emitted by various industrial sources including fossil fuel mining and distribution. Prather et al. (2001) documented a large range in 'bottom-up' estimates for the global source of CH_4. New source estimates published since then are documented in Table 7.6. However, as reported by Bergamaschi et al. (2005), national inventories based on 'bottom-up' studies can grossly underestimate emissions and 'top-down' measurement-based assessments of reported emissions will be required for verification. Keppler et al. (2006) reported the discovery of emissions of CH_4 from living vegetation and estimated that this contributed 10 to 30% of the global CH_4 source. This work extrapolates limited measurements to a global source and has not yet been confirmed by other laboratories, but lends some support to space-borne observations of CH_4 plumes above tropical rainforests reported by Frankenberg et al. (2005). That such a potentially large source of CH_4 could have been missed highlights the large uncertainties involved in current 'bottom-up' estimates of components of the global source (see Section 7.4).

Several wide-ranging hypotheses have been put forward to explain the reduction in the CH_4 growth rate and its variability. For example, Hansen et al. (2000) considered that economic incentives have led to a reduction in anthropogenic CH_4 emissions. The negligible long-term change in its main

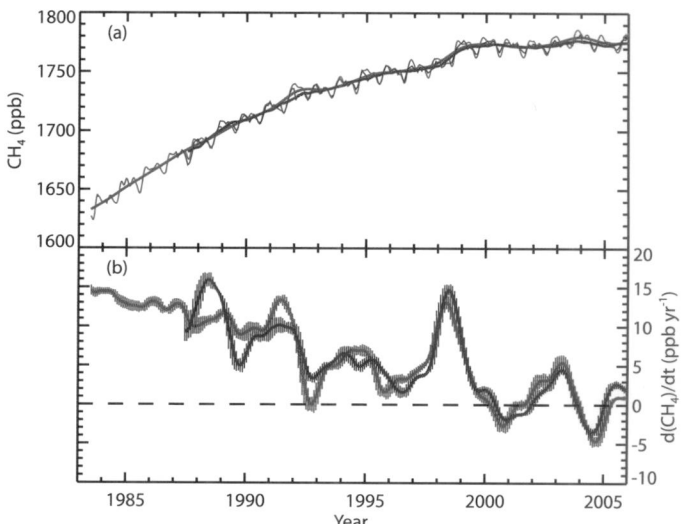

Figure 2.4. *Recent CH_4 concentrations and trends. (a) Time series of global CH_4 abundance mole fraction (in ppb) derived from surface sites operated by NOAA/GMD (blue lines) and AGAGE (red lines). The thinner lines show the CH_4 global averages and the thicker lines are the de-seasonalized global average trends from both networks. (b) Annual growth rate (ppb yr^{-1}) in global atmospheric CH_4 abundance from 1984 through the end of 2005 (NOAA/GMD, blue), and from 1988 to the end of 2005 (AGAGE, red). To derive the growth rates and their uncertainties for each month, a linear least squares method that takes account of the autocorrelation of residuals is used. This follows the methods of Wang et al. (2002) and is applied to the de-seasonalized global mean mole fractions from (a) for values six months before and after the current month. The vertical lines indicate ±2 standard deviation uncertainties (95% confidence interval), and 1 standard deviation uncertainties are between 0.1 and 1.4 ppb yr^{-1} for both AGAGE and NOAA/GMD. Note that the differences between AGAGE and NOAA/GMD calibration scales are determined through occasional intercomparisons.*

sink (OH; see Section 2.3.5 and Figure 2.8) implies that CH_4 emissions are not increasing. Similarly, Dlugokencky et al. (1998) and Francey et al. (1999) suggested that the slowdown in the growth rate reflects a stabilisation of CH_4 emissions, given that the observations are consistent with stable emissions and lifetime since 1982.

Relatively large anomalies occurred in the growth rate during 1991 and 1998, with peak values reaching 15 and 14 ppb yr^{-1}, respectively (about 1% yr^{-1}). The anomaly in 1991 was followed by a dramatic drop in the growth rate in 1992 and has been linked with the Mt. Pinatubo volcanic eruption in June 1991, which injected large amounts of ash and (sulphur dioxide) SO_2 into the lower stratosphere of the tropics with subsequent impacts on tropical photochemistry and the removal of CH_4 by atmospheric OH (Bekki et al., 1994; Dlugokencky et al., 1996). Lelieveld et al. (1998) and Walter et al. (2001a,b) proposed that lower temperatures and lower precipitation in the aftermath of the Mt. Pinatubo eruption could have suppressed CH_4 emissions from wetlands. At this time, and in parallel with the growth rate anomaly in the CH_4 mixing ratio, an anomaly was observed in the $^{13}C/^{12}C$ ratio of CH_4 at surface sites in the SH. This was attributed to a decrease in emissions from an isotopically heavy source such as biomass burning (Lowe et al., 1997; Mak et al., 2000), although these data were not confirmed by lower frequency measurements from the same period made by Francey et al. (1999).

For the relatively large increase in the CH_4 growth rate reported for 1998, Dlugokencky et al. (2001) suggested that wetland and boreal biomass burning sources might have contributed to the anomaly, noting that 1998 was the warmest year globally since surface instrumental temperature records began. Using an inverse method, Chen and Prinn (2006) attributed the same event primarily to increased wetland and rice region emissions and secondarily to biomass burning. The same conclusion was reached by Morimoto et al. (2006), who used carbon isotopic measurements of CH_4 to constrain the relative contributions of biomass burning (one-third) and wetlands (two-thirds) to the increase.

Based on ice core measurements of CH_4 (Etheridge et al., 1998), the pre-industrial global value for CH_4 from 1700 to 1800 was 715 ± 4 ppb (it was also 715 ± 4 ppb in 1750), thus providing the reference level for the RF calculation. This takes into account the inter-polar difference in CH_4 as measured from Greenland and Antarctic ice cores.

The RF due to changes in CH_4 mixing ratio is calculated with the simplified yet still valid expression for CH_4 given in Ramaswamy et al. (2001). The change in the CH_4 mixing ratio from 715 ppb in 1750 to 1,774 ppb (the average mixing ratio from the AGAGE and GMD networks) in 2005 gives an RF of $+0.48 \pm 0.05$ W m^{-2}, ranking CH_4 as the second highest RF of the LLGHGs after CO_2 (Table 2.1). The uncertainty range in mixing ratios for the present day represents intra-annual variability, which is not included in the pre-industrial uncertainty estimate derived solely from ice core sampling precision. The estimate for the RF due to CH_4 is the same as in Ramaswamy et al. (2001) despite the small increase in its mixing ratio. The spectral absorption by CH_4 is overlapped to some extent by N_2O lines (taken into account in the simplified expression). Taking the overlapping lines into account using current N_2O mixing ratios instead of pre-industrial mixing ratios (as in Ramaswamy et al., 2001) reduces the current RF due to CH_4 by 1%.

Collins et al. (2006) confirmed that line-by-line models agree extremely well for the calculation of clear-sky instantaneous RF from CH_4 and N_2O when the same atmospheric background profile is used. However, GCM radiation schemes were found to be in poor agreement with the line-by-line models, and errors of over 50% were possible for CH_4, N_2O and the CFCs. In addition, a small effect from the absorption of solar radiation was found with the line-by-line models, which the GCMs did not include (Section 10.2).

2.3.3 Other Kyoto Protocol Gases

At the time of the TAR, N_2O had the fourth largest RF among the LLGHGs behind CO_2, CH_4 and CFC-12. The TAR quoted an atmospheric N_2O abundance of 314 ppb in 1998, an increase of 44 ppb from its pre-industrial level of around 270 ± 7 ppb, which gave an RF of $+0.15 \pm 0.02$ W m^{-2}. This RF is affected by atmospheric CH_4 levels due to overlapping absorptions. As N_2O is also the major source of ozone-depleting nitric oxide (NO) and nitrogen dioxide (NO_2) in the stratosphere, it is routinely reviewed in the ozone assessments; the most recent assessment (Montzka et al., 2003) recommended an atmospheric lifetime of 114 years for N_2O. The TAR pointed out large uncertainties in the major soil, agricultural, combustion and oceanic sources of N_2O. Given these emission uncertainties, its observed rate of increase of 0.2 to 0.3% yr^{-1} was not inconsistent with its better-quantified major sinks (principally stratospheric destruction). The primary driver for the industrial era increase of N_2O was concluded to be enhanced microbial production in expanding and fertilized agricultural lands.

Ice core data for N_2O have been reported extending back 2,000 years and more before present (MacFarling Meure et al., 2006; Section 6.6). These data, as for CO_2 and CH_4, show relatively little changes in mixing ratios over the first 1,800 years of this record, and then exhibit a relatively rapid rise (see FAQ 2.1, Figure 1). Since 1998, atmospheric N_2O levels have steadily risen to 319 ± 0.12 ppb in 2005, and levels have been increasing approximately linearly (at around 0.26% yr^{-1}) for the past few decades (Figure 2.5). A change in the N_2O mixing ratio

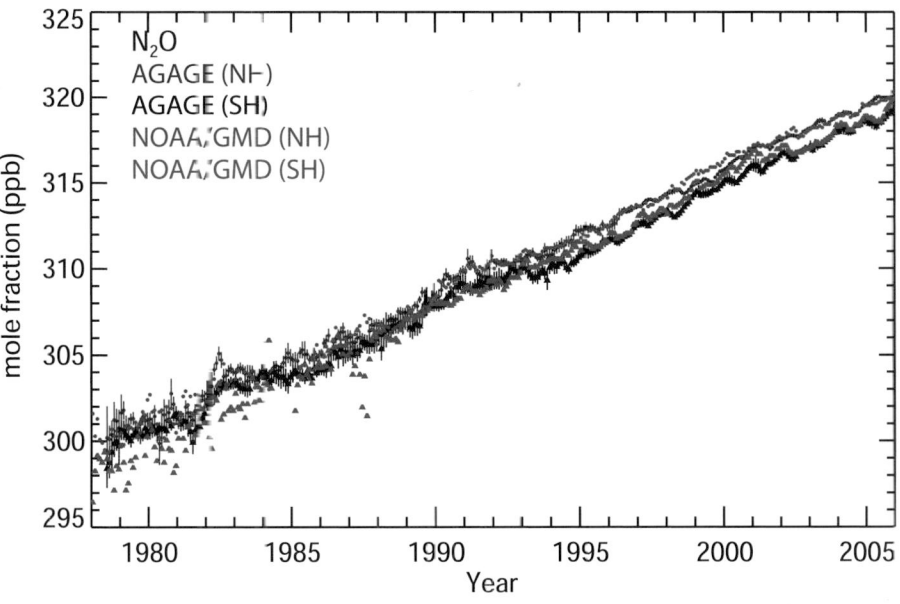

Figure 2.5. *Hemispheric monthly mean N_2O mole fractions (ppb) (crosses for the NH and triangles for the SH). Observations (in situ) of N_2O from the Atmospheric Lifetime Experiment (ALE) and GAGE (through the mid-1990s) and AGAGE (since the mid-1990s) networks (Prinn et al., 2000, 2005b) are shown with monthly standard deviations. Data from NOAA/GMD are shown without these standard deviations (Thompson et al., 2004). The general decrease in the variability of the measurements over time is due mainly to improved instrumental precision. The real signal emerges only in the last decade.*

from 270 ppb in 1750 to 319 ppb in 2005 results in an RF of $+0.16 \pm 0.02$ W m^{-2}, calculated using the simplified expression given in Ramaswamy et al. (2001). The RF has increased by 11% since the time of the TAR (Table 2.1). As CFC-12 levels slowly decline (see Section 2.3.4), N_2O should, with its current trend, take over third place in the LLGHG RF ranking.

Since the TAR, understanding of regional N_2O fluxes has improved. The results of various studies that quantified the global N_2O emissions from coastal upwelling areas, continental shelves, estuaries and rivers suggest that these coastal areas contribute 0.3 to 6.6 TgN yr^{-1} of N_2O or 7 to 61% of the total oceanic emissions (Bange et al., 1996; Nevison et al., 2004b; Kroeze et al., 2005; see also Section 7.4). Using inverse methods and AGAGE Ireland measurements, Manning et al. (2003) estimated EU N_2O emissions of 0.9 ± 0.1 TgN yr^{-1} that agree well with the United Nations Framework Convention on Climate Change (UNFCCC) N_2O inventory (0.8 ± 0.1 TgN yr^{-1}). Melillo et al. (2001) provided evidence from Brazilian land use sequences that the conversion of tropical forest to pasture leads to an initial increase but a later decline in emissions of N_2O relative to the original forest. They also deduced that Brazilian forest soils alone contribute about 10% of total global N_2O production. Estimates of N_2O sources and sinks using observations and inverse methods had earlier implied that a large fraction of global N_2O emissions in 1978 to 1988 were tropical: specifically 20 to 29% from 0° to 30°S and 32 to 39% from 0° to 30°N compared to 11 to 15% from 30°S to 90°S and 22 to 34% from 30°N to 90°N (Prinn et al., 1990). These estimates were uncertain due to their significant sensitivity to assumed troposphere-stratosphere exchange rates that strongly influence inter-hemispheric gradients. Hirsch et al. (2006) used inverse modelling to estimate significantly lower emissions from 30°S to 90°S (0 to 4%) and higher emissions from 0° to 30°N (50 to 64%) than Prinn et al. (1990) during 1998 to 2001, with 26 to 36% from the oceans. The stratosphere is also proposed to play an important role in the seasonal cycles of N_2O (Nevison et al., 2004a). For example, its well-defined seasonal cycle in the SH has been interpreted as resulting from the net effect of seasonal oceanic outgassing of microbially produced N_2O, stratospheric intrusion of low-N_2O air and other processes (Nevison et al., 2005). Nevison et al. also estimated a Southern Ocean (30°S–90°S) N_2O source of 0.9 TgN yr^{-1}, or about 5% of the global total. The complex seasonal cycle in the NH is more difficult to reconcile with seasonal variations in the northern latitude soil sources and stratospheric intrusions (Prinn et al., 2000; T. Liao et al., 2004). The destruction of N_2O in the stratosphere causes enrichment of its heavier isotopomers and isotopologues, providing a potential method to differentiate stratospheric and surface flux influences on tropospheric N_2O (Morgan et al., 2004).

Human-made PFCs, HFCs and SF$_6$ are very effective absorbers of infrared radiation, so that even small amounts of these gases contribute significantly to the RF of the climate system. The observations and global cycles of the major HFCs, PFCs and SF$_6$ were reviewed in Velders et al. (2005), and this section only provides a brief review and an update for these

species. Table 2.1 shows the present mixing ratio and recent trends in the halocarbons and their RFs. Absorption spectra of most halocarbons reviewed here and in the following section are characterised by strongly overlapping spectral lines that are not resolved at tropospheric pressures and temperatures, and there is some uncertainty in cross section measurements. Apart from the uncertainties stemming from the cross sections themselves, differences in the radiative flux calculations can arise from the spectral resolution used, tropopause heights, vertical, spatial and seasonal distributions of the gases, cloud cover and how stratospheric temperature adjustments are performed. IPCC/TEAP (2005) concluded that the discrepancy in the RF calculation for different halocarbons, associated with uncertainties in the radiative transfer calculation and the cross sections, can reach 40%. Studies reviewed in IPCC/TEAP (2005) for the more abundant HFCs show that an agreement better than 12% can be reached for these when the calculation conditions are better constrained (see Section 2.10.2).

The HFCs of industrial importance have lifetimes in the range 1.4 to 270 years. The HFCs with the largest observed mole fractions in 1998 (as reported in the TAR) were, in descending order, HFC-23, HFC-134a and HFC-152a. In 2005, the observed mixing ratios of the major HFCs in the atmosphere were 35 ppt for HFC-134a, 17.5 ppt for HFC-23 (2003 value), 3.7 ppt for HFC-125 and 3.9 ppt for HFC-152a (Table 2.1). Within the uncertainties in calibration and emissions estimates, the observed mixing ratios of the HFCs in the atmosphere can be explained by the anthropogenic emissions. Measurements are available from GMD (Thompson et al., 2004) and AGAGE (Prinn et al., 2000; O'Doherty et al., 2004; Prinn et al., 2005b) networks as well as from University of East Anglia (UEA) studies in Tasmania (updated from Oram et al., 1998; Oram, 1999). These data, summarised in Figure 2.6, show a continuation of positive HFC trends and increasing latitudinal gradients (larger trends in the NH) due to their predominantly NH sources. The air conditioning refrigerant HFC-134a is increasing at a rapid rate in response to growing emissions arising from its role as a replacement for some CFC refrigerants. With a lifetime of about 14 years, its current trends are determined primarily by its emissions and secondarily by its atmospheric destruction. Emissions of HFC-134a estimated from atmospheric measurements are in approximate agreement with industry estimates (Huang and Prinn, 2002; O'Doherty et al., 2004). IPCC/TEAP (2005) reported that global HFC-134a emissions started rapidly increasing in the early 1990s, and that in Europe, sharp increases in emissions are noted for HFC-134a from 1995 to 1998 and for HFC-152a from 1996 to 2000, with some levelling off through 2003. The concentration of the foam blower HFC-152a, with a lifetime of only about 1.5 years, is rising approximately exponentially, with the effects of increasing emissions only partly offset by its rapid atmospheric destruction. Hydrofluorocarbon-23 has a very long atmospheric lifetime (approximately 270 years) and is mainly produced as a by-product of HCFC-22 production. Its concentrations are rising approximately linearly, driven by these emissions, with its destruction being only a minor factor in its budget. There are also smaller but rising concentrations of HFC-125 and HFC-143a, which are both refrigerants.

The PFCs, mainly CF_4 (PFC-14) and C_2F_6 (PFC-116), and SF_6 have very large radiative efficiencies and lifetimes in the range 1,000 to 50,000 years (see Section 2.10, Table 2.14), and make an essentially permanent contribution to RF. The SF_6 and C_2F_6 concentrations and RFs have increased by over 20% since the TAR (Table 2.1 and Figure 2.6), but CF_4 concentrations have not been updated since 1997. Both anthropogenic and natural sources of CF_4 are important to explain its observed atmospheric abundance. These PFCs are produced as by-products of traditional aluminium production, among other activities. The CF_4 concentrations have been increasing linearly since about 1960 and CF_4 has a natural source that accounts for about one-half of its current atmospheric content (Harnisch et al., 1996). Sulphur hexafluoride (SF_6) is produced for use as an electrical insulating fluid in power distribution equipment and also deliberately released as an essentially inert tracer to study atmospheric and oceanic transport processes. Its concentration was 4.2 ppt in 1998 (TAR) and has continued to increase linearly over the past decade, implying that emissions are approximately constant. Its very long lifetime ensures that its emissions accumulate essentially unabated in the atmosphere.

2.3.4 Montreal Protocol Gases

The Montreal Protocol on Substances that Deplete the Ozone Layer regulates many radiatively powerful greenhouse gases for the primary purpose of lowering stratospheric chlorine and bromine concentrations. These gases include the CFCs, HCFCs, chlorocarbons, bromocarbons and halons. Observations and global cycles of these gases were reviewed in detail in Chapter 1 of the 2002 Scientific Assessment of Ozone Depletion (WMO, 2003) and in IPCC/TEAP (2005). The discussion here focuses on developments since these reviews and on those gases that contribute most to RF rather than to halogen loading. Using observed 2005 concentrations, the Montreal Protocol gases have contributed 12% (0.320 W m^{-2}) to the direct RF of all LLGHGs and 95% to the halocarbon RF (Table 2.1). This contribution is dominated by the CFCs. The effect of the Montreal Protocol on these gases has been substantial. IPCC/TEAP (2005) concluded that the combined CO_2-equivalent emissions of CFCs, HCFCs and HFCs decreased from a peak of about 7.5 GtCO$_2$-eq yr^{-1} in the late 1980s to about 2.5 GtCO$_2$-eq yr^{-1} by the year 2000, corresponding to about 10% of that year's CO_2 emissions due to global fossil fuel burning.

Measurements of the CFCs and HCFCs, summarised in Figure 2.6, are available from the AGAGE network (Prinn et al., 2000, 2005b) and the GMD network (Montzka et al., 1999 updated; Thompson et al., 2004). Certain flask measurements are also available from the University of California at Irvine (UCI; Blake et al., 2001 updated) and UEA (Oram et al., 1998; Oram, 1999 updated). Two of the major CFCs (CFC-11 and CFC-113) have both been decreasing in the atmosphere since the mid-1990s. While their emissions have decreased very substantially in response to the Montreal Protocol, their long lifetimes of around 45 and 85 years, respectively, mean that their sinks can reduce their levels by only about 2% and 1% yr^{-1}, respectively. Nevertheless, the effect of the Montreal Protocol has been to substantially reduce the growth of the halocarbon RF, which increased rapidly from 1950 until about 1990. The other major CFC (CFC-12), which is the third most important LLGHG, is finally reaching a plateau in its atmospheric levels (emissions equal loss) and may have peaked in 2003. Its 100-year lifetime means that it can decrease by only about 1% yr^{-1} even when emissions are zero. The levelling off for CFC-12 and approximately linear downward trends for CFC-11 and CFC-113 continue. Latitudinal gradients of all three are very small and decreasing as expected. The combined CFC and HCFC RF has been slowly declining since 2003. Note that the 1998 concentrations of CFC-11 and CFC-12 were overestimated in Table 6.1 of the TAR. This means that the total halocarbon RF quoted for 2005 in Table 2.1 (0.337 W m^{-2}) is slightly smaller than the 0.34 W m^{-2} quoted in the TAR, even though the measurements indicate a small 1% rise in the total halocarbon RF since the time of the TAR (Table 2.1).

The major solvent, methyl chloroform (CH_3CCl_3), is of special importance regarding RFs, not because of its small RF (see Table 2.1 and Figure 2.6), but because this gas is widely

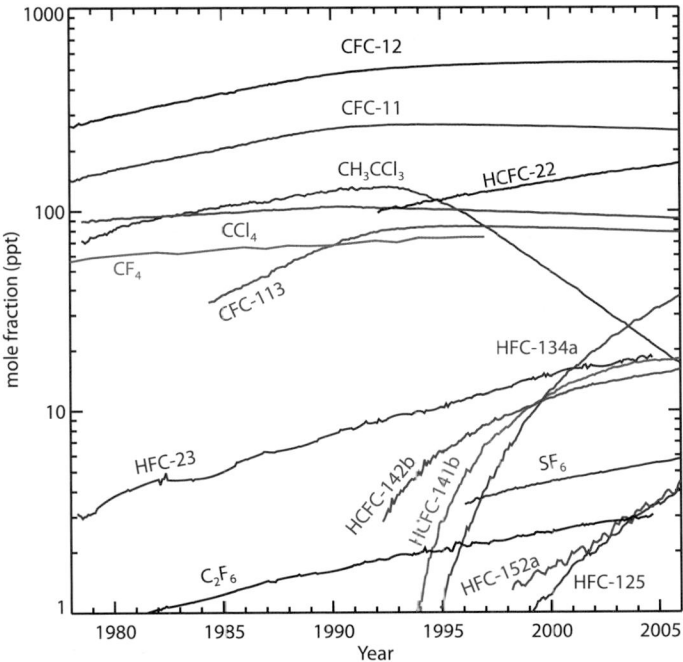

Figure 2.6. *Temporal evolution of the global average dry-air mole fractions (ppt) of the major halogen-containing LLGHGs. These are derived mainly using monthly mean measurements from the AGAGE and NOAA/GMD networks. For clarity, the two network values are averaged with equal weight when both are available. While differences exist, these network measurements agree reasonably well with each other (except for CCl_4 (differences of 2 – 4% between networks) and HCFC-142b (differences of 3 – 6% between networks)), and with other measurements where available (see text for references for each gas).*

used to estimate concentrations of OH, which is the major sink species for CH_4, HFCs, and HCFCs and a major production mechanism for sulphate, nitrate and some organic aerosols as discussed in Section 2.3.5. The global atmospheric methyl chloroform concentration rose steadily from 1978 to reach a maximum in 1992 (Prinn et al., 2001; Montzka et al., 2003). Since then, concentrations have decreased rapidly, driven by a relatively short lifetime of 4.9 years and phase-out under the Montreal Protocol, to levels in 2003 less than 20% of the levels when AGAGE measurements peaked in 1992 (Prinn et al., 2005a). Emissions of methyl chloroform determined from industry data (McCulloch and Midgley, 2001) may be too small in recent years. The 2000 to 2003 emissions from Europe estimated using surface observations (Reimann et al., 2005) show that 1.2 to 2.3 Gg yr^{-1} need to be added over this 4-year period to the above industry estimates for Europe. Estimates of European emissions in 2000 exceeding 20 Gg (Krol et al., 2003) are not supported by analyses of the above extensive surface data (Reimann et al., 2005). From multi-year measurements, Li et al. (2005) estimated 2001 to 2002 emissions from the USA of 2.2 Gg yr^{-1} (or about half of those estimated from more temporally but less geographically limited measurements by Millet and Goldstein, 2004), and suggested that 1996 to 1998 US emissions may be underestimated by an average of about 9.0 Gg yr^{-1} over this 3-year period. East Asian emissions deduced from aircraft data in 2001 are about 1.7 Gg above industry data (Palmer et al., 2003; see also Yokouchi et al., 2005) while recent Australian and Russian emissions are negligible (Prinn et al., 2001; Hurst et al., 2004).

Carbon tetrachloride (CCl_4) is the second most rapidly decreasing atmospheric chlorocarbon after methyl chloroform. Levels peaked in early 1990 and decreased approximately linearly since then (Figure 2.7).

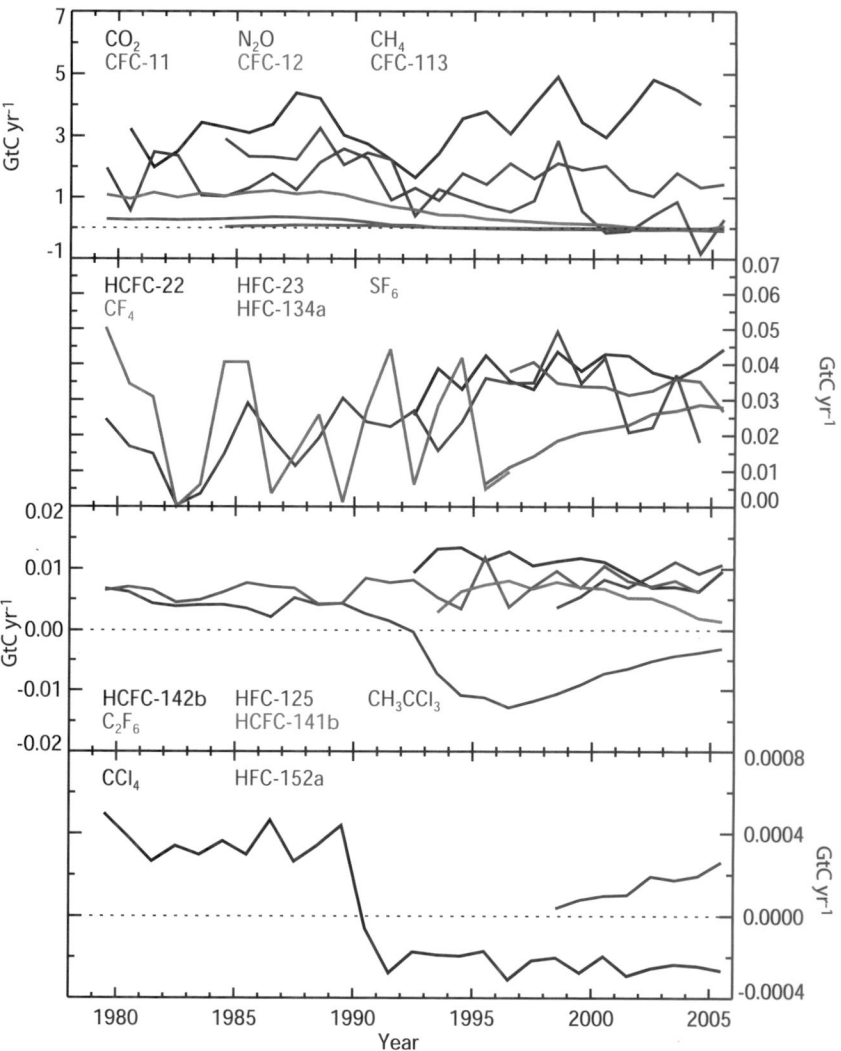

Figure 2.7. *Annual rates of change in the global atmospheric masses of each of the major LLGHGs expressed in common units of GtC yr^{-1}. These rates are computed from their actual annual mass changes in Gt yr^{-1} (as derived from their observed global and annual average mole fractions presented in Figures 2.3 to 2.6 and discussed in Sections 2.3.1 to 2.3.4) by multiplying them by their GWPs for 100-year time horizons and then dividing by the ratio of the CO_2 to carbon (C) masses (44/12). These rates are positive or negative whenever the mole fractions are increasing or decreasing, respectively. Use of these common units provides an approximate way to intercompare the fluxes of LLGHGs, using the same approach employed to intercompare the values of LLGHG emissions under the Kyoto Protocol (see, e.g., Prinn, 2004). Note that the negative indirect RF of CFCs and HCFCs due to stratospheric ozone depletion is not included. The oscillations in the CF_4 curve may result partly from truncation in reported mole fractions.*

Its major use was as a feedstock for CFC manufacturing. Unlike methyl chloroform, a significant inter-hemispheric CCl_4 gradient still exists in 2005 in spite of its moderately long lifetime of 20 to 30 years, resulting from a persistence of significant NH emissions.

HCFCs of industrial importance have lifetimes in the range of 1.3 to 20 years. Global and regional emissions of the CFCs and HCFCs have been derived from observed concentrations and can be used to check emission inventory estimates. Montzka et al. (2003) and IPCC/TEAP (2005) concluded that global emissions of HCFC-22 rose steadily over the period 1975 to 2000, while those of HCFC-141b and HCFC-142b

started increasing quickly in the early 1990s and then began to decrease after 2000.

To provide a direct comparison of the effects on global warming due to the annual changes in each of the non-CO_2 greenhouse gases (discussed in Sections 2.3.2, 2.3.3 and 2.3.4) relative to CO_2, Figure 2.7 shows these annual changes in atmospheric mass multiplied by the GWP (100-year horizon) for each gas (e.g., Prinn, 2004). By expressing them in this way, the observed changes in all non-CO_2 gases in GtC equivalents and the significant roles of CH_4, N_2O and many halocarbons are very evident. This highlights the importance of considering the full suite of greenhouse gases for RF calculations.

2.3.5   Trends in the Hydroxyl Free Radical

The hydroxyl free radical (OH) is the major oxidizing chemical in the atmosphere, destroying about 3.7 Gt of trace gases, including CH_4 and all HFCs and HCFCs, each year (Ehhalt, 1999). It therefore has a very significant role in limiting the LLGHG RF. IPCC/TEAP (2005) concluded that the OH concentration might change in the 21st century by –18 to +5% depending on the emission scenario. The large-scale concentrations and long-term trends in OH can be inferred indirectly using global measurements of trace gases for which emissions are well known and the primary sink is OH. The best trace gas used to date for this purpose is methyl chloroform; long-term measurements of this gas are reviewed in Section 2.3.4. Other gases that are useful OH indicators include ^{14}CO, which is produced primarily by cosmic rays (Lowe and Allan, 2002). While the accuracy of the ^{14}CO cosmic ray and other ^{14}CO source estimates and the frequency and spatial coverage of its measurements do not match those for methyl chloroform, the ^{14}CO lifetime (2 months) is much shorter than that of methyl chloroform (4.9 years). As a result, ^{14}CO provides estimates of average concentrations of OH that are more regional, and is capable of resolving shorter time scales than those estimated from methyl chloroform. The ^{14}CO source variability is better defined than its absolute magnitude so it is better for inferring relative rather than absolute trends. Another useful gas is the industrial chemical HCFC-22. It yields OH concentrations similar to those derived from methyl chloroform, but with less accuracy due to greater uncertainties in emissions and less extensive measurements (Miller et al., 1998). The industrial gases HFC-134a, HCFC-141b and HCFC-142b are potentially useful OH estimators, but the accuracy of their emission estimates needs improvement (Huang and Prinn, 2002; O'Doherty et al., 2004).

Indirect measurements of OH using methyl chloroform have established that the globally weighted average OH concentration in the troposphere is roughly 10^6 radicals per cubic centimetre (Prinn et al., 2001; Krol and Lelieveld, 2003). A similar average concentration is derived using ^{14}CO (Quay et al., 2000), although the spatial weighting here is different. Note that methods to infer global or hemispheric average OH concentrations may be insensitive to compensating regional OH changes such as OH increases over continents and decreases over oceans (Lelieveld et al., 2002). In addition, the quoted absolute OH concentrations (but not their relative trends) depend on the choice of weighting (e.g., Lawrence et al., 2001). While the global average OH concentration appears fairly well defined by these indirect methods, the temporal trends in OH are more difficult to discern since they require long-term measurements, optimal inverse methods and very accurate calibrations, model transports and methyl chloroform emissions data. From AGAGE methyl chloroform measurements, Prinn et al. (2001) inferred that global OH levels grew between 1979 and 1989, but then declined between 1989 and 2000, and also exhibited significant interannual variations. They concluded that these decadal global variations were driven principally by NH OH, with

SH OH decreasing from 1979 to 1989 and staying essentially constant after that. Using the same AGAGE data and identical methyl chloroform emissions, a three-dimensional model analysis (Krol and Lelieveld, 2003) supported qualitatively (but not quantitatively) the earlier result (Prinn et al., 2001) that OH concentrations increased in the 1980s and declined in the 1990s. Prinn et al. (2001) also estimated the emissions required to provide a zero trend in OH. These required methyl chloroform emissions differed substantially from industry estimates by McCulloch and Midgley (2001) particularly for 1996 to 2000. However, Krol and Lelieveld (2003) argued that the combination of possible underestimated recent emissions, especially the >20 Gg European emissions deduced by Krol et al. (2003), and the recent decreasing effectiveness of the stratosphere as a sink for tropospheric methyl chloroform, may be sufficient to yield a zero deduced OH trend. As discussed in Section 2.3.4, estimates of European emissions by Reimann et al. (2005) are an order of magnitude less than those of Krol et al. (2003). In addition, Prinn et al. (2005a) extended the OH estimates through 2004 and showed that the Prinn et al. (2001) decadal and interannual OH estimates remain valid even after accounting for the additional recent methyl chloroform emissions discussed in Section 2.3.4. They also reconfirmed the OH maximum around 1989 and a larger OH minimum around 1998, with OH concentrations then recovering so that in 2003 they were comparable to those in 1979. They noted that the 1997 to 1999 OH minimum coincides with, and is likely caused by, major global wildfires and an intense El Niño at that time. The 1997 Indonesian fires alone have been estimated to have lowered global late-1997 OH levels by 6% due to carbon monoxide (CO) enhancements (Duncan et al., 2003).

Methyl chloroform is also destroyed in the stratosphere. Because its stratospheric loss frequency is less than that in the troposphere, the stratosphere becomes a less effective sink for tropospheric methyl chloroform over time (Krol and Lelieveld, 2003), and even becomes a small source to the troposphere beginning in 1999 in the reference case in the Prinn et al. (2001, 2005a) model. Loss to the ocean has usually been considered irreversible, and its rates and uncertainties have been obtained from observations (Yvon-Lewis and Butler, 2002). However, Wennberg et al. (2004) recently proposed that the polar oceans may have effectively stored methyl chloroform during the pre-1992 years when its atmospheric levels were rising, but began re-emitting it in subsequent years, thus reducing the overall oceanic sink. Prinn et al. (2005a) tried both approaches and found that their inferred interannual and decadal OH variations were present using either formulation, but inferred OH was lower in the pre-1992 years and higher after that using the Wennberg et al. (2004) formulation.

More recently, Bousquet et al. (2005) used an inverse method with a three-dimensional model and methyl chloroform measurements and concluded that substantial year-to-year variations occurred in global average OH concentrations between 1980 and 2000. This conclusion was previously reached by Prinn et al. (2001), but subsequently challenged by Krol and Lelieveld (2003) who argued that these variations were caused by model shortcomings and that models need, in particular, to include observationally-based, interannually varying meteorology to provide accurate annual OH estimates. Neither the two-dimensional Prinn et al. (2001) nor the three-dimensional Krol et al. (2003) inversion models used interannually varying circulation. However, the Bousquet et al. (2005) analysis, which uses observationally based meteorology and estimates OH on monthly time scales, yields interannual OH variations that agree very well with the Prinn et al. (2001) and equivalent Krol and Lelieveld (2003) estimates (see Figure 2.8). However, when Bousquet et al. (2005) estimated both OH concentrations and methyl chloroform emissions (constrained by their uncertainties as reported by McCulloch and Midgley, 2001), the OH variations were reduced by 65% (dashed line in Figure 2.8). The error bars on the Prinn et al. (2001, 2005a)

OH estimates, which account for these emission uncertainties using Monte Carlo ensembles of inversions, also easily allow such a reduction in OH variability (thin vertical bars in Figure 2.8). This implies that these interannual OH variations are real, but only their phasing and not their amplitude, is well defined. Bousquet et al. (2005) also deduced that OH in the SH shows a zero to small negative trend, in qualitative agreement with Prinn et al. (2001). Short-term variations in OH were also recently deduced by Manning et al. (2005) using 13 years of ^{14}CO measurements in New Zealand and Antarctica. They found no significant long-term trend between 1989 and 2003 in SH OH but provided evidence for recurring multi-month OH variations of around 10%. They also deduced even larger (20%) OH decreases in 1991 and 1997, perhaps triggered by the 1991 Mt. Pinatubo eruption and the 1997 Indonesian fires. The similarity

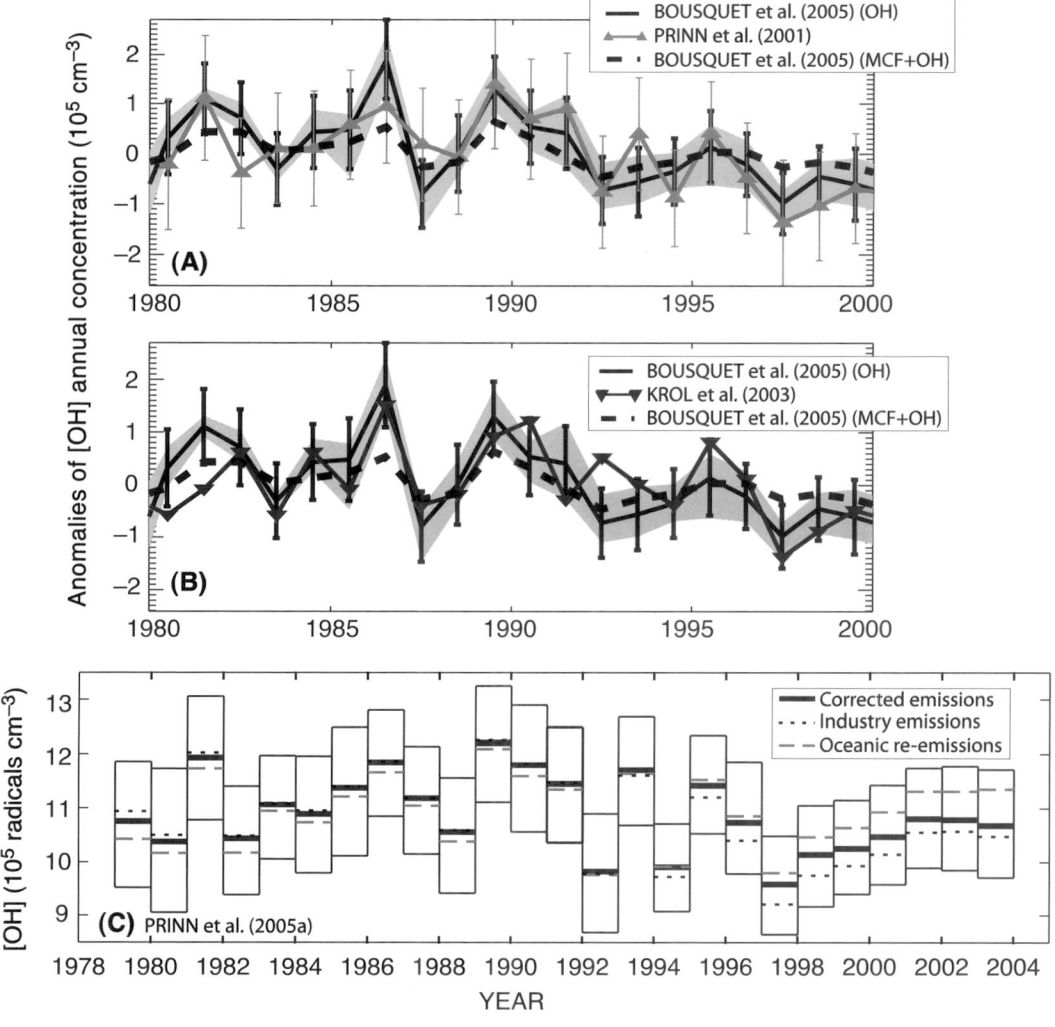

Figure 2.8. *Estimates used to evaluate trends in weighted global average OH concentrations. (A) and (B): comparison of 1980 to 1999 OH anomalies (relative to their long-term means) inferred by Bousquet et al. (2005), Prinn et al. (2001) and Krol et al. (2003) from AGAGE methyl chloroform observations, and by Bousquet et al. (2005) when methyl chloroform emissions as well as OH are inferred; error bars for Bousquet et al. (2005) refer to 1 standard deviation inversion errors while yellow areas refer to the envelope of their 18 OH inversions. (C) OH concentrations for 1979 to 2003 inferred by Prinn et al. (2005a) (utilising industry emissions corrected using recent methyl chloroform observations), showing the recovery of 2003 OH levels to 1979 levels; also shown are results assuming uncorrected emissions and estimates of recent oceanic re-emissions. Error bars in Prinn et al. (2001, 2005a) are 1 standard deviation and include inversion, model, emission and calibration errors from large Monte Carlo ensembles (see Section 2.3.5 for details and references).*

of many of these results to those from methyl chloroform discussed above is very important, given the independence of the two approaches.

RF calculations of the LLGHGs are calculated from observed trends in the LLGHG concentrations and therefore OH concentrations do not directly affect them. Nevertheless OH trends are needed to quantify LLGHG budgets (Section 7.4) and for understanding future trends in the LLGHGs and tropospheric ozone.

2.3.6 Ozone

In the TAR, separate estimates for RF due to changes in tropospheric and stratospheric ozone were given. Stratospheric ozone RF was derived from observations of ozone change from roughly 1979 to 1998. Tropospheric ozone RF was based on chemical model results employing changes in precursor hydrocarbons, CO and nitrogen oxides (NO_x). Over the satellite era (since approximately 1980), stratospheric ozone trends have been primarily caused by the Montreal Protocol gases, and in Ramaswamy et al. (2001) the stratospheric ozone RF was implicitly attributed to these gases. Studies since then have investigated a number of possible causes of ozone change in the stratosphere and troposphere and the attribution of ozone trends to a given precursor is less clear. Nevertheless, stratospheric ozone and tropospheric ozone RFs are still treated separately in this report. However, the RFs are more associated with the vertical location of the ozone change than they are with the agent(s) responsible for the change.

2.3.6.1 Stratospheric Ozone

The TAR reported that ozone depletion in the stratosphere had caused a negative RF of –0.15 W m^{-2} as a best estimate over the period since 1750. A number of recent reports have assessed changes in stratospheric ozone and the research into its causes, including Chapters 3 and 4 of the 2002 Scientific Assessment of Ozone Depletion (WMO, 2003) and Chapter 1 of IPCC/TEAP (2005). This section summarises the material from these reports and updates the key results using more recent research.

Global ozone amounts decreased between the late 1970s and early 1990s, with the lowest values occurring during 1992 to 1993 (roughly 6% below the 1964 to 1980 average), and slightly increasing values thereafter. Global ozone for the period 2000 to 2003 was approximately 4% below the 1964 to 1980 average values. Whether or not recently observed changes in ozone trends (Newchurch et al., 2003; Weatherhead and Andersen, 2006) are already indicative of recovery of the global ozone layer is not yet clear and requires more detailed attribution of the drivers of the changes (Steinbrecht et al., 2004a (see also comment and reply: Cunnold et al., 2004 and Steinbrecht et al., 2004b); Hadjinicolaou et al., 2005; Krizan and Lastovicka, 2005; Weatherhead and Andersen, 2006). The largest ozone changes since 1980 have occurred during the late winter and spring over Antarctica where average total column ozone in September and October is about 40 to 50% below pre-

1980 values (WMO, 2003). Ozone decreases over the Arctic have been less severe than have those over the Antarctic, due to higher temperature in the lower stratosphere and thus fewer polar stratospheric clouds to cause the chemical destruction. Arctic stratospheric ozone levels are more variable due to interannual variability in chemical loss and transport.

The temporally and seasonally non-uniform nature of stratospheric ozone trends has important implications for the resulting RF. Global ozone decreases result primarily from changes in the lower stratospheric extratropics. Total column ozone changes over the mid-latitudes of the SH are significantly larger than over the mid-latitudes of the NH. Averaged over the period 2000 to 2003, SH values are 6% below pre-1980 values, while NH values are 3% lower. There is also significant seasonality in the NH ozone changes, with 4% decreases in winter to spring and 2% decreases in summer, while long-term SH changes are roughly 6% year round (WMO, 2003). Southern Hemisphere mid-latitude ozone shows significant decreases during the mid-1980s and essentially no response to the effects of the Mt. Pinatubo volcanic eruption in June 1991; both of these features remain unexplained. Pyle et al. (2005) and Chipperfield et al. (2003) assessed several studies that show that a substantial fraction (roughly 30%) of NH mid-latitude ozone trends are not directly attributable to anthropogenic chemistry, but are related to dynamical effects, such as tropopause height changes. These dynamical effects are likely to have contributed a larger fraction of the ozone RF in the NH mid-latitudes. The only study to assess this found that 50% of the RF related to stratospheric ozone changes between 20°N to 60°N over the period 1970 to 1997 is attributable to dynamics (Forster and Tourpali, 2001). These dynamical changes may well have an anthropogenic origin and could even be partly caused by stratospheric ozone changes themselves through lower stratospheric temperature changes (Chipperfield et al., 2003; Santer et al., 2004), but are not directly related to chemical ozone loss.

At the time of writing, no study has utilised ozone trend observations after 1998 to update the RF values presented in Ramaswamy et al. (2001). However, Hansen et al. (2005) repeated the RF calculation based on the same trend data set employed by studies assessed in Ramaswamy et al. (2001) and found an RF of roughly –0.06 W m^{-2}. A considerably stronger RF of –0.2 ± 0.1 W m^{-2} previously estimated by the same group affected the Ramaswamy et al. (2001) assessment. The two other studies assessed in Ramaswamy et al. (2001), using similar trend data sets, found RFs of –0.01 W m^{-2} and –0.10 W m^{-2}. Using the three estimates gives a revision of the observationally based RF for 1979 to 1998 to about –0.05 ± 0.05 W m^{-2}.

Gauss et al. (2006) compared results from six chemical transport models that included changes in ozone precursors to simulate both the increase in the ozone in the troposphere and the ozone reduction in the stratosphere over the industrial era. The 1850 to 2000 annually averaged global mean stratospheric ozone column reduction for these models ranged between 14 and 29 Dobson units (DU). The overall pattern of the ozone changes from the models were similar but the magnitude of the ozone changes differed. The models showed a reduction in the ozone at high latitudes, ranging from around 20 to 40% in the SH and smaller changes in the NH. All models have a maximum ozone reduction around 15 km at high latitudes in the SH. Differences between the models were also found in the tropics, with some models simulating about a 10% increase in the lower stratosphere and other models simulating decreases. These differences were especially related to the altitude where the ozone trend switched from an increase in the troposphere to a decrease in the stratosphere, which ranged from close to the tropopause to around 27 km. Several studies have shown that ozone changes in the tropical lower stratosphere are very important for the magnitude and sign of the ozone RF (Ramaswamy et al., 2001). The resulting stratospheric ozone RF ranged between –0.12 and +0.07 W m^{-2}. Note that the models with either a small negative or a positive RF also had a small increase in tropical lower stratospheric ozone, resulting from increases in tropospheric ozone precursors; most of this increase would have occurred before the time of stratospheric ozone destruction by the Montreal Protocol gases. These RF calculations also did not include any negative RF that may have resulted from stratospheric water vapour increases. It has been suggested (Shindell and Faluvegi, 2002) that stratospheric ozone during 1957 to 1975 was lower by about 7 DU relative to the first half of the 20th century as a result of possible stratospheric water vapour increases; however, these long-term increases in stratospheric water vapour are uncertain (see Sections 2.3.7 and 3.4).

The stratospheric ozone RF is assessed to be –0.05 ± 0.10 W m^{-2} between pre-industrial times and 2005. The best estimate is from the observationally based 1979 to 1998 RF of –0.05 ± 0.05 W m^{-2}, with the uncertainty range increased to take into account ozone change prior to 1979, using the model results of Gauss et al. (2006) as a guide. Note that this estimate takes into account causes of stratospheric ozone change in addition to those due to the Montreal Protocol gases. The level of scientific understanding is medium, unchanged from the TAR (see Section 2.9, Table 2.11).

2.3.6.2 Tropospheric Ozone

The TAR identified large regional differences in observed trends in tropospheric ozone from ozonesondes and surface observations. The TAR estimate of RF from tropospheric ozone was +0.35 ± 0.15 W m^{-2}. Due to limited spatial and temporal coverage of observations of tropospheric ozone, the RF estimate is based on model simulations. In the TAR, the models considered only changes in the tropospheric photochemical system, driven by estimated emission changes (NO$_x$, CO, non-methane volatile organic compounds (NMVOCs), and CH$_4$) since pre-industrial times. Since the TAR, there have been major improvements in models. The new generation models include several Chemical Transport Models (CTMs) that couple stratospheric and tropospheric chemistry, as well as GCMs with on-line chemistry (both tropospheric and stratospheric). While the TAR simulations did not consider changes in ozone within the troposphere caused by reduced influx of ozone from

the stratosphere (due to ozone depletion in the stratosphere), the new models include this process (Gauss et al., 2006). This advancement in modelling capabilities and the need to be consistent with how the RF due to changes in stratospheric ozone is derived (based on observed ozone changes) have led to a change in the definition of RF due to tropospheric ozone compared with that in the TAR. Changes in tropospheric ozone due to changes in transport of ozone across the tropopause, which are in turn caused by changes in stratospheric ozone, are now included.

Trends in anthropogenic emissions of ozone precursors for the period 1990 to 2000 have been compiled by the Emission Database for Global Atmospheric Research (EDGAR) consortium (Olivier and Berdowski, 2001 updated). For specific regions, there is significant variability over the period due to variations in the emissions from open biomass burning sources. For all components (NO$_x$, CO and volatile organic compounds (VOCs)) industrialised regions like the USA and Organisation for Economic Co-operation and Development (OECD) Europe show reductions in emissions, while regions dominated by developing countries show significant growth in emissions. Recently, the tropospheric burdens of CO and NO$_2$ were estimated from satellite observations (Edwards et al., 2004; Richter et al., 2005), providing much needed data for model evaluation and very valuable constraints for emission estimates.

Assessment of long-term trends in tropospheric ozone is difficult due to the scarcity of representative observing sites with long records. The long-term tropospheric ozone trends vary both in terms of sign and magnitude and in the possible causes for the change (Oltmans et al., 2006). Trends in tropospheric ozone at northern middle and high latitudes have been estimated based on ozonesonde data by WMO (2003), Naja et al. (2003), Naja and Akimoto (2004), Tarasick et al. (2005) and Oltmans et al. (2006). Over Europe, ozone in the free troposphere increased from the early 20th century until the late 1980s; since then the trend has levelled off or been slightly negative. Naja and Akimoto (2004) analysed 33 years of ozonesonde data from Japanese stations, and showed an increase in ozone in the lower troposphere (750–550 hPa) between the periods 1970 to 1985 and 1986 to 2002 of 12 to 15% at Sapporo and Tsukuba (43°N and 36°N) and 35% at Kagoshima (32°N). Trajectory analysis indicates that the more southerly station, Kagoshima, is significantly more influenced by air originating over China, while Sapporo and Tsukuba are more influenced by air from Eurasia. At Naha (26°N) a positive trend (5% per decade) is found between 700 and 300 hPa (1990–2004), while between the surface and 700 hPa a slightly negative trend is observed (Oltmans et al., 2006). Ozonesondes from Canadian stations show negative trends in tropospheric ozone between 1980 and 1990, and a rebound with positive trends during 1991 to 2001 (Tarasick et al., 2005). Analysis of stratosphere-troposphere exchange processes indicates that the rebound during the 1990s may be partly a result of small changes in atmospheric circulation.

Trends are also derived from surface observations. Jaffe et al. (2003) derived a positive trend of 1.4% yr^{-1} between 1988 and 2003 using measurements from Lassen Volcanic Park in California (1,750 m above sea level), consistent with the trend derived by comparing two aircraft campaigns (Parrish et al., 2004). However, a number of other sites show insignificant changes over the USA over the last 15 years (Oltmans et al., 2006). Over Europe and North America, observations from Whiteface Mountain, Wallops Island, Hohenpeisenberg, Zugspitze and Mace Head (flow from the European sector) show small trends or reductions during summer, while there is an increase during winter (Oltmans et al., 2006). These observations are consistent with reduced NO_x emissions (Jonson et al., 2005). North Atlantic stations (Mace Head, Izana and Bermuda) indicate increased ozone (Oltmans et al., 2006). Over the North Atlantic (40°N–60°N) measurements from ships (Lelieveld et al., 2004) show insignificant trends in ozone, however, at Mace Head a positive trend of 0.49 ± 0.19 ppb yr^{-1} for the period 1987 to 2003 is found, with the largest contribution from air coming from the Atlantic sector (Simmonds et al., 2004).

In the tropics, very few long-term ozonesonde measurements are available. At Irene in South Africa (26°S), Diab et al. (2004) found an increase between the 1990 to 1994 and 1998 to 2002 periods of about 10 ppb close to the surface (except in summer) and in the upper troposphere during winter. Thompson et al. (2001) found no significant trend during 1979 to 1992, based on Total Ozone Mapping Spectrometer (TOMS) satellite data. More recent observations (1994 to 2003, *in situ* data from the Measurement of Ozone by Airbus In-service Aircraft (MOZAIC) program) show significant trends in free-tropospheric ozone (7.7 to 11.3 km altitude) in the tropics: 1.12 ± 0.05 ppb yr^{-1} and 1.03 ± 0.08 ppb yr^{-1} in the NH tropics and SH tropics, respectively (Bortz and Prather, 2006). Ozonesonde measurements over the southwest Pacific indicate an increased frequency of near-zero ozone in the upper troposphere, suggesting a link to an increased frequency of deep convection there since the 1980s (Solomon et al., 2005).

At southern mid-latitudes, surface observations from Cape Point, Cape Grim, the Atlantic Ocean (from ship) and from sondes at Lauder (850–700 hPa) show positive trends in ozone concentrations, in particular during the biomass burning season in the SH (Oltmans et al., 2006). However, the trend is not accompanied by a similar trend in CO, as expected if biomass burning had increased. The increase is largest at Cape Point, reaching 20% per decade (in September). At Lauder, the increase is confined to the lower troposphere.

Changes in tropospheric ozone and the corresponding RF have been estimated in a number of recent model studies (Hauglustaine

and Brasseur, 2001; Mickley et al., 2001; Shindell et al., 2003a; Mickley et al., 2004; Wong et al., 2004; Liao and Seinfeld, 2005; Shindell et al., 2005). In addition, a multi-model experiment including 10 global models was organised through the Atmospheric Composition Change: an European Network (ACCENT; Gauss et al., 2006). Four of the ten ACCENT models have detailed stratospheric chemistry. The adjusted RF for all models was calculated by the same radiative transfer model. The normalised adjusted RF for the ACCENT models was $+0.032 \pm 0.006$ W m^{-2} DU^{-1}, which is significantly lower than the TAR estimate of $+0.042$ W m^{-2} DU^{-1}.

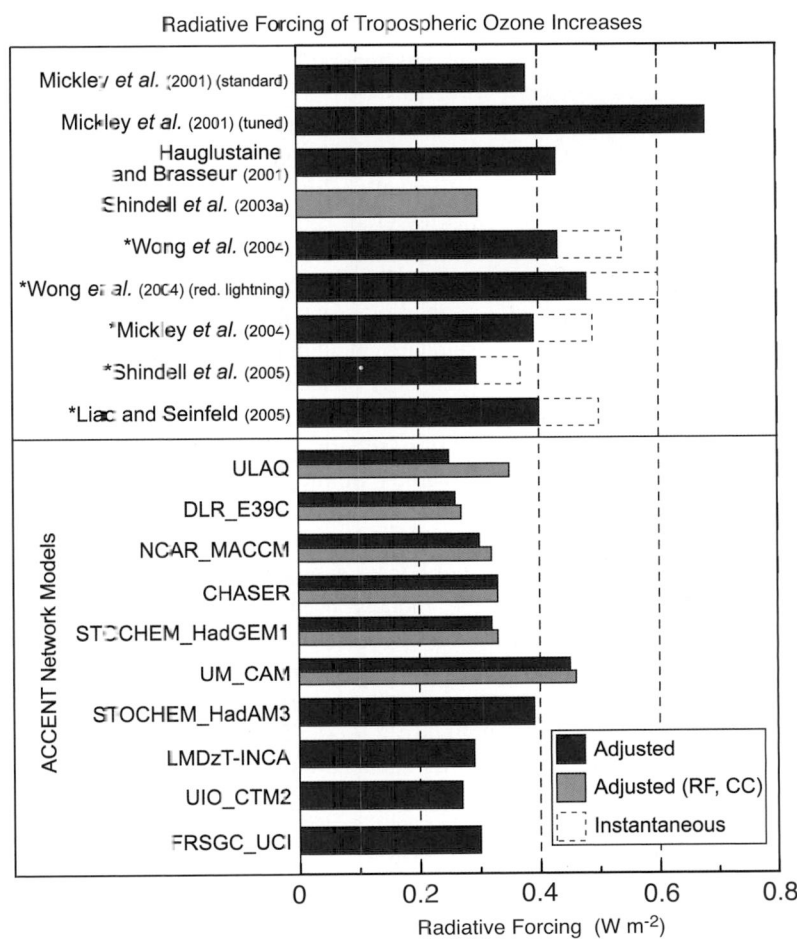

Figure 2.9. *Calculated RF due to tropospheric ozone change since pre-industrial time based on CTM and GCM model simulations published since the TAR. Estimates with GCMs including the effect of climate change since 1750 are given by orange bars (Adjusted RF, CC). Studies denoted with an (*) give only the instantaneous RF in the original publications. Stratospheric-adjusted RFs for these are estimated by reducing the instantaneous RF (indicated by the dashed bars) by 20%. The instantaneous RF from Mickley et al. (2001) is reported as an adjusted RF in Gauss et al. (2006). ACCENT models include ULAQ: University of L'Aquila; DLR_E39C: Deutsches Zentrum für Luft- und Raumfahrt European Centre Hamburg Model; NCAR_MACCM: National Center for Atmospheric Research Middle Atmosphere Community Climate Model; CHASER: Chemical Atmospheric GCM for Study of Atmospheric Environment and Radiative Forcing; STOCHEM_HadGEM1: United Kingdom Meteorological Office global atmospheric chemistry model /Hadley Centre Global Environmental Model 1; UM_CAM: United Kingdom Meteorological Office Unified Model GCM with Cambridge University chemistry; STOCHEM_HadAM3: United Kingdom Meteorological Office global atmospheric chemistry model/ Hadley Centre Atmospheric Model; LMDzT-INCA: Laboratoire de Météorologie Dynamique GCM-INteraction with Chemistry and Aerosols; UIO_CTM2: University of Oslo CTM; FRSGC_UCI: Frontier Research System for Global Change/University of California at Irvine CTM.*

The simulated RFs for tropospheric ozone increases since 1750 are shown in Figure 2.9. Most of the calculations used the same set of assumptions about pre-industrial emissions (zero anthropogenic emissions and biomass burning sources reduced by 90%). Emissions of NO_x from soils and biogenic hydrocarbons were generally assumed to be natural and were thus not changed (see, e.g., Section 7.4). In one study (Hauglustaine and Brasseur, 2001), pre-industrial NO_x emissions from soils were reduced based on changes in the use of fertilizers. Six of the ACCENT models also made coupled climate-chemistry simulations including climate change since pre-industrial times. The difference between the RFs in the coupled climate-chemistry and the chemistry-only simulations, which indicate the possible climate feedback to tropospheric ozone, was positive in all models but generally small (Figure 2.9).

A general feature of the models is their inability to reproduce the low ozone concentrations indicated by the very uncertain semi-quantitative observations (e.g., Pavelin et al., 1999) during the late 19th century. Mickley et al. (2001) tuned their model by reducing pre-industrial lightning and soil sources of NO_x and increasing natural NMVOC emissions to obtain close agreement with the observations. The ozone RF then increased by 50 to 80% compared to their standard calculations. However, there are still several aspects of the early observations that the tuned model did not capture.

The best estimate for the RF of tropospheric ozone increases is $+0.35$ W m^{-2}, taken as the median of the RF values in Figure 2.9 (adjusted and non-climate change values only, i.e., the red bars). The best estimate is unchanged from the TAR. The uncertainties in the estimated RF by tropospheric ozone originate from two factors: the models used (CTM/GCM model formulation, radiative transfer models), and the potential overestimation of pre-industrial ozone levels in the models. The 5 to 95% confidence interval, assumed to be represented by the range of the results in Figure 2.9, is $+0.25$ to $+0.65$ W m^{-2}. A medium level of scientific understanding is adopted, also unchanged from the TAR (see Section 2.9, Table 2.11).

2.3.7 Stratospheric Water Vapour

The TAR noted that several studies had indicated long-term increases in stratospheric water vapour and acknowledged that these trends would contribute a significant radiative impact. However, it only considered the stratospheric water vapour increase expected from CH_4 increases as an RF, and this was estimated to contribute 2 to 5% of the total CH_4 RF (about $+0.02$ W m^{-2}).

Section 3.4 discusses the evidence for stratospheric water vapour trends and presents the current understanding of their possible causes. There are now 14 years of global stratospheric water vapour measurements from Halogen Occultation Experiment (HALOE) and continued balloon-based measurements (since 1980) at Boulder, Colorado. There is some evidence of a sustained long-term increase in stratospheric water vapour of around 0.05 ppm yr^{-1} from 1980 until roughly 2000, since then water vapour concentrations in the lower stratosphere have been decreasing (see Section 3.4 for details and references). As well as CH_4 increases, several other indirect forcing mechanisms have been proposed, including: a) volcanic eruptions (Considine et al., 2001; Joshi and Shine, 2003); b) biomass burning aerosol (Sherwood, 2002); c) tropospheric (SO_2; Notholt et al., 2005) and d) changes in CH_4 oxidation rates from changes in stratospheric chlorine, ozone and OH (Rockmann et al., 2004). These are mechanisms that can be linked to an external forcing agent. Other proposed mechanisms are more associated with climate feedbacks and are related to changes in tropopause temperatures or circulation (Stuber et al., 2001a; Fueglistaler et al., 2004). From these studies, there is little quantification of the stratospheric water vapour change attributable to different causes. It is also likely that different mechanisms are affecting water vapour trends at different altitudes.

Since the TAR, several further calculations of the radiative balance change due to changes in stratospheric water vapour have been performed (Forster and Shine, 1999; Oinas et al., 2001; Shindell, 2001; Smith et al., 2001; Forster and Shine, 2002). Smith et al. (2001) estimated a $+0.12$ to $+0.2$ W m^{-2} per decade range for the RF from the change in stratospheric water vapour, using HALOE satellite data. Shindell (2001) estimated an RF of about $+0.2$ W m^{-2} in a period of two decades, using a GCM to estimate the increase in water vapour in the stratosphere from oxidation of CH_4 and including climate feedback changes associated with an increase in greenhouse gases. Forster and Shine (2002) used a constant 0.05 ppm yr^{-1} trend in water vapour at pressures of 100 to 10 hPa and estimated the RF to be $+0.29$ W m^{-2} for 1980 to 2000. GCM radiation codes can have a factor of two uncertainty in their modelling of this RF (Oinas et al., 2001). For the purposes of this chapter, the above RF estimates are not readily attributable to forcing agent(s) and uncertainty as to the causes of the observed change precludes all but the component due to CH_4 increases being considered a forcing. Two related CTM studies have calculated the RF associated with increases in CH_4 since pre-industrial times (Hansen and Sato, 2001; Hansen et al., 2005), but no dynamical feedbacks were included in those estimates. Hansen et al. (2005) estimated an RF of $+0.07 \pm 0.01$ W m^{-2} for the stratospheric water vapour changes over 1750 to 2000, which is at least a factor of three larger than the TAR value. The RF from direct injection of water vapour by aircraft is believed to be an order of magnitude smaller than this, at about $+0.002$ W m^{-2} (IPCC, 1999). There has been little trend in CH_4 concentration since 2000 (see Section 2.3.2); therefore the best estimate of the stratospheric water vapour RF from CH_4 oxidation ($+0.07$ W m^{-2}) is based on the Hansen et al. (2005) calculation. The 90% confidence range is estimated as ±0.05 W m^{-2}, from the range of the RF studies that included other effects. There is a low level of scientific understanding in this estimate, as there is only a partial understanding of the vertical profile of CH_4-induced stratospheric water vapour change (Section 2.9, Table 2.11). Other human causes of stratospheric water vapour change are unquantified and have a very low level of scientific understanding.

2.3.8 Observations of Long-Lived Greenhouse Gas Radiative Effects

Observations of the clear-sky radiation emerging at the top of the atmosphere and at the surface have been conducted. Such observations, by their nature, do not measure RF as defined here. Instead, they yield a perspective on the influence of various species on the transfer of radiation in the atmosphere. Most importantly, the conditions involved with these observations involve varying thermal and moisture profiles in the atmosphere such that they do not conform to the conditions underlying the RF definition (see Section 2.2). There is a more comprehensive discussion of observations of the Earth's radiative balance in Section 3.4.

Harries et al. (2001) analysed spectra of the outgoing longwave radiation as measured by two satellites in 1970 and 1997 over the tropical Pacific Ocean. The reduced brightness temperature observed in the spectral regions of many of the greenhouse gases is experimental evidence for an increase in the Earth's greenhouse effect. In particular, the spectral signatures were large for CO_2 and CH_4. The halocarbons, with their large change between 1970 and 1997, also had an impact on the brightness temperature. Philipona et al. (2004) found an increase in the measured longwave downward radiation at the surface over the period from 1995 to 2002 at eight stations over the central Alps. A significant increase in the clear-sky longwave downward flux was found to be due to an enhanced greenhouse effect after combining the measurements with model calculations to estimate the contribution from increases in temperature and humidity. While both types of observations attest to the radiative influences of the gases, they should not be interpreted as having a direct linkage to the value of RFs in Section 2.3.

2.4 Aerosols

2.4.1 Introduction and Summary of the Third Assessment Report

The TAR categorised aerosol RFs into direct and indirect effects. The direct effect is the mechanism by which aerosols scatter and absorb shortwave and longwave radiation, thereby altering the radiative balance of the Earth-atmosphere system. Sulphate, fossil fuel organic carbon, fossil fuel black carbon, biomass burning and mineral dust aerosols were all identified as having a significant anthropogenic component and exerting a significant direct RF. Key parameters for determining the direct RF are the aerosol optical properties (the single scattering albedo, ω_0, specific extinction coefficient, k_e and the scattering phase function), which vary as a function of wavelength and relative humidity, and the atmospheric loading and geographic distribution of the aerosols in the horizontal and vertical, which vary as a function of time (e.g., Haywood

and Boucher, 2000; Penner et al., 2001; Ramaswamy et al., 2001). Scattering aerosols exert a net negative direct RF, while partially absorbing aerosols may exert a negative top-of-the-atmosphere (TOA) direct RF over dark surfaces such as oceans or dark forest surfaces, and a positive TOA RF over bright surfaces such as desert, snow and ice, or if the aerosol is above cloud (e.g., Chylek and Wong, 1995; Haywood and Shine, 1995). Both positive and negative TOA direct RF mechanisms reduce the shortwave irradiance at the surface. The longwave direct RF is only substantial if the aerosol particles are large and occur in considerable concentrations at higher altitudes (e.g., Tegen et al., 1996). The direct RF due to tropospheric aerosols is most frequently derived at TOA rather than at the tropopause because shortwave radiative transfer calculations have shown a negligible difference between the two (e.g., Haywood and Shine, 1997; Section 2.2). The surface forcing will be approximately the same as the direct RF at the TOA for scattering aerosols, but for partially absorbing aerosols the surface forcing may be many times stronger than the TOA direct RF (e.g., Ramanathan et al., 2001b and references therein).

The indirect effect is the mechanism by which aerosols modify the microphysical and hence the radiative properties, amount and lifetime of clouds (Figure 2.10). Key parameters for determining the indirect effect are the effectiveness of an aerosol particle to act as a cloud condensation nucleus, which is a function of the size, chemical composition, mixing state and ambient environment (e.g., Penner et al., 2001). The microphysically induced effect on the cloud droplet number concentration and hence the cloud droplet size, with the liquid water content held fixed has been called the 'first indirect effect' (e.g., Ramaswamy et al., 2001), the 'cloud albedo effect' (e.g., Lohmann and Feichter, 2005), or the 'Twomey effect' (e.g., Twomey, 1977). The microphysically induced effect on the liquid water content, cloud height, and lifetime of clouds has been called the 'second indirect effect' (e.g., Ramaswamy et al., 2001), the 'cloud lifetime effect' (e.g., Lohmann and Feichter, 2005) or the 'Albrecht effect' (e.g., Albrecht, 1989). The TAR split the indirect effect into the first indirect effect, and the second indirect effect. Throughout this report, these effects are denoted as 'cloud albedo effect' and 'cloud lifetime effect', respectively, as these terms are more descriptive of the microphysical processes that occur. The cloud albedo effect was considered in the TAR to be an RF because global model calculations could be performed to describe the influence of increased aerosol concentration on the cloud optical properties while holding the liquid water content of the cloud fixed (i.e., in an entirely diagnostic manner where feedback mechanisms do not occur). The TAR considered the cloud albedo effect to be a key uncertainty in the RF of climate but did not assign a best estimate of the RF, and showed a range of RF between 0 and –2 W m^{-2} in the context of liquid water clouds. The other indirect effects were not considered to be RFs because, in suppressing drizzle, increasing the cloud height or the cloud lifetime in atmospheric models (Figure 2.10), the hydrological cycle is invariably altered (i.e., feedbacks occur; see Section

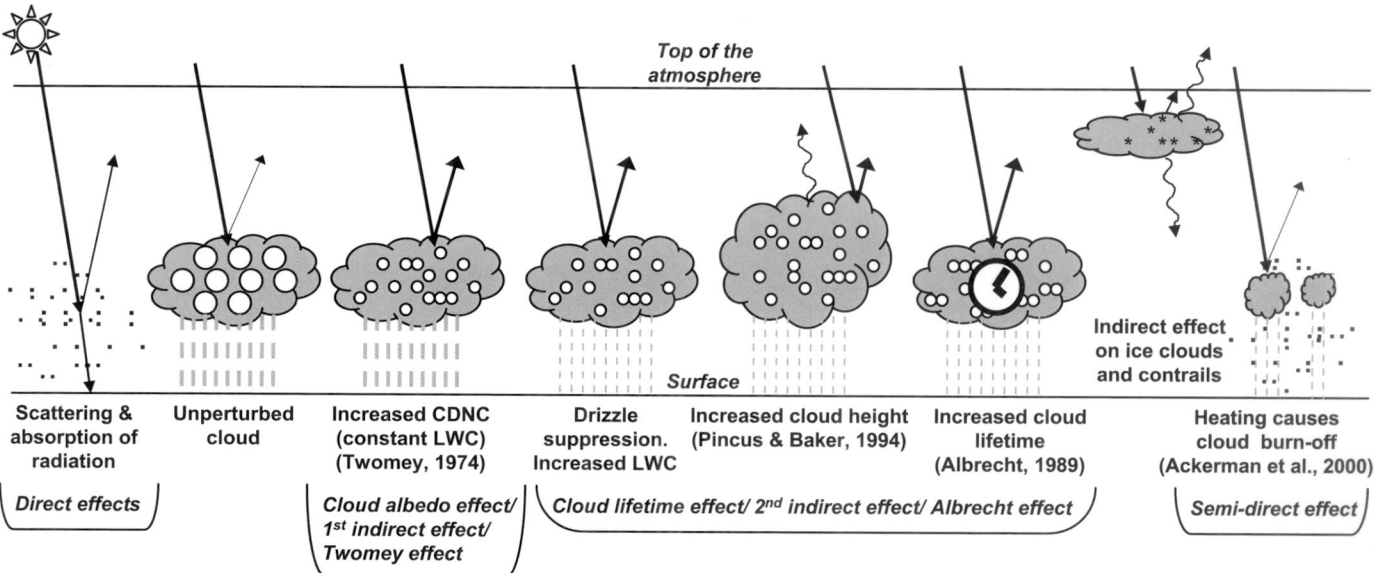

Figure 2.10. *Schematic diagram showing the various radiative mechanisms associated with cloud effects that have been identified as significant in relation to aerosols (modified from Haywood and Boucher, 2000). The small black dots represent aerosol particles; the larger open circles cloud droplets. Straight lines represent the incident and reflected solar radiation, and wavy lines represent terrestrial radiation. The filled white circles indicate cloud droplet number concentration (CDNC). The unperturbed cloud contains larger cloud drops as only natural aerosols are available as cloud condensation nuclei, while the perturbed cloud contains a greater number of smaller cloud drops as both natural and anthropogenic aerosols are available as cloud condensation nuclei (CCN). The vertical grey dashes represent rainfall, and LWC refers to the liquid water content.*

7.5). The TAR also discussed the impact of anthropogenic aerosols on the formation and modification of the physical and radiative properties of ice clouds (Penner et al., 2001), although quantification of an RF from this mechanism was not considered appropriate given the host of uncertainties and unknowns surrounding ice cloud nucleation and physics.

The TAR did not include any assessment of the semi-direct effect (e.g., Hansen et al., 1997; Ackerman et al., 2000a; Jacobson, 2002; Menon et al., 2003; Cook and Highwood, 2004; Johnson et al., 2004), which is the mechanism by which absorption of shortwave radiation by tropospheric aerosols leads to heating of the troposphere that in turn changes the relative humidity and the stability of the troposphere and thereby influences cloud formation and lifetime. In this report, the semi-direct effect is not strictly considered an RF because of modifications to the hydrological cycle, as discussed in Section 7.5 (see also Sections 2.2, 2.8 and 2.4.5).

Since the TAR, there have been substantial developments in observations and modelling of tropospheric aerosols; these are discussed in turn in the following sections.

2.4.2 Developments Related to Aerosol Observations

Surface-based measurements of aerosol properties such as size distribution, chemical composition, scattering and absorption continue to be performed at a number of sites, either at long-term monitoring sites, or specifically as part of intensive field campaigns. These *in situ* measurements provide essential validation for global models, for example, by constraining aerosol concentrations at the surface and by providing high-

quality information about chemical composition and local trends. In addition, they provide key information about variability on various time scales. Comparisons of *in situ* measurements against those from global atmospheric models are complicated by differences in meteorological conditions and because *in situ* measurements are representative of conditions mostly at or near the surface while the direct and indirect RFs depend on the aerosol vertical profile. For example, the spatial resolution of global model grid boxes is typically a few degrees of latitude and longitude and the time steps for the atmospheric dynamics and radiation calculations may be minutes to hours depending on the process to be studied; this poses limitations when comparing with observations conducted over smaller spatial extent and shorter time duration.

Combinations of satellite and surface-based observations provide near-global retrievals of aerosol properties. These are discussed in this subsection; the emissions estimates, trends and *in situ* measurements of the physical and optical properties are discussed with respect to their influence on RF in Section 2.4.4. Further detailed discussions of the recent satellite observations of aerosol properties and a satellite-measurement based assessment of the aerosol direct RF are given by Yu et al. (2006).

2.4.2.1   Satellite Retrievals

Satellite retrievals of aerosol optical depth in cloud-free regions have improved via new generation sensors (Kaufman et al., 2002) and an expanded global validation program (Holben et al., 2001). Advanced aerosol retrieval products such as aerosol fine-mode fraction and effective particle radius have been developed and offer potential for improving estimates of the aerosol direct radiative effect. Additionally, efforts have been made to determine the anthropogenic component of aerosol and associated direct RF, as discussed by Kaufman et al. (2002) and implemented by Bellouin et al. (2005) and Chung et al. (2005). However, validation programs for these advanced products have yet to be developed and initial assessments indicate some systematic errors (Levy et al., 2003; Anderson et al., 2005a; Chu et al., 2005), suggesting that the routine differentiation between natural and anthropogenic aerosols from satellite retrievals remains very challenging.

2.4.2.1.1   *Satellite retrievals of aerosol optical depth*

Figure 2.11 shows an example of aerosol optical depth τ_{aer} (mid-visible wavelength) retrieved over both land and ocean, together with geographical positions of aerosol instrumentation.

Table 2.2 provides a summary of aerosol data currently available from satellite instrumentation, together with acronyms for the instruments. τ_{aer} from the Moderate Resolution Imaging Spectrometer (MODIS) instrument for the January to March 2001 average (Figure 2.11, top panel) clearly differs from that for the August to October 2001 average (Figure 2.11, bottom panel) (Kaufman et al., 1997; Tanré et al., 1997). Seasonal variability in τ_{aer} can be seen; biomass burning aerosol is most strongly evident over the Gulf of Guinea in Figure 2.11 (top panel) but shifts to southern Africa in Figure 2.11 (bottom panel). Likewise, the biomass burning in South America is most evident in Figure 2.11 (bottom panel). In Figure 2.11 (top panel), transport of mineral dust from Africa to South America is discernible while in Figure 2.11 (bottom panel) mineral dust is transported over the West Indies and Central America. Industrial aerosol, which consists of a mixture of sulphates, organic and black carbon,

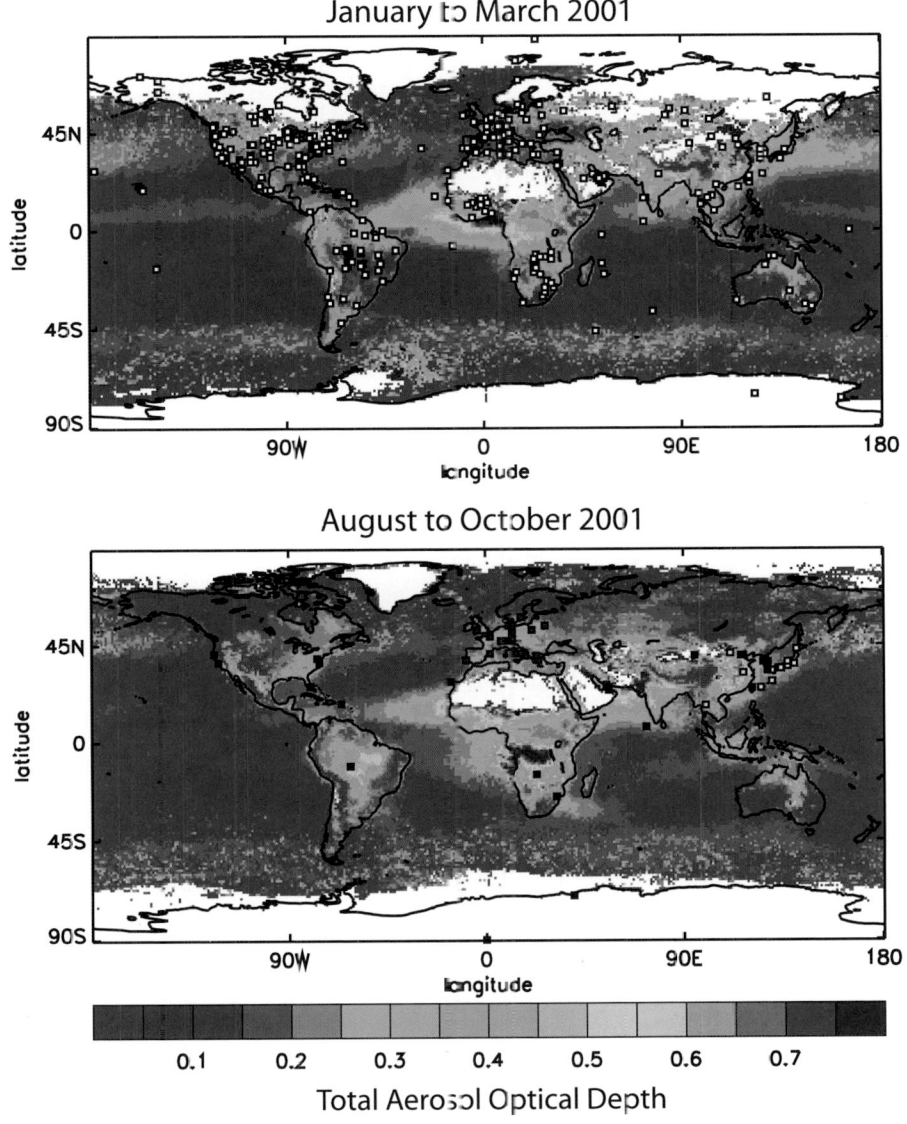

Figure 2.11. *Aerosol optical depth, τ_{aer}, at 0.55 μm (colour bar) as determined by the MODIS instrument for the January to March 2001 mean (top panel) and for the August to October 2001 mean (bottom panel). The top panel also shows the location of AERONET sites (white squares) that have been operated (not necessary continuously) since 1996. The bottom panel also shows the location of different aerosol lidar networks (red: EARLINET, orange: ADNET, black: MPLNET).*

Table 2.2. *Periods of operation, spectral bands and products available from various different satellite sensors that have been used to retrieve aerosol properties.*

Satellite Instrument	Period of Operation	Spectral Bands	Products[a]	Comment and Reference
AVHRR (Advanced Very High Resolution Radiometer) α	1979 to present	5 bands (0.63, 0.87, 3.7, 10.5 and 11.5 μm)	τ_{aer}, α	1-channel retrieval gives $\tau_{\lambda=0.63}$ over ocean (Husar et al., 1997; Ignatov and Stowe, 2002). 2-channel using 0.63 μm and 0.86 μm gives $\tau_{\lambda=0.55}$ and α over ocean assuming mono-modal aerosol size distribution (Mishchenko et al., 1999) 2-channel using 0.63 μm and 0.86 μm gives $\tau_{\lambda=0.55}$ and α over dark forests and lake surfaces (Soufflet et al., 1997) 2-channel using 0.64 μm and 0.83 μm gives $\tau_{\lambda=0.55}$ and α over ocean assuming a bimodal aerosol size distribution (Higurashi and Nakajima, 1999; Higurashi et al., 2000)
TOMS[b] (Total Ozone Mapping Spectrometer)	1979 to present	0.33 μm, 0.36 μm	Aerosol Index, τ_{aer}	Aerosol index to τ_{aer} conversion sensitive to the altitude of the 8 mono-modal aerosol models used in the retrieval (Torres et al., 2002).
POLDER (Polarization and Directionality of the Earth's Reflectances)	Nov 1996 to June 1997; Apr 2003 to Oct 2003; Jan 2005 to present	8 bands (0.44 to 0.91 μm)	τ_{aer}, α, DRE	Multiple view angles and polarization capabilities. 0.67 μm and 0.86 μm radiances used with 12 mono-modal aerosol models over ocean (Goloub et al., 1999; Deuzé et al., 2000). Polarization allows fine particle retrieval over land (Herman et al., 1997; Goloub and Arino, 2000). DRE determined over ocean (Boucher and Tanré, 2000; Bellouin et al., 2003).
OCTS (Ocean Colour and Temperature Scanner)	Nov 1996 to Jun 1997; Apr 2003 to Oct 2003	9 bands (0.41 to 0.86 μm) and 3.9 μm	τ_{aer}, α	0.67 μm and 0.86 μm retrieval gives $\tau_{\lambda=0.50}$ and α over ocean. Bi-modal aerosol size distribution assumed (Nakajima and Higurashi, 1998; Higurashi et al., 2000).
MODIS (Moderate Resolution Imaging Spectrometer)	2000 to present	12 bands (0.41 to 2.1 μm)	τ_{aer}, α, DRE	Retrievals developed over ocean surfaces using bi-modal size distributions (Tanré et al., 1997; Remer et al., 2002). Retrievals developed over land except bright surfaces (Kaufman et al., 1997; Chu et al., 2002). Optical depth speciation and DRE determined over ocean and land (e.g., Bellouin et al., 2005; Kaufman et al., 2005a).
MISR (Multi-angle Imaging Spectro-Radiometer)	2000 to present	4 bands (0.47 to 0.86 μm)	τ_{aer}, α	9 different viewing angles. Five climatological mixing groups composed of four component particles are used in the retrieval algorithm (Kahn et al., 2001; Kahn et al., 2005). Retrievals over bright surfaces are possible (Martonchik et al., 2004).
CERES (Clouds and the Earth's Radiant Energy System)	1998 to present		DRE	DRE determined by a regression of, for example, Visible Infrared Scanner (VIRS; AVHRR-like) τ_{aer} against upwelling irradiance (Loeb and Kato; 2002).
GLAS (Geoscience Laser Altimeter System)	2003 to present	Active lidar (0.53, 1.06 μm)	Aerosol vertical profile	Lidar footprint roughly 70 m at 170 m intervals. 8-day repeat orbiting cycle (Spinhirne et al., 2005).
ATSR-2/AATSR (Along Track Scanning Radiometer/Advanced ATSR)	1996 to present	4 bands (0.56 to 1.65 μm)	τ_{aer}, α	Nadir and 52° forward viewing geometry. 40 aerosol climatological mixtures containing up to six aerosol species are used in the retrievals (Veefkind et al., 1998; Holzer-Popp et al., 2002).
SeaWiFS (Sea-Viewing Wide Field-of-View Sensor)	1997 to present	0.765 and 0.865 μm (ocean) 0.41 to 0.67 μm (land)	τ_{aer}, α	2-channel using 0.765 μm and 0.856 μm gives $\tau_{\lambda=0.856}$ and α over ocean. Bi-modal aerosol size distribution assumed (M. Wang et al., 2005). Retrievals over land and ocean using six visible channels from 0.41 to 0.67 μm (von Hoyningen-Huene, 2003; Lee et al., 2004) also developed.

Notes:

a DRE is the direct radiative effect and includes both natural and anthropogenic sources (see Table 2.3). The Angstrom exponent, α, is the wavelength dependence of τ_{aer} and is defined by $\alpha = -\ln(\tau_{aer\lambda1}/\tau_{aer\lambda2}) / \ln(\lambda_1 / \lambda_2)$ where λ_1 = wavelength 1 and λ_2 = wavelength 2.

b TOMS followed up by the Ozone Monitoring Instrument (OMI) on the Earth Observing System (EOS) Aura satellite, launched July 2004.

nitrates and industrial dust, is evident over many continental regions of the NH. Sea salt aerosol is visible in oceanic regions where the wind speed is high (e.g., south of 45°S). The MODIS aerosol algorithm is currently unable to make routine retrievals over highly reflective surfaces such as deserts, snow cover, ice and areas affected by ocean glint, or over high-latitude regions when the solar insolation is insufficient.

Early retrievals for estimating τ_{aer} include the Advanced Very High Resolution Radiometer (AVHRR) single channel retrieval (e.g., Husar et al., 1997; Ignatov and Stowe, 2002), and the ultraviolet-based retrieval from the TOMS (e.g., Torres et al., 2002). A dual channel AVHRR retrieval has also been developed (e.g., Mishchenko et al., 1999; Geogdzhayev et al., 2002). Retrievals by the AVHRR are generally only performed over ocean surfaces where the surface reflectance characteristics are relatively well known, although retrievals are also possible over dark land surfaces such as boreal forests and lakes (Soufflet et al., 1997). The TOMS retrieval is essentially independent of surface reflectance thereby allowing retrievals over both land and ocean (Torres et al., 2002), but is sensitive to the altitude of the aerosol, and has a relatively low spatial resolution. While these retrievals only use a limited number of spectral bands and lack sophistication compared to those from dedicated satellite instruments, they have the advantage of offering continuous long-term data sets (e.g., Geogdzhayev et al., 2002).

Early retrievals have been superseded by those from dedicated aerosol instruments (e.g., Kaufman et al., 2002). Polarization and Directionality of the Earth's Reflectance (POLDER) uses a combination of spectral channels (0.44–0.91 µm) with several viewing angles, and measures polarization of radiation. Aerosol optical depth and Ångstrom exponent (α) over ocean (Deuzé et al., 2000), τ_{aer} over land (Deuzé et al., 2001) and the direct radiative effect of aerosols (Boucher and Tanré, 2000; Bellouin et al., 2003) have all been developed. Algorithms for aerosol retrievals using MODIS have been developed and validated over both ocean (Tanré et al., 1997) and land surfaces (Kaufman et al., 1997). The uncertainty in these retrievals of τ_{aer} is necessarily higher over land (Chu et al., 2002) than over oceans (Remer et al., 2002) owing to uncertainties in land surface reflectance characteristics, but can be minimised by careful selection of the viewing geometry (Chylek et al., 2003). In addition, new algorithms have been developed for discriminating between sea salt, dust or biomass burning and industrial pollution over oceans (Bellouin et al., 2003, 2005; Kaufman et al., 2005a) that allow for a more comprehensive comparison against aerosol models. Multi-angle Imaging Spectro-Radiometer (MISR) retrievals have been developed using multiple viewing capability to determine aerosol parameters over ocean (Kahn et al., 2001) and land surfaces, including highly reflective surfaces such as deserts (Martonchik et al., 2004). Five typical aerosol climatologies, each containing four aerosol components, are used in the retrievals, and the optimum radiance signature is determined for nine viewing geometries and two different radiances. The results have been validated against those from the Aerosol

RObotic NETwork (AERONET; see Section 2.4.3). Along Track Scanning Radiometer (ATSR) and ATSR-2 retrievals (Veefkind et al., 1998; Holzer-Popp et al., 2002) use a relatively wide spectral range (0.56–1.65 µm), and two viewing directions and aerosol climatologies from the Optical Parameters of Aerosols and Clouds (OPAC) database (Hess et al., 1998) to make τ_{aer} retrievals over both ocean and land (Robles-Gonzalez et al., 2000). The Ocean Colour and Temperature Scanner (OCTS) retrieval has a basis similar to the dual wavelength retrieval from AVHRR and uses wavelengths over the range 0.41 to 0.86 µm to derive τ_{aer} and α over oceans (e.g., Higurashi et al., 2000) using a bi-modal aerosol size distribution. The Sea-Viewing Wide Field-of-View Sensor (SeaWiFS) uses 0.765 µm and 0.856 µm radiances to provide $\tau_{\lambda=0.856}$ and α over ocean using a bi-modal aerosol size distribution (M. Wang et al., 2005). Further SeaWiFS aerosol products have been developed over both land and ocean using six and eight visible channels, respectively (e.g., von Hoyningen-Heune et al., 2003; Lee et al., 2004).

Despite the increased sophistication and realism of the aerosol retrieval algorithms, discrepancies exist between retrievals of τ_{aer} even over ocean regions (e.g., Penner et al., 2002; Myhre et al., 2004a, 2005b; Jeong et al., 2005; Kinne et al., 2006). These discrepancies are due to different assumptions in the cloud clearing algorithms, aerosol models, different wavelengths and viewing geometries used in the retrievals, different parametrizations of ocean surface reflectance, etc. Comparisons of these satellite aerosol retrievals with the surface AERONET observations provide an opportunity to objectively evaluate as well as improve the accuracy of these satellite retrievals. Myhre et al. (2005b) showed that dedicated instruments using multi-channel and multi-view algorithms perform better when compared against AERONET than the simple algorithms that they have replaced, and Zhao et al. (2005) showed that retrievals based on dynamic aerosol models perform better than those based on globally fixed aerosol models. While some systematic biases in specific satellite products exist (e.g., Jeong et al., 2005; Remer et al., 2005), these can be corrected for (e.g., Bellouin et al., 2005; Kaufman et al., 2005b), which then enables an assessment of the direct radiative effect and the direct RF from an observational perspective, as detailed below.

2.4.2.1.2 Satellite retrievals of direct radiative effect

The solar direct radiative effect (DRE) is the sum of the direct effects due to anthropogenic and natural aerosol species while the direct RF only considers the anthropogenic components. Satellite estimates of the global clear-sky DRE over oceans have advanced since the TAR, owing to the development of dedicated aerosol instruments and algorithms, as summarised by Yu et al. (2006) (see Table 2.3). Table 2.3 suggests a reasonable agreement of the global mean, diurnally averaged clear-sky DRE from various studies, with a mean of –5.4 W m^{-2} and a standard deviation of 0.9 W m^{-2}. The clear-sky DRE is converted to an all-sky DRE by Loeb and Manalo-Smith (2005) who estimated an all-sky DRE over oceans of –1.6 to –2.0 W m^{-2} but assumed no aerosol contribution to the DRE from cloudy regions; such an assumption is not valid for optically thin clouds or if partially absorbing aerosols exist above the clouds (see Section 2.4.4.4).

Furthermore, use of a combination of sensors on the same satellite offers the possibility of concurrently deriving τ_{aer} and the DRE (e.g., Zhang and Christopher, 2003; Zhang et al., 2005), which enables estimation of the DRE efficiency, that is, the DRE divided by τ_{aer} (W m^{-2} τ_{aer}^{-1}). Because the DRE efficiency removes the dependence on the geographic distribution of τ_{aer} it is a useful parameter for comparison of models against observations (e.g., Anderson et al., 2005b); however, the DRE efficiency thus derived is not a linear function of τ_{aer} at high τ_{aer} such as those associated with intense mineral dust, biomass burning or pollution events.

2.4.2.1.3. Satellite retrievals of direct radiative forcing

Kaufman et al. (2005a) estimated the anthropogenic-only component of the aerosol fine-mode fraction from the MODIS product to deduce a clear sky RF over ocean of -1.4 W m^{-2}. Christopher et al. (2006) used a combination of the MODIS fine-mode fraction and Clouds and the Earth's Radiant Energy System (CERES) broadband TOA fluxes and estimated an identical value of -1.4 ± 0.9 W m^{-2}. Bellouin et al. (2005) used a combination of MODIS τ_{aer} and fine-mode fraction together with data from AeroCom (see Section 2.4.3) to determine an all-sky RF of aerosols over both land and ocean of -0.8 ± 0.2 W m^{-2}, but this does not include the contribution to the RF and associated uncertainty from cloudy skies. Chung et al. (2005) performed a similar satellite/AERONET/model analysis, but included the contribution from cloudy areas to deduce an RF of -0.35 W m^{-2} or -0.50 W m^{-2} depending upon whether the anthropogenic fraction is determined from a model or from the MODIS fine-mode fraction and suggest an overall uncertainty range of -0.1 to -0.6 W m^{-2}. Yu et al. (2006) used several measurements to estimate a direct RF of -0.5 ± 0.33 W m^{-2}. These estimates of the RF are compared to those obtained from modelling studies in Section 2.4.4.7.

2.4.2.2 Surface-Based Retrievals

A significant advancement since the TAR is the continued deployment and development of surface based remote sensing sun-photometer sites such as AERONET (Holben et al., 1998), and the establishment of networks of aerosol lidar systems such

Table 2.3. The direct aerosol radiative effect (DRE) estimated from satellite remote sensing studies (adapted and updated from Yu et al., 2006).

Reference	Instrument[a]	Data Analysed	Brief Description	Clear Sky DRE (W m^{-2}) ocean
Bellouin et al. (2005)	MODIS; TOMS; SSM/I	2002	MODIS fine and total τ_{aer} with TOMS Aerosol Index and SSM/I to discriminate dust from sea salt.	-6.8
Loeb and Manalo-Smith (2005)	CERES; MODIS	Mar 2000 to Dec 2003	CERES radiances/irradiances and angular distribution models and aerosol properties from either MODIS or from NOAA-NESDIS[b] algorithm used to estimate the direct radiative effect.	-3.8 (NESDIS) to -5.5 (MODIS)
Remer and Kaufman (2006)	MODIS	Aug 2001 to Dec 2003	Best-prescribed aerosol model fitted to MODIS data. τ_{aer} from fine-mode fraction.	-5.7 ± 0.4
Zhang et al. (2005); Christopher and Zhang (2004)	CERES; MODIS	Nov 2000 to Aug 2001	MODIS aerosol properties, CERES radiances/irradiances and angular distribution models used to estimate the direct radiative effect.	-5.3 ± 1.7
Bellouin et al. (2003)	POLDER	Nov 1996 to Jun 1997	Best-prescribed aerosol model fitted to POLDER data	-5.2
Loeb and Kato (2002)	CERES; VIRS	Jan 1998 to Aug 1998; Mar 2000.	τ_{aer} from VIRS regressed against the TOA CERES irradiance (35°N to 35°S)	-4.6 ± 1.0
Chou et al. (2002)	SeaWiFs	1998	Radiative transfer calculations with SeaWiFS τ_{aer} and prescribed optical properties	-5.4
Boucher and Tanré (2000)	POLDER	Nov 1996 to Jun 1997	Best-prescribed aerosol model fitted to POLDER data	-5 to -6
Haywood et al. (1999)	ERBE	Jul 1987 to Dec 1988	DRE diagnosed from GCM-ERBE TOA irradiances	-6.7
Mean (standard deviation)				**-5.4 (0.9)**

Notes:

[a] SSM/I: Special Sensor Microwave/Imager; VIRS: Visible Infrared Scanner; ERBE: Earth Radiation Budget Experiment.

[b] NESDIS: National Environmental Satellite, Data and Information Service.

as the European Aerosol Research Lidar Network (EARLINET, Matthias et al., 2004), the Asian Dust Network (ADNET, Murayama et al., 2001), and the Micro-Pulse Lidar Network (MPLNET, Welton et al., 2001).

The distribution of AERONET sites is also shown in Figure 2.11 (top panel). Currently there are approximately 150 sites operating at any one time, many of which are permanent to enable determination of climatological and interannual column-averaged monthly and seasonal means. In addition to measurements of τ_{aer} as a function of wavelength, new algorithms have been developed that measure sky radiance as a function of scattering angle (Nakajima et al., 1996; Dubovik and King, 2000). From these measurements, the column-averaged size distribution and, if the τ_{aer} is high enough ($\tau_{aer} > 0.5$), the aerosol single scattering albedo, ω_c, and refractive indices may be determined at particular wavelengths (Dubovik et al., 2000), allowing partitioning between scattering and absorption. While these inversion products have not been comprehensively validated, a number of studies show encouraging agreement for both the derived size distribution and ω_0 when compared against *in situ* measurements by instrumented aircraft for different aerosol species (e.g., Dubovik et al., 2002; Haywood et al., 2003a; Reid et al., 2003; Osborne et al., 2004). A climatology of the aerosol DRE based on the AERONET aerosols has also been derived (Zhou et al., 2005).

The MPLNET Lidar network currently consists of 11 lidars worldwide; 9 are co-located with AERONET sites and provide complementary vertical distributions of aerosol backscatter and extinction. Additional temporary MPLNET sites have supported major aerosol field campaigns (e.g., Campbell et al., 2003). The European-wide lidar network EARLINET currently has 15 aerosol lidars making routine retrievals of vertical profiles of aerosol extinction (Mathias et al., 2004), and ADNET is a network of 12 lidars making routine measurements in Asia that have been used to assess the vertical profiles of Asian dust and pollution events (e.g., Husar et al., 2001; Murayama et al., 2001).

2.4.3   Advances in Modelling the Aerosol Direct Effect

Since the TAR, more complete aerosol modules in a larger number of global atmospheric models now provide estimates of the direct RF. Several models have resolutions better than 2° by 2° in the horizontal and more than 20 to 30 vertical levels; this represents a considerable enhancement over the models used in the TAR. Such models now include the most important anthropogenic and natural species. Tables 2.4, 2.5 and 2.6 summarise studies published since the TAR. Some of the more complex models now account explicitly for the dynamics of the aerosol size distribution throughout the aerosol atmospheric lifetime and also parametrize the internal/external mixing of the various aerosol components in a more physically realistic way than in the TAR (e.g., Adams and Seinfeld, 2002; Easter et al., 2004; Stier et al., 2005). Because the most important aerosol species are now included, a comparison of key model output parameters, such as the total τ_{aer}, against satellite retrievals

and surface-based sun photometer and lidar observations is possible (see Sections 2.4.2 and 2.4.4). Progress with respect to modelling the indirect effects due to aerosol-cloud interactions is detailed in Section 2.4.5 and Section 7.5. Several studies have explored the sensitivity of aerosol direct RF to current parametrization uncertainties. These are assessed in the following sections.

Major progress since the TAR has been made in the documentation of the diversity of current aerosol model simulations. Sixteen groups have participated in the Global Aerosol Model Intercomparison (AeroCom) initiative (Kinne et al., 2006). Extensive model outputs are available via a dedicated website (Schulz et al., 2004). Three model experiments (named A, B, and PRE) were analysed. Experiment A models simulate the years 1996, 1997, 2000 and 2001, or a five-year mean encompassing these years. The model emissions and parametrizations are those determined by each research group, but the models are driven by observed meteorological fields to allow detailed comparisons with observations, including those from MODIS, MISR and the AERONET sun photometer network. Experiment B models use prescribed AeroCom aerosol emissions for the year 2000, and experiment PRE models use prescribed aerosol emissions for the year 1750 (Dentener et al., 2006; Schulz et al., 2006). The model diagnostics included information on emission and deposition fluxes, vertical distribution and sizes, thus enabling a better understanding of the differences in lifetimes of the various aerosol components in the models.

This paragraph discusses AeroCom results from Textor et al. (2006). The model comparison study found a wide range in several of the diagnostic parameters; these, in turn, indicate which aerosol parametrizations are poorly constrained and/or understood. For example, coarse aerosol fractions are responsible for a large range in the natural aerosol emission fluxes (dust: ±49% and sea salt: ±200%, where uncertainty is 1 standard deviation of inter-model range), and consequently in the dry deposition fluxes. The complex dependence of the source strength on wind speed adds to the problem of computing natural aerosol emissions. Dust emissions for the same time period can vary by a factor of two or more depending on details of the dust parametrization (Luo et al., 2003; Timmreck and Schulz, 2004; Balkanski et al., 2004; Zender, 2004), and even depend on the reanalysis meteorological data set used (Luo et al., 2003). With respect to anthropogenic and natural emissions of other aerosol components, modelling groups tended to make use of similar best guess information, for example, recently revised emissions information available via the Global Emissions Inventory Activity (GEIA). The vertical aerosol distribution was shown to vary considerably, which is a consequence of important differences in removal and vertical mixing parametrizations. The inter-model range for the fraction of sulphate mass below 2.5 km to that of total sulphate is 45 ± 23%. Since humidification takes place mainly in the boundary layer, this source of inter-model variability increases the range of modelled direct RF. Additionally, differences in the parametrization of the wet deposition/vertical mixing process become more pronounced above 5 km altitude. Some models have a tendency to accumulate insoluble aerosol mass (dust and carbonaceous aerosols) at higher altitudes, while others have much more efficient wet removal schemes. Tropospheric residence times, defined here as the ratio of burden over sinks established for an equilibrated one-year simulation, vary by 20 to 30% for the fine-mode aerosol species. These variations are of interest, since they express the linearity of modelled emissions to aerosol burden and eventually to RF.

Considerable progress has been made in the systematic evaluation of global model results (see references in Tables 2.4 to 2.6). The simulated global τ_{aer} at a wavelength of 0.55 μm in models ranges from 0.11 to 0.14. The values compare favourably to those obtained by remote sensing from the ground (AERONET, about 0.135) and space (satellite composite, about 0.15) (Kinne et al., 2003, 2006), but significant differences exist in regional and temporal distributions. Modelled absorption optical thickness has been suggested to be underestimated by a factor of two to four when compared to observations (Sato et al., 2003) and DRE efficiencies have been shown to be lower in models both for the global average and regionally (Yu et al., 2006) (see Section 2.4.4.7). A merging of modelled and observed fields of aerosol parameters through assimilation methods of different degrees of complexity has also been performed since the TAR (e.g., Yu et al., 2003; Chung et al., 2005). Model results are constrained to obtain present-day aerosol fields consistent with observations. Collins et al. (2001) showed that assimilation of satellite-derived fields of τ_{aer} can reduce the model bias down to 10% with respect to daily mean τ_{aer} measured with a sun photometer at the Indian Ocean Experiment (INDOEX) station Kaashidhoo. Liu et al. (2005) demonstrated similar efficient reduction of errors in τ_{aer}. The magnitude of the global dust cycle has been suggested to range between 1,500 and 2,600 Tg yr^{-1} by minimising the bias between model and multiple dust observations (Cakmur et al., 2006). Bates et al. (2006) focused on three regions downwind of major urban/population centres and performed radiative transfer calculations constrained by intensive and extensive observational parameters to derive 24-hour average clear-sky DRE of -3.3 ± 0.47, -14 ± 2.6 and -6.4 ± 2.1 W m^{-2} for the north Indian Ocean, the northwest Pacific and the northwest Atlantic, respectively. By constraining aerosol models with these observations, the uncertainty associated with the DRE was reduced by approximately a factor of two.

2.4.4   Estimates of Aerosol Direct Radiative Forcing

Unless otherwise stated, this section discusses the TOA direct RF of different aerosol types as a global annual mean quantity inclusive of the effects of clouds. Where possible, statistics from model results are used to assess the uncertainty in the RF. Recently published results and those grouped within AeroCom are assessed. Because the AeroCom results assessed here are based on prescribed emissions, the uncertainty in these results is lowered by having estimates of the uncertainties in the emissions. The quoted uncertainties therefore include the

structural uncertainty (i.e., differences associated with the model formulation and structure) associated with the RF, but do not include the full range of parametric uncertainty (i.e., differences associated with the choice of key model parameters), as the model results are essentially best estimates constrained by observations of emissions, wet and dry deposition, size distributions, optical parameters, hygroscopicity, etc. (Pan et al., 1997). The uncertainties are reported as the 5 to 95% confidence interval to allow the uncertainty in the RF of each species of aerosol to be quantitatively intercompared.

2.4.4.1   Sulphate Aerosol

Atmospheric sulphate aerosol may be considered as consisting of sulphuric acid particles that are partly or totally neutralized by ammonia and that are present as liquid droplets or partly crystallized. Sulphate is formed by aqueous phase reactions within cloud droplets, oxidation of SO_2 via gaseous phase reactions with OH, and by condensational growth onto pre-existing particles (e.g., Penner et al., 2001). Emission estimates are summarised by Haywood and Boucher (2000). The main source of sulphate aerosol is via SO_2 emissions from fossil fuel burning (about 72%), with a small contribution from biomass burning (about 2%), while natural sources are from dimethyl sulphide emissions by marine phytoplankton (about 19%) and by SO_2 emissions from volcanoes (about 7%). Estimates of global SO_2 emissions range from 66.8 to 92.4 TgS yr^{-1} for anthropogenic emissions in the 1990s and from 91.7 to 125.5 TgS yr^{-1} for total emissions. Emissions of SO_2 from 25 countries in Europe were reduced from approximately 18 TgS yr^{-1} in 1980 to 4 TgS yr^{-1} in 2002 (Vestreng et al., 2004). In the USA, the emissions were reduced from about 12 to 8 TgS yr^{-1} in the period 1980 to 2000 (EPA, 2003). However, over the same period SO_2 emissions have been increasing significantly from Asia, which is estimated to currently emit 17 TgS yr^{-1} (Streets et al., 2003), and from developing countries in other regions (e.g., Lefohn et al., 1999; Van Aardenne et al., 2001; Boucher and Pham, 2002). The most recent study (Stern, 2005) suggests a decrease in global anthropogenic emissions from approximately 73 to 54 TgS yr^{-1} over the period 1980 to 2000, with NH emission falling from 64 to 43 TgS yr^{-1} and SH emissions increasing from 9 to 11 TgS yr^{-1}. Smith et al. (2004) suggested a more modest decrease in global emissions, by some 10 TgS yr^{-1} over the same period. The regional shift in the emissions of SO_2 from the USA, Europe, Russia, Northern Atlantic Ocean and parts of Africa to Southeast Asia and the Indian and Pacific Ocean areas will lead to subsequent shifts in the pattern of the RF (e.g., Boucher and Pham, 2002; Smith et al., 2004; Pham et al., 2005). The recently used emission scenarios take into account effective injection heights and their regional and seasonal variability (e.g., Dentener et al., 2006).

The optical parameters of sulphate aerosol have been well documented (see Penner et al., 2001 and references therein). Sulphate is essentially an entirely scattering aerosol across the solar spectrum ($\omega_0 = 1$) but with a small degree of absorption in the near-infrared spectrum. Theoretical and experimental data are available on the relative humidity dependence of the specific extinction coefficient, f_{RH} (e.g., Tang et al., 1995). Measurement campaigns concentrating on industrial pollution, such as the Tropospheric Aerosol Radiative Forcing Experiment (TARFOX; Russell et al., 1999), the Aerosol Characterization Experiment (ACE-2; Raes et al., 2000), INDOEX (Ramanathan et al., 2001b), the Mediterranean Intensive Oxidants Study (MINOS, 2001 campaign), ACE-Asia (2001), Atmospheric Particulate Environment Change Studies (APEX, from 2000 to 2003), the New England Air Quality Study (NEAQS, in 2003) and the Chesapeake Lighthouse and Aircraft Measurements for Satellites (CLAMS; Smith et al., 2005), continue to show that sulphate contributes a significant fraction of the sub-micron aerosol mass, anthropogenic τ_{aer} and RF (e.g., Hegg et al., 1997; Russell and Heintzenberg, 2000; Ramanathan et al., 2001b; Magi et al., 2005; Quinn and Bates, 2005). However, sulphate is invariably internally and externally mixed to varying degrees with other compounds such as biomass burning aerosol (e.g., Formenti et al., 2003), fossil fuel black carbon (e.g., Russell and Heintzenberg, 2000), organic carbon (Novakov et al., 1997; Brock et al., 2004), mineral dust (e.g., Huebert et al., 2003) and nitrate aerosol (e.g., Schaap et al., 2004). This results in a composite aerosol state in terms of effective refractive indices, size distributions, physical state, morphology, hygroscopicity and optical properties.

The TAR reported an RF due to sulphate aerosol of -0.40 W m^{-2} with an uncertainty of a factor of two, based on global modelling studies that were available at that time. Results from model studies since the TAR are summarised in Table 2.4. For models A to L, the RF ranges from approximately -0.21 W m^{-2} (Takemura et al., 2005) to -0.96 W m^{-2} (Adams et al., 2001) with a mean of -0.46 W m^{-2} and a standard deviation of 0.20 W m^{-2}. The range in the RF per unit τ_{aer} is substantial due to differing representations of aerosol mixing state, optical properties, cloud, surface reflectance, hygroscopic growth, sub-grid scale effects, radiative transfer codes, etc. (Ramaswamy et al., 2001). Myhre et al. (2004b) performed several sensitivity studies and found that the uncertainty was particularly linked to the hygroscopic growth and that differences in the model relative humidity fields could cause differences of up to 60% in the RF. The RFs from the models M to U participating in the AeroCom project are slightly weaker than those obtained from the other studies, with a mean of approximately -0.35 W m^{-2} and a standard deviation of 0.15 W m^{-2}; the standard deviation is reduced for the AeroCom models owing to constraints on aerosol emissions, based on updated emission inventories (see Table 2.4). Including the uncertainty in the emissions reported in Haywood and Boucher (2000) increases the standard deviation to 0.2 W m^{-2}. As sulphate aerosol is almost entirely scattering, the surface forcing will be similar or marginally stronger than the RF diagnosed at the TOA. The uncertainty in the RF estimate relative to the mean value remains relatively large compared to the situation for LLGHGs.

The mean and median of the sulphate direct RF from grouping all these studies together are identical at -0.41 W m^{-2}. Disregarding the strongest and weakest direct RF estimates to approximate the 90% confidence interval leads to an estimate of -0.4 ± 0.2 W m^{-2}.

2.4.4.2 *Organic Carbon Aerosol from Fossil Fuels*

Organic aerosols are a complex mixture of chemical compounds containing carbon-carbon bonds produced from fossil fuel and biofuel burning and natural biogenic emissions. Organic aerosols are emitted as primary aerosol particles or formed as secondary aerosol particles from condensation of organic gases considered semi-volatile or having low volatility. Hundreds of different atmospheric organic compounds have been detected in the atmosphere (e.g., Hamilton et al., 2004; Murphy, 2005), which makes definitive modelling of the direct and indirect effects extremely challenging (McFiggans et al., 2006). Emissions of primary organic carbon from fossil fuel burning have been estimated to be 10 to 30 TgC yr^{-1} (Liousse et al., 1996; Cooke et al., 1999; Scholes and Andreae, 2000). More recently, Bond et al. (2004) provided a detailed analysis of primary organic carbon emissions from fossil fuels, biofuels and open burning, and suggested that contained burning (approximately the sum of fossil fuel and biofuel) emissions are in the range of 5 to 17 TgC yr^{-1}, with fossil fuel contributing only 2.4 TgC yr^{-1}. Ito and Penner (2005) estimated global fossil fuel particulate organic matter (POM, which is the sum of the organic carbon and the other associated chemical elements) emissions of around 2.2 Tg(POM) yr^{-1}, and global biofuel emissions of around 7.5 Tg(POM) yr^{-1}. Ito and Penner (2005) estimated that emissions of fossil and biofuel organic carbon increased by a factor of three over the period 1870 to 2000. Subsequent to emission, the hygroscopic, chemical and optical properties of organic carbon particles continue to change because of chemical processing by gas-phase oxidants such as ozone, OH, and the nitrate radical (NO_3) (e.g., Kanakidou et al., 2005). Atmospheric concentrations of organic aerosol are frequently similar to those of industrial sulphate aerosol. Novakov et al. (1997) and Hegg et al. (1997) measured organic carbon in pollution off the East Coast of the USA during the TARFOX campaign, and found organic carbon primarily from fossil fuel burning contributed up to 40% of the total submicron aerosol mass and was frequently the most significant contributor to τ_{aer}. During INDOEX, which studied the industrial plume over the Indian Ocean, Ramanathan et al. (2001b) found that organic carbon was the second largest contributor to τ_{aer} after sulphate aerosol.

Observational evidence suggests that some organic aerosol compounds from fossil fuels are relatively weakly absorbing but do absorb solar radiation at some ultraviolet and visible wavelengths (e.g., Bond et al., 1999; Jacobson, 1999; Bond, 2001) although organic aerosol from high-temperature combustion such as fossil fuel burning (Dubovik et al., 1998; Kirchstetter et al., 2004) appears less absorbing than from low-temperature combustion such as open biomass burning. Observations suggest that a considerable fraction of organic carbon is soluble to some degree, while at low relative humidity more water is often associated with the organic fraction than

Table 2.4. *The direct radiative forcing for sulphate aerosol derived from models published since the TAR and from the AeroCom simulations where different models used identical emissions. Load and aerosol optical depth (τ_{aer}) refer to the anthropogenic sulphate; $\tau_{aer\,ant}$ is the fraction of anthropogenic sulphate to total sulphate τ_{aer} for present day, NRFM is the normalised RF by mass, and NRF is the normalised RF per unit τ_{aer}.*

No	Model[a]	LOAD (mg(SO_4) m^{-2})	τ_{aer} (0.55 µm)	$\tau_{aer\,ant}$ (%)	RF (W m^{-2})	NRFM (W g^{-1})	NRF (W m^{-2} τ_{aer}^{-1})	Reference
Published since IPCC, 2001								
A	CCM3	2.23			−0.56	−251		(Kiehl et al., 2000)
B	GEOSCHEM	1.53	0.018		−0.33	−216	−18	(Martin et al., 2004)
C	GISS	3.30	0.022		−0.65	−206	−32	(Koch, 2001)
D	GISS	3.27			−0.96	−293		(Adams et al., 2001)
E	GISS	2.12			−0.57	−269		(Liao and Seinfeld, 2005)
F	SPRINTARS	1.55	0.015	72	−0.21	−135	−8	(Takemura et al., 2005)
G	LMD	2.76			−0.42	−152		(Boucher and Pham., 2002)
H	LOA	3.03	0.030		−0.41	−135	−14	(Reddy et al., 2005b)
I	GATORG	3.06			−0.32	−105		(Jacobson, 2001a)
J	PNNL	5.50	0.042		−0.44	−80	−10	(Ghan et al., 2001)
K	UIO_CTM	1.79	0.019		−0.37	−207	−19	(Myhre et al., 2004b)
L	UIO_GCM	2.28			−0.29	−127		(Kirkevag and Iversen, 2002)
AeroCom: identical emissions used for year 1750 and 2000								
M	UMI	2.64	0.020	58	−0.58	−220	−28	(Liu and Penner, 2002)
N	UIO_CTM	1.70	0.019	57	−0.35	−208	−19	(Myhre et al., 2003)
O	LOA	3.64	0.035	64	−0.49	−136	−14	(Reddy and Boucher, 2004)
P	LSCE	3.01	0.023	59	−0.42	−138	−18	(Schulz et al., 2006)
Q	ECHAM5-HAM	2.47	0.016	60	−0.46	−186	−29	(Stier et al., 2005)
R	GISS	1.34	0.006	41	−0.19	−139	−31	(Koch, 2001)
S	UIO_GCM	1.72	0.012	59	−0.25	−145	−21	(Iversen and Seland, 2002; Kirkevag and Iversen, 2002)
T	SPRINTARS	1.19	0.013	59	−0.16	−137	−13	(Takemura et al., 2005)
U	ULAQ	1.62	0.020	42	−0.22	−136	−11	(Pitari et al., 2002)
Average A to L		2.80	0.024		−0.46	−176	−17	
Average M to U		2.15	0.018	55	−0.35	−161	−20	
Minimum A to U		1.19	0.006	41	−0.96	−293	−32	
Maximum A to U		5.50	0.042	72	−0.16	−72	−8	
Std. dev. A to L		1.18	0.010		0.20	75	9	
Std. dev. M to U		0.83	0.008	8	0.15	34	7	

Notes:

[a] CCM3: Community Climate Model; GEOSCHEM: Goddard Earth Observing System-Chemistry; GISS: Goddard Institute for Space Studies; SPRINTARS: Spectral Radiation-Transport Model for Aerosol Species; LMD: Laboratoire de Météorologie Dynamique; LOA: Laboratoire d'Optique Atmospherique; GATORG: Gas, Aerosol, Transport, Radiation, and General circulation model; PNNL: Pacific Northwest National Laboratory; UIO_CTM: University of Oslo CTM; UIO_GCM: University of Oslo GCM; UMI: University of Michigan; LSCE: Laboratoire des Sciences du Climat et de l'Environnement; ECHAM5-HAM: European Centre Hamburg with Hamburg Aerosol Module; ULAQ: University of L'Aquila.

with inorganic material. At higher relative humidities, the hygroscopicity of organic carbon is considerably less than that of sulphate aerosol (Kotchenruther and Hobbs, 1998; Kotchenruther et al., 1999).

Based on observations and fundamental chemical kinetic principles, attempts have been made to formulate organic carbon composition by functional group analysis in some main classes of organic chemical species (e.g., Decesari et al., 2000, 2001; Maria et al., 2002; Ming and Russell, 2002), capturing

some general characteristics in terms of refractive indices, hygroscopicity and cloud activation properties. This facilitates improved parametrizations in global models (e.g., Fuzzi et al., 2001; Kanakidou et al., 2005; Ming et al., 2005a).

Organic carbon aerosol from fossil fuel sources is invariably internally and externally mixed to some degree with other combustion products such as sulphate and black carbon (e.g., Novakov et al., 1997; Ramanathan et al., 2001b). Theoretically, coatings of essentially non-absorbing components such as organic carbon or sulphate on strongly absorbing core components such as black carbon can increase the absorption of the composite aerosol (e.g., Fuller et al., 1999; Jacobson, 2001a; Stier et al., 2006a), with results backed up by laboratory studies (e.g., Schnaiter et al., 2003). However, coatings of organic carbon aerosol on hygroscopic aerosol such as sulphate may lead to suppression of the rate of water uptake during cloud activation (Xiong et al., 1998; Chuang, 2003).

Current global models generally treat organic carbon using one or two tracers (e.g., water-insoluble tracer, water-soluble tracer) and highly parametrized schemes have been developed to represent the direct RF. Secondary organic carbon is highly simplified in the global models and in many cases treated as an additional source similar to primary organic carbon. Considerable uncertainties still exist in representing the refractive indices and the water of hydration associated with the particles because the aerosol properties will invariably differ depending on the combustion process, chemical processing in the atmosphere, mixing with the ambient aerosol, etc. (e.g., McFiggans et al., 2006).

The TAR reported an RF of organic carbon aerosols from fossil fuel burning of -0.10 W m^{-2} with a factor of three uncertainty. Many of the modelling studies performed since the TAR have investigated the RF of organic carbon aerosols from both fossil fuel and biomass burning aerosols, and the combined RF of both components. These studies are summarised in Table 2.5. The RF from total organic carbon (POM) from both biomass burning and fossil fuel emissions from recently published models A to K and AeroCom models (L to T) is -0.24 W m^{-2} with a standard deviation of 0.08 W m^{-2} and -0.16 W m^{-2} with a standard deviation of 0.10 W m^{-2}, respectively. Where the RF due to organic carbon from fossil fuels is not explicitly accounted for in the studies, an approximate scaling based on the source apportionment of 0.25:0.75 is applied for fossil fuel organic carbon:biomass burning organic carbon (Bond et al., 2004). The mean RF of the fossil fuel component of organic carbon from those studies other than in AeroCom is -0.06 W m^{-2}, while those from AeroCom produce an RF of -0.03 W m^{-2} with a range of -0.01 W m^{-2} to -0.06 W m^{-2} and a standard deviation of around 0.02 W m^{-2}. Note that these RF estimates, to a large degree, only take into account primary emitted organic carbon. These studies all use optical properties for organic carbon that are either entirely scattering or only weakly absorbing and hence the surface forcing is only slightly stronger than that at the TOA.

The mean and median for the direct RF of fossil fuel organic carbon from grouping all these studies together are identical at -0.05 W m^{-2} with a standard deviation of 0.03 W m^{-2}. The standard deviation is multiplied by 1.645 to approximate the 90% confidence interval.[9] This leads to a direct RF estimate of -0.05 ± 0.05 W m^{-2}.

2.4.4.3   Black Carbon Aerosol from Fossil Fuels

Black carbon (BC) is a primary aerosol emitted directly at the source from incomplete combustion processes such as fossil fuel and biomass burning and therefore much atmospheric BC is of anthropogenic origin. Global, present-day fossil fuel emission estimates range from 5.8 to 8.0 TgC yr^{-1} (Haywood and Boucher, 2000 and references therein). Bond et al. (2004) estimated the total current global emission of BC to be approximately 8 TgC yr^{-1}, with contributions of 4.6 TgC yr^{-1} from fossil fuel and biofuel combustion and 3.3 TgC yr^{-1} from open biomass burning, and estimated an uncertainty of about a factor of two. Ito and Penner (2005) suggested fossil fuel BC emissions for 2000 of around 2.8 TgC yr^{-1}. The trends in emission of fossil fuel BC have been investigated in industrial areas by Novakov et al. (2003) and Ito and Penner (2005). Novakov et al. (2003) reported that significant decreases were recorded in the UK, Germany, the former Soviet Union and the USA over the period 1950 to 2000, while significant increases were reported in India and China. Globally, Novakov et al. (2003) suggested that emissions of fossil fuel BC increased by a factor of three between 1950 and 1990 (2.2 to 6.7 TgC yr^{-1}) owing to the rapid expansion of the USA, European and Asian economies (e.g., Streets et al., 2001, 2003), and have since fallen to around 5.6 TgC yr^{-1} owing to further emission controls. Ito and Penner (2005) determined a similar trend in emissions over the period 1950 to 2000 of approximately a factor of three, but the absolute emissions are smaller than in Novakov et al. (2003) by approximately a factor of 1.7.

Black carbon aerosol strongly absorbs solar radiation. Electron microscope images of BC particles show that they are emitted as complex chain structures (e.g., Posfai et al., 2003), which tend to collapse as the particles age, thereby modifying the optical properties (e.g., Abel et al., 2003). The Indian Ocean Experiment (Ramanathan et al., 2001b and references therein) focussed on emissions of aerosol from the Indian sub-continent, and showed the importance of absorption by aerosol in the atmospheric column. These observations showed that the local surface forcing (-23 W m^{-2}) was significantly stronger than the local RF at the TOA (-7 W m^{-2}). Additionally, the presence of BC in the atmosphere above highly reflective surfaces such as snow and ice, or clouds, may cause a significant positive RF (Ramaswamy et al., 2001). The vertical profile is therefore important, as BC aerosols or mixtures of aerosols containing a relatively large fraction of BC will exert a positive RF when located above clouds. Both microphysical (e.g., hydrophilic-to-hydrophobic nature of emissions into the atmosphere, aging of the aerosols, wet deposition) and meteorological aspects govern the horizontal and vertical distribution patterns of BC aerosols, and the residence time of these aerosols is thus sensitive to these factors (Cooke et al., 2002; Stier et al., 2006b).

The TAR assessed the RF due to fossil fuel BC as being $+0.2$ W m^{-2} with an uncertainty of a factor of two. Those models since the TAR that explicitly model and separate out the RF

---

[9] 1.645 is the factor relating the standard deviation to the 90% confidence interval for a normal distribution.

Table 2.5. Estimates of anthropogenic carbonaceous aerosol forcing derived from models published since the TAR and from the AeroCom simulations where different models used identical emissions. POM: particulate organic matter; BC: black carbon; BCPOM: BC and POM; FFBC: fossil fuel black carbon; FFPOM: fossil fuel particulate organic matter; BB: biomass burning sources included.

No	Model[a]	LOAD POM (mgPOM m⁻²)	τ_{aer} POM	τ_{aer} POM$_{ant}$ (%)	LOAD BC (mg m⁻²)	RF BCPOM (W m⁻²)	RF POM (W m⁻²)	RF BC (W m⁻²)	RF FFPOM (W m⁻²)	RF FFBC (W m⁻²)	RF BB (W m⁻²)	Reference
Published since IPCC, 2001												
A	SPRINT					0.12	-0.24	0.36	-0.05	0.15	-0.01	(Takemura et al., 2001)
B	LOA	2.33	0.016		0.37	0.30	-0.25	0.55	-0.02	0.19	0.14	(Reddy et al., 2005b)
C	GISS	1.86	0.017		0.29	0.35	-0.26	0.61	-0.13	0.49	0.065	(Hansen et al., 2005)
D	GISS	1.86	0.015		0.29	0.05	-0.30	0.35	-0.08b	0.18b	-0.05b	(Koch, 2001)
E	GISS	2.39			0.39	0.32	-0.18	0.50	-0.05b	0.25b	0.12b	(Chung and Seinfeld., 2002)
F	GISS	2.49			0.43	0.30	-0.23	0.53	-0.06b	0.27b	0.09b	(Liao and Seinfeld, 2005)
G	SPRINTARS	2.67	0.029	82	0.53	0.15	-0.27	0.42	-0.07b	0.21b	0.01b	(Takemura et al., 2005)
H	GATORG	2.55			0.39	0.47	-0.06	0.55	-0.01b	0.27b	0.22b	(Jacobson, 2001b)
I	MOZGN	3.03	0.018				-0.34					(Ming et al., 2005a)
J	CCM				0.33			0.34				(Wang, 2004)
K	UIO-GCM				0.30			0.19				(Kirkevag and Iversen, 2002)
AeroCom: identical emissions used for year 1750 and 2000 (Schulz et al., 2006)												
L	UMI	1.16	0.0060	53	0.19	0.02	-0.23	0.25	-0.06b	0.12b	-0.01	(Liu and Penner, 2002)
M	UIO_CTM	1.12	0.0058	55	0.19	0.02	-0.16b	0.22b	-0.04	0.11	-0.05	(Myhre et al., 2003)
N	LOA	1.41	0.0085	52	0.25	0.14	-0.16c	0.32c	-0.04b	0.16b	0.02b	(Reddy and Boucher, 2004)
O	LSCE	1.50	0.0079	46	0.25	0.13	-0.17	0.30	-0.04b	0.15b	0.02b	(Schulz et al., 2006)
P	ECHAM5-HAM	1.00	0.0077		0.16	0.09	-0.10c	0.20c	-0.03b	0.10b	0.01	(Stier et al., 2005)
Q	GISS	1.22	0.0060	51	0.24	0.08	-0.14	0.22	-0.03b	0.11b	0.01b	(Koch, 2001)
R	UIO_GCM	0.88	0.0046	59	0.19	0.24	-0.06	0.36	-0.02b	0.18b	0.08b	(Iversen and Seland, 2002)
S	SPRINTARS	1.84	0.0200	49	0.37	0.22	-0.10	0.32	-0.01	0.13	0.06	(Takemura et al., 2005)
T	ULAQ	1.71	0.0075	58	0.38	-0.01	-0.09	0.08	-0.02b	0.04b	-0.03b	(Pitari et al., 2002)
	Average A–K	2.38	0.019		0.38	0.26	-0.24	0.44	-0.06	0.25	0.07	
	Average L–T	1.32	0.008	53	0.25	0.10	-0.13	0.25	-0.03	0.12	0.01	
	Stddev A–K	0.42	0.006		0.08	0.14	0.08	0.13	0.04	0.11	0.09	
	Stddev L–T	0.32	0.005	4	0.08	0.09	0.05	0.08	0.01	0.04	0.04	

Notes:

[a] MOZGN: MOZART (Model for OZone and Related chemical Tracers-GFDL(Geophysical Fluid Dynamics Laboratory)-NCAR (National Center for Atmospheric Research); for other models see Note (a) in Table 2.4.

[b] Models A to C are used to provide a split in sources derived from total POM and total BC: FFPOM = POM × 0.25; FFBC = BC × 0.5; BB = (BCPOM) – (FFPOM + FFBC); BC = 2 × FFBC; POM = 4 × FFPOM.

[c] Models L, O and Q to T are used to provide a split in components: POM = BCPOM × (-1.16); BC = BCPOM × 2.25.

due to BC from fossil fuels include those from Takemura et al. (2000), Reddy et al. (2005a) and Hansen et al. (2005) as summarised in Table 2.5. The results from a number of studies that continue to group the RF from fossil fuel with that from biomass burning are also shown. Recently published results (A to K) and AeroCom studies (L to T) suggest a combined RF from both sources of +0.44 ± 0.13 W m^{-2} and +0.29 ± 0.15 W m^{-2} respectively. The stronger RF estimates from the models A to K appear to be primarily due to stronger sources and column loadings as the direct RF/column loading is similar at approximately 1.2 to 1.3 W mg^{-1} (Table 2.5). Carbonaceous aerosol emission inventories suggest that approximately 34 to 38% of emissions come from biomass burning sources and the remainder from fossil fuel burning sources. Models that separate fossil fuel from biomass burning suggest an equal split in RF. This is applied to those estimates where the BC emissions are not explicitly separated into emission sources to provide an estimate of the RF due to fossil fuel BC. For the AeroCom results, the fossil fuel BC RF ranges from +0.08 to +0.18 W m^{-2} with a mean of +0.13 W m^{-2} and a standard deviation of 0.03 W m^{-2}. For model results A to K, the RFs range from +0.15 W m^{-2} to approximately +0.27 W m^{-2}, with a mean of +0.25 W m^{-2} and a standard deviation of 0.11 W m^{-2}.

The mean and median of the direct RF for fossil fuel BC from grouping all these studies together are +0.19 and +0.16 W m^{-2}, respectively, with a standard deviation of nearly 0.10 W m^{-2}. The standard deviation is multiplied by 1.645 to approximate the 90% confidence interval and the best estimate is rounded upwards slightly for simplicity, leading to a direct RF estimate of +0.20 ± 0.15 W m^{-2}. This estimate does not include the semi-direct effect or the BC impact on snow and ice surface albedo (see Sections 2.5.4 and 2.8.5.6)

2.4.4.4 Biomass Burning Aerosol

The TAR reported a contribution to the RF of roughly –0.4 W m^{-2} from the scattering components (mainly organic carbon and inorganic compounds) and +0.2 W m^{-2} from the absorbing components (BC) leading to an estimate of the RF of biomass burning aerosols of –0.20 W m^{-2} with a factor of three uncertainty. Note that the estimates of the BC RF from Hansen and Sato (2001), Hansen et al. (2002), Hansen and Nazarenko (2004) and Jacobson (2001a) include the RF component of BC from biomass burning aerosol. Radiative forcing due to biomass burning (primarily organic carbon, BC and inorganic compounds such as nitrate and sulphate) is grouped into a single RF, because biomass burning emissions are essentially uncontrolled. Emission inventories show more significant differences for biomass burning aerosols than for aerosols of fossil fuel origin (Kasischke and Penner, 2004). Furthermore, the pre-industrial levels of biomass burning aerosols are also difficult to quantify (Ito and Penner, 2005; Mouillot et al., 2006).

The Southern African Regional Science Initiative (SAFARI 2000: see Swap et al., 2002, 2003) took place in 2000 and 2001. The main objectives of the aerosol research were to investigate pyrogenic and biogenic emissions of aerosol in southern Africa (Eatough et al., 2003; Formenti et al., 2003; Hély et al., 2003), validate the remote sensing retrievals (Haywood et al., 2003b; Ichoku et al., 2003) and to study the influence of aerosols on the radiation budget via the direct and indirect effects (e.g., Bergstrom et al., 2003; Keil and Haywood, 2003; Myhre et al., 2003; Ross et al., 2003). The physical and optical properties of fresh and aged biomass burning aerosol were characterised by making intensive observations of aerosol size distributions, optical properties, and DRE through *in situ* aircraft measurements (e.g., Abel et al., 2003; Formenti et al., 2003; Haywood et al., 2003b; Magi and Hobbs, 2003; Kirchstetter et al., 2004) and radiometric measurements (e.g., Bergstrom et al., 2003; Eck et al., 2003). The ω_0 at 0.55 μm derived from near-source AERONET sites ranged from 0.85 to 0.89 (Eck et al., 2003), while ω_0 at 0.55 μm for aged aerosol was less absorbing at approximately 0.91 (Haywood et al., 2003b). Abel et al. (2003) showed evidence that ω_0 at 0.55 μm increased from approximately 0.85 to 0.90 over a time period of approximately two hours subsequent to emission, and attributed the result to the condensation of essentially non-absorbing organic gases onto existing aerosol particles. Fresh biomass burning aerosols produced by boreal forest fires appear to have weaker absorption than those from tropical fires, with ω_0 at 0.55 μm greater than 0.9 (Wong and Li 2002). Boreal fires may not exert a significant direct RF because a large proportion of the fires are of natural origin and no significant change over the industrial era is expected. However, Westerling et al. (2006) showed that earlier spring and higher temperatures in USA have increased wildfire activity and duration. The partially absorbing nature of biomass burning aerosol means it exerts an RF that is larger at the surface and in the atmospheric column than at the TOA (see Figure 2.12).

Modelling efforts have used data from measurement campaigns to improve the representation of the physical and optical properties as well as the vertical profile of biomass burning aerosol (Myhre et al., 2003; Penner et al., 2003; Section 2.4.5). These modifications have had important consequences for estimates of the RF due to biomass burning aerosols because the RF is significantly more positive when biomass burning aerosol overlies cloud than previously estimated (Keil and Haywood, 2003; Myhre et al., 2003; Abel et al., 2005). While the RF due to biomass burning aerosol in clear skies is certainly negative, the overall RF of biomass burning aerosol may be positive. In addition to modelling studies, observations of this effect have been made with satellite instruments. Hsu et al. (2003) used SeaWiFs, TOMS and CERES data to show that biomass burning aerosol emitted from Southeast Asia is frequently lifted above the clouds, leading to a reduction in reflected solar radiation over cloudy areas by up to 100 W m^{-2}, and pointed out that this effect could be due to a combination of direct and indirect effects. Similarly, Haywood et al. (2003a) showed that remotely sensed cloud liquid water and effective radius underlying biomass burning aerosol off the coast of Africa are subject to potentially large systematic biases. This may have important consequences for studies that use

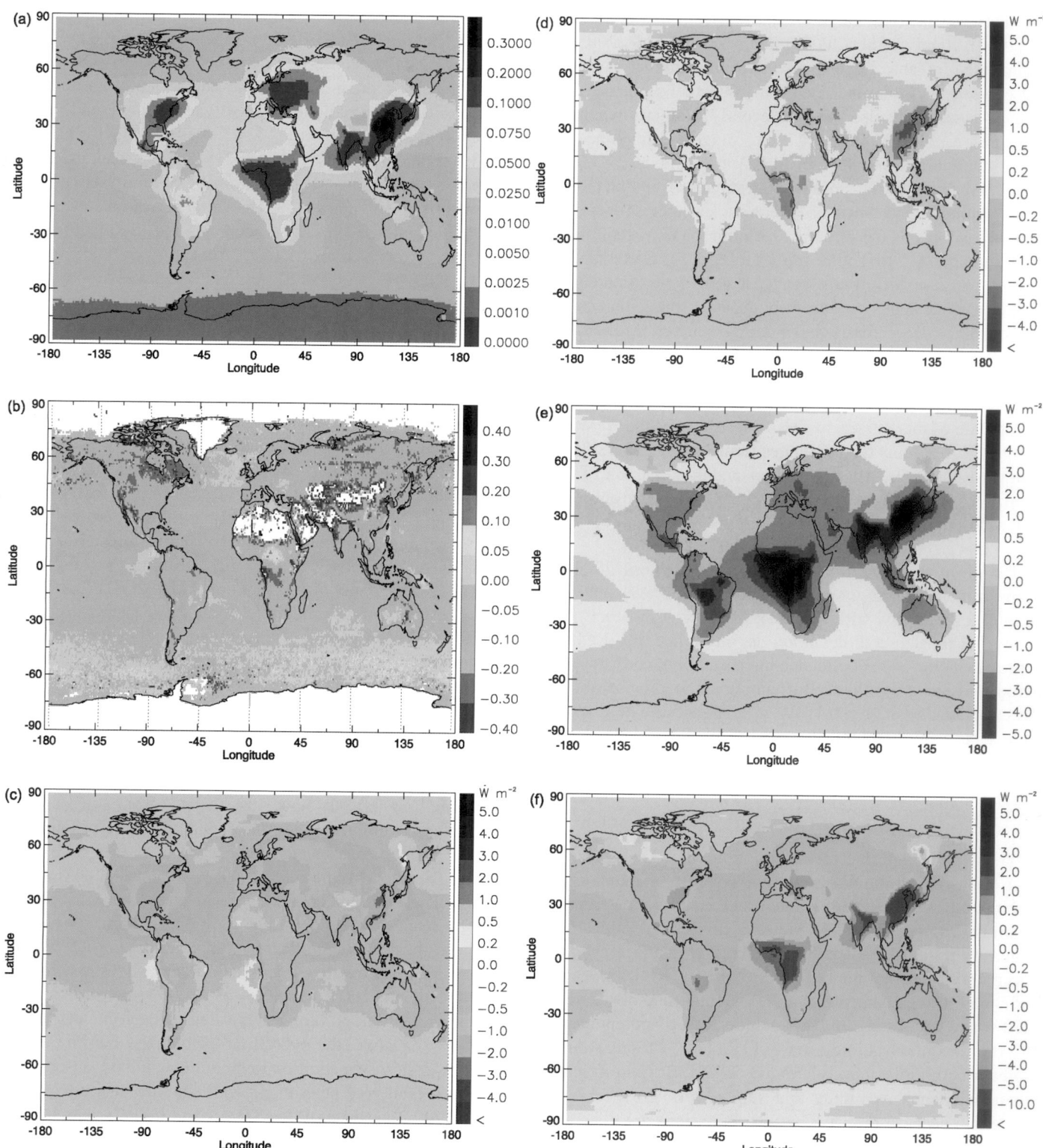

Figure 2.12. *Characteristic aerosol properties related to their radiative effects, derived as the mean of the results from the nine AeroCom models listed in Table 2.5. All panels except (b) relate to the combined anthropogenic aerosol effect. Panel (b) considers the total (natural plus anthropogenic) aerosol optical depth from the models. (a) Aerosol optical depth. (b) Difference in total aerosol optical depth between model and MODIS data. (c) Shortwave RF. (d) Standard deviation of RF from the model results. (e) Shortwave forcing of the atmosphere. (f) Shortwave surface forcing.*

correlations of τ_{aer} and cloud effective radius in estimating the indirect radiative effect of aerosols.

Since the biomass burning aerosols can exert a significant positive RF when above clouds, the aerosol vertical profile is critical in assessing the magnitude and even the sign of the direct RF in cloudy areas. Textor et al. (2006) showed that there are significant differences in aerosol vertical profiles between global aerosol models. These differences are evident in the results from the recently published studies and AeroCom models in Table 2.5. The most negative RF of –0.05 W m^{-2} is from the model of Koch (2001) and from the Myhre et al. (2003) AeroCom submission, while several models have RFs that are slightly positive. Hence, even the sign of the RF due to biomass burning aerosols is in question.

The mean and median of the direct RF for biomass burning aerosol from grouping all these studies together are similar at +0.04 and +0.02 W m^{-2}, respectively, with a standard deviation of 0.07 W m^{-2}. The standard deviation is multiplied by 1.645 to approximate the 90% confidence interval, leading to a direct RF estimate of +0.03 ± 0.12 W m^{-2}. This estimate of the direct RF is more positive than that of the TAR owing to improvements in the models in representing the absorption properties of the aerosol and the effects of biomass burning aerosol overlying clouds.

2.4.4.5   Nitrate Aerosol

Atmospheric ammonium nitrate aerosol forms if sulphate aerosol is fully neutralised and there is excess ammonia. The direct RF due to nitrate aerosol is therefore sensitive to atmospheric concentrations of ammonia as well as NO$_x$ emissions. In addition, the weakening of the RF of sulphate aerosol in many regions due to reduced emissions (Section 2.4.4.1) will be partially balanced by increases in the RF of nitrate aerosol (e.g., Liao and Seinfeld, 2005). The TAR did not quantify the RF due to nitrate aerosol owing to the large discrepancies in the studies available at that time. Van Dorland (1997) and Jacobson (2001a) suggested relatively minor global mean RFs of –0.03 and –0.05 W m^{-2}, respectively, while Adams et al. (2001) suggested a global mean RF as strong as –0.22 W m^{-2}. Subsequent studies include those of Schaap et al. (2004), who estimated that the RF of nitrate over Europe is about 25% of that due to sulphate aerosol, and of Martin et al. (2004), who reported –0.04 to –0.08 W m^{-2} for global mean RF due to nitrate. Further, Liao and Seinfeld (2005) estimated a global mean RF due to nitrate of –0.16 W m^{-2}. In this study, heterogeneous chemistry reactions on particles were included; this strengthens the RF due to nitrate and accounts for 25% of its RF. Feng and Penner (2007) estimated a large, global, fine-mode nitrate burden of 0.58 mg NO$_3$ m^{-2}, which would imply an equivalent of 20% of the mean anthropogenic sulphate burden. Surface observations of fine-mode nitrate particles show that high concentrations are mainly found in highly industrialised regions, while low concentrations are found in rural areas (Malm et al., 2004; Putaud et al., 2004). Atmospheric nitrate is

essentially non-absorbing in the visible spectrum, and laboratory studies have been performed to determine the hygroscopicity of the aerosols (e.g., Tang 1997; Martin et al., 2004 and references therein). In the AeroCom exercise, nitrate aerosols were not included so fewer estimates of this compound exist compared to the other aerosol species considered.

The mean direct RF for nitrate is estimated to be –0.10 W m^{-2} at the TOA, and the conservative scattering nature means a similar flux change at the surface. However, the uncertainty in this estimate is necessarily large owing to the relatively small number of studies that have been performed and the considerable uncertainty in estimates, for example, of the nitrate τ_{aer}. Thus, a direct RF of –0.10 ± 0.10 W m^{-2} is tentatively adopted, but it is acknowledged that the number of studies performed is insufficient for accurate characterisation of the magnitude and uncertainty of the RF.

2.4.4.6   Mineral Dust Aerosol

Mineral dust from anthropogenic sources originates mainly from agricultural practices (harvesting, ploughing, overgrazing), changes in surface water (e.g., Caspian and Aral Sea, Owens Lake) and industrial practices (e.g., cement production, transport) (Prospero et al., 2002). The TAR reported that the RF due to anthropogenic mineral dust lies in the range of +0.4 to –0.6 W m^{-2}, and did not assign a best estimate because of the difficulties in determining the anthropogenic contribution to the total dust loading and the uncertainties in the optical properties of dust and in evaluating the competing shortwave and longwave radiative effects. For the sign and magnitude of the mineral dust RF, the most important factor for the shortwave RF is the single scattering albedo whereas the longwave RF is dependent on the vertical profile of the dust.

Tegen and Fung (1995) estimated the anthropogenic contribution to mineral dust to be 30 to 50% of the total dust burden in the atmosphere. Tegen et al. (2004) provided an updated, alternative estimate by comparing observations of visibility, as a proxy for dust events, from over 2,000 surface stations with model results, and suggested that only 5 to 7% of mineral dust comes from anthropogenic agricultural sources. Yoshioka et al. (2005) suggested that a model simulation best reproduces the North African TOMS aerosol index observations when the cultivation source in the Sahel region contributes 0 to 15% to the total dust emissions in North Africa. A 35-year dust record established from Barbados surface dust and satellite observations from TOMS and the European geostationary meteorological satellite (Meteosat) show the importance of climate control and Sahel drought for interannual and decadal dust variability, with no overall trend yet documented (Chiapello et al., 2005). As further detailed in Section 7.3, climate change and CO$_2$ variations on various time scales can change vegetation cover in semi-arid regions. Such processes dominate over land use changes as defined above, which would give rise to anthropogenic dust emissions (Mahowald and Luo, 2003; Moulin and Chiapello, 2004; Tegen et al., 2004). A best guess of 0 to 20% anthropogenic dust burden from these works is used here, but it is acknowledged that a very large uncertainty remains because the methods used cannot exclude either a reduction of 24% in present-day dust nor a large anthropogenic contribution of up to 50% (Mahowald and Luo, 2003; Mahowald et al., 2004; Tegen et al., 2005). The RF efficiency of anthropogenic dust has not been well differentiated from that of natural dust and it is assumed that they are equal. The RF due to dust emission changes induced by circulation changes between 1750 and the present are difficult to quantify and not included here (see also Section 7.5).

In situ measurements of the optical properties of local Saharan dust (e.g., Haywood et al., 2003c; Tanré et al., 2003), transported Saharan mineral dust (e.g., Kaufman et al., 2001; Moulin et al., 2001; Coen et al., 2004) and Asian mineral dust (Huebert et al., 2003; Clarke et al., 2004; Shi et al., 2005; Mikami et al., 2006) reveal that dust is considerably less absorbing in the solar spectrum than suggested by previous dust models such as that of WMO (1986). These new, spectral, simultaneous remote and *in situ* observations suggest that the single scattering albedo (ω_0) of pure dust at a wavelength of 0.67 µm is predominantly in the range 0.90 to 0.99, with a central global estimate of 0.96. This is in accordance with the bottom-up modelling of ω_0 based on the haematite content in desert dust sources (Claquin et al., 1999; Shi et al., 2005). Analyses of ω_0 from long-term AERONET sites influenced by Saharan dust suggest an average ω_0 of 0.95 at 0.67 µm (Dubovik et al., 2002), while unpolluted Asian dust during the Aeolian Dust Experiment on Climate (ADEC) had an average ω_0 of 0.93 at 0.67 µm (Mikami et al., 2006 and references therein). These high ω_0 values suggest that a positive RF by dust in the solar region of the spectrum is unlikely. However, absorption by particles from source regions with variable mineralogical distributions is generally not represented by global models.

Measurements of the DRE of mineral dust over ocean regions, where natural and anthropogenic contributions are indistinguishably mixed, suggest that the local DRE may be extremely strong: Haywood et al. (2003b) made aircraft-based measurements of the local instantaneous shortwave DRE of as strong as –130 W m^{-2} off the coast of West Africa. Hsu et al. (2000) used Earth Radiation Budget Experiment (ERBE) and TOMS data to determine a peak monthly mean shortwave DRE of around –45 W m^{-2} for July 1985. Interferometer measurements from aircraft and the surface have now measured the spectral signature of mineral dust for a number of cases (e.g., Highwood et al., 2003) indicating an absorption peak in the centre of the 8 to 13 µm atmospheric window. Hsu et al. (2000) determined a longwave DRE over land areas of North Africa of up to +25 W m^{-2} for July 1985; similar results were presented by Haywood et al. (2005) who determined a peak longwave DRE of up to +50 W m^{-2} at the top of the atmosphere for July 2003.

Recent model simulations report the total anthropogenic and natural dust DRE, its components and the net effect as follows (shortwave / longwave = net TOA, in W m^{-2}): H. Liao et al. (2004): –0.21 / +0.31 = +0.1; Reddy et al. (2005a): –0.28 / +0.14 = –0.14; Jacobson (2001a): –0.20 / +0.07 =

–0.13; reference case and [range] of sensitivity experiments in Myhre and Stordal (2001a, except case 6 and 7): –0.53 [–1.4 to +0.2] / +0.13 [+0.0 to +0.8] = –0.4 [–1.4 to +1.0]; and from AeroCom database models, GISS: –0.75 / (+0.19) = (–0.56); UIO-CTM*: –0.56 / (+0.19) = (–0.37); LSCE*: –0.6 / +0.3 = –0.3; UMI*: –0.54 / (+0.19) = (–0.35). (See Table 2.4, Note (a) for model descriptions.) The (*) star marked models use a single scattering albedo (approximately 0.96 at 0.67 µm) that is more representative of recent measurements and show more negative shortwave effects. A mean longwave DRE of 0.19 W m^{-2} is assumed for GISS, UMI and UIO-CTM. The scatter of dust DRE estimates reflects the fact that dust burden and τ_{aer} vary by ±40 and ±44%, respectively, computed as standard deviation from 16 AeroCom A model simulations (Textor et al., 2006; Kinne et al., 2006). Dust emissions from different studies range between 1,000 and 2,150 Tg yr^{-1} (Zender, 2004). Finally, a major effect of dust may be in reducing the burden of anthropogenic species at sub-micron sizes and reducing their residence time (Bauer and Koch, 2005; see Section 2.4.5.7).

The range of the reported dust net DRE (–0.56 to +0.1 W m^{-2}), the revised anthropogenic contribution to dust DRE of 0 to 20% and the revised absorption properties of dust support a small negative value for the anthropogenic direct RF for dust of –0.1 W m^{-2}. The 90% confidence level is estimated to be ±0.2 W m^{-2}, reflecting the uncertainty in total dust emissions and burdens and the range of possible anthropogenic dust fractions. At the limits of this uncertainty range, anthropogenic dust RF is as negative as –0.3 W m^{-2} and as positive as +0.1 W m^{-2}. This range includes all dust DREs reported above, assuming a maximum 20% anthropogenic dust fraction, except the most positive DRE from Myhre and Stordal (2001a).

2.4.4.7 Direct RF for Combined Total Aerosol

The TAR reported RF values associated with several aerosol components but did not provide an estimate of the overall aerosol RF. Improved and intensified *in situ* observations and remote sensing of aerosols suggest that the range of combined aerosol RF is now better constrained. For model results, extensive validation now exists for combined aerosol properties, representing the whole vertical column of the atmosphere, such as τ_{aer}. Using a combined estimate implicitly provides an alternative procedure to estimating the RF uncertainty. This approach may be more robust than propagating uncertainties from all individual aerosol components. Furthermore, a combined RF estimate accounts for nonlinear processes due to aerosol dynamics and interactions between radiation field and aerosols. The role of nonlinear processes of aerosol dynamics in RF has been recently studied in global aerosol models that account for the internally mixed nature of aerosol particles (Jacobson, 2001a; Kirkevåg and Iversen, 2002; Liao and Seinfeld, 2005; Takemura et al., 2005; Stier et al., 2006b). Mixing of aerosol particle populations influences the radiative properties of the combined aerosol, because mixing changes size, chemical composition, state and shape, and this feed backs to the aerosol removal and formation processes itself. Chung and Seinfeld (2002), in reviewing studies where BC is mixed either externally or internally with various other components, showed that BC exerts a stronger positive direct RF when mixed internally. Although the source-related processes for anthropogenic aerosols favour their submicron nature, natural aerosols enter the picture by providing a condensation surface for aerosol precursor gases. Heterogeneous reactions on sea salt and dust can reduce the sub-micron sulphate load by 28% (H. Liao et al., 2004) thereby reducing the direct and indirect RFs. Bauer and Koch (2005) estimated the sulphate RF to weaken from –0.25 to –0.18 W m^{-2} when dust is allowed to interact with the sulphur cycle. It would be useful to identify the RF contribution attributable to different source categories (Section 2.9.3 investigates this). However, few models have separated out the RF from specific emission source categories. Estimating the combined aerosol RF is a first step to quantify the anthropogenic perturbation to the aerosol and climate system caused by individual source categories.

A central model-derived estimate for the aerosol direct RF is based here on a compilation of recent simulation results using multi-component global aerosol models (see Table 2.6). This is a robust method for several reasons. The complexity of multi-component aerosol simulations captures nonlinear effects. Combining model results removes part of the errors in individual model formulations. As shown by Textor et al. (2006), the model-specific treatment of transport and removal processes is partly responsible for the correlated dispersion of the different aerosol components. A less dispersive model with smaller burdens necessarily has fewer scattering and absorbing aerosols interacting with the radiation field. An error in accounting for cloud cover would affect the all-sky RF from all aerosol components. Such errors result in correlated RF efficiencies for major aerosol components within a given model. Directly combining aerosol RF results gives a more realistic aerosol RF uncertainty estimate. The AeroCom compilation suggests significant differences in the modelled local and regional composition of the aerosol (see also Figure 2.12), but an overall reproduction of the total τ_{aer} variability can be performed (Kinne et al., 2006). The scatter in model performance suggests that currently no preference or weighting of individual model results can be used (Kinne et al., 2006). The aerosol RF taken together from several models is more robust than an analysis per component or by just one model. The mean estimate from Table 2.6 of the total aerosol direct RF is –0.2 W m^{-2}, with a standard deviation of ±0.2 W m^{-2}. This is a low-end estimate for both the aerosol RF and uncertainty because nitrate (estimated as –0.1 W m^{-2}, see Section 2.4.4.5) and anthropogenic mineral dust (estimated as –0.1 W m^{-2}, see

Section 2.4.4.6) are missing in most of the model simulations. Adding their contribution yields an overall model-derived aerosol direct RF of –0.4 W m^{-2} (90% confidence interval: 0 to –0.8 W m^{-2}).

Three satellite-based measurement estimates of the aerosol direct RF have become available, which all suggest a more negative aerosol RF than the model studies (see Section 2.4.2.1.3). Bellouin et al. (2005) computed a TOA aerosol RF of –0.8 ± 0.1 W m^{-2}. Chung et al. (2005), based upon similarly extensive calculations, estimated the value to be –0.35 ± 0.25 W m^{-2}, and Yu et al. (2006) estimated it to be –0.5 ± 0.33 W m^{-2}. A central measurement-based estimate would suggest an aerosol direct RF of –0.55 W m^{-2}. Figure 2.13 shows the observationally based aerosol direct RF estimates together with the model estimates published since the TAR.

The discrepancy between measurements and models is also apparent in oceanic clear-sky conditions where the measurement-based estimate of the combined aerosol DRE including natural aerosols is considered unbiased. In these areas, models underestimate the negative aerosol DRE by 20 to 40% (Yu et al., 2006). The anthropogenic fraction of τ_{aer} is similar between model and measurement based studies. Kaufman et al. (2005a) used satellite-observed fine-mode τ_{aer} to estimate the anthropogenic τ_{aer}. Correcting for fine-mode τ_{aer}

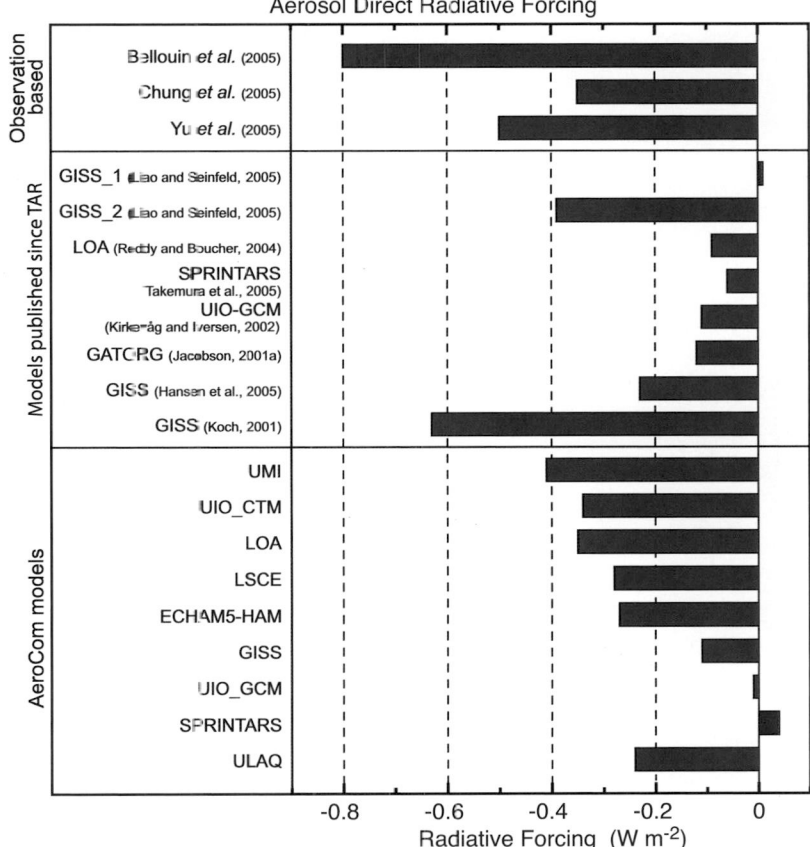

Figure 2.13. Estimates of the direct aerosol RF from observationally based studies, independent modelling studies, and AeroCom results with identical aerosol and aerosol precursor emissions. GISS_1 refers to a study employing an internal mixture of aerosol, and GISS_2 to a study employing an external mixture. See Table 2.4, Note (a) for descriptions of models.

Table 2.6. Quantities related to estimates of the aerosol direct RF. Recent estimates of anthropogenic aerosol load (LOAD), anthropogenic aerosol optical depth ($\tau_{aer\ ant}$), its fraction of the present-day total aerosol optical depth (τ_{aer}), cloud cover in aerosol model, total aerosol direct radiative forcing (RF) for clear sky and all sky conditions, surface forcing and atmospheric all-sky forcing.

No	Model[a]	LOAD (mg m⁻²)	τ_{aer} (0.55 μm)	$\tau_{aer\ ant}$ (0.55 μm) (%)	Cloud Cover (%)	RF top clear sky (W m⁻²)	RF top all sky (W m⁻²)	Surface Forcing all sky (W m⁻²)	Atmospheric Forcing all sky (W m⁻²)	Reference
Published since IPCC, 2001										
A	GISS	5.0			79%		−0.39[b] +0.01[c]	−1.98[b] −2.42[c]	1.59[b] 2.43[c]	(Liao and Seinfeld, 2005)
B	LOA	6.0	0.049	34%	70%	−0.53	−0.09	−1.92	1.86	(Reddy and Boucher, 2004)
C	SPRINTARS	4.8	0.044	50%	63%	−0.77	−0.06			(Takemura et al., 2005)
D	UIO-GCM	2.7			57%		−0.11			(Kirkevag and Iversen, 2002)
E	GATORG	6.4[d]			62%	−0.89	−0.12	−2.5	2.38	(Jacobson, 2001a)
F	GISS	6.7	0.049				−0.23			(Hansen et al., 2005)
G	GISS	5.6	0.040				−0.63			(Koch, 2001)
AeroCom: identical emissions used for year 1750 and 2000 (Schulz et al., 2006)										
H	UMI	4.0	0.028	25%	63%	−0.80	−0.41	−1.24	0.84	(Liu and Penner, 2002)
I	UIO_CTM	3.0	0.026	19%	70%	−0.85	−0.34	−0.95	0.61	(Myhre et al., 2003)
J	LOA	5.3	0.046	28%	70%	−0.80	−0.35	−1.49	1.14	(Reddy and Boucher, 2004)
K	LSCE	4.8	0.033	40%	62%	−0.94	−0.28	−0.93	0.66	(Schulz et al., 2006)
L	ECHAM5	4.3	0.032	30%	62%	−0.64	−0.27	−0.98	0.71	(Stier et al., 2005)
M	GISS	2.8	0.014	11%	57%	−0.29	−0.11	−0.81	0.79	(Koch, 2001)
N	UIO_GCM	2.8	0.017	11%	57%	−0.01	−0.01	−0.84	0.84	(Kirkevag and Iversen, 2002)
O	SPRINTARS	3.2	0.036	44%	62%	−0.35	+0.04	−0.91	0.96	(Takemura et al., 2005)
P	ULAQ	3.7	0.030	23%		−0.79	−0.24			(Pitari et al., 2002)
	Average A–G	5.1	0.046	42%	67%	−0.73	−0.23	−2.21	2.07	
	Average H–P	3.8	0.029	26%	63%	−0.68	−0.22	−1.02	0.82	
	Stddev A–G	1.4	0.004			0.18	0.21			
	Stddev H–P	0.9	0.010	11%	5%	0.24	0.16	0.23	0.17	
	Average A–P	4.3	0.035	29%	64%	−0.70	−0.22	−1.21	1.24	
	Stddev A–P	1.3	0.012	13%	7%	0.26	0.18	0.44	0.65	
	Minimum A–P	2.7	0.014	11%	57%	−0.94	−0.63	−1.98	0.61	
	Maximum A–P	6.7	0.049	50%	79%	−0.29	0.04	−0.81	2.43	

Notes:
a See Note (a) in Table 2.4 for model information.
b External mixture.
c Internal mixture.
d The load excludes that of mineral dust, some of which was considered anthropogenic in Jacobson (2001a).

contributions from dust and sea salt, they found 21% of the total τ_{aer} to be anthropogenic, while Table 2.6 suggests that 29% of τ_{aer} is anthropogenic. Finally, cloud contamination of satellite products, aerosol absorption above clouds, not accounted for in some of the measurement-based estimates, and the complex assumptions about aerosol properties in both methods can contribute to the present discrepancy and increase uncertainty in aerosol RF.

A large source of uncertainty in the aerosol RF estimates is associated with aerosol absorption. Sato et al. (2003) determined the absorption τ_{aer} from AERONET measurements and suggested that aerosol absorption simulated by global aerosol models is underestimated by a factor of two to four. Schuster et al. (2005) estimated the BC loading over continental-scale regions. The results suggest that the model concentrations and absorption τ_{aer} of BC from models are lower than those derived from AERONET. Some of this difference in concentrations could be explained by the assumption that all aerosol absorption is due to BC (Schuster et al., 2005), while a significant fraction may be due to absorption by organic aerosol and mineral dust (see Sections 2.4.4.2, and 2.4.4.6). Furthermore, Reddy et al. (2005a) show that comparison of the aerosol absorption τ_{aer} from models against those from AERONET must be performed very carefully, reducing the discrepancy between their model and AERONET derived aerosol absorption τ_{aer} from a factor of 4 to a factor of 1.2 by careful co-sampling of AERONET and model data. As mentioned above, uncertainty in the vertical position of absorbing aerosol relative to clouds can lead to large uncertainty in the TOA aerosol RF.

The partly absorbing nature of the aerosol is responsible for a heating of the lower-tropospheric column and also results in the surface forcing being considerably more negative than TOA RF, results that have been confirmed through several experimental and observational studies as discussed in earlier sections. Table 2.6 summarises the surface forcing obtained in the different models. Figure 2.12 depicts the regional distribution of several important parameters for assessing the regional impact of aerosol RF. The results are based on a mean model constructed from AeroCom simulation results B and PRE. Anthropogenic τ_{aer} (Figure 2.12a) is shown to have local maxima in industrialised regions and in areas dominated by biomass burning. The difference between simulated and observed τ_{aer} shows that regionally τ_{aer} can be up to 0.1 (Figure 2.12b). Figure 2.12c suggests that there are regions off Southern Africa where the biomass burning aerosol above clouds leads to a local positive RF. Figure 2.12d shows the local variability as the standard deviation from nine models of the overall RF. The largest uncertainties of ±3 W m^{-2} are found in East Asia and in the African biomass burning regions. Figure 2.12e reveals that an average of 0.9 W m^{-2} heating can be expected in the atmospheric column as a consequence of absorption by anthropogenic aerosols. Regionally, this can reach annually averaged values exceeding 5 W m^{-2}. These regional effects and the negative surface forcing in the shortwave (Figure 2.12f) is expected to exert an important effect on climate through alteration of the hydrological cycle.

An uncertainty estimate for the model-derived aerosol direct RF can be based upon two alternative error analyses:

1) An error propagation analysis using the errors given in the sections on sulphate, fossil fuel BC and organic carbon, biomass burning aerosol, nitrate and anthropogenic mineral dust. Assuming linear additivity of the errors, this results in an overall 90% confidence level uncertainty of 0.4 W m^{-2}.

2) The standard deviation of the aerosol direct RF results in Table 2.6, multiplied by 1.645, suggests a 90% confidence level uncertainty of 0.3 W m^{-2}, or 0.4 W m^{-2} when mineral dust and nitrate aerosol are accounted for.

Therefore, the results summarised in Table 2.6 and Figure 2.13, together with the estimates of nitrate and mineral dust RF combined with the measurement-based estimates, provide an estimate for the combined aerosol direct RF of -0.50 ± 0.40 W m^{-2}. The progress in both global modelling and measurements of the direct RF of aerosol leads to a medium-low level of scientific understanding (see Section 2.9, Table 2.11).

2.4.5 Aerosol Influence on Clouds (Cloud Albedo Effect)

As pointed out in Section 2.4.1, aerosol particles affect the formation and properties of clouds. Only a subset of the aerosol population acts as cloud condensation nuclei (CCN) and/or ice nuclei (IN). Increases in ambient concentrations of CCN and IN due to anthropogenic activities can modify the microphysical properties of clouds, thereby affecting the climate system (Penner et al., 2001; Ramanathan et al., 2001a, Jacob et al., 2005). Several mechanisms are involved, as presented schematically in Figure 2.10. As noted in Ramaswamy et al. (2001), enhanced aerosol concentrations can lead to an increase in the albedo of clouds under the assumption of fixed liquid water content (Junge, 1975; Twomey, 1977); this mechanism is referred to in this report as the 'cloud albedo effect'. The aerosol enhancements have also been hypothesised to lead to an increase in the lifetime of clouds (Albrecht, 1989); this mechanism is referred to in this report as the 'cloud lifetime effect' and discussed in Section 7.5.

The interactions between aerosol particles (natural and anthropogenic in origin) and clouds are complex and can be nonlinear (Ramaswamy et al., 2001). The size and chemical composition of the initial nuclei (e.g., anthropogenic sulphates, nitrates, dust, organic carbon and BC) are important in the activation and early growth of the cloud droplets, particularly the water-soluble fraction and presence of compounds that affect surface tension (McFiggans et al., 2006 and references therein). Cloud optical properties are a function of wavelength. They depend on the characteristics of the droplet size distributions and ice crystal concentrations, and on the morphology of the various cloud types.

The interactions of increased concentrations of anthropogenic particles with shallow (stratocumulus and shallow cumulus) and deep convective clouds (with mixed phase) are discussed in this subsection. This section presents new observations and model estimates of the albedo effect. The associated RF in the context of liquid water clouds is assessed. In-depth discussion of the induced changes that are not considered as RFs (e.g., semi-direct and cloud cover and lifetime effects, thermodynamic response and changes in precipitation development) are presented in Section 7.5. The impacts of contrails and aviation-induced cirrus are discussed in Section 2.6 and the indirect impacts of aerosol on snow albedo are discussed in Section 2.5.4.

2.4.5.1 Link Between Aerosol Particles and Cloud Microphysics

The local impact of anthropogenic aerosols has been known for a long time. For example, smoke from sugarcane and forest fires was shown to reduce cloud droplet sizes in early case studies utilising *in situ* aircraft observations (Warner and Twomey, 1967; Eagan et al., 1974). On a regional scale, studies have shown that heavy smoke from forest fires in the Amazon Basin have led to increased cloud droplet number concentrations and to reduced cloud droplet sizes (Reid et al., 1999; Andreae et al., 2004; Mircea et al., 2005). The evidence concerning aerosol modification of clouds provided by the ship track observations reported in the TAR has been further confirmed, to a large extent qualitatively, by results from a number of studies using *in situ* aircraft and satellite data, covering continental cases and regional studies. Twohy et al. (2005) explored the relationship between aerosols and clouds in nine stratocumulus cases, indicating an inverse relationship between particle number and droplet size, but no correlation was found between albedo and particle concentration in the entire data set. Feingold et al. (2003), Kim et al. (2003) and Penner et al. (2004) presented evidence of an increase in the reflectance in continental stratocumulus cases, utilising remote sensing techniques at specific field sites. The estimates in Feingold et al. (2003) confirm that the relationship between aerosol and cloud droplet number concentrations is nonlinear, that is $N_d \approx (N_a)^b$, where N_d is the cloud drop number density and N_a is the aerosol number concentration. The parameter b in this relationship can vary widely, with values ranging from 0.06 to 0.48 (low values of b correspond to low hygroscopicity). This range highlights the sensitivity to aerosol characteristics (primarily size distribution), updraft velocity and the usage of aerosol extinction as a proxy for CCN (Feingold, 2003). Disparity in the estimates of b (or equivalent) based on satellite studies (Nakajima et al., 2001; Breon et al., 2002) suggests that a quantitative estimate of the albedo effect from remote sensors is problematic (Rosenfeld and Feingold 2003), particularly since measurements are not considered for similar liquid water paths.

Many recent studies highlight the importance of aerosol particle composition in the activation process and droplet spectral evolution (indicated in the early laboratory work of Gunn and Philips, 1957), but the picture that emerges is not complete. Airborne aerosol mass spectrometers provide firm evidence that ambient aerosols consist mostly of internal mixtures, for example, biomass burning components, organics and soot are mixed with other aerosol components (McFiggans et al., 2006). Mircea et al. (2005) showed the importance of the organic aerosol fraction in the activation of biomass burning aerosol particles. The presence of internal mixtures (e.g., sea salt and organic compounds) can affect the uptake of water and the resulting optical properties compared to a pure sea salt particle (Randles et al., 2004). Furthermore, the varying contents of water-soluble and insoluble substances in internally mixed particles, the vast diversity of organics, and the resultant effects on cloud droplet sizes, makes the situation even more complex. Earlier observations of fog water (Facchini et al., 1999, 2000) suggested that the presence of organic aerosols would reduce surface tension and lead to a significant increase in the cloud droplet number concentration (Nenes et al., 2002; Rissler et al., 2004; Lohmann and Leck, 2005; Ming et al., 2005a; McFiggans et al., 2006). On the other hand, Feingold and Chuang (2002) and Shantz et al. (2003) indicated that organic coating on CCN delayed activation, leading to a reduction in drop number and a broadening of the cloud droplet spectrum, which had not been previously considered. Ervens et al. (2005) addressed numerous composition effects in unison to show that the effect of composition on droplet number concentration is much less than suggested by studies that address individual composition effects, such as surface tension. The different relationships observed between cloud optical depth and liquid water path in clean and polluted stratocumulus clouds (Penner et al., 2004) have been explained by differences in sub-cloud aerosol particle distributions, while some contribution can be attributed to CCN composition (e.g., internally mixed insoluble dust; Asano et al., 2002). Nevertheless, the review by McFiggans et al. (2006) points to the remaining difficulties in quantitatively explaining the relationship between aerosol size and composition and the resulting droplet size distribution. Dusek et al. (2006) concluded that the ability of a particle to act as a CCN is largely controlled by size rather than composition.

The complexity of the aerosol-cloud interactions and local atmospheric conditions where the clouds are developing are factors in the large variation evidenced for this phenomenon. Advances have been made in the understanding of the regional and/or global impact based on observational studies, particularly for low-level stratiform clouds that constitute a simpler cloud system to study than many of the other cloud types. Column aerosol number concentration and column cloud droplet concentration over the oceans from the AVHRR (Nakajima et al., 2001) indicated a positive correlation, and an increase in shortwave reflectance of low-level, warm clouds with increasing cloud optical thickness, while liquid water path (LWP) remained unmodified. While these results are only applicable over the oceans and are based on data for only four months, the positive correlation between an increase in cloud reflectance and an enhanced ambient aerosol concentration has been confirmed by other studies (Brenguier et al., 2000a,b; Rosenfeld et al., 2002). However, other studies highlight the sensitivity to LWP, linking high pollution entrained into clouds to a decrease in LWP and a reduction in the observed cloud reflectance (Jiang et al., 2002; Brenguier et al., 2003; Twohy et al., 2005). Still others (Han et al., 2002, using AVHRR observations) have reported an absence of LWP changes in response to increases in the column-averaged droplet number concentration, this occurred for one-third of the cloud cases studied for which optical depths ranged between 1 and 15. Results of large-eddy simulations of stratocumulus (Jiang et al., 2002; Ackerman et al., 2004; Lu and Seinfeld, 2005) and cumulus clouds (Jiang and Feingold, 2006; Xue and Feingold, 2006) seem to confirm the lack of increase in LWP due to increases in aerosols; they point to a dependence on precipitation rate and relative humidity above the clouds (Ackerman et al., 2004). The studies above highlight the difficulty of devising observational studies that can isolate the albedo effect from other effects (e.g., meteorological variability, cloud dynamics) that influence LWP and therefore cloud RF.

Results from the POLDER satellite instrument, which retrieves both submicron aerosol loading and cloud droplet size, suggest much larger cloud effective radii in remote oceanic regions than in the highly polluted continental source areas and downwind adjacent oceanic areas, namely from a maximum of 14 μm down to 6 μm (Bréon et al., 2002). This confirms earlier studies of hemispheric differences using AVHRR. Further, the POLDER- and AVHRR-derived correlations between aerosol and cloud parameters are consistent with an aerosol indirect effect (Sekiguchi et al., 2003). These results suggest that the impact of aerosols on cloud microphysics is global. Note that the satellite measurements of aerosol loading and cloud droplet size are not coincident, and an aerosol index is not determined in the presence of clouds. Further, there is a lack of simultaneous measurements of LWP, which makes assessment of the cloud albedo RF difficult.

The albedo effect is also estimated from studies that combined satellite retrievals with a CTM, for example, in the case of two pollution episodes over the mid-latitude Atlantic Ocean. Results indicated a brightening of clouds over a time scale of a few days in instances when LWP did not undergo any significant changes (Hashvardhan et al., 2002; Schwartz et al., 2002; Krüger and Graßl, 2002). There have been fewer studies on aerosol-cloud relationships under more complex meteorological conditions (e.g., simultaneous presence of different cloud types).

The presence of insoluble particles within ice crystals constituting clouds formed at cold temperatures has a significant influence on the radiation transfer. The inclusions of scattering and absorbing particles within large ice crystals (Macke et al., 1996) suggest a significant effect. Hence, when soot particles are embedded, there is an increase in the asymmetry parameter and thus forward scattering. In contrast, inclusions of ammonium sulphate or air bubbles lead to a decrease in the asymmetry parameter of ice clouds. Given the recent observations of partially insoluble nuclei in ice crystals (Cziczo et al., 2004)

and the presence of small crystal populations, there is a need to further develop the solution for radiative transfer through such systems.

2.4.5.2   Estimates of the Radiative Forcing from Models

General Circulation Models constitute an important and useful tool to estimate the global mean RF associated with the cloud albedo effect of anthropogenic aerosols. The model estimates of the changes in cloud reflectance are based on forward calculations, considering emissions of anthropogenic primary particles and secondary particle production from anthropogenic gases. Since the TAR, the cloud albedo effect has been estimated in a more systematic and rigorous way (allowing, for example, for the relaxation of the fixed LWC criterion), and more modelling results are now available. Most climate models use parametrizations to relate the cloud droplet number concentration to aerosol concentration; these vary in complexity from simple empirical fits to more physically based relationships. Some models are run under an increasing greenhouse gas concentration scenario and include estimates of present-day aerosol loadings (including primary and secondary aerosol production from anthropogenic sources). These global modelling studies (Table 2.7) have a limitation arising from the underlying uncertainties in aerosol emissions (e.g., emission rates of primary particles and of secondary particle precursors). Another limitation is the inability to perform a meaningful comparison between the various model results owing to differing formulations of relationships between aerosol particle concentrations and cloud droplet or ice crystal populations; this, in turn, yields differences in the impact of microphysical changes on the optical properties of clouds. Further, even when the relationships used in different models are similar, there are noticeable differences in the spatial distributions of the simulated low-level clouds. Individual models' physics have undergone considerable evolution, and it is difficult to clearly identify all the changes in the models as they have evolved. While GCMs have other well-known limitations, such as coarse spatial resolution, inaccurate representation of convection and hence updraft velocities leading to aerosol activation and cloud formation processes, and microphysical parametrizations, they nevertheless remain an essential tool for quantifying the global cloud albedo effect. In Table 2.7, differences in the treatment of the aerosol mixtures (internal or external, with the latter being the more frequently employed method) are noted. Case studies of droplet activation indicate a clear sensitivity to the aerosol composition (McFiggans et al., 2006); additionally, radiative transfer is sensitive to the aerosol composition and the insoluble fraction present in the cloud droplets.

All models estimate a negative global mean RF associated with the cloud albedo effect, with the range of model results varying widely, from −0.22 to −1.85 W m^{-2}. There are considerable differences in the treatment of aerosol, cloud processes and aerosol-cloud interaction processes in these models. Several models include an interactive sulphur cycle and anthropogenic aerosol particles composed of sulphate, as

Table 2.7. *Published model studies of the RF due to cloud albedo effect, in the context of liquid water clouds, with a listing of the relevant modelling details.*

Model	Model type[a]	Aerosol species[b]	Aerosol mixtures[c]	Cloud types included	Microphysics	Radiative Forcing (W m⁻²)[d]
Lohmann et al. (2000)	AGCM + sulphur cycle (ECHAM4)	S, OC, BC, SS, D	I	warm and mixed phase	Droplet number concentration and LWC, Beheng (1994); Sundqvist et al. (1989). Also, mass and number from field observations	–1.1 (total) **–0.45 (albedo)**
			E			–1.5 (total)
Jones et al. (2001)	AGCM + sulphur cycle, fixed SST (Hadley)	S, SS, D (a crude attempt for D over land, no radiation)	E	stratiform and shallow cumulus	Droplet number concentration and LWC, Wilson and Ballard (1999); Smith (1990); Tripoli and Cotton (1980); Bower et al. (1994). Warm and mixed phase, radiative treatment of anvil cirrus, non-spherical ice particles	–1.89 (total) **–1.34 (albedo)**
Williams et al. (2001b)	GCM with slab ocean + sulphur cycle (Hadley)	S, SS	E	stratiform and shallow cumulus	Jones et al. (2001)	–1.69 (total) **–1.37 (albedo)**
	AGCM, fixed SST					–1.62 (total) **–1.43(albedo)**
Rotstayn and Penner (2001)	AGCM (CSIRO), fixed SST and sulphur loading	S	n.a.	warm and mixed phase	Rotstayn (1997); Rotstayn et al. (2000)	**–1.39 (albedo)**
Rotstayn and Liu (2003)	Interactive sulphur cycle				Inclusion of dispersion	12 to 35% decrease **–1.12 (albedo, mid value decreased)**
Ghan et al. (2001)	AGCM (PNNL) + chemistry (MIRAGE), fixed SST	S, OC, BC, SS, N, D	E (for different modes); I (within modes)	warm and mixed phase	Droplet number concentration and LWC, crystal concentration and ice water content. Different processes affecting the various modes	–1.7 (total) **–0.85 (albedo)**
Chuang et al. (2002)	CCM1 (NCAR) + chemistry (GRANTOUR), fixed SST	S, OC, BC, SS, D	E (for emitted particles); I: when growing by condensation	warm and mixed phase	Modified from Chuang and Penner (1995), no collision/coalescence	**–1.85 (albedo)**
Menon et al. (2002a)	GCM (GISS) + sulphur cycle, fixed SST	S,OC, SS	E	warm	Droplet number concentration and LWC, Del Genio et al. (1996), Sundqvist et al. (1989). Warm and mixed phase, improved vertical distribution of clouds (but only nine layers).Global aerosol burdens poorly constrained	–2.41 (total) **–1.55 (albedo)**
Kristjansson (2002)	CCM3 (NCAR) fixed SST	S, OC, BC, SS, D	E (for nucleation mode and fossil fuel BC); I (for accumulation mode)	warm and mixed phase	Rasch and Kristjánsson (1998). Stratiform and detraining convective clouds	–1.82 (total) **–1.35 (albedo)**
Suzuki et al. (2004)	AGCM (Japan), fixed SST	S, OC, BC, SS	E	stratiform	Berry(1967), Sundqvist(1978)	**0.54 (albedo)**
Quaas et al. (2004)	AGCM (LMDZ) + interactive sulphur cycle, fixed SST	S	n.a.	warm and mixed phase	Aerosol mass and cloud droplet number concentration, Boucher and Lohmann (1995); Boucher et al. (1995)	**–1.3 (albedo)**

Table 2.7 (continued)

Model	Model type[a]	Aerosol species[b]	Aerosol mixtures[c]	Cloud types included	Microphysics	Radiative Forcing (W m^{-2})[d]
Hansen et al. (2005)	GCM (GISS) + 3 different ocean parametrizations	S, OC, BC, SS, N, D (D not included in clouds)	E	warm and shallow (below 720hPa)	Schmidt et al. (2005), 20 vertical layers. Droplet number concentration (Menon and Del Genio, 2007)	−0.77 (albedo)
Kristjansson et al. (2005)	CCM3 (NCAR) + sulphur and carbon cycles slab ocean	S, OC, BC, SS, D	E (for nucleation mode and fossil fuel BC); I (for accumulation mode)	warm and mixed phase	Kristjansson (2002). Stratiform and detraining convective clouds	−1.15 (total, at the surface)
Quaas and Boucher (2005)	AGCM (LMDZ) + interactive sulphur cycle, fixed SST	S, OC, BC, SS, D	E	warm and mixed phase	Aerosol mass and cloud droplet number concentration, Boucher and Lohmann (1995); Boucher et al. (1995) control run fit to POLDER data fit to MODIS data	−0.9 (albedo) −0.5 (albedo)[e] −0.3 (albedo)[e]
Quaas et al. (2005)	AGCM (LMDZ and ECHAM4)	S, OC, BC, SS, D	E	warm and mixed phase	Aerosol mass and cloud droplet number concentration, Boucher and Lohmann, (1995), control runs (ctl) Aerosol mass and cloud droplet number concentration fitted to MODIS data	−0.84 (total LMDZ-ctl) −1.54 (total (ECHAM4-ctl) −0.53 (total LMDZ)[e] 0.20 (total (ECHAM4)[e]
Dufresne et al. (2005)	AGCM (LMDZ) + interactive sulphur cycle, fixed SST	S	n.a.	warm	Aerosol mass and cloud droplet number concentration, Boucher and Lohmann, (1995), fitted to POLDER data	−0.22 (albedo)[a]
Takemura et al. (2005)	AGCM (SPRINTARS) + slab ocean	S, OC, BC, SS, D	E (50% BC from fossil fuel); I (for OC and BC)	warm	Activation based on Kohler theory and updraft velocity	−0.94 (total) −0.52 (albedo)
Chen and Penner (2005)	AGCM (UM) + fixed SST	S, SS, D, OC, BC	I	warm and mixed phase	Aerosol mass and cloud droplet number concentration (lognormal) Control (Abdul-Razzak and Ghan, 2002) Relationship between droplet concentration and dispersion coefficient: High Relationship between droplet concentration and dispersion coefficient: Medium Updraft velocity Relationship between droplet concentration and dispersion coefficient: Low Chuang et al. (1997) Nenes and Seinfeld (2003)	−1.30 (albedo, UM_ctrl)[f] −0.75 (albedo, UM_1)[f] −0.86 (albedo, UM_2)[f] −1.07 (albedo, UM_3)[f] −1.10 (albedo, UM_4)[f] −1.29 (albedo, UM_5)[f] −1.79 (albedo, UM_6)[f]

Table 2.7 (continued)

Model	Model type[a]	Aerosol species[b]	Aerosol mixtures[c]	Cloud types included	Microphysics	Radiative Forcing (W m^{-2})[d]
Ming et al. (2005b)	AGCM (GFDL), fixed SST and sulphur loading	S	n.a.	warm	Rotstayn et al. (2000), Khainroutdinov and Kogan (2000). Aerosols off-line	−2.3 (total) **−1.4 (albedo)**
Penner et al. (2006) results from experiment 1	LMDZ, Oslo and CCSR	S, SS, D, OC, BC	E	warm and mixed phase	Aerosol mass and cloud droplet number concentration; Boucher and Lohmann, (1995); Chen and Penner (2005); Sundqvist (1978)	**−0.65 (albedo Oslo) −0.68 (albedo LMDZ) −0.74 (albedo CCSR)**

Notes:

a   AGCM: Atmospheric GCM; SST: sea surface temperature; CSIRO: Commonwealth Scientific and Industrial Research Organisation; MIRAGE: Model for Integrated Research on Atmospheric Global Exchanges; GRANTOUR: Global Aerosol Transport and Removal model; GFDL: Geophysical Fluid Dynamics Laboratory; CCSR: Centre for Climate System Research; see Table 2.4, Note (a) for listing of other models and modelling centres listed in this column.

b   S: sulphate; SS: sea salt; D: mineral dust; BC: black carbon; OC: organic carbon; N: nitrate.

c   E: external mixtures; I: internal mixtures.

d   Only the bold numbers were used to construct Figure 2.16.

e   These simulations have been constrained by satellite observations, using the same empirical fit to relate aerosol mass and cloud droplet number concentration.

f   The notation UM corresponds to University of Michigan, as listed in Figure 2.14.

well as naturally produced sea salt, dust and continuously outgassing volcanic sulphate aerosols. Lohmann et al. (2000) and Chuang et al. (2002) included internally mixed sulphate, black and organic carbon, sea salt and dust aerosols, resulting in the most negative estimate of the cloud albedo indirect effect. Takemura et al. (2005) used a global aerosol transport-radiation model coupled to a GCM to estimate the direct and indirect effects of aerosols and their associated RF. The model includes a microphysical parametrization to diagnose the cloud droplet number concentration using Köhler theory, which depends on the aerosol particle number concentration, updraft velocity, size distributions and chemical properties of each aerosol species. The results indicate a global decrease in cloud droplet effective radius caused by anthropogenic aerosols, with the global mean RF calculated to be –0.52 W m^{-2}; the land and oceanic contributions are –1.14 and –0.28 W m^{-2}, respectively. Other modelling results also indicate that the mean RF due to the cloud albedo effect is on average somewhat larger over land than over oceans; over oceans there is a more consistent response from the different models, resulting in a smaller inter-model variability (Lohmann and Feichter, 2005).

Chen and Penner (2005), by systematically varying parameters, obtained a less negative RF when the in-cloud updraft velocity was made to depend on the turbulent kinetic energy. Incorporating other cloud nucleation schemes, for example, changing from Abdul-Razzak and Ghan (2002) to the Chuang et al. (1997) parametrization resulted in no RF change, while changing to the Nenes and Seinfeld (2003) parametrization made the RF more negative. Rotstayn and Liu (2003) found a 12 to 35% decrease in the RF when the size dispersion effect was included in the case of sulphate particles. Chen and Penner (2005) further explored the range of parameters used in Rotstayn and Liu (2003) and found the RF to be generally less negative than in the standard integration.

A model intercomparison study (Penner et al., 2006) examined the differences in cloud albedo effect between models through a series of controlled experiments that allowed examination of the uncertainties. This study presented results from three models, which were run with prescribed aerosol mass-number concentration (from Boucher and Lohmann, 1995), aerosol field (from Chen and Penner, 2005) and precipitation efficiency (from Sundqvist, 1978). The cloud albedo RFs in

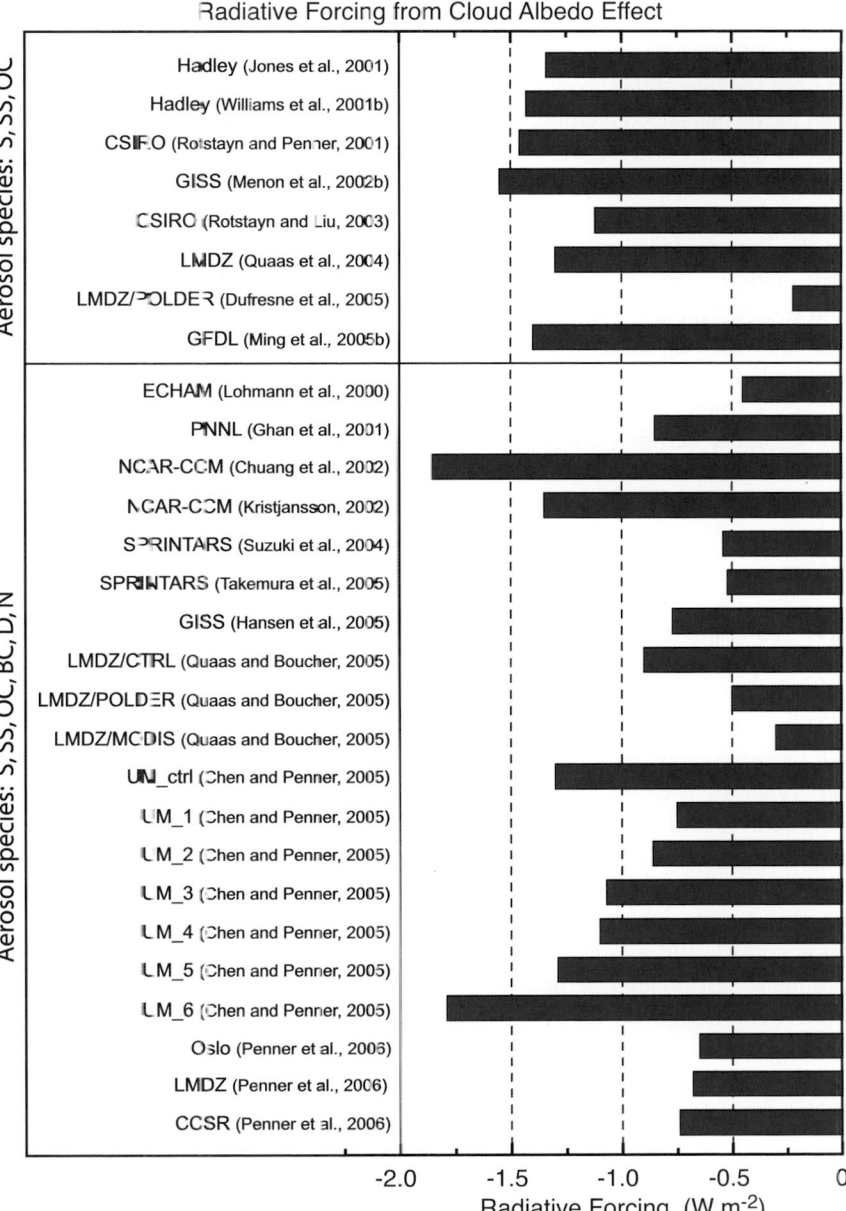

Figure 2.14. Radiative forcing due to the cloud albedo effect, in the context of liquid water clouds, from the global climate models that appear in Table 2.7. The labels next to the bars correspond to the published study; the notes of Table 2.7 explain the species abbreviations listed on the left hand side. Top panel: results for models that consider a limited number of species, primarily anthropogenic sulphate (S). Bottom panel: results from studies that include a variety of aerosol compositions and mixtures; the estimates here cover a larger range than those in the top panel. Chen and Penner (2005) presented a sensitivity study obtained by changing parametrizations in their model, so the results can be considered independent and are thus listed separately. Penner et al. (2006) is an intercomparison study, so the results of the individual models are listed separately.

the three models do not vary widely: –0.65, –0.68 and –0.74 W m^{-2}, respectively. Nevertheless, changes in the autoconversion scheme led to a differing response of the LWP between the models, and this is identified as an uncertainty.

A closer inspection of the treatment of aerosol species in the models leads to a broad separation of the results into two groups: models with only a few aerosol species and those that include a more complex mixture of aerosols of different composition. Thus, in Figure 2.14, RF results are grouped according to the type of aerosol species included in the simulations. In the top panel of Figure 2.14, which shows estimates from models that mainly include anthropogenic sulphate, there is an indication that the results are converging, even though the range of models comes from studies published between 2001 and 2006. These studies show much less scatter than in the TAR, with a mean and standard deviation of -1.37 ± 0.14 W m^{-2}. In contrast, in the bottom panel of Figure 2.14, which shows the studies that include more species, a much larger variability is found. These latter models (see Table 2.7) include 'state of the art' parametrizations of droplet activation for a variety of aerosols, and include both internal and external mixtures.

Some studies have commented on inconsistencies between some of the earlier estimates of the cloud albedo RF from forward and inverse calculations (Anderson et al., 2003). Notwithstanding the fact that these two streams of calculations rely on very different formulations, the results here appear to be within range of the estimates from inverse calculations.

2.4.5.3 Estimates of the Radiative Forcing from Observations and Constrained Models

It is difficult to obtain a best estimate of the cloud albedo RF from pre-industrial times to the present day based solely on observations. The satellite record is not long enough, and other long-term records do not provide the pre-industrial aerosol and cloud microphysical properties needed for such an assessment. Some studies have attempted to estimate the RF by incorporating empirical relationships derived from satellite observations. This approach is valid as long as the observations are robust, but problems still remain, particularly with the use of the aerosol optical depth as proxy for CCN (Feingold et al., 2003), droplet size and cloud optical depth from broken clouds (Marshak et al., 2006), and relative humidity effects (Kapustin et al., 2006) to discriminate between hydrated aerosols and cloud. Radiative forcing estimates constrained by satellite observations need to be considered with these caveats in mind.

By assuming a bimodal lognormal size distribution, Nakajima et al. (2001) determined the Ångstrom exponent from AVHRR data over the oceans (for a period of four months), together with cloud properties, optical thickness and effective radii. The nonlinear relationship between aerosol number concentration and cloud droplet concentration ($N_d \approx (N_a)^b$) obtained is consistent with Twomey's hypothesis; however, the parameter b is smaller than previous estimates (0.5 compared with 0.7 to 0.8; Kaufman et al., 1991), but larger than the 0.26 value obtained by Martin et al. (1994). Using this relationship, Nakajima et al. (2001) provided an estimate of the cloud albedo RF in the range between -0.7 and -1.7 W m^{-2}, with a global average of -1.3 W m^{-2}. Lohmann and Lesins (2002) used POLDER data to estimate aerosol index and cloud droplet radius; they then scaled the results of the simulations with the European Centre Hamburg (ECHAM4) model. The results show that changes in N_a lead to larger changes in N_d in the model than in observations, particularly over land, leading to an overestimate of the cloud albedo effect. The scaled values using

the constraint from POLDER yield a global cloud albedo RF of -0.85 W m^{-2}, an almost 40% reduction from their previous estimate. Sekiguchi et al. (2003) presented results from the analysis of AVHRR data over the oceans, and of POLDER data over land and ocean. Assuming that the aerosol column number concentration increased by 30% from the pre-industrial era, they estimated the effect due to the aerosol influence on clouds as the difference between the forcing under present and pre-industrial conditions. They estimated a global effect due to the total aerosol influence on clouds (sum of cloud albedo and lifetime effects) to be between -0.6 and -1.2 W m^{-2}, somewhat lower than the Nakajima et al. (2001) ocean estimate. When the assumption is made that the liquid water content is constant, the cloud albedo RF estimated from AVHRR data is -0.64 ± 0.16 W m^{-2} and the estimate using POLDER data is -0.37 ± 0.09 W m^{-2}. The results from these two studies are very sensitive to the magnitude of the increase in the aerosol concentration from pre-industrial to current conditions, and the spatial distributions.

Quaas and Boucher (2005) used the POLDER and MODIS data to evaluate the relationship between cloud properties and aerosol concentrations on a global scale in order to incorporate it in a GCM. They derived relationships corresponding to marine stratiform clouds and convective clouds over land that show a decreasing effective radius as the aerosol optical depth increases. These retrievals involve a variety of assumptions that introduce uncertainties in the relationships, in particular the fact that the retrievals for aerosol and cloud properties are not coincident and the assumption that the aerosol optical depth can be linked to the sub-cloud aerosol concentration. When these empirical parametrizations are included in a climate model, the simulated RF due to the cloud albedo effect is reduced by 50% from their baseline simulation. Quaas et al. (2005) also utilised satellite data to establish a relationship between cloud droplet number concentration and fine-mode aerosol optical depth, minimising the dependence on cloud liquid water content but including an adiabatic assumption that may not be realistic in many cases. This relationship is implemented in the ECHAM4 and Laboratoire de Météorologie Dynamique Zoom (LMDZ) climate models and the results indicate that the original parametrizations used in both models overestimated the magnitude of the cloud albedo effect. Even though both models show a consistent weakening of the RF, it should be noted that the original estimates of their respective RFs are very different (by almost a factor of two); the amount of the reduction was 37% in LMDZ and 81% in ECHAM4. Note that the two models have highly different spatial distributions of low clouds, simulated aerosol concentrations and anthropogenic fractions.

When only sulphate aerosols were considered, Dufresne et al. (2005) obtained a weaker cloud albedo RF. Their model used a relationship between aerosol mass concentration and cloud droplet number concentration, modified from that originally proposed by Boucher and Lohmann (1995) and adjusted to POLDER data. Their simulations give a factor of two weaker RF compared to the previous parametrization, but it is noted that the results are highly sensitive to the distribution of clouds over land.

2.4.5.4   Uncertainties in Satellite Estimates

The improvements in the retrievals and satellite instrumentation have provided valuable data to begin observation-motivated assessments of the effect of aerosols on cloud properties, even though satellite measurements cannot unambiguously distinguish natural from anthropogenic aerosols. Nevertheless, an obvious advantage of the satellite data is their global coverage, and such extensive coverage can be analysed to determine the relationships between aerosol and cloud properties at a number of locations around the globe. Using these data some studies (Sekiguchi et al., 2003; Quaas et al., 2004) indicate that the magnitude of the RF is resolution dependent, since the representation of convection and clouds in the GCMs and the simulation of updraft velocity that affects activation themselves are resolution dependent. The rather low spatial and temporal resolution of some of the satellite data sets can introduce biases by failing to distinguish aerosol species with different properties. This, together with the absence of coincident LWP measurements in several instances, handicaps the inferences from such studies, and hinders an accurate analysis and estimate of the RF. Furthermore, the ability to separate meteorological from chemical influences in satellite observations depends on the understanding of how clouds respond to meteorological conditions.

Retrievals involve a variety of assumptions that introduce uncertainties in the relationships. As mentioned above, the retrievals for aerosol and cloud properties are not coincident and the assumption is made that the aerosol optical depth can be linked to the aerosol concentration below the cloud. The POLDER instrument may underestimate the mean cloud-top droplet radius due to uncertainties in the sampling of clouds (Rosenfeld and Feingold, 2003). The retrieval of the aerosol index over land may be less reliable and lead to an underestimate of the cloud albedo effect over land. There is an indication of a systematic bias between MODIS-derived cloud droplet radius and that derived from POLDER (Breon and Doutriaux-Boucher, 2005), as well as differences in the aerosol optical depth retrieved from those instruments (Myhre et al., 2004a) that need to be resolved.

2.4.5.5   Uncertainties Due to Model Biases

One of the large sources of uncertainties is the poor knowledge of the amount and distribution of anthropogenic aerosols used in the model simulations, particularly for pre-industrial conditions. Some studies show a large sensitivity in the RF to the ratio of pre-industrial to present-day aerosol number concentrations.

All climate models discussed above include sulphate particles; some models produce them from gaseous precursors over oceans, where ambient concentrations are low, while some models only condense mass onto pre-existing particles over the continents. Some other climate models also include sea salt and dust particles produced naturally, typically relating particle production to wind speed. Some models include anthropogenic

nitrate, BC and organic compounds, which in turn affect activation. Models also have weaknesses in representing convection processes and aerosol distributions, and simulating updraft velocities and convection-cloud interactions. Even without considering the existing biases in the model-generated clouds, differences in the aerosol chemical composition and the subsequent treatment of activation lead to uncertainties that are difficult to quantify and assess. The presence of organic carbon, owing to its distinct hygroscopic and absorption properties, can be particularly important for the cloud albedo effect in the tropics (Ming et al., 2007).

Modelling the cloud albedo effect from first principles has proven difficult because the representation of aerosol-cloud and convection-cloud interactions in climate models are still crude (Lohmann and Feichter, 2005). Clouds often do not cover a complete grid box and are inhomogeneous in terms of droplet concentration, effective radii and LWP, which introduces added complications in the microphysical and radiative transfer calculations. Model intercomparisons (e.g., Lohmann et al., 2001; Menon et al., 2003) suggest that the predicted cloud distributions vary significantly between models, particularly their horizontal and vertical extents; also, the vertical resolution and parametrization of convective and stratiform clouds are quite different between models (Chen and Penner, 2005). Even high-resolution models have difficulty in accurately estimating the amount of cloud liquid and ice water content in a grid box.

It has proven difficult to compare directly the results from the different models, as uncertainties are not well identified and quantified. All models could be suffering from similar biases, and modelling studies do not often quote the statistical significance of the RF estimates that are presented. Ming et al. (2005b) demonstrated that it is only in the mid-latitude NH that their model yields a RF result at the 95% confidence level when compared to the unforced model variability. There are also large differences in the way that the different models treat the appearance and evolution of aerosol particles and the subsequent cloud droplet formation. Differences in the horizontal and vertical resolution introduce uncertainties in their ability to accurately represent the shallow warm cloud layers over the oceans that are most susceptible to the changes due to anthropogenic aerosol particles. A more fundamental problem is that GCMs do not resolve the small scales (order of hundreds of metres) at which aerosol-cloud interactions occur. Chemical composition and size distribution spectrum are also likely insufficiently understood at a microphysical level, although some modelling studies suggest that the albedo effect is more sensitive to the size than to aerosol composition (Feingold, 2003; Ervens et al., 2005; Dusek et al., 2006). Observations indicate that aerosol particles in nature tend to be composed of several compounds and can be internally or externally mixed. The actual conditions are difficult to simulate and possibly lead to differences among climate models. The calculation of the cloud albedo effect is sensitive to the details of particle chemical composition (activation) and state of the mixture (external or internal). The relationship between ambient aerosol particle concentrations and resulting cloud droplet size distribution is important during the activation process; this is a critical parametrization element in the climate models. It is treated in different ways in different models, ranging from simple empirical functions (Menon et al., 2002a) to more complex physical parametrizations that also tend to be more computationally costly (Abdul-Razzak and Ghan, 2002; Nenes and Seinfeld, 2003; Ming et al., 2006). Finally, comparisons with observations have not yet risen to the same degree of verification as, for example, those for the direct RF estimates; this is not merely due to model limitations, since the observational basis also has not yet reached a sound footing.

Further uncertainties may be due to changes in the droplet spectral shape, typically considered invariant in climate models under clean and polluted conditions, but which can be substantially different in typical atmospheric conditions (e.g., Feingold et al., 1997; Ackerman et al., 2000b; Erlick et al., 2001; Liu and Daum, 2002). Liu and Daum (2002) estimated that a 15% increase in the width of the size distribution can lead to a reduction of between 10 and 80% in the estimated RF of the cloud albedo indirect effect. Peng and Lohmann (2003), Rotstayn and Liu (2003) and Chen and Penner (2005) studied the sensitivity of their estimates to this dispersion effect. These studies confirm that their estimates of the cloud albedo RF, without taking the droplet spectra change into account, are overestimated by about 15 to 35%.

The effects of aerosol particles on heterogeneous ice formation are currently insufficiently understood and present another level of challenge for both observations and modelling. Ice crystal concentrations cannot be easily measured with present *in situ* instrumentation because of the difficulty of detecting small particles (Hirst et al., 2001) and frequent shattering of ice particles on impact with the probes (Korolev and Isaac, 2005). Current GCMs do not have sufficiently rigorous microphysics or sub-grid scale processes to accurately predict cirrus clouds or super-cooled clouds explicitly. Ice particles in clouds are often represented by simple shapes (e.g., spheres), even though it is well known that few ice crystals are like that in reality. The radiative properties of ice particles in GCMs often do not effectively simulate the irregular shapes that are normally found, nor do they simulate the inclusions of crustal material or soot in the crystals.

2.4.5.6 Assessment of the Cloud Albedo Radiative Forcing

As in the TAR, only the aerosol interaction in the context of liquid water clouds is assessed, with knowledge of the interaction with ice clouds deemed insufficient. Since the TAR, the cloud albedo effect has been estimated in a more systematic way, and more modelling results are now available. Models now are more advanced in capturing the complexity of the aerosol-cloud interactions through forward computations. Even though major uncertainties remain, clear progress has been made, leading to a convergence of the estimates from the different modelling efforts. Based on the results from all the modelling studies shown in Figure 2.14, compared to the TAR it is now possible to present a best estimate for the cloud albedo RF of

−0.7 W m^{-2} as the median, with a 5 to 95% range of −0.3 to −1.8 W m^{-2}. The increase in the knowledge of the aerosol-cloud interactions and the reduction in the spread of the cloud albedo RF since the TAR result in an elevation of the level of scientific understanding to low (Section 2.9, Table 2.11).

2.5 Anthropogenic Changes in Surface Albedo and the Surface Energy Budget

2.5.1 Introduction

Anthropogenic changes to the physical properties of the land surface can perturb the climate, both by exerting an RF and by modifying other processes such as the fluxes of latent and sensible heat and the transfer of momentum from the atmosphere. In addition to contributing to changes in greenhouse gas concentrations and aerosol loading, anthropogenic changes in the large-scale character of the vegetation covering the landscape ('land cover') can affect physical properties such as surface albedo. The albedo of agricultural land can be very different from that of a natural landscape, especially if the latter is forest. The albedo of forested land is generally lower than that of open land because the greater leaf area of a forest canopy and multiple reflections within the canopy result in a higher fraction of incident radiation being absorbed. Changes in surface albedo induce an RF by perturbing the shortwave radiation budget (Ramaswamy et al., 2001). The effect is particularly accentuated when snow is present, because open land can become entirely snow-covered and hence highly reflective, while trees can remain exposed above the snow (Betts, 2000). Even a snow-covered canopy exhibits a relatively low albedo as a result of multiple reflections within the canopy (Harding and Pomeroy, 1996). Surface albedo change may therefore provide the dominant influence of mid- and high-latitude land cover change on climate (Betts, 2001; Bounoua et al., 2002). The TAR cited two estimates of RF due to the change in albedo resulting from anthropogenic land cover change relative to potential natural vegetation (PNV), −0.4 W m^{-2} and −0.2 W m^{-2}, and assumed that the RF relative to 1750 was half of that relative to PNV, so gave a central estimate of the RF due to surface albedo change of −0.2 W m^{-2} ± 0.2 W m^{-2}.

Surface albedo can also be modified by the settling of anthropogenic aerosols on the ground, especially in the case of BC on snow (Hansen and Nazarenko, 2004). This mechanism may be considered an RF mechanism because diagnostic calculations may be performed under the strict definition of RF (see Sections 2.2 and 2.8). This mechanism was not discussed in the TAR.

Land cover change can also affect other physical properties such as surface emissivity, the fluxes of moisture through evaporation and transpiration, the ratio of latent to sensible heat fluxes (the Bowen ratio) and the aerodynamic roughness, which exerts frictional drag on the atmosphere and also affects turbulent transfer of heat and moisture. All these processes can affect the air temperature near the ground, and also modify humidity, precipitation and wind speed. Direct human perturbations to the water cycle, such as irrigation, can affect surface moisture fluxes and hence the surface energy balance. Changes in vegetation cover can affect the production of dust, which then exerts an RF. Changes in certain gases, particularly CO_2 and ozone, can also exert an additional influence on climate by affecting the Bowen ratio, through plant responses that affect transpiration. These processes are discussed in detail in Section 7.2. While such processes will act as anthropogenic perturbations to the climate system (Pielke et al., 2002) and will fall at least partly within the 'forcing' component of the forcing-feedback-response conceptual model, it is difficult to unequivocally quantify the pure forcing component as distinct

from feedbacks and responses. The term 'non-radiative forcing' has been proposed (Jacob et al., 2005) and this report adopts the similar term 'non-initial radiative effect', but no quantitative metric separating forcing from feedback and response has yet been implemented for climatic perturbation processes that do not act directly on the radiation budget (see Section 2.2).

Energy consumption by human activities, such as heating buildings, powering electrical appliances and fuel combustion by vehicles, can directly release heat into the environment. This was not discussed in the TAR. Anthropogenic heat release is not an RF, in that it does not directly perturb the radiation budget; the mechanisms are not well identified and so it is here referred to as a non-initial radiative effect. It can, however, be quantified as a direct input of energy to the system in terms of W m^{-2}.

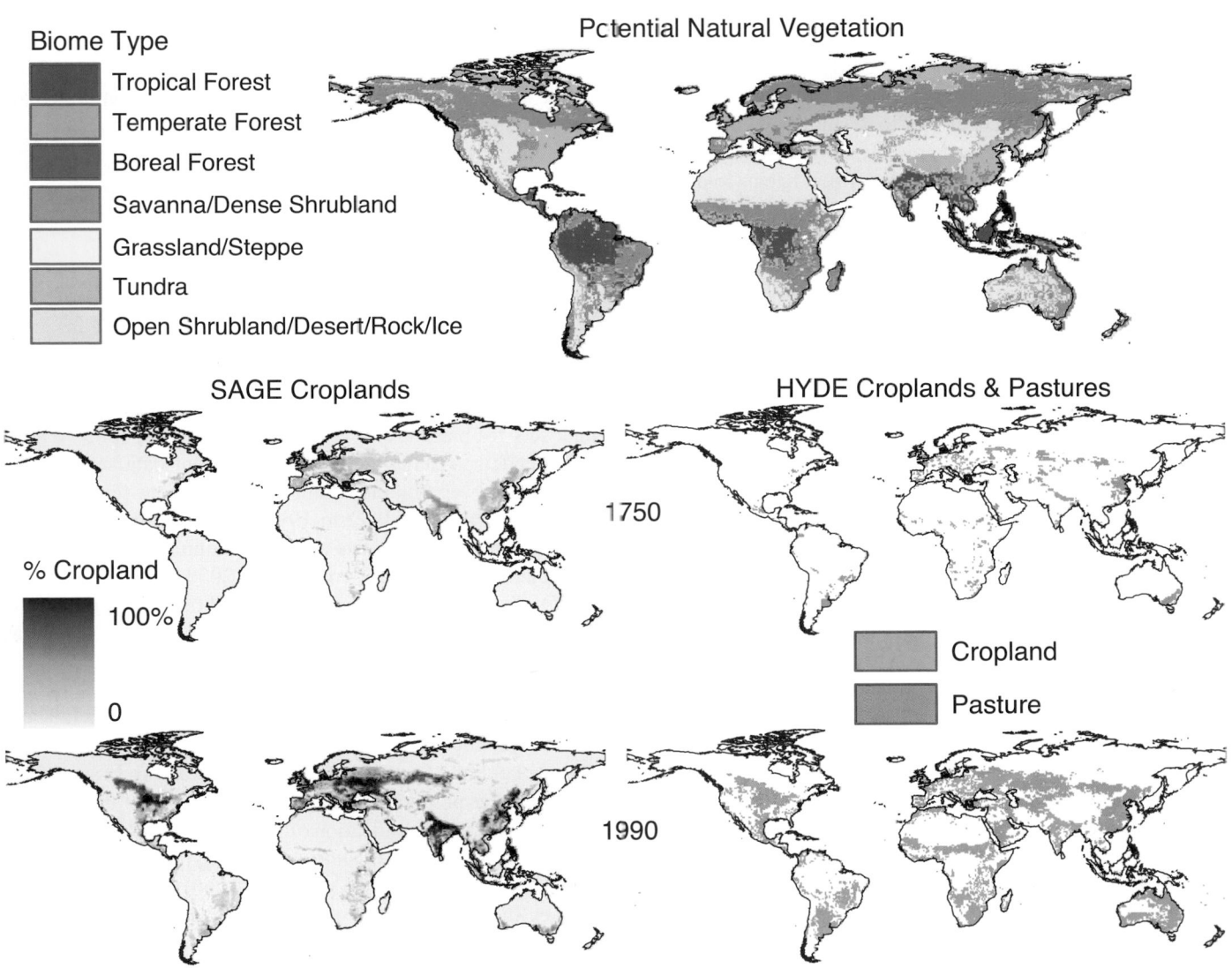

Figure 2.15. *Anthropogenic modifications of land cover up to 1990. Top panel: Reconstructions of potential natural vegetation (Haxeltine and Prentice, 1996). Lower panels: reconstructions of croplands and pasture for 1750 and 1990. Bottom left: fractional cover of croplands from Centre for Sustainability and the Global Environment (SAGE; Ramankutty and Foley, 1999) at 0.5° resolution. Bottom right: reconstructions from the History Database of the Environment (HYDE; Klein Goldewijk, 2001), with one land cover classification per 0.5° grid box.*

2.5.2 Changes in Land Cover Since 1750

In 1750, 7.9 to 9.2 million km² (6 to 7% of the global land surface) were under cultivation or pasture (Figure 2.15), mainly in Europe, the Indo-Gangetic Plain and China (Ramankutty and Foley, 1999; Klein Goldewijk, 2001). Over the next hundred years, croplands and pasture expanded and intensified in these areas, and new agricultural areas emerged in North America. The period 1850 to 1950 saw a more rapid rate of increase in cropland and pasture areas. In the last 50 years, several regions of the world have seen cropland areas stabilise, and even decrease. In the USA, as cultivation shifted from the east to the Midwest, croplands were abandoned along the eastern seaboard around the turn of the century and the eastern forests have regenerated over the last century. Similarly, cropland areas have decreased in China and Europe. Overall, global cropland and pasture expansion was slower after 1950 than before. However, deforestation is occurring more rapidly in the tropics. Latin America, Africa and South and Southeast Asia experienced slow cropland expansion until the 20th century, but have had exponential increases in the last 50 years. By 1990, croplands and pasture covered 45.7 to 51.3 million km² (35% to 39% of global land), and forest cover had decreased by roughly 11 million km² (Ramankutty and Foley, 1999; Klein Goldewijk, 2001; Table 2.8).

Overall, until the mid-20th century most deforestation occurred in the temperate regions (Figure 2.15). In more recent decades, however, land abandonment in Western Europe and North America has been leading to reforestation while deforestation is now progressing rapidly in the tropics. In the 1990s compared to the 1980s, net removal of tropical forest cover had slowed in the Americas but increased in Africa and Asia.

2.5.3 Radiative Forcing by Anthropogenic Surface Albedo Change: Land Use

Since the TAR, a number of estimates of the RF from land use changes over the industrial era have been made (Table 2.8). Unlike the main TAR estimate, most of the more recent studies are 'pure' RF calculations with the only change being land cover; feedbacks such as changes in snow cover are excluded. Brovkin et al. (2006) estimated the global mean RF relative to 1700 to be –0.15 W m⁻², considering only cropland changes (Ramankutty and Foley, 1999) and not pastures. Hansen et al. (2005) also considered only cropland changes (Ramankutty and Foley, 1999) and simulated the RF relative to 1750 to be –0.15 W m⁻². Using historical reconstructions of both croplands (Ramankutty and Foley, 1999) and pasturelands (Klein Goldewijk, 2001), Betts et al. (2007) simulated an RF of –0.18 W m⁻² since 1750. This study also estimated the RF relative to PNV to be –0.24 W m⁻². Other studies since the TAR have also estimated the RF at the present day relative to PNV (Table 2.8). Govindasamy et al. (2001a) estimated this RF as –0.08 W m⁻². Myhre et al. (2005a) used land cover and albedo data from MODIS (Friedl et al., 2002; Schaaf et al., 2002) and

estimated this RF as –0.09 W m⁻². The results of Betts et al. (2007) and Brovkin et al. (2006) suggest that the RF relative to 1750 is approximately 75% of that relative to PNV. Therefore, by employing this factor published RFs relative to PNV can be used to estimate the RF relative to 1750 (Table 2.8).

In all the published studies, the RF showed a very high degree of spatial variability, with some areas showing no RF in 1990 relative to 1750 while values more negative than –5 W m⁻² were typically found in the major agricultural areas of North America and Eurasia. The local RF depends on local albedo changes, which depend on the nature of the PNV replaced by agriculture (see top panel of Figure 2.15). In historical simulations, the spatial patterns of RF relative to the PNV remain generally similar over time, with the regional RFs in 1750 intensifying and expanding in the area covered. The major new areas of land cover change since 1750 are North America and central and eastern Russia.

Changes in the underlying surface albedo could affect the RF due to aerosols if such changes took place in the same regions. Similarly, surface albedo RF may depend on aerosol concentrations. Estimates of the temporal evolution of aerosol RF and surface albedo RF may need to consider changes in each other (Betts et al., 2007).

2.5.3.1 Uncertainties

Uncertainties in estimates of RF due to anthropogenic surface albedo change arise from several factors.

2.5.3.1.1 Uncertainties in the mapping and characterisation of present-day vegetation

The RF estimates reported in the TAR used atlas-based data sets for present-day vegetation (Matthews, 1983; Wilson and Henderson-Sellers, 1985). More recent data sets of land cover have been obtained from satellite remote sensing. Data from the AVHRR in 1992 to 1993 were used to generate two global land cover data sets at 1 km resolution using different methodologies (Hansen and Reed, 2000; Loveland et al., 2000) The International Geosphere-Biosphere Programme Data and Information System (IGBP-DIS) data set is used as the basis for global cropland maps (Ramankutty and Foley, 1999) and historical reconstructions of croplands, pasture and other vegetation types (Ramankutty and Foley, 1999; Klein Goldewijk, 2001) (Table 2.8). The MODIS (Friedl et al., 2002) and Global Land Cover 2000 (Bartholome and Belward, 2005) provide other products. The two interpretations of the AVHRR data agree on the classification of vegetation as either tall (forest and woody savannah) or short (all other land cover) over 84% of the land surface (Hansen and Reed, 2000). However, some of the key disagreements are in regions subject to anthropogenic land cover change so may be important for the estimation of anthropogenic RF. Using the Hadley Centre Atmospheric Model (HadAM3) GCM, Betts et al. (2007) found that the estimate of RF relative to PNV varied from –0.2 W m⁻² with the Wilson and Henderson-Sellers (1985) atlas-based land use data set to –0.24 W m⁻² with a version of the Wilson and Henderson-

Table 2.8. *Estimates of forest area, contribution to CO_2 increase from anthropogenic land cover change, and RF due to the land use change-induced CO_2 increase and surface albedo change, relative to pre-industrial vegetation and PNV. The CO_2 RFs are for 2000 relative to 1850, calculated from the land use change contribution to the total increase in CO_2 from 1850 to 2000 simulated with both land use and fossil fuel emissions by the carbon cycle models. Carbon emissions from land cover change for the 1980s and 1990s are discussed in Section 7.3 and Table 7.2.*

Main Source of Land Cover Data	Forest Area PNV 10^6 km^2	Forest Area circa 1700 10^6 km^2	Forest Area circa 1990 10^6 km^2	Contribution to CO_2 Increase 1850–2000[a] (ppm)	CO_2 RF (W m^{-2})	Albedo RF vs. PNV (W m^{-2})	Albedo RF vs. 1750 (W m^{-2})
Ramankutty and Foley (1999)	55.27	52.77[b]	43.97[c]	16[d]	0.27	−0.24[e] −0.29 to +0.02[f] −0.2[g]	−0.18[e] −0.22 to +0.02[h] −0.14[i,j] −0.15 to −0.28[i,j] −0.15[k] −0.075 to −0.325[i,l]
Klein Goldewijk (2001)	58.6	54.4	41.5	12[d]	0.20	−0.66 to +0.1[f]	−0.50 to +0.08[h] −0.275[i,l]
Houghton (1983[m], 2003)		62.15	50.53[n]	35[d] 26[o]	0.57 0.44		
MODIS (Schaaf et al., 2002)						−0.09[p]	−0.07[h]
Wilson and Henderson-Sellers (1985)						−0.2[q] −0.24[h]	−0.15[h] −0.22[h]
SARB[r]						−0.11 to −0.55[f]	−0.08 to −0.41[h]
Matthews (1983)						−0.12[f] −0.4[s] −0.08[t]	−0.09[h] −0.3[h] −0.06[h]

Notes:

[a] The available literature simulates CO_2 rises with and without land use relative to 1850.

[b] 1750 forest area reported as 51.85×10^6 km^2.

[c] 1992 forest area.

[d] Land use contribution CO_2 rise from Brovkin et al. (2004).

[e] Albedo RF from Betts et al. (2007). Land cover data combined from Ramankutty and Foley (1999), Klein Goldewijk (2001) and Wilson and Henderson-Sellers (1985).

[f] Albedo RF from Myhre and Myhre (2003). Range of estimate for each land cover data set arises from use of different albedo values.

[g] Albedo RF from model of Goosse et al. (2005) in Brovkin et al. (2006).

[h] RF relative to 1750 estimated here as 0.75 of RF relative to PNV following Betts et al. (2007) and Brovkin et al. (2006).

[i] Estimate relative to 1700.

[j] Albedo RF from Matthews et al. (2003).

[k] Albedo RF from Hansen et al. (2005).

[l] Albedo RF from Matthews et al. (2004).

[m] Forest areas aggregated by Richards (1990).

[n] 1980 forest area.

[o] Land use contribution to CO_2 rise from Matthews et al. (2004). Estimate only available relative to 1850 not 1750.

[p] Albedo RF from Myhre et al. (2005a).

[q] Albedo RF from Betts (2001).

[r] Surface and Atmosphere Radiation Budget; http://www-surf.larc.nasa.gov/surf/.

[s] Albedo RF from Hansen et al. (1997).

[t] Albedo RF from Govindasamy et al. (2001a).

Sellers (1985) data set adjusted to agree with the cropland data of Ramankutty and Foley (1999). Myhre and Myhre (2003) found the RF relative to PNV to vary from -0.66 W m^{-2} to 0.29 W m^{-2} according to whether the present-day land cover was from Wilson and Henderson-Sellers (1985), Ramankutty and Foley (1999) or other sources.

2.5.3.1.2 *Uncertainties in the mapping and characterisation of the reference historical state*

Reconstructions of historical land use states require information or assumptions regarding the nature and extent of land under human use and the nature of the PNV. Ramankutty and Foley (1999) reconstructed the fraction of land under crops at 0.5° resolution from 1700 to 1990 (Figure 2.15, Table 2.8) by combining the IGBP Global Land Cover Dataset with historical inventory data, assuming that all areas of past vegetation occur within areas of current vegetation. Klein Goldewijk (2001) reconstructed all land cover types from 1700 to 1990 (Figure 2.15, Table 2.8), combining cropland and pasture inventory data with historical population density maps and PNV. Klein Goldewijk used a Boolean approach, which meant that crops, for example, covered either 100% or 0% of a 0.5° grid box. The total global cropland of Klein Goldewijk is generally 25% less than that reconstructed by Ramankutty and Foley (1999) throughout 1700 to 1990. At local scales, the disagreement is greater due to the high spatial heterogeneity in both data sets. Large-scale PNV (Figure 2.15) is reconstructed either with models or by assuming that small-scale examples of currently undisturbed vegetation are representative of the PNV at the large scale. Matthews et al. (2004) simulated RF relative to 1700 as -0.20 W m^{-2} and -0.28 W m^{-2} with the above land use reconstructions.

2.5.3.1.3 *Uncertainties in the parametrizations of the surface radiation processes*

The albedo for a given land surface or vegetation type may either be prescribed or simulated on the basis of more fundamental characteristics such as vegetation leaf area. But either way, model parameters are set on the basis of observational data that may come from a number of conflicting sources. Both the AVHRR and MODIS (Schaaf et al., 2002; Gao et al., 2005) instruments have been used to quantify surface albedo for the IGBP vegetation classes in different regions and different seasons, and in some cases the albedo for a given vegetation type derived from one source can be twice that derived from the other (e.g., Strugnell et al., 2001; Myhre et al., 2005a). Myhre and Myhre (2003) examined the implications of varying the albedo of different vegetation types either together or separately, and found the RF relative to PNV to vary from -0.65 W m^{-2} to $+0.47$ W m^{-2}; however, the positive RFs occurred in only a few cases and resulted from large reductions in surface albedo in semi-arid regions on conversion to pasture, so were considered unrealistic by the study's authors. The single most important factor for the uncertainty in the study by Myhre and Myhre (2003) was found to be the surface albedo for cropland. In simulations where only the cropland surface albedo was

varied between 0.15, 0.18 and 0.20, the resulting RFs relative to PNV were -0.06, -0.20 and -0.29 W m^{-2}, respectively. Similar results were found by Matthews et al. (2003) considering only cropland changes and not pasture; with cropland surface albedos of 0.17 and 0.20, RFs relative to 1700 were -0.15 and -0.28 W m^{-2}, respectively.

2.5.3.1.3 *Uncertainties in other parts of the model*

When climate models are used to estimate the RF, uncertainties in other parts of the model also affect the estimates. In particular, the simulation of snow cover affects the extent to which land cover changes affect surface albedo. Betts (2000) estimated that the systematic biases in snow cover in HadAM3 introduced errors of up to approximately 10% in the simulation of local RF due to conversion between forest and open land. Such uncertainties could be reduced by the use of an observational snow climatology in a model that just treats the radiative transfer (Myhre and Myhre, 2003). The simulation of cloud cover affects the extent to which the simulated surface albedo changes affect planetary albedo – too much cloud cover could diminish the contribution of surface albedo changes to the planetary albedo change.

On the basis of the studies assessed here, including a number of new estimates since the TAR, the assessment is that the best estimate of RF relative to 1750 due to land-use related surface albedo change should remain at -0.2 ± 0.2 W m^{-2}. In the light of the additional modelling studies, the exclusion of feedbacks, the improved incorporation of large-scale observations and the explicit consideration of land use reconstructions for 1750, the level of scientific understanding is raised to medium-low, compared to low in the TAR (Section 2.9, Table 2.11).

2.5.4 Radiative Forcing by Anthropogenic Surface Albedo Change: Black Carbon in Snow and Ice

The presence of soot particles in snow could cause a decrease in the albedo of snow and affect snowmelt. Initial estimates by Hansen et al. (2000) suggested that BC could thereby exert a positive RF of $+0.2$ W m^{-2}. This estimate was refined by Hansen and Nazarenko (2004), who used measured BC concentrations within snow and ice at a wide range of geographic locations to deduce the perturbation to the surface and planetary albedo, deriving an RF of $+0.15$ W m^{-2}. The uncertainty in this estimate is substantial due to uncertainties in whether BC and snow particles are internally or externally mixed, in BC and snow particle shapes and sizes, in voids within BC particles, and in the BC imaginary refractive index. Jacobson (2004) developed a global model that allows the BC aerosol to enter snow via precipitation and dry deposition, thereby modifying the snow albedo and emissivity. They found modelled concentrations of BC within snow that were in reasonable agreement with those from many observations. The model study found that BC on snow and sea ice caused a decrease in the surface albedo of 0.4% globally and 1% in the NH, although RFs were not reported. Hansen et al. (2005) allowed the albedo change to be proportional to local BC deposition according to Koch (2001) and presented a further revised estimate of 0.08 W m^{-2}. They also suggested that this RF mechanism produces a greater temperature response by a factor of 1.7 than an equivalent CO_2 RF, that is, the 'efficacy' may be higher for this RF mechanism (see Section 2.8.5.7). This report adopts a best estimate for the BC on snow RF of +0.10 ± 0.10 W m^{-2}, with a low level of scientific understanding (Section 2.9, Table 2.11).

2.5.5 Other Effects of Anthropogenic Changes in Land Cover

Anthropogenic land use and land cover change can also modify climate through other mechanisms, some directly perturbing the Earth radiation budget and some perturbing other processes. Impacts of land cover change on emissions of CO_2, CH_4, biomass burning aerosols and dust aerosols are discussed in Sections 2.3 and 2.4. Land cover change itself can also modify the surface energy and moisture budgets through changes in evaporation and the fluxes of latent and sensible heat, directly affecting precipitation and atmospheric circulation as well as temperature. Model results suggest that the combined effects of past tropical deforestation may have exerted regional warmings of approximately 0.2°C relative to PNV, and may have perturbed the global atmospheric circulation affecting regional climates remote from the land cover change (Chase et al., 2000; Zhao et al., 2001; Pielke et al., 2002; Chapters 7, 9 and 11).

Since the dominant aspect of land cover change since 1750 has been deforestation in temperate regions, the overall effect of anthropogenic land cover change on global temperature will depend largely on the relative importance of increased surface albedo in winter and spring (exerting a cooling) and reduced evaporation in summer and in the tropics (exerting a warming) (Bounoua et al., 2002). Estimates of global temperature responses from past deforestation vary from 0.01°C (Zhao et al., 2001) to –0.25°C (Govindasamy et al., 2001a; Brovkin et al., 2006). If cooling by increased surface albedo dominates, then the historical effect of land cover change may still be adequately represented by RF. With tropical deforestation becoming more significant in recent decades, warming due to reduced evaporation may become more significant globally than increased surface albedo. Radiative forcing would then be less useful as a metric of climate change induced by land cover change recently and in the future.

2.5.6 Tropospheric Water Vapour from Anthropogenic Sources

Anthropogenic use of water is less than 1% of natural sources of water vapour and about 70% of the use of water for human activity is from irrigation (Döll, 2002). Several regional studies have indicated an impact of irrigation on temperature, humidity and precipitation (Barnston and Schickedanz, 1984; Lohar and Pal, 1995; de Ridder and Gallée, 1998; Moore and Rojstaczer, 2001; Zhang et al., 2002). Boucher et al. (2004) used a GCM to show that irrigation has a global impact on temperature and

humidity. Over Asia where most of the irrigation takes place, the simulations showed a change in the water vapour content in the lower troposphere of up to 1%, resulting in an RF of +0.03 W m^{-2}. However, the effect of irrigation on surface temperature was dominated by evaporative cooling rather than by the excess greenhouse effect and thus a decrease in surface temperature was found. Irrigation affects the temperature, humidity, clouds and precipitation as well as the natural evaporation through changes in the surface temperature, raising questions about the strict use of RF in this case. Uncertainties in the water vapour flow to the atmosphere from irrigation are significant and Gordon et al. (2005) gave a substantially higher estimate compared to that of Boucher et al. (2004). Most of this uncertainty is likely to be linked to differences between the total withdrawal for irrigation and the amount actually used (Boucher et al., 2004). Furthermore, Gordon et al. (2005) also estimated a reduced water vapour flow to the atmosphere from deforestation, most importantly in tropical areas. This reduced water vapour flow is a factor of three larger than the water vapour increase due to irrigation in Boucher et al. (2004), but so far there are no estimates of the effect of this on the water vapour content of the atmosphere and its RF. Water vapour changes from deforestation will, like irrigation, affect the surface evaporation and temperature and the water cycle in the atmosphere. Radiative forcing from anthropogenic sources of tropospheric water vapour is not evaluated here, since these sources affect surface temperature more significantly through these non-radiative processes, and a strict use of the RF is problematic. The emission of water vapour from fossil fuel combustion is significantly lower than the emission from changes in land use (Boucher et al., 2004).

2.5.7 Anthropogenic Heat Release

Urban heat islands result partly from the physical properties of the urban landscape and partly from the release of heat into the environment by the use of energy for human activities such as heating buildings and powering appliances and vehicles ('human energy production'). The global total heat flux from this is estimated as 0.03 W m^{-2} (Nakicenovic, 1998). If this energy release were concentrated in cities, which are estimated to cover 0.046% of the Earth's surface (Loveland et al., 2000) the mean local heat flux in a city would be 65 W m^{-2}. Daytime values in central Tokyo typically exceed 400 W m^{-2} with a maximum of 1,590 W m^{-2} in winter (Ichinose et al., 1999). Although human energy production is a small influence at the global scale, it may be very important for climate changes in cities (Betts and Best, 2004; Crutzen, 2004).

2.5.8 Effects of Carbon Dioxide Changes on Climate via Plant Physiology: 'Physiological Forcing'

As well as exerting an RF on the climate system, increasing concentrations of atmospheric CO_2 can perturb the climate system through direct effects on plant physiology. Plant stomatal apertures open less under higher CO_2 concentrations (Field et al., 1995), which directly reduces the flux of moisture from the surface to the atmosphere through transpiration (Sellers et al., 1996). A decrease in moisture flux modifies the surface energy balance, increasing the ratio of sensible heat flux to latent heat flux and therefore warming the air near the surface (Sellers et al., 1996; Betts et al., 1997; Cox et al., 1999). Betts et al. (2004) proposed the term 'physiological forcing' for this mechanism. Although no studies have yet explicitly quantified the present-day temperature response to physiological forcing, the presence of this forcing has been detected in global hydrological budgets (Gedney et al., 2006; Section 9.5). This process can be considered a non-initial radiative effect, as distinct from a feedback, since the mechanism involves a direct response to increasing atmospheric CO_2 and not a response to climate change. It is not possible to quantify this with RF. Reduced global transpiration would also be expected to reduce atmospheric water vapour causing a negative forcing, but no estimates of this have been made.

Increased CO_2 concentrations can also 'fertilize' plants by stimulating photosynthesis, which models suggest has contributed to increased vegetation cover and leaf area over the 20th century (Cramer et al., 2001). Increases in the Normalized Difference Vegetation Index, a remote sensing product indicative of leaf area, biomass and potential photosynthesis, have been observed (Zhou et al., 2001), although other causes including climate change itself are also likely to have contributed. Increased vegetation cover and leaf area would decrease surface albedo, which would act to oppose the increase in albedo due to deforestation. The RF due to this process has not been evaluated and there is a very low scientific understanding of these effects.

2.6   Contrails and Aircraft-Induced Cloudiness

2.6.1   Introduction

The IPCC separately evaluated the RF from subsonic and supersonic aircraft operations in the Special Report on Aviation and the Global Atmosphere (IPCC, 1999), hereinafter designated as IPCC-1999. Like many other sectors, subsonic aircraft operations around the globe contribute directly and indirectly to the RF of climate change. This section only assesses the aspects that are unique to the aviation sector, namely the formation of persistent condensation trails (contrails), their impact on cirrus cloudiness, and the effects of aviation aerosols. Persistent contrail formation and induced cloudiness are indirect effects from aircraft operations because they depend on variable humidity and temperature conditions along aircraft flight tracks. Thus, future changes in atmospheric humidity and temperature distributions in the upper troposphere will have consequences for aviation-induced cloudiness. Also noted here is the potential role of aviation aerosols in altering the properties of clouds that form later in air containing aircraft emissions.

2.6.2   Radiative Forcing Estimates for Persistent Line-Shaped Contrails

Aircraft produce persistent contrails in the upper troposphere in ice-supersaturated air masses (IPCC, 1999). Contrails are thin cirrus clouds, which reflect solar radiation and trap outgoing longwave radiation. The latter effect is expected to dominate for thin cirrus (Hartmann et al., 1992; Meerkötter et al., 1999), thereby resulting in a net positive RF value for contrails. Persistent contrail cover has been calculated globally from meteorological data (e.g., Sausen et al., 1998) or by using a modified cirrus cloud parametrization in a GCM (Ponater et al., 2002). Contrail cover calculations are uncertain because the extent of supersaturated regions in the atmosphere is poorly known. The associated contrail RF follows from determining an optical depth for the computed contrail cover. The global RF values for contrail and induced cloudiness are assumed to vary linearly with distances flown by the global fleet if flight ambient conditions remain unchanged. The current best estimate for the RF of persistent linear contrails for aircraft operations in 2000 is +0.010 W m^{-2} (Table 2.9; Sausen et al., 2005). The value is based on independent estimates derived from Myhre and Stordal (2001b) and Marquart et al. (2003) that were updated for increased aircraft traffic in Sausen et al. (2005) to give RF estimates of +0.015 W m^{-2} and +0.006 W m^{-2}, respectively. The uncertainty range is conservatively estimated to be a factor of three. The +0.010 W m^{-2} value is also considered to be the best estimate for 2005 because of the slow overall growth in aviation fuel use in the 2000 to 2005 period. The decrease in the best estimate from the TAR by a factor of two results from reassessments of persistent contrail cover and lower optical depth estimates (Marquart and Mayer, 2002; Meyer et al., 2002; Ponater et al., 2002; Marquart et al., 2003). The new estimates

Table 2.9. *Radiative forcing terms for contrail and cirrus effects caused by global subsonic aircraft operations.*

	Radiative forcing (W m^{-2})[a]		
	1992 IPCC[b]	2000 IPCC[c]	2000[d]
CO_2[d]	0.018	0.025	0.025
Persistent linear contrails	0.020	0.034	0.010 (0.006 to 0.015)
Aviation-induced cloudiness without persistent contrails	0 to 0.040	n.a.	
Aviation-induced cloudiness with persistent contrails			0.030 (0.010 to 0.080)

Notes:

[a] Values for contrails are best estimates. Values in parentheses give the uncertainty range.

[b] Values from IPCC-1999 (IPCC, 1999).

[c] Values interpolated from 1992 and 2015 estimates in IPCC-1999 (Sausen et al., 2005).

[d] Sausen et al. (2005). Values are considered valid (within 10%) for 2005 because of slow growth in aviation fuel use between 2000 and 2005.

include diurnal changes in the solar RF, which decreases the net RF for a given contrail cover by about 20% (Myhre and Stordal, 2001b). The level of scientific understanding of contrail RF is considered low, since important uncertainties remain in the determination of global values (Section 2.9, Table 2.11). For example, unexplained regional differences are found in contrail optical depths between Europe and the USA that have not been fully accounted for in model calculations (Meyer et al., 2002; Ponater et al., 2002; Palikonda et al., 2005).

2.6.3 Radiative Forcing Estimates for Aviation-Induced Cloudiness

Individual persistent contrails are routinely observed to shear and spread, covering large additional areas with cirrus cloud (Minnis et al., 1998). Aviation aerosol could also lead to changes in cirrus cloud (see Section 2.6.4). Aviation-induced cloudiness (AIC) is defined to be the sum of all changes in cloudiness associated with aviation operations. Thus, an AIC estimate includes persistent contrail cover. Because spreading contrails lose their characteristic linear shape, a component of AIC is indistinguishable from background cirrus. This basic ambiguity, which prevented the formulation of a best estimate of AIC amounts and the associated RF in IPCC-1999, still exists for this assessment. Estimates of the ratio of induced cloudiness cover to that of persistent linear contrails range from 1.8 to 10 (Minnis et al., 2004; Mannstein and Schumann, 2005[10]), indicating the uncertainty in estimating AIC amounts. Initial attempts to quantify AIC used trend differences in cirrus cloudiness between regions of high and low aviation fuel consumption (Boucher, 1999). Since IPCC-1999, two studies have also found significant positive trends in cirrus cloudiness in some regions of high air traffic and found lower to negative trends outside air traffic regions (Zerefos et al., 2003; Stordal et al., 2005). Using the International Satellite Cloud Climatology Project (ISCCP) database, these studies derived cirrus cover trends for Europe of 1 to 2% per decade over the last one to two decades. A study with the Television Infrared Observation Satellite (TIROS) Operational Vertical Sounder (TOVS) provides further support for these trends (Stubenrauch and Schumann, 2005). However, cirrus trends that occurred due to natural variability, climate change or other anthropogenic effects could not be accounted for in these studies. Cirrus trends over the USA (but not over Europe) were found to be consistent with changes in contrail cover and frequency (Minnis et al., 2004). Thus, significant uncertainty remains in attributing observed cirrus trends to aviation.

Regional cirrus trends were used as a basis to compute a global mean RF value for AIC in 2000 of +0.030 W m^{-2} with a range of +0.01 to +0.08 W m^{-2} (Stordal et al., 2005). This value is not considered a best estimate because of the uncertainty in the optical properties of AIC and in the assumptions used to derive AIC cover. However, this value is in good agreement with the upper limit estimate for AIC RF in 1992 of +0.026

W m^{-2} derived from surface and satellite cloudiness observations (Minnis et al., 2004). A value of +0.03 W m^{-2} is close to the upper-limit estimate of +0.04 W m^{-2} derived for non-contrail cloudiness in IPCC-1999. Without an AIC best estimate, the best estimate of the total RF value for aviation-induced cloudiness (Section 2.9.2, Table 2.12 and Figure 2.20) includes only that due to persistent linear contrails. Radiative forcing estimates for AIC made using cirrus trend data necessarily cannot distinguish between the components of aviation cloudiness, namely persistent linear contrails, spreading contrails and other aviation aerosol effects. Some aviation effects might be more appropriately considered feedback processes rather than an RF (see Sections 2.2 and 2.4.5). However, the low understanding of the processes involved and the lack of quantitative approaches preclude reliably making the forcing/feedback distinction for all aviation effects in this assessment.

Two issues related to the climate response of aviation cloudiness are worth noting here. First, Minnis et al. (2004, 2005) used their RF estimate for total AIC over the USA in an empirical model, and concluded that the surface temperature response for the period 1973 to 1994 could be as large as the observed surface warming over the USA (around 0.3°C per decade). In response to the Minnis et al. conclusion, contrail RF was examined in two global climate modelling studies (Hansen et al., 2005; Ponater et al., 2005). Both studies concluded that the surface temperature response calculated by Minnis et al. (2004) is too large by one to two orders of magnitude. For the Minnis et al. result to be correct, the climate efficacy or climate sensitivity of contrail RF would need to be much greater than that of other larger RF terms, (e.g., CO_2). Instead, contrail RF is found to have a smaller efficacy than an equivalent CO_2 RF (Hansen et al., 2005; Ponater et al., 2005) (see Section 2.8.5.7), which is consistent with the general ineffectiveness of high clouds in influencing diurnal surface temperatures (Hansen et al., 1995, 2005). Several substantive explanations for the incorrectness of the enhanced response found in the Minnis et al. study have been presented (Hansen et al., 2005; Ponater et al., 2005; Shine, 2005).

The second issue is that the absence of AIC has been proposed as the cause of the increased diurnal temperature range (DTR) found in surface observations made during the short period when all USA air traffic was grounded starting on 11 September 2001 (Travis et al., 2002, 2004). The Travis et al. studies show that during this period: (i) DTR was enhanced across the conterminous USA, with increases in the maximum temperatures that were not matched by increases of similar magnitude in the minimum temperatures, and (ii) the largest DTR changes corresponded to regions with the greatest contrail cover. The Travis et al. conclusions are weak because they are based on a correlation rather than a quantitative model and rely (necessarily) on very limited data (Schumann, 2005). Unusually clear weather across the USA during the shutdown period also has been proposed to account for the observed DTR changes (Kalkstein and Balling, 2004). Thus, more evidence and a

---

[10] A corrigendum to this paper has been submitted for publication by these authors but has not been assessed here.

quantitative physical model are needed before the validity of the proposed relationship between regional contrail cover and DTR can be considered further.

2.6.4 Aviation Aerosols

Global aviation operations emit aerosols and aerosol precursors into the upper troposphere and lower stratosphere (IPCC, 1999; Hendricks et al., 2004). As a result, aerosol number and/or mass are enhanced above background values in these regions. Aviation-induced cloudiness includes the possible influence of aviation aerosol on cirrus cloudiness amounts. The most important aerosols are those composed of sulphate and BC (soot). Sulphate aerosols arise from the emissions of fuel sulphur and BC aerosol results from incomplete combustion of aviation fuel. Aviation operations cause enhancements of sulphate and BC in the background atmosphere (IPCC, 1999; Hendricks et al., 2004). An important concern is that aviation aerosol can act as nuclei in ice cloud formation, thereby altering the microphysical properties of clouds (Jensen and Toon, 1997; Kärcher, 1999; Lohmann et al., 2004) and perhaps cloud cover. A modelling study by Hendricks et al. (2005) showed the potential for significant cirrus modifications by aviation caused by increased numbers of BC particles. The modifications would occur in flight corridors as well as in regions far away from flight corridors because of aerosol transport. In the study, aviation aerosols either increase or decrease ice nuclei in background cirrus clouds, depending on assumptions about the cloud formation process. Results from a cloud chamber experiment showed that a sulphate coating on soot particles reduced their effectiveness as ice nuclei (Möhler et al., 2005). Changes in ice nuclei number or nucleation properties of aerosols can alter the radiative properties of cirrus clouds and, hence, their radiative impact on the climate system, similar to the aerosol-cloud interactions discussed in Sections 2.4.1, 2.4.5 and 7.5. No estimates are yet available for the global or regional RF changes caused by the effect of aviation aerosol on background cloudiness, although some of the RF from AIC, determined by correlation studies (see Section 2.6.3), may be associated with these aerosol effects.

2.7 Natural Forcings

2.7.1 Solar Variability

The estimates of long-term solar irradiance changes used in the TAR (e.g., Hoyt and Schatten, 1993; Lean et al., 1995) have been revised downwards, based on new studies indicating that bright solar faculae likely contributed a smaller irradiance increase since the Maunder Minimum than was originally suggested by the range of brightness in Sun-like stars (Hall and Lockwood, 2004; M. Wang et al., 2005). However, empirical results since the TAR have strengthened the evidence for solar forcing of climate change by identifying detectable tropospheric changes associated with solar variability, including during the solar cycle (Section 9.2; van Loon and Shea, 2000; Douglass and Clader, 2002; Gleisner and Thejll, 2003; Haigh, 2003; Stott et al., 2003; White et al., 2003; Coughlin and Tung, 2004; Labitzke, 2004; Crooks and Gray, 2005). The most likely mechanism is considered to be some combination of direct forcing by changes in total solar irradiance, and indirect effects of ultraviolet (UV) radiation on the stratosphere. Least certain, and under ongoing debate as discussed in the TAR, are indirect effects induced by galactic cosmic rays (e.g., Marsh and Svensmark, 2000a,b; Kristjánsson et al., 2002; Sun and Bradley, 2002).

2.7.1.1 Direct Observations of Solar Irradiance

2.7.1.1.1 Satellite measurements of total solar irradiance

Four independent space-based instruments directly measure total solar irradiance at present, contributing to a database extant since November 1978 (Fröhlich and Lean, 2004). The Variability of Irradiance and Gravity Oscillations (VIRGO) experiment on the Solar Heliospheric Observatory (SOHO) has been operating since 1996, the ACRIM III on the Active Cavity Radiometer Irradiance Monitor Satellite (ACRIMSAT) since 1999 and the Earth Radiation Budget Satellite (ERBS) (intermittently) since 1984. Most recent are the measurements made by the Total Solar Irradiance Monitor (TIM) on the Solar Radiation and Climate Experiment (SORCE) since 2003 (Rottman, 2005).

2.7.1.1.2 Observed decadal trends and variability

Different composite records of total solar irradiance have been constructed from different combinations of the direct radiometric measurements. The Physikalisch-Meteorologisches Observatorium Davos (PMOD) composite (Fröhlich and Lean, 2004), shown in Figure 2.16, combines the observations by the ACRIM I on the Solar Maximum Mission (SMM), the Hickey-Friedan radiometer on Nimbus 7, ACRIM II on the Upper Atmosphere Research Satellite (UARS) and VIRGO on SOHO by analysing the sensitivity drifts in each radiometer prior to determining radiometric offsets. In contrast, the ACRIM composite (Willson and Mordvinov, 2003), also shown in Figure 2.16, utilises ACRIMSAT rather than VIRGO observations in recent times and cross calibrates the reported data assuming that radiometric sensitivity drifts have already been fully accounted for. A third composite, the Space Absolute Radiometric Reference (SARR) composite, uses individual absolute irradiance measurements from the shuttle to cross calibrate satellite records (Dewitte et al., 2005). The gross temporal features of the composite irradiance records are very similar, each showing day-to-week variations associated with the Sun's rotation on its axis, and decadal fluctuations arising from the 11-year solar activity cycle. But the linear slopes differ among the three different composite records, as do levels at solar activity minima (1986 and 1996). These differences are the result of different cross calibrations and drift adjustments applied to individual radiometric sensitivities when constructing the composites (Fröhlich and Lean, 2004).

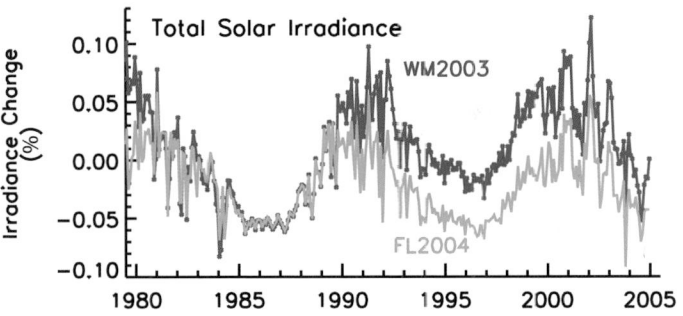

Figure 2.16. *Percentage change in monthly values of the total solar irradiance composites of Willson and Mordvinov (2003; WM2003, violet symbols and line) and Fröhlich and Lean (2004; FL2004, green solid line)*

Solar irradiance levels are comparable in the two most recent cycle minima when absolute uncertainties and sensitivity drifts in the measurements are assessed (Fröhlich and Lean, 2004 and references therein). The increase in excess of 0.04% over the 27-year period of the ACRIM irradiance composite (Willson and Mordvinov, 2003), although incompletely understood, is thought to be more of instrumental rather than solar origin (Fröhlich and Lean, 2004). The irradiance increase in the ACRIM composite is indicative of an episodic increase between 1989 and 1992 that is present in the Nimbus 7 data (Lee et al., 1995; Chapman et al., 1996). Independent, overlapping ERBS observations do not show this increase; nor do they suggest a significant secular trend (Lee et al., 1995). Such a trend is not present in the PMOD composite, in which total irradiance between successive solar minima is nearly constant, to better than 0.01% (Fröhlich and Lean, 2004). Although a long-term trend of order 0.01% is present in the SARR composite between successive solar activity minima (in 1986 and 1996), it is not statistically significant because the estimated uncertainty is ±0.026% (Dewitte et al., 2005).

Current understanding of solar activity and the known sources of irradiance variability suggests comparable irradiance levels during the past two solar minima. The primary known cause of contemporary irradiance variability is the presence on the Sun's disk of sunspots (compact, dark features where radiation is locally depleted) and faculae (extended bright features where radiation is locally enhanced). Models that combine records of the global sunspot darkening calculated directly from white light images and the magnesium (Mg) irradiance index as a proxy for the facular signal do not exhibit a significant secular trend during activity minima (Fröhlich and Lean, 2004; Preminger and Walton, 2005). Nor do the modern instrumental measurements of galactic cosmic rays, 10.7 cm flux and the *aa* geomagnetic index since the 1950s (Benestad, 2005) indicate this feature. While changes in surface emissivity by magnetic sunspot and facular regions are, from a theoretical view, the most effective in altering irradiance (Spruit, 2000), other mechanisms have also been proposed that may cause additional, possibly secular, irradiance changes. Of these, changes in solar diameter have been considered a likely candidate (e.g., Sofia and Li, 2001). But recent analysis of solar imagery, primarily from the Michelson

Doppler Imager (MDI) instrument on SOHO, indicates that solar diameter changes are no more than a few kilometres per year during the solar cycle (Dziembowski et al., 2001), for which associated irradiance changes are 0.001%, two orders of magnitude less than the measured solar irradiance cycle.

2.7.1.1.3 Measurements of solar spectral irradiance

The solar UV spectrum from 120 to 400 nm continues to be monitored from space, with SORCE observations extending those made since 1991 by two instruments on the UARS (Woods et al., 1996). SORCE also monitors, for the first time from space, solar spectral irradiance in the visible and near-infrared spectrum, providing unprecedented spectral coverage that affords a detailed characterisation of solar spectral irradiance variability. Initial results (Harder et al., 2005; Lean et al., 2005) indicate that, as expected, variations occur at all wavelengths, primarily in response to changes in sunspots and faculae. Ultraviolet spectral irradiance variability in the extended database is consistent with that seen in the UARS observations since 1991, as described in the TAR.

Radiation in the visible and infrared spectrum has a notably different temporal character than the spectrum below 300 nm. Maximum energy changes occur at wavelengths from 400 to 500 nm. Fractional changes are greatest at UV wavelengths but the actual energy change is considerably smaller than in the visible spectrum. Over the time scale of the 11-year solar cycle, bolometric facular brightness exceeds sunspot blocking by about a factor of two, and there is an increase in spectral irradiance at most, if not all, wavelengths from the minimum to the maximum of the solar cycle. Estimated solar cycle changes are 0.08% in the total solar irradiance. Broken down by wavelength range these irradiance changes are 1.3% at 200 to 300 nm, 0.2% at 315 to 400 nm, 0.08% at 400 to 700 nm, 0.04% at 700 to 1,000 nm and 0.025% at 1,000 to 1,600 nm.

However, during episodes of strong solar activity, sunspot blocking can dominate facular brightening, causing decreased irradiance at most wavelengths. Spectral irradiance changes on these shorter time scales now being measured by SORCE provide tests of the wavelength-dependent sunspot and facular parametrizations in solar irradiance variability models. The modelled spectral irradiance changes are in good overall agreement with initial SORCE observations but as yet the SORCE observations are too short to provide definitive information about the amplitude of solar spectral irradiance changes during the solar cycle.

2.7.1.2 Estimating Past Solar Radiative Forcing

2.7.1.2.1 Reconstructions of past variations in solar irradiance

Long-term solar irradiance changes over the past 400 years may be less by a factor of two to four than in the reconstructions employed by the TAR for climate change simulations. Irradiance reconstructions such as those of Hoyt and Schatten (1993), Lean et al. (1995), Lean (2000), Lockwood and Stamper (1999) and Solanki and Fligge (1999), used in the TAR, assumed the existence of a long-term variability component in addition to the known 11-year cycle, in which the 17th-century Maunder Minimum total irradiance was reduced in the range of 0.15% to 0.3% below contemporary solar minima. The temporal structure of this long-term component, typically associated with facular evolution, was assumed to track either the smoothed amplitude of the solar activity cycle or the cycle length. The motivation for adopting a long-term irradiance component was three-fold. Firstly, the range of variability in Sun-like stars (Baliunas and Jastrow, 1990), secondly, the long-term trend in geomagnetic activity, and thirdly, solar modulation of cosmogenic isotopes, all suggested that the Sun is capable of a broader range of activity than witnessed during recent solar cycles (i.e., the observational record in Figure 2.16). Various estimates of the increase in total solar irradiance from the 17th-century Maunder Minimum to the current activity minima from these irradiance reconstructions are compared with recent results in Table 2.10.

Each of the above three assumptions for the existence of a significant long-term irradiance component is now questionable. A reassessment of the stellar data was unable to recover the original bimodal separation of lower calcium (Ca) emission in non-cycling stars (assumed to be in Maunder-Minimum type states) compared with higher emission in cycling stars (Hall and Lockwood, 2004), which underpins the Lean et al. (1995) and Lean (2000) irradiance reconstructions. Rather, the current Sun is thought to have 'typical' (rather than high) activity relative to other stars. Plausible lowest brightness levels inferred from stellar observations are higher than the peak of the lower mode of the initial distribution of Baliunas and Jastrow (1990). Other studies raise the possibility of long-term instrumental drifts in historical indices of geomagnetic activity (Svalgaard et al., 2004), which would reduce somewhat the long-term trend in the Lockwood and Stamper (1999) irradiance reconstruction. Furthermore, the relationship between solar irradiance and geomagnetic and cosmogenic indices is complex, and not necessarily linear. Simulations of the transport of magnetic flux on the Sun and propagation of open flux into the heliosphere indicate that 'open' magnetic flux (which modulates geomagnetic activity and cosmogenic isotopes) can accumulate on inter-cycle time scales even when closed flux (such as in sunspots and faculae) does not (Lean et al., 2002; Y. Wang et al., 2005).

A new reconstruction of solar irradiance based on a model of solar magnetic flux variations (Y. Wang et al., 2005), which does not invoke geomagnetic, cosmogenic or stellar proxies, suggests that the amplitude of the background component is significantly less than previously assumed, specifically 0.27 times that of Lean (2000). This estimate results from simulations of the eruption, transport and accumulation of magnetic flux during the past 300 years using a flux transport model with variable meridional flow. Variations in both the total flux and in just the flux that extends into the heliosphere (the open flux) are estimated, arising from the deposition of bipolar magnetic regions (active regions) and smaller-scale bright features (ephemeral regions) on the Sun's surface in strengths and numbers proportional to the sunspot number. The open flux compares reasonably well with the cosmogenic isotopes for

which variations arise, in part, from heliospheric modulation. This gives confidence that the approach is plausible. A small accumulation of total flux (and possibly ephemeral regions) produces a net increase in facular brightness, which, in combination with sunspot blocking, permits the reconstruction of total solar irradiance shown in Figure 2.17. There is a 0.04% increase from the Maunder Minimum to present-day cycle minima.

Prior to direct telescopic measurements of sunspots, which commenced around 1610, knowledge of solar activity is inferred indirectly from the ^{14}C and ^{10}Be cosmogenic isotope records in tree rings and ice cores, respectively, which exhibit solar-related cycles near 90, 200 and 2,300 years. Some studies of cosmogenic isotopes (Jirikowic and Damon, 1994) and spectral analysis of the sunspot record (Rigozo et al., 2001) suggest that solar activity during the 12th-century Medieval Solar Maximum was comparable to the present Modern Solar Maximum. Recent work attempts to account for the chain of physical processes in which solar magnetic fields modulate the heliosphere, in turn altering the penetration of the galactic cosmic rays, the flux of which produces the cosmogenic isotopes that are subsequently deposited in the terrestrial system following additional transport and chemical processes. An initial effort reported exceptionally high levels of solar activity in the past 70 years, relative to the preceding 8,000 years (Solanki et al., 2004). In contrast, when differences among isotopes records are taken into account and the ^{14}C record corrected for fossil fuel burning, current levels of solar activity are found to be historically high, but not exceptionally so (Muscheler et al., 2007).

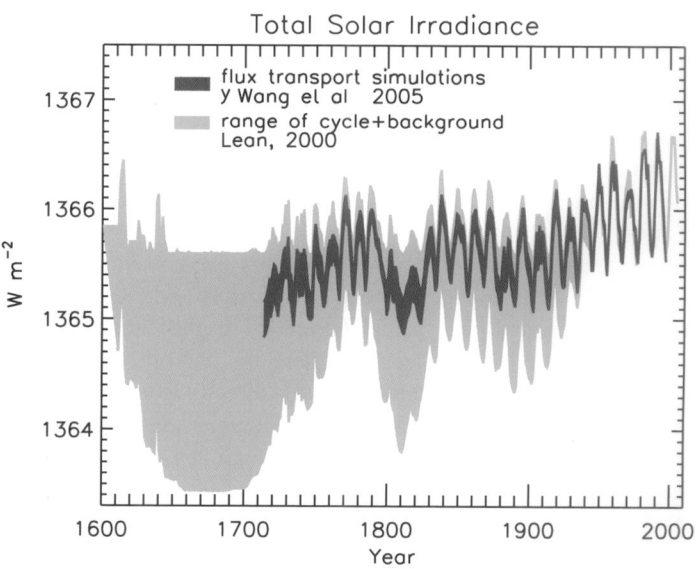

Figure 2.17. *Reconstructions of the total solar irradiance time series starting as early as 1600. The upper envelope of the shaded regions shows irradiance variations arising from the 11-year activity cycle. The lower envelope is the total irradiance reconstructed by Lean (2000), in which the long-term trend was inferred from bright-ness changes in Sun-like stars. In comparison, the recent reconstruction of Y. Wang et al. (2005) is based on solar considerations alone, using a flux transport model to simulate the long-term evolution of the closed flux that generates bright faculae.*

Table 2.10. *Comparison of the estimates of the increase in RF from the 17th-century Maunder Minimum (MM) to contemporary solar minima, documenting new understanding since the TAR.*

Reference	Assumptions and Technique	RF Increase from the Maunder Minimum to Contemporary Minima (W m^{-2})[a]	Comment on Current Understanding
Schatten and Orosz (1990)	Extrapolation of the 11-year irradiance cycle to the MM, using the sunspot record.	~ 0	Irradiance levels at cycle minima remain approximately constant.
Lean et al. (1992)	No spots, plage or network in Ca images assumed during MM.	0.26	Maximum irradiance increase from a non-magnetic sun, due to changes in known bright features on contemporary solar disk.
Lean et al. (1992)	No spots, plage or network and reduced basal emission in cell centres in Ca images to match reduced brightness in non-cycling stars, assumed to be MM analogues.	0.45	New assessment of stellar data (Hall and Lockwood, 2004) does not support original stellar brightness distribution, or the use of the brightness reduction in the Baliunas and Jastrow (1990) 'non-cycling' stars as MM analogues.
Hoyt and Schatten (1993)[b]	Convective restructuring implied by changes in sunspot umbra/penumbra ratios from MM to present; amplitude of increase from MM to present based on brightness of non-cycling stars, from Lean et al. (1992).	0.65	As above
Lean et al. (1995)	Reduced brightness of non-cycling stars, relative to those with active cycles, assumed typical of MM.	0.45	As above
Solanki and Fligge (1999)[b]	Combinations of above.	0.68	As above
Lean (2000)	Reduced brightness of non-cycling stars (revised solar-stellar calibration) assumed typical of MM.	0.38	As above
Foster (2004) Model	Non-magnetic sun estimates by removing bright features from MDI images assumed for MM.	0.28	Similar approach to removal of spots, plage and network by Lean et al. (1992).
Y. Wang et al. (2005)[b]	Flux transport simulations of total magnetic flux evolution from MM to present.	0.1	Solar model suggests that modest accumulation of magnetic flux from one solar cycle to the next produces a modest increase in irradiance levels at solar cycle minima.
Dziembowski et al. (2001)	Helioseismic observations of solar interior oscillations suggest that the historical Sun could not have been any dimmer than current activity minima.	~ 0	

Notes:

[a] The RF is the irradiance change divided by 4 (geometry) and multiplied by 0.7 (albedo). The solar activity cycle, which was negligible during the Maunder Minimum and is of order 1 W m^{-2} (minimum to maximum) during recent cycles, is superimposed on the irradiance changes at cycle minima. When smoothed over 20 years, this cycle increases the net RF in the table by an additional 0.09 W m^{-2}.

[b] These reconstructions extend only to 1713, the end of the Maunder Minimum.

2.7.1.2.2 Implications for solar radiative forcing

In terms of plausible physical understanding, the most likely secular increase in total irradiance from the Maunder Minimum to current cycle minima is 0.04% (an irradiance increase of roughly 0.5 W m^{-2} in 1,365 W m^{-2}), corresponding to an RF[11] of +0.1 W m^{-2}. The larger RF estimates in Table 2.10, in the range of +0.38 to +0.68 W m^{-2}, correspond to assumed changes in solar irradiance at cycle minima derived from brightness fluctuations in Sun-like stars that are no longer valid. Since the 11-year cycle amplitude has increased from the Maunder Minimum to the present, the total irradiance increase to the present-day cycle mean is 0.08%. From 1750 to the present there was a net 0.05% increase in total solar irradiance, according to the 11-year smoothed total solar irradiance time series of Y. Wang et al. (2005), shown in Figure 2.17. This corresponds to an RF of +0.12 W m^{-2}, which is more than a factor of two less than the solar RF estimate in the TAR, also from 1750 to the present. Using the Lean (2000) reconstruction (the lower envelope in Figure 2.17) as an upper limit, there is a 0.12% irradiance increase since 1750, for which the RF is +0.3 W m^{-2}. The lower limit of the irradiance increase from 1750 to the present is 0.026% due to the increase in the 11-year cycle only. The corresponding lower limit of the RF is +0.06 W m^{-2}. As with solar cycle changes, long-term irradiance variations are expected to have significant spectral dependence. For example, the Y. Wang et al. (2005) flux transport estimates imply decreases during the Maunder Minimum relative to contemporary activity cycle minima of 0.43% at 200 to 300 nm, 0.1% at 315 to 400 nm, 0.05% at 400 to 700 nm, 0.03% at 700 to 1,000 nm and 0.02% at 1,000 to 1,600 nm (Lean et al., 2005), compared with 1.4%, 0.32%, 0.17%, 0.1% and 0.06%, respectively, in the earlier model of Lean (2000).

2.7.1.3 Indirect Effects of Solar Variability

Approximately 1% of the Sun's radiant energy is in the UV portion of the spectrum at wavelengths below about 300 nm, which the Earth's atmosphere absorbs. Although of considerably smaller absolute energy than the total irradiance, solar UV radiation is fractionally more variable by at least an order of magnitude. It contributes significantly to changes in total solar irradiance (15% of the total irradiance cycle; Lean et al., 1997) and creates and modifies the ozone layer, but is not considered as a direct RF because it does not reach the troposphere. Since the TAR, new studies have confirmed and advanced the plausibility of indirect effects involving the modification of the stratosphere by solar UV irradiance variations (and possibly by solar-induced variations in the overlying mesosphere and lower thermosphere), with subsequent dynamical and radiative coupling to the troposphere (Section 9.2). Whether solar wind fluctuations

(Boberg and Lundstedt, 2002) or solar-induced heliospheric modulation of galactic cosmic rays (Marsh and Svensmark, 2000b) also contribute indirect forcings remains ambiguous.

As in the troposphere, anthropogenic effects, internal cycles (e.g., the Quasi-Biennial Oscillation) and natural influences all affect the stratosphere. It is now well established from both empirical and model studies that solar cycle changes in UV radiation alter middle atmospheric ozone concentrations (Fioletov et al., 2002; Geller and Smyshlyaev, 2002; Hood, 2003), temperatures and winds (Ramaswamy et al., 2001; Labitzke et al., 2002; Haigh, 2003; Labitzke, 2004; Crooks and Gray, 2005), including the Quasi-Biennial Oscillation (McCormack, 2003; Salby and Callaghan, 2004). In their recent survey of solar influences on climate, Gray et al. (2005) noted that updated observational analyses have confirmed earlier 11-year cycle signals in zonally averaged stratospheric temperature, ozone and circulation with increased statistical confidence. There is a solar-cycle induced increase in global total ozone of 2 to 3% at solar cycle maximum, accompanied by temperature responses that increase with altitude, exceeding 1°C around 50 km. However, the amplitudes and geographical and altitudinal patterns of these variations are only approximately known, and are not linked in an easily discernible manner to the forcing. For example, solar forcing appears to induce a significant lower stratospheric response (Hood, 2003), which may have a dynamical origin caused by changes in temperature affecting planetary wave propagation, but it is not currently reproduced by models.

When solar activity is high, the more complex magnetic configuration of the heliosphere reduces the flux of galactic cosmic rays in the Earth's atmosphere. Various scenarios have been proposed whereby solar-induced galactic cosmic ray fluctuations might influence climate (as surveyed by Gray et al., 2005). Carslaw et al. (2002) suggested that since the plasma produced by cosmic ray ionization in the troposphere is part of an electric circuit that extends from the Earth's surface to the ionosphere, cosmic rays may affect thunderstorm electrification. By altering the population of CCN and hence microphysical cloud properties (droplet number and concentration), cosmic rays may also induce processes analogous to the indirect effect of tropospheric aerosols. The presence of ions, such as produced by cosmic rays, is recognised as influencing several microphysical mechanisms (Harrison and Carslaw, 2003). Aerosols may nucleate preferentially on atmospheric cluster ions. In the case of low gas-phase sulphuric acid concentrations, ion-induced nucleation may dominate over binary sulphuric acid-water nucleation. In addition, increased ion nucleation and increased scavenging rates of aerosols in turbulent regions around clouds seem likely. Because of the difficulty in tracking the influence of one particular modification brought about by

---

[11] To estimate RF, the change in total solar irradiance is multiplied by 0.25 to account for Earth-Sun geometry and then multiplied by 0.7 to account for the planetary albedo (e.g., Ramaswamy et al., 2001). Ideally this resulting RF should also be reduced by 15% to account for solar variations in the UV below 300 nm (see Section 2.7.1.3) and further reduced by about 4% to account for stratospheric absorption of solar radiation above 300 nm and the resulting stratospheric adjustment (Hansen et al., 1997). However, these corrections are not made to the RF estimates in this report because they: 1) represent small adjustments to the RF; 2) may in part be compensated by indirect effects of solar-ozone interaction in the stratosphere (see Section 2.7.1.3); and 3) are not routinely reported in the literature.

ions through the long chain of complex interacting processes, quantitative estimates of galactic cosmic-ray induced changes in aerosol and cloud formation have not been reached.

Many empirical associations have been reported between globally averaged low-level cloud cover and cosmic ray fluxes (e.g., Marsh and Svensmark, 2000a,b). Hypothesised to result from changing ionization of the atmosphere from solar-modulated cosmic ray fluxes, an empirical association of cloud cover variations during 1984 to 1990 and the solar cycle remains controversial because of uncertainties about the reality of the decadal signal itself, the phasing or anti-phasing with solar activity, and its separate dependence for low, middle and high clouds. In particular, the cosmic ray time series does not correspond to global total cloud cover after 1991 or to global low-level cloud cover after 1994 (Kristjánsson and Kristiansen, 2000; Sun and Bradley, 2002) without unproven de-trending (Usoskin et al., 2004). Furthermore, the correlation is significant with low-level cloud cover based only on infrared (not visible) detection. Nor do multi-decadal (1952 to 1997) time series of cloud cover from ship synoptic reports exhibit a relationship to cosmic ray flux. However, there appears to be a small but statistically significant positive correlation between cloud over the UK and galactic cosmic ray flux during 1951 to 2000 (Harrison and Stephenson, 2006). Contrarily, cloud cover anomalies from 1900 to 1987 over the USA do have a signal at 11 years that is anti-phased with the galactic cosmic ray flux (Udelhofen and Cess, 2001). Because the mechanisms are uncertain, the apparent relationship between solar variability and cloud cover has been interpreted to result not only from changing cosmic ray fluxes modulated by solar activity in the heliosphere (Usoskin et al., 2004) and solar-induced changes in ozone (Udelhofen and Cess, 2001), but also from sea surface temperatures altered directly by changing total solar irradiance (Kristjánsson et al., 2002) and by internal variability due to the El Niño-Southern Oscillation (Kernthaler et al., 1999). In reality, different direct and indirect physical processes (such as those described in Section 9.2) may operate simultaneously.

The direct RF due to increase in solar irradiance is reduced from the TAR. The best estimate is +0.12 W m^{-2} (90% confidence interval: +0.06 to +0.30 W m^{-2}). While there have been advances in the direct solar irradiance variation, there remain large uncertainties. The level of scientific understanding is elevated to low relative to TAR for solar forcing due to direct irradiance change, while declared as very low for cosmic ray influences (Section 2.9, Table 2.11).

2.7.2 Explosive Volcanic Activity

2.7.2.1 *Radiative Effects of Volcanic Aerosols*

Volcanic sulphate aerosols are formed as a result of oxidation of the sulphur gases emitted by explosive volcanic eruptions into the stratosphere. The process of gas-to-particle conversion has an e-folding time of roughly 35 days (Bluth et al., 1992; Read et al., 1993). The e-folding time (by mass) for sedimentation of

sulphate aerosols is typically about 12 to 14 months (Lambert et al., 1993; Baran and Foot, 1994; Barnes and Hoffman, 1997; Bluth et al., 1997). Also emitted directly during an eruption are volcanic ash particulates (siliceous material). These are particles usually larger than 2 μm that sediment out of the stratosphere fairly rapidly due to gravity (within three months or so), but could also play a role in the radiative perturbations in the immediate aftermath of an eruption. Stratospheric aerosol data incorporated for climate change simulations tends to be mostly that of the sulphates (Sato et al., 1993; Stenchikov et al., 1998; Ramachandran et al., 2000; Hansen et al., 2002; Tett et al., 2002; Ammann et al., 2003). As noted in the Second Assessment Report (SAR) and the TAR, explosive volcanic events are episodic, but the stratospheric aerosols resulting from them yield substantial transitory perturbations to the radiative energy balance of the planet, with both shortwave and longwave effects sensitive to the microphysical characteristics of the aerosols (e.g., size distribution).

Long-term ground-based and balloon-borne instrumental observations have resulted in an understanding of the optical effects and microphysical evolution of volcanic aerosols (Deshler et al., 2003; Hofmann et al., 2003). Important ground-based observations of aerosol characteristics from pre-satellite era spectral extinction measurements have been analysed by Stothers (2001a,b), but they do not provide global coverage. Global observations of stratospheric aerosol over the last 25 years have been possible owing to a number of satellite platforms, for example, TOMS and TOVS have been used to estimate SO_2 loadings from volcanic eruptions (Krueger et al., 2000; Prata et al., 2003). The Stratospheric Aerosol and Gas Experiment (SAGE) and Stratospheric Aerosol Measurement (SAM) projects (e.g., McCormick, 1987) have provided vertically resolved stratospheric aerosol spectral extinction data for over 20 years, the longest such record. This data set has significant gaps in coverage at the time of the El Chichón eruption in 1982 (the second most important in the 20th century after Mt. Pinatubo in 1991) and when the aerosol cloud is dense; these gaps have been partially filled by lidar measurements and field campaigns (e.g., Antuña et al., 2003; Thomason and Peter, 2006).

Volcanic aerosols transported in the atmosphere to polar regions are preserved in the ice sheets, thus recording the history of the Earth's volcanism for thousands of years (Bigler et al., 2002; Palmer et al., 2002; Mosley-Thompson et al., 2003). However, the atmospheric loadings obtained from ice records suffer from uncertainties due to imprecise knowledge of the latitudinal distribution of the aerosols, depositional noise that can affect the signal for an individual eruption in a single ice core, and poor constraints on aerosol microphysical properties.

The best-documented explosive volcanic event to date, by way of reliable and accurate observations, is the 1991 eruption of Mt. Pinatubo. The growth and decay of aerosols resulting from this eruption have provided a basis for modelling the RF due to explosive volcanoes. There have been no explosive and climatically significant volcanic events since Mt. Pinatubo. As pointed out in Ramaswamy et al. (2001), stratospheric aerosol concentrations are now at the lowest concentrations since the satellite era and global coverage began in about 1980. Altitude-dependent stratospheric optical observations at a few wavelengths, together with columnar optical and physical measurements, have been used to construct the time-dependent global field of stratospheric aerosol size distribution formed in the aftermath of volcanic events. The wavelength-dependent stratospheric aerosol single-scattering characteristics calculated for the solar and longwave spectrum are deployed in climate models to account for the resulting radiative (shortwave plus longwave) perturbations.

Using available satellite- and ground-based observations, Hansen et al. (2002) constructed a volcanic aerosols data set for the 1850 to 1999 period (Sato et al., 1993). This has yielded zonal mean vertically resolved aerosol optical depths for visible wavelengths and column average effective radii. Stenchikov et al. (2006) introduced a slight variation to this data set, employing UARS observations to modify the effective radii relative to Hansen et al. (2002), thus accounting for variations with altitude. Ammann et al. (2003) developed a data set of total aerosol optical depth for the period since 1890 that does not include the Krakatau eruption. The data set

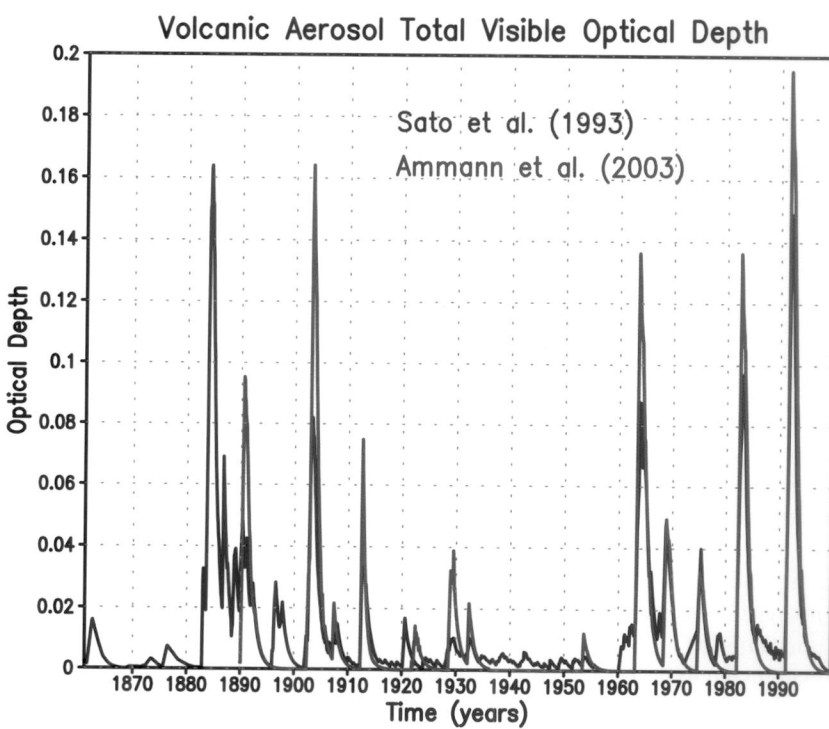

Figure 2.18. *Visible (wavelength 0.55 μm) optical depth estimates of stratospheric sulphate aerosols formed in the aftermath of explosive volcanic eruptions that occurred between 1860 and 2000. Results are shown from two different data sets that have been used in recent climate model integrations. Note that the Ammann et al. (2003) data begins in 1890.*

is based on empirical estimates of atmospheric loadings, which are then globally distributed using a simplified parametrization of atmospheric transport, and employs a fixed aerosol effective radius (0.42 μm) for calculating optical properties. The above data sets have essentially provided the bases for the volcanic aerosols implemented in virtually all of the models that have performed the 20th-century climate integrations (Stenchikov et al., 2006). Relative to Sato et al. (1993), the Ammann et al. (2003) estimate yields a larger value of the optical depth, by 20 to 30% in the second part of the 20th century, and by 50% for eruptions at the end of 19th and beginning of 20th century, for example, the 1902 Santa Maria eruption (Figure 2.18).

The global mean RF calculated using the Sato et al. (1993) data yields a peak in radiative perturbation of about -3 W m^{-2} for the strong (rated in terms of emitted SO$_2$) 1860 and 1991 eruptions of Krakatau and Mt. Pinatubo, respectively. The value is reduced to about -2 W m^{-2} for the relatively less intense El Chichón and Agung eruptions (Hansen et al., 2002). As expected from the arguments above, Ammann's RF is roughly 20 to 30% larger than Sato's RF.

Not all features of the aerosols are well quantified, and extending and improving the data sets remains an important area of research. This includes improved estimates of the aerosol size parameters (Bingen et al., 2004), a new approach for calculating aerosol optical characteristics using SAGE and UARS data (Bauman et al., 2003), and intercomparison of data from different satellites and combining them to fill gaps (Randall et al., 2001). While the aerosol characteristics are

better constrained for the Mt. Pinatubo eruption, and to some extent for the El Chichón and Agung eruptions, the reliability degrades for aerosols from explosive volcanic events further back in time as there are few, if any, observational constraints on their optical depth and size evolution.

The radiative effects due to volcanic aerosols from major eruptions are manifest in the global mean anomaly of reflected solar radiation; this variable affords a good estimate of radiative effects that can actually be tested against observations. However, unlike RF, this variable contains effects due to feedbacks (e.g., changes in cloud distributions) so that it is actually more a signature of the climate response. In the case of the Mt. Pinatubo eruption, with a peak global visible optical depth of about 0.15, simulations yield a large negative perturbation as noted above of about -3 W m^{-2} (Ramachandran et al., 2000; Hansen et al., 2002) (see also Section 9.2). This modelled estimate of reflected solar radiation compares reasonably with ERBS observations (Minnis et al., 1993). However, the ERBS observations were for a relatively short duration, and the model-observation comparisons are likely affected by differing cloud effects in simulations and measurements. It is interesting to note (Stenchikov et al., 2006) that, in the Mt. Pinatubo case, the Goddard Institute for Space Studies (GISS) models that use the Sato et al. (1993) data yield an even greater solar reflection than the National Center for Atmospheric Research (NCAR) model that uses the larger (Ammann et al., 2003) optical depth estimate.

2.7.2.2   Thermal, Dynamical and Chemistry Perturbations Forced by Volcanic Aerosols

Four distinct mechanisms have been invoked with regards to the climate response to volcanic aerosol RF. First, these forcings can directly affect the Earth's radiative balance and thus alter surface temperature. Second, they introduce horizontal and vertical heating gradients; these can alter the stratospheric circulation, in turn affecting the troposphere. Third, the forcings can interact with internal climate system variability (e.g., El Niño-Southern Oscillation, North Atlantic Oscillation, Quasi-Biennial Oscillation) and dynamical noise, thereby triggering, amplifying or shifting these modes (see Section 9.2; Yang and Schlesinger, 2001; Stenchikov et al., 2004). Fourth, volcanic aerosols provide surfaces for heterogeneous chemistry affecting global stratospheric ozone distributions (Chipperfield et al., 2003) and perturbing other trace gases for a considerable period following an eruption. Each of the above mechanisms has its own spatial and temporal response pattern. In addition, the mechanisms could depend on the background state of the climate system, and thus on other forcings (e.g., due to well-mixed gases, Meehl et al., 2004), or interact with each other.

The complexity of radiative-dynamical response forced by volcanic impacts suggests that it is important to calculate aerosol radiative effects interactively within the model rather than prescribe them (Andronova et al., 1999; Broccoli et al., 2003). Despite differences in volcanic aerosol parameters employed, models computing the aerosol radiative effects interactively yield tropical and global mean lower-stratospheric warmings that are fairly consistent with each other and with observations (Ramachandran et al., 2000; Hansen et al., 2002; Yang and Schlesinger, 2002; Stenchikov et al., 2004; Ramaswamy et al., 2006b); however, there is a considerable range in the responses in the polar stratosphere and troposphere. The global mean warming of the lower stratosphere is due mainly to aerosol effects in the longwave spectrum, in contrast to the flux changes at the TOA that are essentially due to aerosol effects in the solar spectrum. The net radiative effects of volcanic aerosols on the thermal and hydrologic balance (e.g., surface temperature and moisture) have been highlighted by recent studies (Free and Angell, 2002; Jones et al., 2003; see Chapter 6; and see Chapter 9 for significance of the simulated responses and model-observation comparisons for 20th-century eruptions). A mechanism closely linked to the optical depth perturbation and ensuing warming of the tropical lower stratosphere is the potential change in the cross-tropopause water vapour flux (Joshi and Shine, 2003; see Section 2.3.7).

Anomalies in the volcanic-aerosol induced global radiative heating distribution can force significant changes in atmospheric circulation, for example, perturbing the equator-to-pole heating gradient (Stenchikov et al., 2002; Ramaswamy et al., 2006a; see Section 9.2) and forcing a positive phase of the Arctic Oscillation that in turn causes a counterintuitive boreal winter warming at middle and high latitudes over Eurasia and North America (Perlwitz and Graf, 2001; Stenchikov et al., 2002,

2004, 2006; Shindell et al., 2003b, 2004; Perlwitz and Harnik, 2003; Rind et al., 2005; Miller et al., 2006).

Stratospheric aerosols affect the chemistry and transport processes in the stratosphere, resulting in the depletion of ozone (Brasseur and Granier, 1992; Tie et al., 1994; Solomon et al., 1996; Chipperfield et al., 2003). Stenchikov et al. (2002) demonstrated a link between ozone depletion and Arctic Oscillation response; this is essentially a secondary radiative mechanism induced by volcanic aerosols through stratospheric chemistry. Stratospheric cooling in the polar region associated with a stronger polar vortex initiated by volcanic effects can increase the probability of formation of polar stratospheric clouds and therefore enhance the rate of heterogeneous chemical destruction of stratospheric ozone, especially in the NH (Tabazadeh et al., 2002). The above studies indicate effects on the stratospheric ozone layer in the wake of a volcanic eruption and under conditions of enhanced anthropogenic halogen loading. Interactive microphysics-chemistry-climate models (Rozanov et al., 2002, 2004; Shindell et al., 2003b; Timmreck et al., 2003; Dameris et al., 2005) indicate that aerosol-induced stratospheric heating affects the dispersion of the volcanic aerosol cloud, thus affecting the spatial RF. However the models' simplified treatment of aerosol microphysics introduces biases; further, they usually overestimate the mixing at the tropopause level and intensity of meridional transport in the stratosphere (Douglass et al., 2003; Schoeberl et al., 2003). For present climate studies, it is practical to utilise simpler approaches that are reliably constrained by aerosol observations.

Because of its episodic and transitory nature, it is difficult to give a best estimate for the volcanic RF, unlike the other agents. Neither a best estimate nor a level of scientific understanding was given in the TAR. For the well-documented case of the explosive 1991 Mt. Pinatubo eruption, there is a good scientific understanding. However, the limited knowledge of the RF associated with prior episodic, explosive events indicates a low level of scientific understanding (Section 2.9, Table 2.11).

2.8   Utility of Radiative Forcing

The TAR and other assessments have concluded that RF is a useful tool for estimating, to a first order, the relative global climate impacts of differing climate change mechanisms (Ramaswamy et al., 2001; Jacob et al., 2005). In particular, RF can be used to estimate the relative equilibrium globally averaged surface temperature change due to different forcing agents. However, RF is not a measure of other aspects of climate change or the role of emissions (see Sections 2.2 and 2.10). Previous GCM studies have indicated that the climate sensitivity parameter was more or less constant (varying by less than 25%) between mechanisms (Ramaswamy et al., 2001; Chipperfield et al., 2003). However, this level of agreement was found not to hold for certain mechanisms such as ozone changes at some altitudes and changes in absorbing aerosol.

Because the climate responses, and in particular the equilibrium climate sensitivities, exhibited by GCMs vary by much more than 25% (see Section 9.6), Ramaswamy et al. (2001) and Jacob et al. (2005) concluded that RF is the most simple and straightforward measure for the quantitative assessment of climate change mechanisms, especially for the LLGHGs. This section discusses the several studies since the TAR that have examined the relationship between RF and climate response. Note that this assessment is entirely based on climate model simulations.

2.8.1 Vertical Forcing Patterns and Surface Energy Balance Changes

The vertical structure of a forcing agent is important both for efficacy (see Section 2.8.5) and for other aspects of climate response, particularly for evaluating regional and vertical patterns of temperature change and also changes in the hydrological cycle. For example, for absorbing aerosol, the surface forcings are arguably a more useful measure of the climate response (particularly for the hydrological cycle) than the RF (Ramanathan et al., 2001a; Menon et al., 2002b). It should be noted that a perturbation to the surface energy budget involves sensible and latent heat fluxes besides solar and longwave irradiance; therefore, it can quantitatively be very different from the RF, which is calculated at the tropopause, and thus is not representative of the energy balance perturbation to the surface-troposphere (climate) system. While the surface forcing adds to the overall description of the total perturbation brought about by an agent, the RF and surface forcing should not be directly compared nor should the surface forcing be considered in isolation for evaluating the climate response (see, e.g., the caveats expressed in Manabe and Wetherald, 1967; Ramanathan, 1981). Therefore, surface forcings are presented as an important and useful diagnostic tool that aids understanding of the climate response (see Sections 2.9.4 and 2.9.5).

2.8.2 Spatial Patterns of Radiative Forcing

Each RF agent has a unique spatial pattern (see, e.g., Figure 6.7 in Ramaswamy et al., 2001). When combining RF agents it is not just the global mean RF that needs to be considered. For example, even with a net global mean RF of zero, significant regional RFs can be present and these can affect the global mean temperature response (see Section 2.8.5). Spatial patterns of RF also affect the pattern of climate response. However, note that, to first order, very different RF patterns can have similar patterns of surface temperature response and the location of maximum RF is rarely coincident with the location of maximum response (Boer and Yu, 2003b). Identification of different patterns of response is particularly important for attributing past climate change to particular mechanisms, and is also important for the prediction of regional patterns of future climate change. This chapter employs RF as the method for ranking the effect of a forcing agent on the equilibrium global temperature change,

and only this aspect of the forcing-response relationship is discussed. However, patterns of RF are presented as a diagnostic in Section 2.9.5.

2.8.3 Alternative Methods of Calculating Radiative Forcing

RFs are increasingly being diagnosed from GCM integrations where the calculations are complex (Stuber et al., 2001b; Tett et al., 2002; Gregory et al., 2004). This chapter also discusses several mechanisms that include some response in the troposphere, such as cloud changes. These mechanisms are not initially radiative in nature, but will eventually lead to a radiative perturbation of the surface-troposphere system that could conceivably be measured at the TOA. Jacob et al. (2005) refer to these mechanisms as non-radiative forcings (see also Section 2.2). Alternatives to the standard stratospherically adjusted RF definition have been proposed that may help account for these processes. Since the TAR, several studies have employed GCMs to diagnose the zero-surface-temperature-change RF (see Figure 2.2 and Section 2.2). These studies have used a number of different methodologies. Shine et al. (2003) fixed both land and sea surface temperatures globally and calculated a radiative energy imbalance: this technique is only feasible in GCMs with relatively simple land surface parametrizations. Hansen et al. (2005) fixed sea surface temperatures and calculated an RF by adding an extra term to the radiative imbalance that took into account how much the land surface temperatures had responded. Sokolov (2006) diagnosed the zero-surface-temperature-change RF by computing surface-only and atmospheric-only components of climate feedback separately in a slab model and then modifying the stratospherically adjusted RF by the atmospheric-only feedback component. Gregory et al. (2004; see also Hansen et al., 2005; Forster and Taylor, 2006) used a regression method with a globally averaged temperature change ordinate to diagnose the zero-surface-temperature-change RF: this method had the largest uncertainties. Shine et al. (2003), Hansen et al. (2005) and Sokolov (2006) all found that that the fixed-surface-temperature RF was a better predictor of the equilibrium global mean surface temperature response than the stratospherically adjusted RF. Further, it was a particularly useful diagnostic for changes in absorbing aerosol where the stratospherically adjusted RF could fail as a predictor of the surface temperature response (see Section 2.8.5.5). Differences between the zero-surface-temperature-change RF and the stratospherically adjusted RF can be caused by semi-direct and cloud-aerosol interaction effects beyond the cloud albedo RF. For most mechanisms, aside from the case of certain aerosol changes, the difference is likely to be small (Shine et al., 2003; Hansen et al., 2005; Sokolov, 2006). These calculations also remove problems associated with defining the tropopause in the stratospherically adjusted RF definition (Shine et al., 2003; Hansen et al., 2005). However, stratospherically adjusted RF has the advantage that it does not depend on relatively uncertain components of a GCM's response, such as cloud changes. For the LLGHGs, the stratospherically adjusted RF also has the advantage that it is readily calculated in detailed off-line radiation codes. For these reasons, the stratospherically adjusted RF is retained as the measure of comparison used in this chapter (see Section 2.2). However, to first order, all methods are comparable and all prove useful for understanding climate response.

2.8.4 Linearity of the Forcing-Response Relationship

Reporting findings from several studies, the TAR concluded that responses to individual RFs could be linearly added to gauge the global mean response, but not necessarily the regional response (Ramaswamy et al., 2001). Since then, studies with several equilibrium and/or transient integrations of several different GCMs have found no evidence of any nonlinearity for changes in greenhouse gases and sulphate aerosol (Boer and Yu, 2003b; Gillett et al., 2004; Matthews et al., 2004; Meehl et al., 2004). Two of these studies also examined realistic changes in many other forcing agents without finding evidence of a nonlinear response (Meehl et al., 2004; Matthews et al., 2004). In all four studies, even the regional changes typically added linearly. However, Meehl et al. (2004) observed that neither precipitation changes nor all regional temperature changes were linearly additive. This linear relationship also breaks down for global mean temperatures when aerosol-cloud interactions beyond the cloud albedo RF are included in GCMs (Feichter et al., 2004; see also Rotstayn and Penner, 2001; Lohmann and Feichter, 2005). Studies that include these effects modify clouds in their models, producing an additional radiative imbalance. Rotstayn and Penner (2001) found that if these aerosol-cloud effects are accounted for as additional forcing terms, the inference of linearity can be restored (see Sections 2.8.3 and 2.8.5). Studies also find nonlinearities for large negative RFs, where static stability changes in the upper troposphere affect the climate feedback (e.g., Hansen et al., 2005). For the magnitude and range of realistic RFs discussed in this chapter, and excluding cloud-aerosol interaction effects, there is high confidence in a linear relationship between global mean RF and global mean surface temperature response.

2.8.5 Efficacy and Effective Radiative Forcing

Efficacy (E) is defined as the ratio of the climate sensitivity parameter for a given forcing agent (λ_i) to the climate sensitivity parameter for CO_2 changes, that is, $E_i = \lambda_i / \lambda_{CO_2}$ (Joshi et al., 2003; Hansen and Nazarenko, 2004). Efficacy can then be used to define an effective RF (= E_i RF_i) (Joshi et al., 2003; Hansen et al., 2005). For the effective RF, the climate sensitivity parameter is independent

of the mechanism, so comparing this forcing is equivalent to comparing the equilibrium global mean surface temperature change. That is, $\Delta T_s = \lambda_{CO_2} \times E_i \times RF_i$ Preliminary studies have found that efficacy values for a number of forcing agents show less model dependency than the climate sensitivity values (Joshi et al., 2003). Effective RFs have been used get one step closer to an estimator of the likely surface temperature response than can be achieved by using RF alone (Sausen and Schumann, 2000; Hansen et al., 2005; Lohmann and Feichter, 2005). Adopting the zero-surface-temperature-change RF, which has efficacies closer to unity, may be another way of achieving similar goals (see Section 2.8.3). This section assesses the efficacy associated with stratospherically adjusted RF, as this is the definition of RF adopted in this chapter (see Section 2.2). Therefore, cloud-aerosol interaction effects beyond the cloud albedo RF are included in the efficacy term. The findings presented in this section are from an assessment of all the studies referenced in the caption of Figure 2.19, which presents a synthesis of efficacy results. As space is limited not all these studies are explicitly discussed in the main text.

2.8.5.1 Generic Understanding

Since the TAR, several GCM studies have calculated efficacies and a general understanding is beginning to emerge as to how and why efficacies vary between mechanisms. The initial climate state, and the sign and magnitude of the RF have less importance but can still affect efficacy (Boer and Yu, 2003a; Joshi et al., 2003; Hansen et al., 2005). These studies have also

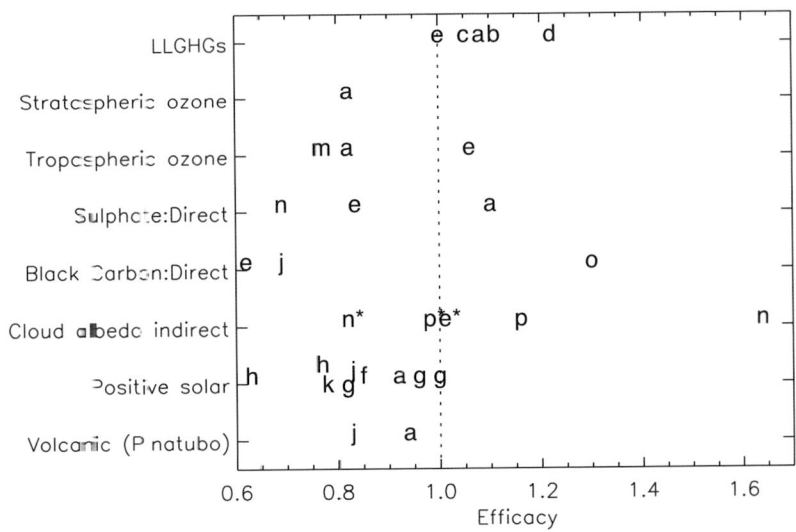

Figure 2.19. Efficacies as calculated by several GCM models for realistic changes in RF agents. Letters are centred on efficacy value and refer to the literature study that the value is taken from (see text of Section 2.8.5 for details and further discussion). In each RF category, only one result is taken per model or model formulation. Cloud-albedo efficacies are evaluated in two ways: the standard letters include cloud lifetime effects in the efficacy term and the letters with asterisks exclude these effects. Studies assessed in the figure are: a) Hansen et al. (2005); b) Wang et al. (1991); c) Wang et al. (1992); d) Govindasamy et al. (2001b); e) Lohmann and Feichter (2005); f) Forster et al. (2000); g) Joshi et al. (2003; see also Stuber et al., 2001a); h) Gregory et al. (2004); j) Sokolov (2006); k) Cook and Highwood (2004); m) Mickley et al. (2004); n) Rotstayn and Penner (2001); o) Roberts and Jones (2004) and p) Williams et al. (2001a).

developed useful conceptual models to help explain variations in efficacy with forcing mechanism. The efficacy primarily depends on the spatial structure of the forcings and the way they project onto the various different feedback mechanisms (Boer and Yu, 2003b). Therefore, different patterns of RF and any nonlinearities in the forcing response relationship affects the efficacy (Boer and Yu, 2003b; Joshi et al., 2003; Hansen et al., 2005; Stuber et al., 2005; Sokolov, 2006). Many of the studies presented in Figure 2.19 find that both the geographical and vertical distribution of the forcing can have the most significant effect on efficacy (in particular see Boer and Yu, 2003b; Joshi et al., 2003; Stuber et al., 2005; Sokolov, 2006). Nearly all studies that examine it find that high-latitude forcings have higher efficacies than tropical forcings. Efficacy has also been shown to vary with the vertical distribution of an applied forcing (Hansen et al., 1997; Christiansen, 1999; Joshi et al., 2003; Cook and Highwood, 2004; Roberts and Jones, 2004; Forster and Joshi, 2005; Stuber et al., 2005; Sokolov, 2006). Forcings that predominately affect the upper troposphere are often found to have smaller efficacies compared to those that affect the surface. However, this is not ubiquitous as climate feedbacks (such as cloud and water vapour) will depend on the static stability of the troposphere and hence the sign of the temperature change in the upper troposphere (Govindasamy et al., 2001b; Joshi et al., 2003; Sokolov, 2006).

2.8.5.2 Long-Lived Greenhouse Gases

The few models that have examined efficacy for combined LLGHG changes generally find efficacies slightly higher than 1.0 (Figure 2.19). Further, the most recent result from the NCAR Community Climate Model (CCM3) GCM (Govindasamy et al., 2001b) indicates an efficacy of over 1.2 with no clear reason of why this changed from earlier versions of the same model. Individual LLGHG efficacies have only been analysed in two or three models. Two GCMs suggest higher efficacies from individual components (over 30% for CFCs in Hansen et al., 2005). In contrast another GCM gives efficacies for CFCs (Forster and Joshi, 2005) and CH_4 (Berntsen et al., 2005) that are slightly less than one. Overall there is medium confidence that the observed changes in the combined LLGHG changes have an efficacy close to 1.0 (within 10%), but there are not enough studies to constrain the efficacies for individual species.

2.8.5.3 Solar

Solar changes, compared to CO_2, have less high-latitude RF and more of the RF realised at the surface. Established but incomplete knowledge suggests that there is partial compensation between these effects, at least in some models, which leads to solar efficacies close to 1.0. All models with a positive solar RF find efficacies of 1.0 or smaller. One study finds a smaller efficacy than other models (0.63: Gregory et al., 2004). However, their unique methodology for calculating climate sensitivity has large uncertainties (see Section 2.8.4). These studies have only examined solar RF from total solar

irradiance change; any indirect solar effects (see Section 2.7.1.3) are not included in this efficacy estimate. Overall, there is medium confidence that the direct solar efficacy is within the 0.7 to 1.0 range.

2.8.5.4 Ozone

Stratospheric ozone efficacies have normally been calculated from idealised ozone increases. Experiments with three models (Stuber et al., 2001a; Joshi et al., 2003; Stuber et al., 2005) found higher efficacies for such changes; these were due to larger than otherwise tropical tropopause temperature changes which led to a positive stratospheric water vapour feedback. However, this mechanism may not operate in the two versions of the GISS model, which found smaller efficacies. Only one study has used realistic stratospheric ozone changes (see Figure 2.19); thus, knowledge is still incomplete. Conclusions are only drawn from the idealised studies where there is (1) medium confidence that the efficacy is within a 0.5 to 2.0 range and (2) established but incomplete physical understanding of how and why the efficacy could be larger than 1.0. There is medium confidence that for realistic tropospheric ozone perturbations the efficacy is within the 0.6 to 1.1 range.

2.8.5.5 Scattering Aerosol

For idealised global perturbations, the efficacy for the direct effect of scattering aerosol is very similar to that for changes in the solar constant (Cook and Highwood, 2004). As for ozone, realistic perturbations of scattering aerosol exhibit larger changes at higher latitudes and thus have a higher efficacy than solar changes (Hansen et al., 2005). Although the number of modelling results is limited, it is expected that efficacies would be similar to other solar effects; thus there is medium confidence that efficacies for scattering aerosol would be in the 0.7 to 1.1 range. Efficacies are likely to be similar for scattering aerosol in the troposphere and stratosphere.

With the formulation of RF employed in this chapter, the efficacy of the cloud albedo RF accounts for cloud lifetime effects (Section 2.8.3). Only two studies contained enough information to calculate efficacy in this way and both found efficacies higher than 1.0. However, the uncertainties in quantifying the cloud lifetime effect make this efficacy very uncertain. If cloud lifetime effects were excluded from the efficacy term, the cloud albedo efficacy would very likely be similar to that of the direct effect (see Figure 2.19).

2.8.5.6 Absorbing Aerosol

For absorbing aerosols, the simple ideas of a linear forcing-response relationship and efficacy can break down (Hansen et al., 1997; Cook and Highwood, 2004; Feichter et al., 2004; Roberts and Jones, 2004; Hansen et al., 2005; Penner et al., 2007). Aerosols within a particular range of single scattering albedos have negative RFs but induce a global mean warming, that is, the efficacy can be negative. The surface albedo and height of the aerosol layer relative to the cloud also affects this relationship (Section 7.5; Penner et al., 2003; Cook and Highwood, 2004; Feichter et al., 2004; Johnson et al., 2004; Roberts and Jones, 2004; Hansen et al., 2005). Studies that increase BC in the planetary boundary layer find efficacies much larger than 1.0 (Cook and Highwood, 2004; Roberts and Jones, 2004; Hansen et al., 2005). These studies also find that efficacies are considerably smaller than 1.0 when BC aerosol is changed above the boundary layer. These changes in efficacy are at least partly attributable to a semi-direct effect whereby absorbing aerosol modifies the background temperature profile and tropospheric cloud (see Section 7.5). Another possible feedback mechanism is the modification of snow albedo by BC aerosol (Hansen and Nazarenko, 2004; Hansen et al., 2005); however, this report does not classify this as part of the response, but rather as a separate RF (see Section 2.5.4 and 2.8.5.7). Most GCMs likely have some representation of the semi-direct effect (Cook and Highwood, 2004) but its magnitude is very uncertain (see Section 7.5) and dependent on aspects of cloud parametrizations within GCMs (Johnson, 2005). Two studies using realistic vertical and horizontal distributions of BC find that overall the efficacy is around 0.7 (Hansen et al., 2005; Lohmann and Feichter, 2005). However, Hansen et al. (2005) acknowledge that they may have underestimated BC within the boundary layer and another study with realistic vertical distribution of BC changes finds an efficacy of 1.3 (Sokolov, 2006). Further, Penner et al. (2007) also modelled BC changes and found efficacies very much larger and very much smaller than 1.0 for biomass and fossil fuel carbon, respectively (Hansen et al. (2005) found similar efficacies for biomass and fossil fuel carbon). In summary there is no consensus as to BC efficacy and this may represent problems with the stratospherically adjusted definition of RF (see Section 2.8.3).

2.8.5.7   Other Forcing Agents

Efficacies for some other effects have been evaluated by one or two modelling groups. Hansen et al. (2005) found that land use albedo RF had an efficacy of roughly 1.0, while the BC-snow albedo RF had an efficacy of 1.7. Ponater et al. (2005) found an efficacy of 0.6 for contrail RF and this agrees with a suggestion from Hansen et al. (2005) that high-cloud changes should have smaller efficacies. The results of Hansen et al. (2005) and Forster and Shine (1999) suggest that stratospheric water vapour efficacies are roughly one.

2.8.6   Efficacy and the Forcing-Response Relationship

Efficacy is a new concept introduced since the TAR and its physical understanding is becoming established (see Section 2.8.5). When employing the stratospherically adjusted RF, there is medium confidence that efficacies are within the 0.75 to 1.25 range for most realistic RF mechanisms aside from aerosol and stratospheric ozone changes. There is medium confidence that realistic aerosol and ozone changes have efficacies within the

0.5 to 2.0 range. Further, zero-surface-temperature-change RFs are very likely to have efficacies significantly closer to 1.0 for all mechanisms. It should be noted that efficacies have only been evaluated in GCMs and actual climate efficacies could be different from those quoted in Section 2.8.5.

2.9   Synthesis

This section begins by synthesizing the discussion of the RF concept. It presents summaries of the global mean RFs assessed in earlier sections and discusses time evolution and spatial patterns of RF. It also presents a brief synthesis of surface forcing diagnostics. It breaks down the analysis of RF in several ways to aid and advance the understanding of the drivers of climate change.

RFs are calculated in various ways depending on the agent: from changes in emissions and/or changes in concentrations; and from observations and other knowledge of climate change drivers. Current RF depends on present-day concentrations of a forcing agent, which in turn depend on the past history of emissions. Some climate response to these RFs is expected to have already occurred. Additionally, as RF is a comparative measure of equilibrium climate change and the Earth's climate is not in an equilibrium state, additional climate change in the future is also expected from present-day RFs (see Sections 2.2 and 10.7). As previously stated in Section 2.2, RF alone is not a suitable metric for weighting emissions; for this purpose, the lifetime of the forcing agent also needs to be considered (see Sections 2.9.4 and 2.10).

RFs are considered external to the climate system (see Section 2.2). Aside from the natural RFs (solar, volcanoes), the other RFs are considered to be anthropogenic (i.e., directly attributable to human activities). For the LLGHGs it is assumed that all changes in their concentrations since pre-industrial times are human-induced (either directly through emissions or from land use changes); these concentration changes are used to calculate the RF. Likewise, stratospheric ozone changes are also taken from satellite observations and changes are primarily attributed to Montreal-Protocol controlled gases, although there may also be a climate feedback contribution to these trends (see Section 2.3.4). For the other RFs, anthropogenic emissions and/or human-induced land use changes are used in conjunction with CTMs and/or GCMs to estimate the anthropogenic RF.

2.9.1   Uncertainties in Radiative Forcing

The TAR assessed uncertainties in global mean RF by attaching an error bar to each RF term that was 'guided by the range of published values and physical understanding'. It also quoted a level of scientific understanding (LOSU) for each RF, which was a subjective judgment of the estimate's reliability.

The concept of LOSU has been slightly modified based on the IPCC Fourth Assessment Report (AR4) uncertainty guidelines. Error bars now represent the 5 to 95% (90%)

confidence range (see Box TS.1). Only 'well-established' RFs are quantified. 'Well established' implies that there is qualitatively both sufficient evidence and sufficient consensus from published results to estimate a central RF estimate and a range. 'Evidence' is assessed by an A to C grade, with an A grade implying strong evidence and C insufficient evidence. Strong evidence implies that observations have verified aspects of the RF mechanism and that there is a sound physical model to explain the RF. 'Consensus' is assessed by assigning a number between 1 and 3, where 1 implies a good deal of consensus and 3 insufficient consensus. This ranks the number of studies, how well studies agree on quantifying the RF and especially how well observation-based studies agree with models. The product of 'Evidence' and 'Consensus' factors give the LOSU rank. These ranks are high, medium, medium-low, low or very low. Ranks of very low are not evaluated. The quoted 90% confidence range of RF quantifies the value uncertainty, as derived from the expert assessment of published values and their ranges. For most RFs, many studies have now been published, which generally makes the sampling of parameter space more complete and the value uncertainty more realistic, compared to the TAR. This is particularly true for both the direct and cloud albedo aerosol RF (see Section 2.4). Table 2.11 summarises the key certainties and uncertainties and indicates the basis for the 90% confidence range estimate. Note that the aerosol terms will have added uncertainties due to the uncertain semi-direct and cloud lifetime effects. These uncertainties in the response to the RF (efficacy) are discussed in Section 2.8.5.

Table 2.11 indicates that there is now stronger evidence for most of the RFs discussed in this chapter. Some effects are not quantified, either because they do not have enough evidence or because their quantification lacks consensus. These include certain mechanisms associated with land use, stratospheric water vapour and cosmic rays. Cloud lifetime and the semi-direct effects are also excluded from this analysis as they are deemed to be part of the climate response (see Section 7.5). The RFs from the LLGHGs have both a high degree of consensus and a very large amount of evidence and, thereby, place understanding of these effects at a considerably higher level than any other effect.

2.9.2 Global Mean Radiative Forcing

The RFs discussed in this chapter, their uncertainty ranges and their efficacies are summarised in Figure 2.20 and Table 2.12. Radiative forcings from forcing agents have been combined into their main groupings. This is particularly useful for aerosol as its total direct RF is considerably better constrained than the RF from individual aerosol types (see Section 2.4.4). Table 2.1 gives a further component breakdown of RF for the LLGHGs. Radiative forcings are the stratospherically adjusted RF and they have not been multiplied by efficacies (see Sections 2.2 and 2.8).

In the TAR, no estimate of the total combined RF from all anthropogenic forcing agents was given because: a) some of the forcing agents did not have central or best estimates; b) a

degree of subjectivity was included in the error estimates; and c) uncertainties associated with the linear additivity assumption and efficacy had not been evaluated. Some of these limitations still apply. However, methods for objectively adding the RF of individual species have been developed (e.g., Schwartz and Andreae, 1996; Boucher and Haywood, 2001). In addition, as efficacies are now better understood and quantified (see Section 2.8.5), and as the linear additivity assumption has been more thoroughly tested (see Section 2.8.4), it becomes scientifically justifiable for RFs from different mechanisms to be combined, with certain exceptions as noted below. Adding together the anthropogenic RF values shown in panel (A) of Figure 2.20 and combining their individual uncertainties gives the probability density functions (PDFs) of RF that are shown in panel (B). Three PDFs are shown: the combined RF from greenhouse gas changes (LLGHGs and ozone); the combined direct aerosol and cloud albedo RFs and the combination of all anthropogenic RFs. The solar RF is not included in any of these distributions. The PDFs are generated by combining the 90% confidence estimates for the RFs, assuming independence and employing a one-million point Monte Carlo simulation to derive the PDFs (see Boucher and Haywood, 2001; and Figure 2.20 caption for details).

The PDFs show that LLGHGs and ozone contribute a positive RF of +2.9 ± 0.3 W m^{-2}. The combined aerosol direct and cloud albedo effect exert an RF that is virtually certain to be negative, with a median RF of –1.3 W m^{-2} and a –2.2 to –0.5 W m^{-2} 90% confidence range. The asymmetry in the combined aerosol PDF is caused by the estimates in Tables 2.6 and 2.7 being non-Gaussian. The combined net RF estimate for all anthropogenic drivers has a value of +1.6 W m^{-2} with a 0.6 to 2.4 W m^{-2} 90% confidence range. Note that the RFs from surface albedo change, stratospheric water vapour change and persistent contrails are only included in the combined anthropogenic PDF and not the other two.

Statistically, the PDF shown in Figure 2.20 indicates just a 0.2% probability that the total RF from anthropogenic agents is negative, which would suggest that it is virtually certain that the combined RF from anthropogenic agents is positive. Additionally, the PDF presented here suggests that it is extremely likely that the total anthropogenic RF is larger than +0.6 W m^{-2}. This combined anthropogenic PDF is better constrained than that shown in Boucher and Haywood (2001) because each of the individual RFs have been quantified to 90% confidence levels, enabling a more definite assessment, and because the uncertainty in some of the RF estimates is considerably reduced. For example, modelling of the total direct RF due to aerosols is better constrained by satellite and surface-based observations (Section 2.4.2), and the current estimate of the cloud albedo indirect effect has a best estimate and uncertainty associated with it, rather than just a range. The LLGHG RF has also increased by 0.20 W m^{-2} since 1998, making a positive RF more likely than in Boucher and Haywood (2001).

Nevertheless, there are some structural uncertainties associated with the assumptions used in the construction of

Table 2.11. *Uncertainty assessment of forcing agents discussed in this chapter. Evidence for the forcing is given a grade (A to C), with A implying strong evidence and C insufficient evidence. The degree of consensus among forcing estimates is given a 1, 2 or 3 grade, where grade 1 implies a good deal of consensus and grade 3 implies an insufficient consensus. From these two factors, a level of scientific understanding is determined (LOSU). Uncertainties are in approximate order of importance with first-order uncertainties listed first.*

	Evidence	Consensus	LOSU	Certainties	Uncertainties	Basis of RF range
LLGHGs	A	1	High	Past and present concentrations; spectroscopy	Pre-industrial concentrations of some species; vertical profile in stratosphere; spectroscopic strength of minor gases	Uncertainty assessment of measured trends from different observed data sets and differences between radiative transfer models
Stratospheric ozone	A	2	Medium	Measured trends and its vertical profile since 1980; cooling of stratosphere; spectroscopy	Changes prior to 1970; trends near tropopause; effect of recent trends	Range of model results weighted to calculations employing trustworthy observed ozone trend data
Tropospheric ozone	A	2	Medium	Present-day concentration at surface and some knowledge of vertical and spatial structure of concentrations and emissions; spectroscopy	Pre-industrial values and role of changes in lightning; vertical structure of trends near tropopause; aspects of emissions and chemistry	Range of published model results, upper bound increased to account for anthropogenic trend in lightning
Stratospheric water vapour from CH_4	A	3	Low	Global trends since 1990; CH_4 contribution to trend; spectroscopy	Global trends prior to 1990; radiative transfer in climate models; CTM models of CH_4 oxidation	Range based on uncertainties in CH_4 contribution to trend and published RF estimates
Direct aerosol	A	2 to 3	Medium to Low	Ground-based and satellite observations; some source regions and modelling	Emission sources and their history vertical structure of aerosol, optical properties, mixing and separation from natural background aerosol	Range of published model results with allowances made for comparisons with satellite data
Cloud albedo effect (all aerosols)	B	3	Low	Observed in case studies – e.g., ship tracks; GCMs model an effect	Lack of direct observational evidence of a global forcing	Range of published model results and published results where models have been constrained by satellite data
Surface albedo (land use)	A	2 to 3	Medium to Low	Some quantification of deforestation and desertification	Separation of anthropogenic changes from natural	Based on range of published estimates and published uncertainty analyses
Surface albedo (BC aerosol on snow)	B	3	Low	Estimates of BC aerosol on snow; some model studies suggest link	Separation of anthropogenic changes from natural; mixing of snow and BC aerosol; quantification of RF	Estimates based on a few published model studies
Persistent linear Contrails	A	3	Low	Cirrus radiative and microphysical properties; aviation emissions; contrail coverage in certain regions	Global contrail coverage and optical properties	Best estimate based on recent work and range from published model results

Table 2.11 (continued)

	Evidence	Consensus	LOSU	Certainties	Uncertainties	Basis of RF range
Solar irradiance	B	3	Low	Measurements over last 25 years; proxy indicators of solar activity	Relationship between proxy data and total solar irradiance; indirect ozone effects	Range from available reconstructions of solar irradiance and their qualitative assessment
Volcanic aerosol	A	3	Low	Observed aerosol changes from Mt. Pinatubo and El Chichón; proxy data for past eruptions; radiative effect of volcanic aerosol	Stratospheric aerosol concentrations from pre-1980 eruptions; atmospheric feedbacks	Past reconstructions/estimates of explosive volcanoes and observations of Mt. Pinatubo aerosol
Stratospheric water vapour from causes other than CH$_4$ oxidation	C	3	Very Low	Empirical and simple model studies suggest link; spectroscopy	Other causes of water vapour trends poorly understood	Not given
Tropospheric water vapour from irrigation	C	3	Very Low	Process understood; spectroscopy; some regional information	Global injection poorly quantified	Not given
Aviation-induced cirrus	C	3	Very Low	Cirrus radiative and microphysical properties; aviation emissions; contrail coverage in certain regions	Transformation of contrails to cirrus; aviation's effect on cirrus clouds	Not given
Cosmic rays	C	3	Very Low	Some empirical evidence and some observations as well as microphysical models suggest link to clouds	General lack/doubt regarding physical mechanism; dependence on correlation studies	Not given
Other surface effects	C	3	Very Low	Some model studies suggest link and some evidence of relevant processes	Quantification of RF and interpretation of results in forcing feedback context difficult	Not given

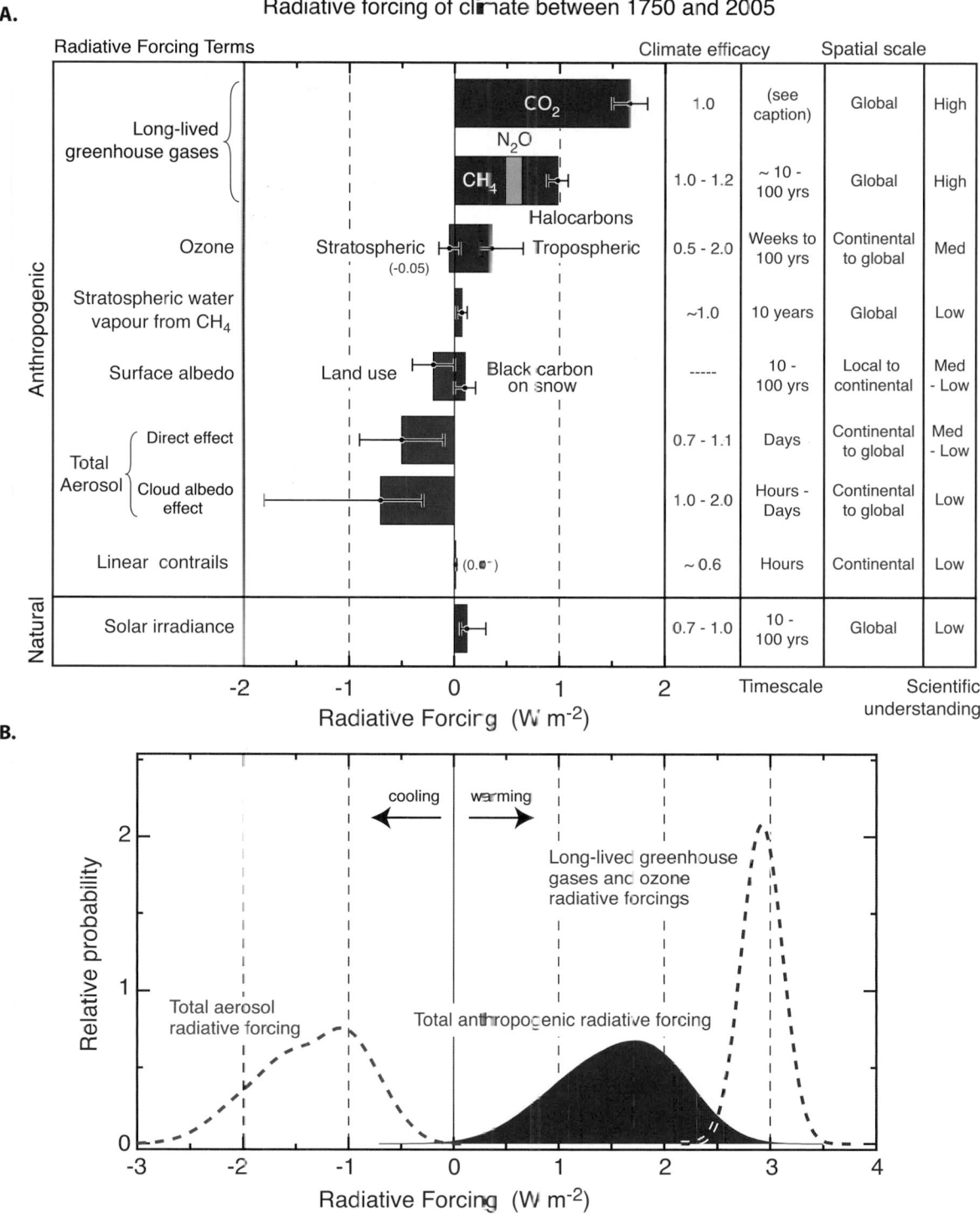

Figure 2.20. *(A) Global mean RFs from the agents and mechanisms discussed in this chapter, grouped by agent type. Anthropogenic RFs and the natural direct solar RF are shown. The plotted RF values correspond to the bold values in Table 2.12. Columns indicate other characteristics of the RF; efficacies are not used to modify the RFs shown. Time scales represent the length of time that a given RF term would persist in the atmosphere after the associated emissions and changes ceased. No CO_2 time scale is given, as its removal from the atmosphere involves a range of processes that can span long time scales, and thus cannot be expressed accurately with a narrow range of lifetime values. The scientific understanding shown for each term is described in Table 2.11. (B) Probability distribution functions (PDFs) from combining anthropogenic RFs in (A). Three cases are shown: the total of all anthropogenic RF terms (block filled red curve; see also Table 2.12); LLGHGs and ozone RFs only (dashed red curve); and aerosol direct and cloud albedo RFs only (dashed blue curve). Surface albedo, contrails and stratospheric water vapour RFs are included in the total curve but not in the others. For all of the contributing forcing agents, the uncertainty is assumed to be represented by a normal distribution (and 90% confidence intervals) with the following exceptions: contrails, for which a lognormal distribution is assumed to account for the fact that the uncertainty is quoted as a factor of three; and tropospheric ozone, the direct aerosol RF (sulphate, fossil fuel organic and black carbon, biomass burning aerosols) and the cloud albedo RF for which discrete values based on Figure 2.9, Table 2.6 and Table 2.7 are randomly sampled. Additional normal distributions are included in the direct aerosol effect for nitrate and mineral dust, as these are not explicitly accounted for in Table 2.6. A one-million point Monte Carlo simulation was performed to derive the PDFs (Boucher and Haywood, 2001). Natural RFs (solar and volcanic) are not included in these three PDFs. Climate efficacies are not accounted for in forming the PDFs.*

Table 2.12. *Global mean radiative forcings since 1750 and comparison with earlier assessments. Bold rows appear on Figure 2.20. The first row shows the combined anthropogenic RF from the probability density function in panel B of Figure 2.20. The sum of the individual RFs and their estimated errors are not quite the same as the numbers presented in this row due to the statistical construction of the probability density function.*

	Global mean radiative forcing (W m^{-2})[a]			Summary comments on changes since the TAR
	SAR (1750–1993)	TAR (1750–1998)	AR4 (1750–2005)	
Combined Anthropogenic RF	**Not evaluated**	**Not evaluated**	**1.6 [–1.0, +0.8]**	**Newly evaluated. Probability density function estimate**
Long-lived Greenhouse gases (Comprising CO$_2$, CH$_4$, N$_2$O, and halocarbons)	**+2.45 [15%] (CO$_2$ 1.56; CH$_4$ 0.47; N$_2$O 0.14; Halocarbons 0.28)**	**+2.43 [10%] (CO$_2$ 1.46; CH$_4$ 0.48; N$_2$O 0.15; Halocarbons 0.34[b])**	**+2.63 [±0.26] (CO$_2$ 1.66 [±0.17]; CH$_4$ 0.48 [±0.05]; N$_2$O 0.16 [±0.02]; Halocarbons 0.34 [±0.03])**	**Total increase in RF, due to upward trends, particularly in CO$_2$. Halocarbon RF trend is positive[b]**
Stratospheric ozone	**–0.1 [2x]**	**–0.15 [67%]**	**–0.05 [±0.10]**	**Re-evaluated to be weaker**
Tropospheric ozone	**+0.40 [50%]**	**+0.35 [43%]**	**+0.35 [–0.1, +0.3]**	**Best estimate unchanged. However, a larger RF could be possible**
Stratospheric water vapour from CH$_4$	**Not evaluated**	**+0.01 to +0.03**	**+0.07 [±0.05]**	**Re-evaluated to be higher**
Total direct aerosol	**Not evaluated**	**Not evaluated**	**–0.50 [±0.40]**	**Newly evaluated**
Direct sulphate aerosol	–0.40 [2x]	–0.40 [2x]	–0.40 [±0.20]	Better constrained
Direct fossil fuel aerosol (organic carbon)	Not evaluated	–0.10 [3x]	–0.05 [±0.05]	Re-evaluated to be weaker
Direct fossil fuel aerosol (BC)	+0.10 [3x]	+0.20 [2x]	+0.20 [±0.15]	Similar best estimate to the TAR. Response affected by semi-direct effects
Direct biomass burning aerosol	–0.20 [3x]	–0.20 [3x]	+0.03 [±0.12]	Re-evaluated and sign changed. Response affected by semi-direct effects
Direct nitrate aerosol	Not evaluated	Not evaluated	–0.10 [±0.10]	Newly evaluated
Direct mineral dust aerosol	Not evaluated	–0.60 to +0.40	–0.10 [±0.20]	Re-evaluated to have a smaller anthropogenic fraction
Cloud albedo effect	**0 to –1.5 (sulphate only)**	**0.0 to –2.0 (all aerosols)**	**–0.70 [–1.1, +0.4] (all aerosols)**	**Best estimate now given**
Surface albedo (land use)	**Not evaluated**	**–0.20 [100%]**	**–0.20 [±0.20]**	**Additional studies**
Surface albedo (BC aerosol on snow)	**Not evaluated**	**Not evaluated**	**+0.10 [±0.10]**	**Newly evaluated**
Persistent linear contrails	**Not evaluated**	**0.02 [3.5x]**	**0.01 [–0.007, +0.02]**	**Re-evaluated to be smaller**
Solar irradiance	**+0.30 [67%]**	**+0.30 [67%]**	**+0.12 [–0.06, +0.18]**	**Re-evaluated to be less than half**

Notes: [a]   For the AR4 column, 90% value uncertainties appear in brackets: when adding these numbers to the best estimate the 5 to 95% confidence range is obtained. When two numbers are quoted for the value uncertainty, the distribution is non-normal. Uncertainties in the SAR and the TAR had a similar basis, but their evaluation was more subjective. [15%] indicates 15% relative uncertainty, [2x], etc. refer to a factor of two, etc. uncertainty and a lognormal distribution of RF estimates.

   [b]   The TAR RF for halocarbons and hence the total LLGHG RF was incorrectly evaluated some 0.01 W m^{-2} too high. The actual trends in these RFs are therefore more positive than suggested by numbers in this table (Table 2.1 shows updated trends).

the PDF and the assumptions describing the component uncertainties. Normal distributions are assumed for most RF mechanisms (with the exceptions noted in the caption); this may not accurately capture extremes. Additionally, as in Boucher and Haywood (2001), all of the individual RF mechanisms are given equal weighting, even though the level of scientific understanding differs between forcing mechanisms. Note also that variation in efficacy and hence the semi-direct and cloud lifetime effects are not accounted for, as these are not considered to be RFs in this report (see Section 2.2). Adding these effects, together with other potential mechanisms that have so far not been defined as RFs and quantified, would introduce further uncertainties but give a fuller picture of the role of anthropogenic drivers. Introducing efficacy would give a broader PDF and a large cloud lifetime effect would reduce the median estimate. Despite these caveats, from the current knowledge of individual forcing mechanisms presented here it remains extremely likely that the combined anthropogenic RF is both positive and substantial (best estimate: +1.6 W m^{-2}).

2.9.3 Global Mean Radiative Forcing by Emission Precursor

The RF due to changes in the concentration of a single forcing agent can have contributions from emissions of several compounds (Shindell et al., 2005). The RF of CH_4, for example, is affected by CH_4 emissions, as well as NO_x emissions. The CH_4 RF quoted in Table 2.12 and shown in Figure 2.20 is a value that combines the effects of both emissions. As an anthropogenic or natural emission can affect several forcing agents, it is useful to assess the current RF caused by each primary emission. For example, emission of NO_x affects CH_4, tropospheric ozone and tropospheric aerosols. Based on a development carried forward from the TAR, this section assesses the RF terms associated with each principal emission including indirect RFs related to perturbations of other forcing agents, with the results shown in Figure 2.21. The following indirect forcing mechanisms are considered:

- fossil carbon from non-CO_2 gaseous compounds, which eventually increase CO_2 in the atmosphere (from CO, CH_4, and NMVOC emissions);
- changes in stratospheric ozone (from N_2O and halocarbon (CFCs, HCFC, halons, etc.) emissions);
- changes in tropospheric ozone (from CH_4, NO_x, CO, and NMVOC emissions);

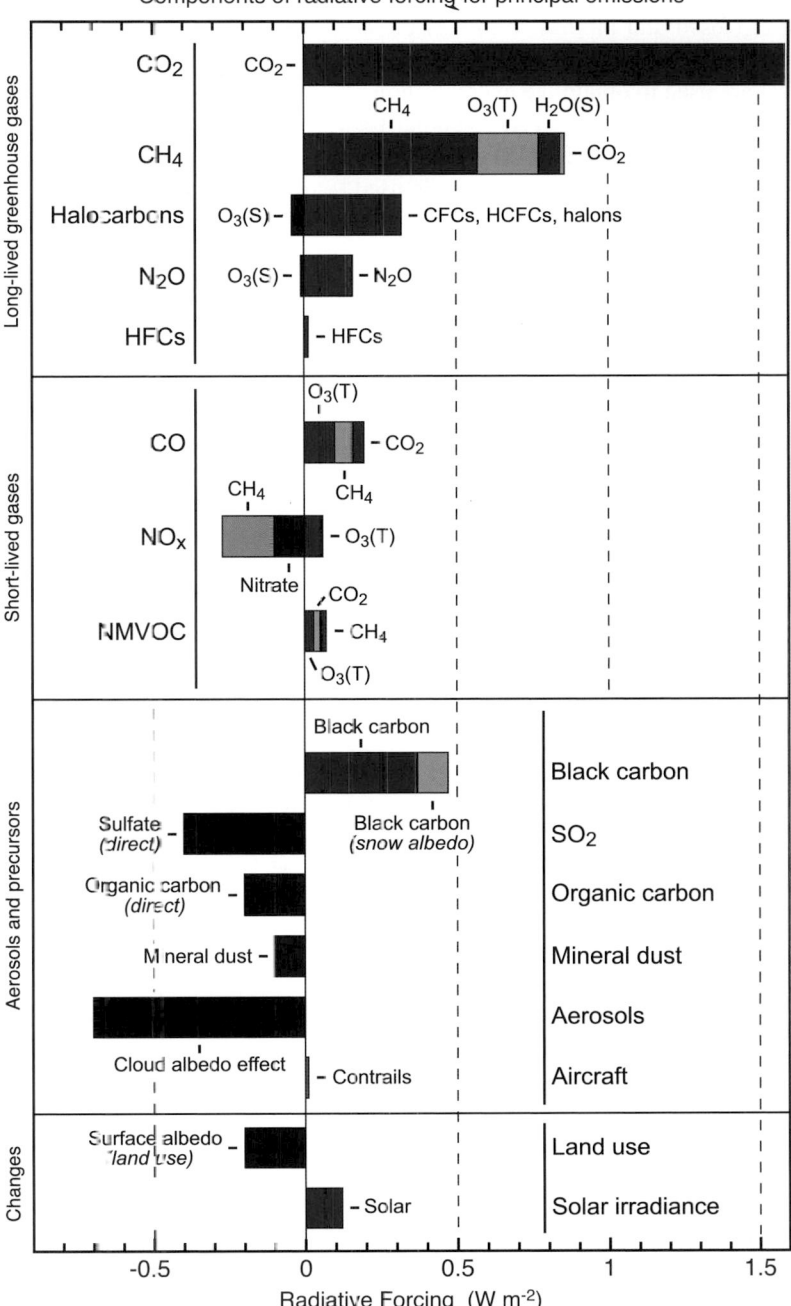

Figure 2.21. *Components of RF for emissions of principal gases, aerosols and aerosol precursors and other changes. Values represent RF in 2005 due to emissions and changes since 1750. (S) and (T) next to gas species represent stratospheric and tropospheric changes, respectively. The uncertainties are given in the footnotes to Table 2.13. Quantitative values are displayed in Table 2.13.*

- changes in OH affecting the lifetime of CH_4 (from CH_4, CO, NO_x, and NMVOC emissions); and
- changing nitrate and sulphate aerosols through changes in NO_x and SO_2 emissions, respectively.

For some of the principal RFs (e.g., BC, land use and mineral dust) there is not enough quantitative information available to assess their indirect effects, thus their RFs are the same as those presented in Table 2.12. Table 2.5 gives the total (fossil and biomass burning) direct RFs for BC and organic carbon aerosols that are used to obtain the average shown in Figure 2.21. Table 2.13 summarises the direct and indirect RFs presented in Figure 2.21, including the methods used for estimating the RFs and the associated uncertainty. Note that for indirect effects through changes in chemically active gases (e.g., OH or ozone), the emission-based RF is not uniquely defined since the effect of one precursor will be affected by the levels of the other precursors. The RFs of indirect effects on CH_4 and ozone by NO_x, CO and VOC emissions are estimated by removing the anthropogenic emissions of one precursor at a time. A sensitivity analysis by Shindell et al. (2005) indicates that the nonlinear effect induced by treating the precursors separately is of the order of 10% or less. Very uncertain indirect effects are not included in Table 2.13 and Figure 2.21. These include ozone changes due to solar effects, changes in secondary organic aerosols through changes in the ozone/OH ratio and apportioning of the cloud albedo changes to each aerosol type (Hansen et al., 2005).

2.9.4 Future Climate Impact of Current Emissions

The changes in concentrations since pre-industrial time of the long-lived components causing the RF shown in Figure 2.20 are strongly influenced by the past history of emissions. A different perspective is obtained by integrating RF over a future time horizon for a one-year 'pulse' of global emissions (e.g., Jacobson (2002) used this approach to compare fossil fuel organic and BC aerosols to CO_2). Comparing the contribution from each forcing agent as shown in Figure 2.22 gives an indication of the future climate impact for current (year 2000) emissions of the different forcing agents. For the aerosols, the integrated RF is obtained based on the lifetimes, burdens and RFs from the AeroCom experiments, as summarised in Tables 2.4 and 2.5. For ozone precursors (CO, NO_x and NMVOCs), data are taken from Derwent et al. (2001), Collins et al. (2002), Stevenson et al. (2004) and Berntsen et al. (2005), while for the long-lived species the radiative efficiencies and lifetimes are used, as well as a response function for CO_2 (see Section 2.10.2, Table 2.14). Uncertainties in the estimates of the integrated RF originate

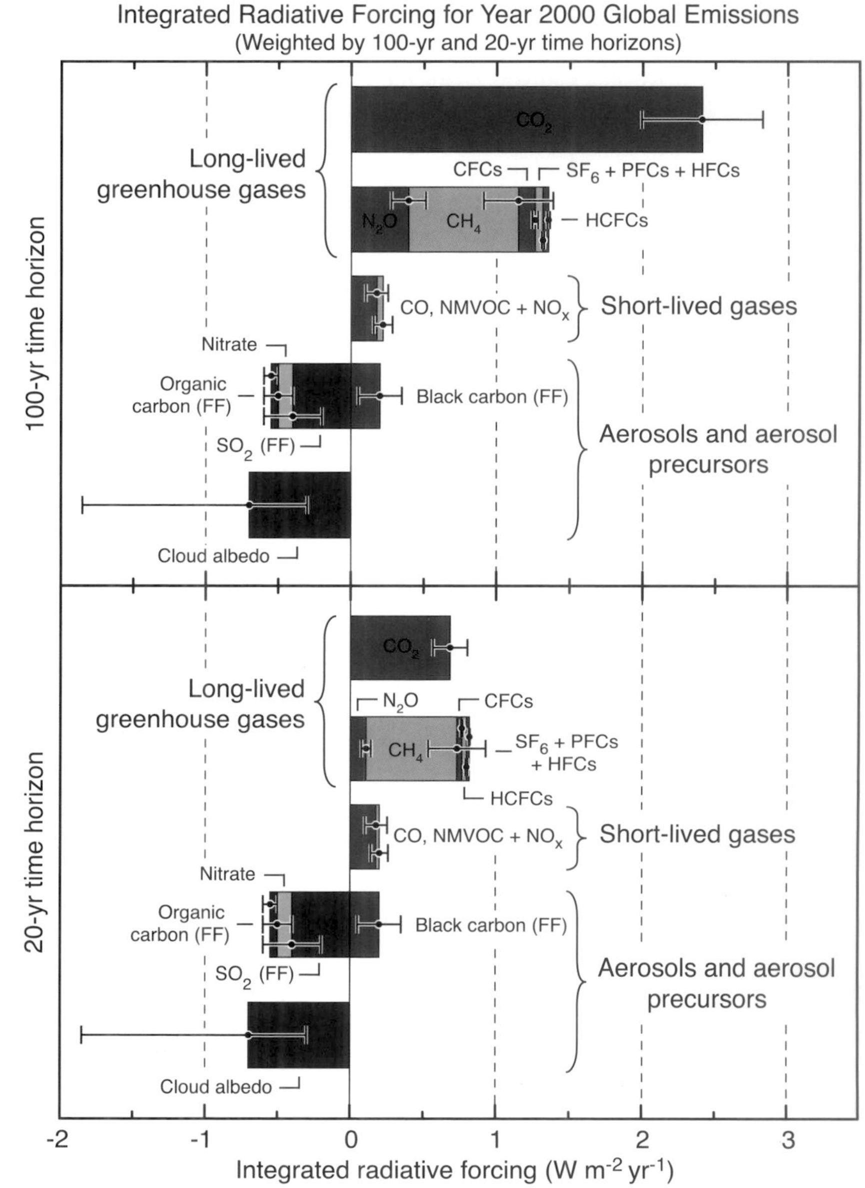

Figure 2.22. *Integrated RF of year 2000 emissions over two time horizons (20 and 100 years). The figure gives an indication of the future climate impact of current emissions. The values for aerosols and aerosol precursors are essentially equal for the two time horizons. It should be noted that the RFs of short-lived gases and aerosol depend critically on both when and where they are emitted; the values given in the figure apply only to total global annual emissions. For organic carbon and BC, both fossil fuel (FF) and biomass burning emissions are included. The uncertainty estimates are based on the uncertainties in emission sources, lifetime and radiative efficiency estimates.*

from uncertainties in lifetimes, optical properties and current global emissions.

Figure 2.22 shows the integrated RF for both a 20- and 100-year time horizon. Choosing the longer time horizon of 100 years, as was done in the GWPs for the long-lived species included in the Kyoto Protocol, reduces the apparent importance of the shorter-lived species. It should be noted that the compounds with long lifetimes and short emission histories will tend to contribute more to the total with this 'forward looking' perspective than in the standard 'IPCC RF bar chart diagram' (Figure 2.20).

Table. 2.13. Emission-based RFs for emitted components with radiative effects other than through changes in their atmospheric abundance. Minor effects where the estimated RF is less than 0.01 W m^{-2} are not included. Effects on sulphate aerosols are not included since SO_2 emission is the only significant factor affecting sulphate aerosols. Method of calculation and uncertainty ranges are given in the footnotes. Values represent RF in 2005 due to emissions and changes since 1750. See Figure 2.21 for graphical presentation of these values.

Component emitted	Atmospheric or surface change directly causing radiative forcing												
	CO_2	CH_4	CFC/HCFC	N_2O	HFC/PFC/SF_6	BC-direct	BC snow albedo	Organic carbon	O_3(T)[a]	O_3(S)[b]	H_2O(S)[c]	Nitrate aerosols	Indirect cloud albedo effect
CO_2	1.56[d]												
CH_4	0.016[d]	0.57[e]							0.2[e]		0.07[f]		
CFC/HCFC/halons			0.32[g]							-0.04[h]			
N_2O				0.15[g]						-0.01[h]			
HFC/PFC/SF_6					0.017[g]								
CO/VOC	0.06[d]	0.08[e]							0.13[e]				
NOx		-0.17[e]							0.06[e]			-0.10[i]	Xi
BC						0.34[k]	0.1[l]						Xi
OC								-0.19[k]					Xi
SO_2													Xi

Notes:

[a] tropospheric ozone.

[b] stratospheric ozone.

[c] stratospheric water vapour.

[d] Derived from the total RF of the observed CO_2 change (Table 2.12), with the contributions from CH_4, CO and VOC emissions from fossil sources subtracted. Historical emissions of CH_4, CO and VOCs from Emission Database for Global Atmospheric Research (EDGAR)-HistorY Database of the Environment (HYDE) (Van Aardenne et al., 2001), CO_2 contribution from these sources calculated with CO_2 model described by Joos et al. (1996).

[e] Derived from the total RF of the observed CH_4 change (Table 2.12). Subtracted from this were the contributions through lifetime changes caused by emissions of NOx, CO and VOC that change OH concentrations. The effects of NOx, CO and VOCs are from Shindell et al. (2005). There are significant uncertainties related to these relations. Following Shindell et al. (2005) the uncertainty estimate is taken to be ±20% for CH_4 emissions, and ±50% for CO, VOC and NOx emissions.

[f] All the radiative forcing from changes in stratospheric water vapour is attributed to CH_4 emissions (Section 2.3.7 and Table 2.12).

[g] RF calculated based on observed concentration change, see Table 2.12 and Section 2.3

[h] 80% of RF from observed ozone depletion in the stratosphere (Table 2.12) is attributed to CFCs/HFCs, remaining 20% to N_2O (Based on Nevison et al., 1999 and WMO, 2003).

[i] RF from Table 2.12, uncertainty ±0.10 W m^{-2}.

[j] Uncertainty too large to apportion the indirect cloud albedo effect to each aerosol type (Hansen et al., 2005).

[k] Mean of all studies in Table 2.5, includes fossil fuel, biofuel and biomass burning. Uncertainty (90% confidence ranges) ±0.25 W m^{-2} (BC) and ±0.20 W m^{-2} (organic carbon) based on range of reported values in Table 2.5.

[l] RF from Table 2.12, uncertainty ±0.10 W m^{-2}.

2.9.5 Time Evolution of Radiative Forcing and Surface Forcing

There is a good understanding of the time evolution of the LLGHG concentrations from *in situ* measurements over the last few decades and extending further back using firn and ice core data (see Section 2.3, FAQ 2.1, Figure 1 and Chapter 6). Increases in RF are clearly dominated by CO_2. Halocarbon RF has grown rapidly since 1950, but the RF growth has been cut dramatically by the Montreal Protocol (see Section 2.3.4). The RF of CFCs is declining; in addition, the combined RF of all ozone-depleting substances (ODS) appears to have peaked at 0.32 W m^{-2} during 2003. However, substitutes for ODS are growing at a slightly faster rate, so halocarbon RF growth is still positive (Table 2.1). Although the trend in halocarbon RF since the time of the TAR has been positive (see Table 2.1), the halocarbon RF in this report, as shown in Table 2.12, is the same as in the TAR; this is due to a re-evaluation of the TAR results.

Radiative forcing time series for the natural (solar, volcanic aerosol) forcings are reasonably well known for the past 25 years; estimates further back are prone to uncertainties (Section 2.7). Determining the time series for aerosol and ozone RF is far more difficult because of uncertainties in the knowledge of past emissions and chemical-microphysical modelling. Several time series for these and other RFs have been constructed (e.g., Myhre et al., 2001; Ramaswamy et al., 2001; Hansen et al., 2002). General Circulation Models develop their own time evolution of many forcings based on the temporal history of the relevant concentrations. As an example, the temporal evolution of the global and annual mean, instantaneous, all-sky RF and surface forcing due to the principal agents simulated by the Model for Interdisciplinary Research on Climate (MIROC) + Spectral Radiation-Transport Model for Aerosol Species (SPRINTARS) GCM (Nozawa et al., 2005; Takemura et al., 2005) is illustrated in Figure 2.23. Although there are differences between models with regards to the temporal reconstructions and thus present-day forcing estimates, they typically have a qualitatively similar temporal evolution since they often base the temporal histories on similar emissions data.

General Circulation Models compute the climate response based on the knowledge of the forcing agents and their temporal evolution. While most current GCMs incorporate the trace gas RFs, aerosol direct effects, solar and volcanoes, a few have in addition incorporated land use change and cloud albedo effect. While LLGHGs have increased rapidly over the past 20 years and contribute the most to the present RF (refer also to Figure 2.20 and FAQ 2.1, Figure 1), Figure 2.23 also indicates that the combined positive RF of the greenhouse gases exceeds the contributions due to all other anthropogenic agents throughout the latter half of the 20th century.

The solar RF has a small positive value. The positive solar irradiance RF is likely to be at least five times smaller than the combined RF due to all anthropogenic agents, and about an order of magnitude less than the total greenhouse gas contribution (Figures 2.20 and 2.23 and Table 2.12; see also the Foukal et al., 2006 review). The combined natural RF consists of the solar

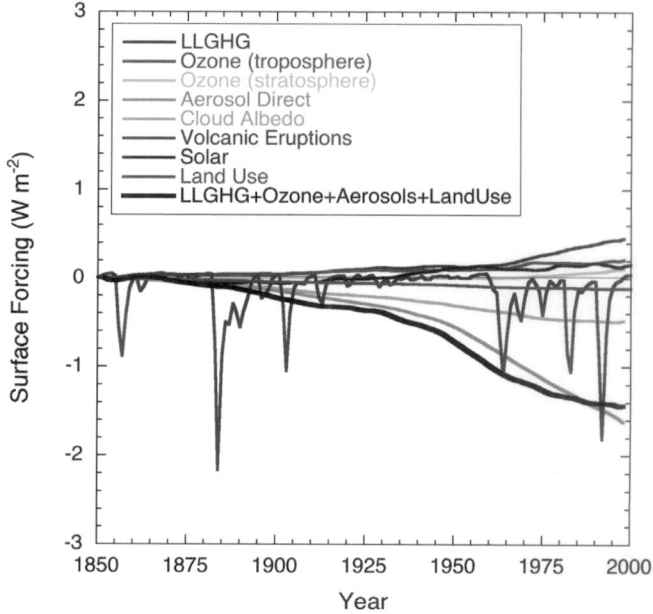

Surface Forcing

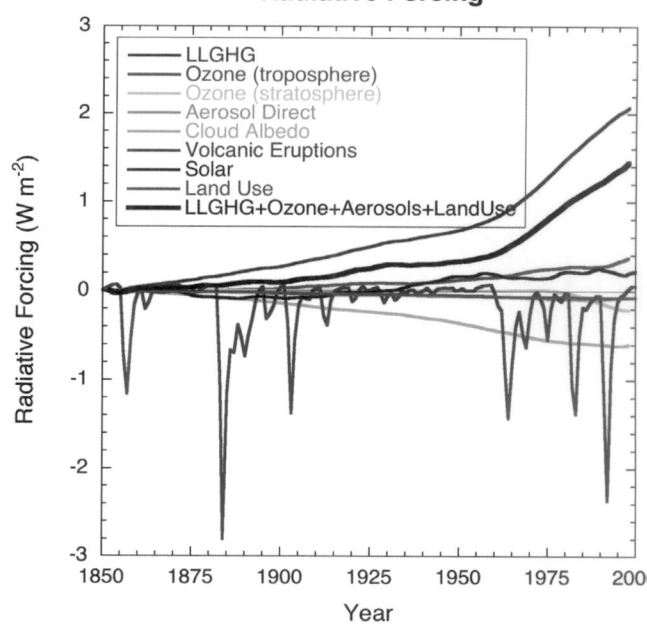

Radiative Forcing

Figure 2.23. *Globally and annually averaged temporal evolution of the instantaneous all-sky RF (top panel) and surface forcing (bottom panel) due to various agents, as simulated in the MIROC+SPRINTARS model (Nozawa et al., 2005; Takemura et al., 2005). This is an illustrative example of the forcings as implemented and computed in one of the climate models participating in the AR4. Note that there could be differences in the RFs among models. Most models simulate roughly similar evolution of the LLGHGs' RF.*

RF plus the large but transitory negative RF from episodic, explosive volcanic eruptions of which there have been several over the past half century (see Figure 2.18). Over particularly the 1950 to 2005 period, the combined natural forcing has been either negative or slightly positive (less than approximately 0.2 W m^{-2}), reaffirming and extending the conclusions in the

TAR. Therefore, it is exceptionally unlikely that natural RFs could have contributed a positive RF of comparable magnitude to the combined anthropogenic RF term over the period 1950 to 2005 (Figure 2.23). Attribution studies with GCMs employ the available knowledge of the evolution of the forcing over the 20th century, and particularly the features distinguishing the anthropogenic from the natural agents (see also Section 9.2).

The surface forcing (Figure 2.23, top panel), in contrast to RF, is dominated by the strongly negative shortwave effect of the aerosols (tropospheric and the episodic volcanic ones), with the LLGHGs exerting a small positive effect. Quantitative values of the RFs and surface forcings by the agents differ across models in view of the differences in model physics and in the formulation of the forcings due to the short-lived species (see Section 10.2, Collins et al. (2006) and Forster and Taylor (2006) for further discussion on uncertainties in GCMs' calculation of RF and surface forcing). As for RF, it is difficult to specify uncertainties in the temporal evolution, as emissions and concentrations for all but the LLGHGs are not well constrained.

2.9.6 Spatial Patterns of Radiative Forcing and Surface Forcing

Figure 6.7 of Ramaswamy et al. (2001) presented examples of the spatial patterns for most of the RF agents discussed in this chapter; these examples still hold. Many of the features seen in Figure 6.7 of Ramaswamy et al. (2001) are generic. However, additional uncertainties exist for the spatial patterns compared to those for the global-mean RF. Spatial patterns of the aerosol RF exhibit some of the largest differences between models, depending on the specification of the aerosols and their properties, and whether or not indirect cloud albedo effects are included. The aerosol direct and cloud albedo effect RF also depend critically on the location of clouds, which differs between the GCMs. Figure 2.24 presents illustrative examples of the spatial pattern of the instantaneous RF between 1860 and present day, due to natural plus anthropogenic agents, from two GCMs. Volcanic aerosols play a negligible role in this calculation owing to the end years considered and their virtual absence during these years. The MIROC+SPRINTARS model includes

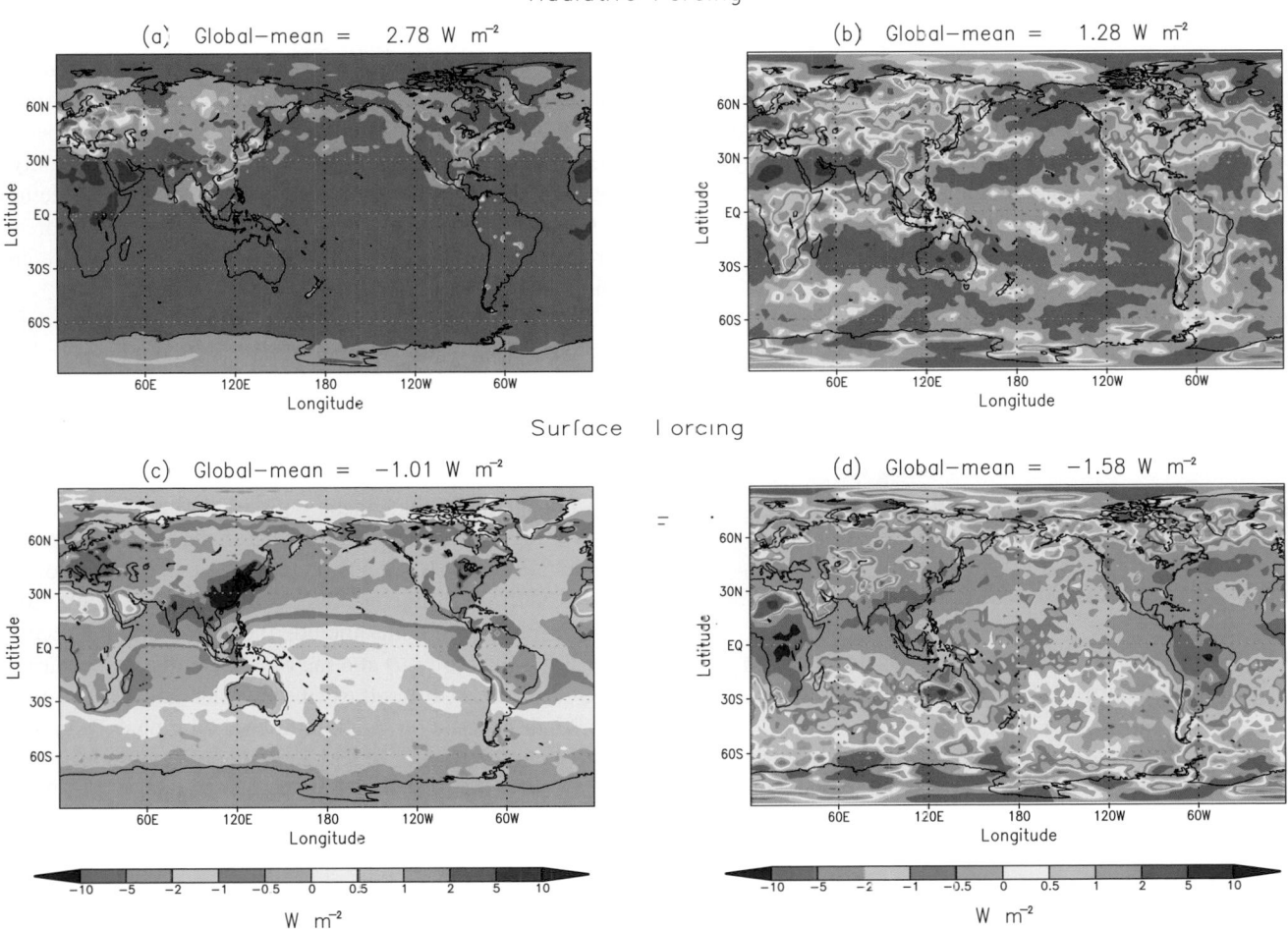

Figure 2.24. *Instantaneous change in the spatial distribution of the net (solar plus longwave) radiative flux (W m⁻²) due to natural plus anthropogenic forcings between the years 1860 and 2000. Results here are intended to be illustrative examples of these quantities in two different climate models. (a) and (c) correspond to tropopause and surface results using the GFDL CM 2.1 model (adapted from Knutson et al., 2006). (b) and (d) correspond to tropopause and surface results using the MIROC+SPRINTARS model (adapted from Nozawa et al., 2005 and Takemura et al., 2005). Note that the MIROC+SPRINTARS model takes into account the aerosol cloud albedo effect while the CM 2.1 model does not.*

an aerosol cloud albedo effect while the Geophysical Fluid Dynamics Laboratory Coupled Climate Model (GFDL CM2.1) (Delworth et al., 2005; Knutson et al., 2006) does not. Radiative forcing over most of the globe is positive and is dominated by the LLGHGs. This is more so for the SH than for the NH, owing to the pronounced aerosol presence in the mid-latitude NH (see also Figure 2.12), with the regions of substantial aerosol RF clearly manifest over the source-rich continental areas. There are quantitative differences between the two GCMs in the global mean RF, which are indicative of the uncertainties in the RF from the non-LLGHG agents, particularly aerosols (see Section 2.4 and Figure 2.12d). The direct effect of aerosols is seen in the total RF of the GFDL model over NH land regions, whereas the cloud albedo effect dominates the MIROC+SPRINTARS model in the stratocumulus low-latitude ocean regions. Note that the spatial pattern of the forcing is not indicative of the climate response pattern.

Wherever aerosol presence is considerable (namely the NH), the surface forcing is negative, relative to pre-industrial times (Figure 2.24). Because of the aerosol influence on the reduction of the shortwave radiation reaching the surface (see also Figure 2.12f), there is a net (sum of shortwave and longwave) negative surface forcing over a large part of the globe (see also Figure 2.23). In the absence of aerosols, LLGHGs increase the atmospheric longwave emission, with an accompanying increase in the longwave radiative flux reaching the surface. At high latitudes and in parts of the SH, there are fewer anthropogenic aerosols and thus the surface forcing has a positive value, owing to the LLGHGs.

These spatial patterns of RF and surface forcing imply different changes in the NH equator-to-pole gradients for the surface and tropopause. These, in turn, imply different changes in the amount of energy absorbed by the troposphere at low and high latitudes. The aerosol influences are also manifest in the difference between the NH and SH in both RF and surface forcing.

2.10 Global Warming Potentials and Other Metrics for Comparing Different Emissions

2.10.1 Definition of an Emission Metric and the Global Warming Potential

Multi-component abatement strategies to limit anthropogenic climate change need a framework and numerical values for the trade-off between emissions of different forcing agents. Global Warming Potentials or other emission metrics provide a tool that can be used to implement comprehensive and cost-effective policies (Article 3 of the UNFCCC) in a decentralised manner so that multi-gas emitters (nations, industries) can compose mitigation measures, according to a specified emission constraint, by allowing for substitution between different climate agents. The metric formulation will differ depending on

whether a long-term climate change constraint has been set (e.g., Manne and Richels, 2001) or no specific long-term constraint has been agreed upon (as in the Kyoto Protocol). Either metric formulation requires knowledge of the contribution to climate change from emissions of various components over time. The metrics assessed in this report are purely physically based. However, it should be noted that many economists have argued that emission metrics need also to account for the economic dimensions of the problem they are intended to address (e.g., Bradford, 2001; Manne and Richels, 2001; Godal, 2003; O'Neill, 2003). Substitution of gases within an international climate policy with a long-term target that includes economic factors is discussed in Chapter 3 of IPCC WGIII AR4. Metrics based on this approach will not be discussed in this report.

A very general formulation of an emission metric is given by (e.g., Kandlikar, 1996):

$$AM_i = \int_0^\infty [(I(\Delta C_{(r+i)}(t)) - I(\Delta C_r(t))) \times g(t)]dt$$

where $I(\Delta C_i(t))$ is a function describing the impact (damage and benefit) of change in climate (ΔC) at time t. The expression $g(t)$ is a weighting function over time (e.g., $g(t) = e^{-kt}$ is a simple discounting giving short-term impacts more weight) (Heal, 1997; Nordhaus, 1997). The subscript r refers to a baseline emission path. For two emission perturbations i and j the absolute metric values AM_i and AM_j can be calculated to provide a quantitative comparison of the two emission scenarios. In the special case where the emission scenarios consist of only one component (as for the assumed pulse emissions in the definition of GWP), the ratio between AM_i and AM_j can be interpreted as a relative emission index for component i versus a reference component j (such as CO_2 in the case of GWP).

There are several problematic issues related to defining a metric based on the general formulation given above (Fuglestvedt et al., 2003). A major problem is to define appropriate impact functions, although there have been some initial attempts to do this for a range of possible climate impacts (Hammitt et al., 1996; Tol, 2002; den Elzen et al., 2005). Given that impact functions can be defined, AM calculations would require regionally resolved climate change data (temperature, precipitation, winds, etc.) that would have to be based on GCM results with their inherent uncertainties (Shine et al., 2005a). Other problematic issues include the definition of the temporal weighting function $g(t)$ and the baseline emission scenarios.

Due to these difficulties, the simpler and purely physical GWP index, based on the time-integrated global mean RF of a pulse emission of 1 kg of some compound (i) relative to that of 1 kg of the reference gas CO_2, was developed (IPCC, 1990) and adopted for use in the Kyoto Protocol. The GWP of component i is defined by

$$GWP_i \equiv \frac{\int_0^{TH} RF_i(t)\,dt}{\int_0^{TH} RF_r(t)\,dt} = \frac{\int_0^{TH} a_i \cdot [C_i(t)]\,dt}{\int_0^{TH} a_r \cdot [C_r(t)]\,dt}$$

where TH is the time horizon, RF_i is the global mean RF of component i, a_i is the RF per unit mass increase in atmospheric abundance of component i (radiative efficiency), $[C_i(t)]$ is the time-dependent abundance of i, and the corresponding quantities for the reference gas (r) in the denominator. The numerator and denominator are called the absolute global warming potential (AGWP) of i and r respectively. All GWPs given in this report use CO_2 as the reference gas. The simplifications made to derive the standard GWP index include (1) setting $g(t) = 1$ (i.e., no discounting) up until the time horizon (TH) and then $g(t) = 0$ thereafter, (2) choosing a 1-kg pulse emission, (3) defining the impact function, $I(\Delta C)$, to be the global mean RF, (4) assuming that the climate response is equal for all RF mechanisms and (5) evaluating the impact relative to a baseline equal to current concentrations (i.e., setting $I(\Delta C_r(t)) = 0$). The criticisms of the GWP metric have focused on all of these simplifications (e.g., O'Neill, 2000; Smith and Wigley, 2000; Bradford, 2001; Godal, 2003). However, as long as there is no consensus on which impact function ($I(\Delta C)$) and temporal weighting functions to use (both involve value judgements), it is difficult to assess the implications of the simplifications objectively (O'Neill, 2000; Fuglestvedt et al., 2003).

The adequacy of the GWP concept has been widely debated since its introduction (O'Neill, 2000; Fuglestvedt et al., 2003). By its definition, two sets of emissions that are equal in terms of their total GWP-weighted emissions will not be equivalent in terms of the temporal evolution of climate response (Fuglestvedt et al., 2000; Smith and Wigley, 2000). Using a 100-year time horizon as in the Kyoto Protocol, the effect of current emissions reductions (e.g., during the first commitment period under the Kyoto Protocol) that contain a significant fraction of short-lived species (e.g., CH_4) will give less temperature reductions towards the end of the time horizon, compared to reductions in CO_2 emissions only. Global Warming Potentials can really only be expected to produce identical changes in one measure of climate change – integrated temperature change following emissions impulses – and only under a particular set of assumptions (O'Neill, 2000). The Global Temperature Potential (GTP) metric (see Section 2.10.4.2) provides an alternative approach by comparing global mean temperature change at the end of a given time horizon. Compared to the GWP, the GTP gives equivalent climate response at a chosen time, while putting much less emphasis on near-term climate fluctuations caused by emissions of short-lived species (e.g., CH_4). However, as long as it has not been determined, neither scientifically, economically nor politically, what the proper time horizon for evaluating 'dangerous anthropogenic interference in the climate system' should be, the lack of temporal equivalence does not invalidate the GWP concept or provide guidance as to how to replace it. Although it has several known shortcomings, a multi-gas strategy using GWPs is very likely to have advantages over a CO_2-only strategy (O'Neill, 2003). Thus, GWPs remain the recommended metric to compare future climate impacts of emissions of long-lived climate gases.

Globally averaged GWPs have been calculated for short-lived species, for example, ozone precursors and absorbing aerosols (Fuglestvedt et al., 1999; Derwent et al., 2001; Collins et al., 2002; Stevenson et al., 2004; Berntsen et al., 2005; Bond and Sun, 2005). There might be substantial co-benefits realised in mitigation actions involving short-lived species affecting climate and air pollutants (Hansen and Sato, 2004); however, the effectiveness of the inclusion of short-lived forcing agents in international agreements is not clear (Rypdal et al., 2005). To assess the possible climate impacts of short-lived species and compare those with the impacts of the LLGHGs, a metric is needed. However, there are serious limitations to the use of global mean GWPs for this purpose. While the GWPs of the LLGHGs do not depend on location and time of emissions, the GWPs for short-lived species will be regionally and temporally dependent. The different response of precipitation to an aerosol RF compared to a LLGHG RF also suggests that the GWP concept may be too simplistic when applied to aerosols.

2.10.2 Direct Global Warming Potentials

All GWPs depend on the AGWP for CO_2 (the denominator in the definition of the GWP). The AGWP of CO_2 again depends on the radiative efficiency for a small perturbation of CO_2 from the current level of about 380 ppm. The radiative efficiency per kilogram of CO_2 has been calculated using the same expression as for the CO_2 RF in Section 2.3.1, with an updated background CO_2 mixing ratio of 378 ppm. For a small perturbation from 378 ppm, the RF is 0.01413 W m^{-2} ppm^{-1} (8.7% lower than the TAR value). The CO_2 response function (see Table 2.14) is based on an updated version of the Bern carbon cycle model (Bern2.5CC; Joos et al. 2001), using a background CO_2 concentration of 378 ppm. The increased background concentration of CO_2 means that the airborne fraction of emitted CO_2 (Section 7.3) is enhanced, contributing to an increase in the AGWP for CO_2. The AGWP values for CO_2 for 20, 100, and 500 year time horizons are 2.47×10^{-14}, 8.69×10^{-14}, and 28.6×10^{-14} W m^{-2} yr (kg CO_2)$^{-1}$, respectively. The uncertainty in the AGWP for CO_2 is estimated to be ±15%, with equal contributions from the CO_2 response function and the RF calculation.

Updated radiative efficiencies for well-mixed greenhouse gases are given in Table 2.14. Since the TAR, radiative efficiencies have been reviewed by Montzka et al. (2003) and Velders et al. (2005). Gohar et al. (2004) and Forster et al. (2005) investigated HFC compounds, with up to 40% differences from earlier published results. Based on a variety of radiative transfer codes, they found that uncertainties could be reduced to around 12% with well-constrained experiments. The HFCs studied were HFC-23, HFC-32, HFC-134a and HFC-227ea. Hurley et al. (2005) studied the infrared spectrum and RF of perfluoromethane ($C\Omega F_4$) and derived a 30% higher GWP value than given in the TAR. The RF calculations for the GWPs for CH_4, N_2O and halogen-containing well-mixed greenhouse gases employ the simplified formulas given in Ramaswamy et al. (2001; see Table 6.2 of the TAR). Table 2.14 gives GWP values for time horizons of 20, 100 and 500 years. The species in Table 2.14 are those for which either significant concentrations or large trends in concentrations have been

Table 2.14. *Lifetimes, radiative efficiencies and direct (except for CH_4) GWPs relative to CO_2. For ozone-depleting substances and their replacements, data are taken from IPCC/TEAP (2005) unless otherwise indicated.*

Industrial Designation or Common Name (years)	Chemical Formula	Lifetime (years)	Radiative Efficiency (W m^{-2} ppb^{-1})	Global Warming Potential for Given Time Horizon			
				SAR[‡] (100-yr)	20-yr	100-yr	500-yr
Carbon dioxide	CO_2	See below[a]	[b]1.4x10^{-5}	1	1	1	1
Methane[c]	CH_4	12[c]	3.7x10^{-4}	21	72	25	7.6
Nitrous oxide	N_2O	114	3.03x10^{-3}	310	289	298	153
Substances controlled by the Montreal Protocol							
CFC-11	CCl_3F	45	0.25	3,800	6,730	4,750	1,620
CFC-12	CCl_2F_2	100	0.32	8,100	11,000	10,900	5,200
CFC-13	$CClF_3$	640	0.25		10,800	14,400	16,400
CFC-113	CCl_2FCClF_2	85	0.3	4,800	6,540	6,130	2,700
CFC-114	$CClF_2CClF_2$	300	0.31		8,040	10,000	8,730
CFC-115	$CClF_2CF_3$	1,700	0.18		5,310	7,370	9,990
Halon-1301	$CBrF_3$	65	0.32	5,400	8,480	7,140	2,760
Halon-1211	$CBrClF_2$	16	0.3		4,750	1,890	575
Halon-2402	$CBrF_2CBrF_2$	20	0.33		3,680	1,640	503
Carbon tetrachloride	CCl_4	26	0.13	1,400	2,700	1,400	435
Methyl bromide	CH_3Br	0.7	0.01		17	5	1
Methyl chloroform	CH_3CCl_3	5	0.06		506	146	45
HCFC-22	$CHClF_2$	12	0.2	1,500	5,160	1,810	549
HCFC-123	$CHCl_2CF_3$	1.3	0.14	90	273	77	24
HCFC-124	$CHClFCF_3$	5.8	0.22	470	2,070	609	185
HCFC-141b	CH_3CCl_2F	9.3	0.14		2,250	725	220
HCFC-142b	CH_3CClF_2	17.9	0.2	1,800	5,490	2,310	705
HCFC-225ca	$CHCl_2CF_2CF_3$	1.9	0.2		429	122	37
HCFC-225cb	$CHClFCF_2CClF_2$	5.8	0.32		2,030	595	181
Hydrofluorocarbons							
HFC-23	CHF_3	270	0.19	11,700	12,000	14,800	12,200
HFC-32	CH_2F_2	4.9	0.11	650	2,330	675	205
HFC-125	CHF_2CF_3	29	0.23	2,800	6,350	3,500	1,100
HFC-134a	CH_2FCF_3	14	0.16	1,300	3,830	1,430	435
HFC-143a	CH_3CF_3	52	0.13	3,800	5,890	4,470	1,590
HFC-152a	CH_3CHF_2	1.4	0.09	140	437	124	38
HFC-227ea	CF_3CHFCF_3	34.2	0.26	2,900	5,310	3,220	1,040
HFC-236fa	$CF_3CH_2CF_3$	240	0.28	6,300	8,100	9,810	7,660
HFC-245fa	$CHF_2CH_2CF_3$	7.6	0.28		3,380	1030	314
HFC-365mfc	$CH_3CF_2CH_2CF_3$	8.6	0.21		2,520	794	241
HFC-43-10mee	$CF_3CHFCHFCF_2CF_3$	15.9	0.4	1,300	4,140	1,640	500
Perfluorinated compounds							
Sulphur hexafluoride	SF_6	3,200	0.52	23,900	16,300	22,800	32,600
Nitrogen trifluoride	NF_3	740	0.21		12,300	17,200	20,700
PFC-14	CF_4	50,000	0.10	6,500	5,210	7,390	11,200
PFC-116	C_2F_6	10,000	0.26	9,200	8,630	12,200	18,200

Table 2.14 (continued)

Industrial Designation or Common Name (years)	Chemical Formula	Lifetime (years)	Radiative Efficiency (W m⁻² ppb⁻¹)	Global Warming Potential for Given Time Horizon			
				SAR‡ (100-yr)	20-yr	100-yr	500-yr
Perfluorinated compounds *(continued)*							
PFC-218	C_3F_8	2,600	0.26	7,000	6,310	8,830	12,500
PFC-318	c-C_4F_8	3,200	0.32	8,700	7,310	10,300	14,700
PFC-3-1-10	C_4F_{10}	2,600	0.33	7,000	6,330	8,860	12,500
PFC-4-1-12	C_5F_{12}	4,100	0.41		6,510	9,160	13,300
PFC-5-1-14	C_6F_{14}	3,200	0.49	7,400	6,600	9,300	13,300
PFC-9-1-18	$C_{10}F_{18}$	>1,000ᵈ	0.56		>5,500	>7,500	>9,500
trifluoromethyl sulphur pentafluoride	SF_5CF_3	800	0.57		13,200	17,700	21,200
Fluorinated ethers							
HFE-125	CHF_2OCF_3	136	0.44		13,800	14,900	8,490
HFE-134	CHF_2OCHF_2	26	0.45		12,200	6,320	1,960
HFE-143a	CH_3OCF_3	4.3	0.27		2,630	756	230
HCFE-235da2	$CHF_2OCHClCF_3$	2.6	0.38		1,230	350	106
HFE-245cb2	$CH_3OCF_2CHF_2$	5.1	0.32		2,440	708	215
HFE-245fa2	$CHF_2OCH_2CF_3$	4.9	0.31		2,280	659	200
HFE-254cb2	$CH_3OCF_2CHF_2$	2.6	0.28		1,260	359	109
HFE-347mcc3	$CH_3OCF_2CF_2CF_3$	5.2	0.34		1,980	575	175
HFE-347pcf2	$CHF_2CF_2OCH_2CF_3$	7.1	0.25		1,900	580	175
HFE-356pcc3	$CH_3OCF_2CF_2CHF_2$	0.33	0.93		386	110	33
HFE-449sl (HFE-7100)	$C_4F_9OCH_3$	3.8	0.31		1,040	297	90
HFE-569sf2 (HFE-7200)	$C_4F_9OC_2H_5$	0.77	0.3		207	59	18
HFE-43-10pccc124 (H-Galden 1040x)	$CHF_2OCF_2OC_2F_4OCHF_2$	6.3	1.37		6,320	1,870	569
HFE-236ca12 (HG-10)	$CHF_2OCF_2OCHF_2$	12.1	0.66		8,000	2,800	860
HFE-338pcc13 (HG-01)	$CHF_2OCF_2CF_2OCHF_2$	6.2	0.87		5,100	1,500	460
Perfluoropolyethers							
PFPMIE	$CF_3OCF(CF_3)CF_2OCF_2OCF_3$	800	0.65		7,620	10,300	12,400
Hydrocarbons and other compounds – Direct Effects							
Dimethylether	CH_3OCH_3	0.015	0.02		1	1	<<1
Methylene chloride	CH_2Cl_2	0.38	0.03		31	8.7	2.7
Methyl chloride	CH_3Cl	1.0	0.01		45	13	4

Notes:

ᵃ The CO_2 response function used in this report is based on the revised version of the Bern Carbon cycle model used in Chapter 10 of this report (Bern2.5CC; Joos et al. 2001) using a background CO_2 concentration value of 378 ppm. The decay of a pulse of CO_2 with time t is given by

$$a_0 + \sum_{i=1}^{3} a_i \cdot e^{-t/\tau_i}$$

Where $a_0 = 0.217$, $a_1 = 0.259$, $a_2 = 0.338$, $a_3 = 0.186$, $\tau_1 = 172.9$ years, $\tau_2 = 18.51$ years, and $\tau_3 = 1.186$ years.

ᵇ The radiative efficiency of CO_2 is calculated using the IPCC (1990) simplified expression as revised in the TAR, with an updated background concentration value of 378 ppm and a perturbation of +1 ppm (see Section 2.10.2).

ᶜ The perturbation lifetime for methane is 12 years as in the TAR (see also Section 7.4). The GWP for methane includes indirect effects from enhancements of ozone and stratospheric water vapour (see Section 2.10.3.1).

ᵈ Shine et al. (2005c), updated by the revised AGWP for CO_2. The assumed lifetime of 1,000 years is a lower limit.

ᵉ Hurley et al. (2005)

ᶠ Robson et al. (2006)

ᵍ Young et al. (2006)

observed or a clear potential for future emissions has been identified. The uncertainties of these direct GWPs are taken to be ±35% for the 5 to 95% (90%) confidence range.

2.10.3 Indirect GWPs

Indirect radiative effects include the direct effects of degradation products or the radiative effects of changes in concentrations of greenhouse gases caused by the presence of the emitted gas or its degradation products. Direct effects of degradation products for the greenhouse gases are not considered to be significant (WMO, 2003). The indirect effects discussed here are linked to ozone formation or destruction, enhancement of stratospheric water vapour, changes in concentrations of the OH radical with the main effect of changing the lifetime of CH_4, and secondary aerosol formation. Uncertainties for the indirect GWPs are generally much higher than for the direct GWPs. The indirect GWP will in many cases depend on the location and time of the emissions. For some species (e.g., NO_x) the indirect effects can be of opposite sign, further increasing the uncertainty of the net GWP. This can be because background levels of reactive species (e.g., NO_x) can affect the chemical response nonlinearly, and/or because the lifetime or the radiative effects of short-lived secondary species formed can be regionally dependent. Thus, the usefulness of the global mean GWPs to inform policy decisions can be limited. However, they are readily calculable and give an indication of the total potential of mitigating climate change by including a certain forcing agent in climate policy. Following the approach taken by the SAR and the TAR, the CO_2 produced from oxidation of CH_4, CO and NMVOCs of fossil origin is not included in the GWP estimates since this carbon has been included in the national CO_2 inventories. This issue may need to be reconsidered as inventory guidelines are revised.

2.10.3.1 Methane

Four indirect radiative effects of CH_4 emissions have been identified (see Prather et al., 2001; Ramaswamy et al., 2001). Methane enhances its own lifetime through changes in the OH concentration: it leads to changes in tropospheric ozone, enhances stratospheric water vapour levels, and produces CO_2. The GWP given in Table 2.14 includes the first three of these effects. The lifetime effect is included by adopting a perturbation lifetime of 12 years (see Section 7.4). The effect of ozone production is still uncertain, and as in the TAR, it is included by enhancing the net of the direct and the lifetime effect by 25%. The estimate of RF caused by an increase in stratospheric water vapour has been increased significantly since the TAR (see Section 2.3.7). This has also been taken into account in the GWP estimate for CH_4 by increasing the enhancement factor from 5% (TAR) to 15%. As a result, the 100-year GWP for CH_4 has increased from 23 in the TAR to 25.

2.10.3.2 Carbon Monoxide

The indirect effects of CO occur through reduced OH levels (leading to enhanced concentrations of CH_4) and enhancement of ozone. The TAR gave a range of 1.0 to 3.0 for the 100-year GWP. Since the TAR, Collins et al. (2002) and Berntsen et al. (2005) have calculated GWPs for CO emissions that range between 1.6 and 2.0, depending on the location of the emissions. Berntsen et al. (2005) found that emissions of CO from Asia had a 25% higher GWP compared to European emissions. Averaging over the TAR values and the new estimates give a mean of 1.9 for the 100-year GWP for CO.

2.10.3.3 Non-methane Volatile Organic Compounds

Collins et al. (2002) calculated indirect GWPs for 10 NMVOCs with a global three-dimensional Lagrangian chemistry-transport model. Impacts on tropospheric ozone, CH_4 (through changes in OH) and CO_2 have been considered, using either an 'anthropogenic' emission distribution or a 'natural' emission distribution depending on the main sources for each gas. The indirect GWP values are given in Table 2.15. Weighting these GWPs by the emissions of the respective compounds gives a weighted average 100-year GWP of 3.4. Due to their short lifetimes and the nonlinear chemistry involved in ozone and OH chemistry, there are significant uncertainties in the calculated GWP values. Collins et al. (2002) estimated an uncertainty range of –50% to +100%.

2.10.3.4 Nitrogen Oxides

The short lifetime and complex nonlinear chemistry, which cause two opposing indirect effects through ozone enhancements and CH_4 reductions, make calculations of GWP for NO_x emissions very uncertain (Shine et al., 2005a). In addition, the effect of nitrate aerosol formation (see Section 2.4.4.5), which has not yet been included in model studies calculating GWPs for NO_x, can be significant. Due to the nonlinear chemistry, the net RF of NO_x emissions will depend strongly on the location of emission and, with a strict definition of a pulse emission for the GWP, also on timing (daily, seasonal) of the emissions (Fuglestvedt et al., 1999; Derwent et al., 2001; Wild et al., 2001; Stevenson et al., 2004; Berntsen et al., 2005, 2006). Due to the lack of agreement even on the sign of the global mean GWP for NO_x among the different studies and the omission of the nitrate aerosol effect, a central estimate for the 100-year GWP for NO_x is not presented.

2.10.3.5 Halocarbons

Chlorine- and bromine-containing halocarbons lead to ozone depletion when the halocarbon molecules are broken down in the stratosphere and chlorine or bromine atoms are released. Indirect GWPs for ozone-depleting halocarbons are estimated in Velders et al. (2005; their Table 2.7). These are based on observed ozone depletion between 1980 and 1990 for 2005 emissions using the Daniel et al. (1995) formulation. Velders et al. (2005) did not quote net GWPs, pointing out that the physical characteristics of the CFC warming effect and ozone cooling effect were very different from each other.

2.10.3.6   Hydrogen

The main loss of hydrogen (H_2) is believed to be through surface deposition, but about 25% is lost through oxidation by OH. In the stratosphere, this enhances the water vapour concentrations and thus also affects the ozone concentrations. In the troposphere, the chemical effects are similar to those of CO, leading to ozone production and CH_4 enhancements (Prather, 2003). Derwent et al. (2001) calculated an indirect 100-year GWP for the tropospheric effects of H_2 of 5.8, which includes the effects of CH_4 lifetime and tropospheric ozone.

2.10.4   New Alternative Metrics for Assessing Emissions

While the GWP is a simple and straightforward index to apply for policy makers to rank emissions of different greenhouse gases, it is not obvious on what basis 'equivalence' between emissions of different species is obtained (Smith and

Wigley, 2000; Fuglestvedt et al., 2003). The GWP metric is also problematic for short-lived gases or aerosols (e.g., NO_x or BC aerosols), as discussed above. One alternative, the RF index (RFI) introduced by IPCC (1999), should not be used as an emission metric since it does not account for the different residence times of different forcing agents.

2.10.4.1   Revised GWP Formulations

2.10.4.1.2   Including the climate efficacy in the GWP

As discussed in Section 2.8.5, the climate efficacy can vary between different forcing agents (within 25% for most realistic RFs). Fuglestvedt et al. (2003) proposed a revised GWP concept that includes the efficacy of a forcing agent. Berntsen et al. (2005) calculated GWP values in this way for NO_x and CO emissions in Europe and in South East Asia. The efficacies are less uncertain than climate sensitivities. However, Berntsen et al. (2005) showed that for ozone produced by NO_x emissions the climate efficacies will also depend on the location of the emissions.

2.10.4.2   The Global Temperature Potential

Shine et al. (2005b) proposed the GTP as a new relative emission metric. The GTP is defined as the ratio between the

Table 2.15. *Indirect GWPs (100-year) for 10 NMVOCs from Collins et al. (2002) and for NO_x emissions (on N-basis) from Derwent et al. (2001), Wild et al. (2001), Berntsen et al. (2005) and Stevenson et al. (2004). The second and third columns respectively represent the methane and ozone contribution to the net GWP and the fourth column represents the net GWP.*

Organic Compound/Study	GWP^{CH_4}	GWP^{O_3}	GWP
Ethane (C_2H_6)	2.9	2.6	5.5
Propane (C_3H_8)	2.7	0.6	3.3
Butane (C_4H_{10})	2.3	1.7	4.0
Ethylene (C_2H_4)	1.5	2.2	3.7
Propylene (C_3H_6)	−2.0	3.8	1.8
Toluene (C_7H_8)	0.2	2.5	2.7
Isoprene (C_5H_8)	1.1	1.6	2.7
Methanol (CH_3OH)	1.6	1.2	2.8
Acetaldehyde (CH_3CHC)	−0.4	1.7	1.3
Acetone (CH_3COCH_3)	0.3	0.2	0.5
Derwent et al. NH surface NO_x[a,b]	−24	11	−12
Derwent et al. SH surface NO_x[a,b]	−64	33	−31
Wild et al., industrial NO_x	−44	32	−12
Berntsen et al., surface NO_x Asia	−31 to −42[c]	55 to 70[c]	25 to 29[c]
Berntsen et al., surface NO_x Europe	−8.6 to −11[c]	8.1 to 12.7	−2.7 to +4.1[c]
Derwent et al., Aircraft NO_x[a,b]	−145	246	100
Wild et al., Aircraft NO_x	−210	340	130
Stevenson et al. Aircraft NO_x	−159	155	−3

Notes:

[a] Corrected values as described in Stevenson et al. (2004).

[b] For January pulse emissions.

[c] Range from two three-dimensional chemistry transport models and two radiative transfer models.

global mean surface temperature change at a given future time horizon (TH) following an emission (pulse or sustained) of a compound x relative to a reference gas r (e.g., CO_2):

$$GTP^{TH}_x = \frac{\Delta T^H_x}{\Delta T^H_r}$$

where ΔT^H_x denotes the global mean surface temperature change after H years following an emission of compound x. The GTPs do not require simulations with AOGCMs, but are given as transparent and simple formulas that employ a small number of input parameters required for calculation. Note that while the GWP is an integral quantity over the time horizon (i.e., the contribution of the RF at the beginning and end of the time horizon is exactly equal), the GTP uses the temperature change at time H (i.e., RF closer to time H contributes relatively more). The GTP metric requires knowledge of the same parameters as the GWP metric (radiative efficiency and lifetimes), but in addition, the response times for the climate system must be known, in particular if the lifetime of component x is very different from the lifetime of the reference gas. Differences in climate efficacies can easily be incorporated into the GTP metric. Due to the inclusion of the response times for the climate system, the GTP values for pulse emissions of gases with shorter lifetimes than the reference gas will be lower than the corresponding GWP values. As noted by Shine et al. (2005b), there is a near equivalence between the GTP for sustained emission changes and the pulse GWP. The GTP metric has the potential advantage over GWP that it is more directly related to surface temperature change.

References

Abdul-Razzak, H., and S.J. Ghan, 2002: A parametrization of aerosol activation: 3. Sectional representation. *J. Geophys. Res.*, **107**(D3), 4026, doi:10.1029/2001JD000483.

Abel, S.J., E.J. Highwood, J.M. Haywood, and M.A. Stringer, 2005: The direct radiative effect of biomass burning aerosol over southern Africa. *Atmos. Chem. Phys. Discuss.*, **5**, 1165–1211.

Abel, S.J., et al., 2003: Evolution of biomass burning aerosol properties from an agricultural fire in southern Africa. *Geophys. Res. Lett.*, **30**(15), 1783, doi:10.1029/2003GL017342.

Ackerman, A.S, M.P. Kirkpatrick, D.E. Stevens, and O.B. Toon, 2004: The impact of humidity above stratiform clouds on indirect aerosol climate forcing. *Nature*, **432**, 1014–1017.

Ackerman, A.S., et al., 2000a: Reduction of tropical cloudiness by soot. *Science*, **288**, 1042–1047.

Ackerman, A.S., et al., 2000b: Effects of aerosols on cloud albedo: evaluation of Twomey's parametrization of cloud susceptibility using measurements of ship tracks. *J. Atmos. Sci.*, **57**, 2684–2695.

Adams, P.J. and J.H. Seinfeld, 2002: Predicting global aerosol size distributions in general circulation models. *J. Geophys. Res.*, **107**(D19), 4370, doi:10.1029/2001JD001010.

Adams, P.J., et al., 2001: General circulation model assessment of direct radiative forcing by the sulfate-nitrate-ammonium-water inorganic aerosol system. *J. Geophys. Res.*, **106**(D1), 1097–1112.

Albrecht, B., 1989: Aerosols, cloud microphysics and fractional cloudiness. *Science*, **245**, 1227–1230.

Allan, W., et al., 2005: Interannual variation of ^{13}C in tropospheric methane: Implications for a possible atomic chlorine sink in the marine boundary layer. *J. Geophys. Res.*, **110**, D11306, doi:10.1029/2004JD005650.

Ammann, C.M., G.A. Meehl, W.M. Washington, and C.S. Zender, 2003: A monthly and latitudinally varying volcanic forcing dataset in simulations of 20th century climate. *Geophys. Res. Lett.*, **30**(12), 1657, doi:10.1029/2003GL016875.

Anderson, T.L., et al., 2003: Climate forcing by aerosols: a hazy picture. *Science*, **300**, 1103–1104.

Anderson, T.L., et al., 2005a. Testing the MODIS satellite retrieval of aerosol fine-mode fraction. *J. Geophys. Res.*, **110**, D18204, doi:10.1029/2005JD005978.

Anderson, T.L., et al., 2005b: An "A-Train" strategy for quantifying direct climate forcing by anthropogenic aerosols. *Bull. Am. Meteorol. Soc.*, **86**, 1795–1809.

Andreae, M.O., and P. Merlet, 2001: Emission of trace gases and aerosols from biomass burning. *Global Biogeochem. Cycles*, **15**(4), 955–966, doi:10.1029/2000GB001382.

Andreae, M.O., et al., 2004: Atmospheric science: smoking rain clouds over the Amazon. *Science*, **303**(5662), 1337–1341.

Andres, R.J., G. Marland, T. Boden, and S. Bischof, 2000: Carbon dioxide emissions from fossil fuel consumption and cement manufacture, 1751-1991, and an estimate of their isotopic composition and latitudinal distribution. In: *The Carbon Cycle* [Wigley, T.M.L., and D.S. Schimel (eds.)]. Cambridge University Press, Cambridge, UK, pp. 53–62.

Andronova, N.G., et al., 1999: Radiative forcing by volcanic aerosols from 1850 to 1994. *J. Geophys. Res.*, **104**(D14), 16807–16826.

Antuña, J.C., et al., 2003: Spatial and temporal variability of the stratospheric aerosol cloud produced by the 1991 Mount Pinatubo eruption. *J. Geophys. Res.*, **108**, 4624, doi:10.1029/2003JD003722.

Asano, S., et al., 2002: Two case studies of winter continental-type water and mixed-phased stratocumuli over the sea. II: Absorption of solar radiation. *J. Geophys. Res.*, **107**(D21), 4570, doi:10.1029/2001JD001108.

Baliunas, S., and R. Jastrow, 1990: Evidence for long-term brightness changes of solar-type stars. *Nature*, **348**, 520–522.

Balkanski, Y., et al., 2004: Global emissions of mineral aerosol: formulation and validation using satellite imagery. In: *Emission of Atmospheric Trace Compounds* [Granier, C., P. Artaxo, and C.E. Reeves (eds.)]. Kluwer, Amsterdam, pp. 239–267.

Bange, H.W., S. Rapsomanikis, and M.O. Andreae, 1996: Nitrous oxide in coastal waters. *Global Biogeochem. Cycles*, **10**(1), 197–207.

Baran, A.J., and J.S. Foot, 1994: A new application of the operational sounder HIRS in determining a climatology of sulfuric acid aerosol from the Pinatubo eruption. *J. Geophys. Res.*, **99**(D12), 25673–25679.

Barnes, J.E., and D.J. Hoffman, 1997: Lidar measurements of stratospheric aerosol over Mauna Loa Observatory. *Geophys. Res. Lett.*, **24**(15), 1923–1926.

Barnston, A.G., and P.T. Schickedanz, 1984: The effect of irrigation on warm season precipitation in the southern Great Plains. *J. Appl. Meteorol.*, **23**, 865–888.

Bartholome, E., and A.S. Belward, 2005: GLC2000: a new approach to global land cover mapping from Earth observation data. *Int. J. Remote Sens.*, **26**, 1959–1977.

Bates, T.S., et al., 2006: Aerosol direct radiative effects over the northwest Atlantic, northwest Pacific, and North Indian Oceans: estimates based on in situ chemical and optical measurements and chemical transport modelling. *Atmos. Chem. Phys. Discuss.*, **6**, 175–362.

Bauer, S.E., and D. Koch, 2005: Impact of heterogeneous sulfate formation at mineral dust surfaces on aerosol loads and radiative forcing in the Goddard Institute for Space Studies general circulation model. *J. Geophys. Res.*, **110**, D17202, doi:10.1029/2005JD005870.

Bauman, J.J., P.B. Russell, M.A. Geller, and P. Hamill, 2003: A stratospheric aerosol climatology from SAGE II and CLAES measurements: 2. Results and comparisons, 1984-1999. *J. Geophys. Res.*, **108**(D13), 4383, doi:10.1029/2002JD002993.

Beheng, K.D., 1994: A parametrization of warm cloud microphysical conversion process. *Atmos. Res.*, **33**, 193–206.

Bekki, S., K. Law, and J. Pyle, 1994: Effect of ozone depletion on atmospheric CH_4 and CO concentrations. *Nature*, **371**, 595–597.

Bellouin, N., O. Boucher, D. Tanré, and O. Dubovik, 2003: Aerosol absorption over the clear-sky oceans deduced from POLDER-1 and AERONET observations. *Geophys. Res. Lett.*, **30**(14), 1748, doi:10.1029/2003GL017121.

Bellouin, N., O. Boucher, J. Haywood, and M.S. Reddy, 2005: Global estimates of aerosol direct radiative forcing from satellite measurements. *Nature*, **438**, 1138–1141.

Benestad, R.E., 2005: A review of the solar cycle length estimates. *Geophys. Res. Lett.*, **32**, L15714, doi:10.1029/2005GL023621.

Bergamaschi, P., et al., 2005: Inverse modelling of national and European CH_4 emissions using the atmospheric zoom model TM5. *Atmos. Chem. Phys.*, **5**, 2431–2460.

Bergstrom, R.W., P. Pilewskie, B. Schmid, and P.B. Russell, 2003: Estimates of the spectral aerosol single scattering albedo and aerosol radiative effects during SAFARI 2000. *J. Geophys. Res.*, **108**(D13), 8474, doi:10.1029/2002JD002435.

Berntsen, T.K., et al., 2005: Climate response to regional emissions of ozone precursors; sensitivities and warming potentials. *Tellus*, **4B**, 283–304.

Berntsen, T.K., et al., 2006: Abatement of greenhouse gases: does location matter? *Clim. Change*, **74**, 377–411.

Berry, E.X., 1967: Cloud droplet growth by collection. *J. Atmos. Sci.*, **24**, 688–701.

Betts, R.A., 2000: Offset of the potential carbon sink from boreal forestation by decreases in surface albedo. *Nature*, **408**(6809), 187–189.

Betts, R.A., 2001: Biogeophysical impacts of land use on present-day climate: near-surface temperature change and radiative forcing. *Atmos. Sci. Lett.*, **2**(1–4), doi:1006/asle.2001.0023.

Betts, R.A., and M.J. Best, 2004: *Relative impact of radiative forcing, landscape effects and local heat sources on simulated climate change in urban areas.* BETWIXT Technical Briefing Note No. 6, Met Office, Exeter, UK, 15 pp.

Betts, R.A., P.M. Cox, S.E. Lee, and F.I. Woodward, 1997: Contrasting physiological and structural vegetation feedbacks in climate change simulations. *Nature*, **387**, 796–799.

Betts, R.A., P.D. Falloon, K.K. Goldewijk, and N. Ramankutty, 2007: Biogeophysical effects of land use on climate: model simulations of radiative forcing and large-scale temperature change. *Agric. For. Meteorol.*, 142, 216-233.

Betts, R.A., et al., 2004: The role of ecosystem-atmosphere interactions in simulated Amazonian precipitation decrease and forest dieback under global climate warming. *Theor. Appl. Climatol.*, **78**, 157–175.

Bigler, M., et al., 2002: Sulphate record from a northeast Greenland ice core over the last 1200 years based on continuous flow analysis. *Ann. Glaciol.*, 35, 250–256.

Bingen, C., D. Fussen, and F. Vanhellemont, 2004: A global climatology of stratospheric aerosol size distribution parameters derived from SAGE II data over the period 1984-2000: 2. Reference data. *J. Geophys. Res.*, **109**, D06202, doi:10.1029/2003JD003511.

Blake, D., and F. Rowland, 1988: Continuing worldwide increase in tropospheric methane, 1978 to 1987. *Science*, **239**, 1129–1131.

Blake, N.J., et al., 2001: Large-scale latitudinal and vertical distributions of NMHCs and selected halocarbons in the troposphere over the Pacific Ocean during the March-April 1999 Pacific Exploratory Mission (PEM-Tropics B). *J. Geophys. Res.*, **106**, 32627–32644.

Bluth, G.J.S., W.I. Rose, I.E. Sprod, and A.J. Krueger, 1997: Stratospheric loading of sulfur from explosive volcanic eruptions. *J. Geol.*, **105**, 671–683.

Bluth, G.J.S., et al., 1992: Global tracking of the SO_2 clouds from the June 1991 Mount Pinatubo eruptions. *Geophys. Res. Lett.*, **19**(2), 151–154, doi:10.1029/91GL02792.

Boberg, F., and H. Lundstedt, 2002: Solar wind variations related to fluctuations of the North Atlantic Oscillation. *Geophys. Res. Lett.*, **29**(15), doi:10.1029/2002GL014903.

Boer, G.J., and B. Yu, 2003a: Climate sensitivity and climate state. *Clim. Dyn.*, **21**, 167–176.

Boer, G.J., and B. Yu, 2003b: Climate sensitivity and response. *Clim. Dyn.*, **20**, 415–429.

Bond, T.C., 2001: Spectral dependence of visible light absorption by carbonaceous particles emitted from coal combustion. *Geophys. Res. Lett.*, **28**(21), 4075–4078.

Bond, T.C., and H. Sun, 2005: Can reducing black carbon emissions counteract global warming? *Environ. Sci. Technol.*, **39**, 5921–5926.

Bond, T.C., et al., 1999: Light absorption by primary particle emissions from a lignite burning plant. *Environ. Sci.Technol.*, **33**, 3887–3891.

Bond, T.C., et al., 2004: A technology-based global inventory of black and organic carbon emissions from combustion. *J. Geophys. Res*, **109**, D14203, doi:10.1029/2003JD003697.

Born, M., H. Dorr, and I. Levin, 1990: Methane consumption in aerated soils of the temperate zone. *Tellus* , **42B**, 2–8.

Bortz, S.E., and M.J. Prather, 2006: Ozone, water vapor, and temperature in the upper tropical troposphere: Variations over a decade of MOZAIC Measurements. *J. Geophys. Res.*, **111**, D05305, doi:10.1029/2005JD006512.

Boucher, O., 1999: Air traffic may increase cirrus cloudiness. *Nature*, **397**, 30–31.

Boucher, O., and U. Lohmann, 1995: The sulfate-CCN-cloud albedo effect: a sensitivity study using two general circulation models. *Tellus*, **47B**, 281–300.

Boucher, O., and D. Tanré, 2000: Estimation of the aerosol perturbation to the Earth's radiative budget over oceans using POLDER satellite aerosol retrievals. *Geophys. Res. Lett.*, **27**(8), 1103–1106.

Boucher, O., and J. Haywood, 2001: On summing the components of radiative forcing of climate change. *Clim. Dyn.*, **18**, 297–302.

Boucher, O., and M. Pham, 2002: History of sulfate aerosol radiative forcings. *Geophys. Res. Lett.*, **29**(9), 22–25.

Boucher, O., H. Le Treut, and M.B. Baker, 1995: Precipitation and radiation modeling in a general circulation model: introduction of cloud microphysical processes. *J. Geophys. Res.*, **100**(D8), 16395–16414.

Boucher, O., G. Myhre, and A. Myhre, 2004: Direct human influence of irrigation on atmospheric water vapour and climate. *Clim. Dyn.*, **22**, 597–604.

Bounoua, L., et al., 2002: Effects of land cover conversion on surface climate. *Clim. Change*, **52**, 29–64.

Bousquet, P., et al., 2005: Two decades of OH variability as inferred by an inversion of atmospheric transport and chemistry of methyl chloroform. *Atmos. Chem. Phys.*, **5**, 1679–1731.

Bower, K.N., et al., 1994: A parametrization of warm clouds for use in atmospheric general circulation models. *J. Atmos. Sci.*, **51**, 2722–2732.

BP, 2006: *Quantifying Energy: BP Statistical Review of World Energy June 2006*. BP p.l.c., London, 45 pp., http://www.bp.com/productlanding. do?categoryId=6842&contentId=7021390.

Bradford, D.F., 2001: Global change: time, money and tradeoffs. *Nature*, **410**, 649–650, doi:10.1038/35070707.

Brasseur, G., and C. Granier, 1992: Mount Pinatubo aerosols, chlorofluorocarbons and ozone depletion. *Science*, **257**, 1239–1242.

Brenguier, J.L., H. Pawlowska, and L. Schuller, 2003: Cloud microphysical and radiative properties for parametrization and satellite monitoring of the indirect effect of aerosol on climate. *J. Geophys. Res.*, **108**(D15), 8632, doi:10.1029/2002JD002682.

Brenguier, J.L., et al., 2000a: Radiative properties of boundary layer clouds: droplet effective radius versus number concentration. *J. Atmos. Sci.*, **57**, 803–821.

Brenguier, J.L., et al., 2000b: An overview of the ACE-2 CLOUDYCOLUMN closure experiment. *Tellus*, **52B**, 815–827.

Bréon, F-M., and M. Doutriaux-Boucher, 2005: A comparison of cloud droplet radii measured from space. *IEEE Trans. Geosci. Remote. Sens.*, **43**, 1796–1805, doi:10.1109/TGRS.2005.852838.

Bréon, F.-M., D. Tanré, and S. Generoso, 2002: Aerosol effect on cloud droplet size monitored from satellite. *Science*, **295**, 834–838.

Broccoli, A.J., et al., 2003: Twentieth-century temperature and precipitation trends in ensemble climate simulations including natural and anthropogenic forcing. *J. Geophys. Res.*, **108**(D24), 4798, doi:10.1029/2003JD003812.

Brock, C.A., et al., 2004: Particle characteristics following cloud-modified transport from Asia to North America. *J. Geophys. Res.*, **109**, 1–17.

Brovkin, V.M., et al., 2004. Role of land cover changes for atmospheric CO_2 increase and climate change during the last 150 years. *Global Change Biol.*, **10**, 1253–1266.

Brovkin, V.M., et al., 2006: Biogeophysical effects of historical land cover changes simulated by six earth system models of intermediate complexity. *Clim. Dyn.*, **26**(6), 587–600.

Cakmur, R.V., et al., 2006: Constraining the magnitude of the global dust cycle by minimizing the difference between a model and observations. *J. Geophys. Res.*, **111**, doi:10.1029/2005JD005791.

Campbell, J.R., et al., 2003: Micropulse Lidar observations of tropospheric aerosols over northeastern South Africa during the ARREX and SAFARI-2000 dry season experiments. *J. Geophys. Res.*, **108**(D13), 8497, doi:10.1029/2002JD002563.

Carslaw, K.S., R.G. Harrison, and J. Kirkby, 2002: Atmospheric science: Cosmic rays, clouds, and climate. *Science*, **298**, 1732–1737.

Chapman, G.A., A.M. Cookson, and J.J. Dobias, 1996: Variations in total solar irradiance during solar cycle 22. *J. Geophys. Res.*, **101**, 13541–13548.

Chase, T.N., et al., 2000: Simulated impacts of historical land cover changes on global climate in northern winter. *Clim. Dyn.*, **16**, 93–105.

Chen, Y., and J.E. Penner, 2005: Uncertainty analysis of the first indirect aerosol effect. *Atmos. Chem. Phys.*, **5**, 2935–2948.

Chen, Y.-H., and R.G. Prinn, 2005: Atmospheric modelling of high- and low-frequency methane observations: Importance of interannually varying transport. *J. Geophys. Res.*, **110**, D10303, doi:10.1029/2004JD005542.

Chen, Y.-H., and R.G. Prinn, 2006: Estimation of atmospheric methane emissions between 1996-2001 using a 3D global chemical transport model. *J. Geophys. Res.*, **111**, D10307, doi:10.1029/2005JD006058.

Chiapello, I., C. Moulin, and J.M. Prospero, 2005: Understanding the long-term variability of African dust transport across the Atlantic as recorded in both Barbados surface concentrations and large-scale Total Ozone Mapping Spectrometer (TOMS) optical thickness. *J. Geophys. Res.*, **110**, D18S10, doi:10.1029/2004JD005132.

Chipperfield, M.P., W.J. Randel, G.E. Bodeker, and P. Johnston, 2003: Global ozone: past and future. In: *Scientific Assessment of Ozone Depletion: 2002* [Ennis, C.A. (ed.)]. World Meteorological Organization, Geneva, pp. 4.1–4.90.

Chou, M.-D., P.-K. Chan, and M. Wang, 2002: Aerosol radiative forcing derived from SeaWiFS-retrieved aerosol optical properties. *J. Atmos. Sci.*, **59**, 748–757.

Christiansen, B., 1999: Radiative forcing and climate sensitivity: the ozone experience. *Q. J. R. Meteorol. Soc.*, **125**, 3011–3035.

Christopher, S.A., and J. Zhang, 2004: Cloud-free shortwave aerosol radiative effect over oceans: strategies for identifying anthropogenic forcing from Terra satellite measurements. *Geophys. Res. Lett.*, **31**, L18101, doi:10.1029/2004GL020510.

Christopher, S.A., J. Zhang, Y.J. Kaufman, and L. Remer, 2006: Satellite-based assessment of the top of the atmosphere anthropogenic aerosol radiative forcing over cloud-free oceans. *Geophys Res. Lett.*, **111**, L15816, doi:10.1029/2005GL025535.

Chu, D.A., et al., 2002: Validation of MODIS aerosol optical depth retrieval over land. *Geophys. Res. Lett.*, **29**(12), doi:10.1029/2001GL013205.

Chu, D.A., et al., 2005: Evaluation of aerosol properties over ocean from Moderate Resolution Imaging Spectroradiometer (MODIS) during ACE-Asia. *J. Geophys. Res.*, **110**, D07308, doi:10.1029/2004JD005208.

Chuang, C.C., and J.E. Penner, 1995: Effects of anthropogenic sulfate on cloud drop nucleation and optical properties. *Tellus*, **47B**, 566–577.

Chuang, C.C., et al., 1997: An assessment of the radiative effects of anthropogenic sulfate. *J. Geophys. Res.*, **102**(D3), 3761–3778.

Chuang, C.C., et al., 2002: Cloud susceptibility and the first aerosol indirect forcing: Sensitivity to black carbon and aerosol concentrations. *J. Geophys. Res.*, **107**(D21), 4564, doi:10.1029/2000JD000215.

Chuang, P.Y., 2003: Measurement of the timescale of hygroscopic growth for atmospheric aerosols. *J. Geophys. Res.*, **108**(D9), 5–13.

Chung, C.E., V. Ramanathan, D. Kim, and I.A. Podgorny, 2005: Global anthropogenic aerosol direct forcing derived from satellite and ground-based observations. *J. Geophys. Res.*, **110**, D24207, doi:10.1029/2005JD006356.

Chung, S.H., and J.H. Seinfeld, 2002: Global distribution and climate forcing of carbonaceous aerosols. *J. Geophys. Res.*, **107**, doi:10.1029/2001JD001397.

Chylek, P., and J. Wong, 1995: Effect of absorbing aerosols on global radiation budget. *Geophys. Res. Lett.*, **22**(8), 929–931.

Chylek, P., B. Henderson, and M. Mishchenko, 2003: Aerosol radiative forcing and the accuracy of satellite aerosol optical depth retrieval. *J. Geophys. Res.*, **108**(D24), 4764, doi:10.1029/2003JD004044.

Clarke, A.D., et al., 2004: Size distributions and mixtures of dust and black carbon aerosol in Asian outflow: Physiochemistry and optical properties. *J. Geophys. Res.*, **109**, D15S09, doi:10.1029/2003JD004378.

Claquin, T., M. Schulz, and Y. Balkanski, 1999: Modeling the mineralogy of atmospheric dust. *J. Geophys. Res.*, **104**(D18), 22243–22256.

Coen, M.C., et al., 2004: Saharan dust events at the Jungfraujoch: detection by wavelength dependence of the single scattering albedo and first climatology analysis. *Atmos. Chem. Phys.*, **4**, 2465–2480.

Collins, W.D., et al., 2001: Simulating aerosols using a chemical transport model with assimilation of satellite aerosol retrievals: Methodology for INDOEX. *J. Geophys. Res.*, **106**(D7), 7313–7336.

Collins, W.D., et al., 2006: Radiative forcing by well-mixed greenhouse gases: Estimates from climate models in the IPCC AR4. *J. Geophys. Res.*, **111**, D14317, doi:10.1029/2005JD006713.

Collins, W.J., R.G. Derwent, C.E. Johnson, and D.S. Stevenson, 2002: The oxidation of organic compounds in the troposphere and their global warming potentials. *Clim. Change*, **52**, 453–479.

Considine, D.B., J.E. Rosenfield, and E.L. Fleming, 2001: An interactive model study of the influence of the Mount Pinatubo aerosol on stratospheric methane and water trends. *J. Geophys. Res.*, **106**(D21), 27711–27727.

Conway, T.J., et al., 1994: Evidence for interannual variability of the carbon cycle from the NOAA/CMDL sampling network. *J. Geophys. Res.*, **99**(D11), 22831–22855.

Cook, J., and E.J. Highwood, 2004: Climate response to tropospheric absorbing aerosols in an intermediate general-circulation model. *Q. J. R. Meteorol. Soc.*, **130**, 175–191.

Cooke, W.F., V. Ramaswamy, and P. Kasibhatla, 2002: A general circulation model study of the global carbonaceous aerosol distribution. *J. Geophys. Res.*, **107**(D16), doi:10.1029/2001JD001274.

Cooke, W.F., C. Liousse, H. Cachier, and J. Feichter, 1999: Construction of a 1° x 1° fossil fuel emission data set for carbonaceous aerosol and implementation and radiative impact in the ECHAM4 model. *J. Geophys. Res.*, **104**(D18), 22137–22162.

Coughlin, K., and K.K. Tung, 2004: Eleven-year solar cycle signal throughout the lower atmosphere. *J. Geophys. Res.*, **109**, D21105, doi:10.1029/2004JD004873.

Cox, P.M., et al., 1999: The impact of new land surface physics on the GCM simulation of climate and climate sensitivity. *Clim. Dyn.* **15**, 183–203.

Cramer, W., et al., 2001: Global response of terrestrial ecosystem structure and function to CO_2 and climate change: Results from six dynamic global vegetation models. *Global Change Biol.*, **7**, 357–373.

Crooks, S.A., and L.J. Gray, 2005: Characterization of the 11-year solar signal using a multiple regression analysis of the ERA-40 dataset. *J. Clim.*, **18**, 996–1015.

Crutzen, P.J., 2004: New directions: the growing urban heat and pollution "island" effect - impact on chemistry and climate. *Atmos. Environ.*, **38**, 3539–3540.

Cunnold, D.M., et al., 2002: In situ measurements of atmospheric methane at GAGE/AGAGE sites during 1985-2000 and resulting source inferences. *J Geophys. Res.*, **107**(D14), doi:10.1029/2001JD001226.

Cunnold, D.M., et al., 2004: Comment on "Enhanced upper stratospheric ozone: Sign of recovery or solar cycle effect?" by W. Steinbrecht et al. *J. Geophys. Res.*, **109**, D14305, doi:10.1029/2004JD004826.

Cziczo, D.J., D.M. Murphy, P.K. Hudson, and D.S. Thompson, 2004: Single particle measurements of the chemical composition of cirrus ice residue from CRYSTAL-FACE. *J. Geophys. Res.*, **109**, D04201, doi:10.1029/2003JD004032.

Dameris, M., et al., 2005: Long-term changes and variability in a transient simulation with a chemistry-climate model employing realistic forcing. *Atmos. Chem. Phys. Discuss.*, **5**, 2121–2145.

Daniel, J.S., S. Solomon, and D.L. Albritton, 1995: On the evaluation of halocarbon radiative forcing and global warming potentials. *J. Geophys. Res.*, **100**(D1), 1271–1286, doi:10.1029/94JD02516.

de Ridder, K., and H. Gallée, 1998: Land surface-induced regional climate change in southern Israel. *J. Appl. Meteorol.*, **37**, 1470–1485.

Decesari, S., M.C. Facchini, S. Fuzzi, and E. Tagliavini, 2000: Characterization of water-soluble organic compounds in atmospheric aerosol: a new approach. *J. Geophys. Res.*, **105**(D1), 1481–1489.

Decesari, S., et al., 2001: Chemical features and seasonal variation of fine aerosol water-soluble organic compounds in the Po Valley, Italy. *Atmos. Environ.*, **35**, 3691–3699.

De Genio, A.D., M.S. Yao, and K.-W. Lo, 1996: A prognostic cloud water parametrization for global climate models. *J. Clim.*, **9**, 270–304.

Delworth, T.L., V. Ramaswamy, and G.L. Stenchikov, 2005: The impact of aerosols on simulated ocean temperature and heat content in the 20th century. *Geophys. Res. Lett.*, **32**, L24709, doi:10.1029/2005GL024457.

den Elzen, M., et al., 2005. Analysing countries' contribution to climate change: Scientific and policy-related choices. *Environ. Sci. Technol.*, **8**(6), 614–636.

Dentener, F., et al., 2006: Emissions of primary aerosol and precursor gases in the years 2000 and 1750 - prescribed data-sets for AeroCom. *Atmos. Chem. Phys. Discuss.*, **6**, 2703–2763.

Derwent, R.G., W.J. Collins, C.E. Johnson, and D.S. Stevenson, 2001: Transient behaviour of tropospheric ozone precursors in a global 3-D CTM and their indirect greenhouse effects. *Clim. Change*, **49**, 463–487.

Deshler, T., et al., 2003: Thirty years of in situ stratospheric aerosol size distribution measurements from Laramie, Wyoming (41N), using balloon-borne instruments. *J. Geophys. Res.*, **108**(D5), 4167, doi:10.1029/2002JD002514.

Deuzé, J.L., et al., 2000: Estimate of the aerosol properties over the ocean with POLDER. *J. Geophys. Res.*, **105**, 15329–15346.

Deuzé, J.L., et al., 2001: Remote sensing of aerosols over land surfaces from POLDER-ADEOS-1 polarized measurements. *J. Geophys. Res.*, **106**, 4913–4926.

Dewitte, S., D. Crommelynck, S. Mekaoui, and A. Joukoff, 2005: Measurement and uncertainty of the long-term total solar irradiance trend. *Sol. Phys.*, **224**, 209–216.

Diab, R.D., et al., 2004: Tropospheric ozone climatology over Irene, South Africa, from 1990 to 1994 and 1998 to 2002. *J. Geophys. Res.*, **109**, D20301, doi:10.1029/2004JD004793.

Dlugokencky, E.J., K.A. Masarie, P.M. Lang, and P.P. Tans, 1998: Continuing decline in the growth rate of atmospheric methane. *Nature*, **393**, 447–450.

Dlugokencky, E.J., et al., 1996: Changes in CH_4 and CO growth rates after the eruption of Mt. Pinatubo and their link with changes in tropical tropospheric UV flux. *Geophys. Res. Lett.*, **23**(20), 2761–2764.

Dlugokencky, E.J., et al., 2001: Measurements of an anomalous global methane increase during 1998. *Geophys. Res. Lett.*, **28**(3), 499–502.

Dlugokencky, E.J., et al., 2003: Atmospheric methane levels off: Temporary pause or a new steady-state? *Geophys. Res. Lett.*, **30**(19), doi:10.1029/2003GL018126.

Dlugokencky, E.J., et al., 2005: Conversion of NOAA CMDL atmospheric dry air CH_4 mole fractions to a gravimetrically prepared standard scale. *J. Geophys. Res.*, **110**, D18306, doi:10.1029/2005JD006035.

Döll, P., 2002: Impact of climate change and variability on irrigation requirements: a global perspective. *Clim. Change*, **54**, 269–293.

Douglass, A., M. Schoeberl, R. Rood, and S. Pawson, 2003: Evaluation of transport in the lower tropical stratosphere in a global chemistry and transport model. *J. Geophys. Res.*, **108**(D9), 4259, doi:10.1029/2002JD002696.

Douglass, D.H., and B.D. Clader, 2002: Climate sensitivity of the Earth to solar irradiance. *Geophys. Res. Lett.*, **29**(16), 33–36.

Dubovik, O., and M.D. King, 2000: A flexible inversion algorithm for retrieval of aerosol optical properties from Sun and sky radiance measurements. *J. Geophys. Res.*, **105**, 20673–20696.

Dubovik, O., et al., 1998: Single-scattering albedo of smoke retrieved from the sky radiance and solar transmittance measured from ground. *J. Geophys. Res.*, **103**(D24), 31903–31923.

Dubovik, O., et al., 2000: Accuracy assessments of aerosol optical properties retrieved from Aerosol Robotic Network (AERONET) Sun and sky radiance measurements. *J. Geophys. Res.*, **105**, 9791–9806.

Dubovik, O., et al., 2002: Variability of absorption and optical properties of key aerosol types observed in worldwide locations. *J. Atmos. Sci.*, **59**, 590–608.

Dufresne, L.-L., et al., 2005: Contrasts in the effects on climate of anthropogenic sulfate aerosols between the 20th and the 21st century. *Geophys. Res. Lett.*, **32**, L21703, doi:10.1029/2005GL023619.

Duncan, B.N., et al., 2003: Indonesian wildfires of 1997: Impact on tropospheric chemistry. *J. Geophys. Res.*, **108**(D15), 4458, doi:10.1029/2002JD003195.

Dusek, U., et al., 2006: Size matters more than chemistry for cloud-nucleating ability of aerosol particles. *Science*, **312**, 1375–1378.

Dziembowski, W.A., P.R. Goode, and J. Schou, 2001: Does the sun shrink with increasing magnetic activity? *Astrophys. J.*, **553**, 897–904.

Eagan, R., P.V. Hobbs, and L. Radke, 1974: Measurements of CCN and cloud droplet size distributions in the vicinity of forest fires. *J. Appl. Meteorol.*, **13**, 537–553.

Easter, R., et al., 2004: MIRAGE: Model description and evaluation of aerosols and trace gases. *J. Geophys. Res.* **109**, D20210, doi:10.1029/2004JD004571.

Eatough, D.J., et al., 2003: Semivolatile particulate organic material in southern Africa during SAFARI 2000. *J. Geophys. Res.*, **108**(D13), 8479, doi:10.1029/2002JD002296.

Eck, T.F., et al., 2003: Variability of biomass burning aerosol optical characteristics in southern Africa during the SAFARI 2000 dry season campaign and a comparison of single scattering albedo estimates from radiometric measurements. *J. Geophys. Res.*, **108**(D13), 8477, doi:10.1029/2002JD002321.

Edwards, D.P., et al., 2004: Observations of carbon monoxide and aerosols from the Terra satellite: Northern Hemisphere variability. *J. Geophys. Res.*, **109**, D24202, doi:10.1029/2004JD004727.

Ehhalt, D.H., 1999: Gas phase chemistry of the troposphere. In: *Global Aspects of Atmospheric Chemistry, Vol. 6* [Baumgaertel, H., W. Gruenbein, and F. Hensel (eds.)]. Springer Verlag, Darmstadt, pp. 21–110.

EPA, 2003: *National Air Quality and Emissions Trends Report, 2003 Special Studies Edition*. Publication No. EPA 454/R-03-005, U. S. Environmental Protection Agency, Washington, DC, 190pp, http://www.epa.gov/air/airtrends/aqtrnd03/

Erlick, C., L.M. Russell, and V. Ramaswamy, 2001: A microphysics-based investigation of the radiative effects of aerosol-cloud interactions for two MAST Experiment case studies. *J. Geophys. Res.*, **106**(D1), 1249–1269.

Ervens, B., G. Feingold, and S.M. Kreidenweis, 2005: The influence of water-soluble organic carbon on cloud drop number concentration. *J. Geophys. Res.*, **110**, D18211, doi:10.1029/2004JD005634.

Etheridge, D.M., L.P. Steele, R.J. Francey, and R.L. Langenfelds, 1998: Atmospheric methane between 1000 A.D. and present: Evidence of anthropogenic emissions and climatic variability. *J. Geophys. Res.*, **103**(D13), 15979–15993.

Etheridge, D.M., et al., 1996: Natural and anthropogenic changes in atmospheric CO_2 over the last 1000 years from air in Antarctic ice and firn. *J. Geophys. Res.*, **101**(D2), 4115–4128.

Facchini, M.C., M. Mircea, S. Fuzzi, and R.J. Charlson, 1999: Cloud albedo enhancement by surface-active organic solutes in growing droplets. *Nature*, **401**, 257–259.

Facchini, M.C., et al., 2000: Surface tension of atmospheric wet aerosol and cloud/fog droplets in relation to their organic carbon content and chemical composition. *Atmos. Environ.*, **34**, 4853–4857.

Feichter, J., E. Roeckner, U. Lohmann, and B. Liepert, 2004: Nonlinear aspects of the climate response to greenhouse gas and aerosol forcing. *J. Clim.*, **17**, 2384–2398.

Feingold, G., 2003: Modelling of the first indirect effect: Analysis of measurement requirements. *Geophys. Res. Lett.*, **30**(19), 1997, doi:10.1029/2003GL017967.

Feingold, G., and P.Y. Chuang, 2002: Analysis of influence of film-forming compounds on droplet growth: Implications for cloud microphysical processes and climate. *J. Atmos. Sci.*, **59**, 2006–2018.

Feingold, G., R. Boers, B. Stevens, and W.R. Cotton, 1997: A modelling study of the effect of drizzle on cloud optical depth and susceptibility. *J. Geophys. Res.*, **102**(D12), 13527–13534.

Feingold, G., W.L. Eberhard, D.E. Veron, and M. Previdi, 2003: First measurements of the Twomey indirect effect using ground-based remote sensors. *Geophys. Res. Lett.*, **30**(6), 1287, doi:10.1029/2002GL016633.

Feng, Y., and J. Penner. 2007: Global modeling of nitrate and ammonium: Interaction of aerosols and tropospheric chemistry. *J. Geophys. Res.*, **112**(D01304), doi:10.1029/2005JD006404.

Ferretti, D.F., D.C. Lowe, R.J. Martin, and G.W. Brailsford, 2000: A new GC-IRMS technique for high precision, N_2O-free analysis of $\delta^{13}C$ and $\delta^{18}O$ in atmospheric CO_2 from small air samples. *J. Geophys. Res.*, **105**(D5), 6709–6718.

Field, C.B., R.B. Jackson, and H.A. Mooney, 1995: Stomatal responses to increased CO_2: implications from the plant to the global scale. *Plant Cell Environ.*, **18**, 1214–1225.

Fioletov, V.E., et al., 2002: Global and zonal total ozone variations estimated from ground-based and satellite measurements: 1964-2000. *J. Geophys. Res.*, **107**(D22), 4647, doi:10.1029/2001JD001350.

Formenti, P., et al., 2003: Inorganic and carbonaceous aerosols during the Southern African Regional Science Initiative (SAFARI 2000) experiment: Chemical characteristics, physical properties, and emission data or smoke from African biomass burning. *J. Geophys. Res.*, **108**(D13), 8488, doi:10.1029/2002JD002408.

Forster, P.M.F., and K.P. Shine, 1999: Stratospheric water vapour changes as a possible contributor to observed stratospheric cooling. *Geophys. Res. Lett.*, **26**(21), 3309–3312.

Forster, P.M.F., and K. Tourpali, 2001: Effect of tropopause height changes on the calculation of ozone trends and their radiative forcing. *J. Geophys. Res.*, **106**(D11), 12241–12251.

Forster, P.M.F., and K.P. Shine, 2002: Assessing the climate impact of trends in stratospheric water vapor. *Geophys. Res. Lett.*, **29**(6), doi:10.1029/2001GL013909.

Forster, P.M.F., and M.J. Joshi, 2005: The role of halocarbons in the climate change of the troposphere and stratosphere. *Clim. Change*, **70**, 249–266.

Forster, P.M.F., and K.E. Taylor, 2006: Climate forcings and climate sensitivities diagnosed from coupled climate model integrations. *J. Clim.*, **19**, 6181–6194.

Forster, P.M.F., M. Blackburn, R. Glover, and K.P. Shine, 2000: An examination of climate sensitivity for idealised climate change experiments in an intermediate general circulation model. *Clim. Dyn.*, **16**, 833–849.

Forster, P.M.F., et al., 2005: Resolving uncertainties in the radiative forcing of HFC-134a. *J. Quant. Spectrosc. Radiative Transfer*, **93**, 447–460.

Foster, S.S., 2004: *Reconstruction of Solar Irradiance Variations for use in Studies of Global Climate Change: Application of Recent SOHO Observations with Historic Data from the Greenwich Observatory.* PhD Thesis, University of Southampton, Faculty of Science, Southampton, 231 p.

Foukal, P., C. Frohlich, H. Spruit, and T.M.L. Wigley, 2006: Variations in solar luminosity and their effect on the Earth's climate. *Nature*, **443**, 161–166.

Francey, R.J., et al., 1995: Changes in oceanic and terrestrial carbon uptake since 1982. *Nature*, **373**, 326–330.

Francey, R.J., et al., 1999: A history of $\delta^{13}C$ in atmospheric CH_4 from the Cape Grim Air Archive and Antarctic firn air. *J. Geophys. Res.*, **104**(D19), 23633–23643.

Frankenberg, C., et al., 2005: Assessing methane emissions from global space-borne observations. *Science*, **308**, 1010–1014.

Free, M., and J. Angell, 2002: Effect of volcanoes on the vertical temperature profile in radiosonde data. *J. Geophys. Res.*, **107**(D10), doi:10.1029/2001JD001128.

Friedl, M.A., et al., 2002: Global land cover mapping from MODIS: algorithms and early results. *Remote Sens. Environ.*, **83**, 287–302.

Fröhlich, C., and J. Lean, 2004: Solar radiative output and its variability: Evidence and mechanisms. *Astron. Astrophys. Rev.*, **12**, 273–320.

Fueglistaler, S., H. Wernli, and T. Peter, 2004: Tropical troposphere-to-stratosphere transport inferred from trajectory calculations. *J. Geophys. Res.*, **109**, D03108, doi:10.1029/2003JD004069.

Fuglestvedt, J.S., T.K. Berntsen, O. Godal, and T. Skodvin, 2000: Climate implications of GWP-based reductions in greenhouse gas emissions. *Geophys. Res. Lett.*, **27**(3), 409–412, doi:10.1029/1999GL010939.

Fuglestvedt, J.S., et al., 1999: Climatic forcing of nitrogen oxides through changes in tropospheric ozone and methane: global 3D model studies. *Atmos. Environ.*, **33**, 961–977.

Fuglestvedt, J.S., et al., 2003: Metrics of climate change: assessing radiative forcing and emission indices. *Clim. Change*, **58**, 267–331.

Fuller, K.A., W.C. Malm, and S.M. Kreidenweis, 1999: Effects of mixing on extinction by carbonaceous particles. *J. Geophys. Res.*, **104**(D13), 15941–15954.

Fuzzi, S., et al., 2001: A simplified model of the water soluble organic component of atmospheric aerosols. *Geophys. Res. Lett.*, **28**(21), 4079–4082.

Gao, F., et al., 2005: MODIS bidirectional reflectance distribution function and albedo Climate Modeling Grid products and the variability of albedo for major global vegetation types. *J. Geophys. Res.*, **110**, D01104, doi:10.1029/2004JD005190.

Gauss, M., et al., 2006: Radiative forcing since preindustrial times due to ozone changes in the troposphere and the lower stratosphere. *Atmos. Chem. Phys.*, **6**, 575–599.

Gedney, N., et al., 2006: Detection of a direct carbon dioxide effect in continental river runoff records. *Nature*, **439**, 835–838, doi:10.1038/nature04504.

Geller, M.A., and S.P. Smyshlyaev, 2002: A model study of total ozone evolution 1979-2000: The role of individual natural and anthropogenic effects. *Geophys. Res. Lett.*, **29**(22), 5–8.

Geogdzhayev, I.V., et al., 2002: Global two-channel AVHRR retrievals of aerosol properties over the ocean for the period of NOAA-9 observations and preliminary retrievals using NOAA-7 and NOAA-11 data. *J. Atmos. Sci.*, **59**, 262–278.

Ghan, S., et al., 2001: Evaluation of aerosol direct radiative forcing in MIRAGE. *J. Geophys. Res.*, **106**(D6), 5295–5316.

Gillett, N.P., M.F. Wehner, S.F.B. Tett, and A.J. Weaver, 2004: Testing the linearity of the response to combined greenhouse gas and sulfate aerosol forcing. *Geophys. Res. Lett.*, **31**, L14201, doi:10.1029/2004GL020111.

Gleisner, H., and P. Thejll, 2003: Patterns of tropospheric response to solar variability. *Geophys. Res. Lett.*, **30**(13), 1711, doi:10.1029/2003GL017129.

Godal, O., 2003: The IPCC assessment of multidisciplinary issues: The choice of greenhouse gas indices. *Clim. Change*, **58**, 243–249.

Gohar, L.K., G. Myhre, and K.P. Shine, 2004: Updated radiative forcing estimates of four halocarbons. *J. Geophys. Res.*, **109**, D01107, doi:10.1029/2003JD004320.

Goloub, P., and O. Arino, 2000: Verification of the consistency of POLDER aerosol index over land with ATSR-2/ERS-2 fire product. *Geophys. Res. Lett.*, **27**(6), 899–902.

Goloub, P., et al., 1999: Validation of the first algorithm applied for deriving the aerosol properties over the ocean using the POLDER/ADEOS measurements. *IEEE Trans. Remote Sens.*, **37**, 1586–1596.

Goosse, H., H. Renssen, A. Timmermann, and R.S. Bradley, 2005: Internal and forced climate variability during the last millennium: a model-data comparison using ensemble simulations. *Quat. Sci. Rev.*, **24**, 1345–1360.

Gordon, L.J., et al., 2005: Human modification of global water vapor flows from the land surface. *Proc. Natl. Acad. Sci. U.S.A.*, **102**, 7612–7617.

Govindasamy, B., P.B. Duffy, and K. Caldeira, 2001a: Land use changes and Northern Hemisphere cooling. *Geophys. Res. Lett.*, **28**(2), 291–294.

Govindasamy, B., et al., 2001b: Limitations of the equivalent CO_2 approximation in climate change simulations. *J. Geophys. Res.*, **106**(D19), 22593–22603.

Gray, L.J., J.D. Haigh, and R.G. Harrison, 2005: *Review of the Influences of Solar Changes on the Earth's Climate.* Hadley Centre Technical Note No. 62, Met Office, Exeter, 82 pp.

Gregory, J.M., et al., 2004: A new method for diagnosing radiative forcing and climate sensitivity. *Geophys. Res. Lett.*, **31**, L03205, doi:10.1029/2003GL018747.

Gunn, R., and B.B. Phillips, 1957: An experimental investigation of the effect of air pollution on the initiation of rain. *J. Meteorol.*, **14**(3), 272–280.

Hadjinicolaou, P., J.A. Pyle, and N.R.P. Harris, 2005: The recent turnaround in stratospheric ozone over northern middle latitudes: a dynamical modeling perspective. *Geophys. Res. Lett.*, **32**, L12821, doi:10.1029/2005GL022476.

Haigh, J.D., 2003: The effects of solar variability on the Earth's climate. *Phil. Trans. R. Soc. London Ser. A*, **361**, 95–111.

Hall, J.C., and G.W. Lockwood, 2004: The chromospheric activity and variability of cycling and flat activity solar-analog stars. *Astrophys. J.*, **614**, 942–946.

Hamilton, J.F., et al., 2004: Partially oxidised organic components in urban aerosol using GCXGC-TOF/MS. *Atmos. Chem. Phys.*, **4**, 1279–1290.

Hammitt, J.K., A.K. Jain, J.L. Adams, and D.J. Wuebbles, 1996: A welfare-based index for assessing environmental effects of greenhouse-gas emissions. *Nature*, **381**, 301–303.

Han, Q., W.B. Rossow, J. Zeng, and R. Welch, 2002: Three different behaviors of liquid water path of water clouds in aerosol-cloud interactions. *J. Atmos. Sci.*, **59**, 726–735.

Hansen, J., and M. Sato, 2001: Trends of measured climate forcing agents. *Proc. Natl. Acad. Sci. U.S.A.*, **98**, 14778–14783.

Hansen, J., and L. Nazarenko, 2004: Soot climate forcing via snow and ice albedos. *Proc. Natl. Acad. Sci. U.S.A.*, **101**, 423–428.

Hansen, J., and M. Sato, 2004: Greenhouse gas growth rates. *Proc. Natl. Acad. Sci. U.S.A.*, **101**, 16109–16114.

Hansen, J., M. Sato, and R. Reudy, 1995: Long-term changes of the diurnal temperature cycle: Implications about mechanisms of global climate change. *Atmos. Res.*, **37**, 175–209.

Hansen, J., M. Sato, and R. Ruedy, 1997: Radiative forcing and climate response. *J. Geophys. Res.*, **102**(D6), 6831–6864.

Hansen, J., et al., 2000: Global warming in the twenty-first century: An alternative scenario. *Proc. Natl. Acad. Sci. U.S.A.*, **97**, 9875–9880.

Hansen, J., et al., 2002: Climate forcings in Goddard Institute for Space Studies SI2000 simulations. *J. Geophys. Res.*, **107**(D18), 4347, doi:10.1029/2001JD001143.

Hansen, J., et al., 2005: Efficacy of climate forcings. *J. Geophys. Res.*, **110**, D18104, doi:10.1029/2005JD005776.

Hansen, M.C., and B. Reed, 2000: A comparison of the IGBP DISCover and University of Maryland 1km global land cover products. *Int. J. Remote Sens.*, **21**, 1365–1373.

Harder, J., et al., 2005: The spectral irradiance monitor I. Scientific requirements, instrument design and operation modes. *Sol. Phys.*, **230**, 141–167.

Harding, R.J., and J.W. Pomeroy, 1996: The energy balance of the winter boreal landscape. *J. Clim.*, **9**, 2778–2787.

Harnisch, J., R.R. Borchers, P.P. Fabian, and M.M. Maiss, 1996: Tropospheric trends for CF_4 and C_2F_6 since 1982 derived from SF_6 dated stratospheric air. *Geophys. Res. Lett.*, **23**(10), 1099–1102.

Harries, J.E., H.E. Brindley, P.J. Sagoo, and R.J. Bantges, 2001: Increases in greenhouse forcing inferred from the outgoing longwave radiation spectra of the Earth in 1970 and 1997. *Nature*, **410**, 355–357.

Harrison, R.G., and K.S. Carslaw, 2003: Ion-aerosol-cloud processes in the lower atmosphere. *Rev. Geophys.*, **41**, 1021.

Harrison, R.G., and D.B. Stephenson, 2006: Empirical evidence for a nonlinear effect of galactic cosmic rays on clouds. *Proc. Roy. Soc. London Ser. A*, **462**, 1221–1233.

Hartmann, D.L., M.E. Ockert-Bell, and M.L. Michelsen, 1992: The effect of cloud type on Earth's energy balance: global analysis. *J. Clim.*, **5**, 1281–1304.

Hashvardhan, S.E. Schwartz, C.E. Benkovitz, and G. Guo, 2002: Aerosol influence on cloud microphysics examined by satellite measurements and chemical transport modelling. *J. Atmos. Sci.*, **59**, 714–725.

Hauglustaine, D.A., and G.P. Brasseur, 2001: Evolution of tropospheric ozone under anthropogenic activities and associated radiative forcing of climate. *J. Geophys. Res.*, **106**(D23), 32337–32360, doi:10.1029/2001JD900175.

Haxeltine, A., and I.C. Prentice, 1996: BIOME3: An equilibrium terrestrial biosphere model based on ecophysiological constraints, resource availability, and competition among plant functional types. *Global Biogeochem. Cycles*, **10**(4), 693–709.

Haywood, J.M., and K.P. Shine, 1995: The effect of anthropogenic sulfate and soot aerosol on the clear sky planetary radiation budget. *Geophys. Res. Lett.*, **22**(5), 603–606.

Haywood, J.M., and K.P. Shine. 1997: Multi-spectral calculations of the direct radiative forcing of tropospheric sulphate and soot aerosols using a column model. *Q. J. R. Meteorol. Soc.* **123**, 1907–1930.

Haywood, J.M., and O. Boucher, 2000: Estimates of the direct and indirect radiative forcing due to tropospheric aerosols: A review. *Rev. Geophys.*, **38**, 513–543.

Haywood, J.M., V. Ramaswamy, and B.J. Soden, 1999: Tropospheric aerosol climate forcing in clear-sky satellite observations over the oceans. *Science*, **283**, 1299–1305.

Haywood, J.M., et al., 2003a: Comparison of aerosol size distributions, radiative properties, and optical depths determined by aircraft observations and Sun photometers during SAFARI 2000. *J. Geophys. Res.*, **108**(D13), 8471, doi:10.1029/2002JD002250.

Haywood, J.M., et al., 2003b: The mean physical and optical properties of regional haze dominated by biomass burning aerosol measured from the C-130 aircraft during SAFARI 2000. *J. Geophys. Res.*, **108**(D13), 8473, doi:10.1029/2002JD002226.

Haywood, J.M, et al., 2003c: Radiative properties and direct radiative effect of Saharan dust measured by the C-130 aircraft during SHADE: 1. Solar spectrum. *J. Geophys. Res.*, **108**(D18), 8577, doi:10.1029/2002JD002687.

Haywood, J.M., et al., 2005: Can desert dust explain the outgoing longwave radiation anomaly over the Sahara during July 2003? *J. Geophys. Res.*, **110**, D05105, doi:10.1029/2004JD005232.

Heal, G., 1997: Discounting and climate change: an editorial comment. *Clim. Change*, **37**, 335–343.

Hegg, D.A., et al., 1997: Chemical apportionment of aerosol column optical depth off the mid-Atlantic coast of the United States. *J. Geophys. Res.*, **102**(D21), 25293–25303.

Hély, C., et al., 2003: Release of gaseous and particulate carbonaceous compound from biomass burning during the SAFARI 2000 dry season field campaign. *J. Geophys. Res.*, **108**(D13), 8470, doi:10.1029/2002JD002482.

Hendricks, J., B. Kärcher, U. Lohmann, and M. Ponater, 2005: Do aircraft black carbon emissions affect cirrus clouds on the global scale? *Geophys. Res. Lett.*, **32**, L12814, doi:10.1029/2005GL022740.

Hendricks, J., et al., 2004: Simulating the global atmospheric black carbon cycle: a revisit to the contribution of aircraft. *Atmos. Chem. Phys.*, **4**, 2521–2541.

Herman, M., et al., 1997: Remote sensing of aerosols over land surfaces including polarization measurements and application to POLDER measurements. *J. Geophys. Res.*, **102**(D14), 17039–17050, doi:10.1029/96JD02109.

Hess, M., P. Koepke, and I. Schult, 1998: Optical properties of aerosols and clouds: the software package OPAC. *Bull. Am. Meteorol. Soc.*, **79**, 831–844.

Highwood, E.J., et al., 2003: Radiative properties and direct effect of Saharan dust measured by the C-130 aircraft during Saharan Dust Experiment (SHADE): 2. Terrestrial spectrum. *J. Geophys. Res.*, **108**(D18), 8577, doi:10.1029/2002JD002552.

Higurashi, A., and T. Nakajima. 1999. Development of a two-channel aerosol retrieval algorithm on a global scale using NOAA AVHRR. *J. Atmos. Sci.*, **56**, 924–941.

Higurashi, A., et al. 2000: A study of global aerosol optical climatology with two-channel AVHRR remote sensing. *J. Clim.*, **13**, 2011–2027.

Hirsch, A.I., et al., 2006: Inverse modeling estimates of the global nitrous oxide surface flux from 1998–2001. *Global Biogeochem. Cycles*, **20**, GB1008, doi:10.1029/2004GB002443.

Hirst, E., et al., 2001: Discrimination of micrometre-sized ice and super-cooled droplets in mixed-phase cloud. *Atmos. Environ.*, **35**, 33–47.

Hofmann, D., 2003: et al., Surface-based observations of volcanic emissions to stratosphere. In: *Volcanism and the Earth's Atmosphere* [Robock, A., and C. Oppenheimer (eds.)]. Geophysical Monograph 139, American Geophysical Union, Washington, DC, pp. 57–73.

Holben, B.N., et al., 1998: AERONET: A federated instrument network and data archive for aerosol characterization. *Remote Sens. Environ.*, **66**, 1–16.

Holben, B.N., et al., 2001: An emerging ground-based aerosol climatology: aerosol optical depth from AERONET. *J. Geophys. Res.*, **106**(D11), 12067–12097.

Holzer-Popp, T., M. Schroedter, and G. Gesell, 2002: Retrieving aerosol optical depth and type in the boundary layer over land and ocean from simultaneous GOME spectrometer and ATSR-2 radiometer measurements - 2. Case study application and validation. *J. Geophys. Res.*, **107**(D21), 4578, doi:10.1029/2002JD002777.

Hood, L.L., 2003: Thermal response of the tropical tropopause region to solar ultraviolet variations. *Geophys. Res. Lett.*, **30**, 2215, doi:10.1029/2003GL018364.

Houghton, R.A., 1999: The annual net flux of carbon to the atmosphere from changes in land use 1850-1990. *Tellus*, **51B**, 298–313.

Houghton, R.A., 2000: A new estimate of global sources and sinks of carbon from land use change. *Eos*, **81**(19), S281.

Houghton, R.A., 2003: Revised estimates of the annual net flux of carbon to the atmosphere from changes in land use and land management. *Tellus*, **55B**, 378–390.

Houghton, R.A., et al., 1983: Changes in the carbon content of terrestrial biota and soils between 1860 and 1980: a net release of CO_2 to the atmosphere. *Ecol. Monogr.*, **53**, 235–262, doi:10.2307/1942531.

Hoyt, D.V., and K.H. Schatten, 1993: A discussion of plausible solar irradiance variations, 1700-1992. *J. Geophys. Res.*, **98**(A11), 18895–18906.

Hsu, N.C., J.R. Herman, and C. Weaver, 2000: Determination of radiative forcing of Saharan dust using combined TOMS and ERBE data. *J. Geophys. Res.*, **105**(D16), 20649–20661.

Hsu, N.C., J.R. Herman, and S.C. Tsay, 2003: Radiative impacts from biomass burning in the presence of clouds during boreal spring in southeast Asia. *Geophys. Res. Lett.*, **30**(5), 28, doi:10.1029/2002GL016485.

Huang, J., and R.G. Prinn, 2002: Critical evaluation of emissions of potential new gases for OH estimation. *J. Geophys. Res.*, **107**(D24), 4784, doi:10.1029/2002JD002394.

Huebert, B.J., et al., 2003: An overview of ACE-Asia: Strategies for quantifying the relationships between Asian aerosols and their climatic impacts. *J. Geophys. Res.*, **108**(D23), 8633, doi:10.1029/2003JD003550.

Hurley, M.D., et al., 2005: IR spectrum and radiative forcing of CF_4 revisited. *J. Geophys. Res.*, **110**, D02102, doi:10.1029/2004JD005201.

Hurst, D.F., et al., 2004: Emissions of ozone-depleting substances in Russia during 2001. *J. Geophys. Res.*, **109**, D14303, doi:10.1029/2004JD004633.

Husar, R.B., J.M. Prospero, and L.L. Stowe, 1997: Characterization of tropospheric aerosols over the oceans with the NOAA advanced very high resolution radiometer optical thickness operational product. *J. Geophys. Res.*, **102**(D14), 16889–16910.

Husar, R.B., et al., 2001: Asian dust events of April 1998. *J. Geophys. Res.*, **106**(D16), 18317–18330.

Ichinose, T., K. Shimodozono, and K. Hanaki, 1999: Impact of anthropogenic heat on urban climate in Tokyo. *Atmos. Environ.*, **33**, 3897–3909.

Ichoku, C., et al., 2003: MODIS observation of aerosols and estimation of aerosol radiative forcing over southern Africa during SAFARI 2000. *J. Geophys. Res.*, **108**(D13), 8499, doi:10.1029/2002JD002366.

Ignatov, A., and L. Stowe, 2002: Aerosol retrievals from individual AVHRR channels. Part I: Retrieval algorithm and transition from Dave to 6S Radiative Transfer Model. *J. Atmos. Sci.*, **59**, 313–334.

IPCC, 1990: *Climate Change: The Intergovernmental Panel on Climate Change Scientific Assessment* [Houghton, J.T., G.J. Jenkins, and J.J. Ephraums (eds.)]. Cambridge University Press, Cambridge, United Kingdom and New York, NY, USA, 364 pp.

IPCC, 1999: *Aviation and the Global Atmosphere: A Special Report of IPCC Working Groups I and III* [Penner, J.E., et al. (eds.)]. Cambridge University Press, Cambridge, United Kingdom and New York, NY, USA, 373 pp.

IPCC, 2001: *Climate Change 2001: The Scientific Basis. Contribution of Working Group I to the Third Assessment Report of the Intergovernmental Panel on Climate Change* [Houghton, J.T., et al. (eds.)]. Cambridge University Press, Cambridge, United Kingdom and New York, NY, USA, 881 pp.

IPCC/TEAP, 2005: *Special Report on Safeguarding the Ozone Layer and the Global Climate System: Issues Related to Hydrofluorocarbons and Perfluorocarbons* [Metz, B., et al. (eds.)]. Cambridge University Press, Cambridge, United Kingdom and New York, NY, USA, 488 pp.

Ito, A., and J.E. Penner, 2005: Historical emissions of carbonaceous aerosols from biomass and fossil fuel burning for the period 1870–2000. *Global Biogeochem. Cycles*, **19**, GB2028, doi:10.1029/2004GB002374.

Iversen, T., and O. Seland, 2002: A scheme for process-tagged SO_4 and BC aerosols in NCAR CCM3: Validation and sensitivity to cloud processes. *J. Geophys. Res.*, **107**(D24), 4751, doi:10.1029/2001JD000885.

Jacob, D.J., et al., 2005: *Radiative Forcing of Climate Change*. The National Academies Press, Washington, DC, 207 pp.

Jacobson, M.Z., 1999: Isolating nitrated and aromatic aerosols and nitrated aromatic gases as sources of ultraviolet light absorption. *J. Geophys. Res.*, **104**(D3), 3527–3542.

Jacobson, M.Z., 2001a: Global direct radiative forcing due to multicomponent anthropogenic and natural aerosols. *J. Geophys. Res.*, **106**(D2), 1551–1568.

Jacobson, M.Z., 2001b: Strong radiative heating due to the mixing state of black carbon in atmospheric aerosols. *Nature*, **409**, 695–697.

Jacobson, M.Z., 2002: Control of fossil-fuel particulate black carbon and organic matter, possibly the most effective method of slowing global warming. *J. Geophys. Res.*, **107**(D19), 4410, doi:10.1029/2001JD001376.

Jacobson, M.Z., 2004: Climate response of fossil fuel and biofuel soot, accounting for soot's feedback to snow and sea ice albedo and emissivity. *J. Geophys. Res.*, **109**, D21201, doi:10.1029/2004JD004945.

Jaffe, D., et al., 2003: Increasing background ozone during spring on the west coast of North America. *Geophys. Res. Lett.*, **30**, 1613, doi:10.1029/2003GL017024.

Jensen, E.J., and O.B. Toon, 1997: The potential impact of soot particles from aircraft exhaust on cirrus clouds. *Geophys. Res. Lett.*, **24**(3), 249–252.

Jeong, M.J., Z.Q. Li, D.A. Chu, and S.C. Tsay, 2005: Quality and compatibility analyses of global aerosol products derived from the advanced very high resolution radiometer and Moderate Resolution Imaging Spectroradiometer. *J. Geophys. Res.*, **110**, D10S09, doi:10.1029/2004JD004648.

Jiang, H., and G. Feingold, 2006: Effect of aerosol on warm convective clouds: Aerosol-clouds-surface flux feedbacks in a new coupled large eddy model. *J. Geophys. Res.*, **111**, D01202, doi:10.1029/2005JD006138.

Jiang, H., G. Feingold, and W.R. Cotton, 2002: Simulations of aerosol-cloud-dynamical feedbacks resulting from of entrainment of aerosols into the marine boundary layer during the Atlantic Stratocumulus Transition Experiment. *J. Geophys. Res.*, **107**(D24), 4813, doi:10.1029/2001JD001502.

Jirikowic, J.L., and P.E. Damon, 1994: The Medieval solar activity maximum. *Clim. Change*, **26**, 309–316.

Johnson, B.T., 2005: The semidirect aerosol effect: comparison of a single-column model with large eddy simulation for marine stratocumulus. *J. Clim.*, **18**, 119–130.

Johnson, B.T., K.P. Shine, and P.M. Forster, 2004: The semi-direct aerosol effect: Impact of absorbing aerosols on marine stratocumulus. *Q. J. R. Meteorol. Soc.*, **130**, 1407–1422.

Jones, A., D.L. Roberts, and M.J. Woodage, 2001: Indirect sulphate aerosol forcing in a climate model with an interactive sulphur cycle. *J. Geophys. Res.*, **106**(D17), 20293–20301.

Jones, P.D., A. Moberg, T.J. Osborn, and K.R. Briffa, 2003: Surface climate responses to explosive volcanic eruptions seen in long European temperature records and mid-to-high latitude tree-ring density around the Northern Hemisphere. In: *Volcanism and the Earth's Atmosphere* [Robock, A., and C. Oppenheimer (eds.)]. Geophysical Monograph 139, American Geophysical Union, Washington, DC, pp. 239–254.

Jonson, J.E., D. Simpson, H. Fagerli, and S. Solberg, 2005: Can we explain the trends in European ozone levels? *Atmos. Chem. Phys. Discuss.*, **5**, 5957–5985.

Joos, F., et al., 1996. An efficient and accurate representation of complex oceanic and biospheric models of anthropogenic carbon uptake. *Tellus*, **48B**, 397–417.

Joos, F., et al., 2001. Global warming feedbacks on terrestrial carbon uptake under the Intergovernmental Panel on Climate Change (IPCC) emission scenarios. *Global Biogeochem. Cycles*, **15**, 891-908, 2001

Joshi, M.M., and K.P. Shine, 2003: A GCM study of volcanic eruptions as a cause of increased stratospheric water vapor. *J. Clim.*, **16**, 3525–3534.

Joshi, M., et al., 2003: A comparison of climate response to different radiative forcings in three general circulation models: Towards an improved metric of climate change. *Clim. Dyn.*, **20**, 843–854.

Junge, C.E., 1975: The possible influence of aerosols on the general circulation and climate and possible approaches for modelling. In: *The Physical Basis of Climate and Climate Modelling: Report of the International Study Conference in Stockholm, 29 July-10 August 1974: organised by WMO and ICSU and supported by UNEP Global Atmospheric Research Programme (GARP), WMO-ICSU Joint Organising Committee.* World Meteorological Organization, Geneva, pp. 244–251.

Kahn, R., P. Banerjee, and D. McDonald, 2001: Sensitivity of multiangle imaging to natural mixtures of aerosols over ocean. *J. Geophys. Res.*, **106**(D16), 18219–18238.

Kahn, R.A., et al., 2005: Multiangle Imaging Spectroradiometer (MISR) global aerosol optical depth validation based on 2 years of coincident Aerosol Robotic Network (AERONET) observation. *J. Geophys. Res.*, **110**, D10S04, doi:10.1029/2004JD004706.

Kalkstein, A.J., and R.C. Balling Jr., 2004: Impact of unusually clear weather on United States daily temperature range following 9/11/2001. *Clim. Res.*, **26**, 1–4.

Kanakidou, M., et al., 2005: Organic aerosol and global climate modelling: a review. *Atmos. Chem. Phys.*, **5**, 1053–1123.

Kandlikar, M., 1996: Indices for comparing greenhouse gas emissions: integrating science and economics. *Energy Econ.*, **18**, 265–282.

Kapustin, V.N., et al., 2006: On the determination of a cloud condensation nuclei from satellite: challenges and possibilities. *J. Geophys. Res.*, **111**, D04202, doi:10.1029/2004JD005527.

Kärcher, B., 1999: Aviation-produced aerosols and contrails. *Surv. Geophys.*, **20**, 113–167.

Kasischke, E.S., and J.E. Penner, 2004: Improving global estimates of atmospheric emissions from biomass burning. *J. Geophys. Res.*, **109**, D14S01, doi:10.1029/2004JD004972.

Kaufman, Y.J., C.J. Tucker, and R.L Mahoney, 1991: Fossil fuel and biomass burning effect on climate: Heating or cooling? *J. Clim.*, **4**, 578–588.

Kaufman, Y.J., D. Tanré, and O. Boucher, 2002: A satellite view of aerosols in the climate system. *Nature*, **419**, 215–223.

Kaufman, Y.J., et al., 1997: Operational remote sensing of tropospheric aerosols over the land from EOS-MODIS. *J. Geophys. Res.*, **102**(D14), 17051–17068.

Kaufman, Y.J., et al., 2001: Absorption of sunlight by dust as inferred from satellite and ground-based remote sensing. *Geophys. Res. Lett.*, **28**(8), 1479–1482.

Kaufman Y.J., et al., 2005a: Aerosol anthropogenic component estimated from satellite data. *Geophys. Res. Lett.*, **32**, L17804, doi:10.1029/2005GL023125.

Kaufman, Y.J., et al., 2005b: A critical examination of the residual cloud contamination and diurnal sampling effects on MODIS estimates of aerosol over ocean. *IEEE Trans. Geosci. Remote*, **43**(12), 2886–2897.

Keeling, C.D., and T.P. Whorf, 2005: Atmospheric CO_2 records from sites in the SIO air sampling network. In: *Trends: A Compendium of Data on Global Change.* Carbon Dioxide Information Analysis Center, Oak Ridge National Laboratory, U.S. Department of Energy, Oak Ridge, TN, http://cdiac.esd.ornl.gov/trends/co2/sio-keel-flask/sio-keel-flask.html.

Keeling, C.D., A.F. Bollenbacher, and T.P. Whorf, 2005: Monthly atmospheric $^{13}C/^{12}C$ isotopic ratios for 10 SIO stations. In: *Trends: A Compendium of Data on Global Change.* Carbon Dioxide Information Analysis Center, Oak Ridge National Laboratory, U.S. Department of Energy, Oak Ridge, TN, http://cdiac.ornl.gov/trends/co2/iso-sio/iso-sio.html.

Keeling, C.D., T.P. Whorf, M. Wahlen, and J. van der Plicht, 1995: Interannual extremes in the rate of rise of atmospheric carbon dioxide since 1980. *Nature*, **375**, 666–670.

Keeling, R.F., and S.R. Shertz, 1992: Seasonal and interannual variations in atmospheric oxygen and implications for the global carbon cycle. *Nature*, **358**, 723–727.

Keeling, R.F., S.C. Piper, and M. Heimann, 1996: Global and hemispheric CO_2 sinks deduced from changes in atmospheric O_2 concentration. *Nature*, **381**, 218–221.

Keil, A., and J.M. Haywood, 2003: Solar radiative forcing by biomass burning aerosol particles during SAFARI 2000: A case study based on measured aerosol and cloud properties. *J. Geophys. Res.*, **108** (D13), 8467, doi:10.1029/2002JD002315.

Keppler, F., J.T.G. Hamilton, M. Brass, and T. Röckmann, 2006: Methane emissions from terrestrial plants under aerobic conditions. *Nature*, **439**, 187–191, doi:10.1038/nature04420.

Kernthaler, S.C., R. Toumi, and J.D. Haigh, 1999: Some doubts concerning a link between cosmic ray fluxes and global cloudiness. *Geophys. Res. Lett.*, **26**(7), 863–866, doi:10.1029/1999GL900121.

Khairoutdinov, M., and Y. Kogan, 2000: A new cloud physics parametrization in a large-eddy simulation model of marine stratocumulus. *Mon. Weather Rev.*, **128**, 229–243.

Khalil, M.A.K., et al., 2003: Atmospheric perfluorocarbons. *Environ. Sci. Technol.*, **37**, 4358–4361.

Kiehl, J.T., et al., 2000: Radiative forcing due to sulfate aerosols from simulations with the National Center for Atmospheric Research Community Climate Model, Version 3. *J. Geophys. Res.*, **105**(D1), 1441–1457.

Kim, B.-G., S.E. Schwartz, and M.A. Miller, 2003: Effective radius of cloud droplets by ground-based remote sensing: Relationship to aerosol. *J. Geophys. Res.*, **108**(D23), doi:10.1029/2003JD003721.

Kinne, S., et al., 2003: Monthly averages of aerosol properties: a global comparison among models, satellite data, and AERONET ground data. *J. Geophys. Res.*, **108** (D20), 4634, doi:10.1029/2001JD001253.

Kinne, S., et al., 2006: An AeroCom initial assessment: optical properties in aerosol component modules of global models. *Atmos. Chem. Phys.*, **6**, 1815–1834.

Kirchstetter, T.W., T. Novakov, and P.V. Hobbs, 2004: Evidence that the spectral dependence of light absorption by aerosols is affected by organic carbon. *J. Geophys. Res.*, **109**, D21208, doi:10.1029/2004JD004999.

Kirkevåg, A., and T. Iversen, 2002: Global direct radiative forcing by process-parametrized aerosol optical properties. *J. Geophys. Res.*, **107**(D20), 4433, doi:10.1029/2001JD000886.

Klein Goldewijk, K., 2001: Estimating global land use change over the past 300 years: The HYDE database. *Global Biogeochem. Cycles*, **15**, 417–433.

Knutson, T.R., et al., 2006: Assessment of twentieth-century regional surface temperature trends using the GFDL CM2 coupled models. *J. Clim.*, **10**(9), 1624–1651.

Koch, D., 2001: Transport and direct radiative forcing of carbonaceous and sulfate aerosols in the GISS GCM. *J. Geophys. Res.*, **106**(D17), 20311–20332.

Korolev, A.V., and G.A. Isaac, 2005: Shattering during sampling OAPs and HVPS. Part I: snow particles. *J. Atmos. Ocean. Technol.*, **22**, 528–542.

Kotchenruther, R.A., and P.V. Hobbs, 1998: Humidification factors of aerosols from biomass burning in Brazil. *J. Geophys. Res.*, **103**(D24), 32081–32089.

Kotchenruther, R.A., P.V. Hobbs, and D.A. Hegg, 1999: Humidification factors for atmospheric aerosols off the mid-Atlantic coast of the United States. *J. Geophys. Res.*, **104**(D2), 2239–2251.

Kristjánsson, J.E., and J. Kristiansen, 2000: Is there a cosmic ray signal in recent variations in global cloudiness and cloud radiative forcing? *J. Geophys. Res.*, **105**(D9), 11851–11863.

Kristjánsson, J.E., A. Staple, J. Kristiansen, and E. Kaas, 2002: A new look at possible connections between solar activity, clouds and climate. *Geophys. Res. Lett.*, **29**, doi:10.1029/2002GL015646.

Kristjánsson, J.E., et al., 2005: Response of the climate system to aerosol direct and indirect forcing: the role of cloud feedbacks. *J. Geophys. Res.*, **110**, D24206, doi:10.1029/2005JD006299.

Krizan P., and J. Lastovicka, 2005: Trends in positive and negative ozone laminae in the Northern Hemisphere. *J. Geophys. Res.*, **110**, D10107, doi:10.1029/2004JD005477.

Kroeze, C., E. Dumont, and S.P. Seitzinger, 2005: New estimates of global emissions of N_2O from rivers, estuaries and continental shelves. *Environ. Sci.*, 2(2–3), 159–165.

Krol, M., and J. Lelieveld, 2003: Can the variability in tropospheric OH be deduced from measurements of 1,1,1-trichloroethane (methyl chloroform)? *J. Geophys. Res.*, **108**(D3), 4125, doi:10.1029/2002JD002423.

Krol, M., et al., 2003: Continuing emissions of methyl chloroform from Europe. *Nature*, **421**, 131–135.

Krueger, A.J., et al., 2000: Ultraviolet remote sensing of volcanic emissions. In: *Remote Sensing of Active Volcanism* [Mouginis-Mark, P.J., J.A. Crisp, and J.H. Fink (eds.)]. Geophysical Monograph 116, American Geophysical Union, Washington, DC, pp. 25–43.

Krüger, O., and H. Graßl, 2002: The indirect aerosol effect over Europe. *Geophys. Res. Lett.*, **29**, doi:10.1029/2001GL14081.

Labitzke, K., 2004: On the signal of the 11-year sunspot cycle in the stratosphere and its modulation by the quasi-biennial oscillation. *J. Atmos. Solar Terr. Phys.*, **66**, 1151–1157.

Labitzke, K., et al., 2002: The global signal of the 11-year solar cycle in the stratosphere: observations and models. *J. Atmos. Solar Terr. Phys.*, **64**, 203–210.

Lambert, A., et al., 1993: Measurements of the evolution of the Mt. Pinatubo aerosol cloud by ISAMS. *Geophys. Res. Lett.*, **20**(12), 1287–1290.

Langenfelds, R.L., et al., 2002: Interannual growth rate variations of atmospheric CO_2 and its delta C-13, H-2, CH_4, and CO between 1992 and 1999 linked to biomass burning. *Global Biogeochem. Cycles*, **16**, doi:10.1029/2001GB001466.

Lawrence, M.G., P. Jöckel, and R. von Kuhlmann, 2001: What does the global mean OH concentration tell us? *Atmos. Chem. Phys.*, **1**, 43–74.

Lean, J., 2000: Evolution of the sun's spectral irradiance since the Maunder Minimum. *Geophys. Res. Lett.*, **27**, 2425–2428.

Lean, J., A. Skumanich, and O. White, 1992: Estimating the sun's radiative output during the Maunder Minimum. *Geophys. Res. Lett.*, **19**(15), 1595–1598.

Lean, J., J. Beer, and R. Bradley, 1995: Reconstruction of solar irradiance since 1610: implications for climate change. *Geophys. Res. Lett.*, **22**, 3195–3198.

Lean, J.L., Y.M. Wang, and N.R. Sheeley, 2002: The effect of increasing solar activity on the Sun's total and open magnetic flux during multiple cycles: Implications for solar forcing of climate. *Geophys. Res. Lett.*, **29**(24), 2224, doi:10.1029/2002GL015880.

Lean, J., G. Rottman, J. Harder, and G. Kopp, 2005: SORCE contribution to new understanding of global change and solar variability. *Solar Phys.*, **230**, 27–53.

Lean, J.L., et al., 1997: Detection and parametrization of variations in solar mid and near ultraviolet radiation (200 to 400 nm). *J. Geophys. Res.*, **102**(D25), 29939–29956.

Lee, K.H., Y.J. Kim, and W. von Hoyningen-Huene, 2004: Estimation of regional aerosol optical thickness from satellite observations during the 2001 ACE-Asia IOP. *J. Geophys. Res.* **109**, D19S16, doi:10.1029/2003JD004126.

Lee, R.B. III, M.A. Gibson, R.S. Wilson, and S. Thomas, 1995: Long-term total solar irradiance variability during sunspot cycle 22. *J. Geophys. Res.*, **100**(A2), 1667–1675.

Lefohn, A.S., J.D. Husar, and R.B. Husar, 1999: Estimating historical anthropogenic global sulfur emission patterns for the period 1850-1990. *Atmos. Environ.*, **33**, 3435–3444.

Lelieveld, J., P.J. Crutzen, and F.J. Dentener, 1998: Changing concentration, lifetime and climate forcing of atmospheric methane. *Tellus*, **50B**, 128–150.

Lelieveld, J., W. Peters, F.J. Dentener, and M.C. Krol, 2002: Stability of tropospheric hydroxyl chemistry. *J. Geophys. Res.*, **107**(D23), 4715, doi:10.1029/2002JD002272.

Lelieveld, J., et al., 2004: Increasing ozone over the Atlantic Ocean. *Science*, **304**, 1483–1487.

Levy, R.C., et al., 2003: Evaluation of the Moderate-Resolution Imaging Spectroradiometer (MODIS) retrievals of dust aerosol over the ocean during PRIDE. *J. Geophys. Res.*, **108**(D19), 8594, doi:10.1029/2002JD002460.

Li, J., et al., 2005: Halocarbon emissions estimated from AGAGE measured pollution events at Trinidad Head, California. *J. Geophys. Res.*, **110**, D14308, doi:10.1029/2004JD005739.

Liao, H., and J.H. Seinfeld, 2005: Global impacts of gas-phase chemistry-aerosol interactions on direct radiative forcing by anthropogenic aerosols and ozone. *J. Geophys. Res.*, **110**, D18208, doi:10.1029/2005JD005907.

Liao, H., J.H. Seinfeld, P.J. Adams, and L.J. Mickley, 2004: Global radiative forcing of coupled tropospheric ozone and aerosols in a unified general circulation model. *J. Geophys. Res.*, **109**, D16207, doi:10.1029/2003JD004456.

Liao, T., C.D. Camp, and Y.L. Yung, 2004: The seasonal cycle of N_2O. *Geophys. Res. Lett.*, **31**, L17108, doi:10.1029/2004GL020345.

Liousse, C., et al., 1996: A global three-dimensional model study of carbonaceous aerosols. *J. Geophys. Res.*, **101**(D14), 19411–19432.

Liu, H.Q., R.T. Pinker, and B.N. Holben, 2005: A global view of aerosols from merged transport models, satellite, and ground observations. *J. Geophys. Res.*, **110**(D10), doi:10.1029/2004JD004695.

Liu, X.H., and J.E. Penner, 2002: Effect of Mount Pinatubo H_2SO_4/H_2O aerosol on ice nucleation in the upper troposphere using a global chemistry and transport model. *J. Geophys. Res.*, **107**(D12), doi:10.1029/2001JD000455.

Liu, Y., and P.H. Daum, 2002: Indirect warming effect from dispersion forcing. *Nature*, **419**, 580–581.

Lockwood, M., and R. Stamper, 1999: Long-term drift of the coronal source magnetic flux and the total solar irradiance. *Geophys. Res. Lett.*, **26**(16), 2461–2464.

Loeb, N.G., and S. Kato, 2002: Top-of-atmosphere direct radiative effect of aerosols over the tropical oceans from the Clouds and the Earth's Radiant Energy System (CERES) satellite instrument. *J. Clim.*, **15**, 1474–1484.

Loeb, N.G., and N. Manalo-Smith, 2005: Top-of-atmosphere direct radiative effect of aerosols over global oceans from merged CERES and MODIS observations. *J. Clim.*, **18**, 3506–3526.

Lohar, D., and B. Pal, 1995: The effect of irrigation on premonsoon season precipitation over south west Bengal, India. *J. Clim.*, **8**, 2567–2570.

Lohmann, U., and G. Lesins, 2002: Stronger constraints on the anthropogenic indirect aerosol effect. *Science*, **298**, 1012–1016.

Lohmann, U., and J. Feichter, 2005: Global indirect aerosol effects: A review. *Atmos. Chem. Phys.*, **5**, 715–737.

Lohmann, U. and C. Leck, 2005: Importance of submicron surface active organic aerosols for pristine Arctic clouds. *Tellus*, **57B**, 261–268.

Lohmann, U., B. Kärcher, and J. Hendrichs, 2004: Sensitivity studies of cirrus clouds formed by heterogeneous freezing in ECHAM GCM. *J. Geophys. Res.*, **109**, D16204, doi:10.1029/2003JD004443.

Lohmann, U., J. Feichter, J.E. Penner, and W.R. Leaitch, 2000: Indirect effect of sulfate and carbonaceous aerosols: a mechanistic treatment. *J. Geophys. Res.*, **105**(D10), 12193–12206.

Lohmann, U., et al., 2001: Vertical distributions of sulfur species simulated by large scale atmospheric models in COSAM: comparison with observations. *Tellus*, **53B**, 646–672.

Loveland, T.R., et al., 2000: Development of a global land cover characteristics database and IGBP DISCover from 1 km AVHR R data. *Int. J. Remote Sens.*, **21**, 1303–1330.

Lowe, D.C., and W. Allan, 2002: A simple procedure for evaluating global cosmogenic ^{14}C production in the atmosphere using neutron monitor data. *Radiocarbon*, **44**, 149–157.

Lowe, D.C., M.R. Manning, G.W. Brailsford, and A.M. Bromley, 1997: The 1991-1992 atmospheric methane anomaly: Southern hemisphere ^{13}C decrease and growth rate fluctuations. *Geophys. Res. Lett.*, **24**(8), 857–860.

Lowe, D.C., et al., 2004: Seasonal cycles of mixing ratio and ^{13}C in atmospheric methane at Suva, Fiji. *J. Geophys. Res.*, **109**, D23308 doi:10.1029/2004JD005166.

Lu, M.-L., and J. Seinfeld, 2005: Study of the aerosol indirect effect by large-eddy simulation of marine stratocumulus. *J. Atmos. Sci.*, **62**, 3909–3932.

Luo, C., N.M. Mahowald, and J. del Corral, 2003: Sensitivity study of meteorological parameters on mineral aerosol mobilization, transport, and distribution, *J. Geophys. Res.*, **108** (D15), 4447, doi:10.1029/2003JD003483.

MacFarling Meure, C., et al., 2006: The Law Dome CO_2, CH_4 and N_2O ice core records extended to 2000 years BP. *Geophys. Res. Lett.*, **33**, L14810, doi:10.1029/2006GL026152.

Macke, A., M. Mishchenko, and B. Cairns, 1996: The influence of inclusions on light scattering by large ice particles. *J. Geophys. Res.*, **101**(D18), 23311–23316.

Magi, B.I., and P.V. Hobbs, 2003: Effects of humidity on aerosols in southern Africa during the biomass burning season. *J. Geophys. Res.*, **108**(D13), 8495, doi:10.1029/2002JD002144.

Magi, B.I., et al., 2005: Properties and chemical apportionment of aerosol optical depth at locations off the U.S. East Coast in July and August 2001. *J. Atmos. Sci.*, **62**(4), 919–933, doi:10.1175/JAS3263.1.

Mahowald, N.M., and C. Luo, 2003: A less dusty future? *Geophys. Res. Lett.*, **30**(17), doi:10.1029/2003GL017880.

Mahowald, N.M., G.C. Rivera, and C. Luo, 2004: Comment on Tegen et al. 2004, on the "Relative importance of climate and land use in determining present and future global soil dust emissions". *Geophys. Res. Lett.*, **31**, L24105, doi:10.1029/2004GL021272.

Mak, J.E., M.R. Manning, and D.C. Lowe, 2000: Aircraft observations of $\delta^{13}C$ of atmospheric methane over the Pacific in August 1991 and 1993: evidence of an enrichment in $^{13}CH_4$ in the Southern Hemisphere. *J. Geophys. Res.*, **105**(D1), 1329–1335.

Malm, W.C., et al., 2004: Spatial and monthly trends in speciated fine particle concentration in the United States. *J. Geophys. Res.*, **109**, D03306, doi:10.1029/2003JD003739.

Manabe, S., and R.T. Wetherald, 1967: Thermal equilibrium of the atmosphere with a given distribution of relative humidity. *J. Atmos. Sci.*, **24**, 241–259.

Manne, A.S., and R.G. Richels, 2001: An alternative approach to establishing trade-offs among greenhouse gases. *Nature*, **410**, 675–676.

Manning, A.C., and R.F. Keeling, 2006: Global oceanic and land biotic carbon sinks from the Scripps atmospheric oxygen flask sampling network. *Tellus*, **58B**, 95–116.

Manning, A.J., et al., 2003: Estimating European emissions of ozone-depleting and greenhouse gases using observations and a modelling back-attribution technique. *J. Geophys. Res.*, **108** (D14), 4405, doi:10.1029/2002JD002312.

Manning, M.R., A. Gomez, and G.W. Brailsford, 1997: Annex B11: The New Zealand CO_2 measurement programme. In: *Report of the Ninth WMO Meeting of Experts on Carbon Dioxide Concentration and Related Tracer Measurement Techniques*. WMO Global Atmosphere Watch No. 132; WMO TD No. 952, Commonwealth Scientific and Industrial Research Organisation, Melbourne, pp. 120–123.

Manning, M.R., et al., 2005: Short term variations in the oxidizing power of the atmosphere. *Nature*, **436**, 1001–1004.

Mannstein, H., and U. Schumann, 2005: Aircraft induced contrail cirrus over Europe. *Meteorol. Z.*, **14**, 549–544.

Maria, S., F. Russell, L.M. Turpin, and R.J. Porcja, 2002: FTIR measurements of functional groups and organic mass in aerosol samples over the Caribbean. *Atmos. Environ.*, **36**, 5185–5196.

Marland, G., T.A. Boden, and R.J. Andres, 2006: Global, regional, and national CO_2 emissions. In: *Trends: A Compendium of Data on Global Change*. Carbon Dioxide Information Analysis Center, Oak Ridge National Laboratory, U.S. Department of Energy, Oak Ridge, TN, http://cdiac.esd.ornl.gov/trends/emis/tre_glob.htm.

Marquart, S., and B. Mayer, 2002: Towards a reliable GCM estimation of contrail radiative forcing. *Geophys. Res. Lett.*, **29**(8), doi:10.1029/2001GL014075.

Marquart, S., M. Ponater, F. Mager, and R. Sausen, 2003: Future development of contrail cover, optical depth, and radiative forcing: Impacts of increasing air traffic and climate change. *J. Clim.*, **16**, 2890–2904.

Marsh, N.D., and H. Svensmark, 2000a: Low cloud properties influenced by cosmic rays. *Phys. Rev. Lett.*, **85**, 5004–5007.

Marsh, N.D., and H. Svensmark, 2000b: Cosmic rays, clouds, and climate. *Space Sci. Rev.*, **94**, 215–230.

Marshak, A., et al., 2006: Impact of three-dimensional radiative effects on satellite retrievals of cloud droplet sizes. *J. Geophys. Res.*, **111**, D09207, doi:10.1029/2005JD006686.

Martin, G.M., D.W. Johnson, and A. Spice, 1994: The measurement and parametrization of effective radius of droplets in warm stratiform clouds. *J. Atmos. Sci.*, **51**, 1823–1842.

Martin, S.T., et al., 2004: Effects of the physical state of tropospheric ammonium-sulfate-nitrate particles on global aerosol direct radiative forcing. *Atmos. Chem. Phys.*, **4**, 183–214.

Martonchik, J.V., et al., 2004: Comparison of MISR and AERONET aerosol optical depths over desert sites. *Geophys. Res. Lett.*, **31**, L16102, doi:10.1029/2004GL019807.

Matthews, E., 1983: Global vegetation and land-use: new high-resolution data-bases for climate studies. *J. Clim. Appl. Meteorol.*, **22**, 474–487.

Matthews, H.D., A.J. Weaver, M. Eby, and K.J. Meissner, 2003: Radiative forcing of climate by historical land cover change. *Geophys. Res. Lett.*, **30**(2), 271–274.

Matthews, H.D., et al., 2004: Natural and anthropogenic climate change: Incorporating historical land cover change, vegetation dynamics and the global carbon cycle. *Clim. Dyn.*, **22**, 461–479.

Matthias, I., et al., 2004: Multiyear aerosol observations with dual wavelength Raman lidar in the framework of EARLINET. *J. Geophys. Res.*, **109**, D13203, doi:10.1029/2004JD004600.

McCormack, J.P., 2003: The influence of the 11-year solar cycle on the quasi-biennial oscillation. *Geophys. Res. Lett.*, **30** (22), 2162, doi:10.1029/2003GL018314.

McCormick, M.P., 1987: SAGE II: An overview. *Adv. Space Res.*, **7**, 219–226.

McCulloch, A., and P.M. Midgley, 2001: The history of methyl chloroform emissions: 1951-2000. *Atmos. Environ.*, **35**, 5311–5319.

McFiggans, G., et al., 2006: The effect of aerosol composition and properties on warm cloud droplet activation. *Atmos. Chem. Phys.*, **6**, 2593–2649.

Meehl, G.A., et al., 2004: Combinations of natural and anthropogenic forcings in twentieth-century climate. *J. Clim.*, **17**, 3721–3727.

Meerkötter, R., et al., 1999: Radiative forcing by contrails. *Ann. Geophys.*, **17**, 1080–1094.

Melillo, J.M., et al., 2001: Nitrous oxide emissions from forests and pastures of various ages in the Brazilian Amazon. *J. Geophys. Res.*, **106**(D24), 34179–34188.

Menon, S., and A. Del Genio, 2007: Evaluating the impacts of carbonaceous aerosols on clouds and climate. In: *Human-Induced Climate Change: An Interdisciplinary Assessment* [Schlesinger, M., et al. (eds.)]. Cambridge University Press, Cambridge, UK, in press.

Menon, S., A.D. Del Genio, D. Koch, and G. Tselioudis, 2002a: GCM simulations of the aerosol indirect effect: sensitivity to cloud parametrization and aerosol burden. *J. Atmos. Sci.*, **59**, 692–713.

Menon, S., J. Hansen, L. Nazarenko, and Y. Luo, 2002b: Climate effects of black carbon aerosols in China and India. *Science*, **297**, 2250–2253.

Menon, S., et al., 2003: Evaluating aerosol/cloud/radiation process parametrizations with single-column models and Second Aerosol Characterization Experiment (ACE-2) cloudy column observations. *J. Geophys. Res.*, **108**(D24), 4762, doi:10.1029/2003JD003902.

Meyer, R., et al., 2002: Regional radiative forcing by line-shaped contrails derived from satellite data. *J. Geophys. Res.*, **107**(D10), 4104, doi:10.1029/2001JD000426.

Mickley, L.J., D.J. Jacob, and D. Rind, 2001: Uncertainty in preindustrial abundance of tropospheric ozone: implications for radiative forcing calculations. *J. Geophys. Res.*, **106**(D4), 3389–3399 doi:10.1029/2000JD900594.

Mickley, L.J., D.J. Jacob, B.D. Field, and D. Rind, 2004: Climate response to the increase in tropospheric ozone since preindustrial times: a comparison between ozone and equivalent CO_2 forcings. *J. Geophys. Res.*, **109**, D05106, doi:10.1029/2003JD003653.

Mikami, M., et al., 2006: Aeolian dust experiment on climate impact: an overview of Japan-China Joint Project ADEC. *Global Planet. Change*, **52**, 142–172, doi:10.1016/j.gloplacha.2006.03.001.

Miller, B.R., et al., 1998: Atmospheric trend and lifetime of chlorodifluoromethane (HCFC-22) and the global tropospheric OH concentration. *J. Geophys. Res.*, **103**(D11), 13237–13248, doi:10.1029/98JD00771.

Miller, R.L., G.A. Schmidt, and D.T. Shindell, 2006: Forced annular variations in the 20th century IPCC AR4 models. *J. Geophys. Res.*, **111**, D18101, doi:10.1029/2005JD006323.

Millet, D.B., and A.H. Goldstein, 2004: Evidence of continuing methylchloroform emissions from the United States. *Geophys. Res. Lett.*, **31**, L17101, doi:10.1029/2004GL020166.

Ming, Y., and L.M. Russell, 2002: Thermodynamic equilibrium of organic-electrolyte mixtures in aerosol particles. *Am. Inst. Chem. Eng. J.*, **48**, 1331–1348.

Ming, Y., V. Ramaswamy, P.A. Ginoux, and L.H. Horowitz, 2005a: Direct radiative forcing of anthropogenic organic aerosol. *J. Geophys. Res.*, **110**, D20208, doi:10.1029/2004JD005573.

Ming, Y., V. Ramaswamy, L.J. Donner, and V.T.J. Phillips, 2006: A new parametrization of cloud droplet activation applicable to general circulation models. *J. Atmos. Sci.*, **63**(4), 1348–1356.

Ming, Y., et al., 2005b: Geophysical Fluid Dynamics Laboratory general circulation model investigation of the indirect radiative effects of anthropogenic sulfate aerosol. *J. Geophys. Res.*, **110**, D22206, doi:10.1029/2005JD006161.

Ming, Y., et al., 2007: Modelling the interactions between aerosols and liquid water clouds with a self-consistent cloud scheme in a general circulation model. *J. Atmos. Sci.*, **64**(4), 1189–1209.

Minnis, P., 2005: Reply. *J. Clim.*, **18**, 2783–2784.

Minnis, P., J.K. Ayers, R. Palikonda, and D. Phan, 2004: Contrails, cirrus trends, and climate. *J. Clim.*, **17**, 1671–1685.

Minnis, P., et al., 1993: Radiative climate forcing by the Mt. Pinatubo eruption. *Science*, **259**, 1411–1415.

Minnis, P., et al., 1998: Transformation of contrails into cirrus during SUCCESS. *Geophys. Res. Lett.*, **25**, 1157–1160.

Mircea, M., et al., 2005: Importance of the organic aerosol fraction for modeling aerosol hygroscopic growth and activation: a case study in the Amazon Basin. *Atmos. Chem. Phys.*, **5**, 3111–3126.

Mishchenko, M.I., et al., 1999: Aerosol retrievals over the ocean by use of channels 1 and 2 AVHRR data: sensitivity analysis and preliminary results. *Appl. Opt.*, **38**, 7325–7341.

Möhler, O., et al., 2005: Effect of sulfuric acid coating on heterogeneous ice nucleation by soot aerosol particles. *J. Geophys. Res.*, **110**, D11210, doi:10.1029/2004JD005169.

Montzka, S.A., et al., 1999: Present and future trends in the atmospheric burden of ozone-depleting halogens. *Nature*, **398**, 690–694.

Montzka, S.A., et al., 2003: Controlled substances and other source gases. In: *Scientific Assessment of Ozone Depletion: 2002*. World Meteorological Organization, Geneva, pp. 1.1–1.83.

Moore, N., and S. Rojstaczer, 2001: Irrigation-induced rainfall and the Great Plains. *J. Appl. Meteorol.*, **40**, 1297–1309.

Morgan, C.G., et al., 2004: Isotopic fractionation of nitrous oxide in the stratosphere: Comparison between model and observations. *J. Geophys. Res.*, **109**, D04305, doi:10.1029/2003JD003402.

Morimoto, S., S. Aoki, T. Nakazawa, and T. Yamanouchi, 2006: Temporal variations of the carbon isotopic ratio of atmospheric methane observed at Ny Ålesund, Svalbard from 1996 to 2004. *Geophys. Res. Lett.*, **33**, L01807, doi:10.1029/2005GL024648.

Mosley-Thompson, E., T.A. Mashiotta, and L.G. Thompson, 2003: High resolution ice core records of Late Holocene volcanism: Current and future contributions from the Greenland PARCA cores. In: *Volcanism and the Earth's Atmosphere* [Robock, A., and C. Oppenheimer (eds.)]. Geophysical Monograph 139, American Geophysical Union, Washington, DC, pp. 153–164.

Mouillot, F., et al., 2006: Global carbon emissions from biomass burning in the 20th century. *Geophys. Res. Lett.*, **33**, L01801, doi:10.1029/2005GL024707.

Moulin, C., and I. Chiapello, 2004: Evidence of the control of summer atmospheric transport of African dust over the Atlantic by Sahel sources from TOMS satellites (1979-2000). *Geophys. Res. Lett.*, **31**, L02107, doi:10.1029/2003GL018931.

Moulin, C., H.R. Gordon, V.F. Banzon, and R.H. Evans, 2001: Assessment of Saharan dust absorption in the visible from SeaWiFS imagery. *J. Geophys. Res.*, **106**(D16), 18239–18249.

Murayama, T., et al., 2001: Ground-based network observation of Asian dust events of April 1998 in East Asia. *J. Geophys. Res.*, **106**, 18345–18360.

Murphy, D.M., 2005: Something in the air. *Science*, **307**, 1888–1890.

Muscheler, R., et al., 2007: Solar activity during the last 1000 yr inferred from radionuclide records. *Quat. Sci. Rev.*, **26**, 82-97, doi:10.1016/j.quascirev.2006.07.012.

Myhre, G., and F. Stordal, 2001a: Global sensitivity experiments of the radiative forcing due to mineral aerosols. *J. Geophys. Res.*, **106**, 18193–18204.

Myhre, G., and F. Stordal, 2001b: On the tradeoff of the solar and thermal infrared radiative impact of contrails. *Geophys. Res. Lett.*, **28**, 3119–3122.

Myhre, G., and A. Myhre, 2003: Uncertainties in radiative forcing due to surface albedo changes caused by land-use changes. *J. Clim.*, **16**, 1511–1524.

Myhre, G., A. Myhre, and F. Stordal, 2001: Historical evolution of radiative forcing of climate. *Atmos. Environ.*, **35**, 2361–2373.

Myhre, G., M.M. Kvalevåg, and C.B. Schaaf, 2005a: Radiative forcing due to anthropogenic vegetation change based on MODIS surface albedo data. *Geophys. Res. Lett.*, **32**, L21410, doi:10.1029/2005GL024004.

Myhre, G., et al., 2003: Modelling the solar radiative impact of aerosols from biomass burning during the Southern African Regional Science Initiative (SAFARI 2000) experiment. *J. Geophys. Res.*, **108**, 8501, doi:10.1029/2002JD002313.

Myhre, G., et al., 2004a: Intercomparison of satellite retrieved aerosol optical depth over ocean. *J. Atmos. Sci.*, **61**, 499–513.

Myhre, G., et al., 2004b: Uncertainties in the radiative forcing due to sulfate aerosols. *J. Atmos. Sci.*, **61**, 485–498.

Myhre, G., et al., 2005b: Intercomparison of satellite retrieved aerosol optical depth over ocean during the period September 1997 to December 2000. *Atmos. Chem. Phys.*, **5**, 1697–1719.

Naja, M., and H. Akimoto, 2004: Contribution of regional pollution and long-range transport to the Asia-Pacific region: Analysis of long-term ozonesonde data over Japan. *J. Geophys. Res.*, **109**, 1306, doi:10.1029/2004JD004687.

Naja, M., H. Akimoto, and J. Staehelin, 2003: Ozone in background and photochemically aged air over central Europe: analysis of long-term ozonesonde data from Hohenpeissenberg and Payerne. *J. Geophys. Res.*, **108**, 4063, doi:10.1029/2002JD002477.

Nakajima, T., and A. Higurashi, 1998: A use of two-channel radiances for an aerosol characterization from space. *Geophys. Res. Lett.*, **25**, 3815–3818.

Nakajima, T., A. Higurashi, K. Kawamoto, and J. Penner, 2001: A possible correlation between satellite-derived cloud and aerosol microphysical parameters. *Geophys. Res. Lett.*, **28**, 1171–1174.

Nakajima, T., et al., 1996: Aerosol optical properties in the Iranian region obtained by ground-based solar radiation measurements in the summer of 1991. *J. Appl. Meteorol.*, **35**, 1265–1278.

Nakazawa, T., S. Moromoto, S. Aoki, and M. Tanaka, 1997: Temporal and spatial variations of the carbon isotopic ratio of atmospheric carbon dioxide in the Western Pacific region. *J. Geophys. Res.*, **102**, 1271–1285.

Nakicenovic, N., A. Grübler, and A. McDonald (eds), 1998: *Global Energy Perspectives*. Cambridge University Press, New York, NY, 299 pp.

Nenes, A., and J.H. Seinfeld, 2003: Parametrization of cloud droplet formation in global climate models. *J. Geophys. Res.*, **108**, doi:10.1029/2002JD002911.

Nenes, A., et al., 2002: Can chemical effects on cloud droplet number rival the first indirect effect? *Geophys. Res. Lett.*, **29**, 1848, doi:10.1029/2002GL015295.

Nevison, C.D., S. Solomon, and R.S. Gao, 1999: Buffering interactions in the modeled response of stratospheric O_3 to increased NO_x and HO_x. *J. Geophys. Res.*, **104**(D3), 3741–3754, 10.1029/1998JD100018.

Nevison, C.D., D.E. Kinnison, and R.F. Weiss, 2004a: Stratospheric influences on the tropospheric seasonal cycles of nitrous oxide and chlorofluorocarbons. *Geophys. Res. Lett.*, **31**, L20103, doi:10.1029/2004GL020398.

Nevison, C., T. Lueker, and R.F. Weiss, 2004b: Quantifying the nitrous oxide source from coastal upwelling. *Global Biogeochem. Cycles*, **18**, GB1018, doi:10.1029/2003GB002110.

Nevison, C.D., et al., 2005: Southern Ocean ventilation inferred from seasonal cycles of atmospheric N_2O and O_2/N_2 at Cape Grim, Tasmania. *Tellus*, **57B**, 218–229.

Newchurch, M.J., et al., 2003: Evidence for slowdown in stratospheric ozone loss: first stage of ozone recovery. *J. Geophys. Res.*, **108**(D16), 4507, doi:10.1029/2003JD003471.

Nordhaus, W.D., 1997: Discounting in economics and climate change: an editorial comment. *Clim. Change*, **37**, 315–328.

Notholt, J., et al., 2005: Influence of tropospheric SO_2 emissions on particle formation and the stratospheric humidity. *Geophys. Res. Lett.*, **32**, L07810, doi:10.1029/2004GL022159.

Novakov, T., D.A. Hegg, and P.V. Hobbs, 1997: Airborne measurements of carbonaceous aerosols on the East coast of the United States. *J. Geophys. Res.*, **102**(D25), 30023–20030.

Novakov, T., et al., 2003: Large historical changes of fossil-fuel black carbon aerosols. *Geophys. Res. Lett.*, **30**(6), 1324, doi:10.1029/2002GL016345.

Nozawa, T., T. Nagashima, H. Shiogama, and S.A. Crooks, 2005: Detecting natural influence on surface air temperature change in the early twentieth century. *Geophys. Res. Lett.*, **32**, L20719, doi:10.1029/2005GL023540.

O'Doherty, S., et al., 2004: Rapid growth of hydrofluorocarbon 134a and hydrochlorofluorocarbons 141b, 142b, and 22 from Advanced Global Atmospheric Gases Experiment (AGAGE) observations at Cape Grim, Tasmania, and Mace Head, Ireland. *J. Geophys. Res.*, **109**, D06310, doi:10.1029/2003JD004277.

Oinas, V., et al., 2001: Radiative cooling by stratospheric water vapor: big differences in GCM results. *Geophys. Res. Lett.*, **28**, 2791–2794.

Olivier, J.G.J., and J.J.M. Berdowski, 2001: Global emissions sources and sinks. In: *The Climate System* [Berdowski, J., R. Guicherit, and B.J. Heij (eds.)]. A.A. Balkema/Swets & Zeitlinger, Lisse, The Netherlands, pp. 33–78, updated at http://www.mnp.nl/edgar/.

Oltmans, S.J., et al., 2006: Long-term changes in tropospheric ozone. *Atmos. Environ.*, **40**, 3156–3173.

O'Neill, B.C., 2000: The jury is still out on global warming potentials. *Clim. Change*, **44**, 427–443.

O'Neill, B., 2003: Economics, natural science, and the costs of global warming potentials. *Clim. Change*, **58**, 251–260.

Oram, D.E., 1999: *Trends of Long-Lived Anthropogenic Halocarbons in the Southern Hemisphere and Model Calculations of Global Emissions*. PhD Thesis, University of East Anglia, Norwich, UK, 249 pp.

Oram, D.E., et al., 1998: Growth of fluoroform (CHF_3, HFC-23) in the background atmosphere. *Geophys. Res. Lett.*, **25**, 35–38, doi:10.1029/97GL03483.

Osborne, S.R., J.M. Haywood, P.N. Francis, and O. Dubovik, 2004: Short-wave radiative effects of biomass burning aerosol during SAFARI2000. *Q. J. R. Meteorol. Soc.*, **130**, 1423–1448.

Palikonda, R., P. Minnis, D.P. Duda, and H. Mannstein, 2005: Contrail coverage derived from 2001 AVHRR data over the continental United States of America and surrounding area. *Meteorol. Z.*, **14**, 525–536.

Palmer, A.S., et al., 2002: Antarctic volcanic flux ratios from Law Dome ice cores. *Ann. Glaciol.*, **35**, 329–332.

Palmer, P.I., et al., 2003: Eastern Asian emissions of anthropogenic halocarbons deduced from aircraft concentration data. *J. Geophys. Res.*, **108**(D24), 4753, doi:10.1029/2003JD003591.

Pan, W., M.A. Tatang, G.J. McRae, and R.G. Prinn, 1997: Uncertainty analysis of direct radiative forcing by anthropogenic sulfate aerosols. *J. Geophys. Res.*, **102**(D18), 21915–21924.

Parrish, D.D., et al., 2004: Changes in the photochemical environment of the temperate North Pacific troposphere in response to increased Asian emissions. *J. Geophys. Res.*, **109**, D23S18, doi:10.1029/2004JD004978.

Pavelin, E.G., C.E. Johnson, S. Rughooputh, and R. Toumi, 1999: Evaluation of pre-industrial surface ozone measurements made using Schönbein's method. *Atmos. Environ.*, **33**, 919–929.

Peng, Y., and U. Lohmann, 2003: Sensitivity study of the spectral dispersion of the cloud droplet size distribution on the indirect aerosol effect. *Geophys. Res. Lett.*, **30**(10), 1507, doi:10.1029/2003GL017192.

Penner, J.E., S.Y. Zhang, and C.C. Chuang, 2003: Soot and smoke aerosol may not warm climate. *J. Geophys. Res.*, **108**(D21), 4657, doi:10.1029/2003JD003409.

Penner, J.E., X. Dong, and Y. Chen, 2004: Observational evidence of a change in radiative forcing due to the indirect aerosol effect. *Nature*, **427**, 231–234.

Penner, J.E., et al., 2001: Aerosols, their direct and indirect effects. In: *Climate Change 2001: The Scientific Basis. Contribution of Working Group I to the Third Assessment Report of the Intergovernmental Panel on Climate Change* [Houghton, J.T., et al. (eds.)]. Cambridge University Press, Cambridge, United Kingdom and New York, NY, USA, pp. 289–348.

Penner, J. E., et al., 2002: A comparison of model- and satellite-derived aerosol optical depth and reflectivity. *J. Atmos. Sci.*, **59**, 441–460.

Penner, J.E., et al., 2006: Model intercomparison of indirect aerosol effects. *Atmos. Chem. Phys. Discuss.*, **6**, 1579–1617.

Penner, J.E., et al., 2007: Effect of black carbon on mid-troposphere and surface temperature trends. In: *Human-Induced Climate Change: An Interdisciplinary Assessment* [Schlesinger, M., et al., (eds.)]. Cambridge University Press, Cambridge, UK, in press.

Perlwitz, J., and H.-F. Graf, 2001: Troposphere-stratosphere dynamic coupling under strong and weak polar vortex conditions. *Geophys. Res. Lett.*, **28**, 271–274.

Perlwitz, J., and N. Harnik, 2003: Observational evidence of a stratospheric influence on the troposphere by planetary wave reflection. *J. Clim.*, **16**, 3011–3026.

Pham, M., O. Boucher, and D. Hauglustaine. 2005: Changes in atmospheric sulfur burdens and concentrations and resulting radiative forcings under IPCC SRES emission scenarios for 1990-2100. *J. Geophys. Res.*, **110**, D06112, doi:10.1029/2004JD005125.

Philipona, R., et al., 2004: Radiative forcing - measured at Earth's surface - corroborate the increasing greenhouse effect. *Geophys. Res. Lett.*, **31**, L03202, doi:10.1029/2003GL018765.

Pielke, R.A. Sr., et al., 2002: The influence of land-use change and landscape dynamics on the climate system - relevance to climate change policy beyond the radiative effect of greenhouse gases. *Philos. Trans. R. Soc. London Ser. A*, **360**, 1705–1719.

Pitari, G., E. Mancini, V. Rizi, and D.T. Shindell, 2002: Impact of future climate and emissions changes on stratospheric aerosols and ozone. *J. Atmos. Sci.*, **59**, 414–440.

Platt, U., W. Allan, and D.C. Lowe, 2004: Hemispheric average Cl atom concentration from $^{13}C/^{12}C$ ratios in atmospheric methane. *Atmos. Chem. Phys.*, **4**, 2393–2399.

Ponater, M., S. Marquart, and R. Sausen, 2002: Contrails in a comprehensive global climate model: Parametrization and radiative forcing results. *J. Geophys. Res.*, **107**(D13), doi:10.1029/2001JD000429.

Ponater, M., S. Marquart, R. Sausen, and U. Schumann, 2005: On contrail climate sensitivity. *Geophys. Res. Lett.*, **32**, L10706, doi:10.1029/2005GL022580.

Posfai, M., et al., 2003: Individual aerosol particles from biomass burning in southern Africa: 1. Compositions and size distributions of carbonaceous particles. *J. Geophys. Res.*, **108**(D13), 8483, doi:10.1029/2002JD002291.

Prata, A., W. Rose, S. Self, and D. O'Brien, 2003: Global, long-term sulphur dioxide measurements from TOVS data: a new tool for studying explosive volcanism and climate. In: *Volcanism and the Earth's Atmosphere* [Robock, A., and C. Oppenheimer (eds).]. Geophysical Monograph 139, American Geophysical Union, Washington, DC, pp. 75–92.

Prather, M.J., 2003: Atmospheric science: an environmental experiment with H_2? *Science*, **302**, 581–583.

Prather, M.J., et al., 2001: Atmospheric chemistry and greenhouse gases. In: *Climate Change 2001: The Scientific Basis. Contribution of Working Group I to the Third Assessment Report of the Intergovernmental Panel on Climate Change* [Houghton, J.T., et al. (eds.)]. Cambridge University Press, Cambridge, United Kingdom and New York, NY, USA, pp. 239–287.

Preminger, D.G., and S.R. Walton, 2005: A new model of total solar irradiance based on sunspot areas. *Geophys. Res. Lett.*, **32**, L14109, doi:10.1029/2005GL022839.

Prentice, I.C., et al., 2001: The carbon cycle and atmospheric carbon dioxide. In: *Climate Change 2001: The Scientific Basis. Contribution of Working Group I to the Third Assessment Report of the Intergovernmental Panel on Climate Change* [Houghton, J.T., et al. (eds.)]. Cambridge University Press, Cambridge, United Kingdom and New York, NY, USA, pp. 184–238.

Prinn, R.G., 2004: Non-CO_2 greenhouse gases. In: *The Global Carbon Cycle* [Field, C., and M. Raupach (eds.)]. Island Press, Washington, DC, pp. 205–216.

Prinn, R.G., et al., 1990: Atmospheric emissions and trends of nitrous oxide deduced from ten years of ALE-GAGE data. *J. Geophys. Res.*, **95**, 18369–18385.

Prinn, R.G., et al., 2000: A history of chemically and radiatively important gases in air deduced from ALE/GAGE/AGAGE. *J. Geophys. Res.*, **105**(D14), 17751–17792.

Prinn, R.G., et al., 2001: Evidence for substantial variations of atmospheric hydroxyl radicals in the past two decades. *Science*, **292**, 1882–1888.

Prinn, R.G., et al., 2005a: Evidence for variability of atmospheric hydroxyl radicals over the past quarter century. *Geophys. Res. Lett.*, **32**, L07809, doi:10.1029/2004GL022228.

Prinn, R.G., et al., 2005b: *The ALE/GAGE/AGAGE Network: DB1001.* Carbon Dioxide Information and Analysis World Data Center, http://cdiac.esd.ornl.gov/ndps/alegage.html.

Prospero, J.M., et al., 2002: Environmental characterization of global sources of atmospheric soil dust identified with the Nimbus 7 Total Ozone Mapping Spectrometer (TOMS) absorbing aerosol product. *Rev. Geophys.*, **40**(1), doi:10.1029/2000RG000095.

Putaud, J.P., et al., 2004: European aerosol phenomenology-2: chemical characteristics of particulate matter at kerbside, urban, rural and background sites in Europe. *Atmos. Environ.*, **38**, 2579–2595.

Pyle, J., et al., 2005: Ozone and climate: a review of interconnections. In: *Special Report on Safeguarding the Ozone Layer and Global Climate System: Issues Related to Hydrofluorocarbons and Perfluorocarbons* [Metz., B., et al. (eds.)]. Cambridge University Press, Cambridge, United Kingdom and New York, NY, USA, pp. 83–132.

Quaas, J., and O. Boucher, 2005: Constraining the first aerosol indirect radiative forcing in the LMDZ GCM using POLDER and MODIS satellite data. *Geophys. Res. Lett.*, **32**, L17814, doi:10.1029/2005GL023850.

Quaas, J., O. Boucher, and F.-M. Breon, 2004: Aerosol indirect effects in POLDER satellite data and the Laboratoire de Météorologie Dynamique-Zoom (LMDZ) general circulation model. *J. Geophys. Res.*, **109**, D08205, doi:10.1029/2003JD004317.

Quaas, J., O. Boucher, and U. Lohmann, 2005: Constraining the total aerosol indirect effect in the LMDZ and ECHAM4 GCMs using MODIS satellite data. *Atmos. Chem. Phys.*, **5**, 9669–9690.

Quay, P., et al., 2000: Atmospheric (CO)-C-14: A tracer of OH concentration and mixing rates. *J. Geophys. Res.*, **105**(D12), 15147–15166.

Quinn, P.K., and T.S. Bates, 2005: Regional aerosol properties: comparisons from ACE 1, ACE 2, Aerosols99, INDOEX, ACE Asia, TARFOX, and NEAQS. *J. Geophys. Res.*, **110**, D14202, doi:10.1029/2004JD004755.

Raes, F., T. Bates, F. McGovern, and M. Van Liedekerke, 2000: The 2nd Aerosol Characterization Experiment (ACE-2): general overview and main results. *Tellus*, **52B**, 111–125.

Ramachandran. S., V. Ramaswamy, G.L. Stenchikov, and A. Robock, 2000: Radiative impact of the Mt. Pinatubo volcanic eruption: Lower stratospheric response. *J. Geophys. Res.*, **105**(D19), 24409–24429.

Ramanathan, V., 1981: The role of ocean-atmosphere interactions in the CO_2 climate problem. *J. Atmos. Sci.*, **38**, 918–930.

Ramanathan, V., P.J. Crutzen, J.T. Kiehl, and D. Rosenfeld, 2001a: Atmosphere: aerosols, climate, and the hydrological cycle. *Science*, **294**, 2119–2124.

Ramanathan, V., et al., 2001b: Indian Ocean experiment: An integrated analysis of the climate forcing and effects of the great Indo-Asian haze. *J. Geophys. Res.*, **106**(D22), 28371–28398.

Ramankutty, N., and J.A. Foley, 1999: Estimating historical changes in global land cover: croplands from 1700 to 1992. *Global Biogeochem. Cycles*, **14**, 997–1027.

Ramaswamy, V., S. Ramachandran, G. Stenchikov, and A. Robock, 2006a: A model study of the effect of Pinatubo volcanic aerosols on the stratospheric temperatures. In: *Frontiers of Climate Modeling* [Kiehl, J.T., and V. Ramanathan (eds.)]. Cambridge University Press, Cambridge, UK, pp. 152–178.

Ramaswamy, V., et al., 2001: Radiative forcing of climate change. In: *Climate Change 2001: The Scientific Basis. Contribution of Working Group I to the Third Assessment Report of the Intergovernmental Panel on Climate Change* [Houghton, J.T., et al. (eds.)]. Cambridge University Press, Cambridge, United Kingdom and New York, NY, USA, pp. 349–416.

Ramaswamy, V., et al., 2006b: Anthropogenic and natural influences in the evolution of lower stratospheric cooling. *Science*, **311**, 1138–1141.

Randall, C.E., R.M. Bevilacqua, J.D. Lumpe, and K.W. Hoppel, 2001: Validation of POAM III aerosols: Comparison to SAGE II and HALOE. *J. Geophys. Res.*, **106**, 27525–27536.

Randles, C.A., L.M. Russell, and V. Ramaswamy, 2004: Hygroscopic and optical properties of organic sea salt aerosol and consequences for climate forcing. *Geophys. Res. Lett.*, **31**, L16108, doi:10.1029/2004GL020628.

Rasch, P.J., and J.E. Kristjánsson, 1998. A comparison of the CCM3 model climate using diagnosed and predicted condensate parametrizations. *J. Clim.*, **11**, 1587–1614.

Read, W.G., L. Froidevaux, and J.W. Waters, 1993: Microwave limb sounder measurements of stratospheric SO_2 from the Mt. Pinatubo volcano. *Geophys. Res. Lett.*, **20**(12), 1299–1302.

Reddy, M.S., and O. Boucher, 2004: A study of the global cycle of carbonaceous aerosols in the LMDZT general circulation model. *J. Geophys. Res.*, **109**, D14202, doi:10.1029/2003JD004048.

Reddy, M.S., O. Boucher, Y. Balanski, and M. Schulz, 2005a: Aerosol optical depths and direct radiative perturbations by species and source type. *Geophys. Res. Lett.*, **32**, L12803, doi:10.1029/2004GL021743.

Reddy, M.S., et al., 2005b: Estimates of global multicomponent aerosol optical depth and direct radiative perturbation in the Laboratoire de Météorologie Dynamique general circulation model. *J. Geophys. Res.*, **110**, D10S16, doi:10.1029/2004JD004757.

Reid, J.S., et al., 1999: Use of the Angstrom exponent to estimate the variability of optical and physical properties of aging smoke particles in Brazil. *J. Geophys. Res.*, **104**(D22), 27473–27490.

Reid, J.S., et al., 2003: Analysis of measurements of Saharan dust by airborne and ground-based remote sensing methods during the Puerto Rico Dust Experiment (PRIDE). *J. Geophys. Res.*, **108**(D19), 8586, doi:10.1029/2002JD002493.

Reimann, S., et al., 2005: Low methyl chloroform emissions inferred from long-term atmospheric measurements. *Nature*, **433**, 506–508, doi:10.1038/nature03220.

Remer, L.A., and Y.J. Kaufman, 2006: Aerosol direct radiative effect at the top of the atmosphere over cloud free oceans derived from four years of MODIS data. *Atmos. Chem. Phys.*, **6**, 237–253.

Remer, L.A., et al., 2002: Validation of MODIS aerosol retrieval over ocean. *Geophys. Res. Lett.*, **29**(12), doi:10.1029/2001GL013204.

Remer, L.A., et al., 2005: The MODIS aerosol algorithm, products, and validation. *J. Atmos. Sci.*, **62**, 947–973.

Richards, J.F, 1990: Land transformation. In: *The Earth as Transformed by Human Action* [Turner, B.L. II, et al. (eds.)]. Cambridge University Press, New York, NY, pp. 163–178.

Richter, A., et al., 2005: Increase in tropospheric nitrogen dioxide over China observed from space. *Nature*, **437**, 129–132.

Rigozo, N.R., E. Echer, L.E.A. Vieira, and D.J.R. Nordemann, 2001: Reconstruction of Wolf sunspot numbers on the basis of spectral characteristics and estimates of associated radio flux and solar wind parameters for the last millennium. *Sol. Phys.*, **203**, 179–191.

Rind, D., J. Perlwitz, and P. Lonergan, 2005: AO/NAO response to climate change: 1. Respective influences of stratospheric and tropospheric climate changes. *J. Geophys. Res.*, **110**, D12107, doi:10.1029/2004JD005103.

Rissler, J., et al., 2004: Physical properties of the sub-micrometer aerosol over the Amazon rain forest during the wet-to-dry season transition – Comparison of modeled and measured CCN concentrations. *Atmos. Chem. Phys.*, **4**, 2119–2143.

Roberts, D.L., and A. Jones, 2004: Climate sensitivity to black carbon aerosol from fossil fuel combustion. *J. Geophys. Res.*, **109**, D16202, doi:10.1029/2004JD004676.

Robles-Gonzalez, C., J.P. Veefkind, and G. de Leeuw, 2000: Mean aerosol optical depth over Europe in August 1997 derived from ATSR-2 data. *Geophys. Res. Lett.* **27**(7), 955–959.

Robson, J.I., et al., 2006: Revised IR spectrum, radiative efficiency and global warming potential of nitrogen trifluoride. *Geophys. Res. Lett.*, **33**, L10817, doi:10.1029/2006GL026210.

Rockmann, T., J. Grooss, and R. Muller, 2004: The impact of anthropogenic chlorine emissions, stratospheric ozone change and chemical feedbacks on stratospheric water. *Atmos. Chem. Phys.*, **4**, 693–699.

Rosenfeld, D., and G. Feingold, 2003: Explanation of discrepancies among satellite observations of the aerosol indirect effects. *Geophys. Res. Lett.*, **30**(14), 1776, doi:10.1029/2003GL017684.

Rosenfeld, D., R. Lahav, A. Khain, and M. Pinsky, 2002: The role of sea spray in cleansing air pollution over the ocean via cloud processes. *Science*, **297**, 1667–1670.

Ross, K.E., et al., 2003: Spatial and seasonal variations in CCN distribution and the aerosol-CCN relationship over southern Africa. *J. Geophys. Res.*, **108**(D13), 8481, doi:10.1029/2002JD002384.

Rotstayn, L.D., 1997: A physically based scheme for the treatment of stratiform clouds and precipitation in large-scale models. I: Description and evaluation of the microphysical processes. *Q. J. R. Meteorol. Soc.*, **123**, 1227–1282.

Rotstayn, L.D., and J.E. Penner, 2001: Indirect aerosol forcing, quasi forcing, and climate response. *J. Clim.*, **14**, 2960–2975.

Rotstayn, L.D., and Y. Liu, 2003: Sensitivity of the first indirect aerosol effect to an increase of the cloud droplet spectral dispersion with droplet number concentration. *J. Clim.*, **16**, 3476–3481.

Rotstayn, L.D., B.F. Ryan, and J. Katzfey, 2000: A scheme for calculation of the liquid fraction in mixed-phase clouds in large scale models. *Mon. Weather Rev.*, **128**, 1070–1088.

Rottman, G., 2005: The SORCE Mission. *Solar Phys.*, **230**, 7–25.

Rozanov, E.V., et al., 2002: Climate/chemistry effect of the Pinatubo volcanic eruption simulated by the UIUC stratosphere/troposphere GCM with interactive photochemistry. *J. Geophys. Res.*, **107**, 4594, doi:10.1029/2001JD000974.

Rozanov, E.V., et al., 2004: Atmospheric response to the observed increase of solar UV radiation from solar minimum to solar maximum simulated by the University of Illinois at Urbana-Champaign climate-chemistry model. *J. Geophys. Res.*, **109**, D01110, doi:10.1029/2003JD003796.

Russell, P.B., and J. Heintzenberg, 2000: An overview of the ACE-2 clear sky column closure experiment (CLEARCOLUMN). *Tellus*, **52B**, 463–483, doi:10.1034/j.1600-0889.2000.00013.x.

Russell, P.B., P.V. Hobbs, and L.L. Stowe, 1999: Aerosol properties and radiative effects in the United States East Coast haze plume: An overview of the Tropospheric Aerosol Radiative Forcing Observational Experiment (TARFOX). *J. Geophys. Res.*, **104**(D2), 2213–2222.

Rypdal, K., et al., 2005: Tropospheric ozone and aerosols in climate agreements: scientific and political challenges. *Environ. Sci. Policy*, **8**, 29–43.

Salby, M., and P. Callaghan, 2004: Evidence of the solar cycle in the general circulation of the stratosphere. *J. Clim.*, **17**, 34–46.

Santer, B.D., et al., 2004: Identification of anthropogenic climate change using a second-generation reanalysis. *J. Geophys. Res.*, **109**, D21104, doi:10.1029/2004JD005075.

Sato, M., J.E. Hansen, M.P. McCormick, and J.B. Pollack, 1993: Stratospheric aerosol optical depths, 1850-1990. *J. Geophys. Res.*, **98**(D12), 22987–22994.

Sato, M., et al., 2003: Global atmospheric black carbon inferred from AERONET. *Proc. Natl. Acad. Sci. U.S.A.*, **100**, 6319–6324.

Sausen, R., and U. Schumann, 2000: Estimates of the climate response to aircraft CO_2 and $NO(x)$ emissions scenarios. *Clim. Change*, **44**, 27–58.

Sausen, R., K. Gierens, M. Ponater, and U. Schumann, 1998: A diagnostic study of the global distribution of contrails part I: Present day climate. *Theor. Appl. Climatol.*, **61**, 127–141.

Sausen, R., et al., 2005: Aviation radiative forcing in 2000: An update on IPCC (1999). *Meteorol. Z.*, **14**, 1–7.

Schaaf, C.B., et al., 2002: First operational BRDF, albedo nadir reflectance products from MODIS. *Remote Sens. Environ.*, **83**, 135–148.

Schaap, M., et al., 2004: Secondary inorganic aerosol simulations for Europe with special attention to nitrate. *Atmos. Chem. Phys.*, **4**, 857–874.

Schatten, K.H., and J.A. Orosz, 1990: Solar constant secular changes. *Sol. Phys.*, **125**, 179–184.

Schmidt, G.A., et al., 2005: Present day atmospheric simulations using GISS ModelE: Comparison to in situ, satellite and reanalysis data. *J. Clim.*, **19**, 153–192.

Schnaiter, M., et al., 2003: UV-VIS-NIR spectral optical properties of soot and soot-containing aerosols. *J. Aerosol Sci.*, **34**(10), 1421–1444.

Schoeberl, M., A. Douglass, Z. Zhu, and S. Pawson, 2003: A comparison of the lower stratospheric age spectra derived from a general circulation model and two data assimilation systems. *J. Geophys. Res.*, **108**, L4113, doi:10.1029/2002JD002652.

Scholes, M., and M.O. Andreae, 2000: Biogenic and pyrogenic emissions from Africa and their impact on the global atmosphere. *Ambio*, **29**, 23–29.

Schulz, M., S. Kinne, C. Textor, and S. Guibert, 2004: *AeroCom Aerosol Model Intercomparison*. http://nansen.ipsl.jussieu.fr/AEROCOM/.

Schulz, M., et al., 2006: Radiative forcing by aerosols as derived from the AeroCom present-day and pre-industrial simulations. *Atmos. Chem. Phys. Discuss.*, **6**, 5095–5136.

Schumann, U., 2005: Formation, properties, and climatic effects of contrails. *Comptes Rendus Physique*, **6**, 549–565.

Schuster, G.L., O. Dubovik, B.N. Holben, and E.E. Clothiaux, 2005: Inferring black carbon content and specific absorption from Aerosol Robotic Network (AERONET) aerosol retrievals. *J. Geophys. Res.*, **110**, D10S17, doi:10.1029/2004JD004548.

Schwartz, S.E., and M.O. Andreae, 1996: Uncertainty in climate change caused by aerosols. *Science*, **272**, 1121–1122.

Schwartz, S.E., D.W. Harshvardhan, and C.M. Benkovitz, 2002: Influence of anthropogenic aerosol on cloud optical depth and albedo shown by satellite measurements and chemical transport modeling. *Proc. Natl. Acad. Sci. U.S.A.*, **99**, 1784–1789.

Sekiguchi, M., et al., 2003: A study of the direct and indirect effects of aerosols using global satellite datasets of aerosol and cloud parameters. *J. Geophys. Res.*, **108**(D22), 4699, doi:10.1029/2002JD003359.

Sellers, P.J., et al., 1996: Comparison of radiative and physiological effects of doubled atmospheric CO_2 on climate. *Science*, **271**, 1402–1406.

Shantz, N.C., W.R. Leaitch, and P. Caffrey, 2003: Effect of organics of low solubility on the growth rate of cloud droplets. *J. Geophys. Res.*, **108**(D5), doi:10.1029/2002JD002540.

Sherwood, S., 2002: A microphysical connection among biomass burning, cumulus clouds, and stratospheric moisture. *Science*, **295**, 1272–1275.

Shi, G.Y., et al., 2005: Sensitivity experiments on the effects of optical properties of dust aerosols on their radiative forcing under clear sky condition. *J. Meteorol. Soc. Japan*, **83A**, 333–346.

Shindell, D.T., 2001: Climate and ozone response to increased stratospheric water vapor. *Geophys. Res. Lett.*, **28**, 1551–1554.

Shindell, D.T., and G. Faluvegi, 2002: An exploration of ozone changes and their radiative forcing prior to the chlorofluorocarbon era. *Atmos. Chem. Phys.*, **2**, 363–374.

Shindell, D.T., G. Faluvegi, and N. Bell, 2003a: Preindustrial-to-present day radiative forcing by tropospheric ozone from improved simulations with the GISS chemistry-climate GCM. *Atmos. Chem. Phys.*, **3**, 1675–1702.

Shindell, D.T., G.A. Schmidt, R.L. Miller, and M. Mann, 2003b: Volcanic and solar forcing of climate change during the preindustrial era. *J. Clim.*, **16**, 4094–4107.

Shindell, D.T., G.A. Schmidt, M. Mann, and G. Faluvegi, 2004: Dynamic winter climate response to large tropical volcanic eruptions since 1600. *J. Geophys. Res.*, **109**, D05104, doi:10.1029/2003JD004151.

Shindell, D.T., G. Faluvegi, N. Bell, and G. Schmidt, 2005: An emissions-based view of climate forcing by methane and tropospheric ozone. *Geophys. Res. Lett.*, **32**, L04803, doi:10.1029/2004GL021900.

Shine, K.P., 2005: Comment on 'Contrails, cirrus, trends, and climate'. *J. Clim.*, **18**, 2781–2782.

Shine, K.P., J. Cook, E.J. Highwood, and M.M. Joshi, 2003: An alternative to radiative forcing for estimating the relative importance of climate change mechanisms. *Geophys. Res. Lett.*, **30** (20), 2047, doi:10.1029/2003GL018141.

Shine, K.P., T.K. Berntsen, J.S. Fuglestvedt, and R. Sausen, 2005a: Scientific issues in the design of metrics for inclusion of oxides of nitrogen in global climate agreements. *Proc. Natl. Acad. Sci. U.S.A.*, **102**, 15768–15773.

Shine, K.P., J.S. Fuglestvedt, K. Hailemariam, and N. Stuber, 2005b: Alternatives to the global warming potential for comparing climate impacts of emissions of greenhouse gases. *Clim. Change*, **68**, 281–302.

Shine, K.P, et al., 2005c: Perfluorodecalin: global warming potential and first detection in the atmosphere. *Atmos. Environ.*, **39**, 1759–1763.

Simmonds, P.G., R. Derwent, A. Manning, and G. Spain, 2004: Significant growth in surface ozone at Mace Head, Ireland ,1987-2003. *Atmos. Environ.*, **38**, 4769–4778.

Simpson, I.J., D.R. Blake, F.S. Rowland, and T.Y. Chen, 2002: Implications of the recent fluctuations in the growth rate of tropospheric methane. *Geophys. Res. Lett.*, **29**(10), doi:10.1029/2001GL014521.

Smith, C.A., J.D. Haigh, and R. Toumi, 2001: Radiative forcing due to trends in stratospheric water vapour. *Geophys. Res. Lett.*, **28**(1), 179–182.

Smith, R.N.B., 1990: A scheme for predicting layer clouds and their water content in a general circulation model. *Q. J. R. Meteorol. Soc.*, **116**, 435–460.

Smith, S.J., and T.M.L. Wigley, 2000: Global warming potentials: 2. Accuracy. *Clim. Change*, **44**, 459–469.

Smith, S.J., E. Conception, R. Andres, and J. Lurz, 2004: *Historical Sulphur Dioxide Emissions 1850-2000: Methods and Results*. Research Report No. PNNL-14537, Joint Global Change Research Institute, College Park, MD, 16 pp.

Smith, W.L. Jr., et al., 2005: EOS terra aerosol and radiative flux validation: an overview of the Chesapeake Lighthouse and Aircraft Measurements for Satellites (CLAMS) experiment. *J. Atmos. Sci.*, **62**(4), 903–918, doi:10.1175/JAS3398.1.

Sofia, S., and L.H. Li, 2001: Solar variability and climate. *J. Geophys. Res.*, **106**(A7), 12969–12974.

Sokolov, A., 2006: Does model sensitivity to changes in CO_2 provide a measure of sensitivity to the forcing of different nature. *J. Clim.*, **19**, 3294–3306.

Solanki, S.K., and M. Fligge, 1999: A reconstruction of total solar irradiance since 1700. *Geophys. Res. Lett.*, **26**(16), 2465–2468.

Solanki, S.K., et al., 2004: Unusual activity of the Sun during recent decades compared to the previous 11,000 years. *Nature*, **431**, 1084–1087.

Solomon, S., et al., 1996: The role of aerosol variations in anthropogenic ozone depletion at northern midlatitudes. *J. Geophys. Res.*, **101**(D3), 5713–6727.

Solomon, S., et al., 2005: On the distribution and variability of ozone in the tropical upper troposphere: Implications for tropical deep convection and chemical-dynamical coupling. *Geophys. Res. Lett.*, **32**, L23813, doi:10.1029/2005GL024323.

Soufflet, V., D. Tanre, A. Royer, and N.T. O'Neill, 1997: Remote sensing of aerosols over boreal forest and lake water from AVHRR data. *Remote Sens. Environ.*, **60**, 22–34.

Spahni, R., et al., 2005: Atmospheric methane and nitrous oxide of the late Pleistocene from Antarctic ice cores. *Science*, **310**, 1317–1321.

Spinhirne, J.D., et al., 2005: Cloud and aerosol measurements from GLAS: overview and initial results. *Geophys. Res. Lett.*, **32**, L22S03, doi:10.1029/2005GL023507.

Spruit, H., 2000: Theory of solar irradiance variations. *Space Sci. Rev.*, **94**, 113–126.

Steinbrecht, W., H. Claude, and P. Winkler, 2004a: Enhanced upper stratospheric ozone: sign of recovery or solar cycle effect? *J. Geophys. Res.*, **109**, D020308, doi:10.1029/2003JD004284.

Steinbrecht, W., H. Claude, and P. Winkler, 2004b: Reply to comment by D. M. Cunnold et al. on "Enhanced upper stratospheric ozone: Sign of recovery or solar cycle effect?". *J. Geophys. Res.*, **109**, D14306, doi:10.1029/2004JD004948.

Stenchikov, G.L., et al., 1998: Radioactive forcing from the 1991 Mount Pinatubo volcanic eruption. *J. Geophys. Res.*, **103**(D12), 13837–13857.

Stenchikov, G.L., et al., 2002: Arctic Oscillation response to the 1991 Mount Pinatubo eruption: effects of volcanic aerosols and ozone depletion. *J. Geophys. Res.*, **107**(D24), 4803, doi:10.1029/2002JD002090.

Stenchikov, G., et al., 2004: Arctic Oscillation response to the 1991 Pinatubo eruption in the SKYHI GCM with a realistic quasi-biennial oscillation. *J. Geophys. Res.*, **109**, D03112, doi:10.1029/2003JD003699.

Stenchikov, G., et al., 2006: Arctic Oscillation response to volcanic eruptions in the IPCC AR4 climate models. *J. Geophys. Res.*, **111**, D07107, doi:10.1029/2005JD006286.

Stern, D.I., 2005: Global sulfur emissions from 1850 to 2000. *Chemosphere*, **58**, 163–175.

Stevenson, D.S., et al., 2004: Radiative forcing from aircraft NO$_x$ emissions: mechanisms and seasonal dependence. *J. Geophys. Res.*, **109**, D17307, doi:10.1029/2004JD004759.

Stier, P., et al., 2005: The aerosol-climate model ECHAM5-HAM. *Atmos. Chem. Phys.*, **5**, 1125–1156.

Stier, P., et al., 2006a: Impact of nonabsorbing anthropogenic aerosols on clear-sky atmospheric absorption. *J. Geophys. Res.*, **111**, D18201, doi:10.1029/2006JD007147.

Stier, P., et al., 2006b: Emission-induced nonlinearities in the global aerosol system: results from the ECHAM5-HAM aerosol-climate model. *J. Clim.*, **19**, 3845–3862.

Stordal, F., et al., 2005: Is there a trend in cirrus cloud cover due to aircraft traffic? *Atmos. Chem. Phys.*, **5**, 2155–2162.

Stothers, R., 2001a: Major optical depth perturbations to the stratosphere from volcanic eruptions: Stellar extinction period, 1961-1978. *J. Geophys. Res.*, **106**(D3), 2993–3003.

Stothers, R., 2001b: A chronology of annual mean radii of stratospheric aerosol from volcanic eruptions during the twentieth century as derived from ground-based spectral extinction measurements. *J. Geophys. Res.*, **106**(D23), 32043–32049.

Stott P.A., G.S. Jones, and J.F.B. Mitchell, 2003: Do models underestimate the solar contribution to recent climate change? *J. Clim.*, **16**, 4079–4093.

Streets, D.G., et al., 2001: Black carbon emissions in China. *Atmos. Environ.*, **35**, 4281–4296.

Streets, D.G., et al., 2003: An inventory of gaseous and primary aerosol emissions in Asia in the year 2000. *J. Geophys. Res.*, **108**(D21), 8809, doi:10.1029/2002JD003093.

Strugnell, N.C., W. Lucht, and C. Schaaf, 2001: A global albedo data set derived from AVHRR data for use in climate simulations. *Geophys. Res. Lett.*, **28**(1), 191–194.

Stubenrauch, C.J., and U. Schumann, 2005: Impact of air traffic on cirrus coverage. *Geophys. Res. Lett.*, **32**, L14813, doi:10.1029/2005GL022707.

Stuber, N., M. Ponater, and R. Sausen, 2001a: Is the climate sensitivity to ozone perturbations enhanced by stratospheric water vapor feedback? *Geophys. Res. Lett.*, **28**(15), 2887–2890.

Stuber, N., R. Sausen, and M. Ponater, 2001b: Stratosphere adjusted radiative forcing calculations in a comprehensive climate model. *Theor. Appl. Climatol.*, **68**, 125–135.

Stuber, N., M. Ponater, and R. Sausen, 2005: Why radiative forcing might fail as a predictor of climate change. *Clim. Dyn.*, **24**, 497–510.

Sun, B., and R.S. Bradley, 2002: Solar influences on cosmic rays and cloud formation: a re-assessment. *J. Geophys. Res.*, **107**(D14), doi:10.1029/2001JD000560.

Sundqvist, H., 1978: A parametrization scheme for non-convective condensation including prediction of cloud water content. *Q. J. R. Meteorol. Soc.*, **104**, 677–690.

Sundqvist, H., E. Berge, and J.E. Kristjánsson, 1989: Condensation and cloud parametrization studies with a mesoscale numerical weather prediction model. *Mon. Weather Rev.*, **117**, 1641–1657.

Suzuki, K., et al., 2004: A study of the aerosol effect on a cloud field with simultaneous use of GCM modeling and satellite observation. *J. Atmos. Sci.*, **61**, 179–194.

Svalgaard, L., E.W. Cliver, and P. Le Sager, 2004: IHV: A new long-term geomagnetic index. *Adv. Space Res.*, **34**, 436–439.

Swap, R.J., et al., 2002: The Southern African Regional Science Initiative (SAFARI 2000): overview of the dry season field campaign. *S. Afr. J. Sci.*, **98**, 125–130.

Swap, R.J., et al., 2003: Africa burning: a thematic analysis of the Southern African Regional Science Initiative (SAFARI 2000). *J. Geophys. Res.*, **108**(D13), 8465, doi:10.1029/2003JD003747.

Tabazadeh, A., et al., 2002: Arctic "ozone hole" in cold volcanic stratosphere. *Proc. Natl. Acad. Sci. U.S.A.*, **99**, 2609–2612.

Takemura, T., T. Nakajima, T. Nozawa, and K. Aoki, 2001: Simulation of future aerosol distribution, radiative forcing, and long-range transport in East Asia. *J. Meteorol. Soc. Japan*, **2**, 79, 1139–1155.

Takemura, T., et al., 2000: Global three-dimensional simulation of aerosol optical thickness distribution of various origins. *J. Geophys. Res.*, **105**(D14), 17853–17874.

Takemura, T., et al., 2005: Simulation of climate response to aerosol direct and indirect effects with aerosol transport-radiation model. *J. Geophys. Res.*, **110**, D02202, doi:10.1029/2004JD005029.

Tang, I.N., 1997: Thermodynamic and optical properties of mixed-salt aerosols of atmospheric importance. *J. Geophys. Res.*, **102**(D2), 1883–1893.

Tang, I.N., K.H. Fung, D.G. Imre, and H.R. Munkelwitz, 1995: Phase transformation and metastability of hygroscopic microparticles. *Aerosol Sci. Tech.*, **23**, 443.

Tanré, D., Y.J. Kaufman, M. Herman, and S. Mattoo, 1997: Remote sensing of aerosol properties over oceans using the MODIS/EOS spectral radiances. *J. Geophys. Res.*, **102**(D14), 16971–16988.

Tanré, D., et al., 2003: Measurement and modeling of the Saharan dust radiative impact: overview of the SaHAran Dust Experiment (SHADE). *J. Geophys. Res.*, **108**(D18), doi:10.1029/2002JD003273.

Tarasick, D.W., et al., 2005: Changes in the vertical distribution of ozone over Canada from ozonesondes: 1980-2001. *J. Geophys. Res.*, **110**, D02304, doi:10.1029/2004JD004643.

Tegen, I., and I. Fung, 1995: Contribution to the atmospheric mineral aerosol load from land surface modification. *J. Geophys. Res.*, **100**, 18707–18726.

Tegen, I., A.A. Lacis, and I. Fung, 1996: The influence on climate forcing of mineral aerosols from disturbed soils. *Nature*, **380**, 419–421.

Tegen, I., M. Werner, S.P. Harrison, and K.E. Kohfeld, 2004: Relative importance of climate and land use in determining present and future global soil dust emission. *Geophys. Res. Lett.*, **31**, L05105, doi:10.1029/2003GL019216.

Tegen, I., M. Werner, S.P. Harrison, and K.E. Kohfeld, 2005: Reply to comment by N. M. Mahowald et al. on "Relative importance of climate and land use in determining present and future global soil dust emission". *Geophys. Res. Lett.*, **32**, doi:10.1029/2004GL021560.

Tett, S.F.B., et al., 2002: Estimation of natural and anthropogenic contributions to twentieth century temperature change. *J. Geophys. Res.*, **107**(D16), 4306, doi:10.1029/2000JD000028.

Textor, C., et al., 2006: AeroCom: The status quo of global aerosol modelling. *Atmos. Chem. Phys.*, **6**, 1777–1813.

Tie, X.X., G.P. Brasseur, B. Breiglib, and C. Granier, 1994: Two dimensional simulation of Pinatubo aerosol and its effect on stratospheric chemistry. *J. Geophys. Res.*, **99**(D10), 20545–20562.

Thomason, L., and T. Peter, 2006: *Assessment of Stratospheric Aerosol Properties (ASAP): Report on the Assessment Kick-Off Workshop, Paris, France, 4-6 November 2001.* SPARC Report No. 4, WCRP-124, WMO/TD No. 1295, http://www.aero.jussieu.fr/~sparc/News18/18_Thomason.html.

Thompson, A.M., et al., 2001: Tropical tropospheric ozone and biomass burning. *Science*, **291**, 2128–2132.

Thompson, T.M., et al., 2004: Halocarbons and other atmospheric trace species. In: *Climate Monitoring and Diagnostics Laboratory, Summary Report No. 27* [Schnell, R.C., A.-M. Buggle, and R.M. Rosson (eds.)]. NOAA CMDL, Boulder, CO, pp. 115–135.

Timmreck, C., and M. Schulz, 2004: Significant dust simulation differences in nudged and climatological operation mode of the AGCM ECHAM. *J. Geophys. Res.*, **109**, D13202, doi:10.1029/2003JD004381.

Timmreck, C., H.-F. Graf, and B. Steil, 2003: Aerosol chemistry interactions after the Mt. Pinatubo Eruption. In: *Volcanism and the Earth's Atmosphere* [Robock, A., and C. Oppenheimer (eds.)]. Geophysical Monograph 139, American Geophysical Union, Washington, DC, pp. 227–236.

Tol, R.S.J., 2002: Estimates of the damage costs of climate change, Part II. Dynamic estimates. *Environ. Resour. Econ.*, **21**, 135–160.

Torres, O., et al., 2002: A long-term record of aerosol optical depth from TOMS: Observations and comparison to AERONET measurements. *J. Atmos. Sci.*, **59**, 398–413.

Travis, D.J., A.M. Carleton, and R.G. Lauritsen, 2002: Contrails reduce daily temperature range. *Nature*, **418**, 601–602.

Travis, D.J., A.M. Carleton, and R.G. Lauritsen, 2004: Regional variations in U.S. diurnal temperature range for the 11-14 September 2001 aircraft groundings: evidence of jet contrail influence on climate. *J. Clim.*, **17**, 1123–1134.

Tripoli, G.J., and W.R. Cotton, 1980: A numerical investigation of several factors contributing to the observed variable intensity of deep convection over South Florida. *J. Appl. Meteorol.*, **19**, 1037–1063.

Twohy, C.H., et al., 2005: Evaluation of the aerosol indirect effect in marine stratocumulus clouds: droplet number, size, liquid water path, and radiative impact. *J. Geophys.Res.*, **110**, D08203, doi:10.1029/2004JD005116.

Twomey, S.A., 1977: The influence of pollution on the shortwave albedo of clouds. *J. Atmos. Sci.*, **34**, 1149–1152.

Udelhofen, P.M., and R.D. Cess, 2001: Cloud cover variations over the United States: An influence of cosmic rays or solar variability? *Geophys. Res. Lett.*, **28**(13), 2617–2620.

Usoskin, I.G., et al., 2004: Latitudinal dependence of low cloud amount on cosmic ray induced ionization. *Geophys. Res. Lett.*, **31**, L16109, doi:10.1029/2004GL019507.

Van Aardenne, J.A., et al., 2001: A 1 x 1 degree resolution dataset of historical anthropogenic trace gas emissions for the period 1890-1990. *Global Biogeochem. Cycles*, **15**, 909–928.

van der Werf, et al., 2004: Continental-scale partitioning of fire emissions during the 1997 to 2001 El Nino/La Nina period. *Science*, **303**, 73–76.

Van Dorland, R., F.J. Dentener, and J. Lelieveld, 1997: Radiative forcing due to tropospheric ozone and sulfate aerosols. *J. Geophys. Res.*, **102**(D23), 28079–28100.

van Loon, H., and D.J. Shea, 2000: The global 11-year solar signal in July-August. *Geophys. Res. Lett.*, **27**(18), 2965–2968.

Veefkind, J.P., G. de Leeuw, and P.A. Durkee, 1998: Retrieval of aerosol optical depth over land using two-angle view satellite radiometry. *Geophys. Res. Lett.* **25**(16), 3135–3138.

Velders, et al., 2005: Chemical and radiative effects of halocarbons and their replacement compounds. In: *Special Report on Safeguarding the Ozone Layer and the Global Climate System: Issues Related to Hydrofluorocarbons and Perfluorocarbons* [Metz, B., et al. (eds.)]. Cambridge University Press, Cambridge, United Kingdom and New York, NY, USA, pp. 133–180.

Vestreng, V., M. Adams, and J. Goodwin, 2004: *Inventory Review 2004: Emission Data Reported to CLRTAP and the NEC Directive*. EMEP/EEA Joint Review Report, Norwegian Meteorological Institute, Norway, 120 pp.

von Hoyningen-Huene, W., M. Freitag, and J.B. Burrows, 2003: Retrieval of aerosol optical thickness over land surfaces from top-of-atmosphere radiance. *J. Geophys. Res.*, **108**(D9), 4260, doi:10.1029/2001JD002018.

Walter, B.P., M. Heimann, and E. Matthews, 2001a: Modeling modern methane emissions from natural wetlands 2. Interannual variations 1982-1993. *J. Geophys. Res.*, **106**(D24), 34207–34220.

Walter, B.P., M. Heimann, and E. Matthews, 2001b: Modeling modern methane emissions from natural wetlands 1. Model description and results. *J. Geophys. Res.*, **106**(D24), 34139–34206.

Wang, C., 2004: A modeling study on the climate impacts of black carbon aerosols. *J. Geophys. Res.*, **109**, D03106, doi:10.1029/2003JD004084.

Wang, H.J., et al., 2002: Assessment of SAGE version 6.1 ozone data quality. *J. Geophys. Res.*, **107**(D23), 4691, doi:10.1029/2002JD002418.

Wang, M.H., K.D. Knobelspiesse, and C.R. McClain, 2005: Study of the Sea-Viewing Wide Field-of-View Sensor (SeaWiFS) aerosol optical property data over ocean in combination with the ocean color products. *J. Geophys. Res.*, **110**, D10S06, doi:10.1029/2004JD004950.

Wang, W.-C., M.P. Dudek, and X.-Z. Liang, 1992: Inadequacy of effective CO_2 as a proxy in assessing the regional climate change due to other radiatively active gases. *Geophys. Res. Lett.*, **19**, 1375–1378.

Wang, W.-C., M.P. Dudek, X.-Z. Liang, and J.T. Kiehl, 1991: Inadequacy of effective CO_2 as a proxy in simulating the greenhouse effect of other radiatively active gases. *Nature*, **350**, 573–577.

Wang, Y.M., J.L. Lean, and N.R. Sheeley, 2005: Modeling the sun's magnetic field and irradiance since 1713. *Astrophys. J.*, **625**, 522–538.

Warner, J., and S.A. Twomey, 1967: The production and cloud nuclei by cane fires and the effect on cloud droplet concentration. *J. Atmos. Sci.*, **24**, 704–706.

Warwick, N.J., et al., 2002: The impact of meteorology on the interannual growth rate of atmospheric methane. *Geophys. Res. Lett.*, **29**(26), doi:10.1029/2002/GL015282.

Weatherhead, E.C., and S. B. Andersen, 2006: The search for signs of recovery of the ozone layer. *Nature*, **441**, 39–45.

Welton, E.J., J.R. Campbell, J.D. Spinhirne, and V.S. Scott, 2001: Global monitoring of clouds and aerosols using a network of micro-pulse lidar systems. In: *Lidar Remote Sensing for Industry and Environmental Monitoring* [Singh, U.N., T. Itabe, and N. Sugimoto (eds.)]. SPIE, Bellingham, WA, pp. 151–158.

Wennberg, P.O., S. Peacock, J.T. Randerson, and R. Bleck, 2004: Recent changes in the air-sea gas exchange of methyl chloroform. *Geophys. Res. Lett.*, **31**, L16112, doi:10.1029/2004GL020476.

Westerling, A.L., H.G. Hidalgo, D.R. Cayan, and T.W. Swetnam, 2006: Warming and earlier spring increase western U.S. forest wildfire activity. *Science*, **313**, 940–943.

White, W.B., M.D. Dettinger, and D.R. Cayan, 2003: Sources of global warming of the upper ocean on decadal period scales. *J. Geophys. Res.*, **108**(C8), doi:10.1029/2002JC001396.

Wild, O., M.J. Prather, and H. Akimoto, 2001: Indirect long-term global radiative cooling from NO_x emissions. *Geophys. Res. Lett.*, **28**(9), 1719–1722.

Williams, K.D., C.A. Senior, and J.F.B. Mitchell, 2001a: Transient climate change in the Hadley Centre models: the role of physical processes. *J. Clim.*, **14**, 2659–2674.

Williams, K.D., et al., 2001b: The response of the climate system to the indirect effects of anthropogenic sulfate aerosols. *Clim. Dyn.*, **17**, 846–856.

Willson, R.C., and A.V. Mordvinov, 2003: Secular total solar irradiance trend during solar cycles 21-23. *Geophys. Res. Lett.*, **30**(5), 3–6.

Wilson, D.R., and S.P. Ballard, 1999: A microphysically based precipitation scheme for the UK Meteorological Office Unified Model. *Q. J. R. Meteorol. Soc.*, **125**, 1607–1636.

Wilson, M.F., and A. Henderson-Sellers, 1985: A global archive of land cover and soils data for use in general-circulation climate models. *J. Climatol.*, **5**, 119–143.

WMO, 1986: *Atmospheric Ozone 1985*. Global Ozone Research and Monitoring Project Report No.16, World Meteorological Organisation, Geneva, Volume 3.

WMO, 2003: *Scientific Assessment of Ozone Depletion: 2002*. Global Ozone Research and Monitoring Project Report No. 47, World Meteorological Organization, Geneva, 498 pp.

Wong, J., and Z. Li, 2002: Retrieval of optical depth for heavy smoke aerosol plumes: uncertainties and sensitivities to the optical properties. *J. Atmos. Sci.*, **59**, 250–261.

Wong, S., et al., 2004: A global climate-chemistry model study of present-day tropospheric chemistry and radiative forcing from changes in tropospheric O_3 since the preindustrial period. *J. Geophys. Res.*, **109**, D11309, doi:10.1029/2003JD003998.

Woods, T.N., et al., 1996: Validation of the UARS solar ultraviolet irradiances: comparison with the ATLAS 1 and 2 measurements. *J. Geophys. Res.*, **101**(D6), 9541–9569.

Xiong, J.Q., et al., 1998: Influence of organic films on the hygroscopicity of ultrafine sulfuric acid aerosol. *Environ. Sci. Technol.*, **32**, 3536–3541.

Xue, H., and G. Feingold, 2006: large eddy simulations of trade-wind cumuli: investigation of aerosol indirect effects. *J. Atmos. Sci.*, **63**, 1605–1622.

Yang, F., and M. Schlesinger, 2001: Identification and separation of Mount Pinatubo and El Nino-Southern Oscillation land surface temperature anomalies. *J. Geophys. Res.*, **106**(D14), 14757–14770.

Yang, F., and M. Schlesinger, 2002: On the surface and atmospheric temperature changes following the 1991 Pinatubo volcanic eruption: a GCM study. *J. Geophys. Res.*, **107**(D8), doi:10.1029/2001JD000373.

Yokouchi, et al., 2005: Estimates of ratios of anthropogenic halocarbon emissions from Japan based on aircraft monitoring over Sagami Bay, Japan. *J. Geophys. Res.*, **110**, D06301, doi:10.1029/2004JD005320.

Yoshioka, M., N. Mahowald, J.L. Dufresne, and C. Luo, 2005: Simulation of absorbing aerosol indices for African dust. *J. Geophys. Res.*, **110**, D18S17, doi:10.1029/2004JD005276.

Yu, H., et al., 2003: Annual cycle of global distributions of aerosol optical depth from integration of MODIS retrievals and GOCART model simulations. *J. Geophys. Res.*, **108**(D3), 4128, doi:10.1029/2002JD002717.

Yu, H., et al., 2006: A review of measurement-based assessments of the aerosol direct radiative effect and forcing. *Atmos. Chem. Phys.*, **6**, 613–666.

Yvon-Lewis, S.A., and J.H. Butler, 2002: Effect of oceanic uptake on atmospheric lifetimes of selected trace gases. *J. Geophys. Res.*, **107**(D20), 4414, doi:10.1029/2001JD001267.

Zender, C.S., 2004: Quantifying mineral dust mass budgets: terminology, constraints, and current estimates. *Eos*, **85**, 509–512.

Zerefos, C.S., et al., 2003: Evidence of impact of aviation on cirrus cloud formation. *Atmos. Chem. Phys.*, **3**, 1633–1644.

Zhang, J., and S. Christopher, 2003: Longwave radiative forcing of Saharan dust aerosols estimated from MODIS, MISR and CERES observations on Terra. *Geophys. Res. Lett.*, **30**(23), doi:10.1029/2003GL018479.

Zhang, J., S.A. Christopher, L.A. Remer, and Y.J. Kaufman, 2005: Shortwave aerosol radiative forcing over cloud-free oceans from Terra: 2. Seasonal and global distributions. *J. Geophys. Res.*, **110**, D10S24, doi:10.1029/2004JD005009.

Zhang, J.P., Z. Yang, D.J. Wang, and X.B. Zhang, 2002: Climate change and causes in the Yuanmou dry-hot valley of Yunnan China. *J. Arid Environ.*, **51**, 153–162.

Zhao, M., A. Pitman, and T.N. Chase, 2001: The impacts of land cover change on the atmospheric circulation. *Clim. Dyn.*, **17**, 467–477.

Zhao, T.X.P., I. Laszlo, P. Minnis, and L. Remer, 2005: Comparison and analysis of two aerosol retrievals over the ocean in the terra/clouds and the earth's radiant energy system: moderate resolution imaging spectroradiometer single scanner footprint data: 1. Global evaluation. *J. Geophys. Res.*, **110**, D21208, doi:10.1029/2005JD005851.

Zhou, L.M., et al., 2001: Variations in northern vegetation activity inferred from satellite data of vegetation index during 1981 to 1999. *J. Geophys. Res.*, **106**(D17), 20069–20083.

Zhou, M., et al., 2005: A normalized description of the direct effect of key aerosol types on solar radiation as estimated from aerosol robotic network aerosols and moderate resolution imaging spectroradiometer albedos. *J. Geophys. Res.*, **110**, D19202, doi:10.1029/2005JD005909.

3

Observations:
Surface and Atmospheric Climate Change

Coordinating Lead Authors:

Kevin E. Trenberth (USA), Philip D. Jones (UK)

Lead Authors:

Peter Ambenje (Kenya), Roxana Bojariu (Romania), David Easterling (USA), Albert Klein Tank (Netherlands), David Parker (UK), Fatemeh Rahimzadeh (Iran), James A. Renwick (New Zealand), Matilde Rusticucci (Argentina), Brian Soden (USA), Panmao Zhai (China)

Contributing Authors:

R. Adler (USA), L. Alexander (UK, Australia, Ireland), H. Alexandersson (Sweden), R. Allan (UK), M.P. Baldwin (USA), M. Beniston (Switzerland), D. Bromwich (USA), I. Camilloni (Argentina), C. Cassou (France), D.R. Cayan (USA), E.K.M. Chang (USA), J. Christy (USA), A. Dai (USA), C. Deser (USA), N. Dotzek (Germany), J. Fasullo (USA), R. Fogt (USA), C. Folland (UK), P. Forster (UK), M. Free (USA), C. Frei (Switzerland), B. Gleason (USA), J. Grieser (Germany), P. Groisman (USA, Russian Federation), S. Gulev (Russian Federation), J. Hurrell (USA), M. Ishii (Japan), S. Josey (UK), P. Kållberg (ECMWF), J. Kennedy (UK), G. Kiladis (USA), R. Kripalani (India), K. Kunkel (USA), C.-Y. Lam (China), J. Lanzante (USA), J. Lawrimore (USA), D. Levinson (USA), B. Liepert (USA), G. Marshall (UK), C. Mears (USA), P. Mote (USA), H. Nakamura (Japan), N. Nicholls (Australia), J. Norris (USA), T. Oki (Japan), F.R. Robertson (USA), K. Rosenlof (USA), F.H. Semazzi (USA), D. Shea (USA), J.M. Shepherd (USA), T.G. Shepherd (Canada), S. Sherwood (USA), P. Siegmund (Netherlands), I. Simmonds (Australia), A. Simmons (ECMWF, UK), C. Thorncroft (USA, UK), P. Thorne (UK), S. Uppala (ECMWF), R. Vose (USA), B. Wang (USA), S. Warren (USA), R. Washington (UK, South Africa), M. Wheeler (Australia), B. Wielicki (USA), T. Wong (USA), D. Wuertz (USA)

Review Editors:

Brian J. Hoskins (UK), Thomas R. Karl (USA), Bubu Jallow (The Gambia)

This chapter should be cited as:

Trenberth, K.E., P.D. Jones, P. Ambenje, R. Bojariu, D. Easterling, A. Klein Tank, D. Parker, F. Rahimzadeh, J.A. Renwick, M. Rusticucci, B. Soden and P. Zhai, 2007: Observations: Surface and Atmospheric Climate Change. In: *Climate Change 2007: The Physical Science Basis*. Contribution of Working Group I to the Fourth Assessment Report of the Intergovernmental Panel on Climate Change [Solomon, S., D. Qin, M. Manning, Z. Chen, M. Marquis, K.B. Averyt, M. Tignor and H.L. Miller (eds.)]. Cambridge University Press, Cambridge, United Kingdom and New York, NY, USA.

Table of Contents

Supplementary Material

The following Supplementary Material is available on CD-ROM and in on-line versions of this report.

Appendix 3.B: *Techniques, Error Estimation and Measurement Systems*

Executive Summary

Global mean surface temperatures have risen by 0.74°C ± 0.18°C when estimated by a linear trend over the last 100 years (1906–2005). The rate of warming over the last 50 years is almost double that over the last 100 years (0.13°C ± 0.03°C vs. 0.07°C ± 0.02°C per decade). Global mean temperatures averaged over land and ocean surfaces, from three different estimates, each of which has been independently adjusted for various homogeneity issues, are consistent within uncertainty estimates over the period 1901 to 2005 and show similar rates of increase in recent decades. The trend is not linear, and the warming from the first 50 years of instrumental record (1850–1899) to the last 5 years (2001–2005) is 0.76°C ± 0.19°C.

2005 was one of the two warmest years on record. The warmest years in the instrumental record of global surface temperatures are 1998 and 2005, with 1998 ranking first in one estimate, but with 2005 slightly higher in the other two estimates. 2002 to 2004 are the 3rd, 4th and 5th warmest years in the series since 1850. Eleven of the last 12 years (1995 to 2006) – the exception being 1996 – rank among the 12 warmest years on record since 1850. Surface temperatures in 1998 were enhanced by the major 1997–1998 El Niño but no such strong anomaly was present in 2005. Temperatures in 2006 were similar to the average of the past 5 years.

Land regions have warmed at a faster rate than the oceans. Warming has occurred in both land and ocean domains, and in both sea surface temperature (SST) and nighttime marine air temperature over the oceans. However, for the globe as a whole, surface air temperatures over land have risen at about double the ocean rate after 1979 (more than 0.27°C per decade vs. 0.13°C per decade), with the greatest warming during winter (December to February) and spring (March to May) in the Northern Hemisphere.

Changes in extremes of temperature are also consistent with warming of the climate. A widespread reduction in the number of frost days in mid-latitude regions, an increase in the number of warm extremes and a reduction in the number of daily cold extremes are observed in 70 to 75% of the land regions where data are available. The most marked changes are for cold (lowest 10%, based on 1961–1990) nights, which have become rarer over the 1951 to 2003 period. Warm (highest 10%) nights have become more frequent. Diurnal temperature range (DTR) decreased by 0.07°C per decade averaged over 1950 to 2004, but had little change from 1979 to 2004, as both maximum and minimum temperatures rose at similar rates. The record-breaking heat wave over western and central Europe in the summer of 2003 is an example of an exceptional recent extreme. That summer (June to August) was the hottest since comparable instrumental records began around 1780 (1.4°C above the previous warmest in 1807) and is very likely to have been the hottest since at least 1500.

Recent warming is strongly evident at all latitudes in SSTs over each of the oceans. There are inter-hemispheric differences in warming in the Atlantic, the Pacific is punctuated by El Niño events and Pacific decadal variability that is more symmetric about the equator, while the Indian Ocean exhibits steadier warming. These characteristics lead to important differences in regional rates of surface ocean warming that affect the atmospheric circulation.

Urban heat island effects are real but local, and have not biased the large-scale trends. A number of recent studies indicate that effects of urbanisation and land use change on the land-based temperature record are negligible (0.006°C per decade) as far as hemispheric- and continental-scale averages are concerned because the very real but local effects are avoided or accounted for in the data sets used. In any case, they are not present in the SST component of the record. Increasing evidence suggests that urban heat island effects extend to changes in precipitation, clouds and DTR, with these detectable as a 'weekend effect' owing to lower pollution and other effects during weekends.

Average arctic temperatures increased at almost twice the global average rate in the past 100 years. Arctic temperatures have high decadal variability. A slightly longer warm period, almost as warm as the present, was also observed from the late 1920s to the early 1950s, but appears to have had a different spatial distribution than the recent warming.

Lower-tropospheric temperatures have slightly greater warming rates than those at the surface over the period 1958 to 2005. The radiosonde record is markedly less spatially complete than the surface record and increasing evidence suggests that it is very likely that a number of records have a cooling bias, especially in the tropics. While there remain disparities among different tropospheric temperature trends estimated from satellite Microwave Sounding Unit (MSU and advanced MSU) measurements since 1979, and all likely still contain residual errors, estimates have been substantially improved (and data set differences reduced) through adjustments for issues of changing satellites, orbit decay and drift in local crossing time (i.e., diurnal cycle effects). It appears that the satellite tropospheric temperature record is broadly consistent with surface temperature trends provided that the stratospheric influence on MSU channel 2 is accounted for. The range (due to different data sets) of global surface warming since 1979 is 0.16°C to 0.18°C per decade compared to 0.12°C to 0.19°C per decade for MSU estimates of tropospheric temperatures. It is likely, however, that there is slightly greater warming in the troposphere than at the surface, and a higher tropopause, with the latter due also to pronounced cooling in the stratosphere.

Lower stratospheric temperatures feature cooling since 1979. Estimates from adjusted radiosondes, satellites (MSU channel 4) and reanalyses are in qualitative agreement, suggesting a lower-stratospheric cooling of between 0.3°C and 0.6°C per decade since 1979. Longer radiosonde records (back to 1958) also indicate cooling but the rate of cooling has been significantly greater since 1979 than between 1958 and 1978. It is likely that radiosonde records overestimate stratospheric cooling, owing to changes in sondes not yet accounted for. Because of the stratospheric warming episodes following major volcanic eruptions, the trends are far from being linear.

Precipitation has generally increased over land north of 30°N over the period 1900 to 2005 but downward trends dominate the tropics since the 1970s. From 10°N to 30°N, precipitation increased markedly from 1900 to the 1950s, but declined after about 1970. Downward trends are present in the deep tropics from 10°N to 10°S, especially after 1976/1977. Tropical values dominate the global mean. It has become significantly wetter in eastern parts of North and South America, northern Europe, and northern and central Asia, but drier in the Sahel, the Mediterranean, southern Africa and parts of southern Asia. Patterns of precipitation change are more spatially and seasonally variable than temperature change, but where significant precipitation changes do occur they are consistent with measured changes in streamflow.

Substantial increases are found in heavy precipitation events. It is likely that there have been increases in the number of heavy precipitation events (e.g., 95th percentile) within many land regions, even in those where there has been a reduction in total precipitation amount, consistent with a warming climate and observed significant increasing amounts of water vapour in the atmosphere. Increases have also been reported for rarer precipitation events (1 in 50 year return period), but only a few regions have sufficient data to assess such trends reliably.

Droughts have become more common, especially in the tropics and subtropics, since the 1970s. Observed marked increases in drought in the past three decades arise from more intense and longer droughts over wider areas, as a critical threshold for delineating drought is exceeded over increasingly widespread areas. Decreased land precipitation and increased temperatures that enhance evapotranspiration and drying are important factors that have contributed to more regions experiencing droughts, as measured by the Palmer Drought Severity Index. The regions where droughts have occurred seem to be determined largely by changes in SSTs, especially in the tropics, through associated changes in the atmospheric circulation and precipitation. In the western USA, diminishing snow pack and subsequent reductions in soil moisture also appear to be factors. In Australia and Europe, direct links to global warming have been inferred through the extreme nature of high temperatures and heat waves accompanying recent droughts.

Tropospheric water vapour is increasing. Surface specific humidity has generally increased after 1976 in close association with higher temperatures over both land and ocean. Total column water vapour has increased over the global oceans by $1.2 \pm 0.3\%$ per decade from 1988 to 2004, consistent in pattern and amount with changes in SST and a fairly constant relative humidity. Strong correlations with SST suggest that total column water vapour has increased by 4% since 1970. Similar upward trends in upper-tropospheric specific humidity, which considerably enhance the greenhouse effect, have also been detected from 1982 to 2004.

'Global dimming' is neither global in extent nor has it continued after 1990. Reported decreases in solar radiation at the Earth's surface from 1970 to 1990 have an urban bias and have reversed in sign. Although records are sparse, pan evaporation is estimated to have decreased in many places due to decreases in surface radiation associated with increases in clouds, changes in cloud properties and/or increases in air pollution (aerosols), especially from 1970 to 1990. However, in many of the same places, actual evapotranspiration inferred from surface water balance exhibits an increase in association with enhanced soil wetness from increased precipitation, as the actual evapotranspiration becomes closer to the potential evaporation measured by the pans. Hence, in determining evapotranspiration there is a trade-off between less solar radiation and increased surface wetness, with the latter generally dominant.

Cloud changes are dominated by the El Niño-Southern Oscillation and appear to be opposite over land and ocean. Widespread (but not ubiquitous) decreases in continental DTR since the 1950s coincide with increases in cloud amounts. Surface and satellite observations disagree about total and low-level cloud changes over the ocean. However, radiation changes at the top of the atmosphere from the 1980s to 1990s, possibly related in part to the El Niño-Southern Oscillation (ENSO) phenomenon, appear to be associated with reductions in tropical upper-level cloud cover, and are linked to changes in the energy budget at the surface and changes in observed ocean heat content.

Changes in the large-scale atmospheric circulation are apparent. Atmospheric circulation variability and change is largely described by relatively few major patterns. The dominant mode of global-scale variability on interannual time scales is ENSO, although there have been times when it is less apparent. The 1976–1977 climate shift, related to the phase change in the Pacific Decadal Oscillation and more frequent El Niños, has affected many areas and most tropical monsoons. For instance, over North America, ENSO and Pacific-North American teleconnection-related changes appear to have led to contrasting changes across the continent, as the west has warmed more than the east, while the latter has become cloudier and wetter. There are substantial multi-decadal variations in the Pacific sector over the 20th century with extended periods of weakened (1900–1924; 1947–1976) as well as strengthened circulation (1925–1946; 1976–2005). Multi-decadal variability is also evident in the Atlantic as the Atlantic Multi-decadal Oscillation in both the atmosphere and the ocean.

Mid-latitude westerly winds have generally increased in both hemispheres. These changes in atmospheric circulation are predominantly observed as 'annular modes', related to the zonally averaged mid-latitude westerlies, which strengthened in most seasons from the 1960s to at least the mid-1990s, with poleward displacements of corresponding Atlantic and southern polar front jet streams and enhanced storm tracks. These were accompanied by a tendency towards stronger winter polar vortices throughout the troposphere and lower stratosphere. On monthly time scales, the southern and northern annular modes (SAM and NAM, respectively) and the North Atlantic Oscillation (NAO) are the dominant patterns of variability in the extratropics and the NAM and NAO are closely related. The westerlies in the Northern Hemisphere, which increased from the 1960s to the 1990s but which have since returned to about normal as part of NAO and NAM changes, alter the flow from oceans to continents and are a major cause of the observed changes in winter storm tracks and related patterns of precipitation and temperature anomalies, especially over Europe. In the Southern Hemisphere, SAM increases from the 1960s to the present are associated with strong warming over the Antarctic Peninsula and, to a lesser extent, cooling over parts of continental Antarctica. Analyses of wind and significant wave height support reanalysis-based evidence for an increase in extratropical storm activity in the Northern Hemisphere in recent decades until the late 1990s.

Intense tropical cyclone activity has increased since about 1970. Variations in tropical cyclones, hurricanes and typhoons are dominated by ENSO and decadal variability, which result in a redistribution of tropical storm numbers and their tracks, so that increases in one basin are often compensated by decreases over other oceans. Trends are apparent in SSTs and other critical variables that influence tropical thunderstorm and tropical storm development. Globally, estimates of the potential destructiveness of hurricanes show a significant upward trend since the mid-1970s, with a trend towards longer lifetimes and greater storm intensity, and such trends are strongly correlated with tropical SST. These relationships have been reinforced by findings of a large increase in numbers and proportion of hurricanes reaching categories 4 and 5 globally since 1970 even as total number of cyclones and cyclone days decreased slightly in most basins. The largest increase was in the North Pacific, Indian and southwest Pacific Oceans. However, numbers of hurricanes in the North Atlantic have also been above normal (based on 1981–2000 averages) in 9 of the last 11 years, culminating in the record-breaking 2005 season. Moreover, the first recorded tropical cyclone in the South Atlantic occurred in March 2004 off the coast of Brazil.

The temperature increases are consistent with observed changes in the cryosphere and oceans. Consistent with observed changes in surface temperature, there has been an almost worldwide reduction in glacier and small ice cap (not including Antarctica and Greenland) mass and extent in the 20th century; snow cover has decreased in many regions of the Northern Hemisphere; sea ice extents have decreased in the Arctic, particularly in spring and summer (Chapter 4); the oceans are warming; and sea level is rising (Chapter 5).

3.1   Introduction

This chapter assesses the observed changes in surface and atmospheric climate, placing new observations and new analyses made during the past six years (since the Third Assessment Report – TAR) in the context of the previous instrumental record. In previous IPCC reports, palaeo-observations from proxy data for the pre-instrumental past and observations from the ocean and ice domains were included within the same chapter. This helped the overall assessment of the consistency among the various variables and their synthesis into a coherent picture of change. However, the amount of information became unwieldy and is now spread over Chapters 3 to 6. Nevertheless, a short synthesis and scrutiny of the consistency of all the observations is included here (see Section 3.9).

In the TAR, surface temperature trends were examined from 1860 to 2000 globally, for 1901 to 2000 as maps and for three sub-periods (1910–1945, 1946–1975 and 1976–2000). The first and third sub-periods had rising temperatures, while the second sub-period had relatively stable global mean temperatures. The 1976 divide is the date of a widely acknowledged 'climate shift' (e.g., Trenberth, 1990) and seems to mark a time (see Chapter 9) when global mean temperatures began a discernible upward trend that has been at least partly attributed to increases in greenhouse gas concentrations in the atmosphere (see the TAR; IPCC, 2001). The picture prior to 1976 has essentially not changed and is therefore not repeated in detail here. However, it is more convenient to document the sub-period after 1979, rather than 1976, owing to the availability of increased and improved satellite data since then (in particular Television InfraRed Observation Satellite (TIROS) Operational Vertical Sounder (TOVS) data) in association with the Global Weather Experiment (GWE) of 1979. The post-1979 period allows, for the first time, a global perspective on many fields of variables, such as precipitation, that was not previously available. For instance, the reanalyses of the global atmosphere from the National Centers for Environmental Prediction/National Center for Atmospheric Research (NCEP/NCAR, referred to as NRA; Kalnay et al., 1996; Kistler et al., 2001) and the European Centre for Medium Range Weather Forecasts (ECMWF, referred to as ERA-40; Uppala et al., 2005) are markedly more reliable after 1979, and spurious discontinuities are present in the analysed record at the end of 1978 (Santer et al., 1999; Bengtsson et al., 2004; Bromwich and Fogt, 2004; Simmons et al., 2004; Trenberth et al., 2005a). Therefore, the availability of high-quality data has led to a focus on the post-1978 period, although physically this new regime seems to have begun in 1976/1977.

Documentation of the climate has traditionally analysed global and hemispheric means, and land and ocean means, and has presented some maps of trends. However, climate varies over all spatial and temporal scales: from the diurnal cycle to El Niño to multi-decadal and millennial variations. Atmospheric waves naturally create regions of temperature and moisture of opposite-signed departures from the zonal mean, as moist warm conditions are favoured in poleward flow while cool dry conditions occur in equatorward flow. Although there is an infinite variety of weather systems, one area of recent substantial progress is recognition that a few preferred patterns (or modes) of variability determine the main seasonal and longer-term climate anomalies (Section 3.6). These patterns arise from the differential effects on the atmosphere of land and ocean, mountains, and anomalous heating, such as occurs during El Niño events. The response is generally felt in regions far removed from the anomalous forcing through atmospheric teleconnections, associated with large-scale waves in the atmosphere. This chapter therefore documents some aspects of temperature and precipitation anomalies associated with these preferred patterns, as they are vitally important for understanding regional climate anomalies (such as observed cooling in parts of the northern North Atlantic from 1901 to 2005; see Section 3.2.2.7, Figure 3.9) and why they differ from global means. Changes in storm tracks, the jet streams, regions of preferred blocking anticyclones and changes in monsoons all occur in conjunction with these preferred patterns and other climate anomalies. Therefore the chapter not only documents changes in variables, but also changes in phenomena (such as El Niño) or patterns, in order to increase understanding of the character of change.

Extremes of climate, such as droughts and wet spells, are very important because of their large impacts on society and the environment, but they are an expression of the variability. Therefore, the nature of variability at different spatial and temporal scales is vital to our understanding of extremes. The global means of temperature and precipitation are most readily linked to global mean radiative forcing and are important because they clearly indicate if unusual change is occurring. However, the local or regional response can be complex and perhaps even counter-intuitive, such as changes in planetary waves in the atmosphere induced by global warming that result in regional cooling. As an indication of the complexity associated with temporal and spatial scales, Table 3.1 provides measures of the magnitude of natural variability of surface temperature in which climate signals are embedded. The measures used are indicators of the range: the mean range of the diurnal and annual cycles, and the estimated 5th to 95th percentiles range of anomalies. These are based on the standard deviation and assumed normal distribution, which is a reasonable approximation in many places for temperature, with the exception of continental interiors in the cold season, which have strongly negatively skewed temperature distributions owing to cold extremes. For the global mean, the variance is somewhat affected by the observed trend, which inflates this estimate of the range slightly. The comparison highlights the large diurnal cycle and daily variability. Daily variability is, however, greatly reduced by either spatial or temporal averaging that effectively averages over synoptic weather systems. Nevertheless, even continental-scale averages contain much greater variability than the global mean in association with planetary-scale waves and events such as El Niño.

Table 3.1. *Typical ranges of surface temperature at different spatial and temporal scales for a sample mid-latitude mid-continental station (Boulder, Colorado; based on 80 years of data) and for monthly mean anomalies (diurnal and annual cycles removed) for the USA as a whole and the globe for the 20th century. For the diurnal and annual cycles, the monthly mean range is given, while other values are the difference between the 5th and 95th percentiles.*

Temporal and Spatial Scale	Temperature Range (°C)
Boulder diurnal cycle	13.1 (December) to 15.1 (September)
Boulder annual cycle	23
Boulder daily anomalies	15
Boulder monthly anomalies	7.0
USA monthly anomalies	3.9
Global mean monthly anomalies	0.79

Throughout the chapter, the authors try to consistently indicate the degree of confidence and uncertainty in trends and other results, as given by Box TS.1 in the Technical Summary. Quantitative estimates of uncertainty include: for the mean, the 5th and 95th percentiles; and for trends, statistical significance at the 0.05 (5%) significance level. This allows assessment of what is unusual. The chapter mainly uses the word 'trend' to designate a generally monotonic change in the level of a variable. Where numerical values are given, they are equivalent linear trends, though more complex changes in the variable will often be clear from the description. The chapter also assesses, if possible, the physical consistency among different variables, which helps to provide additional confidence in trends.

3.2   Changes in Surface Climate: Temperature

3.2.1   Background

Improvements have been made to both land surface air temperature and sea surface temperature (SST) databases during the six years since the TAR was published. Jones and Moberg (2003) revised and updated the Climatic Research Unit (CRU) monthly land-surface air temperature record, improving coverage particularly in the Southern Hemisphere (SH) in the late 19th century. Further revisions by Brohan et al. (2006) include a comprehensive reassessment of errors together with an extension back to 1850. Under the auspices of the World Meteorological Organization (WMO) and the Global Climate Observing System (GCOS), daily temperature (together with precipitation and pressure) data for an increasing number of land stations have also become available, allowing more detailed assessment of extremes (see Section 3.8), as well as potential

urban influences on both large-scale temperature averages and microclimate. A new gridded data set of monthly maximum and minimum temperatures has updated earlier work (Vose et al., 2005a). For the oceans, the International Comprehensive Ocean-Atmosphere Data Set (ICOADS) has been extended by blending the former COADS with the UK's Marine Data Bank and newly digitised data, including the US Maury Collection and Japan's Kobe Collection. As a result, coverage has been improved substantially before 1920, especially over the Pacific, with further modest improvements up to 1950 (Worley et al., 2005; Rayner et al., 2006). Improvements have also been made in the bias reduction of satellite-based infrared (Reynolds et al., 2002) and microwave (Reynolds et al., 2004; Chelton and Wentz, 2005) retrievals of SST for the 1980s onwards. These data represent ocean skin temperature (Section 3.2.2.3), not air temperature or SST, and so must be adjusted to match the latter. Satellite infrared and microwave imagery can now also be used to monitor land surface temperature (Peterson et al., 2000; Jin and Dickinson, 2002; Kwok and Comiso, 2002b). Microwave imagery must allow for variations in surface emissivity and cannot act as a surrogate for air temperature over either snow-covered (Peterson et al., 2000) or sea-ice areas. As satellite-based records are still short in duration, all regional and hemispheric temperature series shown in this section are based on conventional surface-based data sets, except where stated.

Despite these improvements, substantial gaps in data coverage remain, especially in the tropics and the SH, particularly Antarctica. These gaps are largest in the 19th century and during the two world wars. Accordingly, advanced interpolation and averaging techniques have been applied when creating global data sets and hemispheric and global averages (Smith and Reynolds, 2005), and advanced techniques have also been used in the estimation of errors (Brohan et al., 2006), both locally and on a global basis (see Appendix 3.B.1). These errors, as well as the influence of decadal and multi-decadal variability in the climate, have been taken into account when estimating linear trends and their uncertainties (see Appendix 3.A). Estimates of surface temperature from ERA-40 reanalyses have been shown to be of climate quality (i.e., without major time-varying biases) at large scales from 1979 (Simmons et al., 2004). Improvements in ERA-40 over NRA arose from both improved data sources and better assimilation techniques (Uppala et al., 2005). The performance of ERA-40 was degraded prior to the availability of satellite data in the mid-1970s (see Appendix 3.B.5).

3.2.2   Temperature in the Instrumental Record for Land and Oceans

3.2.2.1   Land-Surface Air Temperature

Figure 3.1 shows annual global land-surface air temperatures, relative to the period 1961 to 1990, from the improved analysis (CRU/Hadley Centre gridded land-surface air temperature version 3; CRUTEM3) of Brohan et al. (2006). The long-term variations are in general agreement with those from the operational version of the Global Historical Climatology Network

(GHCN) data set (National Climatic Data Center (NCDC); Smith and Reynolds, 2005; Smith et al. 2005), and with the National Aeronautics and Space Administration's (NASA) Goddard Institute for Space Studies (GISS; Hansen et al., 2001) and Lugina et al. (2005) analyses (Figure 3.1). Most of the differences arise from the diversity of spatial averaging techniques. The global average for CRUTEM3 is a land-area weighted sum ($0.68 \times$ NH + $0.32 \times$ SH). For NCDC it is an area-weighted average of the grid-box anomalies where available worldwide. For GISS it is the average of the anomalies for the zones 90°N to 23.6°N, 23.6°N to 23.6°S and 23.6°S to 90°S with weightings 0.3, 0.4 and 0.3, respectively, proportional to their total areas. For Lugina et al. (2005) it is (NH + $0.866 \times$ SH) / 1.866 because they excluded latitudes south of 60°S. As a result, the recent global trends are largest in CRUTEM3 and NCDC, which give more weight to the NH where recent trends have been greatest (Table 3.2).

Further, small differences arise from the treatment of gaps in the data. The GISS gridding method favours isolated island and coastal sites, thereby reducing recent trends, and Lugina et al. (2005) also obtain reduced recent trends owing to their optimal interpolation method that tends to adjust anomalies towards zero where there are few observations nearby (see, e.g., Hurrell and Trenberth, 1999). The NCDC analysis, which begins in 1880, is higher than CRUTEM3 by between 0.1°C and 0.2°C in the first half of the 20th century and since the late 1990s. This is probably because its anomalies have been interpolated to be spatially complete: an earlier but very similar version (CRUTEM2v; Jones and Moberg, 2003) agreed very closely with NCDC when the global averages were calculated in the same way (Vose et al., 2005b). Differences may also arise because the numbers of stations used by CRUTEM3, NCDC and GISS differ (4,349, 7,230 and >7,200 respectively), although many of the basic station data are in common. Differences in station numbers relate principally to CRUTEM3 requiring series to have sufficient data between 1961 and 1990 to allow the calculation of anomalies (Brohan et al., 2006). Further differences may have arisen from differing homogeneity adjustments (see also Appendix 3.B.2).

Trends and low-frequency variability of large-scale surface air temperature from the ERA-40 reanalysis and from CRUTEM2v (Jones and Moberg, 2003) are in general agreement from the late 1970s onwards (Simmons et al., 2004). When ERA-40 is sub-sampled to match the Jones and Moberg coverage, correlations between monthly hemispheric- and continental-scale averages exceed 0.96, although trends in ERA-40 are then 0.03°C and 0.07°C per decade (NH and SH, respectively) lower than Jones and Moberg (2003). The ERA-40 reanalysis is more homogeneous than previous reanalyses (see Section 3.2.1 and

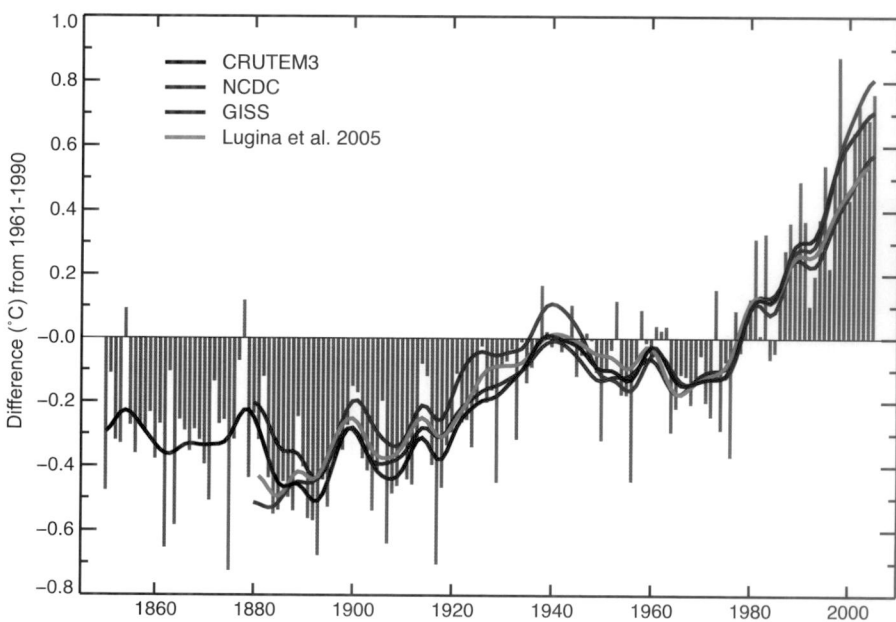

Figure 3.1. *Annual anomalies of global land-surface air temperature (°C), 1850 to 2005, relative to the 1961 to 1990 mean for CRUTEM3 updated from Brohan et al. (2006). The smooth curves show decadal variations (see Appendix 3.A). The black curve from CRUTEM3 is compared with those from NCDC (Smith and Reynolds, 2005; blue), GISS (Hansen et al., 2001; red) and Lugina et al. (2005; green).*

Appendix 3.B.5.4) but is not completely independent of the Jones and Moberg data (Simmons et al., 2004). The warming trends continue to be greatest over the continents of the NH (see maps in Section 3.2.2.7, Figures 3.9 and 3.10), in line with the TAR. Issues of homogeneity of terrestrial air temperatures are discussed in Appendix 3.B.2.

Table 3.2 provides trend estimates from a number of hemispheric and global temperature databases. Warming since 1979 in CRUTEM3 has been 0.27°C per decade for the globe, but 0.33°C and 0.13°C per decade for the NH and SH, respectively. Brohan et al. (2006) and Rayner et al. (2006) (see Section 3.2.2.3) provide uncertainties for annual estimates, incorporating the effects of measurement and sampling error, and uncertainties regarding biases due to urbanisation and earlier methods of measuring SST. These factors are taken into account, although ignoring their serial correlation. In Table 3.2, the effects of persistence on error bars are accommodated using a red noise approximation, which effectively captures the main influences. For more extensive discussion see Appendix 3.A

From 1950 to 2004, the annual trends in minimum and maximum land-surface air temperature averaged over regions with data were 0.20°C per decade and 0.14°C per decade, respectively, with a trend in diurnal temperature range (DTR) of –0.07°C per decade (Vose et al., 2005a; Figure 3.2). This is consistent with the TAR where data extended from 1950 to 1993; spatial coverage is now 71% of the terrestrial surface instead of 54% in the TAR, although tropical areas are still under-represented. Prior to 1950, insufficient data are available to develop global-scale maps of maximum and minimum temperature trends. For 1979 to 2004, the corresponding linear trends for the land areas where data are available were 0.29°C

Table 3.2. *Linear trends in hemispheric and global land-surface air temperatures, SST (shown in table as HadSST2) and Nighttime Marine Air Temperature (NMAT; shown in table as HadMAT1). Annual averages, with estimates of uncertainties for CRU and HadSST2, were used to estimate trends. Trends with 5 to 95% confidence intervals and levels of significance (**bold: <1%**; italic, 1–5%) were estimated by Restricted Maximum Likelihood (REML; see Appendix 3.A), which allows for serial correlation (first order autoregression AR1) in the residuals of the data about the linear trend. The Durbin Watson D-statistic (not shown) for the residuals, after allowing for first-order serial correlation, never indicates significant positive serial correlation.*

	Temperature Trend (°C per decade)		
Dataset	1850–2005	1901–2005	1979–2005
Land: Northern Hemisphere			
CRU (Brohan et al., 2006)	0.063 ± 0.015	0.089 ± 0.025	0.328 ± 0.087
NCDC (Smith and Reynolds, 2005)		0.072 ± 0.026	0.344 ± 0.096
GISS (Hansen et al., 2001)		0.083 ± 0.025	0.294 ± 0.074
Lugina et al. (2006)		0.079 ± 0.029	0.301 ± 0.075
Land: Southern Hemisphere			
CRU (Brohan et al., 2006)	0.036 ± 0.024	0.077 ± 0.029	*0.134 ± 0.070*
NCDC (Smith and Reynolds, 2005)		0.057 ± 0.017	0.220 ± 0.093
GISS (Hansen et al., 2001)		0.056 ± 0.012	*0.085 ± 0.055*
Lugina et al. (2005)		0.058 ± 0.011	0.091 ± 0.048
Land: Globe			
CRU (Brohan et al., 2006)	0.054 ± 0.016	0.084 ± 0.021	0.268 ± 0.069
NCDC (Smith and Reynolds, 2005)		0.068 ± 0.024	0.315 ± 0.088
GISS (Hansen et al., 2001)		0.069 ± 0.017	0.188 ± 0.069
Lugina et al. (2005)		0.069 ± 0.020	0.203 ± 0.058
Ocean: Northern Hemisphere			
UKMO HadSST2 (Rayner et al., 2006)	0.042 ± 0.016	0.071 ± 0.029	*0.190 ± 0.134*
UKMO HadMAT1 (Rayner et al., 2003) from 1861	0.038 ± 0.011	0.065 ± 0.020	0.186 ± 0.060
Ocean: Southern Hemisphere			
UKMO HadSST2 (Rayner et al., 2006)	0.036 ± 0.013	0.068 ± 0.015	0.089 ± 0.041
UKMO HadMAT1 (Rayner et al., 2003) from 1861	0.040 ± 0.012	0.069 ± 0.011	0.092 ± 0.050
Ocean: Globe			
UKMO HadSST2 (Rayner et al., 2006)	0.038 ± 0.011	0.067 ± 0.015	0.133 ± 0.047
UKMO HadMAT1 (Rayner et al., 2003) from 1861	0.039 ± 0.010	0.067 ± 0.013	0.135 ± 0.044

per decade for both maximum and minimum temperature with no trend for DTR. Diurnal temperature range is particularly sensitive to observing techniques, and monitoring it requires adherence to GCOS monitoring principles (GCOS, 2004). A map of the trend of annual DTR over the period 1979 to 2004 (Section 3.2.2.7, Figure 3.11) is discussed later in the chapter.

3.2.2.2 Urban Heat Islands and Land Use Effects

The modified land surface in cities affects the storage and radiative and turbulent transfers of heat and its partition into sensible and latent components (see Section 7.2 and Box 7.2). The relative warmth of a city compared with surrounding rural areas, known as the urban heat island (UHI) effect, arises from these changes and may also be affected by changes in water runoff, pollution and aerosols. Urban heat island effects are often very localised and depend on local climate factors such as windiness and cloudiness (which in turn depend on season), and on proximity to the sea. Section 3.3.2.4 discusses impacts of urbanisation on precipitation.

Many local studies have demonstrated that the microclimate within cities is on average warmer, with a smaller DTR, than if

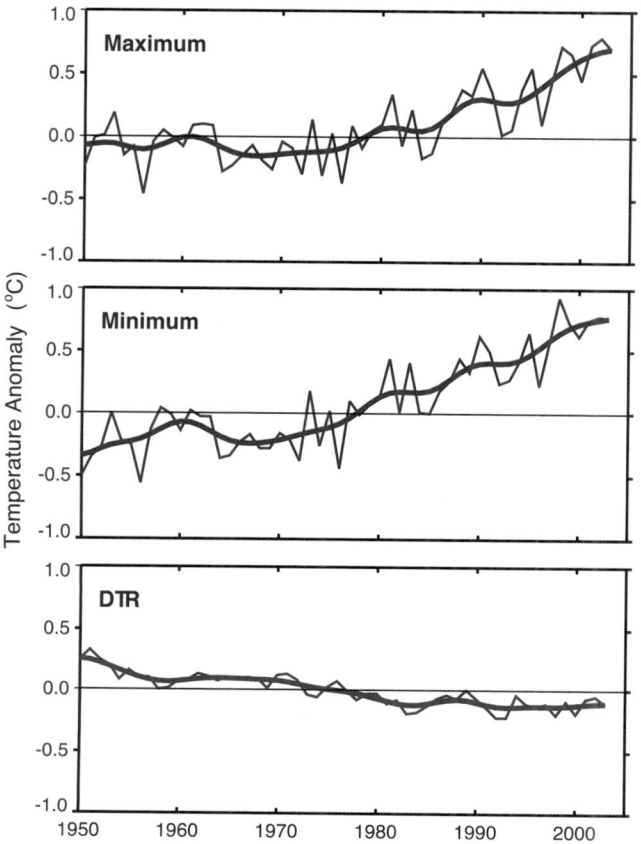

Figure 3.2. *Annual anomalies of maximum and minimum temperatures and DTR (°C) relative to the 1961 to 1990 mean, averaged for the 71% of global land areas where data are available for 1950 to 2004. The smooth curves show decadal variations (see Appendix 3.A). Adapted from Vose et al. (2005a).*

the city were not there. However, the key issue from a climate change standpoint is whether urban-affected temperature records have significantly biased large-scale temporal trends. Studies that have looked at hemispheric and global scales conclude that any urban-related trend is an order of magnitude smaller than decadal and longer time-scale trends evident in the series (e.g., Jones et al., 1990; Peterson et al., 1999). This result could partly be attributed to the omission from the gridded data set of a small number of sites (<1%) with clear urban-related warming trends. In a worldwide set of about 270 stations, Parker (2004, 2006) noted that warming trends in night minimum temperatures over the period 1950 to 2000 were not enhanced on calm nights, which would be the time most likely to be affected by urban warming. Thus, the global land warming trend discussed is very unlikely to be influenced significantly by increasing urbanisation (Parker, 2006). Over the conterminous USA, after adjustment for time-of-observation bias and other changes, rural station trends were almost indistinguishable from series including urban sites (Peterson, 2003; Figure 3.3, and similar considerations apply to China from 1951 to 2001 (Li et al., 2004). One possible reason for the patchiness of UHIs is the location of observing stations in parks where urban influences are reduced (Peterson, 2003). In summary, although some individual sites may be affected, including some small rural locations, the UHI effect is not pervasive, as all global-

scale studies indicate it is a very small component of large-scale averages. Accordingly, this assessment adds the same level of urban warming uncertainty as in the TAR: 0.006°C per decade since 1900 for land, and 0.002°C per decade since 1900 for blended land with ocean, as ocean UHI is zero. These uncertainties are added to the cool side of the estimated temperatures and trends, as explained by Brohan et al. (2006), so that the error bars in Section 3.2.2.4, Figures 3.6 and 3.7 and FAQ 3.1, Figure 1 are slightly asymmetric. The statistical significances of the trends in Table 3.2 and Section 3.2.2.4, Table 3.3 take account of this asymmetry.

McKitrick and Michaels (2004) and De Laat and Maurellis (2006) attempted to demonstrate that geographical patterns of warming trends over land are strongly correlated with geographical patterns of industrial and socioeconomic development, implying that urbanisation and related land surface changes have caused much of the observed warming. However, the locations of greatest socioeconomic development are also those that have been most warmed by atmospheric circulation changes (Sections 3.2.2.7 and 3.6.4), which exhibit large-scale coherence. Hence, the correlation of warming with industrial and socioeconomic development ceases to be statistically significant. In addition, observed warming has been, and transient greenhouse-induced warming is expected to be, greater over land than over the oceans (Chapter 10), owing to the smaller thermal capacity of the land.

Comparing surface temperature estimates from the NRA with raw station time series, Kalnay and Cai (2003) concluded that more than half of the observed decrease in DTR in the eastern USA since 1950 was due to changes in land use, including urbanisation. This conclusion was based on the fact that the reanalysis did not explicitly include these factors, which would affect the observations. However, the reanalysis also did not explicitly include many other natural and anthropogenic

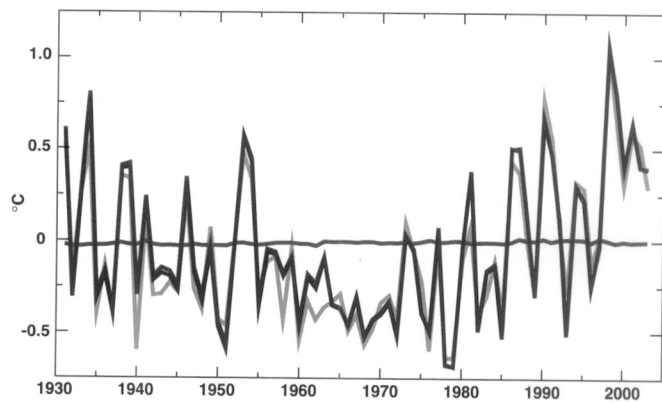

Figure 3.3. *Anomaly (°C) time series relative to the 1961 to 1990 mean of the full US Historical Climatology Network (USHCN) data (red), the USHCN data without the 16% of the stations with populations of over 30,000 within 6 km in the year 2000 (blue), and the 16% of the stations with populations over 30,000 (green). The full USHCN set minus the set without the urban stations is shown in magenta. Both the full data set and the data set without the high-population stations had stations in all of the 2.5° latitude by 3.5° longitude grid boxes during the entire period plotted, but the subset of high-population stations only had data in 56% of these grid boxes. Adapted from Peterson and Owen (2005).*

effects, such as increasing greenhouse gases and observed changes in clouds or soil moisture (Trenberth, 2004). Vose et al. (2004) showed that the adjusted station data for the region (for homogeneity issues, see Appendix 3.B.2) do not support Kalnay and Cai's conclusions. Nor are Kalnay and Cai's results reproduced in the ERA-40 reanalysis (Simmons et al., 2004). Instead, most of the changes appear related to abrupt changes in the type of data assimilated into the reanalysis, rather than to gradual changes arising from land use and urbanisation changes. Current reanalyses may be reliable for estimating trends since 1979 (Simmons et al., 2004) but are in general unsuited for estimating longer-term global trends, as discussed in Appendix 3.B.5.

Nevertheless, changes in land use can be important for DTR at the local-to-regional scale. For instance, land degradation in northern Mexico resulted in an increase in DTR relative to locations across the border in the USA (Balling et al., 1998), and agriculture affects temperatures in the USA (Bonan, 2001; Christy et al., 2006). Desiccation of the Aral Sea since 1960 raised DTR locally (Small et al., 2001). By processing maximum and minimum temperature data as a function of day of the week, Forster and Solomon (2003) found a distinctive 'weekend effect' in DTR at stations examined in the USA, Japan, Mexico and China. The weekly cycle in DTR has a distinctive large-scale pattern with geographically varying sign, and strongly suggests an anthropogenic effect on climate, likely through changes in pollution and aerosols (Jin et al., 2005). Section 7.2 provides fuller discussion of the effects of land use changes.

3.2.2.3   Sea Surface Temperature and Marine Air Temperature

Most analyses of SST estimate the subsurface bulk temperature (i.e., the temperature in the uppermost few metres of the ocean), not the ocean skin temperature measured by satellites. For maximum resolution and data coverage, polar-orbiting infrared satellite data since 1981 can be used so long as the satellite ocean skin temperatures are adjusted to estimate bulk SST values through a calibration procedure (see e.g., Reynolds et al., 2002; Rayner et al., 2003, 2006; Appendix 3.B.3). But satellite SST data alone have not been used as a major resource for estimating climate change because of their strong time-varying biases which are hard to completely remove, for example, as shown in Reynolds et al. (2002) for the Pathfinder polar orbiting satellite SST data set (Kilpatrick et al., 2001). Figures 3.9 and 3.10 (Section 3.2.2.7) do, however, make use of spatial relationships based on adjusted satellite SST estimates after November 1981 to provide nearer-to-global coverage for the 1979 to 2005 period, and O'Carroll et al. (2006) have developed an analysis based on Along-Track Scanning Radiometers (ATSRs) with potential for the future. However, satellite data are unable to fill in estimates of surface temperature over or near sea ice areas.

Recent bulk SSTs estimated using ship and buoy data also have time-varying biases (e.g., Christy et al., 2001; Kent and Kaplan, 2006) that are larger than originally estimated by Folland et al. (1993), but not large enough to prejudice conclusions

about recent warming (see Appendix 3.B.3). As reported in the TAR, a combined physical-empirical method (Folland and Parker, 1995) is mainly used to estimate adjustments to ship SST data obtained up to 1941 to compensate for heat losses from uninsulated (mainly canvas) or partly insulated (mainly wooden) buckets. Details are given in Appendix 3.B.3.

The SST analyses of Rayner et al. (2003) and Smith and Reynolds (2004) are interpolated to fill missing data areas. The main problem for estimating climate variations in the presence of large data gaps is underestimation of change, as most interpolation procedures tend to bias the analysis towards the modern climatologies used in these data sets (Hurrell and Trenberth, 1999). To address non-stationary aspects, Rayner et al. (2003) extracted the leading global covariance pattern, which represents long-term changes, before interpolating using reduced-space optimal interpolation (see Appendix 3.B.1); and Smith and Reynolds removed a smoothed, moving 15-year-average field before interpolating by a related technique.

Figure 3.4a shows annual and decadally smoothed anomalies of global SST from the new, uninterpolated Hadley Centre SST data set version 2 (HadSST2) analysis (Rayner et al., 2006). Figure 3.4a also shows NMAT (referred to as HadMAT: Hadley Centre Marine Air Temperature data set), which is used to avoid daytime heating of ship decks (Bottomley et al., 1990). The global averages are ocean-area weighted sums ($0.44 \times$ NH $+ 0.56 \times$ SH). The HadMAT analysis includes limited optimal interpolation (Rayner et al., 2003) and was chosen because of the demonstration by Folland et al. (2003) of its skill in the sparsely observed South Pacific from the late 19th century onwards, but major gaps (e.g., the Southern Ocean) are not interpolated. Although HadMAT data have been corrected for warm biases during World War II they may still be too warm in the NH and too cool in the SH at that time (Figure 3.4c,d). However, global HadSST2 and HadMAT generally agree well, especially after the 1880s. The SST analysis in the TAR is included in Figure 3.4a. The changes in SST since the TAR are generally fairly small, though the new SST analysis is warmer around 1880 and cooler in the 1950s. The peak warmth in the early 1940s is likely to have arisen partly from closely spaced multiple El Niño events (Brönnimann et al., 2004; see also Section 3.6.2) and also due to the warm phase of the Atlantic Multi-decadal Oscillation (AMO; see Section 3.6.6). The HadMAT data generally confirm the hemispheric SST trends in the 20th century (Figure 3.4c,d and Table 3.2). Overall, the SST data should be regarded as more reliable because averaging of fewer samples is needed for SST than for HadMAT to remove synoptic weather noise. However, the changes in SST relative to NMAT since 1991 in the tropical Pacific may be partly real (Christy et al., 2001). As the atmospheric circulation changes, the relationship between SST and surface air temperature anomalies can change along with surface fluxes. Interannual variations in the heat fluxes to the atmosphere can exceed 100 W m^{-2} locally in individual months, but the main prolonged variations occur with the El Niño-Southern Oscillation (ENSO), where changes in the central tropical Pacific exceed ±50 W m^{-2} for many months during major ENSO events (Trenberth et al., 2002a).

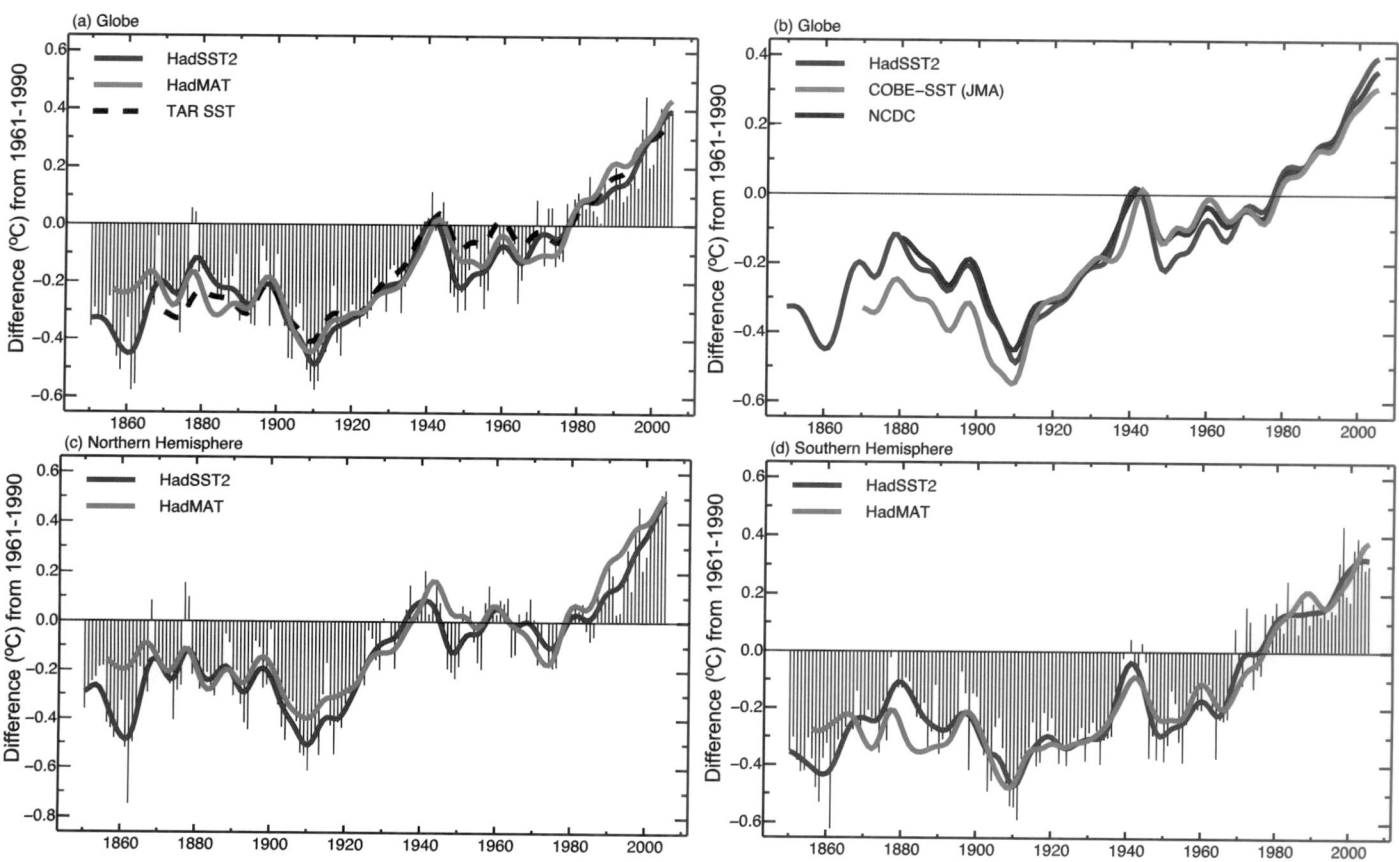

Figure 3.4. *(a) Annual anomalies of global SST (HadSST2; red bars and blue solid curve), 1850 to 2005, and global NMAT (HadMAT, green curve), 1856 to 2005, relative to the 1961 to 1990 mean (°C) from the UK Meteorological Office (UKMO; Rayner et al., 2006). The smooth curves show decadal variations (see Appendix 3.A). The dashed black curve shows equivalent smoothed SST anomalies from the TAR. (b) Smoothed annual global SST anomalies, relative to 1961 to 1990 (°C), from HadSST2 (blue line, as in (a)), from NCDC (Smith et al., 2005; red line) and from COBE-SST (Ishii et al., 2005; green line). The latter two series begin later in the 19th century than HadSST2. (c,d) As in (a) but for the NH and SH showing only the UKMO series.*

Figure 3.4b shows three time series of changes in global SST. Neither the HadSST2 series (as in Figure 3.4a) nor the NCDC series include polar-orbiting satellite data because of possible time-varying biases that remain difficult to correct fully (Rayner et al., 2003). The Japanese series (Ishii et al., 2005; referred to as Centennial *in-situ* Observation-Based Estimates of SSTs (COBE-SST) from the Japan Meteorological Agency (JMA)) is also *in situ* except for the specification of sea ice. The warmest year globally in each SST record was 1998 (0.44°C, 0.38°C and 0.37°C above the 1961 to 1990 average for HadSST2, NCDC and COBE-SST, respectively). The five warmest years in all analyses have occurred after 1995.

Understanding of the variability and trends in different oceans is still developing, but it is already apparent that they are quite different. The Pacific is dominated by ENSO and modulated by the Pacific Decadal Oscillation (PDO), which may provide ways of moving heat from the tropical ocean to higher latitudes and out of the ocean into the atmosphere (Trenberth et al., 2002a), thereby greatly altering how trends are manifested. In the Atlantic, observations reveal the role of the AMO (Folland et al., 1999; Delworth and Mann, 2000; Enfield et al., 2001; Goldenberg et al., 2001; Section 3.6.6 and Figure 3.33). The AMO is likely to be associated with

the Thermohaline Circulation (THC), which transports heat northwards, thereby moderating the tropics and warming the high latitudes. In the Indian Ocean, interannual variability is small compared with the trend. Figure 3.5 presents latitude-time sections from 1900 for SSTs (from HadSST2) for the zonal mean across each ocean, filtered to remove fluctuations of less than about six years, including the ENSO signal. In the Pacific, the long-term warming is clearly evident, but punctuated by cooler episodes centred in the tropics, and no doubt linked to the PDO. The prolonged 1939–1942 El Niño shows up as a warm interval. In the Atlantic, the warming from the 1920s to about 1940 in the NH was focussed on higher latitudes, with the SH remaining cool. This inter-hemispheric contrast is believed to be one signature of the THC (Zhang and Delworth, 2005). The subsequent relative cooling in the NH extratropics and the more recent intense warming in NH mid-latitudes was predominantly a multi-decadal variation of SST; only in the last decade is an overall warming signal clearly emerging. Therefore, the recent strong warming appears to be related in part to the AMO in addition to a global warming signal (Section 3.6.6). The cooling in the northwestern North Atlantic just south of Greenland, reported in the SAR, has now been replaced by strong warming (see also Section 3.2.2.7, Figures 3.9 and 3.10; also Figures

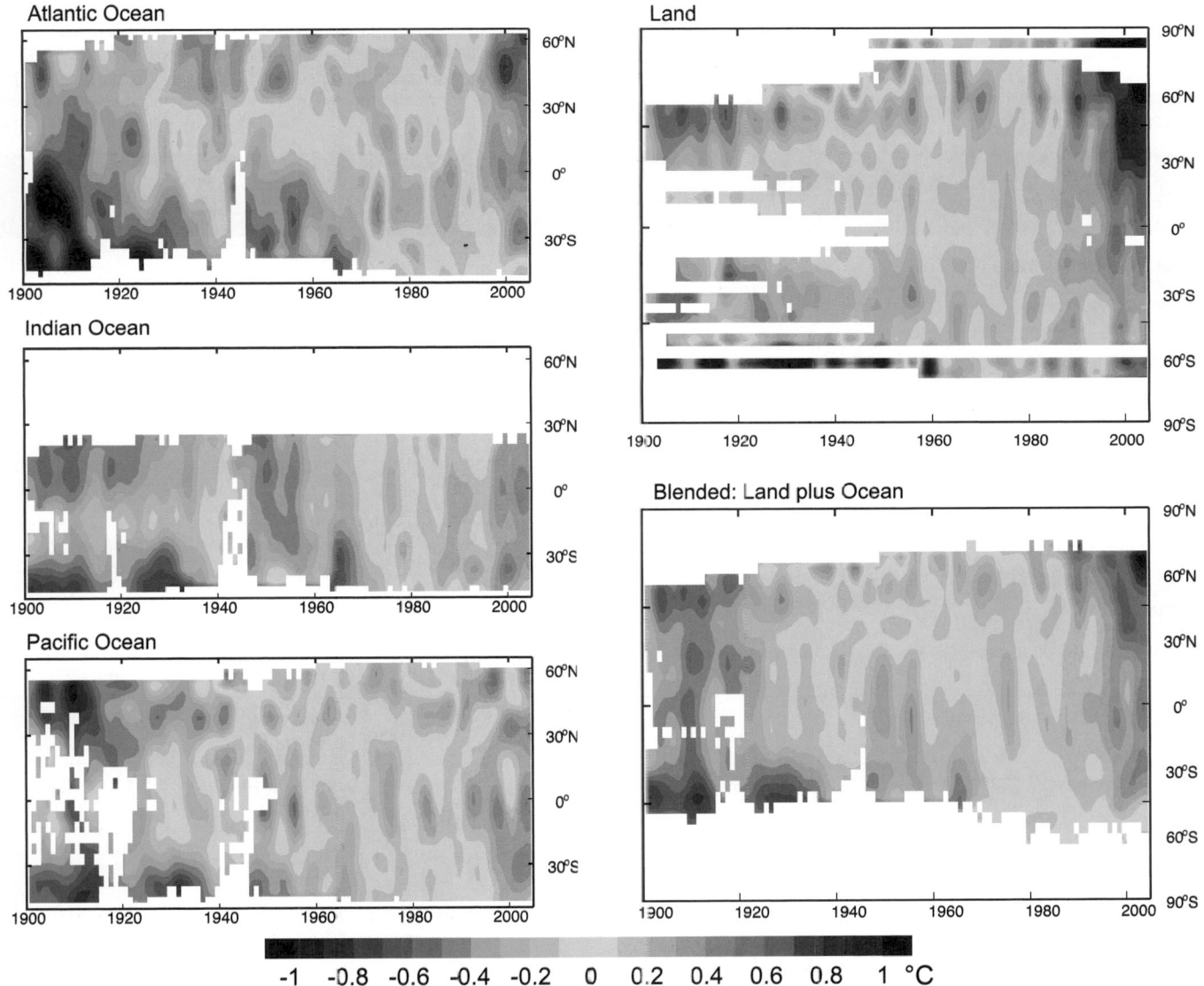

Figure 3.5. *Latitude-time sections of zonal mean temperature anomalies (°C) from 1900 to 2005, relative to the 1961 to 1990 mean. Left panels: SST annual anomalies across each ocean from HadSST2 (Rayner et al., 2006). Right panels: Surface temperature annual anomalies for land (top, CRUTEM3) and land plus ocean (bottom, HadCRUT3). Values are smoothed with the 5-point filter to remove fluctuations of less than about six years (see Appendix 3.A); and white areas indicate missing data.*

5.1 and 5.2 for ocean heat content). The Indian Ocean also reveals a poorly observed warm interval in the early 1940s, and further shows the fairly steady warming in recent years. The multi-decadal variability in the Atlantic has a much longer time scale than that in the Pacific, but it is noteworthy that all oceans exhibit a warm period around the early 1940s.

3.2.2.4　Land and Sea Combined Temperature: Global, Northern Hemisphere, Southern Hemisphere and Zonal Means

Gridded data sets combining land-surface air temperature and SST anomalies have been developed and maintained by three groups: CRU with the UKMO Hadley Centre in the UK (HadCRUT3; Brohan et al., 2006) and NCDC (Smith and Reynolds, 2005) and GISS (Hansen et al., 2001) in the USA. Although the component data sets differ slightly (see Sections 3.2.2.1 and 3.2.2.3) and the combination methods also differ, trends are similar. Table 3.3 provides comparative estimates of linear trends. Overall warming since 1901 has been a little less in the NCDC and GISS analysis than in the HadCRUT3 analysis. All series indicate that the warmest five years have occurred after 1997, although there is slight disagreement about the ordering. The HadCRUT3 data set shows 1998 as warmest, while 2005 is warmest in NCDC and GISS data. Thus the year 2005, with no El Niño, was about as warm globally as 1998 with its major El Niño effects. The GISS analysis of 2005 interpolated the exceptionally warm conditions in the extreme north of Eurasia and North America over the Arctic Ocean (see Figure 3.5). If the GISS data for 2005 are averaged only south of 75°N, then 2005 is cooler than 1998. In addition, there were relatively cool anomalies in 2005 in HadCRUT3 in parts of Antarctica and the Southern Ocean, where sea ice coverage (see Chapter 4) has not declined.

Table 3.3. *Linear trends (°C per decade) in hemispheric and global combined land-surface air temperatures and SST. Annual averages, along with estimates of uncertainties for CRU/UKMO (HadCRUT3), were used to estimate trends. For CRU/UKMO, global annual averages are the simple average of the two hemispheres. For NCDC and GISS the hemispheres are weighted as in Section 3.2.2.1. Trends are estimated and presented as in Table 3.2. R^2 is the squared trend correlation (%). The Durbin Watson D-statistic (not shown) for the residuals, after allowing for first-order serial correlation, never indicated significant positive serial correlation, and plots of the residuals showed virtually no long-range persistence.*

Dataset	Temperature Trend (°C per decade)		
	1850–2005	1901–2005	1979–2005
Northern Hemisphere			
CRU/UKMO (Brohan et al., 2006)	0.047 ± 0.013 R^2=54	0.075 ± 0.023 R^2=63	0.234 ± 0.070 R^2=69
NCDC (Smith and Reynolds, 2005)		0.063 ± 0.022 R^2=55	0.245 ± 0.062 R^2=72
Southern Hemisphere			
CRU/UKMO (Brohan et al., 2006)	0.038 ± 0.014 R^2=51	0.068 ± 0.017 R^2=74	0.092 ± 0.038 R^2=48
NCDC (Smith and Reynolds, 2005)		0.066 ± 0.009 R^2=82	0.096 ± 0.038 R^2=58
Globe			
CRU/UKMO (Brohan et al., 2006)	0.042 ± 0.012 R^2=57	0.071 ± 0.017 R^2=74	0.163 ± 0.046 R^2=67
NCDC (Smith and Reynolds, 2005)		0.064 ± 0.016 R^2=71	0.174 ± 0.051 R^2=72
GISS (Hansen et al., 2001)		0.060 ± 0.014 R^2=70	0.170 ± 0.047 R^2=67

Hemispheric and global series based on Brohan et al. (2006) are shown in Figure 3.6 and tropical and polar series in Figure 3.7. Owing to the sparsity of SST data, the polar series are for land only. The recent warming is strongest in the NH extratropics, while El Niño events are clearly evident in the tropics, particularly the 1997–1998 event that makes 1998 the warmest year in HadCRUT3. The warming over land in the Arctic north of 65°N (Figure 3.7) is more than double the warming in the global mean from the 19th century to the 21st century and also from about the late 1960s to the present. In the arctic series, 2005 is the warmest year. A slightly longer warm period, almost as warm as the present, was observed from the late 1920s to the early 1950s. Although data coverage was limited in the first half of the 20th century, the spatial pattern of the earlier warm period appears to have been different from that of the current warmth. In particular, the current warmth is partly linked to the Northern Annular Mode (NAM; see Section 3.6.4) and affects a broader region (Polyakov et al., 2003). Temperatures over mainland Antarctica (south of 65°S) have not warmed in recent decades (Turner et al., 2005), but it is virtually certain that there has been strong warming over the last 50 years in the Antarctic Peninsula region (Turner et al., 2005; see the discussion of changes in the Southern Annular Mode (SAM) and Figure 3.32 in Section 3.6.5).

3.2.2.5 Consistency between Land and Ocean Surface Temperature Changes

The course of temperature change over the 20th century, revealed by the independent analysis of land air temperatures, SST and NMAT, is generally consistent (Figure 3.8). Warming occurred in two distinct phases (1915–1945 and since 1975), and it has been substantially stronger over land than over the oceans in the later phase, as shown also by the trends in Table 3.2. The land component has also been more variable from year to year (compare Figures 3.1 and 3.4a,c,d). Much of the recent difference between global SST (and NMAT) and global land air temperature trends has arisen from accentuated warming over the continents in the mid-latitude NH (Section 3.2.2.7, Figures 3.9 and 3.10). This is likely to be related to greater evaporation and heat storage in the ocean, and in particular to atmospheric circulation changes in the winter half-year due to the North Atlantic Oscillation (NAO)/NAM (see discussion in Section 3.6.4). Accordingly the differences between NH and SH temperatures follow a course similar to the plot of land air temperature minus SST shown in Figure 3.8.

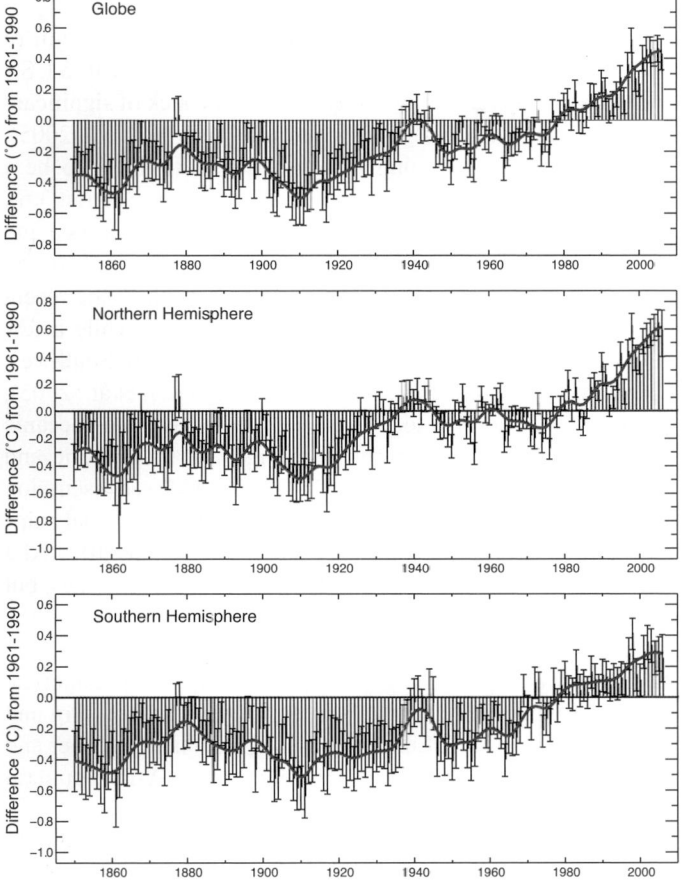

Figure 3.6. *Global and hemispheric annual combined land-surface air temperature and SST anomalies (°C) (red) for 1850 to 2006 relative to the 1961 to 1990 mean, along with 5 to 95% error bar ranges, from HadCRUT3 (adapted from Brohan et al., 2006). The smooth blue curves show decadal variations (see Appendix 3.A).*

3.2.2.6 Temporal Variability of Global Temperatures and Recent Warming

The standard deviation of the HadCRUT3 annual average temperatures for the globe for 1850 to 2005 shown in Figure 3.6 is 0.24°C. The greatest difference between two consecutive years in the global average since 1901 is 0.29°C between 1976 and 1977, demonstrating the importance of the 0.75°C and 0.74°C temperature increases (the HadCRUT3 linear trend estimates for 1901 to 2005 and 1906 to 2005, respectively) in a centennial time-scale context. However, both trends are small compared with interannual variations at one location, and much smaller than day-to-day variations (Table 3.1).

The principal conclusion from the three global analyses is that the global average surface temperature trend has very likely been slightly more than 0.65°C ± 0.2°C over the period from 1901 to 2005 (Table 3.3), a warming greater than any since at least the 11th century (see Chapter 6). A HadCRUT3 linear trend over the 1906 to 2005 period yields a warming of 0.74°C ± 0.18°C, but this rate almost doubles for the last 50 years (0.64°C ± 0.13°C for 1956 to 2005; see FAQ 3.1).

Clearly, the changes are not linear and can also be characterised as level prior to about 1915, a warming to about 1945, levelling out or even a slight decrease until the 1970s, and a fairly linear upward trend since then (Figure 3.6 and FAQ 3.1). Considered this way, the overall warming from the average of the first 50-year period (1850–1899) through 2001 to 2005 is 0.76°C ± 0.19°C. Clearly, the world's surface temperature has continued to increase since the TAR and the trend when computed in the same way as in the TAR remains 0.6°C over the 20th century. In view of Section 3.2.2.2 and the dominance of the globe by ocean, the influence of urbanisation on these estimates is estimated to be very small. The last 12 complete years (1995–2006) now contain 11 of the 12 warmest years since 1850, the earliest year for which comparable records are available. Only 1996 is not in this list – replaced by 1990. 2002 to 2005 are the 3rd, 4th, 5th and 2nd warmest years in the series, with 1998 the warmest in HadCRUT3 but with 2005 and 1998 switching order in GISS and NCDC. The HadCRUT3 surface warming

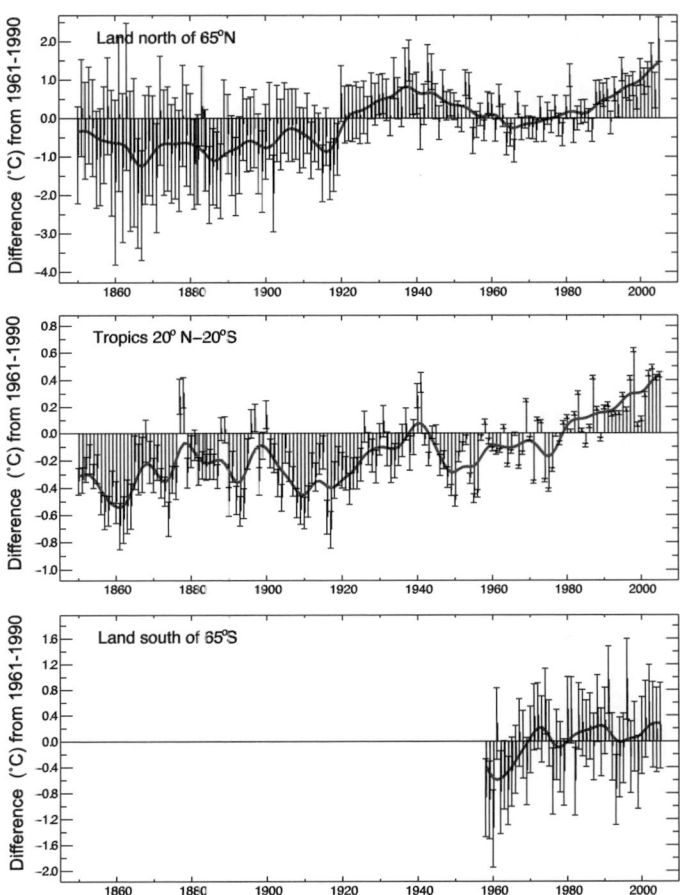

Figure 3.7. *Annual temperature anomalies (°C) up to 2005, relative to the 1961 to 1990 mean (red) with 5 to 95% error bars. The tropical series (middle) is combined land-surface air temperature and SST from HADCRUT3 (adapted from Brohan et al., 2006). The polar series (top and bottom) are land-only from CRUTEM3, because SST data are sparse and unreliable in sea ice zones. The smooth blue curves show decadal variations (see Appendix 3.A).*

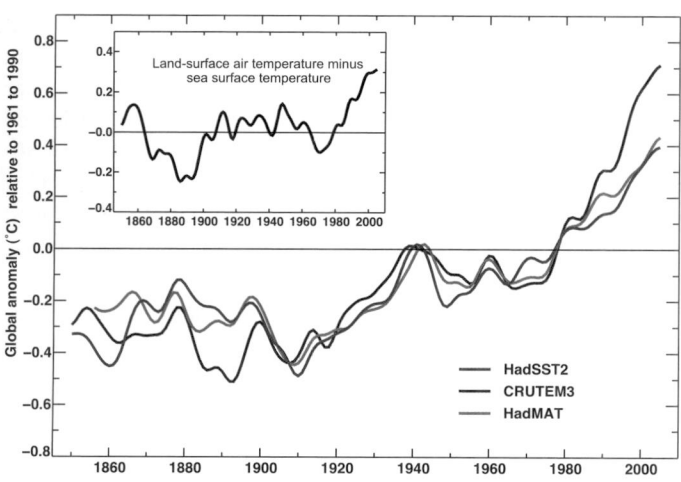

Figure 3.8. *Annual anomalies (°C) of global average SST (blue curve, begins 1850), NMAT (green curve, begins 1856) and land-surface air temperature (red curve, begins 1850) to 2005, relative to their 1961 to 1990 means (Brohan et al., 2006; Rayner et al., 2006). The smooth curves show decadal variations (see Appendix 3.A). Inset shows the smoothed differences between the land-surface air temperature and SST anomalies (i.e., red minus blue).*

trend over 1979 to 2005 was more than 0.16°C per decade, that is, a total warming of 0.44°C ± 0.12°C (the error bars overlap those of NCDC and GISS). During 2001 to 2005, the global average temperature anomaly has been 0.44°C above the 1961–1990 average. The value for 2006 is close to the 2001 to 2005 average.

3.2.2.7 Spatial Trend Patterns

Figure 3.9 illustrates the spatial patterns of annual surface temperature changes for 1901 to 2005 and 1979 to 2005, and Figure 3.10 shows seasonal trends for 1979 to 2005. All maps clearly indicate that differences in trends between locations can be large, particularly for shorter time periods. For the century-

long period, warming is statistically significant over most of the world's surface with the exception of an area south of Greenland and three smaller regions over the southeastern USA and parts of Bolivia and the Congo basin. The lack of significant warming at about 20% of the locations (Karoly and Wu, 2005), and the enhanced warming in other places, is likely to be a result of changes in atmospheric circulation (see Section 3.6). Warming is strongest over the continental interiors of Asia and northwestern North America and over some mid-latitude ocean regions of the SH as well as southeastern Brazil. In the recent period, some regions have warmed substantially while a few have cooled slightly on an annual basis (Figure 3.9). Southwest China has cooled since the mid-20th century (Ren et al., 2005), but most of the cooling locations since 1979 have been oceanic and in the SH, possibly through changes in atmospheric and oceanic circulation related to the PDO and SAM (see discussion in Section 3.6.5). Warming dominates most of the seasonal maps for the period 1979 onwards, but weak cooling has affected a few regions, especially the mid-latitudes of the SH oceans, but also over eastern Canada in spring, possibly in relation to the strengthening NAO (see Section 3.6.4, Figure 3.30). Warming in this period was strongest over western North America, northern Europe and China in winter, Europe and northern and eastern Asia in spring, Europe and North Africa in summer and northern North America, Greenland and eastern Asia in autumn (Figure 3.10).

No single location follows the global average, and the only way to monitor the globe with any confidence is to include observations from as many diverse places as possible. The importance of regions without adequate records is determined from complete model reanalysis fields (Simmons et al., 2004). The importance of the missing areas for hemispheric and global averages is incorporated into the errors bars in Figure 3.6 (see Brohan et al., 2006). Error bars are generally larger in the more data-sparse SH than in the NH; they are larger before the 1950s and largest of all in the 19th century.

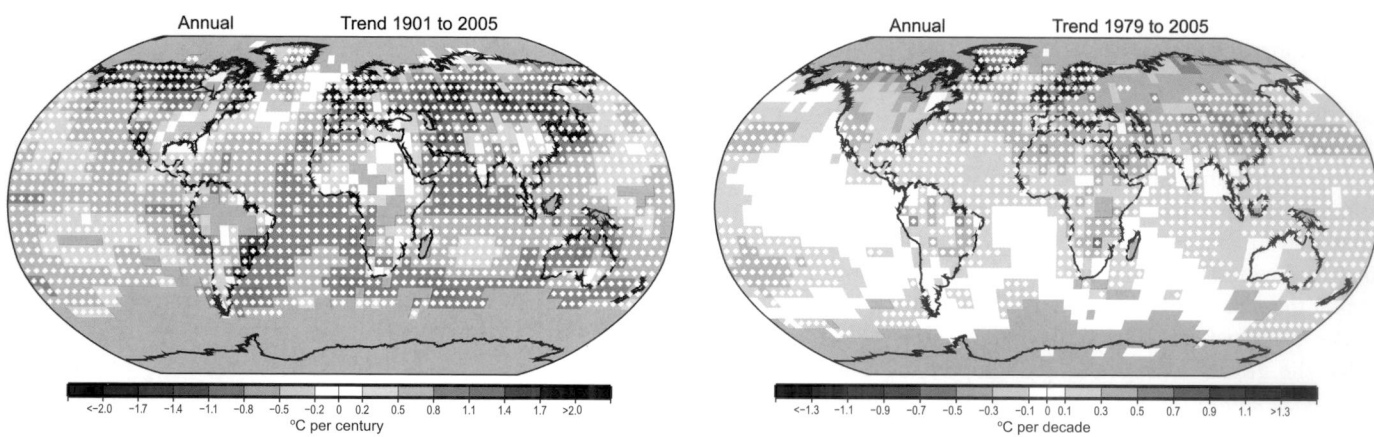

Figure 3.9. *Linear trend of annual temperatures for 1901 to 2005 (left; °C per century) and 1979 to 2005 (right; °C per decade). Areas in grey have insufficient data to produce reliable trends. The minimum number of years needed to calculate a trend value is 66 years for 1901 to 2005 and 18 years for 1979 to 2005. An annual value is available if there are 10 valid monthly temperature anomaly values. The data set used was produced by NCDC from Smith and Reynolds (2005). Trends significant at the 5% level are indicated by white + marks.*

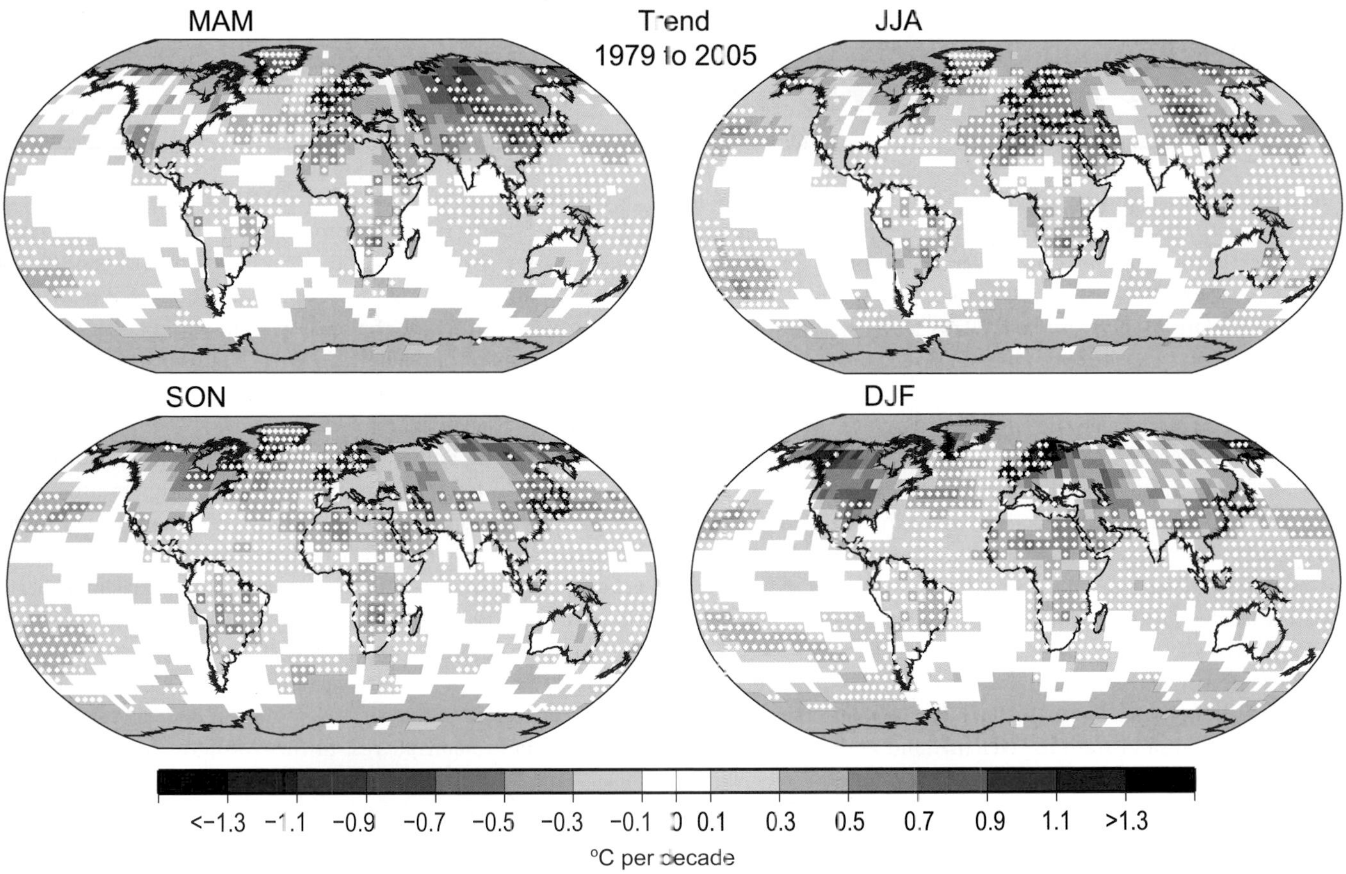

MAM   Trend   JJA
                              1979 to 2005

SON   DJF

<-1.3 -1.1 -0.9 -0.7 -0.5 -0.3 -0.1 0 0.1 0.3 0.5 0.7 0.9 1.1 >1.3

°C per decade

Figure 3.10. *Linear trend of seasonal MAM, JJA, SON and DJF temperature for 1979 to 2005 (°C per decade). Areas in grey have insufficient data to produce reliable trends. The minimum number of years required to calculate a trend value is 18. A seasonal value is available if there are two valid monthly temperature anomaly values. The dataset used was produced by NCDC from Smith and Reynolds (2005). Trends significant at the 5% level are indicated by white + marks.*

Figure 3.11 shows annual trends in DTR from 1979 to 2004. The decline in DTR since 1950 reported in the TAR has now ceased, as confirmed by Figure 3.2. Since 1979, daily minimum temperature increased in most areas except western Australia and southern Argentina, and parts of the western Pacific Ocean; and daily maximum temperature also increased in most regions except northern Peru, northern Argentina, northwestern Australia, and parts of the North Pacific Ocean (Vose et al., 2005a). The changes reported here appear inconsistent with Dai et al. (2006) who reported decreasing DTR in the USA, but this arises partly because Dai et al. (2006) included the high DTR years 1976 to 1978. Furthermore, Figure 3.11 is supported by many other recent regional-scale analyses.

Changes in cloud cover and precipitation explained up to 80% of the variance in historical DTR series for the USA, Australia, mid-latitude Canada and the former Soviet Union during the 20th century (Dai et al., 1999). Cloud cover accounted for nearly half of the change in the DTR in Fennoscandia during the 20th century (Tuomenvirta et al., 2000). Variations in atmospheric circulation also affect DTR. Changes in the frequency of certain synoptic weather types resulted in a decline in DTR during the cold half-year in the Arctic (Przybylak, 2000). A positive phase of the NAM (see Section 3.6.4) is associated with increased DTR in the northeastern USA and Canada (Wettstein and Mearns,

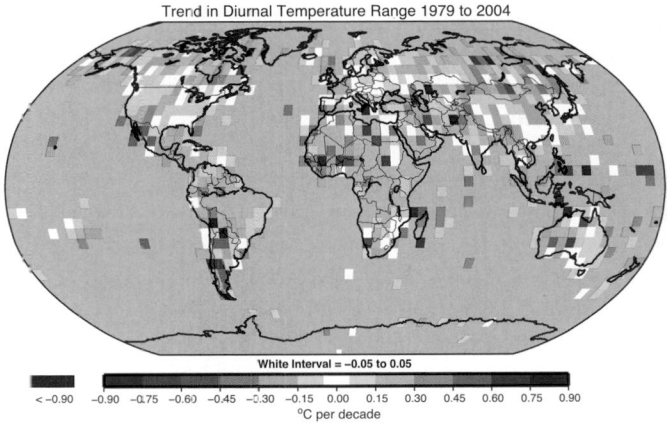

Trend in Diurnal Temperature Range 1979 to 2004

White Interval = −0.05 to 0.05

< −0.90 −0.90 −0.75 −0.60 −0.45 −0.30 −0.15 0.00 0.15 0.30 0.45 0.60 0.75 0.90
°C per decade

Figure 3.11. *Linear trend in annual mean DTR for 1979 to 2004 (°C per decade). Grey regions indicate incomplete or missing data (after Vose et al., 2005a).*

2002). Variations in sea level pressure patterns and associated changes in cloud cover partially accounted for increasing trends in cold-season DTR in the northwestern USA and decreasing trends in the south-central USA (Durre and Wallace, 2001). The relationship between DTR and anthropogenic forcings is complex, as these forcings can affect atmospheric circulation, as well as clouds through both greenhouses gases and aerosols.

Frequently Asked Question 3.1
How are Temperatures on Earth Changing?

Instrumental observations over the past 157 years show that temperatures at the surface have risen globally, with important regional variations. For the global average, warming in the last century has occurred in two phases, from the 1910s to the 1940s (0.35°C), and more strongly from the 1970s to the present (0.55°C). An increasing rate of warming has taken place over the last 25 years, and 11 of the 12 warmest years on record have occurred in the past 12 years. Above the surface, global observations since the late 1950s show that the troposphere (up to about 10 km) has warmed at a slightly greater rate than the surface, while the stratosphere (about 10–30 km) has cooled markedly since 1979. This is in accord with physical expectations and most model results. Confirmation of global warming comes from warming of the oceans, rising sea levels, glaciers melting, sea ice retreating in the Arctic and diminished snow cover in the Northern Hemisphere.

There is no single thermometer measuring the global temperature. Instead, individual thermometer measurements taken every day at several thousand stations over the land areas of the world are combined with thousands more measurements of sea surface temperature taken from ships moving over the oceans to produce an estimate of global average temperature every month. To obtain consistent changes over time, the main analysis is actually of anomalies (departures from the climatological mean at each site) as these are more robust to changes in data availability. It is now possible to use these measurements from 1850 to the present, although coverage is much less than global in the second half of the 19th century, is much better after 1957 when measurements began in Antarctica, and best after about 1980, when satellite measurements began.

Expressed as a global average, surface temperatures have increased by about 0.74°C over the past hundred years (between 1906 and 2005; see Figure 1). However, the warming has been neither steady nor the same in different seasons or in different locations. There was not much overall change from 1850 to about 1915, aside from ups and downs associated with natural variability but which may have also partly arisen from poor sampling. An increase (0.35°C) occurred in the global average temperature from the 1910s to the 1940s, followed by a slight cooling (0.1°C), and then a rapid warming (0.55°C) up to the end of 2006 (Figure 1). The warmest years of the series are 1998 and 2005 (which are statistically indistinguishable), and 11 of the 12 warmest years have occurred in the last 12 years (1995 to 2006). Warming, particularly since the 1970s, has generally been greater over land than over the oceans. Seasonally, warming has been slightly greater in the winter hemisphere. Additional warming occurs in cities and urban areas (often referred to as the urban heat island effect), but is confined in spatial extent, and its effects are allowed for both by excluding as many of the affected sites as possible from the global temperature data and by increasing the error range (the blue band in the figure).

A few areas have cooled since 1901, most notably the northern North Atlantic near southern Greenland. Warming during this time has been strongest over the continental interiors of Asia and northern North America. However, as these are areas with large year-to-year variability, the most evident warming signal has occurred in parts of the middle and lower latitudes, particularly the tropical oceans. In the lower left panel of Figure 1, which shows temperature trends since 1979, the pattern in the Pacific Ocean features warming and cooling regions related to El Niño.

Analysis of long-term changes in daily temperature extremes has recently become possible for many regions of the world (parts of North America and southern South America, Europe, northern and eastern Asia, southern Africa and Australasia). Especially since the 1950s, these records show a decrease in the number of very cold days and nights and an increase in the number of extremely hot days and warm nights (see FAQ 3.3). The length of the frost-free season has increased in most mid- and high-latitude regions of both hemispheres. In the Northern Hemisphere, this is mostly manifest as an earlier start to spring.

In addition to the surface data described above, measurements of temperature above the surface have been made with weather balloons, with reasonable coverage over land since 1958, and from satellite data since 1979. All data are adjusted for changes in instruments and observing practices where necessary. Microwave satellite data have been used to create a 'satellite temperature record' for thick layers of the atmosphere including the troposphere (from the surface up to about 10 km) and the lower stratosphere (about 10 to 30 km). Despite several new analyses with improved cross-calibration of the 13 instruments on different satellites used since 1979 and compensation for changes in observing time and satellite altitude, some uncertainties remain in trends.

For global observations since the late 1950s, the most recent versions of all available data sets show that the troposphere has warmed at a slightly greater rate than the surface, while the stratosphere has cooled markedly since 1979. This is in accord with physical expectations and most model results, which demonstrate the role of increasing greenhouse gases in tropospheric warming and stratospheric cooling; ozone depletion also contributes substantially to stratospheric cooling.

Consistent with observed increases in surface temperature, there have been decreases in the length of river and lake ice seasons. Further, there has been an almost worldwide reduction in glacial mass and extent in the 20th century; melting of the Greenland Ice Sheet has recently become apparent; snow cover has decreased in many Northern Hemisphere regions; sea ice thickness and extent have decreased in the Arctic in all seasons, most dramatically in spring and summer; the oceans are warming; and sea level is rising due to thermal expansion of the oceans and melting of land ice.

(continued)

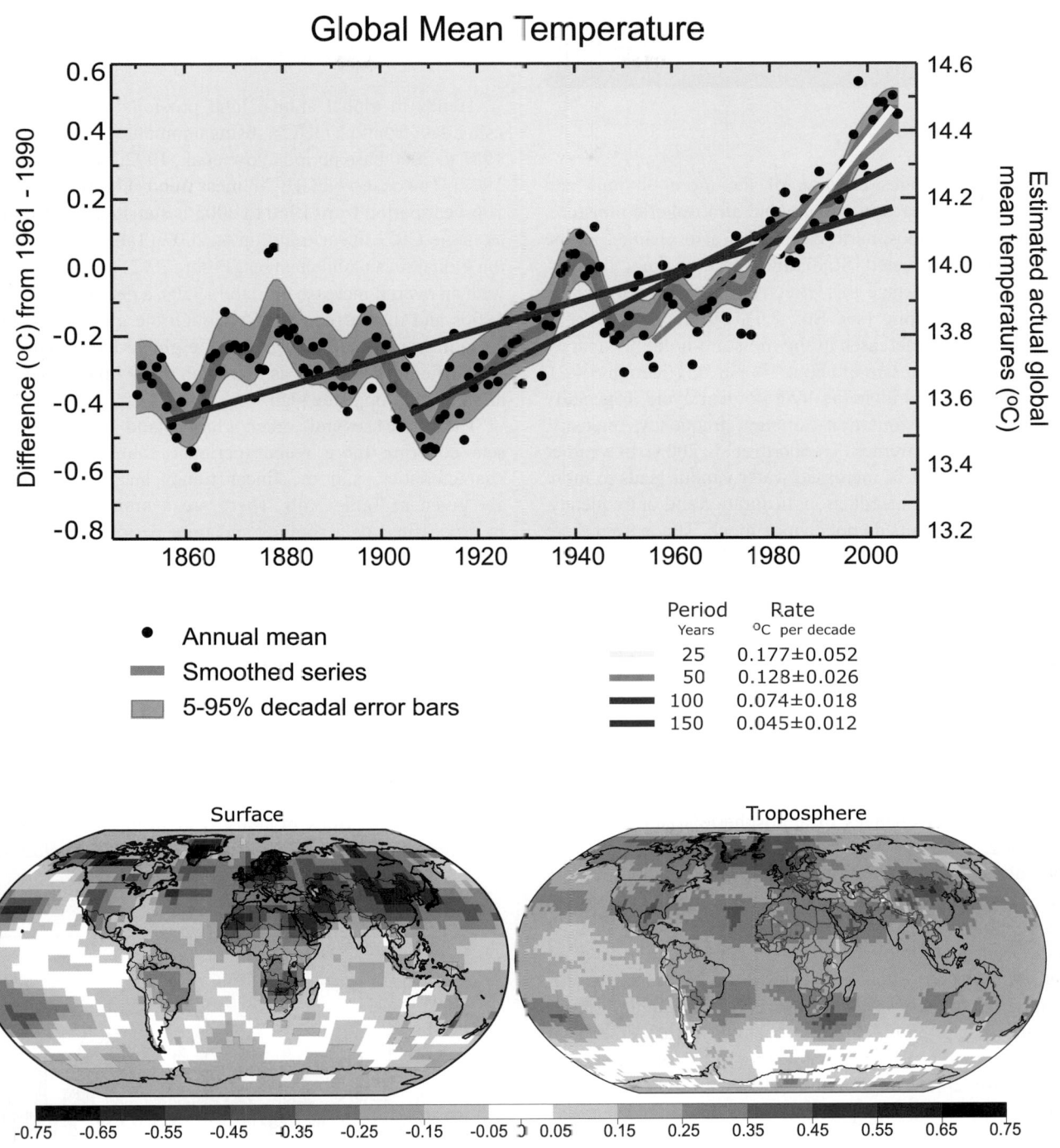

| Period | Rate |
Years	°C per decade
25	0.177±0.052
50	0.128±0.026
100	0.074±0.018
150	0.045±0.012

• Annual mean
▬ Smoothed series
▨ 5-95% decadal error bars

°C per decade

FAQ 3.1, Figure 1. *(Top) Annual global mean observed temperatures[1] (black dots) along with simple fits to the data. The left hand axis shows anomalies relative to the 1961 to 1990 average and the right hand axis shows the estimated actual temperature (°C). Linear trend fits to the last 25 (yellow), 50 (orange), 100 (purple) and 150 years (red) are shown, and correspond to 1981 to 2005, 1956 to 2005, 1906 to 2005, and 1856 to 2005, respectively. Note that for shorter recent periods, the slope is greater, indicating accelerated warming. The blue curve is a smoothed depiction to capture the decadal variations. To give an idea of whether the fluctuations are meaningful, decadal 5% to 95% (light grey) error ranges about that line are given (accordingly, annual values do exceed those limits). Results from climate models driven by estimated radiative forcings for the 20th century (Chapter 9) suggest that there was little change prior to about 1915, and that a substantial fraction of the early 20th-century change was contributed by naturally occurring influences including solar radiation changes, volcanism and natural variability. From about 1940 to 1970 the increasing industrialisation following World War II increased pollution in the Northern Hemisphere, contributing to cooling, and increases in carbon dioxide and other greenhouse gases dominate the observed warming after the mid-1970s. (Bottom) Patterns of linear global temperature trends from 1979 to 2005 estimated at the surface (left), and for the troposphere (right) from the surface to about 10 km altitude, from satellite records. Grey areas indicate incomplete data. Note the more spatially uniform warming in the satellite tropospheric record while the surface temperature changes more clearly relate to land and ocean.*

[1] From the HadCRUT3 data set.

3.3 Changes in Surface Climate: Precipitation, Drought and Surface Hydrology

3.3.1 Background

Temperature changes are one of the more obvious and easily measured changes in climate, but atmospheric moisture, precipitation and atmospheric circulation also change, as the whole system is affected. Radiative forcing alters heating, and at the Earth's surface this directly affects evaporation as well as sensible heating (see Box 7.1). Further, increases in temperature lead to increases in the moisture-holding capacity of the atmosphere at a rate of about 7% per °C (Section 3.4.2). Together these effects alter the hydrological cycle, especially characteristics of precipitation (amount, frequency, intensity, duration, type) and extremes (Trenberth et al., 2003). In weather systems, convergence of increased water vapour leads to more intense precipitation, but reductions in duration and/or frequency, given that total amounts do not change much. The extremes are addressed in Section 3.8.2.2. Expectations for changes in overall precipitation amounts are complicated by aerosols. Because aerosols block the Sun, surface heating is reduced. Absorption of radiation by some, notably carbonaceous, aerosols directly heats the aerosol layer that may otherwise have been heated by latent heat release following surface evaporation, thereby reducing the strength of the hydrological cycle. As aerosol influences tend to be regional, the net expected effect on precipitation over land is especially unclear. This section discusses most aspects of the surface hydrological cycle, except that surface water vapour is included with other changes in atmospheric water vapour in Section 3.4.2.

Difficulties in the measurement of precipitation remain an area of concern in quantifying the extent to which global- and regional-scale precipitation has changed (see Appendix 3.B.4). *In situ* measurements are especially affected by wind effects on the gauge catch, particularly for snow but also for light rain. For remotely sensed measurements (radar and space-based), the greatest problems are that only measurements of instantaneous rate can be made, together with uncertainties in algorithms for converting radiometric measurements (radar, microwave, infrared) into precipitation rates at the surface. Because of measurement problems, and because most historical *in situ*-based precipitation measurements are taken on land leaving the majority of the global surface area under-sampled, it is useful to examine the consistency of changes in a variety of complementary moisture variables. These include both remotely-sensed and gauge-measured precipitation, drought, evaporation, atmospheric moisture, soil moisture and stream flow, although uncertainties exist with all of these variables as well (Huntington, 2006).

3.3.2 Changes in Large-scale Precipitation

3.3.2.1 Global Land Areas

Trends in global annual land precipitation were analysed using data from the GHCN, using anomalies with respect to the 1981 to 2000 base period (Vose et al., 1992; Peterson and Vose, 1997). The observed GHCN linear trend (Figure 3.12) over the 106-year period from 1900 to 2005 is statistically insignificant, as is the CRU linear trend up to 2002 (Table 3.4b). However, the global mean land changes (Figure 3.12) are not at all linear, with an overall increase until the 1950s, a decline until the early 1990s and then a recovery. Although the global land mean is an indicator of a crucial part of the global hydrological cycle, it is difficult to interpret as it is often made up of large regional anomalies of opposite sign.

There are several other global land precipitation data sets covering more recent periods: Table 3.4a gives their characteristics, and the linear trends and their significance are given in Table 3.4b. There are a number of differences in processing, data sources and time periods that lead to the differences in the trend estimates. All but one data set (GHCN) are spatially infilled by either interpolation or the use of satellite estimates of precipitation. The Precipitation Reconstruction over Land (PREC/L) data (Chen et al., 2002) include GHCN data, synoptic data from the National Oceanic and Atmospheric Administration (NOAA) Climate Prediction Center's Climate Anomaly Monitoring System (CAMS), and the Global Precipitation Climatology Project (GPCP) data (Adler et al., 2003), and are a blend of satellite and gauge data. The Global Precipitation Climatology Centre (GPCC; updated from Rudolf et al., 1994) provides monthly data from surface gauges on several grids constructed using GPCC sources (including data

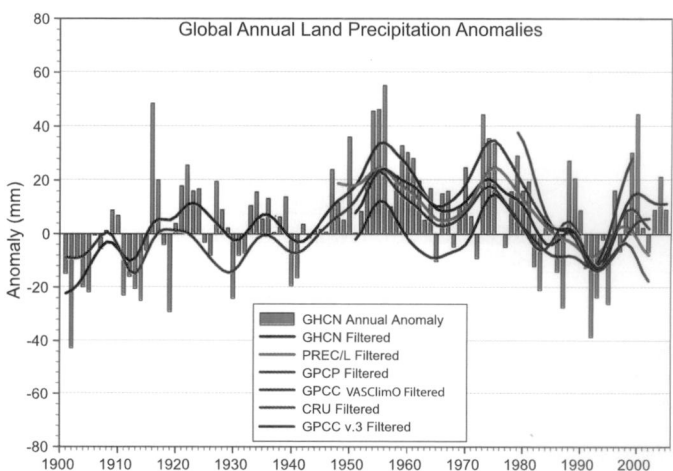

Figure 3.12. *Time series for 1900 to 2005 of annual global land precipitation anomalies (mm) from GHCN with respect to the 1981 to 2000 base period. The smooth curves show decadal variations (see Appendix 3.A) for the GHCN (Peterson and Vose, 1997), PREC/L (Chen et al., 2002), GPCP (Adler et al., 2003), GPCC (Rudolf et al., 1994) and CRU (Mitchell and Jones, 2005) data sets.*

Table 3.4. *(a) Characteristics and references of the six global land-area precipitation data sets used to calculate trends. (b) Global land precipitation trends (mm per decade). Trends with 5 and 95% confidence intervals and levels of significance (italic, 1–5%) were estimated by REML (see Appendix 3.A), which allows for serial correlation in the residuals of the data about the linear trend. All trends are based on annual averages without estimates of intrinsic uncertainties.*

(a)

Series	Period of Record	Gauge only	Satellite and gauge	Spatial infilling	Reference
GHCN	1900–2005	X		No	Vose et al., 1992
PREC/L	1948–2002	X		Yes	Chen et al., 2002
GPCP	1979–2002		X	Yes	Adler et al., 2003
GPCC VASClimO	1951–2000	X		Yes	Beck et al., 2005
GPCC v.3	1951–2002	X		Yes	Rudolf et al., 1994
CRU	1901–2002	X		Yes	Mitchell and Jones, 2005

(b)

Series	Precipitation Trend (mm per decade) 1901–2005	1951–2005	1979–2005
PREC/L		*−5.10 ± 3.25*[a]	−6.38 ± 8.78[a]
CRU	1.10 ± 1.50[a]	−3.87 ± 3.89[a]	−0.90 ± 16.24[a]
GHCN	1.08 ± 1.87	−4.56 ± 4.34	4.16 ± 12.44
GPCC VASClimO		1.82 ± 5.32[b]	12.82 ± 21.45[b]
GPCC v.3		*−6.63 ± 5.18*[a]	−14.64 ± 11.67[a]
GPCP			−15.60 ± 19.84[a]

Notes:
[a] Series ends at 2002
[b] Series ends at 2000

from CRU, GHCN, a Food and Agriculture Organization (FAO) database and many nationally provided data sets). The data set designated GPCC VASClimO (Beck et al., 2005) uses only those quasi-continuous stations whose long-term homogeneity can be assured, while GPCC v.3 has used all available stations to provide more complete spatial coverage. Gridding schemes also vary and include optimal interpolation and grid-box averaging of areally weighted station anomalies. The CRU data set is from Mitchell and Jones (2005).

For 1951 to 2005, trends range from −7 to +2 mm per decade and 5 to 95% error bars range from 3.2 to 5.3 mm per decade. Only the updated PREC/L series (Chen et al., 2002) trend and the GPCC v.3 trend appear to be statistically significant, but the uncertainties, as seen in the different estimates, undermine that result. For 1979 to 2005, GPCP data are added and trends range from −16 to +13 mm per decade but none is significant. Nevertheless, the discrepancies in trends are substantial, and highlight the difficulty of monitoring a variable such as precipitation that has large variability in both space and time. On the other hand, Figure 3.12 also suggests that interannual

fluctuations have some overall reproducibility for land as a whole. The lag-1 autocorrelation of the residuals from the fitted trend (i.e., the de-trended persistence) is in the range 0.3 to 0.5 for the PREC/L, CRU and GHCN series but 0.5 to 0.7 for the two GPCC and the GPCP series. This suggests that either the limited sampling by *in situ* gauge data adds noise, or systematic biases lasting a few years (the lifetime of a satellite) are afflicting the GPCP data, or a combination of the two.

3.3.2.2   Spatial Patterns of Precipitation Trends

The spatial patterns of trends in annual precipitation (% per century or per decade) during the periods 1901 to 2005 and 1979 to 2005 are shown in Figure 3.13. The observed trends over land areas were calculated using GHCN station data interpolated to a 5° × 5° latitude/longitude grid. For most of North America, and especially over high-latitude regions in Canada, annual precipitation has increased during the 105-year period. The primary exception is over the southwest USA, northwest Mexico and the Baja Peninsula, where the trend in annual precipitation has been negative (1 to 2% per decade) as drought has prevailed in recent years. Across South America, increasingly wet conditions were observed over the Amazon Basin and southeastern South America, including Patagonia, while negative trends in annual precipitation were observed over Chile and parts of the western coast of the continent. The largest negative trends in annual precipitation were observed over western Africa and the Sahel. After having concluded that the effect of changing rainfall-gauge networks on Sahel rainfall time series is small, Dai et al. (2004b) noted that Sahel rainfall in the 1990s has recovered considerably from the severe dry years in the early 1980s (see Section 3.7.4 and Figure 3.37). A drying trend is also evident over southern Africa since 1901. Over much of northwestern India the 1901 to 2005 period shows increases of more than 20% per century, but the same area shows a strong decrease in annual precipitation for the 1979 to 2005 period. Northwestern Australia shows areas with moderate to strong increases in annual precipitation over both periods. Over most of Eurasia, increases in precipitation outnumber decreases for both periods.

To assess the expected large regional variations in precipitation trends, Figure 3.14 presents time series of annual precipitation. The regions are 19 of those defined in Table 11.1 (see Section 11.1) and illustrated in Figure 11.26. The GHCN precipitation data set from NCDC was used, and the CRU decadal values allow the reproducibility to be assessed. Based on this, plots for four additional regions (Greenland, Sahara, Antarctica and the Tibetan Plateau) were not included, as precipitation data for these were not considered sufficiently reliable, nor was the first part of the Alaskan series (prior to 1935). Some discrepancies between the decadal variations are still evident at times, mostly owing to different subsets of stations and some stations coming in or dropping out, but overall the confidence in what is presented is quite high. Figure 3.15 presents a latitude-time series of zonal averages over land.

In the tropics, precipitation is highly seasonal, consisting of a dry season and a wet season in association with the summer monsoon. These aspects are discussed in more detail in Section 3.7. Downward trends are strongest in the Sahel (see Section 3.7.4) but occur in both western and eastern Africa in the past 50 years, and are reflected in the zonal means. The downward trends in this zone are also found in southern Asia. The linear trends of rainfall decreases for 1900 to 2005 were 7.5% in both the western Africa and southern Asia regions (significant at <1%). The area of the latter region is much greater than India, whose rainfall features strong variability but little in the way of a century-scale trend. Southern Africa also features a strong overall downward trend, although with strong multi-decadal variability present. Often the change in rainfall in these regions

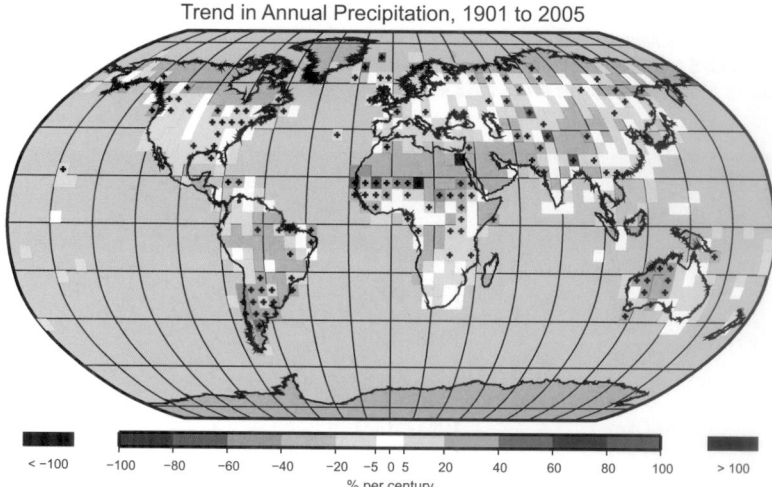

Trend in Annual Precipitation, 1901 to 2005

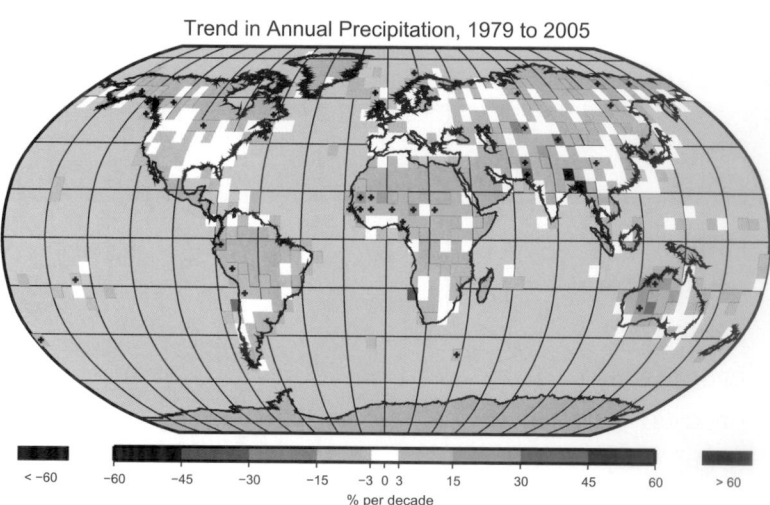

Trend in Annual Precipitation, 1979 to 2005

Figure 3.13. *Trend of annual land precipitation amounts for 1901 to 2005 (top, % per century) and 1979 to 2005 (bottom, % per decade), using the GHCN precipitation data set from NCDC. The percentage is based on the means for the 1961 to 1990 period. Areas in grey have insufficient data to produce reliable trends. The minimum number of years required to calculate a trend value is 66 for 1901 to 2005 and 18 for 1979 to 2005. An annual value is complete for a given year if all 12 monthly percentage anomaly values are present. Note the different colour bars and units in each plot. Trends significant at the 5% level are indicated by black + marks.*

occurs fairly abruptly, and in several cases occurs around the same time in association with the 1976–1977 climate shift (Wang and Ding, 2006). The timing is not the same everywhere, however, and the downward shift occurred earlier in the Sahel (see also Section 3.7.4, Figure 3.37). The main location with different trends at low latitudes is over Australia, but it is clear that large interannual variability, mostly ENSO-related, is dominant (note also the expanded vertical scales for Australia). The apparent upward trend occurs due to two rather wet spells in northern Australia in the early 1970s and 1990s, when it was dry in Southeast Asia (see also Section 3.7.2). Also of note in Australia is the marked downward trend in the far southwest characterised by a downward shift around 1975 (Figure 3.13).

At higher latitudes, especially from 30°N to 85°N, quite distinct upward trends are evident in many regions and these are reflected in the zonal means (Figure 3.15). Central North America, eastern North America, northern Europe, northern

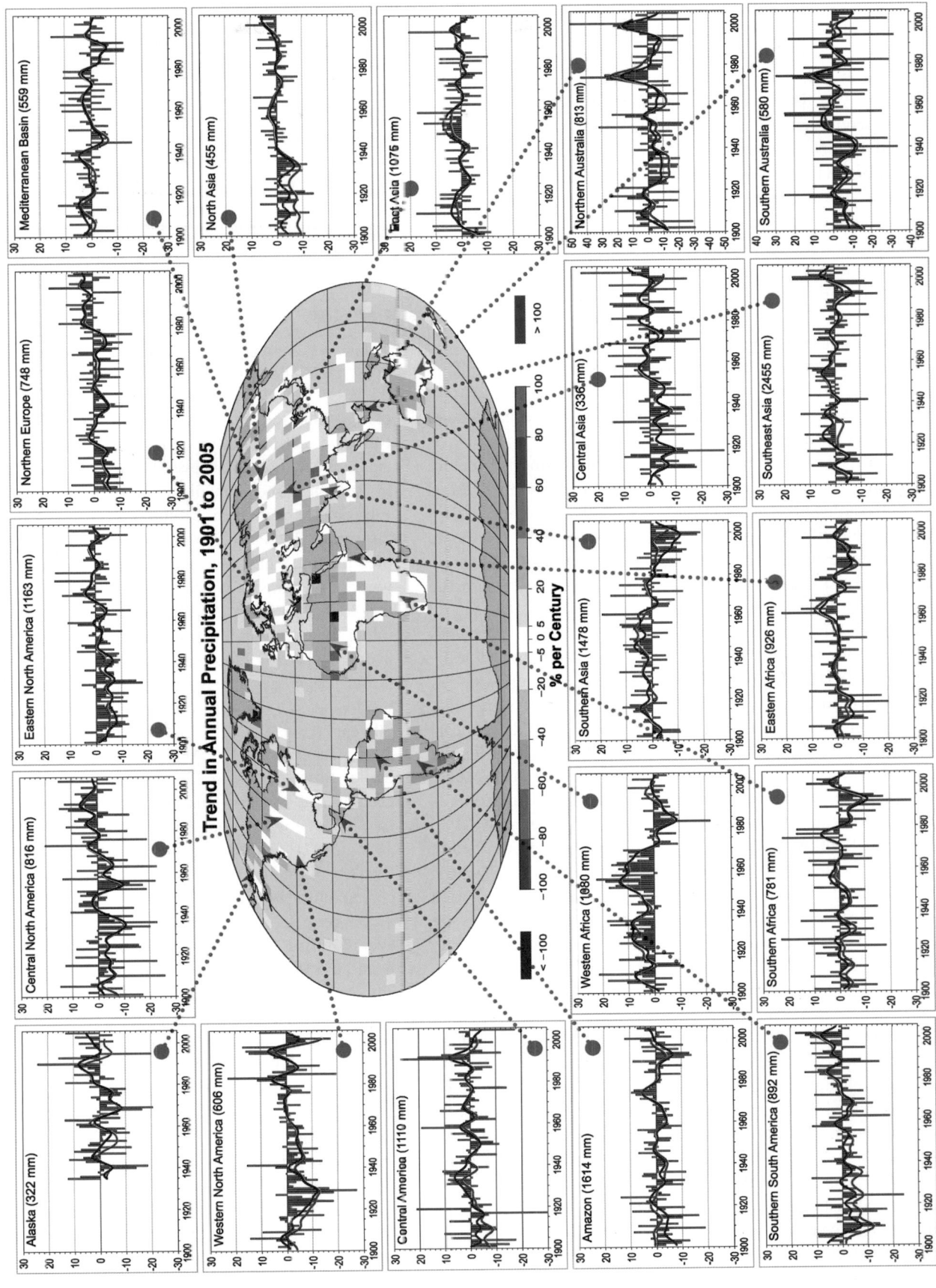

Figure 3.14. *Precipitation for 1900 to 2005. The central map shows the annual mean trends (% per century). Areas in grey have insufficient data to produce reliable trends. The surrounding time series of annual precipitation displayed (% of mean, with the mean given at top for 1961 to 1990) are for the named regions as indicated by the red arrows. The GHCN precipitation from NCDC was used for the annual green bars and black for decadal variations (see Appendix 3.A), and for comparison the CRU decadal variations are in magenta. The range is +30 to –30% except for the two Australian panels. The regions are a subset of those defined in Table 11.1 (Section 11.1) and include: Central North America, Western North America, Alaska, Central America, Eastern North America, Mediterranean, Northern Europe, North Asia, East Asia, Central Asia, Southeast Asia, Southern Asia, Northern Australia, Southern Australia, Eastern Africa, Western Africa, Southern Africa, Southern South America, and the Amazon.*

Asia and central Asia (east of the Caspian Sea) all experienced upward linear trends of between 6 and 8% from 1900 to 2005 (all significant at <5%). These regions all experience snowfall (see also Section 3.3.2.3) and part of the upward trend may arise from changes in efficiency of catching snow, especially in northern Asia. However, there is ample evidence that these trends are real (see Section 3.3.4), and they extend from North America to Europe across the North Atlantic as evidenced by ocean freshening, documented in Sections 5.2.3 and 5.3.2. Western North America shows longer time-scale variability, principally due to the severe drought in the 1930s and lesser events more recently. Note the tendency for inverse variations between northern Europe and the Mediterranean, associated with changes in the NAO (see Section 3.6.4). Southern Europe and parts of central Europe, as well as North Africa, are characterised by a drier winter (DJF) during the positive phase of the NAO, while the reverse is true in the British Isles, Fennoscandia and northwestern Russia.

In the SH, Amazonia and southern South America feature opposite changes, as the South American monsoon features shifted southwards (see Section 3.7.3). This movement was in association with changes in ENSO and the 1976–1977 climate shift. The result is a pronounced upward trend in Argentina and the La Plata River Basin, but not in Chile (where the main declines in precipitation are evident in the austral summer (DJF) and autumn (MAM; Figure 3.13). Decadal-scale variations over Amazonia are also out of phase with the Central American region to the north, which in turn has out-of-phase variations with western North America, again suggestive of latitudinal changes in monsoon features. East and Southeast Asia show hardly any long-term changes, with both having plentiful rains in the 1950s. At interannual time scales there are a number of surprisingly strong correlations: Amazonia is correlated with northern Australia (0.44, significant at <1%) and also Southeast Asia (0.55, <1%), while southern South America is inversely correlated with western Africa (–0.51, <1%). The correlations are surprising because they are based on high-frequency relationships and barely change when the smoothed series are used.

3.3.2.3 Changes in Snowfall

Winter precipitation has increased at high latitudes, although uncertainties exist because of changes in undercatch, especially as snow changes to rain. Snow cover changes are discussed in Section 4.2. Annual precipitation for the circumpolar region north of 50°N has increased during the past 50 years (not shown) by approximately 4% but this increase has not been homogeneous in time and space (Groisman et al., 2003, 2005). Statistically significant increases were documented over Fennoscandia, coastal regions of northern North America (Groisman et al., 2005), most of Canada (particularly northern regions) up until at least 1995 when the analysis ended (Stone

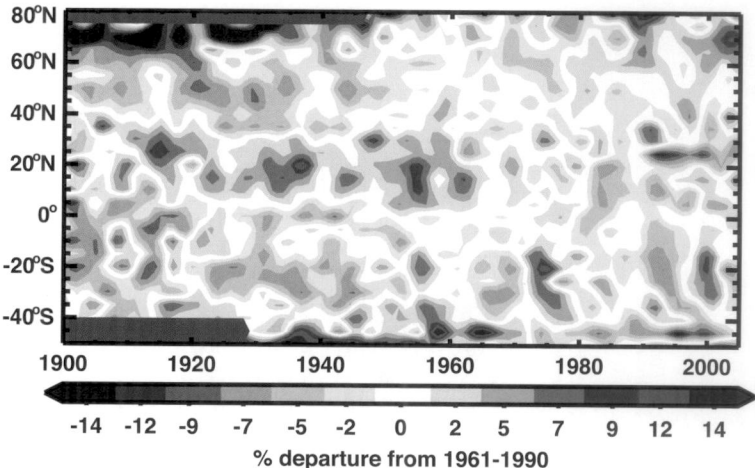

Figure 3.15. *Latitude-time section of zonal average annual anomalies for precipitation (%) over land from 1900 to 2005, relative to their 1961 to 1990 means. Values are smoothed with the 5-point filter to remove fluctuations of less than about six years (see Appendix 3.A). The colour scale is nonlinear and grey areas indicate missing data.*

et al., 2000), the permafrost-free zone of Russia (Groisman and Rankova, 2001) and the entire Great Russian Plain (Groisman et al., 2005, 2007). However, there were no discernible changes in summer and annual precipitation totals over northern Eurasia east of the Ural Mountains (Gruza et al., 1999; Sun and Groisman, 2000; Groisman et al., 2005, 2007). Rainfall (liquid precipitation) has increased during the past 50 years over western portions of North America and Eurasia north of 50°N by about 6%. Rising temperatures have generally resulted in rain rather than snow in locations and seasons where climatological average (1961–1990) temperatures were close to 0°C. The liquid precipitation season has become longer by up to three weeks in some regions of the boreal high latitudes over the last 50 years (Cayan et al., 2001; Groisman et al., 2001; Easterling, 2002; Groisman et al., 2005, 2007) owing, in particular, to an earlier onset of spring. Therefore, in some regions (southern Canada and western Russia), snow has provided a declining fraction of total annual precipitation (Groisman et al., 2003, 2005, 2007). In other regions, particularly north of 55°N, the fraction of annual precipitation falling as snow in winter has changed little.

Berger et al. (2002) found a trend towards fewer snowfall events during winter across the lower Missouri River Basin from 1948 to 2002, but little or no trend in snowfall occurrences within the plains region to the south. In New England, there has been a decrease in the proportion of precipitation occurring as snow at many locations, caused predominantly by a decrease in snowfall, with a lesser contribution from increased rainfall (Huntington et al., 2004). By contrast, Burnett et al. (2003) found large increases in lake-effect snowfall since 1951 for locations near the North American Great Lakes, consistent with the observed decrease in ice cover for most of the Great Lakes since the early 1980s (Assell et al., 2003). In addition to snow data, Burnett et al. (2003) used lake sediment reconstructions for locations south of Lake Ontario to indicate that these increases have been ongoing since the beginning of the 20th century. Ellis and Johnson (2004) found that the increases in snowfall across the regions to the lee of Lakes Erie and Ontario are due to increases in the frequency of snowfall at the expense of rainfall events, an increase in the intensity of snowfall events, and to a lesser extent an increase in the water equivalent of the snow. In Canada, the frequency of heavy snowfall events has decreased since the 1970s in the south and increased in the north (Zhang et al., 2001a).

3.3.2.4   Urban Areas

As noted in Section 3.2.2.2 (see also Box 7.2), the microclimates in cities are clearly different than in neighbouring rural areas. The presence of a city affects runoff, moisture availability and precipitation. Crutzen (2004) pointed out that while human energy production is relatively small globally compared with the Sun, it is locally important in cities, where it can reach 20 to 70 W m^{-2}. Urban effects can lead to increased precipitation (5 to 25% over background values) during the summer months within and 50 to 75 km downwind of the city (Changnon et al., 1981). More frequent or intense storms have been linked to city growth in Phoenix, Arizona (Balling and Brazel, 1987) and Mexico City (Jauregui and Romales, 1996). More recent observational studies (Bornstein and Lin, 2000; Changnon and Westcott, 2002; Shepherd et al., 2002; Diem and Brown, 2003; Dixon and Mote, 2003; Fujibe, 2003; Shepherd and Burian, 2003; Inoue and Kimura, 2004; Shepherd et al., 2004; Burian and Shepherd, 2005) have continued to link urban-induced dynamic processes to precipitation anomalies. Nor is land use change confined to urban areas (see Section 7.2). Other changes in land use also affect precipitation. A notable example arises from deforestation in the Amazon, where Chagnon and Bras (2005) found large changes in local rainfall with increases in deforested areas, associated with local atmospheric circulations that are changed by gradients in vegetation, and also found changes in seasonality.

Suggested mechanisms for urban-induced rainfall include: (1) enhanced convergence due to increased surface roughness in the urban environment (e.g., Changnon et al., 1981; Bornstein and Lin, 2000; Thielen et al., 2000); (2) destabilisation due to UHI thermal perturbation of the boundary layer and resulting downstream translation of the UHI circulation or UHI-generated convective clouds (e.g., Shepherd et al., 2002; Shepherd and Burian, 2003); (3) enhanced aerosols in the urban environment for cloud condensation nuclei sources (e.g., Diem and Brown, 2003; Molders and Olson, 2004); or (4) bifurcating or diverting of precipitating systems by the urban canopy or related processes (e.g., Bornstein and Lin, 2000). The 'weekend effect' noted in Section 3.2.2.2 likely arises from some of these mechanisms. The diurnal cycle in precipitation, which varies over the USA from late afternoon maxima in the Southeast to nocturnal maxima in the Great Plains (Dai and Trenberth, 2004), may be modified in some regions by urban environments. Dixon and Mote (2003) found that a growing UHI effect in Atlanta, Georgia (USA) enhanced and possibly initiated thunderstorms, especially in July (summer) just after midnight. Low-level moisture was found to be a key factor.

3.3.2.5   Ocean Precipitation

Remotely sensed precipitation measurements over the ocean are based on several different sensors in the microwave and infrared that are combined in different ways. Many experimental products exist. Operational merged products seem to perform best in replicating island-observed monthly amounts (Adler et al., 2001). This does not mean they are best for trends or low-frequency variability, because of the changing mixes of input data. The main global data sets available for precipitation, and which therefore include ocean coverage, have been the GPCP (Huffman et al., 1997; Adler et al., 2003) and the NOAA Climate Prediction Center (CPC) Merged Analysis of Precipitation (CMAP; Xie and Arkin, 1997). Comparisons of these data sets and others (Adler et al., 2001; Yin et al., 2004) reveal large discrepancies over the ocean; however, there is better agreement among the passive microwave products even using different algorithms. Over the tropical oceans, mean amounts in CMAP and GPCP differ by 10 to 15%. Calibration using observed rainfall from small atolls in CMAP was extended throughout the tropics in ways that are now recognised as incorrect. However, evaluation of GPCP reveals that it is biased low by 16% at such atolls (Adler et al., 2003), also raising questions about the ocean GPCP values. Differences arise due to sampling and algorithms. Polar-orbiting satellites each obtain only two instantaneous rates per day over any given location, and thus suffer from temporal sampling deficiencies that are offset by using geostationary satellites. However, only less-accurate infrared sensors are available with the latter. Model-based (including reanalysis) products perform poorly in the evaluation of Adler et al. (2001) and are not currently suitable for climate monitoring. Robertson et al. (2001b) examined monthly anomalies from several satellite-derived precipitation data sets (using different algorithms) over the tropical oceans. The expectation in the TAR was that measurements from the Tropical Rainfall Measuring Mission (TRMM) Precipitation Radar (PR) and passive TRMM microwave imager (TMI) would clarify the reasons for the discrepancies, but this has not yet been the case. Robertson et al. (2003) documented poorly correlated behaviour (correlation 0.12) between the monthly, tropical ocean-averaged precipitation anomalies from the PR and TMI sensors. Although the TRMM PR responds directly to precipitation size hydrometeors, it operates with a single attenuating frequency (13.8 GHz) that necessitates significant microphysical assumptions regarding drop size distributions for relating reflectivity, signal attenuation and rainfall, and uncertainties in microphysical assumptions for the primary TRMM algorithm (2A25) remain problematic.

The large regional signals from monsoons and ENSO that emphasise large-scale shifts in precipitation are reasonably well captured in GPCP and CMAP (see Section 3.6.2), but cancel out when area-averaged over the tropics, and the trends and variability of the tropical average are quite different in the two products. Global precipitation from GPCP (updated from Adler et al., 2003, but not shown) has monthly variability with a standard deviation of about 2% of the mean. The variability in the ocean and land areas when examined separately is larger, about 3%, and with variations related to ENSO events (Curtis and Adler, 2003). During El Niño events, area-averaged precipitation increases over the oceans but decreases over land.

Although the trend over 25 years in global total precipitation in the GPCP data set (Adler et al., 2003) is very small, there is a small increase (about 4% over the 25 years) over the oceans in the latitude range 25°S to 25°N, with a partially compensating decrease over land (2%) in the same latitude belt. Northern mid-latitudes show a decrease over land and ocean. Over a slightly longer time frame, precipitation increased over the North Atlantic between 1960 to 1974 and 1975 to 1989 (Josey and Marsh, 2005) and is reflected in changes in salinity in the oceans (Section 5.2.3). The inhomogeneous nature of the data sets and the large ENSO variability limit what can be said about the validity of changes, both globally and regionally.

3.3.3   Evapotranspiration

There are very limited direct measurements of actual evapotranspiration over global land areas. Over oceans, estimates of evaporation depend on bulk flux estimates that contain large errors. Evaporation fields from the ERA-40 and NRA are not considered reliable because they are not well constrained by precipitation and radiation (Betts et al., 2003; Ruiz-Barradas and Nigam, 2005). The physical processes related to changes in evapotranspiration are discussed in Section 7.2 and Section 3.4, Box 3.2.

Decreasing trends during recent decades are found in sparse records of pan evaporation (measured evaporation from an open water surface in a pan) over the USA (Peterson et al., 1995; Golubev et al., 2001; Hobbins et al., 2004), India (Chattopadhyay and Hulme, 1997), Australia (Roderick and Farquhar, 2004), New Zealand (Roderick and Farquhar, 2005), China (Liu et al., 2004a; Qian et al., 2006b) and Thailand (Tebakari et al., 2005). Pan measurements do not represent actual evaporation (Brutsaert and Parlange, 1998), and any trend is more likely caused by decreasing surface solar radiation over the USA and parts of Europe and Russia (Abakumova et al., 1996; Liepert, 2002) and decreased sunshine duration over China (Kaiser and Qian, 2002) that may be related to increases in air pollution and atmospheric aerosols (Liepert et al., 2004; Qian et al., 2006a) and increases in cloud cover (Dai et al., 1999). Whether actual evapotranspiration decreases or not also depends on how surface wetness changes (see Section 3.4, Box 3.2). Changes in evapotranspiration are often calculated using empirical models as a function of precipitation, wind and surface net radiation (Milly and Dunne, 2001), or land surface models (LSMs; e.g., van den Dool et al., 2003; Qian et al., 2006a).

The TAR reported that actual evapotranspiration increased during the second half of the 20th century over most dry regions of the USA and Russia (Golubev et al., 2001), resulting from greater availability of surface moisture due to increased precipitation and larger atmospheric moisture demand due to higher temperature. One outcome is a larger surface latent heat flux (increased evapotranspiration) but decreased sensible heat flux (Trenberth and Shea, 2005). Using observed precipitation, temperature, cloudiness-based surface solar radiation and a comprehensive land surface model, Qian et al. (2006a) found that global land evapotranspiration closely follows variations in land precipitation. Global precipitation values (Figure 3.12) peaked in the early 1970s and then decreased somewhat, but reflect mainly tropical values, and precipitation has increased more generally over land at higher latitudes (Figures 3.13 and 3.14). Changes in evapotranspiration depend not only on moisture supply but also on energy availability and surface wind (see Section 3.4, Box 3.2).

3.3.4   Changes in Soil Moisture, Drought, Runoff and River Discharge

Historical records of soil moisture content measured *in situ* are available for only a few regions and often are very short (Robock et al., 2000). A rare 45-year record of soil moisture over agricultural areas of the Ukraine shows a large upward trend, which was stronger during the first half of the period (Robock et al., 2005). Among over 600 stations from a large variety of climates, including the former Soviet Union, China, Mongolia, India and the USA, Robock et al. (2000) showed an increasing long-term trend in surface (top 1 m) soil moisture content during summer for the stations with the longest records.

One method to examine long-term changes in soil moisture uses calculations based on formulae or LSMs. Since the *in situ* observational record and global estimates of remotely sensed soil moisture data are limited, global soil moisture variations during the 20th century have been estimated by LSM simulations. However, the results depend critically on the 'forcings' used, namely the radiation (clouds), precipitation, winds and other weather variables, which are not sufficiently reliable to determine trends. Consequently the estimates based on simulations disagree. Instead, the primary approach has been to calculate Palmer Drought Severity Index (PDSI; see Box 3.1) values from observed precipitation and temperature (e.g., Dai et al., 2004a). In some locations, much longer proxy extensions have been derived from earlier tree ring data (see Section 6.6.1; e.g., Cook et al., 1999). The longer instrumental-based PDSI estimations are used to look at trends and some recent extreme PDSI events in different regions are placed in a longer-term context (see specific cases in Section 3.8, Box 3.6). As with LSM-based studies, the version of the PDSI used is crucial, and it can partly determine some aspects of the results found (Box 3.1).

Using the PDSI, Dai et al. (2004a) found a large drying trend over NH land since the middle 1950s, with widespread drying over much of Eurasia, northern Africa, Canada and Alaska. In the SH, land surfaces were wet in the 1970s and relatively dry in the 1960s and 1990s, and there was a drying trend from 1974 to 1998 although trends over the entire 1948 to 2002 period were small. Overall patterns of trends in the PDSI are given in FAQ 3.2, Figure 1. Although the long-term (1901–2004) land-based precipitation trend shows a small increase (Figure 3.12), decreases in land precipitation in recent decades are the main

Box 3.1: Drought Terminology and Determination

In general terms, drought is a 'prolonged absence or marked deficiency of precipitation', a 'deficiency of precipitation that results in water shortage for some activity or for some group' or a 'period of abnormally dry weather sufficiently prolonged for the lack of precipitation to cause a serious hydrological imbalance' (Heim, 2002). Drought has been defined in a number of ways. 'Agricultural drought' relates to moisture deficits in the topmost one metre or so of soil (the root zone) that impact crops, 'meteorological drought' is mainly a prolonged deficit of precipitation, and 'hydrologic drought' s related to below-normal streamflow, lake and groundwater levels.

Drought and its severity can be numerically defined using indices that integrate temperature, precipitation and other variables that affect evapotranspiration and soil moisture. Several indices in different countries assess precipitation deficits in various ways, such as the Standardized Precipitation Index. Other indices make use of additional weather variables. An example is the Keetch-Byrum Drought Index (Keetch and Byrum, 1988), which assesses the severity of drought in soils based on rainfall and temperature estimates to assess soil moisture deficiencies. However, the most commonly used index is the PDSI (Palmer, 1965; Heim, 2002) that uses precipitation, temperature and local available water content data to assess soil moisture. Although the PDSI is not an optimal index, since it does not include variables such as wind speed, solar radiation, cloudiness and water vapour, it is widely used and can be calculated across many climates as it requires only precipitation and temperature data for the calculation of potential evapotranspiration (PET) using Thornthwaite's (1948) method. Because these data are readily available for most parts of the globe, the PDSI provides a measure of drought for comparison across many regions.

However, PET is considered to be more reliably calculated using Penman (1948) type approaches that incorporate the effects of wind, water vapour and solar and longwave radiation. In addition, there has been criticism of most Thornthwaite-based estimates of the PDSI because the empirical constants have not been re-computed for each climate (Alley, 1984). Hence, a self-calibrating version of the PDSI has recently been developed to ensure consistency with the climate at any location (Wells et al., 2004). Also, studies that compute changes or trends in the PDSI effectively remove influences of biases in the absolute values. As the effects of temperature anomalies on the PDSI are small compared to precipitation anomalies (Guttman, 1991), the PDSI is largely controlled by precipitation changes.

cause for the drying trends, although large surface warming during the last two to three decades has likely contributed to the drying. Dai et al. (2004a) showed that globally, very dry areas (defined as land areas with a PDSI of less than –3.0) more than doubled (from ~12 to 30%) since the 1970s, with a large jump in the early 1980s due to an ENSO-related precipitation decrease over land and subsequent increases primarily due to surface warming. However, results are dependent on the version of the PDSI model used, since the empirical constants used in a global PDSI model may not be adequately adjusted for the local climate (see Box 3.1).

In Canada, the summer PDSI averaged for the entire country indicates dry conditions during the 1940s and 1950s, generally wet conditions from the 1960s to 1995, but much drier conditions after 1995 (Shabbar and Skinner, 2004) with a relationship between recent increasing summer droughts and the warming trend in SST. Groisman et al. (2007) found increased dryness based on the Keetch-Byrum forest-fire drought index in northern Eurasia, a finding supported by Dai et al. (2004a) using the PDSI. Long European records (van der Schrier et al., 2006) reveal no trend in areas affected by extreme PDSI values (thresholds of either ±2 or ±4) over the 20th century. Nevertheless, recently Europe has suffered prolonged drought, including the 2003 episode associated with the severe summer heat wave (see Section 3.8.4, Box 3.6).

Although there was no significant trend from 1880 to 1998 during summer (JJA) in eastern China, precipitation for 1990 to 1998 was the highest on record for any period of comparable length (Gong and Wang, 2000). Zou et al. (2005) found that

for China as a whole there were no long-term trends in the percentage areas of droughts (defined as PDSI < –1.0) during 1951 to 2003. However, increases in drought areas were found in much of northern China (but not in northwest China; Zou et al., 2005), aggravated by warming and decreasing precipitation (Ma and Fu, 2003; Wang and Zhai, 2003), consistent with Dai et al. (2004a).

A severe drought affecting central and southwest Asia in recent years (see Section 3.8.4, Box 3.6) appears to be the worst since at least 1980 (Barlow et al., 2002). In the Sahel region of Africa, rainfall has recovered somewhat in recent years, after large decreasing rainfall trends from the late 1960s to the late 1980s (Dai et al., 2004b; see also Section 3.3.2.2 and Section 3.7.4, Figure 3.37). Large multi-year oscillations appear to be more frequent and extreme after the late 1960s than previously in the century. A severe drought affected Australia in 2002 and 2003; precipitation deficits were not as severe as during a few episodes earlier in the 20th century, but higher temperatures exacerbated the impacts (see Section 3.8.4, Box 3.6). There have been marked multi-year rainfall deficits and drought since the mid- to late-1990s in several parts of Australia, particularly the far southwest, parts of the southeast and along sections of the east coast.

A multi-decadal period of relative wetness characterised the latter portion of the 20th century in the continental USA, in terms of precipitation (Mauget, 2003a), streamflow (Groisman et al., 2004) and annual moisture surplus (precipitation minus potential evapotranspiration; McCabe and Wolock, 2002). Despite this overall national trend towards wetter conditions,

---

Frequently Asked Question 3.2
How is Precipitation Changing?

Observations show that changes are occurring in the amount, intensity, frequency and type of precipitation. These aspects of precipitation generally exhibit large natural variability, and El Niño and changes in atmospheric circulation patterns such as the North Atlantic Oscillation have a substantial influence. Pronounced long-term trends from 1900 to 2005 have been observed in precipitation amount in some places: significantly wetter in eastern North and South America, northern Europe and northern and central Asia, but drier in the Sahel, southern Africa, the Mediterranean and southern Asia. More precipitation now falls as rain rather than snow in northern regions. Widespread increases in heavy precipitation events have been observed, even in places where total amounts have decreased. These changes are associated with increased water vapour in the atmosphere arising from the warming of the world's oceans, especially at lower latitudes. There are also increases in some regions in the occurrences of both droughts and floods.

Precipitation is the general term for rainfall, snowfall and other forms of frozen or liquid water falling from clouds. Precipitation is intermittent, and the character of the precipitation when it occurs depends greatly on temperature and the weather situation. The latter determines the supply of moisture through winds and surface evaporation, and how it is gathered together in storms as clouds. Precipitation forms as water vapour condenses, usually in rising air that expands and hence cools. The upward motion comes from air rising over mountains, warm air riding over cooler air (warm front), colder air pushing under warmer air (cold front), convection from local heating of the surface, and other weather and cloud systems. Hence, changes in any of these aspects alter precipitation. As precipitation maps tend to be spotty, overall trends in precipitation are indicated by the Palmer Drought Severity Index (see Figure 1), which is a measure of soil moisture using precipitation and crude estimates of changes in evaporation.

A consequence of increased heating from the human-induced enhanced greenhouse effect is increased evaporation, provided that adequate surface moisture is available (as it always is over the oceans and other wet surfaces). Hence, surface moisture effectively acts as an 'air conditioner', as heat used for evaporation acts to moisten the air rather than warm it. An observed consequence of this is that summers often tend to be either warm and dry or cool and wet. In the areas of eastern North and South America where it has become wetter (Figure 1), temperatures have therefore increased less than elsewhere (see FAQ 3.3, Figure 1 for changes in warm days). Over northern continents in winter, however, more precipitation is associated with higher temperatures, as the water holding capacity of the atmosphere increases in the warmer conditions. However, in these regions, where precipitation has generally increased somewhat, increases in temperatures (FAQ 3.1) have increased drying, making the precipitation changes less evident in Figure 1.

As climate changes, several direct influences alter precipitation amount, intensity, frequency and type. Warming accelerates land surface drying and increases the potential incidence and severity of droughts, which has been observed in many places worldwide (Figure 1). However, a well-established physical law (the Clausius-Clapeyron relation) determines that the water-holding capacity of the atmosphere increases by about 7% for every 1°C rise in temperature. Observations of trends in relative humidity are uncertain but suggest that it has remained about the same overall, from the surface throughout the troposphere, and hence increased temperatures will have resulted in increased water vapour. Over the 20th century, based on changes in sea surface temperatures, it is estimated that atmospheric water vapour increased by about 5% in the atmosphere over the oceans. Because precipitation comes mainly from weather systems that feed on the water vapour stored in the atmosphere, this has generally increased precipitation intensity and the risk of heavy rain and snow events. Basic theory, climate model simulations and empirical evidence all confirm that warmer climates, owing to increased water vapour, lead to more intense precipitation events even when the total annual precipitation is reduced slightly, and with prospects for even stronger events when the overall precipitation amounts increase. The warmer climate therefore increases risks of both drought – where it is not raining – and floods – where it is – but at different times and/or places. For instance, the summer of 2002 in Europe brought widespread floods but was followed a year later in 2003 by record-breaking heat waves and drought. The distribution and timing of floods and droughts is most profoundly affected by the cycle of El Niño events, particularly in the tropics and over much of the mid-latitudes of Pacific-rim countries.

In areas where aerosol pollution masks the ground from direct sunlight, decreases in evaporation reduce the overall moisture supply to the atmosphere. Hence, even as the potential for heavier precipitation results from increased water vapour amounts, the duration and frequency of events may be curtailed, as it takes longer to recharge the atmosphere with water vapour.

Local and regional changes in the character of precipitation also depend a great deal on atmospheric circulation patterns determined by El Niño, the North Atlantic Oscillation (NAO; a measure of westerly wind strength over the North Atlantic in winter) and other patterns of variability. Some of these observed circulation changes are associated with climate change. An associated shift in the storm track makes some regions wetter and some – often nearby – drier, making for complex patterns of change. For instance, in the European sector a more positive NAO in the 1990s led to wetter conditions in northern Europe and drier conditions over the Mediterranean and northern African regions (Figure 1). The prolonged drought in the Sahel (see Figure 1), which was pronounced from the late 1960s to the late 1980s,

(continued)

continues although it is not quite as intense as it was; it has been linked, through changes in atmospheric circulation, to changes in tropical sea surface temperature patterns in the Pacific, Indian and Atlantic Basins. Drought has become widespread throughout much of Africa and more common in the tropics and subtropics.

As temperatures rise, the likelihood of precipitation falling as rain rather than snow increases, especially in autumn and spring at the beginning and end of the snow season, and in areas where temperatures are near freezing. Such changes are observed in many places, especially over land in middle and high latitudes of the Northern Hemisphere, leading to increased rains but reduced snowpacks, and consequently diminished water resources in summer, when they are most needed. Nevertheless, the often spotty and intermittent nature of precipitation means observed patterns of change are complex. The long-term record emphasizes that patterns of precipitation vary somewhat from year to year, and even prolonged multi-year droughts are usually punctuated by a year of heavy rains; for instance as El Niño influences are felt. An example may be the wet winter of 2004-2005 in the southwestern USA following a six-year drought and below-normal snowpack.

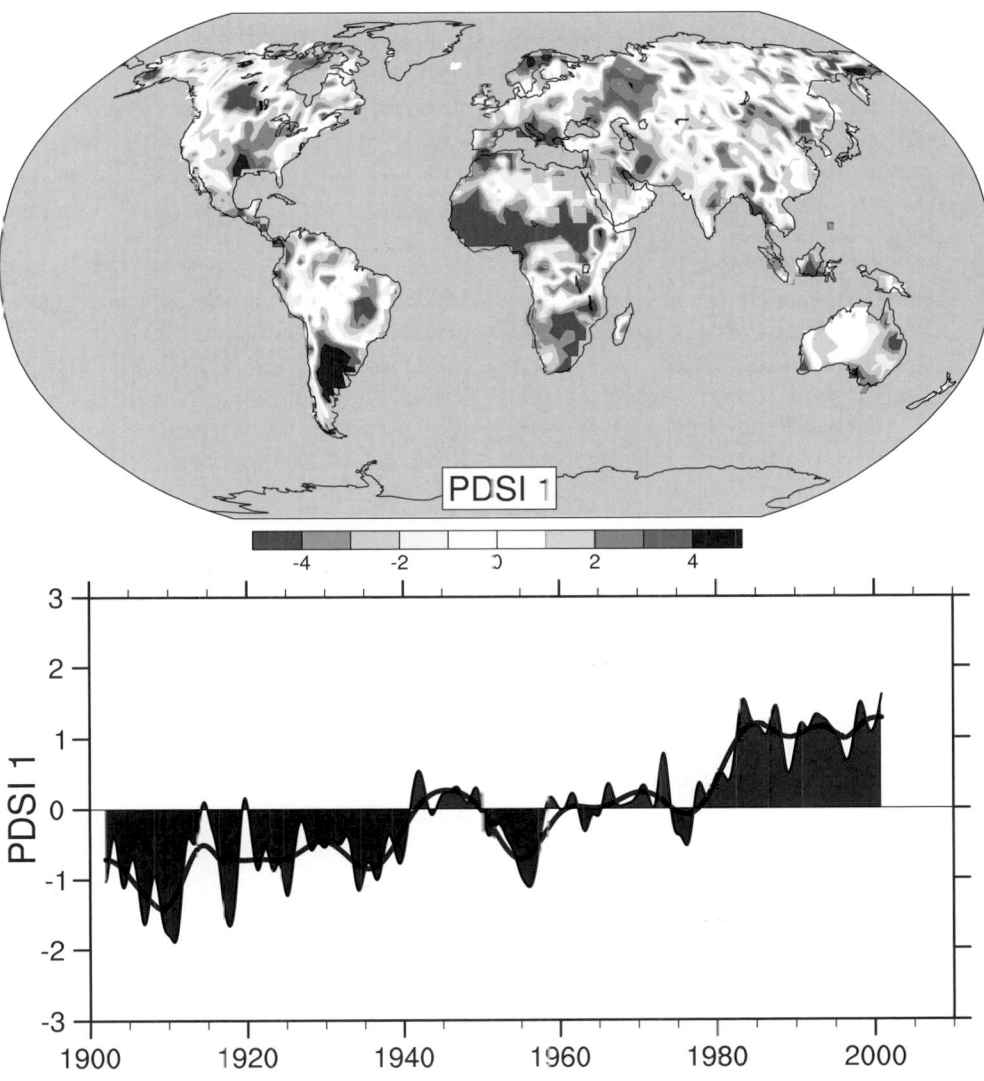

FAQ 3.2, Figure 1. *The most important spatial pattern (top) of the monthly Palmer Drought Severity Index (PDSI) for 1900 to 2002. The PDSI is a prominent index of drought and measures the cumulative deficit (relative to local mean conditions) in surface land moisture by incorporating previous precipitation and estimates of moisture drawn into the atmosphere (based on atmospheric temperatures) into a hydrological accounting system. The lower panel shows how the sign and strength of this pattern has changed since 1900. Red and orange areas are drier (wetter) than average and blue and green areas are wetter (drier) than average when the values shown in the lower plot are positive (negative). The smooth black curve shows decadal variations. The time series approximately corresponds to a trend, and this pattern and its variations account for 67% of the linear trend of PDSI from 1900 to 2002 over the global land area. It therefore features widespread increasing African drought, especially in the Sahel, for instance. Note also the wetter areas, especially in eastern North and South America and northern Eurasia. Adapted from Dai et al. (2004b).*

a severe drought affected the western USA from 1999 to November 2004 (see Section 3.8.4, Box 3.6).

Available streamflow gauge records cover only about two-thirds of the global actively drained land areas and they often have gaps and vary in record length (Dai and Trenberth, 2002). Estimates of total continental river discharge are therefore often based on incomplete gauge records (e.g., Probst and Tardy, 1987, 1989; Guetter and Georgakakos, 1993), reconstructed streamflow time series (Labat et al., 2004) or methods to account for the runoff contribution from the unmonitored areas (Dai and Trenberth, 2002). These estimates show large decadal to multi-decadal variations in continental and global freshwater discharge (excluding groundwater; Guetter and Georgakakos, 1993; Labat et al., 2004).

Streamflow records for the world's major rivers show large decadal to multi-decadal variations, with small secular trends for most rivers (Cluis and Laberge, 2001; Lammers et al., 2001; Mauget, 2003b; Pekárová et al., 2003; Dai et al., 2004a). Increased streamflow during the latter half of the 20th century has been reported over regions with increased precipitation, such as many parts of the USA (Lins and Slack, 1999; Groisman et al., 2004) and southeastern South America (Genta et al., 1998). Decreased streamflow was reported over many Canadian river basins during the last 30 to 50 years (Zhang et al., 2001b), where precipitation has also decreased during the period. Déry and Wood (2005) also found decreases in river discharge into the Arctic and North Atlantic Oceans from high-latitude Canadian rivers, with potential implications for salinity levels in these oceans and possibly the North Atlantic THC. These changes are consistent with observed precipitation decreases in high-latitude Canada from 1963 to 2000. Further, Milly et al. (2002) showed significant trends towards more extreme flood events from streamflow measurements in 29 very large basins, but Kundzewicz et al. (2005) found both increases (in 27 cases) and decreases (in 31 cases) as well as no significant (at the 10% level) long-term changes in annual extreme flows for 137 of the 195 rivers examined worldwide. Recent extreme flood events in central Europe (on the Elbe and some adjacent catchments) are discussed in Section 3.8.4, Box 3.6.

Large changes and trends in seasonal streamflow rates for many of the world's major rivers (Lammers et al., 2001; Cowell and Stoudt, 2002; Ye et al., 2003; Yang et al., 2004) should be interpreted with caution, since many of these streams have been affected by the construction of large dams and reservoirs that increase low flow and reduce peak flow. Nevertheless, there is evidence that the rapid warming since the 1970s has induced earlier snowmelt and associated peak streamflow in the western USA (Cayan et al., 2001) and New England, USA (Hodgkins et al., 2003) and earlier breakup of river ice in Russian Arctic rivers (Smith, 2000) and many Canadian rivers (Zhang et al., 2001b).

River discharges in the La Plata River Basin in southeastern South America exhibit large interannual variability. Consistent evidence linking the Paraná and Uruguay streamflows and ENSO has been found (Bischoff et al., 2000; Camilloni and Barros, 2000, 2003; Robertson et al., 2001a; Berri et al., 2002;

Krepper et al., 2003), indicating that monthly and extreme flows during El Niño are generally larger than those observed during La Niña events. For the Paraguay River, most of the major discharges at the Pantanal wetland outlet occurred in the neutral phases of ENSO, but in the lower reaches of the river the major discharge events occurred during El Niño events (Barros et al., 2004). South Atlantic SST anomalies also modulate regional river discharges through effects on rainfall in southeastern South America (Camilloni and Barros, 2000). The Paraná River shows a positive trend in its annual mean discharge since the 1970s in accordance with the regional rainfall trends (García and Vargas, 1998; Barros et al., 2000a; Liebmann et al., 2004), as do the Paraguay and Uruguay Rivers since 1970 (Figure 3.14).

For 1935 to 1999 in the Lena River Basin in Siberia, Yang et al. (2002) found significant increases in temperature and streamflow and decreases in ice thickness during the cold season. Strong spring warming resulted in earlier snowmelt with a reduced maximum streamflow pulse in June. During the warm season, smaller streamflow increases are related to an observed increase in precipitation. Streamflow in the Yellow River Basin in China decreased significantly during the latter half of the 20th century, even after accounting for increased human water consumption (Yu et al., 2004a). Temperatures have increased over the basin, but precipitation has shown no change, suggesting an increase in evaporation.

In Africa from 1950 to 1995, Jury (2003) found that the Niger and Senegal Rivers show the effects of the Sahel drying trend with a decreasing trend in flow. The Zambezi also exhibits reduced flows, but rainfall over its catchment area appears to be stationary. Other major African rivers, including the Blue and White Nile, Congo and inflow into Lake Malawi show high variability, consistent with interannual variability of SSTs in the Atlantic, Indian and Pacific Oceans. A composite index of streamflow for these rivers shows that the five highest flow years occurred prior to 1979, and the five lowest flow years occurred after 1971.

3.3.5 Consistency and Relationships between Temperature and Precipitation

Observed changes in regional temperature and precipitation can often be physically related to one another. This section assesses the consistencies of these relationships in the observed trends. Significant large-scale correlations between observed monthly mean temperature and precipitation (Madden and Williams, 1978) for North America and Europe have stood up to the test of time and been expanded globally (Trenberth and Shea, 2005). In the warm season over continents, higher temperatures accompany lower precipitation amounts and vice versa. Hence, over land, strong negative correlations dominate, as dry conditions favour more sunshine and less evaporative cooling, while wet summers are cool. However, at latitudes poleward of 40° in winter, positive correlations dominate as the water-holding capacity of the atmosphere limits precipitation amounts in cold conditions and warm air advection in cyclonic storms is accompanied by precipitation. Where ocean conditions drive the atmosphere, higher surface air temperatures are associated with precipitation, as during El Niño events. For South America, Rusticucci and Penalba (2000) showed that warm summers are associated with low precipitation, especially in northeast and central-western Argentina, southern Chile, and Paraguay. Cold season (JJA) correlations are weak but positive to the west of 65°W, as stratiform cloud cover produces a higher minimum temperature. For stations in coastal Chile, the correlation is always positive and significant, as it is adjacent to the ocean, especially in the months of rainfall (May to September), showing that high SSTs favour convection.

This relationship of higher warm-season temperatures with lower precipitation appears to apply also to trends (Trenberth and Shea, 2005). An example is Australia, which exhibits evidence of increased drought severity, consistent with the observed warming during the latter half of the 20th century (Nicholls, 2004). Mean maximum and minimum temperatures during the 2002 Australian drought were much higher than during the previous droughts in 1982 and 1994, suggesting enhanced potential evaporation as well (see Section 3.8.4, Box 3.6). Record-high maximum temperatures also accompanied the dry conditions in 2005.

3.3.6   Summary

Substantial uncertainty remains in trends of hydrological variables because of large regional differences, gaps in spatial coverage and temporal limitations in the data (Huntington, 2006). At present, documenting interannual variations and trends in precipitation over the oceans remains a challenge. Global precipitation averages over land are not very meaningful and mask large regional variations. Precipitation generally increased over the 20th century from 30°N to 85°N over land, and over Argentina, but notable decreases have occurred in the past 30 to 40 years from 10°S to 30°N. Salinity decreases in the North Atlantic and south of 25°S suggest similar precipitation changes over the ocean (Sections 5.3.2 and 5.5.3). Runoff and river discharge generally increased at higher latitudes, along with soil moisture, consistent with precipitation changes. River discharges in many tropical areas of Africa and South America are strongly affected by ENSO, with greater discharges from the Paraná River after the 1976–1977 climate shift but lower discharges from some major African rivers since then.

However, the PDSI suggests there has likely been a large drying trend since the mid-1950s over many land areas, with widespread drying over much of Africa, southern Eurasia, Canada and Alaska. In the SH, there was a drying trend from 1974 to 1998, although trends over the entire 1948 to 2002 period are small. Seasonal decreases in land precipitation since the 1950s are the main cause for some of the drying trends, although large surface warming during the last two to three decades has also likely contributed to the drying. Based on the PDSI data, very dry areas (defined as land areas with a PDSI of less than –3.0) have more than doubled in extent since the 1970s, with a large jump in the early 1980s due to an ENSO-induced precipitation

decrease over land and subsequent increases primarily due to surface warming.

Hence, the observed marked increases in drought in the past three decades arise from more intense and longer droughts over wider areas, as a critical threshold for delineating drought is exceeded over increasingly widespread areas. Overall, consistent with the findings of Huntington (2006), the evidence for increases in both severe droughts and heavy rains (Section 3.8.2) in many regions of the world makes it likely that hydrologic conditions have become more intense.

3.4   Changes in the Free Atmosphere

3.4.1   Temperature of the Upper Air: Troposphere and Stratosphere

Within the community that constructs and actively analyses satellite- and radiosonde-based temperature records there is agreement that the uncertainties about long-term change are substantial. Changes in instrumentation and protocols pervade both sonde and satellite records, obfuscating the modest long-term trends. Historically there is no reference network to anchor the record and establish the uncertainties arising from these changes – many of which are both barely documented and poorly understood. Therefore, investigators have to make seemingly reasonable choices of how to handle these sometimes known but often unknown influences. It is difficult to make quantitatively defensible judgments as to which, if any, of the multiple, independently derived estimates is closer to the true climate evolution. This reflects almost entirely upon the inadequacies of the historical observing network and points to the need for future network design that provides the reference sonde-based ground truth. Karl et al. (2006) provide a comprehensive review of this issue.

3.4.1.1   Radiosondes

Since the TAR, considerable effort has been devoted to assessing and improving the quality of the radiosonde temperature record (see Appendix 3.B.5.1). A particular aim has been to reduce artificial changes arising from instrumental and procedural developments during the seven decades (1940s–2000s) of the radiosonde record (Free and Seidel, 2005; Thorne et al., 2005a; Karl et al., 2006). Comparisons of several adjustment methods showed that they gave disparate results when applied to a common set of radiosonde station data (Free et al., 2002). One approach, based on the physics of heat transfer within the radiosonde, performed poorly when evaluated against satellite temperature records (Durre et al., 2002). Another method, comparison with satellite data (HadRT (Hadley Centre Radiosonde Temperature Data Set); Parker et al., 1997), is limited to the satellite era and to events with available metadata, and causes a reduction in spatial consistency of the data. A comprehensive intercomparison (Seidel et al., 2004) showed that five radiosonde data sets yielded consistent signals for higher-frequency events such as ENSO, the Quasi-Biennial Oscillation (QBO) and volcanic eruptions, but inconsistent signals for long-term trends.

Several approaches have been used to create new adjusted data sets since the TAR. The Lanzante-Klein-Seidel (LKS; Lanzante et al., 2003a,b) data set, using 87 carefully selected stations, has subjectively derived bias adjustments throughout the length of its record but terminates in 1997. It has been updated using the Integrated Global Radiosonde Archive (IGRA; Durre et al., 2006) by applying a different bias adjustment technique (Free et al., 2004b) after 1997, creating a new archive (Radiosonde Atmospheric Temperature Products for Assessing Climate; RATPAC). Another new radiosonde record, HadAT2 (Hadley Centre Atmospheric Temperature Data Set Version 2, successor to HadRT), uses a neighbour comparison approach to build spatial as well as temporal consistency. A third approach (Haimberger, 2005) uses the bias adjustments estimated during data assimilation into model-based reanalyses to identify and reduce inhomogeneities in radiosonde data. Despite the risk of contamination by other biased data or by model bias, the resulting adjustments agree with those estimated by other methods. Rather than adjusting the data, Angell (2003) tried to reduce data quality problems by removing several tropical stations from his radiosonde network.

Despite these efforts to produce homogeneous data sets, recent analyses of radiosonde data indicate that significant problems remain. Sherwood et al. (2005) found substantial changes in the diurnal cycle in unadjusted radiosonde data. These changes are probably a consequence of improved sensors and radiation error adjustments. Relative to nighttime values, they found a daytime warming of sonde temperatures prior to 1971 that is likely to be spurious and then a spurious daytime cooling from 1979 to 1997. They estimated that there was probably a spurious overall downward trend in sonde temperature records during the satellite era (since 1978) throughout the atmosphere of order 0.1°C per decade globally. The assessed spurious cooling is greatest in the tropics (0.16°C per decade for the 850 to 300 hPa layer) and least in the NH extratropics (0.04°C per decade). Randel and Wu (2006) used collocated MSU data to show that cooling biases remain in some of the LKS and RATPAC radiosonde data for the tropical stratosphere and upper troposphere due to changes in instruments and radiation correction adjustments. They also identified problems in night data as well as day, indicating that negative biases are not limited to daytime observations. However, a few stations may have positive biases (Christy and Spencer, 2005).

The radiosonde data set is limited to land areas, and coverage is poor over the tropics and SH. Accordingly, when global estimates based solely on radiosondes are presented, there are

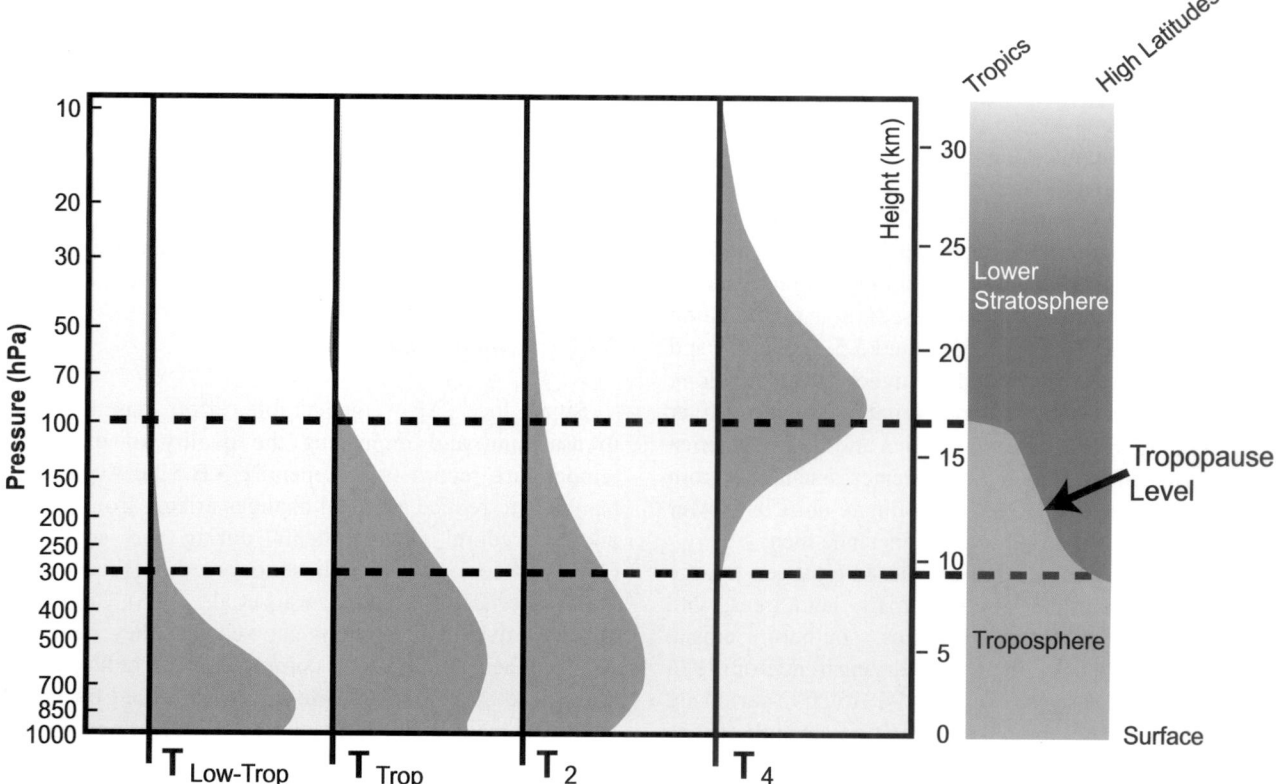

Figure 3.16. *Vertical weighting functions (grey) depicting the layers sampled by satellite MSU measurements and their derivatives, and used also for radiosonde and reanalysis records. The right panel schematically depicts the variation in the tropopause (that separates the stratosphere and troposphere) from the tropics (left) to the high latitudes (right). The fourth panel depicts T4 in the lower stratosphere, the third panel shows T2, the second panel shows the troposphere as a combination of T2 and T4 (Fu et al., 2004a) and the first panel shows T2$_{LT}$ from the UAH for the low troposphere. Adapted from Karl et al. (2006).*

considerable uncertainties (Hurrell et al., 2000; Agudelo and Curry, 2004) and denser networks – which perforce still omit oceanic areas – may not yield more reliable 'global' trends (Free and Seidel, 2005). Radiosonde records have an advantage of starting in the 1940s regionally, and near-globally from about 1958. They monitor the troposphere and lower stratosphere; the layers analysed are described below and in Figure 3.16. Radiosonde-based global mean temperature estimates are given in Figure 3.17, presented later.

3.4.1.2   The Satellite Microwave Sounding Unit Record

3.4.1.2.1   Summary of satellite capabilities and challenges

Satellite-borne microwave sounders estimate the temperature of thick layers of the atmosphere by measuring microwave emissions (radiances) that are proportional to the thermal state of emission of oxygen molecules from a complex of emission lines near 60 GHz. By making measurements at different frequencies near 60 GHz, different atmospheric layers can be sampled. A series of nine instruments called Microwave Sounding Units (MSUs) began making this kind of measurement in late 1978. Beginning in mid-1998, a subsequent series of instruments, the Advanced MSUs (AMSUs), began operation. Unlike infrared sounders, microwave sounders are not affected by most clouds, although some effects are experienced from precipitation and clouds with high liquid water content. Figure 3.16 illustrates the lower troposphere (referred to as $T2_{LT}$), troposphere, and MSU channel 2 (referred to as T2) and channel 4 (lower stratosphere, referred to as T4) layers.

The main advantage of satellite measurements compared to radiosondes is the excellent coverage of the measurements, with complete global coverage every few days. However, like radiosondes, temporal continuity is a major challenge for climate assessment, as data from all the satellites in the series must be merged together. The merging procedure must accurately account for a number of error sources. The most important are: (1) offsets in calibration between satellites; (2) orbital decay and drift and associated long-term changes in the time of day that the measurements are made at a particular location, which combine with the diurnal cycle in atmospheric temperature to produce diurnal drifts in the estimated temperatures; (3) drifts in satellite calibration that are correlated with the temperature of the on-board calibration target. Since the calibration target temperatures vary with the satellite diurnal drift, the satellite calibration and diurnal drift corrections are intricately coupled together (Fu and Johanson 2005). Independent teams of investigators have used different methods to determine and correct for these 'structural' and other sources of error (Thorne et al., 2005b). Appendix 3.B.5.3 discusses adjustments to the data in more detail.

3.4.1.2.2   Progress since the TAR

Since the TAR, several important developments and advances have occurred in the analysis of satellite measurements of atmospheric temperatures. Existing data sets have been scrutinised and problems identified, leading to new versions as described below. A number of new data records have been constructed from the MSU measurements, as well as from global reanalyses (see Section 3.4.1.3). Further, new insights have come from statistical combinations of the MSU records from different channels that have minimised the influence of the stratosphere on the tropospheric records (Fu et al., 2004a,b; Fu and Johanson, 2004, 2005). These new data sets and analyses are very important because the differences highlight assumptions and it becomes possible to estimate the uncertainty in satellite-derived temperature trends that arises from different methods and approaches to the construction of temporally consistent records.

Analyses of MSU channels 2 and 4 have been conducted by the University of Alabama in Huntsville (UAH; Christy et al., 2000, 2003) and by Remote Sensing Systems (RSS; Mears et al., 2003; Mears and Wentz, 2005). Another analysis of channel 2 is that of Vinnikov and Grody (2003; version 1 – VG1), now superseded by Grody et al. (2004) and Vinnikov et al. (2006; version 2 – VG2). MSU channel 2 (T2) measures a thick layer of the atmosphere, with approximately 75 to 80% of the signal coming from the troposphere, 15% from the lower stratosphere, and the remaining 5 to 10% from the surface. MSU channel 4 (T4) is primarily sensitive to temperature in the lower stratosphere (Figure 3.16).

Global time series from each of the MSU records are shown in Figure 3.17 and calculated global trends are depicted in Figure 3.18. These show a global cooling of the stratosphere (T4) of –0.32°C to –0.47 °C per decade and a global warming of the troposphere (T2) of 0.04°C to 0.20°C per decade for the period 1979 to 2004. The large spread in T2 trends stems from differences in the inter-satellite calibration and merging technique, and differences in the corrections for orbital drift, diurnal cycle change and the hot-point calibration temperature (Christy et al., 2003; Mears et al., 2003; Christy and Norris, 2004; Grody et al., 2004; Fu and Johanson, 2005; Mears and Wentz, 2005; Vinnikov et al., 2006; see also Appendix 3.B.5.3)

The RSS results for T2 indicate nearly 0.1°C per decade more warming in the troposphere than UAH (see Figure 3.18) and most of the difference arises from the use of different amounts of data to determine the parameters of the calibration target effect (Appendix 3.B.5.3). The UAH analysis yields parameters for the NOAA-9 satellite (1985–1987) outside of the physical bounds expected by Mears et al. (2003). Hence, the large difference in the calibration parameters for the single instrument mounted on the NOAA-9 satellite accounted for a substantial part of the difference between the UAH and RSS T2 trends. The rest arises from differences in merging parameters for other satellites; differences in the correction for the drift in measurement time, especially for the NOAA-11 satellite (Mears et al., 2003; Christy and Norris, 2004); and differences in the ways the hot-point temperature is corrected for (Grody et al., 2004; Fu and Johanson, 2005). In the tropics, these accounted for differences in T2 trends of about 0.07°C per decade after 1987 and discontinuities were also present in 1992 and 1995 at times of satellite transitions (Fu and Johanson, 2005). The T2 data record of Grody et al. (2004) and Vinnikov et al. (2006) (VG2) shows slightly more warming in the troposphere than the RSS data record (Figure 3.18). See also Appendix 3.B.5.3 for discussion of the VG2 analysis.

Although the T4 from RSS has about 0.1°C per decade less cooling than the UAH product (Figure 3.18), both data sets support the conclusions that the stratosphere has undergone strong cooling since 1979. Because about 15% of the signal for T2 comes from the lower stratosphere, the observed cooling causes the reported T2 trends to underestimate tropospheric warming. By creating a weighted combination of T2 and T4, this effect has been greatly reduced (Fu et al., 2004a; see Figure 3.16). This technique for estimating the global mean temperature implies small negative weights at some stratospheric levels, but because of vertical coherence these merely compensate for other positive weights nearby and it is the integral that matters (Fu and Johanson, 2004). From 1979 to 2001 the stratospheric contribution to the trend in T2 is about –0.08°C per decade. Questions about this technique (Tett and Thorne, 2004) have led to clearer interpretation of its application to the tropics (Fu et al., 2004b). The technique has also been successfully applied to model results (Gillett et al., 2004; Kiehl et al., 2005), although model biases in depicting stratospheric cooling can affect results. In a further development, weighted combinations of T2, MSU channel 3 (T3) and T4 since 1987 have formed tropical series for the upper, lower and whole troposphere (Fu and Johanson, 2005).

By differencing T2 measurements made at different slant angles, the UAH group produced an updated data record weighted for the lower and mid troposphere, $T2_{LT}$ (Christy et al., 2003). This retrieval also has the effect of removing the stratospheric influence on long-term trends, but its uncertainties are augmented by the need to compensate for orbital decay and by computing a small residual from two large values (Wentz and Schabel, 1998). $T2_{LT}$ retrievals include a large signal from the surface and so are adversely affected by changes in surface emissivity, including changes in sea ice cover (Swanson, 2003). Fu and Johanson (2005) found that the $T2_{LT}$ trends were physically inconsistent compared with those of the surface, T2 and T4, even if taken from the UAH record. They also showed that the large trend bias is mainly attributed to the periods when a satellite had substantial drifts in local equator crossing time that caused large changes in calibration target temperatures and large diurnal drifts. Mears and Wentz (2005) further found that the adjustments for diurnal cycle required from satellite drift had the wrong sign in the UAH record in the tropics. Corrections have been made (version 5.2; Christy and Spencer, 2005) and are reflected in Figure 3.18, but the trend in the tropics is still smaller for most periods than both those in the troposphere (using T2 and T4) and those at the surface. Mears and Wentz (2005) computed their own alternative $T2_{LT}$ record and found a $T2_{LT}$ trend nearly 0.1°C per decade larger than the revised UAH trend. After 1987, when MSU channel

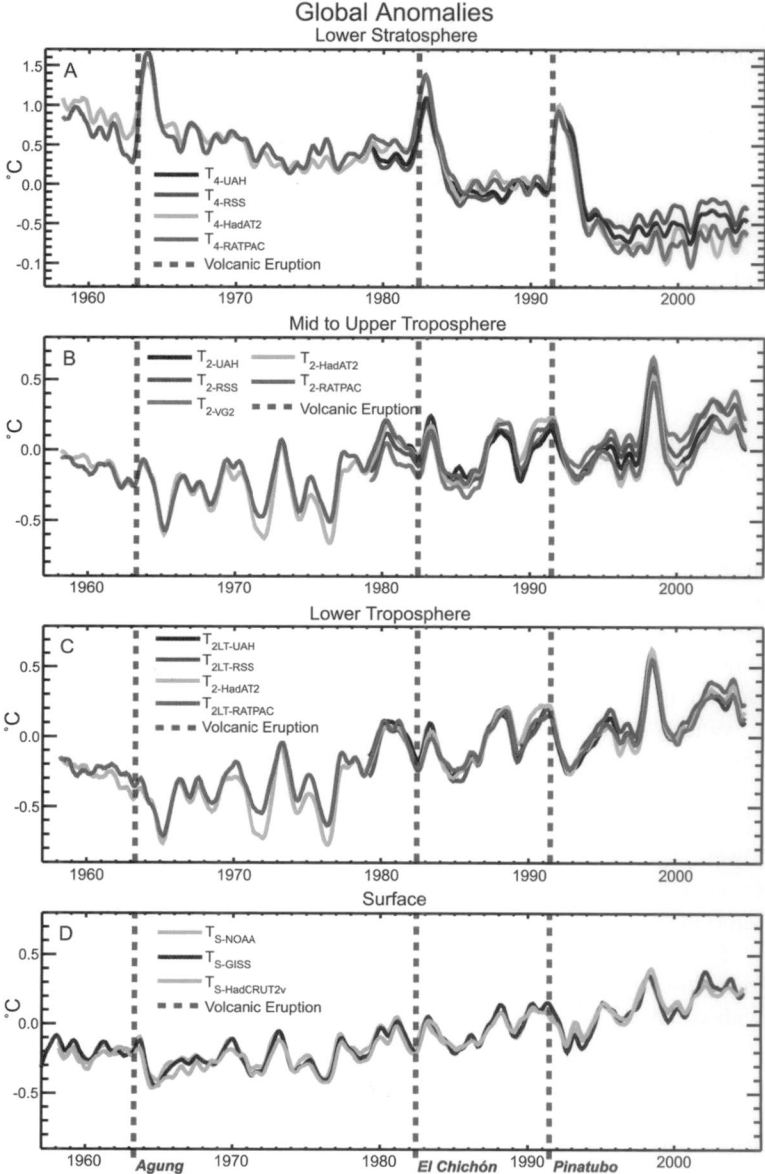

Figure 3.17. *Observed surface and upper-air temperature anomalies (°C). (A) Lower stratospheric T4, (B) Tropospheric T2, (C) Lower tropospheric $T2_{LT}$ from UAH, RSS and VG2 MSU satellite analyses and UKMO HadAT2 and NOAA RATPAC radiosonde observations; and (D) Surface records from NOAA, NASA/GISS and UKMO/CRU (HadCRUT2v). All time series are monthly mean anomalies relative to the period 1979 to 1997 smoothed with a seven-month running mean filter. Major volcanic eruptions are indicated by vertical blue dashed lines. Adapted from Karl et al. (2006).*

3 became available, Fu and Johanson (2005), using RSS data, found a systematic trend of increasing temperature with altitude throughout the tropics.

Comparisons of tropospheric radiosonde station data with collocated satellite data (Christy and Norris, 2004) show considerable scatter, and root mean square differences between UAH satellite data and radiosondes are substantial (Hurrell et al., 2000). Although Christy and Norris (2004) found good agreement between median radiosonde temperature trends and UAH trends, comparisons are more likely to be biased by spurious cooling than by spurious warming in unhomogenised

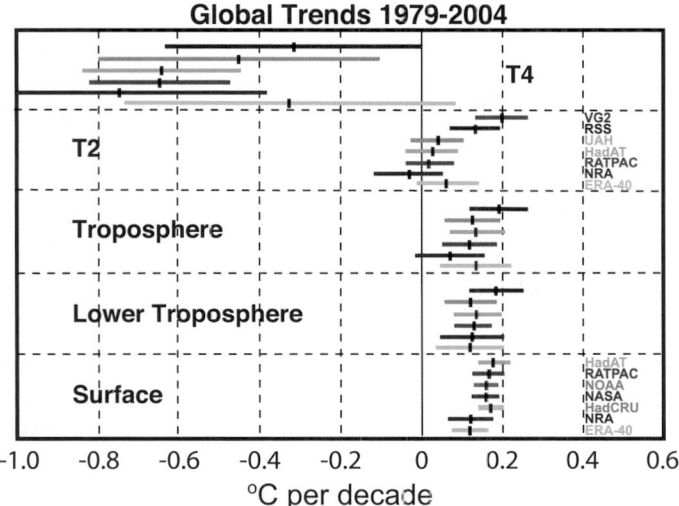

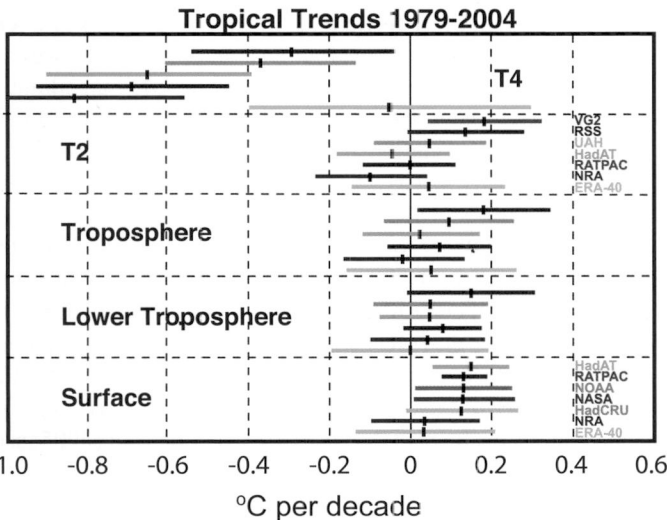

Figure 3.18. Linear temperature trends (°C per decade) for 1979 to 2004 for the globe (top) and tropics (20°N to 20°S; bottom) for the MSU channels T4 (top panel) and T2 (second panel) or equivalent for radiosondes and reanalyses; for the troposphere (third panel) based on T2 with T4 used to statistically remove stratospheric influences (Fu et al., 2004a); for the lower troposphere (fourth panel) based on the UAH retrieval profile; and for the surface (bottom panel). Surface records are from NOAA/NCDC (green), NASA/GISS (blue) and HadCRUT2v (light blue). Satellite records are from UAH (orange), RSS (dark red) and VG2 (magenta); radiosonde-based records are from NOAA RATPAC (brown) and HadAT2 (light green); and atmospheric reanalyses are from NRA (red) and ERA-40 (cyan). The error bars are 5 to 95% confidence limits associated with sampling a finite record with an allowance for autocorrelation. Where the confidence limits exceed –1, the values are truncated. ERA-40 trends are only for 1979 to August 2002. Data from Karl et al. (2006; D. Seidel courtesy of J. Lanzante; and J. Christy).

(Sherwood et al., 2005) and even homogenised (Randel and Wu, 2006) radiosonde data (see Section 3.4.1.1 and Appendix 3.B.5.1). In the stratosphere, radiosonde trends are more negative than both MSU retrievals, especially when compared with RSS, and this is very likely due to changes in sondes and their processing for radiation corrections (Randel and Wu, 2006).

Geographical patterns of the linear trend in tropospheric temperature from 1979 to 2004 (Figure 3.19) are qualitatively

similar in the RSS and UAH MSU data sets. Both show coherent warming over most of the NH, but UAH shows cooling over parts of the tropical Pacific and tropospheric temperature trends differ south of 45°S where UAH indicate more cooling than RSS.

3.4.1.3   Reanalyses

A comprehensive global reanalysis completed since the TAR, ERA-40 (Uppala et al., 2005), extends from September 1957 to August 2002. Reanalysis is designed to prevent changes in the analysis system from contaminating the climate record, as occurs with global analyses from operational numerical weather prediction, and it compensates for some but not all of the effects of changes in the observing system (see Appendix 3.B.5.4). Unlike the earlier NRA that assimilated satellite retrievals, ERA-40 assimilated bias-adjusted radiances including MSU data (Harris and Kelly, 2001; Uppala et al., 2005), and the assimilation procedure itself accounts for orbital drift and change in satellite height – factors that have to be addressed in direct processing of MSU radiances for climate studies (e.g., Christy et al., 2003; Mears et al., 2003; Mears and Wentz, 2005). Onboard calibration biases are treated indirectly via the influence of other data sets. Nonetheless, the veracity of low-frequency variability in atmospheric temperatures is compromised in ERA-40 by residual problems in bias corrections.

Trends and low-frequency variability in large-scale surface air temperature from ERA-40 and from the monthly climate station data analysed by Jones and Moberg (2003) are in generally good agreement from the late 1970s onwards (see also Section 3.2.2.1). Temperatures from ERA-40 vary quite coherently throughout the planetary boundary layer over this period, and earlier for regions with consistently good coverage from both surface and upper-air observations (Simmons et al., 2004).

Processed MSU records of layer temperature have been compared with equivalents derived from the ERA-40 analyses (Santer et al., 2004). The use of deep layers conceals disparate trends at adjacent tropospheric levels in ERA-40.

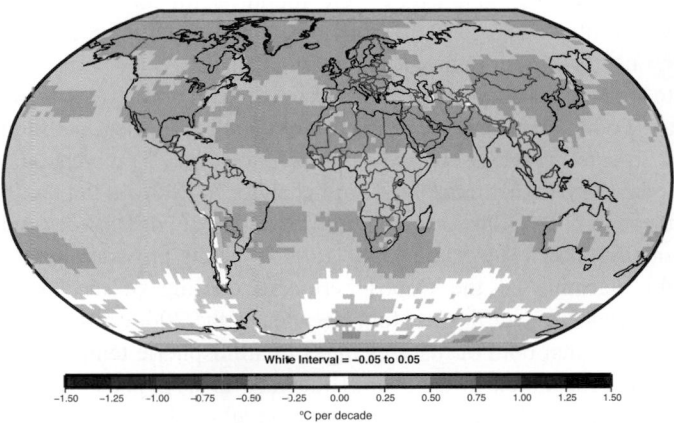

Figure 3.19. Linear tropospheric temperature trends (°C per decade) for 1979 to 2005 from RSS (based on T2 and T4 adjusted as in Fu et al., 2004a). Courtesy Q. Fu.

Relatively cold tropospheric values before the satellite era arose from a combination of scarcity of radiosonde data over the extratropical SH and a cold bias of the assimilating model, giving a tropospheric warming trend that is clearly too large when taken over the full period of the reanalysis (Bengtsson et al., 2004; Simmons et al., 2004; Karl et al., 2006). ERA-40 also exhibits a middle-tropospheric cooling over most of the tropics and subtropics since the 1970s that is certainly too strong owing to a warm bias in the analyses for the early satellite years.

Tropospheric patterns of trends from ERA-40 are similar to Figure 3.19, with coherent warming over the NH, although with discrepancies south of 45°S. These differences are not fully understood, although the treatment of surface emissivity anomalies over snow- and ice-covered surfaces may contribute (Swanson, 2003). At high southern latitudes, ERA-40 shows strong positive temperature trends in JJA in the period 1979 to 2001, in good accord with antarctic radiosonde data (Turner et al., 2006). The large-scale patterns of stratospheric cooling are similar in ERA-40 and the MSU data sets (Santer et al., 2004). However, the ERA-40 analyses in the lower stratosphere are biased cold relative to radiosonde data in the early satellite years, reducing downward trends. Section 3.5 relates the trends to atmospheric circulation changes.

3.4.1.4   The Tropopause

The tropopause marks the division between the troposphere and stratosphere and generally a minimum in the vertical profile of temperature. The height of the tropopause is affected by the heat balance of both the troposphere and the stratosphere. For example, when the stratosphere warms owing to absorption of radiation by volcanic aerosol, the tropopause is lowered. Conversely, a warming of the troposphere raises the tropopause, as does a cooling of the stratosphere. The latter is expected as a result of increasing greenhouse gas concentrations and stratospheric ozone depletion. Accordingly, changes in the height of the tropopause provide a sensitive indicator of human effects on climate. Inaccuracies and spurious trends in NRA preclude their use in determining tropopause trends (Randel et al., 2000) although they were found useful for interannual variability. Over 1979 to 2001, tropopause height increased by nearly 200 m (as a global average) in ERA-40, partly due to tropospheric warming plus stratospheric cooling (Santer et al., 2004). Atmospheric temperature changes in the UAH and RSS satellite MSU data sets (see Section 3.4.1.2) were found to be more highly correlated with changes in ERA-40 than with those in NRA, illustrating the improved quality of ERA-40 and satellite data. The Santer et al. (2004) results provide support for warming of the troposphere and cooling of the lower stratosphere over the last four decades of the 20th century, and indicate that both of these changes in atmospheric temperature have contributed to an overall increase in tropopause height. The radiosonde-based analyses of Randel et al. (2000), Seidel et al. (2001) and Highwood et al. (2000) also show increases in tropical tropopause height.

3.4.1.5   Synthesis and Comparison with the Surface Temperatures

Figure 3.17 presents the radiosonde and satellite global time series and Figure 3.18 gives a summary of the linear trends for 1979 to 2004 for global and tropical (20°N to 20°S) averages. Values at the surface are from NOAA (NCDC), NASA (GISS), UKMO/CRU (HadCRUT2v) and the NRA and ERA-40 reanalyses. Trends aloft are for the lower troposphere corresponding to $T2_{LT}$, T2, T4 and also the linear combination of T2 and T4 to better depict the entire troposphere as given by Fu et al. (2004a). In addition to the reanalyses, the results from the satellite-based methods from UAH, RSS and VG2 are given along with radiosonde estimates from HadAT2 and RATPAC. The ERA-40 trends only extend through August 2002, and VG2 is available only for T2. The error bars plotted here are 5 to 95% confidence limits associated with sampling a finite record where an allowance has been made for temporal autocorrelation in computing degrees of freedom (Appendix 3.A). However, the error bars do not include spatial sampling uncertainty, which increases the noise variance. Noise typically cuts down on temporal autocorrelation and reduces the temporal sampling error bars, which is why the RATPAC error bars are often smaller than the rest. Other sources of 'structural' and 'internal' errors of order 0.08°C for 5 to 95% levels (Mears and Wentz, 2005; see Appendix 3.B.5) are also not explicitly accounted for here. Structural uncertainties and parametric errors (Thorne et al., 2005b) reflect divergence between different data sets after the common climate variability has been accounted for and are better illustrated by use of difference time series, as seen for instance in T2 for RSS vs. UAH in Fu and Johanson (2005; see also Karl et al., 2006).

From Figure 3.17 the first dominant impression is that overall, the records agree remarkably well, especially in the timing and amplitude of interannual variations. This is especially true at the surface, and even the tropospheric records from the two radiosonde data sets agree reasonably well, although HadAT2 has lower values in the 1970s. In the lower stratosphere, all records replicate the dominant variations and the pulses of warming following the volcanic eruptions indicated in the figure. The sonde records differ prior to 1963 in the lower stratosphere when fewer observations were available, and differences also emerge among all data sets after about 1992, with the sonde values lower than the satellite temperatures. The focus on linear trends tends to emphasize these relatively small differences.

A linear trend over the long term is often not a very good approximation of what has occurred (Seidel and Lanzante, 2004; Thorne et al., 2005a,b); alternative interpretations are to factor in the abrupt 1976–1977 climate regime shift (Trenberth, 1990) and episodic stratospheric warming and tropospheric cooling for the two years following major volcanic eruptions. Hence, the confidence limits for linear trends (Figure 3.18) are very large in the lower stratosphere owing to the presence of the large warming perturbations from volcanic eruptions. In the troposphere, the confidence limits are much wider in the tropics than globally, reflecting the strong interannual variability associated with ENSO.

Radiosonde, satellite observations and reanalyses agree that there has been global stratospheric cooling since 1979 (Figures 3.17 and 3.18), although radiosondes almost certainly still overestimate the cooling owing to residual effects of changes in instruments and processing (such as for radiation corrections; Lanzante et al., 2003b; Sherwood et al., 2005; Randel and Wu, 2006) and possibly increased sampling of cold conditions owing to stronger balloons (Parker and Cox, 1995). As the stratosphere is cooling and T2 has a 15% signal from there, it is virtually certain that the troposphere must be warming at a significantly greater rate than indicated by T2 alone. Thus, the tropospheric record adjusted for the stratospheric contribution to T2 has warmed more than T2 in every case. The differences range from 0.06°C per decade for ERA-40 to 0.09°C per decade for both radiosonde and NRA data sets. For UAH and RSS the difference is 0.07°C per decade.

The weakest tropospheric trends occur for NRA. However, unlike ERA-40, the NRA did not allow for changes in greenhouse gas increases over the record (Trenberth, 2004), resulting in errors in radiative forcing and in satellite retrievals in the infrared and making trends unreliable (Randel et al., 2000); indeed, upward trends at high surface mountain stations are stronger than NRA free-atmosphere temperatures at nearby locations (Pepin and Seidel, 2005). The records suggest that since 1979, the global and tropical tropospheric trends are similar to those at the surface although RSS, and by inference VG2, indicate greater tropospheric than surface warming. The reverse is indicated by the UAH and the radiosonde record although these data are subject to significant imperfections discussed above. Amplification occurs in the tropics for the RSS fields, especially after 1987 when there are increasing trends with altitude throughout the troposphere based on T2, T3 and T4 (Fu and Johanson, 2005). In the tropics, the theoretically expected amplification of temperature perturbations with height is borne out by interannual fluctuations (ENSO) in radiosonde, RSS, UAH and model data (Santer et al., 2005), but it is not borne out in the trends of the radiosonde records and UAH data.

The global mean trends since 1979 disguise many regional differences. In particular, in winter much larger temperature trends are present at the surface over northern continents than at higher levels (Karl et al., 2006) (see Figures 3.9 and 3.10; FAQ 3.1, Figure 1). These are associated with weakening of shallow winter temperature inversions and the strong stable surface layers that have little signature in the main troposphere. Such changes are related to changes in surface winds and atmospheric circulation (see Section 3.6.4).

In summary, for the period since 1958, overall global and tropical tropospheric warming estimated from radiosondes has slightly exceeded surface warming (Figure 3.17 and Karl et al., 2006). The climate shift of 1976 appeared to yield greater tropospheric than surface warming (Figure 3.17); such climate variations make differences between the surface and tropospheric temperature trends since 1979 unsurprising. After 1979, there has also been global and tropical tropospheric warming; however, it is uncertain whether tropospheric warming has exceeded that at the surface because the spread of trends among tropospheric data sets encompasses the surface warming trend. The range (due to different data sets, but not including the reanalyses) of global surface warming since 1979 from Figure 3.18 is 0.16°C to 0.18°C per decade compared to 0.12°C to 0.19°C per decade for MSU estimates of tropospheric temperatures. A further complexity is that surface trends have been greater over land than over ocean. Substantial cooling has occurred in the lower stratosphere. Compensation for the effects of stratospheric cooling trends on the T2 record (a cooling of about 0.08°C per decade) has been an important development. However, a linear trend is a poor fit to the data in the stratosphere and the tropics at all levels. The overall global variability is well replicated by all records, although small relative trends exacerbate the differences between records. Inadequacies in the observations and analytical methods result in structural uncertainties that still contribute to the differences between surface and tropospheric temperature trends, and revisions continue to be made. Changes in the height of the tropopause since 1979 are consistent with overall tropospheric warming as well as stratospheric cooling.

3.4.2   Water Vapour

Water vapour is a key climate variable. In the lower troposphere, condensation of water vapour into precipitation provides latent heating which dominates the structure of tropospheric diabatic heating (Trenberth and Stepaniak, 2003a,b). Water vapour is also the most important gaseous source of infrared opacity in the atmosphere, accounting for about 60% of the natural greenhouse effect for clear skies (Kiehl and Trenberth, 1997), and provides the largest positive feedback in model projections of climate change (Held and Soden, 2000).

Water vapour at the land surface has been measured since the late 19th century, but only observations made since the 1950s have been compiled into a database suitable for climate studies. The concentration of surface water vapour is typically reported as the vapour pressure, dew point temperature or relative humidity. Using physical relationships, it is possible to convert from one to the other, but the conversions are exact only for instantaneous values. As the relationships are nonlinearly related to air temperature, errors accumulate as data are averaged to daily and monthly periods. Slightly more comprehensive data exist for oceanic areas, where the dew point temperature is included as part of the ICOADS database, but few analyses have taken place for periods before the 1950s.

The network of radiosonde measurements provides the longest record of water vapour measurements in the atmosphere, dating back to the mid-1940s. However, early radiosonde sensors suffered from significant measurement biases, particularly for the upper troposphere, and changes in instrumentation with time often lead to artificial discontinuities in the data record (e.g., see Elliott et al., 2002). Consequently, most of the analysis of radiosonde humidity has focused on trends for altitudes below 500 hPa and is restricted to those stations and periods for which stable instrumentation and reliable moisture soundings are available.

Additional information on water vapour can be obtained from satellite observations and reanalysis products. Satellite observations provide near-global coverage and thus represent an important source of information over the oceans, where radiosonde observations are scarce, and in the upper troposphere, where radiosonde sensors are often unreliable.

3.4.2.1   Surface and Lower-Tropospheric Water Vapour

Boundary layer moisture strongly determines the longwave (LW) radiative flux from the atmosphere to the surface. It also accounts for a significant proportion of the direct absorption of solar radiation by the atmosphere. The TAR reported widespread increases in surface water vapour in the NH. The overall sign of these trends has been confirmed from analysis of specific humidity over the USA (Robinson, 2000) and over China from 1951 to 1994 (Wang and Gaffen, 2001), particularly for observations made at night. Differences in the spatial, seasonal and diurnal patterns of these changes were found with strong sensitivity of the results to the network choice. Philipona et al. (2004) inferred rapid increases in surface water vapour over central Europe from cloud-cleared LW radiative flux measured over the period 1995 to 2003. Subsequent analyses (Philipona et al., 2005) confirmed that changes in integrated water vapour for this region are strongly coupled to the surface temperature, with regions of warming experiencing increasing moisture and regions of cooling experiencing decreasing moisture. For central Europe, Auer et al. (2007) demonstrated increasing moisture trends. Their vapour pressure series from the Greater Alpine Region closely follow the decadal- to centennial-scale warming at both urban lowland and rural summit sites. In Canada, van Wijngaarden and Vincent (2005) found a decrease in relative humidity of several percent in the spring for 75 stations, after correcting for instrumentation changes, but little change in relative humidity elsewhere or for other seasons. Ishii et al. (2005) reported that globally averaged dew points over the ocean have risen by about 0.25°C between 1950 and 2000. Increasing extremes in summer dew points, and increased humidity during summer heat waves, were found at three stations in northeastern Illinois (Sparks et al., 2002; Changnon et al., 2003) and attributed in part to changes in agricultural practices in the region.

Dai (2006) analysed near-global (60°S–75°N) synoptic data for 1976 to 2005 from ships and buoys and more than 15,000 land stations for specific humidity, temperature and relative humidity. Nighttime relative humidity was found to be greater than daytime by 2 to 15% over most land areas, as temperatures undergo a diurnal cycle, while moisture does not change much. The global trends of near-surface relative humidity are very small. Trends in specific humidity tend to follow surface temperature trends with a global average increase of 0.06 g kg^{-1} per decade (1976–2004). The rise in specific humidity corresponds to about 4.9% per 1°C warming over the globe. Over the ocean, the observed surface specific humidity

increases at 5.7% per 1°C warming, which is consistent with a constant relative humidity. Over land, the rate of increase is slightly smaller (4.3% per 1°C), suggesting a modest reduction in relative humidity as temperatures increase, as expected in water-limited regions.

For the lower troposphere, water vapour information has been available from the TOVS since 1979 and from the Scanning Multichannel Microwave Radiometer (SMMR) from 1979 to 1984. However, the main improvement occurred with the introduction of the Special Sensor Microwave/Imager (SSM/I) in mid-1987 (Wentz and Schabel, 2000). Retrievals of column-integrated water vapour from SSM/I are generally regarded as providing the most reliable measurements of lower-tropospheric water vapour over the oceans, although issues pertaining to the merging of records from successive satellites do arise (Trenberth et al., 2005a; Sohn and Smith, 2003).

Significant interannual variability of column-integrated water vapour has been observed using TOVS, SMMR and SSM/I data. In particular, column water vapour over the tropical oceans increased by 1 to 2 mm during the 1982–1983, 1986–1987 and 1997–1998 El Niño events (Soden and Schroeder, 2000; Allan et al., 2003; Trenberth et al., 2005a) and decreased by a smaller magnitude in response to global cooling following the eruption of Mt. Pinatubo in 1991 (Soden et al., 2002; Trenberth and Smith, 2005; see also Section 8.6.3.1). The linear trend based on monthly SSM/I data over the oceans was 1.2% per decade (0.40 ± 0.09 mm per decade) for 1988 to 2004 (Figure 3.20).

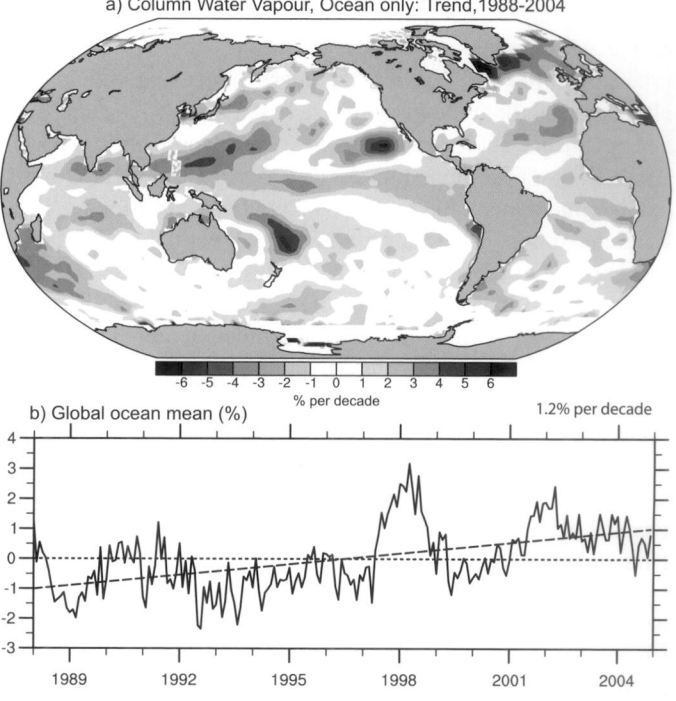

Figure 3.20. *Linear trends in precipitable water (total column water vapour) in % per decade (top) and monthly time series of anomalies relative to the 1988 to 2004 period in % over the global ocean plus linear trend (bottom), from RSS SSM/I (updated from Trenberth et al., 2005a).*

Since the trends are similar in magnitude to the interannual variability, it is likely that the latter affects the magnitude of the linear trends. The trends are overwhelmingly positive in spatial structure, but also suggestive of an ENSO influence. As noted by Trenberth et al. (2005a), most of the patterns associated with the interannual variability and linear trends can be reproduced from the observed SST changes over this period by assuming a constant relative humidity increase in water vapour mixing ratio. Given observed SST increases, this implies an overall increase in water vapour of order 5% over the 20th century and about 4% since 1970.

An independent check on globally vertically integrated water vapour amounts is whether the change in water vapour mass is reflected in the surface pressure field, as this is the only significant influence on the global atmospheric mass to within measurement accuracies. As Trenberth and Smith (2005) showed, such checks indicate considerable problems prior to 1979 in reanalyses, but results are in better agreement thereafter for ERA-40. Evaluations of column integrated water vapour from the NASA Water Vapor Project (NVAP; Randel et al., 1996), and reanalysis data sets from NRA, NCEP-2 and ERA-15/ERA-40 (see Appendix 3.B.5.4) reveal several deficiencies and spurious trends, which limit their utility for climate monitoring (Zveryaev and Chu, 2003; Trenberth et al., 2005a; Uppala et al., 2005). The spatial distributions, trends and interannual variability of water vapour over the tropical oceans are not always well reproduced by reanalyses, even after the 1970s (Allan et al., 2002, 2004; Trenberth et al., 2005a).

To summarise, global, local and regional studies all indicate increases in moisture in the atmosphere near the surface, but highlight differences between regions and between day and night. Satellite observations of oceanic lower-tropospheric water vapour reveal substantial variability during the last two decades. This variability is closely tied to changes in surface temperatures, with the water vapour mass changing at roughly the same rate at which the saturated vapour pressure does. A significant upward trend is observed over the global oceans and some NH land areas, although the calculated trend is likely influenced by large interannual variability in the record.

3.4.2.2   Upper-Tropospheric Water Vapour

Water vapour in the middle and upper troposphere accounts for a large part of the atmospheric greenhouse effect and is believed to be an important amplifier of climate change (Held and Soden, 2000). Changes in upper-tropospheric water vapour in response to a warming climate have been the subject of significant debate.

Due to instrumental limitations, long-term changes in water vapour in the upper troposphere are difficult to assess. Wang et al. (2001) found an increasing trend of 1 to 5% per decade in relative humidity during 1976 to 1995, with the largest increases in the upper troposphere, using 17 radiosonde stations in the tropical west Pacific. Conversely, a combination of Microwave Limb Sounder (MLS) and Halogen Occultation Experiment (HALOE) measurements at 215 hPa suggested increases in

water vapour with increasing temperature (Minschwaner and Dessler, 2004) on interannual time scales, but at a rate smaller than expected from constant relative humidity.

Maistrova et al. (2003) reported an increase in specific humidity at 850 hPa and a decrease from 700 to 300 hPa for 1959 to 2000 in the Arctic, based on data from ships and temporary stations as well as permanent stations. In general, the radiosonde trends are highly suspect owing to the poor quality of, and changes over time in, the humidity sensors (e.g., Wang et al., 2002a). Comparisons of water vapour sensors during recent intensive field campaigns have produced a renewed appreciation of random and systematic errors in radiosonde measurements of upper-tropospheric water vapour and of the difficulty in developing accurate corrections for these measurements (Guichard et al., 2000; Revercombe et al., 2003; Turner et al., 2003; Wang et al., 2003; Miloshevich et al., 2004; Soden et al., 2004).

Information on the decadal variability of upper-tropospheric relative humidity is now provided by 6.7 μm thermal radiance measurements from Meteosat (Picon et al., 2003) and the High-resolution Infrared Radiation Sounder (HIRS) series of instruments flying on NOAA operational polar-orbiting satellites (Bates and Jackson, 2001; Soden et al., 2005). These products rely on the merging of many different satellites to ensure uniform calibration. The HIRS channel 12 (T12) data have been most extensively analysed for variability and show linear trends in relative humidity of order ±1% per decade at various latitudes (Bates and Jackson, 2001), but these trends are difficult to separate from larger interannual fluctuations due to ENSO (McCarthy and Toumi, 2004) and are negligible when averaging over the tropical oceans (Allan et al., 2003).

In the absence of large changes in relative humidity, the observed warming of the troposphere (see Section 3.4.1) implies that the specific humidity in the upper troposphere should have increased. As the upper troposphere moistens, the emission level for T12 increases due to the increasing opacity of water vapour along the satellite line of sight. In contrast, the emission level for the MSU T2 remains constant because it depends primarily on the concentration of oxygen, which does not vary by any appreciable amount. Therefore, if the atmosphere moistens, the brightness temperature difference (T2 − T12) will increase over time due to the divergence of their emission levels (Soden et al., 2005). This radiative signature of upper-tropospheric moistening is evident in the positive trends of T2 − T12 for the period 1982 to 2004 (Figure 3.21). If the specific humidity in the upper troposphere had not increased over this period, the emission level for T12 would have remained unchanged and T2 − T12 would show little trend over this period (dashed line in Figure 3.21).

Clear-sky outgoing longwave radiation (OLR) is also highly sensitive to upper-tropospheric water vapour and a number of scanning instruments have made well-calibrated but non-overlapping measurements since 1985 (see Section 3.4.3). Over this period, the small changes in clear-sky OLR can be explained by the observed temperature changes while maintaining a constant relative humidity (Wong et al., 2000; Allan and Slingo,

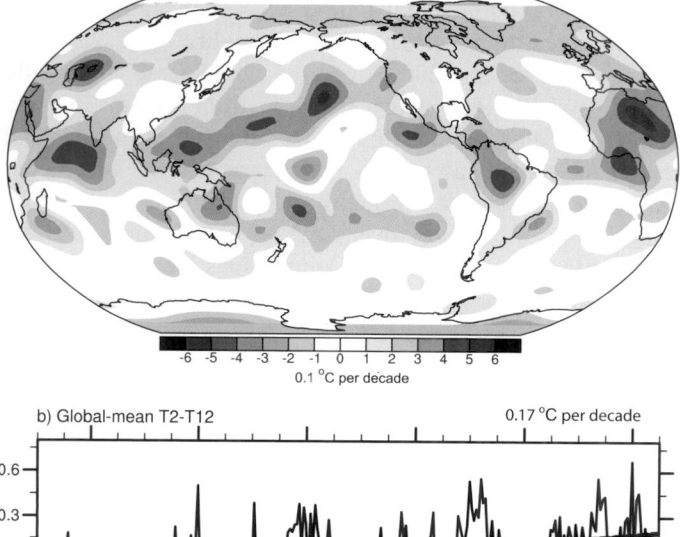

a) Upper Troposphere Moisture: T2-T12 Trend 1982-2004

b) Global-mean T2-T12   0.17 °C per decade

Figure 3.21. *The radiative signature of upper-tropospheric moistening is given by upward linear trends in T2–T12 for 1982 to 2004 (0.1 °C per decade; top) and monthly time series of the global-mean (80°N to 80°S) anomalies relative to 1982 to 2004 (°C) and linear trend (dashed; bottom). Data are from the RSS T2 and HIRS T12 (Soden et al., 2005). The map is smoothed to spectral truncation T31 resolution.*

2002) and changes in well-mixed greenhouse gases (Allan et al., 2003). This again implies a positive relationship between specific humidity and temperature in the upper troposphere.

To summarise, the available data do not indicate a detectable trend in upper-tropospheric relative humidity. However, there is now evidence for global increases in upper-tropospheric specific humidity over the past two decades, which is consistent with the observed increases in tropospheric temperatures and the absence of any change in relative humidity.

3.4.2.4 Stratospheric Water Vapour

The TAR noted an apparent increase of roughly 1% yr^{-1} in stratospheric water vapour content (~0.05 ppm yr^{-1}) during the last half of the 20th century (Kley et al., 2000; Rosenlof et al., 2001). This was based on data taken at mid-latitudes, and from multiple instruments. However, the longest series of data come from just two locations in North America with no temporal overlap. The combination of measurement uncertainties and relatively large variability on time scales from months to years warrants some caution when interpreting the longer-term trends (Kley et al., 2000; Fueglistaler and Haynes, 2005). The moistening is more convincingly documented during the 1980s and most of the 1990s than earlier, due to a longer continuous record (the NOAA Climate Monitoring and Diagnostics Laboratory (CMDL) frost-point balloon record from Boulder,

Colorado; Oltmans et al., 2000) and the availability of satellite observations during much of this period. However, discrepancies between satellite- and balloon-measured variations are apparent at decadal time scales, largely over the latter half of the 1990s (Randel et al., 2004a).

An increase in stratospheric water vapour has important radiative and chemical consequences (see also Section 2.3.8). These may include a contribution to the recent observed cooling of the lower stratosphere and/or warming of the surface (Forster and Shine, 1999, 2002; Smith et al., 2001), although the exact magnitude is difficult to quantify (Oinas et al., 2001; Forster and Shine, 2002). Some efforts to reconcile observed rates of cooling in the stratosphere with those expected based on observed changes in ozone and carbon dioxide (CO_2) since 1979 (Langematz et al., 2003; Shine et al., 2003) have found discrepancies in the lower stratosphere consistent with an additional cooling effect of a stratospheric water vapour increase. However, Shine et al. (2003) noted that because the water vapour observations over the period of consideration are not global in extent, significant uncertainties remain as to whether radiative effects of a water vapour change are a significant contributor to the stratospheric temperature changes. Moreover, other studies which account for uncertainties in the ozone profiles and temperature trends, and natural variability, can reconcile the observed stratospheric temperature changes without the need for sizable water vapour changes (Ramaswamy and Schwarzkopf, 2002; Schwarzkopf and Ramaswamy, 2002).

Although methane oxidation is a major source of water in the stratosphere, and has been increasing over the industrial period, the noted stratospheric trend in water vapour is too large to attribute to methane oxidation alone (Kley et al., 2000; Oltmans et al., 2000). Therefore, other contributors to an increase in stratospheric water vapour are under active investigation. It is likely that different mechanisms are affecting water vapour trends at different altitudes. Aviation emits a small but potentially significant amount of water vapour directly into the stratosphere (IPCC, 1999). Several indirect mechanisms have also been considered including: a) volcanic eruptions (Considine et al., 2001; Joshi and Shine, 2003); b) biomass-burning aerosol (Sherwood, 2002; Andreae et al., 2004); c) tropospheric sulphur dioxide (Notholt et al., 2005); and d) changes to methane oxidation rates from changes in stratospheric chlorine, ozone and the hydroxyl radical (Röckmann et al., 2004). Other proposed mechanisms relate to changes in tropopause temperatures or circulation (Stuber et al., 2001; Zhou et al., 2001; Rosenlof, 2002; Nedoluha et al., 2003; Dessler and Sherwood, 2004; Fueglistaler et al., 2004; Roscoe, 2004).

It has been assumed that temperatures near the tropical tropopause control stratospheric water vapour according to equilibrium thermodynamics, importing more water vapour into the stratosphere when temperatures are warmer. However, tropical tropopause temperatures have cooled slightly over the period of the stratospheric water vapour increase (see Section 3.4.1; Seidel et al., 2001; Zhou et al, 2001). This makes the mid-latitude lower-stratospheric increases harder to explain (Fueglistaler and Haynes, 2005). Satellite observations (Read et al., 2004) show water vapour injected above the tropical tropopause by deep convective clouds, bypassing the traditional control point. Changes in the amount of condensate sublimating in this layer may have contributed to the upward trend, but to what degree is uncertain (Sherwood, 2002). Another suggested source for temperature-independent variability is changes in the efficiency with which air is circulated through the coldest regions before entering the stratosphere (Hatsushika and Yamazaki, 2003; Bonnazola and Haynes, 2004; Dessler and Sherwood, 2004; Fueglistaler et al., 2004). However, it is not yet clear that a circulation-based mechanism can explain the observed trend (Fueglistaler and Haynes, 2005).

The TAR noted a stalling of the upward trend in water vapour during the last few years of observations available at that time. This change in behaviour has persisted, with a near-zero trend in stratospheric water vapour between 1996 and 2000 (Nedoluha et al., 2003; Randel et al., 2004a). The upward trend of methane is also smaller and is currently close to zero (see Section 2.3.2). Further, at the end of 2000 there was a dramatic drop in water vapour in the tropical lower stratosphere as observed by both satellite and CMDL balloon data (Randel et al., 2004a). Temperatures observed near the tropical tropopause also dropped, but the processes producing the tropical tropopause cooling itself are currently not fully understood. The propagation of this recent decrease through the stratosphere should ensure flat or decreasing stratospheric moisture for at least the next few years.

To summarise, water vapour in the stratosphere has shown significant long-term variability and an apparent upward trend over the last half of the 20th century but with no further increases since 1996. It does not appear that this behaviour is a straightforward consequence of known climate changes. Although ideas have been put forward, there is no consensus as to what caused either the upward trend or its recent disappearance.

3.4.3 Clouds

Clouds play an important role in regulating the flow of radiation at the top of the atmosphere and at the surface. They are also integral to the atmospheric hydrological cycle via their integral influence on the balance between radiative and latent heating. The response of cloud cover to increasing greenhouse gases currently represents the largest uncertainty in model predictions of climate sensitivity (see Chapter 8). Surface observations made at weather stations and onboard ships, dating back over a century, provide the longest available records of cloud cover changes. Surface observers report the all-sky conditions, which include the sides as well as bottoms of clouds, but are unable to report upper-level clouds that may be obscured from the observer's view. Although limited by potential inhomogeneities in observation times and methodology, the surface-observed cloud changes are often associated with physically consistent changes in correlative data, strengthening

their credibility. Since the mid-1990s, especially in the USA and Canada, human observations at the surface have been widely replaced with automated ceilometer measurements, which measure only directly overhead low clouds rather than all-sky conditions. In contrast, satellites generally only observe the uppermost level of clouds and have difficulty detecting optically thin clouds. While satellite measurements do provide much better spatial and temporal sampling than can be obtained from the surface, their record is much shorter in length. These disparities in how cloud cover is observed contribute to the lack of consistency between surface- and satellite-measured changes in cloudiness. Condensation trails ('contrails') from aircraft exhaust may expand to form cirrus clouds and these and cosmic ray relations to clouds are addressed in Chapter 2.

3.4.3.1 Surface Cloud Observations

As noted in the TAR and extended with more recent studies, surface observations suggest increased total cloud cover since the middle of the last century over many continental regions including the USA (Sun, 2003; Groisman et al., 2004; Dai et al., 2006), the former USSR (Sun and Groisman, 2000; Sun et al., 2001), Western Europe, mid-latitude Canada, and Australia (Henderson-Sellers, 1992). This increasing cloudiness since 1950 is consistent with an increase in precipitation and a reduction in DTR (Dai et al., 2006). However, decreasing cloudiness over this period has been reported over China (Kaiser, 1998), Italy (Maugeri et al., 2001) and over central Europe (Auer et al.,2007). If the analyses are restricted to after about 1971, changes in continental cloud cover become less coherent. For example, using a worldwide analysis of cloud data (Hahn and Warren, 2003; Minnis et al., 2004) regional reductions were found since the early 1970s over western Asia and Europe but increases over the USA.

Changes in total cloud cover along with an estimate of precipitation over global and hemispheric land (excluding North America) from 1976 to 2003 are shown in Figure 3.22. During this period, secular trends over land are small. The small variability evident in land cloudiness appears to be correlated with precipitation changes, particularly in the SH (Figure 3.22). Note that surface observations from North America are excluded from this figure due to the declining number of human cloud observations since the early 1990s over the USA and Canada, as human observers have been replaced with Automated Surface Observation Systems (ASOS) from which cloud amounts are less reliable and incompatible with previous records (Dai et al., 2006). However, independent human observations from military stations suggest an increasing trend (~1.4% of sky per decade) in total cloud cover over the USA.

The TAR also noted multi-decadal trends in cloud cover over the ocean. An updated analysis of this information (Norris, 2005a) documented substantial decadal variability and decreasing trends in upper-level cloud cover over mid- and low-latitude oceans since 1952. However, there are no direct observations of upper-level clouds from the surface and instead Norris (2005a) infers them from reported total and low cloud cover assuming

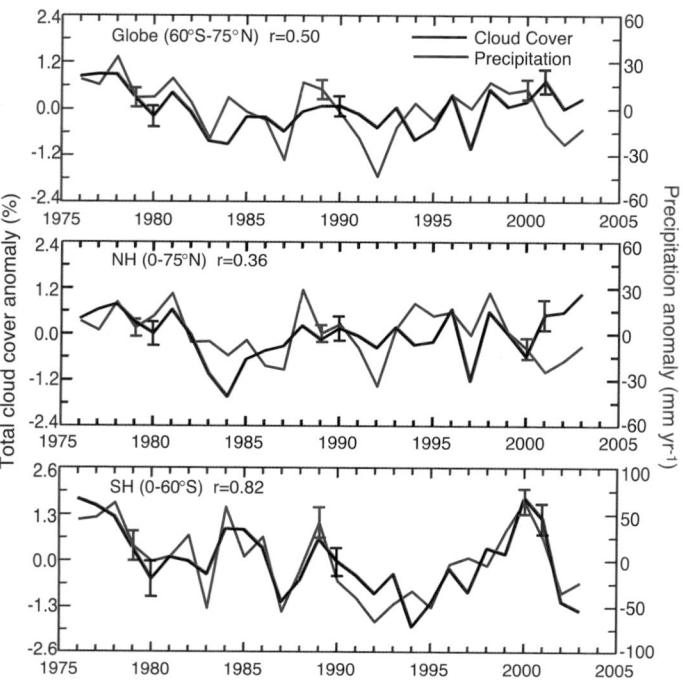

Figure 3.22. *Annual total land (excluding the USA and Canada) cloud cover (black) and precipitation (red) anomalies from 1976 to 2003 over global (60°S–75°N), NH and SH regions, with the correlation coefficient (r) shown at the top. The cloud cover is derived by gridding and area-averaging synoptic observations and the precipitation is updated from Chen et al. (2002). Typical 5 to 95% error bars for each decade are estimates using inter-grid-box variations (from Dai et al., 2006).*

a random overlap. These results partially reverse the finding of increasing trends in mid-level cloud amount in the northern mid-latitude oceans that was reported in the TAR, although the new study does not distinguish between high and middle clouds. Norris (2005b) found that upper-level cloud cover had increased over the equatorial South Pacific between 1952 and 1997 and decreased over the adjacent subtropical regions, the tropical Western Pacific, and the equatorial Indian Ocean. This pattern is consistent with decadal changes in precipitation and atmospheric circulation over these regions noted in the TAR, which further supports their validity. Deser et al. (2004) found similar spatial patterns in inter-decadal variations in total cloud cover, SST and precipitation over the tropical Pacific and Indian Oceans during 1900 to 1995. In contrast, low-cloud cover increased over almost all of the tropical Indian and Pacific Oceans, but this increase bears little resemblance to changes in atmospheric circulation over this period, suggesting that it may be spurious (Norris, 2005b). When averaged globally, oceanic cloud cover appears to have increased over the last 30 years or more (e.g., Ishii et al., 2005).

During El Niño events, cloud cover generally decreases over land throughout much of the tropics and subtropics, but increases over the ocean in association with precipitation changes (Curtis and Adler, 2003). Multi-decadal variations are affected by the 1976–1977 climate shift (Deser et al., 2004), and these dominate the low-latitude trends from 1971 to 1996 found in Hahn and Warren (2003).

3.4.3.2 Satellite Cloud Observations

Since the TAR, there has been considerable effort in the development and analysis of satellite data sets for documenting changes in global cloud cover over the past few decades. The most comprehensive cloud climatology is that of the International Satellite Cloud Climatology Project (ISCCP), begun in July 1983. The ISCCP shows an increase in globally averaged total cloud cover of about 2% from 1983 to 1987, followed by a decline of about 4% from 1987 to 2001 (Rossow and Dueñas, 2004). Cess and Udelhofen (2003) documented decreasing ISCCP total cloud cover in all latitude zones between 40°S and 40°N. Norris (2005a) found that both ISCCP and ship synoptic reports show consistent reductions in middle- or high-level cloud cover from the 1980s to the 1990s over low- and mid-latitude oceans. Minnis et al. (2004) also found consistent trends in high-level cloud cover between ISCCP and surface observations over most areas, except for the North Pacific where they differed by almost 2% per decade. In addition, an analysis of Stratospheric Aerosol and Gas Experiment II (SAGE II) data revealed a decline in cloud frequency above 12 km between 1985 and 1998 (Wang et al., 2002b) that is consistent with the decrease in upper-level cloud cover noted in ISCCP and ocean surface observations. The decline in upper-level cloud cover since 1987 may also be consistent with a decrease in reflected shortwave (SW) radiation during this period as measured by the Earth Radiation Budget Satellite (ERBS; see Section 3.4.4). Radiative transfer calculations, which use the ISCCP cloud properties as input, are able to independently reproduce the decadal changes in outgoing LW and reflected SW radiation reported by ERBS (Zhang et al., 2004c).

Analyses of the spatial trends in ISCCP cloud cover reveal changing biases arising from changes in satellite view angle and coverage that affect the global mean anomaly time series (Norris, 2000; Dai et al., 2006). The ISCCP spurious variability may occur primarily in low-level clouds with the least optical thickness (the ISCCP 'cumulus' category; Norris, 2005a), due to discontinuities in satellite view angles associated with changes in satellites. Such biases likely contribute to ISCCP's negative cloud cover trend, although their magnitude and impact on radiative flux calculations using ISCCP cloud data are not yet known. Additional artefacts, including radiometric noise, navigation and rectification errors are present in the ISCCP data (Norris, 2000), but the effects of known and unknown artefacts on ISCCP cloud and flux data have not yet been quantified.

Other satellite data sets show conflicting decadal changes in total cloud cover. For example, analysis of cloud cover changes from the HIRS shows a slight increase in cloud cover between 1985 and 2001 (Wylie et al., 2005). However, spurious changes have also been identified in the HIRS data set, which may affect its estimates of decadal variability. One important source of uncertainty results from the drift in Equatorial Crossing Time (ECT) of polar-orbiting satellite measurements (e.g., HIRS and the Advanced Very High Resolution Radiometer; AVHRR), which aliases the large diurnal cycle of clouds into spurious lower-frequency variations. After correcting for ECT drift and other small calibration errors in AVHRR measurements of cloudiness, Jacobowitz et al. (2003) found essentially no trend in cloud cover for the tropics from 1981 to 2000.

While the variability in surface-observed upper-level cloud cover has been shown to be consistent with that observed by ISCCP (Norris, 2005a), the variability in total cloud cover is not, implying differences between ISCCP and surface-observed low cloud cover. Norris (2005a) shows that even after taking into account the difference between surface and satellite views of low-level clouds, the decadal changes between the ISCCP and surface data sets still disagree. The extent to which this results from differences in spatial and temporal sampling or differences in viewing perspective is unclear.

In summary, while there is some consistency between ISCCP, ERBS, SAGE II and surface observations of a reduction in high cloud cover during the 1990s relative to the 1980s, there are substantial uncertainties in decadal trends in all data sets and at present there is no clear consensus on changes in total cloudiness over decadal time scales.

3.4.4   Radiation

Measuring the radiation balance accurately is fundamental in quantifying the radiative forcing of the system as well as diagnosing the radiative properties of the atmosphere and surface, which are crucial for understanding radiative feedback processes. At the top of the atmosphere, satellites provide excellent spatial coverage but poorer temporal sampling. The reverse is true at the surface with only a limited number of high-quality point measurements but with excellent temporal coverage.

3.4.4.1   Top-of-Atmosphere Radiation

One important development since the TAR is the apparent unexpectedly large changes in tropical mean radiation flux reported by ERBS (Wielicki et al., 2002a,b). It appears to be related in part to changes in the nature of tropical clouds (Wielicki et al., 2002a), based on the smaller changes in the clear-sky component of the radiative fluxes (Wong et al., 2000; Allan and Slingo, 2002), and appears to be statistically distinct from the spatial signals associated with ENSO (Allan and Slingo, 2002; Chen et al., 2002). A recent reanalysis of the ERBS active-cavity broadband data corrects for a 20 km change in satellite altitude between 1985 and 1999 and changes in the SW filter dome (Wong et al., 2006). Based upon the revised (Edition 3_Rev1) ERBS record (Figure 3.23), outgoing LW radiation over the tropics appears to have increased by about 0.7 W m^{-2} while the reflected SW radiation decreased by roughly 2.1 W m^{-2} from the 1980s to 1990s (Table 3.5).

These conclusions depend upon the calibration stability of the ERBS non-scanner record, which is affected by diurnal sampling issues, satellite altitude drifts and changes in calibration following a three-month period when the sensor was powered off (Trenberth, 2002). Moreover, rather than a trend, the reflected SW radiation change may stem mainly from

a jump in late 1992 in the ERBS record that is also observed in the ISCCP (version FD) record (Zhang et al., 2004c) but not in the AVHRR Pathfinder record (Jacobowitz et al., 2003). However, careful inspection of the sensor calibration revealed no known issues that can explain the decadal shift in the fluxes despite corrections to the ERBS time series relating to diurnal aliasing and satellite altitude changes (Wielicki et al., 2002b; Wong et al., 2006).

As noted in Section 3.4.3, the low-latitude changes in the radiation budget appear consistent with reduced cloud fraction from ISCCP. Detailed radiative transfer computations, using ISCCP cloud products along with additional global data sets, show broad agreement with the ERBS record of tropical radiative fluxes (Hatzianastassiou et al., 2004; Zhang et al., 2004c; Wong et al., 2006). However, the decrease in reflected SW radiation from the 1980s to the 1990s may be inconsistent with the increase in total and low cloud cover over oceans reported by surface observations (Norris, 2005a), which show increased low cloud occurrence. The degree of inconsistency, however, is difficult to ascertain without information on possible changes in low-level cloud albedo.

While the ERBS satellite provides the only continuous long-term top-of-atmosphere (TOA) flux record from broadband active-cavity instruments, narrow spectral band radiometers have made estimates of both reflected SW and outgoing LW radiation trends using regressions to broadband data, or using radiative transfer theory to estimate unmeasured portions of the spectrum of radiation. Table 3.5 shows the 1980s to 1990s TOA tropical mean flux changes for the ERBS Edition 3 data (Wong et al., 2006), the HIRS Pathfinder data (Mehta and Susskind, 1999), the AVHRR Pathfinder data (Jacobowitz et al., 2003) and the ISCCP FD data (Zhang et al., 2004c).

The most accurate of the data sets in Table 3.5 is believed to be the ERBS Edition 3 Rev 1 active-cavity wide field of view data (Wielicki et al., 2005). The ERBS stability is estimated as better than 0.5 W m^{-2} over the 1985 to 1999 period and the spatial and temporal sampling noise is less than 0.5 W m^{-2} on annual time scales (Wong et al., 2006). The outgoing LW radiation changes from ERBS are similar to the decadal changes in the HIRS Pathfinder and ISCCP FD records, but disagree with the AVHRR Pathfinder data (Wong et al., 2006). The AVHRR Pathfinder data also do not support the TOA SW radiation trends. However, calibration issues, conversion from narrow to broadband, and satellite orbit changes are thought to render the AVHRR record less reliable for decadal changes compared to ERBS (Wong et al., 2006). Estimates of the stability of the ISCCP time series for long-term TOA flux records are 3 to 5 W m^{-2} for SW radiative flux and 1 to 2 W m^{-2} for LW radiative flux (Brest et al., 1997), although the time series agreement of the ISCCP and ERBS records are much closer than these estimated calibration drift uncertainties (Zhang et al., 2004c).

The changes in SW radiation measured by ERBS Edition 3 Rev 1 are larger than the clear-sky flux changes due to humidity variations (Wong et al., 2000) or anthropogenic radiative forcing (see Chapter 2). If correct, the large decrease in reflected SW radiation with little change in outgoing LW radiation implies a reduction in tropical low cloud cover over this period. However, specific information on cloud radiative forcing is not available from ERBS after 1989 and, as noted in Section 3.4.3, surface data sets suggest an increase in low cloud cover over this period.

Since most of the net tropical heating of 1.4 W m^{-2} is a decrease in reflected SW radiative flux, the change implies a similar increase in solar insolation at the surface that, if unbalanced by other changes in surface fluxes, would increase the amount of ocean heat storage. Wong et al. (2006) showed that the changes in global net radiation are consistent with a new ocean heat-storage data set from Willis et al. (2004; see Chapter 5 and Figure 5.1). Differences between the two data sets are roughly 0.4 W m^{-2}, in agreement with the estimated annual sampling noise in the ocean heat-storage data.

Using astronomical observations of visible wavelength solar photons reflected from parts of the Earth to the moon and then back to the Earth at a surface-based observatory, Pallé et al. (2004) estimated a dramatic increase of Earth-reflected SW radiative flux of 5.5 W m^{-2} over three years. This is unlikely to be real, as over the same time period (2000–2003), the Clouds and the Earth's Radiant Energy System (CERES) broadband data indicate a decrease in SW radiative flux of almost 1 W m^{-2}, which is much smaller and the opposite sign (Wielicki et al., 2005). In addition, changes in ocean heat storage are more consistent with the CERES data than with the Earthshine indirect observation.

The only long-term time series (1979–2001) of energy divergence in the atmosphere (Trenberth and Stepaniak, 2003b) are based on NRA, which, although not reliable for depicting trends, are reliable on interannual times scales for which they show substantial variability associated with ENSO. Analyses by Trenberth and Stepaniak (2003b) reveal more divergence of energy out of the deep tropics in the 1990s compared with the 1980s due to differences in ENSO, which may account for at least some of the changes discussed above.

In summary, although there is independent evidence for decadal changes in TOA radiative fluxes over the last two decades, the evidence is equivocal. Changes in the planetary and tropical TOA radiative fluxes are consistent with independent global ocean heat-storage data, and are expected to be dominated by changes in cloud radiative forcing. To the extent that they are real, they may simply reflect natural low-frequency variability of the climate system.

3.4.4.2 Surface Radiation

The energy balance at the surface requires net radiative heating to be balanced by turbulent energy fluxes and thus determines the evolution of surface temperature and the cycling of water, which are key parameters of climate change (see Box 7.1). In recent years, several studies have focused on observational evidence of changing surface radiative heating.

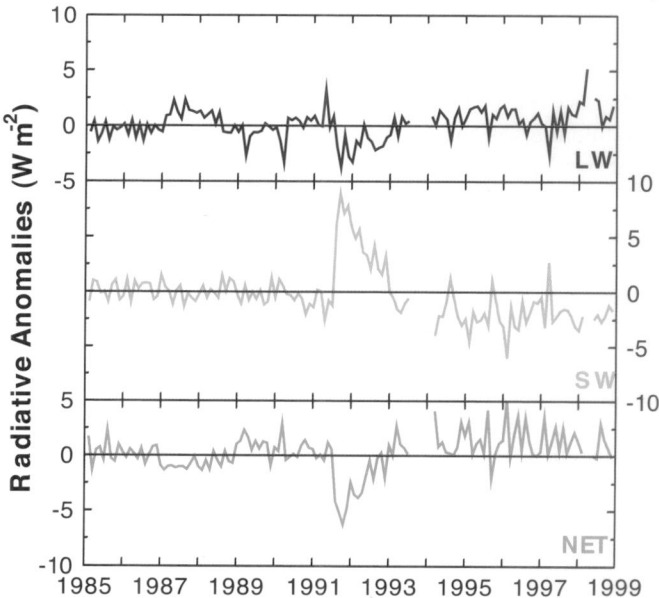

Figure 3.23. *Tropical mean (20°S to 20°N) TOA flux anomalies from 1985 to 1999 (W m^{-2}) for LW, SW, and NET radiative fluxes [NET = –(LW + SW)]. Coloured lines are observations from ERBS Edition 3_Rev1 data from Wong et al. (2006) updated from Wielicki et al. (2002a), including spacecraft altitude and SW dome transmission corrections.*

Reliable SW radiative measurement networks have existed since the 1957–1958 International Geophysical Year.

A reduction in downward solar radiation ('dimming') of about 1.3% per decade or about 7 W m^{-2} was observed from 1961 to 1990 at land stations around the world (Gilgen et al., 1998; Liepert, 2002). Additional studies also found declines in surface solar radiation in the Arctic and Antarctic (Stanhill and Cohen, 2001) as well as at sites in the former Soviet Union (Russak, 1990; Abakumova et al., 1996), around the Mediterranean Sea (Aksoy, 1997; Omran, 2000), China (Ren et al., 2005), the USA (Liepert, 2002) and southern Africa (Power and Mills, 2005). Stanhill and Cohen (2001) claim an overall globally averaged reduction of 2.7% per decade but used only 30 records. However, the stations where these analyses took place are quite limited in domain and dominated by large urban areas, and the dimming is much less at rural sites (Alpert et al., 2005) or even missing altogether over remote areas, except for

Table 3.5. *Top-of-atmosphere (TOA) radiative flux changes from the 1980s to 1990s (W m^{-2}). Values are given as tropical means (20°S to 20°N) for the 1994 to 1997 period minus the 1985 to 1989 period. Dashes are shown where no data are available. From Wong et al. (2006).*

Data Source	Radiative Flux Change (W m^{-2})		
	TOA LW	TOA SW	TOA Net
ERBS Edition 3 Rev 1	0.7	–2.1	1.4
HIRS Pathfinder	0.2	–	–
AVHRR Pathfinder	–1.4	0.7	0.7
ISCCP FD	0.5	–2.4	1.8

Box 3.2: The Dimming of the Planet and Apparent Conflicts in Trends of Evaporation and Pan Evaporation

Several reports have defined the term 'global dimming' (e.g., Cohen et al., 2004). This refers to a widespread reduction of solar radiation received at the surface of the Earth, at least up until about 1990 (Wild et al., 2005). However, recent studies (Alpert et al., 2005; Schwartz, 2005) found that dimming is not global but is rather confined only to large urban areas. At the same time there is considerable confusion in the literature over conflicting trends in pan evaporation and actual evaporation (Ohmura and Wild, 2002; Roderick and Farquhar, 2002, 2004, 2005; Hobbins et al., 2004; Wild et al., 2004, 2005) although the framework for explaining observed changes exists (Brutsaert and Parlange, 1998).

Surface evaporation, or more generally evapotranspiration, depends upon two key components. The first is available energy at the surface, especially solar radiation. The second is the availability of surface moisture, which is not an issue over oceans, but which is related to soil moisture amounts over land. Evaporation pans provide estimates of the potential evaporation that would occur if the surface were wet. Actual evaporation is generally not measured, except at isolated flux towers, but may be computed using bulk flux formulae or estimated as a residual from the surface moisture balance.

The evidence is strong that a key part of the solution to the paradox of conflicting trends in evaporation and pan evaporation lies in changes in the atmospheric circulation and the hydrological cycle. There has been an increase in clouds and precipitation, which reduce solar radiation available for actual and potential evapotranspiration but also increase soil moisture and make the actual evapotranspiration closer to the potential evapotranspiration. An increase in both clouds and precipitation has occurred over many parts of the land surface (Dai et al., 1999, 2004a, 2006), although not in the tropics and subtropics (which dominate the global land mean; Section 3.3.2.2). This reduces solar radiation available for evapotranspiration, as observed since the late 1950s or early 1960s over the USA (Liepert, 2002), parts of Europe and Siberia (Peterson et al., 1995; Abakumova et al., 1996), India (Chattopadhyay and Hulme, 1997), China (Liu et al., 2004a) and over land more generally (Wild et al., 2004). However, increased precipitation also increases soil moisture and thereby increases actual evapotranspiration (Milly and Dunne, 2001). Moreover, increased clouds impose a greenhouse effect and reduce outgoing LW radiation (Philipona and Dürr, 2004), so that changes in net radiation can be quite small or even of reversed sign. Recent re-assessments suggest increasing trends of evapotranspiration over southern Russia during the last 40 years (Golubev et al., 2001) and over the USA during the past 40 or 50 years (Golubev et al., 2001; Walter et al., 2004) in spite of decreases in pan evaporation. Hence, in most, but not all, places the net result has been an increase in actual evaporation but a decrease in pan evaporation. Both are related to observed changes in atmospheric circulation and associated weather.

It is an open question as to how much the changes in cloudiness are associated with other effects, notably impacts of changes in aerosols. Dimming seems to be predominant in large urban areas where pollution plays a role (Alpert et al., 2005). Increases in aerosols are apt to redistribute cloud liquid water over more and smaller droplets, brightening clouds, decreasing the potential for precipitation and perhaps changing the lifetime of clouds (e.g., Rosenfeld, 2000; Ramanathan et al., 2001; Kaufman et al., 2002; see Sections 2.4 and 7.5). Increases in aerosols also reduce direct radiation at the surface under clear skies (e.g., Liepert, 2002), and this appears to be a key part of the explanation in China (Ren et al., 2005).

Another apparent paradox raised by Wild et al. (2004) is that if surface radiation decreases then it should be compensated by a decrease in evaporation from a surface energy balance standpoint, especially given an observed increase in surface air temperature. Of course, back radiation from greenhouse gases and clouds operate in the opposite direction (Philipona and Dürr, 2004). Also, a primary change (not considered by Wild et al., 2004) is in the partitioning of sensible vs. latent heat at the surface and thus in the Bowen ratio. Increased soil moisture means that more heating goes into evapotranspiration at the expense of sensible heating, reducing temperature increases locally (Trenberth and Shea, 2005). Temperatures are affected above the surface where latent heating from precipitation is realised, but then the full dynamics of the atmospheric motions (horizontal advection, adiabatic cooling in rising air and warming in compensating subsiding air) come into play. The net result is a non-local energy balance.

identifiable effects of volcanic eruptions, such as Mt. Pinatubo in 1991 (Schwartz, 2005). At the majority of 421 analysed sites, the decline in surface solar radiation ended around 1990 and a recovery of about 6 W m^{-2} occurred afterwards (Wild et al., 2004; 2005). The increase in surface solar radiation ('brightening') agrees with satellite and surface observations of reduced cloud cover (Wang et al., 2002b; Wielicki et al., 2002a; Rossow and Dueñas, 2004; Norris, 2005b; Pinker et al., 2005), although there is evidence that some of these changes are

spurious (see Section 3.4.3). In addition, the satellite-observed increase in surface radiation noted by Pinker et al. (2005) occured primarily over ocean, whereas the increase observed by Wild et al. (2005) was restricted to land stations.

From 1981 to 2003 over central Europe, Philipona and Dürr (2004) showed that decreases in surface solar radiation from increases in clouds were cancelled by opposite changes in LW radiation and that increases in net radiative flux were dominated by the clear-sky LW radiation component relating to an enhanced water vapour greenhouse effect. Alpert et al. (2005) provided evidence that a significant component of the reductions may relate to increased urbanisation and anthropogenic aerosol concentrations over the period (see also Section 7.5). This has been detected in solar radiation reductions for polluted regions (e.g., China; Luo et al., 2001), but cloudiness changes must also play a major role, as shown for European sites and the USA (Liepert, 2002; Dai et al., 2006). In the USA increasing cloud optical thickness and a shift from cloud-free to more cloudy skies are the dominating factors compared to the aerosol direct effects. Possible causes of the 1990s reversal are reduced cloudiness and increased cloud-free atmospheric transparency due to the reduction of anthropogenic aerosol concentrations and recovery from the effects of the 1991 eruption of Mt. Pinatubo. See Box 3.2 for more discussion and a likely explanation of these aspects.

3.5 Changes in Atmospheric Circulation

Changes in the circulation of the atmosphere and ocean are an integral part of climate variability and change. Accordingly, regional variations in climate can be complex and sometimes counter-intuitive. For example, a rise in global mean temperatures does not mean warming everywhere, but can result in cooling in some places, due to circulation changes.

This section assesses research since the TAR on atmospheric circulation changes, through analysis of global-scale data sets of mean sea level pressure (MSLP), geopotential heights, jet streams and storm tracks. Related quantities at the surface over the ocean, including winds, waves and surface fluxes, are also considered. Many of the results discussed are based on reanalysis data sets. Reanalyses provide a global synthesis of all available observations, but are subject to spurious changes over time as observations change, especially in the late 1970s with the improved satellite and aircraft data and observations from drifting buoys over the SH. See Appendix 3.B.5 for a discussion of the quality of reanalyses from a climate perspective.

3.5.1 Surface or Sea Level Pressure

Maps of MSLP synthesize the atmospheric circulation status. Hurrell and van Loon (1994) noted MSLP changes in the SH beginning in the 1970s while major changes were also occurring over the North Pacific in association with the 1976–1977 climate shift (Trenberth, 1990; Trenberth and Hurrell, 1994). More recently, analyses of sea level pressure from 1948 to 2005 for DJF found decreases over the Arctic, Antarctic and North Pacific, an increase over the subtropical North Atlantic, southern Europe and North Africa (Gillett et al., 2003, 2005), and a weakening of the Siberian High (Gong et al., 2001). The strength of mid-latitude MSLP gradients and associated westerly circulation appears to have increased in both hemispheres, especially during DJF, since at least the late 1970s.

The increase in MSLP gradients in the NH appears to significantly exceed simulated internal and anthropogenically forced variability (Gillett et al., 2003, 2005). However, the significance of changes over the SH is less clear, especially over the oceans prior to satellite observations in the late 1970s, as spurious trends are evident in both major reanalyses (NRA and ERA-40; Marshall, 2003; Bromwich and Fogt, 2004; Trenberth and Smith, 2005; Wang et al., 2006a; see also Appendix 3.B.5). Consistent changes, validated with long-term station-based data, do however seem to be present since the mid-1970s and are often interpreted in terms of time-averaged signatures of weather regimes (Cassou et al., 2004) or annular modes in both hemispheres (Thompson et al., 2000; Marshall, 2003; Bromwich and Fogt, 2004; see Section 3.6).

3.5.2 Geopotential Height, Winds and the Jet Stream

Mean changes in geopotential heights resemble in many ways their MSLP counterparts (Hurrell et al., 2004). Linear trends in 700 hPa height during the solstitial seasons, from ERA-40, are shown in Figure 3.24. The 700 hPa level was used as it is the first atmospheric level to lie largely above the East Antarctic Ice Sheet. The NRA and ERA-40 trends agree closely between 1979 and 2001. Over the NH between 1960 and 2000, winter (DJF) and annual means of geopotential height at 850, 500 and 200 hPa decreased over high latitudes and increased over the mid-latitudes, as for MSLP, albeit shifted westward (Lucarini and Russell, 2002). Using NRA, Frauenfeld and Davis (2003) identified a statistically significant expansion of the NH circumpolar vortex at 700, 500 and 300 hPa from 1949 to 1970. But the vortex has contracted significantly at all levels since then (until 2000) and Angell (2006) found a downward trend in the size of the polar vortex from 1963 to 2001, consistent with warming of the vortex core and analysed increases in 850 to 300 hPa thickness temperatures.

In the NH for 1979 to 2001 during DJF, increases in geopotential height occurred between 30°N and 50°N at many longitudes, notably over the central North Pacific (Figure 3.24). North of 60°N, height changes are consistent with recent occurrences of more neutral phases of the mean polar vortex. Increases in the 700 hPa height outweigh decreases in the northern summer (JJA) during 1979 to 2001. At SH high latitudes, the largest changes are seen in the solstitial seasons (Figure 3.24), with changes of opposite sign in many areas between DJF and JJA. Changes during DJF reflect the increasing strength of the positive phase of the SAM (see Marshall, 2003; Section 3.6.5), with large height decreases over Antarctica and corresponding height increases in the mid-latitudes, through the depth of the troposphere and into the stratosphere. The corresponding enhancement of the near-surface circumpolar westerlies at about 60°S, and associated changes in meridional winds in some sectors, is consistent with a warming trend observed at weather stations over the Antarctic Peninsula and Patagonia (Thompson and Solomon, 2002; see also Sections 3.2.2.4 and 3.6.5). In winter (JJA), there have been height

NH H700 (ERA40), DJF, 1979–2001 SH H700 (ERA40), JJA, 1979–2001

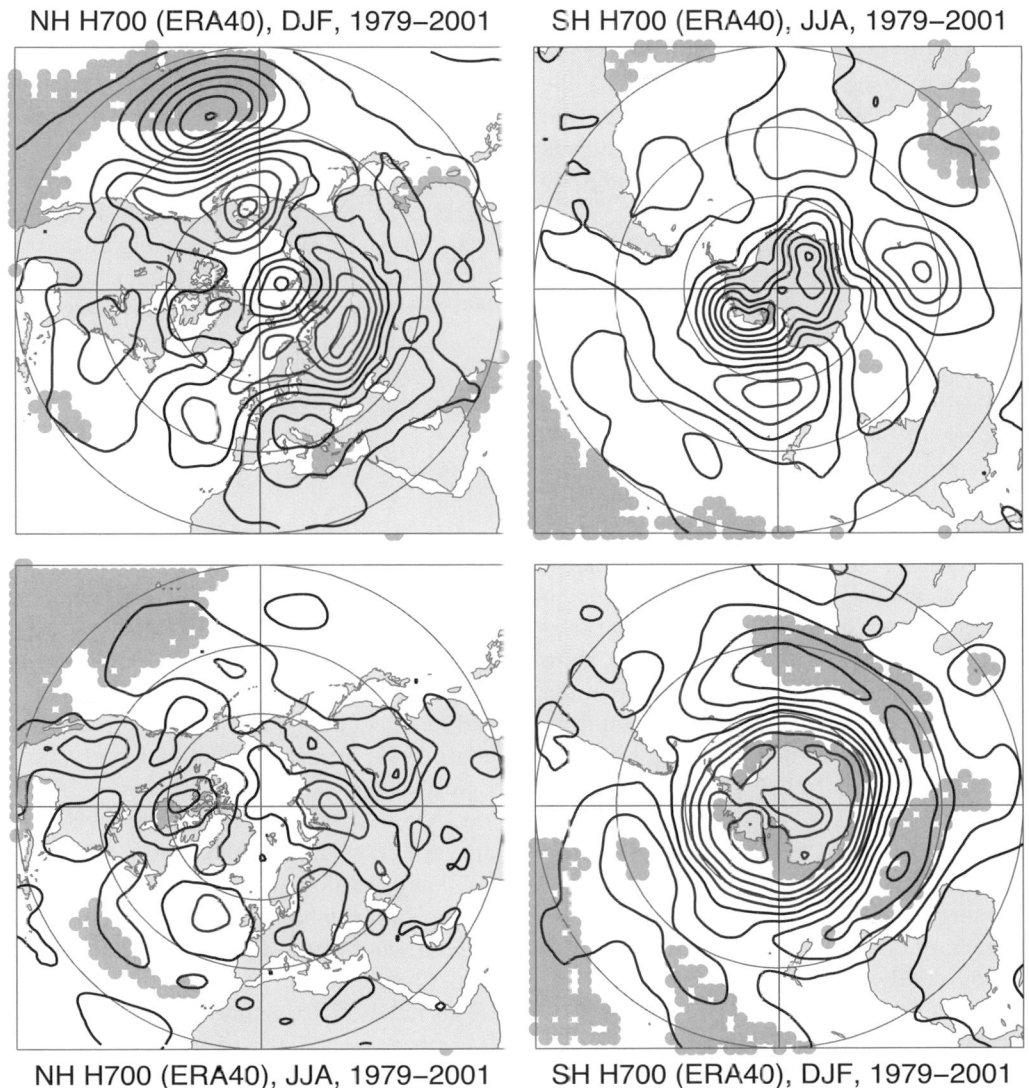

NH H700 (ERA40), JJA, 1979–2001 SH H700 (ERA40), DJF, 1979–2001

Figure 3.24. *Linear trends in ERA-40 700 hPa geopotential height from 1979 to 2001 for DJF (top left and bottom right) and JJA (bottom left and top right), for the NH (left) and SH (right). Trends are contoured in 5 gpm per decade and are calculated from seasonal means of daily 1200 UTC fields. Red contours are positive, blue negative and black zero; the grey background indicates 1% statistical significance using a standard least squares F-test and assuming independence between years.*

increases over Antarctica since 1979, with a zonal wave 3 to wave 4 pattern of rises and falls in southern mid-latitudes. Trends up to 2001 are relatively strong and statistically significant, with annular modes in both hemispheres strongly positive during the 1990s, although less so in recent years. Hence, geopotential height trends in DJF in the SH through 2004 have weakened in magnitude and significance, but with little change in spatial trend patterns.

Hemispheric teleconnections are strongly influenced by jet streams, which alter waves and storm tracks (Branstator, 2002). Using NRA from 1979 to 1995, Nakamura et al. (2002) found a weakening of the North Pacific winter jet since 1987, allowing efficient vertical coupling of upper-level disturbances with the surface temperature gradients (Nakamura and Sampe, 2002; Nakamura et al., 2004). A trend from the 1970s to the 1990s towards a deeper polar vortex and Iceland Low associated with a positive phase of the NAM in winter (Hurrell, 1995;

Thompson et al., 2000; Ostermeier and Wallace, 2003) was accompanied by intensification and poleward displacement of the Atlantic polar frontal jet and associated enhancement of the Atlantic storm track activity (Chang and Fu, 2002; Harnik and Chang, 2003). Analogous trends have also been found in the SH (Gallego et al., 2005).

3.5.3 Storm Tracks

A number of recent studies suggest that cyclone activity over both hemispheres has changed over the second half of the 20th century. General features include a poleward shift in storm track location, increased storm intensity, but a decrease in total storm numbers (e.g., Simmonds and Keay, 2000; Gulev et al., 2001; McCabe et al., 2001). In the NH, McCabe et al. (2001) found that there has been a significant decrease in mid-latitude cyclone activity and an increase in high-latitude cyclone frequency, suggesting a poleward shift of the storm track, with storm intensity increasing over the North Pacific and North Atlantic. In particular, Wang et al. (2006a) found that the North Atlantic storm track has shifted about 180 km northward in winter (JFM) during the past half century. The above findings are corroborated by Paciorek et al. (2002), Simmonds and Keay (2002) and Zhang et al. (2004b).

Several results suggest that cyclone activity in the NH mid-latitudes has increased during the past 40 years. Increases in storm track activity have been found in eddy statistics, based on NRA data. North Pacific storm track activity, identified as poleward eddy heat transport at 850 hPa, was significantly stronger during the late 1980s and early 1990s than during the early 1980s (Nakamura et al., 2002). A striking signal of decadal variability in Pacific storm track activity was its midwinter enhancement since 1987, despite a concurrent weakening of the Pacific jet, concomitant with the sudden weakening of the Siberian High (Nakamura et al., 2002; Chang, 2003). Significant increasing trends over both the Pacific and Atlantic are found in eddy meridional velocity variance at 300 hPa and other statistics (Chang and Fu, 2002; Paciorek et al., 2002). Since 1980, there was an increase in the amount of eddy kinetic energy in the NH due to an increase in the efficiency in the conversion from potential to kinetic energy (Hu et al., 2004). Graham and Diaz (2001) also found an increase in MSLP variance over the Pacific.

There are, however, significant uncertainties with such analyses, with some studies (Bromirski et al., 2003; Chang and Fu, 2003) suggesting that storm track activity during the last part of the 20th century may not be more intense than the activity prior to the 1950s. Eddy meridional velocity variance at 300 hPa in the NRA appears to be biased low prior to the mid-1970s, especially over east Asia and the western USA (Harnik and Chang, 2003). Hence, the increases in eddy variance in the NRA reanalysis data are nearly twice as large as that computed from rawinsonde observations. Better agreement is found for the Atlantic storm track exit region over Europe. Major differences between radiosonde and NRA temperature variance at 500 hPa over Asia (Iskenderian and Rosen, 2000; Paciorek et al., 2002) also cast doubts on the magnitude of the increase in storm track activity, especially over the Pacific.

Station pressure data over the Atlantic-European sector (where records are long and consistent) show a decline of storminess from high levels during the late 19th century to a minimum around 1960 and then a quite rapid increase to a maximum around 1990, followed again by a slight decline (Alexandersson et al., 2000; Bärring and von Storch, 2004; see also Section 3.8.4.1). However, changes in storm tracks are expected to be complex and depend on patterns of variability, and in practice, the noise present in the observations makes the detection of long-term changes in extratropical storm activity difficult. A more relevant approach seems to be the analysis of regional storminess in relation to spatial shifts and strength changes in teleconnection patterns (see Section 3.6).

Significant decreases in cyclone numbers, and increases in mean cyclone radius and depth over the southern extratropics over the last two or three decades (Simmonds and Keay, 2000; Keable et al., 2002; Simmonds, 2003; Simmonds et al., 2003) have been associated with the observed trend in the SAM. Such changes, derived from NRA data, have been related to reductions in mid-latitude winter rainfall (e.g., the drying trend observed in south-western Australia (Karoly, 2003) and to a circumpolar signal of increased precipitation off the coast of Antarctica (Cai et al., 2003). However, there are significant differences between ERA-40 and NRA in the SH: higher strong-cyclone activity and less weak-cyclone activity over all oceanic areas south of 40°S in all seasons, and stronger cyclone activity over the subtropics in the warm season in ERA-40, especially in the early decades (Wang et al., 2006a).

3.5.4 Blocking

Blocking events, associated with persistent high-latitude ridging and a displacement of mid-latitude westerly winds lasting typically a week or two, are an important component of total circulation variability on intra-seasonal time scales. In the NH, the preferred locations for the blocking are over the Atlantic and the Pacific (Tibaldi et al., 1994), with a spring maximum and summer minimum in the Atlantic-European region (Andrea et al., 1998; Trigo et al., 2004). Observations show that in the Euro-Atlantic sector, long-lasting (>10 day) blockings are clearly associated with the negative NAO phase (Quadrelli et al., 2001; Barriopedro et al., 2006), whereas the blockings of 5 to 10 day duration exhibit no such relationship, pointing to the dynamical links between the life cycles of NAO and blocking events (Scherrer et al, 2006; Schwierz et al., 2006). Wiedenmann et al. (2002) did not find any long-term statistically significant trends in NH blocking intensity. However, in the Pacific sector, Barriopedro et al. (2006) found a significant increase from 1948 to 2002 in western Pacific blocking days and events (57 and 62%, respectively). They also found less intense North Atlantic region blocking, with statistically significant decreases in events and days. Wiedenmann et al. (2002) found that blocking events, especially in the North Pacific region, were significantly weaker during El Niño years.

In the SH, blocking occurrence is maximised over the southern Pacific (Renwick and Revell, 1999; Renwick, 2005), with secondary blocking regions over the southern Atlantic and over the southern Indian Ocean and the Great Australian Bight. The frequency of blocking occurrence over the southeast Pacific is strongly ENSO-modulated (Rutllant and Fuenzalida, 1991; Renwick, 1998), while in other regions, much of the interannual variability in occurrence appears to be internally generated (Renwick, 2005). A decreasing trend in blocking frequency and intensity for the SH as a whole from NRA (Wiedenmann et al., 2002) is consistent with observed increases in zonal winds across the southern oceans. However, an overall increasing trend in the frequency of long-lived positive height anomalies is evident in the reanalyses over the SH in the 1970s (Renwick, 2005), apparently related to the introduction of satellite observations. Given data limitations, it may be too early to reliably define trends in SH blocking occurrence.

3.5.5 The Stratosphere

The dynamically stable stratospheric circulation is dominated in mid-latitudes by westerlies in the winter hemisphere and easterlies in the summer hemisphere, and the associated meridional overturning 'Brewer-Dobson' circulation. In the tropics, zonal winds reverse direction approximately every two years, in the downward-propagating QBO (Andrews et al., 1987). Ozone is formed predominantly in the tropics and transported to higher latitudes by the Brewer-Dobson circulation. Climatological stratospheric zonal-mean zonal winds (i.e., the westerly wind averaged over latitude circles) from different data sets show overall good agreement in the extratropics, whereas relatively large differences occur in the tropics (Randel et al., 2004b).

The breaking of vertically propagating waves, originating from the troposphere, decelerates the stratospheric westerlies (see Box 3.3). This sometimes triggers 'sudden warmings' when the westerly polar vortex breaks down with an accompanying warming of the polar stratosphere, which can quickly reverse the latitudinal temperature gradient (Kodera et al., 2000). While no major warming occurred in the NH in nine consecutive winters during 1990 to 1998, seven major warmings occurred during 1999 to 2004 (Manney et al., 2005). As noted by Naujokat et al. (2002), many of the recent stratospheric warmings after 2000 have been atypically early and the cold vortex recovered in March. In September 2002, a major warming was observed for the first time in the SH (e.g., Krüger et al., 2005; Simmons et al., 2005). This major warming followed a relatively weak winter polar vortex (Newman and Nash, 2005).

The analysis of past stratospheric changes relies on a combination of radiosonde information (available since the 1950s), satellite information (available from the 1970s) and global reanalyses. During the mid-1990s, the NH exhibited a number of years when the Arctic winter vortex was colder, stronger (Kodera and Koide, 1997; Pawson and Naujokat, 1999) and more persistent (Waugh et al., 1999; Zhou et al., 2000). Some analyses show a downward trend in the NH wave forcing in the period 1979 to 2000, particularly in January and February (Newman and Nash, 2000; Randel et al., 2002). Trend calculations are, however, very sensitive to the month and period of calculation, so the detection of long-term change from a relatively short stratospheric data series is still problematic (Labitzke and Kunze, 2005).

In the SH, using radiosonde data, Thompson and Solomon (2002) reported a significant decrease of the lower-stratospheric geopotential height averaged over the SH polar cap in October to March and May between 1969 and 1998. The ERA-40 and NRA stratospheric height reanalyses indicate a trend towards a strengthening antarctic vortex since 1980 during summer (DJF; Renwick, 2004; Section 3.5.2), largely related to ozone depletion (Ramaswamy et al., 2001; Gillett and Thompson, 2003). The ozone hole has led to a cooling of the stratospheric polar vortex in late spring (October–November; Randel and Wu, 1999), and to a two- to three-week delay in vortex breakdown (Waugh et al., 1999).

3.5.6 Winds, Waves and Surface Fluxes

Changes in atmospheric circulation imply associated changes in winds, wind waves and surface fluxes. Surface wind and meteorological observations from Voluntary Observing Ships (VOS) became systematic around 150 years ago and are assembled in ICOADS (Worley et al., 2005). Apparent significant trends in scalar wind should be considered with caution as VOS wind observations are influenced by time-dependent biases (Gulev et al., 2007), resulting from the rising proportion of anemometer measurements, increasing anemometer heights, changes in definitions of Beaufort wind estimates (Cardone et al., 1990), growing ship size, inappropriate evaluation of the true wind speed from the relative wind (Gulev and Hasse, 1999) and time-dependent sampling biases (Sterl, 2001; Gulev et al., 2007). Consideration of time series of local surface pressure gradients (Ward and Hoskins, 1996) does not support the existence of any significant globally averaged trends in marine wind speeds, but reveals regional patterns of upward trends in the tropical North Atlantic and extratropical North Pacific and downward trends in the equatorial Atlantic, tropical South Atlantic and subtropical North Pacific (see also Sections 3.5.1 and 3.5.3).

Visual VOS observations of wind waves for more than a century, often measured as significant wave height (SWH, the highest one-third of wave (sea and swell) heights), have been less affected than marine winds by changes in observational practice, although they may suffer from time-dependent sampling uncertainty, which was somewhat higher at the beginning of the record. Local wind speed directly affects only the wind-sea component of SWH, while the swell component is largely influenced by the frequency and intensity of remote storms. Linear trends in the annual mean SWH from ship data (Gulev and Grigorieva, 2004) for 1900 to 2002 were significantly positive almost everywhere in the North Pacific, with a maximum upward trend of 8 to 10 cm per decade (up to 0.5% yr^{-1}). These are supported by buoy records for 1978 to 1999 (Allan and Komar, 2000; Gower, 2002) for annual and winter (October to March) mean SWH and confirmed by the long-term estimates of storminess derived from the tide gauge residuals (Bromirski et al., 2003) and hindcast data (Graham and Diaz, 2001), although Tuller (2004) found primarily negative trends in wind off the west coast of Canada. In the Atlantic, centennial time series (Gulev and Grigorieva, 2004) show weak but statistically significant negative trends along the North Atlantic storm track, with a decrease of 5.2 cm per decade (0.25% yr^{-1}) in the western Atlantic storm formation region. Regional model hindcasts (e.g., Vikebo et al., 2003; Weisse et al., 2005) show growing SWH in the northern North Atlantic over the last 118 years.

Linear trends in SWH for the period 1950 to 2002 (Figure 3.25) are statistically significant and positive over most of the mid-latitudinal North Atlantic and North Pacific, as well as in the western subtropical South Atlantic, the eastern equatorial Indian Ocean and the East China and South China Seas. The largest upward trends (14 cm per decade) occur in the northwest

Box 3.3: Stratospheric-Tropospheric Relations and Downward Propagation

The troposphere influences the stratosphere mainly through planetary-scale waves that propagate upward during the extended winter season when stratospheric winds are westerly. The stratosphere responds to this forcing from below to produce long-lived changes to the strength of the polar vortices. In turn, these fluctuations in the strength of the stratospheric polar vortices are observed to couple downward to surface climate (Baldwin and Dunkerton, 1999, 2001; Kodera et al., 2000; Limpasuvan et al., 2004; Thompson et al., 2005). This relationship occurs in the zonal wind and can be seen clearly in annular modes, which explain a large fraction of the intra-seasonal and interannual variability in the troposphere (Thompson and Wallace, 2000) and most of the variability in the stratosphere (Baldwin and Dunkerton, 1999). Annular modes appear to arise naturally as a result of internal interactions within the troposphere and stratosphere (Limpasuvan and Hartmann, 2000; Lorenz and Hartmann, 2001, 2003).

The relationship between NAM anomalies in the stratosphere and troposphere can be seen in Box 3.3, Figure 1, in which the NAM index at 10 hPa is used to define events when the stratospheric polar vortex was extremely weak (stratospheric warmings). On average, weak vortex conditions in the stratosphere tend to descend to the troposphere and are followed by negative NAM anomalies at the surface for more than two months. Anomalously strong vortex conditions propagate downwards in a similar way.

Long-lived annular mode anomalies in the lowermost stratosphere appear to lengthen the time scale of the surface NAM. The tropospheric annular mode time scale is longest during winter in the NH, but longest during late spring (November–December) in the SH (Baldwin et al., 2003). In both hemispheres, the time scale of the tropospheric annular modes is longest when the variance of the annular modes is greatest in the lower stratosphere.

Downward coupling to the surface depends on having large circulation anomalies in the lowermost stratosphere. In such cases, the stratosphere can be used as a statistical predictor of the monthly mean surface NAM on time scales of up to two months (Baldwin et al., 2003; Scaife et al., 2005). Similarly, SH trends in temperature and geopotential height, associated with the ozone hole, appear to couple downward to affect high-latitude surface climate (Thompson and Solomon, 2002; Gillett and Thompson, 2003). As the stratospheric circulation changes with ozone depletion or increasing greenhouse gases, those changes will likely be reflected in changes to surface climate. Thompson and Solomon (2005) showed that the spring strengthening and cooling of the SH polar stratospheric vortex preceded similarly signed trends in the SH tropospheric circulation by one month in the interval 1973 to 2003. They argued that similar downward coupling is not evident in the NH geopotential trends computed using monthly radiosonde data. An explanation for this difference may be that the stratospheric signal is stronger in the SH, mainly due to ozone depletion, giving a more robust downward coupling.

The dynamical mechanisms by which the stratosphere influences the troposphere are not well understood, but the relatively large surface signal implies that the stratospheric signal is amplified. The processes likely involve planetary waves (Song and Robinson, 2004) and synoptic-scale waves (Wittman et al., 2004), which interact with stratospheric zonal wind anomalies near the tropopause. The altered waves would be expected to affect tropospheric circulation and induce surface pressure changes corresponding to the annular modes (Wittman et al., 2004).

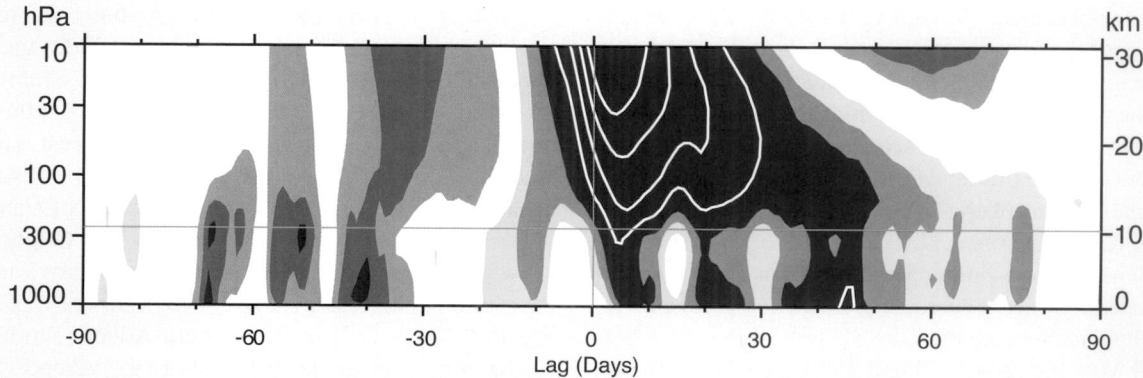

Box 3.3, Figure 1. *Composites of time-height development of the NAM index for 18 weak vortex events. The events are selected by the dates on which the 10 hPa annular mode index crossed –3.0. Day 0 is the start of the weak vortex event. The indices are non-dimensional; the contour interval for the colour shading is 0.25, and 0.5 for the white lines. Values between –0.25 and 0.25 are not shaded. Yellow and red shading indicates negative NAM indices and blue shading indicates positive indices. The thin horizontal lines indicate the approximate boundary between the troposphere and the stratosphere. Modified from Baldwin and Dunkerton (2001).*

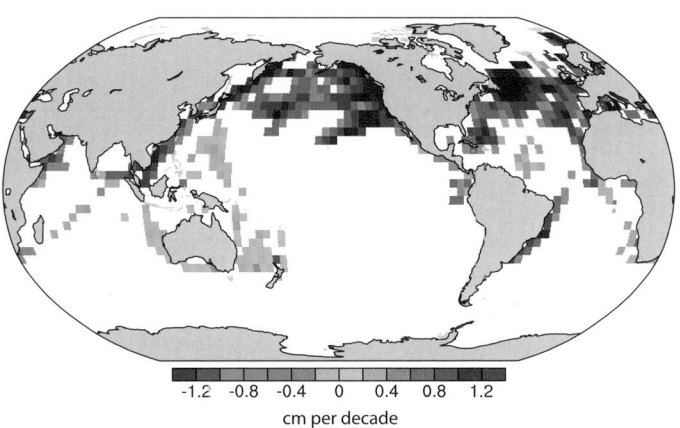

Figure 3.25. *Estimates of linear trends in significant wave height (cm per decade) for regions along the major ship routes of the global ocean for 1950 to 2002. Trends are shown only for locations where they are significant at the 5% level. Adapted from Gulev and Grigorieva (2004).*

Atlantic and the northeast Pacific. Statistically significant negative trends are observed in the western Pacific tropics, the Tasman Sea and the south Indian Ocean (–11 cm per decade). Hindcasts of waves with global and basin-scale models by Wang and Swail (2001, 2002) and Sterl and Caires (2005), based on NRA and ERA-40 winds, respectively, show an increasing mean SWH as well as intensification of SWH extremes during the last 40 years, with the 99% extreme of the winter SWH increasing in the northeast Atlantic by a maximum of 0.4 m per decade. Wave height hindcasts driven with NRA surface winds suggest that worsening wave conditions in the northeastern North Atlantic during the latter half of the 20th century were connected to a northward displacement in the storm track, with decreasing wave heights in the southern North Atlantic (Lozano and Swail, 2002). Increases of SWH in the North Atlantic mid-latitudes are further supported by a 14-year (1988–2002) time series of the merged TOPography EXperiment (TOPEX)/Poseidon and European Remote Sensing (ERS-1/2) satellite altimeter data (Woolf et al., 2002).

Since the TAR, research into surface fluxes has continued to be directed at improving the accuracy of the mean air-sea exchange fields (particularly of heat) with less work on long-term trends. Significant uncertainties remain in global fields of the net heat exchange, stemming from problems in obtaining accurate estimates of the different heat flux components. Estimates of surface flux variability from reanalyses are strongly influenced by inhomogeneous data assimilation input, especially in the Southern Ocean, and Sterl (2004) reported that variability of the surface latent heat flux in the Southern Ocean became much more reliable after 1979, when observations increased. Recent evaluations of heat flux estimates from reanalyses and *in situ* observations indicate some improvements but there are still global biases of several tens of watts per square metre in unconstrained products based on VOS observations (Grist and Josey, 2003). Estimates of the implied ocean heat

transport from the NRA, indirect residual techniques and some coupled models are in reasonable agreement with hydrographic observations (Trenberth and Caron, 2001; Grist and Josey, 2003). However, the hydrographic observations also contain significant uncertainties (see Chapter 5) due to both interannual variability and assumptions made in the computation of the heat transport, and these must be recognised when using them to evaluate the various flux products. For the North Atlantic, there are indications of positive trends in the net heat flux from the ocean of 10 W m^{-2} per decade in the western subpolar gyre and coherent negative changes in the eastern subtropical gyre, closely correlated with the NAO variability in the interval 1948 to 2002 (Marshall et al., 2001; Visbeck et al., 2003; Gulev et al., 2007).

3.5.7 Summary

Changes from the late 1970s to recent years generally reveal decreases in tropospheric geopotential heights over high latitudes of both hemispheres and increases over the mid-latitudes in DJF. The changes amplify with altitude up to the lower stratosphere, but remain similar in shape to lower atmospheric levels and are associated with the intensification and poleward displacement of corresponding Atlantic and southern polar front jet streams and enhanced storm track activity. Based on a variety of measurements at the surface and in the upper troposphere, it is likely that there has been an increase and a poleward shift in NH winter storm-track activity over the second half of the 20th century, but there are still significant uncertainties in the magnitude of the increase due to time-dependent biases in the reanalyses. Analysed decreases in cyclone numbers over the southern extratropics and increases in mean cyclone radius and depth over much of the SH over the last two decades are subject to even larger uncertainties.

The decrease in long-lasting blocking frequency over the North Atlantic-European sector over recent decades is dynamically consistent with NAO variability (see Section 3.6), but given data limitations, it may be too early to define the nature of any trends in SH blocking occurrence, despite observed trends in the SAM. After the late 1990s in the NH, occurrences of major sudden warmings seem to have increased in the polar stratosphere, associated with the occurrence of more neutral states of the tropospheric and stratospheric vortex. In the SH, there has been a strengthening tropospheric antarctic vortex during summer in association with the ozone hole, which has led to a cooling of the stratospheric polar vortex in late spring and to a two- to three-week delay in vortex breakdown. In September 2002, a major warming was observed for the first and only time in the SH. Analysis of observed wind and SWH support the reanalysis-based evidence for an increase in storm activity in the extratropical NH in recent decades (see also Section 3.6) until the late 1990s. For heat flux, there seem to have been NAO-related variations over the Labrador Sea, which is a key region for deep water formation.

3.6 Patterns of Atmospheric Circulation Variability

3.6.1 Teleconnections

The global atmospheric circulation has a number of preferred patterns of variability, all of which have expressions in surface climate. Box 3.4 discusses the main patterns and associated indices. Regional climates in different locations may vary out of phase, owing to the action of such 'teleconnections', which modulate the location and strength of the storm tracks (Section 3.5.3) and poleward fluxes of heat, moisture and momentum. A comprehensive review by Hurrell et al. (2003) has been updated by new analyses, notably from Quadrelli and Wallace (2004) and Trenberth et al. (2005b). Understanding the nature of teleconnections and changes in their behaviour is central to understanding regional climate variability and change. Such seasonal and longer time-scale anomalies have direct impacts on humans, as they are often associated with droughts, floods, heat waves and cold waves and other changes that can severely disrupt agriculture, water supply and fisheries, and can modulate air quality, fire risk, energy demand and supply and human health.

The analysis of teleconnections has typically employed a linear perspective, which assumes a basic spatial pattern with varying amplitude and mirror image positive and negative polarities (Hurrell et al., 2003; Quadrelli and Wallace, 2004). In contrast, nonlinear interpretations would identify preferred climate anomalies as recurrent states of a specific polarity (e.g., Corti et al., 1999; Cassou and Terray, 2001; Monahan et al.,

2001). Climate change may result through changes from one quasi-stationary state to another, as a preference for one polarity of a pattern (Palmer, 1999), or through a change in the nature or number of states (Straus and Molteni, 2004).

In the NH, one-point correlation maps illustrate the Pacific-North American (PNA) pattern and the NAO (Figure 3.26), but in the SH, wave structures do not emerge as readily owing to the dominance of the SAM. Although teleconnections are best defined over a grid, simple indices based on a few key station locations remain attractive as the series can often be carried back in time long before complete gridded fields were available (see Section 3.6.4, Figure 3.31); the disadvantage is increased noise from the reduced spatial sampling. For instance, Hurrell et al. (2003) found that the residence time of the NAO in its positive phase in the early 20th century was not as great as indicated by the positive NAO index for that period.

Many teleconnections have been identified, but combinations of only a small number of patterns can account for much of the interannual variability in the circulation and surface climate. Quadrelli and Wallace (2004) found that many patterns of NH interannual variability can be reconstructed as linear combinations of the first two Empirical Orthogonal Functions (EOFs) of sea level pressure (approximately the NAM and the PNA). Trenberth et al. (2005b) analysed global atmospheric mass and found four key rotated EOF patterns: the two annular modes (SAM and NAM), a global ENSO-related pattern and a fourth closely related to the North Pacific Index and the PDO, which in turn is closely related to ENSO and the PNA pattern.

Teleconnection patterns tend to be most prominent in the winter (especially in the NH), when the mean circulation is

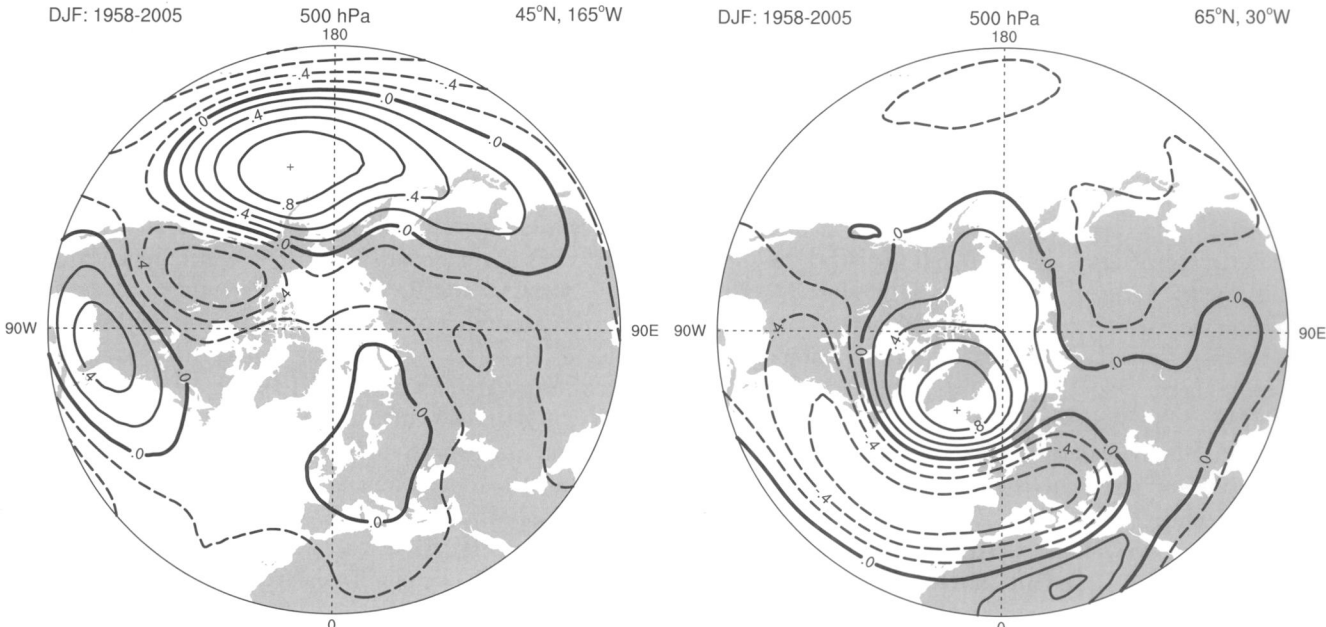

Figure 3.26. *The PNA (left) and NAO (right) teleconnection patterns, shown as one-point correlation maps of 500 hPa geopotential heights for boreal winter (DJF) over 1958 to 2005. In the left panel, the reference point is 45°N, 165°W, corresponding to the primary centre of action of the PNA pattern, given by the + sign. In the right panel, the NAO pattern is illustrated based on a reference point of 65°N, 30°W. Negative correlation coefficients are dashed, and the contour increment is 0.2. Adapted from Hurrell et al. (2003).*

Box 3.4: Defining the Circulation Indices

A teleconnection is made up of a fixed spatial pattern with an associated index time series showing the evolution of its amplitude and phase. Teleconnections are best defined by values over a grid but it is often convenient to devise simplified indices based on key station values. A classic example is the Southern Oscillation (SO), encompassing the entire tropical Pacific, yet encapsulated by a simple SO Index (SOI), based on differences between Tahiti (eastern Pacific) and Darwin (western Pacific) MSLP anomalies.

A number of teleconnections have historically been defined from either station data (SOI, NAO) or from gridded fields (NAM, SAM, PDO/NPI and PNA):

- **Southern Oscillation Index (SOI).** The MSLP anomaly difference of Tahiti minus Darwin, normalised by the long-term mean and standard deviation of the MSLP difference (Troup, 1965; Können et al., 1998). Available from the 1860s. Darwin can be used alone, as its data are more consistent than Tahiti prior to 1935.

- **North Atlantic Oscillation (NAO) Index.** The difference of normalised MSLP anomalies between Lisbon, Portugal and Stykkisholmur, Iceland has become the most widely used NAO index and extends back in time to 1864 (Hurrell, 1995), and to 1821 if Reykjavik is used instead of Stykkisholmur and Gibraltar instead of Lisbon (Jones et al., 1997).

- **Northern Annular Mode (NAM) Index.** The amplitude of the pattern defined by the leading empirical orthogonal function of winter monthly mean NH MSLP anomalies poleward of 20°N (Thompson and Wallace, 1998, 2000). The NAM has also been known as the Arctic Oscillation (AO), and is closely related to the NAO.

- **Southern Annular Mode (SAM) Index.** The difference in average MSLP between SH middle and high latitudes (usually 45°S and 65°S), from gridded or station data (Gong and Wang, 1999; Marshall, 2003), or the amplitude of the leading empirical orthogonal function of monthly mean SH 850 hPa height poleward of 20°S (Thompson and Wallace, 2000). Formerly known as the Antarctic Oscillation (AAO) or High Latitude Mode (HLM).

- **Pacific-North American pattern (PNA) Index.** The mean of normalised 500 hPa height anomalies at 20°N, 160°W and 55°N, 115°W minus those at 45°N, 165°W and 30°N, 85°W (Wallace and Gutzler, 1981).

- **Pacific Decadal Oscillation (PDO) Index and North Pacific Index (NPI).** The NPI is the average MSLP anomaly in the Aleutian Low over the Gulf of Alaska (30°N–65°N, 160°E–140°W; Trenberth and Hurrell, 1994) and is an index of the PDO, which is also defined as the pattern and time series of the first empirical orthogonal function of SST over the North Pacific north of 20°N (Mantua et al., 1997; Deser et al., 2004). The PDO broadened to cover the whole Pacific Basin is known as the Inter-decadal Pacific Oscillation (IPO: Power et al., 1999b). The PDO and IPO exhibit virtually identical temporal evolution (Folland et al., 2002).

strongest. The strength of teleconnections and the way they influence surface climate also vary over long time scales. Both the NAO and ENSO exhibited marked changes in their surface climate expressions on multi-decadal time scales during the 20th century (e.g., Power et al., 1999b; Jones et al., 2003). Multi-decadal changes in influence are often real and not due just to poorer data quality in earlier decades.

3.6.2   El Niño-Southern Oscillation and Tropical/Extratropical Interactions

3.6.2.1   El Niño-Southern Oscillation

El Niño-Southern Oscillation events are a coupled ocean-atmosphere phenomenon. El Niño involves warming of tropical Pacific surface waters from near the International Date Line to the west coast of South America, weakening the usually strong SST gradient across the equatorial Pacific, with associated changes in ocean circulation. Its closely linked atmospheric counterpart, the Southern Oscillation (SO), involves changes in trade winds, tropical circulation and precipitation. Historically, El Niño events occur about every 3 to 7 years and alternate with the opposite phases of below-average temperatures in the eastern tropical Pacific (La Niña). Changes in the trade

winds, atmospheric circulation, precipitation and associated atmospheric heating set up extratropical responses. Wavelike extratropical teleconnections are accompanied by changes in the jet streams and storm tracks in mid-latitudes (Chang and Fu, 2002).

The El Niño-Southern Oscillation has global impacts, manifested most strongly in the northern winter months (November–March). Anomalies in MSLP are much greater in the extratropics while the tropics feature large precipitation variations. Associated patterns of surface temperature and precipitation anomalies around the globe are given in Figure 3.27 (Trenberth and Caron, 2000), and the evolution of these patterns and links to global mean temperature perturbations are given by Trenberth et al. (2002b).

The nature of ENSO has varied considerably over time. Strong ENSO events occurred from the late 19th century through the first 25 years of the 20th century and again after about 1950, but there were few events of note from 1925 to 1950 with the exception of the major 1939–1941 event (Figure 3.27). The 1976–1977 climate shift (Trenberth, 1990; see Figure 3.27 and Section 3.6.3, Figure 3.28) was associated with marked changes in El Niño evolution (Trenberth and Stepaniak, 2001), a shift to generally above-normal SSTs in the eastern and central equatorial Pacific and a tendency towards more prolonged and

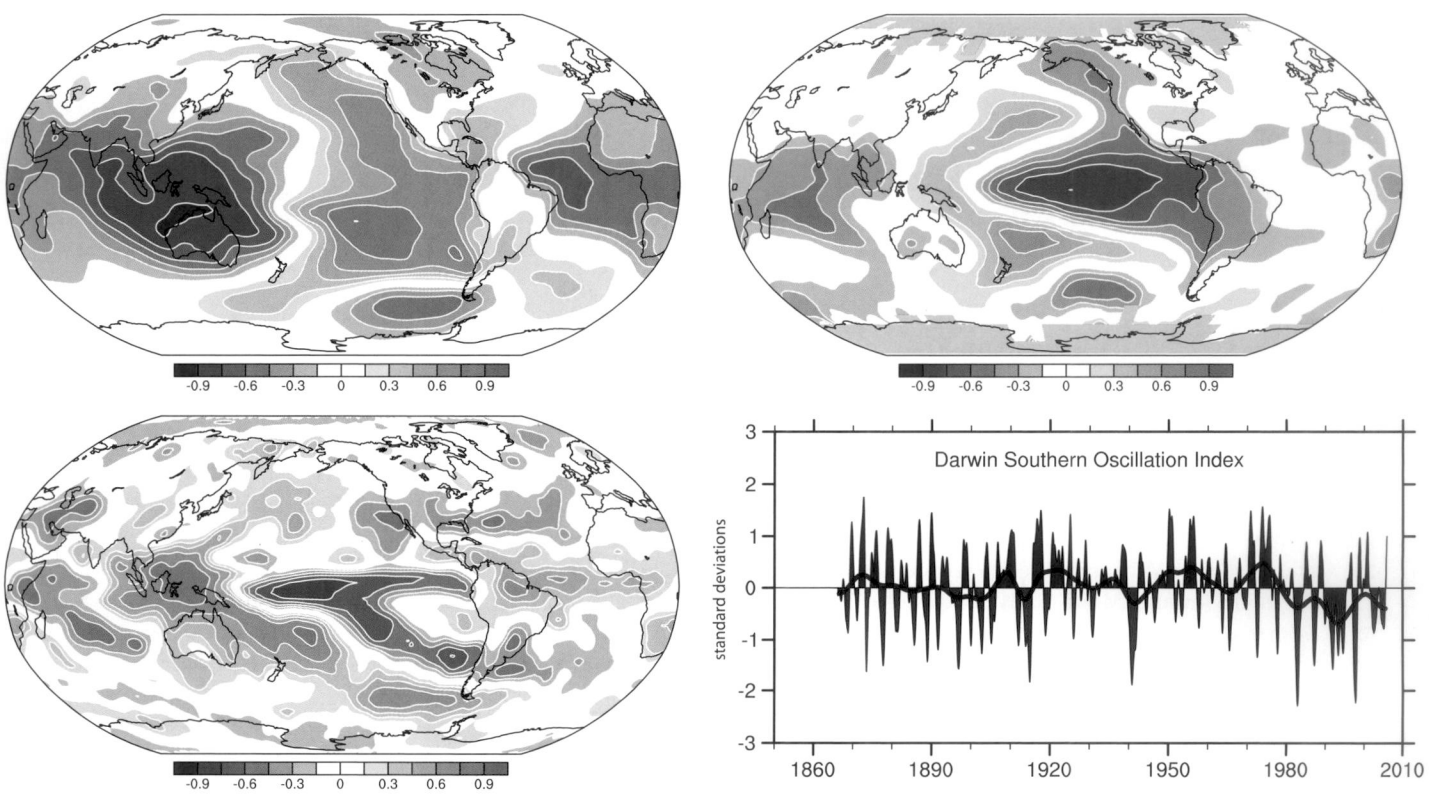

Figure 3.27. *Correlations with the SOI, based on normalised Tahiti minus Darwin sea level pressures, for annual (May to April) means for sea level pressure (top left) and surface temperature (top right) for 1958 to 2004, and GPCP precipitation for 1979 to 2003 (bottom left), updated from Trenberth and Caron (2000). The Darwin-based SOI, in normalized units of standard deviation, from 1866 to 2005 (Können et al., 1998; lower right) features monthly values with an 11-point low-pass filter, which effectively removes fluctuations with periods of less than eight months (Trenberth, 1984). The smooth black curve shows decadal variations (see Appendix 3.A). Red values indicate positive sea level pressure anomalies at Darwin and thus El Niño conditions.*

stronger El Niños. Since the TAR, there has been considerable work on decadal and longer-term variability of ENSO and Pacific climate. Such decadal atmospheric and oceanic variations (Section 3.6.3) are more pronounced in the North Pacific and across North America than in the tropics but are also present in the South Pacific, with evidence suggesting they are at least in part forced from the tropics (Deser et al., 2004).

El Niño-Southern Oscillation events involve large exchanges of heat between the ocean and atmosphere and affect global mean temperatures. The 1997–1998 event was the largest on record in terms of SST anomalies and the global mean temperature in 1998 was the highest on record (at least until 2005). Trenberth et al. (2002b) estimated that global mean surface air temperatures were 0.17°C higher for the year centred on March 1998 owing to the El Niño. Extremes of the hydrological cycle such as floods and droughts are common with ENSO and are apt to be enhanced with global warming (Trenberth et al., 2003). For example, the modest 2002–2003 El Niño was associated with a drought in Australia, made much worse by record-breaking heat (Nicholls, 2004; and see Section 3.8.4, Box 3.6). Thus, whether observed changes in ENSO behaviour are physically linked to global climate change is a research question of great importance.

3.6.2.2 Tropical-Extratropical Teleconnections: PNA and PSA

Circulation variability over the extratropical Pacific features wave-like patterns emanating from the subtropical western Pacific, characteristic of Rossby wave propagation associated with anomalous tropical heating (Horel and Wallace, 1981; Hoskins and Karoly, 1981). These are known as the PNA and Pacific-South American (PSA) patterns and can arise naturally through atmospheric dynamics as well as in response to heating. Over the NH in winter, the PNA pattern lies across North America from the subtropical Pacific, with four centres of action (Figure 3.26). While the PNA pattern can be illustrated by taking a single point correlation, this is not so easy for the PSA pattern (not shown), as its spatial centres of action are not fixed. However, the PSA pattern can be present at all times of year, lying from Australasia over the southern Pacific and Atlantic (Mo and Higgins, 1998; Kidson, 1999; Mo, 2000).

The PNA, or a variant of it (Straus and Shukla, 2002), is associated with modulation of the Aleutian Low, the Asian jet, and the Pacific storm track, affecting precipitation in western North America and the frequency of Alaskan blocking events and associated cold air outbreaks over the western USA in winter (Compo and Sardeshmukh, 2004). The PSA is associated with modulation of the westerlies over the South Pacific, effects of which include significant rainfall variations over New Zealand, changes in the nature and frequency of blocking events across the high-latitude South Pacific, and interannual variations in antarctic sea ice across the Pacific and Atlantic sectors (Renwick and Revell, 1999; Kwok and Comiso, 2002a; Renwick, 2002). While both PNA and PSA activity have varied with decadal modulation of ENSO, no systematic changes in their behaviour have been reported.

3.6.3 Pacific Decadal Variability

Decadal to inter-decadal variability of the atmospheric circulation is most prominent in the North Pacific, where fluctuations in the strength of the winter Aleutian Low pressure system co-vary with North Pacific SST in the PDO. These are linked to decadal variations in atmospheric circulation, SST and ocean circulation throughout the whole Pacific Basin in the Inter-decadal Pacific Oscillation (IPO; Trenberth and Hurrell, 1994; Gershunov and Barnett, 1998; Folland et al., 2002; McPhaden and Zhang, 2002; Deser et al., 2004). Key measures of Pacific decadal variability are the North Pacific Index (NPI; Trenberth and Hurrell, 1994), PDO index (Mantua et al., 1997) and the IPO index (Power et al., 1999b; Folland et al., 2002; see Figures 3.28 and 3.29). Modulation of ENSO by the PDO significantly modifies regional teleconnections around the Pacific Basin (Power et al., 1999b; Salinger et al., 2001), and affects the evolution of the global mean climate.

The PDO/IPO has been described as a long-lived El Niño-like pattern of Indo-Pacific climate variability (Knutson and Manabe, 1998; Evans et al., 2001; Deser et al., 2004; Linsley et al., 2004) or as a low-frequency residual of ENSO variability on multi-decadal time scales (Newman et al., 2003). Indeed, the symmetry of the SST anomaly pattern between the NH and SH may be a reflection of common tropical forcing. However, Folland et al. (2002) showed that the IPO significantly affects the movement of the South Pacific Convergence Zone in a way independent of ENSO (see also Deser et al., 2004). Other results indicate that the extratropical phenomena are generic components of the PDO (Deser et al., 1996, 1999, 2003; Gu and Philander, 1997). The extratropics may also contribute to the tropical SST changes via an 'atmospheric bridge', confounding the simple interpretation of a tropical origin (Barnett et al., 1999; Vimont et al., 2001).

The inter-decadal time scale of tropical Indo-Pacific SST variability is likely due to oceanic processes. Extratropical ocean influences are also likely to play a role as changes in the ocean gyre evolve and heat anomalies are subducted and re-emerge (Deser et al., 1996, 1999, 2003; Gu and Philander, 1997). It is also possible that there is no well-defined coupled ocean-atmosphere 'mode' of variability in the Pacific on decadal to inter-decadal time scales, since instrumental records are too short to provide a robust assessment and palaeoclimate records conflict regarding time scales (Biondi et al., 2001; Gedalof et al., 2002). Schneider and Cornuelle (2005) suggested that the PDO is not itself a mode of variability but is a blend of three

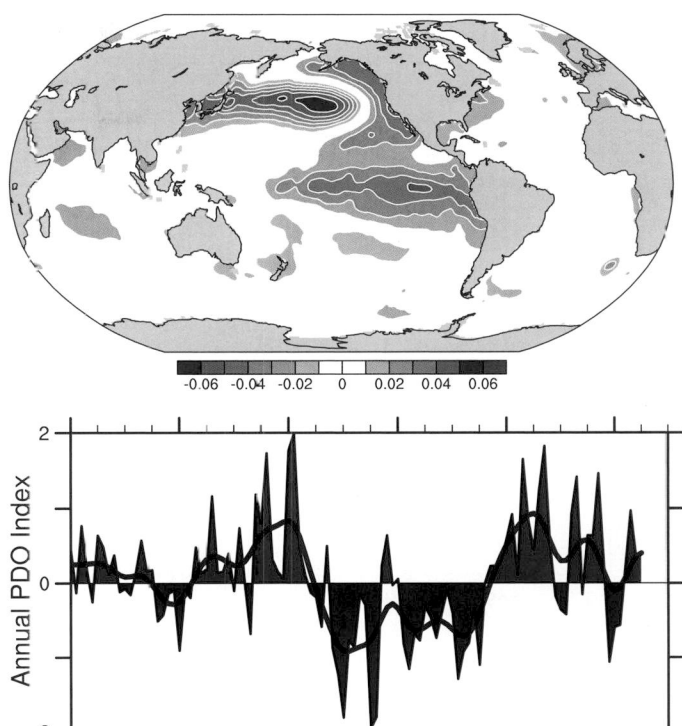

Figure 3.28. *Pacific Decadal Oscillation: (top) SST based on the leading EOF SST pattern for the Pacific basin north of 20°N for 1901 to 2004 (updated; see Mantua et al., 1997; Power et al., 1999b) and projected for the global ocean (units are nondimensional); and (bottom) annual time series (updated from Mantua et al., 1997). The smooth black curve shows decadal variations (see Appendix 3.A).*

phenomena. They showed that the observed PDO pattern and evolution can be recovered from a reconstruction of North Pacific SST anomalies based on a first order autoregressive model and forcing by variability of the Aleutian low, ENSO and oceanic zonal advection in the Kuroshio-Oyashio Extension. The latter results from oceanic Rossby waves that are forced by North Pacific Ekman pumping. The SST response patterns to these processes are not completely independent, but they determine the spatial characteristics of the PDO. Under this hypothesis, the key physical variables for measuring Pacific climate variability are ENSO and NPI (Aleutian Low) indices, rather than the PDO index.

Figure 3.29 (top) shows a time series of the NPI for 1900 to 2005 (Deser et al., 2004). There is substantial low-frequency variability, with extended periods of predominantly high values indicative of a weakened circulation (1900–1924 and 1947–1976) and predominantly low values indicative of a strengthened circulation (1925–1946 and 1977–2005). The well-known decrease in pressure from 1976 to 1977 is analogous to transitions that occurred from 1946 to 1947 and from 1924 to 1925, and these earlier changes were also associated with SST fluctuations in the tropical Indian (Figure 3.29, lower) and Pacific Oceans although not in the upwelling zone of the equatorial eastern Pacific (Minobe, 1997; Deser et al., 2004). In addition, the NPI exhibits variability on shorter time scales, interpreted in part as a bi-decadal rhythm (Minobe, 1999).

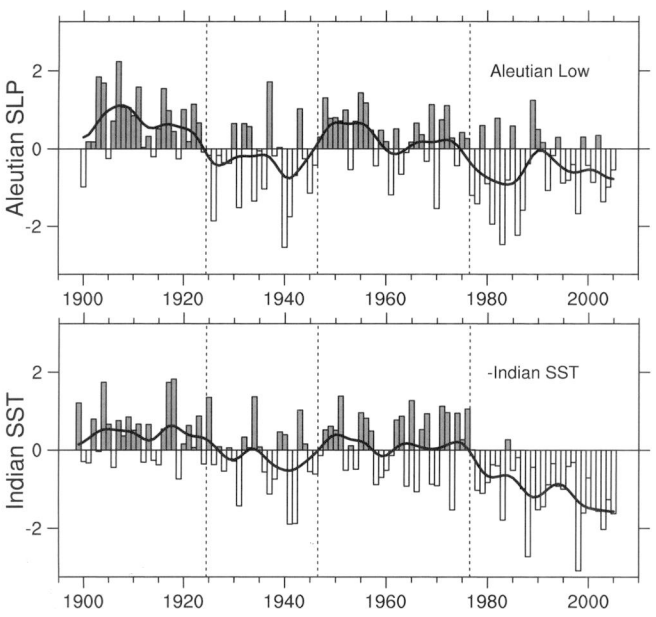

Figure 3.29. *(Top) Time series of the NPI (sea level pressure during December through March averaged over the North Pacific, 30°N to 65°N, 160°E to 140°W) from 1900 to 2005 expressed as normalised departures from the long-term mean (each tick mark on the ordinate represents two standard deviations, or 5.5 hPa). This record reflects the strength of the winter Aleutian Low pressure system, with positive (negative) values indicative of a weak (strong) Aleutian Low. The bars give the winter series and the smooth black curves show decadal variations (see Appendix 3.A). Values were updated and extended to earlier decades from Trenberth and Hurrell (1994). (Bottom) As above but for SSTs averaged over the tropical Indian Ocean (10°S–20°N, 50°E –125°E; each tick mark represents two standard deviations, or 0.36°C). This record has been inverted to facilitate comparison with the top panel. The dashed vertical lines mark years of transition in the Aleutian Low record (1925, 1947, 1977). Updated from Deser et al. (2004).*

There is observational and modelling evidence (Pierce, 2001; Schneider and Cornuelle, 2005) suggesting the PDO/IPO does not excite the climate shifts in the Pacific area, but they share the same forcing. The 1976–1977 climate shift in the Pacific, associated with a phase change in the PDO from negative to positive, was associated with significant changes in ENSO evolution (Trenberth and Stepaniak, 2001) and with changes in ENSO teleconnections and links to precipitation and surface temperatures over North and South America, Asia and Australia (Trenberth, 1990; Trenberth and Hurrell, 1994; Power et al., 1999a; Salinger et al., 2001; Mantua and Hare, 2002; Minobe and Nakanowatari, 2002; Trenberth et al., 2002b; Deser et al., 2004; Marengo, 2004). Schneider and Cornuelle (2005) added extra credence to the hypothesis that the 1976–1977 climate shift is of tropical origin.

3.6.4 The North Atlantic Oscillation and Northern Annular Mode

The only teleconnection pattern prominent throughout the year in the NH is the NAO (Barnston and Livezey, 1987). It is primarily a north-south dipole in sea level pressure characterised by simultaneous out-of-phase pressure and height anomalies between temperate and high latitudes over the Atlantic sector,

and therefore corresponds to changes in the westerlies across the North Atlantic into Europe (Figure 3.30). The NAO has the strongest signature in the winter months (December to March) when its positive (negative) phase exhibits an enhanced (diminished) Iceland Low and Azores High (Hurrell et al., 2003). The NAO is the dominant pattern of near-surface atmospheric circulation variability over the North Atlantic, accounting for one third of the total variance in monthly MSLP in winter. It is closely related to the NAM, which has similar structure over the Atlantic but is more zonally symmetric. The leading winter pattern of variability in the lower stratosphere is also annular, but the MSLP anomaly pattern that is associated with it is confined almost entirely to the Arctic and Atlantic sectors and coincides with the spatial structure of the NAO (Deser, 2000; see also Section 3.5.5 and Box 3.3).

There is considerable debate over whether the NAO or the NAM is more physically relevant to the winter circulation (Deser, 2000; Ambaum et al., 2001, 2002), but the time series are highly correlated in winter (Figure 3.31). As Quadrelli and Wallace (2004) showed, they are near neighbours in terms of their spatial patterns and their temporal evolution. The annular modes are intimately linked to the configuration of the extratropical storm tracks and jet streams. Changes in the phase of the annular modes appear to occur as a result of interactions between the eddies and the mean flow, and external forcing is not required to sustain them (De Weaver and Nigam, 2000). In the NH, stationary waves provide most of the eddy momentum fluxes, although transient eddies are also important. To the extent that the intrinsic excitation of the NAO/NAM pattern is limited to a period less than a few days (Feldstein, 2002), it should not exhibit year-to-year autocorrelation in conditions of constant forcing. Proxy and instrumental data, however, show evidence for intervals with prolonged positive and negative NAO index values in the last few centuries (Cook et al., 2002; Jones et al., 2003). In winter, a reversal occurred from the minimum index values in the late 1960s to strongly positive NAO index values in the mid-1990s. Since then, NAO values have declined to near the long-term mean (Figure 3.31). In summer, Hurrell et al. (2001, 2002) identified significant interannual to multi-decadal fluctuations in the NAO pattern, and the trend towards persistent anticyclonic flow over northern Europe has contributed to anomalously warm and dry conditions in recent decades (Rodwell, 2003).

Feldstein (2002) suggested that the trend and increase in the variance of the NAO/NAM index from 1968 through 1997 was greater than would be expected from internal variability alone, while NAO behaviour during the first 60 years of the 20th century was consistent with atmospheric internal variability. However, the results are not so clear if based on just the period 1975 to 2004 (Overland and Wang, 2005). Although monthly-scale NAO variability is strong (Czaja et al., 2003; Thompson et al., 2003), there may be predictability from stratospheric influences (Thompson et al., 2002; Scaife et al., 2005; see Box 3.3). There is mounting evidence that the recent observed inter-decadal NAO variability comes from tropical and extratropical ocean influences (Hurrell et al., 2003, 2004), land surface

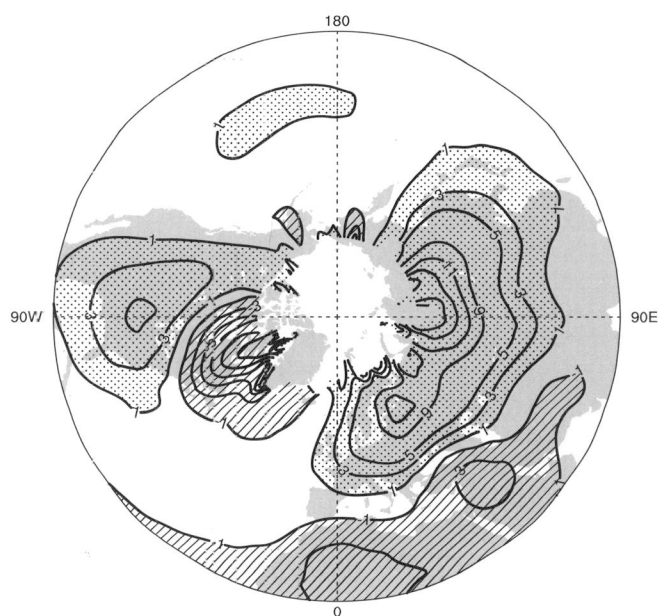

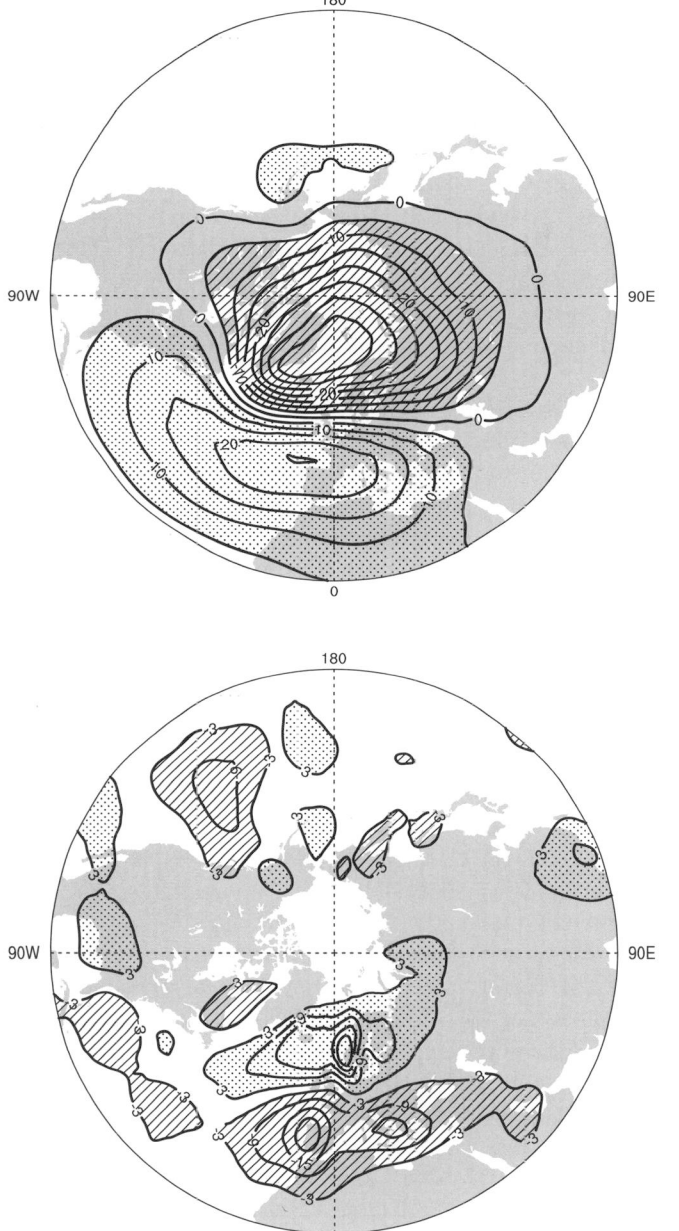

Figure 3.30. *Changes in winter (December–March) surface pressure, temperature, and precipitation corresponding to a unit deviation of the NAO index over 1900 to 2005. (Top left) Mean sea level pressure (0.1 hPa). Values greater than 0.5 hPa are stippled and values less than –0.5 hPa are hatched. (Top right) Land-surface air and sea surface temperatures (0.1°C; contour increment 0.2°C). Temperature changes greater than 0.1°C are indicated by stippling, those less than –0.1°C are indicated by hatching, and regions of insufficient data (e.g., over much of the Arctic) are not contoured. (Bottom left) Precipitation for 1979 to 2003 based on GPCP (0.1 mm per day; contour interval 0.6 mm per day). Stippling indicates values greater than 0.3 mm per day and hatching values less than –0.3 mm per day. Adapted and updated from Hurrell et al. (2003).*

forcing (Gong et al., 2003; Bojariu and Gimeno, 2003) and from other external factors (Gillett et al., 2003).

The NAO exerts a dominant influence on winter surface temperatures across much of the NH (Figure 3.30), and on storminess and precipitation over Europe and North Africa. When the NAO index is positive, enhanced westerly flow across the North Atlantic in winter moves warm moist maritime air over much of Europe and far downstream, with dry conditions over southern Europe and northern Africa and wet conditions in northern Europe, while stronger northerly winds over Greenland and northeastern Canada carry cold air southward and decrease land temperatures and SST over the northwest Atlantic. Temperature variations over North Africa and the Middle East (cooling) and the southeastern USA (warming), associated with the stronger clockwise flow around the subtropical Atlantic high-pressure centre, are also notable.

Following on from Hurrell (1996), Thompson et al. (2000) showed that for JFM from 1968 to 1997, the NAM accounted for 1.6°C of the 3.0°C warming in Eurasian surface temperatures, 4.9 hPa of the 5.7 hPa decrease in sea level pressure from 60°N to 90°N; 37% out of the 45% increase in Norwegian-area precipitation (55°N–65°N, 5°E–10°E), and 33% out of the 49% decrease in Spanish-region rainfall (35°N–45°N, 10°W–0°W). There were also significant effects on ocean heat content, sea ice, ocean currents and ocean heat transport.

Positive NAO index winters are associated with a northeastward shift in the Atlantic storm activity, with enhanced activity from Newfoundland into northern Europe and a modest decrease to the south (Hurrell and van Loon, 1997; Alexandersson et al., 1998). Positive NAO index winters are also typified by more intense and frequent storms in the vicinity of Iceland and the Norwegian Sea (Serreze et al., 1997;

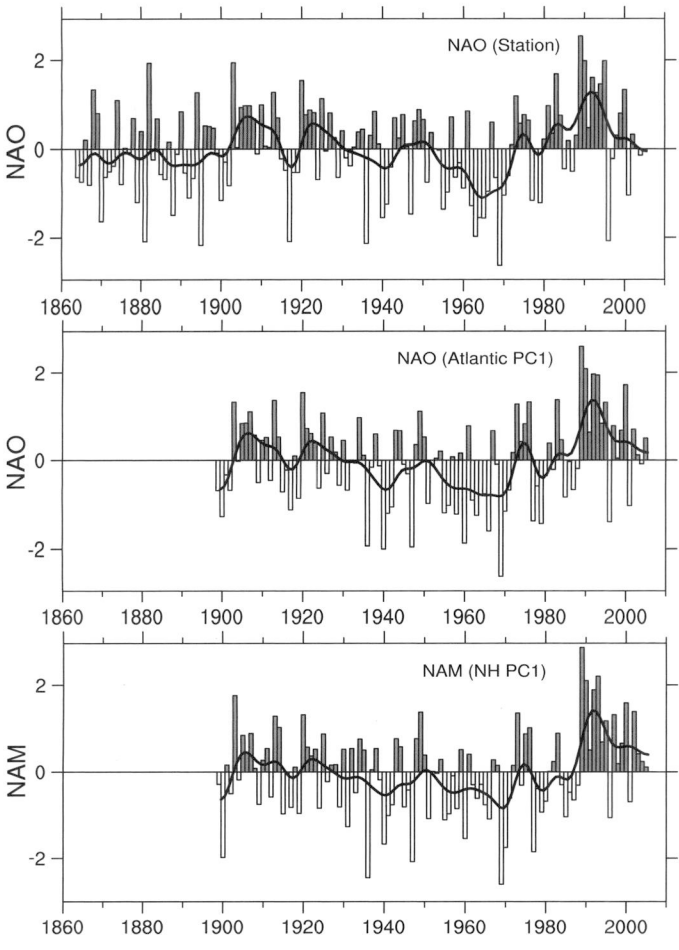

Figure 3.31. *Normalised indices (units of standard deviation) of the mean winter (December–March) NAO developed from sea level pressure data. In the top panel, the index is based on the difference of normalised sea level pressure between Lisbon, Portugal and Stykkisholmur/Reykjavik, Iceland from 1864 to 2005. The average winter sea level pressure data at each station were normalised by dividing each seasonal pressure anomaly by the long-term (1864 to 1983) standard deviation. In the middle panel, the index is the principal component time series of the leading EOF of Atlantic-sector sea level pressure. In the lower panel, the index is the principal component time series of the leading EOF of NH sea level pressure. The smooth black curves show decadal variations (see Appendix 3.A). The individual bar corresponds to the January of the winter season (e.g., 1990 is the winter of 1989/1990). Updated from Hurrell et al. (2003); see http://www.cgd.ucar.edu/cas/jhurrell/indices. html for updated time series.*

Deser et al., 2000). The correlation between the NAO index and cyclone activity is highly negative in eastern Canada and positive in western Canada (Wang et al., 2006b). The upward trend towards more positive NAO index winters from the mid-1960s to the mid-1990s has been associated with increased wave heights over the northeast Atlantic and decreased wave heights south of 40°N (Carter, 1999; Wang and Swail, 2001; see also Section 3.5.6).

The NAO/NAM modulates the transport and convergence of atmospheric moisture and the distribution of evaporation and precipitation (Dickson et al., 2000). Evaporation exceeds precipitation over much of Greenland and the Canadian Arctic and more precipitation than normal falls from Iceland through Scandinavia during winters with a high NAO index, while

the reverse occurs over much of central and southern Europe, the Mediterranean and parts of the Middle East (Dickson et al., 2000). Severe drought has persisted throughout parts of Spain and Portugal as well (Hurrell et al., 2003). As far eastward as Turkey, river runoff is significantly correlated with NAO variability (Cullen and deMenocal, 2000). There are many NAO-related effects on ocean circulation, such as the freshwater balance of the Atlantic Ocean (see Chapter 5), on the cryosphere (see Chapter 4), and on many aspects of the north Atlantic/European biosphere (see the Working Group II contribution to the IPCC Fourth Assessment Report).

3.6.5 The Southern Hemisphere and Southern Annular Mode

The principal mode of variability of the atmospheric circulation in the SH extratropics is now known as the SAM (see Figure 3.32). It is essentially a zonally symmetric structure, but with a zonal wave three pattern superimposed. It is associated with synchronous pressure or height anomalies of opposite sign in mid- and high-latitudes, and therefore reflects changes in the main belt of subpolar westerly winds. Enhanced Southern Ocean westerlies occur in the positive phase of the SAM. The SAM contributes a significant proportion of SH mid-latitude circulation variability on many time scales (Hartmann and Lo, 1998; Kidson, 1999; Thompson and Wallace, 2000; Baldwin, 2001). Trenberth et al. (2005b) showed that the SAM is the leading mode in an EOF analysis of monthly mean global atmospheric mass, accounting for around 10% of total global variance. As with the NAM, the structure and variability of the SAM results mainly from the internal dynamics of the atmosphere and the SAM is an expression of storm track and jet stream variability (e.g., Hartmann and Lo, 1998; Limpasuvan and Hartmann, 2000; Box 3.3). Poleward eddy momentum fluxes interact with the zonal mean flow to sustain latitudinal displacements of the mid-latitude westerlies (Limpasuvan and Hartmann, 2000; Rashid and Simmonds, 2004, 2005).

Gridded reanalysis data sets have been utilised to derive time series of the SAM, particularly the NRA (e.g., Gong and Wang, 1999; Thompson et al., 2000) and more recently ERA-40 (Renwick, 2004; Trenberth et al., 2005b). However, a declining positive bias in pressure at high southern latitudes in both reanalyses before 1979 (Hines et al., 2000; Trenberth and Smith, 2005) means that derived trends in the SAM are too large. Marshall (2003) produced a SAM index based on appropriately located station observations. His index reveals a general increase in the SAM index beginning in the 1960s (Figure 3.32) consistent with a strengthening of the circumpolar vortex and intensification of the circumpolar westerlies, as observed in northern Antarctic Peninsula radiosonde data (Marshall, 2002).

The observed SAM trend has been related to stratospheric ozone depletion (Sexton, 2001; Thompson and Solomon, 2002; Gillett and Thompson, 2003) and to greenhouse gas increases (Hartmann et al., 2000; Marshall et al., 2004; see also Section 9.5.3.3). Jones and Widmann (2004) reconstructed century-

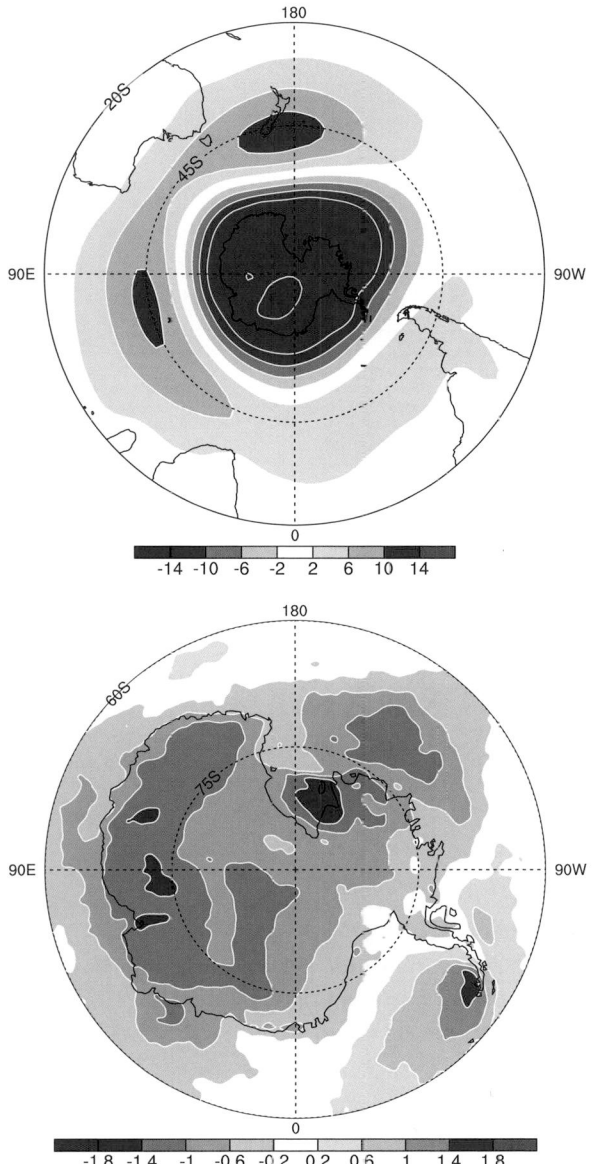

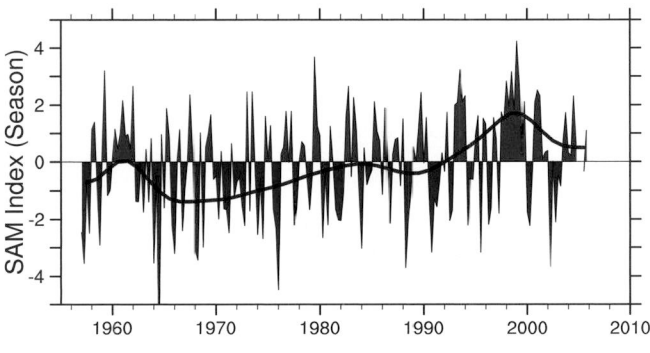

Figure 3.32. *(Bottom) Seasonal values of the SAM index calculated from station data (updated from Marshall, 2003). The smooth black curve shows decadal variations (see Appendix 3.A). (Top) The SAM geopotential height pattern as a regression based on the SAM time series for seasonal anomalies at 850 hPa (see also Thompson and Wallace, 2000). (Middle) The regression of changes in surface temperature (°C) over the 23-year period (1982 to 2004) corresponding to a unit change in the SAM index, plotted south of 60°S. Values exceeding about 0.4°C in magnitude are significant at the 1% significance level (adapted from Kwok and Comiso, 2002b).*

scale records based on proxies of the SAM that indicate that the magnitude of the recent trend may not be unprecedented, even during the 20th century. There is also recent evidence that ENSO variability can influence the SAM in the southern summer (e.g., L'Heureux and Thompson, 2006).

The trend in the SAM, which is statistically significant annually and in summer and autumn (Marshall et al., 2004), has contributed to antarctic temperature trends (Kwok and Comiso, 2002b; Thompson and Solomon, 2002; van den Broeke and van Lipzig, 2003; Schneider et al., 2004); specifically a strong summer warming in the Peninsula region and little change or cooling over much of the rest of the continent (Turner et al., 2005; see Figure 3.32). Through the wave component, the positive SAM is associated with low pressure west of the Peninsula (e.g., Lefebvre et al., 2004) leading to increased poleward flow, warming and reduced sea ice in the region (Liu et al., 2004b). Orr et al. (2004) proposed that this scenario yields a higher frequency of warmer maritime air masses passing over the Peninsula, leading to the marked northeast Peninsula warming observed in summer and autumn (December–May). The positive trend in the SAM has led to more cyclones in the circumpolar trough (Sinclair et al., 1997) and hence a greater contribution to antarctic precipitation from these near-coastal systems that is reflected in $\delta^{18}O$ levels in the snow (Noone and Simmonds, 2002). The SAM also affects spatial patterns of precipitation variability in Antarctica (Genthon et al., 2003) and southern South America (Silvestri and Vera, 2003).

The imprint of SAM variability on the Southern Ocean system is observed as a coherent sea level response around Antarctica (Aoki, 2002; Hughes et al., 2003) and by its regulation of Antarctic Circumpolar Current flow through the Drake Passage (Meredith et al., 2004). Changes in oceanic circulation directly alter the THC (Oke and England, 2004) and may explain recent patterns of observed temperature change at SH high latitudes described by Gille (2002).

3.6.6 Atlantic Multi-decadal Oscillation

Over the instrumental period (since the 1850s), North Atlantic SSTs show a 65 to 75 year variation (0.4°C range), with a warm phase during 1930 to 1960 and cool phases during 1905 to 1925 and 1970 to 1990 (Schlesinger and Ramankutty, 1994), and this feature has been termed the AMO (Kerr, 2000), as shown in Figure 3.33. Evidence (e.g., Enfield et al., 2001; Knight et al., 2005) of a warm phase in the AMO from 1870 to 1900 is revealed as an artefact of the de-trending used (Trenberth and Shea, 2006). The cycle appears to have returned to a warm phase beginning in the mid-1990s, and tropical Atlantic SSTs were at record high levels in 2005. Instrumental observations capture only two full cycles of the AMO, so the robustness of the signal has been addressed using proxies. Similar oscillations in a 60- to 110-year band are seen in North Atlantic palaeoclimatic reconstructions through the last four centuries (Delworth and Mann, 2000; Gray et al., 2004). Both observations and model simulations implicate changes in the strength of the THC as the primary source of the multi-decadal variability, and suggest a

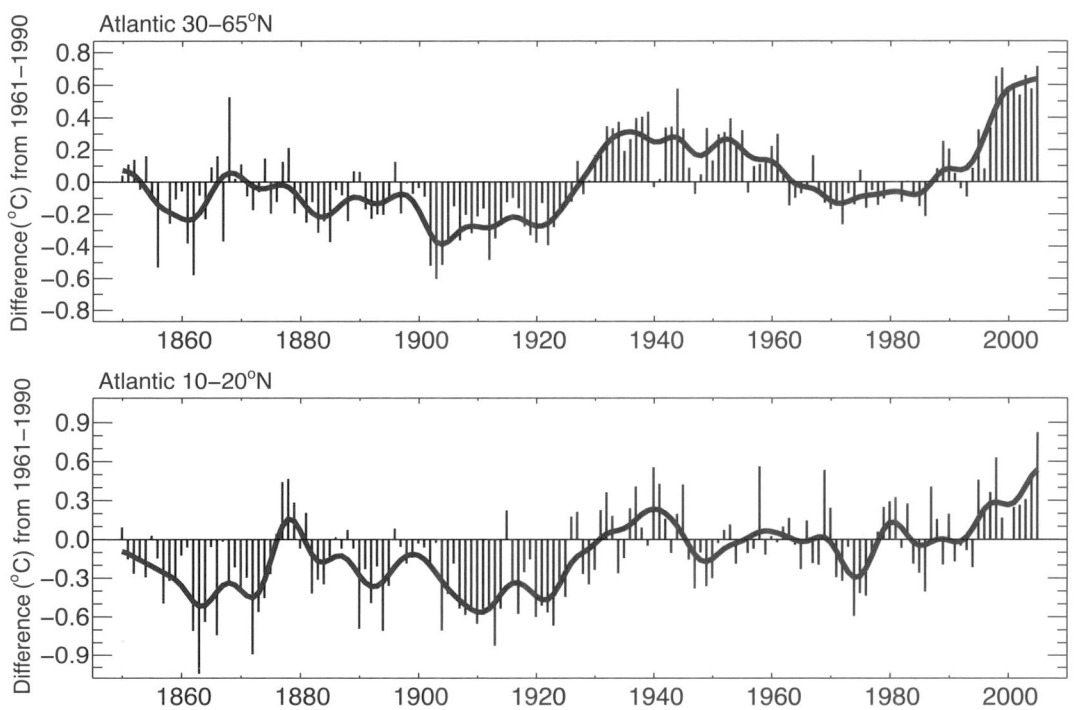

Figure 3.33. *Atlantic Multi-decadal Oscillation index from 1850 to 2005 represented by annual anomalies of SST in the extratropical North Atlantic (30–65°N; top), and in a more muted fashion in the tropical Atlantic (10°N –20°N) SST anomalies (bottom). Both series come from HadSST2 (Rayner et al., 2006) and are relative to the 1961 to 1990 mean (°C). The smooth blue curves show decadal variations (see Appendix 3.A).*

possible oscillatory component to its behaviour (Delworth and Mann, 2000; Latif, 2001; Sutton and Hodson, 2003; Knight et al., 2005). Trenberth and Shea (2006) proposed a revised AMO index, subtracting the global mean SST from the North Atlantic SST. The revised index is about 0.35°C lower than the original after 2000, highlighting the fact that most of the recent warming is global in scale.

The AMO has been linked to multi-year precipitation anomalies over North America, and appears to modulate ENSO teleconnections (Enfield et al., 2001; McCabe et al., 2004; Shabbar and Skinner, 2004). Multi-decadal variability in the North Atlantic also plays a role in Atlantic hurricane formation (Goldenberg et al., 2001; see also Section 3.8.3.2). The revised AMO index (Trenberth and Shea, 2006) indicates that North Atlantic SSTs have recently been about 0.3°C warmer than during 1970 to 1990, emphasizing the role of the AMO in suppressing tropical storm activity during that period. The AMO is likely to be a driver of multi-decadal variations in Sahel droughts, precipitation in the Caribbean, summer climate of both North America and Europe, sea ice concentration in the Greenland Sea and sea level pressure over the southern USA, the North Atlantic and southern Europe (e.g., Venegas and Mysak, 2000; Goldenberg et al., 2001; Sutton and Hodson, 2005; Trenberth and Shea, 2006). Walter and Graf (2002) identified a non-stationary relationship between the NAO and the AMO. During the negative phase of the AMO, the North Atlantic SST is strongly correlated with the NAO index. In contrast, the NAO index is only weakly correlated with the North Atlantic SST during the AMO positive phase. Chelliah and Bell (2004)

defined a tropical multi-decadal pattern related to the AMO, the PDO and winter NAO with coherent variations in tropical convection and surface temperatures in the West African monsoon region, the central tropical Pacific, the Amazon Basin and the tropical Indian Ocean.

3.6.7 Other Indices

As noted earlier, many patterns of variability (sometimes referred to as 'modes') in the climate system have been identified over the years, but few stand out as robust and dynamically significant features in relation to understanding regional climate change. This section discusses two climate signals that have recently drawn the attention of scientific community: the Antarctic Circumpolar Wave and the Indian Ocean Dipole.

3.6.7.1 *Antarctic Circumpolar Wave*

The Antarctic Circumpolar Wave (ACW) is described as a pattern of variability with an approximately four-year period in the southern high-latitude ocean-atmosphere system, characterised by the eastward propagation of anomalies in antarctic sea ice extent, and coupled to anomalies in SST, sea surface height, MSLP and wind (Jacobs and Mitchell, 1996; White and Peterson, 1996; White and Annis, 2004). Since its initial formulation (White and Peterson, 1996), questions have arisen concerning many aspects of the ACW: the robustness of the ACW on inter-decadal time scales (Carril and Navarra, 2001;

Connolley, 2003; Simmonds, 2003), its generating mechanisms (Cai and Baines, 2001; Venegas, 2003; White et al., 2004; White and Simmonds, 2006) and even its very existence (Park et al., 2004).

3.6.7.2 Indian Ocean Dipole

Large interannual variability of SST in the Indian Ocean has been associated with the Indian Ocean Dipole (IOD), also referred to as the Indian Ocean Zonal Mode (IOZM; Saji et al., 1999; Webster et al., 1999). This pattern manifests through a zonal gradient of tropical SST, which in one extreme phase in boreal autumn shows cooling off Sumatra and warming off Somalia in the west, combined with anomalous easterlies along the equator. The magnitude of the secondary rainfall maximum from October to December in East Africa is strongly correlated with positive IOD events (Xie et al., 2002). Several recent IOD events have occurred simultaneously with ENSO events and there is a significant debate on whether the IOD is an Indian Ocean pattern or whether it is triggered by ENSO in the Pacific Ocean (Allan et al., 2001). The strongest IOD episode ever observed occurred in 1997 to 1998 and was associated with catastrophic flooding in East Africa. Trenberth et al. (2002b) showed that Indian Ocean SSTs tend to rise about five months after the peak of ENSO in the Pacific. Monsoon variability and the SAM (Lau and Nath, 2004) are also likely to play a role in triggering or intensifying IOD events. One argument for an independent IOD was the large episode in 1961 when no ENSO event occurred (Saji et al., 1999). Saji and Yamagata (2003), analysing observations from 1958 to 1997, concluded that 11 out of the 19 episodes identified as moderate to strong IOD events occurred independently of ENSO. However, this was disputed by Allan et al. (2001), who found that accounting for varying lag correlations removes the apparent independence from ENSO. Decadal variability in correlations between SST-based indices of the IOD and ENSO has been documented (Clark et al., 2003). At inter-decadal time scales, the SST patterns associated with the inter-decadal variability of ENSO indices are very similar to the SST patterns associated with the Indian monsoon rainfall (Krishnamurthy and Goswami, 2000) and with the North Pacific inter-decadal variability (Deser et al., 2004), raising the issue of coupled mechanisms modulating both ENSO-monsoon system and IOD variability (e.g., Terray et al., 2005).

3.6.8 Summary

Decadal variations in teleconnections considerably complicate the interpretation of climate change. Since the TAR, it has become clear that a small number of teleconnection patterns account for much of the seasonal to interannual variability in the extratropics. On monthly time scales, the SAM, NAM and NAO are dominant in the extratropics. The NAM and NAO are closely related, and are mostly independent from the SAM, except perhaps on decadal time scales. Many other patterns can be explained through combinations of the NAM and PNA in

the NH, and the SAM and PSA in the SH, plus ENSO-related global patterns. Both the NAM/NAO and the SAM have exhibited trends towards their positive phase (strengthened mid-latitude westerlies) over the last three to four decades, although both have returned to near their long-term mean state in the last five years. In the NH, this trend has been associated with the observed winter change in storm tracks, precipitation and temperature patterns. In the SH, SAM changes are related to contrasting trends of strong warming in the Antarctic Peninsula and a cooling over most of interior Antarctica. The increasing positive phase of the SAM has been linked to stratospheric ozone depletion and to greenhouse gas increases. Multi-decadal variability is also evident in the Atlantic, and appears to be related to the THC. Other teleconnection patterns discussed (PNA, PSA) exhibit decadal variations, but have not been shown to have systematic long-term changes.

ENSO has exhibited considerable inter-decadal variability in the past century, in association with the PDO (or IPO). Systematic changes in ENSO behaviour have also been observed, in particular the different evolution of ENSO events and enhanced El Niño activity since the 1976–1977 climate shift. Over North America, ENSO- and PNA-related changes appear to have led to contrasting changes across the continent, as the west has warmed more than the east, while the latter has become cloudier and wetter. Over the Indian Ocean, ENSO, monsoon and SAM variability are related to a zonal gradient of tropical SST associated with anomalous easterlies along the equator, and opposite precipitation and thermal anomalies in East Africa and over the Maritime Continent. The tropical Pacific variability is influenced by interactions with the tropical Atlantic and Indian Oceans, and by the extratropical North and South Pacific. Responses of the extratropical ocean become more important as the time scale is extended, and processes such as subduction, gyre changes and the THC come into play.

3.7 Changes in the Tropics and Subtropics, and in the Monsoons

The global monsoon system embraces an overturning circulation that is intimately associated with the seasonal variation of monsoon precipitation over all major continents and adjacent oceans (Trenberth et al., 2000). It involves the Hadley Circulation, the zonal mean meridional overturning mass flow between the tropics and subtropics entailing the Inter-Tropical Convergence Zone (ITCZ), and the Walker Circulation, which is the zonal east-west overturning. The South Pacific Convergence Zone (SPCZ) is a semi-permanent cloud band extending from around the Coral Sea southeastward towards the extratropical South Pacific, while the South Atlantic Convergence Zone (SACZ) is a more transient feature over and southeast of Brazil that transports moisture originating over the Amazon into the South Atlantic (Liebmann et al., 1999).

Tropical SSTs determine where the upward branch of the Hadley Circulation is located over the oceans, and the dominant variations in the energy transports by the Hadley cell, reflecting its strength, relate to ENSO (Trenberth et al., 2002a; Trenberth and Stepaniak, 2003a). During El Niño, elevated SST causes an increase in convection and relocation of the ITCZ and SPCZ to near the equator over the central and eastern tropical Pacific, with a tendency for drought conditions over Indonesia. There follows a weakening of the Walker Circulation and a strengthening of the Hadley Circulation (Oort and Yienger, 1996; Trenberth and Stepaniak, 2003a), leading to drier conditions over many subtropical regions during El Niño, especially over the Pacific sector. As discussed in Section 3.4.4.1, increased divergence of energy out of the tropics in the 1990s relative to the 1980s (Trenberth and Stepaniak, 2003a) is associated with more frequent El Niño events and especially the major 1997–1998 El Niño event, so these conditions play a role in inter-decadal variability (Gong and Ho, 2002; Mu et al., 2002; Deser et al., 2004). Examination of the Hadley Circulation in several data sets (Mitas and Clement, 2005) suggests some strengthening, although discrepancies among reanalysis data sets and known deficiencies raise questions about the robustness of this strengthening, especially prior to the satellite era (1979).

Monsoons are generally referred to as tropical and subtropical seasonal reversals in both the surface winds and associated precipitation. The strongest monsoons occur over the tropics of southern and eastern Asia and northern Australia, and parts of western and central Africa. Rainfall is the most important monsoon variable because the associated latent heat release drives atmospheric circulations, and because of its critical role in the global hydrological cycle and its vital socioeconomic impacts. Thus, other regions that have an annual reversal in precipitation with an intense rainy summer and a dry winter have been recently recognised as monsoon regions, even though these regions have no explicit seasonal reversal of the surface winds (Wang, 1994; Webster et al., 1998). The latter regions include Mexico and the southwest USA, and parts of South America and South Africa. Owing to the lack of sufficiently reliable and long-term oceanic observations, analyses of observed long-term changes have mainly relied on land-based rain gauge data.

Because the variability of regional monsoons is often the result of interacting circulations from other regions, simple indices of monsoonal strength in adjacent regions may give contradictory indications of strength (Webster and Yang, 1992; Wang and Fan, 1999). Decreasing trends in precipitation over the Indonesian Maritime Continent, equatorial parts of western and central Africa, Central America, Southeast Asia and eastern Australia have been found for 1948 to 2003 (Chen et al., 2004; see Figure 3.13), while increasing trends were evident over the USA and northwestern Australia (see also Section 3.3.2.2 and Figure 3.14), consistent with Dai et al. (1997). Using NRA, Chase et al. (2003) found diminished monsoonal circulations since 1950 and no trends since 1979, but results based on NRA suffer severely from artefacts arising from changes in the observing system (Kinter et al., 2004).

Two precipitation data sets (Chen et al., 2002; GHCN, see Section 3.3) yield very similar patterns of change in the seasonal precipitation contrasts between 1976 to 2003 and 1948 to 1975 (Figure 3.34), despite some differences in details and discrepancies in northwest India. Significant decreases in the annual range (wet minus dry season) were observed over the NH tropical monsoon regions (e.g., Southeast Asia and Central America). Over the East Asian monsoon region, the change over these periods involves increased rainfall in the Yangtze River valley and Korea but decreased rainfall over the lower reaches of the Yellow River and northeast China. In the Indonesian-Australian monsoon region, the change between the two periods is characterised by an increase in northwest Australia and Java but a decrease in northeast Australia and a northeastward movement in the SPCZ (Figure 3.34). However, the average monsoonal rainfall in East Asia, Indonesia-Australia and South America in summer mostly shows no long-term trend

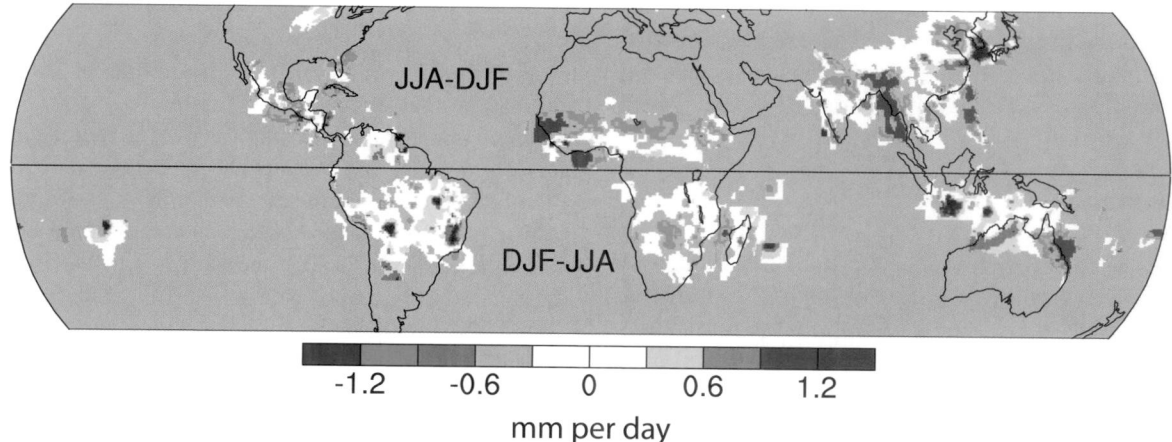

Figure 3.34. *Change in the mean annual range of precipitation: 1976 to 2003 minus 1948 to 1975 periods (mm per day). Blue/green (red/yellow) colour denotes a decreasing (increasing) annual range of the monsoon rainfall. Grey areas indicate missing values (oceans) or areas with insignificant annual changes. Data were from PREC/L (Chen et al., 2002; see Wang and Ding, 2006).*

but significant interannual and inter-decadal variations. In the South African monsoon region there is a slight decrease in the annual range of rainfall (Figure 3.34), and a decreasing trend in area-averaged precipitation (Figure 3.14).

Monsoon variability depends on many factors, from regional air-sea interaction and land processes (e.g., snow cover fluctuations) to teleconnection influences (e.g., ENSO, NAO/ NAM, PDO, IOD). New evidence, relevant to climate change, indicates that increased aerosol loading in the atmosphere may have strong impacts on monsoon evolution (Menon et al., 2002) through changes in local heating of the atmosphere and land surface (see also Box 3.2 and Chapter 2).

3.7.1 Asia

The Asian monsoon can be divided into the East Asian and the South Asian or Indian monsoon systems (Ding et al., 2004). Based on a summer monsoon index derived from MSLP gradients between land and ocean in the East Asian region, Guo et al. (2003) found a systematic reduction in the East Asian summer monsoon during 1951 to 2000, with a stronger monsoon dominant in the first half of the period and a weaker monsoon prevailing in the second half (Figure 3.35). This long-term change in the East Asian monsoon index is consistent with a tendency for a southward shift of the summer rain belt over eastern China (Zhai et al., 2004). However, Figure 3.35, based on the newly developed Hadley Centre MSLP data set version 2 (HadSLP2; Allan and Ansell, 2006), suggests that although there exists a weakening trend starting in the 1920s, it is not reflected in the longer record extending back to the 1850s, which shows marked decadal-scale variability before the 1940s.

There is other evidence that changes in the Asian monsoon occurred about the time of the 1976–1977 climate shift (Wang, 2001) along with changes in ENSO (Huang et al., 2003; Qian et al., 2003), and declines in land precipitation are evident in southern Asia and, to some extent, in Southeast Asia (see Figure 3.14). Gong and Ho (2002) suggested that the change in summer rainfall over the Yangtze River valley was due to

a southward rainfall shift and Ho et al. (2003) noted a sudden change in Korea. These occurred about the same time as a change in the 500 hPa geopotential height and typhoon tracks in summer over the northern Pacific (Gong et al., 2002; see Section 3.6.3) related to the enlargement, intensification and southwestward extension of the northwest Pacific subtropical high. When the equatorial central and eastern Pacific is in a decadal warm period, summer monsoon rainfall is stronger in the Yangtze River valley but weaker in North China. A strong tropospheric cooling trend is found in East Asia during July and August. Accompanying this summer cooling, the upper-level westerly jet stream over East Asia shifts southward and the East Asian summer monsoon weakens, which results in the tendency towards increased droughts in northern China and floods in the Yangtze River valley (Yu et al., 2004b).

Rainfall during the Indian monsoon season, which runs from June to September and accounts for about 70% of annual rainfall, exhibits decadal variability. Observational studies have shown that the impact of El Niño is more severe during the below-normal epochs, while the impact of La Niña is more severe during the above-normal epochs (Kripalani and Kulkarni, 1997a; Kripalani et al., 2001, 2003). Such modulation of ENSO impacts by the decadal monsoon variability was also observed in the rainfall regimes over Southeast Asia (Kripalani and Kulkarni, 1997b). Links between monsoon-related events (rainfall over South Asia, rainfall over East Asia, NH circulation, tropical Pacific circulation) weakened between 1890 and 1930 but strengthened during 1930 to 1970 (Kripalani and Kulkarni, 2001). The strong inverse relationship between El Niño events and Indian monsoon rainfalls that prevailed for more than a century prior to about 1976 has weakened substantially since then (Kumar et al., 1999; Krishnamurthy and Goswami, 2000; Sarkar et al., 2004), involving large-scale changes in atmospheric circulation. Shifts in the Walker Circulation and enhanced land-sea contrasts appear to be countering effects of increased El Niño activity. Ashok et al. (2001) also found that the IOD (see Section 3.6.7.2) plays an important role as a modulator of Indian rainfall. The El Niño-Southern Oscillation is also related to atmospheric fluctuations both in the Indian sector and in northeastern China (Kinter et al., 2002).

3.7.2 Australia

The Australian monsoon covers the northern third of continental Australia and surrounding seas and, considering its closely coincident location and annual evolution, is often studied in conjunction with the monsoon over the islands of Indonesia and Papua New Guinea. The Australian monsoon exhibits large interannual and intra-seasonal variability, largely associated with the effects of ENSO, the Madden-Julian Oscillation (MJO) and tropical cyclone activity (McBride, 1998; Webster et al., 1998; Wheeler and McBride, 2005). Using rain-gauge data, Hennessy et al. (1999) found an increase in calendar-year total rainfall in the Northern Territory of 18% from 1910 to 1995, attributed mostly to enhanced monsoon rainfall in the 1970s and coincident with an almost 20% increase in the number of rain

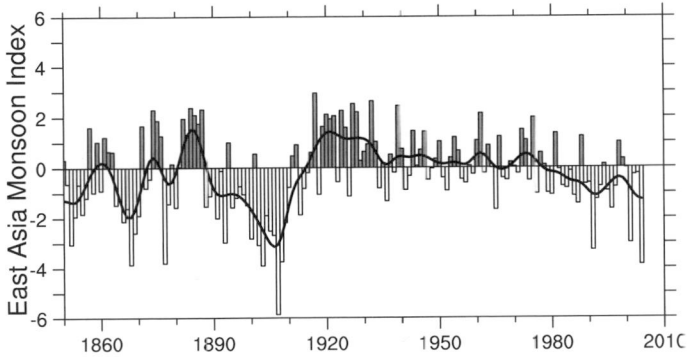

Figure 3.35. *Annual values of the East Asia summer monsoon index derived from MSLP gradients between land and ocean in the East Asia region. The definition of the index is based on Guo et al. (2003) but was recalculated based on the HadSLP2 (Allan and Ansell, 2006) data set. The smooth black curve shows decadal variations (see Appendix 3.A).*

days. With data updated to 2002, Smith (2004) demonstrated that increased monsoon rainfall has become statistically significant over northern, western and central Australia. Northern Australian wet season rainfall (Jones et al., 2004), updated through 2004–2005 (Figure 3.36), shows the positive trend and the contribution of the anomalously wet period of the mid-1970s as well as the more recent anomalously wet period around 2000 (see also Smith, 2004). Wardle and Smith (2004) argued that the upward rainfall trend is consistent with the upward trend in land surface temperatures that has been observed in the south of the continent, independent of changes over the oceans. Strong decadal variations in Australian precipitation have also been noted (Figure 3.36). Using northeastern Australian rainfall, Latif et al. (1997) showed that rainfall was much increased during decades when the tropical Pacific was anomalously cold (the 1950s and 1970s). This strong relationship does not extend to the Australian monsoon as a whole, however, as the rainfall time series (Figure 3.36) has only a weak negative correlation (approximately –0.2) with the IPO. The fact that the long-term trends in rainfall and Pacific SSTs are both positive, the opposite of their interannual relationship (Power et al., 1998), explains only a portion of why the correlation is reduced at decadal time scales.

3.7.3 The Americas

The North American Monsoon System (NAMS) is characterised by ocean-land contrasts including summer heating of higher-elevation mountain and plateau regions of Mexico and the southwestern USA, a large-scale upper-level anticyclonic circulation, a lower-level thermal low and a strong subsidence region to the west in the cool stratus regime of the eastern North Pacific (Vera et al., 2006). The NAMS contains a strong seasonal structure (Higgins and Shi, 2000), with rapid onset of monsoon rains in southwestern Mexico in June, a later northward progression into the southwest USA during its mature phase in July and August and a gradual decay in September and October.

Timing of the start of the northern portion of the NAMS has varied considerably, with some years starting as early as mid-June and others starting as late as early August (Higgins and Shi, 2000). Since part of NAMS variability is governed by larger-scale climate conditions, it is susceptible to interannual and multi-decadal variations. Higgins and Shi (2000) further suggested that the northern portion of the NAMS may be affected by the PDO, wherein anomalous winter precipitation over western North America is correlated with North American monsoon conditions in the subsequent summer.

The South American Monsoon System (SAMS) is evident over South America in the austral summer (Barros et al., 2002; Nogués-Paegle et al., 2002; Vera et al., 2006). It is a key factor for the warm season precipitation regime. In northern Brazil, different precipitation trends (see Figure 3.14 for the Amazon and southern South America regions) have been observed over northern and southern Amazonia, showing a dipole structure (Marengo, 2004) that suggests a southward shift of the SAMS.

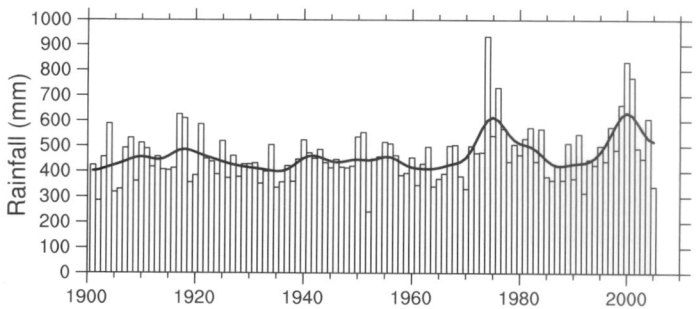

Figure 3.36. *Time series of northern Australian (north of 26°S) wet season (October–April) rainfall (mm) from 1900/1901 to 2004/2005. The individual bar corresponds to the January of the summer season (e.g., 1990 is the summer of 1989/1990). The smooth black curve shows decadal variations (see Appendix 3.A). Data from the Australian Bureau of Meteorology.*

This is consistent with Rusticucci and Penalba (2000), who found a significant positive trend in the amplitude of the annual precipitation cycle, indicating a long-term climate change of the monsoon regime over the semi-arid region of the La Plata Basin. In addition, the mean wind speed of the low-level jet, a component of the SAMS that transports moisture from the Amazon to the south and southwest, showed a positive trend (Marengo et al., 2004). Positive SST anomalies in the western subtropical South Atlantic are associated with positive rainfall anomalies over the SACZ region (Doyle and Barros, 2002; Robertson et al., 2003). Barros et al. (2000b) found that, during summer, the SACZ was displaced northward (southward) and was more intense (weaker) with cold (warm) SST anomalies to its south. The convergence zone is modulated in part by surface features, including the gradient of SST over the equatorial Atlantic (Chang et al., 1999; Nogués-Paegle et al., 2002), and it modulates the interannual variability of seasonal rainfall over eastern Amazonia and northeastern Brazil (Nobre and Shukla, 1996; Folland et al., 2001).

3.7.4 Africa

Since the TAR, a variety of studies have firmly established that ENSO and SSTs in the Indian Ocean are the dominant sources of climate variability over eastern Africa (Goddard and Graham, 1999; Yu and Rienecker, 1999; Indeje et al., 2000; Clark et al., 2003). Further, Schreck and Semazzi (2004) isolated a secondary but significant pattern of regional climate variability based on seasonal (OND) rainfall data. In distinct contrast to the ENSO-related spatial pattern, the trend pattern in their analysis is characterised by positive rainfall anomalies over the northeastern sector of eastern Africa (Ethiopia, Somalia, Kenya and northern Uganda) and opposite conditions over the southwestern sector (Tanzania, southern parts of the Democratic Republic of the Congo and southwestern Uganda). This signal significantly strengthened in recent decades. Warming is associated with an earlier onset of the rainy season over the northeastern Africa region and a late start over the southern sector.

West Africa experiences marked multi-decadal variability in rainfall (e.g., Le Barbe et al., 2002; Dai et al., 2004b). Wet conditions in the 1950s and 1960s gave way to much drier conditions in the 1970s, 1980s and 1990s. The rainfall deficit in this region during 1970 to 1990 was relatively uniform across the region, implying that the deficit was not due to a spatial shift in the peak rainfall (Le Barbe et al., 2002) and was mainly linked to a reduction in the number of significant rainfall events occurring during the peak monsoon period (JAS) in the Sahel and during the first rainy season south of about 9°N. The decreasing rainfall and devastating droughts in the Sahel region during the last three decades of the 20th century (Figure 3.37) are among the largest climate changes anywhere. Dai et al. (2004b) provided an updated analysis of the normalised Sahel rainfall index based on the years 1920 to 2003 (Figure 3.37). Following the major 1982–1983 El Niño event, rainfall reached a minimum of 170 mm below the long-term mean of about 506 mm. Since 1982, there is some evidence for a recovery (see also lower panel of Figure 3.13) but despite this, the mean of the last decade is still well below the pre-1970 level. These authors also noted that large multi-year oscillations appear to be more frequent and extreme after the late 1980s than previously.

ENSO affects the West African monsoon, and the correlation between Sahel rainfall and ENSO during JJA varied between 1945 and 1993 (Janicot et al., 2001). The correlation was always negative but was not significant during the 1960s to the mid-1970s when the role of the tropical Atlantic was relatively more important. Years when ENSO has a larger impact tend to be associated with same-signed rainfall anomalies over the West African region whereas years when the tropical Atlantic is more important tend to have a so-called anomalous 'dipole' pattern, with the Sahel and Guinea Coast having opposite-signed rainfall anomalies (Ward, 1998). Giannini et al. (2003) suggested that both interannual and decadal variability of Sahel rainfall results from the response of the African summer monsoon to oceanic forcing, amplified by land-atmosphere interaction.

While other parts of Africa have experienced statistically significant weakening of the monsoon circulation, analyses of long-term southern African rainfall totals in the wet season

(JFM) have reported no trends (Fauchereau et al., 2003). Decreases in rainfall are evident in analyses of shorter periods, such as the decade from 1986 to 1995 that was the driest of the 20th century. New et al. (2006) reported a decrease in average rainfall intensity and an increase in dry spell length (number of consecutive dry days) for 1961 to 2000.

3.7.5   Summary

Variability at multiple time scales strongly affects monsoon systems. Large interannual variability associated with ENSO dominates the Hadley and Walker Circulations, the ITCZ and monsoons. There is also good evidence for decadal changes associated with monsoonal rainfall changes in many monsoon systems, especially across the 1976–1977 climate shift, but data uncertainties compromise evidence for trends. Some monsoons, especially the East Asian monsoon system, have experienced a dipole change in precipitation with increases in one region and decreases in the other during the last 50 years.

3.8   Changes in Extreme Events

3.8.1   Background

There is increasing concern that extreme events may be changing in frequency and intensity as a result of human influences on climate. Climate change may be perceived most through the impacts of extremes, although these are to a large degree dependent on the system under consideration, including its vulnerability, resiliency and capacity for adaptation and mitigation; see the Working Group II contribution to the IPCC Fourth Assessment Report. Improvements in technology mean that people hear about extremes in most parts of the world within a few hours of their occurrence. Pictures shot by camcorders on the news may foster a belief that weather-related extremes are increasing in frequency, whether they are or not. An extreme weather event becomes a disaster when society and/or ecosystems are unable to cope with it effectively. Growing human vulnerability (due to growing numbers of people living in exposed and marginal areas or due to the development of more high-value property in high-risk zones) is increasing the risk, while human endeavours (such as by local governments) try to mitigate possible effects.

The assessment of extremes in this section is based on long-term observational series of weather elements. As in the TAR, extremes refer to rare events based on a statistical model of particular weather elements, and changes in extremes may relate to changes in the mean and variance in complicated ways. Changes in extremes are assessed at a range of temporal and spatial scales, for example, from extremely warm years globally to peak rainfall intensities locally, and examples are given in Box 3.6. To span this entire range, data are required at a daily (or shorter) time scale. However, the availability of

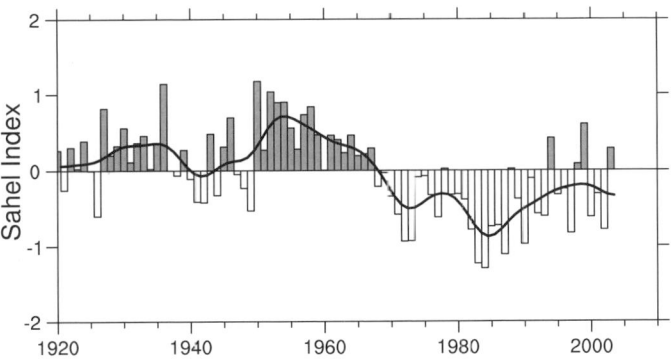

Figure 3.37. *Time series of Sahel (10°N –20°N, 18°W–20°E) regional rainfall (April–October) from 1920 to 2003 derived from gridding normalised station anomalies and then averaging using area weighting (adapted from Dai et al., 2004a). The smooth black curve shows decadal variations (see Appendix 3.A).*

observational data restricts the types of extremes that can be analysed. The rarer the event, the more difficult it is to identify long-term changes, simply because there are fewer cases to evaluate (Frei and Schär, 2001; Klein Tank and Können, 2003). Identification of changes in extremes is also dependent on the analysis technique employed (Zhang et al., 2004a; Trömel and Schönwiese, 2005). To avoid excessive statistical limitations, trend analyses of extremes have traditionally focused on standard and robust statistics that describe moderately extreme events. In percentile terms, these are events occurring between 1 and 10% of the time at a particular location in a particular reference period (generally 1961 to 1990). Unless stated otherwise, this section focuses on changes in these extremes.

Global studies of daily temperature and precipitation extremes over land (e.g., Frich et al., 2002; see also the TAR) suffer from both a scarcity of data and regions with missing data. The main reason is that in various parts of the globe there is a lack of homogeneous observational records with daily resolution covering multiple decades that are part of integrated digitised data sets (GCOS, 2003). In addition, existing records are often inhomogeneous; for instance as a result of changes in observing practices or UHI effects (DeGaetano and Allen, 2002; Vincent et al., 2002; Wijngaard et al., 2003). This affects, in particular, the understanding of extremes, because changes in extremes are often more sensitive to inhomogeneous climate monitoring practices than changes in the mean (see Appendix 3.B.2 and 3.B.4). Consistent observing is also a problem when assessing long-term changes in the frequency and severity of tropical and extratropical storms. Similar difficulties are encountered when trying to find worldwide observational evidence for changes in severe local weather events like tornadoes, hail, thunderstorms and dust storms. Analyses of trends in extremes are also sensitive to the analysis period, for example, the inclusion of the exceptionally hot European summer of 2003 may have a marked influence on results if the period is short.

Since the TAR, the situation with observational data sets has improved, although efforts to update and exchange data must be continued (e.g., GCOS, 2004). Results are now available from newly established regional- and continental-scale daily data sets; from denser networks, from temporally more extended high-quality time series and from many existing national data archives, which have been expanded to cover longer time periods. Moreover, the systematic use and exchange of time series of standard indices of extremes, with common definitions, provides an unprecedented global picture of changes in daily temperature and precipitation extremes (Alexander et al., 2006, updating the results of Frich et al., 2002 presented in the TAR).

As an alternative, but not independent data source, reanalyses can also be analysed for changes in extremes (see Appendix 3.B.5.4). Although spatially and temporally complete, under-representation of certain types of extremes (Kharin and Zwiers, 2000) and spurious trends in the reanalyses (especially in the tropics and in the SH) remain problematic, in particular before the start of the modern satellite era in 1979 (Marshall, 2002, 2003; Sturaro, 2003; Sterl, 2004; Trenberth et al., 2005a).

For instance, Bengtsson et al. (2004) found that analysed global kinetic energy rose by almost 5% in 1979 as a direct consequence of the inclusion of improved satellite information over the oceans, which is expected to significantly affect analysed storm activity over the southern oceans, where ship data are sparse.

In this section, observational evidence for changes in extremes is assessed for temperature, precipitation, tropical and extratropical cyclones and severe local weather events. Most studies of extremes consider the period since about 1950 with even greater emphasis on the last few decades (since 1979), although longer data sets exist for a few regions, enabling more recent events to be placed in a longer context. The section discusses mostly the changes observed in the daily weather elements, where most progress has been made since the TAR. Droughts (although they are considered extremes) are covered in Section 3.3.4 as they are more related to longer periods of anomalous climate.

3.8.2 Evidence for Changes in Variability or Extremes

3.8.2.1 Temperature

For temperature extremes in the 20th century, the TAR highlighted the lengthening of the growing or frost-free season in most mid- and high-latitude regions, a reduction in the frequency of extreme low monthly and seasonal average temperatures and smaller increases in the frequency of extreme high average temperatures. In addition, there was evidence to suggest a decrease in the intra-annual temperature variability with consistent reductions in frost days and increases in warm nighttime temperatures across much of the globe.

Evidence for changes in observed interannual variability (such as standard deviations of seasonal averages) is still sparse. Scherrer et al. (2005) investigated standardised distribution changes for seasonal mean temperature in central Europe and found that temperature variability showed a weak increase (decrease) in summer (winter) for 1961 to 2004, but these changes are not statistically significant at the 10% level. On the daily time scale, regional studies have been completed for southern South America (Vincent et al., 2005), Central America and northern South America (Aguilar et al., 2005), the Caribbean (Peterson et al., 2002), North America (Kunkel et al., 2004; Vincent and Mekis, 2006), the Arctic (Groisman et al., 2003), central and northern Africa (Easterling et al., 2003), southern and western Africa (New et al., 2006), the Middle East (Zhang et al., 2005), Western Europe and east Asia (Kiktev et al., 2003), Australasia and southeast Asia (Griffiths et al., 2005), China (Zhai and Pan, 2003) and central and southern Asia (Klein Tank et al., 2006). They all show patterns of changes in extremes consistent with a general warming, although the observed changes of the tails of the temperature distributions are often more complicated than a simple shift of the entire distribution would suggest (see Figure 3.38). In addition, uneven trends are observed for nighttime and daytime temperature extremes. In southern South America, significant increasing trends were found in the occurrence of warm nights and decreasing trends in the occurrence of cold nights, but no consistent changes in the indices based on daily maximum temperature. In Central America and northern South America, high extremes of both minimum and maximum temperature have increased. Warming of both the nighttime and daytime extremes was also found for the other regions where data have been analysed. For Australasia and Southeast Asia, the dominant distribution change at rural stations for both maximum and minimum temperature involved a change

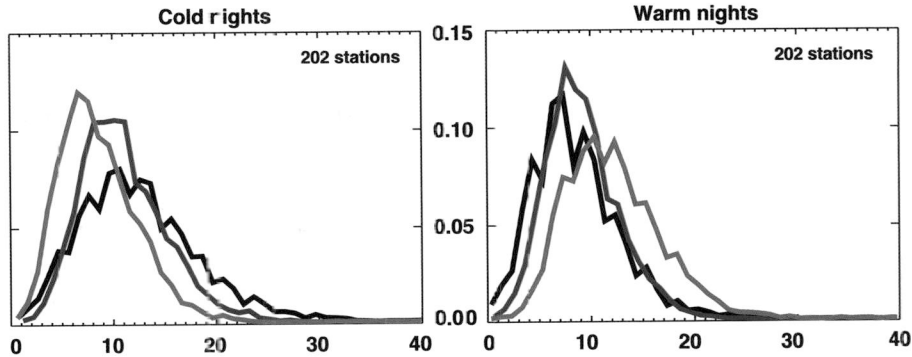

Figure 3.38. *Annual probability distribution functions for temperature indices for 202 global stations with at least 80% complete data between 1901 and 2003 for three time periods: 1901 to 1950 (black), 1951 to 1978 (blue) and 1979 to 2003 (red). The x-axis represents the percentage of time during the year when the indicators were below the 10th percentile for cold nights (left) or above the 90th percentile for warm nights (right). From Alexander et al. (2006).*

in the mean, affecting either one or both distribution tails, with no significant change in standard deviation (Griffiths et al., 2005). For urbanised stations, however, the dominant change also involved a change in the standard deviation. This result was particularly evident for minimum temperature.

Few other studies have considered mutual changes in both the high and low tail of the same daily (minimum, maximum or mean) temperature distribution. Klein Tank and Können (2003) analysed such changes over Europe using standard indices, and found that the annual number of warm extremes (above the 90th percentile for 1961 to 1990) of the daily minimum and maximum temperature distributions increased twice as fast during the last 25 years than expected from the corresponding decrease in the number of cold extremes (lowest 10%). Moberg and Jones (2005) found that both the high and the low tail (defined by the 90th and 10th percentile) of the daily minimum and maximum temperature distribution over Europe in winter increased over the 20th century as a whole, with the low tail of minimum temperature warming significantly in summer. For an even longer period, Yan et al. (2002) found decreasing warm extremes in Europe and China up to the late 19th century, decreasing cold extremes since then, and increasing warm extremes only since 1961, especially in summer (JJA). Brunet et al. (2006) analysed 22 Spanish records for the period 1894 to 2003 and found greater reductions in the number of cold days than increases in hot days. However, since 1973 warm days have been rising dramatically, particularly near the Mediterranean coast. Beniston and Stephenson (2004) showed that changes in extremes of daily temperature in Switzerland were due to changes in both the mean and the variance of the daily temperatures. Vincent and Mekis (2006) found progressively fewer extreme cold nights and cool days but more extreme warm nights and hot days for Canada from 1900 to 2003 and Robeson (2004) found intense warming of the lowest daily minimum temperatures over western and central North America. In Argentina, the strong positive changes in minimum temperature seen during 1959 to 1998 were associated with significant increases in the frequency of warm

nights; there were also decreases in cold days (Rusticucci and Barrucand, 2004).

Alexander et al. (2006) and Caesar et al. (2006) have brought all these and other regional results together, gridding the common indices or data for the period since 1946. Over 74% of the global land area sampled showed a significant decrease in the annual occurrence of cold nights; a significant increase in the annual occurrence of warm nights took place over 73% of the area (Table 3.6, Figure 3.38 and FAQ 3.3). This implies a positive shift in the distribution of daily minimum temperature T_{min} throughout the globe. Changes in the occurrence of cold and warm days show warming as well, but generally less marked. This is consistent with T_{min} increasing more than maximum temperature T_{max}, leading to a reduction in DTR since 1951 (see Sections 3.2.2.1 and 3.2.2.7). The change in the four extremes indices (Table 3.6) also show that the distribution of T_{min} and T_{max} have not only shifted, but also changed in shape. The indices for the number of cold and warm events have changed almost equally, which for a near-Gaussian distributed quantity indicates that the cold tails of the distributions have warmed considerably more than the warm tails over the last 50 years. Considering the last 25 years only, such a change in shape is not seen (Table 3.6).

3.8.2.2 *Precipitation*

The conceptual basis for changes in precipitation has been given by Allen and Ingram (2002) and Trenberth et al. (2003; see Section 3.3 and FAQ 3.2). Issues relate to changes in type, amount, frequency, intensity and duration of precipitation. Observed increases in atmospheric water vapour (see Section 3.4.2) imply increases in intensity, but this will lead to reduced frequency or duration if the total evaporation rate from the Earth's surface (land and ocean) is unchanged. The TAR states that it is likely that there has been a statistically significant 2 to 4% increase in the frequency of heavy and extreme precipitation events when averaged across the middle and high latitudes. Since then a more refined understanding of the observed changes in precipitation extremes has been achieved.

Table 3.6. *Global trends in extremes of temperature and precipitation as measured by the 10th and 90th percentiles (for 1961–1990). Trends with 5 and 95% confidence intervals and levels of significance (**bold**: <1%) were estimated by REML (see Appendix 3.A), which allows for serial correlation in the residuals of the data about the linear trend. All trends are based on annual averages. Values are % per decade. Based on Alexander et al. (2006).*

Series	Trend (% per decade) 1951–2003	1979–2003
TN10: % incidence of T_{min} below coldest decile.	**−1.17 ± 0.20**	**−1.24 ± 0.44**
TN90: % incidence of T_{min} above warmest decile.	**1.43 ± 0.42**	**2.60 ± 0.81**
TX10: % incidence of T_{max} below coldest decile.	**−0.63 ± 0.16**	**−0.91 ± 0.48**
TX90: % incidence of T_{max} above warmest decile.	**0.71 ± 0.35**	**1.74 ± 0.72**
PREC: % contribution of very wet days (above the 95th percentile) to the annual precipitation total.	**0.21 ± 0.10**	0.41 ± 0.38

Many analyses indicate that the evolution of rainfall statistics through the second half of the 20th century is dominated by variations on the interannual to inter-decadal time scale and that trend estimates are spatially incoherent (Manton et al., 2001; Peterson et al., 2002; Griffiths et al., 2003; Herath and Ratnayake, 2004). In Europe, there is a clear majority of stations with increasing trends in the number of moderately and very wet days (defined as wet days (≥1 mm of rain) that exceed the 75th and 95th percentiles, respectively) during the second half of the 20th century (Klein Tank and Können, 2003; Haylock and Goodess, 2004). Similarly, for the contiguous USA, Kunkel et al. (2003) and Groisman et al. (2004) confirmed earlier results and found statistically significant increases in heavy (upper 5%) and very heavy (upper 1%) precipitation of 14 and 20%, respectively. Much of this increase occurred during the last three decades of the 20th century and is most apparent over the eastern parts of the country. In addition, there is new evidence from Europe and the USA that the relative increase in precipitation extremes is larger than the increase in mean precipitation, and this is manifested as an increasing contribution of heavy events to total precipitation (Klein Tank and Können, 2003; Groisman et al., 2004).

Despite a decrease in mean annual rainfall, an increase in the fraction from heavy events was inferred for large parts of the Mediterranean (Alpert et al., 2002; Brunetti et al., 2004;

Maheras et al., 2004). Further, Kostopoulou and Jones (2005) noted contrasting trends of heavy rainfall events between an increase in the central Mediterranean (Italy) and a decrease over the Balkans. In South Africa, Siberia, central Mexico, Japan and the northeastern part of the USA, an increase in heavy precipitation was also observed, while total precipitation and/or the frequency of days with an appreciable amount of precipitation (wet days) was either unchanged or decreasing (Easterling et al., 2000; Faucherau et al., 2003; Sun and Groisman, 2004; Groisman et al., 2005).

A number of recent regional studies have been completed for southern South America (Haylock et al., 2006), Central America and northern South America (Aguilar et al., 2005), southern and western Africa (New et al., 2006), the Middle East (Zhang et al., 2005) and central and southern Asia (Klein Tank et al., 2006). For southern South America, the pattern of trends for extremes between 1960 and 2000 was generally the same as that for total annual rainfall (Haylock et al., 2006). A majority of stations showed a change to wetter conditions, related to the generally lower value of the SOI since 1976/1977, with the exception of southern Peru and southern Chile, where a decrease was observed in many precipitation indices. In the latter region, the change in ENSO has led to a weakening of the continental trough resulting in a southward shift in storm tracks and an important effect on the observed rainfall trends. No significant increases in total precipitation amounts were found over Central America and northern South America (see also Figure 3.14), but rainfall intensities have increased related to changes in SST of tropical Atlantic waters. Over southern and western Africa, and the Middle East, there are no spatially coherent patterns of statistically significant trends in precipitation indices. Averaged over central and southern Asia, a slight indication of disproportionate changes in the precipitation extremes compared with the total amounts is seen. In the Indian sub-continent Sen Roy and Balling (2004) found that about two-thirds of all considered time series exhibit increasing trends in indices of precipitation extremes and that there are coherent regions with increases and decreases.

Alexander et al. (2006) also gridded the extreme indices for precipitation (as for temperature in Section 3.8.2.1). Changes in precipitation extremes are much less coherent than for temperature, but globally averaged over the land area with sufficient data, the percentage contribution to total annual precipitation from very wet days (upper 5%) is greater in recent decades than earlier decades (Figure 3.39, top panel, and Table 3.6, last line). Observed changes in intense precipitation (with geographically varying thresholds between the 90th and 99.9th percentile of daily precipitation events) for more than one half of the global land area indicate an increasing probability of intense precipitation events beyond that expected from changes in the mean for many extratropical regions (Groisman et al., 2005; Figure 3.39, bottom panel). This finding supports the disproportionate changes in the precipitation extremes described in the majority of regional studies above, in particular for the mid-latitudes since about 1950. It is still difficult to draw a consistent picture of changes in the tropics and the subtropics, where many areas are not analysed and data are not readily available.

As well as confirming previous findings, the new analyses provide seasonal detail and insight into longer-term variations for the mid-latitudes. While the increase in the USA is found primarily in the warm season (Groisman et al., 2004), central and northern Europe exhibited changes primarily in winter (DJF) and changes were insignificant in summer (JJA), but the studies did not include the extreme European summers of 2002 (very wet) and 2003 (very dry) (Osborn and Hulme, 2002; Haylock and Goodess, 2004; Schmidli and Frei, 2005). Although data are not as good, the frequencies of precipitation extremes in the USA were at comparable levels from 1895 into the early 1900s and during the 1980s to 1990s (Kunkel et al., 2003). For Canada (excluding the high-latitude Arctic), Zhang et al. (2001a) and Vincent and Mekis (2006) found that the frequency of precipitation days significantly increased during the 20th century, but averaged for the country as a whole, there is no identifiable trend in precipitation extremes. Nevertheless, Groisman et al. (2005) found significant increases in the frequency of heavy and very heavy (between the 95th and 99.7th percentile of daily precipitation events) precipitation in British Columbia south of 55°N for 1910 to 2001, and in other areas (Figure 3.39, bottom panel).

Since the TAR, several regional analyses have been undertaken for statistics with return periods much longer than in the previous studies. For the UK, Fowler and Kilsby (2003a,b), using extreme value statistics, estimated that the recurrence of 10-day precipitation totals with a 50-year return period (based on data for 1961 to 1990) had increased by a factor of two to five by the 1990s in northern England and Scotland. Their results for long return periods are qualitatively similar to changes obtained for traditional (moderate) statistics (Osborn et al., 2000; Osborn and Hulme, 2002), but there are differences in the relative magnitude of the change between seasons (Fowler and Kilsby, 2003b). For the contiguous USA, Kunkel et al. (2003) and Groisman et al. (2004) analysed return periods of 1 to 20 years, and interannual to inter-decadal variations during the 20th century

exhibit a high correlation between all return periods. Similar results were obtained for several extratropical regions (Groisman et al., 2005), including the central USA, the northwestern coast of North America, southern Brazil, Fennoscandia, the East European Plain, South Africa, southeastern Australia and Siberia. In summary, from the available analyses there is

Trend 1951 - 2003 contribution from very wet days

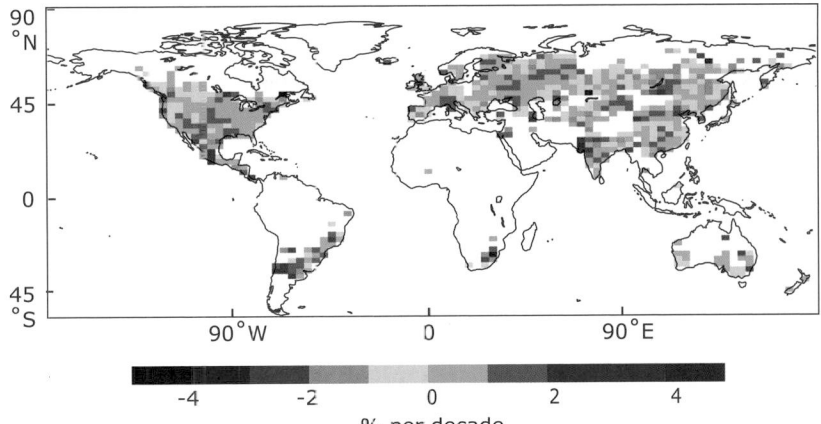

% per decade

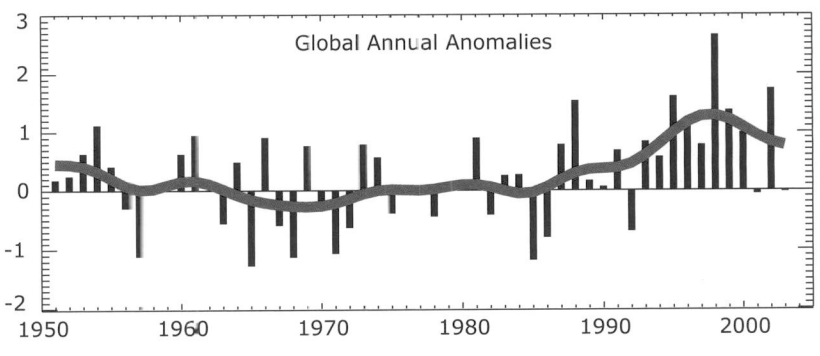

Figure 3.39. *(Top) Observed trends (% per decade) for 1951 to 2003 in the contribution to total annual precipitation from very wet days (95th percentile). Trends were only calculated for grid boxes where both the total and the 95th percentile had at least 40 years of data during this period and had data until at least 1999. (Middle) Anomalies (%) of the global annual time series (with respect to 1961 to 1990) defined as the percentage change of contributions of very wet days from the base period average (22.5%). The smooth orange curve shows decadal variations (see Appendix 3.A). From Alexander et al. (2006). (Bottom) Regions where disproportionate changes in heavy and very heavy precipitation during the past decades were documented as either an increase (+) or decrease (–) compared to the change in the annual and/or seasonal precipitation (updated from Groisman et al., 2005). Thresholds used to define "heavy" and "very heavy" precipitation vary by season and region. However, changes in heavy precipitation frequencies are always greater than changes in precipitation totals and, in some regions, an increase in heavy and/or very heavy precipitation occurred while no change or even a decrease in precipitation totals was observed.*

evidence that the changes at the extreme tail of the distribution (several-decade return periods) are consistent with changes inferred for more robust statistics based on percentiles between the 75th and 95th levels, but practically no regions have sufficient data to assess such trends reliably.

3.8.3 Evidence for Changes in Tropical Storms

The TAR noted that evidence for changes in tropical cyclones (both in number and in intensity) across the various ocean basins is often hampered by classification changes. In addition, considerable inter-decadal variability reduces significance of any long-term trends. Careful interpretation of observational records is therefore required. Traditional measures of tropical cyclones, hurricanes and typhoons have varied in different regions of the globe, and typically have required thresholds of estimated wind speed to be crossed for the system to be called a tropical storm, named storm, cyclone, hurricane or typhoon, or major hurricane or super typhoon. Many other measures or terms exist, such as 'named storm days', 'hurricane days', 'intense hurricanes', 'net tropical cyclone activity', and so on.

The ACE index (see Box 3.5), is essentially a wind energy index, defined as the sum of the squares of the estimated six-hour maximum sustained wind speed (knots) for all named systems while they are at least tropical storm strength. Since this index represents a continuous spectrum of both system duration and intensity, it does not suffer as much from the discontinuities inherent in more widely used measures of activity such as the number of tropical storms, hurricanes or major hurricanes. However, the ACE values reported here are not adjusted for known inhomogeneities in the record (discussed below). The ACE index is also used to define above-, near-, and below-normal hurricane seasons (based on the 1981 to 2000 period). The index has the same meaning in every region. Figure 3.40 shows the ACE index for six regions (adapted from Levinson,

2005, and updated through early 2006). Prior to about 1970, there was no satellite imagery to help estimate the intensity and size of tropical storms, so the estimates of ACE are less reliable, and values are not given prior to about the mid- or late 1970s in the Indian Ocean, South Pacific or Australian regions. Values are given for the Atlantic and two North Pacific regions after 1948, although reliability improves over time, and trends contain unquantified uncertainties.

The Potential Intensity (PI) of tropical cyclones (Emanuel, 2003) can be computed from observational data based primarily on vertical profiles of temperature and humidity (see Box 3.5) and on SSTs. In analysing CAPE (see Box 3.5) from selected radiosonde stations throughout the tropics for the period 1958 to 1997, Gettelman et al. (2002) found mostly positive trends. DeMott and Randall (2004) found more mixed results, although their data may have been contaminated by spurious adjustments (Durre et al., 2002). Further, Free et al. (2004a) found that trends in PI were small and statistically insignificant at a scattering of stations in the tropics. As all of these studies were probably contaminated by problems with tropical radiosondes (Sherwood et al., 2005; Randel and Wu, 2006; see Section 3.4.1 and Appendix 3.B.5), definitive results are not available.

The PDI index of the total power dissipation for the North Atlantic and western North Pacific (Emanuel, 2005a; see also Box 3.5) showed substantial upward trends beginning in the mid-1970s. Because the index depends on wind speed cubed, it is very sensitive to data quality, and the initial Emanuel (2005a) report has been revised to show the PDI increasing by about 75% (vs. about 100%) since the 1970s (Emanuel, 2005b). The

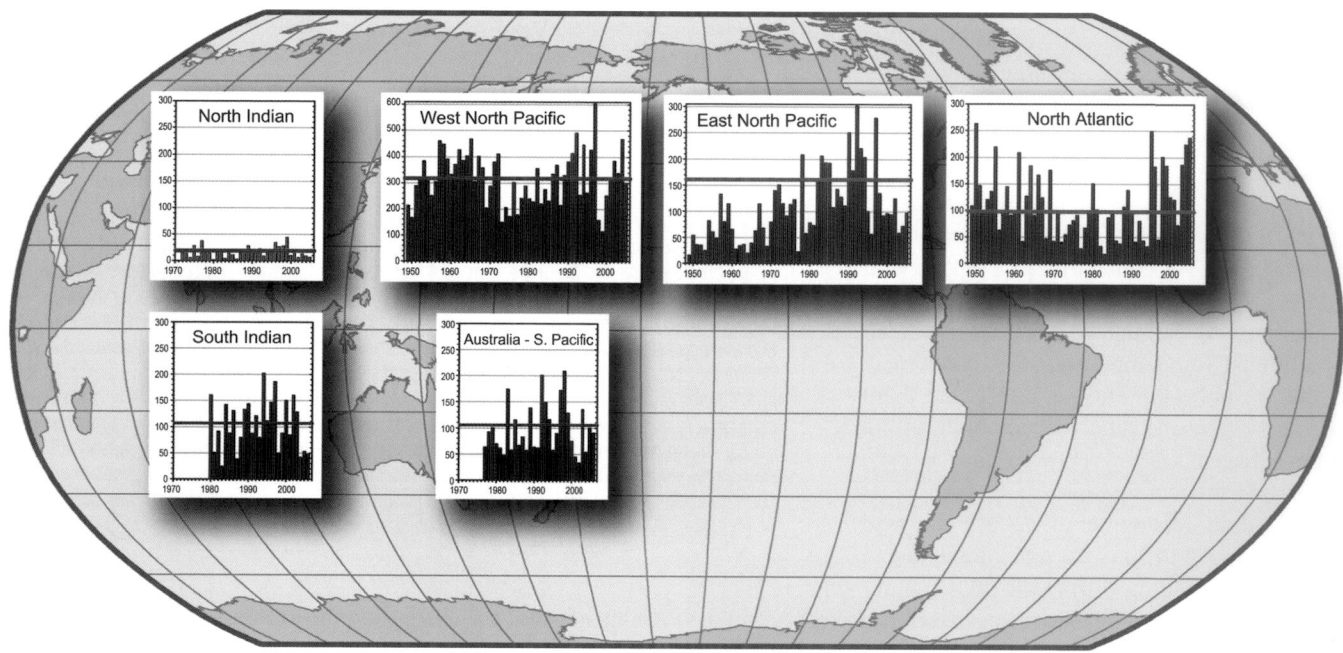

Figure 3.40. *Seasonal values of the ACE index for the North Indian, South Indian, West North Pacific, East North Pacific, North Atlantic and combined Australian-South Pacific regions. The vertical scale in the West North Pacific is twice as large as that of other basins. The SH values are those for the season from July the year before to June of the year plotted. The timeline runs from 1948 or 1970 through 2005 in the NH and through June 2006 in the SH. The ACE index accounts for the combined strength and duration of tropical storms and hurricanes during a given season by computing the sum of squares of the six-hour maximum sustained surface winds in knots while the storm is above tropical storm intensity. Adapted and updated from Levinson (2005).*

Box 3.5: Tropical Cyclones and Changes in Climate

In the summer tropics, outgoing longwave radiative cooling from the surface to space is not effective in the high water vapour, optically thick environment of the tropical oceans. Links to higher latitudes are weakest in the summer tropics, and transports of energy by the atmosphere, such as occur in winter, are also not an effective cooling mechanism, while monsoonal circulations between land and ocean redistribute energy in areas where they are active. However, tropical storms cool the ocean surface through mixing with cooler deeper ocean layers and through evaporation. When the latent heat is realised in precipitation in the storms, the energy is transported high into the troposphere where it can radiate to space, with the system acting somewhat like a Carnot cycle (Emanuel, 2003). Hence, tropical cyclones appear to play a key role in alleviating the heat from the summer Sun over the oceans.

As the climate changes and SSTs continue to increase (see Section 3.2.2.3), the environment in which tropical storms form is changed. Higher SSTs are generally accompanied by increased water vapour in the lower troposphere (see Section 3.4.2.1 and Figure 3.20), thus the moist static energy that fuels convection and thunderstorms is also increased. Hurricanes and typhoons currently form from pre-existing disturbances only where SSTs exceed about 26°C and, as SSTs have increased, it thereby potentially expands the areas over which such storms can form. However, many other environmental factors also influence the generation and tracks of disturbances, and wind shear in the atmosphere greatly influences whether or not these disturbances can develop into tropical storms. The El Niño-Southern Oscillation and variations in monsoons as well as other factors also affect where storms form and track (e.g., Gray, 1984). Whether the large-scale thermodynamic environment and atmospheric static stability (often measured by Convective Available Potential Energy, CAPE) becomes more favourable for tropical storms depends on how changes in atmospheric circulation, especially subsidence, affect the static stability of the atmosphere, and how the wind shear changes. The potential intensity, defined as the maximum wind speed achievable in a given thermodynamic environment (e.g., Emanuel, 2003), similarly depends critically on SSTs and atmospheric structure. The tropospheric lapse rate is maintained mostly by convective transports of heat upwards, in thunderstorms and thunderstorm complexes, including mesoscale disturbances, various waves and tropical storms, while radiative processes serve to cool the troposphere. Increases in greenhouse gases decrease radiative cooling aloft, thus potentially stabilising the atmosphere. In models, the parametrization of sub-grid scale convection plays a critical role in determining whether this stabilisation is realised and whether CAPE is released or not. All of these factors, in addition to SSTs, determine whether convective complexes become organised as rotating storms and form a vortex.

While attention has often been focussed simply on the frequency or number of storms, the intensity, size and duration likely matter more. NOAA's Accumulated Cyclone Energy (ACE) index (Levinson and Waple, 2004) approximates the collective intensity and duration of tropical storms and hurricanes during a given season and is proportional to maximum surface sustained winds squared. The power dissipation of a storm is proportional to the wind speed cubed (Emanuel, 2005a), as the main dissipation is from surface friction and wind stress effects, and is measured by a Power Dissipation Index (PDI). Consequently, the effects of these storms are highly nonlinear and one big storm may have much greater impacts on the environment and climate system than several smaller storms.

From an observational perspective then, key issues are the tropical storm formation regions, the frequency, intensity, duration and tracks of tropical storms, and associated precipitation. For landfalling storms, the damage from winds and flooding, as well as storm surges, are especially of concern, but often depend more on human factors, including whether people place themselves in harm's way, their vulnerability and their resilience through such things as building codes.

increase comes about because of longer storm lifetimes and greater storm intensity, and the index is strongly correlated with tropical SST. These relationships have been reinforced by Webster et al. (2005, 2006) who found a large increase in numbers and proportion of hurricanes reaching categories 4 and 5 globally since 1970 even as the total number of cyclones and cyclone days decreased slightly in most basins. The largest increase was in the North Pacific, Indian and Southwest Pacific Oceans.

These studies have been challenged by several scientists (e.g., Landsea, 2005; Chan, 2006) who have questioned the quality of the data and the start date of the 1970s. In addition, different centres may assign different intensities to the same storm. The historical record typically records the central pressure and the maximum winds, but these turn out not to be physically consistent in older records, mainly prior to about the early 1970s. However, attempts at mutual adjustments result in

increases in some years and decreases in others, with little effect on overall trends. In particular, in the satellite era after about 1970, the trends found by Emanuel (2005a) and Webster et al. (2005) appear to be robust in strong association with higher SSTs (Emanuel, 2005b). There is no doubt that active periods have occurred in the more distant past, notably in the North Atlantic (see below), but the PDI was evidently not as high in the earlier years (Emanuel, 2005a).

There is a clear El Niño connection in most regions, and strong negative correlations between regions in the Pacific and Atlantic, so that the total tropical storm activity is more nearly constant than ACE values in any one basin. During an El Niño event, the incidence of hurricanes typically decreases in the Atlantic (Gray, 1984; Bove et al., 1998) and far western Pacific and Australian regions, while it increases in the central North and South Pacific and especially in the western North Pacific typhoon region (Gray, 1984; Lander, 1994; Kuleshov and de

Hoedt, 2003; Chan and Liu, 2004), emphasizing the change in locations for tropical storms to preferentially form and track with ENSO. Formation and tracks of tropical storms favour either the Australian or South Pacific region depending on the phase of ENSO (Basher and Zheng, 1995; Kuleshov and de Hoedt, 2003), and these two regions have been combined.

The ACE values have been summed over all regions to produce a global value, as given in Klotzbach (2006), beginning in 1986. The highest ACE year through 2005 is 1997, when a major El Niño event began and surface temperatures were subsequently the highest on record (see Section 3.2), and this is followed by 1992, a moderate El Niño year. Such years contain low values in the Atlantic, but much higher values in the Pacific, and they highlight the critical role of SSTs in the distribution and formation of hurricanes. Next in ranking are 1994 and 2004, while 2005 is close to the 1981 to 2000 mean. The PDI also peaks in the late 1990s about the time of the 1997–1998 El Niño for the combined Atlantic and West Pacific regions, although 2004 is almost as high. Webster et al. (2005) found that numbers of intense (category 4 and 5) hurricanes after 1990 are much greater than from 1970 to 1989. Klotzbach (2006) considers ACE values only from 1986 and his record is not long enough to provide reliable trends, given the substantial variability.

3.8.3.1 *Western North Pacific*

In the western North Pacific, long-term trends are masked by strong inter-decadal variability for 1960 to 2004 (Chan and Liu, 2004; Chan, 2006), but results also depend on the statistics used and there are uncertainties in the data prior to the mid-1980s (Klotzbach, 2006). Further increases in activity have occurred in the last few years after Chan and Liu (2004) was completed (Figure 3.40). Tropical cyclones making landfall in China are a small fraction of the total storms, and no obvious long-term trend can be discerned (He et al., 2003; Liu and Chan, 2003; Chan and Liu, 2004). However, Emanuel (2005a) and Webster et al. (2005, 2006) indicated that the typhoons have become more intense in this region, with almost a doubling of PDI values since the 1950s and an increase of about 30% in the number of category 4 and 5 storms from 1990 to 2004 compared with 1975 to 1989. The post-1985 record analysed by Klotzbach (2006) is too short to provide reliable trends.

The main modulating influence on tropical cyclone activity in the western North Pacific appears to be the changes in atmospheric circulation associated with ENSO, rather than local SSTs (Liu and Chan, 2003; Chan and Liu, 2004). In El Niño years, tropical cyclones tend to be more intense and longer-lived than in La Niña years (Camargo and Sobel, 2004) and occur in different locations. In the summer (JJA) and autumn (SON) of strong El Niño years, tropical cyclone numbers increase markedly in the southeastern quadrant of the western North Pacific (0°N–17°N, 140°E–180°E) and decrease in the northwestern quadrant (17°N–30°N, 120°E–140°E; Wang and Chan, 2002). In SON of El Niño years from 1961 to 2000, significantly fewer tropical cyclones made landfall in the

western North Pacific compared with neutral years, although in Japan and the Korean Peninsula no statistically significant change was detected. In contrast, in SON of La Niña years, significantly more landfalls were reported in China (Wu et al., 2004). Overall in 2004, the number of tropical depressions, tropical storms and typhoons was slightly above the 1971 to 2000 median but the number of typhoons (21) was well above the median (17.5) and second highest to 1997, when 23 developed. Moreover, a record number of 10 tropical cyclones or typhoons made landfall in Japan; the previous record was 6 (Levinson, 2005). The ACE index was very close to normal for the 2005 season (Figure 3.40).

3.8.3.2 *North Atlantic*

The North Atlantic hurricane record begins in 1851 and is the longest among cyclone series. Values are considered fairly reliable after about 1950 when measurements from reconnaissance aircraft began. Methods of estimating wind speed from aircraft have evolved over time and, unfortunately, changes were not always well documented. The record is most reliable after the early 1970s (Landsea, 2005). The North Atlantic record shows a fairly active period from the 1930s to the 1960s followed by a less active period in the 1970s and 1980s, similar to the fluctuations of the AMO (Figure 3.33).

Beginning with 1995, all but two Atlantic hurricane seasons have been above normal (relative to the 1981 to 2000 base period). The exceptions are the two El Niño years of 1997 and 2002. As noted in Section 3.8.3, El Niño acts to reduce activity and La Niña acts to increase activity in the North Atlantic. The increased activity after 1995 contrasts sharply with the generally below-normal seasons observed during the previous 25-year period (1970–1994). These multi-decadal fluctuations in hurricane activity result nearly entirely from differences in the number of hurricanes and major hurricanes forming from tropical storms first named in the tropical Atlantic and Caribbean Sea. The change from the negative phase of the AMO in the 1970s and 1980s (see Section 3.6.6) to the post-1995 period has been a contributing factor to the increased hurricane activity (Goldenberg et al., 2001) and is well depicted in Atlantic SSTs (Figure 3.33), including those in the tropics. Nevertheless, it appears likely that most of the warming since the 1970s can be associated with global SST increases rather than the AMO (Trenberth and Shea, 2006; see Section 3.6.6).

During 1995 to 2004, hurricane seasons averaged 13.6 tropical storms, 7.8 hurricanes and 3.8 major hurricanes, and have an average ACE index of 159% of the median. The record-breaking 2005 season is documented in more detail in Section 3.8.4, Box 3.6. In contrast, during the preceding 1970 to 1994 period, hurricane seasons averaged 8.6 tropical storms, 5 hurricanes and 1.5 major hurricanes, and had an average ACE index of only 70% of the median. NOAA classifies 12 (almost one-half) of these 25 seasons as being below normal, and only three as being above normal (1980, 1988, 1989), with the remainder as normal. The positive phase of the AMO was also present during the above-normal hurricane decades of the 1950s and 1960s, as indicated by comparing Atlantic SSTs (Figure 3.33) and seasonal ACE values (Figure 3.40). In 2004, there were 15 named storms, of which 9 were hurricanes, and an unprecedented 4 hit Florida, causing extensive damage (Levinson, 2005). In 2005, record-high SSTs (Figure 3.33) and favourable atmospheric conditions enabled the most active season on record (by many measures), but this was not fully reflected in the ACE index (see also Section 3.8.4, Box 3.6). In 2005, the North Atlantic ACE was the third highest since 1948, while the PDI was the highest on record, exceeding the previous high reached in 2004.

Key factors in the recent increase in Atlantic activity (Chelliah and Bell, 2004) include: (1) warmer SSTs across the tropical Atlantic; (2) an amplified subtropical ridge at upper levels across the central and eastern North Atlantic; (3) reduced vertical wind shear in the deep tropics over the central North Atlantic, which results from an expanded area of easterly winds in the upper atmosphere and weaker easterly trade winds in the lower atmosphere; and (4) a configuration of the African easterly jet that favours hurricane development from tropical disturbances moving westward from the African coast. The vertical shear in the main development region where most Atlantic hurricanes form (Aiyyer and Thorncroft, 2006) fluctuates interannually with ENSO, and with a multi-decadal variation that is correlated with Sahel precipitation. The latter switched sign around 1970 and remained in that phase until the early 1990s, consistent with the AMO variability. It has been argued that the QBO is also a factor in interannual variability (Gray, 1984). The most recent decade has the highest SSTs on record in the tropical North Atlantic (Figure 3.33), apparently as part of global warming and a favourable phase of the AMO. In the Atlantic generally, the changing environmental conditions (Box 3.5) have been more favourable in the past decade for tropical storms to develop.

3.8.3.3   Eastern North Pacific

Tropical cyclone activity (both frequency and intensity) in this region is related especially to SSTs, the phase of ENSO and the phase of the QBO in the tropical lower stratosphere. Above-normal tropical cyclone activity during El Niño years and the lowest activity typically associated with La Niña years is the opposite of the North Atlantic Basin (Landsea et al., 1998). Tropical cyclones tend to attain a higher intensity when the QBO is in its westerly phase at 30 hPa in the tropical lower stratosphere. A well-defined peak in the seasonal ACE occurred in early 1990s, with the largest annual value in 1992 (Figure 3.40), but values are unreliable prior to 1970 in the pre-satellite era. In general, seasonal hurricane activity, including the ACE index, has been below average since 1995, with the exception of the El Niño year of 1997, and is inversely related to the observed increase in activity in the North Atlantic basin over the same time period. This pattern is associated with the AMO (Levinson, 2005) and ENSO. Nevertheless, there has been an increase in category 4 and 5 storms (Webster et al., 2005).

3.8.3.4   Indian Ocean

The North Indian Ocean tropical cyclone season extends from May to December, with peaks in activity during May to June and November when the monsoon trough lies over tropical waters in the basin. Tropical cyclones are usually short-lived and weak, quickly moving into the sub-continent. Tropical storm activity in the northern Indian Ocean has been near normal in recent years (Figure 3.40).

The tropical cyclone season in the South Indian Ocean is normally active from December through April and thus the data are summarised by season in Figure 3.40, rather than by calendar year. The basin extends from the African coastline, where tropical cyclones affect Madagascar, Mozambique and the Mascarene Islands, including Mauritius, to 110°E (tropical cyclones east of 110°E are included in the Australian summary), and from the equator southward, although most cyclones develop south of 10°S. The intensity of tropical cyclones in the South Indian Ocean is reduced during El Niño events (Figure 3.40; Levinson, 2005). Lack of historical record keeping severely hinders trend analysis.

3.8.3.5   Australia and the South Pacific

The tropical cyclone season in the South Pacific-Australia region typically extends over the period November through April, with peak activity from December through March. Tropical cyclone activity in the Australian region (105°E–160°E) apparently declined somewhat over the past decade (Figure 3.40), although this may be partly due to improved analysis and discrimination of weak cyclones that previously were estimated at minimum tropical storm strength (Plummer et al., 1999). Increased cyclone activity in the Australian region has been associated with La Niña years, while below-normal activity has occurred during El Niño years (Plummer et al., 1999; Kuleshov and de Hoedt, 2003). In contrast, in the South Pacific east of 160°E, the opposite signal has been observed, and the most active years have been associated with El Niño events, especially during the strong 1982–1983 and 1997–1998 events (Levinson, 2005), and maximum ACE values occurred from January through March 1998 (Figure 3.40). Webster et al. (2005) found more than a doubling in the numbers of category 4 and 5 hurricanes in the southwest Pacific region between 1975 to 1989 and 1990 to 2004. In the 2005–2006 season, La Niña influences shifted tropical storm activity away from the South Pacific to the Australian region and in March and April 2006, four category 5 typhoons (Floyd, Glenda, Larry and Monica) occurred.

3.8.3.6   South Atlantic

In late March 2004 in the South Atlantic, off the coast of Brazil, the first and only documented hurricane in that region occurred (Pezza and Simmonds, 2005). It came ashore in the Brazilian state of Santa Catarina on 28 March 2004 with winds,

Frequently Asked Question 3.3
Has there been a Change in Extreme Events like Heat Waves, Droughts, Floods and Hurricanes?

Since 1950, the number of heat waves has increased and widespread increases have occurred in the numbers of warm nights. The extent of regions affected by droughts has also increased as precipitation over land has marginally decreased while evaporation has increased due to warmer conditions. Generally, numbers of heavy daily precipitation events that lead to flooding have increased, but not everywhere. Tropical storm and hurricane frequencies vary considerably from year to year, but evidence suggests substantial increases in intensity and duration since the 1970s. In the extratropics, variations in tracks and intensity of storms reflect variations in major features of the atmospheric circulation, such as the North Atlantic Oscillation.

In several regions of the world, indications of changes in various types of extreme climate events have been found. The extremes are commonly considered to be the values exceeded 1, 5 and 10% of the time (at one extreme) or 90, 95 and 99% of the time (at the other extreme). The warm nights or hot days (discussed below) are those exceeding the 90th percentile of temperature, while cold nights or days are those falling below the 10th percentile. Heavy precipitation is defined as daily amounts greater than the 95th (or for 'very heavy', the 99th) percentile.

In the last 50 years for the land areas sampled, there has been a significant decrease in the annual occurrence of cold nights and a significant increase in the annual occurrence of warm nights (Figure 1). Decreases in the occurrence of cold days and increases in hot days, while widespread, are generally less marked. The distributions of minimum and maximum temperatures have not only shifted to higher values, consistent with overall warming, but the cold extremes have warmed more than the warm extremes over the last 50 years (Figure 1). More warm extremes imply an increased frequency of heat waves. Further supporting indications include the observed trend towards fewer frost days associated with the average warming in most mid-latitude regions.

A prominent indication of a change in extremes is the observed evidence of increases in heavy precipitation events over the mid-latitudes in the last 50 years, even in places where mean precipitation amounts are not increasing (see also FAQ 3.2). For very heavy precipitation events, increasing trends are reported as well, but results are available for few areas.

Drought is easier to measure because of its long duration. While there are numerous indices and metrics of drought, many studies use monthly precipitation totals and temperature averages combined into a measure called the Palmer Drought Severity Index (PDSI). The PDSI calculated from the middle of the 20th century shows a large drying trend over many Northern Hemisphere land areas since the mid-1950s, with widespread drying over much of southern Eurasia, northern Africa, Canada and Alaska

(FAQ 3.2, Figure 1), and an opposite trend in eastern North and South America. In the Southern Hemisphere, land surfaces were wet in the 1970s and relatively dry in the 1960s and 1990s, and there was a drying trend from 1974 to 1998. Longer-duration records for Europe for the whole of the 20th century indicate few significant trends. Decreases in precipitation over land since the 1950s are the likely main cause for the drying trends, although large surface warming during the last two to three decades has also likely contributed to the drying. One study shows that very dry land areas across the globe (defined as areas with a PDSI of less than –3.0) have more than doubled in extent since the 1970s, associated with an initial precipitation decrease over land related to the El Niño-Southern Oscillation and with subsequent increases primarily due to surface warming.

Changes in tropical storm and hurricane frequency and intensity are masked by large natural variability. The El Niño-Southern Oscillation greatly affects the location and activity of tropical storms around the world. Globally, estimates of the potential destructiveness of hurricanes show a substantial upward trend since the mid-1970s, with a trend towards longer storm duration and greater storm intensity, and the activity is strongly correlated with tropical sea surface temperature. These relationships have been reinforced by findings of a large increase in numbers and proportion of strong hurricanes globally since 1970 even as total numbers of cyclones and cyclone days decreased slightly in most basins. Specifically, the number of category 4 and 5 hurricanes increased by about 75% since 1970. The largest increases were in the North Pacific, Indian and Southwest Pacific Oceans. However, numbers of hurricanes in the North Atlantic have also been above normal in 9 of the last 11 years, culminating in the record-breaking 2005 season.

Based on a variety of measures at the surface and in the upper troposphere, it is likely that there has been a poleward shift as well as an increase in Northern Hemisphere winter storm track activity over the second half of the 20th century. These changes are part of variations that have occurred related to the North Atlantic Oscillation. Observations from 1979 to the mid-1990s reveal a tendency towards a stronger December to February circumpolar westerly atmospheric circulation throughout the troposphere and lower stratosphere, together with poleward displacements of jet streams and increased storm track activity. Observational evidence for changes in small-scale severe weather phenomena (such as tornadoes, hail and thunderstorms) is mostly local and too scattered to draw general conclusions; increases in many areas arise because of increased public awareness and improved efforts to collect reports of these phenomena.

(continued)

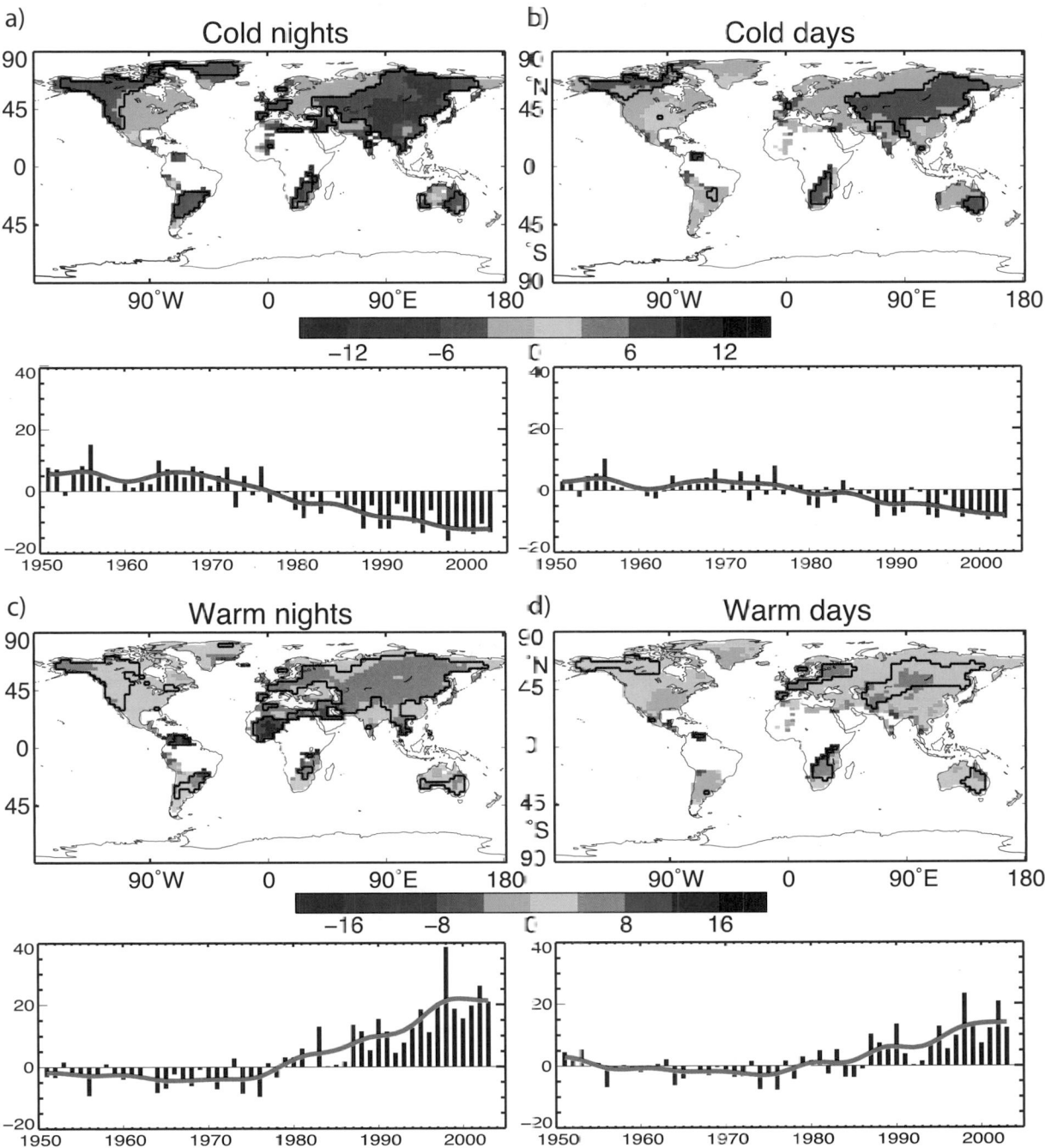

FAQ 3.3, Figure 1. *Observed trends (days per decade) for 1951 to 2003 in the frequency of extreme temperatures, defined based on 1961 to 1990 values, as maps for the 10th percentile: (a) cold nights and (b) cold days; and 90th percentile: (c) warm nights and (d) warm days. Trends were calculated only for grid boxes that had at least 40 years of data during this period and had data until at least 1999. Black lines enclose regions where trends are significant at the 5% level. Below each map are the global annual time series of anomalies (with respect to 1961 to 1990). The orange line shows decadal variations. Trends are significant at the 5% level for all the global indices shown. Adapted from Alexander et al. (2006).*

Box 3.6: Recent Extreme Events

Single extreme events cannot be simply and directly attributed to anthropogenic climate change, as there is always a finite chance the event in question might have occurred naturally. However, when a pattern of extreme weather persists for some time, it may be classed as an extreme climate event, perhaps associated with anomalies in SSTs (such as El Niño). This box provides examples of some recent (post-TAR) notable extreme climate events. A lack of long and homogeneous observational data often makes it difficult to place some of these events in a longer-term context. The odds may have shifted to make some of them more likely than in an unchanging climate, but attribution of the change in odds typically requires extensive model experiments, a topic taken up in Chapter 9. It may be possible, however, to say that the occurrence of recent events is consistent with physically based expectations arising from climate change. Some examples of these recent events are described below (in response to the questions posed to IPCC by the governments) and placed in a long-term perspective.

Drought in Central and Southwest Asia, 1998–2003

Between 1999 and 2003 a severe drought hit much of southwest Asia, including Afghanistan, Kyrgyzstan, Iran, Iraq, Pakistan, Tajikistan, Turkmenistan, Uzbekistan and parts of Kazakhstan (Waple and Lawrimore, 2003; Levinson and Waple, 2004). Most of the area is a semiarid steppe, receiving precipitation only during winter and early spring through orographic capture of eastward-propagating mid-latitude cyclones from the Atlantic Ocean and the Mediterranean Sea (Martyn, 1992). Precipitation between 1998 and 2001 was on average less than 55% of the long-term average, making the drought conditions in 2000 the worst in 50 years (Waple et al., 2002). By June 2000, some parts of Iran had reported no measurable rainfall for 30 consecutive months. In December 2001 and January 2002, snowfall at higher altitudes brought relief for some areas, although the combination of above-average temperatures and early snowmelt, substantial rainfall and hardened ground desiccated by prolonged drought resulted in flash flooding during spring in parts of central and southern Iran, northern Afghanistan and Tajikistan. Other regions in the area continued to experience drought through 2004 (Levinson, 2005). In these years, an anomalous ridge in the upper-level circulation was a persistent feature during the cold season in central and southern Asia. The pattern served to both inhibit the development of baroclinic storm systems and deflect eastward-propagating storms to the north of the drought-affected area. Hoerling and Kumar (2003) linked the drought in certain areas of the mid-latitudes to common global oceanic influences. Both the prolonged duration of the 1998–2002 cold phase ENSO (La Niña) event and the unusually warm ocean waters in the western Pacific and eastern Indian Oceans appear to contribute to the severity of the drought (Nazemosadat and Cordery, 2000; Barlow et al., 2002; Nazemosadat and Ghasemi, 2004).

Drought in Australia, 2002–2003

A severe drought affected Australia during 2002, associated with a moderate El Niño event (Watkins, 2002). However, droughts in 1994 and 1982 were about as dry as the 2002 drought. Earlier droughts in the first half of the 20th century may well have been even drier. The 2002 drought came after several years of good rainfall (averaged across the country), rather than during an extended period of low rainfall such as occurred in the 1940s. If only rainfall is considered, the 2002 drought alone does not provide evidence of Australian droughts becoming more extreme. However, daytime temperatures during the 2002 drought were much higher than during previous droughts. The mean annual maximum temperature for 2002 was 0.5°C warmer than in 1994 and 1.0°C warmer than in 1982. So in this sense, the 2002 drought and associated heat waves were more extreme than the earlier droughts, because the impact of the low rainfall was exacerbated by high potential evaporation (Karoly et al., 2003; Nicholls, 2004). The very high maximum temperatures during 2002 could not simply be attributed to the low rainfall, although there is a strong negative correlation between rainfall and maximum temperature. Severe long-term drought, stemming from at least three years of rainfall deficits, continued during 2005, especially in the eastern third of Australia, although above-normal rainfall in winter and spring 2005 brought some relief. These conditions also have been accompanied by record high maximum temperatures over Australia during 2005 (a comparable national series is only available since 1951).

Drought in Western North America, 1999–2004

The western USA, southern Canada and northwest Mexico experienced a recent pervasive drought (Lawrimore et al., 2002), with dry conditions commencing as early as 1999 and persisting through the end of 2004 (Box 3.6, Figure 1). Drought conditions were recorded by several hydrologic measures, including precipitation, streamflow, lake and reservoir levels and soil moisture (Piechota et al., 2004). The period 2000 through 2004 was the first instance of five consecutive years of below-average flow in the Colorado River since the beginning of modern records in 1922 (Pagano et al., 2004). Cook et al. (2004) provided a longer-term context for this drought. In the western conterminous USA, the area under moderate to extreme drought, as given by the PDSI, rose above 20% in November 1999 and stayed above this level persistently until October 2004. At its peak (August 2002), this drought affected 87% of the West (Rocky Mountains westward), making it the second most extensive and one of the longest droughts in the last 105 years. The impacts of this drought have been exacerbated by depleted or earlier than average melting of the mountain snowpack, due to warm springs,

(continued)

as observed changes in timing from 1948 to 2000 trended earlier by one to two weeks in many parts of the West (Cayan et al., 2001; Regonda et al., 2005; Stewart et al., 2005). Within this episode, the spring of 2004 was unusually warm and dry, resulting in record early snowmelt in several western watersheds (Pagano et al., 2004).

Hoerling and Kumar (2003) attributed the drought to changes in atmospheric circulation associated with warming of the western tropical Pacific and Indian oceans, while McCabe et al. (2004) have produced evidence suggesting that the confluence of both Pacific decadal and Atlantic multi-decadal fluctuations is involved. In the northern winter of 2004 to 2005, the weak El Niño was part of a radical change in atmospheric circulation and storm track across the USA, ameliorating the drought in the Southwest, although lakes remain low.

Floods in Europe, Summer 2002

A catastrophic flood occurred along several central European rivers in August 2002. The floods resulting from extraordinarily high precipitation were enhanced by the fact that the soils were completely saturated and the river water levels were already high because of previous rain (Rudolf

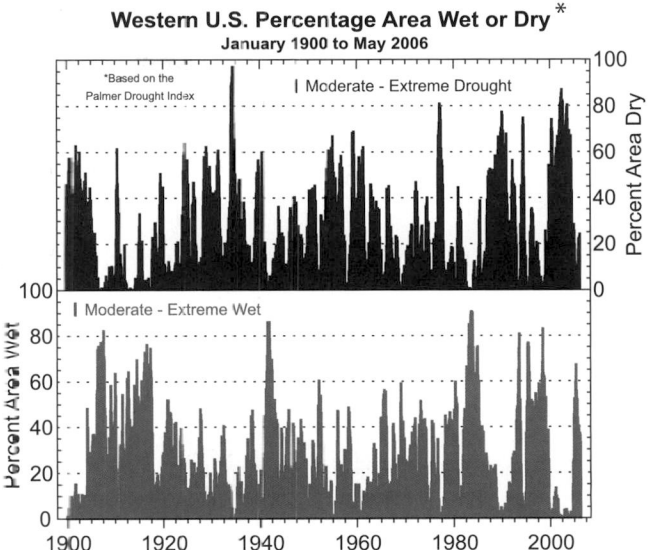

Box 3.6, Figure 1. *Percentage of the USA west of the Rocky Mountains (the 11 states west of and including Montana to New Mexico) that was dry (top) or wet (bottom), based on the Palmer Drought Severity Index for classes of moderate to extreme drought or wet. From NOAA, NCDC.*

and Rapp, 2003; Ulbrich et al., 2003a,b). Hence, it was part of a pattern of weather over an extended period. In the flood, the water levels of the Elbe at Dresden reached a maximum mark of 9.4 m, which is the highest level since records began in 1275 (Ulbrich et al., 2003a). Some small villages in the Ore Mountains (on tributaries of the Elbe) were hit by extraordinary flash floods. The river Vltava inundated the city of Prague before contributing to the Elbe flood. A return period of 500 years was estimated for the flood levels at Prague (Grollmann and Simon, 2002). The central European floods were caused by two heavy precipitation episodes. The first, on 6–7 August, was situated mainly over Lower Austria, the southwestern part of the Czech Republic and southeastern Germany. The second took place on 11–13 August 2002 and most severely affected the Ore Mountains and western parts of the Czech Republic. A persistent low-pressure system moved slowly from the Mediterranean Sea to central Europe on a path over or near the eastern Alps and led to large-scale, strong and quasi-stationary frontal lifting of air with very high liquid water content. Additional to this advective rain were convective precipitation processes (showers and thunderstorms) and a significant orographic lifting (mainly over the Ore Mountains). A maximum 24-hour precipitation total of 353 mm was observed at the German station Zinnwald-Georgenfeld, a new record for Germany. The synoptic situation leading to floods is well known to meteorologists of the region. Similar situations led to the summer floods of the River Oder in 1997 and the River Vistula in 2001 (Ulbrich et al., 2003b). Average summer precipitation trends in the region are negative but barely significant (Schönwiese and Rapp, 1997) and there is no significant trend in flood occurrences of the Elbe within the last 500 years (Mudelsee et al., 2003). However, the observed increase in precipitation variability at a majority of German precipitation stations during the last century (Trömel and Schönwiese, 2005) is indicative of an enhancement of the probability of both floods and droughts.

Heat Wave in Europe, Summer 2003

The heat wave that affected many parts of Europe during the course of summer 2003 produced record-breaking temperatures particularly during June and August (Beniston, 2004; Schär et al., 2004; see Box 3.6, Figure 2). Absolute maximum temperatures exceeded the record highest temperatures observed in the 1940s and early 1950s in many locations in France, Germany, Switzerland, Spain, Italy and the UK, according to the information supplied by national weather agencies (WMO, 2004). Gridded instrumental temperatures (from CRUTEM2v for the region 35°N–50°N, 0–20°E) show that the summer was the hottest since comparable records began in 1780: 3.8°C above the 1961 to 1990 average and 1.4°C hotter than any other summer in this period (the second hottest was 1807). Based on early documentary records, Luterbacher et al. (2004) estimated that 2003 is very likely to have been the hottest summer since at least 1500. The 2003 heat wave was associated with a very robust and persistent blocking high-pressure system that may be a manifestation of an exceptional northward extension of the Hadley Cell (Black et al., 2004; Fink et al., 2004). Already a record month in terms of maximum temperatures, June exhibited high geopotential values that penetrated northwards towards the British Isles, with the

(continued)

greatest northward extension and longest persistence of record-high temperatures observed in August. An exacerbating factor for the temperature extremes was the lack of precipitation in many parts of western and central Europe, leading to much-reduced soil moisture and surface evaporation and evapotranspiration, and thus to a strong positive feedback effect (Beniston and Diaz, 2004).

The 2005 Tropical Storm Season in the North Atlantic

The 2005 North Atlantic hurricane season (1 June to 30 November) was the most active on record by several measures, surpassing the very active season of 2004 (e.g., Levinson, 2005) and causing an unprecedented level of damage. Even before the peak in the seasonal activity, the seven tropical storms in June and July were the most ever, and hurricane Dennis was the strongest on record in July and the earliest ever fourth-named storm. The record 2005 North Atlantic hurricane season featured the largest number of named storms (28; sustained winds greater than 17 m s^{-1}) and is the only time names have ventured into the Greek alphabet. It had the largest number of hurricanes (15; sustained winds greater than 33 m s^{-1}) recorded, and is the only time there have been four category 5 storms (maximum sustained winds greater than 67 m s^{-1}). These included the most intense Atlantic storm on record (Wilma, with recorded surface pressure in the eye of 882

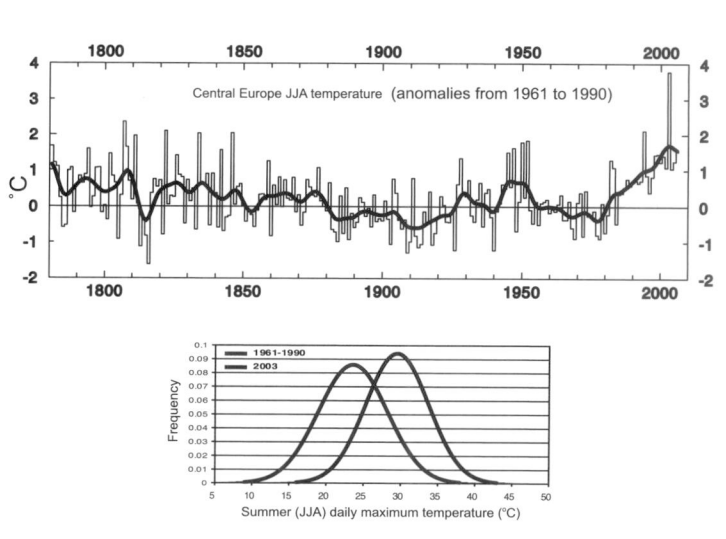

Box 3.6, Figure 2. *Long time series of JJA temperature anomalies in Central Europe relative to the 1961 to 1990 mean (top). The smooth curve shows decadal variations (see Appendix 3.A). In the summer of 2003, the value of 3.8°C far exceeded the next largest anomaly of 2.4°C in 1807, and the highly smoothed Gaussian distribution (bottom) of maximum temperatures (red) compared with normal (blue) at Basel, Switzerland (Beniston and Diaz, 2004) shows how the whole distribution shifted.*

hPa), the most intense storm in the Gulf of Mexico (Rita, 897 hPa), and Katrina. Tropical storm Vince was the first ever to make landfall in Portugal and Spain. In spite of these metrics, the ACE index, although very high and surpassing the 2004 value (Figure 3.40), was not the highest on record, as several storms were quite short lived. Six of the eight most damaging storms on record for the USA occurred from August 2004 to September 2005 (Charlie, Ivan, Francis, Katrina, Rita, Wilma) while another storm in 2005 (Stan) caused severe flooding and mudslides as well as about 2,000 fatalities in central America (Guatemala, El Salvador and southern Mexico).

SSTs in the tropical North Atlantic region critical for hurricanes (10°N to 20°N) were at record-high levels (0.9°C above the 1901 to 1970 normal) in the extended summer (June to October) of 2005 (Figure 3.33), and these high values were a major reason for the very active hurricane season along with favourable atmospheric conditions (see Box 3.5). A substantial component of this warming was the global mean SST increase (Trenberth and Shea, 2006; see Sections 3.2 and 3.6.6).

estimated by the U.S. National Hurricane Center, of near 40 m s^{-1}, causing much damage to property and some loss of life (see Levinson, 2005). The Brazilian meteorologists dubbed it 'Catarina'. This event appears to be unprecedented although records are poor before the satellite era. Pezza and Simmonds (2005) suggest that a key factor in the hurricane development was the more favourable atmospheric circulation regime associated with the positive trend in the SAM (see Section 3.6).

3.8.4 Evidence for Changes in Extratropical Storms and Extreme Events

3.8.4.1 Extratropical Cyclones

Intense extratropical cyclones are often associated with extreme weather, particularly with severe windstorms. Significant increases in the number or strength of intense extratropical cyclone systems have been documented in a number of studies (e.g., Lambert, 1996; Gustafsson, 1997; McCabe et al., 2001; Wang et al., 2006a) with associated changes in the preferred tracks of storms as described in Section 3.5.3. As with tropical cyclones, detection of long-term changes in cyclone measures is hampered by incomplete and changing observing systems. Some earlier results have been questioned because of changes in the observation system (e.g., Graham and Diaz, 2001).

Results from NRA and ERA-40 show that an increase in the number of deep cyclones is apparent over the North Pacific and North Atlantic (Graham and Diaz, 2001; Gulev et al., 2001), with statistically significant winter increases over both ocean basins (Simmonds and Keay, 2002; Wang et al., 2006a). Geng and Sugi (2001) found that cyclone density, deepening rate, central pressure gradient and translation speed have all been increasing in the winter North Atlantic. Caires and Sterl (2005) compared global estimates of 100-year return values of wind speed and SWH in ERA-40, with linear bias corrections based on buoy data, for three different 10-year periods. They showed that the differences in the storm tracks can be attributed to decadal variability in the NH, linked to changes in global circulation patterns, most notably to the NAO (see also Section 3.5.6).

Using NCEP-2 reanalysis data, Lim and Simmonds (2002) showed that for 1979 to 1999, increasing trends in the annual number of explosively developing (deepening by 1 hPa per hour or more) extratropical cyclones are significant in the SH and over the globe (0.56 and 0.78 more systems per year, respectively), while the positive trend did not achieve significance in the NH. Simmonds and Keay (2002) obtained similar results for the change in the number of cyclones in the decile for deepest cyclones averaged over the North Pacific and over the North Atlantic in winter over the period 1958 to 1997.

As noted in Sections 3.5.3 and 3.5.7, the time-dependent biases in the reanalysis cause uncertainties in the trends reported above. Besides reanalyses, station data may also be used to indicate evidence for changes in extratropical cyclone activity. Instead of direct station wind measurements, which may suffer from a lack of consistency of instrumentation, methodology and exposure, values based on pressure gradients have been derived that are more reliable for discerning long-term changes. Alexandersson et al. (2000) used station pressure observations for 21 stations over northwestern Europe back to 1881, from which geostrophic winds were calculated using 'pressure-triangle' methods. They found a decline of storminess expressed by the 95th and 99th percentiles from high levels during the late 19th century to a minimum around 1960 and then a quite rapid increase to a maximum around 1990, followed again by a decline (Figure 3.41). Positive NAO winters are typically associated with more intense and frequent storms (see Section 3.6.4). Similar results were obtained by Schmith et al. (1998) using simpler indices based on pressure tendency. Bärring and von Storch (2004), using both pressure tendencies and the number of very low pressure values, confirmed these results on the basis of two especially long station series in southern Sweden dating back to about 1800. Studies of rapid pressure changes at stations indicate an increase in the frequency, duration and intensity of winter cyclone activity over the lower Canadian Arctic and in the number and intensity of severe storms over the southern

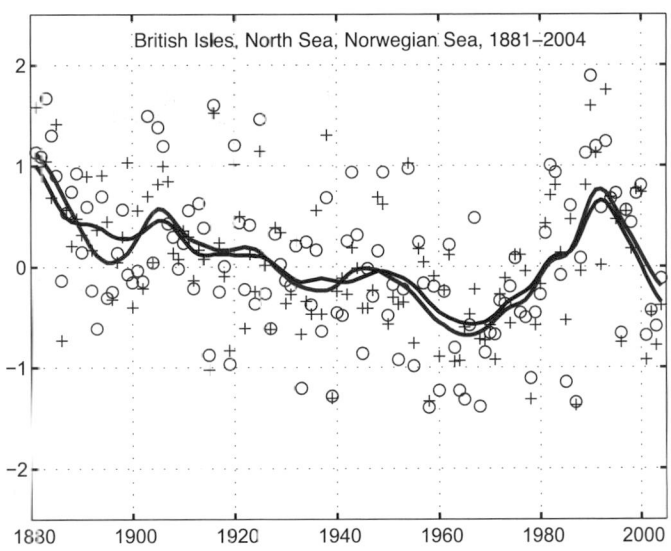

Figure 3.41. Storm index for the British Isles, North Sea and Norwegian Sea, 1881 to 2004. Blue circles are 95th percentiles and red crosses 99th percentiles of standardised geostrophic winds averaged over 10 sets of triangles of stations. The smoothed curves are a decadal filter (updated from Alexandersson et al., 2000).

UK since the 1950s, but a decrease over southern Canada and Iceland (Wang et al., 2006b; Alexander et al., 2005). Besides a northward shift of the storm track (see 3.5.3), the station pressure data for parts of the North Atlantic region show a modest increase in severe storms in recent decades. However, decadal-scale fluctuations of similar magnitude have occurred earlier in the 19th and 20th centuries.

Direct surface wind measurements have, however, been used in a few studies. An analysis of extreme pressure differences and surface winds (Salinger et al., 2005) showed a significant increasing trend over the last 40 years in westerly wind extremes over the southern part of New Zealand and the oceans to the south. The trends are consistent with the increased frequency of El Niño events in recent decades, associated with Pacific decadal variability (see Section 3.6.3). While the zonal pressure gradient and extreme westerly wind frequency have both increased over southern New Zealand, the frequency of extreme easterly winds has also increased there, suggesting more variability in the circulation generally. However, trends in pressure differences (based on the ERA-40, NRA and station data) are not always consistent with changes in surface windiness (e.g., Smits et al., 2005). Based on observed winds at 10 m height over the Netherlands, Smits et al. (2005) found a decline in strong (greater than about 8 on the Beaufort scale) wind events over the last 40 years. Differences cannot entirely be explained by changes in surface aerodynamic roughness, and they concluded that inhomogeneities in the reanalyses are the cause. However, local differences can be important and intensity and severity of storms may not always be synonymous with local extreme surface winds and gusts.

Table 3.7. *Definition of phenomena used to assess extremes in Table 3.8*

PHENOMENON	Definition
Low-temperature days/ nights and frost days	Percentage of days with temperature (maximum for days, minimum for nights) not exceeding some threshold, either fixed (frost days) or varying regionally (cold days/cold nights), based on the 10th percentile of the daily distribution in the reference period (1961–1990).
High-temperature days/nights	See low-temperature days/nights, but now exceeding the 90th percentile.
Cold spells/snaps	Episode of several consecutive low-temperature days/nights.
Warm spells (heat waves)	Episode of several consecutive high-temperature days/nights.
Cool seasons/warm seasons	Seasonal averages (rather than daily temperatures) exceeding some threshold.
Heavy precipitation events (events that occur every year)	Percentage of days (or daily precipitation amount) with precipitation exceeding some threshold, either fixed or varying regionally, based on the 95th or 99th percentile of the daily distribution in the reference period (1961–1990).
Rare precipitation events (with return periods >~10 yr)	As for heavy precipitation events, but for extremes further into the tail of the distribution.
Drought (season/year)	Precipitation deficit; or based on the PDSI (see Box 3.1).
Tropical cyclones (frequency, intensity, track, peak wind, peak precipitation)	Tropical storm with thresholds crossed in terms of estimated wind speed and organisation. Hurricanes in categories 1 to 5, according to the Saffir-Simpson scale, are defined as storms with wind speeds of 33 to 42 m s^{-1}, 43 to 49 m s^{-1}, 50 to 58 m s^{-1}, 59 to 69 m s^{-1}, and >70 m s^{-1}, respectively. NOAA's ACE index is a measure of the total seasonal activity that accounts for the collective intensity and duration of tropical storms and hurricanes during a given tropical cyclone season.
Extreme extratropical storms (frequency, intensity, track, surface wind, wave height)	Intense low-pressure systems that occur throughout the mid-latitudes of both hemispheres fueled by temperature gradients and acting to reduce them.
Small-scale severe weather phenomena	Extreme events, such as tornadoes, hail, thunderstorms, dust storms and other severe local weather.

Table 3.8. *Change in extremes for phenomena over the specified region and period, with the level of confidence and section where the phenomenon is discussed in detail.*

PHENOMENON	Change	Region	Period	Confidence	Section
Low-temperature days/nights and frost days	Decrease, more so for nights than days	Over 70% of global land area	1951–2003 (last 150 years for Europe and China)	*Very likely*	3.8.2.1
High-temperature days/nights	Increase, more so for nights than days	Over 70% of global land area	1951–2003	*Very likely*	3.8.2.1
Cold spells/snaps (episodes of several days)	Insufficient studies, but daily temperature changes imply a decrease				
Warm spells (heat waves) (episodes of several days)	Increase: implicit evidence from changes of daily temperatures	Global	1951–2003	*Likely*	FAQ 3.3
Cool seasons/ warm seasons (seasonal averages)	Some new evidence for changes in inter-seasonal variability	Central Europe	1961–2004	*Likely*	3.8.2.1
Heavy precipitation events (that occur every year)	Increase, generally beyond that expected from changes in the mean (disproportionate)	Many mid-latitude regions (even where reduction in total precipitation)	1951–2003	*Likely*	3.8.2.2
Rare precipitation events (with return periods > ~10 yr)	Increase	Only a few regions have sufficient data for reliable trends (e.g., UK and USA)	Various since 1893	*Likely* (consistent with changes inferred for more robust statistics)	3.8.2.2
Drought (season/year)	Increase in total area affected	Many land regions of the world	Since 1970s	*Likely*	3.3.4 and FAQ 3.3
Tropical cyclones	Trends towards longer lifetimes and greater storm intensity, but no trend in frequency	Tropics	Since 1970s	*Likely;* more confidence in frequency and intensity	3.8.3 and FAQ 3.3
Extreme extratropical storms	Net increase in frequency/intensity and poleward shift in track	NH land	Since about 1950	*Likely*	3.8.4, 3.5, and FAQ 3.3
Small-scale severe weather phenomena	Insufficient studies for assessment				

3.8.4.2 *Tornadoes, Hail, Thunderstorms, Dust Storms and Other Severe Local Weather*

Evidence for changes in the number or intensity of tornadoes relies entirely on local reports. In the USA, databases for tornado reporting are well established, although changes in procedures for evaluating the intensity of tornadoes introduced significant discontinuities in the record. In particular, the apparent decrease in strong tornadoes in the USA from the early period of the official record (1950s–1970s) to the more recent period is, in large part, a result of the way damage from the earlier events was evaluated. Trapp et al. (2005) also questioned the completeness of the tornado record and argued that about 12% of squall-line tornadoes remain unreported. In many European countries, the number of tornado reports has increased considerably over the last decade (Snow, 2003), leading to a much higher estimate of tornado activity (Dotzek, 2003). Bissolli et al. (2007) showed that the increase in Germany between 1950 and 2003 mainly concerns weak tornadoes (F0 and F1 on the Fujita scale), thus paralleling the evolution of tornado reports in the USA after 1950 (see, e.g., Dotzek et al., 2005) and making it likely that the increase in reports in Europe is at least dominated (if not solely caused) by enhanced detection and reporting efficiency. Doswell et al. (2005) highlighted the difficulties encountered when trying to find observational evidence for changes in extreme events at local scales connected to severe thunderstorms. In light of the very strong spatial variability of small-scale severe weather phenomena, the density of surface meteorological observing stations is too coarse to measure all such events. Moreover, homogeneity of existing station series is questionable. While remote sensing techniques allow detection of thunderstorms even in remote areas, they do not always uniquely identify severe weather events from these storms. Another approach links severe thunderstorm occurrence to larger-scale environmental conditions in places where the observations of events are fairly good and then consider the changes in the distribution of those environments (Brooks et al., 2003; Bissolli et al., 2007).

Although a decreasing trend in dust storms was observed from the mid-1950s to the mid-1990s in northern China, the number of dust storm days increased from 1997 to 2002 (Li and Zhai, 2003; Zhou and Zhang, 2003). The decreasing trend appears linked to the reduced cyclone frequency and increasing winter (DJF) temperatures (Qian et al., 2002). The recent increase is associated with vegetation degradation and drought, plus increased surface wind speed (Wang and Zhai, 2004; Zou and Zhai, 2004).

3.8.5 Summary

Even though the archived data sets are not yet sufficient for determining long-term trends in extremes, there are new findings on observed changes for different types of extremes. The definitions of the phenomena are summarised in Table 3.7. A summary of the changes in extremes by phenomena, region and time is given in Table 3.8 along with an assessment of the confidence in these changes.

New analyses since the TAR confirm the picture of a gradual reduction of the number of frost days over most of the mid-latitudes in recent decades. In agreement with this warming trend, the number of warm nights increased between 1951 and 2003, cold nights decreased, and trends in the number of cold and warm days are also consistent with warming, but are less marked than at night.

For precipitation, analysis of updated trends and results for regions that were missing at the time of the TAR show increases in heavy events for the majority of observation stations, with some increase in flooding. This result applies both for areas where total precipitation has increased and for areas where total precipitation has even decreased. Increasing trends are also reported for more rare precipitation events, although results for such extremes are available only for a few areas. Mainly because of lack of data, it remains difficult to draw a consistent picture of changes in extreme precipitation for the tropics and subtropics.

Tropical cyclones, hurricanes and typhoons exhibit large variability from year to year and limitations in the quality of data compromise evaluations of trends. Nonetheless, clear evidence exists for increases in category 4 and 5 storms globally since 1970 along with increases in the PDI due to increases in intensity and duration of storms. The 2005 season in the North Atlantic broke many records. The global view of tropical storm activity highlights the important role of ENSO in all basins, and the most active year was 1997, when a very strong El Niño began, suggesting that observed record high SSTs played a key role.

For extratropical cyclones, positive trends in storm frequency and intensity dominate for recent decades in most regional studies performed. Longer records for the northeastern Atlantic suggest that the recent extreme period may be similar in level to that of the late 19th century.

As noted in Section 3.3.4, the PDSI shows a large drying trend over NH land areas since the mid-1950s and a drying trend in the SH from 1974 to 1998. Decreases in land precipitation, especially since the early 1980s are the main cause for the drying trends, although large surface warming during the last two to three decades has also likely contributed to the drying.

3.9 Synthesis: Consistency Across Observations

This section briefly compares variability and trends within and across different climate variables to see if a physically consistent picture enhances confidence in the realism of apparent recent observed changes. Therefore, this section looks ahead to the subsequent observational chapters on the cryosphere (Chapter 4) and oceans (Chapter 5), which focus on changes in those domains. The emphasis here is on inter-relationships. For example, increases in temperature should enhance the moisture-holding capacity of the atmosphere as a whole and changes in temperature and/or precipitation should be consistent with those evident in circulation indices. Variables treated in this chapter are summarised in the executive summary, with some discussion below. The example of increases in temperature that should also reduce snow seasons and sea ice and cause widespread glacier retreat involves cross-chapter variables. The main sections where more detailed information can be found are given in parentheses following each bullet.

- The observed temperature increases are consistent with the observed nearly worldwide reduction in glacier and small ice cap (not including Antarctica and Greenland) mass and extent in the 20th century. Glaciers and ice caps respond not only to temperatures but also to changes in precipitation, and both winter accumulation and summer melting have increased over the last half century in association with temperature increases. In some regions, moderately increased accumulation observed in recent decades is consistent with changes in atmospheric circulation and associated increases in winter precipitation (e.g., southwestern Norway, parts of coastal Alaska, Patagonia, Karakoram, and Fjordland of the South Island of New Zealand), even though enhanced ablation has led to marked declines in mass balances in Alaska and Patagonia. Tropical glacier changes are synchronous with higher-latitude ones and all have shown declines in recent decades; local temperature records all show a slight warming, but not of the magnitude required to explain the rapid reduction in mass of such glaciers (e.g., on Kilimanjaro). Other factors in recent ablation include changes in cloudiness, water vapour, albedo due to snowfall frequency and the associated radiation balance (Sections 3.2.2, 3.3.3, 3.4.3 and 4.5).

- Snow cover has decreased in many NH regions, particularly in spring, consistent with greater increases in spring as opposed to autumn temperatures in mid-latitude regions, and more precipitation falling as rain instead of snow. These changes are consistent with changes in permafrost: temperatures of the permafrost in the Arctic and subarctic have increased by up to 3°C since the 1980s with permafrost warming also observed on the Tibetan Plateau and in the European mountain permafrost regions. Active layer thickness has increased and seasonally frozen ground depth has decreased over the Eurasian continent (Sections 3.2.2, 3.3.2, 4.2 and 4.8).

- Sea ice extents have decreased in the Arctic, particularly in spring and summer, and patterns of the changes are consistent with regions showing a temperature increase, although changes in winds are also a major factor. Sea ice extents were at record low values in 2005, which was also the warmest year since records began in 1850 for the Arctic north of 65°N. There have also been decreases in sea ice thickness. In contrast to the Arctic, antarctic sea ice does not exhibit any significant trend since the end of the 1970s, which is consistent with the lack of trend in surface temperature south of 65°S over that period. However, along the Antarctic Peninsula, where significant warming has occurred, progressive breakup of ice shelves has occurred beginning in the late 1980s, culminating in the breakup of the Larsen-B ice shelf in 2002. Decreases are found in the length of the freeze season of river and lake ice (Sections 3.2.2, 3.6.5, 4.3 and 4.4).

- Radiation changes at the top of the atmosphere from the 1980s to 1990s, possibly ENSO-related in part, appear to be associated with reductions in tropical cloud cover, and are linked to changes in the energy budget at the surface and in observed ocean heat content in a consistent way (Sections 3.4.3, 3.4.4, 3.6.2 and 5.2.2).

- Reported decreases in solar radiation from 1970 to 1990 at the surface have an urban bias. Although records are sparse, pan evaporation is estimated to have decreased in many places due to decreases in surface radiation associated with increases in clouds, changes in cloud properties and/or increases in air pollution (aerosol), especially from 1970 to 1990. There is evidence to suggest that the solar radiation decrease has reversed in recent years (Sections 2.4.5, 2.4.6, 3.3.3, 3.4.4, 7.2 and 7.5).

- Droughts have increased in spatial extent in various parts of the world. The regions where they have occurred seem to be determined largely by changes in SSTs, especially in the tropics, through changes in the atmospheric circulation and precipitation. Inferred enhanced evapotranspiration and drying associated with warming are additional factors in drought increases, but decreased precipitation is the dominant factor. In the western USA, diminishing snowpack and subsequent summer soil moisture reductions have also been a factor. In Australia and Europe, direct links to warming have been inferred through the extreme nature of high temperatures and heat waves accompanying drought (Sections 3.3.4 and 4.2, FAQ 3.2 and Box 3.6).

- Changes in the freshwater balance of the Atlantic Ocean over the past four decades have been pronounced, as freshening has occurred in the North Atlantic and south of 25°S, while salinity has increased in the tropics and subtropics, especially in the upper 500 m. The implication is that there have been increases in moisture transport by the atmosphere from the subtropics to higher latitudes, in association with changes in atmospheric circulation, including the NAO, thereby producing the observed increases in precipitation over the northern oceans and adjacent land areas (Sections 3.3.2, 3.6.4, 5.3.2 and 5.5.3).

- Sea level likely rose 1.7 ± 0.5 mm yr^{-1} during the 20th century, but the rate increased to 3.1 ± 0.7 mm yr^{-1} from 1993 through 2003, when confidence increases from global altimetry measurements. Increases in ocean heat content and associated ocean expansion are estimated to have contributed 0.4 ± 0.1 mm yr^{-1} from 1961 to 2003, increasing to an estimated value of 1.6 ± 0.5 mm yr^{-1} for 1993 to 2003. In the same interval, glacier and land ice melt has increased ocean mass by approximately 1.2 ± 0.4 mm yr^{-1}. Changes in land water storage are uncertain but may have reduced water in the ocean. The near balance for 1993 to 2003 gives increased confidence that the observed sea level rise is a strong indicator of warming, and an integrator of the cumulative energy imbalance at the top of atmosphere (Sections 4.5, 4.6, 4.8, 5.2 and 5.5).

In summary, global mean temperatures have increased since the 19th century, especially since the mid-1970s. Temperatures have increased nearly everywhere over land, and SSTs and marine air temperatures have also increased, reinforcing the evidence from land. However, temperatures have increased neither monotonically nor in a spatially uniform manner, especially over shorter time intervals. The atmospheric circulation has also changed: in particular, increasing zonal flow is observed in most seasons in both hemispheres, and the mid- to high-latitude annular modes strengthened until the mid-1990s in the NH and up until the present in the SH. In the NH, this brought milder maritime air into Europe and much of high-latitude Asia from the North Atlantic in winter, enhancing warming there. In the SH, where the ozone hole has played a role, it has resulted in cooling over 1971 to 2000 for parts of the interior of Antarctica but large warming in the Antarctic Peninsula region and Patagonia. Temperatures generally have risen more than average where flow has become more poleward, and less than average or even cooled where flow has become more equatorward, reflecting the PDO and other patterns of variability.

Over land, a strong negative correlation is observed between precipitation and surface temperature in summer and at low latitudes throughout the year, and areas that have become wetter, such as the eastern USA and Argentina, have not warmed as much as other land areas (see especially FAQs 3.2 and 3.3). Increased precipitation is associated with increases in cloud and surface wetness, and thus increased evapotranspiration. The inferred increased evapotranspiration and reduced temperature increase is physically consistent with enhanced latent vs. sensible heat fluxes from the surface in wetter conditions.

Consistent with the expectations noted above for a warmer climate, surface specific humidity has generally increased after 1976 in close association with higher temperatures over both land and ocean. Total column water vapour has increased over the global oceans by $1.2 \pm 0.3\%$ per decade from 1988 to 2004, consistent in patterns and amount with changes in SST and a fairly constant relative humidity. Upper-tropospheric water vapour has also increased in ways such that relative humidity remains about constant, providing a major positive feedback to radiative forcing. In turn, widespread observed increases in the fraction of heavy precipitation events are consistent with the increased water vapour amounts.

The three main ocean basins are unique and contain very different wind systems, SST patterns and currents, leading to vastly different variability associated, for instance, with ENSO in the Pacific, and the THC in the Atlantic. Consequently, the oceans have not warmed uniformly, especially at depth. SSTs in the tropics have warmed at different rates and help drive, through coupling with tropical convection and winds, teleconnections around the world. This has changed the atmospheric circulation through ENSO, the PDO, the AMO, monsoons and the Hadley and Walker Circulations. Changes in precipitation and storm tracks are not as well documented but clearly respond to these changes on interannual and decadal time scales. When precipitation increases over the ocean, as it has in recent years in the tropics, it decreases over land, although it has increased over land at higher latitudes. Droughts have increased over many tropical and mid-latitude land areas, in part because of decreased precipitation over land since the 1970s but also from increased evapotranspiration arising from increased atmospheric demand associated with warming.

Changes in the cryosphere (Chapter 4), ocean and land strongly support the view that the world is warming, through observed decreases in snow cover and sea ice, thinner sea ice, shorter freezing seasons of lake and river ice, glacier melt, decreases in permafrost extent, increases in soil temperatures and borehole temperature profiles (see Section 6.6), and sea level rise (Section 5.5).

Acknowledgments

The authors gratefully acknowledge the valuable assistance of Sara Veasey (NCDC, Asheville) and Lisa Butler (NCAR, Boulder) in the development of diagrams and text formatting.

References

Abakumova, G.M., et al., 1996: Evaluation of long-term changes in radiation, cloudiness, and surface temperature on the territory of the former Soviet Union. *J. Clim.*, **9**, 1319–1327.

Adler, R.F., et al., 2001: Intercomparison of global precipitation products: The Third Precipitation Intercomparison Project (PIP-3). *Bull. Am. Meteorol. Soc.*, **82**, 1377–1396.

Adler, R.F., et al., 2003: The version 2 Global Precipitation Climatology Project (GPCP) monthly precipitation analysis (1979–present). *J. Hydrometeorol.*, **4**, 1147–1167.

Agudelo, P.A., and J.A. Curry, 2004: Analysis of spatial distribution in tropospheric temperature trends. *Geophys. Res. Lett.*, **31**, L22207, doi:10.1029/2004GL02818.

Aguilar, E., et al., 2005: Changes in precipitation and temperature extremes in Central America and northern South America, 1961–2003. *J. Geophys. Res.*, **110**, D23107, doi:10.1029/2005JD006119.

Aiyyer, A.R., and C. Thorncroft, 2006: Climatology of vertical wind shear over the tropical Atlantic. *J. Clim.*, **19**, 2969–2983.

Aksoy, B., 1997: Variations and trends in global solar radiation for Turkey. *Theor. Appl. Climatol.*, **58**, 71–77.

Alexander, L.V., S.F.B. Tett, and T. Jónsson, 2005: Recent observed changes in severe storms over the United Kingdom and Iceland. *Geophys. Res. Lett.*, **32**, L13704, doi:10.1029/2005GL022371.

Alexander, L.V., et al., 2006: Global observed changes in daily climate extremes of temperature and precipitation. *J. Geophys. Res.*, **111**, D05109, doi:10.1029/2005JD006290.

Alexandersson, H., et al., 1998: Long-term variations of the storm climate over NW Europe. *Global Atmos. Ocean System*, **6**, 97–120.

Alexandersson, H., et al., 2000: Trends of storms in NW Europe derived from an updated pressure data set. *Clim. Res.*, **14**, 71–73.

Allan, J., and P. Komar, 2000: Are ocean wave heights increasing in the eastern North Pacific? *Eos*, **81**, 561–567.

Allan, R.J., and T. Ansell, 2006: A new globally complete monthly historical gridded mean sea level pressure data set (HadSLP2); 1850–2003. *J. Clim.*, **19**, 5816–5842.

Allan, R.P., and A. Slingo, 2002: Can current climate model forcings explain the spatial and temporal signatures of decadal OLR variations? *Geophys. Res. Lett.*, **29**, 1141, doi:10.1029/2001GL014620.

Allan, R.P., A. Slingo, and V. Ramaswamy, 2002: Analysis of moisture variability in the European Centre for Medium-Range Weather Forecasts 15-year reanalysis over tropical oceans. *J. Geophys. Res.*, **107**, doi:10.1029/2001JD001132.

Allan, R.P., M.A. Ringer, and A. Slingo, 2003: Evaluation of moisture in the Hadley Centre Climate Model using simulations of HIRS water vapour channel radiances. *Q. J. R. Meteorol. Soc.*, **128**, 1–18.

Allan, R.P., et al., 2001: Is there an equatorial Indian Ocean dipole, and is it independent of the El Niño Southern Oscillation? *CLIVAR Exchanges*, **21**, 1–3.

Allan, R.P., et al., 2004: Simulation of the Earth's radiation budget by the European Centre for Medium-Range Weather Forecasts 40-year reanalysis (ERA40). *J. Geophys. Res.*, **109**, D18107, doi:10.1029/2004JD004816.

Allen, M.R., and W.J. Ingram, 2002: Constraints on future changes in climate and the hydrological cycle. *Nature*, **419**, 2224–2232.

Alley, W.M., 1984: The Palmer Drought Severity Index: limitation and assumptions. *J. Clim. Appl. Meteorol.*, **23**, 1100–1109.

Alpert, P., et al., 2002: The paradoxical increase of Mediterranean extreme daily rainfall in spite of decrease in total values. *Geophys. Res. Lett.*, **29**(11), doi:10.1029/2001GL013554.

Alpert, P., et al., 2005: Global dimming or local dimming?: Effect of urbanization on sunlight availability. *Geophys. Res. Lett.*, **32**, L17802, doi:10.1029/2005GL023320.

Ambaum, M.H.P., B.J. Hoskins, and D.B. Stephenson, 2001: Arctic Oscillation or North Atlantic Oscillation? *J. Clim.*, **14**, 3495–3507.

Ambaum, M.H.P., B.J. Hoskins, and D.B. Stephenson, 2002: Corrigendum: Arctic Oscillation or North Atlantic Oscillation? *J. Clim.*, **15**, 553.

Andrea, F., et al., 1998: Northern Hemisphere atmospheric blocking as simulated by 15 atmospheric general circulation models in the period 1979–1988. *Clim. Dyn.*, **14**, 385–407.

Andreae, M.O., et al., 2004: Smoking rain clouds over the Amazon. *Science*, **303**, 1337–1342.

Andrews, D.G., J.R. Holton, and C.B. Leovy, 1987: *Middle Atmosphere Dynamics*. Academic Press, San Diego, CA, 489 pp.

Angell, J.K., 2003: Effect of exclusion of anomalous tropical stations on temperature trends from a 63-station radiosonde network, and comparison with other analyses. *J. Clim.*, **16**, 2288–2295.

Angell, J.K., 2006: Changes in the 300 mb north circumpolar vortex, 1963–2001. *J. Clim.*, **19**, 2984–2994.

Aoki, S., 2002: Coherent sea level response to the Antarctic Oscillation. *Geophys. Res. Lett.*, **29**, 1950, doi:10.1029/2002GL15733.

Ashok, K., Z. Guan, and T. Yamagata, 2001: Impact of the Indian Ocean Dipole on the decadal relationship between the Indian monsoon rainfall and ENSO. *Geophys. Res. Lett.*, **28**, 4499–4502.

Assell, R., K. Cronk, and D. Norton, 2003: Recent trends in Laurentian Great Lakes ice cover. *Clim. Change*, **57**, 185–204.

Auer, I., et al., 2007: HISTALP -- Historical Instrumental Climatological Surface Time Series of the Greater Alpine Region. *Int. J. Climatol.*, **27**, 17-46.

Baldwin, M.P., 2001: Annular modes in global daily surface pressure. *Geophys. Res. Lett.*, **28**, 4115–4118.

Baldwin, M.P., and T. Dunkerton, 1999: Downward propagation of the Arctic Oscillation from the stratosphere to the troposphere. *J. Geophys. Res.*, **104**, 30937–30946.

Baldwin, M.P., and T.J. Dunkerton, 2001: Stratospheric harbingers of anomalous weather regimes. *Science*, **244**, 581–584.

Baldwin, M.P., et al., 2003: Stratospheric memory and extended-range weather forecasts. *Science*, **301**, 636–640.

Balling, R.C. Jr., and S. Brazel, 1987: Recent changes in Phoenix summertime diurnal precipitation patterns. *Theor. Appl. Climatol.*, **38**, 50–54.

Balling, R.C. Jr., et al., 1998: Impacts of land degradation on historical temperature records from the Sonoran Desert. *Clim. Change*, **40**, 669–681.

Barlow, M., H. Cullen, and B. Lyon, 2002: Drought in central and southwest Asia: La Niña, the warm pool, and Indian Ocean precipitation. *J. Clim.*, **15**, 697–700.

Barnett, T.P., et al., 1999: Interdecadal interactions between the tropics and midlatitudes in the Pacific basin. *Geophys. Res. Lett.*, **26**, 615–618.

Barnston, A.G., and R.E. Livezey, 1987: Classification, seasonality and persistence of low frequency atmospheric circulation patterns. *Mon. Weather Rev.*, **115**, 1083–1126.

Bärring, L., and H. von Storch, 2004: Scandinavian storminess since about 1800. *Geophys. Res. Lett.*, **31**, L20202, doi:10.1029/2004GL020441.

Barriopedro, D., et al., 2006: A climatology of Northern Hemisphere blocking. *J. Clim.*, **19**, 1042–1063.

Barros, V.R., M.E. Castañeda, and M.E. Doyle, 2000a: Recent precipitation trends in southern South America east of the Andes: An indication of climatic variability. In: *Southern Hemisphere Paleo- and Neoclimates. Key Sites, Methods, Data and Models* [Smolka, P.P., and W. Volkheimer (eds.)]. Springer Verlag, Berlin, pp. 187–206.

Barros, V.R., et al., 2000b: Influence of the South Atlantic Convergence Zone and South Atlantic sea surface temperature on interannual summer rainfall variability in southeastern South America. *Theor. Appl. Climatol.*, **67**, 123–133.

Barros, V.R., et al., 2002: Climate variability over South America and the South America monsoon: A review. *Meteorologica*, **27**, 33–57.

Barros, V.R., et al., 2004: The major discharge events in the Paraguay River: magnitudes, source regions and climate forcings. *J. Hydrometeorol.*, **6**, 1161–1170.

Basher, R.E., and X. Zheng, 1995: Tropical cyclones in the southwest Pacific: Spatial patterns and relationships to Southern Oscillation and sea surface temperature. *J. Clim.*, **8**, 1249–1260.

Bates, J.J., and D.L. Jackson, 2001: Trends in upper-tropospheric humidity. *Geophys. Res. Lett.*, **28**, 1695–1698.

Beck, C., J. Grieser, and B. Rudolf, 2005: A new monthly precipitation climatology for the global land areas for the period 1951 to 2000. *Climate Status Report, 2004*. German Meteorological Service, pp. 181–190, http://www.dwd.de/de/FundE/Klima/KLIS/prod/KSB/ksb04/28_precipitation.pdf

Bengtsson, L., S. Hagemann, and K.I. Hodges, 2004: Can climate trends be calculated from reanalysis data? *J. Geophys. Res.*, **109**, D11111, doi:10.1029/2004JD004536.

Beniston, M., 2004: The 2003 heat wave in Europe. A shape of things to come? *Geophys. Res. Lett.*, **31**, L02022, doi:10.1029/2003GL018857.

Beniston, M., and H.F. Diaz, 2004: The 2003 heat wave as an example of summers in a greenhouse climate? Observations and climate model simulations for Basel, Switzerland. *Global Planet. Change*, **44**, 73–81.

Beniston, M., and D.B. Stephenson, 2004: Extreme climatic events and their evolution under changing climatic conditions. *Global Planet. Change*, **44**, 1–9.

Berger, C.L., et al., 2002: A climatology of northwest Missouri snowfall events: Long-term trends and interannual variability. *Phys. Geogr.*, **23**, 427–448.

Berri, G.S., M.A. Ghietto, and N.O. García, 2002: The influence of ENSO in the flows of the Upper Paraná River of South America over the past 100 years. *J. Hydrometeorol.*, **3**, 57–65.

Betts, A.K., J.H. Ball, and P. Viterbo, 2003: Evaluation of the ERA-40 surface water budget and surface temperature for the Mackenzie River Basin. *J. Hydrometeorol.*, **4**, 1194–1211.

Biondi, F., A. Gershunov, and D.R. Cayan, 2001: North Pacific decadal climate variability since AD 1661. *J. Clim.*, **14**, 5–10.

Bischoff, S.A., et al., 2000: Climate variability and Uruguay River flows. *Water International*, **25**, 446–456.

Bissolli, P., J. Grieser, N. Dotzek, and M. Welsch, 2007: Tornadoes in Germany 1950-2003 and their relation to particular weather conditions. *Global Planet. Change*, **46**, doi:10.1016/j.gloplacha.2006.11.007

Black, E.M., et al., 2004: Factors contributing to the summer 2003 European heatwave. *Weather*, **59**, 217–223.

Bojariu, R., and L. Gimeno, 2003: The influence of snow cover fluctuations on multiannual NAO persistence. *Geophys. Res. Lett.*, **30**, 1156, doi:10.1029/2002GL015651.

Bonan, G.B., 2001: Observational evidence for reduction of daily maximum temperature by croplands in the midwest United States. *J. Clim.*, **14**, 2430–2442.

Bonnazola, M., and P.H. Haynes, 2004: A trajectory-based study of the tropical tropopause region. *J. Geophys. Res.*, **109**, D020112, doi:10.1029/2003JD004356.

Bornstein, R., and Q. Lin, 2000: Urban heat islands and summertime convective thunderstorms in Atlanta: three cases studies. *Atmos. Environ.*, **34**, 507–516.

Bottomley, M., et al., 1990: *Global Ocean Surface Temperature Atlas "GOSTA"*. HMSO, London, 20 pp.+iv, 313 plates.

Bove, M.C., et al., 1998: Effect of El Niño on U.S. landfalling hurricanes, revisited. *Bull. Am. Meteorol. Soc.*, **79**, 2477–2482.

Branstator, G., 2002: Circum-global teleconnections, the jetstream waveguide, and the North Atlantic Oscillation. *J. Clim.*, **15**, 1893–1910.

Brest, C.L., W.B. Rossow, and M. Roiter, 1997: Update of radiance calibrations for ISCCP. *J. Atmos. Ocean Technol.*, **14**, 1091–1109.

Brohan, P., et al., 2006: Uncertainty estimates in regional and global observed temperature changes: A new dataset from 1850. *J. Geophys. Res.*, **111**, D12106, doi:10.1029/2005JD006548.

Bromirski, P.D., R.E. Flick, and D.R. Cayan, 2003: Storminess variability along the California coast: 1858–2000. *J. Clim.*, **16**, 982–993.

Bromwich, D.H., and R.L. Fogt, 2004: Strong trends in the skill of the ERA-40 and NCEP–NCAR reanalyses in the high and middle latitudes of the Southern Hemisphere, 1958–2001. *J. Clim.*, **17**, 4603–4619.

Brönnimann, S., et al., 2004: Extreme climate of the global troposphere and stratosphere in 1940–42 related to El Niño. *Nature*, **431**, doi:10.1038/nature02982.

Brooks, H.E., J.W. Lee, and J.P. Craven, 2003: The spatial distribution of severe thunderstorm and tornado environments from global reanalysis data. *Atmos. Res.*, **67–68**, 73–94.

Brunet, M., et al., 2006: The development of a new dataset of Spanish daily adjusted temperature series (SDATS) (1850-2003). *Int. J. Climatol.*, **26**, 1777–1802.

Brunetti, M., et al., 2004: Changes in daily precipitation frequency and distribution in Italy over the last 120 years. *J. Geophys. Res.*, **109**, D05102, doi:10.1029/2003JD004296.

Brutsaert, W., and M.B. Parlange, 1998: Hydrologic cycle explains the evaporation paradox. *Nature*, **396**, 30.

Burian, S.J., and J.M. Shepherd, 2005: Effects of urbanization on the diurnal rainfall pattern in Houston. *Hydrological Processes: Special Issue on Rainfall and Hydrological Processes*, **19**, 1089–1103.

Burnett, A.W., et al., 2003: Increasing Great Lake-effect snowfall during the Twentieth Century: A regional response to global warming? *J. Clim.*, **16**, 3535–3541.

Caesar, J., L. Alexander, and R. Vose, 2006: Large-scale changes in observed daily maximum and minimum temperatures: Creation and analysis of a new gridded data set. *J. Geophys. Res.*, **111**, D05101, doi:10.1029/2005JD006280.

Cai, W., and P.G. Baines, 2001: Forcing of the Antarctic Circumpolar Wave by El Niño-Southern Oscillation teleconnections. *J. Geophys. Res.*, **106**, 9019–9038.

Cai, W., P.H. Whetton, and D.J. Karoly, 2003: The response of the Antarctic Oscillation to increasing and stabilized atmospheric CO_2. *J. Clim.*, **16**, 1525–1538.

Caires, S., and A. Sterl, 2005: 100-year return value estimates for wind speed and significant wave height from the ERA-40 data. *J. Clim.*, **18**, 1032–1048.

Camargo, S.J., and A.H. Sobel, 2004: *Western North Pacific Tropical Cyclone Intensity and ENSO*. Technical Report No. 04-03, International Research Institute for Climate Prediction, Palisades, NY, 25 pp.

Camilloni, I.A., and V.R. Barros, 2000: The Paraná River response to El Niño 1982–83 and 1997–98 events. *J. Hydrometeorol.*, **1**, 412–430.

Camilloni, I.A., and V.R. Barros, 2003: Extreme discharge events in the Paraná River and their climate forcing. *J. Hydrol.*, **278**, 94–106.

Cardone, V.J., J.G. Greenwood, and M.A. Cane, 1990: On trends in historical marine wind data. *J. Clim.*, **3**, 113–127.

Carril, A.F., and A. Navarra, 2001: Low-frequency variability of the Antarctic Circumpolar Wave. *Geophys. Res. Lett.*, **28**, 4623–4626.

Carter, D.J.T., 1999: Variability and trends in the wave climate of the North Atlantic: a review. In: *Proceedings of the 9th International Offshore and Polar Engineering Conference, Brest, France,* vol. III. International Society of Offshore and Polar Engineers, Golden, CO, pp. 12–18.

Cassou, C., and L. Terray, 2001: Dual influence of Atlantic and Pacific SST anomalies on the North Atlantic/Europe winter climate. *Geophys. Res. Lett.*, **28**, 3195–3198.

Cassou, C., et al., 2004: North Atlantic winter climate regimes: spatial asymmetry, stationarity with time and oceanic forcing. *J. Clim.*, **17**, 1055–1068.

Cayan, D.R., et al., 2001: Changes in the onset of spring in the western United States. *Bull. Am. Meteorol. Soc.*, **82**, 399–415.

Cess, R.D., and P.M. Udelhofen, 2003: Climate change during 1985–1999: Cloud interactions determined from satellite measurements. *Geophys. Res. Lett.*, **30**, 1019, doi:10.1029/2002GL016128.

Chagnon, F.J.F., and R. L. Bras, 2005: Contemporary climate change in the Amazon. *Geophys. Res. Lett.*, **32**, L13703, doi:10.1029/2005GL022722.

Chan, J.C.L., 2006: Comment on "Changes in tropical cyclone number, duration, and intensity in a warming environment". *Science*, **311**, 1713.

Chan, J.C.L., and K.S. Liu, 2004: Global warming and Western North Pacific typhoon activity from an observational perspective. *J. Clim.*, **17**, 4590–4602.

Chang, C.P., Y. Zhang, and T. Li, 1999: Interannual and interdecadal variations of the East Asian summer monsoon and tropical Pacific SSTs. Pt I: Roles of the subtropical ridge. *J. Clim.*, **13**, 4310–4325.

Chang, E.K.M., 2003: Midwinter suppression of the Pacific storm track activity as seen in aircraft observations. *J. Atmos. Sci.*, **60**, 1345–1358.

Chang, E.K.M., and Y. Fu, 2002: Interdecadal variations in Northern Hemisphere winter storm track intensity. *J. Clim.*, **15**, 642–658.

Chang, E.K.M., and Y. Fu, 2003: Using mean flow change as a proxy to infer interdecadal storm track variability. *J. Clim.*, **16**, 2178–2196.

Changnon, D., M. Sandstrom, and C. Schaffer, 2003: Relating changes in agricultural practices to increasing dew points in extreme Chicago heat waves. *Clim. Res.*, **24**, 243–254.

Changnon, S.A., and N.E. Westcott, 2002: Heavy rainstorms in Chicago: Increasing frequency, altered impacts, and future implications. *J. Am. Water Res. Assoc.*, **38**, 1467–1475.

Changnon, S.A., et al., 1981: *METROMEX: A Review and Summary.* Meteorological Monograph 18, American Meteorological Society, Boston, MA, 81 pp.

Chase, T.N., J.A. Knaff, R.A. Pielke Sr., and E. Kalnay, 2003: Changes in global monsoon circulation since 1950. *Natural Hazards*, **29**, 229–254.

Chattopadhyay, N., and M. Hulme, 1997: Evaporation and potential evapotranspiration in India under conditions of recent and future climate change. *Agric. For. Meteorol.*, **87**, 55–73.

Chelliah, M., and G.D. Bell, 2004: Tropical multidecadal and interannual climate variability in the NCEP–NCAR Reanalysis. *J. Clim.*, **17**, 1777–1803.

Chelton, D.B., and F.J. Wentz, 2005: Global microwave satellite observations of sea surface temperature for numerical weather prediction and climate research. *Bull. Am. Meteorol. Soc.*, **86**, 1097–1115.

Chen, M., P. Xie, and J.E. Janowiak, 2002: Global land precipitation: a 50-yr monthly analysis based on gauge observations. *J. Hydrometeorol.*, **3**, 249–266.

Chen, T.-C., et al., 2004: Variation of the East Asian summer monsoon rainfall. *J. Clim.*, **17**, 744–762.

Christy, J.R., and W.B. Norris, 2004: What may we conclude about tropospheric temperature trends? *Geophys. Res. Lett.*, **31**, L0621, doi:10.1029/2003GL019361.

Christy, J. R., and R.W. Spencer, 2005: Correcting temperature data sets. *Science*, **310**, 972.

Christy, J.R., R.W. Spencer, and W.D. Braswell, 2000: MSU tropospheric temperatures: Dataset construction and radiosonde comparisons. *J. Atmos. Ocean. Technol.*, **17**, 1153–1170.

Christy, J.R., et al., 2001: Differential trends in tropical sea surface and atmospheric temperature since 1979. *Geophys. Res. Lett.*, **28**, 183–186.

Christy, J.R., et al., 2003: Error estimates of version 5.0 of MSU/AMSU bulk atmospheric temperatures. *J. Atmos. Ocean. Technol.*, **20**, 613–629.

Christy, J.R., et al., 2006: Methodology and results of calculating Central California surface temperature trends: Evidence of human-induced climate change? *J. Clim.*, **19**, 548–563.

Clark, C.O., P.J. Webster, and J.E. Cole, 2003: Interdecadal variability of the relationship between the Indian Ocean zonal mode and east African coastal rainfall anomalies. *J. Clim.*, **16**, 548–554.

Cluis, D., and C. Laberge, 2001: Climate change and trend detection in selected rivers within the Asia-Pacific region. *Water International*, **26**, 411–424.

Cohen, S., B. Liepert, and G. Stanhill, 2004: Global dimming comes of age. *Eos*, **85**, 362.

Cohn, T., and H.J. Lins, 2005: Nature's style: Naturally trendy. *Geophys. Res. Lett.*, **32**, L23402, doi:10.1029/2005GL024476.

Compo, G.P., and P.D. Sardeshmukh, 2004: Storm track predictability on seasonal and decadal scales. *J. Clim.*, **17**, 3701–3720.

Connolley, W.M., 2003: Long-term variation of the Antarctic Circumpolar Wave. *J. Geophys. Res.*, **108**, 8076, doi:10.1029/2000JC000380.

Considine, D.B., et al., 2001: An interactive model study of the influence of the Mount Pinatubo aerosol on stratospheric methane and water trends. *J. Geophys. Res.*, **106**, 27711–27728.

Cook, E.R., R.D. D'Arrigo, and M.E. Mann, 2002: A well-verified, multiproxy reconstruction of the winter North Atlantic Oscillation index since A.D. 1400. *J. Clim.*, **15**, 1754–1764.

Cook, E.R., et al., 1999: Drought reconstructions for the continental United States. *J. Clim.*, **12**, 1145–1162.

Cook, E.R., et al., 2004: Long-term aridity changes in the western United States. *Science*, **306**, 1015–1018.

Corti, S., F. Molteni, and T.N. Palmer, 1999: Signature of recent climate change in frequencies of natural atmospheric circulation regimes. *Nature*, **398**, 799–802.

Cowell, C.M., and R.T. Stoudt, 2002: Dam-induced modifications to upper Allegheny River streamflow patterns and their biodiversity implications. *J. Am. Water Res. Assoc.*, **38**, 187–196.

Crutzen, P.J., 2004: New Directions: The growing urban heat and pollution "island" effect—impact on chemistry and climate. *Atmos. Environ.*, **38**, 3539–3540.

Cullen, H., and P.B. deMenocal, 2000: North Atlantic influence on Tigris-Euphrates streamflow. *Int. J. Climatol.*, **20**, 853–863.

Curtis, S., and R.F. Adler, 2003: The evolution of El Niño-precipitation relationships from satellites and gauges. *J. Geophys. Res.*, **108**, 4153, doi:10.1029/2002JD002690.

Czaja, A., A.W. Robertson, and T. Huck: 2003: The role of Atlantic ocean-atmosphere coupling in affecting North Atlantic Oscillation variability. In: *The North Atlantic Oscillation: Climatic Significance and Environmental Impact* [Hurrell, J.W., et al. (eds.)]. Geophysical Monograph 134, American Geophysical Union, Washington, DC, pp. 147–172.

Dai, A., 2006: Recent climatology, variability and trends in global surface humidity. *J. Clim.*, **19**, 3589–3606.

Dai, A., and K.E. Trenberth, 2002: Estimates of freshwater discharge from continents: Latitudinal and seasonal variations. *J. Hydrometeorol.*, **3**, 660–687.

Dai, A., and K.E. Trenberth, 2004: The diurnal cycle and its depiction in the Community Climate System Model. *J. Clim.*, **17**, 930–995.

Dai, A., I.Y. Fung, and A.D. Del Genio, 1997: Surface observed global land precipitation during 1900–1988. *J. Clim.*, **10**, 2943–2962.

Dai, A., K.E. Trenberth, and T.R. Karl, 1999: Effects of clouds, soil moisture, precipitation and water vapor on diurnal temperature range. *J. Clim.*, **12**, 2451–2473.

Dai A., K.E. Trenberth, and T. Qian, 2004a: A global data set of Palmer Drought Severity Index for 1870–2002: Relationship with soil moisture and effects of surface warming. *J. Hydrometeorol.*, **5**, 1117–1130.

Dai, A., et al., 2004b: The recent Sahel drought is real. *Int. J. Climatol.*, **24**, 1323–1331.

Dai, A., et al., 2006: Recent trends in cloudiness over the United States: A tale of monitoring inadequacies. *Bull. Am. Meteorol. Soc*, **87**, 597–606.

De Laat, A.T.J., and A.N. Maurellis, 2006: Evidence for influence of anthropogenic surface processes on lower tropospheric and surface temperature trends. *Int. J. Climatol.*, **26**, 897–913.

DeGaetano, A.T., and R.J. Allen, 2002: Trends in twentieth-century extremes across the United States. *J. Clim.*, **15**, 3188–3205.

Delworth, T.L., and M.E. Mann, 2000: Observed and simulated multidecadal variability in the Northern Hemisphere. *Clim. Dyn.*, **16**, 661–676.

DeMott, C.A., and D.A. Randall, 2004: Observed variations of tropical convective available potential energy. *J. Geophys. Res.*, **109**, D02102, doi:10.1029/2003JD003784.

Déry, S.J., and E.F. Wood, 2005: Decreasing river discharge in northern Canada. *Geophys. Res. Lett.*, **32**, L10401, doi:10.1029/2005GL022845.

Deser, C., 2000: On the teleconnectivity of the Arctic Oscillation. *Geophys. Res. Lett.*, **27**, 779–782.

Deser, C., M.A. Alexander, and M.S. Timlin, 1996: Upper-ocean thermal variations in the North Pacific during 1970–1991. *J. Clim.*, **9**, 1840–1855.

Deser, C., M.A. Alexander, and M.S. Timlin, 1999: Evidence for a wind-driven intensification of the Kuroshio Current Extension from the 1970s to the 1980s. *J. Clim.*, **12**, 1697–1706.

Deser, C., J.E. Walsh, and M.S. Timlin, 2000: Arctic sea ice variability in the context of recent atmospheric circulation trends. *J. Clim.*, **13**, 617–633.

Deser, C., M.A. Alexander, and M. S. Timlin, 2003: Understanding the persistence of sea surface temperature anomalies in midlatitudes. *J. Clim.*, **16**, 57–72.

Deser, C., A.S. Phillips, and J.W. Hurrell, 2004: Pacific interdecadal climate variability: Linkages between the tropics and the north Pacific during boreal winter since 1900. *J. Clim.*, **17**, 3109–3124.

Dessler, A.E., and S.C. Sherwood, 2004: Effect of convection on the summertime extratropical lower stratosphere, *J. Geophys. Res.*, **109**, D23301, doi:10.1029/2004JD005209.

DeWeaver, E., and S. Nigam, 2000: Zonal-eddy dynamics of the North Atlantic Oscillation. *J. Clim.*, **13**, 3893–3914.

Dickson, R.R., et al., 2000: The Arctic Ocean response to the North Atlantic Oscillation. *J. Clim.*, **13**, 2671–2696.

Diem, J.E., and D.P. Brown, 2003: Anthropogenic impacts on summer precipitation in central Arizona, U.S.A. *Professional Geogr.*, **55**, 343–355.

Diggle, P.J., K.Y. Liang, and S.L. Zeger, 1999: *Analysis of Longitudinal Data*. Clarendon Press, Oxford, UK, 253 pp.

Ding, Y.H, C.Y. Li, and Y.J. Liu, 2004: Overview of the South China Seas monsoon experiment. *Adv. Atmos. Sci.*, **21**, 343–360.

Dixon, P.G., and T.L. Mote, 2003: Patterns and causes of Atlanta's urban heat island-initiated precipitation. *J. Appl. Meteorol.*, **42**, 1273–1284.

Doswell, C.A., H.E. Brooks, and M.P. Kay, 2005: Climatological estimates of daily local nontornadic severe thunderstorm probability for the United States. *Weather Forecasting*, **20**, 577–595.

Dotzek, N., 2003: An updated estimate of tornado occurrence in Europe. *Atmos. Res.*, **67–68**, 153–161.

Dotzek, N., et al., 2005: Observational evidence for exponential tornado intensity distributions over specific kinetic energy. *Geophys. Res. Lett.*, **32**, L24813, doi:10.1029/2005GL024583.

Doyle, M.E., and V.R. Barros, 2002: Midsummer low-level circulation and precipitation in subtropical South America and related sea surface temperature anomalies in the South Atlantic. *J. Clim.*, **15**, 3394–3410.

Duchon, C.E., 1979: Lanczos filtering in one and two dimensions. *J Appl. Meteorol.*, **18**, 1016–1022.

Durre, I., and J.M. Wallace, 2001: Factors influencing the cold-season diurnal temperature range in the United States. *J. Clim.*, **14**, 3263–3278.

Durre, I., T. Peterson, and R. Vose, 2002: Evaluation of the effect of the Luers-Eskridge radiation adjustments on radiosonde temperature homogeneity. *J. Clim.*, **15**, 1335–1347.

Durre, I., R.S. Vose, and D.B. Wuertz, 2006: Overview of the integrated global radiosonde archive. *J. Clim.*, **19**, 53–68.

Easterling, D.R., 2002: Recent changes in frost days and the frost-free season in the United States. *Bull. Am. Meteorol. Soc.*, **83**, 1327–1332.

Easterling, D.R., et al., 2000: Observed variability and trends in extreme climate events: A brief review. *Bull. Am. Meteorol. Soc.*, **81**, 417–425.

Easterling, D.R., et al., 2003: CCI/CLIVAR workshop to develop priority climate indices. *Bull. Am. Meteorol. Soc.*, **84**, 1403–1407.

Elliott, W.P., R.J. Ross, and W.H. Blackmore, 2002: Recent changes in NWS upper-air observations with emphasis on changes from VIZ to Vaisala radiosondes. *Bull. Am. Meteorol. Soc.*, **83**, 1003–1017.

Ellis, A.W., and J.J. Johnson, 2004: Hydroclimatic analysis of snowfall trends associated with the North American Great Lakes. *J. Hydrometeorol.*, **5**, 471–486.

Emanuel, K., 2003: Tropical cyclones. *Annu. Rev. Earth. Planet. Sci.*, **31**, 75–104.

Emanuel, K., 2005a: Increasing destructiveness of tropical cyclones over the past 30 years. *Nature*, **436**, 686–688.

Emanuel, K., 2005b: Emanuel replies. *Nature*, **438**, E13, doi:10.1038/nature04427.

Enfield, D.B., A.M. Mestas-Nuñez, and P.J Trimble, 2001: The Atlantic Multidecadal Oscillation and its relation to rainfall and river flows in the continental US. *Geophys. Res. Lett.*, **28**, 2077–2080.

Evans, M.N., et al., 2001: Support for tropically-driven Pacific decadal variability based on paleoproxy evidence. *Geophys. Res. Lett.*, **28**, 3689–3692.

Faucher eau, N., et al., 2003: Rainfall variability and changes in Southern Africa during the 20th century in the global warming context. *Natural Hazards*, **29**, 139–154.

Feldstein, S.B., 2002: The recent trend and variance increase of the Annular Mode. *J. Clim.*, **15**, 88–94.

Fink, A.H., et al., 2004: The 2003 European summer heatwaves and drought – synoptic diagnosis and impacts. *Weather*, **59**, 209–216.

Folland, C.K., and D.E. Parker, 1995: Correction of instrumental biases in historical sea surface temperature data. *Q. J. R. Meteorol. Soc.*, **121**, 319–367.

Folland, C.K., et al., 1993: A study of six operational sea surface temperature analyses. *J. Clim.*, **6**, 96–113.

Folland, C.K., et al., 1999: Large scale modes of ocean surface temperature since the late nineteenth century. In: *Beyond El Niño: Decadal and Interdecadal Climate Variability* [Navarra, A. (ed.)]. Springer-Verlag, Berlin, pp. 73–102.

Folland, C.K., et al., 2001: Predictability of North East Brazil rainfall and real-time forecast skill, 1987-1998. *J. Clim.*, **14**, 1937–1958.

Folland, C.K., et al., 2002: Relative influences of the interdecadal Pacific oscillation and ENSO on the South Pacific convergence zone. *Geophys. Res. Lett.*, **29**(13), doi:10.1029/2001GL014201.

Folland, C.K., et al., 2003: Trends and variations in South Pacific island and ocean surface temperature. *J. Clim.*, **16**, 2859–2874.

Forster, P.M.D., and K.P. Shine, 1999: Stratospheric water vapour changes as a possible contributor to observed stratospheric cooling. *Geophys. Res. Lett.*, **26**, 3309–3312.

Forster, P.M.D., and K.P. Shine, 2002: Assessing the climate impact of trends in stratospheric water vapor. *Geophys. Res. Lett*, **29**, 1086, doi:10.1029/2001GL013909.

Forster, P.M.D., and S. Solomon, 2003: Observations of a "weekend effect" in diurnal temperature range. *Proc. Natl. Acad. Sci. U.S.A.*, **100**, 11225–11230.

Fowler, H.J., and C.G. Kilsby, 2003a: Implications of changes in seasonal and annual extreme rainfall. *Geophys. Res. Lett.*, **30**, 1720, doi:10.1029/2003GL017327.

Fowler, H.J., and C.G. Kilsby, 2003b: A regional frequency analysis of United Kingdom extreme rainfall from 1961 to 2000. *Int. J. Climatol.*, **23**, 1313–1334.

Frauenfeld, O.W., and R.E. Davis, 2003: Northern Hemisphere circumpolar vortex trends and climate change implications. *J. Geophys. Res.*, **108**, 4423, doi:10.1029/2002JD002958.

Free, M., and D. Seidel, 2005: Causes of differing temperature trends in radiosonde upper-air datasets. *J. Geophys. Res.*, **110**, D07101, doi:10.1029/2004JD005481.

Free, M., M. Bister, and K. Emanuel, 2004a: Potential intensity of tropical cyclones: comparison of results from radiosonde and reanalysis data. *J. Clim.*, **17**, 1722–1727.

Free, M., et al., 2002: Creating climate reference datasets: CARDS workshop on adjusting radiosonde temperature data for climate monitoring. *Bull. Am. Meteorol. Soc.*, **83**, 891–899.

Free, M., et al., 2004b: Using first differences to reduce inhomogeneity in radiosonde temperature datasets. *J. Clim.*, **17**, 4171–4179.

Frei, C., and C. Schär, 2001: Detection of probability of trends in rare events: Theory and application to heavy precipitation in the Alpine region. *J. Clim.*, **14**, 1568–1584.

Frich, P., et al., 2002: Observed coherent changes in climatic extremes during the second half of the twentieth century. *Clim. Res.*, **19**, 193–212.

Fu, Q., and C.M. Johanson, 2004: Stratospheric influence on MSU-derived tropospheric temperature trends: A direct error analysis. *J. Clim.*, **17**, 4636–4640.

Fu, Q., and C.M. Johanson, 2005: Satellite-derived vertical dependence of tropical tropospheric temperature trends. *Geophys. Res. Lett.*, **32**, L10703, doi:10.1029/2004GL022266.

Fu, Q., et al., 2004a: Contribution of stratospheric cooling to satellite-inferred tropospheric temperature trends. *Nature*, **429**, 55–58.

Fu, Q., et al., 2004b: Stratospheric cooling and the troposphere (reply). *Nature*, **432**, doi:10.1038/nature03210.

Fueglistaler, S., and P.H. Haynes, 2005: Control of interannual and longer-term variability of stratospheric water vapour. *J. Geophys. Res.*, **110**, D24108, doi:10.1029/2005JD006019.

Fueglistaler, S., H. Wernli, and T. Peter, 2004: Tropical troposphere-to-stratosphere transport inferred from trajectory calculations. *J. Geophys. Res.*, **109**, D03108, doi:10.1029/2003JD004069.

Fujibe, F., 2003: Long-term surface wind changes in the Tokyo metropolitan area in the afternoon of sunny days in the warm season. *J. Meteorol. Soc. Japan*, **81**, 141–149.

Gallego, D., et al., 2005: A new look at the Southern Hemisphere jet stream. *Clim. Dyn.*, **24**, 607–621.

García, N.O., and W.M. Vargas, 1998: The temporal climatic variability in the 'Río de la Plata' basin displayed by the river discharges. *Clim. Change*, **38**, 359–379.

GCOS, 2003: *The Second Report on the Adequacy of the Global Observing Systems for Climate in Support of the UNFCCC*. GCOS-82, WMO/TD No. 1143, Global Climate Observing System, 74 pp.

GCOS, 2004: *GCOS Implementation Plan for the Global Observing System for Climate in support of UNFCCC*. GCOS-92, WMO/TD 1219, Global Climate Observing System, 136 pp.

Gedalof, Z., N.J. Mantua, and D.L. Peterson, 2002: A multi-century perspective of variability in the Pacific Decadal Oscillation: new insights from tree rings and coral. *Geophys. Res. Lett.*, **29**, 2204, doi:10.1029/2002GL015824.

Geng, Q., and M. Sugi, 2001: Variability of the North Atlantic cyclone activity in winter analyzed from NCEP-NCAR reanalysis data. *J. Clim.*, **14**, 3863–3873.

Genta, J.L., G. Perez-Iribarren, and C.R. Mechoso, 1998: A recent increasing trend in the streamflow of rivers in southeastern South America. *J. Clim.*, **11**, 2858–2862.

Genthon, C., G. Krinner, and M. Sacchettini, 2003: Interannual Antarctic tropospheric circulation and precipitation variability. *Clim. Dyn.*, **21**, 289–307.

Gershunov, A., and T.P. Barnett, 1998: Interdecadal modulation of ENSO teleconnections. *Bull. Am. Meteorol. Soc.*, **79**, 2715–2725.

Gettelman, A., et al., 2002: Multi-decadal trends in tropical convective available potential energy. *J. Geophys. Res.*, **107**, 4606, doi:10.1029/2001JD001082.

Giannini, A., R. Saravannan, and P. Chang, 2003: Ocean forcing of Sahel rainfall on interannual to interdecadal time scales. *Science*, **302**, 1027–1030.

Gilgen, H., M. Wild, and A. Ohmura, 1998: Means and trends of shortwave irradiance at the surface estimated from global energy balance archive data. *J. Clim.*, **11**, 2042–2061.

Gille, S.T., 2002: Warming of the Southern Ocean since the 1950s. *Science*, **295**, 1275–1277.

Gillett, N.P., and D. Thompson, 2003: Simulation of recent Southern Hemisphere climate change. *Science*, **302**, 273–275.

Gillett, N.P., B.D. Santer, and A.J. Weaver, 2004: Stratospheric cooling and the troposphere. *Nature*, **432**, doi:10.1038/nature03209.

Gillett, N.P., R.J. Allan, and T.J. Ansell, 2005: Detection of external influence on sea level pressure with a multi-model ensemble. *Geophys. Res. Lett.*, **32**, L19714, doi:10.1029/2005GL023640.

Gillett, N.P., et al., 2003: Detection of human influence on sea-level pressure. *Nature*, **422**, 292–294.

Goddard, L., and N.E. Graham, 1999: Importance of the Indian Ocean for simulating rainfall anomalies over eastern and southern Africa. *J. Geophys. Res.*, **104**, 19099–19116.

Goldenberg, S.B et al., 2001: The recent increase in Atlantic hurricane activity: causes and implications. *Science*, **293**, 474–479.

Golubev, V.S., et al., 2001: Evaporation changes over the contiguous United States and the former USSR: A reassessment. *Geophys. Res. Lett.*, **28**, 2665–2668.

Gong, D.Y., and S.W. Wang, 1999: Definition of Antarctic oscillation index. *Geophys. Res. Lett.*, **26**, 459–462.

Gong D.Y., and S.W. Wang, 2000: Severe summer rainfall in China associated with enhanced global warming. *Clim. Res.*, **16**, 51–59.

Gong, D.Y., and C.-H. Ho, 2002: Shift in the summer rainfall over the Yangtze River valley in the late 1970s. *Geophys. Res. Lett.*, **29**(3), doi:10.1029/2001GL014523.

Gong, D.Y., S.W. Wang, and J.H. Zhu, 2001: East Asian winter monsoon and Arctic Oscillation. *Geophys. Res. Lett.*, **28**, 2073–2076.

Gong, G., D. Entekhabi, and J. Cohen, 2002: A large-ensemble model study of the wintertime AO-NAO and the role of interannual snow perturbations. *J. Clim.*, **15**, 3488–3499.

Gong, G., D. Entekhabi, and J. Cohen, 2003: Modeled Northern Hemisphere winter climate response to realistic Siberian snow anomalies. *J. Clim.*, **16**, 3917–3931.

Gower, J.F.R., 2002: Temperature, wind and wave climatologies, and trends from marine meteorological buoys in the northeast Pacific. *J. Clim.*, **15**, 3709–3718.

Graham, N.E., and H.F. Diaz, 2001: Evidence for intensification of North Pacific winter cyclones since 1948. *Bull. Am. Meteorol. Soc.*, **82**, 1869–1893.

Gray, S.T., et al., 2004: A tree-ring based reconstruction of the Atlantic Multidecadal Oscillation since 1567 A.D. *Geophys. Res. Lett.*, **31**, L12205, doi:10.1029/2004GL019932.

Gray, W.M., 1984: Atlantic seasonal hurricane frequency: Part I: El Niño and 30-mb quasi-biennial oscillation influences. *Mon. Weather Rev.*, **112**, 1649–1668.

Grieser, J., S. Trömel, and C.-D. Schönwiese, 2002: Statistical time series decomposition into significant components and application to European temperature. *Theor. Appl. Climatol.*, **71**, 171–183.

Griffiths, G.M., M.J. Salinger, and I. Leleu, 2003: Trends in extreme daily rainfall in the South Pacific and relations to the South Pacific Convergence Zone. *Int. J. Climatol.*, **23**, 847–869.

Griffiths, G.M., et al., 2005: Change in mean temperature as a predictor of extreme temperature change in the Asia-Pacific region. *Int. J. Climatol.*, **25**, 1301–1330.

Grist, J.P., and S.A. Josey, 2003: Inverse analysis of the SOC air-sea flux climatology using ocean heat transport constraints. *J. Clim.*, **16**, 3274–3295.

Grody, N.C., et al., 2004: Calibration of multi-satellite observations for climatic studies: Microwave sounding unit (MSU). *J. Geophys. Res.*, **109**, D24104, doi:10.1029/2004JD005079.

Groisman, P.Ya., and E.Ya. Rankova, 2001: Precipitation trends over the Russian permafrost-free zone: removing the artifacts of pre-processing. *Int. J. Climatol.*, **21**, 657–678.

Groisman, P.Ya., R.W. Knight, and T.R. Karl, 2001: Heavy precipitation and high streamflow in the contiguous United States: Trends in the 20th century. *Bull. Am. Meteorol. Soc.*, **82**, 219–246.

Groisman, P.Ya., et al., 2003: Contemporary climate changes in high latitudes of the Northern Hemisphere: Daily time resolution. In: *Proceedings of the International Symposium on Climate Change, Beijing, China, 31 March–3 April, 2003*. WMO/TD No. 1172, China Meteorological Press, Beijing, China, pp. 51–55.

Groisman, P.Ya., et al., 2004: Contemporary changes of the hydrological cycle over the contiguous United States: Trends derived from *in situ* observations. *J. Hydrometeorol.*, **5**, 64–85.

Groisman, P.Ya., et al., 2005: Trends in intense precipitation in the climate record. *J. Clim.*, **18**, 1326–1350.

Groisman, P.Ya., et al., 2007: Potential forest fire danger over northern Eurasia: Changes during the 20th century. *Global Planet. Change*, **46**, doi:10.1016/j.gloplacha.2006.07.029.

Grollmann, T., and S. Simon, 2002: Flutkatastrophen – Boten des klimawandels. *Z. Versicher.*, **53**, 682–689.

Gruza, G.V., et al., 1999: Indicators of climatic change for the Russian Federation. *Clim. Change*, **42**, 219–242.

Gu, D.F., and S.G.H. Philander, 1997: Interdecadal climate fluctuations that depend on exchanges between the tropics and extratropics. *Science*, **275**, 805–807.

Guetter, A.K., and K.P. Georgakakos, 1993: River outflow of the conterminous United States, 1939–1988. *Bull. Am. Meteorol. Soc.*, **74**, 1873–1891.

Guichard, F., D. Parsons, and E. Miller, 2000: Thermodynamic and radiative impact of the correction of sounding humidity bias in the tropics. *J. Clim.*, **13**, 3611–3624.

Gulev, S.K., and L. Hasse, 1999: Changes of wind waves in the North Atlantic over the last 30 years. *Int. J. Climatol.*, **19**, 1091–1117.

Gulev, S.K., and V. Grigorieva, 2004: Last century changes in ocean wind wave height from global visual wave data. *Geophys. Res. Lett.*, **31**, L24302, doi:10.1029/2004GL021040.

Gulev, S. K., O. Zolina, and S. Grigoriev, 2001: Extratropical cyclone variability in the Northern Hemisphere winter from the NCEP/NCAR reanalysis data. *Clim. Dyn.*, **17**, 795–809.

Gulev, S.K., T. Jung, and E. Ruprecht, 2007: Estimation of the impact of sampling errors in the VOS observations on air-sea fluxes. Part II. Impact on trends and interannual variability. *J. Clim.*, **20**, 302–315.

Guo, Q.Y., et al., 2003: Interdecadal variability of East-Asian summer monsoon and its impact on the climate of China. *Acta Geogr. Sin.*, **4**, 569–576.

Gustafsson, M.E.R., 1997: Raised levels of marine aerosol deposition owing to increased storm frequency: A cause of forest decline in southern Sweden? *Agric. For. Meteorol.*, **84**, 169–177.

Guttman, N., 1991: A sensitivity analysis of the Palmer Hydrologic Drought Index. *Water Resour. Bull.*, **27**, 797–807.

Hahn, C.J., and S.G. Warren, 2003: *Cloud Climatology for Land Stations Worldwide, 1971–1996.* Report NDP-026D, Carbon Dioxide Information Analysis Center, Oak Ridge, TN, USA, 35 pp., http://cdiac.ornl.gov/epubs/ndp/ndp026d/ndp026d.html.

Haimberger, L., 2005: *Homogenization of Radiosonde Temperature Time Series Using ERA-40 Analysis Feedback Information.* ERA-40 Project Report Series 23, European Centre for Medium Range Weather Forecasts, Reading, UK, 68 pp.

Hansen, J., et al., 2001: A closer look at United States and global surface temperature change. *J. Geophys. Res.*, **106**, 23947–23963.

Harnik, N., and E.K.M. Chang, 2003: Storm track variations as seen in radiosonde observations and reanalysis data. *J. Clim.*, **16**, 480–495.

Harris, B.A., and G.A Kelly, 2001: A satellite radiance bias correction scheme for data assimilation. *Q. J. R. Meteorol. Soc.*, **127**, 1453–1468.

Hartmann, D.L., and F. Lo, 1998: Wave-driven flow vacillation in the Southern Hemisphere. *J. Atmos. Sci.*, **55**, 1303–1315.

Hartmann, D.L., et al., 2000: Can ozone depletion and global warming interact to produce rapid climate change? *Proc. Natl. Acad. Sci. U.S.A.*, **97**, 1412–1417.

Hatsushika, H., and K. Yamazaki, 2003: The stratospheric drain over Indonesia and dehydration within the tropical tropopause layer diagnosed by air parcel trajectories. *J. Geophys. Res.*, **108**, 4610, doi:10.1029/2002JD002986.

Hatzianastassiou, N., et al., 2004: Long-term global distribution of Earth's shortwave radiation budget at the top of atmosphere. *Atmos. Chem. Phys.*, **4**, 1217–1235.

Haylock, M.R., and C.M. Goodess, 2004: Interannual variability of extreme European winter rainfall and links with mean large-scale circulation. *Int. J. Climatol.*, **24**, 759–776.

Haylock M.R., et al., 2006. Trends in total and extreme South American rainfall in1960-2000 and links with sea surface temperature. *J. Clim.*, **19**, 1490–1512.

He, H., et al., 2003: Some climatic features of the tropical cyclones landed onto Guangdong Province during the recent 50 years. *Sci. Meteorol. Sin.*, **23**, 401–409 (in Chinese with English abstract).

Heim, R.R., 2002: A review of twentieth-century drought indices used in the United States. *Bull. Am. Meteorol. Soc.*, **83**, 1149–1165.

Held, I.M., and B.J. Soden, 2000: Water vapor feedback and global warming. *Annu. Rev. Energy Environ.*, **25**, 441–475.

Henderson-Sellers, A., 1992: Continental cloudiness changes this century. *GeoJournal*, **27**, 255–262.

Hennessy, K.J., R. Suppiah, and C.M. Page, 1999: Australian rainfall changes, 1910–1995. *Aust. Meteorol. Mag.*, **48**, 1–13.

Herath, S., and U. Ratnayake, 2004: Monitoring rainfall trends to predict adverse impacts – a case study from Sri Lanka (1964–1993). *Global Environ. Change*, **14**, 71–79.

Higgins, R.W., and W. Shi, 2000: Dominant factors responsible for interannual variability of the summer monsoon in the southwestern United States. *J. Clim.*, **13**, 759–776.

Highwood, E.J., B.J. Hoskins, and P. Berrisford, 2000: Properties of the Arctic tropopause. *Q. J. R. Meteorol. Soc.*, **126**, 1515–1532.

Hines, K.M., D.H. Bromwich, and G.J. Marshall, 2000: Artificial surface pressure trends in the NCEP-NCAR reanalysis over the Southern Ocean and Antarctica. *J. Clim.*, **13**, 3940–3952.

Ho, C.-H., et al., 2003: A sudden change summer rainfall characteristics in Korea during the late 1970s. *Int. J. Climatol.*, **23**, 117–128.

Hobbins, M.T., J.A. Ramirez, and T.C. Brown, 2004: Trends in pan evaporation and actual evapotranspiration across the conterminous U.S.: Paradoxical or complementary? *Geophys. Res. Lett.*, **31**, L13503, doi:10/10029/2004GL019846.

Hodgkins, G.A., R.W. Dudley, and T.G. Huntington, 2003: Changes in the timing of high river flows in New England over the 20th century. *J. Hydrol.*, **278**, 244–252.

Hoerling, M., and A. Kumar, 2003: The perfect ocean for drought. *Science*, **299**, 691–694.

Horel, J.D., and J.M. Wallace, 1981: Planetary-scale atmospheric phenomena associated with the Southern Oscillation. *Mon. Weather Rev.*, **109**, 813–829.

Hoskins, B.J., and D.J. Karoly, 1981: Steady linear response of a spherical atmosphere to thermal and orographic forcing. *J. Atmos. Sci.*, **38**, 1179–1196.

Hu, Q., Y. Tawaye, and S. Feng, 2004: Variations of the Northern Hemisphere atmospheric energetics: 1948–2000. *J. Clim.*, **17**, 1975–1986.

Huang, R.H., L. Zhou, and W. Chen, 2003: The progresses of recent studies on the variabilities of the East Asian monsoon and their causes. *Adv. Atmos. Sci.*, **1**, 55–69.

Huffman, G., et al., 1997: The Global Precipitation Climatology Project (GPCP): combined precipitation dataset. *Bull Am. Meteorol. Soc.*, **78**, 5–20.

Hughes, C.W., et al., 2003: Coherence of Antarctic sea levels, Southern Hemisphere Annular Mode, and flow through the Drake Passage. *Geophys. Res. Lett.*, **30**, 1464, doi:10.1029/2003GL017240.

Huntington, T.G., 2006: Evidence for intensification of the global water cycle: Review and synthesis. *J. Hydrol.*, **319**, 83–95.

Huntington, T.G., et al., 2004: Changes in the proportion of precipitation occurring as snow in New England (1949–2000). *J. Clim.*, **17**, 2626–2636.

Hurrell, J.W., 1995: Decadal trends in the North Atlantic Oscillation and relationships to regional temperature and precipitation. *Science*, **269**, 676–679.

Hurrell, J.W., 1996: Influence of variations in extratropical wintertime teleconnections on Northern Hemisphere temperature. *Geophys. Res. Lett.*, **23**, 665–668.

Hurrell, J.W., and H. van Loon, 1994: A modulation of the atmospheric annual cycle in the Southern Hemisphere. *Tellus*, **46A**, 325–338.

Hurrell, J.W., and H. van Loon, 1997: Decadal variations associated with the North Atlantic Oscillation. *Clim. Change*, **36**, 301–326.

Hurrell, J.W., and K.E. Trenberth, 1999: Global sea surface temperature analyses: multiple problems and their implications for climate analysis, modeling and reanalysis. *Bull. Am. Meteorol. Soc.*, **80**, 2661–2678.

Hurrell, J.W., M.P. Hoerling, and C.K. Folland, 2001: Climatic variability over the North Atlantic. In: *Meteorology at the Millennium* [Pearce, R. (ed.)]. Academic Press, London, pp. 143–151.

Hurrell, J.W., et al., 2000: Comparison of tropospheric temperatures from radiosondes and satellites: 1979–98. *Bull. Am. Meteorol. Soc.*, **81**, 2165–2177.

Hurrell, J.W., et al., 2002: The relationship between tropical Atlantic rainfall and the summer circulation over the North Atlantic. In: *Proc. U. S. CLIVAR Atlantic Conf.*, June 12-14, 2001, Boulder, CO, [Legler, D. (ed.)]. U.S. CLIVAR Office, 193pp. 108–110.

Hurrell, J.W., et al., 2003: An overview of the North Atlantic Oscillation. In: *The North Atlantic Oscillation: Climatic Significance and Environmental Impact* [Hurrell, J.W., et al. (eds.)]. Geophysical Monograph 134, American Geophysical Union, Washington, DC, pp. 1–35.

Hurrell, J.W., et al., 2004: Twentieth century North Atlantic climate change. Pt I: Assessing determinism. *Clim. Dyn.*, **23**, 371–389.

Indeje, M., H.F.M. Semazzi, and L.J.Ogallo, 2000. ENSO signals in East African rainfall seasons. *Int. J. Climatol.*, **20**, 19–46.

Inoue, T., and F. Kimura, 2004: Urban effects on low-level clouds around the Tokyo metropolitan area on clear summer days. *Geophys. Res. Lett.*, **31**, L05103, doi:10.1029/2003GL018908.

IPCC, 1999: *Aviation and the Global Atmosphere* [Penner, J. E., et al. (eds.)]. Cambridge University Press, Cambridge, United Kingdom and New York, NY, USA, 384pp.

IPCC, 2001: *Climate Change 2001: The Scientific Basis. Contribution of Working Group I to the Third Assessment Report of the Intergovernmental Panel on Climate Change* [Houghton, J.T., et al. (eds.)]. Cambridge University Press, Cambridge, United Kingdom and New York, NY, USA, 881 pp.

Ishii, M., et al., 2005: Objective analysis of SST and marine meteorological variables for the 20th Century using ICOADS and the Kobe Collection. *Int. J. Climatol.*, **25**, 865–879.

Iskenderian, H., and R. Rosen, 2000: Low-frequency signals in mid-tropospheric submonthly temperature variance. *J. Clim.*, **13**, 2323–2333.

Jacobowitz, H., et al., 2003: The Advanced Very High Resolution Radiometer Pathfinder Atmosphere (PATMOS) climate dataset: A resource for climate research. *Bull. Am. Meteorol. Soc.*, **84**, 785–793.

Jacobs, G.A., and J.L. Mitchell, 1996: Ocean circulation variations associated with the Antarctic Circumpolar Wave. *Geophys. Res. Lett.*, **23**, 2947–2950.

Janicot, S., S. Trzaska, and I. Poccard, 2001: Summer Sahel-ENSO teleconnection and decadal time scale SST variations. *Clim. Dyn.*, **18**, 303–320.

Jauregui, E., and E. Romales, 1996: Urban effects on convective precipitation in Mexico City. *Atmos. Environ.*, **30**, 3383–3389.

Jin, M., and R.E. Dickinson, 2002: New observational evidence for global warming from satellite. *Geophys. Res. Lett.*, **29**(10), doi:10.1029/2001GL013833.

Jin, M., J.M. Shepherd, and M.D. King, 2005: Urban aerosols and their interaction with clouds and rainfall: A case study for New York and Houston. *J. Geophys. Res.*, **110**, D10S20, doi:10.1029/2004JD005081.

Jones, D., et al., 2004. A new tool for tracking Australia's climate variability and change. *Bull. Aust. Meteorol. Oceanogr. Soc.*, **17**, 65–69.

Jones, J.M., and M. Widmann, 2004: Variability of the Antarctic Oscillation during the 20th century. *Nature*, **432**, 290–291.

Jones, P.D., and A. Moberg, 2003: Hemispheric and large-scale surface air temperature variations: An extensive revision and update to 2001. *J. Clim.*, **16**, 206–223.

Jones, P.D., T. Jónsson, and D. Wheeler, 1997: Extension to the North Atlantic Oscillation using early instrumental pressure observations from Gibraltar and south-west Iceland. *Int. J. Climatol.*, **17**, 1433–1450.

Jones, P.D., T.J. Osborn, and K.R. Briffa, 2003: Pressure-based measures of the North Atlantic Oscillation (NAO): A comparison and an assessment of changes in the strength of the NAO and in its influence on surface climate parameters. In: *The North Atlantic Oscillation: Climatic Significance and Environmental Impact* [Hurrell, J.W., et al. (eds.)]. Geophysical Monograph 134, American Geophysical Union, Washington, DC, pp. 51–62.

Jones, P.D., et al., 1990: Assessment of urbanization effects in time series of surface air temperature over land. *Nature*, **347**, 169–172.

Josey, S.A., and R. Marsh, 2005: Surface freshwater flux variability and recent freshening of the North Atlantic in the eastern subpolar gyre. *J. Geophys. Res.*, **110**, C05008, doi:10.1029/2004JC002521.

Joshi, M.M., and K.P. Shine, 2003: A GCM study of volcanic eruptions as a cause of increased stratospheric water vapour. *J. Clim.*, **16**, 3525–3534.

Jury, M.R., 2003: The coherent variability of African river flows: composite climate structure and the Atlantic Circulation. *Water SA*, **29**, 1–10.

Kaiser, D.P., 1998: Analysis of total cloud amount over China, 1951–1994. *Geophys. Res. Lett*, **25**, 3599–3602.

Kaiser, D.P., and Y. Qian, 2002: Decreasing trends in sunshine duration over China for 1954–1998: Indication of increased haze pollution? *Geophys. Res. Lett.*, **29**, 2042, doi:10.1029/2002GL016057.

Kalnay, E., and M. Cai, 2003: Impact of urbanization and land-use change on climate. *Nature*, **423**, 528–531.

Kalnay, E., et al., 1996: The NCEP/NCAR Reanalysis Project. *Bull. Am. Meteorol. Soc.*, **77**, 437–471.

Karl, T.R., S.J. Hassol, C.D. Miller, and W.L. Murray (eds.), 2006: *Temperature Trends in the Lower Atmosphere: Steps for Understanding and Reconciling Differences*. A report by the Climate Change Science Program and Subcommittee on Global Change Research, Washington, DC, 180pp., http://www.climatescience.gov/Library/sap/sap1-1/finalreport/default.htm.

Karoly, D.J., 2003: Ozone and climate change. *Science*, **302**, 236–237.

Karoly, D.J., and Q. Wu, 2005: Detection of regional surface temperature trends. *J. Clim.*, **18**, 4337–4343.

Karoly, D.J., et al., 2003: Global warming contributes to Australia's worst drought. *Australasian Science*, April, 14–17.

Kaufman, Y.J., D. Tanré, and O. Boucher, 2002: A satellite view of aerosols in the climate system. *Nature*, **419**, 215–223.

Keable, M., I. Simmonds, and K. Keay, 2002: Distribution and temporal variability of 500 hPa cyclone characteristics in the Southern Hemisphere. *Int. J. Climatol.*, **22**, 131–150.

Keetch, J.J., and G.M. Byrum, 1988: *A Drought Index for Forest Fire Control*. Research Paper SE-38, US Department of Agriculture, Asheville, NC, 32 pp., http://www.fl-dof.com/fire_weather/information/se038_keetchbyram_di.pdf.

Kent, E.C., and A. Kaplan, 2006: Toward estimating climatic trends in SST data, part 3: Systematic biases. *J. Atmos. Ocean. Technol.*, **23**, 487–500.

Kerr, R., 2000: A North Atlantic climate pacemaker for the centuries. *Science*, **288**, 1984–1985.

Kharin, V.V., and F.W. Zwiers, 2000: Changes in extremes in an ensemble of transient climate simulations with a coupled atmosphere-ocean GCM. *J. Clim.*, **13**, 3760–3780.

Kidson, J.W., 1999: Principal modes of Southern Hemisphere low frequency variability obtained from NCEP-NCAR reanalyses. *J. Clim.*, **12**, 2808–2830.

Kiehl, J.T., and K.E. Trenberth, 1997: Earth's annual global mean energy budget. *Bull. Am. Meteorol. Soc.*, **78**, 197–208.

Kiehl, J.T., J.M. Caron, and J.J. Hack, 2005: On using global climate model simulations to assess the accuracy of MSU retrieval methods for tropospheric warming trends. *J. Clim.*, **18**, 2533–2539.

Kiktev, D., et al., 2003: Comparison of modeled and observed trends in indices of daily climate extremes. *J. Clim.*, **16**, 3560–3571.

Kilpatrick, K.A., G.P. Podesta, and R. Evans, 2001: Overview of the NOAA/NASA advanced very high resolution radiometer Pathfinder algorithm for sea surface temperature and associated matchup database. *J. Geophys. Res.*, **106**, 9179–9198.

Kinter III, J.L., K. Miyakoda, and S. Yang, 2002: Recent change in the connection from the Asia monsoon to ENSO. *J. Clim.*, **15**, 1203–1215.

Kinter III, J.L., et al., 2004: An evaluation of the apparent interdecadal shift in the tropical divergent circulation in the NCEP–NCAR reanalysis. *J. Clim.*, **17**, 349–361.

Kistler, R., et al., 2001: The NCEP-NCAR 50-year reanalysis: Month means CD-ROM and documentation. *Bull. Am. Meteorol. Soc.*, **82**, 247–268.

Klein Tank, A.M.G., and G.P. Können, 2003: Trends in indices of daily temperature and precipitation extremes in Europe, 1946–1999. *J. Clim.*, **16**, 3665–3680.

Klein Tank, A.M.G., et al., 2006: Changes in daily temperature and precipitation extremes in central and south Asia. *J. Geophys. Res.*, **111**, D16105, doi:10.1029/2005JD006316.

Kley, D., J.M. Russell, and C. Phillips, 2000: *SPARC Assessment of Upper Tropospheric and Stratospheric Water Vapour.* WCRP Report No. 113, WMO/TD Report No. 1043, World Climate Research Programme, Geneva, 325 pp.

Klotzbach, P.J., 2006: Trends in global tropical cyclone activity over the past twenty years (1986–2005). *Geophys. Res. Lett.*, **33**, L10805, doi:10.1029/2006GL025881.

Knight, J., et al., 2005: A signature of persistent natural thermohaline circulation cycles in observed climate. *Geophys. Res. Lett.*, **32**, L20708, doi: 1029/2005GL024233.

Knutson, T.R., and S. Manabe, 1998: Model assessment of decadal variability and trends in the Tropical Pacific Ocean. *J. Clim.*, **11**, 2273–2296

Kodera, K., and H. Koide, 1997: Spatial and seasonal characteristics of recent decadal trends in the northern hemispheric troposphere and stratosphere. *J. Geophys. Res.*, **102**, 19433–19447.

Kodera, K., Y. Kuroda, and S. Pawson, 2000: Stratospheric sudden warmings and slowly propagating zonal-mean zonal wind anomalies. *J. Geophys. Res.*, **105**, 12351–12359.

Können, G.P., et al., 1998: Pre-1866 extensions of the Southern Oscillation index using early Indonesian and Tahitian meteorological readings. *J. Clim.*, **11**, 2325–2339.

Kostopoulou, E., and P.D. Jones, 2005: Assessment of climate extremes in the eastern Mediterranean. *Meteorol. Atmos. Phys.*, **89**, 69–85.

Krepper, C.M., N.O. García, and P.D. Jones, 2003: Interannual variability in the Uruguay River basin. *Int. J. Climatol.*, **23**, 103–115.

Kripalani, R.H., and A. Kulkarni, 1997a: Climatic impact of El Niño / La Niña on the Indian monsoon: A new perspective. *Weather*, **52**, 39–46.

Kripalani, R.H., and A. Kulkarni, 1997b: Rainfall variability over Southeast Asia: Connections with Indian monsoon and ENSO extremes: New perspectives. *Int. J. Climatol.*, **17**, 1155–1168.

Kripalani, R.H., and A. Kulkarni, 2001: Monsoon rainfall variations and teleconnections over South and East Asia. *Int. J. Climatol.*, **21**, 603–616.

Kripalani, R.H., A. Kulkarni, and S.S. Sabade, 2001: El Niño Southern Oscillation, Eurasian snow cover and the Indian monsoon rainfall. *Proc. Indian Nat. Sci. Acad.*, **67A**, 361–368.

Kripalani, R.H., et al., 2003: Indian monsoon variability in a global warming scenario. *Natural Hazards*, **29**, 189–206.

Krishnamurthy, V., and B.N. Goswami, 2000: Indian monsoon-ENSO relationship on interdecadal timescale. *J. Clim.*, **13**, 579–595.

Krüger, K., B. Naujokat, and K. Labitzke, 2005: The unusual midwinter warming in the southern hemisphere stratosphere in 2002: A comparison to northern hemisphere phenomena. *J. Atmos. Sci.*, **62**, 602–613.

Kuleshov, Y., and G. de Hoedt, 2003: Tropical cyclone activity in the Southern Hemisphere. *Bull. Aust. Meteorol. Oceanogr. Soc.*, **16**, 135–137.

Kumar, K.K., B. Rajagopalan, and A.M. Cane, 1999: On the weakening relationship between the Indian monsoon and ENSO. *Science*, **284**, 2156–2159.

Kundzewicz, Z.W., et al., 2005: Trend detection in river flow: 1. Annual maximum flow. *Hydrolog. Sci.*, **50**, 797–810.

Kunkel, K.E., et al., 2003: Temporal variations of extreme precipitation events in the United States: 1895–2000. *Geophys. Res. Lett.*, **30**, 1900, doi:10.1029/2003GL018052.

Kunkel, K.E., et al., 2004: Temporal variations in frost-free season in the United States: 1895–2000. *Geophys. Res. Lett.*, **31**, L03201, doi:10.1029/2003GL018624.

Kwok, R., and J.C. Comiso, 2002a: Southern ocean climate and sea ice anomalies associated with the Southern Oscillation. *J. Clim.*, **15**, 487–501.

Kwok, R., and J.C. Comiso, 2002b: Spatial patterns of variability in Antarctic surface temperature: Connections to the Southern Hemisphere Annular Mode and the Southern Oscillation. *Geophys. Res. Lett.*, **29**, 1705, doi:10.1029/2002GL015415.

Labat, D., et al., 2004: Evidence for global runoff increase related to climate warming. *Adv. Water Resour.*, **27**, 631–642.

Labitzke, K., and M. Kunze, 2005: Stratospheric temperature over the Arctic: Comparison of three data sets. *Meteorol. Z.*, **14**, 65–74.

Lambert, S.J., 1996: Intense extratropical Northern Hemisphere winter cyclone events: 1899–1991. *J. Geophys. Res.*, **101**, 21319–21325.

Lammers, R.B., et al., 2001: Assessment of contemporary Arctic river runoff based on observational discharge records. *J. Geophys. Res.*, **106**, 3321–3334.

Lander, M., 1994: An exploratory analysis of the relationship between tropical storm formation in the Western North Pacific and ENSO. *Mon. Weather Rev.*, **122**, 636–651.

Landsea, C.W., 2005: Hurricanes and global warming: Arising from Emanuel 2005a. *Nature*, **438**, E11–E13, doi:10.1038/nature04477.

Landsea, C.W., et al., 1998: The extremely active 1995 Atlantic hurricane season: Environmental conditions and verification of seasonal forecasts. *Mon. Weather Rev.*, **126**, 1174–1193.

Langematz, U., et al., 2003: Thermal and dynamical changes of the stratosphere since 1979 and their link to ozone and CO_2 changes. *J. Geophys. Res.*, **108**, 4027, doi:10.1029/2002JD002069.

Lanzante, J.R., S.A. Klein, and D.J. Seidel, 2003a: Temporal homogenization of monthly radiosonde temperature data. Pt I: Methodology. *J. Clim.*, **16**, 224–240.

Lanzante, J.R., S.A. Klein, and D.J. Seidel, 2003b: Temporal homogenization of monthly radiosonde temperature data. Pt II: Trends, sensitivities, and MSU comparison. *J. Clim.*, **16**, 241–262.

Latif, M., 2001: Tropical Pacific/Atlantic ocean interactions at multidecadal time scales. *Geophys. Res. Lett.*, **28**, 539–542.

Latif, M., R. Kleeman, and C. Eckert, 1997: Greenhouse warming, decadal variability, or El Nino? An attempt to understand the anomalous 1990s. *J. Clim.*, **10**, 2221–2239.

Lau, N-C., and M.J. Nath, 2004: Coupled GCM simulation of atmosphere-ocean variability associated with zonally asymmetric SST changes in the tropical Indian Ocean. *J. Clim.*, **17**, 245–265.

Lawrimore, J., et al., 2002: Beginning a new era of drought monitoring across North America. *Bull. Am. Meteorol. Soc.*, **83**, 1191–1192.

Le Barbe, L., T. Lebel, and D. Tapsoba, 2002: Rainfall variability in West Africa during the years 1950–1990. *J. Clim.*, **15**, 187–202.

Lefebvre, W., et al., 2004: Influence of the Southern Annular Mode on the sea ice-ocean system. *J. Geophys. Res.*, **109**, C09005, doi:10.1029/2004JC002403.

Levinson, D.H. (ed.), 2005: State of the climate in 2004. *Bull. Am. Meteorol. Soc.*, **86**(6), S1–S84.

Levinson, D.H., and A.M. Waple (eds.), 2004: State of the climate in 2003. *Bull. Am. Meteorol. Soc.*, **85**(6), S1–S72.

L'Heureux, M.L., and D.W.J. Thompson, 2006: Observed relationships between the El Niño–Southern Oscillation and the extratropical zonal-mean circulation. *J. Clim.*, **19**, 276–287.

Li, Q., et al., 2004: Urban heat island effect on annual mean temperature during the last 50 years in China. *Theor. Appl. Climatol.*, **79**, 165–174.

Li, W., and P.M. Zhai, 2003: Variability in occurrence of China's spring dust storm and its relationship with atmospheric general circulation. *Acta Meteorol. Sin.*, **17**(4), 396–405.

Liebmann, B., et al., 1999: Submonthly convective variability over South America and the South Atlantic convergence zone. *J. Clim.*, **12**, 1877–1891.

Liebmann, B., et al., 2004: An observed trend in Central South American precipitation. *J. Clim.*, **22**, 4357–4367.

Liepert, B.G., 2002: Observed reductions of surface solar radiation at sites in the United States and worldwide from 1961 to 1990. *Geophys. Res. Lett.*, **29**, 1421, 10.1029/2002GL014910.

Liepert, B.G., et al., 2004: Can aerosols spin down the water cycle in a warmer and moister world? *Geophys. Res. Lett.*, **31**, doi:10.1029/2003GL019060.

Lim, E.-P., and I. Simmonds, 2002: Explosive cyclone development in the Southern Hemisphere and a comparison with Northern Hemisphere events. *Mon. Weather Rev.*, **130**, 2188–2209.

Limpasuvan, V., and D.L. Hartmann, 2000: Wave-maintained annular modes of climate variability. *J. Clim.*, **13**, 4414–4429.

Limpasuvan, V., D. Thompson and D. Hartmann, 2004: The life cycle of northern hemispheric sudden stratospheric warmings. *J. Clim.*, **17**, 2584–2596.

Lins, H.F., and J.R. Slack, 1999: Streamflow trends in the United States. *Geophys. Res. Lett.*, **26**, 227–230.

Linsley, B.K., et al., 2004: Geochemical evidence from corals for changes in the amplitude and spatial pattern of South Pacific interdecadal climate variability over the last 300 years. *Clim. Dyn.*, **22**, doi:10.1007/s00382-003-0364-y.

Liu, B.H., et al., 2004a: A spatial analysis of pan evaporation trends in China, 1955–2000. *J. Geophys. Res.*, **109**, D15102, doi:10.1029/2004JD004511.

Liu, J., J.A. Curry, and D.G. Martinson, 2004b: Interpretation of recent Antarctic sea ice variability. *Geophys. Res. Lett.*, **31**, L02205, doi:10.1029/2003GL018732.

Liu, K.S., and J.C.L. Chan, 2003: Climatological characteristics and seasonal forecasting of tropical cyclones making landfall along the south China coast. *Mon. Weather Rev.*, **131**, 1650–1662.

Lorenz, D.J., and D.L. Hartmann, 2001: Eddy–zonal flow feedback in the Southern Hemisphere. *J. Atmos. Sci.*, **58**, 3312–3327.

Lorenz, D.J., and D.L. Hartmann, 2003: Eddy–zonal flow feedback in the Northern Hemisphere winter. *J. Clim.*, **16**, 1212–1227.

Lozano, I., and V. Swail, 2002: The link between wave height variability in the North Atlantic and the storm track activity in the last four decades. *Atmos.-Ocean*, **40**, 377–388.

Lucarini, V., and G.L. Russell, 2002: Comparison of mean climate trends in the Northern Hemisphere between National Centers for Environmental Prediction and two atmosphere-ocean model forced runs. *J. Geophys. Res.*, **107**, 4269, doi:10.1029/2001JD001247.

Lugina, K.M., et al., 2005: Monthly surface air temperature time series area-averaged over the 30-degree latitudinal belts of the globe, 1881-2004. In: *Trends: A Compendium of Data on Global Change*. Carbon Dioxide Information Analysis Center, Oak Ridge National Laboratory, US Department of Energy, Oak Ridge, TN, http://cdiac.esd.ornl.gov/trends/temp/lugina/lugina.html.

Luo, Y., et al., 2001: Characteristics of spatial distribution of yearly variation of aerosol optical depth over China in the last 30 years. *J. Geophys. Res.*, **106**(D13), 14501–14513.

Luterbacher, J., et al., 2004: European seasonal and annual temperature variability, trends, and extremes since 1500. *Science*, **303**, 1499–1503.

Ma, Z.G., and C.B. Fu, 2003: Interannual characteristics of the surface hydrological variables over the arid and semi-arid areas of northern China. *Global Planet. Change*, **37**, 189–200.

Madden, R.A., and J. Williams, 1978: The correlation between temperature and precipitation in the United States and Europe. *Mon. Weather Rev.*, **106**, 142–147.

Maheras, P., et al., 2004: On the relationships between circulation types and changes in rainfall variability in Greece. *Int. J. Climatol.*, **24**, 1695–1712.

Maistrova, V.V., et al., 2003: Long-term trends in temperature and specific humidity of free atmosphere in the Northern Polar region. *Dokl. Earth Sci.*, **391**, 755–759.

Mann, M.E., 2004: On smoothing potentially non-stationary climate time series. *Geophys. Res. Lett.*, **31**, L07214, doi:10.1029/2004GL019569.

Manney, G., et al., 2005: The remarkable 2003–2004 winter and other recent warm winters in the Arctic stratosphere since the late 1990s. *J. Geophys. Res.*, **110**, D04107, doi:10.1029/2004JD005367.

Manton, M.J., et al., 2001: Trends in extreme daily rainfall and temperature in Southeast Asia and the South Pacific: 1961–1998. *Int. J. Climatol.*, **21**, 269–284.

Mantua, N.J., and S.J. Hare, 2002: The Pacific Decadal Oscillation. *J. Oceanogr.*, **58**, 35–44.

Mantua, N.J., et al., 1997: A Pacific interdecadal climate oscillation with impacts on salmon production. *Bull. Am. Meteorol. Soc.*, **78**, 1069–1079.

Marengo, J., 2004: Interdecadal variability and trends of rainfall across the Amazon Basin. *Theor. Appl. Climatol.*, **78**, 79–96.

Marengo, J.A., et al., 2004: Climatology of the low-level jet east of the Andes as derived from the NCEP–NCAR Reanalyses: Characteristics and temporal variability. *J. Clim.*, **17**, 2261–2280.

Marshall, G.J., 2002: Analysis of recent circulation and thermal advection change on the northern Antarctic Peninsula. *Int. J. Climatol.*, **22**, 1557–1567.

Marshall, G.J., 2003: Trends in the Southern Annular Mode from observations and reanalyses. *J. Clim.*, **16**, 4134–4143.

Marshall, G.J., et al., 2004: Causes of exceptional atmospheric circulation changes in the Southern Hemisphere. *Geophys. Res. Lett.*, **31**, L14205, doi:10.1029/2004GL019952.

Marshall, J., H. Johnson, and J. Goodman, 2001: A study of the interaction of the North Atlantic Oscillation with the ocean circulation. *J. Clim.*, **14**, 1399–1421.

Martyn, D., 1992: *Climates of the World*. Elsevier, Amsterdam, 436 pp.

Maugeri, M., et al., 2001: Trends in Italian cloud amount 1951-1996. *Geophys. Res. Lett.*, **28**, 4551–4554.

Mauget, S.A., 2003a: Intra- to multidecadal climate variability over the continental United States: 1932–99. *J. Clim.*, **16**, 2215–2231.

Mauget, S.A., 2003b: Multidecadal regime shifts in US streamflow, precipitation, and temperature at the end of the twentieth century. *J. Clim.*, **16**, 3905–3916.

McBride, J.L., 1998: Indonesia, Papua New Guinea, and tropical Australia: The southern hemisphere monsoon. In: *Meteorology of the Southern Hemisphere* [Karoly, D., and D. Vincent (eds.)]. American Meteorological Society, Boston, MA, pp. 89–99.

McCabe, G.J., and D.M. Wolock, 2002: Trends and temperature sensitivity of moisture conditions in the conterminous United States. *Clim. Res.*, **20**, 19–29.

McCabe, G.J., M.P. Clark, and M.C. Serreze, 2001: Trends in Northern Hemisphere surface cyclone frequency and intensity. *J. Clim.*, **14**, 2763–2768.

McCabe, G.J., M. Palecki, and J.L. Betancourt, 2004: Pacific and Atlantic Ocean influences on multi-decadal drought frequency in the United States. *Proc. Natl. Acad. Sci. U.S.A.*, **101**, 4136–4141.

McCarthy, M.P., and R. Toumi, 2004: Observed interannual variability of tropical troposphere relative humidity. *J. Clim.*, **17**, 3181–3191.

McKitrick, R., and P.J. Michaels, 2004: A test of corrections for extraneous signals in gridded surface temperature data. *Clim. Res.*, **26**, 159–173.

McPhaden, M.J., and D. Zhang, 2002: Slowdown of the meridional overturning circulation of the upper Pacific ocean. *Nature*, **415**, 603–608.

Mears, C.A., and F.J. Wentz, 2005: The effect of diurnal correction on satellite-derived lower tropospheric temperature. *Science*, **309**, 1548–1551.

Mears, C.A., M.C. Schabel, and F.J. Wentz, 2003: A reanalysis of the MSU channel 2 tropospheric temperature record. *J. Clim.*, **16**, 3650–3664.

Mehta, A., and J. Susskind, 1999: Outgoing longwave radiation from the TOVS Pathfinder Path A data set. *J. Geophys. Res.*, **104**, 12193–12212.

Menon, S., et al., 2002: Climate effects of black carbon aerosols in China and India. *Science*, **297**, 2250–2253.

Meredith, M.P., et al., 2004: Changes in the ocean transport through Drake Passage during the 1980s and 1990s, forced by changes in the Southern Annular Mode. *Geophys. Res. Lett.*, **31**, L21305, doi:10.1029/2004GL021169.

Milly, P.C.D., and K.A. Dunne, 2001: Trends in evaporation and surface cooling in the Mississippi River basin. *Geophys. Res. Lett.*, **28**, 1219–1222.

Milly, P.C.D., et al., 2002: Increasing risk of great floods in a changing climate. *Nature*, **415**, 514–517.

Miloshevich, L.M., et al., 2004: Development and validation of a time-lag correction for Vaisala radiosonde humidity measurements. *J. Atmos. Ocean. Technol.*, **21**, 1305–1327.

Minnis, P., et al., 2004: Contrails, cirrus trends, and climate. *J. Clim.*, **17**, 1671–1685.

Minschwaner, K., and A.E. Dessler, 2004: Water vapor feedback in the tropical upper troposphere: Model results and observations. *J. Clim.*, **17**, 1272–1282.

Minobe, S., 1997: A 50–70 year climatic oscillation over the North Pacific and North America. *Geophys. Res. Lett.*, **24**, 683–686.

Minobe, S., 1999: Resonance in bidecadal and pentadecadal oscillations over the North Pacific: Role in climate regime shifts. *Geophys. Res. Lett.*, **26**, 855–858.

Minobe, S., and T. Nakanowatari, 2002: Global structure of bidecadal precipitation variability in boreal winter. *Geophys. Res. Lett.*, **29**, 1396, doi:10.1029/2001GL014447.

Mitas, C.M., and A. Clement, 2005: Has the Hadley cell been strengthening in recent decades? *Geophys. Res. Lett.*, **32**, L03809, doi:10.1029/2004GL021765.

Mitchell, T.D., and P.D. Jones, 2005: An improved method of constructing a database of monthly climate observations and associated high-resolution grids. *Int. J. Climatol.*, **25**, 693–712

Mo, K.C., 2000: Relationships between low-frequency variability in the Southern Hemisphere and sea surface temperature anomalies. *J. Clim.*, **13**, 3599–3610.

Mo, K.C., and R.W. Higgins, 1998: The Pacific-South American modes and tropical convection during the Southern Hemisphere winter. *Mon. Weather Rev.*, **126**, 1581–1596.

Moberg, A., and P.D. Jones, 2005: Trends in indices for extremes of daily temperature and precipitation in central and western Europe 1901–1999. *Int. J. Climatol.*, **25**, 1173–1188.

Molders, N., and M.A. Olson, 2004: Impact of urban effects on precipitation in high latitudes. *J. Hydrometeorol.*, **5**, 409–429.

Monahan, A.H., L. Pandolfo, and J.C. Fyfe, 2001: The preferred structure of variability of the Northern Hemisphere atmospheric circulation. *Geophys. Res. Lett.*, **28**, 1019–1022.

Mu, Q.Z., et al., 2002: Simulation study on variation of Western Pacific subtropical high during the last hundred years. *Chin. Sci. Bull.*, **7**, 550–553.

Mudelsee, M., et al., 2003: No upward trends in the occurrence of extreme floods in central Europe. *Nature*, **425**, 166–169.

Nakamura, H., and T. Sampe, 2002: Trapping of synoptic-scale disturbances into the North-Pacific subtropical jet core in midwinter. *Geophys. Res. Lett.*, **29**(16), doi:10.1029/2002GL015535.

Nakamura, H., T. Izumi, and T. Sampe, 2002: Interannual and decadal modulations recently observed in the Pacific storm track activity and East Asia winter monsoon. *J. Clim.*, **15**, 1855–1874.

Nakamura, H., et al., 2004: Observed associations among storm tracks, jet streams and midlatitude oceanic fronts. In: *Earth's Climate: The Ocean-Atmosphere Interaction* [Wang, C., S.-P. Xie, and J. A. Carton (eds.)]. Geophysical Monograph 147, American Geophysical Union, Washington, DC, pp. 329–346.

Naujokat, B., et al., 2002: The early major warming in December 2001 – Exceptional? *Geophys. Res. Lett.*, **29**, 2023, doi:10.1029/2002GL015316.

Nazemosadat, M.J., and I. Cordery, 2000: On the relationships between ENSO and autumn rainfall in Iran. *Int. J. Climatol.*, **20**, 47–61.

Nazemosadat, M.J., and A.R. Ghasemi, 2004: Quantifying the ENSO-related shifts in the intensity and probability of drought and wet periods in Iran. *J. Clim.*, **17**, 4005–4018.

Nedoluha, G.E., et al., 2003: An evaluation of trends in middle atmospheric water vapor as measured by HALOE, WVMS, and POAM. *J. Geophys. Res.*, **108**, 4391, doi:10.1029/2002JD003332.

New, M., et al., 2006: Evidence of trends in daily climate extremes over southern and West Africa. *J. Geophys. Res.*, **111**, D14102, doi:10.1029/2005JD006289.

Newman, M., G. Compo, and M.A. Alexander, 2003: ENSO-forced variability of the Pacific Decadal Oscillation. *J. Clim.*, **23**, 3853–3857.

Newman, P.A., and E.R. Nash, 2000: Quantifying the wave drinking of the stratosphere. *J. Geophys. Res.*, **105**, 12485–12497.

Newman, P.A., and E.R. Nash, 2005: The unusual Southern Hemisphere stratosphere winter of 2002. *J. Atmos. Sci.*, **62**, doi:10.1175/JAS-3323.1.

Nicholls, N., 2004: The changing nature of Australian droughts. *Clim. Change*, **63**, 323–336.

Nobre, P., and J. Shukla, 1996: Variations of sea surface temperature, wind stress, and rainfall over the tropical Atlantic and South America. *J. Clim.*, **9**, 2464–2479.

Nogués-Paegle, J., et al., 2002: Progress in Pan American CLIVAR research: Understanding the South American monsoon. *Meteorologica*, **27**, 3–32.

Noone, D., and I. Simmonds, 2002: Annular variations in moisture transport mechanisms and the abundance of $\delta^{18}O$ in Antarctic snow. *J. Geophys. Res.*, **107**, 4742, doi:10.1029/2002JD002262.

Norris, J.R., 2000: What can cloud observations tell us about climate variability? *Space Sci. Rev.*, **94**, 375–380.

Norris, J.R., 2005a: Multidecadal changes in near-global cloud cover and estimated cloud cover radiative forcing. *J. Geophys. Res.*, **110**, D08206, doi:10.1029/2004JD005600.

Norris, J.R., 2005b: Trends in upper-level cloud cover and atmospheric circulation over the Indo-Pacific region between 1952 and 1997. *J. Geophys. Res.*, **110**, D21110, doi:10.1029/2005JD006183.

Notholt, J., et al., 2005: Influence of tropospheric SO_2 emissions on particle formation and the stratospheric humidity. *Geophys. Res. Lett.*, **32**, L07810, doi:10.1029/2004GL022159.

O'Carroll, A.G., R.W. Saunders, and J.G. Watts, 2006: The measurement of the sea surface temperature climatology by satellites from 1991 to 2005. *J. Atmos. Ocean. Technol.*, **23**,1573-1582.

Ohmura, A., and M. Wild, 2002: Is the hydrological cycle accelerating? *Science*, **298**, 1345–1346.

Oinas, V., et al., 2001: Radiative cooling by stratospheric water vapor: big differences in GCM results. *Geophys. Res. Lett.*, **28**, 2791–2794.

Oke, P.R., and M.H. England, 2004: Oceanic response to changes in the latitude of the Southern Hemisphere subpolar westerly winds. *J. Clim.*, **17**, 1040–1054.

Oltmans, S.J., et al., 2000: The increase in stratospheric water vapor from balloon borne, frostpoint hygrometer measurements at Washington, DC, and Boulder, Colorado. *Geophys. Res. Lett*, **27**, 3453–3456.

Omran, M.A., 2000: Analysis of solar radiation over Egypt. *Theor. Appl. Climatol.*, **67**, 225–240.

Oort, A.H., and J.J.Yienger, 1996: Observed interannual variability in the Hadley circulation and its connection to ENSO. *J. Clim.*, **9**, 2751–2767.

Orr, A., et al., 2004: A 'low-level' explanation for the recent large warming trend over the western Antarctic Peninsula involving blocked winds and changes in zonal circulation. *Geophys. Res. Lett.*, **31**, L06204, doi:10.1029/2003GL019160.

Osborn, T.J., and M. Hulme, 2002: Evidence for trends in heavy rainfall events over the U.K. *Philos. Trans. R. Soc. London Ser. A*, **360**, 1313–1325.

Osborn, T.J., et al., 2000: Observed trends in the daily intensity of United Kingdom precipitation. *Int. J. Climatol.*, **20**, 347–364.

Ostermeier, G.M., and J.M. Wallace, 2003: Trends in the North Atlantic Oscillation – Northern Hemisphere annular mode during the twentieth century. *J. Clim.*, **16**, 336–341.

Overland, J. E., and M. Wang, 2005: The Arctic climate paradox: The recent decrease of the Arctic Oscillation. *Geophys. Res. Lett.*, **32**, L23808, doi:10.1029/2005GL024254.

Paciorek, C.J., et al., 2002: Multiple indices of Northern Hemisphere cyclone activity, winters 1949–99. *J. Clim.*, **15**, 1573–1590.

Pagano, T., et al., 2004: Water year 2004: Western water managers feel the heat. *Eos*, **85**, 392–393.

Pallé, E., et al., 2004: Changes in Earth's reflectance over the past two decades. *Science*, **304**, 1299–1301.

Palmer, T.N., 1999: A nonlinear dynamical perspective on climate prediction. *J. Clim.*, **12**, 575–591.

Palmer, W.C., 1965: *Meteorological Drought*. Research Paper 45, US Department of Commerce, Weather Bureau, Washington, DC, 58 pp. [Available from NOAA Library and Information Services Division, Washington, DC 20852.]

Park, Y., F. Roquet, and F. Vivier, 2004: Quasi-stationary ENSO wave signals versus the Antarctic Circumpolar Wave scenario. *Geophys. Res. Lett.*, **31**, L09315, doi:10.1029/2004GL019806.

Parker, D.E., 2004: Large-scale warming is not urban. *Nature*, **432**, 290–290.

Parker, D.E., 2006: A demonstration that large-scale warming is not urban. *J. Clim.*, **19**, 2882–2895.

Parker, D.E., and D.I. Cox, 1995: Towards a consistent global climatological rawinsonde data-base. *Int. J. Climatol.*, **15**, 473–496.

Parker, D.E., et al., 1997: A new global gridded radiosonde temperature data base and recent temperature trends. *Geophys. Res. Lett.*, **24**, 1499–1502.

Pawson, S., and B. Naujokat, 1999: The cold winters of the middle 1990s in the northern lower stratosphere. *J. Geophys. Res.*, **104**, 14209–14222.

Pekárová, P., P. Miklánek, and J. Pekár, 2003: Spatial and temporal runoff oscillation analysis of the main rivers of the world during the 19th–20th centuries. *J. Hydrol.*, **274**, 62–79.

Penman, H.L., 1948: Natural evaporation from open water, bare soil and grass. *Proc. R. Soc. London Ser. A*, **193**, 120–145.

Pepin, N.C., and D.J. Seidel, 2005: A global comparison of surface and free-air temperatures at high elevations. *J. Geophys. Res.*, **110**, D03104, doi:10.1029/2004JD005047.

Peterson, T.C., 2003: Assessment of urban versus rural *in situ* surface temperatures in the contiguous United States: no difference found. *J. Clim.*, **16**, 2941–2959.

Peterson, T.C., and R.S. Vose, 1997: An overview of the Global Historical Climatology Network temperature database. *Bull. Am. Meteorol. Soc.*, **78**, 2837–2848.

Peterson, T.C., and T.W. Owen, 2005: Urban heat island assessment: Metadata are important. *J. Clim.*, **18**, 2637–2646.

Peterson, T.C., V.S. Golubev, and P.Ya. Groisman, 1995: Evaporation losing its strength. *Nature*, **377**, 687–688.

Peterson, T.C., et al., 1999: Global rural temperature trends. *Geophys. Res. Lett.*, **26**, 329–332.

Peterson, T.C., et al., 2000: A blended satellite – *in situ* near-global surface temperature dataset. *Bull. Am. Meteorol. Soc.*, **81**, 2157–2164.

Peterson, T.C., et al., 2002: Recent changes in climate extremes in the Caribbean region. *J. Geophys. Res.*, **107**, 4601, doi:10.1029/2002JD002251.

Pezza, A.B., and I. Simmonds, 2005: The first South Atlantic hurricane: unprecedented blocking, low shear and climate change. *Geophys. Res. Lett.*, **32**, L15712, doi:10.1029/2005GL023390.

Philipona, R., and B. Dürr, 2004: Greenhouse forcing outweighs decreasing solar radiation driving rapid temperature rise over land. *Geophys. Res. Lett.*, **31**, L22208, doi:10.1029/2004GL020937.

Philipona, R., et al., 2004: Radiative forcing - measured at Earth's surface - corroborate the increasing greenhouse effect. *Geophys. Res. Lett.*, **31**, L15712, doi:10.1029/2003GL018765.

Philipona, R., et al., 2005: Anthropogenic greenhouse forcing and strong water vapor feedback increase temperature in Europe. *Geophys. Res. Lett.*, **32**, L19809, doi:1029/ 2005GL023624.

Picon, L., et al., 2003: A new METEOSAT "water vapor" archive for climate studies. *J. Geophys. Res.*, **108**, 4301, doi:10.1029/2002JD002640.

Piechota, T., et al., 2004: The western drought: How bad is it? *Eos*, **85**(32), 301.

Pierce, D.W., 2001: Distinguishing coupled ocean–atmosphere interactions from background noise in the North Pacific. *Prog. Oceanogr.*, **49**, 331–352.

Pinker, R.T., B. Zhang, and E.G. Dutton, 2005: Do satellites detect trends in surface solar radiation? *Science*, **308**, 850–854.

Plummer, N., et al., 1999: Changes in climate extremes over the Australian region and New Zealand during the Twentieth Century. *Clim. Change*, **42**, 183–202.

Polyakov, I.V., et al., 2003: Variability and trends of air temperature in the Maritime Arctic, 1875–2000. *J. Clim.*, **16**, 2067–2077.

Power, H.C., and D.M. Mills, 2005: Solar radiation climate change over South Africa and an assessment of the radiative impact of volcanic eruptions. *Int. J. Climatol.*, **25**, 295–318.

Power, S., et al., 1998: Australian temperature, Australian rainfall and the Southern Oscillation, 1910–1992: coherent variability and recent changes. *Aust. Meteorol. Mag.*, **47**, 85–101.

Power, S., et al., 1999a: Decadal climate variability in Australia during the twentieth century. *Int. J. Climatol.*, **19**, 169–184.

Power, S., et al., 1999b: Inter-decadal modulation of the impact of ENSO on Australia. *Clim. Dyn.*, **15**, 319–324.

Probst, J.L., and Y. Tardy, 1987: Long-range streamflow and world continental runoff fluctuations since the beginning of this century. *J. Hydrol.*, **94**, 289–311.

Probst, J.L., and Y. Tardy, 1989: Global runoff fluctuations during the last 80 years in relation to world temperature-change. *Am. J. Sci.*, **289**, 267–285.

Przybylak, R., 2000: Diurnal temperature range in the Arctic and its relation to hemispheric and Arctic circulation patterns. *Int. J. Climatol.*, **20**, 231–253.

Qian, T., et al., 2006a: Simulation of global land surface conditions from 1948–2004. Pt I: Forcing data and evaluations. *J. Hydrometeorol.*, **7**, 953–975.

Qian, W.H., L.S. Quan, and S.Y. Shi, 2002: Variations of the dust storm in China and its climatic control. *J. Clim.*, **15**, 1216–1229.

Qian, W.H., et al., 2003: Centennial-scale dry-wet variations in East Asia. *Clim. Dyn.*, **21**, 77–89.

Qian, Y., et al., 2006b: More frequent cloud-free sky and less surface solar radiation in China from 1955 to 2000. *Geophys. Res. Lett.*, **33**, L01812, doi:10.1029/2005GL024586.

Quadrelli, R., and J.M. Wallace, 2004: A simplified linear framework for interpreting patterns of Northern Hemisphere wintertime climate variability. *J. Clim.*, **17**, 3728–3744.

Quadrelli, R., V. Pavan, and F. Molteni, 2001: Wintertime variability of Mediterranean precipitation and its links with large-scale circulation anomalies. *Clim. Dyn.*, **17**, 457–466.

Ramanathan, V., et al., 2001: Aerosols, climate and the hydrological cycle. *Science*, **294**, 2119–2124.

Ramaswamy, V., and M. Schwarzkopf, 2002: Effects of ozone and well-mixed gases on annual-mean stratospheric temperature trends. *Geophys. Res. Lett.*, **29**, 2064, doi:10.1029/2002GL05141.

Ramaswamy, V., et al., 2001: Stratospheric temperature changes: observations and model simulations. *Rev. Geophys.*, **39**, 71–122.

Randel, D.L., et al., 1996: A new global water vapor dataset. *Bull. Am. Meteorol. Soc.*, **77**, 1233–1246.

Randel, W.J., and F. Wu, 1999: Cooling of the Arctic and Antarctic polar stratospheres due to ozone depletion. *J. Clim.*, **12**, 1467–1479.

Randel, W.J., and F. Wu, 2006: Biases in stratospheric temperature trends derived from historical radiosonde data. *J. Clim.*, **19**, 2094–2104.

Randel, W.J., F. Wu, and D.J. Gaffen, 2000: Interannual variability of the tropical tropopause derived from radiosonde data and NCEP reanalyses. *J. Geophys. Res.*, **105**, 15509–15524.

Randel, W.J, F. Wu, and R. Stolarski, 2002: Changes in column ozone correlated with the stratospheric EP flux. *J. Meteorol. Soc. Japan*, **80**, 849–862.

Randel, W.J., et al., 2004a: Interannual changes of stratospheric water vapor and correlations with tropical tropopause temperatures. *J. Atmos. Sci.*, **61**, 2133–2148.

Randel, W.J., et al., 2004b: The SPARC intercomparison of middle-atmosphere climatologies. *J. Clim.*, **17**, 986–1003.

Rashid, H.A., and I. Simmonds, 2004: Eddy-zonal flow interactions associated with the Southern Hemisphere annular mode: Results from NCEP-DOE reanalysis and a quasi-linear model. *J. Atmos. Sci.*, **61**, 873–888.

Rashid, H.A., and I. Simmonds, 2005: Southern Hemisphere annular mode variability and the role of optimal nonmodal growth. *J. Atmos. Sci.*, **62**, 1947–1961

Rayner, N.A., et al., 2003: Global analyses of sea surface temperature, sea ice, and night marine air temperature since the late nineteenth century. *J. Geophys. Res.*, **108**, 4407, doi:10.1029/2002JD002670.

Rayner, N.A., et al., 2006: Improved analyses of changes and uncertainties in sea surface temperature measured *in situ* since the mid-nineteenth century: the HadSST2 dataset. *J. Clim.*, **19**, 446–469.

Read, W.G., et al., 2004: Dehydration in the tropical tropopause layer: Implications from the UARS Microwave Limb Sounder. *J. Geophys. Res.*, **109**, D06110, doi:10.1029/2003JD004056.

Regonda, S.K., et al., 2005: Seasonal cycle shifts in hydroclimatology over the Western U.S. *J. Clim..*, **18**, 372–384.

Ren, G.Y., et al., 2005: Climate changes of mainland China over the past half century, *Acta Meteorol. Sin.*, **63** (6): 942–955 (in Chinese).

Renwick, J.A., 1998: ENSO-related variability in the frequency of South Pacific blocking. *Mon. Weather Rev.*, **126**, 3117–3123.

Renwick, J.A., 2002: Southern Hemisphere circulation and relations with sea ice and sea surface temperature. *J. Clim.*, **15**, 3058–3068.

Renwick, J.A., 2004: Trends in the Southern Hemisphere polar vortex in NCEP and ECMWF reanalyses. *Geophys. Res. Lett.*, **31**, L07209, doi:10.1029/2003GL019302.

Renwick, J.A., 2005: Persistent positive anomalies in the Southern Hemisphere circulation. *Mon. Weather Rev.*, **133**, 977–988.

Renwick, J.A., and M.J. Revell, 1999: Blocking over the South Pacific and Rossby wave propagation. *Mon. Weather Rev.*, **127**, 2233–2247.

Revercombe, H.E., et al., 2003: The ARM program's water vapor intensive observation periods: Overview, initial accomplishments, and future challenges. *Bull. Am. Meteorol. Soc.*, **84**, 217–236.

Reynolds, R.W., C.L. Gentemann, and F. Wentz, 2004: Impact of TRMM SSTs on a climate-scale SST analysis. *J. Clim.*, **17**, 2938–2952.

Reynolds, R.W., et al., 2002: An improved in situ and satellite SST analysis for climate. *J. Clim.*, **15**, 1609–1625.

Robertson, A.W., C.R. Mechoso, and N.O. García, 2001a: Interannual prediction of the Paraná River. *Geophys. Res. Lett.*, **28**, 4235–4238.

Robertson, F.R., R.W. Spencer, and D.E. Fitzjarrald, 2001b: A new satellite deep convective ice index for tropical climate monitoring: possible implications for existing oceanic precipitation data sets, *Geophys. Res. Lett.*, **28**, 251–254.

Robertson F.R., D.E. Fitzjarrald, and C.D. Kummerow, 2003: Effects of uncertainty in TRMM precipitation radar path integrated attenuation on interannual variations of tropical oceanic rainfall. *Geophys. Res. Lett.*, **30**, 1180, doi:10.1029/2002GL016416.

Robeson, S., 2004: Trends in time-varying percentiles of daily minimum and maximum temperature over North America. *Geophys. Res. Lett*, **31**, L04203, doi:10.1029/2003GL019019.

Robinson, P.J., 2000: Temporal trends in United States dew point temperatures. *Int. J. Climatol.*, **20**, 985–1002.

Robock, A., et al., 2000: The global soil moisture data bank. *Bull. Am. Meteorol. Soc.*, **81**, 1281–1299.

Robock, A., et al., 2005: Forty five years of observed soil moisture in Ukraine: No summer desiccation (yet). *Geophys. Res. Lett.*, **32**, L03401, doi:10.0129/2004GL021914.

Röckmann, T., et al., 2004: The impact of anthropogenic chlorine emissions, stratospheric ozone change and chemical feedbacks on stratospheric water. *Atmos. Chem. Phys.*, **4**, 693–699.

Roderick, M.L., and G.D. Farquhar, 2002: The cause of decreased pan evaporation over the past 50 years. *Science*, **298**, 1410–1411.

Roderick, M.L., and G.D. Farquhar, 2004: Changes in Australian pan evaporation from 1970–2002. *Int. J. Climatol.*, **24**, 1077–1090.

Roderick, M.L., and G.D. Farquhar, 2005: Changes in New Zealand pan evaporation since the 1970s. *Int. J. Climatol.*, **25**, 2031–2039.

Rodwell, M.J., 2003: On the predictability of North Atlantic climate. In: *The North Atlantic Oscillation: Climatic significance and environmental impact* [Hurrell, J.W., et al. (eds.)]. Geophysical Monograph 134, American Geophysical Union, Washington, DC, pp. 173–192.

Roscoe, H.K., 2004: A review of stratospheric H_2O and NO_2. *Adv. Space Res.*, **34**, 1747–1754.

Rosenfeld, D., 2000: Suppression of rain and snow by urban and industrial air pollution. *Science*, **287**, 1793–1796.

Rosenlof, K.H., 2002: Transport changes inferred from HALOE water and methane measurements. *J. Meteorol. Soc. Japan*, **80**, 831–848.

Rosenlof, K.H., et al., 2001: Stratospheric water vapor increases over the past half-century. *Geophys. Res. Lett*, **28**, 1195–1198.

Rossow, W.B., and E.N. Dueñas, 2004: The International Satellite Cloud Climatology Project (ISCCP) web site. *Bull. Am. Meteorol. Soc.*, **85**, 167–172.

Rudolf, B., and J. Rapp, 2003: The century flood of the River Elbe in August 2002: Synoptic weather development and climatological aspects. In: *Quart. Rep. German NWP-System Deutscher Wetterdienst*, No. 2, Pt 1, pp. 8–23.

Rudolf, B., et al., 1994: Terrestrial precipitation analysis: Operational method and required density of point measurements. In: *Global Precipitations and Climate Change* [M. Bubois, and F. Désalmand (eds.)]. NATO ASI Series I, **26**, Springer Verlag, Berlin, 173–186.

Ruiz-Barradas, A., and S. Nigam, 2005: Warm-season rainfall variability over the US Great Plains in observations, NCEP and ERA-40 reanalyses, and NCAR and NASA atmospheric model simulations: Intercomparisons for NAME. *J. Clim.*, **18**, 1808–1830.

Russak, V., 1990: Trends of solar radiation, cloudiness and atmospheric transparency during recent decades in Estonia. *Tellus*, **42B**, 206–210.

Rusticucci, M., and O. Penalba, 2000: Precipitation seasonal cycle over southern South America. *Clim. Res.*, **16**, 1–15.

Rusticucci, M., and M. Barrucand, 2004: Observed trends and changes in temperature extremes over Argentina. *J. Clim.*, **17**, 4099–4107.

Rutllant, J., and H. Fuenzalida, 1991: Synoptic aspects of the central Chile rainfall variability associated with the Southern Oscillation. *Int. J. Climatol.*, **11**, 63–76.

Saji, N.H., and T. Yamagata, 2003: Structure of SST and surface wind variability during Indian Ocean dipole mode events: COADS observations. *J. Clim.*, **16**, 2735–2751.

Saji, N.H., et al., 1999: A dipole mode in the tropical Indian Ocean. *Nature*, **401**, 360–363.

Salinger, M.J., J.A. Renwick, and A.B. Mullan, 2001: Interdecadal Pacific Oscillation and South Pacific climate. *Int. J. Climatol.*, **21**, 1705–1721.

Salinger, M.J., G.M. Griffiths, and A. Gosai, 2005: Extreme pressure differences at 0900 NZST and winds across New Zealand. *Int. J. Climatol.*, **25**, 1203–1222.

Santer, B.D., et al., 1999: Uncertainties in observationally based estimates of temperature change in the free atmosphere. *J. Geophys. Res.*, **104**, 6305–6333.

Santer, B.D., et al., 2004: Identification of anthropogenic climate change using a second-generation reanalysis. *J. Geophys. Res.*, **109**, D21104, doi:1029/2004/JD005075.

Santer, B.D., et al., 2005: Amplification of surface temperature trends and variability in the tropical atmosphere. *Science*, **309**, 1551–1556.

Sarkar, S., R.P. Singh, and M. Kafatos, 2004: Further evidences for the weakening relationship of Indian rainfall and ENSO over India. *Geophys. Res. Lett.*, **31**, L13209, doi:10.1029/2004GL020259.

Scaife A.A., et al., 2005: A stratospheric influence on the winter NAO and North Atlantic surface climate, *Geophys. Res. Lett.*, **32**, L18715, doi:10.1029/2005GL023226.

Schär, C., et al., 2004: The role of increasing temperature variability in European summer heat waves. *Nature*, **427**, 332–336.

Scherrer, S.C., et al., 2005: European temperature distribution changes in observations and climate change scenarios. *Geophys. Res. Lett.*, **32**, L19705, doi:10.1029/2005GL024108.

Scherrer, S.C., et al., 2006: Two dimensional indices of atmospheric blocking and their statistical relationship with winter climate patterns in the Euro-Atlantic region. *Int. J. Climatol.*, **20**, 233–249.

Schlesinger, M.E., and N. Ramankutty, 1994: An oscillation in the global climate system of period 65–70 years. *Nature*, **367**, 723–726.

Schmidli, J., and C. Frei, 2005: Trends of heavy precipitation and wet and dry spells in Switzerland during the 20th century. *Int. J. Climatol.*, **25**, 753–771.

Schmith, T., E. Kaas, and T.-S. Li, 1998: Northeast Atlantic winter storminess 1875–1995 re-analysed. *Clim. Dyn.*, **14**, 529–536.

Schneider, D.P., E.J. Steig, and J.C. Comiso, 2004: Recent climate variability in Antarctica from satellite-derived temperature data. *J. Clim.*, **17**, 1569–1583.

Schneider, N., and B.D. Cornuelle, 2005: The forcing of the Pacific Decadal Oscillation. *J. Clim.*, **18**, 4355–4373.

Schönwiese, C.-D., and J. Rapp, 1997: *Climate Trend Atlas of Europe Based on Observations 1891–1990.* Kluwer Academic Press, Dordrecht, 228 pp.

Schreck, C.J. III, and F.H.M. Semazzi, 2004: Variability of the recent climate of Eastern Africa. *Int. J. Climatol.*, **24**, 681–701.

Schwartz, R.D., 2005: Global dimming: clear sky atmospheric transmission from astronomical extinction measurements. *J. Geophys. Res.*, **110**, D14210, doi:10.1029/2005JD005882.

Schwarzkopf, M., and V. Ramaswamy, 2002: Effects of changes in well-mixed gases and ozone on stratospheric seasonal temperatures. *Geophys. Res. Lett.*, **29**, 2184, doi:10.1029/2002GL015759.

Schwierz, C., et al., 2006: Challenges posed by and approaches to the study of seasonal-to-decadal climate variability. *Clim. Change*, **79**, 31–63.

Seidel, D.J., and J. Lanzante, 2004: An assessment of three alternatives to linear trends for characterizing global atmospheric temperature changes. *J. Geophys. Res.*, **109**, D14108, doi:10.1029/2003JD004414.

Seidel, D.J., et al., 2001: Climatological characteristics of the tropical tropopause as revealed by radiosondes. *J. Geophys. Res.*, **106**, 7857–7878.

Seidel, D.J., et al., 2004: Uncertainty in signals of large-scale climate variations in radiosonde and satellite upper-air temperature datasets. *J. Clim.*, **17**, 2225–2240.

Sen Roy, S., and R.C. Balling, 2004: Trends in extreme daily rainfall indices in India. *Intl. J. Climatol.*, **24**, 457–466.

Serreze, M.C., et al., 1997: Icelandic low cyclone activity: climatological features, linkages with the NAO, and relationships with recent changes in the Northern Hemisphere circulation. *J. Clim.*, **10**, 453–464.

Sexton, D.M.H., 2001: The effect of stratospheric ozone depletion on the phase of the Antarctic Oscillation. *Geophys. Res. Lett.*, **28**, 3697–3700.

Shabbar, A., and W. Skinner, 2004: Summer drought patterns in Canada and the relationship to global sea surface temperatures. *J. Clim.*, **17**, 2866–2880.

Shepherd, J.M., and S.J. Burian, 2003: Detection of urban-induced rainfall anomalies in a major coastal city. *Earth Interactions*, **7**, 1–17.

Shepherd, J.M., H. Pierce, and A.J. Negri, 2002: Rainfall modification by major urban areas: Observations from spaceborne rain radar on the TRMM satellite. *J. Appl. Meteorol.*, **41**, 689–701.

Shepherd, J.M., L. Taylor, and C. Garza, 2004: A dynamic multi-criteria technique for siting NASA-Clark Atlanta rain gauge network. *J. Atmos. Ocean. Technol.*, **21**, 1346–1363.

Sherwood, S.C., 2002: A microphysical connection among biomass burning, cumulus clouds, and stratospheric moisture. *Science*, **295**, 1271–1275.

Sherwood, S.C., J. Lanzante, and C. Meyer, 2005: Radiosonde daytime biases and late 20th century warming. *Science*, **309**, 1556–1559.

Shine, K.P., et al., 2003: A comparison of model-simulated trends in stratospheric temperatures. *Q. J. R. Meteorol. Soc.*, **129**, 1565–1588.

Silvestri, G.E., and C.S. Vera, 2003: Antarctic Oscillation signal on precipitation anomalies over southeastern South America. *Geophys. Res. Lett.*, **30**, 2115, doi:10.1029/2003GL018277.

Simmonds, I., 2003: Modes of atmospheric variability over the Southern Ocean. *J. Geophys. Res.*, **108**, 8078, doi:10.1029/2000JC000542.

Simmonds, I., and K. Keay, 2000: Variability of Southern Hemisphere extratropical cyclone behavior 1958–97. *J. Clim.*, **13**, 550–561.

Simmonds, I., and K. Keay, 2002: Surface fluxes of momentum and mechanical energy over the North Pacific and North Atlantic Oceans. *Meteorol. Atmos. Phys.*, **80**, 1–18.

Simmonds, I., K. Keay, and E.-P. Lim, 2003: Synoptic activity in the seas around Antarctica. *Mon. Weather Rev.*, **131**, 272–288.

Simmons, A.J., et al., 2004: Comparison of trends and low-frequency variability in CRU, ERA-40 and NCEP/NCAR analyses of surface air temperature. *J. Geophys. Res.*, **109**, D24115, doi:10.1029/2004JD005306.

Simmons, A.J., et al., 2005: ECMWF analyses and forecasts of stratospheric winter polar vortex breakup: September 2002 in the Southern Hemisphere and related events. *J. Atmos. Sci.*, **62**, 668–689.

Sinclair, M.R., J.A. Renwick, and J.W. Kidson, 1997: Low-frequency variability of Southern Hemisphere sea level pressure and weather system activity. *Mon. Weather Rev.*, **125**, 2531–2543.

Small, E.E., L.C. Sloan, and R. Nychka, 2001: Changes in surface air temperature caused by desiccation of the Aral Sea. *J. Clim.*, **14**, 284–299.

Smith, C.A., J.D. Haigh, and R. Toumi, 2001: Radiative forcing due to trends in stratospheric water vapour. *Geophys. Res. Lett.*, **28**, 179–182.

Smith, I., 2004: An assessment of recent trends in Australian rainfall. *Aust. Meteorol. Mag.*, **53**, 163–173.

Smith, L.C., 2000: Trends in Russian Arctic river-ice formation and breakup, 1917 to 1994. *Phys. Geogr.*, **21**, 46–56.

Smith, T.M., and R.W. Reynolds, 2004: Improved extended reconstruction of SST (1854–1997). *J. Clim.*, **17**, 2466–2477.

Smith, T.M., and R.W. Reynolds, 2005: A global merged land and sea surface temperature reconstruction based on historical observations (1880–1997). *J. Clim.*, **18**, 2021–2036.

Smith, T.M., et al., 2005: New surface temperature analyses for climate monitoring. *Geophys. Res. Lett.*, **32**, L14712, doi:10.1029/2005GL023402.

Smits, A., A.M.G. Klein Tank, and G.P. Können, 2005: Trends in storminess over the Netherlands, 1962–2002. *Int. J. Climatol.*, **25**, 1331–1344.

Snow, J.T. (ed.), 2003: Special Issue: European Conference on Severe Storms. *Atmos. Res.*, **67–68**, 703pp.

Soden, B.J., and S.R. Schroeder, 2000: Decadal variations in tropical water vapor: A comparison of observations and a model simulation. *J. Clim.*, **13**, 3337–3340.

Soden, B.J., et al., 2002: Global cooling after the eruption of Mount Pinatubo: A test of climate feedback by water vapor. *Science*, **296**, 727–730.

Soden, B.J., et al., 2004: An analysis of satellite, radiosonde, and lidar observations of upper tropospheric water vapor from the Atmospheric Radiation Measurement Program. *J. Geophys. Res.*, **109**, D04105, doi:10.1029/2003JD003828.

Soden, B.J., et al., 2005: The radiative signature of upper tropospheric moistening. *Science*, **310**, 841–844.

Sohn, B.-J., and E.A. Smith, 2003: Explaining sources of discrepancy in SSM/I water vapor algorithms. *J. Clim.*, **16**, 3229–3255.

Song, Y., and W.A. Robinson, 2004: Dynamical mechanisms for stratospheric influences on the troposphere. *J. Atmos. Sci.*, **61**, 1711–1725.

Sparks, J., D. Changnon, and J. Starke, 2002: Changes in the frequency of extreme warm-season surface dewpoints in northeastern Illinois: Implications for cooling-system design and operation. *J. Appl. Meteorol.*, **41**, 890–898.

Stanhill, G., and S. Cohen, 2001: Global dimming, a review of the evidence for a widespread and significant reduction in global radiation with a discussion of its probable causes and possible agricultural consequences. *Agric. For. Meteorol.*, **107**, 255–278.

Sterl, A., 2001: On the impact of gap-filling algorithms on variability patterns of reconstructed oceanic surface fields. *J. Geophys. Res.*, **28**, 2473–2476.

Sterl, A., 2004: On the (in)homogeneity of reanalysis products. *J. Clim.*, **17**, 3866–3873.

Sterl, A., and S. Caires, 2005: Climatology, variability and extrema of ocean waves: the Web-based KNMI/ERA-40 wave atlas. *Int. J. Climatol.*, **25**, 963–977.

Stewart, I.T., D.R. Cayan, and M.D. Dettinger, 2005: Changes towards earlier streamflow timing across western North America. *J. Clim.*, **18**, 1136–1155.

Stone, D.A., A.J. Weaver, and F.W. Zwiers, 2000: Trends in Canadian precipitation intensity. *Atmos.-Ocean*, **38**, 321–347.

Straus, D.M., and J. Shukla, 2002: Does ENSO force the PNA? *J. Clim.*, **15**, 2340–2358.

Straus, D.M., and F. Molteni, 2004: Circulation regimes and SST forcing: Results from large GCM ensembles. *J. Clim.*, **17**, 1641–1656.

Stuber, N., et al., 2001: Is the climate sensitivity to ozone perturbations enhanced by stratospheric water vapor feedback? *Geophys. Res. Lett.*, **28**, 2887–2890.

Sturaro, G., 2003: A closer look at the climatological discontinuities present in the NCEP/NCAR reanalysis temperature due to the introduction of satellite data. *Clim. Dyn.*, **21**, 309–316.

Sun, B.M., 2003: Cloudiness over the contiguous United States: Contemporary changes observed using ground-based and ISCCP D2 data. *Geophys. Res. Lett.*, **30**, doi:10.1029/2002GL015887.

Sun, B.M., and P.Ya. Groisman, 2000: Cloudiness variations over the former Soviet Union. *Int. J. Climatol.*, **20**, 1097–1111.

Sun, B.M., and P.Ya. Groisman, 2004: Variations in low cloud cover over the United States during the second half of the twentieth century. *J. Clim.*, **17**, 1883–1888.

Sun, B.M., P.Ya. Groisman, and I.I. Mokhov, 2001: Recent changes in cloud-type frequency and inferred increases in convection over the United States and the former USSR. *J. Clim.*, **14**, 1864–1880.

Sutton, R.T., and D.L.R. Hodson, 2003: Influence of the ocean on North Atlantic climate variability 1871–1999. *J. Clim.*, **16**, 3296–3313.

Sutton, R.T., and D.L.R. Hodson, 2005: Atlantic Ocean forcing of North American and European summer climate. *Science*, **290**, 2133–2137.

Swanson, R.E., 2003: Evidence of possible sea-ice influence on Microwave Sounding Unit tropospheric temperature trends in polar regions. *Geophys. Res. Lett.*, **30**, 2040, doi:10.1029/2003GL017938.

Tebakari, T., J. Yoshitani, and C. Suvanpimol, 2005: Time-space trend analysis in pan evaporation over kingdom of Thailand. *J. Hydrol. Eng.*, **10**, 205–215.

Terray, P., S. Dominiak, and P. Delecluse, 2005: Role of the southern Indian Ocean in the transition of the monsoon-ENSO system during recent decades. *Clim. Dyn.*, **24**, 169–195.

Tett, S.F.B., and P.W. Thorne, 2004: Tropospheric temperature series from satellites. *Nature*, **429**, doi:10.1038/nature03208.

Thielen, J., et al., 2000: The possible influence of urban surfaces on rainfall development: a sensitivity study in 2D in the meso-gamma scale. *Atmos. Res.*, **54**, 15–39.

Thompson, D.W.J., and J.M. Wallace, 1998: The Arctic Oscillation signature in the wintertime geopotential height and temperature fields. *Geophys. Res. Lett.*, **25**, 1297–1300.

Thompson, D.W.J., and J.M. Wallace, 2000: Annular modes in the extratropical circulation, Pt I: Month-to-month variability. *J. Clim.*, **13**, 1000–1016.

Thompson, D.W.J., and S. Solomon, 2002: Interpretation of recent Southern Hemisphere climate change. *Science*, **296**, 895–899.

Thompson, D.W.J., and S. Solomon, 2005: Recent stratospheric climate trends: Global structure and tropospheric linkages. *J. Clim.*, **18**, 4785–4795.

Thompson, D.W.J., J.M. Wallace, and G.C. Hegerl, 2000: Annular modes in the extratropical circulation. Part II: Trends. *J. Clim.*, **13**, 1018–1036.

Thompson, D.W.J., M.P. Baldwin, and J.M. Wallace, 2002: Stratospheric connection to Northern Hemisphere wintertime weather: Implications for prediction. *J. Clim.*, **15**, 1421–1428.

Thompson, D.W.J., S. Lee, and M.P. Baldwin, 2003: Atmospheric processes governing the Northern Hemisphere Annular Mode/North Atlantic Oscillation. In: *The North Atlantic Oscillation: Climatic Significance and Environmental Impact* [Hurrell, J.W., et al. (eds.)]. Geophysical Monograph 134, American Geophysical Union, Washington, DC, pp. 81–112.

Thompson, D.W.J., M.P. Baldwin, and S. Solomon, 2005: Stratosphere/troposphere coupling in the Southern Hemisphere. *J. Atmos. Sci.*, **62**, 708–715.

Thorne, P.W., et al., 2005a: Revisiting radiosonde upper air temperatures from 1958 to 2002. *J. Geophys. Res.*, **110**, D18105, doi:10.1029/2004JD005753.

Thorne, P.W., et al., 2005b: Uncertainties in climate trends: Lessons from upper-air temperature records. *Bull. Am. Meteorol. Soc.*, **86**, 1437–1442.

Thornthwaite, C.W., 1948: An approach toward a rational classification of climate. *Geogr. Rev.*, **38**, 55–94.

Tibaldi, S., et al., 1994: Northern and Southern Hemisphere seasonal variability of blocking frequency and predictability. *Mon. Weather Rev.*, **122**, 1971–2003.

Trapp, R.J., et al., 2005: Tornadoes from squall lines and bow echoes. Pt I: Climatological distribution. *Weather Forecasting*, **20**, 23–34.

Trenberth, K.E., 1984: Signal versus noise in the Southern Oscillation. *Mon. Weather Rev.*, **112**, 326–332.

Trenberth, K.E., 1990: Recent observed interdecadal climate changes in the Northern Hemisphere. *Bull. Am. Meteorol. Soc.*, **71**, 988–993.

Trenberth, K.E., 2002: Changes in tropical clouds and radiation. *Science*, **296**, 2095a (online), http://www.sciencemag.org/cgi/content/full/296/5576/2095a.

Trenberth, K.E., 2004: Rural land-use change and climate. *Nature*, **427**, 213.

Trenberth, K.E., and J.W. Hurrell, 1994: Decadal atmosphere–ocean variations in the Pacific. *Clim. Dyn.*, **9**, 303–319.

Trenberth, K.E., and J.M. Caron, 2000: The Southern Oscillation revisited: Sea level pressures, surface temperatures and precipitation. *J. Clim.*, **13**, 4358–4365.

Trenberth, K.E., and J.M. Caron, 2001: Estimates of meridional atmosphere and ocean heat transports. *J. Clim.*, **14**, 3433–3443.

Trenberth, K.E., and D. P. Stepaniak, 2001: Indices of El Niño evolution. *J. Clim.*, **14**, 1697–1701.

Trenberth, K.E., and D.P. Stepaniak, 2003a: Co-variability of components of poleward atmospheric energy transports on seasonal and interannual timescales. *J. Clim.*, **16**, 3690–3704.

Trenberth, K.E., and D.P. Stepaniak, 2003b: Seamless poleward atmospheric energy transports and implications for the Hadley circulation. *J. Clim.*, **16**, 3705–3721.

Trenberth, K.E., and D.J. Shea, 2005: Relationships between precipitation and surface temperature. *Geophys. Res. Lett.*, **32**, L14703, doi:10.1029/2005GL022760.

Trenberth, K.E., and L. Smith, 2005: The mass of the atmosphere: A constraint on global analyses. *J. Clim.*, **18**, 864–875.

Trenberth, K.E., and D.J. Shea, 2006: Atlantic hurricanes and natural variability in 2005, *Geophys. Res. Lett.*, **33**, L12704, doi:10.1029/2006GL026894.

Trenberth, K.E., D.P. Stepaniak, and J.M. Caron, 2000: The global monsoon as seen through the divergent atmospheric circulation. *J. Clim.*, **13**, 3969–3993.

Trenberth, K.E., D.P. Stepaniak, and J.M. Caron, 2002a: Interannual variations in the atmospheric heat budget. *J. Geophys. Res.*, **107**, 4066, doi:10.1029/2000JD000297.

Trenberth, K.E., J. Fasullo, and L. Smith, 2005a: Trends and variability in column integrated atmospheric water vapor. *Clim. Dyn.*, **24**, 741–758.

Trenberth, K.E., D.P. Stepaniak, and L. Smith, 2005b: Interannual variability of the patterns of atmospheric mass distribution. *J. Clim.*, **18**, 2812–2825.

Trenberth, K.E., et al., 2002b: The evolution of ENSO and global atmospheric temperatures. *J. Geophys. Res.*, **107**, 4065, doi:10.1029/2000JD000298.

Trenberth, K.E., et al., 2003: The changing character of precipitation. *Bull. Am. Meteorol. Soc.*, **84**, 1205–1217.

Trigo, R.M., et al., 2004: Climate impact of the European winter blocking episodes from the NCEP/NCAR Reanalyses. *Clim. Dyn.*, **23**, 17–28.

Trömel, S., and C.-D. Schönwiese, 2005: A generalized method of time series decomposition into significant components including probability assessments of extreme events and application to observed German precipitation data. *Meteorol. Z.*, **14**, 417–427.

Troup, A.J., 1965: The Southern Oscillation. *Q. J. R. Meteorol. Soc.*, **91**, 490–506.

Tuller, S.E., 2004: Measured wind speed trends on the west coast of Canada. *Int. J. Climatol.*, **24**, 1359–1374.

Tuomenvirta, R.H., et al., 2000: Trends in Nordic and Arctic temperature extremes. *J. Clim.*, **13**, 977–990.

Turner, D.D., et al., 2003: Dry bias and variability in Vaisala RS80-H radiosondes: The ARM experience. *J. Atmos. Ocean. Technol.*, **20**, 117–132.

Turner, J., et al., 2005: Antarctic climate change during the last 50 years. *Int. J. Climatol.*, **25**, 279–294.

Turner, J., et al., 2006: Significant warming of the Antarctic winter troposphere. *Science*, **311**, 1914–1917.

Ulbrich, U., et al., 2003a: The central European floods of August 2002: Pt. 1 – Rainfall periods and flood development. *Weather*, **58**, 371–377.

Ulbrich, U., et al., 2003b: The central European floods of August 2002: Pt. 2 – Synoptic causes and considerations with respect to climatic change. *Weather*, **58**, 434–442.

Uppala, S.M., et al., 2005: The ERA-40 reanalysis. *Q. J. R. Meteorol. Soc.*, **131**, 2961–3012.

van den Broeke, M.R., and N.P.M. van Lipzig, 2003: Response of wintertime Antarctic temperatures to the Antarctic Oscillation: Results of a regional climate model. In: *Antarctic Peninsula Climate Variability: Historical and Paleoenvironmental Perspectives* [Domack, E., et al. (eds.)]. Antarctic Research Series 79, American Geophysical Union, Washington, DC, pp. 43–58.

van den Dool, H., J. Huang, and Y. Fan, 2003: Performance and analysis of the constructed analogue method applied to U.S. soil moisture over 1981–2001. *J. Geophys. Res.*, **108**, 8617. doi:10.1029/2002JD003114.

van der Schrier, G., et al., 2006: Summer moisture variability across Europe. *J. Clim.*, **19**, 2818–2834.

van Wijngaarden, W.A., and L.A. Vincent, 2005: Examination of discontinuities in hourly surface relative humidity in Canada during 1953-2003. *J. Geophys. Res.*, **110**, D22102, doi:10.1029/2005JD005925.

Venegas, S.A., 2003: The Antarctic Circumpolar Wave: A combination of two signals? *J. Clim.*, **16**, 2509–2525.

Venegas, S.A., and L.A. Mysak, 2000: Is there a dominant timescale of natural climate variability in the Arctic? *J. Clim.*, **13**, 3412–3434.

Vera, C., et al., 2006: A unified view of the American monsoon systems. *J. Clim.*, **19**, 4977–5000.

Vikebo, F., et al., 2003: Wave height variations in the North Sea and on the Norwegian continental shelf, 1881-1999. *Continental Shelf Res.*, **23**, 251–263.

Vimont, D.J., D.S. Battisti, and A.C. Hirst, 2001: Footprinting: A seasonal connection between the Tropics and midlatitudes. *Geophys. Res. Lett.*, **28**, 3923–2936.

Vincent, L.A., and É. Mekis, 2006: Changes in daily and extreme temperature and precipitation indices for Canada over the 20th century. *Atmos.-Ocean*, **44**, 177–193.

Vincent, L.A., et al., 2002: Homogenization of daily temperatures over Canada. *J. Clim.*, **15**, 1322–1334.

Vincent, L.A., et al., 2005: Observed trends in indices of daily temperature extremes in South America 1960–2000. *J. Clim.*, **18**, 5011–5023.

Vinnikov, K.Y., and N.C. Grody, 2003: Global warming trend of mean tropospheric temperature observed by satellites. *Science*, **302**, 269–272.

Vinnikov, K.Y., et al., 2006: Temperature trends at the surface and in the troposphere. *J. Geophys. Res.*, **111**, D03106, doi:10.1029/2005JD006392.

Visbeck, M., et al., 2003: The ocean's response to North Atlantic Oscillation variability. In: *The North Atlantic Oscillation: Climatic Significance and Environmental Impact* [Hurrell, J.W., et al. (eds.)]. Geophysical Monograph 134, American Geophysical Union, Washington, DC, pp. 113–145.

Vose, R.S., D.R. Easterling, and B. Gleason, 2005a: Maximum and minimum temperature trends for the globe: An update through 2004. *Geophys. Res. Lett.*, **32**, L23822, doi:10.1029/2004GL024379.

Vose, R.S., et al., 1992: *The Global Historical Climatology Network: Long-Term Monthly Temperature, Precipitation, Sea Level Pressure, and Station Pressure Data.* ORNL/CDIAC-53, NDP-041, Carbon Dioxide Information Analysis Center, Oak Ridge National Laboratory, Oak Ridge, TN, 325 pp.

Vose, R.S., et al., 2004: Impact of land-use change on climate. *Nature*, **427**, 213–214.

Vose, R.S., et al., 2005b: An intercomparison of surface air temperature analyses at the global, hemispheric and grid-box scale. *Geophys. Res. Lett.*, **32**, L18718, doi:10.1029/2005GL023502.

Wallace, J.M., and D.S. Gutzler, 1981: Teleconnections in the geopotential height field during the Northern Hemisphere winter. *Mon. Weather Rev.*, **109**, 784–812.

Walter, K., and H.-F. Graf, 2002: On the changing nature of the regional connection between the North Atlantic Oscillation and sea surface temperature. *J. Geophys. Res.*, **107**, 4338, doi:10.1029/2001JD000850.

Walter, M.T., et al., 2004: Increasing evapotranspiration from the conterminous United States. *J. Hydrometeorol.*, **5**, 405–408.

Wang, B., 1994: Climatic regimes of tropical convection and rainfall. *J. Clim.*, **7**, 1109–1118.

Wang, B., and Z. Fan, 1999: Choice of South Asian summer monsoon indices. *Bull. Am. Meteorol. Sci.*, **80**, 629–638.

Wang, B., and J.C.L. Chan, 2002: How strong ENSO events affect tropical storm activity over the western North Pacific. *J. Clim.*, **15**, 1643–1658.

Wang, B., and Q. Ding, 2006: Changes in global monsoon precipitation over the past 56 years. *Geophys. Res. Lett.*, **33**, L06711, doi:10.1029/2005GL025347.

Wang, H.J., 2001: The weakening of the Asian monsoon circulation after the end of 1970's. *Adv. Atmos. Sci.*, **18**, 376–386.

Wang, J., et al., 2003: Performance of operational radiosonde humidity sensors in direct comparison with a chilled mirror dew-point hygrometer and its climate implication. *Geophys. Res. Lett.*, **30**, 1860, 10.129/2003GL016985.

Wang, J.H., H.L. Cole, and D.J. Carlson, 2001: Water vapor variability in the tropical western Pacific from 20-year radiosonde data. *Adv. Atmos. Sci.*, **18**, 752–766.

Wang, J.H., et al., 2002a: Corrections of humidity measurement errors from the Vaisala RS80 radiosonde – Application to TOGA COARE data. *J. Atmos. Ocean. Technol.*, **19**, 981–1002.

Wang, J.X.L., and D.J. Gaffen, 2001: Trends in extremes of surface humidity, temperatures and summertime heat stress in China. *Adv. Atmos. Sci.*, **18**, 742–751.

Wang, P.H., et al., 2002b: Satellite observations of long-term changes in tropical cloud and outgoing longwave radiation from 1985 to 1998. *Geophys. Res. Lett.*, **29**, 1397, doi:10.1029/2001GL014264.

Wang, X.L., and V.R. Swail, 2001: Changes of extreme wave heights in Northern Hemisphere oceans and related atmospheric circulation regimes. *J. Clim.*, **14**, 2204–2201.

Wang, X.L., and V.R. Swail, 2002: Trends of Atlantic wave extremes as simulated in a 40-yr wave hindcast using kinematically reanalyzed wind fields. *J. Clim.*, **15**, 1020–1035.

Wang, X.L., and P.M. Zhai, 2004: Variation of spring dust storms in China and its association with surface winds and sea level pressures. *Acta Meteorol. Sin.*, **62**, 96–103 (in Chinese).

Wang, X.L., V.R. Swail, and F.W. Zwiers, 2006a: Climatology and changes of extratropical storm tracks and cyclone activity: Comparison of ERA-40 with NCEP/NCAR Reanalysis for 1958–2001. *J. Clim.*, **19**, 3145–3166.

Wang, X.L., H. Wan, and V.R. Swail, 2006b: Observed changes in cyclone activity in Canada and their relationships to major circulation regimes. *J. Clim.*, **19**, 896–915.

Wang, Z.W., and P.M. Zhai, 2003: Climate change in drought over northern China during 1950–2000. *Acta Geogr. Sin.*, **58**(supplement), 61–68 (in Chinese).

Waple, A.M., and J.H. Lawrimore, 2003: State of the climate in 2002. *Bull. Am. Meteorol. Soc.*, **84**(6), S1–S68.

Waple, A.M., et al., 2002: Climate assessment for 2001. *Bull. Am. Meteorol. Soc.*, **83**(6), S1–S62.

Ward, M.N., 1998: Diagnosis and short-lead time prediction of summer rainfall in tropical North Africa at interannual and multidecadal timescales. *J. Clim.*, **11**, 3167–3191.

Ward, M.N., and B.J. Hoskins, 1996: Near surface wind over the global ocean 1949–1988. *J. Clim.*, **9**, 1877–1895.

Wardle, R., and I. Smith, 2004: Modeled response of the Australian monsoon to changes in land surface temperatures. *Geophys. Res. Lett.*, **31**, L16205, doi:10.1029/2004GL020157.

Watkins, A., 2002: 2002 Australian climate summary: Dry and warm conditions dominate. *Bull. Aust. Meteorol. Oceanogr. Soc.*, **15**, 109–114.

Waugh, D., et al., 1999: Persistence of the lower stratospheric polar vortices. *J. Geophys. Res.*, **104**, 27191–27201.

Webster, P.J., and S. Yang, 1992: Monsoon and ENSO: selective interactive systems. *Q. J. R. Meteorol. Soc.*, **118**, 877–926.

Webster, P.J., et al., 1998: Monsoons: processes, predictability, and the prospects for prediction. *J. Geophys. Res.*, **103**, 14451–14510.

Webster, P.J., et al., 1999: Coupled ocean-atmosphere dynamics in the Indian Ocean during 1997-98. *Nature*, **401**, 356–360.

Webster, P.J., et al., 2005: Changes in tropical cyclone number, duration and intensity in a warming environment. *Science*, **309**, 1844–1846.

Webster, P. J., et al., 2006: Response to comment on "Changes in tropical cyclone number, duration, and intensity in a warming environment". *Science*, **311**, 1713c.

Weisse, R., H. von Storch, and F. Feser, 2005: Northeast Atlantic and North Sea storminess as simulated by a regional climate model during 1958-2001 and comparison with observations. *J. Clim.*, **18**, 465–479.

Wells, N., S. Goddard, and M.J. Hayes, 2004: A self-calibrating Palmer Drought Severity Index. *J. Clim.*, **17**, 2335–2351.

Wentz, F.J., and M. Schabel, 1998: Effects of satellite orbital decay on MSU lower tropospheric temperature trends. *Nature*, **394**, 661–664.

Wentz, F.J., and M. Schabel, 2000: Precise climate monitoring using complementary satellite data sets. *Nature*, **403**, 414–416.

Wettstein, J.J., and L.O. Mearns, 2002: The influence of the North Atlantic-Arctic Oscillation on mean, variance, and extremes of temperature in the northeastern United States and Canada. *J. Clim.*, **15**, 3586–3600.

Wheeler, M.C., and J.L. McBride, 2005: Australian-Indonesian monsoon. In: *Intraseasonal Variability of the Atmosphere-Ocean Climate System* [Lau, W.K.M., and D.E. Waliser, (eds.)]. Praxis Publishing, Chichester, UK, pp. 125–173.

White, W.B., and R.G. Peterson, 1996: An Antarctic Circumpolar Wave in surface pressure, wind, temperature and sea-ice extent. *Nature*, **380**, 699–702.

White, W.B., and J. Annis, 2004: Influence of the Antarctic Circumpolar Wave on El Niño and its multidecadal changes from 1950 to 2001. *J. Geophys. Res.*, **109**, C06019, doi:10.1029/2002JC001666.

White, W.B., and I. Simmonds, 2006: SST-induced cyclogenesis in the Antarctic Circumpolar Wave. *J. Geophys. Res.*, **111**, C08011, doi:10.1029/2004JC002395.

White, W.B., P. Gloersen, and I. Simmonds, 2004: Tropospheric response in the Antarctic Circumpolar Wave along the sea ice edge around Antarctica. *J. Clim.*, **17**, 2765–2779.

Wiedenmann, J.M., et al., 2002: The climatology of blocking anticyclones for the Northern and Southern Hemispheres: Block intensity as a diagnostic. *J. Clim.*, **15**, 3459–3473.

Wielicki, B.A., et al., 2002a: Evidence for large decadal variability in the tropical mean radiative energy budget. *Science*, **295**, 841–844.

Wielicki, B.A., et al., 2002b: Response. *Science*, **296**, http://www.sciencemag.org/cgi/content/full/296/5576/2095a.

Wielicki, B.A., et al., 2005: Change in Earth's albedo measured by satellite. *Science*, **308**, 825.

Wijngaard, J.B., A.M.G. Klein Tank, and G.P. Können, 2003: Homogeneity of 20th century European daily temperature and precipitation series. *Int. J. Climatol.*, **23**, 679–692.

Wild, M.A., et al., 2004: On the consistency of trends in radiation and temperature records and implications for the global hydrological cycle. *Geophys. Res. Lett.*, **31**, L11201, doi:10.1029/2003GL019188.

Wild, M.A., et al., 2005: From dimming to brightening: Decadal changes in solar radiation at Earth's surface. *Science*, **308**, 847–850.

Willis, J.K., D. Roemmich, and B. Cornuelle, 2004: Interannual variability in upper-ocean heat content, temperature and thermosteric expansion on global scales. *J. Geophys. Res.*, **109**, C12036, doi:10.1029/2003JC002260.

Wittman, M.A.H., et al., 2004: Stratospheric influence on baroclinic lifecycles and its connection to the Arctic Oscillation. *Geophys. Res. Lett.*, **31**, L16113, doi:10.1029/2004GL020503.

WMO, 2004: *World Meteorological Organization Statement on the Status of Global Climate in 2003*. World Meteorological Organization, Geneva, 12 pp.

Wong, T.D.F., M.H. Young, and S. Weckmann, 2000: Validation of the CERES/TRMM ERBE-like monthly mean clear-sky longwave dataset and the effects of the 1998 ENSO event. *J. Clim.*, **13**, 4256–4267.

Wong, T., et al., 2006: Re-examination of the observed decadal variability of Earth Radiation Budget using altitude-corrected ERBE/ERBS nonscanner WFOV data. *J. Clim.*, **19**, 4028–4040.

Woolf, D.K., P.G. Challenor, and P.D. Cotton, 2002: The variability and predictability of North Atlantic wave climate. *J. Geophys. Res.*, **107**, 3145, doi:10.1029/2001JC001124.

Worley, S.J., et al., 2005: ICOADS release 2.1 data and products. *Int. J. Climatol.*, **25**, 823–842.

Wu, M.C., W.L. Chang, and W.M. Leung, 2004: Impacts of El Niño-Southern Oscillation events on tropical cyclone landfalling activity in the western North Pacific. *J. Clim.*, **17**, 1419–1428.

Wylie, D.P., et al., 2005: Trends in global cloud cover in two decades of HIRS observations. *J. Clim.*, **18**, 3021–3031.

Xie, P., and Arkin, P.A. 1997: Global precipitation: A 17-year monthly analysis based on gauge observations, satellite estimates and numerical model outputs. *Bull. Am. Meteorol. Soc.*, **78**, 2539–2558.

Xie, S.P., et al., 2002: Structure and mechanisms of South Indian Ocean climate variability. *J. Clim.*, **15**, 864–878.

Yan, Z., et al., 2002: Trends of extreme temperatures in Europe and China based on daily observations. *Clim. Change*, **53**, 355–392.

Yang, D., B. Ye, and D.L. Kane, 2004: Streamflow changes over Siberian Yenisei River Basin. *J. Hydrol.*, **296**, 59–80.

Yang, D., et al., 2002: Siberian Lena River hydrologic regime and recent change. *J. Geophys. Res.*, **107**, 4694, doi:10.1029/2002JD002542.

Ye, B.S., D.Q. Yang, and D.L. Kane, 2003: Changes in Lena River streamflow hydrology: Human impacts versus natural variations. *Water Resour. Res.*, **39**, 1200, doi:10.1029/2003WR001991.

Yin, X., A. Gruber, and P. Arkin, 2004: Comparison of the GPCP and CMAP merged gauge-satellite monthly precipitation products for the period 1979–2001. *J. Hydrometeorol.*, **5**, 1207–1222.

Yu, L.S., and M.M. Rienecker, 1999: Mechanisms for the Indian Ocean warming during the 1997–1998 El Niño. *Geophys. Res. Lett.*, **26**, 735–738.

Yu, R., B. Wang, and T. Zhou, 2004a: Climate effects of the deep continental stratus clouds generated by the Tibetan Plateau. *J. Clim.*, **17**, 2702–2713.

Yu, R., B. Wang, and T. Zhou, 2004b: Tropospheric cooling and summer monsoon weakening trend over East Asia. *Geophys.Res. Lett.*, **31**, L22212, doi:10.1029/2004GL021270.

Zhai, P.M., and X.H. Pan, 2003: Trends in temperature extremes during 1951–1999 in China. *Geophys. Res. Lett.*, **30**, 1913, doi:10.1029/203GL018004.

Zhai, P.M., et al., 2004: Trends in total precipitation and frequency of daily precipitation extremes over China. *J. Clim.*, **18**, 1096–1108.

Zhang, R., and T.L. Delworth, 2005: Simulated tropical response to a substantial weakening of the Atlantic thermohaline circulation. *J. Clim.*, **18**, 1853–1860.

Zhang, X., W.D. Hogg, and E. Mekis, 2001a: Spatial and temporal characteristics of heavy precipitation events over Canada. *J. Clim.*, **14**, 1923–1936.

Zhang, X., F.W. Zwiers, and G. Li, 2004a: Monte Carlo experiments on the detection of trends in extreme values. *J. Clim.*, **17**, 1945–1952.

Zhang, X., et al., 2001b: Trends in Canadian streamflow. *Water Resour. Res.*, **37**, 987–998.

Zhang, X., et al., 2004b: Climatology and interannual variability of Arctic cyclone activity: 1948–2002. *J. Clim.*, **17**, 2300–2317.

Zhang, X., et al., 2005: Trends in Middle East climate extremes indices from 1950 to 2003. *J. Geophys. Res.*, **110**, D22104, doi:10.1029/2005JD006181.

Zhang, Y., et al., 2004c: Calculation of radiative fluxes from the surface to top of atmosphere based on ISCCP and other global data sets: refinements of the radiative transfer model and the input data. *J. Geophys. Res.*, **109**, D19105, doi:10.1029/2003JD004457.

Zheng, X., and R.E. Basher, 1999: Structural time series models and trend detection in global and regional temperature series. *J. Clim.*, **12**, 2347–2358.

Zhou, S., et al., 2000: An inter-hemisphere comparison of the persistent stratospheric polar vortex. *Geophys. Res. Lett.*, **27**, 1123–1126.

Zhou, X.L., M.A. Geller, and M.H. Zhang, 2001: Cooling trend of the tropical cold point tropopause temperatures and its implications. *J. Geophys. Res.*, **106**, 1511–1522.

Zhou, Z.J., and G.C. Zhang, 2003: Typical severe dust storms in northern China (1954–2002). *Chinese Sci. Bull.*, **48**, 1224–1228 (in Chinese).

Zou, X.K., and P.M. Zhai, 2004: Relationship between vegetation coverage and spring dust storms over northern China. *J. Geophys. Res.*, **109**, D03104, doi:10.1029/2003JD003913.

Zou, X.K., P.M. Zhai, and Q. Zhang, 2005: Variations in droughts over China: 1951–2003. *Geophys. Res. Lett.*, **32**, L04707, doi:10.1029/2004GL021853.

Zveryaev, I.I., and P.S. Chu, 2003: Recent climate changes in precipitable water in the global tropics as revealed in National Centers for Environmental Prediction/National Center for Atmospheric Research reanalysis. *J. Geophys. Res.*, **108**, 4311, doi:10.129/2002JD002476.

Appendix 3.A:
Low-Pass Filters and Linear Trends

The time series used in this report have undergone diverse quality controls that have, for example, led to removal of outliers, thereby building in some smoothing. In order to highlight decadal and longer time-scale variations and trends, it is often desirable to apply some kind of low-pass filter to the monthly, seasonal or annual data. In the literature cited for the many indices used in this chapter, a wide variety of schemes was employed. In this chapter, the same filter was used wherever it was reasonable to do so. The desirable characteristics of such filters are 1) they should be easily understood and transparent; 2) they should avoid introducing spurious effects such as ripples and ringing (Duchon, 1979); 3) they should remove the high frequencies; and 4) they should involve as few weighting coefficients as possible, in order to minimise end effects. The classic low-pass filters widely used have been the binomial set of coefficients that remove $2\Delta t$ fluctuations, where Δt is the sampling interval. However, combinations of binomial filters are usually more efficient, and those have been chosen for use here, for their simplicity and ease of use. Mann (2004) discusses smoothing time series and especially how to treat the ends. This chapter uses the 'minimum slope' constraint at the beginning and end of all time series, which effectively reflects the time series about the boundary. If there is a trend, it will be conservative in the sense that this method will underestimate the anomalies at the end.

The first filter (e.g., Figure 3.5) is used in situations where only the smoothed series is shown, and it is designed to remove interannual fluctuations and those on El Niño time scales. It has 5 weights 1/12 [1-3-4-3-1] and its response function (ratio of amplitude after to before) is 0.0 at 2 and $3\Delta t$, 0.5 at $6\Delta t$, 0.69 at $8\Delta t$, 0.79 at $10\Delta t$, 0.91 at $16\Delta t$, and 1 for zero frequency, so for yearly data ($\Delta t = 1$) the half-amplitude point is for a 6-year period, and the half-power point is near 8.4 years.

The second filter used in conjunction with annual values ($\Delta t = 1$) or for comparisons of multiple curves (e.g., Figure 3.8) is designed to remove fluctuations on less than decadal time scales. It has 13 weights 1/576 [1-6-19-42-71-96-106-96-71-42-19-6-1]. Its response function is 0.0 at 2, 3 and $4\Delta t$, 0.06 at $6\Delta t$, 0.24 at $8\Delta t$, 0.41 at $10\Delta t$, 0.54 at $12\Delta t$, 0.71 at $16\Delta t$, 0.81 at $20\Delta t$, and 1 for zero frequency, so for yearly data the half-amplitude point is about a 12-year period, and the half-power point is 16 years. This filter has a very similar response function to the 21-term binomial filter used in the TAR.

Another low-pass filter, widely used and easily understood, is to fit a linear trend to the time series although there is generally no physical reason why trends should be linear, especially over long periods. The overall change in the time series is often inferred from the linear trend over the given time period, but can be quite misleading. Such measures are typically not stable and are sensitive to beginning and end points, so that adding or subtracting a few points can result in marked differences in the estimated trend. Furthermore, as the climate system exhibits highly nonlinear behaviour, alternative perspectives of overall change are provided by comparing low-pass-filtered values (see above) near the beginning and end of the major series.

The linear trends are estimated by Restricted Maximum Likelihood regression (REML, Diggle et al., 1999), and the estimates of statistical significance assume that the terms have serially uncorrelated errors and that the residuals have an AR1 structure. Brohan et al. (2006) and Rayner et al. (2006) provide annual uncertainties, incorporating effects of measurement and sampling error and uncertainties regarding biases due to urbanisation and earlier methods of measuring SST. These are taken into account, although ignoring their serial correlation. The error bars on the trends, shown as 5 to 95% ranges, are wider and more realistic than those provided by the standard ordinary least squares technique. If, for example, a century-long series has multi-decadal variability as well as a trend, the deviations from the fitted linear trend will be autocorrelated. This will cause the REML technique to widen the error bars, reflecting the greater difficulty in distinguishing a trend when it is superimposed on other long-term variations and the sensitivity of estimated trends to the period of analysis in such circumstances. Clearly, however, even the REML technique cannot widen its error estimates to take account of variations outside the sample period of record. Robust methods for the estimation of linear and nonlinear trends in the presence of episodic components became available recently (Grieser et al., 2002).

As some components of the climate system respond slowly to change, the climate system naturally contains persistence. Hence, the statistical significances of REML AR1-based linear trends could be overestimated (Zheng and Basher, 1999; Cohn and Lins, 2005). Nevertheless, the results depend on the statistical model used, and more complex models are not as transparent and often lack physical realism. Indeed, long-term persistence models (Cohn and Lins, 2005) have not been shown to provide a better fit to the data than simpler models.

Appendix 3.B: Techniques, Error Estimation and Measurement Systems: See Supplementary Material

This material is included in the supplementary material. Please note that the many references that are cited only in Appendix 3.B have not been included in the list above, but are just as valuable in formulating the report.

4

Observations: Changes in Snow, Ice and Frozen Ground

Coordinating Lead Authors:

Peter Lemke (Germany), Jiawen Ren (China)

Lead Authors:

Richard B. Alley (USA), Ian Allison (Australia), Jorge Carrasco (Chile), Gregory Flato (Canada), Yoshiyuki Fujii (Japan), Georg Kaser (Austria, Italy), Philip Mote (USA), Robert H. Thomas (USA, Chile), Tingjun Zhang (USA, China)

Contributing Authors:

J. Box (USA), D. Bromwich (USA), R. Brown (Canada), J.G. Cogley (Canada), J. Comiso (USA), M. Dyurgerov (Sweden, USA), B. Fitzharris (New Zealand), O. Frauenfeld (USA, Austria), H. Fricker (USA), G. H. Gudmundsson (UK, Iceland), C. Haas (Germany), J.O. Hagen (Norway), C. Harris (UK), L. Hinzman (USA), R. Hock (Sweden), M. Hoelzle (Switzerland), P. Huybrechts (Belgium), K. Isaksen (Norway), P. Jansson (Sweden), A. Jenkins (UK), Ian Joughin (USA), C. Kottmeier (Germany), R. Kwok (USA), S. Laxon (UK), S. Liu (China), D. MacAyeal (USA), H. Melling (Canada), A. Ohmura (Switzerland), A. Payne (UK), T. Prowse (Canada), B.H. Raup (USA), C. Raymond (USA), E. Rignot (USA), I. Rigor (USA), D. Robinson (USA), D. Rothrock (USA), S.C. Scherrer (Switzerland), S. Smith (Canada), O. Solomina (Russian Federation), D. Vaughan (UK), J. Walsh (USA), A. Worby (Australia), T. Yamada (Japan), L. Zhao (China)

Review Editors:

Roger Barry (USA), Toshio Koike (Japan)

This chapter should be cited as:

Lemke, P., J. Ren, R.B. Alley, I. Allison, J. Carrasco, G. Flato, Y. Fujii, G. Kaser, P. Mote, R.H. Thomas and T. Zhang, 2007: Observations: Changes in Snow, Ice and Frozen Ground. In: *Climate Change 2007: The Physical Science Basis. Contribution of Working Group I to the Fourth Assessment Report of the Intergovernmental Panel on Climate Change* [Solomon, S., D. Qin, M. Manning, Z. Chen, M. Marquis, K.B. Averyt, M. Tignor and H.L. Miller (eds.)]. Cambridge University Press, Cambridge, United Kingdom and New York, NY, USA.

Table of Contents

Executive Summary

In the climate system, the cryosphere (which consists of snow, river and lake ice, sea ice, glaciers and ice caps, ice shelves and ice sheets, and frozen ground) is intricately linked to the surface energy budget, the water cycle, sea level change and the surface gas exchange. The cryosphere integrates climate variations over a wide range of time scales, making it a natural sensor of climate variability and providing a visible expression of climate change. In the past, the cryosphere has undergone large variations on many time scales associated with ice ages and with shorter-term variations like the Younger Dryas or the Little Ice Age (see Chapter 6). Recent decreases in ice mass are correlated with rising surface air temperatures. This is especially true for the region north of 65°N, where temperatures have increased by about twice the global average from 1965 to 2005.

- Snow cover has decreased in most regions, especially in spring and summer. Northern Hemisphere (NH) snow cover observed by satellite over the 1966 to 2005 period decreased in every month except November and December, with a stepwise drop of 5% in the annual mean in the late 1980s. In the Southern Hemisphere, the few long records or proxies mostly show either decreases or no changes in the past 40 years or more. Where snow cover or snowpack decreased, temperature often dominated; where snow increased, precipitation almost always dominated. For example, NH April snow cover extent is strongly correlated with 40°N to 60°N April temperature, reflecting the feedback between snow and temperature, and declines in the mountains of western North America and in the Swiss Alps have been largest at lower elevations.

- Freeze-up and breakup dates for river and lake ice exhibit considerable spatial variability (with some regions showing trends of opposite sign). Averaged over available data for the NH spanning the past 150 years, freeze-up date has occurred later at a rate of 5.8 ± 1.6 days per century, while the breakup date has occurred earlier at a rate of 6.5 ± 1.2 days per century. (The uncertainty range given throughout this chapter denotes the 5 to 95% confidence interval.)

- Satellite data indicate a continuation of the 2.7 ± 0.6% per decade decline in annual mean arctic sea ice extent since 1978. The decline for summer extent is larger than for winter, with the summer minimum declining at a rate of 7.4 ± 2.4% per decade since 1979. Other data indicate that the summer decline began around 1970. Similar observations in the Antarctic reveal larger interannual variability but no consistent trends.

- Submarine-derived data for the central Arctic indicate that the average sea ice thickness in the central Arctic has *very likely* decreased by up to 1 m from 1987 to 1997. Model-based reconstructions support this, suggesting an arctic-wide reduction of 0.6 to 0.9 m over the same period. Large-scale trends prior to 1987 are ambiguous.

- Mass loss of glaciers and ice caps is estimated to be 0.50 ± 0.18 mm yr^{-1} in sea level equivalent (SLE) between 1961 and 2004, and 0.77 ± 0.22 mm yr^{-1} SLE between 1991 and 2004. The late 20th-century glacier wastage likely has been a response to post-1970 warming. Strongest mass losses per unit area have been observed in Patagonia, Alaska and northwest USA and southwest Canada. Because of the corresponding large areas, the biggest contributions to sea level rise came from Alaska, the Arctic and the Asian high mountains.

- Taken together, the ice sheets in Greenland and Antarctica have *very likely* been contributing to sea level rise over 1993 to 2003. Thickening in central regions of Greenland has been more than offset by increased melting near the coast. Flow speed has increased for some Greenland and Antarctic outlet glaciers, which drain ice from the interior. The corresponding increased ice sheet mass loss has often followed thinning, reduction or loss of ice shelves or loss of floating glacier tongues. Assessment of the data and techniques suggests a mass balance of the Greenland Ice Sheet of between +25 and –60 Gt yr^{-1} (–0.07 to 0.17 mm yr^{-1} SLE) from 1961 to 2003, and –50 to –100 Gt yr^{-1} (0.14 to 0.28 mm yr^{-1} SLE) from 1993 to 2003, with even larger losses in 2005. Estimates for the overall mass balance of the Antarctic Ice Sheet range from +100 to –200 Gt yr^{-1} (–0.28 to 0.55 mm yr^{-1} SLE) for 1961 to 2003, and from +50 to –200 Gt yr^{-1} (–0.14 to 0.55 mm yr^{-1} SLE) for 1993 to 2003. The recent changes in ice flow are *likely* to be sufficient to explain much or all of the estimated antarctic mass imbalance, with changes in ice flow, snowfall and melt water runoff sufficient to explain the mass imbalance of Greenland.

- Temperature at the top of the permafrost layer has increased by up to 3°C since the 1980s in the Arctic. The permafrost base has been thawing at a rate ranging up to 0.04 m yr^{-1} in Alaska since 1992 and 0.02 m yr^{-1} on the Tibetan Plateau since the 1960s. Permafrost degradation is leading to changes in land surface characteristics and drainage systems.

- The maximum extent of seasonally frozen ground has decreased by about 7% in the NH from 1901 to 2002, with a decrease in spring of up to 15%. Its maximum depth has decreased about 0.3 m in Eurasia since the mid-20th century. In addition, maximum seasonal thaw depth over permafrost has increased about 0.2 m in the Russian Arctic from 1956 to 1990. Onset dates of thaw in spring and freeze in autumn advanced five to seven days in Eurasia from 1988 to 2002, leading to an earlier growing season but no change in duration.

- Results summarised here indicate that the total cryospheric contribution to sea level change ranged from 0.2 to 1.2 mm yr^{-1} between 1961 and 2003, and from 0.8 to 1.6 mm yr^{-1} between 1993 and 2003. The rate increased over the 1993 to 2003 period primarily due to increasing losses from mountain glaciers and ice caps, from increasing surface melt on the Greenland Ice Sheet and from faster flow of parts of the Greenland and Antarctic Ice Sheets. Estimates of changes in the ice sheets are highly uncertain, and no best estimates are given for their mass losses or gains. However, strictly for the purpose of considering the possible contributions to the sea level budget, a total cryospheric contribution of 1.2 ± 0.4 mm yr^{-1} SLE is estimated for 1993 to 2003 assuming a midpoint mean plus or minus uncertainties and Gaussian error summation.

4.1   Introduction

The main components of the cryosphere are snow, river and lake ice, sea ice, glaciers and ice caps, ice shelves, ice sheets, and frozen ground (Figure 4.1). In terms of the ice mass and its heat capacity, the cryosphere is the second largest component of the climate system (after the ocean). Its relevance for climate variability and change is based on physical properties, such as its high surface reflectivity (albedo) and the latent heat associated with phase changes, which have a strong impact on the surface energy balance. The presence (absence) of snow or ice in polar regions is associated with an increased (decreased) meridional temperature difference, which affects winds and ocean currents. Because of the positive temperature-ice albedo feedback, some cryospheric components act to amplify both changes and variability. However, some, like glaciers and permafrost, act to average out short-term variability and so are sensitive indicators of climate change. Elements of the cryosphere are found at all latitudes, enabling a near-global assessment of cryosphere-related climate changes.

The cryosphere on land stores about 75% of the world's freshwater. The volumes of the Greenland and Antarctic Ice Sheets are equivalent to approximately 7 m and 57 m of sea level rise, respectively. Changes in the ice mass on land have contributed to recent changes in sea level. On a regional scale, many glaciers and ice caps play a crucial role in freshwater availability.

Presently, ice permanently covers 10% of the land surface, of which only a tiny fraction lies in ice caps and glaciers outside Antarctica and Greenland (Table 4.1). Ice also covers approximately 7% of the oceans in the annual mean. In midwinter, snow covers approximately 49% of the land surface in the Northern Hemisphere (NH). Frozen ground has the largest area of any component of the cryosphere. Changes in the components of the cryosphere occur at different time scales, depending on their dynamic and thermodynamic characteristics (Figure 4.1). All parts of the cryosphere contribute to short-term climate changes, with permafrost, ice shelves and ice sheets also contributing to longer-term changes including the ice age cycles.

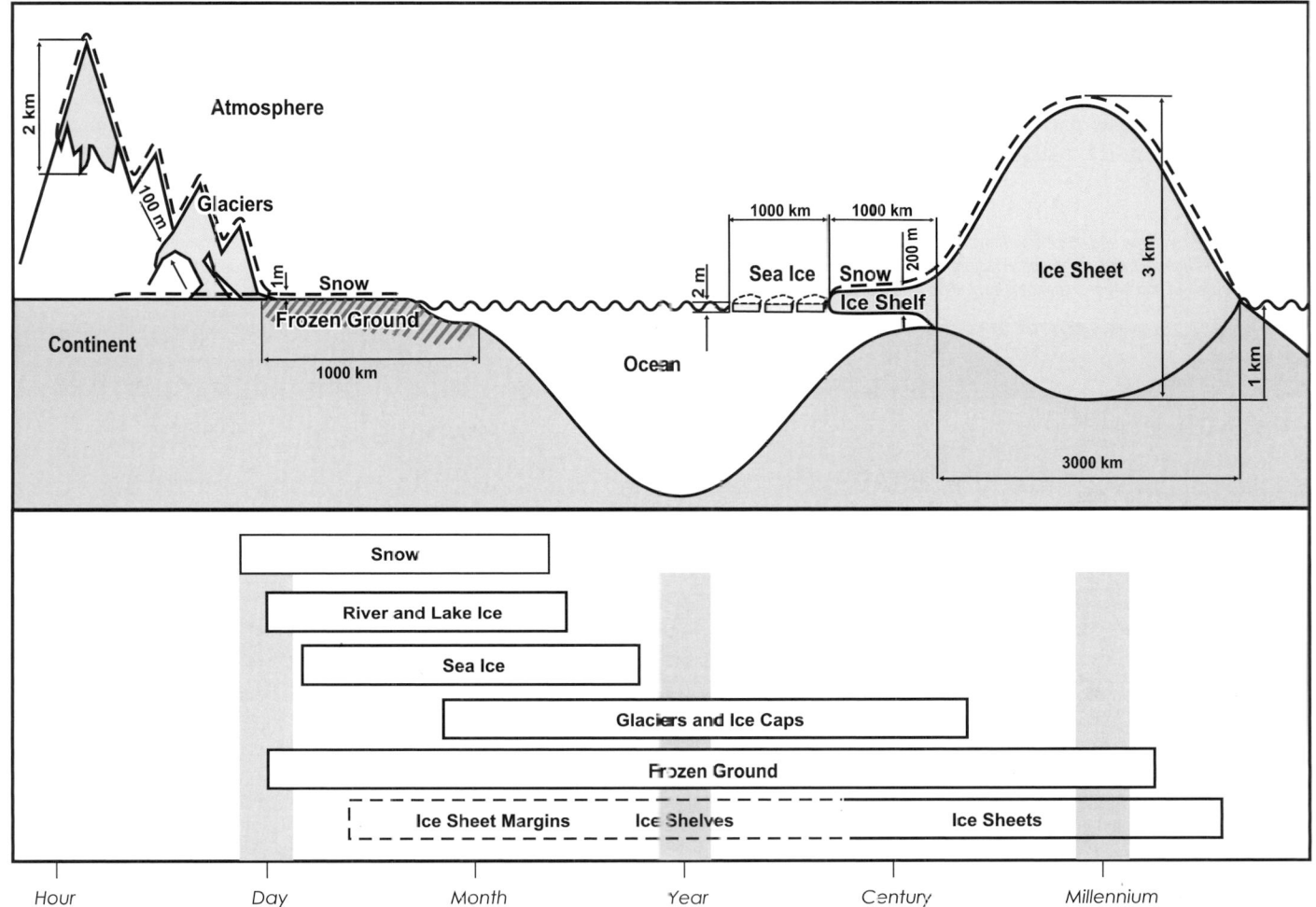

Figure 4.1. *Components of the cryosphere and their time scales.*

Seasonally, the area covered by snow in the NH ranges from a mean maximum in January of 45.2×10^6 km^2 to a mean minimum in August of 1.9×10^6 km^2 (1966–2004). Snow covers more than 33% of lands north of the equator from November to April, reaching 49% coverage in January. The role of snow in the climate system includes strong positive feedbacks related to albedo and other, weaker feedbacks related to moisture storage, latent heat and insulation of the underlying surface (M.P. Clark et al., 1999), which vary with latitude and season.

High-latitude rivers and lakes develop an ice cover in winter. Although the area and volume are small compared to other components of the cryosphere, this ice plays an important role in freshwater ecosystems, winter transportation, bridge and pipeline crossings, etc. Changes in the thickness and duration of these ice covers can therefore have consequences for both the natural environment and human activities. The breakup of river ice is often accompanied by 'ice jams' (blockages formed by accumulation of broken ice); these jams impede the flow of water and may lead to severe flooding.

At maximum extent arctic sea ice covers more than 15×10^6 km^2, reducing to only 7×10^6 km^2 in summer. Antarctic sea ice is considerably more seasonal, ranging from a winter maximum of over 19×10^6 km^2 to a minimum extent of about 3×10^6 km^2. Sea ice less than one year old is termed 'first-year ice' and that which survives more than one year is called 'multi-year ice'. Most sea ice is part of the mobile 'pack ice', which circulates in the polar oceans, driven by winds and surface currents. This pack ice is extremely inhomogeneous, with differences in ice thicknesses and age, snow cover, open

water distribution, etc. occurring at spatial scales from metres to hundreds of kilometres.

Glaciers and ice caps adapt to a change in climate conditions much more rapidly than does a large ice sheet, because they have a higher ratio between annual mass turnover and their total mass. Changes in glaciers and ice caps reflect climate variations, in many cases providing information in remote areas where no direct climate records are available, such as at high latitudes or on the high mountains that penetrate high into the middle troposphere. Glaciers and ice caps contribute to sea level changes and affect the freshwater availability in many mountains and surrounding regions. Formation of large and hazardous lakes is occurring as glacier termini retreat from prominent Little Ice Age moraines, especially in the steep Himalaya and Andes.

The ice sheets of Greenland and Antarctica are the main reservoirs capable of affecting sea level. Ice formed from snowfall spreads under gravity towards the coast, where it melts or calves into the ocean to form icebergs. Until recently (including IPCC, 2001) it was assumed that the spreading velocity would not change rapidly, so that impacts of climate change could be estimated primarily from expected changes in snowfall and surface melting. Observations of rapid ice flow changes since IPCC (2001) have complicated this picture, with strong indications that floating ice shelves 'regulate' the motion of tributary glaciers, which can accelerate manyfold following ice shelf breakup.

Frozen ground includes seasonally frozen ground and permafrost. The permafrost region occupies approximately

Table 4.1: *Area, volume and sea level equivalent (SLE) of cryospheric components. Indicated are the annual minimum and maximum for snow, sea ice and seasonally frozen ground, and the annual mean for the other components. The sea ice area is represented by the extent (area enclosed by the sea ice edge). The values for glaciers and ice caps denote the smallest and largest estimates excluding glaciers and ice caps surrounding Greenland and Antarctica.*

Cryospheric Component	Area (10^6 km^2)	Ice Volume (10^6 km^3)	Potential Sea Level Rise (SLE) (m)[g]
Snow on land (NH)	1.9–45.2	0.0005–0.005	0.001–0.01
Sea ice	19–27	0.019–0.025	~0
Glaciers and ice caps			
Smallest estimate[a]	0.51	0.05	0.15
Largest estimate[b]	0.54	0.13	0.37
Ice shelves[c]	1.5	0.7	~0
Ice sheets	14.0	27.6	63.9
Greenland[d]	1.7	2.9	7.3
Antarctica[c]	12.3	24.7	56.6
Seasonally frozen ground (NH)[e]	5.9–48.1	0.006–0.065	~0
Permafrost (NH)[f]	22.8	0.011–0.037	0.03–0.10

Notes:

[a] Ohmura (2004); glaciers and ice caps surrounding Greenland and Antarctica are excluded.

[b] Dyurgerov and Meier (2005); glaciers and ice caps surrounding Greenland and Antarctica are excluded.

[c] Lythe et al. (2001).

[d] Bamber et al. (2001).

[e] Zhang et al. (2003).

[f] Zhang et al. (1999) , excluding permafrost under ocean, ice sheets and glaciers.

[g] Assuming an oceanic area of 3.62×10^8 km^2, an ice density of 917 kg m^{-3}, a seawater density of 1,028 kg m^{-3}, and seawater replacing grounded ice below sea level.

23×10^6 km² or 24% of the land area in the NH. On average, the long-term maximum areal extent of the seasonally frozen ground, including the active layer over permafrost, is about 48×10^6 km² or 51% of the land area in the NH. In terms of areal extent, frozen ground is the single largest cryospheric component. Permafrost also acts to record air temperature and snow cover variations, and under changing climate can be involved in feedbacks related to moisture and greenhouse gas exchange with the atmosphere.

4.2   Changes in Snow Cover

4.2.1   Background

The high albedo of snow (0.8 to 0.9 for fresh snow) has an important influence on the surface energy budget and on Earth's radiative balance (e.g., Groisman et al., 1994). Snow albedo, and hence the strength of the feedback, depends on a number of factors such as the depth and age of a snow cover, vegetation height, the amount of incoming solar radiation and cloud cover. The albedo of snow may be decreasing because of anthropogenic soot (Hansen and Nazarenko, 2004; see Section 2.5.4 for details).

In addition to the direct snow-albedo feedback, snow may influence climate through indirect feedbacks (i.e., those in which there are more than two causal steps), such as to summer soil moisture. Indirect feedbacks to atmospheric circulation may involve two types of circulation, monsoonal (e.g., Lo and Clark, 2001) and annular (e.g., Saito and Cohen, 2003; see Section 3.6.4), although there are large uncertainties in the physical mechanisms involved (Bamzai, 2003; Robock et al., 2003).

In this section, observations of snow cover extent are updated from IPCC (2001). In addition, several new topics are covered: changes in snow depth and snow water equivalent; relationships of snow to temperature and precipitation; and observations and estimates of changes in snow in the Southern Hemisphere (SH). Changes in the fraction of precipitation falling as snow or other frozen forms are covered in Section 3.3.2.3. This section covers only snow on land; snow on various forms of ice is covered in subsequent sections.

4.2.2   Observations of Snow Cover, Snow Duration and Snow Quantity

4.2.2.1   *Sources of Snow Data*

Daily observations of the depth of snow and of new snowfall have been made by various methods in many countries, dating to the late 1800s in a few countries (e.g., Switzerland, USA, the former Soviet Union and Finland). Measurements of snow depth and snow water equivalent (SWE) became widespread by 1950 in the mountains of western North America and Europe, and a few sites in the mountains of Australia have been monitored since 1960. *In situ* snow data are affected by changes in station location, observing practices and land cover, and are not uniformly distributed.

The premier data set used to evaluate large-scale snow covered area (SCA), which dates to 1966 and is the longest satellite-derived environmental data set of any kind, is the weekly visible wavelength satellite maps of NH snow cover produced by the US National Oceanic and Atmospheric Administration's (NOAA) National Environmental Satellite Data and Information Service (NESDIS; Robinson et al., 1993). Trained meteorologists produce the weekly NESDIS snow product from visual analyses of visible satellite imagery. These maps are well validated against surface observations, although changes in mapping procedures in 1999 affected the continuity of data series at a small number of mountain and coastal grid points. For the SH, mapping of SCA began only in 2000 with the advent of Moderate Resolution Imaging Spectroradiometer (MODIS) satellite data.

Space-borne passive microwave sensors offer the potential for global monitoring since 1978 of not just snow cover, but also snow depth and SWE, unimpeded by cloud cover and winter darkness. In order to generate homogeneous depth or SWE data series, differences between Scanning Multichannel Microwave Radiometer (SMMR; 1978 to 1987) and Special Sensor Microwave/Imager (SSM/I; 1987 to present) in 1987 must be resolved (Derksen et al., 2003). Estimates of SCA from microwave satellite data compare moderately well with visible data except in autumn (when microwave estimates are too low) and over the Tibetan plateau (microwave too high; Armstrong and Brodzik, 2001). Work is ongoing to develop reliable depth and SWE retrievals from passive microwave for areas with heavy forest or deep snowpacks, and the relatively coarse spatial resolution (~10–25 km) still limits applications over mountainous regions.

4.2.2.2   *Variability and Trends in Northern Hemisphere Snow Cover*

In this subsection, following the hemispheric view provided by the large-scale analyses by Brown (2000) and Robinson et al (1993), regional and national-scale studies are discussed. The mean annual NH SCA (1966–2004) is 23.9×10^6 km², not including the Greenland Ice Sheet. Interannual variability of SCA is largest not in winter, when mean SCA is greatest, but in autumn (in absolute terms) or summer (in relative terms). Monthly standard deviations range from 1.0×10^6 km² in August and September to 2.7×10^6 km² in October, and are generally just below 2×10^6 km² in non-summer months.

Since the early 1920s, and especially since the late 1970s, SCA has declined in spring (Figure 4.2) and summer, but not substantially in winter (Table 4.2) despite winter warming (see Section 3.2.2). Recent declines in SCA in the months of

February through August have resulted in (1) a shift in the month of maximum SCA from February to January; (2) a statistically significant decline in annual mean SCA; and (3) a shift towards earlier spring melt by almost two weeks in the 1972 to 2000 period (Dye, 2002). Early in the satellite era, between 1967 and 1987, mean annual SCA was 24.4×10^6 km^2. An abrupt transition occurred between 1986 and 1988, and since 1988 the mean annual extent has been 23.1×10^6 km^2, a statistically significant (T test, p <0.01) reduction of approximately 5% (Robinson and Frei, 2000). Over the longer 1922 to 2005 period (updated from Brown, 2000), the linear trend in March and April NH SCA (Figure 4.2) is a statistically significant reduction of $2.7 \pm 1.5 \times 10^6$ km^2 or $7.5 \pm 3.5\%$.

Temperature variations and trends play a significant role in variability and trends of NH SCA, by determining whether precipitation falls as rain or snow, and by determining snowmelt. In almost every month, SCA is correlated with temperature in the latitude band of greatest variability in SCA, owing to the snow-albedo feedback. For example, temperature in the 40°N to 60°N band and NH SCA are highly correlated in spring (r = –0.68; updated from Brown, 2000) and the largest reductions in March-April average snow cover occurred roughly between the 0°C and 5°C isotherms (Figure 4.3). The snow-albedo feedback also helps determine the longer-term trends (for temperature see Section 3.2.2; see also M.P. Clark et al., 1999; Groisman et al., 1994).

The following paragraphs discuss regional details, including information not available or missing from the satellite data and from Brown's (2000) hemispheric reconstruction.

4.2.2.2.1 North America

From 1915 to 2004, North American SCA increased in November, December and January owing to increases in precipitation (Section 3.3.2; Groisman et al., 2004). Decreases in snow cover are mainly confined to the latter half of the 20th century, and are most apparent in the spring period over western North America (Groisman et al., 2004). Shifts towards earlier melt by about eight days since the mid-1960s were also observed in northern Alaska (Stone et al., 2002).

Another dimension of change in snow is provided by the annual measurements of mountain SWE near April 1 in western North America, which indicate declines since 1950 at about 75% of locations monitored (Mote et al., 2005). The date of maximum mountain SWE appears to have shifted earlier by about two weeks since 1950, as inferred from streamflow measurements (Stewart et al., 2005). That these reductions are predominantly due to warming is shown by regression analysis of streamflow (Stewart et al., 2005) and SWE (Mote, 2006) on temperature and precipitation, and by the dependence of trends in SWE (Mote et al., 2005) on elevation or equivalently mean winter temperature (Figure 4.4a), with the largest percentage changes near the 0°C level.

4.2.2.2.2 Europe and Eurasia

Snow cover trends in mountain regions of Europe are characterised by large regional and altitudinal variations. Recent declines in snow cover have been documented in the mountains of Switzerland (e.g., Scherrer et al., 2004) and Slovakia (Vojtek et al., 2003), but no change was observed in Bulgaria over the 1931 to 2000 period (Petkova et al., 2004). Declines, where observed, were largest at lower elevations, and Scherrer et al.

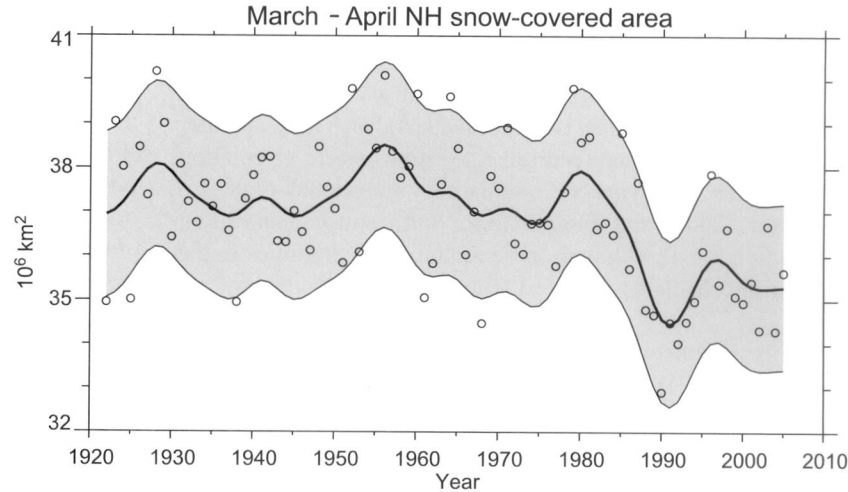

Figure 4.2. *Update of NH March-April average snow-covered area (SCA) from Brown (2000). Values of SCA before 1972 are based on the station-derived snow cover index of Brown (2000); values beginning in 1972 are from the NOAA satellite data set. The smooth curve shows decadal variations (see Appendix 3.A), and the shaded area shows the 5 to 95% range of the data estimated after first subtracting the smooth curve.*

Table 4.2. *Trend (10^6 km^2 per decade) in monthly NH SCA from satellite data (Rutgers-corrected, D. Robinson) over the 1966 to 2005 period and for three months covering the 1922 to 2005 period based on the NH SCA reconstruction of Brown (2000).*

Years	Jan	Feb	Mar	Apr	May	Jun	Jul	Aug	Sep	Oct	Nov	Dec	Ann
1966–2005	–0.11	–0.49	–0.80[a]	–0.74[a]	–0.57	–1.10[a]	–1.17[a]	–0.82[a]	–0.20	–0.36	0.12	0.19	–0.33[a]
1922–2005	n/a	n/a	–0.25[a]	–0.35[a]	n/a	n/a	n/a	n/a	n/a	0.24[a]	n/a	n/a	n/a

Notes:
[a] Statistically significant at the 0.05 level of confidence.
 n/a: not available.

(2004) statistically attributed the declines in the Swiss Alps to warming, as is clear when trends are plotted against winter temperature (Figure 4.4b).

Lowland areas of central Europe are characterised by recent reductions in annual snow cover duration by about 1 day yr⁻¹ (e.g., Falarz, 2002). Trends towards greater maximum snow depth but shorter snow season have been noted in Finland (Hyvärinen, 2003), the former Soviet Union from 1936 to 1995 (Ye and Ellison, 2003), and in the Tibetan Plateau (Zhang et al., 2004) since the late 1970s. Qin et al. (2006) reported no trends in snow depth or snow cover in western China since 1957.

4.2.2.3   Southern Hemisphere

Outside of Antarctica (see Section 4.6), very little land area in the SH experiences snow cover. Long-term records of snow cover, snowfall, snow depth or SWE are scarce. In some cases, proxies for snow line can be used, but the quality of data is much lower than for most NH areas.

4.2.2.3.1   South America

Estimates from microwave satellite observations for mid-latitude alpine regions of South America for the period of

March — April Snow Departure
(1988 - 2004) minus (1967 - 1987)

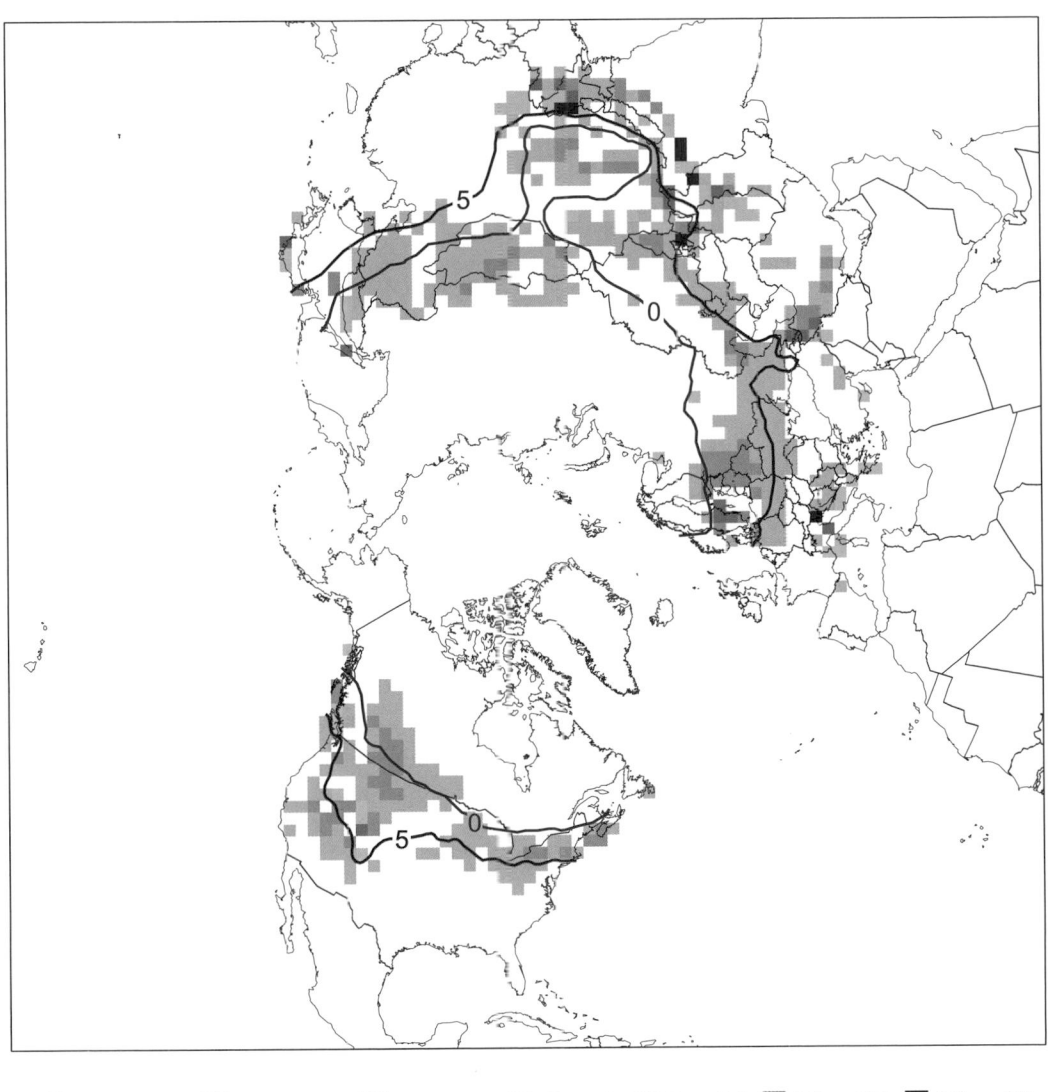

Figure 4.3. *Differences in the distribution of Northern Hemisphere March-April average snow cover between earlier (1967–1987) and later (1988–2004) portions of the satellite era (expressed in % coverage). Negative values indicate greater extent in the earlier portion of the record. Extents are derived from NOAA/NESDIS snow maps. Red curves show the 0°C and 5°C isotherms averaged for March and April 1967 to 2004, from the Climatic Research Unit (CRU) gridded land surface temperature version 2 (CRUTEM2v) data.*

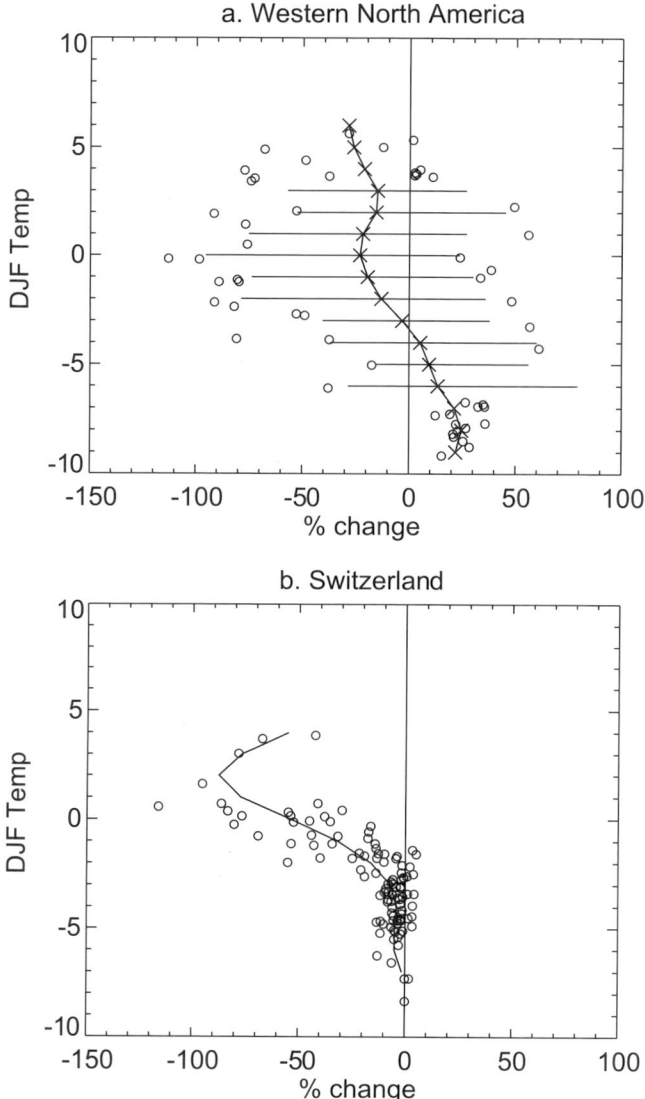

a. Western North America

b. Switzerland

Figure 4.4. *Dependence of trends in snow on mean winter temperature (°C) at each location. (a) Relative trends in 1 April SWE, 1950 to 2000, in the mountains of western North America (British Columbia, Washington, Oregon and California), binned by mean December to February (DJF) temperature. For each 1°C temperature bin, 'x' symbols indicate the mean trend, bars indicate the span of the 5 to 95% confidence interval for bins with at least 10 points, and circles indicate outliers. Total number of data points is 323 (adapted from Mote et al., 2005). (b) Relative trend in days of winter (DJF) snow cover at 109 sites in Switzerland, 1958 to 1999, binned by mean DJF temperature (adapted from Scherrer et al., 2004).*

record 1979 to 2002 show substantial interannual variability with little or no long-term trend. A long-term increasing trend in the number of snow days was found in the eastern side of the central Andes region (33°S) from 1885 to 1996, derived from newspaper reports of Mendoza City (Prieto et al., 2001).

Other approaches suggest some response of snow line to warming in South America. The 0°C isotherm altitude (ZIA), an indication of snow line, has been derived from the daily temperature profile obtained from radiosonde data located at Quintero (32°47'S, 71°33'W, 8 m above sea level; Carrasco et al., 2005), which represents the snow line behaviour in the

western Andes from about 30°S to 36°S. Over the 1975 to 2001 period of record, the linear change in winter ZIA was 121.9 ± 7.7 m, and the positive trend was dominated by atmospheric conditions on dry days (enhancing melt) with no trend on wet days (accumulation zone unchanged).

4.2.2.3.2   Australia and New Zealand

For the mountainous south-eastern area of Australia, studies of late winter (August–September) snow depth have shown some significant declines (as much as 40%) since 1962. Trends in maximum snow depth were more modest. The stronger declines in late winter are attributed to spring season warming, while maximum snow depth is largely determined by winter precipitation, which has declined only slightly (Hennessy et al., 2003; Nicholls, 2005).

In New Zealand, annual observations of end-of-summer snow line on 47 glaciers have been made by airplane since 1977, and reveal large interannual variability primarily associated with atmospheric circulation anomalies (Clare et al., 2002); it is noteworthy, however, that the four years with highest snow line occurred in the 1990s. The only study of seasonal snow cover in the Southern Alps found no trend over the 1930 to 1985 period (Fitzharris and Garr, 1995) and has not been updated.

4.3   Changes in River and Lake Ice

4.3.1   Background

Because of its importance to many human activities, freeze-up and breakup dates of river and lake ice have been recorded for a long time at many locations. These records provide useful climate information, although they must be interpreted with care. In the case of rivers, both freeze-up and breakup at a given location can be strongly affected by conditions far upstream (for example, heavy rains or snowmelt in a distant portion of the watershed). In the case of lakes, the historical observations have typically been made at coastal locations (often protected bays and harbours) and so may not be representative of the lake as a whole, or comparable to more recent satellite-based observations. Nevertheless, these observations represent some of the longest records of cryospheric change available.

Observations of ice thickness are considerably sparser and are generally made using direct drilling methods. Long-term records are available at a few locations; however it should be noted that, just as for sea ice, changes in lake and river ice thickness are a consequence not just of temperature and radiative forcing, but also of changes in snowfall (via the insulating effect of snow).

4.3.2   Changes in Freeze-up and Breakup Dates

Freeze-up is defined conceptually as the time at which a continuous and immobile ice cover forms; however, operational definitions range from local observations of the presence

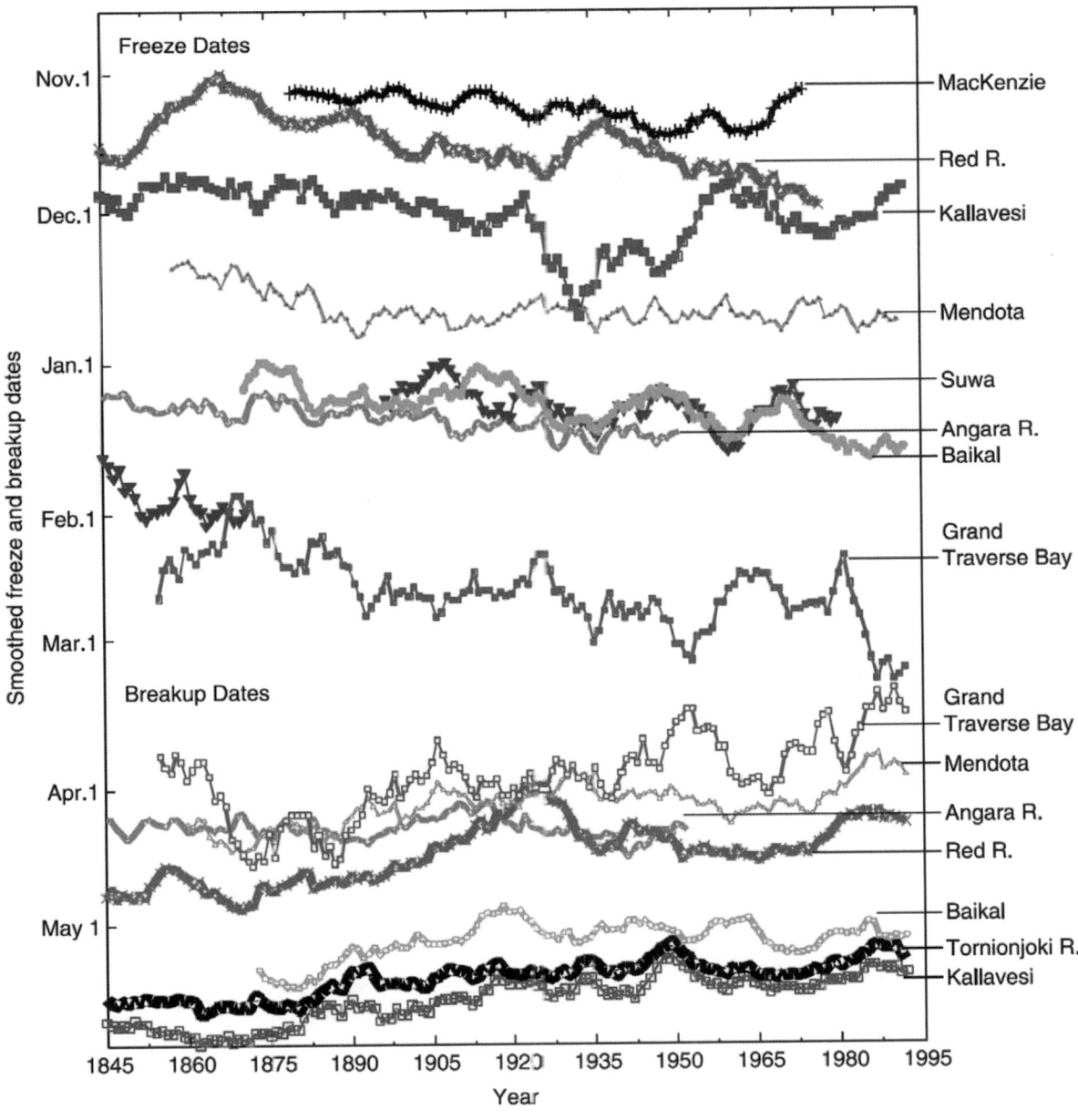

Figure 4.5. *Time series of freeze-up and breakup dates from several northern lakes and rivers (reprinted with permission from Magnuson et al., 2000, copyright AAAS). Dates have been smoothed with a 10-year moving average. See the cited publication for locations and other details.*

or absence of ice to inferences drawn from river discharge measurements. Breakup is typically the time when the ice cover begins to move downstream in a river or when open water becomes extensive at the measurement location for lakes. Here again, there is some ambiguity in the specific date, and in the extent to which local observations reflect conditions elsewhere on a large lake or in a large river basin.

Selected time series from a recent compilation of river and lake freeze-up and breakup records by Magnuson et al. (2000) are shown in Figure 4.5. They limited consideration to records spanning at least 150 years. Eleven out of 15 records showed significant trends towards later freeze-up and 17 out of 25 records showed significant trends towards earlier breakup. When averaged together, the freeze-up date has become later at

a rate of 5.8 ± 1.6 days per century, while the breakup date has occurred earlier at a rate of 6.5 ± 1.2 days per century.

A larger sample of Canadian rivers spanning the last 30 to 50 years was analysed by Zhang et al. (2001). These freeze-up and breakup estimates (based on inferences from streamflow data) exhibit considerable variability, with a trend towards earlier freeze-up and breakup over much of the country. The earlier freeze-up dominates, however, leading to a significant decrease in open water duration at many locations as shown in Figure 4.6. A recent analysis of Russian river data by Smith (2000) revealed a trend towards earlier freeze-up of western Russian rivers and later freeze-up in rivers of eastern Siberia over the last 50 to 70 years. Breakup dates did not exhibit statistically significant trends.

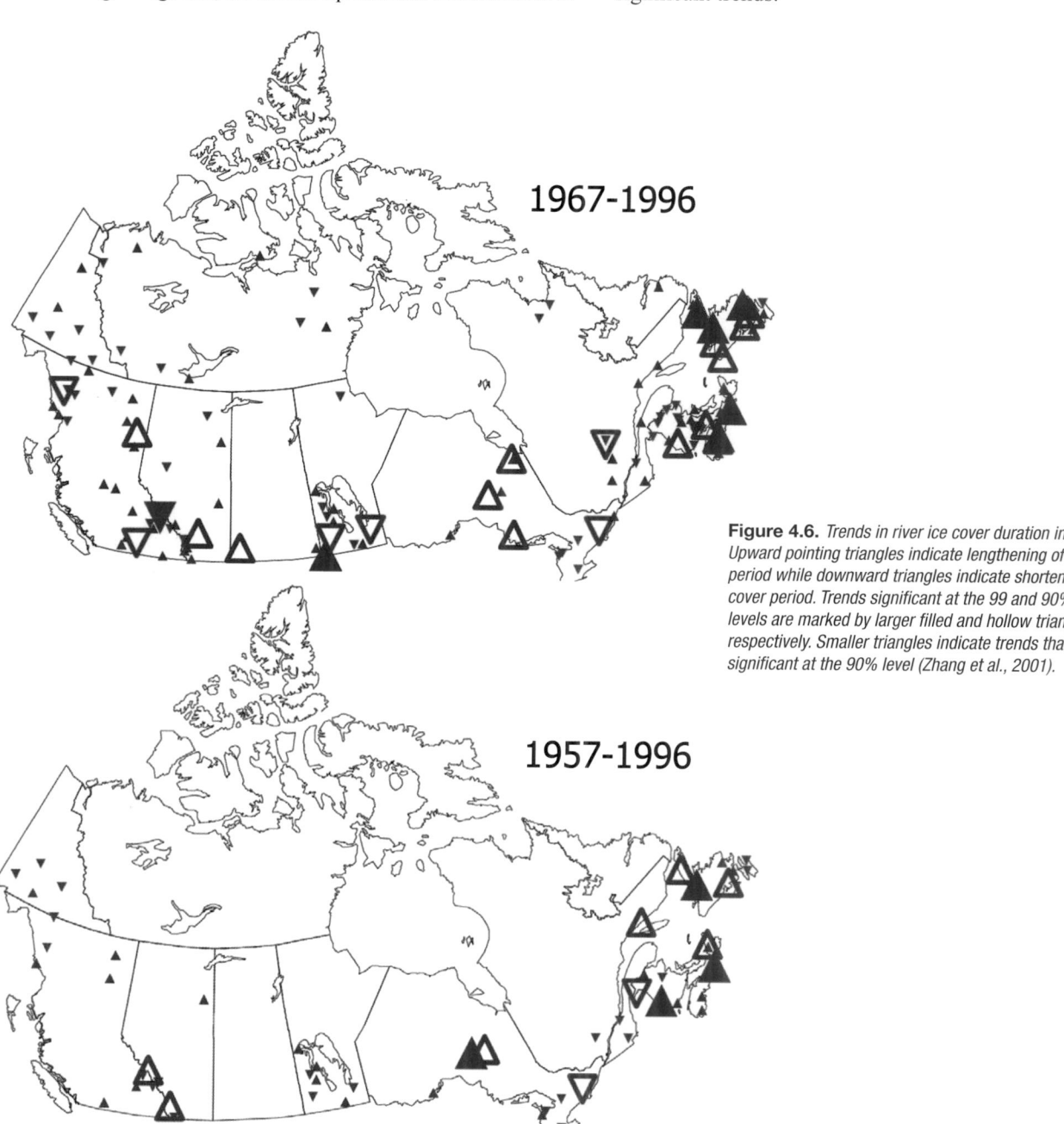

Figure 4.6. *Trends in river ice cover duration in Canada. Upward pointing triangles indicate lengthening of the ice cover period while downward triangles indicate shortening of the ice cover period. Trends significant at the 99 and 90% confidence levels are marked by larger filled and hollow triangles, respectively. Smaller triangles indicate trends that are not significant at the 90% level (Zhang et al., 2001).*

A comparable analysis of freeze-up and breakup dates for Canadian lakes has recently been completed by Duguay et al. (2006). These results (shown in Figure 4.7) indicate a fairly general trend towards earlier breakup (particularly in western Canada), while freeze-up exhibited a mix of early and later dates.

There are insufficient published data on river and lake ice thickness to allow assessment of trends. Modelling studies (e.g., Duguay et al., 2003) indicate that, as with the landfast sea ice case, much of the variability in maximum ice thickness and breakup date is driven by variations in snowfall.

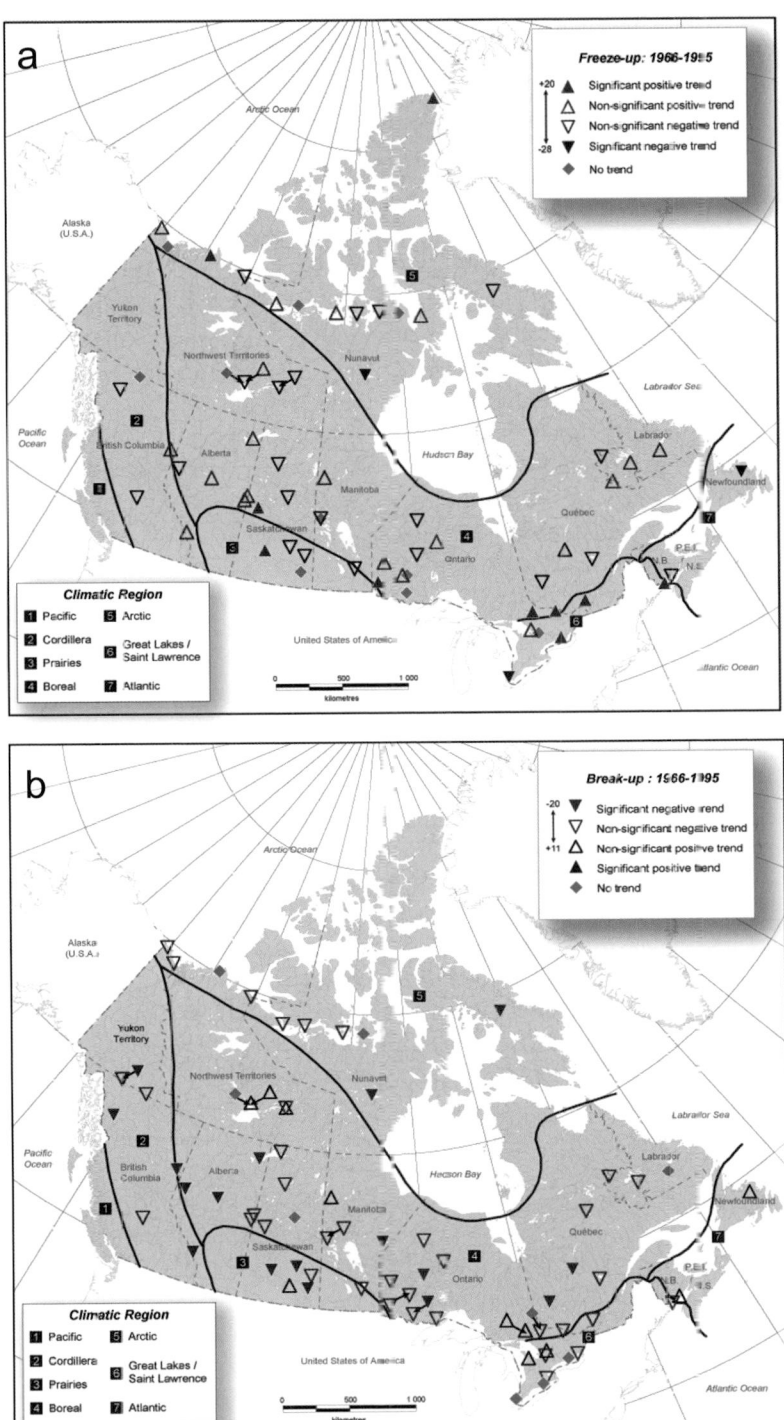

Figure 4.7. *Trends in (a) freeze-up and (b) breakup dates observed at lakes in Canada over the period 1965 to 1995. Downward pointing arrows indicate a trend towards earlier dates; upward pointing arrows, a trend towards later dates. Open symbols indicate that the trend is not significant while solid symbols indicate that the trend is significant at the 90% confidence level (modified from Duguay et al., 2006).*

4.4 Changes in Sea Ice

4.4.1 Background

Sea ice is formed by freezing of seawater in the polar oceans. It is an important, interactive component of the global climate system because: a) it is central to the powerful 'ice-albedo' feedback mechanism that enhances climate response at high latitudes (see Chapter 2); b) it modifies the exchange of heat, gases and momentum between the atmosphere and polar oceans; and c) it redistributes freshwater via the transport and subsequent melt of relatively fresh sea ice, and hence alters ocean buoyancy forcing.

The thickness of sea ice is a consequence of past growth, melt and deformation, and so is an important indicator of climatic conditions. Ice thickness is also closely connected to ice strength, and so changes in thickness are important to navigability by ships, to the stability of the ice as a platform for use by humans and marine mammals, to light transmission through the ice cover, etc. Sea ice increases in thickness as bottom freezing balances heat conduction through the ice to the surface (heat conduction is strongly influenced by the insulating thickness of the ice itself and the snow on it). Most of the inhomogeneity in the pack results from deformation of the ice due to differential movement of individual pieces of ice (called 'floes'). Open water areas created within the ice pack under divergence or shear (called 'leads') are a major contributor to ocean-atmosphere heat exchange (turbulent heat loss from the ocean in winter and shortwave heating in the summer). In some locations, due either to persistent ice divergence or to persistent upwelling of oceanic heat, open water areas within an otherwise ice-covered region can be sustained over much of the winter. These are called 'polynyas' and are important feeding areas for marine mammals and birds.

Under convergence, thin ice sheets may 'raft' on top of each other, doubling the ice thickness, and under strong convergence (for example, when wind drives sea ice against a coast), the ice buckles and crushes to form sinuous 'ridges' of thick ice. In the Arctic, ridges can be tens of metres thick, account for nearly half of the total ice volume and constitute a major impediment to transportation on, through, or under the ice. Although ridging is generally less severe in the Antarctic, ice deformation is still an important process in thickening the ice cover.

Near the shore, in bays and fjords, and among islands like those of the Canadian Arctic Archipelago, sea ice can be attached to land and therefore be immobile. This is termed 'landfast' ice. In the Arctic such ice (and in particular its freeze-up and breakup) is of special importance to local residents as it is used as a platform for hunting and fishing, and is an impediment to shipping.

Some climatically important characteristics of sea ice include: its concentration (that fraction of the ocean covered by ice); its extent (the area enclosed by the ice edge – operationally defined as the 15% concentration contour); the total area of ice within its extent (i.e., extent weighted by concentration); the area of multi-year ice within the total extent; its thickness (and the thickness of the snow cover on it); its velocity; and its growth and melt rates (and hence salt or freshwater flux into the ocean). Ice extent, or ice edge position, is the only sea ice variable for which observations are available for more than a few decades. Expansion or retreat of the ice edge may be amplified by the ice-albedo feedback.

4.4.2 Sea Ice Extent and Concentration

4.4.2.1 Data Sources and Time Periods Covered

The most complete record of sea ice extent is provided by passive microwave data from satellites that are available since the early 1970s. Prior to that, aircraft, ship and coastal observations are available at certain times and in certain locations. Portions of the North Atlantic are unique in having ship observations extending well back into the 19th century. Far fewer historic data exist from the SH, with one notable exception being the record of annual landfast ice duration from the sub-antarctic South Orkney islands starting in 1903 (Murphy et al., 1995).

Estimation of sea ice properties from passive microwave emission requires an algorithm to convert observed radiance into ice concentration (and type). Several such algorithms are available (e.g., Steffen et al., 1992) and their accuracy has been evaluated using high-resolution satellite and aircraft imagery (e.g., Cavalieri, 1992; Kwok, 2002) and operational ice charts (e.g., Agnew and Howell, 2003). The accuracy of satellite-derived ice concentration is usually 5% or better, although errors of 10 to 20% can occur during the melt season. The accuracy of the ice edge (relevant to estimating ice extent) is largely determined by the spatial resolution of the satellite radiometer, and is of the order of 25 km (recently launched instruments provide improved resolution of about 12.5 km). Summer concentration errors do lead to a bias in estimated ice-covered area in both the NH and SH warm seasons (Agnew and Howell, 2003; Worby and Comiso, 2004). This is an important consideration when comparing the satellite period with older proxy records of ice extent.

Distinguishing between first-year and multi-year ice from passive microwave data is more difficult, although algorithms are improving (e.g., Johannessen et al., 1999). However, the summer minimum ice extent, which is by definition the multi-year ice extent at that time of year, is not as prone to algorithm errors (e.g., Comiso, 2002).

4.4.2.2 Hemispheric, Regional and Seasonal Time Series from Passive Microwave

Most analyses of variability and trend in ice extent using the satellite record have focussed on the period after 1978 when the satellite sensors have been relatively constant. Different estimates, obtained using different retrieval algorithms, produce very similar results for hemispheric extent, and all show an asymmetry between changes in the Arctic and Antarctic. As an example, an updated version of the analysis done by Comiso (2003), spanning the period from November 1978 through December 2005, is shown in Figure 4.8. The annual mean ice extent anomalies are shown. There is a significant decreasing trend in arctic sea ice extent of $-33 \pm 7.4 \times 10^3$ km^2 yr^{-1} (equivalent to $-2.7 \pm 0.6\%$ per decade), whereas the antarctic results show a small positive trend of $5.6 \pm 9.2 \times 10^3$ km^2 yr^{-1} ($0.47 \pm 0.8\%$ per decade), which is not statistically significant. The uncertainties represent the 90% confidence interval around the trend estimate and the percentages are based on the 1978 to 2005 mean. In both hemispheres, the trends are larger in summer and smaller in winter. In addition, there is considerable variation in the magnitude, and even the sign, of the trend from region to region within each hemisphere.

The most remarkable change observed in the arctic ice cover has been the decrease in ice that survives the summer, shown in Figure 4.9. The trend in the minimum arctic sea ice extent, between 1979 and 2005, was $-60 \pm 20 \times 10^3$ km^2 yr^{-1} ($-7.4 \pm 2.4\%$ per decade). These trends are superimposed on substantial interannual to decadal variability, which is associated with variability in atmospheric circulation (Belchansky et al., 2005).

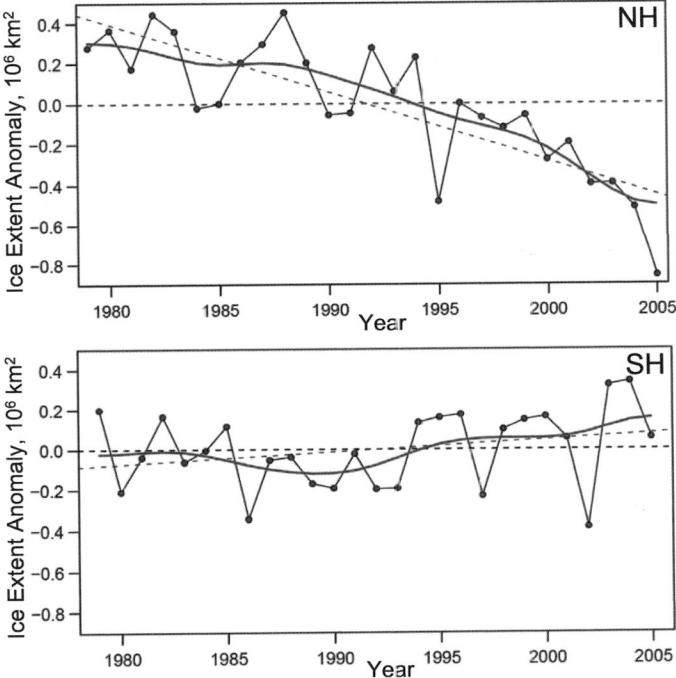

Figure 4.8. *Sea ice extent anomalies (computed relative to the mean of the entire period) for (a) the NH and (b) the SH, based on passive microwave satellite data. Symbols indicate annual mean values while the smooth blue curves show decadal variations (see Appendix 3.A). Linear trend lines are indicated for each hemisphere. For the Arctic, the trend is $-33 \pm 7.4 \times 10^3$ km^2 yr^{-1} (equivalent to approximately -2.7% per decade), whereas the Antarctic results show a small positive trend of $5.6 \pm 9.2 \times 10^3$ km^2 yr^{-1}. The negative trend in the NH is significant at the 90% confidence level whereas the small positive trend in the SH is not significant (updated from Comiso, 2003).*

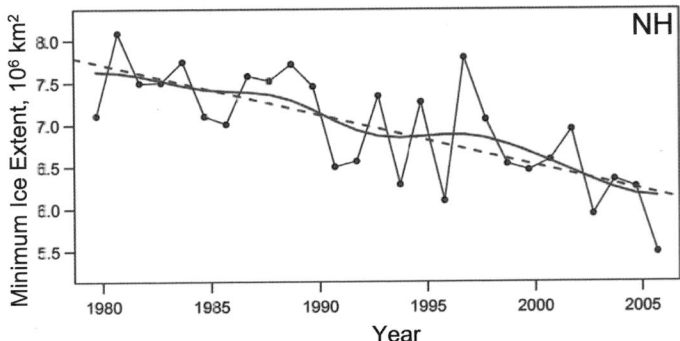

Figure 4.9. *Summer minimum arctic sea ice extent from 1979 to 2005. Symbols indicate annual mean values while the smooth blue curve shows decadal variations (see Appendix 3.A). The dashed line indicates the linear trend, which is $-60 \pm 20 \times 10^3$ km^2 yr^{-1}, or approximately -7.4% per decade (updated from Comiso, 2002).*

4.4.2.3   Longer Records of Hemispheric Extent

The lack of comprehensive sea ice data prior to the satellite era hampers estimates of hemispheric-scale trends over longer time scales. Rayner et al. (2003) compiled a data set of sea ice extent for the 20th century from available sources and accounted for the inhomogeneity between them (Figure 4.10). There is a clear indication of sustained decline in arctic ice extent since about the early 1970s, particularly in summer. On a regional basis, portions of the North Atlantic have sufficient historical data, based largely on ship reports and coastal observations, to permit trend assessments over periods exceeding 100 years. Vinje (2001) compiled information from ship reports in the Nordic Seas to estimate April sea ice extent in this region for the period since about 1860. This time series is also shown in Figure 4.10 and indicates a generally continuous decline from the start of the record to the end. Ice extent data from Russian sources have recently been published (Polyakov et al., 2003), and cover essentially the entire 20th century for the Russian coastal seas (Kara, Laptev, East Siberian and Chukchi). These data, which exhibit large inter-decadal variability, show a declining trend since the 1960s until a reversal in the late 1990s. The Russian data indicate anomalously little ice during the 1940s and 1950s, whereas the Nordic Sea data indicate anomalously large extent at this time, showing the importance of regional variability.

Omstedt and Chen (2001) obtained a proxy record of the annual maximum extent of sea ice in the region of the Baltic Sea over the period 1720 to 1997. This record showed a substantial decline in sea ice that occurred around 1877, and greater variability in sea ice extent in the colder 1720 to 1877 period than in the warmer 1878 to 1997 period. Hill et al. (2002) have examined sea ice information for the Canadian maritime region and deduced that sea ice incursions occurred during the 19th century in the Grand Banks and surrounding areas that are now ice-free. Although there are problems with homogeneity of all these data (with quality declining further back in history), and with the disparity in spatial scales represented by each, they are all consistent in terms of the declining ice extent during the latter decades of the 20th century, with the decline beginning prior to the satellite era. Those data that extend far enough back in time imply, with high confidence, that sea ice was more extensive in the North Atlantic during the 19th century.

Continuous long-term data records for the Antarctic are lacking, as systematic information on the entire Southern Ocean ice cover became available only with the advent of routine microwave satellite reconnaissance in the early 1970s. Parkinson (1990) examined ice edge observations from four exploration voyages in the late 18th and early 19th centuries. Her analysis suggested that the summer antarctic sea ice was more extensive in the eastern Weddell Sea in 1772 and in the Amundsen Sea in 1839 than the present day range from satellite observations. However, many of the early observations are within the present range for the same time of year. An analysis of whaling records by de la Mare (1997) suggested a step decline in antarctic sea ice coverage of 25% (a 2.8° poleward shift in average ice edge latitude) between the mid-

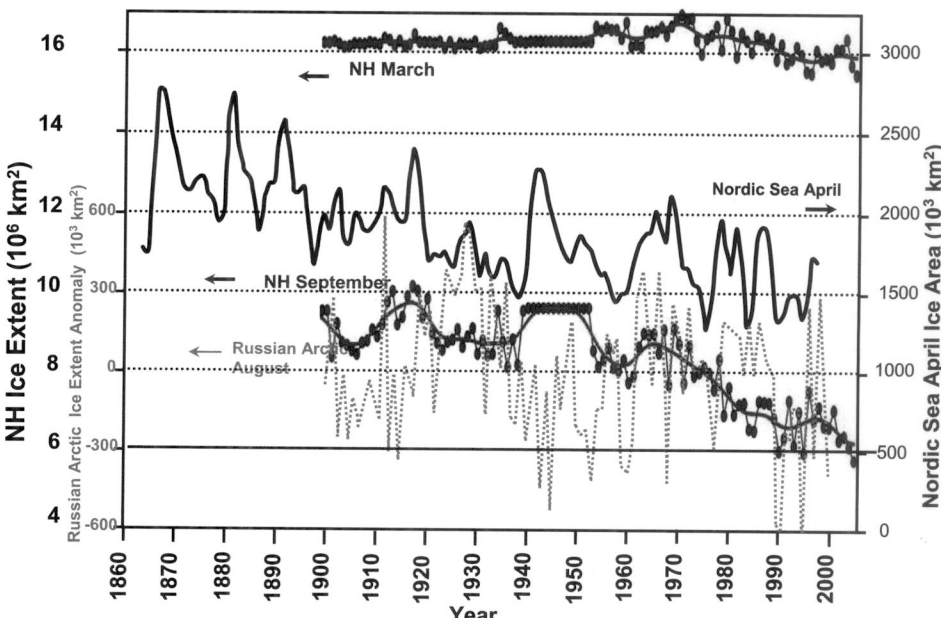

Figure 4.10. *Time series of NH sea ice extent for March and September from the Hadley Centre Sea Ice and Sea Surface Temperature (HadISST) data set (the blue and red curves, updated from Rayner et al., 2003), the April Nordic Sea ice extent (the black curve, redrafted from Vinje, 2001) and the August ice extent anomaly (computed relative to the mean of the entire period) in the Russian Arctic seas – Kara, Laptev, East Siberian and Chukchi (dotted green curve, redrafted from Polyakov et al., 2003). For the NH time series, the symbols indicate yearly values while the curves show the decadal variation (see Appendix 3.A).*

1950s and the early 1970s. A reanalysis by Ackley et al. (2003), which accounted for offsets between satellite-derived ice edge and whaling ship locations, challenged evidence of significant change in ice edge location. Curran et al. (2003) made use of a correlation between methanesulphonic acid concentration (a by-product of marine phytoplankton) in a near-coastal antarctic ice core and the regional sea ice extent in the sector from 80°E to 140°E to infer a quasi-decadal pattern of interannual variability in the ice extent in this region, along with a roughly 20% decline (approximately two degrees of latitude) since the 1950s.

In summary, the antarctic data provide evidence of a decline in sea ice extent in some regions, but there are insufficient data to draw firm conclusions about hemispheric changes prior to the satellite era.

4.4.3 Sea Ice Thickness

4.4.3.1 Sea Ice Thickness Data Sources and Time Periods Covered

Until recently there have been no satellite remote sensing techniques capable of mapping sea ice thickness, and this parameter has primarily been determined by drilling or by under-ice sonar measurement of draft (the submerged portion of sea ice).

Subsea sonar from submarines or moored instruments can be used to measure ice draft over a footprint of 1 to 10 m diameter.

Draft is converted to thickness, assuming an average density for the measured floe, including its snow cover. The principal challenges to accurate observation with sonar are uncertainties in sound speed and atmospheric pressure, and the identification of spurious targets. Upward-looking sonar has been on submarines operating beneath arctic pack ice since 1958. US and UK naval data are now being released for science, and some dedicated arctic submarine missions were made for science during 1993 to 1999. Ice draft measurement by moored ice-profiling sonar, best suited to studies of ice transport or change at fixed sites, began in the Arctic in the late 1980s. Instruments have operated since 1990 in the Beaufort and Greenland Seas and for shorter intervals in other areas, but few records span more than 10 years. In the SH there are no data from submarines and only short time series from moored sonar.

Other techniques, such as electromagnetic induction sounders deployed on the ice surface, ships or aircraft, or airborne laser altimetry to measure freeboard (the portion of sea ice above the waterline), have limited applicability to wide-scale climate analysis of sea ice thickness. Indirect estimates, based on measurement of surface gravity waves, are available in some regions for the 1970s and 1980s (Nagurnyi et al., 1999 as reported in Johannessen et al., 2004), but the accuracy of these estimates is difficult to quantify.

Quantitative data on the thickness of antarctic pack ice only started to become available in the 1980s from sparsely scattered drilling programs covering only small areas and primarily for use in validating other techniques. Visual observations of ice characteristics from ships (Worby and Ackley, 2000) are not adequate for climate monitoring, but are providing one of the first broad pictures of antarctic sea ice thickness.

4.4.3.2   Evidence of Changes in Arctic Pack Ice Thickness from Submarine Sonar

Estimates of thickness change over limited regions are possible when submarine transects are repeated (e.g., Wadhams, 1992). The North Pole is a common waypoint in many submarine cruises and this allowed McLaren et al. (1994) to analyse data from 12 submarine cruises near the pole between 1958 and 1992. They found considerable interannual variability, but no significant trend. Shy and Walsh (1996) examined the same data in relation to ice drift and found that much of the thickness variability was due to the source location and path followed by the ice prior to arrival at the pole.

Rothrock et al. (1999) provided the first 'basin-scale' analysis and found that ice draft in the mid-1990s was less than that measured between 1958 and 1977 at every available location (including the North Pole). The change was least (–0.9 m) in the southern Canada Basin and greatest (–1.7 m) in the Eurasian Basin (with an estimated overall error of less than 0.3 m). The decline averaged about 42% of the average 1958 to 1977 thickness. Their study included very few data within the seasonal sea ice zone and none within 300 km of Canada or Greenland.

Subsequent studies indicate that the reduction in ice thickness was not gradual, but occurred abruptly before 1991. Winsor (2001) found no evidence of thinning along 150°W from six spring cruises during 1991 to 1996, but Tucker et al. (2001), using spring observations from 1976 to 1994 along the same meridian, noted a decrease in ice draft sometime between the mid-1980s and early 1990s, with little subsequent change. The observed change in mean draft resulted from a decrease in the fraction of thick ice (draft of more than 3.5 m) and an increase in the fraction of thin ice, which was probably due to reduced storage of multi-year ice in a smaller Beaufort Gyre and the export of 'surplus' via Fram Strait. Yu et al. (2004) presented evidence of a similar change in ice thickness over a wider area. However, ice thickness varies considerably from year to year at a given location and so the rather sparse temporal sampling provided by submarine data makes inferences regarding long-term change difficult.

4.4.3.3   Other Evidence of Sea Ice Thickness Change in the Arctic and Antarctic

Haas (2004, and references therein) used ground-based electromagnetic induction measurements to show a decrease of approximately 0.5 m between 1991 and 2001 in the modal thickness (i.e., the most commonly observed thickness) of ice floes in the Arctic Trans-Polar Drift. Their survey of 120 km of ice on 146 floes during four cruises is biased by an absence of ice-free and thin-ice fractions and underestimation of ridged ice, but the data are descriptive of floes that are safe to traverse

in summer, and the observed changes are most likely due to thermodynamic forcing.

An emerging new technique, using satellite radar or laser altimetry to estimate ice freeboard from the measured ranges to the ice and sea surface in open leads (and assuming an average floe density and snow depth), offers promise for future monitoring of large-scale sea ice thickness. Laxon et al. (2003) estimated average arctic sea ice thickness over the cold months (October–March) for 1993 to 2001 from satellite-borne radar altimeter measurements. Their data reveal a realistic geographic variation in thickness (increasing from about 2 m near Siberia to 4.5 m off the coasts of Canada and Greenland) and a significant (9%) interannual variability in winter ice thickness, but no indication of a trend over this time.

There are no available data on change in the thickness of antarctic sea ice, much of which is considerably thinner and less ridged than ice in the Arctic Basin.

4.4.3.4   Model-Based Estimates of Change

Physically based sea ice models, forced with winds and temperatures from atmospheric reanalyses and sometimes constrained by observed ice concentration fields, can provide continuous time series of sea ice extent and thickness that can be compared to the sparse observations, and used to interpret the observational record. Models such as those described by Rothrock et al. (2003) and references therein are able to reproduce the observed interannual variations in ice thickness, at least when averaged over fairly large regions. In particular, model studies can elucidate some of the forcing agents responsible for observed changes in ice thickness.

A comparison of various model simulations of historical arctic ice thickness or volume is shown in Figure 4.11. All the models indicate a marked reduction in ice thickness of 0.6 to 0.9 m starting in the late 1980s, but disagree somewhat with respect to trends and/or variations earlier in the century. Most models indicate a maximum in ice thickness in the mid-1960s, with local maxima around 1980 and 1990 as well. There is an emerging suggestion from both models and observations that much of the decrease in thickness occurred between the late 1980s and late 1990s.

It is not possible to attribute the abrupt decrease in thickness inferred from submarine observations entirely to the (rather slow) observed warming in the Arctic, and some of the dramatic decrease may be a consequence of spatial redistribution of ice volume over time (e.g., Holloway and Sou, 2002). Low-frequency, large-scale modes of atmospheric variability (such as interannual changes in circulation connected to the Northern Annular Mode) affect both wind-driving of sea ice and heat transport in the atmosphere, and therefore contribute to interannual variations in ice formation, growth and melt (e.g., Rigor et al., 2002; Dumas et al., 2003).

For the Antarctic, Fichefet et al. (2003) conducted one of the few long-term simulations of ice thickness using observationally based atmospheric forcing covering the period 1958 to 1999. They noted pronounced decadal variability, with area-average

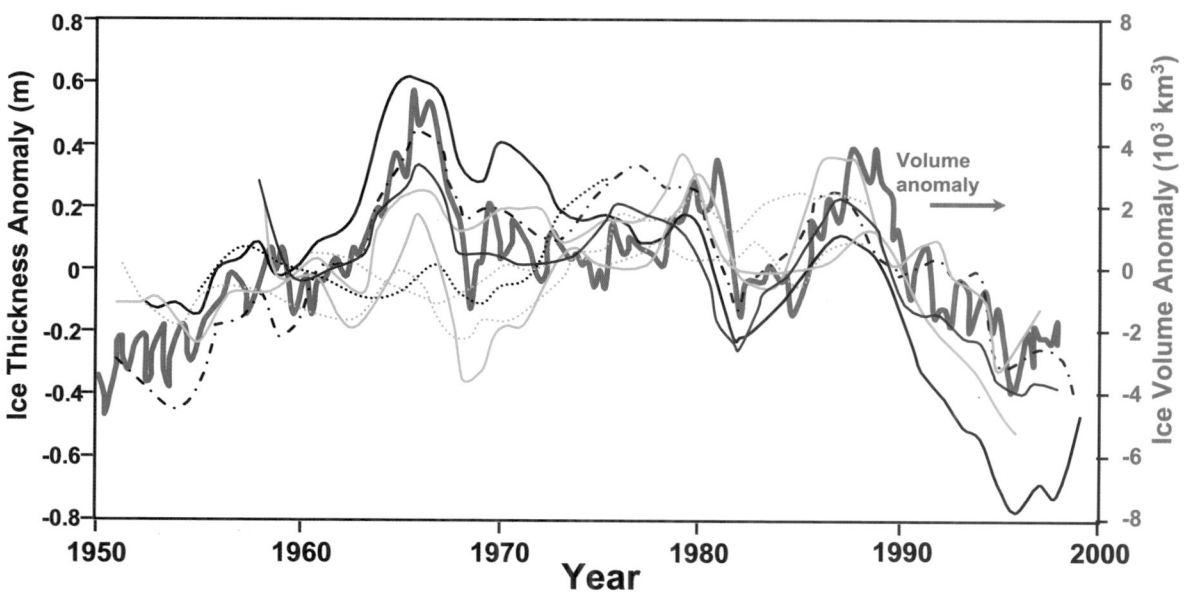

Figure 4.11. *Comparison of model-based time series of the annual mean average sea ice thickness anomaly (computed relative to the mean of the entire period) in the Arctic Basin, obtained from a variety of models (redrafted from Rothrock et al., 2003; see this paper for identification of the individual models and their attributes), along with the sea ice volume anomalies in the Arctic Basin (grey curve and right-hand scale; computed by Koeberle and Gerdes, 2003).*

ice thickness varying by ±0.1 m (over a mean thickness of roughly 0.9 m), but no long-term trend.

4.4.3.5 Landfast Ice Changes

Interannual variations in landfast ice thickness for selected stations in northern Canada were analysed by Brown and Coté (1992). At each of the four sites studied, where ice typically thickens to about 2 m at the end of winter, they detected both positive and negative trends in ice thickness, but no spatially coherent pattern. Interannual variation in ice thickness at the end of the season was determined principally by variation in the amount and timing of snow accumulation, not variation in air temperature. An analysis of several half-century records in Siberian seas has provided evidence that trends in landfast ice thickness over the past century in this area have been small, diverse and generally not statistically significant. Some of the variability is correlated with multi-decadal atmospheric variability (Polyakov et al., 2003).

For the Antarctic, a combined record of the seasonal duration of fast ice in the South Orkney Islands (60.6°S, 45.6°W) has been compiled for observations from two correlated sites for the period 1903 to 1992 (Murphy et al., 1995). The ice duration in these coastal locations is linked to the cycle of pack ice extent in the Weddell Sea, and the duration shows a likely decrease of 7.3 days per decade. This decrease is not linear over the 90-year period and occurs within a strong 7- to 9-year cyclical component of variability over the latter 30 to 40 years of the record. Fast ice thickness measurements have been intermittently made at the coastal sites of Mawson (67.6°S, 62.9°E) and Davis (68.6°S, 78.0°E) for about the last 50 years. Although there is no long-term trend in maximum ice thickness,

at both sites there is a trend for the date of maximum thickness to become later at a rate of about four days per decade (Heil and Allison, 2002).

4.4.3.6 Snow on Sea Ice

Warren et al. (1999) analysed 37 years (1954–1991) of snow depth and density measurements made at Soviet drifting stations on multi-year arctic sea ice. They found a weak negative trend for all months, with the largest trend a decrease of 8 cm (23%) over 37 years in May, the month of maximum snow depth.

There are few data on snow cover and distribution in the Antarctic, and none adequate for detecting any trend in snow cover. Massom et al. (2001) collated available ship observations (between 1981 and 1987) to show that average antarctic snow thickness is typically 0.15 to 0.20 m, and varies widely both seasonally and regionally. An important process in the antarctic sea ice zone is the formation of snow-ice, which occurs when a snow loading depresses thin sea ice below sea level, causing seawater flooding of the near-surface snow and subsequent rapid freezing.

4.4.3.7 Assessment of Changes in Sea Ice Thickness

Sea ice thickness is one of the most difficult geophysical parameters to measure at large scales and, because of the large variability inherent in the sea-ice-climate system, evaluation of ice thickness trends from the available observational data is difficult. Nevertheless, on the basis of submarine sonar data and interpolation of the average sea ice thickness in the Arctic Basin from a variety of physically based sea ice models, it is very likely that the average sea ice thickness in the central Arctic has decreased by up to 1 m since the late 1980s, and that most of this decrease occurred between the late 1980s and the late 1990s. The steady decrease in the area of the summer minimum arctic sea ice cover since the 1980s, resulting in less-thick multi-year ice at the start of the next growth season, is consistent with this. This recent decrease, however, occurs within the context of longer-term decadal variability, with strong maxima in arctic ice thickness in the mid-1960s and around 1980 and 1990, due to both dynamic and thermodynamic forcing of the ice by circulation changes associated with low-frequency modes of atmospheric variability.

There are insufficient data to draw any conclusions about trends in the thickness of antarctic sea ice.

4.4.4 Pack Ice Motion

Pack ice motion influences ice mass locally, through deformation and creation of open water areas; regionally, through advection of ice from one area to another; and globally through export of ice from polar seas to lower latitudes where it melts. The drift of sea ice is primarily forced by the winds and ocean currents. On time scales of days to weeks, winds are responsible for most of the variance in sea ice motion. On longer time scales, the patterns of ice motion follow surface currents and the evolving patterns of wind forcing. Here we consider whether there are trends in the pattern of ice motion.

4.4.4.1   Data Sources and Time Periods Covered

Sea ice motion data are primarily derived from the drift of ships, manned stations and buoys set on or in the pack ice. Although some individual drift trajectories date back to the late 19th century in the Arctic and the early 20th century in the Antarctic, a coordinated observing program did not begin until the International Arctic Buoy Programme (IABP) in the late 1970s. The IABP currently maintains an array of about 25 buoys at any given time and produces gridded fields of ice motion from these using objective analysis (Rigor et al., 2002 and references therein).

Sea ice motion may also be derived from satellite data by estimating the displacement of sea ice features found in two consecutive images from a variety of satellite instruments (e.g., Agnew et al., 1997; Kwok, 2000). The passive microwave sensors provide the longest period of coverage (1979 to present) but their spatial resolution limits the precision of motion estimates. The optimal interpolation of satellite and buoy data (e.g., Kwok et al., 1998) seems to be the most consistent data set to assess interannual variability of sea ice motion.

In the Antarctic, buoy deployments have only been reasonably frequent since the late 1980s. Since 1995, buoy operations have been organised within the World Climate Research Programme (WCRP) International Programme for Antarctic Buoys (IPAB), although spatial and temporal coverage remain poor. A digital atlas of antarctic sea ice has been compiled from two decades of combined passive microwave and IPAB buoy data (Schmitt et al., 2004).

4.4.4.2   Changes in Patterns of Sea Ice Motion and Modes of Climate Variability that Affect Sea Ice Motion

Gudkovich (1961) hypothesised the existence of two regimes of arctic ice motion driven by large-scale variations in atmospheric circulation. Using a coupled atmosphere-ocean-ice model, Proshutinsky and Johnson (1997) showed that the regimes proposed by Gudkovich (1961) alternated on five- to seven-year intervals. Similarly, Rigor et al. (2002) showed that the changes in the patterns of sea ice motion from the 1980s to the 1990s are related to the Northern Annular Mode (NAM). There is, however, no indication of a long-term trend in ice motion.

In the Antarctic, ice motion undergoes an annual cycle caused by stronger winds in winter. Interannual oscillations are found in all regions, most regularly in the Ross, Amundsen and Bellingshausen Seas with periods of about three to six years (Venegas et al., 2001). These wind-driven ice drift oscillations account for the ice extent oscillations seen in the Antarctic Circumpolar Wave (see Section 3.6.6.2). As for the Arctic, no trend in ice motion is apparent based on the limited data available.

4.4.4.3   Ice Export and Advection

The sea ice outflow through Fram Strait is a major component of the ice mass balance of the Arctic Ocean. Approximately 14% of the sea ice mass is exported each year through Fram Strait. Vinje (2001) constructed a time series of ice export during 1950 to 2000 using available moored ice-profiling sonar observations and a parametrization based on geostrophic wind. He found substantial inter-decadal variability in export but no trend.

Kwok and Rothrock (1999) assembled an 18-year time series of ice area and volume flux through Fram Strait based on satellite-derived ice motion and concentration estimates. They found a mean annual area flux of 919×10^3 km^2 yr^{-1} (nearly 10% of the Arctic Ocean area), with large interannual variability that is positively correlated in part with the NAM or North Atlantic Oscillation (NAO) index. Using the thickness data of Vinje et al. (1998), they estimated a mean annual volume flux of 2,366 km^3. Subsequent modelling by Hilmer and Jung (2000) indicated that the correlation between NAO (or nearly equivalently, the NAM) and Fram Strait ice outflow is somewhat transient, with significant correlation during the period 1978 to 1997, but no correlation during 1958 to 1977 (Figure 4.12). This was a consequence of rather subtle shifts in the spatial pattern of surface pressure (and hence wind) anomalies associated with the NAO. A recent update of this time series (Kwok et al., 2004) to 24 years (ending in 2002) shows only minor variations in the mean volume and area flux and the correlation with NAO persists.

Overall, while there is considerable low-frequency variability in the pattern of sea ice motion, there is no evidence of a trend in either hemisphere.

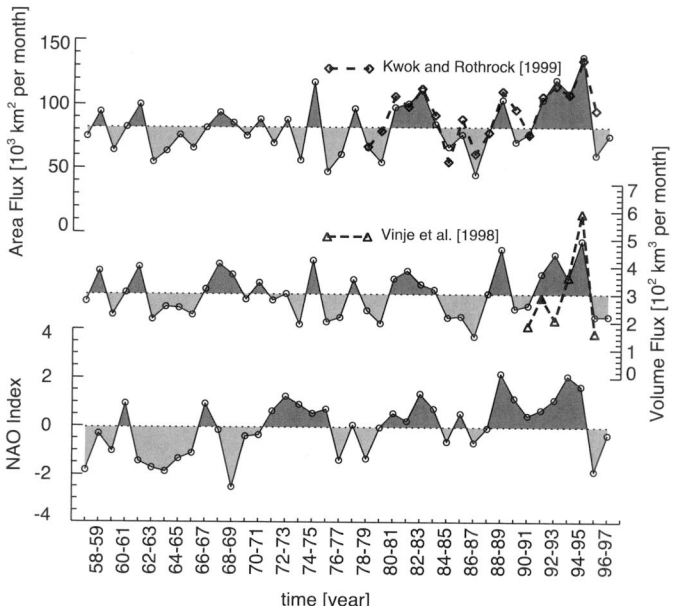

Figure 4.12. *Time series of modelled Fram Strait sea ice area and volume flux, along with the NAO index. Also shown are observational estimates of area flux (Kwok and Rothrock, 1999) and volume flux (Vinje et al., 1998). Reproduced from Hilmer and Jung (2000).*

4.5 Changes in Glaciers and Ice Caps

4.5.1 Background

Those glaciers and ice caps not immediately adjacent to the large ice sheets of Greenland and Antarctica cover an area between 512×10^3 and 546×10^3 km^2 according to inventories from different authors (Table 4.3); volume estimates differ considerably from 51×10^3 to 133×10^3 km^3, representing sea level equivalent (SLE) of between 0.15 and 0.37 m. Including the glaciers and ice caps surrounding the Greenland Ice Sheet and West Antarctica, but excluding those on the Antarctic Peninsula and those surrounding East Antarctica, yields 0.72 ± 0.2 m SLE. These new estimates are about 40% higher than those given in IPCC (2001), but area inventories are still incomplete and volume measurements more so, despite increasing efforts.

Glaciers and ice caps provide among the most visible indications of the effects of climate change. The mass balance at the surface of a glacier (the gain or loss of snow and ice over a hydrological cycle) is determined by the climate. At high and mid-latitudes, the hydrological cycle is determined by the annual cycle of air temperature, with accumulation dominating in winter and ablation in summer. In wide parts of the Himalaya most accumulation and ablation occur during summer (Fujita and Ageta, 2000), in the tropics ablation occurs year round and the seasonality in precipitation controls accumulation (Kaser and Osmaston, 2002). A climate change will affect the magnitude of the accumulation and ablation terms and the length of the mass balance seasons. The glacier will then change its extent towards a size that makes the total mass balance (the mass gain or loss over the entire glacier) zero. However, climate variability and the time lag of the glacier response mean that static equilibrium is never attained. Changes in glacier extent lag behind climate changes by only a few years on the short, steep and shallow glaciers of the tropical mountains with year-round ablation (Kaser et al., 2003), but by up to several centuries on the largest glaciers and ice caps with small slopes and cold ice (Paterson, 2004). Glaciers also lose mass by iceberg calving: this does not have an immediate and straightforward link to climate, but general relations to climate can often be discerned. Mass loss by basal melting is considered negligible at a global or large regional scale.

4.5.2 Large and Global-Scale Analyses

Records of glacier length changes (WGMS(ICSI-IAHS), various years-a) go far back in time – written reports as far back as 1600 in a few cases – and are directly related to low-frequency climate change. From 169 glacier length records, Oerlemans (2005) compiled mean length variations of glacier tongues for large-scale regions between 1700 and 2000 (Figure 4.13).

Table 4.3. *Extents of glaciers and ice caps as given by different authors.*

Reference	Area (10^3 km^2)	Volume (10^3 km^3)	SLE[f] (m)
Raper and Braithwaite, 2005[a,c]	522 ± 42	87 ± 10	0.24 ± 0.03
Ohmura, 2004[a,d]	512	51	0.15
Dyurgerov and Meier, 2005[a,e]	546 ± 30	133 ± 20	0.37 ± 0.06
Dyurgerov and Meier, 2005[b,e]	785 ± 100	260 ± 65	0.72 ± 0.2
IPCC, 2001[b]	680	180 ± 40	0.50 ± 0.1

Notes:
[a] glaciers and ice caps surrounding Greenland and Antarctic Ice Sheets are excluded.
[b] glaciers and ice caps surrounding Greenland and West Antarctic Ice Sheets are included.
[c] volume derived from hypsometry and volume/area scaling within $1° \times 1°$ grid cells.
[d] volume derived from a statistical relationship between glacier volume and area, calibrated with 61 glacier volumes derived from radio-echo-sounding measurements.
[e] volume derived from a statistical relationship between glacier volume and area, calibrated with 144 glacier volumes derived from radio-echo-sounding measurements.
[f] calculated for the ocean surface area of 362×10^6 km^2.

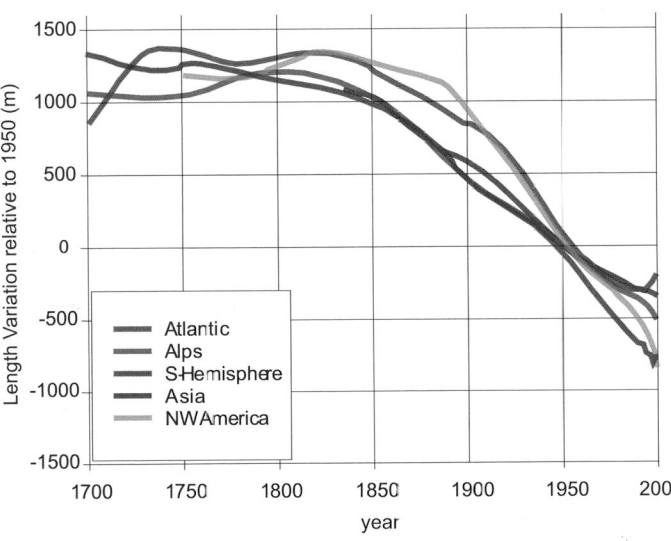

Figure 4.13. *Large-scale regional mean length variations of glacier tongues (Oerlemans, 2005). The raw data are all constrained to pass through zero in 1950. The curves shown are smoothed with the Stineman (1980) method and approximate this. Glaciers are grouped into the following regional classes: SH (tropics, New Zealand, Patagonia), northwest North America (mainly Canadian Rockies), Atlantic (South Greenland, Iceland, Jan Mayen, Svalbard, Scandinavia), European Alps and Asia (Caucasus and central Asia).*

Although much local, regional and high-frequency variability is superimposed, the smoothed series give an apparently homogeneous signal. General retreat of glacier tongues started after 1800, with considerable mean retreat rates in all regions after 1850 lasting throughout the 20th century. A slow down of retreat between about 1970 and 1990 is more evident in the raw data (Oerlemans, 2005). Retreat was again generally rapid in the 1990s; the Atlantic and the SH curves reflect precipitation-driven growth and advances of glaciers in western Scandinavia and New Zealand during the late 1990s (Chinn et al., 2005).

Records of directly measured glacier mass balances are few and stretch back only to the mid-20th century. Because of the very intensive fieldwork required, these records are biased towards logistically and morphologically 'easy' glaciers. Uncertainty in directly measured annual mass balance is typically ± 200 kg m^{-2} yr^{-1} due to measurement and analysis errors (Cogley, 2005). Mass balance data are archived and distributed by the World Glacier Monitoring Service (WGMS(ICSI-IAHS), various years-b). From these and from several other new and historical sources, quality checked time series of the annual mean specific mass balance (the total mass balance of a glacier or ice cap divided by its total surface area) for about 300 individual glaciers have been constructed, analysed and presented in three databases (Ohmura, 2004; Cogley, 2005; Dyurgerov and Meier, 2005). Dyurgerov and Meier (2005) also incorporated recent findings from repeat altimetry of glaciers and ice caps in Alaska (Arendt et al., 2002) and Patagonia (Rignot et al., 2003). Only a few individual series stretch over the entire period. From these statistically small samples, global estimates have been obtained as five-year (pentadal) means by arithmetic averaging (C05a in Figure 4.14), area-weighted averaging (DM05 and O04) and

spatial interpolation (C05i). Although mass balances reported from individual glaciers include the effect of changing glacier area, deficiencies in the inventories do not allow for general consideration of area changes. The effect of this inaccuracy is considered minor. Table 4.4 summarises the data plotted in Figure 4.14.

The time series of globally averaged mean specific mass balance from different authors have very similar shapes despite some offsets in magnitude. Around 1970, mass balances were close to zero or slightly positive in most regions (Figure 4.15) and close to zero in the global mean (Figure 4.14), indicating near-equilibration with climate after the strong earlier mass loss. This gives confidence that the glacier wastage in the late 20th century is essentially a response to post-1970 global warming (Greene, 2005). Strong mass losses are indicated for the 1940s but uncertainty is great since the arithmetic mean values (C05a in Figure 4.14) are from only a few glaciers. The most recent period consists of four years only (2000/2001–2003/2004) and does not cover all regions completely. The shortage of data from Alaska and Patagonia likely causes a positive bias on the area-weighted and interpolated analyses (DM05, O04, C05i) due to the large ice areas in these regions. There is probably also a negative bias in the arithmetic mean (C05a), due to the strongly negative northern mid-latitudes mass balances in 2002/2003, particularly in the European Alps (Zemp et al., 2005). Mass loss rates for 1990/1991 to 2003/2004 are roughly double those for 1960/1961 to 1989/1990 (Table 4.4).

Over the last half century, both global mean winter accumulation and summer melting have increased steadily (Ohmura, 2004; Dyurgerov and Meier, 2005; Greene, 2005). At least in the NH, winter accumulation and summer melting correlate positively with hemispheric air temperature, whereas the mean specific mass balance correlates negatively with hemispheric air temperature (Greene, 2005). Dyurgerov and Dwyer (2000) analysed time series of 21 NH glaciers and found a rather uniformly increased mass turnover rate, qualitatively consistent with moderately increased precipitation and substantially increased low-altitude melting. This general trend is also indicated for Alaska (Arendt et al., 2002), the Canadian Arctic Archipelago (Abdalati et al., 2004) and Patagonia (Rignot et al., 2003).

Regional analyses by Dyurgerov and Meier (2005) show strongest negative mean specific mass balances in Patagonia, the northwest USA and southwest Canada, and Alaska, with losses especially rapid in Patagonia and Alaska after the mid-1990s (Figure 4.15a). A cumulative mean specific mass balance of -10×10^5 kg m^{-2} corresponds to a loss of 10 m of water, or about 11 m of ice, averaged over the glacier area; cumulative losses in Patagonia since 1960 are approximately 40 m of ice thickness averaged over the glaciers. Only Europe showed a mean value close to zero, with strong mass losses in the Alps compensated by mass gains in maritime Scandinavia until the end of the 20th century. High spatial variability in climate and, thus, in glacier variations, also exists in other large regions such as in the high mountains of Asia (Liu et al., 2004; Dyurgerov

Table 4.4. *Global average mass balance of glaciers and ice caps for different periods, showing mean specific mass balance (kg m⁻² yr⁻¹); total mass balance (Gt yr⁻¹); and SLE (mm yr⁻¹) derived from total mass balance and an ocean surface area of 362 × 10⁶ km². Values for glaciers and ice caps excluding those around the ice sheets (total area 546 × 10³ km²) are derived from MB values in Figure 4.14. Values for glaciers and ice caps including those surrounding Greenland and West Antarctica (total area 785.0 × 10³ km²) are modified from Dyurgerov and Meier (2005) by applying pentadal DM05 to MB ratios. Uncertainties are for the 90% confidence level. Sources: Ohmura (2004), Cogley (2005) and Dyurgerov and Meier (2005), all updated to 2003/2004.*

Period	Mean Specific Mass Balance[a] (kg m⁻² yr⁻¹)	Total Mass Balance[a] (Gt yr⁻¹)	Sea Level Equivalent[a] (mm yr⁻¹)	Mean Specific Mass Balance[b] (kg m⁻² yr⁻¹)	Total Mass Balance[b] (Gt yr⁻¹)	Sea Level Equivalent[b] (mm yr⁻¹)
1960/1961–2003/2004	−283 ± 102	−155 ± 55	0.43 ± 0.15	−231 ± 82	−182 ± 64	0.50 ± 0.18
1960/1961–1989/1990	−219 ± 92	−120 ± 50	0.33 ± 0.14	−173 ± 73	−136 ± 57	0.37 ± 0.16
1990/1991–2003/2004	−420 ± 121	−230 ± 66	0.63 ± 0.18	−356 ± 101	−280 ± 79	0.77 ± 0.22

Notes:
[a] Excluding glaciers and ice caps around ice sheets
[b] Including glaciers and ice caps around ice sheets

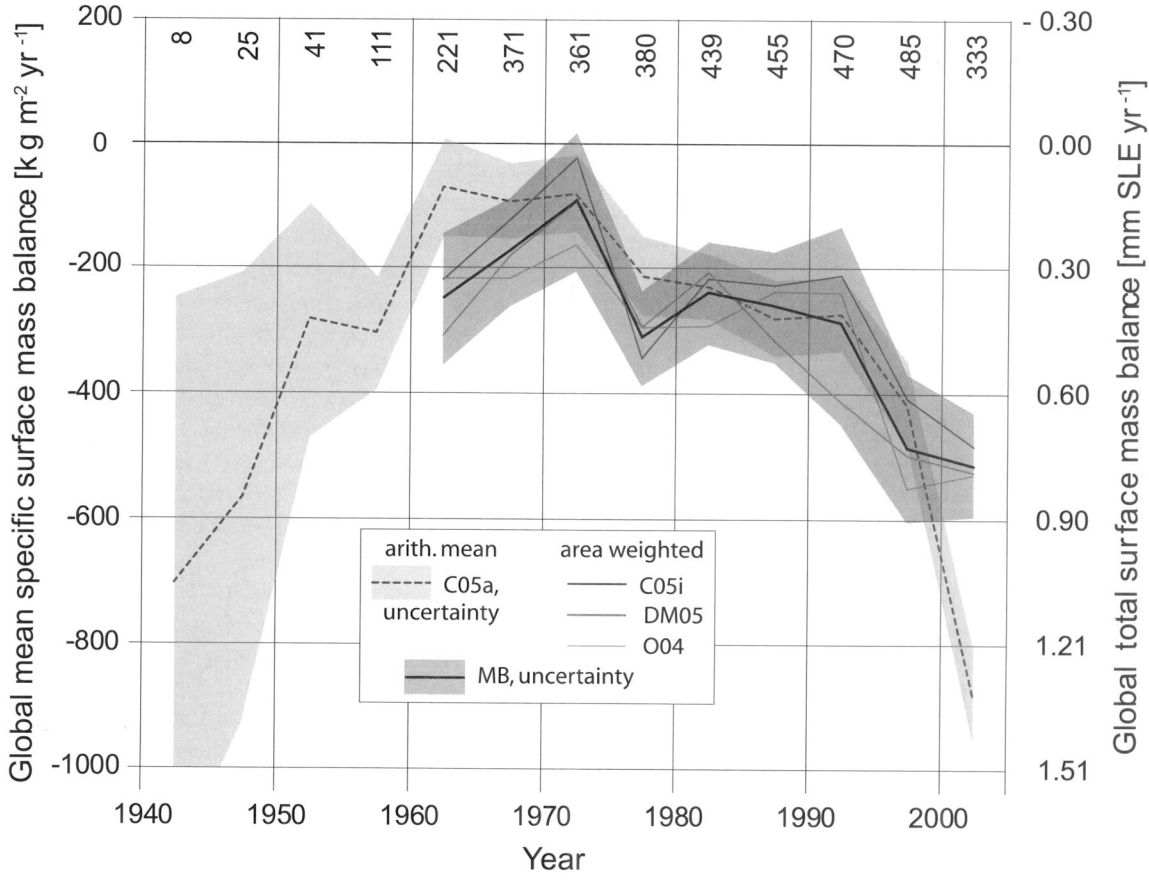

Figure 4.14. *Pentadal (five-year) average mass balance of the world's glaciers and ice caps excluding those around the ice sheets of Greenland and Antarctica. Mean specific mass balance (left axis) is converted to total mass balance and to SLE (right axis) using the total ice surface area of 546 × 10³ km² (Table 4.3) and the ocean surface area of 362 × 10⁶ km². C05a is an arithmetic mean over all annual measurements within each pentade (Cogley, 2005); the grey envelope is the 90% confidence level of the C05a data and represents the spatial variability of the measured mass balances. The number of measurements in each time period is given at the top of the graph. C05i is obtained by spatial interpolation (Cogley, 2005), while DM05 (Dyurgerov and Meier, 2005) and O04 (Ohmura, 2004) are area-weighted global numbers. MB is the arithmetic mean of C05i, DM05 and O04, and its uncertainty (red shading) combines the spatial variability and the structural uncertainty calculated for the 90% confidence level. This does not include uncertainties that derive from uncertainties in the glacier area inventories. The authors performed area weighting and spatial interpolation only after 1960, when up to 100 measured mass balances were available. The most recent period consists of four years only (2000/2001 to 2003/2004).*

and Meier, 2005). Values for Patagonia and Alaska are mainly derived from altimetry evaluations made by Arendt et al. (2002) and Rignot et al. (2003), and authors of both papers note that the observed mass losses cannot be explained by surface mass loss only, but also include increased ice discharge due to enhanced ice velocity. The latter, in turn, has possibly been triggered by previous negative mass balances of glaciers calving icebergs, as well as by increased melt water production that enhances basal sliding. Some glaciers exhibit quasi-periodic internal instabilities (surging), which can affect data from those glaciers (Arendt et al., 2002; Rignot et al., 2003), but these effects are expected to average very close to zero over large regions and many years or decades. Because of a lack of suitable information, the temporal variation of the mass loss of the Patagonian ice fields has been interpolated to match the time series of Alaskan mass balances assuming similar climate regimes (Dyurgerov and Meier, 2005).

The surface mass balance of snow and ice is determined by a complex interaction of energy fluxes towards and away from the surface and the occurrence of solid precipitation. Nevertheless, glacier fluctuations show a strong statistical correlation with air temperature at least at a large spatial scale throughout the 20th century (Greene, 2005), and a strong physical basis exists to explain why warming would cause mass loss (Ohmura, 2001). Changes in snow accumulation also matter, and may dominate in response to strong circulation changes or when temperature is not changing greatly. For example, analyses of glacier mass balances, volume changes, length variations and homogenised temperature records for the western portion of the European Alps (Vincent et al., 2005) clearly indicate the role of precipitation changes in glacier variations in the 18th and 19th centuries. Similarly, Nesje and Dahl (2003) explained glacier advances in southern Norway in the early 18th century as being due to increased winter precipitation rather than colder temperatures.

Total mass balances are the integration of mean specific mass balances (which have a climate signal) over the existing glacier area. Consequently, the biggest mass losses and, thus, contributions to sea level rise are from Alaska with 0.11 mm yr^{-1} SLE from 1960/1961 to 1989/1990 and 0.24 mm yr^{-1} SLE from 1990/1991 to 2002/2003, the Arctic (0.09 and 0.19), and the high mountains of Asia (0.08 and 0.10) (Figure 4.15b).

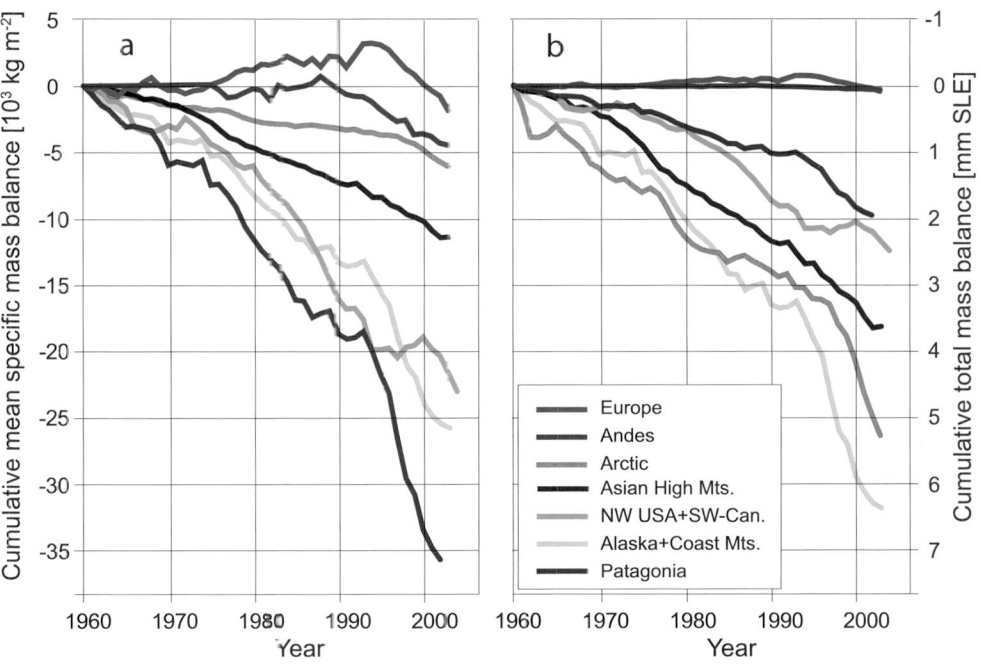

Figure 4.15. *Cumulative mean specific mass balances (a) and cumulative total mass balances (b) of glaciers and ice caps, calculated for large regions (Dyurgerov and Meier, 2005). Mean specific mass balance shows the strength of climate change in the respective region. Total mass balance is the contribution from each region to sea level rise.*

4.5.3 Special Regional Features

Although reports on individual glaciers or limited glacier areas support the global picture of ongoing strong ice shrinkage in almost all regions, some exceptional results indicate the complexity of both regional- to local-scale climate and respective glacier regimes.

For glaciers in the dry and cold Taylor Valley, Antarctica, Fountain et al. (2004) hypothesised that an increase in average air temperature of 2°C alone can explain the observed glacier advance through ice softening.

Altimetric measurements in Svalbard suggested a small ice cap growth (Bamber et al., 2004), however, an alternative evaluation of mass balance processes indicates a slight sea level contribution of 0.01 mm yr^{-1} for the last three decades of the 20th century (Hagen et al., 2003). Svalbard glaciers were recently close to balance, which is exceptional for the Arctic.

In Scandinavia, Norwegian coastal glaciers, which advanced in the 1990s due to increased accumulation in response to a positive phase of the NAO (Nesje et al., 2000), started to shrink around 2000 as a result of a combination of reduced winter accumulation and greater summer melting (Kjøllmoen, 2005). Norwegian glacier tongues farther inland have retreated continuously at a moderate rate. Warming is also indicated by a change in temperature distribution in northern Sweden's Storglaciären where, between 1989 and 2001, 8.3 m of the cold surface layer (or 22% of the long-term average thickness of this cold layer) warmed to the melting point. This is attributed primarily to increased winter temperatures yielding a longer melt season; summer ablation was normal (Pettersson et al., 2003). As with coastal Scandinavia, glaciers in the New Zealand Alps advanced during the 1990s, but have started to shrink since 2000. Increased precipitation may have caused the glacier growth (Chinn et al., 2005).

In the European Alps, exceptional mass loss during 2003 removed an average of 2,500 kg m^{-2} yr^{-1} over nine measured Alpine glaciers, almost 60% higher than the previous record of 1,600 kg m^{-2} yr^{-1} loss in 1996 and four times more than the mean loss from 1980 to 2001 (600 kg m^{-2} yr^{-1}; Zemp et al., 2005). This was caused by extraordinarily high air temperatures over a long period, extremely low precipitation, and albedo feedback from Sahara dust depositions and a previous series of negative mass balance years (see Box 3.6.).

Whereas glaciers in the Asian high mountains have generally shrunk at varying rates (Su and Shi, 2002; Ren et al., 2004; Solomina et al., 2004; Dyurgerov and Meier, 2005), several high glaciers in the central Karakoram are reported to have advanced and/or thickened at their tongues (Hewitt, 2005), probably due to enhanced precipitation.

Tropical glaciers have shrunk from a maximum in the mid-19th century, following the global trend (Figure 4.16). Strong shrinkage rates in the 1940s were followed by relatively stable extents that lasted into the 1970s. Since then, shrinkage has become stronger again; as in other mountain ranges, the smallest glaciers are more strongly affected. Since the publication of IPCC (2001), evidence has increased that changes in the mass balance of tropical glaciers are mainly driven by coupled changes in energy and mass fluxes related to interannual variations of regional-scale hygric seasonality (Wagnon et al., 2001; Francou et al., 2003, 2004). Variations in atmospheric moisture content affect incoming solar radiation, precipitation and albedo, atmospheric longwave emission, and sublimation (Wagnon et al., 2001; Kaser and Osmaston, 2002; Mölg et al., 2003a; Favier et al., 2004; Mölg and Hardy, 2004; Sicart et al., 2005). At a large scale, the mass balance of tropical glaciers strongly correlates with tropical sea surface temperature anomalies and related atmospheric circulation modes (Francou et al., 2003, 2004; Favier et al., 2004). Glaciers on Kilimanjaro behaved exceptionally throughout the 20th century (Figure 4.16). The geometry of the volcano and the dry climate above the freezing level maintain vertical ice walls around the tabular ice on the summit plateau and these retreat at about 0.9 m yr^{-1} (Thompson et al., 2002) forced by solar radiation (Mölg et al., 2003b). Their retreat is responsible for the steady shrinkage of the ice area on the summit plateau (Figure 4.16, insert) (Cullen et al., 2006). In contrast, the slope glaciers, which extend from the plateau rim onto the steep slopes of the volcano, decreased strongly at the beginning of the 20th century, but more slowly recently. This shrinkage is interpreted as an ongoing response to a dramatic change from a wetter to a drier regime, supposedly around 1880, and a subsequent negative trend in mid-troposphere atmospheric moisture content over East Africa (Cullen et al., 2006).

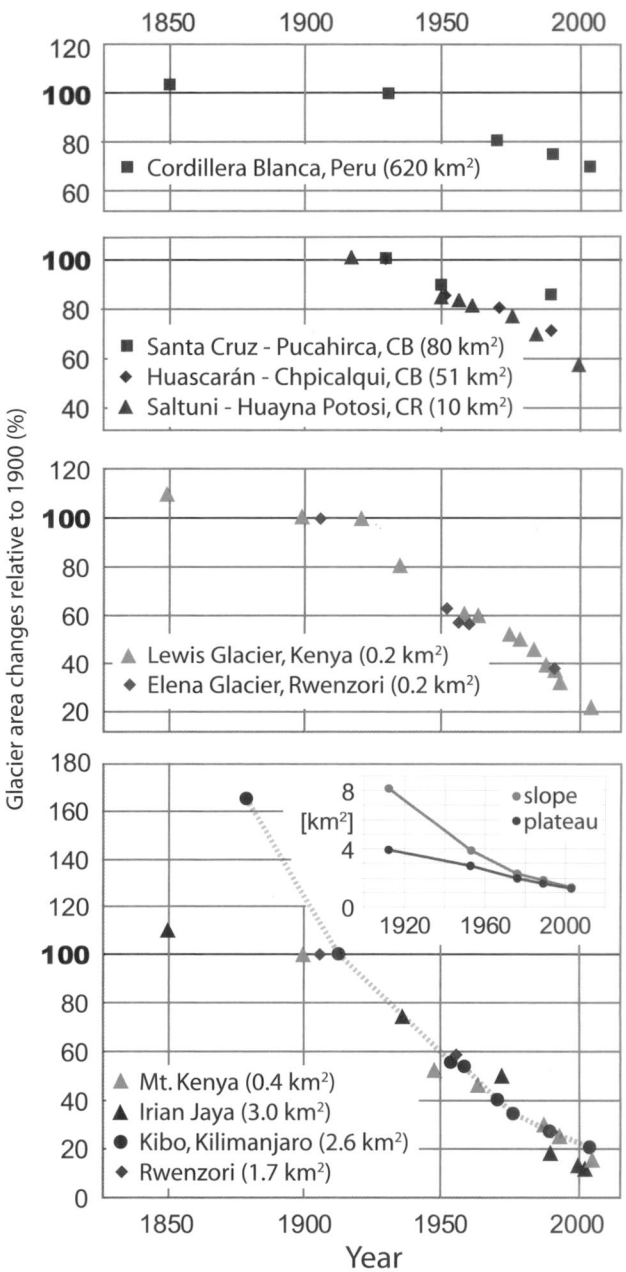

Figure 4.16. *Changes in the surface area of tropical glaciers relative to their extent around 1900, grouped according to different glacier sizes. The sizes are given for 1990 or the closest available date to 1990. The broken red line highlights the retreat of Kilimanjaro glaciers. The insert shows the area change (km²) of the Kilimanjaro plateau (red) and slope (purple) glaciers as separated by the 5,700 m contour line (Kaser and Osmaston, 2002 (updated courtesy of S. Lieb); Mölg et al., 2003b; Georges, 2004; Hastenrath, 2005; Cullen et al., 2006; Klein and Kincaid, 2006).*

4.6 Changes and Stability of Ice Sheets and Ice Shelves

New and improved observational techniques, and extended time series, reveal changes in many parts of the large ice sheets. Greenland has experienced mass loss recently in response to increases in near-coastal melting and in ice flow velocity more than offsetting increases in snowfall. Antarctica appears to be losing mass at least partly in response to recent ice flow acceleration in some near-coastal regions, although with greater uncertainty in overall balance than for Greenland. Shortcomings in forcing, physics and resolution in comprehensive ice flow models have prevented them from fully capturing the ice flow changes.

4.6.1 Background

The ice sheets of Greenland and Antarctica hold enough ice to raise sea level about 64 m if fully melted (Bamber et al., 2001; Lythe et al., 2001). Even a modest change in ice sheet balance could strongly affect future sea level and freshwater flux to the oceans, with possible climatic implications. These ice sheets consist of vast central reservoirs of slow-moving ice drained by rapidly moving, ice-walled ice streams or rock-walled outlet glaciers typically flowing into floating ice shelves or narrower ice tongues, or directly into the ocean. Ice shelves often form in embayments, or run aground on local bedrock highs to produce ice rumples or ice rises, and friction with embayment sides or local grounding points helps restrain the motion of the ice shelves and their tributaries. About half of the ice lost from Greenland is by surface melting and runoff into the sea, but surface melting is much less important to the mass balance of Antarctica. Dynamics of the slow-moving ice and of ice shelves are reasonably well understood and can be modelled adequately, but this is not so for fast-moving ice streams and outlet glaciers. Until recently (including IPCC, 2001), it was assumed that velocities of these outlet glaciers and ice streams cannot change rapidly, and impacts of climate change were estimated primarily as changes in snowfall and surface melting. Recent observations show that outlet glacier and ice stream speeds can change rapidly, for reasons that are still under investigation. Consequently, this assessment will not adequately quantify such effects.

4.6.2 Mass Balance of the Ice Sheets and Ice Shelves

The current state of balance of the Greenland and Antarctic Ice Sheets is discussed here, focussing on the substantial progress made since IPCC (2001). Possible future changes are considered in Chapter 10, and in Chapter 19 of the Working Group II contribution to the IPCC Fourth Assessment Report.

4.6.2.1 Techniques

Several techniques are used to measure the mass balance of large ice masses. The mass budget approach compares input from snow accumulation with output from ice flow and melt water runoff. Repeated altimetry measures surface elevation changes. Temporal variations in gravity over the ice sheets reveal mass changes. Changes in day length and in the direction of the Earth's rotation axis also reveal mass redistribution.

4.6.2.1.1 Mass budget

Snow accumulation is often estimated from annual layering in ice cores, with interpolation between core sites using satellite microwave measurements or radar sounding (Jacka et al., 2004). Increasingly, atmospheric modelling techniques are also applied (e.g., Monaghan et al., 2006). Ice discharge is calculated from radar or seismic measurements of ice thickness, and from *in situ* or remote measurements of ice velocity, usually where the ice begins to float and velocity is nearly depth-independent. A major advance since IPCC (2001) has been widespread application of Interferometric Synthetic Aperture Radar (InSAR) techniques from satellites to measure ice velocity over very large areas of the ice sheets (e.g., Rignot et al., 2005). Calculation of mass discharge also requires estimates for runoff of surface melt water, which is large for low-elevation regions of Greenland and parts of the Antarctic Peninsula but small or zero elsewhere on the ice sheets. Surface melt amounts usually are estimated from modelling driven by atmospheric reanalyses, global models or climatology, and often calibrated against surface observations where available (e.g., Hanna et al., 2005; Box et al., 2006). The typically small mass loss by melting beneath grounded ice is usually estimated from models. Mass loss from melting beneath ice shelves can be large, and is difficult to measure; it is generally calculated as the remainder after accounting for other mass inputs and outputs.

Ice sheet mass inputs and outputs are difficult to estimate with high accuracy. For example, van de Berg et al. (2006) summarised six estimates of net accumulation on the grounded section of Antarctica published between 1999 and 2006, which ranged from 1,811 to 2,076 Gt yr^{-1} or ±7% about the midpoint. Transfer of 360 Gt of grounded (non-floating) ice to the ocean would raise sea level about 1 mm. Uncertainty in the Greenland accumulation rate is probably about 5% (Hanna et al., 2005; Box et al., 2006). Although broad InSAR coverage and progressively improving estimates of grounding-line ice thickness have substantially improved ice discharge estimates, incomplete data coverage implies uncertainties in discharge estimates of a few percent. Uncorrelated errors of 5% on input and output would imply mass budget uncertainties of about 40 Gt yr^{-1} for Greenland and 140 Gt yr^{-1} for Antarctica. Large interannual variability and trends also complicate interpretation. Box et al. (2006) estimated average accumulation on the Greenland Ice Sheet of 543 Gt yr^{-1} from 1988 to 2004, but with an annual minimum of 482 Gt yr^{-1}, a maximum of 613 Gt yr^{-1} and a best-fit linear trend yielding an increase of 68 Gt yr⁻¹ during the period. Glacier velocities can change substantially, sometimes in months or years, adding to the overall uncertainty of mass budget calculations.

4.6.2.1.2   Repeated altimetry

Surface elevation changes reveal ice sheet mass changes after correction for changes in depth-density profiles and in bedrock elevation, or for hydrostatic equilibrium if the ice is floating. Satellite radar altimetry (SRALT) has been widely used to estimate elevation changes (Shepherd et al., 2002; Davis et al., 2005; Johannessen et al., 2005; Zwally et al., 2006), together with laser altimetry from airplanes (Krabill et al., 2004) and from the Ice, Cloud and land Elevation Satellite (ICESat; Thomas et al., 2006). Modelled corrections for isostatic changes in bedrock elevation are small (a few millimetres per year), but with uncertainties nearly as large as the corrections in some cases (Zwally et al., 2006). Corrections for near-surface firn density changes are larger (>10 mm yr⁻¹; Cuffey, 2001) and also uncertain.

Radar altimetry has provided long-term and widespread coverage for more than a decade, but with important challenges (described by Legresy et al., 2006). The available SRALT data are from altimeters with a beam width of 20 km or more, designed and demonstrated to make accurate measurements over the almost flat, horizontal ocean. Data interpretation is more complex over sloping, undulating ice sheet surfaces with spatially and temporally varying dielectric properties and thus penetration into near-surface firn. Empirical corrections are applied for some of these effects, and for inter-satellite biases. The correction for the offset between the European Remote Sensing Satellite (ERS-1 and ERS-2) altimeters is reported by Zwally et al. (2006) to affect mass change estimates for the interval 1992 to 2002 by about 50 Gt yr⁻¹ for Greenland, and to differ from the corresponding correction of Johannessen et al. (2005) by about 20 Gt yr⁻¹, although some of this difference may reflect differences in spatial coverage of the studies combined with spatial dependence of the correction. Changes in surface dielectric properties affect the returned waveform and thus the measured range, so a correction is made for elevation changes correlated to returned-power changes. This effect is small averaged over an ice sheet but often of the same magnitude as the remaining signal at a point, and could remove part or all of the signal if climate change affected both elevation and surface character, hence returned power.

The SRALT tracking algorithms use leading edges of reflected radar waveforms, thus primarily sampling higher-elevation parts of the large footprint. This probably introduces only small errors over most of an ice sheet, where surfaces are nearly flat. However, glaciers and ice streams often flow in surface depressions that can be narrower than the radar footprint, so that SRALT-derived elevation changes are weighted towards slower-moving ice at the glacier sides (Thomas et al., 2006). This is of most concern in Greenland, where other studies show thinning along outlet glaciers just a few kilometres wide

(Abdalati et al., 2001). Elevation-change estimates from SRALT have not been validated against independent data except at higher elevations, where surfaces are nearly flat and horizontal and dielectric properties nearly unchanging (Thomas et al., 2001). Although SRALT coverage is lacking within 900 km of the poles, and some data are lost in steep regions, coverage has now been achieved for about 90% of the Greenland Ice Sheet and 80% of the Antarctic Ice Sheet (Zwally et al., 2006) (Figure 4.19).

Laser altimeters reduce some of the difficulties with SRALT by having negligible penetration of near-surface layers and a smaller footprint (about 1 m for airborne laser, and 60 m for ICESat). However, clouds limit data acquisition, and accuracy is affected by atmospheric conditions and particularly by laser pointing errors. Airborne surveys over Greenland in 1993/1994 and 1998/1999 yielded estimates of elevation change accurate to ±14 mm yr⁻¹ along survey tracks (Krabill et al., 2002). However, the large gaps between flight lines must be filled, often by simple interpolation in regions of weak variability or by interpolation using physical models in more complex regions (Krabill et al., 2004; Figure 4.17).

4.6.2.1.3   Geodetic measurements, including measurement of temporal variations in Earth gravity

Since 2002, the Gravity Recovery and Climate Experiment (GRACE) satellite mission has been providing routine measurement of the Earth's gravity field and its temporal variability. After removing the effects of tides, atmospheric loading, etc., high-latitude data contain information on temporal changes in the mass distribution of the ice sheets and underlying rock (Velicogna and Wahr, 2005). Estimates of ice sheet mass balance are sensitive to modelled estimates of bedrock vertical motion, primarily arising from response to changes in mass loading from the end of the last ice age. Velicogna and Wahr (2005) estimated a correction for Greenland Ice Sheet mass balance of 5 ± 17 Gt yr⁻¹ for the bedrock motion, with an equivalent value of 177 ± 73 Gt yr⁻¹ for Antarctica (Velicogna and Wahr, 2006). (Note that stated uncertainties for ice sheet mass balances referenced to published papers are given here as published. Some papers include error terms that were estimated without formal statistical derivations, and other papers note omission of estimates for certain possible systematic errors, so that these as-published errors generally cannot be interpreted as representing any specific confidence interval such as 5 to 95%.)

Other geodetic data provide constraints on mass changes at high latitudes. These data include the history of changing length of day from eclipse records, the related ongoing changes in the spherical-harmonic coefficients of the geopotential, and true polar wander (changes in the planet's rotation vector; Peltier, 1998; Munk, 2002; Mitrovica et al., 2006). At present, unique solutions are not possible from these techniques, but hypothesised histories of ice sheet changes can be tested against the data for consistency, and progress is rapidly being made.

4.6.2.2   Measured Balance of the Ice Sheets and Ice Shelves

Mass balance of the large ice sheets was summarised by Rignot and Thomas (2002) and Alley et al. (2005a).

4.6.2.2.1   Greenland

Many recent studies have addressed Greenland mass balance. They yield a broad picture (Figure 4.17) of inland thickening (Thomas et al., 2001; Johannessen et al., 2005; Thomas et al., 2006; Zwally et al., 2006), faster near-coastal thinning primarily in the south along fast-moving outlet glaciers (Abdalati et al., 2001; Rignot and Kanagaratnam, 2006), and a recent acceleration in overall shrinkage.

Analysis of GRACE data showed total losses of 75 ± 26 Gt yr⁻¹ between April 2002 and July 2004 (Velicogna and Wahr, 2005). Ramillien et al. (2006), also working from GRACE data, found a mass loss of 129 ± 15 Gt yr⁻¹ for July 2002 to March 2005. Because of the low spatial resolution of GRACE, these include losses from isolated mountain glaciers and ice caps near the coast, whereas the results discussed next do not.

Mass loss from the ice sheet surface (net snow accumulation minus melt water runoff) has increased recently. Box et al. (2006) used calibrated atmospheric modelling and a single approximation for ice flow discharge to estimate average ice sheet mass loss of more than 100 Gt yr⁻¹ during 1988 to 2004; they also found acceleration of surface mass loss during this interval of 43 Gt yr⁻¹. A similar analysis by Hanna et al. (2005) for 1961 to 2003 found somewhat higher net accumulation but similar trends, with ice sheet growth of 22 ± 51 Gt yr⁻¹ from 1961 to 1990, shifting to shrinkage of 14 = 55 Gt yr⁻¹ from 1993 to 1998 and shrinkage of 36 ± 59 Gt yr⁻¹ from 1998 to 2003. Again, ice flow acceleration was not included in these estimates.

In a study especially using SRALT but incorporating laser elevation measurements from aircraft and a correction for the effect of changing temperature on near-surface density, Zwally et al. (2006) estimated slight growth of the ice sheet by 11 ± 3 Gt yr⁻¹ from 1992 to 2002. However, they noted that mass loss of 18 ± 2 Gt yr⁻¹ would be indicated if the thickness changes at higher elevations are largely low-density firn rather than high-density ice, as might apply if the effects of increasing accumulation rate were also taken into account (Hanna et al., 2005; Box et al., 2006). The more spatially limited results of Johannessen et al. (2005) from the same radar data indicated slightly less shrinkage or slightly more growth than found by Zwally et al. (2006) in regions of overlap. Krabill et al. (2000) also found thickening of central regions (~10 mm yr⁻¹) from laser measurements covering 1993/1994 to 1998/1999.

Krabill et al. (2004) used repeat laser altimetry and modelled surface mass balance to estimate mass loss of about 45 Gt yr⁻¹ from 1993/1994 to 1998/1999, with acceleration to a loss of 73 ± 11 Gt yr⁻¹ during the overlapping interval 1997 to 2003. These values may underestimate total losses, because they do not take account of rapid thinning in sparsely surveyed regions such as the southeast, where mass budget studies show large losses (Rignot and Kanagaratnam, 2006). Thomas et al. (2006)

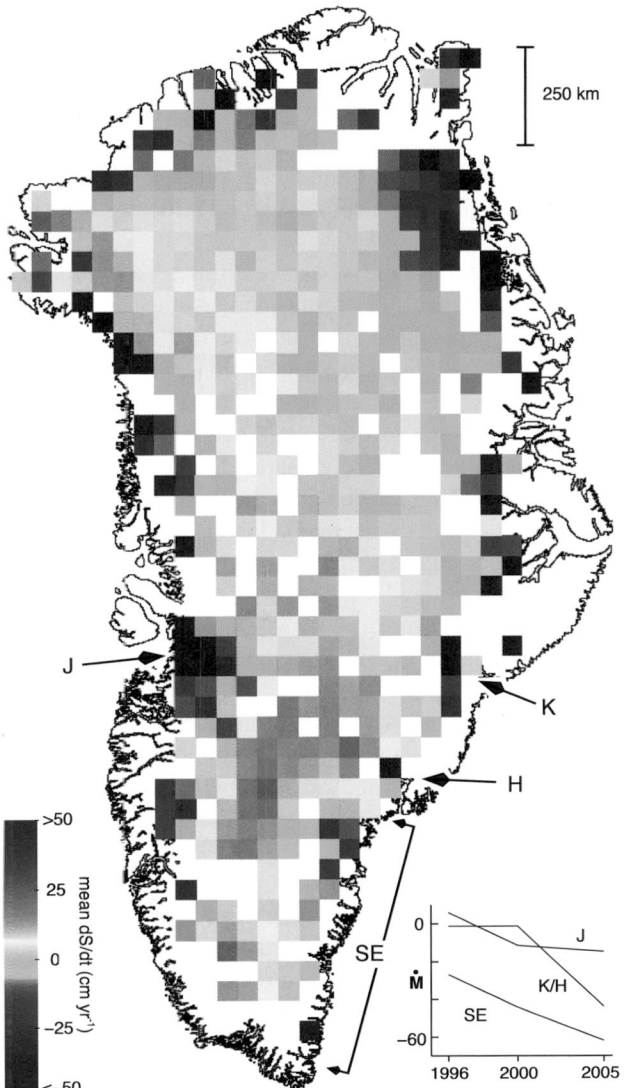

Figure 4.17. *Rates of surface elevation change (dS/dt) derived from laser altimeter measurements at more than 16,000 locations on the Greenland Ice Sheet where ICESat data from 2005 overlay aircraft surveys in 1998/1999 (using methods described by Thomas et al., 2006). Locations of rapidly thinning outlet glaciers at Jakobshavn (J), Kangerdlugssuaq (K), Helheim (H) and along the southeast coast (SE) are shown, together with an inset showing their estimated total mass balance (M, Gt yr⁻¹) between 1996 and 2005 (Rignot and Kanagaratnam, 2006).*

extended these results to 2004 using ICESat data to include approximate corrections for density changes in the near surface. Results showed ice sheet mass loss of 27 ± 23 Gt yr⁻¹ for 1993/1994 to 1998/1999, loss of 55 ± 25 Gt yr⁻¹ for 1997 to 2003, with an updated loss of 81 ± 24 Gt yr⁻¹ from 1998/1999 to 2004.

Rignot and Kanagaratnam (2006) combined several data sets, with special focus on the acceleration in velocity of outlet glaciers measured by Synthetic Aperture Radar (SAR) interferometry. Starting from an estimated excess ice flow discharge of 51 ± 28 Gt yr⁻¹ in 1996, these authors estimated that the ice flow loss increased to 83 ± 27 Gt yr⁻¹ in 2000 and 150 ± 36 Gt yr⁻¹ in 2005. Adding surface mass balance deviations from the long-term average as calculated by Hanna et al. (2005) yielded mass losses of 82 ± 28 Gt yr^{-1} in 1996, 124 ± 28 Gt yr^{-1} in 2000 and 202 ± 37 Gt yr^{-1} in 2005. The more pronounced ice flow accelerations were restricted to regions south of 66°N before 2000 but extended to 70°N by 2005. These estimates of rapid mass loss would be reduced somewhat if ice surface velocities are higher than depth-averaged velocities, which may apply in some places.

Greenland Ice Sheet mass balance estimates are summarised in Figure 4.18 (top). Most results indicate accelerating mass loss from Greenland during the 1990s up to 2005. The different estimates are not fully independent (there is, for example, some commonality in the isostatic corrections used for GRACE and altimetry estimates, and other overlaps can be found), but sufficient independence remains to increase confidence in the result. Different techniques have not fully converged quantitatively, with mismatches larger than formal error estimates suggesting structural uncertainties in the analyses, some of which were discussed above. The SRALT results showing overall near-balance or slight thickening, in contrast to other estimates, may result from the SRALT limitations over narrow glaciers discussed earlier.

Assessment of the data and techniques suggests a mass balance for the Greenland Ice Sheet ranging between growth of 25 Gt yr^{-1} and shrinkage of 60 Gt yr^{-1} for 1961 to 2003, shrinkage of 50 to 100 Gt yr^{-1} for 1993 to 2003 and shrinkage at even higher rates between 2003 and 2005. Lack of agreement between techniques and the small number of estimates preclude assignment of statistically rigorous error bounds. Interannual variability is very large, driven mainly by variability in summer melting, but also by sudden glacier accelerations (Rignot and Kanagaratnam, 2006). Consequently, the short time interval covered by instrumental data is of concern in separating fluctuations from trends.

4.6.2.2.2   Antarctica

Recent estimates of Antarctic Ice Sheet mass balance are summarised in Figure 4.18 (bottom). Rignot and Thomas (2002) combined several data sets including improved estimates of glacier velocities from InSAR to obtain antarctic mass budget estimates. For East Antarctica, growth of 20 ± 21 Gt yr^{-1} was indicated, with estimated losses of 44 ± 13 Gt yr^{-1} from West Antarctica. The balance of the Antarctic Peninsula was not assessed. Combining the East and West Antarctic numbers yielded a loss of 24 ± 25 Gt yr^{-1} for the region monitored. The time interval covered by these estimates is not tightly constrained, because ice input was estimated from data sets of varying length; output data were determined primarily in the few years before 2002.

Zwally et al. (2006) obtained SRALT coverage for about 80% of the ice sheet, including some portions of the Antarctic Peninsula, and interpolated to the rest of the ice sheet. The resulting balance included West Antarctic loss of 47 ± 4 Gt yr^{-1}, East Antarctic gain of 17 ± 11 Gt yr^{-1} and overall loss of 30 ± 12 Gt yr^{-1}. If all the ice thickness changes were low-density firn rather than ice, the loss would be smaller (14 ± 5 Gt yr^{-1}).

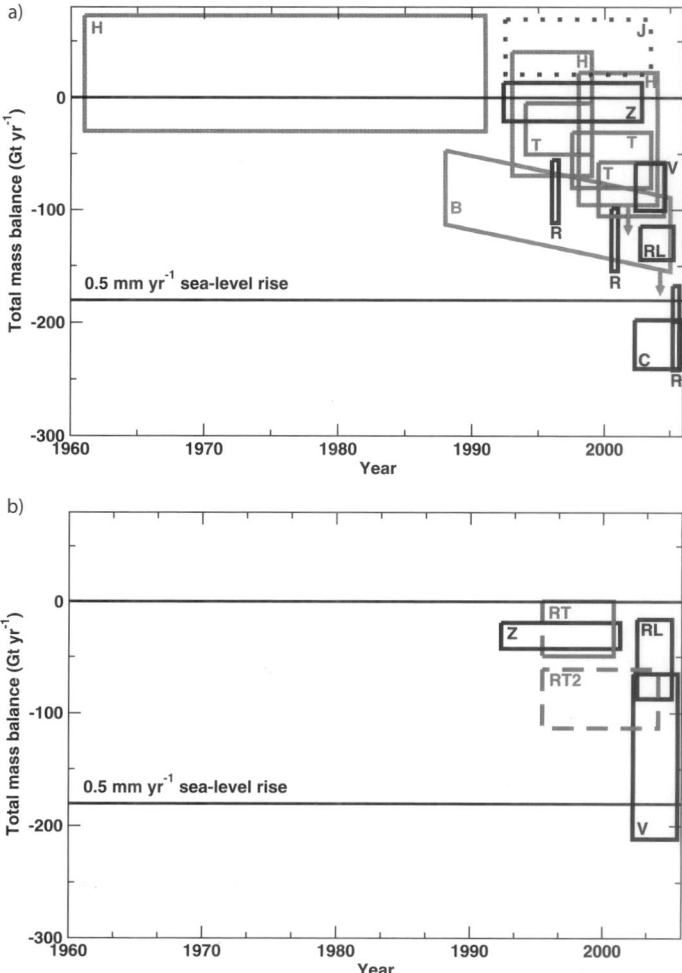

Figure 4.18. *(Top) Mass balance estimates for Greenland. The coloured rectangles, following Thomas et al. (2006), indicate the time span over which the measurements apply and the estimated range, given as (mean + uncertainty) and (mean – uncertainty) as reported in the original papers. Code: B (orange; Box et al, 2006), surface mass balance, using stated trend in accumulation, ice flow discharge (assumed constant), and standard error on regression of accumulation trend, with added arrow indicating additional loss from ice flow acceleration; H (brown; Hanna et al., 2005), surface mass balance, with arrow as for B; T (dark green; Thomas et al., 2006), laser altimetry, showing new results and revision of Krabill et al. (2004) to include firn densification changes; Z (violet; Zwally et al., 2006), primarily radar altimetry, with uncertainty reflecting the difference between a thickness change due to ice everywhere and that due to low-density firn in the accumulation zone; R (red; Rignot and Kanagaratnam, 2006), ice discharge combined with surface mass balance; V (blue; Velicogna and Wahr, 2005) GRACE gravity; RL (blue; Ramillen et al., 2006) GRACE gravity; J (magenta dashed; Johannessen et al., 2005), radar altimetry without firn densification correction and applying only to central regions that are thickening but omitting thinning of coastal regions. (Bottom) Mass balance estimates for grounded ice of Antarctica. Coloured rectangles show age span and error range as in the top panel. Code: Z (violet; Zwally et al., 2006), radar altimetry, with uncertainty reflecting the difference between a thickness change due to ice everywhere and that due to low-density firn everywhere; RT (dark green; Rignot and Thomas, 2002), ice discharge and surface mass balance, with dashed end line because some of the accumulation rate data extend beyond the time limits shown; RT2 (dark green; Rignot and Thomas, 2002), updated to include additional mass losses indicated by Thomas et al. (2004) and Rignot et al. (2005), dashed because the original authors did not produce this as an estimate for the whole ice sheet nor are accumulation rates updated; V (blue; Velicogna and Wahr, 2006), GRACE gravity; RL (blue; Ramillen et al., 2006), GRACE gravity.*

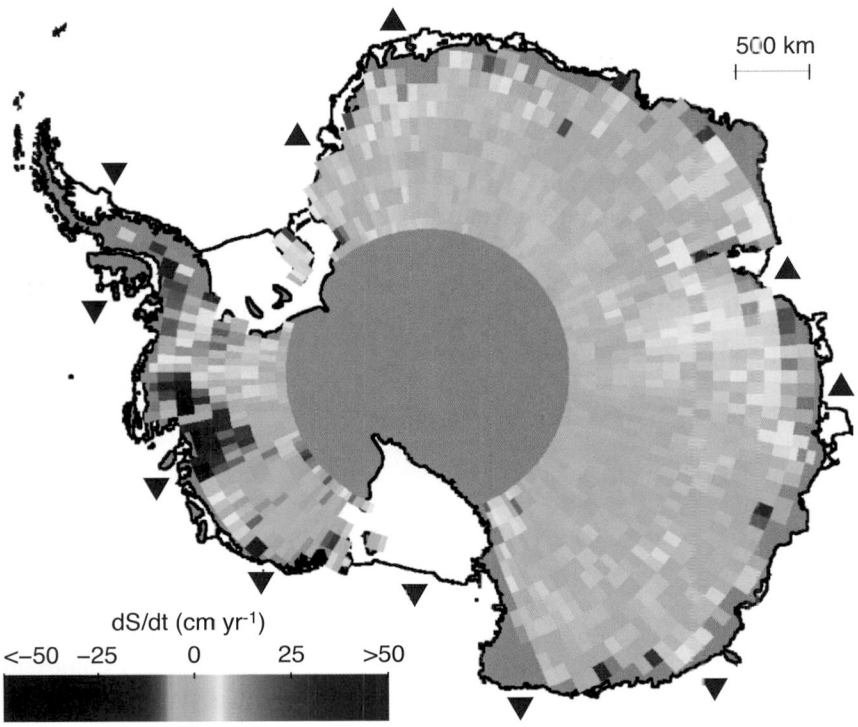

Figure 4.19. *Rates of surface elevation change (dS/dt) derived from ERS radar-altimeter measurements between 1992 and 2003 over the Antarctic Ice Sheet (Davis et al., 2005). Locations of ice shelves estimated to be thickening or thinning by more than 30 cm yr⁻¹ (Zwally et al., 2006) are shown by red triangles (thickening) and purple triangles (thinning).*

although mass loss would be overestimated if snow accumulation has been systematically underestimated (van de Berg et al., 2006).

Taking the Rignot and Thomas (2002), Zwally et al. (2006) and Rignot et al. (2005) results as providing the most complete antarctic coverage suggests ice sheet thinning of about 60 Gt yr⁻¹, with uncertainty of similar magnitude to the signal. Consideration of acceleration of some near-coastal glaciers, discussed below, and the difficulty of SRALT sampling of such regions, might allow slightly faster mass loss. The time interval considered is not uniform; the Rignot et al. (2005) results include changes after the collapse of the Larsen B Ice Shelf in 2002, more recent than data in the other studies, and suggest the possibility of accelerating mass loss. Use of the more spatially restricted Davis et al. (2005) SRALT data rather than the Zwally et al. (2006) results illustrates the persistent uncertainties; depending on the assumed density structure of the changes, Davis et al. (2005) combined with the Rignot et al. (2005) estimate for the Antarctic Peninsula would suggest near-balance or antarctic growth.

Interpretations of GRACE satellite gravity data indicate mass loss from the Antarctic Ice Sheet, including the Antarctic Peninsula and

Davis et al. (2005) compiled SRALT data for about 70% of the ice sheet, and did not interpolate to the rest. The same pattern of East Antarctic thickening and West Antarctic thinning was observed (Figure 4.19). Davis et al. (2005) suggested that the East Antarctic change was primarily from increased snowfall. Assigning all thickness change to low-density firn produces growth of the monitored portions of the ice sheet of 45 ± 8 Gt yr⁻¹; if all thickness changes were ice, this growth would be 105 ± 20 Gt yr⁻¹. Following the suggestion that the East Antarctic changes are from increased snow accumulation and the West Antarctic changes are more likely to be ice-dynamical would yield growth of monitored regions of 33 ± 9 Gt yr⁻¹. Note, however, that Monaghan et al. (2006) did not find the strong increase in snow accumulation suggested by Davis et al. (2005) in arguing for use of low-density firn in East Antarctic changes.

Rignot et al. (2005) documented discharges 84 ± 30% larger than accumulation rates for the glaciers that feed the Wordie Ice Shelf on the west coast of the northern Antarctic Peninsula (which shrank greatly between 1966 and 1989), a region largely absent from the SRALT studies. Consideration of strong imbalances in glaciers feeding the former Larsen B Ice Shelf across the Peninsula, and extrapolation of the results to undocumented basins, suggested mass loss from the ice in the northern part of the Antarctic Peninsula of 42 ± 7 Gt yr⁻¹. Observation of widespread glacier front retreat in the region (Cook et al., 2005) motivates the extrapolation,

small glaciers and ice caps nearby, of 139 ± 73 Gt yr⁻¹ between April 2002 and July 2005 (Velicogna and Wahr, 2006). Near-balance was indicated for East Antarctica, at 0 ± 51 Gt yr⁻¹, with mass loss in West Antarctica of 136 ± 21 Gt yr⁻¹. Independent analyses by Ramillien et al. (2006) found, for July 2002 to March 2005, East Antarctic growth of 67 ± 28 Gt yr⁻¹, West Antarctic shrinkage of 107 ± 23 Gt yr⁻¹ and a net antarctic loss of 40 ± 36 Gt yr⁻¹.

Assessment of the data and techniques suggests overall Antarctic Ice Sheet mass balance ranging from growth of 50 Gt yr⁻¹ to shrinkage of 200 Gt yr⁻¹ from 1993 to 2003. As in the case of Greenland, the small number of measurements, lack of agreement between techniques, and existence of systematic errors that cannot be estimated accurately preclude formal error analyses and confidence limits. There is no implication that the midpoint of the range given provides the best estimate. Lack of older data complicates a similar estimate for the period 1961 to 2003. Acceleration of mass loss is likely to have occurred, but not so dramatically as in Greenland. Considering the lack of estimated strong trends in accumulation rate, assessment of the possible acceleration and the slow time scales affecting central regions of the ice sheets, it is reasonable to estimate that the behaviour from 1961 to 2003 falls between ice sheet growth of 100 Gt yr⁻¹ and shrinkage of 200 Gt yr⁻¹.

Simply summing the 1993 to 2003 contributions from Greenland and Antarctica produces a range from balance

(0 Gt yr^{-1}) to shrinkage of 300 Gt yr^{-1}, or a contribution to sea level rise of 0 to 0.8 mm yr^{-1}. Because it is very unlikely that each of the ice sheets would exhibit the upper limit of its estimated mass balance range, it is very likely that, taken together, the ice sheets in Greenland and Antarctica have been contributing to sea level rise over the period 1993 to 2003. For 1961 to 2003, the same calculation spans growth of 125 Gt yr^{-1} to shrinkage of 260 Gt yr^{-1}, with 1993 to 2003 likely having the fastest mass loss of any decade in the 1961 to 2003 interval. Geodetic data on Earth rotation and polar wander provide additional insight (Peltier, 1998). Although Munk (2002) suggested that the geodetic data did not allow much contribution to sea level rise from ice sheets, subsequent reassessment of the errors involved in some of the data sets and analyses allows an anomalous late 20th century sea level rise of up to about 1 mm yr^{-1} (360 Gt yr^{-1}) from land ice (Mitrovica et al., 2006). Estimated mountain glacier contributions do not supply this, so a contribution from the polar ice sheets is consistent with the geodetic constraints, although little change in polar ice is also consistent.

4.6.2.2.3   Ice shelves

Changes in the mass of ice shelves, which are already floating, do not directly affect sea level, but ice shelf changes can affect flow of adjacent ice that is not floating, and thus affect sea level indirectly. Most ice shelves are in Antarctica, where they cover an area of about 1.5 × 10^6 km^2, or 11% of the entire ice sheet, and where nearly all ice streams and outlet glaciers flow into ice shelves. By contrast, Greenland ice shelves occupy only a few thousand square kilometres, and many are little more than floating glacier tongues. Mass loss by surface melt water runoff is not important for most ice shelf regions, which lose mass primarily by iceberg calving and basal melting, although basal freeze-on occurs in some regions.

Developments since IPCC (2001) include improved velocity and thickness data to estimate fluxes, and interpretation of repeated SRALT surveys over ice shelves to infer thickening or thinning rates. Melting of up to tens of metres per year has been estimated beneath deeper ice near grounding lines (Rignot and Jacobs, 2002; Joughin and Padman, 2003). Significant changes are observed on most ice shelves, with both positive and negative trends, and with faster changes on smaller shelves. Overall, Zwally et al. (2006) estimated mass loss from ice shelves fed by glaciers flowing from West Antarctica of 95 ± 11 Gt yr^{-1}, and mass gain to ice shelves fed by glaciers flowing from East Antarctica of 142 ± 10 Gt yr^{-1}. Rapid thinning of more than 1 m yr^{-1}, and locally more than 5 m yr^{-1}, was observed between 1992 and 2001 for many small ice shelves in the Amundsen Sea and along the Antarctic Peninsula. Thinning of about 1 m yr^{-1} (Shepherd et al., 2003; Zwally et al., 2006) preceded the fragmentation of almost all (3,300 km^2) of the Larsen B Ice Shelf along the Antarctic Peninsula in fewer than five weeks in early 2002 (Scambos et al., 2003).

4.6.3   Causes of Changes

4.6.3.1   Changes in Snowfall and Surface Melting

For Greenland, modelling driven by reanalyses and calibrated against surface observations indicates recent increases in temperature, precipitation minus evaporation, surface melt water runoff and net mass loss from the surface of the ice sheet, as well as areal expansion of melting and reduction in albedo (Hanna et al., 2005, 2006; Box et al., 2006). High interannual variability means that many of the trends are not highly significant, but the trends are supported by the consistency between the various component data sets and results from different groups. Estimated net snowfall minus melt water runoff includes an increase in the Greenland contribution to sea level rise of 58 Gt yr^{-1} between the 1961 to 1990 and 1998 to 2003 intervals (Hanna et al., 2005), or of 43 Gt yr^{-1} from 1998 to 2004 (Box et al., 2006).

For Antarctica, the recent summaries by van de Berg et al. (2006), van den Broeke et al. (2006) and Monaghan et al. (2006) have updated trends in accumulation rate. Contrary to some earlier work, these new studies found no continent-wide significant trends in accumulation over the interval 1980 to 2004 (van de Berg et al., 2006; van den Broeke et al., 2006) or 1985 to 2001 (Monaghan et al., 2006) from atmospheric reanalysis products (National Centers for Environmental Prediction (NCEP), the European Centre for Medium Range Weather Forecasts (ECMWF), Japanese), or from two mesoscale models driven by ECMWF and one by NCEP reanalyses. Strong interannual variability was found, approaching 5% for the continent, and important regional and seasonal trends that fit into larger climatic patterns, including an upward trend in accumulation in the Antarctic Peninsula. Studies of surface temperature (e.g., van den Broeke, 2000; Vaughan et al., 2001; Thompson and Solomon, 2002; Doran et al., 2002; Schneider et al., 2004; Turner et al., 2005) similarly showed regional patterns including strong warming in the Antarctic Peninsula region, and cooling at some other stations. Long-term data are very sparse, precluding confident identification of continent-wide trends.

4.6.3.2   Ongoing Dynamic Ice Sheet Response to Past Forcing

Because some portions of ice sheets respond only slowly to climate changes (decades to thousands of years or longer), past forcing may be influencing ongoing changes (Box 4.1). Some geologic data support recent and perhaps ongoing antarctic mass loss (e.g., Stone et al., 2003). A comprehensive attempt to discern such long-term trends contributing to recently measured imbalances was made by Huybrechts (2002) and Huybrechts et al. (2004). They found little long-term trend in volume of the Greenland Ice Sheet, but a trend in antarctic shrinkage of about 90 Gt yr^{-1}, primarily because of retreat of the West Antarctic grounding line in response to the end of the last ice age. Models project that this trend will largely disappear over

Box 4.1: Ice Sheet Dynamics and Stability

The ice sheets of Antarctica and Greenland could raise sea level greatly. Central parts of these ice sheets have been observed to change only slowly, but near the coast rapid changes over quite large areas have been observed. In these areas, uncertainties about glacier basal conditions, ice deformation and interactions with the surrounding ocean seriously limit the ability to make accurate projections.

Ice sheets are thick, broad masses of ice formed mainly from compaction of snow (Paterson, 2004). They spread under their own weight, transferring mass towards their margins where it is lost primarily by runoff of surface melt water or by calving of icebergs into marginal seas or lakes. Water vapour fluxes (sublimation and condensation), and basal melting or freezing (especially beneath ice shelves) may also be important processes of mass gain and loss.

Ice sheets flow by internal deformation, basal sliding or a combination of both. Deformation in ice occurs through solid-state processes analogous to those involved in polycrystalline metals that are relatively close to their melting points. Deformation rates depend on the gravitational stress (which increases with ice thickness and with the slope of the upper surface), temperature, impurities, and size and orientation of the crystals (which in turn depend in part on the prior deformational history of the ice). While these characteristics are not completely known, model tuning allows slow ice flow by deformation to be simulated with reasonable accuracy.

For basal sliding to be an important component of the total motion, melt water or deformable wet sediment slurries at the base are required for lubrication. While the central regions of ice sheets (typically above 2,000 m elevation) seldom experience surface melting, the basal temperature may be raised to the melting point by heat conducted from the earth's interior, delivered by melt water transport, or from the 'friction' of ice motion. Sliding velocities under a given gravitational stress can differ by orders of magnitude, depending on the presence or absence of unconsolidated sediment, the roughness of the substrate and the supply and distribution of water. Basal conditions are well characterised in few regions, introducing important uncertainties to the modelling of basal motion.

Ice flow is often channelled into fast-moving ice streams (which flow between slower-moving ice walls) or outlet glaciers (with rock walls). Enhanced flow in ice streams arises either from higher gravitational stress linked to thicker ice in bedrock troughs, or from increased basal lubrication.

Ice flowing into a marginal sea or lake may break off immediately to form icebergs, or may remain attached to the ice sheet to become a floating ice shelf. An ice shelf moves forward, spreading and thinning under its own weight, and fed by snowfall on its surface and ice input from the ice sheet. Friction at ice shelf sides and over local shoals slows the flow of the ice shelf and thus the discharge from the ice sheet. An ice shelf loses mass by calving icebergs from the front and by basal melting into the ocean cavity beneath. Estimates based on available data suggest a 1°C ocean warming could increase ice shelf basal melt by 10 m yr^{-1}, but inadequate knowledge of the bathymetry and circulation in the largely inaccessible ice shelf cavities restricts the accuracy of such estimates.

Ice deformation is nonlinear, increasing approximately proportional to the cube of the applied stress. Moreover, an increase in any of the six independent applied stresses (three stretching stresses and three shear stresses) increases the deformation rate for all other stresses. For computational efficiency, most long simulations with comprehensive ice flow models use a simplified stress distribution, but recent changes in ice sheet margins and ice streams cannot be simulated accurately with these models, demonstrating a need for resolving the full stress configuration. Development of such models is still in its infancy, with few results yet available.

Ice sheets respond to environmental forcing over numerous time scales. A surface warming may take more than 10,000 years to penetrate to the bed and change temperatures there, while a crevasse filled with melt water might penetrate to the bed and affect the temperature locally within minutes. Ice velocity over most of an ice sheet changes slowly in response to changes in the ice sheet shape or surface temperature, but large velocity changes may occur rapidly on ice streams and outlet glaciers in response to changing basal conditions or changes in the ice shelves into which they flow.

The palaeo-record of previous ice ages indicates that ice sheets shrink in response to warming and grow in response to cooling. The data also indicate that shrinkage can be far faster than growth. Understanding of the processes suggests that this arises both because surface melting rates can be much larger than the highest snowfall rates, and because ice discharge may be accelerated by strong positive feedbacks (Paterson, 1994; P.U. Clark et al., 1999). Thawing of the bed, loss of restraint from ice shelves or changes in melt water supply and transmission can increase flow speed greatly. The faster flow may then generate additional lubrication from frictional heating and from erosion to produce wet sediment slurries. Surface lowering as the faster flow thins the ice will enhance surface melting, and will reduce basal friction where the thinner ice becomes afloat. Despite competition from stabilising feedbacks, warming-induced changes have led to rapid shrinkage and loss of ice sheets in the past, with possible implications for the future.

the next millennium. In tests of the sensitivity of this result to various model parameters, Huybrechts (2002) found a modern thinning trend in most simulations but an opposite trend in one; in addition, simulated trends for today depend on the poorly known timing of retreat in West Antarctica. Moreover, the ice flow model responds too slowly to some forcings owing to the coarse model grid and lack of some stresses and processes (see Section 4.6.3.3), perhaps causing the modelled long-term trend to end more slowly than it should.

The recent ice flow accelerations discussed in Section 4.6.3.3 are likely to be sufficient to explain much or all of the estimated antarctic mass imbalance, and ice flow and surface mass balance changes are sufficient to explain the mass imbalance in Greenland. This points to little or no contribution from long-term trends to modern ice sheet balance, although with considerable uncertainties.

4.6.3.3 Dynamic Response to Recent Forcing

Numerous papers since IPCC (2001) have documented rapid changes in marginal regions of the ice sheets. Attention has especially focused on increased flow velocity of glaciers along the Antarctic Peninsula (Scambos et al., 2004; Rignot et al., 2004, 2005), the glaciers draining into Pine Island Bay and nearby parts of the Amundsen Sea from West Antarctica (Shepherd et al., 2004; Thomas et al., 2004) and Greenland's Jakobshavn Glacier (Thomas et al., 2003; Joughin et al., 2004) and other glaciers south of about 70°N (Howat et al., 2005; Rignot and Kanagaratnam, 2006). Accelerations may have occurred in some coastal parts of East Antarctica (Zwally et al., 2006), and ice flow deceleration has been observed on Whillans and Bindschadler Ice Streams on the Siple Coast of West Antarctica (Joughin and Tulaczyk, 2002). Rignot and Kanagaratnam (2006) estimated that ice discharge increase in Greenland caused mass loss in 2005 to be about 100 Gt yr^{-1} larger than in 1996; consideration of the changes in the Amundsen Sea and Antarctic Peninsula regions of West Antarctica (and the minor opposing trend on Whillans and Bindschadler Ice Streams) suggests an antarctic signal of similar magnitude, although with greater uncertainty and occurring perhaps over a longer interval (Joughin and Tulaczyk, 2002; Thomas et al., 2004; Rignot et al., 2005; van den Broeke et al., 2006).

Most of the other coastal changes appear to have involved inland acceleration following reduction or loss of ice shelves. Very soon after breakup of the Larsen B Ice Shelf along the Antarctic Peninsula, the speeds of tributary glaciers increased up to eight-fold, but with little change in velocity of adjacent ice still buttressed by the remaining ice shelf (Rignot et al., 2004; Scambos et al., 2004). Thinning and breakup of the floating ice tongue of Jakobshavn Glacier were accompanied by approximate doubling of the ice flow velocity (Thomas et al., 2003; Joughin et al., 2004; Thomas, 2004). Ice shelf thinning has occurred with the acceleration of tributary glaciers entering the Amundsen Sea (Shepherd et al., 2002, 2004; Joughin et al., 2003).

Because of drag between ice shelves and embayment sides or localised re-grounding points on seabed topographic highs, shortening or thinning of ice shelves is expected to accelerate ice flow (Thomas, 1979), with even small ice shelves potentially important (Dupont and Alley, 2006). Targeted models addressing acceleration of particular glaciers in response to ice shelf reduction are capable of simulating the observed time scales (notable changes in years or less) and patterns of change (largest thinning and acceleration near the coast, decreasing inland and following ice streams; Payne et al., 2004; Dupont and Alley, 2005). Comprehensive model runs for ice sheet behaviour over the last century, using known forcings and flow processes but omitting full stress coupling with ice shelves and poorly known details of oceanographic changes beneath the ice shelves, match overall ice sheet trends rather well (Huybrechts et al., 2004) but fail to show these rapid marginal thinning events. This suggests that the changes are in response to processes (either forcings from ocean temperature or ocean circulation changes, or ice flow processes) not included in the comprehensive modelling, or that the coarse spatial resolution of the comprehensive models slows their simulated response rates enough to be important.

The acceleration of Helheim Glacier in Greenland may be akin to changes linked to ice shelves. Enhanced calving may have removed not-quite-floating ice at Helheim, reducing restraint on the remaining ice and allowing faster flow (Howat et al., 2005).

Other ice flow changes have occurred that are not linked to ice shelf reduction. The changes in Siple Coast, Antarctica, likely reflect inherent flow variability rather than recent forcing (Parizek et al., 2003). Zwally et al. (2002) showed for one site near the equilibrium line on the west coast of Greenland that the velocity of comparatively slow-moving ice increased just after the seasonal onset of drainage of surface melt water into the ice sheet, and that greater melt water input produced greater ice flow acceleration. The total acceleration was not large (of the order of 10%), but the effect is not included in most ice flow models. Inclusion in one model (Parizek and Alley, 2004) somewhat increased the sensitivity of the ice sheet to various specified warmings, mostly beyond the year 2100. Much uncertainty remains, especially related to whether fast-moving glaciers and ice streams are similarly affected, and whether access of melt water to the bed through more than 1 km of cold ice would migrate inland if warming caused surface melting to migrate inland (Alley et al., 2005b). This could thaw ice that is frozen to the bed, allowing faster flow through enhanced basal sliding or sub-glacial sediment deformation. Data are not available to assess whether effects of increased surface melting in Greenland have been transmitted to the bed and contributed to ice flow acceleration.

4.6.3.4 *Melting and Calving of Ice Shelves*

Many of the largest and fastest ice sheet changes thus appear to be at least in part responses to ice shelf shrinkage or loss. Although ice shelf shrinkage does not directly contribute to sea level change because shelf ice is already floating, the very tight coupling to inland ice means that ice shelf balance does matter to sea level. The available data suggest that the ice shelf changes have resulted from environmental warming, with both oceanic and atmospheric temperatures important, although changes in oceanic circulation cannot be ruled out as important contributors.

The southward-progressing loss of ice shelves along the Antarctic Peninsula is consistent with a thermal limit to ice shelf viability (Morris and Vaughan, 2003). Cook et al. (2005) found that no ice shelves exist on the warmer side of the –5°C mean annual isotherm, whereas no ice shelves on the colder side of the –9°C isotherm have broken up. Before the 2002 breakup of the Larsen B Ice Shelf, local air temperatures had increased by more than 1.5°C over the previous 50 years (Vaughan et al., 2003), increasing summer melting and formation of large melt ponds on the ice shelf. These likely contributed to breakup by draining into and wedging open surface crevasses that linked to bottom crevasses filled with seawater (Scambos et al., 2000). Large ice flow models do not accurately capture the physical processes involved in such dramatic iceberg calving, or in more common calving behaviour.

Despite an increased ice supply from tributary glaciers, thinning of up to several metres per year has been measured for ice shelves on the Amundsen Sea coastline in the absence of large surface mass balance changes. This suggests that increased basal ice melting is responsible for the thinning (Shepherd et al., 2003, 2004). Similarly, the 15-km floating ice tongue of Jakobshavn Glacier survived air temperatures during the 1950s similar to or even warmer than those associated with thinning and collapse near the end of the 20th century, implicating oceanic heat transport in the more recent changes, although air temperature increases may have contributed (Thomas et al., 2003).

The basal mass balance of an ice shelf depends on temperature and ocean circulation beneath it. Isolation from direct wind forcing means that the main drivers of circulation below an ice shelf are tidal and density (thermohaline) forces. Lack of knowledge of sub-ice bathymetry has hampered the use of three-dimensional models to simulate circulation beneath the thinning ice shelves. Both the west side of the Antarctic Peninsula and the Amundsen Sea coast are exposed to warm Circumpolar Deep Water (CDW; Hellmer et al., 1998), capable of causing rapid ice shelf basal melting. Increased melting in the Amundsen Sea is consistent with observed recent warming by 0.2°C of ocean waters seaward of the continental shelf break (Jacobs et al., 2002; Robertson et al., 2002). Simple regression analysis of available data including those from the Amundsen Sea indicated that 1°C warming of waters below an ice shelf increases basal melt rate by about 10 m yr^{-1} (Shepherd et al., 2004).

4.7 Changes in Frozen Ground

4.7.1 Background

Frozen ground, in a broad sense, includes near-surface soil affected by short-term freeze-thaw cycles, seasonally frozen ground and permafrost. In terms of areal extent, frozen ground is the single largest component of the cryosphere. The presence of frozen ground depends on the ground temperature, which is controlled by the surface energy balance. While the climate is an important factor determining the distribution of frozen ground, local factors are also important, such as vegetation conditions, snow cover, physical and thermal properties of soils and soil moisture conditions. The permafrost temperature regime is a sensitive indicator of decadal to centennial climatic variability (Lachenbruch and Marshall, 1986; Osterkamp, 2005). Thawing of ice-rich permafrost can lead to subsidence of the ground surface as masses of ground ice melt and to the formation of uneven topography know as thermokarst, generating dramatic changes in ecosystems, landscape and infrastructure performance (Nelson et al., 2001; Walsh et al., 2005). Surface soil freezing and thawing processes play a significant role in the land surface energy and moisture balance, hence in climate and hydrologic systems. The primary controls on local hydrological processes in northern regions are the presence or absence of permafrost and the thickness of the active layer (Hinzman et al., 2003). Changes in soil seasonal freeze-thaw processes have a strong influence on spatial patterns, seasonal to interannual variability, and long-term trends in terrestrial carbon budgets and surface-atmosphere trace gas exchange, both directly through biophysical controls on photosynthesis and respiration and indirectly through controls on soil nutrient availability.

4.7.2 Changes in Permafrost

4.7.2.1 *Data Sources*

Although there are some earlier measurements, systematic permafrost temperature monitoring in Russia started in the 1950s at hydrometeorological stations to depths of up to 3.2 m (Zhang et al., 2001) and in boreholes greater than 100 m deep (Pavlov, 1996). Permafrost temperatures in northern Alaska have been measured from deep boreholes (generally >200 m) since the 1940s (Lachenbruch and Marshall, 1986) and from shallow boreholes (generally <80 m) since the mid-1980s (Osterkamp, 2005). Some permafrost temperature measurements on the Tibetan Plateau were conducted in the early 1960s, while continuous permafrost monitoring only started in the late 1980s (Zhao et al., 2003). Monitoring of permafrost temperatures mainly started in the early 1980s in northern Canada (S.L. Smith et al., 2005) and in the 1990s in Europe (Harris et al., 2003).

4.7.2.2 Changes in Permafrost Temperature

Permafrost in the NH has typically warmed in recent decades (Table 4.5), although at a few sites there was little warming or even a cooling trend. For example, measurements (Osterkamp, 2003) and modelling results (see Hinzman et al., 2005; Walsh et al., 2005) indicate that permafrost temperature has increased by up to 2°C to 3°C in northern Alaska since the 1980s. Changes in air temperature alone over the same period cannot account for the permafrost temperature increase, and so changes in the insulation provided by snow may be responsible for some of the change (Zhang, 2005). Data from the northern Mackenzie Valley in the continuous permafrost zone show that permafrost

temperature between depths of 20 to 30 m has increased about 1°C in the 1990s (S.L. Smith et al., 2005), with smaller changes in the central Mackenzie Valley. There is no significant trend in temperatures at the top of permafrost in the southern Mackenzie Valley, where permafrost is thin (less than 10 to 15 m thick) and warmer than –0.3°C (S.L. Smith et al., 2005, Couture et al., 2003). The absence of a trend is likely due to the absorption of latent heat required to melt ice. Similar results are reported for warm permafrost in the southern Yukon Territory (Haeberli and Burn, 2002). Cooling of permafrost was observed from the late 1980s to the early 1990s at a depth of 5 m at Iqaluit in the eastern Canadian Arctic. This cooling, however, was followed by warming of 0.4°C yr^{-1} between 1993 and 2000 (S.L. Smith

Table 4.5. *Recent trends in permafrost temperature (updated from Romanovsky et al., 2002 and Walsh et al., 2005).*

Region	Depth (m)	Period of Record	Permafrost Temperature Change (°C)	Reference
United States				
Northern Alaska	~1	1910s–1980s	2–4	Lachenbruch and Marshall, 1986
Northern Alaska	20	1983–2003	2–3	Osterkamp, 2005
Interior of Alaska	20	1983–2003	0.5–1.5	Osterkamp, 2005
Canada				
Alert, Nunavut	15	1995–2000	0.8	S.L. Smith et al., 2003
Northern Mackenzie Valley	20–30	1990–2002	0.3–0.8	S.L. Smith et al., 2005
Central Mackenzie Valley	10–20	Mid-1980s–2003	0.5	S.L. Smith et al., 2005
Southern Mackenzie Valley & Southern Yukon Territory	~20	Mid-1980s–2003	0	Haeberli and Burn, 2002
Northern Quebec	10	Late 1980s–mid-1990s	<–1	Allard et al., 1995
Northern Quebec	10	1996–2001	1.0	DesJarlais, 2004
Lake Hazen	2.5	1994–2000	1.0	Broll et al., 2003
Iqaluit, Eastern Canadian Arctic	5	1993–2000	2.0	S.L. Smith et al., 2005
Russia				
East Siberia	1.6–3.2	1960–2002	~1.3	Walsh et al., 2005
Northern West Siberia	10	1980–1990	0.3–0.7	Pavlov, 1996
European north of Russia, continuous permafrost zone	6	1973–1992	1.6–2.8	Pavlov, 1996
Northern European Russia	6	1970–1995	1.2–2.8	Oberman and Mazhitova, 2001
Europe				
Juvvasshoe, Southern Norway	~3	Past 30–40 years	0.5–1.0	Isaksen et al., 2001
Janssonhaugen, Svalbard	~2	Past 60–80 years	1–2	Isaksen et al., 2001
Murtel-Corvatsch	11.5	1987–2001	1.0	Vonder Muhll et al., 2004
China				
Tibetan Plateau	~10	1970s–1990s	0.2–0.5	Zhao et al., 2004
Qinghai-Xizang Highway	3–5	1995–2002	Up to 0.5	Wu and Liu, 2003; Zhao et al., 2004
Tianshan Mountains	16–20	1973–2002	0.2–0.4	Qiu et al., 2000; Zhao et al., 2004
Da Hinggan Mountains, Northeastern China	~2	1978–1991	0.7–1.5	Zhou et al., 1996

et al., 2005). This trend is similar to that observed in Northern Quebec, where permafrost cooling was observed between the mid-1980s and mid-1990s at a depth of 10 m (Allard et al., 1995) followed by warming beginning in 1996 (Brown et al., 2000). Warming of permafrost at depths of 15 to 30 m since the mid-1990s has also been observed in the Canadian High Arctic (Smith et al., 2003).

There is also evidence of permafrost warming in the Russian Arctic. Permafrost temperature increased approximately 1°C at depths between 1.6 and 3.2 m from the 1960s to the 1990s in East Siberia, about 0.3°C to 0.7°C at a depth of 10 m in northern West Siberia (Pavlov, 1996) and about 1.2°C to 2.8°C at a depth of 6 m from 1973 through 1992 in northern European Russia (Oberman and Mazhitova, 2001). Fedorov and Konstantinov (2003) reported that permafrost temperatures from three central Siberian stations did not show an apparent trend between 1991 and 2000. Mean annual temperature in Central Mongolia at depths from 10 to 90 m increased 0.05°C to 0.15°C per decade over 30 years (Sharkhuu, 2003).

At the Murtèl-Corvatsch borehole in the Swiss Alps, permafrost temperatures in 2001 and 2003, at a depth of 11.5 m in ice-rich frozen debris, were only slightly below −1°C, and were the highest since readings began in 1987 (Vonder Mühll et al., 2004). Analysis of the long-term thermal record from this site has shown that in addition to summer air temperatures, the depth and duration of snow cover, particularly in early winter, have a major influence on permafrost temperatures (Harris et al., 2003). Results from six years of ground temperature monitoring at Janssonhaugen, Svalbard, indicate that the permafrost has warmed at a rate of about 0.5°C per decade at a depth of 20 m (Isaksen et al., 2001). Results from Juvvasshøe, in southern Norway, indicate that ground temperature has increased by about 0.3°C at a depth of 15 m from 1999 to 2006. At both these sites, wind action prevents snow accumulation in winter and so a close relationship is observed between air, ground surface and ground subsurface temperatures, which makes the geothermal records from Janssonhaugen and Juvvasshøe more direct indicators of climate change.

Permafrost temperature increased about 0.2°C to 0.5°C from the 1970s to 1990s over the hinterland of the Tibetan Plateau (Zhao et al., 2003), up to 0.5°C along the Qinghai-Xizang Highway over a period from 1995 to 2002 (Wu and Liu, 2003; Zhao et al., 2004) and about 0.2°C to 0.4°C from 1973 to 2002 at depths of 16 to 20 m in Tianshan Mountain regions (Qiu et al., 2000; Zhao et al., 2004). Over the Da Hinggan Mountains in north-eastern China, permafrost surface temperature increased about 0.7°C to 1.5°C over a period from 1978 through 1991 from the valley bottom to the north-facing slopes (Zhou et al., 1996). Permafrost temperature at the depth of the zero annual temperature variation increased about 2.1°C on the valley bottom, 0.7°C on the north-facing slopes and 0.8°C on south-facing slopes. In areas of the south-facing slopes where no permafrost exists, soil temperature at depths of 2 to 3 m increased about 2.4°C (Zhou et al., 1996).

4.7.2.3 Permafrost Degradation

Permafrost degradation refers to a naturally or artificially caused decrease in the thickness and/or areal extent of permafrost. Evidence of change in the southern boundary of the discontinuous permafrost zone in the past decades has been reported. In North America, the southern boundary has migrated northward in response to warming since the Little Ice Age, and continues to do so today (Halsey et al., 1995). In recent years, widespread permafrost warming and thawing have occurred on the Tibetan Plateau, China. Based on data from ground penetration radar and *in situ* measurements, the lower limit of permafrost has moved upward about 25 m from 1975 through 2002 on the north-facing slopes of the Kunlun Mountains (Nan et al., 2003). From Amdo to Liangdehe along the Qinghai-Xizang Highway on the Tibetan Plateau, areal extent of permafrost islands decreased approximately 36% over the past three decades (Wang, 2002). Areal extent of taliks (areas of unfrozen ground within permafrost) expanded about 1.2 km on both sides of the Tongtian River (Wang, 2002). Overall, the northern limit of permafrost retreated about 0.5 to 1.0 km southwards and the southern limit moved northwards about 1.0 to 2.0 km along the Qinghai-Xizang (Tibet) Highway (Wang and Zhao, 1997; Wu and Liu, 2003).

When the warming at the top of permafrost eventually penetrates to the base of permafrost and the new surface temperature remains stable, thawing at the base of the ice-bearing permafrost occurs (i.e., basal thawing), especially for thin discontinuous permafrost. At Gulkana, Alaska, permafrost thickness is about 50 to 60 m and the basal thawing of permafrost has averaged 0.04 m yr^{-1} since 1992 (Osterkamp, 2003). Over the Tibetan Plateau, basal thawing of 0.01 to 0.02 m yr^{-1} was observed since the 1960s in permafrost of less than 100 m thickness (Zhao et al., 2003). It is expected that the basal thawing rate will accelerate over the Tibetan Plateau as the permafrost surface continues to warm.

If ice-rich permafrost thaws, the ground surface subsides. This downward displacement of the ground surface is called thaw settlement. Typically, thaw settlement does not occur uniformly and so yields a chaotic surface with small hills and wet depressions known as thermokarst terrain; this is particularly common in areas underlain by ice wedges. On slopes, thawing of ice-rich, near-surface permafrost layers can create mechanical discontinuities in the substrate, leading to active-layer detachment slides (Lewkowicz, 1992), which have the capacity to damage structures similar to other types of rapid mass movements. Thermokarst processes pose a serious threat to arctic biota through either oversaturation or drying (Hinzman et al., 2005; Walsh et al., 2005). Extensive thermokarst development has been discovered near Council, Alaska (Yoshikawa and Hinzman, 2003) and in central Yakutia (Gavrilov and Efremov, 2003). Significant expansion and deepening of thermokarst lakes were observed near Yakutsk (Fedorov and Konstantinov, 2003) between 1992 and 2001. The largest subsidence rates of 17 to 24 cm yr^{-1} were observed in depressions holding young thermokarst lakes. Satellite data reveal that in the continuous permafrost zone of Siberia, total lake area increased by about 12% and lake number rose by 4% during the past three decades (L.C. Smith et al., 2005). Over the discontinuous permafrost zone, total area and lake number decreased by up to 9% and 13%, respectively, probably due to lake water drainage through taliks.

The most sensitive regions of permafrost degradation are coasts with ice-bearing permafrost that are exposed to the Arctic Ocean. Mean annual erosion rates vary from 2.5 to 3.0 m yr⁻¹ for the ice-rich coasts to 1.0 m yr⁻¹ for the ice-poor permafrost coasts along the Russian Arctic Coast (Rachold et al., 2003). Over the Alaskan Beaufort Sea Coast, mean annual erosion rates range from 0.7 to 3.2 m yr⁻¹ with maximum rates up to 16.7 m yr⁻¹ (Jorgenson and Brown, 2005).

4.7.2.4   Subsea Permafrost

Subsea (or offshore) permafrost refers to permafrost occurring beneath the seabed. It exists in continental shelves of the polar regions. Subsea permafrost formed either in response to the negative mean annual sea-bottom temperature or as the result of sea level rise so that terrestrial permafrost was covered by seawater. Although the potential release of methane trapped within subsea permafrost may provide a positive feedback to climate warming, available observations do not permit an assessment of changes that might have occurred.

4.7.3   Changes in Seasonally Frozen Ground

Seasonally frozen ground refers to a soil layer that freezes and thaws annually regardless of whether there is underlying permafrost. It includes both seasonal soil freeze-thaw in non-permafrost regions and the active layer over permafrost. Significant changes in seasonally frozen ground have been observed worldwide.

4.7.3.1   Changes in the Active Layer

The active layer is that portion of the soil above permafrost that thaws and freezes seasonally. It plays an important role in cold regions because most ecological, hydrological, biogeochemical and pedogenic (soil-forming) activity takes place within it (Kane et al., 1991; Hinzman et al., 2003). Changes in active layer thickness are influenced by many factors, including surface

temperature, physical and thermal properties of the surface cover and substrate, vegetation, soil moisture and duration and thickness of snow cover (Brown et al., 2000; Frauenfeld et al., 2004; Zhang et al., 2005). The interannual and spatial variations in thaw depth at point locations can be large, an artefact of year-to-year and microtopographic variations in both surface

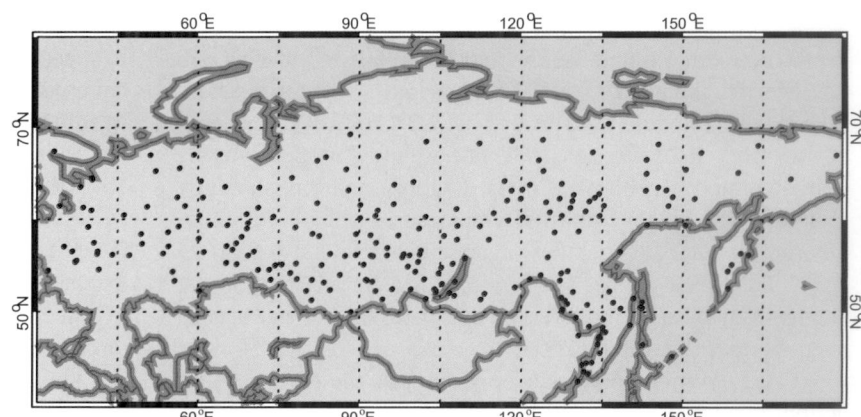

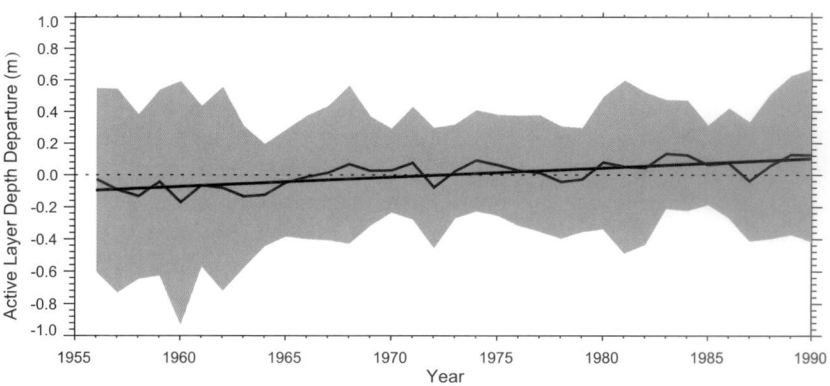

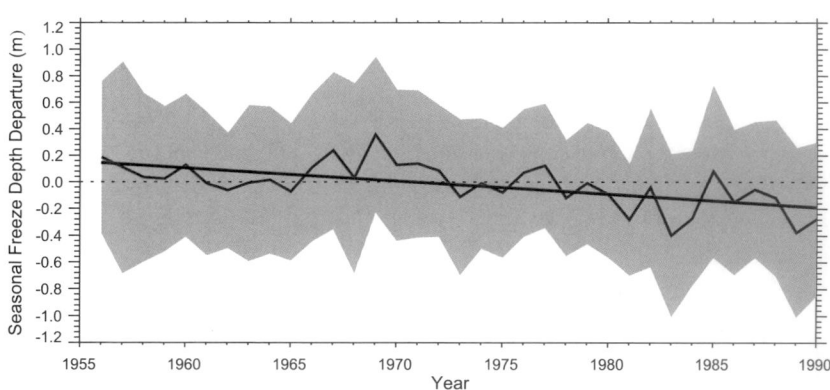

Figure 4.20. *Variations in the thickness of the active layer over permafrost (middle) and maximum soil freeze depth in non-permafrost areas (bottom) in Russia from 1956 through 1990. Active layer thickness has increased by about 20 cm while seasonal freeze depth has decreased by about 34 cm over the period of record (black lines in middle and lower panels). The anomaly in active layer thickness (blue line) is an average of anomalies from 31 stations (blue dots in the top panel) after removing the mean over the period of record for each station. The anomaly in maximum soil freeze depth (red line) is an average of anomalies from 211 stations (red dots in the top panel) after removing the mean over the period of record for each station. The shaded area represents the 5 to 95% confidence interval from the mean for each year, and the dashed line is the zero reference (from Frauenfeld et al., 2004).*

temperature and soil moisture, and so presents monitoring challenges. When the other conditions remain constant, changes in active layer thickness could be expected to increase in response to climate warming, especially in summer.

Long-term monitoring of the active layer has been conducted over the past several decades in Russia. By the early 1990s, there were about 25 stations, each containing 8 to 10 plots and 20 to 30 boreholes to a depth of 10 to 15 m (Pavlov, 1996). Measurements of soil temperature in the active layer and permafrost at depths up to 3.20 m have been carried out in Russia from 31 hydrometeorological stations, most of them started in the 1950s but a few as early as in the 1930s (Figure 4.20). Active layer thickness can be estimated using these daily soil temperature measurements. Over the period 1956 to 1990, the active layer exhibited a statistically significant deepening of about 21 cm. Increases in summer air temperature and winter snow depth are responsible for the increase in active layer thickness.

Monitoring of the active layer was developed at a global scale in the 1990s and currently incorporates more than 125 sites in the Arctic, the Antarctic and several mid-latitude mountain ranges (Brown et al., 2000; Nelson, 2004a,b; Figure 4.21). These sites were designed to observe the response of the active layer and near-surface permafrost to climate change. The results from northern high-latitude sites demonstrate substantial interannual and inter-decadal fluctuations in active layer thickness in response to air temperature variations. During the mid- to late 1990s in Alaska and north-western Canada, maximum and minimum thaw depths were observed in 1998 and in 2000, corresponding to the warmest and coolest summers, respectively. There is evidence of an increase in active layer thickness and thermokarst development, indicating degradation of warmer permafrost (Brown et al., 2000). Evidence from European monitoring sites indicates that active layer thickness has been the greatest in the summers of 2002 and 2003, approximately 20% greater than in previous years (Harris et al., 2003). Active layer thickness has increased by up to 1.0 m along the Qinghai-Xizang Highway over the Tibetan Plateau since the early 1980s (Zhao et al., 2004).

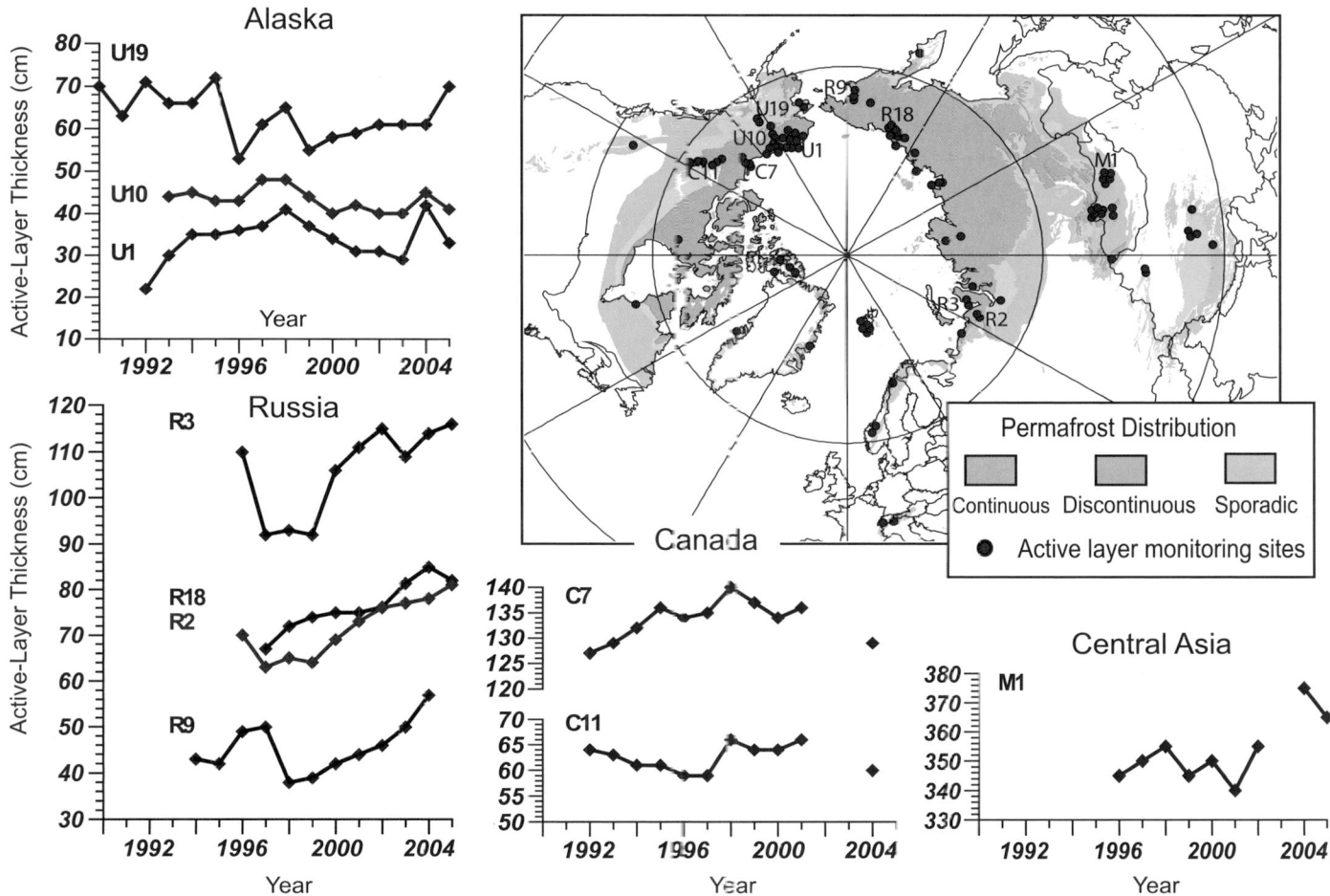

Figure 4.21. *Locations of sites and changes in active layer thickness from selected sites (after Nelson, 2004a,b).*

4.7.3.2   Seasonally Frozen Ground in Non-Permafrost Areas

The thickness of seasonally frozen ground has decreased by more than 0.34 m from 1956 through 1990 in Russia (Figure 4.20), primarily controlled by the increase in winter air temperature and snow depth (Frauenfeld et al., 2004). Over the Tibetan Plateau, the thickness of seasonally frozen ground has decreased by 0.05 to 0.22 m from 1967 through 1997 (Zhao et al., 2004). The driving force for the decrease in thickness of the seasonally frozen ground is the significant warming in cold seasons, while changes in snow depth play a minor role. The duration of seasonally frozen ground decreased by more than 20 days from 1967 through 1997 over the Tibetan Plateau, mainly due to the earlier onset of thaw in spring (Zhao et al., 2004).

The estimated maximum extent of seasonally frozen ground has decreased by about 7% in the NH from 1901 to 2002, with a decrease in spring of up to 15% (Figure 4.22; Zhang et al., 2003). There was little change in the areal extent of seasonally frozen ground during the early and midwinters.

4.7.3.3   Near-Surface Soil Freeze-Thaw Cycle

Satellite remote sensing data have been used to detect the near-surface soil freeze-thaw cycle at regional and hemispheric scales. Evidence from the satellite record indicates that the onset dates of thaw in spring and freeze in autumn advanced five to seven days in Eurasia over the period 1988 to 2002, leading to an earlier start to the growing season but no change in its length (Smith et al., 2004). In North America, a trend towards later freeze dates in autumn by about five days led, in part, to a lengthening of the growing season by eight days. Overall, the timing of seasonal thawing and subsequent initiation of the growing season in early spring has advanced by approximately eight days from 1988 to 2001 for the pan-arctic basin and Alaska (McDonald et al., 2004).

4.8   Synthesis

Observations show a consistent picture of surface warming and reduction in all components of the cryosphere (FAQ 4.1, Figure 1),[1] except antarctic sea ice, which exhibits a small positive but insignificant trend since 1978 (Figure 4.23).

Since IPCC (2001) the cryosphere has undergone significant changes, such as the substantial retreat of arctic sea ice, especially in summer; the continued shrinking of mountain glaciers; the decrease in the extent of snow cover and seasonally frozen ground, particularly in spring; the earlier breakup of river and lake ice; and widespread thinning of antarctic ice shelves along the Amundsen Sea coast, indicating increased basal

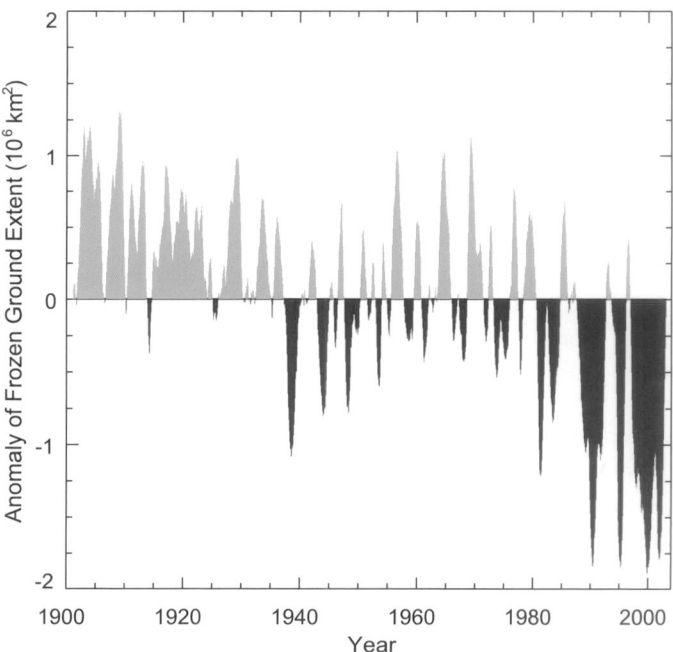

Figure 4.22. *Historical variations in the monthly areal extent (10^6 km^2) of seasonally frozen ground (including the active layer over permafrost) for the period from 1901 through 2002 in the NH. The positive anomaly (blue) represents above-average monthly extent, while the negative anomaly (red) represents below-average extent. The time series is smoothed with a low-pass filter (after Zhang et al., 2003).*

melting due to increased ocean heat fluxes in the cavities below the ice shelves. An additional new feature is the increasingly visible fast dynamic response of ice shelves, for example, the dramatic breakup of the Larsen B Ice Shelf in 2002, and the acceleration of tributary glaciers and ice streams, with possible consequences for the adjacent part of the ice sheets.

One difficulty with using cryospheric quantities as indicators of climate change is the sparse historical database. Although 'extent' of ice (sea ice and glacier margins for example) has been observed for a long time at a few locations, the 'amount' of ice (thickness or depth) is difficult to measure. Therefore, reconstructions of past mass balance are often not possible.

Table 4.6. *Estimates of cryospheric contributions to sea level change.*

Cryospheric component	Sea Level Equivalent (mm yr^-1)	
	1961–2003	1993–2003
Glaciers and Ice Caps	+0.32 to +0.68	+0.55 to +0.99
Greenland	–0.07 to +0.17	+0.14 to +0.28
Antarctica	–0.28 to +0.55	–0.14 to +0.55
Total (adding ranges)	–0.03 to +1.40	+0.55 to +1.82
Total (Gaussian error summation)	+0.22 to +1.15	+0.77 to +1.60

---

[1] Surface air temperature data are updated from Jones and Moberg, 2003; sea ice data are updated from Comiso, 2003; frozen ground data are from Zhang et al., 2003; snow cover data are updated from Brown et al., 2000; glacier mass balance data are from Ohmura, 2004; Cogley, 2005; and Dyurgerov and Meier, 2005.

The most important cryospheric contributions to sea level variations (see Chapter 5) arise from changes in the ice on land (e.g., glaciers, ice caps and ice sheets). In IPCC (2001), the contribution of glaciers and ice caps to sea level rise during the 20th century was estimated as 0.2 to 0.4 mm yr^{-1} (of 1 to 2 mm yr^{-1} total sea level rise). New results presented here indicate that all glaciers contributed about 0.50 ± 0.18 mm yr^{-1} during 1961 to 2003, increasing to 0.77 ± 0.22 mm yr^{-1} from 1993 to 2003 (interpolation from five-year analyses in Table 4.4). Estimates for both ice sheets combined give a contribution ranging from -0.35 to $+0.72$ mm yr^{-1} for 1961 to 2003, increasing to 0 to 0.8 mm yr^{-1} for 1993 to 2003. A conservative error estimate in terms of summing ranges is given in Table 4.6. Assuming a midpoint mean, interpreting the range as uncertainty

and using Gaussian error summation of estimates for glaciers and both ice sheets suggests that the total ice contribution to sea level rise was approximately 0.7 ± 0.5 mm yr^{-1} during 1961 to 2003 and 1.2 ± 0.4 mm yr^{-1} during 1993 to 2003.

The large uncertainties reflect the difficulties in estimating the global ice mass and its variability, because global monitoring of ice thickness is impossible (even the total area of glaciers is not exactly known) and extrapolation from local measurements is therefore necessary. A regional extension of the monitored ice masses and an improvement of measurement and extrapolation techniques are urgently required.

In spite of the large uncertainties, the data that are available portray a rather consistent picture of a cryosphere in decline over the 20th century, increasingly so during 1993 to 2003.

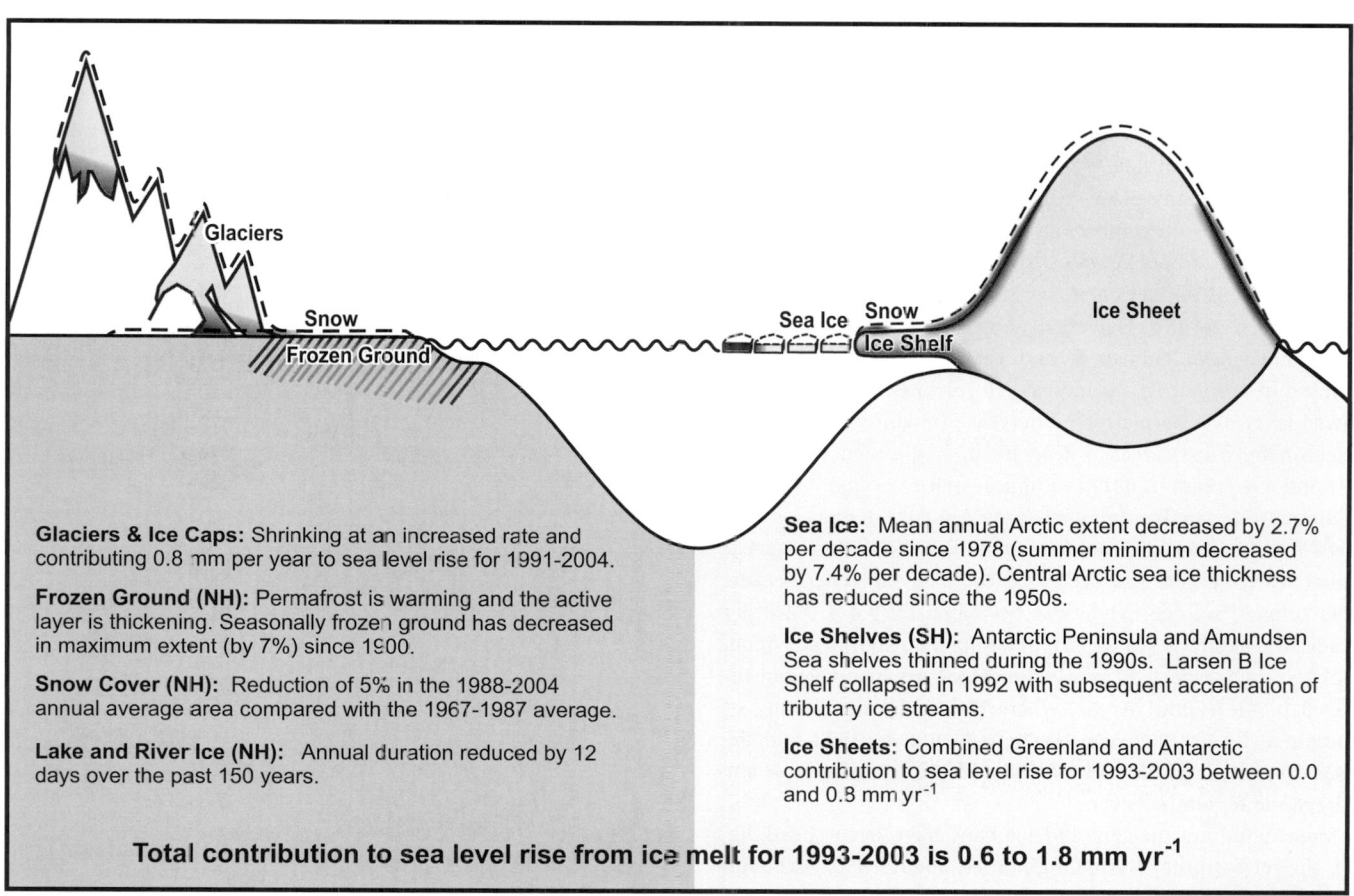

Figure 4.23. *Summary of observed variations in the cryosphere.*

Frequently Asked Question 4.1
Is the Amount of Snow and Ice on the Earth Decreasing?

Yes. Observations show a global-scale decline of snow and ice over many years, especially since 1980 and increasing during the past decade, despite growth in some places and little change in others (Figure 1). Most mountain glaciers are getting smaller. Snow cover is retreating earlier in the spring. Sea ice in the Arctic is shrinking in all seasons, most dramatically in summer. Reductions are reported in permafrost, seasonally frozen ground and river and lake ice. Important coastal regions of the ice sheets on Greenland and West Antarctica, and the glaciers of the Antarctic Peninsula, are thinning and contributing to sea level rise. The total contribution of glacier, ice cap and ice sheet melt to sea level rise is estimated as 1.2 ± 0.4 mm yr^{-1} for the period 1993 to 2003.

Continuous satellite measurements capture most of the Earth's seasonal snow cover on land, and reveal that Northern Hemisphere spring snow cover has declined by about 2% per decade since 1966, although there is little change in autumn or early winter. In many places, the spring decrease has occurred despite increases in precipitation.

Satellite data do not yet allow similarly reliable measurement of ice conditions on lakes and rivers, or in seasonally or permanently frozen ground. However, numerous local and regional reports have been published, and generally seem to indicate warming of permafrost, an increase in thickness of the summer thawed layer over permafrost, a decrease in winter freeze depth in seasonally frozen areas, a decrease in areal extent of permafrost and a decrease in duration of seasonal river and lake ice.

Since 1978, satellite data have provided continuous coverage of sea ice extent in both polar regions. For the Arctic, average annual sea ice extent has decreased by $2.7 \pm 0.6\%$ per decade, while summer sea ice extent has decreased by $7.4 \pm 2.4\%$ per decade. The antarctic sea ice extent exhibits no significant trend. Thickness data, especially from submarines, are available but restricted to the central Arctic, where they indicate thinning of approximately 40% between the period 1958 to 1977 and the 1990s. This is likely an overestimate of the thinning over the entire arctic region however.

Most mountain glaciers and ice caps have been shrinking, with the retreat probably having started about 1850. Although many Northern Hemisphere glaciers had a few years of near-balance around 1970, this was followed by increased shrinkage. Melting of glaciers and ice caps contributed 0.77 ± 0.22 mm yr^{-1} to sea level rise between 1991 and 2004

Taken together, the ice sheets of Greenland and Antarctica are very likely shrinking, with Greenland contributing about 0.2 ± 0.1 mm yr^{-1} and Antarctica contributing 0.2 ± 0.35 mm yr^{-1} to sea level rise over the period 1993 to 2003. There is evidence of accelerated loss through 2005. Thickening of high-altitude, cold regions of Greenland and East Antarctica, perhaps from increased snowfall, has been more than offset by thinning in

coastal regions of Greenland and West Antarctica in response to increased ice outflow and increased Greenland surface melting.

Ice interacts with the surrounding climate in complex ways, so the causes of specific changes are not always clear. Nonetheless, it is an unavoidable fact that ice melts when the local temperature is

(continued)

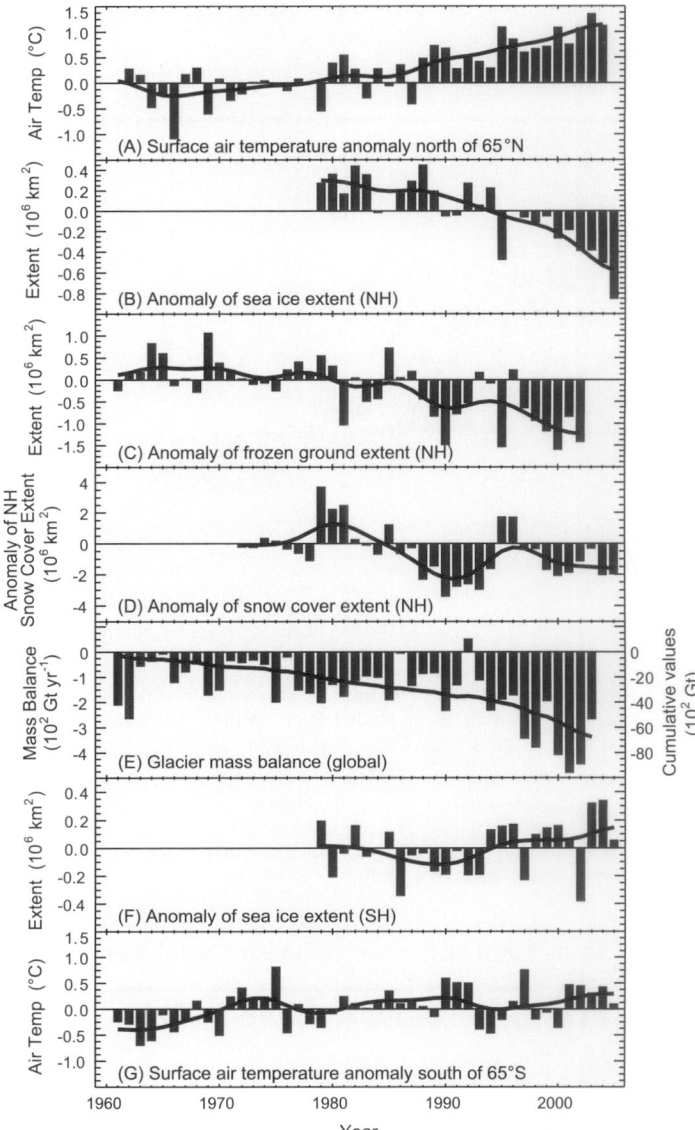

FAQ 4.1, Figure 1. *Anomaly time series (departure from the long-term mean) of polar surface air temperature (A, G), arctic and antarctic sea ice extent (B, F), Northern Hemisphere (NH) frozen ground extent (C), NH snow cover extent (D) and global glacier mass balance (E). The solid red line in E denotes the cumulative global glacier mass balance; in the other panels it shows decadal variations (see Appendix 3.A).*

above the freezing point. Reductions in snow cover and in mountain glaciers have occurred despite increased snowfall in many cases, implicating increased air temperatures. Similarly, although snow cover changes affect frozen ground and lake and river ice, this does not seem sufficient to explain the observed changes, suggesting that increased local air temperatures have been important. Observed arctic sea ice reductions can be simulated fairly well in models driven by historical circulation and temperature changes. The observed increases in snowfall on ice sheets in some cold central regions, surface melting in coastal regions and sub-ice-shelf melting along many coasts are all consistent with warming. The geographically widespread nature of these snow and ice changes suggests that widespread warming is the cause of the Earth's overall loss of ice.

References

Abdalati, W., et al., 2001: Outlet glacier and margin elevation changes: Near-coastal thinning of the Greenland ice sheet. *J. Geophys. Res.*, **106**(D24), 33729–33741.

Abdalati, W., et al., 2004: Elevation changes of ice caps in the Canadian Arctic Archipelago. *J. Geophys. Res.*, **109**, F04007, doi:10.1029/2003JF000045.

Ackley, S., P. Wadhams, J.C. Comiso, and A.P. Worby, 2003: Decadal decrease of Antarctic sea ice extent inferred from whaling records revisited on the basis of historical and modern sea ice records. *Polar Res.*, **22**(1), 19–25.

Agnew, T. and S. Howell, 2003: The use of operational ice charts for evaluating passive microwave ice concentration data. *Atmos.-Ocean*, **41**(4), 317–331.

Agnew, T., H. Le, and T. Hirose, 1997: Estimation of large-scale sea-ice motion from SSM/I 85.5 GHz imagery. *Ann. Glaciol.*, **25**, 305–311.

Allard, M., B. Wang, and J.A. Pilon, 1995: Recent cooling along the southern shore of Hudson Strait Quebec, Canada, documented from permafrost temperature measurements. *Arctic and Alpine Res.*, **27**, 157–166.

Alley, R.B., P.U. Clark, P. Huybrechts, and I. Joughin, 2005a: Ice-sheet and sea-level changes. *Science*, **310**, 456–460.

Alley, R.B., T.K. Dupont, B.R. Parizek, and S. Anandakrishnan, 2005b: Access of surface meltwater to beds of sub-freezing glaciers: preliminary insights. *Ann. Glaciol.*, **40**, 8–14.

Arendt, A.A., et al., 2002: Rapid wastage of Alaska glaciers and their contribution to rising sea level. *Science*, **297**, 382–386.

Armstrong, R.L. and M.J. Brodzik, 2001: Recent Northern Hemisphere snow extent: A comparison of data derived from visible and microwave satellite sensors. *Geophys. Res. Lett.*, **28**, 3673–3676.

Bamber, J.L., R.L. Layberry, and S.P. Gogineni, 2001: A new ice thickness and bed data set for the Greenland ice sheet, 1. Measurement, data reduction, and errors. *J. Geophys. Res.*, **106**, 33733–33780.

Bamber, J.L., W. Krabill, V. Raper, and J. Dowdeswell, 2004: Anomalous recent growth of part of a large Arctic ice cap: Austfonna, Svalbard. *Geophys. Res. Lett.*, **31**(12), L12402, doi: 10.1029/2004GL019667.

Bamzai, A.S., 2003: Relationship between snow cover variability and Arctic Oscillation Index on a hierarchy of time scales. *Int. J. Climatol.*, **23**, 131–142.

Belchansky, G. I., D.C. Douglas, V.A. Eremeev, and N.G. Platonov, 2005: Variations in the Arctic's multiyear sea ice cover: A neural network analysis of SMMR-SSM/I data, 1979-2004. *Geophys. Res. Lett.*, **32**, L09605, doi:10.1029/2005GL022395.

Box, J.E., et al., 2006: Greenland ice-sheet surface mass balance variability (1988-2004) from calibrated Polar MM5 output. *J. Clim.*, **19**(12), 2783–2800.

Broll, G., C. Tarnocai, and J. Gould, 2003: Long-term high Arctic ecosystem monitoring in Quttinirpaaq National Park, Ellesmere Island, Canada. In: *Proceedings of the 8th International Conference on Permafrost, 21-25 July 2003, Zurich, Switzerland* [Phillips, M., S.M. Springman, and L.U. Arenson (eds.)]. A.A. Balkema, Lisse, the Netherlands, pp. 89–94.

Brown, J., K.M. Hinkel, and F.E. Nelson, 2000: The Circumpolar Active Layer Monitoring (CALM) program: research design and initial results. *Polar Geogr.*, **24**(3), 166–258.

Brown, R.D., 2000: Northern hemisphere snow cover variability and change, 1915-97. *J. Clim.*, **13**, 2339–2355.

Brown, R.D., and P. Coté, 1992: Interannual variability of landfast ice thickness in the Canadian high Arctic, 1950-1989. *Arctic*, **45**, 273–284.

Carrasco, J.F., G. Casassa, and J. Quintana, 2005: Changes of the 0°C isotherm and the equilibrium line altitude in central Chile during the last quarter of the XXth century. *Hydrolog. Sci. J.*, **50**, 933–948.

Cavalieri, D.J., 1992: The validation of geophysical products using multisensor data. In: *Microwave Remote Sensing of Sea Ice* [Carsey, F.D. (ed.)]. Geophysical Monograph 68, American Geophysical Union, Washington, DC, pp 233–242.

Chinn, T.J.H., S. Winkler, M.J. Salinger, and N. Haakensen, 2005: Recent glacier advances in Norway and New Zealand: a comparison of their glaciological and meteorological causes. *Geografiska Annaler*, **87A**(1), 141–157.

Clare, G.R., B.B. Fizharris, T.J.H. Chinn, and M.J. Salinger, 2002: Interannual variation in end-of-summer snowlines of the Southern Alps of New Zealand, and relationships with Southern Hemisphere atmospheric circulation and sea surface temperature patterns. *Int. J. Climatol.*, **22**, 107–120.

Clark, M.P., M.C. Serreze, and D.A. Robinson, 1999: Atmospheric controls on Eurasian snow extent. *Int. J. Climatol.*, **19**, 27–40.

Clark, P.U., R.B. Alley, and D. Pollard, 1999: Northern hemisphere ice-sheet influences on global climate change. *Science*, **286**, 1103–1111.

Cogley, J.G., 2005: Mass and energy balances of glaciers and ice sheets. In: *Encyclopedia of Hydrological Sciences* [Anderson, M. (ed.)]. John Wiley & Sons, Ltd, Chichester, pp 2555–2573.

Comiso, J.C., 2002: A rapidly declining perennial sea ice cover in the Arctic. *Geophys. Res. Lett.*, **29**, 1956–1959.

Comiso, J.C., 2003: Large scale characteristics and variability of the global sea ice cover. In: *Sea Ice - An Introduction to its Physics, Biology, Chemistry, and Geology* [Thomas, D. and G.S. Dieckmann (eds.)]. Blackwell Science, Oxford, UK, pp. 112–142.

Cook, A., A. Fox, D. Vaughan, and J. Ferigno, 2005: Retreating glacier fronts on the Antarctic Peninsula over the past half century. *Science*, **308**, 541–544.

Couture, R., et al., 2003: On the hazards to infrastructure in the Canadian North associated with thawing of permafrost. In: *Proceedings of Geohazards 2003, 3rd Canadian Conference on Geotechnique and Natural Hazards*. Canadian Geotechnical Society, Alliston, Ontario, pp. 97–104.

Cuffey, K.M., 2001: Interannual variability of elevation on the Greenland ice sheet: effects of firn densification, and establishment of a benchmark. *Journal of Glaciol.*, **47**(158), 369–377.

Cullen, N.J., et al., 2006: Kilimanjaro glaciers: Recent areal extent from satellite data and new interpretation of observed 20th century retreat rates. *Geophys. Res. Lett.*, **33**, L16502, doi:10.1029/2006GL027084.

Curran, M.A.J., et al., 2003: Ice core evidence for Antarctic sea ice decline since the 1950s. *Science*, **302**(14), 1203–1206.

Davis, C.H., et al., 2005: Snowfall-driven growth in East Antarctic ice sheet mitigates recent sea-level rise. *Science*, **308**, 1898–1901, doi:10.1126/science.1110662.

de la Mare, W.K., 1997: Abrupt mid-20th century decline in Antarctic sea ice extent from whaling records. *Nature*, **389**, 57–61.

Derksen, C., A. Walker, E. LeDrew, and B. Goodison, 2003: Combining SMMR and SSM/I data for time series analysis of central North American snow water equivalent. *J. Hydrometeorol.*, **4**(2), 304–316.

DesJarlais, C., 2004: *S'adapter aux Changements Climatiques*. Ouranos, Montreal, 91pp.

Doran, P.T., et al., 2002: Antarctic climate cooling and terrestrial ecosystem response. *Nature*, **415**, 517–520.

Duguay, C.R., et al., 2003: Ice-cover variability on shallow lakes at high latitudes: model simulations and observations. *Hydrolog. Process.*, **17**, 3465–3483.

Duguay, C.R., et al., 2006: Recent trends in Canadian lake ice cover. *Hydrological Processes*, **20**, 781–801.

Dumas, J.A., G.M. Flato, and A.J. Weaver, 2003: The impact of varying atmospheric forcing on the thickness of arctic multi-year sea ice. *Geophys. Res. Lett.*, **30**(18), 1918–1921.

Dupont, T.K., and R.B. Alley, 2005: Assessment of the importance of ice-shelf buttressing to ice-sheet flow. *Geophys. Res. Lett.*, **32**, L04503, doi:10.1029/2004GL022024.

Dupont, T.K., and R.B. Alley, 2006: The importance of small ice shelves in sea-level rise. *Geophys. Res. Lett.*, **33**, L09503, doi:10.1029/2005GL025665.

Dye, D.G., 2002: Variability and trends in the annual snow-cover cycle in Northern Hemisphere land areas, 1972–2000. *Hydrolog. Process.*, **16**, 3065–3077.

Dyurgerov, M., and J. Dwyer, 2000: The steepening of glacier mass balance gradients with Northern Hemisphere warming. *Zeitschrift für Gletscherkunde und Glazialgeologie*, **36**, 107–118.

Dyurgerov, M., and M.F. Meier, 2005: *Glaciers and the Changing Earth System: A 2004 Snapshot*. Occasional Paper 58, Institute of Arctic and Alpine Research, University of Colorado, Boulder, CO, 118 pp.

Falarz, M., 2002: Long-term variability in reconstructed and observed snow cover over the last 100 winter seasons in Cracow and Zakopane (southern Poland). *Clim. Res.*, **19**(3), 247–256.

Favier, V., P. Wagnon, and P. Ribstein, 2004: Glaciers of the outer and inner tropics: A different behaviour but a common response to climatic forcing. *Geophys. Res. Lett.*, **31**, L16403, doi:10.1029/2004GL020654.

Fedorov, A., and P. Konstantinov, 2003: Observations of surface dynamics with thermokarst initiation, Yukechi site, central Yakutia. In: *Proceedings of the 8th International Conference on Permafrost, 21-25 July 2003, Zurich, Switzerland* [Phillips, M., S.M. Springman, and L.U. Arenson (eds.)]. A.A. Balkema, Lisse, the Netherlands, pp. 239–243.

Fichefet, T., B. Tartinville, and H. Goosse, 2003: Antarctic sea ice variability during 1958-1999: A simulation with a global ice-ocean model. *J. Geophys. Res.*, **108**(C3), 3102–3113.

Fitzharris, B.B., and C.E. Garr, 1995: Simulation of past variability in seasonal snow in the Southern Alps, New Zealand. *Ann. Glaciol.*, **21**, 377–382.

Fountain, A.G., T.A. Neumann, P.L. Glenn, and T. Chinn, 2004: Can warming induce advances of polar glaciers, Taylor Valley, Antarctica. *J. Glaciol.*, **50**(171), 556–564.

Francou, B., et al., 2003: Tropical climate change recorded by a glacier in the central Andes during the last decades of the twentieth century: Chacaltaya, Bolivia, 16°S. *J. Geophys. Res*, **108**(D5), 4154, doi:10.1029/2002JD002959.

Francou, B., M. Vuille, V. Favier, and B. Cáceres, 2004: New evidence for an ENSO impact on low-latitude glaciers: Antizana 15, Andes of Ecuador, 0°28'S. *J. Geophys. Res.*, **109**, D18106, doi:10.1029/2003JD004484.

Frauenfeld, O.W., T. Zhang, R.G. Barry, and D. Gilichinsky, 2004: Interdecadal changes in seasonal freeze and thaw depths in Russia. *J. Geophys. Res.*, **109**, D5101, doi:10.1029/2003JD004245.

Fujita, K., and Y. Ageta, 2000: Effect of summer accumulation on glacier mass balance on the Tibetan Plateau revealed by mass-balance model. *J. Glaciol.*, **46**(153), 244–252.

Gavrilov, P.P. and P.V. Efremov, 2003: Effects of cryogenic processes on Yakutia landscapes under climate warming. In: *Proceedings of the 8th International Conference on Permafrost, 21-25 July 2003, Zurich, Switzerland* [Phillips, M., S.M. Springman, and L.U. Arenson (eds.)]. A.A. Balkema, Lisse, the Netherlands, pp. 277–282.

Georges, C., 2004: The 20th century glacier fluctuations in the Cordillera Blanca (Perú). *Arctic, Antarctic, and Alpine Res.*, **36**(1), 100–107.

Greene, A.M., 2005: A time constant for hemispheric glacier mass balance. *J. Glaciol.*, **51**(174), 353–362.

Groisman, P.Ya, T.R. Karl, and R.W. Knight, 1994: Observed impact of snow cover on the heat balance and the rise of continental spring temperatures. *Science*, **263**, 198–200.

Groisman, P.Ya., et al., 2004: Contemporary changes of the hydrological cycle over the contiguous United States: Trends derived from in situ observations. *J. Hydrometeorol.*, **5**, 64–85.

Gudkovich, Z.M., 1961: Relation of the ice drift in the Arctic Basin to ice conditions in the Soviet Arctic seas. *Tr. Okeanogr. Kom. Akad. Nauk SSSR*, **11**, 14–21 (in Russian).

Haas, C., 2004: Late-summer sea ice thickness variability in the Arctic Transpolar Drift 1991–2001 derived from ground-based electromagnetic sounding. *Geophys. Res. Lett.*, **31**, L09402, doi:10.1029/2003GL019394.

Haeberli, W., and C.R. Burn, 2002: Natural hazards in forests: glacier and permafrost effects as related to climate change. In: *Environmental Change and Geomorphic Hazards in Forests* [Sidle, R.C. (ed.)]. IUFRO Research Series 9, CABI Publishing, Wallingford and New York, pp. 167–202.

Hagen, J.O., K. Melvold, F. Pinglot, and J.A. Dowdeswell, 2003: On the net mass balance of the glaciers and ice caps in Svalbard, Norwegian Arctic. *Arctic, Antarctic, and Alpine Res.*, **35**(2), 264–270.

Halsey, L.A., D.H. Vitt, and S.C. Zoltai, 1995: Disequilibrium response of permafrost in boreal continental western Canada to climate change. *Clim. Change*, **30**, 57–73.

Hanna, E., et al., 2005: Runoff and mass balance of the Greenland ice sheet: 1958-2003. *J. Geophys. Res.*, **110**, D13108, doi:10.1029/2004JD005641.

Hanna, E., et al., 2006: Observed and modeled Greenland Ice Sheet snow accumulation, 1958–2003, and links with regional climate forcing. *J. Clim.*, **19**(3), 344–358.

Hansen, J., and L. Nazarenko, 2004: Soot climate forcing via snow and ice albedos. *Proc. Natl. Acad. Sci. U.S.A.*, **101**(2), 423–428.

Harris, C., et al., 2003: Warming permafrost in European mountains. *Global Planet. Change*, **39**, 215–225.

Hastenrath, S., 2005: The glaciers of Mount Kenya 1899-2004. *Erdkunde*, **59**, 120–125.

Heil, P., and I. Allison, 2002: Long-term fast-ice variability off Davis and Mawson stations, Antarctica. In: *Ice in the Environment: Proceedings of the 16th IAHR International Symposium on Ice (Volume 1), Dunedin, New Zealand, 2-6 December 2002* [Squire, V., and P. Langhorne (eds)]. University of Otago, Dunedin, NZ, pp. 360–367.

Hellmer, H., S. Jacobs, and A. Jenkins, 1998: Oceanic erosion of a floating Antarctic glacier in the Amundsen Sea. In: *Ocean, Ice, and Atmosphere: Interactions at the Antarctic Continental Margin* [Jacobs, S. and R. Weiss (eds.)]. Antarctic Research Series 75, American Geophysical Union, Washington, DC, pp. 83–99.

Hennessy, K.J., et al., 2003: *The Impact of Climate Change on Snow Conditions in Mainland Australia*. CSIRO Atmospheric Research, Aspendale, Australia, 47 pp, http://www.cmar.csiro.au/e-print/open/hennessy_2003a.pdf.

Hewitt, K., 2005: The Karakoram anomaly? Glacier expansion and the "elevation effect", Karakoram Himalaya. *Mountain Research and Development*. **25**(4), 332–340.

Hill, B.T., A. Ruffman, and K. Drinkwater. 2002: Historical record of the incidence of sea ice on the Scotian Shelf and the Gulf of St. Lawrence. In: *Ice in the Environment: Proceedings of the 16th IAHR International Symposium on Ice (Volume 1). Dunedin, New Zealand, 2-6 December 2002* [Squire, V., and P. Langhorne (eds)]. University of Otago, Dunedin, NZ.

Hilmer, M., and T. Jung, 2000: Evidence for a recent change in the link between the North Atlantic Oscillation and Arctic sea ice export. *Geophys. Res. Lett.*, **27**, 989–992.

Hinzman, L.D., et al., 2003: Hydrological variations among watersheds with varying degrees of permafrost. *Proceedings of the 8th International Conference on Permafrost, 21-25 July 2003, Zurich, Switzerland* [Phillips, M., S.M. Springman, and L.U. Arenson (eds.)]. A.A. Balkema, Lisse, the Netherlands, pp. 407–411.

Hinzman, L.D., et al., 2005: Evidence and implications of recent climate change in northern Alaska and other arctic regions. *Clim. Change*, **72**(3), 251–298.

Holloway, G., and T. Sou, 2002: Has Arctic sea ice rapidly thinned? *J. Clim.*, **15**, 1691–1701.

Howat, I.M., I. Joughin, S. Tulaczyk, and S. Gogineni, 2005: Rapid retreat and acceleration of Helheim Glacier, east Greenland. *Geophys. Res. Lett.*, **32**, L22502, doi: 10.1029/2005GL024737.

Huybrechts, P., 2002: Sea-level changes at the LGM from ice-dynamic reconstructions of the Greenland and Antarctic ice sheets during the glacial cycles. *Quat. Sci. Rev.*, **21**(1–3), 203–231.

Huybrechts, P., J. Gregory, I. Janssens, and M. Wild, 2004: Modelling Antarctic and Greenland volume changes during the 20th and 21st centuries forced by GCM time slice integrations. *Global Planet. Change*, **42**(1–4), 83–105.

Hyvärinen, V., 2003: Trends and characteristics of hydrological time series in Finland. *Nord. Hydrol.*, **34**(1–2), 71–90.

IPCC, 2001: *Climate Change 2001: The Scientific Basis. Contribution of Working Group I to the Third Assessment Report of the Intergovernmental Panel on Climate Change* [Houghton, J.T., et al. (eds)]. Cambridge University Press, Cambridge, United Kingdom and New York, NY, USA, 881 pp.

Isaksen, K., P. Holmlund, J.L. Sollid, and C. Harris, 2001: Three deep alpine-permafrost boreholes in Svalbard and Scandinavia. *Permafrost and Periglacial Processes*, **12**, 13–25.

Jacka, T.H., et al., 2004: Recommendations for the collection and synthesis of Antarctic Ice Sheet mass balance data. *Global Planet. Change*, **42**(1–4), 1–15.

Jacobs, S.S., C.F. Giulivi, and P.A. Mele, 2002: Freshening of the Ross Sea during the late 20th century. *Science*, **297**(5580), 386–389.

Johannessen, O.M, E.V. Shalina, and M.W. Miles, 1999: Satellite evidence for an Arctic sea ice cover in transformation. *Science*, 286, 1937–1939.

Johannessen, O.M., K. Khvorostovsky, M.W. Miles, and L.P. Bobylev, 2005: Recent ice-sheet growth in the interior of Greenland. *Science*, **310**, 1013–1016.

Johannessen, O.M., et al., 2004: Arctic climate change – observed and modelled temperature and sea ice. *Tellus*, **56A**, 328–341.

Jones, P.D., and A. Moberg, 2003: Hemispheric and large-scale surface air temperature variations: an extensive revision and an update to 2001. *J Climate*, **16**, 206–223.

Jorgenson, M.T., and J. Brown, 2005: Classification of the Alaskan Beaufort Sea Coast and estimation of carbon and sediment inputs from coastal erosion. *Geo-Marine Lett.*, **25**, 69–80.

Joughin, I., and S. Tulaczyk, 2002: Positive mass balance of the Ross Ice Streams, West Antarctica. *Science*, **295**(5554), 476–480.

Joughin, I., and L. Padman, 2003: Melting and freezing beneath Filchner-Ronne Ice Shelf, Antarctica. *Geophys. Res. Lett.*, **30**(9), 1477, doi:10.1029/2003GL016941.

Joughin, I., W. Abdalati, and M. Fahnestock, 2004: Large fluctuations in speed on Greenland's Jakobshavn Isbræ glacier. *Nature*, **432**, 608–610.

Joughin, I., et al., 2003: Timing of recent accelerations of Pine Island Glacier, Antarctica. *Geophys. Res. Lett.*, **30**(13), 1706, doi:10.1029/2003GL017609.

Kane, D.L., L.D. Hinzman, and J.P. Zarling, 1991: Thermal response of the active layer to climate warming in a permafrost environment. *Cold Regions Sci. Technol.*, **19**, 111–122.

Kaser, G., and H. Osmaston, 2002: *Tropical Glaciers*. UNESCO International Hydrological Series. Cambridge University Press, Cambridge, UK, 207pp.

Kaser, G., et al., 2003: The impact of glaciers on the runoff and the reconstruction of mass balance history from hydrological data in the tropical Cordillera Blanca, Peru. *J. Hydrol.*, **282**, 130–144.

Kjøllmoen, B.E., 2005: *Glaciological Investigations in Norway in 2004*. Norwegian Water Resources and Energy Directorate, Oslo, http://www.nve.no/FileArchive/176/Glac_invest2004.pdf.

Klein, A.G., and J.L. Kincaid, 2006: Retreat of glaciers on Puncak Jaya, Irian Jaya, determined from 2000 and 2002 IKONOS satellite images. *J. Glaciol.*, **52**(176), 65–79.

Koeberle, C., and R. Gerdes, 2003: Mechanisms determining the variability of Arctic sea ice conditions and export. *J. Clim.*, **16**, 2843–2858.

Krabill, W.B., et al., 2000: Greenland Ice Sheet: High elevation balance and peripheral thinning. *Science*, **289**, 428–430.

Krabill, W.B., et al., 2002: Aircraft laser altimetry measurement of elevation changes of the Greenland ice sheet: Technique and accuracy assessment. *J. Geodyn.*, **34**, 357–376.

Krabill, W.B., et al., 2004: Greenland Ice Sheet: Increased coastal thinning. *Geophys. Res. Lett.*, **31**, L24402, doi:10.1029/2004GL021533.

Kwok, R., 2000: Recent changes of the Arctic Ocean sea ice motion associated with the North Atlantic Oscillation. *Geophys. Res. Lett*, **27**(6), 775–778.

Kwok, R., 2002: Sea ice concentration from passive microwave radiometry and openings from SAR ice motion. *Geophys. Res. Lett*, **29**(10), doi:10.1029/2002GL014787.

Kwok, R., and D.A. Rothrock, 1999: Variability of Fram Strait ice flux and North Atlantic Oscillation. *J. Geophys. Res.*, **104**, 5177–5189.

Kwok, R., G.F. Cunningham, and S.S. Pang, 2004: Fram Strait sea ice outflow. *J. Geophys. Res.*, **109**, C01009, doi:10.1029/2003JC001785.

Kwok, R., et al., 1998: Assessment of sea ice motion from sequential passive microwave observations with ERS and buoy ice motions. *J. Geophys. Res.*, **103**(C4), 8191–8213.

Lachenbruch, A.H., and B.V. Marshall, 1986: Changing climate: geothermal evidence from permafrost in the Alaskan Arctic. *Science*, **234**, 689–696.

Laxon, S., N. Peacock, and D. Smith, 2003: High interannual variability of sea ice thickness in the Arctic region. *Nature*, **425**, 947–950.

Legresy, B., F. Remy, and F. Blarel, 2006: Along track repeat altimetry for ice sheets and continental surface studies. In: *Proceedings of the Symposium on 15 years of Progress in Radar Altimetry, Venice, Italy, 13-18 March 2006* [Danesy, D. (ed.)]. ESA-SP614, Paper 181, European Space Agency Publications Division, Noordwijk, The Netherlands, 4 pp., http://earth.esa.int/workshops/venice06/participants/181/paper_181_legrsy.pdf.

Lewkowicz, A.G., 1992: Factors influencing the distribution and initiation of active-layer detachment slides on Ellesmere Island, arctic Canada. In: *Periglacial Geomorphology* [Dixon, J.C. and A.D. Abrahams (eds.)]. Wiley, New York, pp. 223–250.

Liu, S., et al., 2004: Recent progress in glaciological studies in China. *J. Geogr. Sci.*, **14**(4), 401–410.

Lo, F., and M.P. Clark, 2001: Relationships between spring snow mass and summer precipitation in the southwestern US associated with the North American monsoon system. *J. Clim.*, **15**, 1378–1385.

Lythe, M.B., D.G. Vaughan, and the BEDMAP Group, 2001: BEDMAP: A new ice thickness and subglacial topographic model of Antarctica. *J. Geophys. Res.*, **106**(B6), 11335–11351.

Magnuson, J.J., et al., 2000: Historical trends in lake and river ice cover in the Northern Hemisphere. *Science*, **289**, 1743–1746.

Massom, R.A., et al., 2001: Snow on Antarctic sea ice: a review of physical characteristics. *Rev. Geophys.*, **39**(3), 413–445

McDonald, K.C., et al., 2004: Variability in springtime thaw in the terrestrial high latitudes: monitoring a major control on the biospheric assimilation of atmospheric CO_2 with spaceborne microwave remote sensing. *Earth Interactions*, **8**, Paper No. 20.

McLaren, A.S., R.H. Bourke, J.E. Walsh, and R.L. Weaver, 1994: Variability in sea-ice thickness over the North Pole from 1958 to 1992. In: *The Polar Oceans and Their Role in Shaping the Global Environment* [Johannessen, O.M., R.D. Muench, and J.E. Overland (eds.)]. American Geophysical Union, Washington, DC, pp. 363–371.

Mitrovica, J.X., et al., 2006: Reanalysis of ancient eclipse, astronomic and geodetic data: A possible route to resolving the enigma of global sea level rise. *Earth Planet. Sci. Lett.*, **243**, 390–399.

Mölg, T., and D.R. Hardy, 2004: Ablation and associated energy balance of a horizontal glacier surface on Kilimanjaro. *J. Geophys. Res.*, **109**, 1–13, D16104, doi:10.1029/2003JD004338.

Mölg, T., C. Georges, and G. Kaser, 2003a: The contribution of increased incoming shortwave radiation to the retreat of the Rwenzori Glaciers, East Africa, during the 20th century. *Int. J. Climatol.*, **23**, 291–303, doi:10.1002/joc.877.

Mölg, T., D.R. Hardy, and G. Kaser, 2003b: Solar-radiation-maintained glacier recession on Kilimanjaro drawn from combined ice-radiation geometry modeling. *J. Geophys. Res.*, **108**(D23), 4731, doi:10.1029/2003JD003546.

Monaghan, A.J., et al., 2006: Insignificant change in Antarctic snowfall since the International Geophysical Year. *Science*, **313**(5788), 827–831.

Morris, E.M., and D. G. Vaughan, 2003: Glaciological climate relationships spatial and temporal variation of surface temperature on the Antarctic Peninsula and the limit of viability of ice shelves. In: *Antarctic Peninsula Climate Variability: Historical and Paleoenvironmental Perspectives* [Domack, E., et al. (eds.)]. Antarctic Research Series 79, American Geophysical Union, Washington, DC, pp. 61–68.

Mote, P.W., 2006: Climate-driven variability and trends in mountain snowpack in western North America. *J. Clim.*, **19**(23), 6209–6220.

Mote, P.W., A.F. Hamlet, M.P. Clark, and D.P. Lettenmaier, 2005: Declining mountain snowpack in western North America. *Bull. Am. Meteorol. Soc.*, **86**, 39–49, doi:10.1175/BAMS-86-1-39.

Munk, W., 2002: Twentieth century sea level: An enigma. *Proc. Natl. Acad. Sci. U.S.A.*, **99**, 6550–6555.

Murphy, E.J., A. Clarke, C. Symon, and J. Priddle, 1995: Temporal variation in Antarctic sea-ice: analysis of a long term fast-ice record from the South Orkney Islands. *Deep-Sea. Res.*, **42**, 1045–1062.

Nagurnyi, A.P., V.G. Korostolev, and V.V. Ivanov, 1999: Multiyear variability of sea ice thickness in the Arctic basin measured by elastic-gravity waves on the ice surface. *Meteorol. Hydrol.*, **3**, 72–78 (in Russian).

Nan, Z., Z. Gao, S. Li, and T. Wu, 2003: Permafrost changes in the northern limit of permafrost on the Qinghai-Tibet Plateau in the last 30 years. *Acta Geogr. Sin.*, **58**(6), 817–823 (in Chinese).

Nelson, F.E., O.A. Anisimov, and N.I. Shiklomanov, 2001: Subsidence risk from thawing permafrost. *Nature*, **410**, 889–890.

Nelson, F.E. (ed.), 2004a: Circumpolar Active Layer Monitoring (CALM) Workshop. *Permafrost and Periglacial Processes*, **15**(2). 99–188.

Nelson, F.E. (ed.), 2004b: Eurasian contributions to the Circumpolar Active Layer Monitoring (CALM) Workshop. *Polar Geogr.*, **28**(4), 253–340.

Nesje, A., and S.O. Dahl, 2003: The 'Little Ice Age' - only temperature? *The Holocene*, **13**(1), 139–145.

Nesje, A., Ø. Lie, and S.O. Dahl, 2000: Is the North Atlantic Oscillation reflected in Scandinavian glacier mass balance records? *J. Quat. Sci.*, **15**(6), 587–601.

Nicholls, N., 2005: Climate variability, climate change and the Australian snow season. *Aust. Meteorol. Mag.*, **54**, 177–185.

Oberman, N.G., and G.G. Mazhitova, 2001: Permafrost dynamics in the northeast of European Russia at the end of the 20th century. *Norwegian J. Geogr.*, **55**, 241–244.

Oerlemans, J., 2005: Extracting a climate signal from 169 glacier records. *Science*, **308**, 675-677.

Ohmura, A., 2001: Physical basis for the temperature/melt-index method. *J. Appl. Meteorol.*, **40**, 753–761.

Ohmura, A., 2004: Cryosphere during the twentieth century. In: *The State of the Planet: Frontiers and Challenges in Geophysics* [Sparks, R.S.J. and C.J. Hawkesworth (eds.)]. Geophysical Monograph 150, International Union of Geodesy and Geophysics, Boulder, CO and American Geophysical Union, Washington, DC, pp. 239–257.

Omstedt, A., and D. Chen, 2001: Influence of atmospheric circulation on the maximum ice extent in the Baltic Sea. *J. Geophys. Res.*, **106**, 4493–4500.

Osterkamp, T.E., 2003: A thermal history of permafrost in Alaska. In: *Proceedings of the 8th International Conference on Permafrost, 21-25 July 2003, Zurich, Switzerland* [Phillips, M., S.M. Springman, and L.U. Arenson (eds.)]. A.A. Balkema, Lisse, the Netherlands, pp. 863–867.

Osterkamp, T.E., 2005: The recent warming of permafrost in Alaska. *Global Planet. Change*, **49**, 187–202, doi: 10.1016/j.gloplacha.2005.09.001.

Parizek, B.R., and R.B. Alley, 2004: Implications of increased Greenland surface melt under global-warming scenarios: ice-sheet simulations. *Quat. Sci. Rev.*, **23**(9–10), 1013–1027.

Parizek, B.R., R.B. Alley, and C.L. Hulbe, 2003: Subglacial thermal balance permits ongoing grounding-line retreat along the Siple Coast of West Antarctica. *Ann. Glaciol.*, **36**, 251–256.

Parkinson, C.L., 1990: Search for the Little Ice Age in Southern Ocean sea-ice records. *Ann. Glaciol.*, **14**, 221–225.

Paterson, W.S.B., 2004: *The Physics of Glaciers*, Ed. 3.A, Elsevier, Oxford, UK, 496pp.

Pavlov, A.V., 1996: Permafrost-climate monitoring of Russia: analysis of field data and forecast. *Polar Geogr.*, **20**(1), 44–64.

Payne, A.J., et al., 2004: Recent dramatic thinning of largest West Antarctic ice stream triggered by oceans. *Geophys. Res. Lett.*, **31**, L23401, doi:10.1029/2004GL021284.

Peltier, W.R., 1998: Postglacial variations in the level of the sea: implications for climate dynamics and solid-earth geophysics. *Rev. Geophys.*, **36**, 603–689.

Petkova, N., E. Koleva, and V. Alexandrov, 2004: Snow cover variability and change in mountainous regions of Bulgaria, 1931-2000. *Meteorol. Z.*, **13**(1), 19–23.

Pettersson, R., P. Jansson, and P. Holmlund, 2003: Cold surface layer thinning on Storglaciären, Sweden, observed by repeated ground penetrating radar surveys. *J. Geophys. Res.*, **108**(F1), 6004, doi:10.1029/2003JF000024.

Polyakov, I.V., et al., 2003: Long-term ice variability in Arctic marginal seas. *J. Clim.*, **16**, 2078–2085.

Prieto, R. et al., 2001: Interannual oscillations and trend of snow occurrence in the Andes region since 1885. *Aust. Meteorol. Mag.*, **50**(2), 164.

Proshutinsky, A.Y., and M.A. Johnson, 1997: Two circulation regimes of the wind-driven Arctic Ocean. *J. Geophys. Res.*, **102**(C6), 12493–12514.

Qin, D., S. Liu, and P. Li, 2006: Snow cover distribution, variability, and response to climate change in Western China. *J. Clim.*, **19**, 1820–1833.

Qiu, G., Y. Zhou, D. Guo, and Y. Wang, 2000: The map of geocryological regionalization and classification in China. In: *Geocryology in China* [Zhou, Y., D. Guo, G. Qiu, G. Cheng, and S. Li (eds.)]. Science Press, Beijing (in Chinese). The digital version of the map is available at the National Snow and Ice Data Center, University of Colorado at Boulder, Boulder, CO, http://nsidc.org/data/ggd603.html.

Rachold, V., et al., 2003: Modern terrigenous organic carbon input to the Arctic Ocean. In: *Organic Carbon Cycle in the Arctic Ocean: Present and Past* [Stein, R., and R.W. Macdonald (eds.)]. Springer Verlag, Berlin, pp. 33–55.

Ramillien, G., et al., 2006: Interannual variations of the mass balance of the Antarctica and Greenland ice sheets from GRACE. *Global Planet. Change*, **53**, 198–208.

Randerson, J.T., C.B. Field, I.Y. Fung, and C.B. Tans, 1999: Increases in early season ecosystem uptake explain recent changes in the seasonal cycle of the atmospheric CO_2 at high northern latitudes. *Geophys. Res. Lett.*, **26**, 2765–2768.

Raper, S.C.B., and R.J. Braithwaite, 2005: The potential for sea level rise: New estimates from glacier and ice cap area and volume distribution. *Geophys. Res. Lett.*, **32**, L05502, doi:10.1029/2004GL021981.

Rayner, N.A., et al., 2003: Global analyses of sea surface temperature, sea ice and night marine air temperature since the late nineteenth century. *J. Geophys. Res.*, **108**(D14), 4407, doi: 10.1029/2002JD002670.

Ren, J., et al., 2004: Glacier variations and climate warming and drying in the central Himalayas. *Chin. Sci. Bull.*, **49**(1), 65–69.

Rignot, E., and S. Jacobs, 2002: Rapid bottom melting widespread near Antarctic Ice Sheet grounding lines. *Science*, **296**, 2020–2023.

Rignot, E., and R.H. Thomas, 2002: Mass balance of polar ice sheets. *Science*, **297**(5586), 1502–1506.

Rignot, E., and P. Kanagaratnam, 2006: Changes in the velocity structure of the Greenland Ice Sheet. *Science*, **311**, 986–990.

Rignot, E., A. Rivera, and G. Casassa, 2003: Contribution of the Patagonia Icefields of South America to sea level rise. *Science*, **302**, 434–437.

Rignot, E., et al., 2004: Accelerated ice discharge from the Antarctic Peninsula following the collapse of Larsen B ice shelf. *Geophys. Res. Lett.*, **31**(18), L18401, doi:10.1029/2004GL020697.

Rignot, E., et al., 2005: Recent ice loss from the Fleming and other glaciers, Wordie Bay, West Antarctic Peninsula. *Geophys. Res. Lett.*, **32**(7), 1–4.

Rigor, I.G., J.M. Wallace, and R.L. Colony, 2002: Response of sea ice to the arctic oscillation. *J. Clim.*, **15**, 2648– 2663.

Robertson, R., M. Visbek, A. Gordon, and E. Fahrbach, 2002: Long term temperature trends in the deep waters of the Weddell. *Deep-Sea Res.*, **49**, 4791–4802.

Robinson, D.A., and A. Frei, 2000: Seasonal variability of northern hemisphere snow extent using visible satellite data. *Professional Geogr.*, **51**, 307–314.

Robinson, D.A., K.F. Dewey, and R.R. Heim Jr., 1993: Global snow cover monitoring: an update. *Bull. Am. Meteorol. Soc.*, **74**, 1689–1696.

Robock, A., M. Mu, K. Vinnikov, and D. Robinson, 2003: Land surface conditions over Eurasia and Indian summer monsoon rainfall. *J. Geophys. Res.*, **108**(D4), 4131, doi: 10.1029/2002JD002286.

Romanovsky, V.E., et al., 2002: Permafrost temperature records: Indicator of climate change. *Eos*, **83**(50), 589, 593–594.

Rothrock, D.A., Y. Yu, and G.A. Maykut, 1999: Thinning of the Arctic sea-ice cover. *Geophys. Res. Lett.*, **26**(23), 3469.

Rothrock, D.A., J. Zhang, and Y. Yu., 2003: The arctic ice thickness anomaly of the 1990s: A consistent view from observations and models. *J. Geophys. Res.*, **108**(C3), 3083, doi:10.1029/2001JC001208.

Saito, K., and J. Cohen, 2003: The potential role of snow cover in forcing interannual variability of the major Northern Hemisphere mode. *Geophys. Res. Lett.*, **30**, 1302, doi:10.1029/2002GL016341.

Scambos, T., C. Hulbe, and M. Fahnestock, 2003: Climate-induced ice shelf disintegration in the Antarctic Peninsula. In: *Antarctic Peninsula Climate Variability: Historical and Paleoenvironmental Perspectives* [Domack, E., et al. (eds.)]. Antarctic Research Series 79, American Geophysical Union, Washington, DC, pp. 79–92.

Scambos, T., C. Hulbe, M. Fahnestock, and J. Bohlander, 2000: The link between climate warming and break-up of ice shelves in the Antarctic Peninsula. *J. Glaciol.*, **46**, 516–530.

Scambos, T., J. Bohlander, C. Shuman, and P. Skvarca, 2004: Glacier acceleration and thinning after ice shelf collapse in the Larsen B embayment, Antarctica. *Geophys. Res. Lett.*, **31**, L18401, doi:10.1029/2004GL020670.

Scherrer, S.C., C. Appenzeller, and M. Laternser, 2004: Trends in Swiss alpine snow days – the role of local and large scale climate variability. *Geophys. Res. Lett.*, **31**, L13215, doi:10.1029/2004GL020255.

Schmitt, C., Ch. Kottmeier, S. Wassermann, and M. Drinkwater, 2004: *Atlas of Antarctic Sea Ice Drift*. University of Karlsruhe, Karlsruhe, http://imkhp7.physik.uni-karlsruhe.de/~eisatlas/eisatlas_start.html.

Schneider, D.P., E.J. Steig, and J.C. Comiso, 2004: Recent climate variability in Antarctica from satellite-derived temperature data. *J. Clim.*, **17**, 1569–1583.

Sharkhuu, N., 2003: Recent changes in the permafrost of Mongolia. In: *Proceedings of the 8th International Conference on Permafrost, 21-25 July 2003, Zurich, Switzerland* [Phillips, M., S.M. Springman, and L.U. Arenson (eds.)]. A.A. Balkema, Lisse, the Netherlands, pp. 1029–1034.

Shepherd, A., D.J. Wingham, and J.A.D. Mansley, 2002: Inland thinning of the Amundsen Sea sector, West Antarctica. *Geophys. Res. Lett.*, **29**(10), 1364.

Shepherd, A., D. Wingham, and E. Rignot, 2004: Warm ocean is eroding West Antarctic Ice Sheet. *Geophys. Res. Lett.*, **31**(23), 1–4.

Shepherd, A., D. Wingham, T. Payne, and P. Skvarca, 2003: Larsen Ice Shelf has progressively thinned. *Science*, **302**, 856–859.

Shy, T.L., and J.E. Walsh, 1996: North Pole ice thickness and association with ice motion history. *Geophys. Res. Lett.*, **23**(21), 2975–2978.

Sicart, J.E., P. Wagnon, and P. Ribstein, 2005: Atmospheric controls of the heat balance of Zongo Glacier (16°S, Bolivia). *J. Geophys. Res.*, **110**, D12106, doi:10.1029/2004JD005732.

Smith, L.C., 2000: Trends in Russian Arctic river-ice formation and breakup, 1917-1994. *Phys. Geogr.*, **21**, 46–56.

Smith, L.C., Y. Sheng, G.M. MacDonald, and L.D. Hinzman, 2005: Disappearing Arctic lakes. *Science*, **308**, 1429.

Smith, N.V., S.S. Saatchi, and T. Randerson, 2004: Trends in high latitude soil freeze and thaw cycles from 1988 to 2002. *J. Geophys. Res.*, **109**, D12101, doi:10.1029/2003JD004472.

Smith, S.L., M.M. Burgess, and A.E. Taylor, 2003: High Arctic permafrost observatory at Alert, Nunavut – analysis of a 23 year data set. In: *Proceedings of the 8th International Conference on Permafrost, 21-25 July 2003, Zurich, Switzerland* [Phillips, M., S.M. Springman, and L.U. Arenson (eds.)]. A.A. Balkema, Lisse, the Netherlands, pp. 1073–1078.

Smith, S.L., M.M. Burgess, D. Riseborough, and F.M. Nixon, 2005: Recent trends from Canadian permafrost thermal monitoring network sites. *Permafrost and Periglacial Processes*, **16**, 19–30.

Solomina, O., R. Barry, and M. Bodnya, 2004: The retreat of Tien Shan glaciers (Kyrgyzstan) since the Little Ice Age estimated from aerial photographs, lichonometric and historical data. *Geografiska Annaler*, **86A**(2), 205–215.

Steffen, K., et al., 1992: The estimation of geophysical parameters using passive microwave algorithms. In: *Microwave Remote Sensing of Sea Ice* [Carsey, F.D. (ed.)]. Geophysical Monograph 68, American Geophysical Union, Washington, DC, pp 201–231.

Stewart, I.T., D.R. Cayan, and M.D. Dettinger, 2005: Changes towards earlier streamflow timing across western North America. *J. Clim.*, **18**, 1136–1155.

Stineman, R.W., 1980: A consistently well-behaved method of interpolation. *Creative Computing*, July 1980, 54–57.

Stone, J.O., et al., 2003: Holocene deglaciation of Marie Byrd Land, West Antarctica. *Science*, **299**, 99–102.

Stone, R.S., E.G. Dutton, J.M. Harris, and D. Longnecker, 2002: Earlier spring snowmelt in northern Alaska as an indicator of climate change. *J. Geophys. Res.*, **107**(D10), doi:10.1029/2000JD000286.

Su, Z., and Shi, Y., 2002: Response of monsoonal temperate glaciers to global warming since the Littel Ice Age. *Quat. Int.*, **97–98**, 123–131.

Thomas, R.H., 1979: The dynamics of marine ice sheets. *J. Glaciol.*, **24**, 167–177.

Thomas, R., 2004: Force-perturbation analysis of recent thinning and acceleration of Jakobshavn Isbrae, Greenland. *J. Glaciol.*, **50**, 57–66.

Thomas, R., et al., 2001: Mass balance of higher-elevation parts of the Greenland ice sheet. *J. Geophys. Res.*, **106D**, 33707–33716.

Thomas, R., et al., 2003: Investigation of surface melting and dynamic thinning on Jakobshavn Isbrae, Greenland. *J. Glaciol.*, **49**, 231–239.

Thomas, R., et al., 2004: Accelerated sea-level rise from West Antarctica. *Science*, **306**(5694), 255–258.

Thomas, R., et al., 2006: Progressive increase in ice loss from Greenland. *Geophys. Res. Lett.*, **33**, L10503, doi: 10.1029/2006GL026075.

Thompson, D.W.J., and S. Solomon, 2002: Interpretation of recent Southern Hemisphere climate change. *Science*, **296**, 895–899.

Thompson, L.G., et al., 2002: Kilimanjaro ice core records: Evidence of Holocene climate change in tropical Africa. *Science*, **298**, 589–593.

Tucker, W.B. III, et al., 2001: Evidence for the rapid thinning of sea ice in the western Arctic Ocean at the end of the 1980s. *Geophys. Res. Lett.*, **28**(14), 2851–2854.

Turner, J., et al., 2005: Antarctic climate change during the last 50 years. *Int. J. Climatol.*, **25**(3), 279–294.

van de Berg, W.J., M.R. van den Broeke, C.H. Reijmer, and E. van Meijgaard, 2006: Reassessment of the Antarctic surface mass balance using calibrated output of a regional atmospheric climate model. *J. Geophys. Res.*, **111**, D11104, doi:10.1029/2005JD006495.

van den Broeke, M.R., 2000: On the interpretation of Antarctic temperature trends. *J. Clim.*, **13**(21), 3885–3889.

van den Broeke, M.R., W.J. van de Berg, and E. van Meijgaard, 2006: Snowfall in coastal West Antarctica much greater than previously assumed. *Geophys. Res. Lett.*, **33**, L02505, doi:10.1029/2005GL025239.

Vaughan, D.G., et al., 2001: Climate change – Devil in the detail. *Science*, **293**(5536), 1777–1779.

Vaughan, D.G., et al., 2003: Recent rapid regional climate warming on the Antarctic Peninsula. *Clim. Change*, **60**, 243–274.

Velicogna, I., and J. Wahr, 2005: Greenland mass balance from GRACE. *Geophys. Res. Lett.*, **32**, L18505, doi:10.1029/2005GL023955.

Velicogna, I., and J. Wahr, 2006: Measurements of time variable gravity show mass loss in Antarctica. *Science*, **311**(5768), 1754–1756, doi:10.1126/science.1123785.

Venegas, S.A., M.R. Drinkwater, and G. Shaffer, 2001: Coupled oscillations in Antarctic sea ice and atmosphere in the South Pacific sector. *Geophys. Res. Lett.*, **28**,(17), 3301–3304.

Vincent, C., E. Le Meur, D. Six, and M. Funk, 2005: Solving the paradox of the end of the Little Ice Age in the Alps. *Geophys. Res. Lett.*, **32**, L09706, doi:10.1029/2005GL022552.

Vinje, T., 2001: Anomalies and trends of sea ice extent and atmospheric circulation in the Nordic Seas during the period 1864-1998. *J. Clim.*, **14**, 255–267.

Vinje, T., N. Nordlund, and A. Kvambekk, 1998: Monitoring ice thickness in Fram Strait. *J. Geophys. Res.*, **103**(C5), 10437–10450.

Vojtek, M., P. Fasko, and P. St'astny, 2003: Some selected snow climate trends in Slovakia with respect to altitude. *Acta Meteorologica Universitatis Comenianae*, **32**, 17–27.

Vonder Mühll, D., J. Nötzli, K. Makowski, and R. Delaloye, 2004: *Permafrost in Switzerland 2000/2001 and 2001/2002*. Glaciological Report (Permafrost) No. 2/3, Glaciological Commission of the Swiss Academy of Sciences, Zurich, 86 pp.

Wadhams, P., 1992: Sea ice thickness distribution in the Greenland Sea and Eurasian Basin, May 1987. *J. Geophys. Res.*, **97**, 5331–5348.

Wagnon, P., P. Ribstein, B. Francou, and J.E. Sicart, 2001. Anomalous heat and mass budget of Glaciar Zongo, Bolivia during the 1997/98 El Niño year. *J. Glaciol.*, **47**(156), 21–28.

Walsh, J., et al., 2005: Cryosphere and hydrology. In: *Arctic Climate Impact Assessment*. Cambridge University Press, Cambridge and New York, pp. 183–242.

Wang, S., 2002: Permafrost degradation, desertification and CH_4 release. In: *Dynamic Characteristic of Cryosphere in the Central Section of Qinghai-Tibet Plateau* [Yao, T., et al. (eds.)]. Geology Press, Beijing, 234–255 (in Chinese).

Wang, S., and Zhao, X., 1997: Environmental change in patchy permafrost zone in the south section of Qinghai-Tibet Highway. *J. Glaciol. Geocryol.*, **19**, 231–239 (in Chinese).

Warren, S.G., et al., 1999: Snow depth on arctic sea ice. *J. Clim.*, **12**, 1814–1829.

WGMS(ICSI-IAHS), various years-a: *Fluctuations of Glaciers*. World Glacier Monitoring Service, Zurich.

WGMS(ICSI-IAHS), various years-b: *Mass Balance Bulletin*. World Glacier Monitoring Service, Zurich, http://www.wgms.ch/mbb.html.

Winsor, P., 2001: Arctic sea ice thickness remained constant during the 1990s. *Geophys. Res. Lett.*, **28**(6), 1039–1041.

Worby, A.P., and S.F. Ackley, 2000: Antarctic research yields circumpolar sea ice thickness data. *Eos*, **81**(17), 181, 184–185.

Worby, A.P., and J.C. Comiso, 2004: Studies of the Antarctic sea ice edge and ice extent from satellite and ship observations. *Remote Sensing of Environment*, **92**, 98–111.

Wu, Q., and Y. Liu, 2003: Ground temperature monitoring and its recent change in Qinghai–Tibet Plateau. *Cold Regions Sci. Technol.*, **18**, 85–92.

Ye, H.C., and M. Ellison, 2003: Changes in transitional snowfall season length in northern Eurasia. *Geophys. Res. Lett.*, **30**(5), 1252.

Yoshikawa, K., and L.D. Hinzman, 2003: Shrinking thermokarst ponds and groundwater dynamics in discontinuous permafrost. *Permafrost and Periglacial Processes*, **14**(2), 151–160.

Yu, Y., G.A. Maykut, and D.A. Rothrock, 2004: Changes in the thickness distribution of Arctic sea ice between 1958-1970 and 1993-1997. *J. Geophys. Res.*, **109**, C08004, doi:10.1029/2003JC001982.

Zemp, M., R. Frauenfelder, W. Haeberli, and M. Hoelzle, 2005: Worldwide glacier mass balance measurements: general trends and first results of the extraordinary year 2003 in Central Europe. In: *XIII Glaciological Symposium, Shrinkage of the Glaciosphere: Facts and Analyses, St. Petersburg, Russia* [Science, R.A.O. (ed.)]. Data of Glaciological Studies [Materialy glyatsiologicheskikh issledovaniy], Moscow, Russia, pp. 3–12.

Zhang, T., 2005: Influence of the seasonal snow cover on the ground thermal regime: An overview. *Rev. Geophys.*, **43**, RG4002, doi:10.1029/2004RG000157.

Zhang, T., et al., 1999: Statistics and characteristics of permafrost and ground-ice distribution in the Northern Hemisphere. *Polar Geogr.*, **23**(2), 132–154.

Zhang, T., et al., 2003: Distribution of seasonally and perennially frozen ground in the Northern Hemisphere. In: *Proceedings of the 8th International Conference on Permafrost, 21-25 July 2003, Zurich, Switzerland* [Phillips, M., S.M. Springman, and L.U. Arenson (eds.)]. A.A. Balkema, Lisse, the Netherlands, pp. 1289–1294.

Zhang, T., et al., 2005: Spatial and temporal variability in active layer thickness over the Russian Arctic drainage basin. *J. Geophys. Res.*, **110**, D16101, doi:10.1029/2004JD005642.

Zhang, X., L.A. Vincent, W.D. Hogg, and A. Niitsoo, 2000: Temperature and precipitation trends in Canada during the 20th century. *Atmos-Ocean*, **38**(3), 395–429.

Zhang, X., K.D. Harvey, W.D. Hogg, and T.R. Yuzyk, 2001: Trends in Canadian streamflow. *Water Resour. Res.*, **37**(4), 987–998.

Zhang, Y.S., T. Li, and B. Wang, 2004: Decadal change of the spring snow depth over the Tibetan Plateau: The associated circulation and influence on the East Asian summer monsoon. *J. Clim.*, **17**(14), 2780–2793.

Zhao, L., G. Cheng, and Li, S., 2003: Changes of plateau frozen-ground and environmental engineering effects. In: *The Formation Environment and Development of Qinghai-Tibet Plateau* [Zheng, D., et al. (eds)]. Heibei Science and Technology Press, Shijiazhang, pp. 143–150 (in Chinese).

Zhao, L., et al., 2004: Changes of climate and seasonally frozen ground over the past 30 years in Qinghai-Xizang (Tibetan) Plateau, China. *Global Planet. Change*, **43**, 19–31.

Zhou, Y., X. Gao, and Y. Wang, 1996: The ground temperature changes of seasonally freeze-thaw layers and climate warming in Northeast China in the past 40 years. In: *Proceeding of the 5th Chinese Conference on Glaciology and Geocryology (Volume 1)*. Gansu Culture Press, Lanzhou, pp. 3–9 (in Chinese).

Zwally, H.J., et al., 2002: Surface melt-induced acceleration of Greenland ice-sheet flow. *Science*, **297**(5579), 218–222.

Zwally, H.J., et al., 2006: Mass changes of the Greenland and Antarctic ice sheets and shelves and contributions to sea level rise: 1992-2002. *J. Glaciol.*, **51**, 509–527.

5

Observations:
Oceanic Climate Change and Sea Level

Coordinating Lead Authors:
Nathaniel L. Bindoff (Australia), Jürgen Willebrand (Germany)

Lead Authors:
Vincenzo Artale (Italy), Anny Cazenave (France), Jonathan M. Gregory (UK), Sergey Gulev (Russian Federation), Kimio Hanawa (Japan),
Corrine Le Quéré (UK, France, Canada), Sydney Levitus (USA), Yukihiro Nojiri (Japan), C.K. Shum (USA), Lynne D. Talley (USA),
Alakkat S. Unnikrishnan (India)

Contributing Authors:
J. Antonov (USA, Russian Federation), N.R. Bates (Bermuda), T. Boyer (USA), D. Chambers (USA), B. Chao (USA), J. Church (Australia),
R. Curry (USA), S. Emerson (USA), R. Feely (USA), H. Garcia (USA), M. González-Davila (Spain), N. Gruber (USA, Switzerland),
S. Josey (UK), T. Joyce (USA), K. Kim (Republic of Korea), B. King (UK), A. Koertzinger (Germany), K. Lambeck (Australia),
K. Laval (France), N. Lefevre (France), E. Leuliette (USA), R. Marsh (UK), C. Mauritzer (Norway), M. McPhaden (USA), C. Millot (France),
C. Milly (USA), R. Molinari (USA), R.S. Nerem (USA), T. Ono (Japan), M. Pahlow (Canada), T.-H. Peng (USA), A. Proshutinsky (USA),
B. Qiu (USA), D. Quadfasel (Germany), S. Rahmstorf (Germany), S. Rintoul (Australia), M. Rixen (NATO, Belgium), P. Rizzoli (USA, Italy),
C. Sabine (USA), D. Sahagian (USA), F. Schott (Germany), Y. Song (USA), D. Stammer (Germany), T. Suga (Japan), C. Sweeney (USA),
M. Tamisiea (USA), M. Tsimplis (UK, Greece), R. Wanninkhof (USA), J. Willis (USA), A.P.S. Wong (USA, Australia), P. Woodworth (UK),
I. Yashayaev (Canada), I. Yasuda (Japan)

Review Editors:
Laurent Labeyrie (France), David Wratt (New Zealand)

This chapter should be cited as:
Bindoff, N.L., J. Willebrand, V. Artale, A. Cazenave, J. Gregory, S. Gulev, K. Hanawa, C. Le Quéré, S. Levitus, Y. Nojiri, C.K. Shum, L.D. Talley and A. Unnikrishnan, 2007: Observations: Oceanic Climate Change and Sea Level. In: *Climate Change 2007: The Physical Science Basis. Contribution of Working Group I to the Fourth Assessment Report of the Intergovernmental Panel on Climate Change* [Solomon, S., D. Qin, M. Manning, Z. Chen, M. Marquis, K.B. Averyt, M. Tignor and H.L. Miller (eds.)]. Cambridge University Press, Cambridge, United Kingdom and New York, NY, USA.

Table of Contents

Executive Summary

- The oceans are warming. Over the period 1961 to 2003, global ocean temperature has risen by 0.10°C from the surface to a depth of 700 m. Consistent with the Third Assessment Report (TAR), global ocean heat content (0–3,000 m) has increased during the same period, equivalent to absorbing energy at a rate of 0.21 ± 0.04 W m^{-2} globally averaged over the Earth's surface. Two-thirds of this energy is absorbed between the surface and a depth of 700 m. Global ocean heat content observations show considerable interannual and inter-decadal variability superimposed on the longer-term trend. Relative to 1961 to 2003, the period 1993 to 2003 has high rates of warming but since 2003 there has been some cooling.

- Large-scale, coherent trends of salinity are observed for 1955 to 1998, and are characterised by a global freshening in subpolar latitudes and a salinification of shallower parts of the tropical and subtropical oceans. Freshening is pronounced in the Pacific while increasing salinities prevail over most of Atlantic and Indian Oceans. These trends are consistent with changes in precipitation and inferred larger water transport in the atmosphere from low latitudes to high latitudes and from the Atlantic to the Pacific. Observations do not allow for a reliable estimate of the global average change in salinity in the oceans.

- Key oceanic water masses are changing; however, there is no clear evidence for ocean circulation changes. Southern Ocean mode waters and Upper Circumpolar Deep Waters have warmed from the 1960s to about 2000. A similar but weaker pattern of warming in the Gulf Stream and Kuroshio mode waters in the North Atlantic and North Pacific has been observed. Long-term cooling is observed in the North Atlantic subpolar gyre and in the central North Pacific. Since 1995, the upper North Atlantic subpolar gyre has been warming and becoming more saline. It is *very likely* that up to the end of the 20th century, the Atlantic meridional overturning circulation has been changing significantly at interannual to decadal time scales. Over the last 50 years, no coherent evidence for a trend in the strength of the meridional overturning circulation has been found.

- Ocean biogeochemistry is changing. The total inorganic carbon content of the oceans has increased by 118 ± 19 GtC between the end of the pre-industrial period (about 1750) and 1994 and continues to increase. It is *more likely than not* that the fraction of emitted carbon dioxide that was taken up by the oceans has decreased, from $42 \pm 7\%$ during 1750 to 1994 to $37 \pm 7\%$ during 1980 to 2005. This would be consistent with the expected rate at which the oceans can absorb carbon, but the uncertainty in this estimate does not allow firm conclusions. The increase in total inorganic carbon caused a decrease in the depth at which calcium carbonate dissolves, and also caused a decrease in surface ocean pH by an average of 0.1 units since 1750. Direct observations of pH at available time series stations for the last 20 years also show trends of decreasing pH at a rate of 0.02 pH units per decade. There is evidence for decreased oxygen concentrations, likely driven by reduced rates of water renewal, in the thermocline (~100–1,000 m) in most ocean basins from the early 1970s to the late 1990s.

- Global mean sea level has been rising. From 1961 to 2003, the average rate of sea level rise was 1.8 ± 0.5 mm yr^{-1}. For the 20th century, the average rate was 1.7 ± 0.5 mm yr^{-1}, consistent with the TAR estimate of 1 to 2 mm yr^{-1}. There is *high confidence* that the rate of sea level rise has increased between the mid-19th and the mid-20th centuries. Sea level change is highly non-uniform spatially, and in some regions, rates are up to several times the global mean rise, while in other regions sea level is falling. There is evidence for an increase in the occurrence of extreme high water worldwide related to storm surges, and variations in extremes during this period are related to the rise in mean sea level and variations in regional climate.

- The rise in global mean sea level is accompanied by considerable decadal variability. For the period 1993 to 2003, the rate of sea level rise is estimated from observations with satellite altimetry as 3.1 ± 0.7 mm yr^{-1}, significantly higher than the average rate. The tide gauge record indicates that similar large rates have occurred in previous 10-year periods since 1950. It is unknown whether the higher rate in 1993 to 2003 is due to decadal variability or an increase in the longer-term trend.

- There are uncertainties in the estimates of the contributions to sea level change but understanding has significantly improved for recent periods. For the period 1961 to 2003, the average contribution of thermal expansion to sea level rise was 0.4 ± 0.1 mm yr^{-1}. As reported in the TAR, it is *likely* that the sum of all known contributions for this period is smaller than the observed sea level rise, and therefore it is not possible to satisfactorily account for the processes causing sea level rise. However, for the period 1993 to 2003, for which the observing system is much better, the contributions from thermal expansion (1.6 ± 0.5 mm yr^{-1}) and loss of mass from glaciers, ice caps and the Greenland and Antarctic Ice Sheets together give 2.8 ± 0.7 mm yr^{-1}. For the latter period, the climate contributions constitute the main factors in the sea level budget, which is closed to within known errors.

- The patterns of observed changes in global ocean heat content and salinity, sea level, thermal expansion, water mass evolution and biogeochemical parameters described in this chapter are broadly consistent with the observed ocean surface changes and the known characteristics of the large-scale ocean circulation.

5.1   Introduction

The ocean has an important role in climate variability and change. The ocean's heat capacity is about 1,000 times larger than that of the atmosphere, and the oceans net heat uptake since 1960 is around 20 times greater than that of the atmosphere (Levitus et al., 2005a). This large amount of heat, which has been mainly stored in the upper layers of the ocean, plays a crucial role in climate change, in particular variations on seasonal to decadal time scales. The transport of heat and freshwater by ocean currents can have an important effect on regional climates, and the large-scale Meridional Overturning Circulation (MOC; also referred to as thermohaline circulation) influences the climate on a global scale (e.g., Vellinga and Wood, 2002). Life in the sea is dependent on the biogeochemical status of the ocean and is influenced by changes in the physical state and circulation. Changes in ocean biogeochemistry can directly feed back to the climate system, for example, through changes in the uptake or release of radiatively active gases such as carbon dioxide. Changes in sea level are also important for human society, and are linked to changes in ocean circulation. Finally, oceanic parameters can be useful for detecting climate change, in particular temperature and salinity changes in the deeper layers and in different regions where the short-term variability is smaller and the signal-to-noise ratio is higher.

The large-scale, three-dimensional ocean circulation and the formation of water masses that ventilate the main thermocline together create pathways for the transport of heat, freshwater and dissolved gases such as carbon dioxide from the surface ocean into the density-stratified deeper ocean, thereby isolating them from further interaction with the atmosphere. These pathways are also important for the transport of anomalies in these parameters caused by changes in the surface conditions. Furthermore, changes in the storage of heat and in the distribution of ocean salinity cause the ocean to expand or contract and hence change the sea level both regionally and globally.

The ocean varies over a broad range of time scales, from seasonal (e.g., in the surface mixed layer) to decadal (e.g., circulation in the main subtropical gyres) to centennial and longer (associated with the MOC). The main modes of climate variability, which are described in Chapter 3, are the El Niño-Southern Oscillation (ENSO), the Pacific Decadal Oscillation (PDO), the Northern Annular Mode (NAM), which is related to the North Atlantic Oscillation (NAO), and the Southern Annular Mode (SAM). Forcing of the oceans is often related to these modes, which cause changes in ocean circulation through changed patterns of winds and changes in surface ocean density.

The Third Assessment Report (TAR) discussed some aspects of the ocean's role. Folland et al. (2001) concluded that the global ocean has significantly warmed since the late 1950's This assessment provides updated estimates of temperature changes for the oceans. Furthermore, it discusses new evidence for changes in the ocean freshwater budget and the ocean circulation. The TAR estimate of the total inorganic carbon increase in the ocean (Prentice et al., 2001) was based entirely on indirect evidence. This assessment provides updated indirect estimates and reports on new and direct evidence for changes in total carbon increase and for changes in ocean biogeochemistry (including pH and oxygen). Church et al. (2001) determined a range of 1 to 2 mm yr^{-1} for the observed global average sea level rise in the 20th century. This assessment provides new estimates for sea level change and the climate-related contributions to sea level change from thermal expansion and melting of ice sheets, glaciers and ice caps. The focus of this chapter is on observed changes in the global ocean basins, however some regional changes in the ocean state are also considered.

Many ocean observations are poorly sampled in space and time, and regional distributions often are quite heterogeneous. Furthermore, the observational records only cover a relatively short period of time (e.g., the 1950s to the present). Many of the observed changes have significant decadal variability associated with them, and in some cases decadal variability and/or poor sampling may prevent detection of long-term trends. When time series of oceanic parameters are considered, linear trends are often computed in order to quantify the observed long-term changes; however, this does not imply that the original signal is best represented by a linear increase in time. For plotting time series, this chapter generally uses the difference (anomaly) from the average value for the years 1961 to 1990. Wherever possible, error bars are provided to quantify the uncertainty of the observations. As in other parts of this report, 90% confidence intervals are used throughout. If not otherwise stated, values with error bars given as $x \pm e$ should hence be interpreted as a 90% chance that the true value is in the range $x - e$ to $x + e$.

5.2   Changes in Global-Scale Temperature and Salinity

5.2.1   Background

Among the major challenges in understanding the climate system are quantifying the Earth's heat balance and the freshwater balance (hydrological cycle), which both have a substantial contribution from the World Ocean. This chapter presents observational evidence that directly or indirectly helps to quantify changes in these balances.

The TAR included estimates of ocean heat content changes for the upper 3,000 m of the World Ocean. Ocean heat content change is closely proportional to the average temperature change in a volume of seawater, and is defined here as the deviation from a reference period. This section reports on updates of this estimate and presents estimates for the upper 700 m based on additional modern and historical data (Willis et al., 2004; Levitus et al., 2005b; Ishii et al., 2006). The section also presents new estimates of the temporal variability of salinity. The data used for temperature and heat content estimates are based on the World Ocean Database 2001 (e.g., Boyer et al.,

2002; Conkright et al., 2002), which has been updated with more recent data. Temperature data include measurements from reversing thermometers, expendable bathythermographs, mechanical bathythermographs, conductivity-temperature-depth instruments, Argo profiling floats, moored buoys and drifting buoys. The salinity data are described by Locarnini et al. (2002) and Stephens et al. (2002).

5.2.2 Ocean Heat Content

5.2.2.1 Long-Term Temperature Changes

Figure 5.1 shows two time series of ocean heat content for the 0 to 700 m layer of the World Ocean, updated from Ishi et al. (2006) and Levitus et al. (2005a) for 1955 to 2005, and a time series for 0 to 750 m for 1993 to 2005 updated from Willis et al. (2004). Approximately 7.9 million temperature profiles were used in constructing the two longer time series. The three heat content analyses cover different periods but where they overlap in time there is good agreement. The time series shows an overall trend of increasing heat content in the World Ocean with interannual and inter-decadal variations superimposed on this trend. The root mean square difference between the three data sets is 1.5×10^{22} J. These year-to-year differences, which are due to differences in quality control and data used, are small and now approaching the accuracies required to close the Earth's radiation budget (e.g., Carton et al., 2005). On longer time scales, the two longest time series (using independent criteria for selection, quality control, interpolation and analysis

of similar data sets) show good agreement about long-term trends and also on decadal time scales.

For the period 1993 to 2003, the Levitus et al. (2005a) analysis has a linear global ocean trend of 0.42 ± 0.18 W m^{-2}, Willis et al. (2004) has a trend of 0.66 ± 0.18 W m^{-2} and Ishii et al. (2006) a trend of 0.33 ± 0.18 W m^{-2}. Overall, we assess the trend for this period as 0.5 ± 0.18 W m^{-2}. For the 0 to 700 m layer and the period 1955 to 2003 the heat content change is $10.9 \pm 3.1 \times 10^{22}$ J or 0.14 ± 0.04 W m^{-2} (data from Levitus et al., 2005a). All of these estimates are per unit area of Earth surface. Despite the fact that there are differences between these three ocean heat content estimates due to the data used, quality control applied, instrumental biases, temporal and spatial averaging and analysis methods (Appendix 5.A.1), they are consistent with each other giving a high degree of confidence for their use in climate change studies. The global increase in ocean heat content during the period 1993 to 2003 in two ocean models constrained by assimilating altimetric sea level and other observations (Carton et al., 2005; Köhl et al., 2006) is considerably larger than these observational estimates. We assess the heat content change from both of the long time series (0 to 700 m layer and the 1961 to 2003 period) to be $8.11 \pm 0.74 \times 10^{22}$ J, corresponding to an average warming of 0.1°C or 0.14 ± 0.04 W m^{-2}, and conclude that the available heat content estimates from 1961 to 2003 show a significant increasing trend in ocean heat content.

The data used in estimating the Levitus et al. (2005a) ocean temperature fields (for the above heat content estimates) do not include sea surface temperature (SST) observations,

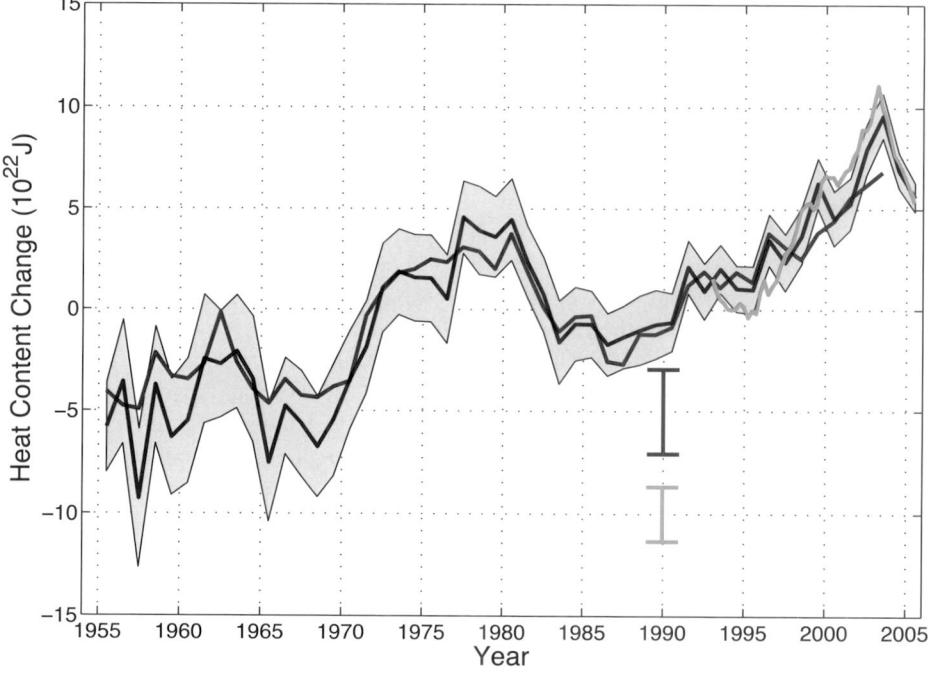

Figure 5.1. *Time series of global annual ocean heat content (10^{22} J) for the 0 to 700 m layer. The black curve is updated from Levitus et al. (2005a), with the shading representing the 90% confidence interval. The red and green curves are updates of the analyses by Ishii et al. (2006) and Willis et al. (2004, over 0 to 750 m) respectively, with the error bars denoting the 90% confidence interval. The black and red curves denote the deviation from the 1961 to 1990 average and the shorter green curve denotes the deviation from the average of the black curve for the period 1993 to 2003.*

which are discussed in Chapter 3. However, comparison of the global, annual mean time series of near-surface temperature (approximately 0 to 5 m depth) from this analysis and the corresponding SST series based on a subset of the International Comprehensive Ocean-Atmosphere Data Set (ICOADS) database (approximately 134 million SST observations; Smith and Reynolds, 2003 and additional data) shows a high correlation (r = 0.96) for the period 1955 to 2005. The consistency between these two data sets gives confidence in the ocean temperature data set used for estimating depth-integrated heat content, and supports the trends in SST reported in Chapter 3.

There is a contribution to the global heat content integral from depths greater than 700 m as documented by Levitus et al. (2000; 2005a). However, due to the lack of data with increasing depth the data must be composited using five-year running pentads in order to have enough data for a meaningful analysis in the deep ocean. Even then, there are not enough deep ocean data to extend the time series for the upper 3,000 m past the 1994–1998 pentad. There is a close correlation between the 0 to 700 and 0 to 3,000 m time series of Levitus et al. (2005a). A comparison of the linear trends from these two series indicates that about 69% of the increase in ocean heat content during 1955 to 1998 (the period when estimates from both time series are available) occurred in the upper 700 m of the World Ocean. Based on the linear trend, for the 0 to 3,000 m layer for the period 1961 to 2003 there has been an increase of ocean heat content of approximately $14.2 \pm 2.4 \times 10^{22}$ J, corresponding to a global ocean volume mean temperature increase of 0.037°C during this period. This increase in ocean heat content corresponds to an average heating rate of 0.21 ± 0.04 W m^{-2} for the Earth's surface.

The geographical distribution of the linear trend of 0 to 700 m heat content for 1955 to 2003 for the World Ocean is shown in Figure 5.2. These trends are non-uniform in space, with some regions showing cooling and others warming. Most of the Atlantic Ocean exhibits warming with a major exception being the subarctic gyre. The Atlantic Ocean accounts for approximately half of the global linear trend of ocean heat content (Levitus et al., 2005a). Much of the Indian Ocean has warmed since 1955 with a major exception being the 5°S to 20°S latitude belt. The Southern Ocean (south of 35°S) in the Atlantic, Indian and Pacific sectors has generally warmed. The Pacific Ocean is characterised by warming with major exceptions along 40°N and the western tropical Pacific.

Figure 5.3 shows the linear trends (1955 to 2003) of zonally averaged temperature anomalies (0 to 1,500 m) for the World Ocean and individual basins based on yearly anomaly fields (Levitus et al., 2005a). The strongest trends in these anomalies are concentrated in the upper ocean. Warming occurs at most latitudes in all three of the ocean basins. The regions that exhibit cooling are mainly in the shallow equatorial areas and in some high-latitude regions. In the Indian Ocean, cooling occurs at subsurface depths centred on 12°S at 150 m depth and in the Pacific centred on the equator and 150 m depth. Cooling also occured in the 32°N to 48°N region of the Pacific Ocean and the 49°N to 60°N region of the Atlantic Ocean. Regional temperature changes are discussed further in Section 5.3.

5.2.2.2   Variability of Heat Content

A major feature of Figure 5.1 is the relatively large increase in global ocean heat content during 1969 to 1980 and a sharp

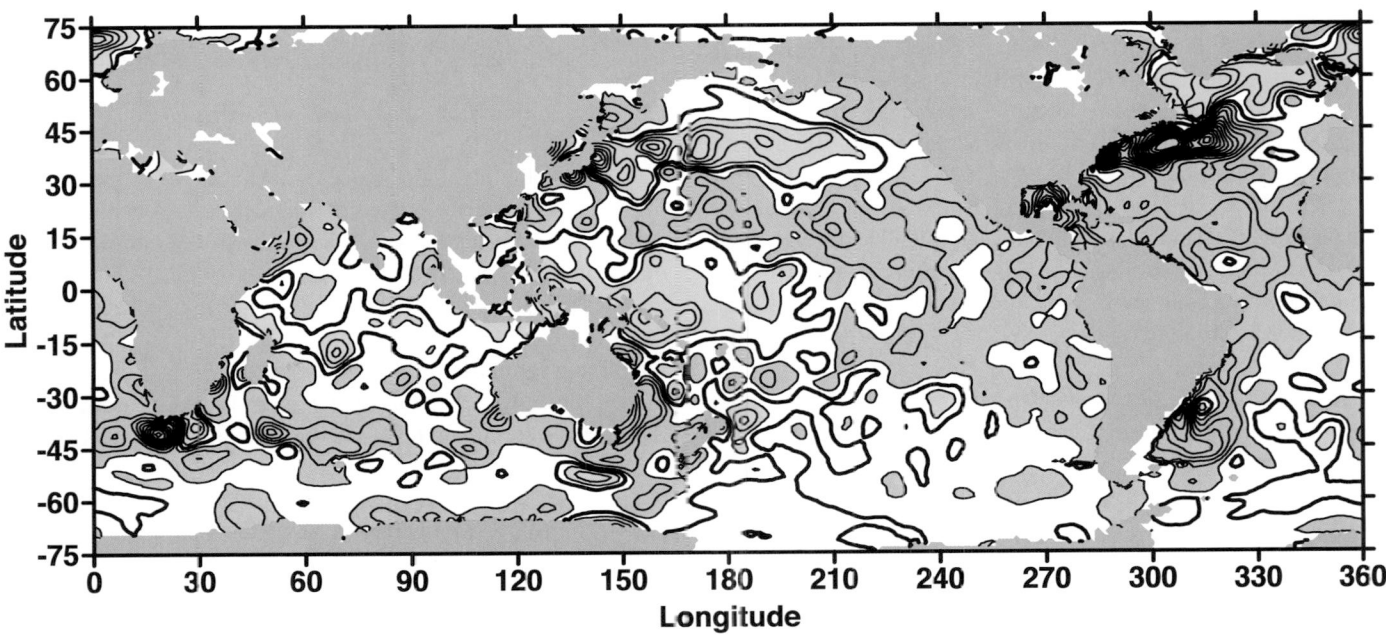

Figure 5.2. *Linear trends (1955–2003) of change in ocean heat content per unit surface area (W m^{-2}) for the 0 to 700 m layer, based on the work of Levitus et al. (2005a). The linear trend is computed at each grid point using a least squares fit to the time series at each grid point. The contour interval is 0.25 W m^{-2}. Red shading indicates values equal to or greater than 0.25 W m^{-2} and blue shading indicates values equal to or less than –0.25 W m^{-2}.*

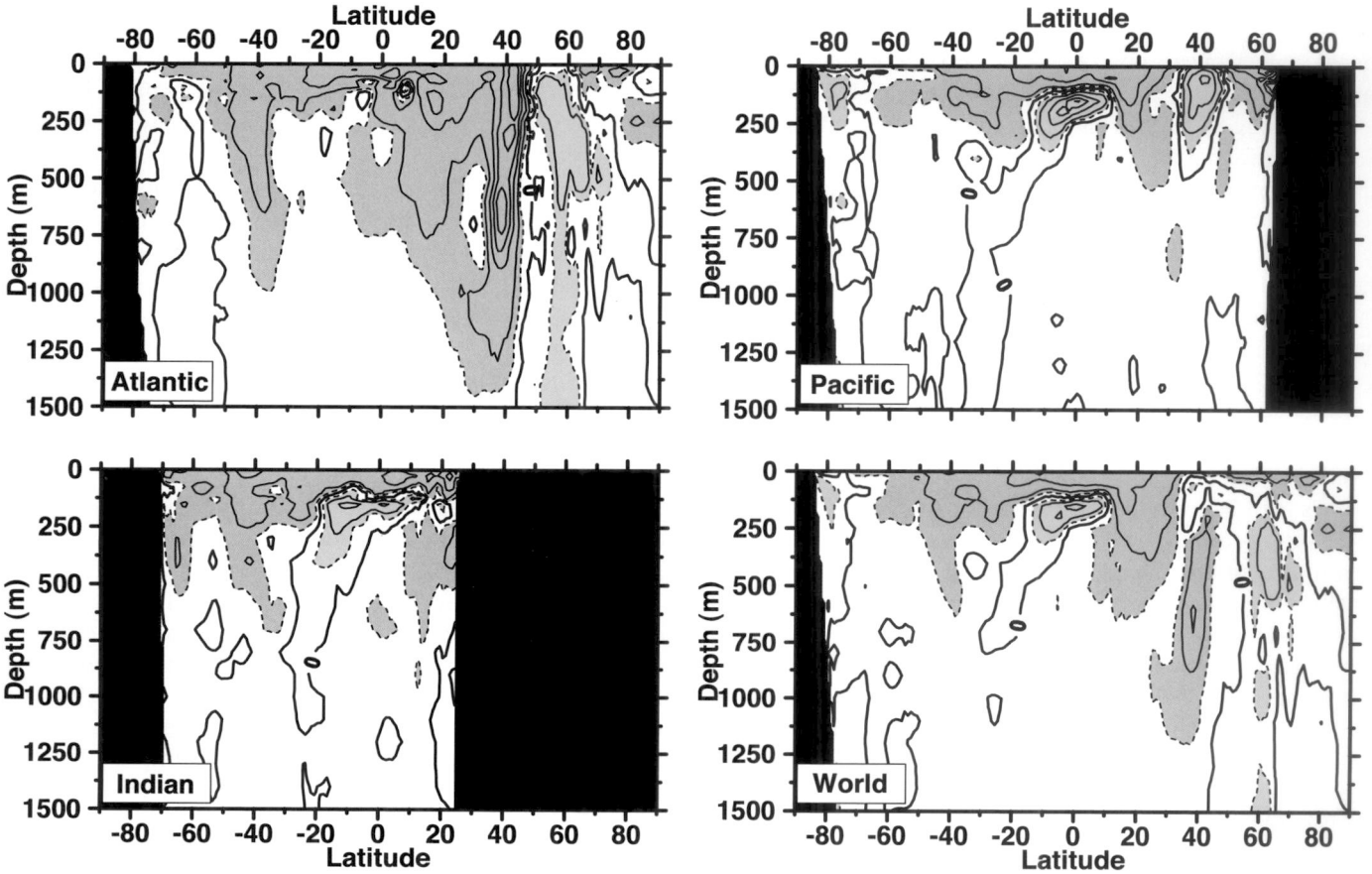

Figure 5.3. *Linear trend (1955–2003) of zonally averaged temperature in the upper 1,500 m of the water column of the Atlantic, Pacific, Indian and World Oceans. The contour interval is 0.05°C per decade, and the dark solid line is the zero contour. Red shading indicates values equal to or greater than 0.025°C per decade and blue shading indicates values equal to or less than –0.025°C per decade. Based on the work of Levitus et al. (2005a).*

decrease during 1980 to 1983. The 0 to 700 m layer cooled at a rate of 1.2 W m^{-2} during this period. Most of this cooling occurred in the Pacific Ocean and may have been associated with the reversal in polarity of the PDO (Stephens et al., 2001; Levitus et al., 2005c, see also Section 3.6.3). Examination of the geographical distribution of the differences in 0 to 700 m heat content between the 1977–1981 and 1965–1969 pentads and the 1986–1990 and 1977–1981 pentads shows that the pattern of heat content change has spatial scales of entire ocean basins and is also found in similar analyses by Ishii et al. (2006). The Pacific Ocean dominates the decadal variations of global heat content during these two periods. The origin of this variability is not well understood.

Based on model experiments, it has been suggested that errors resulting from the highly inhomogeneous distribution of ocean observations in space and time (see Appendix 5.A.1) could lead to spurious variability in the analysis (e.g., Gregory et al., 2004, AchutaRao et al., 2006). As discussed in the appendix, even in periods with overall good coverage in the observing system, large regions in Southern Hemisphere (SH) are not well sampled, and their contribution to global heat content variability is less certain. However, the large-scale nature of heat content variability, the similarity of the Levitus et al. (2005a) and the Ishii et al. (2006) analyses and new results showing a decrease in the

global heat content in a period with much better data coverage (Lyman et al., 2006), gives confidence that there is substantial inter-decadal variability in global ocean heat content.

5.2.2.3   *Implications for Earth's Heat Balance*

To place the changes of ocean heat content in perspective, Figure 5.4 provides updated estimates of the change in heat content of various components of the Earth's climate system for the period 1961 to 2003 (Levitus et al., 2005a). This includes changes in heat content of the lithosphere (Beltrami et al., 2002), the atmosphere (e.g., Trenberth et al., 2001) and the total heat of fusion due to melting of i) glaciers, ice caps and the Antarctic and Greenland Ice Sheets (see Chapter 4) and ii) arctic sea ice (Hilmer and Lemke, 2000). The increase in ocean heat content is much larger than any other store of energy in the Earth's heat balance over the two periods 1961 to 2003 and 1993 to 2003, and accounts for more than 90% of the possible increase in heat content of the Earth system during these periods. Ocean heat content variability is thus a critical variable for detecting the effects of the observed increase in greenhouse gases in the Earth's atmosphere and for resolving the Earth's overall energy balance. It is noteworthy that whereas ice melt from glaciers, ice caps and ice sheets is very important in the sea level budget

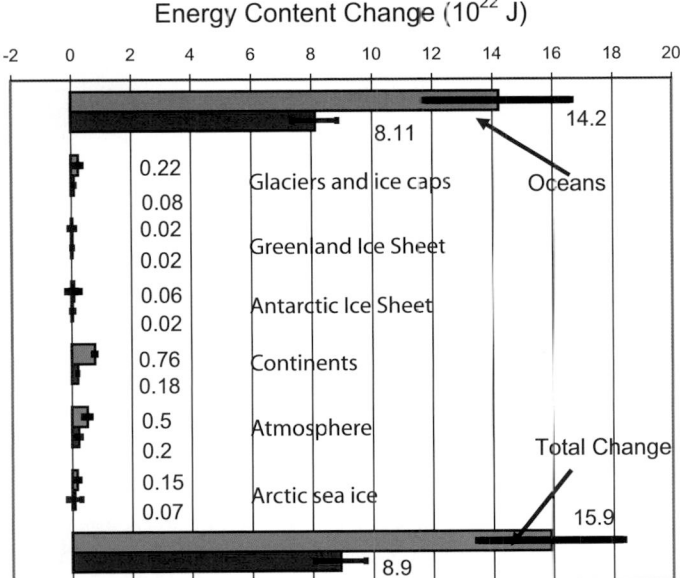

Figure 5.4. *Energy content changes in different components of the Earth system for two periods (1961–2003 and 1993–2003). Blue bars are for 1961 to 2003, burgundy bars for 1993 to 2003. The ocean heat content change is from this section and Levitus et al. (2005c); glaciers, ice caps and Greenland and Antarctic Ice Sheets from Chapter 4; continental heat content from Beltrami et al. (2002); atmospheric energy content based on Trenberth et al. (2001); and arctic sea ice release from Hilmer and Lemke (2000). Positive energy content change means an increase in stored energy (i.e., heat content in oceans, latent heat from reduced ice or sea ice volumes, heat content in the continents excluding latent heat from permafrost changes, and latent and sensible heat and potential and kinetic energy in the atmosphere). All error estimates are 90% confidence intervals. No estimate of confidence is available for the continental heat gain. Some of the results have been scaled from published results for the two respective periods. Ocean heat content change for the period 1961 to 2003 is for the 0 to 3,000 m layer. The period 1993 to 2003 is for the 0 to 700 m (or 750 m) layer and is computed as an average of the trends from Ishii et al. (2006), Levitus et al. (2005a) and Willis et al. (2004).*

(contributing about 40%), the energy associated with ice melt contributes only about 1% to the Earth's energy budget.

5.2.3 Ocean Salinity

Ocean salinity changes are an indirect but potentially sensitive indicator for detecting changes in precipitation, evaporation, river runoff and ice melt. The patterns of salinity change can be used to infer changes in the Earth's hydrological cycle over the oceans (Wong et al., 1999; Curry et al., 2003) and are an important complement to atmospheric measurements. Figure 5.5 shows the linear trends (based on pentadal anomaly fields) of zonally averaged salinity in the upper 500 m of the World Ocean and individual ocean basins (Boyer et al., 2005) from 1955 to 1998. A total of 2.3 million salinity profiles were used in this analysis, about one-third of the amount of data used in the ocean heat content estimates in Section 5.2.2.

Estimates of changes in the freshwater content of the global ocean have suggested that the global ocean is freshening (e.g., Antonov et al., 2002), however, sampling limitations due to data sparsity in some regions, particularly the SH, means that such estimates have an uncertainty that is not possible to quantify.

Between 15°S and 42°N in the Atlantic Ocean there is a salinity increase in the upper 500 m layer. This region includes the North Atlantic subtropical gyre. In the 42°N to 72°N region, including the Labrador, Irminger and Icelandic Seas, there is a freshening trend (discussed further in Section 5.3). The increase in salinity north of 72°N (Arctic Ocean) is highly uncertain because of the paucity of data in this region.

South of 50°S in the polar region of the Southern Ocean, there is a relatively weak freshening signal. Freshening occurs throughout most of the Pacific with the exception of the South Pacific subtropical gyre between 8°S and 32°S and above 300 m where there is an increase in salinity. The near-surface Indian Ocean is characterised mainly by increasing salinity. However, in the latitude band 5°S to 42°S (South Indian gyre) in the depth range of 200 to 1,000 m, there is a freshening of the water column.

The results shown here document that ocean salinity and hence freshwater are changing on gyre and basin scales, with the near-surface waters in the more evaporative regions increasing in salinity in almost all ocean basins. In the high-latitude regions in both hemispheres the surface waters are freshening consistent with these regions having greater precipitation, although higher runoff, ice melting, advection and changes in the MOC (Häkkinen, 2002) may also contribute. In addition to these meridional changes, the Atlantic is becoming saltier over much of the water column (Figure 5.5 and Boyer et al., 2005). Although the South Pacific subtropical region is becoming saltier, on average the whole water column in the Pacific Basin is becoming fresher (Boyer et al., 2005). The increasing difference in volume-averaged salinity between the Atlantic and Pacific Oceans suggests changes in freshwater transport between these two ocean basins.

We are confident that vertically coherent gyre and basin scale changes have occurred in the salinity (freshwater content) of parts of the World Ocean during the past several decades. While the available data and their analyses are insufficient to identify in detail the origin of these changes, the patterns are consistent with a change in the Earth's hydrological cycle, in particular with changes in precipitation and inferred larger water transport in the atmosphere from low latitudes to high latitudes and from the Atlantic to the Pacific (see Section 3.3.2).

5.2.4 Air-Sea Fluxes and Meridional Transports

The global average changes in ocean heat content discussed above are driven by changes in the air-sea net energy flux (see Section 5.2.2.1). At regional scales, few estimates of heat flux changes have been possible. During the last 50 years, net heat fluxes from the ocean to the atmosphere demonstrate locally decreasing values (up to 1 W m^{-2} yr^{-1}) over the southern flank of the Gulf Stream and positive trends (up to 0.5 W m^{-2} yr^{-1}) in the Atlantic central subpolar regions (Gulev et al., 2006). At the global scale, the accuracy of the flux observations is insufficient to permit a direct assessment of changes in heat flux. Air-sea fluxes are discussed in Section 3.5.6.

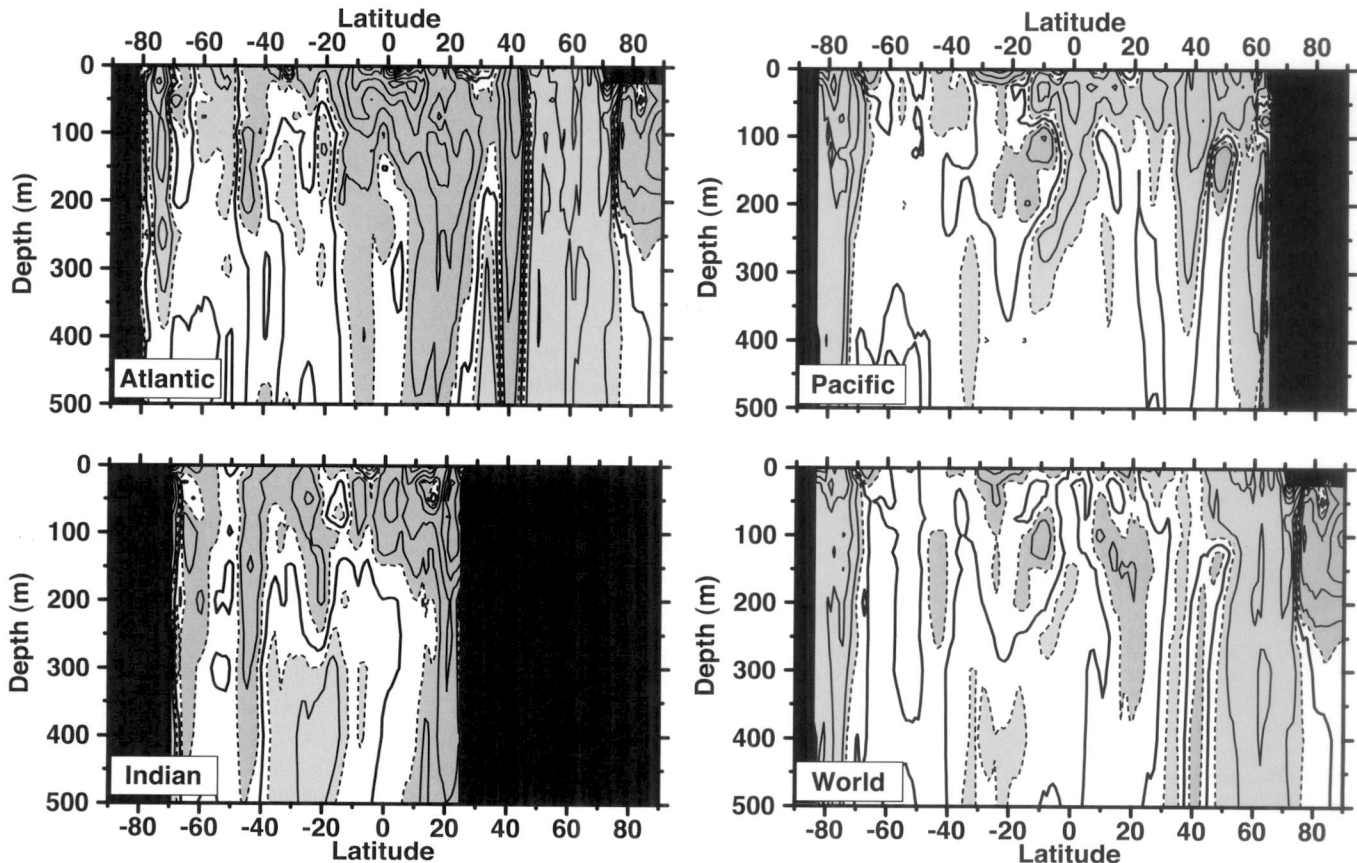

Figure 5.5: *Linear trends (1955–1998) of zonally averaged salinity (psu) in the upper 500 m of the Atlantic, Pacific, Indian and World Oceans. The contour interval is 0.01 psu per decade and dashed contours are ±0.005 psu per decade. The dark solid line is the zero contour. Red shading indicates values equal to or greater than 0.005 psu per decade and blue shading indicates values equal to or less than −0.005 psu per decade. Based on the work of Boyer et al. (2005).*

Estimates of the climatological mean oceanic meridional heat transport derived from atmospheric observations (e.g., Trenberth and Caron, 2001) and from oceanographic cross sections (e.g., Ganachaud and Wunsch, 2003) are in fair agreement, despite considerable uncertainties (see Appendix 5.A.2). The ocean heat transport estimate derived from integration of climatological air-sea heat flux fields (e.g., Grist and Josey, 2003) is in good agreement with an independent oceanographic cross section at 32°S. Estimates of changes in the Atlantic meridional heat transport are discussed in Section 5.3.2.

5.3 Regional Changes in Ocean Circulation and Water Masses

5.3.1 Introduction

Robust long-term trends in global- and basin-scale ocean heat content and basin-scale salinity were shown in Section 5.2. The observed heat and salinity trends are linked to changes in ocean circulation and other manifestations of global change such as oxygen and carbon system parameters (see Section 5.4). Global ocean changes result from regional changes in these properties,

assessed in this section. Evidence for change in temperature, salinity and circulation is described globally and then for each of the major oceans. Two marginal seas with multi-decadal time series are also examined as examples of regional variations.

The upper ocean in all regions is close to the atmospheric forcing and has the largest variability; it is also the best sampled. For these reasons, Section 5.2 mainly assessed upper-ocean observations for long-term trends in heat content and salinity. However, there are important changes in heat and salinity at intermediate and abyssal depths, restricted to regions that are relatively close to the main sources of deep and intermediate waters. These sources are most vigorous in the northern North Atlantic and the Southern Ocean around Antarctica. This is illustrated well in salinity differences shown for the Atlantic (1985–1999 minus 1955–1969) and Pacific (1980s minus 1960s) in Figure 5.6. Striking changes in salinity are found from the surface to the bottom in the northern North Atlantic near water mass formation sites that fill the water column (Section 5.3.2); bottom changes elsewhere are small, being most prevalent at the under-sampled southern ends of both sections. At mid-depth (500 to 2,000 m), the Atlantic and southern end of the Pacific section show widespread change, but the North Pacific signal is weaker and shallower because it has only weak intermediate water formation (and no deep water formation). Changes in intermediate and deep waters can ultimately affect the ocean's vertical stratification and overturning circulation; the topic of the overturning circulation in the North Atlantic is considered in Section 5.3.2.

The observed changes in salinity are of global scale, with similar patterns in different ocean basins (Figure 5.6). The subtropical waters have increased in salinity and the subpolar surface and intermediate waters have freshened in both the Atlantic and Pacific Oceans during the period from the 1960s to the 1990s and in both hemispheres in each ocean. The waters that underlie the near-surface subtropical waters have freshened due to equatorward circulation of the freshened subpolar surface waters; in particular, the fresh intermediate water layer (at ~1,000 m) in the SH has freshened in both the Atlantic and Pacific Oceans. In the Northern Hemisphere (NH), the Pacific intermediate waters have freshened, and the underlying deep waters did not change, consistent with no local bottom water source in the North Pacific. In the central North Atlantic, the intermediate layer (approximately 900–1,200 m) became saltier due to increased salinity in the outflow from the Mediterranean that feeds this layer.

5.3.2 Atlantic and Arctic Oceans

The North Atlantic Ocean has a special role in long-term climate assessment because it is central to one of the two global-scale MOCs (see Box 5.1), the other location being the Southern Ocean. The long-term trends in depth-integrated Atlantic heat content for the period 1955 to 2003 (Figure 5.2) are broadly consistent with the warming tendencies identified from the global analyses of SST (see Section 3.2.2.3). The subtropical gyre warmed and the subpolar gyre cooled over that period, consistent with a predominantly positive phase of the NAO during the last several decades. The warming extended down to below 1,000 m, deeper than anywhere else in the World Ocean (Figure 5.3 Atlantic), and was particularly pronounced under the Gulf Stream and North Atlantic Current near 40°N. Long-term trends in salinity towards freshening in the subpolar regions and increased salinity in the subtropics through the mid-1990s (Figure 5.5 Atlantic and Figure 5.6a) are consistent with the global tendencies for freshening of relatively fresher regions and increased salinity in saltier regions (Section 5.2.3).

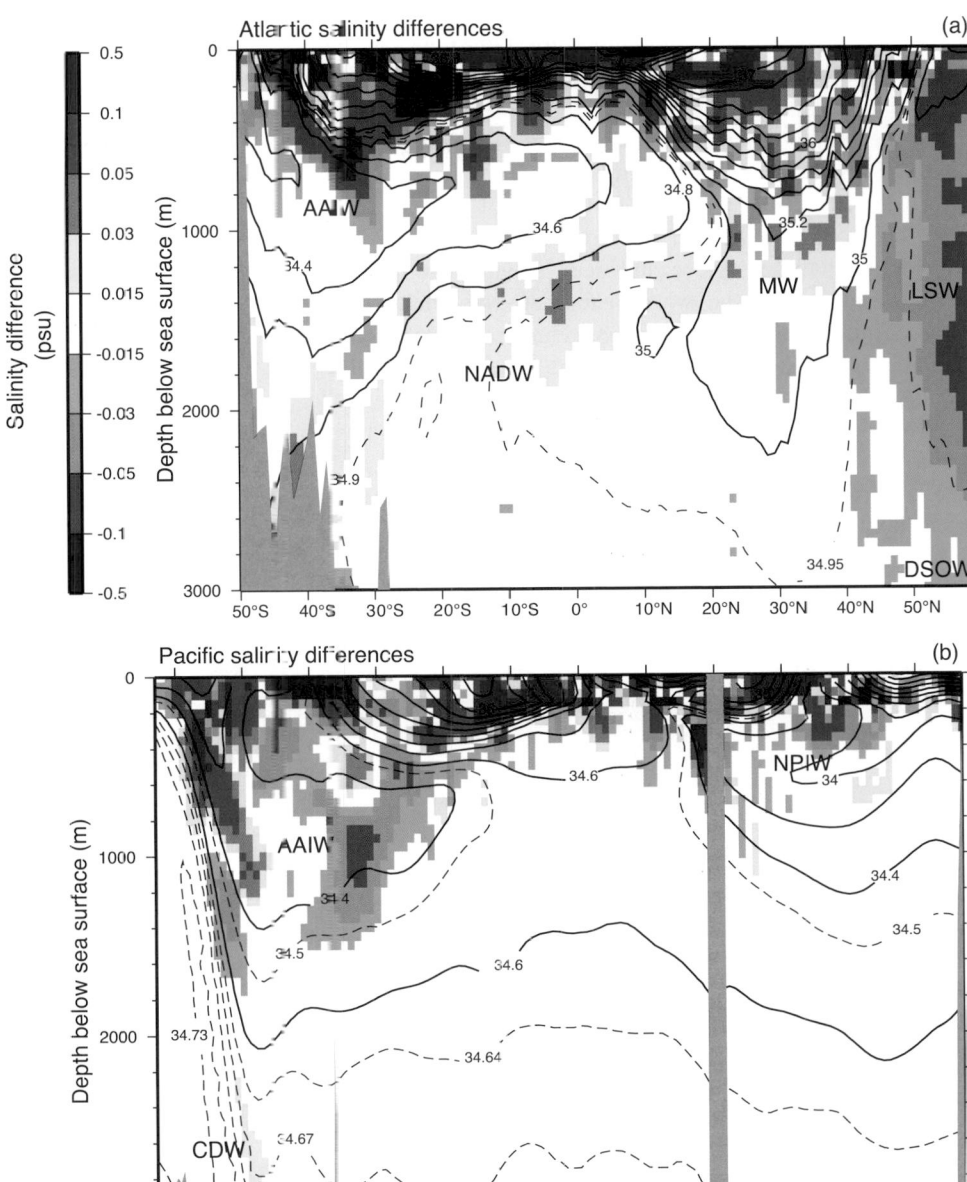

Figure 5.6. *Meridional sections of differences in salinity (psu) of the a) Atlantic Ocean for the period 1985 to 1999 minus 1955 to 1969 and b) Pacific Ocean for the World Ocean Circulation Experiment (WOCE) 150°W section (1991–1992) and historical data from 1968 plus or minus 7.5 years. Contours are the mean salinity fields along each section and show the key features. The salinity differences are differences along isopycnals that have been mapped to pressure surfaces. The Atlantic section is along the western side of the Atlantic Ocean and the Pacific section is along 150°W. The two figures are redrafted from Curry et al. (2003) and Wong et al (2001). Water masses shown include Antarctic Intermediate Water (AAIW), Circumpolar Deep Water (CDW), North Atlantic Deep Water (NADW), Mediterranean Water (MW), Labrador Sea Water (LSW), Denmark Strait Overflow Water (DSOW) and North Pacific Intermediate Water (NPIW). The areas shaded in grey represent the seafloor and oceanic crust.*

5.3.2.1 *North Atlantic Subpolar Gyre, Labrador Sea and Nordic Seas*

In the North Atlantic subpolar gyre, Labrador Sea and Nordic Seas, large salinity changes have been observed that have been associated with changed inputs of fresh water (ice melt, ocean circulation and river runoff) and with the NAO. Advection of these surface and deep salinity anomalies has been traced around the whole subpolar gyre including the Labrador and Nordic Seas. These anomalies are often called 'Great Salinity Anomalies' (GSAs; e.g., Dickson et al., 1988; Belkin, 2004). During a positive phase of the NAO, the subpolar gyre strengthens and expands towards the east, resulting in lower surface salinity in the central subpolar region (Levitus, 1989; Reverdin et al., 1997; Bersch, 2002). Three GSAs have been thoroughly documented: one from 1968 to 1978, one in the 1980s and one in the 1990s. Observational and modelling studies show that the relative influence of local events and advection differ between different GSA events and regions (Houghton and Visbeck, 2002; Josey and Marsh, 2005).

These surface salinity anomalies have affected the Labrador Sea and the production of Labrador Sea Water (LSW), a major component of the North Atlantic Deep Water (NADW) and contributor to the lower limb of the MOC. The LSW appears to alternate between dense, cold types and less dense, warm types (Yashayaev et al., 2003; Kieke et al., 2006) possibly with more production of dense LSW during years of positive-phase NAO (Dickson et al., 1996). Since 1965 to 1970, the LSW has had a significant freshening trend with a superimposed variability consisting of three saltier periods, coinciding with warmer water, and two freshening and cooling periods in the 1970s and 1990s (Figure 5.7). During the period 1988 to 1994, an exceptionally large volume of cold, fresh and dense LSW was produced (Sy et al., 1997; Lazier et al., 2002), unprecedented in the sparse time series that extends back to the 1930s (Talley and McCartney, 1982). The Labrador Sea has now returned to a warmer, more saline state; most of the excess volume of the dense LSW has disappeared, the mid-layers became warmer and saltier, and the production of LSW shifted to the warmer type (e.g., Lazier et al., 2002; Yashayaev et al., 2003; Stramma et al., 2004). This warming and increased salinity and reduction in LSW was associated with the weakening of the North Atlantic subpolar gyre, seen also in satellite altimetry data (Häkkinen and Rhines, 2004).

The eastern half of the subpolar North Atlantic also freshened through the 1980s and into the 1990s, but the upper ocean has been increasing in salinity or remaining steady since then, depending on location. About two-thirds of the freshening in this region has been attributed to an increase in precipitation associated with a climate pattern known as the East Atlantic Pattern (Josey and Marsh, 2005), with the NAO playing a secondary role. From 1965 to 1995, the subpolar freshening amounted to an equivalent freshwater layer of approximately 3 m spread evenly over its total area (Belkin, 2004; Curry and Mauritzen, 2005).

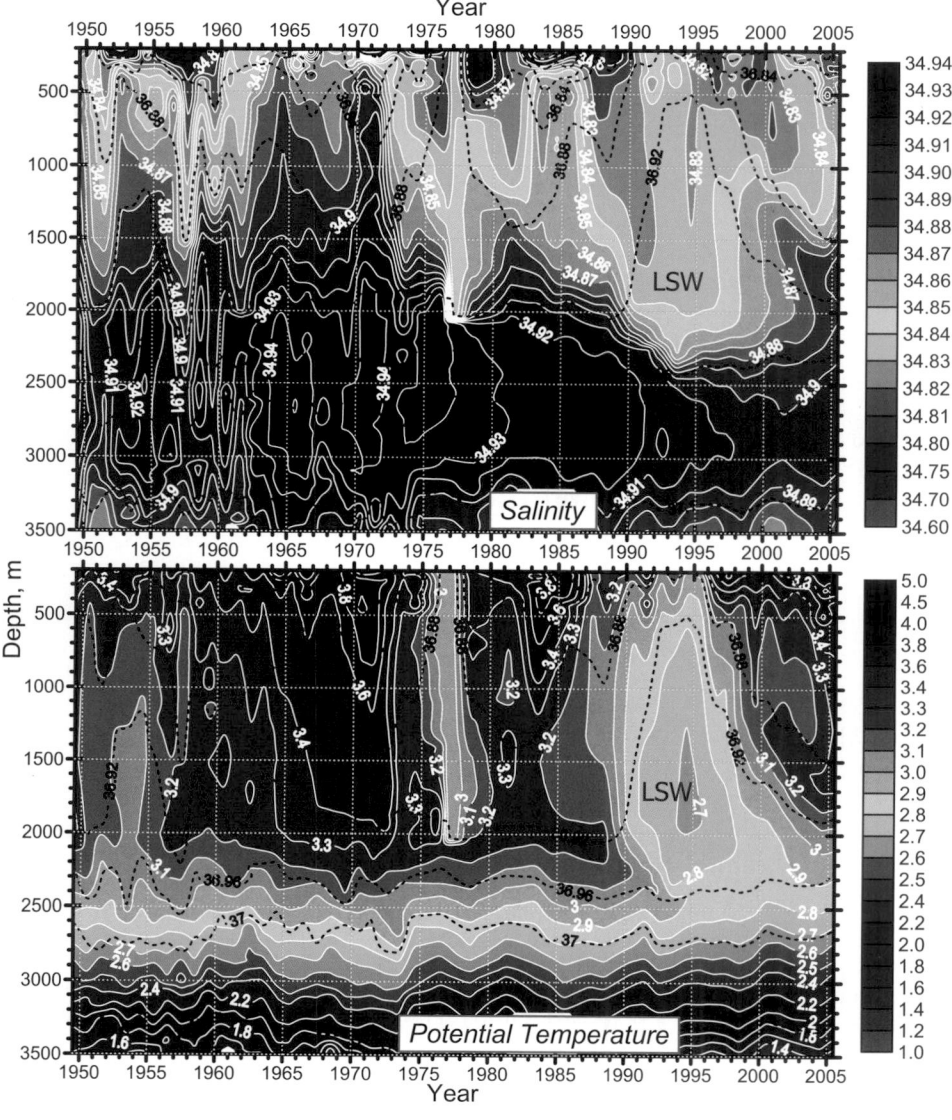

Figure 5.7. *The longest available time series of salinity (psu; upper panel) and potential temperature (°C, lower panel) in the central Labrador Sea from 1949 to 2005 (updated from Yashayaev et al., 2003). The dashed lines are contours of potential density (kg m⁻³, difference from 1,000 kg m⁻³) and are the same on both panels.*

Box 5.1: Has the Meridional Overturning Circulation in the Atlantic Changed?

The global Meridional Overturning Circulation consists primarily of dense waters that sink to the abyssal ocean at high latitudes in the North Atlantic Ocean and near Antarctica. These dense waters then spread across the equator with comparable flows of approximately 17 and 14 Sv (10^6 m³ s⁻¹), respectively (Orsi et al., 2002; Talley et al., 2003a). The North Atlantic overturning circulation (henceforth 'MOC') is characterised by an inflow of warm, saline upper-ocean waters from the south that gradually increase in density from cooling as they move northward through the subtropical and subpolar gyres. They also freshen, which reduces the density increase. The inflows reach the Nordic Seas (Greenland, Iceland and Norwegian Seas) and the Labrador Sea, where they are subject to deep convection, sill overflows and vigorous mixing. Through these processes NADW is formed, constituting the southward-flowing lower limb of the MOC.

Climate models show that the Earth's climate system responds to changes in the MOC (e.g., Vellinga and Wood, 2002), and also suggest that the MOC might gradually decrease in transport in the 21st century as a consequence of anthropogenic warming and additional freshening in the North Atlantic (Bi et al., 2001; Gregory et al., 2005; see also Chapter 10). However, observations of changes in the MOC strength and variability are fragmentary; the best evidence for observational change comes from the North Atlantic.

There is evidence for a link between the MOC and abrupt changes in surface climate during the past 120 kyr, although the exact mechanism is not clear (Clark et al., 2002). At the end of the last glacial period, as the climate warmed and ice sheets melted, there were a number of abrupt oscillations, for example, the Younger Dryas and the 8.2 ka cold event (see Section 6.4), which may have been caused by changes in ocean circulation. The variability of the MOC during the Holocene after the 8.2 ka cooling event is clearly much smaller than during glacial times (Keigwin et al., 1994; see Section 5.4).

Observed changes in MOC transport, water properties and water mass formation are inconclusive about changes in the MOC strength (see Section 5.3.2.2). This is partially due to decadal variability and partially due to inadequate long-term observations. From repeated hydrographic sections in the subtropics, Bryden et al. (2005) concluded that the MOC transport at 25°N had decreased by 30% between 1957 and 2004, but the presence of significant unsampled variability in time and the lack of supporting direct current measurements reduces confidence in this estimate. Direct measurements of the two major sill overflows have shown considerable variability in the dominant Denmark Strait Overflow without enough years of coverage to discern long-term trends (Macrander et al., 2005). The observed freshening of the overflows and the associated reduction in density from 1965 to 2000 (see Section 5.3.2) has so far not led to a significant weakening of the MOC (Dickson et al., 2003; Curry and Mauritzen, 2005). Moreover, large decadal variability observed since 1960 in salinity and temperature of the surface waters, including the recent increase in salinity of the surface waters feeding the MOC, obscures the long-term trend (Hátún et al., 2005; ICES 2005) and hence conclusions about potential MOC changes.

Changes in the MOC can also be caused by changes in Labrador Sea convection, with strong convection corresponding to higher MOC. Convection was strong from the 1970s to 1995, but thereafter the Labrador Sea warmed and re-stratified (Lazier et al., 2002; Yashayaev et al., 2003) and convection has been weaker. Based on observed SST patterns, it was concluded that the MOC transport has increased by about 10% from 1970 to the 1990s (Knight et al., 2005; Latif et al., 2006). From direct current meter observations at the exit of the subpolar North Atlantic, Schott et al. (2004) concluded that the Deep Water outflow, while varying at shorter time scales, had no significant trend during the 1993 to 2001 period.

In summary, it is very likely that up to the end of the 20th century the MOC was changing significantly at interannual to decadal time scales. Given the above evidence from components of the MOC as well as uncertainties in the observational records, over the modern instrumental record no coherent evidence for a trend in the mean strength of the MOC has been found.

Subsurface salinity in the Nordic Seas has also decreased markedly since the 1970s (Dickson et al., 2003), directly affecting the salinity of the Nordic Sea overflow waters that contribute to NADW. This decrease in subsurface salinity was associated with lower salinity of the Atlantic waters entering the Nordic seas and related to the high NAO index and intensification of the subpolar gyre. Since 1994, the salinity of the inflow from the North Atlantic has been increasing, reaching the highest values since 1948, largely due to a weakening of the subpolar gyre circulation that allowed more warm water into the Nordic Seas, associated with a decreasing NAO index (Hátún et al., 2005).

The densest waters contributing to NADW and to the deep limb of the MOC arise as overflows from the upper 1,500 m of the Nordic Seas through the Denmark Strait and Faroe Channel. The marked freshening of the overflow water masses exiting the Arctic was associated with growing sea ice export from the Arctic and precipitation in the Nordic Seas (Dickson et al., 2002, 2003). The transports of the overflow waters, of which the largest component is through Denmark Strait, have varied by about 30% (Macrander et al., 2005), but there has been no clear trend in this location. Overall, the overflows that contribute to NADW from the Nordic Seas have remained constant to within the known variability.

The overall pattern of change in the North Atlantic subpolar gyre is one of a trend towards fresher values over most of the water column from the mid-1960s until the mid-1990s. Since then, there has been a return to warmer and more saline waters

(Figure 5.7), which coincides with the change in NAO and East Atlantic Pattern. However, this return to saltier waters has not been sustained for a long enough period to change the sign of the long-term trends (Figure 5.5 Atlantic).

5.3.2.2 Arctic Ocean

Climate change in the Arctic Ocean and Nordic Seas is closely linked to the North Atlantic subpolar gyre (Østerhus et al., 2005). Within the Arctic Ocean and Nordic Seas, surface temperature has increased since the mid-1980s and continues to increase (Comiso, 2003). In the Atlantic waters entering the Nordic Seas, a temperature increase in the late 1980s and early 1990s (Quadfasel et al., 1991; Carmack et al., 1995) has been associated with the transition in the 1980s towards more positive NAO states. Warm Atlantic waters have also been observed to enter the Arctic as pulses via Fram Strait and then along the slope to the Laptev Sea (Polyakov et al., 2005); the increased heat content and increased transport in the pulses both contribute to net warming of the arctic waters (Schauer et al., 2004). Multi-decadal variability in the temperature of the Atlantic Water core affecting the top 400 m in the Arctic Ocean has been documented (Polyakov et al., 2004). Within the Arctic, salinity increased in the upper layers of the Amundsen and Makarov Basins, while salinity of the upper layers in the Canada Basin decreased (Morison et al., 1998). Compared to the 1980s, the area of upper waters of Pacific origin has decreased (McLaughlin et al., 1996; Steele and Boyd, 1998).

During the 1990s, changed winds caused eastward redirection of river runoff from the Laptev Sea (Lena River, etc.), reducing the low-salinity surface layer in the central Arctic Ocean (Steele and Boyd, 1998), thus allowing greater convection and heat transport into the surface arctic layer from the more saline subsurface Atlantic layer. Thereafter, however, the stratification in the central Arctic (Amundsen Basin) increased and a low-salinity mixed layer was again observed at the North Pole in 2001, possibly due to a circulation change that restored the river water input (Björk et al., 2002). Circulation variability that shifts the balance of fresh and saline surface waters in the Arctic, with associated changes in sea ice, might be associated with the NAM (Proshutinsky and Johnson, 1997; Rigor et al., 2002), however, the long-term decline in arctic sea ice cover appears to be independent of the NAM (Comiso, 2002). While there is significant decadal variability in the Arctic Ocean, no systematic long-term trend in subsurface arctic waters has been identified.

5.3.2.3 Subtropical and Equatorial Atlantic

In the North Atlantic subtropical gyre, circulation, SST, the thickness of near-surface Subtropical Mode Water (STMW, Hanawa and Talley, 2001) and thermocline ventilation are all highly correlated with the NAO, with some time lags. A more positive NAO state, with westerlies shifted northwards, results in a decreased Florida Current transport (Baringer and Larsen, 2001), a likely delayed northward shift of the Gulf Stream position (Joyce et al., 2000; Seager et al., 2001; Molinari,

2004), and decreased subtropical eddy variability (Penduff et al., 2004). In the STMW, low thickness and production and higher temperature result from a high NAO index (e.g., Talley, 1996; e.g., Hazeleger and Drijfhout, 1998; Marsh, 2000). The volume of STMW is likely to lag changes in the NAO by two to three years, and low (high) volumes are associated with high (low) surface layer temperatures because of changes in both convective forcing and location of STMW formation. While quasi-cyclic variability in STMW renewal is apparent over the 1960 to 1980 period, the total volume of STMW has remained low through 2000 since a peak in 1983 to 1984, associated with a relatively persistent positive NAO phase during the late 1980s and early 1990s (Lazier et al., 2002; Kwon and Riser, 2004).

In the subtropics at depths of 1,000 to 2,000 m, the temperature has increased since the late 1950s at Bermuda, at 24°N, and at 52°W and 66°W in the Gulf Stream (Bryden et al., 1996; Joyce and Robbins, 1996; Joyce et al., 1999). These warming trends reflect reduced production of LSW (Lazier, 1995) and increased salinity and temperature of the waters from the Mediterranean (Roether et al., 1996; Potter and Lozier, 2004). After the mid-1990s at greater depths (1,500–2,500 m), temperature and salinity decreased, reversing the previous warming trend, most likely due to delayed appearance of the new colder and fresher Labrador Sea Water produced in the mid-1990s.

Intermediate water (800–1,200 m) in the mid-latitude eastern North Atlantic is strongly influenced by the saline Mediterranean Water (MW; Section 5.3.2.3). This saline layer joins the southward-flowing NADW and becomes part of it in the tropical Atlantic. This layer has warmed and become more saline since at least 1957 (Bryden et al., 1996), continuing during the last decade (1994–2003) at a rate of more than 0.2°C per decade with a rate of 0.4°C per decade at some levels (Vargas-Yáñez et al., 2004). In the Bay of Biscay (44°N; González-Pola et al., 2005) and at Gibraltar (Millot et al., 2006), similar warming was observed through the thermocline and into the core of the MW. From 1955 to 1993, the trend was about 0.1°C per decade in a zone west of Gibraltar (Potter and Lozier, 2004), and of almost the same magnitude even west of the mid-Atlantic Ridge (Curry et al., 2003).

Surface waters in the Southern Ocean, including the high-latitude South Atlantic, set the initial conditions for bottom water in the (SH). This extremely dense Antarctic Bottom Water (AABW), which is formed around the coast of Antarctica (see Section 5.3.5.2), spreads equatorward and enters the Brazil Basin through the narrow Vema Channel of the Rio Grande Rise at 31°S. Ongoing observations of the lowest bottom temperatures there have revealed a slow but consistent increase of the order 0.002°C yr^{-1} in the abyssal layer over the last 30 years (Hogg and Zenk, 1997).

In the tropical Atlantic, the surface water changes are partly associated with the variability of the marine Inter-tropical Convergence Zone, which has strong seasonal variability (Mitchell and Wallace, 1992; Biasutti et al., 2003; Stramma et al., 2003). Tropical Atlantic variability on interannual to decadal time scales can be influenced by a South Atlantic dipole in SST

(Venegas et al., 1998), associated with latent heat fluxes related to changes in the subtropical high (Sterl and Hazeleger, 2003). The South Equatorial Current provides a region for subduction of the water masses (Hazeleger et al., 2003) and may also maintain a propagation pathway for water mass anomalies towards the north (Lazar et al., 2002).

The North Atlantic Oscillation is an important driver of the oceanic water mass variations in the upper North Atlantic subtropical gyre. Its effects are also observed at depths greater than 1,500 m within the subtropical gyre consistent with the large-scale circulation and changes in source waters in the North Atlantic Ocean. While there are coherent changes in the long-term trends in temperature and salinity (Section 5.2), decadal variations are an important climate signal for this region.

5.3.2.4   *Mediterranean Sea*

Marked changes in thermohaline properties have been observed throughout the Mediterranean (Manca et al., 2002). In the western basin, the Western Mediterranean Deep Water (WMDW), formed in the Gulf of Lions, warmed during the last 50 years, interrupted by a short period of cooling in the early 1980s, the latter reflected in cooling of the Levantine Intermediate Water between the late 1970s and mid-1980s (Brankart and Pinardi, 2001). The WMDW warming is in agreement with recent atmospheric temperature changes over the Mediterranean (Luterbacher et al., 2004). The salt content of the WMDW has also been steadily increasing during the last 50 years, mainly attributed to decreasing precipitation over the region since the 1940s (Krahmann and Schott, 1998; Mariotti et al., 2002) and to anthropogenic reduction in the freshwater inflow (Rohling and Bryden, 1992). These changes in water properties and circulation are linked to the long-term variability of surface fluxes (Krahmann and Schott, 1998) with contributions from the NAO (Vignudelli et al., 1999) that produce consistent changes in surface heat fluxes and a net warming of the Mediterranean Sea (Rixen et al., 2005).

These changes in the temperature and salinity within the Mediterranean have affected the outflow of water into the North Atlantic at Gibraltar (see also Section 5.3.2.3). Part of this shift in Mediterranean outflow properties has been traced to the Eastern Mediterranean. During 1987 to 1991, the Eastern Mediterranean Deep Water became warmer and saltier due to the switch of its source water from the Adriatic to the Aegean (Klein et al., 2000; Gertman et al., 2006), most likely related to changes in the heat and freshwater flux anomalies in the Aegean Sea (Tsimplis and Rixen, 2002; Josey, 2003; Rupolo et al., 2003). This 1987 to 1991 switch of source waters has continued and increased its impact, with density of the westward outflow in Sicily Strait now denser (Gasparini et al., 2005). While there are strong natural variations in the Mediterranean, overall there is a discernible trend of increased salinity and warmer temperature in key water masses over the last 50 years and this signal is observable in the North Atlantic.

5.3.3   Pacific Ocean

The upper Pacific Ocean has been warming and freshening overall, as revealed in global heat and freshwater analyses (Section 5.2, Figure 5.5). The subtropical North and South Pacific have been warming. In the SH, the major warming footprint is associated with the thick mode waters north of the Antarctic Circumpolar Current. The North Pacific has cooled along 40°N. Long-term trends are rather difficult to discern in the upper Pacific Ocean because of the strong interannual and decadal variability (ENSO and the PDO) and the relatively short length of the observational records. Changes associated with ENSO are described in Section 3.6.2 and are not included here. Overall, the Pacific is freshening but there are embedded salinity increases in the subtropical upper ocean, where strong evaporation dominates.

5.3.3.1   *Pacific Upper Ocean Changes*

In the North Pacific, the zonally averaged temperature warming trend from 1955 to 2003 (Figure 5.3) is dominated by the PDO increase in the mid-1970s. The strong cooling between 50 and 200 m is due to relaxation and subsequent shallowing of the tropical thermocline, resulting from a decrease in the shallow tropical MOC and a relaxation of the equatorial thermocline (McPhaden and Zhang, 2002), although after 1998 this shallow overturning circulation returned to levels almost as high as in the 1970s (McPhaden and Zhang, 2004).

Warming in the North Pacific subtropics, cooling around 40°N and slight warming farther north is the pattern associated with a positive PDO (strengthened Aleutian Low; Miller and Douglas, 2004; see Figure 3.28). Within the North Pacific Ocean, a positive PDO state such as occurred after 1976 is characterised by a strengthened Kuroshio Extension. After 1976, the Kuroshio Extension and North Pacific Current transport increased by 8% and expanded southward (Parrish et al., 2000). The Kuroshio's advection of temperature anomalies has been shown to be of similar importance to variations in ENSO and the strength of the Aleutian Low in maintaining the positive PDO (Schneider and Cornuelle, 2005). The Oyashio penetrated farther southward along the coast of Japan during the 1980s than during the preceding two decades, consistent with a stronger Aleutian Low (Sekine, 1988; Hanawa, 1995; Sekine, 1999). A shoaling of the halocline in the centre of the western subarctic gyre and a concurrent southward shift of the Oyashio extension front during 1976 to 1998 vs. 1945 to 1975 has been detected (Joyce and Dunworth-Baker, 2003). Similarly, mixed layer depth decreased throughout the eastern subarctic gyre, with a distinct trend over 50 years (Freeland et al., 1997; Li et al., 2005).

Temperature changes in upper-ocean water masses in response to the more positive phase of the PDO after 1976 are well documented. The thick water mass just south of the Kuroshio Extension in the subtropical gyre (Subtropical Mode Water) warmed by 0.8°C from the mid-1970s to the late 1980s, associated with stronger Kuroshio advection, and the thick water mass along the subtropical-subpolar boundary near 40°N

(North Pacific Central Mode Water) cooled by 1°C following the shift in the PDO after 1976 (Yasuda et al., 2000; Hanawa and Kamada, 2001).

Trends towards increased heat content include a major signal in the subtropical South Pacific, within the thick mixed layers just north of the Antarctic Circumpolar Current (Willis et al., 2004; Section 5.3.5). The strength of the South Pacific subtropical gyre circulation increased more than 20% after 1993, peaking in 2003, and subsequently declined. This spin up is linked to an increase of Ekman pumping over the gyre due to an increase in the SAM index (Roemmich et al., 2007).

The marginal seas of the Pacific Ocean are also subject to climate variability and change. Like the Mediterranean in the North Atlantic, the Japan (or East) Sea is nearly completely isolated from the adjacent ocean basin, and forms all of its own waters beneath the shallow pycnocline. Because of this sea's limited size, it responds quickly through its entire depth to surface forcing changes. The warming evident through the global ocean is clearly apparent in this isolated basin, which warmed by 0.1°C at 1,000 m and 0.05°C below 2,500 m since the 1960s. Salinity at these depths also changed, by 0.06 psu per century for depths of 300 to 1,000 m and by –0.02 psu per century below 1,500 m (Kwon et al., 2004). These changes have been attributed to reduced surface heat loss and increased surface salinity, which have changed the mode of ventilation (Kim et al., 2004). Deep water production in the Japan (East) Sea slowed for many decades, with a marked decrease in dissolved oxygen from the 1930s to 2000 at a rate of about 0.8 μmol kg^{-1} yr^{-1} (Gamo et al., 1986; Minami et al., 1998). However, possibly because of weakened vertical stratification at mid-depths associated with the decades-long warming, deep-water production reappeared after the 2000–2001 severe winter (e.g., Kim et al., 2002; Senjyu et al., 2002; Talley et al., 2003b). Nevertheless, the overall trend has continued with lower deep-water production in subsequent years.

5.3.3.2 Intermediate and Deep Circulation and Water Property Changes

Since the 1970s, the major mid-depth water mass in the North Pacific, North Pacific Intermediate Water (NPIW), has been freshening and has become less ventilated, as measured by oxygen content (see Section 5.4.3). The NPIW is formed in the subpolar North Pacific, with most influence from the Okhotsk Sea, and reflects changes in northern North Pacific surface conditions. The salinity of NPIW decreased by 0.1 and 0.02 psu in the subpolar and subtropical gyres, respectively (Wong et al., 2001; Joyce and Dunworth-Baker, 2003). An oxygen decrease and nutrient increase in the NPIW south of Hokkaido from 1970 to 1999 was reported (Ono et al., 2001), along with a subpolar basin-wide oxygen decrease from the mid-1980s to the late 1990s (Watanabe et al., 2001). Warming and freshening occurred in the Okhotsk Sea in the latter half of the 20th century (Hill et al., 2003). The Okhotsk Sea intermediate water thickness was reduced and its density decreased in the 1990s (Yasuda et al., 2001).

In the southwest Pacific, in the deepest waters originating from the North Atlantic and Antarctica, cooling and freshening of 0.07°C and 0.01 psu from 1968 to 1991 was observed (Johnson and Orsi, 1997) and attributed to a change in the relative importance of Antarctic and North Atlantic source waters and weakening bottom transport. Bottom waters in the North Pacific are farther from the surface sources than any other of the world's deep waters. They are also the most uniform, in terms of spatial temperature and salinity variations. A large-scale, significant warming of the bottom 1,000 m across the entire North Pacific of the order of 0.002°C occurred between 1985 and 1999, measurable because of the high accuracy of modern instruments (Fukasawa et al., 2004). The cause of this warming is uncertain, but could have resulted from warming of the deep waters in the South Pacific and Southern Ocean, where mid-depth changes since the 1950s are as high as 0.17°C (Gille, 2002; see Figure 5.8), and/or from the declining bottom water transport into the deep North Pacific (Johnson et al., 1994).

5.3.4 Indian Ocean

The upper Indian Ocean has been warming everywhere except in a band centred at about 12°S (South Equatorial Current), as seen in Section 5.2 (Figure 5.3). In the tropical and eastern subtropical Indian Ocean (north of 10°S), warming in the upper 100 m (Qian et al., 2003) is consistent with the significant warming of the sea surface from 1900 to 1999 (see Section 3.2.2 and Figure 3.9). The surface warming trend during the period 1900 to 1970 was relatively weak, but increased significantly in the 1970 to 1999 period, with some regions exceeding 0.2°C per decade.

The global-scale circulation includes transport of warm, relatively fresh waters from the Pacific passing through the Indonesian Seas to the Indian Ocean and then onward into the South Atlantic. Much of this throughflow occurs in the tropics south of the equator, and is strongly affected by ENSO and the Indian Ocean Dipole (see Section 3.6.7.2). The latter causes pronounced thermocline variability (Qian et al., 2003) and includes propagation of upper-layer thickness anomalies by Rossby waves (Xie et al., 2002; Feng and Meyers, 2003; Yamagata et al., 2004) in the 3°S to 15°S latitude band that includes the westward-flowing throughflow water.

Long-term trends in transport and properties of the throughflow have not been reported. The mean transport into the Indonesian Seas measured at Makassar Strait from 1996 to 1998 was 9 to 10 Sv (Vranes et al., 2002), matching transports exiting the Indonesian Seas (e.g., Sprintall et al., 2004). Large variability in this transport is associated with varying tropical Pacific and Indian winds (Wijffels and Meyers, 2004), including a strong ENSO response (e.g., Meyers, 1996), and may be associated with changes in SST in the tropical Indian Ocean.

Models suggest that upper-ocean warming in the south Indian Ocean can be attributed to a reduction in the southeast trade winds and associated decrease in the southward transport of heat from the tropics to the subtropics (Lee, 2004). The export of heat from the northern Indian Ocean to the south across the equator is accomplished by a wind-driven, shallow cross-equatorial cell: data assimilation analysis has shown a significant decadal reduction in the mass exchange during 1950 to 1990 but little change in heat transport (Schoenefeldt and Schott, 2006).

Changes in Indian subtropical gyre circulation since the 1960s include a 20% slowdown from 1962 to 1987 (Bindoff and McDougall, 2000) and a 20% speedup from 1987 to 2002 (Bryden et al., 2003; McDonagh et al., 2005), with the speedup mainly between 1995 and 2002 (Palmer et al., 2004). The upper thermocline warmed during the slowdown, and then cooled during speedup. Simulations of this region and the analysis of climate change scenarios show that the slowdown and speedup were part of an oscillatory pattern in the upper part of this gyre over periods of decades (Murray et al., 2007; Stark et al., 2006). On the other hand, the lower thermocline (<10°C) freshened and warmed from 1936 to 2002 (Bryden et al., 2003), consistent with heat content increases discussed in Section 5.2 and earlier results.

5.3.5   Southern Ocean

The Southern Ocean, which is the region south of 30°S, connects the Atlantic, Indian and Pacific Oceans together, allowing inter-ocean exchange. This region is active in the formation and subduction of waters that contributed strongly to the storage of anthropogenic carbon and heat (see Section 5.2). It is also the location of the densest part of the global overturning circulation, through formation of bottom waters around Antarctica, fed by deep waters from all of the oceans to the north. Note that some observed changes found in the Atlantic, Indian and Pacific Oceans are related to changes in the Southern Ocean waters but have largely been described in those sections.

5.3.5.1   *Upper-Ocean Property Changes*

The upper ocean in the SH has warmed since the 1960s, dominated by changes in the thick near-surface layers called Subantarctic Mode Water (SAMW), located just north of the Antarctic Circumpolar Current (ACC) that encircles Antarctica. The observed warming of SAMW is consistent with the subduction of warmer surface waters from south of the ACC (Wong et al., 2001; Aoki et al., 2003). In the Upper Circumpolar Deep Water (UCDW) in the Indian and Pacific sectors of the Southern Ocean, temperature and salinity have been increasing (on density surfaces) and oxygen has been decreasing between the Subantarctic Front near 45°S and the Antarctic Divergence near 60°S (Aoki et al., 2005a). These changes just below the mixed layer (~100 to 300 m) are consistent with the mixing of warmer and fresher surface waters with UCDW, suggesting an increase in stratification in the surface layer of this polar region.

Mid-depth waters of the Southern Ocean have also warmed in recent decades. As shown in Figure 5.8, temperatures increased near 900 m depth between the 1950s and the 1980s throughout most of the Southern Ocean (Aoki et al., 2003; Gille, 2004).

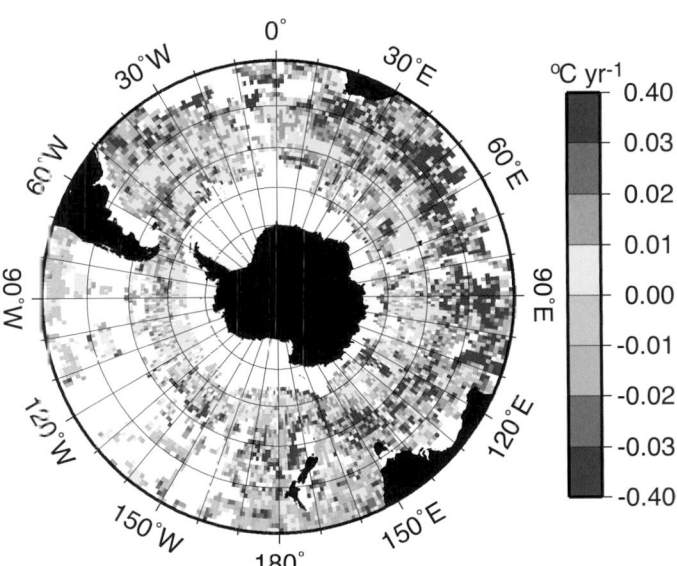

Figure 5.8. *Temperature trends (°C yr⁻¹) at 900 m depth using data collected from the 1930s to 2000, including shipboard profile and Autonomous LAgrangian Current Explorer float data. The largest warming occurs in subantarctic regions, and a slight cooling occurs to the north. From Gille (2002).*

The largest changes are found near the Antarctic Circumpolar Current, where the warming at 900 m depth is similar in magnitude to the increase in regional surface air temperatures. Analysis of altimeter and Argo float profile data suggests that, over the last 10 years, the zonally averaged warming in the upper 400 m of the ocean near 40°S (Willis et al. 2004) is much larger than that seen in long-term trends (see Section 5.2, Figure 5.3 World). The warming results from these analyses have been attributed to a southward shift and increased intensity of the SH westerlies, which would shift the ACC slightly southward and intensify the subtropical gyres (e.g., Cai, 2006).

The major mid-depth water mass in the SH, Antarctic Intermediate Water (AAIW), has also been freshening since the 1960s (Wong et al., 1999; Bindoff and McDougall, 2000; see Figure 5.6). The Atlantic freshening of AAIW is also supported by direct observations of a freshening of southern surface waters (Curry et al., 2003).

5.3.5.2.   *Antarctic Regions and Antarctic Circumpolar Current*

The ACC, the longest current system in the world, has a transport through Drake Passage of about 130 Sv, with significant interannual variability. Measurements over 25 years across Drake Passage show no evidence for a systematic trend in total volume transport between the 1970s and the present (Cunningham et al., 2003), although continuous subsurface pressure measurements suggest that trends in seasonality of transport are highly correlated with similar trends in the SAM index (Meredith and King, 2005).

There is growing evidence for the changes in the AABW and intermediate depth waters around Antarctica. In the Weddell Sea, the deep and bottom water properties varied in the 1990s

(Robertson et al., 2002; Fahrbach et al., 2004). Changes in bottom water properties have also been observed downstream of these source regions (Hogg, 2001; Andrie et al., 2003) and in the South Atlantic (Section 5.3.2.3). The upper ocean adjacent to the West Antarctic Peninsula warmed by more than 1°C and became more saline by 0.25 psu from 1951 to 1994 (Meredith and King, 2005). The warming is likely to have resulted from large regional atmospheric warming (Vaughan et al., 2003) and reduced winter sea ice observed in this region.

In the Ross Sea and near the Ross Ice Shelf, significant decreases in salinity of 0.003 psu yr^{-1} (and density decreases) over the last four decades (Jacobs et al., 2002) have been observed. Downstream of the Ross Ice Shelf in the Australian-Antarctic Basin, AABW has also cooled and freshened (Aoki et al., 2005b). These observed decreases are significantly greater than earlier reports of AABW variability (Whitworth, 2002) and suggest that changes in the antarctic shelf waters can be quite quickly communicated to deep waters. Jacobs et al. (2002) concluded that the freshening appears to have resulted from a combination of factors including increased precipitation, reduced sea ice production and increased melting of the West Antarctic Ice Sheet.

5.3.6 Relation of Regional to Global Changes

5.3.6.1 Changes in Global Water Mass Properties

The regional analyses described in the previous sections have global organisation, as described partially in Section 5.3.1 (Figure 5.6), and as reflected in the global trend analyses in Section 5.2. The data sets used for the largest-scale descriptions over the last 30 to 50 years are reliable; different types of data and widely varying methods yield similar results, increasing confidence in the reality of the changes found in both the global and regional analyses.

The regional and global analyses of ocean warming generally show a pattern of increased ocean temperature in the regions of very thick surface mixed layer (mode water) formation. This is clearest in the North Atlantic and North Pacific and in all sectors of the Southern Ocean (Figure 5.3). There are also regions of decreased ocean temperature in both the global and regional analyses in parts of the subpolar and equatorial regions.

Both the global and regional analyses show long-term freshening in the subpolar waters in the North Atlantic and North Pacific and a salinity increase in the upper ocean (<100 m deep) at low to mid-latitudes. This is consistent with an increase in the atmospheric hydrological cycle over the oceans and could result in changes in ocean advection (Section 5.3.2). In the North Atlantic, the subpolar freshening occurred throughout the entire water column, from the 1960s to the mid-1990s (Figure 5.5 and Figure 5.6a). Increased salinity and temperature in the upper water column in the subpolar North Atlantic after 1994 are not apparent from the linear trend applied to the full time series in Figure 5.5, but are clear in all regional time series (Section 5.3.2). Freshening in the North Pacific subpolar gyre north of 45°N is apparent in both regional analyses (Section 5.3.3.2) and

global analyses (Figure 5.5). Freshening of intermediate depth waters (>300 m) from Southern Ocean sources (Section 5.3.5) is apparent in both the global and regional analyses (e.g., Figure 5.5 World).

Many of the observed changes in the temperature and salinity fields have been linked to atmospheric forcing through correlations with atmospheric indices associated with the NAO, PDO and SAM. Indeed, most of the few time series of ocean measurements or repeat measurements of long sections (see Sections 5.3.2 and 5.3.4) show evidence of decadal variability. Because of the long time scales of these natural climate patterns, it is difficult to discern if observed decadal oceanic variability is natural or a climate change signal; indeed, changes in these natural patterns themselves might be related to climate change. In the North Atlantic, freshening at high latitudes and increased evaporation at subtropical latitudes prior to the mid-1990s might have been associated with an increasing NAO index, and the reversal towards higher salinity at high latitudes thereafter with a decreasing NAO index after 1990 (see Figure 3.31). Likewise in the Pacific, freshening at high latitudes and increased evaporation in the subtropics, cooling in the central North Pacific, warming in the eastern and tropical Pacific and reduced ventilation in the Kuroshio region, Japan and Okhotsk Seas could be associated with the extended positive phase of the PDO. The few detection and attribution studies of ocean changes are discussed in Section 9.5.1.

At a global scale, the observed long-term patterns of zonal temperature and salinity changes tend to be approximately symmetric around the equator (Figure 5.6) and occur simultaneously in different ocean basins (Figures 5.3 and 5.5). The scale of these patterns, which extends beyond the regions of influence normally associated with the NAO, PDO and SAM, suggests that these coherent changes between both hemispheres are associated with a global phenomenon.

5.3.6.2 Consistency with the Large-Scale Ocean Circulation

The observed changes are broadly consistent with scientific understanding of the circulation of the global oceans. The North Atlantic and antarctic regions, where the oceans ventilate the deep waters over short time scales (<50 years), show strong evidence of change over the instrumental record. For example, the North Atlantic shows evidence of a deep warming and freshening. There is evidence of change in the Southern Ocean bottom waters consistent with the sinking of fresher antarctic shelf waters. Deep waters that are far from the North Atlantic and Antarctic, remote from interaction with the atmosphere, and with replenishment rates that are long compared with the instrumental record, typically show no significant changes. Mode waters, key global water masses found in every ocean basin equatorward of major oceanic frontal systems or separated boundary currents, have a relatively rapid formation and ventilation rate (<20 years) and provide a pathway for heat (and salinity) to be transported into the main subtropical gyres of the global oceans as observed.

5.4 Ocean Biogeochemical Changes

5.4.1 Introduction

The observed increase in atmospheric carbon dioxide (CO_2; see Chapter 2) and the observed changes in the physical properties of the ocean reported in this chapter can affect marine biogeochemical cycles (here mainly carbon, oxygen, and nutrients). The increase in atmospheric CO_2 causes additional CO_2 to dissolve in the ocean. Changes in temperature and salinity affect the solubility and chemical equilibration of gases. Changes in circulation affect the supply of carbon and nutrients from below, the ventilation of oxygen-depleted waters and the downward penetration of anthropogenic carbon. The combined physical and biogeochemical changes also affect biological activity, with further consequences for the biogeochemical cycles.

The increase in surface ocean CO_2 has consequences for the chemical equilibrium of the ocean. As CO_2 increases, surface waters become more acidic and the concentration of carbonate ions decreases. This change in chemical equilibrium causes a reduction of the capacity of the ocean to take up additional CO_2. However, the response of marine organisms to ocean acidification is poorly known and could cause further changes in the marine carbon cycle with consequences that are difficult to estimate (see Section 7.3.4 and Chapter 4 of the Working Group II contribution to the IPCC Fourth Assessment Report).

Dissolved oxygen (O_2) in the ocean is affected by the same physical processes that affect CO_2, but in contrast to CO_2, O_2 is not affected by changes in its atmospheric concentration (which are only of the order of 10^{-4} of its mean concentration). Changes in oceanic O_2 concentration thus provide information on the changes in the physical or biological processes that occur within the ocean, such as ventilation (here used to describe the rate of renewal of thermocline waters), mode water formation, upwelling or biological export and respiration. Furthermore, changes in the oceanic O_2 content are needed to estimate the CO_2 budget from atmospheric O_2/molecular nitrogen (N_2) ratio measurements. However, the method currently estimates the change in air-sea fluxes of O_2 indirectly based on heat flux changes (see Section 7.3.2).

This section reports observed changes in biogeochemical cycles and assesses their consistency with observed changes in physical properties. Changes in oceanic nitrous oxide (N_2O) and methane (CH_4) have not been assessed because of the lack of large-scale observations. Observations of the mean fluxes of N_2O and CH_4 (including CH_4 hydrates) are discussed in Chapter 7.

5.4.2 Carbon

5.4.2.1 Total Change in Dissolved Inorganic Carbon and Air-Sea Carbon Dioxide Flux

Direct observations of oceanic dissolved inorganic carbon (DIC; i.e., the sum of CO_2 plus carbonate and bicarbonate)

reflect changes in both the natural carbon cycle and the uptake of anthropogenic CO_2 from the atmosphere. Links between the main modes of climate variability and the marine carbon cycle have been observed on interannual time scales in several regions of the world (see Section 7.3.2.4 for quantitative estimates). In the equatorial Pacific, the reduced upwelling associated with El Niño events decreases the regional outgas of natural CO_2 to the atmosphere (Feely et al., 1999). In the subtropical North Atlantic, reduced mode water formation and reduced deep winter mixing during the positive NAO phase increase the storage of carbon in the intermediate ocean (Bates et al., 2002). These observations show that variability in the content of natural DIC in the ocean has occurred in association with climate variability.

Longer observations exist for the partial pressure of CO_2 (pCO_2) at the surface only. Over more than two decades, the oceanic pCO_2 increase has generally followed the atmospheric CO_2 within the given uncertainty, although regional differences have been observed (Feely et al., 1999; Takahashi et al., 2006). The three stations with the longest time series, all in the northern subtropics, show pCO_2 increases at a rate varying between 1.6 and 1.9 µatm yr^{-1} (Figure 5.9), indistinguishable from the atmospheric increase of 1.5 to 1.9 µatm yr^{-1}. Variability on the order of 20 µatm over periods of five years was observed in the three time series, as well as in other data sets, and has been associated with regional changes in the natural carbon cycle driven by changes in ocean circulation and by climate variability (Gruber et al., 2002; Dore et al., 2003) or with variations in biological activity (Lefèvre et al., 2004).

Direct surface pCO_2 observations have been used to compute a global air-sea CO_2 flux of 1.6 ± 1 GtC yr^{-1} for the year 1995 (Takahashi et al., 2002; Section 7.3.2.3.2, Figure 7.8). It is not yet possible to detect large-scale changes in the global air-sea CO_2 flux from direct observations because of the large influence of climate variability. However, estimates from inverse methods of the air-sea CO_2 flux from the spatio-temporal distribution of atmospheric CO_2 suggest that the global air-sea CO_2 flux increased by 0.1 to 0.6 GtC yr^{-1} between the 1980s and 1990s, consistent with results from ocean models (Le Quéré et al., 2003).

5.4.2.2 Anthropogenic Carbon Change

The recent uptake of anthropogenic carbon in the ocean is well constrained by observations to a decadal mean of 2.2 ± 0.4 GtC yr^{-1} for the 1990s (see Section 7.3.2, Table 7.1). The uptake of anthropogenic carbon over longer time scales can be estimated from oceanic measurements. Changes in DIC between two time periods reflect the anthropogenic carbon uptake plus the changes in DIC concentration due to changes in water masses and biological activity. To estimate the contribution of anthropogenic carbon alone, several corrections must be applied. From observed DIC changes between surveys in the 1970s and the 1990s, an increase in anthropogenic carbon has been inferred down to depths of 1,100 m in the North Pacific (Peng et al., 2003; Sabine et al., 2004a), 200 to 1,200 m in the Indian Ocean (Peng et al., 1998; Sabine et al., 1999) and 1,900 m in the Southern Ocean (McNeil et al., 2003).

An indirect method was used to estimate anthropogenic carbon from observations made at a single time period based on well-known processes that control the distribution of natural DIC in the ocean. The method corrects the observed DIC concentration for organic matter decomposition and dissolution of carbonate minerals, and removes an estimate of the DIC concentration of the water when it was last in contact with the atmosphere (Gruber et al., 1996). With this method, a global DIC increase of 118 ± 19 GtC between pre-industrial times (roughly 1750) and 1994 has been estimated, using 9,618 profiles from the 1990s (Sabine et al., 2004b; see Figure 5.10). The uncertainty of ±19 GtC in this estimate is based on uncertainties in the anthropogenic DIC estimates and mapping errors, which have characteristics of random error, and on an estimate of potential biases, which are not necessarily centred on the mean value. Potential biases of up to 7% in the technique have been identified, mostly caused by assumptions about the time evolution of CO_2, the age or the identification of water masses (Matsumoto and Gruber, 2005), and the recent changes in surface warming and stratification (Keeling, 2005). Potential biases from assumptions of constant carbon and nutrient uptake ratios for biological activity have not been assessed. While the magnitude and direction of all potential biases are not yet clear, the given uncertainty of ±16% appears realistic compared to the biases already identified.

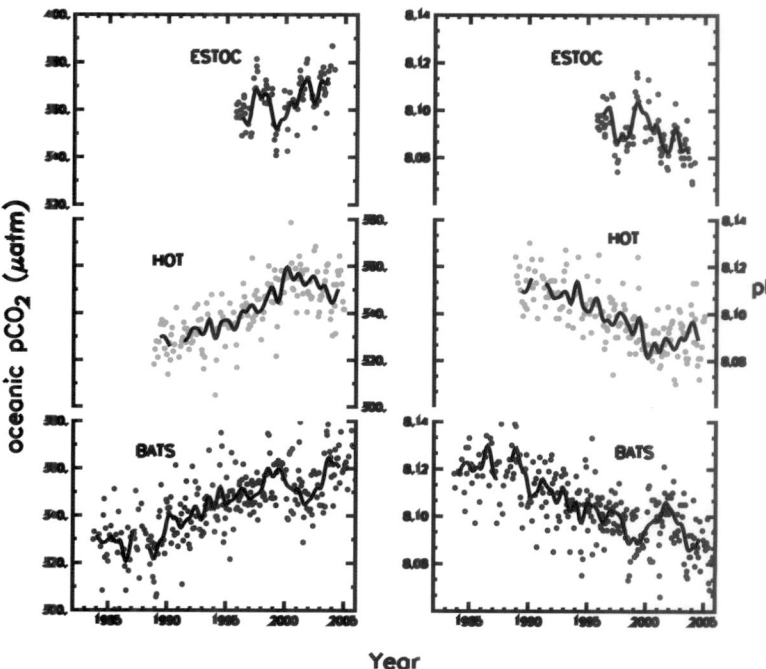

Figure 5.9. *Changes in surface oceanic pCO_2 (left; in µatm) and pH (right) from three time series stations: Blue: European Station for Time-series in the Ocean (ESTOC, 29°N, 15°W; Gonzalez-Dávila et al., 2003); green: Hawaii Ocean Time-Series (HOT, 23°N, 158°W; Dore et al., 2003); red: Bermuda Atlantic Time-series Study (BATS, 31/32°N, 64°W; Bates et al., 2002; Gruber et al., 2002). Values of pCO_2 and pH were calculated from DIC and alkalinity at HOT and BATS; pH was directly measured at ESTOC and pCO_2 was calculated from pH and alkalinity. The mean seasonal cycle was removed from all data. The thick black line is smoothed and does not contain variability less than 0.5 years period.*

Because of the limited rate of vertical transport in the ocean, more than half of the anthropogenic carbon can still be found in the upper 400 m, and it is undetectable in most of the deep ocean (Figure 5.11). The vertical penetration of anthropogenic carbon is consistent with the DIC changes observed between two cruises (Peng et al., 1998, 2003). Anthropogenic carbon has penetrated deeper in the North Atlantic and subantarctic Southern Ocean compared to other basins, due to a combination of: i) high surface alkalinity (in the Atlantic) which favours the uptake of CO_2, and ii) more active vertical exchanges caused by intense winter mixing and by the formation of deep waters (Sabine et al., 2004b). The deeper penetration of anthropogenic carbon in these regions is consistent with similar features observed in the oceanic distribution of chlorofluorocarbons

(CFCs) of atmospheric origin (Willey et al., 2004), confirming that it takes decades to many centuries to transport carbon from the surface into the thermocline and the deep ocean. Deeper penetration in the North Atlantic and subantarctic Southern Ocean is also observed in the changes in heat content shown in Figure 5.3. The large storage of anthropogenic carbon observed in the subtropical gyres is caused by the lateral transport of carbon from the region of mode water formation towards the lower latitudes (Figure 5.10).

The fraction of the net CO_2 emissions taken up by the ocean (the uptake fraction) was possibly lower during 1980 to 2005 (37% ± 7%) compared to 1750 to 1994 (42% ± 7%); however the uncertainty in the estimates is larger than the difference between the estimates (Table 5.1). The net CO_2 emissions

Table 5.1. *Fraction of CO_2 emissions taken up by the ocean for different time periods.*

Time Period	Oceanic Increase (GtC)	Net CO₂ Emissions[a] (GtC)	Uptake Fraction (%)	Reference
1750–1994	118 ± 19	283 ± 19	42 ± 7	Sabine et al., 2004b
1980–2005[b]	53 ± 9	143 ± 10	37 ± 7	Chapter 7[c]

Notes:

[a] Sum of emissions from fossil fuel burning, cement production, land use change and the terrestrial biosphere response.

[b] The longest possible time period was used for the recent decades to minimise the effect of the variability in atmospheric CO_2.

[c] Sum of the estimates for the 1980s, 1990s and 2000 to 2005 from Table 7.1.

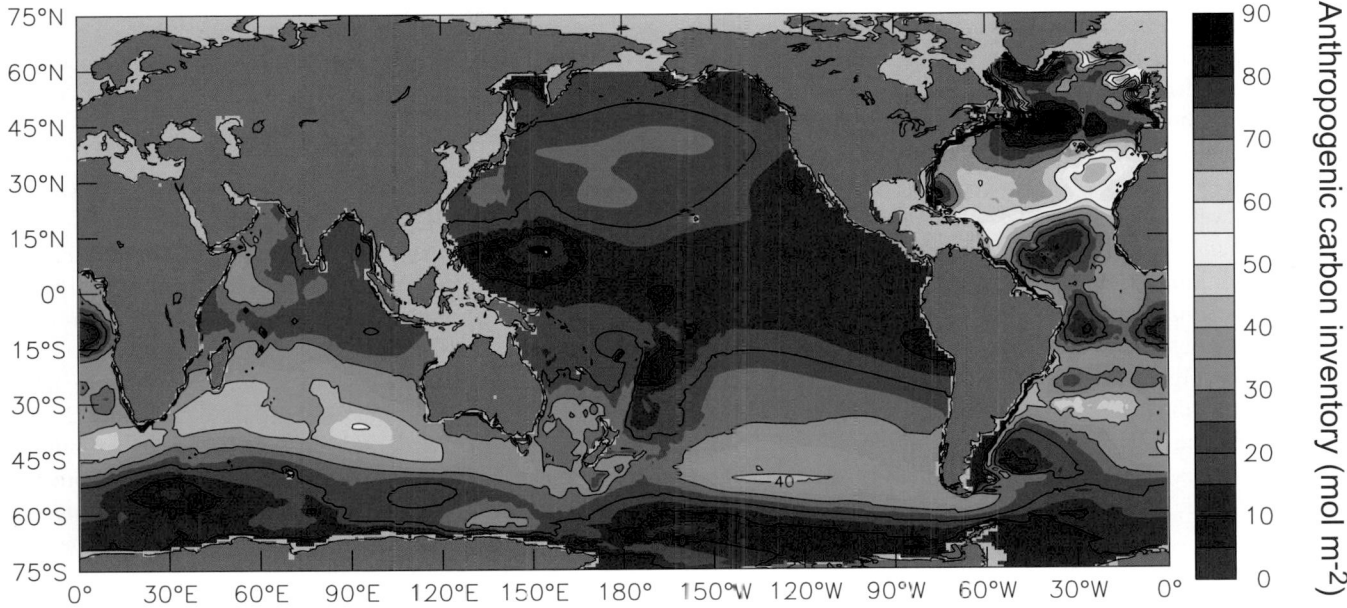

Figure 5.10. *Column inventory of anthropogenic carbon (mol m^{-2}) as of 1994 from Sabine et al. (2004b). Anthropogenic carbon is estimated indirectly by correcting the measured DIC for the contributions of organic matter decomposition and dissolution of carbonate minerals, and taking into account the DIC concentration the water had in the pre-industrial ocean when it was last in contact with the atmosphere. The global inventory of anthropogenic carbon taken up by the ocean between 1750 and 1994 is estimated to be 118 ± 19 GtC.*

include all emissions that have an influence on the atmospheric CO_2 concentration (i.e., emissions from fossil fuel burning, cement production, land use change and the terrestrial biosphere response). It is equivalent to the sum of the atmospheric and oceanic CO_2 increase. Because the atmospheric CO_2 is well constrained by observations, the uncertainty in the net CO_2 emissions is nearly equal to the uncertainty in the oceanic CO_2 increase. The decrease in oceanic uptake fraction would be consistent with the understanding that the ocean CO_2 sink is limited by the transport rate of anthropogenic carbon from the surface to the deep ocean, and also with the nonlinearity in carbon chemistry that reduces the CO_2 uptake capacity of water as its CO_2 concentration increases (Sarmiento et al., 1995).

5.4.2.3   Ocean Acidification by Carbon Dioxide

The uptake of anthropogenic carbon by the ocean changes the chemical equilibrium of the ocean. Dissolved CO_2 forms a weak acid.[1] As CO_2 increases, pH decreases, that is, the ocean becomes more acidic. Ocean pH can be computed from measurements of DIC and alkalinity. A decrease in surface pH of 0.1 over the global ocean was calculated from the estimated uptake of anthropogenic carbon between 1750 and 1994 (Sabine et al., 2004b; Raven et al., 2005), with the lowest decrease (0.06) in the tropics and subtropics, and the highest decrease (0.12) at high latitudes, consistent with the lower buffer capacity of the high latitudes compared to the low latitudes. The mean pH of surface waters ranges between 7.9 and 8.3 in the open

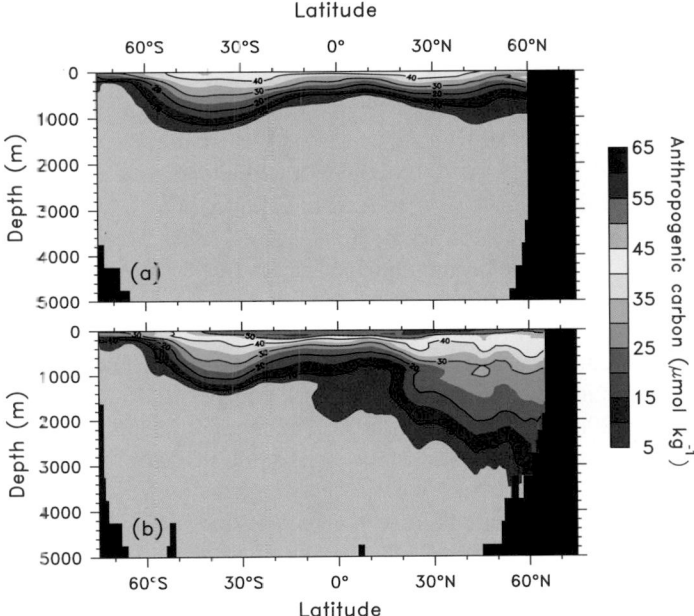

Figure 5.11. *Mean concentration of anthropogenic carbon as of 1994 in μmol kg^{-1} from Sabine et al. (2004b) averaged over (a) the Pacific and Indian Oceans and (b) the Atlantic Ocean. The calculation of anthropogenic carbon is described in the caption of Figure 5.10 and in the text (Section 5.4).*

---

[1] Acidity is a measure of the concentration of H$^+$ ions and is reported in pH units, where pH $= -\log_{10}($H$^+$). A pH decrease of 1 unit means a 10-fold increase in the concentration of H$^+$, or acidity.

ocean, so the ocean remains alkaline (pH > 7) even after these decreases. For comparison, pH was higher by 0.1 unit during glaciations, and there is no evidence of pH values more than 0.6 units below the pre-industrial pH during the past 300 million years (Caldeira and Wickett, 2003). A decrease in ocean pH of 0.1 units corresponds to a 30% increase in the concentration of H^+ in seawater, assuming that alkalinity and temperature remain constant. Changes in surface temperature may have induced an additional decrease in pH of <0.01. The calculated anthropogenic impact on pH is consistent with results from time series stations where a decrease in pH of 0.02 per decade was observed (Figure 5.9). Results from time series stations include not only the increase in anthropogenic carbon, but also other changes due to local physical and biological variability. The consequences of changes in pH on marine organisms are poorly known (see Section 7.3.4 and Box 7.3).

5.4.2.4   Change in Carbonate Species

The uptake of anthropogenic carbon occurs through the injection of CO_2 and causes a shift in the distribution of carbon species (i.e., the balance between CO_2, carbonate and bicarbonate). The availability of carbonate is particularly important because it controls the maximum amount of CO_2 that the ocean is able to absorb. Marine organisms use carbonate to produce shells of calcite and aragonite (both consisting of calcium carbonate; $CaCO_3$). Currently, the surface ocean is super-saturated with respect to both calcite and aragonite, but undersaturated below a depth called the 'saturation horizon'. The undersaturation starts at a depth varying between 200 m in parts of the high-latitude and the Indian Ocean and 3,500 m in the Atlantic. Calcium carbonate dissolves either when it sinks below the calcite or aragonite saturation horizons or under the action of biological activity.

Shoaling of the aragonite saturation horizon has been observed in all ocean basins based on alkalinity, DIC and oxygen measurements (Feely and Chen, 1982; Feely et al., 2002; Sabine et al., 2002; Sarma et al., 2002). The amplitude and direction of the signal was everywhere consistent with the uptake of anthropogenic carbon, with potentially smaller contributions from changes in circulation, temperature and biology. Feely et al. (2004) calculated that the uptake of anthropogenic carbon alone has caused a shoaling of the aragonite saturation horizon between 1750 and 1994 by 30 to 200 m in the eastern Atlantic (50°S–15°N), the North Pacific and the North Indian Ocean, and a shoaling of the calcite saturation horizon by 40 to 100 m in the Pacific (north of 20°N). This calculation is based on the anthropogenic DIC increase estimated by Sabine et al. (2004a), on a global compilation of biogeochemical data and on carbonate chemistry equations. Furthermore, an increase in total alkalinity (primarily controlled by carbonate and bicarbonate) at the depth of the aragonite saturation horizon between 1970 and 1990 has been reported (Sarma et al., 2002). These results are consistent with the calculated increase in $CaCO_3$ dissolution as a result of the shoaling of the aragonite saturation horizon, but with large uncertainty. Carbonate decreases at high latitudes

and particularly in the Southern Ocean may have consequences for marine ecosystems because the current saturation horizon is closer to the surface than in other basins (Orr et al., 2005; see Section 7.3.4).

5.4.3   Oxygen

In the thermocline (~100 to 1,000 m), a decrease in the O_2 concentration has been observed between about the early 1970s and the late 1990s or later in several repeated hydrographic sections in the North and South Pacific, North Atlantic, and Southern Indian Oceans (Figure 5.12; see summary table in Emerson et al., 2004, and Section 5.3). Section 5.3 reports on a number of O_2 decreases that fit the overall message of Section 5.4. The reported O_2 decreases range from 0.1 to 6 μmol kg^{-1} yr^{-1}, superposed on decadal variations of ±2 μmol kg^{-1} yr^{-1} (Ono et al., 2001; Andreev and Watanabe, 2002). In all published studies, the observed O_2 decrease appeared to be driven primarily by changes in ocean circulation, and less by changes in the rate of O_2 demand from downward settling of organic matter. A few studies have quantified the contribution of the change in ocean circulation using estimates of changes in apparent CFC ages (Doney et al., 1998; Watanabe et al., 2001; Mecking et al., 2006). In nearly all cases, the decrease in O_2 could entirely be accounted for by the increased apparent CFC age that resulted from reduced rate of renewal of intermediate waters. Changes in biological processes were only significant at the coast of California and may result from assumptions in the method (Mecking et al., 2006).

It is unclear whether the recent changes in O_2 are indicative of trends or of variability. Recent data in the Indian Ocean have shown a reversal of the O_2 decrease between 1987 and 2002 in the South Indian Ocean of similar amplitude to the decrease observed during the previous decades (McDonagh et al., 2005). Variability has been observed on decadal time scales in the North Atlantic large enough to mask any potential trends (Johnson and Gruber, 2007).

In the upper 100 m of the global ocean surface, decadal variations of ±0.5 μmol kg^{-1} in O_2 concentration were observed for the period 1956 to 1998 based on a global analysis of 530,000 oxygen profiles, with no clear trends (Garcia et al., 2005). However, the near-surface changes in O_2 concentration are difficult to interpret. They can be caused by changes in biological activity, by changes in the physical transport of O_2 from intermediate waters or by changes in temperature and salinity. Because there is less confidence in the early measurements and the reported changes cannot be explained by known processes, it cannot be said whether the absence of a long-term trend in surface O_2 is realistic or not.

5.4.4   Nutrients

Changes in nutrient concentrations can provide information on changes in the physical and biological processes that affect the carbon cycle and could potentially be used as indicators for large-scale changes in marine biology. However, only a

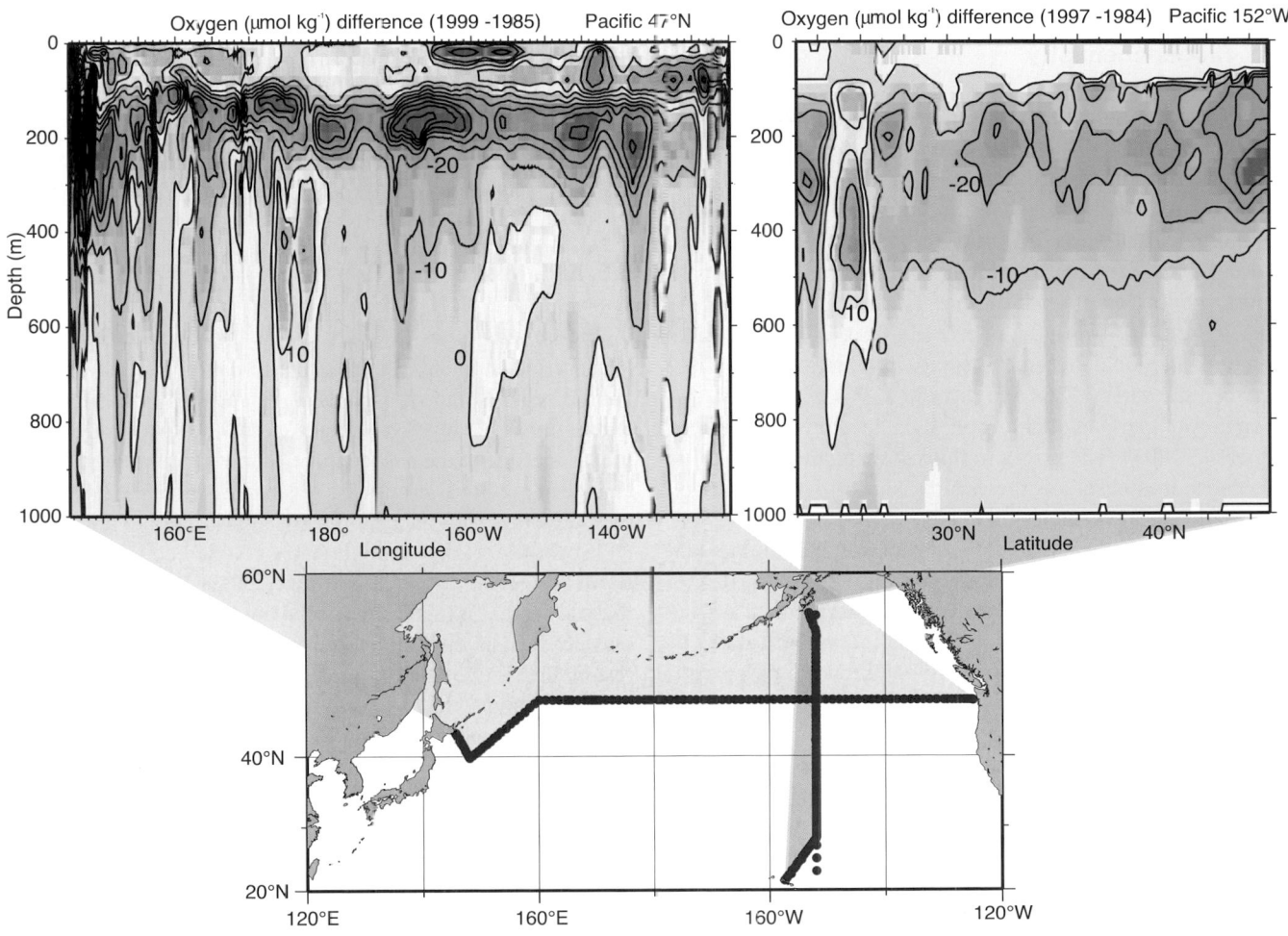

Figure 5.12. *Changes in oxygen concentration (μmol kg⁻¹) along two sections in the North Pacific (see map, bottom panel). Top left panel: Difference (1999 minus 1985) along 47°N. Top right panel: Difference (1997 minus 1984) at 152°W. Blue colours indicate a decrease and yellow colours indicate an increase in oxygen over time. The differences were calculated using density as the vertical coordinate. After Deutsch et al. (2005).*

few studies reported decadal changes in inorganic nutrient concentrations. In the North Pacific, the concentration of nitrate plus nitrite (N) and phosphate decreased at the surface (Freeland et al., 1997; Watanabe et al., 2005) and increased below the surface (Emerson et al., 2001; Ono et al., 2001; Keller et al., 2002) in the past two decades. Nutrient changes were observed in the deep ocean of all basins but no clear pattern emerges from available observations. Pahlow and Riebesell (2000) found changes in the ratio of nutrients in the North Pacific and Atlantic Oceans, and no significant changes in the South Pacific. In the North Pacific, Keller et al. (2002) observed a decrease in N associated with the increase in O_2 between 1970 and 1990 at 1,050 m, opposite to the results of Pahlow and Riebesell's longer study. Using the same data set extended to the world, large regional changes in nutrient ratios were observed (Li and Peng, 2002) but no consistent basin-scale patterns. Uncertainties in deep ocean nutrient observations may be responsible for the lack of coherence in the nutrient changes. Sources of inaccuracy include the limited number of observations and the lack of compatibility between measurements from different laboratories at different times.

In some cases, the observed trends in nutrients can be explained by either a change in thermocline ventilation or a change in biological activity (Pahlow and Riebesell, 2000; Emerson et al., 2001), but in other cases are mostly consistent with a reduction in thermocline ventilation (Freeland et al., 1997; Ono et al., 2001; Watanabe et al., 2005). Thus, all of the reported trends are consistent with a physical explanation of the observed changes, although changes in biological activity cannot be ruled out.

The concentration of surface nutrients can also be influenced by surface mixing, as a reduction in mixing leads to a decreased concentration of surface nutrients. The observed changes in surface temperature and salinity (see Sections 5.2.3 and 5.3) are indicative of changes in the surface mixing (see Section 7.3.4.3). In most of the Pacific Ocean, surface warming and freshening act in the same direction and contribute to reduced mixing (Figures 5.2 and 5.5), consistent with regional observations (Freeland et al., 1997; Watanabe et al., 2005). In the Atlantic and Indian Oceans, temperature and salinity trends generally act in opposite directions and changes in mixing have not been quantified regionally.

5.4.5 Biological Changes Relevant to Ocean Biogeochemistry

Changes in biological activity are an important part of the carbon cycle but are difficult to quantify at the global scale. Marine export production (the fraction of primary production that is not respired at the ocean surface and thus sinks to depth) is the biological process that has the largest influence on element cycles. There are no global observations on changes in export production or respiration. However, estimates of changes in primary production provide partial information. A reduction in global oceanic primary production by about 6% between the early 1980s and the late 1990s was estimated based on the comparison of chlorophyll data from two satellites (Gregg et al., 2003). The errors in this estimate are potentially large because it is based on the comparison of data from two different sensors. Nevertheless, a change in biological fluxes of this order of magnitude is plausible considering that biological production is controlled primarily by nutrient input from intermediate waters, and that a decrease in intermediate water renewal has been observed during that period as indicated by the decrease in O_2. Shifts and trends in plankton biomass have been observed for instance in the North Atlantic (Beaugrand and Reid, 2003), the North Pacific (Karl, 1999; Chavez et al., 2003) and in the Southern Indian Ocean (Hirawake et al., 2005), but the spatial and temporal coverage is limited. The potential impacts of changes in marine ecosystems or dissolved organic matter on climate are discussed in Section 7.3.4, and the impact of climate on marine ecosystems in Chapter 4 of the Working Group II contribution to the IPCC Fourth Assessment Report.

5.4.6 Consistency with Physical Changes

It is clearly established that climate variability affects the oceanic content of natural and anthropogenic DIC and the air-sea flux of CO_2, although the amplitude and physical processes responsible for the changes are less well known. Variability in the marine carbon cycle has been observed in response to physical changes associated with the dominant modes of climate variability such as El Niño events and the PDO (Feely et al., 1999; Takahashi et al., 2006), and the NAO (Bates et al., 2002; Johnson and Gruber, 2007). The regional patterns of anthropogenic CO_2 storage are consistent with those of CFCs and with changes in heat content. The observed trends in CO_2, DIC, pH and carbonate species can be primarily explained by the response of the ocean to the increase in atmospheric CO_2.

Large-scale changes in the O_2 content of the thermocline have been observed between the 1970s and the late 1990s. These changes are everywhere consistent with the local changes in ocean ventilation as identified either by changes in density gradients or by changes in apparent CFC ages. Nevertheless, an influence of changes in marine biology cannot be ruled out. The available data are insufficient to say if the changes in O_2 are caused by natural variability or are trends that are likely to persist in the future, but they do indicate that large-scale changes

in ocean physics influence natural biogeochemical cycles, and thus the cycles of O_2 and CO_2 are likely to undergo changes if ocean circulation changes persist in the future.

5.5 Changes in Sea Level

5.5.1 Introductory Remarks

Present-day sea level change is of considerable interest because of its potential impact on human populations living in coastal regions and on islands. This section focuses on global and regional sea level variations, over time spans ranging from the last decade to the past century; a brief discussion of sea level change in previous centuries is given in Section 5.5.2.4. Changes over previous millennia are discussed in Section 6.4.3.

Processes in several nonlinearly coupled components of the Earth system contribute to sea level change, and understanding these processes is therefore a highly interdisciplinary endeavour. On decadal and longer time scales, global mean sea level change results from two major processes, mostly related to recent climate change, that alter the volume of water in the global ocean: i) thermal expansion (Section 5.5.3), and ii) the exchange of water between oceans and other reservoirs (glaciers and ice caps, ice sheets, other land water reservoirs - including through anthropogenic change in land hydrology, and the atmosphere; Section 5.5.5). All these processes cause geographically non-uniform sea level change (Section 5.5.4) as well as changes in the global mean; some oceanographic factors (e.g., changes in ocean circulation or atmospheric pressure) also affect sea level at the regional scale, while contributing negligibly to changes in the global mean. Vertical land movements such as resulting from glacial isostatic adjustment (GIA), tectonics, subsidence and sedimentation influence local sea level measurements but do not alter ocean water volume; nonetheless, they affect global mean sea level through their alteration of the shape and hence the volume of the ocean basins containing the water.

Measurements of present-day sea level change rely on two different techniques: tide gauges and satellite altimetry (Section 5.5.2). Tide gauges provide sea level variations with respect to the land on which they lie. To extract the signal of sea level change due to ocean water volume and other oceanographic change, land motions need to be removed from the tide gauge measurement. Land motions related to GIA can be simulated in global geodynamic models. The estimation of other land motions is not generally possible unless there are adequate nearby geodetic or geological data, which is usually not the case. However, careful selection of tide gauge sites such that records reflecting major tectonic activity are rejected, and averaging over all selected gauges, results in a small uncertainty for global sea level estimates (Appendix 5.A.4). Sea level change based on satellite altimetry is measured with respect to the Earth's centre of mass, and thus is not distorted by land motions, except for a small component due to large-scale deformation of ocean basins from GIA.

Frequently Asked Question 5.1
Is Sea Level Rising?

Yes, there is strong evidence that global sea level gradually rose in the 20th century and is currently rising at an increased rate, after a period of little change between AD 0 and AD 1900. Sea level is projected to rise at an even greater rate in this century. The two major causes of global sea level rise are thermal expansion of the oceans (water expands as it warms) and the loss of land-based ice due to increased melting.

Global sea level rose by about 120 m during the several millennia that followed the end of the last ice age (approximately 21,000 years ago), and stabilised between 3,000 and 2,000 years ago. Sea level indicators suggest that global sea level did not change significantly from then until the late 19th century. The instrumental record of modern sea level change shows evidence for onset of sea level rise during the 19th century. Estimates for the 20th century show that global average sea level rose at a rate of about 1.7 mm yr^{-1}.

Satellite observations available since the early 1990s provide more accurate sea level data with nearly global coverage. This decade-long satellite altimetry data set shows that since 1993, sea level has been rising at a rate of around 3 mm yr^{-1}, significantly higher than the average during the previous half century. Coastal tide gauge measurements confirm this observation, and indicate that similar rates have occurred in some earlier decades.

In agreement with climate models, satellite data and hydrographic observations show that sea level is not rising uniformly around the world. In some regions, rates are up to several times the global mean rise, while in other regions sea level is falling. Substantial spatial variation in rates of sea level change is also inferred from hydrographic observations. Spatial variability of the rates of sea level rise is mostly due to non-uniform changes in temperature and salinity and related to changes in the ocean circulation.

Near-global ocean temperature data sets made available in recent years allow a direct calculation of thermal expansion. It is believed that on average, over the period from 1961 to 2003, thermal expansion contributed about one-quarter of the observed sea level rise, while melting of land ice accounted for less than half. Thus, the full magnitude of the observed sea level rise during that period was not satisfactorily explained by those data sets, as reported in the IPCC Third Assessment Report.

During recent years (1993–2003), for which the observing system is much better, thermal expansion and melting of land ice each account for about half of the observed sea level rise, although there is some uncertainty in the estimates.

The reasonable agreement in recent years between the observed rate of sea level rise and the sum of thermal expansion and loss of land ice suggests an upper limit for the magnitude of change in land-based water storage, which is relatively poorly known. Model results suggest no net trend in the storage of water over land due to climate-driven changes but there are large interannual and decadal fluctuations. However, for the recent period 1993 to 2003, the small discrepancy between observed sea level rise and the sum of known contributions might be due to unquantified human-induced processes (e.g., groundwater extraction, impoundment in reservoirs, wetland drainage and deforestation).

Global sea level is projected to rise during the 21st century at a greater rate than during 1961 to 2003. Under the IPCC Special Report on Emission Scenarios (SRES) A1B scenario by the mid-2090s, for instance, global sea level reaches 0.22 to 0.44 m above 1990 levels, and is rising at about 4 mm yr^{-1}. As in the past, sea level change in the future will not be geographically uniform, with regional sea level change varying within about ±0.15 m of the mean in a typical model projection. Thermal expansion is projected to contribute more than half of the average rise, but land ice will lose mass increasingly rapidly as the century progresses. An important uncertainty relates to whether discharge of ice from the ice sheets will continue to increase as a consequence of accelerated ice flow, as has been observed in recent years. This would add to the amount of sea level rise, but quantitative projections of how much it would add cannot be made with confidence, owing to limited understanding of the relevant processes.

Figure 1 shows the evolution of global mean sea level in the past and as projected for the 21st century for the SRES A1B scenario.

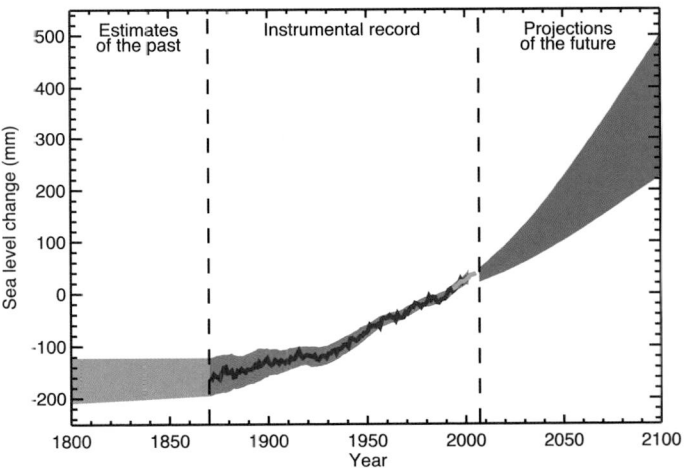

FAQ 5.1, Figure 1. *Time series of global mean sea level (deviation from the 1980-1999 mean) in the past and as projected for the future. For the period before 1870, global measurements of sea level are not available. The grey shading shows the uncertainty in the estimated long-term rate of sea level change (Section 6.4.3). The red line is a reconstruction of global mean sea level from tide gauges (Section 5.5.2.1), and the red shading denotes the range of variations from a smooth curve. The green line shows global mean sea level observed from satellite altimetry. The blue shading represents the range of model projections for the SRES A1B scenario for the 21st century, relative to the 1980 to 1999 mean, and has been calculated independently from the observations. Beyond 2100, the projections are increasingly dependent on the emissions scenario (see Chapter 10 for a discussion of sea level rise projections for other scenarios considered in this report). Over many centuries or millennia, sea level could rise by several metres (Section 10.7.4).*

The TAR chapter on sea level change provided estimates of climate and other anthropogenic contributions to 20th-century sea level rise, based mostly on models (Church et al., 2001). The sum of these contributions ranged from –0.8 to 2.2 mm yr^{-1}, with a mean value of 0.7 mm yr^{-1}, and a large part of this uncertainty was due to the lack of information on anthropogenic land water change. For observed 20th-century sea level rise, based on tide gauge records, Church et al. (2001) adopted as a best estimate a value in the range of 1 to 2 mm yr^{-1}, which was more than twice as large as the TAR's estimate of climate-related contributions. It thus appeared that either the processes causing sea level rise had been underestimated or the rate of sea level rise observed with tide gauges was biased towards higher values.

Since the TAR, a number of new results have been published. The global coverage of satellite altimetry since the early 1990s (TOPography EXperiment (TOPEX)/Poseidon and Jason) has improved the estimate of global sea level rise and has revealed the complex geographical patterns of sea level change in open oceans. Near-global ocean temperature data for the last 50 years have been recently made available, allowing the first observationally based estimate of the thermal expansion contribution to sea level rise in past decades. For recent years, better estimates of the land ice contribution to sea level are available from various observations of glaciers, ice caps and ice sheets.

In this section, we summarise the current knowledge of present-day sea level rise. The observational results are assessed, followed by our current interpretation of these observations in terms of climate change and other processes, and ending with a discussion of the sea level budget (Section 5.5.6).

5.5.2 Observations of Sea Level Changes

5.5.2.1 20th-Century Sea Level Rise from Tide Gauges

Table 11.9 of the TAR listed several estimates for global and regional 20th-century sea level trends based on the Permanent Service for Mean Sea Level (PSMSL) data set (Woodworth and Player, 2003). The concerns about geographical bias in the PSMSL data set remain, with most long sea level records stemming from the NH, and most from continental coastlines rather than ocean interiors. Based on a small number (~25) of high-quality tide gauge records from stable land regions, the rate of sea level rise has been estimated as 1.8 mm yr^{-1} for the past 70 years (Douglas, 2001; Peltier, 2001), and Miller and Douglas (2004) find a range of 1.5 to 2.0 mm yr^{-1} for the 20th century from 9 stable tide gauge sites. Holgate and Woodworth (2004) estimated a rate of 1.7 ± 0.4 mm yr^{-1} sea level change averaged along the global coastline during the period 1948 to 2002, based on data from 177 stations divided into 13 regions. Church et al. (2004) (discussed further below) determined

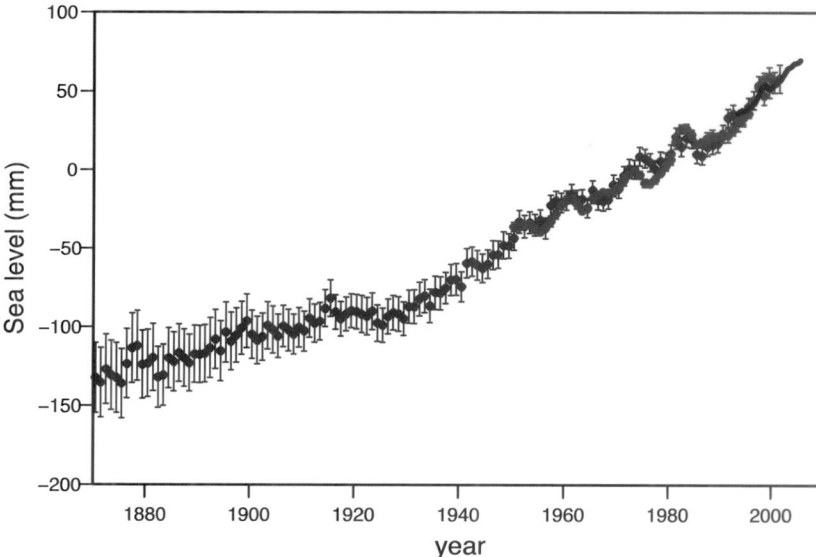

Figure 5.13. *Annual averages of the global mean sea level (mm). The red curve shows reconstructed sea level fields since 1870 (updated from Church and White, 2006); the blue curve shows coastal tide gauge measurements since 1950 (from Holgate and Woodworth, 2004) and the black curve is based on satellite altimetry (Leuliette et al., 2004). The red and blue curves are deviations from their averages for 1961 to 1990, and the black curve is the deviation from the average of the red curve for the period 1993 to 2001. Error bars show 90% confidence intervals.*

a global rise of 1.8 ± 0.3 mm yr^{-1} during 1950 to 2000, and Church and White (2006) determined a change of 1.7 ± 0.3 mm yr^{-1} for the 20th century. Changes in global sea level as derived from analyses of tide gauges are displayed in Figure 5.13. Considering the above results, and allowing for the ongoing higher trend in recent years shown by altimetry (see Section 5.5.2.2), we assess the rate for 1961 to 2003 as 1.8 ± 0.5 mm yr^{-1} and for the 20th century as 1.7 ± 0.5 mm yr^{-1}.

While the recently published estimates of sea level rise over the last decades remain within the range of the TAR values (i.e., 1–2 mm yr^{-1}), there is an increasing opinion that the best estimate lies closer to 2 mm yr^{-1} than to 1 mm yr^{-1}. The lower bound reported in the TAR resulted from local and regional studies; local and regional rates may differ from the global mean, as discussed below (see Section 5.5.2.5).

A critical issue concerns how the records are adjusted for vertical movements of the land upon which the tide gauges are located and of the oceans. Trends in tide gauge records are corrected for GIA using models, but not for other land motions. The GIA correction ranges from about 1 mm yr^{-1} (or more) near to former ice sheets to a few tenths of a millimetre per year in the far field (e.g., Peltier, 2001); the error in tide-gauge based global average sea level change resulting from GIA is assessed as 0.15 mm yr^{-1}. The TAR mentioned the developing geodetic technologies (especially the Global Positioning System; GPS) that hold the promise of measuring rates of vertical land movement at tide gauges, no matter if those movements are due to GIA or to other geological processes. Although there has been some model validation, especially for GIA models, systematic problems with such techniques, including short data spans, have yet to be fully resolved.

5.5.2.2 Sea Level Change during the Last Decade from Satellite Altimetry

Since 1992, global mean sea level can be computed at 10-day intervals by averaging the altimetric measurements from the TOPEX/Poseidon (T/P) and Jason satellites over the area of coverage (66°S to 66°N) (Nerem and Mitchum, 2001). Each 10-day estimate of global mean sea level has an accuracy of approximately 5 mm. Numerous papers on the altimetry results (see Cazenave and Nerem, 2004, for a review) show a current rate of sea level rise of 3.1 ± 0.7 mm yr^{-1} over 1993 to 2003 (Cazenave and Nerem, 2004; Leuliette et al., 2004; Figure 5.14). A significant fraction of the 3 mm yr^{-1} rate of change has been shown to arise from changes in the Southern Ocean (Cabanes et al., 2001).

The accuracy needed to compute mean sea level change pushes the altimeter measurement system to its performance limits, and thus care must be taken to ensure that the instrument is precisely calibrated (see Appendix 5.A.4.1). The tide gauge calibration method (Mitchum, 2000) provides diagnoses of problems in the altimeter instrument, the orbits, the measurement corrections and ultimately the final sea level data. Errors in determining the altimeter instrument drift using the tide gauge calibration, currently estimated to be about 0.4 mm yr^{-1}, are almost entirely driven by errors in knowledge of vertical land motion at the gauges (Mitchum, 2000).

Altimetry-based sea level measurements include variations in the global ocean basin volume due to GIA. Averaged over the oceanic regions sampled by the altimeter satellites, this effect yields a value close to –0.3 mm yr^{-1} in sea level (Peltier, 2001), with possible uncertainty of 0.15 mm yr^{-1}. This number is subtracted from altimetry-derived global mean sea level in order to obtain the contribution due to ocean (water) volume change.

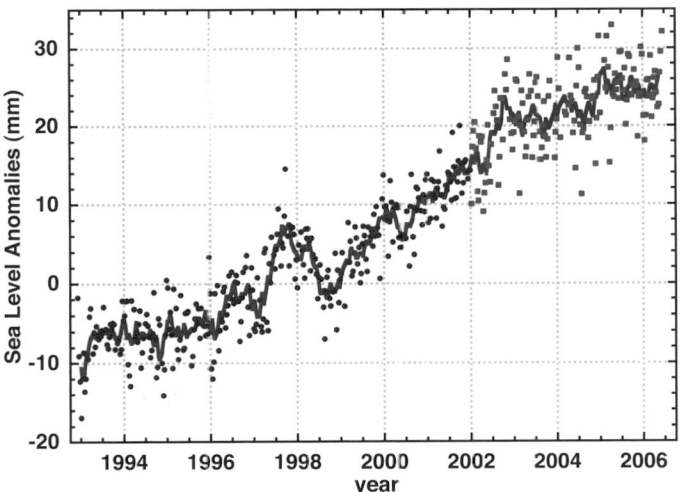

Figure 5.14. *Variations in global mean sea level (difference to the mean 1993 to mid-2001) computed from satellite altimetry from January 1993 to October 2005, averaged over 65°S to 65°N. Dots are 10-day estimates (from the TOPEX/Poseidon satellite in red and from the Jason satellite in green). The blue solid curve corresponds to 60-day smoothing. Updated from Cazenave and Nerem (2004) and Leuliette et al. (2004).*

Altimetry from T/P allows the mapping of the geographical distribution of sea level change (Figure 5.15a). Although regional variability in coastal sea level change had been reported from tide gauge analyses (e.g., Douglas, 1992; Lambeck, 2002), the global coverage of satellite altimetry provides unambiguous evidence of non-uniform sea level change in open oceans, with some regions exhibiting rates of sea level change about five times the global mean. For the past decade, sea level rise shows the highest magnitude in the western Pacific and eastern Indian oceans, regions that exhibit large interannual variability associated with ENSO. Except for the Gulf Stream region, most of the Atlantic Ocean shows sea level rise during the past decade. Despite the global mean rise, Figure 5.15a shows that sea level has been dropping in some regions (eastern Pacific and western Indian Oceans). These spatial patterns likely reflect decadal fluctuations rather than long-term trends. Empirical Orthogonal Functions (EOF) analyses of altimetry-based sea level maps over 1993 to 2003 show a strong influence of the 1997–1998 El Niño, with the geographical patterns of the dominant mode being very similar to those of the sea level trend map (e.g., Nerem et al., 1999).

5.5.2.3 Reconstructions of Sea Level Change during the Last 50 Years Based on Satellite Altimetry and Tide Gauges

Attempts have been made to reconstruct historical sea level fields by combining the near-global coverage from satellite altimeter data with the longer but spatially sparse tide gauge records (Chambers et al., 2002; Church et al., 2004). These sea level reconstructions use the short altimeter record to determine the principal EOF of sea level variability, and the tide gauge data to estimate the evolution of the amplitude of the EOFs over time. The method assumes that the geographical patterns of decadal sea level trends can be represented by a superposition of the patterns of variability that are manifest in interannual variability. The sea level for the period 1870 to 2000 (Church and White, 2006) shown in Figure 5.13 is based on this approach. As a caveat, note that variability on different time scales may have different characteristic patterns (see Section 5.5.4.1).

The trends in the EOF amplitudes (and the implied global correlations) allow the reconstruction of a spatially variable rate of sea level rise. Figure 5.16a (updated from Church et al., 2004) shows the geographical distribution of linear sea level trends for 1955 to 2003 based on this reconstruction technique. Comparison with the altimetry-based trend map for the shorter period (1993 to 2003) indicates quite different geographical patterns. These differences mainly arise from thermal expansion changes through time (see Section 5.5.3)

Changes in spatial sea level patterns through time may help reconcile apparently inconsistent estimates of regional variations in tide-gauge based sea level rise. For example, the minimum in rise along the northwest Australian coast is consistent with the results of Lambeck (2002) in having smaller rates of sea level rise and indeed sea level fall off north-western Australia over the last few decades. In addition, for the North Atlantic Ocean,

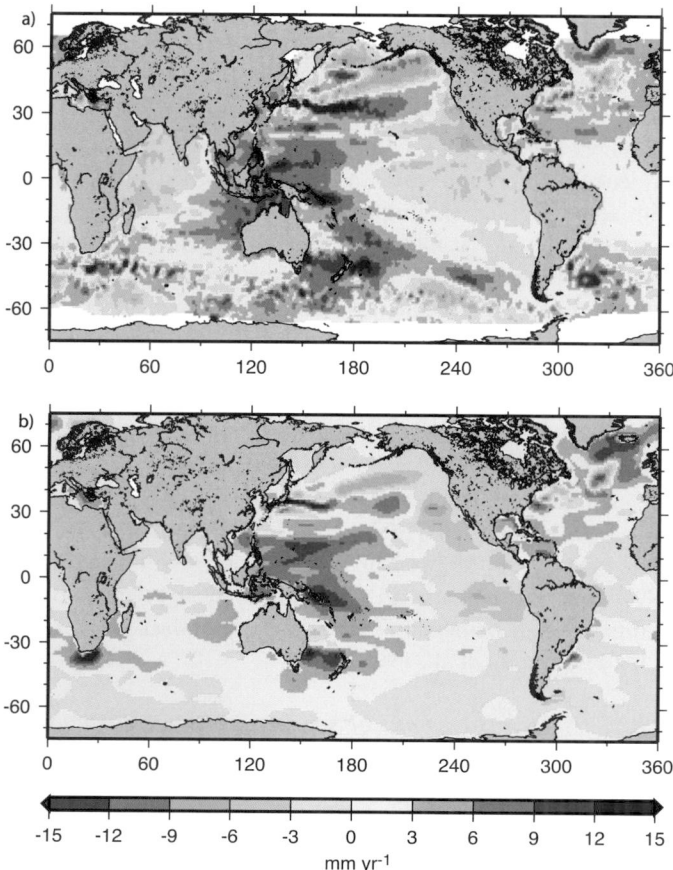

Figure 5.15. *(a) Geographic distribution of short-term linear trends in mean sea level (mm yr⁻¹) for 1993 to 2003 based on TOPEX/Poseidon satellite altimetry (updated from Cazenave and Nerem, 2004) and (b) geographic distribution of linear trends in thermal expansion (mm yr⁻¹) for 1993 to 2003 (based on temperature data down to 700 m from Ishii et al., 2006).*

the rate of rise reaches a maximum (over 2 mm yr⁻¹) in a band running east-northeast from the US east coast. The trends are lower in the eastern than in the western Atlantic (Lambeck et al., 1998; Woodworth et al., 1999; Mitrovica et al., 2001).

5.5.2.4 *Interannual and Decadal Variability and Long-Term Changes in Sea Level*

Sea level records contain a considerable amount of interannual and decadal variability, the existence of which is coherent throughout extended parts of the ocean. For example, the global sea level curve in Figure 5.13 shows an approximately 10 mm rise and fall of global mean sea level accompanying the 1997–1998 ENSO event. Over the past few decades, the time series of the first EOF of Church et al. (2004) represents ENSO variability, as shown by a significant (negative) correlation with the Southern Oscillation Index. The signature of the 1997–1998 El Niño is also clear in the altimetric maps of sea level anomalies (see Section 5.5.2.2). Model results suggest that large volcanic eruptions produce interannual to decadal fluctuations in the global mean sea level (see Section 9.5.2).

Holgate and Woodworth (2004) concluded that the 1990s had one of the fastest recorded rates of sea level rise averaged along the global coastline (~4 mm yr⁻¹), slightly higher than the altimetry-based open ocean sea level rise (3 mm yr⁻¹). However, their analysis also shows that some previous decades had comparably large rates of coastal sea level rise (e.g., around 1980; Figure 5.17). White et al. (2005) confirmed the larger sea level rise during the 1990s around coastlines compared to the open ocean but found that in some previous periods the coastal rate was smaller than the open ocean rate, and concluded that over the last 50 years the coastal and open ocean rates of change were the same on average. The global reconstruction of Church et al. (2004) and Church and White (2006) also exhibits large decadal variability in the rate of global mean sea level rise, and the 1993 to 2003 rate has been exceeded in some previous decades (Figure 5.17). The variability is smaller in the global reconstruction (standard deviation of overlapping 10-year rates is 1.1 mm yr⁻¹) than in the Holgate and Woodworth (2004) coastal time series (standard deviation 1.7 mm yr⁻¹). The rather

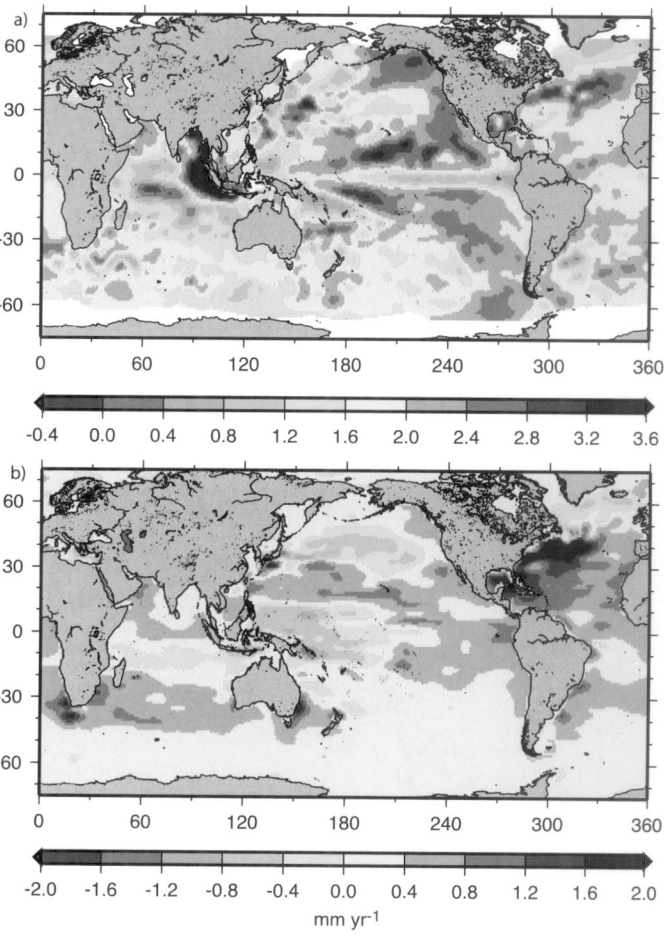

Figure 5.16. *(a) Geographic distribution of long-term linear trends in mean sea level (mm yr⁻¹) for 1955 to 2003 based on the past sea level reconstruction with tide gauges and altimetry data (updated from Church et al., 2004) and (b) geographic distribution of linear trends in thermal expansion (mm yr⁻¹) for 1955 to 2003 (based on temperature data down to 700 m from Ishii et al., 2006). Note that colours in (a) denote 1.6 mm yr⁻¹ higher values than those in (b).*

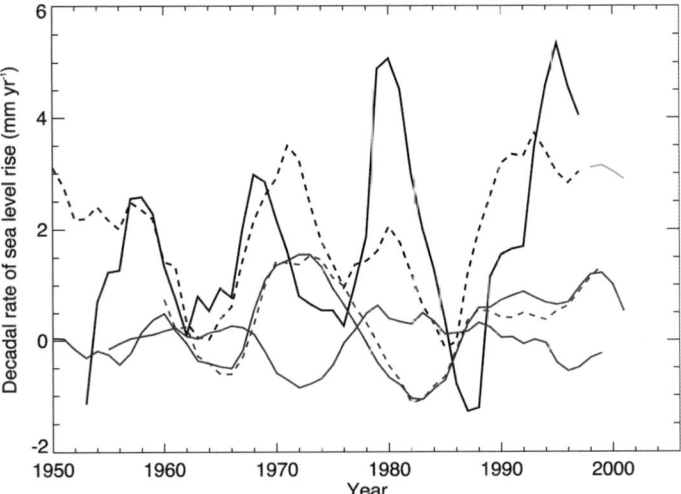

Figure 5.17. *Overlapping 10-year rates of global sea level change from tide gauge data sets (Holgate and Woodworth, 2004, in solid black; Church and White, 2006, in dashed black) and satellite altimetry (updated from Cazenave and Nerem, 2004, in green), and contributions to global sea level change from thermal expansion (Ishii et al., 2006, in solid red; Antonov et al., 2005, in dashed red) and climate-driven land water storage (Ngo-Duc et al., 2005, in blue). Each rate is plotted against the middle of its 10-year period.*

low temporal correlation (r = 0.44) between the two time series suggests that the statistical uncertainty in the linear trends calculated from either data set probably underestimates the systematic uncertainty in the results (Section 5.5.6).

Interannual or longer variability is a major reason why no long-term acceleration of sea level has been identified using 20th-century data alone (Woodworth, 1990; Douglas, 1992). Another possibility is that the sparse tide gauge network may have been inadequate to detect it if present (Gregory et al., 2001). The longest records available from Europe and North America contain accelerations of the order of 0.4 mm yr[-1] per century between the 19th and 20th century (Ekman, 1988; Woodworth et al., 1999). For the reconstruction shown in Figure 5.13, Church and White (2006) found an acceleration of 1.3 ± 0.5 mm yr[-1] per century over the period 1870 to 2000. These data support an inference that the onset of acceleration occurred during the 19th century (see Section 9.5.2).

Geological observations indicate that during the last 2,000 years (i.e., before the recent rise recorded by tide gauges), sea level change was small, with an average rate of only 0.0 to 0.2 mm yr[-1] (see Section 6.4.3). The use of proxy sea level data from archaeological sources is well established in the Mediterranean. Oscillations in sea level from 2,000 to 100 yr before present did not exceed ±0.25 m, based on the Roman-Byzantine-Crusader well data (Sivan et al., 2004). Many Roman and Greek constructions are relatable to the level of the sea. Based on sea level data derived from Roman fish ponds, which are considered to be a particularly reliable source of such information, together with nearby tide gauge records, Lambeck et al. (2004) concluded that the onset of the modern sea level rise occurred between 1850 and 1950. Donnelly et al. (2004) and Gehrels et al. (2004), employing geological data from Connecticut, Maine and Nova Scotia salt-marshes together with

nearby tide gauge records, demonstrated that the sea level rise observed during the 20th century was in excess of that averaged over the previous several centuries.

The joint interpretation of the geological observations, the longest instrumental records and the current rate of sea level rise for the 20th century gives a clear indication that the rate of sea level rise has increased between the mid-19th and the mid-20th centuries.

5.5.2.5 Regional Sea Level Change

Two regions are discussed here to give examples of local variability in sea level: the northeast Atlantic and small Pacific Islands.

Interannual variability in northeast Atlantic sea level records exhibits a clear relationship to the air pressure and wind changes associated with the NAO, with the magnitude and sign of the response depending primarily upon latitude (Andersson, 2002; Wakelin et al., 2003; Woolf et al., 2003). The signal of the NAO can also be observed to some extent in ocean temperature records, suggesting a possible, smaller NAO influence on regional mean sea level via steric (density) changes (Tsimplis et al., 2006). In the Russian Arctic Ocean, sea level time series for recent decades also have pronounced decadal variability that correlates with the NAO index. In this region, wind stress and atmospheric pressure loading contribute nearly half of the observed sea level rise of 1.85 mm yr[-1] (Proshutinsky et al., 2004).

Small Pacific Islands are the subject of much concern in view of their vulnerability to sea level rise. The Pacific Ocean region is the centre of the strongest interannual variability of the climate system, the coupled ocean-atmosphere ENSO mode. There are only a few Pacific Island sea level records extending back to before 1950. Mitchell et al. (2001) calculated rates of relative sea level rise for the stations in the Pacific region. Using their results (from their Table 1) and focusing on only the island stations with more than 50 years of data (only 4 locations), the average rate of sea level rise (relative to the Earth's crust) is 1.6 mm yr[-1]. For island stations with record lengths greater than 25 years (22 locations), the average rate of relative sea level rise is 0.7 mm yr[-1]. However, these data sets contain a large range of rates of relative sea level change, presumably as a result of poorly quantified vertical land motions.

An example of the large interannual variability in sea level is Kwajalein (8°44'N, 167°44'E) (Marshall Archipelago). As shown in Figure 5.18, the local tide gauge data, the sea level reconstructions of Church et al. (2004) and Church and White (2006) and the shorter satellite altimeter record all agree and indicate that interannual variations associated with ENSO events are greater than 0.2 m. The Kwajalein data also suggest increased variability in sea level after the mid-1970s, consistent with the trend towards more frequent, persistent and intense ENSO events since the mid-1970s (Folland et al., 2001). For the Kwajalein record, the rate of sea level rise, after correction for GIA land motions and isostatic response to atmospheric pressure changes, is 1.9 ± 0.7 mm yr[-1]. However,

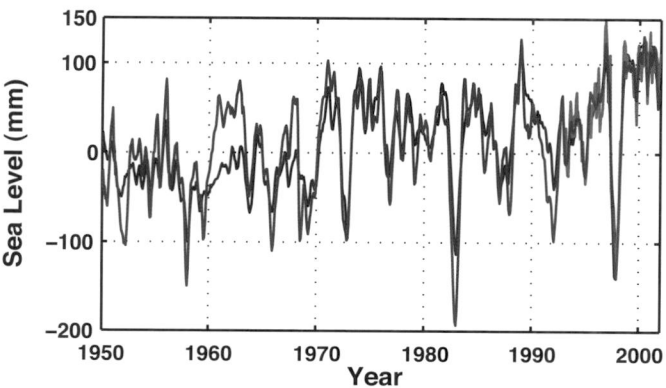

Figure 5.18. *Monthly mean sea level curve for 1950 to 2000 at Kwajalein (8°44'N, 167°44'E). The observed sea level (from tide gauge measurements) is in blue, the reconstructed sea level in red and the satellite altimetry record in green. Annual and semi-annual signals have been removed from each time series and the tide gauge data have been smoothed. The figure was drawn using techniques in Church et al. (2004) and Church and White (2006).*

the uncertainties in rates of sea level change increase rapidly with decreasing record length and can be several mm yr^{-1} for decade-long records (depending on the magnitude of the interannual variability). Sea level change on the atolls of Tuvalu (western Pacific) has been the subject of intense interest as a result of their low-lying nature and increasing incidence of flooding. There are two records available at Funafuti, Tuvalu; the first record commences in 1977 and the second (with rigorous datum control) in 1993. After allowing for subsidence affecting the first record, Church et al. (2006) estimate sea level rise at Tuvalu to be 2.0 ± 1.7 mm yr^{-1}, in agreement with the reconstructed rate of sea level rise.

5.5.2.6 *Changes in Extreme Sea Level*

Societal impacts of sea level change primarily occur via the extreme levels rather than as a direct consequence of mean sea level changes. Apart from non-climatic events such as tsunamis, extreme sea levels occur mainly in the form of storm surges generated by tropical or extra tropical cyclones. Secular changes and decadal variability in storminess are discussed in Chapter 3. Studies of variations in extreme sea levels during the 20th century based on tide gauge data are fewer than studies of changes in mean sea level for several reasons. A study on changes in extremes, which are caused by changes in mean sea level as well as changes in surges, is more complex than the study of mean sea level changes. Moreover, the hourly sampling interval normally used in tide gauge records is not always sufficient to accurately capture the true extreme. Among the different parameters often used to describe extremes, annual maximum surge is a good indicator of climatic trends. For study of long records extending back to the 19th century or before, annual maximum surge-at-high-water (defined as the maximum of the difference between observed high water and the predicted tide at high water) is a better-suited parameter because during that period high waters and not the full tidal curve were recorded.

Studies of the longest records of extremes are inevitably restricted to a small number of locations. From observed sea level extremes at Liverpool since 1768, Woodworth and Blackman (2002) concluded that the annual maximum surge-at-high-water was larger in the late 18th, late 19th and late 20th centuries than for most of the 20th century, qualitatively consistent with the long-term variability in storminess from meteorological data. From the tide gauge record at Brest from 1860 to 1994, Bouligand and Pirazzoli (1999) found an increasing trend in annual maxima and 99th percentile of surges; however, a decreasing trend was found during the period 1953 to 1994. From non-tidal residuals ('surges') at San Francisco since 1858, Bromirski et al. (2003) concluded that extreme winter residuals have exhibited a significant increasing trend since about 1950, a trend that is attributed to an increase in storminess during this period. Zhang et al. (2000) concluded from records at 10 stations along the east coast of the USA since 1900 that the rise in extreme sea level closely followed the rise in mean sea level. A similar conclusion can be drawn from a recent study of Firing and Merrifield (2004), who found long-term increases in the number and height of daily extremes at Honolulu (interestingly, the highest-ever value being due an anticyclonic oceanic eddy system in 2003), but no evidence for an increase relative to the underlying upward mean sea level trend.

An analysis of 99th percentiles of hourly sea level at 141 stations over the globe for recent decades (Woodworth and Blackman, 2004) showed that there is evidence for an increase in extreme high sea level worldwide since 1975. In many cases, the secular changes in extremes were found to be similar to those in mean sea level. Likewise, interannual variability in extremes was found to be correlated with regional mean sea level, as well as to indices of regional climate patterns.

5.5.3 Ocean Density Changes

Sea level will rise if the ocean warms and fall if it cools, since the density of the water column will change. If the thermal expansivity were constant, global sea level change would parallel the global ocean heat content discussed in Section 5.2. However, since warm water expands more than cold water (with the same input of heat), and water at higher pressure expands more than at lower pressure, the global sea level change depends on the three-dimensional distribution of ocean temperature change.

Analysis of the last half century of temperature observations indicates that the ocean has warmed in all basins (see Section 5.2). The average rate of thermosteric sea level rise caused by heating of the global ocean is estimated to be 0.40 ± 0.09 mm yr^{-1} over 1955 to 1995 (Antonov et al., 2005), based on five-year mean temperature data down to 3,000 m. For the 0 to 700 m layer and the 1955 to 2003 period, the averaged thermosteric trend, based on annual mean temperature data from Levitus et al. (2005a), is 0.33 ± 0.07 mm yr^{-1} (Antonov et al., 2005). For the same period and depth range, the mean thermosteric rate based on monthly ocean temperature data from Ishii et al. (2006) is 0.36 ± 0.12 mm yr^{-1}. Figure 5.19

shows the thermosteric sea level curve over 1955 to 2003 for both the Levitus and Ishii data sets. The rate of thermosteric sea level rise is clearly not constant in time and shows considerable fluctuations (Figure 5.17). A rise of more than 20 mm occurred from the late 1960s to the late 1970s (giving peak 10-year rates in the early 1970s) with a smaller drop afterwards. Another large rise began in the 1990s, but after 2003, the steric sea level is decreasing in both estimates (peak rates in the late 1990s). Overlapping 10-year rates from these two estimates have a very high temporal correlation (r = 0.97) and the standard deviation of the rates is 0.7 mm yr^{-1}.

The Levitus and Ishii data sets both give 0.32 ± 0.09 mm yr^{-1} for the upper 700 m during 1961 to 2003, but the Levitus data set of temperature down to 3,000 m ends in 1998. From the results of Antonov et al. (2005) for thermal expansion, the difference between the trends in the upper 3,000 m and the upper 700 m for 1961 to 1998 is about 0.1 mm yr^{-1}. Assuming that the ocean below 700 m continues to contribute beyond 1998 at a similar rate, with an uncertainty similar to that of the upper-ocean contribution, we assess the thermal expansion of the ocean down to 3,000 m during 1961 to 2003 as 0.42 ± 0.12 mm yr^{-1}.

For the recent period 1993 to 2003, a value of 1.2 ± 0.5 mm yr^{-1} for thermal expansion in the upper 700 m is estimated both by Antonov et al. (2005) and Ishii et al. (2006). Willis et al. (2004) estimate thermal expansion to be 1.6 ± 0.5 mm yr^{-1}, based on combined in situ temperature profiles down to 750 m and satellite measurements of altimetric height. Including the satellite data reduces the error caused by the inadequate sampling of the profile data. Error bars were estimated to be about 2 mm for individual years in the time series, with most of the remaining error due to inadequate profile availability. A close result (1.8 ± 0.4 mm yr^{-1} steric sea level rise for 1993 to 2003) was recently obtained by Lombard et al. (2006), based on a combined analysis of in situ hydrographic data and satellite sea surface height and

SST data (Guinehut et al., 2004). It is presently unclear why the latter two estimates are significantly larger than the thermosteric rates based on temperature data alone. It is possible that the in situ data underestimate thermal expansion because of poor coverage in Southern Oceans, and it is interesting to note that a model based on assimilation of hydrographic data yields a somewhat higher estimate of 2.3 mm yr^{-1} (Carton et al., 2005). Published estimates of the steric sea level rates for 1955 to 2003 and 1993 to 2003 are shown in Table 5.2.

We assess the thermal expansion of the upper 700 m during 1993 to 2003 as 1.5 ± 0.5 mm yr^{-1}, and that of the upper 3,000 m as 1.6 ± 0.5 mm yr^{-1}, allowing for the ocean below 700 m as for the earlier period (see also Section 5.5.6, Table 5.3).

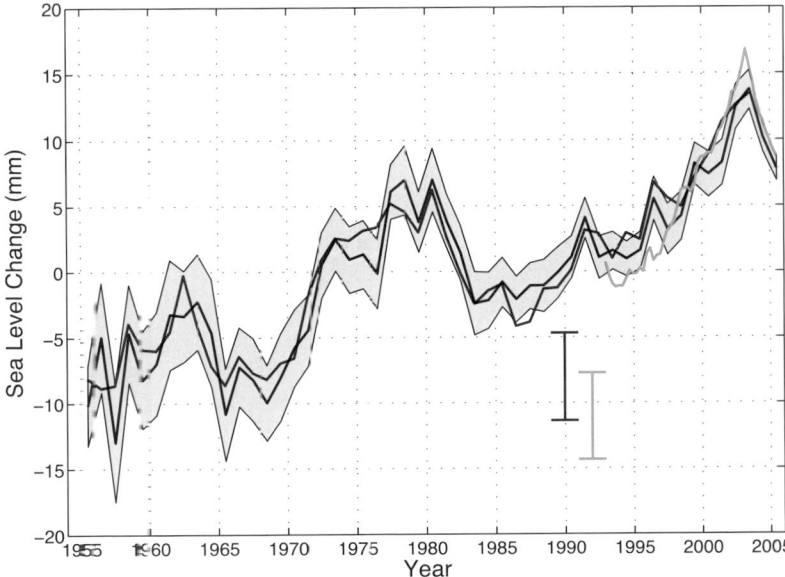

Figure 5.19. Global sea level change due to thermal expansion for 1955 to 2003, based on Levitus et al. (2005a, black line) and Ishii et al. (2006; red line) for the 0 to 700 m layer, and based on Willis et al. (2004, green line) for the upper 750 m. The shaded area and the vertical red and green error bars represent the 90% confidence interval. The black and red curves denote the deviation from their 1961 to 1990 average, the shorter green curve the deviation from the average of the black curve for the period 1993 to 2003.

Table 5.2. Recent estimates for steric sea level trends from different studies.

Reference	Steric sea level change with errors (mm yr^{-1})	Period	Depth range (m)	Data Source
Antonov et al. (2005)	0.40 ± 0.09	1955–1998	0–3,000	Levitus et al. (2005b)
Antonov et al. (2005)	0.33 ± 0.07	1955–2003	0–700	Levitus et al. (2005b)
Ishii et al. (2006)	0.36 ± 0.06	1955–2003	0–700	Ishii et al. (2006)
Antonov et al. (2005)	1.2 ± 0.5	1993–2003	0–700	Levitus et al. (2005b)
Ishii et al. (2006)	1.2 ± 0.5	1993–2003	0–700	Ishii et al. (2006)
Willis et al. (2004)	1.6 ± 0.5	1993–2003	0–750	Willis et al. (2004)
Lombard et al. (2006)	1.8 ± 0.4	1993-2003	0–700	Guinehut et al. (2004)

Antonov et al. (2002) attributed about 10% of the global average steric sea level rise during recent decades to halosteric expansion (i.e., the volume increase caused by freshening of the water column). A similar result was obtained by Ishii et al. (2006) who estimated a halosteric contribution to 1955 to 2003 sea level rise of 0.04 ± 0.02 mm yr^{-1}. While it is of interest to quantify this effect, only about 1% of the halosteric expansion contributes to the global sea level rise budget. This is because the halosteric expansion is nearly compensated by a decrease in volume of the added freshwater when its salinity is raised (by mixing) to the mean ocean value; the compensation would be exact for a linear state equation (Gille, 2004; Lowe and Gregory, 2006). Hence, for global sums of sea level change, halosteric expansion cannot be counted separately from the volume of added land freshwater (which Antonov et al., 2002, also calculate; see Section 5.5.5.1). However, for regional changes in sea level, thermosteric and halosteric contributions can be comparably important (see, e.g., Section 5.5.4.1).

5.5.4 Interpretation of Regional Variations in the Rate of Sea Level Change

Sea level observations show that whatever the time span considered, rates of sea level change display considerable regional variability (see Sections 5.5.2.2 and 5.5.2.3). A number of processes can cause regional sea level variations.

5.5.4.1 Steric Sea Level Changes

Like the sea level trends observed by satellite altimetry (see Section 5.5.2.3), the global distribution of thermosteric sea level trends is not spatially uniform. This is illustrated by Figure 5.15b and Figure 5.16b, which show the geographical distribution of thermosteric sea level trends over two different periods, 1993 to 2003 and 1955 to 2003 respectively (updated from Lombard et al., 2005). Some regions experienced sea level rise while others experienced a fall, often with rates that are several times the global mean. However, the patterns of thermosteric sea level rise over the approximately 50-year period are different from those seen in the 1990s. This occurs because the spatial patterns, like the global average, are also subject to decadal variability. In other words, variability on different time scales may have different characteristic patterns.

An EOF analysis of gridded thermosteric sea level time series since 1955 (updated from Lombard et al., 2005) displays a spatial pattern that is similar to the spatial distribution of thermosteric sea level trends over the same time span (compare Figure 5.20 with Figure 5.16b). In addition, the first principal component is negatively correlated with the Southern Oscillation Index. Thus, it appears that ENSO-related ocean variability accounts for the

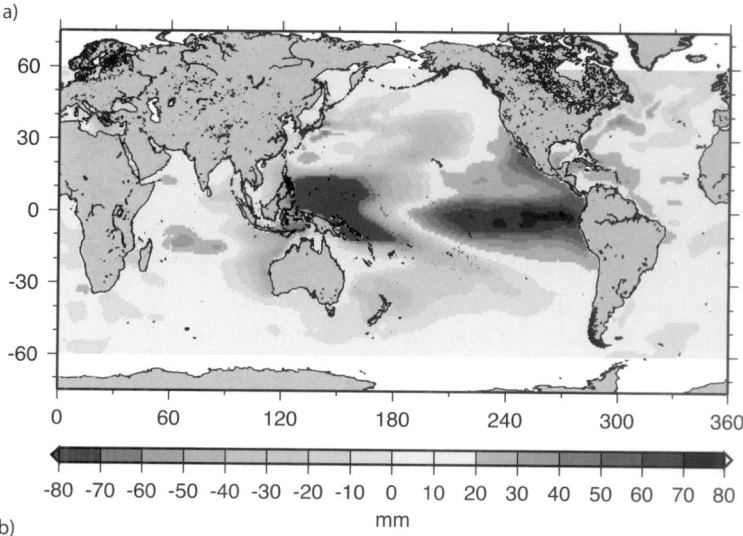

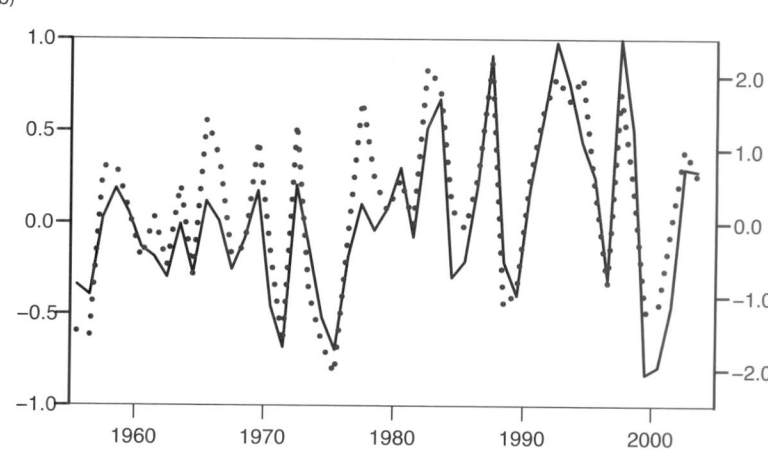

Figure 5.20. *(a) First mode of the EOF decomposition of the gridded thermosteric sea level time series of yearly temperature data down to 700 m from Ishii et al. (2006). (b) The normalised principal component (black solid curve) is highly correlated with the negative Southern Oscillation Index (dotted red curve).*

largest fraction of variance in spatial patterns of thermosteric sea level. Similarly, decadal thermosteric sea level in the North Pacific and North Atlantic appears strongly influenced by the PDO and NAO respectively.

For the recent years (1993–2003), the geographic distribution of observed sea level trends (Figure 5.15a) shows correlation with the spatial patterns of thermosteric sea level change (Figure 5.15b). This suggests that at least part of the non-uniform pattern of sea level rise observed in the altimeter data over the past decade can be attributed to changes in the ocean's thermal structure, which is itself driven by surface heating effects and ocean circulation. Note that the steric changes due to salinity changes have not been included in these figures due to insufficient salinity data in parts of the World Ocean.

Ocean salinity changes, while unimportant for sea level at the global scale, can have an effect on regional sea level (e.g., Antonov et al., 2002; Ishii et al., 2006; Section 5.5.3). For example, in the subpolar gyre of the North Atlantic, especially in the Labrador Sea, the halosteric contribution nearly counteracts the thermosteric contribution. This observational result is supported by results from data assimilation into models (e.g., Stammer et al., 2003). Since density changes can result not only from surface buoyancy fluxes but also from the wind, a simple attribution of density changes to buoyancy forcing is not possible.

While much of the non-uniform pattern of sea level change can be attributed to thermosteric volume changes, the difference between observed and thermosteric spatial trends show a high residual signal in a number of regions, especially in the southern oceans. Part of these residuals is likely due to the lack of ocean temperature coverage in remote oceans as well as in deep layers (below 700 m), and to regional salinity change.

5.5.4.2   Ocean Circulation Changes

The highly non-uniform geographical distribution of steric sea level trends is closely connected, through geostrophic balance, with changes in ocean surface circulation. Density and circulation changes result from changes in atmospheric forcing that is primarily by surface wind stress and buoyancy flux (i.e., heat and freshwater fluxes). The wind alone can therefore cause local (but not global) changes in steric sea level. Ocean general circulation models based on the assimilation of ocean data satisfactorily reproduce the spatial structure of sea level trends for the past decade, and show in particular that the tropical Pacific pattern results from decadal fluctuations in the depth of the tropical thermocline and change in equatorial trade winds (Carton et al., 2005; Köhl et al., 2006). The similarity of the patterns of steric and actual sea level change indicates that density changes are the dominant influence. Discrepancies may indicate a significant contribution from changes in the wind-driven barotropic circulation, especially at high latitudes.

5.5.4.3   Surface Atmospheric Pressure Changes

Surface atmospheric pressure also causes regional sea level variations. Over time scales longer than a few days, the ocean adjusts nearly isostatically to changes in atmospheric pressure (inverted barometer effect), that is, for each 1 hPa sea level pressure increase the ocean is depressed by approximately 10 mm, shifting the underlying mass sideways to other regions. For the temporal average, regional changes in sea level caused by atmospheric pressure loading reach about 0.2 m (e.g., between the subtropical Atlantic and the subpolar Atlantic). Such effects are generally corrected for in tide gauge and altimetry-based sea level analyses. The inverted barometer effect has a negligible effect on global mean sea level, because water is nearly incompressible, but is significant when averaged over the area of T/P and Jason-1 altimetry, which does not cover the whole World Ocean (Ponte, 2006). For that reason, the altimetry-based mean sea level curve is corrected for the inverted barometer effect.

5.5.4.4   Solid Earth and Geoid Changes

Geodynamical processes related to the solid Earth's elastic and viscoelastic response to spatially variable ice melt loading (due to the last deglaciation and present-day land ice melt) also cause non-uniform sea level change (e.g., Mitrovica et al., 2001; Peltier, 2001, 2004; Plag, 2006). The solid Earth and oceans continue to respond to the ice and complementary water loads associated with the late Pleistocene and early Holocene glacial cycles through GIA. This process not only drives large crustal uplift near the location of former ice complexes, but also produces a worldwide signature in sea level that results from gravitational, deformational and rotational effects: as the viscous mantle material flows to restore isostasy during and after the last deglaciation, uplift occurs under the former centres of the ice sheets while the surrounding peripheral bulges experience a subsidence. The return of the melt water to the oceans produces an ongoing geoid change resulting in subsidence of the ocean basins and an upward warping of the continents, while the flow of water into the subsiding peripheral bulges contributes a broad scale sea level fall in the far field of the ice complexes. The combined gravitational and deformational effects also perturb the rotation vector of the planet, and this perturbation feeds back into variations in the position of the crust and the geoid (an equipotential surface of the Earth's gravity field that coincides with the mean surface of the oceans). Corrections for GIA effects are made to both tide gauge and altimeter estimates of global sea level change (see Sections 5.5.2.1 and 5.5.2.2).

Self-gravitation and deformation of the Earth's surface in response to the ongoing change in loading by glaciers and ice sheets is another cause of regional sea level variations. Model predictions show quite different patterns of non-uniform sea level change depending on the source of the ice melt (Mitrovica et al., 2001; Plag, 2006), and associated regional sea level variations reach up to a few 0.1 mm yr^{-1}.

5.5.5   Ocean Mass Change

Global mean sea level will rise if water is added to the ocean from other reservoirs in the climate system. Water storage in the atmosphere is equivalent to only about 35 mm of global mean sea level, and the observed atmospheric storage trend of about 0.04 mm yr^{-1} in recent decades (Section 3.4.2.1) is unimportant compared with changes in ice and water stored on land, described in this subsection. Variations in land water storage result from variations in climatic conditions, direct human intervention in the water cycle and human modification of the land surface.

5.5.5.1   Ocean Mass Change Estimated from Salinity Change

Global salinity changes can be caused by changes in the global sea ice volume (which do not influence sea level) and by ocean mass changes (which do). Thus in principle, global salinity changes can be used to estimate the global average sea level change due to fresh water input (Antonov et al., 2002; Munk, 2003; Wadhams and Munk, 2004). However, the accuracy of these estimates depends on the accuracy of the estimates for both sea ice volume (Hilmer and Lemke, 2000; Wadhams and Munk, 2004; see also Section 4.4) and global salinity change (Section 5.2.3). We assess that the error in estimates of ocean mass changes derived from salinity changes and sea ice melt is too large to provide useful constraints on the sea level change budget (Section 5.5.6).

5.5.5.2   Land Ice

During the 20th century, glaciers and ice caps have experienced considerable mass losses, with strong retreats in response to global warming after 1970. For 1961 to 2003, their contribution to sea level is assessed as 0.50 ± 0.18 mm yr^{-1} and for 1993 to 2003 as 0.77 ± 0.22 mm yr^{-1} (see Section 4.5.2).

As discussed in Section 4.6.2.2 and Table 4.6, the Greenland Ice Sheet has also been losing mass in recent years, contributing 0.05 ± 0.12 mm yr^{-1} to sea level rise during 1961 to 2003 and 0.21 ± 0.07 mm yr^{-1} during 1993 to 2003. Assessments of contributions to sea level rise from the Antarctic Ice Sheet are less certain, especially before the advent of satellite measurements, and are 0.14 ± 0.41 mm yr^{-1} for 1961 to 2003 and 0.21 ± 0.35 mm yr^{-1} for 1993 to 2003. Geodetic data on Earth rotation and polar wander allow a late-20th century sea level contribution of up to about 1 mm yr^{-1} from land ice (Mitrovica et al., 2006). However, recent estimates of ice sheet mass change exclude the large contribution inferred for Greenland by Mitrovica et al. (2001) from the geographical pattern of sea level change, confirming the lower rates reported above.

5.5.5.3   Climate-Driven Change in Land Water Storage

Continental water storage includes water (both liquid and solid) stored in subsurface saturated (groundwater) and unsaturated (soil water) zones, in the snowpack, and in surface water bodies (lakes, artificial reservoirs, rivers, floodplains and wetlands). Changes in concentrated stores, most notably very large lakes, are relatively well known from direct observation. In contrast, global estimates of changes in distributed surface stores (soil water, groundwater, snowpack and small areas of surface water) rely on computations with detailed hydrological models coupled to global ocean-atmosphere circulation models or forced by observations. Such models estimate the variation in land water storage by solving the water balance equation. The Land Dynamics (LaD) model developed by Milly and Shmakin (2002) provides global 1° by 1° monthly gridded time series of root zone soil water, groundwater and snowpack for the last two decades. With these data, the contributions of time-varying land water storage to sea level rise in response to climate change have been estimated, resulting in a small positive sea level trend of about 0.12 mm yr^{-1} for the last two decades, with larger interannual and decadal fluctuations (Milly et al., 2003).

From a land surface model forced by a global climatic data set based on standard reanalysis products and on observations, land water changes during the past five decades were found to have low-frequency (decadal) variability of about 2 mm in amplitude but no significant trend (Ngo-Duc et al., 2005). These decadal variations are related to groundwater and are caused by precipitation variations. They are strongly negatively correlated with the de-trended thermosteric sea level (Figure 5.17). This suggests that the land water contribution to sea level and thermal expansion partly compensate each other on decadal time scales. However, this conclusion depends on the accuracy of the precipitation in reanalysis products.

5.5.5.4   Anthropogenic Change in Land Water Storage

The amount of anthropogenic change in land water storage systems cannot be estimated with much confidence, as already discussed by Church et al. (2001). A number of factors can contribute to sea level rise. First, natural groundwater systems typically are in a condition of dynamic equilibrium where, over long time periods, recharge and discharge are in balance. When the rate of groundwater pumping greatly exceeds the rate of recharge, as is often the case in arid or even semi-arid regions, water is removed permanently from storage. The water that is lost from groundwater storage eventually reaches the ocean through the atmosphere or surface flow, resulting in sea level rise. Second, wetlands contain standing water, soil moisture and water in plants equivalent to water roughly 1 m deep. Hence, wetland destruction contributes to sea level rise. Over time scales shorter than a few years, diversion of surface waters for irrigation in the internally draining basins of arid regions results in increased evaporation. The water lost from the basin hydrologic system eventually reaches the ocean. Third, forests store water in living tissue both above and below ground. When a forest is removed, transpiration is eliminated so that runoff is favoured in the hydrologic budget.

On the other hand, impoundment of water behind dams removes water from the ocean and lowers sea level. Dams have led to a sea level drop over the past few decades of -0.5 to -0.7 mm yr^{-1} (Chao, 1994; Sahagian et al., 1994). Infiltration from dams and irrigation may raise the water table, storing more water. Gornitz (2001) estimated -0.33 to -0.27 mm yr^{-1} sea level change equivalent held by dams (not counting additional potential storage due to subsurface infiltration).

It is very difficult to provide accurate estimates of the net anthropogenic contribution, given the lack of worldwide information on each factor, although the effect caused by dams is possibly better known than other effects. According to Sahagian (2000), the sum of the above effects could be of the order of 0.05 mm yr^{-1} sea level rise over the past 50 years, with an uncertainty several times as large.

In summary, our assessment of the land hydrology contribution to sea level change has not led to a reduction in the uncertainty compared to the TAR, which estimated the rather wide ranges of -1.1 to $+0.4$ mm yr^{-1} for 1910 to 1990 and –1.9 to +1.0 mm yr^{-1} for 1990. However, indirect evidence from considering other contributions to the sea level budget (see Section 5.5.6) suggests that the land contribution either is small (<0.5 mm yr^{-1}) or is compensated for by unaccounted or underestimated contributions.

5.5.6 Total Budget of the Global Mean Sea Level Change

The various contributions to the budget of sea level change are summarised in Table 5.3 and Figure 5.21 for 1961 to 2003 and 1993 to 2003. Some terms known to be small have been omitted, including changes in atmospheric water vapour and climate-driven change in land water storage (Section 5.5.5), permafrost and sedimentation (see, e.g., Church et al., 2001), which very likely total less than 0.2 mm yr^{-1}. The poorly known anthropogenic contribution from terrestrial water storage (see Section 5.5.5.4) is also omitted.

For 1961 to 2003, thermal expansion accounts for only 23 ± 9% of the observed rate of sea level rise. Miller and Douglas (2004) reached a similar conclusion by computing steric sea level change over the past 50 years in three oceanic regions (northeast Pacific, northeast Atlantic and western Atlantic); they found it to be too small by about a factor of three to account for the observed sea level rise based on nine tide gauges in these regions. They concluded that sea level rise in the second half of the 20th century was mostly due to water mass added to the oceans. However, Table 5.3 shows that the sum of thermal expansion and contributions from land ice is smaller by 0.7 ± 0.7 mm yr^{-1} than the observed global average sea level rise. This is likely to be a significant difference. The assessment of Church et al. (2001) could allow this difference to be explained by positive anthropogenic terms (especially groundwater mining) but these are expected to have been

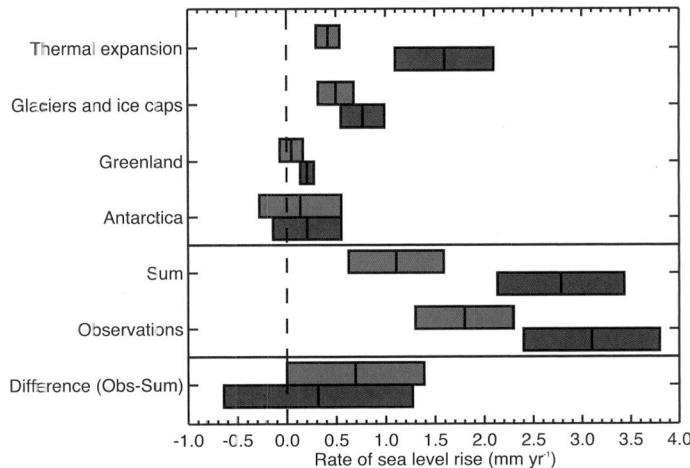

Figure 5.21. *Estimates of the various contributions to the budget of the global mean sea level change (upper four entries), the sum of these contributions and the observed rate of rise (middle two), and the observed rate minus the sum of contributions (lower), all for 1961 to 2003 (blue) and 1993 to 2003 (brown). The bars represent the 90% error range. For the sum, the error has been calculated as the square root of the sum of squared errors of the contributions. Likewise the errors of the sum and the observed rate have been combined to obtain the error for the difference.*

outweighed by negative terms (especially impoundment). We conclude that the budget has not yet been closed satisfactorily.

Given the large temporal variability in the rate of sea level rise evaluated from tide gauges (Section 5.5.2.4 and Figure 5.17), the budget is rather problematic on decadal time scales. The thermosteric contribution has smaller variability (though still substantial; Section 5.5.3) and there is only moderate temporal correlation between the thermosteric rate and the tide gauge rate. The difference between them has to be explained by ocean mass change. Because the thermosteric and climate-driven land water contributions are negatively correlated (Section 5.5.5.3.),

Table 5.3. *Estimates of the various contributions to the budget of global mean sea level change for 1961 to 2003 and 1993 to 2003 compared with the observed rate of rise. Ice sheet mass loss of 100 Gt yr^{-1} is equivalent to 0.28 mm yr^{-1} of sea level rise. A GIA correction has been applied to observations from tide gauges and altimetry. For the sum, the error has been calculated as the square root of the sum of squared errors of the contributions. The thermosteric sea level changes are for the 0 to 3,000 m layer of the ocean.*

Source	Sea Level Rise (mm yr^{-1})		Reference
	1961–2003	1993–2003	
Thermal Expansion	0.42 ± 0.12	1.6 ± 0.5	Section 5.5.3
Glaciers and Ice Caps	0.50 ± 0.18	0.77 ± 0.22	Section 4.5
Greenland Ice Sheet	0.05 ± 0.12	0.21 ± 0.07	Section 4.6.2
Antarctic Ice Sheet	0.14 ± 0.41	0.21 ± 0.35	Section 4.6.2
Sum	1.1 ± 0.5	2.8 ± 0.7	
Observed	1.8 ± 0.5		Section 5.5.2.1
		3.1 ± 0.7	Section 5.5.2.2
Difference (Observed –Sum)	0.7 ± 0.7	0.3 ± 1.0	

the apparent difference implies contributions during some 10-year periods from land ice, the only remaining term, exceeding 2 mm yr⁻¹ (Figure 5.17). Since it is unlikely that the land ice contributions of 1993 to 2003 were exceeded in earlier decades (Figure 4.14 and Section 4.6.2.2), we conclude that the maximum 10-year rates of global sea level rise are likely overestimated from tide gauges, indicating that the estimated variability is excessive.

For 1993 to 2003, thermal expansion is much larger and land ice contributes 1.2 ± 0.4 mm yr⁻¹. These increases may partly reflect decadal variability rather than an acceleration (Section 5.5.3; attribution of changes in rates and comparison with model results are discussed in Section 9.5.2). The sum is still less than the observed trend but the discrepancy of 0.3 ± 1.0 mm yr⁻¹ is consistent with zero. It is interesting to note that the difference between the observed total and thermal expansion (assumed to be due to ocean mass change) is about the same in the two periods. The more satisfactory assessment for recent years, during which individual terms are better known and satellite altimetry is available, indicates progress since the TAR.

5.6   Synthesis

The patterns of observed changes in global heat content and salinity, sea level, steric sea level, water mass evolution and biogeochemical cycles described in the previous four sections are broadly consistent with known characteristics of the large-scale ocean circulation (e.g., ENSO, NAO and SAM).

There is compelling evidence that the heat content of the World Ocean has increased since 1955 (Section 5.2). In the North Atlantic, the warming is penetrating deeper than in the Pacific, Indian and Southern Oceans (Figure 5.3), consistent with the strong convection, subduction and deep overturning circulation cell that occurs in the North Atlantic Ocean. The overturning cell in the North Atlantic region (carrying heat and water downwards through the water column) also suggests that there should be a higher anthropogenic carbon content as observed (Figure 5.11). Subduction of SAMW (and to a lesser extent AAIW) also carries anthropogenic carbon into the ocean, which is observed to be higher in the formation areas of these subantarctic water masses (Figure 5.10). The transfer of heat into the ocean also leads to sea level rise through thermal expansion, and the geographical pattern of sea level change since 1955 is largely consistent with thermal expansion and with the change in heat content (Figure 5.2).

Although salinity measurements are relatively sparse compared with temperature measurements, the salinity data also show significant changes. In global analyses, the waters at high latitudes (poleward of 50°N and 70°S) are fresher in the upper 500 m (Figure 5.5 World). In the upper 500 m, the subtropical latitudes in both hemispheres are characterised by an increase in salinity. The regional analyses of salinity also

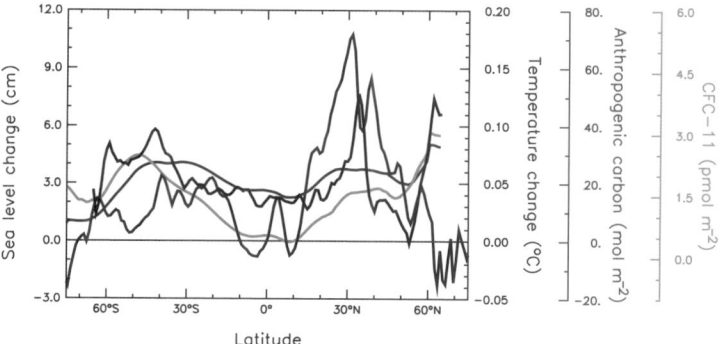

Figure 5.22. *Averages of temperature change (blue, from Levitus et al., 2005a), anthropogenic carbon (red, from Sabine et al., 2004b) and CFC-11 (green, from Willey et al., 2004) along lines of constant latitude over the top 700-m layer of the upper ocean. Also shown is sea level change averaged along lines of constant latitude (black, from Cazenave and Nerem, 2004). The temperature changes are for the 1955 to 2003 period, the anthropogenic carbon is since pre-industrial times (i.e., 1750), CFC-11 concentrations are for the period 1930 to 1994 and sea level for the period 1993 to 2003.*

show a similar distributional change with a freshening of key high-latitude water masses such as LSW, AAIW and NPIW, and increased salinity in some of the subtropical gyres such as that at 24°N. The North Atlantic (and other key ocean water masses) also shows significant decadal variations, such as the recent increase in surface salinity in the North Atlantic subpolar gyre. At high latitudes (particularly in the NH), there is an observed increase in melting of perennial sea ice, precipitation, and glacial melt water (see Chapter 4), all of which act to freshen high-latitude surface waters. At mid-latitudes it is likely that evaporation minus precipitation has increased (i.e., the transport of freshwater from the ocean to the atmosphere has increased). The pattern of salinity change suggests an intensification in the Earth's hydrological cycle over the last 50 years. These trends are consistent with changes in precipitation and inferred greater water transport in the atmosphere from low latitudes to high latitudes and from the Atlantic to the Pacific.

Figure 5.22 shows zonal means of changes in temperature, anthropogenic carbon, sea level rise and a passive tracer (CFC). It is remarkable that these independent variables (albeit with widely varying reference periods) show a common pattern of change in the ocean. Specifically, the close similarity of higher levels of warming, sea level rise, anthropogenic carbon and CFC-11 at mid-latitudes and near the equator strongly suggests that these changes are the result of changes in ocean ventilation and circulation. Warming of the upper ocean should lead to a decrease in ocean ventilation and subduction rates, for which there is some evidence from observed decreases in O_2 concentrations.

In the equatorial Pacific, the pattern of steric sea level rise also shows that strong west to east gradients in the Pacific have weakened (i.e., it is now cooler in the western Pacific and warmer in the eastern Pacific). This decrease in the equatorial temperature gradient is consistent with a tendency towards more prolonged and stronger El Niños over this same period (see Section 3.6.2).

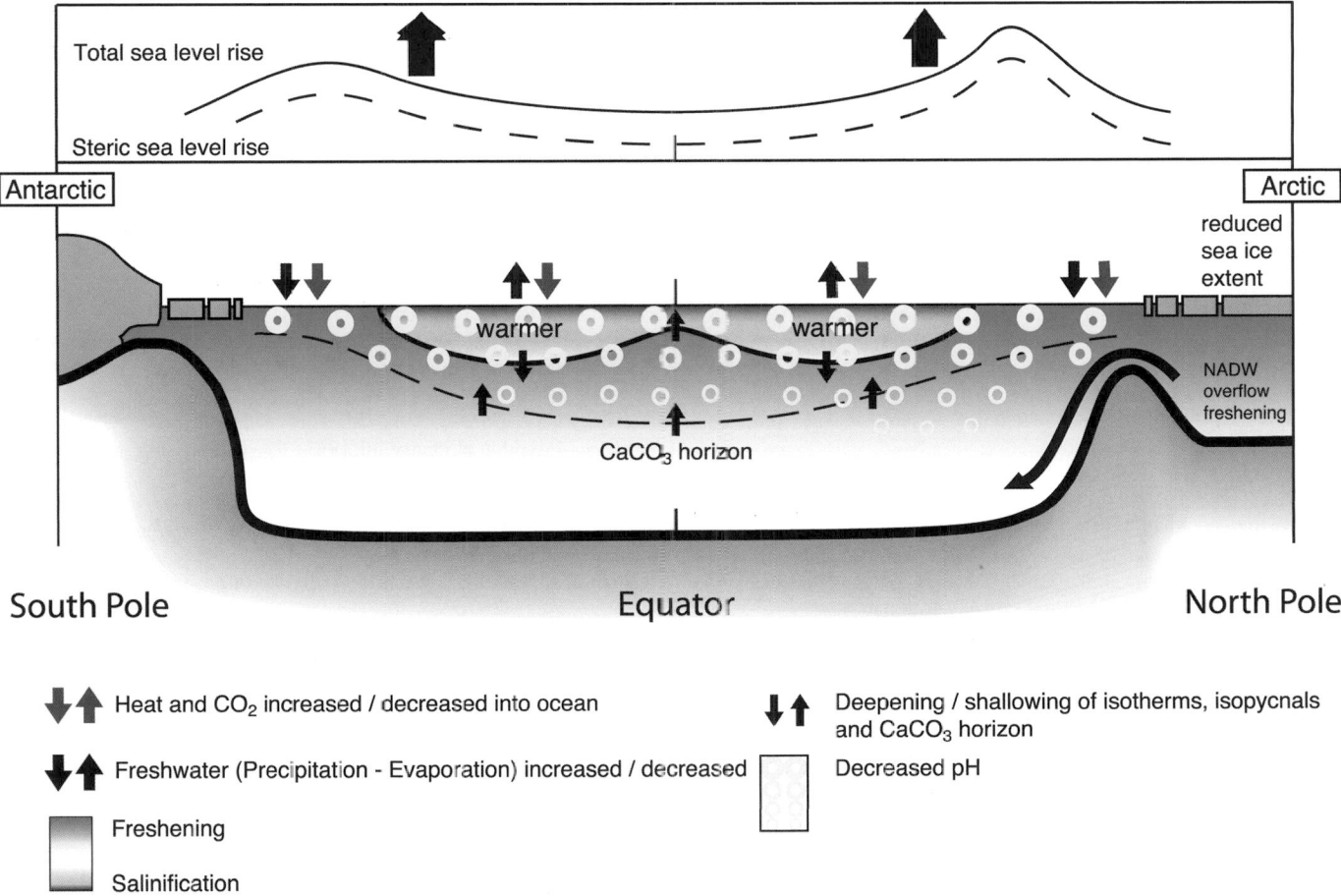

Figure 5.23. *Schematic of the observed changes in the ocean state, including ocean temperature, ocean salinity, sea level, sea ice and biogeochemical cycles. The legend identifies the direction of the changes in these variables.*

The subduction of carbon into the ocean has resulted in calcite and aragonite saturation horizons generally becoming shallower and pH decreasing primarily in the surface and near-surface ocean causing the ocean to become more acidic.

Since the TAR, the capability to measure most of the processes that contribute to sea level has been developed. In the 1990s, the observed sea level rise that was not explained through steric sea level rise could largely be explained by the transfer of mass from glaciers, ice sheets and river runoff (see Section 5.5). Figure 5.23 is a schematic that summarises the observed changes.

All of these observations taken together give high confidence that the ocean state has changed, that the spatial distribution of the changes is consistent with the large-scale ocean circulation and that these changes are in response to changed ocean surface conditions.

While there are many robust findings regarding the changed ocean state, key uncertainties still remain. Limitations in ocean sampling (particularly in the SH) mean that decadal variations in global heat content, regional salinity patterns, and rates of global sea level rise can only be evaluated with moderate confidence. Furthermore, there is low confidence in the evidence for trends

in the MOC and the global ocean freshwater budget. Finally, the global average sea level rise for the last 50 years is likely to be larger than can be explained by thermal expansion and loss of land ice due to increased melting, and thus for this period it is not possible to satisfactorily quantify the known processes causing sea level rise.

References

AchutaRao, K.M. et al., 2006: Variability of ocean heat uptake: Reconciling observations and models, *J. Geophys. Res.*, **111**, C05019, doi:10.1029/2005JC003136.

Andersson, H.C., 2002: Influence of long-term regional and large-scale atmospheric circulation on the Baltic sea level. *Tellus*, **A54**, 76–88.

Andreev, A., and S. Watanabe, 2002: Temporal changes in dissolved oxygen of the intermediate water in the subarctic North Pacific. *Geophys. Res. Lett.*, **29**(14), 1680, doi:10.1029/2002GL015021.

Andrie, C., et al., 2003: Variability of AABW properties in the equatorial channel at 35 degrees W. *Geophys. Res. Lett.*, **30**(5), 8007, doi:10.1029/2002GL015766.

Antonov, J.I., S. Levitus, and T.P. Boyer, 2002: Steric sea level variations during 1957-1994: Importance of salinity. *J. Geophys. Res.*, **107**(C12), 8013, doi:10.1029/2001JC000964.

Antonov, J.I., S. Levitus, and T.P. Boyer, 2005: Steric variability of the world ocean, 1955-2003. *Geophys. Res. Lett.*, **32**(12), L12602, doi:10.1029/2005GL023112.

Aoki, S., M. Yoritaka, and A. Masuyama, 2003: Multidecadal warming of subsurface temperature in the Indian sector of the Southern Ocean. *J. Geophys. Res.*, **108**(C4), 8081, doi:10.1029/JC000307.

Aoki, S., N.L. Bindoff, and J.A. Church, 2005a: Interdecadal water mass changes in the Southern Ocean between 30E and 160E. *Geophys. Res. Lett.*, **32**, L07607, doi:10.1029/2004GL022220.

Aoki, S., S.R. Rintoul, S. Ushio, and S. Watanabe, 2005b: Freshening of the Adélie Land Bottom water near 140°E. *Geophys. Res. Lett.*, **32**, L23601, doi:10.1029/2005GL024246.

Baringer, M.O., and J.C. Larsen, 2001: Sixteen years of Florida Current transport at 27° N. *Geophys. Res. Lett.*, **28**(16), 3179–3182.

Bates, N.R., A.C. Pequignet, R.J. Johnson, and N. Gruber, 2002: A short-term sink for atmospheric CO_2 in subtropical mode water of the North Atlantic Ocean. *Nature*, **420**(6915), 489–493.

Beaugrand, G., and P.C. Reid, 2003: Long-term changes in phytoplankton, zooplankton and salmon related to climate. *Global Change Biol.*, **9**(6), 801–817.

Belkin, I.M., 2004: Propagation of the "Great Salinity Anomaly" of the 1990s around the northern North. *Geophys. Res. Lett.*, **31**, L08306, doi:10.1029/2003GL019334.

Beltrami, H., J.E. Smerdon, H.N. Pollack, and S. Huang, 2002: Continental heat gain in the global climate system. *Geophys. Res. Lett.*, **29**, doi:10.1029/2001GL014310.

Bersch, M., 2002: North Atlantic Oscillation-induced changes of the upper layer circulation in the northern North Atlantic Ocean. *J. Geophys. Res.*, **107**(C10), 3156, doi:10.1029/JC000901.

Bi, D.H., W.F. Budd, A.C. Hirst, and X.R. Wu, 2001: Collapse and reorganisation of the Southern Ocean overturning under global warming in a coupled model. *Geophys. Res. Lett.*, **28**(20), 3927–3930.

Biasutti, M., D.S. Battisti, and E.S. Sarachik, 2003: The annual cycle over the tropical Atlantic, South America, and Africa. *J. Clim.*, **16**(15), 2491–2508.

Bindoff, N.L., and T.J. McDougall, 2000: Decadal changes along an Indian Ocean section at 32 degrees S and their interpretation. *J. Phys. Oceanogr.*, **30**(6), 1207–1222.

Björk, G., et al., 2002: Return of the cold halocline layer to the Amundsen Basin of the Arctic Ocean: Implications for the sea ice mass balance. *Geophys. Res. Lett.*, **29**(11), 1513, doi:10.1029/2001GL014157.

Bouligand, R., and P.A. Pirazzoli, 1999: Les surcotes et les décotes marines à Brest, étude statistique et évolution. *Oceanol. Acta*, **22**(2), 153–166.

Boyer, T.P., J.I. Antonov, S. Levitus, and R. Locarnini, 2005: Linear trends of salinity for the world ocean, 1955-1998. *Geophys. Res. Lett.*, **32**, L01604, doi:1029/2004GL021791.

Boyer, T.P., et al., 2002: World ocean database 2001, Volume 2: Temporal distribution of bathythermograph profiles. In: *NOAA Atlas NESDIS 43* [Levitus, S. (ed.)]. Vol. 2. U.S. Government Printing Office, Washington, DC, 119 pp, CD-ROMs.

Brankart, J.M., and N. Pinardi, 2001: Abrupt cooling of the Mediterranean Levantine Intermediate Water at the beginning of the 1980s: Observational evidence and model simulation. *J. Phys. Oceanogr.*, **31**(8), 2307–2320.

Bromirski, P.D., R.E. Flick, and D.R. Cayan, 2003: Storminess variability along the California coast: 1858-2000. *J. Clim.*, **16**, 982–993.

Bryden, H.L., E.L. McDonagh, and B.A. King, 2003: Changes in ocean water mass properties: Oscillations or trends? *Science*, **300**, 2086–2088.

Bryden, H.L., H.R. Longworth, and S.A. Cunningham, 2005: Slowing of the Atlantic meridional overturning circulation at 25°N. *Nature*, **438**, 655–657, doi:10.1038/nature04385.

Bryden, H.L., et al., 1996: Decadal changes in water mass characteristics at 24 degrees N in the subtropical North Atlantic Ocean. *J. Clim.*, **9**(12), 3162–3186.

Cabanes, C., A. Cazenave, and C. Le Provost, 2001: Sea level change from Topex-Poseidon altimetry for 1993-1999 and possible warming of the southern oceans. *Geophys. Res. Lett.*, **28**(1), 9–12.

Cai, W., 2006: Antarctic ozone depletion causes an intensification of the Southern Ocean super-gyre circulation. *Geophys. Res. Lett.*, **33**, doi:1029/2005GL024911.

Caldeira, K., and M.E. Wickett, 2003: Anthropogenic carbon and ocean pH. *Nature*, **425**(6956), 365.

Carmack, E.C., et al., 1995: Evidence for warming of Atlantic Water in the southern Canadian Basin of the Arctic-Ocean - Results from the Larsen-93 Expedition. *Geophys. Res. Lett.*, **22**(9), 1061–1064.

Carton, J., B. Giese, and S. Grodsky, 2005: Sea level rise and the warming of the oceans in the Simple Ocean Data Assimilation (SODA) ocean reanalysis. *J. Geophys. Res.*, **110**, C09006, doi:1029/2004JC002817.

Cazenave, A., and R.S. Nerem, 2004: Present-day sea level change: observations and causes. *Rev. Geophys.*, **42**(3), RG3001, doi:10.1029/2003RG000139.

Chambers, D., J.C. Ries, C.K. Shum, and B.D. Tapley, 1998: On the use of tide gauges to determine altimeter drift. *J. Geophys. Res.*, **103**(C6), 12885–12890.

Chambers, D.P., et al., 2002: Low-frequency variations in global mean sea level: 1950-2000. *J. Geophys. Res.*, **107**(C4), doi:10.1029/2001JC001089.

Chao, B., 1994: Man-made lakes and global sea level. *Nature*, **370**, 258.

Chavez, F.P., J. Ryan, S.E. Lluch-Cota, and M. Niquen, 2003: From anchovies to sardines and back: Multidecadal change in the Pacific Ocean. *Science*, **299**(5604), 217–221.

Chelton, B., et al., 2001: Satellite altimetry. In: *Satellite Altimetry and Earth Sciences: A Handbook of Techniques and Applications* [Fu, L.-L., and A. Cazenave (eds.)]. Academic Press, San Diego, pp. 1–131.

Church, J.A., and N.J. White, 2006: A 20th century acceleration in global sea-level rise. *Geophys. Res. Lett.*, **33**, L01602, doi:10.1029/2005GL024826.

Church, J.A., N.J. White, and J.R. Hunter, 2006: Sea-level rise at tropical Pacific and Indian Ocean islands. *Global Planet. Change*, **53**, 155–168.

Church, J.A., et al., 2001: Changes in sea level. In: *Climate Change 2001: The Scientific Basis. Contribution of Working Group 1 to the Third Assessment Report of the Intergovernmental Panel on Climate Change* [Houghton, J.T., et al. (eds.)]. Cambridge University Press, Cambridge, United Kingdom and New York, NY, USA, pp. 639–693.

Church, J.A., et al., 2004: Estimates of the regional distribution of sea-level rise over the 1950 to 2000 period. *J. Clim.*, **17**(13), 2609–2625.

Clark, P.U., N.G. Pisias, T.F. Stocker, and A.J. Weaver, 2002: The role of the thermohaline circulation in abrupt climate change. *Nature*, **415**, 863–869.

Comiso, J.C., 2002: A rapidly declining perennial sea ice cover in the Arctic. *Geophys. Res. Lett.*, **29**, 1956–1959.

Comiso, J.C., 2003: Warming trends in the Arctic from clear sky satellite observations. *J. Clim.*, **16**(21), 3498–3510.

Conkright, M.E., et al., 2002: World ocean database 2001, Volume 1: Introduction. In: *NOAA Atlas NESDIS 42* [Levitus, S. (ed.)]. Vol. 1. U.S. Government Printing Office, Washington, DC, 159 pp, CD-ROMs.

Cunningham, S.A., S.G. Alderson, B.A. King, and M.A. Brandon, 2003: Transport and variability of the Antarctic Circumpolar Current in Drake Passage. *J. Geophys. Res.*, **108**(C5), 8084, doi:10.1029/2001JC001147.

Curry, R., and C. Mauritzen, 2005: Dilution of the northern North Atlantic Ocean in recent decades. *Science*, **308**(5729), 1772–1774.

Curry, R., B. Dickson, and I. Yashayaev, 2003: A change in the freshwater balance of the Atlantic Ocean over the past four decades. *Nature*, **426**(6968), 826–829.

Deutsch, C., S. Emerson, and L. Thompson, 2005: Fingerprints of climate change in North Pacific oxygen. *Geophys. Res. Lett.*, **32**, L16604, doi:10.1029/2005GL023190.

Dickson, B., et al., 2002: Rapid freshening of the deep North Atlantic Ocean over the past four decades. *Nature*, **416**(6883), 832–837.

Dickson, R., et al., 1996: Long-term coordinated changes in the convective activity of the North Atlantic. *Prog. Oceanogr.*, **38**, 241–295.

Dickson, R.R., R. Curry, and I. Yashayaev, 2003: Recent changes in the North Atlantic. *Philos. Trans. R. Soc. London Ser. A*, **361**(1810), 1917–1933.

Dickson, R.R., J. Meincke, S.A. Malmberg, and A.J. Lee, 1988: The Great Salinity Anomaly in the Northern North Atlantic 1968-1982. *Prog. Oceanogr.*, **20**(2), 103–151.

Doney, S.C., J.L. Bullister, and R. Wanninkhof, 1998: Climatic variability in upper ocean ventilation rates diagnosed using chlorofluorocarbons. *Geophys. Res. Lett.*, **25**(9), 1399–1402.

Donnelly, J.P., P. Cleary, P. Newby, and R. Ettinger, 2004: Coupling instrumental and geological records of sea-level change: Evidence from southern New England of an increase in the rate of sea-level rise in the late 19th century. *Geophys. Res. Lett.*, **31**(5), L05203, doi:10.1029/2003GL018933.

Dore, J.E., R. Lukas, D.W. Sadler, and D.M. Karl, 2003: Climate-driven changes to the atmospheric CO_2 sink in the subtropical North Pacific Ocean. *Nature*, **424**(6950), 754–757.

Douglas, B.C., 1992: Global sea level acceleration. *J. Geophys. Res.*, **97**(C8), 12699–12706.

Douglas, B.C., 2001: Sea level change in the era of the recording tide gauges. In: *Sea Level Rise: History and Consequences* [Douglas, B.C., Kearney, M.S., and S.P. Leatherman (eds.)]. Academic Press, New York, pp. 37–64.

Ekman, M., 1988: The world's longest continued series of sea level observations. *Pure Appl. Geophys.*, **127**, 73–77.

Emerson, S., S. Mecking, and J. Abell, 2001: The biological pump in the subtropical North Pacific Ocean: Nutrient sources, Redfield ratios, and recent changes. *Global Biogeochem. Cycles*, **15**(3), 535–554.

Emerson, S., Y.W. Watanabe, T. Ono, and S. Mecking, 2004: Temporal trends in apparent oxygen utilization in the upper pycnocline of the North Pacific: 1980-2000. *J. Oceanogr.*, **60**(1), 139–147.

Fahrbach, E., et al., 2004: Decadal-scale variations of water mass properties in the deep Weddell Sea. *Ocean Dyn.*, **54**(1), 77–91.

Feely, R.A., and C.T.A. Chen, 1982: The effect of excess CO_2 on the calculated calcite and aragonite saturation horizons in the northeast Pacific. *Geophys. Res. Lett.*, **9**(11), 1294–1297.

Feely, R.A., R. Wanninkhof, T. Takahashi, and P. Tans, 1999: Influence of El Niño on the equatorial Pacific contribution to atmospheric CO_2 accumulation. *Nature*, **398**(6728), 597–601.

Feely, R.A., et al., 2002: In situ calcium carbonate dissolution in the Pacific Ocean. *Global Biogeochem. Cycles*, **16**(4), 1144, doi:10.1029/2002GB001866.

Feely, R.A., et al., 2004: Impact of anthropogenic CO_2 on the $CaCO_3$ system in the oceans. *Science*, **305**(5682), 362–366.

Feng, M., and G. Meyers, 2003: Interannual variability in the tropical Indian Ocean: A two-year time-scale of Indian Ocean Dipole. *Deep-Sea Res. II*, **50**, 2263–2284.

Firing, Y.L., and M.A. Merrifield, 2004: Extreme sea level events at Hawaii: the influence of mesoscale eddies. *Geophys. Res. Lett.*, **31**(24), L24306, doi:10.1029/2004GL021539.

Folland, C.K., et al., 2001: Observed climate variability and change. In: *Climate Change 2001: The Scientific Basis. Contribution of Working Group I to the Third Assessment Report of the Intergovernmental Panel on Climate Change* [Houghton, J.T., et al. (eds.)]. Cambridge University Press, Cambridge, United Kingdom and New York, NY, pp. 99–181.

Freeland, H., et al., 1997: Evidence of change in the winter mixed layer in the Northeast Pacific Ocean. *Deep-Sea Res. I*, **44**(12), 2117–2129.

Fu, L.L., and A. Cazenave, 2001: *Satellite Altimetry and Earth Sciences: A Handbook of Techniques and Applications*. International Geophysics Series Vol. 69, Academic Press, San Diego, 457 pp.

Fukasawa, M., et al., 2004: Bottom water warming in the North Pacific Ocean. *Nature*, **427**(6977), 825–827.

Gamo, T., et al., 1986: Spatial and temporal variations of water characteristics in the Japan Sea bottom layer. *J. Mar. Res.*, **44**(4), 781–793.

Ganachaud, A., and C. Wunsch, 2003: Large-scale ocean heat and freshwater transports during the World Ocean Circulation Experiment. *J. Clim.*, **16**(4), 696–705.

Garcia, H.E., et al., 2005: On the variability of dissolved oxygen and apparent oxygen utilization content for the upper world ocean: 1955 to 1998. *Geophys. Res. Lett.*, **32**, L09604, doi:10.1029/GL022286.

Gasparini, G.P., et al., 2005: The effect of the Eastern Mediterranean Transient on the hydrographic characteristics in the Strait of Sicily and in the Tyrrhenian Sea. *Deep-Sea Res. I*, **52**(6), 915–935.

Gehrels, W.R., et al., 2004: Late Holocene sea-level changes and isostatic crustal movements in Atlantic Canada. *Quat. Int.*, **120**, 79–89.

Gertman, I., N. Pinardi, Y. Popov, and A. Hecht, 2006: Aegean sea water masses during the early stages of the eastern Mediterranean climatic transient (1988-1990). *J. Phys. Oceanogr.*, **36**(9), 1841–1859.

Gille, S.T., 2002: Warming of the Southern Ocean since the 1950s. *Science*, **295**(5558), 1275–1277.

Gille, S.T., 2004: How nonlinearities in the equation of state of seawater can confound estimates of steric sea level change. *J. Geophys. Res.*, **109**(3), C03005, doi:10.1029/2003JC002012.

Gonzalez-Dávila, M., et al., 2003: Seasonal and interannual variability of sea-surface carbon dioxide species at the European Station for Time Series in the Ocean at the Canary Islands (ESTOC) between 1996 and 2000. *Global Biogeochem. Cycles*, **17**(3), 1076, doi:10.1029/2002GB001993.

González-Pola, C., A. Lavin, and M. Vargas-Yanez, 2005: Intense warming and salinity modification of intermediate water masses in the southeastern corner of the Bay of Biscay for the period 1992-2003. *J. Geophys. Res.*, **110**, C05020, doi:10.1029/2004JC002367.

Gornitz, V., 2001: Impoundment, groundwater mining, and other hydrologic transformations: Impacts on global sea level rise. In: *Sea Level Rise: History and Consequences* [Douglas, B.C., M.S. Kearney, and S.P. Leatherman (eds.)]. Academic Press, San Diego, pp. 97–119.

Gregg, W.W., et al., 2003: Ocean primary production and climate: Global decadal changes. *Geophys. Res. Lett.*, **30**(15), 1809, doi:10.1029/2003GL016889.

Gregory, J.M., et al., 2001: Comparison of results from several AOGCMs for global and regional sea-level changes 1900-2100. *Clim. Dyn.*, **18**, 225–240.

Gregory, J.M., et al., 2004: Simulated and observed decadal variability in ocean heat content. *Geophys. Res. Lett.*, **31**, L15312, doi:10.1029/2004GL020258.

Gregory, J.M., et al., 2005: A model intercomparison of changes in the Atlantic thermohaline circulation in response to increasing atmospheric CO_2 concentration. *Geophys. Res. Lett.*, **32**, L12703, doi:10.1029/2005GL023209.

Grist, J.P., and S.A. Josey, 2003: Inverse analysis adjustment of the SOC air-sea flux climatology using ocean heat transport constraints. *J. Clim.*, **16**(20), 3274–3295.

Gruber, N., J.L. Sarmiento, and T.F. Stocker, 1996: An improved method for detecting anthropogenic CO_2 in the oceans. *Global Biogeochem. Cycles*, **10**(4), 809–837.

423

Gruber, N., C.D. Keeling, and N.R. Bates, 2002: Interannual variability in the North Atlantic Ocean carbon sink. *Science*, **298**(5602), 2374–2378.

Guinehut, S., P.-Y. Le Traon, G. Larnicol, and S. Phillips, 2004: Combining ARGO and remote-sensing data to estimate the ocean three-dimensional temperature fields. *J. Mar. Syst.*, **46**, 85–98.

Gulev, S.K., T. Jung, and E. Ruprecht, 2006: Estimation of the sampling errors in global surface flux fields based on VOS data. *J. Clim.*, **20**(2), 279-301.

Häkkinen, S., 2002: Surface salinity variability in the northern North Atlantic during recent decades. *J. Geophys. Res.*, **107**(C12), doi:10.1029/2001JC000812.

Häkkinen, S., and P.B. Rhines, 2004: Decline of subpolar North Atlantic circulation during the 1990s. *Science*, **304**, 555–559.

Hanawa, K., 1995: Southward penetration of the Oyashio water system and the wintertime condition of midlatitude westerlies over the North Pacific. *Bull. Hokkaido Natl. Fish. Res. Inst.*, **59**, 103–119.

Hanawa, K., and J. Kamada, 2001: Variability of core layer temperature (CLT) of the North Pacific subtropical mode water. *Geophys. Res. Lett.*, **28**(11), 2229–2232.

Hanawa, K., and L.D. Talley, 2001: Mode waters. In: *Ocean Circulation and Climate* [Siedler, G., J.A. Church, and J. Gould (eds.)]. Academic Press, San Diego, pp. 373–386.

Harrison, D.E., and M. Carson, 2006: Is the World Ocean warming? Upper ocean temperature trends, 1950-2000. *J. Phys. Oceanogr.*, **37** (2), 174–187.

Hátún, H., et al., 2005: Influence of the Atlantic Subpolar Gyre on the thermohaline circulation. *Science*, **309**, 1841–1844.

Hazeleger, W., and S.S. Drijfhout, 1998: Mode water variability in a model of the subtropical gyre: Response to anomalous forcings. *J. Phys. Oceanogr.*, **28**, 266–288.

Hazeleger, W., P. de Vries, and Y. Friocourt, 2003: Sources of the Equatorial Undercurrent in the Atlantic in a high-resolution ocean model. *J. Phys. Oceanogr.*, **33**, 677–693.

Hill, K.L., A.J. Weaver, H.J. Freeland, and A. Bychkov, 2003: Evidence of change in the Sea of Okhotsk: Implications for the North Pacific. *Atmos.-Ocean*, **41**(1), 49–63.

Hilmer, M., and P. Lemke, 2000: On the decrease of Arctic sea ice volume. *Geophys. Res. Lett.*, **27**(22), 3751–3754.

Hirawake, T., T. Odate, and M. Fukuchi, 2005: Long-term variation of surface phytoplankton chlorophyll a in the Southern Ocean during 1965-2002. *Geophys. Res. Lett.*, **32**(5), L05606, doi:10.1029/2004GL021394.

Hogg, N.G., 2001: Quantification of the deep circulation. In: *Ocean Circulation and Climate* [Siedler, G., J.A. Church, and J. Gould (eds.)]. Academic Press, San Diego, pp. 259–270.

Hogg, N.G., and W. Zenk, 1997: Long-period changes in the bottom water flowing through Vema Channel. *J. Geophys. Res.*, **102**, 15639–15646.

Holgate, S.J., and P.L. Woodworth, 2004: Evidence for enhanced coastal sea level rise during the 1990s. *Geophys. Res. Lett.*, **31**, L07305, doi:10.1029/2004GL019626.

Houghton, R.W., and M. Visbeck, 2002: Quasi-decadal salinity fluctuations in the Labrador Sea. *J. Phys. Oceanogr.*, **32**, 687–701.

ICES, 2005: *The Annual ICES Ocean Climate Status Summary 2004/2005.* ICES Cooperative Research Report No.275, International Council for the Exploration of the Sea, Copenhagen, Denmark, 37 pp.

IOC, 2002: *Manual on Sea-Level Measurement and Interpretation. Volume 3 - Reappraisals and Recommendations as of the Year 2000.* Manuals and Guides No. 14, Intergovernmental Oceanographic Commission, Paris, 47 pp.

IPCC, 2001: *Climate Change 2001: The Scientific Basis. Contribution of Working Group I to the Third Assessment Report of the Intergovernmental Panel on Climate Change* [Houghton, J.T., et al. (eds.)]. Cambridge University Press, Cambridge, United Kingdom and New York, NY, USA, 881 pp.

Ishii, M., M. Kimoto, K. Sakamoto, and S.I. Iwasaki, 2006: Steric sea level changes estimated from historical ocean subsurface temperature and salinity analyses. *J. Oceanogr.*, **62**(2), 155–170.

Jacobs, S.S., C.F. Giulivi, and P.A. Mele, 2002: Freshening of the Ross Sea during the late 20th century. *Science*, **297**(5580), 386–389.

Johnson, G.C., and A.H. Orsi, 1997: Southwest Pacific Ocean water-mass changes between 1968/69 and 1990/91. *J. Clim.*, **10**(2), 306–316.

Johnson, G.C., and N. Gruber, 2007: Decadal water mass variations along 20°W in the northeastern Atlantic Ocean. *Prog. Oceanogr.*, in press.

Johnson, G.C., D.L. Rudnick, and B.A. Taft, 1994: Bottom water variability in the Samoa Passage. *J. Mar. Res.*, **52**, 177–196.

Josey, S.A., 2003: Changes in the heat and freshwater forcing of the eastern Mediterranean and their influence on deep water formation. *J. Geophys. Res.*, **108**(C7), 3237, doi:10.1029/2003JC001778.

Josey, S.A., and R. Marsh, 2005: Surface freshwater flux variability and recent freshening of the North Atlantic in the eastern Subpolar Gyre. *J. Geophys. Res.*, **110**, C05008, doi:10.1029/2004JC002521.

Joyce, T.M., and P.E. Robbins, 1996: The long-term hydrographic record at Bermuda. *J. Clim.*, **9**, 3121–3131.

Joyce, T.M., and J. Dunworth-Baker, 2003: Long-term hydrographic variability in the Northwest Pacific Ocean. *Geophys. Res. Lett.*, **30**(2), 1043, doi:10.1029/2002GL015225.

Joyce, T.M., R.S. Pickart, and R.C. Millard, 1999: Long-term hydrographic changes at 52 and 66 degrees W in the North Atlantic Subtropical Gyre & Caribbean. *Deep-Sea Res. II*, **46**(1–2), 245–278.

Joyce, T.M., C. Deser, and M.A. Spall, 2000: The relation between decadal variability of subtropical mode water and the North Atlantic Oscillation. *J. Clim.*, **13**(14), 2550–2569.

Karl, D.M., 1999: A sea of change: Biogeochemical variability in the North Pacific Subtropical Gyre. *Ecosystems*, **2**(3), 181–214.

Keeling, R.F., 2005: Comment on "The ocean sink for anthropogenic CO_2". *Science*, **308**(5729), 1743c.

Keigwin, L.D., W.B. Curry, S.J. Lehman, and S. Johnsen, 1994: The role of the deep ocean in North Atlantic climate change between 70 and 130 kyr ago. *Nature*, **371**, 323–326.

Keller, K., R.D. Slater, M. Bender, and R.M. Key, 2002: Possible biological or physical explanations for decadal scale trends in North Pacific nutrient concentrations and oxygen utilization. *Deep-Sea Res. II*, **49**(1–3), 345–362.

Kieke, D., et al., 2006: Changes in the CFC inventories and formation rates of Upper Labrador Sea water. *J. Phys. Oceanogr.*, **36**, 64–86.

Kim, K., et al., 2004: Water masses and decadal variability in the East Sea (Sea of Japan). *Prog. Oceanogr.*, **61**(2–4), 157–174.

Kim, K.R., et al., 2002: A sudden bottom-water formation during the severe winter 2000-2001: The case of the East/Japan Sea. *Geophys. Res. Lett.*, **29**(8), doi:10.1029/2001GL014498.

Klein, B., et al., 2000: Is the Adriatic returning to dominate the production of Eastern Mediterranean Deep Water? *Geophys. Res. Lett.*, **27**(20), 3377–3380.

Knight, J.R., et al., 2005: A signature of persistent natural thermohaline circulation cycles in observed climate. *Geophys. Res. Lett.*, **32**, L20708, doi:1029/2005GL024233.

Köhl, A., D. Stammer, and B. Cornuelle, 2006: Interannual to decadal changes in the ECCO global synthesis. *J. Phys. Oceanogr.*, **37**(2), 313-337.

Krahmann, G., and F. Schott, 1998: Long-term increases in Western Mediterranean salinities and temperatures: anthropogenic and climatic sources. *Geophys. Res. Lett.*, **25**(22), 4209–4212.

Kwon, Y.O., and S.C. Riser, 2004: North Atlantic Subtropical Mode Water: A history of ocean-atmosphere interaction 1961-2000. *Geophys. Res. Lett.*, **31**(19), L19307, doi:10.1029/2004GL021116.

Kwon, Y.O., K. Kim, Y.G. Kim, and K.R. Kim, 2004: Diagnosing long-term trends of the water mass properties in the East Sea (Sea of Japan). *Geophys. Res. Lett.*, **31**(20), L20306, doi:10.1029/2004GL020881.

Lambeck, K., 2002: Sea-level change from mid-Holocene to recent time: An Australian example with global implications. In: *Ice Sheets, Sea Level and the Dynamic Earth* [Mitrovica, J.X., and B. L.A. Vermeersen (eds.)]. Geodynamics Series Vol. 29, American Geophysical Union, Washington, DC, doi:10.1029/029GD03. 33–50.

Lambeck, K., C. Smither, and M. Ekman, 1998: Tests of glacial rebound models for Fennoscandinavia based on instrumental sea- and lake-level records. *Geophys. J. Int.*, **135**, 375–387.

Lambeck, K., et al., 2004: Sea level in Roman time in the Central Mediterranean and implications for recent change. *Earth Planet. Sci. Lett.*, **224**, 563–575.

Latif, M., et al., 2006: Is the thermohaline circulation changing? *J. Clim.*, **19**, 4631–4637.

Lazar, A., et al., 2002: Seasonality of the ventilation of the tropical Atlantic thermocline in an ocean general circulation mode. *J. Geophys. Res.*, **107**(C8), doi:10.1029/2000JC000667.

Lazier, J.R.N., 1995: The salinity decrease in the Labrador Sea over the past thirty years. In: *Natural Climate Variability on Decade-to-Century Time Scales* [Martinson, D.G., et al. (eds.)]. National Academy Press, Washington, DC, pp. 295–304.

Lazier, J.R.N., et al., 2002: Convection and restratification in the Labrador Sea, 1990-2000. *Deep-Sea Res. I*, **49**(10), 1819–1835.

Le Quéré, C., et al., 2003: Two decades of ocean CO_2 sink and variability. *Tellus*, **B55**(2), 649–656.

Lee, T., 2004: Decadal weakening of the shallow overturning circulation of the South Indian Ocean. *Geophys. Res. Lett.*, **31**, L18305, doi:10.1029/2004GL020884.

Lefèvre, N., et al., 2004: A decrease in the sink for atmospheric CO_2 in the North Atlantic. *Geophys. Res. Lett.*, **31**(7), L07306, doi:10.1029/2003GL018957.

Leuliette, E.W., R.S. Nerem, and G.T. Mitchum, 2004: Calibration of TOPEX/Poseidon and Jason altimeter data to construct a continuous record of mean sea level change. *Mar. Geodesy*, **27**(1–2), 79–94.

Levitus, S., 1989: Interpentadal variability of salinity in the upper 150m of the North-Atlantic Ocean, 1970-1974 versus 1955-1959. *J. Geophys. Res.*, **94**(C7), 9679–9685.

Levitus, S., J.I. Antonov, and T.P. Boyer, 2005a: Warming of the World Ocean, 1955-2003. *Geophys. Res. Lett.*, **32**, L02604, doi:10.1029/2004GL021592.

Levitus, S., J. Antonov, T.P. Boyer, and C. Stephens, 2000: Warming of the World Ocean. *Science*, **287**, 2225–2229.

Levitus, S., et al., 2005b: *Building Ocean Profile-Plankton Databases for Climate and Ecosystem System Research.* NOAA Technical Report NESDIS 117, U.S. Government Printing Office, Washington, DC, 29 pp.

Levitus, S., et al., 2005c: EOF analysis of upper ocean heat content, 1956-2003. *Geophys. Res. Lett.*, **32**, L18607, doi:10.1029/2005GL023606.

Li, M., P.G. Myers, and H. Freeland, 2005: An examination of historical mixed layer depths along Line-P in the Gulf of Alaska. *Geophys. Res. Lett.*, **32**, L05613, doi:10.1029/2004GL021911.

Li, Y.-H., and T.-H. Peng, 2002: Latitudinal change of remineralization ratios in the oceans and its implication for nutrient cycles. *Global Biogeochem. Cycles*, **16**(4), 1130, doi:10.1029/2001GB001828.

Locarnini, R.A., et al., 2002: World ocean database 2001. In: *NOAA Atlas NESDIS 45. Vol. 4: Temporal Distribution of Temperature, Salinity and Oxygen Profiles* [Levitus, S. (ed.)]. U.S. Government Printing Office, Washington, DC, 332 pp, CD-ROMs.

Lombard, A., et al., 2005: Thermosteric sea level rise for the past 50 years; comparison with tide gauges and inference on water mass contribution. *Global Planet. Change*, **48**, 303–312.

Lombard, A., et al., 2006: Perspectives on present-day sea level change: a tribute to Christian le Provost. *Ocean Dyn.*, **56**(5-6), doi:10.1007/s10236-005-0046-x.

Lowe, J.A., and J.M. Gregory, 2006: Understanding projections of sea level rise in a Hadley Centre coupled climate model. *J. Geophys. Res.*, **111**, C11014, doi:10.1029/2005JC003421.

Luterbacher, J., et al., 2004: European seasonal and annual temperature variability, trends, and extremes since 1500. *Science*, **303**(5663), 1499–1503.

Lyman, J.M., J.K. Willis, and G.C. Johnson, 2006: Recent cooling of the upper ocean. *Geophys. Res. Lett.*, **33**, L18604, doi:10.1029/2006GL027033.

Marander, A., et al., 2005: Interannual changes in the overflow from the Nordic Seas into the Atlantic Ocean through Denmark Strait. *Geophys. Res. Lett.*, **32**, L06606, doi:10.1029/2004GL021463.

Manca, B.B., V. Kovacevic, M. Gacic, and D. Viezzoli, 2002: Dense water formation in the Southern Adriatic Sea and spreading into the Ionian Sea in the period 1997-1999. *J. Mar. Syst.*, **33**, 133–154.

Mariotti, A., et al., 2002: The hydrological cycle in the Mediterranean region and implications for the water budget of the Mediterranean Sea. *J. Clim.*, **15**(13), 1674.

Marsh, R., 2000: Recent variability of the North Atlantic thermohaline circulation inferred from surface heat and freshwater fluxes. *J. Clim.*, **13**(18), 3239–3260.

Matsumoto, K., and N. Gruber, 2005: How accurate is the estimation of anthropogenic carbon in the ocean? An evaluation of the delta C* method. *Global Biogeochem. Cycles*, **19**, GB3014, doi:10.1029/2004GB002397.

McDonagh, E.L., et al., 2005: Decadal changes in the south Indian Ocean thermocline. *J. Clim.*, **18**, 1575–1590.

McLaughlin, F.A., E.C. Carmack, R.W. Macdonald, and J.K.B. Bishop, 1996: Physical and geochemical properties across the Atlantic Pacific water mass front in the southern Canadian Basin. *J. Geophys. Res.*, **101**(C1), 1183–1197.

McNeil, B.I., et al., 2003: Anthropogenic CO_2 uptake by the ocean based on the global chlorofluorocarbon data set. *Science*, **299**(5604), 235–239.

McPhaden, M.J., and D.X. Zhang, 2002: Slowdown of the meridional overturning circulation in the upper Pacific Ocean. *Nature*, **415**(6872), 603–608.

McPhaden, M.J., and D.X. Zhang, 2004: Pacific Ocean circulation rebounds. *Geophys. Res. Lett.*, **31**(18), L18301, doi:10.1029/2004GL020727.

Mecking, S., M.J. Warner, and J.L. Bullister, 2006: Temporal changes in pCFC-12 ages and AOU along two hydrographic sections in the eastern subtropical North Pacific. *Deep-Sea Res.*, **53**(1), 169–187.

Meredith, M.P., and J.C. King, 2005: Rapid climate change in the ocean west of the Antarctic Peninsula during the second half of the 20th century. *Geophys. Res. Lett.*, **32**, L19604, doi:1029/2005GL024042.

Meyers, G., 1996: Variation of the Indonesian throughflow and the El Nino-Southern Oscillation. *J. Geophys. Res.*, **101**, 12255–12263.

Miller, L., and B.C. Douglas, 2004: Mass and volume contributions to 20th century global sea level rise. *Nature*, **428**, 406–409.

Millot, C., J.-L. Fuda, J. Candela, and Y. Tber, 2006: Large warming and salinification of the Mediterranean outflow due to changes in its composition. *Deep-Sea Res. I*, **53**, 656–666.

Milly, P.C.D., and A.B. Shmakin, 2002: Global modeling of land water and energy balances: 1. The land dynamics (LaD) model. *J. Hydrometeorol.*, **3**, 283–299.

Milly, P.C.D., A. Cazenave, and M.C. Gennero, 2003: Contribution of climate-driven change in continental water storage to recent sea-level rise. *Proc. Natl. Acad. Sci. U.S.A.*, **100**(213), 13158–13161.

Minami, H., Y. Kano, and K. Ogawa, 1998: Long-term variations of potential temperature and dissolved oxygen of the Japan Sea Water. *J. Oceanogr.*, **55**, 197–205.

Mitchell, T.P., and J.M. Wallace, 1992: The annual cycle in equatorial convection and sea-surface temperature. *J. Clim.*, **5**(10), 1140–1156.

Mitchell, W., J. Chittleborough, B. Ronai, and G.W. Lennon, 2001: Sea level rise in Australia and the Pacific. In: *Pacific Islands Conference on Climate Change, Climate Variability and Sea Level Rise, National Tidal Facility Australia, Rarotonga, Cook Islands, 3-7 April 2000.* Flinders Press, Adelaide, Australia, pp. 47–57.

Mitchum, G.T., 1994: Comparison of Topex sea surface heights and tide gauge sea levels. *J. Geophys. Res.*, **99**(C12), 24541–24554.

Mitchum, G.T., 2000: An improved calibration of satellite altimetric heights using tide gauge sea levels with adjustment for land motion. *Mar. Geodesy*, **23**, 145–166.

Mitrovica, J.X., M. Tamisiea, J.L. Davis, and G.A. Milne, 2001: Recent mass balance of polar ice sheets inferred from patterns of global sea-level change. *Nature*, **409**, 1026–1029.

Mitrovica, J.X., et al., 2006: Reanalysis of ancient eclipses, astronomic and geodetic data: a possible route to resolving the enigma of global sea level rise. *Earth Planet. Sci. Lett.*, **243**, 390–399.

Molinari, R.L., 2004: Annual and decadal variability in the western subtropical North Atlantic: signal characteristics and sampling methodologies. *Prog. Oceanogr.*, **62**(1), 33–66.

Morison, J., M. Steele, and R. Andersen, 1998: Hydrography of the upper Arctic Ocean measured from the nuclear submarine USS Pargo. *Deep-Sea Res. I*, **45**(1), 15–38.

Munk, W., 2003: Ocean freshening, sea level rising. *Science*, **300**, 2041–2043.

Murray, R.J., N.L. Bindoff, and C.J.C. Reason, 2007: Modelling decadal changes on the Indian Ocean Section I5 at 32°S. *J. Clim.*, in press.

Nerem, R.S., and G.T. Mitchum, 2001: Observations of sea level change from satellite altimetry. In: *Sea Level Rise: History and Consequences* [Douglas, B.C., M.S. Kearney, and S.P. Leatherman (eds.)]. Academic Press, San Diego, pp. 121–163.

Nerem, R.S., et al., 1999: Variations in global mean sea level associated with the 1997-1998 ENSO event: Implications for measuring long term sea level change. *Geophys. Res. Lett.*, **26**, 3005–3008.

Ngo-Duc, T., et al., 2005: Effects of land water storage on the global mean sea level over the last half century. *Geophys. Res. Lett.*, **32**, L09704, doi:10.1029/2005GL022719.

Ono, T., et al., 2001: Temporal increases of phosphate and apparent oxygen utilization in the subsurface waters of western subarctic Pacific from 1968 to 1998. *Geophys. Res. Lett.*, **28**(17), 3285–3288.

Orr, J.C., et al., 2005: 21st century decline in ocean carbonate and high latitude aragonitic organisms. *Nature*, **437**, 681–686.

Orsi, A.H., W.M. Smethie, and J.L. Bullister, 2002: On the total input of Antarctic waters to the deep ocean: A preliminary estimate from chlorofluorocarbon measurements. *J. Geophys. Res.*, **107**(C8), 3122, doi:10.1029/2001JC000976.

Østerhus, S., W.R. Turrell, S. Jónsson, and B. Hansen, 2005: Measured volume, heat, and salt fluxes from the Atlantic to the Arctic Mediterranean. *Geophys. Res. Lett.*, **32**, L07603, doi:10.1029/2004GL022188.

Pahlow, M., and U. Riebesell, 2000: Temporal trends in deep ocean Redfield ratios. *Science*, **287**(5454), 831–833.

Palmer, M.H., H.L. Bryden, J.L. Hirschi, and J. Marotzke, 2004: Observed changes in the South Indian Ocean gyre circulation, 1987-2002. *Geophys. Res. Lett.*, **31**(15), L15303, doi:15310.11029/12004GL020506.

Parrish, R.H., F.B. Schwing, and R. Mendelssohn, 2000: Midlatitude wind stress: the energy source for climatic regimes in the North Pacific Ocean. *Fish. Oceanogr.*, **9**, 224–238.

Peltier, W.R., 2001: Global glacial isostatic adjustment and modern instrumental records of relative sea level history. In: *Sea Level Rise: History and Consequences* [Douglas, B.C., M.S. Kearney, and S.P. Leatherman (eds.)]. Academic Press, San Diego, pp. 65–95.

Peltier, W.R., 2004: Global glacial isostasy and the surface of the ice-age earth: the ICE-5G (VM2) model and GRACE. *Annu. Rev. Earth Planet. Sci.*, **32**, 111–149.

Penduff, T., B. Barnier, W.K. Dewar, and J.J. O'Brien, 2004: Dynamical response of the oceanic eddy field to the North Atlantic Oscillation: A model-data comparison. *J. Phys. Oceanogr.*, **34**, 2615–2629.

Peng, T.-H., R. Wanninkhof, and R.A. Feely, 2003: Increase of anthropogenic CO_2 in the Pacific Ocean over the last two decades. *Deep-Sea Res. II*, **50**, 3065–3082.

Peng, T.-H., et al., 1998: Quantification of decadal anthropogenic CO_2 uptake in the ocean based on dissolved inorganic carbon measurements. *Nature*, **396**(6711), 560–563.

Plag, H.-P., 2006: Recent relative sea level trends: an attempt to quantify the forcing factors. *Philos. Trans. R. Soc. London A*, **364**(1841), 821–844.

Polyakov, I.V., et al., 2004: Variability of the intermediate Atlantic water of the Arctic Ocean over the last 100 years. *J. Clim.*, **17**(23), 4485–4497.

Polyakov, I.V., et al., 2005: One more step toward a warmer Arctic. *Geophys. Res. Lett.*, **32**, L17605, doi:10.1029/2005GL023740.

Ponte, R.M., 2006: Low frequency sea level variability and the inverted barometer effect. *J. Atmos. Ocean. Technol.*, **23**(4), 619–629.

Potter, R.A., and M.S. Lozier, 2004: On the warming and salinification of the Mediterranean outflow waters in the North Atlantic. *Geophys. Res. Lett.*, **31**(1), L01202, doi:10.1029/2003GL018161.

Prentice, I.C., et al., 2001: The carbon cycle and atmospheric carbon dioxide. In: *Climate Change 2001: The Scientific Basis. Contribution of Working Group I to the Third Assessment Report of the Intergovernmental Panel on Climate Change* [Houghton, J.T., et al. (eds.)]. Cambridge University Press, Cambridge, United Kingdom and New York, NY, USA, pp. 183–237.

Proshutinsky, A.Y., and M.A. Johnson, 1997: Two circulation regimes of the wind-driven Arctic Ocean. *J. Geophys. Res.*, **102**(C6), 12493–12514.

Proshutinsky, A., et al., 2004: Secular sea level change in the Russian sector of the Arctic Ocean. *J. Geophys. Res.*, **109**(C3), C03042, doi:10.1029/2003JC002007.

Qian, H., Y. Yin, and Y. Ni, 2003: Tropical Indian Ocean subsurface dipole mode and diagnostic analysis of dipole event in 1997-1998. *J. Appl. Meteorol. Sci.*, **14**, 129–139 (in Chinese).

Quadfasel, D., A. Sy, D. Wells, and A. Tunik, 1991: Warming in the Arctic. *Nature*, **350**(6317), 385.

Raven, J., et al., 2005: *Ocean Acidification due to Increasing Atmospheric Carbon Dioxide.* The Royal Society, London, 59 pp.

Reverdin, G., D. Cayan, and Y. Kushnir, 1997: Decadal variability of hydrography in the upper northern North Atlantic in 1948-1990. *J. Geophys. Res.*, **102**(C4), 8505–8531.

Rigor, I.G., J.M. Wallace, and R.L. Colony, 2002: Response of sea ice to the Arctic Oscillation. *J. Clim.*, **15**, 2648–2663.

Rixen, M., et al., 2005: The Western Mediterranean Deep Water: A new proxy for global climate change. *Geophys. Res. Lett.*, **32**, L12608, doi:10.1029/2005GL022702.

Robertson, R., M. Visbeck, A.L. Gordon, and E. Fahrbach, 2002: Long-term temperature trends in the deep waters of the Weddell Sea. *Deep-Sea Res. II*, **49**(21), 4791–4806.

Roemmich, et al., 2007: Decadal spin-up of the South Pacific Subtropical Gyre. *J. Phys. Oceanogr.*, 37, 162-173.

Roether, W., et al., 1996: Recent changes in eastern Mediterranean deep waters. *Science*, **271**(5247), 333–335.

Rohling, E.J., and H.L. Bryden, 1992: Man-induced salinity and temperature increases in western Mediterranean deep-water. *J. Geophys. Res.*, **97**(C7), 11191–11198.

Rupolo, V., S. Marullo, and D. Iudicone, 2003: Eastern Mediterranean transient studied with Lagrangian diagnostics applied to a Mediterranean OGCM forced by satellite SST and ECMWF wind stress for the years 1988-1993. *J. Geophys. Res.*, **108**(C9), 8121.

Sabine, C.L., R.M. Key, R.A. Feely, and D. Greeley, 2002: Inorganic carbon in the Indian Ocean: Distribution and dissolution processes. *Global Biogeochem. Cycles*, **16**(4), 1067, doi:10.1029/2002GB001869.

Sabine, C.L., R.A. Feely, Y.W. Watanabe, and M. Lamb, 2004a: Temporal evolution of the North Pacific CO_2 uptake rate. *J. Oceanogr.*, **60**(1), 5–15.

Sabine, C.L., et al., 1999: Anthropogenic CO_2 inventory of the Indian Ocean. *Global Biogeochem. Cycles*, **13**, 179–198.

Sabine, C.L., et al., 2004b: The oceanic sink for anthropogenic CO_2. *Science*, **305**(5682), 367–371.

Sahagian, D.L., 2000: Global physical effects of anthropogenic hydrological alterations: sea level and water redistribution. *Global Planet. Change*, **25**, 39–48.

Sahagian, D.L., F.W. Schwartz, and D.K. Jacobs, 1994: Direct anthropogenic contributions to sea level rise in the twentieth century. *Nature*, **367**, 54–56.

Sarma, V.V.S.S., T. Ono, and T. Saino, 2002: Increase of total alkalinity due to shoaling of aragonite saturation horizon in the Pacific and Indian Oceans: Influence of anthropogenic carbon inputs. *Geophys. Res. Lett.*, **29**(20), 1971, doi:10.1029/2002GL015135.

Sarmiento, J.L., C. Le Quéré, and S.W. Pacala, 1995: Limiting future atmospheric carbon dioxide. *Global Biogeochem. Cycles*, **9**(1), 121–137.

Schauer, U., E. Fahrbach, and S. Østerhus, 2004: Arctic warming through the Fram Strait - Oceanic heat transport from three years of measurements. *J. Geophys. Res.*, **109**, C06026, doi:10.1029/2003JC001823.

Schneider, N., and B.D. Cornuelle, 2005: The forcing of the Pacific Decadal Oscillation. *J. Clim.*, **18**(21), 4355–4373.

Schoenefeldt, R., and F. Schott, 2006: Decadal variability of the Indian Ocean cross-equatorial exchange in SODA. *Geophys. Res. Lett.*, **33**, L08602, doi:10.1029/2006GL025891.

Schott, F.A., et al., 2004: Circulation and deep-water export at the western exit of the subpolar North Atlantic. *J. Phys. Oceanogr.*, **34**, 817–843.

Seager, R., et al., 2001: Wind-driven shifts in the latitude of the Kuroshio-Oyashio Extension and generation of SST anomalies on decadal timescales. *J. Clim.*, **14**(22), 4249–4265.

Sekine, Y., 1988: Anomalous southward intrusion of the Oyashio east of Japan.1. Influence of the seasonal and interannual variations in the wind stress over the North Pacific. *J. Geophys. Res.*, **93**(C3), 2247–2255.

Sekine, Y., 1999: On variations in the subarctic circulation in the North Pacific. *Prog. Oceanogr.*, **43**(2–4), 193–203.

Senjyu, T., et al., 2002: Renewal of the bottom water after the winter 2000-2001 may spin-up the thermohaline circulation in the Japan Sea. *Geophys. Res. Lett.*, **29**(7), 1149, doi:10.1029/2001GL014093.

Sivan, D., et al., 2004: Ancient coastal wells of Caesarea Maritima, Israel, an indicator for sea level changes during the last 2000 years. *Earth Planet. Sci. Lett.*, **222**, 315–330.

Smith, T.M., and R.W. Reynolds, 2003: Extended reconstructions of global sea surface temperatures based on COADS Data (1854-1997). *J. Clim.*, **16**, 1495–1510.

Sprintall, J., et al., 2004: INSTANT: A new international array to measure the Indonesian Throughflow. *EOS*, **85**(39), 369.

Stammer, D., et al., 2003: Volume, heat and freshwater transports of the global ocean circulation 1993-2000. *J. Geophys. Res.*, **108**(C1), doi:10.1029/2001JC001115.

Stark, S., R.A. Wood, and H.T. Banks, 2006: Re-evaluating the causes of observed changes in Indian Ocean water masses. *J. Clim.*, **19**(16), 4075–4086.

Steele, M., and T. Boyd, 1998: Retreat of the cold halocline layer in the Arctic Ocean. *J. Geophys. Res.*, **103**(C5), 10419–10435.

Stephens, C., S. Levitus, J. Antonov, and T. Boyer, 2001: On the Pacific Ocean regime shift. *Geophys. Res. Lett.*, **28**, 3721–3724.

Stephens, C., et al., 2002: World ocean database 2001, Volume 3: Temporal distribution of conductivity-temperature-depth profiles. In: *NOAA Atlas NESDIS 44* [Levitus, S. (ed.)]. U.S. Government Printing Office, Washington, DC, pp. 47, CD-ROMs.

Sterl, A., and W. Hazeleger, 2003: Coupled variability and air-sea interaction in the South Atlantic Ocean. *J. Clim.*, **21**, 559–571.

Stramma, L., J. Fischer, P. Brandt, and F. Schott, 2003: Circulation, variability and near-equatorial meridional flow in the central tropical Atlantic. In: *Interhemispheric Water Exchange in the Atlantic Ocean* [Goni, G., and P. Malanotte-Rizzoli (eds.)]. Elsevier, Amsterdam, pp. 1–22.

Stramma, L., et al., 2004: Deep water changes at the western boundary of the subpolar North Atlantic during 1996 to 2001. *Deep-Sea Res.*, **51A**, 1033–1056.

Sy, A., et al., 1997: Surprisingly rapid spreading of newly formed intermediate waters across the North Atlantic Ocean. *Nature*, **386**(6626), 675–679.

Takahashi, T., S.C. Sutherland, R.A. Feely, and R. Wanninkhof, 2006: Decadal change of the surface water pCO_2 in the North Pacific: A synthesis of 35 years of observations. *J. Geophys. Res.*, **111**, C07S05, doi:10.1029/2005JC003074.

Takahashi, T., et al., 2002: Global sea-air CO_2 flux based on climatological surface ocean pCO_2, and seasonal biological and temperature effects. *Deep-Sea Res. II*, **49**(9–10), 1601–1622.

Talley, L.D., 1996: North Atlantic circulation and variability, reviewed for the CNLS conference. *Physica D*, **98**(2–4), 625–646.

Talley, L.D., and M.S. McCartney, 1982: Distribution and circulation of Labrador Sea-water. *J. Phys. Oceanogr.*, **12**(11), 1189–1205.

Talley, L.D., J.L. Reid, and P.E. Robbins, 2003a: Data-based meridional overturning streamfunctions for the global ocean. *J. Clim.*, **16**, 3213–3226.

Talley, L.D., et al., 2003b: Deep convection and brine rejection in the Japan Sea. *Geophys. Res. Lett.*, **30**(4), 1159, doi:10.1029/2002GL0165451.

Trenberth, K.E., and J.M. Caron, 2001: Estimates of meridional atmosphere and ocean heat transports. *J. Clim.*, **14**(16), 3433–3443.

Trenberth, K.E., J.M. Caron, and D.P. Stepaniak, 2001: The atmospheric energy budget and implications for surface fluxes and ocean heat transports. *Clim. Dyn.*, **17**, 259–276.

Tsimplis, M.N., and M. Rixen, 2002: Sea level in the Mediterranean Sea: The contribution of temperature and salinity changes. *Geophys. Res. Lett.*, **29**(23), 2136, doi:10.1029/2002GL015870.

Tsimplis, M.N., A.G.P. Shaw, R.A. Flather, and D.K. Woolf, 2006: The influence of the North Atlantic Oscillation on the sea level around the northern European coasts reconsidered: the thermosteric effects. *Phil. Trans. R. Soc.London A*, **364**(1841), 845–856, doi:10.1098/rsta.2006.1740.

Vargas-Yáñez, M., et al., 2004: Temperature and salinity increase in the eastern North Atlantic along the 24.5°N in the last ten years. *Geophys. Res. Lett.*, **31**, L06210, doi:10.1029/2003GL019308.

Vaughan, D., et al., 2003: Recent rapid regional climate warming on the Antarctic Peninsula. *Clim. Change*, **60**, 243–274.

Vellinga, M., and R.A. Wood, 2002: Global climatic impacts of a collapse of the Atlantic thermohaline circulation. *Clim. Change*, **54**, 251–267.

Venegas, S.A., L.A. Mysak, and D.N. Straub, 1998: An interdecadal climate cycle in the South Atlantic and its links to other ocean basins. *J. Geophys. Res.*, **103**(C11), 24723–24736.

Vignudelli, S., G.P. Gasparini, M. Astraldi, and M.E. Schiano, 1999: A possible influence of the North Atlantic Oscillation on the circulation of the Western Mediterranean Sea. *Geophys. Res. Lett.*, **26**(5), 623–626.

Vranes, K., A.L. Gordon, and A. Ffield, 2002: The heat transport of the Indonesian Throughflow and implications for the Indian Ocean heat budget. *Deep-Sea Res. I*, **49**, 1391–1410.

Wadhams, P., and W. Munk, 2004: Ocean freshening, sea level rising, sea ice melting. *Geophys. Res. Lett.*, **31**(11), L11311, doi:10.1029/2004GL020039.

Wakelin, S.L., P.L. Woodworth, R.A. Flather, and J.A. Williams, 2003: Sea-level dependence on the NAO over the NW European continental shelf. *Geophys. Res. Lett.*, **30**(7), 1403, doi:10.1029/2003GL017041.

Watanabe, Y.W., H. Ishida, T. Nakano, and N. Nagai, 2005: Spatiotemporal decreases of nutrients and chlorophyll-a in the surface mixed layer of the western North Pacific from 1971 to 2000. *J. Oceanogr.*, **61**, 1011–1016.

Watanabe, Y.W., et al., 2001: Probability of a reduction in the formation rate of the subsurface water in the North Pacific during the 1980s and 1990s. *Geophys. Res. Lett.*, **28**(17), 3289–3292.

White, N.J., J.A. Church, and J.M. Gregory, 2005: Coastal and global averaged sea-level rise for 1950 to 2000. *Geophys. Res. Lett.*, **32**(1), L01601, doi:10.1029/2004GL021391.

Whitworth, T., 2002: Two modes of bottom water in the Australian-Antarctic Basin. *Geophys. Res. Lett.*, **29**(5), 1973, doi:10.1029/2001GL014282.

Wijffels, S., and G.A. Meyers, 2004: An intersection of oceanic wave guides: Variability in the Indonesian Throughflow region. *J. Phys. Oceanogr.*, **34**, 1232–1253.

Willey, D.A., et al., 2004: Global oceanic chlorofluorocarbon inventory. *Geophys. Res. Lett.*, **31**, L01303, doi:10.1029/2003GL018816.

Willis, J.K., D. Roemmich, and B. Cornuelle, 2004: Interannual variability in upper-ocean heat content, temperature and thermosteric expansion on global scales. *J. Geophys. Res.*, **109**, C12036, doi:10.1029/2003JC002260.

Wong, A.P.S., N.L. Bindoff, and J.A. Church, 1999: Large-scale freshening of intermediate waters in the Pacific and Indian oceans. *Nature*, **400**(6743), 440–443.

Wong, A.P.S., N.L. Bindoff, and J.A. Church, 2001: Freshwater and heat changes in the North and South Pacific Oceans between the 1960s and 1985-94. *J. Clim.*, **14**(7), 1613–1633.

Woodworth, P.L., 1990: A search for accelerations in records of European mean sea level. *Int. J. Climatol.*, **10**, 129–143.

Woodworth, P.L., and D.L. Blackman, 2002: Changes in extreme high waters at Liverpool since 1768. *Int. J. Climatol.*, **22**, 697–714.

Woodworth, P.L., and R. Player, 2003: The Permanent Service for Mean Sea Level: An update to the 21st century. *J. Coastal Res.*, **19**, 287–295.

Woodworth, P.L., and D.L. Blackman, 2004: Evidence for systematic changes in extreme high waters since the mid-1970s. *J. Clim.*, **17**, 1190–1197.

Woodworth, P.L., M.N. Tsimplis, R.A. Flather, and I. Shennan, 1999: A review of the trends observed in British Isles mean sea level data measured by tide gauges. *Geophys. J. Int.*, **136**, 651–670.

Woolf, D., A. Shaw, and M.N. Tsimplis, 2003: The influence of the North Atlantic Oscillation on sea level variability in the North Atlantic Region. *Global Atmos. Ocean System*, **9**(4), 145–167.

Xie, S.P., H. Annamalai, F.A. Schott, and J.P. McCreary, 2002: Structure and mechanisms of South Indian Ocean climate variability. *J. Clim.*, **15**(8), 864–878.

Yamagata, T., et al., 2004: Coupled ocean-atmosphere variability in the tropical Indian Ocean. In: *Earth Climate: The Ocean-Atmosphere Interaction* [Wang, C., S.-P. Xie, and J.A. Carton (eds.)]. American Geophysical Union, Washington, DC, pp. 189–212.

Yashayaev, I., J.R.N. Lazier, and R.A. Clarke, 2003: Temperature and salinity in the central Labrador Sea. *ICES Marine Symposia Series*, **219**, 32–39.

Yasuda, I., T. Tozuka, M. Noto, and S. Kouketsu, 2000: Heat balance and regime shifts of the mixed layer in the Kuroshio Extension. *Prog. Oceanogr.*, **47**(2–4), 257–278.

Yasuda, I., et al., 2001: Hydrographic structure and transport of the Oyashio south of Hokkaido and the formation of North Pacific Intermediate Water. *J. Geophys. Res.*, **106**(C4), 6931–6942.

Zhang, K., B.C. Douglas, and S.P. Leatherman, 2000: Twentieth-century storm activity along the U.S. east coast. *J. Clim.*, **13**, 1748–1761.

Appendix 5.A:
Techniques, Error Estimation and
Measurement Systems

5.A.1   Ocean Temperature and Salinity

Sections 5.2 and 5.3 report on the changes in the oceans using two different approaches to the oceanic part of the climate system. Section 5.2 documents the changes found in the most comprehensive ocean data sets that exist for temperature and salinity. These data sets are collected from a wide range of organisations and are a composite of heterogeneous measurement systems, including mechanical and expendable bathythermographs, research ship measurements, voluntary observing ships, moored and drifting buoys and Argo floats for recent years. The advantage of these composite data sets is the greater spatial and temporal coverage that they offer for climate studies. The main disadvantage of these composite data sets, relative to the research data sets used in Section 5.3, is that they can have more problems related to the quality and heterogeneity of the measurements systems. This heterogeneity can lead to subtle biases and artificial noise and consequently difficulties in estimating trends at small regional scales (Harrison and Carson, 2006). On the other hand, Section 5.3 described the changes found in detailed analyses of very specific research voyages that consist mainly of very tightly calibrated and monitored temperature and salinity measurements (and other variables). The internal consistency of these research data sets is much higher than the composite data sets, and as a consequence they have significant advantages in their ease of interpretation and analysis. However, research quality oceanographic data sets are only collected occasionally and are focussed more frequently on regional rather than global issues. This means that in the poorly sampled oceans, such as the Indian, South Pacific and Southern Oceans, observational records only cover a relatively short period of time (e.g., the 1960s to present) with some decades poorly covered and highly heterogeneous in space (see Figure 5.A.1).

An example of the distribution of ocean temperature observations in both space and time is shown in Figure 5.A.1. This figure shows the *in situ* temperature data distribution for two five-year periods used to create estimates of global heat content change (e.g., Figure 5.1), one with a low (a) and one with a high (b) density of observations. It is clear that parts of the ocean, in particular in the SH, are not well sampled even in periods of high observation density. Hence, sampling errors resulting from the lack of data are potentially important but cannot easily be quantified.

Several different objective analysis techniques have been used to produce the gridded fields of temperature anomalies used to compute ocean heat content and steric sea level rise presented in this chapter. The technique used by Levitus et al. (2005b), Garcia et al. (2005) and Antonov et al. (2005) in their estimates of temperature (heat content), oxygen and the

thermosteric component of sea level change is based on the construction of gridded (1° latitude by 1° longitude grid) fields at standard depth measurement levels. The objective analysis procedure used for interpolation (filling in data-void areas and smoothing the entire field) is described by Boyer et al. (2002). At each standard depth level, all data are averaged within each 1° square, and the deviation from climatology yields the observed anomaly. From all observations within the surrounding region of diameter 888 km, the analysed value is computed. Features with a wavelength of less than 500 to 600 km are substantially reduced in amplitude; in regions without sufficient data, it is essentially the climatological information that is used. Ishii et al. (2006) employed similar techniques, with a smaller de-correlation length scale of 300 km and a least-squares technique for estimating corrections to the climatological field. Willis et al. (2004) used a two-scale covariance function, but also used altimetric data in areas where ocean observations were lacking.

There are some differences in the data used in these studies. In addition to ocean temperature profile data, Ishii et al. (2006) also used the product of climatological mixed layer depth and individual SST measurements in their estimates of ocean heat content. Southern Hemisphere World Ocean Circulation Experiment profiling float temperature profiles for the 1990s were used by Willis et al. (2004) that were not used by Levitus et al. (2005a) and Ishii et al. (2006). The similarity of the three independently estimated heat content time series shown in Figure 5.1 to within confidence intervals indicates that the differences between analysis techniques and data sources do not substantially influence the estimates of the three global ocean heat content time series.

All analyses are subject to statistical errors and sampling errors. Statistical errors are estimated in a straightforward way. For example, for the Levitus et al. (2005a) fields, the uncertainty at any grid point is estimated from the variability of observations that contributed to the analysed value. In this way, 90% errors for all analysed variables are computed as a function of depth and horizontal position, and correspondingly for integrated variables such as heat content. Both Ishii et al. (2006) and Willis et al. (2004) used the interannual variability of heat content as the basis for error analyses.

5.A.2   Heat Transports

Estimates of meridional heat transport (MHT) derived from the surface heat balance involve the integration of the zonally averaged balances in the longitudinal direction. This integration also implies the integration of uncertainties in the zonally averaged estimates. For instance, an uncertainty in zonally averaged estimates of ±10 W m^{-2} results in an uncertainty of 0.5×10^{15} W in MHT in the Atlantic and nearly twice that value in the Pacific. Thus, all climatological estimates of MHT based on the surface heat balance have considerable uncertainties, and estimates of MHT variability are unlikely to be significant when derived from the surface heat balance.

In addition to the uncertainties in diagnostic computations of transports from vertical sections, estimates of MHT based on oceanic cross sections are largely influenced by sparse sampling of these sections during continuous time periods. As a result, there is no way to discriminate between the long-term signals and interannual variability using the estimates of MHT for individual years.

5.A.3 Estimates of Oxygen Changes

Estimates of changes in O_2 in the surface 100 m of the ocean between 1955 and 1998 were made for each pentad using a total of 530,000 O_2 profiles (Garcia et al., 2005). The measurement method was not reported for all the cruises. Only the Winkler titration was reported, with only manual titrations prior to 1990. The Carpenter method to improve accuracy was reported for some cruises after 1970. An automated titration gives a significant improvement for measurement reproducibility but is not the essential solution for accuracy. Problems of O_2 leakage were reported from the older samples using Nansen bottles (generally before 1970). The Niskin bottles more widely used after 1970 are thought to be more reliable. There are no agreed standards for O_2 measurements because of reagent impurity and the difficulty in preparing a stable solution, which limits accuracy of these measurements to typically less than 10 μmol kg^{-1} for modern methods.

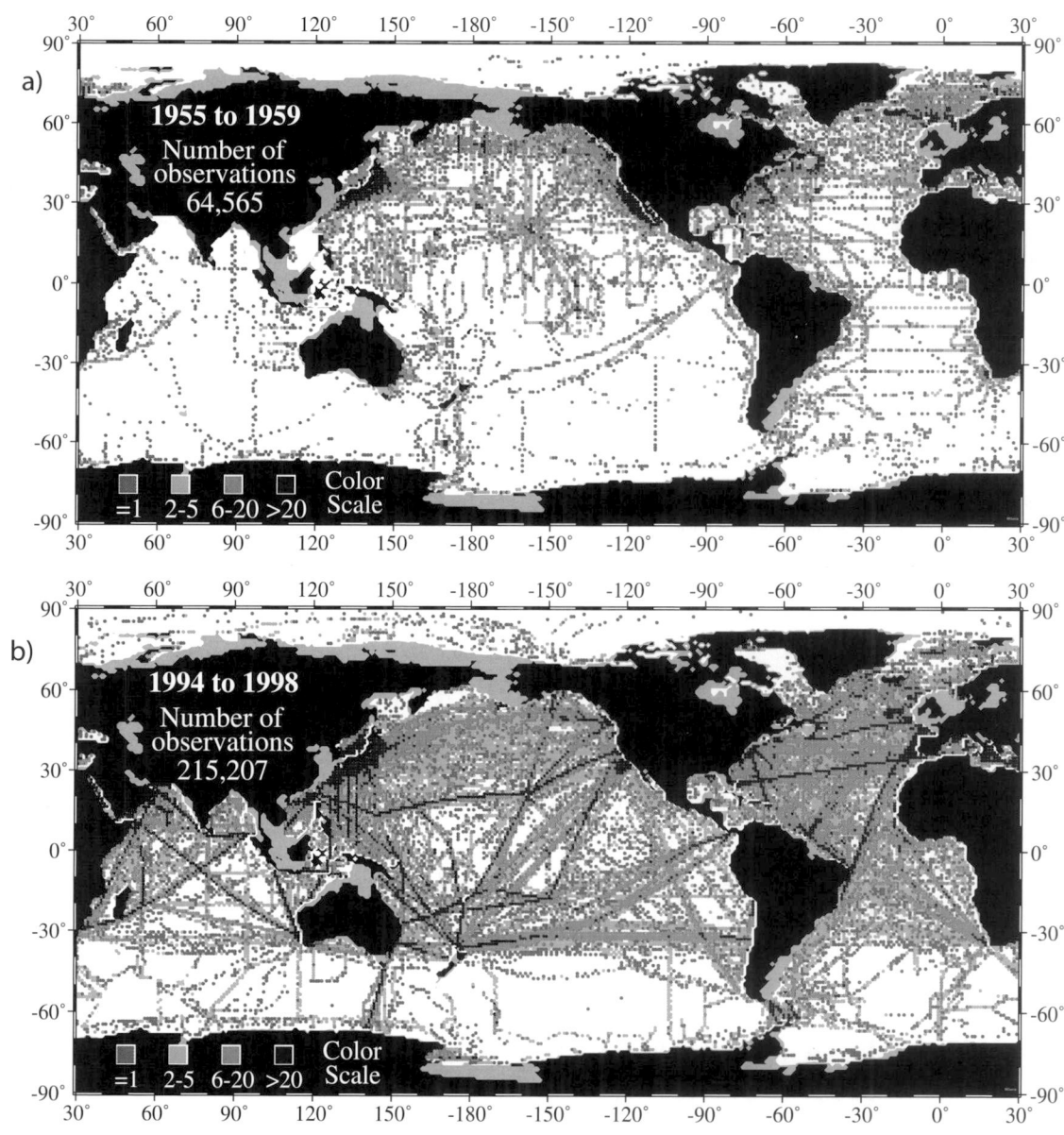

Figure 5.A.1 *The number of ocean temperature observations in each 1° grid box at 250 m depth for two periods: (a) 1955 to 1959, with a low density of observations, and (b) 1994 to 1998, with a high density of observations. A blue dot indicates a 1° grid box containing 1 observation, a green dot 2 to 5 observations, an orange dot 6 to 20 observations, and a red dot more than 20 observations.*

5.A.4   Estimation of Sea Level Change

5.A.4.1   Satellite Altimetry: Measurement Principle and Associated Errors

The concept of satellite altimetry measurement is rather straightforward. The onboard radar altimeter transmits a short pulse of microwave radiation with known power towards the nadir. Part of the incident radiation reflects back to the altimeter. Measurement of the round-trip travel time provides the height of the satellite above the instantaneous sea surface. The quantity of interest in oceanography is the height of the instantaneous sea surface above a fixed reference surface, which is computed as the difference between the altitude of the satellite above the reference ellipsoid and the altimeter range. The satellite position is computed through precise orbit determination, combining accurate modelling of the satellite motion and tracking measurements between the satellite and observing stations on Earth or other observing satellites. A number of corrections must be applied to obtain the correct sea surface height. These include instrumental corrections, ionospheric correction, dry and wet tropospheric corrections, electromagnetic bias correction, ocean and solid Earth tidal corrections, ocean loading correction, pole tide correction and an inverted barometer correction that has to be applied since the altimeter does not cover the global ocean completely. The total measurement accuracy for the TOPEX/Poseidon altimetry-based sea surface height is about 80 mm (95% error) for a single measurement based on one-second along-track averages (Chelton et al., 2001).

The above error estimates concern instantaneous sea surface height measurements. For estimating the mean sea level variations, the procedure consists of simply averaging over the ocean the point-to-point measurements collected by the satellite during a complete orbital cycle (10 days for TOPEX/Poseidon and Jason-1), accounting for the spatial distribution of the data using an equal area weighting. In effect, during this time interval, the satellite realises an almost complete coverage of the oceanic domain. The 95% error associated with a 10-day mean sea level estimate is approximately 8 mm.

When computing global mean sea level variations through time, proper account of instrumental bias and drifts (including the terrestrial reference frame) is of considerable importance. These effects (e.g., the radiometer drift onboard TOPEX/ Poseidon used to correct for the wet tropospheric delay) are of the same order of magnitude as the sea level signal. Studies by Chambers et al. (1998) and Mitchum (1994; 2000) have demonstrated that comparing the altimeter sea level measurements to tide gauge sea level measurements produces the most robust way of correcting for instrumental bias and drifts. This approach uses a network of high-quality tide gauges, well distributed over the ocean domain. Current results indicate that the residual error in the mean sea level variation using the tide gauge calibration is about 0.8 mm yr^{-1} (a value resulting mainly from the uncertainties in vertical land

motion at the tide gauges). The current altimeter-inferred sea level measurements do not include modelling of the geocenter or mitigating the effect resulting from the potential drift of the terrestrial reference frame.

Detailed information about satellite altimetry, uncertainty and applications can be found in Fu and Cazenave (2001).

5.A.4.2   Sea Level from Tide Gauge Observations

Tide gauges are based on a number of different technologies (float, pressure, acoustic, radar), each of which has its advantages in particular applications. The Global Sea Level Observing System (GLOSS) specifies that a gauge must be capable of measuring sea level to centimetre accuracy (or better) in all weather conditions (i.e., in all wave conditions). The most important consideration is the need to maintain the gauge datum relative to the level of the Tide Gauge Bench Mark (TGBM), which provides the land reference level for the sea level measurements. The specifications for GLOSS require that local levelling must be repeated at least annually between the reference mark of the gauge, TGBM and a set of approximately five ancillary marks in the area, in order to maintain the geodetic integrity of the measurements. In practice, this objective is easier to meet if the area around the gauge is hard rock, rather than reclaimed land, for example. The question of whether the TGBM is moving vertically within a global reference frame (for whatever reason) is being addressed by advanced geodetic methods (GPS, Determination d'Orbite et Radiopositionnement Intégrés par Satellite (DORIS), Absolute Gravity). With typical rates of sea and land level change of order of 1 mm yr^{-1}, it is necessary to maintain the accuracy of the overall gauge system at the centimetre level over many decades. This demanding requirement has been met in many countries for many years (see IOC, 2002 for more information). The tide gauge observation system for three periods is shown in Figure 5.A.2, together with the evolution over time of the number of stations in both hemispheres. The distribution of tide gauge stations was particularly sparse in space at the beginning of the 20th century, but rapidly improved in the 1950s through to the current network of GLOSS standard instruments. This distribution of instruments through time means that confidence in the estimates of sea level rise has been improving and this certainty is reflected in the shrinking confidence intervals (Figure 5.13).

Figure 5.A.2. *(a) Number of tide gauge stations in the Northern Hemisphere (NH) and Southern Hemisphere (SH) used to derive the global sea level curve (red and blue curves in Figure 5.13) as a function of time. Lower panels show the spatial distribution of tide gauge stations (denoted by red dots) for the periods (b) 1900 to 1909, (c) 1950 to 1959 and (d) 1980 to 1989.*

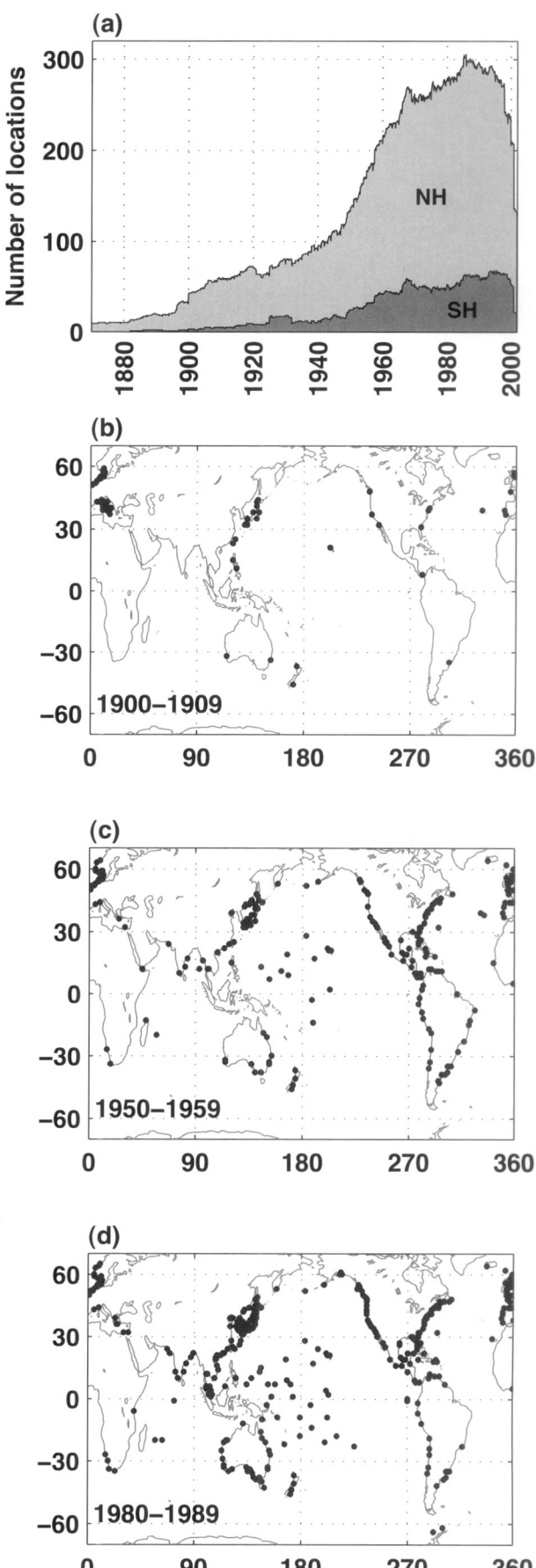

6

Palaeoclimate

Coordinating Lead Authors:

Eystein Jansen (Norway), Jonathan Overpeck (USA)

Lead Authors:

Keith R. Briffa (UK), Jean-Claude Duplessy (France), Fortunat Joos (Switzerland), Valérie Masson-Delmotte (France), Daniel Olago (Kenya), Bette Otto-Bliesner (USA), W. Richard Peltier (Canada), Stefan Rahmstorf (Germany), Rengaswamy Ramesh (India), Dominique Raynaud (France), David Rind (USA), Olga Solomina (Russian Federation), Ricardo Villalba (Argentina), De'er Zhang (China)

Contributing Authors:

J.-M. Barnola (France), E. Bauer (Germany), E. Brady (USA), M. Chandler (USA), J. Cole (USA), E. Cook (USA), E. Cortijo (France), T. Dokken (Norway), D. Fleitmann (Switzerland, Germany), M. Kageyama (France), M. Khodri (France), L. Labeyrie (France), A. Laine (France), A. Levermann (Germany), Ø. Lie (Norway), M.-F. Loutre (Belgium), K. Matsumoto (USA), E. Monnin (Switzerland), E. Mosley-Thompson (USA), D. Muhs (USA), R. Muscheler (USA), T. Osborn (UK), Ø. Paasche (Norway), F. Parrenin (France), G.-K. Plattner (Switzerland), H. Pollack (USA), R. Spahni (Switzerland), L.D. Stott (USA), L. Thompson (USA), C. Waelbroeck (France), G. Wiles (USA), J. Zachos (USA), G. Zhengteng (China)

Review Editors:

Jean Jouzel (France), John Mitchell (UK)

This chapter should be cited as:

Jansen, E., J. Overpeck, K.R. Briffa, J.-C. Duplessy, F. Joos, V. Masson-Delmotte, D. Olago, B. Otto-Bliesner, W.R. Peltier, S. Rahmstorf, R. Ramesh, D. Raynaud, D. Rind, O. Solomina, R. Villalba and D. Zhang, 2007: Palaeoclimate. In: *Climate Change 2007: The Physical Science Basis. Contribution of Working Group I to the Fourth Assessment Report of the Intergovernmental Panel on Climate Change* [Solomon, S., D. Qin, M. Manning, Z. Chen, M. Marquis, K.B. Averyt, M. Tignor and H.L. Miller (eds.)]. Cambridge University Press, Cambridge, United Kingdom and New York, NY, USA.

Table of Contents

Supplementary Material

The following supplementary material is available on CD-ROM and in on-line versions of this report.

Appendix 6.A: *Glossary of Terms Specific to Chapter 6*

Executive Summary

What is the relationship between past greenhouse gas concentrations and climate?

- The sustained rate of increase over the past century in the combined radiative forcing from the three well-mixed greenhouse gases carbon dioxide (CO_2), methane (CH_4), and nitrous oxide (N_2O) is *very likely* unprecedented in at least the past 16 kyr. Pre-industrial variations of atmospheric greenhouse gas concentrations observed during the last 10 kyr were small compared to industrial era greenhouse gas increases, and were *likely* mostly due to natural processes.

- It is *very likely* that the current atmospheric concentrations of CO_2 (379 ppm) and CH_4 (1,774 ppb) exceed by far the natural range of the last 650 kyr. Ice core data indicate that CO_2 varied within a range of 180 to 300 ppm and CH_4 within 320 to 790 ppb over this period. Over the same period, antarctic temperature and CO_2 concentrations co-vary, indicating a close relationship between climate and the carbon cycle.

- It is *very likely* that glacial-interglacial CO_2 variations have strongly amplified climate variations, but it is *unlikely* that CO_2 variations have triggered the end of glacial periods. Antarctic temperature started to rise several centuries before atmospheric CO_2 during past glacial terminations.

- It is *likely* that earlier periods with higher than present atmospheric CO_2 concentrations were warmer than present. This is the case both for climate states over millions of years (e.g., in the Pliocene, about 5 to 3 Ma) and for warm events lasting a few hundred thousand years (i.e., the Palaeocene-Eocene Thermal Maximum, 55 Ma). In each of these two cases, warming was *likely* strongly amplified at high northern latitudes relative to lower latitudes.

What is the significance of glacial-interglacial climate variability?

- Climate models indicate that the Last Glacial Maximum (about 21 ka) was 3°C to 5°C cooler than the present due to changes in greenhouse gas forcing and ice sheet conditions. Including the effects of atmospheric dust content and vegetation changes gives an additional 1°C to 2°C global cooling, although scientific understanding of these effects is very low. It is *very likely* that the global warming of 4°C to 7°C since the Last Glacial Maximum occurred at an average rate about 10 times slower than the warming of the 20th century.

- For the Last Glacial Maximum, proxy records for the ocean indicate cooling of tropical sea surface temperatures (average *likely* between 2°C and 3°C) and much greater cooling and expanded sea ice over the high-latitude oceans. Climate models are able to simulate the magnitude of these latitudinal ocean changes in response to the estimated Earth orbital, greenhouse gas and land surface changes for this period, and thus indicate that they adequately represent many of the major processes that determine this past climate state.

- Last Glacial Maximum land data indicate significant cooling in the tropics (up to 5°C) and greater magnitudes at high latitudes. Climate models vary in their capability to simulate these responses.

- It is *virtually certain* that global temperatures during coming centuries will not be significantly influenced by a natural orbitally induced cooling. It is *very unlikely* that the Earth would naturally enter another ice age for at least 30 kyr.

- During the last glacial period, abrupt regional warmings (likely up to 16°C within decades over Greenland) and coolings occurred repeatedly over the North Atlantic region. They *likely* had global linkages, such as with major shifts in tropical rainfall patterns. It is *unlikely* that these events were associated with large changes in global mean surface temperature, but instead *likely* involved a redistribution of heat within the climate system associated with changes in the Atlantic Ocean circulation.

- Global sea level was *likely* between 4 and 6 m higher during the last interglacial period, about 125 ka, than in the 20th century. In agreement with palaeoclimatic evidence, climate models simulate arctic summer warming of up to 5°C during the last interglacial. The inferred warming was largest over Eurasia and northern Greenland, whereas the summit of Greenland was simulated to be 2°C to 5°C higher than present. This is consistent with ice sheet modelling suggestions that large-scale retreat of the south Greenland Ice Sheet and other arctic ice fields *likely* contributed a maximum of 2 to 4 m of sea level rise during the last interglacial, with most of any remainder *likely* coming from the Antarctic Ice Sheet.

What does the study of the current interglacial climate show?

- Centennial-resolution palaeoclimatic records provide evidence for regional and transient pre-industrial warm periods over the last 10 kyr, but it is *unlikely* that any of these commonly cited periods were globally synchronous. Similarly, although individual decadal-resolution interglacial palaeoclimatic records support the existence of regional quasi-periodic climate variability, it is *unlikely* that any of these regional signals were coherent at the global scale, or are capable of explaining the majority of global warming of the last 100 years.

- Glaciers in several mountain regions of the Northern Hemisphere retreated in response to orbitally forced regional warmth between 11 and 5 ka, and were smaller (or even absent) at times prior to 5 ka than at the end of the 20th century. The present day near-global retreat of mountain glaciers cannot be attributed to the same natural causes, because the decrease of summer insolation during the past few millennia in the Northern Hemisphere should be favourable to the growth of the glaciers.

- For the mid-Holocene (about 6 ka), GCMs are able to simulate many of the robust qualitative large-scale features of observed climate change, including mid-latitude warming with little change in global mean temperature (<0.4°C), as well as altered monsoons, consistent with the understanding of orbital forcing. For the few well-documented areas, models tend to underestimate hydrological change. Coupled climate models perform generally better than atmosphere-only models, and reveal the amplifying roles of ocean and land surface feedbacks in climate change.

- Climate and vegetation models simulate past northward shifts of the boreal treeline under warming conditions. Palaeoclimatic results also indicated that these treeline shifts *likely* result in significant positive climate feedback. Such models are also capable of simulating changes in the vegetation structure and terrestrial carbon storage in association with large changes in climate boundary conditions and forcings (i.e., ice sheets, orbital variations).

- Palaeoclimatic observations indicate that abrupt decadal-to centennial-scale changes in the regional frequency of tropical cyclones, floods, decadal droughts and the intensity of the African-Asian summer monsoon *very likely* occurred during the past 10 kyr. However, the mechanisms behind these abrupt shifts are not well understood, nor have they been thoroughly investigated using current climate models.

How does the 20th-century climate change compare with the climate of the past 2,000 years?

- It is *very likely* that the average rates of increase in CO_2, as well as in the combined radiative forcing from CO_2, CH_4 and N_2O concentration increases, have been at least five times faster over the period from 1960 to 1999 than over any other 40-year period during the past two millennia prior to the industrial era.

- Ice core data from Greenland and Northern Hemisphere mid-latitudes show a *very likely* rapid post-industrial era increase in sulphate concentrations above the pre-industrial background.

- Some of the studies conducted since the Third Assessment Report (TAR) indicate greater multi-centennial Northern Hemisphere temperature variability over the last 1 kyr than was shown in the TAR, demonstrating a sensitivity to the particular proxies used, and the specific statistical methods of processing and/or scaling them to represent past temperatures. The additional variability shown in some new studies implies mainly cooler temperatures (predominantly in the 12th to 14th, 17th and 19th centuries), and only one new reconstruction suggests slightly warmer conditions (in the 11th century, but well within the uncertainty range indicated in the TAR).

- The TAR pointed to the 'exceptional warmth of the late 20th century, relative to the past 1,000 years'. Subsequent evidence has strengthened this conclusion. It is *very likely* that average Northern Hemisphere temperatures during the second half of the 20th century were higher than for any other 50-year period in the last 500 years. It is also *likely* that this 50-year period was the warmest Northern Hemisphere period in the last 1.3 kyr, and that this warmth was more widespread than during any other 50-year period in the last 1.3 kyr. These conclusions are most robust for summer in extratropical land areas, and for more recent periods because of poor early data coverage.

- The small variations in pre-industrial CO_2 and CH_4 concentrations over the past millennium are consistent with millennial-length proxy Northern Hemisphere temperature reconstructions; climate variations larger than indicated by the reconstructions would *likely* yield larger concentration changes. The small pre-industrial greenhouse gas variations also provide indirect evidence for a limited range of decadal- to centennial-scale variations in global temperature.

- Palaeoclimate model simulations are broadly consistent with the reconstructed NH temperatures over the past 1 kyr. The rise in surface temperatures since 1950 *very likely* cannot be reproduced without including anthropogenic greenhouse gases in the model forcings, and it is *very unlikely* that this warming was merely a recovery from a pre-20th century cold period.

- Knowledge of climate variability over the last 1 kyr in the Southern Hemisphere and tropics is very limited by the low density of palaeoclimatic records.

- Climate reconstructions over the past millennium indicate with *high confidence* more varied spatial climate teleconnections related to the El Niño-Southern Oscillation than are represented in the instrumental record of the 20th century.

- The palaeoclimate records of northern and eastern Africa, as well as the Americas, indicate with *high confidence* that droughts lasting decades or longer were a recurrent feature of climate in these regions over the last 2 kyr.

What does the palaeoclimatic record reveal about feedback, biogeochemical and biogeophysical processes?

- The widely accepted orbital theory suggests that glacial-interglacial cycles occurred in response to orbital forcing. The large response of the climate system implies a strong positive amplification of this forcing. This amplification has *very likely* been influenced mainly by changes in greenhouse gas concentrations and ice sheet growth and decay, but also by ocean circulation and sea ice changes, biophysical feedbacks and aerosol (dust) loading.

- It is *virtually certain* that millennial-scale changes in atmospheric CO_2 associated with individual antarctic warm events were less than 25 ppm during the last glacial period. This suggests that the associated changes in North Atlantic Deep Water formation and in the large-scale deposition of wind-borne iron in the Southern Ocean had limited impact on CO_2.

- It is *very likely* that marine carbon cycle processes were primarily responsible for the glacial-interglacial CO_2 variations. The quantification of individual marine processes remains a difficult problem.

- Palaeoenvironmental data indicate that regional vegetation composition and structure are *very likely* sensitive to climate change, and in some cases can respond to climate change within decades.

6.1   Introduction

This chapter assesses palaeoclimatic data and knowledge of how the climate system changes over interannual to millennial time scales, and how well these variations can be simulated with climate models. Additional palaeoclimatic perspectives are included in other chapters.

Palaeoclimate science has made significant advances since the 1970s, when a primary focus was on the origin of the ice ages, the possibility of an imminent future ice age, and the first explorations of the so-called Little Ice Age and Medieval Warm Period. Even in the first IPCC assessment (IPCC, 1990), many climatic variations prior to the instrumental record were not that well known or understood. Fifteen years later, understanding is much improved, more quantitative and better integrated with respect to observations and modelling.

After a brief overview of palaeoclimatic methods, including their strengths and weaknesses, this chapter examines the palaeoclimatic record in chronological order, from oldest to youngest. This approach was selected because the climate system varies and changes over all time scales, and it is instructive to understand the contributions that lower-frequency patterns of climate change might make in influencing higher-frequency patterns of variability and change. In addition, an examination of how the climate system has responded to large changes in climate forcing in the past is useful in assessing how the same climate system might respond to the large anticipated forcing changes in the future.

Cutting across this chronologically based presentation are assessments of climate forcing and response, and of the ability of state-of-the-art climate models to simulate the responses. Perspectives from palaeoclimatic observations, theory and modelling are integrated wherever possible to reduce uncertainty in the assessment. Several sections also assess the latest developments in the rapidly advancing area of abrupt climate change, that is, forced or unforced climatic change that involves crossing a threshold to a new climate regime (e.g., new mean state or character of variability), often where the transition time to the new regime is short relative to the duration of the regime (Rahmstorf, 2001; Alley et al., 2003; Overpeck and Trenberth, 2004).

6.2   Palaeoclimatic Methods

6.2.1   Methods – Observations of Forcing and Response

The field of palaeoclimatology has seen significant methodological advances since the Third Assessment Report (TAR), and the purpose of this section is to emphasize these advances while giving an overview of the methods underlying the data used in this chapter. Many critical methodological details are presented in subsequent sections where needed.

Thus, this methods section is designed to be more general, and to give readers more insight to and confidence in the findings of the chapter. Readers are referred to several useful books and special issues of journals for additional methodological detail (Bradley, 1999; Cronin, 1999; Fischer and Wefer, 1999; Ruddiman and Thomson, 2001; Alverson et al., 2003; Mackay et al., 2003; Kucera et al., 2005; NRC, 2006).

6.2.1.1   How are Past Climate Forcings Known?

Time series of astronomically driven insolation change are well known and can be calculated from celestial mechanics (see Section 6.4, Box 6.1). The methods behind reconstructions of past solar and volcanic forcing continue to improve, although important uncertainties still exist (see Section 6.6).

6.2.1.2   How are Past Changes in Global Atmospheric Composition Known?

Perhaps one of the most important aspects of modern palaeoclimatology is that it is possible to derive time series of atmospheric trace gases and aerosols for the period from about 650 kyr to the present from air trapped in polar ice and from the ice itself (see Sections 6.4 to 6.6 for more methodological citations). As is common in palaeoclimatic studies of the Late Quaternary, the quality of forcing and response series are verified against recent (i.e., post-1950) measurements made by direct instrumental sampling. Section 6.3 cites several papers that reveal how atmospheric CO_2 concentrations can be inferred back millions of years, with much lower precision than the ice core estimates. As is common across all aspects of the field, palaeoclimatologists seldom rely on one method or proxy, but rather on several. This provides a richer and more encompassing view of climatic change than would be available from a single proxy. In this way, results can be cross-checked and uncertainties understood. In the case of pre-Quaternary carbon dioxide (CO_2), multiple geochemical and biological methods provide reasonable constraints on past CO_2 variations, but, as pointed out in Section 6.3, the quality of the estimates is somewhat limited.

6.2.1.3   How Precisely Can Palaeoclimatic Records of Forcing and Response be Dated?

Much has been researched and written on the dating methods associated with palaeoclimatic records, and readers are referred to the background books cited above for more detail. In general, dating accuracy gets weaker farther back in time and dating methods often have specific ranges where they can be applied. Tree ring records are generally the most accurate, and are accurate to the year, or season of a year (even back thousands of years). There are a host of other proxies that also have annual layers or bands (e.g., corals, varved sediments, some cave deposits, some ice cores) but the age models associated with these are not always exact to a specific year. Palaeoclimatologists strive to generate age information from multiple sources to reduce age uncertainty, and palaeoclimatic interpretations must take into account uncertainties in time control.

There continue to be significant advances in radiometric dating. Each radiometric system has ranges over which the system is useful, and palaeoclimatic studies almost always publish analytical uncertainties. Because there can be additional uncertainties, methods have been developed for checking assumptions and cross verifying with independent methods. For example, secular variations in the radiocarbon clock over the last 12 kyr are well known, and fairly well understood over the last 35 kyr. These variations, and the quality of the radiocarbon clock, have both been well demonstrated via comparisons with age models derived from precise tree ring and varved sediment records, as well as with age determinations derived from independent radiometric systems such as uranium series. However, for each proxy record, the quality of the radiocarbon chronology also depends on the density of dates, the material available for dating and knowledge about the radiocarbon age of the carbon that was incorporated into the dated material.

6.2.1.4 *How Can Palaeoclimatic Proxy Methods Be Used to Reconstruct Past Climate Dynamics?*

Most of the methods behind the palaeoclimatic reconstructions assessed in this chapter are described in some detail in the aforementioned books, as well as in the citations of each chapter section. In some sections, important methodological background and controversies are discussed where such discussions help assess palaeoclimatic uncertainties.

Palaeoclimatic reconstruction methods have matured greatly in the past decades, and range from direct measurements of past change (e.g., ground temperature variations, gas content in ice core air bubbles, ocean sediment pore-water change and glacier extent changes) to proxy measurements involving the change in chemical, physical and biological parameters that reflect – often in a quantitative and well-understood manner – past change in the environment where the proxy carrier grew or existed. In addition to these methods, palaeoclimatologists also use documentary data (e.g., in the form of specific observations, logs and crop harvest data) for reconstructions of past climates. While a number of uncertainties remain, it is now well accepted and verified that many organisms (e.g., trees, corals, plankton, insects and other organisms) alter their growth and/or population dynamics in response to changing climate, and that these climate-induced changes are well recorded in the past growth of living and dead (fossil) specimens or assemblages of organisms. Tree rings, ocean and lake plankton and pollen are some of the best-known and best-developed proxy sources of past climate going back centuries and millennia. Networks of tree ring width and density chronologies are used to infer past temperature and moisture changes based on comprehensive calibration with temporally overlapping instrumental data. Past distributions of pollen and plankton from sediment cores can be used to derive quantitative estimates of past climate (e.g., temperatures, salinity and precipitation) via statistical methods calibrated against their modern distribution and associated climate

parameters. The chemistry of several biological and physical entities reflects well-understood thermodynamic processes that can be transformed into estimates of climate parameters such as temperature. Key examples include: oxygen (O) isotope ratios in coral and foraminiferal carbonate to infer past temperature and salinity; magnesium/calcium (Mg/Ca) and strontium/calcium (Sr/Ca) ratios in carbonate for temperature estimates; alkenone saturation indices from marine organic molecules to infer past sea surface temperature (SST); and O and hydrogen isotopes and combined nitrogen and argon isotope studies in ice cores to infer temperature and atmospheric transport. Lastly, many physical systems (e.g., sediments and aeolian deposits) change in predictable ways that can be used to infer past climate change. There is ongoing work on further development and refinement of methods, and there are remaining research issues concerning the degree to which the methods have spatial and seasonal biases. Therefore, in many recent palaeoclimatic studies, a combination of methods is applied since multi-proxy series provide more rigorous estimates than a single proxy approach, and the multi-proxy approach may identify possible seasonal biases in the estimates. No palaeoclimatic method is foolproof, and knowledge of the underlying methods and processes is required when using palaeoclimatic data.

The field of palaeoclimatology depends heavily on replication and cross-verification between palaeoclimate records from independent sources in order to build confidence in inferences about past climate variability and change. In this chapter, the most weight is placed on those inferences that have been made with particularly robust or replicated methodologies.

6.2.2 Methods – Palaeoclimate Modelling

Climate models are used to simulate episodes of past climate (e.g., the Last Glacial Maximum, the last interglacial period or abrupt climate events) to help understand the mechanisms of past climate changes. Models are key to testing physical hypotheses, such as the Milankovitch theory (Section 6.4, Box 6.1), quantitatively. Models allow the linkage of cause and effect in past climate change to be investigated. Models also help to fill the gap between the local and global scale in palaeoclimate, as palaeoclimatic information is often sparse, patchy and seasonal. For example, long ice core records show a strong correlation between local temperature in Antarctica and the globally mixed gases CO_2 and methane, but the causal connections between these variables are best explored with the help of models. Developing a quantitative understanding of mechanisms is the most effective way to learn from past climate for the future, since there are probably no direct analogues of the future in the past.

At the same time, palaeoclimate reconstructions offer the possibility of testing climate models, particularly if the climate forcing can be appropriately specified, and the response is sufficiently well constrained. For earlier climates (i.e., before the current 'Holocene' interglacial), forcing and responses cover a much larger range, but data are more sparse and uncertain, whereas for recent millennia more records are available, but forcing and response are much smaller. Testing models with palaeoclimatic data is important, as not all aspects of climate models can be tested against instrumental climate data. For example, good performance for present climate is not a conclusive test for a realistic sensitivity to CO_2 – to test this, simulation of a climate with a very different CO_2 level can be used. In addition, many parametrizations describing sub-grid scale processes (e.g., cloud parameters, turbulent mixing) have been developed using present-day observations; hence climate states not used in model development provide an independent benchmark for testing models. Palaeoclimate data are key to evaluating the ability of climate models to simulate realistic climate change.

In principle the same climate models that are used to simulate present-day climate, or scenarios for the future, are also used to simulate episodes of past climate, using differences in prescribed forcing and (for the deep past) in configuration of oceans and continents. The full spectrum of models (see Chapter 8) is used (Claussen et al., 2002), ranging from simple conceptual models, through Earth System Models of Intermediate Complexity (EMICs) and coupled General Circulation Models (GCMs). Since long simulations (thousands of years) can be required for some palaeoclimatic applications, and computer power is still a limiting factor, relatively 'fast' coupled models are often used. Additional components that are not standard in models used for simulating present climate are also increasingly added for palaeoclimate applications, for example, continental ice sheet models or components that track the stable isotopes in the climate system (LeGrande et al., 2006). Vegetation modules as well as terrestrial and marine ecosystem modules are increasingly included, both to capture biophysical and biogeochemical feedbacks to climate, and to allow for validation of models against proxy palaeoecological (e.g., pollen) data. The representation of biogeochemical tracers and processes is a particularly important new advance for palaeoclimatic model simulations, as a rich body of information on past climate has emerged from palaeoenvironmental records that are intrinsically linked to the cycling of carbon and other nutrients.

6.3 The Pre-Quaternary Climates

6.3.1 What is the Relationship Between Carbon Dioxide and Temperature in this Time Period?

Pre-Quaternary climates prior to 2.6 Ma (e.g., Figure 6.1) were mostly warmer than today and associated with higher CO_2 levels. In that sense, they have certain similarities with the anticipated future climate change (although the global biology and geography were increasingly different further back in time). In general, they verify that warmer climates are to be expected with increased greenhouse gas concentrations. Looking back in time beyond the reach of ice cores, that is, prior to about

1 Ma, data on greenhouse gas concentrations in the atmosphere become much more uncertain. However, there are ongoing efforts to obtain quantitative reconstructions of the warm climates over the past 65 Myr and the following subsections discuss two particularly relevant climate events of this period.

How accurately is the relationship between CO_2 and temperature known? There are four primary proxies used for pre-Quaternary CO_2 levels (Jasper and Hayes, 1990; Royer et al., 2001; Royer, 2003). Two proxies apply the fact that biological entities in soils and seawater have carbon isotope ratios that are distinct from the atmosphere (Cerling, 1991; Freeman and Hayes, 1992; Yapp and Poths, 1992; Pagani et al., 2005). The third proxy uses the ratio of boron isotopes (Pearson and Palmer, 2000), while the fourth uses the empirical relationship between stomatal pores on tree leaves and atmospheric CO_2 content (McElwain and Chaloner, 1995; Royer, 2003). As shown in Figure 6.1 (bottom panel), while there is a wide range of reconstructed CO_2 values, magnitudes are generally higher than the interglacial, pre-industrial values seen in ice core data. Changes in CO_2 on these long time scales are thought to be driven by changes in tectonic processes (e.g., volcanic activity source and silicate weathering drawdown; e.g., Ruddiman, 1997). Temperature reconstructions, such as that shown in Figure 6.1 (middle panel), are derived from O isotopes (corrected for variations in the global ice volume), as well as Mg/Ca in forams and alkenones. Indicators for the presence of continental ice on Earth show that the planet was mostly ice-free during geologic history, another indication of the general warmth. Major expansion of antarctic glaciations starting around 35 to 40 Ma was likely a response, in part, to declining atmospheric CO_2 levels from their peak in the Cretaceous (~100 Ma) (DeConto and Pollard, 2003). The relationship between CO_2 and temperature can be traced further back in time as indicated in Figure 6.1 (top panel), which shows that the warmth of the Mesozoic Era (230–65 Ma) was likely associated with high levels of CO_2 and that the major glaciations around 300 Ma likely coincided with low CO_2 concentrations relative to surrounding periods.

6.3.2 What Does the Record of the Mid-Pliocene Show?

The Mid-Pliocene (about 3.3 to 3.0 Ma) is the most recent time in Earth's history when mean global temperatures were substantially warmer for a sustained period (estimated by GCMs to be about 2°C to 3°C above pre-industrial temperatures; Chandler et al., 1994; Sloan et al., 1996; Haywood et al., 2000; Jiang et al., 2005), providing an accessible example of a world that is similar in many respects to what models estimate could be the Earth of the late 21st century. The Pliocene is also recent enough that the continents and ocean basins had nearly reached their present geographic configuration. Taken together, the average of the warmest times during the middle Pliocene presents a view of the equilibrium state of a globally warmer world, in which atmospheric CO_2 concentrations (estimated

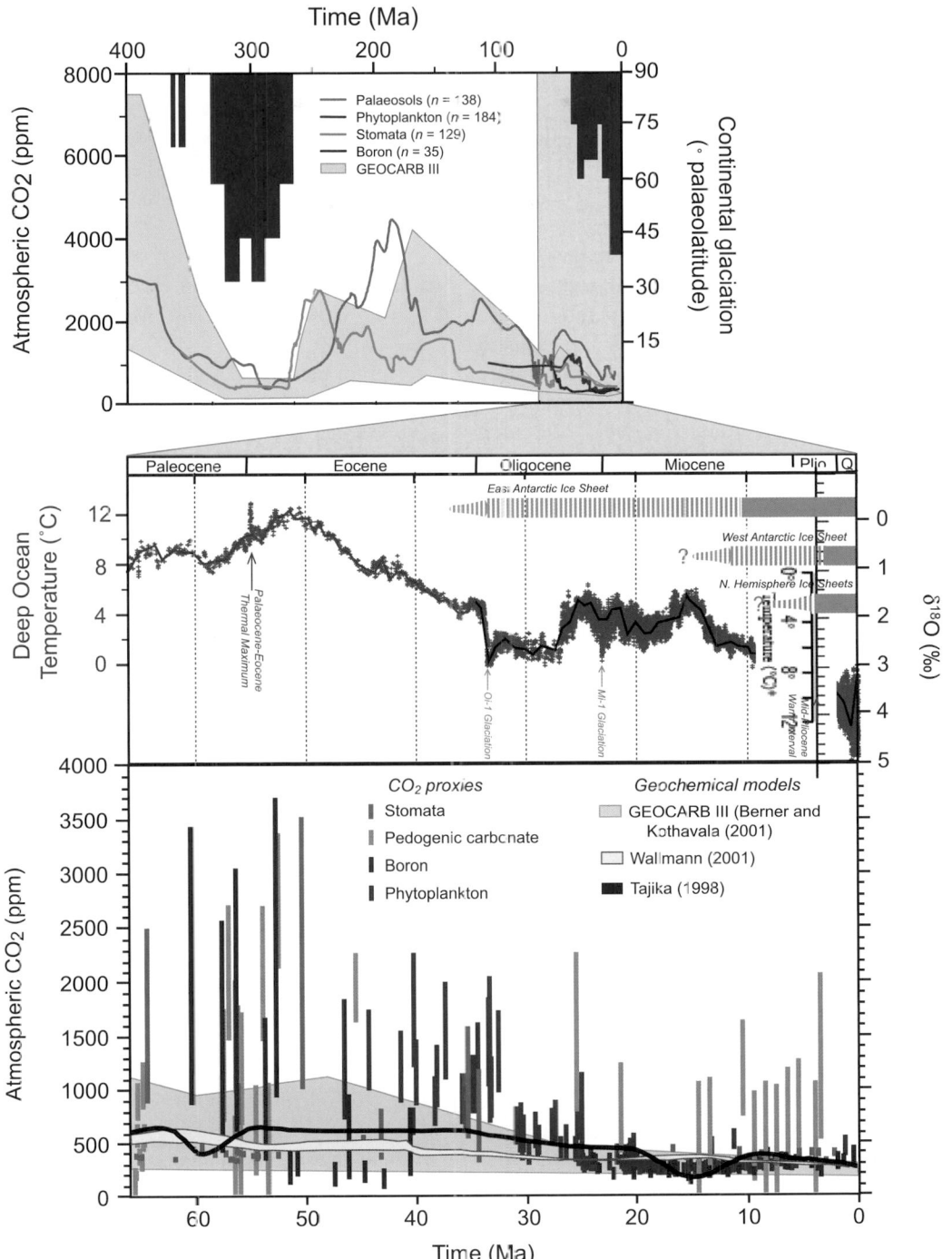

Figure 6.1. *(Top) Atmospheric CO_2 and continental glaciation 400 Ma to present. Vertical blue bars mark the timing and palaeolatitudinal extent of ice sheets (after Crowley, 1998). Plotted CO_2 records represent five-point running averages from each of the four major proxies (see Royer, 2006 for details of compilation). Also plotted are the plausible ranges of CO_2 from the geochemical carbon cycle model GEOCARB III (Berner and Kothavala, 2001). All data have been adjusted to the Gradstein et al. (2004) time scale. (Middle) Global compilation of deep-sea benthic foraminifera ^{18}O isotope records from 40 Deep Sea Drilling Program and Ocean Drilling Program sites (Zachos et al., 2001) updated with high-resolution records for the Eocene through Miocene interval (Billups et al., 2002; Bohaty and Zachos, 2003; Lear et al., 2004). Most data were derived from analyses of two common and long-lived benthic taxa, Cibicidoides and Nuttallides. To correct for genus-specific isotope vital effects, the ^{18}O values were adjusted by +0.64 and +0.4 (Shackleton et al., 1984), respectively. The ages are relative to the geomagnetic polarity time scale of Berggren et al. (1995). The raw data were smoothed using a five-point running mean, and curve-fitted with a locally weighted mean. The ^{18}O temperature values assume an ice-free ocean (–1.0‰ Standard Mean Ocean Water), and thus only apply to the time preceding large-scale antarctic glaciation (~35 Ma). After the early Oligocene much of the variability (~70%) in the ^{18}O record reflects changes in antarctic and Northern Hemisphere ice volume, which is represented by light blue horizontal bars (e.g., Hambrey et al., 1991; Wise et al., 1991; Ehrmann and Mackensen, 1992). Where the bars are dashed, they represent periods of ephemeral ice or ice sheets smaller than present, while the solid bars represent ice sheets of modern or greater size. The evolution and stability of the West Antarctic Ice Sheet (e.g., Lemasurier and Rocchi, 2005) remains an important area of uncertainty that could affect estimates of future sea level rise. (Bottom) Detailed record of CO_2 for the last 65 Myr. Individual records of CO_2 and associated errors are colour-coded by proxy method; when possible, records are based on replicate samples (see Royer, 2006 for details and data references). Dating errors are typically less than ±1 Myr. The range of error for each CO_2 proxy varies considerably, with estimates based on soil nodules yielding the greatest uncertainty. Also plotted are the plausible ranges of CO_2 from three geochemical carbon cycle models.*

to be between 360 to 400 ppm) were likely higher than pre-industrial values (Raymo and Rau, 1992; Raymo et al., 1996), and in which geologic evidence and isotopes agree that sea level was at least 15 to 25 m above modern levels (Dowsett and Cronin, 1990; Shackleton et al., 1995), with correspondingly reduced ice sheets and lower continental aridity (Guo et al., 2004).

Both terrestrial and marine palaeoclimate proxies (Thompson, 1991; Dowsett et al., 1996; Thompson and Fleming, 1996) show that high latitudes were significantly warmer, but that tropical SSTs and surface air temperatures were little different from the present. The result was a substantial decrease in the lower-tropospheric latitudinal temperature gradient. For example, atmospheric GCM simulations driven by reconstructed SSTs from the Pliocene Research Interpretations and Synoptic Mapping Group (Dowsett et al., 1996; Dowsett et al., 2005) produced winter surface air temperature warming of 10°C to 20°C at high northern latitudes with 5°C to 10°C increases over the northern North Atlantic (~60°N), whereas there was essentially no tropical surface air temperature change (or even slight cooling) (Chandler et al., 1994; Sloan et al., 1996; Haywood et al., 2000, Jiang et al., 2005). In contrast, a coupled atmosphere-ocean experiment with an atmospheric CO_2 concentration of 400 ppm produced warming relative to pre-industrial times of 3°C to 5°C in the northern North Atlantic, and 1°C to 3°C in the tropics (Haywood et al., 2005), generally similar to the response to higher CO_2 discussed in Chapter 10.

The estimated lack of tropical warming is a result of basing tropical SST reconstructions on marine microfaunal evidence. As in the case of the Last Glacial Maximum (see Section 6.4), it is uncertain whether tropical sensitivity is really as small as such reconstructions suggest. Haywood et al. (2005) found that alkenone estimates of tropical and subtropical temperatures do indicate warming in these regions, in better agreement with GCM simulations from increased CO_2 forcing (see Chapter 10). As in the study noted above, climate models cannot produce a response to increased CO_2 with large high-latitude warming, and yet minimal tropical temperature change, without strong increases in ocean heat transport (Rind and Chandler, 1991).

The substantial high-latitude response is shown by both marine and terrestrial palaeodata, and it may indicate that high latitudes are more sensitive to increased CO_2 than model simulations suggest for the 21st century. Alternatively, it may be the result of increased ocean heat transports due to either an enhanced thermohaline circulation (Raymo et al., 1989; Rind and Chandler, 1991) or increased flow of surface ocean currents due to greater wind stresses (Ravelo et al., 1997; Haywood et al., 2000), or associated with the reduced extent of land and sea ice (Jansen et al., 2000; Knies et al., 2002; Haywood et al., 2005). Currently available proxy data are equivocal concerning a possible increase in the intensity of the meridional overturning cell for either transient or equilibrium climate states during the Pliocene, although an increase would contrast with the North Atlantic transient deep-water production decreases that are found in most coupled model simulations for the 21st century

(see Chapter 10). The transient response is likely to be different from an equilibrium response as climate warms. Data are just beginning to emerge that describe the deep ocean state during the Pliocene (Cronin et al., 2005). Understanding the climate distribution and forcing for the Pliocene period may help improve predictions of the likely response to increased CO_2 in the future, including the ultimate role of the ocean circulation in a globally warmer world.

6.3.3 What Does the Record of the Palaeocene-Eocene Thermal Maximum Show?

Approximately 55 Ma, an abrupt warming (in this case of the order of 1 to 10 kyr) by several degrees celsius is indicated by changes in ^{18}O isotope and Mg/Ca records (Kennett and Stott, 1991; Zachos et al., 2003; Tripati and Elderfield, 2004). The warming and associated environmental impact was felt at all latitudes, and in both the surface and deep ocean. The warmth lasted approximately 100 kyr. Evidence for shifts in global precipitation patterns is present in a variety of fossil records including vegetation (Wing et al., 2005). The climate anomaly, along with an accompanying carbon isotope excursion, occurred at the boundary between the Palaeocene and Eocene epochs, and is therefore often referred to as the Palaeocene-Eocene Thermal Maximum (PETM). The thermal maximum clearly stands out in high-resolution records of that time (Figure 6.2). At the same time, ^{13}C isotopes in marine and continental records show that a large mass of carbon with low ^{13}C concentration must have been released into the atmosphere and ocean. The mass of carbon was sufficiently large to lower the pH of the ocean and drive widespread dissolution of seafloor carbonates (Zachos et al., 2005). Possible sources for this carbon could have been methane (CH_4) from decomposition of clathrates on the sea floor, CO_2 from volcanic activity, or oxidation of sediments rich in organic matter (Dickens et al., 1997; Kurtz et al., 2003; Svensen et al., 2004). The PETM, which altered ecosystems worldwide (Koch et al., 1992; Bowen et al., 2002; Bralower, 2002; Crouch et al., 2003; Thomas, 2003; Bowen et al., 2004; Harrington et al., 2004), is being intensively studied as it has some similarity with the ongoing rapid release of carbon into the atmosphere by humans. The estimated magnitude of carbon release for this time period is of the order of 1 to 2×10^{18} g of carbon (Dickens et al., 1997), a similar magnitude to that associated with greenhouse gas releases during the coming century. Moreover, the period of recovery through natural carbon sequestration processes, about 100 kyr, is similar to that forecast for the future. As in the case of the Pliocene, the high-latitude warming during this event was substantial (~20°C; Moran et al., 2006) and considerably higher than produced by GCM simulations for the event (Sluijs et al., 2006) or in general for increased greenhouse gas experiments (Chapter 10). Although there is still too much uncertainty in the data to derive a quantitative estimate of climate sensitivity from the PETM, the event is a striking example of massive carbon release and related extreme climatic warming.

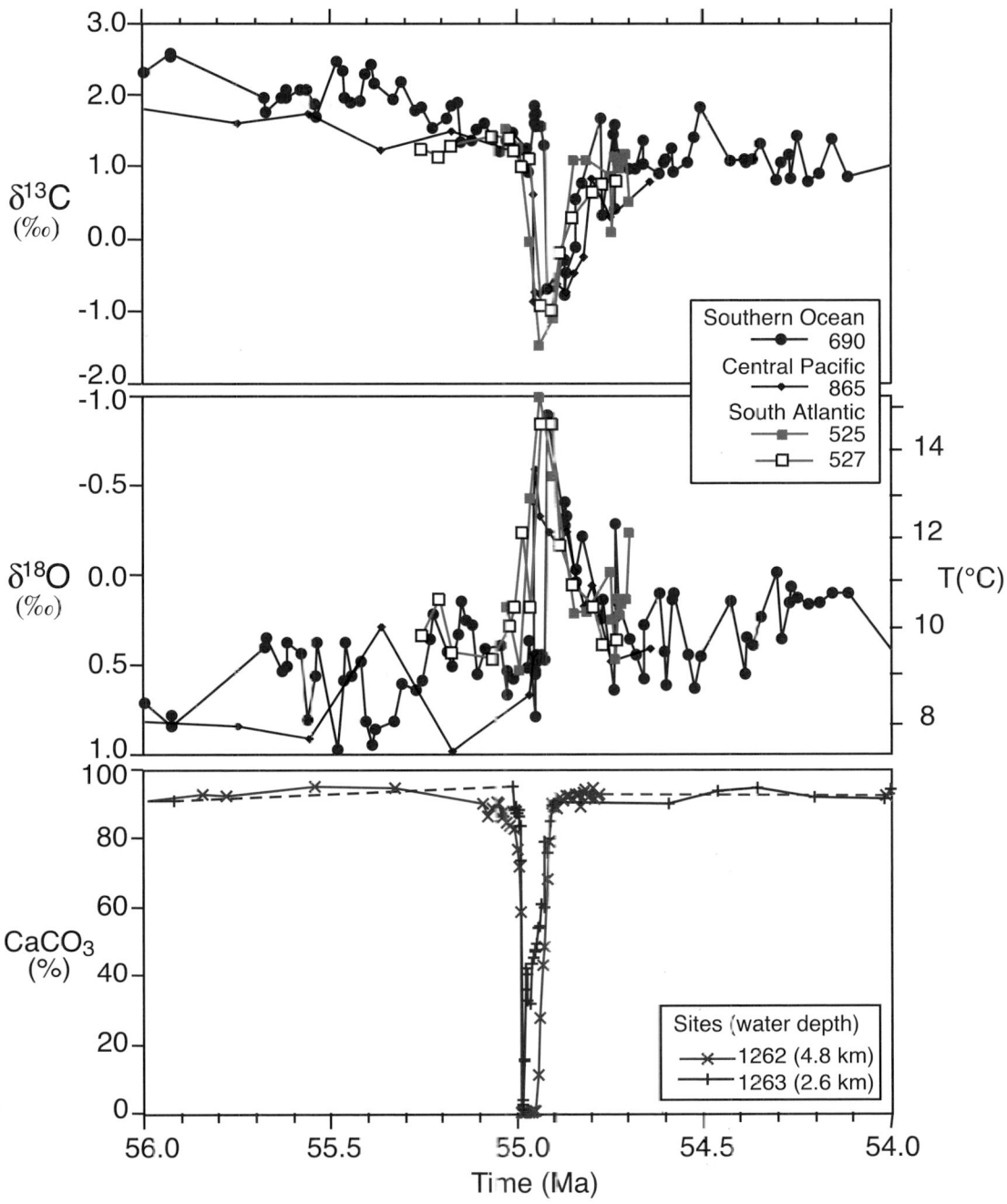

Figure 6.2. *The Palaeocene-Eocene Thermal Maximum as recorded in benthic (bottom dwelling) foraminifer (Nuttallides truempyi) isotopic records from sites in the Antarctic, south Atlantic and Pacific (see Zachos et al., 2003 for details). The rapid decrease in carbon isotope ratios in the top panel is indicative of a large increase in atmospheric greenhouse gases CO₂ and CH₄ that was coincident with an approximately 5°C global warming (centre panel). Using the carbon isotope records, numerical models show that CH₄ released by the rapid decomposition of marine hydrates might have been a major component (~2,000 GtC) of the carbon flux (Dickens and Owen, 1996). Testing of this and other models requires an independent constraint on the carbon fluxes. In theory, much of the additional greenhouse carbon would have been absorbed by the ocean, thereby lowering seawater pH and causing widespread dissolution of seafloor carbonates. Such a response is evident in the lower panel, which shows a transient reduction in the carbonate (CaCO₃) content of sediments in two cores from the south Atlantic (Zachos et al., 2004, 2005). The observed patterns indicate that the ocean's carbonate saturation horizon rapidly shoaled more than 2 km, and then gradually recovered as buffering processes slowly restored the chemical balance of the ocean. Initially, most of the carbonate dissolution is of sediment deposited prior to the event, a process that offsets the apparent timing of the dissolution horizon relative to the base of the benthic foraminifer carbon isotope excursion. Model simulations show that the recovery of the carbonate saturation horizon should precede the recovery in the carbon isotopes by as much as 100 kyr (Dickens and Owen, 1996), another feature that is evident in the sediment records.*

6.4 Glacial-Interglacial Variability and Dynamics

6.4.1 Climate Forcings and Responses Over Glacial-Interglacial Cycles

Palaeoclimatic records document a sequence of glacial-interglacial cycles covering the last 740 kyr in ice cores (EPICA community members, 2004), and several million years in deep oceanic sediments (Lisiecki and Raymo, 2005) and loess (Ding et al., 2002). The last 430 kyr, which are the best documented, are characterised by 100-kyr glacial-interglacial cycles of very large amplitude, as well as large climate changes corresponding to other orbital periods (Hays et al., 1976; Box 6.1), and at millennial time scales (McManus et al.; 2002; NorthGRIP, 2004). A minor proportion (20% on average) of each glacial-interglacial cycle was spent in the warm interglacial mode, which normally lasted for 10 to 30 kyr (Figure 6.3). There is

evidence for longer interglacial periods between 430 and 740 ka, but these were apparently colder than the typical interglacials of the latest Quaternary (EPICA community members, 2004). The Holocene, the latest of these interglacials, extends to the present.

The ice core record indicates that greenhouse gases co-varied with antarctic temperature over glacial-interglacial cycles, suggesting a close link between natural atmospheric greenhouse gas variations and temperature (Box 6.2). Variations in CO_2 over the last 420 kyr broadly followed antarctic temperature, typically by several centuries to a millennium (Mudelsee, 2001). The sequence of climatic forcings and responses during deglaciations (transitions from full glacial conditions to warm interglacials) are well documented. High-resolution ice core records of temperature proxies and CO_2 during deglaciation indicates that antarctic temperature starts to rise several hundred years before CO_2 (Monnin et al., 2001; Caillon et al., 2003). During the last deglaciation, and likely also the three previous ones, the onset of warming at both high southern and northern

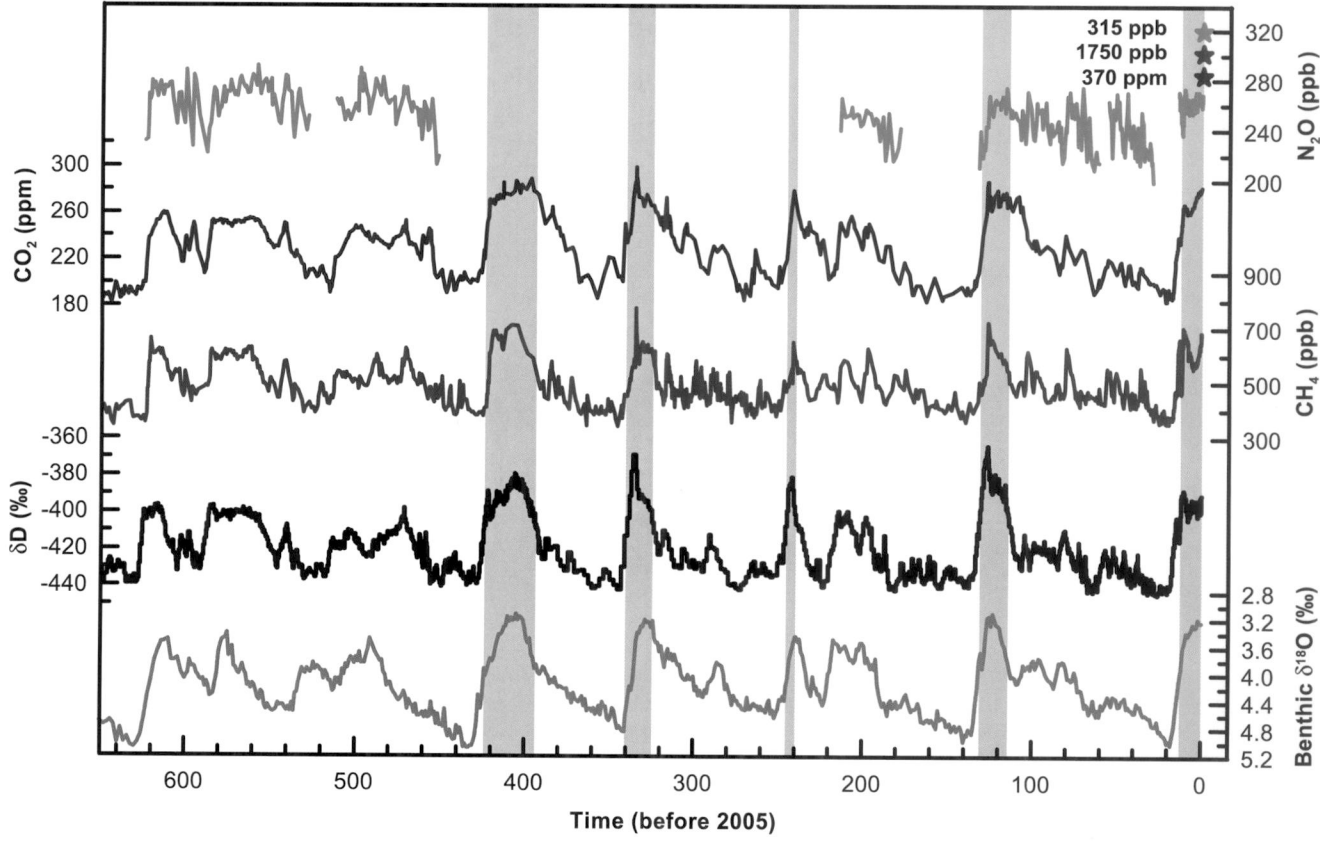

Figure 6.3. *Variations of deuterium (δD; black), a proxy for local temperature, and the atmospheric concentrations of the greenhouse gases CO_2 (red), CH_4 (blue), and nitrous oxide (N_2O; green) derived from air trapped within ice cores from Antarctica and from recent atmospheric measurements (Petit et al., 1999; Indermühle et al., 2000; EPICA community members, 2004; Spahni et al., 2005; Siegenthaler et al., 2005a,b). The shading indicates the last interglacial warm periods. Interglacial periods also existed prior to 450 ka, but these were apparently colder than the typical interglacials of the latest Quaternary. The length of the current interglacial is not unusual in the context of the last 650 kyr. The stack of 57 globally distributed benthic $\delta^{18}O$ marine records (dark grey), a proxy for global ice volume fluctuations (Lisiecki and Raymo, 2005), is displayed for comparison with the ice core data. Downward trends in the benthic $\delta^{18}O$ curve reflect increasing ice volumes on land. Note that the shaded vertical bars are based on the ice core age model (EPICA community members, 2004), and that the marine record is plotted on its original time scale based on tuning to the orbital parameters (Lisiecki and Raymo, 2005). The stars and labels indicate atmospheric concentrations at year 2000.*

Box 6.1: Orbital Forcing

It is well known from astronomical calculations (Berger, 1978) that periodic changes in parameters of the orbit of the Earth around the Sun modify the seasonal and latitudinal distribution of incoming solar radiation at the top of the atmosphere (hereafter called 'insolation'). Past and future changes in insolation can be calculated over several millions of years with a high degree of confidence (Berger and Loutre, 1991; Laskar et al., 2004). This box focuses on the time period from the past 800 kyr to the next 200 kyr.

Over this time interval, the obliquity (tilt) of the Earth axis varies between 22.05° and 24.50° with a strong quasi-periodicity around 41 kyr. Changes in obliquity have an impact on seasonal contrasts. This parameter also modulates annual mean insolation changes with opposite effects in low vs. high latitudes (and therefore no effect on global average insolation). Local annual mean insolation changes remain below 6 W m^{-2}.

The eccentricity of the Earth's orbit around the Sun has longer quasi-periodicities at 400 and around 100 kyr, and varies between values of about 0.002 and 0.050 during the time period from 800 ka to 200 kyr in the future. Changes in eccentricity alone modulate the Sun-Earth distance and have limited impacts on global and annual mean insolation. However, changes in eccentricity affect the intra-annual changes in the Sun-Earth distance and thereby modulate significantly the seasonal and latitudinal effects induced by obliquity and climatic precession.

Associated with the general precession of the equinoxes and the longitude of perihelion, periodic shifts in the position of solstices and equinoxes on the orbit relative to the perihelion occur, and these modulate the seasonal cycle of insolation with periodicities of about 19 and about 23 kyr. As a result, changes in the position of the seasons on the orbit strongly modulate the latitudinal and seasonal distribution of insolation. When averaged over a season, insolation changes can reach 60 W m^{-2} (Box 6.1, Figure 1). During periods of low eccentricity, such as about 400 ka and during the next 100 kyr, seasonal insolation changes induced by precession are less strong than during periods of larger eccentricity (Box 6.1, Figure 1). High-frequency variations of orbital variations appear to be associated with very small insolation changes (Bertrand et al., 2002a).

The Milankovitch theory proposes that ice ages are triggered by minima in summer insolation near 65°N, enabling winter snowfall to persist all year and therefore accumulate to build NH glacial ice sheets. For example, the onset of the last ice age, about 116 ± 1 ka (Stirling et al., 1998), corresponds to a 65°N mid-June insolation about 40 W m^{-2} lower than today (Box 6.1, Figure 1).

Studies of the link between orbital parameters and past climate changes include spectral analysis of palaeoclimatic records and the identification of orbital periodicities; precise dating of specific climatic transitions; and modelling of the climate response to orbital forcing, which highlights the role of climatic and biogeochemical feedbacks. Sections 6.4 and 6.5 describe some aspects of the state-of-the-art understanding of the relationships between orbital forcing, climate feedbacks and past climate changes.

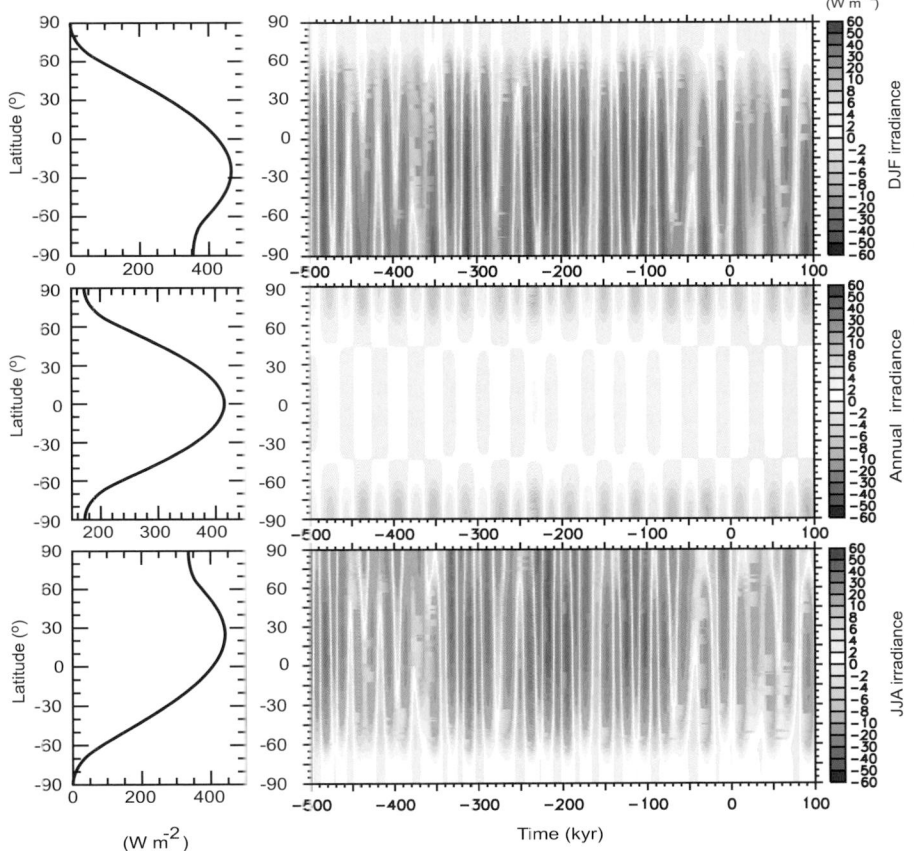

Box 6.1, Figure 1. *(Left) December to February (top), annual mean (middle) and June to August (bottom) latitudinal distribution of present-day (year 1950) incoming mean solar radiation (W m^{-2}). (Right) Deviations with respect to the present of December to February (top), annual mean (middle) and June to August (bottom) latitudinal distribution of incoming mean solar radiation (W m^{-2}) from the past 500 kyr to the future 100 kyr (Berger and Loutre, 1991; Loutre et al., 2004).*

Box 6.2: What Caused the Low Atmospheric Carbon Dioxide Concentrations During Glacial Times?

Ice core records show that atmospheric CO_2 varied in the range of 180 to 300 ppm over the glacial-interglacial cycles of the last 650 kyr (Figure 6.3; Petit et al., 1999; Siegenthaler et al., 2005a). The quantitative and mechanistic explanation of these CO_2 variations remains one of the major unsolved questions in climate research. Processes in the atmosphere, in the ocean, in marine sediments and on land, and the dynamics of sea ice and ice sheets must be considered. A number of hypotheses for the low glacial CO_2 concentrations have emerged over the past 20 years, and a rich body of literature is available (Webb et al., 1997; Broecker and Henderson, 1998; Archer et al., 2000; Sigman and Boyle, 2000; Kohfeld et al., 2005). Many processes have been identified that could potentially regulate atmospheric CO_2 on glacial-interglacial time scales. However, the existing proxy data with which to test hypotheses are relatively scarce, uncertain, and their interpretation is partly conflicting.

Most explanations propose changes in oceanic processes as the cause for low glacial CO_2 concentrations. The ocean is by far the largest of the relatively fast-exchanging (<1 kyr) carbon reservoirs, and terrestrial changes cannot explain the low glacial values because terrestrial storage was also low at the Last Glacial Maximum (see Section 6.4.1). On glacial-interglacial time scales, atmospheric CO_2 is mainly governed by the interplay between ocean circulation, marine biological activity, ocean-sediment interactions, seawater carbonate chemistry and air-sea exchange. Upon dissolution in seawater, CO_2 maintains an acid/base equilibrium with bicarbonate and carbonate ions that depends on the acid-titrating capacity of seawater (i.e., alkalinity). Atmospheric CO_2 would be higher if the ocean lacked biological activity. CO_2 is more soluble in colder than in warmer waters; therefore, changes in surface and deep ocean temperature have the potential to alter atmospheric CO_2. Most hypotheses focus on the Southern Ocean, where large volume-fractions of the cold deep-water masses of the world ocean are currently formed, and large amounts of biological nutrients (phosphate and nitrate) upwelling to the surface remain unused. A strong argument for the importance of SH processes is the co-evolution of antarctic temperature and atmospheric CO_2.

One family of hypotheses regarding low glacial atmospheric CO_2 values invokes an increase or redistribution in the ocean alkalinity as a primary cause. Potential mechanisms are (i) the increase of calcium carbonate ($CaCO_3$) weathering on land, (ii) a decrease of coral reef growth in the shallow ocean, or (iii) a change in the export ratio of $CaCO_3$ and organic material to the deep ocean. These mechanisms require large changes in the deposition pattern of $CaCO_3$ to explain the full amplitude of the glacial-interglacial CO_2 difference through a mechanism called carbonate compensation (Archer et al., 2000). The available sediment data do not support a dominant role for carbonate compensation in explaining low glacial CO_2 levels. Furthermore, carbonate compensation may only explain slow CO_2 variation, as its time scale is multi-millennial.

Another family of hypotheses invokes changes in the sinking of marine plankton. Possible mechanisms include (iv) fertilization of phytoplankton growth in the Southern Ocean by increased deposition of iron-containing dust from the atmosphere after being carried by winds from colder, drier continental areas, and a subsequent redistribution of limiting nutrients; (v) an increase in the whole ocean nutrient content (e.g., through input of material exposed on shelves or nitrogen fixation); and (vi) an increase in the ratio between carbon and other nutrients assimilated in organic material, resulting in a higher carbon export per unit of limiting nutrient exported. As with the first family of hypotheses, this family of mechanisms also suffers from the inability to account for the full amplitude of the reconstructed CO_2 variations when constrained by the available information. For example, periods of enhanced biological production and increased dustiness (iron supply) are coincident with CO_2 concentration changes of 20 to 50 ppm (see Section 6.4.2, Figure 6.7). Model simulations consistently suggest a limited role for iron in regulating past atmospheric CO_2 concentration (Bopp et al., 2002).

Physical processes also likely contributed to the observed CO_2 variations. Possible mechanisms include (vii) changes in ocean temperature (and salinity), (viii) suppression of air-sea gas exchange by sea ice, and (ix) increased stratification in the Southern Ocean. The combined changes in temperature and salinity increased the solubility of CO_2, causing a depletion in atmospheric CO_2 of perhaps 30 ppm. Simulations with general circulation ocean models do not fully support the gas exchange-sea ice hypothesis. One explanation (ix) conceived in the 1980s invokes more stratification, less upwelling of carbon and nutrient-rich waters to the surface of the Southern Ocean and increased carbon storage at depth during glacial times. The stratification may have caused a depletion of nutrients and carbon at the surface, but proxy evidence for surface nutrient utilisation is controversial. Qualitatively, the slow ventilation is consistent with very saline and very cold deep waters reconstructed for the last glacial maximum (Adkins et al., 2002), as well as low glacial stable carbon isotope ratios ($^{13}C/^{12}C$) in the deep South Atlantic.

In conclusion, the explanation of glacial-interglacial CO_2 variations remains a difficult attribution problem. It appears likely that a range of mechanisms have acted in concert (e.g., Köhler et al., 2005). The future challenge is not only to explain the amplitude of glacial-interglacial CO_2 variations, but the complex temporal evolution of atmospheric CO_2 and climate consistently.

latitudes preceded by several thousand years the first signals of significant sea level increase resulting from the melting of the northern ice sheets linked with the rapid warming at high northern latitudes (Petit et al., 1999; Shackleton, 2000; Pépin et al., 2001). Current data are not accurate enough to identify whether warming started earlier in the Southern Hemisphere (SH) or Northern Hemisphere (NH), but a major deglacial feature is the difference between North and South in terms of the magnitude and timing of strong reversals in the warming trend, which are not in phase between the hemispheres and are more pronounced in the NH (Blunier and Brook, 2001).

Greenhouse gas (especially CO_2) feedbacks contributed greatly to the global radiative perturbation corresponding to the transitions from glacial to interglacial modes (see Section 6.4.1.2). The relationship between antarctic temperature and CO_2 did not change significantly during the past 650 kyr, indicating a rather stable coupling between climate and the carbon cycle during the late Pleistocene (Siegenthaler et al., 2005a). The rate of change in atmospheric CO_2 varied considerably over time. For example, the CO_2 increase from about 180 ppm at the Last Glacial Maximum to about 265 ppm in the early Holocene occurred with distinct rates over different periods (Monnin et al., 2001; Figure 6.4).

6.4.1.1　How Do Glacial-Interglacial Variations in the Greenhouse Gases Carbon Dioxide, Methane and Nitrous Oxide Compare with the Industrial Era Greenhouse Gas Increase?

The present atmospheric concentrations of CO_2, CH_4 and nitrous oxide (N_2O) are higher than ever measured in the ice core record of the past 650 kyr (Figures 6.3 and 6.4). The measured concentrations of the three greenhouse gases fluctuated only slightly (within 4% for CO_2 and N_2O and within 7% for CH_4) over the past millennium prior to the industrial era, and also varied within a restricted range over the late Quaternary. Within the last 200 years, the late Quaternary natural range has been exceeded by at least 25% for CO_2, 120% for CH_4 and 9% for N_2O. All three records show effects of the large and increasing growth in anthropogenic emissions during the industrial era.

Variations in atmospheric CO_2 dominate the radiative forcing by all three gases (Figure 6.4). The industrial era increase in CO_2, and in the radiative forcing (Section 2.3) by all three gases, is similar in magnitude to the increase over the transitions from glacial to interglacial periods, but started from an interglacial level and occurred one to two orders of magnitude faster (Stocker and Monnin, 2003). There is no indication in the ice core record that an increase comparable in magnitude and rate to the industrial era has occurred in the past 650 kyr. The data resolution is sufficient to exclude with very high confidence a peak similar to the anthropogenic rise for the past 50 kyr for CO_2, for the past 80 kyr for CH_4 and for the past 16 kyr for N_2O. The ice core records show that during the industrial era, the average rate of increase in the radiative forcing from CO_2, CH_4 and N_2O is greater than at any time during the past 16

kyr (Figure 6.4). The smoothing of the atmospheric signal (Schwander et al., 1993; Spahni et al., 2003) is small at Law Dome, a high-accumulation site in Antarctica, and decadal-scale rates of change can be computed from the Law Dome record spanning the past two millennia (Etheridge et al., 1996; Ferretti et al., 2005; MacFarling Meure et al., 2006). The average rate of increase in atmospheric CO_2 was at least five times larger over the period from 1960 to 1999 than over any other 40-year period during the two millennia before the industrial era. The average rate of increase in atmospheric CH_4 was at least six times larger, and that for N_2O at least two times larger over the past four decades, than at any time during the two millennia before the industrial era. Correspondingly, the recent average rate of increase in the combined radiative forcing by all three greenhouse gases was at least six times larger than at any time during the period AD 1 to AD 1800 (Figure 6.4d).

6.4.1.2　What Do the Last Glacial Maximum and the Last Deglaciation Show?

Past glacial cold periods, sometimes referred to as 'ice ages', provide a means for evaluating the understanding and modelling of the response of the climate system to large radiative perturbations. The most recent glacial period started about 116 ka, in response to orbital forcing (Box 6.1), with the growth of ice sheets and fall of sea level culminating in the Last Glacial Maximum (LGM), around 21 ka. The LGM and the subsequent deglaciation have been widely studied because the radiative forcings, boundary conditions and climate response are relatively well known.

The response of the climate system at the LGM included feedbacks in the atmosphere and on land amplifying the orbital forcing. Concentrations of well-mixed greenhouse gases at the LGM were reduced relative to pre-industrial values (Figures 6.3 and 6.4), amounting to a global radiative perturbation of –2.8 W m^{-2} – approximately equal to, but opposite from, the radiative forcing of these gases for the year 2000 (see Section 2.3). Land ice covered large parts of North America and Europe at the LGM, lowering sea level and exposing new land. The radiative perturbation of the ice sheets and lowered sea level, specified as a boundary condition for some LGM simulations, has been estimated to be about –3.2 W m^{-2}, but with uncertainties associated with the coverage and height of LGM continental ice (Mangerud et al., 2002; Peltier, 2004; Toracinta et al., 2004; Masson-Delmotte et al., 2006) and the parametrization of ice albedo in climate models (Taylor et al., 2000). The distribution of vegetation was altered, with tundra expanded over the northern continents and tropical rain forest reduced (Prentice et al., 2000), and atmospheric aerosols (primarily dust) were increased (Kohfeld and Harrison, 2001), partly as a consequence of reduced vegetation cover (Mahowald et al., 1999). Vegetation and atmospheric aerosols are treated as specified conditions in some LGM simulations, each contributing about –1 W m^{-2} of radiative perturbation, but with very low scientific understanding of their radiative influence at the LGM (Claquin et al., 2003;

Crucifix and Hewitt, 2005). Changes in biogeochemical cycles thus played an important role and contributed, through changes in greenhouse gas concentration, dust loading and vegetation cover, more than half of the known radiative perturbation during the LGM. Overall, the radiative perturbation for the changed greenhouse gas and aerosol concentrations and land surface was approximately –8 W m^{-2} for the LGM, although with

significant uncertainty in the estimates for the contributions of aerosol and land surface changes (Figure 6.5).

Understanding of the magnitude of tropical cooling over land at the LGM has improved since the TAR with more records, as well as better dating and interpretation of the climate signal associated with snow line elevation and vegetation change. Reconstructions of terrestrial climate show strong spatial

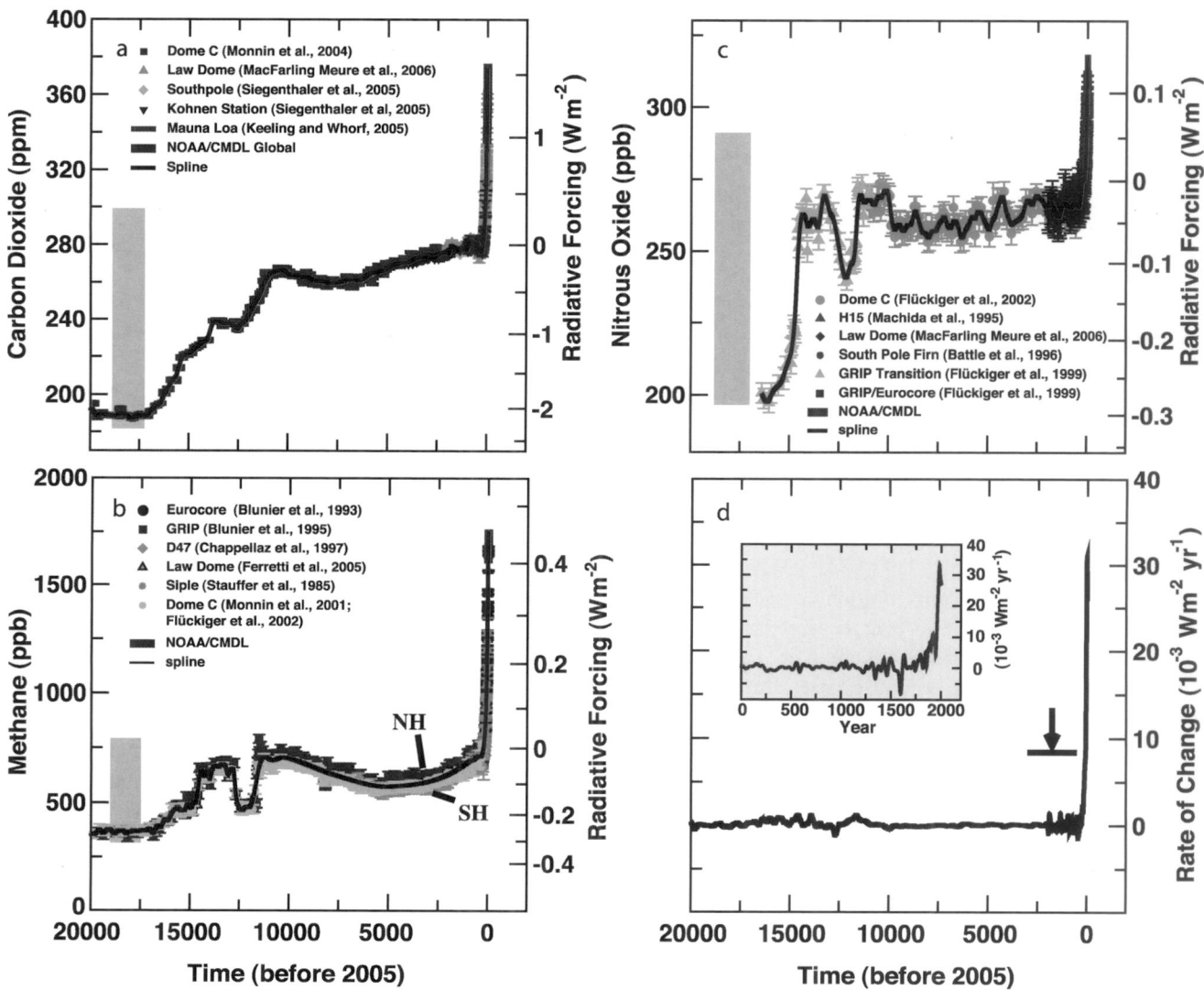

Figure 6.4. *The concentrations and radiative forcing by (a) CO$_2$, (b) CH$_4$ and (c) nitrous oxide (N$_2$O), and (d) the rate of change in their combined radiative forcing over the last 20 kyr reconstructed from antarctic and Greenland ice and firn data (symbols) and direct atmospheric measurements (red and magenta lines). The grey bars show the reconstructed ranges of natural variability for the past 650 kyr (Siegenthaler et al., 2005a; Spahni et al., 2005). Radiative forcing was computed with the simplified expressions of Chapter 2 (Myhre et al., 1998). The rate of change in radiative forcing (black line) was computed from spline fits (Enting, 1987) of the concentration data (black lines in panels a to c). The width of the age distribution of the bubbles in ice varies from about 20 years for sites with a high accumulation of snow such as Law Dome, Antarctica, to about 200 years for low-accumulation sites such as Dome C, Antarctica. The Law Dome ice and firn data, covering the past two millennia, and recent instrumental data have been splined with a cut-off period of 40 years, with the resulting rate of change in radiative forcing shown by the inset in (d). The arrow shows the peak in the rate of change in radiative forcing after the anthropogenic signals of CO$_2$, CH$_4$ and N$_2$O have been smoothed with a model describing the enclosure process of air in ice (Spahni et al., 2003) applied for conditions at the low accumulation Dome C site for the last glacial transition. The CO$_2$ data are from Etheridge et al. (1996); Monnin et al. (2001); Monnin et al. (2004); Siegenthaler et al. (2005b; South Pole); Siegenthaler et al. (2005a; Kohnen Station); and MacFarling Meure et al. (2006). The CH$_4$ data are from Stauffer et al. (1985); Steele et al. (1992); Blunier et al. (1993); Dlugokencky et al. (1994); Blunier et al. (1995); Chappellaz et al. (1997); Monnin et al. (2001); Flückiger et al. (2002); and Ferretti et al. (2005). The N$_2$O data are from Machida et al. (1995); Battle et al. (1996); Flückiger et al. (1999, 2002); and MacFarling Meure et al. (2006). Atmospheric data are from the National Oceanic and Atmospheric Administration's global air sampling network, representing global average concentrations (dry air mole fraction; Steele et al., 1992; Dlugokencky et al., 1994; Tans and Conway, 2005), and from Mauna Loa, Hawaii (Keeling and Whorf, 2005). The globally averaged data are available from http://www.cmdl.noaa.gov/.*

Frequently Asked Question 6.1

What Caused the Ice Ages and Other Important Climate Changes Before the Industrial Era?

Climate on Earth has changed on all time scales, including long before human activity could have played a role. Great progress has been made in understanding the causes and mechanisms of these climate changes. Changes in Earth's radiation balance were the principal driver of past climate changes, but the causes of such changes are varied. For each case – be it the Ice Ages, the warmth at the time of the dinosaurs or the fluctuations of the past millennium – the specific causes must be established individually. In many cases, this can now be done with good confidence, and many past climate changes can be reproduced with quantitative models.

Global climate is determined by the radiation balance of the planet (see FAQ 1.1). There are three fundamental ways the Earth's radiation balance can change, thereby causing a climate change: (1) changing the incoming solar radiation (e.g., by changes in the Earth's orbit or in the Sun itself), (2) changing the fraction of solar radiation that is reflected (this fraction is called the albedo – it can be changed, for example, by changes in cloud cover, small particles called aerosols or land cover), and (3) altering the long-wave energy radiated back to space (e.g., by changes in greenhouse gas concentrations). In addition, local climate also depends on how heat is distributed by winds and ocean currents. All of these factors have played a role in past climate changes.

Starting with the ice ages that have come and gone in regular cycles for the past nearly three million years, there is strong evidence that these are linked to regular variations in the Earth's orbit around the Sun, the so-called Milankovitch cycles (Figure 1). These cycles change the amount of solar radiation received at each latitude in each season (but hardly affect the global annual mean), and they can be calculated with astronomical precision. There is still some discussion about how exactly this starts and ends ice ages, but many studies suggest that the amount of summer sunshine on northern continents is crucial: if it drops below a critical value, snow from the past winter does not melt away in summer and an ice sheet starts to grow as more and more snow accumulates. Climate model simulations confirm that an Ice Age can indeed be started in this way, while simple conceptual models have been used to successfully 'hindcast' the onset of past glaciations based on the orbital changes. The next large reduction in northern summer insolation, similar to those that started past Ice Ages, is due to begin in 30,000 years.

Although it is not their primary cause, atmospheric carbon dioxide (CO_2) also plays an important role in the ice ages. Antarctic ice core data show that CO_2 concentration is low in the cold glacial times (~190 ppm), and high in the warm interglacials (~280 ppm); atmospheric CO_2 follows temperature changes in Antarctica with a lag of some hundreds of years. Because the climate changes at the beginning and end of ice ages take several thousand years,

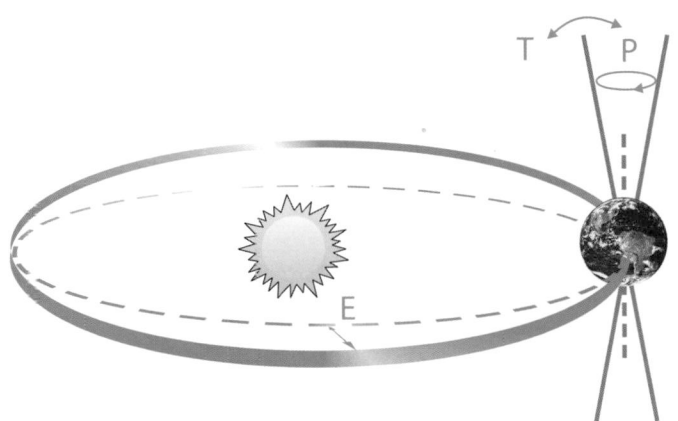

FAQ 6.1, Figure 1. *Schematic of the Earth's orbital changes (Milankovitch cycles) that drive the ice age cycles. 'T' denotes changes in the tilt (or obliquity) of the Earth's axis, 'E' denotes changes in the eccentricity of the orbit (due to variations in the minor axis of the ellipse), and 'P' denotes precession, that is, changes in the direction of the axis tilt at a given point of the orbit. Source: Rahmstorf and Schellnhuber (2006).*

most of these changes are affected by a positive CO_2 feedback; that is, a small initial cooling due to the Milankovitch cycles is subsequently amplified as the CO_2 concentration falls. Model simulations of ice age climate (see discussion in Section 6.4.1) yield realistic results only if the role of CO_2 is accounted for.

During the last ice age, over 20 abrupt and dramatic climate shifts occurred that are particularly prominent in records around the northern Atlantic (see Section 6.4). These differ from the glacial-interglacial cycles in that they probably do not involve large changes in global mean temperature: changes are not synchronous in Greenland and Antarctica, and they are in the opposite direction in the South and North Atlantic. This means that a major change in global radiation balance would not have been needed to cause these shifts; a redistribution of heat within the climate system would have sufficed. There is indeed strong evidence that changes in ocean circulation and heat transport can explain many features of these abrupt events; sediment data and model simulations show that some of these changes could have been triggered by instabilities in the ice sheets surrounding the Atlantic at the time, and the associated freshwater release into the ocean.

Much warmer times have also occurred in climate history – during most of the past 500 million years, Earth was probably completely free of ice sheets (geologists can tell from the marks ice leaves on rock), unlike today, when Greenland and Antarctica are ice-covered. Data on greenhouse gas abundances going back beyond a million years, that is, beyond the reach of antarctic ice cores, are still rather uncertain, but analysis of geological

(continued)

samples suggests that the warm ice-free periods coincide with high atmospheric CO_2 levels. On million-year time scales, CO_2 levels change due to tectonic activity, which affects the rates of CO_2 exchange of ocean and atmosphere with the solid Earth. See Section 6.3 for more about these ancient climates.

Another likely cause of past climatic changes is variations in the energy output of the Sun. Measurements over recent decades show that the solar output varies slightly (by close to 0.1%) in an 11-year cycle. Sunspot observations (going back to the 17th century), as well as data from isotopes generated by cosmic radiation, provide evidence for longer-term changes in solar activity. Data correlation and model simulations indicate that solar variability and volcanic activity are likely to be leading reasons for climate variations during the past millennium, before the start of the industrial era.

These examples illustrate that different climate changes in the past had different causes. The fact that natural factors caused climate changes in the past does not mean that the current climate change is natural. By analogy, the fact that forest fires have long been caused naturally by lightning strikes does not mean that fires cannot also be caused by a careless camper. FAQ 2.1 addresses the question of how human influences compare with natural ones in their contributions to recent climate change.

differentiation, regionally and with elevation. Pollen records with their extensive spatial coverage indicate that tropical lowlands were on average 2°C to 3°C cooler than present, with strong cooling (5°C–6°C) in Central America and northern South America and weak cooling (<2°C) in the western Pacific Rim (Farrera et al., 1999). Tropical highland cooling estimates derived from snow-line and pollen-based inferences show similar spatial variations in cooling although involving substantial uncertainties from dating and mapping, multiple climatic causes of treeline and snow line changes during glacial periods (Porter, 2001; Kageyama et al., 2004), and temporal asynchroneity between different regions of the tropics (Smith et al., 2005). These new studies give a much richer regional picture of tropical land cooling, and stress the need to use more than a few widely scattered proxy records as a measure of low-latitude climate sensitivity (Harrison, 2005).

The Climate: Long-range Investigation, Mapping, and Prediction (CLIMAP) reconstruction of ocean surface temperatures produced in the early 1980s indicated about 3°C cooling in the tropical Atlantic, and little or no cooling in the tropical Pacific. More pronounced tropical cooling for the LGM tropical oceans has since been proposed, including 4°C to 5°C based on coral skeleton records from off Barbados (Guilderson et al., 1994) and up to 6°C in the cold tongue off western South America based on foraminiferal assemblages (Mix et al., 1999). New data syntheses from multiple proxy types using carefully defined chronostratigraphies and new calibration data sets are now available from the Glacial Ocean Mapping (GLAMAP) and Multiproxy Approach for the Reconstruction of the Glacial Ocean surface (MARGO) projects, although with caveats including selective species dissolution, dating precision, non-analogue situations, and environmental preferences of the organisms (Sarnthein et al., 2003b; Kucera et al., 2005; and references therein). These recent reconstructions confirm moderate cooling, generally 0°C to 3.5°C, of tropical SST at the LGM, although with significant regional variation, as well as greater cooling in eastern boundary currents and equatorial upwelling regions. Estimates of cooling show notable differences among the different proxies. Faunal-based proxies argue for an intensification of the eastern equatorial Pacific cold tongue in contrast to Mg/Ca-based SST estimates that suggest a relaxation of SST gradients within the cold tongue (Mix et al., 1999; Koutavas et al., 2002; Rosenthal and Broccoli, 2004). Using a Bayesian approach to combine different proxies, Ballantyne et al. (2005) estimated a LGM cooling of tropical SSTs of 2.7°C ± 0.5°C (1 standard deviation).

These ocean proxy synthesis projects also indicate a colder glacial winter North Atlantic with more extensive sea ice than present, whereas summer sea ice only covered the glacial Arctic Ocean and Fram Strait with the northern North Atlantic and Nordic Seas largely ice free and more meridional ocean surface circulation in the eastern parts of the Nordic Seas (Sarnthein et al., 2003a; Meland et al., 2005; de Vernal et al., 2006). Sea ice around Antarctica at the LGM also responded with a large expansion of winter sea ice and substantial seasonal variation (Gersonde et al., 2005). Over mid- and high-latitude northern continents, strong reductions in temperatures produced southward displacement and major reductions in forest area (Bigelow et al., 2003), expansion of permafrost limits over northwest Europe (Renssen and Vandenberghe, 2003), fragmentation of temperate forests (Prentice et al., 2000; Williams et al., 2000) and predominance of steppe-tundra in Western Europe (Peyron et al., 2005). Temperature reconstructions from polar ice cores indicate strong cooling at high latitudes of about 9°C in Antarctica (Stenni et al., 2001) and about 21°C in Greenland (Dahl-Jensen et al., 1998).

The strength and depth extent of the LGM Atlantic overturning circulation have been examined through the application of a variety of new marine proxy indicators (Rutberg et al., 2000; Duplessy et al., 2002; Marchitto et al., 2002; McManus et al., 2004). These tracers indicate that the boundary between North Atlantic Deep Water (NADW) and Antarctic Bottom Water was much shallower during the LGM, with a reinforced pycnocline between intermediate and particularly cold and salty deep water (Adkins et al., 2002). Most of the deglaciation occurred over the period about 17 to 10 ka, the same period of maximum deglacial atmospheric CO_2 increase (Figure 6.4). It is thus very likely that the global warming of 4°C to 7°C since the LGM occurred at an average rate about 10 times slower than the warming of the 20th century.

In summary, significant progress has been made in the understanding of regional changes at the LGM with the development of new proxies, many new records, improved understanding of the relationship of the various proxies to climate variables and syntheses of proxy records into reconstructions with stricter dating and common calibrations.

6.4.1.3   How Realistic Are Results from Climate Model Simulations of the Last Glacial Maximum?

Model intercomparisons from the first phase of the Paleoclimate Modelling Intercomparison Project (PMIP-1), using atmospheric models (either with prescribed SST or with simple slab ocean models), were featured in the TAR. There are now six simulations of the LGM from the second phase (PMIP-2) using Atmosphere-Ocean General Circulation Models (AOGCMs) and EMICs, although only a few regional comparisons were completed in time for this assessment. The radiative perturbation for the PMIP-2 LGM simulations available for this assessment, which do not yet include the effects of vegetation or aerosol changes, is –4 to –7 W m⁻².

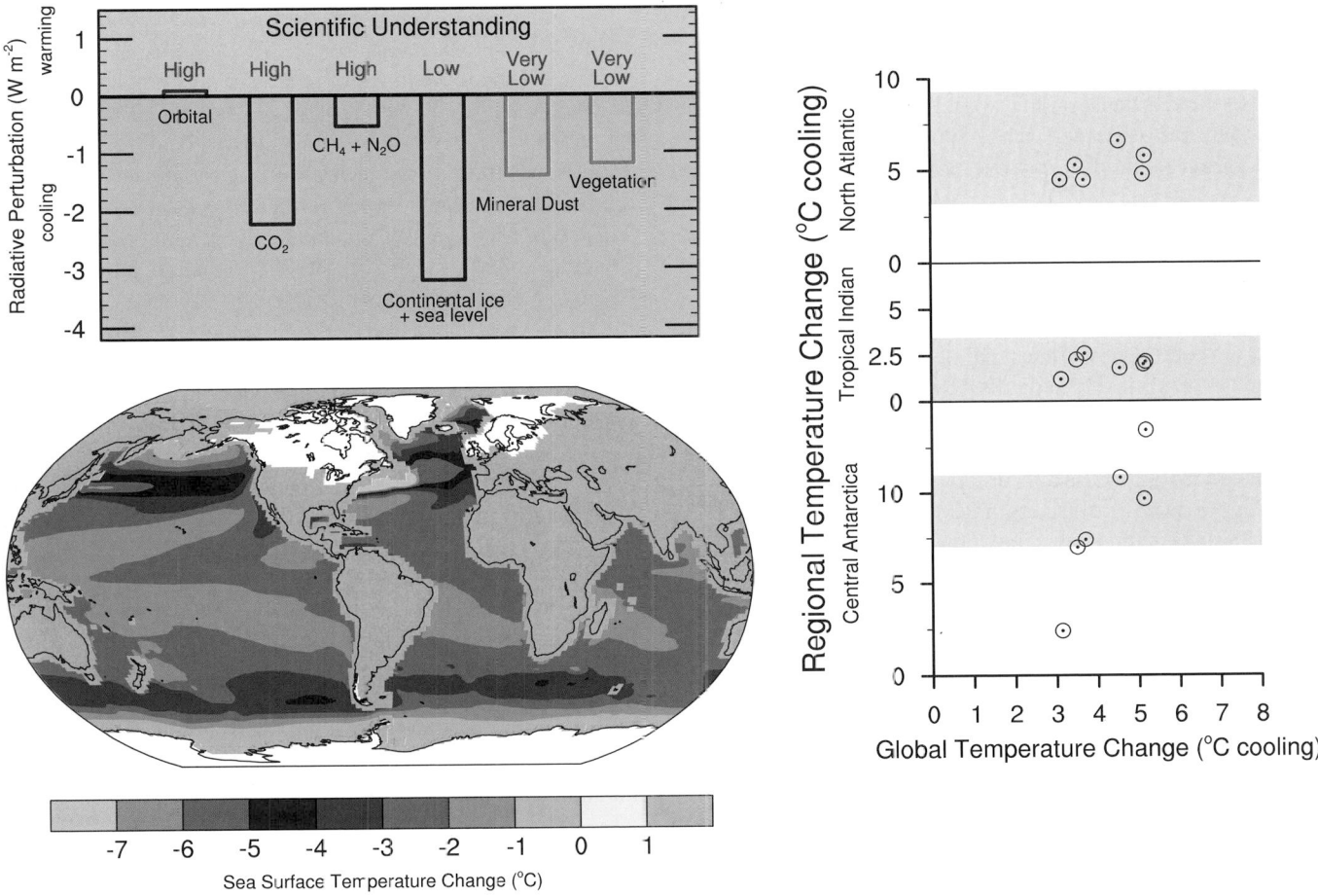

Figure 6.5. *The Last Glacial Maximum climate (approximately 21 ka) relative to the pre-industrial (1750) climate. (Top left) Global annual mean radiative influences (W m⁻²) of LGM climate change agents, generally feedbacks in glacial-interglacial cycles, but also specified in most Atmosphere-Ocean General Circulation Model (AOGCM) simulations for the LGM. The heights of the rectangular bars denote best estimate values guided by published values of the climate change agents and conversion to radiative perturbations using simplified expressions for the greenhouse gas concentrations and model calculations for the ice sheets, vegetation and mineral dust. References are included in the text. A judgment of each estimate's reliability is given as a level of scientific understanding based on uncertainties in the climate change agents and physical understanding of their radiative effects. Paleoclimate Modelling Intercomparison Project 2 (PMIP-2) simulations shown in bottom left and right panels do not include the radiative influences of LGM changes in mineral dust or vegetation. (Bottom left) Multi-model average SST change for LGM PMIP-2 simulations by five AOGCMs (Community Climate System Model (CCSM), Flexible Global Ocean-Atmosphere-Land System (FGOALS), Hadley Centre Coupled Model (HadCM), Institut Pierre Simon Laplace Climate System Model (IPSL-CM), Model for Interdisciplinary Research on Climate (MIROC)). Ice extent over continents is shown in white. (Right) LGM regional cooling compared to LGM global cooling as simulated in PMIP-2, with AOGCM results shown as red circles and EMIC (ECBilt-CLIO) results shown as blue circles. Regional averages are defined as: Antarctica, annual for inland ice cores; tropical Indian Ocean, annual for 15°S to 15°N, 50°E to 100°E; and North Atlantic Ocean, July to September for 42°N to 57°N, 35°W to 20°E. Grey shading indicates the range of observed proxy estimates of regional cooling: Antarctica (Stenni et al., 2001; Masson-Delmotte et al., 2006), tropical Indian Ocean (Rosell-Mele et al., 2004; Barrows and Juggins, 2005), and North Atlantic Ocean (Rosell-Mele et al., 2004; Kucera et al., 2005; de Vernal et al., 2006; Kageyama et al., 2006).*

These simulations allow an assessment of the response of a subset of the models presented in Chapters 8 and 10 to very different conditions at the LGM.

The PMIP-2 multi-model LGM SST change shows a modest cooling in the tropics, and greatest cooling at mid- to high latitudes in association with increases in sea ice and changes in ocean circulation (Figure 6.5). The PMIP-2 modelled strengthening of the SST meridional gradient in the LGM North Atlantic, as well as cooling and expanded sea ice, agrees with proxy indicators (Kageyama et al., 2006). Polar amplification of global cooling, as recorded in ice cores, is reproduced for Antarctica (Figure 6.5), but the strong LGM cooling over Greenland is underestimated, although with caveats about the heights of these ice caps in the PMIP-2 simulations (Masson-Delmotte et al., 2006).

The PMIP-2 AOGCMs give a range of tropical ocean cooling between 15°S to 15°N of 1.7°C to 2.4°C. Sensitivity simulations with models indicate that this tropical cooling can be explained by the reduced glacial greenhouse gas concentrations, which had direct effects on the tropical radiative forcing (Shin et al., 2003; Otto-Bliesner et al., 2006b) and indirect effects through LGM cooling by positive sea ice-albedo feedback in the Southern Ocean contributing to enhanced ocean ventilation of the tropical thermocline and the intermediate waters (Liu et al., 2002). Regional variations in simulated tropical cooling are much smaller than indicated by MARGO data, partly related to models at current resolutions being unable to simulate the intensity of coastal upwelling and eastern boundary currents. Simulated cooling in the Indian Ocean (Figure 6.5), a region with important present-day teleconnections to Africa and the North Atlantic, compares favourably to proxy estimates from alkenones (Rosell-Mele et al., 2004) and foraminifera assemblages (Barrows and Juggins, 2005).

Considering changes in vegetation appears to improve the realism of simulations of the LGM, and points to important climate-vegetation feedbacks (Wyputta and McAvaney, 2001; Crucifix and Hewitt, 2005). For example, extension of the tundra in Asia during the LGM contributes to the local surface cooling, while the tropics warm where savannah replaces tropical forest (Wyputta and McAvaney, 2001). Feedbacks between climate and vegetation occur locally, with a decrease in the tree fraction in central Africa reducing precipitation, and remotely with cooling in Siberia (tundra replacing trees) altering (diminishing) the Asian summer monsoon. The physiological effect of CO_2 concentration on vegetation needs to be included to properly represent changes in global forest (Harrison and Prentice, 2003), as well as to widen the climatic range where grasses and shrubs dominate. The biome distribution simulated with dynamic global vegetation models reproduces the broad features observed in palaeodata (e.g., Harrison and Prentice, 2003).

In summary, the PMIP-2 LGM simulations confirm that current AOGCMs are able to simulate the broad-scale spatial patterns of regional climate change recorded by palaeodata in response to the radiative forcing and continental ice sheets of the LGM, and thus indicate that they adequately represent

the primary feedbacks that determine the climate sensitivity of this past climate state to these changes. The PMIP-2 AOGCM simulations using glacial-interglacial changes in greenhouse gas forcing and ice sheet conditions give a radiative perturbation in reference to pre-industrial conditions of –4.6 to –7.2 W m^{-2} and mean global temperature change of –3.3°C to –5.1°C, similar to the range reported in the TAR for PMIP-1 (IPCC, 2001). The climate sensitivity inferred from the PMIP-2 LGM simulations is 2.3°C to 3.7°C for a doubling of atmospheric CO_2 (see Section 9.6.3.2). When the radiative perturbations of dust content and vegetation changes are estimated, climate models yield an additional cooling of 1°C to 2°C (Crucifix and Hewitt, 2005; Schneider et al., 2006), although scientific understanding of these effects is very low.

6.4.1.4   How Realistic Are Simulations of Terrestrial Carbon Storage at the Last Glacial Maximum?

There is evidence that terrestrial carbon storage was reduced during the LGM compared to today. Mass balance calculations based on ^{13}C measurements on shells of benthic foraminifera yield a reduction in the terrestrial biosphere carbon inventory (soil and living vegetation) of about 300 to 700 GtC (Shackleton, 1977; Bird et al., 1994) compared to the pre-industrial inventory of about 3,000 GtC. Estimates of terrestrial carbon storage based on ecosystem reconstructions suggest an even larger difference (e.g., Crowley, 1995). Simulations with carbon cycle models yield a reduction in global terrestrial carbon stocks of 600 to 1,000 GtC at the LGM compared to pre-industrial time (Francois et al., 1998; Beerling, 1999; Francois et al., 1999; Kaplan et al., 2002; Liu et al., 2002; Kaplan et al., 2003; Joos et al., 2004). The majority of this simulated difference is due to reduced simulated growth resulting from lower atmospheric CO_2. A major regulating role for CO_2 is consistent with the model-data analysis of Bond et al. (2003), who suggested that low atmospheric CO_2 could have been a significant factor in the reduction of trees during glacial times, because of their slower regrowth after disturbances such as fire. In summary, results of terrestrial models, also used to project future CO_2 concentrations, are broadly compatible with the range of reconstructed differences in glacial-interglacial carbon storage on land.

6.4.1.5   How Long Did the Previous Interglacials Last?

The four interglacials of the last 450 kyr preceding the Holocene (Marine Isotope Stages 5, 7, 9 and 11) were all different in multiple aspects, including duration (Figure 6.3). The shortest (Stage 7) lasted a few thousand years, and the longest (Stage 11; ~420 to 395 ka) lasted almost 30 kyr. Evidence for an unusually long Stage 11 has been recently reinforced by new ice core and marine sediment data. The European Programme for Ice Coring in Antarctica (EPICA) Dome C antarctic ice core record suggests that antarctic temperature remained approximately as warm as the Holocene for 28 kyr (EPICA community members, 2004). A new stack of 57 globally distributed benthic $\delta^{18}O$ records presents age estimates at Stage 11 nearly identical to those provided by the EPICA results (Lisiecki and Raymo, 2005).

It has been suggested that Stage 11 was an extraordinarily long interglacial period because of its low orbital eccentricity, which reduces the effect of climatic precession on insolation (Box 6.1) (Berger and Loutre, 2003). In addition, the EPICA Dome C and the recently revisited Vostok records show CO_2 concentrations similar to pre-industrial Holocene values throughout Stage 11 (Raynaud et al., 2005). Thus, both the orbital forcing and the CO_2 feedback were providing favourable conditions for an unusually long interglacial. Moreover, the length of Stage 11 has been simulated by conceptual models of the Quaternary climate, based on threshold mechanisms (Paillard, 1998). For Stage 11, these conceptual models show that the deglaciation was triggered by the insolation maximum at about 427 ka, but that the next insolation minimum was not sufficiently low to start another glaciation. The interglacial thus lasts an additional precessional cycle, yielding a total duration of 28 kyr.

6.4.1.6   How Much Did the Earth Warm During the Previous Interglacial?

Globally, there was less glacial ice on Earth during the Last Interglacial, also referred to as "Last Interglaciation" (LIG, 130 ± 1 to 116 ± 1 ka; Stirling et al., 1998) than now. This suggests significant reduction in the size of the Greenland and possibly Antarctic Ice Sheets (see Section 6.4.3). The climate of the LIG has been inferred to be warmer than present (Kukla et al., 2002), although the evidence is regional and not necessarily synchronous globally, consistent with understanding of the primary forcing. For the first half of this interglacial (~130–123 ka), orbital forcing (Box 6.1) produced a large increase in NH summer insolation. Proxy data indicate warmer-than-present coastal waters in parts of the Pacific, Atlantic, and Indian Oceans as well as in the Mediterranean Sea, greatly reduced sea ice in the coastal waters around Alaska, extension of boreal forest into areas now occupied by tundra in interior Alaska and Siberia and a generally warmer Arctic (Brigham-Grette and Hopkins, 1995; Lozhkin and Anderson, 1995; Muhs et al., 2001, CAPE Last Interglacial Project Members, 2006). Ice core data indicate a large response over Greenland and Antarctica with early LIG temperatures 3°C to 5°C warmer than present (Watanabe et al., 2003; NGRIP, 2004; Landais et al., 2006). Palaeofauna evidence from New Zealand indicates LIG warmth during the late LIG consistent with the latitudinal dependence of orbital forcing (Marra, 2003).

There are AOGCM simulations available for the LIG, but no standardised intercomparison simulations have been performed. When forced with orbital forcing of 130 to 125 ka (Box 6.1), with over 10% more summer insolation in the NH than today, AOGCMs produce a summer arctic warming of up to 5°C, with greatest warming over Eurasia and in the Baffin Island/northern Greenland region (Figure 6.6) (Montoya et al., 2000; Kaspar et al., 2005; Otto-Bliesner et al., 2006a). Simulations generally match proxy reconstructions of the maximum arctic summer warmth (Kaspar and Cubasch, 2006; CAPE Last Interglacial Project Members, 2006) although may still underestimate warmth in Siberia because vegetation feedbacks are not included in current simulations. Simulated LIG annual average global temperature is not notably higher than present, consistent with the orbital forcing.

6.4.1.7   What Is Known About the Mechanisms of Transitions Into Ice Ages?

Successful simulation of glacial inception has been a key target for models simulating climate change. The Milankovitch theory proposes that ice ages were triggered by reduced summer insolation at high latitudes in the NH, enabling winter snowfall to persist all year and accumulate to build NH glacial ice sheets (Box 6.1). Continental ice sheet growth and associated sea level lowering took place at about 116 ka (Waelbroeck et al., 2002) when the summer incoming solar radiation in the NH at high latitudes reached minimum values. The inception took place while the continental ice volume was minimal and stable, and low and mid-latitudes of the North Atlantic continuously warm (Cortijo et al., 1999; Goni et al., 1999; McManus et al., 2002; Risebrobakken et al., 2005). When forced with orbital insolation changes, atmosphere-only models failed in the past to find the proper magnitude of response to allow for perennial snow cover. Models and data now show that shifts in the northern treeline, expansion of sea ice at high latitudes and warmer low-latitude oceans as a source of moisture for the ice sheets provide feedbacks that amplify the local insolation forcing over the high-latitude continents and allow for growth of ice sheets (Pons et al., 1992; Cortijo et al., 1999; Goni et al., 1999; Crucifix and Loutre, 2002; McManus et al., 2002; Jackson and Broccoli, 2003; Khodri et al., 2003; Meissner et al., 2003; Vettoretti and Peltier, 2003; Khodri et al., 2005; Risebrobakken et al., 2005). The rapid growth of ice sheets after inception is captured by EMICs that include models for continental ice, with increased Atlantic Meridional Overturning Circulation (MOC) allowing for increased snowfall. Increasing ice sheet altitude and extent is also important, although the ice volume-equivalent drop in sea level found in data records (Waelbroeck et al., 2002; Cutler et al., 2003) is not well reproduced in some EMIC simulations (Wang and Mysak, 2002; Kageyama et al., 2004; Calov et al., 2005).

6.4.1.8   When Will the Current Interglacial End?

There is no evidence of mechanisms that could mitigate the current global warming by a natural cooling trend. Only a strong reduction in summer insolation at high northern latitudes, along with associated feedbacks, can end the current interglacial. Given that current low orbital eccentricity will persist over the next tens of thousand years, the effects of precession are minimised, and extremely cold northern summer orbital configurations like that of the last glacial initiation at 116 ka will not take place for at least 30 kyr (Box 6.1). Under a natural CO_2 regime (i.e., with the global temperature-CO_2 correlation continuing as in the Vostok and EPICA Dome C ice cores), the next glacial period would not be expected to start within the next 30 kyr (Loutre and Berger, 2000; Berger and Loutre, 2002; EPICA Community Members, 2004). Sustained high atmospheric greenhouse gas concentrations, comparable to a mid-range CO_2 stabilisation scenario, may lead to a complete melting of the Greenland Ice Sheet (Church et al., 2001) and further delay the onset of the next glacial period (Loutre and Berger, 2000; Archer and Ganopolski, 2005).

6.4.2 Abrupt Climatic Changes in the Glacial-Interglacial Record

6.4.2.1 *What Is the Evidence for Past Abrupt Climate Changes?*

Abrupt climate changes have been variously defined either simply as large changes within less than 30 years (Clark et al., 2002), or in a physical sense, as a threshold transition or a response that is rapid compared to forcing (Rahmstorf, 2001; Alley et al., 2003). Overpeck and Trenberth (2004) noted that not all abrupt changes need to be externally forced. Numerous terrestrial, ice and oceanic climatic records show that large, widespread, abrupt climate changes have occurred repeatedly throughout the past glacial interval (see review by Rahmstorf,

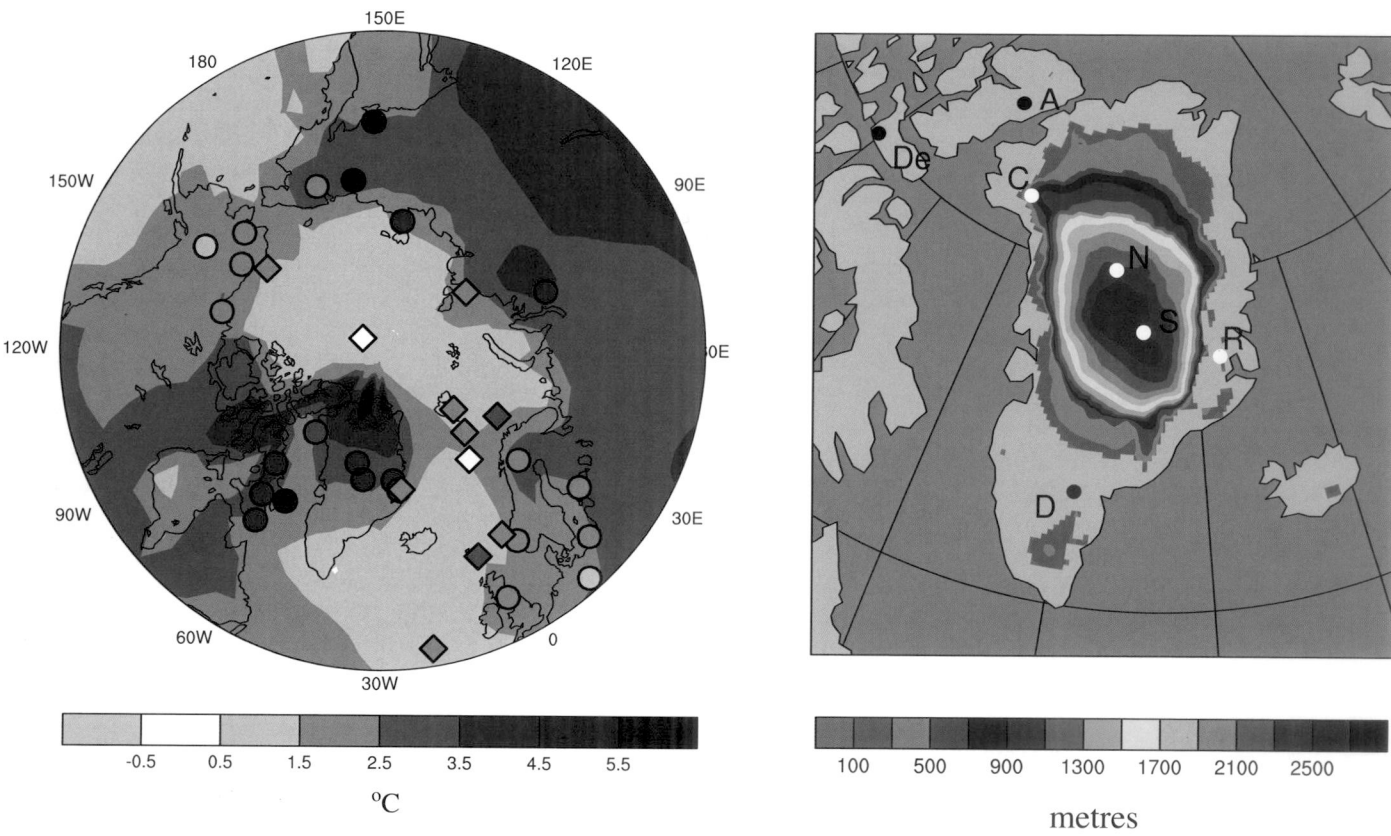

Figure 6.6. *Summer surface air temperature change over the Arctic (left) and annual minimum ice thickness and extent for Greenland and western arctic glaciers (right) for the LIG from a multi-model and a multi-proxy synthesis. The multi-model summer warming simulated by the National Center for Atmospheric Research (NCAR) Community Climate System Model (CCSM), 130 ka minus present (Otto-Bliesner et al., 2006b), and the ECHAM4 HOPE-G (ECHO-G) model, 125 ka minus pre-industrial (Kaspar et al., 2005), is contoured in the left panel and is overlain by proxy estimates of maximum summer warming from terrestrial (circles) and marine (diamonds) sites as compiled in the syntheses published by the CAPE Project Members (2006) and Kaspar et al. (2005). Extents and thicknesses of the Greenland Ice Sheet and eastern Canadian and Iceland glaciers are shown at their minimum extent for the LIG as a multi-model average from three ice models (Tarasov and Peltier, 2003; Lhomme et al., 2005a; Otto-Bliesner et al., 2006a). Ice core observations (Koerner, 1989; NGRIP, 2004) indicate LIG ice (white dots) at Renland (R), North Greenland Ice Core Project (N), Summit (S, Greenland Ice Core Project and Greenland Ice Sheet Project 2) and possibly Camp Century (C), but no LIG ice (black dots) at Devon (De) and Agassiz (A) in the eastern Canadian Arctic. Evidence for LIG ice at Dye-3 (D) in southern Greenland is equivocal (grey dot; see text for detail).*

2002). High-latitude records show that ice age abrupt temperature events were larger and more widespread than were those of the Holocene. The most dramatic of these abrupt climate changes were the Dansgaard-Oeschger (D-O) events, characterised by a warming in Greenland of 8°C to 15°C within a few decades (see Severinghaus and Brook, 1999; Masson-Delmotte et al., 2005a for a review) followed by much slower cooling over centuries. Another type of abrupt change were the Heinrich events; characterised by large discharges of icebergs into the northern Atlantic leaving diagnostic drop-stones in the ocean sediments (Hemming, 2004). In the North Atlantic, Heinrich events were accompanied by a strong reduction in sea surface salinity (Bond et al., 1993), as well as a sea surface cooling on a centennial time scale. Such ice age cold periods lasted hundreds to thousands of years, and the warming that ended them took place within decades (Figure 6.7; Cortijo et al., 1997; Voelker, 2002). At the end of the last glacial, as the climate warmed and ice sheets melted, climate went through a number of abrupt cold phases, notably the Younger Dryas and the 8.2 ka event.

The effects of these abrupt climate changes were global, although out-of-phase responses in the two hemispheres (Blunier et al., 1998; Landais et al., 2006) suggest that they were not primarily changes in global mean temperature. The highest amplitude of the changes, in terms of temperature, appears centred around the North Atlantic. Strong and fast changes are found in the global CH_4 concentration (of the order of 100 to 150 ppb within decades), which may point to changes in the extent or productivity of tropical wetlands (see Chappellaz et al., 1993; Brook et al., 2000 for a review; Masson-Delmotte et al., 2005a), and in the Asian monsoon (Wang et al., 2001). The NH cold phases were linked with a reduced northward flow of warm waters in the Nordic Seas (Figure 6.7), southward shift of the Inter-Tropical Convergence Zone (ITCZ) and thus the location of the tropical rainfall belts (Peterson et al., 2000; Lea et al., 2003). Cold, dry and windy conditions with low CH_4 and high dust aerosol concentrations generally occurred together in the NH cold events. The accompanying changes in atmospheric CO_2 content were relatively small (less than 25 ppm; Figure 6.7) and parallel to the antarctic counterparts of Greenland D-O events. The record in N_2O is less complete and shows an increase of about 50 ppb and a decrease of about 30 ppb during warm and cold periods, respectively (Flückiger et al., 2004).

A southward shift of the boreal treeline and other rapid vegetation responses were associated with past cold events (Peteet, 1995; Shuman et al., 2002; Williams et al., 2002). Decadal-scale changes in vegetation have been recorded in annually laminated sequences at the beginning and the end of the Younger Dryas and the 8.2 ka event (Birks and Ammann, 2000; Tinner and Lotter, 2001; Veski et al., 2004). Marine pollen records with a typical sampling resolution of 200 years provide unequivocal evidence of the immediate response of vegetation in Southern Europe to the climate fluctuations during glacial times (Sánchez Goñi et al., 2002; Tzedakis, 2005). The same holds true for the vegetation response in northern South America during the last deglaciation (Hughen et al., 2004).

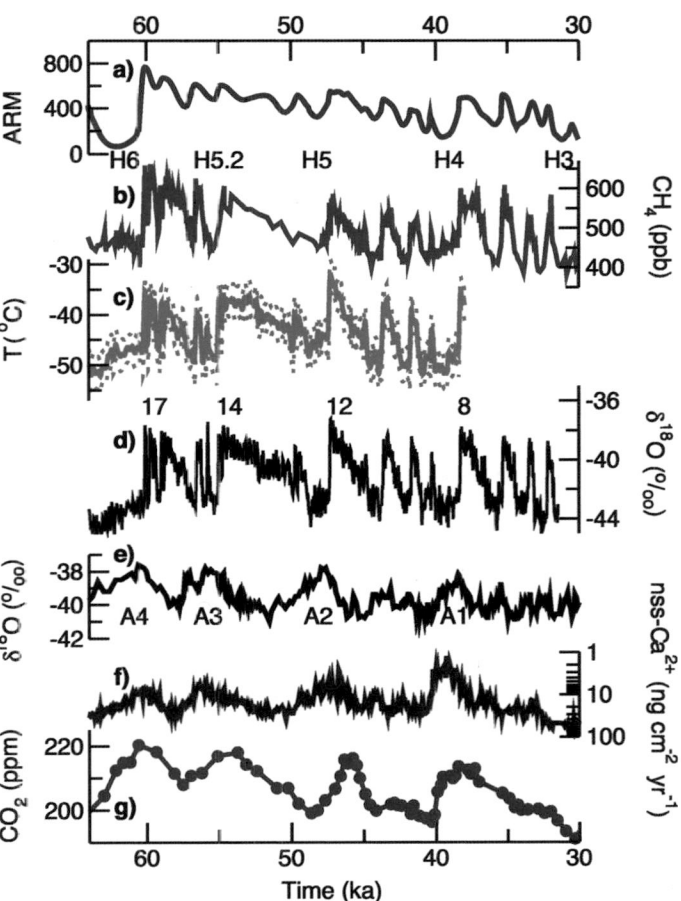

Figure 6.7. *The evolution of climate indicators from the NH (panels a to d), and from Antarctica (panels e to g), over the period 64 to 30 ka. (a) Anhysteretic remanent magnetisation (ARM), here a proxy of the northward extent of Atlantic MOC, from an ocean sediment core from the Nordic Seas (Dokken and Jansen, 1999); (b) CH_4 as recorded in Greenland ice cores at the Greenland Ice Core Project (GRIP), Greenland Ice Sheet Project (GISP) and North GRIP (NGRIP) sites (Blunier and Brook, 2001; Flückiger et al., 2004; Huber et al., 2006); CH_4 data for the period 40 to 30 ka were selected for the GRIP site and for 64 to 40 ka for the GISP site when sample resolution is highest in the cores; (c) surface temperature estimated from nitrogen isotope ratios that are influenced by thermal diffusion (Huber et al., 2006); (d) $\delta^{18}O$, a proxy for surface temperature, from NGRIP (2004) with the D-O NH warm events 8, 12, 14 and 17 indicated; (e) $\delta^{18}O$ from Byrd, Antarctica (Blunier and Brook, 2001) with A1 to A4 denoting antarctic warm events; (f) nss-Ca^{2+}, a proxy of dust and iron deposition, from Dome C, Antarctica (Röthlisberger et al., 2004); and (g) CO_2 as recorded in ice from Taylor Dome, Antarctica (Indermühle et al., 2000). The Heinrich events (periods of massive ice-rafted debris recorded in marine sediments) H3, H4, H5, H5.2, and H6, are shown. All data are plotted on the Greenland SS09sea time scale (Johnsen et al., 2001). CO_2 and CH_4 are well mixed in the atmosphere. CH_4 variations are synchronous within the resolution of ±50 years with variations in Greenland temperature, but a detailed analysis suggests that CH_4 rises lag temperature increases at the onset of the D-O events by 25 to 70 years (Huber et al., 2006). CO_2 co-varied with the antarctic temperature, but the exact synchronisation between Taylor Dome and Byrd is uncertain, thus making the determination of leads or lags between temperature and CO_2 elusive. The evolution of Greenland and antarctic temperature is consistent with a reorganisation of the heat transport and the MOC in the Atlantic (Knutti et al., 2004).*

6.4.2.2   What Is Known About the Mechanism of these Abrupt Changes?

There is good evidence now from sediment data for a link between these glacial-age abrupt changes in surface climate and ocean circulation changes (Clark et al., 2002). Proxy data show that the South Atlantic cooled when the north warmed (with a possible lag), and vice versa (Voelker, 2002), a seesaw of NH and SH temperatures that indicates an ocean heat transport change (Crowley, 1992; Stocker and Johnsen 2003). During D-O warming, salinity in the Irminger Sea increased strongly (Elliot et al., 1998; van Kreveld et al., 2000) and northward flow of temperate waters increased in the Nordic Seas (Dokken and Jansen, 1999), indicative of saline Atlantic waters advancing northward. Abrupt changes in deep water properties of the Atlantic have been documented from proxy data (e.g., ^{13}C, $^{231}Pa/^{230}Th$), which reconstruct the ventilation of the deep water masses and changes in the overturning rate and flow speed of the deep waters (Vidal et al., 1998; Dokken and Jansen, 1999; McManus et al., 2004; Gherardi et al., 2005). Despite this evidence, many features of the abrupt changes are still not well constrained due to a lack of precise temporal control of the sequencing and phasing of events between the surface, the deep ocean and ice sheets.

Heinrich events are thought to have been caused by ice sheet instability (MacAyeal, 1993). Iceberg discharge would have provided a large freshwater forcing to the Atlantic, which can be estimated from changes in the abundance of the isotope ^{18}O. These yield a volume of freshwater addition typically corresponding to a few (up to 15) metres of global sea level rise occurring over several centuries (250–750 years), that is, a flux of the order of 0.1 Sv (Hemming, 2004). For Heinrich event 4, Roche et al. (2004) have constrained the freshwater amount to 2 ± 1 m of sea level equivalent provided by the Laurentide Ice Sheet, and the duration of the event to 250 ± 150 years. Volume and timing of freshwater release is still controversial, however.

Freshwater influx is the likely cause for the cold events at the end of the last ice age (i.e., the Younger Dryas and the 8.2 ka event). Rather than sliding ice, it is the inflow of melt water from melting ice due to the climatic warming at this time that could have interfered with the MOC and heat transport in the Atlantic – a discharge into the Arctic Ocean of the order 0.1 Sv may have triggered the Younger Dryas (Tarasov and Peltier, 2005), while the 8.2 ka event was probably linked to one or more floods equal to 11 to 42 cm of sea level rise within a few years (Clarke et al., 2004; see Section 6.5.2). This is an important difference relative to the D-O events, for which no large forcing of the ocean is known; model simulations suggest that a small forcing may be sufficient if the ocean circulation is close to a threshold (Ganopolski and Rahmstorf, 2001). The exact cause and nature of these ocean circulation changes, however, are not universally agreed. Some authors have argued that some of the abrupt climate shifts discussed could have been triggered from the tropics (e.g., Clement and Cane, 1999), but a more specific and quantitative explanation for D-O events building on this idea is yet to emerge.

Atmospheric CO_2 changes during the glacial antarctic warm events, linked to changes in NADW (Knutti et al., 2004), were small (less than 25 ppm; Figure 6.7). A relatively small positive feedback between atmospheric CO_2 and changes in the rate of NADW formation is found in palaeoclimate and global warming simulations (Joos et al., 1999; Marchal et al., 1999). Thus, palaeodata and available model simulations agree that possible future changes in the NADW formation rate would have only modest effects on atmospheric CO_2. This finding does not, however, preclude the possibility that circulation changes in other ocean regions, in particular in the Southern Ocean, could have a larger impact on atmospheric CO_2 (Greenblatt and Sarmiento, 2004).

6.4.2.3   Can Climate Models Simulate these Abrupt Changes?

Modelling the ice sheet instabilities that are the likely cause of Heinrich events is a difficult problem because the physics are not sufficiently understood, although recent results show some promise (Calov et al., 2002). Many model studies have been performed in which an influx of freshwater from an ice sheet instability (Heinrich event) or a melt water release (8.2 ka event; see Section 6.5.2) has been assumed and prescribed, and its effects on ocean circulation and climate have been simulated. These experiments suggest that freshwater input of the order of magnitude deduced from palaeoclimatic data could indeed have caused the Atlantic MOC to shut down, and that this is a physically viable explanation for many of the climatic repercussions found in the data (e.g., the high-latitude northern cooling, the shift in the ITCZ and the hemispheric seesaw; Vellinga and Wood, 2002; Dahl et al., 2005; Zhang and Delworth, 2005). The phase relation between temperature in Greenland and Antarctic has been explained by a reduction in the NADW formation rate and oceanic heat transport into the North Atlantic region, producing cooling in the North Atlantic and a lagged warming in the SH (Ganopolski and Rahmstorf, 2001; Stocker and Johnsen, 2003). In freshwater simulations where the North Atlantic MOC is forced to collapse, the consequences also include an increase in nutrient-rich water in the deep Atlantic Ocean, higher $^{231}Pa/^{230}Th$ ratios in North Atlantic sediments (Marchal et al., 2000), a retreat of the northern treeline (Scholze et al., 2003; Higgins, 2004; Köhler et al., 2005), a small (10 ppm) temporary increase in atmospheric CO_2 in response to a reorganisation of the marine carbon cycle (Marchal et al., 1999) and CO_2 changes of a few parts per million due to carbon stock changes in the land biosphere (Köhler et al., 2005). A 10 ppb reduction in atmospheric N_2O is found in one ocean-atmosphere model (Goldstein et al., 2003), suggesting that part of the measured N_2O variation (up to 50 ppb) is of terrestrial origin. In summary, model simulations broadly reproduce the observed variations during abrupt events of this type.

Dansgaard-Oeschger events appear to be associated with latitudinal shifts in oceanic convection between the Nordic Seas and the open mid-latitude Atlantic (Alley and Clark, 1999). Models suggest that the temperature evolution in Greenland, the seesaw response in the South Atlantic, the observed Irminger Sea salinity changes and other observed features of the events may be explained by such a mechanism (Ganopolski and Rahmstorf, 2001), although the trigger for the ocean circulation changes remains undetermined. Alley et al. (2001) showed evidence for a stochastic resonance process at work in the timing of these events, which means that a regular cycle together with random 'noise' could have triggered them. This can be reproduced in models (e.g., the above), as long as a threshold mechanism is involved in causing the events.

Some authors have argued that climate models tend to underestimate the size and extent of past abrupt climate changes (Alley et al., 2003), and hence may underestimate the risk of future ones. However, such a general conclusion is probably too simple, and a case-by-case evaluation is required to understand which effects may be misinterpreted in the palaeoclimatic record and which mechanisms may be underestimated in current models. This issue is important for an assessment of risks for the future: the expected rapid warming in the coming centuries could approach the amount of warming at the end of the last glacial, and would occur at a much faster rate. Hence, melt water input from ice sheets could again become an important factor influencing the ocean circulation, as for the Younger Dryas and 8.2 ka events. A melting of the Greenland Ice Sheet (equivalent to 7 m of global sea level) over 1 kyr would contribute an average freshwater flux of 0.1 Sv; this is a comparable magnitude to the estimated freshwater fluxes associated with past abrupt climate events. Most climate models used for future scenarios have thus far not included melt water runoff from melting ice sheets. Intercomparison experiments subjecting different models to freshwater influx have revealed that while responses are qualitatively similar, the amount of freshwater needed for a shutdown of the Atlantic circulation can differ greatly between models; the reasons for this model dependency are not yet fully understood (Rahmstorf et al., 2005; Stouffer et al., 2006). Given present knowledge, future abrupt climate changes due to ocean circulation changes cannot be ruled out.

6.4.3   Sea Level Variations Over the Last Glacial-Interglacial Cycle

6.4.3.1   *What Is the Influence of Past Ice Volume Change on Modern Sea Level Change?*

Palaeorecords of sea level history provide a crucial basis for understanding the background variations upon which the sea level rise related to modern processes is superimposed. Even if no anthropogenic effect were currently operating in the climate system, measurable and significant changes in relative sea level (RSL) would still be occurring. The primary cause of this natural variability in sea level has to do with the planet's memory of the last deglaciation event. Through the so-called glacial isostatic adjustment (GIA) process, gravitational equilibrium is restored

following deglaciation, not only by crustal 'rebound', but also through the horizontal redistribution of water in the ocean basins required to maintain the ocean surface at gravitational equipotential.

Models of the global GIA process have enabled isolation of a contribution to the modern rate of global sea level rise being measured by the TOPography EXperiment (TOPEX)/Poseidon (T/P) satellite of -0.28 mm yr^{-1} for the ICE-4G(VM2) model of Peltier (1996) and -0.36 mm yr^{-1} for the ICE-5G(VM2) model of Peltier (2004). These analyses (Peltier, 2001) imply that the impact of modern climate change on the global rate of sea level rise is larger than implied by the uncorrected T/P measurements (see also Chapter 5).

By employing the same theory to predict the impact upon Earth's rotational state due to both the Late Pleistocene glacial cycle and the influence of present-day melting of the great polar ice sheets on Greenland and Antarctica, it has also proven possible to estimate the extent to which these ice sheets may have been losing mass over the past century. In Peltier (1998), such analysis led to an upper-bound estimate of approximately 0.5 mm yr^{-1} for the rate of global sea level rise equivalent to the mass loss. This suggests the plausibility of the notion that polar ice sheet and glacier melting may provide the required closure of the global sea level rise budget (see Chapters 4 and 5).

6.4.3.2   *What Was the Magnitude of Glacial-Interglacial Sea Level Change?*

Model-based palaeo-sea level analysis also helps to refine estimates of the eustatic (globally averaged) sea level rise that occurred during the most recent glacial-interglacial transition from the LGM to the Holocene. The extended coral-based RSL curve from the island of Barbados in the Caribbean Sea (Fairbanks, 1989; Peltier and Fairbanks, 2006) is especially important, as the RSL history from this site has been shown to provide a good approximation to the ice-equivalent eustatic curve itself (Peltier, 2002). The fit of the prediction of the ICE-5G(VM2) model to the Fairbanks data set, as shown in Figure 6.8b, constrains the net ice-equivalent eustatic rise subsequent to 21 ka to a value of 118.7 m, very close to the value of approximately 120 m conventionally inferred (e.g., Shackleton, 2000) on the basis of deep-sea O isotopic information (Figure 6.8b). Waelbroeck et al. (2002) produced a sea level reconstruction based upon coral records and deep-sea O isotopes corrected for the influence of abyssal ocean temperature changes for the entire glacial-interglacial cycle. This record (Figure 6.8a) is characterised by a best estimate of the LGM depression of ice-equivalent eustatic sea level that is also near 120 m. The analysis of the Red Sea O isotopic record by Siddal et al. (2003) further supports the validity of the interpretation of the extended Barbados record by Peltier and Fairbanks (2006).

The ice-equivalent eustatic sea level curve of Lambeck and Chappell (2001), based upon data from a variety of different

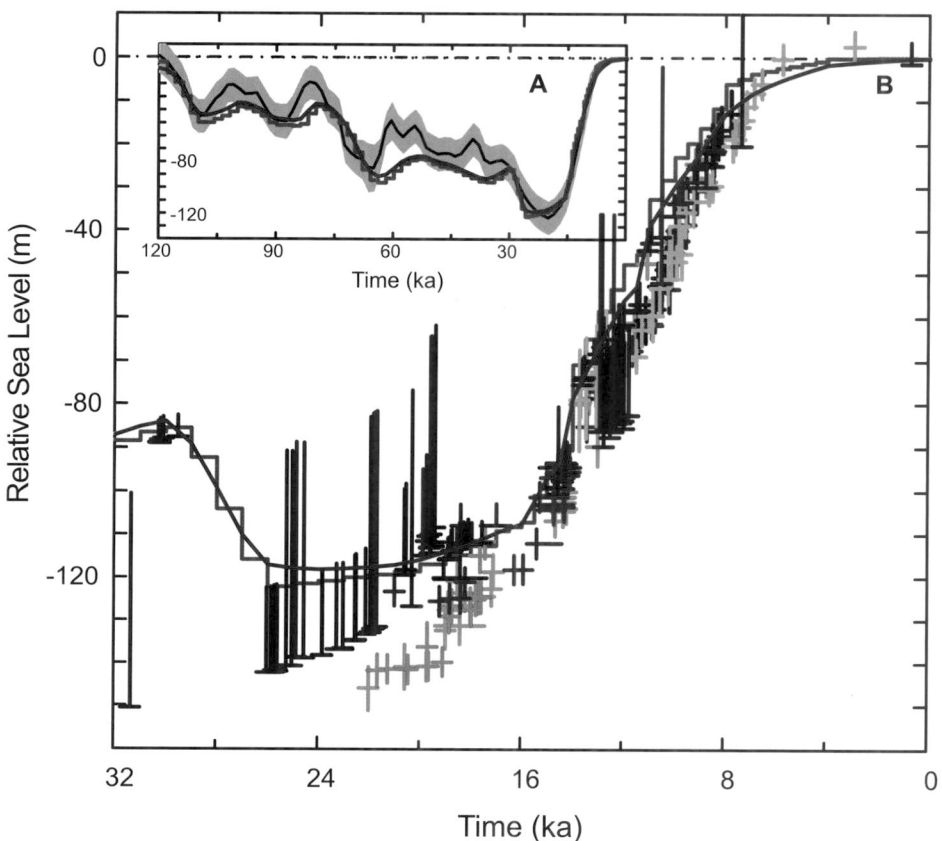

Figure 6.8. *(A) The ice-equivalent eustatic sea level history over the last glacial-interglacial cycle according to the analysis of Waelbroeck et al. (2002). The smooth black line defines the mid-point of their estimates for each age and the surrounding hatched region provides an estimate of error. The red line is the prediction of the ICE-5G(VM2) model for the Barbados location for which the RSL observations themselves provide an excellent approximation to the ice-equivalent eustatic sea level curve. (B) The fit of the ICE-5G(VM2) model prediction (red line) to the extended coral-based record of RSL history from the island of Barbados in the Caribbean Sea (Fairbanks, 1989; Peltier and Fairbanks, 2006) over the age range from 32 ka to present. The actual ice-equivalent eustatic sea level curve for this model is shown as the step-discontinuous brown line. The individual coral-based estimates of RSL (blue) have an attached error bar that depends upon the coral species. The estimates denoted by the short error bars are derived from the* Acropora palmata *species, which provide the tightest constraints upon relative sea level as this species is found to live within approximately 5 m of sea level in the modern ecology. The estimates denoted by the longer error bars are derived either from the* Montastrea annularis *species of coral (error bars of intermediate 20 m length) or from further species that are found over a wide range of depths with respect to sea level (longest error bars). These additional data are most useful in providing a lower bound for the sea level depression. The data denoted by the coloured crosses are from the ice-equivalent eustatic sea level reconstruction of Lambeck and Chappell (2001) for Barbados (cyan), Tahiti (grey), Huon (black), Bonaparte Gulf (orange) and Sunda Shelf (magenta).*

sources, including the Barbados coral record, measurements from the Sunda Shelf of Indonesia (Hanebuth et al., 2000) and observations from the Bonaparte Gulf of northern Australia (Yokoyama et al., 2000), is also shown in Figure 6.8b. This suggests an ice-equivalent eustatic sea level history that conflicts with that based upon the extended Barbados record. First, the depth of the low stand of the sea at the LGM is approximately 140 m below present sea level rather than the value of approximately 120 m required by the Barbados data set. Second, the Barbados data appear to rule out the possibility of the sharp rise of sea level at 19 ka suggested by Yokoyama et al. (2000). That the predicted RSL history at Barbados using the ICE-5G(VM2) model is essentially identical to the ice-equivalent eustatic curve for the same model is shown explicitly in Figure 6.8, where the red curve is the model prediction and the step-discontinuous purple curve is the ice-equivalent eustatic curve.

6.4.3.3 What Is the Significance of Higher than Present Sea Levels During the Last Interglacial Period?

The record of eustatic sea level change can be extended into the time of the LIG. Direct sea level measurements based upon coastal sedimentary deposits and tropical coral sequences (e.g., in tectonically stable settings) have clearly established that eustatic sea level was higher than present during this last interglacial by approximately 4 to 6 m (e.g., Rostami et al., 2000; Muhs et al., 2002). The undisturbed ice core record of the North Greenland Ice Core Project (NGRIP) to 123 ka, and older but disturbed LIG ice in the Greenland Ice Core Project (GRIP) and Greenland Ice Sheet Project 2 (GISP2) cores, indicate that the Greenland Summit region remained ice-covered during the LIG (Raynaud et al., 1997; NGRIP, 2004). Similar isotopic value differences found in the Camp Century and Renland cores (Johnsen et al., 2001) suggest that relative elevation differences during the LIG in northern Greenland were not large

(NGRIP, 2004). Interpretation of the Dye-3 ice core in southern Greenland is equivocal. The presence of isotopically enriched ice, possibly LIG ice, at the bottom of the Dye-3 core has been interpreted as substantial reduction in southern Greenland ice thickness during the LIG (NGRIP, 2004). Equally plausible interpretations suggest that the Greenland Ice Sheet's southern dome did not survive the peak interglacial warmth and that Dye-3 is recording the growth of late-LIG ice when the ice sheet re-established itself in southern Greenland (Koerner and Fisher, 2002), or ice that flowed into the region from central Greenland or from a surviving but isolated southern dome (Lhomme et al., 2005a). The absence of pre-LIG ice in the larger ice caps in the eastern Canadian Arctic indicates that they melted completely during the LIG (Koerner, 1989).

Most of the global sea level rise during the LIG must have been the result of polar ice sheet melting. Greenland Ice Sheet models forced with Greenland temperature scenarios derived from data (Cuffey and Marshall, 2000; Tarasov and Peltier, 2003; Lhomme et al., 2005a), or temperatures and precipitation produced by an AOGCM (Otto-Bliesner et al., 2006a), simulate the minimal LIG Greenland Ice Sheet as a steep-sided ice sheet in central and northern Greenland (Figure 6.6). This inferred ice sheet, combined with the change in other arctic ice fields, likely generated no more than 2 to 4 m of early LIG sea level rise over several millennia. The simulated contribution of Greenland to this sea level rise was likely driven by orbitally forced summer warming in the Arctic (see Section 6.4.1). The evidence that sea level was 4 to 6 m above present implies there may also have been a contribution from Antarctica (Scherer et al., 1998; Overpeck et al., 2006). Overpeck et al. (2006) argued that since the circum-arctic LIG warming was very similar to that expected in a future doubled CO_2 climate, significant retreat of the Greenland Ice Sheet can be expected to occur under this future condition. Since not all of the LIG increment of sea level appears to be explained by the melt-back of the Greenland Ice Sheet, it is possible that parts of the Antarctic Ice Sheet might also retreat under this future condition (see also Scherer et al., 1998; Tarasov and Peltier, 2003; Domack et al., 2005 and Oppenheimer and Alley, 2005).

6.4.3.4   *What Is the Long-Term Contribution of Polar Ice-sheet Derived Melt Water to the Observed Globally Averaged Rate of Sea Level Rise?*

Models of postglacial RSL history together with Holocene observations can be employed to assess whether or not a significant fraction of the observed globally averaged rate of sea level rise of about 2 mm yr^{-1} during the 20th century can be explained as a long term continuing influence of the most recent partial deglaciation of the polar ice sheets. Based upon post-TAR estimates derived from geological observations of Holocene sea level from 16 equatorial Pacific islands (Peltier, 2002; Peltier et al., 2002), it appears likely that the average rate of sea level rise due to this hypothetical source over the last 2 kyr was zero and at most in the range 0 to 0.2 mm yr^{-1} (Lambeck, 2002).

6.5   The Current Interglacial

A variety of proxy records provide detailed temporal and spatial information concerning climate change during the current interglacial, the Holocene, an approximately 11.6 kyr period of increasingly intense anthropogenic modifications of the local (e.g., land use) to global (e.g., atmospheric composition) environment. The well-dated reconstructions of the past 2 kyr are covered in Section 6.6. In the context of both climate forcing and response, the Holocene is far better documented in terms of spatial coverage, dating and temporal resolution than previous interglacials. The evidence is clear that significant changes in climate forcing during the Holocene induced significant and complex climate responses, including long-term and abrupt changes in temperature, precipitation, monsoon dynamics and the El Niño-Southern Oscillation (ENSO). For selected periods such as the mid-Holocene, about 6 ka, intensive efforts have been dedicated to the synthesis of palaeoclimatic observations and modelling intercomparisons. Such extensive data coverage provides a sound basis to evaluate the capacity of climate models to capture the response of the climate system to the orbital forcing.

6.5.1   Climate Forcing and Response During the Current Interglacial

6.5.1.1   *What Were the Main Climate Forcings During the Holocene?*

During the current interglacial, changes in the Earth's orbit modulated the latitudinal and seasonal distribution of insolation (Box 6.1). Ongoing efforts to quantify Holocene changes in stratospheric aerosol content recorded in the chemical composition of ice cores from both poles (Zielinski, 2000; Castellano et al., 2005) confirm that volcanic forcing amplitude and occurrence varied significantly during the Holocene (see also Section 6.6.3). Fluctuations of cosmogenic isotopes (ice core ^{10}Be and tree ring ^{14}C) have been used as proxies for Holocene changes in solar activity (e.g., Bond et al., 2001), although the quantitative link to solar irradiance remains uncertain and substantial work is needed to disentangle solar from non-solar influences on these proxies over the full Holocene (Muscheler et al., 2006). Residual continental ice sheets formed during the last ice age were retreating during the first half of the current interglacial period (Figure 6.8). The associated ice sheet albedo is thought to have locally modulated the regional climate response to the orbital forcing (e.g., Davis et al., 2003).

The evolution of atmospheric trace gases during the Holocene is well known from ice core analyses (Figure 6.4). A first decrease in atmospheric CO_2 of about 7 ppm from 11 to 8 ka was followed by a 20 ppm CO_2 increase until the onset of the industrial revolution (Monnin et al., 2004). Atmospheric CH_4 decreased from a NH value of about 730 ppb around 10 ka to about 580 ppb around 6 ka, and increased again slowly to 730 ppb in pre-industrial times (Chappellaz et al., 1997; Flückiger et al., 2002). Atmospheric N_2O largely followed the evolution of atmospheric CO_2 and shows an early Holocene decrease of about 10 ppb and an increase of the same magnitude between 8 and 2 ka (Flückiger et al., 2002). Implied radiative forcing changes from Holocene greenhouse gas variations are 0.4 W m^{-2} (CO_2) and 0.1 W m^{-2} (N_2O and CH_4), relative to pre-industrial forcing.

6.5.1.2   Why Did Holocene Atmospheric Greenhouse Gas Concentrations Vary Before the Industrial Period?

Recent transient carbon cycle-climate model simulations with a predictive global vegetation model have attributed the early Holocene CO_2 decrease to forest regrowth in areas of the waning Laurentide Ice Sheet, partly counteracted by ocean sediment carbonate compensation (Joos et al., 2004). Carbonate compensation of terrestrial carbon uptake during the glacial-interglacial transition and the early Holocene, as well as coral reef buildup during the Holocene, likely contributed to the subsequent CO_2 rise (Broecker and Clark, 2003; Ridgwell et al., 2003; Joos et al., 2004), whereas recent carbon isotope data (Eyer, 2004) and model results (Brovkin et al., 2002; Kaplan et al., 2002; Joos et al., 2004) suggest that the global terrestrial carbon inventory has been rather stable over the 7 kyr preceding industrialisation. Variations in carbon storage in northern peatlands may have contributed to the observed atmospheric CO_2 changes. Such natural mechanisms cannot account for the much more significant industrial trace gas increases; atmospheric CO_2 would be expected to remain well below 290 ppm in the absence of anthropogenic emissions (Gerber et al., 2003).

It has been hypothesised, based on Vostok ice core CO_2 data (Petit et al., 1999), that atmospheric CO_2 would have dropped naturally by 20 ppm during the past 8 kyr (in contrast with the observed 20 ppm increase) if prehistoric agriculture had not caused a release of terrestrial carbon and CH_4 during the Holocene (Ruddiman, 2003; Ruddiman et al., 2005). This hypothesis also suggests that incipient late-Holocene high-latitude glaciation was prevented by these pre-industrial greenhouse gas emissions. However, this hypothesis conflicts with several, independent lines of evidence, including the lack of orbital similarity of the three previous interglacials with the Holocene and the recent finding that CO_2 concentrations were high during the entire Stage 11 (Siegenthaler et al., 2005a; Figure 6.3), a long (~28 kyr) interglacial (see Section 6.4.1.5). This hypothesis also requires much larger changes in the Holocene atmospheric stable carbon isotope ratio ($^{13}C/^{12}C$) than found in ice cores (Eyer, 2004), as well as a carbon release by anthropogenic land use that is larger than estimated by comparing carbon storage for natural vegetation and present day land cover (Joos et al., 2004).

6.5.1.3   Was Any Part of the Current Interglacial Period Warmer than the Late 20th Century?

The temperature evolution over the Holocene has been established for many different regions, often with centennial-resolution proxy records more sensitive to specific seasons (see Section 6.1). At high latitudes of the North Atlantic and adjacent Arctic, there was a tendency for summer temperature maxima to occur in the early Holocene (10 to 8 ka), pointing to the direct influence of the summer insolation maximum on sea ice extent (Kim et al., 2004; Kaplan and Wolfe, 2006). Climate reconstructions for the mid-northern latitudes exhibit a long-term decline in SST from the warmer early to mid-Holocene to the cooler pre-industrial period of the late Holocene (Johnsen et al., 2001; Marchal et al., 2002; Andersen et al., 2004; Kim et al., 2004; Kaplan and Wolfe 2006), most likely in response to annual mean and summer orbital forcings at these latitudes (Renssen et al., 2005). Near ice sheet remnants in northern Europe or North America, peak warmth was locally delayed, probably as a result of the interplay between ice elevation, albedo, atmospheric and oceanic heat transport and orbital forcing (MacDonald et al., 2000; Davis et al., 2003; Kaufman et al., 2004). The warmest period in northern Europe and north-western North America occurs from 7 to 5 ka (Davis et al., 2003; Kaufman et al., 2004). During the mid-Holocene, global pollen-based reconstructions (Prentice and Webb, 1998; Prentice et al., 2000) and macrofossils (MacDonald et al., 2000) show a northward expansion of northern temperate forest (Bigelow et al., 2003; Kaplan et al., 2003), as well as substantial glacier retreat (see Box 6.3). Warmer conditions at mid- and high latitudes of the NH in the early to mid-Holocene are consistent with deep borehole temperature profiles (Huang et al., 1997). Other early warm periods were identified in the equatorial west Pacific (Stott et al., 2004), China (He et al., 2004), New Zealand (Williams et al., 2004), southern Africa (Holmgren et al., 2003) and Antarctica (Masson et al., 2000). At high southern latitudes, the early warm period cannot be explained by a linear response to local summer insolation changes (see Box 6.1), suggesting large-scale reorganisation of latitudinal heat transport. In contrast, tropical temperature reconstructions, only available from marine records, show that Mediterranean, tropical Atlantic, Pacific and Indian Ocean SSTs exhibit a progressive warming from the beginning of the current interglacial onwards (Kim et al., 2004; Rimbu et al., 2004; Stott et al., 2004), possibly a reflection of tropical annual mean insolation increase (Box 6.1, Figure 1).

Extratropical centennial-resolution records therefore provide evidence for local multi-centennial periods warmer than the last decades by up to several degrees in the early to mid-Holocene. These local warm periods were very likely not globally synchronous and occurred at times when there is evidence that some areas of the tropical oceans were cooler than today (Figure 6.9) (Lorenz et al., 2006). When forced by 6 ka orbital parameters, state-of-the-art coupled climate models and EMICs capture reconstructed regional temperature and precipitation

Box 6.3: Holocene Glacier Variability

The near-global retreat of mountain glaciers is among the most visible evidence of 20th- and 21st-century climate change (see Chapter 4), and the question arises as to the significance of this current retreat within a longer time perspective. The climatic conditions that cause an advance or a retreat may be different for glaciers located in different climate regimes (see Chapter 4). This distinction is crucial if reconstructions of past glacier activity are to be understood properly.

Records of Holocene glacier fluctuations provide a necessary backdrop for evaluating the current global retreat. However, in most mountain regions, records documenting past glacier variations exist as discontinuous low-resolution series (see Box 6.3, Figure 1), whereas continuous records providing the most coherent information for the whole Holocene are available so far only in Scandinavia (e.g., Nesje et al., 2005; see Box 6.3, Figure 1).

What do glaciers reveal about climate change during the Holocene?

Most archives from the NH and the tropics indicate short, or in places even absent, glaciers between 11 and 5 ka, whereas during the second half of the Holocene, glaciers reformed and expanded. This tendency is most probably related to changes in summer insolation due to the configuration of orbital forcing (see Box 6.1). Long-term changes in solar insolation, however, cannot explain the shorter, regionally diverse glacier responses, driven by complex glacier and climate (mainly precipitation and temperature) interactions. On these shorter time scales, climate phenomena such as the North Atlantic Oscillation (NAO) and ENSO affected glaciers' mass balance, explaining some of the discrepancies found between regions. This is exemplified in the anti-phasing between glacier mass balance variations from the Alps and Scandinavia (Reichert et al., 2001; Six et al., 2001). Comparing the ongoing retreat of glaciers with the reconstruction of glacier variations during the Holocene, no period analogous to the present with a globally homogenous trend of retreating glaciers over centennial and shorter time scales could be identified in the past, although account must be taken of the large gaps in the data coverage on retreated glaciers in most regions. This is in line with model experiments suggesting that present-day glacier retreat exceed any variations simulated by the GCM control experiments and must have an external cause, with anthropogenic forcing the most likely candidate (Reichert et al., 2002).

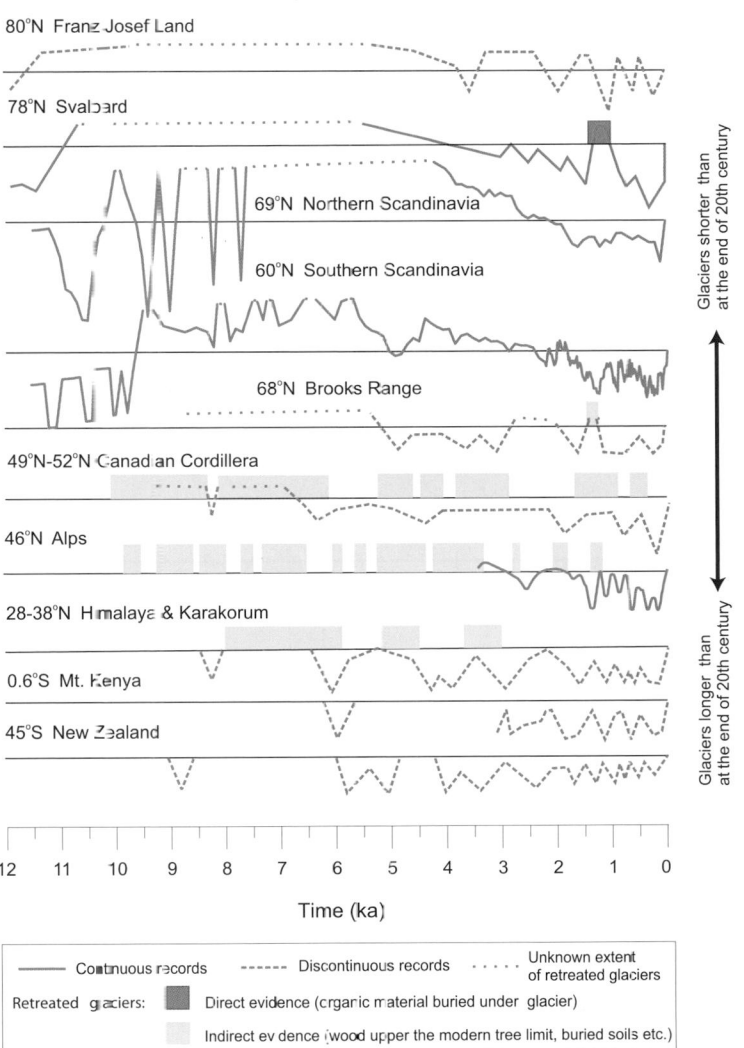

Box 6.3, Figure 1. *Timing and relative scale of selected glacier records from both hemispheres. The different records show that Holocene glacier patterns are complex and that they should be interpreted regionally in terms of precipitation and temperature. In most cases, the scale of glacier retreat is unknown and indicated on a relative scale. Lines upper the horizontal line indicate glaciers smaller than at the end of the 20th century and lines below the horizontal line denote periods with larger glaciers than at the end of the 20th century. The radiocarbon dates are calibrated and all curves are presented in calendar years. Franz Josef Land (Lubinski et al., 1999), Svalbard from Svendsen and Mangerud (1997) corrected with Humlum et al. (2005), Northern Scandinavia (Bakke et al., 2005a,b; Nesje et al., 2005), Southern Scandinavia (Dahl and Nesje, 1996; Matthews et al., 2000, 2005; Lie et al., 2004), Brooks Range (Ellis and Calkin, 1984), Canadian Cordillera (Luckman and Kearney, 1986; Osborn and Luckman, 1988; Koch et al., 2004; Menounos et al., 2004), Alps (Holzhauser et al., 2005; Jörin et al., 2006), Himalaya and Karakorum (Röthlisberger and Geyh, 1985; Bao et al., 2003), Mt. Kenya (Karlén et al., 1999), New Zealand (Gellatly et al., 1988).*

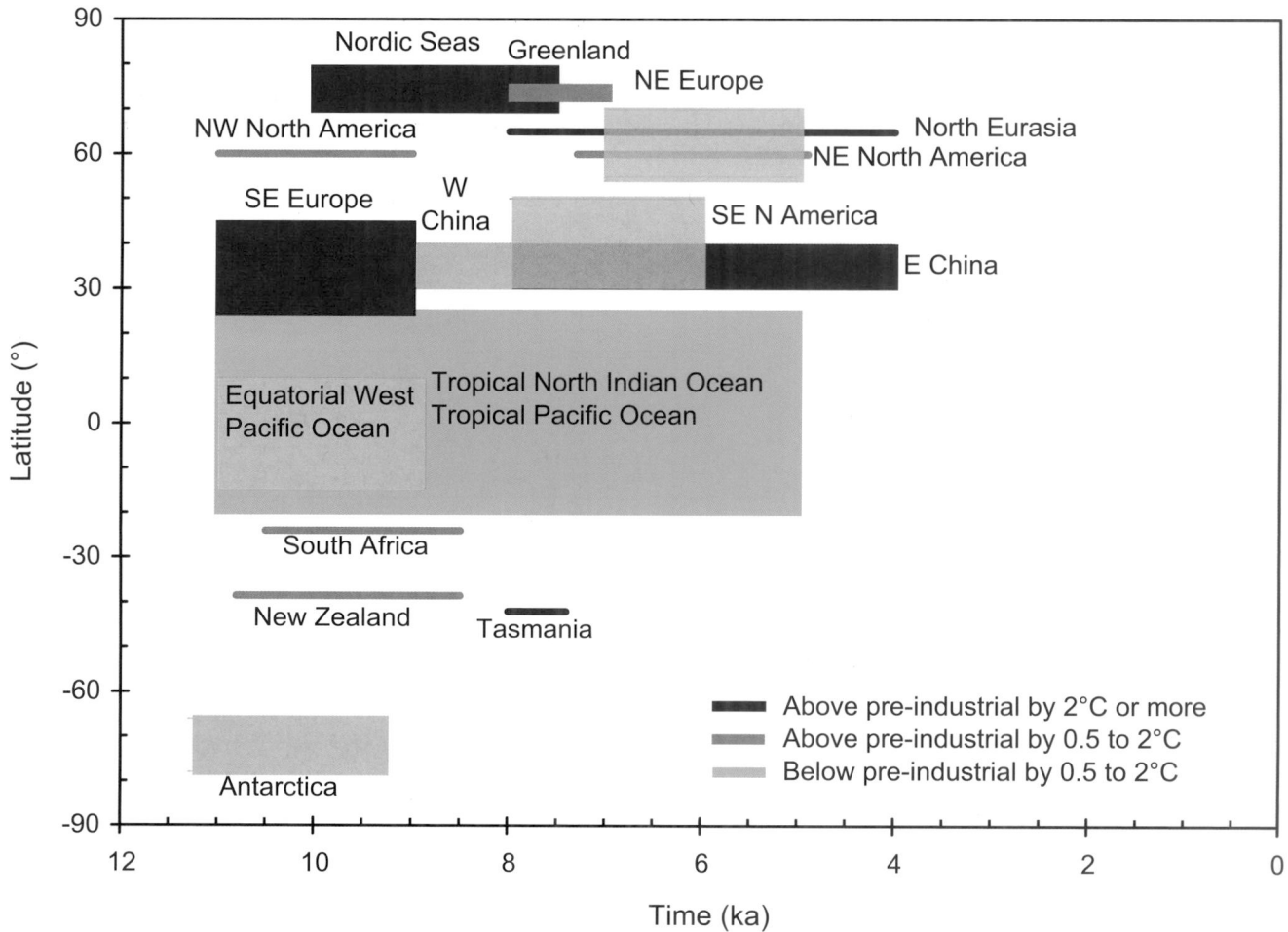

Figure 6.9. *Timing and intensity of maximum temperature deviation from pre-industrial levels, as a function of latitude (vertical axis) and time (horizontal axis, in thousands of years before present). Temperatures above pre-industrial levels by 0.5°C to 2°C appear in orange (above 2°C in red). Temperatures below pre-industrial levels by 0.5°C to 2°C appear in blue. References for data sets are: Barents Sea (Duplessy et al., 2001), Greenland (Johnsen et al., 2001), Europe (Davis et al., 2003), northwest and northeast America (MacDonald et al., 2000; Kaufman et al., 2004), China (He et al., 2004), tropical oceans (Rimbu et al., 2004; Stott et al., 2004; Lorentz et al., 2006), north Atlantic (Marchal et al., 2002; Kim et al., 2004), Tasmania (Xia et al., 2001), East Antarctica (Masson et al., 2000), southern Africa (Holmgren et al., 2003) and New Zealand (Williams et al., 2004).*

changes (Sections 6.5.1.4 and 6.5.1.5), whereas simulated global mean temperatures remain essentially unchanged (<0.4°C; Masson-Delmotte et al., 2005b), just as expected from the seasonality of the orbital forcing (see Box 6.1). Due to different regional temperature responses from the tropics to high latitudes, as well as between hemispheres, commonly used concepts such as 'mid-Holocene thermal optimum', 'altithermal', etc. are not globally relevant and should only be applied in a well-articulated regional context. Current spatial coverage, temporal resolution and age control of available Holocene proxy data limit the ability to determine if there were multi-decadal periods of global warmth comparable to the last half of 20th century.

6.5.1.4   *What Are the Links Between Orbital Forcing and Mid-Holocene Monsoon Intensification?*

Lake levels and vegetation changes reconstructed for the early to mid-Holocene indicate large precipitation increases in North Africa (Jolly et al., 1998). Simulating this intensification

of the African monsoon is widely used as a benchmark for climate models within PMIP. When forced by mid-Holocene insolation resulting from changes in the Earth's orbit (see Box 6.1), but fixed present-day vegetation and ocean temperatures, atmospheric models simulate NH summer continental warming and a limited enhancement of summer monsoons but underestimate the reconstructed precipitation increase and extent over the Sahara (Joussaume et al., 1999; Coe and Harrison., 2002; Braconnot et al., 2004). Differences among simulations appear related to atmospheric model characteristics together with the mean tropical temperature of the control simulation (Braconnot et al., 2002). As already noted in the TAR, the vegetation and surface albedo feedbacks play a major role in the enhancement of the African monsoon (e.g., Claussen and Gayler, 1997; de Noblet-Ducoudre et al., 2000; Levis et al., 2004). New coupled ocean-atmosphere simulations show that the ocean feedback strengthens the inland monsoon flow and the length of the monsoon season, due to robust changes in late summer dipole SST patterns and in mixed layer depth (Braconnot et al., 2004; Zhao et al., 2005). When combined, vegetation, soil characteristics and ocean feedbacks produce nonlinear interactions resulting in simulated precipitation in closer agreement with data (Braconnot et al., 2000; Levis et al., 2004). Transient simulations of Holocene climate performed with EMICs have further shown that land surface feedbacks are possibly involved in abrupt monsoon fluctuations (see Section 6.5.2). The mid-Holocene intensification of the North Australian, Indian and southwest American monsoons is captured by coupled ocean-atmosphere climate models in response to orbital forcing, again with amplifying ocean feedbacks (Harrison et al., 2003; Liu et al., 2004; Zhao et al., 2005).

6.5.1.5 *What Are the Links Between Orbital Forcing and mid-Holocene Climate at Middle and High Latitudes?*

Terrestrial records of the mid-Holocene indicate an expansion of forest at the expense of tundra at mid- to high latitudes of the NH (MacDonald et al., 2000; Prentice et al., 2000). Since the TAR, coupled atmosphere-ocean models, including the recent PMIP-2 simulations, have investigated the response of the climate system to orbital forcing at 6 ka during the mid-Holocene (Section 6.6.1, Box 6.1). Fully coupled atmosphere-ocean-vegetation models do produce the northward shift in the position of the northern limit of boreal forest, in response to simulated summer warming, and the northward expansion of temperate forest belts in North America, in response to simulated winter warming (Wohlfahrt et al., 2004). At high latitudes, the vegetation-snow albedo and ocean feedbacks enhance the warming in spring and autumn, respectively and transform the seasonal orbital forcing into an annual response (Crucifix et al., 2002; Wohlfahrt et al., 2004). Ocean changes simulated for this period are generally small and difficult to quantify from data due to uncertainties in the way proxy methods respond to the seasonality and stratification of the surface waters (Waelbroeck et al., 2005). Simulations with atmosphere and slab ocean models indicate that a change in the mean tropical Pacific SSTs in the mid-Holocene to conditions more like La Niña conditions can explain North American drought conditions at mid-Holocene (Shin et al., 2006). Based on proxies of SST in the North Atlantic, it has been suggested that trends from early to late Holocene are consistent with a shift from a more meridional regime over northern Europe to a positive North Atlantic Oscillation (NAO)-like mean state in the early to mid-Holocene (Rimbu et al., 2004). A PMIP2 intercomparison shows that three of nine models support a positive NAO-like atmospheric circulation in the mean state for the mid-Holocene as compared to the pre-industrial period, without significant changes in simulated NAO variability (Gladstone et al., 2005).

6.5.1.6 *Are There Long-Term Modes of Climate Variability Identified During the Holocene that Could Be Involved in the Observed Current Warming?*

An increasing number of Holocene proxy records are of sufficiently high resolution to describe the climate variability on centennial to millennial time scales, and to identify possible natural quasi-periodic modes of climate variability at these time scales (Haug et al., 2001; Gupta et al., 2003). Although earlier studies suggested that Holocene millennial variability could display similar frequency characteristics as the glacial variability in the North Atlantic (Bond et al., 1997), this assumption is being increasingly questioned (Risebrobakken et al., 2003; Schulz et al., 2004). In many records, there is no apparent consistent pacing at specific centennial to millennial frequencies through the Holocene period, but rather shifts between different frequencies (Moros et al., 2006). The suggested synchroneity of tropical and North Atlantic centennial to millennial variability (de Menocal et al., 2000; Mayewski et al., 2004; Y.J. Wang et al., 2005) is not common to the SH (Masson et al., 2000; Holmgren et al., 2003), suggesting that millennial scale variability cannot account for the observed 20th-century warming trend. Based on the correlation between changes in cosmogenic isotopes (^{10}Be or ^{14}C) – related to solar activity changes – and climate proxy records, some authors argue that solar activity may be a driver for centennial to millennial variability (Karlén and Kuylenstierna, 1996; Bond et al., 2001; Fleitmann et al., 2003; Y.J. Wang et al., 2005). The possible importance of (forced or unforced) modes of variability within the climate system, for instance related to the deep ocean circulation, have also been highlighted (Bianchi and McCave, 1999; Duplessy et al., 2001; Marchal et al., 2002; Oppo et al., 2003). The current lack of consistency between various data sets makes it difficult, based on current knowledge, to attribute the millennial time scale large-scale climate variations to external forcings (solar activity, episodes of intense volcanism), or to variability internal to the climate system.

6.5.2 Abrupt Climate Change During the Current Interglacial

6.5.2.1 *What Do Abrupt Changes in Oceanic and Atmospheric Circulation at Mid- and High-Latitudes Show?*

An abrupt cooling of 2°C to 6°C identified as a prominent feature of Greenland ice cores at 8.2 ka (Alley et al., 1997; Alley and Agustsdottir, 2005) is documented in Europe and North America by high-resolution continental proxy records (Klitgaard-Kristensen et al., 1998; von Grafenstein et al., 1998; Barber et al., 1999; Nesje et al., 2000; Rohling and Palike, 2005). A large decrease in atmospheric CH_4 concentrations (several tens of parts per billion; Spahni et al., 2003) reveals the widespread signature of the abrupt '8.2 ka event' associated with large-scale atmospheric circulation change recorded from the Arctic to the tropics with associated dry episodes (Hughen et al., 1996; Stager and Mayewski, 1997; Haug et al., 2001; Fleitmann et al., 2003; Rohling and Palike, 2005). The 8.2 ka event is interpreted as resulting from a brief reorganisation of the North Atlantic MOC (Bianchi and McCave, 1999; Risebrobakken et al., 2003; McManus et al., 2004), however, without a clear signature identified in deep water formation records. Significant volumes of freshwater were released in the North Atlantic and Arctic at the beginning of the Holocene by the decay of the residual continental ice (Nesje et al., 2004). A likely cause for the 8.2 ka event is an outburst flood during which pro-glacial Lake Agassiz drained about 10^{14} m^3 of freshwater into Hudson Bay extremely rapidly (possibly 5 Sv over 0.5 year; Clarke et al., 2004). Climate models have been used to test this hypothesis and assess the vulnerability of the ocean and atmospheric circulation to different amounts of freshwater release (see Alley and Agustsdottir, 2005 for a review; Section 6.4.2.2). Ensemble simulations conducted with EMICs (Renssen et al., 2002; Bauer et al., 2004) and coupled ocean-atmosphere GCMs (Alley and Agustsdottir, 2005; LeGrande et al., 2006) with different boundary conditions and freshwater forcings show that climate models are capable of simulating the broad features of the observed 8.2 ka event (including shifts in the ITCZ).

The end of the first half of the Holocene – between about 5 and 4 ka – was punctuated by rapid events at various latitudes, such as an abrupt increase in NH sea ice cover (Jennings et al., 2001); a decrease in Greenland deuterium excess, reflecting a change in the hydrological cycle (Masson-Delmotte et al., 2005b); abrupt cooling events in European climate (Seppa and Birks, 2001; Lauritzen, 2003); widespread North American drought for centuries (Booth et al., 2005); and changes in South American climate (Marchant and Hooghiemstra, 2004). The processes behind these observed abrupt shifts are not well understood. As these particular events took place at the end of a local warm period caused by orbital forcing (see Box 6.1 and Section 6.5.1), these observations suggest that under gradual climate forcings (e.g., orbital) the climate system can change abruptly.

6.5.2.2   What Is the Understanding of Abrupt Changes in Monsoons?

In the tropics, precipitation-sensitive records and models indicate that summer monsoons in Africa, India and Southeast Asia were enhanced in the early to mid-Holocene due to orbital forcing, a resulting increase in land-sea temperature gradients and displacement of the ITCZ. All high-resolution precipitation-sensitive records reveal that the local transitions from wetter conditions in the early Holocene to drier modern conditions occurred in one or more steps (Guo et al., 2000; Fleitmann et al., 2003; Morrill et al., 2003; Y.J.Wang et al., 2005). In the early Holocene, large increases in monsoon-related northern African runoff and/or wetter conditions over the Mediterranean

are associated with dramatic changes in Mediterranean Sea ventilation, as evidenced by sapropel layers (Ariztegui et al., 2000).

Transient simulations of the Holocene, although usually after the final disappearance of ice sheets, have been performed with EMICs and forced by orbital parameters (Box 6.1). These models have pointed to the operation of mechanisms that can generate rapid events in response to orbital forcing, such as changes in African monsoon intensity due to nonlinear interactions between vegetation and monsoon dynamics (Claussen et al., 1999; Renssen et al., 2003).

6.5.3   How and Why Has the El Niño-Southern Oscillation Changed Over the Present Interglacial?

High-resolution palaeoclimate records from diverse sources (corals, archaeological middens, lake and ocean sediments) consistently indicate that the early to mid-Holocene likely experienced weak ENSO variability, with a transition to a stronger modern regime occurring in the past few thousand years (Shulmeister and Lees, 1995; Gagan et al., 1998; Rodbell et al., 1999; Tudhope et al., 2001; Moy et al., 2002; McGregor and Gagan, 2004). Most data sources are discontinuous, providing only snapshots of mean conditions or interannual variability, and making it difficult to precisely characterise the rate and timing of the transition to the modern regime.

A simple model of the coupled Pacific Ocean and atmosphere, forced with orbital insolation variations, suggests that seasonal changes in insolation can produce systematic changes in ENSO behaviour (Clement et al., 1996, 2000; Cane, 2005). This model simulates a progressive, somewhat irregular increase in both event frequency and amplitude throughout the Holocene, due to the Bjerknes feedback mechanism (Bjerknes, 1969) and ocean dynamical thermostat (Clement and Cane, 1999; Clement et al., 2001; Cane, 2005). Snapshot experiments conducted with some coupled GCMs also reproduce an intensification of ENSO between the early Holocene and the present, although with some disagreement as to the magnitude of change. Both model results and data syntheses suggest that before the mid-Holocene, the tropical Pacific exhibited a more La Niña-like background state (Clement et al., 2000; Liu et al., 2000; Kitoh and Murakami, 2002; Otto-Bliesner et al., 2003; Liu, 2004). In palaeoclimate simulations with GCMs, ENSO teleconnections robust in the modern system show signs of weakening under mid-Holocene orbital forcing (Otto-Bliesner, 1999; Otto-Bliesner et al., 2003).

Frequently Asked Question 6.2

Is the Current Climate Change Unusual Compared to Earlier Changes in Earth's History?

Climate has changed on all time scales throughout Earth's history. Some aspects of the current climate change are not unusual, but others are. The concentration of CO_2 in the atmosphere has reached a record high relative to more than the past half-million years, and has done so at an exceptionally fast rate. Current global temperatures are warmer than they have ever been during at least the past five centuries, probably even for more than a millennium. If warming continues unabated, the resulting climate change within this century would be extremely unusual in geological terms. Another unusual aspect of recent climate change is its cause: past climate changes were natural in origin (see FAQ 6.1), whereas most of the warming of the past 50 years is attributable to human activities.

When comparing the current climate change to earlier, natural ones, three distinctions must be made. First, it must be clear which variable is being compared: is it greenhouse gas concentration or temperature (or some other climate parameter), and is it their absolute value or their rate of change? Second, local changes must not be confused with global changes. Local climate changes are often much larger than global ones, since local factors (e.g., changes in oceanic or atmospheric circulation) can shift the delivery of heat or moisture from one place to another and local feedbacks operate (e.g., sea ice feedback). Large changes in global mean temperature, in contrast, require some global forcing (such as a change in greenhouse gas concentration or solar activity). Third, it is necessary to distinguish between time scales. Climate changes over millions of years can be much larger and have different causes (e.g., continental drift) compared to climate changes on a centennial time scale.

The main reason for the current concern about climate change is the rise in atmospheric carbon dioxide (CO_2) concentration (and some other greenhouse gases), which is very unusual for the Quaternary (about the last two million years). The concentration of CO_2 is now known accurately for the past 650,000 years from antarctic ice cores. During this time, CO_2 concentration varied between a low of 180 ppm during cold glacial times and a high of 300 ppm during warm interglacials. Over the past century, it rapidly increased well out of this range, and is now 379 ppm (see Chapter 2). For comparison, the approximately 80-ppm rise in CO_2 concentration at the end of the past ice ages generally took over 5,000 years. Higher values than at present have only occurred many millions of years ago (see FAQ 6.1).

Temperature is a more difficult variable to reconstruct than CO_2 (a globally well-mixed gas), as it does not have the same value all over the globe, so that a single record (e.g., an ice core) is only of limited value. Local temperature fluctuations, even those over just a few decades, can be several degrees celsius, which is larger than the global warming signal of the past century of about 0.7°C.

More meaningful for global changes is an analysis of large-scale (global or hemispheric) averages, where much of the local varia-

tion averages out and variability is smaller. Sufficient coverage of instrumental records goes back only about 150 years. Further back in time, compilations of proxy data from tree rings, ice cores, etc., go back more than a thousand years with decreasing spatial coverage for earlier periods (see Section 6.5). While there are differences among those reconstructions and significant uncertainties remain, all published reconstructions find that temperatures were warm during medieval times, cooled to low values in the 17th, 18th and 19th centuries, and warmed rapidly after that. The medieval level of warmth is uncertain, but may have been reached again in the mid-20th century, only to have likely been exceeded since then. These conclusions are supported by climate modelling as well. Before 2,000 years ago, temperature variations have not been systematically compiled into large-scale averages, but they do not provide evidence for warmer-than-present global annual mean temperatures going back through the Holocene (the last 11,600 years; see Section 6.4). There are strong indications that a warmer climate, with greatly reduced global ice cover and higher sea level, prevailed until around 3 million years ago. Hence, current warmth appears unusual in the context of the past millennia, but not unusual on longer time scales for which changes in tectonic activity (which can drive natural, slow variations in greenhouse gas concentration) become relevant (see Box 6.1).

A different matter is the current rate of warming. Are more rapid global climate changes recorded in proxy data? The largest temperature changes of the past million years are the glacial cycles, during which the global mean temperature changed by 4°C to 7°C between ice ages and warm interglacial periods (local changes were much larger, for example near the continental ice sheets). However, the data indicate that the global warming at the end of an ice age was a gradual process taking about 5,000 years (see Section 6.3). It is thus clear that the current rate of global climate change is much more rapid and very unusual in the context of past changes. The much-discussed abrupt climate shifts during glacial times (see Section 6.3) are not counter-examples, since they were probably due to changes in ocean heat transport, which would be unlikely to affect the global mean temperature.

Further back in time, beyond ice core data, the time resolution of sediment cores and other archives does not resolve changes as rapid as the present warming. Hence, although large climate changes have occurred in the past, there is no evidence that these took place at a faster rate than present warming. If projections of approximately 5°C warming in this century (the upper end of the range) are realised, then the Earth will have experienced about the same amount of global mean warming as it did at the end of the last ice age; there is no evidence that this rate of possible future global change was matched by any comparable global temperature increase of the last 50 million years.

6.6 The Last 2,000 Years

6.6.1 Northern Hemisphere Temperature Variability

6.6.1.1 *What Do Reconstructions Based on Palaeoclimatic Proxies Show?*

Figure 6.10 shows the various instrumental and proxy climate evidence of the variations in average large-scale surface temperatures over the last 1.3 kyr. Figure 6.10a shows two instrumental compilations representing the mean annual surface temperature of the NH since 1850, one based on land data only, and one using land and surface ocean data combined (see Chapter 3). The uncertainties associated with one of these series are also shown (30-year smoothed combined land and marine). These arise primarily from the incomplete spatial coverage of instrumentation through time (Jones et al., 1997) and, whereas these uncertainties are larger in the 19th compared to the 20th century, the prominence of the recent warming, especially in the last two to three decades of the record, is clearly apparent in this 150-year context. The land-only record shows similar variability, although the rate of warming is greater than in the combined record after about 1980. The land-only series can be extended back beyond the 19th century, and is shown plotted from 1781 onwards. The early section is based on a much sparser network of available station data, with at least 23 European stations, but only one North American station, spanning the first two decades, and the first Asian station beginning only in the 1820s. Four European records (Central England, De Bilt, Berlin and Uppsala) provide an even longer, though regionally restricted, indication of the context for the warming observed in the last approximately 20 to 30 years, which is even greater in this area than is observed over the NH land as a whole.

The instrumental temperature data that exist before 1850, although increasingly biased towards Europe in earlier periods, show that the warming observed after 1980 is unprecedented compared to the levels measured in the previous 280 years, even allowing for the greater variance expected in an average of so few early data compared to the much greater number in the 20th century. Recent analyses of instrumental, documentary and proxy climate records, focussing on European temperatures, have also pointed to the unprecedented warmth of the 20th century and shown that the extreme summer of 2003 was very likely warmer than any that has occurred in at least 500 years (Luterbacher et al., 2004; Guiot et al., 2005; see Box 3.6).

If the behaviour of recent temperature change is to be understood, and the mechanisms and causes correctly attributed, parallel efforts are needed to reconstruct the longer and more widespread pre-instrumental history of climate variability, as well as the detailed changes in various factors that might influence climate (Bradley et al., 2003b; Jones and Mann, 2004).

The TAR discussed various attempts to use proxy data to reconstruct changes in the average temperature of the NH for the period after AD 1000, but focused on three reconstructions

(included in Figure 6.10), all with yearly resolution. The first (Mann et al., 1999) represents mean annual temperatures, and is based on a range of proxy types, including data extracted from tree rings, ice cores and documentary sources; this reconstruction also incorporates a number of instrumental (temperature and precipitation) records from the 18th century onwards. For 900 years, this series exhibits multi-decadal fluctuations with amplitudes up to 0.3°C superimposed on a negative trend of 0.15°C, followed by an abrupt warming (~0.4°C) matching that observed in the instrumental data during the first half of the 20th century. Of the other two reconstructions, one (Jones et al., 1998) was based on a much smaller number of proxies, whereas the other (Briffa et al., 2001) was based solely on tree ring density series from an expansive area of the extratropics, but reached back only to AD 1400. These two reconstructions emphasise warm season rather than annual temperatures, with a geographical focus on extratropical land areas. They indicate a greater range of variability on centennial time scales prior to the 20th century, and also suggest slightly cooler conditions during the 17th century than those portrayed in the Mann et al. (1998, 1999) series.

The 'hockey stick' reconstruction of Mann et al. (1999) has been the subject of several critical studies. Soon and Baliunas (2003) challenged the conclusion that the 20th century was the warmest at a hemispheric average scale. They surveyed regionally diverse proxy climate data, noting evidence for relatively warm (or cold), or alternatively dry (or wet) conditions occurring at any time within pre-defined periods assumed to bracket the so-called 'Medieval Warm Period' (and 'Little Ice Age'). Their qualitative approach precluded any quantitative summary of the evidence at precise times, limiting the value of their review as a basis for comparison of the relative magnitude of mean hemispheric 20th-century warmth (Mann and Jones, 2003; Osborn and Briffa, 2006). Box 6.4 provides more information on the 'Medieval Warm Period'.

McIntyre and McKitrick (2003) reported that they were unable to replicate the results of Mann et al. (1998). Wahl and Ammann (2007) showed that this was a consequence of differences in the way McIntyre and McKitrick (2003) had implemented the method of Mann et al. (1998) and that the original reconstruction could be closely duplicated using the original proxy data. McIntyre and McKitrick (2005a,b) raised further concerns about the details of the Mann et al. (1998) method, principally relating to the independent verification of the reconstruction against 19th-century instrumental temperature data and to the extraction of the dominant modes of variability present in a network of western North American tree ring chronologies, using Principal Components Analysis. The latter may have some theoretical foundation, but Wahl and Amman (2006) also show that the impact on the amplitude of the final reconstruction is very small (~0.05°C; for further discussion of these issues see also Huybers, 2005; McIntyre and McKitrick, 2005c,d; von Storch and Zorita, 2005).

Since the TAR, a number of additional proxy data syntheses based on annually or near-annually resolved data, variously representing mean NH temperature changes over the last

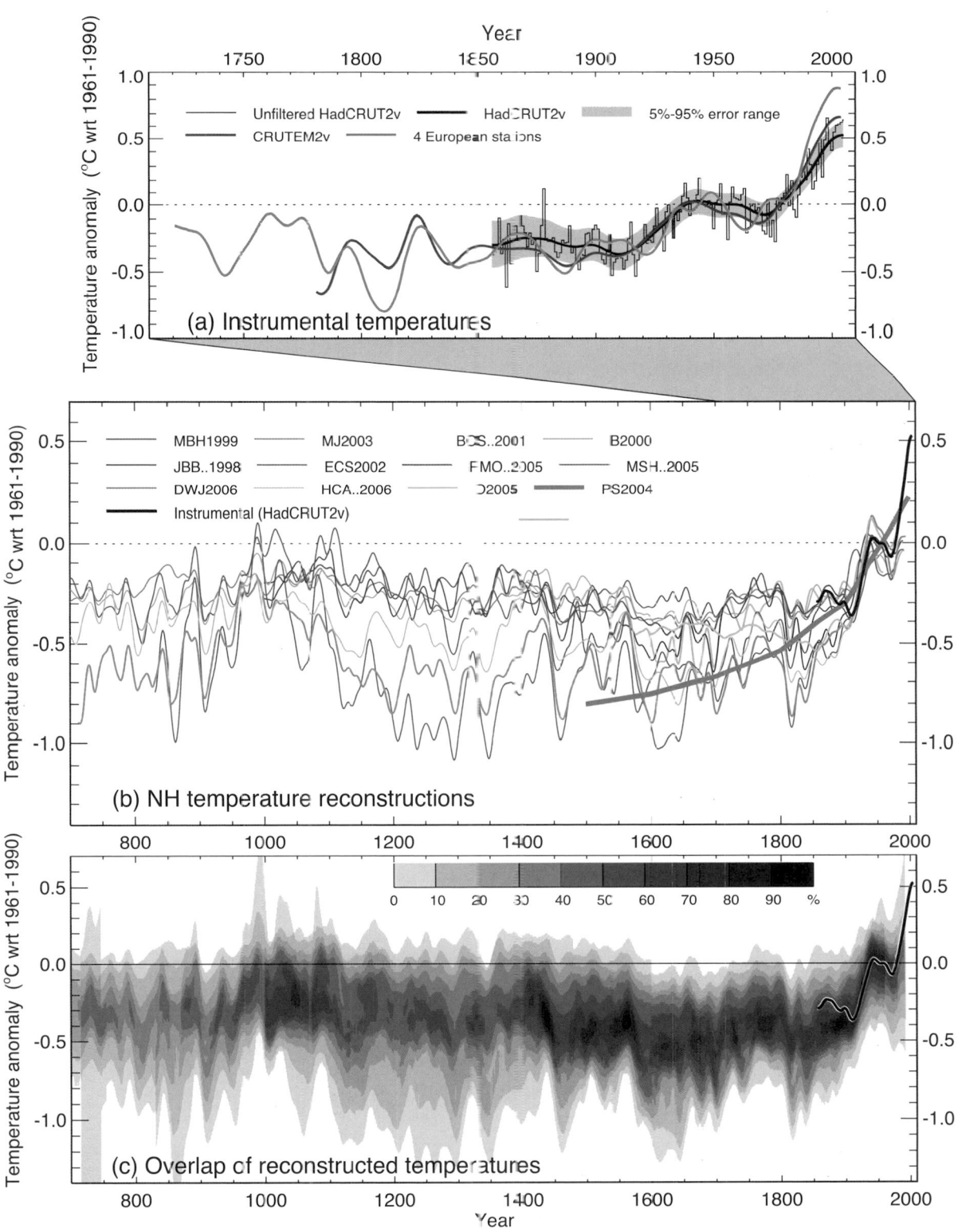

Figure 6.10. *Records of NH temperature variation during the last 1.3 kyr. (a) Annual mean instrumental temperature records, identified in Table 6.1. (b) Reconstructions using multiple climate proxy records, identified in Table 6.1, including three records (JBB..1998, MBH..1999 and BOS..2001) shown in the TAR, and the HadCRUT2v instrumental temperature record in black. (c) Overlap of the published multi-decadal time scale uncertainty ranges of all temperature reconstructions identified in Table 6.1 (except for RMO..2005 and PS2004), with temperatures within ±1 standard error (SE) of a reconstruction 'scoring' 10%, and regions within the 5 to 95% range 'scoring' 5% (the maximum 100% is obtained only for temperatures that fall within ±1 SE of all 10 reconstructions). The HadCRUT2v instrumental temperature record is shown in black. All series have been smoothed with a Gaussian-weighted filter to remove fluctuations on time scales less than 30 years; smoothed values are obtained up to both ends of each record by extending the records with the mean of the adjacent existing values. All temperatures represent anomalies (°C) from the 1961 to 1990 mean.*

Box 6.4: Hemispheric Temperatures in the 'Medieval Warm Period'

At least as early as the beginning of the 20th century, different authors were already examining the evidence for climate changes during the last two millennia, particularly in relation to North America, Scandinavia and Eastern Europe (Brooks, 1922). With regard to Iceland and Greenland, Pettersson (1914) cited evidence for considerable areas of Iceland being cultivated in the 10th century. At the same time, Norse settlers colonised areas of Greenland, while a general absence of sea ice allowed regular voyages at latitudes far to the north of what was possible in the colder 14th century. Brooks (1922) described how, after some amelioration in the 15th and 16th centuries, conditions worsened considerably in the 17th century; in Iceland, previously cultivated land was covered by ice. Hence, at least for the area of the northern North Atlantic, a picture was already emerging of generally warmer conditions around the centuries leading up to the end of the first millennium, but framed largely by comparison with strong evidence of much cooler conditions in later centuries, particularly the 17th century.

Lamb (1965) seems to have been the first to coin the phrase 'Medieval Warm Epoch' or 'Little Optimum' to describe the totality of multiple strands of evidence principally drawn from western Europe, for a period of widespread and generally warmer temperatures which he put at between AD 1000 and 1200 (Lamb, 1982). It is important to note that Lamb also considered the warmest conditions to have occurred at different times in different areas: between 950 and 1200 in European Russia and Greenland, but somewhat later, between 1150 and 1300 (though with notable warmth also in the later 900s) in most of Europe (Lamb, 1977).

Much of the evidence used by Lamb was drawn from a very diverse mixture of sources such as historical information, evidence of treeline and vegetation changes, or records of the cultivation of cereals and vines. He also drew inferences from very preliminary analyses of some Greenland ice core data and European tree ring records. Much of this evidence was difficult to interpret in terms of accurate quantitative temperature influences. Much was not precisely dated, representing physical or biological systems that involve complex lags between forcing and response, as is the case for vegetation and glacier changes. Lamb's analyses also predate any formal statistical calibration of much of the evidence he considered. He concluded that 'High Medieval' temperatures were probably 1.0°C to 2.0°C above early 20th-century levels at various European locations (Lamb, 1977; Bradley et al., 2003a).

A later study, based on examination of more quantitative evidence, in which efforts were made to control for accurate dating and specific temperature response, concluded that it was not possible to say anything other than '… in some areas of the Globe, for some part of the year, relatively warm conditions may have prevailed' (Hughes and Diaz, 1994).

In medieval times, as now, climate was unlikely to have changed in the same direction, or by the same magnitude, everywhere (Box 6.4, Figure 1). At some times, some regions may have experienced even warmer conditions than those that prevailed throughout the 20th century (e.g., see Bradley et al., 2003a). Regionally restricted evidence by itself, especially when the dating is imprecise, is of little practical relevance to the question of whether climate in medieval times was globally as warm or warmer than today. Local climate variations can be dominated by internal climate variability, often the result of the redistribution of heat by regional climate processes. Only very large-scale climate averages can be expected to reflect global forcings over recent millennia (Mann and Jones, 2003; Goosse

(continued)

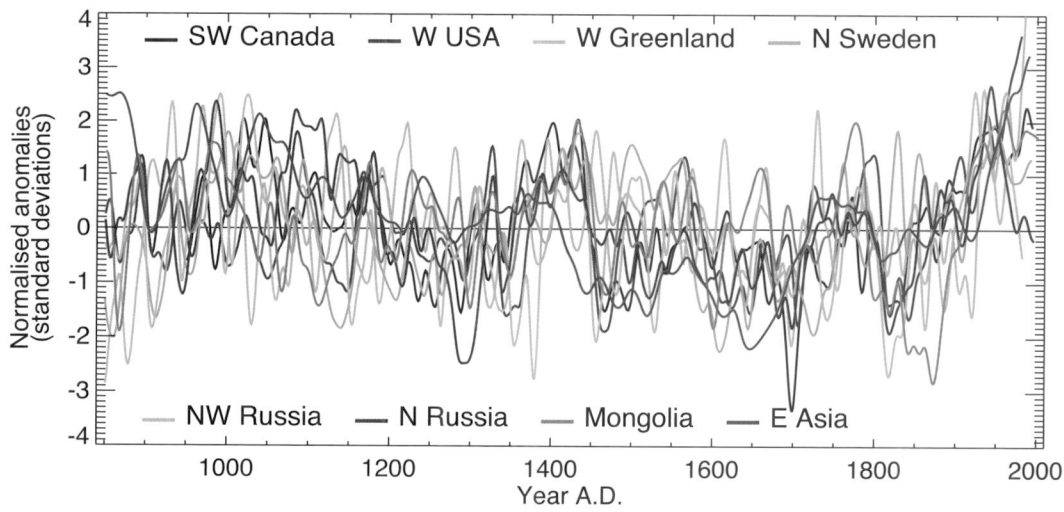

Box 6.4, Figure 1. *The heterogeneous nature of climate during the 'Medieval Warm Period' is illustrated by the wide spread of values exhibited by the individual records that have been used to reconstruct NH mean temperature. These consist of individual, or small regional averages of, proxy records collated from those used by Mann and Jones (2003), Esper et al. (2002) and Luckman and Wilson (2005), but exclude shorter series or those with no evidence of sensitivity to local temperature. These records have not been calibrated here, but each has been smoothed with a 20-year filter and scaled to have zero mean and unit standard deviation over the period 1001 to 1980.*

et al., 2005a). To define medieval warmth in a way that has more relevance for exploring the magnitude and causes of recent large-scale warming, widespread and continuous palaeoclimatic evidence must be assimilated in a homogeneous way and scaled against recent measured temperatures to allow a meaningful quantitative comparison against 20th-century warmth (Figure 6.10).

A number of studies that have attempted to produce very large spatial-scale reconstructions have come to the same conclusion: that medieval warmth was heterogeneous in terms of its precise timing and regional expression (Crowley and Lowery, 2000; Folland et al., 2001; Esper et al., 2002; Bradley et al., 2003a; Jones and Mann, 2004; D'Arrigo et al., 2006).

The uncertainty associated with present palaeoclimate estimates of NH mean temperatures is significant, especially for the period prior to 1600 when data are scarce (Mann et al., 1999; Briffa and Osborn, 2002; Cook et al., 2004a). However, Figure 6.10 shows that the warmest period prior to the 20th century very likely occurred between 950 and 1100, but temperatures were probably between 0.1°C and 0.2°C below the 1961 to 1990 mean and significantly below the level shown by instrumental data after 1980.

In order to reduce the uncertainty, further work is necessary to update existing records, many of which were assembled up to 20 years ago, and to produce many more, especially early, palaeoclimate series with much wider geographic coverage. There are far from sufficient data to make any meaningful estimates of *global* medieval warmth (Figure 6.11). There are very few long records with high temporal resolution data from the oceans, the tropics or the SH.

The evidence currently available indicates that NH mean temperatures during medieval times (950–1100) were indeed warm in a 2-kyr context and even warmer in relation to the less sparse but still limited evidence of widespread average cool conditions in the 17th century (Osborn and Briffa, 2006). However, the evidence is not sufficient to support a conclusion that hemispheric mean temperatures were as warm, or the extent of warm regions as expansive, as those in the 20th century as a whole, during any period in medieval times (Jones et al., 2001; Bradley et al., 2003a,b; Osborn and Briffa, 2006).

Table 6.1. *Records of Northern Hemisphere temperature shown in Figure 6.10.*

Instrumental temperatures			
Series	**Period**	**Description**	**Reference**
HadCRUT2v[a]	1856–2005	Land and marine temperatures for the NH	Jones and Moberg, 2003; errors from Jones et al., 1997
CRUTEM2v[b]	1781–2004	Land-only temperatures for the NH	Jones and Moberg, 2003; extended using data from Jones et al., 2003
4 European Stations	1721–2004	Average of central England, De Bilt, Berlin and Uppsala	Jones et al., 2003

Proxy-based reconstructions of temperature								
		Reconstructed		**Location Of Proxies[c]**				
Series	**Period**	**Season**	**Region**	**H**	**M**	**L**	**O**	**Reference**
JBB..1998	1000–1991	Summer	Land, 20°N–90°N	◢	◢	□	□	Jones et al., 1998; calibrated by Jones et al., 2001
MBH1999	1000–1980	Annual	Land + marine, 0–90°N	■	■	◢	◢	Mann et al., 1999
BOS..2001	1402–1960	Summer	Land, 20°N–90°N	■	◢	□	□	Briffa et al., 2001
ECS2002	831–1992	Annual	Land, 20°N–90°N	◢	◢	□	□	Esper et al., 2002; recalibrated by Cook et al., 2004a
B2000	1–1993	Summer	Land, 20°N–90°N	◢	□	□	□	Briffa, 2000; calibrated by Briffa et al., 2004
MJ2003	200–1980	Annual	Land + marine, 0–90°N	◢	◢	□	□	Mann and Jones, 2003
RMO..2005	1400–1960	Annual	Land + marine, 0–90°N	■	■	◢	◢	Rutherford et al., 2005
MSH..2005	1–1979	Annual	Land + marine, 0–90°N	◢	◢	◢	◢	Moberg et al., 2005
DWJ2006	713–1995	Annual	Land, 20°N–90°N	■	◢	□	□	D'Arrigo et al., 2006
HCA..2006	558–1960	Annual	Land, 20°N–90°N	◢	◢	□	□	Hegerl et al., 2006
PS2004	1500–2000	Annual	Land, 0–90°N	◢	■	□	□	Pollack and Smerdon, 2004; reference level adjusted following Moberg et al., 2005
O2005	1600–1990	Summer	Global land	◢	■	□	□	Oerlemans, 2005

Notes:
[a] Hadley Centre/Climatic Research Unit gridded surface temperature data set, version 2 variance adjusted.
[b] Climatic Research Unit gridded land surface air temperature, version 2 variance corrected.
[c] Location of proxies from H = high-latitude land, M = mid-latitude land, L = low-latitude land, O = oceans is indicated by □ (none or very few), ◢ (limited coverage) or ■ (moderate or good coverage).

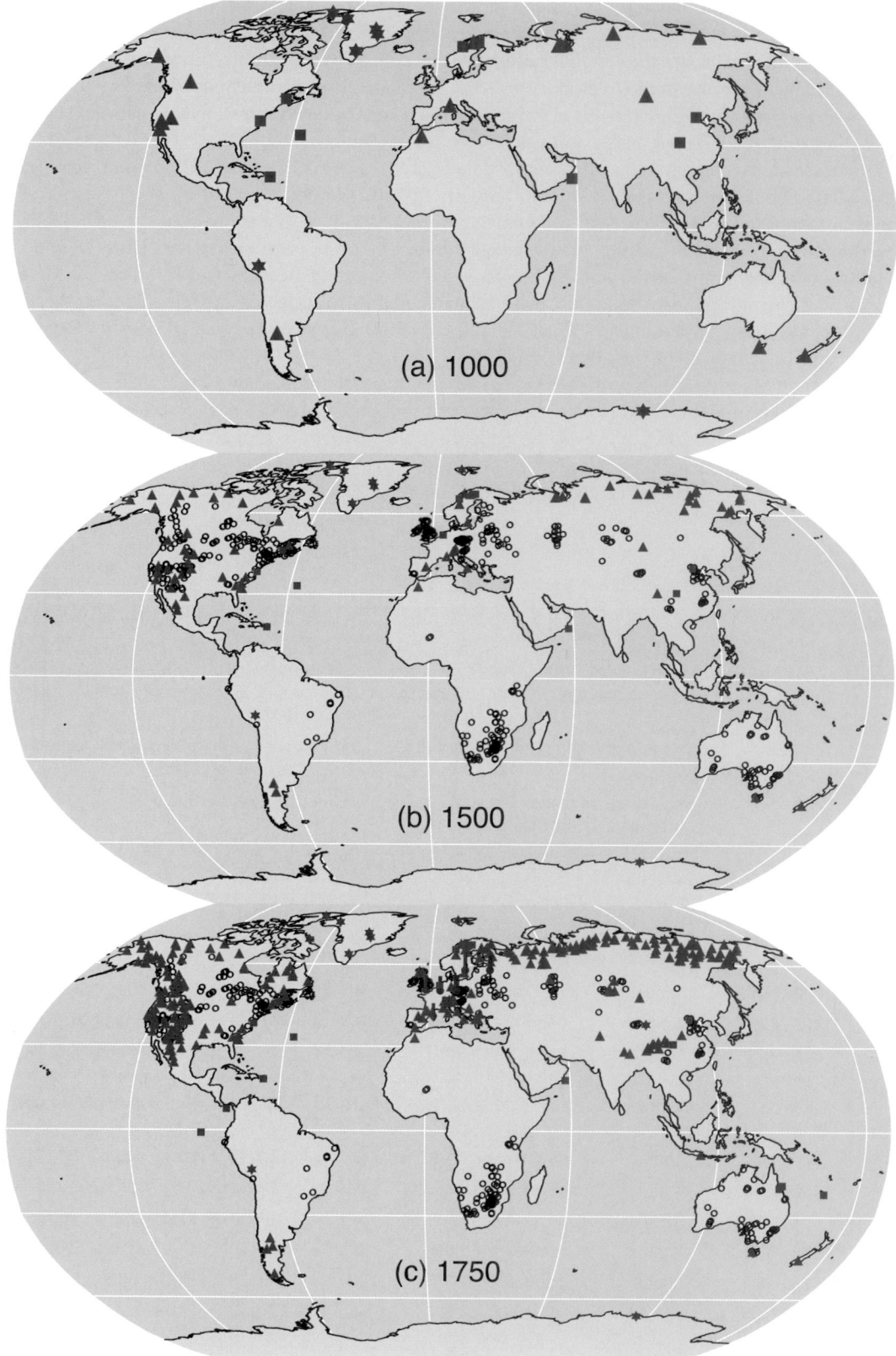

Figure 6.11. *Locations of proxy records with data back to AD 1000, 1500 and 1750 (instrumental: red thermometers; tree ring: brown triangles; borehole: black circles; ice core/ice borehole: blue stars; other including low-resolution records: purple squares) that have been used to reconstruct NH or SH temperatures by studies shown in Figure 6.10 (see Table 6.1, excluding O2005) or used to indicate SH regional temperatures (Figure 6.12).*

1 or 2 kyr, have been published (Esper et al., 2002; Crowley et al., 2003; Mann and Jones, 2003; Cook et al., 2004a; Moberg et al., 2005; Rutherford et al., 2005; D'Arrigo et al., 2006). These are shown, plotted from AD 700 in Figure 6.10b, along with the three series from the TAR. As with the original TAR series, these new records are not entirely independent reconstructions inasmuch as there are some predictors (most often tree ring data and particularly in the early centuries) that are common between them, but in general, they represent some expansion in the length and geographical coverage of the previously available data (Figures 6.10 and 6.11).

Briffa (2000) produced an extended history of interannual tree ring growth incorporating records from sites across northern Fennoscandia and northern Siberia, using a statistical technique to construct the tree ring chronologies that is capable of preserving multi-centennial time scale variability. Although ostensibly representative of northern Eurasian summer conditions, these data were later scaled using simple linear regression against a mean NH land series to provide estimates of summer temperature over the past 2 kyr (Briffa et al., 2004). Esper et al. (2002) took tree ring data from 14 sites in Eurasia and North America, and applied a variant of the same statistical technique designed to produce ring width chronologies in which evidence of long time scale climate forcing is better represented compared with earlier tree ring processing methods. The resulting series were averaged, smoothed and then scaled so that the multi-decadal variance matched that in the Mann et al. (1998) reconstruction over the period 1900 to 1977. This produced a reconstruction with markedly cooler temperatures during the 12th to the end of the 14th century than are apparent in any other series. The relative amplitude of this reconstruction is reduced somewhat when recalibrated directly against smoothed instrumental temperatures (Cook et al., 2004a) or by using annually resolved temperature data (Briffa and Osborn, 2002), but even then, this reconstruction remains at the coldest end of the range defined by all currently available reconstructions.

Mann and Jones (2003) selected only eight normalised series (all screened for temperature sensitivity) to represent annual mean NH temperature change over the last 1.8 kyr. Four of these eight represent integrations of multiple proxy site records or reconstructions, including some O isotope records from ice cores and documentary information as well tree ring records. A weighted average of these decadally smoothed series was scaled so that its mean and standard deviation matched those of the NH decadal mean land and marine record over the period 1856 to 1980. Moberg et al. (2005) used a mixture of tree ring and other proxy-based climate reconstructions to represent changes at short and longer time scales, respectively, across the NH. Seven tree ring series provided information on time scales shorter than 80 years, while 11 far less accurately dated records with lower resolution (including ice melt series, lake diatoms and pollen data, chemistry of marine shells and foraminifera, and one borehole temperature record from the Greenland Ice Sheet) were combined and scaled to match the mean and standard deviation of the instrumental record between 1856 and

1979. This reconstruction displays the warmest temperatures of any reconstruction during the 10th and early 11th centuries, although still below the level of warmth observed since 1980.

Many of the individual annually resolved proxy series used in the various reconstruction studies cited above have been combined in a new reconstruction (only back to AD 1400) based on a climate field reconstruction technique (Rutherford et al., 2005). This study also involved a methodological exploration of the sensitivity of the results to the precise specification of the predictor set, as well as the predictand target region and seasonal window. It concluded that the reconstructions were reasonably robust to differences in the choice of proxy data and statistical reconstruction technique.

D'Arrigo et al. (2006) used only tree ring data, but these include a substantial number not used in other reconstructions, particularly in northern North America. Their reconstruction, similar to that of Esper et al. (2002), displays a large amplitude of change during the past 1 kyr, associated with notably cool excursions during most of the 9th, 13th and 14th centuries, clearly below those of most other reconstructions. Hegerl et al. (2006) used a mixture of 14 regional series, of which only 3 were not made up from tree ring data (a Greenland ice O isotope record and two composite series, from China and Europe, including a mixture of instrumental, documentary and other data). Many of these are common to the earlier reconstructions. However, these series were combined and scaled using a regression approach (total least squares) intended to prevent the loss of low-frequency variance inherent in some other regression approaches. The reconstruction produced lies close to the centre of the range defined by the other reconstructions.

Various statistical methods are used to convert the various sets of original palaeoclimatic proxies into the different estimates of mean NH temperatures shown in Figure 6.10 (see discussions in Jones and Mann, 2004; Rutherford et al., 2005). These range from simple averaging of regional data and scaling of the resulting series so that its mean and standard deviation match those of the observed record over some period of overlap (Jones et al., 1998; Crowley and Lowery, 2000), to complex climate field reconstruction, where large-scale modes of spatial climate variability are linked to patterns of variability in the proxy network via a multivariate transfer function that explicitly provides estimates of the spatio-temporal changes in past temperatures, and from which large-scale average temperature changes are derived by averaging the climate estimates across the required region (Mann et al., 1998; Rutherford et al., 2003, 2005). Other reconstructions can be considered to represent what are essentially intermediate applications between these two approaches, in that they involve regionalisation of much of the data prior to the use of a statistical transfer function, and so involve fewer, but potentially more robust, regional predictors (Briffa et al., 2001; Mann and Jones, 2003; D'Arrigo et al., 2006). Some of these studies explicitly or implicitly reconstruct tropical temperatures based on data largely from the extratropics, and assume stability in the patterns of climate association between these regions. This assumption has been questioned on the basis of both observational and model-simulated data suggesting that tropical to extratropical climate variability can be decoupled (Rind et al., 2005), and also that extratropical teleconnections associated with ENSO may vary through time (see Section 6.5.6).

Oerlemans (2005) constructed a temperature history for the globe based on 169 glacier length records. He used simplified glacier dynamics that incorporate specific response time and climate sensitivity estimates for each glacier. The reconstruction suggests that moderate global warming occurred after the middle of the 19th century, with about 0.6°C warming by the middle of the 20th century. Following a 25-year cooling, temperatures rose again after 1970, though much regional and high-frequency variability is superimposed on this overall interpretation. However, this approach does not allow for changing glacier sensitivity over time, which may limit the information before 1900. For example, analyses of glacier mass balances, volume changes and length variations along with temperature records in the western European Alps (Vincent et al., 2005) indicate that between 1760 and 1830, glacier advance was driven by precipitation that was 25% above the 20th century average, while there was little difference in average temperatures. Glacier retreat after 1830 was related to reduced winter precipitation and the influence of summer warming only became effective at the beginning of the 20th century. In southern Norway, early 18th-century glacier advances can be attributed to increased winter precipitation rather than cold temperatures (Nesje and Dahl, 2003).

Changes in proxy records, either physical (such as the isotopic composition of various elements in ice) or biological (such as the width of a tree ring or the chemical composition of a growth band in coral), do not respond precisely or solely to changes in any specific climate parameter (such as mean temperature or total rainfall), or to the changes in that parameter as measured over a specific 'season' (such as June to August or January to December). For this reason, the proxies must be 'calibrated' empirically, by comparing their measured variability over a number of years with available instrumental records to identify some optimal climate association, and to quantify the statistical uncertainty associated with scaling proxies to represent this specific climate parameter. All reconstructions, therefore, involve a degree of compromise with regard to the specific choice of 'target' or dependent variable. Differences between the temperature reconstructions shown in Figure 6.10b are to some extent related to this, as well as to the choice of different predictor series (including differences in the way these have been processed). The use of different statistical scaling approaches (including whether the data are smoothed prior to scaling, and differences in the period over which this scaling is carried out) also influences the apparent spread between the various reconstructions. Discussions of these issues can also be found in Harris and Chapman (2001), Beltrami (2002), Briffa and Osborn (2002), Esper et al. (2002), Trenberth and Otto-Bliesner (2003), Zorita et al. (2003), Jones and Mann (2004), Pollack and Smerdon (2004), Esper et al. (2005) and Rutherford et al. (2005).

The considerable uncertainty associated with individual reconstructions (2-standard-error range at the multi-decadal time scale is of the order of ±0.5°C) is shown in several publications, calculated on the basis of analyses of regression residuals (Mann et al., 1998; Briffa et al., 2001; Jones et al., 2001; Gerber et al., 2003; Mann and Jones, 2003; Rutherford et al., 2005; D'Arrigo et al., 2006). These are often calculated from the error apparent in the calibration of the proxies. Hence, they are likely to be minimum uncertainties, as they do not take into account other sources of error not apparent in the calibration period, such as any reduction in the statistical robustness of the proxy series in earlier times (Briffa and Osborn, 1999; Esper et al., 2002; Bradley et al., 2003b; Osborn and Briffa, 2006).

All of the large-scale temperature reconstructions discussed in this section, with the exception of the borehole and glacier interpretations, include tree ring data among their predictors so it is pertinent to note several issues associated with them. The construction of ring width and ring density chronologies involves statistical processing designed to remove non-climate trends that could obscure the evidence of climate that they contain. In certain situations, this process may restrict the extent to which a chronology portrays the evidence of long time scale changes in the underlying variability of climate that affected the growth of the trees; in effect providing a high-pass filtered version of past climate. However, this is generally not the case for chronologies used in the reconstructions illustrated in Figure 6.10. Virtually all of these used chronologies or tree ring climate reconstructions produced using methods that preserve multi-decadal and centennial time scale variability. As with all biological proxies, the calibration of tree ring records using linear regression against some specific climate variable represents a simplification of what is inevitably a more complex and possibly time-varying relationship between climate and tree growth. That this is a defensible simplification, however, is shown by the general strength of many such calibrated relationships, and their significant verification using independent instrumental data. There is always a possibility that non-climate factors, such as changing atmospheric CO_2 or soil chemistry, might compromise the assumption of uniformity implicit in the interpretation of regression-based climate reconstructions, but there remains no evidence that this is true for any of the reconstructions referred to in this assessment. A group of high-elevation ring width chronologies from the western USA that show a marked growth increase during the last 100 years, attributed by LaMarche et al. (1984) to the fertilizing effect of increasing atmospheric CO_2, were included among the proxy data used by Mann et al. (1998, 1999). However, their tree ring data from the western USA were adjusted specifically in an attempt to mitigate this effect. Several analyses of ring width and ring density chronologies, with otherwise well-established sensitivity to temperature, have shown that they do not emulate the general warming trend evident in instrumental temperature records over recent decades, although they do track the warming that occurred during the early part of the 20th century and they continue to maintain a good correlation with observed temperatures over the full instrumental period at the interannual time scale (Briffa et al., 2004; D'Arrigo, 2006). This 'divergence' is apparently restricted to some northern, high-latitude regions, but it is certainly not ubiquitous even there. In their large-scale reconstructions based on tree ring density data, Briffa et al. (2001) specifically excluded the post-1960 data in their calibration against instrumental records, to avoid biasing the estimation of the earlier reconstructions (hence they are not shown in Figure 6.10), implicitly assuming that the 'divergence' was a uniquely recent phenomenon, as has also been argued by Cook et al. (2004a). Others, however, argue for a breakdown in the assumed linear tree growth response to continued warming, invoking a possible threshold exceedance beyond which moisture stress now limits further growth (D'Arrigo et al., 2004). If true, this would imply a similar limit on the potential to reconstruct possible warm periods in earlier times at such sites. At this time there is no consensus on these issues (for further references see NRC, 2006) and the possibility of investigating them further is restricted by the lack of recent tree ring data at most of the sites from which tree ring data discussed in this chapter were acquired.

Figure 6.10b illustrates how, when viewed together, the currently available reconstructions indicate generally greater variability in centennial time scale trends over the last 1 kyr than was apparent in the TAR. It should be stressed that each of the reconstructions included in Figure 6.10b is shown scaled as it was originally published, despite the fact that some represent seasonal and others mean annual temperatures. Except for the borehole curve (Pollack and Smerdon, 2004) and the interpretation of glacier length changes (Oerlemans, 2005), they were originally also calibrated against different instrumental data, using a variety of statistical scaling approaches. For all these reasons, these reconstructions would be expected to show some variation in relative amplitude.

Figure 6.10c is a schematic representation of the most likely course of hemispheric mean temperature change during the last 1.3 kyr based on all of the reconstructions shown in Figure 6.10b, and taking into account their associated statistical uncertainty. The envelopes that enclose the two standard error confidence limits bracketing each reconstruction have been overlain (with greater emphasis placed on the area within the 1 standard error limits) to show where there is most agreement between the various reconstructions. The result is a picture of relatively cool conditions in the 17th and early 19th centuries and warmth in the 11th and early 15th centuries, but the warmest conditions are apparent in the 20th century. Given that the confidence levels surrounding all of the reconstructions are wide, virtually all reconstructions are effectively encompassed within the uncertainty previously indicated in the TAR. The major differences between the various proxy reconstructions relate to the magnitude of past cool excursions, principally during the 12th to 14th, 17th and 19th centuries. Several reconstructions exhibit a short-lived maximum just prior to AD 1000 but only one (Moberg et al., 2005) indicates persistent hemispheric-scale conditions (i.e., during AD 990 to 1050 and AD 1080 to 1120) that were as warm as those in the 1940s and 50s. However, the long time scale variability in this reconstruction is determined by low-resolution proxy records that cannot be rigorously calibrated against recent instrumental temperature data (Mann et al., 2005b). None of the reconstructions in Fig. 6.10 show pre-20th century temperatures reaching the levels seen in the instrumental temperature record for the last two decades of the 20th century.

It is important to recognise that in the NH as a whole there are few long and well-dated climate proxies, particularly for the period prior to the 17th century (Figure 6.11). Those that do exist are concentrated in extratropical, terrestrial locations, and many have greatest sensitivity to summer rather than winter (or annual) conditions. Changes in seasonality probably limit the conclusions that can be drawn regarding annual temperatures derived from predominantly summer-sensitive proxies (Jones et al., 2003). There are very few strongly temperature-sensitive proxies from tropical latitudes. Stable isotope data from high-elevation ice cores provide long records and have been interpreted in terms of past temperature variability (Thompson, 2000), but recent calibration and modelling studies in South America and southern Tibet (Hoffmann et al., 2003; Vuille and Werner, 2005; Vuille et al., 2005) indicate a dominant sensitivity to precipitation changes, at least on seasonal to decadal time scales, in these regions. Very rapid and apparently unprecedented melting of tropical ice caps has been observed in recent decades (Thompson et al., 2000; Thompson, 2001; see Box 6.3), likely associated with enhanced warming at high elevations (Gaffen et al., 2000; see Chapter 4). Coral O isotopes and Sr/Ca ratios reflect SSTs, although the former are also influenced by salinity changes associated with precipitation variability (Lough, 2004). Unfortunately, these records are invariably short, of the order of centuries at best, and can be associated with age uncertainties of 1 or 2%. Virtually all coral records currently available from the tropical Indo-Pacific indicate unusual warmth in the 20th century (Cole, 2003), and in the tropical Indian Ocean many isotope records show a trend towards warmer conditions (Charles et al., 1997; Kuhnert et al., 1999; Cole et al., 2000). In most multi-centennial length coral series, the late 20th century is warmer than any time in the last 100 to 300 years.

Using pseudo-proxy networks extracted from GCM simulations of global climate for the last millennium, von Storch et al. (2004) suggested that temperature reconstructions may not fully represent variance on long time scales. This would represent a bias, as distinct from the random error represented by published reconstruction uncertainty ranges. At present, the extent of any such biases in specific reconstructions and as indicated by pseudo-proxy studies is uncertain (being dependent on the choice of statistical regression model and climate model simulation used to provide the pseudo-proxies). It is very unlikely, however, that any bias would be as large as the factor of two suggested by von Storch et al. (2004) with regard to the reconstruction by Mann et al. (1998), as discussed by Burger and Cubash (2005) and Wahl et al. (2006). However, the bias will depend on the degree to which past climate departs from the range of temperatures encompassed within the calibration period data (Mann et al., 2005b; Osborn and Briffa, 2006) and on the proportions of temperature variability occurring on short and long time scales (Osborn and Briffa, 2004). In any case, this bias would act to damp the amplitude of reconstructed departures that are further from the calibration period mean, so that temperatures during cooler periods may have been colder than estimated by some reconstructions, while periods with comparable temperatures (e.g., possible portions of the period between AD 950 and 1150, Figure 6.10) would be largely unbiased. As only one reconstruction (Moberg et al., 2005) shows an early period that is noticeably warmer than the mean for the calibration period, the possibility of a bias does not affect the general conclusion about the relative warmth of the 20th century based on these data.

The weight of current multi-proxy evidence, therefore, suggests greater 20th-century warmth, in comparison with temperature levels of the previous 400 years, than was shown in the TAR. On the evidence of the previous and four new reconstructions that reach back more than 1 kyr, it is likely that the 20th century was the warmest in at least the past 1.3 kyr. Considering the recent instrumental and longer proxy evidence together, it is very likely that average NH temperatures during the second half of the 20th century were higher than for any other 50-year period in the last 500 years. Greater uncertainty associated with proxy-based temperature estimates for individual years means that it is more difficult to gauge the significance, or precedence, of the extreme warm years observed in the recent instrumental record, such as 1998 and 2005, in the context of the last millennium.

6.6.1.2 What Do Large-Scale Temperature Histories from Subsurface Temperature Measurements Show?

Hemispheric or global ground surface temperature (GST) histories reconstructed from measurements of subsurface temperatures in continental boreholes have been presented by several geothermal research groups (Huang et al., 2000; Harris and Chapman, 2001; Beltrami, 2002; Beltrami and Bourlon, 2004; Pollack and Smerdon, 2004); see Pollack and Huang (2000) for a review of this methodology. These borehole reconstructions have been derived using the contents of a publicly available database of borehole temperatures and climate reconstructions (Huang and Pollack, 1998) that in 2004 included 695 sites in the NH and 166 in the SH (Figure 6.11). Because the solid Earth acts as a low-pass filter on downward-propagating temperature signals, borehole reconstructions lack annual resolution; accordingly they typically portray only multi-decadal to centennial changes. These geothermal reconstructions provide independent estimates of surface temperature history with which to compare multi-proxy reconstructions. Figure 6.10b shows a reconstruction of average NH GST by Pollack and Smerdon (2004). This reconstruction, very similar to that presented by Huang et al. (2000), shows an overall warming of the ground surface of about 1.0°C over the past five centuries. The two standard error uncertainties for their series (not shown here) are 0.20°C (in 1500), 0.10°C (1800) and 0.04°C (1900). These are errors associated with various scales of areal weighting

and consequent suppression of site-specific noise through aggregation (Pollack and Smerdon, 2004). The reconstruction is similar to the cooler multi-proxy reconstructions in the 16th and 17th centuries, but sits in the middle of the multi-proxy range in the 19th and early 20th centuries. A geospatial analysis by Mann et al. (2003; see correction by Rutherford and Mann, 2004) of the results of Huang et al. (2000) argued for significantly less overall warming, a conclusion contested by Pollack and Smerdon (2004) and Beltrami and Bourlon (2004). Geothermal reconstructions based on the publicly available database generally yield somewhat muted estimates of the 20th-century trend, because of a relatively sparse representation of borehole data north of 60°N. About half of the borehole sites at the time of measurement had not yet been exposed to the significant warming of the last two decades of the 20th century (Taylor et al., 2006; Majorowicz et al., 2004).

The assumption that the reconstructed GST history is a good representation of the surface air temperature (SAT) history has been examined with both observational data and model studies. Observations of SAT and GST display differences at daily and seasonal time scales, and indicate that the coupling of SAT and GST over a single year is complex (Sokratov and Barry, 2002; Stieglitz et al., 2003; Bartlett et al., 2004; Smerdon et al., 2006). The mean annual GST differs from the mean annual SAT in regions where there is snow cover and/or seasonal freezing and thawing (Gosnold et al., 1997; Smerdon et al., 2004; Taylor et al., 2006), as well as in regions without those effects (Smerdon et al., 2006). Observational time series of ground temperatures are not long enough to establish whether the mean annual differences are stable over long time scales. The long-term coupling between SAT and GST has been addressed by simulating both air and soil temperatures in global three-dimensional coupled climate models. Mann and Schmidt (2003), in a 50-year experiment using the GISS Model E, suggested that GST reconstructions may be biased by seasonal influences and snow cover variability, an interpretation contested by Chapman et al (2004). Thousand-year simulations by González-Rouco et al. (2003, 2006) using the ECHO-G model suggest that seasonal differences in coupling are of little significance over long time scales. They also indicate that deep soil temperature is a good proxy for the annual SAT on continents and that the spatial array of borehole locations is adequate to reconstruct the NH mean SAT. Neither of these climate models included time-varying vegetation cover.

6.6.2 Southern Hemisphere Temperature Variability

There are markedly fewer well-dated proxy records for the SH compared to the NH (Figure 6.11), and consequently little evidence of how large-scale average surface temperatures have changed over the past few thousand years. Mann and Jones (2003) used only three series to represent annual mean SH temperature change over the last 1.5 kyr. A weighted combination of the individual standardised series was scaled to match (at decadal time scales) the mean and the standard deviation of SH annual mean land and marine temperatures over the period 1856 to 1980. The recent proxy-based temperature estimates, up to the end of the reconstruction in 1980, do not capture the full magnitude of the warming seen in the instrumental temperature record. Earlier periods, around AD 700 and 1000, are reconstructed as warmer than the estimated level in the 20th century, and may have been as warm as the measured values in the last 20 years. The paucity of SH proxy data also means that uncertainties associated with hemispheric temperature estimates are much greater than for the NH, and it is more appropriate at this time to consider the evidence in terms of limited regional indicators of temperature change (Figure 6.12).

The long-term oscillations in warm-season temperatures shown in a tree ring reconstruction for Tasmania (Cook et al., 2000) suggest that the last 30 years was the warmest multi-decadal period in the last 1 kyr, but only by a marginal degree. Conditions were generally warm over a longer period from 1300 to 1500 (Figure 6.12). Another tree ring reconstruction, of austral summer temperatures based on data from South Island, New Zealand, spans the past 1.1 kyr and is the longest yet produced for the region (Cook et al., 2002a). Disturbance at the site from which the trees were sampled restricts the calibration of this record to the 70 years up until 1950, but both tree rings and instrumental data indicate that the 20th century was not anomalously warm when compared to several warm periods reconstructed in the last 1 kyr (around the mid-12th and early 13th centuries and around 1500).

Tree-ring based temperature reconstructions across the Southern Andes (37°S to 55°S) of South America indicate that the annual temperatures during the 20th century have been anomalously high in the context of the past four centuries. The mean annual temperatures for northern and southern Patagonia during the interval 1900 to 1990 are 0.53°C and 0.86°C above the 1640 to 1899 means, respectively (Figure 6.12). In Northern Patagonia, the highest temperatures occurred in the 1940s. In Southern Patagonia, the year 1998 was the warmest of the past four centuries. The rate of temperature increase from 1850 to 1920 was the highest over the past 360 years (Villalba et al., 2003).

Figure 6.12 also shows the evidence of GST changes over the last 500 years, provided by regionally aggregated borehole

temperature inversions (Figure 6.11) from southern Africa (92 records) and Australia (57 records) described in Huang et al. (2000). The instrumental records for these areas show warmer conditions that postdate the time when the boreholes were logged; thus, the most recent warming is not registered in these borehole curves. A more detailed analysis of the Australian geothermal reconstruction (Pollack et al., 2006), indicates that the warming of Australia in the past five centuries was apparently only half that experienced over the continents of the NH during the same period and shows good correspondence with the tree-ring based reconstructions for Tasmania and New Zealand (Cook et al., 2000, 2002a). Contrasting evidence of past temperature variations at Law Dome, Antarctica has been derived from ice core isotope measurements and from the inversion of a subsurface temperature profile (Dahl-Jensen et al., 1999; Goosse et al., 2004; Jones and Mann, 2004). The borehole analysis indicates colder intervals at around 1250 and 1850, followed by a gradual warming of 0.7°C to the present. The isotope record indicates a relatively cold 20th century and warmer conditions throughout the period 1000 to 1750.

Taken together, the very sparse evidence for SH temperatures prior to the period of instrumental records indicates that unusual warming is occurring in some regions. However, more proxy data are required to verify the apparent warm trend.

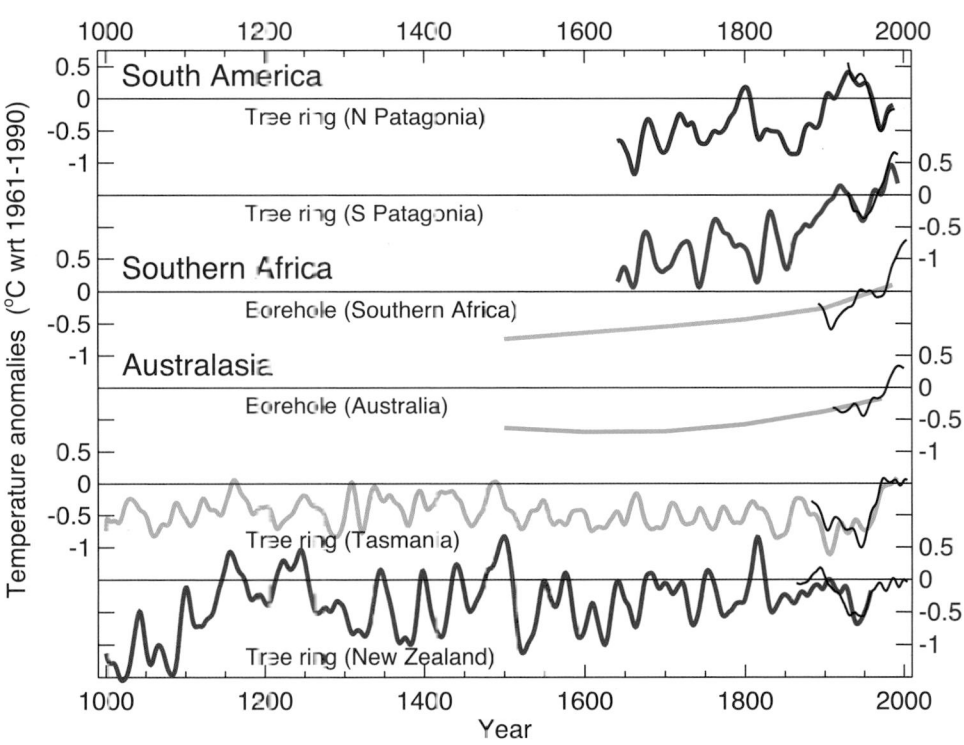

Figure 6.12. *Temperature reconstructions for regions in the SH: two annual temperature series from South American tree ring data (Villalba et al., 2003); annual temperature estimates from borehole inversions for southern Africa and Australia (Huang et al., 2000); summer temperature series from Tasmania and New Zealand tree ring data (Cook et al., 2000, 2002a). The black curves show summer or annual instrumental temperatures for each region. All tree ring and instrumental series were smoothed with a 25-year filter and represent anomalies (°C) from the 1961 to 1990 mean (indicated by the horizontal lines).*

6.6.3   Comparisons of Millennial Simulations with Palaeodata

A range of increasingly complex climate models has been used to simulate NH temperatures over the last 500 to 1,000 years using both natural and anthropogenic forcings (Figure 6.13). These models include an energy balance formulation (Crowley et al., 2003, Gerber et al., 2003), two- and three-dimensional reduced complexity models (Bertrand et al., 2002b; Bauer et al., 2003), and three fully coupled AOGCMs (Ammann et al., 2003; Von Storch et al., 2004; Tett et al., 2007).

Comparison and evaluation of the output from palaeoclimate simulations is complicated by their use of different historical forcings, as well as by the way indirect evidence of the history of various forcings is translated into geographically and seasonally specific radiative inputs within the models. Some factors, such as orbital variations of the Earth in relation to the Sun, can be calculated accurately (e.g., Berger, 1977; Bradley et al., 2003b) and directly implemented in terms of latitudinal and seasonal changes in incoming shortwave radiation at the top of the atmosphere. For the last 2 kyr, although this forcing is incorporated in most models, its impact on climate can be neglected compared to the other forcings (Bertrand et al., 2002b).

Over recent millennia, the analysis of the gas bubbles in ice cores with high deposition rates provides good evidence of greenhouse gas changes at near-decadal resolution (Figure 6.4). Other factors, such as land use changes (Ramankutty and Foley, 1999) and the concentrations and distribution of tropospheric aerosols and ozone, are not as well known (Mickley et al.,

2001). However, because of their magnitude, uncertainties in the history of solar irradiance and volcanic effects are more significant for the pre-industrial period.

6.6.3.1   Solar Forcing

The direct measurement of solar irradiance by satellite began less than 30 years ago, and over this period only very small changes are apparent (0.1% between the peak and trough of recent sunspot cycles, which equates to only about 0.2 W m^{-2} change in radiative forcing; Fröhlich and Lean (2004); see Section 2.7). Earlier extensions of irradiance change used in most model simulations are estimated by assuming a direct correlation with evidence of changing sunspot numbers and cosmogenic isotope production as recorded in ice cores (^{10}Be) and tree rings (^{14}C) (Lean et al., 1995; Crowley, 2000).

There is general agreement in the evolution of the different proxy records of solar activity such as cosmogenic isotopes, sunspot numbers or aurora observations, and the annually resolved records clearly depict the well-known 11-year solar cycle (Muscheler et al., 2006). For example, palaeoclimatic ^{10}Be and ^{14}C values are higher during times of low or absent sunspot numbers. During these periods, their production is high as the shielding of the Earth's atmosphere from cosmic rays provided by the Sun's open magnetic field is weak (Beer et al., 1998). However, the relationship between the isotopic records indicative of the Sun's open magnetic field, sunspot numbers and the Sun's closed magnetic field or energy output are not fully understood (Wang and Sheeley, 2003).

Table 6.2. *Climate model simulations shown in Figure 6.13.*

Series	Model[a]	Model type	Forcings[b]	Reference
GSZ2003	ECHO-G	GCM	SV -G - - - -	González-Rouco et al., 2003
ORB2006	ECHO-G/MAGICC	GCM adj. using EBM[c]	SV -G -A -Z	Osborn et al., 2006
TBC..2006	HadCM3	GCM	SVOG -ALZ	Tett et al., 2007
AJS..2006	NCAR CSM	GCM	SV -G -A -Z	Mann et al., 2005b
BLC..2002	MoBiDiC	EMIC	SV -G -AL -	Bertrand et al., 2002b
CBK..2003	-	EBM[c]	SV -G -A - -	Crowley et al., 2003
GRT..2005	ECBilt-CLIO	EMIC	SV -G -A - -	Goosse et al., 2005b
GJB..2003	Bern CC	EBM[c]	SV -G -A -Z	Gerber et al., 2003
B..03-14C	Climber2	EMIC (solar from ^{14}C)	SV - -C -L -	Bauer et al., 2003
B..03-10Be	Climber2	EMIC (solar from ^{10}Be)	SV - -C -L -	Bauer et al., 2003
GBZ..2006	ECHO-G	GCM	SV -G - - - -	González-Rouco et al., 2006
SMC2006	ECHAM4/OPYC3	GCM	SV -G -A -Z	Stendel et al., 2006

Notes:

[a] Models: ECHO-G = ECHAM4 atmospheric GCM/HOPE-G ocean GCM, MAGICC = Model for the Assessment of Greenhouse-gas Induced Climate Change, HadCM3 = Hadley Centre Coupled Model 3; NCAR CSM = National Center for Atmospheric Research Climate System Model, MoBiDiC = Modèle Bidimensionnel du Climat , ECBilt-CLIO = ECBilt-Coupled Large-scale Ice Ocean, Bern CC = Bern Carbon Cycle-Climate Model, CLIMBER2 = Climate Biosphere Model 2, ECHAM4/OPYC3 = ECHAM4 atmospheric GCM/Ocean Isopycnal GCM 3.

[b] Forcings: S = solar, V = volcanic, O = orbital, G = well-mixed greenhouse gases, C = CO_2 but not other greenhouse gases, A = tropospheric sulphate aerosol, L = land use change, Z=tropospheric and/or stratospheric ozone changes and/or halocarbons.

[c] EBM = Energy Balance Model.

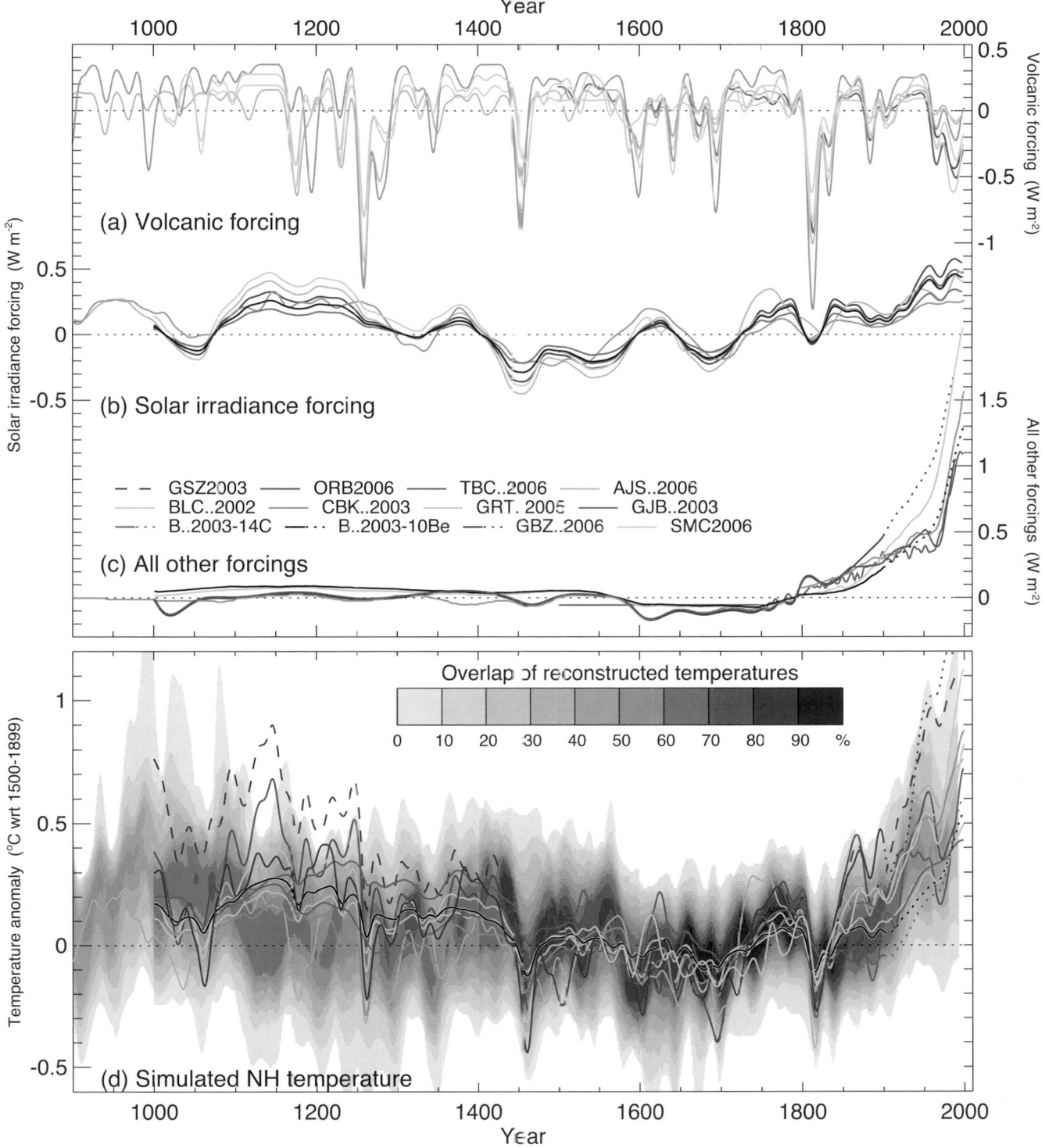

Figure 6.13. *Radiative forcings and simulated temperatures during the last 1.1 kyr. Global mean radiative forcing (W m⁻²) used to drive climate model simulations due to (a) volcanic activity, (b) solar irradiance variations and (c) all other forcings (which vary between models, but always include greenhouse gases, and, except for those with dotted lines after 1900, tropospheric sulphate aerosols). (d) Annual mean NH temperature (°C) simulated under the range of forcings shown in (a) to (c), compared with the concentration of overlapping NH temperature reconstructions (shown by grey shading, modified from Figure 6.10c to account for the 1500 to 1899 reference period used here). All forcings and temperatures are expressed as anomalies from their 1500 to 1899 means and then smoothed with a Gaussian-weighted filter to remove fluctuations on time scales less than 30 years; smoothed values are obtained up to both ends of each record by extending the records with the mean of the adjacent existing values. The individual series are identified in Table 6.2.*

The cosmogenic isotope records have been linearly scaled to estimate solar energy output (Bard et al., 2000) in many climate simulations. More recent studies utilise physics-based models to estimate solar activity from the production rate of cosmogenic isotopes taking into account nonlinearities between isotope production and the Sun's open magnetic flux and variations in the geomagnetic field (Solanki et al., 2004; Muscheler et al., 2005). Following this approach, Solanki et al. (2004) suggested that the current level of solar activity has been without precedent over the last 8 kyr. This is contradicted by a more recent analysis linking the isotope proxy records to instrumental data that identifies, for the last millennium, three periods (around AD 1785, 1600 and 1140) when solar activity was as high, or higher, than in the satellite era (Muscheler et al., 2006).

The magnitude of the long-term trend in solar irradiance remains uncertain. A reassessment of the stellar data (Hall and Lockwood, 2004) has been unable to confirm or refute the analysis by Baliunas and Jastrow (1990) that implied significant long-term solar irradiance changes, and also underpinned some of the earlier reconstructions (see Section 2.7). Several new studies (Lean et al., 2002; Foster, 2004; Foukal et al., 2004; Y.M. Wang et al., 2005) suggest that long-term irradiance changes were notably less than in earlier reconstructions (Hoyt and Schatten, 1993; Lean et al., 1995; Lockwood and Stamper, 1999; Bard et al., 2000; Fligge and Solanki, 2000; Lean, 2000) that were employed in a number of TAR climate change simulations and in many of the simulations shown in Figure 6.13d.

In the previous reconstructions, the 17th-century 'Maunder Minimum' total irradiance was 0.15 to 0.65% (irradiance change about 2.0 to 8.7 W m^{-2}; radiative forcing about 0.36 to 1.55 W m^{-2}) below the present-day mean (Figure 6.13b). Most of the recent studies (with the exception of Solanki and Krivova, 2003) calculate a reduction of only around 0.1% (irradiance change of the order of –1 W m^{-2}, radiative forcing of –0.2 W m^{-2}; section 2.7). Following these results, the magnitude of the radiative forcing used in Chapter 9 for the Maunder Minimum period is relatively small (–0.2 W m^{-2} relative to today).

6.6.3.2   Volcanic Forcing

There is also uncertainty in the estimates of volcanic forcing during recent millennia because of the necessity to infer atmospheric optical depth changes (including geographic details as well as temporal accuracy and persistence), where there is only indirect evidence in the form of levels of acidity and sulphate measured in ice cores (Figures 6.14 and 6.15). All of the volcanic histories used in current model-based palaeoclimate simulations are based on analyses of polar ice cores containing minor dating uncertainty and obvious geographical bias.

The considerable difficulties in calculating hemispheric and regional volcanic forcing changes (Robock and Free, 1995; Robertson et al., 2001; Crowley et al., 2003) result from sensitivity to the choice of which ice cores are considered,

assumptions as to the extent of stratosphere penetration by eruption products, and the radiative properties of different volcanic aerosols and their residence time in the stratosphere. Even after producing some record of volcanic activity, there are major differences in the way models implement this. Some use a direct reduction in global radiative forcing with no spatial discrimination (von Storch et al., 2004), while other models prescribe geographical changes in radiative forcing (Crowley et al., 2003; Goosse et al., 2005a; Stendel et al., 2006). Models with more sophisticated radiative schemes are able to incorporate prescribed aerosol optical depth changes, and interactively calculate the perturbed (longwave and shortwave) radiation budgets (Tett et al., 2007). The effective level of (prescribed or diagnosed) volcanic forcing therefore varies considerably between the simulations (Figure 6.13a).

6.6.3.3   Industrial Era Sulphate Aerosols

Ice core data from Greenland and the mid-latitudes of the NH (Schwikowski et al., 1999; Bigler et al., 2002) provide evidence of the rapid increase in sulphur dioxide emissions (Stern, 2005) and tropospheric sulphate aerosol loading, above the pre-industrial background, during the modern industrial era but they also show a very recent decline in these emissions (Figure 6.15). Data from ice cores show that sulphate aerosol deposition has not changed on Antarctica, remote from anthropogenic sulphur dioxide sources. The ice records are indicative of the regional-to-hemispheric scale atmospheric loading of sulphate aerosols that varies regionally as aerosols have a typical lifetime of only weeks in the troposphere. In recent years, sulphur dioxide emissions have decreased globally and in many regions of the NH (Stern, 2005; see Chapter 2). In general, tropospheric sulphate aerosols exert a negative temperature forcing that will be less if sulphur dioxide emissions and the sulphate loading in the atmosphere continue to decrease.

6.6.3.4   Comparing Simulations of Northern Hemisphere Mean Temperatures with Palaeoclimatic Observations

Various simulations of NH (mean land and marine) surface temperatures produced by a range of climate models, and the forcings that were used to drive them, are shown in Figure 6.13. Despite differences in the detail and implementation of the different forcing histories, there is generally good qualitative agreement between the simulations regarding the major features: warmth during much of the 12th through 14th centuries, with lower temperatures being sustained during the 17th, mid 15th and early 19th centuries, and the subsequent sharp rise to unprecedented levels of warmth at the end of the 20th century. The spread of this multi-model ensemble is constrained to be small during the 1500 to 1899 reference period (selected following Osborn et al., 2006), but the model spread also remains small back to 1000, with the exception of the ECHO-G simulation (Von Storch et al., 2004). The implications of the greater model spread in the rates of warming after 1840 will be clear only after determining the extent to which it can be attributed to differences in prescribed forcings and individual model sensitivities (Goosse et al., 2005b). The ECHO-G simulation (dashed red line in Figure 6.13d) is atypical compared to the ensemble as a whole, being notably warmer in the pre-1300 and post-1900 periods. Osborn et al. (2006) showed that these anomalies are likely the result of a large initial disequilibrium and the lack of anthropogenic tropospheric aerosols in that simulation (see Figure 6.13c). One other simulation (González-Rouco et al., 2006) also exhibits greater early 20th-century warming in comparison to the other simulations but, similarly, does not include tropospheric aerosols among the forcings. All of these simulations, therefore, appear to be consistent with the reconstructions of past NH temperatures, for which the evidence (taken from Figure 6.10c)

is shown by the grey shading underlying the simulations in Figure 6.13d.

It is important to note that many of the simulated temperature variations during the pre-industrial period shown in Figure 6.13 have been driven by assumed solar forcing, the magnitude of which is currently in doubt. Therefore, although the data and simulations appear consistent at this hemispheric scale, they are not a powerful test of the models because of the large uncertainty in both the reconstructed NH changes and the total radiative forcing. The influence of solar irradiance variability and anthropogenic forcings on simulated NH surface temperature is further illustrated in Figure 6.14. A range of EMICs (Petoukhov et al., 2000; Plattner et al., 2001; Montoya et al., 2005) were forced with two different reconstructions of solar irradiance (Bard et al., 2000; Y.M. Wang et al., 2005) to compare the

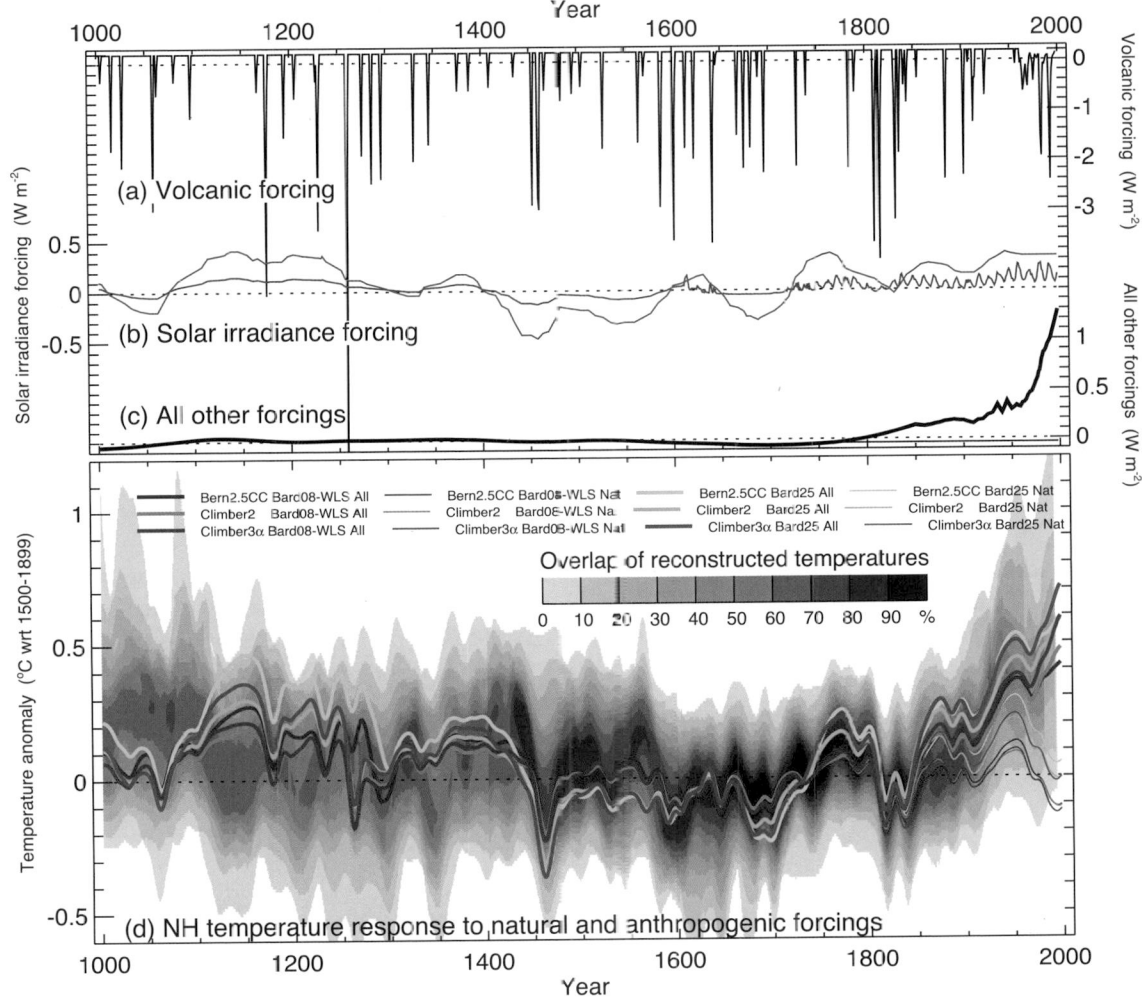

Figure 6.14. *Simulated temperatures during the last 1 kyr with and without anthropogenic forcing, and also with weak or strong solar irradiance variations. Global mean radiative forcing (W m^{-2}) used to drive climate model simulations due to (a) volcanic activity, (b) strong (blue) and weak (brown) solar irradiance variations, and (c) all other forcings, including greenhouse gases and tropospheric sulphate aerosols (the thin flat line after 1765 indicates the fixed anthropogenic forcing used in the 'Nat' simulations). (d) Annual mean NH temperature (°C) simulated by three climate models under the forcings shown in (a) to (c), compared with the concentration of overlapping NH temperature reconstructions (shown by grey shading, modified from Figure 6.10c to account for the 1500 to 1899 reference period used here). 'All' (thick lines) used anthropogenic and natural forcings; 'Nat' (thin lines) used only natural forcings. All forcings and temperatures are expressed as anomalies from their 1500 to 1899 means; the temperatures were then smoothed with a Gaussian-weighted filter to remove fluctuations on time scales less than 30 years. Note the different vertical scale used for the volcanic forcing compared with the other forcings. The individual series are identified in Table 6.3.*

influence of large versus small changes in the long-term strength of solar irradiance over the last 1 kyr (Figure 6.14b). Radiative forcing related to explosive volcanism (Crowley, 2000), atmospheric CO_2 and other anthropogenic agents (Joos et al., 2001) were identically prescribed within each model simulation. Additional simulations, in which anthropogenic forcings were not included, enable a comparison to be made between 'natural' versus 'all' (i.e., natural plus anthropogenic) forcings on the evolution of hemispheric temperatures before and during the 20th century.

The alternative solar irradiance histories used in the simulations differ in their low-frequency amplitudes by a factor of about three. The 'high-amplitude' case (strong solar irradiance forcing) corresponds roughly with the level of irradiance change assumed in many of the simulations shown in Figure 6.13b, whereas the 'low-amplitude'

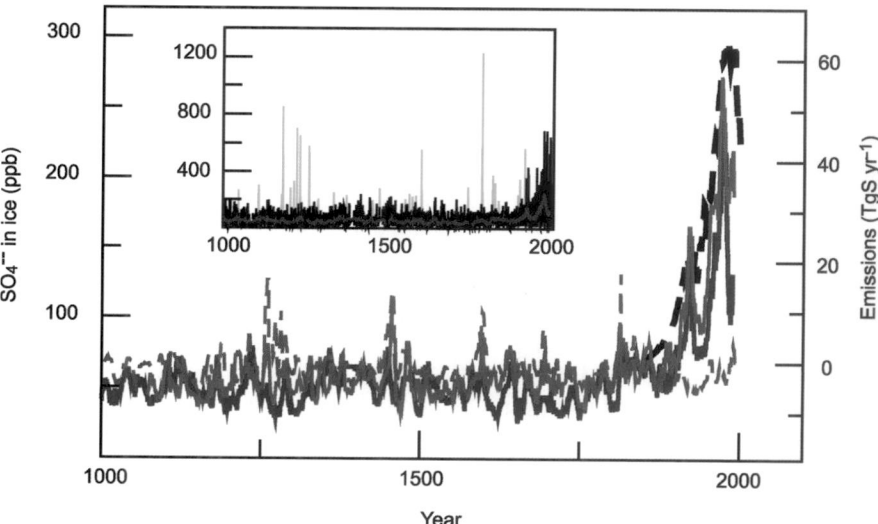

Figure 6.15. *Sulphate (SO_4^{2-}) concentrations in Greenland (Bigler et al., 2002, red line; Mieding, 2005, blue) and antarctic (Traufetter et al., 2004, dash, violet) ice cores during the last millennium. Also shown are the estimated anthropogenic sulphur (S) emissions for the NH (Stern, 2005; dashed black). The ice core data have been smoothed with a 10-year running median filter, thereby removing the peaks of major volcanic eruptions. The inset illustrates the influence of volcanic emissions over the last millennium and shows monthly sulphate data in ppm as measured (green), with identified volcanic spikes removed (black, most recent volcanic events were not assigned nor removed), and results from the 10-year filter (red) (Bigler et al., 2002). The records represent illustrative examples and can be influenced by local deposition events.*

case (weaker solar irradiance forcing) is representative of the more recent reconstructions of solar irradiance changes (as discussed in Section 6.6.3). The high-amplitude forcing history ('Bard25', Table 6.3) is based on an ice core record of ¹⁰Be scaled to give an average reduction in solar irradiance of 0.25% during the Maunder Minimum, as compared to today (Bard et al., 2000). The low-amplitude history ('Bard08-WLS') is estimated using sunspot data and a model of the Sun's closed magnetic flux for the period from 1610 to the present (Y.M. Wang et al., 2005), with an earlier extension based on the Bard et al. (2000) record scaled to a Maunder Minimum reduction of 0.08% compared to today. The low-frequency evolution of these two reconstructions is very similar (Figure 6.14) even

though they are based on completely independent sources of observational data (sunspots versus cosmogenic isotopes) and are produced differently (simple linear scaling versus modelled Sun's magnetic flux) after 1610.

The EMIC simulations shown in Figure 6.14, like those in Figure 6.13d, fall within the range of proxy-based NH temperature reconstructions shown in Figure 6.10c and are compatible with reconstructed and observed 20th-century warming only when anthropogenic forcings are incorporated. The standard deviation of multi-decadal variability in NH SAT is greater by 0.04°C to 0.07°C for the stronger solar forcing (Bard25, Table 6.3) compared to the weaker solar forcing (Bard08-WLS). The uncertainty associated with the proxy-

Table 6.3. *Simulations with intermediate complexity climate models shown in Figure 6.14.*

Modelsᵃ:	
Bern2.5CC	Plattner et al., 2001
Climber2	Petoukhov et al., 2000
Climber3α	Montoya et al., 2005
Forcings:	
Volcanic	Forcing from Crowley (2000) used in all runs
Solar	'Bard25' runs used strong solar irradiance changes, based on ¹⁰Be record scaled to give a Maunder Minimum irradiance 0.25% lower than today, from Bard et al. (2000) 'Bard08-WLS' runs used weak solar irradiance changes, using sunspot records and a model of the Sun's magnetic flux for the period since 1610, from Y.M. Wang et al. (2005), and extended before this by the ¹⁰Be record scaled to give a Maunder Minimum irradiance 0.08% lower than today
Anthropogenic	'All' runs included anthropogenic forcings after 1765, from Joos et al. (2001) 'Nat' runs did not include any anthropogenic forcings

Notes:
ᵃ Models: Bern2.5CC = Bern 2.5D Carbon Cycle-Climate Model, CLIMBER = Climate Biosphere Model.

based temperature reconstructions and climate sensitivity of the models is too large to establish, based on these simulations, which of the two solar irradiance histories is the most likely. However, in the simulations that do not include anthropogenic forcing, NH temperatures reach a peak in the middle of the 20th century, and decrease afterwards, for both the strong and weak solar irradiance cases. This suggests that the contribution of natural forcing to observed 20th-century warming is small, and that solar and volcanic forcings are not responsible for the degree of warmth that occurred in the second half of the 20th century, consistent with the evidence of earlier work based on simple and more complex climate models (Crowley and Lowery, 2000; Bertrand et al., 2002b; Gerber et al., 2003; Hegerl et al., 2006; Tett et al., 2007; see also Chapter 9).

An overall conclusion can be drawn from the available instrumental and proxy evidence for the history of hemispheric average temperature change over the last 500 to 2,000 years, as well as the modelling studies exploring the possible roles of various causal factors: that is, greenhouse gases must be included among the forcings in order to simulate hemispheric mean temperatures that are compatible with the evidence of unusual warmth observed in the second half of the 20th century. It is very unlikely that this warming was merely a recovery from a pre-20th century cold period.

6.6.4   Consistency Between Temperature, Greenhouse Gas and Forcing Records; and Compatibility of Coupled Carbon Cycle-Climate Models with the Proxy Records

It is difficult to constrain the climate sensitivity from the proxy records of the last millennium (see Chapter 9). As noted above, the evidence for hemispheric temperature change as interpreted from the different proxy records, and for atmospheric trace greenhouse gases, inferred solar forcing and reconstructed volcanic forcing, is to varying degrees uncertain. The available temperature reconstructions suggest that decadally averaged NH temperatures varied within 1°C or less during the two millennia preceding the 20th century (Figure 6.10), but the magnitude of the reconstructed low-frequency variations differs by up to about a factor of two for different reconstructions. The reconstructions of natural forcings (solar and volcanic) are uncertain for this period. If they produced substantial negative energy balances (reduced solar, increased volcanic activity), then low-to-medium estimates of climate sensitivity are compatible with the reconstructed temperature variations (Figure 6.10); however, if solar and volcanic forcing varied only weakly, then moderate-to-high climate sensitivity would be consistent with the temperature reconstructions, especially those showing larger cooling (see also Chapter 9), assuming that the sensitivity of the climate system to solar irradiance changes and explosive volcanism is not different than the sensitivity to changes in greenhouse gases or other forcing agents.

The greenhouse gas record provides indirect evidence for a limited range of low-frequency hemispheric-scale climate variations over the last two millennia prior to the industrial period (AD 1–1750). The greenhouse gas histories of CO_2, CH_4 and N_2O show only small changes over this time period (MacFarling Meure et al., 2006; Figure 6.4), although there is evidence from the ice core record (Figures 6.3 and 6.7), as well as from models, that greenhouse gas concentrations react sensitively to climatic changes.

The sensitivity of atmospheric CO_2 to climatic changes as simulated by coupled carbon cycle-climate models is broadly consistent with the ice core CO_2 record and the amplitudes of the pre-industrial, decadal-scale NH temperature changes in the proxy-based reconstructions (Joos and Prentice, 2004). The CO_2 climate sensitivity can be formally defined as the change in atmospheric CO_2 relative to a nominal change in NH temperature in units of ppm per °C. Its strength depends on several factors, including the change in solubility of CO_2 in seawater, and the responses of productivity and heterotrophic respiration on land to temperature and precipitation (see Section 7.3). The sensitivity was estimated for modest (NH temperature change less than about 1°C) temperature variations from simulations with the Bern Carbon Cycle-Climate model driven with solar and volcanic forcing over the last millennium (Gerber et al., 2003) and from simulations with the range of models participating in the Coupled Carbon Cycle-Climate Model Intercomparison Project (C^4MIP) over the industrial period (Friedlingstein et al., 2006). The range of the CO_2 climate sensitivity is 4 to 16 ppm per °C for the 10 models participating in the C^4MIP intercomparison (evaluated as the difference in atmospheric CO_2 for the 1990 decade between a simulation with, and without, climate change, divided by the increase in NH temperature from the 1860 decade to the 1990 decade). This is comparable to a range of 10 to 17 ppm per °C obtained for CO_2 variations in the range of 6 to 10 ppm (Etheridge et al., 1996; Siegenthaler et al., 2005b) and the illustrative assumption that decadally averaged NH temperature varied within 0.6°C.

6.6.5   Regional Variability in Quantities Other than Temperature

6.6.5.1   *Changes in the El Niño-Southern Oscillation System*

Considerable interest in the ENSO system has encouraged numerous attempts at its palaeoclimatic reconstruction. These include a boreal winter (December–February) reconstruction of the Southern Oscillation Index (SOI) based on ENSO-sensitive tree ring indicators (Stahle et al., 1998), two multi-proxy reconstructions of annual and October to March Niño 3 index (average SST anomalies over 5°N to 5°S, 150°W to 90°W; Mann et al., 2005a,b), and a tropical coral-based Niño 3.4 SST reconstruction (Evans et al., 2002). Fossil coral records from Palmyra Island in the tropical Pacific also provide 30- to 150-year windows of ENSO variability within the last 1.1 kyr (Cobb et al., 2003). Finally, a new 600-year reconstruction of December to February Niño-3 SST has recently been developed (D'Arrigo et al., 2005), which is considerably longer than previous series. Although not totally independent (i.e., the reconstructions share a number of common predictors), these palaeorecords display significant common variance (typically more than 30% during their respective cross-validation periods), suggesting a relatively consistent history of El Niño in past centuries (Jones and Mann, 2004). In most coral records from the western Pacific and the Indian Ocean, late 20th-century warmth is unprecedented over the past 100 to 300 years (Bradley et al., 2003b). However, reliable and consistent interpretation of geochemical records from corals is still problematic (Lough, 2004). Reconstructions of extratropical temperatures and atmospheric circulation features (e.g., the North Pacific Index) correlate significantly with tropical estimates, supporting evidence for tropical/high-latitude Pacific links during the past three to four centuries (Evans et al., 2002; Linsley et al., 2004; D'Arrigo et al., 2006).

The El Niño-Southern Oscillation may have responded to radiative forcing induced by solar and volcanic variations over the past millennium (Adams et al., 2003; Mann et al., 2005a). Model simulations support a statistically significant response of ENSO to radiative changes such that during higher radiative inputs, La Niña-like conditions result from an intensified zonal SST gradient that drives stronger trade winds, and vice versa (Mann et al., 2005a). Comparing data and model results over the past millennium suggests that warmer background conditions are associated with higher variability (Cane, 2005). Numerical experiments suggest that the dynamics of ENSO may have played an important role in the climatic response to past changes in radiative forcing (Mann et al., 2005b). Indeed, the low-frequency changes in both amplitude of variability and mean state indicated by ENSO reconstructions from Palmyra corals (Cobb et al., 2003) were found to correspond well with the model responses to changes in tropical volcanic radiative forcing over the past 1 kyr, with solar forcing playing a secondary role.

Proxy records suggest that ENSO's global climate imprint evolves over time, complicating predictions. Comparisons of ENSO and drought indices clearly show changes in the linkage between ENSO and moisture balance in the USA over the past 150 years. Significant ENSO-drought correlations occur consistently in the southwest USA, but the strength of moisture penetration into the continent varies substantially over time (Cole and Cook, 1998; Cook et al., 2000). Comparing reconstructed Niño 3 SST with global temperature patterns suggests that some features are robust through time, such as the warming in the eastern tropical Pacific and western coasts of North and South America, whereas teleconnections into North America, the Atlantic and Eurasia are variable (Mann et al., 2000). The spatial correlation pattern for the period 1801 to 1850 provides striking evidence of non-stationarity in ENSO teleconnections, showing a distinct absence of the typical pattern of tropical Pacific warming (Mann et al., 2000).

6.6.5.2   The Record of Past Atlantic Variability

Climate variations over the North Atlantic are related to changes in the NAO (Hurrell, 1995) and the Atlantic Multidecadal Oscillation (Delworth and Mann, 2000; Sutton

and Hodson, 2005). From 1980 to 1995, the NAO tended to remain in one extreme phase and accounted for a substantial part of the winter warming over Europe and northern Eurasia. The North Atlantic region has a unique combination of long instrumental observations, many documentary records and multiple sources of proxy records. However, it remains difficult to document past variations in the dominant modes of climate variability in the region, including the NAO, due to problems of establishing proxies for atmospheric pressure, as well as the lack of stationarity in the NAO frequency and in storm tracks. Several reconstructions of NAO have been proposed (Cook et al., 2002b; Cullen et al., 2002; Luterbacher et al., 2002). Although the reconstructions differ in many aspects, there is a general tendency for more negative NAO indices during the 17th and 18th centuries than in the 20th century, thus indicating that the colder mean climate was characterised by a more zonal atmospheric pattern than in the 20th century. The coldest reconstructed European winter in 1708/1709, and the strong warming trend between 1684 and 1738 (+0.32°C per decade), have been related to a negative NAO index and the NAO response to increasing radiative forcing, respectively (Luterbacher et al., 2004). Some spatially resolved simulations employing GCMs indicate that solar and volcanic forcings lead to continental warming associated with a shift towards a high NAO index (Shindell et al., 2001, 2003, 2004; Stendel et al., 2006). Increased solar irradiance at the end of the 17th century and through the first half of the 18th century might have induced such a shift towards a high NAO index (Shindell et al., 2001; Luterbacher et al., 2004; Xoplanki et al., 2005).

It is well known that the NAO exerts a dominant influence on winter temperature and precipitation over Europe, but the strength of the relationship can change over time and region (Jones et al., 2003). The strong trend towards a more positive NAO index in the early part of the 18th century in the Luterbacher et al. (2002) NAO reconstruction appears connected with positive winter precipitation anomalies over northwest Europe and marked expansions of maritime glaciers in a manner similar to the effect of positive winter precipitation anomalies over recent decades for the same glaciers (Nesje and Dahl, 2003; Pauling et al., 2006).

6.6.5.3   Asian Monsoon Variability

Fifteen severe (three years or longer) droughts have occurred in a region of China dominated by the East Asian Monsoon over the last 1 kyr (Zhang, 2005). These palaeodroughts were generally more severe than droughts in the same region within the last 50 years. In contrast, the South Asian (Indian) monsoon, in the drier areas of its influence, has recently reversed its millennia-long orbitally driven low-frequency trend towards less rainfall. This recent reversal in monsoon rainfall also appears to coincide with a synchronous increase in inferred monsoon winds over the western Arabian Sea (Anderson et al., 2002), a change that could be related to increased summer heating over and around the Tibetan Plateau (Braüning and Mantwill, 2004; Morrill et al., 2006).

6.6.5.4   Northern and Eastern Africa Hydrologic Variability

Lake sediment and historical documentary evidence indicate that northern Africa and the Sahel region have for a long time experienced substantial droughts lasting from decades to centuries (Kadomura, 1992; Verschuren, 2001; Russell et al., 2003; Stager et al., 2003; Nguetsop et al., 2004; Brooks et al., 2005; Stager et al., 2005). Although there have been attempts to link these dry periods to solar variations, the evidence is not conclusive (Stager et al., 2005), particularly given that the relationship between hypothesised solar proxies and variation in total solar irradiance remains unclear (see Section 6.6.3). The palaeoclimate record indicates that persistent droughts have been a common feature of climate in northern and eastern Africa. However, it has not been demonstrated that these droughts can be simulated with coupled ocean-atmosphere models.

6.6.5.5   The Record of Hydrologic Variability and Change in the Americas

Multiple proxies, including tree rings, sediments, historical documents and lake sediment records make it clear that the past 2 kyr included periods with more frequent, longer and/or geographically more extensive droughts in North America than during the 20th century (Stahle and Cleaveland, 1992; Stahle et al., 1998; Woodhouse and Overpeck, 1998; Forman et al., 2001; Cook et al., 2004b; Hodell et al., 2005; MacDonald and Case, 2005). Past droughts, including decadal-length 'megadroughts' (Woodhouse and Overpeck, 1998), are most likely due to extended periods of anomalous SST (Hoerling and Kumar, 2003; Schubert et al., 2004; MacDonald and Case, 2005; Seager et al., 2005), but remain difficult to simulate with coupled ocean-atmosphere models. Thus, the palaeoclimatic record suggests that multi-year, decadal and even centennial-scale drier periods are likely to remain a feature of future North American climate, particularly in the area west of the Mississippi River.

There is some evidence that North American drought was more regionally extensive, severe and frequent during past intervals that were characterised by warmer than average NH summer temperatures (e.g., during medieval times and the mid-Holocene; Forman et al., 2001; Cook et al., 2004b). There is evidence that changes in the North American hydrologic regime can occur abruptly relative to the rate of change in climate forcing and duration of the subsequent climate regime. Abrupt shifts in drought frequency and duration have been found in palaeohydrologic records from western North America (Cumming et al., 2002; Laird et al., 2003; Cook et al., 2004b). Similarly, the upper Mississippi River Basin and elsewhere have seen abrupt shifts in the frequency and size of the largest flood events (Knox, 2000). Recent investigations of past large-hurricane activity in the southeast USA suggest that changes in the regional frequency of large hurricanes can shift abruptly in response to more gradual forcing (Liu, 2004). Although the palaeoclimatic record indicates that hydrologic shifts in drought, floods and tropical storms have occurred abruptly (i.e., within years), this past abrupt change has not been simulated with coupled atmosphere ocean models. Decadal variability of Central Chilean precipitation was greater before the 20th century, with more intense and prolonged dry episodes in the past. Tree-ring based precipitation reconstructions for the past eight centuries reveal multi-year drought episodes in the 14th and 16th to 18th centuries that exceed the estimates of decadal drought during the 20th century (LeQuesne et al., 2006).

6.7   Concluding Remarks on Key Uncertainties

Each palaeoclimatic time scale covered in this chapter contributes to the understanding of how the climate system varies naturally and also responds to changes in climate forcing. The existing body of knowledge is sufficient to support the assertions of this chapter. At the same time, key uncertainties remain, and greater confidence would result if these uncertainties were reduced.

Even though a great deal is known about glacial-interglacial variations in climate and greenhouse gases, a comprehensive mechanistic explanation of these variations remains to be articulated. Similarly, the mechanisms of abrupt climate change (for example, in ocean circulation and drought frequency) are not well enough understood, nor are the key climate thresholds that, when crossed, could trigger an acceleration in sea level rise or regional climate change. Furthermore, the ability of climate models to simulate realistic abrupt change in ocean circulation, drought frequency, flood frequency, ENSO behaviour and monsoon strength is uncertain. Neither the rates nor the processes by which ice sheets grew and disintegrated in the past are known well enough.

Knowledge of climate variability over the last 1 to 2 kyr in the SH and tropics is severely limited by the lack of palaeoclimatic records. In the NH, the situation is better, but there are important limitations due to a lack of tropical records and ocean records. Differing amplitudes and variability observed in available millennial-length NH temperature reconstructions, and the extent to which these differences relate to choice of proxy data and statistical calibration methods, need to be reconciled. Similarly, the understanding of how climatic extremes (i.e., in temperature and hydro-climatic variables) varied in the past is incomplete. Lastly, this assessment would be improved with extensive networks of proxy data that run right up to the present day. This would help measure how the proxies responded to the rapid global warming observed in the last 20 years, and it would also improve the ability to investigate the extent to which other, non-temperature, environmental changes may have biased the climate response of proxies in recent decades.

References

Adams, J.B., M.E. Mann, and C.M. Ammann, 2003: Proxy evidence for an El Niño-like response to volcanic forcing. *Nature*, **426**(6964), 274–278.

Adkins, J.F., K. McIntyre, and D.P. Schrag, 2002: The salinity, temperature, and $\delta^{18}O$ of the glacial deep ocean. *Science*, **298**, 1769–1773.

Alley, R.B., and P.U. Clark, 1999: The deglaciation of the northern hemisphere: A global perspective. *Annu. Rev. Earth Planet. Sci.*, **27**, 149–182.

Alley, R.B., and A.M. Agustsdottir, 2005: The 8k event: cause and consequences of a major Holocene abrupt climate change. *Quat. Sci. Rev.*, **24**, 1123–1149.

Alley, R.B., S. Anandakrishnan, and P. Jung, 2001: Stochastic resonance in the North Atlantic. *Paleoceanography*, **16**, 190–198.

Alley, R.B., et al., 1997: Holocene climatic instability: A large, widespread event 8200 years ago. *Geology*, **25**, 483–486.

Alley, R.B., et al., 2003: Abrupt climate change. *Science*, **299**(5615), 2005–2010.

Alverson, K.D., R.S. Bradley, and T.F. Pedersen (eds.), 2003: *Paleoclimate, Global Change and the Future*. International Geosphere Biosphere Programme Book Series, Springer-Verlag, Berlin, 221 pp.

Ammann, C.M., G.A. Meehl, W.M. Washington, and C.S. Zender, 2003: A monthly and latitudinally varying volcanic forcing dataset in simulations of 20th century climate. *Geophys. Res. Lett.*, **30**(12), 1657, doi:10.1029/2003GL016875.

Andersen, C., N. Koç, A. Jennings, and J.T. Andrews, 2004: Non uniform response of the major surface currents in the Nordic Seas to insolation forcing: implications for the Holocene climate variability. *Paleoceanography*, **19**, 1–16.

Anderson, D.M., J.T. Overpeck, and A.K. Gupta, 2002: Increase in the Asian southwest monsoon during the past four centuries. *Science*, **297**(5581), 596–599.

Archer, D., and A. Ganopolski, 2005: A movable trigger: Fossil fuel CO_2 and the onset of the next glaciation. *Geochem. Geophys. Geosystems*, **6**, Q05003.

Archer, D.A., A. Winguth, D. Lea, and N. Mahowald, 2000: What caused the glacial/interglacial atmospheric pCO_2 cycles? *Rev. Geophys.*, **12**, 159–189.

Ariztegui, D., et al., 2000: Paleoclimate and the formation of sapropel S1: inferences from Late Quaternary lacustrine and marine sequences in the central Mediterranean region. *Palaeogeogr. Palaeoclimatol. Palaeoecol.*, **158**, 215–240.

Bakke, J., S.O. Dahl, and A. Nesje, 2005a: Late glacial and early Holocene palaeoclimatic reconstruction based on glacier fluctuations and equilibrium-line altitudes at northern Folgefonna, Hardanger, western Norway. *J. Quat. Sci.*, **20**(2), 179–198.

Bakke, J., et al., 2005b: Glacier fluctuations, equilibrium-line altitudes and palaeoclimate in Lyngen, northern Norway, during the Late glacial and Holocene. *The Holocene*, **15**(4), 518–540.

Baliunas, S., and R. Jastrow, 1990: Evidence for long-term brightness changes of solar-type stars. *Nature*, **348**, 520–522.

Ballantyne, A.P., et al., 2005: Meta-analysis of tropical surface temperatures during the Last Glacial Maximum. *Geophys. Res. Lett.*, **32**, L05712, doi:10.1029/2004GL021217.

Bao, Y., A. Brauning, and S. Yafeng, 2003: Late Holocene temperature fluctuations on the Tibetan Plateau. *Quat. Sci. Rev.*, **22**(21), 2335–2344.

Barber, D.C., et al., 1999: Forcing of the cold event of 8,200 years ago by catastrophic drainage of Laurentide lakes. *Nature*, **400**, 344–347.

Bard, E., G. Raisbeck, F. Yiou, and J. Jouzel, 2000: Solar irradiance during the last millennium based on cosmogenic nuclides. *Tellus*, **52B**, 985–992.

Barrows, T.T., and S. Juggins, 2005: Sea-surface temperatures around the Australian margin and Indian Ocean during the Last Glacial Maximum. *Quat. Sci. Rev.*, **24**, 1017–1047.

Bartlett, M.G., D.S. Chapman, and R.N. Harris, 2004: Snow and the ground temperature record of climate change. *J. Geophys Res.*, **109**, F04008, doi:10.1029/2004JF000224.

Battle, M., et al., 1996: Atmospheric gas concentrations over the past century measured in air from firn at the South Pole. *Nature*, **383**(6597), 231–235.

Bauer, E., A. Ganopolski, and M. Montoya, 2004: Simulation of the cold climate event 8200 years ago by meltwater outburst from Lake Agassiz. *Paleoceanography*, **19**, PA3014, doi:10.1029/2004PA001030.

Bauer, E., M. Claussen, V. Brovkin, and A. Huenerbein, 2003: Assessing climate forcings of the Earth system for the past millennium. *Geophys. Res. Lett.*, **30**(6), 1276, doi:10.1029/2002GL016639.

Beer, J., S. Tobias, and N. Weiss, 1998: An active sun throughout the Maunder Minimum. *Sol. Phys.*, **181**(1), 237–249.

Beerling, D.J., 1999: New estimates of carbon transfer to terrestrial ecosystems between the last glacial maximum and the Holocene. *Terra Nova*, **11**(4), 162–167.

Beltrami, H., 2002: Paleoclimate: Earth's long-term memory. *Science*, **297**(5579), 206–207.

Beltrami, H., and E. Bourlon, 2004: Ground warming patterns in the Northern Hemisphere during the last five centuries. *Earth Planet. Sci. Lett.*, **227**(3–4), 169–177.

Berger, A., 1977: Long-term variations of earth's orbital elements. *Celestial Mechanics*, **15**(1), 53–74.

Berger, A., 1978: Long-term variation of caloric solar radiation resulting from the earth's orbital elements. *Quat. Res.*, **9**, 139–167.

Berger, A.L., and M.F. Loutre, 1991: Insolation values for the climate of the last 10 million years. *Quat. Sci. Rev.*, **10**, 297–317.

Berger, A.L., and M.F. Loutre, 2002: An exceptionally long interglacial ahead? *Science*, **297**, 1287–1288.

Berger, A.L., and M.F. Loutre, 2003: Climate 400,000 years ago, a key to the future? In: *Earth's Climate and Orbital Eccentricity* [Droxler, A.W., R.Z. Poore, and L.H. Burckle (eds.)]. American Geophysical Union, Washington, DC, pp. 17–26.

Berggren, W.A., D.V. Kent, C.C.I. Swisher, and M.P. Aubry, 1995: *Geochronology, Time Scales and Global Stratigraphic Correlation* [W.A. Berggren (ed)]. Special Publication No. 54, Society for Sedimentary Geology, Tulsa, OK, 386 pp.

Berner, R.A., and Z. Kothavala, 2001: GEOCARB III: A revised model of atmospheric CO_2 over phanerozoic time. *Am. J. Sci.*, **301**(2), 182–204.

Bertrand, C., M.F. Loutre, and A. Berger, 2002a: High frequency variations of the Earth's orbital parameters and climate change. *Geophys. Res. Lett.*, **29**, doi:10.1029/2002GL015622.

Bertrand, C., M.F. Loutre, M. Crucifix, and A. Berger, 2002b: Climate of the last millennium: a sensitivity study. *Tellus*, **54A**(3), 221–244.

Bianchi, G., and I.N. McCave, 1999: Holocene periodicity in north Atlantic climate and deep ocean flow south of Iceland. *Nature*, **397**, 515–518.

Bigelow, N., et al., 2003: Climate change and Arctic ecosystems: 1. Vegetation changes north of 55 degrees N between the last glacial maximum, mid-Holocene, and present. *J. Geophys. Res.*, **108**, doi:10.1029/2002JD002558.

Bigler, M., et al., 2002: Sulphate record from a northeast Greenland ice core over the last 1200 years based on continuous flow analysis. *Ann. Glaciol.*, **35**, 250–256.

Billups, K., J.E.T. Channell, and J. Zachos, 2002: Late Oligocene to early Miocene geochronology and paleoceanography from the subantarctic South Atlantic. *Paleoceanography*, **17**(1), 1004, doi:10.1029/2000PA000568.

Bird, M.I., J. Lloyd, and G.D. Farquhar, 1994: Terrestrial carbon storage at the LGM. *Nature*, **371**(6498), 566–566.

Birks, H.H., and B. Ammann, 2000: Two terrestrial records of rapid climatic change during the glacial-Holocene transition (14,000-9,000 calendar years B.P.) from Europe. *Proc. Natl. Acad. Sci. U.S.A.*, **97**, 1390–1394.

Bjerknes, J., 1969: Atmospheric teleconnections from equatorial pacific. *Mon. Weather Rev.*, **97**(3), 163–172

Blunier, T., and E.J. Brook, 2001: Timing of millennial-scale climate change in Antarctica and Greenland during the last glacial period. *Science*, **291**, 109–112.

Blunier, T., et al., 1993: Atmospheric methane record from a Greenland ice core over the last 1000 years. *Geophys. Res. Lett.*, **20**(20), 2219–2222.

Blunier, T., et al., 1995: Variations in atmospheric methane concentration during the Holocene epoch. *Nature*, **374**(6517), 46–49.

Blunier, T., et al., 1998: Asynchrony of Antarctic and Greenland climate change during the last glacial period. *Nature*, **394**, 739–743.

Bohaty, S.M., and J.C. Zachos, 2003: Significant Southern Ocean warming event in the late middle Eocene. *Geology*, **31**(11), 1017–1020.

Bond, G., et al., 1993: Correlations between climate records from North Atlantic sediments and Greenland ice. *Nature*, **365**, 143–147.

Bond, G., et al., 1997: A pervasive millennial-scale cycle in the North Atlantic Holocene and glacial climates. *Science*, **278**, 1257–1266.

Bond, G., et al., 2001: Persistent solar influence on North Atlantic climate during the Holocene. *Science*, **294**, 2130–2136.

Bond, W.J., G.F. Midgley, and F.I. Woodward, 2003: The importance of low atmospheric CO_2 and fire in promoting the spread of grasslands and savannas. *Global Change Biol.*, **9**(7), 973–982.

Booth, R.K., et al., 2005: A severe centennial-scale drought in mid-continental North America 4200 years ago and apparent global linkages. *The Holocene*, **15**, 321–328.

Bopp, L., K.E. Kohlfeld, C. Le Quéré, and O.O. Aumont, 2002: Dust impact on marine biota and atmospheric CO_2 in glacial periods. *Geochim. Cosmochim. Acta*, **66**(15A), A91, Suppl. 1, Aug. 2002.

Bowen, G.J., et al., 2002: Mammalian dispersal at the Paleocene/Eocene boundary. *Science*, **295**(5562), 2062–2065.

Bowen, G.J., et al., 2004: A humid climate state during the Palaeocene/Eocene thermal maximum. *Nature*, **432**(7016), 495–499.

Braconnot, P., O. Marti, S. Joussaume, and Y. Leclaninche, 2000: Ocean feedbacks in response to 6 kyr insolation. *J. Clim.*, **13**, 1537–1553.

Braconnot, P., et al., 2002: How the simulated change in monsoon at 6 ka BP is related to the simulation of the modern climate: results from the Paleoclimate Modeling Intercomparison Project. *Clim. Dyn.*, **19**(2), 107–121.

Braconnot, P., et al., 2004: Evaluation of PMIP coupled ocean-atmosphere simulations of the Mid-Holocene. In: *Past Climate Variability through Europe and Africa*, Vol. 6 [Battarbee, R.W., F. Gasse, and C.E. Stickley (eds)], Springer, Dordrecht, The Netherlands, 515–534.

Bradley, R.S., 1999: Climatic variability in sixteenth-century Europe and its social dimension - Preface. *Clim. Change*, **43**(1), 1–2.

Bradley, R.S., M.K. Hughes, and H.F. Diaz, 2003a: Climate in Medieval time. *Science*, **302**(5644), 404–405.

Bradley, R.S., K.R. Briffa, J. Cole, and T.J. Osborn, 2003b: The climate of the last millennium. In: *Paleoclimate, Global Change and the Future* [Alverson, K.D., R.S. Bradley, and T.F. Pedersen (eds.)]. Springer, Berlin, pp. 105–141.

Bralower, T.J., 2002: Evidence of surface water oligotrophy during the Paleocene-Eocene thermal maximum: Nannofossil assemblage data from Ocean Drilling Program Site 690, Maud Rise, Weddell Sea. *Paleoceanography*, **17**(2), 1023, doi:10.1029/2001PA000662.

Bräuning, A., and B. Mantwill, 2004: Summer temperature and summer monsoon history on the Tibetan plateau during the last 400 years recorded by tree rings. *Geophys. Res. Lett.*, **31**(24), L24205, doi:10.1029/2004GL020793.

Briffa, K.R., 2000: Annual climate variability in the Holocene: interpreting the message of ancient trees. *Quat. Sci. Rev.*, **19**(1–5), 87–105.

Briffa, K.R., and T.J. Osborn, 1999: Perspectives: Climate warming - Seeing the wood from the trees. *Science*, **284**(5416), 926–927.

Briffa, K.R., and T.J. Osborn, 2002: Paleoclimate - Blowing hot and cold. *Science*, **295**(5563), 2227–2228.

Briffa, K.R., T.J. Osborn, and F.H. Schweingruber, 2004: Large-scale temperature inferences from tree rings: a review. *Global Planet. Change*, **40**(1–2), 11–26.

Briffa, K.R., et al., 2001: Low-frequency temperature variations from a northern tree ring density network. *J. Geophys. Res.*, **106**(D3), 2929–2941.

Brigham-Grette, J., and D.M. Hopkins, 1995: Emergent marine record and paleoclimate of the last interglaciation along the northwest Alaskan coast. *Quat. Res.*, **43**, 159–173.

Broecker, W.S., and G.M. Henderson, 1998: The sequence of events surrounding Termination II and their implications for the cause of glacial-interglacial CO_2 changes. *Paleoceanography*, **13**, 352–364.

Broecker, W.S., and E. Clark, 2003: Holocene atmospheric CO_2 increase as viewed from the seafloor. *Global Biogeochem. Cycles*, **17**(2), doi:10.1029/2002GB001985.

Brook, E.J., et al., 2000: On the origin and timing of rapid changes in atmospheric methane during the last glacial period. *Global Biogeochem. Cycles*, **14**(2), 559–572.

Brooks, C.E.P., 1922: *The Evolution of Climate*. [Preface by Simpson, G.C.] Benn Brothers, London, 173 pp.

Brooks, K., et al., 2005: Late-Quaternary lowstands of Lake Bosumtwi, Ghana: evidence from high-resolution seismic-reflection and sediment-core data. *Palaeogeogr. Palaeoclimatol. Palaeoecol.*, **216**(3–4), 235–249.

Brovkin, V., et al., 2002: Carbon cycle, vegetation and climatic dynamics in the Holocene: Experiments with the CLIMBER-2 model. *Global Biogeochem. Cycles*, **16**, 1139, doi:10.1029/2001GB001662.

Burger, G., and U. Cubasch, 2005: Are multiproxy climate reconstructions robust? *Geophys. Res. Lett.*, **32**(23), doi:10.1029/2005GL024155.

Caillon, N., et al., 2003: Timing of atmospheric CO_2 and Antarctic temperature changes across Termination III. *Science*, **299**, 1728–1731.

Calov, R., A. Ganopolski, V. Petoukhov, and M. Claussen, 2002: Large-scale instabilities of the Laurentide ice sheet simulated in a fully coupled climate-system model. *Geophys. Res. Lett.*, **29**, 2216, doi:10.1029/2002GL016078.

Calov, R., et al., 2005: Transient simulation of the last glacial inception. Part II: Sensitivity and feedback analysis. *Clim. Dyn.*, **24**, 563–576.

Cane, M.A., 2005: The evolution of El Niño, past and future. *Earth Planet. Sci. Lett.*, **230**(3–4), 227–240.

CAPE Last Interglacial Project Members, 2006: Last Interglacial Arctic warmth confirms polar amplification of climate change. *Quat. Sci. Rev.*, **25**(13–14), 1383–1400.

Castellano, E., et al., 2005: Holocene volcanic history as recorded in the sulfate stratigraphy of the European Project for Ice Coring in Antarctica Dome C (EDC96) ice core. *J. Geophys. Res.*, **110**, D06114, doi:10.1029/2004JD005259.

Cerling, T.E., 1991: Carbon dioxide in the atmosphere: Evidence from Cenozoic and Mesozoic paleosols. *Am. J. Sci.*, **291**, 377–400.

Chandler, M.A., D. Rind, and R.S. Thompson, 1994: Joint investigations of the middle Pliocene climate II: GISS GCM Northern Hemisphere results. *Global Planet. Change*, **9**, 197–219.

Chapman, D.S., M.G. Bartlett, and R.N. Harris, 2004: Comment on "Ground vs. surface air temperature trends: Implications for borehole surface temperature reconstructions" by M. E. Mann and G. Schmidt. *Geophys. Res. Lett.*, **31**(7), L07205, doi:10.1029/2003GL019054.

Chappellaz, J.A., I.Y. Fung, and A.M. Thompson, 1993: The atmospheric CH_4 increase since the last glacial maximum. *Tellus*, **B45**(3), 228–241.

Chappellaz, J., et al., 1997: Changes in the atmospheric CH_4 gradient between Greenland and Antarctica during the Holocene. *J. Geophys. Res.*, **102**(D13), 15987–15997.

Charles, C.D., D.E. Hunter, and R.G. Fairbanks, 1997: Interaction between the ENSO and the Asian monsoon in a coral record of tropical climate. *Science*, **277**(5328), 925–928.

Church, J.A., et al., 2001: Changes in sea level. In: *Climate Change 2001: The Scientific Basis. Contribution of Working Group I to the Third Assessment Report of the Intergovernmental Panel on Climate Change* [Houghton, J.T. et al. (eds.)]. Cambridge University Press, Cambridge, United Kingdom and New York, NY, USA, pp. 639–693.

Claquin, T., et al., 2003: Radiative forcing of climate by ice-age atmospheric dust. *Clim. Dyn.*, **20**, 193–202.

Clark, P.U., N.G. Pisias, T.F. Stocker, and A.J. Weaver, 2002: The role of the thermohaline circulation in abrupt climate change. *Nature*, **415**, 863–869.

Clarke, G.K.C., D.W. Leverington, J.T. Teller, and A.S. Dyke, 2004: Paleohydraulics of the last outburst flood from glacial Lake Agassiz and the 8200 BP cold event. *Quat. Sci. Rev.*, **23**, 389–407.

Claussen, M., and Gayler, V., 1997: The greening of Sahara during the mid-Holocene: results of an interactive atmosphere-biome model. *Global Ecol. Biogeogr. Lett.*, **6**, 369–377.

Claussen, M., et al., 1999: Simulation of an abrupt change in Saharan vegetation in the mid-Holocene. *Geophys. Res. Lett.*, **26**(14), 2037–2040.

Claussen, M., et al., 2002: Earth system models of intermediate complexity: closing the gap in the spectrum of climate system models. *Clim. Dyn.*, **18**(7), 579–586.

Clement, A.C., and M.A. Cane, 1999: A role for the tropical Pacific coupled ocean-atmosphere system on Milankovitch and millennial timescales. Part I: A modeling study of tropical Pacific variability. In: *Mechanisms of Global Climate Change at Millennial Time Scales* [Clark, P.U., R.S. Webb, and L.D. Keigwin (eds.)]. American Geophysical Union, Washington, DC, pp. 363–371.

Clement, A.C., R. Seager, and M.A. Cane, 2000: Suppression of El Niño during the mid-Holocene by changes in the earth's orbit. *Paleoceanography*, **15**(6), 731–737.

Clement, A.C., M.A. Cane, and R. Seager, 2001: An orbitally driven tropical source for abrupt climate change. *J. Clim.*, **14**(11), 2369–2375.

Clement, A.C., R. Seager, M.A. Cane, and S.E. Zebiak, 1996: An ocean dynamical thermostat. *J. Clim.*, **9**(9), 2190–2196.

Cobb, K.M., C.D. Charles, H. Cheng, and R.L. Edwards, 2003: El Niño/Southern Oscillation and tropical Pacific climate during the last millennium. *Nature*, **424**(6946), 271–276.

Coe, M.T., and S.P. Harrison 2002: The water balance of northern Africa during the mid-Holocene: an evaluation of the 6 ka BPPMIP simulations. *Clim. Dyn.*, **19**(2), 155–166.

Cole, J., 2003: Global change - Dishing the dirt on coral reefs. *Nature*, **421**(6924), 705–706.

Cole, J.E., and E.R. Cook, 1998: The changing relationship between ENSO variability and moisture balance in the continental United States. *Geophys. Res. Lett.*, **25**(24), 4529–4532.

Cole, J.E., R.B. Dunbar, T.R. McClanahan, and N.A. Muthiga, 2000: Tropical Pacific forcing of decadal SST variability in the western Indian Ocean over the past two centuries. *Science*, **287**(5453), 617–619.

Cook, E.R., J.G. Palmer, and R.D. D'Arrigo, 2002a: Evidence for a 'Medieval Warm Period' in a 1,100 year tree-ring reconstruction of past austral summer temperatures in New Zealand. *Geophys. Res. Lett.*, **29**(14), 1667, doi:10.1029/2001GL014580.

Cook, E.R., R.D. D'Arrigo, and M.E. Mann, 2002b: A well-verified, multiproxy reconstruction of the winter North Atlantic Oscillation index since AD 1400. *J. Clim.*, **15**(13), 1754–1764.

Cook, E.R., J. Esper, and R.D. D'Arrigo, 2004a: Extra-tropical Northern Hemisphere land temperature variability over the past 1000 years. *Quat. Sci. Rev.*, **23**(20–22), 2063–2074.

Cook, E.R., B.M. Buckley, R.D. D'Arrigo, and M.J. Peterson, 2000: Warm-season temperatures since 1600 BC reconstructed from Tasmanian tree rings and their relationship to large-scale sea surface temperature anomalies. *Clim. Dyn.*, **16**(2–3), 79–91.

Cook, E.R., et al., 2004b: Long-term aridity changes in the western United States. *Science*, **306**(5698), 1015–1018.

Cortijo, E., et al., 1997: Changes in the sea surface hydrology associated with Heinrich event 4 in the North Atlantic Ocean (40-60°N). *Earth Planet. Sci. Lett.*, **146**, 29–45.

Cortijo, E., et al., 1999: Changes in meridional temperature and salinity gradients in the North Atlantic Ocean (30°-72°N) during the last interglacial period. *Paleoceanography*, **14**(1), 23–33.

Cronin, T.M., 1999: *Principles of Paleoclimatology.* Perspectives in Paleobiology and Earth History. Columbia University Press, New York, NY, 560 pp.

Cronin, T.M., et al., 2005: Mid-Pliocene deep-sea bottom-water temperatures based on ostracode Mg/Ca ratios. *Mar. Micropaleontol.*, **54**(3–4), 249–261.

Crouch, E.M., et al., 2003: The Apectodinium acme and terrestrial discharge during the Paleocene-Eocene thermal maximum: new palynological, geochemical and calcareous nannoplankton observations at Tawanui, New Zealand. *Palaeogeogr. Palaeoclimatol. Palaeoecol.*, **194**(4), 387–403.

Crowley, T.J., 1992: North Atlantic deep water cools the Southern Hemisphere. *Paleoceanography*, **7**, 489–497.

Crowley, T.J., 1995: Ice-age terrestrial carbon changes revisited. *Global Biogeochem. Cycles*, **9**(3), 377–389.

Crowley, T.J., 1998: Significance of tectonic boundary conditions for paleoclimate simulations. In: *Tectonic Boundary Conditions for Climate Reconstructions* [Crowley, T.J., and K.C. Burke (eds.)]. Oxford University Press, New York, pp. 3–17.

Crowley, T.J., 2000: Causes of climate change over the past 1000 years. *Science*, **289**(5477), 270–277.

Crowley, T.J., and T.S. Lowery, 2000: How warm was the medieval warm period? *Ambio*, **29**(1), 51–54.

Crowley, T.J., et al., 2003: Modeling ocean heat content changes during the last millennium. *Geophys. Res. Lett.*, **30**(18), 1932, doi:10.1029/2003GL017801.

Crucifix, M., and M.F. Loutre, 2002: Transient simulations over the last interglacial period (126-115 kyr BP). *Clim. Dyn.*, **19**, 417–433.

Crucifix, M., and C.D. Hewitt, 2005: Impact of vegetation changes on the dynamics of the atmosphere at the Last Glacial Maximum. *Clim. Dyn.*, **25**(5), 447–459.

Crucifix, M., et al., 2002: Climate evolution during the Holocene, a study with an Earth System model of intermediate complexity. *Clim. Dyn.*, **19**, 43–60.

Cuffey, K.M., and S.J. Marshall, 2000: Substantial contribution to sea-level rise during the last interglacial from the Greenland ice sheet. *Nature*, **404**, 591–594.

Cullen, H.M., A. Kaplan, P.A. Arkin, and P.B. Demenocal, 2002: Impact of the North Atlantic Oscillation on Middle Eastern climate and streamflow. *Clim. Change*, **55**(3), 315–338.

Cumming, B.F., et al., 2002: Persistent millennial-scale shifts in moisture regimes in western Canada during the past six millennia. *Proc. Natl. Acad. Sci. U.S.A.*, **99**(25), 16117–16121.

Cutler, K.B., et al., 2003: Rapid sea-level fall and deep-ocean temperature change since the last interglacial period, *Earth Planet. Sci. Lett.*, **206**, 253–271.

Dahl, K., A. Broccoli, and R. Stouffer, 2005: Assessing the role of North Atlantic freshwater forcing in millennial scale climate variability: a tropical Atlantic perspective. *Clim. Dyn.*, **24**(4), 325–346.

Dahl, S.O., and A. Nesje, 1996: A new approach to calculating Holocene winter precipitation by combining glacier equilibrium-line altitudes and pine-tree limits: A case study from Hardangerjøkulen, central southern Norway. *The Holocene*, **6**(4), 381–398.

Dahl-Jensen, D., V.I. Morgan, and A. Elcheikh, 1999: Monte Carlo inverse modelling of the Law Dome (Antarctica) temperature profile. *Ann. Glaciol.*, **29**, 145–150.

Dahl-Jensen, D., et al., 1998: Past temperature directly from the Greenland Ice Sheet. *Science*, **282**, 268–271.

D'Arrigo, R., R. Wilson, and G. Jacoby, 2006: On the long-term context for late twentieth century warming. *J. Geophys. Res.*, **111**(D3), doi:10.1029/2005JD006352.

D'Arrigo, R.D., et al., 2004: Thresholds for warming-induced growth decline at elevational tree line in the Yukon Territory, Canada. *Global Biogeochem. Cycles*, **18**(3), GB3021, doi:10.1029/2004GB002249.

D'Arrigo, R., et al., 2005: On the variability of ENSO over the past six centuries. *Geophys. Res. Lett.*, **32**(3), L03711, doi:10.1029/2004GL022055.

Davis, B.A.S., et al., 2003: The temperature of Europe during the Holocene reconstructed from pollen data. *Quat. Sci. Rev.*, **22**, 1701–1716.

de Menocal, P., J. Ortiz, T. Guilderson, and M. Sarnthein, 2000: Coherent high- and low-latitude climate variability during the Holocene warm period. *Science*, **288**(5474), 2198–2202.

de Noblet-Ducoudre, N., R. Claussen, and C. Prentice, 2000: Mid-Holocene greening of the Sahara: first results of the GAIM 6000 year BP experiment with two asynchronously coupled atmosphere/biome models. *Clim. Dyn.*, **16**(9), 643–659.

de Vernal, A., et al., 2006: Comparing proxies for the reconstruction of LGM sea-surface conditions in the northern North Atlantic. *Quat. Sci. Rev.*, **25**(21–22), 2820–2834.

DeConto, R.M., and D. Pollard, 2003: Rapid Cenozoic glaciation of Antarctica induced by declining atmospheric CO_2. *Nature*, **421**(6920), 245–249.

Delworth, T.L., and M.E. Mann, 2000: Observed and simulated multidecadal variability in the Northern Hemisphere. *Clim. Dyn.*, **16**(9), 661–676.

Dickens, G.R., and R.M. Owen, 1996: Sediment geochemical evidence for an early-middle Gilbert (early Pliocene) productivity peak in the North Pacific Red Clay Province. *Mar. Micropaleontol.*, **27**(1–4), 107–120.

Dickens, G.R., M.M. Castillo, and J.C.G. Walker, 1997: A blast of gas in the latest Paleocene: Simulating first-order effects of massive dissociation of oceanic methane hydrate. *Geology*, **25**(3), 259–262.

Ding, Z.L., et al., 2002: Stacked 2.6-Ma grain size record from the Chinese loess based on five sections and correlation with the deep-sea $\delta^{18}O$ record. *Paleoceanography*, **17**(3), 1033, doi:10.1029/2001PA000725.

Dlugokencky, E.J., L.P. Steele, P.M. Lang, and K.A. Masarie, 1994: The growth rate and distribution of atmospheric methane. *J. Geophys. Res.*, **99**, 17021–17043.

Dokken, T.M., and E. Jansen, 1999: Rapid changes in the mechanism of ocean convection during the last glacial period. *Nature*, **401**, 458–461.

Domack, E., et al., 2005: Stability of the Larsen B ice shelf on the Antarctic Peninsula during the Holocene epoch. *Nature*, **436**, 681–685.

Dowsett, H.J., and T.M. Cronin, 1990: High eustatic sea level during the middle Pliocene: evidence from southeastern U.S. Atlantic coastal plain. *Geology*, **18**, 435–438.

Dowsett, H., J. Barron, and R. Poore, 1996: Middle Pliocene sea surface temperatures: A global reconstruction. *Mar. Micropaleontol.*, **27**(1–4), 13–25.

Dowsett, H.J., M.A. Chandler, T.M. Cronin, and G.S. Dwyer, 2005: Middle Pliocene sea surface temperature variability. *Paleoceanography*, **20**(2), doi:10.1029/2005PA001133.

Duplessy, J.C., L. Labeyrie, and C. Waelbroeck, 2002: Constraints on the ocean oxygen isotopic enrichment between the Last Glacial Maximum and the Holocene: Paleoceanographic implications. *Quat. Sci. Rev.*, **21**, 315–330.

Duplessy, J.C., et al., 2001: Holocene paleoceanography of the northern Barents Sea and variations of the northward heat transport by the Atlantic Ocean. *Boreas*, **30**, 2–16.

Ehrmann, W.U., and A. Mackensen, 1992: Sedimentological evidence for the formation of an East Antarctic ice-sheet in Eocene Oligocene time. *Palaeogeogr. Palaeoclimatol. Palaeoecol.*, **93**(1–2), 85–112.

Elliot, M., et al., 1998: Millennial scale iceberg discharges in the Irminger Basin during the last glacial period: relationship with the Heinrich events and environmental settings. *Paleoceanography*, **13**, 433–446.

Ellis, J.M., and Calkin, P.E., 1984: Chronology of Holocene glaciation, central Brooks Range, Alaska. *Geol. Soc. Am. Bull.*, **95**, 897–912.

Enting, I.G., 1987: On the use of smoothing splines to filter CO_2 data. *J. Geophys. Res.*, **92**, 10977–10984.

EPICA community members, 2004: Eight glacial cycles from an Antarctic ice core. *Nature*, **429**(6992), 623–628.

Esper, J., E.R. Cook, and F.H. Schweingruber, 2002: Low-frequency signals in long tree-ring chronologies for reconstructing past temperature variability. *Science*, **295**(5563), 2250–2253.

Esper, J., D.C. Frank, R.J.S. Wilson, and K.R. Briffa, 2005: Effect of scaling and regression on reconstructed temperature amplitude for the past millennium. *Geophys. Res. Lett.*, **32**(7), doi:10.1029/2004GL021236.

Etheridge, D.M., et al., 1996: Natural and anthropogenic changes in atmospheric CO_2 over the last 1000 years from air in Antarctic ice and firn. *J. Geophys. Res.*, **101**(D2), 4115–4128.

Evans, M.N., A. Kaplan, and M.A. Cane, 2002: Pacific sea surface temperature field reconstruction from coral $\delta^{18}O$ data using reduced space objective analysis. *Paleoceanography*, **17**(1), 1007, doi:10.1029/2000PA000590.

Eyer, M., 2004: *Highly Resolved $\delta^{13}C$ Measurements on CO_2 in Air from Antarctic Ice Cores*. PhD Thesis, University of Bern, 113 pp.

Fairbanks, R.G., 1989: A 17,000 year glacio-eustatic sea level record: Influence of glacial melting rates on the Younger Dryas event and deep-ocean circulation. *Paleoceanography*, **342**, 637–642.

Farrera, I., et al., 1999: Tropical climates at the Last Glacial Maximum: a new synthesis of terrestrial palaeoclimate data. I. Vegetation, lake-levels and geochemistry. *Clim. Dyn.*, **15**, 823–856.

Ferretti, D.F., et al., 2005: Unexpected changes to the global methane budget over the past 2000 years. *Science*, **309**, 1714–1717.

Fischer, G., and G. Wefer (eds.), 1999: *Use of Proxies in Paleoceanography: Examples from the South Atlantic*. Springer, Berlin, 735 pp.

Fleitmann, D., et al., 2003: Holocene forcing of the Indian monsoon recorded in a stalagmite from southern Oman. *Science*, **300**, 1737–1740.

Flagge, M., and S.K. Solanki, 2000: The solar spectral irradiance since 1700. *Geophys. Res. Lett.*, **27**, 2157–2160.

Flückiger, J., et al., 1999: Variations in atmospheric N_2O concentration during abrupt climatic changes. *Science*, **285**(5425), 227–230.

Flückiger, J., et al., 2002: High resolution Holocene N_2O ice core record and its relationship with CH_4 and CO_2. *Global Biogeochem. Cycles*, **16**, doi:10.1029/2001GB001417.

Flückiger, J., et al., 2004: N_2O and CH_4 variations during the last glacial epoch: Insight into global processes. *Global Biogeochem. Cycles*, **18**, doi:10.1029/2003GB002122.

Folland, C.K., et al., 2001: Observed climate variability and change. In: *Climate Change 2001: The Scientific Basis. Contribution of Working Group I to the Third Assessment Report of the Intergovernmental Panel on Climate Change* [Houghton, J.T. et al. (eds.)]. Cambridge University Press, Cambridge, United Kingdom and New York, NY, USA, pp. 99–181.

Forman, S.L., R. Oglesby, and R.S. Webb, 2001: Temporal and spatial patterns of Holocene dune activity on the Great Plains of North America: megadroughts and climate links. *Global Planet. Change*, **29**(1–2), 1–29.

Foster, S., 2004: *Reconstruction of Solar Irradiance Variations for Use in Studies of Global Climate Change: Application of Recent SOHO Observations with Historic Data from the Greenwich Observatory*. Ph.D. Thesis, University of Southampton, Southampton, UK.

Foukal, P., G. North, and T. Wigley, 2004: A stellar view on solar variations and climate. *Science*, **306**(5693), 68–69.

Francois, L.M., C. Delire, P. Warnant, and G. Munhoven, 1998: Modelling the glacial-interglacial changes in the continental biosphere. *Global Planet. Change*, **17**, 37–52.

Francois, L.M., et al., 1999: Carbon stocks and isotopic budgets of the terrestrial biosphere at mid-Holocene and last glacial maximum times. *Chem. Geol.*, **159**, 163–189.

Freeman, K.H., and J.M. Hayes, 1992: Fractionation of carbon isotopes by phytoplankton and estimates of ancient CO2 levels. *Global Biogeochem. Cycles*, **6**, 185–198.

Friedlingstein, P., et al., 2006: Climate-carbon cycle feedback analysis, results from the C^{4}MIP model intercomparison. *J. Clim.*, **19** (14), 3337–3353.

Fröhlich, C., and J. Lean, 2004: Solar radiative output and its variability: evidence and mechanisms. *Astron. Astrophys. Rev.*, **12**, 273–320.

Gaffen, D.J., et al., 2000: Multidecadal changes in the vertical temperature structure of the tropical troposphere. *Science*, **287**(5456), 1242–1245.

Gagan, M.K., et al., 1998: Temperature and surface-ocean water balance of the mid-Holocene tropical western Pacific. *Science*, **279**, 1014–1018.

Ganopolski, A., and S. Rahmstorf, 2001: Rapid changes of glacial climate simulated in a coupled climate model. *Nature*, **409**, 153–158.

Gellatly, A.F., T.J. Chinn, and F. Röthlisberger, 1988: Holocene glacier variations in New Zealand: a review. *Quat. Sci. Rev.*, **7**, 227–242.

Gerber, S., et al., 2003: Constraining temperature variations over the last millennium by comparing simulated and observed atmospheric CO_2. *Clim. Dyn.*, **20**(2–3), 281–299.

Gersonde, R., X. Crosta, A. Abelmann, and L. Armand, 2005: Sea-surface temperature and sea ice distribution of the Southern Ocean at the EPILOG Last Glacial Maximum - a circum-Antarctic view based on siliceous microfossil records. *Quat. Sci. Rev.*, **24** (7–9), 869–896

Gherardi, J.M., et al., 2005: Evidence from the North Eastern Atlantic Basin for variability of the Meridional Overturning Circulation through the last deglaciation. *Earth Planet. Sci. Lett.*, **240**, 710–723.

Gladstone, R.M., et al., 2005: Mid-Holocene NAO: a PMIP2 model intercomparison. *Geophys. Res. Lett.*, **32**, L16707, doi:10.1029/2005GL023596.

Goldstein, B., F. Joos, and T.F. Stocker, 2003: A modeling study of oceanic nitrous oxide during the Younger Dryas cold period. *Geophys. Res. Lett.*, **30**, doi:10.1029/2002GL016418.

Goni, M.F.S., F. Eynaud, J.L. Turon, and N.J. Shackleton, 1999: High resolution palynological record off the Iberian margin: direct land-sea correlation for the Last Interglacial complex. *Earth Planet. Sci. Lett.*, **171**(1), 123–137.

González-Rouco, F., H. von Storch, and E. Zorita, 2003: Deep soil temperature as proxy for surface air-temperature in a coupled model simulation of the last thousand years. *Geophys. Res. Lett.*, **30**(21), 2116, doi:10.1029/2003GL018264.

González-Rouco, J.F., H. Beltrami, E. Zorita, and H. von Storch, 2006: Simulation and inversion of borehole temperature profiles in surrogate climates: Spatial distribution and surface coupling. *Geophys. Res. Lett.*, **33**(1), L01703, doi:10.1029/2005GL024693.

Goosse, H., H. Renssen, A. Timmermann, and R.S. Bradley, 2005a: Internal and forced climate variability during the last millennium: a model-data comparison using ensemble simulations. *Quat. Sci. Rev.*, **24**, 1345–1360.

Goosse, H., et al., 2004: A late medieval warm period in the Southern Ocean as a delayed response to external forcing? *Geophys. Res. Lett.*, **31**, L06203, doi:10.1029/2003GL019140.

Goosse, H., et al., 2005b: Modelling the climate of the last millennium: what causes the differences between simulations? *Geophys. Res. Lett.*, **32**, L06710, doi:10.1029/2005GL022368.

Gosnold, W.D., P.E. Todhunter, and W. Schmidt, 1997: The borehole temperature record of climate warming in the mid- continent of North America. *Global Planet. Change*, **15**(1–2), 33–45.

Gradstein, F.M., J.G. Ogg, and A.G. Smith (eds.), 2004: *A Geologic Time Scale*. Cambridge University Press, Cambridge, 589 pp.

Greenblatt, J.B., and J.L. Sarmiento, 2004: Variability and climate feedback mechanisms in ocean uptake of CO_2. In: *The Global Carbon Cycle* [Field, C.B., and M.R. Raupach (eds)]. Island Press, Washington, DC, pp. 257–275.

Guilderson, T.P., R.G. Fairbanks, and J.L. Rubenstone, 1994: Tropical temperature variations since 20,000 years ago: modulating interhemispheric climate change. *Science*, **263**, 663–665.

Guiot, J., et al., 2005: Last-millennium summer-temperature variations in Western Europe based on proxy data. *The Holocene*, **15**(4), 489–500.

Guo, Z.T., N. Petit-Maire, and S. Kropelin, 2000: Holocene non-orbital climatic events in present-day arid areas of Northern Africa and China. *Global Planet. Change*, **26**(1–3), 97–103.

Guo, Z.T., et al., 2004: Late Miocene-Pliocene development of Asian aridification as recorded in the Red-Earth formation in northern China. *Global Planet. Change*, **41**(3–4), 135–145.

Gupta, A.K., D.M. Anderson, and J.T. Overpeck, 2003: Abrupt changes in the Asian southwest monsoon during the Holocene and their links to the North Atlantic Ocean. *Nature*, **421**, 354–357.

Hall, J.C., and G.M. Lockwood, 2004: The chromospheric activity and variability of cycling and flat activity solar-analog stars. *Astrophys. J.*, **614**, 942–946.

Hambrey, M.J., W.U. Ehrmann, and B. Larsen, 1991: Cenozoic glacial record of the Prydz Bay continental shelf, East Antarctica. In: *Proceedings of the Ocean Drilling Program: Scientific Results*, Vol. 119. Ocean Drilling Program, College Station, TX, pp. 77–131.

Hanebuth, T., K. Stattegger, and P.M. Grootes, 2000: Rapid flooding of the Sunda Shelf: A late-glacial sea-level record. *Science*, **288**(5468), 1033–1035.

Harrington, G.J., S.J. Kemp, and P.L. Koch, 2004: Palaeocene-Eocene paratropical floral change in North America: responses to climate change and plant immigration. *J. Geol. Soc. London*, **161**, 173–184.

Harris, R.N., and D.S. Chapman, 2001: Mid-latitude (30°-60° N) climatic warming inferred by combining borehole temperatures with surface air temperatures. *Geophys. Res. Lett.*, **28**(5), 747–750.

Harrison, S.P., 2005: Snowlines at the last glacial maximum and tropical cooling. *Quat. Int.*, **138**, 5–7.

Harrison, S.P., and I.C. Prentice, 2003: Climate and CO_2 controls on global vegetation distribution at the last glacial maximum: analysis based on palaeovegetation data, biome modelling and paleoclimate simulations. *Global Change Biol.*, **9**, 983–1004.

Harrison, S.P., et al., 2003: Mid-Holocene climates of the Americas: a dynamical response to changed seasonality. *Clim. Dyn.*, **20**(7–8), 663–688.

Haug, G.H., et al., 2001: Southward migration of the Intertropical Convergence Zone through the Holocene. *Science*, **17**(293), 1304–1308.

Hays, J.D., J. Imbrie, and N.J. Shackleton, 1976: Variations in the earth's orbit: pacemaker of the ice ages. *Science*, **194**, 1121–1132.

Haywood, A.M., P.J. Valdes, and B.W. Sellwood, 2000: Global scale paleoclimate reconstruction of the middle Pliocene climate using the UKMO GCM: initial results. *Global Planet. Change*, **25**, 239–256.

Haywood, A.M., P. Dekens, A.C. Ravelo, and M. Williams, 2005: Warmer tropics during the mid-Pliocene? Evidence from alkenone paleothermometry and a fully coupled ocean-atmosphere GCM. *Geochem. Geophys. Geosystems*, **6**, Q03010, doi:10.1029/2004GC000799.

He, Y., et al., 2004: Asynchronous Holocene climatic change across China. *Quat. Res.*, **61**, 52–63.

Hegerl, G.C., T.J. Crowley, W.T. Hyde, and D.J. Frame, 2006: Climate sensitivity constrained by temperature reconstructions over the past seven centuries. *Nature*, **440**, 1029–1032.

Hemming, S.R., 2004: Heinrich events: Massive late Pleistocene detritus layers of the North Atlantic and their global climate imprint. *Rev. Geophys.*, **42**(1), RG1005, doi:10.1029/2003RG000128.

Higgins, P.A.T., 2004: Biogeochemical and biophysical responses of the land surface to a sustained thermohaline circulation weakening. *J. Clim.*, **17**, 4135–4142.

Hodell, D.A., M. Brenner, and J.H. Curtis, 2005: Terminal classic drought in the northern Maya lowlands inferred from multiple sediment cores in Lake Chichancanab (Mexico). *Quat. Sci. Rev.*, **24**(12–13), 1413–1427.

Hoerling, M., and A. Kumar, 2003: The perfect ocean for drought. *Science*, **299**(5607), 691–694.

Hoffmann, G., et al., 2003: Coherent isotope history of Andean ice cores over the last century. *Geophys. Res. Lett.*, **30**(4), doi:10.1029/2002GL014870.

Holmgren, K., et al., 2003: Persistent millennial-scale climate variability over the past 25,000 years in Southern Africa. *Quat. Sci. Rev.*, **22**, 2311–2326.

Holzhauser, H., M.J. Magny, and H.J. Zumbuhl, 2005: Glacier and lake-level variations in west-central Europe over the last 3500 years. *The Holocene*, **15**(6), 789–801.

Hoyt, D.V., and K.H. Schatten, 1993: A discussion of plausible solar irradiance variations. *J. Geophys. Res.*, **98**, 18895–18906.

Huang, S.P., and H.N. Pollack, 1998: *Global Borehole Temperature Database for Climate Reconstruction*. IGBP PAGES/World Data Center-A for Paleoclimatology Data Contribution Series #1998-044, NOAA/NGDC Paleoclimatology Program, Boulder, CO.

Huang, S.P., H.N. Pollack, and P.Y. Shen, 1997: Late Quaternary temperature changes seen in the world-wide continental heat flow measurements. *Geophys. Res. Lett.*, 24, 1947–1950.

Huang, S.P., H.N. Pollack, and P.Y. Shen, 2000: Temperature trends over the past five centuries reconstructed from borehole temperatures. *Nature*, 403(6771), 756–758.

Huber, C., et al., 2006: Isotope calibrated Greenland temperature record over Marine Isotope Stage 3 and its relation to CH$_4$. *Earth Planet. Sci. Lett.*, 243(3–4), 504–519.

Hughen, K.A., J.T. Overpeck, L.C. Peterson, and S. Trumbore, 1996: Rapid climate changes in the tropical Atlantic region during the last deglaciation. *Nature*, 380(6569), 51–54.

Hughen, K.A., T.I. Eglinton, L. Xu, and M. Makou, 2004: Abrupt tropical vegetation response to rapid climate changes. *Science*, 304(5679), 1955–1959.

Hughes, M.K., and H.F. Diaz, 1994: Was there a Medieval Warm Period, and if so, where and when? *Clim. Change*, 26(2–3), 109–142.

Humlum, O., et al., 2005: Late-Holocene glacier growth in Svalbard, documented by subglacial relict vegetation and living soil microbes. *The Holocene*, 15(3), 396–407.

Hurrell, J.W., 1995: Decadal trends in the North-Atlantic Oscillation - regional temperatures and precipitation. *Science*, 269(5224), 676–679.

Huybers, P., 2005: Comment on "Hockey sticks, principal components, and spurious significance" by S. McIntyre and R. McKitrick. *Geophys. Res. Lett.*, 32(20), doi:10.1029/2005GL023395.

Indermühle, A., et al., 2000: Atmospheric CO$_2$ concentration from 60 to 20 kyr BP from the Taylor Dome ice core, Antarctica. *Geophys. Res. Lett.*, 27(5), 735–738.

IPCC, 1990: *Climate Change: The IPCC Scientific Assessment* [Houghton, J.T., G.J. Jenkins, and J.J. Ephraums (eds.)]. Cambridge University Press, Cambridge, United Kingdom and New York, NY, USA, 362 pp.

IPCC, 2001: *Climate Change 2001: The Scientific Basis. Contribution of Working Group I to the Third Assessment Report of the Intergovernmental Panel on Climate Change* [Houghton, J.T., et al. (eds.)]. Cambridge University Press, Cambridge, United Kingdom and New York, NY, USA, 881 pp.

Jackson, S.C., and A.J. Broccoli, 2003: Orbital forcing of Arctic climate: mechanisms of climate response and implications for continental glaciation. *Clim. Dyn.*, 21, 539–557.

Jansen, E., T. Fronval, F. Rack, and J.E.T. Channell, 2000: Pliocene-Pleistocene ice rafting history and cyclicity in the Nordic Seas during the last 3.5 Myr. *Paleoceanography*, 15(6), 709–721.

Jasper, J.P., and J.M. Hayes, 1990: A carbon isotope record of CO$_2$ levels during the late Quaternary. *Nature*, 347, 462–464.

Jennings, A.E., et al., 2001: A mid-Holocene shift in Arctic sea-ice variability on the East Greenland Shelf. *The Holocene*, 12, 49–58.

Jiang, D., et al., 2005: Modeling the middle Pliocene climate with a global atmospheric general circulation model. *J. Geophys. Res.*, 110, D14107, doi:10.1029/2004JD005639.

Jörin, U.E., T.F. Stocker, and C. Schlüchter, 2006: Multi-century glacier fluctuations in the Swiss Alps during the Holocene. *The Holocene*, 16(5), 697–704.

Johnsen, S.J., et al., 2001: Oxygen isotope and palaeotemperature records from six Greenland ice-core stations: Camp Century, Dye-3, GRIP, GISP2, Renland and NorthGRIP. *J. Quat. Sci.*, 16, 299–307.

Jolly, D., S.P. Harrison, B. Damnati, and R. Bonnefille, 1998: Simulated climate and biomes of Africa during the Late Quaternary: comparison with pollen and lake status data. *Quat. Sci. Rev.*, 17, 629–657.

Jones, P.D., and A. Moberg, 2003: Hemispheric and large-scale surface air temperature variations: An extensive revision and an update to 2001. *J. Clim.*, 16(2), 206–223.

Jones, P.D., and M.E. Mann, 2004: Climate over past millennia. *Rev. Geophys.*, 42(2), RG2002, doi:10.1029/2003RG000143.

Jones, P.D., T.J. Osborn, and K.R. Briffa, 1997: Estimating sampling errors in large-scale temperature averages. *J. Clim.*, 10(10), 2548–2568.

Jones, P.D., T.J. Osborn, and K.R. Briffa, 2001: The evolution of climate over the last millennium. *Science*, 292(5517), 662–667.

Jones, P.D., K.R. Briffa, and T.J. Osborn, 2003: Changes in the Northern Hemisphere annual cycle: Implications for paleoclimatology? *J. Geophys. Res.*, 108(D18), 4588, doi:10.1029/2003JD003695.

Jones, P.D., K.R. Briffa, T.P. Barnett, and S.F.B. Tett, 1998: High-resolution palaeoclimatic records for the last millennium: interpretation, integration and comparison with General Circulation Model control-run temperatures. *The Holocene*, 8(4), 455–471.

Joos, F., and I.C. Prentice, 2004: A paleo-perspective on changes in atmospheric CO$_2$ and climate. In: *The Global Carbon Cycle: Integrating Humans, Climate and the Natural World* [Field, C.B., and M.R. Raupach (eds.)]. Island Press, Washington DC, pp. 165–186.

Joos, F., et al., 1999: Global warming and marine carbon cycle feedbacks on future atmospheric CO2. *Science*, 284, 464–467.

Joos, F., et al., 2001: Global warming feedbacks on terrestrial carbon uptake under the Intergovernmental Panel on Climate Change (IPCC) emission scenarios. *Global Biogeochem. Cycles*, 15(4), 891–907.

Joos, F., et al., 2004: Transient simulations of Holocene atmospheric carbon dioxide and terrestrial carbon since the Last Glacial Maximum. *Global Biogeochem. Cycles*, 18, doi:10.1029/2003GB002156.

Joussaume, S., et al., 1999: Monsoon changes for 6000 years ago: Results of 18 simulations from the Paleoclimate Modeling Intercomparison Project (PMIP). *Geophys. Res. Lett.*, 26(7), 859–862.

Kadomura, H., 1992: Climate change in the West African Sahel-Sudan zone since the Little Ice Age. In: *Symposium on the Little Ice Age* [Mikami, T. (ed.)]. Tokyo Metropolitan University, Tokyo, pp. 40–45.

Kageyama, M., et al., 2004: Quantifying ice-sheet feedbacks during the last glacial inception. *Geophys. Res. Lett.*, 31, doi:10.1029/2004GL021339.

Kageyama, M., et al., 2006: Last Glacial Maximum temperatures over the North Atlantic, Europe, and western Siberia: a comparison between PMIP models, MARGO sea-surface temperatures and pollen-base reconstructions. *Quat. Sci. Rev.*, 25, 2082–2102.

Kaplan, J.O., I.C. Prentice, W. Knorr, and P.J. Valdes, 2002: Modeling the dynamics of terrestrial carbon storage since the Last Glacial Maximum. *Geophys. Res. Lett.*, 29, doi:10.1029/2002GL015230.

Kaplan, J.O, et al., 2003: Climate change and Arctic ecosystems: 2. Modeling, paleodata-model comparisons, and future projections. *J. Geophys. Res.*, 108, doi:10.1029/2002JD002559.

Kaplan, M.R., and A.P. Wolfe, 2006: Spatial and temporal variability of Holocene temperature in the North Atlantic region. *Quat. Res.*, 65, 223–231.

Karlén, W. and J. Kuylenstierna, 1996: On solar forcing of Holocene climate: evidence from Scandinavia. *The Holocene*, 6, 359–365.

Karlén, W., et al., 1999: Glacier fluctuations on Mount Kenya since ca 6000 cal. years BP: implications for Holocene climatic change in Africa. *Ambio*, 28(5), 409–418.

Kaspar, F., and U. Cubasch, 2006: Simulations of the Eemian interglacial and the subsequent glacial inception with a coupled ocean-atmosphere general circulation model. In: *The Climate of Past Interglacials* [Sirocko, F., M. Claussen, M.F. Sánchez-Goñi and T. Litt (eds.)], Elsevier Science, Amsterdam, pp. 499-516.

Kaspar, F., N. Kuhl, U. Cubasch, and T. Litt, 2005: A model-data comparison of European temperatures in the Eemian interglacial. *Geophys. Res. Lett.*, 32, L11703, doi:10.1029/2005GL022456.

Kaufman, D.S., et al., 2004: Holocene thermal maximum in the western Arctic (0-180°W). *Quat. Sci. Rev.*, 23, 529–560.

Keeling, C.D., and T.P. Whorf, 2005: Atmospheric CO$_2$ records from sites in the SiO air sampling network. In: *Trends: A Compendium of Data on Global Change*. Carbon Dioxide Information Analysis Center, Oak Ridge National Laboratory, U.S. Department of Energy, Oak Ridge, TN.

Kennett, J.P., and L.D. Stott, 1991: Abrupt deep-sea warming, palaeoceanographic changes and benthic extinctions at the end of the Palaeocene. *Nature*, **353**, 225–229.

Khodri, M., G. Ramstein, N. De Noblet, and M. Kageyama, 2003: Sensitivity of the northern extratropics hydrological cycle to the changing insolation forcing at 126 and 115 ky BP. *Clim. Dyn.*, **21**, 273–287.

Khodri, M., et al., 2005: The impact of precession changes on the Arctic climate during the last interglacial glacial transition. *Earth Planet. Sci. Lett.*, **236**, 285–304.

Kim, J.H., et al., 2004: North Pacific and North Atlantic sea-surface temperature variability during the Holocene. *Quat. Sci. Rev.*, **23**, 2141–2154.

Kitoh, A., and S. Murakami, 2002: Tropical Pacific climate at the mid-Holocene and the Last Glacial Maximum simulated by a coupled ocean-atmosphere general circulation model. *Paleoceanography*, **17**, 1047, doi:10.1029/2001PA000724.

Klitgaard-Kristensen, D., et al., 1998: The short cold period 8,200 years ago documented in oxygen isotope records of precipitation in Europe and Greenland. *J. Quat. Sci.*, **13**(2), 165–169.

Knies, J., J. Matthiessen, C. Vogt, and R. Stein, 2002: Evidence of 'Mid-Pliocene (similar to 3 Ma) global warmth' in the eastern Arctic Ocean and implications for the Svalbard/Barents Sea ice sheet during the late Pliocene and early Pleistocene (similar to 3-1.7 Ma). *Boreas*, **31**(1), 82–93.

Knox, J.C., 2000: Sensitivity of modern and Holocene floods to climate change. *Quat. Sci. Rev.*, **19**(1–5), 439–457.

Knutti, R., J. Flückiger, T.F. Stocker, and A. Timmermann, 2004: Strong hemispheric coupling of glacial climate through freshwater discharge and ocean circulation. *Nature*, **430**(7002), 851–856.

Koch, J., B. Menounos, J. Clague, and G.D. Osborn, 2004: Environmental change in Garibaldi Provincial Park, Southern Coast Mountains, British Columbia. *Geoscience Canada*, **31**(3), 127–135.

Koch, P.L., J.C. Zachos, and P.D. Gingerich, 1992: Correlation between isotope records in marine and continental carbon reservoirs near the Paleocene Eocene boundary. *Nature*, **358**(6384), 319–322.

Koerner, R.M., 1989: Ice core evidence for extensive melting of the Greenland Ice-sheet in the last interglacial. *Science*, **244**(4907), 964–968.

Koerner, R.M., and D.A. Fisher, 2002: Ice-core evidence for widespread Arctic glacier retreat in the Last Interglacial and the early Holocene. *Ann. Glaciol.*, **35**, 19–24.

Kohfeld, K., and S.P. Harrison, 2001: DIRTMAP: the geological record of dust. *Earth Sci. Rev.*, **54**, 81–114.

Kohfeld, K.E., C. LeQuéré, S.P. Harrison, and R.F. Anderson, 2005: Role of marine biology in glacial-interglacial CO_2 cycles. *Science*, **308**, 74–78.

Köhler, P., F. Joos, S. Gerber, and R. Knutti, 2005: Simulating changes in vegetation distribution, land carbon storage, and atmospheric CO_2 in response to a collapse of the North Atlantic thermohaline circulation. *Clim. Dyn.*, **25** (7–8), 689–708.

Koutavas, A., J. Lynch-Stieglitz, T.M. Marchitto Jr., and J.P. Sachs, 2002: El Niño-like pattern in ice age tropical Pacific sea surface temperature. *Science*, **297**, 226–230.

Kucera, M., et al., 2005: Multiproxy approach for the reconstruction of the glacial ocean surface (MARGO). *Quat. Sci. Rev.*, **24**, 813–819.

Kuhnert, H., et al., 1999: A 200-year coral stable oxygen isotope record from a high-latitude reef off western Australia. *Coral Reefs*, **18**(1), 1–12.

Kukla, G.J., et al., 2002: Last interglacial climates. *Quat. Res.*, **58**, 2–13.

Kurtz, A.C., et al., 2003: Early Cenozoic decoupling of the global carbon and sulphur cycles. *Paleoceanography*, **18**(4), doi:10.1029/2003PA000908.

Laird, K.R., et al., 2003: Lake sediments record large-scale shifts in moisture regimes across the northern prairies of North America during the past two millennia. *Proc. Natl. Acad. Sci. U.S.A.*, **100**(5), 2483–2488.

LaMarche, V.C., D.A. Graybill, H.C. Fritts, and M.R. Rose, 1984: Increasing atmospheric carbon dioxide: Tree ring evidence for growth enhancement in natural vegetation. *Science*, **225**, 1019–1021.

Lamb, H.H., 1965. The early medieval warm epoch and its sequel. *Palaeogeogr. Palaeoclimatol. Palaeoecol.*, **1**(13), 13–37.

Lamb, H.H., 1977: *Climates of the Past, Present and Future*. Vol. I and II. Metheun, London.

Lamb, H.H., 1982: *Climate History and the Modern World*. Routledge, London and New York, 433 pp.

Lambeck, K., 2002: Sea-level change from mid-Holocene to recent time: An Australian example with global implications. In: *Ice Sheets, Sea Level and the Dynamic Earth* [Mitrovica, J.X., and L.A. Vermeersen (eds.)]. Geodynamic Series Vol. 29, American Geophysical Union, Washington, DC, pp. 33–50.

Lambeck, K., and J. Chappell, 2001: Sea level change through the last glacial cycle. *Science*, **292**(5517), 679–686.

Landais, A., et al., 2006: The glacial inception as recorded in the NorthGRIP Greenland ice core: timing, structure and associated abrupt temperature changes. *Clim. Dyn.*, **26**(2–3), 273–284.

Laskar, J., et al., 2004: A long-term numerical solution for the insolation quantities of the Earth. *Astron. Astrophys.*, **428**(1), 261–285.

Lauritzen, S.E., 2003: Reconstruction of Holocene climate records from speleothems. In: *Global Change in the Holocene* [Mackay, A., R. Battarbee, J. Birks, and F. Oldfield (eds)]. Arnold, London, pp. 242–263.

Le Quesne, C., et al., 2006: Ancient *Austrocedrus* tree-ring chronologies used to reconstruct Central Chile precipitation variability from AD 1200 to 2000. *J. Clim.*, **19**(22), 5731–5744.

Lea, D.W., D.K. Pak, L.C. Peterson, and K.A. Hughen, 2003: Synchroneity of tropical and high-latitude Atlantic temperatures over the last glacial termination. *Science*, **301**(5638), 1361–1364.

Lean, J., 2000: Evolution of the sun's spectral irradiance since the Maunder Minimum. *Geophys. Res. Lett.*, **27**(16), 2425–2428.

Lean, J.L., Y.M. Wang, and N.R. Sheeley, 2002: The effect of increasing solar activity on the sun's total and open magnetic flux during multiple cycles: Implications for solar forcing of climate. *Geophys. Res. Lett.*, **29**(24), 2224, doi:10.1029/2002GL015880.

Lean, J.L., et al., 1995: Correlated brightness variations in solar radiative output from the photosphere to the corona. *Geophys. Res. Lett.*, **22**(5), 655–658.

Lear, C.H., Y. Rosenthal, H.K. Coxall, and P.A. Wilson, 2004: Late Eocene to early Miocene ice sheet dynamics and the global carbon cycle. *Paleoceanography*, **19**(4), PA4015, doi:10.1029/2004PA001039.

LeGrande, A.N., et al., 2006: Consistent simulations of multiple proxy responses to an abrupt climate change event. *Proc. Natl. Acad. Sci. U.S.A.*, **103**(4), 837–842.

Lemasurier, W.E., and S. Rocchi, 2005: Terrestrial record of post-Eocene climate history in Marie Byrd Land, West Antarctica. *Geografiska Annaler*, **87A**(1), 51–66.

Levis, S., G. B. Bonan, and C. Bonfils 2004: Soil feedback drives the mid-Holocene North African monsoon northward in fully coupled CCSM2 simulations with a dynamic vegetation model. *Clim. Dyn.*, **23**, 791–802.

Lhomme, N., G.K.C. Clarke, and S.J. Marshall, 2005: Tracer transport in the Greenland Ice Sheet: constraints on ice cores and glacial history. *Quat. Sci. Rev.*, **24**, 173–194.

Lie, Ø., et al., 2004: Holocene fluctuations of a polythermal glacier in high-alpine eastern Jotunheimen, central-southern Norway. *Quat. Sci. Rev.*, **23**(18–19), 1925–1945.

Linsley, B.K., et al., 2004: Geochemical evidence from corals for changes in the amplitude and spatial pattern of South Pacific interdecadal climate variability over the last 300 years. *Clim. Dyn.*, **22**(1), 1–11.

Lisiecki, L.E., and M.E. Raymo, 2005: A Pliocene-Pleistocene stack of 57 globally distributed benthic $\delta^{18}O$ records. *Paleoceanography*, **20**, PA1003, doi:10.1029/2004PA001071.

Liu, K.B., 2004: Paleotempestology: Principles, methods, and examples from Gulf coast lake-sediments. In: *Hurricanes and Typhoons: Past, Present and Future* [Murnane, R., and K. Liu (eds.)]. Columbia University Press, New York, pp. 13–57.

Liu, Z., J.E. Kutzbach, and L. Wu, 2000: Modeling climate shift of El Niño variability in the Holocene. *Geophys. Res. Lett.*, **27**, 2265–2268.

Liu, Z., S.P. Harrison, J.E. Kutzbach, and B. Otto-Bleisner, 2004: Global monsoons in the mid-Holocene and oceanic feedback. *Clim. Dyn.*, **22**, 157–182.

Liu, Z., et al., 2002: Tropical cooling at the last glacial maximum and extratropical ocean ventilation. *Geophys. Res. Lett.*, **29**, 1409, doi:10.1029/2001GL013938.

Lockwood, M., and R. Stamper, 1999: Long-term drift of the coronal source magnetic flux and the total solar irradiance. *Geophys. Res. Lett.*, **26**, 2461–2464.

Lorentz, S.J., et al., 2006: Orbitally driven insolation forcing on Holocene climate trends: evidence from alkenone data and climate modeling. *Paleoceanography*, **21**, doi:10.1029/2005PA001152.

Lough, J.M., 2004: A strategy to improve the contribution of coral data to high-resolution paleoclimatology. *Palaeogeogr. Paleoclimatol. Palaeoecol.*, **204**, 115–143.

Loutre, M.F., and A.L. Berger, 2000: Future climate changes: Are we entering an exceptionally long interglacial? *Clim. Change*, **46**, 61–90.

Loutre, M.F., D. Paillard, F. Vimeux, and E. Cortijo, 2004: Does mean annual insolation have the potential to change the climate? *Earth Planet. Sci. Lett.*, **221**(1–4), 1–14.

Lozhkin, A.V., and P.M. Anderson, 1995: The last interglaciation in northeast Siberia. *Quat. Res.*, **43**, 147–158.

Lubinski, D.J., S.L. Forman, and G.H. Miller, 1999: Holocene glacier and climate fluctuations on Franz Josef Land, Arctic Russia, 80° N. *Quat. Sci. Rev.*, **18**(1), 85–108.

Luckman, B.H., and M.S. Kearney, 1986: Reconstruction of Holocene changes in alpine vegetation and climate in the Maligne Range, Jasper National Park, Alberta. *Quat. Res.*, **26**(2), 244–261.

Luckman, B.H., and R.J.S. Wilson, 2005: Summer temperatures in the Canadian Rockies during the last millennium: a revised record. *Clim. Dyn.*, **24**(2–3), 131–144.

Luterbacher, J., et al., 2002: Reconstruction of sea level pressure fields over the Eastern North Atlantic and Europe back to 1500. *Clim. Dyn.*, **18**(7), 545–561.

Luterbacher, J., et al., 2004: European seasonal and annual temperature variability, trends, and extremes since 1500. *Science*, **303**(5663), 1499–1503.

MacAyeal, D.R., 1993: Binge/Purge oscillations of the Laurentide Ice-Sheet as a cause of the North-Atlantic Heinrich Events. *Paleoceanography*, **8**(6), 775–784.

MacDonald, G.M., and R.A. Case, 2005: Variations in the Pacific Decadal Oscillation over the past millennium. *Geophys. Res. Lett.*, **32**(8), doi:10.1029/2005GL022478.

MacDonald, G.M., et al., 2000: Holocene treeline history and climate change across northern Eurasia. *Quat. Res.*, **53**, 302–311.

MacFarling Meure, C., et al., 2006: The Law Dome CO_2, CH_4 and N_2O ice core records extended to 2000 years BP. *Geophys. Res. Lett.*, **33**, L14810, doi:10.1029/2006GL026152.

Machida, T., et al., 1995: Increase in the atmospheric nitrous oxide concentration during the last 250 years. *Geophys. Res. Lett.*, **22**, 2921–2924.

Mackay, A., R. Battarbee, J. Birks, and F.E. Oldfield (eds.), 2003: *Global Change in the Holocene*. Hodder Arnold, London, 480 pp.

Mahowald, N., et al., 1999: Dust sources and deposition during the Last Glacial Maximum and current climate: A comparison of model results with paleodata from ice cores and marine sediments. *J. Geophys. Res.*, **104**, 15859–15916.

Majorowicz, J.A., W.R. Skinner, and J. Safanda, 2004: Large ground warming in the Canadian Arctic inferred from inversions of temperature logs. *Earth Planet. Sci. Lett.*, **221**, 15–25.

Mangerud, J., V. Astakhov, and J.I. Svendsen, 2002: The extent of the Barents-Kara Ice Sheet during the Last Glacial Maximum. *Quat. Sci. Rev.*, **21**, 111–119.

Mann, M.E., and P.D. Jones, 2003: Global surface temperatures over the past two millennia. *Geophys. Res. Lett.*, **30**(15), 1820, doi:10.1029/2003GL017814.

Mann, M.E., and G.A. Schmidt, 2003: Ground vs. surface air temperature trends: Implications for borehole surface temperature reconstructions. *Geophys. Res. Lett.*, **30**(12), 1607, doi:10.1029/2003GL017170.

Mann, M.E., R.S. Bradley, and M.K. Hughes, 1998: Global-scale temperature patterns and climate forcing over the past six centuries. *Nature*, **392**(6678), 779–787.

Mann, M.E., R.S. Bradley, and M.K. Hughes, 1999: Northern hemisphere temperatures during the past millennium: Inferences, uncertainties, and limitations. *Geophys. Res. Lett.*, **26**(6), 759–762.

Mann, M.E., R. Bradley, and M.K. Hughes, 2000: Long-term variability in the El Niño/Southern Oscillation and associated teleconnections. In: *El Niño and the Southern Oscillation: Multiscale Variability and Global and Regional Impacts* [Diaz, H.F., and V. Markgraf (eds.)]. Cambridge University Press, Cambridge, pp. 357–412.

Mann, M.E., M.A. Cane, S.E. Zebiak, and A. Clement, 2005a: Volcanic and solar forcing of the tropical Pacific over the past 1000 years. *J. Clim.*, **18**(3), 447–456.

Mann, M.E., S. Rutherford, E. Wahl, and C.M. Ammann, 2005b: Testing the fidelity of methods used in 'proxy-based' reconstructions of past climate. *J. Clim.*, **18**(20), 4097–4107.

Mann, M.E., et al., 2003: Optimal surface temperature reconstructions using terrestrial borehole data. *J. Geophys. Res.*, **108**(D7), doi:10.1029/2002JD002532.

Marchal, O., R. Francois, T.F. Stocker, and F. Joos, 2000: Ocean thermohaline circulation and sedimentary $^{231}Pa/^{230}Th$ ratio. *Paleoceanography*, **6**, 625–641.

Marchal, O., et al., 1999: Modelling the concentration of atmospheric CO_2 during the Younger Dryas climate event. *Clim. Dyn.*, **15**, 341–354.

Marchal, O., et al., 2002: Apparent long-term cooling of the sea surface in the northeast Atlantic and Mediterranean during the Holocene. *Quat. Sci. Rev.*, **21** (4–6), 455–483.

Marchant, R., and H. Hooghiemstra, 2004: Rapid environmental change in African and South American tropics around 4000 years before present : a review. *Earth Sci. Rev.*, **66**, 217–260.

Marchitto, T.N.J., D.W. Oppo, and W.B. Curry, 2002: Paired benthic foraminiferal Cd/Ca and Zn/Ca evidence for a greatly increased presence of Southern Ocean Water in the glacial North Atlantic. *Paleoceanography*, **17**, 1038, doi:10.1029/2000PA000598.

Marra, M.J., 2003: Last interglacial beetle fauna from New Zealand. *Quart. Res.*, **59**, 122–131.

Masson, V., et al., 2000: Holocene climate variability in Antarctica based on 11 ice cores isotopic records. *Quat. Res.*, **54**, 348–358.

Masson-Delmotte, V., et al., 2005a: Rapid climate variability during warm and cold periods in polar regions and Europe. *Comptes Rendus Geoscience*, **337**(10–11), 935–946.

Masson-Delmotte, V., et al., 2005b: GRIP deuterium excess reveals rapid and orbital-scale changes in Greenland moisture origin. *Science*, **309**(5731), 118–121.

Masson-Delmotte, V., et al., 2006: Past and future polar amplification of climate change: climate model intercomparisons and ice-core constraints. *Clim. Dyn.*, **26** (5), 513–529.

Matthews, J.A., et al., 2000: Holocene glacier variations in central Jotunheimen, southern Norway based on distal glaciolacustrine sediment cores. *Quat. Sci. Rev.*, **19**, 1625–1647.

Matthews, J.A., et al., 2005: Holocene glacier history of Bjørnbreen and climatic reconstruction in central Jotunheimen, Norway, based on proximal glaciofluvial stream-bank mires. *Quat. Sci. Rev.*, **24**(1–2), 67–90.

Mayewski, P.A., et al., 2004: Holocene climate variability. *Quat. Res.*, **62** (3), 243–255.

McElwain, J.C., and W.G. Chaloner, 1995: Stomatal density and index of fossil plants track atmospheric carbon dioxide in the Palaeozoic. *Ann. Bot.(London)*, **76**, 389–395.

McGregor, H.V., and M.K. Gagan, 2004: Western Pacific coral δ^{18}O records of anomalous Holocene variability in the El Niño-Southern Oscillation. *Geophys. Res. Lett.*, **31**(11), doi:10.1029/2004GL019972.

McIntyre, S., and R. McKitrick, 2003: Corrections to the Mann et al. (1998) proxy database and northern hemispheric average temperature series. *Energy Environ.*, **14**, 751–771.

McIntyre, S., and R. McKitrick, 2005a: Hockey sticks, principal components, and spurious significance. *Geophys. Res. Lett.*, **32**(3), L03710, doi:10.1029/2004GL021750.

McIntyre, S., and R. McKitrick, 2005b: The M&M critique of the MBH98 Northern Hemisphere climate index: Update and implications. *Energy Environ.*, **16**, 69–99.

McIntyre, S., and R. McKitrick, 2005c: Reply to comment by von Storch and Zorita on "Hockey sticks, principal components, and spurious significance". *Geophys. Res. Lett.*, **32**(20), L20714, doi:10.1029/2005GL023089.

McIntyre, S., and R. McKitrick, 2005d: Reply to comment by von Huybers on "Hockey sticks, principal components, and spurious significance". *Geophys. Res. Lett.*, **32**(20), L20713, doi:10.1029/2005GL023586.

McManus, J.F., et al., 2002: Thermohaline circulation and prolonged interglacial warmth in the North Atlantic. *Quat. Res.*, **58**, 17–21.

McManus, J.F., et al., 2004: Collapse and rapid resumption of Atlantic meridional circulation linked to deglacial climate changes. *Nature*, **428**, 834–837.

Meissner, K.J., A.J. Weaver, H.D. Matthews, and P.M. Cox, 2003: The role of land surface dynamics in glacial inception: a study with the UVic Earth System Model. *Clim. Dyn.*, **21**, 7–8.

Meland, M.Y., E. Jansen, and H. Elderfield, 2005: Constraints on SST estimates for the northern North Atlantic/Nordic Seas during the LGM. *Quat. Sci. Rev.*, **24**(7–9), 835–852.

Menounos, B., et al., 2004: Early Holocene glacier advance, southern Coast Mountains, British Columbia, Canada. *Quat. Sci. Rev.*, **23**(14–15), 1543–1550.

Mickley, L.J., D.J. Jacob, and D. Rind, 2001: Uncertainty in preindustrial abundance of tropospheric ozone: Implications for radiative forcing calculations. *J. Geophys. Res.*, **106**(D4), 3389–3399.

Mieding, B., 2005: *Reconstruction of Millennial Aerosol-Chemical Ice Core Records from the Northeast Greenland: Quantification of Temporal Changes in the Atmospheric Circulation, Emission and Deposition.* Reports on Polar and Marine Research No. 513, Alfred Wegener Institute for Polar and Marine Research, Bremerhaven, 119 pp.

Mix, A.C., A.E. Morey, N.G. Pisias, and S.W. Hostetler, 1999: Foraminiferal faunal estimates of paleotemperature: Circumventing the no-analog problem yields cool ice age tropics. *Paleoceanography*, **14**, 350–359.

Moberg, A., et al., 2005: Highly variable Northern Hemisphere temperatures reconstructed from low- and high-resolution proxy data. *Nature*, **433**(7026), 613–617.

Monnin, E., et al., 2001: Atmospheric CO_2 concentrations over the last glacial termination. *Science*, **291**(5501), 112–114.

Monnin, E., et al., 2004: Evidence for substantial accumulation rate variability in Antarctica during the Holocene, through synchronization of CO_2 in the Taylor Dome, Dome C and DML ice cores. *Earth Planet. Sci. Lett.*, **224**(1–2), 45–54.

Montoya, M., H. von Storch, and T.J. Crowley, 2000: Climate simulation for 125 kyr BP with a coupled ocean-atmosphere general circulation model. *J. Clim.*, **13**, 1057–1072.

Montoya, M., et al., 2005: The Earth System Model of Intermediate Complexity CLIMBER-3α Part I: description and performance for present day conditions. *Clim. Dyn.*, **25**, 237–263.

Moran, K., et al., 2006: The Cenozoic palaeoenvironment of the Arctic Ocean. *Nature*, **441**, 601–605.

Moros, M., J.T. Andrews, D.E. Eberl, and E. Jansen, 2006: Holocene history of drift ice in the northern North Atlantic: Evidence for different spatial and temporal modes. *Paleoceanography*, **21**, PA2017, doi:10.1029/2005PA001214.

Morrill, C., J.T. Overpeck, and J.E. Cole, 2003: A synthesis of abrupt changes in the Asian summer monsoon since the last deglaciation. *The Holocene*, **13**, 465–476.

Morrill, C., et al., 2006: Holocene variations in the Asian monsoon inferred from the geochemistry of lake sediments in central Tibet. *Quat. Res.*, **65**(2), 232–243.

Moy, C.M., G.O. Seltzer, D.T. Rodbell, and D.M. Anderson, 2002: Variability of El Niño/Southern Oscillation activity at millennial timescales during the Holocene epoch. *Nature*, **420**, 162–165.

Mudelsee, M., 2001: The phase relations among atmospheric CO_2 content, temperature and global ice volume over the past 420 ka. *Quat. Sci. Rev.*, **20**, 583–589.

Muhs, D.R., T.A. Ager, and J.E. Beget, 2001: Vegetation and paleoclimate of the last interglacial period, central Alaska. *Quat. Sci. Rev.*, **20**, 41–61.

Muhs, D.R., K.R. Simmons, and B. Steinke, 2002: Timing and warmth of the last interglacial period: New U-series evidence from Hawaii and Bermuda and a new fossil compilation for North America. *Quat. Sci. Rev.*, **21**, 1355–1383.

Muscheler, R., F. Joos, S.A. Müller, and I. Snowball, 2005: Climate - How unusual is today's solar activity? *Nature*, **436**(7050), E3–E4.

Muscheler, R., et al., 2007: Solar activity during the last 1000 years inferred from radionuclide records. *Quat. Sci. Rev.*, **26**, 82-97.

Myhre, G., E.J. Highwood, K.P. Shine, and F. Stordal, 1998: New estimates of radiative forcing due to well mixed greenhouse gases. *Geophys. Res. Lett.*, **25**, 2715–1718.

Nesje, A., and S.O. Dahl, 2003: The 'Little Ice Age' - only temperature? *The Holocene*, **13**(1), 139–145.

Nesje, A., S.O. Dahl, and J. Bakke, 2004: Were abrupt late glacial and early-Holocene climate changes in northwest Europe linked to freshwater outbursts to the North Atlantic and Arctic oceans? *The Holocene*, **14**, 299–310.

Nesje, A., S.O. Dahl, C. Andersson, and J.A. Matthews, 2000: The lacustrine sedimentary sequence in Sygneskardvatnet, western Norway: a continuous, high-resolution record of the Jostedalsbreen ice cap during the Holocene. *Quat. Sci. Rev.*, **19**, 1047–1065.

Nesje, A., et al., 2005: Holocene climate variability in the Northern North Atlantic region: A review of terrestrial and marine evidence. In: *The Nordic Seas: An Integrated Perspective* [Drange, H., et al. (eds.)]. Geophysical Monographs Vol. 158, American Geophysical Union, Washington, DC, pp. 289–322.

NGRIP (North Greenland Ice Core Project), 2004: High-resolution record of Northern Hemisphere climate extending into the last interglacial period. *Nature*, **431**, 147–151.

Nguetsop, V.F., S. Servant-Vildary, and M. Servant, 2004: Late Holocene climate changes in west Africa, a high resolution diatom record from equatorial Cameroon. *Quat. Sci. Rev.*, **23**(5–6), 591–609.

NRC (National Research Council), 2006: *Surface Temperature Reconstructions for the Last 2,000 Years.* National Academies Press, Washington, DC, 196 pp.

Oerlemans, J., 2005: Extracting a climate signal from 169 glacier records. *Science*, **308**(5722), 675–677.

Oppenheimer, M., and R.B. Alley, 2005: Ice sheets, global warming, and Article 2 of the UNFCCC. *Clim. Change*, **68**(3), 257–267.

Oppo, D.W., J.F. McManus, and J.L. Cullen, 2003: Deepwater variability in Holocene epoch. *Nature*, **422**, 277–278.

Osborn, G., and B.H. Luckman, 1988: Holocene glacier fluctuations in the Canadian Cordillera (Alberta and British Columbia). *Quat. Sci. Rev.*, **7**(2), 115–128.

Osborn, T.J., and K.R. Briffa, 2004: The real color of climate change? *Science*, **306**(5296), 621–622.

Osborn, T.J., and K.R. Briffa, 2006: The spatial extent of 20th-century warmth in the context of the past 1200 years. *Science*, **311**(5762), 841–844.

Osborn, T.J., S.C.B. Raper, and K.R. Briffa, 2006: Simulated climate change during the last 1,000 years: comparing the ECHO-G general circulation model with the MAGICC simple climate model. *Clim. Dyn.*, **27** (2–3), 185–197, doi:10.1007/s00382-006-0129-5.

Otto-Bliesner, B.L., 1999: El Niño La Niña and Sahel precipitation during the middle Holocene. *Geophys. Res. Lett.*, **26**, 87–90.

Otto-Bliesner, B.L., S.J. Marshall, J.T. Overpeck, and G. Miller, 2006a: Simulating Arctic climate warmth and icefield retreat in the Last Interglaciation. *Science*, **311**, 1751–1753.

Otto-Bliesner, B.L., et al., 2003: Modeling El Niño and its tropical teleconnections during the last glacial-interglacial cycle. *Geophys. Res. Lett.*, **30**(23), doi:10.1029/2003GL018553.

Otto-Bliesner, B.L., et al., 2006b: Last Glacial Maximum and Holocene climate in CCSM3. *J. Clim.*, **19**, 2567–2583.

Overpeck, J., and K.E. Trenberth, 2004: *A Multi-Millennia Perspective on Drought and Implications for the Future.* Proceedings of a joint CLIVAR/PAGES/IPCC Workshop, 18-21 Nov. 2003, Tucson, AZ. University Corporation for Atmospheric Research, Boulder CO, 30 pp.

Overpeck, J.T., et al., 2006: Paleoclimatic evidence for future ice sheet instability and rapid sea level rise. *Science*, **311**(5768), 1747–1750.

Pagani, M., et al., 2005: Marked decline in atmospheric carbon dioxide concentrations during the Paleogene. *Science*, **309**(5734), 600–603.

Paillard, D., 1998: The timing of Pleistocene glaciations from a simple multiple-state climate model. *Nature*, **391**, 378–381.

Pauling, A., J. Luterbacher, C. Casty, and H. Wanner, 2006: 500 years of gridded high-resolution precipitation reconstructions over Europe and the connection to large-scale circulation. *Clim. Dyn.*, **26**, 387–405.

Pearson, P.N., and M.R. Palmer, 2000: Atmospheric carbon dioxide concentrations over the past 60 million years. *Nature*, **406**, 695–699.

Peltier, W.R., 1996: Mantle viscosity and ice age ice sheet topography. *Science*, **273**, 1359–1364.

Peltier, W.R., 1998: Postglacial variations in the level of the sea: Implications for climate dynamics and solid-earth geophysics. *Rev. Geophys.*, **36**(4), 603–689.

Peltier, W.R., 2001: Global glacial isostatic adjustment and modern instrumental records of relative sea level history. In: *Sea Level Rise: History and Consequences* [Douglas, B.C., M.S. Kearney, and S.P. Leatherman (eds.)]. Academic Press, San Diego, CA, pp. 65–95.

Peltier, W.R., 2002: On eustatic sea level history: Last Glacial Maximum to Holocene. *Quat. Sci. Rev.*, **21**(1–3), 377–396.

Peltier, W.R., 2004: Global glacial isostasy and the surface of the ice-age Earth: The ICE-5G (VM2) model and GRACE. *Annu. Rev. Earth Planet. Sci.*, **32**, 111–149.

Peltier, W.R., and R.G. Fairbanks, 2006: Global glacial ice volume and last glacial maximum duration from an extended Barbados sea level record. *Quat. Sci. Rev.*, **25**, 3322-3337.

Peltier, W.R., I. Shennan, R. Drummond, and B. Horton, 2002: On the postglacial isostatic adjustment of the British Isles and the shallow viscoelastic structure of the earth. *Geophys. J. Int.*, **148**(3), 443–475.

Pépin, L., D. Raynaud, J.-M. Barnola, and M.F. Loutre, 2001: Hemispheric roles of climate forcings during glacial-interglacial transitions, as deduced from the Vostok record and LLN-2D model experiments. *J. Geophys. Res.*, **106**(D23), 31885–31892.

Peteet, D., 1995: Global Younger Dryas. *Quat. Int.*, **28**, 93–104.

Peterson, L.C., G.H. Haug, K.A. Hughen, and U. Röhl, 2000: Rapid changes in the hydrologic cycle of the tropical Atlantic during the last glacial. *Science*, **290**, 1947–1951.

Petit, J.R., et al., 1999: Climate and atmospheric history of the past 420,000 years from the Vostok ice core, Antarctica. *Nature*, **399**, 429–436.

Petoukhov, V., et al., 2000: CLIMBER-2: a climate system model of intermediate complexity. Part I: model description and performance for present climate. *Clim. Dyn.*, **16**(1), 1–17.

Pettersson, O., 1914: Climate variations in historic and prehistoric time. *Svenska Hydrogr. - Biol. Komm. Skriften*, **5**, 1–26.

Peyron, O., et al., 2005: Lateglacial climate in the Jura Mountains (France) based on different quantitative reconstruction approaches from pollen, lake-levels, and chironomids. *Quat. Res.*, **62**(2), 197–211.

Plattner, G.K., F. Joos, T.F. Stocker, and O. Marchal, 2001: Feedback mechanisms and sensitivities of ocean carbon uptake under global warming. *Tellus*, **53B**(5), 564–592.

Pollack, H.N., and S.P. Huang, 2000: Climate reconstruction from subsurface temperatures. *Annu. Rev. Earth Planet. Sci.*, **28**, 339–365.

Pollack, H.N., and J.E. Smerdon, 2004: Borehole climate reconstructions: Spatial structure and hemispheric averages. *J. Geophys. Res.*, **109**(D11), D11106, doi:10.1029/2003JD004163.

Pollack, H.N., S. Huang, and J.E. Smerdon, 2006: Five centuries of climate change in Australia: The view from underground. *J. Quat. Sci*, **21**(7), 701–706.

Pons, A., J. Guiot, J.L. Debeaulieu, and M. Reille, 1992: Recent contributions to the climatology of the last glacial interglacial cycle based on French pollen sequences. *Quat. Sci. Rev.*, **11**(4), 439–448.

Porter, S.C., 2001: Snowline depression in the tropics during the last glaciation. *Quat. Sci. Rev.*, **20**, 1067–1091.

Prentice, I.C., and T. Webb, 1998: BIOME 6000: reconstructing global mid-Holocene vegetation patterns from palaeoecological records. *J. Biogeogr.*, **25** (6), 997–1005.

Prentice, I.C., D. Jolly, and BIOME 6000 participants, 2000: Mid-Holocene and glacial-maximum vegetation geography of the northern continents and Africa. *J. Biogeogr.*, **27**, 507–519.

Rahmstorf, S., 2001: Abrupt climate change. In: *Encyclopedia of Ocean Sciences*, Vol.1 [Steele, J., S. Thorpe, and K. Turekian (eds.)]. Academic Press, London, pp. 1–6.

Rahmstorf, S., 2002: Ocean circulation and climate during the past 120,000 years. *Nature*, **419**, 207–214.

Rahmstorf, S., and H.J. Schellnhuber, 2006: *Der Klimawandel*. Beck Verlag, Munich, 144 pp.

Rahmstorf, S., et al., 2005: Thermohaline circulation hysteresis: A model intercomparison. *Geophys. Res. Lett.*, **32**(23), doi:10.1029/2005GL023655.

Ramankutty, N., and J.A. Foley, 1999: Estimating historical changes in global land cover: Croplands from 1700 to 1992. *Global Biogeochem. Cycles*, **13**(4), 997–1027.

Ravelo, A.C., et al., 1997: Pliocene carbonate accumulation along the California margin. *Paleoceanography*, **12**, 729–741.

Raymo, M.E., and G.H. Rau, 1992: Plio-Pleistocene atmospheric CO_2 levels inferred from POM δ^{13}C at DSDP Site 607. *Eos*, **73**, 95.

Raymo, M.E., B. Grant, M. Horowitz, and G.H. Rau, 1996: Mid-Pliocene warmth: Stronger greenhouse and stronger conveyor. *Mar. Micropaleontol.*, **27**(1–4), 313–326.

Raymo, M.E., et al., 1989: Late Pliocene variation in northern hemisphere ice sheets and North Atlantic deep water circulation. *Paleoceanography*, **4**, 413–446.

Raynaud, D., J. Chappellaz, C. Ritz, and P. Martinerie, 1997: Air content along the Greenland Ice Core Project core: A record of surface climatic parameters and elevation in central Greenland. *J. Geophys. Res.*, **102**, 26607–26614.

Raynaud, D., et al., 2005: The record for marine isotopic stage 11. *Nature*, **436**(7047), 39–40.

Reichert, B.K., L. Bengtsson, and J. Oerlemans, 2001: Midlatitude forcing mechanisms for glacier mass balance investigated using general circulation models. *J. Clim.*, **14**(17), 3767–3784.

Reichert, B.K., L. Bengtsson, and J. Oerlemans, 2002: Recent glacier retreat exceeds internal variability. *J. Clim.*, **15**(21), 3069–3081.

Renssen, H., and J. Vandenberghe, 2003: Investigation of the relationship between permafrost distribution in NW Europe and extensive winter sea-ice cover in the North Atlantic Ocean during the cold phases of the last glaciation. *Quat. Sci. Rev.*, **22**, 209–223.

Renssen, H., H. Goosse, and T. Fichefet, 2002: Modeling the effect of freshwater pulses on the early Holocene climate: the influence of high frequency climate variability. *Paleoceanography*, **17**, 1020, doi:10.1029/2001PA000649.

Renssen, H., V. Brovkin, T. Fichefet, and H. Goosse, 2003: Holocene climate instability during the termination of the African humid period. *Geophys. Res. Lett.*, **30**, 1184, doi:10.1029/2002GL016636.

Renssen, H., et al., 2005: Simulating the Holocene climate evolution at northern high latitudes using a coupled atmosphere-sea ice-ocean-vegetation model. *Clim. Dyn.*, **24**(1), 23–43.

Ridgwell, A.J., A.J. Watson, M.A. Maslin, and J.O. Kaplan, 2003: Implications of coral reef buildup for the controls on atmospheric CO_2 since the Last Glacial Maximum *Paleoceanography*, **18**(4), doi:10.1029/2003PA000893.

Rimbu, N., et al., 2004: Holocene climate variability as derived from alkenone sea surface temperature and coupled ocean-atmosphere model experiments. *Clim. Dyn.*, **23**, 215–227.

Rind, D., and M.A. Chandler, 1991: Increased ocean heat transports and warmer climate. *J. Geophys. Res.*, **96**, 7437–7461.

Rind, D., J. Perlwitz, and P. Lonergan, 2005: AO/NAO response to climate change: 1. Respective influences of stratospheric and tropospheric climate changes. *J. Geophys. Res.*, **110**(D12), doi:10.1029/2004JD005103.

Risebrobakken, B., T.M. Dokken, and E. Jansen, 2005: Extent and variability of the meridional Atlantic circulation in the eastern Nordic seas during marine isotope stage 5 and its influence on the inception of the last glacial. In: *The Nordic Seas: An Integrated Perspective* [Drange, H., et al. (eds.)]. Geophysical Monographs Vol. 158, American Geophysical Union, Washington, DC, pp. 323–340.

Risebrobakken, B., et al., 2003: A high resolution study of Holocene paleoclimatic and paleoceanographic changes in the Nordic Seas. *Paleoceanography*, **18**, 1–14.

Robertson, A., et al., 2001: Hypothesized climate forcing time series for the last 500 years. *J. Geophys. Res.*, **106**(D14), 14783–14803.

Robock, A., and M.P. Free, 1995: Ice cores as an index of global volcanism from 1850 to the present. *J. Geophys. Res.*, **100**(D6), 11549–11567.

Roche, D., D. Paillard, and E. Cortijo, 2004: Constraints on the duration and freshwater release of Heinrich event 4 through isotope modelling. *Nature*, **432**, 379–382.

Rodbell, D.T., et al., 1999: An ~15,000-year record of El Niño-driven alluviation in southwestern Ecuador. *Science*, **283**, 516–520.

Rohling, E.J., and H. Palike, 2005: Centennial-scale climate cooling with a sudden cold event around 8,200 years ago. *Nature*, **434**, 975–979.

Rosell-Mele, A., et al., 2004: Sea surface temperature anomalies in the oceans at the LGM estimated from the alkenone-$U^{K'}_{37}$ index: comparison with GCMs. *Geophys. Res. Lett.*, **31**, L03208, doi:10.1029/2003GL018151.

Rosenthal, Y., and A.J. Broccoli, 2004: In search of Paleo-ENSO. *Science*, **304**, 219–221.

Rostami, K., W.R. Peltier, and A. Mangini, 2000: Quaternary marine terraces, sea-level changes and uplift history of Patagonia, Argentina. Comparisons with predictions of ICE-4G (VM2) model of the global process of glacial isostatic adjustment. *Quat. Sci. Rev.*, **19**, 1495–1525.

Röthlisberger, F., and M.A. Geyh, 1985: Glacier variations in Himalayas and Karakorum. *Z. Gletscherkunde Glazialgeologie*, **21**, 237–249.

Röthlisberger, R., et al., 2004: Ice core evidence for the extent of past atmospheric CO_2 change due to iron fertilisation. *Geophys. Res. Lett.*, **31**(16), L16207, doi:10.1029/2004GL020338.

Royer, D., 2003: Estimating latest Cretaceous and Tertiary atmospheric CO_2 from stomatal indices. In: *Causes and Consequences of Globally Warm Climates in the Early Paleogene* [Wing, S.L., P.D. Gingerich, B. Schmitz, and E. Thomas (eds.)]. Special Paper Vol. 369, Geological Society of America, Boulder, CO, pp. 79–93.

Royer, D.L., 2006: CO_2-forced climate thresholds during the Phanerozoic. *Geochim. Cosmochim. Acta*, **70**(23), 5665–5675.

Royer, D.L., et al., 2001: Paleobotanical evidence for near present-day levels of atmospheric CO_2 during part of the tertiary. *Science*, **292**(5525), 2310–2313.

Ruddiman, W.F. (ed.), 1997: *Tectonic Uplift and Climate Change.* Plenum Press, New York, 535 pp.

Ruddiman, W.F., 2003: Orbital insolation, ice volume and greenhouse gases. *Quat. Sci. Rev.*, **15–17**, 1597–1629.

Ruddiman, W.F., and J.S. Thomson, 2001: The case for human causes of increased atmospheric CH_4. *Quat. Sci. Rev.*, **20**(18), 1769–1777.

Ruddiman, W.F., S.J. Vavrus, and J.E. Kutzbach, 2005: A test of the overdue-glaciation hypothesis. *Quat. Sci. Rev.*, **24**, 1–10.

Russell, J.M., T.C. Johnson, and M.R. Talbot, 2003: A 725 yr cycle in the climate of central Africa during the late Holocene. *Geology*, **31**(8), 677–680.

Rutberg, R.L., S.R. Hemming, and S.L. Goldstein, 2000: Reduced North Atlantic deep water flux to the glacial Southern Ocean inferred from neodymium isotope ratios. *Nature*, **405**, 935–938.

Rutherford, S., and M.E. Mann, 2004: Correction to "Optimal surface temperature reconstructions using terrestrial borehole data" by Mann et al. *J. Geophys. Res.*, **109**, D11107, doi:10.1029/2003JD004163.

Rutherford, S., M.E. Mann, T.L. Delworth, and R.J. Stouffer, 2003: Climate field reconstruction under stationary and nonstationary forcing. *J. Clim.*, **16**(3), 462–479.

Rutherford, S., et al., 2005: Proxy-based Northern Hemisphere surface temperature reconstructions: Sensitivity to method, predictor network, target season, and target domain. *J. Clim.*, **18**(13), 2308–2329.

Sànchez Goñi, M.F., et al., 2002: Synchroneity between marine and terrestrial responses to millennial scale climatic variability during the last glacial period in the Mediterranean region. *Clim. Dyn.*, **19**, 95–105.

Sarnthein, M., U. Pflaumann, and M. Weinelt, 2003a: Past extent of sea ice in the northern North Atlantic inferred from foraminiferal paleotemperature estimates. *Paleoceanography*, **18**, doi:10.1029/2002PA000771.

Sarnthein, M., et al., 2003b: Overview of the Glacial Atlantic Ocean Mapping (GLAMAP 2000). *Paleoceanography*, **18**, 1030, doi:10.1029/2002PA000769.

Scherer, R.P., et al., 1998: Pleistocene collapse of the West Antarctic ice sheet. *Science*, **281**(5373), 82–85.

Schneider von Deimling, T., A. Ganopolski, H. Held, and S. Rahmstorf, 2006: How cold was the Last Glacial Maximum? *Geophys. Res. Lett.*, **33**, doi: 10.1029/2006GL026484.

Scholze, M., W. Knorr, and M. Heimann, 2003: Modelling terrestrial vegetation dynamics and carbon cycling for an abrupt climate change event. *The Holocene*, **13**, 327–333.

Schubert, S.D., et al., 2004: Causes of long-term drought in the US Great Plains. *J. Clim.*, **17**(3), 485–503.

Schulz, M., A. Paul, and A. Timmermann, 2004: Glacial-interglacial contrast in climate variability at centennial-to-millennial timescales: observations and conceptual model. *Quat. Sci. Rev.*, **23**, 2219–2230.

Schwander, J., et al., 1993: The age of the air in the firn and the ice at Summit, Greenland. *J. Geophys. Res.*, **98**(D2), 2831–2838.

Schwikowski, M., A. Döscher, H.W. Gäggeler, and U. Schotterer, 1999: Anthropogenic versus natural sources of atmospheric sulphate from an Alpine ice core. *Tellus*, **51B**, 938–951.

Seager, R., et al., 2005: Modeling of tropical forcing of persistent droughts and pluvials over western North America: 1856-2000. *J. Clim.*, **18**(19), 4065–4088.

Seppa, H., and H.J.B. Birks, 2001: July mean temperature and annual precipitation trends during Holocene in the Fennoscandian tree-line area: pollen-based climate reconstructions. *The Holocene*, **11**, 527–539.

Severinghaus, J.P., and E.J. Brook, 1999: Abrupt climate change at the end of the last glacial period inferred from trapped air in polar ice. *Science*, **286**(5441), 930–934.

Shackleton, N.J., 1977: Carbon-13 in Uvigerina: Tropical rainforest history and the equatorial Pacific carbonate dissolution cycles. In: *The Fate of Fossil Fuel CO_2 in the Ocean* [Andersen, N., and A. Malahoff (eds.)]. Plenum, New York, pp. 401–428.

Shackleton, N.J., 2000: The 100,000-year ice-age cycle identified and found to lag temperature, carbon dioxide, and orbital eccentricity. *Science*, **289**, 1897–1902.

Shackleton, N.J., M.A. Hall, and A. Boersma, 1984: Oxygen and carbon isotope data from Leg-74 foraminifers. In: *Initial Reports of the Deep Sea Drilling Project*, Vol. 74. Ocean Drilling Program, College Station, TX, pp. 599–612.

Shackleton, N.J., J.C. Hall, and D. Pate, 1995: Pliocene stable isotope stratigraphy of ODP Site 846. In: *Proceedings of the Ocean Drilling Program, Scientific Results*. Vol. 138. Ocean Drilling Program, College Station, TX, pp. 337–356.

Shin, S.I., et al., 2003: A simulation of the Last Glacial Maximum climate using the NCAR CSM. *Clim. Dyn.*, **20**, 127–151.

Shin, S.I., et al., 2006: Understanding the mid-Holocene climate. *J. Clim.*, **19**(12), 2801–2818.

Shindell, D.T., G.A. Schmidt, R.L. Miller, and M.E. Mann, 2003: Volcanic and solar forcing of climate change during the preindustrial era. *J. Clim.*, **16**(24), 4094–4107.

Shindell, D.T., G.A. Schmidt, M.E. Mann, and G. Faluvegi, 2004: Dynamic winter climate response to large tropical volcanic eruptions since 1600. *J. Geophys. Res.*, **109**(D5), D05104, doi:10.1029/2003JD004151.

Shindell, D.T., et al., 2001: Solar forcing of regional climate change during the Maunder Minimum. *Science*, **294**(5549), 2149–2152.

Shulmeister, J., and B.G. Lees, 1995: Pollen evidence from tropical Australia for the onset of an ENSO-dominated climate at c. 4000 BP. *The Holocene*, **5**, 10–18.

Shuman, B., W. Thompson, P. Bartlein, and J.W. Williams, 2002: The anatomy of a climatic oscillation: vegetation change in eastern North America during the Younger Dryas chronozone. *Quat. Sci. Rev.*, **21**(16–17), 1777–1791.

Siddall, M., et al., 2003: Sea-level fluctuations during the last glacial cycle. *Nature*, **423**, 853–858.

Siegenthaler, U., et al., 2005a: Stable carbon cycle-climate relationship during the late Pleistocene. *Science*, **310**(5752), 1313–1317.

Siegenthaler, U., et al., 2005b: Supporting evidence from the EPICA Dronning Maud Land ice core for atmospheric CO_2 changes during the past millennium. *Tellus*, **57B**(1), 51–57.

Sigman, D.M., and E.A. Boyle, 2000: Glacial/interglacial variations in atmospheric carbon dioxide. *Nature*, **407**, 859–869.

Six, D., L. Reynaud, and A. Letréguilly, 2001: Bilans de masse des glaciers alpins et scandinaves, leurs relations avec l oscillation du climat de l Atlantique nord. *C. R. Acad. Sci. Paris, Sciences de la Terre et des planètes/Earth and Planetary Sciences*, **333**, 693–698.

Sloan, L.C., T.J. Crowley, and D. Pollard, 1996: Modeling of middle Pliocene climate with the NCAR GENESIS general circulation model. *Mar. Micropaleontol.*, **27**, 51–61.

Sluijs, A., et al., 2006: Subtropical Arctic Ocean temperatures during the Palaeocene/Eocene thermal maximum. *Nature*, **441**, 610–613.

Smerdon, J.E., et al., 2004: Air-ground temperature coupling and subsurface propagation of annual temperature signals. *J. Geophys. Res.*, **109**(D21), D21107, doi:10.1029/2004JD005056.

Smerdon, J.E., et al., 2006: Daily, seasonal and annual relationships between air and subsurface temperatures. *J. Geophys. Res*, **111**, D07101, doi:10.1029/2004JD005578.

Smith, J.A., et al., 2005: Early local last glacial maximum in the tropical Andes. *Science*, **308**, 678–681.

Sokratov, S.A., and R.G. Barry, 2002: Intraseasonal variation in the thermoinsulation effect of snow cover on soil temperatures and energy balance. *J. Geophys. Res.*, **107**(D19), 4374, doi:10.1029/2001JD000489.

Solanki, S.K., and N.S. Krivova, 2003: Can solar variability explain global warming since 1970? *J. Geophys. Res.*, **108**, 1200, doi:10.1029/2002JA009753.

Solanki, S.K., et al., 2004: Unusual activity of the sun during recent decades compared to the previous 11,000 years. *Nature*, **431**, 1084–1087.

Soon, W., and S. Baliunas, 2003: Proxy climatic and environmental changes of the past 1000 years. *Clim. Res*, **23**(2), 89–110.

Spahni, R., et al., 2003: The attenuation of fast atmospheric CH_4 variations recorded in polar ice cores. *Geophys. Res. Lett.*, **30**(11), doi:10.1029/2003GL017093.

Spahni, R., et al., 2005: Atmospheric methane and nitrous oxide of the late Pleistocene from Antarctic ice cores. *Science*, **310**(5752), 1317–1321.

Stager, J.C., and P.A. Mayewski, 1997: Abrupt early to Mid-Holocene climatic transition registered at the equator and the poles. *Science*, **276**, 1834–1836.

Stager, J.C., B.F. Cumming, and L.D. Meeker, 2003: A 10,000-year high-resolution diatom record from Pilkington Bay, Lake Victoria, East Africa. *Quat. Res.*, **59**(2), 172–181.

Stager, J.C., et al., 2005: Solar variability and the levels of Lake Victoria, East Africa, during the last millennium. *J. Paleolimnol.*, **33**(2), 243–251.

Stahle, D.W., and M.K. Cleaveland, 1992: Reconstruction and analysis of spring rainfall over southeastern U.S. for the past 1000 years. *Bull. Am. Meteorol. Soc.*, **73**, 1947–1961.

Stahle, D.W., et al., 1998: Experimental dendroclimatic reconstruction of the Southern Oscillation. *Bull. Am. Meteorol. Soc.*, **79**(10), 2137–2152.

Stauffer, B., G. Fischer, A. Neftel, and H. Oeschger, 1985: Increase of atmospheric methane recorded in Antarctic ice core. *Science*, **229**(4720), 1386–1388.

Steele, L.P., et al., 1992: Slowing down of the accumulation of atmospheric methane during the 1980s. *Nature*, **358**, 313–316.

Stendel, M., I.A. Mogensen, and J.H. Christensen, 2006: Influence of various forcings on global climate in historical times using a coupled atmosphere-ocean general circulation model. *Clim. Dyn.*, **26**(1), 1–15.

Stenni, B., et al., 2001: An oceanic cold reversal during the last deglaciation. *Science*, **293**, 2074–2077.

Stern, D.I., 2005: Global sulfur emissions from 1850 to 2000. *Chemosphere*, **58**, 163–175.

Stieglitz, M., S.J. Dery, V.E. Romanovsky, and T.E. Osterkamp, 2003: The role of snow cover in the warming of arctic permafrost. *Geophys. Res. Lett.*, **30**(13), 1721, doi:10.1029/2003GL017337.

Stirling C.H., T.M. Esat, K. Lambeck, and M.T. McCulloch, 1998: Timing and duration of the last interglacial: evidence for a restricted interval of widespread coral reef growth. *Earth Planet. Sci. Lett.*, **160**, 745–762.

Stocker, T.F., and S.J. Johnsen, 2003: A minimum thermodynamic model for the bipolar seesaw. *Paleoceanography*, **18**(4), doi:10.1029/2003PA000920.

Stocker, T.F., and E. Monnin, 2003: Past rates of carbon dioxide changes and their relevance for future climate. *Pages News*, **11**(1), 6–8.

Stott, L., et al., 2004: Decline in surface temperature and salinity in the western tropical Pacific Ocean in Holocene epoch. *Nature*, **431**, 56–59.

Stouffer, R.J., et al., 2006: Investigating the causes of the response of the thermohaline circulation to past and future climate changes. *J. Clim.*, **19**(3), 1365–1386.

Sutton, R.T., and D.L.R. Hodson, 2005: Atlantic Ocean forcing of North American and European summer climate. *Science*, **309**(5731), 115–118.

Svendsen, J.I., and J. Mangerud, 1997: Holocene glacial and climatic variations on Spitsbergen, Svalbard. *The Holocene*, **7**, 45–57.

Svensen, H., et al., 2004: Release of methane from a volcanic basin as a mechanism for initial Eocene global warming. *Nature*, **429**, 542–545.

Tajika, E., 1998: Climate change during the last 150 million years: reconstruction from a carbon cycle. *Earth Planet. Sci. Lett.*, **160**(3–4), 695–707.

Tans, P.P., and T.J. Conway, 2005: Monthly atmospheric CO_2 mixing ratios from the NOAA CMDL Carbon Cycle Cooperative Global Air Sampling Network, 1968–2002. In: *Trends: A Compendium of Data on Global Change*. Carbon Dioxide Information Analysis Center, Oak Ridge National Laboratory, U.S. Department of Energy, Oak Ridge, TN.

Tarasov, L., and W.R. Peltier, 2003: Greenland glacial history, borehole constraints, and Eemian extent. *J. Geophys. Res.*, **108**, 2143, doi:10.1029/2001JB001731.

Tarasov, L., and W.R. Peltier, 2005: Arctic freshwater forcing of the Younger Dryas cold reversal. *Nature*, **435**(7042), 662–665.

Taylor, A.E., et al., 2006: Canadian arctic permafrost observatories: detecting contemporary climate change through inversion of subsurface temperature time-series. *J. Geophys. Res.*, **111**, B02411, doi:10.1029/2004JB003208.

Taylor, K.E., et al., 2000: Analysis of forcing, response, and feedbacks in a paleoclimate modeling experiment. In: *Proceedings of the Third Paleoclimate Modelling Intercomparison Project (PMIP) Workshop, 4-8 Oct. 1999, La Huardière, Canada* [Braconnot, P. (ed.)]. WCRP-111, WMO/TD-No. 1007, World Meteorological Organization, Geneva, pp. 43–49.

Tett, S.F.B., et al., 2007: The impact of natural and anthropogenic forcings on climate and hydrology since 1550. *Clim. Dyn.*, **28**(1), 3–34.

Thomas, E., 2003: Extinction and food at the sea floor: a high-resolution benthic foraminiferal record across the Initial Eocene Thermal Maximum, Southern Ocean Site 690. In: *Causes and Consequences of Globally Warm Climates of the Paleogene* [Wing, S., Gingerich, P., Schmitz, B., and Thomas, E., (eds.)]. Special Paper Vol. 369, Geological Society of America, Boulder, CO, pp. 319–332.

Thompson, L.G., 2000: Ice core evidence for climate change in the Tropics: implications for our future. *Quat. Sci. Rev.*, **19**(1–5), 19–35.

Thompson, L.G., 2001: Stable isotopes and their relationship to temperature as recorded in low latitude ice cores. In: *Geological Perspectives of Global Climate Change* [Gerhard, L.C., W.E. Harrison, and B.M. Hanson (eds.)]. Studies in Geology No. 47, American Association of Petroleum Geologists, Tulsa, OK, pp. 99–119.

Thompson, L.G., et al., 2000: A high-resolution millennial record of the South Asian Monsoon from Himalayan ice cores. *Science*, **289**(5486), 1916–1919.

Thompson, R.S., 1991: Pliocene environments and climates in the Western United States. *Quat. Sci. Rev.*, **10**, 115–132.

Thompson, R.S., and R.F. Fleming, 1996: Middle Pliocene vegetation: Reconstructions, paleoclimate inferences, and boundary conditions for climate modeling. *Mar. Micropaleontol.*, **27**, 27–49.

Tinner, W., and A.F. Lotter, 2001: Central European vegetation response to abrupt climate change at 8.2 ka. *Geology*, **29**, 551–554.

Toracinta, E.R., R.J. Oglesby, and D.H. Bromwich, 2004: Atmospheric response to modified CLIMAP ocean boundary conditions during the Last Glacial Maximum. *J. Clim.*, **17**, 504–522.

Traufetter, F., et al., 2004: Spatio-temporal variability in volcanic sulphate deposition over the past 2 kyr in snow pits and firn cores from Amundsenisen, Dronning Maud Land, Antarctica. *J. Glaciol.*, **50**, 137–146.

Trenberth, K.E., and B.L. Otto-Bliesner, 2003: Toward integrated reconstruction of past climates. *Science*, **300**(5619), 589–591.

Tripati, A.K., and H. Elderfield, 2004: Abrupt hydrographic changes in the equatorial Pacific and subtropical Atlantic from foraminiferal Mg/Ca indicate greenhouse origin for the thermal maximum at the Paleocene-Eocene Boundary. *Geochem. Geophys. Geosystems*, **5**, doi:10.1029/2003GC000631.

Tudhope, A.W., et al., 2001: Variability in the El Niño-Southern Oscillation through a glacial-interglacial cycle. *Science*, **291**, 1511–1517.

Tzedakis, P.C., 2005: Towards an understanding of the response of southern European vegetation to orbital and suborbital climate variability. *Quat. Sci. Rev.*, **24**, 1585–1599.

van Kreveld, S., et al., 2000: Potential links between surging ice sheets, circulation changes, and the Dansgaard-Oeschger cycles in the Irminger Sea, 60-18 kyr. *Paleoceanography*, **15**, 425–442.

Vellinga, M., and R.A. Wood, 2002: Global climatic impacts of a collapse of the Atlantic thermohaline circulation. *Clim. Change*, **54**(3), 251–267.

Verschuren, D., 2001: Reconstructing fluctuations of a shallow East African lake during the past 1800 yrs from sediment stratigraphy in a submerged crater basin. *J. Paleolimnol.*, **25**(3), 297–311.

Veski, S., H. Seppa, and A.E.K. Ojala, 2004: Cold event at 8200 yr BP recorded in annually laminated lake sediments in eastern Europe. *Geology*, **32**(8), 681–684.

Vettoretti, G., and W.R. Peltier, 2003: Post-Eemian glacial inception. Part II: Elements of a cryospheric moisture pump. *J. Clim.*, **16**(6), 912–927.

Vidal, L., L. Labeyrie, and T.C.E. van Weering, 1998: Benthic δ^{18}O records in the North Atlantic over the last glacial period (60-10 kyr): Evidence for brine formation. *Paleoceanography*, **13**(3), 245–251.

Villalba, R., et al., 2003: Large-scale temperature changes across the southern Andes: 20th-century variations in the context of the past 400 years. *Clim. Change*, **59**(1–2), 177–232.

Vincent, C., et al., 2005: Glacier fluctuations in the Alps and in the tropical Andes. *Comptes Rendus Geoscience*, **337**(1–2), 97–106.

Voelker, A.H.L., 2002: Global distribution of centennial-scale records for Marine Isotope Stage (MIS) 3: a database. *Quat. Sci. Rev.*, **21**(10), 1185–1212.

von Grafenstein, U., et al., 1998: The cold event 8,200 years ago documented in oxygen isotope records of precipitation in Europe and Greenland. *Clim. Dyn.*, **14**, 73–81.

von Storch, H., and E. Zorita, 2005: Comment on "Hockey sticks, principal components, and spurious significance" by S. McIntyre and R. McKitrick. *Geophys. Res. Lett.*, **32**(20), doi:10.1029/2005GL022753.

von Storch, H., et al., 2004: Reconstructing past climate from noisy data. *Science*, **306**(5296), 679–682.

Vuille, M., and M. Werner, 2005: Stable isotopes in precipitation recording South American summer monsoon and ENSO variability: observations and model results. *Clim. Dyn.*, **25**(4), 401–413.

Vuille, M., M. Werner, R.S. Bradley, and F. Keimig, 2005: Stable isotopes in precipitation in the Asian monsoon region. *J. Geophys. Res.*, **110**(D23), doi:10.1029/2005JD006022.

Waelbroeck, C., et al., 2002: Sea-level and deep water temperature changes derived from benthic foraminifera isotopic records. *Quat. Sci. Rev.*, **21**(1–3), 295–305.

Waelbroeck, C., et al., 2005: A global compilation of late Holocene planktonic foraminiferal δ^{18}O: Relationship between surface water temperature and δ^{18}O. *Quat. Sci. Rev.*, **24**, 853–858.

Wahl, E.R., and C.M. Ammann, 2007: Robustness of the Mann, Bradley, Hughes reconstruction of Northern Hemisphere surface temperatures: Examination of criticisms based on the nature and processing of proxy climate evidence. *Clim. Change*, in press.

Wahl, E.R., D.M. Ritson and C.M. Ammann, 2006: Comment on "Reconstructing past climate from noisy data". *Science*, **312**, 529.

Wallmann, K., 2001: Controls on the Cretaceous and Cenozoic evolution of seawater composition, atmospheric CO_2 and climate. *Geochim. Cosmochim. Acta*, **65**(18), 3005–3025.

Wang, Y.J., et al., 2001: A high-resolution absolute-dated late Pleistocene monsoon record from Hulu Cave, China. *Science*, **294**, 2345–2348.

Wang, Y.J., et al., 2005: Holocene Asian monsoon: Links to solar changes and North Atlantic climate. *Science*, **308**, 854–857.

Wang, Y.M., and N.R. Sheeley, 2003: Modeling the sun's large-scale magnetic field during the Maunder minimum. *Astrophys. J.*, **591**(2), 1248–1256.

Wang, Y.M., J.L. Lean, and N.R. Sheeley, 2005: Modeling the sun's magnetic field and irradiance since 1713. *Astrophys. J.*, **625**, 522–538.

Wang, Z., and L.A. Mysak, 2002: Simulation of the last glacial inception and rapid ice sheet growth in the McGill paleoclimate model. *Geophys. Res. Lett.*, **29**, doi:10.1029/2002GL015120.

Watanabe, O., et al., 2003: Homogeneous climate variability across East Antarctica over the past three glacial cycles. *Nature*, **422**, 509–512.

Webb, R.S., et al., 1997: Influence of ocean heat transport on the climate of the Last Glacial Maximum. *Nature*, **385**, 695–699.

Williams, J.W., T.I. Webb, P.H. Richard, and P. Newby, 2000: Late Quaternary biomes of Canada and the eastern United States. *J. Biogeogr.*, **27**, 585–607.

Williams, J.W., et al., 2002: Rapid and widespread vegetation responses to past climate change in the North Atlantic region. *Geology*, **30**(11), 971–974.

Williams, P.W., D.N.T. King, J.-X. Zhao, and K.D. Collerson, 2004: Speleotherm master chronologies : combined Holocene [18]O and ^{13}C records from the north Island of New Zealand and their palaeoenvironmental interpretation. *The Holocene*, **14**, 194–208.

Wing, S.L., et al., 2005: Transient floral change and rapid global warming at the Paleocene-Eocene boundary. *Science*, **310**(5750), 993–996.

Wise, S.W.J., J.R. Breza, D.M. Harwood, and W. Wei, 1991: Paleogene glacial history of Antarctica. In: *Controversies in Modern Geology: Evolution of Geological Theories in Sedimentology, Earth History and Tectonics* [Müller, D.W., J.A. McKenzie, and H. Weissert (eds.)]. Cambridge University Press, Cambridge, pp. 133–171.

Wohlfahrt, J., S.P. Harrison, and P. Braconnot, 2004: Synergistic feedbacks between ocean and vegetation on mid- and high- latitude climates during the mid-Holocene. *Clim. Dyn.*, **22**, 223–238.

Woodhouse, C.A., and J.T. Overpeck, 1998: 2000 years of drought variability in the central United States. *Bull. Am. Meteorol. Soc.*, **79**(12), 2693–2714.

Wyputta, U., and B.J. McAvaney, 2001: Influence of vegetation changes during the Last Glacial Maximum using the BMRC atmospheric general circulation model. *Clim. Dyn.*, **17**, 923–932.

Xia, Q.K., H.X. Zhao, and K.D. Collerson, 2001: Early-Mid Holocene climatic variations in Tasmania, Australia: multi-proxy records in a stalagmite from Lynds Cave. *Earth Planet. Sci. Lett.*, **194**(1–2), 177–187.

Xoplaki, E., et al., 2005: European spring and autumn temperature variability and change of extremes over the last half millennium. *Geophys. Res. Lett.*, **32**, L15713, doi:10.1029/2005GL023424.

Yapp, C.J., and H. Poths, 1992: Ancient atmospheric CO_2 pressures inferred from natural goethites. *Nature*, **355**, 342–344.

Yokoyama, Y., et al., 2000: Timing of the Last Glacial Maximum from observed sea-level minima. *Nature*, **406**(6797), 713–716.

Zachos, J., et al., 2001: Trends, rhythms, and aberrations in global climate 65 Ma to present. *Science*, **292**(5517), 686–693.

Zachos, J.C., et al., 2003: A transient rise in tropical sea surface temperature during the Paleocene-Eocene Thermal Maximum. *Science*, **302**(5650), 1551–1554.

Zachos, J.C., et al., 2004*: Early Cenozoic Extreme Climates: The Walvis Ridge Transect, Sites 1262-1267.* Proceedings of the Ocean Drilling Program, Initial Reports Vol. 208, Ocean Drilling Program, College Station, TX.

Zachos, J.C., et al., 2005: Rapid acidification of the ocean during the Paleocene-Eocene thermal maximum. *Science*, **308**(5728), 1611–1615.

Zhang, D.E., 2005: Severe drought events as revealed in the climate record of China and their temperature situations over the last 1000 years. *Acta Meteorol. Sin.*, **19**(4), 485–491.

Zhang, R., and T.L. Delworth, 2005: Simulated tropical response to a substantial weakening of the Atlantic thermohaline circulation. *J. Clim.*, **18**(12), 1853–1860.

Zhao, Y., et al., 2005: A multi-model analysis of the role of the ocean on the African and Indian monsoon during the mid-Holocene. *Clim. Dyn.*, **25**, (7–8), 777–800.

Zielinski, G.A., 2000: Use of paleo-records in determining variability within the volcanism-climate system. *Quat. Sci. Rev.*, **19**(1), 417–438.

Zorita, E., F. Gonzalez-Rouco, and S. Legutke, 2003: Testing the Mann et al. (1998) approach to paleoclimate reconstructions in the context of a 1000-yr control simulation with the ECHO-G coupled climate model. *J. Clim.*, **16**(9), 1378–1390.

7

Couplings Between Changes in the Climate System and Biogeochemistry

Coordinating Lead Authors:

Kenneth L. Denman (Canada), Guy Brasseur (USA, Germany)

Lead Authors:

Amnat Chidthaisong (Thailand), Philippe Ciais (France), Peter M. Cox (UK), Robert E. Dickinson (USA), Didier Hauglustaine (France), Christoph Heinze (Norway, Germany), Elisabeth Holland (USA), Danie Jacob (USA, France), Ulrike Lohmann (Switzerland), Srikanthan Ramachandran (India), Pedro Leite da Silva Dias (Brazil), Steven C. Wofsy (USA), Xiaoye Zhang (China)

Contributing Authors:

D. Archer (USA), V. Arora (Canada), J. Austin (USA), D. Baker (USA), J.A. Berry (USA), R. Betts (UK), G. Bonan (USA), P. Bousquet (France), J. Canadell (Australia), J. Christian (Canada), D.A. Clark (USA), M. Dameris (Germany), F. Dentener (EU), D. Easterling (USA), V. Eyring (Germany), J. Feichter (Germany), P. Friedlingstein (France, Belgium), I. Fung (USA), S. Fuzzi (Italy), S. Gong (Canada), N. Gruber (USA, Switzerland), A. Guenther (USA), K. Gurney (USA), A. Henderson-Sellers (Switzerland), J. House (UK), A. Jones (UK), C. Jones (UK), B. Kärcher (Germany), M. Kawamiya (Japan), K. Lassey (New Zealand), C. Le Quéré (UK, France, Canada), C. Leck (Sweden), J. Lee-Taylor (USA, UK), Y. Malhi (UK), K. Masarie (USA), G. McFiggans (UK), S. Menon (USA), J.B. Miller (USA), P. Peylin (France), A. Pitman (Australia), J. Quaas (Germany), M. Rauzach (Australia), P. Rayner (France), G. Rehder (Germany), U. Riebesell (Germany), C. Rödenbeck (Germany), L. Rotstayn (Australia), N. Roulet (Canada), C. Sabine (USA), M.G. Schultz (Germany), M. Schulz (France, Germany), S.E. Schwartz (USA), W. Steffen (Australia), D. Stevenson (UK), Y. Tian (USA, China), K.E. Trenberth (USA), T. Van Noije (Netherlands), O. Wild (Japan, UK), T. Zhang (USA, China), L. Zhou (USA, China)

Review Editors:

Kansri Boonpragob (Thailand), Martin Heimann (Germany, Switzerland), Mario Molina (USA, Mexico)

This chapter should be cited as:

Denman, K.L., G. Brasseur, A. Chidthaisong, P. Ciais, P.M. Cox, R.E. Dickinson, D. Hauglustaine, C. Heinze, E. Holland, D. Jacob, U. Lohmann, S Ramachandran, P.L. da Silva Dias, S.C. Wofsy and X. Zhang, 2007: Couplings Between Changes in the Climate System and Biogeochemistry. In: *Climate Change 2007: The Physical Science Basis. Contribution of Working Group I to the Fourth Assessment Report of the Intergovernmental Panel on Climate Change* [Solomon, S., D. Qin, M. Manning, Z. Chen, M. Marquis, K.B. Averyt, M.Tignor and H.L. Miller (eds.)]. Cambridge University Press, Cambridge, United Kingdom and New York, NY, USA.

Table of Contents

Executive Summary

Emissions of carbon dioxide, methane, nitrous oxide and of reactive gases such as sulphur dioxide, nitrogen oxides, carbon monoxide and hydrocarbons, which lead to the formation of secondary pollutants including aerosol particles and tropospheric ozone, have increased substantially in response to human activities. As a result, biogeochemical cycles have been perturbed significantly. Nonlinear interactions between the climate and biogeochemical systems could amplify (positive feedbacks) or attenuate (negative feedbacks) the disturbances produced by human activities.

The Land Surface and Climate

- Changes in the land surface (vegetation, soils, water) resulting from human activities can affect regional climate through shifts in radiation, cloudiness and surface temperature.

- Changes in vegetation cover affect surface energy and water balances at the regional scale, from boreal to tropical forests. Models indicate increased boreal forest reduces the effects of snow albedo and causes regional warming. Observations and models of tropical forests also show effects of changing surface energy and water balance.

- The impact of land use change on the energy and water balance may be very significant for climate at regional scales over time periods of decades or longer.

The Carbon Cycle and Climate

- Atmospheric carbon dioxide (CO_2) concentration has continued to increase and is now almost 100 ppm above its pre-industrial level. The annual mean CO_2 growth rate was significantly higher for the period from 2000 to 2005 (4.1 ± 0.1 GtC yr^{-1}) than it was in the 1990s (3.2 ± 0.1 GtC yr^{-1}). Annual emissions of CO_2 from fossil fuel burning and cement production increased from a mean of 6.4 ± 0.4 GtC yr^{-1} in the 1990s to 7.2 ± 0.3 GtC yr^{-1} for 2000 to 2005.[1]

- Carbon dioxide cycles between the atmosphere, oceans and land biosphere. Its removal from the atmosphere involves a range of processes with different time scales. About 50% of a CO_2 increase will be removed from the atmosphere within 30 years, and a further 30% will be removed within a few centuries. The remaining 20% may stay in the atmosphere for many thousands of years.

- Improved estimates of ocean uptake of CO_2 suggest little change in the ocean carbon sink of 2.2 ± 0.5 GtC yr^{-1}

between the 1990s and the first five years of the 21st century. Models indicate that the fraction of fossil fuel and cement emissions of CO_2 taken up by the ocean will decline if atmospheric CO_2 continues to increase.

- Interannual and inter-decadal variability in the growth rate of atmospheric CO_2 is dominated by the response of the land biosphere to climate variations. Evidence of decadal changes is observed in the net land carbon sink, with estimates of 0.3 ± 0.9, 1.0 ± 0.6, and 0.9 ± 0.6 GtC yr^{-1} for the 1980s, 1990s and 2000 to 2005 time periods, respectively.

- A combination of techniques gives an estimate of the flux of CO_2 to the atmosphere from land use change of 1.6 (0.5 to 2.7) GtC yr^{-1} for the 1990s. A revision of the Third Assessment Report (TAR) estimate for the 1980s downwards to 1.4 (0.4 to 2.3) GtC yr^{-1} suggests little change between the 1980s and 1990s, and continuing uncertainty in the net CO_2 emissions due to land use change.

- Fires, from natural causes and human activities, release to the atmosphere considerable amounts of radiatively and photochemically active trace gases and aerosols. If fire frequency and extent increase with a changing climate, a net increase in CO_2 emissions is expected during this fire regime shift.

- There is yet no statistically significant trend in the CO_2 growth rate as a fraction of fossil fuel plus cement emissions since routine atmospheric CO_2 measurements began in 1958. This 'airborne fraction' has shown little variation over this period.

- Ocean CO_2 uptake has lowered the average ocean pH (increased acidity) by approximately 0.1 since 1750. Consequences for marine ecosystems may include reduced calcification by shell-forming organisms, and in the longer term, the dissolution of carbonate sediments.

- The first-generation coupled climate-carbon cycle models indicate that global warming will increase the fraction of anthropogenic CO_2 that remains in the atmosphere. This positive climate-carbon cycle feedback leads to an additional increase in atmospheric CO_2 concentration of 20 to 224 ppm by 2100, in models run under the IPCC (2000) Special Report on Emission Scenarios (SRES) A2 emissions scenario.

Reactive Gases and Climate

- Observed increases in atmospheric methane concentration, compared with pre-industrial estimates, are directly linked to human activity, including agriculture, energy production,

---

[1] The uncertainty ranges given here and especially in Tables 7.1 and 7.2 are the authors' estimates of the *likely* (66%) range for each term based on their assessment of the currently available studies. There are not enough comparable studies to enable estimation of a *very likely* (90%) range for all the main terms in the carbon cycle budget.

waste management and biomass burning. Constraints from methyl chloroform observations show that there have been no significant trends in hydroxyl radical (OH) concentrations, and hence in methane removal rates, over the past few decades (see Chapter 2). The recent slowdown in the growth rate of atmospheric methane since about 1993 is thus *likely* due to the atmosphere approaching an equilibrium during a period of near-constant total emissions. However, future methane emissions from wetlands are *likely* to increase in a warmer and wetter climate, and to decrease in a warmer and drier climate.

- No long-term trends in the tropospheric concentration of OH are expected over the next few decades due to offsetting effects from changes in nitric oxides (NO_x), carbon monoxide, organic emissions and climate change. Interannual variability of OH may continue to affect the variability of methane.

- New model estimates of the global tropospheric ozone budget indicate that input of ozone from the stratosphere (approximately 500 Tg yr^{-1}) is smaller than estimated in the TAR (770 Tg yr^{-1}), while the photochemical production and destruction rates (approximately 5,000 and 4,500 Tg yr^{-1} respectively) are higher than estimated in the TAR (3,400 and 3,500 Tg yr^{-1}). This implies greater sensitivity of ozone to changes in tropospheric chemistry and emissions.

- Observed increases in NO_x and nitric oxide emissions, compared with pre-industrial estimates, are *very likely* directly linked to 'acceleration' of the nitrogen cycle driven by human activity, including increased fertilizer use, intensification of agriculture and fossil fuel combustion.

- Future climate change may cause either an increase or a decrease in background tropospheric ozone, due to the competing effects of higher water vapour and higher stratospheric input; increases in regional ozone pollution are expected due to higher temperatures and weaker circulation.

- Future climate change may cause significant air quality degradation by changing the dispersion rate of pollutants, the chemical environment for ozone and aerosol generation and the strength of emissions from the biosphere, fires and dust. The sign and magnitude of these effects are highly uncertain and will vary regionally.

- The future evolution of stratospheric ozone, and therefore its recovery following its destruction by industrially manufactured halocarbons, will be influenced by stratospheric cooling and changes in the atmospheric circulation resulting from enhanced CO_2 concentrations. With a possible exception in the polar lower stratosphere where colder temperatures favour ozone destruction by chlorine activated on polar stratospheric cloud particles, the expected cooling of the stratosphere should reduce ozone depletion and therefore enhance the ozone column amounts.

Aerosol Particles and Climate

- Sulphate aerosol particles are responsible for globally averaged temperatures being lower than expected from greenhouse gas concentrations alone.

- Aerosols affect radiative fluxes by scattering and absorbing solar radiation (direct effect, see Chapter 2). They also interact with clouds and the hydrological cycle by acting as cloud condensation nuclei (CCN) and ice nuclei. For a given cloud liquid water content, a larger number of CCN increases cloud albedo (indirect cloud albedo effect) and reduces the precipitation efficiency (indirect cloud lifetime effect), both of which are *likely* to result in a reduction of the global, annual mean net radiation at the top of the atmosphere. However, these effects may be partly offset by evaporation of cloud droplets due to absorbing aerosols (semi-direct effect) and/or by more ice nuclei (glaciation effect).

- The estimated total aerosol effect is lower than in TAR mainly due to improvements in cloud parametrizations, but large uncertainties remain.

- The radiative forcing resulting from the indirect cloud albedo effect was estimated in Chapter 2 as –0.7 W m^{-2} with a 90% confidence range of –0.3 to –1.8 W m^{-2}. Feedbacks due to the cloud lifetime effect, semi-direct effect or aerosol-ice cloud effects can either enhance or reduce the cloud albedo effect. Climate models estimate the sum of all aerosol effects (total indirect plus direct) to be –1.2 W m^{-2} with a range from –0.2 to –2.3 W m^{-2} in the change in top-of-the-atmosphere net radiation since pre-industrial times, whereas inverse estimates constrain the indirect aerosol effect to be between –0.1 and –1.7 W m^{-2} (see Chapter 9).

- The magnitude of the total aerosol effect on precipitation is more uncertain, with model results ranging from almost no change to a decrease of 0.13 mm day^{-1}. Decreases in precipitation are larger when the atmospheric General Circulation Models are coupled to mixed-layer ocean models where the sea surface temperature and, hence, the evaporation is allowed to vary.

- Deposition of dust particles containing limiting nutrients can enhance photosynthetic carbon fixation on land and in the oceans. Climate change is likely to affect dust sources.

- Since the TAR, advances have been made to link the marine and terrestrial biospheres with the climate system via the aerosol cycle. Emissions of aerosol precursors from vegetation and from the marine biosphere are expected to respond to climate change.

The Earth's climate is determined by a number of complex connected physical, chemical and biological processes occurring in the atmosphere, land and ocean. The radiative properties of the atmosphere, a major controlling factor of the Earth's climate, are strongly affected by the biophysical state of the Earth's surface and by the atmospheric abundance of a variety of trace constituents. These constituents include long-lived greenhouse gases (LLGHGs) such as carbon dioxide (CO_2), methane (CH_4) and nitrous oxide (N_2O), as well as other radiatively active constituents such as ozone and different types of aerosol particles. The composition of the atmosphere is determined by processes such as natural and anthropogenic emissions of gases and aerosols, transport at a variety of scales, chemical and microphysical transformations, wet scavenging and surface uptake by the land and terrestrial ecosystems, and by the ocean and its ecosystems. These processes and, more generally the rates of biogeochemical cycling, are affected by climate change, and involve interactions between and within the different components of the Earth system. These interactions are generally nonlinear and may produce negative or positive feedbacks to the climate system.

An important aspect of climate research is to identify potential feedbacks and assess if such feedbacks could produce large and undesired responses to perturbations resulting from human activities. Studies of past climate evolution on different time scales can elucidate mechanisms that could trigger nonlinear responses to external forcing. The purpose of this chapter is to identify the major biogeochemical feedbacks of significance to the climate system, and to assess current knowledge of their magnitudes and trends. Specifically, this chapter will examine the relationships between the physical climate system and the land surface, the carbon cycle, chemically reactive atmospheric gases and aerosol particles. It also presents the current state of knowledge on budgets of important trace gases. Large uncertainties remain in many issues discussed in this chapter, so that quantitative estimates of the importance of the coupling mechanisms discussed in the following sections are not always available. In addition, regional differences in the role of some cycles and the complex interactions between them limit our present ability to provide a simple quantitative description of the interactions between biogeochemical processes and climate change.

7.1.1   Terrestrial Ecosystems and Climate

The terrestrial biosphere interacts strongly with the climate, providing both positive and negative feedbacks due to biogeophysical and biogeochemical processes. Some of these feedbacks, at least on a regional basis, can be large. Surface climate is determined by the balance of fluxes, which can be changed by radiative (e.g., albedo) or non-radiative (e.g., water cycle related processes) terms. Both radiative and non-radiative terms are controlled by details of vegetation. High-latitude climate is strongly influenced by snow albedo feedback, which

is drastically reduced by the darkening effect of vegetation. In semi-arid tropical systems, such as the Sahel or northeast Brazil, vegetation exerts both radiative and hydrological feedbacks. Surface climate interacts with vegetation cover, biomes, productivity, respiration of vegetation and soil, and fires, all of which are important for the carbon cycle. Various processes in terrestrial ecosystems influence the flux of carbon between land and the atmosphere. Terrestrial ecosystem photosynthetic productivity changes in response to changes in temperature, precipitation, CO_2 and nutrients. If climate becomes more favourable for growth (e.g., increased rainfall in a semi-arid system), productivity increases, and carbon uptake from the atmosphere is enhanced. Organic carbon compounds in soils, originally derived from plant material, are respired (i.e., oxidized by microbial communities) at different rates depending on the nature of the compound and on the microbial communities; the aggregate rate of respiration depends on soil temperature and moisture. Shifts in ecosystem structure in response to a changing climate can alter the partitioning of carbon between the atmosphere and the land surface. Migration of boreal forest northward into tundra would initially lead to an increase in carbon storage in the ecosystem due to the larger biomass of trees than of herbs and shrubs, but over a longer time (e.g., centuries), changes in soil carbon would need to be considered to determine the net effect. A shift from tropical rainforest to savannah, on the other hand, would result in a net flux of carbon from the land surface to the atmosphere.

7.1.2   Ocean Ecosystems and Climate

The functioning of ocean ecosystems depends strongly on climatic conditions including near-surface density stratification, ocean circulation, temperature, salinity, the wind field and sea ice cover. In turn, ocean ecosystems affect the chemical composition of the atmosphere (e.g. CO_2, N_2O, oxygen (O_2), dimethyl sulphide (DMS) and sulphate aerosol). Most of these components are expected to change with a changing climate and high atmospheric CO_2 conditions. Marine biota also influence the near-surface radiation budget through changes in the marine albedo and absorption of solar radiation (bio-optical heating). Feedbacks between marine ecosystems and climate change are complex because most involve the ocean's physical responses and feedbacks to climate change. Increased surface temperatures and stratification should lead to increased photosynthetic fixation of CO_2, but associated reductions in vertical mixing and overturning circulation may decrease the return of required nutrients to the surface ocean and alter the vertical export of carbon to the deeper ocean. The sign of the cumulative feedback to climate of all these processes is still unclear. Changes in the supply of micronutrients required for photosynthesis, in particular iron, through dust deposition to the ocean surface can modify marine biological production patterns. Ocean acidification due to uptake of anthropogenic CO_2 may lead to shifts in ocean ecosystem structure and dynamics, which may alter the biological production and export from the surface ocean of organic carbon and calcium carbonate ($CaCO_3$).

7.1.3 Atmospheric Chemistry and Climate

Interactions between climate and atmospheric oxidants, including ozone, provide important coupling mechanisms in the Earth system. The concentration of tropospheric ozone has increased substantially since the pre-industrial era, especially in polluted areas of the world, and has contributed to radiative warming. Emissions of chemical ozone precursors (carbon monoxide, CH_4, non-methane hydrocarbons, nitrogen oxides) have increased as a result of larger use of fossil fuel, more frequent biomass burning and more intense agricultural practices. The atmospheric concentration of pre-industrial tropospheric ozone is not accurately known, so that the resulting radiative forcing cannot be accurately determined, and must be estimated from models. The decrease in concentration of stratospheric ozone in the 1980s and 1990s due to manufactured halocarbons (which produced a slight cooling) has slowed down since the late 1990s. Model projections suggest a slow steady increase over the next century, but continued recovery could be affected by future climate change. Recent changes in the growth rate of atmospheric CH_4 and in its apparent lifetime are not well understood, but indications are that there have been changes in source strengths. Nitrous oxide continues to increase in the atmosphere, primarily as a result of agricultural activities. Changes in atmospheric chemical composition that could result from climate changes are even less well quantified. Photochemical production of the hydroxyl radical (OH), which efficiently destroys many atmospheric compounds, occurs in the presence of ozone and water vapour, and should be enhanced in an atmosphere with increased water vapour, as projected under future global warming. Other chemistry-related processes affected by climate change include the frequency of lightning flashes in thunderstorms (which produce nitrogen oxides), scavenging mechanisms that remove soluble species from the atmosphere, the intensity and frequency of convective transport events, the natural emissions of chemical compounds (e.g., biogenic hydrocarbons by the vegetation, nitrous and nitric oxide by soils, DMS from the ocean) and the surface deposition on molecules on the vegetation and soils. Changes in the circulation and specifically the more frequent occurrence of stagnant air events in urban or industrial areas could enhance the intensity of air pollution events. The importance of these effects is not yet well quantified.

7.1.4 Aerosol Particles and Climate

Atmospheric aerosol particles modify Earth's radiation budget by absorbing and scattering incoming solar radiation. Even though some particle types may have a warming effect, most aerosol particles, such as sulphate (SO_4) aerosol particles, tend to cool the Earth surface by scattering some of the incoming solar radiation back to space. In addition, by acting as cloud condensation nuclei, aerosol particles affect radiative properties of clouds and their lifetimes, which contribute to additional surface cooling. A significant natural source of sulphate is DMS, an organic compound whose production by phytoplankton and release to the atmosphere depends on climatic factors. In many areas of the Earth, large amounts of SO_4 particles are produced as a result of human activities (e.g., coal burning). With an elevated atmospheric aerosol load, principally in the Northern Hemisphere (NH), it is likely that the temperature increase during the last century has been smaller than the increase that would have resulted from radiative forcing by greenhouse gases alone. Other indirect effects of aerosols on climate include the evaporation of cloud particles through absorption of solar radiation by soot, which in this case provides a positive warming effect. Aerosols (i.e., dust) also deliver nitrogen (N), phosphorus and iron to the Earth's surface; these nutrients could increase uptake of CO_2 by marine and terrestrial ecosystems.

7.1.5 Coupling the Biogeochemical Cycles with the Climate System

Models that attempt to perform reliable projections of future climate changes should account explicitly for the feedbacks between climate and the processes that determine the atmospheric concentrations of greenhouse gases, reactive gases and aerosol particles. An example is provided by the interaction between the carbon cycle and climate. It is well established that the level of atmospheric CO_2, which directly influences the Earth's temperature, depends critically on the rates of carbon uptake by the ocean and the land, which are also dependent on climate. Climate models that include the dynamics of the carbon cycle suggest that the overall effect of carbon-climate interactions is a positive feedback. Hence predicted future atmospheric CO_2 concentrations are therefore higher (and consequently the climate warmer) than in models that do not include these couplings. As understanding of the role of the biogeochemical cycles in the climate system improves, they should be explicitly represented in climate models. The present chapter assesses the current understanding of the processes involved and highlights the role of biogeochemical processes in the climate system.

7.2 The Changing Land Climate System

7.2.1 Introduction to Land Climate

The land surface relevant to climate consists of the terrestrial biosphere, that is, the fabric of soils, vegetation and other biological components, the processes that connect them and the carbon, water and energy they store. This section addresses from a climate perspective the current state of understanding of the land surface, setting the stage for consideration of carbon and other biogenic processes linked to climate. The land climate consists of 'internal' variables and 'external' drivers, including the various surface energy, carbon and moisture stores, and their response to precipitation, incoming radiation and near-surface atmospheric variables. The drivers and response variables change over various temporal and spatial scales.

This variation in time and space can be at least as important as averaged quantities. The response variables and drivers for the terrestrial system can be divided into biophysical, biological, biogeochemical and human processes. The present biophysical viewpoint emphasizes the response variables that involve the stores of energy and water and the mechanisms coupling these terms to the atmosphere. The exchanges of energy and moisture between the atmosphere and land surface (Boxes 7.1 and 7.2) are driven by radiation, precipitation and the temperature, humidity and winds of the overlying atmosphere. Determining how much detail to include to achieve an understanding of the system is not easy: many choices can be made and more detail becomes necessary when more processes are to be addressed.

7.2.2 Dependence of Land Processes and Climate on Scale

7.2.2.1 *Multiple Scales are Important*

Temporal variability ranges from the daily and weather time scales to annual, interannual, and decadal or longer scales: the amplitudes of shorter time scales change with long-term changes from global warming. The land climate system has controls on amplitudes of variables on all these time scales, varying with season and geography. For example, Trenberth and Shea (2005)

evaluate from climatic observations the correlation between surface air temperature and precipitation, and find a strong $r > 0.3$) positive correlation over most winter land areas (i.e., poleward of 40°N) but a strong ($|r| > 0.3$) negative correlation over much of summer and tropical land. These differences result from competing feedbacks with the water cycle. On scales large enough that surface temperatures control atmospheric temperatures, the atmosphere will hold more water vapour and may provide more precipitation with warmer temperatures. Low clouds strongly control surface temperatures, especially in cold regions where they make the surface warmer. In warm regions without precipitation, the land surface can become warmer because of lack of evaporation, or lack of clouds. Although a drier surface will become warmer from lack of evaporative cooling, more water can evaporate from a moist surface if the temperature is warmer (see Box 7.1).

7.2.2.2 *Spatial Dependence*

Drivers of the land climate system have larger effects at regional and local scales than on global climate, which is controlled primarily by processes of global radiation balance. Myhre et al. (2005) point out that the albedo of agricultural systems may be only slightly higher than that of forests and estimate that the impact since pre-agricultural times of land use

Box 7.1: Surface Energy and Water Balance

The land surface on average is heated by net radiation balanced by exchanges with the atmosphere of sensible and latent heat, known as the 'surface energy balance'. Sensible heat is the energy carried by the atmosphere in its temperature and latent heat is the energy lost from the surface by evaporation of surface water. The latent heat of the water vapour is converted to sensible heat in the atmosphere through vapour condensation and this condensed water is returned to the surface through precipitation.

The surface also has a 'surface water balance'. Water coming to the surface from precipitation is eventually lost either through water vapour flux or by runoff. The latent heat flux (or equivalently water vapour flux) under some conditions can be determined from the energy balance. For a fixed amount of net surface radiation, if the sensible heat flux goes up, the latent flux will go down by the same amount. Thus, if the ratio of sensible to latent heat flux depends only on air temperature, relative humidity and other known factors, the flux of water vapour from the surface can be found from the net radiative energy at the surface. Such a relationship is most readily obtained when water removal (evaporation from soil or transpiration by plants) is not limited by availability of water. Under these conditions, the increase of water vapour concentration with temperature increases the relative amount of the water flux as does low relative humidity. Vegetation can prolong the availability of soil water through the extent of its roots and so increase the latent heat flux but also can resist movement through its leaves, and so shift the surface energy fluxes to a larger fraction carried by the sensible heat flux. Fluxes to the atmosphere modify atmospheric temperatures and humidity and such changes feed back to the fluxes. Storage and the surface can also be important at short time scales, and horizontal transports can be important at smaller spatial scales.

If a surface is too dry to exchange much water with the atmosphere, the water returned to the atmosphere should be on average not far below the incident precipitation, and radiative energy beyond that needed for evaporating this water will heat the surface. Under these circumstances, less precipitation and hence less water vapour flux will make the surface warmer. Reduction of cloudiness from the consequently warmer and drier atmosphere may act as a positive feedback to provide more solar radiation. A locally moist area (such as an oasis or pond), however, would still evaporate according to energy balance with no water limitation and thus should increase its evaporation under such warmer and drier conditions.

Various feedbacks coupling the surface to the atmosphere may work in opposite directions and their relative importance may depend on season and location as well as on temporal and spatial scales. A moister atmosphere will commonly be cloudier making the surface warmer in a cold climate and cooler in a warm climate. The warming of the atmosphere by the surface may reduce its relative humidity and reduce precipitation as happens over deserts. However, it can also increase the total water held by the atmosphere, which may lead to increased precipitation as happens over the tropical oceans.

conversion to agriculture on global radiative forcing has been only –0.09 W m^{-2}, that is, about 5% of the warming contributed by CO_2 since pre-industrial times (see Chapter 2 for a more comprehensive review of recent estimates of land surface albedo change). Land comprises only about 30% of the Earth's surface, but it can have the largest effects on the reflection of global solar radiation in conjunction with changes in ice and snow cover, and the shading of the latter by vegetation.

At a regional scale and at the surface, additional more localised and shorter time-scale processes besides radiative forcing can affect climate in other ways, and possibly be of comparable importance to the effects of the greenhouse gases. Changes over land that modify its evaporative cooling can cause large changes in surface temperature, both locally and regionally (see Boxes 7.1, 7.2). How this change feeds back to precipitation remains a major research question. Land has a strong control on the vertical distribution of atmospheric heating. It determines how much of the radiation delivered to land goes into warming the near-surface atmosphere compared with how much is released as latent heat fuelling precipitation at higher levels. Low clouds are normally closely coupled to the surface and over land can be significantly changed by modifications of surface temperature or moisture resulting from changes in land properties. For example, Chagnon et al. (2004) find a large increase in boundary layer clouds in the Amazon in areas of partial deforestation (also, e.g., Durieux et al., 2003; Ek and Holtslag, 2004). Details of surface properties at scales as small as a few kilometres can be important for larger scales. Over some fraction of moist soils, water tables can be high enough to be hydrologically connected to the rooting zone, or reach the surface as in wetlands (e.g., Koster et al., 2000; Marani et al., 2001; Milly and Shmakin, 2002; Liang et al., 2003; Gedney and Cox, 2003).

The consequences of changes in atmospheric heating from land changes at a regional scale are similar to those from ocean temperature changes such as from El Niño, potentially producing patterns of reduced or increased cloudiness and precipitation elsewhere to maintain global energy balance. Attempts have been made to find remote adjustments (e.g., Avissar and Werth, 2005). Such adjustments may occur in multiple ways, and are part of the dynamics of climate models. The locally warmer temperatures can lead to more rapid vertical decreases of atmospheric temperature so that at some level overlying temperature is lower and radiates less. The net effect of such compensations is that averages over larger areas or longer time scales commonly will give smaller estimates of change. Thus, such regional changes are better described by local and regional metrics or at larger scales by measures of change in spatial and temporal variability rather than simply in terms of a mean global quantity.

7.2.2.3 Daily and Seasonal Variability

Diurnal and seasonal variability result directly from the temporal variation of the solar radiation driver. Large-scale changes in climate variables are of interest as part of the

Box 7.2: Urban Effects on Climate

If the properties of the land surface are changed locally, the surface net radiation and the partitioning between latent and sensible fluxes (Box 7.1) may also change, with consequences for temperatures and moisture storage of the surface and near-surface air. Such changes commonly occur to meet human needs for agriculture, housing, or commerce and industry. The consequences of urban development may be especially significant for local climates. However, urban development may have different features in different parts of an urban area and between geographical regions.

Some common modifications are the replacement of vegetation by impervious surfaces such as roads or the converse development of dry surfaces into vegetated surfaces by irrigation, such as lawns and golf courses. Buildings cover a relatively small area but in urban cores may strongly modify local wind flow and surface energy balance (Box 7.1). Besides the near-surface effects, urban areas can provide high concentrations of aerosols with local or downwind impacts on clouds and precipitation. Change to dark dry surfaces such as roads will generally increase daytime temperatures and lower humidity while irrigation will do the opposite. Changes at night may depend on the retention of heat by buildings and can be exacerbated by the thinness of the layer of atmosphere connected to the surface by mixing of air. Chapter 3 further addresses urban effects.

observational record of climate changes (Chapter 3). Daytime during the warm season produces a thick layer of mixed air with temperature relatively insensitive to perturbations in daytime radiative forcing. Nighttime and high-latitude winter surface temperatures, on the other hand, are coupled by mixing to only a thin layer of atmosphere, and can be more readily altered by changes in atmospheric downward thermal radiation. Thus, land is more sensitive to changes in radiative drivers under cold stable conditions and weak winds than under warm unstable conditions. Winter or nighttime temperatures (hence diurnal temperature range) are strongly correlated with downward longwave radiation (e.g., Betts, 2006; Dickinson et al., 2006); consequently, average surface temperatures may change (e.g., Pielke and Matsui, 2005) with a change in downward longwave radiation.

Modification of downward longwave radiation by changes in clouds can affect land surface temperatures. Qian and Giorgi (2000) discussed regional aerosol effects, and noted a reduction in the diurnal temperature range of –0.26°C per decade over Sichuan China. Huang et al. (2006) model the growth of sulphate aerosols and their interactions with clouds in the context of a RCM, and find over southern China a decrease in the diurnal temperature range comparable with that observed by Zhou et al. (2004) and Qian and Giorgi. They show the nighttime temperature change to be a result of increased nighttime cloudiness and hence downward longwave radiation connected to the increase in aerosols.

In moist warm regions, large changes are possible in the fraction of energy going into water fluxes, for example, by changes in vegetation cover or precipitation, and hence in soil moisture. Bonan (2001) and Oleson et al. (2004) indicate that conversion of mid-latitude forests to agriculture could cause a daytime cooling. This cooling is apparently a result of higher albedo and increased transpiration. Changes in reflected solar radiation due to changing vegetation, hence feedbacks, are most pronounced in areas with vegetation underlain by snow or light-coloured soil. Seasonal and diurnal precipitation cycles can be pronounced. Climate models simulate the diurnal precipitation cycle but apparently not yet very well (e.g., Collier and Bowman, 2004). Betts (2004) reviews how the diurnal cycle of tropical continental precipitation is linked to land surface fluxes and argues that errors in a model can feed back to model dynamics with global impacts.

7.2.2.4 Coupling of Precipitation Intensities to Leaf Water – An Issue Involving both Temporal and Spatial Scales

The bulk of the water exchanged with the atmosphere is stored in the soil until taken up by plant roots, typically weeks later. However, the rapidity of evaporation of the near-surface stores allows plant uptake and evaporation to be of comparable importance for surface water and energy balances. (Dickinson et al., 2003, conclude that feedbacks between surface moisture and precipitation may act differently on different time scales). Evaporation from the fast reservoirs acts primarily as a surface energy removal mechanism. Leaves initially intercept much of the precipitation over vegetation, and a significant fraction of this leaf water re-evaporates in an hour or less. This loss reduces the amount of water stored in the soil for use by plants. Its magnitude depends inversely on the intensity of the precipitation, which can be larger at smaller temporal and spatial scales. Modelling results can be wrong either through neglect of or through exaggeration of the magnitude of the fast time-scale moisture stores.

Leaf water evaporation may have little effect on the determination of monthly evapotranspiration (e.g., as found in the analysis of Desborough, 1999) but may still produce important changes in temperature and precipitation. Pitman et al. (2004), in a coupled study with land configurations of different complexity, were unable to find any impacts on atmospheric variability, but Bagnoud et al. (2005) found that precipitation and temperature extremes were affected. Some studies that change the intensity of precipitation find a very large impact from leaf water. For example, Wang and Eltahir (2000) studied the effect of including more realistic precipitation intensity compared to the uniform intensity of a climate model. Hahmann (2003) used another model to study this effect. Figure 7.1 compares their tropical results (Wang and Eltahir over equatorial Africa and Hahmann over equatorial Amazon). The model of Wang and Eltahir shows that more realistic precipitation greatly

increases runoff whereas Hahmann shows that it reduces runoff. It has not been determined whether these contradictory results are more a consequence of model differences or of differences between the climates of the two continents, as Hahmann suggests.

7.2.3 Observational Basis for the Effects of Land Surface on Climate

7.2.3.1 Vegetative Controls on Soil Water and its Return Flux to the Atmosphere

Scanlon et al. (2005) provide an example of how soil moisture can depend on vegetation. They monitored soil moisture in the Nevada desert with lysimeters either including or excluding vegetation and for a multi-year period that included times of anomalously strong precipitation. Without vegetation, much of the moisture penetrated deeply, had a long lifetime and became available for recharge of deep groundwater, whereas for the vegetated plot, the soil moisture was all transpired. In the absence of leaves, forests in early spring also appear as especially dry surfaces with consequent large sensible fluxes that mix the atmosphere to a great depth (e.g., Betts et al., 2001). Increased water fluxes with spring green-up are observed in terms of a reduction in temperature. Trees in the Amazon can have the largest water fluxes in the dry season by development of deep roots (Da Rocha et al., 2004; Quesada et al., 2004). Forests can also retard fluxes through control by their leaves. Such control by vegetation of water fluxes is most pronounced for taller or sparser vegetation in cooler or drier climates, and from leaves that are sparse or exert the strongest resistance to water movement. The boreal forest, in particular, has been characterised as a 'green desert' because of its small release of water to the atmosphere (Gamon et al., 2003).

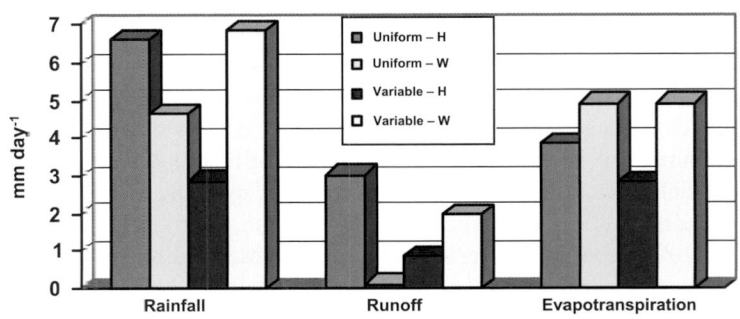

Figure 7.1. Rainfall, runoff and evapotranspiration derived from climate simulation results of Hahmann (H; 2004) and Wang and Eltahir (W; 2000). Hahmann's results are for the Amazon centred on the equator, and Wang and Eltahir's for Africa at the equator. Both studies examined the differences between 'uniform' precipitation over a model grid square and 'variable' precipitation (added to about 10% of the grid square). Large differences are seen between the two cases in the two studies: a large reduction in precipitation is seen in the Hahmann variable case relative to the uniform case, whereas an increase is seen for the Wang and Eltahir variable case. The differences are even greater for runoff: Hahmann's uniform case runoff is three times as large as the variable case, whereas Wang and Eltahir have almost no runoff for their uniform case.

7.2.3.2 Land Feedback to Precipitation

Findell and Eltahir (2003) examine the correlation between early morning near-surface humidity over the USA and an index of the likelihood of precipitation occurrence. They identify different geographical regions with positive, negative or little correlation. Koster et al. (2003) and Koster and Suarez (2004) show during summer over the USA, and all land 30°N to 60°N, respectively, a significant correlation of monthly precipitation with that of prior months. They further show that their model only reproduces this correlation if soil moisture feedback is allowed to affect precipitation. Additional observational evidence for such feedback is noted by D'Odorico and Porporato (2004) in support of a simplified model of precipitation soil moisture coupling (see, e.g., Salvucci et al., 2002, for support of the null hypothesis of no coupling). Liebmann and Marengo (2001) point out that the interannual variation of precipitation over the Amazon is largely controlled by the timing of the onset and end of the rainy season. Li and Fu (2004) provide evidence that onset time of the rainy season is strongly dependent on transpiration by vegetation during the dry season. Previous modelling and observational studies have also suggested that Amazon deforestation should lead to a longer dry season. Fu and Li (2004) further argue from observations that removal of tropical forest reduces surface moisture fluxes, and that such land use changes should contribute to a lengthening of the Amazon dry season. Durieux et al. (2003) find more rainfall in the deforested area in the wet season and a reduction of the dry season precipitation over deforested regions compared with forested areas. Negri et al. (2004) obtain an opposite result (although their result is consistent with Durieux during the wet season).

7.2.3.3 Properties Affecting Radiation

Albedo (the fraction of reflected solar radiation) and emissivity (the ratio of thermal radiation to that of a black body) are important variables for the radiative balance. Surfaces that have more or taller vegetation are commonly darker than are those with sparse or shorter vegetation. With sparse vegetation, the net surface albedo also depends on the albedo of the underlying surfaces, especially if snow or a light-coloured soil. A large-scale transformation of tundra to shrubs, possibly connected to warmer temperatures over the last few decades, has been observed (e.g., Chapin et al., 2005). Sturm et al. (2005) report on winter and melt season observations of how varying extents of such shrubs can modify surface albedo. New satellite data show the importance of radiation heterogeneities at the plot scale for the determination of albedo and the solar radiation used for photosynthesis, and appropriate modelling concepts to incorporate the new data are being advanced (e.g., Yang and Friedl, 2003; Niu and Yang, 2004; Wang, 2005; Pinty et al., 2006).

7.2.3.4 Improved Global and Regional Data

Specification of land surface properties has improved through new, more accurate global satellite observations. In particular, satellite observations have provided albedos of soils in non-vegetated regions (e.g., Tsvetsinskaya et al., 2002; Ogawa and Schmugge, 2004; Z. Wang, et al., 2004; Zhou et al., 2005) and their emissivities (Zhou et al., 2003a,b). They also constrain model-calculated albedos in the presence of vegetation (Oleson et al., 2003) and vegetation underlain by snow (Jin et al., 2002), and help to define the influence of leaf area on albedo (Tian et al., 2004). Precipitation data sets combining rain gauge and satellite observations (Chen et al., 2002; Adler et al., 2003) are providing diagnostic constraints for climate modelling, as are observations of runoff (Dai and Trenberth, 2002; Fekete et al., 2002).

7.2.3.5 Field Observational Programs

New and improved local site observational constraints collectively describe the land processes that need to be modelled. The largest recent such activity has been the Large-Scale Biosphere-Atmosphere Experiment in Amazonia (LBA) project (Malhi et al., 2002; Silva Dias et al., 2002). Studies within LBA have included physical climate at all scales, carbon and nutrient dynamics and trace gas fluxes. The physical climate aspects are reviewed here. Goncalves et al. (2004) discuss the importance of incorporating land cover heterogeneity. Da Rocha et al. (2004) and Quesada et al. (2004) quantify water and energy budgets for a forested and a savannah site, respectively. Dry season evapotranspiration for the savannah averaged 1.6 mm day^{-1} compared with 4.9 mm day^{-1} for the forest. Both ecosystems depend on deep rooting to sustain evapotranspiration during the dry season, which may help control the length of the dry season (see, e.g., Section 7.2.3.2). Da Rocha et al. (2004) also observed that hydraulic lift recharged the forest upper soil profiles each night. At Tapajós, the forest showed no signs of drought stress allowing uniformly high carbon uptake throughout the dry season (July through December 2000; Da Rocha et al., 2004; Goulden et al., 2004). Tibet, another key region, continues to be better characterised from observational studies (e.g., Gao et al., 2004; Hong et al., 2004). With its high elevation, hence low air densities, heating of the atmosphere by land mixes air to a much higher altitude than elsewhere, with implications for vertical exchange of energy. However, the daytime water vapour mixing ratio in this region decreases rapidly with increasing altitude (Yang et al., 2004), indicating a strong insertion of dry air from above or by lateral transport.

7.2.3.6 Connecting Changing Vegetation to Changing Climate

Only large-scale patterns are assessed here. Analysis of satellite-sensed vegetation greenness and meteorological station data suggest an enhanced plant growth and lengthened growing season duration at northern high latitudes since the 1980s (Zhou et al., 2001, 2003c). This effect is further supported by modelling linked to observed climate data (Lucht et al., 2002). Nemani et al. (2002, 2003) suggest that increased rainfall and humidity spurred plant growth in the USA and that climate changes may have eased several critical climatic constraints to plant growth and thus increased terrestrial net primary production.

7.2.4 Modelling the Coupling of Vegetation, Moisture Availability, Precipitation and Surface Temperature

7.2.4.1 How do Models of Vegetation Control Surface Water Fluxes?

Box 7.1 provides a general description of water fluxes from surface to atmosphere. The most important factors affected by vegetation are soil water availability, leaf area and surface roughness. Whether water has been intercepted on the surface of the leaves or its loss is only from the leaf interior as controlled by stomata makes a large difference. Shorter vegetation with more leaves has the most latent heat flux and the least sensible flux. Replacement of forests with shorter vegetation together with the normally assumed higher albedo could then cool the surface. However, if the replacement vegetation has much less foliage or cannot access soil water as successfully, a warming may occur. Thus, deforestation can modify surface temperatures by up to several degrees celsius in either direction depending on what type of vegetation replaces the forest and the climate regime. Drier air can increase evapotranspiration, but leaves may decrease their stomatal conductance to counter this effect.

7.2.4.2 Feedbacks Demonstrated Through Simple Models

In semi-arid systems, the occurrence and amounts of precipitation can be highly variable from year to year. Are there mechanisms whereby the growth of vegetation in times of adequate precipitation can act to maintain the precipitation? Various analyses with simple models have demonstrated how this might happen (Zeng et al., 2002; Foley et al., 2003; G. Wang, et al., 2004; X. Zeng et al., 2004). Such models demonstrate how assumed feedbacks between precipitation and surface fluxes generated by dynamic vegetation may lead to the possibility of transitions between multiple equilibria for two soil moisture and precipitation regimes. That is, the extraction of water by roots and shading of soil by plants can increase precipitation and maintain the vegetation, but if the vegetation is removed, it may not be able to be restored for a long period. The Sahel region between the deserts of North Africa and the African equatorial forests appears to most readily generate such an alternating precipitation regime.

7.2.4.3 Consequences of Changing Moisture Availability and Land Cover

Soil moisture control of the partitioning of energy between sensible and latent heat flux is very important for local and

regional temperatures, and possibly their coupling to precipitation. Oglesby et al. (2002) carried out a study starting with dry soil where the dryness of the soil over the US Great Plains for at least the first several summer months of their integration produced a warming of about 10°C to 20°C. Williamson et al. (2005), have shown that flaws in model formulation of thunderstorms can cause excessive evapotranspiration that lowers temperatures by more than 1°C. Many modelling studies have demonstrated that changing land cover can have local and regional climate impacts that are comparable in magnitude to temperature and precipitation changes observed over the last several decades as reported in Chapter 3. However, since such regional changes can be of both signs, the global average impact is expected to be small. Current literature has large disparities in conclusions. For example, Snyder et al. (2004) found that removal of northern temperate forests gave a summer warming of 1.3°C and a reduction in precipitation of 1.5 mm day^{-1}. Conversely, Oleson et al. (2004) found that removal of temperate forests in the USA would cool summer temperatures by 0.4°C to 1.5°C and probably increase precipitation, depending on the details of the model and prescription of vegetation. The discrepancy between these two studies may be largely an artefact of different assumptions. The first study assumes conversion of forest to desert and the second to crops. Such studies collectively demonstrate a potentially important impact of human activities on climate through land use modification.

Other recent such studies illustrate various aspects of this issue. Maynard and Royer (2004) address the sensitivity to different parameter changes in African deforestation experiments and find that changes in roughness, soil depth, vegetation cover, stomatal resistance, albedo and leaf area index all could make significant contributions. Voldoire and Royer (2004) find that such changes may affect temperature and precipitation extremes more than means, in particular the daytime maximum temperature and the drying and temperature responses associated with El Niño events. Guillevic et al. (2002) address the importance of interannual leaf area variability as inferred from Advanced Very High Resolution Radiometer (AVHRR) satellite data, and infer a sensitivity of climate to this variation. In contrast, Lawrence and Slingo (2004) find little difference in climate simulations that use annual mean vegetation characteristics compared with those that use a prescribed seasonal cycle. However, they do suggest model modifications that would give a much larger sensitivity. Osborne et al. (2004) examine effects of changing tropical soils and vegetation: variations in vegetation produce variability in surface fluxes and their coupling to precipitation. Thus, interactive vegetation can promote additional variability of surface temperature and precipitation as analysed by Crucifix et al. (2005). Marengo and Nobre (2001) found that removal of vegetation led to a decrease in precipitation and evapotranspiration and a decrease in moisture convergence in central and northern Amazonia. Oyama and Nobre (2004) show that removal of vegetation in northeast Brazil would substantially decrease precipitation.

7.2.4.4 Mechanisms for Modification of Precipitation by Spatial Heterogeneity

Clark et al. (2004) show an example of a 'squall-line' simulation where soil moisture variation at the scale of the rainfall modifies the rainfall pattern. Pielke (2001), Weaver et al. (2002) and S. Roy et al. (2003) also address various aspects of small-scale precipitation coupling to land surface heterogeneity. If deforestation occurs in patches rather than uniformly, the consequences for precipitation could be different. Avissar et al. (2002) and Silva Dias et al. (2002) suggest that there may be a small increase in precipitation (of the order of 10%) resulting from partial deforestation as a consequence of the mesoscale circulations triggered by the deforestation.

7.2.4.5 Interactive Vegetation Response Variables

Prognostic approaches estimate leaf cover based on physiological processes (e.g., Arora and Boer, 2005). Levis and Bonan (2004) discuss how spring leaf emergence in mid-latitude forests provides a negative feedback to rapid increases in temperature. The parametrization of water uptake by roots contributes to the computed soil water profile (Feddes et al., 2001; Barlage and Zeng, 2004), and efforts are being made to make the roots interactive (e.g., Arora and Boer, 2003). Dynamic vegetation models have advanced and now explicitly simulate competition between plant functional types (e.g., Bonan et al., 2003; Sitch et al., 2003; Arora and Boer, 2006). New coupled climate-carbon models (Betts et al., 2004; Huntingford et al., 2004) demonstrate the possibility of large feedbacks between future climate change and vegetation change, discussed further in Section 7.3.5 (i.e., a die back of Amazon vegetation and reductions in Amazon precipitation). They also indicate that the physiological forcing of stomatal closure by rising atmospheric CO_2 levels could contribute 20% to the rainfall reduction. Levis et al. (2004) demonstrate how African rainfall and dynamic vegetation could change each other.

7.2.5 Evaluation of Models Through Intercomparison

Intercomparison of vegetation models usually involves comparing surface fluxes and their feedbacks. Henderson-Sellers et al. (2003), in comparing the surface fluxes among 20 models, report over an order of magnitude range among sensible fluxes of different models. However, recently developed models cluster more tightly. Irannejad et al. (2003) developed a statistical methodology to fit monthly fluxes from a large number of climate models to a simple linear statistical model, depending on factors such as monthly net radiation and surface relative humidity. Both the land and atmosphere models are major sources of uncertainty for feedbacks. Irannejad et al. find that coupled models agree more closely due to offsetting differences in the atmospheric and land models. Modelling studies have long reported that soil moisture can influence precipitation. Only recently, however, have there been attempts to quantify this coupling from a statistical viewpoint (Dirmeyer, 2001; Koster and Suarez, 2001; Koster et al., 2002; Reale and Dirmeyer, 2002; Reale et al., 2002; Koster et al., 2003; Koster and Suarez, 2004). Koster et al. (2004, 2006) and Guo et al. (2006) report on a new model intercomparison activity, the Global Land Atmosphere Coupling Experiment (GLACE), which compares among climate models differences in precipitation variability caused by interaction with soil moisture. Using an experimental protocol to generate ensembles of simulations with soil moisture that is either prescribed or interactive as it evolves in time, they report a wide range of differences between models (Figure 7.2). Lawrence and Slingo (2005) show that the relatively weak coupling strength of the Hadley Centre model results from its atmospheric component. There is yet little confidence in this feedback component of climate models and therefore its possible contribution to global warming (see Chapter 8).

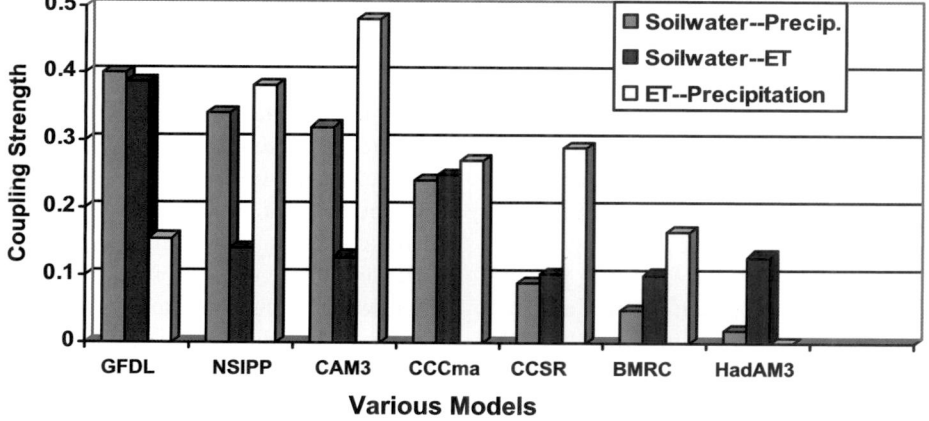

Figure 7.2. *Coupling strength (a nondimensional pattern similarity diagnostic defined in Koster et al., 2006) between summer rainfall and soil water in models assessed by the GLACE study (Guo et al., 2006), divided into how strongly soil water causes evaporation (including from plants) and how strongly this evaporation causes rainfall. The soil water-precipitation coupling is scaled up by a factor of 10, and the two indices for evaporation to precipitation coupling given in the study are averaged. Models include the Geophysical Fluid Dynamics Laboratory (GFDL) model, the National Aeronautics and Space Administration (NASA) Seasonal to Interannual Prediction Program (NSIPP) model, the National Center for Atmospheric Research Community Atmosphere Model (CAM3), the Canadian Centre for Climate Modelling and Analysis (CCCma) model, the Centre for Climate System Research (CCSR) model, the Bureau of Meteorology Research Centre (BMRC) model and the Hadley Centre Atmospheric Model version 3 (HadAM3).*

7.2.6   Linking Biophysical to Biogeochemical and Ecohydrological Components

Soil moisture and surface temperatures work together in response to precipitation and radiative inputs. Vegetation influences these terms through its controls on energy and water fluxes, and through these fluxes, precipitation. It also affects the radiative heating. Clouds and precipitation are affected through modifications of the temperature and water vapour content of near-surface air. How the feedbacks of land to the atmosphere work remains difficult to quantify from either observations or modelling (as addressed in Sections 7.2.3.2 and 7.2.5.1). Radiation feedbacks depend on vegetation or cloud cover that has changed because of changing surface temperatures or moisture conditions. How such conditions may promote or discourage the growth of vegetation is established by various ecological studies. The question of how vegetation will change its distribution at large scales and the consequent changes in absorbed radiation is quantified through remote sensing studies. At desert margins, radiation and precipitation feedbacks may act jointly with vegetation. Radiation feedbacks connected to vegetation may be most pronounced at the margins between boreal forests and tundra and involve changes in the timing of snowmelt. How energy is transferred from the vegetation to underlying snow surfaces is understood in general terms but remains problematic in modelling and process details. Dynamic vegetation models (see Section 7.2.4.5) synthesize current understanding.

Changing soil temperatures and snow cover affect soil microbiota and their processing of soil organic matter. How are nutrient supplies modified by these surface changes or delivery from the atmosphere? In particular, the treatment of carbon fluxes (addressed in more detail in Section 7.3) may require comparable or more detail in the treatment of N cycling (as attempted by S. Wang, et al., 2002; Dickinson et al., 2003). The challenge is to establish better process understanding at local scales and appropriately incorporate this understanding into global models. The Coupled Carbon-Cycle Climate Model Intercomparison Project (C^4MIP) simulations described in Section 7.3.5 are a first such effort.

Biomass burning is a major mechanism for changing vegetation cover and generation of atmospheric aerosols and is directly coupled to the land climate variables of moisture and near-surface winds, as addressed for the tropics by Hoffman et al. (2002). The aerosol plume produced by biomass burning at the end of the dry season contains black carbon that absorbs radiation. The combination of a cooler surface due to lack of solar radiation and a warmer boundary layer due to absorption of solar radiation increases the thermal stability and reduces cloud formation, and thus can reduce rainfall. Freitas et al. (2005) indicate the possibility of rainfall decrease in the Plata Basin as a response to the radiative effect of the aerosol load transported from biomass burning in the Cerrado and Amazon regions. Aerosols and clouds reduce the availability of visible light needed by plants for photosynthesis. However, leaves in full sun may be light saturated, that is, they do not develop sufficient enzymes to utilise that level of light. Leaves that are shaded, however, are generally light limited. They are only illuminated by diffuse light scattered by overlying leaves or by atmospheric constituents. Thus, an increase in diffuse light at the expense of direct light may promote leaf carbon assimilation and transpiration (Roderick et al., 2001; Cohan et al., 2002; Gu et al., 2002, 2003). Yamasoe et al. (2006) report the first observational tower evidence for this effect in the tropics. Diffuse radiation resulting from the Mt. Pinatubo eruption may have created an enhanced terrestrial carbon sink (Roderick et al., 2001; Gu et al., 2003). Angert et al. (2004) provide an analysis that rejects this hypothesis relative to other possible mechanisms.

7.3   The Carbon Cycle and the Climate System

7.3.1   Overview of the Global Carbon Cycle

7.3.1.1   The Natural Carbon Cycle

Over millions of years, CO_2 is removed from the atmosphere through weathering by silicate rocks and through burial in marine sediments of carbon fixed by marine plants (e.g., Berner, 1998). Burning fossil fuels returns carbon captured by plants in Earth's geological history to the atmosphere. New ice core records show that the Earth system has not experienced current atmospheric concentrations of CO_2, or indeed of CH_4, for at least 650 kyr – six glacial-interglacial cycles. During that period the atmospheric CO_2 concentration remained between 180 ppm (glacial maxima) and 300 ppm (warm interglacial periods) (Siegenthaler et al., 2005). It is generally accepted that during glacial maxima, the CO_2 removed from the atmosphere was stored in the ocean. Several causal mechanisms have been identified that connect astronomical changes, climate, CO_2 and other greenhouse gases, ocean circulation and temperature, biological productivity and nutrient supply, and interaction with ocean sediments (see Box 6.2).

Prior to 1750, the atmospheric concentration of CO_2 had been relatively stable between 260 and 280 ppm for 10 kyr (Box 6.2). Perturbations of the carbon cycle from human activities were insignificant relative to natural variability. Since 1750, the concentration of CO_2 in the atmosphere has risen, at an increasing rate, from around 280 ppm to nearly 380 ppm in 2005 (see Figure 2.3 and FAQ 2.1, Figure 1). The increase in atmospheric CO_2 concentration results from human activities: primarily burning of fossil fuels and deforestation, but also cement production and other changes in land use and management such as biomass burning, crop production and conversion of grasslands to croplands (see FAQ 7.1). While human activities contribute to climate change in many direct and indirect ways, CO_2 emissions from human activities are considered the single largest anthropogenic factor contributing to climate change (see FAQ 2.1, Figure 2). Atmospheric CH_4

Frequently Asked Question 7.1

Are the Increases in Atmospheric Carbon Dioxide and Other Greenhouse Gases During the Industrial Era Caused by Human Activities?

Yes, the increases in atmospheric carbon dioxide (CO_2) and other greenhouse gases during the industrial era are caused by human activities. In fact, the observed increase in atmospheric CO_2 concentrations does not reveal the full extent of human emissions in that it accounts for only 55% of the CO_2 released by human activity since 1959. The rest has been taken up by plants on land and by the oceans. In all cases, atmospheric concentrations of greenhouse gases, and their increases, are determined by the balance between sources (emissions of the gas from human activities and natural systems) and sinks (the removal of the gas from the atmosphere by conversion to a different chemical compound). Fossil fuel combustion (plus a smaller contribution from cement manufacture) is responsible for more than 75% of human-caused CO_2 emissions. Land use change (primarily deforestation) is responsible for the remainder. For methane, another important greenhouse gas, emissions generated by human activities exceeded natural emissions over the last 25 years. For nitrous oxide, emissions generated by human activities are equal to natural emissions to the atmosphere. Most of the long-lived halogen-containing gases (such as chlorofluorcarbons) are manufactured by humans, and were not present in the atmosphere before the industrial era. On average, present-day tropospheric ozone has increased 38% since pre-industrial times, and the increase results from atmospheric reactions of short-lived pollutants emitted by human activity. The concentration of CO_2 is now 379 parts per million (ppm) and methane is greater than 1,774 parts per billion (ppb), both very likely much higher than any time in at least 650 kyr (during which CO_2 remained between 180 and 300 ppm and methane between 320 and 790 ppb). The recent rate of change is dramatic and unprecedented; increases in CO_2 never exceeded 30 ppm in 1 kyr – yet now CO_2 has risen by 30 ppm in just the last 17 years.

Carbon Dioxide

Emissions of CO_2 (Figure 1a) from fossil fuel combustion, with contributions from cement manufacture, are responsible for more than 75% of the increase in atmospheric CO_2 concentration since pre-industrial times. The remainder of the increase comes from land use changes dominated by deforestation (and associated biomass burning) with contributions from changing agricultural practices. All these increases are caused by human activity. The natural carbon cycle cannot explain the observed atmospheric increase of 3.2 to 4.1 GtC yr^{-1} in the form of CO_2 over the last 25 years. (One GtC equals 10^{15} grams of carbon, i.e., one billion tonnes.)

Natural processes such as photosynthesis, respiration, decay and sea surface gas exchange lead to massive exchanges, sources and sinks of CO_2 between the land and atmosphere (estimated at

~120 GtC yr^{-1}) and the ocean and atmosphere (estimated at ~90 GtC yr^{-1}; see figure 7.3). The natural sinks of carbon produce a small net uptake of CO_2 of approximately 3.3 GtC yr^{-1} over the last 15 years, partially offsetting the human-caused emissions. Were it not for the natural sinks taking up nearly half the human-produced CO_2 over the past 15 years, atmospheric concentrations would have grown even more dramatically.

The increase in atmospheric CO_2 concentration is known to be caused by human activities because the character of CO_2 in the atmosphere, in particular the ratio of its heavy to light carbon atoms, has changed in a way that can be attributed to addition of fossil fuel carbon. In addition, the ratio of oxygen to nitrogen in the atmosphere has declined as CO_2 has increased; this is as expected because oxygen is depleted when fossil fuels are burned. A heavy form of carbon, the carbon-13 isotope, is less abundant in vegetation and in fossil fuels that were formed from past vegetation, and is more abundant in carbon in the oceans and in volcanic or geothermal emissions. The relative amount of the carbon-13 isotope in the atmosphere has been declining, showing that the added carbon comes from fossil fuels and vegetation. Carbon also has a rare radioactive isotope, carbon-14, which is present in atmospheric CO_2 but absent in fossil fuels. Prior to atmospheric testing of nuclear weapons, decreases in the relative amount of carbon-14 showed that fossil fuel carbon was being added to the atmosphere.

Halogen-Containing Gases

Human activities are responsible for the bulk of long-lived atmospheric halogen-containing gas concentrations. Before industrialisation, there were only a few naturally occurring halogen-containing gases, for example, methyl bromide and methyl chloride. The development of new techniques for chemical synthesis resulted in a proliferation of chemically manufactured halogen-containing gases during the last 50 years of the 20th century. Emissions of key halogen-containing gases produced by humans are shown in Figure 1b. Atmospheric lifetimes range from 45 to 100 years for the chlorofluorocarbons (CFCs) plotted here, from 1 to 18 years for the hydrochlorofluorocarbons (HCFCs), and from 1 to 270 years for the hydrofluorocarbons (HFCs). The perfluorocarbons (PFCs, not plotted) persist in the atmosphere for thousands of years. Concentrations of several important halogen-containing gases, including CFCs, are now stabilising or decreasing at the Earth's surface as a result of the Montreal Protocol on Substances that Deplete the Ozone Layer and its Amendments. Concentrations of HCFCs, production of which is to be phased out by 2030, and of the Kyoto Protocol gases HFCs and PFCs, are currently increasing. *(continued)*

Methane

Methane (CH_4) sources to the atmosphere generated by human activities exceed CH_4 sources from natural systems (Figure 1c). Between 1960 and 1999, CH_4 concentrations grew an average of at least six times faster than over any 40-year period of the two millennia before 1800, despite a near-zero growth rate since 1980. The main natural source of CH_4 to the atmosphere is wetlands. Additional natural sources include termites, oceans, vegetation and CH_4 hydrates. The human activities that produce CH_4 include energy production from coal and natural gas, waste disposal in landfills, raising ruminant animals (e.g., cattle and sheep), rice agriculture and biomass burning. Once emitted, CH_4 remains in the atmosphere for approximately 8.4 years before removal, mainly by chemical oxidation in the troposphere. Minor sinks for CH_4 include uptake by soils and eventual destruction in the stratosphere.

Nitrous Oxide

Nitrous oxide (N_2O) sources to the atmosphere from human activities are approximately equal to N_2O sources from natural systems (Figure 1d). Between 1960 and 1999, N_2O concentrations grew an average of at least two times faster than over any 40-year period of the two millennia before 1800. Natural sources of N_2O include oceans, chemical oxidation of ammonia in the atmosphere, and soils. Tropical soils are a particularly important source of N_2O to the atmosphere. Human activities that emit N_2O include transformation of fertilizer nitrogen into N_2O and its subsequent emission from agricultural soils, biomass burning, raising cattle and some industrial activities, including nylon manufacture. Once emitted, N_2O remains in the atmosphere for approximately 114 years before removal, mainly by destruction in the stratosphere.

Tropospheric Ozone

Tropospheric ozone is produced by photochemical reactions in the atmosphere involving forerunner chemicals such as carbon monoxide, CH_4, volatile organic compounds and nitrogen oxides. These chemicals are emitted by natural biological processes and by human activities including land use change and fuel combustion. Because tropospheric ozone is relatively short-lived, lasting for a few days to weeks in the atmosphere, its distributions are highly variable and tied to the abundance of its forerunner compounds, water vapour and sunlight.

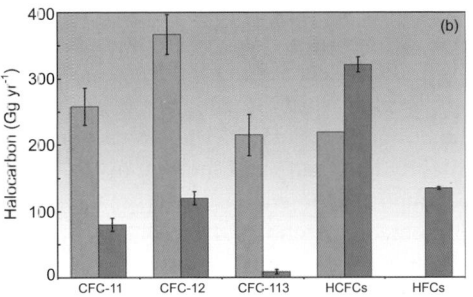

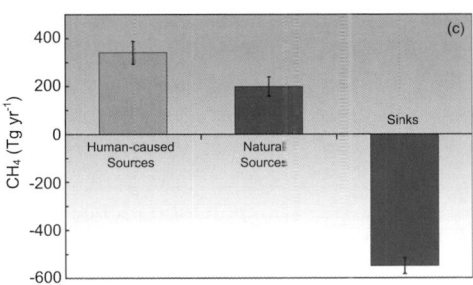

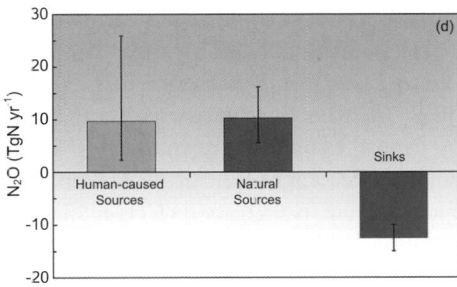

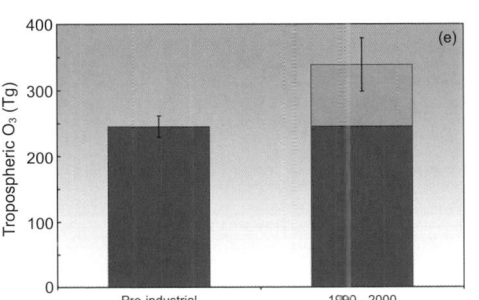

FAQ 7.1, Figure 1. *Breakdown of contributions to the changes in atmospheric greenhouse gas concentrations, based on information detailed in Chapters 4 and 7. In (a) through (d), human-caused sources are shown in orange, while natural sources and sinks are shown in teal. In (e), human-caused tropospheric ozone amounts are in orange while natural ozone amounts are in green. (a) Sources and sinks of CO_2 (GtC). Each year CO_2 is released to the atmosphere from human activities including fossil fuel combustion and land use change. Only 57 to 60% of the CO_2 emitted from human activity remains in the atmosphere. Some is dissolved into the oceans and some is incorporated into plants as they grow. Land-related fluxes are for the 1990s; fossil fuel and cement fluxes and net ocean uptake are for the period 2000 to 2005. All values and uncertainty ranges are from Table 7.1. (b) Global emissions of CFCs and other halogen-containing compounds for 1990 (light orange) and 2002 (dark orange). These chemicals are exclusively human-produced. Here, 'HCFCs' comprise HCFC-22, -141b and -142b, while 'HFCs' comprise HFC-23, -125, -134a and -152a. One Gg = 10^9 g (1,000 tonnes). Most data are from reports listed in Chapter 2. (c) Sources and sinks of CH_4 for the period 1983 to 2004. Human-caused sources of CH_4 include energy production, landfills, ruminant animals (e.g., cattle and sheep), rice agriculture and biomass burning. One Tg = 10^{12} g (1 million tonnes). Values and uncertainties are the means and standard deviations for CH_4 of the corresponding aggregate values from Table 7.6. (d) Sources and sinks of N_2O. Human-caused sources of N_2O include the transformation of fertilizer nitrogen into N_2O and its subsequent emission from agricultural soils, biomass burning, cattle and some industrial activities including nylon manufacture. Source values and uncertainties are the midpoints and range limits from Table 7.7. N_2O losses are from Chapter 7.4. (e) Tropospheric ozone in the 19th and early 20th centuries and the 1990 to 2000 period. The increase in tropospheric ozone formation is human-induced, resulting from atmospheric chemical reactions of pollutants emitted by burning of fossil fuels or biofuels. The pre-industrial value and uncertainty range are from Table 4.9 of the IPCC Third Assessment Report (TAR), estimated from reconstructed observations. The present-day total and its uncertainty range are the average and standard deviation of model results quoted in Table 7.9 of this report, excluding those from the TAR.*

Tropospheric ozone concentrations are significantly higher in urban air, downwind of urban areas and in regions of biomass burning. The increase of 38% (20–50%) in tropospheric ozone since the pre-industrial era (Figure 1e) is human-caused.

It is very likely that the increase in the combined radiative forcing from CO_2, CH_4 and N_2O was at least six times faster between 1960 and 1999 than over any 40-year period during the two millennia prior to the year 1800.

concentrations have similarly experienced a rapid rise from about 700 ppb in 1750 (Flückiger et al., 2002) to about 1,775 ppb in 2005 (see Section 2.3.2): sources include fossil fuels, landfills and waste treatment, peatlands/wetlands, ruminant animals and rice paddies. The increase in CH_4 radiative forcing is slightly less than one-third that of CO_2, making it the second most important greenhouse gas (see Chapter 2). The CH_4 cycle is presented in Section 7.4.1.

Both CO_2 and CH_4 play roles in the natural cycle of carbon, involving continuous flows of large amounts of carbon among the ocean, the terrestrial biosphere and the atmosphere, that maintained stable atmospheric concentrations of these gases for 10 kyr prior to 1750. Carbon is converted to plant biomass by photosynthesis. Terrestrial plants capture CO_2 from the atmosphere; plant, soil and animal respiration (including decomposition of dead biomass) returns carbon to the atmosphere as CO_2, or as CH_4 under anaerobic conditions. Vegetation fires can be a significant source of CO_2 and CH_4 to the atmosphere on annual time scales, but much of the CO_2 is recaptured by the terrestrial biosphere on decadal time scales if the vegetation regrows.

Carbon dioxide is continuously exchanged between the atmosphere and the ocean. Carbon dioxide entering the surface ocean immediately reacts with water to form bicarbonate (HCO_3^-) and carbonate (CO_3^{2-}) ions. Carbon dioxide, HCO_3^- and CO_3^{2-} are collectively known as dissolved inorganic carbon (DIC). The residence time of CO_2 (as DIC) in the surface ocean, relative to exchange with the atmosphere and physical exchange with the intermediate layers of the ocean below, is less than a decade. In winter, cold waters at high latitudes, heavy and enriched with CO_2 (as DIC) because of their high solubility, sink from the surface layer to the depths of the ocean. This localised sinking, associated with the Meridional Overturning Circulation (MOC; Box 5.1) is termed the 'solubility pump'. Over time, it is roughly balanced by a distributed diffuse upward transport of DIC primarily into warm surface waters.

Phytoplankton take up carbon through photosynthesis. Some of that sinks from the surface layer as dead organisms and particles (the 'biological pump'), or is transformed into dissolved organic carbon (DOC). Most of the carbon in sinking particles is respired (through the action of bacteria) in the surface and intermediate layers and is eventually recirculated to the surface as DIC. The remaining particle flux reaches abyssal depths and a small fraction reaches the deep ocean sediments, some of which is re-suspended and some of which is buried. Intermediate waters mix on a time scale of decades to centuries, while deep waters mix on millennial time scales. Several mixing times are required to bring the full buffering capacity of the ocean into effect (see Section 5.4 for long-term observations of the ocean carbon cycle and their consistency with ocean physics).

Together the solubility and biological pumps maintain a vertical gradient in CO_2 (as DIC) between the surface ocean (low) and the deeper ocean layers (high), and hence regulate exchange of CO_2 between the atmosphere and the ocean. The strength of the solubility pump depends globally on the strength of the MOC, surface ocean temperature, salinity, stratification and ice cover. The efficiency of the biological pump depends on the fraction of photosynthesis exported from the surface ocean as sinking particles, which can be affected by changes in ocean circulation, nutrient supply and plankton community composition and physiology.

In Figure 7.3 the natural or unperturbed exchanges (estimated to be those prior to 1750) among oceans, atmosphere and land are shown by the black arrows. The gross natural fluxes between the terrestrial biosphere and the atmosphere and between the oceans and the atmosphere are (circa 1995) about 120 and 90 GtC yr^{-1}, respectively. Just under 1 GtC yr^{-1} of carbon is transported from the land to the oceans via rivers either dissolved or as suspended particles (e.g., Richey, 2004). While these fluxes vary from year to year, they are approximately in balance when averaged over longer time periods. Additional small natural fluxes that are important on longer geological time scales include conversion of labile organic matter from terrestrial plants into inert organic carbon in soils, rock weathering and sediment accumulation ('reverse weathering'), and release from volcanic activity. The net fluxes in the 10 kyr prior to 1750, when averaged over decades or longer, are assumed to have been less than about 0.1 GtC yr^{-1}. For more background on the carbon cycle, see Prentice et al. (2001), Field and Raupach (2004) and Sarmiento and Gruber (2006).

7.3.1.2 Perturbations of the Natural Carbon Cycle from Human Activities

The additional burden of CO_2 added to the atmosphere by human activities, often referred to as 'anthropogenic CO_2' leads to the current 'perturbed' global carbon cycle. Figure 7.3 shows that these 'anthropogenic emissions' consist of two fractions: (i) CO_2 from fossil fuel burning and cement production, newly released from hundreds of millions of years of geological storage (see Section 2.3) and (ii) CO_2 from deforestation and agricultural development, which has been stored for decades to centuries. Mass balance estimates and studies with other gases indicate that the net land-atmosphere and ocean-atmosphere fluxes have become significantly different from zero, as indicated by the red arrows in Figure 7.3 (see also Section 7.3.2). Although the anthropogenic fluxes of CO_2 between the atmosphere and both the land and ocean are just a few percent of the gross natural fluxes, they have resulted in measurable changes in the carbon content of the reservoirs since pre-industrial times as shown in red. These perturbations to the natural carbon cycle are the dominant driver of climate change because of their persistent effect on the atmosphere. Consistent with the response function to a CO_2 pulse from the Bern Carbon Cycle Model (see footnote (a) of Table 2.14), about 50% of an increase in atmospheric CO_2 will be removed within 30 years, a further 30% will be removed within a few centuries and the remaining 20% may remain in the atmosphere for many thousands of years (Prentice et al., 2001; Archer, 2005; see also Sections 7.3.4.2 and 10.4)

About 80% of anthropogenic CO_2 emissions during the 1990s resulted from fossil fuel burning, with about 20% from land use

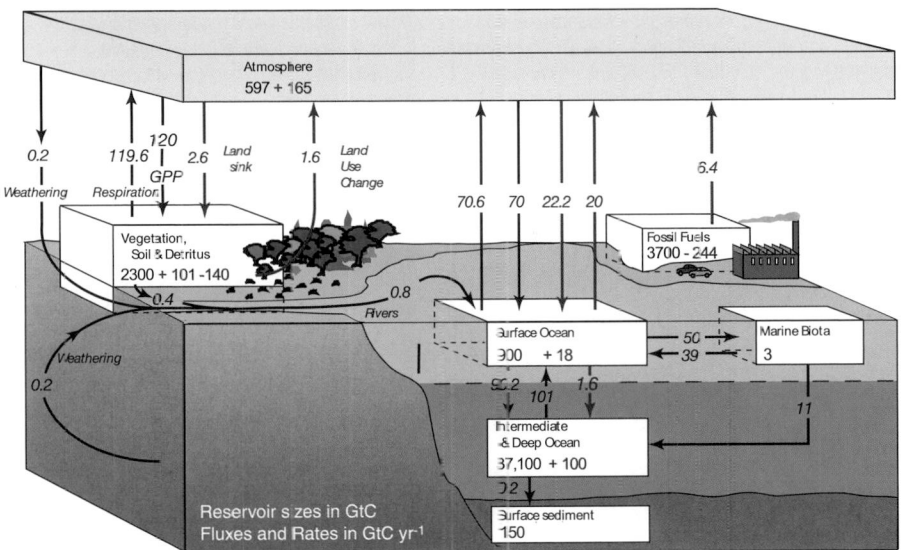

Figure 7.3. *The global carbon cycle for the 1990s, showing the main annual fluxes in GtC yr⁻¹: pre-industrial 'natural' fluxes in black and 'anthropogenic' fluxes in red (modified from Sarmiento and Gruber, 2006, with changes in pool sizes from Sabine et al., 2004a). The net terrestrial loss of –39 GtC is inferred from cumulative fossil fuel emissions minus atmospheric increase minus ocean storage. The loss of –140 GtC from the 'vegetation, soil and detritus' compartment represents the cumulative emissions from land use change (Houghton, 2003), and requires a terrestrial biosphere sink of 101 GtC (in Sabine et al., given only as ranges of –140 to –80 GtC and 61 to 141 GtC, respectively; other uncertainties given in their Table 1). Net anthropogenic exchanges with the atmosphere are from Column 5 'AR4' in Table 7.1. Gross fluxes generally have uncertainties of more than ±20% but fractional amounts have been retained to achieve overall balance when including estimates in fractions of GtC yr⁻¹ for riverine transport, weathering, deep ocean burial, etc. 'GPP' is annual gross (terrestrial) primary production. Atmospheric carbon content and all cumulative fluxes since 1750 are as of end 1994.*

change (primarily deforestation) (Table 7.1). Almost 45% of combined anthropogenic CO_2 emissions (fossil fuel plus land use) have remained in the atmosphere. Oceans are estimated to have taken up approximately 30% (about 118 ± 19 GtC: Sabine et al., 2004a; Figure 7.3), an amount that can be accounted for by increased atmospheric concentration of CO_2 without any change in ocean circulation or biology. Terrestrial ecosystems have taken up the rest through growth of replacement vegetation on cleared land, land management practices and the fertilizing effects of elevated CO_2 and N deposition (see Section 7.3.3).

Because CO_2 does not limit photosynthesis significantly in the ocean, the biological pump does not take up and store anthropogenic carbon directly. Rather, marine biological cycling of carbon may undergo changes due to high CO_2 concentrations, via feedbacks in response to a changing climate. The speed with which anthropogenic CO_2 is taken up effectively by the ocean, however, depends on how quickly surface waters are transported and mixed into the intermediate and deep layers of the ocean. A considerable amount of anthropogenic CO_2 can be buffered or neutralized by dissolution of $CaCO_3$ from surface sediments in the deep sea, but this process requires many thousands of years.

The increase in the atmospheric CO_2 concentration relative to the emissions from fossil fuels and cement production only is defined here as the 'airborne fraction'.[2] Land emissions, although significant, are not included in this definition due to the difficulty of quantifying their contribution, and to the complication that much land emission from logging and clearing of forests may be compensated a few years later by

uptake associated with regrowth. The 'airborne fraction of total emissions' is thus defined as the atmospheric CO_2 increase as a fraction of total anthropogenic CO_2 emissions, including the net land use fluxes. The airborne fraction varies from year to year mainly due to the effect of interannual variability in land uptake (see Section 7.3.2).

7.3.1.3   New Developments in Knowledge of the Carbon Cycle Since the Third Assessment Report

Sections 7.3.2 to 7.3.5 describe where knowledge and understanding have advanced significantly since the Third Assessment Report (TAR). In particular, the budget of anthropogenic CO_2 (shown by the red fluxes in Figure 7.3) can be calculated with improved accuracy. In the ocean, newly available high-quality data on the ocean carbon system have been used to construct robust estimates of the cumulative ocean burden of anthropogenic carbon (Sabine et al., 2004a) and associated changes in the carbonate system (Feely et al., 2004). The pH in the surface ocean is decreasing, indicating the need to understand both its interaction with a changing climate and the potential impact on organisms in the ocean (e.g., Orr et al., 2005; Royal Society, 2005). On land, there is a better understanding of the contribution to the buildup of CO_2 in the atmosphere since 1750 associated with land use and of how the land surface and the terrestrial biosphere interact with a changing climate. Globally, inverse techniques used to infer the magnitude and location of major fluxes in the global carbon

---

[2] This definition follows the usage of C. Keeling, distinct from that of Oeschger et al. (1980).

Table 7.1. *The global carbon budget (GtC yr⁻¹); errors represent ±1 standard deviation uncertainty estimates and not interannual variability, which is larger. The atmospheric increase (first line) results from fluxes to and from the atmosphere: positive fluxes are inputs to the atmosphere (emissions); negative fluxes are losses from the atmosphere (sinks); and numbers in parentheses are ranges. Note that the total sink of anthropogenic CO_2 is well constrained. Thus, the ocean-to-atmosphere and land-to-atmosphere fluxes are negatively correlated: if one is larger, the other must be smaller to match the total sink, and vice versa.*

| | 1980s | | 1990s | | 2000–2005c |
	TAR	TAR revised[a]	TAR	AR4	AR4
Atmospheric Increase[b]	3.3 ± 0.1	3.3 ± 0.1	3.2 ± 0.1	3.2 ± 0.1	4.1 ± 0.1
Emissions (fossil + cement)[c]	5.4 ± 0.3	5.4 ± 0.3	6.4 ± 0.4	6.4 ± 0.4	7.2 ± 0.3
Net ocean-to-atmosphere flux[d]	−1.9 ± 0.6	−1.8 ± 0.8	−1.7 ± 0.5	−2.2 ± 0.4	−2.2 ± 0.5
Net land-to-atmosphere flux[e]	−0.2 ± 0.7	−0.3 ± 0.9	−1.4 ± 0.7	−1.0 ± 0.6	−0.9 ± 0.6
Partitioned as follows					
Land use change flux	1.7 (0.6 to 2.5)	1.4 (0.4 to 2.3)	n.a.	1.6 (0.5 to 2.7)	n.a.
Residual terrestrial sink	−1.9 (−3.8 to −0.3)	−1.7 (−3.4 to 0.2)	n.a.	−2.6 (−4.3 to −0.9)	n.a.

Notes:

[a] TAR values revised according to an ocean heat content correction for ocean oxygen fluxes (Bopp et al., 2002) and using the Fourth Assessment Report (AR4) best estimate for the land use change flux given in Table 7.2.

[b] Determined from atmospheric CO_2 measurements (Keeling and Whorf, 2005, updated by S. Piper until 2006) at Mauna Loa (19°N) and South Pole (90°S) stations, consistent with the data shown in Figure 7.4, using a conversion factor of 2.12 GtC yr⁻¹ = 1 ppm.

[c] Fossil fuel and cement emission data are available only until 2003 (Marland et al., 2006). Mean emissions for 2004 and 2005 were extrapolated from energy use data with a trend of 0.2 GtC yr⁻¹.

[d] For the 1980s, the ocean-to-atmosphere and land-to-atmosphere fluxes were estimated using atmospheric $O_2:N_2$ and CO_2 trends, as in the TAR. For the 1990s, the ocean-to-atmosphere flux alone is estimated using ocean observations and model results (see Section 7.3.2.2.1), giving results identical to the atmospheric $O_2:N_2$ method (Manning and Keeling, 2006), but with less uncertainty. The net land-to-atmosphere flux then is obtained by subtracting the ocean-to-atmosphere flux from the total sink (and its errors estimated by propagation). For 2000 to 2005, the change in ocean-to-atmosphere flux was modelled (Le Quéré et al., 2005) and added to the mean ocean-to-atmosphere flux of the 1990s. The error was estimated based on the quadratic sum of the error of the mean ocean flux during the 1990s and the root mean square of the five-year variability from three inversions and one ocean model presented in Le Quéré et al. (2003).

[e] Balance of emissions due to land use change and a residual land sink. These two terms cannot be separated based on current observations.

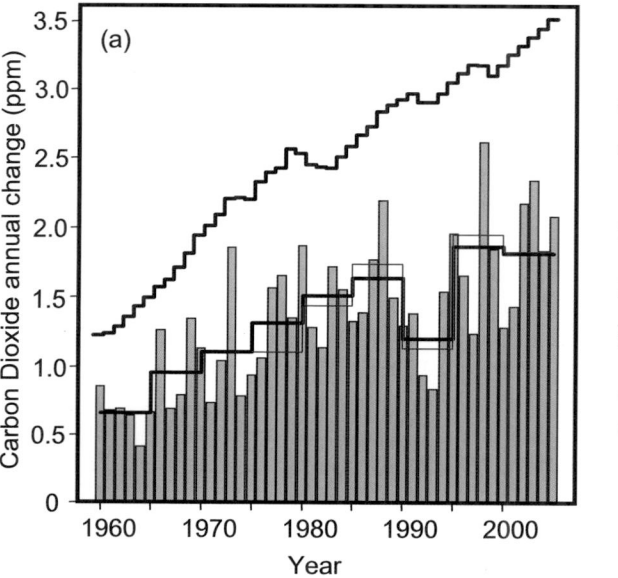

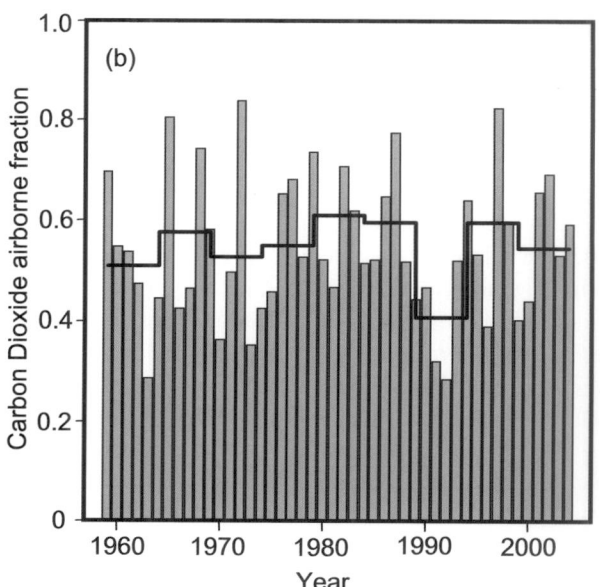

Figure 7.4. *Changes in global atmospheric CO_2 concentrations. (a) Annual (bars) and five-year mean (lower black line) changes in global CO_2 concentrations, from Scripps Institution of Oceanography observations (mean of South Pole and Mauna Loa; Keeling and Whorf, 2005, updated). The upper stepped line shows annual increases that would occur if 100% of fossil fuel emissions (Marland et al., 2006, updated as described in Chapter 2) remained in the atmosphere, and the red line shows five-year mean annual increases from National Oceanic and Atmospheric Administration (NOAA) data (mean of Samoa and Mauna Loa; Tans and Conway, 2005, updated). (b) Fraction of fossil fuel emissions remaining in the atmosphere ('airborne fraction') each year (bars), and five-year means (solid black line) (Scripps data) (mean since 1958 is 0.55). Note the anomalously low airborne fraction in the early 1990s.*

cycle have continued to mature, reflecting both refinement of the techniques and the availability of new observations. During preparation of the TAR, inclusion of the carbon cycle in climate models was new. Now, results from the first C[4]MIP are available: when the carbon cycle is included, the models consistently simulate climate feedbacks to land and ocean carbon cycles that tend to reduce uptake of CO_2 by land and ocean from 1850 to 2100 (see Section 7.3.5).

7.3.2 The Contemporary Carbon Budget

7.3.2.1 Atmospheric Increase

The atmospheric CO_2 increase is measured with great accuracy at various monitoring stations (see Chapter 2; and Keeling and Whorf, 2005 updated by S. Piper through 2006). The mean yearly increase in atmospheric CO_2 (the CO_2 'growth rate') is reported in Table 7.1. Atmospheric CO_2 has continued to increase since the TAR (Figure 7.4), and the rate of increase appears to be higher, with the average annual increment rising from 3.2 ± 0.1 GtC yr[-1] in the 1990s to 4.1 ± 0.1 GtC yr[-1] in the period 2000 to 2005. The annual increase represents the net effect of several processes that regulate global land-atmosphere and ocean-atmosphere fluxes, examined below. The 'airborne fraction' (atmospheric increase in CO_2 concentration/fossil fuel emissions) provides a basic benchmark for assessing short- and long-term changes in these processes. From 1959 to the present, the airborne fraction has averaged 0.55, with remarkably little variation when block-averaged into five-year bins (Figure 7.4). Thus, the terrestrial biosphere and the oceans together have consistently removed 45% of fossil CO_2 for the last 45 years, and the recent higher rate of atmospheric CO_2 increase largely reflects increased fossil fuel emissions. Year-to-year fluctuations in the airborne fraction are associated with major climatic events (see Section 7.3.2.4). The annual increase in 1998, 2.5 ppm, was the highest ever observed, but the airborne fraction (0.82) was no higher than values observed several times in prior decades. The airborne fraction dropped significantly below the average in the early 1990s, and preliminary data suggest it may have risen above the average in 2000 to 2005.

The inter-hemispheric gradient of CO_2 provides additional evidence that the increase in atmospheric CO_2 is caused primarily by NH sources. The excess atmospheric CO_2 in the NH compared with the Southern Hemisphere (SH), ΔCO_2^{N-S}, has increased in proportion to fossil fuel emission rates (which are predominantly in the NH) at about 0.5 ppm per (GtC yr[-1]) (Figure 7.5). The intercept of the best-fit line indicates that, without anthropogenic emissions, atmospheric CO_2 would be 0.8 ppm higher in the SH than in the NH, presumably due to transport of CO_2 by the ocean circulation. The consistency of the airborne fraction and the relationship between ΔCO_2^{N-S} and fossil fuel emissions suggest broad consistency in the functioning of the carbon cycle over the period. There are interannual fluctuations in ΔCO_2^{N-S} as large as ± 0.4 ppm, at least some of which may be attributed to changes in atmospheric circulation (Dargaville et al., 2000), while others may be due to shifts in sources and sinks, such as large forest fires.

7.3.2.1.1 Fossil fuel and cement emissions

Fossil fuel and cement emissions rose from 5.4 ± 0.3 GtC yr[-1] in the 1980s to 6.4 ± 0.4 GtC yr[-1] in the 1990s (Marland et al., 2006). They have continued to increase between the 1990s and 2000 to 2005, climbing to 7.2 ± 0.3 GtC yr[-1]. These numbers are estimated based upon international energy statistics for the 1980 to 2003 period (Marland et al., 2006) with extrapolated trends for 2004 to 2005 (see Table 7.1). The error (± 1 standard deviation) for fossil fuel and cement emissions is of the order of 5% globally. Cement emissions are small compared with fossil fuel emissions (roughly 3% of the total).

7.3.2.1.2 Land use change

During the past two decades, the CO_2 flux caused by land use changes has been dominated by tropical deforestation. Agriculture and exploitation of forest resources have reached into formerly remote areas of old growth forest in the tropics, in contrast to mid-latitudes where exploitation previously eliminated most old growth forests. The land use change fluxes reported in this section include explicitly some accumulation of carbon by regrowing vegetation (e.g., Houghton et al., 2000). In the TAR, the global land use flux, adapted from Houghton (1999), was estimated to be 1.7 (0.6–2.5) GtC yr[-1] for the 1980s. No estimate was available at the time for the 1990s. This estimate is based on a 'bookkeeping' carbon model prescribed with deforestation statistics (Houghton, 1999). A markedly lower estimate of the land use flux in the 1980s (Table 7.2) was obtained by McGuire et al. (2001) from four process-driven

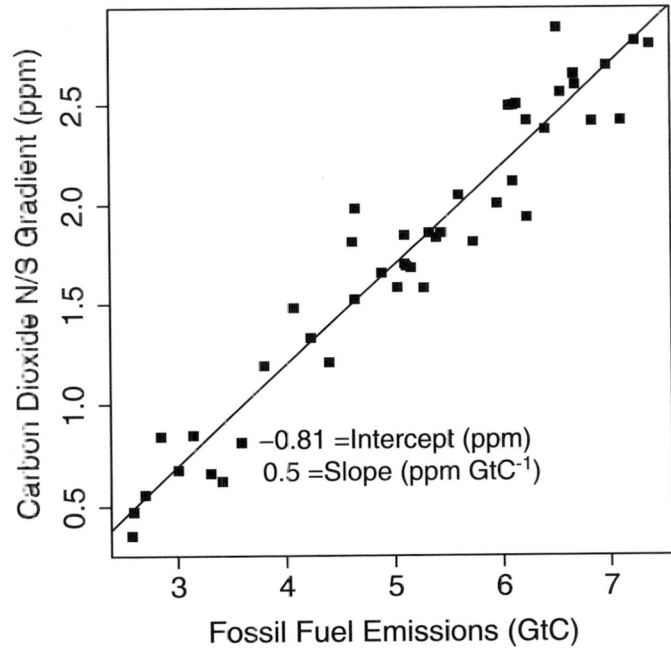

Figure 7.5. *The difference between CO_2 concentration in the NH and SH (y axis), computed as the difference between annual mean concentrations (ppm) at Mauna Loa and the South Pole (Keeling and Whorf, 2005, updated), compared with annual fossil fuel emissions (x axis; GtC; Marland, et al., 2006), with a line showing the best fit. The observations show that the north-south difference in CO_2 increases proportionally with fossil fuel use, verifying the global impact of human-caused emissions.*

terrestrial carbon models, prescribed with changes in cropland area from Ramankutty and Foley (1999). The higher land use emissions of Houghton (2003a) may reflect both the additional inclusion of conversion of forest to pasture and the use of a larger cropland expansion rate than the one of Ramankutty and Foley (1999), as noted by Jain and Yang (2005). Houghton (2003a) updated the land use flux to 2.0 ± 0.8 GtC yr^{-1} for the 1980s and 2.2 ± 0.8 GtC yr^{-1} for the 1990s (see Table 7.2). This update gives higher carbon losses from tropical deforestation than those in the TAR (Houghton 2003b).

In addition, DeFries et al. (2002) estimated a tropical land use flux of 0.7 (0.4–1.0) GtC yr^{-1} for the 1980s and 1.0 (0.5–1.6) GtC yr^{-1} for the 1990s, using the same bookkeeping approach as Houghton (1999) but driven by remotely sensed data on deforested areas. A similar estimate was independently produced by Achard et al. (2004) for the 1990s, also based on

remote sensing. These different land use emissions estimates are reported in Table 7.2. Although the two recent satellite-based estimates point to a smaller source than that of Houghton (2003a), it is premature to say that Houghton's numbers are overestimated. The land use carbon source has the largest uncertainties in the global carbon budget. If a high value for the land use source is adopted in the global budget, then the residual land uptake over undisturbed ecosystems should be a large sink, and vice versa. For evaluating the global carbon budget, the mean of DeFries et al. (2002) and Houghton (2003a), which both cover the 1980s and the 1990s (Table 7.2), was chosen and the full range of uncertainty is reported. The fraction of carbon emitted by fossil fuel burning, cement production and land use changes that does not accumulate in the atmosphere must be taken up by land ecosystems and by the oceans.

Table 7.2. *Land to atmosphere emissions resulting from land use changes during the 1990s and the 1980s (GtC yr^{-1}). The Fourth Assessment Report (AR4) estimates used in the global carbon budget (Table 7.1) are shown in bold. Positive values indicate carbon losses from land ecosystems. Uncertainties are reported as ±1 standard deviation. Numbers in parentheses are ranges of uncertainty.*

	Tropical Americas	Tropical Africa	Tropical Asia	Pan-Tropical	Non-tropics	Total Globe
1990s						
Houghton (2003a)[a]	0.8 ± 0.3	0.4 ± 0.2	1.1 ± 0.5	2.2 ± 0.6	-0.02 ± 0.5	2.2 ± 0.8
DeFries et al. (2002)[b]	0.5 (0.2 to 0.7)	0.1 (0.1 to 0.2)	0.4 (0.2 to 0.6)	1.0 (0.5 to 1.6)	n.a.	n.a.
Achard et al. (2004)[c]	0.3 (0.3 to 0.4)	0.2 (0.1 to 0.2)	0.4 (0.3 to 0.5)	0.9 (0.5 to 1.4)	n.a.	n.a.
AR4[d]	0.7 (0.4 to 0.9)	0.3 (0.2 to 0.4)	0.8 (0.4 to 1.1)	1.6 (1.0 to 2.2)	−0.02 (−0.5 to +0.5)	**1.6 (0.5 to 2.7)**
1980s						
Houghton (2003a)[a]	0.8 ± 0.3	0.3 ± 0.2	0.9 ± 0.5	1.9 ± 0.6	0.06 ± 0.5	2.0 ± 0.8
DeFries et al. (2002)[b]	0.4 (0.2 to 0.5)	0.1 (0.08 to 0.14)	0.2 (0.1 to 0.3)	0.7 (0.4 to 1.0)	n.a.	n.a.
McGuire et al. (2001)[e]				0.6 to 1.2	−0.1 to +0.4	(0.6 to 1.0)
Jain and Yang (2005)[f]	0.22 to 0.24	0.08 to 0.48	0.58 to 0.34	-	-	1.33 to 2.06
TAR[g]						1.7 (0.6 to 2.5)
AR4[d]	0.6 (0.3 to 0.8)	0.2 (0.1 to 0.3)	0.6 (0.3 to 0.9)	1.3 (0.9 to 1.8)	0.06 (−0.4 to +0.6)	**1.4 (0.4 to 2.3)**

Notes:

[a] His Table 2.

[b] Their Table 3.

[c] Their Table 2 for mean estimates with the range indicated in parentheses corresponding to their reported minimum and maximum estimates.

[d] Best estimate calculated from the mean of Houghton (2003a) and DeFries et al. (2002), the only two studies covering both the 1980s and the 1990s. For non-tropical regions where DeFries et al. have no estimate, Houghton has been used.

[e] Their Table 5; range is obtained from four terrestrial carbon models.

[f] The range indicated in parentheses corresponds to two simulations using the same model, but forced with different land cover change datasets from Houghton (2003a) and DeFries et al. (2002).

[g] In the TAR estimate, no values were available for the 1990s.

7.3.2.2   Uptake of CO_2 by Natural Reservoirs and Global Carbon Budget

7.3.2.2.1   Ocean-atmosphere flux

To assess the mean ocean sink, seven methods have been used. The methods are based on: (1) observations of the partial pressure of CO_2 at the ocean surface and gas-exchange estimates (Takahashi et al., 2002); (2) atmospheric inversions based upon diverse observations of atmospheric CO_2 and atmospheric transport modelling (see Section 7.2.3.4); (3) observations of carbon, oxygen, nutrients and chlorofluorocarbons (CFCs) in seawater, from which the concentration of anthropogenic CO_2 is estimated (Sabine et al., 2004a) combined with estimates of oceanic transport (Gloor et al., 2003; Mikaloff Fletcher et al., 2006); (4) estimates of the distribution of water age based on CFC observations combined with the atmospheric CO_2 history (McNeil et al., 2003); (5) the simultaneous observations of the increase in atmospheric CO_2 and decrease in atmospheric O_2 (Manning and Keeling, 2006); (6) various methods using observations of change in ^{13}C in the atmosphere (Ciais et al., 1995) or the oceans (Gruber and Keeling, 2001; Quay et al., 2003); and (7) ocean General Circulation Models (Orr et al., 2001). The ocean uptake estimates obtained with methods (1) and (2) include in part a flux component due to the outgassing of river-supplied inorganic and organic carbon (Sarmiento and Sundquist, 1992). The magnitude of this necessary correction to obtain the oceanic uptake flux of anthropogenic CO_2 is not well known, as these estimates pertain to the open ocean, whereas a substantial fraction of the river-induced outgassing likely occurs in coastal regions. These estimates of the net oceanic sink are shown in Figure 7.3.

With these corrections, estimates from all methods are consistent, resulting in a well-constrained global oceanic sink for anthropogenic CO_2 (see Table 7.1). The uncertainty around the different estimates is more difficult to judge and varies considerably with the method. Four estimates appear better constrained than the others. The estimate for the ocean uptake of atmospheric CO_2 of -2.2 ± 0.5 GtC yr^{-1} centred around 1998 based on the atmospheric O_2/N_2 ratio needs to be corrected for the oceanic O_2 changes (Manning and Keeling, 2006). The estimate of -2.0 ± 0.4 GtC yr^{-1} centred around 1995 based on CFC observations provides a constraint from observed physical transport in the ocean. These estimates of the ocean sink are shown in Figure 7.6. The mean estimates of -2.2 ± 0.25 and -2.2 ± 0.2 GtC yr^{-1} centred around 1995 and 1994 provide constraints based on a large number of ocean carbon observations. These well-constrained estimates all point to a decadal mean ocean CO_2 sink of -2.2 ± 0.4 GtC yr^{-1} centred around 1996, where the uncertainty is the root mean square of all errors. See Section 5.4 for a discussion of changes in the ocean CO_2 sink.

7.3.2.2.2   Land-atmosphere flux

The land-atmosphere CO_2 flux is the sum of the land use change CO_2 flux (see Section 7.3.2.1) plus sources and sinks due for instance to legacies of prior land use, climate, rising CO_2 or N deposition (see Section 7.3.3 for a review of

processes). For assessing the global land-atmosphere flux, more than just direct terrestrial observations must be used, because observations of land ecosystem carbon fluxes are too sparse and the ecosystems are too heterogeneous to allow global assessment of the net land flux with sufficient accuracy. For instance, large-scale biomass inventories (Goodale et al., 2002; UN-ECE/FAO, 2000) are limited to forests with commercial value, and they do not adequately survey tropical forests. Direct flux observations by the eddy covariance technique are only available at point locations, most do not yet have long-term coverage and they require considerable upscaling to obtain global estimates (Baldocchi et al., 2001). As a result, two methods can be used to quantify the net global land-atmosphere flux: (1) deducing that quantity as a residual between the fossil fuel and cement emissions and the sum of ocean uptake and atmospheric increase (Table 7.1), or (2) inferring the land-atmosphere flux simultaneously with the ocean sink by inverse analysis or mass balance computations using atmospheric CO_2 data, with terrestrial and marine processes distinguished using O_2/N_2 and/or ^{13}C observations. Individual estimates of the land-atmosphere flux deduced using either method 1 or method 2

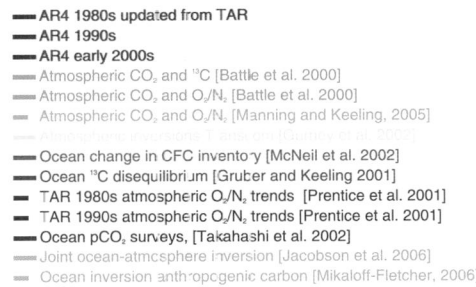

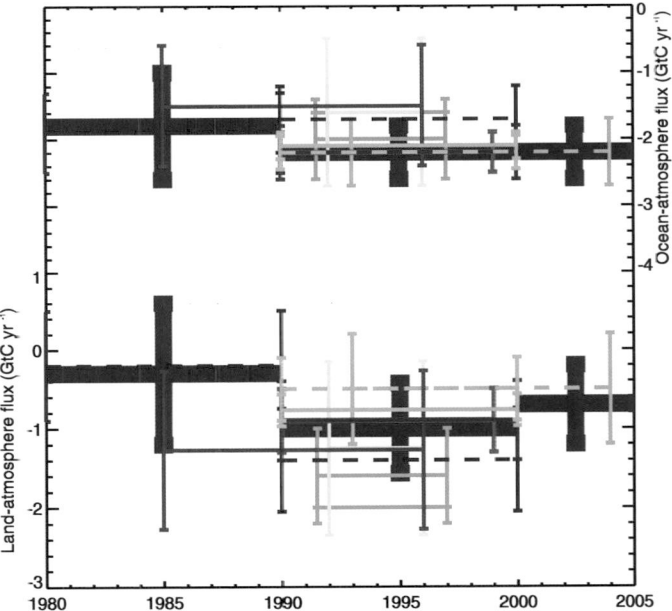

Figure 7.6. *Individual estimates of the ocean-atmosphere flux reported in Chapter 5 and of the related land-atmosphere flux required to close the global carbon budget. The dark thick lines are the revised budget estimates in the AR4 for the 1980s, the 1990s and the early 2000s, respectively.*

are shown in Figure 7.6. Method 2 was used in the TAR, based upon O_2/N_2 data (Langenfelds et al., 1999; Battle et al., 2000). Corrections have been made to the results of method 2 to account for the effects of thermal O_2 fluxes by the ocean (Le Quéré et al., 2003). This chapter includes these corrections to update the 1980s budget, resulting in a land net flux of -0.3 ± 0.9 GtC yr^{-1} during the 1980s. For the 1990s and after, method 1 was adopted for assessing the ocean sink and the land-atmosphere flux. Unlike in the TAR, method 1 is preferred for the 1990s and thereafter (i.e., estimating first the ocean uptake, and then deducing the land net flux) because the ocean uptake is now more robustly determined by various oceanographic approaches (see 7.3.2.2.1) than by the atmospheric O_2 trends. The numbers are reported in Table 7.1. The land-atmosphere flux evolved from a small sink in the 1980s of -0.3 ± 0.9 GtC yr^{-1} to a large sink during the 1990s of -1.0 ± 0.6 GtC yr^{-1}, and returned to an intermediate value of -0.9 ± 0.6 GtC yr^{-1} over the past five years. A recent weakening of the land-atmosphere uptake has also been suggested by other independent studies of the flux variability over the past decades (Jones and Cox, 2005). The global CO_2 budget is summarised in Table 7.1.

7.3.2.2.3 Residual land sink

In the context of land use change, deforestation dominates over forest regrowth (see Section 7.3.2.1), and the observed net uptake of CO_2 by the land biosphere implies that there must be an uptake by terrestrial ecosystems elsewhere, called the 'residual land sink' (formerly the 'missing sink'). Estimates of the residual land sink necessarily depend on the land use change flux, and its uncertainty reflects predominantly the (large) errors associated with the land use change term. With the high land use source of Houghton (2003a), the residual land sink equals -2.3 (-4.0 to -0.3) and -3.2 (-4.5 to -1.9) GtC yr^{-1} respectively for the 1980s and the 1990s. With the smaller land use source of DeFries et al. (2002), the residual land sink is -0.9 (-2.0 to -0.3) and -1.9 (-2.9 to -1.0) GtC yr^{-1} for the 1980s and the 1990s. Using the mean value of the land use source from Houghton (2003a) and DeFries et al. (2002) as reported in Table 7.2, a mean residual land sink of -1.7 (-3.4 to 0.2) and -2.6 (-4.3 to -0.9) GtC yr^{-1} for the 1980s and 1990s respectively is obtained. Houghton (2003a) and DeFries et al. (2002) give different estimates of the land use source, but they robustly indicate that deforestation emissions were 0.2 to 0.3 GtC yr^{-1} higher in the 1990s than in the 1980s (see Table 7.2). To compensate for that increase and to match the larger land-atmosphere uptake during the 1990s, the inferred residual land sink must have increased by 1 GtC yr^{-1} between the 1980s and the 1990s. This finding is insensitive to the method used to determine the land use flux, and shows considerable decadal variability in the residual land sink.

7.3.2.2.4 Undisturbed tropical forests: are they a carbon dioxide sink?

Despite expanding areas of deforestation and degradation, there are still large areas of tropical forests that are among the world's great wilderness areas, with fairly light human impact,

especially in Amazonia. A major uncertainty in the carbon budget relates to possible net change in the carbon stocks in these forests. Old growth tropical forests contain huge stores of organic matter, and are very dynamic, accounting for a major fraction of global net primary productivity (and about 46% of global biomass; Brown and Lugo, 1982). Changes in the carbon balance of these regions could have significant effects on global CO_2.

Recent studies of the carbon balance of study plots in mature, undisturbed tropical forests (Phillips et al., 1998; Baker et al., 2004) report accumulation of carbon at a mean rate of 0.7 ± 0.2 MgC ha^{-1} yr^{-1}, implying net carbon uptake into global Neotropical biomass of 0.6 ± 0.3 GtC yr^{-1}. An intriguing possibility is that rising CO_2 levels could stimulate this uptake by accelerating photosynthesis, with ecosystem respiration lagging behind. Atmospheric CO_2 concentration has increased by about 1.5 ppm (0.4%) yr^{-1}, suggesting incremental stimulation of photosynthesis of about 0.25% (e.g., next year's photosynthesis should be 1.0025 times this year's) (Lin et al., 1999; Farquhar et al., 2001). For a mean turnover rate of about 10 years for organic matter in tropical forests, the present imbalance between uptake of CO_2 and respiration might be 2.5% (1.0025^{10}), consistent with the reported rates of live biomass increase (~3%).

But the recent pan-tropical warming, about 0.26°C per decade (Malhi and Wright, 2004), could increase water stress and respiration, and stimulation by CO_2 might be limited by nutrients (Chambers and Silver, 2004; Koerner, 2004; Lewis et al., 2005; see below), architectural constraints on how much biomass a forest can hold, light competition, or ecological shifts favouring short lived trees or agents of disturbance (insects, lianas) (Koerner, 2004). Indeed, Baker et al. (2004) note higher mortality rates and increased prevalence of lianas, and, since dead organic pools were not measured, effects of increased disturbance may give the opposite sign of the imbalance inferred from live biomass only (see, e.g., Rice et al., 2004). Methodological bias associated with small plots, which under-sample natural disturbance and recovery, might also lead to erroneous inference of net growth (Koerner, 2004). Indeed, studies involving large-area plots (9–50 ha) have indicated either no net long-term change or a long-term net decline in above ground live biomass (Chave et al., 2003; Baker et al., 2004; Clark, 2004; Laurance et al., 2004), and a five-year study of a 20 ha plot in Tapajos, Brazil show increasing live biomass offset by decaying necromass (Fearnside, 2000; Saleska et al., 2003).

Koerner (2004) argues that accurate assessment of trends in forest carbon balance requires long-term monitoring of many replicate plots or very large plots; lacking these studies, the net carbon balance of undisturbed tropical forests cannot be authoritatively assessed based on *in situ* studies. If the results from the plots are extrapolated for illustration, the mean above ground carbon sink would be 0.89 ± 0.32 MgC ha^{-1} yr^{-1} (Baker et al., 2004), or 0.54 ± 0.19 GtC yr^{-1} (Malhi and Phillips 2004) extrapolated to all Neotropical moist forest area (6.0×10^6 km^2). If the uncompiled data from the African and Asian tropics (50% of global moist tropical forest area) were to show a similar trend, the associated tropical live biomass sink would be about

1.2 ± 0.4 GtC yr^{-1}, close to balancing the net source due to deforestation inferred by DeFries et al. (2002) and Achard et al. (2004) (Table 7.2).

7.3.2.2.5   New findings on the carbon budget

The revised carbon budget in Table 7.1 shows new estimates of two key numbers. First, the flux of CO_2 released to the atmosphere from land use change is estimated to be 1.6 (0.5 to 2.7) GtC yr^{-1} for the 1990s. A revision of the TAR estimate for the 1980s (see TAR, Chapter 3) downwards to 1.4 (0.4 to 2.3) GtC yr^{-1} suggests little change between the 1980s and 1990s, but there continues to be considerable uncertainty in these estimates. Second, the net residual terrestrial sink seems to have been larger in the 1990s than in the periods before and after. Thus, a transient increase in terrestrial uptake during the 1990s explains the lower airborne fraction observed during that period. The ocean uptake has increased by 22% between the 1980s and the 1990s, but the fraction of emissions (fossil plus land use) taken up by the ocean has remained constant.

7.3.2.3   Regional Fluxes

Quantifying present-day regional carbon sources and sinks and understanding the underlying carbon mechanisms are needed to inform policy decisions. Furthermore, by analysing spatial and temporal detail, mechanisms can be isolated.

7.3.2.3.1   The top-down view: atmospheric inversions

The atmosphere mixes and integrates surface fluxes that vary spatially and temporally. The distribution of regional fluxes over land and oceans can be retrieved using observations of atmospheric CO_2 and related tracers within models of atmospheric transport. This is called the 'top-down' approach to estimating fluxes. Atmospheric inversions belong to that approach, and determine an optimal set of fluxes that minimise the mismatch between modelled and observed concentrations, accounting for measurement and model errors. Fossil fuel emissions have small uncertainties that are often ignored and, when considered (e.g., Enting et al., 1995; Rodenbeck et al., 2003a), are found to have little influence on the inversion. Fossil fuel emissions are generally considered perfectly known in inversions, so that their effect can be easily modelled and subtracted from atmospheric CO_2 data to solve for regional land-atmosphere and ocean-atmosphere fluxes, although making such an assumption biases the results (Gurney et al., 2005). Input data for inversions come from a global network of about 100 CO_2 concentration measurement sites,[3] with mostly discrete flask sampling, and a smaller number of in situ continuous measurement sites. Generally, regional fluxes derived from inverse models have smaller uncertainties upwind of regions with denser data coverage. Measurement and modelling errors and uneven and sparse coverage of the network generate random errors in inversion results. In addition, inverse methodological details, such as the choice of transport model, can introduce

systematic errors. A number of new inversion ensembles, with different methodological details, have been produced since the TAR (Gurney et al., 2003; Rödenbeck et al., 2003a,b; Peylin et al., 2005; Baker et al., 2006). Generally, confidence in the long-term mean inverted regional fluxes is lower than confidence in the year-to-year anomalies (see Section 7.3.2.4). For individual regions, continents or ocean basins, the errors of inversions increase and the significance can be lost. Because of this, Figure 7.7 reports the oceans and land fluxes aggregated into large latitude bands, as well as a breakdown of five land and ocean regions in the NH, which is constrained by denser atmospheric stations. Both random and systematic errors are reported in Figure 7.7.

7.3.2.3.2   The bottom-up view: land and ocean observations and models

The range of carbon flux and inventory data enables quantification of the distribution and variability of CO_2 fluxes between the Earth's surface and the atmosphere. This is called the 'bottom-up' approach. The fluxes can be determined by measuring carbon stock changes at repeated intervals, from which time-integrated fluxes can be deduced, or by direct observations of the fluxes. The stock change approach includes basin-scale in situ measurements of dissolved and particulate organic and inorganic carbon or tracers in the ocean (e.g., Sabine et al., 2004a), extensive forest biomass inventories (e.g., UN-ECE/FAO, 2000; Fang et al., 2001; Goodale et al., 2002; Nabuurs et al., 2003; Shvidenko and Nilsson, 2003) and soil carbon inventories and models (e.g., Ogle et al., 2003; Bellamy et al., 2005; van Wesemael et al., 2005; Falloon et al., 2006). The direct flux measurement approach includes surveys of ocean CO_2 partial pressure (pCO$_2$) from ship-based measurements, drifters and time series (e.g., Lefèvre et al., 1999; Takahashi et al., 2002), and ecosystem flux measurements via eddy covariance flux networks (e.g., Valentini et al., 2000; Baldocchi et al., 2001).

The air-sea CO_2 fluxes consist of a superposition of natural and anthropogenic CO_2 fluxes, with the former being globally nearly balanced (except for a small net outgassing associated with the input of carbon by rivers). Takahashi et al. (2002) present both surface ocean pCO$_2$ and estimated atmosphere-ocean CO_2 fluxes (used as prior knowledge in many atmospheric inversions) normalised to 1995 using National Centers for Environmental Prediction (NCEP)/National Center for Atmospheric Research (NCAR) 41-year mean monthly winds. Large annual CO_2 fluxes to the ocean occur in the Southern Ocean subpolar regions (40°S–60°S), in the North Atlantic poleward of 30°N and in the North Pacific poleward of 30°N (see Figure 7.8). Ocean inversions calculate natural and anthropogenic air-sea fluxes (Gloor et al., 2003; Mikaloff Fletcher et al., 2006), by optimising ocean carbon model results against vertical profiles of DIC data. These studies indicate that the Southern Ocean is the largest sink of anthropogenic CO_2, together with mid- to high-latitude regions in the North Atlantic. This is consistent with global ocean hydrographic surveys (Sabine et al., 2004a

---

[3] Data can be accessed for instance via the World Data Centre for Greenhouse Gases (http://gaw.kishou.go.jp/wdcgg.html) or the NOAA ESRL Global Monitoring Division (http://www.cmdl.noaa.gov/ccgg/index.html)

and Figure 5.10). However, only half of the anthropogenic CO_2 absorbed by the Southern Ocean is stored there, due to strong northward transport (Mikaloff Fletcher et al., 2006). The tropical Pacific is a broad area of natural CO_2 outgassing to the atmosphere, but this region is a sink of anthropogenic CO_2.

Models are used to extrapolate flux observations into regional estimates, using remote-sensing properties and knowledge of the processes controlling the CO_2 fluxes and their variability. Rayner et al. (2005) use inverse process-based models, where observations are 'assimilated' to infer optimised fluxes. Since the TAR, the global air-sea flux synthesis has been updated (Takahashi et al., 2002 and Figure 7.8), and new syntheses have been made of continental-scale carbon budgets of the NH continents (Pacala et al., 2001; Goodale et al., 2002; Janssens et al., 2003; Shvidenko and Nilsson, 2003; Ciais et al., 2005a), and of tropical forests (Malhi and Grace, 2000). These estimates are shown in Figure 7.7 and compared with inversion results.

Comparing bottom-up regional fluxes with inversion results is not straightforward because: (1) inversion fluxes may contain a certain amount of prior knowledge of bottom-up fluxes so that the two approaches are not fully independent; (2) the time period for which inversion models and bottom-up estimates are compared is often not consistent, in the presence of interannual variations in fluxes[4] (see Section 7.3.2.4); and (3) inversions of CO_2 data produce estimates of CO_2 fluxes, so the results will differ from budgets for carbon fluxes (due to the emission of reduced carbon compounds that get oxidized into CO_2 in the atmosphere and are subject to transport and chemistry) and carbon storage changes (due to lateral carbon transport, e.g., by rivers) (Sarmiento and Sundquist, 1992). Some of these effects can be included by 'off-line' conversion of inversion results (Enting and Mansbridge, 1991; Suntharalingam et al., 2005). Reduced carbon compounds such as volatile organic compounds (VOCs), carbon monoxide (CO) and CH_4 emitted by ecosystems and human activities are transported and oxidized into CO_2 in the atmosphere (Folberth et al., 2005). Trade of forest and crop products displaces carbon from ecosystems (Imhoff et al., 2004). Rivers displace dissolved and particulate inorganic and organic carbon from land to ocean (e.g., Aumont et al., 2001). A summary of the main results of inversion and bottom-up estimates of regional CO_2 fluxes is given below.

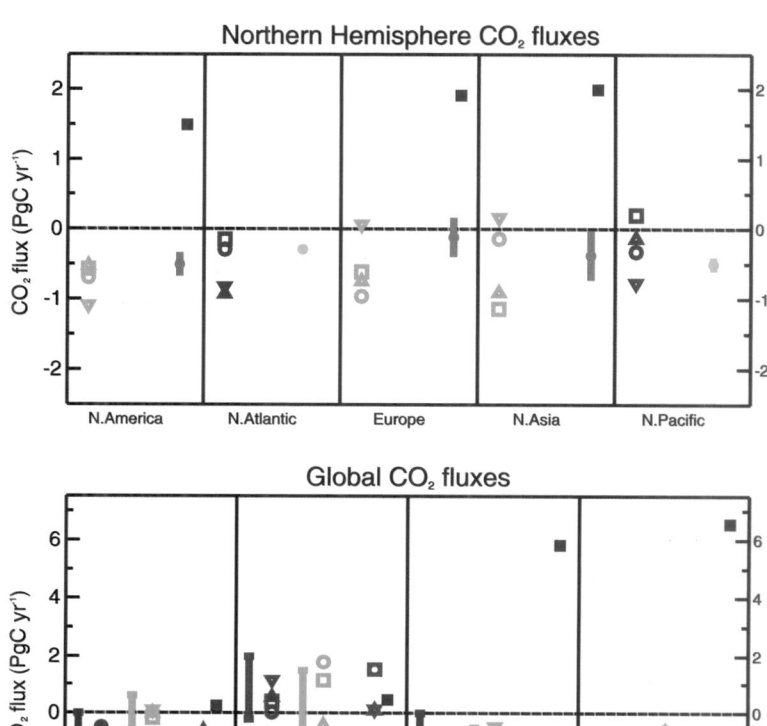

Figure 7.7. *Regional ocean-atmosphere and land-atmosphere CO_2 fluxes for the NH (top) and the globe (bottom) from inversion ensembles and bottom-up studies. Fluxes to the atmosphere are positive and uptake has a negative sign. Inversion results all correspond to the post-Pinatubo period 1992 to 1996. Orange: Bottom-up terrestrial fluxes from Pacala et al. (2001) and Kurz and Apps (1999) for North America, from Janssens et al. (2003) for Europe and from Shvidenko and Nilsson (2003) plus Fang et al. (2001) for North Asia (Asian Russia and China). Cyan (filled circles): Bottom-up ocean flux estimates from Takahashi, et al. (2002). Blue: ocean fluxes from atmospheric inversions. Green: terrestrial fluxes from inversion models. Magenta: total inversion fluxes. Red: fossil fuel emissions. The mean flux of different inversion ensembles is reported. Inversion errors for regional fluxes are not reported here; their values usually range between 0.5 and 1 GtC yr^{-1}. Error bar: range of atmospheric inversion fluxes from the TAR. Squares: Gurney et al. (2002) inversions using annual mean CO_2 observations and 16 transport models. Circles: Gurney et al. (2003) inversions using monthly CO_2 observations and 13 transport models. Triangles: Peylin et al. (2005) inversions with three transport models, three regional breakdowns and three inversion settings. Inverted triangles: Rödenbeck et al. (2003a) inversions where the fluxes are solved on the model grid using monthly flask data.*

7.3.2.3.3 Robust findings of regional land-atmosphere flux

- Tropical lands are found in inversions to be either carbon neutral or sink regions, despite widespread deforestation, as is apparent in Figure 7.7, where emissions from land include deforestation. This implies carbon uptake by undisturbed tropical ecosystems, in agreement with limited forest inventory data in the Amazon (Phillips et al., 1998; Malhi and Grace, 2000).

- Inversions place a substantial land carbon sink in the NH. The inversion estimate is −1.7 (−0.4 to −2.3) GtC yr^{-1} (from data in Figure 7.7). A bottom-up value of the NH land sink of −0.98 (−0.38 to −1.6) GtC yr^{-1} was also estimated,

---

[4] For instance, the chosen 1992 to 1996 time period for assessing inversion fluxes, dictated by the availability of the Atmospheric Tracer Transport Model Intercomparison Project (TransCom 3) intercomparison results (Gurney et al., 2002, 2003, 2004), corresponds to a low growth rate and to a stronger terrestrial carbon sink, likely due to the eruption of Mt. Pinatubo.

based upon regional synthesis studies (Kurz and Apps, 1999; Fang et al., 2001; Pacala et al., 2001; Janssens et al., 2003; Nilsson et al., 2003; Shvidenko and Nilsson, 2003). The inversion sink value is on average higher than the bottom-up value. Part of this discrepancy could be explained by lateral transport of carbon via rivers, crop trade and emission of reduced carbon compounds.

- The longitudinal partitioning of the northern land sink between North America, Europe and Northern Asia has large uncertainties (see Figure 7.7). Inversions give a very large spread over Europe (–0.9 to +0.2 GtC yr⁻¹), and Northern Asia (–1.2 to +0.3 GtC yr⁻¹) and a large spread over North America (–0.6 to –1.1 GtC yr⁻¹). Within the uncertainties of each approach, continental-scale carbon fluxes from bottom-up and top-down methods over Europe, North America and Northern Asia are mutually consistent (Pacala et al., 2001; Janssens et al., 2003). The North American carbon sink estimated by recent inversions is on average lower

than an earlier widely cited study by Fan et al. (1998). Nevertheless, the Fan et al. (1998) estimate remains within the range of inversion uncertainties. In addition, the fluxes calculated in Fan et al. (1998) coincide with the low growth rate post-Pinatubo period, and hence are not necessarily representative of long-term behaviour.

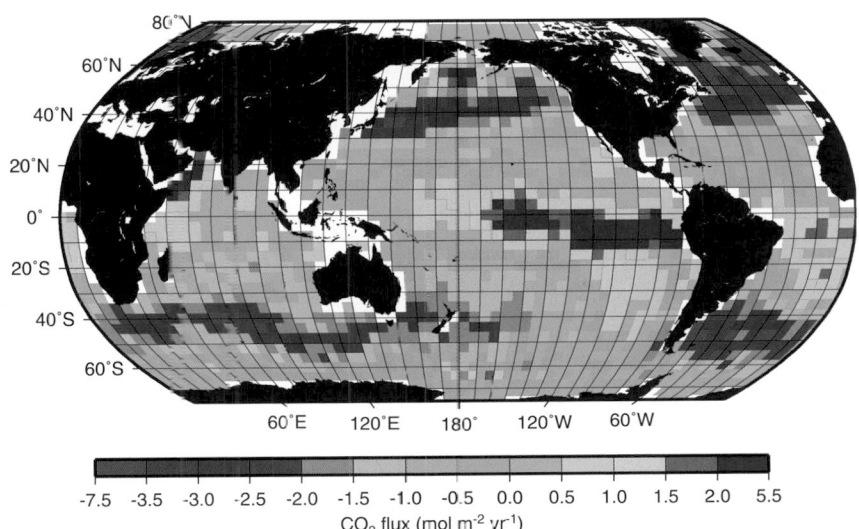

Figure 7.8. *Estimates (4° × 5°) of sea-to-air flux of CO$_2$, computed using 940,000 measurements of surface water pCO$_2$ collected since 1956 and averaged monthly, together with NCEP/NCAR 41-year mean monthly wind speeds and a (10-m wind speed)² dependence on the gas transfer rate (Wanninkhof, 1992). The fluxes were normalised to the year 1995 using techniques described in Takahashi et al. (2002), who used wind speeds taken at the 0.995 standard deviation level (about 40 m above the sea surface). The annual flux of CO$_2$ for 1995 with 10-m winds is –1.6 GtC yr⁻¹, with an approximate uncertainty (see Footnote 1) of ±1 GtC yr⁻¹, mainly due to uncertainty in the gas exchange velocity and limited data coverage. This estimated global flux consists of an uptake of anthropogenic CO$_2$ of –2.2 GtC yr⁻¹ (see text) plus an outgassing of 0.6 GtC yr⁻¹, corresponding primarily to oxidation of organic carbon borne by rivers (Figure 7.3). The monthly flux values with 10-m winds used here are available from T. Takahashi at http://www.ldeo. columbia.edu/res/pi/CO2/carbondioxide/pages/air_sea_flux_rev1.html.*

7.3.2.3.4   Robust findings of regional ocean-atmosphere flux

- The regional air-sea CO$_2$ fluxes consist of a superposition of natural and anthropogenic CO$_2$ fluxes, with the former being globally nearly balanced (except for a small net outgassing associated with the input of carbon by rivers), and the latter having a global integral uptake of 2.2 ± 0.5 GtC yr⁻¹ (see Table 7.1).
- The tropical oceans are outgassing CO$_2$ to the atmosphere (see Figure 7.8), with a mean flux of the order of 0.7 GtC yr⁻¹, estimated from an oceanic inversion (Gloor et al., 2003), in good agreement with atmospheric inversions (0 to 1.5 GtC yr⁻¹), and estimates based on oceanic pCO$_2$ observations (0.8 GtC yr⁻¹; Takahashi et al., 2002).
- The extratropical NH ocean is a net sink for anthropogenic and natural CO$_2$, with a magnitude of the order of 1.2 GtC yr⁻¹, consistent among various estimates.
- The Southern Ocean is a large sink of atmospheric CO$_2$ (Takahashi et al., 2002; Gurney et al., 2002) and of anthropogenic CO$_2$ (Gloor et al., 2003; Mikaloff Fletcher et al., 2006). Its magnitude has been estimated to be about 1.5 GtC yr⁻¹. This estimate is consistent among the different methods at the scale of the entire Southern Ocean. However,

differences persist with regard to the Southern Ocean flux distribution between subpolar and polar latitudes (T. Roy et al., 2003). Atmospheric inversions and oceanic inversions indicate a larger sink in subpolar regions (Gurney et al., 2002; Gloor et al., 2003), consistent with the distribution of CO$_2$ fluxes based on available ΔpCO$_2$ observations (Figure 7.8 and Takahashi, 2002).

7.3.2.4   Interannual Changes in the Carbon Cycle

7.3.2.4.1   Interannual changes in global fluxes

The atmospheric CO$_2$ growth rate exhibits large interannual variations (see Figure 3.3, the TAR and http://lgmacweb.env. uea.ac.uk/lequere/co2/carbon_budget). The variability of fossil fuel emissions and the estimated variability in net ocean uptake are too small to account for this signal, which must be caused by year-to-year fluctuations in land-atmosphere fluxes. Over the past two decades, higher than decadal-mean CO$_2$ growth rates occurred in 1983, 1987, 1994 to 1995, 1997 to 1998 and 2002 to 2003. During such episodes, the net uptake of anthropogenic CO$_2$ (sum of land and ocean sinks) is temporarily weakened. Conversely, small growth rates occurred in 1981, 1992 to 1993 and 1996 to 1997, associated with enhanced uptake. Generally, high CO$_2$ growth rates correspond to El Niño climate conditions, and low growth rates to La Niña (Bacastow and Keeling, 1981; Lintner, 2002). However, two episodes of CO$_2$ growth rate variations during the past two decades did not reflect such an El Niño forcing. In 1992 to 1993, a marked reduction in growth rate occurred, coincident with the cooling and radiation anomaly caused by the eruption of Mt. Pinatubo in June 1991. In 2002 to 2003, an increase in growth rate occurred, larger than expected based on the very weak El Niño event (Jones and Cox, 2005). It coincided with droughts in Europe (Ciais et al., 2005b), in North America (Breshears et al., 2005) and in Asian Russia (IFFN, 2003).

Since the TAR, many studies have confirmed that the variability of CO_2 fluxes is mostly due to land fluxes, and that tropical lands contribute strongly to this signal (Figure 7.9). A predominantly terrestrial origin of the growth rate variability can be inferred from (1) atmospheric inversions assimilating time series of CO_2 concentrations from different stations (Bousquet et al., 2000; Rödenbeck et al., 2003b; Baker et al., 2006), (2) consistent relationships between $\delta^{13}C$ and CO_2 (Rayner et al., 1999), (3) ocean model simulations (e.g., Le Quéré et al., 2003; McKinley et al., 2004a) and (4) terrestrial carbon cycle and coupled model simulations (e.g., C. Jones et al., 2001; McGuire et al., 2001; Peylin et al., 2005; Zeng et al., 2005). Currently, there is no evidence for basin-scale interannual variability of the air-sea CO_2 flux exceeding ±0.4 GtC yr^{-1}, but there are large ocean regions, such as the Southern Ocean, where interannual variability has not been well observed.

7.3.2.4.2 Interannual variability in regional fluxes, atmospheric inversions and bottom-up models

Year-to-year flux anomalies can be more robustly inferred by atmospheric inversions than mean fluxes. Yet, at the scale of continents or ocean basins, the inversion errors increase and the statistical significance of the inferred regional fluxes decreases.[5] This is why Figure 7.9 shows the land-atmosphere and ocean-atmosphere flux anomalies over broad latitude bands only for the inversion ensembles of Baker et al. (2006), Bousquet et al. (2000) and Rödenbeck et al. (2003b). An important finding of these studies is that differences in transport models have little impact on the interannual variability of fluxes. Interannual variability of global land-atmosphere fluxes (±4 GtC yr^{-1} between extremes) is larger than that of air-sea fluxes and dominates the global fluxes. This result is also true over large latitude bands (Figure 7.9). Tropical land fluxes exhibit on average a larger variability than temperate and boreal fluxes. Inversions give tropical land flux anomalies of the order of ±1.5 to 2 GtC yr^{-1}, which compare well in timing and magnitude with terrestrial model results (Tian et al., 1998; Peylin et al., 2005; Zeng et al., 2005). In these studies, enhanced sources occur during El Niño episodes and abnormal sinks during La Niña. In addition to the influence of these climate variations on ecosystem processes (Gérard et al., 1999; C. Jones et al., 2001), regional droughts during El Niño events promote large biomass fires, which appear to contribute to high CO_2 growth rates during the El Niño episodes (Barbosa et al., 1999; Langenfelds et al., 2002; Page et al., 2002; van der Werf et al., 2003, 2004; Patra et al., 2005).

Inversions robustly attribute little variability to ocean-atmosphere CO_2 flux (±0.5 GtC yr^{-1} between extremes), except for the recent work of Patra et al. (2005). This is in agreement with ocean model and ocean observations (Lee et al., 1998; Le Quéré et al., 2003; Obata and Kitamura, 2003; McKinley et al., 2004b). However, inversions and ocean models differ on the dominant geographic contributions to the variability. Inversions estimate similar variability in both hemispheres, whereas ocean models estimate more variability in the Southern Ocean (Bousquet et al., 2000; Rödenbeck et al., 2003b; Baker et al., 2006). Over the North Atlantic, Gruber et al. (2002) suggest a regional CO_2 flux variability (extremes of ±0.3 GtC yr^{-1}) by extrapolating data from a single ocean station, but McKinley et al. (2004a,b) model a small variability (extremes of ±0.1 GtC yr^{-1}). The equatorial Pacific is the ocean region of the world where the variability is constrained with repeated ΔpCO_2 observations (variations of about ±0.4 GtC yr^{-1}; Feely et al., 2002), with a reduced source of CO_2 during El Niño associated with decreased upwelling of CO_2-rich waters. Over this region, some inversion results (e.g., Bousquet et al., 2000) compare well in magnitude and timing with ocean and coupled model results (Le Quéré et al., 2000; C. Jones et al., 2001; McKinley et al., 2004a,b) and with ΔpCO_2 observations (Feely et al., 1999, 2002).

7.3.2.4.3 Slowdown in carbon dioxide growth rates during the early 1990s

The early 1990s had anomalously strong global sinks for atmospheric CO_2, compared with the decadal mean (Table 7.1). Although a weak El Niño from 1991 to 1995 may have helped to enhance ocean uptake at that time, inversions and $O_2:N_2$ and $\delta^{13}C$-CO_2 atmospheric data (Battle et al., 2000) indicate that the enhanced uptake was of predominantly terrestrial origin. The regions where the 1992 to 1993 abnormal sink is projected to be are not robustly estimated by inversions. Both Bousquet et al. (2000) and Rödenbeck et al. (2003b) project a large fraction of that sink in temperate North America, while Baker et al. (2006) place it predominantly in the tropics. Model results suggest that cooler temperatures caused by the Mt. Pinatubo eruption reduced soil respiration and enhanced NH carbon uptake (Jones and Cox, 2001b; Lucht et al., 2002), despite lower productivity as indicated by remote sensing of vegetation activity. In addition, aerosols from the volcanic eruption scattered sunlight and increased its diffuse fraction, which is used more efficiently by plant canopies in photosynthesis than direct light (Gu et al., 2003). It has been hypothesised that a transient increase in the diffuse fraction of radiation enhanced CO_2 uptake by land ecosystems in 1992 to 1993, but the global significance and magnitude of this effect remains unresolved (Roderick et al., 2001; Krakauer and Randerson, 2003; Angert et al., 2004; Robock, 2005).

---

[5] In other words, the model bias has only a small influence on inversions of interannual variability. These interannual inversion studies all report a random error and a systematic error range derived from sensitivity tests with different settings. Bousquet et al. (2000) used large regions and different inversion settings for the period 1980 to 1998. Rödenbeck et al. (2003) used one transport model and inverted fluxes at the resolution of the model grid for the period 1982 to 2002, with different inversion settings. Baker et al. (2006) used large regions but 13 different transport models for the period 1988 to 2002.

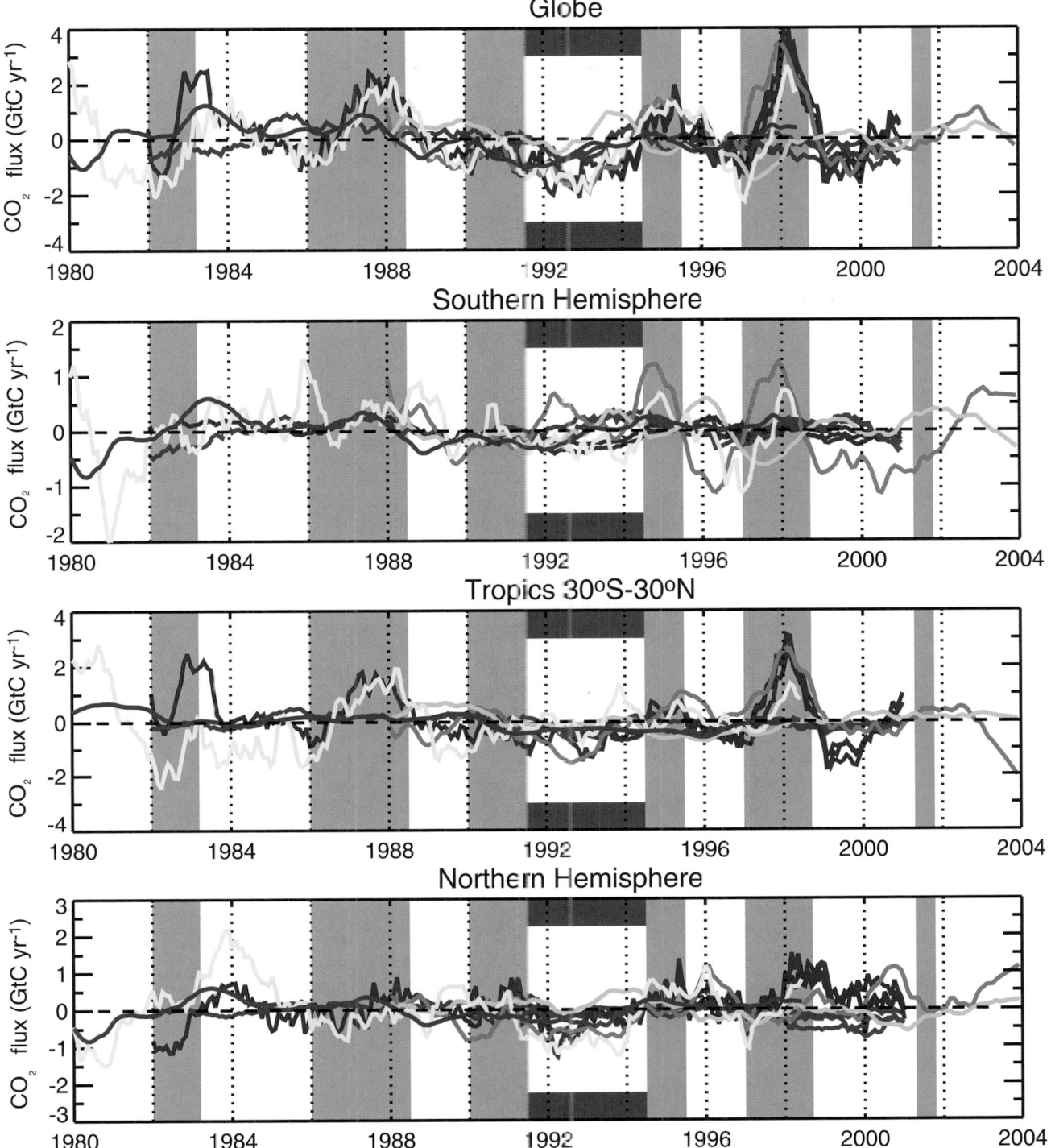

Baker et al. 2005 (orange = land ; cyan = ocean)
Rodenbeck et al. 2003 (red = land ; blue = ocean)
Bousquet et al. 2000 + (yellow = land ; purple = ocean)

Figure 7.9. *Year-to-year anomalies in ocean-atmosphere and land-atmosphere CO_2 fluxes (GtC yr^{-1}) from interannual inversion ensembles covering the past 20 years or so, grouped into large latitude bands, and over the globe. Three different inversion ensembles from Bousquet et al. (2000), Rödenbeck et al. (2003a) and Baker et al. (2006) are shown. For each flux and each region, the anomalies were obtained by subtracting the long-term mean flux and removing the seasonal signal. Grey shaded regions indicate El Niño episodes, and the black bars indicate the cooling period following the Mt. Pinatubo eruption.*

7.3.2.4.4 Speed-up in carbon dioxide growth rates during the late 1990s

The high CO_2 growth in 1998 coincided with a global increase in CO concentrations attributable to wildfires (Yurganov et al., 2005) in Southeast Asia (60%), South America (30%) and Siberia (van der Werf et al., 2004). Langenfelds et al. (2002) analyse the correlations in the interannual growth rate of CO_2 and other species at 10 stations and link the 1997 to 1998 (and the 1994 to 1995) anomalies to high fire emissions as a single process. Achard et al. (2004) estimate a source of 0.88 ± 0.07 GtC emitted from the burning of 2.4×10^6 ha of peatland in the Indonesian forest fires in 1997 to 1998, and Page et al. (2002) estimate a source of +0.8 to +2.6 GtC. During the 1997 to 1998 high CO_2 growth rate episode, inversions place an abnormal source over tropical Southeast Asia, in good agreement with such bottom-up evidence. The relationship between El Niño and CO_2 emissions from fires is not uniform: fire emissions from low productivity ecosystems in Africa and northern Australia are limited by fuel load density and thus decrease during drier periods, in contrast to the response in tropical forests (Barbosa et al., 1999; Randerson et al., 2005). In addition, co-varying processes such as reduced productivity caused by drought in tropical forests during El Niño episodes may be superimposed on fire emissions. From 1998 to 2003, extensive drought in mid-latitudes of the NH (Hoerling and Kumar, 2003), accompanied by more wildfires in some regions (Balzter et al., 2005; Yurganov et al., 2005) may have led to decreased photosynthesis and carbon uptake (Angert et al., 2005; Ciais et al., 2005b), helping to increase the atmospheric CO_2 growth rate.

7.3.3 Terrestrial Carbon Cycle Processes and Feedbacks to Climate

The net exchange of carbon between the terrestrial biosphere and the atmosphere is the difference between carbon uptake by photosynthesis and release by plant respiration, soil respiration and disturbance processes (fire, windthrow, insect attack and herbivory in unmanaged systems, together with deforestation, afforestation, land management and harvest in managed systems). Over at least the last 30 years, the net result of all these processes has been uptake of atmospheric CO_2 by terrestrial ecosystems (Table 7.1, 'land-atmosphere flux' row). It is critical to understand the reasons for this uptake and its likely future course. Will uptake by the terrestrial biosphere grow or diminish with time, or even reverse so that the terrestrial biosphere becomes a net source of CO_2 to the atmosphere? To answer this question it is necessary to understand the underlying processes and their dependence on the key drivers of climate, atmospheric composition and human land management.

Drivers that affect the carbon cycle in terrestrial ecosystems can be classified as (1) direct climate effects (changes in precipitation, temperature and radiation regime); (2) atmospheric composition effects (CO_2 fertilization, nutrient deposition, damage by pollution); and (3) land use change effects (deforestation, afforestation, agricultural practices, and

their legacies over time). This section first summarises current knowledge of the processes by which each of these drivers influence the terrestrial carbon balance, and then examines knowledge of the integrative consequences of all these processes in the key case of tropical forests.

7.3.3.1 Processes Driven by Climate, Atmospheric Composition and Land Use Change

7.3.3.1.1 Climatic regulation of terrestrial carbon exchange

Ecosystem responses to environmental drivers (sunlight, temperature, soil moisture) and to ecological factors (e.g., forest age, nutrient supply, organic substrate availability; see, e.g., Clark, 2002; Ciais et al., 2005b; Dunn et al., 2007) are complex. For example, elevated temperature and higher soil water content enhance rates for heterotrophic respiration in well-aerated soils, but depress these rates in wet soils. Soil warming experiments typically show marked soil respiration increases at elevated temperature (Oechel et al., 2000; Rustad et al., 2001; Melillo et al., 2002), but CO_2 fluxes return to initial levels in a few years as pools of organic substrate re-equilibrate with inputs (Knorr et al., 2005). However, in dry soils, decomposition may be limited by moisture and not respond to temperature (Luo et al., 2001). Carbon cycle simulations need to capture both the short- and long-term responses to changing climate to predict carbon cycle responses.

Current models of terrestrial carbon balance have difficulty simulating measured carbon fluxes over the full range of temporal and spatial scales, including instantaneous carbon exchanges at the leaf, plot or ecosystem level, seasonal and annual carbon fluxes at the stand level and decadal to centennial accumulation of biomass and organic matter at stand or regional scales (Melillo et al., 1995; Thornton et al., 2002). Moreover, projections of changes in land carbon storage are tied not only to ecosystem responses to climate change, but also to the modelled projections of climate change itself. As there are strong feedbacks between these components of the Earth system (see Section 7.3.5), future projections must be considered cautiously.

7.3.3.1.2 Effects of elevated carbon dioxide

On physiological grounds, almost all models predict stimulation of carbon assimilation and sequestration in response to rising CO_2, called 'CO_2 fertilization' (Cramer et al., 2001; Oren et al., 2001; Luo et al., 2004; DeLucia et al., 2005). Free Air CO_2 Enrichment (FACE) and chamber studies have been used to examine the response of ecosystems to large (usually about 50%) step increases in CO_2 concentration. The results have been variable (e.g., Oren et al., 2001; Nowak et al., 2004; Norby et al., 2005). On average, net CO_2 uptake has been stimulated, but not as much as predicted by some models. Other factors (e.g., nutrients or genetic limitations on growth) can limit plant growth and reduce response to CO_2. Eleven FACE experiments, encompassing bogs, grasslands, desert and young temperate tree stands report an average increased net primary productivity (NPP) of 12% when compared to ambient CO_2 levels (Nowak et al., 2004). There is a large range of responses, with woody plants consistently showing NPP increases of 23 to 25% (Norby et al., 2005), but much smaller increases for grain crops (Ainsworth and Long, 2005), reflecting differential allocation of the incremental organic matter to shorter- vs. longer-lived compartments. Overall, about two-thirds of the experiments show positive response to increased CO_2 (Ainsworth and Long, 2005; Luo et al., 2005). Since saturation of CO_2 stimulation due to nutrient or other limitations is common (Dukes et al., 2005; Koerner et al., 2005), it is not yet clear how strong the CO_2 fertilization effect actually is.

7.3.3.1.3   Nutrient and ozone limitations to carbon sequestration

The basic biochemistry of photosynthesis implies that stimulation of growth will saturate under high CO_2 concentrations and be further limited by nutrient availability (Dukes et al., 2005; Koerner et al., 2005) and by possible acclimation of plants to high CO_2 levels (Ainsworth and Long, 2005). Carbon storage by terrestrial plants requires net assimilation of nutrients, especially N, a primary limiting nutrient at middle and high latitudes and an important nutrient at lower latitudes (Vitousek et al., 1998). Hungate et al. (2003) argue that 'soil C sequestration under elevated CO_2 is constrained both directly by N availability and indirectly by nutrients needed to support N_2 fixation', and Reich et al. (2006) conclude that 'soil N supply is probably an important constraint on global terrestrial responses to elevated CO_2'. This view appears to be consistent with other recent studies (e.g., Finzie et al., 2006; Norby et al., 2006; van Groenigen et al., 2006) and with at least some of the FACE data, further complicating estimation of the current effects of rising CO_2 on carbon sequestration globally.

Additional N supplied through atmospheric deposition or direct fertilization can stimulate plant growth (Vitousek, 2004) and in principle could relieve the nutrient constraint on CO_2 fertilization. Direct canopy uptake of atmospheric N may be particularly effective (Sievering et al., 2000). Overall, the effectiveness of N inputs appears to be limited by immobilisation and other mechanisms. For example, when labelled nitrogen (^{15}N) was added to soil and litter in a forest over seven years, only a small fraction became available for tree growth (Nadelhoffer et al., 2004). Moreover, atmospheric N deposition is spatially correlated with air pollution, including elevated atmospheric ozone. Ozone and other pollutants may have detrimental effects on plant growth, possibly further limiting the stimulation of carbon uptake by anthropogenic N emissions (Ollinger and Aber, 2002; Holland and Carroll, 2003). Indeed, Felzer et al. (2004) estimate that surface ozone increases since 1950 may have reduced CO_2 sequestration in the USA by 18 to 20 TgC yr^{-1}. The current generation of coupled carbon-climate models (see Section 7.3.5) does not include nutrient limitations or air pollution effects.

7.3.3.1.4   Fire

Fire is a major agent for conversion of biomass and soil organic matter to CO_2 (Randerson et al., 2002a–d; Cochrane, 2003; Nepstad et al., 2004; Jones and Cox, 2005; Kasischke et al., 2005; Randerson et al., 2005). Globally, wildfires (savannah and forest fires, excluding biomass burning for fuel and land clearing) oxidize 1.7 to 4.1 GtC yr^{-1} (Mack et al., 1996; Andreae and Merlet, 2001), or about 3 to 8% of total terrestrial NPP. There is an additional large enhancement of CO_2 emissions associated with fires stimulated by human activities, such as deforestation and tropical agricultural development. Thus, there is a large potential for future alteration in the terrestrial carbon balance through altered fire regimes. A striking example occurred during the 1997 to 1998 El Niño, when large fires in the Southeast Asian archipelago are estimated to have released 0.8 to 2.6 GtC (see Section 7.3.2.4). Fire frequency and intensity are strongly sensitive to climate change and variability, and to land use practices. Over the last century, trends in burned area have been largely driven by land use practices, through fire suppression policies in mid-latitude temperate regions and increased use of fire to clear forest in tropical regions (Mouillot and Field, 2005). However, there is also evidence that climate change has contributed to an increase in fire frequency in Canada (Gillett et al., 2004). The decrease in fire frequency in regions like the USA and Europe has contributed to the land carbon sink there, while increased fire frequency in regions like Amazonia, Southeast Asia and Canada has contributed to the carbon source. At high latitudes, the role of fire appears to have increased in recent decades: fire disturbance in boreal forests was higher in the 1980s than in any previous decade on record (Kurz et al., 1995; Kurz and Apps, 1999; Moulliot and Field, 2005). Flannigan et al. (2005) estimate that in the future, the CO_2 source from fire will increase.

7.3.3.1.5   Direct effects of land use and land management

Evolution of landscape structure, including woody thickening: Changes in the structure and distribution of ecosystems are driven in part by changes in climate and atmospheric CO_2, but also by human alterations of landscapes through land management and the introduction of invasive species and exotic pathogens. The single most important process in the latter category is woody encroachment or vegetation thickening, the increase in woody biomass occurring in (mainly semi-arid) grazing lands. In many regions, this increase arises from fire suppression and associated grazing management practices, but there is also a possibility that increases in CO_2 are giving C_3 woody plants a competitive advantage over C_4 grasses (Bond et al., 2003). Woody encroachment could account for as much as 22 to 40% of the regional carbon sink in the USA (Pacala et al., 2001), and a high proportion in northeast Australia (Burrows et al., 2002). Comprehensive data are lacking to define this effect accurately.

Deforestation: Forest clearing (mainly in the tropics) is a large contributor to the land use change component of the current atmospheric CO_2 budget, accounting for up to one-third of total anthropogenic emissions (see Table 7.2; Section 7.3.2.1; also Table 7.1, row 'land use change flux'). The future evolution of this term in the CO_2 budget is therefore of critical importance. Deforestation in Africa, Asia and the tropical Americas is expected to decrease towards the end of the 21st century to a small fraction of the levels in 1990 (IPCC, 2000). The declines in Asia and Africa are driven by the depletion of forests, while trends in the Americas have the highest uncertainty given the extent of the forest resource.

Afforestation: Recent (since 1970) afforestation and reforestation as direct human-induced activities have not yet had much impact on the global terrestrial carbon sink. However, regional sinks have been created in areas such as China, where afforestation since the 1970s has sequestered 0.45 GtC (Fang et al., 2001). The largest effect of afforestation is not immediate but through its legacy.

Agricultural practices: Improvement of agricultural practices on carbon-depleted soils has created a carbon sink. For instance, the introduction of conservation tillage in the USA is estimated to have increased soil organic matter (SOM) stocks by about 1.4 GtC over the last 30 years. However, yearly increases in SOM can be sustained only for 50 to 100 years, after which the system reaches a new equilibrium (Cole et al., 1996; Smith et al., 1997). Moreover, modern conservation tillage often entails large inputs of chemicals and fertilizer, which are made using fossil fuels, reducing the CO_2 benefit from carbon sequestration in agricultural soils. The increase in soil carbon stocks under low-tillage systems may also be mostly a topsoil effect with little increase in total profile carbon storage observed, confounded by the fact that most studies of low-tillage systems have only sampled the uppermost soil layers.

7.3.3.1.6 Forest regrowth

Some studies suggest that forest regrowth could be a major contributor to the global land carbon sink (e.g., Pacala et al., 2001; Schimel et al., 2001; Hurtt et al., 2002). Forest areas generally increased during the 20th century at middle and high latitudes (unlike in the tropics). This surprising trend reflects the intensification of agriculture and forestry. Globally, more food is being grown on less land, reflecting mechanisation of agriculture, increased fertilizer use and adoption of high-yield cultivars, although in parts of Africa and Asia the opposite is occurring. Likewise, intensive forest management and agroforestry produce more fibre on less land; improved forest management favours more rapid regrowth of forests after harvest. These trends have led to carbon sequestration by regrowing forests. It should be noted, however, that industrialised agriculture and forestry require high inputs of fossil energy, so it is difficult to assess the net global effects of agricultural intensification on atmospheric greenhouse gases and radiative forcing.

Regional studies have confirmed the plausibility of strong mid-latitude sinks due to forest regrowth. Data from the eddy flux tower network show that forests on long-abandoned former agricultural lands (Curtis et al., 2002) and in industrial managed forests (Hollinger et al., 2002) take up significant amounts of carbon every year. Analysis of forest inventory data shows that, in aggregate, current forest lands are significant sinks for atmospheric CO_2 (Pacala et al., 2001). Few old growth forests remain at mid-latitudes (most forests are less than 70 years old), in part due to forest management. Therefore, forests in

these areas are accumulating biomass because of their ages and stages of succession. Within wide error bands (see Section 7.3.2.3), the uptake rates inferred from flux towers are generally consistent with those inferred from inverse methods (e.g., Hurtt et al., 2002). Stocks of soil carbon are also likely increasing due to replenishment of soil organic matter and necromass depleted during the agricultural phase, and changes in soil microclimate associated with reforestation; these effects might add 30 to 50% to the quantity of CO_2 sequestered (e.g., Barford et al., 2001). It is important to note that at least some of this sequestration is 'refilling' the deficits in biomass and soil organic matter, accumulated in previous epochs (see Figure 7.3), and the associated CO_2 uptake should be expected to decline in the coming decades unless sustained by careful management strategies designed to accomplish that purpose.

7.3.4 Ocean Carbon Cycle Processes and Feedbacks to Climate

7.3.4.1 Overview of the Ocean Carbon Cycle

Oceanic carbon exists in several forms: as DIC, DOC, and particulate organic carbon (POC) (living and dead) in an approximate ratio DIC:DOC:POC = 2000:38:1 (about 37,000 GtC DIC: Falkowski et al., 2000 and Sarmiento and Gruber, 2006; 685 GtC DOC: Hansell and Carlson, 1998; and 13 to 23 GtC POC: Eglinton and Repeta, 2004). Before the industrial revolution, the ocean contained about 60 times as much carbon as the atmosphere and 20 times as much carbon as the terrestrial biosphere/soil compartment.

Seawater can, through inorganic processes, absorb large amounts of CO_2 from the atmosphere, because CO_2 is a weakly acidic gas and the minerals dissolved in the ocean have over geologic time created a slightly alkaline ocean (surface pH 7.9 to 8.25: Degens et al., 1984; Royal Society, 2005). The air-sea exchange of CO_2 is determined largely by the air-sea gradient in pCO_2 between atmosphere and ocean. Equilibration of surface ocean and atmosphere occurs on a time scale of roughly one year. Gas exchange rates increase with wind speed (Wanninkhof and McGillis, 1999; Nightingale et al., 2000) and depend on other factors such as precipitation, heat flux, sea ice and surfactants. The magnitudes and uncertainties in local gas exchange rates are maximal at high wind speeds. In contrast, the equilibrium values for partitioning of CO_2 between air and seawater and associated seawater pH values are well established (Zeebe and Wolf-Gladrow, 2001; see Box 7.3).

In addition to changes in advection and mixing, the ocean can alter atmospheric CO_2 concentration through three mechanisms (Volk and Hoffert, 1985), illustrated in Figure 7.10: (1) absorption or release of CO_2 due to changes in solubility of gaseous CO_2 ('solubility pump'); (2) changes in carbon fixation to POC in surface waters by photosynthesis and export of this carbon through sinking of organic particles out of the surface layer ('organic carbon pump') – this process is limited to first order by availability of light and nutrients (phosphate, nitrate, silicic acid and micronutrients such as iron); and (3) changes in

Box 7.3: Marine Carbon Chemistry and Ocean Acidification

The marine carbonate buffer system allows the ocean to take up CO_2 far in excess of its potential uptake capacity based on solubility alone, and in doing so controls the pH of the ocean. This control is achieved by a series of reactions that transform carbon added as CO_2 into HCO_3^- and CO_3^{2-}. These three dissolved forms (collectively known as DIC) are found in the approximate ratio CO_2:HCO_3^-:CO_3^{2-} of 1:100:10 (Equation (7.1)). CO_2 is a weak acid and when it dissolves, it reacts with water to form carbonic acid, which dissociates into a hydrogen ion (H^+) and a HCO_3^- ion, with some of the H^+ then reacting with CO_3^{2-} to form a second HCO_3^- ion (Equation (7.2)).

$$CO_2 + H_2O \rightarrow H^+ + HCO_3^- \rightarrow 2H^+ + CO_3^{2-}$$ (7.1)

$$CO_2 + H_2O + CO_3^{2-} \rightarrow HCO_3^- + H^+ + CO_3^{2-} \rightarrow 2HCO_3^-$$ (7.2)

Therefore, the net result of adding CO_2 to seawater is an increase in H^+ and HCO_3^-, but a reduction in CO_3^{2-}. The decrease in the CO_3^{2-} ion reduces the overall buffering capacity as CO_2 increases, with the result that proportionally more H^+ ions remain in solution and increase acidity.

This ocean acidification is leading to a decrease in the saturation state of $CaCO_3$ in the ocean. Two primary effects are expected: (1) the biological production of corals as well as calcifying phytoplankton and zooplankton within the water column may be inhibited or slowed down (Royal Society, 2005), and (2) the dissolution of $CaCO_3$ at the ocean floor will be enhanced (Archer, 2005). Aragonite, the meta-stable form of $CaCO_3$ produced by corals and pteropods (planktonic snails; Lalli and Gilmer, 1989), will be particularly susceptible to a pH reduction (Kleypas et al., 1999b; Hughes et al., 2003; Orr et al., 2005). Laboratory experiments under high ambient CO_2 with the coccolithophore species *Emiliania huxleyi* and *Gephyrocapsa oceanica* produce a significant reduction in $CaCO_3$ production and a stimulation of POC production (Riebesell et al., 2000; Zondervan et al., 2001). Other species and growth under other conditions may show different responses, so that no conclusive quantification of the $CaCO_3$ feedback is possible at present (Tortell et al., 2002; Sciandra et al., 2003).

The sinking speed of marine particle aggregates depends on their composition: $CaCO_3$ may act as an efficient ballast component, leading to high sinking speeds of aggregates (Armstrong et al., 2002; Klaas and Archer, 2002). The relatively small negative feedback of reduced $CaCO_3$ production to atmospheric pCO_2 may be compensated for by a change in the ballast for settling biogenic particles and the associated shallowing of re-mineralization depth levels in the water column for organic carbon (Heinze, 2004). On the other hand, production of extracellular organic carbon could increase under high CO_2 levels and lead to an increase in export (Engel et al., 2004).

Ecological changes due to expected ocean acidification may be severe for corals in tropical and cold waters (Gattuso et al., 1999; Kleypas et al., 1999a; Langdon et al., 2003; Buddemeier et al., 2004; Roberts et al., 2006) and for pelagic ecosystems (Tortell et al., 2002; Royal Society, 2005). Acidification can influence the marine food web at higher trophic levels (Langenbuch and Pörtner, 2003; Ishimatsu et al., 2004).

Since the beginning of the industrial revolution, sea surface pH has dropped by about 0.1 pH units (corresponding to a 30% increase in the H ion concentration). The expected continued decrease may lead within a few centuries to an ocean pH estimated to have occurred most recently a few hundred million years before present (Caldeira and Wickett, 2003; Key et al., 2004; Box 7.3, Figure 1).

According to a model experiment based on the IPCC Scenarios 1992a (IS92a) emission scenario, bio-calcification will be reduced by 2100, in particular within the Southern Ocean (Orr et al., 2005), and by 2050 for aragonite-producing organisms (see also Figure 10.24). It is important to note that ocean acidification is not a direct consequence of climate change but a consequence of fossil fuel CO_2 emissions, which are the main driver of the anticipated climate change.

Box 7.3, Figure 1. *(a) Atmospheric CO_2 emissions, historical atmospheric CO_2 levels and predicted CO_2 concentrations from the given emission time series, together with changes in ocean pH based on horizontally averaged chemistry. The emission time series is based on the mid-range IS92a emission scenario (solid line) prior to 2100 and then assumes that emissions continue until fossil fuel reserves decline. (b) Estimated maximum change in surface ocean pH as a function of final atmospheric CO_2 pressure, and the transition time over which this CO_2 pressure is linearly approached from 280 ppm. A: Glacial-interglacial CO_2 changes; B: slow changes over the past 300 Myr; C: historical changes in ocean surface waters; D: unabated fossil fuel burning over the next few centuries. Source: Caldeira and Wickett (2003). Reprinted with permission from Macmillan Publishers Ltd: Nature, Caldeira and Wickett (2003), copyright (2003).*

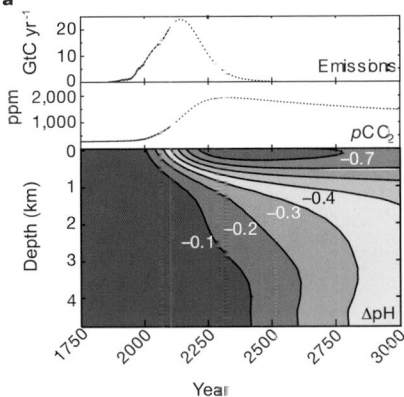

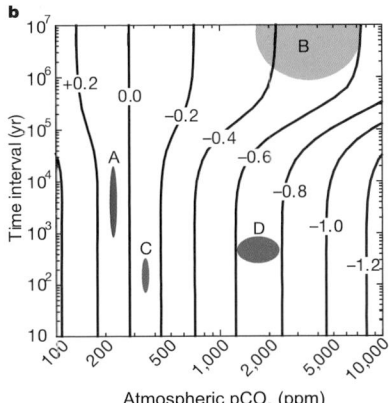

the release of CO_2 in surface waters during formation of $CaCO_3$ shell material by plankton ('$CaCO_3$ counter pump').

Organic particles are re-mineralized (oxidized to DIC and other inorganic compounds through the action of bacteria) primarily in the upper 1,000 m of the oceanic water column, with an accompanying decrease in dissolved O. On the average, $CaCO_3$ particles sink deeper before they undergo dissolution: deep waters are undersaturated with respect to $CaCO_3$. The remainder of the particle flux enters marine sediments and is subject to either re-dissolution within the water column or accumulation within the sediments. Although the POC reservoir is small, it plays an important role in keeping DIC concentrations low in surface waters and high in deep waters. The loop is closed through the three-dimensional ocean circulation: upwelling water brings inorganic carbon and nutrients to the surface again, leading to outgassing and biogenic particle production. Dissolved organic carbon enters the ocean water column from rivers and marine metabolic processes. A large fraction of DOC has a long ocean residence time (1–10 kyr), while other fractions are more short-lived (days to hundreds of years; Loh

et al., 2004). The composition of dissolved organic matter is still largely unknown.

In conjunction with the global ocean mixing or overturning time of the order of 1 kyr (Broecker and Peng, 1982), small changes in the large ocean carbon reservoir can induce significant changes in atmospheric CO_2 concentration. Likewise, perturbations in the atmospheric pCO_2 can be buffered by the ocean. Glacial-interglacial changes in the atmospheric CO_2 content can potentially be attributed to a change in functioning of the marine carbon pump (see Chapter 6). The key role for the timing of the anthropogenic carbon uptake by the ocean is played by the downward transport of surface water, with a high burden of anthropogenic carbon, into the ocean's interior. The organic carbon cycle and the $CaCO_3$ counter pump modulate, but do not dominate, the net marine uptake of anthropogenic carbon.

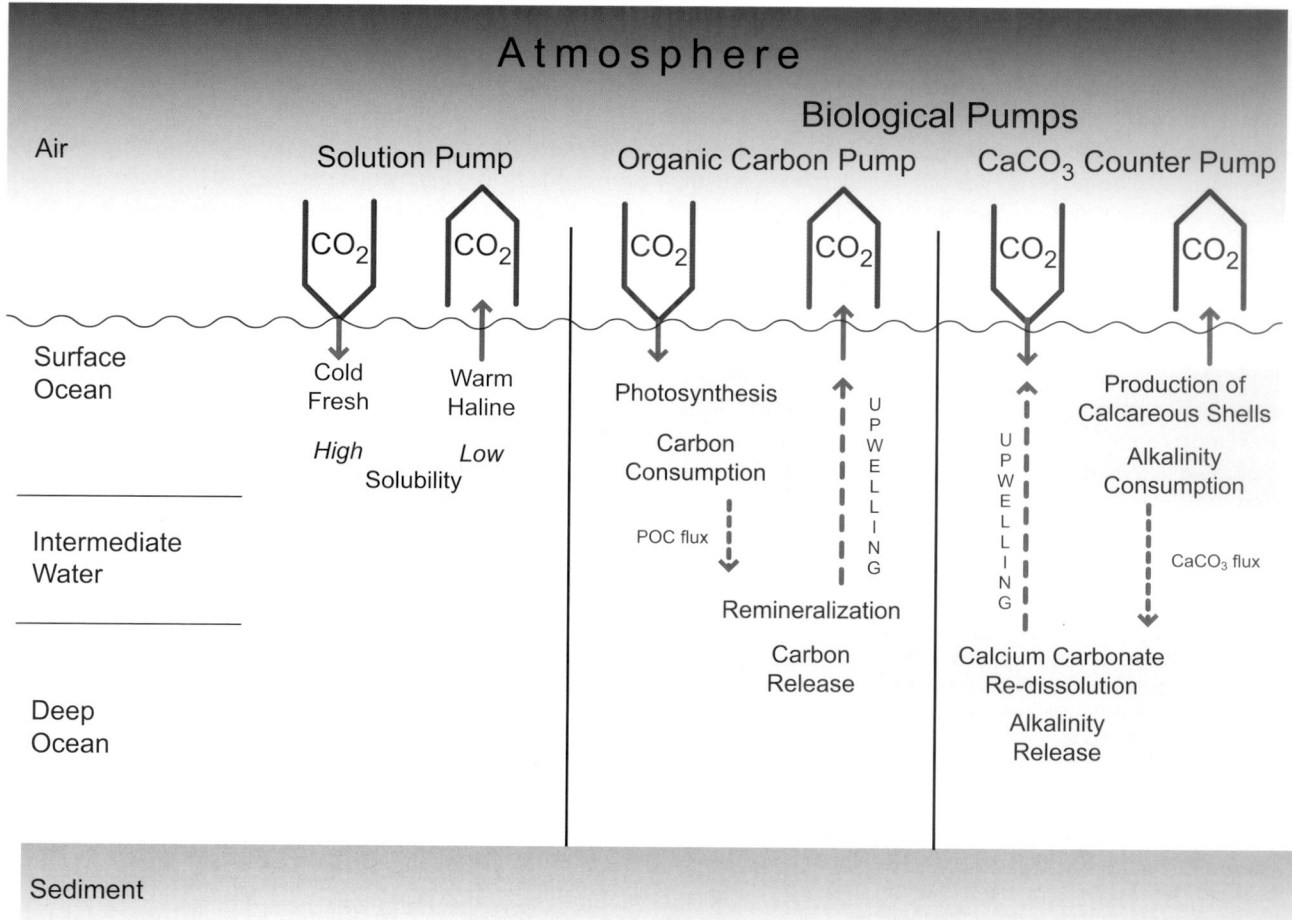

Figure 7.10. *Three main ocean carbon pumps govern the regulation of natural atmospheric CO_2 changes by the ocean (Heinze et al., 1991): the solubility pump, the organic carbon pump and the $CaCO_3$ 'counter pump'. The oceanic uptake of anthropogenic CO_2 is dominated by inorganic carbon uptake at the ocean surface and physical transport of anthropogenic carbon from the surface to deeper layers. For a constant ocean circulation, to first order, the biological carbon pumps remain unaffected because nutrient cycling does not change. If the ocean circulation slows down, anthropogenic carbon uptake is dominated by inorganic buffering and physical transport as before, but the marine particle flux can reach greater depths if its sinking speed does not change, leading to a biologically induced negative feedback that is expected to be smaller than the positive feedback associated with a slower physical downward mixing of anthropogenic carbon. Reprinted with permission, copyright 1991 American Geophysical Union.*

7.3.4.2   *Carbon Cycle Feedbacks to Changes in Atmospheric Carbon Dioxide*

Chemical buffering of anthropogenic CO_2 is the quantitatively most important oceanic process acting as a carbon sink. Carbon dioxide entering the ocean is buffered due to scavenging by the CO_3^{2-} ions and conversion to HCO_3^{-}, that is, the resulting increase in gaseous seawater CO_2 concentration is smaller than the amount of CO_2 added per unit of seawater volume. Carbon dioxide buffering in seawater is quantified by the Revelle factor ('buffer factor', Equation (7.3)), relating the fractional change in seawater pCO_2 to the fractional change in total DIC after re-equilibration (Revelle and Suess, 1957; Zeebe and Wolf-Gladrow, 2001):

Revelle factor (or buffer factor) =
$$(\Delta[CO_2] / [CO_2]) / (\Delta[DIC] / [DIC]) \qquad (7.3)$$

The lower the Revelle factor, the larger the buffer capacity of seawater. Variability of the buffer factor in the ocean depends mainly on changes in pCO_2 and the ratio of DIC to total alkalinity. In the present-day ocean, the buffer factor varies between 8 and 13 (Sabine et al., 2004a; Figure 7.11). With respect to atmospheric pCO_2 alone, the inorganic carbon system of the ocean reacts in two ways: (1) seawater re-equilibrates, buffering a significant amount of CO_2 from the atmosphere depending on the water volume exposed to equilibration; and (2) the Revelle factor increases with pCO_2 (positive feedback; Figure 7.11). Both processes are quantitatively important. While the first is generally considered as a system response, the latter is a feedback process.

The ocean will become less alkaline (seawater pH will decrease) due to CO_2 uptake from the atmosphere (see Box 7.3).

The ocean's capacity to buffer increasing atmospheric CO_2 will decline in the future as ocean surface pCO_2 increases (Figure 7.11a). This anticipated change is certain, with potentially severe consequences.

Increased carbon storage in the deep ocean leads to the dissolution of calcareous sediments below their saturation depth (Broecker and Takahashi, 1978; Feely et al., 2004). The feedback of $CaCO_3$ sediment dissolution to atmospheric pCO_2 increase is negative and quantitatively significant on a 1 to 100 kyr time scale, where $CaCO_3$ dissolution will account for a 60 to 70% absorption of the anthropogenic CO_2 emissions, while the ocean water column will account for 22 to 33% on a time scale of 0.1 to 1 kyr. In addition, the remaining 7 to 8% may be compensated by long-term terrestrial weathering cycles involving silicate carbonates (Archer et al., 1998). Due to the slow $CaCO_3$ buffering mechanism (and the slow silicate weathering), atmospheric pCO_2 will approach a new equilibrium asymptotically only after several tens of thousands of years (Archer, 2005; Figure 7.12).

Elevated ambient CO_2 levels appear to also influence the production rate of POC by marine calcifying planktonic organisms (e.g., Zondervan et al., 2001). This increased carbon fixation under higher CO_2 levels was also observed for three diatom (siliceous phytoplankton) species (Riebesell et al., 1993). It is critical to know whether these increased carbon fixation rates translate into increased export production rates (i.e., removal of carbon to greater depths). Studies of the nutrient to carbon ratio in marine phytoplankton have not yet shown any significant changes related to CO_2 concentration of the nutrient utilisation efficiency (expressed through the 'Redfield ratio' – carbon:nitrogen:phosphorus:silicon) in organic tissue (Burkhardt et al., 1999).

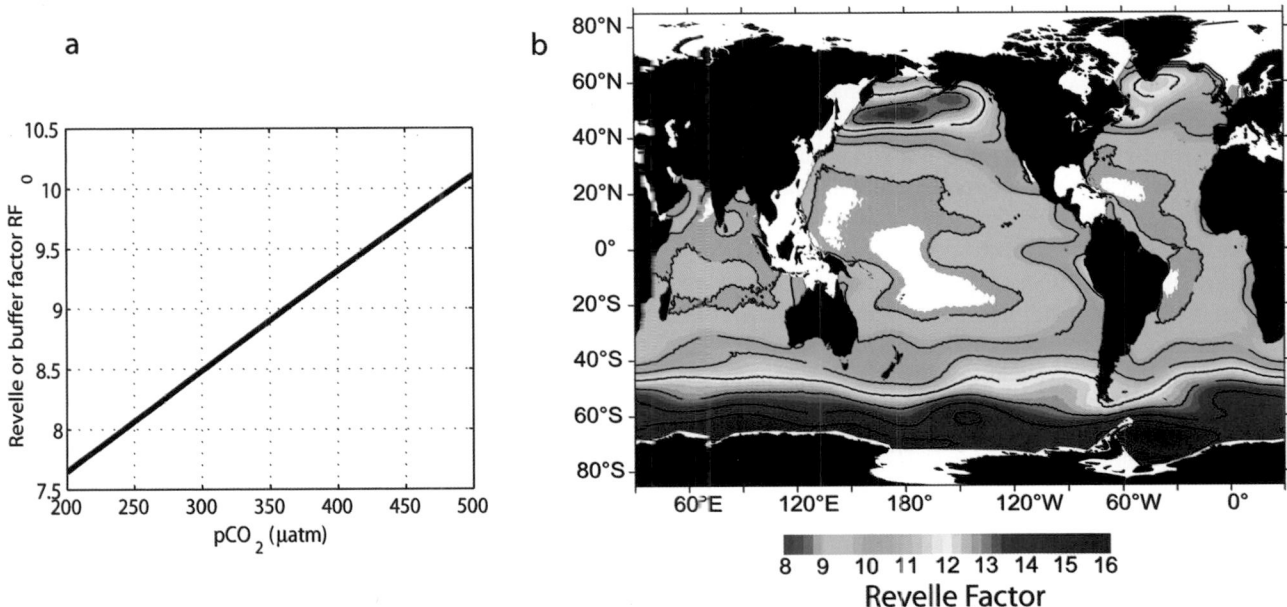

Figure 7.11. *(a) The Revelle factor (or buffer factor) as a function of CO_2 partial pressure (for temperature 25°C, salinity 35 psu, and total alkalinity 2,300 µmol kg⁻¹) (Zeebe and Wolf-Gladrow, 2001, page 73; reprinted with permission, copyright 2001 Elsevier). (b) The geographical distribution of the buffer factor in ocean surface waters in 1994 (Sabine et al., 2004a; reprinted with permission, copyright 2004 American Association for the Advancement of Science). High values indicate a low buffer capacity of the surface waters.*

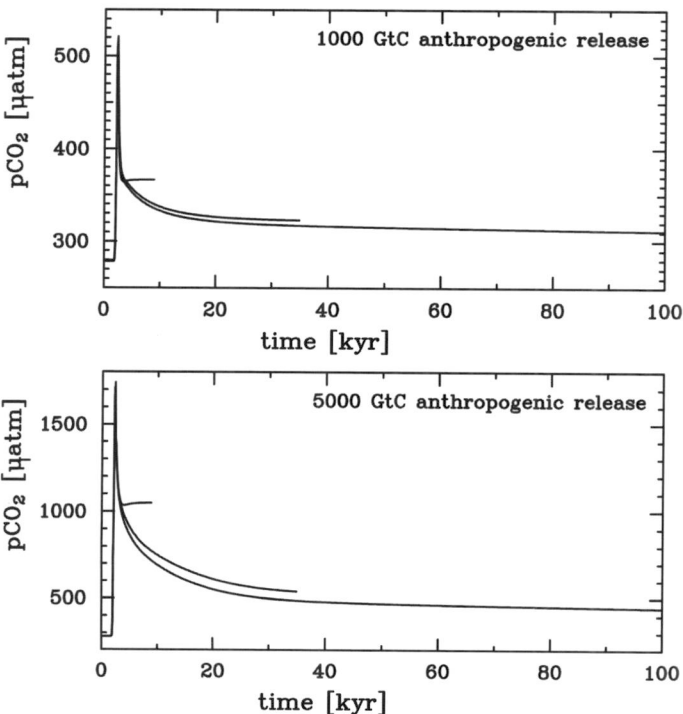

Figure 7.12. *Model projections of the neutralization of anthropogenic CO_2 for an ocean-only model, a model including dissolution of $CaCO_3$ sediment and a model including weathering of silicate rocks, (top) for a total of 1,000 GtC of anthropogenic CO_2 emissions and (bottom) for a total of 5,000 GtC of anthropogenic CO_2. Note that the y-axis is different for the two diagrams. Without $CaCO_3$ dissolution from the sea-floor, the buffering of anthropogenic CO_2 is limited. Even after 100 kyr, the remaining pCO_2 is substantially higher than the pre-industrial value. Source: Archer (2005).*

7.3.4.3 Carbon Cycle Feedbacks to Changes in Physical Forcing

A more sluggish ocean circulation and increased density stratification, both expected in a warmer climate, would slow down the vertical transport of carbon, alkalinity and nutrients, and the replenishment of the ocean surface with water that has not yet been in contact with anthropogenic CO_2. This narrowing of the 'bottleneck' for anthropogenic CO_2 invasion into the ocean would provide a significant positive feedback to atmospheric greenhouse gas concentrations (Bolin and Eriksson, 1959; see also the carbon cycle climate model simulations by Cox et al., 2000; Friedlingstein et al., 2001, 2006). As long as the vertical transfer rates for marine biogenic particles remain unchanged, in a more sluggish ocean the biological carbon pump will be more efficient (Boyle, 1988; Heinze et al., 1991), thus inducing a negative feedback, which is expected to be smaller than the physical transport feedback (Broecker, 1991; Maier-Reimer et al., 1996; Plattner et al., 2001; see Figure 7.10). However, a modelling study by Bopp et al. (2005) predicts a decrease in vertical particle transfer and hence shallower depths of re-mineralization of particulate organic carbon resulting in a positive CO_2 feedback. Further changes in plankton community structure including the role of N_2-fixing organisms can feed

back to the carbon cycle (Sarmiento et al., 2004; Mahaffey et al., 2005). Changes in ocean circulation can affect the regional circulation of shelf and coastal seas, leading either to increased export of nutrients plus carbon from the shallow seas into the open ocean or to increased upwelling of nutrients plus carbon onto the shelf and towards coastal areas (Walsh, 1991; Smith and Hollibaugh, 1993; Chen et al., 2003; Borges, 2005). A reduction in sea ice cover may increase the uptake area for anthropogenic CO_2 and act as a minor negative carbon feedback (ACIA, 2005). The physical 'bottleneck' feedback dominates over biological feedbacks induced by circulation change, resulting in an anticipated overall positive feedback to climate change. Both feedbacks depend on details of the future ocean circulation and model projections show a large range.

The solubility of CO_2 gas in seawater and the two dissociation constants of carbonic acid in seawater depend on temperature and salinity (Weiss, 1974; Millero et al., 2002). A 1°C increase in sea surface temperature produces an increase in pCO_2 of 6.9 to 10.2 ppm after 100 to 1,000 years (Heinze et al., 2003; see also Broecker and Peng, 1986; Plattner et al., 2001). Warming may increase the biological uptake rate of nutrients and carbon from surface waters, but the net effect on export and DIC is uncertain. Laws et al. (2000) proposed that export efficiency increases with net photosynthesis at low temperatures, which implies a positive feedback to warming. In addition, DOC may be degraded more quickly at higher temperatures.

7.3.4.4 Carbon Cycle Feedbacks Induced by Nutrient Cycling and Land Ocean Coupling

Rivers deliver carbon (DIC, DOC) and nutrients to the ocean. Rising CO_2 levels in the atmosphere and land use may lead to increased chemical and physical weathering, resulting in increased carbon and alkalinity loads in rivers (Clair et al., 1999; Hejzlar et al., 2003; Raymond and Cole, 2003; Freeman et al., 2004). Depending on the lithology and soil composition of the catchment areas, increased levels of alkalinity, DIC or DOC can lead to local positive or negative feedbacks. Mobilisation of silicate carbonates from soils and transfer to the ocean would lead to a negative feedback to atmospheric CO_2 on long time scales (Dupre et al., 2003). Variations in nutrient supply can lead to species shifts and to deviations from the large-scale average Redfield ratios mainly in coastal waters, but also in the open ocean (Pahlow and Riebesell, 2000). Nutrient supply to the ocean has been changed through increased nitrate release from land due to fertilizer use as well as nitrogen deposition from the atmosphere in highly polluted areas (De Leeuw et al., 2001; Green et al., 2004).

Dust deposition to the ocean provides an important source of micronutrients (iron, zinc and others, e.g., Frew et al., 2001; Boyd et al., 2004) and ballast material to the ocean. Areas where iron is not supplied by aeolian dust transport in sufficient amounts tend to be iron-limited. A warmer climate may result on the average in a decrease of dust mobilisation and transport (Werner et al., 2002; Mahowald and Luo, 2003) although increased dust loads may result as well due to changes in land use (Tegen et al., 2004) and in vegetation cover (Woodward et al., 2005). A decrease in dust loads could result in a net positive feedback, further increasing CO_2 through a weakening of marine biological production and export of aggregates due to clay ballast (Haake and Ittekkot, 1990; Ittekkot, 1993). Changes in plankton species composition and regional shifts of high production zones due to a changing climate could lead to a series of further feedbacks. Light absorption due to changes in bio-optical heating may change and induce a respective temperature change in ocean surface water (Sathyendranath et al., 1991; Wetzel et al., 2006). An increase in blooms involving calcifying organisms as indicated for the high northern latitudes (Broerse et al., 2003; Smyth et al., 2004) can temporarily increase surface ocean albedo, though the effect on the radiation budget is small (Tyrell et al., 1999).

7.3.4.5 Summary of Marine Carbon Cycle Climate Couplings

Couplings between the marine carbon cycle and climate are summarised in Table 7.3 and below.

7.3.4.5.1 Robust findings

- A potential slowing down of the ocean circulation and the decrease of seawater buffering with rising CO_2 concentration will suppress oceanic uptake of anthropogenic CO_2.
- Ocean CO_2 uptake has lowered the average ocean pH (increased acidity) by approximately 0.1 since 1750. Ocean acidification will continue and is directly and inescapably coupled to the uptake of anthropogenic CO_2 by the ocean.
- Inorganic chemical buffering and dissolution of marine $CaCO_3$ sediments are the main oceanic processes for neutralizing anthropogenic CO_2. These processes cannot prevent a temporary buildup of a large atmospheric CO_2 pool because of the slow large-scale overturning circulation.

7.3.4.5.2 Key uncertainties

- Future changes in ocean circulation and density stratification are still highly uncertain. Both the physical uptake of CO_2 by the ocean and changes in the biological cycling of carbon depend on these factors.
- The overall reaction of marine biological carbon cycling (including processes such as nutrient cycling as well as ecosystem changes including the role of bacteria and viruses) to a warm and high-CO_2 world is not yet well understood. Several small feedback mechanisms may add up to a significant one.
- The response of marine biota to ocean acidification is not yet clear, both for the physiology of individual organisms and for ecosystem functioning as a whole. Potential impacts are expected especially for organisms that build $CaCO_3$ shell material ('bio-calcification'). Extinction thresholds will likely be crossed for some organisms in some regions in the coming century.

7.3.5 Coupling Between the Carbon Cycle and Climate

7.3.5.1 Introduction

Atmospheric CO_2 is increasing at only about half the rate implied by fossil fuel plus land use emissions, with the remainder being taken up by the ocean, and vegetation and soil on land. Therefore, the land and ocean carbon cycles are currently helping to mitigate CO_2-induced climate change. However, these carbon cycle processes are also sensitive to climate. The glacial-interglacial cycles are an example of tight coupling between climate and the carbon cycle over long time scales, but there is also clear evidence of the carbon cycle responding to short-term climatic anomalies such as the El Niño-Southern Oscillation (ENSO) and Arctic Oscillation (Rayner et al., 1999; Bousquet et al., 2000; C. Jones et al., 2001; Lintner, 2002; Russell and Wallace, 2004) and the climate perturbation arising from the Mt. Pinatubo volcanic eruption (Jones and Cox, 2001a; Lucht et al., 2002; Angert et al., 2004).

Previous IPCC reports have used simplified or 'reduced-form' models to estimate the impact of climate change on the carbon cycle. However, detailed climate projections carried out with Atmosphere-Ocean General Circulation Models (AOGCMs) have typically used a prescribed CO_2 concentration scenario, neglecting two-way coupling between climate and the carbon cycle. This section discusses the first generation of coupled climate-carbon cycle AOGCM simulations, using the results to highlight a number of critical issues in the interaction between climate change and the carbon cycle.

7.3.5.2 Coupled Climate-Carbon Cycle Projections

The TAR reported two initial climate projections using AOGCMs with interactive carbon cycles. Both indicated positive feedback due largely to the impacts of climate warming on land carbon storage (Cox et al., 2000; Friedlingstein et al., 2001), but the magnitude of the feedback varied markedly between the models (Friedlingstein et al., 2003). Since the TAR a number of other climate modelling groups have completed climate-carbon cycle projections (Brovkin et al., 2004; Thompson et al., 2004; N. Zeng et al., 2004, Fung et al., 2005; Kawamiya et al., 2005; Matthews et al., 2005; Sitch et al., 2005) as part of C^4MIP. The 11 models involved in C^4MIP differ in the complexity of their components (Friedlingstein et al., 2006), including both Earth System Models of Intermediate Complexity and AOGCMs.

The models were forced by historical and Special Report on Emission Scenarios (SRES; IPCC, 2000) A2 anthropogenic CO_2 emissions for the 1850 to 2100 time period. Each modelling group carried out at least two simulations: one 'coupled' in which climate change affects the carbon cycle, and one 'uncoupled' in which atmospheric CO_2 increases do not influence climate (so that the carbon cycle experiences no CO_2-induced climate change). A comparison of the runs defines the climate-carbon cycle feedback, quantified by the feedback factor:

Table 7.3. Couplings between climate change (increased atmospheric pCO_2, warming) and ocean carbon cycle processes. The response in terms of direct radiative forcing is considered (furthering or counteracting uptake of anthropogenic CO_2 from the atmosphere). The two quantitatively most important marine processes for neutralization of anthropogenic CO_2 work on long time scales only and are virtually certain to be in effect.

Marine Carbon Cycle Process	Major Forcing Factors	Response + = positive feedback − = negative feedback and Quantitative Potential	Start	Re-equilibration Time Scale (kyr)	Likelihood	Comment
Biological export production of organic carbon and changes in organic carbon cycling	Warming, ocean circulation, nutrient supply, radiation, atmospheric CO_2, pH value	(Sum of effects not clear) +/− medium	immediate	0.001–10	Likely	Complex feedback chain, reactions can be fast for surface ocean, nutrient supply from land works on longer time scales, patterns of biodiversity and ecosystem functioning may be affected
Biological export production of calcium carbonate	Warming, atmospheric CO_2, pH value	(Sum of effects not clear) +/− small	immediate	0.001–1	Likely	Complex feedback chain, extinction of species likely, patterns of biodiversity and ecosystem functioning may be affected
Seawater buffering	Atmospheric CO_2, ocean circulation	− high	immediate	5–10	Virtually certain	System response, leads to ocean acidification
Changes in inorganic carbon chemistry (solubility, dissociation, buffer factor)	Warming, atmospheric CO_2, ocean circulation	+ medium	immediate	5–10	Virtually certain	Positive feedback dependent on 'bottleneck' ocean mixing
Dissolution of calcium carbonate sediments	pH value, ocean circulation	− high	immediate	40	Virtually certain	Patterns of biodiversity and ecosystem functioning in deep sea may be affected
Weathering of silicate carbonates	Atmospheric CO_2, warming	− medium	immediate	100	Likely	Very long-term negative feedback

Table 7.4. *Impact of carbon cycle feedbacks in the C⁴MIP models. Column 2 shows the impact of climate change on the CO₂ concentration by 2100, and column 3 shows the related amplification of the atmospheric CO₂ increase (i.e., the climate-carbon cycle feedback factor). Columns 4 to 8 list effective sensitivity parameters of the models: transient sensitivity of mean global temperature to CO₂, and the sensitivities of land and ocean carbon storage to CO₂ and climate (Friedlingstein et al., 2006). These parameters were calculated by comparison of the coupled and uncoupled runs over the entire period of the simulations (typically 1860 to 2100). Model details are given in Friedlingstein et al. (2006).*

Model[a]	Impact of Climate Change on the CO₂ Concentration by 2100 (ppm)	Climate-Carbon Feedback Factor	Transient Climate Sensitivity to Doubling CO₂ (°C)	Land Carbon Storage Sensitivity to CO₂ (GtC ppm⁻¹)	Ocean Carbon Storage Sensitivity to CO₂ (GtC ppm⁻¹)	Land Carbon Storage Sensitivity to Climate (GtC °C⁻¹)	Ocean Carbon Storage Sensitivity to Climate (GtC °C⁻¹)
A. HadCM3LC	224	1.44	2.3	1.3	0.9	−175	−24
B. IPSL-CM2C	74	1.18	2.3	1.6	1.6	−97	−30
C. MPI-M	83	1.18	2.6	1.4	1.1	−64	−22
D. LLNL	51	1.13	2.5	2.5	0.9	−81	−14
E. NCAR CSM-1	20	1.04	1.2	1.1	0.9	−24	−17
F. FRCGC	128	1.26	2.3	1.4	1.2	−111	−47
G. Uvic-2.7	129	1.25	2.3	1.2	1.1	−97	−43
H. UMD	98	1.17	2.0	0.2	1.5	−36	−60
I. BERN-CC	65	1.15	1.5	1.6	1.3	−104	−38
J. CLIMBER2-LPJ	59	1.11	1.9	1.2	0.9	−64	−22
K. IPSL-CM4-LOOP	32	1.07	2.7	1.2	1.1	−19	−17
Mean	87	1.18	2.1	1.4	1.1	−79	−30
Standard Deviation	±57	±0.11	±0.4	±0.5	±0.3	±45	±15

Notes:

[a] HadCM3LC: Hadley Centre coupled climate-carbon cycle general circulation model; IPSL-CM2C: Institut Pierre-Simon Laplace; MPI-M: Max Planck Institute for Meteorology; LLNL: Lawrence Livermore National Laboratory; NCAR CSM-1: NCAR Climate System Model version 1; FRCGC: Frontier Research Center for Global Change; Uvic-2.7: University of Victoria Earth System Climate Model; UMD: University of Maryland; BERN-CC: Bern Carbon Cycle Model; CLIMBER2-LPJ: Climate Biosphere Model 2 - Lund Potsdam Jena Terrestrial Carbon Model; IPSL-CM4-LOOP: Institute Pierre-Simon Laplace.

$F = \Delta C_A{}^c / \Delta C_A{}^u$, where $\Delta C_A{}^c$ is the change in CO₂ in the coupled run, and $\Delta C_A{}^u$ is the change in CO₂ in the uncoupled run. All of the eleven C⁴MIP models produce a positive climate-carbon cycle feedback, but with feedback factors varying from 1.04 (Model E) to 1.44 (Model A). This translates into an additional CO₂ concentration of between 20 and 224 ppm by 2100, with a mean of 87 ppm (Table 7.4).

All C⁴MIP models predict that an increasing fraction of total anthropogenic CO₂ emissions will remain airborne through the 21st century. Figure 7.13 shows the simulated partitioning of anthropogenic CO₂ for the entire simulation period to 2100 from each of the coupled models, and compares this with the partitioning simulated by the same models over the historical period to 1999. The dashed box shows observational constraints on the historical CO₂ partitioning, based on estimates of changes in ocean carbon storage (Sabine et al., 2004a) and total anthropogenic CO₂ emissions. The area of this box is largely due to uncertainties in the net land use emissions. The majority of the models sit within or very close to the historical constraints, but they differ in the magnitude of the changes projected for the 21st century. However, all models produce an increase in the fraction of total emissions that remain in the atmosphere, and most also indicate a decline in the fraction of emissions absorbed by the ocean (9 out of 11 models) and the land (10 out of 11 models).

In the case of the oceanic uptake, this is largely a consequence of the reduced buffering capacity as CO₂ increases, and therefore also occurs in the uncoupled C⁴MIP models.

7.3.5.3 Sensitivity Analysis

The coupled and uncoupled model experiments can be used to separate the effects of climate change and CO₂ increase on land and ocean carbon storage (Friedlingstein et al., 2003). Table 7.4 also shows the linear sensitivity parameters diagnosed from each of the C⁴MIP models (Friedlingstein et al., 2006).

7.3.5.3.1 Increase in ocean carbon uptake with increasing atmospheric carbon dioxide

The ocean takes up CO₂ at a rate that depends on the difference between pCO₂ in the atmosphere and in the surface ocean. Model estimates of uptake differ primarily because of differences in the rate at which carbon is exported from the surface ocean to depth by the large-scale circulation (Doney et al., 2004; Section 7.3.4.1; Box 7.3) and the biological pump (Sarmiento et al., 2004). Ocean carbon cycle model intercomparisons have shown that the simulated circulation in the Southern Ocean can have a large impact on the efficiency with which CO₂, and other anthropogenic tracers such as CFCs,

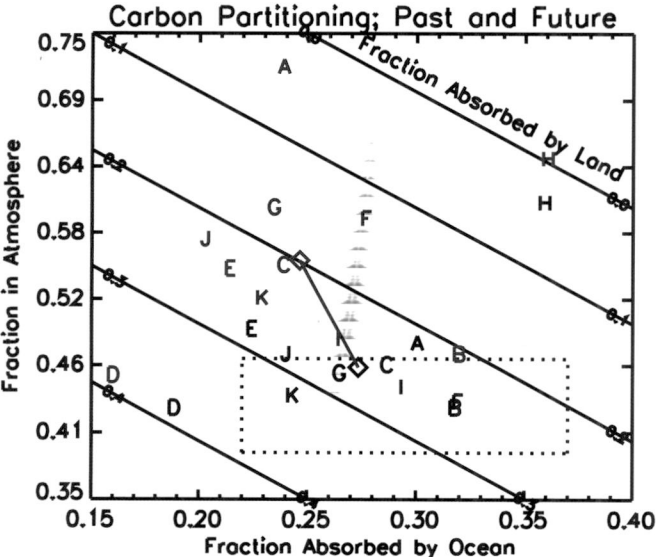

Figure 7.13. *Predicted increase in the fraction of total emissions that add to atmospheric CO_2. Changes in the mean partitioning of emissions as simulated by the C^4MIP models up to 2000 (black symbols) and for the entire simulation period to 2100 (red symbols). The letters represent the models as given in Table 7.4. The box shown by the dotted line is a constraint on the historical carbon balance based on records of atmospheric CO_2 increase, and estimates of total emissions (fossil fuel plus land use emissions) and the oceanic uptake of anthropogenic CO_2 (Sabine et al., 2004a). The black and red diamonds show the model-mean carbon partitioning for the historical period and the entire simulation period, respectively. The red line shows the mean tendency towards an increasing airborne fraction through the 21st century, which is common to all models.*

are drawn down (Orr et al., 2001; Dutay et al., 2002). The C^4MIP models show ocean carbon storage increases ranging from 0.9 to 1.6 GtC ppm^{-1}, which is equivalent to ocean uptake increasing at between 42 and 75% of the rate of atmospheric CO_2 increase. Basic ocean carbonate chemistry suggests that the ocean-borne fraction of emissions will fall in the future, even in the absence of climate change, because of an increasing ocean buffer factor (Section 7.3.4.2).

7.3.5.3.2 Increase in land carbon uptake with increasing atmospheric carbon dioxide

In the absence of land use change and forest fires, land carbon storage depends on the balance between the input of carbon as NPP, and the loss of carbon as heterotrophic (soil) respiration (Section 7.3.3). There is an ongoing debate concerning the importance of CO_2 fertilization at the patch scale where other constraints such as N limitation may dominate; recent surveys indicate a wide range of possible responses to a CO_2 increase of around 50%, with average increases of 12 to 23% (Norby et al., 2005; see Section 7.3.3.1).

The C^4MIP models show increases in global NPP of between 6 and 33% when CO_2 increases over the same range. These figures are not directly comparable: some C^4MIP models include vegetation dynamics, which are likely to increase the vegetation cover as well as the NPP per unit of vegetation area, and therefore lead to higher overall sensitivity of global NPP to CO_2. The FACE experiments also typically involve an

instantaneous increase in CO_2. However, most C^4MIP models are within the range of the CO_2 sensitivities measured.

The overall response of land carbon storage to CO_2 is given by the fifth column of Table 7.4. The C^4MIP models show time-mean land carbon storage increases ranging from 0.2 to 2.5 GtC ppm^{-1}, with all but two models between 1.1 and 1.6 GtC ppm^{-1}. This response is driven by the CO_2 fertilization of NPP in each model, with a counteracting tendency for the mean soil carbon turnover rate (i.e., the heterotrophic respiration by unit soil carbon) to increase even in the absence of climate change. This somewhat surprising effect of CO_2 is seen to varying degrees in all C^4MIP models. It appears to arise because CO_2 fertilization of NPP acts particularly to increase vegetation carbon, and therefore litter fall and soil carbon, in productive tropical regions that have high intrinsic decomposition rates. This increases the average turnover rate of the global soil carbon pool even though local turnover rates are unchanged. In some models (e.g., model C) this acts to offset a significant fraction of the land carbon increase arising from CO_2 fertilization. Models with large responses of ocean or land carbon storage to CO_2 tend to have weaker climate-carbon cycle feedbacks because a significant fraction of any carbon released through climate change effects is reabsorbed through direct CO_2 effects (Thompson et al., 2004).

7.3.5.3.3 Transient climate sensitivity to carbon dioxide

The strength of the climate-carbon cycle feedback loop depends on both the sensitivity of the carbon cycle to climate, and the sensitivity of climate to CO_2. The equilibrium climate sensitivity to a doubling of atmospheric CO_2 concentration remains a critical uncertainty in projections of future climate change, but also has a significant bearing on future CO_2 concentrations, with higher climate sensitivities leading to larger climate-carbon cycle feedbacks (Andreae et al., 2005). The fourth column of Table 7.4 shows the transient global climate sensitivity (i.e., the global climate warming that results when the transient simulation passes doubled atmospheric CO_2) for each of the C^4MIP models. All but two models (models E and I) have transient climate sensitivities in the range 1.9°C to 2.7°C. However, differences in carbon cycle responses are likely to occur because of potentially large differences in regional climate change, especially where this affects water availability on the land.

7.3.5.3.4 Dependence of ocean carbon uptake on climate.

Climate change can reduce ocean uptake through reductions in CO_2 solubility, suppression of vertical mixing by thermal stratification and decreases in surface salinity. On longer time scales (>70 years) the ocean carbon sink may also be affected by climate-driven changes in large-scale circulation (e.g., a slowing down of the thermohaline circulation). The last column of Table 7.4 shows the sensitivity of ocean carbon storage to climate change as diagnosed from the C^4MIP models. All models indicate a reduction in the ocean carbon sink by climate change of between –14 and –60 GtC °C^{-1}, implying a positive climate-CO_2 feedback.

7.3.5.3.5   Dependence of land carbon storage on climate.

The major land-atmosphere fluxes of CO_2 are strongly climate dependent. Heterotrophic respiration and NPP are both very sensitive to water availability and ambient temperatures. Changes in water availability depend critically on uncertain regional aspects of climate change projections and are therefore likely to remain a dominant source of uncertainty (see Chapter 11). The overall sensitivity of land carbon storage to climate (Table 7.4, seventh column) is negative in all models, implying a positive climate-CO_2 feedback, but the range is large: –19 to –175 GtC $°C^{-1}$. These values are determined by the combined effects of climate change on NPP and the soil carbon turnover (or decomposition) rate, as shown in Table 7.5.

The C4MIP models utilise different representations of soil carbon turnover, ranging from single-pool models (model A) to nine-pool models (model E). However, most soil models assume a similar acceleration of decay with temperature, approximately equivalent to a doubling of the specific respiration rate for every 10°C warming. This temperature sensitivity is broadly consistent with a long history of lab and field measurements of soil efflux (Raich and Schlesinger, 1992), although there is an ongoing difficulty in separating root and soil respiration. Note, however, that the expected dependence on temperature was not found at the whole ecosystem level for decadal time scales, in forest soils (Giardina and Ryan, 2000; Melillo et al., 2002), grasslands (Luo et al., 2001) or boreal forests (Dunn et al., 2007). These apparent discrepancies may reflect the rapid depletion of labile pools of organic matter, with strong temperature responses likely so long as litter inputs are maintained (Knorr

et al., 2005). Nevertheless, the temperature sensitivity of the slow carbon pools is still poorly known.

Table 7.5 shows that all C4MIP models simulate an overall increase in soil carbon turnover rate as the climate warms, ranging from 2 to 10% per °C. The use of a single soil carbon pool in the Hadley model (A) cannot completely account for the relatively large sensitivity of soil respiration to temperature in this model (Jones et al., 2005), as evidenced by the lower effective sensitivity diagnosed from the UVic model (model G), which uses the same soil-vegetation component. It seems more likely that differences in soil moisture simulations are playing the key part in determining the effective sensitivity of soil turnover rate to climate. Table 7.5 also shows the effective sensitivities of NPP to climate, ranging from a significant reduction of 6% per °C to smaller climate-change driven increases of 2% per °C under climate change. This variation may reflect different time scales for boreal forest response to warming (leading to a positive impact on global NPP), as well as different regional patterns of climate change (Fung et al., 2005). The models with the largest negative responses of NPP to climate (models A, B and C) also show the tendency for tropical regions to dry under climate change, in some cases significantly (Cox et al., 2004).

7.3.5.4   Summary of Coupling Between the Carbon Cycle and Climate

7.3.5.4.1   Robust findings

Results from the coupled climate-carbon cycle models participating in the C4MIP project support the following statements:

Table 7.5. *Effective sensitivities of land processes in the C4MIP models: percent change of vegetation NPP to a doubling of atmospheric CO_2 concentration (Column 2), and sensitivities of vegetation NPP and specific heterotrophic soil respiration to a 1°C global temperature increase (Columns 3 and 4).*

Model[a]	Sensitivity of Vegetation NPP to CO_2: % change for a CO_2 doubling	Sensitivity of Vegetation NPP to Climate: % change for a 1°C increase	Sensitivity of Specific Heterotrophic Respiration Rate to Climate: % change for a 1°C increase
A. HadCM3LC	57	–5.8	10.2
B. IPSL-CM2C	50	–4.5	2.3
C. MPI-M	76	–4.0	2.8
D. LLNL	73	–0.4	7.0
E. NCAR CSM-1	34	0.8	6.2
F. FRCGC	21	1.2	7.2
G. UVic-2.7	47	–2.3	6.5
H. UMC	12	–1.6	4.8
I. BERN-CC	46	1.2	8.7
J. CLIMBER2-LPJ	44	1.9	9.4
K. IPSL-CM4-LOOP	64	–0.3	2.9
Mean	48	–1.3	6.2
Std Dev	±20	±2.6	±2.7

Notes:

[a] See Table 7.4 for model descriptions.

- All C⁴MIP models project an increase in the airborne fraction of total anthropogenic CO_2 emissions through the 21st century.
- The CO_2 increase alone will lead to continued uptake by the land and the ocean, although the efficiency of this uptake will decrease through the carbonate buffering mechanism in the ocean, and through saturation of the land carbon sink.
- Climate change alone will tend to suppress both land and ocean carbon uptake, increasing the fraction of anthropogenic CO_2 emissions that remain airborne and producing a positive feedback to climate change. The magnitude of this feedback varies among the C⁴MIP models, ranging from a 4 to 44% increase in the rate of increase of CO_2, with a mean (± standard deviation) of 18 ± 11%.

7.3.5.4.2   Key uncertainties

The C⁴MIP models also exhibit uncertainties in the evolution of atmospheric CO_2 for a given anthropogenic emissions scenario. Figure 7.14 shows how uncertainties in the sensitivities of ocean and land carbon processes contribute to uncertainties in the fraction of emissions that remain in the atmosphere. The confidence limits were produced by spanning the range of sensitivities diagnosed from the 11 C⁴MIP models

(Tables 7.4 and 7.5). In the absence of climate change effects (lowest three bars), models simulate increased uptake by ocean and land (primarily as a result of CO_2 enhancement of NPP), with a slight offset of the land uptake by enhancement of the specific heterotrophic respiration rate (see Section 7.3.5.3.2). However, there is a wide range of response to CO_2, even in the absence of climate change effects on the carbon cycle. Climate change increases the fraction of emissions that remain airborne by suppressing ocean uptake, enhancing soil respiration and reducing plant NPP. The sensitivity of NPP to climate change is especially uncertain because it depends on changing soil water availability, which varies significantly between General Circulation Models (GCMs), with some models suggesting major drying and reduced productivity in tropical ecosystems (Cox et al., 2004). The transient climate sensitivity to CO_2 is also a major contributor to the overall uncertainty in the climate-carbon cycle feedback (top bar).

Other potentially important climate-carbon cycle interactions were not included in these first generation C⁴MIP experiments. The ocean ecosystem models used in C⁴MIP are at an early stage of development. These models have simple representations of the biological fluxes, which include the fundamental response to changes in internal nutrients, temperature and light availability, but for most models do not include the more complex responses

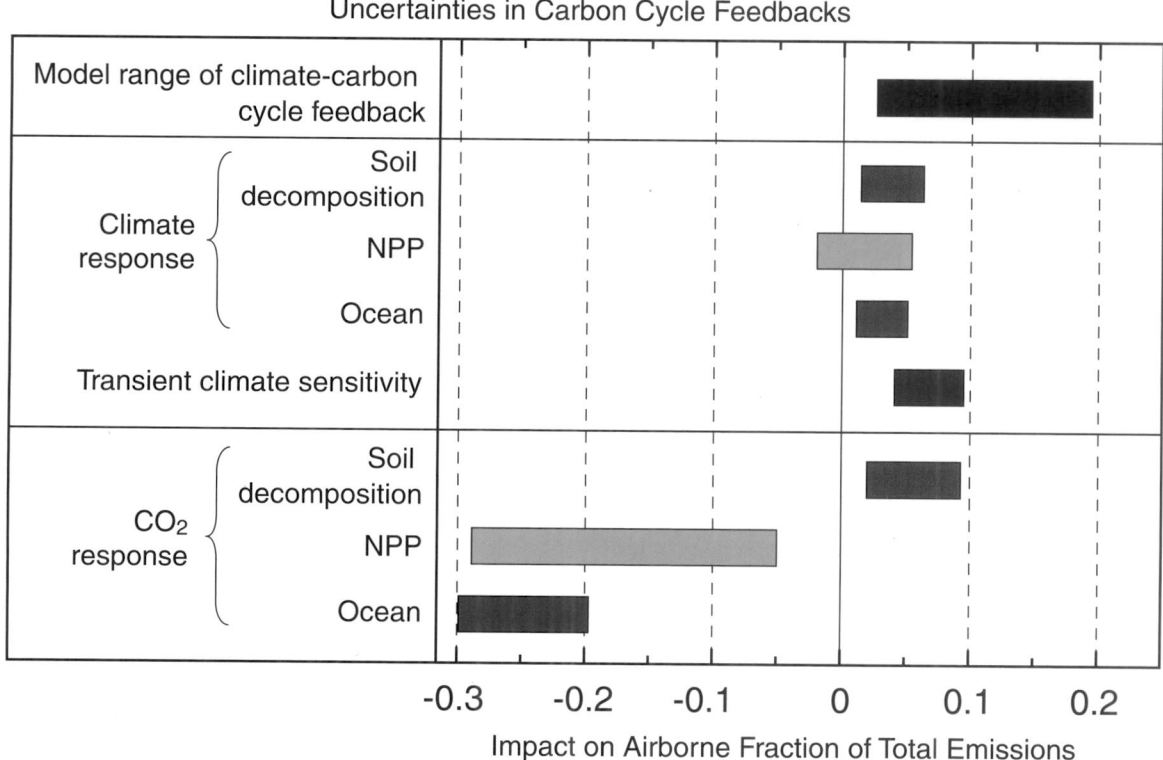

Figure 7.14. *Uncertainties in carbon cycle feedbacks estimated from analysis of the results from the C⁴MIP models. Each effect is given in terms of its impact on the mean airborne fraction over the simulation period (typically 1860 to 2100), with bars showing the uncertainty range based on the ranges of effective sensitivity parameters given in Tables 7.4 and 7.5. The lower three bars are the direct response to increasing atmospheric CO_2 (see Section 7.3.5 for details), the middle four bars show the impacts of climate change on the carbon cycle, and the top black bar shows the range of climate-carbon cycle feedbacks given by the C⁴MIP models.*

to changes in ecosystem structure. Changes in ecosystem structure can occur when specific organisms respond to surface warming, acidification, changes in nutrient ratios resulting from changes in external sources of nutrients (atmosphere or rivers) and changes in upper trophic levels (fisheries). Shifts in the structure of ocean ecosystems can influence the rate of CO_2 uptake by the ocean (Bopp et al., 2005).

The first-generation C4MIP models also currently exclude, by design, the effects of forest fires and prior land use change. Forest regrowth may account for a large part of the land carbon sink in some regions (e.g., Pacala et al., 2001; Schimel et al., 2001; Hurtt et al., 2002; Sitch et al., 2005), while combustion of vegetation and soil organic matter may be responsible for a significant fraction of the interannual variability in CO_2 (Cochrane, 2003; Nepstad et al., 2004; Kasischke et al., 2005; Randerson et al., 2005). Other important processes were excluded in part because modelling these processes is even less straightforward. Among these are N cycling on the land (which could enhance or suppress CO_2 uptake by plants) and the impacts of increasing ozone concentrations on plants (which could suppress CO_2 uptake).

7.4 Reactive Gases and the Climate System

The atmospheric concentration of many reactive gases has increased substantially during the industrial era as a result of human activities. Some of these compounds (CH_4, N_2O, halocarbons, ozone, etc.) interact with longwave (infrared) solar radiation and, as a result, contribute to 'greenhouse warming'. Ozone also efficiently absorbs shortwave (ultraviolet and visible) solar energy, so that it protects the biosphere (including humans) from harmful radiation and plays a key role in the energy budget of the middle atmosphere. Many atmospheric chemical species are emitted at the surface as a result of biological processes (soils, vegetation, oceans) or anthropogenic activities (fossil fuel consumption, land use changes) before being photochemically destroyed in the atmosphere and converted to compounds that are eventually removed by wet and dry deposition. The oxidizing power (or capacity) of the atmosphere is determined primarily by the atmospheric concentration of the OH radical (daytime) and to a lesser extent the concentrations of the nitrate radical (NO_3; nighttime), ozone and hydrogen peroxide (H_2O_2). The coupling between chemical processes in the atmosphere and the climate system (Figure 7.15) are complex because they involve a large number of physical, chemical and biological processes that are not always very well quantified. An important issue is to determine to what extent predicted climate change could affect air quality (see Box 7.4). The goal of this section is assess recent progress made in the understanding of the two-way interactions between reactive gases and the climate system.

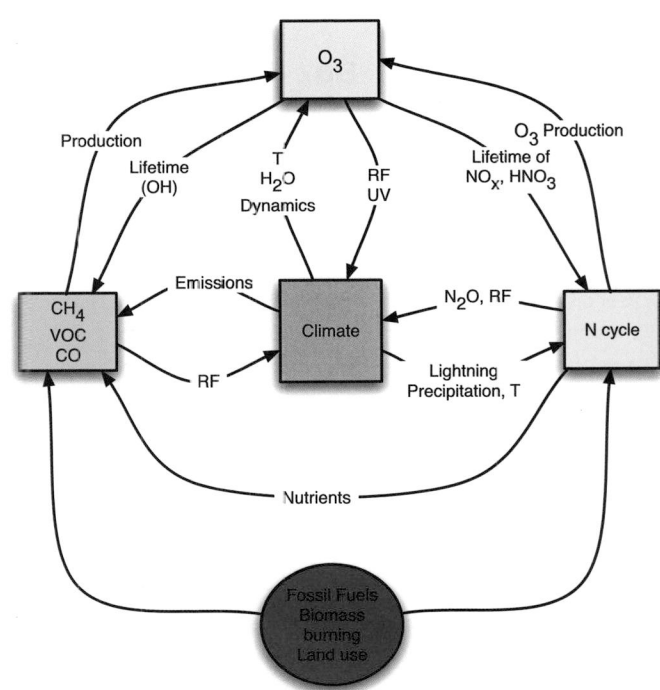

Figure 7.15. *Schematic representation of the multiple interactions between tropospheric chemical processes, biogeochemical cycles and the climate system. RF represents radiative forcing, UV ultraviolet radiation, T temperature and HNO_3 nitric acid.*

7.4.1 Methane

7.4.1.1 Biogeochemistry and Budgets of Methane

Atmospheric CH_4 originates from both non-biogenic and biogenic sources. Non-biogenic CH_4 includes emissions from fossil fuel mining and burning (natural gas, petroleum and coal), biomass burning, waste treatment and geological sources (fossil CH_4 from natural gas seepage in sedimentary basins and geothermal/volcanic CH_4). However, emissions from biogenic sources account for more than 70% of the global total. These sources include wetlands, rice agriculture, livestock, landfills, forests, oceans and termites. Emissions of CH_4 from most of these sources involve ecosystem processes that result from complex sequences of events beginning with primary fermentation of organic macromolecules to acetic acid (CH_3COOH), other carboxylic acids, alcohols, CO_2 and hydrogen (H_2), followed by secondary fermentation of the alcohols and carboxylic acids to acetate, H_2 and CO_2, which are finally converted to CH_4 by the so-called methanogenic Archaea: $CH_3COOH \rightarrow CH_4 + CO_2$ and $CO_2 + 4H_2 \rightarrow CH_4 + 2H_2O$ (Conrad, 1996). Alternatively, CH_4 sources can be divided into anthropogenic and natural. The anthropogenic sources include rice agriculture, livestock, landfills and waste treatment, some biomass burning, and fossil fuel combustion. Natural CH_4 is emitted from sources such as wetlands, oceans, forests, fire, termites and geological sources (Table 7.6).

Box 7.4: Effects of Climate Change on Air Quality

Weather is a key variable affecting air quality. Surface air concentrations of pollutants are highly sensitive to boundary layer ventilation, winds, temperature, humidity and precipitation. Anomalously hot and stagnant conditions in the summer of 1988 were responsible for the highest ozone year on record in the north-eastern USA (Lin et al., 2001). The summer heat wave in Europe in 2003 was associated with exceptionally high ozone (Ordonez et al., 2005). Such high interannual variability of surface ozone correlated with temperature demonstrates the potential air quality implications of climate change over the next century.

A few GCM studies have investigated how air pollution meteorology might respond to future climate change. Rind et al. (2001) found that increased continental ventilation as a result of more vigorous convection should decrease surface concentrations, while Holzer and Boer (2001) found that weaker winds should result in slower dilution of pollution plumes and hence higher concentrations. A focused study by Mickley et al. (2004) for the eastern USA found an increase in the severity and persistence of regional pollution episodes due to the reduced frequency of ventilation by cyclones tracking across Canada. This effect more than offsets

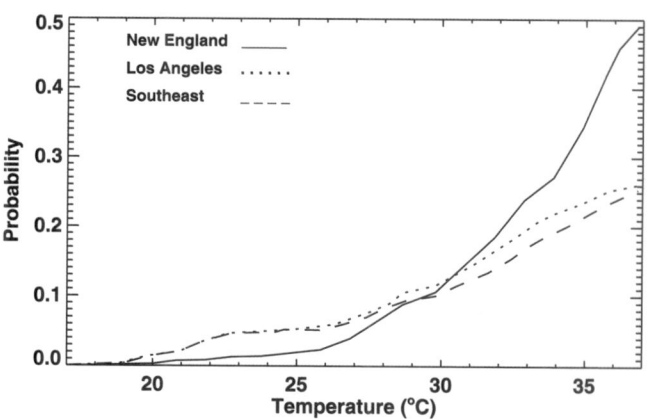

Box 7.4, Figure 1. *Probability that the daily maximum eight-hour average ozone concentration will exceed the US National Ambient Air Quality Standard of 0.08 ppm for a given daily maximum temperature based on 1980 to 1998 data. Values are shown for New England (bounded by 36°N, 44°N, 67.5°W and 87.5°W), the Los Angeles Basin (bounded by 32°N, 40°N, 112.5°W and 122.5°W) and the southeastern USA (bounded by 32°N, 36°N, 72.5°W and 92.5°W). Redrawn from Lin et al. (2001).*

the dilution associated with the small rise in mixing depths. A decrease in cyclone frequency at northern mid-latitudes and a shift to higher latitudes has been noted in observations from the past few decades (McCabe et al., 2001). An urban air quality model study by Jacobson (1999) pointed out that decreasing soil moisture or increasing surface temperature would decrease mixing depths and reduce near-surface pollutant concentrations.

A number of studies in the USA have shown that summer daytime ozone concentrations correlate strongly with temperature (NRC, 1991). This correlation appears to reflect contributions of comparable magnitude from (1) temperature-dependent biogenic VOC emissions, (2) thermal decomposition of peroxyacetylnitrate, which acts as a reservoir for NO_x and (3) association of high temperatures with regional stagnation (Jacob et al., 1993; Sillman and Samson, 1995; Hauglustaine et al., 2005). Empirical relationships between ozone air quality standard exceedances and temperature, as shown in Figure 1, integrate all of these effects and could be used to estimate how future regional changes in temperature would affect ozone air quality. Changes in the global ozone background would also have to be accounted for (Stevenson et al., 2005).

A few GCM studies have examined more specifically the effect of changing climate on regional ozone air quality, assuming constant emissions. Knowlton et al. (2004) use a GCM coupled to a Regional Climate Model (RCM) to investigate the impact of 2050 climate change (compared with 1990) on ozone concentrations in the New York City metropolitan area. They found a significant ozone increase that they translated into a 4.5% increase in ozone-related acute mortality. Langner et al. (2005) use an RCM driven by two different GCMs to examine changes in the Accumulated Ozone concentration above a Threshold of 40 ppb (AOT40) statistic (ozone-hours above 40 ppb) over Europe in 2050 to 2070 relative to the present. They found an increase in southern and central Europe and a decrease in northern Europe that they attributed to different regional trends in cloudiness and precipitation. Dentener et al. (2006) synthesise the results of 10 global model simulations for 2030 driven by future compared with present climate. They find that climate change caused mean decreases in surface ozone of 0.5 to 1 ppb over continents and 1 to 2 ppb over the oceans, although some continental regions such as the Eastern USA experienced slight increases.

There has been less work on the sensitivity of aerosols to meteorological conditions. Regional model simulations by Aw and Kleeman (2003) find that increasing temperatures should increase surface aerosol concentrations due to increased production of aerosol precursors (in particular semi-volatile organic compounds and HNO_3) although this is partly compensated by the increasing vapour pressure of these compounds at higher temperatures. Perturbations of precipitation frequencies and patterns might be expected to have a major impact on aerosol concentrations, but the GCM study by Mickley et al. (2004) for 2000 to 2050 climate change finds little effect in the USA.

The net rate of CH_4 emissions is generally estimated from three approaches: (1) extrapolation from direct flux measurements and observations, (2) process-based modelling (bottom-up approach) and (3) inverse modelling that relies on spatially distributed, temporally continuous observations of concentration, and in some cases isotopic composition in the atmosphere (top-down approach). The top-down method also includes aircraft and satellite observations (Xiao et al., 2004; Frankenberg et al., 2005, 2006). When the bottom-up approach is used to extrapolate the emissions to larger scales, uncertainty results from the inherent large temporal and spatial variations of fluxes and the limited range of observational conditions. The top-down approach helps to overcome the weaknesses in bottom-up methods. However, obstacles to extensive application of the top-down approach include inadequate observations, and insufficient capabilities of the models to account for error amplification in the inversion process and to simulate complex topography and meteorology (Dentener et al., 2003a; Mikaloff Fletcher et al., 2004a, 2004b; Chen and Prinn, 2005, 2006). Measurements of isotopes of CH_4 (^{13}C, ^{14}C, and ^{2}H) provide additional constraints on CH_4 budgets and specific sources, but such data are even more limited (Bergamaschi et al., 2000; Lassey et al., 2000; Mikaloff Fletcher et al., 2004a, 2004b).

Since the TAR, availability of new data from various measurement networks and from national reporting documents has enabled re-estimates of CH_4 source magnitudes and insights into individual source strengths. Total global pre-industrial emissions of CH_4 are estimated to be 200 to 250 $Tg(CH_4)$ yr^{-1} (Chappellaz et al., 1993; Etheridge et al., 1998; Houweling et al., 2000; Ferretti et al., 2005; Valdes et al., 2005). Of this, natural CH_4 sources emitted between 190 and 220 $Tg(CH_4)$ yr^{-1}, and anthropogenic sources (rice agriculture, livestock, biomass burning and waste) accounted for the rest (Houweling et al., 2000; Ruddiman and Thomson, 2001). In contrast, anthropogenic emissions dominate present-day CH_4 budgets, accounting for more than 60% of the total global budget (Table 7.6).

The single largest CH_4 source is natural wetlands. Recent estimates combine bottom-up and top-down fluxes, and global observations of atmospheric CH_4 concentrations in a three-dimensional Atmospheric Transport and Chemical Model (ATCM) simulation (Chen and Prinn, 2005, 2006). In these estimates, southern and tropical regions account for more than 70% of total global wetland emissions. Other top-down studies that include both direct observations and $^{13}C/^{12}C$ ratios of CH_4 suggest greater emissions in tropical regions compared with previously estimates (Mikaloff Fletcher et al., 2004a, 2004b; Xiao et al., 2004; Frankenberg et al., 2006). However, several bottom-up studies indicate fewer emissions from tropical rice agriculture (Li et al., 2002; Yan et al., 2003; Khalil and Shearer 2006). Frankenberg et al. (2005, 2006) and Keppler et al. (2006) suggest that tropical trees emit CH_4 via an unidentified process. The first estimate of this source was 10 to 30% (62–236 $Tg(CH_4)$ yr^{-1}) of the global total, but Kirschbaum et al. (2006) revise this estimate downwards to 10 to 60 $Tg(CH_4)$ yr^{-1}. Representative $^{13}C/^{12}C$ ratios ($\delta^{13}C$ values) of CH_4 emitted from individual sources are included in Table 7.6. Due to isotope fractionation associated with CH_4 production and consumption processes, CH_4 emitted from each source exhibits a measurably different $\delta^{13}C$ value. Therefore, it is possible, using mixing models, to constrain further the sources of atmospheric CH_4.

Geological sources of CH_4 are not included in Table 7.6. However, several studies suggest that significant amounts of CH_4, produced within the Earth's crust (mainly by bacterial and thermogenic processes), are released into the atmosphere through faults and fractured rocks, mud volcanoes on land and the seafloor, submarine gas seepage, microseepage over dry lands and geothermal seeps (Etiope and Klusman, 2002; Etiope, 2004; Kvenvolden and Rogers, 2005). Emissions from these sources are estimated to be as large as 40 to 60 $Tg(CH_4)$ yr^{-1}.

The major CH_4 sinks are oxidation by OH in the troposphere, biological CH_4 oxidation in drier soil, and loss to the stratosphere (Table 7.6). Oxidation by chlorine (Cl) atoms in the marine atmospheric boundary layer is suggested as an additional sink for CH_4, possibly constituting an additional loss of about 19 $Tg(CH_4)$ yr^{-1} (Gupta et al., 1997; Tyler et al., 2000; Platt et al., 2004; Allan et al., 2005). However, the decline in the growth rate of atmospheric CH_4 concentration since the TAR shows no clear correlation with change in sink strengths over the same period (Prinn et al., 2001, 2005; Allan et al., 2005). This trend has continued since 1993, and the reduction in the CH_4 growth rate has been suggested to be a consequence of source stabilisation and the approach of the global CH_4 budget towards steady state (Dlugokencky et al., 1998, 2003). Thus, total emissions are likely not increasing but partitioning among the different sources may have changed (see Section 2.3). Consequently, in the Fourth Assessment Report (AR4) the sink strength is treated as in the TAR (576 $Tg(CH_4)$ yr^{-1}). However, the AR4 estimate has been increased by 1% (to 581 $Tg(CH_4)$ yr^{-1}) to take into account the recalibration of the CH_4 scale explained in Chapter 2. The main difference between TAR and AR4 estimates is the source-sink imbalance inferred from the annual increment in concentration. The TAR used 8 ppb yr^{-1} for a period centred on 1998 when there was clearly an anomalously high growth rate. The present assessment uses 0.2 ppb yr^{-1}, the average over 2000 to 2005 (see Section 2.3 and Figure 2.4). Thus, using the CH_4 growth rate for a single anomalous year, as in the TAR, gives an anomalously high top-down value relative to the longer-term average source. For a conversion factor of 2.78 $Tg(CH_4)$ per ppb and an atmospheric concentration of 1,774 ppb, the atmospheric burden of CH_4 in 2005 was 4,932 Tg, with an annual average increase (2000–2005) of about 0.6 Tg yr^{-1}. Total average annual emissions during the period considered here are approximately 582 $Tg(CH_4)$ yr^{-1}.

Uncertainty in this estimate may arise from several sources. Uncertainty in the atmospheric concentration measurement, given in Chapter 2 as 1,774 ± 1.8 ppb in 2005, is small (about 0.1%). Uncertainty ranges for individual sink estimates are ±103 $Tg(CH_4)$ (20%), ±15 $Tg(CH_4)$ (50%), ±8 $Tg(CH_4)$ (20%) for OH, soil and stratospheric loss, respectively (as reported in the Second Assessment Report). The use of a different lifetime for CH_4 (8.7 ± 1.3 years) leads to an uncertainty in overall sink strength of ±15%. Thus, the top-down method used in AR4 is

Table 7.6 Sources, sinks and atmospheric budgets of CH_4 $(Tg(CH_4)\ yr^{-1})$.[a]

References	Indicative ^{13}C, ‰[b]	Hein et al., 1997[c]	Houweling et al., 2000[c]	Olivier et al., 2005	Wuebbles and Hayhoe, 2002	Scheehle et al., 2002	J. Wang et al., 2004[c]	Mikaloff Fletcher et al., 2004a[c]	Chen and Prinn, 2006[c]	TAR	AR4
Base year		1983–1989		2000		1990	1994	1999	1996–2001	1998	2000–2004
Natural sources			222		145		200	260	168		
Wetlands	−58	231	163		100		176	231	145		
Termites	−70		20		20		20	29	23		
Ocean	−60		15		4						
Hydrates	−60				5		4				
Geological sources	−40		4		14						
Wild animals	−60		15								
Wildfires	−25		5		2						
Anthropogenic sources		361		320	358	264	307	350	428		
Energy						74	77				
Coal mining	−37	32		34	46			30	48[d]		
Gas, oil, industry	−44	68		64	60			52	36[e]		
Landfills & waste	−55	43		66	61	69	49	35			
Ruminants	−60	92		80	81	76	83	91	189[f]		
Rice agriculture	−63	83		39	60	31	57	54	112		
Biomass burning	−25	43			50	14	41	88	43[e]		
C3 vegetation	−25			27							
C4 vegetation	−12			9							
Total sources		592			503		507	610	596	598	582
Imbalance		+33								+22	+1
Sinks											
Soils	−18	26			30		34	30		30	30[g]
Tropospheric OH	−3.9	488			445		428	507		506	511[g]
Stratospheric loss		45			40		30	40		40	40[g]
Total sink		559			515		492	577		576	581[g]

Notes:

[a] Table shows the best estimate values.

[b] Indicative ^{13}C values for sources are taken mainly from Mikaloff Fletcher et al. (2004a). Entries for sinks are the fractionation, $(k_{13}/k_{12}-1)$ where k_n is the removal rate of $^nCH_4$; the fractionation for OH is taken from Saueressig et al. (2001) and that for the soil sink from Snover and Quay (2000) as the most recent determinations.

[c] Estimates from global inverse modelling (top-down method).

[d] Includes natural gas emissions.

[e] Biofuel emissions are included under Industry.

[f] Includes emissions from landfills and wastes.

[g] Numbers are increased by 1% from the TAR according to recalibration described in Chapter 2.

constrained mainly by uncertainty in sink estimates and the choice of lifetime used in the mass balance calculation.

7.4.1.2   *Effects of Climate*

Effects of climate on CH_4 biogeochemistry are investigated by examining records of the past and from model simulations under various climate change scenarios. Ice core records going back 650 ka (Petit et al., 1999; Spahni et al., 2005) reveal that the atmospheric concentration of CH_4 is closely tied to atmospheric temperature, falling and rising in phase with temperature at the inception and termination of glacial episodes (Wuebbles and Hayhoe, 2002). Brook et al. (2000) show that, following each transition, temperature increased more rapidly than CH_4 concentration. Since biogenic CH_4 production and emission from major sources (wetlands, landfills, rice agriculture and biomass burning) are influenced by climate variables such as temperature and moisture, the effect of climate on emissions from these sources is significant.

Several studies indicate a high sensitivity of wetland CH_4 emissions to temperature and water table. Before the 1990s, elevated surface temperature and emissions from wetlands were believed to contribute to the increase in global CH_4 emissions (Walter and Heimann, 2001a,b; Christensen et al., 2003; Zhuang et al., 2004). Observations indicate substantial increases in CH_4 released from northern peatlands that are experiencing permafrost melt (Christensen et al., 2004; Wickland et al., 2006). Based on the relationship between emissions and temperature at two wetland sites in Scotland, Chapman and Thurlow (1996) predicted that CH_4 emissions would increase by 17, 30 and 60% for warmings of 1.5°C, 2.5°C and 4.5°C (warming above the site's mean temperature during 1951 to 1980), respectively. A model simulation by Cao et al. (1998) yielded a 19% emission increase under a uniform 2°C warming. The combined effects of a 2°C warming and a 10% increase in precipitation yielded an increase of 21% in emissions. In most cases, the net emission depends on how an increase in temperature affects net ecosystem production (NEP), as this is the source of methanogenic substrates (Christensen et al., 2003), and on the moisture regime of wetlands, which determines if decomposition is aerobic or anaerobic. Emissions increase under a scenario where an increase in temperature is associated with increases in precipitation and NEP, but emissions decrease if elevated temperature results in either reduced precipitation or reduced NEP.

For a doubling in atmospheric CO_2 concentration, the GCM of Shindell et al. (2004) simulates a 3.4°C warming. Changes in the hydrological cycle due to this CO_2 doubling cause CH_4 emissions from wetlands to increase by 78%. Gedney et al. (2004) also simulate an increase in CH_4 emissions from northern wetlands due to an increase in wetland area and an increase in CH_4 production due to higher temperatures. Zhuang et al. (2004) use a terrestrial ecosystem model based on emission data for the 1990s to study how rates of CH_4 emission and consumption in high-latitude soils of the NH (north of 45°N) have changed over the past century (1900–2000) in response to observed change in

the region's climate. They estimate that average net emissions of CH_4 increased by 0.08 Tg yr^{-1} over the 20th century. Their decadal net CH_4 emission rate correlates with soil temperature and water table depth.

In rice agriculture, climate factors that will likely influence CH_4 emission are those associated with plant growth. Plant growth controls net emissions by determining how much substrate will be available for either methanogenesis or methanotrophy (Matthews and Wassmann, 2003). Sass et al. (2002) show that CH_4 emissions correlate strongly with plant growth (height) in a Texas rice field. Any climate change scenario that results in an increase in plant biomass in rice agriculture is likely to increase CH_4 emissions (Xu et al., 2004). However, the magnitude of increased emission depends largely on water management. For example, field drainage could significantly reduce emission due to aeration of the soil (i.e., influx of air into anaerobic zones that subsequently suppresses methanogenesis, Li et al., 2002).

Past observations indicate large interannual variations in CH_4 growth rates (Dlugokencky et al., 2001). The mechanisms causing these variations are poorly understood and the role of climate is not well known. Emissions from wetlands and biomass burning may have contributed to emission peaks in 1993 to 1994 and 1997 to 1998 (Langenfelds et al., 2002; Butler et al., 2004). Unusually warm and dry conditions in the NH during ENSO periods increase biomass burning. Kasischke and Bruhwiler (2002) attribute CH_4 releases of 3 to 5 Tg in 1998 to boreal forest fires in Eastern Siberia resulting from unusually warm and dry conditions.

Meteorological conditions can affect global mean removal rates (Warwick et al., 2002; Dentener et al., 2003a). Dentener et al. find that over the period 1979 to 1993, the primary effect resulted from changes in OH distribution caused by variations in tropical tropospheric water vapour. Johnson et al. (2001) studied predictions of the CH_4 evolution over the 21st century and found that there is also a substantial increase in CH_4 destruction due to increases in the $CH_4 + OH$ rate coefficient in a warming climate. There also appear to be significant interannual variations in the active Cl sink, but a climate influence has yet to be identified (Allan et al., 2005). On the other hand, several model studies indicate that CH_4 oxidation in soil is relatively insensitive to temperature increase (Ridgwell at al., 1999; Zhuang et al., 2004). A doubling of atmospheric CO_2 would likely change the sink strength only marginally (in the range of –1 to +3 $Tg(CH_4)$ yr^{-1}; Ridgwell et al., 1999). However, any change in climate that alters the amount and pattern of precipitation may significantly affect the CH_4 oxidation capacity of soils. A process-based model simulation indicated that CH_4 oxidation strongly depends on soil gas diffusivity, which is a function of soil bulk density and soil moisture content (Bogner et al., 2000; Del Grosso et al., 2000).

Climate also affects the stability of CH_4 hydrates beneath the ocean, where large amounts of CH_4 are stored (~4 ×10^6 Tg; Buffett and Archer, 2004). The $\delta^{13}C$ values of ancient seafloor carbonates reveal several hydrate dissociation events that appear to have occurred in connection with rapid warming episodes in the Earth's history (Dickens et al., 1997; Dickens, 2001). Model results indicate that these hydrate decomposition events occurred too fast to be controlled by the propagation of the temperature change into the sediments (Katz et al., 1999; Paull et al., 2003). Additional studies infer other indirect and inherently more rapid mechanisms such as enhanced migration of free gas, or reordering of gas hydrates due to slump slides (Hesselbo et al., 2000; Jahren et al., 2001; Kirschvink et al., 2003; Ryskin, 2003). Recent modelling suggests that today's seafloor CH_4 inventory would be diminished by 85% with a warming of bottom water temperatures by 3°C (Buffett and Archer, 2004). Based on this inventory, the time-dependent feedback of hydrate destabilisation to global warming has been addressed using different assumptions for the time constant of destabilisation: an anthropogenic release of 2,000 GtC to the atmosphere could cause an additional release of CH_4 from gas hydrates of a similar magnitude (~2,000 Gt(CH_4)) over a period of 1 to 100 kyr (Archer and Buffett, 2005). Thus, gas hydrate decomposition represents an important positive CH_4 feedback to be considered in global warming scenarios on longer time scales.

In summary, advances have been made since the TAR in constraining estimates of CH_4 source strengths and in understanding emission variations. These improvements are attributed to increasing availability of worldwide observations and improved modelling techniques. Emissions from anthropogenic sources remain the major contributor to atmospheric CH_4 budgets. Global emissions are likely not to have increased since the time of the TAR, as nearly zero growth rates in atmospheric CH_4 concentrations have been observed with no significant change in the sink strengths.

7.4.2   Nitrogen Compounds

The N cycle is integral to functioning of the Earth system and to climate (Vitousek et al., 1997; Holland et al., 2005a). Over the last century, human activities have dramatically increased emissions and removal of reactive N to the global atmosphere by as much as three to five fold. Perturbations of the N cycle affect the atmosphere climate system through production of three key N-containing trace gases: N_2O, ammonia (NH_3) and NO_x (nitric oxide (NO) + nitrogen dioxide (NO_2)). Nitrous oxide is the fourth largest single contributor to positive radiative forcing, and serves as the only long-lived atmospheric tracer of human perturbations of the global N cycle (Holland et al., 2005a). Nitrogen oxides have short atmospheric lifetimes of hours to days (Prather et al., 2001). The dominant impact of NO_x emissions on the climate is through the formation of tropospheric ozone, the third largest single contributor to positive radiative forcing (Sections 2.3.6, 7.4.4). Emissions of NO_x generate indirect negative radiative forcing by shortening the atmospheric lifetime of CH_4 (Prather 2002). Ammonia contributes to the formation of sulphate and nitrate aerosols, thereby contributing to aerosol cooling and the aerosol indirect effect (Section 7.5), and to increased nutrient supply for the carbon cycle (Section 7.5). Ammonium and NO_x are removed from the atmosphere by deposition, thus affecting the carbon cycle through increased nutrient supply (Section 7.3.3.1.3).

Atmospheric concentrations of N_2O have risen 16%, from about 270 ppb during the pre-industrial era to 319 ppb in 2005 (Figure 7.16a). The average annual growth rate for 1999 to 2000 was 0.85 to 1.1 ppb yr^{-1}, or about 0.3% per year (WMO, 2003). The main change in the global N_2O budget since the TAR is quantification of the substantial human-driven emission of N_2O (Table 7.7; Naqvi et al., 2000; Nevison et al., 2004; Kroeze et al., 2005; Hirsch et al., 2006). The annual source of N_2O from the Earth's surface has increased by about 40 to 50% over pre-industrial levels as a result of human activity (Hirsch et al., 2006). Human activity has increased N supply to coastal and open oceans, resulting in decreased O_2 availability and N_2O emissions (Naqvi et al., 2000; Nevison et al., 2004).

Since the TAR, both top-down and bottom-up estimates of N_2O have been refined. Agriculture remains the single biggest anthropogenic N_2O source (Bouwman et al., 2002; Smith and Conen, 2004; Del Grosso et al., 2005). Land use change continues to affect N_2O and NO emissions (Neill et al., 2005): logging is estimated to increase N_2O and NO emissions by 30 to 350% depending on conditions (Keller et al., 2005). Both studies underscore the importance of N supply, temperature and moisture as regulators of trace gas emissions. The inclusion of several minor sources (human excreta, landfills and atmospheric deposition) has increased the total bottom-up budget to 20.6 TgN yr^{-1} (Bouwman et al., 2002). Sources of N_2O now estimated since the TAR include coastal N_2O fluxes of 0.2 TgN yr^{-1} (±70%; Nevison et al., 2004) and river and estuarine N_2O fluxes of 1.5 TgN yr^{-1} (Kroeze et al., 2005). Box model calculations show the additional river and estuarine sources to be consistent with the observed rise in atmospheric N_2O (Kroeze et al., 2005).

Top-down estimates of surface sources use observed concentrations to constrain total sources and their spatial distributions. A simple calculation, using the present-day N_2O burden divided by its atmospheric lifetime, yields a global stratospheric loss of about 12.5 ± 2.5 TgN yr^{-1}. Combined with the atmospheric increase, this loss yields a surface source of 16 TgN yr^{-1}. An inverse modelling study of the surface flux of N_2O yields a global source of 17.2 to 17.4 TgN yr^{-1} with an estimated uncertainty of 1.4 (1 standard deviation; Hirsch et al., 2006). The largest sources of N_2O are from land at tropical latitudes, the majority located north of the equator. The Hirsch et al. inversion results further suggest that N_2O source estimates from agriculture and fertilizer may have increased markedly over the last three decades when compared with an earlier inverse model estimate (Prinn et al., 1990). Bottom-up estimates, which sum individual source estimates, are more evenly distributed with latitude and lack temporal variability. However, there is clear consistency between top-down and bottom-up global source estimates, which are 17.3 (15.8–18.4) and 17.7 (8.5–27.7) TgN yr^{-1}, respectively.

Concentrations of NO_x and reduced nitrogen (NH_x = NH_3 + ammonium ion (NH_4^+)) are difficult to measure because the atmospheric lifetimes of hours to days instead of years generate pronounced spatial and temporal variations in their distributions. Atmospheric concentrations of NO_x and NH_x

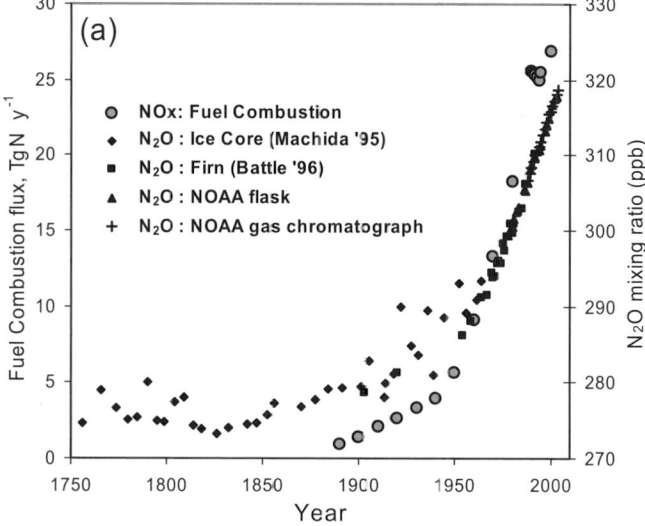

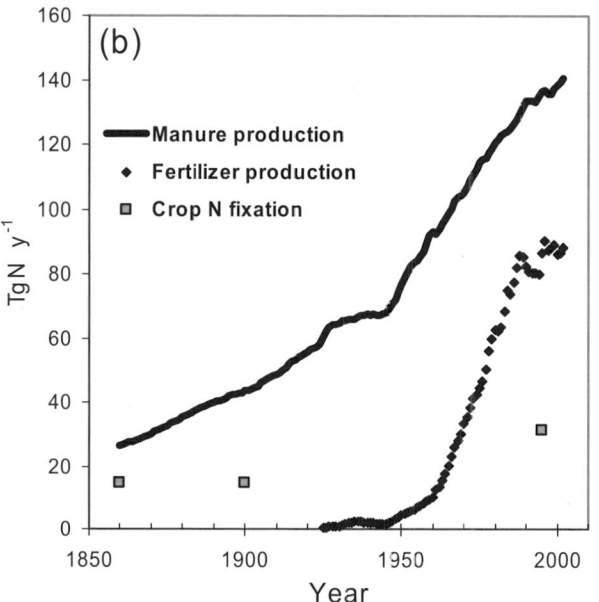

Figure 7.16. *(a) Changes in the emissions of fuel combustion NO_x and atmospheric N_2O mixing ratios since 1750. Mixing ratios of N_2O provide the atmospheric measurement constraint on global changes in the N cycle. (b) Changes in the indices of the global agricultural N cycle since 1850: the production of manure, fertilizer and estimates of crop N fixation. For data sources see http://www-eosdis.ornl.gov/ (Holland et al., 2005b) and http://www.cmdl.noaa.gov/. Figure adapted from Holland et al. (2005c).*

vary more regionally and temporally than concentrations of N_2O. Total global NO_x emissions have increased from an estimated pre-industrial value of 12 TgN yr^{-1} (Holland et al., 1999; Galloway et al., 2004) to between 42 and 47 TgN yr^{-1} in 2000 (Table 7.7). Lamarque et al. (2005a) forecast them to be 105 to 131 TgN yr^{-1} by 2100. The range of surface NO_x emissions (excluding lightning and aircraft) used in the current generation of global models is 33 to 45 TgN yr^{-1} with small ranges for individual sources. The agreement reflects the use

of similar inventories and parametrizations. Current estimates of NO_x emissions from fossil fuel combustion are smaller than in the TAR.

Since the TAR, estimates of tropospheric NO_2 columns from space by the Global Ozone Monitoring Experiment (GOME, launched in 1995) and the SCanning Imaging Absorption SpectroMeter for Atmospheric CHartographY (SCIAMACHY, launched in 2002) (Richter and Burrows, 2002; Heue et al., 2005) provide constraints on estimates of NO_x emissions (Leue et al., 2001). Martin et al. (2003a) use GOME data to estimate a global surface source of NO_x of 38 TgN yr^{-1} for 1996 to 1997 with an uncertainty factor of 1.6. Jaeglé et al. (2005) partition the surface NO_x source inferred from GOME into 25.6 TgN yr^{-1} from fuels, 5.9 TgN yr^{-1} from biomass burning and 8.9 TgN yr^{-1} from soils. Interactions between soil emissions and scavenging by plant canopies have a significant impact on soil NO_x emissions to the free troposphere: the impact may be greatest in subtropical and tropical regions where emissions from fuel combustion are rising (Ganzeveld et al., 2002). Boersma et al. (2005) find that GOME data constrain the global lightning NO_x source for 1997 to the range 1.1 to 6.4 TgN yr^{-1}. Comparison of the tropospheric NO_2 column of three state-of-the-art retrievals from GOME for the year 2000 with model results from 17 global atmospheric chemistry models highlights significant differences among the various models and among the three GOME retrievals (Figure 7.17, van Noije et al., 2006). The discrepancies among the retrievals (10 to 50% in the annual mean over polluted regions) indicate that the previously estimated retrieval uncertainties have a large systematic component. Top-down estimates of NO_x emissions from satellite retrievals of tropospheric NO_2 are strongly dependent on the choice of model and retrieval.

Knowledge of the spatial distribution of NO_x emissions has evolved significantly since the TAR. An Asian increase in emissions has been compensated by a European decrease over the past decade (Naja et al., 2003). Richter et al. (2005; see also Irie et al., 2005) use trends for 1996 to 2004 observed by GOME and SCIAMACHY to deduce a 50% increase in NO_x emissions over industrial areas of China. Observations of NO_2 in shipping lanes from GOME (Beirle et al., 2004) and SCIAMACHY (Richter et al., 2004) give values at the low end of emission inventories. Data from GOME and SCIAMACHY further reveal large pulses of soil NO_x emissions associated with rain (Jaeglé et al., 2004) and fertilizer application (Bertram et al., 2005).

All indices show an increase since pre-industrial times in the intensity of agricultural nitrogen cycling, the primary source of NH_3 emissions (Figure 7.16b and Table 7.7; Bouwman et al., 2002). Total global NH_3 emissions have increased from an estimated pre-industrial value of 11 TgN yr^{-1} to 54 TgN yr^{-1} for 2000 (Holland et al., 1999; Galloway et al., 2004), and are projected to increase to 116 TgN yr^{-1} by 2050.

The primary sink for NH_x and NO_x and their reaction products is wet and dry deposition. Estimates of the removal rates of both NH_x and NO_x are provided by measurements of wet deposition over the USA and Western Europe to quantify acid rain inputs (Hauglustaine et al., 2004; Holland et al., 2005a; Lamarque et al., 2005a). Chemical transport models represent the wet and dry deposition of NO_x and NH_x and their reaction products. A study of 29 simulations with 6 different tropospheric chemistry models, focusing on present-day and 2100 conditions for NO_x and its reaction products, projects an average increase in N deposition over land by a factor of 2.5 by 2100 (Lamarque et al., 2005b), mostly due to increases in NO_x

emissions. Nitrogen deposition rates over Asia are projected to increase by a factor of 1.4 to 2 by 2030. Climate contributions to the changes in oxidized N deposition are limited by the models' ability to represent changes in precipitation patterns. An intercomparison of 26 global atmospheric chemistry models demonstrates that current scenarios and projections are not sufficient to stabilise or reduce N deposition or ozone pollution before 2030 (Dentener et al., 2006).

Table 7.7. *Global sources (TgN yr^{-1}) of NO_x, NH_3 and N_2O for the 1990s.*

Source	NO_x TAR[a]	NO_x AR4[b]	NH_3 TAR[a]	NH_3 AR4[a]	N_2O TAR[c]	N_2O AR4
Anthropogenic sources						
Fossil fuel combustion & industrial processes	33 (20–24)	25.6 (21–28)	0.3 (0.1–0.5)	2.5[d]	1.3/0.7 (0.2–1.8)	0.7 (0.2–1.8)[d]
Aircraft	0.7 (0.2–0.9)	– [e] (0.5–0.8)	-	-	-	-
Agriculture	2.3[f] (0–4)	1.6[g]	34.2 (16–48)	35[g] (16–48)	6.3/2.9 (0.9–17.9)	2.8 (1.7–4.8)[g]
Biomass and biofuel burning	7.1 (2–12)	5.9 (6–12)	5.7 (3–8)	5.4[d] (3–8)	0.5 (0.2–1.0)	0.7 (0.2–1.0)[g]
Human excreta	–	–	2.6 (1.3–3.9)	2.6[g] (1.3–3.9)	–	0.2[g] (0.1–0.3)[h]
Rivers, estuaries, coastal zones	–	–	–	–	–	1.7 (0.5–2.9)[i]
Atmospheric deposition	–	0.3[g]	–	–	–	0.6[i] (0.3–0.9)[h]
Anthropogenic total	**43.1**	**33.4**	**42.8**	**45.5**	**8.1/4.1**	**6.7**
Natural sources						
Soils under natural vegetation	3.3[f] (3–8)	7.3[j] (5–8)	2.4 (1–10)	2.4[g] (1–10)	6.0/6.6 (3.3–9.9)	6.6 (3.3–9.0)[g]
Oceans	–	–	8.2 (3–16)	8.2[g] (3–6)	3.0/3.6 (1.0–5.7)	3.8 (1.8–5.8)[k]
Lightning	5 (2–12)	1.1–6.4 (3–7)	–	–	–	–
Atmospheric chemistry	<0.5	–	–	–	0.6 (0.3–1.2)	0.6 (0.3–1.2)[c]
Natural total	**8.8**	**8.4–13.7**	**10.6**	**10.6**	**9.6/10.8**	**11.0**
Total sources	**51.9** (27.2–60.9)	**41.8–47.1** (37.4–57.7)	**53.4** (40–70)	**56.1** (26.8–78.4)	**17.7/14.9** (5.9–37.5)	**17.7** (8.5–27.7)

Notes:

[a] Values from the TAR: NO_x from Table 4.8 with ranges from Tables 4.8 and 5.2; NH_3 from Table 5.2, unless noted.

[b] Parentheses show the range of emissions used in the model runs described in Table 7.9. See text for explanation. Where possible, the best estimate NO_x emission is based on satellite observations. None of the model studies includes the NO_x source from oxidation of NH_3, which could contribute up to 3 TgN yr^{-1}. The source of NO_x from stratosphere-troposphere exchange is less than 1 TgN yr^{-1} in all models, which is well constrained from observations of N_2O-NO_x correlations in the lower stratosphere (Olsen et al., 2001).

[c] Values are from the TAR, Table 4.4; Mosier et al. (1998); Kroeze et al. (1999)/Olivier et al. (1998): a single value indicates agreement between the sources and methodologies of the different studies.

[d] Van Aardenne et al. (2001), range from the TAR.

[e] The aircraft source is included in the total for industrial processes. The parentheses indicate values used in model runs.

[f] The total soil NO_x emissions estimate of 5.6 provided in Table 4.8 of the TAR was distributed between agriculture and soil NO_x according to the proportions provided in the TAR, Table 5.2.

[g] Bouwman et al. (2001, Table 1); Bouwman et al. (2002) for the 1990s; range from the TAR or calculated as ±50%.

[h] Estimated as ±50%.

[i] Kroeze et al. (2005); Nevison et al. (2004); estimated uncertainty is ±70% from Nevison et al. (2004).

[j] All soils, minus the fertilized agricultural soils indicated above.

[k] Nevison et al. (2003, 2004), combining the uncertainties in ocean production and oceanic exchange.

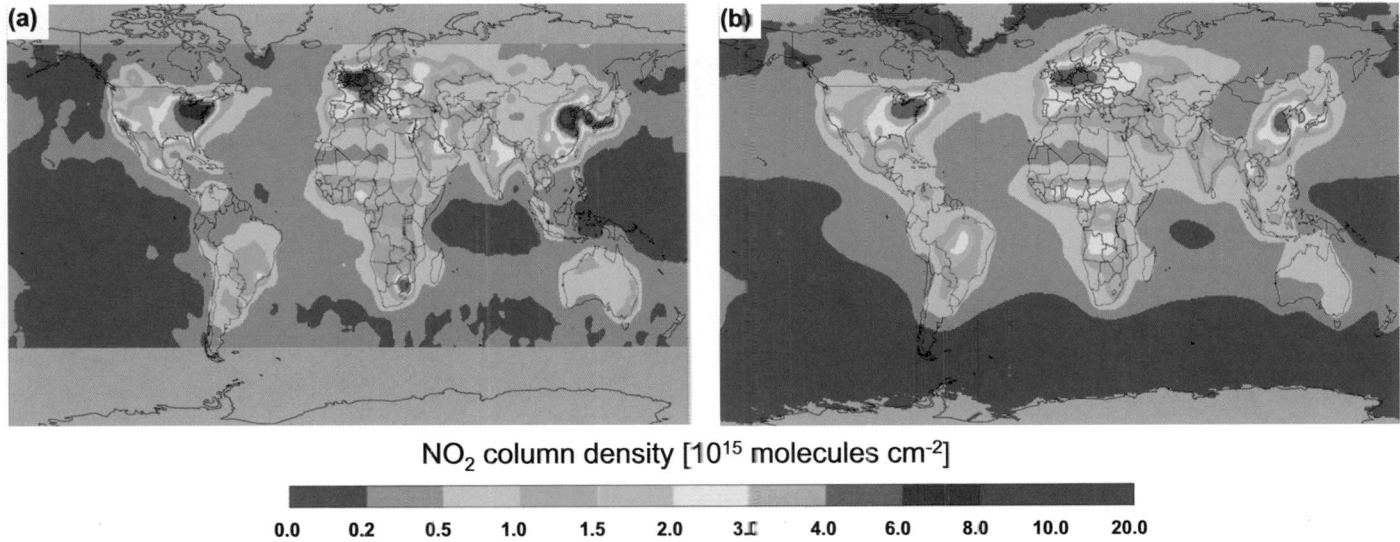

NO_2 column density [10^{15} molecules cm^{-2}]

0.0   0.2   0.5   1.0   1.5   2.0   3.0   4.0   6.0   8.0   10.0   20.0

Figure 7.17. *Tropospheric column NO_2 from (a) satellite measurements and (b) atmospheric chemistry models. The maps represent ensemble average annual mean tropospheric NO_2 column density maps for the year 2000. The satellite retrieval ensemble comprises three state-of-the-art retrievals from GOME; the model ensemble includes 17 global atmospheric chemistry models. These maps were obtained after smoothing the data to a common horizontal resolution of 5° × 5° (adapted from van Noije et al., 2006).*

7.4.3. Molecular Hydrogen

Increased interest in atmospheric H_2 is due to its potential role as an indirect greenhouse gas (Derwent et al., 2001) and expected perturbations of its budget in a prospective 'hydrogen economy' (Schultz et al., 2003; Tromp et al., 2003; Warwick et al., 2004). Potential consequences of increased H_2 emissions include a reduction of global oxidizing capacity (presently H_2 constitutes 5 to 10% of the global average OH sink, Schultz et al., 2003) and increased formation of water vapour, which could lead to increased cirrus formation in the troposphere and increased polar stratospheric clouds (PSCs) and additional cooling in the stratosphere, thereby leading to more efficient ozone depletion (Tromp et al., 2003).

Studies of the global tropospheric H_2 budget (see Table 7.8) generally agree on a total source strength of between 70 and 90 Tg(H_2) yr^{-1}, which is approximately balanced by its sinks. About half of the H_2 is produced in the atmosphere via photolysis of formaldehyde (CH_2O), which itself originates from the oxidation of CH_4 and other volatile organic compounds. The other half stems mostly from the combustion of fossil fuels (e.g., car exhaust) and biomass burning. About 10% of the global H_2 source is due to ocean biochemistry and N fixation in soils. Presently, about 50 Tg(H_2) yr^{-1} are produced in the industrial sector, mostly for the petrochemical industry (e.g., refineries) (Lovins, 2003). Evaporative losses of industrial H_2 are generally assumed to be negligible (Zittel and Altmann, 1996). The dominant sink of atmospheric H_2 is deposition with catalytic destruction by soil microorganisms and possibly enzymes (Conrad and Seiler, 1981). The seasonal cycle of observed H_2 concentrations implies an atmospheric lifetime of about 2 years (Novelli et al., 1999; Simmonds et al., 2000; Hauglustaine and Ehhalt, 2002), whereas the lifetime with respect to OH oxidation is 9 to 10 years, which implies

that the deposition sink is about three to four times as large as the oxidation. Loss of H_2 to the stratosphere and its subsequent escape to space is negligible for the tropospheric H_2 budget, because the budgets of the troposphere and stratosphere are largely decoupled (Warneck, 1988).

Estimates of H_2 required to fuel a future carbon-free energy system are highly uncertain and depend on the technology as well as the fraction of energy that might be provided by H_2. In the future, H_2 emissions could at most double: the impact on global oxidizing capacity and stratospheric temperatures and ozone concentrations is estimated to be small (Schultz et al., 2003; Warwick et al., 2004). According to Schultz et al. (2003), the side effects of a global H_2 economy could have a stronger impact on global climate and air pollution. Global oxidizing capacity is predominantly controlled by the concentration of NO_x. Large-scale introduction of H_2-powered vehicles would lead to a significant decrease in global NO_x emissions, leading to a reduction in OH of the order of 5 to 10%. Reduced NO_x levels could also significantly reduce tropospheric ozone concentrations in urban areas. Despite the expected large-scale use of natural gas for H_2 production, the impact of a H_2 economy on the global CH_4 budget is likely to be small, except for the feedback between reduced oxidizing capacity (via NO_x reduction) and CH_4 lifetime.

7.4.4   Global Tropospheric Ozone

7.4.4.1   *Present-Day Budgets of Ozone and its Precursors*

Tropospheric ozone is (after CO_2 and CH_4) the third most important contributor to greenhouse radiative forcing. Trends over the 20th century are discussed in Chapter 2. Ozone is produced in the troposphere by photochemical oxidation of CO, CH_4 and non-methane VOCs (NMVOCs) in the presence

Table 7.8. *Summary of global budget studies of atmospheric H_2 ($Tg(H_2)$ yr^{-1}).*

	Sanderson et al. (2003a)	Hauglustaine and Ehhalt (2002)	Novelli et al. (1999)	Ehhalt (1999)	Warneck (1988)	Seiler and Conrad (1987)
Sources						
Oxidation of CH_4 and VOC	30.2	31	40 ± 16	35 ± 15	50	40 ± 15
Fossil fuel combustion	20	16	15 ± 10	15 ± 10	17	20 ± 10
Biomass burning	20	13	16 ± 11	16 ± 5	15	20 ± 10
N_2 fixation	4	5	3 ± 1	3 ± 2	3	3 ± 2
Ocean release	4	5	3 ± 2	3 ± 2	4	4 ± 2
Volcanoes	–	–	–	–	0.2	–
Total	78.2	70	77 ± 16	71 ± 20	89	87
Sinks						
Deposition	58.3	55	56 ± 41	40 ± 30	78	90 ± 20
Oxidation by OH	17.1	15	19 ± 5	25 ± 5	11	8 ± 3
Total	74.4	70	75 ± 41	65 ± 30	89	98

of NO_x. Stratosphere-troposphere exchange (STE) is another source of ozone to the troposphere. Loss of tropospheric ozone takes place through chemical reactions and dry deposition. Understanding of tropospheric ozone and its relationship to sources requires three-dimensional tropospheric chemistry models that describe the complex nonlinear chemistry involved and its coupling to transport.

The past decade has seen considerable development in global models of tropospheric ozone, and the current generation of models can reproduce most climatological features of ozone observations. The TAR reported global tropospheric ozone budgets from 11 models in the 1996 to 2000 literature. Table 7.9 presents an update to the post-2000 literature, including a recent intercomparison of 25 models (Stevenson et al., 2006). Models concur that chemical production and loss are the principal terms in the global budget. Although STE is only a minor term in the global budget, it delivers ozone to the upper troposphere where its lifetime is particularly long (about one month, limited by transport to the lower troposphere) and where it is of most importance from a radiative forcing perspective.

The post-2000 model budgets in Table 7.9 show major differences relative to the older generation TAR models: on average a 34% weaker STE, a 35% stronger chemical production, a 10% larger tropospheric ozone burden, a 16% higher deposition velocity and a 10% shorter chemical lifetime. It is now well established that many of the older studies overestimated STE, as observational constraints in the lower stratosphere impose an STE ozone flux of 540 ± 140 Tg yr^{-1} (Gettelman et al., 1997; Olsen et al., 2001). Overestimation of the STE flux appears to be most serious in models using assimilated meteorological data, due to the effect of assimilation on vertical motions (Douglass et al., 2003; Schoeberl et al., 2003; Tan et al., 2004; Van Noije et al., 2004). The newer models correct for this effect by using dynamic flux boundary conditions in the

tropopause region (McLinden et al., 2000) or by relaxing model results to observed climatology (Horowitz et al., 2003). Such corrections, although matching the global STE flux constraints, may still induce errors in the location of the transport (Hudman et al., 2004) with implications for the degree of stratospheric influence on tropospheric concentrations (Fusco and Logan, 2003).

The faster chemical production and loss of ozone in the current generation of models could reflect improved treatment of NMVOC sources and chemistry (Houweling et al., 1998), ultraviolet (UV) actinic fluxes (Bey et al., 2001) and deep convection (Horowitz et al., 2003), as well as higher NO_x emissions (Stevenson et al., 2006). Subtracting ozone chemical production and loss terms in Table 7.9 indicates that the current generation of models has net production of ozone in the troposphere, while the TAR models had net loss, reflecting the decrease in STE. Net production is not a useful quantity in analysing the ozone budget because (1) it represents only a small residual between production and loss and (2) it reflects a balance between STE and dry deposition, both of which are usually parametrized in models.

Detailed budgets of ozone precursors were presented in the TAR. The most important precursors are CH_4 and NO_x (Wang et al., 1998; Grenfell et al., 2003; Dentener et al., 2005). Methane is in general not simulated explicitly in ozone models and is instead constrained from observations. Nitrogen oxides are explicitly simulated and proper representation of sources and chemistry is critical for the ozone simulation. The lightning source is particularly uncertain (Nesbitt et al., 2000; Tie et al., 2002), yet is of great importance because of the high production efficiency of ozone in the tropical upper troposphere. The range of the global lightning NO_x source presently used in models (3–7 TgN yr^{-1}) is adjusted to match atmospheric observations of ozone and NO_x, although large model uncertainties in deep

Table 7.9. *Global budgets of tropospheric ozone (Tg yr^{-1}) for the present-day atmosphere[a].*

Reference	Model[b]	Stratosphere-Troposphere Exchange	Chemical Production[c]	Chemical Loss[c]	Dry Deposition	Burden (Tg)	Lifetime[d] (days)
TAR[e]	11 models	770 ± 400	3420 ± 770	3470 ± 520	770 ± 180	300 ± 30	24 ± 2
Lelieveld and Dentener (2000)	TM3	570	3310	3170	710	350	33
Bey et al. (2001)	GEOS-Chem	470	4900	4300	1070	320	22
Sudo et al. (2002b)	CHASER	593	4895	4498	990	322	25
Horowitz et al. (2003)	MOZART-2	340	5260	4750	860	360	23
Von Kuhlmann et al. (2003)	MATCH-MPIC	540	4560	4290	820	290	21
Shindell et al. (2003)	GISS	417	NR[f]	NR	1470	349	NR
Hauglustaine et al. (2004)	LMDz-INCA	523	4486	3918	1090	296	28
Park et al. (2004)	UMD-CTM	480	NR	NR	1290	340	NR
Rotman et al. (2004)	IMPACT	660	NR	NR	830	NR	NR
Wong et al. (2004)	SUNY/UiO GCCM	600	NR	NR	1100	376	NR
Stevenson et al. (2004)	STOCHEM	395	4980	4420	950	273	19
Wild et al. (2004)	FRSGC/UCI	520	4090	3850	760	283	22
Folberth et al. (2006)	LMDz-INCA	715	4436	3890	1261	303	28
Stevenson et al. (2006)	25 models	520 ± 200	5060 ± 570	4560 ± 720	1010 ± 220	340 ± 40	22 ± 2

Notes:

[a] From global model simulations describing the atmosphere of the last decade of the 20th century.

[b] TM3: Royal Netherlands Meteorological Institute (KNMI) chemistry transport model; GEOS-Chem: atmospheric composition model driven by observations from the Goddard Earth Observing System; CHASER: Chemical AGCM for Study of Atmospheric Environment and Radiative Forcing; MOZART-2: Model for (tropospheric) Ozone and Related Tracers; MATCH-MPIC: Model of Atmospheric Transport and Chemistry – Max Planck Institute for Chemistry; GISS: Goddard Institute for Space Studies chemical transport model; LMDz-INCA: Laboratoire de Météorologie Dynamique GCM-Interactive Chemistry and Aerosols model; UMD-CTM: University of Maryland Chemical Transport Model; IMPACT: Integrated Massively Parallel Atmospheric Chemistry Transport model; SUNY/UiO GCCM: State University of New York/University of Oslo Global Tropospheric Climate-Chemistry Model; STOCHEM: Hadley Centre global atmospheric chemistry model; FRSGC/UCI: Frontier Research System for Global Change/University of California at Irvine chemical transport model.

[c] Chemical production and loss rates are calculated for the odd oxygen family, usually defined as O_x = ozone + O + NO_2 + $2NO_3$ + 3 dinitrogen pentoxide (N_2O_5) + pernitric acid (HNO_4) + peroxyacylnitrates (and sometimes nitric acid; HNO_3), to avoid accounting for rapid cycling of ozone with short-lived species that have little implication for its budget. Chemical production is mainly contributed by reactions of NO with peroxy radicals, while chemical loss is mainly contributed by the oxygen radical in the 1D excited state (O(^{1}D)) plus water (H_2O) reaction and by the reactions of ozone with the hydroperoxyl radical (HO_2), OH, and alkenes.

[d] Calculated as the ratio of the burden to the sum of chemical and deposition losses.

[e] Means and standard deviations for 11 global model budgets from the 1996 to 2000 literature reported in the TAR. The mean budget does not balance exactly because only nine chemical transport models reported their chemical production and loss statistics.

[f] Not reported.

convection and lightning vertical distributions detract from the strength of this constraint. Process-based models tend to predict higher lightning emissions (5–20 TgN yr^{-1}; Price et al., 1997).

Other significant precursors for tropospheric ozone are CO and NMVOCs, the most important of which is biogenic isoprene. Satellite measurements of CO from the Measurements of Pollution in the Troposphere (MOPITT) instrument launched in 1999 (Edwards et al., 2004) have provided important new constraints for CO emissions, pointing in particular to an underestimate of Asian sources in current inventories (Kasibhatla et al., 2002; Arellano et al., 2004; Heald et al., 2004; Petron et al., 2004), as confirmed also by aircraft observations of Asian outflow (Palmer et al., 2003a; Allen et al., 2004). Satellite measurements of formaldehyde columns from

the GOME instrument (Chance et al., 2000) have been used to place independent constraints on isoprene emissions and indicate values generally consistent with current inventories, although with significant regional discrepancies (Palmer et al., 2003b; Shim et al., 2005).

A few recent studies have examined the effect of aerosols on global tropospheric ozone involving both heterogeneous chemistry and perturbations to actinic fluxes. Jacob (2000) reviewed the heterogeneous chemistry involved. Hydrolysis of dinitrogen pentoxide (N_2O_5) in aerosols is a well-known sink for NO_x, but other processes involving reactive uptake of the hydroperoxyl radical (HO_2), NO_2 and ozone itself could also be significant. Martin et al. (2003b) find that including these processes along with effects of aerosols on UV radiation in a global Chemical Transport Model (CTM) reduced ozone production rates by 6% globally, with larger effects over aerosol source regions.

Although the current generation of tropospheric ozone models is generally successful in describing the principal features of the present-day global ozone distribution, there is much less confidence in the ability to reproduce the changes in ozone associated with perturbations of emissions or climate. There are major discrepancies with observed long-term trends in ozone concentrations over the 20th century (Hauglustaine and Brasseur, 2001; Mickley et al., 2001; Shindell and Favulegi, 2002; Shindell et al., 2003; Lamarque et al., 2005c), including after 1970 when the reliability of observed ozone trends is high (Fusco and Logan, 2003). Resolving these discrepancies is needed to establish confidence in the models.

7.4.4.2   Effects of Climate Change

Climate change can affect tropospheric ozone by modifying emissions of precursors, chemistry, transport and removal (European Commission, 2003). These and other effects are discussed below. They could represent positive or negative feedbacks to climate change.

7.4.4.2.1   Effects on emissions

Climate change affects the sources of ozone precursors through physical response (lightning), biological response (soils, vegetation, biomass burning) and human response (energy generation, land use, agriculture). It is generally expected that lightning will increase in a warmer climate (Price and Rind, 1994a; Brasseur et al., 2005; Hauglustaine et al., 2005), although a GCM study by Stevenson et al. (2006) for the 2030 climate finds no global increase but instead a shift from the tropics to mid-latitudes. Perturbations to lightning could have a large effect on ozone in the upper troposphere (Toumi et al., 1996; Thompson et al., 2000; Martin et al., 2002; Wong et al., 2004). Mickley et al. (2001) find that observed long-term trends in ozone over the past century might be explainable by an increase in lightning.

Biomass burning in the tropics and at high latitudes is likely to increase with climate change, both as a result of increased lightning and as a result of increasing temperatures and dryness (Price and Rind, 1994b; Stocks et al., 1998; A. Williams et al., 2001; Brown et al., 2004). Biomass burning is known to make a large contribution to the budget of ozone in the tropical troposphere (Thompson et al., 1996), and there is evidence that boreal forest fires can enhance ozone throughout the extratropical NH (Jaffe et al., 2004). With climate warming, it is likely that boreal fires will increase due to a shorter duration of the seasonal snowpack and decreased soil moisture (Kasischke et al., 1995).

Biogenic VOC emissions may be highly sensitive to climate change. The most important global ozone precursors are CH_4 and isoprene. The effect of climate change on CH_4 is discussed in Section 7.4.1. The effect on NMVOCs was examined by Constable et al. (1999), Sanderson et al. (2003b), and Lathière

et al. (2005). Although biogenic NMVOC emissions increase with increasing temperature, all three studies concur that climate-driven changes in vegetation types unfavourable to isoprene emissions (notably the recession of tropical forests) would partly compensate for the effect of warming in terms of ozone generation.

7.4.4.2.2   Effects on chemistry

Changes in temperature, humidity and UV radiation intensity brought about by climate change could affect ozone significantly. Simulations with GCMs by Stevenson et al. (2000) and Grewe et al. (2001) for the 21st century indicate a decrease in the lifetime of tropospheric ozone as increasing water vapour enhances the dominant ozone sink from the oxygen radical in the 1D excited state ($O(^1D)$) plus water (H_2O) reaction. Stevenson et al. (2006) find similar results in an intercomparison of nine models for 2030 compared with 2000 climate. However, regional ozone pollution may increase in the future climate as a result of higher temperatures (see Section 7.6, Box 7.4).

7.4.4.2.3   Effects on transport

Changes in atmospheric circulation could have a major effect on tropospheric ozone. Studies using GCMs concur that STE should increase in the future climate because of the stronger Brewer-Dobson stratospheric circulation (Sudo et al., 2002a; Collins et al., 2003; Zeng and Pyle, 2003; Hauglustaine et al., 2005; Stevenson et al., 2005). Changes in vertical transport within the troposphere are also important, in view of the rapid increase in both ozone production efficiency and ozone lifetime with altitude. Convection is expected to intensify as climate warms (Rind et al., 2001), although this might not be the case in the tropics (Stevenson et al., 2005). The implications are complex, as recently discussed by Pickering et al. (2001), Lawrence et al. (2003), Olivié et al. (2004), Doherty et al. (2005) and Li et al. (2005). On the one hand, convection brings down ozone-rich air from the upper troposphere to the lower troposphere where it is rapidly destroyed, and replaces it with low-ozone air. On the other hand, injection of NO_x to the upper troposphere greatly increases its ozone production efficiency.

7.4.5   The Hydroxyl Radical

The hydroxyl radical (OH) is the primary cleansing agent of the lower atmosphere, providing the dominant sink for many greenhouse gases (e.g., CH_4, hydrochlorofluorocarbons (HCFCs), hydrofluorocarbons) and pollutants (e.g., CO, non-methane hydrocarbons). Steady-state lifetimes of these trace gases are determined by the morphology of their atmospheric distribution, the kinetics of their reaction with OH and the OH distribution. Local abundance of OH is controlled mainly by local abundances of NO_x, CO, CH_4 and higher hydrocarbons, ozone, water vapour, as well as the intensity of solar UV radiation at wavelengths shorter than 0.310 μm. New laboratory and field work also shows significant formation of $O(^1D)$ from ozone photolysis in the wavelength range between

0.310 μm and 0.350 μm (Matsumi et al., 2002; Hofzumahaus et al., 2004). The primary source of tropospheric OH is a pair of reactions starting with the photodissociation of ozone by solar UV radiation.

Additionally, in the remote, and in particular upper, troposphere, photodissociation of oxygenated volatile organic chemicals such as peroxides, acetone and other ketones, alcohols, and aldehydes may be the dominant sources of OH radical (e.g., Müller and Brasseur, 1999; Collins et al., 1999; Jaeglé et al., 2001; Tie et al., 2003; Singh et al., 2004). Over continents, measurements in the lower troposphere suggest that processing of unsaturated hydrocarbons or photolysis of carbonyls can also sustain a large pool of radicals (e.g., Handisides et al., 2003; Heard et al., 2004). Furthermore, the net formation of OH by photolysis of nitrous acid (HONO) was found to be the dominant OH radical source in urban atmospheres (e.g., Ren et al., 2003) and in a forest canopy (Kleffmann et al., 2005). The hydroxyl radical reacts with many atmospheric trace gases, in most cases as the first and rate-determining step of a reaction chain that leads to more or less complete oxidation of the compound. These chains often lead to formation of HO_2, which then reacts with ozone or NO to recycle back to OH. Tropospheric OH and HO_2 are lost through radical-radical reactions leading to the formation of peroxides or with NO_2 to form nitric acid (HNO_3). Sources and sinks of OH involve most of the fast photochemistry of the troposphere.

7.4.5.1 Changes in the Hydroxyl Radical Over Time

7.4.5.1.1 Impact of emissions

Because of its dependence on CH_4 and other pollutants, tropospheric OH is also expected to have changed since the pre-industrial era and to change in the future. Pre-industrial OH is likely to have been different than today, but because of the counteracting effects of higher CO and CH_4 (decreasing OH) and increased NO_x and ozone (increasing OH) there is still little consensus on the magnitude of this change. Several model studies suggest a decline in weighted global mean OH from pre-industrial time to the present of less than 10% (Shindell et al., 2001; Lelieveld et al., 2002a; Lamarque et al., 2005a). Other studies have reported larger decreases in global OH of 16% (Mickley et al., 1999), 25% (Wong et al., 2004) and 33% (Hauglustaine and Brasseur, 2001). The model study by Lelieveld et al. (2002b) suggests that during the past century, OH concentration decreased substantially in the marine troposphere through reaction with CH_4 and CO. However, on a global scale it has been compensated by an increase over the continents associated with strong emissions of NO_x.

Karlsdottir and Isaksen (2000) used a three-dimensional CTM accounting for varying NO_x, CO and NMVOC emissions and found a positive trend in OH of 0.43% yr^{-1} over the period 1980 to 1996. Dentener et al. (2003a,b), with a three-dimensional CTM accounting for varying emissions of ozone precursors and CH_4, meteorology and column ozone, derive a positive trend of 0.26% yr^{-1} over the 1979 to 1993 period. J. Wang et al. (2004) also use a three-dimensional CTM accounting for interannual variations in CH_4 and CO emissions, transport and column ozone to analyse the trend in CH_4 from 1988 to 1997. They do not account for interannual variability of a number of other variables that affect OH such as concentrations of NO_x, tropospheric ozone and NMVOCs. They also derive a positive trend in OH over the period considered of 0.63% yr^{-1}. Their calculated trend in OH is associated primarily with the negative trend in the overhead column ozone over the period considered and the trend is reduced to 0.16% yr^{-1} when the total ozone column is held constant.

Future changes in OH depend on relative changes in hydrocarbons compared with NO_x abundances. In the TAR, Prather et al. (2001), using scenarios reported in the IPCC SRES (IPCC, 2000) and on the basis of a comparison of results from 14 models, predicted that global OH could decrease by 10 to 18% by 2100 for five emission scenarios and increase by 5% for one scenario (which assumes large decreases in CH_4 and other ozone precursor emissions). Based on a different emission scenario for future emissions, Wang and Prinn (1999) also predicted an OH decrease of $16 \pm 3\%$ in 2100.

7.4.5.1.2 Effects of climate change

In addition to the emission changes, future increases in greenhouse gases could also induce changes in OH, arising through direct participation in OH-controlling chemistry and indirectly through stratospheric ozone changes that could increase solar UV radiation in the troposphere. OH will also be affected by changes in temperature, humidity and clouds or climate change effects on biogenic emissions of CH_4 and other ozone precursors. Changes in tropospheric water could have important chemical repercussions. The reaction between water vapour and electronically excited oxygen atoms constitutes the major source of tropospheric OH. So, in a warmer climate characterised by increased specific humidity, the abundance of OH is expected to increase. This effect was proposed by Pinto and Khalil (1991) to explain the variation of OH during the cold dry Last Glacial Maximum (LGM). It was quantified by Martinerie et al. (1995) who calculated that the global mean OH concentration during the LGM was 7% lower than at present because the atmospheric water vapour concentration was lower during that period. Valdes et al. (2005) estimate that the cold and dry LGM climate was responsible for a 7% decrease in global OH. Brasseur et al. (1998) and Johnson et al. (1999) estimated that in a warmer (doubled atmospheric CO_2) climate, the global and annual mean OH concentration would increase by 7% and 12.5%, respectively. More recently, Hauglustaine et al. (2005) use a climate-chemistry three-dimensional model to estimate a 16% reduction in global OH from the present day to 2100 accounting solely for changes in surface emissions. The effect of climate change and mainly of increased water vapour in this model is to increase global OH by 13%. In this study, the competing effects of emissions and climate change maintain the future global average OH concentration close to its present-day value. The importance of the water vapour distribution to global OH is illustrated by Lamarque et al. (2005a), who show that under reduced aerosol emissions, a warmer and moister climate significantly increases global OH concentration.

Changes in lightning NO_x emissions in a warmer climate may also affect OH. Labrador et al. (2004) show that global OH is sensitive to the magnitude of lightning NO_x emissions, and increases by 10% and 23% when global lightning is increased by a factor of 2 and 4, respectively, from a 5 TgN yr^{-1} best estimate. Similar sensitivity of global OH to the lightning source was estimated by Wang et al. (1998), who calculated a 10.6% increase in OH for a doubling of the source (from 3 to 6 TgN yr^{-1}). Regarding the large uncertainty about lightning emissions and the sensitivity of OH to the total amount of N emitted, an improved understanding of this source appears important for the ability to simulate OH accurately over time.

7.4.5.2 Consequences for Lifetimes

7.4.5.2.1 Lifetime definition

The global instantaneous atmospheric lifetime of a trace gas in the atmosphere is obtained by integrating the loss frequency l over the atmospheric domain considered. The integral must be weighted by the distribution of the trace gas on which the sink processes act. Considering a distribution of the trace gas $C(x,y,z,t)$, a global instantaneous lifetime derived from the budget can be defined as:

$$\tau_{global} = \int C \, dv \, / \int C \, l \, dv \qquad (7.4)$$

where dv is an atmospheric volume element. This expression can be averaged over one year to determine the global and annual mean lifetime. The global atmospheric lifetime (also called 'burden lifetime' or 'turnover lifetime') characterises the time required to turn over the global atmospheric burden.

The global atmospheric lifetime characterises the time to achieve an e-fold decrease of the global atmospheric burden. Unfortunately τ_{global} is a constant only in very limited circumstances. In the case where the loss rate depends on the burden, the perturbation or pulse decay lifetime (τ_{pert}) is introduced (see Velders et al., 2005). The perturbation lifetime is used to determine how a one-time pulse emission may decay as a function of time as needed for the calculation of Global Warming Potentials (GWPs). The perturbation lifetime can be distinctly different from the global atmospheric lifetime. For example, if the CH_4 abundance increases above its present-day value due to a one-time emission, the time it takes for CH_4 to decay back to its background value is longer than its global unperturbed atmospheric lifetime. This delay occurs because the added CH_4 will cause a suppression of OH, in turn increasing the background CH_4. Such feedbacks cause the decay time of a perturbation (τ_{pert}) to differ from the global atmospheric lifetime (τ_{global}). In the limit of small perturbations, the relation between the perturbation lifetime of a gas and its global atmospheric lifetime can be derived from a simple budget relationship as $\tau_{pert} = \tau_{global} / (1 - f)$, where the sensitivity coefficient $f = d\ln(\tau_{global}) / d\ln(B)$. Prather et al. (2001) estimated the feedback of CH_4 to tropospheric OH and its lifetime and determined a sensitivity coefficient $f = 0.28$, giving a ratio $\tau_{pert} / \tau_{global}$ of 1.4. Stevenson et al. (2006), from 25 CTMs,

calculate an ensemble mean and 1 standard deviation uncertainty in present-day CH_4 global lifetime τ_{global} of 8.7 ± 1.3 years, which is the AR4 updated value. The corresponding perturbation lifetime that should be used in the GWP calculation is 12 ± 1.8 years.

Perturbation lifetimes can be estimated from global models by simulating the injection of a pulse of gas and tracking the decay of the added amount. The pulse of added CO, HCFCs or hydrocarbons, by causing the concentration of OH to decrease and thus the lifetime of CH_4 to increase temporarily, causes a buildup of CH_4 while the added burden of the gas persists. Thus, changes in the emissions of short-lived gases can generate long-lived perturbations as shown in global models (Derwent et al., 2001; Wild et al., 2001; Collins et al., 2002). Changes in tropospheric ozone accompany the CH_4 decay on a 12-year time scale as an inherent component of this mode, a key example of chemical coupling in the troposphere. Any chemically reactive gas, whether a greenhouse gas or not, will produce some level of indirect greenhouse effect through its impact on atmospheric chemistry.

7.4.5.2.2 Changes in lifetime

Since OH is the primary oxidant in the atmosphere of many greenhouse gases including CH_4 and hydrogenated halogen species, changes in OH will directly affect their lifetime in the atmosphere and hence their impact on the climate system. Recent studies show that interannual variations in the chemical removal of CH_4 by OH have an important impact on the variability of the CH_4 growth rate (Johnson et al., 2002; Warwick et al., 2002; J. Wang et al., 2004). Variations in CH_4 oxidation by OH contribute to a significant fraction of the observed variations in the annual accumulation rate of CH_4 in the atmosphere. In particular, the 1992 to 1993 anomaly in the CH_4 growth rate can be explained by fluctuations in OH and wetland emissions after the eruption of Mt. Pinatubo (J. Wang et al., 2004). CH_4 variability simulated by Johnson et al. (2002), resulting only from OH sink processes, also indicates that the ENSO cycle is the largest component of that variability. These findings are consistent with the variability of global OH reconstructed by Prinn et al. (2005), Manning and Keeling (2006) and Bousquet et al. (2005), which is strongly affected by large-scale wildfires as in 1997 to 1998, by El Niño events and by the Mt. Pinatubo eruption.

The effect of climate change on tropospheric chemistry has been investigated in several studies. In most cases, the future CH_4 lifetime increases when emissions increase and climate change is ignored (Brasseur et al., 1998; Stevenson et al., 2000; Hauglustaine and Brasseur, 2001; Prather et al., 2001; Hauglustaine et al., 2005). This reflects the fact that increased levels of CH_4 and CO depress OH, reducing the CH_4 sink. However, climate warming increases the temperature-dependent CH_4 oxidation rate coefficient (Johnson et al., 1999), and increases in water vapour and NO_x concentrations tend to increase OH. In most cases, these effects partly offset or exceed the CH_4 lifetime increase due to emissions. As a consequence, the future CH_4 lifetime calculated by Brasseur et al. (1998), Stevenson et al. (2000) and Hauglustaine et al.

(2005) remains relatively constant (within a few percent) over the 21st century. In their transient simulation over the period 1990 to 2100, Johnson et al. (2001) find a dominant effect of climate change on OH in the free troposphere so that the global CH_4 lifetime declines from about 9 years in 1990 to about 8.3 years by 2025 but does not change significantly thereafter. Hence the evolution of the CH_4 lifetime depends on the relative timing of NO_x and hydrocarbon emission changes in the emission scenarios, causing the calculated CH_4 increase in 2100 to be reduced by 27% when climate change is considered. Stevenson et al. (2006) reach a similar conclusion about the relatively constant CH_4 lifetime. As a result of future changes in emissions, the CH_4 steady-state lifetime simulated by 25 state-of-the-art CTMs increases by $2.7 \pm 2.3\%$ in 2030 from an ensemble mean of 8.7 ± 1.3 years for the present day (mean $\pm$ 1 standard deviation) for a current legislation scenario of future emissions of ozone precursors. Under the 2030 warmer climate scenario, the lifetime is reduced by $4.0 \pm 1.8\%$: the total effect of both emission and climate changes reduces the CH_4 lifetime by only 1.3%.

7.4.6 Stratospheric Ozone and Climate

From about 1980 to the mid-1990s a negative trend in globally averaged total ozone occurred, due primarily to an increase in Cl and bromine loading (Montzka et al., 1999). A reduction in halogen loading appears to have occurred recently (Montzka et al., 2003) as well as the beginning of ozone recovery (e.g., Newchurch et al., 2003; Huck et al., 2005; Reinsel et al., 2005; Yang et al., 2005). Evidence suggests that a sustainable recovery of ozone is not expected before the end of the current decade (e.g., Steinbrecht et al., 2004; Dameris et al., 2006). Atmospheric concentrations of LLGHGs have increased (see Chapter 2) and are expected to continue to increase, with consequences for the ozone layer. This section assesses current understanding of interactions and feedbacks between stratospheric ozone and climate. More detailed discussions can be found in recent reports (European Commission, 2003; IPCC/TEAP, 2005).

7.4.6.1   Interactions

Stratospheric ozone is affected by climate change through changes in dynamics and in the chemical composition of the troposphere and stratosphere. An increase in the concentrations of LLGHGs, especially CO_2, cools the stratosphere, allowing the possibility of more PSCs, and alters the ozone distribution (Rosenlof et al., 2001; Rosenfield et al., 2002; Randel et al., 2004, 2006; Fueglistaler and Haynes, 2005). With the possible exception of the polar lower stratosphere, a decrease in temperature reduces ozone depletion leading to higher ozone column amounts and a positive correction to the LLGHG-induced radiative cooling of the stratosphere. Moreover, ozone itself is a greenhouse gas and absorbs UV radiation in the stratosphere. Absorption of UV radiation provides the heating responsible for the observed temperature increase with height above the tropopause. Changes in stratospheric temperatures, induced by changes in ozone or LLGHG concentration, alter the Brewer-Dobson circulation (Butchart and Scaife, 2001; Butchart et al., 2006), controlling the rate at which long-lived molecules, such as LLGHGs, CFCs, HCFCs and halogens are transported from the troposphere to various levels in the stratosphere. Furthermore, increases in the Brewer-Dobson circulation increase temperatures adiabatically in the polar regions and decrease temperatures adiabatically in the tropics.

Climate is affected by changes in stratospheric ozone, which radiates infrared radiation down to the troposphere. For a given percentage change in the vertical structure of ozone, the largest dependence of the radiative forcing is in the upper troposphere and ozone layer regions (e.g., TAR, Figure 6.1). Past ozone depletion has induced surface cooling (Chapter 2). The observed decrease in stratospheric ozone and the resultant increase in UV irradiance (e.g., Zerefos et al., 1998; McKenzie et al., 1999) have affected the biosphere and biogenic emissions (Larsen, 2005). Such UV radiation increases lead to an enhanced OH production, reducing the lifetime of CH_4 and influencing tropospheric ozone, both important greenhouse gases (European Commission, 2003). In addition to global mean equilibrium surface temperature changes, local surface temperature changes have been identified by Gillett and Thompson (2003) as a result of ozone loss from the lower stratosphere. Observational (e.g., Baldwin and Dunkerton, 1999, 2001; Thompson et al., 2005) and modelling (Polvani and Kushner, 2002; Norton, 2003; Song and Robinson, 2004; Thompson et al., 2005) evidence exists for month-to-month changes to the stratospheric flow feedback to the troposphere, affecting its circulation. Model results show that trends in the SH stratosphere can affect high-latitude surface climate (Gillett and Thompson, 2003).

7.4.6.2   Past Changes in Stratospheric Ozone

Ozone losses have been largest in the polar lower stratosphere during later winter and spring. For example, the ozone hole over Antarctica has occurred every spring since the early 1980s (Fioletov et al., 2002). Antarctic ozone destruction is driven by climatologically low temperatures combined with high Cl and bromine amounts produced from photochemical breakdown of primarily anthropogenic CFCs and halons. Similar losses, smaller in magnitude, have occurred over the Arctic due to the same processes during cold winters. During warm winters, arctic ozone has been relatively unaffected (Tilmes et al., 2004). The antarctic lower stratosphere is nearly always cold enough to produce substantial ozone loss, but in the year 2002, a sudden stratospheric warming split the early ozone hole into two separate regions (e.g., Simmons et al., 2005). Temperatures were subsequently too high to produce further ozone loss. Following the later merging of the two separate regions back into a single vortex, the dynamical conditions were unsuitable for further ozone loss. This is not an indication of recovery in ozone amounts, but rather the result of a dynamical disturbance (e.g., Newman et al., 2004). A summary of recent stratospheric ozone changes is given in Chapter 2.

7.4.6.3   Future Changes in Stratospheric Ozone

The evolution of stratospheric ozone over the next few decades will depend on natural, including solar and volcanic activity (e.g., Steinbrecht et al., 2004; Dameris et al., 2005), and human-caused factors such as stratospheric halogen loading, which is expected to decrease over future decades (WMO, 2003; IPCC/TEAP, 2005). The evolution of ozone will also depend on changes in many stratospheric constituents: it is expected that the reduction of ozone-depleting substances in the 21st century will cause ozone to increase via chemical processes (Austin et al., 2003). However, this increase could be strongly affected by temperature changes (due to LLGHGs), other chemical changes (e.g., due to water vapour) and transport changes. Coupled Chemistry-Climate Models (CCMs) provide tools to simulate future atmospheric composition and climate. For this purpose, a set of consistent model forcings has been prescribed as part of the CCM Validation Activity for Stratospheric Processes and their Role in Climate (SPARC CCMVal; Eyring et al., 2005). Forcings include natural and anthropogenic emissions based on existing scenarios, atmospheric observations and the Kyoto and Montreal Protocols and Amendments. The simulations follow the IPCC SRES scenario A1B (IPCC, 2000) and changes in halocarbons as prescribed in Table 4B-2 of WMO (2003). Figure 7.18 shows the late winter minimum total column ozone poleward of 60° for various transient CCM reference simulations compared with observations. Antarctic ozone follows mainly the behaviour of Cl and bromine in the models. The peak depletion simulated by the CCMs occurs around the year 2000 followed by a slow increase with minimum values remaining constant between 2000 and 2010 in many models. Most models predict that antarctic ozone amounts will increase to 1980 values close to the time when modelled halogen amounts decrease to 1980 values, lagging the recovery in mid-latitudes due to the delay associated with transport of stratospheric air to polar regions. The late return to pre-1980 values by about 2065 in the Atmospheric Model with Transport and Chemistry (AMTRAC) model (Austin and Wilson, 2006) is consistent with an empirical model study based on observations (Newman et al., 2006). Moreover, increased atmospheric fluxes of CFCs have recently been reported (Hurst et al., 2006), which may point to a still later recovery. The CCMs do not predict consistent values for minimum arctic column ozone, with some models showing large discrepancies with observations. In all CCMs that have been run long enough, arctic ozone increases to 1980 values before antarctic ozone does, by as much as 30 years (e.g., Austin and Wilson 2006). This delay in the Antarctic arises from an increased Brewer-Dobson circulation (Butchart and Scaife, 2001; Butchart et al., 2006) combined with a reduction in stratospheric temperatures.

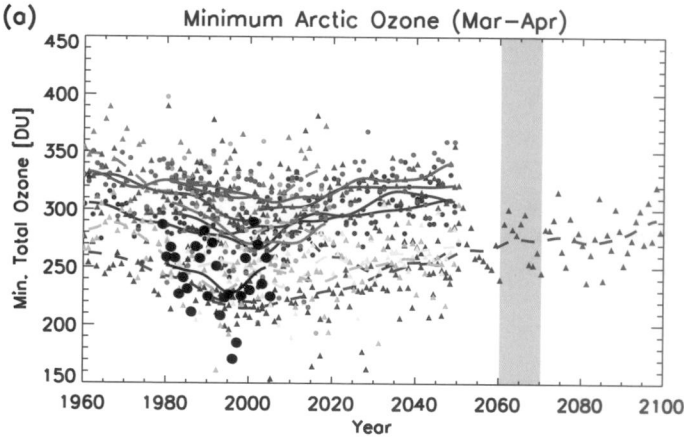

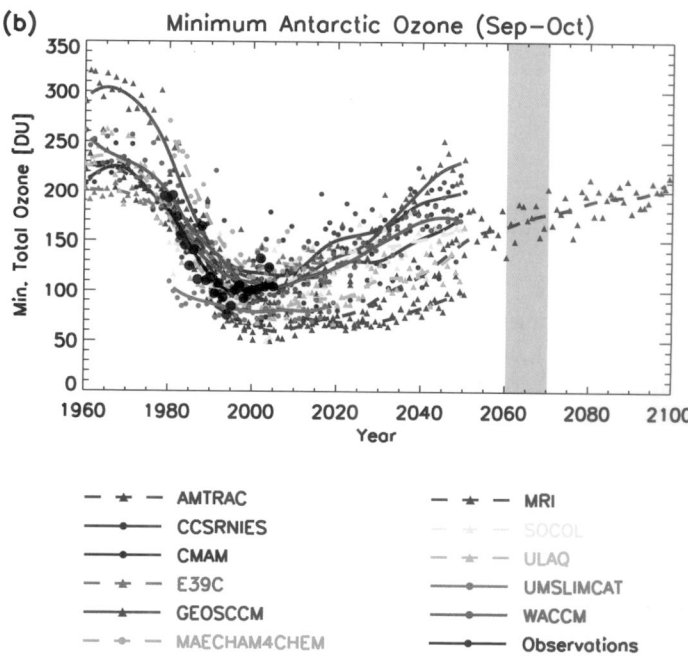

Figure 7.18. *(a) Minimum arctic total column ozone for March to April and (b) minimum antarctic total column ozone for September to October (both poleward of 60°) in Dobson Units (DU). Simulations of future evolution of ozone were performed by 11 CCMs analysed as part of the CCM Validation Activity for SPARC (http://www. pa.op.dlr.de/CCMVal/). Model results are compared with values calculated from the National Institute of Water and Atmospheric Research (NIWA) assimilated total column ozone database shown as black dots (Bodeker et al., 2005). The light grey shading between 2060 and 2070 shows the period when halogen amounts in the polar lower stratosphere are expected to return to 1980 values. Models include AMTRAC: Atmospheric Model with Transport and Chemistry; CCSRNIES: Center for Climate System Research - National Institute for Environmental Studies; CMAM: Canadian Middle Atmosphere Model; E39C: German Aerospace Center (DLR) version of ECHAM4 with chemistry and 39 levels; GEOSCCM: Goddard Earth Observing System Chemistry-Climate Model; MAECHAM4/CHEM: Middle Atmosphere ECHAM4 with Chemistry; MRI: Meteorological Research Institute; SOCOL: Solar Climate Ozone Links; ULAQ: University of L'Aquila; UMSLIMCAT: Unified Model SLIMCAT; WACCM: Whole Atmosphere Community Climate Model.*

7.4.6.4   Uncertainties Due to Atmospheric Dynamics

Changes in atmospheric dynamics could affect ozone. For example, sub-grid scale processes such as gravity wave propagation (e.g., Warner and McIntyre, 2001), prescribed for past and present conditions, may change in the future. Tropospheric climate changes will also alter planetary-scale waves. Together with changes in orographic gravity waves, these waves give rise to the increase in the Brewer-Dobson circulation seen in most models (Butchart et al., 2006). The magnitude of this effect varies from model to model and leads to increased adiabatic heating of the polar regions, compensating in part the increased radiative cooling from CO_2 increases. Hence, the net heating or cooling is subject to large uncertainty, and available model simulations do not give a consistent picture of future development of ozone, particularly in the Arctic (Figure 7.18).

7.5   Aerosol Particles and the Climate System

Aerosols are an integral part of the atmospheric hydrological cycle and the atmosphere's radiation budget, with many possible feedback mechanisms that are not yet fully understood. This section assesses (1) the impact of meteorological (climatic) factors like wind, temperature and precipitation on the natural aerosol burden and (2) possible effects of aerosols on climate parameters and biogeochemistry. The most easily understood interaction between aerosols and climate is the direct effect (scattering and absorption of shortwave and thermal radiation), which is discussed in detail in Chapter 2. Interactions with the hydrological cycle, and additional impacts on the radiation budget, occur through the role of aerosols in cloud microphysical processes, as aerosol particles act as cloud condensation nuclei (CCN) and ice nuclei (IN). The suite of possible impacts of aerosols through the modification of cloud properties is called 'indirect effects'. The forcing aspect of the indirect effect at the top of the atmosphere is discussed in Chapter 2, while the processes that involve feedbacks or interactions, like the 'cloud lifetime effect'[6], the 'semi-direct effect' and aerosol impacts on the large-scale circulation, convection, the biosphere through nutrient supply and the carbon cycle, are discussed here.

7.5.1   Aerosol Emissions and Burdens Affected by Climatic Factors

Most natural aerosol sources are controlled by climatic parameters like wind, moisture and temperature. Hence, human-induced climate change is also expected to affect the natural aerosol burden. The sections below give a systematic overview of the major natural aerosol sources and their relations to climate parameters while anthropogenic aerosol emissions and combined aerosols are the subject of Chapter 2.

7.5.1.1   Dust

Estimates of the global source strength of bulk dust aerosols with diameters below 10 µm of between 1,000 and 3,000 Tg yr^{-1} agree well with a wide range of observations (Duce, 1995; Textor et al., 2005; Cakmur et al., 2006). Seven to twenty percent of the dust emissions are less than 1 µm in diameter (Cakmur et al., 2006; Schulz et al., 1998). Zhang et al. (1997) estimated that about 800 Tg yr^{-1} of Asian dust emissions are injected into the atmosphere annually, about 30% of which is re-deposited onto the deserts and 20% is transported over regional scales, while the remaining approximately 50% is subject to long-range transport to the Pacific Ocean and beyond. Asian dust appears to be a continuous source that dominates background dust aerosol concentrations on the west coast of the USA (Duce, 1995; Perry et al., 2004). Uncertainties in the estimates of global dust emissions are greater than a factor of two (Zender et al., 2004) due to problems in validating and modelling the global emissions. The representation of the high wind tail of the wind speed distribution alone, responsible for most of the dust flux, leads to differences in emissions of more then 30% (Timmreck and Schulz, 2004). Observations suggest that annual mean African dust may have varied by a factor of four during 1960 to 2000 (Prospero and Lamb, 2003), possibly due to rainfall variability in the Sahel zone. Likewise, simulations of dust emissions in 2100 are highly uncertain, ranging from a 60% decrease to a factor of 3.8 increase as compared to present-day dust emissions (Mahowald and Luo, 2003; Tegen et al., 2004; Woodward et al., 2005; Stier et al., 2006a). Reasons for these discrepancies include different treatments of climate-biosphere interactions and the climate model used to drive the vegetation and dust models. The potentially large impact of climate change on dust emissions shows up in particular when comparing present-day with LGM conditions for dust erosion (e.g., Werner et al., 2002).

The radiative effect of dust, which, for example, could intensify the African Easterly Waves, may be a feedback mechanism between climate and dust (Jones et al., 2004). It also alters the atmospheric circulation, which feeds back to dust emission from natural sources (see Section 7.5.4). Perlwitz et al. (2001) estimate that this feedback reduces the global dust load by roughly 15%, as dust radiative forcing reduces the downward mixing of momentum within the planetary boundary layer, the surface wind speed, and thus dust emission (Miller et al., 2004a). In addition to natural dust production, human activities have created another potential source for dust mobilisation through desertification. The contribution to global dust emission of desertification through human activities is uncertain: estimates vary from 50% (Tegen et al., 1996; Mahowald et al., 2004) to less than 10% (Tegen et al., 2004) to insignificant values (Ginoux et al., 2001; Prospero et al., 2002). A 43-year estimate of Asian dust emissions reveals that meteorology and climate have a greater influence on Asian

---

6   The processes involved are more complex than can be encompassed in a single expression. The term 'cloud lifetime effect' thus should be understood to mean that aerosols can change precipitation efficiency in addition to increasing cloud albedo.

dust emissions and associated Asian dust storm occurrences than does desertification (Figure 7.19; Zhang et al., 2003).

In addition, aerosol deposition affects global ecosystems. Deposition of mineral dust plays an important role in the biogeochemical cycle of the oceans, by providing the nutrient iron, which affects ocean biogeochemistry with feedbacks to climate and dust production (Jickells et al., 2005; Section 7.3.4.4). Conversely, water-soluble particulate iron over the Pacific Ocean is linked to elemental carbon emissions resulting from anthropogenic activity in Asia (Chuang et al., 2005). The input of trace elements by dust deposition is also of importance to terrestrial ecosystems. For example, it has been proposed that the vegetation of the Amazon basin is highly dependent on Saharan dust deposition, which provides phosphorus, necessary for maintenance of long-term productivity (Okin et al., 2004; Section 7.3). The Hawaiian Islands also depend on phosphorus from Asian dust transport (Chadwick et al., 1999). Moreover, mineral dust can act as a sink for acidic trace gases, such as sulphur dioxide (SO_2) and HNO_3, and thereby interact with the sulphur and N cycles (e.g., Dentener et al., 1996; Umann et al., 2005). Coatings with soluble substances, such as sulphate or nitrate, will change the ability of mineral dust aerosols to nucleate cloud droplets (Levin et al., 1996; Section 7.5.2.1).

7.5.1.2 Sea Salt

Sea salt aerosol is a key aerosol constituent of the marine atmosphere. Sea salt aerosol particles affect the formation of clouds and rain. They serve as sinks for reactive gases and small particles and possibly suppress new particle formation. Sea salt is also responsible for a large fraction of the non-sea salt sulphate formation (e.g., Sievering et al., 1992). The major meteorological and environmental factors that affect sea salt formation are wind speed, atmospheric stability and wind friction velocity, sea surface and air temperatures, present and prior rain or snow and the amount and nature of surface-active materials in the near-surface ocean waters (Lewis and Schwartz, 2005). The average annual global sea salt flux from 12 models is estimated to be 16,300 Tg ± 200% (Textor et al., 2005) of which 15% is emitted into the submicron mode.

7.5.1.3 Natural Organic Carbon

Biogenic organic material is both directly emitted into the atmosphere and produced by VOCs. Primary emissions from the continents have been thought to be a relatively minor source but some studies suggest that these emissions could be much higher

a)

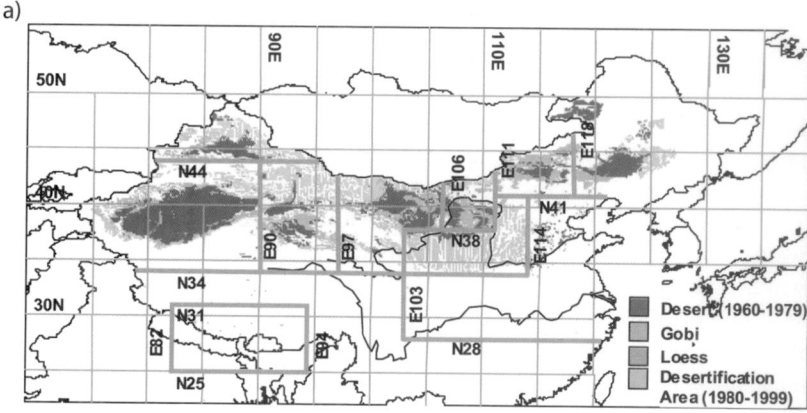

b)

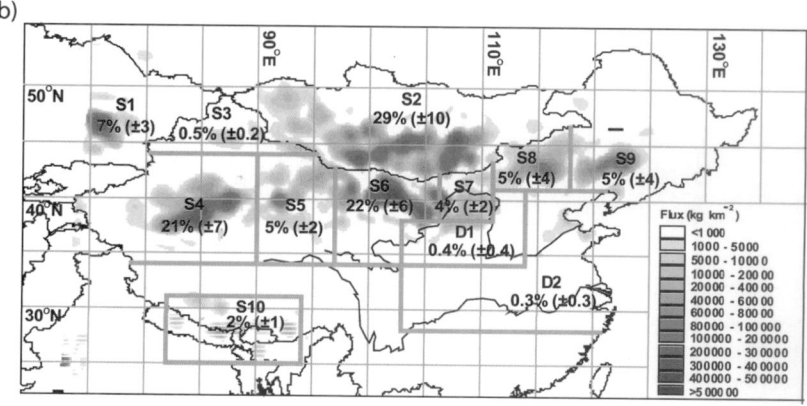

Figure 7.19. *(a) Chinese desert distributions from 1960 to 1979 and desert plus desertification areas from 1980 to 1999. (b) Sources (S1 to S10) and typical depositional areas (D1 and D2) for Asian dust indicated by spring average dust emission flux (kg km⁻² per month) averaged over 1960 to 2002. The percentages with standard deviations in parentheses denote the average amount of dust production in each source region and the total amount of emissions between 1960 and 2002. The deserts in Mongolia (S2) and in western (S4) and northern (S6) China (mainly the Taklimakan and Badain Juran, respectively) can be considered the major sources of Asian dust emissions. Several areas with more expansions of deserts (S7, S8, S9 and S5) are not key sources. Adapted from Zhang et al. (2003).*

than previously estimated (Folberth et al., 2005; Jaenicke, 2005). Kanakidou et al. (2005) estimate a global biogenic secondary organic aerosol production of about 30 Tg yr⁻¹ and recognise the potentially large, but uncertain, flux of primary biogenic particles. Annual global biogenic VOC emission estimates range from 500 to 1,200 Tg yr⁻¹ (Guenther et al., 1995). There is a large range (less than 5 to greater than 90%) of organic aerosol yield for individual compounds and atmospheric conditions resulting in estimates of global annual secondary organic aerosol production from biogenic VOCs that range from 2.5 to 44.5 Tg of organic matter per year (Tsigaridis and Kanakidou, 2003). All biogenic VOC emissions are highly sensitive to changes in temperature, and some emissions respond to changes in solar radiation and precipitation (Guenther et al., 1995). In addition to the direct response to climatic changes, biogenic VOC emissions are also highly sensitive to climate-induced changes in plant species composition and biomass distributions.

Global biogenic VOC emissions respond to climate change (e.g., Turner et al., 1991; Adams et al., 2001; Penner et al., 2001; Sanderson et al., 2003b). These model studies predict that solar radiation and climate-induced vegetation change can affect emissions, but they do not agree on the sign of the change. Emissions are predicted to increase by 10% per °C (Guenther et al., 1993). There is evidence of physiological adaptation to higher temperatures that would lead to a greater response for long-term temperature changes (Guenther et al., 1999). The response of biogenic secondary organic carbon aerosol production to a temperature change, however, could be considerably lower than the response of biogenic VOC emissions since aerosol yields can decrease with increasing temperature. A potentially important feedback among forest ecosystems, greenhouse gases, aerosols and climate exists through increased photosynthesis and forest growth due to increasing temperatures and CO_2 fertilization (Kulmala et al., 2004). Increased forest biomass would increase VOC emissions and thereby organic aerosol production. This couples the climate effect of CO_2 with that of aerosols.

New evidence shows that the ocean also acts as a source of organic matter from biogenic origin (O'Dowd et al., 2004; Leck and Bigg, 2005b). O'Dowd et al. (2004) show that during phytoplankton blooms (summer conditions), the organic aerosols can constitute up to 63% of the total aerosol. Surface-active organic matter of biogenic origin (such as lipidic and proteinaceous material and humic substances), enriched in the oceanic surface layer and transferred to the atmosphere by bubble-bursting processes, are the most likely candidates to contribute to the observed organic fraction in marine aerosol. Insoluble heat-resistant organic sub-micrometre particles (peaking at 40 to 50 nm in diameter), mostly combined into chains or aggregated balls of 'marine microcolloids' linked by an amorphous electron-transparent material with properties entirely consistent with exopolymer secretions (Decho, 1990; Verdugo et al., 2004), are found in near-surface water of lower-latitude oceans (Benner et al., 1992; Wells and Goldberg, 1994), in leads between ice floes (Bigg et al., 2004), above the arctic pack ice (Leck and Bigg, 2005a) and over lower-latitude oceans (Leck and Bigg, 2005b). This aerosol formation pathway may constitute an ice (microorganisms)-ocean-aerosol-cloud feedback.

7.5.1.4   *Aerosols from Dimethyl Sulphide*

Dimethyl sulphide produced by phytoplankton is the most abundant form in which the ocean releases gaseous sulphur. Sea-air fluxes of DMS vary by orders of magnitude depending mainly on DMS sea surface concentration and on wind speed. Estimates of the global DMS flux vary widely depending mainly on the DMS sea surface climatology utilised, sea-air exchange parametrization and wind speed data, and range from 16 to 54 Tg yr^{-1} of sulphur (see Kettle and Andreae, 2000 for a review). According to model studies (Gondwe et al., 2003; Kloster et al., 2006), 18 to 27% of the DMS is converted into sulphate aerosols. Penner et al. (2001) show a small increase in DMS emissions between 2000 and 2100 (from 26.0 to 27.7

Tg yr^{-1} of sulphur) using constant DMS sea surface concentrations together with a constant monthly climatological ice cover. Gabric et al. (2004) predict an increase of the globally integrated DMS flux perturbation of 14% for a tripling of the pre-industrial atmospheric CO_2 concentration.

Bopp et al. (2004) estimate the feedback of DMS to cloud albedo with a coupled atmosphere-ocean-biogeochemical climate model that includes phytoplankton species in the ocean and a sulphur cycle in the atmospheric climate model. They obtain an increase in the sea-air DMS flux of 3% for doubled atmospheric CO_2 conditions, with large spatial heterogeneities (–15 to +30%). The mechanisms affecting those fluxes are marine biology, relative abundance of phytoplankton types and wind intensity. The simulated increase in fluxes causes an increase in sulphate aerosols and, hence, cloud droplets resulting in a radiative perturbation of cloud albedo of –0.05 W m^{-2}, which represents a small negative climate feedback to global warming.

7.5.1.5   *Aerosols from Iodine Compounds*

Intense new aerosol particle formation has been frequently observed in the coastal environment (O'Dowd et al., 2002a). Simultaneous coastal observations of reactive iodine species (Saiz-Lopez et al., 2005), chamber studies using iodocarbon precursors and laboratory characterisation of iodine oxide particles formed from exposure of *Laminaria* macroalgae to ozone (McFiggans et al., 2004) have demonstrated that coastal particle formation is linked to iodine compound precursor released from abundant infralittoral beds of macroalgae. The particle bursts overwhelmingly occur during daytime low tides (O'Dowd et al., 2002b; Saiz-Lopez et al., 2005). Tidal exposure of kelp leads to the well-documented release of significant fluxes of iodocarbons (Carpenter et al., 2003), the most photolabile of which, di-iodomethane (CH_2I_2), may yield a high iodine atom flux. However, the iodine monoxide (IO) and iodine dioxide (OIO) radicals, and new particles are thought more likely to result from emissions of molecular iodine (McFiggans et al., 2004), which will yield a much greater iodine atom flux (Saiz-Lopez and Plane, 2004). It is unclear whether such particles grow sufficiently to act as CCN (O'Dowd, 2002; Saiz-Lopez et al., 2005). Thus, a hitherto undiscovered remote ocean source of iodine atoms (such as molecular iodine) must be present if iodine-mediated particle formation is to be important in the remote marine boundary layer (McFiggans, 2005).

7.5.1.6   *Climatic Factors Controlling Aerosol Burdens and Cycling*

As discussed above, near-surface wind speed determines the source strength for primary aerosols (sea salt, dust, primary organic particles) and precursors of secondary aerosols (mainly DMS). Progress has been made in the development of source functions (in terms of wind speed) for sea salt and desert dust (e.g., Tegen et al., 2002; Gong, 2003; Balkanski et al., 2004).

Wind speed also affects dry deposition velocities and hence the lifetime of aerosols. In addition, biogenic emissions are strongly dependent on temperature (together with humidity/moisture; e.g., Guenther et al., 1995). Temperature also is a key factor in the gas-aerosol partitioning of semi-volatile secondary organics (Kanakidou et al., 2005).

Precipitation directly affects the wet removal and hence the lifetime of atmospheric aerosols. More aerosols decrease the precipitation formation rate, which in turn increases the lifetime of aerosols and results in more long-range aerosol transport to remote regions where wet removal is less efficient. At the same time, precipitating boundary layer clouds maintain themselves by keeping aerosol concentrations low (e.g., Baker and Charlson, 1990; Stevens et al., 2005; Sharon et al., 2006). Precipitation also affects soil moisture, with impacts on dust source strength and on stomatal opening/closure of plant leaves, hence affecting biogenic emissions. Cloud processing is an important pathway in the gas-to-particle conversion. It is the most important oxidation pathway for sulphate aerosols and shifts the aerosol size distribution to larger sizes, such that aerosols are more easily activated in subsequent cloud events

(e.g., Hoppel et al., 1990; Kerkweg et al., 2003; Yin et al., 2005). It is also important in the conversion of hydrophobic to hydrophilic carbon.

Aerosol burden and lifetime are also affected by microphysical interactions among the different aerosol compounds as well as by changes in the spatial and seasonal distribution of the emissions. Sea salt aerosols, for example, provide surfaces for conversion of SO_2 into sulphate aerosols (Sievering et al., 1992) with consequences for cloud formation (Gong and Barrie, 2003; Section 7.5.2.1). A future reduction in SO_2 emissions and the associated reduced conversion of hydrophobic to hydrophilic soot could lead to a prolonged residence time of soot (Cooke et al., 2002; Stier et al., 2006b) and increased ammonium nitrate (Liao and Seinfeld, 2005). However, in a transient AOGCM climate simulation with an embedded microphysical aerosol module, Stier et al. (2006a) show that the effect on the hydrophobic to hydrophilic conversion can be outweighed by a general shift to low-latitude dry-season soot emissions. Consequently, soot lifetime increases in a future climate despite an enhanced conversion of hydrophobic to hydrophilic soot.

Table 7.10a. *Overview of the different aerosol indirect effects and their sign of the net radiative flux change at the top of the atmosphere (TOA).*

Effect	Cloud Types Affected	Process	Sign of Change in TOA Radiation	Potential Magnitude	Scientific Understanding
Cloud albedo effect	All clouds	For the same cloud water or ice content more but smaller cloud particles reflect more solar radiation	Negative	Medium	Low
Cloud lifetime effect	All clouds	Smaller cloud particles decrease the precipitation efficiency thereby presumably prolonging cloud lifetime	Negative	Medium	Very low
Semi-direct effect	All clouds	Absorption of solar radiation by absorbing aerosols affects static stability and the surface energy budget, and may lead to an evaporation of cloud particles	Positive or negative	Small	Very low
Glaciation indirect effect	Mixed-phase clouds	An increase in IN increases the precipitation efficiency	Positive	Medium	Very low
Thermodynamic effect	Mixed-phase clouds	Smaller cloud droplets delay freezing causing super-cooled clouds to extend to colder temperatures	Positive or negative	Medium	Very low

Table 7.10b. *Overview of the different aerosol indirect effects and their implications for the global mean net shortwave radiation at the surface, F_{sfc} (Columns 2-4) and for precipitation (Columns 5-7).*

Effect	Sign of Change in F_{sfc}	Potential Magnitude	Scientific Understanding	Sign of Change in Precipitation	Potential Magnitude	Scientific Understanding
Cloud albedo effect	Negative	Medium	Low	n.a.	n.a.	n.a.
Cloud lifetime effect	Negative	Medium	Very low	Negative	Small	Very low
Semi-direct effect	Negative	Large	Very low	Negative	Large	Very low
Glaciation indirect effect	Positive	Medium	Very low	Positive	Medium	Very low
Thermodynamic effect	Positive or negative	Medium	Very low	Positive or negative	Medium	Very low

7.5.2 Indirect Effects of Aerosols on Clouds and Precipitation

Aerosols can interact with clouds and precipitation in many ways, acting either as CCN or IN, or as absorbing particles, redistributing solar energy as thermal energy inside cloud layers. These indirect effects (in contrast to the direct interaction with radiation, see Chapter 2) are the subject of this subsection. They can be subdivided into different contributing processes, as summarised in Table 7.10 and shown in Figure 7.20. Cloud feedbacks remain the largest source of uncertainty in climate sensitivity estimates and the relatively poor simulation of boundary layer clouds in the present climate is a reason for some concern (see Chapter 8). Therefore the results discussed below need to be considered with caution.

The cloud-albedo effect, that is, the distribution of the same cloud liquid water content over more, hence smaller, cloud droplets leading to higher cloud reflectivity, is a purely radiative forcing and is therefore treated in Chapter 2. The other effects involve feedbacks in the climate system and are discussed here. The albedo effect cannot be easily separated from the other effects; in fact, the processes that decrease the cloud droplet size per given liquid water content also decrease precipitation formation, presumably prolonging cloud lifetime (cloud lifetime effect, Section 7.5.2.1 and Figure 7.20). In turn, an increase in cloud lifetime also contributes to a change in the time-averaged cloud albedo. The

Cloud albedo and lifetime effect (negative radiative effect for warm clouds at TOA; less precipitation and less solar radiation at the surface)

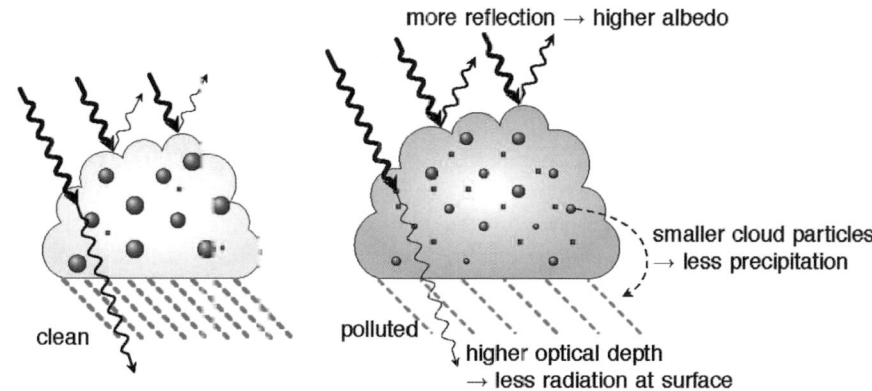

Semi-direct effect (positive radiative effect at TOA for soot inside clouds, negative for soot above clouds)

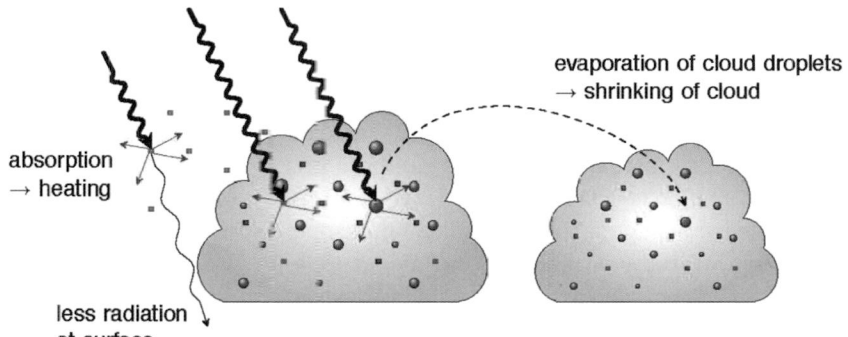

Glaciation effect (positive radiative effect at TOA and more precipitation), thermodynamic effect (sign of radiative effect and change in precipitation not yet known)

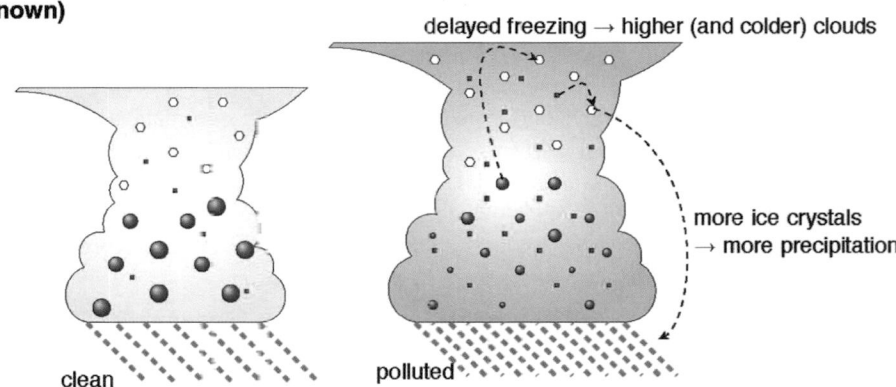

Figure 7.20. *Schematic diagram of the aerosol effects discussed in Table 7.10. TOA refers to the top-of-the-atmosphere.*

semi-direct effect refers to the absorption of solar radiation by soot, re-emitted as thermal radiation, hence heating the air mass and increasing static stability relative to the surface. It may also cause evaporation of cloud droplets (see Sections 2.4 and 7.5.4.1 and Figure 7.20). The glaciation effect refers to an increase in IN resulting in a rapid glaciation of a super-cooled liquid water cloud due to the difference in vapour pressure over ice and water. Unlike cloud droplets, these ice crystals grow in an environment of high super-saturation with respect to ice,

quickly reaching precipitation size, with the potential to turn a non-precipitating cloud into a precipitating cloud (Section 7.5.2.2 and Figure 7.20). The thermodynamic effect refers to a delay in freezing by the smaller droplets causing super-cooled clouds to extend to colder temperatures (Section 7.5.2.2 and Figure 7.20). In addition to aerosol-induced changes at the top of the atmosphere (TOA), aerosols affect the surface energy budget (Table 7.10b; Section 7.5.2) with consequences for convection, evaporation and precipitation (Figure 7.20).

7.5.2.1 Aerosol Effects on Water Clouds and Warm Precipitation

Aerosols are hypothesised to increase the lifetime of clouds because increased concentrations of smaller droplets lead to decreased drizzle production and reduced precipitation efficiency (Albrecht, 1989). It is difficult to devise observational studies that can separate the cloud lifetime from the cloud albedo effect (see Section 2.4). Thus, observational studies usually provide estimates of the combined effects. Similarly, climate models cannot easily separate the cloud lifetime indirect effect once the aerosol scheme is fully coupled to a cloud microphysics scheme, but also predict the combined cloud albedo, lifetime and semi-direct effect.

Evidence for the absence of a drizzle mode due to anthropogenic emissions of aerosols and their precursors comes, for instance, from ship tracks perturbing marine stratus cloud decks off the coast of California (Ferek et al., 1998) as well as from analysing polluted compared with clean clouds off the Atlantic coast of Canada (Peng et al., 2002). One problem is that most climate models suggest an increase in liquid water when adding anthropogenic aerosols, whereas newer ship track studies show that polluted marine water clouds can have less liquid water than clean clouds (Platnick et al., 2000; Coakley and Walsh, 2002). Ackerman et al. (2004) attribute this effect to enhanced entrainment of dry air in polluted clouds in these instances with subsequent evaporation of cloud droplets. Similarly, when cloud lifetime is analysed, an increase in aerosol concentration from very clean to very polluted does not increase cloud lifetime, even though precipitation is suppressed (Jiang et al., 2006). This effect is due to competition between precipitation suppression and enhanced evaporation of the more numerous smaller cloud droplets in polluted clouds. Observed lower aerosol concentrations in pockets of open cells (Stevens et al., 2005) and in rifts of broken clouds surrounded by solid decks of stratocumulus with higher aerosol concentrations (Sharon et al., 2006) are manifestations of two stable aerosol regimes (Baker and Charlson, 1990). The low aerosol concentration regimes maintain themselves by higher drizzle rates. However, it is hard to disentangle cause and effect from these studies.

Smoke from burning vegetation reduces cloud droplet sizes and delays the onset of precipitation (Warner and Twomey, 1967; Rosenfeld, 1999; Andreae et al., 2004). In addition, desert dust suppresses precipitation in thin low-altitude clouds (Rosenfeld et al., 2001; Mahowald and Kiehl, 2003). Contradictory results have been found regarding the suppression of precipitation by aerosols downwind of urban areas (Givati and Rosenfeld, 2004; Jin et al., 2005) and in Australia (Rosenfeld, 2000; Ayers, 2005).

Models suggest that anthropogenic aerosols suppress precipitation in the absence of giant CCN and aerosol-induced changes in ice microphysics (e.g., Lohmann, 2002; Menon and DelGenio, 2007) as well as in mixed-phase clouds where the ice phase only plays a minor role (Phillips et al., 2002). A reduction in precipitation formation leads to increased cloud processing

of aerosols. Feingold et al. (1998) and Wurzler et al. (2000) showed that cloud processing could either lead to an increase or decrease in precipitation formation in subsequent cloud cycles, depending on the size and concentration of activated CCN. Giant sea salt nuclei, on the other hand, may override the precipitation suppression effect of the large number of small pollution nuclei (Johnson, 1982; Feingold et al., 1999; Rosenfeld et al., 2002). Likewise, Gong and Barrie (2003) predict a reduction of 20 to 60% in marine cloud droplet number concentrations and an increase in precipitation when interactions of sulphate with sea salt aerosols are considered. When aerosol effects on warm convective clouds are included in addition to their effect on warm stratiform clouds, the overall indirect aerosol effect and the change in surface precipitation can be larger or smaller than if just the aerosol effect on stratiform clouds is considered (Nober et al., 2003; Menon and Rotstayn, 2006). Besides changes in the distribution of precipitation, the frequency of extreme events may also be reduced by the presence of aerosols (Paeth and Feichter, 2006).

Observations show that aerosols can decrease or increase cloud cover. Kaufman et al. (2005) conclude from satellite observations that the aerosol indirect effect is likely primarily due to an increase in cloud cover, rather than an increase in cloud albedo. In contrast, model results of Lohmann et al. (2006) associate the increase in cloud cover with differing dynamic regimes and higher relative humidities that maintain higher aerosol optical depths. On the other hand, the semi-direct effect of absorbing aerosols can cause evaporation of cloud droplets and/or inhibit cloud formation. In a large area with absorbing biomass-burning aerosol, few low-lying clouds were observed when the aerosol optical depth exceeded 1.2 (Koren et al., 2004). Increasing emissions of absorbing aerosols from the late 1980s to the late 1990s in China also reduced cloud amount leading to a decrease in local planetary albedo, as deduced from satellite data (Krüger and Grassl, 2004). When the combined effect of pollution and smoke aerosols is considered from ground-based observations, the net effect seems to be an increase in cloud cover with increasing aerosol column concentrations (Kaufman and Koren, 2006).

7.5.2.2 Aerosol Impacts on Mixed-Phase Clouds

As satellite observations of aerosol effects on mixed-phase clouds are not conclusive (Mahowald and Kiehl, 2003), this section only refers to model results and field studies. Studies with GCMs suggest that if, in addition to mineral dust, hydrophilic black carbon aerosols are assumed to act as IN at temperatures between 0°C and −35°C, then increases in aerosol concentration from pre-industrial to present times may cause a glaciation indirect effect (Lohmann, 2002). Increases in IN can result in more frequent glaciation of super-cooled stratiform clouds and increase the amount of precipitation via the ice phase, which could decrease the global mean cloud cover leading to more absorption of solar radiation. Whether the glaciation effect or warm cloud lifetime effect is larger depends on the chemical nature of the dust (Lohmann and Diehl, 2006). Likewise, the number and size of ice particles in convective mixed-phase clouds is sensitive to the chemical composition of the insoluble fraction (e.g., dust, soot, biological particles) of the aerosol particles (Diehl and Wurzler, 2004).

Rosenfeld (1999) and Rosenfeld and Woodley (2000) analysed aircraft data together with satellite data suggesting that pollution aerosols suppress deep convective precipitation by decreasing cloud droplet size and delaying the onset of freezing. This hypothesis was supported by a cloud-resolving model study (Khain et al., 2001) showing that super-cooled cloud droplets down to $-37.5°C$ could only be simulated if the cloud droplets were small and numerous. Precipitation from single-cell mixed-phase convective clouds is reduced under continental and maritime conditions when aerosol concentrations are increased (Yin et al., 2000; Khain et al., 2004; Seifert and Beheng, 2006). In the modelling study by Cui et al. (2006), this is caused by drops evaporating more rapidly in the high aerosol case (see also Jiang et al., 2006), which eventually reduces ice mass and hence precipitation. Khain et al. (2005) postulate that smaller cloud droplets, such as those originating from human activity, would change the thermodynamics of convective clouds. More, smaller droplets would reduce the production of rain in convective clouds. When these droplets freeze, the associated latent heat release would then result in more vigorous convection and more precipitation. In a clean cloud, on the other hand, rain would have depleted the cloud so that less latent heat is released when the cloud glaciates, resulting in less vigorous convection and less precipitation. Similar results were obtained by Koren et al. (2005), Zhang et al. (2005) and for the multi-cell cloud systems studied by Seifert and Beheng (2006). For a thunderstorm in Florida in the presence of Saharan dust, the simulated precipitation enhancement only lasted two hours after which precipitation decreased as compared with clean conditions (van den Heever et al., 2006). Cloud processing of dust particles, sulphate particles and trace gases can lead to an acceleration of precipitation formation in continental mixed-phase clouds, whereas in maritime clouds, which already form on rather large CCN, the simulated effect on precipitation is small (Yin et al., 2002). This highlights the complexity of the system and indicates that the sign of the global change in precipitation due to aerosols is not yet known. Note that microphysical processes can only change the temporal and spatial distribution of precipitation while the total amount of precipitation can only change if evaporation from the surface changes.

7.5.2.3   Aerosol Impacts on Cirrus Clouds

Cirrus clouds can form by homogeneous and heterogeneous ice nucleation mechanisms at temperatures below 235 K. While homogeneous freezing of super-cooled aqueous phase aerosol particles is rather well understood, understanding of heterogeneous ice nucleation is still in its infancy. A change in the number of ice crystals in cirrus clouds could exert a cloud albedo effect in the same way that the cloud albedo effect acts for water

clouds. In addition, a change in the cloud ice water content could exert a radiative effect in the infrared. The magnitude of these effects in the global mean has not yet been fully established, but the development of physically based parametrization schemes of cirrus formation for use in global models led to significant progress in understanding underlying mechanisms of aerosol-induced cloud modifications (Kärcher and Lohmann, 2002; Liu and Penner, 2005; Kärcher et al., 2006).

A global climate model study concluded that a cloud albedo effect based solely on ubiquitous homogeneous freezing is small globally (Lohmann and Kärcher, 2002). This is expected to also hold in the presence of heterogeneous IN that cause cloud droplets to freeze at relative humidities over ice close to homogeneous values (above 130–140%) (Kärcher and Lohmann, 2003). Efficient heterogeneous IN, however, would be expected to lower the relative humidity over ice, so that the climate effect may be larger (Liu and Penner, 2005). *In situ* measurements reveal that organic-containing aerosols are less abundant than sulphate aerosols in ice cloud particles, suggesting that organics do not freeze preferentially (Cziczo et al., 2004). A model study explains this finding by the disparate water uptake of organic aerosols, and suggests that organics are unlikely to significantly modify cirrus formation unless they are present in very high concentrations (compared with sulphate-rich particles) at low temperatures (Kärcher and Koop, 2004).

With regard to aerosol effects on cirrus clouds, a strong link has been established between gravity wave induced, mesoscale variability in vertical velocities and climate forcing by cirrus (Kärcher and Ström, 2003; Hoyle et al., 2005). Hemispheric-scale studies of aerosol-cirrus interactions using ensemble trajectories suggest that changes in upper-tropospheric cooling rates and ice-forming aerosols in a future climate may induce changes in cirrus occurrence and optical properties that are comparable in magnitude with observed decadal trends in global cirrus cover (Haag and Kärcher, 2004). Optically thin and sub-visible cirrus are particularly susceptible to IN and therefore likely affected by anthropogenic activities.

Radiative forcing estimates and observed trends of aviation-induced cloudiness are discussed in Section 2.6. In terms of indirect effects of aircraft-induced aerosols on cirrus clouds, Lohmann and Kärcher (2002) show that the impact of aircraft sulphur emissions on cirrus properties via homogeneous freezing is small. The contribution from air traffic to the global atmospheric black carbon cycle was assessed by Hendricks et al. (2004). Assuming that black carbon particles from aviation serve as efficient IN, maximum increases or decreases in ice crystal number concentrations of more than 40% are simulated in a climate model study assuming that the 'background' (no aviation impact) cirrus cloud formation is dominated by heterogeneous or homogeneous nucleation, respectively (Hendricks et al., 2005). Progress in assessing the impact of aircraft black carbon on cirrus is hampered by the poor knowledge of natural freezing modes in cirrus conditions and the inability to describe the full complexity of cirrus processes in global models.

7.5.2.4 Global Climate Model Estimates of the Total Anthropogenic Aerosol Effect

The total anthropogenic aerosol effect as defined here includes estimates of the direct effect, semi-direct effect, indirect cloud albedo and cloud lifetime effect for warm clouds from several climate models. The total anthropogenic aerosol effect is obtained as the difference between a multi-year simulation with present-day aerosol emissions and a simulation representative for pre-industrial conditions, where anthropogenic emissions are turned off. It should be noted that the representation of the cloud lifetime effect in GCMs is essentially one of changing the auto-conversion of cloud water to rainwater.

The global mean total anthropogenic aerosol effect on net radiation at TOA from pre-industrial times to the present day is shown in Figure 7.21. Whereas Chapter 2 only considers the radiative forcing of the cloud albedo effect, here feedbacks are included in the radiative flux change. In most simulations shown in Figures 7.21 to 7.23, the total aerosol effect is restricted to warm clouds except for the simulations by Jacobson (2006) and Lohmann and Diehl (2006), who also include aerosol effects on mixed-phase and ice clouds. The total aerosol effect ranges from –0.2 W m^{-2} in the combined GCM plus satellite simulations (Quaas et al., 2006) to –2.3 W m^{-2} in the simulations by Ming et al. (2005), with an average forcing of –1.2 W m^{-2}. The total aerosol effect is larger when sulphate aerosols are used as surrogates for all anthropogenic aerosols than if multiple

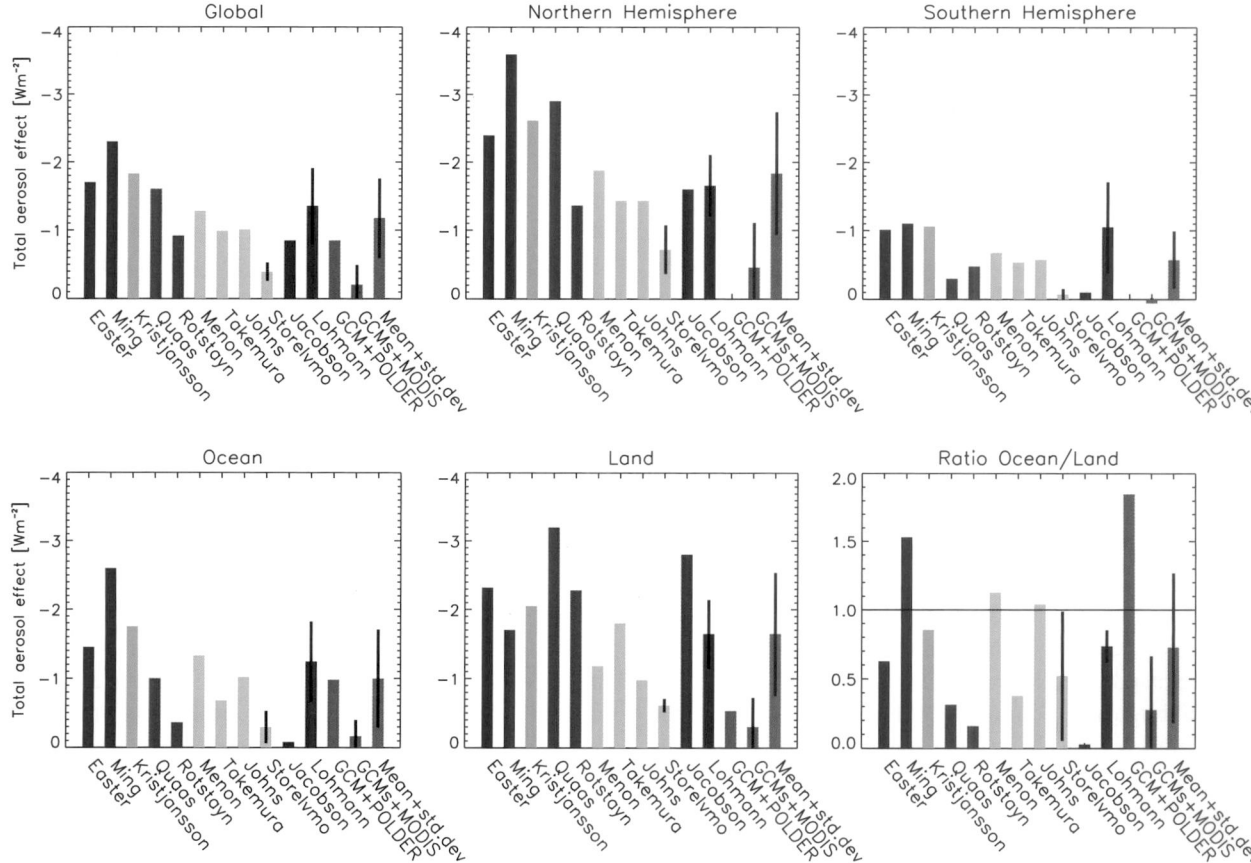

Figure 7.21. *Global mean total anthropogenic aerosol effect (direct, semi-direct and indirect cloud albedo and lifetime effects) defined as the response in net radiation at TOA from pre-industrial times to the present day and its contribution over the NH and SH, over oceans and over land, and the ratio over oceans/land. Red bars refer to anthropogenic sulphate (Easter et al., 2004; Ming et al., 2005+), green bars refer to anthropogenic sulphate and black carbon (Kristjánsson, 2002*,+), blue bars to anthropogenic sulphate and organic carbon (Quaas et al., 2004; Rotstayn and Liu, 2005+), cyan bars to anthropogenic sulphate and black and organic carbon (Menon and Del Genio, 2005; Takemura et al., 2005; Johns et al., 2006; Storelvmo et al., 2006), dark purple bars to anthropogenic sulphate and black and organic carbon effects on water and ice clouds (Jacobson, 2006; Lohmann and Diehl, 2006), teal bars refer to a combination of GCM and satellite results (European Centre for Medium Range Weather Forecasts/Max-Planck Institute for Meteorology Atmospheric GCM (ECHAM) plus Polarisation and Directionality of the Earth's Reflectance (POLDER), Lohmann and Lesins, 2002; Laboratoire de Météorologie Dynamique GCM (LMDZ)/ECHAM plus Moderate Resolution Imaging Spectroradiometer (MODIS), Quaas et al., 2006) and olive bars to the mean and standard deviation from all simulations. Vertical black lines for individual results refer to ±1 standard deviation in cases of multiple simulations and/or results.*

* refers to estimates of the aerosol effect deduced from the shortwave radiative flux only
+ refers to estimates solely from the indirect effects

aerosol types are considered (Figure 7.21). Although most model estimates also include the direct and semi-direct effects, their contribution to the TOA radiation is generally small compared with the indirect effect, ranging from +0.1 to –0.5 W m^{-2} due to variations in the locations of black carbon with respect to the cloud (Lohmann and Feichter, 2005). The simulated cloud lifetime effect in a subset of models displayed in Figure 7.21 varies between –0.3 and –1.4 W m^{-2} (Lohmann and Feichter, 2005), which highlights some of the differences among models. The importance of the cloud albedo effect compared with the cloud lifetime effect varies even when the models use the same aerosol fields (Penner et al., 2006). Other differences among the simulations include an empirical treatment of the relationship between aerosol mass and cloud droplet number concentration vs. a mechanistic relationship, the dependence of the indirect aerosol effect on the assumed background aerosol or cloud droplet number concentration, and the competition between natural and anthropogenic aerosols as CCN (Ghan et al., 1998; O'Dowd et al., 1999). Likewise, differences in the cloud microphysics scheme, especially in the auto-conversion rate, cause different cloud responses (e.g., A. Jones et al., 2001; Menon et al., 2002a, 2003; Penner et al., 2006).

All models agree that the total aerosol effect is larger over the NH than over the SH (Figure 7.21). The values of the NH total aerosol effect vary between –0.5 and –3.6 W m^{-2} and in the SH between slightly positive and –1.1 W m^{-2}, with an average SH to NH ratio of 0.3. Estimates of the ocean/land partitioning of the total indirect effect vary from 0.03 to 1.8 with an average value of 0.7. While the combined European Centre for Medium Range Weather Forecasts/Max-Planck Institute for Meteorology Atmospheric GCM (ECHAM4) plus Polarisation and Directionality of the Earth's Reflectance (POLDER) satellite estimate suggests that the total aerosol effect should be larger over oceans (Lohmann and Lesins, 2002), combined estimates of the Laboratoire de Météorologie Dynamique (LMD) and ECHAM4 GCMs with Moderate Resolution Imaging Spectroradiometer (MODIS) satellite data reach the opposite conclusion (Quaas et al., 2006). The average total aerosol effect over the ocean of –1 W m^{-2} agrees with estimates of between –1 and –1.6 W m^{-2} from the Advanced Very High Resolution Radiometer (AVHRR)/POLDER (Sekiguchi et al., 2003). Estimates from GCMs of the total aerosol effect are generally larger than those from inverse models (Anderson et al., 2003 and Chapter 9).

As compared with the estimates of the total aerosol effect in Lohmann and Feichter (2005), some new estimates (Chen and Penner, 2005; Rotstayn and Liu, 2005; Lohmann and Diehl, 2006) now also include the influence of aerosols on the cloud droplet size distribution (dispersion effect; Liu and Daum, 2002). The dispersion effect refers to a widening of the size distribution in the polluted clouds that partly counteracts the reduction in the effective cloud droplet radius in these clouds. Thus, if the dispersion effect is taken into account, the indirect cloud albedo aerosol effect is reduced by 12 to 42% (Peng and Lohmann, 2003; Rotstayn and Liu, 2003; Chen and Penner, 2005). The global mean total indirect aerosol effect in the

simulation by Rotstayn and Liu (2005) has also been reduced due to a smaller cloud lifetime effect resulting from a new treatment of auto-conversion.

Global climate model estimates of the change in global mean precipitation due to the total aerosol effects are summarised in Figure 7.22. Consistent with the conflicting results from detailed cloud system studies, the change in global mean precipitation varies between 0 and –0.13 mm day^{-1}. These differences are amplified over the SH, ranging from –0.06 mm day^{-1} to 0.12 mm day^{-1}. In general, the decreases in precipitation are larger when the atmospheric GCMs are coupled to mixed-layer ocean models (green bars), where the sea surface temperature and, hence, evaporation are allowed to vary.

7.5.3 Effects of Aerosols and Clouds on Solar Radiation at the Earth's Surface

By increasing aerosol and cloud optical depth, anthropogenic emissions of aerosols and their precursors contribute to a reduction of solar radiation at the surface. As such, worsening air quality contributes to regional aerosol effects. The partially conflicting observations on solar dimming/brightening are discussed in detail in Section 3.4 and Box 3.2. This section focuses on the possible contribution by aerosols. The decline in solar radiation from 1961 to 1990 affects the partitioning between direct and diffuse solar radiation: Liepert and Tegen (2002) concluded that over Germany, both aerosol absorption and scattering must have declined from 1975 to 1990 in order to explain the simultaneously weakened aerosol forcing and increased direct/diffuse solar radiation ratio. The direct/diffuse solar radiation ratio over the USA also increased from 1975 to 1990, likely due to increases in absorbing aerosols. Increasing aerosol optical depth associated with scattering aerosols alone in otherwise clear skies produces a larger fraction of diffuse radiation at the surface, which results in larger carbon assimilation into vegetation (and therefore greater transpiration) without a substantial reduction in the total surface solar radiation (Niyogi et al., 2004; Section 7.2.6.2).

For the tropical Indian Ocean, Ramanathan et al. (2001) estimate an indirect aerosol effect of –5 W m^{-2} at TOA and –6 W m^{-2} at the surface. While the direct effect is negligible at TOA, its surface forcing amounts to –14 W m^{-2} as a consequence of large atmospheric absorption in this region. In South Asia, absorbing aerosols may have masked up to 50% of the surface warming due to the global increase in greenhouse gases (Ramanathan et al., 2005). Global climate model estimates of the mean decrease in surface shortwave radiation in response to all aerosol effects vary between –1.3 and –3.3 W m^{-2} (Figure 7.23). It is larger than the TOA radiation flux change because some aerosols like black carbon absorb solar radiation within the atmosphere (see also Jacobson, 2001; Lohmann and Feichter, 2001; Ramanathan et al., 2001; Liepert et al., 2004). As for the TOA net radiation, the decrease is largest over land, with values approaching –9 W m^{-2}. Consistent with the above-mentioned regional studies, most models predict larger decreases over land than over the oceans.

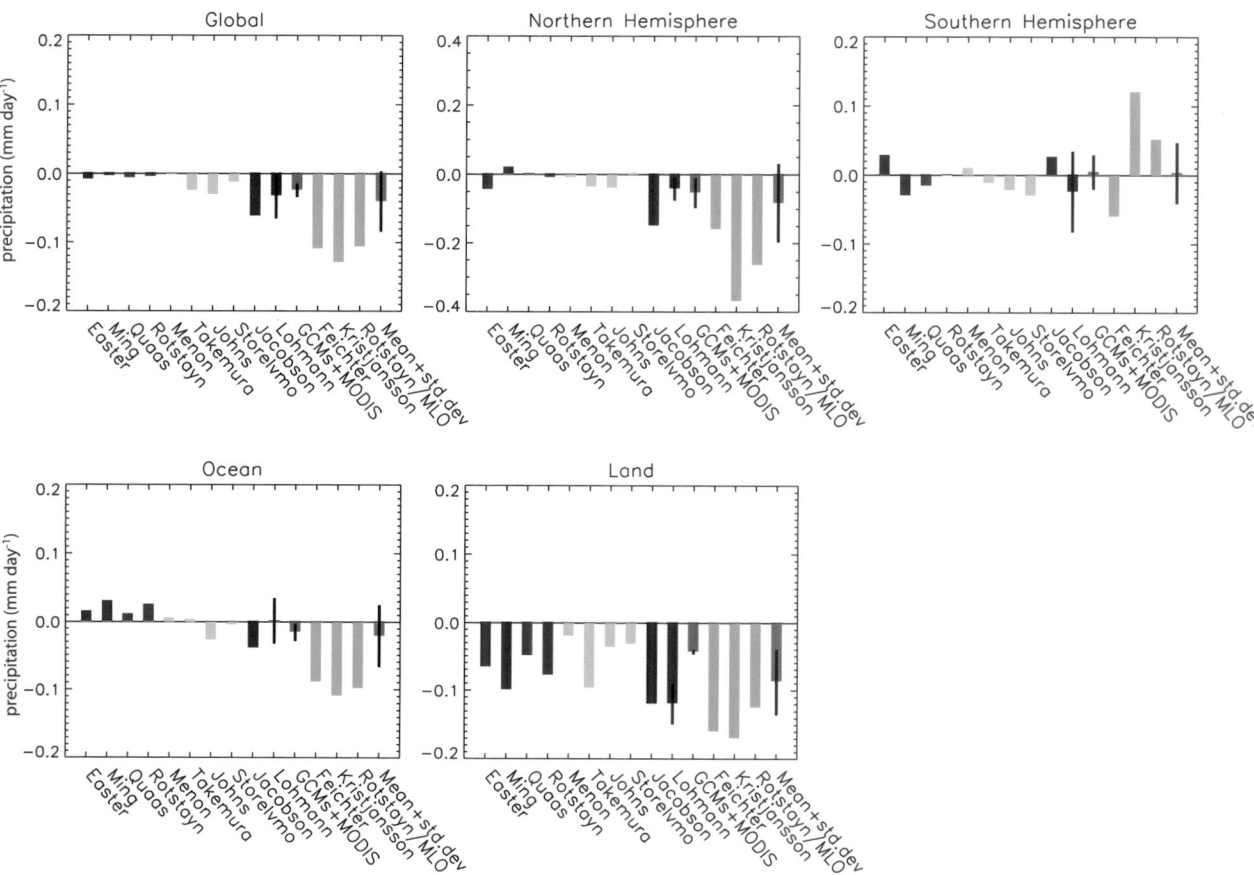

Figure 7.22. *Global mean change in precipitation due to the total anthropogenic aerosol effect (direct, semi-direct and indirect cloud albedo and lifetime effects) from pre-industrial times to the present day and its contribution over the NH and SH, over oceans and over land. Red bars refer to anthropogenic sulphate (Easter et al., 2004; Ming et al., 2005+), blue bars to anthropogenic sulphate and organic carbon (Quaas et al., 2004; Rotstayn and Liu, 2005+), cyan bars to anthropogenic sulphate, and black and organic carbon (Menon and Del Genio, 2005; Takemura et al., 2005; Johns et al., 2006; Storelvmo et al., 2006), dark purple bars to anthropogenic sulphate and black and organic carbon effects on water and ice clouds (Jacobson, 2006; Lohmann and Diehl, 2006), teal bars refer to a combination of GCM and satellite results (LMDZ/ECHAM plus MODIS, Quaas et al., 2006), green bars refer to results from coupled atmosphere/mixed-layer ocean (MLO) experiments (Feichter et al., 2004: sulphate and black and organic carbon; Kristjansson et al., 2005: sulphate and black carbon; Rotstayn and Lohmann, 2002+: sulphate only) and olive bars to the mean from all simulations. Vertical black lines refer to ±1 standard deviation.*

+ refers to estimates solely from the indirect effects

Transient simulations (Roeckner et al., 1999) and coupled GCM-mixed-layer ocean equilibrium simulations (Feichter et al., 2004; Liepert et al., 2004) suggest that the decrease in solar radiation at the surface resulting from increases in optical depth due to the direct and indirect anthropogenic aerosol effects is more important for controlling the surface energy budget than the greenhouse-gas induced increase in surface temperature. There is a slight increase in downwelling longwave radiation due to aerosols, which in the global mean is small compared to the decrease in shortwave radiation at the surface. The other components of the surface energy budget (thermal radiative flux, sensible and latent heat fluxes) decrease in response to the reduced input of solar radiation. As global mean evaporation must equal precipitation, a reduction in the latent heat flux in the model leads to a reduction in precipitation (Liepert et al., 2004). This is in contrast to the observed precipitation evolution in the last century (see Section 3.3) and points to an overestimation of aerosol influences on precipitation. The simulated decrease

in global mean precipitation from pre-industrial times to the present may reverse into an increase of about 1% in 2031 to 2050 as compared to 1981 to 2000, because the increased warming due to black carbon and greenhouse gases then dominates over the sulphate cooling (Roeckner et al., 2006).

7.5.4 Effects of Aerosols on Circulation Patterns

7.5.4.1 Effects on Stability

Changes in the atmospheric lapse rate modify the longwave emission and affect the water vapour feedback (Hu, 1996) and the formation of clouds (see, e.g., Section 8.6). Observations and model studies show that an increase in the lapse rate produces an amplification of the water vapour feedback (Sinha, 1995). As aerosols cool the Earth's surface and warm the aerosol layer, the lapse rate will decrease globally and suppress the water vapour feedback (e.g., Feichter et al., 2004). The local change

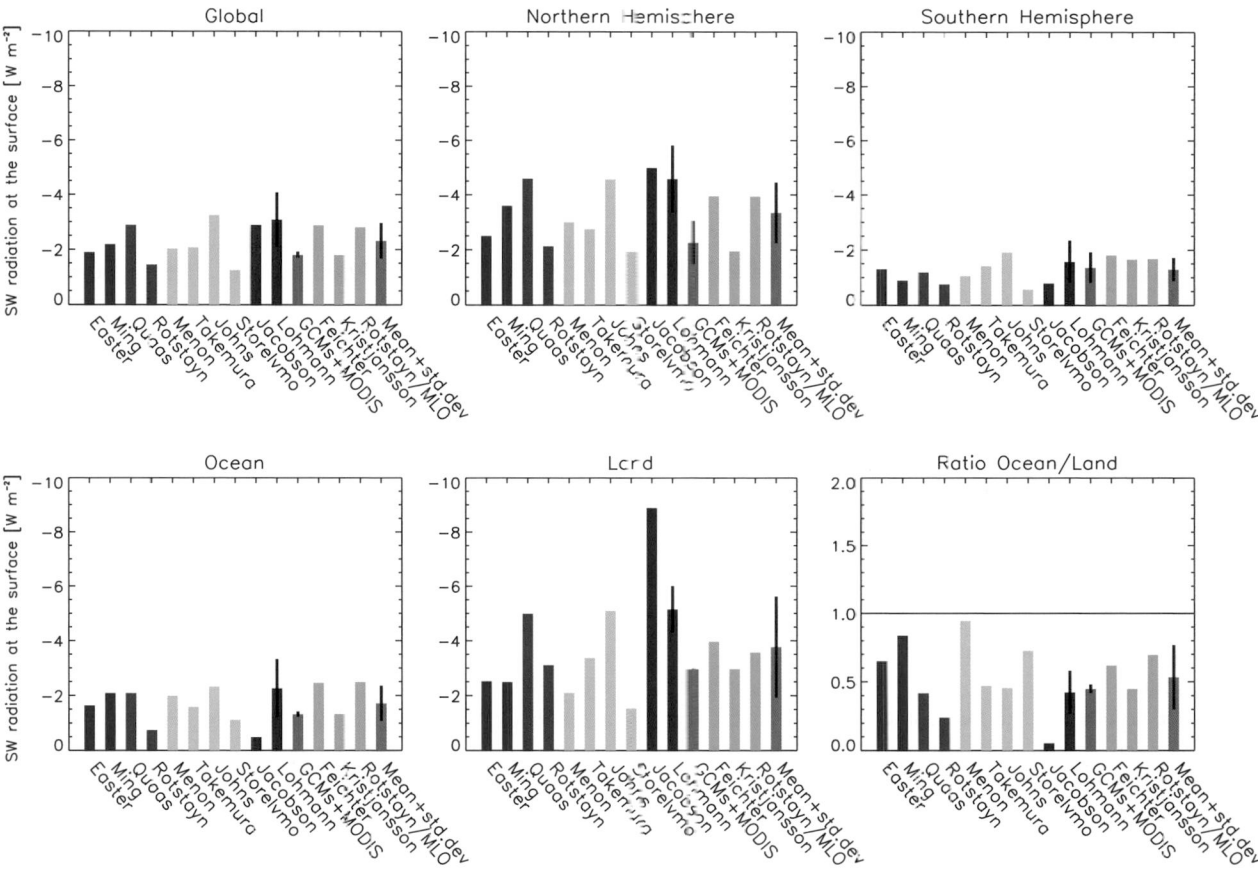

Figure 7.23. *Global mean change in net solar radiation at the surface due to the total anthropogenic aerosol effect (direct, semi-direct and indirect cloud albedo and lifetime effects) from pre-industrial times to the present day and its contribution over the NH and SH, over oceans and over land and the ratio over oceans/land. Red bars refer to anthropogenic sulphate (Easter et al., 2004; Ming et al., 2005+), blue bars to anthropogenic sulphate and organic carbon (Quaas et al., 2004; Rotstayn and Liu, 2005+), cyan bars to anthropogenic sulphate and black and organic carbon (Menon and Del Genio, 2005; Takemura et al., 2005; Johns et al., 2006; Storelvmo et al., 2006), dark purple bars to anthropogenic sulphate and black and organic carbon effects on water and ice clouds (Jacobson, 2006; Lohmann and Diehl, 2006), teal bars refer to a combination of GCM and satellite results (LMDZ/ECHAM plus MODIS, Quaas et al., 2006), green bars refer to results from coupled atmosphere/mixed-layer ocean (MLO) experiments (Feichter et al., 2004: sulphate and black and organic carbon; Kristjansson et al., 2005: sulphate and black carbon; Rotstayn and Lohmann, 2002+: sulphate only) and olive bars to the mean from all simulations. Vertical black lines refer to ±1 standard deviation.*

+ refers to estimates solely from the indirect effects

in atmospheric stability strongly depends on the altitude of the black carbon heating (Penner et al., 2003).

Absorption of solar radiation by aerosols can change the cloud amount (semi-direct effect; Grassl, 1975; Hansen et al., 1997; Ackerman et al., 2000; Ramanathan et al., 2001; Jacobson, 2006; Figure 7.20). The semi-direct effect has been simulated with GCMs and high-resolution cloud-resolving models, since it is implicitly accounted for whenever absorbing aerosols coupled to the radiation scheme are included (Hansen et al., 1997; Lohmann and Feichter, 2001; Jacobson, 2002; Menon et al., 2002b; Penner et al., 2003; Cook and Highwood, 2004; Hansen et al., 2005). Aerosol heating within cloud layers reduces cloud fractions, whereas aerosol heating above the cloud layer tends to increase cloud fractions. When diagnosed within a GCM framework, the semi-direct effect can also include cloud changes due to circulation effects and/or surface albedo effects. Moreover, the semi-direct effect is not exclusive to absorbing aerosol, as potentially any radiative heating of

the mid-troposphere can produce a similar response in a GCM (Hansen et al., 2005; see also Section 2.8). Cloud-resolving models of cumulus and stratocumulus case studies also diagnose semi-direct effects indicating a similar relationship between the height of the aerosol layer relative to the cloud and the sign of the semi-direct effect (Ackerman et al., 2000; Ramanathan et al., 2001; Johnson et al., 2004; Johnson, 2005). Using a large eddy simulation, Feingold et al. (2005) show that the reduction in net surface radiation and in surface latent and sensible heat fluxes is the most simple explanation of the reduction in cloudiness associated with absorbing aerosols.

7.5.4.2 Effects on the Large-Scale Circulation

Several studies have considered the response of a GCM with a mixed-layer ocean to indirect aerosol effects (Rotstayn et al., 2000; K. Williams et al., 2001; Rotstayn and Lohmann, 2002) or to a combination of direct and indirect aerosol effects

(Feichter et al., 2004; Kristjansson et al., 2005; Takemura et al., 2005). All of these, and recent transient simulations (Held et al., 2005; Paeth and Feichter, 2006), found a substantial cooling that was strongest in the NH, with a consequent southward shift of the Inter-Tropical Convergence Zone (ITCZ) and the associated tropical rainfall belt. Rotstayn and Lohmann (2002) even suggest that aerosol effects might have contributed to the Sahelian droughts of the 1970s and 1980s (see Sections 9.5 and 11.2). If in turn the NH is warmed, for instance due to the direct forcing by black carbon aerosols, the ITCZ is found to shift northward (Chung and Seinfeld, 2005).

Menon et al. (2002b) and Wang (2004) found that circulation changes could be caused by aerosols in southeast China. In India and China, where absorbing aerosols have been added, increased rising motions are seen as well as increased subsidence to the south and north (Menon et al., 2002b). However, Ramanathan et al. (2005) found that convection was suppressed due to increased stability resulting from black carbon heating. Drier conditions resulting from suppressed rainfall can induce more dust and smoke due to the burning of drier vegetation (Ramanathan et al., 2001), thus affecting both regional and global hydrological cycles (Wang, 2004). Heating of a lofted dust layer could increase the occurrence of deep convection (Stephens et al., 2004). It can also strengthen the Asian summer monsoon circulation and cause a local increase in precipitation, despite the global reduction of evaporation that compensates aerosol radiative heating at the surface (Miller et al., 2004b). The dust-induced thermal contrast changes between the Eurasian continent and the surrounding oceans are found to trigger or modulate a rapidly varying or unstable Asian winter monsoon circulation, with a feedback to reduce the dust emission from its sources (Zhang et al., 2002).

In summary, an increase in atmospheric aerosol load decreases air quality and reduces the amount of solar radiation reaching the surface. This negative radiative forcing competes with the greenhouse gas warming for determining the change in evaporation and precipitation. At present, no transient climate simulation accounts for all aerosol-cloud interactions, so that the net aerosol effect on clouds deduced from models is not conclusive.

7.6 Concluding Remarks

Biogeochemical cycles interact closely with the climate system over a variety of temporal and spatial scales. On geological time scales, this interaction is illustrated by the Vostok ice core record, which provides dramatic evidence of the coupling between the carbon cycle and the climate system. The dynamics of the Earth system inferred from this record result from a combination of external forcing (in this case long-term periodic changes in the orbital parameters of the Earth and hence solar forcing) and an array of feedback mechanisms within the Earth environment (see Chapter 6). On shorter time scales, a range of forcings originating from human activities (conversion and fragmentation of natural ecosystems, emissions of greenhouse gases, nitrogen fixation, degradation of air quality, stratospheric ozone depletion) is expected to produce planet-wide effects and perturb numerous feedback mechanisms that characterise the dynamics of the Earth system.

A number of feedbacks that amplify or attenuate the climate response to radiative forcing have been identified. In addition to the well-known positive water vapour and ice-albedo feedbacks, a feedback between the carbon cycle and the climate system could produce substantial effects on climate. The reduction in surface carbon uptake expected in future climate should produce an additional increase in the atmospheric CO_2 concentration and therefore enhance climate forcing. Large differences between models, however, make the quantitative estimate of this feedback uncertain. Other feedbacks (involving, for example, atmospheric chemical and aerosol processes) are even less well understood. Their magnitude and even their sign remain uncertain. Potentially important aerosol-cloud interactions such as changes in cloud lifetime and aerosol effects on ice clouds can influence the hydrologic cycle and the radiative budget; however, the scientific understanding of these processes is low. The response of the climate system to anthropogenic forcing is expected to be more complex than simple cause and effect relationships would suggest; rather, it could exhibit chaotic behaviour with cascades of effects across the different scales and with the potential for abrupt and perhaps irreversible transitions.

This chapter has assessed how processes related to vegetation dynamics, carbon exchanges, gas-phase chemistry and aerosol microphysics could affect the climate system. These processes, however, cannot be considered in isolation because of the potential interactions that exist between them. Air quality and climate change, for example, are intimately coupled (Dentener et al., 2006). Brasseur and Roeckner (2005) estimate that the hypothetical removal from the atmosphere of the entire burden of anthropogenic sulphate aerosol particles (in an effort to improve air quality) would produce a rather immediate increase of about 0.8°C in the globally averaged temperature, with geographical patterns that bear a resemblance to the temperature changes found in greenhouse gas scenario experiments (Figure 7.24). Thus, environmental strategies aimed at maintaining 'global warming' below a prescribed threshold must therefore account not only for CO_2 emissions but also for measures implemented to improve air quality. To cope with the complexity of Earth system processes and their interactions, and particularly to evaluate sophisticated models of the Earth system, observations and long-term monitoring of climate and biogeochemical quantities will be essential. Climate models will have to reproduce accurately the important processes and feedback mechanisms discussed in this chapter.

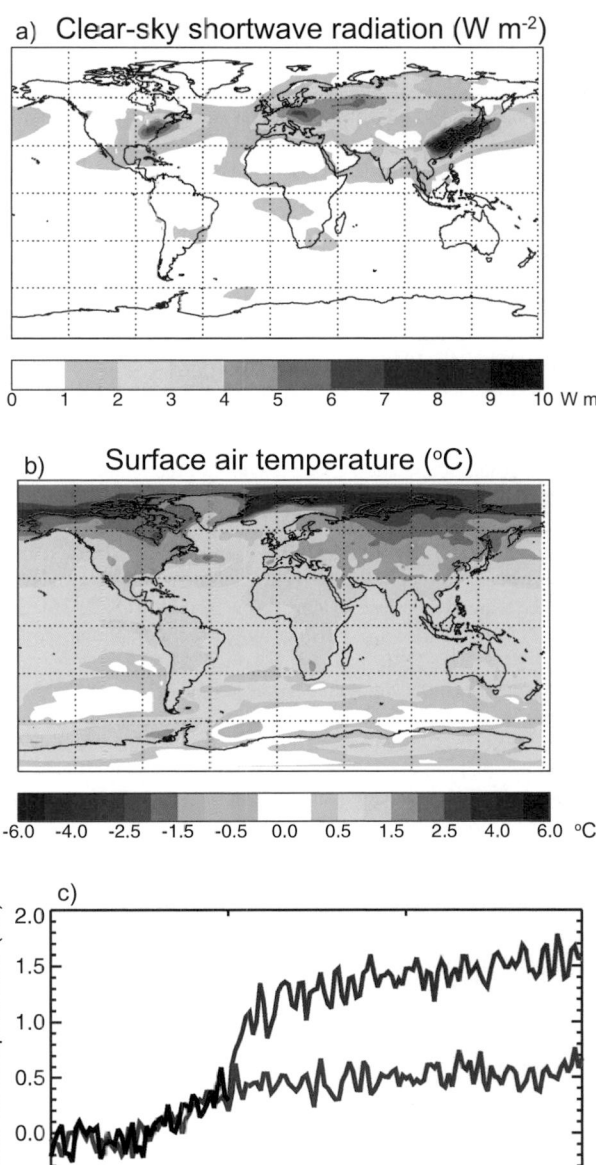

Figure 7.24. *Effect of removing the entire burden of sulphate aerosols in the year 2000 on (a) the annual mean clear sky TOA shortwave radiation (W m⁻²) calculated by Brasseur and Roeckner (2005) for the time period 2071 to 2100 and (b) on the annual mean surface air temperature (°C) calculated for the same time period. (c) temporal evolution of global and annual mean surface air temperature anomalies (°C) with respect to the mean 1961 to 1990 values. The evolution prior to the year 2000 is driven by observed atmospheric concentrations of greenhouse gases and aerosols as adopted by IPCC (see Chapter 10). After 2000, the concentration of greenhouse gases remains constant while the aerosol burden is unchanged (blue line) or set to zero (red line). The black curve shows observations (A. Jones et al., 2001; Jones et al., 2006).*

References

Achard, F., et al., 2004: Improved estimates of net carbon emissions from land cover change in the tropics for the 1990s. *Global Biogeochem. Cycles*, **18**, GB2008, doi:10.1029/2003GB002142.

ACIA, 2005: *Arctic Climate Impact Assessment*. Cambridge University Press, Cambridge, 1042 pp.

Ackerman, A.S., M.P. Kirkpatrick, D.E. Stevens, and O.B. Toon, 2004: The impact of humidity above stratiform clouds on indirect climate forcing. *Nature*, **432**, 1014–1017.

Ackerman, A.S., et al., 2000: Reduction of tropical cloudiness by soot. *Science*, **288**, 1042–1047.

Adams, J., J. Constable, A. Guenther, and P. Zimmerman, 2001: An estimate of natural volatile organic compound emissions from vegetation since the last glacial maximum. *Chemosphere*, **3**, 73–91.

Adler, R.F., et al., 2003: The version-2 Global Precipitation Climatology Project (GPCP) monthly precipitation analysis (1979-present). *J. Hydrometeorol.*, **4**, 1147–1167.

Ainsworth, E.A., and S.P. Long, 2005: What have we learned from 15 years of free-air CO_2 enrichment (FACE)? A meta-analytic review of the responses of photosynthesis, canopy. *New Phytol.*, **165**(2), 351–371.

Albrecht, B., 1989: Aerosols, cloud microphysics, and fractional cloudiness. *Science*, **245**, 1227–1230.

Allan, W., et al., 2005: Interannual variations of ^{13}C in tropospheric methane: Implications for a possible atomic chlorine sink in the marine boundary layer. *J. Geophys. Res.*, **110**, doi:10.1029/2004JD005650.

Allen, D., K. Pickering, and M. Fox-Rabinovitz, 2004: Evaluation of pollutant outflow and CO sources during TRACE-P using model-calculated, aircraft-based, and Measurements of Pollution in the Troposphere (MOPITT)-derived CO concentrations. *J. Geophys. Res.*, **109**, D15S03, doi:10.1029/2003JD004250.

Anderson, T.L., et al., 2003: Climate forcing by aerosols - a hazy picture. *Science*, **300**, 1103–1104.

Andreae, M.O., and P. Merlet, 2001: Emission of trace gases and aerosols from biomass burning. *Global Biogeochem. Cycles*, **15**, 955–966.

Andreae, M.O., C.D. Jones, and P.M. Cox, 2005: Strong present-day aerosol cooling implies a hot future. *Nature*, **435**(7046), 1187–1190.

Andreae, M.O., et al., 2004: Smoking rain clouds over the Amazon. *Science*, **303**, 1337–1342.

Angert, A., et al., 2004: CO_2 seasonality indicates origins of post-Pinatubo sink. *Geophys. Res. Lett.*, **31**(11), L11103, doi:10.1029/2004GL019760.

Angert, A., et al., 2005: Drier summers cancel out the CO_2 uptake enhancement induced by warmer springs. *Proc. Natl. Acad. Sci. U.S.A.*, **102**, 10823–10827.

Archer, D., 2005: The fate of fossil fuel CO_2 in geologic time. *J. Geophys. Res.*, **110**(C9), C09S05, doi:10.1029/2004JC002625.

Archer, D., and B. Buffett, 2005: Time-dependent response of the global ocean clathrate reservoir to climatic and anthropogenic forcing. *Geochem. Geophys. Geosystems*, **6**, Q03002, doi:10.1029/2004GC000854.

Archer, D., H. Kheshgi, and E. Maier-Reimer, 1998: Dynamics of fossil fuel CO_2 neutralization by marine $CaCO_3$. *Global Biogeochem. Cycles*, **12**(2), 259–276.

Arellano, A.F. Jr., et al., 2004: Top-down estimates of global CO sources using MOPITT measurements. *Geophys. Res. Lett.*, **31**, L01104, doi:10.1029/ 2003GL018609.

Armstrong, R.A., et al., 2002: A new, mechanistic model for organic carbon fluxes in the ocean based on the quantitative association of POC with ballast minerals. *Deep-Sea Res. II*, **49**, 219–236.

Arora, V.K., and G.J. Boer, 2003: A representation of variable root distribution in dynamic vegetation models. *Earth Interactions*, **7**(6), 1–19.

Arora, V.K., and G.J. Boer, 2005: A parameterization of leaf phenology for the terrestrial ecosystem component of climate models. *Global Change Biol.*, **11**(1), 39–59.

Arora, V.K., and G.J. Boer, 2006: Simulating competition and coexistence between plant functional types in a dynamic vegetation model. *Earth Interactions*, **10**, Paper 10, 30 pp., doi:10.1175/EI170.1.

Aumont, O., et al., 2001: Riverine-driven interhemispheric transport of carbon. *Global Biogeochem. Cycles*, **15**, 393–405.

Austin, J., and R.J. Wilson, 2006: Ensemble simulations of the decline and recovery of stratospheric ozone. *J. Geophys. Res.*, **111**, D16314, doi:10.1029/2005JD006907.

Austin, J., et al., 2003: Uncertainties and assessments of chemistry-climate models of the stratosphere. *Atmos. Chem. Phys.*, **3**, 1–27.

Avissar, R., and D. Werth, 2005: Global hydroclimatological teleconnections resulting from tropical deforestation. *J. Hydrometeorol.*, **6**(2), 134–145.

Avissar, R., P.L. Silva Dias, M.A.F. Silva Dias, and C. Nobre, 2002: The Large-scale Biosphere-Atmosphere Experiment in Amazonia (LBA): Insights and future research needs. *J. Geophys. Res.*, **107**(D20), 8034, doi:10.1029/2002JD002507.

Aw, J., and M.J. Kleeman, 2003: Evaluating the first-order effect of intraannual air pollution on urban air pollution. *J. Geophys. Res.*, **108**, 4365, doi:10.1029/2002JD002688.

Ayers, G.P., 2005: "Air pollution and climate change: has air pollution suppressed rainfall over Australia?" *Clean Air and Environmental Quality*, **39**, 51–57.

Bacastow, R.B., and C.D. Keeling, 1981: Atmospheric carbon dioxide concentration and the observed airborne fraction. In: *Carbon Cycle Modelling* [Bolin, B. (ed.)]. SCOPE 16. John Wiley and Sons, New York, pp. 103–112.

Bagnoud, N., A.J. Pitman, B.J. McAvaney, and N.J. Holbrook, 2005: The contribution of the land surface energy balance complexity to differences in means, variances and extremes using the AMIP-II methodology. *Clim. Dyn.*, **25**, 171–188. doi:10.1007/S00382-005-0004-9.

Baker, D.F., et al., 2006: TransCom 3 inversion intercomparison: impact of transport model errors on the interannual variability of regional CO_2 fluxes, 1988-2003. *Global Biogeochem. Cycles*, **20**, GB1002, doi:1010.1029/2004GB002439.

Baker, M., and R.J. Charlson, 1990: Bistability of CCN concentrations and thermodynamics in the cloud-topped boundary layer. *Nature*, **345**, 142–145.

Baker, T.R., et al., 2004: Increasing biomass in Amazonian forest plots. *Philos. Trans. R. Soc. London Ser. B*, **359**, 353–365.

Baldocchi, D., et al., 2001: FLUXNET: A new tool to study the temporal and spatial variability of ecosystem-scale carbon dioxide, water vapor, and energy flux densities. *Bull. Am. Meteorol. Soc.*, **82**, 2415–2434.

Baldwin, M.P., and T.J. Dunkerton, 1999: Downward propagation of the Arctic Oscillation from the stratosphere to the troposphere. *J. Geophys. Res.*, **104**, 30937–30946.

Baldwin, M.P., and T.J. Dunkerton, 2001: Stratospheric harbingers of anomalous weather regimes. *Science*, **244**, 581–584.

Balkanski, Y., et al., 2004: Global emissions of mineral aerosol: formulation and validation using satellite imagery. In: *Emission of Atmospheric Trace Compounds* [Granier, C., P. Artaxo, and C.E. Reeves (eds.)]. Kluwer, Dordrecht, pp. 239–267.

Balzter H., et al., 2005: Impact of the Arctic Oscillation pattern on interannual forest fire variability in Central Siberia. *Geophys. Res. Lett.*, **32**, L14709, doi:10.1029/2005GL022526.

Barbosa, P.M., et al., 1999: An assessment of vegetation fire in Africa (1981–1991): Burned areas, burned biomass, and atmospheric emissions. *Global Biogeochem. Cycles*, **13**, 933–950.

Barford, C.C., et al., 2001: Factors controlling long and short term sequestration of atmospheric CO_2 in a mid-latitude forest. *Science*, **294**(5547), 1688–1691.

Barlage, M., and X. Zeng, 2004: Impact of observed vegetation root distribution on seasonal global simulations of land surface processes. *J. Geophys. Res.*, **109**, D09101, doi:10.1029/2003JD003847.

Battle, M., et al., 2000: Global carbon sinks and their variability inferred from atmospheric O_2 and $\delta^{13}C$. *Science*, **287**(5462), 2467–2470.

Beirle, S., et al., 2004: Estimate of nitrogen oxide emissions from shipping by satellite remote sensing. *Geophys. Res. Lett.*, **31**, L18102, doi:10.1029/2004GL020312.

Bellamy, P.H., et al., 2005: Carbon losses from all soils across England and Wales 1978–2003. *Nature*, **437**, 245248.

Benner R., et al., 1992: Bulk chemical characteristics of dissolved organic matter in the ocean. *Science*, **255**, 1561–1564.

Bergamaschi, P., M. Braeunlich, T. Marik, and C.A.M. Brenninkmeijer. 2000: Measurements of the carbon and hydrogen isotopes of atmospheric methane at Izana, Tenerife: Seasonal cycles and synoptic-scale variations. *J. Geophys. Res.*, **105**, 14531–14546.

Berner, R.A., 1998: The carbon cycle and CO_2 over Phanerozoic time: the role of land plants. *Philos. Trans. R. Soc. London Ser. B*, **353**(1365), 75–81.

Bertram, T.H., et al., 2005: Satellite measurements of daily variations in soil NO_x emissions. *Geophys. Res. Lett.*, **32**, L24812, doi:10.1029/2005GL024640.

Betts, A.K., 2004: Understanding hydrometeorology using global models. *Bull. Am. Meteorol. Soc.*, **85**, 1673–1688.

Betts, A.K., 2006: Radiative scaling of the nocturnal boundary layer and the diurnal temperature range. *J. Geophys. Res.*, **111**, D07105, doi:10.1029/2005JD006560.

Betts, A.K., J. Ball, and J. McCaughey, 2001: Near-surface climate in the boreal forest. *J. Geophys. Res.*, **106**, 33529–33541.

Betts, R., et al., 2004: The role of ecosystem-atmosphere interactions in simulated Amazonian precipitation decrease and forest dieback under global change warming. *Theor. Appl. Climatol.*, **78**(1–3), 157–175.

Bey, I., et al., 2001: Global modeling of tropospheric chemistry with assimilated meteorology: model description and evaluation. *J. Geophys. Res.*, **106**(D19), 23073–23096.

Bigg, E.K., C. Leck, and L. Tranvik, 2004: Particulates of the surface microlayer of open water in the central Arctic Ocean in summer. *Mar. Chem.*, **91**(1–4), 131–141.

Bodeker, G.E., H. Shiona, and H. Eskes, 2005: Indicators of Antarctic ozone depletion. *Atmos. Chem. Phys.*, **5**, 2603–2615.

Boersma, K.F., H.J. Eskes, E.W. Meijer, and H.M. Keider, 2005: Estimates of lightning NO_x production from GOME satellite observations. *Atmos. Chem. Phys. Discussions*, **5**, 3047–3104.

Bogner, J.E., R.L. Sass, and B.P. Walter, 2000: Model comparisons of methane oxidation across a management gradient: Wetlands, rice production systems, and landfill. *Global Biogeochem. Cycles.*, **14**, 1021–1033.

Bolin, B., and E. Eriksson, 1959: Changes in the carbon dioxide content of the atmosphere and sea due to fossil fuel combustion. In: *The Atmosphere and Sea in Motion* [Bolin, B. (ed.)]. Rossby Memorial Volume. Rockefeller Institute, New York, NY, pp. 130–142.

Bonan, G.B., 2001: Observational evidence for reduction of daily maximum temperature by croplands in the midwest United States. *J. Clim.*, **14**, 2430–2442.

Bonan, G.B., et al., 2003: A dynamic global vegetation model for use with climate models: concepts and description of simulated vegetation dynamics. *Global Change Biol.*, **9**, 1543–1566.

Bond, W.J., G.F. Midgley, and F.I. Woodward, 2003: The importance of low atmospheric CO_2 and fire in promoting the spread of grasslands and savannas. *Global Change Biol.*, **9**, 973–982.

Bopp, L., et al., 2002: Climate-induced oceanic oxygen fluxes: Implications for the contemporary carbon budget. *Global Biogeochem. Cycles*, **16**, doi:10.1029/2001GB001445.

Bopp, L., et al., 2004: Will marine dimethyl sulfide emissions amplify or alleviate global warming? A model study. *Can. J. Fish Aquat. Sci.*, **61**(5), 826–835.

Bopp, L., et al., 2005: Response of diatoms distribution to global warming and potential implications – a global model study. *Geophys. Res. Lett.*, **32**(19), L19606, doi:10.1029/2005GL023653.

Borges, A.V., 2005: Do we have enough pieces of the jigsaw to integrate CO_2 fluxes in the coastal ocean? *Estuaries*, **28**, 3–27.

Bousquet, P., et al., 2000: Regional changes in carbon dioxide fluxes of land and oceans since 1980. *Science*, **290**(5495), 1342–1346.

Bousquet, P., et al., 2005: Two decades of OH variability as inferred by an inversion of atmospheric transport and chemistry of methyl chloroform, *Atmos. Chem. Phys.*, **5**, 2635–2656.

Bouwman, A.F., L.J.M. Boumans, and N.H. Batjes, 2001: *Global Estimates of Gaseous Emission of NH_3, NO and N_2O from Agricultural Land*. Food and Agriculture Organisation, Rome, 57 pp.

Bouwman, A.F., L.J.M. Boumans, and N.H. Batjes, 2002: Modeling global annual N_2O and NO emissions from fertilized fields. *Global Biogeochem. Cycles*, **16**(4), 1080, doi:10.1029/2001GB001812.

Boyd, P.W., et al., 2004: The decline and fate of an iron-induced subarctic phytoplankton bloom. *Nature*, **428**, 549–553.

Boyle, E.D., 1988: The role of vertical chemical fractionation in controlling late quaternary atmospheric carbon dioxide. *J. Geophys. Res.*, **93**(C12), 15701–15714.

Brasseur, G.P., and E. Roeckner, 2005: Impact of improved air quality on the future evolution of climate. *Geophys. Res. Lett.*, **32**, L23704, doi:10.1029/2005GL023902.

Brasseur, G.P., et al., 1998: Past and future changes in global tropospheric ozone: impact on radiative forcing. *Geophys. Res. Lett.*, **25**(20), 3807–3810.

Brasseur, G.P., et al., 2005: Impact of climate change on the future chemical composition of the global troposphere. *J. Clim.*, **19**, 3932–3951

Breshears, D.D., et al., 2005: Regional vegetation die-off in response to global-change-type drought. *Proc. Natl. Acad. Sci. U.S.A.*, **102**(42), 15144–15148.

Broecker, W.S., 1991: Keeping global change honest. *Global Biogeochem. Cycles*, **5**, 191–195.

Broecker, W.S., and T. Takahashi, 1978: Neutralization of fossil fuel CO_2 by marine calcium carbonate. In: *The Fate of Fossil Fuel CO_2 in the Ocean* [Andersen, N.R., and A. Malahoff (eds.)]. Plenum Press, New York, NY, pp. 213–248.

Broecker, W.S., and T.-H. Peng, 1982: *Tracers in the Sea*. ELDIGIO Press, New York, NY, 689 pp.

Broecker, W.S., and T.-H. Peng, 1986: Carbon cycle: 1985 – glacial to interglacial changes in the operation of the global carbon cycle. *Radiocarbon*, **28**, 309–327.

Broerse, A.T.C., et al., 2003: The cause of bright waters in the Bering Sea in winter. *Continental Shelf Res.*, **23**, 1579–1596.

Brook, E., et al., 2000: On the origin and timing of rapid changes in atmospheric methane during the last glacial period. *Global Biogeochem. Cycles*, **14**, 559–572.

Brovkin, V., et al., 2004: Role of land cover changes for atmospheric CO_2 increase and climate change during the last 150 years. *Global Change Biol.*, **10**, 1253–1266, doi:10.1111/j.1365-2486.2004.00812.

Brown, S., and A.E. Lugo, 1982: The storage and production of organic-matter in tropical forests and their role in the global carbon-cycle. *Biotropica*, **14**(3), 161–187.

Brown, T.J., B.L. Hall, and A.L. Westerling, 2004: The impact of twenty-first century climate change on wildland fire danger in the western United States: an applications perspective. *Clim. Change*, **62**, 365–388.

Buddemeier, R.W., J.A. Kleypas, and R.B. Aronson, 2004: *Coral Reefs and Global Climate Change*. Pew Centre on Global Climate Change, Arlington, VA, 44 pp.

Buffett, B., and D. Archer, 2004: Global inventory of methane clathrate: sensitivity to changes in the deep ocean. *Earth Planet. Sci. Lett.*, **227**, 185–199.

Burkhardt, S., I. Zondervan, and U. Riebesell, 1999: Effect of CO_2 concentration on C:N:P ratio in marine phytoplankton: a species comparison. *Limnol. Oceanogr.*, **44**(3), 683–690.

Burrows, W.H., et al., 2002: Growth and carbon stock change in eucalypt woodlands in northeast Australia: ecological and greenhouse sink implications. *Global Change Biol.*, **8**, 769–784.

Butchart, N., and A.A. Scaife, 2001: Removal of chlorofluorocarbons by increased mass exchange between the stratosphere and troposphere in a changing climate. *Nature*, **410**, 799–802.

Butchart, N., et al., 2006: Simulations of anthropogenic change in the strength of the Brewer–Dobson circulation. *Clim. Dyn.*, **27**, doi:10.1007/s00382-006-0162-4.

Butler, T.M., I. Simmonds, and P.J. Rayner, 2004: Mass balance inverse modeling of methane in the 1990s using a chemistry transport model. *Atmos. Chem. Phys.*, **4**, 2561–2580.

Cakmur, R.V., et al., 2006: Constraining the magnitude of the global dust cycle by minimizing the difference between a model and observations. *J. Geophys. Res.*, **111**, doi:10.1029/2005JD005791

Caldeira, K., and M.E. Wickett, 2003: Anthropogenic carbon and ocean pH. *Nature*, **425**(6956), 365–368.

Cao, M., K. Gregson, and S. Marshall, 1998: Global methane emission from wetlands and its sensitivity to climate change. *Atmos. Environ.*, **32**, 3291–3299.

Carpenter, L.J., 2003: Iodine in the marine boundary layer. *Chem. Rev.*, 103, 4953–4962.

Chadwick, O.A., et al., 1999: Changing sources of nutrients during four million years of ecosystem development. *Nature*, **397**(6719), 491.

Chagnon, F.J.F., R.L. Bras, and J. Wang, 2004: Climatic shift in patterns of shallow clouds over the Amazon. *Geophys. Res. Lett.*, **31**(24), L24212, doi:10.1029/2004GL021188.

Chambers, J.Q., and S.E. Trumbore, 1999: An age-old problem. *Trends Plant Sci.*, **4**(10), 385–386.

Chambers, J.Q., and W.L. Silver, 2004: Some aspects of ecophysiological and biogeochemical responses of tropical forests to atmospheric change. *Philos. Trans. R. Soc. London Ser. B*, **359**(1443), 463–476.

Chance, K., et al., 2000: Satellite observations of formaldehyde over North America from GOME. *Geophys. Res. Lett.*, **27**, 3461–3464.

Chapin, F.S. III, et al., 2005: Role of land-surface changes in arctic summer warming. *Science*, **310**, 657–660.

Chapman, S.J., and M. Thurlow, 1996: The influence of climate on CO_2 and CH_4 emissions from organic soils. *J. Agric. For. Meteorol.*, **79**, 205–217.

Chappellaz, J.A., I.Y. Fung, and A.M. Thompson, 1993: The atmospheric CH_4 increase since the last Glacial Maximum (1) Source estimates. *Tellus*, **45B**, 228–241.

Chave, J., et al., 2003: Spatial and temporal variation of biomass in a tropical forest: results from a large census plot in Panama. *J. Ecol.*, **91**, 240–252.

Chen, C.-T.A., K.-K. Liu, and R. MacDonald, 2003: Continental margin exchanges. In: *Ocean Biogeochemistry* [Fasham, M.J.R. (ed.)]. Springer-Verlag, Berlin, pp. 53–97.

Chen, M., P. Xie, and J.E. Janowiak, 2002: Global land precipitation: a 50-yr monthly analysis based on gauge observations. *J. Hydrometeorol.*, **3**, 249–266.

Chen, Y., and J.E. Penner, 2005: Uncertainty analysis for estimates of the first indirect effect. *Atmos. Chem. Phys.*, **5**, 2935–2948.

Chen, Y-H., and R.G. Prinn, 2005: Atmospheric modeling of high- and low-frequency methane observations: Importance of interannually varying transport. *J. Geophys. Res.*, **110**, D10303, doi:10.1029/2004JD005542.

Chen, Y.-H., and R.G. Prinn, 2006: Estimation of atmospheric methane emission between 1996-2001 using a 3-D global chemical transport model. *J. Geophys. Res.*, **111**, D10307, doi:10.1029/2005JD006058.

Christensen, T.R., A. Ekberg, L. Ström, and M. Mastepanov, 2003: Factors controlling large scale variations in methane emission from wetlands. *Geophys. Res. Lett.*, **30**, 1414, doi:10.1029/2002GL016848.

Christensen, T.R., et al., 2004: Thawing sub-arctic permafrost: Effects on vegetation and methane emissions. *Geophys. Res. Lett.*, **31**, doi:10.1029/2003GL018680.

Chuang, P.Y., R.M. Duvall, M.M. Shafer, and J.J. Schauer, 2005: The origin of water soluble particulate iron in the Asian atmospheric outflow. *Geophys. Res. Lett.*, **32**, doi:10.1029/2004GL021946.

Chung, S.H., and J.H. Seinfeld, 2005: Climate response of direct radiative forcing of anthropogenic black carbon. *J. Geophys. Res.*, **110**, D11102, doi:10.1029/2004JD005441.

Ciais, P., et al., 1995: Partitioning of ocean and land uptake of CO_2 as inferred by $\delta^{13}C$ measurements from the NOAA Climate Monitoring and Diagnostics Laboratory Global Air Sampling Network. *J. Geophys. Res.*, **100**(D3), 5051–5070.

Ciais, P., et al., 2005a: The potential for rising CO_2 to account for the observed uptake of carbon by tropical, temperate, and boreal forest biomes. In: *The Carbon Balance of Forest Biomes* [Griffiths, H., and P. G. Jarvis (eds.)]. Taylor and Francis, New York, pp. 109–150.

Ciais, P., et al., 2005b: Europe-wide reduction in primary productivity caused by the heat and drought in 2003. *Nature*, **437**(7058), 529–533.

Clair, T.A., J.M. Ehrman, and K. Higuchi, 1999: Changes in freshwater carbon exports from Canadian terrestrial basins to lakes and estuaries under $2xCO_2$ atmospheric scenario. *Global Biogeochem. Cycles*, **13**(4), 1091–1097.

Clark, D.A., 2002: Are tropical forests an important carbon sink? Reanalysis of the long-term plot data. *Ecol. Appl.*, **12**, 3–7.

Clark, D.A., 2004: Sources or sinks? The responses of tropical forests to current and future climate and atmospheric composition. *Philos. Trans. R. Soc. London Ser. B*, **359**, 477–491.

Clark, D.B., C.M. Taylor, and A.J. Thorpe, 2004: Feedback between the land surface and rainfall at convective length scales. *J. Hydrometeorol.*, **5**(4), 625–639.

Coakley, J.A. Jr., and C.D. Walsh, 2002: Limits to the aerosol indirect radiative forcing derived from observations of ship tracks. *J. Atmos. Sci.*, **59**, 668–680.

Cochrane, M.A., 2003: Fire science for rainforests. *Nature*, **421**(6926), 913–919.

Cohan, D.S., et al., 2002: Impact of atmospheric aerosol light scattering and absorption on terrestrial net primary productivity. *Global Biogeochem. Cycles*, **16**(4), 25–34, 1090, doi:10.1029/2001GB001441.

Cole, V., et al., 1996: Agricultural options for mitigation of greenhouse gas emissions. In: *Climate Change 1995. Impacts, Adaptations and Mitigation of Climate Change: Scientific-Technical Analyses* [Watson, R.T, M.C. Zinyowera, R.H. Moss, and D.J. Dokken (eds)]. Cambridge University Press, Cambridge, United Kingdom and New York, NY, USA, pp 745–771.

Collier, J.C., and K.P. Bowman, 2004: Diurnal cycle of tropical precipitation in a general circulation model. *J. Geophys. Res.*, **109**, D17105, doi:10.1029/2004JD004818.

Collins, W.J., D.S. Stevenson, C.E. Johnson, and R.G. Derwent, 1999: Role of convection in determining the budget of odd hydrogen in the upper troposphere. *J. Geophys. Res.*, **104**(D21), 26927–26942.

Collins, W.J., R.G. Derwent, C.E. Johnson, and D.S. Stevenson, 2002: The oxidation of organic compounds in the troposphere and their global warming potentials. *Clim. Change*, **52**(4), 453–479.

Collins, W.J., et al., 2003: Effect of stratosphere-troposphere exchange on the future tropospheric ozone trend. *J. Geophys.Res.*, **108**(D12), 8528, doi:10.1029/2002JD002617.

Conrad, R., 1996: Soil microorganisms as controllers of atmospheric trace gases (H_2, CO, CH_4, OCS, N_2O, and NO). *Microbiol. Rev.*, **60**, 609–640.

Conrad, R., and W. Seiler, 1981: Decomposition of atmospheric hydrogen by soil-microorganisms and soil enzymes. *Soil Biol. Biochem.*, **13**, 43–49.

Constable, J.V.H., A.B. Guenther, D.S. Schimel, and R.K. Monson, 1999: Modeling changes in VOC emission in response to climate change in the continental United States. *Global Change Biol.*, **5**, 791–806.

Cook, J., and E.J. Highwood, 2004: Climate response to tropospheric absorbing aerosols in an intermediate general circulation model. *Q. J. R. Meteorol. Soc.*, **130**, 175–191.

Cooke, W.F., V. Ramaswamy, and P. Kasibhatla, 2002: A general circulation model study of the global carbonaceous aerosol distribution. *J. Geophys. Res.*, **107**, 4279, doi:10.1029/2001JD001274.

Cox, P.M., et al., 2000: Acceleration of global warming due to carbon-cycle feedbacks in a coupled climate model. *Nature*, **408**(6809), doi:10.1038/35041539.

Cox, P.M., et al., 2004: Amazonian forest dieback under climate-carbon cycle projections for the 21st century. *Theor. Appl. Climatol.*, **78**, 137–156.

Cramer, W., et al., 2001: Global response of terrestrial ecosystem structure and function to CO_2 and climate change: results from six dynamic global vegetation models. *Global Change Biol.*, **7**(4), 357–374.

Crucifix, M., R.A. Betts, and P.M. Cox, 2005: Vegetation and climate variability: a GCM modeling study. *Clim. Dyn.*, **24**, 457–467, doi:10.1007/S00382-004-0504-z.

Cui, Z.Q., K.S. Carslaw, Y. Yin, and S. Davies, 2006: A numerical study of aerosol effects on the dynamics and microphysics of a deep convective cloud in a continental environment. *J. Geophys. Res.*, **111**, D05201, doi:10.1029/2005JD005981.

Curtis, P.S., et al., 2002: Biometric and eddy-covariance based estimates of annual carbon storage in five eastern North American deciduous forests. *Agric. For. Meteorol.*, **113**, 3–19.

Cziczo, D.J., et al., 2004: Observations of organic species and atmospheric ice formation. *Geophys. Res. Lett.*, **31**, doi:10.1029/2004GL019822.

Da Rocha, H.R., et al., 2004: Seasonality of water and heat fluxes over a tropical forest in eastern Amazonia. *Ecol. Appl.*, **14**, S114–S126.

Dai, A., and K.E. Trenberth, 2002: Estimates of freshwater discharge from continents: latitudinal and seasonal variations. *J. Hydrometeorol.*, **3**, 660–687.

Dameris, M., et al., 2005: Long-term changes and variability in a transient simulation with a chemistry-climate model employing realistic forcing. *Atmos. Chem. Phys.*, **5**, 2121–2145.

Dameris, M., et al., 2006: Impact of solar cycle for onset of ozone recovery. *Geophys. Res. Lett.*, **33**, L03806, doi:10.1029/2005GL024741.

Dargaville, R.J., et al., 2000: Implications of interannual variability in atmospheric circulation on modeled CO_2 concentrations and source estimates. *Global Biogeochem. Cycles*, **14**, 931–943.

De Leeuw, G., et al., 2001: Atmospheric input of nitrogen into the North Sea: ANICE project overview. *Continental Shelf Res.*, **21**(18–19), 2073–2094.

Decho, A.W., 1990: Microbial exopolymer secretions in ocean environments: their role(s) in food webs and marine processes. *Oceanogr. Mar. Biol. Annu. Rev.*, **28**, 73–153.

DeFries, R.S., et al., 2002: Carbon emissions from tropical deforestation and regrowth based on satellite observations for the 1980s and 1990s. *Proc. Natl. Acad. Sci. U.S.A.*, **99**(22), 14256–14261.

Degens, E.T., S. Kempe, and A. Spitzy, 1984: Carbon dioxide: A biogeochemical portrait. In: *The Handbook of Environmental Chemistry* [Hutzinger, O. (ed.)]. Vol. 1, Part C, Springer-Verlag, Berlin, Heidelberg, pp. 127–215.

Del Grosso, S.J., A.R. Mosier, W.J. Parton, and D.S. Ojima, 2005: DAYCENT model analysis of past and contemporary soil N_2O and net greenhouse gas flux for major crops in the USA. *Soil Tillage Res.*, **83**(1), 9–24.

Del Grosso, S.J., et al., 2000: General CH_4 oxidation model and comparison of CH_4 oxidation in natural and managed systems. *Global Biogeochem. Cycles*, **14**, 999–1019.

DeLucia, E.H., D.J. Moore, and R.J. Norby, 2005: Contrasting responses of forest ecosystems to rising atmospheric CO_2: implications for the global C cycle. *Global Biogeochem. Cycles*, **19**, G3006, doi:10.1029/2004GB002346.

Dentener, F., et al., 1996: Role of mineral aerosol as a reactive surface in the global troposphere. *J. Geophys. Res.*, **101**, 22869–22889.

Dentener, F., et al., 2003a: Interannual variability and trend of CH_4 lifetime as a measure for OH changes in the 1979–1993 time period. *J. Geophys. Res.*, **108**(D15), 4442, doi:10.1029/2002JD002916.

Dentener, F., et al., 2003b: Trends and inter-annual variability of methane emissions derived from 1979–1993 global CTM simulations. *Atmos. Chem. Phys.*, **3**, 73–88.

Dentener, F., et al., 2005: The impact of air pollutant and methane emission controls on tropospheric ozone and radiative forcing: CTM calculations for the period 1990–2030. *Atmos. Chem. Phys.*, **5**, 1731–1755.

Dentener, F., et al., 2006: The global atmospheric environment for the next generation. *Environ. Sci. Technol.*, **40**(11), 3586–3594.

Derwent, R.G., W.J. Collins, C.E. Johnson, and D.S. Stevenson, 2001: Transient behaviour of tropospheric ozone precursors in a global 3-D CTM and their indirect greenhouse effects. *Clim. Change*, **49**(4), 463–487.

Desborough, C.E., 1999: Surface energy balance complexity in GCM land surface models. *Clim. Dyn.*, **15**, 389–403.

Dickens, G.R., 2001: Modeling the global carbon cycle with gas hydrate capacitor: Significance for the latest Paleocene thermal maximum. In: *Natural Gas Hydrates: Occurrence, Distribution, and Detection* [Paull, C.K., and W.P. Dillon (eds.)]. Geophysical Monographs Vol. 124, American Geophysical Union, Washington, DC, pp. 19–38.

Dickens, G.R., M.M. Castillo, and J.G.C. Walker, 1997: A blast of gas in the latest Paleocene: Simulating first-order effects of massive dissociation of oceanic methane hydrate. *Geology*, **25**, 259–262.

Dickinson, R.E., G. Wang, X. Zeng, and Q.-C. Zeng, 2003: How does the partitioning of evapotranspiration and runoff between different processes affect the variability and predictability of soil moisture and precipitation? *Adv. Atmos. Sci.*, **20**(3), 475–478.

Dickinson, R.E., et al., 2006: The community land model and its climate statistics as a component of the community climate system model. *J. Clim.*, **19**, 2302–2324.

Diehl, K., and S. Wurzler, 2004: Heterogeneous drop freezing in the immersion mode: Model calculations considering soluble and insoluble particles in the drops. *J. Atmos. Sci.*, **61**, 2063–2072.

Dirmeyer, P.A., 2001: An evaluation of the strength of land-atmosphere coupling. *J. Hydrometeorol.*, **2**(4), 329–344.

Dlugokencky, E.J., K.A. Masarie, P.M. Lang, and P.P. Tans, 1998: Continuing decline in the growth rate of the atmospheric methane burden. *Nature*, **393**, 447–450.

Dlugokencky, E.J., et al., 2001: Measurements of an anomalous global methane increase during 1998. *Geophys. Res. Lett.*, **28**, 499–502.

Dlugokencky, E.J., et al., 2003: Atmospheric methane levels off: Temporary pause or a new steady state. *Geophys. Res. Lett.*, **30**, doi:10.1029/2003GL018126.

D'Odorico, P., and A. Porporato, 2004: Preferential states in soil moisture and climate dynamics. *Proc. Natl. Acad. Sci. U.S.A.*, **101**(24), 8848–8851.

Doherty, R.M., D.S. Stevenson, W.J. Collins, and M.G. Sanderson, 2005: Influence of convective transport on tropospheric ozone and its precursors in a chemistry-climate model. *Atmos. Chem. Phys.*, **5**, 3747–3771.

Doney, S.C., et al., 2004: Evaluating global ocean carbon models: the importance of realistic physics. *Global Biogeochem. Cycles*, **18**(3), GB3017, doi:10.1029/2003GB002150.

Douglass, A.R., M.R. Schoeberl, R.B. Rood, and S. Pawson, 2003: Evaluation of transport in the lower tropical stratosphere in a global chemistry and transport model. *J. Geophys. Res.*, **108**(D9), 4259, doi:10.1029/2002JD002696.

Duce, R.A., 1995: Sources, distributions and fluxes of mineral aerosols and their relationship to climate. In: *Aerosol Forcing of Climate* [Charlson, R.J. and J. Heintzenberg (eds.)]. John Wiley & Sons Ltd., Chichester, New York, pp. 43–72.

Dukes, J.S., et al., 2005: Responses of grassland production to single and multiple global environmental changes. *PLoS Biol.*, **3**(10), 1829–1836.

Dunn, A.L, et al., 2007: A long-term record of carbon exchange in a boreal black spruce forest: means, responses to interannual variability, and decadal trends. *Global Change Biol.*, **13**, 577-590, doi:10.1111/j.1365-2486.2006.01221.x.

Dupre, B., et al., 2003: Rivers, chemical weathering and Earth's climate. *Comptes Rendus Geoscience*, **335**(16), 1141–1160.

Durieux, L., L.A.T. Machado, and H. Laurent, 2003: The impact of deforestation on cloud cover over the Amazon arc of deforestation. *Remote Sens. Environ.*, **86**(1), 132–140.

Dutay, J.C., et al., 2002: Evaluation of ocean model ventilation with CFC-11: comparison of 13 global ocean models. *Ocean Modelling*, **4**(2), 89–102.

Easter, R.C., et al., 2004: MIRAGE: Model description and evaluation of aerosols and trace gases. *J. Geophys. Res.*, **109**, doi:10.1029/2004JD004571.

Edwards, D.P., et al., 2004: Observations of carbon monoxide and aerosols from the Terra satellite: Northern Hemisphere variability. *J. Geophys. Res.*, **109**, D24202, doi:10.1029/2004JD004727.

Eglinton, T.I., and D.J. Repeta, 2004, Organic matter in the contemporary ocean. In: *Treatise on Geochemistry* [Holland, H.D., and K.K. Turekian (eds.)]. Volume 6, The Oceans and Marine Geochemistry, Elsevier Pergamon, Amsterdam, pp. 145–180.

Ehhalt, D.H., 1999: Gas phase chemistry of the troposphere. In: *Global Aspects of Atmospheric Chemistry* [Baumgärtl, H., W. Grünbein, and F. Hensel (eds.)]. Dr. Dietrich Steinkopf Verlag, Darmstadt, Germany, pp. 21–110.

Ek, M.B., and A.A.M. Holtslag, 2004: Influence of soil moisture on boundary layer cloud development. *J. Hydrometeorol.*, **5**, 86–99.

Engel, A., et al., 2004: Polysaccharide aggregation as a potential sink of marine dissolved organic carbon. *Nature*, **428**, 929–932.

Enting, I.G., and J.V. Mansbridge, 1991: Latitudinal distribution of sources and sinks of CO_2 - Results of an inversion study. *Tellus*, **43B**, 156–170.

Enting, I.G., C.M. Trudinger, and R.J. Francey, 1995: A synthesis inversion of the concentration and ^{13}C of atmospheric CO_2. *Tellus*, **47B**, 35–52.

Etheridge, D.M., L.P. Steel, R.J. Francey, and R.L. Langenfelds, 1998: Atmospheric methane between 1000 A.D. and present: Evidence of anthropogenic emissions and climatic variability. *J. Geophys. Res.*, **103**, 15979–15993.

Etiope, G., 2004: GEM-Geologic Emission of Methane, the missing source in the atmospheric methane budget. *Atmos. Environ.*, **38**, 3099–3100.

Etiope, G., and R.W. Klusman, 2002: Geologic emissions of methane to the atmosphere. *Chemosphere*, **49**, 777–789.

European Commission, 2003: *Ozone-Climate Interactions*. Air Pollution Research Report 81, EUR 20623, European Commission, Luxembourg, 143 pp.

Eyring, V., et al., 2005: A strategy for process-oriented validation of coupled chemistry-climate models. *Bull. Am. Meteorol. Soc.*, **86**, 1117–1133.

Falkowski, P., et al., 2000: The global carbon cycle: A test of our knowledge of Earth as a system. *Science*, **290**(5490), 291–296.

Falloon, P., et al., 2006: RothCUK – a dynamic modelling system for estimating changes in soil C at 1km scale in the UK. *Soil Use Management*, **22**, 274–288.

Fan, S., et al., 1998: A large terrestrial carbon sink in North America implied by atmospheric and oceanic carbon dioxide data and models. *Science*, **282**, 442–446.

Fang, J., et al., 2001: Changes in forest biomass carbon storage in China between 1949 and 1998. *Science*, **292**, 2320–2322.

Farquhar, G.D., S. von Caemmerer, and J.A. Berry, 2001: Models of photosynthesis. *Plant Physiol.*, **125**(1), 42–45.

Fearnside, P.M., 2000: Global warming and tropical land-use change: greenhouse gas emissions from biomass burning, decomposition and soils in forest conversion, shifting cultivation and secondary vegetation. *Clim. Change*, **46**, 115–158.

Feddes, R.A., et al., 2001: Modeling root water uptake in hydrological and climate models. *Bull. Am. Meteorol. Soc.*, **82**(12), 2797–2809.

Feely, R.A., R. Wanninkhof, T. Takahashi, and P. Tans, 1999: Influence of El Nino on the equatorial Pacific contribution to atmospheric CO_2 accumulation. *Nature*, **398**(6728), 597–601.

Feely, R.A., et al., 2002: Seasonal and interannual variability of CO_2 in the equatorial Pacific. *Deep-Sea Res. II*, **49**, 2443–2469.

Feely, R.A., et al., 2004: Impact of anthropogenic CO_2 on the $CaCO_3$ system in the oceans. *Science*, **305**, 362–366.

Feichter, J., E. Roeckner, U. Lohmann, and B. Liepert, 2004: Nonlinear aspects of the climate response to greenhouse gas and aerosol forcing. *J. Clim.*, **17**(12), 2384–2398.

Feingold, G., S.M. Kreidenweis, and Y.P. Zhang, 1998: Stratocumulus processing of gases and cloud condensation nuclei - 1. Trajectory ensemble model. *J. Geophys. Res.*, **103**(D16), 19527–19542.

Feingold, G., H. Jiang, and J. Y. Harrington, 2005: On smoke suppression of clouds in Amazonia. *Geophys. Res. Lett.*, **32**, L02804, doi:10.1029/2004GL021369.

Feingold, G., W.R. Cotton, S.M. Kreidenweis, and J.T. Davis, 1999: The impact of giant cloud condensation nuclei on drizzle formation in stratocumulus: Implications for cloud radiative properties. *J. Atmos. Sci.*, **56**, 4100–4117.

Fekete, B.M., C.J. Vorosmarty, and W. Grabs, 2002: High-resolution fields of global runoff combining observed river discharge and simulated water balances. *Global Biogeochem. Cycles*, **16**, doi:10.1029/1999GB001254.

Felzer, B., et al., 2004: Effects of ozone on net primary production and carbon sequestration in the conterminous United States using a biogeochemistry model. *Tellus*, **56B**, 230–248.

Ferek, R.J., et al., 1998: Measurements of ship-induced tracks in clouds off the Washington coast. *J. Geophys. Res.*, **103**, 23199–23206.

Ferretti, D.F., et al., 2005: Unexpected changes to the global methane budget over the past 2000 years. *Science*, **309**, 1714–1717.

Field, C.B., and M.R. Raupach (eds.), 2004: *The Global Carbon Cycle: Integrating Humans, Climate, and the Natural World*. SCOPE 62, Island Press, Washington, DC, 526 pp.

Finzi, A.C., et al., 2006: Progressive nitrogen limitation of ecosystem processes under elevated CO_2 in a warm-temperate forest. *Ecology*, 87, 15–25.

Findell, K.L., and E.A.B. Eltahir, 2003: Atmospheric controls on soil moisture–boundary layer interactions. Part II: Feedbacks within the continental United States. *J. Hydrometeorol.*, **4**, 570–583.

Fioletov, V.E., et al., 2002: Global and zonal total ozone variations estimated from ground-based and satellite measurements: 1964–2000. *J. Geophys. Res.*, **107**(D22), 4647, doi:10.1029/2001JD001350.

Flannigan, M.D., B.J. Stocks, and B.M. Wotton, 2000: Climate change and forest fires. *Sci. Total Environ.*, **262**, 221–229.

Flückiger, J., et al., 2002: High resolution Holocene N_2O ice core record and its relationship with CH_4 and CO_2. *Global Biogeochem. Cycles*, **16**, doi: 10.1029/2001GB001417.

Folberth, G., D.A. Hauglustaine, P. Ciais, and J. Lathière, 2005: On the role of atmospheric chemistry in the global CO_2 budget. *Geophys. Res. Lett.*, **32**, L08801, doi:10.1029/2004GL021812.

Folberth, G.A., D.A. Hauglustaine, J. Lathière, and F. Brocheton, 2006: Interactive chemistry in the Laboratoire de Météorologie Dynamique general circulation model: model description and impact of biogenic hydrocarbons on tropospheric chemistry. *Atmos. Chem. Phys.*, **6**, 2273–2319.

Foley, J.A., et al., 2003: Green Surprise? How terrestrial ecosystems could affect Earth's climate. *Frontiers Ecol. Environ.*, **1**(1), 38–44.

Frankenberg, C., et al., 2005: Assessing methane emission from global space-borne observation. *Science*, **308**, 1010–1014.

Frankenberg, C., et al., 2006: Satellite chartography of atmospheric methane from SCIAMACHY on board EMVISAT: Analysis of the years 2003 and 2004. *J. Geophys. Res.*, **111**, doi:10.1029/2005JD006235.

Freeman, C., et al., 2004: Export of dissolved organic carbon from peatlands under elevated carbon dioxide levels. *Nature*, **430**, 195–198.

Freitas, S.R., et al., 2005: Monitoring the transport of biomass burning emissions in South America. *Environ. Fluid Mech.*, **5**, 135–167.

Frew, R., A. Bowie, P. Croot, and S. Pickmere, 2001: Macronutrient and trace-metal geochemistry of an in situ iron-induced Southern Ocean bloom. *Deep-Sea Res. II*, **48**(11–12), 2467–2481.

Friedlingstein, P., J.-L. Dufresne, P.M. Cox, and P. Rayner, 2003: How positive is the feedback between climate change and the carbon cycle? *Tellus*, **55B**(2), 692–700.

Friedlingstein, P., et al., 2001: Positive feedback between future climate change and the carbon cycle. *Geophys. Res. Lett.*, **28**, 1543–1546, doi:10.1029/2000GL012015.

Friedlingstein, P., et al., 2006: Climate-carbon cycle feedback analysis: results from the C4MIP model intercomparison. *J. Clim.*, **19**, 3337–3353.

Fu, R., and W. Li, 2004: The influence of the land surface on the transition from dry to wet season in Amazonia. *Theor. Appl. Climatol.*, **78**, 97–110, doi:10.1007/s00704-004-0046-7.

Fueglistaler, S., and P.H. Haynes, 2005: Control of interannual and longer-term variability of stratospheric water vapor. *J. Geophys. Res.*, **110**, D24108, doi:10.1029/2005JD006019.

Fung, I., S.C. Doney, K. Lindsay, and J. John, 2005: Evolution of carbon sinks in a changing climate. *Proc. Natl. Acad. Sci. U.S.A.*, **102**(32), 11201–11206.

Fusco, A.C., and J.A. Logan, 2003: Analysis of 1970-1995 trends in tropospheric ozone at northern hemisphere midlatitudes with the GEOS-CHEM model. *J. Geophys. Res.*, **108**(D15), 4449, doi:10.1029/2002JD002742.

Gabric, A.J., et al., 2004: Modeling estimates of the global emission of dimethylsulfide under enhanced greenhouse conditions. *Global Biogeochem. Cycles*, **18**(2), GB2014, doi:10.1029/2003GB002183.

Galloway, J.N., et al., 2004: Nitrogen cycles: past, present, and future. *Biogeochemistry*, **70**(2), 153–226.

Gamon, J.A., et al., 2003. Remote sensing in BOREAS: Lessons learned. *Remote Sens. Environ.*, **89**, 139–162.

Ganzeveld, L.N., et al., 2002: Global soil-biogenic NO_x emissions and the role of canopy processes. *J. Geophys. Res.*, **107**(D16), 4298, doi:10.1029/2001JD001289.

Gao, Z., et al., 2004: Modeling of surface energy partitioning, surface temperature, and soil wetness in the Tibetan prairie using the Simple Biosphere Model 2 (SiB2). *J. Geophys. Res.*, **109**, D06102, doi:10.1029/2003JD004089.

Gattuso, J.-P., D. Allemand, and M. Frankignoulle, 1999: Photosynthesis and calcification at cellular, organismal and community levels in coral reefs: a review on interactions and control by carbonate chemistry. *Am. Zool.*, **39**, 160–183.

Gedney, N., and P. Cox, 2003: The sensitivity of global climate model simulations to the representation of soil moisture heterogeneity. *J. Hydrometeorol.*, **4**, 1265–1275.

Gedney, N., P.M. Cox, and C. Huntingford, 2004: Climate feedback from wetland methane emissions. *Geophys. Res. Lett.*, **31**, L20503, doi:10.1029/2004GL020919.

Gérard, J.C., et al., 1999: The interannual change of atmospheric CO_2: contribution of subtropical ecosystems. *Geophys. Res. Lett.*, **26**, 243–246.

Gettelman, A., J.R. Holton, and K.H. Rosenlof, 1997: Mass fluxes of O_3, CH_4, N_2O, and CF_2Cl_2 in the lower stratosphere calculated from observational data. *J. Geophys. Res.*, **102**, 19149–19159.

Ghan, S.J., G. Guzman, and H. Abdul-Razzak, 1998: Competition between sea salt and sulphate particles as cloud condensation nuclei. *J. Atmos. Sci.*, **55**, 3340–3347.

Giardina, C.P., and M.G. Ryan, 2000: Evidence that decomposition rates of organic carbon in mineral soil do not vary with temperature. *Nature*, **404**, 858–861.

Gillett, N.P., and D.W.J. Thompson, 2003: Simulation of recent Southern Hemisphere climate change. *Science*, **302**, 273–275.

Gillett, N.P., A.J. Weaver, F.W. Zwiers, and M.D. Flannigan, 2004: Detecting the effect of climate change on Canadian forest fires. *Geophys. Res. Lett.*, **31**(18), L18211, doi:10.1029/2004GL020876.

Ginoux, P., et al., 2001: Sources and distributions of dust aerosols simulated with the GOCART model. *J. Geophys. Res.*, **16**, 20255–20274.

Givati, A., and D. Rosenfeld, 2004: Quantifying precipitation suppression due to air pollution. *J. Appl. Meteorol.*, **43**(7), 1038–1056.

Gloor, M., et al., 2003: A first estimate of present and preindustrial air-sea CO_2 flux patterns based on ocean interior carbon measurements and models. *Geophys. Res. Lett.*, **30**(1), 1010, doi:10.1029/2002GL015594.

Goncalves, L.G.G., E.J. Burke, and W.J. Shuttleworth, 2004: Application of improved ecosystem aerodynamics in regional weather forecasts. *Ecol. Appl.*, **14**, S17–S21.

Gondwe, M., et al., 2003: Correction to "The contribution of ocean-leaving DMS to the global atmospheric burdens of DMS, MSA, SO, and NSS SO=". *Global Biogeochem. Cycles*, **17**, 1106, doi:10.1029/2003GB002153.

Gong, S.L., 2003: A parameterization of sea-salt aerosol source function for sub- and super-micron particles. *Global Biogeochem. Cycles*, **17**(4), 1097, doi:10.1029/2003GB002079.

Gong, S.L., and L.A. Barrie, 2003: Simulating the impact of sea salt on global nss sulphate aerosols. *J. Geophys. Res.*, **108**(D16), 4516, doi:10.1029/2002JD003181.

Goodale, C.L., et al., 2002: Forest carbon sinks in the northern hemisphere. *Ecol. Appl.*, **12**(3), 891–899.

Goulden, M.L., et al., 2004: Diel and seasonal patterns of tropical forest CO_2 exchange. *Ecol. Appl.*, **14**, S42–S54.

Grassl, H., 1975: Albedo reduction and radiative heating of clouds by absorbing aerosol particles. *Contrib. Atmos. Phys.*, **48**, 199–210.

Green, P.A., et al., 2004: Pre-industrial and contemporary fluxes of nitrogen through rivers: a global assessment based on typology. *Biogeochemistry*, **68**(1), 71–105.

Grenfell, J.L., D.T. Shindell, and V. Grewe, 2003: Sensitivity studies of oxidative changes in the troposphere in 2100 using the GISS GCM. *Atmos. Chem. Phys.*, **3**, 1267–1283.

Grewe, V., et al., 2001: Future changes of the atmospheric composition and the impact on climate change. *Tellus*, **53B**(2), 103–121.

Gruber, N., and C.D. Keeling, 2001: An improved estimate of the isotopic air-sea disequilibrium of CO_2: Implications for the oceanic uptake of anthropogenic CO_2. *Geophys. Res. Lett.*, **28**, 555–558.

Gruber, N., N. Bates, and C.D. Keeling, 2002: Interannual variability in the North Atlantic Ocean carbon sink. *Science*, **298**(5602), 2374–2378.

Gu, L., et al., 2002: Advantages of diffuse radiation for terrestrial ecosystem productivity. *J. Geophys. Res.*, **107**(6), 4050, doi:10.1029/2001JD001242.

Gu, L., et al., 2003: Response of a deciduous forest to the Mt. Pinatubo eruption: enhanced photosynthesis. *Science*, **299**(5615), 2035–2038.

Guenther, A.B., et al., 1993: Isoprene and monoterpene emission rate variability - model evaluations and sensitivity analyses. *J. Geophys. Res.*, **98**(D7), 12609–12617.

Guenther, A.B., et al., 1995: A global-model of natural volatile organic-compound emissions. *J. Geophys. Res.*, **100**(D5), 8873–8892.

Guenther, A.B., et al., 1999: Isoprene emission estimates and uncertainties for the Central African EXPRESSO study domain. *J. Geophys. Res.*, **104**(D23), 30625–30639.

Guillevic, P., et al., 2002: Influence of the interannual variability of vegetation on the surface energy balance - a global sensitivity study. *J. Hydrometeorol.*, **3**, 617–629.

Guo, Z., et al., 2006. GLACE: The Global Land-Atmosphere Coupling Experiment. 2. Analysis. *J. Hydrometeorol.*, **7**, 611–625.

Gupta, M., et al., 1997: $^{12}C/^{13}C$ kinetic isotope effects in the reactions of CH_4 with OH and Cl. *Geophys. Res. Lett.*, **24**, 2761–2764.

Gurney, K.R., et al., 2002: Towards robust regional estimates of CO_2 sources and sinks using atmospheric transport models. *Nature*, **415**(6872), 626–630.

Gurney, K.R., et al., 2003: TransCom 3 CO_2 inversion intercomparison: 1. Annual mean control results and sensitivity to transport and prior flux information. *Tellus*, **55B**(2), 555–579.

Gurney, K.R., et al., 2004: Transcom 3 inversion intercomparison: model mean results for the estimation of seasonal carbon sources and sinks. *Global Biogeochem. Cycles*, **18**(1), GB1010, doi:10.1029/2003GB002111.

Gurney, K.R., et al., 2005: Sensitivity of atmospheric CO_2 inversions to seasonal and interannual variations in fossil fuel emissions. *J. Geophys. Res.*, **110**, D10308, doi:10.1029/2004JD005373.

Haag, W., and B. Kärcher, 2004: The impact of aerosols and gravity waves on cirrus clouds at midlatitudes. *J. Geophys. Res.*, **109**, doi:10.1029/2004JD004579.

Haake, B., and V. Ittekkot, 1990: The wind-driven biological pump and carbon removal in the ocean. *Naturwissenschaften*, **77**(2), 75–79.

Hahmann, A.N., 2003: Representing spatial sub-grid precipitation variability in a GCM. *J. Hydrometeorol.*, **4**(5), 891–900.

Handisides, G.M., et al., 2003: Hohenpeissenberg photochemical experiment (HOPE 2000): measurements and photostationary state calculations of OH and peroxy radicals. *Atmos. Chem. Phys.*, **3**, 1565–1588.

Hansell, D.A., and C.A. Carlson, 1998: Deep-ocean gradients in the concentration of dissolved organic carbon. *Nature*, **395**, 263–266.

Hansen, J., M. Sato, and R. Ruedy, 1997: Radiative forcing and climate response. *J. Geophys. Res.*, **102**, 6831–6864.

Hansen, J., et al., 2005: Efficacy of climate forcings. *J. Geophys. Res.*, **110**(D18), D18104, doi:10.1029/2005JD005776.

Hauglustaine, D.A., and G.P. Brasseur, 2001: Evolution of tropospheric ozone under anthropogenic activities and associated radiative forcing of climate. *J. Geophys. Res.*, **106**(D23), 32337–32360.

Hauglustaine, D., and D.H. Ehhalt, 2002: A three-dimensional model of molecular hydrogen in the troposphere. *J. Geophys. Res.*, **107**(D17), doi:10.1029/2001JD001156.

Hauglustaine, D.A., J. Lathière, S. Szopa, and G. Folberth, 2005: Future tropospheric ozone simulated with a climate-chemistry-biosphere model. *Geophys. Res. Lett.*, **32**, L24807, doi:10.1029/2005GL024031.

Hauglustaine, D.A., et al., 2004: Interactive chemistry in the laboratoire de meteorologie dynamique general circulation model: description and background tropospheric chemistry. *J. Geophys. Res.*, **109**, D04314, doi:10.1029/2003JD003957.

Heald, C.L., et al., 2004: Comparative inverse analysis of satellite (MOPITT) and aircraft (TRACE-P) observations to estimate Asian sources of carbon monoxide. *J. Geophys. Res.*, **109**(D23), D23306, doi:10.1029/2004JD005185.

Heard, D.E., et al., 2004: High levels of the hydroxyl radical in the winter urban troposphere. *Geophys. Res. Lett.*, **31**, L18112, doi:10.1029/2004GL020544.

Hein, R., P.J. Crutzen, and M. Heimann, 1997: An inverse modeling approach to investigate the global atmospheric methane cycle. *Global Biogeochem. Cycles*, **11**, 43–76.

Heinze, C., 2004: Simulating oceanic $CaCO_3$ export production in the greenhouse. *Geophys. Res. Lett.*, **31**, L16308, doi:10.1029/2004GL020613.

Heinze, C., E. Maier-Reimer, and K. Winn, 1991: Glacial pCO_2 reduction by the World Ocean: experiments with the Hamburg carbon cycle model. *Paleoceanography*, **6**(4), 395–430.

Heinze, C., et al., 2003: Sensitivity of the marine biospheric Si cycle for biogeochemical parameter variations. *Global Biogeochem. Cycles*, **17**(3), 1086, doi:10.1029/2002GB001943.

Hejzlar, J., M. Dubrovsky, J. Buchtele, and M. Ruzicka, 2003: The apparent and potential effects of climate change on the inferred concentration of dissolved organic matter in a temperate stream (the Malse River, South Bohemia). *Sci. Total Environ.*, **310**(1–3), 143–152.

Held, I.M., et al., 2005: Simulation of Sahel drought in the 20th and 21st centuries. *Proc. Natl. Acad. Sci. U.S.A.*, **102**(50), 17891–17896

Henderson-Sellers, A., P. Irannejad, K. McGuffie, and A.J. Pitman, 2003: Predicting land-surface climates - better skill or moving targets? *Geophys. Res. Lett.*, **30**(14), 1777, doi:10.1029/2003GL017387.

Hendricks, J., B. Kärcher, M. Ponater, and U. Lohmann, 2005: Do aircraft black carbon emissions affect cirrus clouds on a global scale? *Geophys. Res. Lett.*, **32**, L12814, doi:10.1029/2005GL022740.

Hendricks, J., et al., 2004: Simulating the global atmospheric black carbon cycle: A revisit to the contribution of aircraft emissions. *Atmos. Chem. Phys.* **4**, 2521–2541.

Hesselbo, S.P., et al., 2000: Massive dissociation of gas hydrate during a Jurassic oceanic anoxic event. *Nature*, **406**, 392–395.

Heue, K.-P., et al., 2005: Validation of SCIAMACHY tropospheric NO_2 columns with AMAXDOAS measurements. *Atmos. Chem. Phys.*, **5**, 1039–1051.

Hirsch, A.I., et al., 2006: Inverse modeling estimates of the global nitrous oxide surface flux from 1998–2001. *Global Biogeochem. Cycles*, **20**, GB1008, doi:10.1029/2004GB002443.

Hoerling, M., and A. Kumar, 2003: The perfect ocean for drought. *Science*, **299**(5607), 691–694.

Hoffman, W.A., W. Schroeder, and R.B. Jackson, 2002: Positive feedbacks of fire, climate, and vegetation and the conversion of tropical savanna. *Geophys. Res. Lett.*, **15**, doi:10.1029/2002G0152.

Hofzumahaus, A., et al., 2004: Photolysis frequency of O_3 to $O(1D)$: Measurements and modeling during the International Photolysis Frequency Measurement and Modeling Intercomparison (IPMMI). *J. Geophys. Res.*, **109**, D08S90, doi:10.1029/2003JD004333.

Holland, E.A., and M.A. Carroll, 2003: Atmospheric chemistry and the bio-atmospheric carbon and nitrogen cycles. In: *Interactions of the Major Biogeochemical Cycles, Global Change and Human Impacts* [Melillo, J.M., C.B. Field, and B. Moldan (eds.)]. SCOPE 61, Island Press, Washington, DC, pp. 273–294.

Holland, E.A., F.J. Dentener, B.H. Braswell, and J.M. Sulzman, 1999: Contemporary and pre-industrial reactive nitrogen budgets. *Biogeochemistry*, **46**, 7–43.

Holland, E.A., B.H. Braswell, J. Sulzman, and J.F. Lamarque, 2005a: Nitrogen deposition onto the United States and Western Europe: synthesis of observations and models. *Ecol. Appl.*, **15**, 38–57.

Holland, E.A., J. Lee-Taylor, C. Nevison, and J. Sulzman, 2005b: Global N Cycle: Fluxes and N_2O mixing ratios originating from human activity. Data set. Available online from Oak Ridge National Laboratory Distributed Active Archive Center, Oak Ridge, TN, http://www.daac.ornl.gov.

Holland, E.A., et al., 2005c: U.S. nitrogen science plan focuses collaborative efforts. *Eos*, **86**(27), 253–260.

Hollinger, D.Y., et al., 1999: Seasonal patterns and environmental control of carbon dioxide and water vapour exchange in an ecotonal boreal forest. *Global Change Biol.*, **5**, 891–902.

Holzer, M., and G.J. Boer, 2001: Simulated changes in atmospheric transport climate. *J. Clim.*, **14**, 4398–4420.

Hong, J., T. Choi, H. Ishikawa, and J. Kim, 2004: Turbulence structures in the near-neutral surface layer on the Tibetan Plateau. *Geophys. Res. Lett.*, **31**, L15106, doi:10.1029/2004GL019935.

Hoppel, W.A., J.W. Fitzgerald, G.M. Frick, and R.E. Larson, 1990: Aerosol size distributions and optical properties found in the marine boundary layer over the Atlantic Ocean. *J. Geophys. Res.*, **95**, 3659–3686.

Horowitz, L.W., et al., 2003: A global simulation of tropospheric ozone and related tracers: description and evaluation of MOZART, version 2. *J. Geophys. Res.*, **108**, 4784, doi:10.1029/2002JD002853.

Houghton, R.A., 1999: The annual net flux of carbon to the atmosphere from changes in land use 1850-1990. *Tellus*, **51B**, 298–313.

Houghton, R.A., 2003a: Revised estimates of the annual net flux of carbon to the atmosphere from changes in land use and land management 1850-2000. *Tellus*, **55B**(2), 378–390.

Houghton, R.A., 2003b: Why are estimates of the terrestrial carbon balance so different? *Global Change Biol.*, **9**, 500–509.

Houghton, R.A., et al., 2000: Annual fluxes of carbon from deforestation and regrowth in the Brazilian Amazon. *Nature*, **403**, 301–304.

Houweling, S., F. Dentener, and J. Lelieveld, 1998: The impact of nonmethane hydrocarbon compounds on tropospheric photochemistry. *J. Geophys. Res.*, **103**, 10673–10696.

Houweling, S., F. Dentener, and J. Lelieveld, 2000: Simulation of preindustrial atmospheric methane to constrain the global source strength of natural wetlands. *J. Geophys. Res.*, **105**, 17243–17255.

Hoyle, C.R., B.P. Luo, and T. Peter, 2005: The origin of high ice crystal number densities in cirrus clouds. *J. Atmos. Sci.*, **62**, 2568–2579.

Hu, H., 1996: Water vapour and temperature lapse rate feedbacks in the mid-latitude seasonal cycle. *Geophys. Res. Lett.*, **23**, 1761–1764.

Huang, Y., R.E. Dickinson, and W.L. Chameides, 2006: Impact of aerosol indirect effect on climate over East Asia. *Proc. Natl. Acad. Sci. U.S.A.*, **103**, 4371–4376.

Huck, P.E., A.J. McDonald, G.E. Bodeker, and H. Struthers, 2005: Interannual variability in Antarctic ozone depletion controlled by planetary waves and polar temperatures. *Geophys. Res. Lett.*, **32**, L13819, doi:10.1029/2005GL022943.

Hudman, R.C., et al., 2004: Ozone production in transpacific Asian pollution plumes and implications for ozone air quality in California. *J. Geophys. Res.*, **109**, D23S10, doi:10.1029/2004JD004974.

Hughes, T.P., et al., 2003: Climate change, human impacts, and the resilience of coral reefs. *Science*, **301**, 929–933.

Hungate, B., et al., 2003: Nitrogen and climate change. *Science*, **302**(5650), 1512–1513.

Huntingford, C., et al., 2004: Using a GCM analogue model to investigate the potential for Amazonian forest dieback. *Theor. Appl. Climatol.*, **78**(1–3), 177–185.

Hurst, D.F., et al., 2006: Continuing global significance of emissions of Montreal Protocol-restricted halocarbons in the USA and Canada. *J. Geophys. Res.*, **111**, D15302, doi:10.1029/2005JD006785.

Hurtt, G.C., et al., 2002: Projecting the future of the U.S. carbon sink. *Proc. Natl. Acad. Sci. U.S.A.*, **99**(3), 1389–1394.

IFFN, 2003: Russian Federation Fire 2002 Special. Part III: The 2002 fire season in the Asian part of the Russian Federation: A view from space. *International Forest Fire News (IFFN)*, **28**, 18–28.

Imhoff, M.L., et al., 2004: Global patterns in human consumption of net primary production. *Nature*, **429**(6994), 870–873.

IPCC, 2000: *Special Report on Emission Scenarios. A Special Report of Working Group III of the Intergovernmental Panel on Climate Change* [Nakićenović, N., et al. (eds.)]. Cambridge University Press, Cambridge, United Kingdom and New York, NY, USA, 599 pp.

IPCC/TEAP, 2005: *IPCC/TEAP Special Report on Safeguarding the Ozone Layer and the Global Climate System: Issues related to Hydrofluorocarbons and Perfluorocarbons. Prepared by Working Group I and III of the Intergovernmental Panel on Climate Change and the Technology and Economic Assessment Panel* [Metz, B., et al. (eds.)]. Cambridge University Press, Cambridge, United Kingdom and New York, NY, USA, 488 pp.

Irannejad, P., A. Henderson-Sellers, and S. Sharmeen, 2003: Importance of land-surface parameterisation for latent heat simulation in global atmospheric models. *Geophys. Res. Lett.*, **30**(17), 1904, doi:10.1029/2003GL018044.

Irie, H., et al., 2005: Evaluation of long-term tropospheric NO_2 data obtained by GOME over East Asia in 1996-2002. *Geophys. Res. Lett.*, **32**, L11810, doi:10.1029/2005GL022770.

Ishimatsu, A., et al., 2004: Effects of CO_2 on marine fish: larvae and adults. *J. Oceanogr.*, **60**, 731–741.

Ittekkot, V., 1993: The abiotically driven biological pump in the ocean and short-term fluctuations in atmospheric CO_2 contents. *Global Planet. Change*, **8**(1–2), 17–25.

Jacob, D.J., 2000: Heterogeneous chemistry and tropospheric ozone. *Atmos. Environ.*, **34**, 2131–2159.

Jacob, D.J., et al., 1993: Factors regulating ozone over the United States and its export to the global atmosphere. *J. Geophys. Res.*, **98**, 14817–14826.

Jacobson, M.Z., 1999: Effects of soil moisture on temperatures, winds, and pollutant concentrations in Los Angeles. *J. Appl. Meteorol.*, **38**(5), 607–616.

Jacobson, M.Z., 2001: Global direct radiative forcing due to multicomponent anthropogenic and natural aerosols. *J. Geophys. Res.*, **106**, 1551–1568.

Jacobson, M.Z., 2002: Control of fossil-fuel particulate black carbon and organic matter, possibly the most effective method of slowing global warming. *J. Geophys. Res.*, **107**, doi:10.1029/2001JD001376.

Jacobson, M.Z., 2006: Effects of externally-through-internally-mixed soot inclusions within clouds and precipitation on global climate. *J. Phys. Chem. A*, **110**, 6860–6873.

Jaeglé, L., D.J. Jacob, W.H. Brune, and P.O. Wenberg, 2001: Chemistry of HO_x radicals in the upper troposphere. *Atmos. Environ.*, **35**, 469–489.

Jaeglé, L., L. Steinberger, R.V. Martin, and K. Chance, 2005: Global partitioning of NO_x sources using satellite observations: Relative roles of fossil fuel combustion, biomass burning and soil emissions. *Faraday Discuss.*, **130**, 407–423.

Jaeglé, L., et al., 2004: Satellite mapping of rain-induced nitric oxide emissions from soils. *J. Geophys. Res.*, **109**, D21310, doi:10.1029/2004JD004787.

Jaenicke, R., 2005: Abundance of cellular material and proteins in the atmosphere. *Science*, **308**(5718), doi:10.1126/science.1106335.

Jaffe, D., et al., 2004: Long-range transport of Siberian biomass burning emissions and impact on surface ozone in western North America. *Geophys. Res. Lett.*, **31**, L16106, doi:10.1029/2004GL020093.

Jahren, A.H., et al., 2001: Terrestrial record of methane hydrate dissociation in the Early Cretaceous. *Geology*, **29**(2), 159–162.

Jain, A.K., and X. Yang, 2005: Modeling the effects of two different land cover change data sets on the carbon stocks of plants and soils in concert in CO_2 and climate change. *Global Biogeochem. Cycles*, **19**, doi:10.1029/2004GB002349.

Janssens, I.A., et al., 2003: Europe's terrestrial biosphere absorbs 7 to 12% of European anthropogenic CO_2 emissions. *Science*, **300**(5625), 1538–1542.

Jiang, H., et al., 2006: Aerosol effects on the lifetime of shallow cumulus. *Geophys. Res. Lett.*, **33**, doi:10.1029/2006GL026024.

Jickells, T.D., et al., 2005: Global iron connections between desert dust, ocean biogeochemistry, and climate. *Science*, **308**(5718), 67–71.

Jin, M.L., J.M. Shepherd, and M.D. King, 2005: Urban aerosols and their variations with clouds and rainfall: a case study for New York and Houston. *J. Geophys. Res.*, **110**, doi:10.1029/2004JD005081.

Jin, Y., et al., 2002: How does snow impact the albedo of vegetated land surfaces as analyzed with MODIS data? *Geophys. Res. Lett.*, **29**, doi:10.1029/2001GLO14132.

Johns, T.C., et al., 2006: The new Hadley Centre climate model HadGEM1: Evaluation of coupled simulations. *J. Clim.*, **19**, 1327–1353.

Johnson, B.T., 2005: The semidirect aerosol effect: Comparison of a single-column model with large eddy simulation for marine stratocumulus. *J. Clim.*, **18**, 119–130.

Johnson, B.T., K.P. Shine, and P.M. Forster, 2004: The semi-direct aerosol effect: Impact of absorbing aerosols on marine stratocumulus. *Q. J. R. Meteorol. Soc.*, **130**, 1407–1422.

Johnson, C.E., W.J. Collins, D.S. Stevenson, and R.G. Derwent, 1999: Relative roles of climate and emissions changes on future tropospheric oxidant concentrations. *J. Geophys. Res.*, **104**(D15), 18631–18645.

Johnson, C.E., D.S. Stevenson, W.J. Collins, and R.G. Derwent, 2001: Role of climate feedback on methane and ozone studied with a coupled ocean-atmosphere-chemistry model. *Geophys. Res. Lett.*, **28**(9), 1723–1726.

Johnson, C.E., D.S. Stevenson, W.J. Collins, and R.G. Derwent, 2002: Interannual variability in methane growth rate simulated with a coupled ocean-atmosphere-chemistry model. *Geophys. Res. Lett.*, **29**(19), 1903, doi:10.1029/2002GL015269.

Johnson, D.B., 1982: The role of giant and ultragiant aerosol particles in warm rain initiation. *J. Atmos. Sci.*, **39**, 448–460.

Jones, A., D.L. Roberts, M.J. Woodage, and C.E. Johnson, 2001: Indirect sulphate aerosol forcing in a climate model with an interactive sulphur cycle. *J. Geophys. Res.*, **106**, 20293–20310.

Jones, C., N. Mahowald, and C. Luo, 2004: Observational evidence of African desert dust intensification of easterly waves. *Geophys. Res. Lett.*, **31**, doi:10.1029/2004GL020107.

Jones, C.D., and P.M. Cox, 2001a: Modelling the volcanic signal in the atmospheric CO_2 record. *Global Biogeochem. Cycles*, **15**(2), 453–466.

Jones, C.D., and P.M. Cox, 2001b: Constraints on the temperature sensitivity of global soil respiration from the observed interannual variability in atmospheric CO_2. *Atmos. Sci. Lett.*, **1**, doi:10.1006/asle.2001.0041.

Jones, C.D., and P.M. Cox, 2005: On the significance of atmospheric CO_2 growth rate anomalies in 2002-2003. *Geophys. Res. Lett.*, **32**, L14816, doi:10.1029/2005GL023027.

Jones, C.D., M. Collins, P.M. Cox, and S.A. Spall, 2001: The carbon cycle response to ENSO: a coupled climate-carbon cycle model study. *J. Clim.*, **14**, 4113–4129.

Jones, C.D., et al., 2005: Global climate change and soil carbon stocks: predictions from two contrasting models for the turnover of organic carbon in soil. *Global Change Biol.*, **11**(1), 154–166.

Jones, P.D., D.E. Parker, T.J. Osborn, and K.R. Briffa, 2006: Global and hemispheric temperature anomalies--land and marine instrumental records. In: *Trends: A Compendium of Data on Global Change*. Carbon Dioxide Information Analysis Center, Oak Ridge National Laboratory, U.S. Department of Energy, Oak Ridge, TN.

Kanakidou, M., et al., 2005: Organic aerosol and global climate modelling: a review. *Atmos. Chem. Phys.*, **5**, 1053–1123.

Kärcher, B., and U. Lohmann, 2002: A parameterization of cirrus cloud formation: homogeneous freezing of supercooled aerosols. *J. Geophys. Res.*, **107**, doi:10.1029/2001JD000470.

Kärcher, B., and U. Lohmann, 2003: A parameterization of cirrus cloud formation: heterogeneous freezing. *J. Geophys. Res.*, **108**, doi:10.1029/2002JD003220.

Kärcher, B., and J. Ström, 2003: The roles of dynamical variability and aerosols in cirrus cloud formation. *Atmos. Chem. Phys.*, **3**, 823–838.

Kärcher, B., and T. Koop, 2004: The role of organic aerosols in homogeneous ice formation. *Atmos. Chem. Phys.*, **4**, 6719–6745.

Kärcher, B., J. Hendricks, and U. Lohmann, 2006: Physically-based parameterization of cirrus cloud formation for use in global atmospheric models. *J. Geophys. Res.*, **111**, doi:10.1029/2005JD006219.

Karlsdottir, S., and I.S.A. Isaksen, 2000: Changing methane lifetime: Possible cause for reduced growth. *Geophys. Res. Lett.*, **27**(1), 93–96.

Kasibhatla, P., et al., 2002: Top-down estimate of a large source of atmospheric carbon monoxide associated with fuel combustion in Asia. *Geophys. Res. Lett.*, **29**(19), 1900, doi:10.1029/2002GL015581.

Kasischke, E.S., and L.P. Bruhwiler, 2002: Emissions of carbon dioxide, carbon monoxide, and methane from boreal forest fires in 1998. *J. Geophys. Res.*, **107**, 8146, doi:10.1029/2001JD000461.

Kasischke, E.S., N.L. Christensen, and B.J. Stocks, 1995: Fire, global warming and the carbon balance of boreal forests. *Ecol. Appl.*, **5**(2), 437–451.

Kasischke, E.S., et al., 2005: Influences of boreal fire emissions on Northern Hemisphere atmospheric carbon and carbon monoxide. *Global Biogeochem. Cycles*, **19**(1), GB1012, doi:10.1029/2004GB002300.

Katz, M.E., D.K. Pak, G.R. Dickens, and K.G. Miller. 1999: The source and fate of massive carbon input during the Latest Paleocene Thermal Maximum. *Science*, **286**, 1531–1533.

Kaufman, Y.J., and I. Koren, 2006: Smoke and pollution aerosol effect on cloud cover. *Science*, **313**, 655–658, doi:10.1126/science.1126232.

Kaufman, Y.J., et al., 2005: The effect of smoke, dust, and pollution aerosol on shallow cloud development over the Atlantic Ocean. *Proc. Natl. Acad. Sci. U.S.A.*, **102**(32), 11207–11212.

Kawamiya, M., et al., 2005: Development of an integrated Earth system model on the Earth Simulator. *J. Earth Simulator*, **4**, 18–30.

Keeling, C.D., and T.P. Whorf, 2005: Atmospheric CO_2 records from sites in the SIO air sampling network. In: *Trends: A Compendium of Data on Global Change*. Carbon Dioxide Information Analysis Center, Oak Ridge National Laboratory, U.S. Department of Energy, Oak Ridge, TN, http://cdiac.esd.ornl.gov/trends/co2/sio-keel-flask/sio-keel-flask.html.

Keller, M., et al., 2005: Soil-atmosphere exchange for nitrous oxide, nitric oxide, methane, and carbon dioxide in logged and undisturbed forest in the Tapajos National Forest, Brazil. *Earth Interactions*, **9**, 1–28, doi:10.1175/EI125.1.

Keppler, F., J.T.G. Hamilton, M. Brass, and T. Roeckmann, 2006: Methane emissions from terrestrial plants under aerobic conditions. *Nature*, **439**, 187–191.

Kerkweg, A., S. Wurzler, T. Reisin, and A. Bott, 2003: On the cloud processing of aerosol particles: An entraining air-parcel model with two-dimensional spectral cloud microphysics and a new formulation of the collection kernel. *Q. J. R. Meteorol. Soc.*, **129**(587), 1–18.

Kettle, A., and M. Andreae, 2000: Flux of the dimethylsulfide from the oceans: A comparison of updated data sets and flux models. *J. Geophys. Res.*, **105**, 26793–26808.

Key, R.M., et al., 2004: A global ocean carbon climatology: Results from Global Data Analysis Project (GLODAP). *Global Biogeochem. Cycles*, **18**(4), GB4031, doi:10.1029/2004GB002247.

Khain, A.P., D. Rosenfeld, and A. Pokrovsky, 2001: Simulating convective clouds with sustained supercooled liquid water down to –37.5°C using a spectral microphysics model. *Geophys. Res. Lett.*, **28**, 3887–3890.

Khain, A.P., D. Rosenfeld, and A. Pokrovsky, 2005: Aerosol impact on the dynamics and microphysics of convective clouds. *Q. J. R. Meteorol. Soc.*, **131**(611), 2639–2663.

Khain, A.P., et al., 2004: Simulation of effects of atmospheric aerosols on deep turbulent convective using a spectral microphysics mixed-phase cumulus cloud model. 1. Model description and possible applications. *J. Atmos. Sci.*, **61**, 2963–2982.

Khalil, M.A.K., and M.J. Shearer, 2006: Decreasing emissions of methane from rice agriculture. *Int. Congress Ser.*, **1293**, 33–41.

Kirschbaum, M.U.F., et al., 2006: A comment on the quantitative significance of aerobic methane release by plants. *Funct. Plant Biol.*, **33**, 521–530.

Kirschvink, J.L., and T.D. Raub, 2003: A methane fuse for the Cambrian explosion: carbon cycles and true polar wander. *Comptes Rendus Geoscience*, **335**, 65–78.

Klaas, C., and D.E. Archer, 2002: Association of sinking organic matter with various types of mineral ballast in the deep sea: Implications for the rain ratio. *Global Biogeochem. Cycles*, **16**(4), 1116, doi:10.1029/2001GB001765.

Kleffmann, J., et al., 2005: Daytime formation of nitrous acid: a major source of OH radicals in a forest. *Geophys. Res. Lett.*, **32**, L05818, doi:10.1029/2005GL022524.

Kleypas, J.A., J. McManus, and L. Menez, 1999a: Using environmental data to define reef habitat: where do we draw the line? *Am. Zool.*, **39**, 146–159.

Kleypas, J.A., et al., 1999b: Geochemical consequences of increased atmospheric carbon dioxide on coral reefs. *Science*, **284**, 118–120.

Kloster, S., et al., 2006: DMS cycle in the marine ocean-atmosphere system - a global model study. *Biogeosciences*, **3**, 29–51.

Knorr, W., I.C. Prentice, J.I. House, and E.A. Holland, 2005: Long-term sensitivity of soil carbon turnover to warming. *Nature*, **433**, 298–301.

Knowlton, K., et al., 2004: Assessing ozone-related health impacts under a changing climate. *Environ. Health Perspect.*, **112**, 1557–1563.

Koerner, C., 2004: Through enhanced tree dynamics carbon dioxide enrichment may cause tropical forests to lose carbon. *Philos. Trans. R. Soc. London Ser. B*, **359**, 493–498.

Koerner, C., et al., 2005: Carbon flux and growth in mature deciduous forest trees exposed to elevated CO_2. *Science*, **309**(5739), 1360–1362.

Koren, I., Y.J. Kaufman, L.A. Remer, and J.V. Martins, 2004: Measurements of the effect of smoke aerosol on inhibition of cloud formation. *Science*, **303**, 1342–1345.

Koren, I., et al., 2005: Aerosol invigoration and restructuring of Atlantic convective clouds. *Geophys. Res. Lett.*, **32**(14), L14828, doi:10.1029/2005GL023187.

Koster, R.D., and M.J. Suarez, 2001: Soil moisture memory in climate models. *J. Hydrometeorol.*, **2**(6), 558–570.

Koster, R.D., and M.J. Suarez, 2004: Suggestions in the observational record of land-atmosphere feedback operating at seasonal time scales. *J. Hydrometeorol.*, **5**(3), doi: 10.1175/1525.

Koster, R.D., M.J. Suarez, R.W. Higgins, and H.M. Van den Dool, 2003: Observational evidence that soil moisture variations affect precipitation. *Geophys. Res. Lett.*, **30**(5), 1241. doi:10.1029/2002GL016571.

Koster, R.D., et al., 2000: A catchment-based approach to modeling land surface processes in a general circulation model. 1. Model structure. *J. Geophys. Res.*, **105**, 24809–24822.

Koster, R.D., et al., 2002: Comparing the degree of land-atmosphere interaction in four atmospheric general circulation models. *J. Hydrometeorol.*, **3**(3), 363–375.

Koster, R.D., et al., 2004: Regions of strong coupling between soil moisture and precipitation. *Science*, **305**, 1138–1140.

Koster, R.D., et al., 2006: GLACE: The Global Land-Atmosphere Coupling Experiment. 1. Overview. *J. Hydrometeorol.*, **7**, 590–610.

Krakauer, N.Y., and J.T. Randerson, 2003: Do volcanic eruptions enhance or diminish net primary production? Evidence from tree rings. *Global Biogeochem. Cycles*, **17**(4), 1118, doi:10.1029/2003GB002076.

Kristjánsson, J.E., 2002: Studies of the aerosol indirect effect from sulphate and black carbon aerosols. *J. Geophys. Res.*, **107**, doi:10.1029/2001JD000887.

Kristjánsson, J.E., et al., 2005: Response of the climate system to aerosol direct and indirect forcing: Role of cloud feedbacks. *J. Geophys. Res.*, **110**, D24206, doi:10.1029/2005JD006299.

Kroeze, C., A. Mosier, and L. Bouwman, 1999: Closing the N_2O budget: A retrospective analysis. *Global Biogeochem. Cycles*, **13**, 1–8.

Kroeze, C., E. Dumont, and S.P. Seitzinger, 2005: New estimates of global emissions of N_2O from rivers and estuaries. *Environ. Sci.*, **2**, 159–165.

Krüger, O., and H. Grassl, 2004: Albedo reduction by absorbing aerosols over China. *Geophys. Res. Lett.*, **31**, doi:10.1029/2003GL019111.

Kulmala, M., et al., 2004: A new feedback mechanism linking forests, aerosols, and climate. *Atmos. Chem. Phys.*, **4**, 557–562.

Kurz, W.A., and M. Apps, 1999: A 70-years retrospective analysis of carbon fluxes in the Canadian forest sector. *Ecol. Appl.*, **9**, 526–547.

Kurz, W.A., M.J. Apps, S.J. Beukema, and T. Lekstrum, 1995: 20th-century carbon budget of Canadian forests. *Tellus*, **47B**(1–2), 170–177.

Kvenvolden, K.A., and B.W. Rogers, 2005: Gaia's breath - global methane exhalations. *Mar. Petrol. Geol.*, **22**, 579–590.

Labrador, L.J., R. von Kuhlmann, and M.G. Lawrence, 2004: Strong sensitivity of the global mean OH concentration and the tropospheric oxidizing efficiency to the source of NO_x from lightning. *Geophys. Res. Lett.*, **31**, L06102, doi:10.1029/2003GL019229.

Lalli, C.M., and R.W. Gilmer, 1989: *Pelagic Snails: The Biology of Holoplanktonic Gastropod Mollusks*. Stanford University Press, Palo Alto, CA, 259 pp.

Lamarque, J.-F., et al., 2005a: Coupled chemistry-climate response to changes in aerosol emissions: global impact on the hydrological cycle and the tropospheric burdens of OH, ozone and NO_x. *Geophys. Res. Lett.*, **32**, L16809, doi:10.1029/2005GL023419.

Lamarque J.-F., et al., 2005b: Assessing future nitrogen deposition and carbon cycle feedback using a multimodel approach: Part 1. Analysis of nitrogen deposition. *J. Geophys. Res.*, **110**, D19303, doi:10.1029/2005JD005825.

Lamarque, J.-F., et al., 2005c: Tropospheric ozone evolution between 1890 and 1990. *J. Geophys. Res.*, **110**, D08304, doi:10.1029/2004JD005537.

Langdon, C., et al., 2003: Effect of elevated CO_2 on the community metabolism of an experimental coral reef. *Global Biogeochem. Cycles*, **17**, 1011, doi:10.1029/2002GB001941.

Langenbuch, H., and H.O. Pörtner, 2003: Energy budget of heptocytes from Antarctic fish (*Pachycara brachycephalum* and *Lepidonotothen kempi*) as a function of ambient CO_2: pH-dependent limitations of cellular protein biosynthesis? *J. Exp. Biol.*, **206**, 3895–3903.

Langenfelds, R.L., et al., 1999: Partitioning of the global fossil CO_2 sink using a 19-year trend in atmospheric O_2. *Geophys. Res. Lett.*, **26**, 1897–1900.

Langenfelds, R.L., et al., 2002: Interannual growth rate variations of atmospheric CO_2 and its ^{13}C, H_2, CH_4, and CO between 1992 and 1999 linked to biomass burning. *Global Biogeochem. Cycles*, **16**(3), 1048, doi:10.1029/2001GB001466.

Langner, J., R. Bergstrom, and V. Foltescu, 2005: Impact of climate change on surface ozone and deposition of sulphur and nitrogen in Europe. *Atmos. Environ.*, **39**, 1129–1141.

Larsen, S.H., 2005: Solar variability, dimethyl sulphide, clouds, and climate. *Global Biogeochem. Cycles*, **19**, GB1014, doi:10.1029/2004GB002333.

Lassey, K.R., D.C Lowe, and M.R. Manning, 2000: The trend in atmospheric methane $\delta^{13}C$ and implications for isotopic constraints on the global methane budget. *Global Biogeochem. Cycles*, **14**, 41–49.

Lathière, J., et al., 2005: Past and future changes in biogenic volatile organic compound emissions simulated with a global dynamic vegetation model. *Geophys. Res. Lett.*, **32**, L20818, doi:10.1029/2005GL024164.

Laurance, W.F., et al., 2004: Pervasive alteration of tree communities in undisturbed Amazonian forests. *Nature*, **428**, 171–175.

Lawrence, D.M., and J.M. Slingo, 2004: An annual cycle of vegetation in a GCM. Part I: implementation and impact on evaporation. *Clim. Dyn.*, **22**, doi:10.1007/s0038200303669.

Lawrence, D.M., and J.M. Slingo, 2005: Weak land-atmosphere coupling strength in HadAM3: The role of soil moisture variability. *J. Hydrometeorol.*, **6**(5), 670–680, doi:10.1775/JHM445.1.

Lawrence, M.G., R. von Kuhlmann, M. Salzmann, and P.J. Rasch, 2003: The balance of effects of deep convective mixing on tropospheric ozone. *Geophys. Res. Lett.*, **30**(18), 1940, doi:10.1029/2003GL017644.

Laws, E.A., et al., 2000: Temperature effects on export production in the open ocean. *Global Biogeochem. Cycles*, **14**, 1231–1246.

Le Quéré, C., et al., 2000: Interannual variability of the oceanic sink of CO_2 from 1979 to 1997. *Global Biogeochem. Cycles*, **14**, 1247–1265.

Le Quéré, C., et al., 2003: Two decades of ocean CO_2 sink and variability. *Tellus*, **55B**(2), 649–656.

Le Quéré, C., et al., 2005: Ecosystem dynamics based on plankton functional types for global ocean biogeochemistry models. *Global Change Biol.*, **11**, doi:10.1111/j.1365-2486.2005.001004.x.

Leck, C., and E.K. Bigg. 2005a: Biogenic particles in the surface microlayer and overlaying atmosphere in the central Arctic Ocean during summer. *Tellus*, **57B**(4), 305–316.

Leck, C., and E.K. Bigg. 2005b: Source and evolution of the marine aerosol - A new perspective. *Geophys. Res. Lett.*, **32**, L19803, doi:10.1029/2005GL023651.

Lee, K., et al., 1998: Low interannual variability in recent oceanic uptake of atmospheric carbon dioxide. *Nature*, **396**, 155–159.

Lefèvre, N., et al., 1999: Assessing the seasonality of the oceanic sink for CO_2 in the northern hemisphere. *Glob. Biogeochem. Cycles*, **13**, 273–286.

Lelieveld, J., and F.J. Dentener, 2000: What controls tropospheric ozone? *J. Geophys. Res.*, **105**, 3531–3551.

Lelieveld, J., W. Peters, F.J. Dentener, and M.C. Krol, 2002a: Stability of tropospheric hydroxyl chemistry. *J. Geophys. Res.*, **107**(D23), 4715, doi:10.1029/2002JD002272.

Lelieveld, J., et al., 2002b: Global air pollution crossroads over the Mediterranean. *Science*, **298**, 794–799.

Leue, C., et al. 2001: Quantitative analysis of NO_x emissions from GOME satellite image sequences. *J. Geophys. Res.*, **106**, 5493–5505.

Levin, Z., E. Ganor, and V. Gladstein, 1996: The effects of desert particles coated with sulfate on rain formation in the eastern Mediterranean. *J. Appl. Meteorol.*, **35**, 1511–1523.

Levis, S., and G.B. Bonan, 2004: Simulating springtime temperature patterns in the community atmosphere model coupled to the community land model using prognostic leaf area. *J. Clim.*, **17**, 4531–4540.

Levis, S., G.B. Bonan, and C. Bonfils, 2004: Soil feedback drives the mid-Holocene North African monsoon northward in fully coupled CCSM2 simulations with a dynamic vegetation model. *Clim. Dyn.*, **23**, doi:10.1007/s00382-004-0477-y.

Lewis, E.R., and S.E. Schwartz, 2005: *Sea Salt Aerosol Production: Mechanisms. Methods, Measurements, and Models: A Critical Review*. Geophysical Monograph Vol. 152, American Geophysical Union, Washington, DC, 413 pp.

Lewis, S.L., Y. Malhi, and O.L. Phillips, 2005: Fingerprinting the impacts of global change on tropical forests. *Philos. Trans. R. Soc. London Ser. B*, **359**, doi:10.1098/rstb.2003.1432.

Li, C., et al., 2002: Reduced methane emissions from large-scale changes in water management of China's rice paddies during 1980-2000. *Geophys. Res. Lett.*, **29**, doi: 10.1029/2002GL015370.

Li, Q., et al., 2005: North American pollution outflow and the trapping of convectively lifted pollution by upper-level anticyclone. *J. Geophys. Res.*, **110**, D10301, doi:10.1029/2004JD005039.

Li, W., and R. Fu, 2004: Transition of the large-scale atmospheric and land surface conditions from the dry to the wet season over Amazonia as diagnosed by the ECMWF re-analysis. *J. Clim.*, **17**, 2637–2651.

Liang X., Z. Xie, and M. Huang, 2003: A new parameterization for surface and groundwater interactions and its impact on water budgets with the variable infiltration capacity (VIC) land surface model. *J. Geophys. Res.*, **108**, 8613, doi:19.1029/2002JD003090.

Liao, H., and J.H. Seinfeld, 2005: Global impacts of gas-phase chemistry-aerosol interactions on direct radiative forcing by anthropogenic aerosols and ozone. *J. Geophys. Res.*, **110**, D18208, doi:10.1029/2005JD005907.

Liebmann, B., and J.A. Marengo, 2001: Interannual variability of the rainy season and rainfall in the Brazilian Amazon basin. *J. Clim.*, **14**(22), 4308–4318.

Liepert, B.G., and I. Tegen, 2002: Multidecadal solar radiation trends in the United States and Germany and direct tropospheric aerosol forcing. *J. Geophys. Res.*, **107**, doi:10.1029/2001JD000760.

Liepert, B.G., J. Feichter, U. Lohmann, and E. Roeckner, 2004: Can aerosols spin down the water cycle in a warmer and moister world. *Geophys. Res. Lett.*, **31**, L06207, doi:10.1029/2003GL019060.

Lin, C.-Y.C., D.J. Jacob, and A.M. Fiore, 2001: Trends in exceedances of the ozone air quality standard in the continental United States, 1980-1998. *Atmos. Environ.*, **35**, 3217–3228.

Lin, G.H., et al., 1999: Ecosystem carbon exchange in two terrestrial ecosystem mesocosms under changing atmospheric CO_2 concentrations. *Oecologia*, **119**(1), 97–108.

Lintner, B.R., 2002: Characterizing global CO_2 interannual variability with empirical orthogonal function/principal component (EOF/PC) analysis. *Geophys. Res. Lett.*, **29**(19), 1921, doi:10.1029/2001GL014419.

Liu, X., and J.E. Penner, 2005: Ice nucleation parameterization for a global model. *Meteorol. Z.*, **14**(4), 499–514.

Liu, Y., and P.H. Daum, 2002: Indirect warming effect from dispersion forcing. *Nature*, **419**, 580–581.

Loh, A.N., J.E. Bauer, and E.R.M. Druffel, 2004: Variable ageing and storage of dissolved organic components in the open ocean. *Nature*, **430**, 877–881.

Lohmann, U., 2002: A glaciation indirect aerosol effect caused by soot aerosols. *Geophys. Res. Lett.*, **29**, doi:10.1029/2001GL014357.

Lohmann, U., and J. Feichter, 2001: Can the direct and semi-direct aerosol effect compete with the indirect effect on a global scale? *Geophys. Res. Lett.*, **28**(1), 159–161, doi:10.1029/2000GL012051.

Lohmann, U., and B. Kärcher, 2002: First interactive simulations of cirrus clouds formed by homogeneous freezing in the ECHAM GCM. *J. Geophys. Res.*, **107**, doi:10.1029/2001JD000767.

Lohmann, U., and G. Lesins, 2002: Stronger constraints on the anthropogenic indirect aerosol effect. *Science*, **298**, 1012–1016.

Lohmann, U., and J. Feichter, 2005: Global indirect aerosol effects: a review. *Atmos. Chem. Phys.*, **5**, 715–737.

Lohmann, U., and K. Diehl, 2006: Sensitivity studies of the importance of dust ice nuclei for the indirect aerosol effect on stratiform mixed-phase clouds. *J. Atmos. Sci.*, **63**, 1338–1347.

Lohmann, U., I. Koren, and Y.J. Kaufman, 2006: Disentangling the role of microphysical and dynamical effects in determining cloud properties over the Atlantic. *Geophys. Res. Lett.*, **33**, L09802, doi:10.1029/2005GL024625.

Lovins, A.B., 2003: Hydrogen primer. *RMI Solutions Newsletter*, **19**(2), 1–4; 36–39.

Lucht, W., et al., 2002: Climatic control of the high-latitude vegetation greening trend and Pinatubo effect. *Science*, **296**(5573), 1687–1689.

Luo, Y., S.Q. Wan, D.F. Hui, and L.L. Wallace, 2001: Acclimatization of soil respiration to warming in a tall grass prairie. *Nature*, **413**, 622–625.

Luo, Y., et al., 2004: Progressive nitrogen limitation of ecosystem responses to rising atmospheric carbon dioxide. *Bioscience*, **54**, 731–739.

Mack, F., J. Hoffstadt, G. Esser, and J.G. Goldammer, 1996: Modeling the influence of vegetation fires on the global carbon cycle. In: *Biomass Burning and Global Change* [Levine, J.S. (ed)]. MIT Press, Cambridge, MA, pp. 149–159.

Mahaffey, C., A.F. Michaels, and D.G. Capone, 2005: The conundrum of marine N_2 fixation. *Am. J. Sci.*, **305**(6–8): 546–595.

Mahowald, N.M., and L.M. Kiehl, 2003: Mineral aerosol and cloud interactions. *Geophys. Res. Lett.*, **30**, doi:10.1029/2002GL016762.

Mahowald, N.M., and C. Luo, 2003: A less dusty future? *Geophys. Res. Lett.*, **30**(7), 1903, doi:10.1029/2003GL017880.

Mahowald, N.M., G.D.R. Rivera, and C. Luo, 2004: Comment on "Relative importance of climate and land use in determining present and future global soil dust emission" by I. Tegen et al. *Geophys. Res. Lett.*, **31**(24), L24105, doi:10.1029/2004GL021272.

Maier-Reimer, E., U. Mikolajewicz, and A. Winguth, 1996: Future ocean uptake of CO_2: interaction between ocean circulation and biology. *Clim. Dyn.*, **12**, 711–721.

Malhi, Y., and J. Grace, 2000: Tropical forests and atmospheric carbon dioxide. *Trends Ecol. Evol.*, **15**(8), 332–337.

Malhi, Y., and O.L. Phillips, 2004: Tropical forests and global atmospheric change: a synthesis. *Philos. Trans. R. Soc. London Ser. B*, **359**, doi:10.1098/rstb.2003.1449.

Malhi, Y., and J. Wright, 2004: Spatial patterns and recent trends in the climate of tropical rainforest regions. *Philos. Trans. R. Soc. London Ser. B*, **359**, doi:10.1098/rstb.2003.1433.

Malhi, Y., et al., 2002: An international network to understand the biomass and dynamics of Amazonian forests (RAINFOR). *J. Veg. Sci.*, **13**, 439–450.

Manning, A.C., and R.F. Keeling, 2006: Global oceanic and land biotic carbon sinks from the Scripps atmospheric oxygen flask sampling network. *Tellus*, **58B**(2), 95–116.

Marani, M., E. Eltahir, and A. Rinaldo, 2001: Geomorphic controls on regional base flow. *Water Resour. Res.*, **37**, 2619–2630.

Marengo, J., and C.A. Nobre, 2001: The hydroclimatological framework in Amazonia. In: *Biogeochemistry of the Amazon Basin* [McClaine, M., R. Victoria, and J. Richey (eds.)]. Oxford University Press, Oxford, UK, pp. 17–42.

Marland, G., T.A. Boden, and R.J. Andres, 2006: Global, regional, and national CO_2 emissions. In: *Trends: A Compendium of Data on Global Change*. Carbon Dioxide Information Analysis Center, Oak Ridge National Laboratory, U.S. Department of Energy, Oak Ridge, TN, http://cdiac.esd.ornl.gov/trends/emis/tre_glob.htm.

Martin, R.V., et al., 2002: Interpretation of TOMS observations of tropical tropospheric ozone with a global model and in-situ observations. *J. Geophys. Res.*, **107**(D18), 4351, doi:10.1029/2001JD001480.

Martin, R.V., et al., 2003a: Global inventory of nitrogen oxide emissions constrained by space-based observations of NO_2 columns. *J. Geophys. Res.*, **108**(D17), 4537, doi:10.1029/2003JD003453.

Martin, R.V., et al., 2003b: Global and regional decreases in tropospheric oxidants from photochemical effects of aerosols. *J. Geophys. Res.*, **108**(D3), 4097, doi:10.1029/2002JD002622.

Martinerie, P., G.P. Brasseur, and C. Granier, 1995: The chemical composition of ancient atmospheres: a model study constrained by ice core data. *J. Geophys. Res.*, **100**, 14291–14304.

Matsumi, Y., et al., 2002: Quantum yields for production of O(1D) in the ultraviolet photolysis of ozone: Recommendation based on evaluation of laboratory data. *J. Geophys. Res.*, **104**(D3), doi:10.1029/2001JD000510.

Matthews, H.D., M. Eby, A.J. Weaver, and B.J. Hawkins, 2005: Primary productivity control of simulated carbon cycle-climate feedbacks. *Geophys. Res. Lett.*, **32**, L14708, doi:10.1029/2005GL022941.

Matthews, R., and R. Wassmann, 2003: Modelling the impacts of climate change and methane emission reductions on rice production: a review. *Eur. J. Agron.*, **19**, 573–598.

Maynard, K., and J.-F. Royer, 2004: Sensitivity of a general circulation model to land surface parameters in African tropical deforestation experiments. *Clim. Dyn.*, **22**, doi:10.1007/s0038200403989.

McCabe, G.J., M.P. Clark, and M.C. Serreze, 2001: Trends in northern hemisphere surface cyclone frequency and intensity. *J. Clim.*, **14**, 2763–2768.

McFiggans, G., 2005: Marine aerosols and iodine emissions. *Nature*, **433**, E13.

McFiggans, G., et al., 2004: Direct evidence for coastal iodine particles from Laminaria macroalgae - linkage to emissions of molecular iodine. *Atmos. Chem. Phys.*, **4**, 701–713.

McGuire, A.D. III, et al., 2001: Carbon balance of the terrestrial biosphere in the twentieth century: Analyses of CO_2, climate and land use effects with four process-based ecosystem models. *Global Biogeochem. Cycles*, **15**, 183–206.

McKenzie, R.L., B.J. Connor, and G.E. Bodeker, 1999: Increased summertime UV observed in New Zealand in response to ozone loss. *Science*, **285**, 1709–1711.

McKinley, G.A., M.J. Follows, and J. Marshall, 2004a: Mechanisms of air-sea CO_2 flux variability in the equatorial Pacific and the North Atlantic. *Global Biogeochem. Cycles*, **18**, GB2011, doi:10.1029/2003GB002179.

McKinley, G.A., et al., 2004b: Pacific dominance to global air-sea CO_2 flux variability: A novel atmospheric inversion agrees with ocean models. *Geophys. Res. Lett.*, **31**, L22308, doi:10.1029/2004GL021069.

McLinden, C., et al., 2000: Stratospheric ozone in 3-D models: a simple chemistry and the cross-tropopause flux. *J. Geophys. Res.*, **105**, 14653–14665.

McNeil, B.I., et al., 2003: Anthropogenic CO_2 uptake by the ocean based on the global chlorofluorocarbon data set. *Science*, **299**(5604), 235–239.

Melillo, J.M., et al., 1995: Vegetation/ecosystem modeling and analysis project: Comparing biogeography and biogeochemistry models in a continental-scale study of terrestrial ecosystem responses to climate change and CO_2 doubling. *Global Biogeochem. Cycles*, **9**, 407–437.

Melillo, J.M., et al., 2002: Soil warming and carbon-cycle feedbacks to the climate system. *Science*, **298**, 2173–2176.

Menon, S., and A. Del Genio, 2007: Evaluating the impacts of carbonaceous aerosols on clouds and climate. In: *An Interdisciplinary Assessment: Human-Induced Climate Change* [Schlesinger, M., et al. (eds.)]. Cambridge University Press, Cambridge, UK, in press.

Menon, S., and L. Rotstayn, 2006: The radiative influence of aerosol effects on liquid-phase cumulus and stratiform clouds based on sensitivity studies with two climate models. *Clim. Dyn.*, **27**, 345–356.

Menon, S., A.D. Del Genio, D. Koch, and G. Tselioudis, 2002a: GCM Simulations of the aerosol indirect effect: sensitivity to cloud parameterization and aerosol burden. *J. Atmos. Sci.*, **59**, 692–713.

Menon, S., J. Hansen, L. Nazarenko, and Y. Luo, 2002b: Climate effects of black carbon aerosols in China and India. *Science*, **297**, 2250–2252.

Menon, S., et al., 2003: Evaluating aerosol/cloud/radiation process parameterizations with single-column models and Second Aerosol Characterization Experiment (ACE-2) cloudy column observations. *J. Geophys. Res.*, **108**, doi:10.1029/2003JD003902.

Mickley, L.J., D.J. Jacob, and D. Rind, 2001: Uncertainty in preindustrial abundance of tropospheric ozone: implications for radiative forcing calculations. *J. Geophys. Res.*, **106**, 3389–3399.

Mickley, L.J., D.J. Jacob, B.D. Field, and D. Rind, 2004: Effects of future climate change on regional air pollution episodes in the United States. *Geophys. Res. Lett.*, **30**, 1862, doi:10.1029/2003GL017933.

Mickley, L.J., et al., 1999: Radiative forcing from tropospheric ozone calculated with a unified chemistry–climate model. *J. Geophys. Res.*, **104**(D23), 30153–30172.

Mikaloff Fletcher, S.E., et al., 2004a: CH_4 sources estimated from atmospheric observations of CH_4 and its $^{13}C/^{12}C$ isotopic ratios: 1. Inverse modeling of source processes. *Global Biogeochem. Cycles*, **18**, GB4004, doi:10.1029/2004GB002223.

Mikaloff Fletcher, S.E., et al., 2004b: CH_4 sources estimated from atmospheric observations of CH_4 and its $^{13}C/^{12}C$ isotopic ratios: 2. Inverse modeling of CH_4 fluxes from geographical regions. *Global Biogeochem. Cycles*, **18**, doi:10.1029/2004GB002224.

Mikaloff Fletcher, S.E., et al., 2006: Inverse estimates of anthropogenic CO_2 uptake, transport, and storage by the ocean. *Global Biogeochem. Cycles*, **18**, doi:10.1029/2005GB002530.

Miller, R.L., J. Perlwitz, and I. Tegen, 2004a: Feedback upon dust emission by dust radiative forcing through the planetary boundary layer. *J. Geophys. Res.*, **109**, D24209, doi:10.1029/2004JD004912.

Miller, R.L., I. Tegen, and J. Perlwitz, 2004b: Surface radiative forcing by soil dust aerosols and the hydrologic cycle. *J. Geophys. Res.*, **109**, D04203, doi:10.1029/2003JD004085.

Millero, F.J., et al., 2002: Dissociation constants for carbonic acid determined from field measurements. *Deep-Sea Res. I*, **49**, 1705–1723.

Milly, P.C.D, and A.B. Schmakin, 2002: Global modeling of land water and energy balances, Part I: The Land Dynamics (LaD) model. *J. Hydrometeorol.*, **3**, 301–310.

Ming, Y., et al., 2005: Geophysical Fluid Dynamics Laboratory general circulation model investigation of the indirect radiative effects of anthropogenic sulfate aerosol. *J. Geophys. Res.*, **110**, D22206, doi:10.1029/2005JD006161.

Montzka, S.A., et al., 1999: Present and future trends in the atmospheric burden of ozone depleting halogens. *Nature*, **398**, 690–694.

Montzka, S.A., et al., 2003: A decline in tropospheric bromine. *Geophys. Res. Lett.*, **30**(15), 1826, doi:10.1029/2003GL017745.

Mosier, A., et al., 1998: Closing the global N_2O budget: N_2O emissions through the agricultural nitrogen cycle-OECD/IPCC.IEA phase II development of IPCC guidelines for national greenhouse gas inventory methodology. *Nutrient Cycling in Agroecosystems*, **52**, 225–248.

Mouillot, F., and C.B. Field, 2005: Fire history and the global carbon budget: a 1° x 1° fire history reconstruction for the 20th century. *Global Change Biol.*, **11**, 398–420.

Müller, J., and G. Brasseur, 1999: Sources of upper tropospheric HO_x: A three-dimensional study. *J. Geophys. Res.*, **104**(D1), 1705–1716.

Myhre, G., M.M. Kvalevag, and C.B. Schaaf, 2005: Radiative forcing due to anthropogenic vegetation change based on MODIS surface albedo data. *Geophys. Res. Lett.*, **32**, L21410, doi: 10.1029/2005GLO24004.

Nabuurs, G.J., et al., 2003: Temporal evolution of the European forest sector carbon sink from 1950 to 1999. *Global Change Biol.*, **9**, 152–160.

Nadelhoffer, K., et al., 2004: Decadal-scale fates of N-15 tracers added to oak and pine stands under ambient and elevated N inputs at the Harvard Forest (USA). *For. Ecol. Manage.*, **196**, 89–107.

Neja, M., H. Akimoto, and J. Staehelin, 2003: Ozone in background and photochemically aged air over central Europe: analysis of long-term ozonesonde data from Hohenpeissenberg and Payerne. *J. Geophys. Res.*, **108**(D2), 4063, doi:10.1029/2002JD002477.

Naqvi, S.W.A., et al., 2000: Increased marine production of N_2O due to intensifying anoxia on the Indian continental shelf. *Nature*, **408**(6810), 346–349.

Negri, A.J., R.F. Adler, L.M. Xu, and J. Surratt, 2004: The impact of Amazonian deforestation on dry season rainfall. *J. Clim.*, **17**(6), 1306–1319.

Neill, C., et al., 2005: Rates and controls of nitrous oxide and nitric oxide emissions following conversion of forest to pasture in Rondônia. *Nutrient Cycling in Agroecosystems*, **71**, 1–15.

Nemani, R., et al., 2002: Recent trends in hydrologic balance have enhanced the terrestrial carbon sink in the United States. *Geophys. Res. Lett.*, **29**(10), 1468, doi:10.1029/2002GL014867.

Nemani, R.R., et al., 2003: Climate-driven increases in global terrestrial net primary production from 1982 to 1999. *Science*, **300**, 1560–1563.

Nepstad, D., et al., 2004: Amazon drought and its implications for forest flammability and tree growth: a basin-wide analysis. *Global Change Biol.*, **10**(5), 704–717.

Nesbitt, S.W., R.Y. Zhang, and R.E. Orville, 2000: Seasonal and global NO_x production by lightning estimated from the Optical Transient Detector (OTD). *Tellus*, **52B**, 1206–1215.

Nevison, C.D., J.H. Butler, and J.W. Elkins, 2003: Global distribution of N_2O and the N_2O/AOU yield in the subsurface ocean. *Global Biogeochem. Cycles*, **17**(4), 1119, doi:10.1029/2003GB002068.

Nevison, C.D., T. Lueker, and R.F. Weiss, 2004: Quantifying the nitrous oxide source from coastal upwelling. *Global Biogeochem. Cycles*, **18**, GB1018, doi:10.1029/2003GB002110.

Newchurch, M.J., et al., 2003: Evidence for slowdown in stratospheric ozone loss: first stage of ozone recovery. *J. Geophys. Res.*, **108**(D16), 4507, doi:10.1029/2003jd003471.

Newman, P.A., S.R. Kawa, and E.R. Nash, 2004: On the size of the Antarctic ozone hole. *Geophys. Res. Lett.*, **31**, L21104, doi:10.1029/2004GL020596.

Newman, P.A., et al., 2006: When will the Antarctic ozone hole recover? *Geophys. Res. Lett.*, **33**, L12814, doi:10.1029/2005GL025232.

Nightingale, P.D., et al., 2000: In situ evaluation of air-sea gas exchange parameterisations using novel conservative and volatile tracers. *Global Biogeochem. Cycles*, **14**(1), 373–387.

Nilsson, S., et al., 2003: The missing "missing sink". *Forestry Chronicle*, **79**(6), 1071–1074.

Niu, G.-Y., and Z.-L. Yang, 2004: Effects of vegetation canopy processes on snow surface energy and mass balances. *J. Geophys. Res.*, **109**, D23111, doi:10.1029/2004JD004884.

Niyogi, D., et al., 2004: Direct observations of the effects of aerosol loading on net ecosystem CO_2 exchanges over different landscapes. *Geophys. Res. Lett.*, **31**, L20506, doi:10.1029/2004GL020915.

Nober, F.J., H.-F. Graf, and D. Rosenfeld, 2003: Sensitivity of the global circulation to the suppression of precipitation by anthropogenic aerosols. *Global Planet. Change*, **37**, 57–80.

Norby, R.J., and C.M. Iversen, 2006: Nitrogen uptake, distribution, turnover, and efficiency of use in a CO_2-enriched sweetgum forest. *Ecology*, **87**, 5–14.

Norton, W.A., 2003: Sensitivity of northern hemisphere surface climate to simulation of the stratospheric polar vortex. *Geophys. Res. Lett.*, **30**(12), doi:10.1029/2003GL016958.

Novelli, P.C., et al., 1999: Molecular hydrogen in the troposphere: global distribution and budget. *J. Geophys. Res.*, **104**(D23), 30427–30444.

Nowak, R.S., D.S. Ellsworth, and S.D. Smith, 2004: Functional responses of plants to elevated atmospheric CO_2 – do photosynthetic and productivity data from FACE experiments support early predictions? *New Phytol.*, **162**, 253–280.

NRC (National Research Council), 1991: *Rethinking the Ozone Problem in Urban and Regional Air Pollution*. National Academy Press, Washington, DC, 524 pp.

Obata, A., and Y. Kitamura, 2003: Interannual variability of the sea-air exchange of CO_2 from 1961 to 1998 simulated with a global ocean circulation-biogeochemistry model. *J. Geophys. Res.*, **108**, 3337, doi:10.1029/2001JC001088.

O'Dowd, C.D., 2002: On the spatial extent and evolution of coastal aerosol plumes. *J. Geophys. Res.*, **107**(D19), 8105, doi:10.1029/2001JD000422.

O'Dowd, C.D., J.A. Lowe, and M.H. Smith, 1999: Coupling sea-salt and sulphate interactions and its impact on cloud droplet concentration predictions. *Geophys. Res. Lett.*, **26**, 1311–1314.

O'Dowd, C.D., et al., 2002a: Coastal new particle formation: Environmental conditions and aerosol physicochemical characteristics during nucleation bursts. *J. Geophys. Res.*, **107**(D19), 8107, doi:10.1029/2000JD000206.

O'Dowd, C.D., et al., 2002b: A dedicated study of New Particle Formation and Fate in the Coastal Environment (PARFORCE): Overview of objectives and achievements. *J. Geophys. Res.*, **107**(D19), 8108, doi:10.1029/2001JD000555.

O'Dowd, C.D., et al., 2004: Biogenically driven organic contribution to marine aerosols. *Nature*, **431**, 676–680.

Oechel, W.C, et al., 2000: Acclimation of ecosystem CO_2 exchange in the Alaskan Arctic in response to decadal climate warming. *Nature*, **406**, 978–981.

Ogawa, K., and T. Schmugge, 2004: Mapping surface broadband emissivity of the Sahara desert using ASTER and MODIS data. *Earth Interactions*, **8**(7), 1–14.

Ogle, S.M., M.D. Eve, F.J. Breidt, and K. Paustian, 2003: Uncertainty in estimating land use and management impacts on soil organic carbon storage for US agroecosystems between 1982 and 1997. *Global Change Biol.*, **9**, 1521–1542.

Oglesby, R.J., et al., 2002: Thresholds in atmosphere-soil moisture interactions: results from climate model studies. *J. Geophys. Res.*, **107**(14), doi:10.1029/2001JD001045.

Okin, G.S., N. Mahowald, O.A. Chadwick, and P. Artaxo, 2004: Impact of desert dust on the biogeochemistry of phosphorus in terrestrial ecosystems. *Global Biogeochem. Cycles*, **18**, GB2005, doi:10.1029/2003GB002145.

Oleson, K.W., G.B. Bonan, S. Levis, and M. Vertenstein, 2004: Effects of land use change on North American climate: impact of surface datasets and model biogeophysics. *Clim. Dyn.*, **23**, 117–132, doi:10.1007/s00382-004-0426-9.

Oleson, K.W., et al., 2003: Assessment of global climate model land surface albedo using MODIS data. *Geophys. Res. Lett.*, **30**(8), 1443, doi:10.1029/2002GL016749.

Olivié, D.J.L., P.F.J. van Velthoven, A.C.M. Beliaars, and H.M. Kelder, 2004: Comparison between archived and off-line diagnosed convective mass fluxes in the chemistry transport model TM3. *J. Geophys. Res.*, **109**, D11303, doi:10.1029/2003JD004036.

Olivier, J.G.J., A.F. Bouwman, K.W. Van Der Hoek, and J.J.M. Berdowski, 1998: Global air emission inventories for anthropogenic sources of NO_x, NH_3 and N_2O in 1990. *Environ. Pollut.*, **102**(1, S1), 135–148.

Olivier, J.G.J., et al., 2005: Recent trends in global greenhouse emissions: regional trends 1970–2000 and spatial distribution of key sources in 2000. *Environ. Sci.*, **2**, 81–99.

Ollinger, S.V., and J.D. Aber, 2002: The interactive effects of land use, carbon dioxide, ozone, and N deposition. *Global Change Biol.*, **8**, 545–562

Olsen, S.C., C.A. McLinden, and M.J. Prather, 2001: Stratospheric N_2O-NO_y system: testing uncertainties in a three-dimensional framework. *J. Geophys. Res.*, **106**, 28771–28784.

Ordóñez, C., et al., 2005: Changes of daily surface ozone maxima in Switzerland in all seasons from 1992 to 2002 and discussion of summer 2003. *Atmos. Chem. Phys.*, **5**, 1187–1203.

Oren, R., et al., 2001: Soil fertility limits carbon sequestration by forest ecosystems in a CO_2-enriched atmosphere. *Nature*, **411**(6836), 469–472.

Orr, J.C., et al., 2001: Estimates of anthropogenic carbon uptake from four three-dimensional global ocean models. *Global Biogeochem. Cycles*, **15**(1), 43–60, doi:10.1029/1999GB001256.

Orr, J.C., et al., 2005: Anthropogenic ocean acidification over the twenty-first century and its impact on calcifying organisms. *Nature*, **437**(7059), 681–686.

Osborne, T.M., et al., 2004: Influence of vegetation on the local climate and hydrology in the tropics: sensitivity to soil parameters. *Clim. Dyn.*, **23**, 45–61.

Oyama, M.D., and C.A. Nobre, 2004: Climatic consequences of a large-scale desertification in northeast Brazil: a GCM simulation study. *J. Clim.*, **17**(16), 3203–3213.

Pacala, S.W., et al., 2001: Consistent land- and atmosphere-based US carbon sink estimates. *Science*, **292**, 2316–2320.

Paeth, H., and J. Feichter, 2006: Greenhouse-gas versus aerosol forcing and African climate response. *Clim Dyn.*, **26**(1), 35–54.

Page, S., et al., 2002: The amount of carbon released from peat and forest fires in Indonesia during 1997. *Nature*, **320**, 61–65.

Pahlow, M., and U. Riebesell, 2000, Temporal trends in deep ocean Redfield ratios. *Science*, **287**, 831–833.

Palmer, P.I., et al., 2003a: Inverting for emissions of carbon monoxide from Asia using aircraft observations over the western Pacific. *J. Geophys. Res.*, **108**(D21), 8828, doi:10.1029/2003JD003397.

Palmer, P.I., et al., 2003b: Mapping isoprene emissions over North America using formaldehyde column observations from space. *J. Geophys. Res.*, **108**, 4180, doi:10.1029/2002JD002153.

Park, R.J., et al., 2004: Global simulation of tropospheric ozone using the University of Maryland Chemical Transport Model (UMD-CTM): 1. Model description and evaluation. *J. Geophys. Res.*, **109**, D09301, doi:10.1029/2003JD004266.

Patra, P.K., et al., 2005: Role of biomass burning and climate anomalies for land-atmosphere carbon fluxes based on inverse modeling of atmospheric CO_2. *Global Biogeochem. Cycles*, **19**, GB3005, doi:10.1029/2004GB002258 .

Paull, C.K., et al., 2003: An experiment demonstrating that marine slumping is a mechanism to transfer methane from seafloor gas-hydrate deposits into the upper ocean and atmosphere. *Geo.-Marine Lett.*, **22**, 198–203.

Peng, Y., and U. Lohmann, 2003: Sensitivity study of the spectral dispersion of the cloud droplet size distribution on the indirect aerosol effect. *Geophys. Res. Lett.*, **30**(10), 1507, doi:10.1029/2003GL017192.

Peng, Y., et al., 2002: The cloud albedo-cloud droplet effective radius relationship for clean and polluted clouds from ACE and FIRE. *J. Geophys. Res.*, **107**(D11), doi:10.1029/2002JD000281.

Penner, J., et al., 2001: Aerosols, their direct and indirect effects. In: *Climate Change 2001: The Scientific Basis. Contribution of Working Group I to the Third Assessment Report of the Intergovernmental Panel on Climate Change* [Houghton, J.T., et al. (eds.)]. Cambridge University Press, Cambridge, United Kingdom and New York, NY, USA, pp. 289–348.

Penner, J.E., S.Y. Zhang, and C.C. Chuang. 2003: Soot and smoke aerosol may not warm climate. *J. Geophys. Res.*, **108**(21), 4657, doi:10.1029/2003JD003409.

Penner., J.E., et al., 2006: Model intercomparison of indirect aerosol effects. *Atmos. Chem. Phys. Discuss.*, **6**, 1579–1617.

Perlwitz, J., I. Tegen, and R.L. Miller, 2001: Interactive soil dust aerosol model in the GISS GCM 1. Sensitivity of the soil dust cycle to radiative properties of soil dust aerosols. *J. Geophys. Res.*, **106**(D16), doi:10.1029/2000JD900668.

Perry, K.D., S.S. Cliff, and M.P. Jimenez-Cruz, 2004: Evidence for hygroscopic mineral dust particles from the Intercontinental Transport and Chemical Transformation Experiment. *J. Geophys. Res.*, **109**, D23S28, doi:10.1029/2004JD004979.

Petit, J., et al., 1999: Climate and atmospheric history of the past 420,000 years from the Vostok ice core, Antarctica. *Nature*, **399**, 429–436.

Pétron, G., et al., 2004: Monthly CO surface sources inventory based on the 2000-2001 MOPITT data. *Geophys. Res. Lett.*, **31**, L21107, doi:10.1029/2004GL020560.

Peylin, P., et al., 2005: Multiple constraints on regional CO_2 flux variations over land and oceans. *Global Biogeochem. Cycles*, **19**, GB1011, doi:10.1029/2003GB002214.

Phillips, O.L., et al., 1998: Changes in the carbon balance of tropical forests: evidence from long-term plots. *Science*, **282**(5388), 439–442.

Phillips, V.T.J., T.W. Choularton, A.M. Blyth, and J. Latham, 2002: The influence of aerosol concentrations on the glaciation and precipitation of a cumulus cloud. *Q. J. R. Meteorol. Soc.*, **128**(581), 951–971.

Pickering, K.E., et al., 2001: Trace gas transport and scavenging in PEM-Tropics B South Pacific convergence zone convection. *J. Geophys. Res.*, **106**(D23), doi:10.1029/2001JD000328.

Pielke, R.A. Sr., 2001: Influence of the spatial distribution of vegetation and soils on the prediction of cumulus convective rainfall. *Rev. Geophys.*, **39**(2), 151–177.

Pielke, R.A. Sr., and T. Matsui, 2005: Should light wind and windy nights have the same temperature trends at individual levels even if the boundary layer averaged heat content change is the same? *Geophys. Res. Lett.*, **32**, L21813, doi:10.1029/2005GL024407.

Pinto, J.P., and M.A.K. Khalil, 1991: The stability of tropospheric OH during ice ages, inter-glacial epochs and modern times. *Tellus*, **43B**, 347–352.

Pinty, B., et al., 2006: Simplifying the interaction of land surfaces with radiation for relating remote sensing products to climate models. *J. Geophys. Res.*, **111**, D02116, doi:10.1029/2005JD005952.

Pitman, A.J., B.J. McAvaney, N. Bagnound, and B. Chemint, 2004: Are inter-model differences in AMIP-II near surface air temperature means and extremes explained by land surface energy balance complexity. *Geophys. Res. Lett.*, **31**, L05205, doi:10.1029/2003GL019233.

Platnick, S., et al., 2000: The role of background cloud microphysics in the radiative formation of ship tracks. *J. Atmos. Sci.*, **57**, 2607–2624.

Platt, U., W. Allan, and D. Lowe. 2004: Hemispheric average Cl atom concentration from $^{12}C/^{13}C$ ratios in atmospheric methane. *Atmos. Chem. Phys.*, **4**, 2393–2399.

Plattner, G.-K., F. Joos, T.F. Stocker, and O. Marchal, 2001: Feedback mechanisms and sensitivities of ocean carbon uptake under global warming. *Tellus*, **53B**, 564–592.

Polvani, L.M. and P.J. Kushner, 2002: Tropospheric response to stratospheric perturbations in a relatively simple general circulation model. *Geophys. Res. Lett.*, **29**, doi:10.1029/2001GL014284.

Prather, M.J., 2002: Lifetimes of atmospheric species: integrating environmental impacts. *Geophys. Res. Lett.*, **29**(22), 2063, doi:10.1029/2002GL016299

Prather, M.J., et al., 2001: Atmospheric chemistry and greenhouse gases. In: *Climate Change 2001: The Scientific Basis. Contribution of Working Group I to the Third Assessment Report of the Intergovernmental Panel on Climate Change* [Houghton, J.T., et al. (eds.)]. Cambridge University Press, Cambridge, United Kingdom and New York, NY, USA, pp. 239–287.

Prentice, I.C., et al., 2001: The Carbon Cycle and Atmospheric Carbon Dioxide. In: *Climate Change 2001: The Scientific Basis. Contribution of Working Group I to the Third Assessment Report of the Intergovernmental Panel on Climate Change* [Houghton, J.T., et al. (eds.)]. Cambridge University Press, Cambridge, United Kingdom and New York, NY, USA, pp. 99–181.

Price, C., and D. Rind, 1994a: Possible implications of global climate change on global lightning distributions and frequencies. *J. Geophys. Res.*, **99**(D5), doi:10.1029/94JD00019.

Price, C., and D. Rind, 1994b: The impact of a $2xCO_2$ climate on lightning-caused fires. *J. Clim.*, **7**, 1484–1494.

Price, C., J. Penner, and M. Prather, 1997: NO_x from lightning 1. Global distribution based on lightning physics. *J. Geophys. Res.*, **102**(D5), doi:10.1029/96JD03504.

Prinn, R.G., et al., 1990: Atmospheric emissions and trends of nitrous-oxide deduced from 10 years of ALE-gauge data. *J. Geophys. Res.*, **95**(D11), 18369–18385.

Prinn, R.G., et al., 2001: Evidence for substantial variations of atmospheric hydroxyl radicals in the past two decades. *Science*, **292**(5523), 1882–1888.

Prinn, R.G., et al., 2005: Evidence for variability of atmospheric hydroxyl radicals over the past quarter century. *Geophys.Res.Lett.*, **32**, L07809, doi:10.1029/2004GL022228.

Prospero, J.M., and P.J. Lamb, 2003: African droughts and dust transport to the Caribbean: Climate change implications. *Science*, **302**, 1024–1027.

Prospero, J.M., et al., 2002: Environmental characterization of global sources of atmospheric soil dust identified with the NIMBUS 7 total ozone mapping spectrometer (TOMS) absorbing aerosol product. *Rev. Geophys.*, **40**, doi:10.1029/2000RG000095.

Qian, Y., and F. Giorgi, 2000: Regional climatic effects of anthropogenic aerosols? The case of Southwestern China. *Geophys. Res. Lett.*, **27**(21), doi:10.1029/2000GL011942.

Quaas, J., O. Boucher, and F.-M. Bréon, 2004: Aerosol indirect effects in POLDER satellite data and the Laboratoire de Météorologie Dynamique-Zoom (LMDZ) general circulation model. *J. Geophys. Res.*, **109**, doi:10.1029/2003JD004317.

Quaas, J., O. Boucher, and U. Lohmann, 2006: Constraining the total aerosol indirect effect in the LMDZ and ECHAM4 GCMs using MODIS satellite data. *Atmos. Chem. Phys.*, **6**, 947-955.

Quay, P., et al., 2003: Changes in the $^{13}C/^{12}C$ of dissolved inorganic carbon in the ocean as a tracer of anthropogenic CO_2 uptake. *Global Biogeochem. Cycles*, **17**(1), 1004, doi:10.1029/2001GB001817.

Quesada, C.A., et al., 2004: Seasonal and depth variation of soil moisture in a burned open savanna (campo sujo) in central Brazil. *Ecol. Appl.*, **14**, S33–41.

Raich, J., and W. Schlesinger, 1992: The global carbon dioxide flux in soil respiration and its relationship to vegetation and climate. *Tellus*, **44B**, 81–99.

Ramanathan, V., P.J. Crutzen, J.T. Kiehl, and D. Rosenfeld, 2001: Aerosols, climate, and the hydrological cycle. *Science*, **294**, 2119–2123.

Ramanathan, V., et al., 2005: Atmospheric brown clouds: impacts on South Asian climate and hydrological cycle. *Proc. Natl. Acad. Sci. U.S.A.*, **102**, 5326–5333.

Ramankutty, N., and J.A. Foley, 1999: Estimating historical changes in global land cover: Croplands from 1700 to 1992. *Global Biogeochem. Cycles*, **13**, 997–1028.

Randel, W.J., et al., 2004: Interannual changes of stratospheric water vapor and correlations with tropical tropopause temperatures. *J. Atmos. Sci.*, **61**, 2133–2148.

Randel, W.J., et al., 2006: Decreases in stratospheric water vapor after 2001: Links to changes in the tropical tropopause and the Brewer-Dobson circulation. *J. Geophys. Res.*, **111**, D12312, doi:10.1029/2005JD006744.

Randerson, J.T., et al., 2002a: Net ecosystem production: A comprehensive measure of net carbon accumulation by ecosystems. *Ecol. Appl.*, **12**(4), 937–947.

Randerson, J.T., et al., 2002b: Seasonal and latitudinal variability of troposphere $\Delta^{14}CO_2$: Post bomb contributions from fossil fuels, oceans, the stratosphere, and the terrestrial biosphere. *Global Biogeochem. Cycles*, **16**(4), 1112, doi:10.1029/2002GB001876.

Randerson, J.T., et al., 2002c: A possible global covariance between terrestrial gross primary production and ^{13}C discrimination: Consequences for the atmospheric ^{13}C budget and its response to ENSO. *Global Biogeochem. Cycles*, **16**(4), 1136, doi:10.1029/2001GB001845.

Randerson, J.T., et al., 2002d: Carbon isotope discrimination of arctic and boreal biomes inferred from remote atmospheric measurements and a biosphere-atmosphere model. *Global Biogeochem. Cycles*, **16**(3), doi:10.1029/2001GB001435.

Randerson, J.T., et al., 2005: Fire emissions from C-3 and C-4 vegetation and their influence on interannual variability of atmospheric CO_2 and $\Delta^{13}CO_2$. *Global Biogeochem. Cycles*, **19**(2), GB2019, doi:10.1029/2004GB002366.

Raymond, P.A., and J.J. Cole, 2003: Increase in the export of alkalinity from North America's largest river. *Science*, **301**, 88–91.

Rayner, P.J., I.G. Enting, R.J. Francey, and R. Langenfelds, 1999: Reconstructing the recent carbon cycle from atmospheric CO_2, $\delta^{13}C$ and O_2/N_2 observations. *Tellus*, **51B**(2), 213–232.

Rayner, P.J., et al., 2005: Two decades of terrestrial carbon fluxes from a Carbon Cycle Data Assimilation System (CCDAS). *Global Biogeochem. Cycles*, **19**, doi:10.1029/2004GB002254.

Reale, O., and P. Dirmeyer, 2002: Modeling the effect of land surface evaporation variability on precipitation variability. I: General response. *J. Hydrometeorol.*, **3**(4), 433–450.

Reale, O., P. Dirmeyer, and A. Schlosser, 2002: Modeling the effect of land surface evaporation variability on precipitation variability. II: Time- and space-scale structure. *J. Hydrometeorol.*, **3**(4), 451–466.

Reich, P.B., et al., 2006: Nitrogen limitation constrains sustainability of ecosystem response to CO_2. *Nature*, **440**, 922–925, doi:10.1038/nature04486.

Reinsel, G.C., et al., 2005: Trend analysis of total ozone data for turnaround and dynamical contributions. *J. Geophys. Res.*, **110**, D16306, doi:10.1029/2004JD004662.

Ren, X., et al., 2003: OH and HO_2 chemistry in the urban atmosphere of New York City. *Atmos. Environ.*, **37**, 3639–3651.

Revelle, R., and H.E. Suess, 1957: Carbon dioxide exchange between atmosphere and ocean and the question of an increase of atmospheric CO_2 during past decades. *Tellus*, **9**, 18–27.

Rice, A.H., et al., 2004: Carbon balance and vegetation dynamics in an old-growth Amazonian Forest. *Ecol. Appl.*, **14**(4), S55–S71.

Richey, J.E., 2004: Pathways of atmospheric CO_2 through fluvial systems. In: *The Global Carbon Cycle: Integrating Humans, Climate, and the Natural World* [Field, C., and M. Raupach (eds)]. SCOPE 62, Island Press, Washington, DC, pp. 329–340.

Richter, A., and J.P. Burrows, 2002: Tropospheric NO_2 from GOME measurements. *Adv. Space Res.*, **29**, 1673–1683.

Richter, A., et al., 2004: Satellite measurements of NO_2 from international shipping emissions. *Geophys. Res. Lett.*, **31**, doi:10.1029/2004GL020822.

Richter, A., et al., 2005: Increase in tropospheric nitrogen dioxide over China observed from space. *Nature*, **437**, 129–132.

Ridgwell, A.J., S.J. Marshall, and K. Gregson, 1999: Consumption of atmospheric methane by soils: A process-based model. *Global Biogeochem. Cycles*, **13**, 59–70.

Riebesell, U., D.A. Wolf-Gladrow, and V. Smetacek, 1993: Carbon dioxide limitation of marine phytoplankton growth rates. *Nature*, **361**, 249–251.

Riebesell, U., et al., 2000: Reduced calcification of marine plankton in response to increased atmospheric CO_2. *Nature*, **407**, 364–367.

Rind, D., J. Lerner, and C. McLinden, 2001: Changes of tracer distribution in the doubled CO_2 climate. *J. Geophys. Res.*, **106**(D22), doi:10.1029/2001JD000439.

Roberts, J.M., A.J. Wheeler, and A. Freiwald, 2006: Reefs of the deep: The biology and geology of cold-water coral ecosystems. *Science*, **312**, 543–547.

Robock, A., 2005: Cooling following large volcanic eruptions corrected for the effect of diffuse radiation on tree rings. *Geophys. Res. Lett.*, **32**, L06702, doi:10.1029/2004GL022116.

Rödenbeck, C., S. Houweling, M. Gloor, and M. Heimann, 2003a: Time-dependent atmospheric CO_2 inversions based on interannually varying tracer transport. *Tellus*, **55B**, 488–497.

Rödenbeck, C., S. Houweling, M. Gloor, and M. Heimann, 2003b: CO_2 flux history 1982–2001 inferred from atmospheric data using a global inversion of atmospheric transport. *Atmos. Chem. Phys.*, **3**, 2575–2659.

Roderick, M.L., G.D. Farquhar, S.L. Berry, and I.R. Noble, 2001: On the direct effect of clouds and atmospheric particles on the productivity and structure of vegetation. *Oecologia*, **129**, 21–30.

Roeckner, E., et al., 1999: Transient climate change simulations with a coupled atmosphere-ocean GCM including the tropospheric sulphur cycle. *J. Clim.*, **12**, 3004–3032.

Roeckner, E., et al., 2006: Impact of carbonaceous aerosol emissions on regional climate change. *Clim. Dyn.*, **27**, 553–571.

Rosenfeld, D., 1999: TRMM observed first direct evidence of smoke from forest fires inhibiting rainfall. *Geophys. Res. Lett.*, **26**(20), doi:10.1029/1999GL006066.

Rosenfeld, D., 2000: Suppression of rain and snow by urban and industrial air pollution. *Science*, **287**, 1793–1796.

Rosenfeld, D., and W.L. Woodley, 2000: Deep convective clouds with sustained supercooled liquid water down to –37.5 °C. *Nature*, **405**, 440–442.

Rosenfeld, D., Y. Rudich, and R. Lahav, 2001: Desert dust suppressing precipitation: a possible desertification feedback loop. *Proc. Natl. Acad. Sci. U.S.A.*, **98**, 5975–5980.

Rosenfeld, D., R. Lahav, A. Khain, and M. Pinsky, 2002: The role of sea spray in cleansing air pollution over ocean via cloud processes. *Science*, **297**, 1667–1670.

Rosenfield, J.E., A.R. Douglass, and D.B. Considine, 2002: The impact of increasing carbon dioxide on ozone recovery. *J. Geophys. Res.*, **107**(D6), 4049, doi:10.1029/2001JD000824.

Rosenlof, K.H., et al., 2001. Stratospheric water vapor increase over the past half-century. *Geophys. Res. Lett.*, **28**, 1195–1198.

Rotman, D.A., et al., 2004: IMPACT, the LLNL 3-D global atmospheric chemical transport model for the combined troposphere and stratosphere: Model description and analysis of ozone and other trace gases. *J. Geophys. Res.*, **109**, D04303, doi:10.1029/2002JD003155.

Rotstayn, L.D., and U. Lohmann, 2002: Tropical rainfall trends and the indirect aerosol effect. *J. Clim.*, **15**, 2103–2116.

Rotstayn, L.D., and Y. Liu, 2003: Sensitivity of the first indirect aerosol effect to an increase of cloud droplet spectral dispersion with droplet number concentration. *J. Clim.*, **16**, 3476–3481.

Rotstayn, L.D., and Y. Liu, 2005: A smaller global estimate of the second indirect aerosol effect. *Geophys. Res. Lett.*, **32**, L05708, doi:10.1029/2004GL021922.

Rotstayn, L.D., B.F. Ryan, and J.E. Penner, 2000: Precipitation changes in a GCM resulting from the indirect effects of anthropogenic aerosols. *Geophys. Res. Lett.*, **27**, 3045–3048.

Roy, S.B., G.C. Hurtt, C.P. Weaver, and S.W. Pacala, 2003: Impact of historical land cover change on the July climate of the United States. *J. Geophys. Res.*, **108**(D24), 4793, doi:10.1029/2003JD003565.

Roy, T., P. Rayner, R. Matear, and R. Francey, 2003: Southern hemisphere ocean CO_2 uptake: reconciling atmospheric and oceanic estimates. *Tellus*, **55B**(2), 701–710.

Royal Society, 2005: *Ocean Acidification Due to Increasing Atmospheric Carbon Dioxide*. Policy document 12/05, June 2005, The Royal Society, London, 60 pp., http://www.royalsoc.ac.uk/document.asp?tip=0&id=3249.

Ruddiman, W.F., and J.S. Thomson, 2001: The case for human causes of increased atmospheric CH_4 over the last 5000 years. *Quat. Sci. Rev.*, **20**, 1769–1777.

Russell, J.L., and J.M. Wallace, 2004: Annual carbon dioxide drawdown and the Northern Annular Mode. *Global Biogeochem. Cycles*, **18**(1), GB1012, doi:10.1029/2003GB002044.

Rustad, L.E., et al., 2001: A meta-analysis of the response of soil respiration, net nitrogen mineralization, and above ground plant growth to experimental ecosystem warming. *Oecologia*, **126**, 543–562.

Ryskin, G., 2003: Methane-driven oceanic eruptions and mass extinctions. *Geology*, **31**(9), 741–744.

Sabine, C.L., et al., 2004a: The oceanic sink for anthropogenic CO_2. *Science*, **305**(5682), 367–371.

Sabine, C.L., et al., 2004b: Current status and past trends of the global carbon cycle. In: *The Global Carbon Cycle: Integrating Humans, Climate and the Natural World* [Field, C., and M. Raupach (eds.)]. SCOPE 62, Island Press, Washington, DC, pp. 17–44.

Saiz-Lopez, A., and J.M.C. Plane, 2004: Novel iodine chemistry in the marine boundary layer. *Geophys. Res. Lett.*, **31**, L04112, doi:10.1029/2003GL019215.

Saiz-Lopez, A., et al., 2005: Modelling molecular iodine emissions in a coastal marine environment: the link to new particle formation. *Atmos. Chem. Phys. Discuss.*, **5**, 5405–5439.

Saleska, S.R., et al., 2003: Carbon in Amazon forests: Unexpected seasonal fluxes and disturbance–induced losses. *Science*, **302**(5650), 1554–1557.

Salvucci, G.D., J.A. Saleem, and R. Kaufmann, 2002: Investigating soil moisture feedbacks on precipitation with tests of Granger causality. *Adv. Water Resour.*, **25**, 1305–1312.

Sanderson, M.G., W.J. Collins, R.G. Derwent, and C.E. Johnson, 2003a: Simulation of global hydrogen levels using a Lagrangian three-dimensional model. *J. Atmos. Chem.*, **46**(1), 15–28.

Sanderson, M.G., et al., 2003b: Effect of climate change on isoprene emissions and surface ozone levels. *Geophys. Res. Lett.*, **30**(18), 1936, doi:10.1029/2003GL017642.

Sarmiento, J.L., and E.T. Sundquist, 1992: Revised budget for the oceanic uptake of anthropogenic carbon dioxide. *Nature*, **356**, 589–593.

Sarmiento, J.L., and N. Gruber, 2006: *Ocean Biogeochemical Dynamics*. Princeton University Press, Princeton, NJ, 503 pp.

Sarmiento, J.L., et al., 2004: Response of ocean ecosystems to climate warming. *Global Biogeochem. Cycles*, **18**(3), GB3003, doi:10.1029/2003GB002134.

Sass, R.L., J.A. Andrews, A.J. Ding, and F.M. Fisher, 2002: Spatial and temporal variability in methane emissions from rice paddies: implications for assessing regional methane budgets. *Nutrient Cycling in Agroecosystems*, **64**(1–2), 3–7.

Sathyendranath, S., et al., 1991: Biological control of surface temperature in the Arabian Sea. *Nature*, **349**, 54–56.

Saueressig, G., et al., 2001: Carbon 13 and D kinetic isotope effects in the reaction of CH_4 with $O(^1D)$ and OH: New laboratory measurements and their implications for the isotopic composition of stratospheric methane. *J. Geophys. Res.*, **106**, 23127–23138.

Scanlon, B.R., et al., 2005: Ecological controls on water-cycle response to climate variability in deserts. *Proc. Natl. Acad. Sci. U.S.A.*, **102**, 6033–6038.

Scheehle, E.A., W.N. Irving, and D. Kruger, 2002: Global anthropogenic methane emission. In: *Non-CO_2 Greenhouse Gases* [Van Ham, J., A.P. Baede, R. Guicherit, and J.Williams-Jacobse (eds)]. Millpress, Rotterdam, pp. 257–262.

Schimel, D.S., et al., 2001: Recent patterns and mechanisms of carbon exchange by terrestrial ecosystems. *Nature*, **414**, 169–172.

Schoeberl, M.R., A.R. Douglass, Z. Zhu, and S. Pawson, 2003: A comparison of the lower stratospheric age-spectra derived from a general circulation model and two data assimilation systems. *J. Geophys. Res.*, **108**(D3), doi:10.1029/2002JD002652.

Schultz, M.G., T. Diehl, G.P. Brasseur, and W. Zittel, 2003: Air pollution and climate-forcing impacts of a global hydrogen economy. *Science*, **302**, 624–627.

Schulz, M., Y. Balkanski, F. Dulac, and W. Guelle, 1998: Role of aerosol size distribution and source location in a three-dimensional simulation of a Saharan dust episode tested against satellite-derived optical thickness. *J. Geophys Res*, **103**, 10579–10592.

Sciandra, A., et al., 2003: Response of the coccolithophorid *Emiliana huxleyi* to elevated partial pressure of CO_2 under nitrogen limitation. *Mar. Ecol. Prog. Ser.*, **261**, 111–122.

Seifert, A., and K.D. Beheng, 2006: A two-moment cloud microphysics parameterization for mixed-phase clouds. Part II: Deep convective storms. *Meteorol. Atmos. Phys.*, **92**, doi:10.1007/s00703-005-0113-3.

Seiler, W., and R. Conrad, 1987: Contribution of tropical ecosystems to the global budget of trace gases, especially CH_4, H_2, CO, and N_2O. In: *The Geophysiology of Amazonia: Vegetation and Climate Interactions* [Dickinson, R.E. (ed.)]. John Wiley, New York, pp. 33–62.

Sekiguchi, M., et al., 2003: A study of the direct and indirect effects of aerosols using global satellite data sets of aerosol and cloud parameters. *J. Geophys. Res.*, **108**. 4699, doi:10.1029/2002JD003359.

Sharon, T.M., et al., 2006: Aerosol and cloud microphysical characteristics of rifts and gradients in maritime stratocumulus clouds. *J. Atmos. Sci.*, **63**(3), 983–997.

Shim, C., et al., 2005: Constraining global isoprene emissions with Global Ozone Monitoring Experiment (GOME) formaldehyde column measurements. *J. Geophys. Res.*, **110**, D24301, doi:10.1029/2004JD005629.

Shindell, D.T., and G. Faluvegi, 2002: An exploration of ozone changes and their radiative forcing prior to the chlorofluorocarbon era. *Atmos. Chem. Phys.*, **2**, 363–374.

Shindell, D.T., G. Faluvegi, and N. Bell, 2003: Preindustrial-to-present-day radiative forcing by tropospheric ozone from improved simulations with the GISS chemistry-climate GCM. *Atmos. Chem. Phys.*, **3**, 1675–1702.

Shindell, D.T., B.P. Walter, and G. Faluvegi, 2004: Impacts of climate change on methane emissions from wetlands. *Geophys. Res. Lett.*, **31**, L21202, doi:10.1029/2004GL021009.

Shindell, D.T., et al., 2001: Chemistry-climate interactions in the Goddard Institute for Space Studies general circulation model 1. Tropospheric chemistry model description and evaluation. *J. Geophys. Res.*, **106**(D8), doi:10.1029/2000JD900704.

Shvidenko, A.Z., and S. Nilsson, 2003: A synthesis of the impact of Russian forests on the global carbon budget for 1961-1998. *Tellus*, **55B**, 391–415.

Siegenthaler, U., et al., 2005: Stable carbon cycle-climate relationship during the late Pleistocene. *Science*, **310**(5752), 1313–1317.

Sievering, H., et al., 1992: Removal of sulphur from the marine boundary layer by ozone oxidation in sea-salt aerosols. *Nature*, **360**, 571–573.

Sievering, H., et al., 2000: Forest canopy uptake of atmospheric nitrogen deposition at eastern U.S. conifer sites: Carbon storage implications. *Global Biogeochem. Cycles*, **14**(4), doi:10.1029/1999GB001250.

Sillman, S., and P.J. Samson, 1995: Impact of temperature on oxidant photochemistry in urban, polluted rural, and remote environments. *J. Geophys. Res.*, **100**(D6), 11497–11508, doi:10.1029/94JD02146.

Silva Dias, M.A.F., et al., 2002: Clouds and rain processes in a biosphere atmosphere interaction context. *J. Geophys. Res.*, **107**(D20), 8072, doi:10.1029/2001JD000335.

Simmonds, P.G., et al., 2000: Continuous high-frequency observations of hydrogen at the Mace Head baseline atmospheric monitoring station over the 1994-1998 period. *J. Geophys. Res.*, **105**(D10), 12105–12121, doi:10.1029/2000JD900007.

Simmons, A.J., et al., 2005: ECMWF analyses and forecasts of stratospheric winter polar vortex breakup: September 2002 in the southern hemisphere and related events. *J. Atmos. Sci.*, **62**, 668–689.

Singh, H.B., et al., 2004: Analysis of the atmospheric distribution, sources, and sinks of oxygenated volatile organic chemicals based on measurements over the Pacific during TRACE-P. *J. Geophys. Res.*, **109**, D15S07, doi:10.1029/2003JD003883.

Sinha, A., 1995: Relative influence of lapse rate and water vapour on the greenhouse effect. *J. Geophys. Res.*, **100**(D3), 5095–5103, doi:10.1029/94JD03248.

Sitch, S., et al., 2003: Evaluation of ecosystem dynamics, plant geography and terrestrial carbon cycling in the LPJ dynamic global vegetation model. *Global Change Biol.*, **9**, 161–185.

Sitch, S., et al., 2005: Impacts of future land cover changes on atmospheric CO_2 and climate. *Global Biogeochem. Cycles*, **19**, doi:10.1029/2004GB002311.

Smith, K.A., and F. Conen, 2004: Impacts of land management on fluxes of trace greenhouse gases. *Soil Use Management*, **20**, 255–263.

Smith, P., D.S. Powlson, M.J. Glendining, and J.U. Smith, 1997: Potential for carbon sequestration in European soils: preliminary estimates for five scenarios using results from long-term experiments. *Global Change Biol.*, **3**, 67–79.

Smith, S.V., and J.T. Hollibaugh, 1993: Coastal metabolism and the oceanic organic carbon balance. *Rev. Geophys.*, **31**(1), 75–89.

Smyth, T.J., T. Tyrrell, and B. Tarrant, 2004: Time series coccolithophore activity in the Barents Sea, from twenty years of satellite imagery. *Geophys. Res. Lett.*, **31**, L11302, doi:10.1029/2004GL019735.

Snover, A.K., and P.D. Quay, 2000: Hydrogen and carbon kinetic effects during soil uptake of atmospheric methane. *Global Biogeochem. Cycles*, **14**, 25–39.

Snyder, P.K., C. Delire, and J.A. Foley, 2004: Evaluating the influence of different vegetation biomes on the global climate. *Clim. Dyn.*, **23**, 279–302, doi:10.1007/s00382-004-0430-0.

Song, Y., and W.A. Robinson, 2004: Dynamical mechanisms for stratospheric influences on the troposphere. *J. Atmos. Sci.*, **61**, 1711–1725.

Spahni, R., et al., 2005: Atmospheric methane and nitrous oxide of the Late Pleistocene from Antarctic ice cores. *Science*, **310**(5752), 1317–1321.

Steinbrecht, W., H. Claude, and P. Winkler, 2004: Enhanced upper stratospheric ozone: Sign of recovery or solar cycle effect? *J. Geophys. Res.*, **109**, D02308, doi:10.1029/2003JD004284.

Stephens, G.L., N.B. Wood, and L.A. Pakula, 2004: On the radiative effects of dust on tropical convection. *Geophys. Res. Lett.*, **31**, L23112, doi:10.1029/2004GL021342.

Stevens, B., et al., 2005: Pockets of open cells and drizzle in marine stratocumulus. *Bull. Am. Meteorol. Soc.*, **86**, 51–57.

Stevenson, D.S., et al., 2000: Future estimates of tropospheric ozone radiative forcing and methane turnover - the impact of climate change. *Geophys. Res. Lett.*, **105**(14), doi:10.1029/1999GL010887.

Stevenson, D.S., et al., 2004: Radiative forcing from aircraft NO_x emissions: mechanisms and seasonal dependence. *J. Geophys. Res.*, **109**, D17307, doi:10.1029/2004JD004759.

Stevenson, D.S., et al., 2005: Impacts of climate change and variability on tropospheric ozone and its precursors. *Faraday Discuss.*, **130**, doi:10.1039/b417412g.

Stevenson, D.S., et al., 2006: Multi-model ensemble of present-day and near-future tropospheric ozone. *J. Geophys. Res.*, **111**, D8301, doi:10.1029/2005JD006338.

Stier, P., et al., 2006a: The evolution of the global aerosol system in a transient climate simulation from 1860 to 2100. *Atmos. Chem. Phys.*, **6**, 3059–3076.

Stier, P., et al., 2006b: Emission-induced nonlinearities in the global aerosol system - results from the ECHAM5-HAM aerosol-climate model. *J. Clim.*, **19**, 3845–3862.

Stocks, B.J., et al., 1998: Climate change and forest fire potential in Russian and Canadian boreal forests. *Clim. Change*, **38**, 1–13.

Storelvmo T., et al., 2006: Predicting cloud droplet number concentration in Community Atmosphere Model (CAM)-Oslo, *J. Geophys. Res.*, **111**, D24208, doi:10.1029/2005JD006300.

Sturm, M., T. Douglas, C. Racine, and G. Liston, 2005: Changing snow and shrub conditions affect albedo with global implications. *J. Geophys. Res.*, **110**, G01004, doi:10.1029/2005JG000013.

Sudo, K., M. Takahashi, and H. Akimoto, 2002a: CHASER: A global chemical model of the troposphere 2. Model results and evaluation. *J. Geophys. Res.*, **107**, 4586, doi:10.1029/2001JD001114.

Sudo, K., M. Takahashi, J. Kurokawa, and H. Akimoto, 2002b: CHASER: A global chemical model of the troposphere 1. Model description. *J. Geophys. Res.*, **107**, 4339, doi:10.1029/2001JD001113.

Suntharalingam, P., et al., 2005: Influence of reduced carbon emissions and oxidation on the distribution of atmospheric CO_2: Implications for inversion analyses. *Global Biogeochem. Cycles*, **19**, GB4003, doi:10.1029/2005GB002466.

Takahashi, T., et al., 2002: Global sea-air CO_2 flux based on climatological surface ocean pCO_2, and seasonal biological and temperature effects. *Deep-Sea Res. II*, **49**(9–10), 1601–1622.

Takemura, T., et al., 2005: Simulation of climate response to aerosol direct and indirect effects with aerosol transport-radiation model. *J. Geophys. Res*, **110**, doi:10.1029/2004JD00502.

Tan, W.W., M.A. Geller, S. Pawson, and A. da Silva, 2004: A case study of excessive subtropical transport in the stratosphere of a data assimilation system. *J. Geophys. Res.*, **109**, D11102, doi:10.1029/2003JD004057.

Tans, P. P., and T.J. Conway, 2005: Monthly atmospheric CO_2 mixing ratios from the NOAA CMDL Carbon Cycle Cooperative Global Air Sampling Network, 1968-2002. In: *Trends: A Compendium of Data on Global Change*. Carbon Dioxide Information Analysis Center, Oak Ridge National Laboratory, U.S. Department of Energy, Oak Ridge, TN, http://cdiac.ornl.gov/trends/co2/cmdl-flask/cmdl-flask.html.

Tegen, I., A.A. Lacis, and I. Fung, 1996: The influence of mineral aerosols from disturbed soils on the global radiation budget. *Nature*, **380**, 419–422.

Tegen, I., M. Werner, S.P. Harrison, and K.E. Kohfeld, 2004: Relative importance of climate and land use in determining present and future global soil dust emission. *Geophys. Res. Lett.*, **31**, L05105, doi:10.1029/2003GL019216.

Tegen, I., et al., 2002. Impact of vegetation and preferential source areas on global dust aerosol: results from a model study. *J. Geophys. Res.*, **107**(D21), 4576, doi:10.1029/2001JD000963.

Textor, C., et al., 2005: Analysis and quantification of the diversities of aerosol life cycles within AEROCOM. *Atmos. Chem. Phys. Discuss.*, **5**, 8331–8420.

Thompson, A.M., et al., 1996: Where did tropospheric ozone over southern Africa and the tropical Atlantic come from in October 1992? Insights from TOMS, GTE TRACE A, and SAFARI 1992. *J. Geophys. Res.*, **101**(D19), doi:10.1029/96JD01463.

Thompson, A.M., et al., 2000: A tropical Atlantic paradox: shipboard and satellite views of a tropospheric ozone maximum and wave-one in January–February 1999. *Geophys. Res. Lett.*, **27**(20), doi:10.1029/1999GL011273.

Thompson, D.W.J., M.P. Baldwin, and S. Solomon, 2005: Stratosphere/troposphere coupling in the Southern Hemisphere. *J. Atmos. Sci.*, **62**, 708–715.

Thompson, S.L., et al., 2004: Quantifying the effects of CO_2-fertilized vegetation on future global climate and carbon dynamics. *Geophys. Res. Lett.*, **31**, L23211, doi:10.1029/2004GL021239.

Thornton, P.E., et al., 2002: Modeling and measuring the effects of disturbance history and climate on carbon and water budgets in evergreen needleleaf forests. *Agric. For. Meteorol.*, **113**, 185–222.

Tian, H., et al., 1998: Effect of interannual climate variability on carbon storage in Amazonian ecosystems. *Nature*, **396**, 664–667.

Tian, Y., et al., 2004: Comparison of seasonal and spatial variations of leaf area index and fraction of absorbed photosynthetically active radiation from Moderate Resolution Imaging Spectroradiometer (MODIS) and common land model. *J. Geophys. Res.*, **109**, doi:10.1029/2003JD003777.

Tie, X.X., A. Guenther, and E. Holland, 2003: Biogenic methanol and its impact on tropospheric oxidants. *Geophys. Res. Lett.*, **30**(17), 1881, doi:10.1029/2003GL017167.

Tie, X.X., R.Y. Zhang, G. Brasseur, and W.F. Lei, 2002: Global NO_x production by lightning. *J. Atmos. Chem.*, **43**(1), 61–74.

Tilmes, S., R. Müller, J.-U. Grooß, and J.M. Russell III, 2004: Ozone loss and chlorine activation in the Arctic winters 1991–2003 derived with the TRAC method. *Atmos. Chem. Phys.*, **4**, 2181–2213.

Timmreck, C., and M. Schulz, 2004: Significant dust simulation differences in nudged and climatological operation mode of the AGCM ECHAM. *J. Geophys. Res.*, **109**, D13202, doi:10.1029/2003JD004381.

Tortell, P.D., G.R. DiTullio, D.M. Sigman, and F.M.M. Morel, 2002: CO_2 effects on taxonomic composition and nutrient utilization in an Equatorial Pacific phytoplankton assemblage. *Mar. Ecol. Prog. Ser.*, **236**, 37–43.

Toumi, R., J.D. Haigh, and K.S. Law, 1996: A tropospheric ozone-lightning climate feedback. *Geophys. Res. Lett.*, **23**(9), doi:10.1029/96GL00944.

Trenberth, K.E., and D.J. Shea, 2005: Relationships between precipitation and surface temperature. *Geophys. Res. Lett.*, **32**, L14703, doi:10.1029/2005GL022760.

Tromp, T.K., et al., 2003: Potential environmental impact of a hydrogen economy on the stratosphere. *Science*, **300**, 1740–1742.

Tsigaridis, K., and M. Kanakidou, 2003: Global modelling of secondary organic aerosol in the troposphere: a sensitivity analysis. *Atmos. Chem. Phys.*, **3**, 1849–1869.

Tsvetsinskaya, E.A., et al., 2002: Relating MODIS-derived surface albedo to soils and rock types over Northern Africa and the Arabia Peninsula. *Geophys. Res. Lett.*, **29**(9), doi:10.1029/2001GL014096.

Turner, D.P., et al., 1991: Climate change and isoprene emissions from vegetation. *Chemosphere*, **23**, 37–56.

Tyler, S.C., et al., 2000: Experimentally determined kinetic isotope effects in the reaction of CH_4 with Cl: Implications for atmospheric CH_4. *Geophys. Res. Lett.*, **27**, 1715–1718.

Tyrrell, T., P.M. Holligan, and C.D. Mobley, 1999: Optical impacts of oceanic coccolithophore blooms. *J. Geophys. Res.*, **104**(C2), 3223–3241.

Umann, B., et al., 2005: Interaction of mineral dust with gas phase nitric acid and sulfur dioxide during the MINATROC II field campaign: First estimate of the uptake coefficient gamma(HNO_3) from atmospheric data. *J. Geophys. Res.*, *110*, D22306, doi:10.1029/2005JD005906.

UN-ECE/FAO (ed.), 2000: *Forest Resources of Europe, CIS, North America, Australia, Japan and New Zealand*. UN-ECE/FAO Contribution to the Global Forest Resources Assessment 2000, United Nations, New York and Geneva, 445 pp.

Valdes, P.J., D.J. Beerling, and C.E. Johnson, 2005: The ice age methane budget. *Geophys. Res. Lett.*, **32**, doi:10.1029/2004GL021004.

Valentini, R., et al., 2000: Respiration as the main determinant of carbon balance in European forests. *Nature*, **404**(6780), 861–865.

Van Aardenne, J.A., et al., 2001: A 1°×1° resolution data set of historical anthropogenic trace gas emissions for the period 1890–1990. *Global Biogeochem. Cycles*, **15**, 909–928.

Van den Heever, S.C., et al., 2006: Impacts of nucleating aerosol on Florida storms, Part I: Mesoscale simulations. *J. Atmos. Sci.*, **63**, 1752–1775.

van der Werf, G.R., J.T. Randerson, G.J. Collatz, and L. Giglio, 2003: Carbon emissions from fires in tropical and subtropical ecosystems. *Global Change Biol.*, **9**, 547–562.

van der Werf, G.R., et al., 2004: Continental-scale partitioning of fire emissions during the 1997 to 2001 El Niño/La Niña period. *Science*, **303**(5654), 73–76.

van Groenigen, K.J., et al., 2006: Element interactions limit soil carbon storage. *Proc. Natl. Acad. Sci. U.S.A.*, **103**, 6571–6574.

van Noije, T.P.C., H.J. Eskes, M. Van Weele, and P.F.J. van Velthoven, 2004: Implications of enhanced Brewer-Dobson circulation in European Centre for Medium-Range Weather Forecasts reanalysis for the stratosphere-troposphere exchange of ozone in global chemistry transport models. *J. Geophys. Res.*, **109**, D19308, doi:10.1029/2004JD004586.

van Noije, T.P.C., et al., 2006: Multi-model ensemble simulations of tropospheric NO_2 compared with GOME retrievals for the year 2000. *Atmos. Chem. Phys.*, **6**, 2943–2979.

van Wesemael, B., S. Lettens, C. Roelandt, and J. Van Orshoven, 2005: Modelling the evolution of regional carbon stocks in Belgian 19 cropland soils. *Can. J. Soil Sci.*, **85**(4), 511–521.

Velders, G.J.M., et al., 2005: Chemical and radiative effects of halocarbons and their replacement compounds. In: *IPCC/TEAP Special Report on Safeguarding the Ozone Layer and the Global Climate System: Issues related to Hydrofluorocarbons and Perfluorocarbons. Prepared by Working Group I and III of the Intergovernmental Panel on Climate Change and the Technology and Economic Assessment Panel* [Metz, B., et al. (eds.)]. Cambridge University Press, Cambridge, United Kingdom and New York, NY. USA, pp. 133–180.

Verdugo, P., et al., 2004: The oceanic gel phase: a bridge in the DOM-POM continuum. *Mar. Chem.*, **92**, 67–85.

Vitousek, P.M., 2004: *Nutrient Cycling and Limitations: Hawai'i as a Model Ecosystem*. Princeton University Press, Princeton, NJ, 232 pp.

Vitousek, P.M., et al., 1997: Human alteration of the global nitrogen cycle: sources and consequences. *Ecol. Appl.*, **7**, 737–750.

Vitousek, P.M., et al., 1998: Within-system element cycles, input-output budgets, and nutrient limitations. In: *Successes, Limitations, and Frontiers in Ecosystem Science* [Pace, M., and P. Groffman (eds.)]. Springer-Verlag, New York, pp. 432–451.

Voldoire, A., and J.-F. Royer, 2004: Tropical deforestation and climate variability. *Clim. Dyn.*, **22**, 857–874, doi:10.1007/s00382-004-0423-z.

Volk, T., and M.I. Hoffert, 1985: Ocean carbon pumps: Analysis of relative strengths and efficiencies in ocean-driven atmospheric CO_2 changes. In: *The Carbon Cycle and Atmospheric CO_2: Natural Variations Archean to Present* [Sundquist, E.T., and W.S. Broecker (eds.)]. Geophysical Monograph Vol. 32, American Geophysical Union, Washington, DC, pp. 99–110.

Von Kuhlmann, R., M.G. Lawrence, P.J. Crutzen, and P.J. Rasch, 2003: A model for studies of tropospheric ozone and nonmethane hydrocarbons: model description and ozone results. *J. Geophys. Res.*, **108**, 4294, doi:10.1029/2002JD002893.

Walsh, J.J., 1991: Importance of continental margins in the marine biogeochemical cycling of carbon and nitrogen. *Nature*, **350**, 53–55.

Walter, B.P., and M. Heimann, 2001a: Modeling modern methane emission from natural wetlands, 1. Model description and results. *J. Geophys. Res.*, **106**, 34189–34206.

Walter, B.P., and M. Heimann, 2001b: Modeling modern methane emission from natural wetlands, 2. Interannual variations 1982-1993. *J. Geophys. Res.*, **106**, 34207–37219.

Wang, C., 2004: A modeling study on the climate impacts of black carbon aerosols. *J. Geophys. Res.*, **109**, doi:10.1029/2003JD004084.

Wang, C., and R. Prinn, 1999: Impact of emissions, chemistry and climate on atmospheric carbon monoxide: 100 year predictions from a global chemistry model. *Chemosphere*, **1**, 73–81.

Wang, G., and E. Eltahir, 2000: Modeling the biosphere-atmosphere system: the impact of the subgrid variability in rainfall interception. *J. Clim.*, **13**, 2887–3078.

Wang, G., et al., 2004: Decadal variability of rainfall in the Sahel: results from the coupled GENESIS-IBIS atmosphere-biosphere model. *Clim. Dyn.*, **22**, doi:10.1007/s00382-004-0411-3.

Wang, J.S., M.B. McElroy, C.M. Spivakovsky, and D.B.A. Jones, 2002: On the contribution of anthropogenic Cl to the increase in $\delta^{13}C$ of atmospheric methane. *Global Biogeochem. Cycles*, **16**, doi:10.1029/2001GB001572.

Wang, J.S., et al., 2004: A 3-D model analysis of the slowdown and interannual variability in the methane growth rate from 1988 to 1997. *Global Biogeochem. Cycles*, **18**, GB3011, doi:10.1029/3003GB002180.

Wang, S.S., 2005: Dynamics of surface albedo of a boreal forest and its simulation. *Ecol. Model.*, **183**, 477–494.

Wang, S.S., et al., 2002: Modelling carbon dynamics of boreal forest ecosystems using the Canadian Land Surface Scheme. *Clim. Change*, **55**(4), 451–477.

Wang, Y., D.J. Jacob, and J.A. Logan, 1998: Global simulation of tropospheric O_3-NO_x-hydrocarbon chemistry, 3. Origin of tropospheric ozone and effects of non-methane hydrocarbons. *J. Geophys. Res.*, **103**(D9), 10757–10768.

Wang, Z., et al., 2004: Using MODIS BRDF and albedo data to evaluate global model land surface albedo. *J. Hydrometeorol.*, **5**, 3–14.

Wanninkhof, R., and W.R. McGillis, 1999: A cubic relationship between air-sea CO_2 exchange and wind speed. *Geophys. Res. Lett.*, **26**(13), 1889–1892.

Warneck, P., 1988: *Chemistry of the Natural Atmosphere*. Academic Press, London, 757 pp.

Warner, C.D., and M.E. McIntyre, 2001: An ultrasimple spectral parameterization for nonorographic gravity waves. *J. Atmos Sci.*, **58**, 1837–1857.

Warner, J., and S. Twomey, 1967: The production of cloud nuclei by cane fires and the effect on cloud droplet concentration. *J. Atmos Sci.*, **24**, 704–706.

Warwick, N.J., S. Bekki, E.G. Nisbet, and J.A. Pyle, 2004: Impact of a hydrogen economy on the stratosphere and troposphere studied in a 2-D model. *Geophys. Res. Lett.*, **31**, L05107, doi:10.1029/2003GL019224.

Warwick, N.J., et al., 2002: The impact of meteorology on the interannual growth rate of atmospheric methane. *Geophys. Res. Lett.*, **29**(20), 1947, doi:10.1029/2002GL015282.

Weaver, C.P., S.B. Roy, and R. Avissar, 2002: Sensitivity of simulated mesoscale atmospheric circulations resulting from landscape heterogeneity to aspects of model configuration. *J. Geophys. Res.*, **107**(D20), 8041, doi:10.1029/2001JD000376.

Weiss, R.F., 1974: Carbon dioxide in water and seawater: the solubility of a non-ideal gas. *Mar. Chem.*, **2**, 203–215.

Wells, M.L., and E.D. Goldberg, 1994: The distribution of colloids in the North Atlantic and Southern Oceans. *Limnol. Oceanogr.*, **39**, 286–302.

Werner, M., et al., 2002: Seasonal and interannual variability of the mineral dust cycle under present and glacial climate conditions. *J. Geophys. Res.*, **107**(D24), 4744, doi:10.1029/2002JD002365.

Wetzel, P., et al., 2006. Effects of ocean biology on the penetrative radiation in a coupled climate model. *J. Clim.*, **19**, 3973–3987.

Wickland, K., R. Striegl, J. Neff, and T. Sachs, 2006: Effects of permafrost melting on CO_2 and CH_4 exchange of a poorly drained black spruce lowland. *J. Geophys. Res.*, **111**, G02011, doi:10.1029/2005JG000099.

Wild, O., M.J. Prather, and H. Akimoto, 2001: Indirect long-term global radiative cooling from NO_x emissions. *Geophys. Res. Lett.*, **28**(9), 1719–1722.

Wild, O., P. Pochanart, and H. Akimoto, 2004: Trans-Eurasian transport of ozone and its precursors. *J. Geophys. Res.*, **109**, D11302, doi:10.1029/2003JD004501.

Williams, A.A.J., D.J. Karoly, and N. Tapper, 2001: The sensitivity of Australian fire danger to climate change. *Clim. Change*, **49**, 171–191.

Williams, K.D., et al., 2001: The response of the climate system to the indirect effects of anthropogenic sulphate aerosols. *Clim. Dyn.*, **17**, 845–856.

Williamson, D., et al., 2005: Moisture and temperature at the Atmospheric Radiation Measurement Southern Great Plains site in forecasts with the Community Atmosphere Model (CAM2). *J. Geophys. Res.*, **110**, doi:10.1029/2004JD005109.

WMO, 2003: *Scientific Assessment of Ozone Depletion: 2002*. Global Ozone Research and Monitoring Project Report No. 47, World Meteorological Organization, Geneva, 498 pp.

Wong, S., et al., 2004: A global climate-chemistry model study of present-day tropospheric chemistry and radiative forcing from changes in tropospheric O_3 since the preindustrial period. *J. Geophys. Res.*, **109**, D11309, doi:10.1029/2003JD003998.

Woodward, S., D.L. Roberts, and R.A. Betts, 2005: A simulation of the effect of climate change-induced desertification on mineral dust aerosol. *Geophys. Res. Lett.*, **32**, L18810, doi:10.1029/2005GL023482.

Wuebbles, D.J., and K. Hayhoe, 2002: Atmospheric methane and global change. *Earth Sci. Rev.*, **57**, 177–210.

Wurzler, S., T.G. Reisin, and Z. Levin, 2000: Modification of mineral dust particles by cloud processing and subsequent effects on drop size distributions. *J. Geophys. Res.*, **105**(D4), 4501–4512.

Xiao, Y., et al., 2004: Constraints on Asian and European sources of methane from CH_4-C_2H_6-Co correlation in Asian outflow. *J. Geophys. Res.*, **109**, doi:10.1029/2003JD004475.

Xu, Z., et al., 2004: Effects of elevated CO_2 and N fertilization of CH_4 emissions from paddy rice fields. *Global Biogeochem. Cycles*, **18**, GB3009, doi:10.1029/2004GB002233.

Yamasoe, M.A., et al., 2006: Effect of smoke and clouds on the transmissivity of photosynthetically active radiation inside the canopy. *Atmos. Chem. Phys.*, **6**, 1645–1656.

Yan, X., T. Ohara, and H. Akimoto, 2003: Development of region-specific emission factors and estimation of methane emission from rice fields in the East, Southeast, and South Asian countries. *Global Change Biol.*, **9**, 237–254.

Yang, E.-S., D.M. Cunnold, M.J. Newchurch, and R.J. Salawitch, 2005: Change in ozone trends at southern high latitudes. *Geophys. Res. Lett.*, **32**, L12812, doi:10.1029/2004GL022296.

Yang, K., et al., 2004: The daytime evolution of the atmospheric boundary layer and convection over the Tibetan Plateau: observations and simulations. *J. Meteorol. Soc. Japan*, **82**(6), 1777–1792.

Yang, R., and M.A. Friedl, 2003: Modeling the effects of 3-D vegetation structure on surface radiation and energy balance in boreal forests. *J. Geophys. Res.*, **108**(D16), 8615, doi:10.1029/2002JD003109.

Yin, Y., K.S. Carslaw, and G. Feingold, 2005: Vertical transport and processing of aerosols in a mixed-phase convective cloud and the feedback on cloud development. *Q. J. R. Meteorol. Soc.*, **131**, 221–246.

Yin, Y., A. Levin, T.G. Reisin, and S. Tzivion, 2000: The effect of giant cloud condensation nuclei on the development of precipitation in convective clouds - a numerical study. *Atmos. Res.*, **53**, 91–116.

Yin, Y., S. Wurzler, A. Levin, and T.G. Reisin, 2002: Interactions of mineral dust particles and clouds: Effects on precipitation and cloud optical properties. *J. Geophys. Res.*, **107**, doi:10.1029/2001JD001544.

Yurganov, L.N., et al., 2005: Increased Northern Hemispheric carbon monoxide burden in the troposphere in 2002 and 2003 detected from the ground and from space. *Atm. Chem. Phys.*, **5**, 563–573.

Zeebe, R.E., and D. Wolf-Gladrow, 2001: *CO_2 in Seawater: Equilibrium, Kinetics, Isotopes*. Elsevier Oceanography Series 65, Elsevier, Amsterdam, 346 pp.

Zender, C., R.L. Miller, and I. Tegen, 2004: Quantifying mineral dust mass budgets: systematic terminology, constraints, and current estimates. *Eos*, **85**(48), 509, 512.

Zeng, G., and J.A. Pyle, 2003: Changes in tropospheric ozone between 2000 and 2100 modeled in a chemistry-climate model. *Geophys. Res. Lett.*, **30**, 1392, doi:10.1029/2002GL016708.

Zeng, N., K. Hales, and J.D. Neelin, 2002: Nonlinear dynamics in a coupled vegetation-atmosphere system and implications for desert-forest gradient. *J. Clim.*, **15**, 3474–3485.

Zeng, N., A. Mariotti, and P. Wetzel, 2005: Terrestrial mechanisms of interannual CO_2 variability, *Global Biogeochem. Cycles*, **19**, GB1016, doi:10.1029/2004GB002273.

Zeng, N., H. Qian, E. Munoz, and R. Iacono, 2004: How strong is carbon cycle-climate feedback under global warming? *Geophys. Res. Lett.*, **31**, L20203, doi:10.1029/2004GL020904.

Zeng, X.-D., S.S.P. Shen, X. Zeng, and R.E. Dickinson, 2004: Multiple equilibrium states and the abrupt transitions in a dynamical system of soil water interacting with vegetation. *Geophys. Res. Lett.*, **31**, L05501, doi:10.1029/2003GL018910.

Zerefos, C., et al., 1998: Quasi-biennial and longer-term changes in clear sky UV-B solar irradiance. *Geophys. Res. Lett.*, **25**, 4345–4348.

Zhang, J., U. Lohmann, and P. Stier, 2005: A microphysical parameterization for convective clouds in the ECHAM5 climate model, 1. Single column results evaluated at the Oklahoma ARM site. *J. Geophys. Res.*, **110**, doi:10.1029/2004JD005128.

Zhang, X.Y., R. Arimoto, and Z.S. An, 1997: Dust emission from Chinese desert sources linked to variations in atmospheric circulation. *J. Geophys. Res.*, **102**, 28041–28047.

Zhang, X.Y., H.Y. Lu, R. Arimoto, and S.L. Gong, 2002: Atmospheric dust loadings and their relationship to rapid oscillations of the Asian winter monsoon climate: two 250-kyr loess records. *Earth Planet. Sci. Lett.*, **202**, 637–643.

Zhang, X.Y., et al., 2003: Sources of Asian dust and role of climate change versus desertification in Asian dust emission. *Geophys. Res. Lett.*, **30**, 2272, doi:10.1029/2003GL018206.

Zhou, L., R.E. Dickinson, and Y. Tian, 2005: Derivation of a soil albedo dataset from MODIS using principal component analysis: Northern Africa and the Arabian Peninsula. *Geophys. Res. Lett.*, **32**, L21407, doi:10.1029/205GL024448.

Zhou, L., et al., 2001: Variations in northern vegetation activity inferred from satellite data of vegetation index during 1981 to 1999. *J. Geophys. Res.*, **106**, 20069–20083.

Zhou, L., et al., 2003a: A sensitivity study of climate and energy balance simulations with use of satellite-derived emissivity data over Northern Africa and the Arabian Peninsula. *J. Geophys. Res.*, **108**(D24), 4795, doi:10.1029/2003JD004083.

Zhou, L., et al., 2003b: Relations between albedos and emissivities from MODIS and ASTER data over North African desert. *Geophys. Res. Lett.*, **30**(20), 2026, doi:10/1029/2003GL018069.

Zhou, L., et al., 2003c: Relation between interannual variations in satellite measures of northern forest greenness and climate between 1982 and 1999. *J. Geophys. Res.*, **108**(D1), 4004, doi:10.1029/2002JD002510.

Zhou, L., et al., 2004: Evidence for a significant urbanization effect on climate in China. *Proc. Natl. Acad. Sci. U.S.A.*, **101**(26), 9540–9544.

Zhuang, Q., et al., 2004: Methane fluxes between terrestrial ecosystem and the atmosphere at northern high latitudes during the past century: A retrospective analysis with a process-based biogeochemistry model. *Global Biogeochem. Cycles*, **18**, doi:10.1029/2008GB002239.

Zittel, W., and M. Altmann, 1996: Molecular hydrogen and water vapour emissions in a global hydrogen energy economy. In: *Proceedings of the 11th World Hydrogen Energy Conference, Stuttgart, Germany, June 1996*. Schön & Wetzel, Frankfurt, Germany, pp. 71–82.

Zondervan, I., R.E. Zeebe, B. Rost, and U. Riebesell, 2001: Decreasing marine biogenic calcification: a negative feedback on rising pCO_2. *Global Biogeochem. Cycles*, **15**, 507–516.

8

Climate Models and Their Evaluation

Coordinating Lead Authors:

David A. Randall (USA), Richard A. Wood (UK)

Lead Authors:

Sandrine Bony (France), Robert Colman (Australia), Thierry Fichefet (Belgium), John Fyfe (Canada), Vladimir Kattsov (Russian Federation), Andrew Pitman (Australia), Jagadish Shukla (USA), Jayaraman Srinivasan (India), Ronald J. Stouffer (USA), Akimasa Sumi (Japan), Karl E. Taylor (USA)

Contributing Authors:

K. AchutaRao (USA), R. Allan (UK), A. Berger (Belgium), H. Blatter (Switzerland), C. Bonfils (USA, France), A. Boone (France, USA), C. Bretherton (USA), A. Broccoli (USA), V. Brovkin (Germany, Russian Federation), W. Cai (Australia), M. Claussen (Germany), P. Dirmeyer (USA), C. Doutriaux (USA, France), H. Drange (Norway), J.-L. Dufresne (France), S. Emori (Japan), P. Forster (UK), A. Frei (USA), A. Ganopolski (Germany), P. Gent (USA), P. Gleckler (USA), H. Goosse (Belgium), R. Graham (UK), J.M. Gregory (UK), R. Gudgel (USA), A. Hall (USA), S. Hallegatte (USA, France), H. Hasumi (Japan), A. Henderson-Sellers (Switzerland), H. Hendon (Australia), K. Hodges (UK), M. Holland (USA), A.A.M. Holtslag (Netherlands), E. Hunke (USA), P. Huybrechts (Belgium), W. Ingram (UK), F. Joos (Switzerland), B. Kirtman (USA), S. Klein (USA), R. Koster (USA), P. Kushner (Canada), J. Lanzante (USA), M. Latif (Germany), N.-C. Lau (USA), M. Meinshausen (Germany), A. Monahan (Canada), J.M. Murphy (UK), T. Osborn (UK), T. Pavlova (Russian Federationi), V. Petoukhov (Germany), T. Phillips (USA), S. Power (Australia), S. Rahmstorf (Germany), S.C.B. Raper (UK), H. Renssen (Netherlands), D. Rind (USA), M. Roberts (UK), A. Rosati (USA), C. Schär (Switzerland), A. Schmittner (USA, Germany), J. Scinocca (Canada), D. Seidov (USA), A.G. Slater (USA, Australia), J. Slingo (UK), D. Smith (UK), B. Soden (USA), W. Stern (USA), D.A. Stone (UK), K.Sudo (Japan), T. Takemura (Japan), G. Tselioudis (USA, Greece), M. Webb (UK), M. Wild (Switzerland)

Review Editors:

Elisa Manzini (Italy), Taroh Matsuno (Japan), Bryant McAvaney (Australia)

This chapter should be cited as:

Randall, D.A., R.A. Wood, S. Bony, R. Colman, T. Fichefet, J. Fyfe, V. Kattsov, A. Pitman, J. Shukla, J. Srinivasan, R.J. Stouffer, A. Sumi and K.E. Taylor, 2007: Cilmate Models and Their Evaluation. In: *Climate Change 2007: The Physical Science Basis. Contribution of Working Group I to the Fourth Assessment Report of the Intergovernmental Panel on Climate Change* [Solomon, S., D. Qin, M. Manning, Z. Chen, M. Marquis, K.B. Averyt, M.Tignor and H.L. Miller (eds.)]. Cambridge University Press, Cambridge, United Kingdom and New York, NY, USA.

Table of Contents

Supplementary Material

The following supplementary material is available on CD-ROM and in on-line versions of this report.

Figures S8.1–S8.15: *Model Simulations for Different Climate Variables*

Table S8.1: *MAGICC Parameter Values*

Executive Summary

This chapter assesses the capacity of the global climate models used elsewhere in this report for projecting future climate change. Confidence in model estimates of future climate evolution has been enhanced via a range of advances since the IPCC Third Assessment Report (TAR).

Climate models are based on well-established physical principles and have been demonstrated to reproduce observed features of recent climate (see Chapters 8 and 9) and past climate changes (see Chapter 6). There is considerable confidence that Atmosphere-Ocean General Circulation Models (AOGCMs) provide credible quantitative estimates of future climate change, particularly at continental and larger scales. Confidence in these estimates is higher for some climate variables (e.g., temperature) than for others (e.g., precipitation). This summary highlights areas of progress since the TAR:

- Enhanced scrutiny of models and expanded diagnostic analysis of model behaviour have been increasingly facilitated by internationally coordinated efforts to collect and disseminate output from model experiments performed under common conditions. This has encouraged a more comprehensive and open evaluation of models. The expanded evaluation effort, encompassing a diversity of perspectives, makes it less likely that significant model errors are being overlooked.

- Climate models are being subjected to more comprehensive tests, including, for example, evaluations of forecasts on time scales from days to a year. This more diverse set of tests increases confidence in the fidelity with which models represent processes that affect climate projections.

- Substantial progress has been made in understanding the inter-model differences in equilibrium climate sensitivity. Cloud feedbacks have been confirmed as a primary source of these differences, with low clouds making the largest contribution. New observational and modelling evidence strongly supports a combined water vapour-lapse rate feedback of a strength comparable to that found in General Circulation Models (approximately 1 W m^{-2} °C^{-1}, corresponding to around a 50% amplification of global mean warming). The magnitude of cryospheric feedbacks remains uncertain, contributing to the range of model climate responses at mid- to high latitudes.

- There have been ongoing improvements to resolution, computational methods and parametrizations, and additional processes (e.g., interactive aerosols) have been included in more of the climate models.

- Most AOGCMs no longer use flux adjustments, which were previously required to maintain a stable climate.

At the same time, there have been improvements in the simulation of many aspects of present climate. The uncertainty associated with the use of flux adjustments has therefore decreased, although biases and long-term trends remain in AOGCM control simulations.

- Progress in the simulation of important modes of climate variability has increased the overall confidence in the models' representation of important climate processes. As a result of steady progress, some AOGCMs can now simulate important aspects of the El Niño-Southern Oscillation (ENSO). Simulation of the Madden-Julian Oscillation (MJO) remains unsatisfactory.

- The ability of AOGCMs to simulate extreme events, especially hot and cold spells, has improved. The frequency and amount of precipitation falling in intense events are underestimated.

- Simulation of extratropical cyclones has improved. Some models used for projections of tropical cyclone changes can simulate successfully the observed frequency and distribution of tropical cyclones.

- Systematic biases have been found in most models' simulation of the Southern Ocean. Since the Southern Ocean is important for ocean heat uptake, this results in some uncertainty in transient climate response.

- The possibility that metrics based on observations might be used to constrain model projections of climate change has been explored for the first time, through the analysis of ensembles of model simulations. Nevertheless, a proven set of model metrics that might be used to narrow the range of plausible climate projections has yet to be developed.

- To explore the potential importance of carbon cycle feedbacks in the climate system, explicit treatment of the carbon cycle has been introduced in a few climate AOGCMs and some Earth System Models of Intermediate Complexity (EMICs).

- Earth System Models of Intermediate Complexity have been evaluated in greater depth than previously. Coordinated intercomparisons have demonstrated that these models are useful in addressing questions involving long time scales or requiring a large number of ensemble simulations or sensitivity experiments.

Developments in model formulation

Improvements in atmospheric models include reformulated dynamics and transport schemes, and increased horizontal and vertical resolution. Interactive aerosol modules have been incorporated into some models, and through these, the direct and the indirect effects of aerosols are now more widely included.

Significant developments have occurred in the representation of terrestrial processes. Individual components continue to be improved via systematic evaluation against observations and against more comprehensive models. The terrestrial processes that might significantly affect large-scale climate over the next few decades are included in current climate models. Some processes important on longer time scales are not yet included.

Development of the oceanic component of AOGCMs has continued. Resolution has increased and models have generally abandoned the 'rigid lid' treatment of the ocean surface. New physical parametrizations and numerics include true freshwater fluxes, improved river and estuary mixing schemes and the use of positive definite advection schemes. Adiabatic isopycnal mixing schemes are now widely used. Some of these improvements have led to a reduction in the uncertainty associated with the use of less sophisticated parametrizations (e.g., virtual salt flux).

Progress in developing AOGCM cryospheric components is clearest for sea ice. Almost all state-of-the-art AOGCMs now include more elaborate sea ice dynamics and some now include several sea ice thickness categories and relatively advanced thermodynamics. Parametrizations of terrestrial snow processes in AOGCMs vary considerably in formulation. Systematic evaluation of snow suggests that sub-grid scale heterogeneity is important for simulating observations of seasonal snow cover. Few AOGCMs include ice sheet dynamics; in all of the AOGCMs evaluated in this chapter and used in Chapter 10 for projecting climate change in the 21st century, the land ice cover is prescribed.

There is currently no consensus on the optimal way to divide computer resources among: finer numerical grids, which allow for better simulations; greater numbers of ensemble members, which allow for better statistical estimates of uncertainty; and inclusion of a more complete set of processes (e.g., carbon feedbacks, atmospheric chemistry interactions).

Developments in model climate simulation

The large-scale patterns of seasonal variation in several important atmospheric fields are now better simulated by AOGCMs than they were at the time of the TAR. Notably, errors in simulating the monthly mean, global distribution of precipitation, sea level pressure and surface air temperature have all decreased. In some models, simulation of marine low-level clouds, which are important for correctly simulating sea surface temperature and cloud feedback in a changing climate, has also improved. Nevertheless, important deficiencies remain in the simulation of clouds and tropical precipitation (with their important regional and global impacts).

Some common model biases in the Southern Ocean have been identified, resulting in some uncertainty in oceanic heat uptake and transient climate response. Simulations of the thermocline, which was too thick, and the Atlantic overturning and heat transport, which were both too weak, have been substantially improved in many models.

Despite notable progress in improving sea ice formulations, AOGCMs have typically achieved only modest progress in simulations of observed sea ice since the TAR. The relatively slow progress can partially be explained by the fact that improving sea ice simulation requires improvements in both the atmosphere and ocean components in addition to the sea ice component itself.

Since the TAR, developments in AOGCM formulation have improved the representation of large-scale variability over a wide range of time scales. The models capture the dominant extratropical patterns of variability including the Northern and Southern Annular Modes, the Pacific Decadal Oscillation, the Pacific-North American and Cold Ocean-Warm Land Patterns. AOGCMs simulate Atlantic multi-decadal variability, although the relative roles of high- and low-latitude processes appear to differ between models. In the tropics, there has been an overall improvement in the AOGCM simulation of the spatial pattern and frequency of ENSO, but problems remain in simulating its seasonal phase locking and the asymmetry between El Niño and La Niña episodes. Variability with some characteristics of the MJO is simulated by most AOGCMs, but the events are typically too infrequent and too weak.

Atmosphere-Ocean General Circulation Models are able to simulate extreme warm temperatures, cold air outbreaks and frost days reasonably well. Models used in this report for projecting tropical cyclone changes are able to simulate present-day frequency and distribution of cyclones, but intensity is less well simulated. Simulation of extreme precipitation is dependent on resolution, parametrization and the thresholds chosen. In general, models tend to produce too many days with weak precipitation (<10 mm day^{-1}) and too little precipitation overall in intense events (>10 mm day^{-1}).

Earth system Models of Intermediate Complexity have been developed to investigate issues in past and future climate change that cannot be addressed by comprehensive AOGCMs because of their large computational cost. Owing to the reduced resolution of EMICs and their simplified representation of some physical processes, these models only allow inferences about very large scales. Since the TAR, EMICs have been evaluated via several coordinated model intercomparisons which have revealed that, at large scales, EMIC results compare well with observational data and AOGCM results. This lends support to the view that EMICS can be used to gain understanding of processes and interactions within the climate system that evolve on time scales beyond those generally accessible to current AOGCMs. The uncertainties in long-term climate change projections can also be explored more comprehensively by using large ensembles of EMIC runs.

Developments in analysis methods

Since the TAR, an unprecedented effort has been initiated to make available new model results for scrutiny by scientists outside the modelling centres. Eighteen modelling groups performed a set of coordinated, standard experiments, and the resulting model output, analysed by hundreds of researchers worldwide, forms the basis for much of the current IPCC assessment of model results. The benefits of coordinated model intercomparison include increased communication among modelling groups, more rapid identification and correction of errors, the creation of standardised benchmark calculations and a more complete and systematic record of modelling progress.

A few climate models have been tested for (and shown) capability in initial value predictions, on time scales from weather forecasting (a few days) to seasonal forecasting (annual). The capability demonstrated by models under these conditions increases confidence that they simulate some of the key processes and teleconnections in the climate system.

Developments in evaluation of climate feedbacks

Water vapour feedback is the most important feedback enhancing climate sensitivity. Although the strength of this feedback varies somewhat among models, its overall impact on the spread of model climate sensitivities is reduced by lapse rate feedback, which tends to be anti-correlated. Several new studies indicate that modelled lower- and upper-tropospheric humidity respond to seasonal and interannual variability, volcanically induced cooling and climate trends in a way that is consistent with observations. Recent observational and modelling evidence thus provides strong additional support for the combined water vapour-lapse rate feedback being around the strength found in AOGCMs.

Recent studies reaffirm that the spread of climate sensitivity estimates among models arises primarily from inter-model differences in cloud feedbacks. The shortwave impact of changes in boundary-layer clouds, and to a lesser extent mid-level clouds, constitutes the largest contributor to inter-model differences in global cloud feedbacks. The relatively poor simulation of these clouds in the present climate is a reason for some concern. The response to global warming of deep convective clouds is also a substantial source of uncertainty in projections since current models predict different responses of these clouds. Observationally based evaluation of cloud feedbacks indicates that climate models exhibit different strengths and weaknesses, and it is not yet possible to determine which estimates of the climate change cloud feedbacks are the most reliable.

Despite advances since the TAR, substantial uncertainty remains in the magnitude of cryospheric feedbacks within AOGCMs. This contributes to a spread of modelled climate response, particularly at high latitudes. At the global scale, the surface albedo feedback is positive in all the models, and varies between models much less than cloud feedbacks. Understanding and evaluating sea ice feedbacks is complicated by the strong coupling to polar cloud processes and ocean heat and freshwater transport. Scarcity of observations in polar regions also hampers evaluation. New techniques that evaluate surface albedo feedbacks have recently been developed. Model performance in reproducing the observed seasonal cycle of land snow cover may provide an indirect evaluation of the simulated snow-albedo feedback under climate change.

Systematic model comparisons have helped establish the key processes responsible for differences among models in the response of the ocean to climate change. The importance of feedbacks from surface flux changes to the meridional overturning circulation has been established in many models. At present, these feedbacks are not tightly constrained by available observations.

The analysis of processes contributing to climate feedbacks in models and recent studies based on large ensembles of models suggest that in the future it may be possible to use observations to narrow the current spread in model projections of climate change.

8.1 Introduction and Overview

The goal of this chapter is to evaluate the capabilities and limitations of the global climate models used elsewhere in this assessment. A number of model evaluation activities are described in various chapters of this report. This section provides a context for those studies and a guide to direct the reader to the appropriate chapters.

8.1.1 What is Meant by Evaluation?

A specific prediction based on a model can often be demonstrated to be right or wrong, but the model itself should always be viewed critically. This is true for both weather prediction and climate prediction. Weather forecasts are produced on a regular basis, and can be quickly tested against what actually happened. Over time, statistics can be accumulated that give information on the performance of a particular model or forecast system. In climate change simulations, on the other hand, models are used to make projections of possible future changes over time scales of many decades and for which there are no precise past analogues. Confidence in a model can be gained through simulations of the historical record, or of palaeoclimate, but such opportunities are much more limited than are those available through weather prediction. These and other approaches are discussed below.

8.1.2 Methods of Evaluation

A climate model is a very complex system, with many components. The model must of course be tested at the system level, that is, by running the full model and comparing the results with observations. Such tests can reveal problems, but their source is often hidden by the model's complexity. For this reason, it is also important to test the model at the component level, that is, by isolating particular components and testing them independent of the complete model.

Component-level evaluation of climate models is common. Numerical methods are tested in standardised tests, organised through activities such as the quasi-biennial Workshops on Partial Differential Equations on the Sphere. Physical parametrizations used in climate models are being tested through numerous case studies (some based on observations and some idealised), organised through programs such as the Atmospheric Radiation Measurement (ARM) program, EUROpean Cloud Systems (EUROCS) and the Global Energy and Water cycle Experiment (GEWEX) Cloud System Study (GCSS). These activities have been ongoing for a decade or more, and a large body of results has been published (e.g., Randall et al., 2003).

System-level evaluation is focused on the outputs of the full model (i.e., model simulations of particular observed climate variables) and particular methods are discussed in more detail below.

8.1.2.1 Model Intercomparisons and Ensembles

The global model intercomparison activities that began in the late 1980s (e.g., Cess et al., 1989), and continued with the Atmospheric Model Intercomparison Project (AMIP), have now proliferated to include several dozen model intercomparison projects covering virtually all climate model components and various coupled model configurations (see http://www.clivar.org/science/mips.php for a summary). By far the most ambitious organised effort to collect and analyse Atmosphere-Ocean General Circulation Model (AOGCM) output from standardised experiments was undertaken in the last few years (see http://www-pcmdi.llnl.gov/ipcc/about_ipcc.php). It differed from previous model intercomparisons in that a more complete set of experiments was performed, including unforced control simulations, simulations attempting to reproduce observed climate change over the instrumental period and simulations of future climate change. It also differed in that, for each experiment, multiple simulations were performed by some individual models to make it easier to separate climate change signals from internal variability within the climate system. Perhaps the most important change from earlier efforts was the collection of a more comprehensive set of model output, hosted centrally at the Program for Climate Model Diagnosis and Intercomparison (PCMDI). This archive, referred to here as 'The Multi-Model Data set (MMD) at PCMDI', has allowed hundreds of researchers from outside the modelling groups to scrutinise the models from a variety of perspectives.

The enhancement in diagnostic analysis of climate model results represents an important step forward since the Third Assessment Report (TAR). Overall, the vigorous, ongoing intercomparison activities have increased communication among modelling groups, allowed rapid identification and correction of modelling errors and encouraged the creation of standardised benchmark calculations, as well as a more complete and systematic record of modelling progress.

Ensembles of models represent a new resource for studying the range of plausible climate responses to a given forcing. Such ensembles can be generated either by collecting results from a range of models from different modelling centres ('multi-model ensembles' as described above), or by generating multiple model versions within a particular model structure, by varying internal model parameters within plausible ranges ('perturbed physics ensembles'). The approaches are discussed in more detail in Section 10.5.

8.1.2.2 Metrics of Model Reliability

What does the accuracy of a climate model's simulation of past or contemporary climate say about the accuracy of its projections of climate change? This question is just beginning to be addressed, exploiting the newly available ensembles of models. A number of different observationally based metrics have been used to weight the reliability of contributing models when making probabilistic projections (see Section 10.5.4).

For any given metric, it is important to assess how good a test it is of model results for making projections of future climate change. This cannot be tested directly, since there are no observed periods with forcing changes exactly analogous to those expected over the 21st century. However, relationships between observable metrics and the predicted quantity of interest (e.g., climate sensitivity) can be explored across model ensembles. Shukla et al. (2006) correlated a measure of the fidelity of the simulated surface temperature in the 20th century with simulated 21st-century temperature change in a multi-model ensemble. They found that the models with the smallest 20th-century error produced relatively large surface temperature increases in the 21st century. Knutti et al. (2006), using a different, perturbed physics ensemble, showed that models with a strong seasonal cycle in surface temperature tended to have larger climate sensitivity. More complex metrics have also been developed based on multiple observables in present day climate, and have been shown to have the potential to narrow the uncertainty in climate sensitivity across a given model ensemble (Murphy et al., 2004; Piani et al., 2005). The above studies show promise that quantitative metrics for the likelihood of model projections may be developed, but because the development of robust metrics is still at an early stage, the model evaluations presented in this chapter are based primarily on experience and physical reasoning, as has been the norm in the past.

An important area of progress since the TAR has been in establishing and quantifying the feedback processes that determine climate change response. Knowledge of these processes underpins both the traditional and the metric-based approaches to model evaluation. For example, Hall and Qu (2006) developed a metric for the feedback between temperature and albedo in snow-covered regions, based on the simulation of the seasonal cycle. They found that models with a strong feedback based on the seasonal cycle also had a strong feedback under increased greenhouse gas forcing. Comparison with observed estimates of the seasonal cycle suggested that most models in the MMD underestimate the strength of this feedback. Section 8.6 discusses the various feedbacks that operate in the atmosphere-land surface-sea ice system to determine climate sensitivity, and Section 8.3.2 discusses some processes that are important for ocean heat uptake (and hence transient climate response).

8.1.2.3 Testing Models Against Past and Present Climate

Testing models' ability to simulate 'present climate' (including variability and extremes) is an important part of model evaluation (see Sections 8.3 to 8.5, and Chapter 11 for specific regional evaluations). In doing this, certain practical choices are needed, for example, between a long time series or mean from a 'control' run with fixed radiative forcing (often pre-industrial rather than present day), or a shorter, transient time series from a '20th-century' simulation including historical variations in forcing. Such decisions are made by individual researchers, dependent on the particular problem being studied.

Differences between model and observations should be considered insignificant if they are within:

1. unpredictable internal variability (e.g., the observational period contained an unusual number of El Niño events);
2. expected differences in forcing (e.g., observations for the 1990s compared with a 'pre-industrial' model control run); or
3. uncertainties in the observed fields.

While space does not allow a discussion of the above issues in detail for each climate variable, they are taken into account in the overall evaluation. Model simulation of present-day climate at a global to sub-continental scale is discussed in this chapter, while more regional detail can be found in Chapter 11.

Models have been extensively used to simulate observed climate change during the 20th century. Since forcing changes are not perfectly known over that period (see Chapter 2), such tests do not fully constrain future response to forcing changes. Knutti et al. (2002) showed that in a perturbed physics ensemble of Earth System Models of Intermediate Complexity (EMICs), simulations from models with a range of climate sensitivities are consistent with the observed surface air temperature and ocean heat content records, if aerosol forcing is allowed to vary within its range of uncertainty. Despite this fundamental limitation, testing of 20th-century simulations against historical observations does place some constraints on future climate response (e.g., Knutti et al., 2002). These topics are discussed in detail in Chapter 9.

8.1.2.4 Other Methods of Evaluation

Simulations of climate states from the more distant past allow models to be evaluated in regimes that are significantly different from the present. Such tests complement the 'present climate' and 'instrumental period climate' evaluations, since 20th-century climate variations have been small compared with the anticipated future changes under forcing scenarios derived from the IPCC Special Report on Emission Scenarios (SRES). The limitations of palaeoclimate tests are that uncertainties in both forcing and actual climate variables (usually derived from proxies) tend to be greater than in the instrumental period, and that the number of climate variables for which there are good palaeo-proxies is limited. Further, climate states may have been so different (e.g., ice sheets at last glacial maximum) that processes determining quantities such as climate sensitivity were different from those likely to operate in the 21st century. Finally, the time scales of change were so long that there are difficulties in experimental design, at least for General Circulation Models (GCMs). These issues are discussed in depth in Chapter 6.

Climate models can be tested through forecasts based on initial conditions. Climate models are closely related to the models that are used routinely for numerical weather prediction, and increasingly for extended range forecasting on seasonal to interannual time scales. Typically, however, models used for numerical weather prediction are run at higher resolution than is possible for climate simulations. Evaluation of such forecasts tests the models' representation of some key processes in the atmosphere and ocean, although the links between these processes and long-term climate response have not always been established. It must be remembered that the quality of an initial value prediction is dependent on several factors beyond the numerical model itself (e.g., data assimilation techniques, ensemble generation method), and these factors may be less relevant to projecting the long-term, forced response of the climate system to changes in radiative forcing. There is a large body of literature on this topic, but to maintain focus on the goal of this chapter, discussions here are confined to the relatively few studies that have been conducted using models that are very closely related to the climate models used for projections (see Section 8.4.11).

8.1.3 How Are Models Constructed?

The fundamental basis on which climate models are constructed has not changed since the TAR, although there have been many specific developments (see Section 8.2). Climate models are derived from fundamental physical laws (such as Newton's laws of motion), which are then subjected to physical approximations appropriate for the large-scale climate system, and then further approximated through mathematical discretization. Computational constraints restrict the resolution that is possible in the discretized equations, and some representation of the large-scale impacts of unresolved processes is required (the parametrization problem).

8.1.3.1   *Parameter Choices and 'Tuning'*

Parametrizations are typically based in part on simplified physical models of the unresolved processes (e.g., entraining plume models in some convection schemes). The parametrizations also involve numerical parameters that must be specified as input. Some of these parameters can be measured, at least in principle, while others cannot. It is therefore common to adjust parameter values (possibly chosen from some prior distribution) in order to optimise model simulation of particular variables or to improve global heat balance. This process is often known as 'tuning'. It is justifiable to the extent that two conditions are met:

1. Observationally based constraints on parameter ranges are not exceeded. Note that in some cases this may not provide a tight constraint on parameter values (e.g., Heymsfield and Donner, 1990).

2. The number of degrees of freedom in the tuneable parameters is less than the number of degrees of freedom in the observational constraints used in model evaluation. This is believed to be true for most GCMs – for example, climate models are not explicitly tuned to give a good representation of North Atlantic Oscillation (NAO) variability – but no

studies are available that formally address the question. If the model has been tuned to give a good representation of a particular observed quantity, then agreement with that observation cannot be used to build confidence in that model. However, a model that has been tuned to give a good representation of certain key observations may have a greater likelihood of giving a good prediction than a similar model (perhaps another member of a 'perturbed physics' ensemble) that is less closely tuned (as discussed in Section 8.1.2.2 and Chapter 10).

Given sufficient computer time, the tuning procedure can in principle be automated using various data assimilation procedures. To date, however, this has only been feasible for EMICs (Hargreaves et al., 2004) and low-resolution GCMs (Annan et al., 2005b; Jones et al., 2005; Severijns and Hazeleger, 2005). Ensemble methods (Murphy et al., 2004; Annan et al., 2005a; Stainforth et al., 2005) do not always produce a unique 'best' parameter setting for a given error measure.

8.1.3.2   *Model Spectra or Hierarchies*

The value of using a range of models (a 'spectrum' or 'hierarchy') of differing complexity is discussed in the TAR (Section 8.3), and here in Section 8.8. Computationally cheaper models such as EMICs allow a more thorough exploration of parameter space, and are simpler to analyse to gain understanding of particular model responses. Models of reduced complexity have been used more extensively in this report than in the TAR, and their evaluation is discussed in Section 8.8. Regional climate models can also be viewed as forming part of a climate modelling hierarchy.

8.2   Advances in Modelling

Many modelling advances have occurred since the TAR. Space does not permit a comprehensive discussion of all major changes made over the past several years to the 23 AOGCMs used widely in this report (see Table 8.1). Model improvements can, however, be grouped into three categories. First, the dynamical cores (advection, etc.) have been improved, and the horizontal and vertical resolutions of many models have been increased. Second, more processes have been incorporated into the models, in particular in the modelling of aerosols, and of land surface and sea ice processes. Third, the parametrizations of physical processes have been improved. For example, as discussed further in Section 8.2.7, most of the models no longer use flux adjustments (Manabe and Stouffer, 1988; Sausen et al., 1988) to reduce climate drift. These various improvements, developed across the broader modelling community, are well represented in the climate models used in this report.

Despite the many improvements, numerous issues remain. Many of the important processes that determine a model's response to changes in radiative forcing are not resolved by

Table 8.1. *Selected model features. Salient features of the AOGCMs participating in the MMD at PCMDI are listed by IPCC identification (ID) along with the calendar year ('vintage') of the first publication of results from each model. Also listed are the respective sponsoring institutions, the pressure at the top of the atmospheric model, the horizontal and vertical resolution of the model atmosphere and ocean models, as well as the oceanic vertical coordinate type (Z: see Griffies (2004) for definitions) and upper boundary condition (BC: free surface or rigid lid). Also listed are the characteristics of sea ice dynamics/structure (e.g., rheology vs 'free drift' assumption and inclusion of ice leads), and whether adjustments of surface momentum, heat or freshwater fluxes are applied in coupling the atmosphere, ocean and sea ice components. Land features such as the representation of soil moisture (single-layer 'bucket' vs multi-layered scheme) and the presence of a vegetation canopy or a river routing scheme also are noted. Relevant references describing details of these aspects of the models are cited.*

Model ID, Vintage	Sponsor(s), Country	Atmosphere Top Resolution[a] References	Ocean[b] Resolution, Z Coord, Top BC References	Sea Ice Dynamics, Leads References	Coupling Flux Adjustments References	Land Soil, Plants, Routing References
1: BCC-CM1, 2005	Beijing Climate Center, China	top = 25 hPa T63 (1.9° x 1.9°) L16 Dong et al., 2000; CSMD, 2005; Xu et al., 2005	1.9° x 1.9° L30 depth, free surface Jin et al., 1999	no rheology or leads Xu et al., 2005	heat, momentum Yu and Zhang, 2000; CSMD, 2005	layers, canopy, routing CSMD, 2005
2: BCCR-BCM2.0, 2005	Bjerknes Centre for Climate Research, Norway	top = 10 hPa T63 (1.9° x 1.9°) L31 Déqué et al., 1994	0.5°–1.5° x 1.5° L35 density, free surface Bleck et al., 1992	rheology, leads Hibler, 1979; Harder, 1996	no adjustments Furevik et al., 2003	Layers, canopy, routing Mahfouf et al., 1995; Douville et al., 1995; Oki and Sud, 1998
3: CCSM3, 2005	National Center for Atmospheric Research, USA	top = 2.2 hPa T85 (1.4° x 1.4°) L26 Collins et al., 2004	0.3°–1° x 1° L40 depth, free surface Smith and Gent, 2002	rheology, leads Briegleb et al., 2004	no adjustments Collins et al., 2006	layers, canopy, routing Oleson et al., 2004; Branstetter, 2001
4: CGCM3.1(T47), 2005	Canadian Centre for Climate Modelling and Analysis, Canada	top = 1 hPa T47 (~2.8° x 2.8°) L31 McFarlane et al., 1992; Flato, 2005	1.9° x 1.9° L29 depth, rigid lid Pacanowski et al., 1993	rheology, leads Hibler, 1979; Flato and Hibler, 1992	heat, freshwater Flato, 2005	layers, canopy, routing Verseghy et al., 1993
5: CGCM3.1(T63), 2005		top = 1 hPa T63 (1.0° x 1.0°) L31 McFarlane et al., 1992; Flato 2005	0.9° x 1.4° L29 depth, rigid lid Flato and Boer, 2001; Kim et al., 2002	rheology, leads Hibler, 1979; Flato and Hibler, 1992	heat, freshwater Flato, 2005	layers, canopy, routing Verseghy et al., 1993
6: CNRM-CM3, 2004	Météo-France/Centre National de Recherches Météorologiques, France	top = 0.05 hPa T63 (~1.9° x 1.9°) L45 Déqué et al., 1994	0.5°–2° x 2° L31 depth, rigid lid Madec et al., 1998	rheology, leads Hunke-Dukowicz, 1997; Salas-Mélia, 2002	no adjustments Terray et al., 1998	layers, canopy, routing Mahfouf et al., 1995; Douville et al., 1995; Oki and Sud, 1998
7: CSIRO-MK3.0, 2001	Commonwealth Scientific and Industrial Research Organisation (CSIRO) Atmospheric Research, Australia	top = 4.5 hPa T63 (~1.9° x 1.9°) L18 Gordon et al., 2002	0.8° x 1.9° L31 depth, rigid lid Gordon et al., 2002	rheology, leads O'Farrell, 1998	no adjustments Gordon et al., 2002	layers, canopy Gordon et al., 2002
8: ECHAM5/MPI-OM, 2005	Max Planck Institute for Meteorology, Germany	top = 10 hPa T63 (~1.9° x 1.9°) L31 Roeckner et al., 2003	1.5° x 1.5° L40 depth, free surface Marsland et al., 2003	rheology, leads Hibler, 1979; Semtner, 1976	no adjustments Jungclaus et al., 2005	bucket, canopy, routing Hagemann, 2002; Hagemann and Dümenil-Gates, 2001
9: ECHO-G, 1999	Meteorological Institute of the University of Bonn, Meteorological Research Institute of the Korea Meteorological Administration (KMA), and Model and Data Group, Germany/Korea	top = 10 hPa T30 (~3.9° x 3.9°) L19 Roeckner et al., 1996	0.5°–2.8° x 2.8° L20 depth, free surface Wolff et al., 1997	rheology, leads Wolff et al., 1997	heat, freshwater Min et al., 2005	bucket, canopy, routing Roeckner et al., 1996; Dümenil and Todini, 1992

Table 8.1 (continued)

Model ID, Vintage	Sponsor(s), Country	Atmosphere Top Resolution[a] References	Ocean Resolution[b] Z Coord, Top BC References	Sea Ice Dynamics, Leads References	Coupling Flux Adjustments References	Land Soil, Plants, Routing References
10: FGOALS-g1.0, 2004	National Key Laboratory of Numerical Modeling for Atmospheric Sciences and Geophysical Fluid Dynamics (LASG)/Institute of Atmospheric Physics, China	top = 2.2 hPa T42 (~2.8° x 2.8°) L26 Wang et al., 2004	1.0° x 1.0° L16 eta, free surface Jin et al., 1999; Liu et al., 2004	rheology, leads Briegleb et al., 2004	no adjustments Yu et al., 2002, 2004	layers, canopy, routing Bonan et al., 2002
11: GFDL-CM2.0, 2005	U.S. Department of Commerce/ National Oceanic and Atmospheric Administration (NOAA)/Geophysical Fluid Dynamics Laboratory (GFDL), USA	top = 3 hPa 2.0° x 2.5° L24 GFDL GAMDT, 2004	0.3°–1.0° x 1.0° depth, free surface Gnanadesikan et al., 2004	rheology, leads Winton, 2000; Delworth et al., 2006	no adjustments Delworth et al., 2006	bucket, canopy, routing Milly and Shmakin, 2002; GFDL GAMDT, 2004
12: GFDL-CM2.1, 2005	Geophysical Fluid Dynamics Laboratory (GFDL), USA	top = 3 hPa 2.0° x 2.5° L24 GFDL GAMDT, 2004 with semi-Lagrangian transports	0.3°–1.0° x 1.0° depth, free surface Gnanadesikan et al., 2004	rheology, leads Winton, 2000; Delworth et al., 2006	no adjustments Delworth et al., 2006	bucket, canopy, routing Milly and Shmakin, 2002; GFDL GAMDT, 2004
13: GISS-AOM, 2004	National Aeronautics and Space Administration (NASA)/ Goddard Institute for Space Studies (GISS), USA	top = 10 hPa 3° x 4° L12 Russell et al., 1995; Russell, 2005	3° x 4° L16 mass/area, free surface Russell et al., 1995; Russell, 2005	rheology, leads Flato and Hibler, 1992; Russell, 2005	no adjustments Russell, 2005	layers, canopy, routing Abramopoulos et al., 1988; Miller et al., 1994
14: GISS-EH, 2004	NASA/GISS, USA	top = 0.1 hPa 4° x 5° L20 Schmidt et al., 2006	2° x 2° L16 density, free surface Bleck, 2002	rheology, leads Liu et al., 2003; Schmidt et al., 2004	no adjustments Schmidt et al., 2006	layers, canopy, routing Friend and Kiang, 2005
15: GISS-ER, 2004	NASA/GISS, USA	top = 0.1 hPa 4° x 5° L20 Schmidt et al., 2006	4° x 5° L13 mass/area, free surface Russell et al., 1995	rheology, leads Liu et al., 2003; Schmidt et al., 2004	no adjustments Schmidt et al., 2006	layers, canopy, routing Friend and Kiang, 2005
16: INM-CM3.0, 2004	Institute for Numerical Mathematics, Russia	top = 10 hPa 4° x 5° L21 Alekseev et al., 1998; Galin et al., 2003	2° x 2.5° L33 sigma, rigid lid Diansky et al., 2002	no rheology or leads Diansky et al., 2002	regional freshwater Diansky and Volodin, 2002; Volodin and Diansky, 2004	layers, canopy, no routing Alekseev et al., 1998; Volodin and Lykosoff, 1998
17: IPSL-CM4, 2005	Institut Pierre Simon Laplace, France	top = 4 hPa 2.5° x 3.75° L19 Hourdin et al., 2006	2° x 2° L31 depth, free surface Madec et al., 1998	rheology, leads Fichefet and Morales Maqueda, 1997; Goosse and Fichefet, 1999	no adjustments Marti et al., 2005	layers, canopy, routing Krinner et al., 2005
18: MIROC3.2(hires), 2004	Center for Climate System Research (University of Tokyo), National Institute for Environmental Studies, and Frontier Research Center for Global Change (JAMSTEC), Japan	top = 40 km T106 (~1.1° x 1.1°) L56 K-1 Developers, 2004	0.2° x 0.3° L47 sigma/depth, free surface K-1 Developers, 2004	rheology, leads K-1 Developers, 2004	no adjustments K-1 Developers, 2004	layers, canopy, routing K-1 Developers, 2004; Oki and Sud, 1998
19: MIROC3.2(medres), 2004	Frontier Research Center for Global Change (JAMSTEC), Japan	top = 30 km T42 (-2.8° x 2.8°) L20 K-1 Developers, 2004	0.5°–1.4° x 1.4° L43 sigma/depth, free surface K-1 Developers, 2004	rheology, leads K-1 Developers, 2004	no adjustments K-1 Developers, 2004	layers, canopy, routing K-1 Developers, 2004; Oki and Sud, 1998

Table 8.1 (continued)

Model ID, Vintage	Sponsor(s), Country	Atmosphere Top Resolution[a] References	Ocean Resolution[b] Z Coord, Top BC References	Sea Ice Dynamics, Leads References	Coupling Flux Adjustments References	Land Soil, Plants, Routing References
20: MRI-CGCM2.3.2, 2003	Meteorological Research Institute, Japan	top = 0.4 hPa T42 (~2.8° x 2.8°) L30 Shibata et al., 1999	0.5°–2.0° x 2.5° L23 depth, rigid lid Yukimoto et al., 2001	free drift, leads Mellor and Kantha, 1989	heat, freshwater, momentum (12°S–12°N) Yukimoto et al., 2001; Yukimoto and Noda, 2003	layers, canopy, routing Sellers et al., 1986; Sato et al., 1989
21: PCM, 1998	National Center for Atmospheric Research, USA	top = 2.2 hPa T42 (~2.8° x 2.8°) L26 Kiehl et al., 1998	0.5°–0.7° x 1.1° L40 depth, free surface Maltrud et al., 1998	rheology, leads Hunke and Dukowicz 1997, 2003; Zhang et al., 1999	no adjustments Washington et al., 2000	layers, canopy, no routing Bonan, 1998
22: UKMO-HadCM3, 1997	Hadley Centre for Climate Prediction and Research/Met Office, UK	top = 5 hPa 2.5° x 3.75° L19 Pope et al., 2000	1.25° x 1.25° L20 depth, rigid lid Gordon et al., 2000	free drift, leads Cattle and Crossley, 1995	no adjustments Gordon et al., 2000	layers, canopy, routing Cox et al., 1999
23: UKMO-HadGEM1, 2004	Office, UK	top = 39.2 km ~1.3° x 1.9° L38 Martin et al., 2004	0.3°–1.0° x 1.0° L40 depth, free surface Roberts, 2004	rheology, leads Hunke and Dukowicz, 1997; Semtner, 1976; Lipscomb, 2001	no adjustments Johns et al., 2006	layers, canopy, routing Essery et al., 2001; Oki and Sud, 1998

Notes:

[a] Horizontal resolution is expressed either as degrees latitude by longitude or as a triangular (T) spectral truncation with a rough translation to degrees latitude and longitude. Vertical resolution (L) is the number of vertical levels.

[b] Horizontal resolution is expressed as degrees latitude by longitude, while vertical resolution (L) is the number of vertical levels.

Frequently Asked Question 8.1
How Reliable Are the Models Used to Make Projections of Future Climate Change?

There is considerable confidence that climate models provide credible quantitative estimates of future climate change, particularly at continental scales and above. This confidence comes from the foundation of the models in accepted physical principles and from their ability to reproduce observed features of current climate and past climate changes. Confidence in model estimates is higher for some climate variables (e.g., temperature) than for others (e.g., precipitation). Over several decades of development, models have consistently provided a robust and unambiguous picture of significant climate warming in response to increasing greenhouse gases.

Climate models are mathematical representations of the climate system, expressed as computer codes and run on powerful computers. One source of confidence in models comes from the fact that model fundamentals are based on established physical laws, such as conservation of mass, energy and momentum, along with a wealth of observations.

A second source of confidence comes from the ability of models to simulate important aspects of the current climate. Models are routinely and extensively assessed by comparing their simulations with observations of the atmosphere, ocean, cryosphere and land surface. Unprecedented levels of evaluation have taken place over the last decade in the form of organised multi-model 'intercomparisons'. Models show significant and

increasing skill in representing many important mean climate features, such as the large-scale distributions of atmospheric temperature, precipitation, radiation and wind, and of oceanic temperatures, currents and sea ice cover. Models can also simulate essential aspects of many of the patterns of climate variability observed across a range of time scales. Examples include the advance and retreat of the major monsoon systems, the seasonal shifts of temperatures, storm tracks and rain belts, and the hemispheric-scale seesawing of extratropical surface pressures (the Northern and Southern 'annular modes'). Some climate models, or closely related variants, have also been tested by using them to predict weather and make seasonal forecasts. These models demonstrate skill in such forecasts, showing they can represent important features of the general circulation across shorter time scales, as well as aspects of seasonal and interannual variability. Models' ability to represent these and other important climate features increases our confidence that they represent the essential physical processes important for the simulation of future climate change. (Note that the limitations in climate models' ability to forecast weather beyond a few days do not limit their ability to predict long-term climate changes, as these are very different types of prediction – see FAQ 1.2.)

(continued)

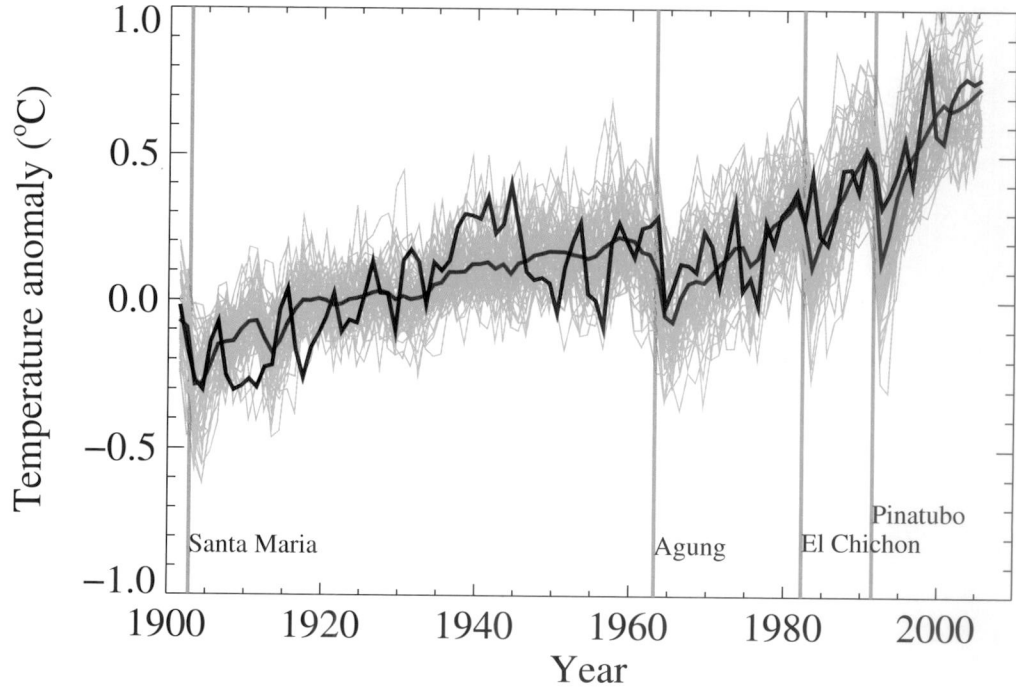

FAQ 8.1, Figure 1. *Global mean near-surface temperatures over the 20th century from observations (black) and as obtained from 58 simulations produced by 14 different climate models driven by both natural and human-caused factors that influence climate (yellow). The mean of all these runs is also shown (thick red line). Temperature anomalies are shown relative to the 1901 to 1950 mean. Vertical grey lines indicate the timing of major volcanic eruptions. (Figure adapted from Chapter 9, Figure 9.5. Refer to corresponding caption for further details.)*

A third source of confidence comes from the ability of models to reproduce features of past climates and climate changes. Models have been used to simulate ancient climates, such as the warm mid-Holocene of 6,000 years ago or the last glacial maximum of 21,000 years ago (see Chapter 6). They can reproduce many features (allowing for uncertainties in reconstructing past climates) such as the magnitude and broad-scale pattern of oceanic cooling during the last ice age. Models can also simulate many observed aspects of climate change over the instrumental record. One example is that the global temperature trend over the past century (shown in Figure 1) can be modelled with high skill when both human and natural factors that influence climate are included. Models also reproduce other observed changes, such as the faster increase in nighttime than in daytime temperatures, the larger degree of warming in the Arctic and the small, short-term global cooling (and subsequent recovery) which has followed major volcanic eruptions, such as that of Mt. Pinatubo in 1991 (see FAQ 8.1, Figure 1). Model global temperature projections made over the last two decades have also been in overall agreement with subsequent observations over that period (Chapter 1).

Nevertheless, models still show significant errors. Although these are generally greater at smaller scales, important large-scale problems also remain. For example, deficiencies remain in the simulation of tropical precipitation, the El Niño-Southern Oscillation and the Madden-Julian Oscillation (an observed variation in tropical winds and rainfall with a time scale of 30 to 90 days). The ultimate source of most such errors is that many important small-scale processes cannot be represented explicitly in models, and so must be included in approximate form as they interact with larger-scale features. This is partly due to limitations in computing power, but also results from limitations in scientific understanding or in the availability of detailed observations of some physical processes. Significant uncertainties, in particular, are associated with the representation of clouds, and in the resulting cloud responses to climate change. Consequently, models continue to display a substantial range of global temperature change in response to specified greenhouse gas forcing (see Chapter 10). Despite such uncertainties, however, models are unanimous in their prediction of substantial climate warming under greenhouse gas increases, and this warming is of a magnitude consistent with independent estimates derived from other sources, such as from observed climate changes and past climate reconstructions.

Since confidence in the changes projected by global models decreases at smaller scales, other techniques, such as the use of regional climate models, or downscaling methods, have been specifically developed for the study of regional- and local-scale climate change (see FAQ 11.1). However, as global models continue to develop, and their resolution continues to improve, they are becoming increasingly useful for investigating important smaller-scale features, such as changes in extreme weather events, and further improvements in regional-scale representation are expected with increased computing power. Models are also becoming more comprehensive in their treatment of the climate system, thus explicitly representing more physical and biophysical processes and interactions considered potentially important for climate change, particularly at longer time scales. Examples are the recent inclusion of plant responses, ocean biological and chemical interactions, and ice sheet dynamics in some global climate models.

In summary, confidence in models comes from their physical basis, and their skill in representing observed climate and past climate changes. Models have proven to be extremely important tools for simulating and understanding climate, and there is considerable confidence that they are able to provide credible quantitative estimates of future climate change, particularly at larger scales. Models continue to have significant limitations, such as in their representation of clouds, which lead to uncertainties in the magnitude and timing, as well as regional details, of predicted climate change. Nevertheless, over several decades of model development, they have consistently provided a robust and unambiguous picture of significant climate warming in response to increasing greenhouse gases.

the model's grid. Instead, sub-grid scale parametrizations are used to parametrize the unresolved processes, such as cloud formation and the mixing due to oceanic eddies. It continues to be the case that multi-model ensemble simulations generally provide more robust information than runs of any single model. Table 8.1 summarises the formulations of each of the AOGCMs used in this report.

There is currently no consensus on the optimal way to divide computer resources among finer numerical grids, which allow for better simulations; greater numbers of ensemble members, which allow for better statistical estimates of uncertainty; and inclusion of a more complete set of processes (e.g., carbon feedbacks, atmospheric chemistry interactions).

8.2.1 Atmospheric Processes

8.2.1.1 Numerics

In the TAR, more than half of the participating atmospheric models used spectral advection. Since the TAR, semi-Lagrangian advection schemes have been adopted in several atmospheric models. These schemes allow long time steps and maintain positive values of advected tracers such as water vapour, but they are diffusive, and some versions do not formally conserve mass. In this report, various models use spectral, semi-Lagrangian, and Eulerian finite-volume and finite-difference advection schemes, although there is still no consensus on which type of scheme is best.

8.2.1.2 Horizontal and Vertical Resolution

The horizontal and vertical resolutions of AOGCMs have increased relative to the TAR. For example, HadGEM1 has eight times as many grid cells as HadCM3 (the number of cells has doubled in all three dimensions). At the National Center for Atmospheric Research (NCAR), a T85 version of the Climate System Model (CSM) is now routinely used, while a T42 version was standard at the time of the TAR. The Center for Climate System Research (CCSR), National Institute for Environmental Studies (NIES) and Frontier Research Center for Global Change (FRCGC) have developed a high-resolution climate model (MIROC-hi, which consists of a T106 L56 Atmospheric GCM (AGCM) and a 1/4° by 1/6° L48 Ocean GCM (OGCM)), and The Meteorological Research Institute (MRI) of the Japan Meteorological Agency (JMA) has developed a TL959 L60 spectral AGCM (Oouchi et al., 2006), which is being used in time-slice mode. The projections made with these models are presented in Chapter 10.

Due to the increased horizontal and vertical resolution, both regional- and global-scale climate features are better simulated. For example, a far-reaching effect of the Hawaiian Islands in the Pacific Ocean (Xie et al., 2001) has been well simulated (Sakamoto et al., 2004) and the frequency distribution of precipitation associated with the Baiu front has been improved (Kimoto et al., 2005).

8.2.1.3 Parametrizations

The climate system includes a variety of physical processes, such as cloud processes, radiative processes and boundary-layer processes, which interact with each other on many temporal and spatial scales. Due to the limited resolutions of the models, many of these processes are not resolved adequately by the model grid and must therefore be parametrized. The differences between parametrizations are an important reason why climate model results differ. For example, a new boundary-layer parametrization (Lock et al., 2000; Lock, 2001) had a strong positive impact on the simulations of marine stratocumulus cloud produced by the Geophysical Fluid Dynamics Laboratory (GFDL) and the Hadley Centre climate models, but the same parametrization had less positive impact when implemented in an earlier version of the Hadley Centre model (Martin et al., 2006). Clearly, parametrizations must be understood in the context of their host models.

Cloud processes affect the climate system by regulating the flow of radiation at the top of the atmosphere, by producing precipitation, by accomplishing rapid and sometimes deep redistributions of atmospheric mass and through additional mechanisms too numerous to list here (Arakawa and Schubert, 1974; Arakawa, 2004). Cloud parametrizations are based on physical theories that aim to describe the statistics of the cloud field (e.g., the fractional cloudiness or the area-averaged precipitation rate) without describing the individual cloud elements. In an increasing number of climate models, microphysical parametrizations that represent such processes as cloud particle and raindrop formation are used to predict the distributions of liquid and ice clouds. These parametrizations improve the simulation of the present climate, and affect climate sensitivity (Iacobellis et al., 2003). Realistic parametrizations of cloud processes are a prerequisite for reliable current and future climate simulation (see Section 8.6).

Data from field experiments such as the Global Atmospheric Research Program (GARP) Atlantic Tropical Experiment (GATE, 1974), the Monsoon Experiment (MONEX, 1979), ARM (1993) and the Tropical Ocean Global Atmosphere (TOGA) Coupled Ocean-Atmosphere Response Experiment (COARE, 1993) have been used to test and improve parametrizations of clouds and convection (e.g., Emanuel and Zivkovic-Rothmann, 1999; Sud and Walker, 1999; Bony and Emanuel, 2001). Systematic research such as that conducted by the GCSS (Randall et al., 2003) has been organised to test parametrizations by comparing results with both observations and the results of a cloud-resolving model. These efforts have influenced the development of many of the recent models. For example, the boundary-layer cloud parametrization of Lock et al. (2000) and Lock (2001) was tested via the GCSS. Parametrizations of radiative processes have been improved and tested by comparing results of radiation parametrizations used in AOGCMs with those of much more detailed 'line-by-line' radiation codes (Collins et al., 2006). Since the TAR, improvements have been made in several models to the physical coupling between cloud and convection parametrizations, for example, in the Max Planck Institute (MPI) AOGCM using

Tompkins (2002), in the IPSL-CM4 AOGCM using Bony and Emanuel (2001) and in the GFDL model using Tiedtke (1993). These are examples of component-level testing.

In parallel with improvement in parametrizations, a non-hydrostatic model has been used for downscaling. A model with a 5 km grid on a domain of 4,000 x 3,000 x 22 km centred over Japan has been run by MRI/JMA, using the time-slice method for the Fourth Assessment Report (AR4) (Yoshizaki et al., 2005).

Aerosols play an important role in the climate system. Interactive aerosol parametrizations are now used in some models (HADGEM1, MIROC-hi, MIROC-med). Both the 'direct' and 'indirect' aerosol effects (Chapter 2) have been incorporated in some cases (e.g., IPSL-CM4). In addition to sulphates, other types of aerosols such as black and organic carbon, sea salt and mineral dust are being introduced as prognostic variables (Takemura et al., 2005; see Chapter 2). Further details are given in Section 8.2.5.

8.2.2   Ocean Processes

8.2.2.1   Numerics

Recently, isopycnic or hybrid vertical coordinates have been adopted in some ocean models (GISS-EH and BCCR-BCM2.0). Tests show that such models can produce solutions for complex regional flows that are as realistic as those obtained with the more common depth coordinate (e.g., Drange et al., 2005). Issues remain over the proper treatment of thermobaricity (nonlinear relationship of temperature, salinity and pressure to density), which means that in some isopycnic coordinate models the relative densities of, for example, Mediterranean and Antarctic Bottom Water masses are distorted. The merits of these vertical coordinate systems are still being established.

An explicit representation of the sea surface height is being used in many models, and real freshwater flux is used to force those models instead of a 'virtual' salt flux. The virtual salt flux method induces a systematic error in sea surface salinity prediction and causes a serious problem at large river basin mouths (Hasumi, 2002a,b; Griffies, 2004).

Generalised curvilinear horizontal coordinates with bipolar or tripolar grids (Murray, 1996) have become widely used in the oceanic component of AOGCMs. These are strategies used to deal with the North Pole coordinate singularity, as alternatives to the previously common polar filter or spherical coordinate rotation. The newer grids have the advantage that the singular points can be shifted onto land while keeping grid points aligned on the equator. The older methods of representing the ocean surface, surface water flux and North Pole are still in use in several AOGCMs.

8.2.2.2   Horizontal and Vertical Resolution

There has been a general increase in resolution since the TAR, with a horizontal resolution of order one to two degrees now commonly used in the ocean component of most climate models.

To better resolve the equatorial waveguide, several models use enhanced meridional resolution in the tropics. Resolution high enough to allow oceanic eddies, eddy permitting, has not been used in a full suite of climate scenario integrations due to computational cost, but since the TAR it has been used in some idealised and scenario-based climate experiments as discussed below. A limited set of integrations using the eddy-permitting MIROC3.2 (hires) model is used here and in Chapter 10. Some modelling centres have also increased vertical resolution since the TAR.

A few coupled climate models with eddy-permitting ocean resolution (1/6° to 1/3°) have been developed (Roberts et al., 2004; Suzuki et al., 2005), and large-scale climatic features induced by local air-sea coupling have been successfully simulated (e.g., Sakamoto et al., 2004).

Roberts et al. (2004) found that increasing the ocean resolution of the HadCM3 model from about 1° to 0.33° by 0.33° by 40 levels (while leaving the atmospheric component unchanged) resulted in many improvements in the simulation of ocean circulation features. However, the impact on the atmospheric simulation was relatively small and localised. The climate change response was similar to the standard resolution model, with a slightly faster rate of warming in the Northern Europe-Atlantic region due to differences in the Atlantic Meridional Overturning Circulation (MOC) response. The adjustment time scale of the Atlantic Basin freshwater budget decreased from being of order 400 years to being of order 150 years with the higher resolution ocean, suggesting possible differences in transient MOC response on those time scales, but the mechanisms and the relative roles of horizontal and vertical resolution are not clear.

The Atlantic MOC is influenced by freshwater as well as thermal forcing. Besides atmospheric freshwater forcing, freshwater transport by the ocean itself is also important. For the Atlantic MOC, the fresh Pacific water coming through the Bering Strait could be poorly simulated on its transit to the Canadian Archipelago and the Labrador Sea (Komuro and Hasumi, 2005). These aspects have been improved since the TAR in many of the models evaluated here.

Changes around continental margins are very important for regional climate change. Over these areas, climate is influenced by the atmosphere and open ocean circulation. High-resolution climate models contribute to the improvement of regional climate simulation. For example, the location of the Kuroshio separation from the Japan islands is well simulated in the MIROC3.2 (hires) model (see Figure 8.1), which makes it possible to study a change in the Kuroshio axis in the future climate (Sakamoto et al., 2005).

Guilyardi et al. (2004) suggested that ocean resolution may play only a secondary role in setting the time scale of model El Niño-Southern Oscillation (ENSO) variability, with the dominant time scales being set by the atmospheric model provided the basic speeds of the equatorial ocean wave modes are adequately represented.

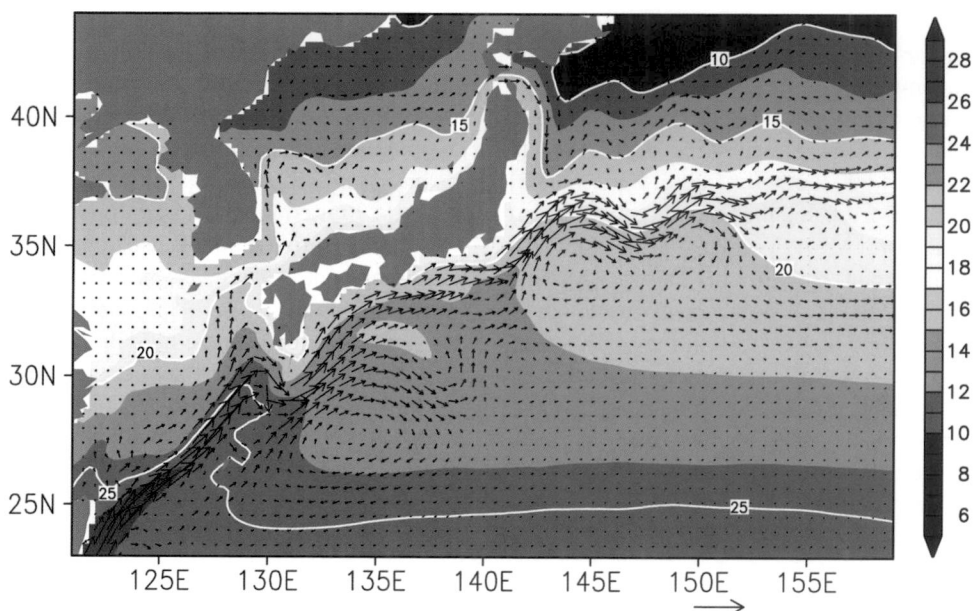

Figure 8.1. *Long-term mean ocean current velocities at 100 m depth (vectors, unit: m s⁻¹) and sea surface temperature (colours, °C) around the Kuroshio and the Kuroshio Extension obtained from a control experiment forced by pre-industrial conditions (CO₂ concentration 295.9 ppm) using MIROC3.2 (hires).*

8.2.2.3   *Parametrizations*

In the tracer equations, isopycnal diffusion (Redi, 1982) with isopycnal layer thickness diffusion (Gent et al., 1995), including its modification by Visbeck et al. (1997), has become a widespread choice instead of a simple horizontal diffusion. This has led to improvements in the thermocline structure and meridional overturning (Böning et al., 1995; see Section 8.3.2). For vertical mixing of tracers, a wide variety of parametrizations is currently used, such as turbulence closures (e.g., Mellor and Yamada, 1982), non-local diffusivity profiles (Large et al., 1994) and bulk mixed-layer models (e.g., Kraus and Turner, 1967). Representation of the surface mixed layer has been much improved due to developments in these parametrizations (see Section 8.3.2). Observations have shown that deep ocean vertical mixing is enhanced over rough bottoms, steep slopes and where stratification is weak (Kraus, 1990; Polzin et al., 1997; Moum et al., 2002). While there have been modelling studies indicating the significance of such inhomogeneous mixing for the MOC (e.g., Marotzke, 1997; Hasumi and Suginohara, 1999; Otterå et al., 2004; Oliver et al., 2005, Saenko and Merryfield 2005), comprehensive parametrizations of the effects and their application in coupled climate models are yet to be seen.

Many of the dense waters formed by oceanic convection, which are integral to the global MOC, must flow over ocean ridges or down continental slopes. The entrainment of ambient water around these topographic features is an important process determining the final properties and quantity of the deep waters. Parametrizations for such bottom boundary-layer processes have come into use in some AOGCMs (e.g., Winton et al., 1998; Nakano and Suginohara, 2002). However, the impact of the bottom boundary-layer representation on the coupled system is not fully understood (Tang and Roberts, 2005). Thorpe et al.

(2004) studied the impact of the very simple scheme used in the HadCM3 model to control mixing of overflow waters from the Nordic Seas into the North Atlantic. Although the scheme does result in a change of the subpolar water mass properties, it appears to have little impact on the simulation of the strength of the large-scale MOC, or its response to global warming.

8.2.3   Terrestrial Processes

Few multi-model analyses of terrestrial processes included in the models in Table 8.1 have been conducted. However, significant advances since the TAR have been reported based on climate models that are similar to these models. Analysis of these models provides insight on how well terrestrial processes are included in the AR4 models.

8.2.3.1   *Surface Processes*

The addition of the terrestrial biosphere models that simulate changes in terrestrial carbon sources and sinks into fully coupled climate models is at the cutting edge of climate science. The major advance in this area since the TAR is the inclusion of carbon cycle dynamics including vegetation and soil carbon cycling, although these are not yet incorporated routinely into the AOGCMs used for climate projection (see Chapter 10). The inclusion of the terrestrial carbon cycle introduces a new and potentially important feedback into the climate system on time scales of decades to centuries (see Chapters 7 and 10). These feedbacks include the responses of the terrestrial biosphere to increasing carbon dioxide (CO₂), climate change and changes in climate variability (see Chapter 7). However, many issues remain to be resolved. The magnitude of the sink remains uncertain (Cox et al., 2000; Friedlingstein et al., 2001;

Dufresne et al., 2002) because it depends on climate sensitivity as well as on the response of vegetation and soil carbon to increasing CO_2 (Friedlingstein et al., 2003). The rate at which CO_2 fertilization saturates in terrestrial systems dominates the present uncertainty in the role of biospheric feedbacks. A series of studies have been conducted to explore the present modelling capacity of the response of the terrestrial biosphere rather than the response of just one or two of its components (Friedlingstein et al., 2006). This work has built on systematic efforts to evaluate the capacity of terrestrial biosphere models to simulate the terrestrial carbon cycle (Cramer et al., 2001) via intercomparison exercises. For example, Friedlingstein et al. (2006) found that in all models examined, the sink decreases in the future as the climate warms.

Other individual components of land surface processes have been improved since the TAR, such as root parametrization (Arora and Boer, 2003; Kleidon, 2004) and higher-resolution river routing (Ducharne et al., 2003). Cold land processes have received considerable attention with multi-layer snowpack models now more common (e.g., Oleson et al., 2004) as is the inclusion of soil freezing and thawing (e.g., Boone et al., 2000; Warrach et al., 2001). Sub-grid scale snow parametrizations (Liston, 2004), snow-vegetation interactions (Essery et al., 2003) and the wind redistribution of snow (Essery and Pomeroy, 2004) are more commonly considered. High-latitude organic soils are included in some models (Wang et al., 2002). A recent advance is the coupling of groundwater models into land surface schemes (Liang et al., 2003; Maxwell and Miller, 2005; Yeh and Eltahir, 2005). These have only been evaluated locally but may be adaptable to global scales. There is also evidence emerging that regional-scale projection of warming is sensitive to the simulation of processes that operate at finer scales than current climate models resolve (Pan et al., 2004). In general, the improvements in land surface models since the TAR are based on detailed comparisons with observational data. For example, Boone et al. (2004) used the Rhone Basin to investigate how land surface models simulate the water balance for several annual cycles compared to data from a dense observation network. They found that most land surface schemes simulate very similar total runoff and evapotranspiration but the partitioning between the various components of both runoff and evaporation varies greatly, resulting in different soil water equilibrium states and simulated discharge. More sophisticated snow parametrizations led to superior simulations of basin-scale runoff.

An analysis of results from the second phase of AMIP (AMIP-2) explored the land surface contribution to climate simulation. Henderson-Sellers et al. (2003) found a clear chronological sequence of land surface schemes (early models that excluded an explicit canopy, more recent biophysically based models and very recent biophysically based models). Statistically significant differences in annually averaged evaporation were identified that could be associated with the parametrization of canopy processes. Further improvements in land surface models depends on enhanced surface observations, for example, the use of stable isotopes (e.g., Henderson-Sellers et al., 2004) that allow several components of evaporation to be

evaluated separately. Pitman et al. (2004) explored the impact of the level of complexity used to parametrize the surface energy balance on differences found among the AMIP-2 results. They found that quite large variations in surface energy balance complexity did not lead to systematic differences in the simulated mean, minimum or maximum temperature variance at the global scale, or in the zonal averages, indicating that these variables are not limited by uncertainties in how to parametrize the surface energy balance. This adds confidence to the use of the models in Table 8.1, as most include surface energy balance modules of more complexity than the minimum identified by Pitman et al. (2004).

While little work has been performed to assess the capability of the land surface models used in coupled climate models, the upgrading of the land surface models is gradually taking place and the inclusion of carbon in these models is a major conceptual advance. In the simulation of the present-day climate, the limitations of the standard bucket hydrology model are increasingly clear (Milly and Shmakin, 2002; Henderson-Sellers et al., 2004; Pitman et al., 2004), including evidence that it overestimates the likelihood of drought (Seneviratne et al., 2002). Relatively small improvements to the land surface model, for example, the inclusion of spatially variable water-holding capacity and a simple canopy conductance, lead to significant improvements (Milly and Shmakin, 2002). Since most models in Table 8.1 represent the continental-scale land surface more realistically than the standard bucket hydrology scheme, and include spatially variable water-holding capacity, canopy conductance, etc. (Table 8.1), most of these models likely capture the key contribution made by the land surface to current large-scale climate simulations. However, it is not clear how well current climate models can capture the impact of future warming on the terrestrial carbon balance. A systematic evaluation of AOGCMs with the carbon cycle represented would help increase confidence in the contribution of the terrestrial surface resulting from future warming.

8.2.3.2   Soil Moisture Feedbacks in Climate Models

A key role of the land surface is to store soil moisture and control its evaporation. An important process, the soil moisture-precipitation feedback, has been explored extensively since the TAR, building on regionally specific studies that demonstrated links between soil moisture and rainfall. Recent studies (e.g., Gutowski et al., 2004; Pan et al., 2004) suggest that summer precipitation strongly depends on surface processes, notably in the simulation of regional extremes. Douville (2001) showed that soil moisture anomalies affect the African monsoon while Schär et al. (2004) suggested that an active soil moisture-precipitation feedback was linked to the anomalously hot European summer in 2003.

The soil moisture-precipitation feedback in climate models had not been systematically assessed at the time of the TAR. It is associated with the strength of coupling between the land and atmosphere, which is not directly measurable at the large scale in nature and has only recently been quantified in models

(Dirmeyer, 2001). Koster et al. (2004) provided an assessment of where the soil moisture-precipitation feedback is regionally important during the Northern Hemisphere (NH) summer by quantifying the coupling strength in 12 atmospheric GCMs. Some similarity was seen among the model responses, enough to produce a multi-model average estimate of where the global precipitation pattern during the NH summer was most strongly affected by soil moisture variations. These 'hot spots' of strong coupling are found in transition regions between humid and dry areas. The models, however, also show strong disagreement in the strength of land-atmosphere coupling. A few studies have explored the differences in coupling strength. Seneviratne et al. (2002) highlighted the importance of differing water-holding capacities among the models while Lawrence and Slingo (2005) explored the role of soil moisture variability and suggested that frequent soil moisture saturation and low soil moisture variability could partially explain the weak coupling strength in the HadAM3 model (note that 'weak' does not imply 'wrong' since the real strength of the coupling is unknown).

Overall, the uncertainty in surface-atmosphere coupling has implications for the reliability of the simulated soil moisture-atmosphere feedback. It tempers our interpretation of the response of the hydrologic cycle to simulated climate change in 'hot spot' regions. Note that no assessment has been attempted for seasons other than NH summer.

Since the TAR, there have been few assessments of the capacity of climate models to simulate observed soil moisture. Despite the tremendous effort to collect and homogenise soil moisture measurements at global scales (Robock et al., 2000), discrepancies between large-scale estimates of observed soil moisture remain. The challenge of modelling soil moisture, which naturally varies at small scales, linked to landscape characteristics, soil processes, groundwater recharge, vegetation type, etc., within climate models in a way that facilitates comparison with observed data is considerable. It is not clear how to compare climate-model simulated soil moisture with point-based or remotely sensed soil moisture. This makes assessing how well climate models simulate soil moisture, or the change in soil moisture, difficult.

8.2.4 Cryospheric Processes

8.2.4.1 Terrestrial Cryosphere

Ice sheet models are used in calculations of long-term warming and sea level scenarios, though they have not generally been incorporated in the AOGCMs used in Chapter 10. The models are generally run in 'off-line' mode (i.e., forced by atmospheric fields derived from high-resolution time-slice experiments), although Huybrechts et al. (2002) and Fichefet et al. (2003) reported early efforts at coupling ice sheet models to AOGCMs. Ice sheet models are also included in some EMICs (e.g., Calov et al., 2002). Ridley et al. (2005) pointed out that the time scale of projected melting of the Greenland Ice Sheet may be different in coupled and off-line simulations. Presently available thermomechanical ice sheet models do not

include processes associated with ice streams or grounding line migration, which may permit rapid dynamical changes in the ice sheets. Glaciers and ice caps, due to their relatively small scales and low likelihood of significant climate feedback at large scales, are not currently included interactively in any AOGCMs. See Chapters 4 and 10 for further detail. For a discussion of terrestrial snow, see Section 8.3.4.1.

8.2.4.2 Sea Ice

Sea ice components of current AOGCMs usually predict ice thickness (or volume), fractional cover, snow depth, surface and internal temperatures (or energy) and horizontal velocity. Some models now include prognostic sea ice salinity (Schmidt et al., 2004). Sea ice albedo is typically prescribed, with only crude dependence on ice thickness, snow cover and puddling effects.

Since the TAR, most AOGCMs have started to employ complex sea ice dynamic components. The complexity of sea ice dynamics in current AOGCMs varies from the relatively simple 'cavitating fluid' model (Flato and Hibler, 1992) to the viscous-plastic model (Hibler, 1979), which is computationally expensive, particularly for global climate simulations. The elastic-viscous-plastic model (Hunke and Dukowicz, 1997) is being increasingly employed, particularly due to its efficiency for parallel computers. New numerical approaches for solving the ice dynamics equations include more accurate representations on curvilinear model grids (Hunke and Dukowicz, 2002; Marsland et al., 2003; Zhang and Rothrock, 2003) and Lagrangian methods for solving the viscous-plastic equations (Lindsay and Stern, 2004; Wang and Ikeda, 2004).

Treatment of sea ice thermodynamics in AOGCMs has progressed more slowly: it typically includes constant conductivity and heat capacities for ice and snow (if represented), a heat reservoir simulating the effect of brine pockets in the ice, and several layers, the upper one representing snow. More sophisticated thermodynamic schemes are being developed, such as the model of Bitz and Lipscomb (1999), which introduces salinity-dependent conductivity and heat capacities, modelling brine pockets in an energy-conserving way as part of a variable-layer thermodynamic model (e.g., Saenko et al., 2002). Some AOGCMs include snow ice formation, which occurs when an ice floe is submerged by the weight of the overlying snow cover and the flooded snow layer refreezes. The latter process is particularly important in the antarctic sea ice system.

Even with fine grid scales, many sea ice models incorporate sub-grid scale ice thickness distributions (Thorndike et al., 1975) with several thickness 'categories', rather than considering the ice as a uniform slab with inclusions of open water. An ice thickness distribution enables more accurate simulation of thermodynamic variations in growth and melt rates within a single grid cell, which can have significant consequences for ice-ocean albedo feedback processes (e.g., Bitz et al., 2001; Zhang and Rothrock, 2001). A well-resolved ice thickness distribution enables a more physical formulation for ice ridging and rafting events, based on energetic principles. Although parametrizations of ridging mechanics and their relationship with the ice thickness distribution have improved (Babko et al., 2002; Amundrud et al., 2004; Toyota et al., 2004), inclusion of advanced ridging parametrizations has lagged other aspects of sea ice dynamics (rheology, in particular) in AOGCMs. Better numerical algorithms used for the ice thickness distribution (Lipscomb, 2001) and ice strength (Hutchings et al., 2004) have also been developed for AOGCMs.

8.2.5 Aerosol Modelling and Atmospheric Chemistry

Climate simulations including atmospheric aerosols with chemical transport have greatly improved since the TAR. Simulated global aerosol distributions are better compared with observations, especially satellite data (e.g., Advanced Very High Resolution Radar (AVHRR), Moderate Resolution Imaging Spectroradiometer (MODIS), Multi-angle Imaging Spectroradiometer (MISR), Polarization and Directionality of the Earth's Reflectance (POLDER), Total Ozone Mapping Spectrometer (TOMS)), the ground-based network (Aerosol Robotic Network; AERONET) and many measurement campaigns (e.g., Chin et al., 2002; Takemura et al., 2002). The global Aerosol Model Intercomparison project, AeroCom, has also been initiated in order to improve understanding of uncertainties of model estimates, and to reduce them (Kinne et al., 2003). These comparisons, combined with cloud observations, should result in improved confidence in the estimation of the aerosol direct and indirect radiative forcing (e.g., Ghan et al., 2001a,b; Lohmann and Lesins, 2002; Takemura et al., 2005). Interactive aerosol sub-component models have been incorporated in some of the climate models used in Chapter 10 (HadGEM1 and MIROC). Some models also include indirect aerosol effects (e.g., Takemura et al., 2005); however, the formulation of these processes is still the subject of much research.

Interactive atmospheric chemistry components are not generally included in the models used in this report. However, CCSM3 includes the modification of greenhouse gas concentrations by chemical processes and conversion of sulphur dioxide and dimethyl sulphide to sulphur aerosols.

8.2.6 Coupling Advances

In an advance since the TAR, a number of groups have developed software allowing easier coupling of the various components of a climate model (e.g., Valcke et al., 2006). An example, the Ocean Atmosphere Sea Ice Soil (OASIS) coupler, developed at the Centre Europeen de Recherche et de Formation Avancee en Calcul Scientific (CERFACS) (Terray et al., 1998), has been used by many modelling centres to synchronise the different models and for the interpolation of the coupling fields between the atmosphere and ocean grids. The schemes for interpolation between the ocean and the atmosphere grids have been revised. The new schemes ensure both a global and local conservation of the various fluxes at the air-sea interface, and track terrestrial, ocean and sea ice fluxes individually.

Coupling frequency is an important issue, because fluxes are averaged during a coupling interval. Typically, most AOGCMs evaluated here pass fluxes and other variables between the component parts once per day. The K-Profile Parametrization ocean vertical scheme (Large et al., 1994), used in several models, is very sensitive to the wind energy available for mixing. If the models are coupled at a frequency lower than once per ocean time step, nonlinear quantities such as wind mixing power (which depends on the cube of the wind speed) must be accumulated over every time step before passing to the ocean. Improper averaging therefore could lead to too little mixing energy and hence shallower mixed-layer depths, assuming the parametrization is not re-tuned. However, high coupling frequency can bring new technical issues. In the MIROC model, the coupling interval is three hours, and in this case, a poorly resolved internal gravity wave is excited in the ocean so some smoothing is necessary to damp this numerical problem. It should also be noted that the AOGCMs used here have relatively thick top oceanic grid boxes (typically 10 m or more), limiting the sea surface temperature (SST) response to frequent coupling (Bernie et al., 2005).

8.2.7 Flux Adjustments and Initialisation

Since the TAR, more climate models have been developed that do not adjust the surface heat, water and momentum fluxes artificially to maintain a stable control climate. As noted by Stouffer and Dixon (1998), the use of such flux adjustments required relatively long integrations of the component models before coupling. In these models, normally the initial conditions for the coupled integrations were obtained from long spin ups of the component models.

In AOGCMs that do not use flux adjustments (see Table 8.1), the initialisation methods tend to be more varied. The oceanic components of many models are initialised using values obtained either directly from an observationally based, gridded data set (Levitus and Boyer, 1994; Levitus and Antonov, 1997; Levitus et al., 1998) or from short ocean-only integrations that used an observational analysis for their initial conditions. The initial atmospheric component data are usually obtained from atmosphere-only integrations using prescribed SSTs.

To obtain initial data for the pre-industrial control integrations discussed in Chapter 10, most AOGCMs use variants of the Stouffer et al. (2004) scheme. In this scheme, the coupled model is initialised as discussed above. The radiative forcing is then set back to pre-industrial conditions. The model is integrated for a few centuries using constant pre-industrial radiative forcing, allowing the coupled system to partially adjust to this forcing. The degree of equilibration in the real pre-industrial climate to the pre-industrial radiative forcing is not known. Therefore, it seems unnecessary to have the pre-industrial control fully equilibrated. After this spin-up integration, the pre-industrial control is started and perturbation integrations can begin. An important next step, once the start of the control integration is determined, is the assessment of the control integration climate drift. Large climate drifts can distort both the natural variability (e.g., Inness et al., 2003) and the climate response to changes in radiative forcing (Spelman and Manabe, 1984).

In earlier IPCC reports, the initialisation methods were quite varied. In some cases, the perturbation integrations were initialised using data from control integrations where the SSTs were near present-day values and not pre-industrial. Given that many climate models now use some variant of the Stouffer et al. (2004) method, this situation has improved.

8.3 Evaluation of Contemporary Climate as Simulated by Coupled Global Models

Due to nonlinearities in the processes governing climate, the climate system response to perturbations depends to some extent on its basic state (Spelman and Manabe, 1984). Consequently, for models to predict future climatic conditions reliably, they must simulate the current climatic state with some as yet unknown degree of fidelity. Poor model skill in simulating present climate could indicate that certain physical or dynamical processes have been misrepresented. The better a model simulates the complex spatial patterns and seasonal and diurnal cycles of present climate, the more confidence there is that all the important processes have been adequately represented. Thus, when new models are constructed, considerable effort is devoted to evaluating their ability to simulate today's climate (e.g., Collins et al., 2006; Delworth et al., 2006).

Some of the assessment of model performance presented here is based on the 20th-century simulations that constitute a part of the MMD archived at PCMDI. In these simulations, modelling groups initiated the models from pre-industrial (circa 1860) 'control' simulations and then imposed the natural and anthropogenic forcing thought to be important for simulating the climate of the last 140 years or so. The 23 models considered here (see Table 8.1) are those relied on in Chapters 9 and 10 to investigate historical and future climate changes. Some figures in this section are based on results from a subset of the models because the data set is incomplete.

In order to identify errors that are systematic across models, the mean of fields available in the MMD, referred to here as the 'multi-model mean field', will often be shown. The multi-model mean field results are augmented by results from individual models available as Supplementary Material (see Figures S8.1 to S8.15). The multi-model averaging serves to filter out biases of individual models and only retains errors that are generally pervasive. There is some evidence that the multi-model mean field is often in better agreement with observations than any of the fields simulated by the individual models (see Section 8.3.1.1.2), which supports continued reliance on a diversity of modelling approaches in projecting future climate change and provides some further interest in evaluating the multi-model mean results.

Faced with the rich variety of climate characteristics that could potentially be evaluated here, this section focuses on those elements that can critically affect societies and natural ecosystems and that are most likely to respond to changes in radiative forcing.

8.3.1 Atmosphere

8.3.1.1 Surface Temperature and the Climate System's Energy Budget

For models to simulate accurately the global distribution of the annual and diurnal cycles of surface temperature, they must, in the absence of compensating errors, correctly represent a variety of processes. The large-scale distribution of annual mean surface temperature is largely determined by the distribution of insolation, which is moderated by clouds, other surface heat fluxes and transport of energy by the atmosphere and to a lesser extent by the ocean. Similarly, the annual and diurnal cycles of surface temperature are governed by seasonal and diurnal changes in these factors, respectively, but they are also damped by storage of energy in the upper layers of the ocean and to a lesser degree the surface soil layers.

8.3.1.1.1 Temperature

Figure 8.2a shows the observed time mean surface temperature as a composite of surface air temperature over regions of land and SST elsewhere. Also shown is the difference between the multi-model mean field and the observed field. With few exceptions, the absolute error (outside polar regions and other data-poor regions) is less than 2°C. Individual models typically have larger errors, but in most cases still less than 3°C, except at high latitudes (see Figure 8.2b and Supplementary Material, Figure S8.1). Some of the larger errors occur in regions of sharp elevation changes and may result simply from mismatches between the model topography (typically smoothed) and the actual topography. There is also a tendency for a slight, but general, cold bias. Outside the polar regions, relatively large errors are evident in the eastern parts of the tropical ocean basins, a likely symptom of problems in the simulation of low clouds. The extent to which these systematic model errors affect a model's response to external perturbations is unknown, but may be significant (see Section 8.6).

In spite of the discrepancies discussed here, the fact is that models account for a very large fraction of the global temperature pattern: the correlation coefficient between the simulated and observed spatial patterns of annual mean temperature is typically about 0.98 for individual models. This supports the view that major processes governing surface temperature climatology are represented with a reasonable degree of fidelity by the models.

An additional opportunity for evaluating models is afforded by the observed annual cycle of surface temperature. Figure 8.3 shows the standard deviation of monthly mean surface temperatures, which is dominated by contributions from the amplitudes of the annual and semi-annual components of the annual cycle. The difference between the mean of the model results and the observations is also shown. The absolute

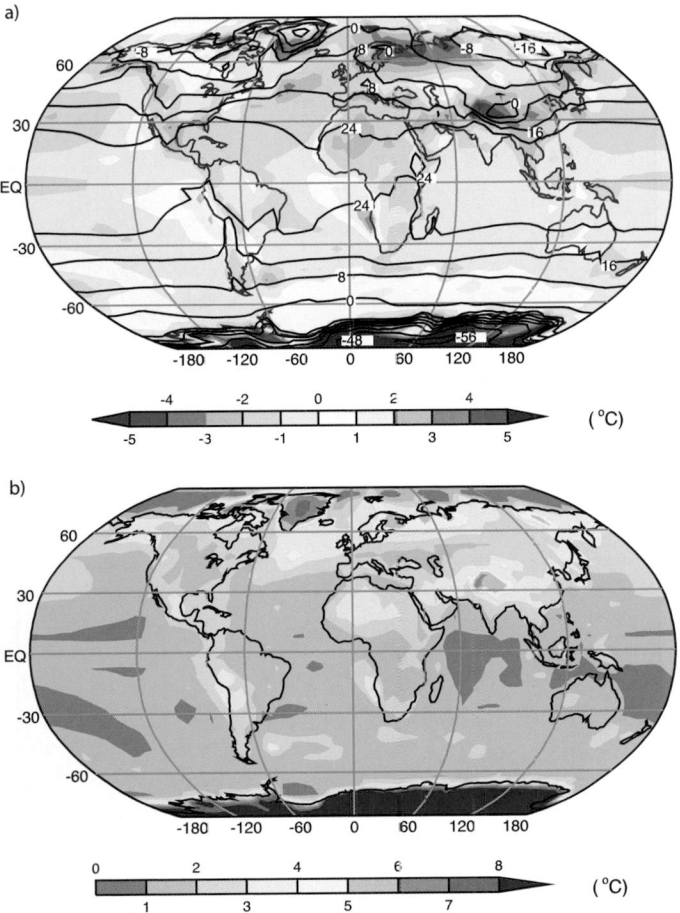

Like the annual range of temperature, the diurnal range (the difference between daily maximum and minimum surface air temperature) is much smaller over oceans than over land, where it is also better observed, so the discussion here is restricted to continental regions. The diurnal temperature range, zonally and annually averaged over the continents, is generally too small in the models, in many regions by as much as 50% (see Supplementary Material, Figure S8.3). Nevertheless, the models simulate the general pattern of this field, with relatively high values over the clearer, drier regions. It is not yet known why models generally underestimate the diurnal temperature range; it is possible that in some models it is in part due to shortcomings of the boundary-layer parametrizations or in the simulation of freezing and thawing soil, and it is also known that the diurnal cycle of convective cloud, which interacts strongly with surface temperature, is rather poorly simulated.

Surface temperature is strongly coupled with the atmosphere above it. This is especially evident at mid-latitudes, where migrating cold fronts and warm fronts can cause relatively large swings in surface temperature. Given the strong interactions between the surface temperature and the temperature of the air above, it is of special interest to evaluate how well models simulate the vertical profile of atmospheric temperature. The multi-model mean absolute error in the zonal mean, annual mean air temperature is almost everywhere less than 2°C (compared with the observed range of temperatures, which spans more than 100°C when the entire troposphere is considered; see Supplementary Material, Figure S8.4). It is notable, however, that near the tropopause at high latitudes the models are generally biased cold. This bias is a problem that has persisted for many years, but in general is now less severe than in earlier models. In a few of the models, the bias has been eliminated entirely, but

Figure 8.2. *(a) Observed climatological annual mean SST and, over land, surface air temperature (labelled contours) and the multi-model mean error in these temperatures, simulated minus observed (colour-shaded contours). (b) Size of the typical model error, as gauged by the root-mean-square error in this temperature, computed over all AOGCM simulations available in the MMD at PCMDI. The Hadley Centre Sea Ice and Sea Surface Temperature (HadISST; Rayner et al., 2003) climatology of SST for 1980 to 1999 and the Climatic Research Unit (CRU; Jones et al., 1999) climatology of surface air temperature over land for 1961 to 1990 are shown here. The model results are for the same period in the 20th-century simulations. In the presence of sea ice, the SST is assumed to be at the approximate freezing point of seawater (−1.8°C). Results for individual models can be seen in the Supplementary Material, Figure S8.1.*

differences are in most regions less than 1°C. Even over extensive land areas of the NH where the standard deviation generally exceeds 10°C, the models agree with observations within 2°C almost everywhere. The models, as a group, clearly capture the differences between marine and continental environments and the larger magnitude of the annual cycle found at higher latitudes, but there is a general tendency to underestimate the annual temperature range over eastern Siberia. In general, the largest fractional errors are found over the oceans (e.g., over much of tropical South America and off the east coasts of North America and Asia). These exceptions to the overall good agreement illustrate a general characteristic of current climate models: the largest-scale features of climate are simulated more accurately than regional- and smaller-scale features.

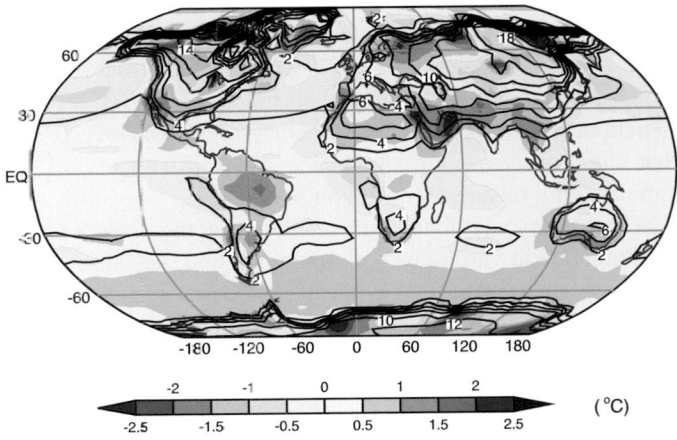

Figure 8.3. *Observed standard deviation (labelled contours) of SST and, over land, surface air temperature, computed over the climatological monthly mean annual cycle, and the multi-model mean error in the standard deviations, simulated minus observed (colour-shaded contours). In most regions, the standard deviation provides a measure of the amplitude of the seasonal range of temperature. The observational data sets, the model results and the climatological periods are as described in Figure 8.2. Results for individual models can be seen in the Supplementary Material, Figure S8.2.*

compensating errors may be responsible. It is known that the tropopause cold bias is sensitive to several factors, including horizontal and vertical resolution, non-conservation of moist entropy, and the treatment of sub-grid scale vertical convergence of momentum ('gravity wave drag'). Although the impact of the tropopause temperature bias on the model's response to radiative forcing changes has not been definitively quantified, it is almost certainly small relative to other uncertainties.

8.3.1.1.2 The balance of radiation at the top of the atmosphere

The primary driver of latitudinal and seasonal variations in temperature is the seasonally varying pattern of incident sunlight, and the fundamental driver of the circulation of the atmosphere and ocean is the local imbalance between the shortwave (SW) and longwave (LW) radiation at the top of the atmosphere. The impact on temperature of the distribution of insolation can be strongly modified by the distribution of clouds and surface characteristics.

Considering first the annual mean SW flux at the 'top' of the atmosphere (TOA)[1], the insolation is determined by well-known orbital parameters that ensure good agreement between models and observations. The annual mean insolation is strongest in the tropics, decreasing to about half as much at the poles. This largely drives the strong equator-to-pole temperature gradient. As for outgoing SW radiation, the Earth, on average, reflects about the same amount of sunlight (~100 W m^{-2} in the annual mean) at all latitudes. At most latitudes, the difference between the multi-model mean zonally averaged outgoing SW radiation and observations is in the annual mean less than 6 W m^{-2} (i.e., an error of about 6%; see Supplementary Material, Figure S8.5). Given that clouds are responsible for about half the outgoing SW radiation, these errors are not surprising, for it is known that cloud processes are among the most difficult to simulate with models (see Section 8.6.3.2.3).

There are additional errors in outgoing SW radiation due to variations with longitude and season, and these can be quantified by means of the root-mean-square (RMS) error, calculated for each latitude over all longitudes and months and plotted in Figure 8.4a (see also Supplementary Material, Figure S8.6). Errors in the complete two-dimensional fields (see Supplementary Material, Figure S8.6) tend to be substantially larger than the zonal mean errors of about 6 W m^{-2}, an example of the common result that model errors tend to increase as smaller spatial scales and shorter time scales are considered. Figure 8.4a also illustrates a common result that the errors in the multi-model average of monthly mean fields are often smaller than the errors in the individual

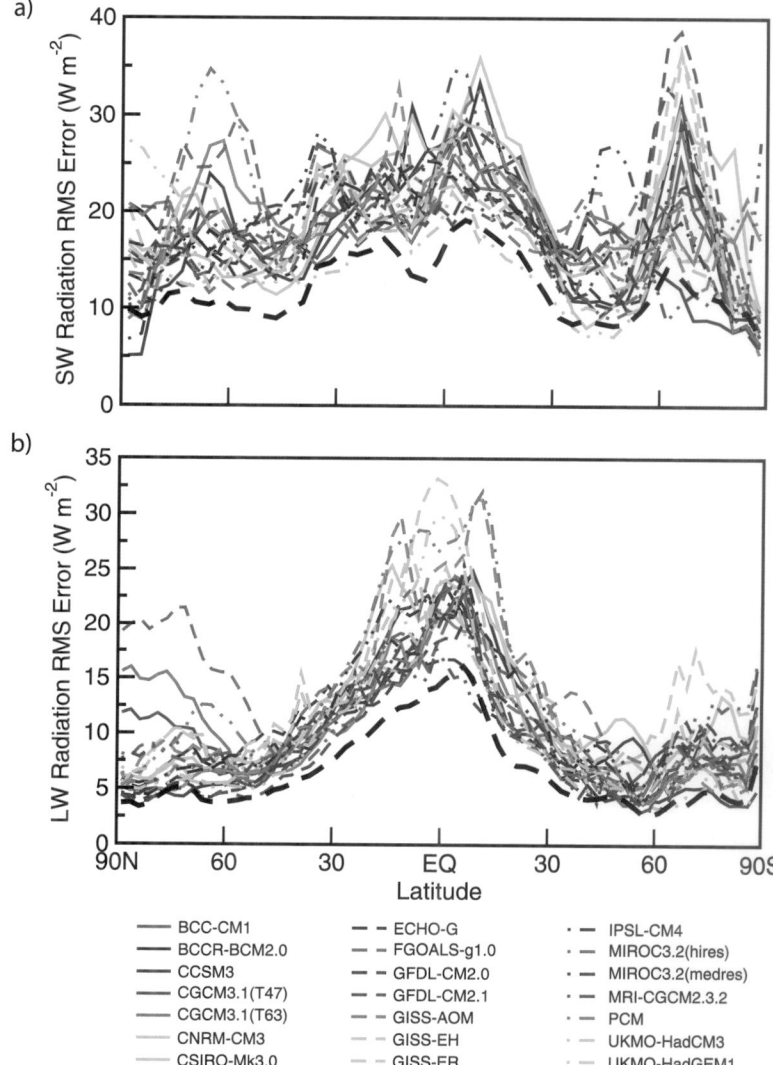

Figure 8.4. Root-mean-square (RMS) model error, as a function of latitude, in simulation of (a) outgoing SW radiation reflected to space and (b) outgoing LW radiation. The RMS error is calculated over all longitudes and over all 12 months of a climatology formed from several years of data. The RMS statistic labelled 'Mean Model' is computed by first calculating the multi-model monthly mean fields, and then calculating the RMS error (i.e., it is not the mean of the individual model RMS errors). The Earth Radiation Budget Experiment (ERBE; Barkstrom et al., 1989) observational estimates used here are for the period 1985 to 1989 from satellite-based radiometers, and the model results are for the same period in the 20th-century simulations in the MMD at PCMDI. See Table 8.1 for model descriptions. Results for individual models can be seen in the Supplementary Material, Figures S8.5 to S8.8.

model fields. In the case of outgoing SW radiation, this is true at nearly all latitudes. Calculation of the global mean RMS error, based on the monthly mean fields and area-weighted over all grid cells, indicates that the individual model errors are in the range 15 to 22 W m^{-2}, whereas the error in the multi-model mean climatology is only 13.1 W m^{-2}. Why the multi-model mean field turns out to be closer to the observed than the fields in any of the individual models is the subject of ongoing research; a superficial explanation is that at each location and

---

[1] The atmosphere clearly has no identifiable 'top', but the term is used here to refer to an altitude above which the absorption of SW and LW radiation is negligibly small.

for each month, the model estimates tend to scatter around the correct value (more or less symmetrically), with no single model consistently closest to the observations. This, however, does not explain why the results should scatter in this way.

At the TOA, the net SW radiation is everywhere partially compensated by outgoing LW radiation (i.e., infrared emissions) emanating from the surface and the atmosphere. Globally and annually averaged, this compensation is nearly exact. The pattern of LW radiation emitted by earth to space depends most critically on atmospheric temperature, humidity, clouds and surface temperature. With a few exceptions, the models can simulate the observed zonal mean of the annual mean outgoing LW within 10 W m^{-2} (an error of around 5%; see Supplementary Material, Figure S8.7). The models reproduce the relative minimum in this field near the equator where the relatively high humidity and extensive cloud cover in the tropics raises the effective height (and lowers the effective temperature) at which LW radiation emanates to space.

The seasonal cycle of the outgoing LW radiation pattern is also reasonably well simulated by models (see Figure 8.4b). The RMS error for most individual models varies from about 3% of the outgoing LW radiation (OLR) near the poles to somewhat less than 10% in the tropics. The errors for the multi-model mean simulation, ranging from about 2 to 6% across all latitudes, are again generally smaller than those in the individual models.

For a climate in equilibrium, any local annual mean imbalance in the net TOA radiative flux (SW plus LW) must be balanced by a vertically integrated net horizontal divergence of energy carried by the ocean and atmosphere. The fact that the TOA SW and LW fluxes are well simulated implies that the models must also be properly accounting for poleward transport of total energy by the atmosphere and ocean. This proves to be the case, with most models correctly simulating poleward energy transport within about 10%. Although superficially this would seem to provide an important check on models, it is likely that in current models compensating errors improve the agreement of the simulations with observations. There are theoretical and model studies that suggest that if the atmosphere fails to transport the observed portion of energy, the ocean will tend to largely compensate (e.g., Shaffrey and Sutton, 2004).

8.3.1.2   *Moisture and Precipitation*

Water is fundamental to life, and if regional seasonal precipitation patterns were to change, the potential impacts could be profound. Consequently, it is of real practical interest to evaluate how well models can simulate precipitation, not only at global scales, but also regionally. Unlike seasonal variation in temperature, which at large scales is strongly determined by the insolation pattern and configuration of the continents, precipitation variations are also strongly influenced by vertical movement of air due to atmospheric instabilities of various kinds and by the flow of air over orographic features. For models to simulate accurately the seasonally varying pattern of precipitation, they must correctly simulate a number of processes (e.g., evapotranspiration, condensation, transport) that are difficult to evaluate at a global scale. Some of these are discussed further in Sections 8.2 and 8.6. In this subsection, the focus is on the distribution of precipitation and water vapour.

Figure 8.5a shows observation-based estimates of annual mean precipitation and Figure 8.5b shows the multi-model mean field. At the largest scales, the lower precipitation rates at higher latitudes reflect both reduced local evaporation at lower temperatures and a lower saturation vapour pressure of cooler air, which tends to inhibit the transport of vapour from other regions. In addition to this large-scale pattern, captured well by models, is a local minimum in precipitation near the equator in the Pacific, due to a tendency for the Inter-Tropical Convergence Zone (ITCZ)[2] to reside off the equator. There are local maxima at mid-latitudes, reflecting the tendency for subsidence to suppress precipitation in the subtropics and for storm systems to enhance precipitation at mid-latitudes. The models capture these large-scale zonal mean precipitation differences, suggesting that they can adequately represent these features of atmospheric circulation. Moreover, there is some evidence provided in Section 8.3.5 that models have improved over the last several years in simulating the annual cycle of the precipitation patterns.

Models also simulate some of the major regional characteristics of the precipitation field, including the major convergence zones and the maxima over tropical rain forests, although there is a tendency to underestimate rainfall over the Amazon. When considered in more detail, however, there are deficiencies in the multi-model mean precipitation field. There is a distinct tendency for models to orient the South Pacific convergence zone parallel to latitudes and to extend it too far eastward. In the tropical Atlantic, the precipitation maximum is too weak in most models with too much rain south of the equator. There are also systematic east-west positional errors in the precipitation distribution over the Indo-Pacific Warm Pool in most models, with an excess of precipitation over the western Indian Ocean and over the Maritime Continent. These lead to systematic biases in the location of the major rising branches of the Walker Circulation and can compromise major teleconnection[3] pathways, in particular those associated with El Niño (e.g., Turner et al., 2005). Systematic dry biases over the Bay of Bengal are related to errors in the monsoon simulations.

Despite the apparent skill suggested by the multi-model mean (Figure 8.5), many models individually display substantial precipitation biases, especially in the tropics, which often approach the magnitude of the mean observed climatology (e.g., Johns et al., 2006; see also the Supplementary Material, Figures S8.9 and S8.10). Although some of these biases can be attributed to errors in the SST field of the coupled model, even

---

[2] The ITCZ is manifested as a band of relatively intense convective precipitation, accompanied by surface convergence of moisture, which tends to locate seasonally over the warmest surface temperatures and circumnavigates the earth in the tropics (though not continuously).

[3] Teleconnection describes the process through which changes in one part of the climate system affect a remote location via changes in atmospheric circulation patterns.

a)

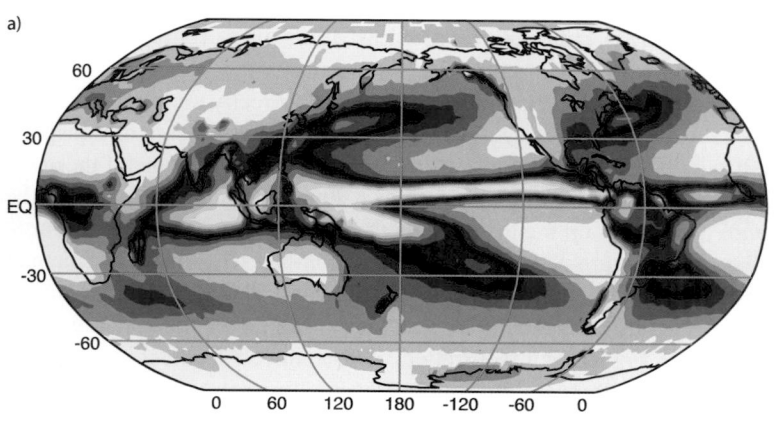

b)

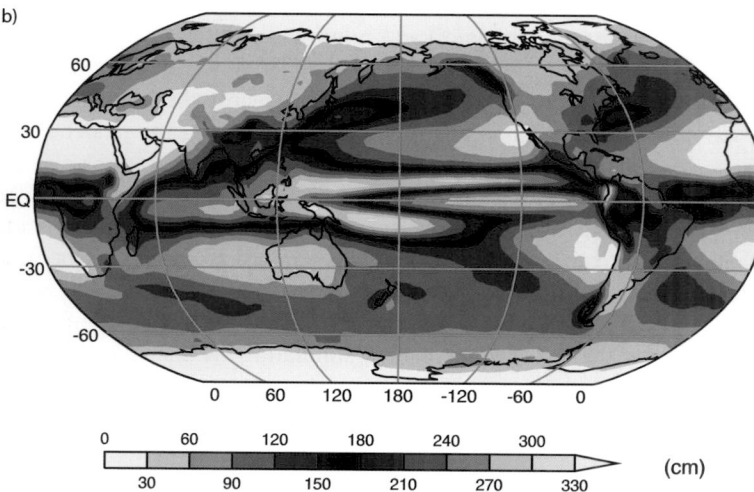

Figure 8.5. *Annual mean precipitation (cm), observed (a) and simulated (b), based on the multi-model mean. The Climate Prediction Center Merged Analysis of Precipitation (CMAP; Xie and Arkin, 1997) observation-based climatology for 1980 to 1999 is shown, and the model results are for the same period in the 20th-century simulations in the MMD at PCMDI. In (a), observations were not available for the grey regions. Results for individual models can be seen in Supplementary Material, Figure S8.9.*

atmosphere-only versions of the models show similarly large errors (e.g., Slingo et al., 2003). This may be one factor leading to a lack of consensus among models even as to the sign of future regional precipitation changes predicted in parts of the tropics (see Chapter 10).

At the heart of understanding what determines the regional distribution of precipitation over land and oceans in the tropics is atmospheric convection and its interaction with large-scale circulation. Convection occurs on a wide range of spatial and temporal scales, and there is increasing evidence that interactions across all scales may be crucial for determining the mean tropical climate and its regional rainfall distributions (e.g., Khairoutdinov et al., 2005). Over tropical land, the diurnal cycle dominates, and yet many models have difficulty simulating the early evening maximum in rainfall. Instead, they systematically tend to simulate rain before noon (Yang and Slingo, 2001; Dai, 2006), which compromises the energy budget of the land surface. Similarly, the land-sea breezes around the complex system of islands in Indonesia have been implicated

in the failure of models to capture the regional rainfall patterns across the Indo-Pacific Warm Pool (Neale and Slingo, 2003). Over the oceans, the precipitation distribution along the ITCZ results from organised convection associated with weather systems occurring on synoptic and intra-seasonal time scales (e.g., the Madden-Julian Oscillation (MJO); see Section 8.4.8). These systems are frequently linked to convectively coupled equatorial wave structures (e.g., Yang et al., 2003), but these are poorly represented in models (e.g., Lin et al., 2006; Ringer et al., 2006). Thus the rain-bearing systems, which establish the mean precipitation climatology, are not well simulated, contributing also to the poor temporal characteristics of daily rainfall (e.g., Dai, 2006) in which many models simulate rain too frequently but with reduced intensity.

Precipitation patterns are intimately linked to atmospheric humidity, evaporation, condensation and transport processes. Good observational estimates of the global pattern of evaporation are not available, and condensation and vertical transport of water vapour can often be dominated by sub-grid scale convective processes which are difficult to evaluate globally. The best prospect for assessing water vapour transport processes in humid regions, especially at annual and longer time scales, may be to compare modelled and observed streamflow, which must nearly balance atmospheric transport since terrestrial water storage variations on longer time scales are small (Milly et al., 2005; see Section 8.3.4.2).

Although an analysis of runoff in the MMD at PCMDI has not yet been performed, the net result of evaporation, transport and condensation processes can be seen in the atmospheric humidity distribution. Models reproduce the large-scale decrease in humidity with both latitude and altitude (see Supplementary Material, Figure S8.11), although this is not truly an independent check of models, since it is almost a direct consequence of their reasonably realistic simulation of temperature. The multi-model mean bias in humidity, zonally and annually averaged, is less than 10% throughout most of the lower troposphere compared with reanalyses, but model evaluation in the upper troposphere is considerably hampered by observational uncertainty.

Any errors in the water vapour distribution should affect the outgoing LW radiation (see Section 8.3.1.1.2), which was seen to be free of systematic zonal mean biases. In fact, the observed differences in outgoing LW radiation between the moist and dry regions are reproduced by the models, providing some confidence that any errors in humidity are not critically affecting the net fluxes at the TOA. However, the strength of water vapour feedback, which strongly affects global climate sensitivity, is primarily determined by fractional changes in water vapour in response to warming, and the ability of models to correctly represent this feedback is perhaps better assessed with process studies (see Section 8.6).

8.3.1.3 Extratropical Storms

The impact of extratropical cyclones on global climate derives primarily from their role in transporting heat, momentum and humidity. Regionally and individually, these mid-latitude storms often provide beneficial precipitation, but also occasionally produce destructive flooding and high winds. For these reasons, the effect of climate change on extratropical cyclones is of considerable importance and interest.

Among the several approaches used to characterise cyclone activity (e.g., Paciorek et al., 2002), analysis methods that identify and track extratropical cyclones can provide the most direct information concerning their frequency and movement (Hoskins and Hodges, 2002, 2005). Climatologies for the distribution and properties of cyclones found in models can be compared with reanalysis products (Chapter 3), which provide the best observation-constrained data.

Results from a systematic analysis of AMIP-2 simulations (Hodges, 2004; Stratton and Pope, 2004) indicate that models run with observed SSTs are capable of producing storm tracks located in about the right locations, but nearly all show some deficiency in the distribution and level of cyclone activity. In particular, simulated storm tracks are often more zonally oriented than is observed. A study by Lambert and Fyfe (2006), based on the MMD at PCMDI, finds that as a group, the recent models, which include interactive oceans, tend to underestimate slightly the total number of cyclones in both hemispheres. However, the number of intense storms is slightly overestimated in the NH, but underestimated in the Southern Hemisphere (SH), although observations are less certain there.

Increases in model resolution (characteristic of models over the last several years) appear to improve some aspects of extratropical cyclone climatology (Bengtsson et al., 2006), particularly in the NH where observations are most reliable (Hodges et al., 2003; Hanson et al., 2004; Wang et al., 2006). Improvements to the dynamical core and physics of models have also led to better agreement with reanalyses (Ringer et al., 2006; Watterson, 2006).

Our assessment is that although problems remain, climate models are improving in their simulation of extratropical cyclones.

8.3.2 Ocean

As noted earlier, this chapter focuses only on those variables important in determining the transient response of climate models (see Section 8.6). Due to space limitations, much of the analysis performed for this section is found in the Supplementary Material (Figures S8.12 to S8.15). An assessment of the modes of natural, internally generated variability can be found in Section 8.4. Comparisons of the type performed here need to be made with an appreciation of the uncertainties in the historical estimates of radiative forcing and various sampling issues in the observations (see Chapters 2 and 5). Unless otherwise noted, all results discussed here are based on the MMD at PCMDI.

8.3.2.1 Simulation of Mean Temperature and Salinity Structure

Before discussing the oceanic variables directly involved in determining the climatic response, it is important to discuss the fluxes between the ocean and atmosphere. Modelling experience shows that the surface fluxes play a large part in determining the fidelity of the oceanic simulation. Since the atmosphere and ocean are coupled, the fidelity of the oceanic simulation feeds back to the atmospheric simulation, affecting the surface fluxes.

Unfortunately, the total surface heat and water fluxes (see Supplementary Material, Figure S8.14) are not well observed. Normally, they are inferred from observations of other fields, such as surface temperature and winds. Consequently, the uncertainty in the observational estimate is large – of the order of tens of watts per square metre for the heat flux, even in the zonal mean. An alternative way of assessing the surface fluxes is by looking at the horizontal transports in the ocean. In the long-term average, the heat and water storage in the ocean are small so that the horizontal transports have to balance the surface fluxes. Since the heat transport seems better constrained by the available observations, it is presented here.

North of 45°N, most model simulations transport too much heat northward when compared to the observational estimates used here (Figure 8.6), but there is uncertainty in the observations. At 45°N, for example, the model simulations lie much closer to the estimate of 0.6×10^{15} W obtained by Ganachaud and Wunsch (2003). From 45°N to the equator, most model estimates lie near or between the observational estimates shown. In the tropics and subtropical zone of the SH, most models underestimate the southward heat transport away from the equator. At middle

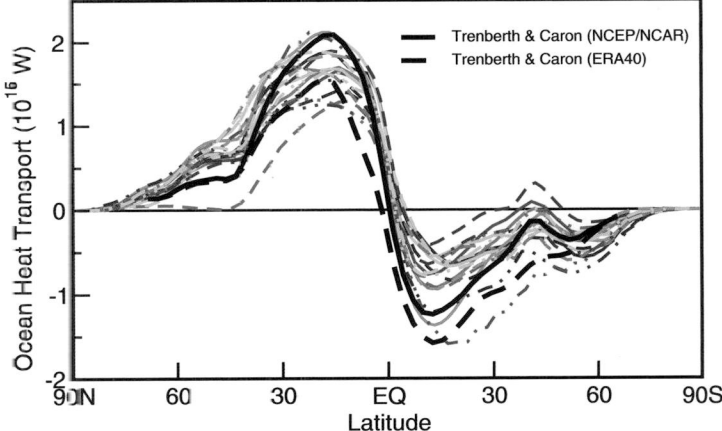

Figure 8.6. *Annual mean, zonally averaged oceanic heat transport implied by net heat flux imbalances at the sea surface, under an assumption of negligible changes in oceanic heat content. The observationally based estimate, taken from Trenberth and Caron (2001) for the period February 1985 to April 1989, derives from reanalysis products from the National Centers for Environmental Prediction (NCEP)/NCAR (Kalnay et al., 1996) and European Centre for Medium Range Weather Forecasts 40-year reanalysis (ERA40; Uppala et al., 2005). The model climatologies are derived from the years 1980 to 1999 in the 20th-century simulations in the MMD at PCMDI. The legend identifying individual models appears in Figure 8.4.*

and high latitudes of the SH, the observational estimates are more uncertain and the model-simulated heat transports tend to surround the observational estimates.

The oceanic heat fluxes have large seasonal variations which lead to large variations in the seasonal storage of heat by the oceans, especially in mid-latitudes. The oceanic heat storage tends to damp and delay the seasonal cycle of surface temperature. The model simulations evaluated here agree well with the observations of seasonal heat storage by the oceans (Gleckler et al., 2006a). The most notable problem area for the models is in the tropics, where many models continue to have biases in representing the flow of heat from the tropics into middle and high latitudes.

The annually averaged zonal component of surface wind stress, zonally averaged over the oceans, is reasonably well simulated by the models (Figure 8.7). At most latitudes, the reanalysis estimates (based on atmospheric models constrained by observations) lie within the range of model results. At middle to low latitudes, the model spread is relatively small and all the model results lie fairly close to the reanalysis. At middle to high latitudes, the model-simulated wind stress maximum tends to lie equatorward of the reanalysis. This error is particularly large in the SH, a region where there is more uncertainty in the reanalysis. Almost all model simulations place the SH wind stress maximum north of the reanalysis estimate. The Southern Ocean wind stress errors in the control integrations may adversely affect other aspects of the simulation and possibly the oceanic heat uptake under climate change, as discussed below.

The largest individual model errors in the zonally averaged SST (Figure 8.8) are found at middle and high latitudes, particularly the mid-latitudes of the NH where the model-simulated temperatures are too cold. Almost every model has some tendency for this cold bias. This error seems to be associated with poor simulation of the path of the North Atlantic Current and seems to be due to an ocean component problem rather than a problem with the surface fluxes. In the zonal averages near 60°S, there is a warm bias in the multi-model mean results. Many models suffer from a too-warm bias in the Southern Ocean SSTs.

In the individual model SST error maps (see Supplementary Material, Figure S8.1), it is apparent that most models have a large warm bias in the eastern parts of the tropical ocean basins, near the continental boundaries. This is also evident in the multi-model mean result (Figure 8.2a) and is associated with insufficient resolution, which leads to problems in the simulation of the local wind stress, oceanic upwelling and under-prediction of the low cloud amounts (see Sections 8.2 and 8.3.1). These are also regions where there is a relatively large spread among the model simulations, indicating a relatively wide range in the magnitude of these errors. Another area where the model error spread is relatively large is found in the North Atlantic Ocean. As noted above, this is an area where many models have problems properly locating the North Atlantic Current, a region of large SST gradients.

In spite of the errors, the model simulation of the SST field is fairly realistic overall. Over all latitudes, the multi-model mean

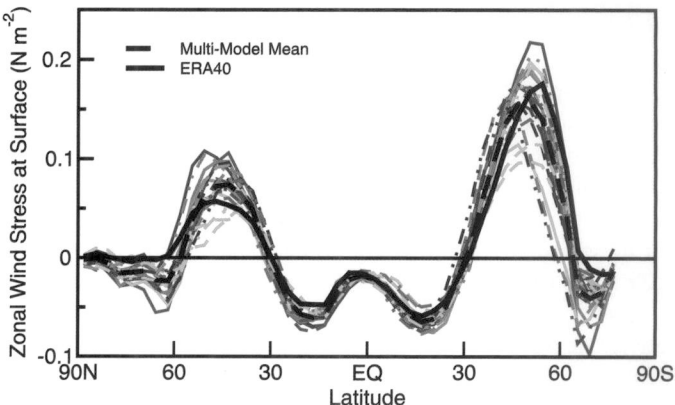

Figure 8.7. *Annual mean east-west component of wind stress zonally averaged over the oceans. The observationally constrained estimate is from the years 1980 to 1999 in the European Centre for Medium Range Weather Forecasts 40-year re-analysis (ERA40; Uppala et al., 2005), and the model climatologies are calculated for the same period in the 20th-century simulations in the MMD at PCMDI. The legend identifying individual models appears in Figure 8.4.*

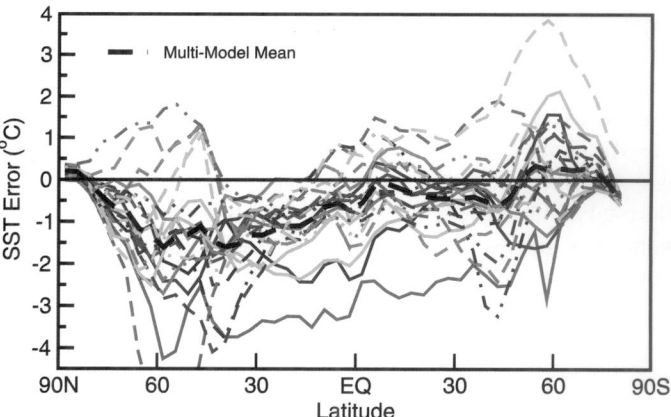

Figure 8.8. *Annual mean, zonally averaged SST error, simulated minus observed climatology. The Hadley Centre Sea Ice and Sea Surface Temperature (HadISST; Rayner et al., 2003) observational climatology for 1980 to 1999 is the reference used here, and the model results are for the same period in the 20th-century simulations in the MMD at PCMDI. In the presence of sea ice, the SST is assumed to be at the freezing point of seawater. The legend identifying individual models appears in Figure 8.4.*

zonally averaged SST error is less than 2°C, which is fairly small considering that most models do not use flux adjustments in these simulations. The model mean local SST errors are also less than 2°C over most regions, with only relatively small areas exceeding this value. Even relatively small SST errors, however, can adversely affect the simulation of variability and teleconnections (Section 8.4).

Over most latitudes, at depths ranging from 200 to 3,000 m, the multi-model mean zonally averaged ocean temperature is too warm (see Figure 8.9). The maximum warm bias (about 2°C) is located in the region of the North Atlantic Deep Water (NADW) formation. Above 200 m, however, the multi-model mean is too cold, with maximum cold bias (more than 1°C) near the surface at mid-latitudes of the NH, as discussed above. Most models generally have an error pattern similar to the

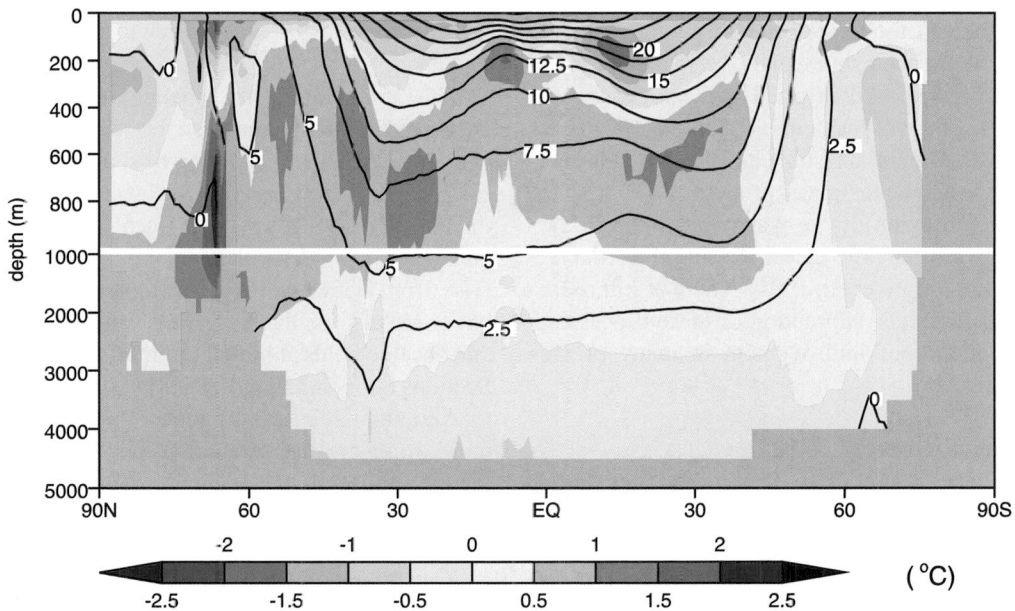

Figure 8.9. *Time-mean observed potential temperature (°C), zonally averaged over all ocean basins (labelled contours) and multi-model mean error in this field, simulated minus observed (colour-filled contours). The observations are from the 2004 World Ocean Atlas compiled by Levitus et al. (2005) for the period 1957 to 1990, and the model results are for the same period in the 20th-century simulations in the MMD at PCMDI. Results for individual models can be seen in the Supplementary Material, Figure S8.12.*

multi-model mean (see Supplementary Material, Figure S8.12) except for CNRM-CM3 and MRI-CGCM2.3.2, which are too cold throughout most of the mid- and low-latitude ocean (see Supplementary Material, Figure S8.12). The GISS-EH model is much too cold throughout the subtropical thermocline and only the NH part of the FGOALS-g1.0 error pattern is similar to the model mean error described here. The magnitude of these errors, especially in the deeper parts of the ocean, depends on the AOGCM initialisation method (Section 8.2.7).

The error pattern, in which the upper 200 m of the ocean tend to be too cold while the layers below are too warm, indicates that the thermocline in the multi-model mean is too diffuse. This error, which was also present at the time of the TAR, seems partly related to the wind stress errors in the SH noted above and possibly to errors in formation and mixing of NADW. The multi-model mean errors in temperature (too warm) and salinity (too salty; see Supplementary Material, Figure S8.13) at middle and low latitudes near the base of the thermocline tend to cancel in terms of a density error and appear to be associated with the problems in the formation of Antarctic Intermediate Water (AAIW), as discussed below.

8.3.2.2 Simulation of Circulation Features Important for Climate Response

8.3.2.2.1 Meridional overturning circulation

The MOC is an important component of present-day climate and many models indicate that it will change in the future (Chapter 10). Unfortunately, many aspects of this circulation are not well observed. The MOC transports large amounts of heat and salt into high latitudes of the North Atlantic Ocean, where the relatively warm, salty surface waters are cooled by the atmosphere, making the water dense enough to sink to

depth. These waters then flow southward towards the Southern Ocean where they mix with the rest of the World Ocean waters (see Supplementary Material, Figure S8.15).

The models simulate this major aspect of the MOC and also simulate a number of distinct wind-driven surface cells (see Supplementary Material, Figure S8.15). In the tropics and subtropics, these cells are quite shallow, but at the latitude of the Drake Passage (55°S) the wind-driven cell extends to a much greater depth (2 to 3 km). Most models in the multi-model data set have some manifestation of the wind-driven cells. The strength and pattern of the overturning circulation varies greatly from model to model (see Supplementary Material, Figure S8.15). The GISS-AOM exhibits the strongest overturning circulation, almost 40 to 50 Sv (10^6 m^3 s^{-1}). The CGCM (T47 and T63) and FGOALS have the weakest overturning circulations, about 10 Sv. The observed value is about 18 Sv (Ganachaud and Wunsch 2000).

In the Atlantic, the MOC, extending to considerable depth, is responsible for a large fraction of the northward oceanic heat transport in both observations and models (e.g., Hall and Bryden, 1982; Gordon et al., 2000). Figure 10.15 contains an index of the Atlantic MOC at 30°N for the suite of AOGCM 20th-century simulations. While the majority of models show an MOC strength that is within observational uncertainty, some show higher and lower values and a few show substantial drifts which could make interpretation of MOC projections using those models very difficult.

Overall, some aspects of the simulation of the MOC have improved since the TAR. This is due in part to improvements in mixing schemes, the use of higher resolution ocean models (see Section 8.2) and better simulation of the surface fluxes. This improvement can be seen in the individual model MOC sections (see Supplementary Material, Figure S8.15) by the fact that (1) the location of the deep-water formation is more realistic, with more sinking occurring in the Greenland-Iceland-Norwegian and Labrador Seas as evidenced by the larger stream function values north of the sill located at 60°N (e.g., Wood et al., 1999) and (2) deep waters are subjected to less spurious mixing, resulting in better water mass properties (Thorpe et al., 2004) and a larger fraction of the water that sinks in the northern part of the North Atlantic Ocean exiting the Atlantic Ocean near 30°S (Danabasoglu et al., 1995). There is still room for improvement in the models' simulation of these processes, but there is clear evidence of improvement in many of the models analysed here.

8.3.2.2.2 Southern Ocean circulation

The Southern Ocean wind stress error has a particularly large detrimental impact on the Southern Ocean simulation by the models. Partly due to the wind stress error identified above, the simulated location of the Antarctic Circumpolar Current (ACC) is also too far north in most models (Russell et al., 2006). Since the AAIW is formed on the north side of the ACC, the water mass properties of the AAIW are distorted (typically too warm and salty: Russell et al., 2006). The relatively poor AAIW simulation contributes to the multi-model mean error identified above where the thermocline is too diffuse, because the waters near the base of thermocline are too warm and salty.

It is likely that the relatively poor Southern Ocean simulation will influence the transient climate response to increasing greenhouse gases by affecting the oceanic heat uptake. When forced by increases in radiative forcing, models with too little Southern Ocean mixing will probably underestimate the ocean heat uptake; models with too much mixing will likely exaggerate it. These errors in oceanic heat uptake will also have a large impact on the reliability of the sea level rise projections. See Chapter 10 for more discussion of this subject.

8.3.2.3 Summary of Oceanic Component Simulation

Overall, the improvements in the simulation of the observed time mean ocean state noted in the TAR (McAvaney et al., 2001) have continued in the models evaluated here. It is notable that this improvement has continued in spite of the fact that nearly all models no longer use flux adjustments. This suggests that the improvements in the physical parametrizations, increased resolution (see Section 8.2) and improved surface fluxes are together having a positive impact on model simulations. The temperature and salinity errors in the thermocline, while still large, have been reduced in many models. In the NH, many models still suffer from a cold bias in the upper ocean which is at a maximum near the surface and may distort the ice-albedo feedback in some models (see Section 8.3.3). In the Southern Ocean, the equatorward bias of the westerly wind stress maximum found in most model simulations is a problem that may affect the models' response to increasing radiative forcing.

8.3.3 Sea Ice

The magnitude and spatial distribution of the high-latitude climate changes can be strongly affected by sea ice characteristics, but evaluation of sea ice in models is hampered by insufficient observations of some key variables (e.g., ice thickness) (see Section 4.4). Even when sea ice errors can be quantified, it is difficult to isolate their causes, which might arise from deficiencies in the representation of sea ice itself, but could also be due to flawed simulation of the atmospheric and oceanic fields at high latitudes that drive ice movement (see Sections 8.3.1, 8.3.2 and 11.3.8).

Although sea ice treatment in AOGCMs has become more sophisticated, including better representation of both the dynamics and thermodynamics (see Section 8.2.4), improvement in simulating sea ice in these models, as a group, is not obvious (compare Figure 8.10 with TAR Figure 8.10; or Kattsov and Källén, 2005, Figure 4.11). In some models, however, the geographic distribution and seasonality of sea ice is now better reproduced.

For the purposes of model evaluation, the most reliably measured characteristic of sea ice is its seasonally varying extent (i.e., the area enclosed by the ice edge, operationally defined as the 15% contour; see Section 4.4). Despite the wide differences among the models, the multi-model mean of sea ice extent is in reasonable agreement with observations. Based on 14 of the 15 AOGCMs available at the time of analysis (one model was excluded because of unrealistically large ice extents; Arzel et al., 2006), the mean extent of simulated sea ice exceeds that observed in the NH by up to roughly 1×10^6 km^2 throughout the year, whereas in the SH the annual cycle is exaggerated, with too much sea ice in September ($\sim 2 \times 10^6$ km^2) and too little in March by a lesser amount. In many models the regional distribution of sea ice is poorly simulated, even if the hemispheric areal extent is approximately correct (Arzel et al., 2006; Holland and Raphael, 2006; Zhang and Walsh, 2006). The spread of simulated sea ice extents, measured as the multi-model standard deviation from the model mean, is generally narrower in the NH than in the SH (Arzel et al., 2006). Even in the best case (NH winter), the range of simulated sea ice extent exceeds 50% of the mean, and ice thickness also varies considerably, suggesting that projected decreases in sea ice cover remain rather uncertain. The model sea ice biases may influence global climate sensitivity (see Section 8.6). There is a tendency for models with relatively large sea ice extent in the present climate to have higher sensitivity. This is apparently especially true of models with low to moderate polar amplification (Holland and Bitz, 2003).

Among the primary causes of biases in simulated sea ice (especially its distribution) are biases in the simulation of high-latitude winds (Bitz et al., 2002; Walsh et al., 2002; Chapman and Walsh, 2007), as well as vertical and horizontal mixing in the ocean (Arzel et al., 2006). Also important are surface heat flux errors, which in particular may result from inadequate parametrizations of the atmospheric boundary layer (under stable conditions commonly occurring at night and in the winter over sea ice) and generally from poor simulation

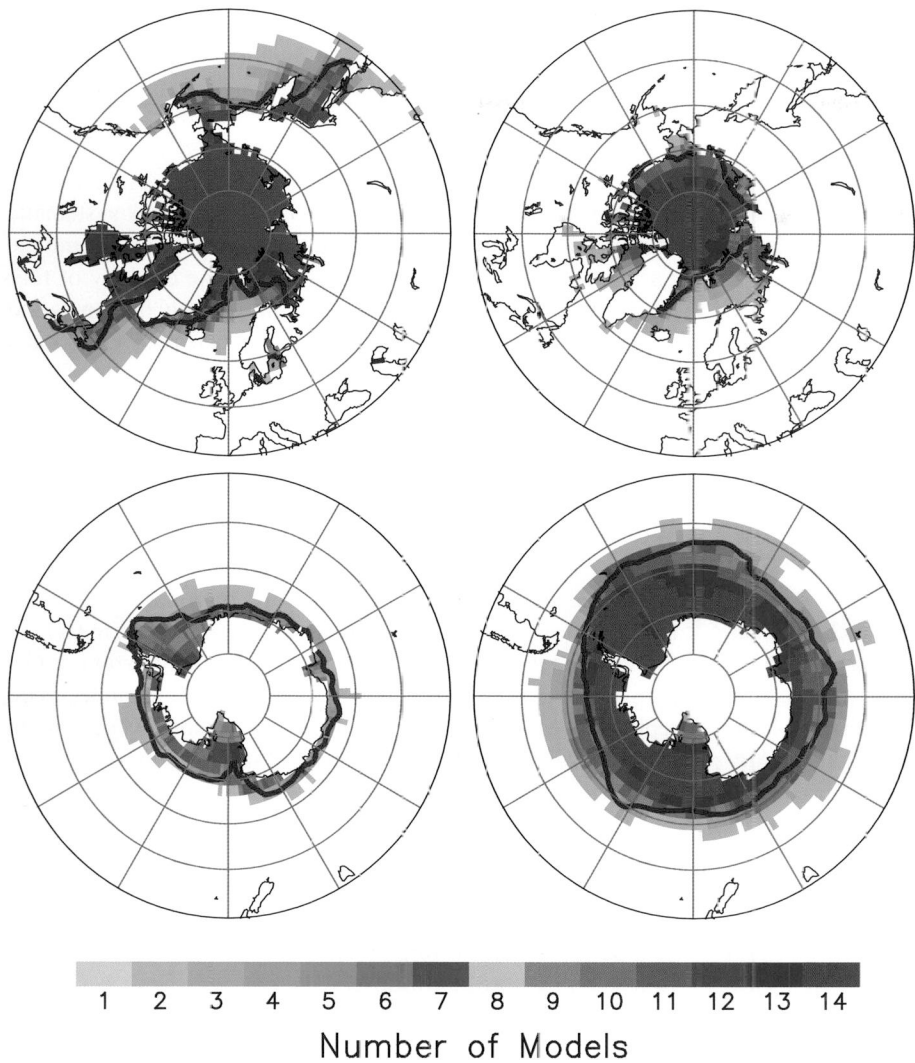

Number of Models

Figure 8.10. *Baseline climate (1980–1999) sea ice distribution in the Northern Hemisphere (upper panels) and Southern Hemisphere (lower panels) simulated by 14 of the AOGCMs listed in Table 8.1 for March (left) and September (right), adapted from Arzel et al. (2006). For each 2.5° x 2.5° longitude-latitude grid cell, the figure indicates the number of models that simulate at least 15% of the area covered by sea ice. The observed 15% concentration boundaries (red line) are based on the Hadley Centre Sea Ice and Sea Surface Temperature (HadISST; Rayner et al., 2003) data set.*

of high-latitude cloudiness, which is evident from the striking inter-model scatter (e.g., Kattsov and Källén, 2005).

8.3.4 Land Surface

Evaluation of the land surface component in coupled models is severely limited by the lack of suitable observations. The terrestrial surface plays key climatic roles in influencing the partitioning of available energy between sensible and latent heat fluxes, determining whether water drains or remains available for evaporation, determining the surface albedo and whether snow melts or remains frozen, and influencing surface fluxes of carbon and momentum. Few of these can be evaluated at large spatial or long temporal scales. This section therefore evaluates those quantities for which some observational data exist.

8.3.4.1 Snow Cover

Analysis and comparison of AMIP-2 results, available at the time of the TAR, and more recent AOGCM results in the present MMD at PCMDI, show that models are now more consistent in their simulation of snow cover. Problems remain, however, and Roesch (2006) showed that the recent models predict excessive snow water equivalent (SWE) in spring, likely because of excessive winter precipitation. Frei et al. (2005) found that AMIP-2 models simulate the seasonal timing and the relative spatial patterns of SWE over North America fairly well, but identified a tendency to overestimate ablation during spring. At the continental scale, the highest monthly SWE integrated over the North American continent in AMIP-2 models varies within ±50% of the observed value of about 1,500 km³. The magnitude of these model errors is large enough to affect continental water balances. Snow cover area (SCA) is well captured by the recent models, but interannual variability is too low during melt. Frei et al. (2003) showed where observations were within the inter-quartile range of AMIP-2 models for all months at the hemispheric and continental scale. Encouragingly, there was significant improvement over earlier AMIP-1 simulations for seasonal and interannual variability of SCA (Frei et al., 2005). Both the recent AOGCMs and AMIP models reproduced the observed decline in annual SCA over the period 1979 to 1995 and most models captured the observed decadal-scale variability over the 20th century. Despite these improvements, a minority of models still exaggerate SCA.

Large discrepancies remain in albedo for forested areas under snowy conditions, due to difficulties in determining the extent of masking of snow by vegetation (Roesch, 2006). The ability of terrestrial models to simulate snow under observed meteorological forcing has been evaluated via several intercomparisons. At the scale of individual grid cells, for mid-latitude (Slater et al., 2001) and alpine (Etchevers et al., 2004) locations, the spread of model simulations usually encompasses observations. However, grid-box scale simulations of snow over high-latitude river basins identified significant limitations (Nijssen et al., 2003), due to difficulties relating to calculating net radiation, fractional snow cover and interactions with vegetation.

8.3.4.2 Land Hydrology

The evaluation of the hydrological component of climate models has mainly been conducted uncoupled from AOGCMs (Bowling et al., 2003; Nijssen et al., 2003; Boone et al., 2004). This is due in part to the difficulties of evaluating runoff simulations across a range of climate models due to variations in rainfall, snowmelt and net radiation. Some attempts have, however, been made. Arora (2001) used the AMIP-2 framework to show that the Canadian Climate Model's simulation of the global hydrological cycle compared well to observations, but regional variations in rainfall and runoff led to differences at the basin scale. Gerten et al. (2004) evaluated the hydrological performance of the Lund-Potsdam-Jena (LPJ) model and showed that the model performed well in the simulation of runoff and evapotranspiration compared to other global hydrological models, although the version of LPJ assessed had been enhanced to improve the simulation of hydrology over the versions used by Sitch et al. (2003).

Milly et al. (2005) made use of the MMD, which contains results from recent models, to investigate whether observed 20th-century trends in regional land hydrology could be attributed to variations in atmospheric composition and solar irradiance. Their analysis, based on an ensemble of 26 integrations of 20th-century climate from nine climate models, showed that at regional scales these models simulated observed streamflow measurements with good qualitative skill. Further, the models demonstrated highly significant quantitative skill in identifying the regional runoff trends indicated by 165 long-term stream gauges. They concluded that the impact of changes in atmospheric composition and solar irradiance on observed streamflow was, at least in part, predictable. This is an important scientific advance: it suggests that despite limitations in the hydrological parametrizations included in climate models, these models can capture observed changes in 20th-century streamflow associated with atmospheric composition and solar irradiance changes. This enhances confidence in the use of these models for future projection.

8.3.4.3 Surface Fluxes

Despite considerable effort since the TAR, uncertainties remain in the representation of solar radiation in climate models (Potter and Cess, 2004). The AMIP-2 results and the recent model results in the MMD provide an opportunity for a major systematic evaluation of model ability to simulate solar radiation. Wild (2005) and Wild et al. (2006) evaluated these models and found considerable differences in the global annual mean solar radiation absorbed at the Earth's surface. In comparison to global surface observations, Wild (2005) concluded that many climate models overestimate surface absorption of solar radiation partly due to problems in the parametrizations of atmospheric absorption, clouds and aerosols. Similar uncertainties exist in the simulation of downwelling infrared radiation (Wild et al., 2001). Difficulties in simulating absorbed solar and infrared radiation at the surface leads inevitably to uncertainty in the simulation of surface sensible and latent heat fluxes.

8.3.4.4 Carbon

A major advance since the TAR is some systematic assessments of the capability of land surface models to simulate carbon. Dargaville et al. (2002) evaluated the capacity of four global vegetation models to simulate the seasonal dynamics and interannual variability of atmospheric CO_2 between 1980 and 1991. Using off-line forcing, they evaluated the capacity of these models to simulate carbon fluxes, via an atmospheric transport model, using observed atmospheric CO_2 concentration. They found that the terrestrial models tended to underestimate the amplitude of the seasonal cycle and simulated the spring uptake of CO_2 approximately one to two months too early. Of the four models, none was clearly superior in its capacity to simulate the global carbon budget, but all four reproduced the main features of the observed seasonal cycle in atmospheric CO_2. A further off-line evaluation of the LPJ global vegetation model by Sitch et al. (2003) provided confidence that the model could replicate the observed vegetation pattern, seasonal variability in net ecosystem exchange and local soil moisture measurements when forced by observed climatologies.

The only systematic evaluation of carbon models that were interactively coupled to climate models occurred as part of the Coupled Climate-Carbon Cycle Model Intercomparison Project (C⁴MIP), where Friedlingstein et al. (2006) compared the ability of a suite of models to simulate historical atmospheric CO_2 concentration forced by observed emissions. Issues relating to the magnitude of the fertilization effect and the partitioning between land and ocean uptake were identified in individual models, but it is only under increasing CO_2 in the future (see Chapter 10) that the differences become large. Several other groups have evaluated the impact of coupling specific models of carbon to climate models but clear results are difficult to obtain because of inevitable biases in both the terrestrial and atmospheric modules (e.g., Delire et al., 2003).

8.3.5 Changes in Model Performance

Standard experiments, agreed upon by the climate modelling community to facilitate model intercomparison (see Section 8.1.2.2), have produced archives of model output that make it easier to track historical changes in model performance. Most of the modelling groups that contributed output to the current MMD at PCMDI also archived simulations from their earlier models (circa 2000) as part of the Coupled Model Intercomparison Project (CMIP1&2). The TAR largely relied on the earlier generation of models in its assessment.

Based on the archived model output, it is possible to quantify changes in performance of evolving models.[4] This can be done most straightforwardly by only considering the 14 modelling

---

[4] One modelling group participating in CMIP1&2 did not contribute to the MMD, and four groups providing output to the MMD did not do so for CMIP1&2. Results from these five groups are therefore not considered in this subsection. Some modelling groups contributed results from more than one version of their model (sometimes, simply running it at two different resolutions), and in these cases the mean of the two model results is considered here.

groups that contributed output from both their earlier and more recent models. One important aspect of model skill is how well the models simulate the seasonally varying global pattern of climatically important fields. The only monthly mean fields available in the CMIP1&2 archive are surface air temperature, precipitation and mean sea level pressure, so these are the focus of this analysis. Although the simulation conditions in the MMD 20th-century simulations were not identical to those in the CMIP1&2 control runs, the differences do not alter the conclusions summarised below because the large-scale climatological features dominate, not the relatively small perturbations resulting from climate change.

A summary of the ability of AOGCMs to simulate the seasonally varying climate state is provided by Figure 8.11, which displays error measures that gauge how well recent models simulate precipitation, sea level pressure and surface temperature, compared with their predecessors. The normalised RMS error shown is a so-called space-time statistic, computed from squared errors, summed over all 12 climatological months and over the entire globe, with grid cell values weighted by the corresponding grid cell area. This statistic can be used to assess the combined contributions of both spatial pattern errors and seasonal cycle errors. The RMS error is divided by the corresponding observed standard deviation of the field to provide a relative measure of the error. In Figure 8.11 this scaling implies that pressure is better simulated than precipitation, and that surface temperature is simulated best of all.

The models in Figure 8.11 are categorised based on whether or not flux adjustments were applied (see Section 8.2.7). Of the earlier generation models, 8 of the 14 models were flux adjusted, but only two of these groups continue this practice. Several conclusions can be drawn from the figure: 1) although flux-adjusted models on average have smaller errors than those without (in both generations), the smallest errors in simulating sea level pressure and surface temperature are found in models without flux adjustment; 2) despite the elimination of flux adjustment in all but two of the recent models, the mean error obtained from the recent suite of 14 models is smaller than errors found in the corresponding earlier suite of models; and 3) models without flux adjustment have improved on average, as have the flux-adjusted models. An exception to this last statement is the slight increase in mean RMS error for sea level pressure found in non-flux-adjusted models. Despite no apparent improvement in the mean in this case, three of the recent generation models have smaller sea level pressure errors than any of the earlier models.

These results demonstrate that the models now being used in applications by major climate modelling groups better simulate seasonally varying patterns of precipitation, mean sea level pressure and surface air temperature than the models relied on by these same groups at the time of the TAR.

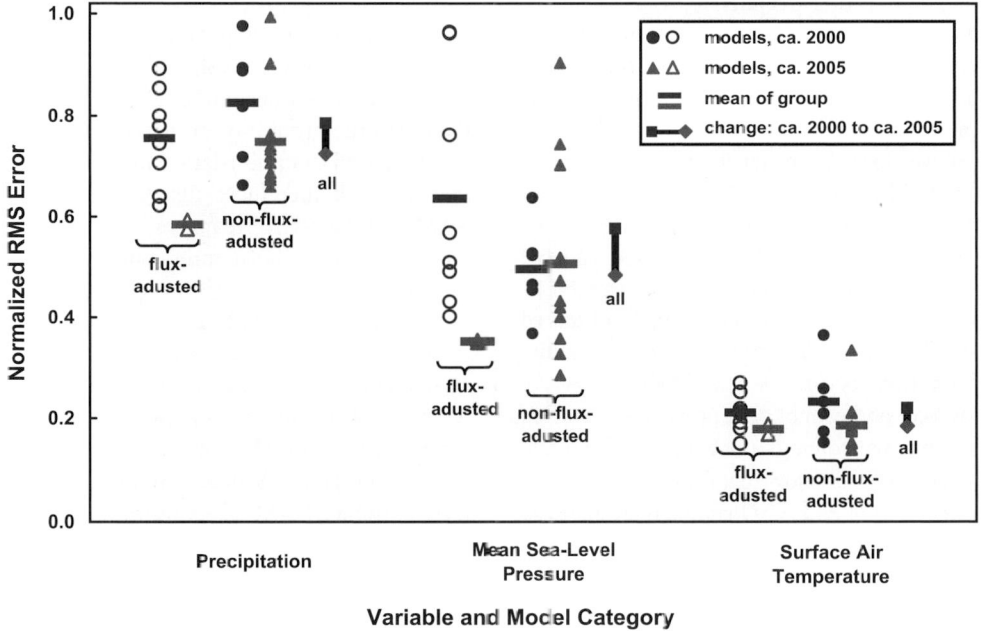

Figure 8.11. *Normalised RMS error in simulation of climatological patterns of monthly precipitation, mean sea level pressure and surface air temperature. Recent AOGCMs (circa 2005) are compared to their predecessors (circa 2000 and earlier). Models are categorised based on whether or not any flux adjustments were applied. The models are gauged against the following observation-based datasets: Climate Prediction Center Merged Analysis of Precipitation (CMAP; Xie and Arkin, 1997) for precipitation (1980–1999), European Centre for Medium Range Weather Forecasts 40-year reanalysis (ERA40; Uppala et al., 2005) for sea level pressure (1980–1999) and Climatic Research Unit (CRU; Jones et al., 1999) for surface temperature (1961–1990). Before computing the errors, both the observed and simulated fields were mapped to a uniform 4° x 5° latitude-longitude grid. For the earlier generation of models, results are based on the archived output from control runs (specifically, the first 30 years, in the case of temperature, and the first 20 years for the other fields), and for the recent generation models, results are based on the 20th-century simulations with climatological periods selected to correspond with observations. (In both groups of models, results are insensitive to the period selected.)*

8.4 Evaluation of Large-Scale Climate Variability as Simulated by Coupled Global Models

The atmosphere-ocean coupled climate system shows various modes of variability that range widely from intra-seasonal to inter-decadal time scales. Successful simulation and prediction over a wide range of these phenomena increase confidence in the AOGCMs used for climate predictions of the future.

8.4.1 Northern and Southern Annular Modes

There is evidence (e.g., Fyfe et al., 1999; Shindell et al., 1999) that the simulated response to greenhouse gas forcing in AOGCMs has a pattern that resembles the models' Northern Annular Mode (NAM), and thus it would appear important that the NAM (see Chapters 3 and 9) is realistically simulated. Analyses of individual AOGCMs (e.g., Fyfe et al., 1999; Shindell et al., 1999) have demonstrated that they are capable of simulating many aspects of the NAM and NAO patterns including linkages between circulation and temperature. Multi-model comparisons of winter atmospheric pressure (Osborn, 2004), winter temperature (Stephenson and Pavan, 2003) and atmospheric pressure across all months of the year (AchutaRao et al., 2004), including assessments of the MMD at PCMDI (Miller et al., 2006) confirm the overall skill of AOGCMs but also identify that teleconnections between the Atlantic and Pacific Oceans are stronger in many models than is observed (Osborn, 2004). In some models this is related to a bias towards a strong polar vortex in all winters so that their simulations nearly always reflect behaviour that is only observed at times with strong vortices (when a stronger Atlantic-Pacific correlation is observed; Castanheira and Graf, 2003).

Most AOGCMs organise too much sea level-pressure variability into the NAM and NAO (Miller et al., 2006). The year-to-year variance of the NAM or NAO is correctly simulated by some AOGCMs, while other simulations are significantly too variable (Osborn, 2004); for the models that simulate stronger variability, the persistence of anomalous states is greater than is observed (AchutaRao et al., 2004). The magnitude of multi-decadal variability (relative to sub-decadal variability) is lower in AOGCM control simulations than is observed, and cannot be reproduced in current model simulations with external forcings (Osborn, 2004; Gillett, 2005). However, Scaife et al. (2005) show that the observed multi-decadal trend in the surface NAM and NAO can be reproduced in an AOGCM if observed trends in the lower stratospheric circulation are prescribed in the model. Troposphere-stratosphere coupling processes may therefore need to be included in models to fully simulate NAM variability. The response of the NAM and NAO to volcanic aerosols (Stenchikov et al., 2002), sea surface temperature variability (Hurrell et al., 2004) and sea ice anomalies (Alexander et al., 2004) demonstrate some compatibility with observed variations, though the difficulties in determining cause and effect in the coupled system limit the conclusions that can be drawn with regards to the trustworthiness of model behaviour.

Like its NH counterpart, the NAM, the Southern Annular Mode (SAM; see Chapters 3 and 9) has signatures in the tropospheric circulation, the stratospheric polar vortex, mid-latitude storm tracks, ocean circulation and sea ice. AOGCMs generally simulate the SAM realistically (Fyfe et al., 1999; Cai et al., 2003; Miller et al., 2006). For example, Figure 8.12 compares the austral winter SAM simulated in the MMD at PCMDI to the observed SAM as represented in the National Centers for Environmental Prediction (NCEP) reanalysis. The main elements of the pattern, the low-pressure anomaly over Antarctica and the high-pressure anomalies equatorward of 60°S are captured well by the AOGCMs. In all but two AOGCMs, the spatial correlation between the observed and simulated SAM is greater than 0.95. Further analysis shows that the SAM signature in surface temperature, such as the surface warm anomaly over the Antarctic Peninsula associated with a positive SAM event, is also captured by some AOGCMs (e.g., Delworth et al., 2006; Otto-Bliesner et al., 2006). This follows from the realistic simulation of the SAM-related circulation shown in Figure 8.12, because the surface temperature signatures of the SAM typically reflect advection of the climatological temperature distribution by the SAM-related circulation (Thompson and Wallace, 2000).

Although the spatial structure of the SAM is well simulated by the AOGCMs in the MMD at PCMDI, other features of the SAM, such as the amplitude, the detailed zonal structure and the temporal spectra, do not always compare well with the NCEP reanalysis SAM (Miller et al., 2006; Raphael and Holland, 2006). For example, Figure 8.12 shows that the simulated SAM variance (the square of the SAM amplitude) ranges between 0.9 and 2.4 times the NCEP reanalysis SAM variance. However, such features vary considerably among different realisations of multiple-member ensembles (Raphael and Holland, 2006), and the temporal variability of the NCEP reanalysis SAM does not compare well to station data (Marshall, 2003). Thus, it is difficult to assess whether these discrepancies between the simulated SAM and the NCEP reanalysis SAM point to shortcomings in the models or to shortcomings in the observed analysis.

Resolving these issues may require a better understanding of SAM dynamics. Although the SAM exhibits clear signatures in the ocean and stratosphere, its tropospheric structure can be simulated, for example, in atmospheric GCMs with a poorly resolved stratosphere and driven by prescribed SSTs (e.g., Limpasuvan and Hartmann, 2000; Cai et al., 2003). Even much simpler atmospheric models with one or two vertical levels produce SAM-like variability (Vallis et al., 2004). These relatively simple models capture the dynamics that underlie SAM variability – namely, interactions between the tropospheric jet stream and extratropical weather systems (Limpasuvan and Hartmann, 2000; Lorenz and Hartmann, 2001). Nevertheless, the ocean and stratosphere might still influence SAM variability in important ways. For example, AOGCM simulations suggest strong SAM-related impacts on ocean temperature, ocean heat transport and sea ice distribution (Watterson, 2001; Hall and Visbeck, 2002), suggesting a potential for air-sea interactions to influence SAM dynamics. Furthermore, observational

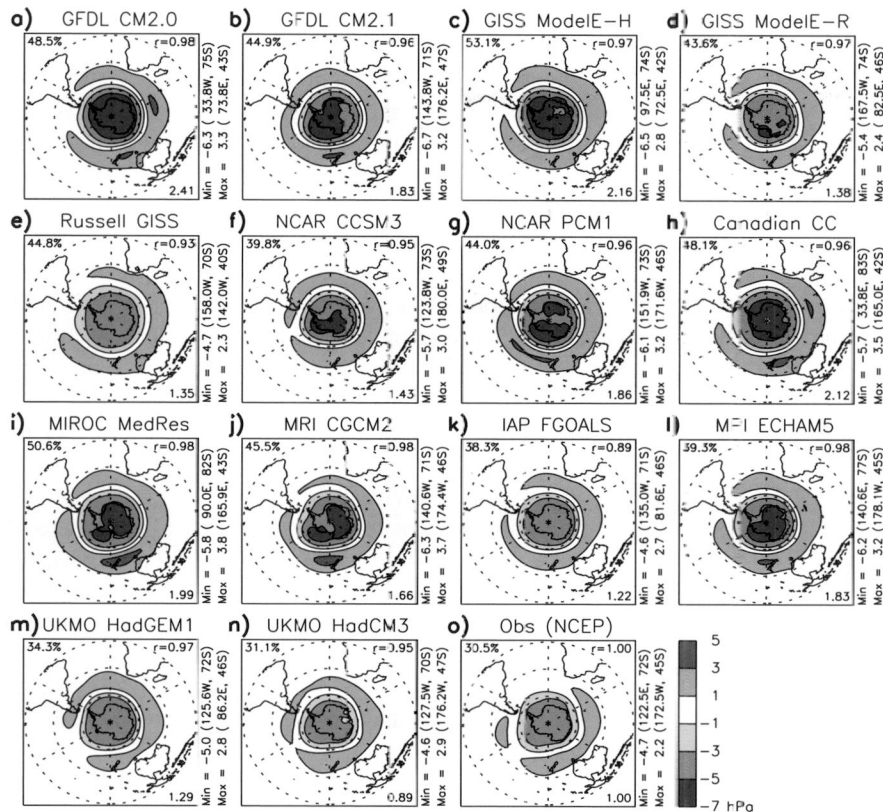

Figure 8.12. *Ensemble mean leading Empirical Orthogonal Function (EOF) of summer (November through February) Southern Hemisphere sea level pressure (hPa) for 1950 to 1999. The EOFs are scaled so that the associated principal component has unit variance over this period. The percentage of variance accounted for by the leading mode is listed at the upper left corner of each panel. The spatial correlation (r) with the observed pattern is given at the upper right corner. At the lower right is the ratio of the EOF spatial variance to the observed value. "Canadian CC" refers to CGCM3.1 (T47), and "Russell GISS" refers to the GISS AOM. Adapted from Miller et al. (2006).*

and modelling studies (e.g., Thompson and Solomon, 2002; Baldwin et al., 2003; Gillett and Thompson, 2003) suggest that the stratosphere might also influence the tropospheric SAM, at least in austral spring and summer. Thus, an accurate simulation of stratosphere-troposphere and ocean-atmosphere coupling may still be necessary to accurately simulate the SAM.

8.4.2 Pacific Decadal Variability

Recent work suggests that the Pacific Decadal Oscillation (PDO, see Chapters 3 and 9) is the North Pacific expression of a near-global ENSO-like pattern of variability called the Inter-decadal Pacific Oscillation or IPO (Power et al., 1999; Deser et al., 2004). The appearance of the IPO as the leading Empirical Orthogonal Function (EOF) of SST in AOGCMs that do not include inter-decadal variability in natural or external forcing indicates that the IPO is an internally generated, natural form of variability. Note, however, that some AOGCMs exhibit an El Niño-like response to global warming (Cubasch et al., 2001) that can take decades to emerge (Cai and Whetton, 2000). Therefore some, though certainly not all, of the variability seen in the IPO and PDO indices might be anthropogenic in origin (Shiogama et al., 2005). The IPO and PDO can be partially understood as the residual of random inter-decadal changes in ENSO activity (e.g., Power et al., 2006), with their spectra reddened (i.e., increasing energy at lower frequencies) by the integrating effect of the upper ocean mixed layer (Newman et al., 2003; Power and Colman, 2006) and the excitation of low frequency off-equatorial Rossby waves (Power and Colman,

2006). Some of the inter-decadal variability in the tropics also has an extratropical origin (e.g., Barnett et al., 1999; Hazeleger et al., 2001) and this might give the IPO a predictable component (Power et al., 2006).

Atmosphere-Ocean General Circulation Models do not seem to have difficulty in simulating IPO-like variability (e.g., Yeh and Kirtman, 2004; Meehl and Hu, 2006), even AOGCMs that are too coarse to properly resolve equatorially trapped waves important for ENSO dynamics. Some studies have provided objective measures of the realism of the modelled decadal variability. For example, Pierce et al. (2000) found that the ENSO-like decadal SST mode in the Pacific Ocean of their AOGCM had a pattern that gave a correlation of 0.56 with its observed counterpart. This compared with a correlation coefficient of 0.79 between the modelled and observed interannual ENSO mode. The reduced agreement on decadal time scales was attributed to lower than observed variability in the North Pacific subpolar gyre, over the southwest Pacific and along the western coast of North America. The latter was attributed to poor resolution of the coastal waveguide in the AOGCM. The importance of properly resolving coastally trapped waves in the context of simulating decadal variability in the Pacific has been raised in a number studies (e.g., Meehl and Hu, 2006). Finally, there has been little work evaluating the amplitude of Pacific decadal variability in AOGCMs. Manabe and Stouffer (1996) showed that the variability has roughly the right magnitude in their AOGCM, but a more detailed investigation using recent AOGCMs with a specific focus on IPO-like variability would be useful.

8.4.3 Pacific-North American Pattern

The Pacific-North American (PNA) pattern (see Chapter 3) is commonly associated with the response to anomalous boundary forcing. However, PNA-like patterns have been simulated in atmospheric GCM experiments subjected to constant boundary conditions. Hence, both external and internal processes may contribute to the formation of this pattern. Particular attention has been paid to the external influences due to SST anomalies related to ENSO episodes in the tropical Pacific, as well as those situated in the extratropical North Pacific. Internal mechanisms that might play a role in the formation of the PNA pattern include interactions between the slowly varying component of the circulation and high-frequency transient disturbances, and instability of the climatological flow pattern. Trenberth et al. (1998) reviewed the myriad observational and modelling studies on various processes contributing to the PNA pattern.

The ability of GCMs to replicate various aspects of the PNA pattern has been tested in coordinated experiments. Until several years ago, such experiments were conducted by prescribing observed SST anomalies as lower boundary conditions for atmospheric GCMs. Particularly noteworthy are the ensembles of model runs performed under the auspices of the European Prediction of Climate Variations on Seasonal to Interannual Time Scales (PROVOST) and the US Dynamical Seasonal Prediction (DSP) projects. The skill of seasonal hindcasts of the participating models' atmospheric anomalies in different regions of the globe (including the PNA sector) was summarised in a series of articles edited by Palmer and Shukla (2000). These results demonstrate that the prescribed SST forcing exerts a notable impact on the model atmospheres. The hindcast skill for the winter extratropical NH is particularly high during the largest El Niño and La Niña episodes. However, these experiments indicate considerable variability of the responses in individual models, and among ensemble members of a given model. This large scatter of model responses suggests that atmospheric changes in the extratropics are only weakly constrained by tropical SST forcing.

The performance of the dynamical seasonal forecast system at the US NCEP in predicting the atmospheric anomalies given prescribed anomalous SST forcing (in the PNA sector) was assessed by Kanamitsu et al. (2002). During the large El Niño event of 1997 to 1998, the forecasts based on this system with one-month lead time are in good agreement with the observed changes in the PNA sector, with anomaly correlation scores of 0.8 to 0.9 (for 200 mb height), 0.6 to 0.8 (surface temperature) and 0.4 to 0.5 (precipitation). More recently, hindcast experiments have been launched using AOGCMs. The European effort was supported by the Development of a European Multimodel Ensemble System for Seasonal to Interannual Prediction (DEMETER) programme (Palmer et al., 2004). For the boreal winter season, and with hindcasts initiated in November, the model-generated PNA indices exhibit statistically significant temporal correlations with the corresponding observations. The fidelity of the PNA simulations is evident in both the multi-model ensemble means, as well as in the output from individual member models. However,

the strength of the ensemble mean signal remains low when compared with the statistical spread due to sampling fluctuations among different models, and among different realisations of a given model. The model skill is notably lower for other seasons and longer lead times. Empirical Orthogonal Function analyses of the geopotential height data produced by individual member models confirm that the PNA pattern is a leading spatial mode of atmospheric variability in these models.

Multi-century integrations have also been conducted at various institutions using the current generation of AOGCMs. Unlike the hindcasting or forecasting experiments mentioned above, these climate simulations are not aimed at reproducing specific ENSO events in the observed system. Diagnosis of the output from one such AOGCM integration indicates that the modelled ENSO events are linked to a PNA-like pattern in the upper troposphere (Wittenberg et al., 2006). The centres of action of the simulated patterns are systematically displaced 20 to 30 degrees of longitude west of the observed positions. This discrepancy is evidently linked to a corresponding spatial shift in the ENSO-related SST and precipitation anomaly centres simulated in the tropical Pacific. This finding illustrates that the spatial configuration of the PNA pattern in AOGCMs is crucially dependent on the accuracy of ENSO simulations in the tropics.

8.4.4 Cold Ocean-Warm Land Pattern

The Cold Ocean-Warm Land (COWL) pattern indicates that the oceans are relatively cold and the continents are relatively warm poleward of 40°N when the NH is relatively warm. The COWL pattern results from the contrast in thermal inertia between the continents and oceans, which allows continental temperature anomalies to have greater amplitude, and thus more strongly influence hemispheric mean temperature. The COWL pattern has been simulated in climate models of varying degrees of complexity (e.g., Broccoli et al., 1998), and similar patterns have been obtained from cluster analysis (Wu and Straus, 2004a) and EOF analysis (Wu and Straus, 2004b) of reanalysis data. In a number of studies, cold season trends in NH temperature and sea level pressure during the late 20th century have been associated with secular trends in indices of the COWL pattern (Wallace et al., 1996; Lu et al., 2004).

In their analysis of AOGCM simulations, Broccoli et al. (1998) found that the original method for extracting the COWL pattern could yield potentially misleading results when applied to a simulation forced by past and future variations in anthropogenic forcing (as is the case with most other patterns, or modes, of climate variability). The resulting spatial pattern was a mixture of the patterns associated with unforced climate variability and the anthropogenic fingerprint. Broccoli et al. (1998) also noted that temperature anomalies in the two continental centres of the COWL pattern are virtually uncorrelated, suggesting that different atmospheric teleconnections are involved in producing this pattern. Quadrelli and Wallace (2004) recently showed that the COWL pattern can be reconstructed as a linear combination of the first two EOFs of monthly mean December to March sea level pressure.

These two EOFs are the NAM and a mode closely resembling the PNA pattern. A linear combination of these two fundamental patterns can also account for a substantial fraction of the winter trend in NH sea level pressure during the late 20th century.

8.4.5 Atmospheric Regimes and Blocking

Weather, or climate, regimes are important factors in determining climate at various locations around the world and they can have a large impact on day-to-day variability (e.g., Plaut and Simonnet, 2001; Trigo et al., 2004; Yiou and Nogaj, 2004). General Circulation Models have been found to simulate hemispheric climate regimes quite similar to those found in observations (Robertson, 2001; Achatz and Opsteegh, 2003; Selten and Branstator, 2004). Simulated regional climate regimes over the North Atlantic strongly similar to the observed regimes were reported by Cassou et al. (2004), while the North Pacific regimes simulated by Farrara et al. (2000) were broadly consistent with those in observations. Since the TAR, agreement between different studies has improved regarding the number and structure of both hemispheric and sectoral atmospheric regimes, although this remains a subject of research (e.g., Wu and Straus, 2004a) and the statistical significance of the regimes has been discussed and remains an unresolved issue (e.g., Hannachi and O'Neill, 2001; Hsu and Zwiers, 2001; Stephenson et al., 2004; Molteni et al., 2006).

Blocking events are an important class of sectoral weather regimes (see Chapter 3), associated with local reversals of the mid-latitude westerlies. The most recent systematic intercomparison of atmospheric GCM simulations of NH blocking (D'Andrea et al., 1998) was reported in the TAR. Consistent with the conclusions of this earlier study, recent studies have found that GCMs tend to simulate the location of NH blocking more accurately than frequency or duration: simulated events are generally shorter and rarer than observed events (e.g., Pelly and Hoskins, 2003b). An analysis of one of the AOGCMs from the MMD at the PCMDI found that increased horizontal resolution combined with better physical parametrizations has led to improvements in simulations of NH blocking and synoptic weather regimes over Europe. Finally, both GCM simulations and analyses of long data sets suggest the existence of considerable interannual to inter-decadal variability in blocking frequency (e.g., Stein, 2000; Pelly and Hoskins, 2003a), highlighting the need for caution when assessing blocking climatologies derived from short records (either observed or simulated). Blocking events also occur in the SH mid-latitudes (Sinclair, 1996); no systematic intercomparison of observed and simulated SH blocking climatologies has been carried out. There is also evidence of connections between North and South Pacific blocking and ENSO variability (e.g., Renwick, 1998; Chen and Yoon, 2002), and between North Atlantic blocks and sudden stratospheric warmings (e.g., Kodera and Chiba, 1995; Monahan et al., 2003) but these connections have not been systematically explored in AOGCMs.

8.4.6 Atlantic Multi-decadal Variability

The Atlantic Ocean exhibits considerable multi-decadal variability with time scales of about 50 to 100 years (see Chapter 3). This multi-decadal variability appears to be a robust feature of the surface climate in the Atlantic region, as shown by tree ring reconstructions for the last few centuries (e.g., Mann et al., 1998). Atlantic multi-decadal variability has a unique spatial pattern in the SST anomaly field, with opposite changes in the North and South Atlantic (e.g., Mestas-Nunez and Enfield, 1999; Latif et al., 2004), and this dipole pattern has been shown to be significantly correlated with decadal changes in Sahelian rainfall (Folland et al., 1986). Decadal variations in hurricane activity have also been linked to the multi-decadal SST variability in the Atlantic (Goldenberg et al., 2001). Atmosphere-Ocean General Circulation Models simulate Atlantic multi-decadal variability (e.g., Delworth et al., 1993; Latif, 1998 and references therein; Knight et al., 2005), and the simulated space-time structure is consistent with that observed (Delworth and Mann, 2000). The multi-decadal variability simulated by the AOGCMs originates from variations in the MOC (see Section 8.3). The mechanisms, however, that control the variations in the MOC are fairly different across the ensemble of AOGCMs. In most AOGCMs, the variability can be understood as a damped oceanic eigenmode that is stochastically excited by the atmosphere. In a few other AOGCMs, however, coupled interactions between the ocean and the atmosphere appear to be more important. The relative roles of high- and low-latitude processes differ also from model to model. The variations in the Atlantic SST associated with the multi-decadal variability appear to be predictable a few decades ahead, which has been shown by potential (diagnostic) and classical (prognostic) predictability studies. Atmospheric quantities do not exhibit predictability at decadal time scales in these studies, which supports the picture of stochastically forced variability.

8.4.7 El Niño-Southern Oscillation

During the last decade, there has been steady progress in simulating and predicting ENSO (see Chapters 3 and 9) and the related global variability using AOGCMs (Latif et al., 2001; Davey et al., 2002; AchutaRao and Sperber, 2002). Over the last several years the parametrized physics have become more comprehensive (Gregory et al., 2000; Collins et al., 2001; Kiehl and Gent, 2004), the horizontal and vertical resolutions, particularly in the atmospheric component models, have markedly increased (Guilyardi et al., 2004) and the application of observations in initialising forecasts has become more sophisticated (Alves et al., 2004). These improvements in model formulation have led to a better representation of the spatial pattern of the SST anomalies in the eastern Pacific (AchutaRao and Sperber, 2006). In fact, as an indication of recent model improvements, some IPCC class models are being used for ENSO prediction (Wittenberg et al., 2006). Despite this progress, serious systematic errors in both the simulated mean climate and the natural variability persist. For example, the

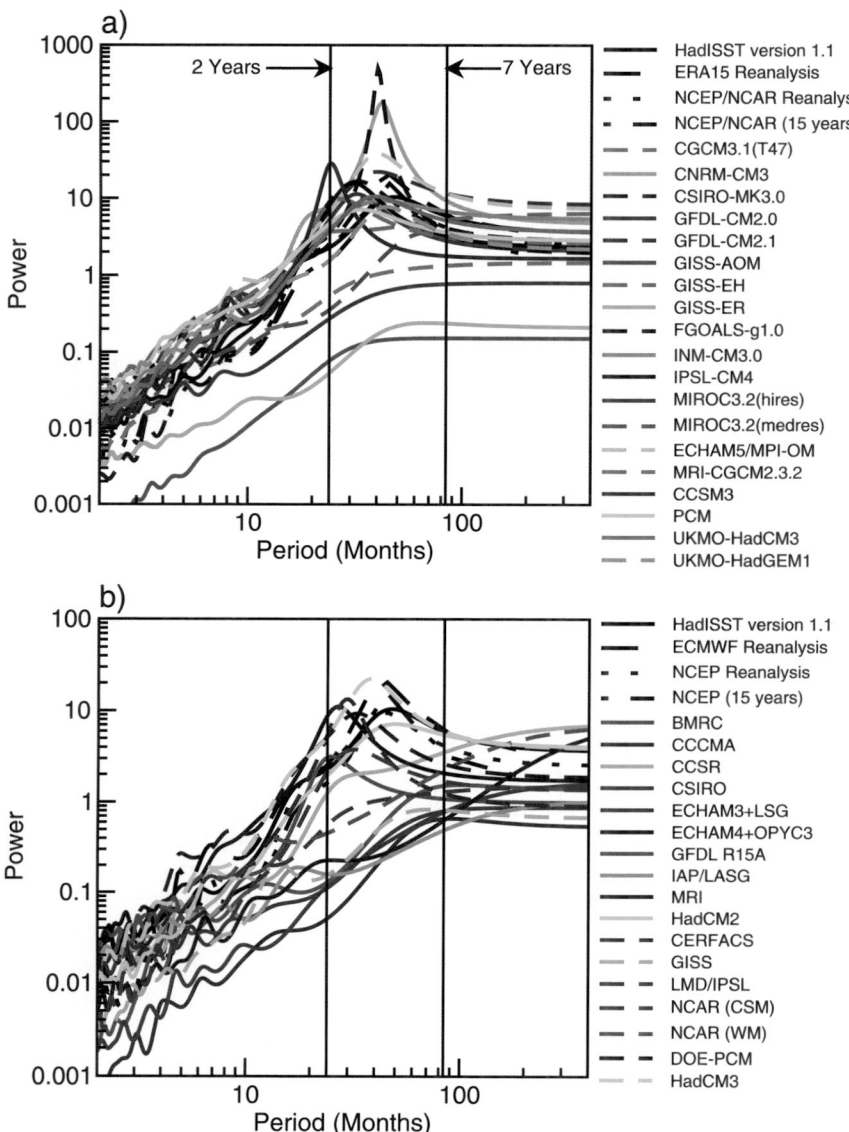

Figure 8.13. *Maximum entropy power spectra of surface air temperature averaged over the NINO₃ region (i.e., 5°N to 5°S, 150°W to 90° W) for (a) the MMD at the PCMDI and (b) the CMIP2 models. Note the differing scales on the vertical axes and that ECMWF reanalysis in (b) refers to the European Centre for Medium Range Weather Forecasts (ECMWF) 15-year reanalysis (ERA15) as in (a). The vertical lines correspond to periods of two and seven years. The power spectra from the reanalyses and for SST from the Hadley Centre Sea Ice and Sea Surface Temperature (HadISST) version 1.1 data set are given by the series of solid, dashed and dotted black curves. Adapted from AchutaRao and Sperber (2006).*

considerably faster than observed (AchutaRao and Sperber, 2002), although there has been some notable progress in this regard over the last decade (AchutaRao and Sperber, 2006) in that more models are consistent with the observed time scale for ENSO (see Figure 8.13). The models also have difficulty capturing the correct phase locking between the annual cycle and ENSO. Further, some AOGCMs fail to represent the spatial and temporal structure of the El Niño-La Niña asymmetry (Monahan and Dai, 2004). Other weaknesses in the simulated amplitude and structure of ENSO variability are discussed in Davey et al. (2002) and van Oldenborgh et al. (2005).

Current research points to some promise in addressing some of the above problems. For example, increasing the atmospheric resolution in both the horizontal (Guilyardi et al., 2004) and vertical (NCEP Coupled Forecast System) may improve the simulated spectral characteristics of the variability, ocean parametrized physics have also been shown to significantly influence the coupled variability (Meehl et al., 2001) and continued methodical numerical experimentation into the sources of model error (e.g., Schneider, 2001) will ultimately suggest model improvement strategies.

In terms of ENSO prediction, the two biggest recent advances are: (i) the recognition that forecasts must include quantitative information regarding uncertainty (i.e., probabilistic prediction) and that verification must include skill measures for probability forecasts (Kirtman, 2003); and (ii) that a multi-model ensemble strategy may be the best current approach for adequately dealing with forecast uncertainty, for example, Palmer et al. (2004), in which Figure 2 demonstrates that a multi-model ensemble forecast has better skill than a comparable ensemble

so-called 'double ITCZ' problem noted by Mechoso et al. (1995; see Section 8.3.1) remains a major source of error in simulating the annual cycle in the tropics in most AOGCMs, which ultimately affects the fidelity of the simulated ENSO. Along the equator in the Pacific the models fail to adequately capture the zonal SST gradient, the equatorial cold tongue structure is equatorially confined and extends too far too to the west (Cai et al., 2003), and the simulations typically have thermoclines that are far too diffuse (Davey et al., 2002). Most AOGCMs fail to capture the meridional extent of the anomalies in the eastern Pacific and tend to produce anomalies that extend too far into the western tropical Pacific. Most, but not all, AOGCMs produce ENSO variability that occurs on time scales

based on a single model. Improvements in the use of data, particularly in the ocean, for initialising forecasts continues to yield enhancements in forecast skill (Alves et al., 2004); moreover, other research indicates that forecast initialisation strategies that are implemented within the framework of the coupled system as opposed to the individual component models may also lead to substantial improvements in skill (Chen et al., 1995). However, basic questions regarding the predictability of SST in the tropical Pacific remain open challenges in the forecast community. For instance, it is unclear how westerly wind bursts, intra-seasonal variability or atmospheric weather noise in general limit the predictability of ENSO (e.g., Thompson and Battisti, 2001; Kleeman et al., 2003; Flugel et al., 2004; Kirtman et al., 2005). There are also apparent decadal variations in ENSO forecast skill (Balmaseda et al., 1995; Ji et al., 1996; Kirtman and Schopf, 1998), and the sources of these variations are the subject of some debate. Finally, it remains unclear how changes in the mean climate will ultimately affect ENSO predictability (Collins et al., 2002).

8.4.8 Madden-Julian Oscillation

The MJO (Madden and Julian, 1971) refers to the dominant mode of intra-seasonal variability in the tropical troposphere. It is characterised by large-scale regions of enhanced and suppressed convection, coupled to a deep baroclinic, primarily zonal circulation anomaly. Together, they propagate slowly eastward along the equator from the western Indian Ocean to the central Pacific and exhibit local periodicity in a broad 30- to 90-day range. Simulation of the MJO in contemporary coupled and uncoupled climate models remains unsatisfactory (e.g., Zhang, 2005; Lin et al., 2006), partly because more is now demanded from the model simulations, as understanding of the role of the MJO in the coupled atmosphere-ocean climate system expands. For instance, simulations of the MJO in models at the time of the TAR were judged using gross metrics (e.g., Slingo et al., 1996). The spatial phasing of the associated surface fluxes, for instance, are now recognised as critical for the development of the MJO and its interaction with the underlying ocean (e.g., Hendon, 2005; Zhang, 2005). Thus, while a model may simulate some gross characteristics of the MJO, the simulation may be deemed unsuccessful when the detailed structure of the surface fluxes is examined (e.g., Hendon, 2000).

Variability with MJO characteristics (e.g., convection and wind anomalies of the correct spatial scale that propagate coherently eastward with realistic phase speeds) is simulated in many contemporary models (e.g., Sperber et al., 2005; Zhang, 2005), but this variability is typically not simulated to occur often enough or with sufficient strength so that the MJO stands out realistically above the broadband background variability (Lin et al., 2006). This underestimation of the strength and coherence of convection and wind variability at MJO temporal and spatial scales means that contemporary climate models still simulate poorly many of the important climatic effects of the MJO (e.g., its impact on rainfall variability in the monsoons or the modulation of tropical cyclone development). Simulation of the spatial structure of the MJO as it evolves through its life cycle is also problematic, with tendencies for the convective anomaly to split into double ITCZs in the Pacific and for erroneously strong convective signals to sometimes develop in the eastern Pacific ITCZ (e.g., Inness and Slingo, 2003). It has also been suggested that inadequate representation in climate models of cloud-radiative interactions and/or convection-moisture interactions may explain some of the difficulties in simulating the MJO (e.g., Lee et al., 2001; Bony and Emanuel, 2005).

Even though the MJO is probably not fundamentally a coupled ocean-atmosphere mode (e.g., Waliser et al., 1999), air-sea coupling does appear to promote more coherent eastward, and, in northern summer, northward propagation at

MJO temporal and spatial scales. The interaction with an active ocean is important especially in the suppressed convective phase when SSTs are warming and the atmospheric boundary layer is recovering (e.g., Hendon, 2005). Thus, the most realistic simulation of the MJO is anticipated to be with AOGCMs. However, coupling, in general, has not been a panacea. While coupling in some models improves some aspects of the MJO, especially eastward propagation and coherence of convective anomalies across the Indian and western Pacific Oceans (e.g., Kemball-Cook et al., 2002; Inness and Slingo, 2003), problems with the horizontal structure and seasonality remain. Typically, models that show the most beneficial impact of coupling on the propagation characteristics of the MJO are also the models that possess the most unrealistic seasonal variation of MJO activity (e.g., Zhang, 2005). Unrealistic simulation of the seasonal variation of MJO activity implies that the simulated MJO will improperly interact with climate phenomena that are tied to the seasonal cycle (e.g., the monsoons and ENSO).

Simulation of the MJO is also adversely affected by biases in the mean state (see Section 8.4.7). These biases include the tendency for coupled models to exaggerate the double ITCZ in the Indian and western Pacific Oceans, under-predict the eastward extent of surface monsoonal westerlies into the western Pacific, and over-predict the westward extension of the Pacific cold tongue. Together, these flaws limit development, maintenance and the eastward extent of convection associated with the MJO, thereby reducing its overall strength and coherence (e.g., Inness et al., 2003). To date, simulation of the MJO has proven to be most sensitive to the convective parametrization employed in climate models (e.g., Wang and Schlesinger, 1999; Maloney and Hartmann, 2001; Slingo et al., 2005). A consensus, although with exceptions (e.g., Liu et al., 2005), appears to be emerging that convective schemes based on local vertical stability and that include some triggering threshold produce more realistic MJO variability than those that convect too readily. However, some sophisticated models, with arguably the most physically based convective parametrizations, are unable to simulate reasonable MJO activity (e.g., Slingo et al., 2005).

8.4.9 Quasi-Biennial Oscillation

The Quasi-Biennial Oscillation (QBO; see Chapter 3) is a quasi-periodic wave-driven zonal mean wind reversal that dominates the low-frequency variability of the lower equatorial stratosphere (3 to 100 hPa) and affects a variety of extratropical phenomena including the strength and stability of the winter polar vortex (e.g., Baldwin et al., 2001). Theory and observations indicate that a broad spectrum of vertically propagating waves in the equatorial atmosphere must be considered to explain the QBO. Realistic simulation of the QBO in GCMs therefore depends on three important conditions: (i) sufficient vertical resolution in the stratosphere to allow the representation of equatorial waves at the horizontally resolved scales of a GCM, (ii) a realistic excitation of resolved equatorial waves by simulated tropical weather and (iii) parametrization of the effects of unresolved gravity waves. Due to the computational cost associated with the requirement of a well-resolved stratosphere, the models employed for the current assessment do not generally include the QBO.

The inability of resolved wave driving to induce a spontaneous QBO in GCMs has been a long-standing issue (Boville and Randel, 1992). Only recently (Takahashi, 1996, 1999; Horinouchi and Yoden, 1998; Hamilton et al., 2001) have two necessary conditions been identified that allow resolved waves to induce a QBO: high vertical resolution in the lower stratosphere (roughly 0.5 km), and a parametrization of deep cumulus convection with sufficiently large temporal variability. However, recent analysis of satellite and radar observations of deep tropical convection (Horinouchi, 2002) indicates that the forcing of a QBO by resolved waves alone requires a parametrization of deep convection with an unrealistically large amount of temporal variability. Consequently, it is currently thought that a combination of resolved and parametrized waves is required to properly model the QBO. The utility of parametrized non-orographic gravity wave drag to force a QBO has now been demonstrated by a number of studies (Scaife et al., 2000; Giorgetta et al., 2002, 2006). Often an enhancement of input momentum flux in the tropics relative to that needed in the extratropics is required. Such an enhancement, however, depends implicitly on the amount of resolved waves and in turn, the spatial and temporal properties of parametrized deep convection employed in each model (Horinouchi et al., 2003; Scinocca and McFarlane, 2004).

8.4.10 Monsoon Variability

Monsoon variability (see Chapters 3, 9 and 11) occurs over a range of temporal scales from intra-seasonal to inter-decadal. Since the TAR, the ability of AOGCMs to simulate monsoon variability on intra-seasonal as well as interannual time scales has been examined. Lambert and Boer (2001) compared the AOGCMs that participated in CMIP, finding large errors in the simulated precipitation in the equatorial regions and in the Asian monsoon region. Lin et al. (2006) evaluated the intra-seasonal variation of precipitation in the MMD at PCMDI. They found that the intra-seasonal variance of precipitation simulated by most AOGCMs was smaller than observed. The space-time spectra of most model simulations have much less power than is observed, especially at periods shorter than six days. The speed of the equatorial waves is too fast, and the persistence of the precipitation is too long, in most of the AOGCM simulations. Annamalai et al (2004) examined the fidelity of precipitation simulation in the Asian monsoon region in the MMD at PCMDI. They found that just 6 of the 18 AOGCMs considered realistically simulated climatological monsoon precipitation for the 20th century. For the former set of models, the spatial correlation of the patterns of monsoon precipitation between the models exceeded 0.6, and the seasonal cycle of monsoon rainfall was simulated well. Among these models, only four exhibited a robust ENSO-monsoon contemporaneous teleconnection. Cook and Vizy (2006) evaluated the simulation of the 20th-century

climate in North Africa in the MMD at PCMDI. They found that the simulation of North African summer precipitation was less realistic than the simulation of summer precipitation over North America or Europe. In short, most AOGCMs do not simulate the spatial or intra-seasonal variation of monsoon precipitation accurately. See Chapter 11 for a more detailed regional evaluation of simulated monsoon variability.

8.4.11 Shorter-Term Predictions Using Climate Models

This subsection focuses on the few results of initial value predictions made using models that are identical, or very close to, the models used in other chapters of this report for understanding and predicting climate change.

Weather prediction

Since the TAR, it has been shown that climate models can be integrated as weather prediction models if they are initialised appropriately (Phillips et al., 2004). This advance appears to be due to: (i) improvements in the forecast model analyses and (ii) increases in the climate model spatial resolution. An advantage of testing a model's ability to predict weather is that some of the sub-grid scale physical processes that are parametrized in models (e.g., cloud formation, convection) can be evaluated on time scales characteristic of those processes, without the complication of feedbacks from these processes altering the underlying state of the atmosphere (Pope and Stratton, 2002; Boyle et al., 2005; Williamson et al., 2005; Martin et al., 2006). Full use can be made of the plentiful meteorological data sets and observations from specialised field experiments. According to these studies, some of the biases found in climate simulations are also evident in the analysis of their weather forecasts. This suggests that ongoing improvements in model formulation driven primarily by the needs of weather forecasting may lead also to more reliable climate predictions.

Seasonal prediction

Verification of seasonal-range predictions provides a direct test of a model's ability to represent the physical and dynamical processes controlling (unforced) fluctuations in the climate system. Satisfactory prediction of variations in key climate signals such as ENSO and its global teleconnections provides evidence that such features are realistically represented in long-term forced climate simulations.

A version of the HadCM3 AOGCM (known as GloSea) has been assessed for skill in predicting observed seasonal climate variations (Davey et al., 2002; Graham et al., 2005). Graham et al. (2005) analysed 43 years of retrospective six-month forecasts ('hindcasts') with GloSea, run from observed ocean-land-atmosphere initial conditions. A nine-member ensemble was used to sample uncertainty in the initial conditions. Conclusions relevant to HadCM3 include: (i) the model is able to reproduce observed large-scale lagged responses to ENSO events in the tropical Atlantic and Indian Ocean SSTs; and (ii) the model can realistically predict anomaly patterns in North

Atlantic SSTs, shown to have important links with the NAO and seasonal temperature anomalies over Europe.

The GFDL-CM2.0 AOGCM has also been assessed for seasonal prediction. Twelve-month retrospective and contemporaneous forecasts were produced using a six-member ensemble over 15 years starting in 1991. The forecasts were initialised using global ocean data assimilation (Derber and Rosati, 1989; Rosati et al., 1997) and observed atmospheric forcing, combined with atmospheric initial conditions derived from the atmospheric component of the model forced with observed SSTs. Results indicated considerable model skill out to 12 months for ENSO prediction (see http://www.gfdl.noaa. gov/~rgg/si_workdir/Forecasts.html). Global teleconnections, as diagnosed from the NCEP reanalysis (GFDL GAMDT, 2004), were evident throughout the 12-month forecasts.

8.5 Model Simulations of Extremes

Society's perception of climate variability and climate change is largely formed by the frequency and the severity of extremes. This is especially true if the extreme events have large and negative impacts on lives and property. As climate models' resolution and the treatment of physical processes have improved, the simulation of extremes has also improved. Mainly because of increased data availability (e.g., daily data, various indices, etc.), the modelling community has now examined the model simulations in greater detail and presented a comprehensive description of extreme events in the coupled models used for climate change projections.

Some extreme events, by their very nature of being smaller in scale and shorter in duration, are manifestations of either a rapid amplification, or an equilibration at a higher amplitude, of naturally occurring local instabilities. Large-scale and long-duration extreme events are generally due to persistence of weather patterns associated with air-sea and air-land interactions. A reasonable hypothesis might be that the coarse-resolution AOGCMs might not be able to simulate local short-duration extreme events, but that is not the case. Our assessment of the recent scientific literature shows, perhaps surprisingly, that the global statistics of the extreme events in the current climate, especially temperature, are generally well simulated by the current models (see Section 8.5.1). These models have been more successful in simulating temperature extremes than precipitation extremes.

The assessment of extremes, especially for temperature, has been done by examining the amplitude, frequency and persistence of the following quantities: daily maximum and minimum temperature (e.g., hot days, cold days, frost days), daily precipitation intensity and frequency, seasonal mean temperature and precipitation and frequency and tracks of tropical cyclones. For precipitation, the assessment has been done either in terms of return values or extremely high rates of precipitation.

8.5.1 Extreme Temperature

Kiktev et al. (2003) compared station observations of extreme events with the simulations of an atmosphere-only GCM (Hadley Centre Atmospheric Model version 3; HadAM3) forced by prescribed oceanic forcing and anthropogenic radiative forcing during 1950 to 1995. The indices of extreme events they used were those proposed by Frich et al. (2002). They found that inclusion of anthropogenic radiative forcing was required to reproduce observed changes in temperature extremes, particularly at large spatial scales. The decrease in the number of frost days in Southern Australia simulated by HadAM3 with anthropogenic forcing is in good agreement with the observations. The increase in the number of warm nights over Eurasia is poorly simulated when anthropogenic forcing is not included, but the inclusion of anthropogenic forcing improves the modelled trend patterns over western Russia and reproduces the general increase in the occurrence of warm nights over much of the NH.

Meehl et al. (2004) compared the number of frost days simulated by the PCM model with observations. The 20th-century simulations include the variations in solar, volcano, sulphate aerosol, ozone and greenhouse gas forcing. Both model simulations and observations show that the number of frost days decreased by two days per decade in the western USA during the 20th century. The model simulations do not agree with observations in the southeastern USA, where the model simulates a decrease in the number of frost days in this region in the 20th century, while observations indicate an increase in this region. Meehl et al. (2004) argue that this discrepancy could be due to the model's inability to simulate the impact of El Niño events on the number of frost days in the southeastern USA. Meehl and Tebaldi (2004) compared the heat waves simulated by the PCM with observations. They defined a heat wave as the three consecutive warmest nights during the year. During the period 1961 to 1990, there is good agreement between the model and observations (NCEP reanalysis).

Kharin et al. (2005) examined the simulations of temperature and precipitation extremes for AMIP-2 models, some of which are atmospheric components of coupled models used in this assessment. They found that models simulate the temperature extremes, especially the warm extremes, reasonably well. Models have serious deficiencies in simulating precipitation extremes, particularly in the tropics. Vavrus et al. (2006) used daily values of 20th-century integrations from seven models. They defined a cold air outbreak as 'an occurrence of two or more consecutive days during which the local mean daily surface air temperature is at least two standard deviations below the local winter mean temperature'. They found that the climate models reproduce the location and magnitude of cold air outbreaks in the current climate.

Researchers have also established relationships between large-scale circulation features and cold air outbreaks or heat waves. For example, Vavrus et al. (2006) found that 'the favored regions of cold air outbreaks are located near and downstream from preferred locations of atmosphere blocking'. Likewise, Meehl and Tebaldi (2004) found that heat waves over Europe and North America were associated with changes in the 500 hPa circulation pattern.

8.5.2   Extreme Precipitation

Sun et al. (2006) investigated the intensity of daily precipitation simulated by 18 AOGCMs, including several used in this report. They found that most of the models produce light precipitation (<10 mm day^{-1}) more often than observed, too few heavy precipitation events and too little precipitation in heavy events (>10 mm day^{-1}). The errors tend to cancel, so that the seasonal mean precipitation is fairly realistic (see Section 8.3).

Since the TAR, many simulations have been made with high-resolution GCMs. Iorio et al. (2004) examined the impact of model resolution on the simulation of precipitation in the USA using the Community Climate Model version 3 (CCM3). They found that the high-resolution simulation produces more realistic daily precipitation statistics. The coarse-resolution model had too many days with weak precipitation and not enough with intense precipitation. This tendency was partially eliminated in the high-resolution simulation, but, in the simulation at the highest resolution (T239), the high-percentile daily precipitation was still too low. This problem was eliminated when a cloud-resolving model was embedded in every grid point of the GCM.

Kimoto et al. (2005) compared the daily precipitation over Japan in an AOGCM with two different resolutions (high res. and med res. of MIROC 3.2) and found more realistic precipitation distributions with the higher resolution. Emori et al. (2005) showed that a high-resolution AGCM (the atmospheric part of high res. MIROC 3.2) can simulate the extreme daily precipitation realistically if there is provision in the model to suppress convection when the ambient relative humidity is below 80%, suggesting that modelled extreme precipitation can be strongly parametrization dependent. Kiktev et al. (2003) compared station observations of rainfall with the simulations of the atmosphere-only GCM HadAM3 forced by prescribed oceanic forcing and anthropogenic radiative forcing. They found that this model shows little skill in simulating changing precipitation extremes. May (2004) examined the variability and extremes of daily rainfall in the simulation of present day climate by the ECHAM4 GCM. He found that this model simulates the variability and extremes of rainfall quite well over most of India when compared to satellite-derived rainfall, but has a tendency to overestimate heavy rainfall events in central India. Durman et al. (2001) compared the extreme daily European precipitation simulated by the HadCM2 GCM with station observations. They found that the GCM's ability to simulate daily precipitation events exceeding 15 mm per day was good but its ability to simulate events exceeding 30 mm per day was poor. Kiktev et al. (2003) showed that HadAM3 was able to simulate the natural variability of the precipitation intensity index (annual mean precipitation divided by number of days with precipitation less than 1 mm) but was not able to simulate accurately the variability

in the number of wet days (the number of days in a year with precipitation greater than 10 mm).

Using the Palmer Drought Severity Index (PDSI), Dai at al. (2004) concluded that globally very dry or wet areas (PDSI above +3 or below –3) have increased from 20% to 38% since 1972. In addition to simulating the short-duration events like heat waves, frost days and cold air outbreaks, models have also shown success in simulating long time-scale anomalies. For example, Burke et al. (2006) showed that the HadCM3 model, on a global basis and at decadal time scales, 'reproduces the observed drying trend' as defined by the PDSI if the anthropogenic forcing is included, although the model does not always simulate correctly the regional distributions of wet and dry areas.

8.5.3   Tropical Cyclones

The spatial resolution of the coupled ocean-atmosphere models used in the IPCC assessment is generally not high enough to resolve tropical cyclones, and especially to simulate their intensity. A common approach to investigate the effects of global warming on tropical cyclones has been to utilise the SST boundary conditions from a global change scenario run to force a high-resolution AGCM. That model run is then compared with a control run using the high-resolution AGCM forced with specified observed SST for the current climate (Sugi et al., 2002; Camargo et al., 2005; McDonald et al., 2005; Bengtsson et al., 2006; Oouchi et al., 2006; Yoshimura et al., 2006). There are also several idealised model experiments in which a high-resolution AGCM is integrated with and without a fixed global warming or cooling of SST. Another method is to embed a high-resolution regional model in the lower-resolution climate model (Knutson and Tuleya, 1999; Walsh et al., 2004). Projections using these methods are discussed in Chapter 10.

Bengtsson et al. (2006) showed that the global metrics of tropical cyclones (tropical or hemispheric averages) are broadly reproduced by the ECHAM5 model, even as a function of intensity. However, varying degrees of errors (in some cases substantial) in simulated tropical storm frequency and intensity have been noted in some models (e.g., GFDL GAMDT, 2004; Knutson and Tuleya, 2004; Camargo et al., 2005). The tropical cyclone simulation has been shown to be sensitive to the choice of convection parametrization in some cases.

Oouchi et al. (2006) used one of the highest-resolution (20 km) atmospheric models to simulate the frequency, distribution and intensity of tropical cyclones in the current climate. Although there were some deficiencies in simulating the geographical distribution of tropical cyclones (over-prediction of tropical cyclones between 0° to 10°S in the Indian Ocean, and under-prediction between 0° to 10°N in the western Pacific), the overall simulation of geographical distribution and frequency was remarkably good. The model could not simulate the strongest observed maximum wind speeds, and central pressures were not as low as observed, suggesting that even higher resolution may be required to simulate the most intense tropical cyclones.

8.5.4   Summary

Because most AOGCMs have coarse resolution and large-scale systematic errors, and extreme events tend to be short lived and have smaller spatial scales, it is somewhat surprising how well the models simulate the statistics of extreme events in the current climate, including the trends during the 20th century (see Chapter 9 for more detail). This is especially true for the temperature extremes, but intensity, frequency and distribution of extreme precipitation are less well simulated. The higher-resolution models used for projections of tropical cyclone changes (Chapter 10) produce generally good simulation of the frequency and distribution of tropical cyclones, but less good simulation of their intensity. Improvements in the simulation of the intensity of precipitation and tropical cyclones with increases in the resolution of AGCMs (Oouchi et al., 2006) suggest that when climate models have sufficient resolution to explicitly resolve at least the large convective systems without using parametrizations for deep convection, it is likely that simulation of precipitation and intensity of tropical cyclones will improve.

8.6   Climate Sensitivity and Feedbacks

8.6.1   Introduction

Climate sensitivity is a metric used to characterise the response of the global climate system to a given forcing. It is broadly defined as the equilibrium global mean surface temperature change following a doubling of atmospheric CO_2 concentration (see Box 10.2). Spread in model climate sensitivity is a major factor contributing to the range in projections of future climate changes (see Chapter 10) along with uncertainties in future emission scenarios and rates of oceanic heat uptake. Consequently, differences in climate sensitivity between models have received close scrutiny in all four IPCC reports. Climate sensitivity is largely determined by internal feedback processes that amplify or dampen the influence of radiative forcing on climate. To assess the reliability of model estimates of climate sensitivity, the ability of climate models to reproduce different climate changes induced by specific forcings may be evaluated. These include the Last Glacial Maximum and the evolution of climate over the last millennium and the 20th century (see Section 9.6). The compilation and comparison of climate sensitivity estimates derived from models and from observations are presented in Box 10.2. An alternative approach, which is followed here, is to assess the reliability of key climate feedback processes known to play a critical role in the models' estimate of climate sensitivity.

This section explains why the estimates of climate sensitivity and of climate feedbacks differ among current models (Section 8.6.2), summarises understanding of the role of key radiative feedback processes associated with water vapour and lapse rate, clouds, snow and sea ice in climate sensitivity, and assesses the

treatment of these processes in the global climate models used to make projections of future climate change (Section 8.6.3). Finally we discuss how we can assess our relative confidence in the different climate sensitivity estimates derived from climate models (Section 8.6.4). Note that climate feedbacks associated with chemical or biochemical processes are not discussed in this section (they are addressed in Chapters 7 and 10), nor are local-scale feedbacks (e.g., between soil moisture and precipitation; see Section 8.2.3.2).

8.6.2   Interpreting the Range of Climate Sensitivity Estimates Among General Circulation Models

8.6.2.1   Definition of Climate Sensitivity

As defined in previous assessments (Cubasch et al., 2001) and in the Glossary, the global annual mean surface air temperature change experienced by the climate system after it has attained a new equilibrium in response to a doubling of atmospheric CO_2 concentration is referred to as the 'equilibrium climate sensitivity' (unit is °C), and is often simply termed the 'climate sensitivity'. It has long been estimated from numerical experiments in which an AGCM is coupled to a simple non-dynamic model of the upper ocean with prescribed ocean heat transports (usually referred to as 'mixed-layer' or 'slab' ocean models) and the atmospheric CO_2 concentration is doubled. In AOGCMs and non-steady-state (or transient) simulations, the 'transient climate response' (TCR; Cubasch et al., 2001) is defined as the global annual mean surface air temperature change (with respect to a 'control' run) averaged over a 20-year period centred at the time of CO_2 doubling in a 1% yr^{-1} compound CO_2 increase scenario. That response depends both on the sensitivity and on the ocean heat uptake. An estimate of the equilibrium climate sensitivity in transient climate change integrations is obtained from the 'effective climate sensitivity' (Murphy, 1995). It corresponds to the global temperature response that would occur if the AOGCM was run to equilibrium with feedback strengths held fixed at the values diagnosed at some point of the transient climate evolution. It is computed from the oceanic heat storage, the radiative forcing and the surface temperature change (Cubasch et al., 2001; Gregory et al., 2002).

The climate sensitivity depends on the type of forcing agents applied to the climate system and on their geographical and vertical distributions (Allen and Ingram, 2002; Sausen et al., 2002; Joshi et al., 2003). As it is influenced by the nature and the magnitude of the feedbacks at work in the climate response, it also depends on the mean climate state (Boer and Yu, 2003). Some differences in climate sensitivity will also result simply from differences in the particular radiative forcing calculated by different radiation codes (see Sections 10.2.1 and 8.6.2.3). The global annual mean surface temperature change thus presents limitations regarding the description and the understanding of the climate response to an external forcing. Indeed, the regional temperature response to a uniform forcing (and even more to a vertically or geographically distributed forcing) is highly inhomogeneous. In addition, climate sensitivity only considers the surface mean temperature and gives no indication of the occurrence of abrupt changes or extreme events. Despite its limitations, however, the climate sensitivity remains a useful concept because many aspects of a climate model scale well with global average temperature (although not necessarily across models), because the global mean temperature of the Earth is fairly well measured, and because it provides a simple way to quantify and compare the climate response simulated by different models to a specified perturbation. By focusing on the global scale, climate sensitivity can also help separate the climate response from regional variability.

8.6.2.2 Why Have the Model Estimates Changed Since the TAR?

The current generation of GCMs[5] covers a range of equilibrium climate sensitivity from 2.1°C to 4.4°C (with a mean value of 3.2°C; see Table 8.2 and Box 10.2), which is quite similar to the TAR. Yet most climate models have undergone substantial developments since the TAR (probably more than between the Second Assessment Report and the TAR) that generally involve improved parametrizations of specific processes such as clouds, boundary layer or convection (see Section 8.2). In some cases, developments have also concerned numerics, dynamical cores or the coupling to new components (ocean, carbon cycle, etc.). Developing new versions of a model to improve the physical basis of parametrizations or the simulation of the current climate is at the heart of modelling group activities. The rationale for these changes is generally based upon a combination of process-level tests against observations or against cloud-resolving or large-eddy simulation models (see Section 8.2), and on the overall quality of the model simulation (see Sections 8.3 and 8.4). These developments can, and do, affect the climate sensitivity of models.

The equilibrium climate sensitivity estimates from the latest model version used by modelling groups have increased (e.g., CCSM3 vs CSM1.0, ECHAM5/MPI-OM vs ECHAM3/LSG, IPSL-CM4 vs IPSL-CM2, MRI-CGCM2.3.2 vs MRI2, UKMO-HadGEM1 vs UKMO-HadCM3), decreased (e.g., CSIRO-MK3.0 vs CSIRO-MK2, GFDL-CM2.0 vs GFDL_R30_c, GISS-EH and GISS-ER vs GISS2, MIROC3.2(hires) and MIROC3.2(medres) vs CCSR/NIES2) or remained roughly unchanged (e.g., CGCM3.1(T47) vs CGCM1, GFDL-CM2.1 vs GFDL_R30_c) compared to the TAR. In some models, changes in climate sensitivity are primarily ascribed to changes in the cloud parametrization or in the representation of cloud-radiative properties (e.g., CCSM3, MRI-CGCM2.3.2, MIROC3.2(medres) and MIROC3.2(hires)). However, in most models the change in climate sensitivity cannot be attributed to a specific change in the model. For instance, Johns et al. (2006) showed that most of the individual changes made during the development of HadGEM1 have a small impact on the climate sensitivity, and that the global effects of the individual changes

largely cancel each other. In addition, the parametrization changes can interact nonlinearly with each other so that the sum of change A and change B does not produce the same as the change in A plus B (e.g., Stainforth et al., 2005). Finally, the interaction among the different parametrizations of a model explains why the influence on climate sensitivity of a given change is often model dependent (see Section 8.2). For instance, the introduction of the Lock boundary-layer scheme (Lock et al., 2000) to HadCM3 had a minimal impact on the climate sensitivity, in contrast to the introduction of the scheme to the GFDL atmospheric model (Soden et al., 2004; Johns et al., 2006).

8.6.2.3 What Explains the Current Spread in Models' Climate Sensitivity Estimates?

As discussed in Chapter 10 and throughout the last three IPCC assessments, climate models exhibit a wide range of climate sensitivity estimates (Table 8.2). Webb et al. (2006), investigating a selection of the slab versions of models in Table 8.1, found that differences in feedbacks contribute almost three times more to the range in equilibrium climate sensitivity estimates than differences in the models' radiative forcings (the spread of models' forcing is discussed in Section 10.2).

Several methods have been used to diagnose climate feedbacks in GCMs, whose strengths and weaknesses are reviewed in Stephens (2005) and Bony et al. (2006). These methods include the 'partial radiative perturbation' approach and its variants (e.g., Colman, 2003a; Soden and Held, 2006), the use of radiative-convective models and the 'cloud radiative forcing' method (e.g., Webb et al., 2006). Since the TAR, there has been progress in comparing the feedbacks produced by climate models in doubled atmospheric CO_2 equilibrium experiments (Colman, 2003a; Webb et al., 2006) and in transient climate change integrations (Soden and Held, 2006). Water vapour, lapse rate, cloud and surface albedo feedback parameters, as estimated by Colman (2003a), Soden and Held (2006) and Winton (2006a) are shown in Figure 8.14.

In AOGCMs, the water vapour feedback constitutes by far the strongest feedback, with a multi-model mean and standard deviation for the MMD at PCMDI of 1.80 ± 0.18 W m^{-2} °C^{-1}, followed by the (negative) lapse rate feedback (-0.84 ± 0.26 W m^{-2} °C^{-1}) and the surface albedo feedback (0.26 ± 0.08 W m^{-2} °C^{-1}). The cloud feedback mean is 0.69 W m^{-2} °C^{-1} with a very large inter-model spread of ± 0.38 W m^{-2} °C^{-1} (Soden and Held, 2006).

A substantial spread is apparent in the strength of water vapour feedback that is smaller in Soden and Held (2006) than in Colman (2003a). It is not known whether this smaller spread indicates a closer consensus among current AOGCMs than among older models, differences in the methodology or differences in the nature of climate change integrations between the two studies. In both studies, the lapse rate feedback also shows a substantial spread among models, which is explained

---

[5] Unless explicitly stated, GCM here refers both to AOGCM (used to estimate TCR) and AGCM coupled to a slab ocean (used to estimate equilibrium climate sensitivity).

Table 8.2. *Climate sensitivity estimates from the AOGCMs assessed in this report (see Table 8.1 for model details). Transient climate response (TCR) and equilibrium climate sensitivity (ECS) were calculated by the modelling groups (using atmosphere models coupled to slab ocean for equilibrium climate sensitivity), except those in italics, which were calculated from simulations in the MMD at PCMDI. The ocean heat uptake efficiency (W m^{-2} $°C^{-1}$), discussed in Chapter 10, may be roughly estimated as $F_{2x} \times (TCR^{-1} - ECS^{-1})$, where F_{2x} is the radiative forcing for doubled atmospheric CO_2 concentration (see Supplementary Material, Table 8.SM.1)*

AOGCM	Equilibrium climate sensitivity (°C)	Transient climate response (°C)
1: BCC-CM1	n.a.	n.a.
2: BCCR-BCM2.0	n.a.	n.a.
3: CCSM3	2.7	1.5
4: CGCM3.1(T47)	3.4	*1.9*
5: CGCM3.1(T63)	*3.4*	n.a.
6: CNRM-CM3	n.a.	1.6
7: CSIRO-MK3.0	3.1	1.4
8: ECHAM5/MPI-OM	3.4	2.2
9: ECHO-G	3.2	1.7
10: FGOALS-g1.0	*2.3*	*1.2*
11: GFDL-CM2.0	2.9	1.6
12: GFDL-CM2.1	3.4	1.5
13: GISS-AOM	n.a.	n.a.
14: GISS-EH	2.7	1.6
15: GISS-ER	2.7	1.5
16: INM-CM3.0	2.1	1.6
17: IPSL-CM4	4.4	2.1
18: MIROC3.2(hires)	4.3	2.6
19: MIROC3.2(medres)	4.0	2.1
20: MRI-CGCM2.3.2	3.2	2.2
21: PCM	2.1	1.3
22: UKMO-HadCM3	3.3	2.0
23: UKMO-HadGEM1	4.4	1.9

the difference in mean lapse rate feedback between the two studies is unclear, but may relate to inappropriate inclusion of stratospheric temperature response in some feedback analyses (Soden and Held, 2006).

The three studies, using different methodologies to estimate the global surface albedo feedback associated with snow and sea ice changes, all suggest that this feedback is positive in all the models, and that its range is much smaller than that of cloud feedbacks. Winton (2006a) suggests that about three-quarters of the global surface albedo feedback arises from the NH (see Section 8.6.3.3).

The diagnosis of global radiative feedbacks allows better understanding of the spread of equilibrium climate sensitivity estimates among current GCMs. In the idealised situation that the climate response to a doubling of atmospheric CO_2 consisted of a uniform temperature change only, with no feedbacks operating (but allowing for the enhanced radiative cooling resulting from the temperature increase), the global warming from GCMs would be around 1.2°C (Hansen et al., 1984; Bony et al., 2006). The water vapour feedback, operating alone on top of this, would at least double the response.[6] The water vapour feedback is, however, closely related to the lapse rate feedback (see above), and the two combined result in a feedback parameter of approximately 1 W m^{-2} $°C^{-1}$, corresponding to an amplification of the basic temperature response by approximately 50%. The

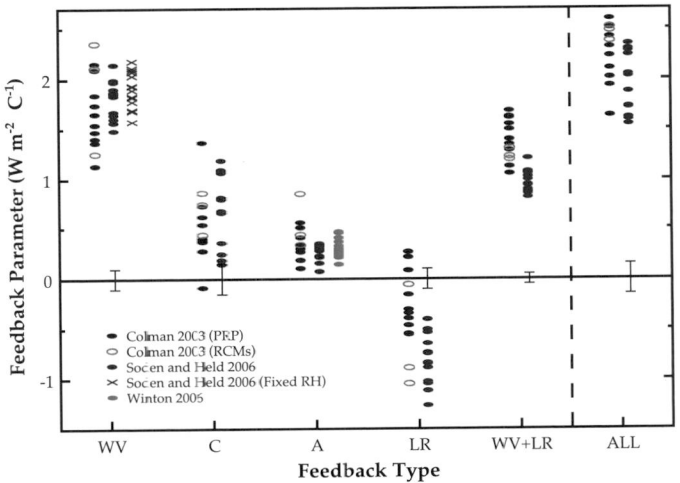

Figure 8.14. *Comparison of GCM climate feedback parameters for water vapour (WV), cloud (C), surface albedo (A), lapse rate (LR) and the combined water vapour plus lapse rate (WV + LR) in units of W m^{-2} $°C^{-1}$. 'ALL' represents the sum of all feedbacks. Results are taken from Colman (2003a; blue, black), Soden and Held (2006; red) and Winton (2006a, green). Closed blue and open black symbols from Colman (2003a) represent calculations determined using the partial radiative perturbation (PRP) and the radiative-convective method (RCM) approaches respectively. Crosses represent the water vapour feedback computed for each model from Soden and Held (2006) assuming no change in relative humidity. Vertical bars depict the estimated uncertainty in the calculation of the feedbacks from Soden and Held (2006).*

by inter-model differences in the relative surface warming of low and high latitudes (Soden and Held, 2006). Because the water vapour and temperature responses are tightly coupled in the troposphere (see Section 8.6.3.1), models with a larger (negative) lapse rate feedback also have a larger (positive) water vapour feedback. These act to offset each other (see Box 8.1). As a result, it is more reasonable to consider the sum of water vapour and lapse rate feedbacks as a single quantity when analysing the causes of inter-model variability in climate sensitivity. This makes inter-model differences in the combination of water vapour and lapse rate feedbacks a substantially smaller contributor to the spread in climate sensitivity estimates than differences in cloud feedback (Figure 8.14). The source of

---

[6] Under these simplifying assumptions the amplification of the global warming from a feedback parameter λ (in W m^{-2} $°C^{-1}$) with no other feedbacks operating is $\dfrac{1}{1 + \lambda / \lambda_p}$, where λ_p is the 'uniform temperature' radiative cooling response (of value approximately -3.2 W m^{-2} $°C^{-1}$; Bony et al., 2006). If *n* independent feedbacks operate, λ is replaced by $(\lambda_1 + \lambda_2 + ... \lambda_n)$.

Box 8.1: Upper-Tropospheric Humidity and Water Vapour Feedback

Water vapour is the most important greenhouse gas in the atmosphere. Tropospheric water vapour concentration diminishes rapidly with height, since it is ultimately limited by saturation-specific humidity, which strongly decreases as temperature decreases. Nevertheless, these relatively low upper-tropospheric concentrations contribute disproportionately to the 'natural' greenhouse effect, both because temperature contrast with the surface increases with height, and because lower down the atmosphere is nearly opaque at wavelengths of strong water vapour absorption.

In the stratosphere, there are potentially important radiative impacts due to anthropogenic sources of water vapour, such as from methane oxidation (see Section 2.3.7). In the troposphere, the radiative forcing due to direct anthropogenic sources of water vapour (mainly from irrigation) is negligible (see Section 2.5.6). Rather, it is the response of tropospheric water vapour to warming itself – the water vapour feedback – that matters for climate change. In GCMs, water vapour provides the largest positive radiative feedback (see Section 8.6.2.3): alone, it roughly doubles the warming in response to forcing (such as from greenhouse gas increases). There are also possible stratospheric water vapour feedback effects due to tropical tropopause temperature changes and/or changes in deep convection (see Sections 3.4.2 and 8.6.3.1.1).

The radiative effect of absorption by water vapour is roughly proportional to the logarithm of its concentration, so it is the fractional change in water vapour concentration, not the absolute change, that governs its strength as a feedback mechanism. Calculations with GCMs suggest that water vapour remains at an approximately constant fraction of its saturated value (close to unchanged relative humidity (RH)) under global-scale warming (see Section 8.6.3.1). Under such a response, for uniform warming, the largest fractional change in water vapour, and thus the largest contribution to the feedback, occurs in the upper troposphere. In addition, GCMs find enhanced warming in the tropical upper troposphere, due to changes in the lapse rate (see Section 9.4.4). This further enhances moisture changes in this region, but also introduces a partially offsetting radiative response from the temperature increase, and the net effect of the combined water vapour/lapse rate feedback is to amplify the warming in response to forcing by around 50% (Section 8.6.2.3). The close link between these processes means that water vapour and lapse rate feedbacks are commonly considered together. The strength of the combined feedback is found to be robust across GCMs, despite significant inter-model differences, for example, in the mean climatology of water vapour (see Section 8.6.2.3).

Confidence in modelled water vapour feedback is thus affected by uncertainties in the physical processes controlling upper-tropospheric humidity, and confidence in their representation in GCMs. One important question is what the relative contribution of large-scale advective processes (in which confidence in GCMs' representation is high) is compared with microphysical processes (in which confidence is much lower) for determining the distribution and variation in water vapour. Although advection has been shown to establish the general distribution of tropical upper-tropospheric humidity in the present climate (see Section 8.6.3.1), a significant role for microphysics in humidity response to climate change cannot yet be ruled out.

Difficulties in observing water vapour in the upper troposphere have long hampered both observational and modelling studies, and significant limitations remain in coverage and reliability of observational humidity data sets (see Section 3.4.2). To reduce the impact of these problems, in recent years there has been increased emphasis on the use of satellite data (such as 6.3 to 6.7 µm thermal radiance measurements) for inferring variations or trends in humidity, and on direct simulation of satellite radiances in models as a basis for model evaluation (see Sections 3.4.2 and 8.6.3.1.1).

Variations in upper-tropospheric water vapour have been observed across time scales from seasonal and interannual to decadal, as well as in response to external forcing (see Section 3.4.2.2). At tropics-wide scales, they correspond to roughly unchanged RH (see Section 8.6.3.1), and GCMs are generally able to reproduce these observed variations. Both column-integrated (see Section 3.4.2.1) and upper-tropospheric (see Section 3.4.2.2) specific humidity have increased over the past two decades, also consistent with roughly unchanged RH. There remains substantial disagreement between different observational estimates of lapse rate changes over recent decades, but some of these are consistent with GCM simulations (see Sections 3.4.1 and 9.4.4).

Overall, since the TAR, confidence has increased in the conventional view that the distribution of RH changes little as climate warms, particularly in the upper troposphere. Confidence has also increased in the ability of GCMs to represent upper-tropospheric humidity and its variations, both free and forced. Together, upper-tropospheric observational and modelling evidence provide strong support for a combined water vapour/lapse rate feedback of around the strength found in GCMs (see Section 8.6.3.1.2).

surface albedo feedback amplifies the basic response by about 10%, and the cloud feedback does so by 10 to 50% depending on the GCM. Note, however, that because of the inherently nonlinear nature of the response to feedbacks, the final impact on sensitivity is not simply the sum of these responses. The effect of multiple positive feedbacks is that they mutually amplify each other's impact on climate sensitivity.

Using feedback parameters from Figure 8.14, it can be estimated that in the presence of water vapour, lapse rate and surface albedo feedbacks, but in the absence of cloud feedbacks, current GCMs would predict a climate sensitivity (±1 standard deviation) of roughly 1.9°C ± 0.15°C (ignoring spread from radiative forcing differences). The mean and standard deviation of climate sensitivity estimates derived from current GCMs are larger (3.2°C ± 0.7°C) essentially because the GCMs all predict a positive cloud feedback (Figure 8.14) but strongly disagree on its magnitude.

The large spread in cloud radiative feedbacks leads to the conclusion that differences in cloud response are the primary source of inter-model differences in climate sensitivity (see discussion in Section 8.6.3.2.2). However, the contributions of water vapour/lapse rate and surface albedo feedbacks to sensitivity spread are non-negligible, particularly since their impact is reinforced by the mean model cloud feedback being positive and quite strong.

8.6.3 Key Physical Processes Involved in Climate Sensitivity

The traditional approach in assessing model sensitivity has been to consider water vapour, lapse rate, surface albedo and cloud feedbacks separately. Although this division can be regarded as somewhat artificial because, for example, water vapour, clouds and temperature interact strongly, it remains conceptually useful, and is consistent in approach with previous assessments. Accordingly, and because of the relationship between lapse rate and water vapour feedbacks, this subsection separately addresses the water vapour/lapse rate feedbacks and then the cloud and surface albedo feedbacks.

8.6.3.1   Water Vapour and Lapse Rate

Absorption of LW radiation increases approximately with the logarithm of water vapour concentration, while the Clausius-Clapeyron equation dictates a near-exponential increase in moisture-holding capacity with temperature. Since tropospheric and surface temperatures are closely coupled (see Section 3.4.1), these constraints predict a strongly positive water vapour feedback if relative humidity (RH) is close to unchanged. Furthermore, the combined water vapour-lapse rate feedback is relatively insensitive to changes in lapse rate for unchanged RH (Cess, 1975) due to the compensating effects of water vapour and temperature on the OLR (see Box 8.1). Understanding processes determining the distribution and variability in RH is therefore central to understanding of the water vapour-lapse

rate feedback. To a first approximation, GCM simulations indeed maintain a roughly unchanged distribution of RH under greenhouse gas forcing. More precisely, a small but widespread RH decrease in GCM simulations typically reduces feedback strength slightly compared with a constant RH response (Colman, 2004; Soden and Held, 2006; Figure 8.14).

In the planetary boundary layer, humidity is controlled by strong coupling with the surface, and a broad-scale quasi-unchanged RH response is uncontroversial (Wentz and Schabel, 2000; Trenberth et al., 2005; Dai, 2006). Confidence in GCMs' water vapour feedback is also relatively high in the extratropics, because large-scale eddies, responsible for much of the moistening throughout the troposphere, are explicitly resolved, and keep much of the atmosphere at a substantial fraction of saturation throughout the year (Stocker et al., 2001). Humidity changes in the tropical middle and upper troposphere, however, are less well understood and have more TOA radiative impact than do other regions of the atmosphere (e.g., Held and Soden, 2000; Colman, 2001). Therefore, much of the research since the TAR has focused on the RH response in the tropics with emphasis on the upper troposphere (see Bony et al., 2006 for a review), and confidence in the humidity response of this region is central to confidence in modelled water vapour feedback.

The humidity distribution within the tropical free troposphere is determined by many factors, including the detrainment of vapour and condensed water from convective systems and the large-scale atmospheric circulation. The relatively dry regions of large-scale descent play a major role in tropical LW cooling, and changes in their area or humidity could potentially have a significant impact on water vapour feedback strength (Pierrehumbert, 1999; Lindzen et al., 2001; Peters and Bretherton, 2005). Given the complexity of processes controlling tropical humidity, however, simple convincing physical arguments about changes under global-scale warming are difficult to sustain, and a combination of modelling and observational studies are needed to assess the reliability of model water vapour feedback.

In contrast to cloud feedback, a strong positive water vapour feedback is a robust feature of GCMs (Stocker et al., 2001), being found across models with many different schemes for advection, convection and condensation of water vapour. High-resolution mesoscale (Larson and Hartmann, 2003) and cloud-resolving models (Tompkins and Craig, 1999) run on limited tropical domains also display humidity responses consistent with strong positive feedback, although with differences in the details of upper-tropospheric RH (UTRH) trends with temperature. Experiments with GCMs have found water vapour feedback strength to be insensitive to large changes in vertical resolution, as well as convective parametrization and advection schemes (Ingram, 2002). These modelling studies provide evidence that the free-tropospheric RH response of global coupled models under climate warming is not simply an artefact of GCMs or of coarse GCM resolution, since broadly similar changes are found in a range of models of different complexity and scope. Indirect supporting evidence for model water vapour feedback strength also comes from experiments which show that suppressing humidity variation from the radiation code in an AOGCM produces unrealistically low interannual variability (Hall and Manabe, 1999).

Confidence in modelled water vapour feedback is dependent upon understanding of the physical processes important for controlling UTRH, and confidence in their representation in GCMs. The TAR noted a sensitivity of UTRH to the representation of cloud microphysical processes in several simple modelling studies. However, other evidence suggests that the role of microphysics is limited. The observed RH field in much of the tropics can be well simulated without microphysics, but simply by observed winds while imposing an upper limit of 100% RH on parcels (Pierrehumbert and Roca, 1998; Gettelman et al., 2000; Dessler and Sherwood, 2000), or by determining a detrainment profile from clear-sky radiative cooling (Folkins et al., 2002). Evaporation of detrained cirrus condensate also does not play a major part in moistening the tropical upper troposphere (Soden, 2004; Luo and Rossow, 2004), although cirrus might be important as a water vapour sink (Luo and Rossow, 2004). Overall, these studies increase confidence in GCM water vapour feedback, since they emphasise the importance of large-scale advective processes, or radiation, in which confidence in representation by GCMs is high, compared with microphysical processes, in which confidence is much lower. However, a significant role for microphysics in determining the distribution of changes in water vapour under climate warming cannot yet be ruled out.

Observations provide ample evidence of regional-scale increases and decreases in tropical UTRH in response to changes in convection (Zhu et al., 2000; Bates and Jackson, 2001; Blankenship and Wilheit, 2001; Wang et al., 2001; Chen et al., 2002; Chung et al., 2004; Sohn and Schmetz, 2004). Such changes, however, provide little insight into large-scale thermodynamic relationships (most important for the water vapour feedback) unless considered over entire circulation systems. Recent observational studies of the tropical mean UTRH response to temperature have found results consistent with that of near-unchanged RH at a variety of time scales (see Section 3.4.2.2). These include responses from interannual variability (Bauer et al., 2002; Allan et al., 2003; McCarthy and Toumi, 2004), volcanic forcing (Soden et al., 2002; Forster and Collins, 2004) and decadal trends (Soden et al., 2005), although modest RH decreases are noted at high levels on interannual time scales (Minschwaner and Dessler, 2004; Section 3.4.2.3). Seasonal variations in observed global LW radiation trapping are also consistent with a strong positive water vapour feedback (Inamdar and Ramanathan, 1998; Tsushima et al., 2005). Note, however, that humidity responses to variability or shorter time-scale forcing must be interpreted cautiously, as they are not direct analogues to that from greenhouse gas increases, because of differences in patterns of warming and circulation changes.

8.6.3.1.1 Evaluation of water vapour/lapse rate feedback processes in models

Evaluation of the humidity distribution and its variability in GCMs, while not directly testing their climate change feedbacks, can assess their ability to represent key physical processes controlling water vapour and therefore affect confidence in their water vapour feedback. Limitations in coverage or accuracy of radiosonde measurements or reanalyses have long posed a problem for UTRH evaluation in models (Trenberth et al., 2001; Allan et al., 2004), and recent emphasis has been on assessments using satellite measurements, along with increasing efforts to directly simulate satellite radiances in models (so as to reduce errors in converting to model-level RH) (e.g., Soden et al., 2002; Allan et al., 2003; Iacono et al., 2003; Brogniez et al., 2005; Huang et al., 2005).

Major features of the mean humidity distribution are reasonably simulated by GCMs, along with the consequent distribution of OLR (see Section 8.3.1). In the important subtropical subsidence regions, models show a range of skill in representing the mean UTRH. Some large regional biases have been found (Iacono et al., 2003; Chung et al., 2004), although good agreement of distribution and variability with satellite data has also been noted in some models (Allan et al., 2003; Brogniez et al., 2005). Uncertainties in satellite-derived data sets further complicate such comparisons, however. Skill in the reproduction of 'bimodality' in the humidity distribution at different time scales has also been found to differ between models (Zhang et al., 2003; Pierrehumbert et al., 2007), possibly associated with mixing processes and resolution. Note, however, that given the near-logarithmic dependence of LW radiation on humidity, errors in the control climate humidity have little direct effect on climate sensitivity: it is the fractional change of humidity as climate changes that matters (Held and Soden, 2000).

A number of new tests of large-scale variability of UTRH have been applied to GCMs since the TAR, and have generally found skill in model simulations. Allan et al. (2003) found that an AGCM forced by observed SSTs simulated interannual changes in tropical mean 6.7 μm radiance (sensitive to UTRH and temperature) in broad agreement with High Resolution Infrared Radiation Sounder (HIRS) observations over the last two decades. Minschwaner et al. (2006) analysed the interannual response of tropical mean 250 hPa RH to the mean SST of the most convectively active region in 16 AOGCMs from the MMD at PCMDI. The mean model response (a small decrease in RH) was statistically consistent with the 215 hPa response inferred from satellite observations, when uncertainties from observations and model spread were taken into account. AGCMs have been able to reproduce global or tropical mean variations in clear sky OLR (sensitive to water vapour and temperature distributions) over seasonal (Tsushima et al., 2005) as well as interannual and decadal (Soden, 2000; Allan and Slingo, 2002) time scales (although aerosol or greenhouse gas uncertainties and sampling differences can affect these latter comparisons; Allan et al., 2003). In the lower troposphere, GCMs can simulate global-scale interannual moisture variability well (e.g., Allan et al., 2003). At a smaller scale, a number of GCMs have also shown skill in reproducing regional changes in UTRH in response to circulation changes such as from seasonal or interannual variability (e.g., Soden, 1997; Allan et al., 2003; Brogniez et al., 2005).

A further test of the response of free tropospheric temperature and humidity to surface temperature in models is how well they can reproduce interannual correlations between surface temperature and vertical humidity profiles. Although GCMs are only partially successful in reproducing regional (Ross et al., 2002) and mean tropical (Bauer et al., 2002) correlations, the marked disagreement found in previous studies (Sun and Held, 1996; Sun et al., 2001) has been shown to be in large part an artefact of sampling techniques (Bauer et al., 2002).

There have also been efforts since the TAR to test GCMs' water vapour response against that from global-scale temperature changes of recent decades. One recent study used a long period of satellite data (1982–2004) to infer trends in UTRH, and found that an AGCM, forced by observed SSTs, was able to capture the observed global and zonal humidity trends well (Soden et al., 2005). A second approach uses the cooling following the eruption of Mt Pinatubo. Using estimated aerosol forcing, Soden et al. (2002) found a model-simulated response of HIRS 6.7 μm radiance consistent with satellite observations. They also found a model global temperature response similar to that observed, but not if the water vapour feedback was switched off (although the study neglected changes in cloud cover and potential heat uptake by the deep ocean). Using radiation calculations based on humidity observations, Forster and Collins (2004) found consistency in inferred water vapour feedback strength with an ensemble of coupled model integrations, although the latitude-height pattern of the observed humidity response did not closely match any single realisation. They deduced a water vapour feedback of 0.9 to 2.5 W m^{-2} °C^{-1}, a range which covers that of models under greenhouse gas forcing (see Figure 8.14). An important caveat to these studies is that the climate perturbation from Mt Pinatubo was small, not sitting clearly above natural variability (Forster and Collins, 2004). Caution is also required when comparing with feedbacks from increased greenhouse gases, because radiative forcing from volcanic aerosol is differently distributed and occurs over shorter time scales, which can induce different changes in circulation and bias the relative land/ocean response (although a recent AOGCM study found similar global LW radiation clear sky feedbacks between the two forcings; Yokohata et al., 2005). Nevertheless, comparing observed and modelled water vapour response to the eruption of Mt. Pinatubo constitutes one way to test model ability to simulate humidity changes induced by an external global-scale forcing.

At low latitudes, GCMs show negative lapse rate feedback because of their tendency towards a moist adiabatic lapse rate, producing amplified warming aloft. At middle to high latitudes, enhanced low-level warming, particularly in winter, contributes a positive feedback (e.g., Colman, 2003b), and global feedback strength is dependent upon the meridional warming

gradient (Soden and Held, 2006). There has been extensive testing of GCM tropospheric temperature response against observational trends for climate change detection purposes (see Section 9.4.4). Although some recent studies have suggested consistency between modelled and observed changes (e.g., Fu et al., 2004; Santer et al., 2005), debate continues as to the level of agreement, particularly in the tropics (Section 9.4.4). Regardless, if RH remains close to unchanged, the combined lapse rate and water vapour feedback is relatively insensitive to differences in lapse rate response (Cess, 1975; Allan et al., 2002; Colman, 2003a).

In the stratosphere, GCM water vapour response is sensitive to the location of initial radiative forcing (Joshi et al., 2003; Stuber et al., 2005). Forcing concentrated in the lower stratosphere, such as from ozone changes, invoked a positive feedback involving increased stratospheric water vapour and tropical cold point temperatures in one study (Stuber et al., 2005). However, for more homogenous forcing, such as from CO_2, the stratospheric water vapour contribution to model sensitivity appears weak (Colman, 2001; Stuber et al., 2001, 2005). There is observational evidence of possible long-term increases in stratospheric water vapour (Section 3.4.2.3), although it is not yet clear whether this is a feedback process. If there is a significant global mean trend associated with feedback mechanisms, however, this could imply a significant stratospheric water vapour feedback (Forster and Shine, 2002).

8.6.3.1.2   Summary of water vapour and lapse rate feedbacks

Significant progress has been made since the TAR in understanding and evaluating water vapour and lapse rate feedbacks. New tests have been applied to GCMs, and have generally found skill in the representation of large-scale free tropospheric humidity responses to seasonal and interannual variability, volcano-induced cooling and climate trends. New evidence from both observations and models has reinforced the conventional view of a roughly unchanged RH response to warming. It has also increased confidence in the ability of GCMs to simulate important features of humidity and temperature response under a range of different climate perturbations. Taken together, the evidence strongly favours a combined water vapour-lapse rate feedback of around the strength found in global climate models.

8.6.3.2   Clouds

By reflecting solar radiation back to space (the albedo effect of clouds) and by trapping infrared radiation emitted by the surface and the lower troposphere (the greenhouse effect of clouds), clouds exert two competing effects on the Earth's radiation budget. These two effects are usually referred to as the SW and LW components of the cloud radiative forcing (CRF). The balance between these two components depends on many factors, including macrophysical and microphysical cloud properties. In the current climate, clouds exert a cooling effect on climate (the global mean CRF is negative). In response to global warming, the cooling effect of clouds on climate might be enhanced or weakened, thereby producing a radiative feedback to climate warming (Randall et al., 2006; NRC, 2003; Zhang, 2004; Stephens, 2005; Bony et al., 2006).

In many climate models, details in the representation of clouds can substantially affect the model estimates of cloud feedback and climate sensitivity (e.g., Senior and Mitchell, 1993; Le Treut et al., 1994; Yao and Del Genio, 2002; Zhang, 2004; Stainforth et al., 2005; Yokohata et al., 2005). Moreover, the spread of climate sensitivity estimates among current models arises primarily from inter-model differences in cloud feedbacks (Colman, 2003a; Soden and Held, 2006; Webb et al., 2006; Section 8.6.2, Figure 8.14). Therefore, cloud feedbacks remain the largest source of uncertainty in climate sensitivity estimates.

This section assesses the evolution since the TAR in the understanding of the physical processes involved in cloud feedbacks (see Section 8.6.3.2.1), in the interpretation of the range of cloud feedback estimates among current climate models (see Section 8.6.3.2.2) and in the evaluation of model cloud feedbacks using observations (see Section 8.6.3.2.3).

8.6.3.2.1 Understanding of the physical processes involved in cloud feedbacks

The Earth's cloudiness is associated with a large spectrum of cloud types, ranging from low-level boundary-layer clouds to deep convective clouds and anvils. Understanding cloud feedbacks requires an understanding of how a change in climate may affect the spectrum and the radiative properties of these different clouds, and an estimate of the impact of these changes on the Earth's radiation budget. Moreover, since cloudy regions are also moist regions, a change in the cloud fraction matters for both the water vapour and the cloud feedbacks (Pierrehumbert, 1995; Lindzen et al., 2001). Since the TAR, there have been some advances in the analysis of physical processes involved in cloud feedbacks, thanks to the combined analysis of observations, simple conceptual models, cloud-resolving models, mesoscale models and GCMs (reviewed in Bony et al., 2006). Major issues are presented below.

Several climate feedback mechanisms involving convective anvil clouds have been examined. Hartmann and Larson (2002) proposed that the emission temperature of tropical anvil clouds is essentially independent of the surface temperature (Fixed Anvil Temperature hypothesis), and that it will thus remain unchanged during climate change. This suggestion is consistent with cloud-resolving model simulations showing that in a warmer climate, the vertical profiles of mid- and upper-tropospheric cloud fraction, condensate and RH all tend to be displaced upward in height together with the temperature (Tompkins and Craig, 1999). However, this hypothesis has not yet been tested with observations or with cloud-resolving model simulations having a fine vertical resolution in the upper troposphere. The response of the anvil cloud fraction to a change in temperature remains a subject of debate. Assuming that an increase with temperature in the precipitation efficiency of convective clouds could decrease the amount of water detrained in the upper troposphere,

Lindzen et al. (2001) speculated that the tropical area covered by anvil clouds could decrease with rising temperature, and that would lead to a negative climate feedback (iris hypothesis). Numerous objections have been raised about various aspects of the observational evidence provided so far (Chambers et al., 2002; Del Genio and Kovari, 2002; Fu et al., 2002; Harrison, 2002; Hartmann and Michelsen, 2002; Lin et al., 2002, 2004), leading to a vigorous debate with the authors of the hypothesis (Bell et al., 2002; Chou et al., 2002; Lindzen et al., 2002). Other observational studies (Del Genio and Kovari, 2002; Del Genio et al., 2005b) suggest an increase in the convective cloud cover with surface temperature.

Boundary-layer clouds have a strong impact on the net radiation budget (e.g., Harrison et al., 1990; Hartmann et al., 1992) and cover a large fraction of the global ocean (e.g., Norris, 1998a,b). Understanding how they may change in a perturbed climate is thus a vital part of the cloud feedback problem. The observed relationship between low-level cloud amount and a particular measure of lower tropospheric stability (Klein and Hartmann, 1993), which has been used in some simple climate models and in some GCMs' parametrizations of boundary-layer cloud amount (e.g., CCSM3, FGOALS), led to the suggestion that a global climate warming might be associated with an increased low-level cloud cover, which would produce a negative cloud feedback (e.g., Miller, 1997; Zhang, 2004). However, variants of the lower-tropospheric stability measure, which may predict boundary-layer cloud amount as well as the Klein and Hartmann (1993) measure, would not necessarily predict an increase in low-level clouds in a warmer climate (e.g., Williams et al., 2006). Moreover, observations indicate that in regions covered by low-level clouds, the cloud optical depth decreases and the SW CRF weakens as temperature rises (Tselioudis and Rossow, 1994; Greenwald et al., 1995; Bony et al., 1997; Del Genio and Wolf, 2000; Bony and Dufresne, 2005), but the different factors that may explain these observations are not well established. Therefore, understanding of the physical processes that control the response of boundary-layer clouds and their radiative properties to a change in climate remains very limited.

At mid-latitudes, the atmosphere is organised in synoptic weather systems, with prevailing thick, high-top frontal clouds in regions of synoptic ascent and low-level or no clouds in regions of synoptic descent. In the NH, several climate models report a decrease in overall extratropical storm frequency and an increase in storm intensity in response to climate warming (e.g., Carnell and Senior, 1998; Geng and Sugi, 2003) and a poleward shift of the storm tracks (Yin, 2005). Using observations and reanalyses to investigate the impact that dynamical changes such as those found by Carnell and Senior (1998) would have on the NH radiation budget, Tselioudis and Rossow (2006) suggested that the increase in storm strength would have a larger radiative impact than the decrease in storm frequency, and that this would produce increased reflection of SW radiation and decreased emission of LW radiation. However, the poleward shift of the storm tracks may decrease the amount of SW radiation reflected (Tsushima et al., 2006). In addition, several studies have used observations to investigate the dependence of mid-latitude cloud radiative properties on temperature. Del Genio and Wolf (2000) showed that the physical thickness of low-level continental clouds decreases with rising temperature, resulting in a decrease in the cloud water path and optical thickness as temperature rises, and Norris and Iacobellis (2005) suggested that over the NH ocean, a uniform change in surface temperature would result in decreased cloud amount and optical thickness for a large range of dynamical conditions. The sign of the climate change radiative feedback associated with the combined effects of dynamical and temperature changes on extratropical clouds is still unknown.

The role of polar cloud feedbacks in climate sensitivity has been emphasized by Holland and Bitz (2003) and Vavrus (2004). However, these feedbacks remain poorly understood.

8.6.3.2.2 Interpretation of the range of cloud feedbacks among climate models

In doubled atmospheric CO_2 equilibrium experiments performed by mixed-layer ocean-atmosphere models as well as in transient climate change integrations performed by fully coupled ocean-atmosphere models, models exhibit a large range of global cloud feedbacks, with roughly half of the climate models predicting a more negative CRF in response to global warming, and half predicting the opposite (Soden and Held, 2006; Webb et al., 2006). Several studies suggest that the sign of cloud feedbacks may not be necessarily that of CRF changes (Zhang et al., 1994; Colman, 2003a; Soden et al., 2004), due to the contribution of clear-sky radiation changes (i.e., of water vapour, temperature and surface albedo changes) to the change in CRF. The Partial Radiative Perturbation (PRP) method, that excludes clear-sky changes from the definition of cloud feedbacks, diagnoses a positive global net cloud feedback in virtually all the models (Colman, 2003a; Soden and Held, 2006). However, the cloud feedback estimates diagnosed from either the change in CRF or the PRP method are well correlated (i.e., their relative ranking is similar), and they exhibit a similar spread among GCMs.

By decomposing the GCM feedbacks into regional components or dynamical regimes, substantial progress has been made in the interpretation of the range of climate change cloud feedbacks. The comparison of coupled AOGCMs used for the climate projections presented in Chapter 10 (Bony and Dufresne, 2005), of atmospheric or slab ocean versions of current GCMs (Webb et al., 2006; Williams et al., 2006; Wyant et al., 2006), or of slightly older models (Williams et al., 2003; Bony et al., 2004; Volodin, 2004; Stowasser et al.; 2006) show that inter-model differences in cloud feedbacks are mostly attributable to the

SW cloud feedback component, and that the responses to global warming of both deep convective clouds and low-level clouds differ among GCMs. Recent analyses suggest that the response of boundary-layer clouds constitutes the largest contributor to the range of climate change cloud feedbacks among current GCMs (Bony and Dufresne, 2005; Webb et al., 2006; Wyant et al., 2006). It is due both to large discrepancies in the radiative response simulated by models in regions dominated by low-level cloud cover (Figure 8.15), and to the large areas of the globe covered by these regions. However, the response of other cloud types is also important because for each model it either reinforces or partially cancels the radiative response from low-level clouds. The spread of model cloud feedbacks is substantial at all latitudes, and tends to be larger in the tropics (Bony et al., 2006; Webb et al., 2006). Differences in the representation of mixed-phase clouds and in the degree of latitudinal shift of the storm tracks predicted by the models also contribute to inter-model differences in the CRF response to climate change, particularly in the extratropics (Tsushima et al., 2006).

8.6.3.2.3 Evaluation of cloud feedbacks produced by climate models

The evaluation of clouds in climate models has long been based on comparisons of observed and simulated climatologies of TOA radiative fluxes and total cloud amount (see Section 8.3.1).

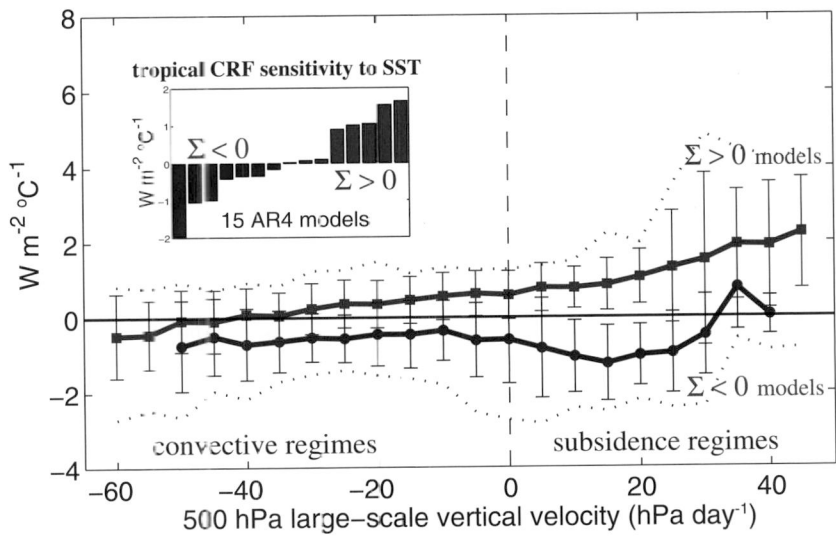

Figure 8.15. *Sensitivity (in W m⁻² °C⁻¹) of the tropical net cloud radiative forcing (CRF) to SST changes associated with global warming (simulations in which CO₂ increases by 1% yr⁻¹). The inset shows the tropically averaged sensitivity Σ predicted by 15 AOGCMs used in this report: 7 models predict Σ < 0 and 8 models predict Σ > 0. The main panel compares the CRF sensitivity to SST predicted by the two groups of models in different regimes of the large-scale tropical circulation (the 500 hPa vertical pressure velocity is used as a proxy for large-scale motions, with negative values corresponding to large-scale ascending motions, and positive values to sinking motions). Thick lines and vertical lines represent the mean and the standard deviation of model sensitivities within each group; dotted lines represent the minimum and maximum values of model sensitivities within each dynamical regime. The discrepancy between the two groups of models is greatest in regimes of large-scale subsidence. These regimes, which have a large statistical weight in the tropics, are primarily covered by boundary-layer clouds. As a result, the spread of tropical cloud feedbacks among the models (inset) primarily arises from inter-model differences in the radiative response of low-level clouds in regimes of large-scale subsidence. Adapted from Bony and Dufresne (2005).*

However, a good agreement with these observed quantities may result from compensating errors. Since the TAR, and partly due to the use of an International Satellite Cloud Climatology Project (ISCCP) simulator (Klein and Jakob, 1999; Webb et al., 2001), the evaluation of simulated cloud fields is increasingly done in terms of cloud types and cloud optical properties (Klein and Jakob, 1999; Webb et al., 2001; Williams et al., 2003; Lin and Zhang, 2004; Weare, 2004; Zhang et al., 2005; Wyant et al., 2006). It has thus become more powerful and constrains the models more. In addition, a new class of observational tests has been applied to GCMs, using clustering or compositing techniques, to diagnose errors in the simulation of particular cloud regimes or in specific dynamical conditions (Tselioudis et al., 2000; Norris and Weaver, 2001; Jakob and Tselioudis, 2003; Williams et al., 2003; Bony et al., 2004; Lin and Zhang, 2004; Ringer and Allan, 2004; Bony and Dufresne, 2005; Del Genio et al., 2005a; Gordon et al., 2005; Bauer and Del Genio, 2006; Williams et al., 2006; Wyant et al., 2006). An observational test focused on the global response of clouds to seasonal variations has been proposed to evaluate model cloud feedbacks (Tsushima et al., 2005), but has not yet been applied to current models.

These studies highlight some common biases in the simulation of clouds by current models (e.g., Zhang et al., 2005). This includes the over-prediction of optically thick clouds and the under-prediction of optically thin low and middle-top clouds. However, uncertainties remain in the observational determination of the relative amounts of the different cloud types (Chang and Li, 2005). For mid-latitudes, these biases have been interpreted as the consequence of the coarse resolution of climate GCMs and their resulting inability to simulate the right strength of ageostrophic circulations (Bauer and Del Genio, 2006) and the right amount of sub-grid scale variability (Gordon et al., 2005). Although the errors in the simulation of the different cloud types may eventually compensate and lead to a prediction of the mean CRF in agreement with observations (see Section 8.3), they cast doubts on the reliability of the model cloud feedbacks. For instance, given the nonlinear dependence of cloud albedo on cloud optical depth, the overestimate of the cloud optical thickness implies that a change in cloud optical depth, even of the right sign and magnitude, would produce a too small radiative signature. Similarly, the under-prediction of low- and mid-level clouds presumably affects the magnitude of the radiative response to climate warming in the widespread regions of subsidence. Modelling assumptions controlling the cloud water phase (liquid, ice or mixed) are known to be critical for the prediction of climate sensitivity. However, the evaluation of these assumptions is just beginning (Doutriaux-Boucher and Quaas, 2004; Naud et al., 2006). Tsushima et al. (2006) suggested that observations of the distribution of each phase of cloud water in the current climate would provide a substantial constraint on the model cloud feedbacks at middle and high latitudes.

As an attempt to assess some components of the cloud response to a change in climate, several studies have investigated the ability of GCMs to simulate the sensitivity of clouds and CRF to interannual changes in environmental conditions. When examining atmosphere-mixed-layer ocean models, Williams

et al. (2006) found for instance that by considering the CRF response to a change in large-scale vertical velocity and in lower-tropospheric stability, a component of the local mean climate change cloud response can be related to the present-day variability, and thus evaluated using observations. Bony and Dufresne (2005) and Stowasser and Hamilton (2006) examined the ability of the AOGCMs of Chapter 10 to simulate the change in tropical CRF to a change in SST, in large-scale vertical velocity and in lower-tropospheric RH. They showed that the models are most different and least realistic in regions of subsidence, and to a lesser extent in regimes of deep convective activity. This emphasizes the necessity to improve the representation and the evaluation of cloud processes in climate models, and especially those of boundary-layer clouds.

8.6.3.2.4   Conclusion on cloud feedbacks

Despite some advances in the understanding of the physical processes that control the cloud response to climate change and in the evaluation of some components of cloud feedbacks in current models, it is not yet possible to assess which of the model estimates of cloud feedback is the most reliable. However, progress has been made in the identification of the cloud types, the dynamical regimes and the regions of the globe responsible for the large spread of cloud feedback estimates among current models. This is likely to foster more specific observational analyses and model evaluations that will improve future assessments of climate change cloud feedbacks.

8.6.3.3   Cryosphere Feedbacks

A number of feedbacks that significantly contribute to the global climate sensitivity are due to the cryosphere. A robust feature of the response of climate models to increases in atmospheric concentrations of greenhouse gases is the poleward retreat of terrestrial snow and sea ice, and the polar amplification of increases in lower-tropospheric temperature. At the same time, the high-latitude response to increased greenhouse gas concentrations is highly variable among climate models (e.g., Holland and Bitz, 2003) and does not show substantial convergence in the latest generation of AOGCMs (Chapman and Walsh, 2007; see also Section 11.8). The possibility of threshold behaviour also contributes to the uncertainty of how the cryosphere may evolve in future climate scenarios.

Arguably, the most important simulated feedback associated with the cryosphere is an increase in absorbed solar radiation resulting from a retreat of highly reflective snow or ice cover in a warmer climate. Since the TAR, some progress has been made in quantifying the surface albedo feedback associated with the cryosphere. Hall (2004) found that the albedo feedback was responsible for about half the high-latitude response to a doubling of atmospheric CO_2. However, an analysis of long control simulations showed that it accounted for surprisingly little internal variability. Hall and Qu (2006) show that biases of a number of MMD models in reproducing the observed seasonal cycle of land snow cover (especially the spring melt) are tightly related to the large variations in snow albedo feedback strength

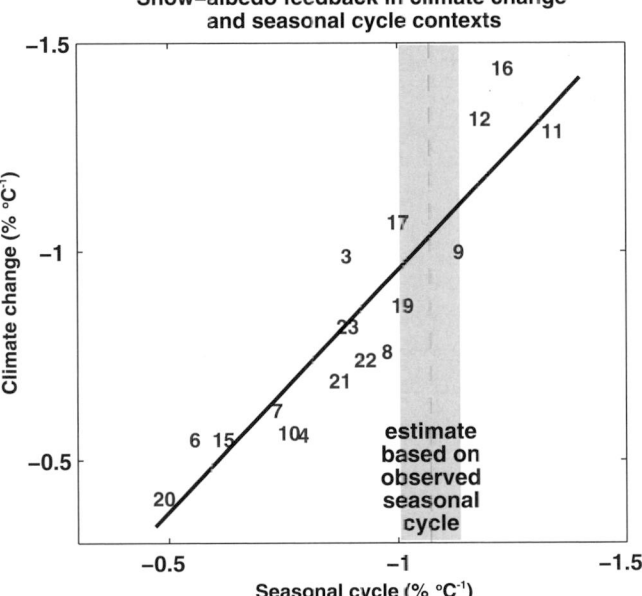

Figure 8.16. Scatter plot of simulated springtime $\Delta\alpha_s/\Delta T_s$ values in climate change (ordinate) vs simulated springtime $\Delta\alpha_s/\Delta T_s$ values in the seasonal cycle (abscissa) in transient climate change experiments with 17 AOGCMs used in this report ($\Delta\alpha_s$ and T_s are surface albedo and surface air temperature, respectively). The climate change $\Delta\alpha_s/\Delta T_s$ values are the reduction in springtime surface albedo averaged over Northern Hemisphere continents between the 20th and 22nd centuries divided by the increase in surface air temperature in the region over the same time period. Seasonal cycle $\Delta\alpha_s/\Delta T_s$ values are the difference between 20th-century mean April and May α_s averaged over Northern Hemisphere continents divided by the difference between April and May T_s averaged over the same area and time period. A least-squares fit regression line for the simulations (solid line) and the observed seasonal cycle $\Delta\alpha_s/\Delta T_s$ value based on ISCCP and ERA40 reanalysis (dashed vertical line) are also shown. The grey bar gives an estimate of statistical error, according to a standard formula for error in the estimate of the mean of a time series (in this case the observed time series of $\Delta\alpha_s/\Delta T_s$) given the time series' length and variance. If this statistical error only is taken into account, the probability that the actual observed value lies outside the grey bar is 5%. Each number corresponds to a particular AOGCM (see Table 8.1). Adapted from Hall and Qu (2006).

simulated by the same models in climate change scenarios. Addressing the seasonal cycle biases would therefore provide a constraint that would reduce divergence in simulations of snow albedo feedback under climate change. However, possible use of seasonal snow albedo feedback to evaluate snow albedo feedback under climate change conditions is of course dependent upon the realism of the correlation between the two feedbacks suggested by GCMs (Figure 8.16). A new result found independently by Winton (2006a) and Qu and Hall (2005) is that surface processes are the main source of divergence in climate simulations of surface albedo feedback, rather than simulated differences in cloud fields in cryospheric regions.

Understanding of other feedbacks associated with the cryosphere (e.g., ice insulating feedback, MOC/SST-sea ice feedback, ice thickness/ice growth feedback) has improved since the TAR (NRC, 2003; Bony et al., 2006). However, the relative influence on climate sensitivity of these feedbacks has not been quantified.

Understanding and evaluating sea ice feedbacks is complicated by their strong coupling to processes in the high-latitude atmosphere and ocean, particularly to polar cloud processes and ocean heat and freshwater transport. Additionally, while impressive advances have occurred in developing sea ice components of the AOGCMs since the TAR, particularly by the inclusion of more sophisticated dynamics in most of them (see Section 8.2.4), evaluation of cryospheric feedbacks through the testing of model parametrizations against observations is hampered by the scarcity of observational data in the polar regions. In particular, the lack of sea ice thickness observations is a considerable problem.

The role of sea ice dynamics in climate sensitivity has remained uncertain for years. Some recent results with AGCMs coupled to slab ocean models (Hewitt et al., 2001; Vavrus and Harrison, 2003) support the hypothesis that a representation of sea ice dynamics in climate models has a moderating impact on climate sensitivity. However, experiments with full AOGCMs (Holland and Bitz, 2003) show no compelling relationship between the transient climate response and the presence or absence of ice dynamics, with numerous model differences presumably overwhelming whatever signal might be due to ice dynamics. A substantial connection between the initial (i.e., control) simulation of sea ice and the response to greenhouse gas forcing (Holland and Bitz, 2003; Flato, 2004) further hampers 'clean' experiments aimed at identifying or quantifying the role of sea ice dynamics.

A number of processes, other than surface albedo feedback, have been shown to also contribute to the polar amplification of warming in models (Alexeev, 2003, 2005; Holland and Bitz, 2003; Vavrus, 2004; Cai, 2005; Winton, 2006b). An important one is additional poleward energy transport, but contributions from local high-latitude water vapour, cloud and temperature feedbacks have also been found. The processes and their interactions are complex, however, with substantial variation between models (Winton, 2006b), and their relative importance contributing to or dampening high-latitude amplification has not yet been properly resolved.

8.3.4   How to Assess Our Relative Confidence in Feedbacks Simulated by Different Models?

Assessments of our relative confidence in climate projections from different models should ideally be based on a comprehensive set of observational tests that would allow us to quantify model errors in simulating a wide variety of climate statistics, including simulations of the mean climate and variability and of particular climate processes. The collection of measures that quantify how well a model performs in an ensemble of tests of this kind are referred to as 'climate metrics'. To have the ability to constrain future climate projections, they would ideally have strong connections with one or several aspects of climate change: climate sensitivity, large-scale patterns of climate change (inter-hemispheric symmetry, polar amplification, vertical patterns of temperature change, land-sea contrasts), regional patterns or transient aspects of climate change. For example, to assess confidence in model projections of the Australian climate, the metrics would need to include some measures of the quality of ENSO simulation because the Australian climate depends much on this variability (see Section 11.7).

To better assess confidence in the different model estimates of climate sensitivity, two kinds of observational tests are available: tests related to the global climate response associated with specified external forcings (discussed in Chapters 6, 9 and 10; Box 10.2) and tests focused on the simulation of key feedback processes.

Based on the understanding of both the physical processes that control key climate feedbacks (see Section 8.6.3), and also the origin of inter-model differences in the simulation of feedbacks (see Section 8.6.2), the following climate characteristics appear to be particularly important: (i) for the water vapour and lapse rate feedbacks, the response of upper-tropospheric RH and lapse rate to interannual or decadal changes in climate; (ii) for cloud feedbacks, the response of boundary-layer clouds and anvil clouds to a change in surface or atmospheric conditions and the change in cloud radiative properties associated with a change in extratropical synoptic weather systems; (iii) for snow albedo feedbacks, the relationship between surface air temperature and snow melt over northern land areas during spring and (iv) for sea ice feedbacks, the simulation of sea ice thickness.

A number of diagnostic tests have been proposed since the TAR (see Section 8.6.3), but few of them have been applied to a majority of the models currently in use. Moreover, it is not yet clear which tests are critical for constraining future projections. Consequently, a set of model metrics that might be used to narrow the range of plausible climate change feedbacks and climate sensitivity has yet to be developed.

8.7 Mechanisms Producing Thresholds and Abrupt Climate Change

8.7.1 Introduction

This discussion of thresholds and abrupt climate change is based on the definitions of 'threshold' and 'abrupt' proposed by Alley et al. (2002). The climate system tends to respond to changes in a gradual way until it crosses some threshold: thereafter any change that is defined as abrupt is one where the change in the response is much larger than the change in the forcing. The changes at the threshold are therefore abrupt relative to the changes that occur before or after the threshold and can lead to a transition to a new state. The spatial scales for these changes can range from global to local. In this definition, the magnitude of the forcing and response are important. In addition to the magnitude, the time scale being considered is also important. This section focuses mainly on decadal to centennial time scales.

Because of the somewhat subjective nature of the definitions of threshold and abrupt, there have been efforts to develop

quantitative measures to identify these points in a time series of a given variable (e.g., Lanzante, 1996; Seidel and Lanzante, 2004; Tomé and Miranda, 2004). The most common way to identify thresholds and abrupt changes is by linearly de-trending the input time series and looking for large deviations from the trend line. More statistically rigorous methods are usually based on Bayesian statistics.

This section explores the potential causes and mechanisms for producing thresholds and abrupt climate change and addresses the issue of how well climate models can simulate these changes. The following discussion is split into two main areas: forcing changes that can result in abrupt changes and abrupt climate changes that result from large natural variability on long time scales. Formally, the latter abrupt changes do not fit the definition of thresholds and abrupt changes, because the forcing (at least radiative forcing – the external boundary condition) is not changing in time. However these changes have been discussed in the literature and popular press and are worthy of assessment here.

8.7.2 Forced Abrupt Climate Change

8.7.2.1 Meridional Overturning Circulation Changes

As the radiative forcing of the planet changes, the climate system responds on many different time scales. For the physical climate system typically simulated in coupled models (atmosphere, ocean, land, sea ice), the longest response time scales are found in the ocean (Stouffer, 2004). In terms of thresholds and abrupt climate changes on decadal and longer time scales, the ocean has also been a focus of attention. In particular, the ocean's Atlantic MOC (see Box 5.1 for definition and description) is a main area of study.

The MOC transports large amounts of heat (order of 10^{15} Watts) and salt into high latitudes of the North Atlantic. There, the heat is released to the atmosphere, cooling the surface waters. The cold, relatively salty waters sink to depth and flow southward out of the Atlantic Basin. The complete set of climatic drivers of this circulation remains unclear but it is likely that both density (e.g., Stommel 1961; Rooth 1982) and wind stress forcings (e.g., Wunsch, 2002; Timmermann and Goosse, 2004) are important. Both palaeoclimate studies (e.g., Broecker, 1997; Clark et al., 2002) and modelling studies (e.g., Manabe and Stouffer, 1988, 1997; Vellinga and Wood, 2002) suggest that disruptions in the MOC can produce abrupt climate changes. A systematic model intercomparison study (Rahmstorf et al., 2005) found that all 11 participating EMICs had a threshold where the MOC shuts down (see Section 8.8.3). Due to the high computational cost, such a search for thresholds has not yet been performed with AOGCMs.

It is important to note the distinction between the equilibrium and transient or time-dependent responses of the MOC to changes in forcing. Due to the long response time scales found in the ocean (some longer than 1 kyr), it is possible that the short-term response to a given forcing change may be very different from the equilibrium response. Such behaviour of the coupled system has been documented in at least one AOGCM (Stouffer and Manabe, 2003) and suggested in the results of a few other AOGCM studies (e.g., Hirst, 1999; Senior and Mitchell, 2000; Bryan et al., 2006). In these AOGCM experiments, the MOC weakens as the greenhouse gases increase in the atmosphere. When the CO_2 concentration is stabilised, the MOC slowly returns to its unperturbed value.

As discussed in section 10.3.4, the MOC typically weakens as greenhouse gases increase due to the changes in surface heat and freshwater fluxes at high latitudes (Manabe et al., 1991). The surface flux changes reduce the surface density, hindering the vertical movement of water and slowing the MOC. As the MOC slows, it could approach a threshold where the circulation can no longer sustain itself. Once the MOC crosses this threshold, it could rapidly change states, causing abrupt climate change where the North Atlantic and surrounding land areas would cool relative to the case where the MOC is active. This cooling is the result of the loss of heat transport from low latitudes in the Atlantic and the feedbacks associated with the reduction in the vertical mixing of high-latitude waters.

A common misunderstanding is that the MOC weakening could cause the onset of an ice age. However, no model has supported this speculation when forced with realistic estimates of future climate forcings (see Section 10.3.4). In addition, in idealised modelling studies where the MOC was forced to shut down through very large sources of freshwater (not changes in greenhouse gases), the surface temperature changes do not support the idea that an ice age could result from a MOC shut down, although the impacts on climate would be large (Manabe and Stouffer, 1988, 1997; Schiller et al., 1997; Vellinga and Wood, 2002; Stouffer et al., 2006). In a recent intercomparison involving 11 coupled atmosphere-ocean models (Gregory et al., 2005), the MOC decreases by only 10 to 50% during a 140-year period (as atmospheric CO_2 quadruples), and in no model is there a land cooling anywhere (as the global-scale heating due to increasing CO_2 overwhelms the local cooling effect due to reduced MOC).

Because of the large amount of heat and salt transported northward and its sensitivity to surface fluxes, the changes in the MOC are able to produce abrupt climate change on decadal to centennial time scales (e.g., Manabe and Stouffer, 1995; Stouffer et al., 2006). Idealised studies using present-day simulations have shown that models can simulate many of the variations seen in the palaeoclimate record on decadal to centennial time scales when forced by fluxes of freshwater water at the ocean surface. However, the quantitative response to freshwater inputs varies widely among models (Stouffer et al., 2006), which led the CMIP and Paleoclimate Modelling Intercomparison Project (PMIP) panels to design and support a set of coordinated experiments to study this issue (http://www.gfdl.noaa.gov/~kd/CMIP.html and http://www.pmip2.cnrs-gif.fr/pmip2/design/experiments/waterhosing.shtml).

In addition to the amount of the freshwater input, the exact location of that input may also be important (Rahmstorf 1996, Manabe and Stouffer, 1997; Rind et al., 2001). Designing experiments and determining the realistic past forcings needed to test the models' response on decadal to centennial time scales remains to be accomplished.

The processes determining MOC response to increasing greenhouse gases have been studied in a number of models. In many models, initial MOC response to increasing greenhouse gases is dominated by thermal effects. In most models, this is enhanced by changes in salinity driven by, among other things, the expected strengthening of the hydrological cycle (Gregory et al., 2005; Chapter 10). Melt water runoff from a melting of the Greenland Ice Sheet is a potentially major source of freshening not yet included in the models found in the MMD (see Section 8.7.2.2). More complex feedbacks, associated with wind and hydrological changes, are also important in many models. These include local surface flux anomalies in deep-water formation regions (Gent, 2001) and oceanic teleconnections driven by changes to the freshwater budget of the tropical and South Atlantic (e.g., Latif et al., 2000; Thorpe et al., 2001; Vellinga et al., 2002; Hu et al., 2004). The magnitudes of the climate factors causing the MOC to weaken, along with the feedbacks and the associated restoring factors, are all uncertain at this time. Evaluation of these processes in AOGCMs is mainly restricted by lack of observations, but some early progress has been made in individual studies (e.g., Schmittner et al., 2000; Pardaens et al., 2003; Wu et al., 2005; Chapter 9). Model intercomparison studies (e.g., Gregory et al., 2005; Rahmstorf et al., 2005; Stouffer et al., 2006) were developed to identify and understand the causes for the wide range of MOC responses in the coupled models used here (see Chapters 4, 6 and 10).

8.7.2.2 Rapid West Antarctic and/or Greenland Ice Sheet Collapse and Meridional Overturning Circulation Changes

Increased influx of freshwater to the ocean from the ice sheets is a potential forcing for abrupt climate changes. For Antarctica in the present climate, these fluxes chiefly arise from melting below the ice shelves and from melting of icebergs transported by the ocean; both fluxes could increase significantly in a warmer climate. Ice sheet runoff and iceberg calving, in roughly equal shares, currently dominate the freshwater flux from the Greenland Ice Sheet (Church et al., 2001; Chapter 4). In a warming climate, runoff is expected to quickly increase and become much larger than the calving rate, the latter of which in turn is likely to decrease as less and thinner ice borders the ocean; basal melting from below the grounded ice will remain several orders of magnitude smaller than the other fluxes (Huybrechts et al., 2002). For a discussion of the likelihood of these ice sheet changes and the effects on sea level, see the discussion in Chapter 10.

Changes in the surface forcing near the deep-water production areas seem to be most capable of producing rapid climate changes on decadal and longer time scales due to changes in the ocean circulation and mixing. If there are large changes in the ice volume over Greenland, it is likely that much of this melt water will freshen the surface waters in the high-latitude North Atlantic, slowing down the MOC (see Section 8.7.2.1; Chapter 10). Rind et al. (2001) found that changes in the NADW formation rate could instigate changes in the deep-water formation around Antarctica.

The response of the Atlantic MOC to changes in the Antarctic Ice Sheet is less well understood. Experiments with ocean-only models where the melt water changes are imposed as surface salinity changes indicate that the Atlantic MOC will intensify as the waters around Antarctica become less dense (Seidov et al., 2001). Weaver et al. (2003) showed that by adding freshwater in the Southern Ocean, the MOC could change from an 'off' state to a state similar to present day. However, in an experiment with an AOGCM, Seidov et al. (2005) found that an external source of freshwater in the Southern Ocean resulted in a surface freshening throughout the world ocean, weakening the Atlantic MOC. In these model results, the SH MOC associated with Antarctic Bottom Water (AABW) formation weakened, causing a cooling around Antarctica. See Chapters 4, 6 and 10 for more discussion about the likelihood of large melt water fluxes from the ice sheets affecting the climate.

In summary, there is a potential for rapid ice sheet changes to produce rapid climate change both through sea level changes and ocean circulation changes. The ocean circulation changes result from increased freshwater flux over the particularly sensitive deep-water production sites. In general, the possible climate changes associated with future evolution of the Greenland Ice Sheet are better understood than are those associated with changes in the Antarctic Ice Sheets.

8.7.2.3 Volcanoes

Volcanoes produce abrupt climate responses on short time scales. The surface cooling effect of the stratospheric aerosols, the main climatic forcing factor, decays in one to three years after an eruption due to the lifetime of the aerosols in the stratosphere. It is possible for one large volcano or a series of large volcanic eruptions to produce climate responses on longer time scales, especially in the subsurface region of the ocean (Delworth et al., 2005; Gleckler et al., 2006b).

The models' ability to simulate any possible abrupt response of the climate system to volcanic eruptions seems conceptually similar to their ability to simulate the climate response to future changes in greenhouse gases in that both produce changes in the radiative forcing of the planet. However, mechanisms involved in the exchange of heat between the atmosphere and ocean may be different in response to volcanic forcing when compared to the response to increase greenhouse gases. Therefore, the feedbacks involved may be different (see Section 9.6.2.2 for more discussion).

8.7.2.4 Methane Hydrate Instability/Permafrost Methane

Methane hydrates are stored on the seabed along continental margins where they are stabilised by high pressures and low temperatures, implying that ocean warming may cause hydrate instability and release of methane into the atmosphere (see

Section 4.7.2.4). Methane is also stored in the soils in areas of permafrost and warming increases the likelihood of a positive feedback in the climate system via permafrost melting and the release of trapped methane into the atmosphere. The likelihood of methane release from methane hydrates found in the oceans or methane trapped in permafrost layers is assessed in Chapter 7.

This subsection considers the potential usefulness of models in determining if those releases could trigger an abrupt climate change. Both forms of methane release represent a potential threshold in the climate system. As the climate warms, the likelihood of the system crossing a threshold for a sudden release increases (see Chapters 4, 7 and 10). Since these changes produce changes in the radiative forcing through changes in the greenhouse gas concentrations, the climatic impacts of such a release are the same as an increase in the rate of change in the radiative forcing. Therefore, the models' ability to simulate any abrupt climate change should be similar to their ability to simulate future abrupt climate changes due to changes in the greenhouse gas forcing.

8.7.2.5 Biogeochemical

Two questions concerning biogeochemical aspects of the climate system are addressed here. First, can biogeochemical changes lead to abrupt climate change? Second, can abrupt changes in the MOC further affect radiative forcing through biogeochemical feedbacks?

Abrupt changes in biogeochemical systems of relevance to our capacity to simulate the climate of the 21st century are not well understood (Friedlingstein et al., 2003). The potential for major abrupt change exists in the uptake and storage of carbon by terrestrial systems. While abrupt change within the climate system is beginning to be seriously considered (Rial et al., 2004; Schneider, 2004), the potential for abrupt change in terrestrial systems, such as loss of soil carbon (Cox et al., 2000) or die back of the Amazon forests (Cox et al., 2004) remains uncertain. In part this is due to lack of understanding of processes (see Friedlingstein et al., 2003; Chapter 7) and in part it results from the impact of differences in the projected climate sensitivities in the host climate models (Joos et al., 2001; Govindasamy et al., 2005; Chapter 10) where changes in the physical climate system affect the biological response.

There is some evidence of multiple equilibria within vegetation-soil-climate systems. These include North Africa and Central East Asia where Claussen (1998), using an EMIC with a land vegetation component, showed two stable equilibria for rainfall, dependent on initial land surface conditions. Kleidon et al. (2000), Wang and Eltahir (2000) and Renssen et al. (2003) also found evidence for multiple equilibria. These are preliminary assessments using relatively simple physical climate models that highlight the possibility of irreversible change in the Earth system but require extensive further research to assess the reliability of the phenomena found.

There have only been a few preliminary studies of the impact of abrupt climate changes such as the shutdown of the MOC on the carbon cycle. The findings of these studies indicate that the shutdown of the MOC would tend to increase the amount of greenhouse gases in the atmosphere (Joos et al., 1999; Plattner et al., 2001; Chapter 6). In both of these studies, only the effect of the oceanic component of the carbon cycle changes was considered.

8.7.3 Unforced Abrupt Climate Change

Formally, as noted above, the changes discussed here do not fall into the definition of abrupt climate change. In the literature, unforced abrupt climate change falls into two general categories. One is just a red noise time series, where there is power at decadal and longer time scales. A second category is a bimodal or multi-modal distribution. In practice, it can be very difficult to distinguish between the two categories unless the time series are very long – long enough to eliminate sampling as an issue – and the forcings are fairly constant in time. In observations, neither of these conditions is normally met.

Models, both AOGCMs and less complex models, have produced examples of large abrupt climate change (e.g., Hall and Stouffer 2001; Goosse et al., 2002) without any changes in forcing. Typically, these events are associated with changes in the ocean circulation, mainly in the North Atlantic. An abrupt event can last for several years to a few centuries. They bear some similarities with the conditions observed during a relatively cold period in the recent past in the Arctic (Goosse et al., 2003)

Unfortunately, the probability of such an event is difficult to estimate as it requires a very long experiment and is certainly dependent on the mean state simulated by the model. Furthermore, comparison with observations is nearly impossible since it would require a very long period with constant forcing which does not exist in nature. Nevertheless, if an event such as the one of those mentioned above were to occur in the future, it would make the detection and attribution of climate changes very difficult.

8.8 Representing the Global System with Simpler Models

8.8.1 Why Lower Complexity?

An important concept in climate system modelling is that of a spectrum of models of differing levels of complexity, each being optimum for answering specific questions. It is not meaningful to judge one level as being better or worse than another independently of the context of analysis. What is important is that each model be asked questions appropriate for its level of complexity and quality of its simulation.

The most comprehensive models available are AOGCMs. These models, which include more and more components of the climate system (see Section 8.2), are designed to provide the best representation of the system and its dynamics, thereby serving as the most realistic laboratory of nature. Their major limitation

is their high computational cost. To date, unless modest-resolution models are executed on an exceptionally large-scale distributed computed system, as in the climate*prediction.net* project (http://climateprediction.net; Stainforth et al., 2005), only a limited number of multi-decadal experiments can be performed with AOGCMs, which hinders a systematic exploration of uncertainties in climate change projections and prevents studies of the long-term evolution of climate.

At the other end of the spectrum of climate system model complexity are the so-called simple climate models (see Harvey et al., 1997 for a review of these models). The most advanced simple climate models contain modules that calculate in a highly parametrized way (1) the abundances of atmospheric greenhouse gases for given future emissions, (2) the radiative forcing resulting from the modelled greenhouse gas concentrations and aerosol precursor emissions, (3) the global mean surface temperature response to the computed radiative forcing and (4) the global mean sea level rise due to thermal expansion of sea water and the response of glaciers and ice sheets. These models are much more computationally efficient than AOGCMs and thus can be utilised to investigate future climate change in response to a large number of different scenarios of greenhouse gas emissions. Uncertainties from the modules can also be concatenated, potentially allowing the climate and sea level results to be expressed as probabilistic distributions, which is harder to do with AOGCMs because of their computational expense. A characteristic of simple climate models is that climate sensitivity and other subsystem properties must be specified based on the results of AOGCMs or observations. Therefore, simple climate models can be tuned to individual AOGCMs and employed as a tool to emulate and extend their results (e.g., Cubasch et al., 2001; Raper et al., 2001). They are useful mainly for examining global-scale questions.

To bridge the gap between AOGCMs and simple climate models, EMICs have been developed. Given that this gap is quite large, there is a wide range of EMICs (see the reviews of Saltzman, 1978 and Claussen et al., 2002). Typically, EMICs use a simplified atmospheric component coupled to an OGCM or simplified atmospheric and oceanic components. The degree of simplification of the component models varies among EMICs.

Earth System Models of Intermediate Complexity are reduced-resolution models that incorporate most of the processes represented by AOGCMs, albeit in a more parametrized form. They explicitly simulate the interactions between various components of the climate system. Similar to AOGCMs, but in contrast to simple climate models, the number of degrees of freedom of an EMIC exceeds the number of adjustable parameters by several orders of magnitude. However, these models are simple enough to permit climate simulations over several thousand of years or even glacial cycles (with a period of some 100 kyr), although not all are suitable for this purpose. Moreover, like simple climate models, EMICs can explore the parameter space with some completeness and are thus appropriate for assessing uncertainty. They can also be utilised to screen the phase space of climate or the history of climate in order to identify interesting time slices, thereby providing guidance for more detailed studies to be undertaken with AOGCMs. In addition, EMICs are invaluable tools for understanding large-scale processes and feedbacks acting within the climate system. Certainly, it would not be sensible to apply an EMIC to studies that require high spatial and temporal resolution. Furthermore, model assumptions and restrictions, hence the limit of applicability of individual EMICs, must be carefully studied. Some EMICs include a zonally averaged atmosphere or zonally averaged oceanic basins. In a number of EMICs, cloudiness and/or wind fields are prescribed and do not evolve with changing climate. In still other EMICs, the atmospheric synoptic variability is not resolved explicitly, but diagnosed by using a statistical-dynamical approach. *A priori*, it is not obvious how the reduction in resolution or dynamics/ physics affects the simulated climate. As shown in Section 8.8.3 and in Chapters 6, 9 and 10, at large scales most EMIC results compare well with observational or proxy data and AOGCM results. Therefore, it is argued that there is a clear advantage in having available a spectrum of climate system models.

8.8.2   Simple Climate Models

As in the TAR, a simple climate model is utilised in this report to emulate the projections of future climate change conducted with state-of-the-art AOGCMs, thus allowing the investigation of the temperature and sea level implications of all relevant emission scenarios (see Chapter 10). This model is an updated version of the Model for the Assessment of Greenhouse-Gas Induced Climate Change (MAGICC) model (Wigley and Raper, 1992, 2001; Raper et al., 1996). The calculation of the radiative forcings from emission scenarios closely follows that described in Chapter 2, and the feedback between climate and the carbon cycle is treated consistently with Chapter 7. The atmosphere-ocean module consists of an atmospheric energy balance model coupled to an upwelling-diffusion ocean model. The atmospheric energy balance model has land and ocean boxes in each hemisphere, and the upwelling-diffusion ocean model in each hemisphere has 40 layers with inter-hemispheric heat exchange in the mixed layer.

This simple climate model has been tuned to outputs from 19 of the AOGCMs described in Table 8.1, with resulting parameter values as given in the Supplementary Material, Table S8.1. The applied tuning procedure involves an iterative optimisation to derive least-square optimal fits between the simple model results and the AOGCM outputs for temperature time series and net oceanic heat uptake. This procedure attempts to match not only the global mean temperature but also the hemispheric land and ocean surface temperature changes of the AOGCM results by adjusting the equilibrium land-ocean warming ratio. Where data availability allowed, the tuning procedure took simultaneous account of low-pass filtered AOGCM data for two scenarios, namely a 1% per year compounded increase in atmospheric CO_2 concentration to twice and quadruple the pre-industrial level, with subsequent stabilisation. Before tuning, the AOGCM temperature and heat uptake data was de-drifted by subtracting the respective low-pass filtered pre-industrial control run segments. The three tuned parameters in the simple

climate model are the effective climate sensitivity, the ocean effective vertical diffusivity, and the equilibrium land-ocean warming ratio. Values specific to each AOGCM for the radiative forcing for CO_2 doubling were used in the tuning procedure where available (from Forster and Taylor, 2006, supplemented with values provided directly from the modelling groups). Otherwise, a default value of 3.71 W m^{-2} was chosen (Myhre et al., 1998). Default values of 1 W m^{-2} °C^{-1}, 1 W m^{-2} °C^{-1} and 8°C were used for the land-ocean heat exchange coefficient, the inter-hemispheric heat exchange coefficient and the magnitude of the warming that would result in a collapse of the MOC, respectively (see Appendix 9.1 of the TAR).

The obtained best-fit climate sensitivity estimates differ for various reasons from other estimates that were derived with alternative methods. Such alternative methods include, for example, regression estimates that use a global energy balance equation around the year of atmospheric CO_2 doubling or the analysis of slab ocean equilibrium warmings. The resulting differences in climate sensitivity estimates can be partially explained by the non-time constant effective climate sensitivities in many of the AOGCM runs. Furthermore, tuning results of a simple climate model will be affected by the model structure, although simple, and other default parameter settings that affect the simple model transient response.

8.8.3   Earth System Models of Intermediate Complexity

Pictorially, EMICs can be defined in terms of the components of a three-dimensional vector (Claussen et al., 2002): the number of interacting components of the climate system explicitly represented in the model, the number of processes explicitly simulated and the detail of description. Some basic information on the EMICs used in Chapter 10 of this report is presented in Table 8.3. A comprehensive description of all EMICs in operation can be found in Claussen (2005). Actually, there is a broad range of EMICs, reflecting the differences in scope. In some EMICs, the number of processes and the detail of description are reduced to simulate feedbacks between as many components of the climate system as feasible. Others, with fewer interacting components, are utilised in long-term ensemble experiments to investigate specific aspects of climate variability. The gap between some of the most complicated EMICs and AOGCMs is not very large. In fact, this particular class of EMICs is derived from AOGCMs. On the other hand, EMICs and simple climate models differ much more. For instance, EMICs as well as AOGCMs realistically represent the large-scale geographical structures of the Earth, like the shape of continents and ocean basins, which is certainly not the case for simple climate models.

Since the TAR, EMICs have intensively been used to study past and future climate changes (see Chapters 6, 9 and 10). Furthermore, a great deal of effort has been devoted to the evaluation of those models through coordinated intercomparisons.

Figure 8.17 compares the results from some of the EMICs utilised in Chapter 10 (see Table 8.3) with observation-based estimates and results of GCMs that took part in AMIP and CMIP1 (Gates et al., 1999; Lambert and Boer, 2001). The EMIC results refer to simulations in which climate is in equilibrium with an atmospheric CO_2 concentration of 280 ppm. Figures 8.17a and 8.17b show that the simulated latitudinal distributions of the zonally averaged surface air temperature for boreal winter and boreal summer are in good agreement with observations, except at northern and southern high latitudes. Interestingly, the GCM results also exhibit a larger scatter in these regions, and they somewhat deviate from data there. Figures 8.17c and 8.17d indicate that EMICs satisfactorily reproduce the general structure of the observed zonally averaged precipitation. Here again, at most latitudes, the scatter in the EMIC results seems to be as large as the scatter in the GCM results, and both EMIC and GCM results agree with observational estimates. When these EMICs are allowed to adjust to a doubling of atmospheric CO_2 concentration, they all simulate an increase in globally averaged annual mean surface temperature and precipitation that falls largely within the range of GCM results (Petoukhov et al., 2005).

The responses of the North Atlantic MOC to increasing atmospheric CO_2 concentration and idealised freshwater perturbations as simulated by EMICs have also been compared to those obtained by AOGCMs (Gregory et al., 2005; Petoukhov et al., 2005; Stouffer et al., 2006). These studies reveal no systematic difference in model behaviour, which gives added confidence to the use of EMICs.

In a further intercomparison, Rahmstorf et al. (2005) compared results from 11 EMICs in which the North Atlantic Ocean was subjected to a slowly varying change in freshwater input. All the models analysed show a characteristic hysteresis response of the North Atlantic MOC to freshwater forcing, which can be explained by Stommel's (1961) salt advection feedback. The width of the hysteresis curve varies between 0.2 and 0.5 Sv in the models. Major differences are found in the location of the present-day climate on the hysteresis diagram. In seven of the models, the present-day climate for standard parameter choices is found in the bi-stable regime, while in the other four models, this climate is situated in the mono-stable regime. The proximity of the present-day climate to Stommel's

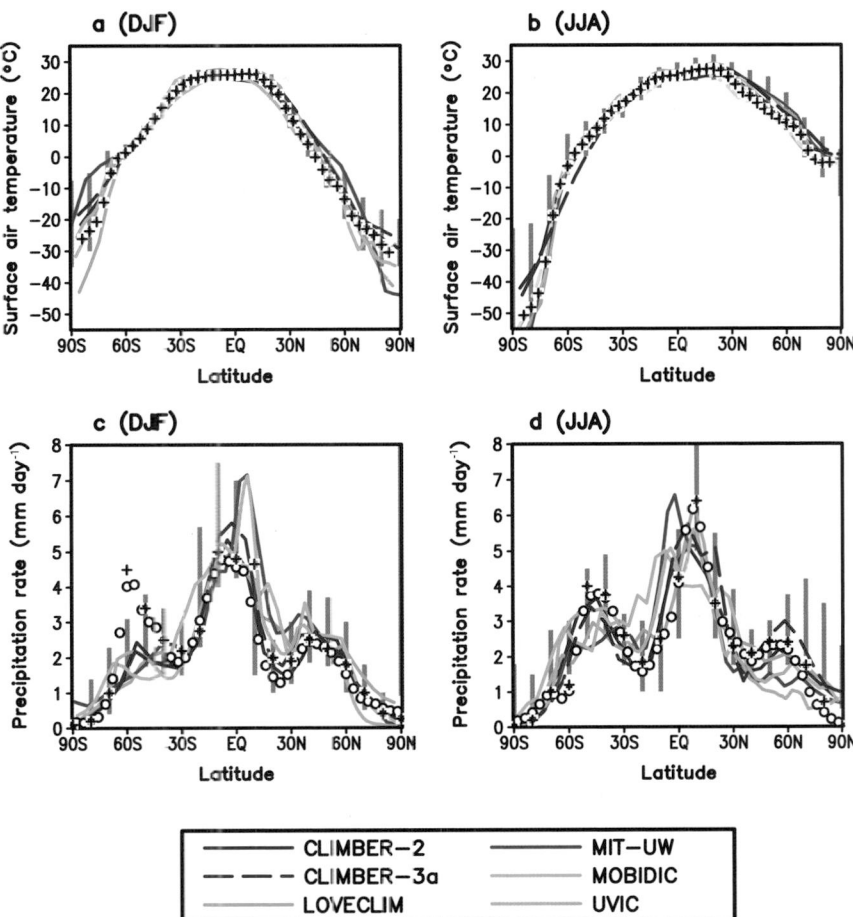

Figure 8.17. *Latitudinal distributions of the zonally averaged surface air temperature (a, b) and precipitation rate (c, d) for boreal winter (DJF) (a, c) and boreal summer (JJA) (b, d) as simulated at equilibrium by some of the EMICs used in Chapter 10 (see Table 8.3) for an atmospheric CO_2 concentration of 280 ppm. In (a) and (b), observational data merged from Jennings (1975), Jones (1988), Schubert et al. (1992), da Silva et al. (1994) and Fiorino (1997) are shown by crosses. In (c) and (d), observation-based estimates from Jaeger (1976; crosses) and Xie and Arkin (1997; open circles) are shown. The vertical grey bars indicate the range of GCM results from AMIP and CMIP1 (see text). Note that the model versions used in this intercomparison have no interactive biosphere and ice sheet components. The MIT-UW model is an earlier version of MIT-IGSM2.3. Adapted from Petoukhov et al., 2005.*

bifurcation point, beyond which NADW formation cannot be sustained, varies from less than 0.1 Sv to over 0.5 Sv.

A final example of EMIC intercomparison is discussed in Brovkin et al. (2006). Earth System Models of Intermediate Complexity that explicitly simulate the interactions between atmosphere, ocean and land surface were forced by a reconstruction of land cover changes during the last millennium. In response to historical deforestation of about 18×10^6 km², all models exhibited a decrease in globally averaged annual mean surface temperature in the range of 0.13°C to 0.25°C, mainly due to the increase in land surface albedo. Further experiments with the models forced by the historical atmospheric CO_2 trend reveal that, for the whole last millennium, the biogeophysical cooling due to land cover changes is less pronounced than the warming induced by the elevated atmospheric CO_2 level (0.27°C–0.62°C). During the 19th century, the cooling effect of deforestation appears to counterbalance, albeit not completely, the warming effect of increasing CO_2 concentration.

Table 8.3. *Description of the EMICs used in Chapter 10. The naming convention for the models is as agreed by all modelling groups involved. An asterisk after a component or parametrization means that this component or parametrization was not activated in the experiments discussed in Chapter 10.*

Name	Atmosphere[a]	Ocean[b]	Sea Ice[c]	Coupling/Flux Adjustments[d]	Land Surface[e]	Biosphere[f]	Ice Sheets[g]
E1: BERN2.5CC (Plattner et al., 2001; Joos et al., 2001)	EMBM, 1-D (φ), NCL, 7.5° × 15° (Schmittner and Stocker, 1999)	FG with parametrized zonal pressure gradient, 2-D (φ, z), 3 basins, RL, ISO, MESO, 7.5°×15°, L14 (Wright and Stocker, 1992)	0-LT, 2-LIT (Wright and Stocker, 1993)	PM, NH, NW (Stocker et al., 1992; Schmittner and Stocker, 1999)	NST, NSM (Schmittner and Stocker, 1999)	BO (Marchal et al., 1998), BT (Sitch et al., 2003; Gerber et al., 2003), BV (Sitch et al., 2003; Gerber et al., 2003)	
E2: C-GOLDSTEIN (Edwards and Marsh, 2005)	EMBM, 2-D(φ,λ), NCL, 5° × 10° (Edwards and Marsh, 2005)	FG, 3-D, RL, ISO, MESO, 5° × 10°, L8 (Edwards and Marsh, 2005)	0-LT, DOC, 2-LIT (Edwards and Marsh, 2005)	GM, NH, RW (Edwards and Marsh, 2005)	NST, NSM, RIV (Edwards and Marsh, 2005)		
E3: CLIMBER-2 (Petoukhov et al., 2000)	SD, 3-D, CRAD, ICL, 10° × 51°, L10 (Petoukhov et al., 2000)	FG with parametrized zonal pressure gradient, 2-D (φ, z), 3 basins, RL, 2.5°, L21 (Wright and Stocker, 1992)	0-LT, DOC, 2-LIT (Petoukhov et al., 2000)	NM, NH, NW (Petoukhov et al., 2000)	1-LST, CSM, RIV (Petoukhov et al., 2000)	BO (Brovkin et al., 2002), BT (Brovkin et al., 2002), BV (Brovkin et al., 2002)	TM, 3-D, 0.75° × 1.5° (Calov et al., 2005), L20*
E4: CLIMBER-3a (Montoya et al., 2005)	SD, 3-D, CRAD, ICL, 7.5° × 22.5°, L10 (Petoukhov et al., 2000)	PE, 3-D, FS, ISO, MESO, TCS, DC*, 3.75° × 3.75°, L24 (Montoya et al., 2005)	2-LT, R, 2-LIT (Fichefet and Morales Maqueda, 1997)	AM, NH, RW (Montoya et al., 2005)	1-LST, CSM, RIV (Petoukhov et al., 2000)	BO* (Six and Maier-Reimer, 1996), BT* (Brovkin et al., 2002), BV* (Brovkin et al., 2002)	
E5: LOVECLIM (Driesschaert, 2005)	QG, 3-D, LRAD, NCL, T21 (5.6° × 5.6°), L3 (Opsteegh et al., 1998)	PE, 3-D, FS, ISO, MESO, TCS, DC, 3° × 3°, L30 (Goosse and Fichefet, 1999)	3-LT, R, 2-LIT (Fichefet and Morales Maqueda, 1997)	NM, NH, RW (Driesschaert, 2005)	1-LST, BSM, RIV (Opsteegh et al., 1998)	BO (Mouchet and François, 1996), BT (Brovkin et al., 2002), BV (Brovkin et al., 2002)	TM, 3-D, 10 km x 10 km, L30 (Huybrechts, 2002)
E6: MIT-IGSM2.3 (Sokolov et al., 2005)	SD, 2-D(φ,z), CRAD, ICL, 4°, L11 (Sokolov and Stone, 1998), CHEM* (Mayer et al., 2000)	PE, 3-D, FS, ISO, MESO, 4° × 4°, L15 (Marshall et al., 1997)	3-LT, 2-LIT (Winton, 2000)	AM, GH, GW (Sokolov et al., 2005)	10-LST, CSM (Bonan et al., 2002)	BO (Parekh et al., 2005), BT (Felzer et al., 2005), BV* (Felzer et al., 2005)	
E7: MOBIDIC (Crucifix et al., 2002)	QG, 2-D (φ, z), CRAD, NCL, 5°, L2 (Gallée et al., 1991)	PE with parametrized zonal pressure gradient, 2-D (φ, z), 3 basins, RL, DC, 5°, L15 (Hovine and Fichefet, 1994)	0-LT, PD, 2-LIT (Crucifix et al., 2002)	NM, NH, NW (Crucifix et al., 2002)	1-LST, BSM (Gallée et al., 1991)	BO* (Crucifix, 2005), BT* (Brovkin et al., 2002), BV (Brovkin et al., 2002)	M, 1-D (φ), 0.5° (Crucifix and Berger, 2002)
E8: UVIC (Weaver et al., 2001)	DEMBM, 2-D (φ, λ), NCL, 1.8° × 3.6° (Weaver et al., 2001)	PE, 3-D, RG, ISO, MESO, 1.8° × 3.6° (Weaver et al., 2001)	0-LT, R, 2-LIT (Weaver et al., 2001)	AM, NH, NW (Weaver et al., 2001)	1-LST, CSM, RIV (Meissner et al., 2003)	BO (Weaver et al., 2001), BT (Cox, 2001), BV (Cox, 2001)	M, 2-D (φ, λ), 1.8° x 3.6° (Weaver et al., 2001)

Notes:

a   EMBM = energy-moisture balance model; DEMBM = energy-moisture balance model including some dynamics; SD = statistical-dynamical model; QG = quasi-geostrophic model; 1-D (φ) = zonally averaged; 2-D(φ, λ) = vertically averaged; 2-D(φ, z) = zonally averaged; 3-D = three-dimensional; LRAD = linearized radiation scheme; CRAD = comprehensive radiation scheme; NCL = non-interactive cloudiness; ICL = interactive cloudiness; CHEM = chemistry module; horizontal and vertical resolutions: the horizontal resolution is expressed either as degrees latitude x longitude or as spectral truncation with a rough translation to degrees latitude x longitude; the vertical resolution is expressed as 'Lm', where m is the number of vertical levels.

b   FG = frictional geostrophic model; PE = primitive equation model; 2-D (φ, z) = zonally averaged; 3-D (φ, z) = zonally averaged; 3-D = three-dimensional; RL = rigid lid; FS = free surface; ISO = isopycnal diffusion; MESO = parametrization of the effect of mesoscale eddies on tracer distribution; TCS = complex turbulence closure scheme; DC = parametrization of density-driven down-sloping currents; horizontal and vertical resolutions: the horizontal resolution is expressed as 'Lm', where m is the number of vertical levels.

c   n-LT = n-layer thermodynamic scheme; PD = prescribed drift; DOC = drift with oceanic currents; R = viscous-plastic or elastic-viscous-plastic rheology; 2-LIT = two-level ice thickness distribution (level ice and leads).

Notes (continued):

d   PM = prescribed momentum flux; GM = global momentum flux adjustment; AM = momentum flux anomalies relative to the control run are computed and added to climatological data; NM = no momentum flux adjustment; GH = global heat flux adjustment; NH = no heat flux adjustment; GW = global freshwater flux adjustment; RW = regional freshwater flux adjustment; NW = no freshwater flux adjustment.

e   NST = no explicit computation of soil temperature; n-LST = n-layer soil temperature scheme; NSM = no moisture storage in soil; BSM = bucket model for soil moisture; CSM = complex model for soil moisture; RIV = river routing scheme.

f   BO = model of oceanic carbon dynamics; BT = model of terrestrial carbon dynamics; BV = dynamical vegetation model.

g   TM = thermomechanical model; M = mechanical model (isothermal); 1-D (φ) = vertically averaged with east-west parabolic profile; 2-D (φ, λ) = vertically averaged; 3-D = three-dimensional; horizontal and vertical resolutions: the horizontal resolution is expressed either as degrees latitude x longitude or kilometres x kilometres; the vertical resolution is expressed as 'Lm', where m is the number of vertical levels.

References

Abramopoulos, F., C. Rosenzweig, and B. Choudhury, 1988: Improved ground hydrology calculations for global climate models (GCMs): Soil water movement and evapotranspiration. *J. Clim.*, **1**, 921–941.

Achatz, U., and J.D. Opsteegh, 2003: Primitive-equation-based low-order models with seasonal cycle, Part II: Application to complexity and nonlinearity of large-scale atmospheric dynamics. *J. Atmos. Sci.*, **60**, 478–490.

AchutaRao, K., and K.R. Sperber, 2002: Simulation of the El Niño Southern Oscillation: Results from the coupled model intercomparison project. *Clim. Dyn.*, **19**, 191–209.

AchutaRao, K., and K.R. Sperber, 2006: ENSO simulation in coupled ocean-atmosphere models: Are the current models better? *Clim. Dyn.*, **27**, 1–15.

AchutaRao, K., et al., 2004: *An Appraisal of Coupled Climate Model Simulations*. UCRL-TR-202550, Lawrence Livermore National Laboratory, Livermore, CA, 197 pp.

Alexander, M.A., et al., 2004: The atmospheric response to realistic Arctic sea ice anomalies in an AGCM during winter. *J. Clim.*, **17**, 890–905.

Alexeev, V.A., 2003: Sensitivity to CO_2 doubling of an atmospheric GCM coupled to an oceanic mixed layer: a linear analysis. *Clim. Dyn.*, **20**, 775–787.

Alexeev, V.A., P.L. Langen, and J.R. Bates, 2005: Polar amplification of surface warming on an aquaplanet in "ghost forcing" experiments without sea ice feedbacks. *Clim. Dyn.*, **24**, 655–666.

Alexeev, V.A., et al., 1998: *Modelling of the present-day climate by the INM RAS atmospheric model "DNM GCM"*. Institute of Numerical Mathematics, Moscow, Russia, 200 pp.

Allan, R.P., and A. Slingo, 2002: Can current climate forcings explain the spatial and temporal signatures of decadal OLR variations? *Geophys. Res. Lett.*, **29**(7), 1141, doi:10.1029/2001GL014620.

Allan, R.P., V. Ramaswamy, and A. Slingo, 2002: A diagnostic analysis of atmospheric moisture and clear-sky radiative feedback in the Hadley Centre and Geophysical Fluid Dynamics Laboratory (GFDL) climate models. *J. Geophys. Res.*, **107**(D17), 4329, doi:10.1029/2001JD001131.

Allan, R.P., M.A. Ringer, and A. Slingo, 2003: Evaluation of moisture in the Hadley Centre Climate Model using simulations of HIRS water vapour channel radiances. *Q. J. R. Meteorol. Soc.*, **129**, 3371–3389.

Allan, R.P., M.A. Ringer, J.A. Pamment, and A. Slingo, 2004: Simulation of the Earth's radiation budget by the European Centre for Medium Range Weather Forecasts 40-year Reanalysis (ERA40). *J. Geophys. Res.*, **109**, D18107, doi:10.1029/2004JD004816.

Allen, M.R., and W.J. Ingram, 2002: Constraints on future changes in climate and the hydrologic cycle. *Nature*, **419**, 224–231.

Alley, R.B., et al., 2002: *Abrupt Climate Changes: Inevitable Surprises*. National Research Council, National Academy Press, Washington, DC, 221 pp.

Alves, O., M.A. Balmaseda, D. Anderson, and T. Stockdale, 2004: Sensitivity of dynamical seasonal forecast to ocean initial conditions. *Q. J. R. Meteorol. Soc.*, **130**, 647–667.

Amundrud, T.L., H. Mailing, and R.G. Ingram, 2004: Geometrical constraints on the evolution of ridged sea ice. *J. Geophys. Res.*, **109**, C06005, doi:10.1029/2003JC002251.

Annamalai, H., K. Hamilton, and K.R. Sperber, 2007: South Asian summer monsoon and its relationship with ENSO in the IPCC AR4 simulations. *J. Clim.*, **20**, 1071-1083.

Annan, J.D., J.C. Hargreaves, N.R. Edwards, and R. Marsh, 2005a: Parameter estimation in an intermediate complexity Earth System Model using an ensemble Kalman filter. *Ocean Modelling*, **8**, 135–154.

Annan, J.D., et al., 2005b: Efficiently constraining climate sensitivity with palaeoclimate observations. *Scientific Online Letters on the Atmosphere*, **1**, 181–184.

Arakawa, A., 2004: The cumulus parameterization problem: Past, present, and future. *J. Clim.*, **17**, 2493–2525.

Arakawa, A., and W.H. Schubert, 1974: Interaction of a cumulus cloud ensemble with the large-scale environment, Part I. *J. Atmos. Sci.*, **31**, 674–701.

Arora, V.K., 2001: Assessment of simulated water balance for continental-scale river basins in an AMIP 2 simulation. *J. Geophys. Res.*, **106**, 14827–14842.

Arora, V.K., and G.J. Boer, 2003: A representation of variable root distribution in dynamic vegetation models. *Earth Interactions*, **7**, 1–19.

Arzel, O., T. Fichefet, and H. Goosse, 2006: Sea ice evolution over the 20th and 21st centuries as simulated by the current AOGCMs. *Ocean Modelling*, **12**, 401–415.

Babko, O., D.A. Rothrock, and G.A. Maykut, 2002: Role of rafting in the mechanical redistribution of sea ice thickness. *J. Geophys. Res.*, **107**, 3113, doi:10.1029/1999JC000190.

Baldwin, M.P., et al., 2001: The quasi-biennial oscillation. *Rev. Geophys.*, **39**, 179–229.

Baldwin, M.P., et al., 2003: Stratospheric memory and skill of extended-range weather forecasts. *Science*, **301**, 636–640.

Balmaseda, M.A., M.K. Davey, and D.L.T. Anderson, 1995: Decadal and seasonal dependence of ENSO prediction skill. *J. Clim.*, **8**, 2705–2715.

Barkstrom, B., et al., 1989: Earth Radiation Budget Experiment (ERBE) archival and April 1985 results. *Bull. Am. Meteorol. Soc.*, **70**, 1254–1262.

Barnett, T.P., et al., 1999: Origins of midlatitude Pacific decadal variability. *Geophys. Res. Lett.*, **26**, 1453–1456.

Bates, J.J., and D.L. Jackson, 2001: Trends in upper-tropospheric humidity. *Geophys. Res. Lett.*, **28**, 1695–1698.

Bauer, M., and A.D. Del Genio, 2006: Composite analysis of winter cyclones in a GCM: Influence on climatological humidity. *J. Clim.*, **19**, 1652–1672. .

Bauer, M., A.D. Del Genio, and J.R. Lanzante, 2002: Observed and simulated temperature humidity relationships: sensitivity to sampling and analysis. *J. Clim.*, **15**, 203–215.

Bell, T.L., M.-D. Chou, R.S. Lindzen, and A.Y. Hou, 2002: Comments on "Does the Earth have an adaptive infrared iris?" Reply. *Bull. Am. Meteorol. Soc.*, **83**, 598–600.

Bengtsson, L.K., I. Hodges, and E. Roeckner, 2006: Storm tracks and climate change. *J. Clim.*, **19**, 3518–3543.

Bernie, D., S.J. Woolnough, J.M. Slingo, and E. Guilyardi, 2005: Modelling diurnal and intraseasonal variability of the ocean mixed layer. *J. Clim.*, **15**, 1190–1202.

Bitz, C.M., and W.H. Lipscomb, 1999: An energy-conserving thermodynamic sea ice model for climate study. *J. Geophys. Res.*, **104**, 15669–15677.

Bitz, C.M., G. Flato, and J. Fyfe, 2002: Sea ice response to wind forcing from AMIP models. *J. Clim.*, **15**, 523–535.

Bitz, C.M., M.M., Holland, A.J. Weaver, and M. Eby, 2001: Simulating the ice-thickness distribution in a coupled climate model. *J. Geophys. Res.*, **106**, 2441–2463.

Blankenship, C.B., and T.T Wilheit, 2001: SSM/T-2 measurements of regional changes in three-dimensional water vapour fields during ENSO events. *J. Geophys. Res.*, **106**, 5239–5254.

Bleck, R., 2002: An oceanic general circulation model framed in hybrid isopycnic-Cartesian coordinates. *Ocean Modelling*, **4**, 55–88.

Bleck, R., C. Rooth, D. Hu, and L.T. Smith, 1992: Salinity-driven thermocline transients in a wind- and thermohaline-forced isopycnic coordinate model of the North Atlantic. *J. Phys. Oceanogr.*, **22**, 1486–1505.

Boer, G.J., and B. Yu, 2003: Climate sensitivity and climate state. *Clim. Dyn.*, **21**, 167–176.

Bonan, G.B., 1998: The land surface climatology of the NCAR land surface model (LSM 1.0) coupled to the NCAR Community Climate Model (CCM3). *J. Clim.*, **11**, 1307–1326.

Bonan, G.B., K.W. Oleson, M. Vertenstein, and S. Levis, 2002: The land surface climatology of the Community Land Model coupled to the NCAR Community Climate Model. *J. Clim.*, **15**, 3123–3149.

Böning, C.W., et al., 1995: An overlooked problem in model simulations of the thermohaline circulation and heat transports in the Atlantic Ocean. *J. Clim.*, **8**, 515–523.

Bony, S., and K.A. Emanuel, 2001: A parameterization of the cloudiness associated with cumulus convection: Evaluation using TOGA COARE data. *J. Atmos. Sci.*, **58**, 3158–3183.

Bony, S., and J.-L. Dufresne, 2005: Marine boundary-layer clouds at the heart of tropical cloud feedback uncertainties in climate models. *Geophys. Res. Lett.*, **32**(20), L20806, doi:10.1029/2005GL023851.

Bony, S., and K.A. Emanuel, 2005: On the role of moist processes in tropical intraseasonal variability: cloud-radiation and moisture-convection feedbacks. *J. Atmos. Sci.*, **62**, 2770–2789.

Bony, S., K.-M. Lau, and Y.C. Sud, 1997: Sea surface temperature and large-scale circulation influences on tropical greenhouse effect and cloud radiative forcing. *J. Clim.*, **10**, 2055–2077.

Bony, S., et al., 2004: On dynamic and thermodynamic components of cloud changes. *Clim. Dyn.*, **22**, 71–86.

Bony, S., et al., 2006: How well do we understand and evaluate climate change feedback processes? *J. Clim.*, **19**, 3445–3482.

Boone, A., V. Masson, T. Meyers, and J. Noilhan, 2000: The influence of the inclusion of soil freezing on simulations by a soil-vegetation-atmosphere transfer scheme. *J. Appl. Meteorol.*, **39**(9), 1544–1569.

Boone, A., et al., 2004: The Rhone-Aggregation land surface scheme intercomparison project: An overview. *J. Clim.*, **17**, 187–208.

Boville, B.A., and W.J. Randel, 1992: Equatorial waves in a stratospheric GCM: Effects of resolution. *J. Atmos. Sci.*, **49**, 785–801.

Bowling, L.C., et al., 2003: Simulation of high latitude hydrological processes in the Torne-Kalix basin: PILPS Phase 2(e) 1: Experiment description and summary intercomparisons. *Global Planet. Change*, **38**, 1–30.

Boyle, J.S., et al., 2005: Diagnosis of Community Atmospheric Model 2 (CAM2) in numerical weather forecast configuration at Atmospheric Radiation Measurement (ARM) sites. *J. Geophys. Res.*, **110**, doi:10.1029/2004JD005042.

Branstetter, M.L., 2001: *Development of a Parallel River Transport Algorithm and Application to Climate Studies*. PhD Dissertation, University of Texas. Austin, TX.

Briegleb, B.P., et al., 2004: *Scientific Description of the Sea Ice Component in the Community Climate System Model, Version Three*. Technical Note TN-463STR, NTIS #PB2004-106574, National Center for Atmospheric Research, Boulder, CO, 75 pp.

Broccoli, A.J., N.-C. Lau, and M.J. Nath, 1998: The cold ocean-warm land pattern: Model simulation and relevance to climate change detection. *J. Clim.*, **11**, 2743–2763.

Broecker, W.S., 1997: Thermohaline circulation, the Achilles heel of our climate system: will man-made CO_2 upset the current balance? *Science*, **278**, 1582–1588.

Brogniez, H., R. Roca, and L. Picon, 2005: Evaluation of the distribution of subtropical free tropospheric humidity in AMIP-2 simulations using METEOSAT water vapour channel data. *Geophys. Res. Lett.*, **32**, L19708, doi:10.1029/2005GL024341.

Brovkin, V., et al., 2002: Carbon cycle, vegetation and climate dynamics in the Holocene: Experiments with the CLIMBER-2 model. *Global Biogeochem. Cycles*, **16**(4), 1139, doi:10.1029/2001GB001662.

Brovkin, V., et al., 2006: Biogeophysical effects of historical land cover changes simulated by six Earth system models of intermediate complexity. *Clim. Dyn.*, **26**, 587–600, doi:10.1007/s00382-005-0092-6.

Bryan, F.O., et al., 2006: Response of the North Atlantic thermohaline circulation and ventilation to increasing carbon dioxide in CCSM3. *J. Clim.*, **19**, 2382–2397.

Burke, E.J., S.J. Brown, and N. Christidis, 2006: Modelling the recent evolution of global drought and projections for the 21st century with the Hadley Centre climate model. *J. Hydrometeorol.*, **7**, 1113–1125.

Cai, M., 2005: Dynamical amplification of polar warming. *Geophys. Res. Lett.*, **32**, L22710, doi:10.1029/2005GL024481.

Cai, W.J., and P.H. Whetton, 2000: Evidence for a time-varying pattern of greenhouse warming in the Pacific Ocean. *Geophys. Res. Lett.*, **27**(16), 2577–2580.

Cai, W.J., P.H. Whetton, and D.J. Karoly, 2003: The response of the Antarctic Oscillation to increasing and stabilized atmospheric CO_2. *J. Clim.*, **16**, 1525–1538.

Calov, R., et al., 2002: Large-scale instabilities of the Laurentide ice sheet simulated in a fully coupled climate-system model. *Geophys. Res. Lett.*, **29**(24), 2216, doi:10.1029/2002GL016078.

Calov, R., et al., 2005: Transient simulation of the last glacial inception. Part I: Glacial inception as a bifurcation of the climate system. *Clim. Dyn.*, **24**(6), 545–561.

Camargo, S., A.G. Barnston, and S.E. Zebiak, 2005: A statistical assessment of tropical cyclone activity in atmospheric general circulation models. *Tellus*, **57A**, 589–604.

Carnell, R., and C. Senior, 1998: Changes in mid-latitude variability due to increasing greenhouse gases and sulphate aerosols. *Clim. Dyn.*, **14**, 369–383.

Cassou, C., L. Terray, J.W. Hurrell, and C. Deser, 2004: North Atlantic winter climate regimes: Spatial asymmetry, stationarity with time, and oceanic forcing. *J. Clim.*, **17**, 1055–1068.

Castanheira, J.M., and H.-F. Graf, 2003: North Pacific–North Atlantic relationships under stratospheric control? *J. Geophys. Res.*, **108**, 4036, doi:10.1029/2002JD002754.

Cattle, H., and J. Crossley, 1995: Modelling Arctic climate change. *Philos. Trans. R. Soc. London Ser. A*, **352**, 201–213.

Cess, R.D., 1975: Global climate change: an investigation of atmospheric feedback mechanisms. *Tellus*, **27**, 193–198.

Cess, R.D., et al., 1989: Interpretation of cloud-climate feedback as produced by 14 atmospheric general circulation models. *Science*, **245**, 513–516.

Chambers, L.H., B. Lin, and D.F. Young, 2002: Examination of new CERES data for evidence of tropical Iris feedback. *J. Clim.*, **15**, 3719–3726.

Chang, F.-L., and Z. Li, 2005: A comparison of the global surveys of high, mid and low clouds from satellite and general circulation models. In: *Proceedings of the Fifteenth Atmospheric Radiation Measurement (ARM) Science Team Meeting, Daytona Beach, Florida, 14–18 March 2005.* Atmospheric Radiation Measurement Program, US Department of Energy, Washington, DC, http://www.arm.gov/publications/proceedings/conf15/

Chapman, W.L., and J. E. Walsh, 2007: Simulations of arctic temperature and pressure by global coupled models. *J. Clim.*, **20**, 609-632.

Chen, D., S.E. Zebiak, A.J. Busalacchi, and M.A. Cane, 1995: An improved procedure for El Niño forecasting. *Science*, **269**, 1699–1702.

Chen, J., B.E. Carlson, and A.D. Del Genio, 2002: Evidence for strengthening of the tropical general circulation in the 1990s. *Science*, **295**, 838–841.

Chen, T.-C., and J.-H. Yoon, 2002: Interdecadal variation of the North Pacific wintertime blocking. *Mon. Weather Rev.*, **130**, 3136–3143.

Chin, M., et al., 2002: Tropospheric aerosol optical thickness from GOCART model and comparisons with satellite and sun photometer measurements. *J. Atmos. Sci.*, **59**, 461–483.

Chou, M.-D., R.S. Lindzen, and A.Y. Hou, 2002: Reply to: "Tropical cirrus and water vapor: An effective Earth infrared iris feedback?". *Atmos. Chem. Phys.*, **2**, 99–101.

Chung, E.S., B.J. Sohn, and V. Ramanathan, 2004: Moistening processes in the upper troposphere by deep convection: a case study over the tropical Indian Ocean. *J. Meteorol. Soc. Japan*, **82**, 959–965.

Church, J.A., et al., 2001: Changes in sea level. In: *Climate Change 2001: The Scientific Basis. Contribution of Working Group I to the Third Assessment Report of the Intergovernmental Panel on Climate Change* [Houghton, J.T., et al. (eds.)]. Cambridge University Press, Cambridge, United Kingdom and New York, NY, USA, pp. 663–693.

Clark, P.U., N.G. Pisias, T.F. Stocker, and A.J. Weaver, 2002: The role of the thermohaline circulation in abrupt climate change. *Nature*, **415**, 863–869.

Claussen, M., 1998: On multiple solutions of the atmosphere-vegetation system in present-day climate. *Global Change Biol.*, **4**, 549–559.

Claussen, M., 2005: *Table of EMICs (Earth System Models of Intermediate Complexity)*. PIK Report 98, Potsdam-Institut für Klimafolgenforschung, Potsdam, Germany, 55 pp, http://www.pik-potsdam.de/emics.

Claussen, M., et al., 2002: Earth system models of intermediate complexity: closing the gap in the spectrum of climate system models. *Clim. Dyn.*, **18**, 579–586.

Collins, M., S.F.B. Tett, and C. Cooper, 2001: The internal climate variability of HadCM3, a version of the Hadley Centre coupled model without flux adjustments. *Clim. Dyn.*, **17**, 61–81.

Collins, M., D. Frame, B. Sinha, and C. Wilson, 2002: How far ahead could we predict El Niño? *Geophys. Res. Lett.*, **29**(10), 1492, doi:10.1029/2001GL013919.

Collins, W.D., et al., 2004: *Description of the NCAR Community Atmosphere Model (CAM3.0)*. Technical Note TN-464+STR, National Center for Atmospheric Research, Boulder, CO, 214 pp.

Collins, W.D., et al., 2006: The Community Climate System Model: CCSM3. *J. Clim.*, **19**, 2122–2143.

Colman, R.A., 2001: On the vertical extent of atmospheric feedbacks. *Clim. Dyn.*, **17**, 391–405.

Colman, R.A., 2003a: A comparison of climate feedbacks in general circulation models. *Clim. Dyn.*, **20**, 865–873.

Colman, R.A., 2003b: Seasonal contributions to climate feedbacks. *Clim. Dyn.*, **20**, 825–841.

Colman, R.A., 2004: On the structure of water vapour feedbacks in climate models. *Geophys. Res. Lett.*, **31**, L21109, doi:10.1029/2004GL020708.

Cook, K.H., and E.K. Vizy, 2006: Coupled model simulations of the West African monsoon system: 20th century simulations and 21st century predictions. *J. Clim.*, **19**, 3681–3703.

Cox, P., 2001: *Description of the "TRIFFID" Dynamic Global Vegetation Model*. Technical Note 24, Hadley Centre, United Kingdom Meteorological Office, Bracknell, UK.

Cox, P.M., et al., 1999: The impact of new land surface physics on the GCM simulation of climate and climate sensitivity. *Clim. Dyn.*, **15**, 183–203.

Cox, P.M., et al., 2000: Acceleration of global warming due to carbon-cycle feedbacks in a coupled climate model. *Nature*, **408**, 184–187.

Cox, P.M., et al., 2004: Amazonian forest dieback under climate-carbon cycle projections for the 21st century. *Theor. Appl. Climatol.*, **78**, 137–156, doi:10.1007/s00704-004-0049-4.

Cramer, W., et al., 2001: Global response of terrestrial ecosystem structure and function to CO_2 and climate change: results from six dynamic global vegetation models. *Global Change Biol.*, **7**, 357–373.

Crucifix, M., 2005: Carbon isotopes in the glacial ocean: A model study. *Paleoceanography*, **20**, PA4020, doi:10.1029/2005PA001131.

Crucifix, M., and A. Berger, 2002: Simulation of ocean–ice sheet interactions during the last deglaciation. *Paleoceanography*, **17**(4), 1054, doi:10.1029/2001PA000702.

Crucifix, M., et al., 2002: Climate evolution during the Holocene: A study with an Earth system model of intermediate complexity. *Clim. Dyn.*, **19**, 43–60, doi:10.10007/s00382-001-0208-6.

CSMD (Climate System Modeling Division), 2005: An introduction to the first general operational climate model at the National Climate Center. *Advances in Climate System Modeling*, 1, National Climate Center, China Meteorological Administration, 14 pp (in English and Chinese).

Cubasch, U., et al., 2001: Projections of future climate changes. In: *Climate Change 2001: The Scientific Basis. Contribution of Working Group I to the Third Assessment Report of the Intergovernmental Panel on Climate Change* [Houghton, J.T., et al. (eds.)]. Cambridge University Press, Cambridge, United Kingdom and New York, NY, USA, pp. 525–582.

da Silva, A.M., C.C. Young, and S. Levitus, 1994: *Atlas of Surface Marine Data 1994, NOAA Atlas NESDIS 6*. NOAA/NESDIS E/OC21 (6 Volumes). US Department of Commerce, National Oceanographic Data Center, User Services Branch, Washington, DC.

Dai, A., 2006: Precipitation characteristics in eighteen coupled climate models. *J. Clim.*, **19**, 4605–4630.

Dai, A., K.E. Trenberth, and T. Qian, 2004: A global data set of Palmer Drought Severity Index for 1870-2002: Relationship with soil moisture and effects of surface warming. *J. Hydrometeorol.*, **5**, 1117–1130. PDSI data: http://www.cgd.ucar.edu/cas/catalog/climind/pdsi.html.

Danabasoglu, G., J.C. McWilliams, and P.R. Gent, 1995: The role of mesoscale tracer transports in the global ocean circulation. *Science*, **264**, 1123–1126.

D'Andrea, F., et al., 1998: Northern Hemisphere atmospheric blocking as simulated by 15 atmospheric general circulation models in the period 1979–1988. *Clim. Dyn.*, **14**(6), 385–407.

Dargaville, R.J., et al., 2002: Evaluation of terrestrial carbon cycle models with atmospheric CO_2 measurements: Results from transient simulations considering increasing CO_2, climate, and land-use effects. *Global Biogeochem. Cycles*, **16**, 1092, doi:10.1029/2001GB001426.

Davey, M., et al., 2002: STOIC: A study of coupled GCM climatology and variability in tropical ocean regions. *Clim. Dyn.*, **18**, 403–420, doi:10.1007/s00382-001-0188-6.

Del Genio, A.D., and A.B. Wolf, 2000: The temperature dependence of the liquid water path of low clouds in the southern great plains. *J. Clim.*, **13**, 3465 3486.

Del Genio, A.D., and W. Kovari, 2002: Climatic properties of tropical precipitating convection under varying environmental conditions. *J. Clim.*, **15**, 2597–2615.

Del Genio, A.D., A. Wolf, and M.-S. Yao, 2005a: Evaluation of regional cloud feedbacks using single-column models. *J. Geophys. Res.*, **110**, D15S13, doi:10.1029/2004JD005011.

Del Genio, A.D., W. Kovari, M.-S. Yao, and J. Jonas, 2005b: Cumulus microphysics and climate sensitivity. *J. Clim.*, **18**, 2376–2387, doi:10.1175/JCLI3413.1.

Delire, C., J.A. Foley, and S. Thompson, 2003: Evaluating the carbon cycle of a coupled atmosphere–biosphere model. *Global Biogeochem. Cycles*, **17**, 1012, doi:10.1029/2002GB001870.

Delworth, T.L., and M.E. Mann, 2000: Observed and simulated multidecadal variability in the Northern Hemisphere. *Clim. Dyn.*, **16**(9), 661–676.

Delworth, T., S. Manabe, and R.J. Stouffer, 1993: Interdecadal variations of the thermohaline circulation in a coupled ocean-atmosphere model. *J. Clim.*, **6**, 1993–2011.

Delworth, T.L., V. Ramaswamy, and G.L. Stenchikov, 2005: The impact of aerosols on simulated ocean temperature and heat content in the 20th century. *Geophys. Res. Lett.*, **32**, L24709, doi:10.1029/2005GL024457.

Delworth, T., et al., 2006: GFDL's CM2 global coupled climate models – Part 1: Formulation and simulation characteristics. *J. Clim.*, **19**, 643–674.

Déqué, M., C. Dreveton, A. Braun, and D. Cariolle, 1994: The ARPEGE/IFS atmosphere model: A contribution to the French community climate modeling. *Clim. Dyn.*, **10**, 249–266.

Derber, J., and A. Rosati, 1989: A global oceanic data assimilation system. *J. Phys. Oceanogr.*, **19**(9), 1333–1347.

Deser, C., A.S. Phillips, and J.W. Hurrell, 2004: Pacific interdecadal climate variability: Linkages between the tropics and North Pacific during boreal winter since 1900. *J. Clim.*, **17**, 3109–3124.

Dessler, A.E., and S.C. Sherwood, 2000: Simulations of tropical upper tropospheric humidity. *J. Geophys. Res.*, **105**, 20155–20163.

Diansky, N.A., and E.M. Volodin, 2002: Simulation of the present-day climate with a coupled atmosphere-ocean general circulation model. *Izv. Atmos. Ocean. Phys.*, **38**, 732–747 (English translation).

Diansky, N.A., A.V. Bagno, and V.B. Zalesny, 2002: Sigma model of global ocean circulation and its sensitivity to variations in wind stress. *Izv. Atmos. Ocean. Phys.*, **38**, 477–494 (English translation).

Dirmeyer, P.A., 2001: An evaluation of the strength of land-atmosphere coupling. *J. Hydrometeorol.*, **2**, 329–344.

Dong, M., et al., 2000: Developments and implications of the atmospheric general circulation model. In: *Investigations on the Model System of the Short-Term Climate Predictions* [Ding, Y., et al. (eds.)]. China Meteorological Press, Beijing, China, pp. 63–69 (in Chinese).

Doutriaux-Boucher, M., and J. Quaas, 2004: Evaluation of cloud thermodynamic phase parametrizations in the LMDZ GCM by using POLDER satellite data. *Geophys. Res. Lett.*, **31**, L06126, doi:10.1029/2003GL019095.

Douville, H., 2001: Influence of soil moisture on the Asian and African Monsoons. Part II: interannual variability. *J. Clim.*, **15**, 701–720.

Douville, H., J.-F. Royer, and J.-F. Mahfouf, 1995: A new snow parameterization for the Meteo-France climate model. *Clim. Dyn.*, **12**, 21–35.

Drange, H., et al., 2005: Ocean general circulation modelling of the Nordic Seas. In: *The Nordic Seas: An Integrated Perspective* [Drange, H., et al. (eds.)]. Geophysical Monograph 158, American Geophysical Union, Washington, DC, pp. 199–220.

Driesschaert, E., 2005: *Climate Change over the Next Millennia Using LOVECLIM, a New Earth System Model Including Polar Ice Sheets.* PhD Thesis, Université Catholique de Louvain, Louvain-la-Neuve, Belgium, 214 pp., http://edoc.bib.ucl.ac.be:81/ETD-db/collection/available/BelnUcetd-10172005-185914/.

Ducharne, A., et al., 2003: Development of a high resolution runoff routing model, calibration and application to assess runoff from the LMD GCM. *J. Hydrol.*, **280**, 207–228.

Dufresne, J.-L., et al., 2002: On the magnitude of positive feedback between future climate change and the carbon cycle. *Geophys. Res. Lett.*, **29**(10), doi:10.1029/2001GL013777.

Dümenil, L., and E. Todini, 1992: A rainfall-runoff scheme for use in the Hamburg climate model. In: *Advances in Theoretical Hydrology: A Tribute to James Dooge. European Geophysical Society Series on Hydrological Sciences, Vol. 1* [O'Kane, J.P. (ed.)]. Elsevier Press, Amsterdam, pp. 129–157.

Durman, C.F., et al., 2001: A comparison of extreme European daily precipitation simulated by a global model and regional climate model for present and future climates. *Q. J. R. Meteorol. Soc.*, **127**, 1005–1015.

Edwards, N.R., and R.J. Marsh, 2005: Uncertainties due to transport-parameter sensitivity in an efficient 3-D ocean-climate model. *Clim. Dyn.*, **24**, 415–433, doi:10.1007/s00382-004-0508-8.

Emanuel, K.A., and M. Zivkovic-Rothman, 1999: Development and evaluation of a convection scheme for use in climate models. *J. Atmos. Sci.*, **56**, 1766–1782.

Emori, S., A. Hasegawa, T. Suzuki, and K. Dairaku, 2005: Validation, parameterization dependence and future projection of daily precipitation simulated with an atmospheric GCM. *Geophys. Res. Lett.*, **32**, L06708, doi:10.1029/2004GL022306.

Essery, R.H., and J. Pomeroy, 2004: Vegetation and topographic control of wind-blown snow distributions in distributed and aggregated simulations. *J. Hydrometeorol.*, **5**(5), 735–744.

Essery, R., M. Best, and P. Cox, 2001: *MOSES 2.2 Technical Documentation.* Hadley Centre Technical Note No. 30, Hadley Centre for Climate Prediction and Research, UK Met Office, Exeter, UK, http://www.metoffice.gov.uk/research/hadleycentre/pubs/HCTN/index.html.

Essery, R.H., J. Pomeroy, J. Parvianen, and P. Storck, 2003: Sublimation of snow from boreal forests in a climate model. *J. Clim.*, **16**, 1855–1864.

Etchevers, P., et al., 2004: Validation of the energy budget of an alpine snowpack simulated by several snow models (SnowMIP project). *Ann. Glaciol.*, **38**, 150–158.

Farrara, J.D., C.R. Mechoso, and A.W. Robertson, 2000: Ensembles of AGCM two-tier predictions and simulations of the circulation anomalies during winter 1997–1998. *Mon. Weather Rev.*, **128**, 3589–3604.

Felzer, B., et al., 2005: Global and future implications of ozone on net primary production and carbon sequestration using a biogeochemical model. *Clim. Change*, **73**, 345–373.

Fichefet, T., and M.A. Morales Maqueda, 1997: Sensitivity of a global sea ice model to the treatment of ice thermodynamics and dynamics. *J. Geophys. Res.*, **102**, 12609–12646.

Fichefet, T., et al., 2003: Implications of changes in freshwater flux from the Greenland ice sheet for the climate of the 21st century. *Geophys. Res. Lett.*, **30**(17), 1911, doi:10.1029/2003GL017826.

Fiorino, M., 1997: *PCMDI IPCC '95 AMIP Analysis: Observations used in the analysis.* PCMDI Web. Rep., Program for Climate Model Diagnosis and Intercomparison, Lawrence Livermore National Laboratory, Livermore, CA, http://www-pcmdi.llnl.gov/obs/ipcc/ipcc.obs.dat.htm.

Flato, G.M., 2004: Sea-ice and its response to CO_2 forcing as simulated by global climate models. *Clim. Dyn.*, **23**, 229–241, doi:10.1007/s00382-004-0436-7.

Flato, G.M., 2005: *The Third Generation Coupled Global Climate Model (CGCM3)* (and included links to the description of the AGCM3 atmospheric model). http://www.cccma.bc.ec.gc.ca/models/cgcm3.shtml.

Flato, G.M., and W.D. Hibler, 1992: Modeling pack ice as a cavitating fluid. *J. Phys. Oceanogr.*, **22**, 626–651.

Flato, G.M., and G.J. Boer, 2001: Warming asymmetry in climate change simulations. *Geophys. Res. Lett.*, **28**, 195–198.

Flugel, M., P. Chang, and C. Penland, 2004: The role of stochastic forcing in modulating ENSO predictability. *J. Clim.*, **17**(16), 3125–3140.

Folkins, I., K.K. Kelly, and E.M. Weinstock, 2002: A simple explanation of the increase in relative humidity between 11 and 14 km in the tropics. *J. Geophys. Res.*, **107**, doi:10.1029/2002JD002185.

Folland, C.K., T.K. Palmer, and D.E. Parker, 1986: Sahel rainfall and worldwide sea temperatures. *Nature*, **320**, 602–607.

Forster, P.M. de F., and K.P. Shine, 2002: Assessing the climate impact of trends in stratospheric water vapour. *Geophys. Res. Lett.*, **6**, doi:10.1029/2001GL013909.

Forster, P.M. de F., and M. Collins, 2004: Quantifying the water vapour feedback associated with post-Pinatubo cooling. *Clim. Dyn.*, **23**, 207–214.

Forster, P.M. de F., and K.E. Taylor, 2006: Climate forcings and climate sensitivities diagnosed from coupled climate model integrations. *J. Clim.*, **19**, 6181–6194.

Frei, A., J. Miller, and D. Robinson, 2003: Improved simulations of snow extent in the second phase of the Atmospheric Model Intercomparison Project (AMIP-2). *J. Geophys. Res.*, **108**(D12), 4369, doi:10.1029/2002JD003030.

Frei, A., J.A. Miller, R. Brown, and D.A. Robinson, 2005: Snow mass over North America: observations and results from the second phase of the Atmospheric Model Intercomparison Project (AMIP-2). *J. Hydrometeorol.*, **6**, 681–695.

Frich, P., et al., 2002: Observed coherent changes in climatic extremes during the second half of the twentieth century. *Clim. Res.*, **19**, 193–212.

Friedlingstein, P., et al., 2001: Positive feedback between future climate change and the carbon cycle. *Geophys. Res. Lett.*, **28**(8), 1543–1546.

Friedlingstein, P., J.-L. Dufresne, P.M Cox, and P. Rayner, 2003: How positive is the feedback between climate change and the carbon cycle? *Tellus*, **55B**, 692–700.

Friedlingstein, P., et al., 2006: Climate–carbon cycle feedback analysis, results from the C4MIP model intercomparison. *J. Clim.*, **19**, 3337–3353.

Friend, A.D., and N.Y. Kiang, 2005: Land surface model development for the GISS GCM: Effects of improved canopy physiology on simulated climate. *J. Clim.*, **18**, 2883–2902.

Fu, Q., M. Baker, and D.L. Hartmann, 2002: Tropical cirrus and water vapour: an effective Earth infrared iris? *Atmos. Chem. Phys.*, **2**, 31–37.

Fu, Q., C.M. Johanson, S.G. Warren, and D.J. Seidel, 2004: Contribution of stratospheric cooling to satellite-inferred tropospheric temperature trends. *Nature*, **429**, 55–58.

Furevik, T., et al., 2003: Description and evaluation of the Bergen climate model: ARPEGE coupled with MICOM. *Clim. Dyn.*, **21**, 27–51.

Fyfe, J.C., G.J. Boer, and G.M. Flato, 1999: The Arctic and Antarctic Oscillations and their projected changes under global warming. *Geophys. Res. Lett.*, **11**, 1601–1604.

Galin, V. Ya., E.M. Volodin, and S.P. Smyshliaev, 2003: Atmospheric general circulation model of INM RAS with ozone dynamics. *Russ. Meteorol. Hydrol.*, **5**, 13–22.

Gallée, H., et al., 1991: Simulation of the last glacial cycle by a coupled, sectorally averaged climate–ice sheet model. Part I: The climate model. *J. Geophys. Res.*, **96**, 13139–13161.

Ganachaud, A., and C. Wunsch, 2000: Improved estimates of global ocean circulation, heat transport and mixing from hydrographic data. *Nature*, **408**, 453–457.

Ganachaud, A., and C. Wunsch, 2003: Large-scale ocean heat and freshwater transports during the World Ocean Circulation Experiment. *J. Clim.*, **16**, 696–705.

Gates, W.L., et al., 1999: An overview of the results of the Atmospheric Model Intercomparison Project (AMIP I). *Bull. Am. Meteorol. Soc.*, **80**, 29–55.

Geng, Q., and M. Sugi, 2003: Possible change of extratropical cyclone activity due to enhanced greenhouse gases and sulfate aerosols–Study with a high-resolution AGCM. *J. Clim.*, **16**, 2262–2274.

Gent, P.R., 2001: Will the North Atlantic Ocean thermohaline circulation weaken during the 21st century? *Geophys. Res. Lett.*, **28**, 1023–1026.

Gent, P.R., J. Willebrand, T.J. McDougall, and J.C. McWilliams, 1995: Parameterizing eddy-induced tracer transports in ocean circulation models. *J. Phys. Oceanogr.*, **25**, 463–474.

Gerber, S., et al., 2003: Constraining temperature variations over the last millennium by comparing simulated and observed atmospheric CO_2. *Clim. Dyn.*, **20**, 281–299.

Gerten, D., et al., 2004: Terrestrial vegetation and water balance–hydrological evaluation of a dynamic global vegetation model. *J. Hydrol.*, **286**, 249–270.

Gettelman, A., J.R. Holton, and A.R. Douglass, 2000: Simulations of water vapor in the lower stratosphere and upper troposphere. *J. Geophys. Res.*, **105**, 9003–9023.

GFDL GAMDT (The GFDL Global Atmospheric Model Development Team), 2004: The new GFDL global atmosphere and land model AM2-LM2: Evaluation with prescribed SST simulations. *J. Clim.*, **17**, 4641–4673.

Ghan, S.J., R. Easter, J. Hudson, and F.-M. Bréon, 2001a: Evaluation of aerosol indirect radiative forcing in MIRAGE. *J. Geophys. Res.*, **106**, 5317–5334.

Ghan, S.J., et al., 2001b: Evaluation of aerosol direct radiative forcing in MIRAGE. *J. Geophys. Res.*, **106**, 5295–5316.

Gillett, N.P., 2005: Northern Hemisphere circulation. *Nature*, **437**, 496.

Gillett, N.P., and D.W.J. Thompson, 2003: Simulation of recent Southern Hemisphere climate change. *Science*, **302**, 273–275.

Giorgetta, M.A., E. Manzini, and E. Roeckner, 2002: Forcing of the quasi-biennial oscillation from a broad spectrum of atmospheric waves. *Geophys. Res. Lett.*, **29**, 1245, doi:10.1029/2002GL014756.

Giorgetta M.A., et al., 2006: Climatology and forcing of the quasi-biennial oscillation in the MAECHAM5 model. *J. Clim.*, **19**, 3882–3901.

Gleckler, P.J., K.R. Sperber, and K. AchutaRao, 2006a: The annual cycle of global ocean heat content: observed and simulated. *J. Geophys. Res.*, **111**, C06008, doi:10.1029/2005JC003223.

Gleckler, P.J., et al., 2006b: Krakatoa's signature persists in the ocean. *Nature*, **439**, 675, doi:10.1038/439675a.

Gnanadesikan, A., et al., 2004: GFDL's CM2 global coupled climate models–Part 2: The baseline ocean simulation. *J. Clim.*, **19**, 675–697.

Goldenberg, S.B., C.W. Landsea, A.M. Mestas-Nunez, and W.M. Gray, 2001: The recent increase in Atlantic hurricane activity: Causes and implications. *Science*, **293**, 474–479.

Goosse, H., and T. Fichefet, 1999: Importance of ice-ocean interactions for the global ocean circulation: A model study. *J. Geophys. Res.*, **104**, 23337–23355.

Goosse, H., F.M. Selten, R.J. Haarsma, and J.D. Opsteegh, 2003: Large sea-ice volume anomalies simulated in a coupled climate model. *Clim. Dyn.*, **20**, 523–536, doi:10.1007/s00382-002-0290-4.

Goosse, H., et al., 2002: Potential causes of abrupt climate events: a numerical study with a three-dimensional climate model. *Geophys. Res. Lett.*, **29**(18), 1860, doi:10.1029/2002GL014993.

Gordon, C., et al., 2000: The simulation of SST, sea ice extents and ocean heat transports in a version of the Hadley Centre coupled model without flux adjustments. *Clim. Dyn.*, **16**, 147–168.

Gordon, H.B., et al., 2002: *The CSIRO Mk3 Climate System Model*. CSIRO Atmospheric Research Technical Paper No. 60, Commonwealth Scientific and Industrial Research Organisation Atmospheric Research, Aspendale, Victoria, Australia, 130 pp, http://www.cmar.csiro.au/e-print/open/gordon_2002a.pdf.

Gordon, N.D., J.R. Norris, C.P. Weaver, and S.A. Klein, 2005: Cluster analysis of cloud regimes and characteristic dynamics of midlatitude synoptic systems in observations and a model. *J. Geophys. Res.*, **110**, D15S17, doi:10.1029/2004JD005027.

Govindasamy, B., et al., 2005: Increase of the carbon cycle feedback with climate sensitivity: results from a coupled and carbon climate and carbon cycle model. *Tellus*, **57B**, 153–163.

Graham, R.J., et al., 2005: A performance comparison of coupled and uncoupled versions of the Met Office seasonal prediction general circulation model. *Tellus*, **57A**, 320–339.

Greenwald, T.J., G.L. Stephens, S.A. Christopher, and T.H.V. Haar, 1995: Observations of the global characteristics and regional radiative effects of marine cloud liquid water. *J. Clim.*, **8**, 2928–2946.

Gregory, D., et al., 2000: Revision of convection, radiation and cloud schemes in the ECMWF Integrated Forecasting System. *Q. J. R. Meteorol. Soc.*, **126**, 1685–1710.

Gregory, J.M., et al., 2002: An observationally based estimate of the climate sensitivity. *J. Clim.*, **15**, 3117–3121.

Gregory, J.M., et al., 2005: A model intercomparison of changes in the Atlantic thermohaline circulation in response to increasing atmospheric CO_2 concentration. *Geophys. Res. Lett.*, **32**, L12703, doi:10.1029/2005GL023209.

Griffies, S.M., 2004: *Fundamentals of Ocean Climate Models*. Princeton University Press, Princeton, NJ, 496 pp.

Guilyardi, E., et al., 2004: Representing El Niño in coupled ocean-atmosphere GCMs: the dominant role of the atmospheric component. *J. Clim.*, **17**, 4623–4629.

Gutowski, W.J., et al., 2004: Diagnosis and attribution of a seasonal precipitation deficit in a US regional climate simulation. *J. Hydrometeorol.*, **5**(1), 230–242.

Hagemann, S., 2002: *An Improved Land Surface Parameter Dataset for Global and Regional Climate Models*. Max Planck Institute for Meteorology Report 162, MPI for Meteorology, Hamburg, Germany, 21 pp.

Hagemann, S., and L. Dümenil-Gates, 2001: Validation of the hydrological cycle of ECMWF and NCEP reanalyses using the MPI hydrological discharge model. *J. Geophys. Res.*, **106**, 1503–1510.

Hall, A., 2004: The role of surface albedo feedback in climate. *J. Clim.*, **17**, 1550–1568.

Hall, A., and S. Manabe, 1999: The role of water vapour feedback in unperturbed climate variability and global warming. *J. Clim.*, **12**, 2327–2346.

Hall, A., and R.J. Stouffer, 2001: An abrupt climate event in a coupled ocean-atmosphere simulation without external forcing. *Nature*, **409**(6817), 171–174.

Hall, A., and M. Visbeck, 2002: Synchronous variability in the Southern Hemisphere atmosphere, sea ice and ocean resulting from the annular mode. *J. Clim.*, **15**, 3043–3057.

Hall, A., and X. Qu, 2006: Using the current seasonal cycle to constrain snow albedo feedback in future climate change. *Geophys. Res. Lett.*, **33**, L03502, doi:10.1029/2005GL025127.

Hall, M.M., and H.L. Bryden, 1982: Direct estimates and mechanisms of ocean heat transport. *Deep Sea Res.*, **29**, 339–359.

Hamilton, K., R.J. Wilson, and R.S. Hemler, 2001: Spontaneous stratospheric QBO-like oscillations simulated by the GFDL SKYHI general circulation model. *J. Atmos. Sci.*, **58**, 3271–3292.

Hannachi, A., and A. O'Neill, 2001: Atmospheric multiple equilibria and non-Gaussian behaviour in model simulations. *Q. J. R. Meteorol. Soc.*, **127**, 939–958.

Hansen, J., et al., 1984: Climate sensitivity: analysis of feedback mechanisms. *Meteorol. Monogr.*, **29**, 130–163.

Hanson, C.E., J.P. Palutikof, and T.D. Davies, 2004: Objective cyclone climatologies of the North Atlantic - a comparison between the ECMWF and NCEP Reanalyses. *Clim. Dyn.*, **22**, 757–769.

Harder, M., 1996: *Dynamik, Rauhigkeit und Alter des Meereises in der Arktis*. PhD Thesis, Alfred-Wegener-Institut für Polar und Meeresforschung, Bremerhaven, Germany, 124 pp.

Hargreaves, J.C., J.D. Annan, N.R. Edwards, and R. Marsh, 2004: An efficient climate forecasting method using an intermediate complexity Earth System Model and the ensemble Kalman filter. *Clim. Dyn.*, **23**, 745–760.

Harrison, E.F., et al., 1990: Seasonal variation of cloud radiative forcing derived from the Earth Radiation Budget Experiment. *J. Geophys. Res.*, **95**, 18687–18703.

Harrison, H., 2002: Comments on "Does the Earth have an adaptive infrared iris?". *Bull. Am. Meteorol. Soc.*, **83**, 597.

Hartmann, D.L., and K. Larson, 2002: An important constraint on tropical cloud-climate feedback. *Geophys. Res. Lett.*, **29**(20), 1951–1954.

Hartmann, D.L., and M.L. Michelsen, 2002: No evidence for iris. *Bull. Am. Meteorol. Soc.*, **83**, 249–254.

Hartmann, D.L., M.E. Ockert-Bell, and M.L. Michelsen, 1992: The effect of cloud type on Earth's energy balance: Global analysis. *J. Clim.*, **5**, 1281–1304.

Harvey, D., et al., 1997: *An Introduction to Simple Climate Models Used in the IPCC Second Assessment Report*. IPCC Technical Paper 2 [Houghton, J.T., L.G. Meira Filho, D.J. Griggs, and K. Maskell (eds.)]. IPCC, Geneva, Switzerland, 51 pp.

Hasumi, H., 2002a: Sensitivity of the global thermohaline circulation to interbasin freshwater transport by the atmosphere and the Bering Strait throughflow. *J. Clim.*, **15**, 2516–2526.

Hasumi, H., 2002b: Modeling the global thermohaline circulation. *J. Oceanogr.*, **58**, 25–33.

Hasumi, H., and N. Suginohara, 1999: Effects of locally enhanced vertical diffusivity over rough bathymetry on the world ocean circulation. *J. Geophys. Res.*, **104**, 23367–23374.

Hazeleger, W., et al., 2001: Decadal upper ocean temperature variability in the tropical Pacific. *J. Geophys. Res.*, **106**(C5), 8971–8988.

Held, I.M., and B.J. Soden, 2000: Water vapour feedback and global warming. *Annu. Rev. Energy Environ.*, **25**, 441–475.

Henderson-Sellers, A., P. Irannejad, K. McGuffie, and A.J. Pitman, 2003: Predicting land-surface climates - better skill or moving targets? *Geophys. Res. Lett.*, **30**(14), 1777–1780.

Henderson-Sellers, A., K. McGuffie, D. Noone, and P. Irannejad, 2004: Using stable water isotopes to evaluate basin-scale simulations of surface water budgets. *J. Hydrometeorol.*, **5**(5), 805–822.

Hendon, H.H., 2000: Impact of air–sea coupling on the Madden–Julian oscillation in a general circulation model. *J. Atmos. Sci.*, **57**, 3939–3952.

Hendon, H.H., 2005: Air sea interaction. In: *Intraseasonal Variability in the Atmosphere-Ocean Climate System* [Lau, W.K.M., and D.E. Waliser (eds.)]. Praxis Publishing, 436 pp.

Hewitt, C.D., C.S. Senior, and J.F.B. Mitchell, 2001: The impact of dynamic sea-ice on the climate sensitvity of a GCM: a study of past, present and future climates. *Clim. Dyn.*, **17**, 655–668.

Heymsfield, A.J., and L. Donner, 1990: A scheme for parameterizing ice-cloud water content in general circulation models. *J. Atmos. Sci.*, **47**, 1865–1877.

Hibler, W.D., 1979: A dynamic thermodynamic sea ice model. *J. Phys. Oceanogr.*, **9**, 817–846.

Hirst, A.C., 1999: The Southern Ocean response to global warming in the CSIRO coupled ocean-atmosphere model. *Environ. Model. Software*, **14**, 227–241.

Hodges, K.I., B.J. Hoskins, J. Boyle, and C. Thorncroft, 2003: A comparison of recent reanalysis data sets using objective feature tracking: storm tracks and tropical easterly waves. *Mon. Weather Rev.*, **131**, 2012–2037.

Hodges, K.: Feature based diagnostics from ECMWF/NCEP Analyses and AMIP II: Model Climatologies. In: *The Second Phase of the Atmospheric Model Intercomparison Project (AMIP2)* [Gleckler, P. (ed.)]. Proceedings of the WCRP/WGNE Workshop, Toulouse, France, pp. 201-204.

Holland, M.M., and C.M. Bitz, 2003: Polar amplification of climate change in coupled models. *Clim. Dyn.*, **21**, 221–232, doi:10.1007/s00382-003-0332-6.

Holland, M.M., and M. Raphael, 2006: Twentieth century simulation of the Southern Hemisphere climate in coupled models. Part II: sea ice conditions and variability. *Clim. Dyn.*, **26**, 229–245, doi:10.1007/s00382-005-0087-3.

Horinouchi, T., 2002: Mesoscale variability of tropical precipitation: Validation of satellite estimates of wave forcing using TOGA COARE radar data. *J. Atmos. Sci.*, **59**, 2428–2437.

Horinouchi, T., and S. Yoden, 1998: Wave-mean flow interaction associated with a QBO-like oscillation simulated in a simplified GCM. *J. Atmos. Sci.*, **55**, 502–526.

Horinouchi, T., et al., 2003: Tropical cumulus convection and upward-propagating waves in middle-atmospheric GCMs. *J. Atmos. Sci.*, **60**, 2765–2782.

Hoskins, B.J., and K.I. Hodges, 2002: New perspectives on the Northern Hemisphere winter storm tracks. *J. Atmos. Sci.*, **59**, 1041–1061.

Hoskins, B.J., and K.I. Hodges, 2005: New perspectives on the Southern Hemisphere storm tracks. *J. Clim.*, **18**, 4108–4129.

Hourdin, F., et al., 2006: The LMDZ4 general circulation model: Climate performance and sensitivity to parameterized physics with emphasis on tropical convection. *Clim. Dyn.*, **27**, 787–813.

Hovine, S., and T. Fichefet, 1994: A zonally averaged, three-basin ocean circulation model for climate studies. *Clim. Dyn.*, **15**, 1405–1413.

Hsu, C.J., and F. Zwiers, 2001: Climate change in recurrent regimes and modes of atmospheric variability. *J. Geophys. Res.*, **106**, 20145–20160.

Hu, A.X., G.A. Meehl, W.M. Washington, and A. Dai, 2004: Response of the Atlantic thermohaline circulation to increased atmospheric CO_2 in a coupled model. *J. Clim.*, **17**, 4267–4279.

Huang, X., B.J. Soden, and D.L. Jackson, 2005: Interannual co-variability of tropical temperature and humidity: A comparison of model, reanalysis data and satellite observation. *Geophys. Res. Lett.*, **32**, L17808, doi:10.1029/2005GL023375.

Hunke, E.C., and J.K. Dukowicz, 1997: An elastic-viscous-plastic model for sea ice dynamics. *J. Phys. Oceanogr.*, **27**, 1849–1867.

Hunke, E.C., and J.K. Dukowicz, 2002: The Elastic-Viscous-Plastic sea ice dynamics model in general orthogonal curvilinear coordinates on a sphere–Effect of metric terms. *Mon. Weather Rev.*, **130**, 1848–1865.

Hunke, E.C., and J.K. Dukowicz, 2003: *The Sea Ice Momentum Equation in the Free Drift Regime*. Technical Report LA-UR-03-2219, Los Alamos National Laboratory, Los Alamos, NM.

Hurrell, J.W., M.P. Hoerling, A.S. Phillips, and T. Xu, 2004: Twentieth century North Atlantic climate change. Part I: assessing determinism. *Clim. Dyn.*, **23**, 371–389.

Hutchings, J.K., H. Jasak, and S.W. Laxon, 2004: A strength implicit correction scheme for the viscous-plastic sea ice model. *Ocean Modelling*, **7**, 111–133.

Huybrechts, P., 2002: Sea-level changes at the LGM from ice-dynamics reconstructions of the Greenland and Antarctic ice sheets during the glacial cycles. *Quat. Sci. Rev.*, **21**, 203–231.

Huybrechts, P., I. Janssens, C. Poncin, and T. Fichefet, 2002: The response of the Greenland ice sheet to climate changes in the 21st century by interactive coupling of an AOGCM with a thermomechanical ice sheet model. *Ann. Glaciol.*, **35**, 409–415.

Iacobellis, S.F., G.M. McFarquhar, D.L. Mitchell, and R.C.J. Somerville, 2003: The sensitivity of radiative fluxes to parameterized cloud microphysics. *J. Clim.*, **16**, 2979–2996.

Iacono, M.J., J.S. Delamere, E.J. Mlawer, and S.A. Clough, 2003: Evaluation of upper tropospheric water vapor in the NCAR Community Climate Model, CCM3, using modeled and observed HIRS radiances. *J. Geophys. Res.*, **108**(D2), 4037, doi:10.1029/2002JD002539.

Inamdar, A.K., and V. Ramanathan, 1998: Tropical and global scale interactions among water vapour, atmospheric greenhouse effect, and surface temperature. *J. Geophys. Res.*, **103**, 32177–32194.

Ingram, W.J., 2002: On the robustness of the water vapor feedback: GCM vertical resolution and formulation. *J. Clim.*, **15**, 917–921.

Inness, P.M., and J.M. Slingo, 2003: Simulation of the MJO in a coupled GCM. I: Comparison with observations and atmosphere-only GCM. *J. Clim.*, **16**, 345–364.

Inness, P.M., J.M. Slingo, E. Guilyardi, and J. Cole, 2003: Simulation of the MJO in a coupled GCM. II: The role of the basic state. *J. Clim.*, **16**, 365–382.

Iorio, J.P., et al., 2004: Effects of model resolution and subgrid scale physics on the simulation of precipitation in the continental United States. *Clim. Dyn.*, **23**, 243–258, doi:10.1007/s00382-004-0440-y.

Jaeger, L., 1976: *Monatskarten des Niederschlags für die Ganze Erde*. Ber. Deutsche Wetterdienstes 139, Germany, 38 pp.

Jakob, C., and G. Tselioudis, 2003: Objective identification of cloud regimes in the tropical western pacific. *Geophys. Res. Lett.*, **30**, doi:10.1029/2003GL018367.

Jennings, R.L., 1975: *Data Sets for Meteorological Research*. NCAR-TN/1A, National Center for Atmospheric Research, Boulder, CO, 156 pp.

Ji, M., A. Leetmaa, and V.E. Kousky, 1996: Coupled model predictions of ENSO during the 1980s and the 1990s at the National Centers for Environmental Prediction. *J. Clim.*, **9**, 3105–3120.

Jin, X.Z., X.H. Zhang, and T.J. Zhou, 1999: Fundamental framework and experiments of the third generation of the IAP/LASG World Ocean General Circulation Model. *Adv. Atmos. Sci.*, **16**, 197–215.

Johns, T.C., et al., 2006: The new Hadley Centre climate model HadGEM1: Evaluation of coupled simulations. *J. Clim.*, **19**, 1327–1353.

Jones, C.D., et al., 2005: Systematic optimisation and climate simulation of FAMOUS, a fast version of HadCM3. *Clim. Dyn.*, **25**, 189–204.

Jones, P.D., 1988: Hemispheric surface air temperature variations: Recent trends and an update to 1987. *J. Clim.*, **1**, 654–660.

Jones, P.D., et al., 1999: Surface air temperature and its variations over the last 150 years. *Rev. Geophys.*, **37**, 173–199.

Joos, F., et al., 1999: Global warming and marine carbon cycle feedbacks on future atmospheric CO_2. *Science*, **284**, 464–467.

Joos, F., et al., 2001: Global warming feedbacks on terrestrial carbon uptake under the IPCC emission scenarios. *Global Biogeochem. Cycles*, **15**, 891–907.

Joshi, M., et al., 2003: A comparison of climate response to different radiative forcings in three general circulation models: towards an improved metric of climate change. *Clim. Dyn.*, **20**, 843–854.

Jungclaus, J.H., et al., 2006: Ocean circulation and tropical variability in the AOGCM ECHAM5/MPI-OM. *J. Clim.*, **19**, 3952–3972.

K-1 Model Developers, 2004: *K-1 Coupled Model (MIROC) Description*. K-1 Technical Report 1 [Hasumi, H., and S. Emori (eds.)]. Center for Climate System Research, University of Tokyo, Tokyo, Japan, 34 pp., http://www.ccsr.u-tokyo.ac.jp/kyosei/hasumi/MIROC/tech-repo.pdf.

Kalnay, E., et al., 1996: The NCEP/NCAR 40-year reanalysis project. *Bull. Am. Meteorol. Soc.*, **77**, 437–471.

Kanamitsu, M., et al., 2002: NCEP dynamical seasonal forecast system 2000. *Bull. Am. Meteorol. Soc.*, **83**, 1019–1037.

Kattsov, V., and E. Källén, 2005: Future climate change: Modeling and scenarios for the Arctic. In: *Arctic Climate Impact Assessment (ACIA)*. Cambridge University Press, Cambridge, UK, pp. 99–150.

Kemball-Cook, S., B. Wang, and X. Fu, 2002: Simulation of the intraseasonal oscillation in ECHAM-4 model: The impact of coupling with an ocean model. *J. Atmos. Sci.*, **59**, 1433–1453.

Khairoutdinov, M., D. Randall, and C. DeMott, 2005: Simulations of the atmospheric general circulation using a cloud-resolving model as a superparameterization of physical processes. *J. Atmos. Sci.*, **62**, 2136–2154.

Kharin, V.V., F.W. Zwiers, and X. Zhang, 2005: Intercomparison of near surface temperature and precipitation extremes in AMIP-2 simulations, reanalyses and observations. *J. Clim.*, **18**(24), 5201–5223.

Kiehl, J.T., and P.R. Gent, 2004: The Community Climate System Model, Version 2. *J. Clim.*, **17**, 3666–3682.

Kiehl, J.T., et al., 1998: The National Center for Atmospheric Research Community Climate Model: CCM3. *J. Clim.*, **11**, 1131–1149.

Kiktev, D., D.M.H. Sexton, L. Alexander, and C.K. Folland, 2003: Comparison of modeled and observed trends in indices of daily climate extremes. *J. Clim.*, **16**(22), 3560–3571.

Kim, S.-J., G.M. Flato, G.J. Boer, and N.A. McFarlane, 2002: A coupled climate model simulation of the Last Glacial Maximum, Part 1: Transient multi-decadal response. *Clim. Dyn.*, **19**, 515–537.

Kimoto, M., N. Yasutomi, C. Yokoyama, and S. Emori, 2005: Projected changes in precipitation characteristics near Japan under the global warming. *Scientific Online Letters on the Atmosphere*, **1**, 85–88, doi:10.2151/sola.2005-023.

Kinne, S., et al., 2003: Monthly averages of aerosol properties: A global comparison among models, satellite, and AERONET ground data. *J. Geophys. Res.*, **108**(D20), 4634, doi:10.1029/2001JD001253.

Kirtman, B.P., 2003: The COLA anomaly coupled model: Ensemble ENSO prediction. *Mon. Weather Rev.*, **131**, 2324–2341.

Kirtman, B.P., and P.S. Schopf, 1998: Decadal variability in ENSO predictability and prediction. *J. Clim.*, **11**, 2804–2822.

Kirtman, B.P., K. Pegion, and S. Kinter, 2005: Internal atmospheric dynamics and tropical indo-pacific climate variability. *J. Atmos. Sci.*, **62**, 2220–2233.

Kleeman, R., Y. Tang, and A.M. Moore, 2003: The calculation of climatically relevant singular vectors in the presence of weather noise as applied to the ENSO problem. *J. Atmos. Sci.*, **60**, 2856–2868.

Kleidon, A., 2004: Global datasets of rooting zone depth inferred from inverse methods. *J. Clim.*, **17**, 2714–2722.

Kleidon, A., K. Fraedrich, and M. Heimann, 2000: A green planet versus a desert world: estimating the maximum effect of vegetation on the land surface climate. *Clim. Change*, **44**, 471–493.

Klein, S.A., and D.L. Hartmann, 1993: The seasonal cycle of low stratiform clouds. *J. Clim.*, **6**, 1587–1606.

Klein, S.A., and C. Jakob, 1999: Validation and sensitivities of frontal clouds simulated by the ECMWF model. *Mon. Weather Rev.*, **127**, 2514–2531.

Knight, J.R, et al., 2005: A signature of persistent natural thermohaline circulation cycles in observed climate. *Geophys. Res. Lett.*, **32**, L20708, doi:10.1029/2005GL024233.

Knutson, T.R., and R.E. Tuleya, 1999: Increased hurricane intensities with CO_2-induced global warming as simulated using the GFDL hurricane prediction system. *Clim. Dyn.*, **15**(7), 503–519.

Knutson, T.R., and R.E. Tuleya, 2004: Impact of CO_2-induced warming on simulated hurricane intensity and precipitation: Sensitivity to the choice of climate model and convective parameterization. *J. Clim.*, **17**, 3477–3495.

Knutti, R., T.F. Stocker, F. Joos, and G.K. Plattner, 2002: Constraints on radiative forcing and future climate change from observations and climate model ensembles. *Nature*, **416**, 719–723.

Knutti, R., G.A. Meehl, M.R. Allen and D.A. Stainforth, 2006: Constraining climate sensitivity from the seasonal cycle in surface temperature. *J. Clim.*, **19**, 4224–4233.

Kodera, K., and M. Chiba, 1995: Tropospheric circulation changes associated with stratospheric sudden warmings: A case study. *J. Geophys. Res.*, **100**, 11055–11068.

Komuro, Y., and H. Hasumi, 2005: Intensification of the Atlantic deep circulation by the Canadian Archipelago throughflow. *J. Phys. Oceanogr.*, **35**, 775–789.

Koster, R.D., et al., 2004: Regions of coupling between soil moisture and precipitation. *Science*, **305**, 1138–1140.

Kraus, E.B., 1990: Diapycnal mixing. In: *Climate-Ocean Interaction* [Schlesinger, M.E. (ed.)]. Kluwer, Amsterdam, pp. 269–293.

Kraus, E.B., and J.S. Turner, 1967: A one-dimensional model of the seasonal thermocline. II. The general theory and its consequences. *Tellus*, **19**, 98–105.

Krinner, G., et al., 2005: A dynamic global vegetation model for studies of the coupled atmosphere-biosphere system. *Global Biogeochem. Cycles*, **19**, GB1015, doi:10.1029/2003GB002199.

Lambert, S.J., and G.J. Boer, 2001: CMIP1 evaluation and intercomparison of coupled climate models. *Clim. Dyn.*, **17**, 83–106.

Lambert, S.J., and J. Fyfe, 2006: Changes in winter cyclone frequencies and strengths simulated in enhanced greenhouse gas simulations: Results from the models participating in the IPCC diagnostic exercise. *Clim. Dyn.*, **26**, 713–728.

Lanzante, J.R., 1996: Resistant, robust and nonparametric techniques for analysis of climate data: Theory and examples, including applications to historical radiosonde station data. *Int. J. Climatol.*, **16**, 1197–1226.

Large, W.G., J.C. McWilliams, and S.C. Doney, 1994: Oceanic vertical mixing: a review and a model with a nonlocal boundary layer parameterization. *Rev. Geophys.*, **32**, 363–403.

Larson, K., and D.L. Hartmann, 2003: Interactions among cloud, water vapour, radiation and large-scale circulation in the tropical climate. Part 1: sensitivity to uniform sea surface temperature changes. *J. Clim.*, **15**, 1425–1440.

Latif, M., 1998: Dynamics of interdecadal variability in coupled ocean-atmosphere models. *J. Clim.*, **11**, 602–624.

Latif, M., E. Roeckner, U. Mikolajewicz, and R. Voss, 2000: Tropical stabilisation of the thermohaline circulation in a greenhouse warming simulation. *J. Clim.*, **13**, 1809–1813.

Latif, M., et al., 2001: ENSIP: The El Niño simulation intercomparison project. *Clim. Dyn.*, **18**, 255–276.

Latif, M., et al., 2004: Reconstructing, monitoring, and predicting multidecadal scale changes in the North Atlantic thermohaline circulation with sea surface temperatures. *J. Clim.*, **17**, 1605–1614.

Lawrence, D.M., and J.M. Slingo, 2005: Weak land–atmosphere coupling strength in HadAM3: The role of soil moisture variability. *J. Hydrometeorol.*, **6**, 670–680.

Le Treut, H., Z.X. Li, and M. Forichon, 1994: Sensitivity of the LMD general circulation model to greenhouse forcing associated with two different cloud water parametrizations. *J. Clim.*, **7**, 1827–1841.

Lee, M.-I., I.-S. Kang, J.-K. Kim, and B. E. Mapes, 2001: Influence of cloud-radiation interaction on simulating tropical intraseasonal oscillation with an atmospheric general circulation model. *J. Geophys. Res.*, **106**, 14219–14233.

Levitus, S., and T.P. Boyer, 1994: *World Ocean Atlas 1994, Volume 4: Temperature*. NOAA NESDIS E/OC21, Washington, DC, 117 pp.

Levitus, S., and J. Antonov, 1997: *Variability of Heat Storage of and the Rate of Heat Storage of the World Ocean*. NOAA NESDIS Atlas 16, US Government Printing Office, Washington, DC, 6 pp., 186 figures.

Levitus, S., J. Antonov, and T. Boyer, 2005: Warming of the world ocean, 1955-2003. *Geophys. Res. Lett.*, **32**, L02604, doi:10.1029/2004GLO21592.

Levitus, S., et al., 1998: *World Ocean Database 1998, Volume 1: Introduction*. NOAA Atlas NESDIS 18, US Government Printing Office, Washington, DC.

Liang, X., Z. Xie, and M. Huang, 2003: A new parameterization for surface and groundwater interactions and its impact on water budgets with the variable infiltration capacity (VIC) land surface model. *J. Geophys. Res.*, **108**, 8613, doi:10.1029/2002JD003090.

Limpasuvan, V., and D.L. Hartmann, 2000: Wave-maintained annular modes of climate variability. *J. Clim.*, **13**, 4414–4429.

Lin, B., T. Wong, B.A. Wielicki, and Y. Hu, 2004: Examination of the decadal tropical mean ERBS nonscanner radiation data for the iris hypothesis. *J. Clim.*, **17**, 1239–1246.

Lin, B., et al., 2002: The iris hypothesis: A negative or positive cloud feedback? *J. Clim.*, **15**, 3–7.

Lin, J.L., et al., 2006: Tropical intraseasonal variability in 14 IPCC AR4 climate models. Part I: Convective signals. *J. Clim.*, **19**, 2665–2690.

Lin, W.Y., and M.H. Zhang, 2004: Evaluation of clouds and their radiative effects simulated by the NCAR Community Atmospheric Model against satellite observations. *J. Clim.*, **17**, 3302–3318.

Lindsay, R.W., and H.L. Stern, 2004: A new Lagrangian model of Arctic sea ice. *J. Phys. Oceanogr.*, **34**, 272–283.

Lindzen, R.S., M.-D. Chou, and A.Y. Hou, 2001: Does the Earth have an adaptative infrared iris? *Bull. Am. Meteorol. Soc.*, **82**, 417–432.

Lindzen, R.S., M.-D. Chou, and A.Y. Hou, 2002: Comment on "No evidence for iris". *Bull. Am. Meteorol. Soc.*, **83**, 1345–1349.

Lipscomb, W.H., 2001: Remapping the thickness distribution in sea ice models. *J. Geophys. Res.*, **106**, 13989–14000.

Liston, G., 2004: Representing subgrid snow cover heterogeneities in regional and global models. *J. Clim.*, **17**, 1381–1397.

Liu, H., et al., 2004: An eddy-permitting oceanic general circulation model and its preliminary evaluations. *Adv. Atmos. Sci.*, **21**, 675–690.

Liu, J., et al., 2003: Sensitivity of sea ice to physical parameterizations in the GISS global climate model. *J. Geophys. Res.*, **108**, 3053, doi:10.1029/2001JC001167.

Liu, P., et al., 2005: MJO in the NCAR CAM2 with the Tiedtke convective scheme. *J. Clim.*, **18**, 3007–3020.

Lock, A.P., 2001: The numerical representation of entrainment in parameterizations of boundary layer turbulent mixing. *Mon. Weather Rev.*, **129**, 1148–1163.

Lock, A.P., et al., 2000: A new boundary layer mixing scheme. Part I: Scheme description and SCM tests. *Mon. Weather Rev.*, **128**, 3187–3199.

Lohmann, U., and G. Lesins, 2002: Stronger constraints on the anthropogenic indirect aerosol effect. *Science*, **298**, 1012–1015.

Lorenz, D.J., and D.L. Hartmann, 2001: Eddy–zonal flow feedback in the Southern Hemisphere. *J. Atmos. Sci.*, **58**, 3312–3327.

Lu, J., R.J. Greatbatch, and K.A. Peterson, 2004: Trend in Northern Hemisphere winter atmospheric circulation during the last half of the twentieth century. *J. Clim.*, **17**, 3745–3760.

Luo, Z., and W.B. Rossow, 2004: Characterising tropical cirrus life cycle, evolution and interaction with upper tropospheric water vapour using a Lagrangian trajectory analysis of satellite observations. *J. Clim.*, **17**, 4541–4563.

Madden, R.A., and P.R. Julian, 1971: Detection of a 40-50 day oscillation in the zonal wind in the tropical Pacific. *J. Atmos. Sci.*, **28**, 702–708.

Madec, G., P. Delecluse, M. Imbard, and C. Lévy, 1998: *OPA Version 8.1 Ocean General Circulation Model Reference Manual*. Notes du Pôle de Modélisation No. 11, Institut Pierre-Simon Laplace, Paris, 91 pp., http://www.lodyc.jussieu.fr/opa/Docu_Free/Doc_models/Doc_OPA8.1.pdf.

Mahfouf, J.-F., et al., 1995: The land surface scheme ISBA within the Meteo-France climate model ARPEGE. Part 1: Implementation and preliminary results. *J. Clim.*, **8**, 2039–2057.

Maloney, E.D., and D.L. Hartmann, 2001: The sensitivity of the intraseasonal variability in the NCAR CCM3 to changes in convective parameterization. *J. Clim.*, **14**, 2015–2034.

Maltrud, M.E., R.D. Smith, A.J. Semtner, and R.C. Malone, 1998: Global eddy-resolving ocean simulations driven by 1985–1995 atmospheric winds. *J. Geophys. Res.*, **103**, 30825–30853.

Manabe, S., and R.J. Stouffer, 1988: Two stable equilibria of a coupled ocean-atmosphere model. *J. Clim.*, **1**(9), 841–866.

Manabe, S., and R.J. Stouffer, 1995: Simulation of abrupt climate change induced by fresh water input to the North Atlantic Ocean. *Nature*, **378**, 165–167.

Manabe, S., and R.J. Stouffer, 1996: Low-frequency variability of surface air temperature in a 1000-year integration of a coupled atmosphere-ocean-land surface model. *J. Clim.*, **9**, 376–393.

Manabe, S., and R.J. Stouffer, 1997: Coupled ocean-atmosphere model response to freshwater input: Comparison to Younger Dryas event. *Paleoceanography*, **12**, 321–336.

Manabe, S., R.J. Stouffer, M.J. Spelman, and K. Bryan, 1991: Transient responses of a coupled ocean atmosphere model to gradual changes of atmospheric CO_2. I: Annual mean response. *J. Clim.*, **4**, 785–818.

Mann, M.E., R.S. Bradley, and M.K. Hughes, 1998: Global-scale temperature patterns and climate forcing over the past six centuries. *Nature*, **392**, 779–787.

Marchal, O., T.F. Stocker, and F. Joos, 1998: A latitude-depth, circulation-biogeochemical ocean model for paleoclimate studies. *Tellus*, **50B**, 290–316.

Marotzke, J., 1997: Boundary mixing and the dynamics of three-dimensional thermohaline circulation. *J. Phys. Oceanogr.*, **27**, 1713–1728.

Marshall, G.J., 2003: Trends in the Southern Annular Mode from observations and reanalyses. *J. Clim.*, **16**, 4134–4143.

Marshall, J.C., C. Hill, L. Perelman, and A. Adcroft, 1997: Hydrostatic, quasi-hydrostatic and non-hydrostatic ocean modeling. *J. Geophys. Res.*, **102**, 5733–5752.

Marsland, S.J., et al., 2003: The Max-Planck-Institute global ocean/sea ice model with orthogonal curvilinear coordinates. *Ocean Modelling*, **5**, 91–127.

Marti, O., et al., 2005: *The New IPSL Climate System Model: IPSL-CM4.* Note du Pôle de Modélisation No. 26, Institut Pierre Simon Laplace des Sciences de l'Environnement Global, Paris, http://dods.ipsl.jussieu.fr/omamce/IPSLCM4/DocIPSLCM4/FILES/DocIPSLCM4.pdf.

Martin, G.M., et al., 2004: *Evaluation of the Atmospheric Performance of HadGAM/GEM1.* Hadley Centre Technical Note No. 54, Hadley Centre for Climate Prediction and Research/Met Office, Exeter, UK, http://www.metoffice.gov.uk/research/hadleycentre/pubs/HCTN/index.html.

Martin, G.M., et al., 2006: The physical properties of the atmosphere in the new Hadley Centre Global Environmental Model, HadGEM1. Part I: Model description and global climatology. *J. Clim.*, **19**, 1274–1301.

Maxwell, R.M., and N.L. Miller, 2005: Development of a coupled land surface and groundwater model. *J. Hydrometeorol.*, **6**, 233–247.

May, W., 2004: Simulation of the variability and extremes of daily rainfall during the Indian summer monsoon for present and future times in a global time-slice experiment. *Clim. Dyn.*, **22**, 183–204.

Mayer, M., C. Wang, M. Webster, and R. Prinn, 2000: Linking air pollution to global chemistry and climate. *J. Geophys. Res.*, **105**, 22869–22896.

McAvaney, B.J., et al., 2001: Model evaluation. In: *Climate Change 2001: The Scientific Basis. Contribution of Working Group I to the Third Assessment Report of the Intergovernmental Panel on Climate Change* [Houghton, J.T., et al. (eds.)]. Cambridge University Press, Cambridge, United Kingdom and New York, NY, USA, pp. 471–523.

McCarthy, M.P., and R. Toumi, 2004: Observed interannual variability of tropical troposphere relative humidity. *J. Clim.*, **17**, 3181–3191.

McDonald, R.E., et al., 2005: Tropical storms: representation and diagnosis in climate models and the impacts of climate change. *Clim. Dyn.*, **25**, 19–36.

McFarlane, N.A., G.J. Boer, J.-P. Blanchet, and M. Lazare, 1992: The Canadian Climate Centre second-generation general circulation model and its equilibrium climate. *J. Clim.*, **5**, 1013–1044.

Mechoso, C.R., et al., 1995: The seasonal cycle over the tropical Pacific in general circulation model. *Mon. Weather Rev.*, **123**, 2825–2838.

Meehl, G.A., and C. Tebaldi, 2004: More intense, more frequent, and longer lasting heat waves in the 21st century. *Science*, **305**, 994–997.

Meehl, G.A., and A. Hu, 2006: Mega droughts in the Indian monsoon and southwest North America and a mechanism for associated multi-decadal sea surface temperature anomalies. *J. Clim.*, **19**, 1605–1623.

Meehl, G.A., C. Tebaldi, and D. Nychka, 2004: Changes in frost days in simulations of twenty-first century climate. *Clim. Dyn.*, **23**, 495–511.

Meehl, G.A., et al., 2001: Factors that affect the amplitude of El Niño in global coupled climate models. *Clim. Dyn.*, **17**, 515–526.

Meissner, K.J., A.J. Weaver, H.D. Matthews, and P.M. Cox, 2003: The role of land surface dynamics in glacial inception: A study with the UVic Earth System Model. *Clim. Dyn.*, **21**, 515–537, doi:10.1007/s00382-003-0352-2.

Mellor, G.L., and T. Yamada, 1982: Development of a turbulence closure model for geophysical fluid problems. *Rev. Geophys.*, **20**, 851–875.

Mellor, G.L., and L. Kantha, 1989: An ice-ocean coupled model. *J. Geophys. Res.*, **94**, 10937–10954.

Mestas-Nunez, A.M., and D.B. Enfield, 1999: Rotated global modes of non-ENSO sea surface temperature variability. *J. Clim.*, **12**, 2734–2745.

Miller, J.R., G.L. Russell, and G. Caliri, 1994: Continental-scale river flow in climate models. *J. Clim.*, **7**, 914–928.

Miller, R.L., 1997: Tropical thermostats and low cloud cover. *J. Clim.*, **10**, 409–440.

Miller, R.L., G.A. Schmidt, and D.T. Shindell, 2006: Forced variations of annular modes in the 20th century IPCC AR4 simulations. *J. Geophys. Res.*, **111**, D18101, doi:10.1029/2005JD006323.

Milly, P.C.D., and A.B. Shmakin, 2002: Global modeling of land water and energy balances, Part I: The Land Dynamics (LaD) model. *J. Hydrometeorol.*, **3**, 283–299.

Milly, P.C.D., K.A. Dunne, and A.V. Vecchia, 2005: Global pattern of trends in streamflow and water availability in a changing climate. *Nature*, **438**, 347–350, doi:10.1038/nature04312.

Min, S.-K., S. Legutke, A. Hense, and W.-T. Kwon, 2005: Climatology and internal variability in a 1000-year control simulation with the coupled climate model ECHO-G—I. Near-surface temperature, precipitation and mean sea level pressure. *Tellus*, **57A**, 605–621.

Minschwaner, K., and A.E. Dessler, 2004: Water vapor feedback in the tropical upper troposphere: model results and observations. *J. Clim.*, **17**, 1272–1282.

Minschwaner, K., A.E. Dessler, and S. Parnchai, 2006: Multi-model analysis of the water vapour feedback in the tropical upper troposphere. *J. Clim.*, **19**, 5455–5464.

Mitchell, T.D., and P.D. Jones, 2005: An improved method of constructing a database of monthly climate observations and associated high-resolution grids. *Int. J. Climatol.*, **25**, 693712.

Molteni, F., Kuchraski, F., and Corti, S., 2006: On the predictability of flow-regime properties on interannual to interdecadal timescales. In: *Predictability of Weather and Climate* [Palmer, T. and R. Hagedorn (eds.)]. Cambridge University Press, Cambridge, UK.

Monahan, A.H., and A. Dai, 2004: The spatial and temporal structure of ENSO nonlinearity. *J. Clim.*, **17**, 3026–3036.

Monahan, A.H., J.C. Fyfe, and L. Pandolfo, 2003: The vertical structure of wintertime climate regimes of the Northern Hemisphere extratropical atmosphere. *J. Clim.*, **16**, 2005–2021.

Montoya, M., et al., 2005: The Earth System Model of Intermediate Complexity CLIMBER-3α. Part I: Description and performance for present day conditions. *Clim. Dyn.*, **25**, 237–263, doi:10.1007/s00382-005-0044-1.

Mouchet, A., and L. François, 1996: Sensitivity of a global oceanic carbon cycle model to the circulation and to the fate of organic matter: Preliminary results. *Phys. Chem. Earth*, **21**, 511–516.

Moum, J.N., D.R. Caldwell, J.D. Nash, and G.D. Gunderson, 2002: Observations of boundary mixing over the continental slope. *J. Phys. Oceanogr.*, **32**, 2113–2130.

Murphy, J.M., 1995: Transient response of the Hadley Centre coupled ocean-atmosphere model to increasing carbon dioxide. Part III: analysis of global-mean response using simple models. *J. Clim.*, **8**, 496–514.

Murphy, J.M., et al., 2004: Quantification of modelling uncertainties in a large ensemble of climate change simulations. *Nature*, **430**, 768–772.

Murray, R.J., 1996: Explicit generation of orthogonal grids for ocean models. *J. Comput. Phys.*, **126**, 251–273.

Myhre, G., E.J. Highwood, K.P. Shine, and F. Stordal, 1998: New estimates of radiative forcing due to well mixed greenhouse gases. *Geophys. Res. Lett.*, **25**, 2715–2718.

Nakano, H., and N. Suginohara, 2002: Effects of bottom boundary layer parameterization on reproducing deep and bottom waters in a World Ocean model. *J. Phys. Oceanogr.*, **32**, 1209–1227.

Naud, C.M., A.D. Del Genio, and M. Bauer, 2006: Observational constraints on cloud thermodynamic phase in midlatitude storms. *J. Clim.*, **19**, 5273–5288.

Neale, R., and J. Slingo, 2003: The maritime continent and its role in the global climate: A GCM study. *J. Clim.*, **16**, 834–848.

Newman, M., G.P. Compo, and M.A. Alexander, 2003: ENSO-forced variability of the PDO. *J. Clim.*, **16**, 3853–3857.

Nijssen, B., et al., 2003: Simulation of high latitude hydrological processes in the Torne-Kalix basin: PILPS Phase 2(e) 2: Comparison of model results with observations. *Global Planet. Change*, **38**, 31–53.

Norris, J.R., 1998a: Low cloud type over the ocean from surface observations. Part I: relationship to surface meteorology and the vertical distribution of temperature and moisture. *J. Clim.*, **11**, 369–382.

Norris, J.R., 1998b: Low cloud type over the ocean from surface observations. Part II: geographical and seasonal variations. *J. Clim.*, **11**, 383–403.

Norris, J.R., and C.P. Weaver, 2001: Improved techniques for evaluating GCM cloudiness applied to the NCAR CCM3. *J. Clim.*, **14**, 2540–2550.

Norris, J.R., and S.F. Iacobellis, 2005: North pacific cloud feedbacks inferred from synoptic-scale dynamic and thermodynamic relationships. *J. Clim.*, **18**, 4862–4878.

NRC (National Research Council), 2003: *Understanding Climate Change Feedbacks*. National Academies Press, Washington, DC, 152 pp.

O'Farrell, S.P., 1998: Investigation of the dynamic sea ice component of a coupled atmosphere sea-ice general circulation model. *J. Geophys. Res.*, **103**, 15751–15782.

Oki, T., and Y.C. Sud, 1998: Design of total runoff integrating pathways (TRIP)—A global river channel network. *Earth Interactions*, **2**, 1–37.

Oleson, K.W., et al., 2004: *Technical Description of the Community Land Model (CLM)*. NCAR Technical Note NCAR/TN-461+STR, National Center for Atmospheric Research, Boulder, CO, 173 pp.

Oliver, K.I.C., A.J. Watson, and D.P. Stevens, 2005: Can limited ocean mixing buffer rapid climate change? *Tellus*, **57A**, 676–690.

Oouchi, K., et al., 2006: Tropical cyclone climatology in a global-warming climate as simulated in a 20 km-mesh global atmospheric model: Frequency and wind intensity analyses. *J. Meteorol. Soc. Japan*, **84**, 259–276.

Opsteegh, J.D., R.J. Haarsma, F.M. Selten, and A. Kattenberg, 1998: ECBILT: A dynamic alternative to mixed boundary conditions in ocean models. *Tellus*, **50A**, 348–367.

Osborn, T.J., 2004: Simulating the winter North Atlantic Oscillation: the roles of internal variability and greenhouse gas forcing. *Clim. Dyn.*, **22**, 605–623.

Otterå, O.H., et al., 2004: Transient response of the Atlantic meridional overturning circulation to enhanced freshwater input to the Nordic Seas-Arctic Ocean in the Bergen Climate Model. *Tellus*, **56A**, 342–361.

Otto-Bliesner, B.L., et al., 2006: Climate sensitivity of moderate- and low-resolution versions of CCSM3 to preindustrial forcings. *J. Clim.*, **19**, 2567–2583.

Pacanowski, R.C., K. Dixon, and A. Rosati, 1993: *The GFDL Modular Ocean Model Users Guide, Version 1.0*. GFDL Ocean Group Technical Report No. 2, Geophysical Fluid Dynamics Laboratory, Princeton, NJ.

Paciorek, C.J., J.S. Risbey, V. Ventura, and R.D. Rosen, 2002: Multiple indices of Northern Hemisphere cyclone activity, winters 1949-99. *J. Clim.*, **15**, 1573–1590.

Palmer, T.N., and J. Shukla, 2000: Editorial (for special issue on DSP/PROVOST). *Q. J. R. Meteorol. Soc.*, **126**, 1989–1990.

Palmer, T.N., et al., 2004: Development of a European multimodel ensemble system for seasonal to interannual prediction (DEMETER). *Bull. Am. Meteorol. Soc.*, **85**, 853–872.

Pan, Z., et al., 2004: Evaluation of uncertainties in regional climate change simulations. *J. Geophys. Res.*, **106**, 17735–17752.

Pardaens, A.K., H.T. Banks, J.M. Gregory, and P.R. Rowntree, 2003: Freshwater transports in HadCM3. *Clim. Dyn.*, **21**, 177–195.

Parekh, P., M.J. Follows, and E. Boyle, 2005: Decoupling of iron and phosphate in the global ocean. *Global Biogeochem. Cycles*, **19**, doi:10.1029/2004GB002280.

Pelly, J.L., and B.J. Hoskins, 2003a: A new perspective on blocking. *J. Atmos. Sci.*, **60**, 743–755.

Pelly, J.L., and B.J. Hoskins, 2003b: How well does the ECMWF Ensemble Prediction System predict blocking? *Q. J. R. Meteorol. Soc.*, **129**, 1683–1702.

Peters, M.E., and C.S. Bretherton, 2005: A simplified model of the Walker circulation with an interactive ocean mixed layer and cloud-radiative feedbacks. *J. Clim.*, **18**, 4216–4234.

Petoukhov, V., et al., 2000: CLIMBER-2: A climate system model of intermediate complexity. Part I: Model description and performance for present climate. *Clim. Dyn.*, **16**, 1–17.

Petoukhov, V., et al., 2005: EMIC Intercomparison Project (EMIP-CO$_2$): Comparative analysis of EMIC simulations of current climate and equilibrium and transient responses to atmospheric CO$_2$ doubling. *Clim. Dyn.*, **25**, 363–385, doi:10.1007/s00382-005-0042-3.

Phillips, T.J., et al., 2004: Evaluating parameterizations in general circulation models: Climate simulation meets weather prediction. *Bull. Am. Meteorol. Soc.*, **85**, 1903–1915.

Piani, C., D.J. Frame, D.A. Stainforth, and M.R. Allen, 2005: Constraints on climate change from a multi-thousand member ensemble of simulations. *Geophys. Res. Lett.*, **32**, L23825, doi:10.1029/2005GL024452.

Pierce, D.W., T.P. Barnett, and M. Latif, 2000: Connections between the Pacific Ocean tropics and midlatitudes on decadal time scales. *J. Clim.*, **13**, 1173–1194.

Pierrehumbert, R.T., 1995: Thermostats, radiator fins, and the local runaway greenhouse. *J. Atmos. Sci.*, **52**, 1784–180.

Pierrehumbert, R.T., 1999: Subtropical water vapour as a mediator of rapid global climate change. In: *Mechanisms of Global Climate Change at Millennial Timescales*. Geophysical Monograph 112, American Geophysical Union, Washington, DC, pp. 339–361.

Pierrehumbert, R.T., and R. Roca, 1998: Evidence for control of Atlantic subtropical humidity by large scale advection. *Geophys. Res. Lett.*, **25**, 4537–4540.

Pierrehumbert, R.T., H. Brogniez, and R. Roca, 2007: On the relative humidity of the Earth's atmosphere. In: *The General Circulation* [Schneider, T., and A. Sobel (eds.)]. Princeton University Press, Princeton, NJ, in press.

Pitman, A.J., B.J. McAvaney, N. Bagnoud, and B. Cheminat, 2004: Are inter-model differences in AMIP-II near surface air temperature means and extremes explained by land surface energy balance complexity? *Geophys. Res. Lett.*, **31**, L05205, doi:10.1029/2003GL019233.

Plattner, G.-K., F. Joos, T.F. Stocker, and O. Marchal, 2001: Feedback mechanisms and sensitivities of ocean carbon uptake under global warming. *Tellus*, **53B**, 564–592.

Plaut, G., and E. Simonnet, 2001: Large-scale circulation classification, weather regimes, and local climate over France, the Alps, and Western Europe. *Clim. Res.*, **17**, 303–324.

Polzin, K.L., J.M. Toole, J.R. Redwell, and R.W. Schmitt, 1997: Spatial variability of turbulent mixing in the abyssal ocean. *Science*, **276**, 93–96.

Pope, V.D., and R.A. Stratton, 2002: The processes governing horizontal resolution sensitivity in a climate model. *Clim. Dyn.*, **19**, 211–236.

Pope, V.D., M.L. Gallani, P.R. Rowntree, and R.A. Stratton, 2000: The impact of new physical parametrizations in the Hadley Centre climate model: HadAM3. *Clim. Dyn.*, **16**, 123–146.

Potter, G.L., and R.D. Cess, 2004: Testing the impact of clouds on the radiation budgets of 19 atmospheric general circulation models. *J. Geophys. Res.*, **109**, doi:10.1029/2003JD004018.

Power, S.B., and R. Colman, 2006: Multi-decadal predictability in a coupled GCM. *Clim. Dyn.*, **26**, 247–272.

Power, S.B., M.H. Haylock, R. Colman, and X. Wang, 2006: The predictability of interdecadal changes in ENSO activity and ENSO teleconnections. *J. Clim.*, **19**, 4755–4771.

Power, S., et al., 1999: Interdecadal modulation of the impact of ENSO on Australia. *Clim. Dyn.*, **15**, 319–324.

Qu, X., and A. Hall, 2005: Surface contribution to planetary albedo variability in cryosphere regions. *J. Clim.*, **18**, 5239–5252.

Quadrelli, R., and J.M. Wallace, 2004: A simplified linear framework for interpreting patterns of northern hemisphere wintertime climate variability. *J. Clim.*, **17**, 3728–3744.

Rahmstorf, S., 1996: On the freshwater forcing and transport of the Atlantic thermohaline circulation. *Clim. Dyn.*, **12**, 799–811.

Rahmstorf, S., et al., 2005: Thermohaline circulation hysteresis: A model intercomparison. *Geophys. Res. Lett.*, **32**, L23605, doi:10.1029/2005GL023655.

657

Randall, D.A., et al., 2003: Confronting models with data: The GEWEX Cloud Systems Study. *Bull. Am. Meteorol. Soc.*, **84**, 455–469.

Randall, D.A., et al., 2006: Cloud feedbacks. In: *Frontiers in the Science of Climate Modeling* [Kiehl, J.T., and V. Ramanathan (eds.)]. Proceedings of a symposium in honor of Professor Robert D. Cess.

Raper, S.C.B., T.M.L. Wigley, and R.A. Warrick, 1996: Global sea-level rise: past and future. In: *Sea-Level Rise and Coastal Subsidence: Causes, Consequences and Strategies* [Milliman, J.D., and B.U. Haq (eds.)]. Kluwer Academic Publishers, Dordrecht, The Netherlands, pp. 11–46.

Raper, S.C.B., J.M. Gregory, and T.J. Osborn, 2001: Use of an upwelling-diffusion energy balance model to simulate and diagnose A/OGCM results. *Clim. Dyn.*, **17**, 601–613.

Raphael, M.N., and M.M. Holland, 2006: Twentieth century simulation of the Southern Hemisphere climate in coupled models. Part 1: Large scale circulation variability. *Clim. Dyn.*, **26**, 217–228, doi:10.1007/s00382-005-0082-8.

Rayner, N.A., et al., 2003: Global analyses of sea surface temperature, sea ice, and night marine air temperature since the late nineteenth century. *J. Geophys. Res.*, **108**(D14), doi:10.1029/2002JD002670.

Redi, M.H., 1982: Oceanic isopycnal mixing by coordinate rotation. *J. Phys. Oceanogr.*, **12**, 1154–1158.

Renssen, H., V. Brovkin, T. Fichefet, and H. Goosse, 2003: Holocene climate instability during the termination of the African Humid Period. *Geophys. Res. Lett.*, **30**(4), 1184, doi:10.1029/2002GL016636.

Renwick, J.A., 1998: ENSO-related variability in the frequency of South Pacific blocking. *Mon. Weather Rev.*, **126**, 3117–3123.

Rial, J.A., 2004: Abrupt climate change: chaos and order at orbital and millennial scales. *Global Planet. Change*, **41**, 95–109.

Ridley, J.K., P. Huybrechts, J.M. Gregory, and J.A. Lowe, 2005: Elimination of the Greenland ice sheet in a high CO_2 climate. *J. Clim.*, **18**, 3409–3427.

Rind, D.G., et al., 2001: Effects of glacial meltwater in the GISS Coupled Atmosphere-Ocean model: Part II. A bipolar seesaw in deep water production. *J. Geophys. Res.*, **106**, 27355–27365.

Ringer, M.A., and R.P. Allan, 2004: Evaluating climate model simulations of tropical clouds. *Tellus*, **56A**, 308–327.

Ringer, M.A., et al., 2006: The physical properties of the atmosphere in the new Hadley Centre Global Environmental Model (HadGEM1). Part II: Aspects of variability and regional climate. *J. Clim.*, **19**, 1302–1326.

Roberts, M.J., 2004: *The Ocean Component of HadGEM1*. GMR Report Annex IV.D.3, Met Office, Exeter, UK.

Roberts, M., et al., 2004: Impact of an eddy-permitting ocean resolution on control and climate change simulations with a global coupled GCM. *J. Clim.*, **17**, 3–20.

Robertson, A.W., 2001: Influence of ocean-atmosphere interaction on the Arctic Oscillation in two general circulation models. *J. Clim.*, **14**, 3240–3254.

Robock, A., et al., 2000: The global soil moisture data bank. *Bull. Am. Meteorol. Soc.*, **81**, 1281–1299.

Roeckner, E., et al., 1996: *The Atmospheric General Circulation Model ECHAM4: Model Description and Simulation of Present-Day Climate*. MPI Report No. 218, Max-Planck-Institut für Meteorologie, Hamburg, Germany, 90 pp.

Roeckner, E., et al., 2003: *The Atmospheric General Circulation Model ECHAM5. Part I: Model Description*. MPI Report 349, Max Planck Institute for Meteorology, Hamburg, Germany, 127 pp.

Roesch, A., 2006: Evaluation of surface albedo and snow cover in AR4 coupled climate models. *J. Geophys. Res.*, **111**, D15111, doi:10.1029/2005JD006473.

Rooth, C., 1982: Hydrology and ocean circulation. *Prog. Oceanogr.*, **11**, 131–149.

Rosati, A., K. Miyakoda, and R. Gudgel, 1997: The impact of ocean initial conditions on ENSO forecasting with a coupled model. *Mon. Weather Rev.*, **125**(5), 754–772.

Ross, R.J., W.P. Elliott, D.J. Seidel, and participating AMIP-II modelling groups, 2002: Lower tropospheric humidity-temperature relationships in radiosonde observations and atmospheric general circulation models. *J. Hydrometeorol.*, **3**, 26–38.

Russell, G.L., 2005: *4x3 Atmosphere-Ocean Model Documentation*. http://aom.giss.nasa.gov/doc4x3.html.

Russell, G.L., J.R. Miller, and D. Rind, 1995: A coupled atmosphere-ocean model for transient climate change studies. *Atmos.-Ocean*, **33**, 683–730.

Russell, J.L., R.J. Stouffer, and K.W. Dixon, 2006: Intercomparison of the Southern Ocean circulations in IPCC coupled model control simulations. *J. Clim.*, **19**, 4560–4575.

Saenko, O.A., and W.J. Merryfield, 2005: On the effect of topographically-enhanced mixing on the global ocean circulation. *J. Phys. Oceanogr.*, **35**, 826–834.

Saenko, O.A., G.M. Flato, and A.J. Weaver, 2002: Improved representation of sea-ice processes in climate models. *Atmos.-Ocean*, **40**, 21–43.

Sakamoto, T.T., et al., 2004: Far-reaching effects of the Hawaiian Islands in the CCSR/NIES/FRCGC high-resolution climate model. *Geophys. Res. Lett.*, **31**, doi:10.1029/2004GL020907.

Sakamoto, T., et al., 2005: Responses of the Kuroshio and the Kuroshio Extension to global warming in a high-resolution climate model. *Geophys. Res. Lett.*, **32**, L14617, doi:10.1029/2005GL023384.

Salas-Mélia, D., 2002: A global coupled sea ice-ocean model. *Ocean Modelling*, **4**, 137–172.

Saltzman, B., 1978: A survey of statistical-dynamical models of the terrestrial climate. *Adv. Geophys.*, **20**, 183–295.

Santer, B.D., et al., 2005: Amplification of surface temperature trends and variability in the tropical atmosphere. *Science*, **309**, 1551–1556.

Sato, N., et al., 1989: Effects of implementing the simple biosphere model in a general circulation model. *J. Atmos. Sci.*, **46**, 2757–2782.

Sausen, R., K. Barthel, and K. Hasselmann, 1988: Coupled ocean-atmosphere models with flux correction. *Clim. Dyn.*, **2**, 145–163.

Sausen, R., et al., 2002: Climate response to inhomogeneously distributed forcing agents. In: *Non-CO_2 Greenhouse Gases: Scientific Understanding, Control Options and Policy Aspects* [van Ham, J., A.P.M. Baede, R. Guicherit, and J.G.F.M. Williams-Jacobse (eds.)]. Millpress, Rotterdam, Netherlands, pp. 377–381.

Schär, C., et al., 2004: The role of increasing temperature variability for European summer heat waves. *Nature*, **427**, 332–336, doi:10.1038/nature02300.

Scaife, A.A., J.R. Knight, C.K. Folland, and G.K. Vallis, 2005: A stratospheric influence on the winter NAO and North Atlantic surface climate. *Geophys. Res. Lett.*, **32**, L18715.

Scaife, A.A., et al., 2000: Realistic quasi-biennial oscillations in a simulation of the global climate. *Geophys. Res. Lett.*, **27**, 3481–3484.

Schiller, A., U. Mikolajewicz, and R. Voss, 1997: The stability of the North Atlantic thermohaline circulation in a coupled ocean-atmosphere general circulation model. *Clim. Dyn.*, **13**, 325–347.

Schmidt, G.A., C.M. Bitz, U. Mikolajewicz, and L.B. Tremblay, 2004: Ice-ocean boundary conditions for coupled models. *Ocean Modelling*, **7**, 59–74.

Schmidt, G.A., et al., 2006: Present day atmospheric simulations using GISS ModelE: Comparison to in-situ, satellite and reanalysis data. *J. Clim.*, **19**, 153–192, http://www.giss.nasa.gov/tools/modelE/.

Schmittner, A., and T.F. Stocker, 1999: The stability of the thermohaline circulation in global warming experiments. *J. Clim.*, **12**, 1117–1133.

Schmittner, A., C. Appenzeller, and T.F. Stocker, 2000: Enhanced Atlantic freshwater export during El Niño. *Geophys. Res. Lett.*, **27**, 1163–1166.

Schneider, E.K., 2001: Causes of differences between the equatorial Pacific as simulated by two coupled GCM's. *J. Clim.*, **15**, 2301–2320.

Schneider, S.H., 2004: Abrupt non-linear climate change, irreversibility and surprise. *Global Environ. Change*, **14**, 245–258.

Schubert, S., et al., 1992: *Monthly Means of Selected Climate Variables for 1985–1989.* NASA Technical Memorandum, Goddard Space Flight Center, Greenbelt, MD, 376 pp. Available from the NASA Technical Report Server, Accession Number: 92N29653; Document ID: 19920020410; Report Number: NAS 1.15104565, NASA-TM-104565, REPT-92B00088.

Scinocca, J.F., and N.A. McFarlane, 2004: The variability of modelled tropical precipitation. *J. Atmos. Sci.*, **61**, 1993–2015.

Seidel, D.J., and J.R. Lanzante, 2004: An assessment of three alternatives to linear trends for characterizing global atmospheric temperature changes. *J. Geophys. Res.*, **109**, D14108, doi:10.1029/2003JD004414.

Seidov, D., E.J. Barron, and B.J. Haupt, 2001: Meltwater and the global ocean conveyor: Northern versus southern connections. *Global Planet. Change*, **30**, 253–266.

Seidov, D., R.J. Stouffer, and B.J. Haupt, 2005: Is there a simple bi-polar ocean seesaw? *Global Planet. Change*, **49**, 19–27.

Sellers, P.J., Y. Mintz, Y.C. Sud, and A. Dalcher, 1986: A simple biosphere model (SiB) for use within general circulation models. *J. Atmos. Sci.*, **43**, 505–531.

Selten, F.M., and G. Branstator, 2004: Preferred regime transition routes and evidence for an unstable periodic orbit in a baroclinic model. *J. Atmos. Sci.*, **61**, 2267–2268.

Semtner, A.J., 1976: A model for the thermodynamic growth of sea ice in numerical investigations of climate. *J. Phys. Oceanogr.*, **6**, 379–389.

Seneviratne, S.I., J.S. Pal, E.A.B. Eltahir, and C. Schär, 2002: Summer dryness in a warmer climate: A process study with a regional climate model. *Clim. Dyn.*, **20**, 69–85.

Senior, C.A., and J.F.B. Mitchell, 1993: Carbon dioxide and climate: The impact of cloud parameterization. *J. Clim.*, **6**, 393–418.

Senior, C.A., and J.F.B. Mitchell, 2000: The time dependence of climate sensitivity. *Geophys. Res. Lett.*, **27**, 2685–2688.

Severijns, C.A., and W. Hazeleger, 2005: Optimising parameters in an atmospheric general circulation model. *J. Clim.*, **18**, 3527–3535.

Shaffrey, L., and R. Sutton, 2004: The interannual variability of energy transports within and over the Atlantic Ocean in a coupled climate model. *J. Clim.*, **17**, 1433–1448.

Shibata, K., et al., 1999: A simulation of troposphere, stratosphere and mesosphere with an MRI/JMA98 GCM. *Papers in Meteorology and Geophysics*, **50**, 15–53.

Shindell, D.T., R.L. Miller, G.A. Schmidt, and L. Pandolfo, 1999: Simulation of recent northern winter climate trends by greenhouse-gas forcing. *Nature*, **399**, 452–455.

Shiogama, H., M. Watanabe, M. Kimoto, and T. Nozawa, 2005: Anthropogenic and natural forcing impacts on the Pacific Decadal Oscillation during the second half of the 20th century. *Geophys. Res. Lett.*, **32**, L21714, doi:10.1029/2005GL023871.

Shukla, J., et al., 2006: Climate model fidelity and projections of climate change. *Geophys. Res. Lett.*, **33**, L07702, doi:10.1029/2005GL025579.

Sinclair, M.R., 1996: A climatology of anticyclones and blocking for the Southern Hemisphere. *Mon. Weather Rev.*, **124**, 245–263.

Sitch, S., et al., 2003: Evaluation of ecosystem dynamics, plant geography and terrestrial carbon cycling in the LPJ dynamic global vegetation model. *Global Change Biol.*, **9**, 161–185.

Six, K.D., and E. Maier-Reimer, 1996: Effects of plankton dynamics on seasonal carbon fluxes in an ocean general circulation model. *Global Biogeochem. Cycles*, **10**, 559–583.

Slater, A.G., et al., 2001: The representation of snow in land-surface schemes: Results from PILPS 2(d). *J. Hydrometeorol.*, **2**, 7–25.

Slingo, J.M., P.M. Inness, and K.R. Sperber, 2005: Modelling the Madden Julian Oscillation. In: *Intraseasonal Variability of the Atmosphere-Ocean Climate System* [Lau, W.K.-M., and D.E. Waliser (eds.)]. Praxis Publishing.

Slingo, J.M., et al., 1996: Intraseasonal oscillations in 15 atmospheric general circulation models: Results from an AMIP Diagnostic Subproject. *Clim. Dyn.*, **12**, 325–357.

Slingo, J., et al., 2003: Scale interactions on diurnal to seasonal timescales and their relevance to model systematic errors. *Ann. Geophys.*, **46**, 139–155.

Smith, R.D., and P.R. Gent, 2002: *Reference Manual for the Parallel Ocean Program (POP), Ocean Component of the Community Climate System Model (CCSM2.0 and 3.0).* Technical Report LA-UR-02-2484, Los Alamos National Laboratory, Los Alamos, NM, http://www.ccsm.ucar.edu/models/ccsm3.0/pop/.

Soden, B.J., 1997: Variations in the tropical greenhouse effect during El Niño. *J. Clim.*, **10**(5), 1050–1055.

Soden, B.J., 2000: The sensitivity of the tropical hydrological cycle to ENSO. *J. Clim.*, **13**, 538–549.

Soden, B.J., 2004: The impact of tropical convection and cirrus on upper tropospheric humidity: A Lagrangian analysis of satellite measurements. *Geophys. Res. Lett.*, **31**, L20104, doi:10.1029/2004GL020980.

Soden, B.J., and I.M. Held, 2006: An assessment of climate feedbacks in coupled ocean-atmosphere models. *J. Clim.*, **19**, 3354–3360.

Soden, B.J., A.J. Broccoli, and R.S. Hemler, 2004: On the use of cloud forcing to estimate cloud feedback. *J. Clim.*, **17**, 3661–3665.

Soden, B.J., R.T. Wetherald, G.L. Stenchikov, and A. Robock, 2002: Global cooling after the eruption of Mount Pinatubo: A test of climate feedback by water vapour. *Science*, **296**, 727–730.

Soden, B.J., et al., 2005: The radiative signature of upper tropospheric moistening. *Science*, **310**(5749), 841–844.

Sohn, B.-J., and J. Schmetz, 2004: Water vapor-induced OLR variations associated with high cloud changes over the tropics: a study from Meteosat-5 observations. *J. Clim.*, **17**, 1987–1996.

Sokolov, A., and P. Stone, 1998: A flexible climate model for use in integrated assessments. *Clim. Dyn.*, **14**, 291–303.

Sokolov, A.P., et al., 2005: *The MIT Integrated Global System Model (IGSM), Version 2: Model Description And Baseline Evaluation.* Report No. 124, Joint Program on the Science and Policy of Global Change, Massachusetts Institute of Technology, Cambridge, MA, http://web.mit.edu/globalchange/www/MITJPSPGC_Rpt124.pdf.

Spelman, M.J., and S. Manabe, 1984: Influence of oceanic heat transport upon the sensitivity of a model climate. *J. Geophys. Res.*, **89**, 571–586.

Sperber, K.R., S. Gualdi, S. Legutke, and V. Gayler, 2005: The Madden-Julian Oscillation in ECHAM4 coupled and uncoupled GCMs. *Clim. Dyn.*, **25**, doi:10.1007/s00382-005-0026-3.

Stainforth, D.A., et al., 2005: Uncertainty in predictions of the climate response to rising levels of greenhouse gases. *Nature*, **433**, 403–406.

Stein, O., 2000: The variability of Atlantic-European blocking as derived from long SLP time series. *Tellus*, **52A**, 225–236.

Stenchikov, G., et al., 2002: Arctic Oscillation response to the 1991 Mount Pinatubo eruption: Effects of volcanic aerosols and ozone depletion. *J. Geophys. Res.*, **107**(D24), 4803.

Stephens, G.L., 2005: Cloud feedbacks in the climate system: a critical review. *J. Clim.*, **18**, 237–273.

Stephenson, D.B., and V. Pavan, 2003: The North Atlantic Oscillation in coupled climate models: a CMIP1 evaluation. *Clim. Dyn.*, **20**, 381–399.

Stephenson, D.B., A. Hannachi, and A. O'Neill, 2004: On the existence of multiple climate regimes. *Q. J. R. Meteorol. Soc.*, **130**, 583–605.

Stocker, T.F., D.G. Wright, and L.A. Mysak, 1992: A zonally averaged, coupled atmosphere-ocean model for paleoclimate studies. *J. Clim.*, **5**, 773–797.

Stocker, T.F., et al., 2001: Physical climate processes and feedbacks. In: *Climate Change 2001: The Scientific Basis. Contribution of Working Group I to the Third Assessment Report of the Intergovernmental Panel on Climate Change* [Houghton, J.T., et al. (eds.)]. Cambridge University Press, Cambridge, United Kingdom and New York, NY, USA, pp. 419–470.

Stommel, H., 1961: Thermohaline convection with two stable regimes of flow. *Tellus*, **13**, 224–230.

Stouffer, R.J., 2004: Time scales of climate response. *J. Clim.*, **17**(1), 209–217.

Stouffer, R.J., and K.W. Dixon, 1998: *Initialization of Coupled Models for Use in Climate Studies: A Review*. Research Activities in Atmospheric and Oceanic Modelling, Report No. 27, WMO/TD-No. 865, World Meteorological Organization, Geneva, Switzerland, I.1–I.8.

Stouffer, R.J., and S. Manabe, 2003: Equilibrium response of thermohaline circulation to large changes in atmospheric CO_2 concentration. *Clim. Dyn.*, **20**(7/8), 759–773.

Stouffer, R.J., A.J. Weaver, and M. Eby, 2004: A method for obtaining pre-twentieth century initial conditions for use in climate change studies. *Clim. Dyn.*, **23**, 327–339.

Stouffer, R.J., et al., 2006: Investigating the causes of the response of the thermohaline circulation to past and future climate changes. *J. Clim.*, **19**, 1365–1387.

Stowasser, M., and K. Hamilton, 2006: Relationship between shortwave cloud radiative forcing and local meteorological variables compared in observations and several global climate models. *J. Clim.*, **19**, 4344–4359.

Stowasser, M., K. Hamilton, and G.J. Boer, 2006: Local and global climate feedbacks in models with differing climate sensitivity. *J. Clim.*, **19**, 193–209.

Stratton, R.A., and V.D. Pope, 2004: Modelling the climatology of storm tracks - Sensitivity to resolution. In: *The Second Phase of the Atmospheric Model Intercomparison Project (AMIP2)* [Gleckler, P. (ed.)]. Proceedings of the WCRP/WGNE Workshop, Toulouse, pp. 207-210.

Stuber, N., M. Ponater, and R. Sausen, 2001: Is the climate sensitivity to ozone perturbations enhanced by stratospheric water vapor feedback? *Geophys. Res. Lett.*, **28**, doi:10.1029/2001GL013000.

Stuber, N., M. Ponater, and R. Sausen, 2005: Why radiative forcing might fail as a predictor of climate change. *Clim. Dyn.*, **24**, 497–510.

Sud, Y.C., and G.K. Walker, 1999: Microphysics of clouds with the relaxed Arakawa-Schubert Cumulus Scheme (McRAS). Part I: Design and evaluation with GATE Phase III data. *J. Atmos. Sci.*, **56**, 3196–3220.

Sugi, M., A. Noda, and N. Sato, 2002: Influence of the global warming on tropical cyclone climatology: An experiment with the JMA global model. *J. Meteorol. Soc. Japan*, **80**, 249–272.

Sun, D.-Z., and I.M. Held, 1996: A comparison of modeled and observed relationships between interannual variations of water vapor and temperature. *J. Clim.*, **9**, 665–675.

Sun, D.-Z., C. Covey, and R.S. Lindzen, 2001: Vertical correlations of water vapor in GCMs. *Geophys. Res. Lett.*, **28**, 259–262.

Sun, Y., S. Solomon, A. Dai, and R. Portmann, 2006: How often does it rain? *J. Clim.*, **19**, 916–934.

Suzuki, T., et al., 2005: Projection of future sea level and its variability in a high-resolution climate model: Ocean processes and Greenland and Antarctic ice-melt contributions. *Geophys. Res. Lett.*, **32**, L19706, doi:10.1029/2005GL023677.

Takahashi, M., 1996: Simulation of the stratospheric quasi-biennial oscillation using a general circulation model. *Geophys. Res. Lett.*, **23**, 661–664.

Takahashi, M., 1999: The first realistic simulation of the stratospheric quasi-biennial oscillation in a general circulation model. *Geophys. Res. Lett.*, **26**, 1307–1310.

Takemura, T., et al., 2002: Single scattering albedo and radiative forcing of various aerosol species with a global three-dimensional model. *J. Clim.*, **15**, 333–352.

Takemura, T., et al., 2005: Simulation of climate response to aerosol direct and indirect effects with aerosol transport-radiation model. *J. Geophys. Res.*, **110**, D02202, doi:10.1029/2004JD005029.

Tang, Y.M., and M.J. Roberts, 2005: The impact of a bottom boundary layer scheme on the North Atlantic Ocean in a global coupled climate model. *J. Phys. Oceanogr.*, **35**(2), 202–217.

Terray, L., S. Valcke, and A. Piacentini, 1998: *OASIS 2.2 Guide and Reference Manual*. Technical Report TR/CMGC/98-05, Centre Europeen de Recherche et de Formation Avancée en Calcul Scientifique, Toulouse, France.

Thompson, C.J., and D.S. Battisti, 2001: A linear stochastic dynamical model of ENSO. Part II: Analysis. *J. Clim.*, **14**, 445–466.

Thompson, D.W.J., and J.M. Wallace, 2000: Annular modes in the extratropical circulation. Part I: Month-to-month variability. *J. Clim.*, **13**, 1000–1016.

Thompson, D.W.J., and S. Solomon, 2002: Interpretation of recent Southern Hemisphere climate change. *Science*, **296**, 895–899.

Thorndike, A.S., D.A. Rothrock, G.A. Maykut, and R. Colony, 1975: The thickness distribution of sea ice. *J. Geophys. Res.*, **80**, 4501–4513.

Thorpe, R.B., R.A. Wood, and J.F.B. Mitchell, 2004: The sensitivity of the thermohaline circulation response to preindustrial and anthropogenic greenhouse gas forcing to the parameterisation of mixing across the Greenland-Scotland ridge. *Ocean Modelling*, **7**, 259–268.

Thorpe, R.B., et al., 2001: Mechanisms determining the Atlantic thermohaline circulation response to greenhouse gas forcing in a non-flux-adjusted coupled climate model. *J. Clim.*, **14**, 3102–3116.

Tiedtke, M., 1993: Representation of clouds in large-scale models. *Mon. Weather Rev.*, **121**, 3040–3061

Timmermann, A., and H. Goosse, 2004: Is the wind stress forcing essential for the meridional overturning circulation? *Geophys. Res. Lett.*, **31**(4), L04303, doi:10.1029/2003GL018777.

Tomé, A., and P.M.A. Miranda, 2004: Piecewise linear fitting and trend changing points of climate parameters. *Geophys. Res. Lett.*, **31**, L02207, doi:10.1029/2003GL019100.

Tompkins, A., 2002: A prognostic parameterization for the subgrid-scale variability of water vapor and clouds in large-scale models and its use to diagnose cloud cover. *J. Atmos. Sci.*, **59**, 1917–1942.

Tompkins, A.M., and G.C. Craig, 1999: Sensitivity of tropical convection to sea surface temperature in the absence of large-scale flow. *J. Clim.*, **12**, 462–476.

Toyota, T., et al., 2004: Thickness distribution, texture and stratigraphy, and a simple probabilistic model for dynamical thickening of sea ice in the southern Sea of Okhotsk. *J. Geophys. Res.*, **109**, C06001, doi:10.1029/2003JC002090.

Trenberth, K.E., and J.M. Caron, 2001: Estimates of meridional atmosphere and ocean heat transports. *J. Clim.*, **14**, 3433–3443.

Trenberth, K.E., J. Fasullo, and L. Smith, 2005: Trends and variability in column-integrated atmospheric water vapour. *Clim. Dyn.*, **24**, 741–758.

Trenberth, K.E., D.P. Stepaniak, J.W. Hurrel, and M. Fiorino, 2001: Quality of re-analyses in the tropics. *J. Clim.*, **14**, 1499–1510.

Trenberth, K.E., et al., 1998: Progress during TOGA in understanding and modeling global teleconnection associated with tropical sea surface temperatures. *J. Geophys. Res.*, **103**, 14291–14324.

Trigo, R.M., I.F. Trigo, C.C. DaCamra, and T.J. Osborn, 2004: Climate impact of the European winter blocking episodes from the NCEP/NCAR reanalyses. *Clim. Dyn.*, **23**, 17–28.

Tselioudis, G., and W.B. Rossow, 1994: Global, multiyear variations of optical-thickness with temperature in low and cirrus clouds. *Geophys. Res. Lett.*, **21**, 2211–2214

Tselioudis, G., and W.B. Rossow, 2006: Climate feedback implied by observed radiation and precipitation changes with midlatitude storm strength and frequency. *Geophys. Res. Lett.*, **33**, L02704, doi:10.1029/2005GL024513.

Tselioudis, G., Y.-C. Zhang, and W.R. Rossow, 2000: Cloud and radiation variations associated with northern midlatitude low and high sea level pressure regimes. *J. Clim.*, **13**, 312–327.

Tsushima, Y., A. Abe-Ouchi, and S. Manabe, 2005: Radiative damping of annual variation in global mean surface temperature: Comparison between observed and simulated feedback. *Clim. Dyn.*, **24**, 591–597, doi:10.1007/s00382-005-0002-y.

Tsushima, Y., et al., 2006: Importance of the mixed-phase cloud distribution in the control climate for assessing the response of clouds to carbon dioxide increase: a multi-model study. *Clim. Dyn.*, **27**, 113–126, doi:10.1007/s00382-006-0127-7.

Turner, A.G., P.M. Inness and J.M. Slingo, 2005: The role of the basic state in monsoon prediction. *Q. J. R. Meteorol. Soc.*, **131**, 781–804.

Uppala, S.M., et al., 2005: The ERA-40 Reanalysis. *Q. J. R. Meteorol. Soc.*, **131**, 2961–3012, doi:10.1256/qj.04.176.

Valcke, S., E. Guilyardi, and C. Larsson, 2006: PRISM and ENES: A European approach to Earth system modelling. *Concurrency and Computation: Practice and Experience*, **18**(2), 247–262.

Van Oldenborgh, G.J., S.Y. Philip, and M. Collins, 2005: El Nino in a changing climate: a multi-model study. *Ocean Sci.*, **1**, 81–95.

Vallis, G.K., E.P. Gerber, P.J. Kushner, and B.A. Cash, 2004: A mechanism and simple dynamical model of the North Atlantic Oscillation and Annular Modes. *J. Atmos. Sci.*, **61**, 264–280.

Vavrus, S., 2004: The impact of cloud feedbacks on Arctic climate under greenhouse forcing. *J. Clim.*, **17**, 603–615.

Vavrus, S., and S.P. Harrison, 2003: The impact of sea-ice dynamics on the Arctic climate system. *Clim. Dyn.*, **20**, 741–757.

Vavrus, S., J.E. Walsh, W.L. Chapman, and D. Portis, 2006: The behavior of extreme cold air outbreaks under greenhouse warming. *Int. J. Climatol.*, **26**, 1133–1147.

Vellinga, M., and R.A. Wood, 2002: Global climate impacts of a collapse of the Atlantic thermohaline circulation. *Clim. Change*, **54**, 251–267.

Vellinga, M., R.A. Wood, and J.M. Gregory, 2002: Processes governing the recovery of a perturbed thermohaline circulation in HadCM3. *J. Clim.*, **15**, 764–780.

Verseghy, D.L., N.A. McFarlane, and M. Lazare, 1993: A Canadian land surface scheme for GCMs: II. Vegetation model and coupled runs. *Int. J. Climatol.*, **13**, 347–370.

Visbeck, M., J. Marshall, T. Haine, and M. Spall, 1997: Specification of eddy transfer coefficients in coarse-resolution ocean circulation models. *J. Phys. Oceanogr.*, **27**, 381–402.

Volodin, E.M., 2004: Relation between the global-warming parameter and the heat balance on the Earth's surface at increased contents of carbon dioxide. *Izv. Atmos. Ocean. Phys.*, **40**, 269–275.

Volodin, E.M., and V.N. Lykossov, 1998: Parameterization of heat and moisture processes in the soil-vegetation system: 1. Formulation and simulations based on local observational data. *Izv. Atmos. Ocean. Phys.*, **34**(4), 453–465.

Volodin, E.M., and N.A. Diansky, 2004: El-Niño reproduction in a coupled general circulation model of atmosphere and ocean. *Russ. Meteorol. Hydrol.*, **12**, 5–14.

Waliser, D.E., K.M. Lau, and J.H. Lim, 1999: The influence of coupled sea surface temperatures on the Madden–Julian oscillation: A model perturbation experiment. *J. Atmos. Sci.*, **56**, 333–358.

Wallace, J.M., Y. Zhang, and L. Bajuk, 1996: Interpretation of interdecadal trends in Northern Hemisphere surface air temperature. *J. Clim.*, **9**, 249–259.

Walsh, J.E., et al., 2002: Comparison of Arctic climate simulations by uncoupled and coupled global models. *J. Clim.*, **15**, 1429–1446.

Walsh, K.J.E., K.C. Nguyen and J.L. McGregor, 2004: Fine-resolution regional climate model simulations of the impact of climate change on tropical cyclones near Australia. *Clim. Dyn.*, **22**, 47–56.

Wang, B., et al., 2004: Design of a new dynamical core for global atmospheric models based on some efficient numerical methods. *Science in China, Ser. A*, **47** Suppl., 4–21.

Wang, G.L., and E.A.B. Eltahir, 2000: Ecosystem dynamics and the Sahel drought. *Geophys. Res. Lett.*, **27**, 795–798.

Wang, J., H.L. Cole, and D.J. Carlson, 2001: Water vapor variability in the tropical western Pacific from 20-year radiosonde data. *Adv. Atmos. Sci.*, **18**(5), 752–766.

Wang, L.R., and M. Ikeda, 2004: A Lagrangian description of sea ice dynamics using the finite element method. *Ocean Modelling*, **7**, 21–38.

Wang, S., R.F. Grant, D.L. Verseghy, and T.A. Black, 2002: Modelling carbon dynamics of boreal forest ecosystems using the Canadian land surface scheme. *Clim. Change*, **55**, 451–477.

Wang, W., and M. Schlesinger, 1999: The dependence on convection parameterization of the tropical intraseasonal oscillation simulated by the UIUC 11-layer atmospheric GCM. *J. Clim.*, **12**, 1423–1457.

Wang, X.L.L., V.R. Swai, and F.W. Zwiers, 2006: Climatology and changes of extratropical cyclone activity: Comparison of ERA-40 with NCEP-NCAR reanalysis for 1958-2001. *J. Clim.*, **19**, 3145–3166.

Warrach, K., H.T. Mengelkamp, and E. Raschke, 2001: Treatment of frozen soil and snow cover in the land surface model SEWAB. *Theor. Appl. Climatol.*, **69**(1–2), 23–37.

Washington, W.M., et al., 2000: Parallel Climate Model (PCM) control and transient simulations. *Clim. Dyn.*, **16**, 755–774.

Watterson, I.G., 2001: Zonal wind vacillation and its interaction with the ocean: Implications for interannual variability and predictability. *J. Geophys. Res.*, **106**, 23965–23975.

Watterson, I.G., 2006: The intensity of precipitation during extratropical cyclones in global warming simulations: a link to cyclone intensity? *Tellus*, **58A**, 82–97.

Weare, B.C., 2004: A comparison of AMIP II model cloud layer properties with ISCCP D2 estimates. *Clim. Dyn.*, **22**, 281–292.

Weaver, A.J., O.A. Saenko, P.U. Clark, and J.X. Mitrovica, 2003: Meltwater pulse 1A from Antarctica as a trigger of the Bølling-Allerød warm interval. *Science*, **299**, 1709–1713.

Weaver, A.J., et al., 2001: The UVic Earth System Climate Model: Model description, climatology and application to past, present and future climates. *Atmos.-Ocean*, **39**, 361–428.

Webb, M., C. Senior, S. Bony, and J.-J. Morcrette, 2001: Combining ERBE and ISCCP data to assess clouds in the Hadley Centre ECMWF and LMD atmospheric climate models. *Clim. Dyn.*, **17**, 905–922.

Webb, M.J., et al., 2006: On the contribution of local feedback mechanisms to the range of climate sensitivity in two GCM ensembles. *Clim. Dyn.*, **27**, 17–38.

Wentz, F.J., and M. Schabel, 2000: Precise climate monitoring using complementary satellite data sets. *Nature*, **403**, 414–416.

Wigley, T.M.L., and S.C.B. Raper, 1992: Implications for climate and sea level of revised IPCC emissions scenarios. *Nature*, **357**, 293–300.

Wigley, T.M.L., and S.C.B. Raper, 2001: Interpretation of high projections for global-mean warming. *Science*, **293**, 451–454.

Wild, M., 2005: Solar radiation budgets in atmospheric model intercomparisons from a surface perspective. *Geophys. Res. Lett.*, **32**, doi:10.1029/2005GL022421.

Wild, M., C.N. Long, and A. Ohmura, 2006: Evaluation of clear-sky solar fluxes in GCMs participating in AMIP and IPCC-AR4 from a surface perspective. *J. Geophys. Res.*, **111**, D01104, doi:10.1029/2005JD006118.

Wild, M., et al., 2001: Downward longwave radiation in General Circulation Models. *J. Clim.*, **14**, 3227–3239.

Williams, K.D., M.A. Ringer, and C.A. Senior, 2003: Evaluating the cloud response to climate change and current climate variability. *Clim. Dyn.*, **20**(7–8), 705–721.

Williams, K.D., et al., 2006: Evaluation of a component of the cloud response to climate change in an intercomparison of climate models. *Clim. Dyn.*, **26**, 145–165.

Williamson, D.L., et al., 2005: Moisture and temperature balances at the Atmospheric Radiation Measurement Southern Great Plains Site in forecasts with the Community Atmosphere Model (CAM2). *J. Geophys. Res.*, **110**, D15S16, doi:10.1029/2004JD00510.

Winton, M., 2000: A reformulated three-layer sea ice model. *J. Atmos. Ocean. Technol.*, **17**(4), 525–531.

Winton, M., 2006a: Surface albedo feedback estimates for the AR4 climate models. *J. Clim.*, **19**, 359–365.

Winton, M., 2006b: Amplified Arctic climate change: what does surface albedo feedback have to do with it? *Geophys. Res. Lett.*, **33**, L03701, doi:10.1029/2005GL025244.

Winton, M., R. Hallberg, and A. Gnanadesikan, 1998: Simulation of density-driven frictional downslope flow in z-coordinate ocean models. *J. Phys. Oceanogr.*, **28**, 2163–2174.

Wittenberg, A.T., A. Rosati, N.-C. Lau, and J.J. Ploshay, 2006: GFDL's CM2 global coupled climate models, Part 3: Tropical Pacific climate and ENSO. *J. Clim.*, **19**, 698–722.

Wolff, J.-O., E. Maier-Reimer, and S. Lebutke, 1997: *The Hamburg Ocean Primitive Equation Model*. DKRZ Technical Report No. 13, Deutsches KlimaRechenZentrum, Hamburg, Germany, 100 pp., http://www.mad.zmaw.de/Pingo/reports/ReportNo.13.pdf.

Wood, R.A., A.B. Keen, J.F.B. Mitchell, and J.M. Gregory, 1999: Changing spatial structure of the thermohaline circulation in response to atmospheric CO_2 forcing in a climate model. *Nature*, **399**, 572–575.

Wright, D.G., and T.F. Stocker, 1992: Sensitivities of a zonally averaged global ocean circulation model. *J. Geophys. Res.*, **97**, 12707–12730.

Wright, D.G., and T.F. Stocker, 1993: Younger Dryas experiments. In: *Ice in the Climate System, NATO ASI Series, 112* [Peltier, R. (ed.)]. Springer-Verlag, London, pp. 395–416.

Wu, P., R.A. Wood, and P. Stott, 2005: Human influence on increasing Arctic river discharges. *Geophys. Res. Lett.*, **32**, L02703, doi:10.1029/2004GL021570.

Wu, Q., and D.M. Straus, 2004a: On the existence of hemisphere-wide climate variations. *J. Geophys. Res.*, **109**, D06118, doi:10.1029/2003JD004230.

Wu, Q., and D.M. Straus, 2004b: AO, COWL, and observed climate trends. *J. Clim.*, **17**, 2139–2156.

Wunsch, C., 2002: What is the thermohaline circulation? *Science*, **298**, 1179–1180.

Wyant, M.C., et al., 2006: A comparison of low-latitude cloud properties and their response to climate change in three US AGCMs sorted into regimes using mid-tropospheric vertical velocity. *Clim. Dyn.*, **27**, 261–279.

Xie, P., and P.A. Arkin, 1997: Global precipitation: A 17-year monthly analysis based on gauge observations, satellite estimates, and numerical model outputs. *Bull. Am. Meteorol. Soc.*, **78**, 2539–2558.

Xie, S.-P., W.T. Liu, Q. Liu and M. Nonaka, 2001: Far-reaching effects of the Hawaiian Islands on the Pacific ocean-atmosphere system. *Science*, **292**, 2057–2060.

Xu, Y., et al., 2005: Detection of climate change in the 20th century by the NCC T63. *Acta Meteorol. Sin.*, Special Report on Climate Change, **4**, 1–15.

Yang, G.Y., and J. Slingo, 2001: The diurnal cycle in the tropics. *Mon. Weather Rev.*, **129**, 784–801.

Yang, G.Y., B. Hoskins, and J. Slingo, 2003: Convectively coupled equatorial waves: A new methodology for identifying wave structures in observational data. *J. Atmos. Sci.*, **60**, 1637–1654.

Yao, M.-S., and A. Del Genio, 2002: Effects of cloud parameterization on the simulation of climate changes in the GISS GCM. Part II: Sea surface temperature and cloud feedbacks. *J. Clim.*, **15**, 2491–2503.

Yeh, P. J.-F., and E.A.B. Eltahir, 2005: Representation of water table dynamics in a land surface scheme. Part 1. Model development. *J. Clim.*, **18**, 1861–1880.

Yeh, S.-W., and B.P. Kirtman, 2004: Decadal North Pacific sea surface temperature variability and the associated global climate anomalies in a coupled GCM. *J. Geophys. Res.*, **109**, D20113, doi:10.1029/2004JD004785.

Yin, H., 2005: A consistent poleward shift of the storm tracks in simulations of 21st century climate. *Geophys. Res. Lett.*, **32**, L18701, doi:10.1029/2005GL023684.

Yiou, P., and M. Nogaj, 2004: Extreme climatic events and weather regimes over the North Atlantic: When and where? *Geophys. Res. Lett.*, **31**, doi:10.1029/2003GL019119.

Yokohata, T., et al., 2005: Climate response to volcanic forcing: Validation of climate sensitivity of a coupled atmosphere-ocean general circulation model. *Geophys. Res. Lett.*, **32**, L21710, doi:10.1029/2005GL023542.

Yoshimura, J., M. Sugi, and A. Noda, 2006: Influence of greenhouse warming on tropical cyclone frequency. *J. Meteorol. Soc. Japan*, **84**, 405–428.

Yoshizaki, M., et al., 2005: Changes of Baui (Mei-yu) frontal activity in the global warming climate simulated by a non-hydrostatic regional model. *Scientific Online Letters on the Atmosphere*, **1**, 25–28.

Yu, Y., and X. Zhang, 2000: Coupled schemes of flux adjustments of the air and sea. In: *Investigations on the Model System of the Short-Term Climate Predictions* [Ding, Y., et al. (eds.)]. China Meteorological Press, Beijing, China, pp. 201–207 (in Chinese).

Yu, Y., Z. Zhang, and Y. Guo, 2004: Global coupled ocean-atmosphere general circulation models in LASG/IAP. *Adv. Atmos. Sci.*, **21**, 444–455.

Yu, Y., R. Yu, X. Zhang, and H. Liu, 2002: A flexible global coupled climate model. *Adv. Atmos. Sci.*, **19**(1), 169–190.

Yukimoto, S., and A. Noda, 2003: *Improvements of the Meteorological Research Institute Global Ocean-Atmosphere Coupled GCM (MRI-GCM2) and its Climate Sensitivity*. CGER's Supercomputing Activity Report, National Institute for Environmental Studies, Ibaraki, Japan.

Yukimoto, S., et al., 2001: The new Meteorological Research Institute global ocean-atmosphere coupled GCM (MRI-CGCM2)--Model climate and variability. *Papers in Meteorology and Geophysics*, **51**, 47–88.

Zhang, C., 2005: Madden-Julian Oscillation. *Rev. Geophys.*, **43**, RG2003, doi:10.1029/2004RG000158.

Zhang, C., B. Mapes, and B.J. Soden, 2003: Bimodality of water vapour. *Q. J. R. Meteorol. Soc.*, **129**, 2847–2866.

Zhang, J., and D. Rothrock, 2001: A thickness and enthalpy distribution sea-ice model. *J. Phys. Oceanogr.*, **31**, 2986–3001.

Zhang, J., and D. Rothrock, 2003: Modeling global sea ice with a thickness and enthalpy distribution model in generalized curvilinear coordinates. *Mon. Weather Rev.*, **131**, 845–861.

Zhang, M., 2004: Cloud-climate feedback: how much do we know? In: *Observation, Theory, and Modeling of Atmospheric Variability, World Scientific Series on Meteorology of East Asia, Vol. 3* [Zhu et al. (eds.)]. World Scientific Publishing Co., Singapore, 632 pp.

Zhang, M.H., R.D. Cess, J.J. Hack, and J.T. Kiehl, 1994: Diagnostic study of climate feedback processed in atmospheric general circulation models. *J. Geophys. Res.*, **99**, 5525–5537.

Zhang, M.H., et al., 2005: Comparing clouds and their seasonal variations in 10 atmospheric general circulation models with satellite measurements. *J. Geophys. Res.*, **110**, D15S02, doi:10.1029/2004JD005021.

Zhang, X., and J.E. Walsh, 2006: Toward a seasonally ice-covered Arctic Ocean: scenarios from the IPCC AR4 model simulations. *J. Clim.*, **19**, 1730–1747.

Zhang, Y., W. Maslowski, and A.J. Semtner, 1999: Impacts of mesoscale ocean currents on sea ice in high-resolution Arctic ice and ocean simulations. *J. Geophys. Res.*, **104**(C8), 18409–18429.

Zhu, Y., R.E. Newell, and W.G. Read, 2000: Factors controlling upper-troposphere water vapour. *J. Clim.*, **13**, 836–848.

9

Understanding and Attributing Climate Change

Coordinating Lead Authors:

Gabriele C. Hegerl (USA, Germany), Francis W. Zwiers (Canada)

Lead Authors:

Pascale Braconnot (France), Nathan P. Gillett (UK), Yong Luo (China), Jose A. Marengo Orsini (Brazil, Peru), Neville Nicholls (Australia), Joyce E. Penner (USA), Peter A. Stott (UK)

Contributing Authors:

M. Allen (UK), C. Ammann (USA), N. Andronova (USA), R.A. Betts (UK), A. Clement (USA), W.D. Collins (USA), S. Crooks (UK), T.L. Delworth (USA), C. Forest (USA), P. Forster (UK), H. Goosse (Belgium), J.M. Gregory (UK), D. Harvey (Canada), G.S. Jones (UK), F. Joos (Switzerland), J. Kenyon (USA), J. Kettleborough (UK), V. Kharin (Canada), R. Knutti (Switzerland), F.H. Lambert (UK), M. Lavine (USA), T.C.K. Lee (Canada), D. Levinson (USA), V. Masson-Delmotte (France), T. Nozawa (Japan), B. Otto-Bliesner (USA), D. Pierce (USA), S. Power (Australia), D. Rind (USA), L. Rotstayn (Australia), B. D. Santer (USA), C. Senior (UK), D. Sexton (UK), S. Stark (UK), D.A. Stone (UK), S. Tett (UK), P. Thorne (UK), R. van Dorland (The Netherlands), M. Wang (USA), B. Wielicki (USA), T. Wong (USA), L. Xu (USA, China), X. Zhang (Canada), E. Zorita (Germany, Spain)

Review Editors:

David J. Karoly (USA, Australia), Laban Ogallo (Kenya), Serge Planton (France)

This chapter should be cited as:

Hegerl, G.C., F. W. Zwiers, P. Braconnot, N.P. Gillett, Y. Luo, J.A. Marengo Orsini, N. Nicholls, J.E. Penner and P.A. Stott, 2007: Understanding and Attributing Climate Change. In: *Climate Change 2007: The Physical Science Basis. Contribution of Working Group I to the Fourth Assessment Report of the Intergovernmental Panel on Climate Change* [Solomon, S., D. Qin, M. Manning, Z. Chen, M. Marquis, K.B. Averyt, M. Tignor and H.L. Miller (eds.)]. Cambridge University Press, Cambridge, United Kingdom and New York, NY, USA.

Table of Contents

Supplementary Material

*The following supplementary material is available on CD-ROM and
in on-line versions of this report.*

Appendix 9.B: *Methods Used to Estimate Climate Sensitivity and Aerosol
Forcing*

Appendix 9.C: *Notes and technical details on Figures displayed in Chapter 9*

Appendix 9.D: *Additional Figures and Tables*

References for Appendices 9.B to 9.D

Executive Summary

Evidence of the effect of external influences on the climate system has continued to accumulate since the Third Assessment Report (TAR). The evidence now available is substantially stronger and is based on analyses of widespread temperature increases throughout the climate system and changes in other climate variables.

Human-induced warming of the climate system is widespread. Anthropogenic warming of the climate system can be detected in temperature observations taken at the surface, in the troposphere and in the oceans. Multi-signal detection and attribution analyses, which quantify the contributions of different natural and anthropogenic forcings to observed changes, show that greenhouse gas forcing alone during the past half century would *likely* have resulted in greater than the observed warming if there had not been an offsetting cooling effect from aerosol and other forcings.

It is *extremely unlikely* (<5%) that the global pattern of warming during the past half century can be explained without external forcing, and *very unlikely* that it is due to known natural external causes alone. The warming occurred in both the ocean and the atmosphere and took place at a time when natural external forcing factors would *likely* have produced cooling.

Greenhouse gas forcing has *very likely* caused most of the observed global warming over the last 50 years. This conclusion takes into account observational and forcing uncertainty, and the possibility that the response to solar forcing could be underestimated by climate models. It is also robust to the use of different climate models, different methods for estimating the responses to external forcing and variations in the analysis technique.

Further evidence has accumulated of an anthropogenic influence on the temperature of the free atmosphere as measured by radiosondes and satellite-based instruments. The observed pattern of tropospheric warming and stratospheric cooling is *very likely* due to the influence of anthropogenic forcing, particularly greenhouse gases and stratospheric ozone depletion. The combination of a warming troposphere and a cooling stratosphere has *likely* led to an increase in the height of the tropopause. It is *likely* that anthropogenic forcing has contributed to the general warming observed in the upper several hundred meters of the ocean during the latter half of the 20th century. Anthropogenic forcing, resulting in thermal expansion from ocean warming and glacier mass loss, has *very likely* contributed to sea level rise during the latter half of the 20th century. It is difficult to quantify the contribution of anthropogenic forcing to ocean heat content increase and glacier melting with presently available detection and attribution studies.

It is *likely* that there has been a substantial anthropogenic contribution to surface temperature increases in every continent except Antarctica since the middle of the 20th century. Anthropogenic influence has been detected in every continent except Antarctica (which has insufficient observational coverage to make an assessment), and in some sub-continental land areas. The ability of coupled climate models to simulate the temperature evolution on continental scales and the detection of anthropogenic effects on each of six continents provides stronger evidence of human influence on the global climate than was available at the time of the TAR. No climate model that has used natural forcing only has reproduced the observed global mean warming trend or the continental mean warming trends in all individual continents (except Antarctica) over the second half of the 20th century.

Difficulties remain in attributing temperature changes on smaller than continental scales and over time scales of less than 50 years. Attribution at these scales, with limited exceptions, has not yet been established. Averaging over smaller regions reduces the natural variability less than does averaging over large regions, making it more difficult to distinguish between changes expected from different external forcings, or between external forcing and variability. In addition, temperature changes associated with some modes of variability are poorly simulated by models in some regions and seasons. Furthermore, the small-scale details of external forcing, and the response simulated by models are less credible than large-scale features.

Surface temperature extremes have *likely* been affected by anthropogenic forcing. Many indicators of climate extremes and variability, including the annual numbers of frost days, warm and cold days, and warm and cold nights, show changes that are consistent with warming. An anthropogenic influence has been detected in some of these indices, and there is evidence that anthropogenic forcing may have substantially increased the risk of extremely warm summer conditions regionally, such as the 2003 European heat wave.

There is evidence of anthropogenic influence in other parts of the climate system. Anthropogenic forcing has *likely* contributed to recent decreases in arctic sea ice extent and to glacier retreat. The observed decrease in global snow cover extent and the widespread retreat of glaciers are consistent with warming, and there is evidence that this melting has *likely* contributed to sea level rise.

Trends over recent decades in the Northern and Southern Annular Modes, which correspond to sea level pressure reductions over the poles, are *likely* related in part to human activity, affecting storm tracks, winds and temperature patterns in both hemispheres. Models reproduce the sign of the Northern Annular Mode trend, but the simulated response is smaller than observed. Models including both greenhouse gas and stratospheric ozone changes simulate a realistic trend in the Southern Annular Mode, leading to a detectable human influence on global sea level pressure patterns.

The response to volcanic forcing simulated by some models is detectable in global annual mean land precipitation during the latter half of the 20th century. The latitudinal pattern of change in land precipitation and observed increases in heavy precipitation over the 20th century appear to be consistent with the anticipated response to anthropogenic forcing. It is *more likely than not* that anthropogenic influence has contributed to increases in the frequency of the most intense tropical cyclones. Stronger attribution to anthropogenic factors is not possible at present because the observed increase in the proportion of such storms appears to be larger than suggested by either theoretical or modelling studies and because of inadequate process knowledge, insufficient understanding of natural variability, uncertainty in modelling intense cyclones and uncertainties in historical tropical cyclone data.

Analyses of palaeoclimate data have increased confidence in the role of external influences on climate. Coupled climate models used to predict future climate have been used to understand past climatic conditions of the Last Glacial Maximum and the mid-Holocene. While many aspects of these past climates are still uncertain, key features have been reproduced by climate models using boundary conditions and radiative forcing for those periods. A substantial fraction of the reconstructed Northern Hemisphere inter-decadal temperature variability of the seven centuries prior to 1950 is *very likely* attributable to natural external forcing, and it is *likely* that anthropogenic forcing contributed to the early 20th-century warming evident in these records.

Estimates of the climate sensitivity are now better constrained by observations. Estimates based on observational constraints indicate that it is *very likely* that the equilibrium climate sensitivity is larger than 1.5°C with a most likely value between 2°C and 3°C. The upper 95% limit remains difficult to constrain from observations. This supports the overall assessment based on modelling and observational studies that the equilibrium climate sensitivity is *likely* 2°C to 4.5°C with a most likely value of approximately 3°C (Box 10.2). The transient climate response, based on observational constraints, is *very likely* larger than 1°C and *very unlikely* to be greater than 3.5°C at the time of atmospheric CO_2 doubling in response to a 1% yr^{-1} increase in CO_2, supporting the overall assessment that the transient climate response is *very unlikely* greater than 3°C (Chapter 10).

Overall consistency of evidence. Many observed changes in surface and free atmospheric temperature, ocean temperature and sea ice extent, and some large-scale changes in the atmospheric circulation over the 20th century are distinct from internal variability and consistent with the expected response to anthropogenic forcing. The simultaneous increase in energy content of all the major components of the climate system as well as the magnitude and pattern of warming within and across the different components supports the conclusion that the cause of the warming is *extremely unlikely* (<5%) to be the result of internal processes. Qualitative consistency is also apparent in some other observations, including snow cover, glacier retreat and heavy precipitation.

Remaining uncertainties. Further improvements in models and analysis techniques have led to increased confidence in the understanding of the influence of external forcing on climate since the TAR. However, estimates of some radiative forcings remain uncertain, including aerosol forcing and inter-decadal variations in solar forcing. The net aerosol forcing over the 20th century from inverse estimates based on the observed warming likely ranges between –1.7 and –0.1 W m^{-2}. The consistency of this result with forward estimates of total aerosol forcing (Chapter 2) strengthens confidence in estimates of total aerosol forcing, despite remaining uncertainties. Nevertheless, the robustness of surface temperature attribution results to forcing and response uncertainty has been evaluated with a range of models, forcing representations and analysis procedures. The potential impact of the remaining uncertainties has been considered, to the extent possible, in the overall assessment of every line of evidence listed above. There is less confidence in the understanding of forced changes in other variables, such as surface pressure and precipitation, and on smaller spatial scales.

Better understanding of instrumental and proxy climate records, and climate model improvements, have increased confidence in climate model-simulated internal variability. However, uncertainties remain. For example, there are apparent discrepancies between estimates of ocean heat content variability from models and observations. While reduced relative to the situation at the time of the TAR, uncertainties in the radiosonde and satellite records still affect confidence in estimates of the anthropogenic contribution to tropospheric temperature change. Incomplete global data sets and remaining model uncertainties still restrict understanding of changes in extremes and attribution of changes to causes, although understanding of changes in the intensity, frequency and risk of extremes has improved.

9.1   Introduction

The objective of this chapter is to assess scientific understanding about the extent to which the observed climate changes that are reported in Chapters 3 to 6 are expressions of natural internal climate variability and/or externally forced climate change. The scope of this chapter includes 'detection and attribution' but is wider than that of previous detection and attribution chapters in the Second Assessment Report (SAR; Santer et al., 1996a) and the Third Assessment Report (TAR; Mitchell et al., 2001). Climate models, physical understanding of the climate system and statistical tools, including formal climate change detection and attribution methods, are used to interpret observed changes where possible. The detection and attribution research discussed in this chapter includes research on regional scales, extremes and variables other than temperature. This new work is placed in the context of a broader understanding of a changing climate. However, the ability to interpret some changes, particularly for non-temperature variables, is limited by uncertainties in the observations, physical understanding of the climate system, climate models and external forcing estimates. Research on the impacts of these observed climate changes is assessed by Working Group II of the IPCC.

9.1.1   What are Climate Change and Climate Variability?

'Climate change' refers to a change in the state of the climate that can be identified (e.g., using statistical tests) by changes in the mean and/or the variability of its properties, and that persists for an extended period, typically decades or longer (see Glossary). Climate change may be due to internal processes and/or external forcings. Some external influences, such as changes in solar radiation and volcanism, occur naturally and contribute to the total natural variability of the climate system. Other external changes, such as the change in composition of the atmosphere that began with the industrial revolution, are the result of human activity. A key objective of this chapter is to understand climate changes that result from anthropogenic and natural external forcings, and how they may be distinguished from changes and variability that result from internal climate system processes.

Internal variability is present on all time scales. Atmospheric processes that generate internal variability are known to operate on time scales ranging from virtually instantaneous (e.g., condensation of water vapour in clouds) up to years (e.g., troposphere-stratosphere or inter-hemispheric exchange). Other components of the climate system, such as the ocean and the large ice sheets, tend to operate on longer time scales. These components produce internal variability of their own accord and also integrate variability from the rapidly varying atmosphere (Hasselmann, 1976). In addition, internal variability is produced by coupled interactions between components, such as is the case with the El-Niño Southern Oscillation (ENSO; see Chapters 3 and 8).

Distinguishing between the effects of external influences and internal climate variability requires careful comparison between observed changes and those that are expected to result from external forcing. These expectations are based on physical understanding of the climate system. Physical understanding is based on physical principles. This understanding can take the form of conceptual models or it might be quantified with climate models that are driven with physically based forcing histories. An array of climate models is used to quantify expectations in this way, ranging from simple energy balance models to models of intermediate complexity to comprehensive coupled climate models (Chapter 8) such as those that contributed to the multi-model data set (MMD) archive at the Program for Climate Model Diagnosis and Intercomparison (PCMDI). The latter have been extensively evaluated by their developers and a broad investigator community. The extent to which a model is able to reproduce key features of the climate system and its variations, for example the seasonal cycle, increases its credibility for simulating changes in climate.

The comparison between observed changes and those that are expected is performed in a number of ways. Formal detection and attribution (Section 9.1.2) uses objective statistical tests to assess whether observations contain evidence of the expected responses to external forcing that is distinct from variation generated within the climate system (internal variability). These methods generally do not rely on simple linear trend analysis. Instead, they attempt to identify in observations the responses to one or several forcings by exploiting the time and/or spatial pattern of the expected responses. The response to forcing does not necessarily evolve over time as a linear trend, either because the forcing itself may not evolve in that way, or because the response to forcing is not necessarily linear.

The comparison between model-simulated and observed changes, for example, in detection and attribution methods (Section 9.1.2), also carefully accounts for the effects of changes over time in the availability of climate observations to ensure that a detected change is not an artefact of a changing observing system. This is usually done by evaluating climate model data only where and when observations are available, in order to mimic the observational system and avoid possible biases introduced by changing observational coverage.

9.1.2   What are Climate Change Detection and Attribution?

The concepts of climate change 'detection' and 'attribution' used in this chapter remain as they were defined in the TAR (IPCC, 2001; Mitchell et al., 2001). 'Detection' is the process of demonstrating that climate has changed in some defined statistical sense, without providing a reason for that change (see Glossary). In this chapter, the methods used to identify change in observations are based on the expected responses to external forcing (Section 9.1.1), either from physical understanding or as simulated by climate models. An identified change is 'detected' in observations if its likelihood of occurrence by chance due to internal variability alone is determined to be small. A failure to detect a particular response might occur for a number of reasons, including the possibility that the response is weak relative to internal variability, or that the metric used to measure change is insensitive to the expected change. For example, the annual global mean precipitation may not be a sensitive indicator of the influence of increasing greenhouse concentrations given the expectation that greenhouse forcing would result in moistening at some latitudes that is partially offset by drying elsewhere (Chapter 10; see also Section 9.5.4.2). Furthermore, because detection studies are statistical in nature, there is always some small possibility of spurious detection. The risk of such a possibility is reduced when corroborating lines of evidence provide a physically consistent view of the likely cause for the detected changes and render them less consistent with internal variability (see, for example, Section 9.7).

Many studies use climate models to predict the expected responses to external forcing, and these predictions are usually represented as patterns of variation in space, time or both (see Chapter 8 for model evaluation). Such patterns, or 'fingerprints', are usually derived from changes simulated by a climate model in response to forcing. Physical understanding can also be used to develop conceptual models of the anticipated pattern of response to external forcing and the consistency between responses in different variables and different parts of the climate system. For example, precipitation and temperature are ordinarily inversely correlated in some regions, with increases in temperature corresponding to drying conditions. Thus, a warming trend in such a region that is not associated with rainfall change may indicate an external influence on the climate of that region (Nicholls et al., 2005; Section 9.4.2.3). Purely diagnostic approaches can also be used. For example, Schneider and Held (2001) use a technique that discriminates between slow changes in climate and shorter time-scale variability to identify in observations a pattern of surface temperature change that is consistent with the expected pattern of change from anthropogenic forcing.

The spatial and temporal scales used to analyse climate change are carefully chosen so as to focus on the spatio-temporal scale of the response, filter out as much internal variability as possible (often by using a metric that reduces the influence of internal variability, see Appendix 9.A) and enable the separation of the responses to different forcings. For example, it is expected that greenhouse gas forcing would cause a large-scale pattern of warming that evolves slowly over time, and thus analysts often smooth data to remove small-scale variations. Similarly, when fingerprints from Atmosphere-Ocean General Circulation Models (AOGCMs) are used, averaging over an ensemble of coupled model simulations helps separate the model's response to forcing from its simulated internal variability.

Detection does not imply attribution of the detected change to the assumed cause. 'Attribution' of causes of climate change is the process of establishing the most likely causes for the detected change with some defined level of confidence (see Glossary). As noted in the SAR (IPCC, 1996) and the TAR (IPCC, 2001), unequivocal attribution would require controlled experimentation with the climate system. Since that is not possible, in practice attribution of anthropogenic climate change is understood to mean demonstration that a detected change is 'consistent with the estimated responses to the given combination of anthropogenic and natural forcing' and 'not consistent with alternative, physically plausible explanations of recent climate change that exclude important elements of the given combination of forcings' (IPCC, 2001).

The consistency between an observed change and the estimated response to a hypothesised forcing is often determined by estimating the amplitude of the hypothesised pattern of change from observations and then assessing whether this estimate is statistically consistent with the expected amplitude of the pattern. Attribution studies additionally assess whether the response to a key forcing, such as greenhouse gas increases, is distinguishable from that due to other forcings (Appendix 9.A). These questions are typically investigated using a multiple regression of observations onto several fingerprints representing climate responses to different forcings that, ideally, are clearly distinct from each other (i.e., as distinct spatial patterns or distinct evolutions over time; see Section 9.2.2). If the response to this key forcing can be distinguished, and if even rescaled combinations of the responses to other forcings do not sufficiently explain the observed climate change, then the evidence for a causal connection is substantially increased. For example, the attribution of recent warming to greenhouse gas forcing becomes more reliable if the influences of other external forcings, for example solar forcing, are explicitly accounted for in the analysis. This is an area of research with considerable challenges because different forcing factors may lead to similar large-scale spatial patterns of response (Section 9.2.2). Note that another key element in attribution studies is the consideration of the physical consistency of multiple lines of evidence.

Both detection and attribution require knowledge of the internal climate variability on the time scales considered, usually decades or longer. The residual variability that remains in instrumental observations after the estimated effects of external forcing have been removed is sometimes used to estimate internal variability. However, these estimates are uncertain because the instrumental record is too short to give a well-constrained estimate of internal variability, and because of uncertainties in the forcings and the estimated responses. Thus, internal climate variability is usually estimated from long control simulations from coupled climate models. Subsequently, an assessment is usually made of the consistency between the residual variability referred to above and the model-based estimates of internal variability; analyses that yield implausibly large residuals are not considered credible (for example, this might happen if an important forcing is missing, or if the internal variability from the model is too small). Confidence is further increased by systematic intercomparison of the ability of models to simulate the various modes of observed variability (Chapter 8), by comparisons between variability in observations and climate model data (Section 9.4) and by comparisons between proxy reconstructions and climate simulations of the last millennium (Chapter 6 and Section 9.3).

Studies where the estimated pattern amplitude is substantially different from that simulated by models can still provide some understanding of climate change but need to be treated with caution (examples are given in Section 9.5). If this occurs for variables where confidence in the climate models is limited, such a result may simply reflect weaknesses in models. On the other hand, if this occurs for variables where confidence in the models is higher, it may raise questions about the forcings, such as whether all important forcings have been included or whether they have the correct amplitude, or questions about uncertainty in the observations.

Model and forcing uncertainties are important considerations in attribution research. Ideally, the assessment of model uncertainty should include uncertainties in model parameters (e.g., as explored by multi-model ensembles), and in the representation of physical processes in models (structural uncertainty). Such a complete assessment is not yet available, although model intercomparison studies (Chapter 8) improve the understanding of these uncertainties. The effects of forcing uncertainties, which can be considerable for some forcing agents such as solar and aerosol forcing (Section 9.2), also remain difficult to evaluate despite advances in research. Detection and attribution results based on several models or several forcing histories do provide information on the effects of model and forcing uncertainty. Such studies suggest that while model uncertainty is important, key results, such as attribution of a human influence on temperature change during the latter half of the 20th century, are robust.

Detection of anthropogenic influence is not yet possible for all climate variables for a variety of reasons. Some variables respond less strongly to external forcing, or are less reliably modelled or observed. In these cases, research that describes observed changes and offers physical explanations, for example, by demonstrating links to sea surface temperature changes, contributes substantially to the understanding of climate change and is therefore discussed in this chapter.

The approaches used in detection and attribution research described above cannot fully account for all uncertainties, and thus ultimately expert judgement is required to give a calibrated assessment of whether a specific cause is responsible for a given climate change. The assessment approach used in this chapter is to consider results from multiple studies using a variety of observational data sets, models, forcings and analysis techniques. The assessment based on these results typically takes into account the number of studies, the extent to which there is consensus among studies on the significance of detection results, the extent to which there is consensus on the consistency between the observed change and the change expected from forcing, the degree of consistency with other types of evidence, the extent to which known uncertainties are accounted for in and between studies, and whether there might be other physically plausible explanations for the given climate change. Having determined a particular likelihood assessment, this was then further downweighted to take into account any remaining uncertainties, such as, for example, structural uncertainties or a limited exploration of possible forcing histories of uncertain

forcings. The overall assessment also considers whether several independent lines of evidence strengthen a result.

While the approach used in most detection studies assessed in this chapter is to determine whether observations exhibit the expected response to external forcing, for many decision makers a question posed in a different way may be more relevant. For instance, they may ask, 'Are the continuing drier-than-normal conditions in the Sahel due to human causes?' Such questions are difficult to respond to because of a statistical phenomenon known as 'selection bias'. The fact that the questions are 'self selected' from the observations (only large observed climate anomalies in a historical context would be likely to be the subject of such a question) makes it difficult to assess their statistical significance from the same observations (see, e.g., von Storch and Zwiers, 1999). Nevertheless, there is a need for answers to such questions, and examples of studies that attempt to do so are discussed in this chapter (e.g., see Section 9.4.3.3).

9.1.3   The Basis from which We Begin

Evidence of a human influence on the recent evolution of the climate has accumulated steadily during the past two decades. The first IPCC Assessment Report (IPCC, 1990) contained little observational evidence of a detectable anthropogenic influence on climate. However, six years later the IPCC Working Group I SAR (IPCC, 1996) concluded that 'the balance of evidence' suggested there had been a 'discernible' human influence on the climate of the 20th century. Considerably more evidence accumulated during the subsequent five years, such that the TAR (IPCC, 2001) was able to draw a much stronger conclusion, not just on the detectability of a human influence, but on its contribution to climate change during the 20th century.

The evidence that was available at the time of the TAR was considerable. Using results from a range of detection studies of the instrumental record, which was assessed using fingerprints and estimates of internal climate variability from several climate models, it was found that the warming over the 20th century was 'very unlikely to be due to internal variability alone as estimated by current models'.

Simulations of global mean 20th-century temperature change that accounted for anthropogenic greenhouse gases and sulphate aerosols as well as solar and volcanic forcing were found to be generally consistent with observations. In contrast, a limited number of simulations of the response to known natural forcings alone indicated that these may have contributed to the observed warming in the first half of the 20th century, but could not provide an adequate explanation of the warming in the second half of the 20th century, nor the observed changes in the vertical structure of the atmosphere.

Attribution studies had begun to use techniques to determine whether there was evidence that the responses to several different forcing agents were simultaneously present in observations, mainly of surface temperature and of temperature in the free atmosphere. A distinct greenhouse gas signal was found to be detectable whether or not other external influences were explicitly considered, and the amplitude of the simulated greenhouse gas response was generally found to be consistent with observationally based estimates on the scales that were considered. Also, in most studies, the estimated rate and magnitude of warming over the second half of the 20th century due to increasing greenhouse gas concentrations alone was comparable with, or larger than, the observed warming. This result was found to be robust to attempts to account for uncertainties, such as observational uncertainty and sampling error in estimates of the response to external forcing, as well as differences in assumptions and analysis techniques.

The TAR also reported on a range of evidence of qualitative consistencies between observed climate changes and model responses to anthropogenic forcing, including global temperature rise, increasing land-ocean temperature contrast, diminishing arctic sea ice extent, glacial retreat and increases in precipitation at high northern latitudes.

A number of uncertainties remained at the time of the TAR. For example, large uncertainties remained in estimates of internal climate variability. However, even substantially inflated (doubled or more) estimates of model-simulated internal variance were found unlikely to be large enough to nullify the detection of an anthropogenic influence on climate. Uncertainties in external forcing were also reported, particularly in anthropogenic aerosol, solar and volcanic forcing, and in the magnitude of the corresponding climate responses. These uncertainties contributed to uncertainties in detection and attribution studies. Particularly, estimates of the contribution to the 20th-century warming by natural forcings and anthropogenic forcings other than greenhouse gases showed some discrepancies with climate simulations and were model dependent. These results made it difficult to attribute the observed climate change to one specific combination of external influences.

Based on the available studies and understanding of the uncertainties, the TAR concluded that 'in the light of new evidence and taking into account the remaining uncertainties, most of the observed warming over the last 50 years is likely to have been due to the increase in greenhouse gas concentrations'. Since the TAR, a larger number of model simulations using more complete forcings have become available, evidence on a wider range of variables has been analysed and many important uncertainties have been further explored and in many cases reduced. These advances are assessed in this chapter.

9.2   Radiative Forcing and Climate Response

This section briefly summarises the understanding of radiative forcing based on the assessment in Chapter 2, and of the climate response to forcing. Uncertainties in the forcing and estimates of climate response, and their implications for understanding and attributing climate change are also discussed. The discussion of radiative forcing focuses primarily on the period since 1750, with a brief reference to periods in the more distant past that are also assessed in the chapter, such as the last millennium, the Last Glacial Maximum and the mid-Holocene.

Two basic types of calculations have been used in detection and attribution studies. The first uses best estimates of forcing together with best estimates of modelled climate processes to calculate the effects of external changes in the climate system (forcings) on the climate (the response). These 'forward calculations' can then be directly compared to the observed changes in the climate system. Uncertainties in these simulations result from uncertainties in the radiative forcings that are used, and from model uncertainties that affect the simulated response to the forcings. Forward calculations are explored in this chapter and compared to observed climate change.

Results from forward calculations are used for formal detection and attribution analyses. In such studies, a climate model is used to calculate response patterns ('fingerprints') for individual forcings or sets of forcings, which are then combined linearly to provide the best fit to the observations. This procedure assumes that the amplitude of the large-scale pattern of response scales linearly with the forcing, and that patterns from different forcings can be added to obtain the total response. This assumption may not hold for every forcing, particularly not at smaller spatial scales, and may be violated when forcings interact nonlinearly (e.g., black carbon absorption decreases cloudiness and thereby decreases the indirect effects of sulphate aerosols). Generally, however, the assumption is expected to hold for most forcings (e.g., Penner et al., 1997; Meehl et al., 2004). Errors or uncertainties in the magnitude of the forcing or the magnitude of a model's response to the forcing should not affect detection results provided that the space-time pattern of the response is correct. However, for the linear combination of responses to be considered consistent with the observations, the scaling factors for individual response patterns should indicate that the model does not need to be rescaled to match the observations (Sections 9.1.2, 9.4.1.4 and Appendix 9.A) given uncertainty in the amplitude of forcing, model response and estimate due to internal climate variability. For detection studies, if the space-time pattern of response is incorrect, then the scaling, and hence detection and attribution results, will be affected.

In the second type of calculation, the so-called 'inverse' calculations, the magnitude of uncertain parameters in the forward model (including the forcing that is applied) is varied in order to provide a best fit to the observational record. In general, the greater the degree of *a priori* uncertainty in the parameters of the model, the more the model is allowed to adjust. Probabilistic posterior estimates for model parameters and uncertain forcings are obtained by comparing the agreement between simulations and observations, and taking into account prior uncertainties (including those in observations; see Sections 9.2.1.2, 9.6 and Supplementary Material, Appendix 9.B).

9.2.1 Radiative Forcing Estimates Used to Simulate Climate Change

9.2.1.1 Summary of 'Forward' Estimates of Forcing for the Instrumental Period

Estimates of the radiative forcing (see Section 2.2 for a definition) since 1750 from forward model calculations and observations are reviewed in detail in Chapter 2 and provided in Table 2.12. Chapter 2 describes estimated forcing resulting from increases in long-lived greenhouse gases (carbon dioxide (CO_2), methane, nitrous oxide, halocarbons), decreases in stratospheric ozone, increases in tropospheric ozone, sulphate aerosols, nitrate aerosols, black carbon and organic matter from fossil fuel burning, biomass burning aerosols, mineral dust aerosols, land use change, indirect aerosol effects on clouds. aircraft cloud effects, solar variability, and stratospheric and tropospheric water vapour increases from methane and irrigation. An example of one model's implemented set of forcings is given in Figure 2.23. While some members of the MMD at PCMDI have included a nearly complete list of these forcings for the purpose of simulating the 20th-century climate (see Supplementary Material, Table S9.1), most detection studies to date have used model runs with a more limited set of forcings. The combined anthropogenic forcing from the estimates in Section 2.9.2 since 1750 is 1.6 W m^{-2}, with a 90% range of 0.6 to 2.4 W m^{-2}, indicating that it is extremely likely that humans have exerted a substantial warming influence on climate over that time period. The combined forcing by greenhouse gases plus ozone is 2.9 ± 0.3 W m^{-2} and the total aerosol forcing (combined direct and indirect 'cloud albedo' effect) is virtually certain to be negative and estimated to be –1.3 (90% uncertainty range of –2.2 to –0.5 W m^{-2}; see Section 2.9). In contrast, the direct radiative forcing due to increases in solar irradiance is estimated to be +0.12 (90% range from 0.06 to 0.3) W m^{-2}. In addition, Chapter 2 concludes that it is exceptionally unlikely that the combined natural (solar and volcanic) radiative forcing has had a warming influence comparable to that of the combined anthropogenic forcing over the period 1950 to 2005. As noted in Chapter 2, the estimated global average surface temperature response from these forcings may differ for a particular magnitude of forcing since all forcings do not have the same 'efficacy' (i.e., effectiveness at changing the surface temperature compared to CO_2; see Section 2.8). Thus, summing these forcings does not necessarily give an adequate estimate of the response in global average surface temperature.

9.2.1.2 Summary of 'Inverse' Estimates of Net Aerosol Forcing

Forward model approaches to estimating aerosol forcing are based on estimates of emissions and models of aerosol physics and chemistry. They directly resolve the separate contributions by various aerosol components and forcing mechanisms. This must be borne in mind when comparing results to those from inverse calculations (see Section 9.6 and Supplementary Material, Appendix 9.B for details), which, for example, infer the net aerosol forcing required to match climate model simulations with observations. These methods can be applied using a global average forcing and response, or using the spatial and temporal patterns of the climate response in order to increase the ability to distinguish between responses to different external forcings. Inverse methods have been used to constrain one or several uncertain radiative forcings (e.g., by aerosols), as well as climate sensitivity (Section 9.6) and other uncertain climate parameters (Wigley, 1989; Schlesinger and Ramankutty, 1992; Wigley et al., 1997; Andronova and Schlesinger, 2001; Forest et al., 2001, 2002; Harvey and Kaufmann, 2002; Knutti et al., 2002, 2003; Andronova et al., 2007; Forest et al., 2006; see Table 9.1 – Stott et al., 2006c). The reliability of the spatial and temporal patterns used is discussed in Sections 9.2.2.1 and 9.2.2.2.

In the past, forward calculations have been unable to rule out a total net negative radiative forcing over the 20th century (Boucher and Haywood, 2001). However, Section 2.9 updates the Boucher and Haywood analysis for current radiative forcing estimates since 1750 and shows that it is extemely likely that the combined anthropogenic RF is both positive and substantial (best estimate: +1.6 W m^{-2}). A net forcing close to zero would imply a very high value of climate sensitivity, and would be very difficult to reconcile with the observed increase in temperature (Sections 9.6 and 9.7). Inverse calculations yield only the 'net forcing', which includes all forcings that project on the fingerprint of the forcing that is estimated. For example, the response to tropospheric ozone forcing could project onto that for sulphate aerosol forcing. Therefore, differences between forward estimates and inverse estimates may have one of several causes, including (1) the magnitude of the forward model calculation is incorrect due to inadequate physics and/or chemistry, (2) the forward calculation has not evaluated all forcings and feedbacks or (3) other forcings project on the fingerprint of the forcing that is estimated in the inverse calculation.

Studies providing inverse estimates of aerosol forcing are compared in Table 9.1. One type of inverse method uses the ranges of climate change fingerprint scaling factors derived from detection and attribution analyses that attempt to separate the climate response to greenhouse gas forcing from the response to aerosol forcing and often from natural forcing as well (Gregory et al., 2002a; Stott et al., 2006c; see also Section 9.4.1.4). These provide the range of fingerprint magnitudes (e.g., for the combined temperature response to different aerosol forcings) that are consistent with observed climate change, and can therefore be used to infer the likely range of forcing that is consistent with the observed record. The separation between greenhouse gas and aerosol fingerprints exploits the fact that the forcing from well-mixed greenhouse gases is well known, and that errors in the model's transient sensitivity can therefore be separated from errors in aerosol forcing in the model (assuming that there are similar errors in a model's sensitivity to greenhouse gas and aerosol

Table 9.1. *Inverse estimates of aerosol forcing from detection and attribution studies and studies estimating equilibrium climate sensitivity (see Section 9.6 and Table 9.3 for details on studies). The 5 to 95% estimates for the range of aerosol forcing relate to total or net fossil-fuel related aerosol forcing (in W m⁻²).*

	Forest et al. (2006)	Andronova and Schlesinger (2001)	Knutti et al. (2002, 2003)	Gregory et al. (2002a)	Stott et al. (2006c)	Harvey and Kaufmann (2002)
Observational data used to constrain aerosol forcing	Upper air, surface and deep ocean space-time temperature, latter half of 20th century	Global mean and hemispheric difference in surface air temperature 1856 to 1997	Global mean ocean heat uptake 1955 to 1995, global mean surface air temperature increase 1860 to 2000	Surface air temperature space-time patterns, one AOGCM	Surface air temperature space-time patterns, three AOGCMs	Global mean and hemispheric difference in surface air temperature 1856 to 2000
Forcings considered[a]	G, Sul, Sol, Vol, OzS, land surface changes	G, OzT, Sul, Sol, Vol	G, Sul, Suli, OzT, OzS, BC+OM, stratospheric water vapour, Vol, Sol	G, Sul, Suli, Sol, Vol	G, Sul, Suli, OzT, OzS, Sol, Vol	G, Sul, biomass aerosol, Sol, Vol
Year[b]	1980s	1990	2000	2000	2000	1990
Aerosol forcing (W m⁻²)[c]	−0.14 to −0.74 −0.07 to −0.65 with expert prior	−0.54 to −1.3	0 to −1.2 indirect aerosol −0.6 to −1.7 total aerosol	−0.4 to −1.6 total aerosol	−0.4 to −1.4 total aerosol	Fossil fuel aerosol unlikely < −1, biomass plus dust unlikely < −0.5[d]

Notes:

[a] G: greenhouse gases; Sul: direct sulphate aerosol effect; Suli: (first) indirect sulphate aerosol effect; OzT: tropospheric ozone; OzS: stratospheric ozone; Vol: volcanic forcing; Sol: solar forcing; BC+OM: black carbon and organic matter from fossil fuel and biomass burning.

[b] Year(s) for which aerosol forcing is calculated, relative to pre-industrial conditions.

[c] 5 to 95% inverse estimate of the total aerosol forcing in the year given relative to pre-industrial forcing. The aerosol range refers to the net fossil-fuel related aerosol range, which tends to be all forcings not directly accounted for that project onto the pattern associated with fossil fuel aerosols, and includes all unknown forcings and those not explicitly considered (for example, OzT and BC+OM in several of the studies).

[d] Explores IPCC TAR range of climate sensitivity (i.e., 1.5°C to 4.5°C), while other studies explore wider ranges

forcing; see Gregory et al., 2002a; Table 9.1). By scaling spatio-temporal patterns of response up or down, this technique takes account of gross model errors in climate sensitivity and net aerosol forcing but does not fully account for modelling uncertainty in the patterns of temperature response to uncertain forcings.

Another approach uses the response of climate models, most often simple climate models or Earth System Models of Intermediate Complexity (EMICs, Table 8.3) to explore the range of forcings and climate parameters that yield results consistent with observations (Andronova and Schlesinger, 2001; Forest et al., 2002; Harvey and Kaufmann, 2002; Knutti et al., 2002, 2003; Forest et al., 2006). Like detection methods, these approaches seek to fit the space-time patterns, or spatial means in time, of observed surface, atmospheric or ocean temperatures. They determine the probability of combinations of climate sensitivity and net aerosol forcing based on the fit between simulations and observations (see Section 9.6 and Supplementary Material, Appendix 9.B for further discussion). These are often based on Bayesian approaches, where prior assumptions about

ranges of external forcing are used to constrain the estimated net aerosol forcing and climate sensitivity. Some of these studies use the difference between Northern and Southern Hemisphere mean temperature to separate the greenhouse gas and aerosol forcing effects (e.g., Andronova and Schlesinger, 2001; Harvey and Kaufmann, 2002). In these analyses, it is necessary to accurately account for hemispheric asymmetry in tropospheric ozone forcing in order to infer the hemispheric aerosol forcing. Additionally, aerosols from biomass burning could cause an important fraction of the total aerosol forcing although this forcing shows little hemispheric asymmetry. Since it therefore projects on the greenhouse gas forcing, it is difficult to separate in an inverse calculation. Overall, results will be only as good as the spatial or temporal pattern that is assumed in the analysis. Missing forcings or lack of knowledge about uncertainties, and the highly parametrized spatial distribution of response in some of these models may hamper the interpretation of results.

Aerosol forcing appears to have grown rapidly during the period from 1945 to 1980, while greenhouse gas forcing grew more slowly (Ramaswamy et al., 2001). Global sulphur emissions (and thus sulphate aerosol forcing) appear to have decreased after 1980 (Stern, 2005), further rendering the temporal evolution of aerosols and greenhouse gases distinct. As long as the temporal pattern of variation in aerosol forcing is approximately correct, the need to achieve a reasonable fit to the temporal variation in global mean temperature and the difference between Northern and Southern Hemisphere temperatures can provide a useful constraint on the net aerosol radiative forcing (as demonstrated, e.g., by Harvey and Kaufmann, 2002; Stott et al., 2006c).

The inverse estimates summarised in Table 9.1 suggest that to be consistent with observed warming, the net aerosol forcing over the 20th century should be negative with likely ranges between -1.7 and -0.1 W m^{-2}. This assessment accounts for the probability of other forcings projecting onto the fingerprints. These results typically provide a somewhat smaller upper limit for the total aerosol forcing than the estimates given in Chapter 2, which are derived from forward calculations and range between -2.2 and -0.5 W m^{-2} (5 to 95% range, median -1.3 W m^{-2}). Note that the uncertainty ranges from inverse and forward calculations are different due to the use of different information, and that they are affected by different uncertainties. Nevertheless, the similarity between results from inverse and forward estimates of aerosol forcing strengthens confidence in estimates of total aerosol forcing, despite remaining uncertainties. Harvey and Kaufmann (2002), who use an approach that focuses on the TAR range of climate sensitivity, further conclude that global mean forcing from fossil-fuel related aerosols was probably less than -1.0 W m^{-2} in 1990 and that global mean forcing from biomass burning and anthropogenically enhanced soil dust aerosols is 'unlikely' to have exceeded -0.5 W m^{-2} in 1990.

9.2.1.3   Radiative Forcing of Pre-Industrial Climate Change

Here we briefly discuss the radiative forcing estimates used for understanding climate during the last millennium, the mid-Holocene and the Last Glacial Maximum (LGM) (Section 9.3) and in estimates of climate sensitivity based on palaeoclimatic records (Section 9.6.3).

Regular variation in the Earth's orbital parameters has been identified as the pacemaker of climate change on the glacial to interglacial time scale (see Berger, 1988 for a review). These orbital variations, which can be calculated from astronomical laws (Berger, 1978), force climate variations by changing the seasonal and latitudinal distribution of solar radiation (Chapter 6).

Insolation at the time of the LGM (21 ka) was similar to today. Nonetheless, the LGM climate remained cold due to the presence of large ice sheets in the Northern Hemisphere (Peltier, 1994, 2004) and reduced atmospheric CO_2 concentration (185 ppm according to recent ice core estimates, see Monnin et al., 2001). Most modelling studies of this period do not treat ice sheet extent and elevation or CO_2 concentration prognostically,

but specify them as boundary conditions. The LGM radiative forcing from the reduced atmospheric concentrations of well-mixed greenhouse gases is likely to have been about -2.8 W m^{-2} (see Figure 6.5). Ice sheet albedo forcing is estimated to have caused a global mean forcing of about -3.2 W m^{-2} (based on a range of several LGM simulations) and radiative forcing from increased atmospheric aerosols (primarily dust and vegetation) is estimated to have been about -1 W m^{-2} each. Therefore, the total annual and global mean radiative forcing during the LGM is likely to have been approximately -8 W m^{-2} relative to 1750, with large seasonal and geographical variations and significant uncertainties (see Section 6.4.1).

The major mid-Holocene forcing relative to the present was due to orbital perturbations that led to large changes in the seasonal cycle of insolation. The Northern Hemisphere (NH) seasonal cycle was about 27 W m^{-2} greater, whereas there was only a negligible change in NH annual mean solar forcing. For the Southern Hemisphere (SH), the seasonal forcing was -6.5 W m^{-2}. In contrast, the global and annual mean net forcing was only 0.011 W m^{-2}.

Changes in the Earth's orbit have had little impact on annual mean insolation over the past millennium. Summer insolation decreased by 0.33 W m^{-2} at 45°N over the millennium, winter insolation increased by 0.83 W m^{-2} (Goosse et al., 2005), and the magnitude of the mean seasonal cycle of insolation in the NH decreased by 0.4 W m^{-2}. Changes in insolation are also thought to have arisen from small variations in solar irradiance, although both timing and magnitude of past solar radiation fluctuations are highly uncertain (see Chapters 2 and 6; Lean et al., 2002; Gray et al., 2005; Foukal et al., 2006). For example, sunspots were generally missing from approximately 1675 to 1715 (the so-called Maunder Minimum) and thus solar irradiance is thought to have been reduced during this period. The estimated difference between the present-day solar irradiance cycle mean and the Maunder Minimum is 0.08% (see Section 2.7.1.2.2), which corresponds to a radiative forcing of about 0.2 W m^{-2}, which is substantially lower than estimates used in the TAR (Chapter 2).

Natural external forcing also results from explosive volcanism that introduces aerosols into the stratosphere (Section 2.7.2), leading to a global negative forcing during the year following the eruption. Several reconstructions are available for the last two millennia and have been used to force climate models (Section 6.6.3). There is close agreement on the timing of large eruptions in the various compilations of historic volcanic activity, but large uncertainty in the magnitude of individual eruptions (Figure 6.13). Different reconstructions identify similar periods when eruptions happened more frequently. The uncertainty in the overall amplitude of the reconstruction of volcanic forcing is also important for quantifying the influence of volcanism on temperature reconstructions over longer periods, but is difficult to quantify and may be a substantial fraction of the best estimate (e.g., Hegerl et al., 2006a).

9.2.2 Spatial and Temporal Patterns of the Response to Different Forcings and their Uncertainties

9.2.2.1 Spatial and Temporal Patterns of Response

The ability to distinguish between climate responses to different external forcing factors in observations depends on the extent to which those responses are distinct (see, e.g., Section 9.4.1.4 and Appendix 9.A). Figure 9.1 illustrates the zonal average temperature response in the PCM model (see Table 8.1 for model details) to several different forcing agents over the last 100 years, while Figure 9.2 illustrates the zonal average temperature response in the Commonwealth Scientific and Industrial Research Organisation (CSIRO) atmospheric model (when coupled to a simple mixed layer ocean model) to fossil fuel black carbon and organic matter, and to the combined effect of these forcings together with biomass burning aerosols (Penner et al., 2007). These figures indicate that the modelled vertical and zonal average signature of the temperature response should depend on the forcings. The major features shown in Figure 9.1 are robust to using different climate models. On the other hand, the response to black carbon forcing has not been widely examined and therefore the features in Figure 9.2 may be model dependent. Nevertheless, the response to black carbon forcings appears to be small.

Greenhouse gas forcing is expected to produce warming in the troposphere, cooling in the stratosphere, and, for transient simulations, somewhat more warming near the surface in the NH due to its larger land fraction, which has a shorter surface response time to the warming than do ocean regions (Figure 9.1c). The spatial pattern of the transient surface temperature response to greenhouse gas forcing also typically exhibits a land-sea pattern of stronger warming over land, for the same reason (e.g., Cubasch et al., 2001). Sulphate aerosol forcing results in cooling throughout most of the globe, with greater cooling in the NH due to its higher aerosol loading (Figure 9.1e; see Chapter 2), thereby partially offsetting the greater NH greenhouse-gas induced warming. The combined effect of tropospheric and stratospheric ozone forcing (Figure 9.1d) is expected to warm the troposphere, due to increases in tropospheric ozone, and cool the stratosphere, particularly at high latitudes where stratospheric ozone loss has been greatest. Greenhouse gas forcing is also expected to change the hydrological cycle worldwide, leading to disproportionately greater increases in heavy precipitation (Chapter 10 and Section 9.5.4), while aerosol forcing can influence rainfall regionally (Section 9.5.4).

The simulated responses to natural forcing are distinct from those due to the anthropogenic forcings described above. Solar forcing results in a general warming of the atmosphere (Figure 9.1a) with a pattern of surface warming that is similar to that expected from greenhouse gas warming, but in contrast to the response to greenhouse warming, the simulated solar-forced warming extends throughout the atmosphere (see, e.g., Cubasch

et al., 1997). A number of independent analyses have identified tropospheric changes that appear to be associated with the solar cycle (van Loon and Shea, 2000; Gleisner and Thejll, 2003; Haigh, 2003; White et al., 2003; Coughlin and Tung, 2004; Labitzke, 2004; Crooks and Gray, 2005), suggesting an overall warmer and moister troposphere during solar maximum. The peak-to-trough amplitude of the response to the solar cycle globally is estimated to be approximately 0.1°C near the surface. Such variations over the 11-year solar cycle make it is necessary to use several decades of data in detection and attribution studies. The solar cycle also affects atmospheric ozone concentrations with possible impacts on temperatures and winds in the stratosphere, and has been hypothesised to influence clouds through cosmic rays (Section 2.7.1.3). Note that there is substantial uncertainty in the identification of climate response to solar cycle variations because the satellite period is short relative to the solar cycle length, and because the response is difficult to separate from internal climate variations and the response to volcanic eruptions (Gray et al., 2005).

Volcanic sulphur dioxide (SO_2) emissions ejected into the stratosphere form sulphate aerosols and lead to a forcing that causes a surface and tropospheric cooling and a stratospheric warming that peak several months after a volcanic eruption and last for several years. Volcanic forcing also likely leads to a response in the atmospheric circulation in boreal winter (discussed below) and a reduction in land precipitation (Robock and Liu, 1994; Broccoli et al., 2003; Gillett et al., 2004b). The response to volcanic forcing causes a net cooling over the 20th century because of variations in the frequency and intensity of volcanic eruptions. This results in stronger volcanic forcing towards the end of the 20th century than early in the 20th century. In the PCM, this increase results in a small warming in the lower stratosphere and near the surface at high latitudes, with cooling elsewhere (Figure 9.1b).

The net effect of all forcings combined is a pattern of NH temperature change near the surface that is dominated by the positive forcings (primarily greenhouse gases), and cooling in the stratosphere that results predominantly from greenhouse gas and stratospheric ozone forcing (Figure 9.1f). Results obtained with the CSIRO model (Figure 9.2) suggest that black carbon, organic matter and biomass aerosols would slightly enhance the NH warming shown in Figure 9.1f. On the other hand, indirect aerosol forcing from fossil fuel aerosols may be larger than the direct effects that are represented in the CSIRO and PCM models, in which case the NH warming could be somewhat diminished. Also, while land use change may cause substantial forcing regionally and seasonally, its forcing and response are expected to have only a small impact at large spatial scales (Sections 9.3.3.3 and 7.2.2; Figures 2.20 and 2.23).

The spatial signature of a climate model's response is seldom very similar to that of the forcing, due in part to the strength of the feedbacks relative to the initial forcing. This comes about because climate system feedbacks vary spatially and because the atmospheric and ocean circulation cause a redistribution of energy over the globe. For example, sea ice albedo feedbacks

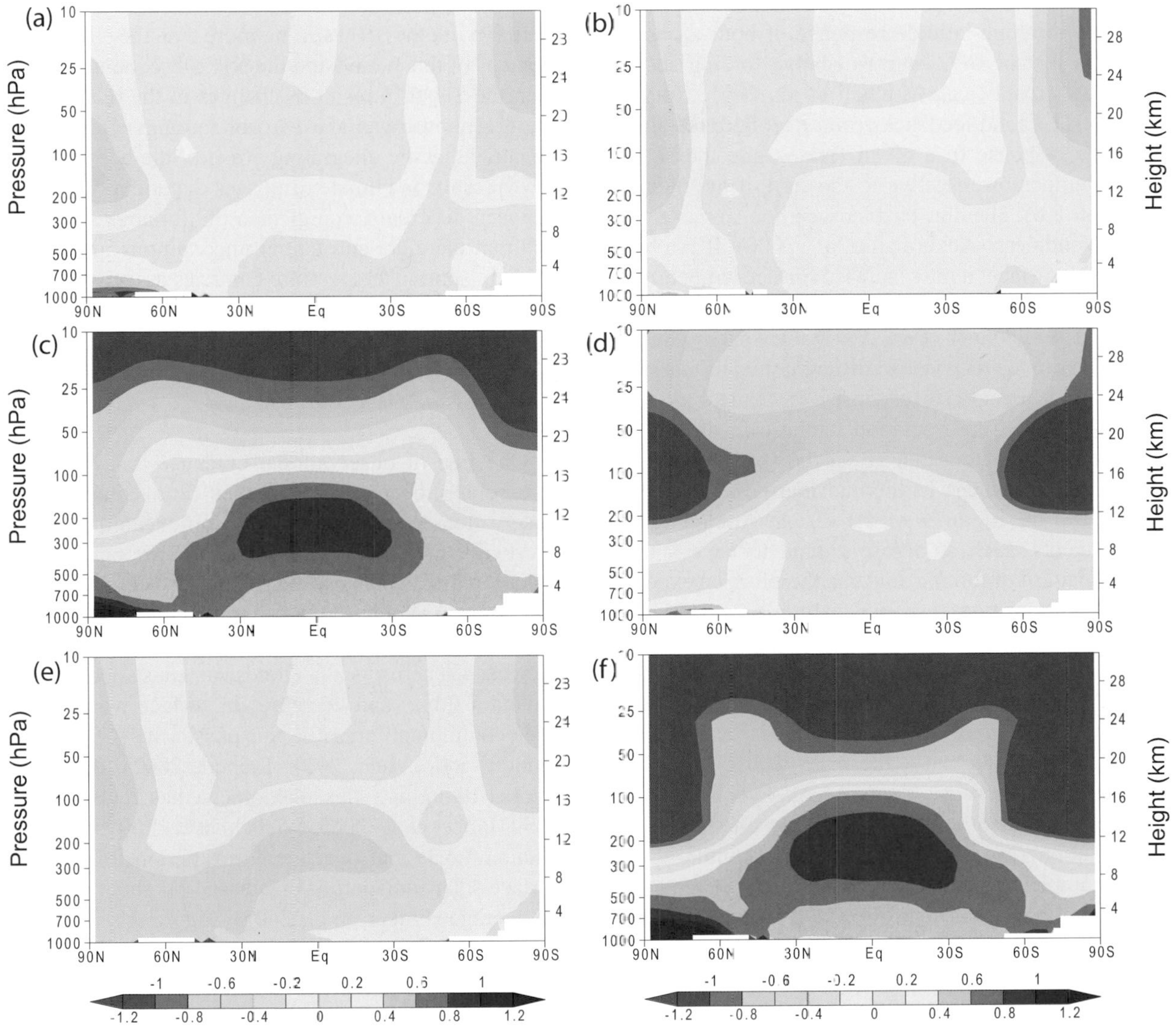

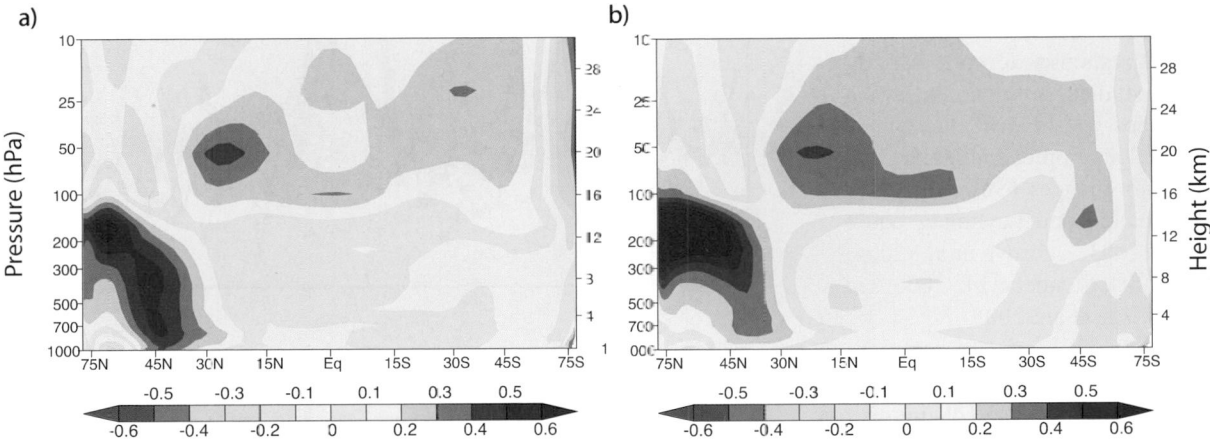

Figure 9.1. *Zonal mean atmospheric temperature change from 1890 to 1999 (°C per century) as simulated by the PCM model from (a) solar forcing, (b) volcanoes, (c) well-mixed greenhouse gases, (d) tropospheric and stratospheric ozone changes, (e) direct sulphate aerosol forcing and (f) the sum of all forcings. Plot is from 1,000 hPa to 10 hPa (shown on left scale) and from 0 km to 30 km (shown on right). See Appendix 9.C for additional information. Based on Santer et al. (2003a).*

Figure 9.2. *The zonal mean equilibrium temperature change (°C) between a present day minus a pre-industrial simulation by the CSIRO atmospheric model coupled to a mixed-layer ocean model from (a) direct forcing from fossil fuel black carbon and organic matter (BC+OM) and (b) the sum of fossil fuel BC+OM and biomass burning. Plot is from 1,000 hPa to 10 hPa (shown on left scale) and from 0 km to 30 km (shown on right). Note the difference in colour scale from Figure 9.1. See Supplementary Material, Appendix 9.C for additional information. Based on Penner et al. (2007).*

tend to enhance the high-latitude response of both a positive forcing, such as that of CO_2, and a negative forcing such as that of sulphate aerosol (e.g., Mitchell et al., 2001; Rotstayn and Penner, 2001). Cloud feedbacks can affect both the spatial signature of the response to a given forcing and the sign of the change in temperature relative to the sign of the radiative forcing (Section 8.6). Heating by black carbon, for example, can decrease cloudiness (Ackerman et al., 2000). If the black carbon is near the surface, it may increase surface temperatures, while at higher altitudes it may reduce surface temperatures (Hansen et al., 1997; Penner et al., 2003). Feedbacks can also lead to differences in the response of different models to a given forcing agent, since the spatial response of a climate model to forcing depends on its representation of these feedbacks and processes. Additional factors that affect the spatial pattern of response include differences in thermal inertia between land and sea areas, and the lifetimes of the various forcing agents. Shorter-lived agents, such as aerosols, tend to have a more distinct spatial pattern of forcing, and can therefore be expected to have some locally distinct response features.

The pattern of response to a radiative forcing can also be altered quite substantially if the atmospheric circulation is affected by the forcing. Modelling studies and data comparisons suggest that volcanic aerosols (e.g., Kirchner et al., 1999; Shindell et al., 1999; Yang and Schlesinger, 2001; Stenchikov et al., 2006) and greenhouse gas changes (e.g., Fyfe et al., 1999; Shindell et al., 1999; Rauthe et al., 2004) can alter the North Atlantic Oscillation (NAO) or the Northern Annular Mode (NAM). For example, volcanic eruptions, with the exception of high-latitude eruptions, are often followed by a positive phase of the NAM or NAO (e.g., Stenchikov et al., 2006) leading to Eurasian winter warming that may reduce the overall cooling effect of volcanic eruptions on annual averages, particularly over Eurasia (Perlwitz and Graf, 2001; Stenchikov et al., 2002; Shindell et al., 2003; Stenchikov et al., 2004; Oman et al., 2005; Rind et al., 2005a; Miller et al., 2006; Stenchikov et al., 2006). In contrast, NAM or NAO responses to solar forcing vary between studies, some indicating a response, perhaps with dependence of the response on season or other conditions, and some finding no changes (Shindell et al., 2001a,b; Ruzmaikin and Feynman, 2002; Tourpali et al., 2003; Egorova et al., 2004; Palmer et al., 2004; Stendel et al., 2006; see also review in Gray et al., 2005).

In addition to the spatial pattern, the temporal evolution of the different forcings (Figure 2.23) generally helps to distinguish between the responses to different forcings. For example, Santer et al. (1996b,c) point out that a temporal pattern in the hemispheric temperature contrast would be expected in the second half of the 20th

century with the SH warming more than the NH for the first two decades of this period and the NH subsequently warming more than the SH, as a result of changes in the relative strengths of the greenhouse gas and aerosol forcings. However, it should be noted that the integrating effect of the oceans (Hasselmann, 1976) results in climate responses that are more similar in time between different forcings than the forcings are to each other, and that there are substantial uncertainties in the evolution of the hemispheric temperature contrasts associated with sulphate aerosol forcing.

9.2.2.2 Aerosol Scattering and Cloud Feedback in Models and Observations

One line of observational evidence that reflective aerosol forcing has been changing over time comes from satellite observations of changes in top-of-atmosphere outgoing shortwave radiation flux. Increases in the outgoing shortwave radiation flux can be caused by increases in reflecting aerosols, increases in clouds or a change in the vertical distribution of clouds and water vapour, or increases in surface albedo. Increases in aerosols and clouds can cause decreases in surface radiation fluxes and decreases in surface warming. There has been continuing interest in this possibility (Gilgen et al., 1998; Stanhill and Cohen, 2001; Liepert, 2002). Sometimes called 'global dimming', this phenomena has reversed since about 1990 (Pinker et al., 2005; Wielicki et al., 2005; Wild et al., 2005; Section 3.4.3), but over the entire period from 1984 to 2001, surface solar radiation has increased by about 0.16 W m^{-2} yr^{-1} on average (Pinker et al., 2005). Figure 9.3 shows the top-of-

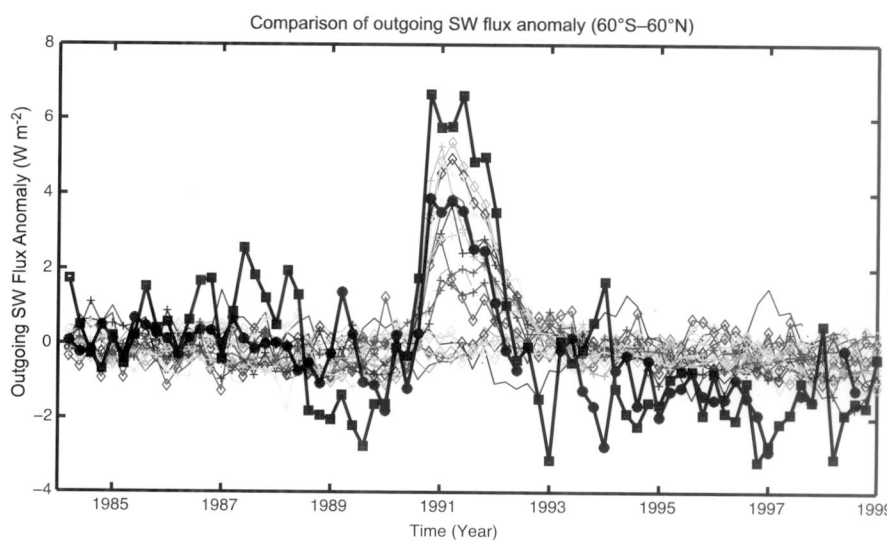

Figure 9.3. *Comparison of outgoing shortwave radiation flux anomalies (in W m^{-2}, calculated relative to the entire time period) from several models in the MMD archive at PCMDI (coloured curves) with ERBS satellite data (black with stars; Wong et al., 2006) and with the ISCCP flux data set (black with squares; Zhang et al., 2004). Models shown are CCSM3, CGCM3.1(T47), CGCM3.1(T63), CNRM-CM3, CSIRO-MK3.0, FGOALS-g1.0, GFDL-CM2.0, GFDL-CM2.1, GISS-AOM, GISS-EH, GISS-ER, INM-CM3.0, IPSL-CM4, and MRI-CGCM2.3.2 (see Table 8.1 for model details). The comparison is restricted to 60°S to 60°N because the ERBS data are considered more accurate in this region. Note that not all models included the volcanic forcing from Mt. Pinatubo (1991–1993) and so do not predict the observed increase in outgoing solar radiation. See Supplementary Material, Appendix 9.C for additional information.*

atmosphere outgoing shortwave radiation flux anomalies from the MMD at PCMDI, compared to that measured by the Earth Radiation Budget Satellite (ERBS; Wong et al., 2006) and inferred from International Satellite Cloud Climatology Project (ISCCP) flux data (FD) (Zhang et al., 2004). The downward trend in outgoing solar radiation is consistent with the long-term upward trend in surface radiation found by Pinker et al. (2005). The effect of the eruption of Mt. Pinatubo in 1991 results in an increase in the outgoing shortwave radiation flux (and a corresponding dimming at the surface) and its effect has been included in most (but not all) models in the MMD. The ISCCP flux anomaly for the Mt. Pinatubo signal is almost 2 W m^{-2} larger than that for ERBS, possibly due to the aliasing of the stratospheric aerosol signal into the ISCCP cloud properties. Overall, the trends from the ISCCP FD (–0.18 with 95% confidence limits of ±0.11 W m^{-2} yr^{-1}) and the ERBS data (–0.13 ± 0.08 W m^{-2} yr^{-1}) from 1984 to 1999 are not significantly different from each other at the 5% significance level, and are in even better agreement if only tropical latitudes are considered (Wong et al., 2006). These observations suggest an overall decrease in aerosols and/or clouds, while estimates of changes in cloudiness are uncertain (see Section 3.4.3). The model-predicted trends are also negative over this time period, but are smaller in most models than in the ERBS observations (which are considered more accurate than the ISCCP FD). Wielicki et al. (2002) explain the observed downward trend by decreases in cloudiness, which are not well represented in the models on these decadal time scales (Chen et al., 2002; Wielicki et al., 2002).

9.2.2.3 Uncertainty in the Spatial Pattern of Response

Most detection methods identify the magnitude of the space-time patterns of response to forcing (sometimes called 'fingerprints') that provide the best fit to the observations. The fingerprints are typically estimated from ensembles of climate model simulations forced with reconstructions of past forcing. Using different forcing reconstructions and climate models in such studies provides some indication of forcing and model uncertainty. However, few studies have examined how uncertainties in the spatial pattern of forcing explicitly contribute to uncertainties in the spatial pattern of the response. For short-lived components, uncertainties in the spatial pattern of forcing are related to uncertainties in emissions patterns, uncertainties in the transport within the climate model or chemical transport model and, especially for aerosols, uncertainties in the representation of relative humidities or clouds. These uncertainties affect the spatial pattern of the forcing. For example, the ratio of the SH to NH indirect aerosol forcing associated with the total aerosol forcing ranges from –0.12 to 0.63 (best guess 0.29) in different studies, and that between ocean and land forcing ranges from 0.03 to 1.85 (see Figure 7.21; Rotstayn and Penner, 2001; Chuang et al., 2002; Kristjansson, 2002; Lohmann and Lesins, 2002; Menon et al., 2002a; Rotstayn and Liu, 2003; Lohmann and Feichter, 2005).

9.2.2.4 Uncertainty in the Temporal Pattern of Response

Climate model studies have also not systematically explored the effect of uncertainties in the temporal evolution of forcings. These uncertainties depend mainly on the uncertainty in the spatio-temporal expression of emissions, and, for some forcings, fundamental understanding of the possible change over time.

The increasing forcing by greenhouse gases is relatively well known. In addition, the global temporal history of SO$_2$ emissions, which have a larger overall forcing than the other short-lived aerosol components, is quite well constrained. Seven different reconstructions of the temporal history of global anthropogenic sulphur emissions up to 1990 have a relative standard deviation of less than 20% between 1890 and 1990, with better agreement in more recent years. This robust temporal history increases confidence in results from detection and attribution studies that attempt to separate the effects of sulphate aerosol and greenhouse gas forcing (Section 9.4.1).

In contrast, there are large uncertainties related to the anthropogenic emissions of other short-lived compounds and their effects on forcing. For example, estimates of historical emissions from fossil fuel combustion do not account for changes in emission factors (the ratio of the emitted gas or aerosol to the fuel burned) of short-lived species associated with concerns over urban air pollution (e.g., van Aardenne et al., 2001). Changes in these emission factors would have slowed the emissions of nitrogen oxides as well as carbon monoxide after about 1970 and slowed the accompanying increase in tropospheric ozone compared to that represented by a single emission factor for fossil fuel use. In addition, changes in the height of SO$_2$ emissions associated with the implementation of tall stacks would have changed the lifetime of sulphate aerosols and the relationship between emissions and effects. Another example relates to the emissions of black carbon associated with the burning of fossil fuels. The spatial and temporal emissions of black carbon by continent reconstructed by Ito and Penner (2005) are significantly different from those reconstructed using the methodology of Novakov et al. (2003). For example, the emissions in Asia grow significantly faster in the inventory based on Novakov et al. (2003) compared to those based on Ito and Penner (2005). In addition, before 1988 the growth in emissions in Eastern Europe using the Ito and Penner (2005) inventory is faster than the growth based on the methodology of Novakov et al. (2003). Such spatial and temporal uncertainties will contribute to both spatial and temporal uncertainties in the net forcing and to spatial and temporal uncertainties in the distribution of forcing and response.

There are also large uncertainties in the magnitude of low-frequency changes in forcing associated with changes in total solar radiation as well as its spectral variation, particularly on time scales longer than the 11-year cycle. Previous estimates of change in total solar radiation have used sunspot numbers to calculate these slow changes in solar irradiance over the last few centuries, but these earlier estimates are not necessarily supported by current understanding and the estimated magnitude of low-frequency changes has been substantially reduced since the TAR (Lean et al., 2002; Foukal et al., 2004, 2006; Sections 6.6.3.1 and 2.7.1.2). In addition, the magnitude of radiative forcing associated with major volcanic eruptions is uncertain and differs between reconstructions (Sato et al., 1993; Andronova et al., 1999; Ammann et al., 2003), although the timing of the eruptions is well documented.

9.2.3 Implications for Understanding 20th-Century Climate Change

Any assessment of observed climate change that compares simulated and observed responses will be affected by errors and uncertainties in the forcings prescribed in a climate model and its corresponding responses. As noted above, detection studies scale the response patterns to different forcings to obtain the best match to observations. Thus, errors in the magnitude of the forcing or in the magnitude of the model response to a forcing (which is approximately, although not exactly, a function of climate sensitivity), should not affect detection results provided that the large-scale space-time pattern of the response is correct. Attribution studies evaluate the consistency between the model-simulated amplitude of response and that inferred from observations. In the case of uncertain forcings, scaling factors provide information about the strength of the forcing (and response) needed to reproduce the observations, or about the possibility that the simulated pattern or strength of response is incorrect. However, for a model simulation to be considered consistent with the observations given forcing uncertainty, the forcing used in the model should remain consistent with the uncertainty bounds from forward model estimates of forcing.

Detection and attribution approaches that try to distinguish the response to several external forcings simultaneously may be affected by similarities in the pattern of response to different forcings and by uncertainties in forcing and response. Similarities between the responses to different forcings, particularly in the spatial patterns of response, make it more difficult to distinguish between responses to different external forcings, but also imply that the response patterns will be relatively insensitive to modest errors in the magnitude and distribution of the forcing. Differences between the temporal histories of different kinds of forcing (e.g., greenhouse gas versus sulphate aerosol) ameliorate the problem of the similarity between the spatial patterns of response considerably. For example, the spatial response of surface temperature to solar forcing resembles that due to anthropogenic greenhouse gas forcing (Weatherall and Manabe, 1975; Nesme-Ribes et al., 1993; Cubasch et al., 1997; Rind et al., 2004; Zorita et al., 2005). Distinct features of the vertical structure of the responses in the atmosphere to different types of forcing further help to distinguish between the different sources of forcing. Studies that interpret observed climate in subsequent sections use such strategies, and the overall assessment in this chapter uses results from a range of climate variables and observations.

Many detection studies attempt to identify in observations both temporal and spatial aspects of the temperature response to a given set of forcings because the combined space-time responses tend to be more distinct than either the space-only or the time-only patterns of response. Because the emissions and burdens of different forcing agents change with time, the net forcing and its rate of change vary with time. Although explicit accounting for uncertainties in the net forcing is not available (see discussion in Sections 9.2.2.3 and 9.2.2.4), models often employ different implementations of external forcing. Detection and attribution studies using such simulations suggest that results are not very sensitive to moderate forcing uncertainties. A further problem arises due to spurious temporal correlations between the responses to different forcings that arise from sampling variability. For example, spurious correlation between the climate responses to solar and volcanic forcing over parts of the 20th century (North and Stevens, 1998) can lead to misidentification of one as the other, as in Douglass and Clader (2002).

The spatial pattern of the temperature response to aerosol forcing is quite distinct from the spatial response pattern to CO_2 in some models and diagnostics (Hegerl et al., 1997), but less so in others (Reader and Boer, 1998; Tett et al., 1999; Hegerl et al., 2000; Harvey, 2004). If it is not possible to distinguish the spatial pattern of greenhouse warming from that of fossil-fuel related aerosol cooling, the observed warming over the last century could be explained by large greenhouse warming balanced by large aerosol cooling or alternatively by small greenhouse warming with very little or no aerosol cooling. Nevertheless, estimates of the amplitude of the response to greenhouse forcing in the 20th century from detection studies are quite similar, even though the simulated responses to aerosol forcing are model dependent (Gillett et al., 2002a; Hegerl and Allen, 2002). Considering three different climate models, Stott et al. (2006c) conclude that an important constraint on the possible range of responses to aerosol forcing is the temporal evolution of the global mean and hemispheric temperature contrast as was suggested by Santer et al. (1996a; see also Section 9.4.1.5).

9.2.4 Summary

The uncertainty in the magnitude and spatial pattern of forcing differs considerably between forcings. For example, well-mixed greenhouse gas forcing is relatively well constrained and spatially homogeneous. In contrast, uncertainties are large for many non-greenhouse gas forcings. Inverse model studies, which use methods closely related to those used in climate change detection research, indicate that the magnitude of the total net aerosol forcing has a likely range of -1.7 to -0.1 W m^{-2}. As summarised in Chapter 2, forward calculations of aerosol radiative forcing, which do not depend on knowledge of observed climate change or the ability of climate models to simulate the transient response to forcings, provide results (-2.2 to -0.5 W m^{-2}; 5 to 95%) that are quite consistent with inverse estimates; the uncertainty ranges from inverse and forward calculations are different due to the use of different information. The large uncertainty in total aerosol forcing makes it more difficult to accurately infer the climate sensitivity from observations (Section 9.6). It also increases uncertainties in results that attribute cause to observed climate change (Section 9.4.1.4), and is in part responsible for differences in probabilistic projections of future climate change (Chapter 10). Forcings from black carbon, fossil fuel organic matter and biomass burning aerosols, which have not been considered in most detection studies performed to date, are likely small but with large uncertainties relative to the magnitudes of the forcings.

Uncertainties also differ between natural forcings and sometimes between different time scales for the same forcing. For example, while the 11-year solar forcing cycle is well documented, lower-frequency variations in solar forcing are highly uncertain. Furthermore, the physics of the response to solar forcing and some feedbacks are still poorly understood. In contrast, the timing and duration of forcing due to aerosols ejected into the stratosphere by large volcanic eruptions is well known during the instrumental period, although the magnitude of that forcing is uncertain.

Differences in the temporal evolution and sometimes the spatial pattern of climate response to external forcing make it possible, with limitations, to separate the response to these forcings in observations, such as the responses to greenhouse gas and sulphate aerosol forcing. In contrast, the climate response and temporal evolution of other anthropogenic forcings is more uncertain, making the simulation of the climate response and its detection in observations more difficult. The temporal evolution, and to some extent the spatial and vertical pattern, of the climate response to natural forcings is also quite different from that of anthropogenic forcing. This makes it possible to separate the climate response to solar and volcanic forcing from the response to anthropogenic forcing despite the uncertainty in the history of solar forcing noted above.

9.3 Understanding Pre-Industrial Climate Change

9.3.1 Why Consider Pre-Industrial Climate Change?

The Earth system has experienced large-scale climate changes in the past (Chapter 6) that hold important lessons for the understanding of present and future climate change. These changes resulted from natural external forcings that, in some instances, triggered strong feedbacks as in the case of the LGM (see Chapter 6). Past periods offer the potential to provide information not available from the instrumental record, which is affected by anthropogenic as well as natural external forcings and is too short to fully understand climate variability and major climate system feedbacks on inter-decadal and longer time scales. Indirect indicators ('proxy data' such as tree ring width and density) must be used to infer climate variations (Chapter 6) prior to the instrumental era (Chapter 3). A complete description of these data and of their uncertainties can be found in Chapter 6.

The discussion here is restricted to several periods in the past for which modelling and observational evidence can be compared to test understanding of the climate response to external forcings. One such period is the last millennium, which places the recent instrumental record in a broader context (e.g., Mitchell et al., 2001). The analysis of the past 1 kyr focuses mainly on the climate response to natural forcings (changes in solar radiation and volcanism) and on the role of anthropogenic forcing during the most recent part of the record. Two time periods analysed in the Paleoclimate Modelling Intercomparison Project (PMIP, Joussaume and Taylor, 1995; PMIP2, Harrison et al., 2002) are also considered, the mid-Holocene (6 ka) and the LGM (21 ka). Both periods had a substantially different climate compared to the present, and there is relatively good information from data synthesis and model simulation experiments (Braconnot et al., 2004; Cane et al., 2006). An increased number of simulations using EMICs or Atmosphere-Ocean General Circulation Models (AOGCMs) that are the same as, or related to, the models used in simulations of the climates of the 20th and 21st centuries are available for these periods.

9.3.2 What Can be Learned from the Last Glacial Maximum and the Mid-Holocene?

Relatively high-quality global terrestrial climate reconstructions exist for the LGM and the mid-Holocene and as part of the Global Palaeovegetation Mapping (BIOME 6000) project (Prentice and Webb, 1998; Prentice and Jolly, 2000). The Climate: Long-range Investigation, Mapping and Prediction (CLIMAP, 1981) reconstruction of LGM sea surface temperatures has also been improved (Chapter 6). The LGM climate was colder and drier than at present as is indicated by the extensive tundra and steppe vegetation that existed during this period. Most LGM proxy data suggest that the tropical oceans were colder by about 2°C than at present, and that the frontal zones in the SH and NH were shifted equatorward (Kucera et al., 2005), even though large differences are found between temperature estimates from the different proxies in the North Atlantic.

Several new AOGCM simulations of the LGM have been produced since the TAR. These simulations show a global cooling of approximately 3.5°C to 5.2°C when LGM greenhouse gas and ice sheet boundary conditions are specified (Chapter 6), which is within the range (–1.8°C to –6.5°C) of PMIP results from simpler models that were discussed in the TAR (McAvaney et al., 2001). Only one simulation exhibits a very strong response with a cooling of approximately 10°C (Kim et al., 2002). All of these simulations exhibit a strongly damped hydrological cycle relative to that of the modern climate, with less evaporation over the oceans and continental-scale drying over land. Changes in greenhouse gas concentrations may account for about half of the simulated tropical cooling (Shin et al., 2003), and for the production of colder and saltier water found at depth in the Southern Ocean (Liu et al., 2005). Most LGM simulations with coupled models shift the deep-water formation in the North

Atlantic southward, but large differences exist between models in the intensity of the Atlantic meridional overturning circulation. Including vegetation changes appears to improve the realism of LGM simulations (Wyputta and McAvaney, 2001). Furthermore, including the physiological effect of the atmospheric CO_2 concentration on vegetation has a non-negligible impact (Levis et al., 1999) and is necessary to properly represent changes in global forest (Harrison and Prentice, 2003) and terrestrial carbon storage (e.g., Kaplan et al., 2002; Joos et al., 2004; see also Chapter 6). To summarise, despite large uncertainties, LGM simulations capture the broad features found in palaeoclimate data, and better agreement is obtained with new coupled simulations using more recent models and more complete feedbacks from ocean, sea ice and land surface characteristics such as vegetation and soil moisture (Chapter 6).

Closer to the present, during the mid-Holocene, one of the most noticeable indications of climate change is the northward extension of northern temperate forest (Bigelow et al., 2003), which reflects warmer summers than at present. In the tropics the more vegetated conditions inferred from pollen records in the now dry sub-Saharan regions indicate wetter conditions due to enhanced summer monsoons (see Braconnot et al., 2004 for a review). Simulations of the mid-Holocene with AOGCMs (see Section 9.2.1.3 for forcing) produce an amplification of the mean seasonal cycle of temperature of approximately 0.5°C to 0.7°C. This range is slightly smaller than that obtained using atmosphere-only models in PMIP1 (~0.5°C to ~1.2°C) due to the thermal response of the ocean (Braconnot et al., 2000). Simulated changes in the ocean circulation have strong seasonal features with an amplification of the sea surface temperature (SST) seasonal cycle of 1°C to 2°C in most places within the tropics (Zhao et al., 2005), influencing the Indian and African monsoons. Over West Africa, AOGCM-simulated changes in annual mean precipitation are about 5 to 10% larger than for atmosphere-only simulations, and in better agreement with data reconstructions (Braconnot et al., 2004). Results for the Indian and Southwest Asian monsoon are less consistent between models.

As noted in the TAR (McAvaney et al., 2001), vegetation change during the mid-Holocene likely triggered changes in the hydrological cycle, explaining the wet conditions that prevailed in the Sahel region that were further enhanced by ocean feedbacks (Ganopolski et al., 1998; Braconnot et al., 1999), although soil moisture may have counteracted some of these feedbacks (Levis et al., 2004). Wohlfahrt et al. (2004) show that at middle and high latitudes the vegetation and ocean feedbacks enhanced the warming in spring and autumn by about 0.8°C. However, models have a tendency to overestimate the mid-continental drying in Eurasia, which is further amplified when vegetation feedbacks are included (Wohlfahrt et al., 2004).

A wide range of proxies containing information about ENSO variability during the mid-Holocene is now also available (Section 6.5.3). These data suggest that ENSO variability was weaker than today prior to approximately 5 kyr before present (Moy et al., 2002 and references therein; Tudhope and Collins, 2003). Several studies have attempted to analyse these changes

in interannual variability from model simulations. Even though some results are controversial, a consistent picture has emerged for the mid-Holocene, for which simulations produce reduced variability in precipitation over most ocean regions in the tropics (Liu et al., 2000; Braconnot et al., 2004; Zhao et al., 2005). Results obtained with the Cane-Zebiak model suggest that the Bjerknes (1969) feedback mechanism may be a key element of the ENSO response in that model. The increased mid-Holocene solar heating in boreal summer leads to more warming in the western than in the eastern Pacific, which strengthens the trade winds and inhibits the development of ENSO (Clement et al., 2000, 2004). Atmosphere-Ocean General Circulation Models also tend to simulate less intense ENSO events, in qualitative agreement with data, although there are large differences in magnitude and proposed mechanisms, and inconsistent responses of the associated teleconnections (Otto-Bliesner, 1999; Liu et al., 2000; Kitoh and Murakami, 2002; Otto-Bliesner et al., 2003).

9.3.3 What Can be Learned from the Past 1,000 Years?

External forcing relative to the present is generally small for the last millennium when compared to that for the mid-Holocene and LGM. Nonetheless, there is evidence that climatic responses to forcing, together with natural internal variability of the climate system, produced several well-defined climatic events, such as the cool conditions during the 17th century or relatively warm periods early in the millennium.

9.3.3.1 Evidence of External Influence on the Climate Over the Past 1,000 Years

A substantial number of proxy reconstructions of annual or decadal NH mean surface temperature are now available (see Figure 6.11, and the reviews by Jones et al., 2001 and Jones and Mann, 2004). Several new reconstructions have been published, some of which suggest larger variations over the last millennium than assessed in the TAR, but uncertainty remains in the magnitude of inter-decadal to inter-centennial variability. This uncertainty arises because different studies rely on different proxy data or use different reconstruction methods (Section 6.6.1). Nonetheless, NH mean temperatures in the second half of the 20th century were likely warmer than in any other 50-year period in the last 1.3 kyr (Chapter 6), and very likely warmer than any such period in the last 500 years. Temperatures subsequently decreased, and then rose rapidly during the most recent 100 years. This long-term tendency is punctuated by substantial shorter-term variability (Figure 6.10). For example, cooler conditions with temperatures 0.5°C to 1°C below the 20th-century mean value are found in the 17th and early 18th centuries.

A number of simulations of the last millennium (Figure 6.13) have been performed using a range of models, including some simulations with AOGCMs (e.g., Crowley, 2000; Goosse and Renssen, 2001; Bertrand et al., 2002; Bauer et al., 2003; Gerber et al., 2003; see also Gonzalez-Rouco et al., 2003; Jones and Mann, 2004; Zorita et al., 2004; Weber, 2005; Tett et al., 2007). These simulations use different reconstructions of external forcing, particularly solar, volcanic and greenhouse gas forcing, and often include land use changes (e.g., Bertrand et al., 2002; Stendel et al., 2006; Tett et al., 2007). While the use of different models and forcing reconstructions leads to differences, the simulated evolution of the NH annual mean surface temperature displays some common characteristics between models that are consistent with the broad features of the data (Figures 6.13 and 9.4). For example, all simulations show relatively cold conditions during the period around 1675 to 1715 in response to natural forcing, which is in qualitative agreement with the proxy reconstructions. In all simulations shown in Figure 6.13, the late 20th century is warmer than any other multi-decadal period during the last millennium. In addition, there is significant correlation between simulated and reconstructed variability (e.g., Yoshimori et al., 2005). By comparing simulated and observed atmospheric CO_2 concentration during the last 1 kyr, Gerber et al. (2003) suggest that the amplitude of the temperature evolution simulated by simple climate models and EMICs is consistent with the observed evolution of CO_2. Since reconstructions of external forcing are virtually independent from the reconstructions of past temperatures, this broad consistency increases confidence in the broad features of the reconstructions and the understanding of the role of external forcing in recent climate variability. The simulations also show that it is not possible to reproduce the large 20th-century warming without anthropogenic forcing regardless of which solar or volcanic forcing reconstruction is used (Crowley, 2000; Bertrand et al., 2002; Bauer et al., 2003; Hegerl et al., 2003, 2007), stressing the impact of human activity on the recent warming.

While there is broad qualitative agreement between simulated and reconstructed temperatures, it is difficult to fully assess model-simulated variability because of uncertainty in the magnitude of historical variations in the reconstructions and differences in the sensitivity to external forcing (Table 8.2). The role of internal variability has been found to be smaller than that of the forced variability for hemispheric temperature means at decadal or longer time scales (Crowley, 2000; Hegerl et al., 2003; Goosse et al., 2004; Weber et al., 2004; Hegerl et al., 2007; Tett et al., 2007), and thus internal variability is a relatively small contributor to differences between different simulations of NH mean temperature. Other sources of uncertainty in simulations include model ocean initial conditions, which, for example, explain the warm conditions found in the Zorita et al. (2004) simulation during the first part of the millennium (Goosse et al., 2005; Osborn et al., 2006).

9.3.3.2   Role of Volcanism and Solar Irradiance

Volcanic eruptions cause rapid decreases in hemispheric and global mean temperatures followed by gradual recovery over several years (Section 9.2.2.1) in climate simulations driven by volcanic forcing (Figure 6.13; Crowley, 2000; Bertrand

et al., 2002; Weber, 2005; Yoshimori et al., 2005; Tett et al., 2007). These simulated changes appear to correspond to cool episodes in proxy reconstructions (Figure 6.13). This suggestive correspondence has been confirmed in comparisons between composites of temperatures following multiple volcanic eruptions in simulations and reconstructions (Hegerl et al., 2003; Weber, 2005). In addition, changes in the frequency of large eruptions result in climate variability on decadal and possibly longer time scales (Crowley, 2000; Briffa et al., 2001; Bertrand et al., 2002; Bauer et al., 2003; Weber, 2005). Hegerl et al. (2003; 2007), using a multi-regression approach based on Energy Balance Model (EBM) simulated fingerprints of solar, volcanic and greenhouse gas forcing (Appendix 9.A.1; see also Section 9.4.1.4 for the 20th century), simultaneously detect the responses to volcanic and greenhouse gas forcing in a number of proxy reconstructions of average NH mean annual and growing season temperatures (Figure 9.4) with high significance. They find that a high percentage of decadal variance in the reconstructions used can be explained by external forcing (between 49 and 70% of decadal variance depending upon the reconstruction).

There is more uncertainty regarding the influence of solar forcing. In addition to substantial uncertainty in the timing and amplitude of solar variations on time scales of several decades to centuries, which has increased since the TAR although the estimate of solar forcing has been revised downwards (Sections 9.2.1.3 and 2.7.1), uncertainty also arises because the spatial response of surface temperature to solar forcing resembles that due to greenhouse gas forcing (Section 9.2.3). Analyses that make use of differences in the temporal evolution of solar and volcanic forcings are better able to distinguish between the two (Section 9.2.3; see also Section 9.4.1.5 for the 20th century). In such an analysis, solar forcing can only be detected and distinguished from the effect of volcanic and greenhouse gas forcing over some periods in some reconstructions (Hegerl et al., 2003, 2007), although the effect of solar forcing has been detected over parts of the 20th century in some time-space analyses (Section 9.4.1.5) and there are similarities between regressions of solar forcing on model simulations and several proxy reconstructions (Weber, 2005; see also Waple, 2002). A model simulation (Shindell et al., 2003) suggests that solar forcing may play a substantial role in regional anomalies due to dynamical feedbacks. These uncertainties in the contribution of different forcings to climatic events during the last millennium reflect substantial uncertainty in knowledge about past solar and volcanic forcing, as well as differences in the way these effects are taken into account in model simulations.

Overall, modelling and detection and attribution studies confirm a role of volcanic, greenhouse gas and probably solar forcing in explaining the broad temperature evolution of the last millennium, although the role of solar forcing has recently been questioned (Foukal et al., 2006). The variability that remains in proxy reconstructions after estimates of the responses to external forcing have been removed is broadly consistent with AOGCM-simulated internal variability (e.g., Hegerl et al., 2003, 2007), providing a useful check on AOGCMs even though

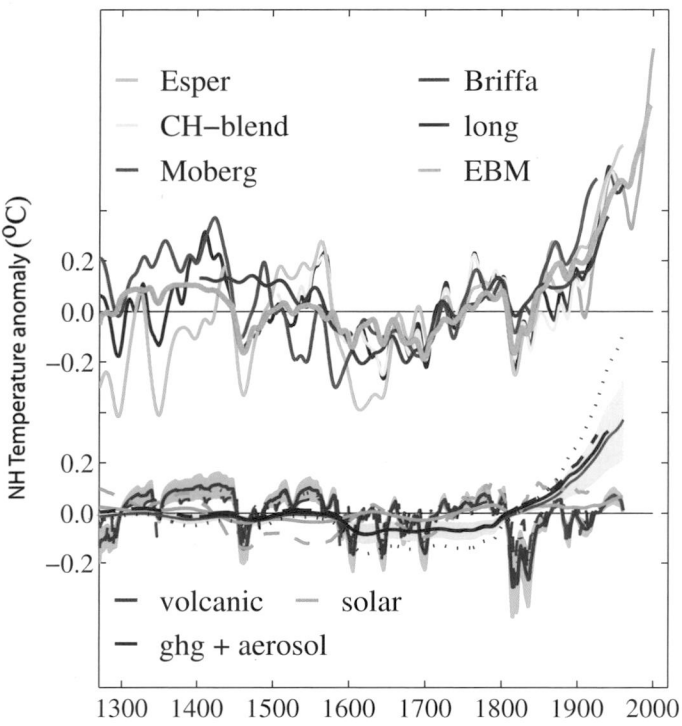

Figure 9.4. *Contribution of external forcing to several high-variance reconstructions of NH temperature anomalies, (Esper et al., 2002; Briffa et al., 2001; Hegerl et al., 2007, termed CH-blend and CH-blend long; and Moberg et al., 2005). The top panel compares reconstructions to an EBM simulation (equilibrium climate sensitivity of 2.5°C) of NH 30°N to 90°N average temperature, forced with volcanic, solar and anthropogenic forcing. All timeseries are centered on the 1500-1925 average. Instrumental temperature data are shown by a green line (centered to agree with CH-blend average over the period 1880-1960). The displayed data are low-pass filtered (20-year cutoff) for clarity. The bottom panel shows the estimated contribution of the response to volcanic (blue lines with blue uncertainty shade), solar (green) and greenhouse gas (GHG) and aerosol forcing (red line with yellow shades, aerosol only in 20th century) to each reconstruction (all timeseries are centered over the analysis period). The estimates are based on multiple regression of the reconstructions on fingerprints for individual forcings. The contributions to different reconstructions are indicated by different line styles (Briffa et al.: solid, fat; Esper et al.: dotted; Moberg: dashed; CH-blend: solid, thin; with shaded 90% confidence limits around best estimates for each detectable signal). All reconstructions show a highly significant volcanic signal, and all but Moberg et al. (which ends in 1925) show a detectable greenhouse gas signal at the 5% significance level. The latter shows a detectable greenhouse gas signal with less significance. Only Moberg et al. contains a detectable solar signal (only shown for these data and CH-blend, where it is not detectable). All data are decadally averaged. The reconstructions represent slightly different regions and seasons: Esper et al. (2002) is calibrated to 30°N to 90°N land temperature, CH-blend and CH-blend long (Hegerl et al., 2007) to 30°N to 90°N mean temperature and Moberg et al. (2005) to 0° to 90°N temperature. From Hegerl et al. (2007).*

uncertainties are large. Such studies also help to explain episodes during the climate of the last millennium. For example, several modelling studies suggest that volcanic activity has a dominant role in explaining the cold conditions that prevailed from 1675 to 1715 (Andronova et al., 2007; Yoshimori et al., 2005). In contrast, Rind et al. (2004) estimate from model simulations that the cooling relative to today was primarily associated with reduced greenhouse gas forcing, with a substantial contribution from solar forcing.

There is also some evidence from proxy data that the response to external forcing may influence modes of climate variability. For example, Cobb et al. (2003), using fossil corals, attempt to extend the ENSO record back through the last millennium. They find that ENSO events may have been as frequent and intense during the mid-17th century as during the instrumental period, with events possibly rivalling the strong 1997–1998 event. On the other hand, there are periods during the 12th and 14th centuries when there may have been significantly less ENSO variability, a period during which there were also cooler conditions in the northeast Pacific (MacDonald and Case, 2005) and evidence of droughts in central North America (Cook et al., 2004). Cobb et al. (2003) find that fluctuations in reconstructed ENSO variability do not appear to be correlated in an obvious way with mean state changes in the tropical Pacific or global mean climate, while Adams et al. (2003) find statistical evidence for an El Niño-like anomaly during the first few years following explosive tropical volcanic eruptions. The Cane-Zebiak model simulates changes similar to those in the Cobb et al. (2003) data when volcanism and solar forcing are accounted for, supporting the link with volcanic forcing over the past millennium (Mann et al., 2005). However, additional studies with different models are needed to fully assess this relationship, since previous work was less conclusive (Robock, 2000).

Extratropical variability also appears to respond to volcanic forcing. During the winter following a large volcanic eruption, the zonal circulation may be more intense, causing a relative warming over the continents during the cold season that could partly offset the direct cooling due to the volcanic aerosols (Sections 9.2.2.1 and 8.4.1; Robock, 2000; Shindell et al., 2003). A tendency towards the negative NAO state during periods of reduced solar input is found in some reconstructions of this pattern for the NH (Shindell et al., 2001b; Luterbacher et al., 2002, 2004; Stendel et al., 2006), possibly implying a solar forcing role in some long-term regional changes, such as the cooling over the NH continents around 1700 (Shindell et al., 2001b; Section 9.2.2). Indications of changes in ENSO variability during the low solar irradiance period of the 17th to early 18th centuries are controversial (e.g., D'Arrigo et al., 2005).

9.3.3.3   *Other Forcings and Sources of Uncertainties*

In addition to forcing uncertainties discussed above, a number of other uncertainties affect the understanding of pre-industrial climate change. For example, land cover change may have influenced the pre-industrial climate (Bertrand et al., 2002; Bauer et al., 2003), leading to a regional cooling of 1°C to 2°C in winter and spring over the major agricultural regions of North America and Eurasia in some model simulations, when pre-agriculture vegetation was replaced by present-day vegetation (Betts, 2001). The largest anthropogenic land cover changes involve deforestation (Chapter 2). The greatest proportion of deforestation has occurred in the temperate regions of the NH (Ramankutty and Foley, 1999; Goldewijk, 2001). Europe had cleared about 80% of its agricultural area by 1860, but over half of the forest removal in North America took place after 1860 (Betts, 2001), mainly in the late 19th century (Stendel et al., 2006). During the past two decades, the CO_2 flux caused by land use changes has been dominated by tropical deforestation (Section 7.3.2.1.2). Climate model simulations suggest that the effect of land use change was likely small at hemispheric and global scales, estimated variously as –0.02°C relative to natural pre-agricultural vegetation (Betts, 2001), less than –0.1°C since 1700 (Stendel et al., 2006) and about –0.05°C over the 20th century and too small to be detected statistically in observed trends (Matthews et al., 2004). However, the latter authors did find a larger cooling effect since 1700 of between –0.06°C and –0.22°C when they explored the sensitivity to different representations of land cover change.

Oceanic processes and ocean-atmosphere interaction may also have played a role in the climate evolution during the last millennium (Delworth and Knutson, 2000; Weber et al., 2004; van der Schrier and Barkmeijer, 2005). Climate models generally simulate a weak to moderate increase in the intensity of the oceanic meridional overturning circulation in response to a decrease in solar irradiance (Cubasch et al., 1997; Goosse and Renssen, 2004; Weber et al., 2004). A delayed response to natural forcing due to the storage and transport of heat anomalies by the deep ocean has been proposed to explain the warm Southern Ocean around the 14th to 15th centuries (Goosse et al., 2004).

9.3.4 Summary

Considerable progress has been made since the TAR in understanding the response of the climate system to external forcings. Periods like the mid-Holocene and the LGM are now used as benchmarks for climate models that are used to simulate future climate (Chapter 6). While considerable uncertainties remain in the climate reconstructions for these periods, and in the boundary conditions used to force climate models, comparisons between simulated and reconstructed conditions in the LGM and mid-Holocene demonstrate that models capture the broad features of changes in the temperature and precipitation patterns. These studies have also increased understanding of the roles of ocean and vegetation feedbacks in determining the response to solar and greenhouse gas forcing. Moreover, although proxy data on palaeoclimatic interannual to multi-decadal variability during these periods remain very uncertain, there is an increased appreciation that external forcing may, in the past, have affected climatic variability such as that associated with ENSO.

The understanding of climate variability and change, and its causes during the past 1 kyr, has also improved since the TAR (IPCC, 2001). There is consensus across all millennial reconstructions on the timing of major climatic events, although their magnitude remains somewhat uncertain. Nonetheless, the collection of reconstructions from palaeodata, which is larger and more closely scrutinised than that available for the TAR, indicates that it is likely that NH average temperatures during the second half of the 20th century were warmer than any other 50-year period during the past 1.3 kyr (Chapter 6). While

uncertainties remain in temperature and forcing reconstructions, and in the models used to estimate the responses to external forcings, the available detection studies, modelling and other evidence support the conclusion that volcanic and possibly solar forcings have very likely affected NH mean temperature over the past millennium and that external influences explain a substantial fraction of inter-decadal temperature variability in the past. The available evidence also indicates that natural forcing may have influenced the climatic conditions of individual periods, such as the cooler conditions around 1700. The climate response to greenhouse gas increases can be detected in a range of proxy reconstructions by the end of the records.

When driven with estimates of external forcing for the last millennium, AOGCMs simulate changes in hemispheric mean temperature that are in broad agreement with proxy reconstructions (given their uncertainties), increasing confidence in the forcing reconstructions, proxy climate reconstructions and models. In addition, the residual variability in the proxy climate reconstructions that is not explained by forcing is broadly consistent with AOGCM-simulated internal variability. Overall, the information on temperature change over the last millennium is broadly consistent with the understanding of climate change in the instrumental era.

9.4 Understanding of Air Temperature Change During the Industrial Era

9.4.1 Global-Scale Surface Temperature Change

9.4.1.1 Observed Changes

Six additional years of observations since the TAR (Chapter 3) show that temperatures are continuing to warm near the surface of the planet. The annual global mean temperature for every year since the TAR has been among the 10 warmest years since the beginning of the instrumental record. The global mean temperature averaged over land and ocean surfaces warmed by 0.76°C ± 0.19°C between the first 50 years of the instrumental record (1850–1899) and the last 5 years (2001–2005) (Chapter 3; with a linear warming trend of 0.74°C ± 0.18°C over the last 100 years (1906–2005)). The rate of warming over the last 50 years is almost double that over the last 100 years (0.13°C ± 0.03°C vs 0.07°C = 0.02°C per decade; Chapter 3). The larger number of proxy reconstructions from palaeodata than were available for the TAR indicate that it is very likely that average NH temperatures during the second half of the 20th century were warmer than any other 50-year period in the last 500 years and it is likely that this was the warmest period in the past 1.3 kyr (Chapter 6). Global mean temperature has not increased smoothly since 1900 as would be expected if it were influenced only by forcing from increasing greenhouse gas concentrations (i.e., if natural variability and other forcings did not have a role; see Section 9.2.1; Chapter 2). A rise in near-surface temperatures also occurred over several decades during the first half of the 20th century, followed by a period of more than three decades when temperatures showed no pronounced trend (Figure 3.6). Since the mid-1970s, land regions have warmed at a faster rate than oceans in both hemispheres (Figure 3.8) and warming over the SH was smaller than that over the NH during this period (Figure 3.6), while warming rates during the early 20th century were similar over land and ocean.

9.4.1.2   Simulations of the 20th Century

There are now a greater number of climate simulations from AOGCMs for the period of the global surface instrumental record than were available for the TAR, including a greater variety of forcings in a greater variety of combinations. These simulations used models with different climate sensitivities, rates of ocean heat uptake and magnitudes and types of forcings (Supplementary Material, Table S9.1). Figure 9.5 shows that simulations that incorporate anthropogenic forcings, including increasing greenhouse gas concentrations and the effects of aerosols, and that also incorporate natural external forcings provide a consistent explanation of the observed temperature record, whereas simulations that include only natural forcings do not simulate the warming observed over the last three decades. A variety of different forcings is used in these simulations. For example, some anthropogenically forced simulations include both the direct and indirect effects of sulphate aerosols whereas others include just the direct effect, and the aerosol forcing that is calculated within models differs due to differences in the representation of physics. Similarly, the effects of tropospheric and stratospheric ozone changes are included in some simulations but not others, and a few simulations include the effects of carbonaceous aerosols and land use changes, while the naturally forced simulations include different representations of changing solar and volcanic forcing. Despite this additional uncertainty, there is a clear separation in Figure 9.5 between the simulations with anthropogenic forcings and those without.

Global mean and hemispheric-scale temperatures on multi-decadal time scales are largely controlled by external forcings (Stott et al., 2000). This external control is demonstrated by ensembles of model simulations with identical forcings (whether anthropogenic or natural) whose members exhibit very similar simulations of global mean temperature on multi-decadal time scales (e.g., Stott et al., 2000; Broccoli et al., 2003; Meehl et al., 2004). Larger interannual variations are seen in the observations than in the ensemble mean model simulation of the 20th century because the ensemble averaging process filters out much of the natural internal interannual variability that is simulated by the models. The interannual variability in the individual simulations that is evident in Figure 9.5 suggests that current models generally simulate large-scale natural internal variability quite well, and also capture the cooling associated with volcanic eruptions on shorter time scales. Section 9.4.1.3 assesses the variability of near surface temperature observations and simulations.

The fact that climate models are only able to reproduce observed global mean temperature changes over the 20th century when they include anthropogenic forcings, and that they fail to do so when they exclude anthropogenic forcings, is evidence for the influence of humans on global climate. Further evidence is provided by spatial patterns of temperature change. Figure 9.6 compares observed near-surface temperature trends over the

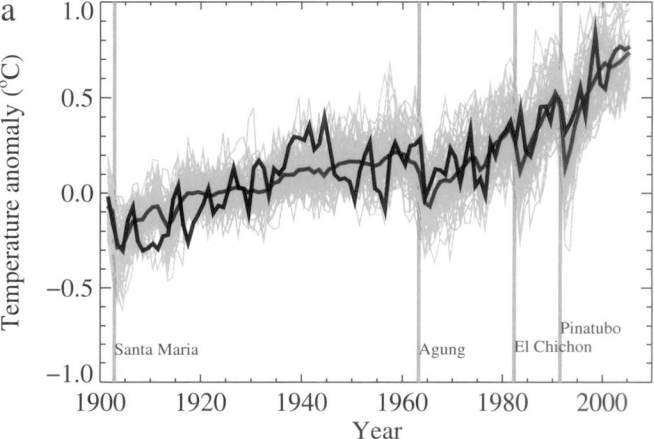

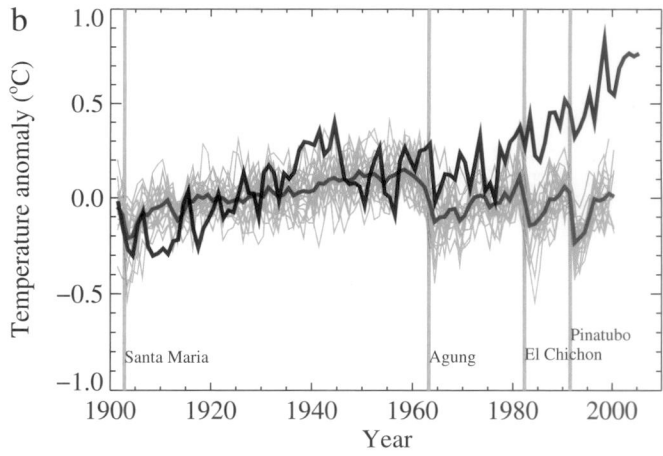

Figure 9.5. *Comparison between global mean surface temperature anomalies (°C) from observations (black) and AOGCM simulations forced with (a) both anthropogenic and natural forcings and (b) natural forcings only. All data are shown as global mean temperature anomalies relative to the period 1901 to 1950, as observed (black, Hadley Centre/Climatic Research Unit gridded surface temperature data set (HadCRUT3); Brohan et al., 2006) and, in (a) as obtained from 58 simulations produced by 14 models with both anthropogenic and natural forcings. The multi-model ensemble mean is shown as a thick red curve and individual simulations are shown as thin yellow curves. Vertical grey lines indicate the timing of major volcanic events. Those simulations that ended before 2005 were extended to 2005 by using the first few years of the IPCC Special Report on Emission Scenarios (SRES) A1B scenario simulations that continued from the respective 20th-century simulations, where available. The simulated global mean temperature anomalies in (b) are from 19 simulations produced by five models with natural forcings only. The multi-model ensemble mean is shown as a thick blue curve and individual simulations are shown as thin blue curves. Simulations are selected that do not exhibit excessive drift in their control simulations (no more than 0.2°C per century). Each simulation was sampled so that coverage corresponds to that of the observations. Further details of the models included and the methodology for producing this figure are given in the Supplementary Material, Appendix 9.C. After Stott et al. (2006b).*

globe (top row) with those simulated by climate models when they include anthropogenic and natural forcing (second row) and the same trends simulated by climate models when only natural forcings are included (third row). The observed trend over the entire 20th century (Figure 9.6, top left panel) shows warming almost everywhere with the exception of the southeastern USA, northern North Atlantic, and isolated grid boxes in Africa and South America (see also Figure 3.9). Such a pattern of warming is not associated with known modes of internal climate variability. For example, while El Niño or El Niño-like decadal variability results in unusually warm annual temperatures, the spatial pattern associated with such a warming is more structured, with cooling in the North Pacific and South Pacific (see, e.g., Zhang et al., 1997). In contrast, the trends in climate model simulations that include anthropogenic and natural forcing (Figure 9.6, second row) show a pattern of spatially near-uniform warming similar to that observed. There is much greater similarity between the general evolution of the warming in observations and that simulated by models when anthropogenic and natural forcings are included than when only natural forcing is included (Figure 9.6, third row). Figure 9.6 (fourth row) shows that climate models are only able to reproduce the observed patterns of zonal mean near-surface temperature trends over the 1901 to 2005 and 1979 to 2005 periods when they include anthropogenic forcings and fail to do so when they exclude anthropogenic forcings. Although there is less warming at low latitudes than at high northern latitudes, there is also less internal variability at low latitudes, which results in a greater separation of the climate simulations with and without anthropogenic forcings.

Climate simulations are consistent in showing that the global mean warming observed since 1970 can only be reproduced when models are forced with combinations of external forcings

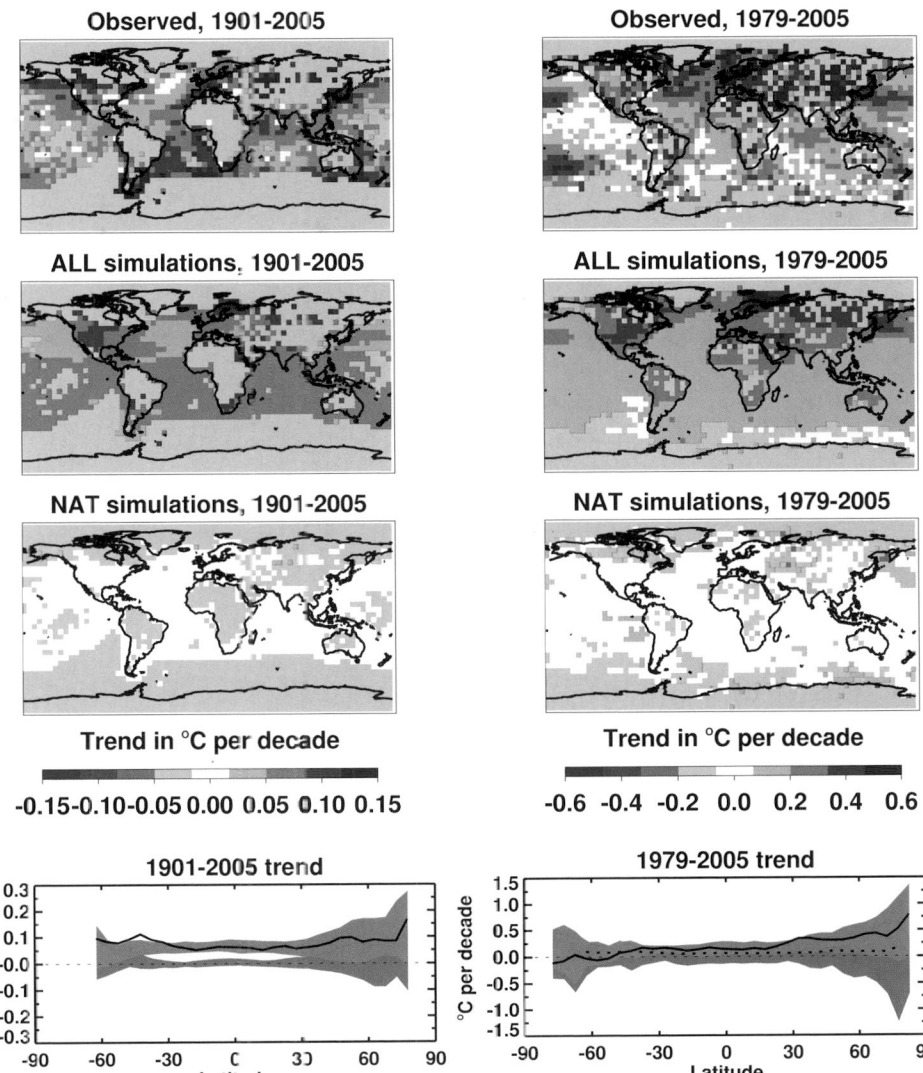

Figure 9.6. *Trends in observed and simulated temperature changes (°C) over the 1901 to 2005 (left column) and 1979 to 2005 (right column) periods. First row: trends in observed temperature changes (Hadley Centre/Climatic Research Unit gridded surface temperature data set (HadCRUT3), Brohan et al., 2006). Second row: average trends in 58 historical simulations from 14 climate models including both anthropogenic and natural forcings. Third row: average trends in 19 historical simulations from five climate models including natural forcings only. Grey shading in top three rows indicates regions where there are insufficient observed data to calculate a trend for that grid box (see Supplementary Material, Appendix 9.C for further details of data exclusion criteria). Fourth row: average trends for each latitude; observed trends are indicated by solid black curves. Red shading indicates the middle 90% range of trend estimates from the 58 simulations including both anthropogenic and natural forcings (estimated as the range between 4th and 55th of the 58 ranked simulations); blue shading indicates the middle 90% range of trend estimates from the 19 simulations with natural forcings only (estimated as the range between 2nd and 18th of the 19 ranked simulations); for comparison, the dotted black curve in the right-hand plot shows the observed 1901 to 2005 trend. Note that scales are different between columns. The 'ALL' simulations were extended to 2005 by adding their IPCC Special Report on Emission Scenarios (SRES) A1B continuation runs where available. Where not available, and in the case of the 'NAT' simulations, the mean for the 1996 to 2005 decade was estimated using model output from 1996 to the end of the available runs. In all plots, each climate simulation was sampled so that coverage corresponds to that of the observations. Further details of the models included and the methodology for producing this figure are given in the Supplementary Material, Appendix 9.C.*

that include anthropogenic forcings (Figure 9.5). This conclusion holds despite a variety of different anthropogenic forcings and processes being included in these models (e.g., Tett et al., 2002; Broccoli et al., 2003; Meehl et al., 2004; Knutson et al., 2006). In all cases, the response to forcing from well-mixed greenhouse gases dominates the anthropogenic warming in the model. No climate model using natural forcings alone has reproduced the observed global warming trend in the second half of the 20th century. Therefore, modelling studies suggest that late 20th-century warming is much more likely to be anthropogenic than natural in origin, a finding which is confirmed by studies relying on formal detection and attribution methods (Section 9.4.1.4).

Modelling studies are also in moderately good agreement with observations during the first half of the 20th century when both anthropogenic and natural forcings are considered, although assessments of which forcings are important differ, with some studies finding that solar forcing is more important (Meehl et al., 2004) while other studies find that volcanic forcing (Broccoli et al., 2003) or internal variability (Delworth and Knutson, 2000) could be more important. Differences between simulations including greenhouse gas forcing only and those that also include the cooling effects of sulphate aerosols (e.g., Tett et al., 2002) indicate that the cooling effects of sulphate aerosols may account for some of the lack of observational warming between 1950 and 1970, despite increasing greenhouse gas concentrations, as was proposed by Schwartz (1993). In contrast, Nagashima et al. (2006) find that carbonaceous aerosols are required for the MIROC model (see Table 8.1 for a description) to provide a statistically consistent representation of observed changes in near-surface temperature in the middle part of the 20th century. The mid-century cooling that the model simulates in some regions is also observed, and is caused in the model by regional negative surface forcing from organic and black carbon associated with biomass burning. Variations in the Atlantic Multi-decadal Oscillation (see Section 3.6.6 for a more detailed discussion) could account for some of the evolution of global and hemispheric mean temperatures during the instrumental period (Schlesinger and Ramankutty, 1994; Andronova and Schlesinger, 2000; Delworth and Mann, 2000); Knight et al. (2005) estimate that variations in the Atlantic Multi-decadal Oscillation could account for up to 0.2°C peak-to-trough variability in NH mean decadal temperatures.

9.4.1.3 Variability of Temperature from Observations and Models

Year-to-year variability of global mean temperatures simulated by the most recent models compares reasonably well with that of observations, as can be seen by comparing observed and modelled variations in Figure 9.5a. A more quantitative evaluation of modelled variability can be carried out by comparing the power spectra of observed and modelled global mean temperatures. Figure 9.7 compares the power spectrum of observations with the power spectra of transient simulations of the instrumental period. This avoids the need to compare variability estimated from long control runs of models with observed variability, which is difficult because observations are likely to contain a response to external forcings that cannot be reliably removed by subtracting a simple linear trend. The simulations considered contain both anthropogenic and natural forcings, and include most 20th Century Climate in Coupled Models (20C3M) simulations in the MMD at PCMDI. Figure

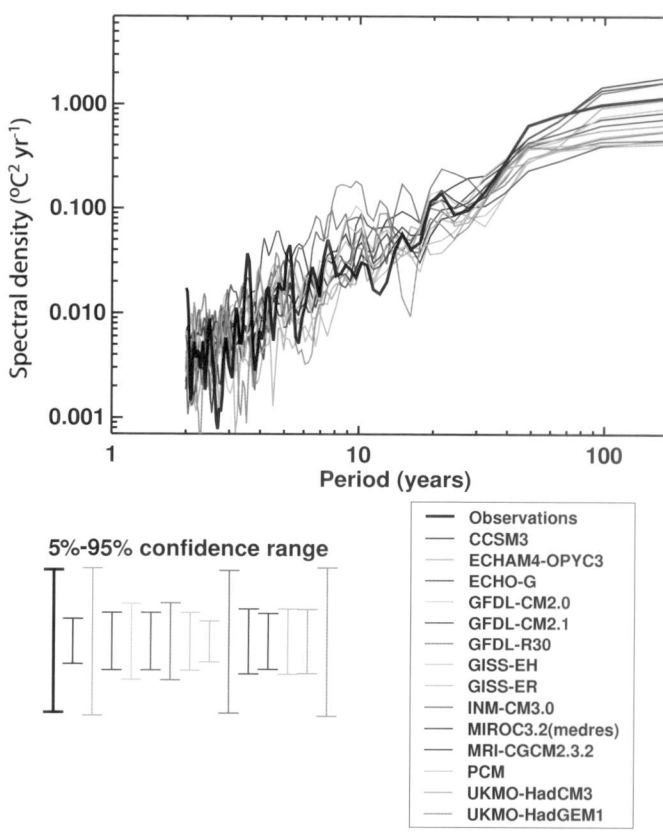

Figure 9.7. *Comparison of variability as a function of time scale of annual global mean temperatures (°C² yr⁻¹) from the observed record (Hadley Centre/Climatic Research Unit gridded surface temperature data set (HadCRUT3), Brohan et al., 2006) and from AOGCM simulations including both anthropogenic and natural forcings. All power spectra are estimated using a Tukey-Hanning filter of width 97 years. The model spectra displayed are the averages of the individual spectra estimated from individual ensemble members. The same 58 simulations and 14 models are used as in Figure 9.5a. All models simulate variability on decadal time scales and longer that is consistent with observations at the 10% significance level. Further details of the method of calculating the spectra are given in the Supplementary Material, Appendix 9.C.*

9.7 shows that the models have variance at global scales that is consistent with the observed variance at the 5% significance level on the decadal to inter-decadal time scales important for detection and attribution. Figure 9.8 shows that this is also generally the case at continental scales, although model uncertainty is larger at smaller scales (Section 9.4.2.2).

Detection and attribution studies routinely assess if the residual variability unexplained by forcing is consistent with the estimate of internal variability (e.g., Allen and Tett, 1999; Tett et al., 1999; Stott et al., 2001; Zwiers and Zhang, 2003). Furthermore, there is no evidence that the variability in palaeoclimatic reconstructions that is not explained by forcing is stronger than that in models, and simulations of the last 1 kyr show similar variability to reconstructions (Section 9.3.3.2). Chapter 8 discusses the simulation of major modes of variability and the extent to which they are simulated by models (including on decadal to inter-decadal time scales).

9.4.1.4   The Influence of Greenhouse Gas and Total Anthropogenic Forcing on Global Surface Temperature

Since the TAR, a large number of studies based on the longer observational record, improved models and stronger signal-to-noise ratio have increased confidence in the detection of an anthropogenic signal in the instrumental record (see, e.g., the recent review by IDAG, 2005). Many more detection and attribution studies are now available than were available for the TAR, and these have used more recent climate data than previous studies and a much greater variety of climate simulations with more sophisticated treatments of a greater number of both anthropogenic and natural forcings.

Fingerprint studies that use climate change signals estimated from an array of climate models indicate that detection of an anthropogenic contribution to the observed warming is a result that is robust to a wide range of model uncertainty, forcing uncertainties and analysis techniques (Hegerl et al., 2001; Gillett et al., 2002c; Tett et al., 2002; Zwiers and Zhang, 2003; IDAG, 2005; Stone and Allen, 2005b; Stone et al., 2007a,b; Stott et al., 2006b,c; Zhang et al., 2006). These studies account for the possibility that the agreement between simulated and observed global mean temperature changes could be fortuitous as a result of, for example, balancing too great (or too small) a model sensitivity with a too large (or too small) negative aerosol forcing (Schwartz, 2004; Hansen et al., 2005) or a too small (or too large) warming due to solar changes. Multi-signal detection and attribution analyses do not rely on such agreement because they seek to explain the observed temperature changes in terms of the responses to individual forcings, using model-derived patterns of response and a noise-reducing metric (Appendix 9.A) but determining their amplitudes from observations. As discussed in Section 9.2.2.1, these approaches make use of differences in the temporal and spatial responses to forcings to separate their effect in observations.

Since the TAR, there has also been an increased emphasis on quantifying the greenhouse gas contribution to observed warming, and distinguishing this contribution from other factors, both anthropogenic, such as the cooling effects of aerosols, and natural, such as from volcanic eruptions and changes in solar radiation.

A comparison of results using four different models (Figure 9.9) shows that there is a robust identification of a significant greenhouse warming contribution to observed warming that is likely greater than the observed warming over the last 50 years

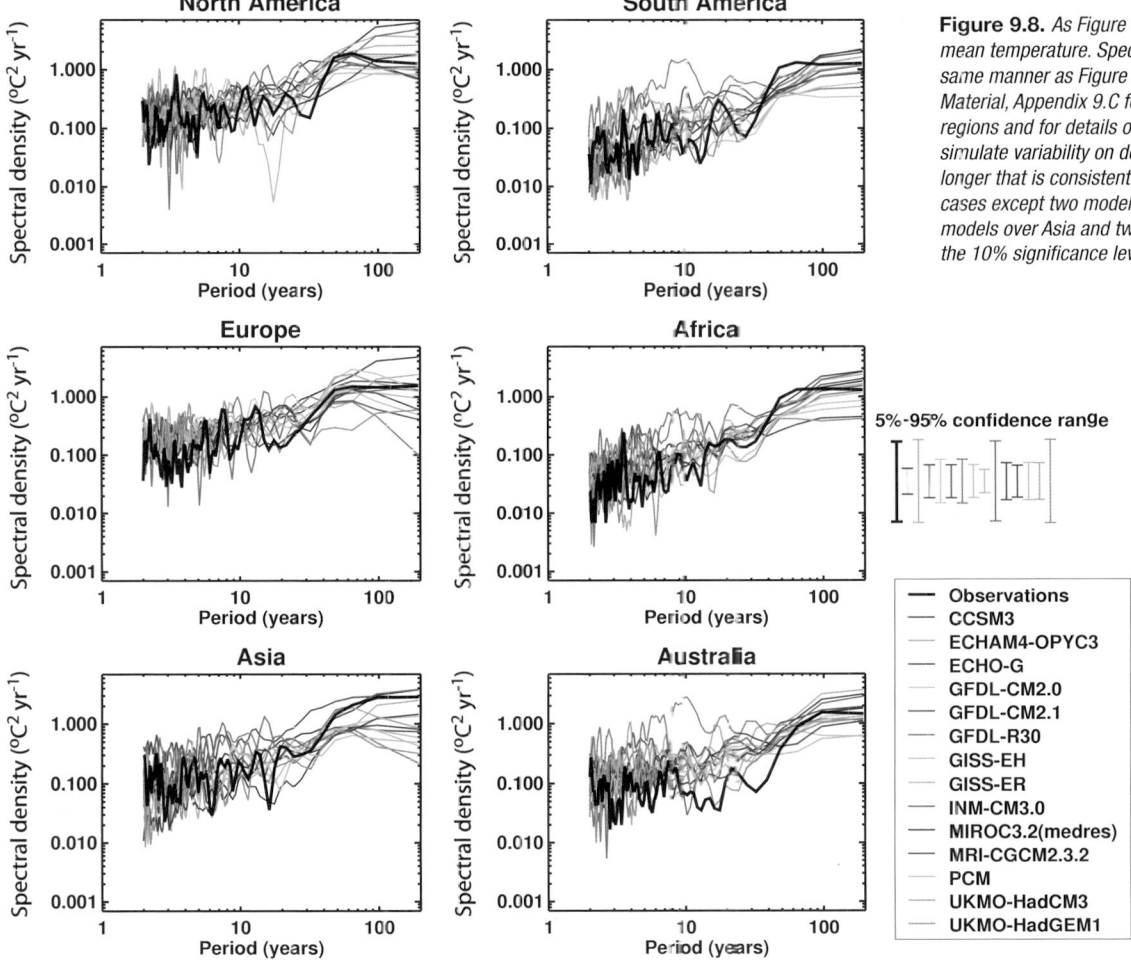

Figure 9.8. *As Figure 9.7, except for continental mean temperature. Spectra are calculated in the same manner as Figure 9.7. See the Supplementary Material, Appendix 9.C for a description of the regions and for details of the method used. Models simulate variability on decadal time scales and longer that is consistent with observations in all cases except two models over South America, five models over Asia and two models over Australia (at the 10% significance level).*

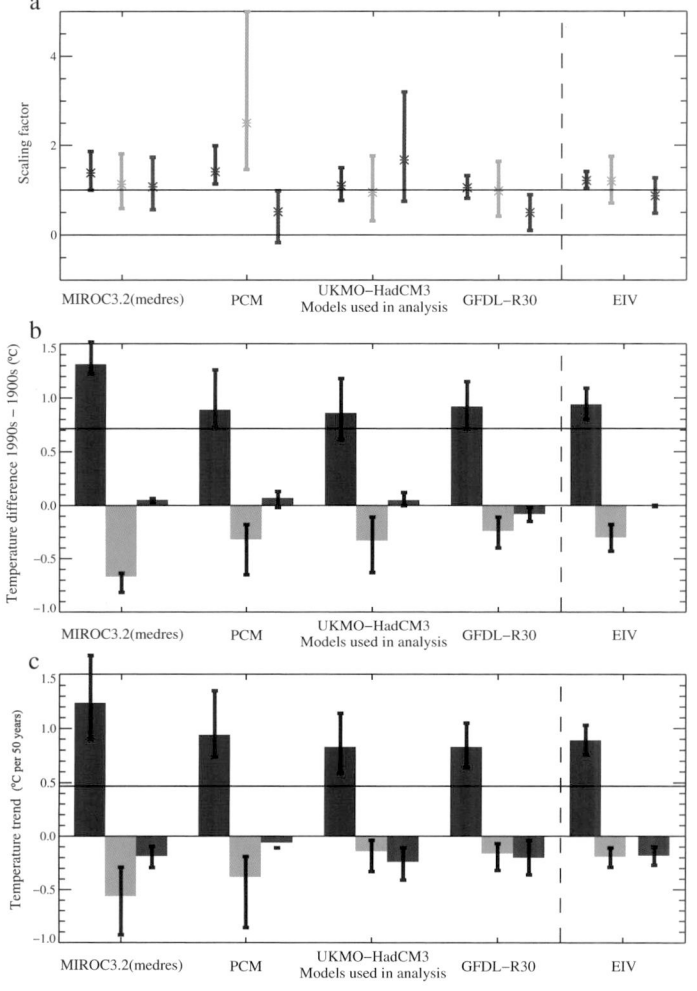

Figure 9.9. *Estimated contribution from greenhouse gas (red), other anthropogenic (green) and natural (blue) components to observed global mean surface temperature changes, based on 'optimal' detection analyses (Appendix 9.A). (a) 5 to 95% uncertainty limits on scaling factors (dimensionless) based on an analysis over the 20th century, (b) the estimated contribution of forced changes to temperature changes over the 20th century, expressed as the difference between 1990 to 1999 mean temperature and 1900 to 1909 mean temperature (°C) and (c) estimated contribution to temperature trends over 1950 to 1999 (°C per 50 years). The horizontal black lines in (b) and (c) show the observed temperature changes from the Hadley Centre/Climatic Research Unit gridded surface temperature data set (HadCRUT2v; Parker et al., 2004). The results of full space-time optimal detection analyses (Nozawa et al., 2005; Stott et al., 2006c) using a total least squares algorithm (Allen and Stott, 2003) from ensembles of simulations containing each set of forcings separately are shown for four models, MIROC3.2(medres), PCM, UKMO-HadCM3 and GFDL-R30. Also shown, labelled 'EIV', is an optimal detection analysis using the combined spatio-temporal patterns of response from three models (PCM, UKMO-HadCM3 and GFDL-R30) for each of the three forcings separately, thus incorporating inter-model uncertainty (Huntingford et al., 2006).*

with a significant net cooling from other anthropogenic forcings over that period, dominated by aerosols. Stott et al. (2006c) compare results over the 20th century obtained using the UKMO-HadCM3, PCM (see Table 8.1 for model descriptions) and Geophysical Fluid Dynamics Laboratory (GFDL) R30 models. They find consistent estimates for the greenhouse gas attributable warming over the century, expressed as the

difference between temperatures in the last and first decades of the century, of 0.6°C to 1.3°C (5 to 95%) offset by cooling from other anthropogenic factors associated mainly with cooling from aerosols of 0.1°C to 0.7°C and a small net contribution from natural factors over the century of –0.1°C to 0.1°C (Figure 9.9b). Scaling factors for the model response to three forcings are shown in Figure 9.9a. A similar analysis for the MIROC3.2 model (see Table 8.1 for a description) finds a somewhat larger warming contribution from greenhouse gases of 1.2°C to 1.5°C offset by a cooling of 0.6°C to 0.8°C from other anthropogenic factors and a very small net natural contribution (Figure 9.9b). In all cases, the fifth percentile of the warming attributable to greenhouse gases is greater than the observed warming over the last 50 years of the 20th century (Figure 9.9c).

The detection and estimation of a greenhouse gas signal is also robust to accounting more fully for model uncertainty. An analysis that combines results from three climate models and thereby incorporates uncertainty in the response of these three models (by including an estimate of the inter-model covariance structure in the regression method; Huntingford et al., 2006), supports the results from each of the models individually that it is likely that greenhouse gases would have caused more warming than was observed over the 1950 to 1999 period (Figure 9.9, results labelled 'EIV'). These results are consistent with the results of an earlier analysis, which calculated the mean response patterns from five models and included a simpler estimate of model uncertainty (obtained by a simple rescaling of the variability estimated from a long control run, thereby assuming that inter-model uncertainty has the same covariance structure as internal variability; Gillett et al., 2002c). Both the results of Gillett et al. (2002c) and Huntingford et al. (2006) indicate that inter-model differences do not greatly increase detection and attribution uncertainties and that averaging fingerprints improves detection results.

A robust anthropogenic signal is also found in a wide range of climate models that do not have the full range of simulations required to directly estimate the responses to individual forcings required for the full multi-signal detection and attribution analyses (Stone et al., 2007a,b). In these cases, an estimate of the model's pattern of response to each individual forcing can be diagnosed by fitting a series of EBMs, one for each forcing, to the mean coupled model response to all the forcings to diagnose the time-dependent response in the global mean for each individual forcing. The magnitude of these time-only signals can then be inferred from observations using detection methods (Stone et al., 2007a,b). When applied to 13 different climate models that had transient simulations of 1901 to 2005 temperature change, Stone et al. (2007a) find a robust detection across the models of greenhouse gas warming over this period, although uncertainties in attributable temperature changes due to the different forcings are larger than when considering spatio-temporal patterns. By tuning an EBM to the observations, and using an AOGCM solely to estimate internal variability, Stone and Allen (2005b) detect the effects of greenhouse gases and tropospheric sulphate aerosols in the observed 1900 to 2004 record, but not the effects of volcanic and solar forcing.

The detection of an anthropogenic signal is also robust to using different methods. For example, Bayesian detection analyses (Appendix 9.A.2) robustly detect anthropogenic influence on near-surface temperature changes (Smith et al., 2003; Schnur and Hasselmann, 2005; Min and Hense, 2006a,b). In these studies, Bayes Factors (ratios of posterior to prior odds) are used to assess evidence supporting competing hypotheses (Kass and Raftery, 1995; see Appendix 9.A.2). A Bayesian analysis of seven climate models (Schnur and Hasselman, 2005) and Bayesian analyses of MMD 20C3M simulations (Min and Hense, 2006a,b) find decisive evidence for the influence of anthropogenic forcings. Lee et al. (2005), using an approach suggested by Berliner et al. (2000), evaluate the evidence for the presence of the combined greenhouse gas and sulphate aerosol (GS) signal, estimated from CGCM1 and CGCM2 (Table 8.1; McAvaney et al., 2001), in observations for several five-decade windows, beginning with 1900 to 1949 and ending with 1950 to 1999. Very strong evidence was found in support of detection of the forced response during both halves of the 20th century regardless of the choice of prior distribution. However, evidence for attribution in that approach is based on the extent to which observed data narrow the prior uncertainty on the size of the anthropogenic signal. That evidence was not found to be very strong, although Lee et al. (2005) estimate that

strong evidence for attribution as defined in their approach may emerge within the next two decades as the anthropogenic signal strengthens.

In a further study, Lee et al. (2006) assess whether anthropogenic forcing has enhanced the predictability of decadal global-scale temperature changes; a forcing-related enhancement in predictability would give a further indication of its role in the evolution of the 20th-century climate. Using an ensemble of simulations of the 20th century with GS forcing, they use Bayesian tools similar to those of Lee et al. (2005) to produce, for each decade beginning with 1930 to 1939, a forecast of the probability of above-normal temperatures where 'normal' is defined as the mean temperature of the preceding three decades. These hindcasts become skilful during the last two decades of the 20th century as indicated both by their Brier skill scores, a standard measure of the skill of probabilistic forecasts, and by the confidence bounds on hindcasts of global mean temperature anomalies (Figure 9.10). This indicates that greenhouse gas forcing contributes to predictability of decadal temperature changes during the latter part of the 20th century.

Another type of analysis is a Granger causality analysis of the lagged covariance structure of observed hemispheric temperatures (Kaufmann and Stern, 2002), which also provides evidence for an anthropogenic signal, although such

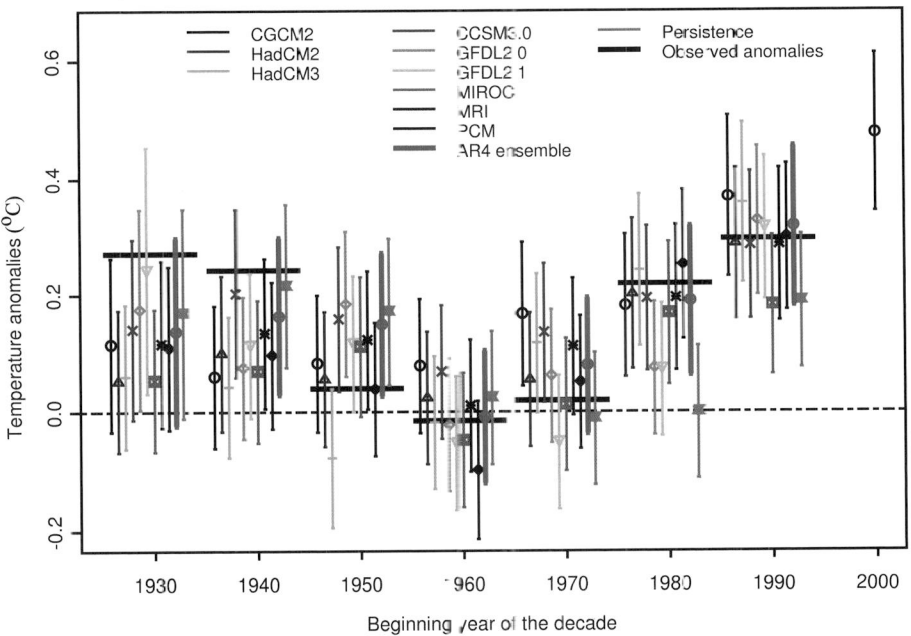

Figure 9.10. *Observed and hindcast decadal mean surface temperature anomalies (°C) expressed, for each decade, relative to the preceding three decades. Observed anomalies are represented by horizontal black lines. Hindcast decadal anomalies and their uncertainties (5 to 95% confidence bounds) are displayed as vertical bars. Hindcasts are based on a Bayesian detection analysis using the estimated response to historical external forcing. Hindcasts made with CGCM2, HadCM2 (see Table 8.1 of the TAR) and HadCM3 (see Table 8.1, this report) use the estimated response to anthropogenic forcing only (left hand column of legend) while those made with selected MMD 20C3M models used anthropogenic and natural forcings (centre column of legend; see Table 8.1 for model descriptions). Hindcasts made with the ensemble mean of the selected 20C3M models are indicated by the thick green line. A hindcast based on persisting anomalies from the previous decade is also shown. The hindcasts agree well with observations from the 1950s onward. Hindcasts for the decades of the 1930s and 1940s are sensitive to the details of the hindcast procedure. A forecast for the decadal global mean anomaly for the decade 2000 to 2009, relative to the 1970 to 1999 climatology, based on simulations performed with the Canadian Centre for Climate Modelling and Analysis Coupled Global Climate Model (CGCM2) is also displayed. From Lee et al. (2006).*

evidence may not be conclusive on its own without additional information from climate models (Triacca, 2001). Consistently, a neural network model is unable to reconstruct the observed global temperature record from 1860 to 2000 if anthropogenic forcings are not taken into account (Pasini et al., 2006). Further, an assessment of recent climate change relative to the long-term persistence of NH mean temperature as diagnosed from a range of reconstructed temperature records (Rybski et al., 2006) suggests that the recent warming cannot be explained solely in terms of natural factors, regardless of the reconstruction used. Similarly, Fomby and Vogelsang (2002), using a test of trend that accounts for the effects of serial correlation, find that the increase in global mean temperature over the 20th century is statistically significant even if it is assumed that natural climate variability has strong serial correlation.

9.4.1.5   The Influence of Other Anthropogenic and Natural Forcings

A significant cooling due to other anthropogenic factors, dominated by aerosols, is a robust feature of a wide range of detection analyses. These analyses indicate that it is likely that greenhouse gases alone would have caused more than the observed warming over the last 50 years of the 20th century, with some warming offset by cooling from natural and other anthropogenic factors, notably aerosols, which have a very short residence time in the atmosphere relative to that of well-mixed greenhouse gases (Schwartz, 1993). A key factor in identifying the aerosol fingerprint, and therefore the amount of aerosol cooling counteracting greenhouse warming, is the change through time of the hemispheric temperature contrast, which is affected by the different evolution of aerosol forcing in the two hemispheres as well as the greater thermal inertia of the larger ocean area in the SH (Santer et al., 1996b,c; Hegerl et al., 2001; Stott et al., 2006c). Regional and seasonal aspects of the temperature response may help to distinguish further the response to greenhouse gas increases from the response to aerosols (e.g., Ramanathan et al., 2005; Nagashima et al., 2006).

Results on the importance and contribution from anthropogenic forcings other than greenhouse gases vary more between different approaches. For example, Bayesian analyses differ in the strength of evidence they find for an aerosol effect. Schnur and Hasselman (2005), for example, fail to find decisive evidence for the influence of aerosols. They postulate that this could be due to taking account of modelling uncertainty in the response to aerosols. However, two other studies using frequentist methods that also include modelling uncertainty find a clear detection of sulphate aerosols, suggesting that the use of multiple models helps to reduce uncertainties and improves detection of a sulphate aerosol effect (Gillett et al., 2002c; Huntingford et al., 2006). Similarly, a Bayesian study of hemispheric mean temperatures from 1900 to 1996 finds decisive evidence for an aerosol cooling effect (Smith et al., 2003). Differences in the separate detection of sulphate aerosol influences in multi-signal approaches can also reflect differences

in the diagnostics applied (e.g., the space-time analysis of Tett et al. (1999) versus the space-only analysis of Hegerl et al. (1997, 2000)) as was shown by Gillett et al. (2002a).

Recent estimates (Figure 9.9) indicate a relatively small combined effect of natural forcings on the global mean temperature evolution of the second half of the 20th century, with a small net cooling from the combined effects of solar and volcanic forcings. Coupled models simulate much less warming over the 20th century in response to solar forcing alone than to greenhouse gas forcing (Cubasch et al., 1997; Broccoli et al., 2003; Meehl et al., 2004), independent of which solar forcing reconstruction is used (Chapter 2). Several studies have attempted to estimate the individual contributions from solar and volcanic forcings separately, thus allowing for the possibility of enhancement of the solar response in observations due to processes not represented in models. Optimal detection studies that attempt to separate the responses to solar and other forcings in observations can also account for gross errors in the overall magnitude of past solar forcing, which remains uncertain (Chapter 2), by scaling the space-time patterns of response (Section 9.2.2.1). Using such a method, Tett et al. (1999) estimate that the net anthropogenic warming in the second half of the 20th century was much greater than any possible solar warming, even when using the solar forcing reconstruction by Hoyt and Schatten (1993), which indicates larger solar forcing and a different evolution over time than more recent reconstructions (Section 2.7.1). However, Stott et al. (2003b), using the same solar reconstruction but a different model, are not able to completely rule out the possibility that solar forcing might have caused more warming than greenhouse gas forcing over the 20th century due to difficulties in distinguishing between the patterns of response to solar and greenhouse forcing. This was not the case when using the response to solar forcing based on the alternative reconstruction of Lean et al. (1995), in which case they find a very small likelihood (less than 1%, as opposed to approximately 10%) that solar warming could be greater than greenhouse warming since 1950. Note that recent solar forcing reconstructions show a substantially decreased magnitude of low-frequency variations in solar forcing (Section 2.7.1) compared to Lean et al. (1995) and particularly Hoyt and Schatten (1993).

The conclusion that greenhouse warming dominates over solar warming is supported further by a detection and attribution analysis using 13 models from the MMD at PCMDI (Stone et al., 2007a) and an analysis of the National Center for Atmospheric Research (NCAR) Community Climate System Model (CCSM1.4; Stone et al., 2007b). In both these analyses, the response to solar forcing in the model was inferred by fitting a series of EBMs to the mean coupled model response to the combined effects of anthropogenic and natural forcings. In addition, a combined analysis of the response at the surface and through the depth of the atmosphere using HadCM3 and the solar reconstruction of Lean et al. (1995) concluded that the near-surface temperature response to solar forcing over 1960 to 1999 is much smaller than the response to greenhouse gases (Jones et al., 2003). This conclusion is also supported by the vertical pattern of climate change, which is more consistent with the response to greenhouse gas than to solar forcing (Figure 9.1). Further evidence against a dominant solar role arises from older analyses targeted at detecting the solar response (e.g., North and Stevens, 1998). Based on these detection results, which allow for possible amplification of the solar influence by processes not represented in climate models, we conclude that it is very likely that greenhouse gases caused more global warming over the last 50 years than changes in solar irradiance.

Detection and attribution as well as modelling studies indicate more uncertainty regarding the causes of early 20th-century warming than the recent warming. A number of studies detect a significant natural contribution to early 20th-century warming (Tett et al., 2002; Stott et al., 2003b; Nozawa et al., 2005; Shiogama et al., 2006). Some studies find a greater role for solar forcing than other forcings before 1950 (Stott et al., 2003b), although one detection study finds a roughly equal role for solar and volcanic forcing (Shiogama et al., 2006), and others find that volcanic forcing (Hegerl et al., 2003, 2007) or a substantial contribution from natural internal variability (Tett et al., 2002; Hegerl et al., 2007) could be important. There could also be an early expression of greenhouse warming in the early 20th century (Tett et al., 2002; Hegerl et al., 2003, 2007).

9.4.1.6 Implications for Transient Climate Response

Quantification of the likely contributions of greenhouse gases and other forcing factors to past temperature change (Section 9.4.1.4) in turn provides observational constraints on the transient climate response, which determines the rapidity and strength of a global temperature response to external forcing (see Glossary and Sections 9.6.2.3 and 8.6.2.1 for detailed definitions) and therefore helps to constrain likely future rates of warming. Scaling factors derived from detection analyses can be used to scale predictions of future change by assuming that the fractional error in model predictions of global mean temperature change is constant (Allen et al., 2000, 2002; Allen and Stainforth, 2002; Stott and Kettleborough, 2002). This linear relationship between past and future fractional error in temperature change has been found to be sufficiently robust over a number of realistic forcing scenarios to introduce little additional uncertainty (Kettleborough et al., 2007). In this approach based on detection and attribution methods, which is compared with other approaches for producing probabilistic projections in Section 10.5.4.5, different scaling factors are applied to the greenhouse gases and to the response to other anthropogenic forcings (notably aerosols); these separate scaling factors are used to account for possible errors in the models and aerosol forcing. Uncertainties calculated in this way are likely to be more reliable than uncertainty ranges derived from simulations by coupled AOGCMs that happen to be available. Such ensembles could provide a misleading estimate of forecast uncertainty because they do not systematically explore modelling uncertainty (Allen et al., 2002; Allen and Stainforth, 2002). Stott et al. (2006c) compare observationally constrained predictions from three coupled climate models with a range of sensitivities and show that predictions made in this way are relatively insensitive to the particular choice of model used to produce them. The robustness to choice of model of such observationally constrained predictions was also demonstrated by Stone et al. (2007a) for the MMD ensemble. The observationally constrained transient climate response at the time of doubling of atmospheric CO_2 following a 1% per year increase in CO_2 was estimated by Stott et al. (2006c) to lie between 1.5°C and 2.8°C (Section 9.6.2, Figure 9.21). Such approaches have also been used to provide observationally constrained predictions of global mean (Stott and Kettleborough, 2002; Stone et al., 2007a) and continental-scale temperatures (Stott et al., 2006a) following the IPCC Special Report on Emission Scenarios (SRES) emissions scenarios, and these are discussed in Sections 10.5.4.5 and 11.10.

9.4.1.7 Studies of Indices of Temperature Change

Another method for identifying fingerprints of climate change in the observational record is to use simple indices of surface air temperature patterns that reflect features of the anticipated response to anthropogenic forcing (Karoly and Braganza, 2001; Braganza et al., 2003). By comparing modelled and observed changes in such indices, which include the global mean surface temperature, the land-ocean temperature contrast, the temperature contrast between the NH and SH, the mean magnitude of the annual cycle in temperature over land and the mean meridional temperature gradient in the NH mid-latitudes, Braganza et al. (2004) estimate that anthropogenic forcing accounts for almost all of the warming observed between 1946 and 1995 whereas warming between 1896 and 1945 is explained by a combination of anthropogenic and natural forcing and internal variability. These results are consistent with the results from studies using space-time detection techniques (Section 9.4.1.4).

Diurnal temperature range (DTR) has decreased over land by about 0.4°C over the last 50 years, with most of that change occurring prior to 1980 (Section 3.2.2.1). This decreasing trend has been shown to be outside the range of natural internal variability estimated from models. Hansen et al. (1995) demonstrate that tropospheric aerosols plus increases in continental cloud cover, possibly associated with aerosols, could account for the observed decrease in DTR. However, although models simulate a decrease in DTR when they include anthropogenic changes in greenhouse gases and aerosols, the observed decrease is larger than the model-simulated decrease (Stone and Weaver, 2002, 2003; Braganza et al., 2004). This discrepancy is associated with simulated increases in daily maximum temperature being larger than observed, and could be associated with simulated increases in cloud cover being smaller than observed (Braganza et al., 2004; see Section 3.4.3.1 for observations), a result supported by other analyses (Dai et al., 1999; Stone and Weaver, 2002, 2003).

9.4.1.8   Remaining Uncertainties

A much larger range of forcing combinations and climate model simulations has been analysed in detection studies than was available for the TAR (Supplementary Material, Table S9.1). Detection and attribution analyses show robust evidence for an anthropogenic influence on climate. However, some forcings are still omitted by many models and uncertainties remain in the treatment of those forcings that are included by the majority of models.

Most studies omit two forcings that could have significant effects, particularly at regional scales, namely carbonaceous aerosols and land use changes. However, detection and attribution analyses based on climate simulations that include these forcings, (e.g., Stott et al., 2006b), continue to detect a significant anthropogenic influence in 20th-century temperature observations even though the near-surface patterns of response to black carbon aerosols and sulphate aerosols could be so similar at large spatial scales (although opposite in sign) that detection analyses may be unable to distinguish between them (Jones et al., 2005). Forcing from surface albedo changes due to land use change is expected to be negative globally (Sections 2.5.3, 7.3.3 and 9.3.3.3) although tropical deforestation could increase evaporation and warm the climate (Section 2.5.5), counteracting cooling from albedo change. However, the albedo-induced cooling effect is expected to be small and was not detected in observed trends in the study by Matthews et al. (2004).

For those forcings that have been included in attribution analyses, uncertainties associated with the temporal and spatial pattern of the forcing and the modelled response can affect the results. Large uncertainties associated with estimates of past solar forcing (Section 2.7.1) and omission of some chemical and dynamical response mechanisms (Gray et al., 2005) make it difficult to reliably estimate the contribution of solar forcing to warming over the 20th century. Nevertheless, as discussed above, results generally indicate that the contribution is small even if allowance is made for amplification of the response in observations, and simulations used in attribution analyses use several different estimates of solar forcing changes over the 20th century (Supplementary Material, Table S9.1). A number of different volcanic reconstructions are included in the modelling studies described in Section 9.4.1.2 (e.g., Sato et al., 1993; Andronova et al., 1999; Ammann et al., 2003; Supplementary Material, Table S9.1). Some models include volcanic effects by simply perturbing the incoming shortwave radiation at the top of the atmosphere, while others simulate explicitly the radiative effects of the aerosols in the stratosphere. In addition, some models include the indirect effects of tropospheric sulphate aerosols on clouds (e.g., Tett et al., 2002), whereas others consider only the direct radiative effect (e.g., Meehl et al., 2004). In models that include indirect effects, different treatments of the indirect effect are used, including changing the albedo of clouds according to an off-line calculation (e.g., Tett et al., 2002) and a fully interactive treatment of the effects of aerosols on clouds (e.g., Stott et al., 2006b). The overall level of consistency between attribution results derived from different models (as shown in Figure 9.9), and the ability of climate models to simulate large-scale temperature changes during the 20th century (Figures 9.5 and 9.6), indicate that such model differences are likely to have a relatively small impact on attribution results of large-scale temperature change at the surface.

There have also been methodological developments that have resulted in attribution analyses taking uncertainties more fully into account. Attribution analyses normally directly account for errors in the magnitude of the model's pattern of response to different forcings by the inclusion of factors that scale the model responses up or down to best match observed climate changes. These scaling factors compensate for under- or overestimates of the amplitude of the model response to forcing that may result from factors such as errors in the model's climate sensitivity, ocean heat uptake efficiency or errors in the imposed external forcing. Older analyses (e.g., Tett et al., 2002) did not take account of uncertainty due to sampling signal estimates from finite-member ensembles. This can lead to a low bias, particularly for weak forcings, in the scaling factor estimates (Appendix 9.A.1; Allen and Stott, 2003; Stott et al., 2003a). However, taking account of sampling uncertainty (as most more recent detection and attribution studies do, including those shown in Figure 9.9) makes relatively little difference to estimates of attributable warming rates, particularly those due to greenhouse gases; the largest differences occur in estimates of upper bounds for small signals, such as the response to solar forcing (Allen and Stott, 2003; Stott et al., 2003a). Studies that compare results between models and analysis techniques (e.g., Hegerl et al., 2000; Gillett et al., 2002a; Hegerl and Allen, 2002), and more recently, that use multiple models to determine fingerprints of climate change (Gillett et al., 2002c; Huntingford et al., 2006; Stott et al., 2006c; Zhang et al., 2006) find a robust detection of an anthropogenic signal in past temperature change.

A common aspect of detection analyses is that they assume the response in models to combinations of forcings to be additive. This was shown to be the case for near-surface temperatures in the PCM (Meehl et al., 2004), in the Hadley Centre Climate Model version 2 (HadCM2; Gillett et al., 2004c) and in the GFDL CM2.1 (see Table 8.1) model (Knutson et al., 2006), although none of these studies considered the indirect effects of sulphate aerosols. Sexton et al. (2003) did find some evidence for a nonlinear interaction between the effects of greenhouse gases and the indirect effect of sulphate aerosols in the atmosphere-only version of HadCM3 forced by observed SSTs; the additional effect of combining greenhouse gases and indirect aerosol effects together was much smaller than each term separately but was found to be comparable to the warming due to increasing tropospheric ozone. In addition, Meehl et al. (2003) found that additivity does not hold so well for regional responses to solar and greenhouse forcing in the PCM. Linear additivity was found to hold in the PCM model for changes in tropopause height and synthetic satellite-borne Microwave Sounding Unit (MSU) temperatures (Christy et al., 2000; Mears et al., 2003; Santer et al., 2003b).

A further source of uncertainty derives from the estimates of internal variability that are required for all detection analyses. These estimates are generally model-based because of difficulties in obtaining reliable internal variability estimates from the observational record on the spatial and temporal scales considered in detection studies. However, models would need to underestimate variability by factors of over two in their standard deviation to nullify detection of greenhouse gases in near-surface temperature data (Tett et al., 2002), which appears unlikely given the quality of agreement between models and observations at global and continental scales (Figures 9.7 and 9.8) and agreement with inferences on temperature variability from NH temperature reconstructions of the last millennium. The detection of the effects of other forcings, including aerosols, is likely to be more sensitive (e.g., an increase of 40% in the estimate of internal variability is enough to nullify detection of aerosol and natural forcings in HadCM3; Tett et al., 2002)

Few detection studies have explicitly considered the influence of observational uncertainty on near-surface temperature changes. However, Hegerl et al. (2001) show that inclusion of observational sampling uncertainty has relatively little effect on detection results and that random instrumental error has even less effect. Systematic instrumental errors, such as changes in measurement practices or urbanisation, could be more important, especially earlier in the record (Chapter 3), although these errors are calculated to be relatively small at large spatial scales. Urbanisation effects appear to have negligible effects on continental and hemispheric average temperatures (Chapter 3). Observational uncertainties are likely to be more important for surface temperature changes averaged over small regions (Section 9.4.2) and for analyses of free atmosphere temperature changes (Section 9.4.4).

9.4.2 Continental and Sub-continental Surface Temperature Change

9.4.2.1 *Observed Changes*

Over the 1901 to 2005 period there has been warming over most of the Earth's surface with the exception of an area south of Greenland and parts of North and South America (Figure 3.9 and Section 3.2.2.7, see also Figure 9.6). Warming has been strongest over the continental interiors of Asia and north-western North America and some mid-latitude ocean regions of the SH as well as south-eastern Brazil. Since 1979, almost all land areas with observational data coverage show warming (Figure 9.6). Warming is smaller in the SH than in the NH, with cooling over parts of the mid-latitude oceans. There have been widespread decreases in continental DTR since the 1950s which coincide with increases in cloud amounts (Section 3.4.3.1).

9.4.2.2 *Studies Based on Space-Time Patterns*

Global-scale analyses using space-time detection techniques (Section 9.4.1.4) have robustly identified the influence of anthropogenic forcing on the 20th-century global climate. A number of studies have now extended these analyses to consider sub-global scales. Two approaches have been used; one to assess the extent to which global studies can provide information at sub-global scales, the other to assess the influence of external forcing on the climate in specific regions. Limitations and problems in using smaller spatial scales are discussed at the end of this section.

The approach taken by IDAG (2005) was to compare analyses of full space-time fields with results obtained after removing the globally averaged warming trend, or after removing the annual global mean from each year in the analysis. They find that the detection of anthropogenic climate change is driven by the pattern of the observed warming in space and time, not just by consistent global mean temperature trends between models and observations. These results suggest that greenhouse warming should also be detectable at sub-global scales (see also Barnett et al., 1999). It was also shown by IDAG (2005) that uncertainties increase, as expected, when global mean information, which has a high signal-to-noise ratio, is disregarded (see also North et al., 1995).

Another approach for assessing the regional influence of external forcing is to apply detection and attribution analyses to observations in specific continental- or sub-continental scale regions. A number of studies using a range of models and examining various continental- or sub-continental scale land areas find a detectable human influence on 20th-century temperature changes, either by considering the 100-year period from 1900 or the 50-year period from 1950. Stott (2003) detects the warming effects of increasing greenhouse gas concentrations in six continental-scale regions over the 1900 to 2000 period, using HadCM3 simulations. In most regions, he finds that cooling from sulphate aerosols counteracts some of the greenhouse warming. However, the separate detection of a sulphate aerosol signal in regional analyses remains difficult because of lower signal-to-noise ratios, loss of large-scale spatial features of response such as hemispheric asymmetry that help to distinguish different signals, and greater modelling and forcing uncertainty at smaller scales. Zwiers and Zhang (2003) also detect human influence using two models (CGCM1 and CGCM2; see Table 8.1, McAvaney et al., 2001) over the 1950 to 2000 period in a series of nested regions, beginning with the full global domain and descending to separate continental domains for North America and Eurasia. Zhang et al. (2006) update this study using additional models (HadCM2 and HadCM3). They find evidence that climates in both continental domains have been influenced by anthropogenic emissions during 1950 to 2000, and generally also in the sub-continental domains (Figure 9.11). This finding is robust to the exclusion of NAO/Arctic Oscillation (AO) related variability, which is associated with part of the warming in Central Asia and could itself be related to anthropogenic forcing (Section 9.5.3). As the spatial scales considered become smaller, the uncertainty in estimated signal amplitudes (as demonstrated by the size of the vertical bars in Figure 9.11) becomes larger, reducing the signal-to-noise ratio (see also Stott and Tett, 1998). The signal-to-noise ratio, however, also depends on the strength of the climate change

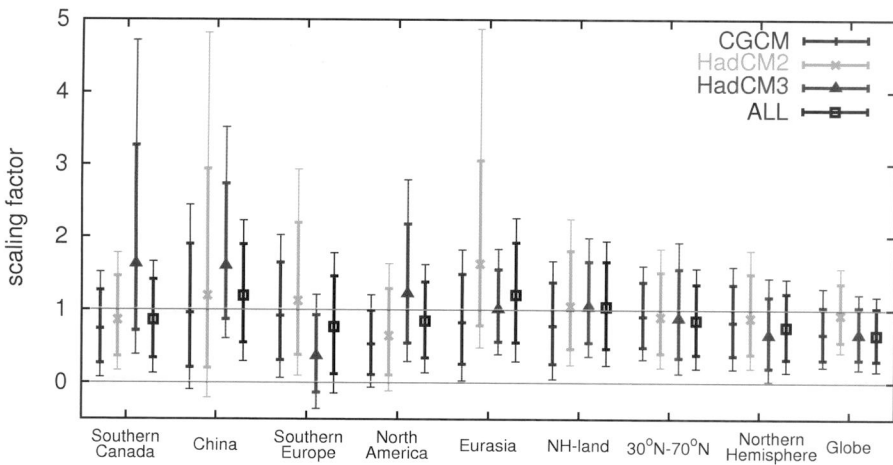

Figure 9.11. *Scaling factors indicating the match between observed and simulated decadal near-surface air temperature change (1950–1999) when greenhouse gas and aerosol forcing responses (GS) are taken into account in 'optimal' detection analyses (Appendix 9.A), at a range of spatial scales from global to sub-continental. Thick bars indicate 90% confidence intervals on the scaling factors, and the thin extensions indicate the increased width of these confidence intervals when estimates of the variance due to internal variability are doubled. Scaling factors and uncertainties are provided for different spatial domains including Canada (Canadian land area south of 70°N), China, Southern Europe (European land area bounded by 10°W to 40°E, 35° to 50°N), North America (North American land area between 30°N and 70°N), Eurasia (Eurasian land area between 30°N and 70°N), mid-latitude land area between 30°N and 70°N (labelled NH-land), the NH mid-latitudes (30°N to 70°N including land and ocean), the NH, and the globe. The GS signals are obtained from CGCM1 and CGCM2 combined (labelled CGCM, see Table 8.1 of the TAR), HadCM2 (see Table 8.1 of the TAR), and HadCM3 (see Table 8.1, this report), and these four models combined ('ALL'). After Zhang et al. (2006) and Hegerl et al. (2006b).*

and the local level of natural variability, and therefore differs between regions. Most of the results noted above hold even if the estimate of internal climate variability from the control simulation is doubled.

The ability of models to simulate many features of the observed temperature changes and variability at continental and sub-continental scales and the detection of anthropogenic effects on each of six continents provides stronger evidence of human influence on climate than was available to the TAR. A comparison between a large ensemble of 20th-century simulations of regional temperature changes made with the MMD at PCMDI (using the same simulations for which the global mean temperatures are plotted in Figure 9.5) shows that the spread of the multi-model ensembles encompasses the observed changes in regional temperature changes in almost all sub-continental regions (Figure 9.12; see also FAQ 9.2, Figure 1 and related figures in Chapter 11). In many of the regions, there is a clear separation between the ensembles of simulations that include only natural forcings and those that contain both anthropogenic and natural forcings. A more detailed analysis of one particular model, HadCM3, shows that it reproduces many features of the observed temperature changes and variability in the different regions (IDAG, 2005). The GFDL-CM2 model (see Table 8.1) is also able to reproduce many features of the evolution of temperature change in a number of regions of the globe (Knutson et al., 2006). Other studies show success at simulating regional temperatures when models include anthropogenic and natural forcings. Wang et al. (2007) showed that all MMD 20C3M simulations replicated the late 20th-century arctic warming to various degrees, while both

forced and control simulations reproduce multi-year arctic warm anomalies similar in magnitude to the observed mid 20th-century warming event.

There is some evidence that an anthropogenic signal can now be detected in some sub-continental scale areas using formal detection methods (Appendix 9.A.1), although this evidence is weaker than at continental scales. Zhang et al. (2006) detect anthropogenic fingerprints in China and southern Canada. Spagnoli et al. (2002) find some evidence for a human influence on 30-year trends of summer daily minimum temperatures in France, but they use a fingerprint estimated from a simulation of future climate change and do not detect an anthropogenic influence on the other indices they consider, including summer maximum daily temperatures and winter temperatures. Min et al. (2005) find an anthropogenic influence on East Asian temperature changes in a Bayesian framework, but they do not consider anthropogenic aerosols or natural forcings in their analysis. Atmosphere-only general circulation model (AGCM) simulations forced with observed SSTs can potentially detect anthropogenic influence at smaller spatial and temporal scales than coupled model analyses, but have the weakness that they do not explain the observed SST changes (Sexton et al., 2003). Two studies have applied attribution analysis to sub-continental temperatures to make inferences about changes in related variables. Stott et al. (2004) detect an anthropogenic influence on southern European summer mean temperature changes of the past 50 years and then infer the likelihood of exceeding an extreme temperature threshold (Section 9.4.3.3). Gillett et al. (2004a) detect an anthropogenic contribution to summer season warming in Canada and demonstrate a statistical link with area burned in forest fires. However, the robustness of these results to factors such as the choice of model or analysis method remains to be established given the limited number of studies at sub-continental scales.

Knutson et al. (2006) assess temperature changes in regions of the world covering between 0.3 and 7.4% of the area of the globe and including tropical and extratropical land and ocean regions. They find much better agreement between climate simulations and observations when the models include rather than exclude anthropogenic forcings, which suggests a detectable anthropogenic warming signal over many of the regions they examine. This would indicate the potential for formal detection studies to detect anthropogenic warming in many of these regions, although Knutson et al. (2006) also note that in some regions the climate simulations they examined were not very realistic and showed that some of these discrepancies are associated with modes of variability such as the AO.

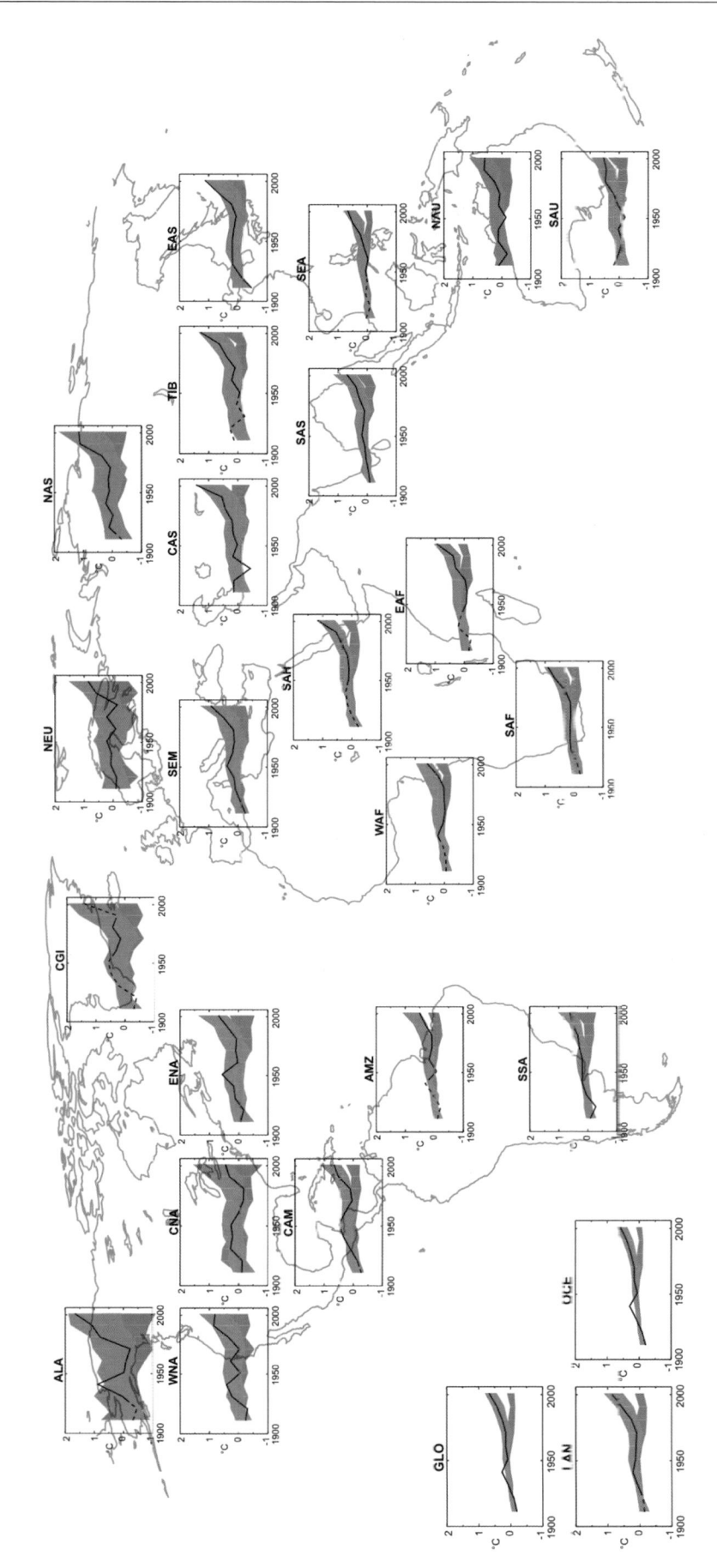

Figure 9.12. *Comparison of multi-model data set 20C3M model simulations containing all forcings (red shaded regions) and containing natural forcings only (blue shaded regions) with observed decadal mean temperature changes (°C) from 1906 to 2005 from the Hadley Centre/Climatic Research Unit gridded surface temperature data set (HadCRUT3; Brohan et al., 2006). The panel labelled GLO shows comparison for global mean; LAN, global land; and OCE, global ocean data. Remaining panels display results for 22 sub-continental scale regions (see the Supplementary Material, Appendix 9.C for a description of the regions). This figure is produced identically to FAQ 9.2, Figure 1 except sub-continental regions were used; a full description of the procedures for producing FAQ 9.2, Figure 1 is given in the Supplementary Material, Appendix 9.C. Shaded bands represent the middle 90% range estimated from the multi-model ensemble. Note that the model simulations have not been scaled in any way. The same simulations are used as in Figure 9.5 (58 simulations using all forcings from 14 models, and 19 simulations using natural forcings only from 5 models). Each simulation was sampled so that coverage corresponds to that of the observations, and was centred relative to the 1901 to 1950 mean obtained by that simulation in the region of interest. Observations in each region were centred relative to the same period. The observations in each region are generally consistent with model simulations that include anthropogenic and natural forcings, whereas in many regions the observations are inconsistent with model simulations that include natural forcings only. Lines are dashed where spatial coverage is less than 50%.*

Frequently Asked Question 9.1
Can Individual Extreme Events
be Explained by Greenhouse Warming?

Changes in climate extremes are expected as the climate warms in response to increasing atmospheric greenhouse gases resulting from human activities, such as the use of fossil fuels. However, determining whether a specific, single extreme event is due to a specific cause, such as increasing greenhouse gases, is difficult, if not impossible, for two reasons: 1) extreme events are usually caused by a combination of factors and 2) a wide range of extreme events is a normal occurrence even in an un-changing climate. Nevertheless, analysis of the warming observed over the past century suggests that the likelihood of some extreme events, such as heat waves, has increased due to greenhouse warming, and that the likelihood of others, such as frost or extremely cold nights, has decreased. For example, a recent study estimates that human influences have more than doubled the risk of a very hot European summer like that of 2003.

People affected by an extreme weather event often ask whether human influences on the climate could be held to some extent responsible. Recent years have seen many extreme events that some commentators have linked to increasing greenhouse gases. These include the prolonged drought in Australia, the extremely hot summer in Europe in 2003 (see Figure 1), the intense North Atlantic hurricane seasons of 2004 and 2005 and the extreme rainfall events in Mumbai, India in July 2005. Could a human influence such as increased concentrations of greenhouse gases in the atmosphere have 'caused' any of these events?

Extreme events usually result from a combination of factors. For example, several factors contributed to the extremely hot European summer of 2003, including a persistent high-pressure system that was associated with very clear skies and dry soil, which left more solar energy available to heat the land because less energy was consumed to evaporate moisture from the soil. Similarly, the formation of a hurricane requires warm sea surface temperatures and specific atmospheric circulation conditions. Because some factors may be strongly affected by human activities, such as sea surface temperatures, but others may not, it is not simple to detect a human influence on a single, specific extreme event.

Nevertheless, it may be possible to use climate models to determine whether human influences have changed the likelihood of certain types of extreme

events. For example, in the case of the 2003 European heat wave, a climate model was run including only historical changes in natural factors that affect the climate, such as volcanic activity and changes in solar output. Next, the model was run again including both human and natural factors, which produced a simulation of the evolution of the European climate that was much closer to that which had actually occurred. Based on these experiments, it was estimated that over the 20th century, human influences more than doubled the risk of having a summer in Europe as hot as that of 2003, and that in the absence of human influences, the risk would probably have been one in many hundred years. More detailed modelling work will be required to estimate the change in risk for specific high-impact events, such as the occurrence of a series of very warm nights in an urban area such as Paris.

The value of such a probability-based approach – 'Does human influence change the likelihood of an event?' – is that it can be used to estimate the influence of external factors, such as increases in greenhouse gases, on the frequency of specific types of events, such as heat waves or frost. Nevertheless, careful statistical analyses are required, since the likelihood of individual extremes, such as a late-spring frost, could change due to changes in climate variability as well as changes in average climate conditions. Such analyses rely on climate-model based estimates of climate variability, and thus the climate models used should adequately represent that variability.

The same likelihood-based approach can be used to examine changes in the frequency of heavy rainfall or floods. Climate models predict that human influences will cause an increase in many types of extreme events, including extreme rainfall. There is already evidence that, in recent decades, extreme rainfall has increased in some regions, leading to an increase in flooding.

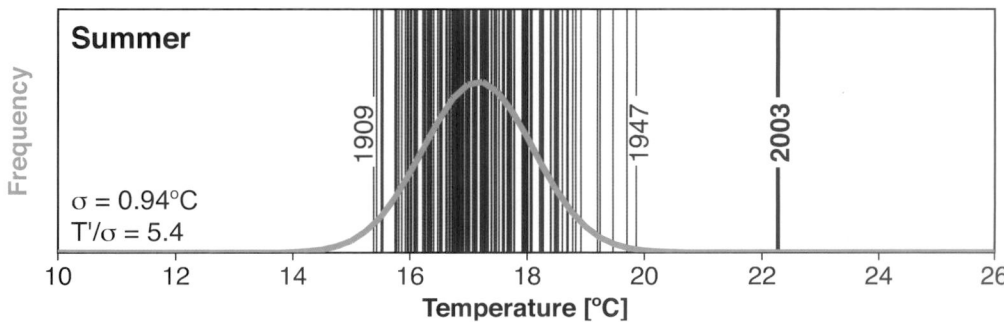

FAQ 9.1, Figure 1. *Summer temperatures in Switzerland from 1864 to 2003 are, on average, about 17°C, as shown by the green curve. During the extremely hot summer of 2003, average temperatures exceeded 22°C, as indicated by the red bar (a vertical line is shown for each year in the 137-year record). The fitted Gaussian distribution is indicated in green. The years 1909, 1947 and 2003 are labelled because they represent extreme years in the record. The values in the lower left corner indicate the standard deviation (σ) and the 2003 anomaly normalised by the 1864 to 2000 standard deviation (T'/σ). From Schär et al. (2004).*

Karoly and Wu (2005) compare observed temperature trends in 5° × 5° grid boxes globally over 30-, 50- and 100-year periods ending in 2002 with 1) internal variability as simulated by three models (GFDL R30, HadCM2, PCM) and 2) the simulated response to greenhouse gas and sulphate aerosol forcing in those models (see also Knutson et al., 1999). They find that a much higher percentage of grid boxes show trends that are inconsistent with model-estimated internal variability than would be expected by chance and that a large fraction of grid boxes show changes that are consistent with the forced simulations, particularly over the two shorter periods. This assessment is essentially a global-scale detection result because its interpretation relies upon a global composite of grid-box scale statistics. As discussed in the paper, this result does not rule out the possibility that individual grid box trends may be explained by different external forcing combinations, particularly since natural forcings and forcings that could be important at small spatial scales, such as land use change or black carbon aerosols, are missing from these models. The demonstration of local consistency between models and observations in this study does not necessarily imply that observed changes can be attributed to anthropogenic forcing in a specific grid box, and it does not allow confident estimates of the anthropogenic contribution to change at those scales.

Models do not reproduce the observed temperature changes equally well in all regions. Areas where temperature changes are not particularly well simulated by some models include parts of North America (Knutson et al., 2006) and mid-Asia (IDAG, 2005). This could be due to a regional trend or variation that was caused by internal variability (a result that models would not be expected to reproduce), uncertain forcings that are locally important, or model errors. Examples of uncertain forcings that play a small role globally, but could be more important regionally, are the effects of land use changes (Sections 9.2 and 9.3) or atmospheric brown clouds. The latter could be important in explaining observed temperature trends in South Asia and the northern Indian Ocean (Ramanathan et al., 2005; see Chapter 2).

An analysis of the MMD 20C3M experiments indicates that multi-decadal internal variability could be responsible for some of the rapid warming seen in the central USA between 1901 and 1940 and rapid cooling between 1940 and 1979 (Kunkel et al., 2006). Also, regional temperature is more strongly influenced by variability and changes in climate dynamics, such as temperature changes associated with the NAO, which may itself show an anthropogenic influence (Section 9.5.3.2), or the Atlantic Multi-decadal Oscillation (AMO), which could in some regions and seasons be poorly simulated by models and could be confounded with the expected temperature response to external forcings. Thus the anthropogenic signal is likely to be more easy to identify in some regions than in others, with temperature changes in those regions most affected by multi-decadal scale variability being the most difficult to attribute, even if those changes are inconsistent with model estimated internal variability and therefore detectable.

The extent to which temperature changes at sub-continental scales can be attributed to anthropogenic forcings, and the extent to which it is possible to estimate the contribution of greenhouse gas forcing to regional temperature trends, remains a topic for further research. Idealised studies (e.g., Stott and Tett, 1998) suggest that surface temperature changes are detectable mainly at large spatial scales of the order of several thousand kilometres (although they also show that as the signal of climate change strengthens in the 21st century, surface temperature changes are expected to become detectable at increasingly smaller scales). Robust detection and attribution are inhibited at the grid box scales because it becomes difficult to separate the effects of the relatively well understood large-scale external influences on climate, such as greenhouse gas, aerosols, solar and volcanic forcing, from each other and from local influences that may not be related to these large-scale forcings. This occurs because the contribution from internal climate variability increases at smaller scales, because the spatial details that can help to distinguish between different forcings at large scales are not available or unreliable at smaller scales, and because forcings that could be important at small spatial scales, such as land use change or black carbon aerosols, are uncertain and may not have been included in the models used for detection. Although models do not typically underestimate natural internal variability of temperature at continental scales over land (Figure 9.8), even at a grid box scale (Karoly and Wu, 2005), the credibility of small-scale details of climate simulated by models is lower than for large-scale features. While the large-scale coherence of temperatures means that temperatures at a particular grid box should adequately represent a substantial part of the variability of temperatures averaged over the area of that grid box, the remaining variability from local-scale processes and the upward cascades from smaller to larger scales via nonlinear interactions may not be well represented in models at the grid box scale. Similarly, the analysis of shorter temporal scales also decreases the signal-to-noise ratio and the ability to use temporal information to distinguish between different forcings. This is why most detection and attribution studies use temporal scales of 50 or more years.

9.4.2.3 Studies Based on Indices of Temperature Change and Temperature-Precipitation Relationships

Studies based on indices of temperature change support the robust detection of human influence on continental-scale land areas. Observed trends in indices of North American continental-scale temperature change, (including the regional mean, the mean land-ocean temperature contrast and the annual cycle) were found by Karoly et al. (2003) to be generally consistent with simulated trends under historical forcing from greenhouse gases and sulphate aerosols during the second half of the 20th century. In contrast, they find only a small likelihood of agreement with trends driven by natural forcing only during this period. An analysis of changes in Australian mean, daily maximum and daily minimum temperatures and diurnal temperature range using six coupled climate models showed that it is likely that there has been a significant contribution to observed warming in Australia from increasing greenhouse gases and sulphate aerosols (Karoly and Braganza, 2005a). An anomalous warming has been found over all Australia (Nicholls, 2003) and in New South Wales (Nicholls et al., 2005) since the early 1970s, associated with a changed relationship between annual mean maximum temperature and rainfall. Whereas interannual rainfall and temperature variations in this region are strongly inversely correlated, in recent decades temperatures have tended to be higher for a given rainfall than in previous decades. By removing the rainfall-related component of Australian temperature variations, thereby enhancing the signal-to-noise ratio, Karoly and Braganza (2005b) detect an anthropogenic warming signal in south-eastern Australia, although their results are affected by some uncertainty associated with their removal of rainfall-related temperature variability. A similar technique applied to the Sudan and Sahel region improved the agreement between model simulations and observations of temperature change over the last 60 years in this region (Douville, 2006) and could improve the detectability of regional temperature signals over other regions where precipitation is likely to affect the surface energy budget (Trenberth and Shea, 2005).

9.4.3   Surface Temperature Extremes

9.4.3.1   *Observed Changes*

Observed changes in temperature extremes are consistent with the observed warming of the climate (Alexander et al., 2006) and are summarised in Section 3.8.2.1. There has been a widespread reduction in the number of frost days in mid-latitude regions in recent decades, an increase in the number of warm extremes, particularly warm nights, and a reduction in the number of cold extremes, particularly cold nights. A number of regional studies all show patterns of changes in extremes consistent with a general warming, although the observed changes in the tails of the temperature distributions are generally not consistent with a simple shift of the entire distribution alone.

9.4.3.2   *Global Assessments*

Evidence for observed changes in short-duration extremes generally depends on the region considered and the analysis method (IPCC, 2001). Global analyses have been restricted by the limited availability of quality-controlled and homogenised daily station data. Indices of temperature extremes have been calculated from station data, including some indices from regions where daily station data are not released (Frich et al., 2002; Klein Tank and Können, 2003; Alexander et al., 2006). Kiktev et al. (2003) analyse a subset of such indices by using fingerprints from atmospheric model simulations driven by prescribed SSTs. They find significant decreases in the number of frost days and increases in the number of very warm nights over much of the NH. Comparisons of observed and modelled trend estimates show that inclusion of anthropogenic effects in the model integrations improves the simulation of

these changing temperature extremes, indicating that human influences are probably an important contributor to changes in the number of frost days and warm nights. Tebaldi et al. (2006) find that changes simulated by eight MMD models agreed well with observed trends in heat waves, warm nights and frost days over the last four decades.

Christidis et al. (2005) analyse a new gridded data set of daily temperature data (Caesar et al., 2006) using the indices shown by Hegerl et al. (2004) to have a potential for attribution, namely the average temperature of the most extreme 1, 5, 10 and 30 days of the year. Christidis et al. (2005) detect robust anthropogenic changes in indices of extremely warm nights using signals estimated with the HadCM3 model, although with some indications that the model overestimates the observed warming of warm nights. They also detect human influence on cold days and nights, but in this case the model underestimates the observed changes, significantly so in the case of the coldest day of the year. Anthropogenic influence was not detected in observed changes in extremely warm days.

9.4.3.3   *Attributable Changes in the Risk of Extremes*

Many important impacts of climate change may manifest themselves through a change in the frequency or likelihood of occurrence of extreme events. While individual extreme events cannot be attributed to external influences, a change in the probability of such events might be attributable to external influences (Palmer, 1999; Palmer and Räisänen, 2002). One study estimates that anthropogenic forcings have significantly increased the risk of extremely warm summer conditions over southern Europe, as was observed during the 2003 European heat wave. Stott et al. (2004) apply a methodology for making quantitative statements about change in the likelihood of such specific types of climatic events (Allen, 2003; Stone and Allen, 2005a), by expressing the contribution of external forcing to the risk of an event exceeding a specific magnitude. If P_1 is the probability of a climatic event (such as a heat wave) occurring in the presence of anthropogenic forcing of the climate system, and P_0 is the probability of it occurring if anthropogenic forcing had not been present, then the fraction of the current risk that is attributable to past greenhouse gas emissions (fraction of attributable risk; FAR) is given by FAR = $1 - P_0 / P_1$ (Allen, 2003). Stott et al. (2004) apply the FAR concept to mean summer temperatures of a large part of continental Europe and the Mediterranean. Using a detection and attribution analysis, they determine that regional summer mean temperature has likely increased due to anthropogenic forcing, and that the observed change is inconsistent with natural forcing. They then use the HadCM3 model to estimate the FAR associated with a particular extreme threshold of regional summer mean temperature that was exceeded in 2003, but in no other year since the beginning of the record in 1851. Stott et al. (2004) estimate that it is very likely that human influence has more than doubled the risk of the regional summer mean temperature exceeding this threshold (Figure 9.13).

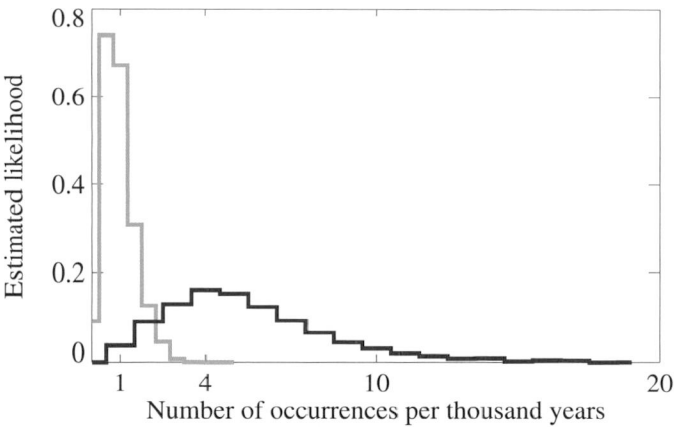

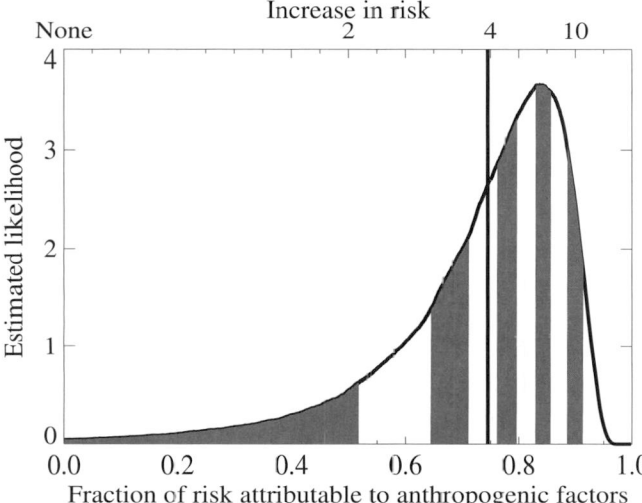

Figure 9.13. *Change in risk of mean European summer temperatures exceeding a threshold of 1.6°C above 1961 to 1990 mean temperatures, a threshold that was exceeded in 2003 but in no other year since the start of the instrumental record in 1851. (Top) Frequency histograms of the estimated likelihood of the risk (probability) of exceeding a 1.6°C threshold (relative to the 1961–1990 mean) in the 1990s in the presence (red curve) and absence (green curve) of anthropogenic change, expressed as an occurrence rate. (Bottom) Fraction of attributable risk (FAR). The vertical line indicates the 'best estimate' FAR, the mean risk attributable to anthropogenic factors averaged over the distribution. The alternation between gray and white bands indicates the deciles of the estimated FAR distribution. The shift from the green to the red distribution in (a) implies a FAR distribution with mean 0.75, corresponding to a four-fold increase in the risk of such an event (b). From Stott et al. (2004).*

This study considered only continental mean seasonally averaged temperatures. Consideration of shorter-term and smaller-scale heat waves will require higher resolution modelling and will need to take complexities such as land surface processes into account (Schär and Jendritzky, 2004). Also, Stott et al. (2004) assume no change in internal variability in the region they consider (which was the case in HadCM3 21st-century climate projections for summer mean temperatures in the region they consider), thereby ascribing the increase in risk only to an increase in mean temperatures (i.e., as shown in Box TS.5, Figure 1, which illustrates how a shift in the mean of a distribution can cause a large increase in the frequency of extremes). However, there is some evidence for a weak

increase in European temperature variability in summer (and a decrease in winter) for the period 1961 to 2004 (Scherrer et al., 2005), which could contribute to an increase in the likelihood of extremes. Schär et al. (2004) show that the central European heat wave of 2003 could also be consistent with model-predicted increases in temperature variability due to soil moisture and vegetation feedbacks. In addition, multi-decadal scale variability, associated with basin-scale changes in the Atlantic Ocean related to the Meridional Overturning Circulation (MOC) could have contributed to changes in European summer temperatures (Sutton and Hodson, 2005), although Klein Tank et al. (2005) show evidence that patterns of change in European temperature variance in spring and summer are not consistent with patterns of change in temperature variance expected from natural variability. Meteorological aspects of the summer 2003 European heat wave are discussed in Box 3.6.

9.4.4 Free Atmosphere Temperature

9.4.4.1 Observed Changes

Observed free atmosphere temperature changes are discussed in Section 3.4.1 and Karl et al. (2006) provide a comprehensive review. Radiosonde-based observations (with near global coverage since 1958) and satellite-based temperature measurements (beginning in late 1978) show warming trends in the troposphere and cooling trends in the stratosphere. All data sets show that the global mean and tropical troposphere has warmed from 1958 to the present, with the warming trend in the troposphere slightly greater than at the surface. Since 1979, it is likely that there is slightly greater warming in the troposphere than at the surface, although uncertainties remain in observed tropospheric warming trends and whether these are greater or less than the surface trend. The range (due to different data sets) of the global mean tropospheric temperature trend since 1979 is 0.12°C to 0.19°C per decade based on satellite-based estimates (Chapter 3) compared to a range of 0.16°C to 0.18°C per decade for the global surface warming. While all data sets show that the stratosphere has cooled considerably from 1958 and from 1979 to present, there are large differences in the linear trends estimated from different data sets. However, a linear trend is a poor fit to the data in the stratosphere and the tropics at all levels (Section 3.4.1). The uncertainties in the observational records are discussed in detail in Section 3.4.1 and by Karl et al. (2006). Uncertainties remain in homogenised radiosonde data sets which could result in a spurious inference of net cooling in the tropical troposphere. Differences between temperature trends measured from different versions of tropospheric satellite data result primarily from differences in how data from different satellites are merged.

9.4.4.2 Changes in Tropopause Height

The height of the lapse rate tropopause (the boundary between the stratosphere and the troposphere) is sensitive to bulk changes in the thermal structure of the stratosphere and the troposphere, and may also be affected by changes in surface temperature gradients

(Schneider, 2004). Analyses of radiosonde data have documented increases in tropopause height over the past 3 to 4 decades (Highwood et al., 2000; Seidel et al., 2001). Similar increases have been inferred from three different reanalysis products, the European Centre for Medium Range Weather Forecasts (ECMWF) 15- and 40-year reanalyses (ERA-15 and ERA-40) and the NCAR- National Center for Environmental Prediction (NCEP) reanalysis (Kalnay et al., 1996; Gibson et al., 1997; Simmons and Gibson, 2000; Kistler et al., 2001), and from model simulations with combined anthropogenic and natural forcing (Santer et al., 2003a,b, 2004; see Figure 9.14). In both models and reanalyses, changes in tropopause height over the satellite and radiosonde eras are smallest in the tropics and largest over Antarctica (Santer et al., 2003a,b, 2004). Model simulations with individual forcings indicate that the major drivers of the model tropopause height increases are ozone-induced stratospheric cooling and the tropospheric warming caused by greenhouse gas increases (Santer et al., 2003a). However, earlier model studies have found that it is difficult to alter tropopause height through stratospheric ozone changes alone (Thuburn and Craig, 2000). Santer et al. (2003c) found that the model-simulated response to combined anthropogenic and natural forcing is robustly detectable in different reanalysis products, and that solar and volcanic forcing alone could not explain the tropopause height increases (Figure 9.14). Climate data from reanalyses, especially the 'first generation' reanalysis analysed by Santer et al. (2003a), are subject to some deficiencies, notably inhomogeneities related to changes over time in the availability and quality of input data, and are subject to a number of specific technical choices in the reanalysis scheme (see Santer et al., 2004, for a discussion). Also, the NCEP reanalysis detection results could be due to compensating errors because of excessive stratospheric cooling in the reanalysis (Santer et al., 2004), since the stratosphere cools more relative to the troposphere in the NCEP reanalysis while models warm the troposphere. In contrast, the finding of a significant anthropogenic influence on tropopause height in the 'second generation' ERA-40 reanalysis is driven by similar large-scale changes in both models and the reanalysis. Detection results there are robust to removing global mean tropopause height increases.

9.4.4.3 Overall Atmospheric Temperature Change

Anthropogenic influence on free atmosphere temperatures has been detected in analyses of satellite data since 1979, although this finding has been found to be sensitive to which analysis of satellite data is used. Satellite-borne MSUs, beginning in 1978, estimate the temperature of thick layers of the atmosphere. The main layers represent the lower troposphere ($T2_{LT}$), the mid-troposphere (T2) and the lower stratosphere (T4) (Section 3.4.1.2.1). Santer et al. (2003c) compare T2 and T4 temperature changes simulated by the PCM model including anthropogenic and natural forcings with the University of Alabama in Huntsville (UAH; Christy et al., 2000) and Remote Sensing Systems (RSS; Mears and Wentz, 2005) satellite data sets (Section 3.4.1.2.2). They find that the model fingerprint of the T4 response to combined anthropogenic and natural forcing is consistently detected in both satellite data sets, whereas the T2 response is detected only in the RSS data set. However, when the global mean changes are removed, the T2 fingerprint is detected in both data sets, suggesting a common spatial pattern of response overlain by a systematic global mean difference.

Anthropogenic influence on free atmosphere temperatures has been robustly detected in a number of different studies analysing various versions of the Hadley Centre Radiosonde Temperature (HadRT2) data set (Parker et al., 1997) by means of a variety of different diagnostics and fingerprints estimated with the HadCM2 and HadCM3 models (Tett et al., 2002; Thorne et al., 2002, 2003; Jones et al., 2003). Whereas an analysis of spatial

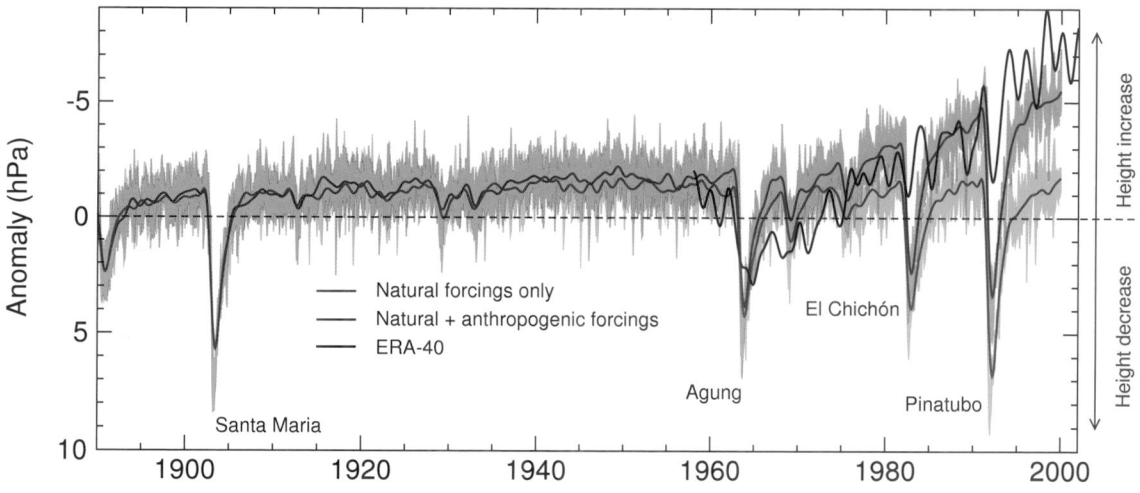

Figure 9.14. *Comparison between reanalysis and climate-model simulated global monthly mean anomalies in tropopause height. Model results are from two different PCM (Table 8.1) ensemble experiments using either natural forcings, or natural and anthropogenic forcings (ALL). There are four realisations of each experiment. Both the low-pass filtered ensemble mean and the unfiltered range between the highest and lowest values of the realisations are shown. All model anomalies are defined relative to climatological monthly means computed over 1890 to 1999. Reanalysis-based tropopause height anomalies estimated from ERA-40 were filtered in the same way as model data. The ERA-40 record spans 1957 to 2002 and was forced to have the same mean as ALL over 1960 to 1999. After Santer et al. (2003a) and Santer et al. (2004).*

patterns of zonal mean free atmosphere temperature changes was unable to detect the response to natural forcings (Tett et al., 2002), an analysis of spatio-temporal patterns detected the influence of volcanic aerosols and (less convincingly) solar irradiance changes, in addition to detecting the effects of greenhouse gases and sulphate aerosols (Jones et al., 2003). In addition, Crooks (2004) detects a solar signal in atmospheric temperature changes as seen in the HadRT2.1s radiosonde data set when a diagnostic chosen to extract the solar signal from other signals is used. The models used in these studies have poor vertical resolution in the stratosphere and they significantly underestimate stratospheric variability, thus possibly overestimating the significance of these detected signals (Tett et al., 2002). However, a sensitivity study (Thorne et al., 2002) showed that detection of human influence on free atmosphere temperature changes does not depend on the inclusion of stratospheric temperatures. An analysis of spatial patterns of temperature change, represented by large-scale area averages at the surface, in broad atmospheric layers and in lapse rates between layers, showed robust detection of an anthropogenic influence on climate when a range of uncertainties were explored relating to the choice of fingerprints and the radiosonde and model data sets (Thorne et al., 2003). However, Thorne et al. were not able to attribute recent observed tropospheric temperature changes to any particular combination of external forcing influences because the models analysed (HadCM2 and HadCM3) overestimate free atmosphere warming as estimated by the radiosonde data sets, an effect also seen by Douglass et al. (2004) during the satellite era. However, there is evidence that radiosonde data during the satellite era are contaminated by spurious cooling trends (Sherwood et al., 2005; Randel and Wu, 2006; Section 3.4.1), and since structural uncertainty arising from the choice of techniques used to analyse radiosonde data has not yet been quantified (Thorne et al., 2005), it is difficult to assess, based on these analyses alone, whether model-data discrepancies are due to model or observational deficiencies. However further information is provided by an analysis of modelled and observed tropospheric lapse rates, discussed in Section 9.4.4.4.

A different approach is to assess detectability of observed temperature changes through the depth of the atmosphere with AGCM simulations forced with observed SSTs, although the vertical profile of the atmospheric temperature change signal estimated in this way can be quite different from the same signal estimated by coupled models with the same external forcings (Hansen et al., 2002; Sun and Hansen, 2003; Santer et al., 2005). Sexton et al. (2001) find that inclusion of anthropogenic effects improves the simulation of zonally averaged upper air temperature changes from the HadRTt1.2 data set such that an anthropogenic signal is detected at the 5% significance level in patterns of seasonal mean temperature change calculated as overlapping eight-year means over the 1976 to 1994 period and expressed as anomalies relative to the 1961 to 1975 base period. In addition, analysing patterns of annual mean temperature change for individual years shows that an anthropogenic signal is also detected on interannual time scales for a number of years towards the end of the analysis period.

9.4.4.4   Differential Temperature Trends

Subtracting temperature trends at the surface from those in the free atmosphere removes much of the common variability between these layers and tests whether the model-predicted trends in tropospheric lapse rate are consistent with those observed by radiosondes and satellites (Karl et al., 2006). Since 1979, globally averaged modelled trends in tropospheric lapse rates are consistent with those observed. However, this is not the case in the tropics, where most models have more warming aloft than at the surface while most observational estimates show more warming at the surface than in the troposphere (Karl et al., 2006). Karl et al. (2006) carried out a systematic review of this issue. There is greater consistency between simulated and observed differential warming in the tropics in some satellite measurements of tropospheric temperature change, particularly when the effect of the cooling stratosphere on tropospheric retrievals is taken into account (Karl et al., 2006). External forcing other than greenhouse gas changes can also help to reconcile some of the differential warming, since both volcanic eruptions and stratospheric ozone depletion are expected to have cooled the troposphere more than the surface over the last several decades (Santer et al., 2000, 2001; IPCC, 2001; Free and Angell, 2002; Karl et al., 2006). There are, however, uncertainties in quantifying the differential cooling caused by these forcings, both in models and observations, arising from uncertainties in the forcings and model response to the forcings. Differential effects of natural modes of variability, such as ENSO and the NAM, on observed surface and tropospheric temperatures, which arise from differences in the amplitudes and spatial expression of these modes at the surface and in the troposphere, make only minor contributions to the overall differences in observed surface and tropospheric warming rates (Santer et al., 2001; Hegerl and Wallace, 2002; Karl et al., 2006).

A systematic intercomparison between radiosonde-based (Radiosonde Atmospheric Temperature Products for Assessing Climate (RATPAC); Free et al., 2005, and Hadley Centre Atmospheric Temperature (HadAT), Thorne et al., 2005) and satellite-based (RSS, UAH) observational estimates of tropical lapse rate trends with those simulated by 19 MMD models shows that on monthly and annual time scales, variations in temperature at the surface are amplified aloft in both models and observations by consistent amounts (Santer et al., 2005; Karl et al., 2006). It is only on longer time scales that disagreement between modelled and observed lapse rates arises (Hegerl and Wallace, 2002), that is, on the time scales over which discrepancies would arise from inhomogeneities in the observational record. Only one observational data set (RSS) was found to be consistent with the models on both short and long time scales. While Vinnikov et al. (2006) have not produced a lower-tropospheric retrieval, their estimate of the T2 temperature trend (Figure 3.18) is consistent with model simulations (Karl et al., 2006). One possibility is that amplification effects are controlled by different physical mechanisms on short and long time scales, although a more probable explanation is that some observational records are contaminated by errors that affect their long-term trends (Section 3.4.1; Karl et al., 2006).

Frequently Asked Question 9.2
Can the Warming of the 20th Century be Explained by Natural Variability?

It is very unlikely that the 20th-century warming can be explained by natural causes. The late 20th century has been unusually warm. Palaeoclimatic reconstructions show that the second half of the 20th century was likely the warmest 50-year period in the Northern Hemisphere in the last 1300 years. This rapid warming is consistent with the scientific understanding of how the climate should respond to a rapid increase in greenhouse gases like that which has occurred over the past century, and the warming is inconsistent with the scientific understanding of how the climate should respond to natural external factors such as variability in solar output and volcanic activity. Climate models provide a suitable tool to study the various influences on the Earth's climate. When the effects of increasing levels of greenhouse gases are included in the models, as well as natural external factors, the models produce good simulations of the warming that has occurred over the past century. The models fail to reproduce the observed warming when run using only natural factors. When human factors are included, the models also simulate a geographic pattern of temperature change around the globe similar to that which has occurred in recent decades. This spatial pattern, which has features such as a greater warming at high northern latitudes, differs from the most important patterns of natural climate variability that are associated with internal climate processes, such as El Niño.

Variations in the Earth's climate over time are caused by natural internal processes, such as El Niño, as well as changes in external influences. These external influences can be natural in origin, such as volcanic activity and variations in solar output, or caused by human activity, such as greenhouse gas emissions, human-sourced aerosols, ozone depletion and land use change. The role of natural internal processes can be estimated by studying observed variations in climate and by running climate models without changing any of the external factors that affect climate. The effect of external influences can be estimated with models by changing these factors, and by using physical understanding of the processes involved. The combined effects of natural internal variability and natural external factors can also be estimated from climate information recorded in tree rings, ice cores and other types of natural 'thermometers' prior to the industrial age.

The natural external factors that affect climate include volcanic activity and variations in solar output. Explosive volcanic eruptions occasionally eject large amounts of dust and sulphate aerosol high into the atmosphere, temporarily shielding the Earth and reflecting sunlight back to space. Solar output has an 11-year cycle and may also have longer-term variations. Human activities over the last 100 years, particularly the burning of fossil fuels, have caused a rapid increase in carbon dioxide and other greenhouse gases in the atmosphere. Before

the industrial age, these gases had remained at near stable concentrations for thousands of years. Human activities have also caused increased concentrations of fine reflective particles, or 'aerosols', in the atmosphere, particularly during the 1950s and 1960s.

Although natural internal climate processes, such as El Niño, can cause variations in global mean temperature for relatively short periods, analysis indicates that a large portion is due to external factors. Brief periods of global cooling have followed major volcanic eruptions, such as Mt. Pinatubo in 1991. In the early part of the 20th century, global average temperature rose, during which time greenhouse gas concentrations started to rise, solar output was probably increasing and there was little volcanic activity. During the 1950s and 1960s, average global temperatures levelled off, as increases in aerosols from fossil fuels and other sources cooled the planet. The eruption of Mt. Agung in 1963 also put large quantities of reflective dust into the upper atmosphere. The rapid warming observed since the 1970s has occurred in a period when the increase in greenhouse gases has dominated over all other factors.

Numerous experiments have been conducted using climate models to determine the likely causes of the 20th-century climate change. These experiments indicate that models cannot reproduce the rapid warming observed in recent decades when they only take into account variations in solar output and volcanic activity. However, as shown in Figure 1, models are able to simulate the observed 20th-century changes in temperature when they include all of the most important external factors, including human influences from sources such as greenhouse gases and natural external factors. The model-estimated responses to these external factors are detectable in the 20th-century climate globally and in each individual continent except Antarctica, where there are insufficient observations. The human influence on climate very likely dominates over all other causes of change in global average surface temperature during the past half century.

An important source of uncertainty arises from the incomplete knowledge of some external factors, such as human-sourced aerosols. In addition, the climate models themselves are imperfect. Nevertheless, all models simulate a pattern of response to greenhouse gas increases from human activities that is similar to the observed pattern of change. This pattern includes more warming over land than over the oceans. This pattern of change, which differs from the principal patterns of temperature change associated with natural internal variability, such as El Niño, helps to distinguish the response to greenhouse gases from that of natural external factors. Models and observations also both show warming in the lower part of

(continued)

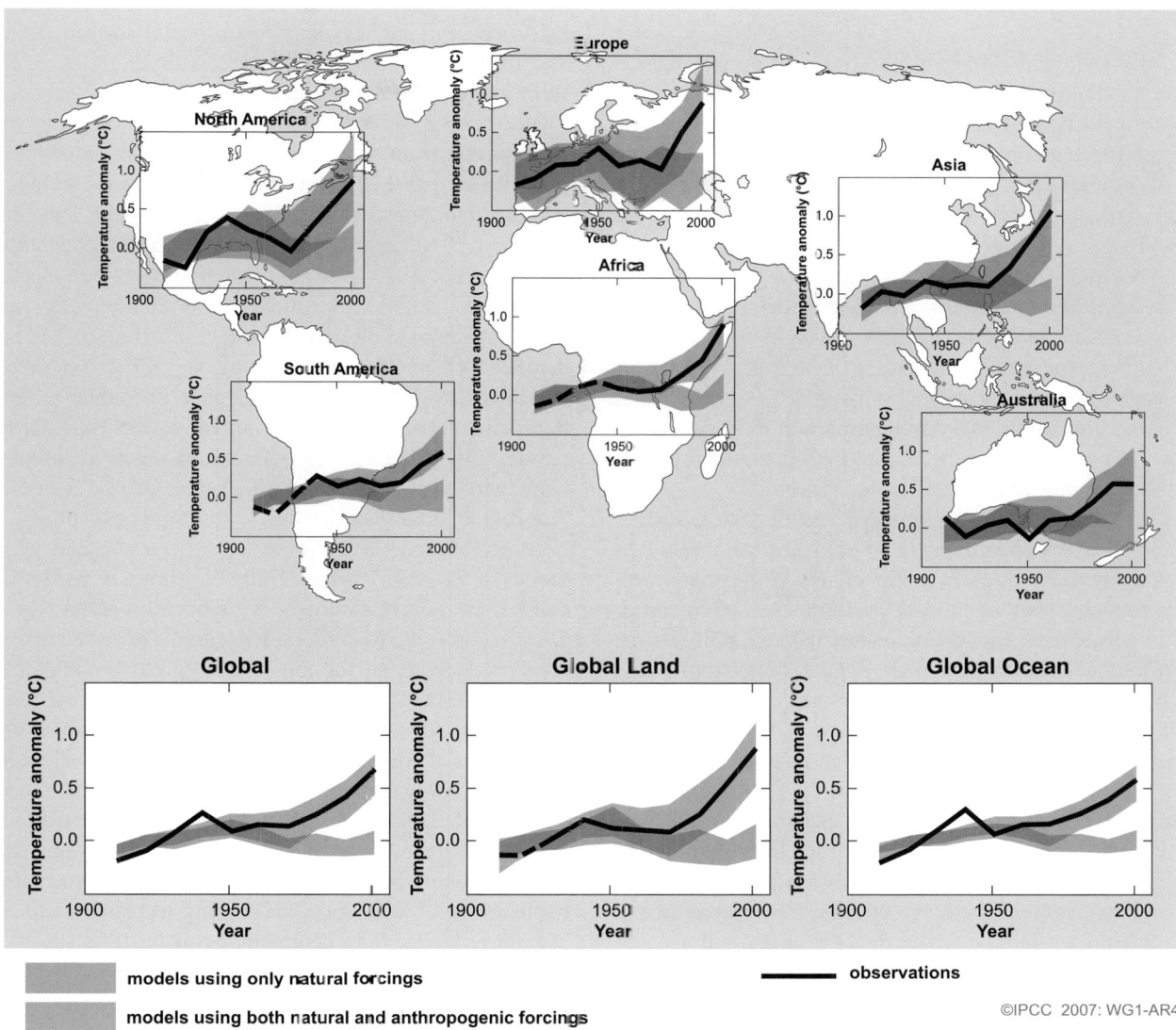

FAQ 9.2, Figure 1. *Temperature changes relative to the corresponding average for 1901-1950 (°C) from decade to decade from 1906 to 2005 over the Earth's continents, as well as the entire globe, global land area and the global ocean (lower graphs). The black line indicates observed temperature change, while the coloured bands show the combined range covered by 90% of recent model simulations. Red indicates simulations that include natural and human factors, while blue indicates simulations that include only natural factors. Dashed black lines indicate decades and continental regions for which there are substantially fewer observations. Detailed descriptions of this figure and the methodology used in its production are given in the Supplementary Material, Appendix 9.C.*

the atmosphere (the troposphere) and cooling higher up in the stratosphere. This is another 'fingerprint' of change that reveals the effect of human influence on the climate. If, for example an increase in solar output had been responsible for the recent climate warming, both the troposphere and the stratosphere would have warmed. In addition, differences in the timing of the human and natural external influences help to distinguish the climate responses to these factors. Such considerations increase confidence that human rather than natural factors were the dominant cause of the global warming observed over the last 50 years.

Estimates of Northern Hemisphere temperatures over the last one to two millennia, based on natural 'thermometers' such as tree rings that vary in width or density as temperatures change, and historical weather records, provide additional evidence that

the 20th-century warming cannot be explained by only natural internal variability and natural external forcing factors. Confidence in these estimates is increased because prior to the industrial era, much of the variation they show in Northern Hemisphere average temperatures can be explained by episodic cooling caused by large volcanic eruptions and by changes in the Sun's output. The remaining variation is generally consistent with the variability simulated by climate models in the absence of natural and human-induced external factors. While there is uncertainty in the estimates of past temperatures, they show that it is likely that the second half of the 20th century was the warmest 50-year period in the last 1300 years. The estimated climate variability caused by natural factors is small compared to the strong 20th-century warming.

9.4.5 Summary

Since the TAR, the evidence has strengthened that human influence has increased global temperatures near the surface of the Earth. Every year since the publication of the TAR has been in the top ten warmest years in the instrumental global record of near-surface temperatures. Many climate models are now available which simulate global mean temperature changes that are consistent with those observed over the last century when they include the most important forcings of the climate system. The fact that no coupled model simulation so far has reproduced global temperature changes over the 20th century without anthropogenic forcing is strong evidence for the influence of humans on global climate. This conclusion is robust to variations in model formulation and uncertainties in forcings as far as they have been explored in the large multi-model ensemble now available (Figure 9.5).

Many studies have detected a human influence on near-surface temperature changes, applying a variety of statistical techniques and using many different climate simulations. Comparison with observations shows that the models used in these studies appear to have an adequate representation of internal variability on the decadal to inter-decadal time scales important for detection (Figure 9.7). When evaluated in a Bayesian framework, very strong evidence is found for a human influence on global temperature change regardless of the choice of prior distribution.

Since the TAR, there has been an increased emphasis on partitioning the observed warming into contributions from greenhouse gas increases and other anthropogenic and natural factors. These studies lead to the conclusion that greenhouse gas forcing has very likely been the dominant cause of the observed global warming over the last 50 years, and account for the possibility that the agreement between simulated and observed temperature changes could be reproduced by different combinations of external forcing. This is because, in addition to detecting the presence of model-simulated spatio-temporal response patterns in observations, such analyses also require consistency between the model-simulated and observational amplitudes of these patterns.

Detection and attribution analyses indicate that over the past century there has likely been a cooling influence from aerosols and natural forcings counteracting some of the warming influence of the increasing concentrations of greenhouse gases (Figure 9.9). Spatial information is required in addition to temporal information to reliably detect the influence of aerosols and distinguish them from the influence of increased greenhouse gases. In particular, aerosols are expected to cause differential warming and cooling rates between the NH and SH that change with time depending on the evolution of the aerosol forcing, and this spatio-temporal fingerprint can help to constrain the possible range of cooling from aerosols over the century. Despite continuing uncertainties in aerosol forcing and the climate response, it is likely that greenhouse gases alone would have caused more warming than observed during the last 50 years, with some warming offset by cooling from aerosols

and other natural and anthropogenic factors. The overall evidence from studies using instrumental surface temperature and free atmospheric temperature data, along with evidence from analysis of temperature over the last few hundred years (Section 9.3.3.2), indicates that it is very unlikely that the contribution from solar forcing to the warming of the last 50 years was larger than that from greenhouse gas forcing.

An important development since the TAR has been the detection of an anthropogenic signal in surface temperature changes since 1950 over continental and sub-continental scale land areas. The ability of models to simulate many aspects of the temperature evolution at these scales (Figure 9.12) and the detection of significant anthropogenic effects on each of six continents provides stronger evidence of human influence on the global climate than was available to the TAR. Difficulties remain in attributing temperature changes at smaller than continental scales and over time scales of less than 50 years. Attribution at these scales has, with limited exceptions, not yet been established. Temperature changes associated with some modes of variability, which could be wholly or partly naturally caused, are poorly simulated by models in some regions and seasons and could be confounded with the expected temperature response to external forcings. Averaging over smaller regions reduces the natural variability less than averaging over large regions, making it more difficult to distinguish changes expected from external forcing. In addition, the small-scale details of external forcing and the response simulated by models are less credible than large-scale features. Overall, uncertainties in observed and model-simulated climate variability and change at smaller spatial scales make it difficult at present to estimate the contribution of anthropogenic forcing to temperature changes at scales smaller than continental and on time scales shorter than 50 years.

There is now some evidence that anthropogenic forcing has affected extreme temperatures. There has been a significant decrease in the frequency of frost days and an increase in the incidence of warm nights. A detection and attribution analysis has shown a significant human influence on patterns of changes in extremely warm nights and evidence for a human-induced warming of the coldest nights and days of the year. Many important impacts of climate change are likely to manifest themselves through an increase in the frequency of heat waves in some regions and a decrease in the frequency of extremely cold events in others. Based on a single study, and assuming a model-based estimate of temperature variability, past human influence may have more than doubled the risk of European mean summer temperatures as high as those recorded in 2003 (Figure 9.13).

Since the TAR, further evidence has accumulated that there has been a significant anthropogenic influence on free atmosphere temperature since widespread measurements became available from radiosondes in the late 1950s. The influence of greenhouse gases on tropospheric temperatures has been detected, as has the influence of stratospheric ozone depletion on stratospheric temperatures. The combination of a warming troposphere and a cooling stratosphere has likely led to an increase in the height of the tropopause and model-data comparisons show that greenhouse gases and stratospheric ozone changes are likely largely responsible (Figure 9.14).

Whereas, on monthly and annual time scales, variations of temperature in the tropics at the surface are amplified aloft in both the MMD simulations and observations by consistent amounts, on longer time scales, simulations of differential tropical warming rates between the surface and the free atmosphere are inconsistent with some observational records. One possible explanation for the discrepancies on multi-annual but not shorter time scales is that amplification effects are controlled by different physical mechanisms, but a more probable explanation is that some observational records are contaminated by errors that affect their long-term trends.

9.5　Understanding of Change in Other Variables during the Industrial Era

The objective of this section is to assess large-scale climate change in variables other than air temperature, including changes in ocean climate, atmospheric circulation, precipitation, the cryosphere and sea level. This section draws heavily on Chapters 3, 4, 5 and 8. Where possible, it attempts to identify links between changes in different variables, such as those that associate some aspects of SST change with precipitation change. It also discusses the role of external forcing, drawing where possible on formal detection studies.

9.5.1　Ocean Climate Change

9.5.1.1　Ocean Heat Content Changes

Since the TAR, evidence of climate change has accumulated within the ocean, both at regional and global scales (Chapter 5). The overall heat content in the World Ocean is estimated to have increased by 14.2×10^{22} J during the period 1961 to 2003 (Section 5.2.2). This overall increase has been superimposed on strong interannual and inter-decadal variations. The fact that the entire ocean, which is by far the system's largest heat reservoir (Levitus et al., 2005; see also Figure 5.4) gained heat during the latter half of the 20th century is consistent with a net positive radiative forcing of the climate system. Late 20th-century ocean heat content changes were at least one order of magnitude larger than the increase in energy content of any other component of the Earth's ocean-atmosphere-cryosphere system (Figure 5.4, Levitus et al., 2005).

All analyses indicate a large anthropogenic component of the positive trend in global ocean heat content. Levitus et al. (2001) and Gregory et al (2004) analyse simulations from the GFDL R30 and HadCM3 models respectively and show that climate simulations agree best with observed changes when the models include anthropogenic forcings from increasing greenhouse gas concentrations and sulphate aerosols. Gent and Danabasoglu

(2004) show that the observed trend cannot be explained by natural internal variability as simulated by a long control run of the Community Climate System Model (CCSM2). Barnett et al. (2001) and Reichert et al. (2002b) use detection analyses similar to those described in Section 9.4 to detect model-simulated ocean climate change signals in the observed spatio-temporal patterns of ocean heat content across the ocean basins.

Barnett et al. (2005) extend previous detection and attribution analyses of ocean heat content changes to a basin by basin analysis of the temporal evolution of temperature changes in the upper 700 m of the ocean (see also Pierce et al., 2006). They report that whereas the observed change is not consistent with internal variability and the response to natural external forcing as simulated by two climate models (PCM and HadCM3), the simulated ocean warming due to anthropogenic factors (including well-mixed greenhouse gases and sulphate aerosols) is consistent with the observed changes and reproduces many of the different responses seen in the individual ocean basins (Figure 9.15), indicating a human-induced warming of the world's oceans with a complex vertical and geographical structure that is simulated quite well by the two AOGCMs. Barnett et al. (2005) find that the earlier conclusions of Barnett et al. (2001) were not affected by the Levitus et al. (2005) revisions to the Levitus et al. (2000) ocean heat content data.

In contrast, changes in solar forcing can potentially explain only a small fraction of the observationally based estimates of the increase in ocean heat content (Crowley et al., 2003), and the cooling influence of natural (volcanic) and anthropogenic aerosols would have slowed ocean warming over the last half century. Delworth et al. (2005) find a delay of several decades and a reduction in the magnitude of the warming of approximately two-thirds in simulations with the GFDL-CM2 model that included these forcings compared to the response to increasing greenhouse gases alone, consistent with results based on an upwelling diffusion EBM (Crowley et al., 2003). Reductions in ocean heat content are found following volcanic eruptions in climate simulations (Church et al., 2005), including a persistent centennial time-scale signal of ocean cooling at depth following the eruption of Krakatoa (Gleckler et al., 2006).

Although the heat uptake in the ocean cannot be explained without invoking anthropogenic forcing, there is some evidence that the models have overestimated how rapidly heat has penetrated below the ocean's mixed layer (Forest et al., 2006; see also Figure 9.15). In simulations that include natural forcings in addition to anthropogenic forcings, eight coupled climate models simulate heat uptake of 0.26 ± 0.06 W m^{-2} (± 1 standard deviation) for 1961 to 2003, whereas observations of ocean temperature changes indicate a heat uptake of 0.21 ± 0.04 W m^{-2} (Section 5.2.2.1). These could be consistent within their uncertainties but might indicate a tendency of climate models to overestimate ocean heat uptake.

In addition, the interannual to decadal variability seen in Levitus et al. (2000, 2005) (Section 5.2.2) is underestimated by models; Gregory et al. (2004) show significant differences between observed and modelled interannual deviations from a linear trend in five-year running means of world ocean heat content above 3,000 m for 1957 to 1994. While some studies

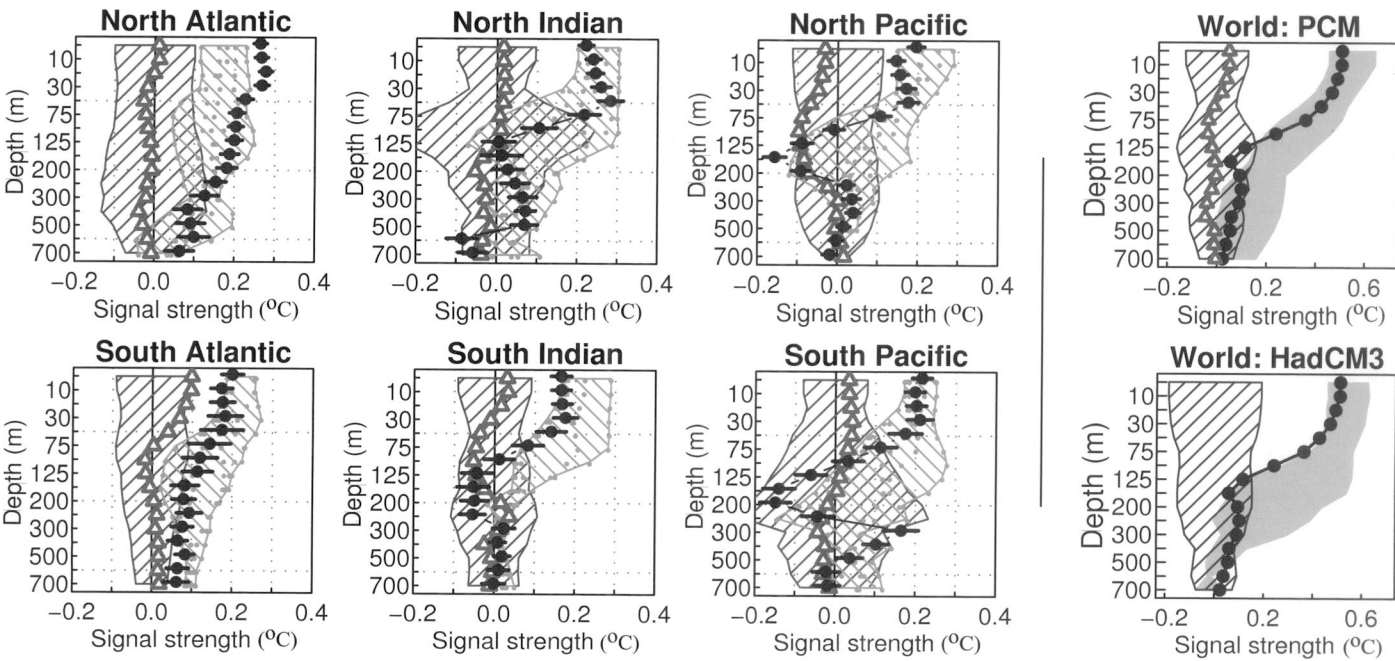

Figure 9.15. *Strength of observed and model-simulated warming signal by depth for the World Ocean and for each ocean basin individually (in ºC, see Barnett et al., 2005 and Pierce et al., 2006 for calculation of signal strength). For ocean basins, the signal is estimated from PCM (Table 8.1) while for the World Ocean it is estimated from both PCM and HadCM3 (Table 8.1). Red dots represent the projection of the observed temperature changes onto the normalised model-based pattern of warming. They show substantial basin-to-basin differences in how the oceans have warmed over the past 40 years, although all oceans have experienced net warming over that interval. The red bars represent the ±2 standard deviation limits associated with sampling uncertainty. The blue crosshatched swaths represent the 90% confidence limits of the natural internal variability strength. The green crosshatched swaths represent the range of the anthropogenically forced signal estimates from different realisations of identically forced simulations with the PCM model for each ocean basin (the smaller dots within the green swaths are the individual realisations) and the green shaded regions represent the range of anthropogenically forced signal estimates from different realisations of identically forced simulations with the PCM and HadCM3 models for the World Ocean (note that PCM and HadCM3 use different representations of anthropogenic forcing). The ensemble-averaged strength of the warming signal in four PCM simulations with solar and volcanic forcing is also shown (grey triangles). From Barnett et al. (2005) and Pierce et al. (2006).*

note the potential importance of the choice of infilling method in poorly sampled regions (Gregory et al., 2004; AchutaRao et al., 2006), the consistency of the differently processed data from the Levitus et al. (2005), Ishii et al. (2006) and Willis et al. (2004) analyses adds confidence to their use for analysing trends in climate change studies (Chapter 5). Gregory et al. (2004) show that agreement between models and observations is better in the well-observed upper ocean (above 300 m) in the NH and that there is large sensitivity to the method of infilling the observational data set outside this well-observed region. They find a strong maximum in variability in the Levitus data set at around 500 m depth that is not seen in HadCM3 simulations, a possible indication of model deficiency or an artefact in the Levitus data. AchutaRao et al. (2006) also find that observational estimates of temperature variability over much of the oceans may be substantially affected by sparse observational coverage and the method of infilling.

9.5.1.2   Water Mass Properties

Interior water masses, which are directly ventilated at the ocean surface, act to integrate highly variable surface changes in heat and freshwater, and could therefore provide indicators of global change (Stark et al., 2006). Some studies have

attempted to investigate changes in three-dimensional water mass properties (Section 5.3). Sub-Antarctic Mode Water (SAMW) and the subtropical gyres have warmed in the Indian and Pacific basins since the 1960s, waters at high latitudes have freshened in the upper 500 m and salinity has increased in some of the subtropical gyres. These changes are consistent with an increase in meridional moisture flux over the oceans over the last 50 years leading to increased precipitation at high latitudes (Section 5.2.3; Wong et al., 1999) and a reduction in the difference between precipitation and evaporation at mid-latitudes (Section 5.6). This suggests that the ocean might integrate rainfall changes to produce detectable salinity changes. Boyer et al. (2005) estimated linear trends in salinity for the global ocean from 1955 to 1998 that indicate salinification in the Antarctic Polar Frontal Zone around 40°S and in the subtropical North Atlantic, and freshening in the sub-polar Atlantic (Figures 5.5 and 5.7). However, variations in other terms (e.g., ocean freshwater transport) may be contributing substantially to the observed salinity changes and have not been quantified.

An observed freshening of SAMW in the South Indian Ocean between the 1960s and 1990s has been shown to be consistent with anthropogenically forced simulations from HadCM3 (Banks et al., 2000) but care should be taken in interpreting sparse hydrographic data, since apparent trends could reflect natural variability or the aliased effect of changing observational coverage. Although SAMW was fresher along isopycnals in 1987 than in the 1960s, in 2002 the salinity was again near the 1960s values (Bindoff and McDougall, 2000; Bryden et al., 2003). An analysis of an ocean model forced by observed atmospheric fluxes and SSTs indicates that this is likely associated with natural variability (Murray et al., 2007), a result supported by an analysis of 20th-century simulations with HadCM3 which shows that it is not possible to reject the null hypothesis that the observed differences are due to internal variability (Stark et al., 2006), although this model does project a long-term freshening trend in the 21st century due to the large-scale response to surface heating and hydrological changes (Banks et al., 2000).

9.5.1.3   Changes in the Meridional Overturning Circulation

It is possible that anthropogenic and natural forcing may have influenced the MOC in the Atlantic (see also Box 5.1). One possible oceanic consequence of climate change is a slowing down or even halting of the MOC. An estimate of the overturning circulation and associated heat transport based on a trans-Atlantic section along latitude 25°N indicates that the Atlantic MOC has slowed by about 30% over five samples taken between 1957 and 2004 (Bryden et al., 2005), although given the infrequent sampling and considerable variability it is not clear whether the trend estimate is robust (Box 5.1). Freshening of North East Atlantic Deep Water has been observed (Dickson et al., 2002; Curry et al., 2003; Figure 5.6) and has been interpreted as being consistent with an enhanced difference between precipitation and evaporation at high latitudes and a possible slowing down of the MOC. Wu et al. (2004) show that the observed freshening trend is well reproduced by an ensemble of HadCM3 simulations that includes both anthropogenic and natural forcings, but this freshening coincides with a strengthening rather than a weakening trend in the MOC. Therefore, this analysis is not consistent with an interpretation of the observed freshening trends in the North Atlantic as an early signal of a slowdown of the thermohaline circulation. Dickson et al. (2002) propose a possible role for the Arctic in

driving the observed freshening of the subpolar North Atlantic. Wu et al. (2005) show that observed increases in arctic river flow (Peterson et al., 2002) are well simulated by HadCM3 including anthropogenic and natural forcings and propose that this increase is anthropogenic, since it is not seen in HadCM3 simulations including just natural forcing factors. However, the relationship between this increased source of freshwater and freshening in the Labrador Sea is not clear in the HadCM3 simulations, since Wu et al. (2007) find that recent freshening in the Labrador Sea is simulated by the model when it is driven by natural rather than anthropogenic forcings. Importantly, freshening is also associated with decadal and multi-decadal variability, with links to the NAO (Box 5.1) and the AMO (Box 5.1; Vellinga and Wu, 2004; Knight et al., 2005).

9.5.2   Sea Level

A precondition for attributing changes in sea level rise to anthropogenic forcing is that model-based estimates of historical global mean sea level rise should be consistent with observational estimates. Although AOGCM simulations of global mean surface air temperature trends are generally consistent with observations (Section 9.4.1, Figure 9.5), consistency with surface air temperature alone does not guarantee a realistic simulation of thermal expansion, as there may be compensating errors among climate sensitivity, ocean heat uptake and radiative forcing (see, e.g., Raper et al., 2002, see also Section 9.6). Model simulations also offer the possibility of attributing past sea level changes to particular forcing factors. The observational budget for sea level (Section 5.5.6) assesses the periods 1961 to 2003 and 1993 to 2003. Table 9.2 evaluates the same terms from 20C3M simulations in the MMD at PCMDI, although most 20C3M simulations end earlier (between 1999 and 2002), so the comparison is not quite exact.

Simulations including natural as well as anthropogenic forcings (the 'ALL' models in Table 9.2) generally have smaller ocean heat uptake during the period 1961 to 2003 than those without volcanic forcing, since several large volcanic eruptions cooled the climate during this period (Gleckler et al., 2006). This leads to a better agreement of those simulations with

Table 9.2. *Components of the rate of global mean sea level rise (mm yr⁻¹) from models and observations. All ranges are 5 to 95% confidence intervals. The observational components and the observed rate of sea level rise ('Obs' column) are repeated from Section 5.5.6 and Table 5.3. The 'ALL' column is computed (following the methods of Gregory and Huybrechts, 2006 and Section 10.6.3.1) from eight 20C3M simulations that include both natural and anthropogenic forcings (models 3, 9, 11, 12, 14, 15, 19 and 21; see Table 8.1), and the 'ALL/ANT' column from 16 simulations: the eight ALL and eight others that have anthropogenic forcings only (models 4, 6, 7, 8, 13, 16, 20 and 22; see Table 8.1).*

	1961–2003			1993–2003		
	Obs	**ALL**	**ALL/ANT**	**Obs**	**ALL**	**ALL/ANT**
Thermal expansion	0.42 ± 0.12	0.5 ± 0.2	0.7 ± 0.4	1.60 ± 0.50	1.5 ± 0.7	1.2 ± 0.9
Glaciers and ice caps	0.50 ± 0.18	0.5 ± 0.2	0.5 ± 0.3	0.77 ± 0.22	0.7 ± 0.3	0.8 ± 0.3
Ice sheets (observed)	0.19 ± 0.43			0.41 ± 0.35		
Sum of components	1.1 ± 0.5	1.2 ± 0.5	1.4 ± 0.7	2.8 ± 0.7	2.6 ± 0.8	2.4 ± 1.0
Observed rate of rise	1.8 ± 0.5			3.1 ± 0.7		

thermal expansion estimates based on observed ocean warming (Section 5.5.3) than for the complete set of model simulations ('ALL/ANT' in Table 9.2). For 1993 to 2003, the models that include natural forcings agree well with observations. Although this result is somewhat uncertain because the simulations end at various dates from 1999 onwards, it accords with results obtained by Church et al. (2005) using the PCM and Gregory et al. (2006) using HadCM3, which suggest that 0.5 mm yr^{-1} of the trend in the last decade may result from warming as a recovery from the Mt. Pinatubo eruption of 1991. Comparison of the results for 1961 to 2003 and 1993 to 2003 shows that volcanoes influence the ocean differently over shorter and longer periods. The rapid expansion of 1993 to 2003 was caused, in part, by rapid warming of the upper ocean following the cooling due to the Mt. Pinatubo eruption, whereas the multi-decadal response is affected by the much longer persistence in the deep ocean of cool anomalies caused by volcanic eruptions (Delworth et al., 2005; Gleckler et al., 2006; Gregory et al., 2006).

Both observations and model results indicate that the global average mass balance of glaciers and ice caps depends linearly on global average temperature change, but observations of accelerated mass loss in recent years suggest a greater sensitivity than simulated by models. The global average temperature change simulated by AOGCMs gives a good match to the observational estimates of the contribution of glaciers and ice caps to sea level change in 1961 to 2003 and 1993 to 2003 (Table 9.2) with the assumptions that the global average mass balance sensitivity is 0.80 mm yr^{-1} °C^{-1} (sea level equivalent) and that the climate of 1900 to 1929 was 0.16°C warmer than the temperature required to maintain the steady state for glaciers (see discussion in Section 10.6.3.1 and Appendix 10.A).

Calculations of ice sheet surface mass balance changes due to climate change (following the methods of Gregory and Huybrechts, 2006 and Section 10.6.3.1) indicate small but uncertain contributions during 1993 to 2003 of 0.1 ± 0.1 mm yr^{-1} (5 to 95% range) from Greenland and –0.2 ± 0.4 mm yr^{-1} from Antarctica, the latter being negative because rising temperature in AOGCM simulations leads to greater snow accumulation (but negligible melting) at present. The observational estimates (Sections 4.6.2 and 5.5.6) are 0.21 ± 0.07 mm yr^{-1} for Greenland and 0.21 ± 0.35 mm yr^{-1} for Antarctica. For both ice sheets, there is a significant contribution from recent accelerations in ice flow leading to greater discharge of ice into the sea, an effect that is not included in the models because its causes and mechanisms are not yet properly understood (see Sections 4.6.2 and 10.6.4 for discussion). Hence, the surface mass balance model underestimates the sea level contribution from ice sheet melting. Model-based and observational estimates may also differ because the model-based estimates are obtained using estimates of the correlation between global mean climate change and local climate change over the ice sheets in the 21st century under SRES scenarios. This relationship may not represent recent changes over the ice sheets.

Summing the modelled thermal expansion, global glacier and ice cap contributions and the observational estimates of the ice sheet contributions results in totals that lie below the observed rates of global mean sea level rise during 1961 to 2003 and 1993 to 2003. As shown by Table 9.2, the terms are reasonably well reproduced by the models. Nevertheless, the discrepancy in the total, especially for 1961 to 2003, indicates the lack of a satisfactory explanation of sea level rise. This is also a difficulty for the observational budget (discussed in Section 5.5.6).

A discrepancy between model and observations could also be partly explained by the internally generated variability of the climate system, which control simulations suggest could give a standard deviation in the thermal expansion component of ~0.2 mm yr^{-1} in 10-year trends. This variability may be underestimated by models, since observations give a standard deviation in 10-year trends of 0.7 mm yr^{-1} in thermal expansion (see Sections 5.5.3 and Section 9.5.1.1; Gregory et al., 2006).

Since recent warming and thermal expansion are likely largely anthropogenic (Section 9.5.1.1), the model results suggest that the greater rate of rise in 1993 to 2003 than in 1961 to 2003 could have been caused by rising anthropogenic forcing. However, tide gauge estimates suggests larger variability than models in 10-year trends, and that rates as large as that observed during 1993 to 2003 occurred in previous decades (Section 5.5.2.4).

Overall, it is very likely that the response to anthropogenic forcing contributed to sea level rise during the latter half of the 20th century. Models including anthropogenic and natural forcing simulate the observed thermal expansion since 1961 reasonably well. Anthropogenic forcing dominates the surface temperature change simulated by models, and has likely contributed to the observed warming of the upper ocean and widespread glacier retreat. It is very unlikely that the warming during the past half century is due only to known natural causes. Lack of studies quantifying the contribution of anthropogenic forcing to ocean heat content increase and glacier melting, and the fact that the observational budget is not closed, make it difficult to estimate the anthropogenic contribution. Nevertheless, an expert assessment based on modelling and ocean heat content studies suggests that anthropogenic forcing has likely contributed at least one-quarter to one-half of the sea level rise during the second half of the 20th century (see also Woodworth et al., 2004).

Anthropogenic forcing is also expected to produce an accelerating rate of sea level rise (Woodworth et al., 2004). On the other hand, natural forcings could have increased the rate of sea level rise in the early 20th century and decreased it later in the 20th century, thus producing a steadier rate of rise during the 20th century when combined with anthropogenic forcing (Crowley et al., 2003; Gregory et al., 2006). Observational evidence for acceleration during the 20th century is equivocal, but the rate of sea level rise was greater in the 20th than in the 19th century (Section 5.5.2.4). An onset of higher rates of rise in the early 19th century could have been caused by natural factors, in particular the recovery from the Tambora eruption of 1815 (Crowley et al., 2003; Gregory et al., 2006), with anthropogenic forcing becoming important later in the 19th century.

9.5.3   Atmospheric Circulation Changes

Natural low-frequency variability of the climate system is dominated by a small number of large-scale circulation patterns such as ENSO, the Pacific Decadal Oscillation (PDO), and the NAM and Southern Annular Mode (SAM) (Section 3.6 and Box 3.4). The impact of these modes on terrestrial climate on annual to decadal time scales can be profound, but the extent to which they can be excited or altered by external forcing remains uncertain. While some modes might be expected to change as a result of anthropogenic effects such as the enhanced greenhouse effect, there is little *a priori* expectation about the direction or magnitude of such changes.

9.5.3.1   El Niño-Southern Oscillation/Pacific Decadal Oscillation

The El Niño-Southern Oscillation is the leading mode of variability in the tropical Pacific, and it has impacts on climate around the globe (Section 3.6.2). There have been multi-decadal oscillations in the ENSO index (conventionally defined as a mean SST anomaly in the eastern equatorial Pacific) throughout the 20th century, with more intense El Niño events since the late 1970s, which may reflect in part a mean warming of the eastern equatorial Pacific (Mendelssohn et al., 2005). Model projections of future climate change generally show a mean state shift towards more El-Niño-like conditions, with enhanced warming in the eastern tropical Pacific and a weakened Walker Circulation (Section 10.3.5.3); there is some evidence that such a weakening has been observed over the past 140 years (Vecchi et al., 2006). While some simulations of the response to anthropogenic influence have shown an increase in ENSO variability in response to greenhouse gas increases (Timmermann, 1999; Timmermann et al., 1999; Collins, 2000b), others have shown no change (e.g., Collins, 2000a) or a decrease in variability (Knutson et al., 1997). A recent survey of the simulated response to atmospheric CO_2 doubling in 15 MMD AOGCMs (Merryfield, 2006) finds that three of the models exhibited significant increases in ENSO variability, five exhibited significant decreases and seven exhibited no significant change. Thus, as yet there is no detectable change in ENSO variability in the observations, and no consistent picture of how it might be expected to change in response to anthropogenic forcing (Section 10.3.5.3).

Decadal variability in the North Pacific is characterised by variations in the strength of the Aleutian Low coupled to changes in North Pacific SST (Sections 3.6.3 and 8.4.2). The leading mode of decadal variability in the North Pacific is usually referred to as the PDO, and has a spatial structure in the atmosphere and upper North Pacific Ocean similar to the pattern that is associated with ENSO. One recent study showed a consistent tendency towards the positive phase of the PDO in observations and simulations with the MIROC model that included anthropogenic forcing (Shiogama et al., 2005),

although differences between the observed and simulated PDO patterns, and the lack of additional studies, limit confidence in these findings.

9.5.3.2   North Atlantic Oscillation/Northern Annular Mode

The NAM is an approximately zonally symmetric mode of variability in the NH (Thompson and Wallace, 1998), and the NAO (Hurrell, 1996) may be viewed as its Atlantic counterpart (Section 3.6.4). The NAM index exhibited a pronounced trend towards its positive phase between the 1960s and the 1990s, corresponding to a decrease in surface pressure over the Arctic and an increase over the subtropical North Atlantic (see Section 3.6.4; see also Hurrell, 1996; Thompson et al., 2000; Gillett et al., 2003a). Several studies have shown this trend to be inconsistent with simulated internal variability (Osborn et al., 1999; Gillett et al., 2000, 2002b; Osborn, 2004; Gillett, 2005). Although the NAM index has decreased somewhat since its peak in the mid-1990s, the trend calculated over recent decades remains significant at the 5% significance level compared to simulated internal variability in most models (Osborn, 2004; Gillett, 2005), although one study found that the NAO index trend was marginally consistent with internal variability in one model (Selten et al., 2004).

Most climate models simulate some increase in the NAM index in response to increased concentrations of greenhouse gases (Fyfe et al., 1999; Paeth et al., 1999; Shindell et al., 1999; Gillett et al., 2003a,b; Osborn, 2004; Rauthe et al., 2004), although the simulated trend is generally smaller than that observed (Gillett et al., 2002b, 2003b; Osborn, 2004; Gillett, 2005; and see Figure 9.16). Simulated sea level pressure changes are generally found to project more strongly onto the hemispheric NAM index than onto a two-station NAO index (Gillett et al., 2002b; Osborn, 2004; Rauthe et al., 2004). Some studies have postulated an influence of ozone depletion (Volodin and Galin, 1999; Shindell et al., 2001a), changes in solar irradiance (Shindell et al., 2001a) and volcanic eruptions (Kirchner et al., 1999; Shindell et al., 2001a; Stenchikov et al., 2006) on the NAM. Stenchikov et al. (2006) examine changes in sea level pressure following nine volcanic eruptions in the MMD 20C3M ensemble of 20th-century simulations, and find that the models simulated a positive NAM response to the volcanoes, albeit one that was smaller than that observed. Nevertheless, ozone, solar and volcanic forcing changes are generally not found to have made a large contribution to the observed NAM trend over recent decades (Shindell et al., 2001a; Gillett et al., 2003a). Simulations incorporating all the major anthropogenic and natural forcings from the MMD 20C3M ensemble generally showed some increase in the NAM over the latter part of the 20th century (Gillett, 2005; Miller et al., 2006; and see Figure 9.16), although the simulated trend is in all cases smaller than that observed, indicating inconsistency between simulated and observed trends at the 5% significance level (Gillett, 2005).

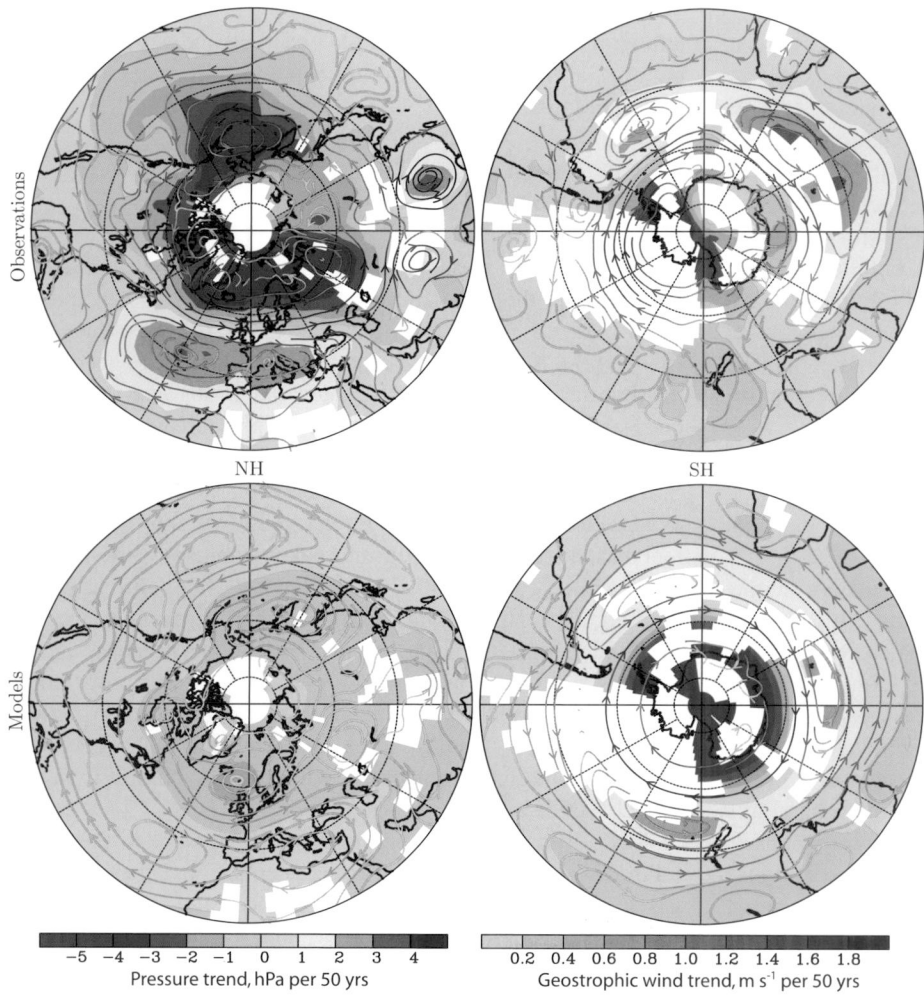

NH

SH

Observations

Models

−5 −4 −3 −2 −1 0 1 2 3 4
Pressure trend, hPa per 50 yrs

0.2 0.4 0.6 0.8 1.0 1.2 1.4 1.6 1.8
Geostrophic wind trend, m s⁻¹ per 50 yrs

Figure 9.16. *Comparison between observed (top) and model-simulated (bottom) December to February sea-level pressure trends (hPa per 50 years) in the NH (left panels) and SH (right panels) based on decadal means for the period 1955 to 2005. Observed trends are based on the Hadley Centre Mean Sea Level Pressure data set (HadSLP2r, an infilled observational data set; Allan and Ansell, 2006). Model-simulated trends are the mean simulated response to greenhouse gas, sulphate aerosol, stratospheric ozone, volcanic aerosol and solar irradiance changes from eight coupled models (CCSM3, GFDL-CM2.0, GFDL-CM2.1, GISS-EH, GISS-ER, MIROC3.2(medres), PCM, UKMO-HadCM3; see Table 8.1 for model descriptions). Streamlines indicate the direction of the trends (m s⁻¹ per 50 years) in the geostrophic wind velocity derived from the trends in sea level pressure, and the shading of the streamlines indicates the magnitude of the change, with darker streamlines corresponding to larger changes in geostrophic wind. White areas in all panels indicate regions with insufficient station-based measurements to constrain analysis. Further explanation of the construction of this figure is provided in the Supplementary Material, Appendix 9.C. Updated after Gillett et al. (2005).*

Over the period 1968 to 1997, the trend in the NAM was associated with approximately 50% of the winter surface warming in Eurasia, due to increased advection of maritime air onto the continent, but only a small fraction (16%) of the NH extratropical annual mean warming trend (Thompson et al., 2000; Section 3.6.4 and Figure 3.30). It was also associated with a decrease in winter precipitation over southern Europe and an increase over northern Europe, due the northward displacement of the storm track (Thompson et al., 2000).

9.5.3.3 Southern Annular Mode

The SAM is more zonally symmetric than its NH counterpart (Thompson and Wallace, 2000; Section 3.6.5). It too has exhibited a pronounced upward trend over the past 30 years, corresponding to a decrease in surface pressure over the Antarctic and an increase over the southern mid-latitudes (Figure 9.16), although the mean SAM index since 2000 has been below the mean in the late 1990s, but above the long term mean (Figure 3.32). An upward trend in the SAM has occurred in all seasons, but the largest trend has been observed during the southern summer (Thompson et al., 2000; Marshall, 2003). Marshall et al. (2004) show that observed trends in the SAM are not consistent with simulated internal variability in HadCM3, suggesting an external cause. On the other hand, Jones and Widmann (2004) develop a 95-year reconstruction of the summer SAM index based largely on mid-latitude pressure measurements, and find that their reconstructed SAM index was as high in the early 1960s as in the late 1990s. However, a more reliable reconstruction from 1958, using more Antarctic data and a different method, indicates that the summer SAM index was higher at the end of the 1990s than at any other time in the observed record (Marshall et al., 2004).

Based on an analysis of the structure and seasonality of the observed trends in SH circulation, Thompson and Solomon (2002) suggest that they have been largely induced by stratospheric ozone depletion. Several modelling studies simulate an upward trend in the SAM in response to stratospheric ozone depletion (Sexton, 2001; Gillett and Thompson, 2003; Marshall et al., 2004; Shindell and Schmidt, 2004; Arblaster and Meehl, 2006; Miller et al., 2006), particularly in the southern summer. Stratospheric ozone depletion cools and strengthens

The mechanisms underlying NH circulation changes remain open to debate. Simulations in which observed SST changes, which may in part be externally forced, were prescribed either globally or in the tropics alone were able to capture around half of the recent trend towards the positive phase of the NAO (Hoerling et al., 2005; Hurrell et al., 2005), suggesting that the trend may in part relate to SST changes, particularly over the Indian Ocean (Hoerling et al., 2005). Another simulation in which a realistic trend in stratospheric winds was prescribed was able to reproduce the observed trend in the NAO (Scaife et al., 2005). Rind et al. (2005a,b) find that both stratospheric changes and changes in SST can force changes in the NAM and NAO, with changes in SSTs being the dominant forcing mechanism.

the antarctic stratospheric vortex in spring, and observations and models indicate that this strengthening of the stratospheric westerlies can be communicated downwards into the troposphere (Thompson and Solomon, 2002; Gillett and Thompson, 2003). While ozone depletion may be the dominant cause of the trends, other studies have indicated that greenhouse gas increases have also likely contributed (Fyfe et al., 1999; Kushner et al., 2001; Stone et al., 2001; Cai et al., 2003; Marshall et al., 2004; Shindell and Schmidt, 2004; Stone and Fyfe, 2005; Arblaster and Meehl, 2006). During the southern summer, the trend in the SAM has been associated with the observed increase of about 3 m s^{-1} in the circumpolar westerly winds over the Southern Ocean. This circulation change is estimated to explain most of the summer surface cooling over the Antarctic Plateau, and about one-third to one-half of the warming of the Antarctic Peninsula (Thompson and Solomon, 2002; Carril et al., 2005; Section 3.6.5), with the largest influence on the eastern side of the Peninsula (Marshall et al., 2006), although other factors are also likely to have contributed to this warming (Vaughan et al., 2001).

### 9.5.3.4	Sea Level Pressure Detection and Attribution

Global December to February sea level pressure changes observed over the past 50 years have been shown to be inconsistent with simulated internal variability (Gillett et al., 2003b, 2005), but are consistent with the simulated response to greenhouse gas, stratospheric ozone, sulphate aerosol, volcanic aerosol and solar irradiance changes based on 20C3M simulations by eight MMD coupled models (Gillett et al., 2005; Figure 9.16). This result is dominated by the SH, where the inclusion of stratospheric ozone depletion leads to consistency between simulated and observed sea level pressure changes. In the NH, simulated sea level pressure trends are much smaller than those observed (Gillett, 2005). Global mean sea level pressure changes associated with increases in atmospheric water vapour are small in comparison to the spatial variations in the observed change in sea level pressure, and are hard to detect because of large observational uncertainties (Trenberth and Smith, 2005).

### 9.5.3.5	Monsoon Circulation

The current understanding of climate change in the monsoon regions remains one of considerable uncertainty with respect to circulation and precipitation (Sections 3.7, 8.4.10 and 10.3.5.2). The Asian monsoon circulation in the MMD models was found to decrease by 15% by the late 21st century under the SRES A1B scenario (Tanaka et al., 2005; Ueda et al., 2006), but trends during the 20th century were not examined. Ramanathan et al. (2005) simulate a pronounced weakening of the Asian monsoon circulation between 1985 and 2000 in response to black carbon aerosol increases. Chase et al. (2003) examine changes in several indices of four major tropical monsoonal circulations (Southeastern Asia, western Africa, eastern Africa and the Australia/Maritime Continent) for the period 1950 to

1998. They find significantly diminished monsoonal circulation in each region, although this result is uncertain due to changes in the observing system affecting the NCEP reanalysis (Section 3.7). These results are consistent with simulations (Ramanathan et al., 2005; Tanaka et al., 2005) of weakening monsoons due to anthropogenic factors, but further model and empirical studies are required to confirm this.

### 9.5.3.6	Tropical Cyclones

Several recent events, including the active North Atlantic hurricane seasons of 2004 and 2005, the unusual development of a cyclonic system in the subtropical South Atlantic that hit the coast of southern Brazil in March 2004 (e.g., Pezza and Simmonds, 2005) and a hurricane close to the Iberian Peninsula in October 2005, have raised public and media interest in the possible effects of climate change on tropical cyclone activity. The TAR concluded that there was 'no compelling evidence to indicate that the characteristics of tropical and extratropical storms have changed', but that an increase in tropical peak wind intensities was likely to occur in some areas with an enhanced greenhouse effect (see also Box 3.5 and Trenberth, 2005). The spatial resolution of most climate models limits their ability to realistically simulate tropical cyclones (Section 8.5.3), therefore, most studies of projected changes in hurricanes have either used time slice experiments with high-resolution atmosphere models and prescribed SSTs, or embedded hurricane models in lower-resolution General Circulation Models (GCMs) (Section 10.3.6.3). While results vary somewhat, these studies generally indicate a reduced frequency of tropical cyclones in response to enhanced greenhouse gas forcing, but an increase in the intensity of the most intense cyclones (Section 10.3.6.3). It has been suggested that the simulated frequency reduction may result from a decrease in radiative cooling associated with increased CO_2 concentration (Sugi and Yoshimura, 2004; Yoshimura and Sugi, 2005; Section 10.3.6.3; Box 3.5), while the enhanced atmospheric water vapour concentration under greenhouse warming increases available potential energy and thus cyclone intensity (Trenberth, 2005).

There continues to be little evidence of any trend in the observed total frequency of global tropical cyclones, at least up until the late 1990s (e.g., Solow and Moore, 2002; Elsner et al., 2004; Pielke et al., 2005; Webster et al., 2005). However, there is some evidence that tropical cyclone intensity may have increased. Globally, Webster et al. (2005) find a strong increase in the number and proportion of the most intense tropical cyclones over the past 35 years. Emanuel (2005) reports a marked increase since the mid-1970s in the Power Dissipation Index (PDI), an index of the destructiveness of tropical cyclones (essentially an integral, over the lifetime of the cyclone, of the cube of the maximum wind speed), in the western North Pacific and North Atlantic, reflecting the apparent increases in both the duration of cyclones and their peak intensity. Several studies have shown that tropical cyclone activity was also high in the 1950 to 1970 period in the North Atlantic (Landsea, 2005) and North Pacific (Chan, 2006), although recent values of the PDI may be higher than those recorded previously (Emanuel, 2005; Section 3.8.3). Emanuel (2005) and Elsner et al. (2006) report a strong correlation between the PDI and tropical Atlantic SSTs, although Chan and Liu (2004) find no analogous relationship in the western North Pacific. While changes in Atlantic SSTs have been linked in part to the AMO, the recent warming appears to be mainly associated with increasing global temperatures (Section 3.8.3.2; Mann and Emanuel, 2006; Trenberth and Shea, 2006). Tropical cyclone development is also strongly influenced by vertical wind shear and static stability (Box 3.5). While increasing greenhouse gas concentrations have likely contributed to a warming of SSTs, effects on static stability and wind shear may have partly opposed this influence on tropical cyclone formation (Box 3.5). Thus, detection and attribution of observed changes in hurricane intensity or frequency due to external influences remains difficult because of deficiencies in theoretical understanding of tropical cyclones, their modelling and their long-term monitoring (e.g., Emanuel, 2005; Landsea, 2005; Pielke, 2005). These deficiencies preclude a stronger conclusion than an assessment that anthropogenic factors more likely than not have contributed to an increase in tropical cyclone intensity.

9.5.3.7 *Extratropical Cyclones*

Simulations of 21st-century climate change in the MMD 20C3M model ensemble generally exhibit a decrease in the total number of extratropical cyclones in both hemispheres, but an increase in the number of the most intense events (Lambert and Fyfe, 2006), although this behaviour is not reproduced by all models (Bengtsson et al., 2006; Section 10.3.6.4). Many 21st-century simulations also show a poleward shift in the storm tracks in both hemispheres (Bengtsson et al., 2006; Section 10.3.6.4). Recent observational studies of winter NH storms have found a poleward shift in storm tracks and increased storm intensity, but a decrease in total storm numbers, in the second half of the 20th century (Section 3.5.3). Analysis of observed wind and significant wave height suggests an increase in storm activity in the NH. In the SH, the storm track has also shifted poleward, with increases in the radius and depth of storms, but decreases in their frequency. These features appear to be associated with the observed trends in the SAM and NAM. Thus, simulated and observed changes in extratropical cyclones are broadly consistent, but an anthropogenic influence has not yet been detected, owing to large internal variability and problems due to changes in observing systems (Section 3.5.3).

9.5.4 Precipitation

9.5.4.1 *Changes in Atmospheric Water Vapour*

The amount of moisture in the atmosphere is expected to increase in a warming climate (Trenberth et al., 2005) because saturation vapour pressure increases with temperature according to the Clausius-Clapeyron equation. Satellite-borne Special Sensor Microwave/Imager (SSM/I) measurements of water vapour since 1988 are of higher quality than either

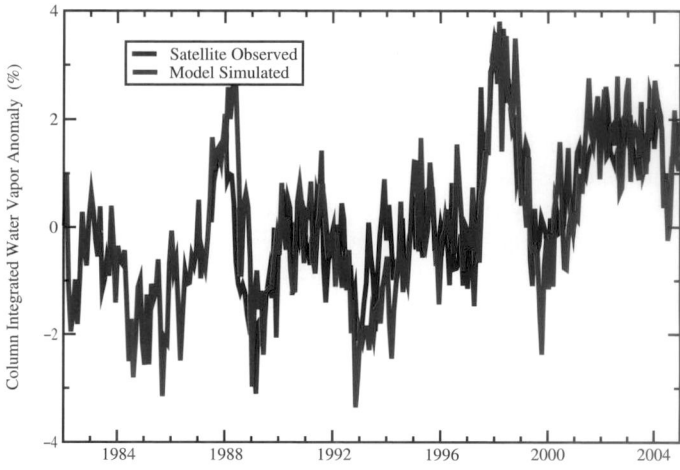

Figure 9.17. *Global mean (ocean-only) anomalies relative to 1987 to 2000 in column-integrated water vapour (%) from simulations with the GFDL AM2-LM2 AGCM forced with observed SSTs (red), and satellite observations from SSM/I (black, Wentz and Schabel, 2000). From Soden et al. (2005).*

radiosonde or reanalysis data (Trenberth et al., 2005) and show a statistically significant upward trend in precipitable (column-integrated) water of 1.2 ± 0.3 % per decade averaged over the global oceans (Section 3.4.2.1). Soden et al. (2005) demonstrate that the observed changes, including the upward trend, are well simulated in the GFDL atmospheric model when observed SSTs are prescribed (Figure 9.17). The simulation and observations show common low-frequency variability, which is largely associated with ENSO. Soden et al. (2005) also demonstrate that upper-tropospheric changes in water vapour are realistically simulated by the model. Observed warming over the global oceans is likely largely anthropogenic (Figure 9.12), suggesting that anthropogenic influence has contributed to the observed increase in atmospheric water vapour over the oceans.

9.5.4.2 *Global Precipitation Changes*

The increased atmospheric moisture content associated with warming might be expected to lead to increased global mean precipitation (Section 9.5.4.1). Global annual land mean precipitation showed a small, but uncertain, upward trend over the 20th century of approximately 1.1 mm per decade (Section 3.3.2.1 and Table 3.4). However, the record is characterised by large inter-decadal variability, and global annual land mean precipitation shows a non-significant decrease since 1950 (Figure 9.18; see also Table 3.4).

9.5.4.2.1 *Detection of external influence on precipitation*
Mitchell et al. (1987) argue that global mean precipitation changes should be controlled primarily by the energy budget of the troposphere where the latent heat of condensation is balanced by radiative cooling. Warming the troposphere enhances the cooling rate, thereby increasing precipitation, but this may be partly offset by a decrease in the efficiency of

radiative cooling due to an increase in atmospheric CO_2 (Allen and Ingram, 2002; Yang et al., 2003; Lambert et al., 2004; Sugi and Yoshimura, 2004). This suggests that global mean precipitation should respond more to changes in shortwave forcing than CO_2 forcing, since shortwave forcings, such as volcanic aerosol, alter the temperature of the troposphere without affecting the efficiency of radiative cooling. This is consistent with a simulated decrease in precipitation following large volcanic eruptions (Robock and Liu, 1994; Broccoli et al., 2003), and may explain why anthropogenic influence has not been detected in measurements of global land mean precipitation

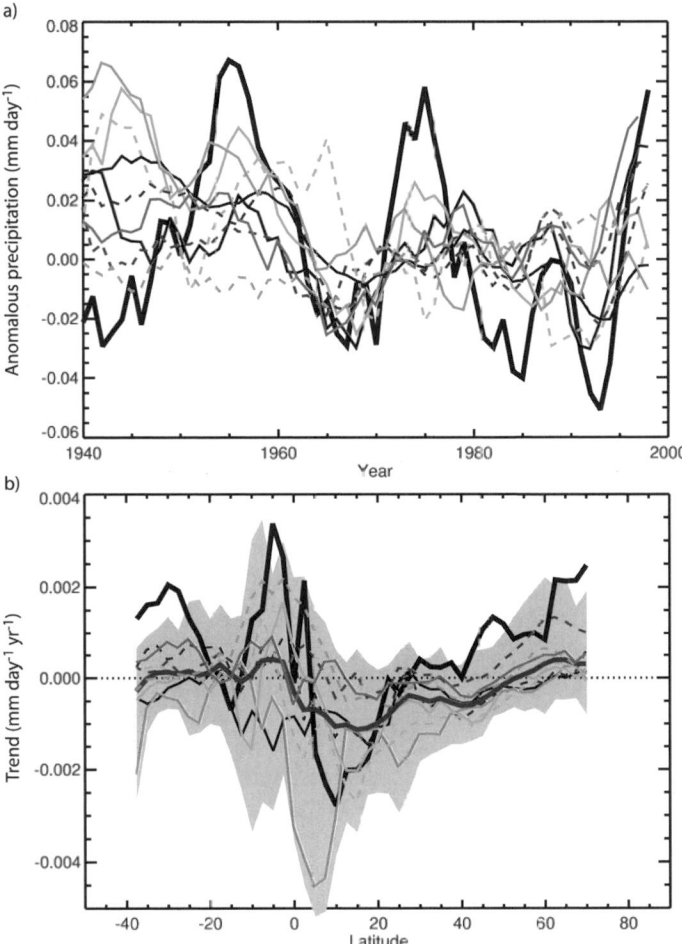

Figure 9.18. *Simulated and observed anomalies (with respect to 1961-1990) in terrestrial mean precipitation (a), and zonal mean precipitation trends 1901-1998 (b). Observations (thick black line) are based on a gridded data set of terrestrial rain gauge measurements (Hulme et al., 1998). Model data are from 20th-century MMD integrations with anthropogenic, solar and volcanic forcing from the following coupled climate models (see Table 8.1 for model details): UKMO-HadCM3 (brown), CCSM3 (dark blue), GFDL-CM2.0 (pale green), GFDL-CM2.1 (pale blue), GISS-EH (red), GISS-ER (thin black), MIROC3.2(medres) (orange), MRI-CGCM2.3.2 (dashed green) and PCM (pink). Coloured curves are ensemble means from individual models. In (a), a five-year running mean was applied to suppress other sources of natural variability, such as ENSO. In (b), the grey band indicates the range of trends simulated by individual ensemble members, and the thick dark blue line indicates the multi-model ensemble mean. External influence in observations on global terrestrial mean precipitation is detected with those precipitation simulations shown by continuous lines in the top panel. Adapted from Lambert et al. (2005).*

(Ziegler et al., 2003; Gillett et al., 2004b), although Lambert et al. (2004) urge caution in applying the energy budget argument to land-only data. Greenhouse-gas induced increases in global precipitation may have also been offset by decreases due to anthropogenic aerosols (Ramanathan et al., 2001).

Several studies have demonstrated that simulated land mean precipitation in climate model integrations including both natural and anthropogenic forcings is significantly correlated with that observed (Allen and Ingram, 2002; Gillett et al., 2004b; Lambert et al., 2004), thereby detecting external influence in observations of precipitation (see Section 8.3.1.2 for an evaluation of model-simulated precipitation). Lambert et al. (2005) examine precipitation changes in simulations of nine MMD 20C3M models including anthropogenic and natural forcing (Figure 9.18a), and find that the responses to combined anthropogenic and natural forcing simulated by five of the nine models are detectable in observed land mean precipitation (Figure 9.18a). Lambert et al. (2004) detect the response to shortwave forcing, but not longwave forcing, in land mean precipitation using HadCM3, and Gillett et al. (2004b) similarly detect the response to volcanic forcing using the PCM. Climate models appear to underestimate the variance of land mean precipitation compared to that observed (Gillett et al., 2004b; Lambert et al., 2004, 2005), but it is unclear whether this discrepancy results from an underestimated response to shortwave forcing (Gillett et al., 2004b), underestimated internal variability, errors in the observations, or a combination of these.

Greenhouse gas increases are also expected to cause enhanced horizontal transport of water vapour that is expected to lead to a drying of the subtropics and parts of the tropics (Kumar et al., 2004; Neelin et al., 2006), and a further increase in precipitation in the equatorial region and at high latitudes (Emori and Brown, 2005; Held and Soden, 2006). Simulations of 20th-century zonal mean land precipitation generally show an increase at high latitudes and near the equator, and a decrease in the subtropics of the NH (Hulme et al., 1998; Held and Soden, 2006; Figure 9.18b). Projections for the 21st century show a similar effect (Figure 10.12). This simulated drying of the northern subtropics and southward shift of the Inter-Tropical Convergence Zone may relate in part to the effects of sulphate aerosol (Rotstayn and Lohmann, 2002), although simulations without aerosol effects also show drying in the northern subtropics (Hulme et al., 1998). This pattern of zonal mean precipitation changes is broadly consistent with that observed over the 20th century (Figure 9.18b; Hulme et al., 1998; Allen and Ingram, 2002; Rotstayn and Lohmann, 2002), although the observed record is characterised by large inter-decadal variability (Figure 3.15). The agreement between the simulated and observed zonal mean precipitation trends is not sensitive to the inclusion of forcing by volcanic eruptions in the simulations, suggesting that anthropogenic influence may be evident in this diagnostic.

Changes in runoff have been observed in many parts of the world, with increases or decreases corresponding to changes in precipitation (Section 3.3.4). Climate models suggest that runoff will increase in regions where precipitation increases faster than evaporation, such as at high northern latitudes (Section 10.3.2.3 and Figure 10.12; see also Milly et al., 2005; Wu et al., 2005). Gedney et al. (2006) attribute increased continental runoff in the latter decades of the 20th century in part to suppression of transpiration due to CO_2-induced stomatal closure. They find that observed climate changes (including precipitation changes) alone are insufficient to explain the increased runoff, although their result is subject to considerable uncertainty in the runoff data. In addition, Qian et al. (2006) simulate observed runoff changes in response to observed temperature and precipitation alone, and Milly et al. (2005) demonstrate that 20th-century runoff trends simulated by the MMD models are significantly correlated with observed runoff trends. Wu et al. (2005) demonstrate that observed increases in arctic river discharge are reproduced in coupled model simulations with anthropogenic forcing, but not in simulations with natural forcings only.

Mid-latitude summer drying is another anticipated response to greenhouse gas forcing (Section 10.3.6.1), and drying trends have been observed in the both the NH and SH since the 1950s (Section 3.3.4). Burke et al. (2006), using the HadCM3 model with all natural and anthropogenic external forcings and a global Palmer Drought Severity Index data set compiled from observations by Dai et al. (2004), are able to formally detect the observed global trend towards increased drought in the second half of the 20th century, although the model trend is weaker than observed and the relative contributions of natural external forcings and anthropogenic forcings are not assessed. The model also simulates some aspects of the spatial pattern of observed drought trends, such as the trends across much of Africa and southern Asia, but not others, such as the trend to wetter conditions in Brazil and northwest Australia.

9.5.4.2.2 Changes in extreme precipitation

Allen and Ingram (2002) suggest that while global annual mean precipitation is constrained by the energy budget of the troposphere, extreme precipitation is constrained by the atmospheric moisture content, as predicted by the Clausius-Clapeyron equation. For a given change in temperature, they therefore predict a larger change in extreme precipitation than in mean precipitation, which is consistent with the HadCM3 response. Consistent with these findings, Emori and Brown (2005) discuss physical mechanisms governing changes in the dynamic and thermodynamic components of mean and extreme precipitation and conclude that changes related to the dynamic component (i.e., that due to circulation change) are secondary factors in explaining the greater percentage increase in extreme precipitation than in mean precipitation that is seen in models. Meehl et al. (2005) demonstrate that tropical precipitation intensity increases are related to water vapour increases, while mid-latitude intensity increases are related to circulation changes that affect the distribution of increased water vapour.

Climatological data show that the most intense precipitation occurs in warm regions (Easterling et al., 2000) and diagnostic analyses have shown that even without any change in total precipitation, higher temperatures lead to a greater proportion of

total precipitation in heavy and very heavy precipitation events (Karl and Trenberth, 2003). In addition, Groisman et al. (1999) demonstrate empirically, and Katz (1999) theoretically, that as total precipitation increases a greater proportion falls in heavy and very heavy events if the frequency remains constant. Similar characteristics are anticipated under global warming (Cubasch et al., 2001; Semenov and Bengtsson, 2002; Trenberth et al., 2003). Trenberth et al. (2005) point out that since the amount of moisture in the atmosphere is likely to rise much faster as a consequence of rising temperatures than the total precipitation, this should lead to an increase in the intensity of storms, offset by decreases in duration or frequency of events.

Model results also suggest that future changes in precipitation extremes will likely be greater than changes in mean precipitation (Section 10.3.6.1; see Section 8.5.2 for an evaluation of model-simulated precipitation extremes). Simulated changes in globally averaged annual mean and extreme precipitation appear to be quite consistent between models. The greater and spatially more uniform increases in heavy precipitation as compared to mean precipitation may allow extreme precipitation change to be more robustly detectable (Hegerl et al., 2004).

Evidence for changes in observations of short-duration precipitation extremes varies with the region considered (Alexander et al., 2006) and the analysis method employed (Folland et al., 2001; Section 3.8.2.2). Significant increases in observed extreme precipitation have been reported over some parts of the world, for example over the USA, where the increase is similar to changes expected under greenhouse warming (e.g., Karl and Knight, 1998; Semenov and Bengtsson, 2002; Groisman et al., 2005). However, a quantitative comparison between area-based extreme events simulated in models and station data remains difficult because of the different scales involved (Osborn and Hulme, 1997). A first attempt based on Frich et al. (2002) indices used fingerprints from atmospheric model simulations with prescribed SST (Kiktev et al., 2003) and found little similarity between patterns of simulated and observed rainfall extremes, in contrast to the qualitative similarity found in other studies (Semenov and Bengtsson, 2002; Groisman et al., 2005). Tebaldi et al. (2006) report that eight MMD 20C3M models show a general tendency towards a greater frequency of heavy precipitation events over the past four decades, most coherently at high latitudes of the NH, broadly consistent with observed changes (Groisman et al., 2005).

9.5.4.3 Regional Precipitation Changes

Observed trends in annual precipitation during the period 1901 to 2003 are shown in Figure 3.13 for regions in which data is available. Responses to external forcing in regional precipitation trends are expected to exhibit low signal-to-noise ratios and are likely to exhibit strong spatial variations because of the dependence of precipitation on atmospheric circulation and on geographic factors such as orography. There have been some suggestions, for specific regions, of a possible anthropogenic influence on precipitation, which are discussed below.

9.5.4.3.1   Sahel drought

Rainfall decreased substantially across the Sahel from the 1950s until at least the late 1980s (Dai et al., 2004; Figure 9.19, see also Figure 3.37). There has been a partial recovery since about 1990, although rainfall has not returned to levels typical of the period 1920 to 1965. Zeng (2003) note that two main hypotheses have been proposed as a cause of the extended drought: overgrazing and conversion of woodland to agriculture increasing surface albedo and reducing moisture supply to the atmosphere, and large-scale atmospheric circulation changes related to decadal global SST changes that could be of anthropogenic or natural origin (Nicholson, 2001). Black carbon has also been suggested as a contributor (Menon et al., 2002b). Taylor et al. (2002) examine the impact of land use change with an atmospheric GCM forced only by estimates of Sahelian land use change since 1961. They simulate a small decrease in Sahel rainfall (around 5% by 1996) and conclude that the impacts of recent land use changes are not large enough to have been the principal cause of the drought.

Several recent studies have demonstrated that simulations with a range of atmospheric models using prescribed observed SSTs are able to reproduce observed decadal variations in Sahel rainfall (Bader and Latif, 2003; Giannini et al., 2003; Rowell, 2003; Haarsma et al., 2005; Held et al., 2005; Lu and Delworth, 2005; see also Figure 9.19; Hoerling et al., 2006), consistent with earlier findings (Folland, 1986; Rowell, 1996). Hoerling et al. (2006) show that AGCMs with observed SST changes typically underestimate the magnitude of the observed precipitation changes, although the models and observations are not inconsistent. These studies differ somewhat in terms of which ocean SSTs they find to be most important: Giannini et al. (2003) and Bader and Latif (2003) emphasize the role of tropical Indian Ocean warming, Hoerling et al. (2006) attribute the drying trend to a progressive warming of the South Atlantic relative to the North Atlantic, and Rowell (2003) finds that Mediterranean SSTs are an additional important contributor to decadal variations in Sahel rainfall. Based on a multi-model ensemble of coupled model simulations Hoerling et al. (2006) conclude that the observed drying trend in the Sahel is not consistent with simulated internal variability alone.

Thus, recent research indicates that changes in SSTs are probably the dominant influence on rainfall in the Sahel, although land use changes possibly also contribute (Taylor et al., 2002). But what has caused the differential SST changes? Rotstayn and Lohmann (2002) propose that spatially varying, anthropogenic sulphate aerosol forcing (both direct and indirect) can alter low-latitude atmospheric circulation leading to a decline in Sahel rainfall. They find a southward shift of tropical rainfall due to a hemispheric asymmetry in the SST response to changes in cloud albedo and lifetime in a climate simulation forced with recent anthropogenic changes in sulphate aerosol. Williams et al. (2001) also find a southward shift of tropical rainfall as a response to the indirect effect of sulphate aerosol. These results suggest that sulphate aerosol changes may have led to reduced warming of the northern tropical oceans, which in turn led to the decrease in Sahel rainfall, possibly enhanced

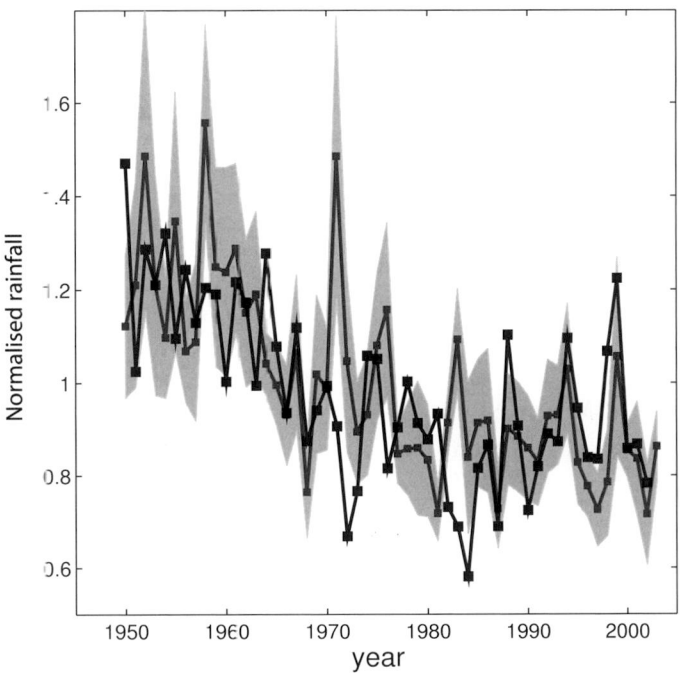

Figure 9.19. *Observed (Climatic Research Unit TS 2.1; Mitchell and Jones, 2005) Sahel July to September rainfall for each year (black), compared to an ensemble mean of 10 simulations of the atmospheric/land component of the GFDL-CM2.0 model (see Table 8.1 for model details) forced with observed SSTs (red). Both model and observations are normalized to unit mean over 1950-2000. The grey band represents ±1 standard deviation of intra-ensemble variability. After Held et al. (2005), based on results in Lu and Delworth (2005).*

through land-atmosphere interaction, although a full attribution analysis has yet to be conducted. Held et al. (2005) show that historical climate simulations with the both the GFDL-CM2.0 and CM2.1 models (see Table 8.1 for details) exhibit drying trends over the Sahel in the second half of the 20th century, which they ascribe to a combination of greenhouse gas and sulphate aerosol changes. The spatial pattern of the trends in simulated rainfall also shows some agreement with observations. However, Hoerling et al. (2006) find that eight other coupled climate models with prescribed anthropogenic forcing do not simulate significant trends in Sahel rainfall over the 1950 to 1999 period.

9.5.4.3.2   Southwest Australian drought

Early winter (May–July) rainfall in the far southwest of Australia declined by about 15% in the mid-1970s (IOCI, 2002) and remained low subsequently. The rainfall decrease was accompanied by a change in large-scale atmospheric circulation in the surrounding region (Timbal, 2004). The circulation and precipitation changes are somewhat consistent with, but larger than, those simulated by climate models in response to greenhouse gas increases. The Indian Ocean Climate Initiative (IOCI, 2005) concludes that land cover change could not be the primary cause of the rainfall decrease because of the link between the rainfall decline and changes in large-scale atmospheric circulation, and re-affirms the conclusion of IOCI

(2002) that both natural variability and greenhouse forcing likely contributed. Timbal et al. (2005) demonstrate that climate change signals downscaled from the PCM show some similarity to observed trends, although the significance of this finding is uncertain.

Some authors (e.g., Karoly, 2003) have suggested that the decrease in rainfall is related to anthropogenic changes in the SAM (see Section 9.5.3.3). However, the influence of changes in circulation on southwest Australian drought remains unclear as the largest SAM trend has occurred during the SH summer (December–March; Thompson et al., 2000; Marshall et al., 2004), while the largest rainfall decrease has occurred in early winter (May–July).

9.5.4.3.3 Monsoon precipitation

Decreasing trends in precipitation over the Indonesian Maritime Continent, equatorial western and central Africa, Central America, Southeast Asia and eastern Australia have been observed over the period 1948 to 2003, while increasing trends were found over the USA and north-western Australia (Section 3.7). The TAR (IPCC, 2001, pp. 568) concluded that an increase in Southeast Asian summer monsoon precipitation is simulated in response to greenhouse gas increases in climate models, but that this effect is reduced by an increase in sulphate aerosols, which tend to decrease monsoon precipitation. Since then, additional modelling studies have come to conflicting conclusions regarding changes in monsoon precipitation (Lal and Singh, 2001; Douville et al., 2002; Maynard et al., 2002; May, 2004; Wardle and Smith, 2004; see also Section 9.5.3.5). Ramanathan et al. (2005) were able to simulate realistic changes in Indian monsoon rainfall, particularly a decrease that occurred between 1950 and 1970, by including the effects of black carbon aerosol. In both the observations and model, these changes were associated with a decreased SST gradient over the Indian Ocean and an increase in tropospheric stability, and they were not reproduced in simulations with greenhouse gas and sulphate aerosol changes only.

9.5.5 Cryosphere Changes

9.5.5.1 Sea Ice

Widespread warming would, in the absence of other countervailing effects, lead to declines in sea ice, snow, and glacier and ice sheet extent and thickness. The annual mean area of arctic sea ice cover has decreased in recent decades, with stronger declines in summer than in winter, and some thinning (Section 4.4). Gregory et al. (2002b) show that a four-member ensemble of HadCM3 integrations with all major anthropogenic and natural forcing factors simulates a decline in arctic sea ice extent of about 2.5% per decade over the period 1970 to 1999, which is close to the observed decline of 2.7% per decade over the satellite period 1978 to 2004. This decline is inconsistent with simulated internal climate variability and the response to natural forcings alone (Vinnikov et al., 1999; Gregory et al., 2002b; Johannssen et al., 2004), indicating that

anthropogenic forcing has likely contributed to the trend in NH sea ice extent. Models such as those described by Rothrock et al. (2003) and references therein are able to reproduce the observed interannual variations in ice thickness, at least when averaged over fairly large regions. Simulations of historical arctic ice thickness or volume (Goeberle and Gerdes, 2003; Rothrock et al., 2003) show a marked reduction in ice thickness starting in the late 1980s, but disagree somewhat with respect to trends and/or variations earlier in the century. Although some of the dramatic change inferred may be a consequence of a spatial redistribution of ice volume over time (e.g., Holloway and Sou, 2002), thermodynamic changes are also believed to be important. Low-frequency atmospheric variability (such as interannual changes in circulation connected to the NAM) appears to be important in flushing ice out of the Arctic Basin, thus increasing the amount of summer open water and enhancing thermodynamic thinning through the ice-albedo feedback (e.g., Lindsay and Zhang, 2005). Large-scale modes of variability affect both wind driving and heat transport in the atmosphere, and therefore contribute to interannual variations in ice formation, growth and melt (e.g., Rigor et al., 2002; Dumas et al., 2003). Thus, the decline in arctic sea ice extent and its thinning appears to be largely, but not wholly, due to greenhouse gas forcing.

Unlike in the Arctic, a strong decline in sea ice extent has not been observed in the Antarctic during the period of satellite observations (Section 4.4.2.2). Fichefet et al. (2003) conducted a simulation of Antarctic ice thickness using observationally based atmospheric forcing covering the period 1958 to 1999. They note pronounced decadal variability, with area average ice thickness varying by ±0.1 m (compared to a mean thickness of roughly 0.9 m), but no long-term trend. However, Gregory et al. (2002b) find a decline in antarctic sea ice extent in their model, contrary to observations. They suggest that the lack of consistency between the observed and modelled changes in sea ice extent might reflect an unrealistic simulation of regional warming around Antarctica, rather than a deficiency in the ice model. Holland and Raphael (2006) examine sea ice variability in six MMD 20C3M simulations that include stratospheric ozone depletion. They conclude that the observed weak increase in antarctic sea ice extent is not inconsistent with simulated internal variability, with some simulations reproducing the observed trend over 1979 to 2000, although the models exhibit larger interannual variability in sea ice extent than satellite observations.

9.5.5.2 Snow and Frozen Ground

Snow cover in the NH, as measured from satellites, has declined substantially in the past 30 years, particularly from early spring through summer (Section 4.2). Trends in snow depth and cover can be driven by precipitation or temperature trends. The trends in recent decades have generally been driven by warming at lower and middle elevations. Evidence for this includes: (a) interannual variations in NH April snow-covered area are strongly correlated (r = –0.68) with April

40°N to 60°N temperature; (b) interannual variations in snow (water equivalent, depth or duration) are strongly correlated with temperature at lower- and middle-elevation sites in North America (Mote et al., 2005), Switzerland (Scherrer et al., 2004) and Australia (Nicholls, 2005); (c) trends in snow water equivalent or snow depth show strong dependence on elevation or equivalently mean winter temperature, both in western North America and Switzerland (with stronger decreases at lower, warmer elevations where a warming is more likely to affect snowfall and snowmelt); and (d) the trends in North America, Switzerland and Australia have been shown to be well explained by warming and cannot be explained by changes in precipitation. In some very cold places, increases in snow depth have been observed and have been linked to higher precipitation.

Widespread permafrost warming and degradation appear to be the result of increased summer air temperatures and changes in the depth and duration of snow cover (Section 4.7.2). The thickness of seasonally frozen ground has decreased in response to winter warming and increases in snow depth (Section 4.7.3).

9.5.5.3   Glaciers, Ice Sheets and Ice Shelves

During the 20th century, glaciers generally lost mass with the strongest retreats in the 1930s and 1940s and after 1990 (Section 4.5). The widespread shrinkage appears to imply widespread warming as the probable cause (Oerlemans, 2005), although in the tropics changes in atmospheric moisture might be contributing (Section 4.5.3). Over the last half century, both global mean winter accumulation and summer melting have increased steadily (Ohmura, 2004; Dyurgerov and Meier, 2005; Greene, 2005), and at least in the NH, winter accumulation and summer melting correlate positively with hemispheric air temperature (Greene, 2005); the negative correlation of net balance with temperature indicates the primary role of temperature in forcing the respective glacier fluctuations.

There have been a few studies for glaciers in specific regions examining likely causes of trends. Mass balances for glaciers in western North America are strongly correlated with global mean winter (October–April) temperatures and the decline in glacier mass balance has paralleled the increase in temperature since 1968 (Meier et al., 2003). Reichert et al. (2002a) forced a glacier mass balance model for the Nigardsbreen and Rhône glaciers with downscaled data from an AOGCM control simulation and conclude that the rate of glacier advance during the 'Little Ice Age' could be explained by internal climate variability for both glaciers, but that the recent retreat cannot, implying that the recent retreat of both glaciers is probably due to externally forced climate change. As well, the thinning and acceleration of some polar glaciers (e.g., Thomas et al., 2004) appear to be the result of ice sheet calving driven by oceanic and atmospheric warming (Section 4.6.3.4).

Taken together, the ice sheets of Greenland and Antarctica are shrinking. Slight thickening in inland Greenland is more than compensated for by thinning near the coast (Section 4.6.2.2). Warming is expected to increase low-altitude melting and high-

altitude precipitation in Greenland; altimetry data suggest that the former effect is dominant. However, because some portions of ice sheets respond only slowly to climate changes, past forcing may be influencing ongoing changes, complicating attribution of recent trends (Section 4.6.3.2).

9.5.6   Summary

In the TAR, quantitative evidence for human influence on climate was based almost exclusively on atmospheric and surface temperature. Since then, anthropogenic influence has also been identified in a range of other climate variables, such as ocean heat content, atmospheric pressure and sea ice extent, thereby contributing further evidence of an anthropogenic influence on climate, and improving confidence in climate models.

Observed changes in ocean heat content have now been shown to be inconsistent with simulated natural climate variability, but consistent with a combination of natural and anthropogenic influences both on a global scale, and in individual ocean basins. Models suggest a substantial anthropogenic contribution to sea level rise, but underestimate the actual rise observed. While some studies suggest that an anthropogenic increase in high-latitude rainfall may have contributed to a freshening of the Arctic Ocean and North Atlantic deep water, these results are still uncertain.

There is no evidence that 20th-century ENSO behaviour is distinguishable from natural variability. By contrast, there has been a detectable human influence on global sea level pressure. Both the NAM and SAM have shown significant trends. Models reproduce the sign but not magnitude of the NAM trend, and models including both greenhouse gas and ozone simulate a realistic trend in the SAM. Anthropogenic influence on either tropical or extratropical cyclones has not been detected, although the apparent increased frequency of intense tropical cyclones, and its relationship to ocean warming, is suggestive of an anthropogenic influence.

Simulations and observations of total atmospheric water vapour averaged over oceans agree closely when the simulations are constrained by observed SSTs, suggesting that anthropogenic influence has contributed to an increase in total atmospheric water vapour. However, global mean precipitation is controlled not by the availability of water vapour, but by a balance between the latent heat of condensation and radiative cooling in the troposphere. This may explain why human influence has not been detected in global precipitation, while the influence of volcanic aerosols has been detected. However, observed changes in the latitudinal distribution of land precipitation are suggestive of a possible human influence as is the observed increased incidence of drought as measured by the Palmer Drought Severity Index. Observational evidence indicates that the frequency of the heaviest rainfall events has likely increased within many land regions in general agreement with model simulations that indicate that rainfall in the heaviest events is likely to increase in line with atmospheric water vapour concentration. Many AGCMs capture the observed decrease in Sahel rainfall when constrained by observed SSTs, although this decrease is not simulated by most AOGCMs. One study found that an observed decrease in Asian monsoon rainfall could only be simulated in response to black carbon aerosol, although conclusions regarding the monsoon response to anthropogenic forcing differ.

Observed decreases in arctic sea ice extent have been shown to be inconsistent with simulated internal variability, and consistent with the simulated response to human influence, but SH sea ice extent has not declined. The decreasing trend in global snow cover and widespread melting of glaciers is consistent with a widespread warming. Anthropogenic forcing has likely contributed substantially to widespread glacier retreat during the 20th century.

9.6 Observational Constraints on Climate Sensitivity

This section assesses recent research that infers equilibrium climate sensitivity and transient climate response from observed changes in climate. 'Equilibrium climate sensitivity' (ECS) is the equilibrium annual global mean temperature response to a doubling of equivalent atmospheric CO_2 from pre-industrial levels and is thus a measure of the strength of the climate system's eventual response to greenhouse gas forcing. 'Transient climate response' (TCR) is the annual global mean temperature change at the time of CO_2 doubling in a climate simulation with a 1% yr^{-1} compounded increase in CO_2 concentration (see Glossary and Section 8.6.2.1 for detailed definitions). TCR is a measure of the strength and rapidity of the climate response to greenhouse gas forcing, and depends in part on the rate at which the ocean takes up heat. While the direct temperature change that results from greenhouse gas forcing can be calculated in a relatively straightforward manner, uncertain atmospheric feedbacks (Section 8.6) lead to uncertainties in estimates of future climate change. The objective here is to assess estimates of ECS and TCR that are based on observed climate changes, while Chapter 8 assesses feedbacks individually. Inferences about climate sensitivity from observed climate *changes* complement approaches in which uncertain parameters in climate models are varied and assessed by evaluating the resulting skill in reproducing observed *mean* climate (Section 10.5.4.4). While observed climate changes have the advantage of being most clearly related to future climate change, the constraints they provide on climate sensitivity are not yet very strong, in part because of uncertainties in both climate forcing and the estimated response (Section 9.2). An overall summary assessment of ECS and TCR, based on the ability of models to simulate climate change and mean climate and on other approaches, is given in Box 10.2. Note also that this section does not assess regional climate sensitivity or sensitivity to forcings other than CO_2.

9.6.1 Methods to Estimate Climate Sensitivity

The most straightforward approach to estimating climate sensitivity would be to relate an observed climate change to a known change in radiative forcing. Such an approach is strictly correct only for changes between equilibrium climate states. Climatic states that were reasonably close to equilibrium in the past are often associated with substantially different climates than the pre-industrial or present climate, which is probably not in equilibrium (Hansen et al., 2005). An example is the climate of the LGM (Chapter 6 and Section 9.3). However, the climate's sensitivity to external forcing will depend on the mean climate state and the nature of the forcing, both of which affect feedback mechanisms (Chapter 8). Thus, an estimate of the sensitivity directly derived from the ratio of response to forcing cannot be readily compared to the sensitivity of climate to a doubling of CO_2 under idealised conditions. An alternative approach, which has been pursued in most work reported here, is based on varying parameters in climate models that influence the ECS in those models, and then attaching probabilities to the different ECS values based on the realism of the corresponding climate change simulations. This ameliorates the problem of feedbacks being dependent on the climatic state, but depends on the assumption that feedbacks are realistically represented in models and that uncertainties in all parameters relevant for feedbacks are varied. Despite uncertainties, results from simulations of climates of the past and recent climate change (Sections 9.3 to 9.5) increase confidence in this assumption.

The ECS and TCR estimates discussed here are generally based on large ensembles of simulations using climate models of varying complexity, where uncertain parameters influencing the model's sensitivity to forcing are varied. Studies vary key climate and forcing parameters in those models, such as the ECS, the rate of ocean heat uptake, and in some instances, the strength of aerosol forcing, within plausible ranges. The ECS can be varied directly in simple climate models and in some EMICs (see Chapter 8), and indirectly in more complex EMICs and AOGCMs by varying model parameters that influence the strength of atmospheric feedbacks, for example, in cloud parametrizations. Since studies estimating ECS and TCR from observed climate changes require very large ensembles of simulations of past climate change (ranging from several hundreds to thousands of members), they are often, but not always, performed with EMICs or EBMs.

The idea underlying this approach is that the plausibility of a given combination of parameter settings can be determined from the agreement of the resulting simulation of historical climate with observations. This is typically evaluated by means of Bayesian methods (see Supplementary Material, Appendix 9.B for methods). Bayesian approaches constrain parameter values by combining prior distributions that account for uncertainty in the knowledge of parameter values with information about the parameters estimated from data (Kennedy and O'Hagan, 2001). The uniform distribution has been used widely as a prior distribution, which enables comparison of constraints obtained from the data in different approaches. ECS ranges encompassed by the uniform prior distribution must be limited due to computer time limiting the size of model ensembles, but generally cover the range considered possible by experts, such as from 0°C to 10°C. Note that uniform prior distributions for ECS, which only require an expert assessment of possible range, generally assign a higher prior belief to high sensitivity than, for example, non-uniform prior distributions that depend more heavily on expert assessments (e.g., Forest et al., 2006). In addition, Frame et al. (2005) point out that care must be taken when specifying the uniform prior distribution. For example, a uniform prior distribution for the climate feedback parameter (see Glossary) implies a non-uniform prior distribution for ECS due to the nonlinear relationship between the two parameters.

Since observational constraints on the upper bound of ECS are still weak (as shown below), these prior assumptions influence the resulting estimates. Frame et al. (2005) advocate sampling a flat prior distribution in ECS if this is the target of the estimate, or in TCR if future temperature trends are to be constrained. In contrast, statistical research on the design and interpretation of computer experiments suggests the use of prior distributions for model input parameters (e.g., see Kennedy and O'Hagan, 2001; Goldstein and Rougier, 2004). In such Bayesian studies, it is generally good practice to explore the sensitivity of results to different prior beliefs (see, for example, Tol and Vos, 1998; O'Hagan and Forster, 2004). Furthermore, as demonstrated by Annan and Hargreaves (2005) and Hegerl et al. (2006a), multiple and independent lines of evidence about climate sensitivity from, for example, analysis of climate change at different times, can be combined by using information from one line of evidence as prior information for the analysis of another line of evidence. The extent to which the different lines of evidence provide complete information on the underlying physical mechanisms and feedbacks that determine the climate sensitivity is still an area of active research. In the following, uniform prior distributions for the target of the estimate are used unless otherwise specified.

Methods that incorporate a more comprehensive treatment of uncertainty generally produce wider uncertainty ranges for the inferred climate parameters. Methods that do not vary uncertain parameters, such as ocean diffusivity, in the course of the uncertainty analysis will yield probability distributions for climate sensitivity that are conditional on these values, and therefore are likely to underestimate the uncertainty in climate sensitivity. On the other hand, approaches that do not use all available evidence will produce wider uncertainty ranges than estimates that are able to use observations more comprehensively.

9.6.2 Estimates of Climate Sensitivity Based on Instrumental Observations

9.6.2.1 Estimates of Climate Sensitivity Based on 20th-Century Warming

A number of recent studies have used instrumental records of surface, ocean and atmospheric temperature changes to estimate climate sensitivity. Most studies use the observed surface temperature changes over the 20th century or the last 150 years (Chapter 3). In addition, some studies also use the estimated ocean heat uptake since 1955 based on Levitus et al. (2000, 2005) (Chapter 5), and temperature changes in the free atmosphere (Chapter 3; see also Table 9.3). For example, Frame et al. (2005) and Andronova and Schlesinger (2000) use surface air temperature alone, while Forest et al. (2002, 2006), Knutti et al. (2002, 2003) and Gregory et al. (2002a) use both surface air temperature and ocean temperature change to constrain climate sensitivity. Forest et al. (2002, 2006) and Lindzen and Giannitsis (2002) use free atmospheric temperature data from radiosondes in addition to surface air temperature. Note that studies using radiosonde data may be affected by recently discovered inhomogeneities (Section 3.4.1.1), although Forest et al. (2006) illustrate that the impact of the radiosonde atmospheric temperature data on their climate sensitivity estimate is smaller than that of surface and ocean warming data. A further recent study uses Earth Radiation Budget Experiment (ERBE) data (Forster and Gregory, 2006) in addition to surface temperature changes to estimate climate feedbacks (and thus ECS) from observed changes in forcing and climate.

Wigley et al. (1997) pointed out that uncertainties in forcing and response made it impossible to use observed global temperature changes to constrain ECS more tightly than the range explored by climate models at the time (1.5°C to 4.5°C), and particularly the upper end of the range, a conclusion confirmed by subsequent studies. A number of subsequent publications qualitatively describe parameter values that allow models to reproduce features of observed changes, but without directly estimating a climate sensitivity probability density function (PDF). For example, Harvey and Kaufmann (2002) find a best-fit ECS of 2.0°C out of a range of 1°C to 5°C, and constrain fossil fuel and biomass aerosol forcing (Section 9.2.1.2). Lindzen and Giannitsis (2002) pose the hypothesis that the rapid change in tropospheric (850–300 hPa) temperatures around 1976 triggered a delayed response in surface temperature that is best modelled with a climate sensitivity of less than 1°C. However, their estimate does not account for substantial uncertainties in the analysis of such a short time period, most notably those associated with the role of internal climate variability in the rapid tropospheric warming of 1976. The 1976–1977 climate shift occurred along with a phase shift of the PDO, and a concurrent change in the ocean (Section 3.6.3) that appears to contradict the Lindzen and Giannitsis (2002) assumption that the change was initiated by tropospheric forcing. In addition, the authors do not account for uncertainties in the simple model whose sensitivity is fitted. The finding of Lindzen and Giannitsis is in contrast with that of Forest et al. (2002, 2006) who consider the joint evolution of surface and upper air temperatures on much longer time scales.

Several recent studies have derived probability estimates for ECS using a range of models and diagnostics. The diagnostics, which are used to compare model-simulated and observed changes, are often simple temperature indices such as the global mean surface temperature and ocean mean warming

(Knutti et al., 2002, 2003) or the differential warming between the SH and NH (together with the global mean; Andronova and Schlesinger, 2001). Results that use more detailed information about the space-time evolution of climate may be able to provide tighter constraints than those that use simpler indices. Forest et al. (2002, 2006) use a so-called 'optimal' detection method (Section 9.4.1.4 and Appendix 9.A.1) to diagnose the fit between model-simulated and observed patterns of zonal mean temperature change. Frame et al. (2005) use detection results from an analysis based on several multi-model AOGCM fingerprints (Section 9.4.1.4) that separate the greenhouse gas response from that to other anthropogenic and natural forcings (Stott et al., 2006c). Similarly, Gregory et al. (2002a) apply an inverse estimate of the range of aerosol forcing based on fingerprint detection results. Note that while results from fingerprint detection approaches will be affected by uncertainty in separation between greenhouse gas and aerosol forcing, the resulting uncertainty in estimates of the near-surface temperature response to greenhouse gas forcing is relatively small (Sections 9.2.3 and 9.4.1.4).

A further consideration in assessing these results is the extent to which realistic forcing estimates were used, and whether forcing uncertainty was included. Most studies consider a range of anthropogenic forcing factors, including greenhouse gases and sulphate aerosol forcing, sometimes directly including the indirect forcing effect, such as Knutti et al. (2002, 2003), and sometimes indirectly accounting for the indirect effect by using a wide range of direct forcing (e.g., Andronova and Schlesinger, 2001; Forest et al., 2002, 2006). Many studies also consider tropospheric ozone (e.g., Andronova and Schlesinger, 2001; Knutti et al., 2002, 2003). Forest et al. (2006) demonstrate that the inclusion of natural forcing affects the estimated PDF of climate sensitivity since net negative natural forcing in the second half of the 20th century favours higher sensitivities than earlier results that disregarded natural forcing (Forest et al., 2002; see Figure 9.20), particularly if the same ocean warming estimates were used. Note that some of the changes due to inclusion of natural forcing are offset by using recently revised ocean warming data (Levitus et

al., 2005), which favour somewhat smaller ocean heat uptakes than earlier data (Levitus et al., 2001; Forest et al., 2006). Only a few estimates account for uncertainty in forcings other than from aerosols (e.g., Gregory et al., 2002a; Knutti et al., 2002, 2003); some other studies perform some sensitivity testing to assess the effect of forcing uncertainty not accounted for, for example, in natural forcing (e.g., Forest et al., 2006; see Table 9.1 for an overview).

The treatment of uncertainty in the ocean's uptake of heat varies, from assuming a fixed value for a model's ocean diffusivity (Andronova and Schlesinger, 2001) to trying to allow for a wide range of ocean mixing parameters (Knutti et al., 2002, 2003) or systematically varying the ocean's effective diffusivity (e.g., Forest et al., 2002, 2006; Frame et al., 2005). Furthermore, all approaches that use the climate's time evolution attempt to account for uncertainty due to internal climate variability, either by bootstrapping (Andronova and Schlesinger, 2001), by using a noise model in fingerprint studies whose results are used (Frame et al., 2005) or directly (Forest et al., 2002, 2006).

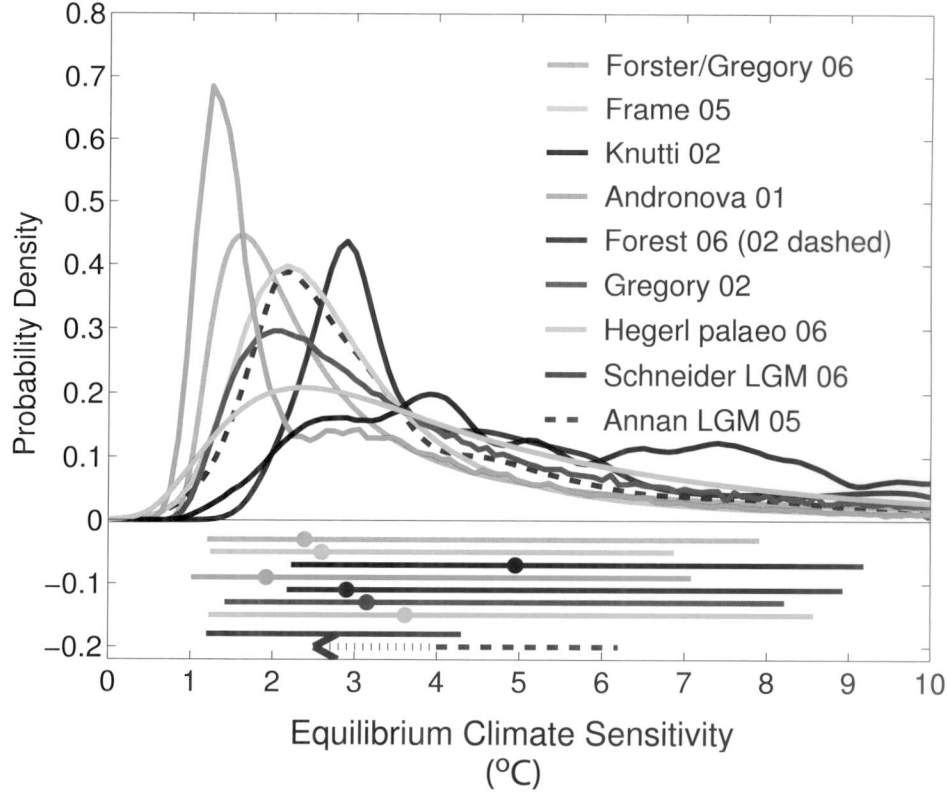

Figure 9.20. *Comparison between different estimates of the PDF (or relative likelihood) for ECS (°C). All PDFs/ likelihoods have been scaled to integrate to unity between 0°C and 10°C ECS. The bars show the respective 5 to 95% ranges, dots the median estimate. The PDFs/likelihoods based on instrumental data are from Andronova and Schlesinger (2001), Forest et al. (2002; dashed line, considering anthropogenic forcings only), Forest et al. (2006; solid, anthropogenic and natural forcings), Gregory et al. (2002a), Knutti et al. (2002), Frame et al. (2005), and Forster and Gregory (2006), transformed to a uniform prior distribution in ECS using the method after Frame et al. (2005). Hegerl et al. (2006a) is based on multiple palaeoclimatic reconstructions of NH mean temperatures over the last 700 years. Also shown are the 5 to 95% approximate ranges for two estimates from the LGM (dashed, Annan et al., 2005; solid, Schneider von Deimling et al., 2006) which are based on models with different structural properties. Note that ranges extending beyond the published range in Annan et al. (2005), and beyond that sampled by the climate model used there, are indicated by dots and an arrow, since Annan et al. only provide an upper limit. For details of the likelihood estimates, see Table 9.3. After Hegerl et al. (2006a).*

Figure 9.20 compares results from many of these studies. All PDFs shown are based on a uniform prior distribution of ECS and have been rescaled to integrate to unity for all positive sensitivities up to 10°C to enable comparisons of results using different ranges of uniform prior distributions (this affects both median and upper 95th percentiles if original estimates were based on a wider uniform range). Thus, zero prior probability is assumed for sensitivities exceeding 10°C, since many results do not consider those, and for negative sensitivities. Negative climate sensitivity would lead to cooling in response to a positive forcing and is inconsistent with understanding of the energy balance of the system (Stouffer et al., 2000; Gregory et al., 2002a; Lindzen and Giannitsis, 2002). This figure shows that best estimates of the ECS (mode of the estimated PDFs) typically range between 1.2°C and 4°C when inferred from constraints provided by historical instrumental data, in agreement with estimates derived from more comprehensive climate models. Most studies suggest a 5th percentile for climate sensitivity of 1°C or above. The upper 95th percentile is not well constrained, particularly in studies that account conservatively for uncertainty in, for example, 20th-century radiative forcing and ocean heat uptake. The upper tail is particularly long in studies using diagnostics based on large-scale mean data because separation of the greenhouse gas response from that to aerosols or climate variability is more difficult with such diagnostics (Andronova and Schlesinger, 2001; Gregory et al., 2002a; Knutti et al., 2002, 2003). Forest et al. (2006) find a 5 to 95% range of 2.1°C to 8.9°C for climate sensitivity (Table 9.3), which is a wider range than their earlier result based on anthropogenic forcing only (Forest et al., 2002). Frame et al. (2005) infer a 5 to 95% uncertainty range for the ECS of 1.2°C to 11.8°C, using a uniform prior distribution that extends well beyond 10°C sensitivity. Studies generally do not find meaningful constraints on the rate at which the climate system mixes heat into the deep ocean (e.g., Forest et al., 2002, 2006). However, Forest et al. (2006) find that many coupled AOGCMs mix heat too rapidly into the deep ocean, which is broadly consistent with comparisons based on heat uptake (Section 9.5.1.1,). However the relevance of this finding is unclear because most MMD AOGCMs were not included in the Forest et al. comparison, and because they used a relatively simple ocean model. Knutti et al. (2002) also determine that strongly negative aerosol forcing, as has been suggested by several observational studies (Anderson et al., 2003), is incompatible with the observed warming trend over the last century (Section 9.2.1.2 and Table 9.1).

Table 9.3. *Results from key studies on observational estimates of ECS (in °C) from instrumental data, individual volcanic eruptions, data for the last millennium, and simulations of the LGM . The final three rows list some studies using non-uniform prior distributions, while the other studies use uniform prior distributions of ECS.*

Study	Observational Data Used to Constrain Study[a]	Model[b]	External Forcings Included[c]	Treatment of uncertainties[d]	Estimated ECS Range 5 to 95% (°C)
From Instrumental Data					
Forest et al. (2006)	Upper air, surface and deep ocean space-time 20th-century temperatures Prior 0°C to 10°C	2-D EMIC (~E6)	G, Su , Sol, Vol, OzS, land surface changes (2002: G, Sul, OzS)	ε_{obs}, noise, κ, ε_{aer}, sensitivity tests for solar/volcanic. forcing uncertainty	2.1 to 8.9 (1.4 to 7.7 without natural forcings)
Andronova and Schlesinger (2001)	Global mean and hemispheric difference in surface air temperature 1856 to 1997	EBM	G, OzT, Sul, Sol, Vol	Noise (bootstrap residual), choice of radiative forcing factors	1.0 to 9.3 prob ~ 54% that ECS outside 1.5 to 4.5
Knutti et al. (2002; 2003)	Global mean ocean heat uptake 1955 to 1995, mean surface air temperature 1860 to 2000 Prior 0°C to 10°C	EMIC (~E1) plus neural net	G, OzT, OzS, fossil fuel and biomass burning BC+OM, stratospheric water vapour, Vol, Sol, Sul, Suli	ε_{obs}, ε_{forc} for multiple forcings from IPCC (2001), κ, different ocean mixing schemes	2.2 to 9.2 prob ~ 50% that ECS outside 1.5 to 4.5
Gregory et al. (2002a)	Global mean change in surface air temperature and ocean heat change between 1861 to 1900 and 1957 to 1994	1-Box	G, Sul and Suli (top down via Stott et al., 2001), Sol, Vol	ε_{obs}, ε_{forc}	1.1 to ∞
Frame et al. (2005)	Global change in surface temperature	EBM	G, accounted for other anthropogenic and natural forcing by fingerprints, Sul, Nat	Noise, uncertainty in amplitude but not pattern of natural and anthropogenic. forcings and response (scaling factors), κ (range consistent with ocean warming)	1.2 to 11.8

(continued)

Table 9.3 (continued)

Study	Observational Data Used to Constrain Study[a]	Model[b]	External Forcings Included[c]	Treatment of uncertainties[d]	Estimated ECS Range 5 to 95% (°C)
Forster and Gregory (2006)	1985 to 1996 ERBE data 60°N to 60°S, global surface temperature Prior 0°C to 18.5°C, transformed after Frame et al. (2005)	1-Box	G, Vol, Sol, Sul	ε_{obs}, ε_{forc}	1.2 to 14.2
From individual volcanic eruptions					
Wigley et al. (2005a)	Global mean surface temperature	EBM	From volcanic forcing only	El Niño	Agung: 1.3 to 6.3; El Chichon: 0.3 to 7.7; Mt. Pinatubo: 1.8 to 5.2
From last millenium					
Hegerl et al. (2006a)	NH mean surface air temperature pre-industrial (1270/1505 to 1850) from multiple reconstructions Prior 0°C to 10°C	2D EBM	G, Sul, Sol, Vol	Noise (from residual), κ; uncertainty in magnitude of reconstructions and solar and volcanic forcing	1.2 to 8.6
From LGM					
Schneider von Deimling et al. (2006)	LGM tropical SSTs and other LGM data	EMIC (~E3)	LGM forcing: greenhouse gases, dust, ice sheets, vegetation, insolation	uncertainty of proxy-based ice age SSTs (one type of data); attempt to account for structural uncertainty, estimate of forcing uncertainty	1.2 to 4.3 (based on encompassing several ranges given)
Annan et al. (2005)	LGM tropical SSTs, present-day seasonal cycle of a number of variables for sampling prior distribution of model parameters	AGCM with mixed-layer ocean	PMIP2 LGM forcing	Observational uncertainty in tropical SST estimates (one type of data)	<7% chance of sensitivity >6
Using non-uniform prior distributions					
Forest et al. (2002, 2006)	Expert prior, 20th-century temperature change (see above)	See Forest et al.	see above	See individual estimates	1.9 to 4.7
Annan et al. (2006)	Estimates from LGM, 20th-century change, volcanism combined	See Annan et al.	see above	See individual estimates	> 1.7 to 4.5
Hegerl et al. (2006a)	1950 to 2000 surface temperature change (Frame et al., 2005), NH mean pre-industrial surface air temperature from last millennium	See Hegerl et al. and Frame et al. (2005)	see above	See individual estimates	1.5 to 6.2

Notes:

[a] Range covered by uniform prior distribution if narrower than 0°C to 20°C.

[b] Energy Balance Model (EBM), often with upwelling-diffusive ocean; 1-box energy balance models; EMIC (numbers refer to related EMICs described in Table 8.3).

[c] G: greenhouse gases; Sul: direct sulphate aerosol effect; Suli: (first) indirect sulphate effect; OzT: tropospheric ozone; OzS: stratospheric ozone; Vol: volcanism; Sol: solar; BC+OM: black carbon and organic matter.).

[d] Uncertainties taken into account (e.g., uncertainty in ocean diffusivity $\mathcal{K}$, or total aerosol forcing ε_{forc}). Ideally, studies account for model uncertainty, forcing uncertainty (for example, in aerosol forcing ε_{aer} or natural forcing ε_{nat}), uncertainty in observations, ε_{obs}, and internal climate variability ('noise').

Some studies have further attempted to use non-uniform prior distributions. Forest et al. (2002, 2006) obtained narrower uncertainty ranges when using expert prior distributions (Table 9.3). While they reflect credible prior ranges of ECS, expert priors may also be influenced by knowledge about observed climate change, and thus may yield overly confident estimates when combined with the same data (Supplementary Material, Appendix 9.B). Frame et al. (2005) find that sampling uniformly in TCR results in an estimated ECS of 1.2°C to 5.2°C with a

median value of 2.3°C. In addition, several approaches have been based on a uniform prior distribution of climate feedback. Translating these results into ECS estimates is equivalent to using a prior distribution that favours smaller sensitivities, and hence tends to result in narrower ECS ranges (Frame et al., 2005). Forster and Gregory (2006) estimate ECS based on radiation budget data from the ERBE combined with surface temperature observations based on a regression approach, using the observation that there was little change in aerosol forcing over that time. They find a climate feedback parameter of 2.3 ± 1.4 W m^{-2} °C^{-1}, which corresponds to a 5 to 95% ECS range of 1.0°C to 4.1°C if using a prior distribution that puts more emphasis on lower sensitivities as discussed above, and a wider range if the prior distribution is reformulated so that it is uniform in sensitivity (Table 9.3). The climate feedback parameter estimated from the MMD AOGCMs ranges from about 0.7 to 2.0 W m^{-2} °C^{-1} (Supplementary Material, Table S8.1).

9.6.2.2   Estimates Based on Individual Volcanic Eruptions

Some recent analyses have attempted to derive insights into ECS from the well-observed forcing and response to the eruption of Mt. Pinatubo, or from other major eruptions during the 20th century. Such events allow for the study of physical mechanisms and feedbacks and are discussed in detail in Section 8.6. For example, Soden et al. (2002) demonstrate agreement between observed and simulated responses based on an AGCM with a climate sensitivity of 3.0°C coupled to a mixed-layer ocean, and that the agreement breaks down if the water vapour feedback in the model is switched off. Yokohata et al. (2005) find that a version of the MIROC climate model with a sensitivity of 4.0°C yields a much better simulation of the Mt. Pinatubo eruption than a model version with sensitivity of 6.3°C, concluding that the cloud feedback in the latter model appears inconsistent with data. Note that both results may be specific to the model analysed.

Constraining ECS from the observed responses to individual volcanic eruptions is difficult because the response to short-term volcanic forcing is strongly nonlinear in ECS, yielding only slightly enhanced peak responses and substantially extended response times for very high sensitivities (Frame et al., 2005; Wigley et al., 2005a). The latter are difficult to distinguish from a noisy background climate. A further difficulty arises from uncertainty in the rate of heat taken up by the ocean in response to a short, strong forcing. Wigley et al. (2005a) find that the lower boundary and best estimate obtained by comparing observed and simulated responses to major eruptions in the 20th century are consistent with the TAR range of 1.5°C to 4.5°C, and that the response to the eruption of Mt. Pinatubo suggests a best fit sensitivity of 3.0°C and an upper 95% limit of 5.2°C. However, as pointed out by the authors, this estimate does not account for forcing uncertainties. In contrast, an analysis by Douglass and Knox (2005) based on a box model suggests a very low climate sensitivity (under 1°C) and negative climate feedbacks based on the eruption of Mt. Pinatubo. Wigley et al. (2005b) demonstrate that the analysis method of Douglass and Knox (2005) severely underestimates (by a factor of three) climate sensitivity if applied to a model with known sensitivity. Furthermore, as pointed out by Frame et al. (2005), the effect of noise on the estimate of the climatic background level can lead to a substantial underestimate of uncertainties if not taken into account.

In summary, the responses to individual volcanic eruptions provide a useful test for feedbacks in climate models (Section 8.6). However, due to the physics involved in the response,

such individual events cannot provide tight constraints on ECS. Estimates of the most likely sensitivity from most such studies are, however, consistent with those based on other analyses.

9.6.2.3   Constraints on Transient Climate Response

While ECS is the equilibrium global mean temperature change that eventually results from atmospheric CO_2 doubling, the smaller TCR refers to the global mean temperature change that is realised at the time of CO_2 doubling under an idealised scenario in which CO_2 concentrations increase by 1% yr^{-1} (Cubasch et al., 2001; see also Section 8.6.2.1). The TCR is therefore indicative of the temperature trend associated with external forcing, and can be constrained by an observable quantity, the observed warming trend that is attributable to greenhouse gas forcing. Since external forcing is likely to continue to increase through the coming century, TCR may be more relevant to determining near-term climate change than ECS.

Stott et al. (2006c) estimate TCR based on scaling factors for the response to greenhouse gases only (separated from aerosol and natural forcing in a three-pattern optimal detection analysis) using fingerprints from three different model simulations (Figure 9.21) and find a relatively tight constraint. Using three model simulations together, their estimated median TCR is 2.1°C at the time of CO_2 doubling (based on a 1% yr^{-1} increase in CO_2), with a 5 to 95% range of 1.5°C to 2.8°C. Note that since TCR scales linearly with the errors in the estimated scaling factors, estimates do not show a tendency for a long upper tail, as is the case for ECS. However, the separation of greenhouse gas response from the responses to other external forcing in a multi-fingerprint analysis introduces a small uncertainty, illustrated by small differences in results between three models (Figure 9.21). The TCR does not scale linearly with ECS because the transient response is strongly influenced by the speed with which the ocean transports heat into its interior, while the equilibrium sensitivity is governed by feedback strengths (discussion in Frame et al., 2005).

Estimates of a likely range for TCR can also be inferred directly from estimates of attributable greenhouse warming obtained in optimal detection analyses since there is a direct linear relationship between the two (Frame et al., 2005). The attributable greenhouse warming rates inferred from Figure 9.9 generally support the TCR range shown in Figure 9.21, although the lowest 5th percentile (1.3°C) and the highest 95th percentile (3.3°C) estimated in this way from detection and attribution analyses based on individual models lie outside the 5% to 95% range of 1.5°C to 2.8°C obtained from Figure 9.21.

Choosing lower and upper limits that encompass the range of these results and deflating significance levels in order to account for structural uncertainty in the estimate leads to the conclusion that it is very unlikely that TCR is less than 1°C and very unlikely that TCR is greater than 3.5°C. Information based on the models discussed in Chapter 10 provides additional information that can help constrain TCR further (Section 10.5.4.5).

9.6.3 Estimates of Climate Sensitivity Based on Palaeoclimatic Data

The palaeoclimate record offers a range of opportunities to assess the response of climate models to changes in external forcing. This section discusses estimates from both the palaeoclimatic record of the last millennium, and from the climate of the LGM. The latter gives a different perspective on feedbacks than anticipated with greenhouse warming, and thus provides a test bed for the physics in climate models. There also appears to be a likely positive relationship between temperature and CO_2 prior to the 650 kyr period covered by ice core measurements of CO_2 (Section 6.3).

As with analyses of the instrumental record discussed in Section 9.6.2, some studies using palaeoclimatic data have also estimated PDFs for ECS by varying model parameters. Inferences about ECS made through direct comparisons between radiative forcing and climate response, without using climate models, show large uncertainties since climate feedbacks, and thus sensitivity, may be different for different climatic background states and for different seasonal characteristics of forcing (e.g., Montoya et al., 2000). Thus, sensitivity to forcing during these periods cannot be directly compared to that for atmospheric CO_2 doubling.

9.6.3.1 *Estimates of Climate Sensitivity Based on Data for the Last Millennium*

The relationship between forcing and response based on a long time horizon can be studied using palaeoclimatic reconstructions of temperature and radiative forcing, particularly volcanism and solar forcing, for the last millennium. However, both forcing and temperature reconstructions are subject to large uncertainties (Chapter 6). To account for the uncertainty in reconstructions, Hegerl et al. (2006a) use several proxy data reconstructions of NH extratropical temperature for the past millennium (Briffa et al., 2001; Esper et al., 2002; Mann and Jones, 2003; Hegerl et al., 2007) to constrain ECS estimates for the pre-industrial period up to 1850. This study used a large ensemble of simulations of the last millennium performed with an energy balance model forced with reconstructions of volcanic (Crowley, 2000, updated), solar (Lean et al., 2002) and greenhouse gas forcing (see Section 9.3.3 for results on the detection of these external influences). Their estimated PDFs for ECS incorporate an estimate of uncertainty in the overall amplitude (including an attempt to account for uncertainty

in efficacy), but not the time evolution, of volcanic and solar forcing. They also attempt to account for uncertainty in the amplitude of reconstructed temperatures in one reconstruction (Hegerl et al., 2007), and assess the sensitivity of their results to changes in amplitude for others. All reconstructions combined yield a median climate sensitivity of 3.4°C and a 5 to 95% range of 1.2°C to 8.6°C (Figure 9.20). Reconstructions with a higher amplitude of past climate variations (e.g., Esper et al., 2002; Hegerl et al., 2007) are found to support higher ECS estimates than reconstructions with lower amplitude (e.g., Mann and Jones, 2003). Note that the constraint on ECS originates mainly from low-frequency temperature variations associated with changes in the frequency and intensity of volcanism which lead to a highly significant detection of volcanic response (Section 9.3.3) in all records used in the study.

The results of Andronova et al. (2004) are broadly consistent with these estimates. Andronova et al. (2004) demonstrate that climate sensitivities in the range of 2.3°C to 3.4°C yield reasonable simulations of both the NH mean temperature from 1500 onward when compared to the Mann and Jones (2003) reconstruction, and for the instrumental period. The agreement is less good for reconstructed SH temperature, where reconstructions are substantially more uncertain (Chapter 6).

Rind et al. (2004) studied the period from about 1675 to 1715 to attempt a direct estimate of climate sensitivity. This period has reduced radiative forcing relative to the present due to decreased solar radiation, decreased greenhouse gas and possibly increased volcanic forcing (Section 9.2.1.3). Different NH temperature reconstructions (Figure 6.10) have a wide range of cooling estimates relative to the late 20th century that

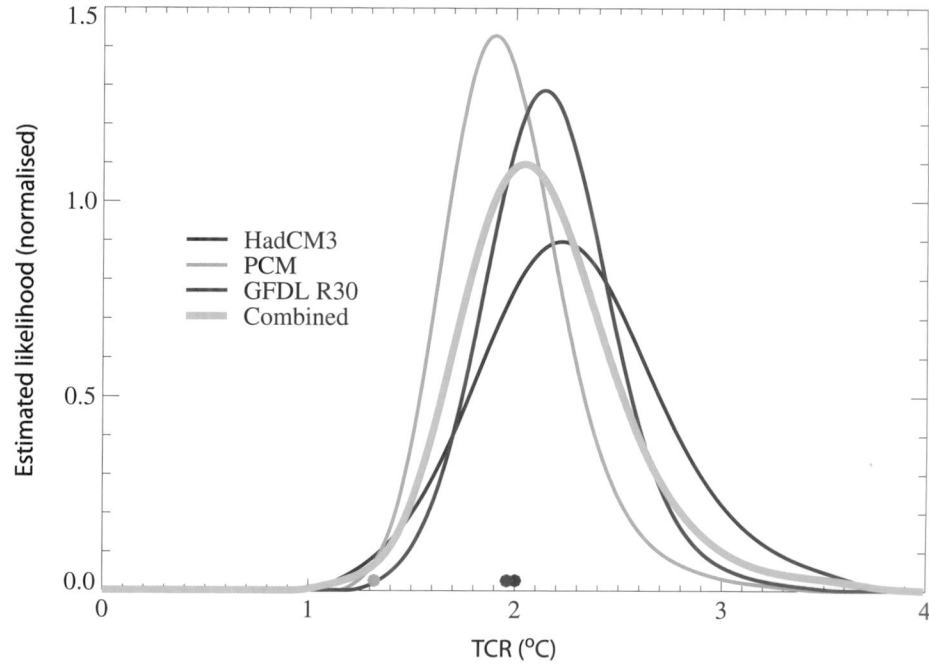

Figure 9.21. *Probability distributions of TCR (expressed as warming at the time of CO_2 doubling), as constrained by observed 20th-century temperature change, for the HadCM3 (Table 8.1, red), PCM (Table 8.1, green) and GFDL R30 (Delworth et al., 2002, blue) models. The average of the PDFs derived from each model is shown in cyan. Coloured circles show each model's TCR. (After Stott et al., 2006c).*

is broadly reproduced by climate model simulations. While climate in this cold period may have been close to radiative balance (Rind et al., 2004), some of the forcing during the present period is not yet realised in the system (estimated as 0.85 W m^{-2}; Hansen et al., 2005). Thus, ECS estimates based on a comparison between radiative forcing and climate response are subject to large uncertainties, but are broadly similar to estimates discussed above. Again, reconstructions with stronger cooling in this period imply higher climate sensitivities than those with weaker cooling (results updated from Rind et al., 2004).

9.6.3.2 Inferences About Climate Sensitivity Based on the Last Glacial Maximum

The LGM is one of the key periods used to estimate ECS (Hansen et al., 1984; Lorius et al., 1990; Hoffert and Covey, 1992), since it represents a quasi-equilibrium climate response to substantially altered boundary conditions. When forced with changes in greenhouse gas concentrations and the extent and height of ice sheet boundary conditions, AOGCMs or EMICs identical or similar to those used for 20th- and 21st-century simulations produce a 3.3°C to 5.1°C cooling for this period in response to radiative perturbations of 4.6 to 7.2 W m^{-2} (Sections 6.4.1.3; see also Section 9.3.2; see also Masson-Delmotte et al., 2006). The simulated cooling in the tropics ranges from 1.7°C to 2.4°C. The ECS of the models used in PMIP2 ranges from 2.3°C to 3.7°C (Table 8.2), and there is some tendency for models with larger sensitivity to produce larger tropical cooling for the LGM, but this relationship is not very tight. Comparison between simulated climate change and reconstructed climate is affected by substantial uncertainties in forcing and data (Chapter 6 and Section 9.2.1.3). For example, the PMIP2 forcing does not account for changes in mineral dust, since the level of scientific understanding for this forcing is very low (Figure 6.5). The range of simulated temperature changes is also affected by differences in the radiative influence of the ice-covered regions in different models (Taylor et al., 2000). Nevertheless, the PMIP2 models simulate LGM climate changes that are approximately consistent with proxy information (Chapter 6).

Recent studies (Annan et al., 2005; Schneider von Deimling et al., 2006) attempt to estimate the PDF of ECS from ensemble simulations of the LGM by systematically exploring model uncertainty. Both studies investigate the relationship between climate sensitivity and LGM tropical SSTs, which are influenced strongly by CO_2 changes. In a perturbed physics ensemble, Schneider von Deimling et al. (2006) vary 11 ocean and atmospheric parameters in a 1,000-member ensemble simulation of the LGM with the CLIMBER-2 EMIC (Table 8.3). They find a close relationship between ECS and tropical SST cooling in their model, implying a 5 to 95% range of ECS of 1.2°C to 4.3°C when attempting to account for model parameter, forcing and palaeoclimate data uncertainties. Similar constraints on climate sensitivity are found when proxy reconstructions of LGM antarctic temperatures are used instead of tropical SSTs (Schneider von Deimling et al., 2006). In contrast, Annan

et al. (2005) use a perturbed physics ensemble based on a low-resolution version of the atmospheric component of the MIROC3.2 model, perturbing a range of model parameters over prior distributions determined from the ability of the model to reproduce seasonal mean climate in a range of climate variables. They find a best-fit sensitivity of about 4.5°C, and their results suggest that sensitivities in excess of 6°C are unlikely given observational estimates of LGM tropical cooling and the relationship between tropical SST and sensitivity in their model. Since the perturbed physics ensemble based on that atmospheric model does not produce sensitivities less than 4°C, this result cannot provide a lower limit or a PDF for ECS.

The discrepancy between the inferred upper limits in the two studies probably arises from both different radiative forcing and structural differences between the models used. Forcing from changes in vegetation cover and dust is not included in the simulations done by Annan et al. (2005), which according to Schneider von Deimling et al. (2006) would reduce the Annan et al. ECS estimates and yield better agreement between the results of the two studies. However, the effect of these forcings and their interaction with other LGM forcings is very uncertain, limiting confidence in such estimates of their effect (Figure 6.5). Structural differences in models are also likely to play a role. The Annan et al. (2005) estimate shows a weaker association between simulated tropical SST changes and ECS than the Schneider von Deimling et al. (2006) result. Since Annan et al. use a mixed-layer ocean model, and Schneider von Deimling a simplified ocean model, both models may not capture the full ocean response affecting tropical SSTs. The atmospheric model used in Schneider von Deimling is substantially simpler than that used in the Annan et al. (2005) study. Overall, estimates of climate sensitivity from the LGM are broadly consistent with other estimates of climate sensitivity derived, for example, from the instrumental period.

9.6.4 Summary of Observational Constraints for Climate Sensitivity

Any constraint of climate sensitivity obtained from observations must be interpreted in light of the underlying assumptions. These assumptions include (i) the choice of prior distribution for each of the model parameters (Section 9.6.1 and Supplementary Material, Appendix 9.B), including the parameter range explored, (ii) the treatment of other parameters that influence the estimate, such as effective ocean diffusivity, and (iii) the methods used to account for uncertainties, such as structural and forcing uncertainties, that are not represented by the prior distributions. Neglecting important sources of uncertainty in these estimates will result in overly narrow ranges that overstate the certainty with which the ECS or TCR is known. Errors in assumptions about forcing or model response will also result in unrealistic features of model simulations, which can result in erroneous modes (peak probabilities) and shapes of the PDF. On the other hand, using less than all available information will yield results that are less constrained than they could be under optimal use of available data.

While a variety of important uncertainties (e.g., radiative forcing, mixing of heat into the ocean) have been taken into account in most studies (Table 9.3), some caveats remain. Some processes and feedbacks might be poorly represented or missing, particularly in simple and many intermediate complexity models. Structural uncertainties in the models, for example, in the representation of cloud feedback processes (Chapter 8) or the physics of ocean mixing, will affect results for climate sensitivity and are very difficult to quantify. In addition, differences in efficacy between forcings are not directly represented in simple models, so they may affect the estimate (e.g., Tett et al., 2007), although this uncertainty may be folded into forcing uncertainty (e.g., Hegerl et al., 2003, 2007). The use of a single value for the ECS further assumes that it is constant in time. However, some authors (e.g., Senior and Mitchell, 2000; Boer and Yu, 2003) have shown that ECS varies in time in the climates simulated by their models. Since results from instrumental data and the last millennium are dominated primarily by decadal- to centennial-scale changes, they will therefore only represent climate sensitivity at an equilibrium that is not too far from the present climate. There is also a small uncertainty in the radiative forcing due to atmospheric CO_2 doubling (<10%; see Chapter 2), which is not accounted for in most studies that derive observational constraints on climate sensitivity.

Despite these uncertainties, which are accounted for to differing degrees in the various studies, confidence is increased by the similarities between individual ECS estimates (Figure 9.20). Most studies find a lower 5% limit of between 1°C and 2.2°C, and studies that use information in a relatively complete manner generally find a most likely value between 2°C and 3°C (Figure 9.20). Constraints on the upper end of the likely range of climate sensitivities are also important, particularly for probabilistic forecasts of future climate with constant radiative forcing. The upper 95% limit for ECS ranges from 5°C to 10°C, or greater in different studies depending upon the approach taken, the number of uncertainties included and specific details of the prior distribution that was used. This wide range is largely caused by uncertainties and nonlinearities in forcings and response. For example, a high sensitivity is difficult to rule out because a high aerosol forcing could nearly cancel greenhouse gas forcing over the 20th century. This problem can be addressed, at least to some extent, if the differences in the spatial and temporal patterns of response between aerosol and greenhouse gas forcing are used for separating these two responses in observations (as, for example, in Gregory et al., 2002a; Harvey and Kaufmann, 2002; Frame et al., 2005). In addition, nonlinearities in the response to transient forcing make it more difficult to constrain the upper limit on ECS based on observed transient forcing responses (Frame et al., 2005). The TCR, which may be more relevant for near-term climate change, is easier to constrain since it relates more linearly to observables. For the pre-instrumental part of the last millennium, uncertainties in temperature and forcing reconstructions, and the nonlinear connection between ECS and the response to volcanism, prohibit tighter constraints.

Estimates of climate sensitivity based on the ability of climate models to reproduce climatic conditions of the LGM broadly support the ranges found from the instrumental period, although a tight constraint is also difficult to obtain from this period alone because of uncertainties in tropical temperature changes, forcing uncertainties and the effect of structural model uncertainties. In addition, the number of studies providing estimates of PDFs from palaeoclimatic data, using independent approaches and complementary sources of proxy data, are limited.

Thus, most studies that use a simple uniform prior distribution of ECS are not able to exclude values beyond the traditional IPCC First Assessment Report range of 1.5°C to 4.5°C (IPCC, 1990). However, considering all available evidence on ECS together provides a stronger constraint than individual lines of evidence. Bayesian methods can be used to incorporate multiple lines of evidence to sharpen the posterior distribution of ECS, as in Annan and Hargreaves (2006) and Hegerl et al. (2006a). Annan and Hargreaves (2006) demonstrate that using three lines of evidence, namely 20th-century warming, the response to individual volcanic eruptions and the LGM response, results in a tighter estimate of ECS, with a probability of less than 5% that ECS exceeds 4.5°C. The authors find a similar constraint using five lines of evidence under more conservative assumptions about uncertainties (adding cooling during the Little Ice Age and studies based on varying model parameters to match climatological means, see Box 10.2). However, as discussed in Annan and Hargreaves (2006), combining multiple lines of evidence may produce overly confident estimates unless every single line of evidence is entirely independent of others, or dependence is explicitly taken into account. Hegerl et al. (2006a) argue that instrumental temperature change during the second half of the 20th century is essentially independent of the palaeoclimate record of the last millennium and of the instrumental data from the first half of the 20th century that is used to calibrate the palaeoclimate records. Hegerl et al. (2006a) therefore base their prior probability distribution for the climate sensitivity on results from the late 20th century (Frame et al., 2005), which reduces the 5 to 95% ECS range from all proxy reconstructions analysed to 1.5°C to 6.2°C compared to the previous range of 1.2°C to 8.6°C. Both results demonstrate that independent estimates, when properly combined in a Bayesian analysis, can provide a tighter constraint on climate sensitivity, even if they individually provide only weak constraints. These studies also find a 5% lower limit of 1.5°C or above, consistent with several studies based on the 20th-century climate change alone (Knutti et al., 2002; Forest et al., 2006) and estimates that greenhouse warming contributes substantially to observed temperature changes (Section 9.4.1.4).

Overall, several lines of evidence strengthen confidence in present estimates of ECS, and new results based on objective analyses make it possible to assign probabilities to ranges of climate sensitivity previously assessed from expert opinion alone. This represents a significant advance. Results from studies of observed climate change and the consistency of estimates from different time periods indicate that ECS is very likely larger than 1.5°C with a most likely value between 2°C and 3°C. The lower bound is consistent with the view that the sum of all atmospheric feedbacks affecting climate sensitivity is positive. Although upper limits can be obtained by combining multiple lines of evidence, remaining uncertainties that are not accounted for in individual estimates (such as structural model uncertainties) and possible dependencies between individual lines of evidence make the upper 95% limit of ECS uncertain at present. Nevertheless, constraints from observed climate change support the overall assessment that the ECS is likely to lie between 2°C and 4.5°C with a most likely value of approximately 3°C (Box 10.2).

9.7   Combining Evidence of Anthropogenic Climate Change

The widespread change detected in temperature observations of the surface (Sections 9.4.1, 9.4.2, 9.4.3), free atmosphere (Section 9.4.4) and ocean (Section 9.5.1), together with consistent evidence of change in other parts of the climate system (Section 9.5), strengthens the conclusion that greenhouse gas forcing is the dominant cause of warming during the past several decades. This combined evidence, which is summarised in Table 9.4, is substantially stronger than the evidence that is available from observed changes in global surface temperature alone (Figure 3.6).

The evidence from surface temperature observations is strong: The observed warming is highly significant relative to estimates of internal climate variability which, while obtained from models, are consistent with estimates obtained from both instrumental data and palaeoclimate reconstructions. It is extremely unlikely (<5%) that recent global warming is due to internal variability alone such as might arise from El Niño (Section 9.4.1). The widespread nature of the warming (Figures 3.9 and 9.6) reduces the possibility that the warming could have resulted from internal variability. No known mode of internal variability leads to such widespread, near universal warming as has been observed in the past few decades. Although modes of internal variability such as El Niño can lead to global average warming for limited periods of time, such warming is regionally variable, with some areas of cooling (Figures 3.27 and 3.28). In addition, palaeoclimatic evidence indicates that El Niño variability during the 20th century is not unusual relative to earlier periods (Section 9.3.3.2; Chapter 6). Palaeoclimatic evidence suggests that such a widespread warming has not been observed in the NH in at least the past 1.3 kyr (Osborn and Briffa, 2006), further strengthening the evidence that the recent warming is not due to natural internal variability. Moreover, the response to anthropogenic forcing is detectable on all continents individually except Antarctica, and in some sub-continental regions. Climate models only reproduce the observed 20th-century global mean surface warming when both anthropogenic and natural forcings are included (Figure 9.5). No model that has used natural forcing only has reproduced the observed

global mean warming trend or the continental mean warming trends in all individual continents (except Antarctica) over the second half of the 20th century. Detection and attribution of external influences on 20th-century and palaeoclimatic reconstructions, from both natural and anthropogenic sources (Figure 9.4 and Table 9.4), further strengthens the conclusion that the observed changes are very unusual relative to internal climate variability.

The energy content change associated with the observed widespread warming of the atmosphere is small relative to the energy content change of the ocean, and also smaller than that associated with other components such as the cryosphere. In addition, the solid Earth also shows evidence for warming in boreholes (Huang et al., 2000; Beltrami et al., 2002; Pollack and Smerdon, 2004). It is theoretically feasible that the warming of the near surface could have occurred due to a reduction in the heat content of another component of the system. However, all parts of the cryosphere (glaciers, small ice caps, ice sheets and sea ice) have decreased in extent over the past half century, consistent with anthropogenic forcing (Section 9.5.5, Table 9.4), implying that the cryosphere consumed heat and thus indicating that it could not have provided heat for atmospheric warming. More importantly, the heat content of the ocean (the largest reservoir of heat in the climate system) also increased, much more substantially than that of the other components of the climate system (Figure 5.4; Hansen et al., 2005; Levitus et al., 2005). The warming of the upper ocean during the latter half of the 20th century was likely due to anthropogenic forcing (Barnett et al., 2005; Section 9.5.1.1; Table 9.4). While the statistical evidence in this research is very strong that the warming cannot be explained by ocean internal variability as estimated by two different climate models, uncertainty arises since there are discrepancies between estimates of ocean heat content variability from models and observations, although poor sampling of parts of the World Ocean may explain this discrepancy. However, the spatial pattern of ocean warming with depth is very consistent with heating of the ocean resulting from net positive radiative forcing, since the warming proceeds downwards from the upper layers of the ocean and there is deeper penetration of heat at middle to high latitudes and shallower penetration at low latitudes (Barnett et al., 2005; Hansen et al., 2005). This observed ocean warming pattern is inconsistent with a redistribution of heat between the surface and the deep ocean.

Thus, the evidence appears to be inconsistent with the ocean or land being the source of the warming at the surface. In addition, simulations forced with observed SST changes cannot fully explain the warming in the troposphere without increases in greenhouse gases (e.g., Sexton et al., 2001), further strengthening the evidence that the warming does not originate from the ocean. Further evidence for forced changes arises from widespread melting of the cryosphere (Section 9.5.5), increases in water vapour in the atmosphere (Section 9.5.4.1) and changes in top-of-the atmosphere radiation that are consistent with changes in forcing.

The simultaneous increase in energy content of all the major components of the climate system and the pattern and amplitude of warming in the different components, together with evidence that the second half of the 20th century was likely the warmest in 1.3 kyr (Chapter 6) indicate that the cause of the warming is extremely unlikely to be the result of internal processes alone. The consistency across different lines of evidence makes a strong case for a significant human influence on observed warming at the surface. The observed rates of surface temperature and ocean heat content change are consistent with the understanding of the likely range of climate sensitivity and net climate forcings. Only with a net positive forcing, consistent with observational and model estimates of the likely net forcing of the climate system (as used in Figure 9.5), is it possible to explain the large increase in heat content of the climate system that has been observed (Figure 5.4).

Table 9.4. *A synthesis of climate change detection results: (a) surface and atmospheric temperature evidence and (b) evidence from other variables. Note that our likelihood assessments are reduced compared to individual detection studies in order to take into account remaining uncertainties (see Section 9.1.2), such as forcing and model uncertainty not directly accounted for in the studies. The likelihood assessment is indicated in percentage terms, in parentheses where the term is not from the standard IPCC likelihood levels.*

a)

Result	Region	Likelihood	Factors contributing to likelihood assessment
Surface temperature			
Warming during the past half century cannot be explained without external radiative forcing	Global	Extremely likely (>95%)	Anthropogenic change has been detected in surface temperature with very high significance levels (less than 1% error probability). This conclusion is strengthened by detection of anthropogenic change in the upper ocean with high significance level. Upper ocean warming argues against the surface warming being due to natural internal processes. Observed change is very large relative to climate-model simulated internal variability. Surface temperature variability simulated by models is consistent with variability estimated from instrumental and palaeorecords. Main uncertainty from forcing and internal variability estimates (Sections 9.4.1.2, 9.4.1.4, 9.5.1.1, 9.3.3.2, 9.7).
Warming during the past half century is not solely due to known natural causes	Global	Very Likely	This warming took place at a time when non-anthropogenic external factors would likely have produced cooling. The combined effect of known sources of forcing would have been extremely likely to produce a warming. No climate model that has used natural forcing only has reproduced the observed global warming trend over the 2nd half of the 20th century. Main uncertainties arise from forcing, including solar, model-simulated responses and internal variability estimates (Sections 2.9.2, 9.2.1, 9.4.1.2, 9.4.1.4; Figures 9.5, 9.6, 9.9).
Greenhouse gas forcing has been the dominant cause of the observed global warming over the last 50 years.	Global	Very likely	All multi-signal detection and attribution studies attribute more warming to greenhouse gas forcing than to a combination of all other sources considered, including internal variability, with a very high significance. This conclusion accounts for observational, model and forcing uncertainty, and the possibility that the response to solar forcing could be underestimated by models. Main uncertainty from forcing and internal variability estimates (Section 9.4.1.4; Figure 9.9).
Increases in greenhouse gas concentrations alone would have caused more warming than observed over the last 50 years because volcanic and anthropogenic aerosols have offset some warming that would otherwise have taken place.	Global	Likely	Estimates from different analyses using different models show consistently more warming than observed over the last 50 years at the 5% significance level. However, separation of the response to non-greenhouse gas (particularly aerosol) forcing from greenhouse gas forcing varies between models (Section 9.4.1.4; Figure 9.9).
There has been a substantial anthropogenic contribution to surface temperature increases in every continent except Antarctica since the middle of the 20th century	Africa, Asia, Australia, Europe, North America and South America	Likely	Anthropogenic change has been estimated using detection and attribution methods on every individual continent (except Antarctica). Greater variability compared to other continental regions makes detection more marginal in Europe. No climate model that used natural forcing only reproduced the observed continental mean warming trend over the second half of the 20th century. Uncertainties arise because sampling effects result in lower signal-to-noise ratio at continental than at global scales. Separation of the response to different forcings is more difficult at these spatial scales (Section 9.4.2; FAQ 9.2, Figure 1).
Early 20th-century warming is due in part to external forcing.	Global	Very Likely	A number of studies detect the influence of external forcings on early 20th-century warming, including a warming from anthropogenic forcing. Both natural forcing and response are uncertain, and different studies find different forcings dominant. Some studies indicate that internal variability could have made a large contribution to early 20th-century warming. Some observational uncertainty in early 20th-century trend (Sections 9.3.3.2, 9.4.1.4; Figures 9.4, 9.5).

(continued)

Table 9.4 (continued)

Result	Region	Likelihood	Factors contributing to likelihood assessment
Surface temperature			
Pre-industrial temperatures were influenced by natural external forcing (period studied is past 7 centuries)	NH (mostly extratropics)	Very Likely	Detection studies indicate that external forcing explains a substantial fraction of inter-decadal variability in NH temperature reconstructions. Simulations in response to estimates of pre-industrial forcing reproduce broad features of reconstructions. Substantial uncertainties in reconstructions and past forcings are unlikely to lead to a spurious agreement between temperature reconstructions and forcing reconstructions as they are derived from independent proxies (Section 9.3.3; Figures 9.4, 6.13).
Temperature extremes have changed due to anthropogenic forcing	NH land areas and Australia combined.	Likely	A range of observational evidence indicates that temperature extremes are changing. An anthropogenic influence on the temperatures of the 1, 5, 10 and 30 warmest nights, coldest days and coldest nights annually has been formally detected and attributed in one study, but observed change in the temperature of the warmest day annually is inconsistent with simulated change. The detection of changes in temperature extremes is supported by other comparisons between models and observations. Model uncertainties in changes in temperature extremes are greater than for mean temperatures and there is limited observational coverage and substantial observational uncertainty (Section 9.4.3).
Free atmosphere changes			
Tropopause height increases are detectable and attributable to anthropogenic forcing (latter half of the 20th century)	Global	Likely	There has been robust detection of anthropogenic influence on increasing tropopause height. Simulated tropopause height increases result mainly from greenhouse gas increases and stratospheric ozone decreases. Detection and attribution studies rely on reanalysis data, which are subject to inhomogeneities related to differing availability and quality of input data, although tropopause height increases have also been identified in radiosonde observations. Overall tropopause height increases in recent model and one reanalysis (ERA-40) appear to be driven by similar large-scale changes in atmospheric temperature, although errors in tropospheric warming and stratospheric cooling could lead to partly spurious agreement in other data sets (Section 9.4.2; Figure 9.14).
Tropospheric warming is detectable and attributable to anthropogenic forcing (latter half of the 20th century)	Global	Likely	There has been robust detection and attribution of anthropogenic influence on tropospheric warming, which does not depend on including stratospheric cooling in the fingerprint pattern of response. There are observational uncertainties in radiosonde and satellite records. Models generally predict a relative warming of the free troposphere compared to the surface in the tropics since 1979, which is not seen in one version of the radiosonde record (possibly due to uncertainties in the radiosonde record) but is seen in one version of the satellite record, although not others (Section 9.4.4).
Simultaneous tropospheric warming and stratospheric cooling due to the influence of anthropogenic forcing has been observed (latter half of the 20th century)	Global	Very Likely	Simultaneous warming of the troposphere and cooling of the stratosphere due to natural factors is less likely than warming of the troposphere or cooling of the stratosphere alone. Cooling of the stratosphere is in part related to decreases in stratospheric ozone. Modelled and observational uncertainties as discussed under entries for tropospheric warming with additional uncertainties due to stratospheric observing systems and the relatively poor representations of stratospheric processes and variability in climate models (Section 9.4.4).

b)

Result	Region	Likelihood	Factors contributing to likelihood assessment
Ocean changes			
Anthropogenic forcing has warmed the upper several hundred metres of the ocean during the latter half of the 20th century	Global (but with limited sampling in some regions)	Likely	Robust detection and attribution of anthropogenic fingerprint from three different models in ocean temperature changes, and in ocean heat content data, suggests high likelihood, but observational and modelling uncertainty remains. 20th-century simulations with MMD models simulate comparable ocean warming to observations only if anthropogenic forcing is included. Simulated and observed variability appear inconsistent, either due to sampling errors in the observations or under-simulated internal variability in the models. Limited geographical coverage in some ocean basins (Section 9.5.1.1; Figure 9.15).
Anthropogenic forcing contributed to sea level rise during the latter half 20th century	Global	Very likely	Natural factors alone do not satisfactorily explain either the observed thermal expansion of the ocean or the observed sea level rise. Models including anthropogenic and natural forcing simulate the observed thermal expansion since 1961 reasonably well. Anthropogenic forcing dominates the surface temperature change simulated by models, and has likely contributed to the observed warming of the upper ocean and widespread glacier retreat. It is very unlikely that the warming during the past half century is due only to known natural causes. It is therefore very likely that anthropogenic forcing contributed to sea level rise associated with ocean thermal expansion and glacier retreat. However, it remains difficult to estimate the anthropogenic contribution to sea level rise because suitable studies quantifying the anthropogenic contribution to sea level rise and glacier retreat are not available, and because the observed sea level rise budget is not closed (Table 9.2; Section 9.5.2).
Circulation			
Sea level pressure shows a detectable anthropogenic signature during the latter half of the 20th century	Global	Likely	Changes of similar nature are observed in both hemispheres and are qualitatively, but not quantitatively consistent with model simulations. Uncertainty in models and observations. Models underestimate the observed NH changes for reasons that are not understood, based on a small number of studies. Simulated response to 20th century forcings is consistent with observations in SH if effect of stratospheric ozone depletion is included (Section 9.5.3.4; Figure 9.16).
Anthropogenic forcing contributed to the increase in frequency of the most intense tropical cyclones since the 1970s	Tropical regions	More likely than not (>50%)	Recent observational evidence suggests an increase in frequency of intense storms. Increase in intensity is consistent with theoretical expectations. Large uncertainties due to models and observations. Modelling studies generally indicate a reduced frequency of tropical cyclones in response to enhanced greenhouse gas forcing, but an increase in the intensity of the most intense cyclones. Observational evidence, which is affected by substantial inhomogeneities in tropical cyclone data sets for which corrections have been attempted, suggests that increases in cyclone intensity since the 1970s are associated with SST and atmospheric water vapour increases (Section 3.8.3, Box 3.5 and Section 9.5.2.6).
Precipitation, Drought, Runoff			
Volcanic forcing influences total rainfall	Global land areas	More likely than not (>50%)	Model response detectable in observations for some models and result supported by theoretical understanding. However, uncertainties in models, forcings and observations. Limited observational sampling, particularly in the SH (Section 9.5.4.2; Figure 9.18).
Increases in heavy rainfall are consistent with anthropogenic forcing during latter half 20th century	Global land areas (limited sampling)	More likely than not (>50%)	Observed increases in heavy precipitation appear to be consistent with expectations of response to anthropogenic forcing. Models may not represent heavy rainfall well; observations suffer from sampling inadequacies (Section 9.5.4.2).

(continued)

Table 9.4 (continued)

Result	Region	Likelihood	Factors contributing to likelihood assessment
Precipitation, Drought, Runoff			
Increased risk of drought due to anthropogenic forcing during latter half 20th century	Global land areas	More likely than not (>50%)	One detection study has identified an anthropogenic fingerprint in a global Palmer Drought Severity Index data set with high significance, but the simulated response to anthropogenic and natural forcing combined is weaker than observed, and the model appears to have less inter-decadal variability than observed. Studies of some regions indicate that droughts in those regions are linked either to SST changes that, in some instances, may be linked to anthropogenic aerosol forcing (e.g., Sahel) or to a circulation response to anthropogenic forcing (e.g., southwest Australia). Models, observations and forcing all contribute uncertainty (Section 9.5.3.2).
Cryosphere			
Anthropogenic forcing has contributed to reductions in NH sea ice extent during the latter half of the 20th century	Arctic	Likely	The observed change is qualitatively consistent with model-simulated changes for most models and expectation of sea ice melting under arctic warming. Sea ice extent change detected in one study. The model used has some deficiencies in arctic sea ice annual cycle and extent. The conclusion is supported by physical expectations and simulations with another climate model. Change in SH sea ice probably within range explained by internal variability (Section 9.5.5.1).
Anthropogenic forcing has contributed to widespread glacier retreat during the 20th century	Global	Likely	Observed changes are qualitatively consistent with theoretical expectations and temperature detection. Anthropogenic contribution to volume change difficult to estimate. Few detection and attribution studies, but retreat in vast majority of glaciers consistent with expected reaction to widespread warming (Section 9.5.5.3).

References

AchutaRao, K.M., et al., 2006: Variability of ocean heat uptake: Reconciling observations and models. *J. Geophys. Res.*, **111**, C05019.

Ackerman, A.S., et al., 2000: Reduction of tropical cloudiness by soot. *Science*, **288**, 1042–1047.

Adams, J.B., M.E. Mann, and C.M. Ammann, 2003: Proxy evidence for an El Nino-like response to volcanic forcing. *Nature*, **426**(6964), 274–278.

Alexander, L.V., et al., 2006: Global observed changes in daily climate extremes of temperature and precipitation. *J. Geophys. Res.*, **111**, D05109, doi:10.1029/2005JD006290.

Allan, R.J., and T.J. Ansell, 2006: A new globally-complete monthly historical gridded mean sea level pressure data set (HadSLP2): 1850-2004. *J. Clim.*, **19**, 5816–5842.

Allen, M.R., 2003: Liability for climate change. *Nature*, **421**, 891–892.

Allen, M.R., and S.F.B. Tett, 1999: Checking for model consistency in optimal fingerprinting. *Clim. Dyn.*, **15**, 419–434.

Allen, M.R., and W.J. Ingram, 2002: Constraints on future changes in climate and the hydrologic cycle. *Nature*, **419**, 224–232.

Allen, M.R., and D.A. Stainforth, 2002: Towards objective probabilistic climate forecasting. *Nature*, **419**, 228–228.

Allen, M.R., and P.A. Stott, 2003: Estimating signal amplitudes in optimal fingerprinting, Part I: Theory. *Clim. Dyn*, **21**, 477–491.

Allen, M.R., J.A. Kettleborough, and D.A. Stainforth, 2002: Model error in weather and climate forecasting. In: *ECMWF Predictability of Weather and Climate Seminar* [Palmer, T.N. (ed.)]. European Centre for Medium Range Weather Forecasts, Reading, UK, http://www.ecmwf. int/publications/library/do/references/list/209.

Allen, M.R., et al., 2000: Quantifying the uncertainty in forecasts of anthropogenic climate change. *Nature*, **407**, 617–620.

Ammann, C.M., G.A. Meehl, W.M. Washington, and C. Zender, 2003: A monthly and latitudinally varying volcanic forcing dataset in simulations of 20th century climate. *Geophys. Res. Lett.*, **30**(12), 1657.

Anderson, T.L., et al., 2003: Climate forcing by aerosols: A hazy picture. *Science*, **300**, 1103–1104.

Andronova, N.G., and M.E. Schlesinger, 2000: Causes of global temperature changes during the 19th and 20th centuries. *Geophys. Res. Lett.*, **27**(14), 2137–2140.

Andronova, N.G., and M.E. Schlesinger, 2001: Objective estimation of the probability density function for climate sensitivity. *J. Geophys. Res.*, **106**(D19), 22605–22611.

Andronova, N.G., M.E. Schlesinger, and M.E. Mann, 2004: Are reconstructed pre-instrumental hemispheric temperatures consistent with instrumental hemispheric temperatures? *Geophys. Res. Lett.*, **31**, L12202, doi:10.1029/2004GL019658.

Andronova, N.G., et al., 1999: Radiative forcing by volcanic aerosols from 1850 to 1994. *J. Geophys. Res.*, **104**, 16807–16826.

Andronova, N.G., et al., 2007: The concept of climate sensitivity History and development. In: *Human-Induced Climate Change: An Interdisciplinary Assessment* [Schlesinger, M., et al. (eds.)]. Cambridge University Press, Cambridge, UK, in press.

Annan, J.D., and J.C. Hargreaves, 2006: Using multiple observationally-based constraints to estimate climate sensitivity. *Geophys. Res. Lett.*, **33**, L06704, doi:10.1029/2005GL025259.

Annan, J.D., et al., 2005: Efficiently constraining climate sensitivity with paleoclimate simulations. *Scientific Online Letters on the Atmosphere*, **1**, 181–184.

Arblaster, J.M., and G.A. Meehl, 2006: Contributions of external forcing to Southern Annular Mode trends. *J. Clim.*, **19**, 2896–2905.

Bader, J., and M. Latif, 2003: The impact of decadal-scale Indian Ocean sea surface temperature anomalies on Sahelian rainfall and the North Atlantic Oscillation. *Geophys. Res. Lett.*, **30**(22), 2169.

Banks, H.T., et al., 2000: Are observed decadal changes in intermediate water masses a signature of anthropogenic climate change? *Geophys. Res. Lett.*, **27**, 2961–2964.

Barnett, T.P., D.W. Pierce, and R. Schnur, 2001: Detection of anthropogenic climate change in the world's oceans. *Science*, **292**, 270–274.

Barnett, T.P., et al., 1999: Detection and attribution of recent climate change. *Bull. Am. Meteorol. Soc.*, **80**, 2631–2659.

Barnett, T.P., et al., 2005: Penetration of a warming signal in the world's oceans: human impacts. *Science*, **309**, 284–287.

Bauer, E., M. Claussen, V. Brovkin, and A. Huenerbein, 2003: Assessing climate forcings of the Earth system for the past millennium. *Geophys. Res. Lett.*, **30**(6), 1276.

Betrami, H., J.E. Smerdon, H.N. Pollack, and S. Huang, 2002: Continental heat gain in the global climate system. *Geophys. Res. Lett.*, **29**, 1167.

Bengtsson, L., K.I. Hodges, and E. Roechner, 2006: Storm tracks and climate change. *J. Clim.*, **19**, 3518–3543.

Berger, A., 1978: Long-term variations of caloric solar radiation resulting from the earth's orbital elements. *Quat. Res.*, **9**, 139–167.

Berger, A., 1988: Milankovitch theory and climate. *Rev. Geophys.*, **26**, 524–657.

Berliner, L.M., R.A. Levine, and D.J. Shea, 2000: Bayesian climate change assessment. *J. Clim.*, **13**, 3805–3820.

Bertrand, C., M.F. Loutre, M. Crucifix, and A. Berger, 2002: Climate of the last millennium: a sensitivity study. *Tellus*, **54A**(3), 221–244.

Betts, R.A., 2001: Biogeophysical impacts of land use on present-day climate: near surface temperature and radiative forcing. *Atmos. Sci. Lett.*, **2**, 39–51.

Bigelow, N.H., et al., 2003: Climate change and Arctic ecosystems: 1. Vegetation changes north of 55 degrees N between the last glacial maximum, mid-Holocene, and present. *J. Geophys. Res.*, **108**(D19), 8170, doi:10.1029/2002JD002558.

Bindoff, N.L., and T.J. McDougall, 2000: Decadal changes along an Indian Ocean section at 32S and their interpretation. *J. Phys. Oceanogr.*, **30**(6), 1207–1222.

Bjerknes, J., 1969: Atmospheric teleconnections from the equatorial Pacific. *Mon. Weather Rev.*, **97**, 163–172.

Boer, G.J., and B. Yu, 2003: Climate sensitivity and climate state. *Clim. Dyn.*, **21**, 167–176.

Boucher, O., and J. Haywood, 2001: On summing the components of radiative forcing of climate change. *Clim. Dyn.*, **18**, 297–302.

Boyer, T.P., et al., 2005: Linear trends in salinity for the World Ocean, 1955-1998. *Geophys. Res. Lett.*, **32**, L01604.

Braconnot, P., S. Joussaume, O. Marti, and N. de Noblet, 1999: Synergistic feedbacks from ocean and vegetation on the African monsoon response to mid-Holocene insolation. *Geophys. Res. Lett.*, **26**, 2481–2484.

Braconnot, P., O. Marti, S. Joussaume, and Y. Leclainche, 2000: Ocean feedback in response to 6 kyr BP insolation. *J. Clim.*, **13**(9), 1537–1553.

Braconnot, P., et al., 2004: Evaluation of PMIP coupled ocean-atmosphere simulations of the Mid-Holocene. In: *Past Climate Variability through Europe and Africa* [Battarbee, R.W., F. Gasse, and C.E. Stickley (eds.)]. Springer, London, UK, pp. 515-533.

Braganza, K., et al., 2003: Simple indices of global climate variability and change: Part I - Variability and correlation structure. *Clim. Dyn.*, **20**, 491–502.

Braganza, K., et al., 2004: Simple indices of global climate variability and change: Part II - Attribution of climate change during the 20th century. *Clim. Dyn.*, **22**, 823–838.

Briffa, K.R., et al., 2001: Low-frequency temperature variations from a northern tree ring density network. *J. Geophys. Res.*, **106**(D3), 2929–2941.

Broccoli, A.J., et al., 2003: Twentieth-century temperature and precipitation trends in ensemble climate simulations including natural and anthropogenic forcing. *J. Geophys. Res.*, **108**(D24), 4798.

Brohan, P., et al., 2006: Uncertainty estimates in regional and global observed temperature changes: a new dataset from 1850. *J. Geophys. Res.*, **111**, D12106, doi:10.1029/2005JD006548.

Bryden, H.L., E. McDonagh, and B.A. King, 2003: Changes in ocean water mass properties: oscillations of trends? *Science*, **300**, 2086–2088.

Bryden, H.L., H.R. Longworth, and S.A. Cunningham, 2005: Slowing of the Atlantic meridional overturning circulation at 25° N. *Nature*, **438**, 655–657.

Burke, E.J., S.J. Brown, and N. Christidis, 2006: Modelling the recent evolution of global drought and projections for the 21st century with the Hadley Centre climate model. *J. Hydrometeorol.*, **7**, 1113–1125.

Caesar, J., L. Alexander, and R. Vose, 2006: Large-scale changes in observed daily maximum and minimum temperatures, 1946-2000. *J. Geophys. Res.*, **111**, D05101, doi:10.1029/2005JD006280.

Cai, W., P.H. Whetton, and D.J. Karoly, 2003: The response of the Antarctic Oscillation to increasing and stabilized atmospheric CO_2. *J. Clim.*, **16**, 1525–1538.

Cane, M., et al., 2006: Progress in paleoclimate modeling. *J. Clim.*, **19**, 5031–5057.

Carril, A.F., C.G. Menéndez, and A. Navarra, 2005: Climate response associated with the Southern Annular Mode in the surroundings of Antarctic Peninsula: A multimodel ensemble analysis. *Geophys. Res. Lett.*, **32**, L16713, doi:10.1029/2005GL023581.

Chan, J.C.L., 2006: Comment on "Changes in tropical cyclone number, duration, and intensity in a warming environment". *Science*, **311**, 1713.

Chan, J.C.L., and K.S. Liu, 2004: Global warming and western North Pacific typhoon activity from an observational perspective. *J. Clim.*, **17**, 4590–4602.

Chase, T.N., J.A. Knaff, R.A. Pielke, and E. Kalnay, 2003: Changes in global monsoon circulations since 1950. *Natural Hazards*, **29**, 229–254.

Chen, J., B.E. Carlson, and A.D. Del Genio, 2002: Evidence for strengthening of the tropical general circulation in the 1990s. *Science*, **295**, 838–841.

Christidis, N., et al., 2005: Detection of changes in temperature extremes during the second half of the 20th century. *Geophys. Res. Lett.*, **32**, L20716, doi:10.1029/2005GL023885.

Christy, J.R., R.W. Spencer, and W.D. Braswell, 2000: MSU tropospheric temperatures: Dataset construction and radiosonde comparison. *J. Atmos. Ocean. Technol.*, **17**, 1153–1170.

Chuang, C.C., et al., 2002: Cloud susceptibility and the first aerosol indirect forcing: Sensitivity to black carbon and aerosol concentrations. *J. Geophys. Res.*, **107**(D21), 4564, doi:10.1029/2000JD000215.

Church, J.A., N.J. White, and J.M. Arblaster, 2005: Volcanic eruptions: their impact on sea level and oceanic heat content. *Nature*, **438**, 74–77.

Clement, A.C., R. Seager, and M.A. Cane, 2000: Suppression of El Nino during the mid-Holocene by changes in the Earth's orbit. *Paleoceanography*, **15**(6), 731–737.

Clement, A.C., A. Hall, and A.J. Broccoli, 2004: The importance of precessional signals in the tropical climate. *Clim. Dyn.*, **22**, 327–341.

CLIMAP (Climate: Long-range Investigation, Mapping and Prediction), 1981: *Seasonal Reconstructions of the Earth's Surface at the Last Glacial Maximum.* Map Series Technical Report MC-36, Geological Society of America, Boulder, CO.

Cobb, K.M., C.D. Charles, H. Cheng, and R.L. Edwards, 2003: El Nino/Southern Oscillation and tropical Pacific climate during the last millennium. *Nature*, **424**(6946), 271–276.

Collins, M., 2000a: The El-Nino Southern Oscillation in the second Hadley Centre coupled model and its response to greenhouse warming. *J. Clim.*, **13**, 1299–1312.

Collins, M., 2000b: Understanding uncertainties in the response of ENSO to greenhouse warming. *Geophys. Res. Lett.*, **27**, 3509–3513.

Cook, E.R., et al., 2004: Long-term aridity changes in the western United States. *Science*, **306**(5698), 1015–1018.

Coughlin, K., and K.K. Tung, 2004: Eleven-year solar cycle signal throughout the lower atmosphere. *J. Geophys. Res.*, **109**, D21105, doi:10.1029/2004JD004873.

Crooks, S., 2004: *Solar Influence On Climate.* PhD Thesis, University of Oxford.

Crooks, S.A., and L.J. Gray, 2005: Characterization of the 11-year solar signal using a multiple regression analysis of the ERA-40 dataset. *J. Clim.*, **18**(7), 996–1015.

Crowley, T.J., 2000: Causes of climate change over the past 1000 years. *Science*, **289**(5477), 270–277.

Crowley, T.J., et al., 2003: Modeling ocean heat content changes during the last millennium. *Geophys. Res. Lett.*, **30**(18), 1932.

Cubasch, U., et al., 1997: Simulation of the influence of solar radiation variations on the global climate with an ocean-atmosphere general circulation model. *Clim. Dyn.*, **13**(11), 757–767.

Cubasch, U., et al., 2001: Projections of future climate change. In: *Climate Change 2001: The Scientific Basis. Contribution of Working Group I to the Third Assessment Report of the Intergovernmental Panel on Climate Change* [Houghton, J.T., et al. (eds.)]. Cambridge University Press, Cambridge, United Kingdom and New York, NY, USA, pp. 99–181.

Curry, R., B. Dickson, and I. Yashayaev, 2003: A change in the freshwater balance of the Atlantic Ocean over the past four decades. *Nature*, **426**, 826–829.

Dai, A., K.E. Trenberth, and T.R. Karl, 1999: Effects of clouds, soil, moisture, precipitation and water vapour on diurnal temperature range. *J. Clim.*, **12**, 2451–2473.

Dai, A., et al., 2004: The recent Sahel drought is real. *Int. J. Climatol.*, **24**, 1323–1331.

D'Arrigo, R., et al., 2005: On the variability of ENSO over the past six centuries. *Geophys. Res. Lett.*, **32**(3), L03711, doi:10.1029/2004GL022055.

Delworth, T.L., and T.R. Knutson, 2000: Simulation of early 20th century global warming. *Science*, **287**, 2246–2250.

Delworth, T.L., and M.E. Mann, 2000: Observed and simulated multidecadal variability in the Northern Hemisphere. *Clim. Dyn.*, **16**(9), 661–676.

Delworth, T.L., V. Ramaswamy, and G.L. Stenchikov, 2005: The impact of aerosols on simulated ocean temperature and heat content in the 20th century. *Geophys. Res. Lett.*, **32**, L24709, doi:10.1029/2005GL024457.

Delworth, T., et al., 2002: Review of simulations of climate variability and change with the GFDL R30 coupled climate model. *Clim. Dyn.*, **19**, 555–574.

Dickson, R.R., et al., 2002: Rapid freshening of the deep North Atlantic Ocean over the past four decades. *Nature*, **416**, 832–837.

Douglass, D.H., and B.D. Clader, 2002: Climate sensitivity of the Earth to solar irradiance. *Geophys. Res. Lett.*, **29**(16), 1786.

Douglass, D.H., and R.S. Knox, 2005: Climate forcing by volcanic eruption of Mount Pinatubo. *Geophys. Res. Lett.*, **32**, L05710, doi:10.1029/2004GL022119.

Douglass, D.H., B.D. Pearson, and S.F. Singer, 2004: Altitude dependence of atmospheric temperature trends: Climate models versus observation. *Geophys. Res. Lett.*, **31**, doi:10.1029/2004GL020103.

Douville, H., 2006: Detection-attribution of global warming at the regional scale: How to deal with precipitation variability. *Geophys. Res. Lett.*, **33**, L02701, doi:10.1029/2005GL024967.

Douville, H., et al., 2002: Sensitivity of the hydrological cycle to increasing amounts of greenhouse gases and aerosols. *Clim. Dyn.*, **20**, 45–68.

Dumas, J.A., G.M. Flato, and A.J. Weaver, 2003: The impact of varying atmospheric forcing on the thickness of Arctic multi-year sea ice. *Geophys. Res. Lett.*, **30**, 1918.

Dyurgerov, M.B., and M.F. Meier, 2005: *Glaciers and the Changing Earth System: A 2004 Snapshot.* Institute of Arctic and Alpine Research, University of Colorado, Boulder, CO, 117 pp.

Easterling, D.R., et al., 2000: Climate extremes: Observations, modeling and impacts. *Science*, **289**, 2068–2074.

Egorova, T., et al., 2004: Chemical and dynamical response to the 11-year variability of the solar irradiance simulated with a chemistry-climate model. *Geophys. Res. Lett.*, **31**, L06119, doi:10.1029/2003GL019294.

Elsner, J.B., X. Niu, and T.H. Jagger, 2004: Detecting shifts in hurricane rates using a Markov chain Monte Carlo approach. *J. Clim.*, **17**, 2652–2666.

Elsner, J.B., A.A. Tsonis, and T.H. Jagger, 2006: High-frequency variability in hurricane power dissipation and its relationship to global temperature. *Bull. Am. Meteorol. Soc.*, **87**, 763–768.

Emanuel, K., 2005: Increasing destructiveness of tropical cyclones over the past 30 years. *Nature*, **436**, 686–688.

Emori, S., and S.J. Brown, 2005: Dynamic and thermodynamic changes in mean and extreme precipitation under changed climate. *Geophys. Res. Lett.*, **32**, L17706, doi:10.1029/2005GL023272.

Esper, J., E.R. Cook, and F.H. Schweingruber, 2002: Low-frequency signals in long tree-ring chronologies for reconstructing past temperature variability. *Science*, **295**(5563), 2250–2253.

Fichefet, T., B. Tartinville, and H. Goosse, 2003: Antarctic sea ice variability during 1958-1999: A simulation with a global ice-ocean model. *J. Geophys. Res.*, **108**(C3), 3102–3113.

Folland, C.K., T. N. Palmer, and D. E. Parker, 1986: Sahel rainfall and worldwide sea temperatures 1901-85. *Nature*, **320**, 602–607.

Folland, C.K., et al., 2001: Observed variability and change. In: *Climate Change 2001: The Scientific Basis. Contribution of Working Group I to the Third Assessment Report of the Intergovernmental Panel on Climate Change* [Houghton, J.T., et al. (eds.)]. Cambridge University Press, Cambridge, United Kingdom and New York, NY, USA, pp. 881pp.

Fomby, T.B., and T.J. Vogelsang, 2002: The application of size-robust trend statistics to global-warming temperature series. *J. Clim.*, **15**, 117–123.

Forest, C.E., M.R. Allen, A.P. Sokolov, and P.H. Stone, 2001: Constraining climate model properties using optimal fingerprint detection methods. *Clim. Dyn.*, **18**, 277–295.

Forest, C.E., et al., 2002: Quantifying uncertainties in climate system properties with the use of recent observations. *Science*, **295**, 113.

Forest, D.J., P.H. Stone, and A.P. Sokolov, 2006: Estimated PDFs of climate system properties including natural and anthropogenic forcings. *Geophys. Res. Lett.*, **33**, L01705, doi:10.1029/2005GL023977.

Forster, P.M.D.F., and J.M. Gregory, 2006: The climate sensitivity and its components diagnosed from Earth radiation budget data. *J. Clim.*, **19**, 39–52.

Foukal, P., G. North, and T. Wigley, 2004: A stellar view on solar variations and climate. *Science*, **306**, 68–69.

Foukal, P., C. Froehlich, H. Sruit, and T.M.L. Wigley, 2006: Variations in solar luminosity and their effect on Earth's climate. *Nature*, **443**, 161–166, doi:10.1038/nature05072.

Frame, D.J., et al., 2005: Constraining climate forecasts: The role of prior assumptions. *Geophys. Res. Lett.*, **32**, L09702, doi:10.1029/2004GL022241.

Free, M., and J.K. Angell, 2002: Effect of volcanoes on the vertical temperature profile in radiosonde data. *J. Geophys. Res.*, **107**, doi:10.1029/2001JD001128.

Free, M., et al., 2005: Radiosonde Atmospheric Temperature Products for Assessing Climate (RATPAC): A new dataset of large-area anomaly time series. *J. Geophys. Res.*, **110**, D22101, doi:10.1029/2005JD006169.

Frich, P., et al., 2002: Observed coherent changes in climatic extremes during the second half of the twentieth century. *Clim. Res.*, **19**, 193–212.

Fyfe, J.C., G.J. Boer, and G.M. Flato, 1999: The Arctic and Antarctic Oscillations and their projected changes under global warming. *Geophys. Res. Lett.*, **26**, 1601–1604.

Ganopolski, A., et al., 1998: The influence of vegetation-atmosphere-ocean interaction on climate during the mid-Holocene. *Science*, **280**, 1916–1919.

Gedney, N., et al., 2006: Detection of a direct carbon dioxide effect in continental river runoff records. *Nature*, **439**, 835–838.

Gent, P.R., and G. Danabasoglu, 2004: Heat uptake and the thermohaline circulation in the Coummunity Climate System Model, Version 2. *J. Clim.*, **17**, 4058–4069.

Gerber, S., et al., 2003: Constraining temperature variations over the last millennium by comparing simulated and observed atmospheric CO_2. *Clim. Dyn.*, **20**(2–3), 281–299.

Giannini, A., R. Saravanan, and P. Chang, 2003: Oceanic forcing of Sahel rainfall on interannual to interdecadal time scales. *Science*, **302**, 1027–1030.

Gibson, J.K., et al., 1997: *ERA Description*. ECMWF Reanalysis Project Report Series Vol. 1. European Centre for Medium-Range Weather Forecasts, Reading, UK, 66 pp.

Gilgen, H., M. Wild, and A. Ohmura, 1998: Means and trends of shortwave irradiance at the surface estimated from global energy balance archive data. *J. Clim.*, **11**, 2042–2061.

Gillett, N.P., 2005: Northern Hemisphere circulation. *Nature*, **437**, 496.

Gillett, N.P., and D.W.J. Thompson, 2003: Simulation of recent Southern Hemisphere climate change. *Science*, **302**, 273–275.

Gillett, N.P., H.F. Graf, and T.J. Osborn, 2003a: Climate change and the North Atlantic Oscillation. In: *The North Atlantic Oscillation: Climate Significance and Environmental Impact* [Hurrell, Y.K.J., G. Ottersen, and M. Visbeck (eds.)]. Geophysical Monograph Vol. 134, American Geophysical Union, Washington, DC, pp. 193-209.

Gillett, N.P., R.J. Allan, and T.J. Ansell, 2005: Detection of external influence on sea level pressure with a multi-model ensemble. *Geophys. Res. Lett.*, **32**(19), L19714, doi:10.1029/2005GL023640.

Gillett, N.P., G.C. Hegerl, M.R. Allen, and P.A. Stott, 2000: Implications of changes in the Northern Hemispheric circulation for the detection of anthropogenic climate change. *Geophys. Res. Lett.*, **27**, 993–996.

Gillett, N.P., F.W. Zwiers, A.J. Weaver, and P.A. Stott, 2003b: Detection of human influence on sea level pressure. *Nature*, **422**, 292–294.

Gillett, N.P., A.J. Weaver, F.W. Zwiers, and M.D. Flannigan, 2004a: Detecting the effect of climate change on Canadian forest fires. *Geophys. Res. Lett.*, **31**(18), L18211, doi:10.1029/2004GL020876.

Gillett, N.P., A.J. Weaver, F.W. Zwiers, and M.F. Wehner, 2004b: Detection of volcanic influence on global precipitation. *Geophys. Res. Lett.*, **31**(12), L12217, doi:10.1029/2004GL020044.

Gillett, N.P., M.F. Wehner, S.F.B. Tett, and A.J. Weaver, 2004c: Testing the linearity of the response to combined greenhouse gas and sulfate aerosol forcing. *Geophys. Res. Lett.*, **31**, L14201, doi:10.1029/2004GL020111.

Gillett, N.P., et al., 2002a: Reconciling two approaches to the detection of anthropogenic influence on climate. *J. Clim.*, **15**, 326–329.

Gillett, N.P., et al., 2002b: How linear is the Arctic Oscillation response to greenhouse gases? *J. Geophys. Res.*, **107**, doi: 10.1029/2001JD000589.

Gillett, N.P., et al., 2002c: Detecting anthropogenic influence with a multi-model ensemble. *Geophys. Res. Lett.*, **29**, doi:10.1029/2002GL015836.

Gleckler, P.J., et al., 2006: Krakatoa's signature persists in the ocean. *Nature*, **439**, 675.

Gleisner, H., and P. Thejll, 2003: Patterns of tropospheric response to solar variability. *Geophys. Res. Lett.*, **30**, 44–47.

Goeberle, C., and R. Gerdes, 2003: Mechanisms determining the variability of Arctic sea ice conditions and export. *J. Clim.*, **16**, 2843–2858.

Goldewijk, K.K., 2001: Estimating global land use change over the past 300 years: The HYDE Database. *Global Biogeochem. Cycles*, **15**(2), 417–433.

Goldstein, M., and J. Rougier, 2004: Probabilistic formulations for transferring inferences from mathematical models to physical systems. *SIAM J. Sci. Computing*, **26**(2), 467–487.

Gonzalez-Rouco, F., H. von Storch, and E. Zorita, 2003: Deep soil temperature as proxy for surface air-temperature in a coupled model simulation of the last thousand years. *Geophys. Res. Lett.*, **30**(21), 2116, doi:10.1029/2003GL018264.

Goosse, H., and H. Renssen, 2001: A two-phase response of the Southern Ocean to an increase in greenhouse gas concentrations. *Geophys. Res. Lett.*, **28**(18), 3469–3472.

Goosse, H., and H. Renssen, 2004: Exciting natural modes of variability by solar and volcanic forcing: idealized and realistic experiments. *Clim. Dyn.*, **23**(2), 153–163.

Goosse, H., et al., 2004: A late medieval warm period in the Southern Ocean as a delayed response to external forcing? *Geophys. Res. Lett.*, **31**(6), L06203, doi:10.1029/2003GL19140.

Goosse, H., et al., 2005: Modelling the climate of the last millennium: What causes the differences between simulations? *Geophys. Res. Lett.*, **32**(6), L06710, doi:10.1029/2005GL022368.

Gray, L.J., R.G. Harrison, and J.D. Haigh, 2005: *The Influence of Solar Changes on the Earth's Climate*. Hadley Centre Technical Note 62, The UK Met Office.

Greene, A.M., 2005: A time constant for hemispheric glacier mass balance. *J. Glaciol.*, **51**(174), 353–362.

Gregory, J.M., and P. Huybrechts, 2006: Ice-sheet contributions to future sea-level change. *Philos. Trans. R. Soc. London Ser. A*, **364**, 1709–1731.

Gregory, J.M., J.A. Lowe, and S.F.B. Tett, 2006: Simulated global-mean sea-level changes over the last half-millennium. *J. Clim.*, **19**, 4576–4591.

Gregory, J.M., et al., 2002a: An observationally based estimate of the climate sensitivity. *J. Clim.*, **15**(22), 3117–3121.

Gregory, J.M., et al., 2002b: Recent and future changes in Arctic sea ice simulated by the HadCM3 AOGCM. *Geophys. Res. Lett.*, **29**, 2175.

Gregory, J.M., et al., 2004: Simulated and observed decadal variability in ocean heat content. *Geophys. Res. Lett.*, **31**, L15312.

Groisman, P.Y., et al., 1999: Changes in the probability of heavy precipitation: Important indicators of climatic change. *Clim. Change*, **42**, 243–283.

Groisman, P.Y., et al., 2005: Trends in intense precipitation in the climate record. *J. Clim.*, **18**, 1326–1350.

Haarsma, R.J., F. Selten, N. Weber, and M. Kliphuis, 2005: Sahel rainfall variability and response to greenhouse warming. *Geophys. Res. Lett.*, **32**, L17702, doi:10.1029/2005GL023232.

Haigh, J.D., 2003: The effects of solar variability on the Earth's climate. *Philos. Trans. R. Soc. London Ser. A*, **361**, 95–111.

Hansen, J.E., M. Sato, and R. Ruedy, 1995: Long-term changes of the diurnal temperature cycle: implications about mechanisms of global climate change. *Atmos. Res.*, **37**, 175–209.

Hansen, J.E., M. Sato, and R. Ruedy, 1997: Radiative forcing and climate response. *J. Geophys. Res.*, **102**, 6831–6864.

Hansen, J., et al., 1984: Climate sensitivity: Analysis of feedback mechanisms. In: *Climate Processes and Climate Sensitivity* [Hansen, J.E., and T. Takahashi (eds.)]. Geophysical Monographs Vol. 29, American Geophysical Union, Washington, DC, pp. 130–163.

Hansen, J., et al., 2002: Climate forcings in Goddard Institute for Space Studies SI2000 simulations. *J. Geophys. Res.*, **107**(D18), 4347.

Hansen, J., et al., 2005: Earth's energy imbalance: Confirmation and implications. *Science*, **308**, 1431–1435.

Harrison, S., and C. Prentice, 2003: Climate and CO_2 controls on global vegetation distribution at the last glacial maximum: analysis based on palaeovegetation data, biome modelling and palaeoclimate simulations. *Global Change Biol.*, **9**, 983–1004.

Harrison, S., P. Braconnot, C. Hewitt, and R.J. Stouffer, 2002: Fourth international workshop of The Palaeoclimate Modelling Intercomparison Project (PMIP): launching PMIP Phase II. *Eos*, **83**, 447.

Harvey, L.D.D., 2004: Characterizing the annual-mean climatic effect of anthropogenic CO_2 and aerosol emissions in eight coupled atmosphere-ocean GCMs. *Clim. Dyn.*, **23**, 569–599.

Harvey, L.D.D., and R.K. Kaufmann, 2002: Simultaneously constraining climate sensitivity and aerosol radiative forcing. *J. Clim.*, **15** (20), 2837–2861.

Hasselmann, K., 1976: Stochastic climate models. Part 1. Theory. *Tellus*, **28**, 473–485.

Hasselmann, K., 1979: On the signal-to-noise problem in atmospheric response studies. In: *Meteorology of Tropical Oceans* [Shaw, D.B. (ed.)]. Royal Meteorological Society, Bracknell, UK, pp. 251–259.

Hasselmann, K., 1997: Multi-pattern fingerprint method for detection and attribution of climate change. *Climate Dyn.*, **13**, 601–612.

Hasselmann, K., 1998: Conventional and Bayesian approach to climate-change detection and attribution. *Q. J. R. Meteorol. Soc.*, **124**, 2541–2565.

Hegerl, G.C., and M.R. Allen, 2002: Origins of model-data discrepancies in optimal fingerprinting. *J. Clim.*, **15**, 1348–1356.

Hegerl, G.C., and J.M. Wallace, 2002: Influence of patterns of climate variability on the difference between satellite and surface temperature trends. *J. Clim.*, **15**, 2412–2428.

Hegerl, G.C., P.D. Jones, and T.P. Barnett, 2001: Effect of observational sampling error on the detection and attribution of anthropogenic climate change. *J. Clim.*, **14**, 198–207.

Hegerl, G.C., F.W. Zwiers, V.V. Kharin, and P.A. Stott, 2004: Detectability of anthropogenic changes in temperature and precipitation extremes. *J. Clim.*, **17**, 3683–3700.

Hegerl, G.C., T. Crowley, W.T. Hyde, and D. Frame, 2006a: Constraints on climate sensitivity from temperature reconstructions of the past seven centuries. *Nature*, **440**, doi:10.1038/nature04679.

Hegerl, G.C., et al., 1996: Detecting greenhouse gas induced climate change with an optimal fingerprint method. *J. Clim.*, **9**, 2281–2306.

Hegerl, G.C., et al., 1997: Multi-fingerprint detection and attribution of greenhouse-gas and aerosol-forced climate change. *Clim. Dyn.*, **13**, 613–634.

Hegerl, G.C., et al., 2000: Detection and attribution of climate change: Sensitivity of results to climate model differences. *Clim. Dyn.*, **16**, 737–754.

Hegerl, G.C., et al., 2003: Detection of volcanic, solar and greenhouse gas signals in paleo-reconstructions of Northern Hemispheric temperature. *Geophys. Res. Lett.*, **30**(5), 1242.

Hegerl, G.C., et al., 2006b: Climate change detection and attribution: beyond mean temperature signals. *J. Clim.*, **19**, 5058–5077.

Hegerl, G.C., et al., 2007: Detection of human influence on a new 1500yr climate reconstruction. *J. Clim.*, **20**, 650-666.

Held, I.M., and B.J. Soden, 2006: Robust responses of the hydrological cycle to global warming. *J. Clim.*, **19**, 5686–5699.

Held, I.M., et al., 2005: Simulation of Sahel drought in the 20th and 21st centuries. *Proc. Natl. Acad. Sci. U.S.A.*, **102**(50), 17891–17896.

Highwood, E.J., B.J. Hoskins, and P. Berrisford, 2000: Properties of the Arctic tropopause. *Q. J. R. Meteorol. Soc.*, **126**, 1515–1532.

Hoerling, M.P., J.W. Hurrell, J. Eischeid, and A. Phillips, 2006: Detection and attribution of twentieth-century northern and southern African rainfall change. *J. Clim.*, **19**, 3989–4008.

Hoerling, M.P., et al., 2005: Twentieth century North Atlantic climate change. Part II: Understanding the effect of Indian Ocean warming. *Clim. Dyn.*, **23**, 391–405.

Hoffert, M.I., and C. Covey, 1992: Deriving global climate sensitivity from paleoclimate reconstructions. *Nature*, **360**, 573–576.

Holland, M.M., and M.N. Raphael, 2006: Twentieth century simulation of the southern hemisphere climate in coupled models. Part II: sea ice conditions and variability. *Clim. Dyn.*, **26**, 229–245.

Holloway, G., and T. Sou, 2002: Has Arctic sea ice rapidly thinned? *J. Clim.*, **15**, 1691–1701.

Hoyt, D.V., and K.H. Schatten, 1993: A discussion of plausible solar irradiance variations, 1700-1992. *J. Geophys. Res.*, **98**, 18895–18906.

Huang, S.P., H.N. Pollack, and P.Y. Shen, 2000: Temperature trends ever the past five centuries reconstructed from borehole temperatures. *Nature*, **403**(6771), 756–758.

Hulme, M., T.J. Osborn, and T.C. Johns, 1998: Precipitation sensitivity to global warming: Comparison of observations with HadCM2 simulations. *Geophys. Res. Lett.*, **25**, 3379–3382.

Huntingford, C., P.A. Stott, M.R. Allen, and F.H. Lambert, 2006: Incorporating model uncertainty into attribution of observed temperature change. *Geophys. Res. Lett.*, **33**, L05710, doi:10.1029/2005GL024831.

Hurrell, J.W., 1996: Influence of variations in extratropical wintertime teleconnections on Northern Hemisphere temperature. *Geophys. Res. Lett.*, **23**, 665–668.

Hurrell, J.W., M.P. Hoerling, A.S. Phillips, and T. Xu, 2005: Twentieth century North Atlantic climate change. Part I: Assessing determinism. *Clim. Dyn.*, **23**, 371–389.

IDAG (International Ad Hoc Detection and Attribution Group), 2005: Detecting and attributing external influences on the climate system: A review of recent advances. *J. Clim.*, **18**, 1291–1314.

IOCI, 2002: *Climate Variability And Change In South West Western Australia, September 2002*. Indian Ocean Climate Initiative, Perth, Australia, 34 pp.

IOCI, 2005: *Indian Ocean Climate Initiative Stage 2: Report of Phase 1 Activity*. Indian Ocean Climate Initiative, Perth, Australia, 42 pp.

IPCC, 1990: *Climate Change: The Intergovernmental Panel on Climate Change Scientific Assessment* [Houghton, J.T., G.J. Jenkins, and J.J. Ephraums (eds.)]. Cambridge University Press, Cambridge, United Kingdom and New York, NY, USA, 365 pp.

IPCC, 1996: *Climate Change 1995: The Science of Climate Change. Contribution of the Working Group I to the Second Assessment Report of the Intergovernmental Panel on Climate Change* [Houghton, J.T., et al. (eds.)]. Cambridge University Press, Cambridge, United Kingdom and New York, NY, USA, 572 pp.

IPCC, 2001: *Climate Change 2001: The Scientific Basis. Contribution of Working Group I to the Third Assessment Report of the Intergovernmental Panel on Climate Change* [Houghton, J.T., et al. (eds.)]. Cambridge University Press, Cambridge, United Kingdom and New York, NY, USA, 881 pp.

Ishii, M., M. Kimoto, K. Sakamoto, and S.-I. Iwasaki, 2006: Steric sea level changes estimated from historical ocean subsurface temperature and salinity analyses. *J. Oceanogr.*, **62**, 155–170.

Ito, A., and J.E. Penner, 2005: Historical emissions of carbonaceous aerosols from biomass and fossil fuel burning for the period 1870-2000. *Global Biogeochem. Cycles*, **19**(2), GB2028, doi:10.1029/2004GB002374.

Johannssen, O.M., et al., 2004: Arctic climate change: observed and modeled temperature and sea-ice variability. *Tellus*, **56A**, 328–341.

Jones, G.S., S.F.B. Tett, and P.A. Stott, 2003: Causes of atmospheric temperature change 1960-2000: A combined attribution analysis. *Geophys. Res. Lett.*, **30**, 1228.

Jones, G.S., et al., 2005: Sensitivity of global scale attribution results to inclusion of climatic response to black carbon. *Geophys. Res. Lett.*, **32**, L14701, doi:10.1029/2005GL023370.

Jones, J.M., and M. Widmann, 2004: Early peak in Antarctic Oscillation index. *Nature*, **432**, 290–291.

Jones, P.D., and M.E. Mann, 2004: Climate over past millennia. *Rev. Geophys.*, **42**(2), RG2002, doi:10.1029/2003RG000143.

Jones, P.D., T.J. Osborn, and K.R. Briffa, 2001: The evolution of climate over the last millennium. *Science*, **292**(5517), 662–667.

Joos, F., et al., 2004: Transient simulations of Holocene atmospheric carbon dioxide and terrestrial carbon since the Last Glacial Maximum. *Global Biogeochem. Cycles*, **18**, 1–18.

Joussaume, S., and K.E. Taylor, 1995: Status of the Paleoclimate Modeling Intercomparison Project. In: *Proceedings of the First International AMIP Scientific Conference, WCRP-92, Monterey, USA*. WMO/TD-No. 732, Geneva, Switzerland, pp. 425–430.

Kalnay, E., et al., 1996: The NCEP/NCAR Reanalysis Project. *Bull. Am. Meteorol. Soc.*, **77**, 437–471.

Kaplan, J.O., I.C. Prentice, W. Knorr, and P.J. Valdes, 2002: Modeling the dynamics of terrestrial carbon storage since the Last Glacial Maximum. *Geophys. Res. Lett.*, **29**(22), 2074.

Karl, T.R., and R.W. Knight, 1998: Secular trends of precipitation amount, frequency, and intensity in the USA. *Bull. Am. Meteorol. Soc.*, **79**, 231–241.

Karl, T.R., and K.E. Trenberth, 2003: Modern global climate change. *Science*, **302**, 1719–1723.

Karl, T.R., S.J. Hassol, C.D. Miller, and W.L. Murray (eds.), 2006: *Temperature Trends in the Lower Atmosphere: Steps for Understanding and Reconciling Differences*. A Report by the Climate Change Science Program and Subcommittee on Global Change Research, Washington, DC, 180pp, http://www.climatescience.gov/Library/sap/sap1-1/finalreport/default.htm.

Karoly, D.J., 2003: Ozone and climate change. *Science*, **302**, 236–237.

Karoly, D.J., and K. Braganza, 2001: Identifying global climate change using simple indices. *Geophys. Res. Lett.*, **28**, 2205–2208.

Karoly, D.J., and K. Braganza, 2005a: Attribution of recent temperature changes in the Australian region. *J. Clim.*, **18**, 457–464.

Karoly, D.J., and K. Braganza, 2005b: A new approach to detection of anthropogenic temperature changes in the Australian region. *Meteorol. Atmos. Phys.*, **89**, 57–67.

Karoly, D.J., and Q. Wu, 2005: Detection of regional surface temperature trends. *J. Clim.*, **18**, 4337–4343.

Karoly, D.J., et al., 2003: Detection of a human influence on North American climate. *Science*, **302**, 1200–1203.

Kass, R.E., and A.E. Raftery, 1995: Bayes Factors. *J. Am. Stat. Assoc.*, **90**, 773–795.

Katz, R.W., 1999: Extreme value theory for precipitation: Sensitivity analysis for climate change. *Adv. Water Resour.*, **23**, 133–139.

Kaufmann, R.K., and D.L. Stern, 2002: Cointegration analysis of hemispheric temperature relations. *J. Geophys. Res.*, **107**, 4012.

Kennedy, M.C., and A. O'Hagan, 2001: Bayesian calibration of computer models. *J. Roy. Stat. Soc. Ser. B*, **63**(3), 425–464.

Kettleborough, J.A., B.B.B. Booth, P.A. Stott, and M.R. Allen, 2007: Estimates of uncertainty in predictions of global mean surface temperature. *J. Clim.*, **20**, 843-855.

Kiktev, D., D. Sexton, L. Alexander, and C. Folland, 2003: Comparison of modelled and observed trends in indices of daily climate extremes. *J. Clim.*, **16**, 3560–3571.

Kim, S.J., G.M. Flato, G.J. Boer, and N.A. McFarlane, 2002: A coupled climate model simulation of the Last Glacial Maximum, part 1: Transient multi-decadal response. *Clim. Dyn.*, **19**(5–6), 515–537.

Kirchner, I., et al., 1999: Climate model simulation of winter warming and summer cooling following the 1991 Mount Pinatubo volcanic eruption. *J. Geophys. Res.*, **104**, 19039–19055.

Kistler, R., et al., 2001: The NCEP-NCAR 50-year reanalysis: Monthly means CD-ROM and documentation. *Bull. Am. Meteorol. Soc.*, **82**, 247–267.

Kitoh, A., and S. Murakami, 2002: Tropical Pacific climate at the mid-Holocene and the Last Glacial Maximum simulated by a coupled ocean-atmosphere general circulation model. *Paleoceanography*, **17**(3), 1047, doi:10.1029/2001PA000724.

Klein Tank, A.M.G., and G.P. Können, 2003: Trends in indices of daily temperature and precipitation extremes in Europe, 1946-99. *J. Clim.*, **16**, 3665–3680.

Klein Tank, A.M.G., G.P. Können, and F.M. Selten, 2005: Signals of anthropogenic influence on European warming as seen in the trend patterns of daily temperature variance. *Int. J. Climatol.*, **25**, 1–16.

Knight, J.R., et al., 2005: A signature of persistent natural thermohaline circulation cycles in observed climate. *Geophys. Res. Lett.*, **32**, L20708, doi:10.1029/2005GL024233.

Knutson, T.R., S. Manabe, and D. Gu, 1997: Simulated ENSO in a global coupled ocean-atmosphere model: Multidecadal amplitude modulation and CO_2 sensitivity. *J. Clim.*, **10**(1), 138–161.

Knutson, T.R., T.L. Delworth, K.W. Dixon, and R.J. Stouffer, 1999: Model assessment of regional surface temperature trends (1949-1997). *J. Geophys. Res.*, **104**, 30981–30996.

Knutson, T.R., et al., 2006: Assessment of twentieth-century regional surface temperature trends using the GFDL CM2 coupled models. *J. Clim.*, **19**, 1624–1651.

Knutti, R., T.F. Stocker, F. Joos, and G.-K. Plattner, 2002: Constraints on radiative forcing and future climate change from observations and climate model ensembles. *Nature*, **416**, 719–723.

Knutti, R., T.F. Stocker, F. Joos, and G.-K. Plattner, 2003: Probabilistic climate change projections using neural networks. *Clim. Dyn.*, **21**, 257–272.

Kristjansson, J.E., 2002: Studies of the aerosol indirect effect from sulfate and black carbon aerosols. *J. Geophys. Res.*, **107**, doi:10.1029/2001JD000887.

Kucera, M., et al., 2005: Reconstruction of sea-surface temperatures from assemblages of planktonic foraminifera: multi-technique approach based on geographically constrained calibration data sets and its application to glacial Atlantic and Pacific Oceans. *Quat. Sci. Rev.*, **24**(7–9), 951–998.

Kumar, A., F. Yang, L. Goddard, and S. Schubert, 2004: Differing trends in the tropical surface temperatures and precipitation over land and oceans. *J. Clim.*, **17**, 653–664.

Kunkel, K.E., X.-Z. Liang, J. Zhu, and Y. Lin, 2006: Can CGCMS simulate the twentieth century "warming hole" in the central United States? *J. Clim.*, **19**, 4137–4153.

Kushner, P.J., I.M. Held, and T.L. Delworth, 2001: Southern Hemisphere atmospheric circulation response to global warming. *J. Clim.*, **14**, 2238–3349.

Labitzke, K., 2004: On the signal of the 11-year sunspot cycle in the stratosphere and its modulation by the quasi, biennial oscillation. *J. Atmos. Solar Terr. Phys.*, **66**, 1151–1157.

Lal, M., and S.K. Singh, 2001: Global warming and monsoon climate. *Mausam*, **52**, 245–262.

Lambert, F.H., P.A. Stott, M.R. Allen, and M.A. Palmer, 2004: Detection and attribution of changes in 20th century land precipitation. *Geophys. Res. Lett.*, **31**(10), L10203, doi:10.1029/2004GL019545.

Lambert, F.H., N.P. Gillett, D.A. Stone, and C. Huntingford, 2005: Attribution studies of observed land precipitation changes with nine coupled models. *Geophys. Res. Lett.*, **32**, L18704, doi:10.1029/2005GL023654.

Lambert, S.J., and J.C. Fyfe, 2006: Changes in winter cyclone frequencies and strengths simulated in enhanced greenhouse warming experiments: Results from the models participating in the IPCC diagnostic exercise. *Clim. Dyn.*, **26**, 713–728.

Landsea, C.W., 2005: Hurricanes and global warming. *Nature*, **438**, E11–E12.

Lean, J.L., J. Beer, and R. Bradley, 1995: Reconstruction of solar irradiance changes since 1610: Implications for climate change. *Geophys. Res. Lett.*, **22**, 3195.

Lean, J.L., Y.M. Wang, and N.R. Sheeley, 2002: The effect of increasing solar activity on the Sun's total and open magnetic flux during multiple cycles: Implications for solar forcing of climate. *Geophys. Res. Lett.*, **29**(24), 2224, doi:10.1029/2002GL015880.

Lee, T.C.K., F.W. Zwiers, X. Zhang, and M. Tsao, 2006: Evidence of decadal climate prediction skill resulting from changes in anthropogenic forcing. *J. Clim.*, **19**, 5305–5318.

Lee, T.C.K., et al., 2005: A Bayesian approach to climate change detection and attribution. *J. Clim.*, **18**, 2429–2440.

Leroy, S.S., 1998: Detecting climate signals: Some Bayesian aspects. *J. Clim.*, **11**, 640–651.

Levis, S., J.A. Foley, and D. Pollard, 1999: CO_2, climate, and vegetation feedbacks at the Last Glacial Maximum. *J. Geophys. Res.*, **104**(D24), 31191–31198.

Levis, S., G.B. Bonan, and C. Bonfils, 2004: Soil feedback drives the mid-Holocene North African monsoon northward in fully coupled CCSM2 simulations with a dynamic vegetation model. *Clim. Dyn.*, **23**(7–8), 791–802.

Levitus, S., J. Antonov, and T. Boyer, 2005: Warming of the world ocean, 1955–2003. *Geophys. Res. Lett.*, **32**, L02604, doi:10.1029/2004GL021592.

Levitus, S., J. Antonov, T.P. Boyer, and C. Stephens, 2000: Warming of the world ocean. *Science*, **287**, 2225–2229.

Levitus, S., et al., 2001: Anthropogenic warming of the Earth's climate system. *Science*, **292**, 267–270.

Liepert, B., 2002: Observed reductions of surface solar radiation at sites in the United States and worldwide from 1961 to 1990. *Geophys. Res. Lett.*, **29**, 1421.

Lindsay, R.W., and J. Zhang, 2005: The thinning of arctic sea ice, 1988-2003: Have we passed a tipping point? *J. Clim.*, **18**, 4879–4894.

Lindzen, R.S., and C. Giannitsis, 2002: Reconciling observations of global temperature change. *Geophys. Res. Lett.*, **29**, doi:10.1029/2001GL014074.

Liu, Z.Y., J. Kutzbach, and L.X. Wu, 2000: Modeling climate shift of El Nino variability in the Holocene. *Geophys. Res. Lett.*, **27**(15), 2265–2268.

Liu, Z.Y., et al., 2005: Atmospheric CO_2 forcing on glacial thermohaline circulation and climate. *Geophys. Res. Lett.*, **32**(2), L02706, doi:10.1029/2004GL021929.

Lohmann, U., and G. Lesins, 2002: Stronger constraints on the anthropogenic indirect aerosol effect. *Science*, **298**, 1012–1016.

Lohmann, U., and J. Feichter, 2005: Global indirect aerosol effects: A review. *Atmos. Chem. Phys.*, **5**, 715–737.

Lorius, C., et al., 1990: The ice-core record: climate sensitivity and future greenhouse warming. *Nature*, **347**, 139–145.

Lu, J., and T.L. Delworth, 2005: Oceanic forcing of the late 20th century Sahel drought. *Geophys. Res. Lett.*, **32**, L22706, doi:10.1029/2005GL023316.

Luterbacher, J., et al., 2002: Extending North Atlantic Oscillation reconstructions back to 1500. *Atmos. Sci. Lett.*, **2**(114–124).

Luterbacher, J., et al., 2004: European seasonal and annual temperature variability, trends, and extremes since 1500. *Science*, **303**(5663), 1499–1503.

MacDonald, G.M., and R.A. Case, 2005: Variations in the Pacific Decadal Oscillation over the past millennium. *Geophys. Res. Lett.*, **32**(8), L08703, doi:10.1029/2005GL022478.

Mann, M.E., and P.D. Jones, 2003: Global surface temperature over the past two millennia. *Geophys. Res. Lett.*, **30**, 1820.

Mann, M.E., and K.A. Emanuel, 2006: Atlantic hurricane trends linked to climate change. *Eos*, **87**, 233–241.

Mann, M.E., M.A. Cane, S.E. Zebiak, and A. Clement, 2005: Volcanic and solar forcing of the tropical Pacific over the past 1000 years. *J. Clim.*, **18**(3), 447–456.

Marshall, G.J., 2003: Trends in the Southern Annular Mode from observations and reanalyses. *J. Clim.*, **16**, 4134–4143.

Marshall, G.J., A. Orr, N.P.M. van Lipzig, and J.C. King, 2006: The impact of a changing Southern Hemisphere Annular Mode on Antarctic Peninsula summer temperatures. *J. Clim.*, **19**, 5388–5404.

Marshall, G.J., et al., 2004: Causes of exceptional atmospheric circulation changes in the Southern Hemisphere. *Geophys. Res. Lett.*, **31**, L14205, doi:10.1029/2004GL019952.

Masson-Delmotte, V., et al., 2006: Past and future polar amplification of climate change: climate model intercomparisons and ice-core constraints. *Clim. Dyn.*, **26**, 513–529.

Matthews, H.D., et al., 2004: Natural and anthropogenic climate change: incorporating historical land cover change, vegetation dynamics and the global carbon cycle. *Clim. Dyn.*, **22**(5), 461–479.

May, W., 2004: Potential future changes in the Indian summer monsoon due to greenhouse warming: analysis of mechanisms in a global time-slice experiment. *Clim. Dyn.*, **22**, 389–414.

Maynard, K., J.F. Royer, and F. Chauvin, 2002: Impact of greenhouse warming on the West African summer monsoon. *Clim. Dyn.*, **19**, 499–514.

McAvaney, B.J., et al., 2001: Model evaluation. In: *Climate Change 2001: The Scientific Basis. Contribution of Working Group I to the Third Assessment Report of the Intergovernmental Panel on Climate Change* [Houghton, J.T., et al. (eds.)]. Cambridge University Press, Cambridge, United Kingdom and New York, NY, USA, pp. 471–525.

Mears, C.A., and F.J. Wentz, 2005: The effect of diurnal correction on satellite-derived lower tropospheric temperature. *Science*, **309**, 1548–1551.

Mears, C.A., M.C. Schabel, and F.J. Wentz, 2003: A reanalysis of the MSU channel 2 tropospheric temperature record. *J. Clim.*, **16**, 3650–3664.

Meehl, G.A., J.M. Arblaster, and C. Tebaldi, 2005: Understanding future patterns of precipitation extremes in climate model simulations. *Geophys. Res. Lett.*, **32**, L18719, doi:10.1029/2005GL023680.

Meehl, G.A., et al., 2003: Solar and greenhouse gas forcing and climate response in the 20th century. *J. Clim.*, **16**, 426–444.

Meehl, G.A., et al., 2004: Combinations of natural and anthropogenic forcings in 20th century climate. *J. Clim.*, **17**, 3721–3727.

Meier, M.F., M.B. Dyurgerov, and G.J. McCabe, 2003: The health of glaciers: Recent changes in glacier regime. *Clim. Change*, **59**, 123–135.

Mendelssohn, R., S.J. Bograd, F.B. Schwing, and D.M. Palacios, 2005: Teaching old indices new tricks: A state-space analysis of El Niño related climate indices. *Geophys. Res. Lett.*, **32**, L07709, doi:10.1029/2005GL022350.

Menon, S., A.D. Del Genio, D. Koch, and G. Tselioudis, 2002a: GCM Simulations of the aerosol indirect effect: Sensitivity to cloud parameterization and aerosol burden. *J. Atmos. Sci.*, **59**, 692–713.

Menon, S., J.E. Hansen, L. Nazarenko, and Y. Luo, 2002b: Climate effects of black carbon aerosols in China and India. *Science*, **297**, 2250–2253.

Merryfield, W.J., 2006: Changes to ENSO under CO_2 doubling in a multimodel ensemble. *J. Clim.*, **19**, 4009–4027.

Miller, R.L., G.A. Schmidt, and D.T. Shindell, 2006: Forced variations in the annular modes in the 20th century IPCC AR4 simulations. *J. Geophys. Res.*, **111**, D18101, doi:10.1029/2005JD006323.

Milly, P.C.D., K.A. Dunne, and A.V. Vecchia, 2005: Global patterns of trends in streamflow and water availability in a changing climate. *Nature*, **438**, 347–350.

Min, S.-K., and A. Hense, 2006a: A Bayesian approach to climate model evaluation and multi-model averaging with an application to global mean surface temperatures from IPCC AR4 coupled climate models. *Geophys. Res. Lett.*, **33**, L08708, doi:10.1029/2006GL025779.

Min, S.-K., and A. Hense, 2006b: A Bayesian assessment of climate change using multi-model ensembles. Part I: Global mean surface temperature. *J. Clim.*, **19**, 3237–3256.

Min, S.-K., A. Hense, and W.-T. Kwon, 2005: Regional-scale climate change detection using a Bayesian decision method. *Geophys. Res. Lett.*, **32**, L03706, doi:10.1029/2004GL021028.

Min, S.-K., A. Hense, H. Paeth, and W.-T. Kwon, 2004: A Bayesian decision method for climate change signal analysis. *Meteorol. Z.*, **13**, 421–436.

Mitchell, J.F.B., C.A. Wilson, and W.M. Cunningham, 1987: On CO_2 climate sensitivity and model dependence of results. *Q. J. R. Meteorol. Soc.*, **113**, 293–322.

Mitchell, J.F.B., et al., 2001: Detection of climate change and attribution of causes. In: *Climate Change 2001: The Scientific Basis. Contribution of Working Group I to the Third Assessment Report of the Intergovernmental Panel on Climate Change* [Houghton, J.T., et al. (eds.)]. Cambridge University Press, Cambridge, United Kingdom and New York, NY, USA, pp. 695–738.

Mitchell, T.D., and P.D. Jones, 2005: An improved method of constructing a database of monthly climatological observations and associated high-resolution grids. *Int. J. Climatol.*, **25**, 693–712.

Moberg, A., et al., 2005: Highly variable Northern Hemisphere temperatures reconstructed from low- and high-resolution proxy data. *Nature*, **433**, 613–617.

Monnin, E., et al., 2001: Atmospheric CO_2 concentrations over the last glacial termination. *Science*, **291**(5501), 112–114.

Montoya, M., H. von Storch, and T.J. Crowley, 2000: Climate simulation for 125,000 years ago with a coupled ocean-atmosphere General Circulation Model. *J. Clim.*, **13**, 1057–1070.

Mote, P.W., A.F. Hamlet, M.P. Clark, and D.P. Lettenmaier, 2005: Declining mountain snowpack in western North America. *Bull. Am. Meteorol. Soc.*, **86**, 39–49.

Moy, C.M., G.O. Seltzer, D.T. Rodbell, and D.M. Anderson, 2002: Variability of El Nino/Southern Oscillation activity at millennial timescales during the Holocene epoch. *Nature*, **420**(6912), 162–165.

Murray, R.J., N.L. Bindoff, and C.J.C. Reason, 2007: Modelling decadal changes on the Indian Ocean Section I5 at 32°S. *J. Clim.*, accepted.

Nagashima, T., et al., 2006: The effect of carbonaceous aerosols on surface temperature in the mid twentieth century. *Geophys. Res. Lett.*, **33**, L04702, doi:10.1029/2005GL024887.

Neelin, J.D., et al., 2006: Tropical drying trends in global warming models and observations. *Proc. Natl. Acad. Sci. U.S.A.*, **103**, 6110–6115.

Nesme-Ribes, E., et al., 1993: Solar dynamics and its impact on solar irradiance and the terrestrial climate. *J. Geophys. Res.*, **98**, 18923–18935.

New, M.G., M. Hulme, and P.D. Jones, 2000: Representing twentieth-century space-time climate variability. Part II: development of 1901-96 monthly grids of terrestrial surface climate. *J. Clim.*, **13**, 2217–2238.

Nicholls, N., 2003: Continued anomalous warming in Australia. *Geophys. Res. Lett.*, **30**, doi:10.1029/2003GL017037.

Nicholls, N., 2005: Climate variability, climate change, and the Australian snow season. *Aust. Meteorol. Mag.*, **54**, 177–185.

Nicholls, N., P. Della-Marta, and D. Collins, 2005: 20th century changes in temperature and rainfall in New South Wales. *Aust. Meteorol. Mag.*, **53**, 263–268.

Nicholson, S.E., 2001: Climatic and environmental change in Africa during the last two centuries. *Clim. Res.*, **17**, 123–144.

North, G.R., and M. Stevens, 1998: Detecting climate signals in the surface temperature record. *J. Clim.*, **11**, 563–577.

North, G.R., K.-Y. Kim, S.S.P. Shen, and J.W. Hardin, 1995: Detection of forced climate signals. Part 1: Filter theory. *J. Climate*, **8**, 401–408.

Novakov, T., et al., 2003: Large historical changes of fossil-fuel black carbon aerosols. *Geophys. Res. Lett.*, **30**(6), 1324.

Nozawa, T., T. Nagashima, H. Shiogama, and S. Crooks, 2005: Detecting natural influence on surface air temperature in the early twentieth century. *Geophys. Res. Lett.*, **32**, L20719, doi:10.1029/2005GL023540.

Oerlemans, J., 2005: Extracting a climate signal from 169 glacier records. *Science*, **308**, 675–677.

O'Hagan, A., and J. Forster, 2004: *Kendall's Advanced Theory of Statistics. Volume 2b, Bayesian Inference*. Arnold, London, 480 pp.

Ohmura, A., 2004: Cryosphere during the twentieth century, the state of the planet. In: *The State of the Planet: Frontiers and Challenges in Geophysics* [Sparks, R.S.J., and C.J. Hawkesworth (eds.)]. International Union of Geodesy and Geophysics, Washington, DC, pp. 239–257.

Oman, L., et al., 2005: Climatic response to high latitude volcanic eruptions. *J. Geophys. Res.*, **110**, D13103, doi:10.1029/2004JD005487.

Osborn, T.J., 2004: Simulating the winter North Atlantic Oscillation: the roles of internal variability and greenhouse gas forcing. *Clim. Dyn.*, **22**, 605–623.

Osborn, T.J., and M. Hulme, 1997: Development of a relationship between station and grid-box rainday frequencies for climate model evaluation. *J. Clim.*, **10**, 1885–1908.

Osborn, T.J., and K.R. Briffa, 2006: The spatial extent of 20th-century warmth in the context of the past 1200 years. *Science*, **311**, 841–844.

Osborn, T.J., S. Raper, and K.R. Briffa, 2006: Simulated climate change during the last 1000 years: comparing the ECHO-G general circulation model with the MAGICC simple climate model. *Clim. Dyn.*, **27**, 185–197.

Osborn, T.J., et al., 1999: Evaluation of the North Atlantic Oscillation as simulated by a coupled climate model. *Clim. Dyn.*, **15**, 685–702.

Otto-Bliesner, B.L., 1999: El Nino La Nina and Sahel precipitation during the middle Holocene. *Geophys. Res. Lett.*, **26**(1), 87–90.

Otto-Bliesner, B.L., et al., 2003: Modeling El Nino and its tropical teleconnections during the last glacial-interglacial cycle. *Geophys. Res. Lett.*, **30**(23), 2198, doi:10.1029/2003GL018553.

Paeth, H., A. Hense, R. Glowienka-Hense, and R. Voss, 1999: The North Atlantic Oscillation as an indicator for greenhouse-gas induced regional climate change. *Clim. Dyn.*, **15**, 953–960.

Palmer, M.A., L.J. Gray, M.R. Allen, and W.A. Norton, 2004: Solar forcing of climate: model results. *Adv. Space Res.*, **34**, 343–348.

Palmer, T.N., 1999: Predicting uncertainty in forecasts of weather and climate. *Rep. Prog. Phys.*, **63**, 71–116.

Palmer, T.N., and J. Räisänen, 2002: Quantifying the risk of extreme seasonal precipitation events in a changing climate. *Nature*, **415**, 512–514.

Parker, D.E., L.V. Alexander, and J. Kennedy, 2004: Global and regional climate in 2003. *Weather*, **59**, 145–152.

Parker, D.E., et al., 1997: A new global gridded radiosonde temperature database and recent temperature trends. *Geophys. Res. Lett.*, **24**, 1499–1452.

Pasini, A., M. Lorè, and F. Ameli, 2006: Neural network modelling for the analysis of forcings/temperatures relationships at different scales in the climate system. *Ecol. Model.*, **191**, 58–67.

Peltier, W.R., 1994: Ice age paleotopography. *Science*, **265**, 195–201.

Peltier, W.R., 2004: Global glacial isostasy and the surface of the ice-age Earth: the ICE-5G(VM2) model and GRACE. *Annu. Rev. Earth Planet. Sci.*, **32**, 111–149.

Penner, J.E., S.Y. Zhang, and C.C. Chuang, 2003: Soot and smoke aerosol may not warm climate. *J. Geophys. Res.*, **108**(D21), 4657, doi:10.1029/2003JD003409.

Penner, J.E., et al., 1997: Anthropogenic aerosols and climate change: A method for calibrating forcing. In: *Assessing Climate Change: Results from the Model Evaluation Consortium for Climate Assessment* [Howe, W., and A. Henderson-Sellers (eds.)]. Gordon & Breach Science Publishers, Sydney, Australia, pp. 91–111.

Penner, J.E., et al., 2007: Effect of black carbon on mid-troposphere and surface temperature trends. In: *Human-Induced Climate Change: An Interdisciplinary Assessment* [Schlesinger, M., et al. (eds.)]. Cambridge University Press, Cambridge, UK, in press.

Perlwitz, J., and H.-F. Graf, 2001: Troposphere-stratosphere dynamic coupling under strong and weak polar vortex conditions. *Geophys. Res. Lett.*, **28**, 271–274.

Peterson, B.J., et al., 2002: Increasing river discharge to the Arctic Ocean. *Science*, **298**, 2171–2173.

Pezza, A.B., and I. Simmonds, 2005: The first South Atlantic hurricane: Unprecedented blocking, low shear and climate change. *Geophys. Res. Lett.*, **32**, L15712, doi:10.1029/2005GL023390.

Pielke, R.A. Jr., 2005: Are there trends in hurricane destruction? *Nature*, **438**, E11.

Pielke, R.A. Jr., et al., 2005: Hurricanes and global warming. *Bull. Am. Meteorol. Soc.*, **86**, 1571–1575.

Pierce, D.W., et al., 2006: Anthropogenic warming of the oceans: observations and model results. *J. Clim.*, **19**, 1873–1900.

Pinker, R.T., B. Zhang, and E.G. Dutton, 2005: Do satellites detect trends in surface solar radiation? *Science*, **308**, 850–854.

Pollack, H.N., and J.E. Smerdon, 2004: Borehole climate reconstructions: Spatial structure and hemispheric averages. *J. Geophys. Res.*, **109**, D11106, doi:10.1029/2003JD004163.

Prentice, I.C., and T. Webb, 1998: BIOME 6000: reconstructing global mid-Holocene vegetation patterns from palaeoecological records. *J. Biogeogr.*, **25**(6), 997–1005.

Prentice, I.C., and D. Jolly, 2000: Mid-Holocene and glacial-maximum vegetation geography of the northern continents and Africa. *J. Biogeogr.*, **27**(3), 507–519.

Qian, T., A. Dai, K.E. Trenberth, and K.W. Oleson, 2006: Simulation of global land surface conditions from 1948 to 2002: Part I: Forcing data and evaluations. *J. Hydrometeorol.*, **7**, 953–975.

Ramanathan, V., P.J. Crutzen, J.T. Kiehl, and D. Rosenfeld, 2001: Aerosols, climate, and the hydrological cycle. *Science*, **294**, 2119–2124.

Ramanathan, V., et al., 2005: Atmospheric brown clouds: Impacts on South Asian climate and hydrological cycle. *Proc. Natl. Acad. Sci. U.S.A.*, **102**, 5326–5333.

Ramankutty, N., and J.A. Foley, 1999: Estimating historical changes in global land cover: Croplands from 1700 to 1992. *Global Biogeochem. Cycles*, **13**(4), 997–1027.

Ramaswamy, V., et al., 2001: Radiative forcing of climate change. In: *Climate Change 2001: The Scientific Basis. Contribution of Working Group I to the Third Assessment Report of the Intergovernmental Panel on Climate Change* [Houghton, J.T., et al. (eds.)]. Cambridge University Press, Cambridge, United Kingdom and New York, NY, USA, pp. 349–416.

Randel, W.J., and F. Wu, 2006: Biases in stratospheric temperature trends derived from historical radiosonde data. *J. Clim.*, **19**, 2094–2104.

Raper, S.C.B., J.M. Gregory, and R.J. Stouffer, 2002: The role of climate sensitivity and ocean heat uptake on AOGCM transient temperature response. *J. Clim.*, **15**, 124–130.

Rauthe, M., A. Hense, and H. Paeth, 2004: A model intercomparison study of climate change-signals in extratropical circulation. *Int. J. Climatol.*, **24**, 643–662.

Reader, M., and G. Boer, 1998: The modification of greenhouse gas warming by the direct effect of sulphate aerosols. *Clim. Dyn.*, **14**, 593–607.

Reichert, B.K., L. Bengtsson, and J. Oerlemans, 2002a: Recent glacier retreat exceeds internal variability. *J. Clim.*, **15**, 3069–3081.

Reichert, B.K., R. Schnur, and L. Bengtsson, 2002b: Global ocean warming tied to anthropogenic forcing. *Geophys. Res. Lett.*, **29**(11), 1525.

Rigor, I.G., J.M. Wallace, and R.L. Colony, 2002: Response of sea ice to the Arctic Oscillation. *J. Clim.*, **15**, 2648–2668.

Rind, D., J. Perlwitz, and P. Lonergan, 2005a: AO/NAO response to climate change. Part I: The respective influences of stratospheric and tropospheric climate changes. *J. Geophys. Res.*, **110**, D12107, doi:10.1029/2004JD005103.

Rind, D., J. Perlwitz, and P. Lonergan, 2005b: AO/NAO response to climate change. Part II: The relative importance of low and high latitude temperature changes. *J. Geophys. Res.*, **110**, D12108, doi:10.1029/2004JD005686.

Rind, D., et al., 2004: The relative importance of solar and anthropogenic forcing of climate change between the Maunder Minimum and the present. *J. Clim.*, **17**(5), 906–929.

Robock, A., 2000: Volcanic eruptions and climate. *Rev. Geophys.*, **38**(2), 191–219.

Robock, A., and Y. Liu, 1994: The volcanic signal in Goddard Institute for Space Studies three-dimensional model simulations. *J. Clim.*, **7**, 44–55.

Rothrock, D.A., J. Zhang, and Y. Yu, 2003: The arctic ice thickness anomaly of the 1990s: A consistent view from observations and models. *J. Geophys. Res.*, **108**(C3), 3083, doi:10.1029/2001JC001208.

Rotstayn, L.D., and J.E. Penner, 2001: Forcing, quasi-forcing and climate response. *J. Clim.*, **14**, 2960–2975.

Rotstayn, L.D., and U. Lohmann, 2002: Tropical rainfall trends and the indirect aerosol effect. *J. Clim.*, **15**, 2103–2116.

Rotstayn, L.D., and Y. Liu, 2003: Sensitivity of the first indirect aerosol effect to an increase of cloud droplet spectral dispersion with droplet number concentration. *J. Clim.*, **16**, 3476–3481.

Rowell, D.P., 1996: Reply to comments by Y.C. Sud and W.K.-M. Lau. *Q. J. R. Meteorol. Soc.*, **122**, 1007–1013.

Rowell, D.P., 2003: The Impact of Mediterranean SSTs on the Sahelian rainfall season. *J. Clim.*, **16**, 849–862.

Ruzmaikin, A., and J. Feynman, 2002: Solar influence on a major mode of atmospheric variability. *J. Geophys. Res.*, **107**(D14), doi:10.1029/2001JD001239.

Rybski, D., A. Bunde, S. Havlin, and H. von Storch, 2006: Long-term persistence in climate and the detection problem. *Geophys. Res. Lett.*, **33**, L06718, doi:10.1029/2005GL025591.

Santer, B.D., T.M.L. Wigley, T. Barnett, and E. Anyamba, 1996a: Detection of climate change and attribution of causes. In: *Climate Change 1995: The Science of Climate Change. Contribution of Working Group I to the Second Assessment Report of the Intergovernmental Panel on Climate Change* [Houghton, J.T. et al. (eds.)]. Cambridge University Press, Cambridge, United Kingdom and New York, NY, USA, pp. 407–444.

Santer, B.D., et al., 1996b: A search for human influences on the thermal structure of the atmosphere. *Nature*, **382**, 39–46.

Santer, B.D., et al., 1996c: Reply to "Human effect on global climate?" *Nature*, **384**, 522–525.

Santer, B.D., et al., 2000: Interpreting differential temperature trends at the surface and in the lower troposphere. *Science*, **287**, 1227–1231.

Santer, B.D., et al., 2001: Accounting for the effects of volcanoes and ENSO in comparisons of modeled and observed temperature trends. *J. Geophys. Res.*, **106**, 28033–28059.

Santer, B.D., et al., 2003a: Contributions of anthropogenic and natural forcing to recent tropopause height changes. *Science*, **301**, 479–483.

Santer, B.D., et al., 2003b: Behavior of tropopause height and atmospheric temperature in models, reanalyses, and observations: Decadal changes. *J. Geophys. Res.*, **108**(D1), 4002.

Santer, B.D., et al., 2003c: Influence of satellite data uncertainties on the detection of externally-forced climate change. *Science*, **300**, 1280–1284.

Santer, B.D., et al., 2004: Identification of anthropogenic climate change using a second-generation reanalysis. *J. Geophys. Res.*, **109**, doi:10.1029/2004JD005075.

Santer, B.D., et al., 2005: Amplification of surface temperature trends and variability in the tropical atmosphere. *Science*, **309**, 1551–1556.

Sato, M., J.E. Hansen, M.P. McCormick, and J.B. Pollack, 1993: Stratospheric aerosol optical depths, 1850-1990. *J. Geophys. Res.*, **98**, 22987–22994.

Scaife, A.A., J.R. Knight, G.K. Vallis, and C.K. Folland, 2005: A stratospheric influence on the winter NAO and North Atlantic surface climate. *Geophys. Res. Lett.*, **32**, L18715, doi:10.1029/2005GL023226.

Schär, C., and G. Jendritzky, 2004: Hot news for summer 2003. *Nature*, **432**, 559–560.

Schär, C., et al., 2004: The role of increasing temperature variability in European summer heat waves. *Nature*, **427**, 332–336

Scherrer, S.C., C. Appenzeller, and M. Laternser, 2004: Trends in Swiss alpine snow days – the role of local and large scale climate variability. *Geophys. Res. Lett.*, **31**, doi:10.1029/2004GL020255.

Scherrer, S.C., C. Appenzeller, M. A. Linger and C. Schär, 2005: European temperature distribution changes in observations and climate change scenarios. *Geophys. Res. Lett.*, **32**, doi:10.1029/2005GL024108.

Schlesinger, M.E., and N. Ramankutty, 1992: Implications for global warming of intercycle solar irradiance variations. *Nature*, **360**, 330–333.

Schlesinger, M.E., and N. Ramankutty, 1994: An oscillation in the global climate system of period 65-70 years. *Nature*, **367**, 723–726.

Schneider, T., 2004: The tropopause and the thermal stratification in the extratropics of a dry atmosphere. *J. Atmos. Sci.*, **61**, 1317–1340.

Schneider, T., and I.M. Held, 2001: Discriminants of twentieth-century changes in Earth surface temperatures. *J. Clim.*, **14**, 249–254.

Schneider von Deimling, T., H. Held, A. Ganopolski, and S. Rahmstorf, 2006: Climate sensitivity estimated from ensemble simulations of glacial climate. *Clim. Dyn.*, **27**, 149–163.

Schnur, R., and K. Hasselmann, 2005: Optimal filtering for Bayesian detection of climate change. *Clim. Dyn.*, **24**, 45–55.

Schwartz, S.E., 1993: Does fossil fuel combustion lead to global warming? *Energy Int. J.*, **18**, 1229–1248.

Schwartz, S.E., 2004: Uncertainty requirements in radiative forcing of climate change. *J. Air Waste Manage. Assoc.*, **54**, 1351–1359.

Seidel, D.J., R.J. Ross, J.K. Angell, and G.C. Reid, 2001: Climatological characteristics of the tropical tropopause as revealed by radiosondes. *J. Geophys. Res.*, **106**, 7857–7878.

Selten, F.M., G.W. Branstator, H.A. Dijkstra, and M. Kliphuis, 2004: Tropical origins for recent and future Northern Hemisphere climate change. *Geophys. Res. Lett.*, **31**, L21205, doi:10.1029/2004GL020739.

Semenov, V.A., and L. Bengtsson, 2002: Secular trends in daily precipitation characteristics: Greenhouse gas simulation with a coupled AOGCM. *Clim. Dyn.*, **19**, 123–140.

Senior, C.A., and J.F.B. Mitchell, 2000: The time dependence of climate sensitivity. *Geophys. Res. Lett.*, **27**, 2685–2689.

Sexton, D.M.H., 2001: The effect of stratospheric ozone depletion on the phase of the Antarctic Oscillation. *Geophys. Res Lett.*, **28**, 3697–3700.

Sexton, D.M.H., D.P. Rowell, C.K. Folland, and D.J. Karoly, 2001: Detection of anthropogenic climate change using an atmospheric GCM. *Clim. Dyn.*, **17**, 669–685.

Sexton, D.M.H., H. Grubb, K.P. Shine, and C.K. Folland, 2003: Design and analysis of climate model experiments for the efficient estimation of anthropogenic signals. *J. Clim.*, **16**, 1320–1336.

Sherwood, S., J. Lanzante, and C. Meyer, 2005: Radiosonde daytime biases and late-20th century warming. *Science*, **309**, 1156–1159.

Shin, S.I., et al., 2003: A simulation of the last glacial maximum climate using the NCAR-CCSM. *Clim. Dyn.*, **20**(2–3), 127–151.

Shindell, D.T., and G.A. Schmidt, 2004: Southern Hemisphere climate response to ozone changes and greenhouse gas increases. *Geophys. Res. Lett.*, **31**, L18209, doi:10.1029/2004GL020724.

Shindell, D.T., R.L. Miller, G.A. Schmidt, and L. Pandolfo, 1999: Simulation of recent northern winter climate trends by greenhouse-gas forcing. *Nature*, **399**, 452–455.

Shindell, D.T., G.A. Schmidt, R.L. Miller, and D. Rind, 2001a: Northern Hemispheric climate response to greenhouse gas, ozone, solar and volcanic forcing. *J. Geophys. Res.*, **106**, 7193–7210.

Shindell, D.T., G.A. Schmidt, R.L. Miller, and M.E. Mann, 2003: Volcanic and solar forcing of climate change during the preindustrial era. *J. Clim.*, **16**(24), 4094–4107.

Shindell, D.T., et al., 2001b: Solar forcing of regional climate change during the Maunder Minimum. *Science*, **294**(5549), 2149–2152.

Shiogama, H., M. Watanabe, M. Kimoto, and T. Nozawa, 2005: Anthropogenic and natural forcing impacts on ENSO-like decadal variability during the second half of the 20th century. *Geophys. Res. Lett.*, **32**, L21714, doi 10.1029/2005GL023871.

Shiogama, H., et al., 2006: Influence of volcanic activity and changes in solar irradiance on surface air temperatures in the early twentieth century. *Geophys. Res. Lett.*, **33**, L09702, doi:10.1029/2005GL025622.

Simmons, A.J., and J.K. Gibson, 2000: *The ERA-40 Project Plan*. ERA-40 Project Report Series, Vol. 1, European Centre for Medium-Range Weather Forecasts, Reading, UK, 62 pp.

Smith, R.L., T.M.L. Wigley, and B.D. Santer, 2003: A bivariate time series approach to anthropogenic trend detection in hemispheric mean temperatures. *J. Clim.*, **16**, 1228–1240.

Soden, B.J., R.T. Wetherald, G.L. Stenchikov, and A. Robock, 2002: Global cooling after the eruption of Mount Pinatubo: A test of climate feedback by water vapor. *Science*, **296**(5568), 727–730.

Soden, B.J., et al., 2005: The radiative signature of upper tropospheric moistening. *Science*, **310**(5749), 841–844.

Solow, A.R., and L.J. Moore, 2002: Testing for trend in North Atlantic hurricane activity, 1900-98. *J. Clim.*, **15**, 3111–3114.

Spagnoli, B., et al., 2002: Detecting climate change at a regional scale: the case of France. *Geophys. Res. Lett.*, **29**, doi:10.1029/2001GL014619.

Stanhill, G., and S. Cohen, 2001: Global dimming, a review of the evidence for a widespread and significant reduction in global radiation with a discussion of its probable causes and possible agricultural consequences. *Agric. Forest Meteorol.*, **107**, 255–278.

Stark, S., R.A. Wood, and H.T. Banks, 2006: Reevaluating the causes of observed changes in Indian Ocean water masses. *J. Clim.*, **19**, 4075–4086.

Stenchikov, G.L., et al., 2002: Arctic Oscillation response to the 1991 Mount Pinatubo eruption: Effects of volcanic aerosols and ozone depletion. *J. Geophys. Res.*, **107**, 4803.

Stenchikov, G., et al., 2004: Arctic Oscillation response to the 1991 Pinatubo eruption in the SKYHI GCM with a realistic Quasi-Biennial Oscillation. *J. Geophys. Res.*, **109**, D03112, doi:10.1029/2003JD003699.

Stenchikov, G., et al., 2006: Arctic Oscillation response to volcanic eruptions in the IPCC AR4 climate models. *J. Geophys. Res.*, **111**, D07107, doi:10.1029/2005JD006286.

Stendel, M., I.A. Mogensen, and J.H. Christensen, 2006: Influence of various forcings on global climate in historical times using a coupled atmosphere–ocean general circulation model. *Clim. Dyn.*, **26**, 1–15.

Stern, D.I., 2005: Global sulfur emissions from 1850 to 2000. *Chemosphere*, **58**, 163–175.

Stone, D.A., and A.J. Weaver, 2002: Daily minimum and maximum temperature trends in a climate model. *Geophys. Res. Lett.*, **29**, doi:10.1029/2001GL014556.

Stone, D.A., and A.J. Weaver, 2003: Factors contributing to diurnal temperature trends in twentieth and twenty-first century simulations of the CCCma coupled model. *Clim. Dyn.*, **20**, 435–445.

Stone, D.A., and M.R. Allen, 2005a: The end-to-end attribution problem: From emissions to impacts. *Clim. Change*, **71**, 303–318.

Stone, D.A., and M.R. Allen, 2005b: Attribution of global surface warming without dynamical models. *Geophys. Res. Lett.*, **32**, L18711, doi:10.1029/2005GL023682.

Stone, D.A., and J.C. Fyfe, 2005: The effect of ocean mixing parameterisation on the enhanced CO_2 response of the Southern Hemisphere mid-latitude jet. *Geophys. Res. Lett.*, **32**, L06811, doi:10.1029/2004GL022007.

Stone, D.A., A.J. Weaver, and R.J. Stouffer, 2001: Projection of climate change onto modes of atmospheric variability. *J. Clim.*, **14**, 3551–3565.

Stone, D.A., M.R. Allen, and P.A. Stott, 2007a: A multi-model update on the detection and attribution of global surface warming. *J. Clim.*, **20**, 517-530.

Stone, D.A., M.R. Allen, F. Selten, and M. Kilphuis, 2007b: The detection and attribution of climate change using an ensemble of opportunity. *J. Clim.*, **20**, 504-516.

Stott, P.A., 2003: Attribution of regional-scale temperature changes to anthropogenic and natural causes. *Geophys. Res. Lett.*, **30**, doi:10.1029/2003GL017324.

Stott, P.A., and S.F.B. Tett, 1998: Scale-dependent detection of climate change. *J. Clim.*, **11**, 3282–3294.

Stott, P.A., and J.A. Kettleborough, 2002: Origins and estimates of uncertainty in predictions of 21st century temperature rise. *Nature*, **416**, 723–726.

Stott, P.A., M.R. Allen, and G.S. Jones, 2003a: Estimating signal amplitudes in optimal fingerprinting, Part II: Application to general circulation models. *Clim. Dyn.*, **21**, doi:10.1007/s00382-003-0314-8.

Stott, P.A., G.S. Jones, and J.F.B. Mitchell, 2003b: Do models underestimate the solar contribution to recent climate change? *J. Clim.*, **16**, 4079–4093.

Stott, P.A., D.A. Stone, and M.R. Allen, 2004: Human contribution to the European heatwave of 2003. *Nature*, **432**, 610–614.

Stott, P.A., J.A. Kettleborough, and M.R. Allen, 2006a: Uncertainty in predictions of continental scale temperature rise. *Geophys. Res. Lett.*, **33**, doi:10.1029/GL024423.

Stott, P.A., et al., 2000: External control of 20th century temperature by natural and anthropogenic forcings. *Science*, **290**, 2133–2137.

Stott, P.A., et al., 2001: Attribution of twentieth century temperature change to natural and anthropogenic causes. *Clim. Dyn.*, **17**, 1–21.

Stott, P.A., et al., 2006b: Transient climate simulations with the HadGEM1 model: causes of past warming and future climate change. *J. Clim.*, **19**, 2763–2782.

Stott, P.A., et al., 2006c: Observational constraints on past attributable warming and predictions of future global warming. *J. Clim.*, **19**, 3055–3069.

Stouffer, R.J., G.C. Hegerl, and S.F.B. Tett, 2000: A comparison of surface air temperature variability in three 1000-year coupled ocean-atmosphere model integrations. *J. Clim.*, **13**, 513–547.

Sugi, M., and J. Yoshimura, 2004: A mechanism of tropical precipitation change due to CO_2 increase. *J. Clim.*, **17**, 238–243.

Sun, S., and J. Hansen, 2003: Climate simulations for 1951-2050 with a coupled atmosphere-ocean model. *J. Clim.*, **16**, 2807–2826.

Sutton, R.T., and L.R. Hodson, 2005: Atlantic ocean forcing of North American and European summer climate. *Science*, **309**, 115–118.

Tanaka, H.L., N. Ishizaki, and N. Nohara, 2005: Intercomparison of the intensities and trends of Hadley, Walker and Monsoon Circulations in the global warming predictions. *Scientific Online Letters on the Atmosphere*, **1**, 77–80.

Taylor, C.M., et al., 2002: The influence of land use change on climate in the Sahel. *J. Clim.*, **15**, 3615–3629.

Taylor, K.E., et al., 2000: Analysis of forcing, response, and feedbacks in a paleoclimate modeling experiment. In: *Paleoclimate Modeling Intercomparison Project (PMIP): Proceedings of the Third PMIP Workshop, La Huardière, Canada, 4-8 October 1999* [P. Braconnot (ed.)]. WCRP-111, WMO/TD-1007, World Meteorological Organization, Geneva, Switzerland, pp. 271.

Tebaldi, C., K. Hayhoe, J.M. Arblaster, and G.A. Meehl, 2006: Going to extremes: An intercomparison of model-simulated historical and future changes in extreme events. *Clim. Change*, **79**, 185–211.

Tett, S.F.B., et al., 1999: Causes of twentieth-century temperature change near the Earth's surface. *Nature*, **399**, 569–572.

Tett, S.F.B., et al., 2002: Estimation of natural and anthropogenic contributions to twentieth century temperature change. *J. Geophys. Res.*, **107**(D16), 4306, doi:10.1029/2000JD000028.

Tett, S.F.B., et al., 2007: The impact of natural and anthropogenic forcings on climate and hydrology since 1550. *Clim. Dyn.*, **28**, 3–34, doi:10.1007/s00382-006-0165-1.

Thomas, R., et al., 2004: Accelerated sea level rise from West Antarctica. *Science*, **306**, 255–258.

Thompson, D.W.J., and J.M. Wallace, 1998: The Arctic Oscillation signature in the wintertime geopotential height and temperature fields. *Geophys. Res. Lett.*, **25**, 1297–1300.

Thompson, D.W.J., and J.M. Wallace, 2000: Annular modes in the extratropical circulation. Part I: Month-to-month variability. *J. Clim.*, **13**, 1000–1016.

Thompson, D.W.J., and S. Solomon, 2002: Interpretation of recent Southern Hemisphere climate change. *Science*, **296**, 895–899.

Thompson, D.W.J., J.M. Wallace, and G.C. Hegerl, 2000: Annular modes in the extratropical circulation: Part II, Trends. *J. Clim.*, **13**, 1018–1036.

Thorne, P.W., et al., 2002: Assessing the robustness of zonal mean climate change detection. *Geophys. Res. Lett.*, **29**, doi:10.1029/2002GL015717.

Thorne, P.W., et al., 2003: Probable causes of late twentieth century tropospheric temperature trends. *Clim. Dyn.*, **21**, 573–591.

Thorne, P.W., et al., 2005: Revisiting radiosonde upper air temperatures from 1958 to 2002. *J. Geophys. Res.*, **110**, D18105, doi:10.1029/2004JD005753.

Thuburn, J., and G.C. Craig, 2000: Stratospheric influence on tropopause height: The radiative constraint. *J. Atmos. Sci.*, **57**, 17–28.

Timbal, B., 2004: Southwest Australia past and future rainfall trends. *Clim. Res.*, **26**, 233–249.

Timbal, B., J.M. Arblaster, and S. Power, 2005: Attribution of late 20th century rainfall decline in South-West Australia. *J. Clim.*, **19**, 2046–2062.

Timmermann, A., 1999: Detecting the nonstationary response of ENSO to greenhouse warming. *J. Atmos. Sci.*, **56**, 2313–2325.

Timmermann, A., et al., 1999: Increased El Niño frequency in a climate model forced by future greenhouse warming. *Nature*, **398**, 694–696.

Tol, R.S.J., and A.F. De Vos, 1998: A Bayesian statistical analysis of the enhanced greenhouse effect. *Clim. Change*, **38**, 87–112.

Tourpali, K., et al., 2003: Stratospheric and tropospheric response to enhanced solar UV radiation: A model study. *Geophys. Res. Lett.*, **30**(5), doi:10.1029/2002GL016650.

Trenberth, K.E., 2005: Uncertainty in hurricanes and global warming. *Science*, **308**, 1753–1754.

Trenberth, K.E., and D.J. Shea, 2005: Relationships between precipitation and surface temperature. *Geophys. Res. Lett.*, **32**, L14703, doi:10.1029/2005GL022760.

Trenberth, K.E., and L. Smith, 2005: The mass of the atmosphere: A constraint on global analyses. *J. Clim.*, **18**, 864–875.

Trenberth, K.E., and D. Shea, 2006: Atlantic hurricanes and natural variability in 2005. *Geophys. Res. Lett.*, **33**, L12704, doi:10.1029/2006GL026894.

Trenberth, K.E., J. Fasullo, and L. Smith, 2005: Trends and variability in column-integrated water vapour. *Clim. Dyn.*, **24**, 741–758.

Trenberth, K.E., A. Dai, R.M. Rasmussen, and D.B. Parsons, 2003: The changing character of precipitation. *Bull. Am. Meteorol. Soc.*, **84**, 1205–1217.

Triacca, U., 2001: On the use of Granger causality to investigate the human influence on climate. *Theor. Appl. Climatol.*, **69**, 137–138.

Tudhope, S., and M. Collins, 2003: Global change – The past and future of El Nino. *Nature*, **424**(6946), 261–262.

Ueda, H., A. Iwai, K. Kuwako, and M.E. Hori, 2006: Impact of anthropogenic forcing on the Asian summer monsoon simulated by 8 GCMs. *Geophys. Res. Lett.*, **33**, L06703, doi:10.1029/2005GL025336.

van Aardenne, J.A., et al., 2001: A 1° x 1° resolution data set of historical anthropogenic trace gas emissions for the period 1890-1990. *Global Biogeochem. Cycles*, **15**, 909–928.

van der Schrier, G., and J. Barkmeijer, 2005: Bjerknes' hypothesis on the coldness during AD 1790-1820 revisited. *Clim. Dyn.*, **24**(4), 355–371.

van Loon, H., and D.J. Shea, 2000: The global 11-year solar signal in July-August. *Geophys. Res. Lett.*, **27**, 2965–2968.

Vaughan, D.G., et al., 2001: Devil in the detail. *Science*, **293**, 1777–1779.

Vecchi, G.A., et al., 2006: Weakening of tropical Pacific atmospheric circulation due to anthropogenic forcing. *Nature*, **44**, 73–76.

Vellinga, M., and P. Wu, 2004: Low-latitude freshwater influence on centennial variability of the thermohaline circulation. *J. Clim.*, **17**, 4498–4511.

Vinnikov, K.Y., et al., 1999: Global warming and Northern Hemisphere sea ice extent. *Science*, **286**(5446), 1934–1937.

Vinnikov, K.Y., et al., 2006: Trends at the surface and in the troposphere. *J. Geophys. Res.*, **111**, D03106, doi:10.1029/2005JD006392.

Volodin, E.M., and V.Y. Galin, 1999: Interpretation of winter warming on Northern Hemisphere continents in 1977-1994. *J. Clim.*, **12**, 2947–2955.

von Storch, H., and F.W. Zwiers, 1999: *Statistical Analysis in Climate Research.* Cambridge University Press, Cambridge, UK, 484 pp.

Wang, M., et al., 2007: Intrinsic versus forced variation in coupled climate model simulations over the Arctic during the 20th century. *J. Clim.*, **20**, 1084-1098.

Waple, A.M., M.E. Mann, and R.S. Bradley, 2002: Long-term patterns of solar irradiance forcing in model experiments and proxy based surface temperature reconstructions. *Clim. Dyn.*, **18**, 563–578.

Wardle, R., and I. Smith, 2004: Modeled response of the Australian monsoon to changes in land surface temperatures. *Geophys. Res. Lett.*, **31**, L16205, doi:10.1029/2004GL020157.

Weatherall, R.T., and S. Manabe, 1975: The effects of changing the solar constant on the climate of a general circulation model. *J. Atmos. Sci.*, **32**, 2044–2059.

Weber, S.L., 2005: A timescale analysis of the NH temperature response to volcanic and solar forcing in the past millennium. *Climate of the Past*, **1**, 9–17.

Weber, S.L., T.J. Crowley, and G. van der Schrier, 2004: Solar irradiance forcing of centennial climate variability during the Holocene. *Clim. Dyn.*, **22**(5), 539–553.

Webster, P.J., G.J. Holland, J.A. Curry, and H.-R. Chang, 2005: Changes in tropical cyclone number, duration, and intensity in a warming environment. *Science*, **309**(5742), 1844–1846.

Wentz, F.J., and M. Schabel, 2000: Precise climate monitoring using complementary satellite data sets. *Nature*, **403**, 414–416.

White, W.B., M.D. Dettinger, and D.R. Cayan, 2003: Sources of global warming of the upper ocean on decadal period scales. *J. Geophys. Res.*, **108**, 3248, doi:10.1029/2002JC001396.

Wielicki, B.A., et al., 2002: Evidence of large decadal variability in tropical mean radiative energy budget. *Science*, **295**, 841–844.

Wielicki, B.A., et al., 2005: Changes in Earth's albedo measured by satellite. *Science*, **308**, 825.

Wigley, T.M.L., 1989: Climate variability on the 10-100-year time scale: Observations and possible causes. In: *Global Changes of the Past: Papers Arising from the 1989 OIES Global Change Institute* [Bradley, R.S. (ed.)]. University Corporation for Atmospheric Research, Boulder, CO, pp. 83–101.

Wigley, T.M.L., P.D. Jones, and S.C.B. Raper, 1997: The observed global warming record: What does it tell us? *Proc. Natl. Acad. Sci. U.S.A.*, **94**, 8314–8320.

Wigley, T.M.L., C.M. Ammann, B.D. Santer, and S.C.B. Raper, 2005a: Effect of climate sensitivity on the response to volcanic forcing. *J. Geophys. Res.*, **110**, D09107, doi:10.1029/2004JD005557.

Wigley, T.M.L., C.M. Ammann, B.D. Santer, and K.E. Taylor, 2005b: Comment on "Climate forcing by the volcanic eruption of Mount Pinatubo" by David H. Douglass and Robert S. Knox. *Geophys. Res. Lett.*, **32**, L20709, doi:10.1029/2005GL023312.

Wild, M., et al., 2005: From dimming to brightening: Decadal changes in solar radiation at Earth's surface. *Science*, **308**, 847–850.

Williams, K.D., et al., 2001: The response of the climate system to indirect effects of anthropogenic aerosol. *Clim. Dyn.*, **17**, 845–856.

Willis, J.K., D. Roemmich, and B. Cornuelle, 2004: Interannual variability in upper ocean heat content, temperature and thermosteric expansion on global scales. *J. Geophys. Res.*, **109**, C12036, doi:10.1029/2003JC002260.

Wohlfahrt, J., S.P. Harrison, and P. Braconnot, 2004: Synergistic feedbacks between ocean and vegetation on mid- and high-latitude climates during the mid-Holocene. *Clim. Dyn.*, **22**(2–3), 223–238.

Wong, A.P.S., N.L. Bindoff, and J. Church, 1999: Large-scale freshening of intermediate waters in the Pacific and Indian Oceans. *Nature*, **400**, 440–444.

Wong, T., et al., 2006: Reexamination of the observed decadal variability of the Earth Radiation Budget using altitude-corrected ERBE/ERBS nonscanner WFOV data. *J. Clim.*, **19**, 4028–4040.

Woodworth, P.L., J.M. Gregory, and R.J. Nicholls, 2004: Long term sea level changes and their impacts. In: *The Sea* [Robinson, A.R. and K.H. Brink (eds.)]. Harvard University Press, Cambridge, MA, pp. 715–753.

Wu, P., R. Wood, and P.A. Stott, 2004: Does the recent freshening trend in the North Atlantic indicate a weakening thermohaline circulation? *Geophys. Res. Lett.*, **31**, doi:10.1029/2003GL018584.

Wu, P., R. Wood, and P.A. Stott, 2005: Human influence on increasing Arctic river discharges. *Geophys. Res. Lett.*, **32**, L02703, doi:10.1029/2004GL021570.

Wu, P., R. Wood, P.A. Stott, and G.S. Jones, 2007: Deep North Atlantic freshening simulated in a coupled model. *Progr. Oceanogr.*, accepted.

Wyputta, U., and B.J. McAvaney, 2001: Influence of vegetation changes during the Last Glacial Maximum using the BMRC atmospheric general circulation model. *Clim. Dyn.*, **17**(12), 923–932.

Yang, F., and M. Schlesinger, 2001: Identification and separation of Mount Pinatubo and El Nino-Southern Oscillation land surface temperature anomalies. *J. Geophys. Res.*, **106**, 14757–14770.

Yang, F., A. Kumar, M.E. Schlesinger, and W. Wang, 2003: Intensity of hydrological cycles in warmer climates. *J. Clim.*, **16**, 2419–2423.

Yokohata, T., et al., 2005: Climate response to volcanic forcing: Validation of climate sensitivity of a coupled atmosphere-ocean general circulation model. *Geophys. Res. Lett.*, **32**, L21710, doi:10.1029/2005GL023542.

Yoshimori, M., T. Stocker, C.C. Raible, and M. Renold, 2005: Externally forced and internal variability in ensemble climate simulations of the Maunder Minimum. *J. Clim.*, **18**, 4253–4270.

Yoshimura, J., and M. Sugi, 2005: Tropical cyclone climatology in a high-resolution AGCM -Impacts of SST warming and CO2 increase. *Scientific Online Letters on the Atmosphere*, **1**, 133–136.

Zeng, N., 2003: Drought in the Sahel. *Science*, **302**, 999–1000.

Zhang, X., F.W. Zwiers, and P.A. Stott, 2006: Multi-model multi-signal climate change detection at regional scale. *J. Clim.*, **19**, 4294–4307.

Zhang, Y., J.M. Wallace, and D.S. Battisti, 1997: ENSO-like interdecadal variability: 1900-93. *J. Clim.*, **10**, 1004–1020.

Zhang, Y.-C., et al., 2004: Calculation of radiative flux profiles from the surface to top-of-atmosphere based on ISCCP and other global data sets: Refinements of the radiative transfer model and the input data. *J. Geophys. Res.*, **109**, D19105, doi:10.1029/2003JD004457.

Zhao, Y., et al., 2005: A multi-model analysis of the role of the ocean on the African and Indian monsoon during the mid-Holocene. *Clim. Dyn.*, **25**, 777–800.

Ziegler, A.D., et al., 2003: Detection of intensification in global- and continental-scale hydrological cycles: Temporal scale of evaluation. *J. Clim.*, **16**, 535–547.

Zorita, E., et al., 2004: Climate evolution in the last five centuries simulated by an atmosphere-ocean model: Global temperatures, the North Atlantic Oscillation and the Late Maunder Minimum. *Meteorol. Z.*, **13**(4), 271–289.

Zorita, E., et al., 2005: Natural and anthropogenic modes of surface temperature variations in the last thousand years. *Geophys. Res. Lett.*, **32**(8), L08707, doi:10.1029/2004GL021563.

Zwiers, F.W., and X. Zhang, 2003: Toward regional scale climate change detection. *J. Clim.*, **16**, 793–797.

Appendix 9.A: Methods Used to Detect Externally Forced Signals

This appendix very briefly reviews the statistical methods that have been used in most recent detection and attribution work. Standard 'frequentist' methods (methods based on the relative frequency concept of probability) are most often used, but there is also increasing use of Bayesian methods of statistical inference. The following sections briefly describe the optimal fingerprinting technique followed by a short discussion on the differences between the standard and Bayesian approaches to statistical inferences that are relevant to detection and attribution.

9.A.1   Optimal Fingerprinting

Optimal fingerprinting is generalised multivariate regression adapted to the detection of climate change and the attribution of change to externally forced climate change signals (Hasselmann, 1979, 1997; Allen and Tett, 1999). The regression model has the form $y = Xa + u$, where vector y is a filtered version of the observed record, matrix X contains the estimated response patterns to the external forcings (signals) that are under investigation, a is a vector of scaling factors that adjusts the amplitudes of those patterns and u represents internal climate variability. Vector u is usually assumed to be a Gaussian random vector with covariance matrix C. Vector a is estimated with $\hat{a} = (X^T C^{-1} X)^{-1} X^T C^{-1} y$, which is equivalent to $(\tilde{X}^T \tilde{X})^{-1} \tilde{X}^T \tilde{y}$, where matrix $\tilde{X}$ and vector $\tilde{y}$ represent the signal patterns and observations after normalisation by the climate's internal variability. This normalisation, standard in linear regression, is used in most detection and attribution approaches to improve the signal-to-noise ratio (see, e.g., Hasselmann, 1979; Allen and Tett, 1999; Mitchell et al., 2001).

The matrix X typically contains signals that are estimated with either an AOGCM, an AGCM (see Sexton et al., 2001, 2003) or a simplified climate model such as an EBM. Because AOGCMs simulate natural internal variability as well as the response to specified anomalous external forcing, AOGCM-simulated climate signals are typically estimated by averaging across an ensemble of simulations (for a discussion of optimal ensemble size and composition, see Sexton et al., 2003). If an observed response is to be attributed to anthropogenic influence, X should at a minimum contain separate natural and anthropogenic responses. In order to relax the assumption that the relative magnitudes of the responses to individual forcings are correctly simulated, X may contain separate responses to all the main forcings, including greenhouse gases, sulphate aerosol, solar irradiance changes and volcanic aerosol. The vector a accounts for possible errors in the amplitude of the external forcing and the amplitude of the climate model's response by scaling the signal patterns to best match the observations.

Fitting the regression model requires an estimate of the covariance matrix C (i.e., the internal variability), which is usually obtained from unforced variation simulated by AOGCMs (e.g., from long control simulations) because the instrumental record is too short to provide a reliable estimate and may be affected by external forcing. Atmosphere-Ocean General Circulation Models may not simulate natural internal climate variability accurately, particularly at small spatial scales, and thus a residual consistency test (Allen and Tett, 1999) is typically used to assess the model-simulated variability at the scales that are retained in the analysis. To avoid bias (Hegerl et al., 1996, 1997), uncertainty in the estimate of the vector of scaling factors a is usually assessed with a second, statistically independent estimate of the covariance matrix C which is ordinarily obtained from an additional, independent sample of simulated unforced variation.

Signal estimates are obtained by averaging across an ensemble of forced climate change simulations, but contain remnants of the climate's natural internal variability because the ensembles are finite. When ensembles are small or signals weak, these remnants may bias ordinary least-squares estimates of a downward. This is avoided by estimating a with the total least-squares algorithm (Allen and Stott 2003).

9.A.2   Methods of Inference

Detection and attribution questions are assessed through a combination of physical reasoning (to determine, for example, by assessing consistency of possible responses, whether other mechanisms of change not included in the climate model could plausibly explain the observed change) and by evaluating specific hypotheses about the scaling factors contained in a. Most studies evaluate these hypotheses using standard frequentist methods (Hasselmann, 1979, 1997; Hegerl et al., 1997; Allen and Tett, 1999). Several recent studies have also used Bayesian methods (Hasselmann, 1998; Leroy, 1998; Min et al., 2004, 2005; Lee et al., 2005, 2006; Schnur and Hasselmann, 2005; Min and Hense, 2006a,b).

In the standard approach, detection of a postulated climate change signal occurs when its amplitude in observations is shown to be significantly different from zero (i.e., when the null hypothesis $H_D : a = 0$ where 0 is a vector of zeros, is rejected) with departure from zero in the physically plausible direction. Subsequently, the second attribution requirement (consistency with a combination of external forcings and natural internal variability) is assessed with the 'attribution consistency test' (Hasselmann, 1997; see also Allen and Tett, 1999) that evaluates the null hypothesis $H_A : a = 1$ where 1 denotes a vector of units. This test does not constitute a complete attribution assessment, but contributes important evidence to such assessments (see Mitchell et al., 2001). Attribution studies usually also test whether the response to a key forcing, such as greenhouse gas increases, is distinguishable from that to other forcings, usually based on the results of multiple regression (see above) using the most important forcings simultaneously in X. If the response to a key forcing (e.g., due to greenhouse gas increases) is detected by rejecting the hypothesis that its amplitude $a_{GHG} = 0$ in such a multiple regression, this provides strong attribution information because it demonstrates that the observed climate change is 'not consistent with alternative, physically plausible explanations of recent climate change that exclude important elements of the given combination of forcings' (Mitchell et al., 2001).

Bayesian approaches are of interest because they can be used to integrate information from multiple lines of evidence, and can incorporate independent prior information into the analysis. Essentially two approaches (described below) have been taken to date. In both cases, inferences are based on a posterior distribution that blends evidence from the observations with the independent prior information, which may include information on the uncertainty of external forcing estimates, climate models and their responses to forcing. In this way, all information that enters into the analysis is declared explicitly.

Schnur and Hasselmann (2005) approach the problem by developing a filtering technique that optimises the impact of the data on the prior distribution in a manner similar to the way in which optimal fingerprints maximise the ratio of the anthropogenic signal to natural variability noise in the conventional approach. The optimal filter in the Bayesian approach depends on the properties of both the natural climate variability and model errors. Inferences are made by comparing evidence, as measured by Bayes Factors (Kass and Raftery, 1995), for competing hypotheses. Other studies using similar approaches include Min et al. (2004) and Min and Hense (2006a,b). In contrast, Berliner et al. (2000) and Lee et al. (2005) use Bayesian methods only to make inferences about the estimate of **a** that is obtained from conventional optimal fingerprinting.

10

Global Climate Projections

Coordinating Lead Authors:

Gerald A. Meehl (USA), Thomas F. Stocker (Switzerland)

Lead Authors:

William D. Collins (USA), Pierre Friedlingstein (France, Belgium), Amadou T. Gaye (Senegal), Jonathan M. Gregory (UK),
Akio Kitoh (Japan), Reto Knutti (Switzerland), James M. Murphy (UK), Akira Noda (Japan), Sarah C.B. Raper (UK),
Ian G. Watterson (Australia), Andrew J. Weaver (Canada), Zong-Ci Zhao (China)

Contributing Authors:

R.B. Alley (USA), J. Annan (Japan, UK), J. Arblaster (USA, Australia), C. Bitz (USA), P. Brockmann (France),
V. Brovkin (Germany, Russian Federation), L. Buja (USA), P. Cadule (France), G. Clarke (Canada), M. Collier (Australia), M. Collins (UK),
E. Driesschaert (Belgium), N.A. Diansky (Russian Federation), M. Dix (Australia), K. Dixon (USA), J.-L. Dufresne (France),
M. Dyurgerov (Sweden, USA), M. Eby (Canada), N.R. Edwards (UK), S. Emori (Japan), P. Forster (UK), R. Furrer (USA, Switzerland),
P. Gleckler (USA), J. Hansen (USA), G. Harris (UK, New Zealand), G.C. Hegerl (USA, Germany), M. Holland (USA), A. Hu (USA, China),
P. Huybrechts (Belgium), C. Jones (UK), F. Joos (Switzerland), J.H. Jungclaus (Germany), J. Kettleborough (UK), M. Kimoto (Japan),
T. Knutson (USA), M. Krynytzky (USA), D. Lawrence (USA), A. Le Brocq (UK), M.-F. Loutre (Belgium), J. Lowe (UK),
H.D. Matthews (Canada), M. Meinshausen (Germany), S.A. Müller (Switzerland), S. Nawrath (Germany), J. Oerlemans (Netherlands),
M. Oppenheimer (USA), J. Orr (Monaco, USA), J. Overpeck (USA), T. Palmer (ECMWF, UK), A. Payne (UK), G.-K. Plattner (Switzerland),
J. Räisänen (Finland), A. Rinke (Germany), E. Roeckner (Germany), G.L. Russell (USA), D. Salas y Melia (France), B. Santer (USA),
G. Schmidt (USA, UK), A. Schmittner (USA, Germany), B. Schneider (Germany), A. Shepherd (UK), A. Sokolov (USA, Russian Federation),
D. Stainforth (UK), P.A. Stott (UK), R.J. Stouffer (USA), K.E. Taylor (USA), C. Tebaldi (USA), H. Teng (USA, China), L. Terray (France),
R. van de Wal (Netherlands), D. Vaughan (UK), E. M. Volodin (Russian Federation), B. Wang (China), T. M. L. Wigley (USA),
M. Wild (Switzerland), J. Yoshimura (Japan), R. Yu (China), S. Yukimoto (Japan)

Review Editors:

Myles Allen (UK), Govind Ballabh Pant (India)

This chapter should be cited as:

Meehl, G.A., T.F. Stocker, W.D. Collins, P. Friedlingstein, A.T. Gaye, J.M. Gregory, A. Kitoh, R. Knutti, J.M. Murphy, A. Noda, S.C.B. Raper,
I.G. Watterson, A.J. Weaver and Z.-C. Zhao, 2007: Global Climate Projections. In: *Climate Change 2007: The Physical Science Basis.
Contribution of Working Group I to the Fourth Assessment Report of the Intergovernmental Panel on Climate Change* [Solomon, S.,
D. Qin, M. Manning, Z. Chen, M. Marquis, K.B. Averyt, M. Tignor and H.L. Miller (eds.)]. Cambridge University Press, Cambridge, United
Kingdom and New York, NY, USA.

Table of Contents

Supplementary Material

The following Supplementary Material is available on CD-ROM and in on-line versions of this report.

Supplementary Figures S10.1 to S10.4

Figures Showing Individual Model Results for Different Climate Variables

Executive Summary

The future climate change results assessed in this chapter are based on a hierarchy of models, ranging from Atmosphere-Ocean General Circulation Models (AOGCMs) and Earth System Models of Intermediate Complexity (EMICs) to Simple Climate Models (SCMs). These models are forced with concentrations of greenhouse gases and other constituents derived from various emissions scenarios ranging from non-mitigation scenarios to idealised long-term scenarios. In general, we assess non-mitigated projections of future climate change at scales from global to hundreds of kilometres. Further assessments of regional and local climate changes are provided in Chapter 11. Due to an unprecedented, joint effort by many modelling groups worldwide, climate change projections are now based on multi-model means, differences between models can be assessed quantitatively and in some instances, estimates of the probability of change of important climate system parameters complement expert judgement. New results corroborate those given in the Third Assessment Report (TAR). Continued greenhouse gas emissions at or above current rates will cause further warming and induce many changes in the global climate system during the 21st century that would *very likely* be larger than those observed during the 20th century.

Mean Temperature

All models assessed here, for all the non-mitigation scenarios considered, project increases in global mean surface air temperature (SAT) continuing over the 21st century, driven mainly by increases in anthropogenic greenhouse gas concentrations, with the warming proportional to the associated radiative forcing. There is close agreement of globally averaged SAT multi-model mean warming for the early 21st century for concentrations derived from the three non-mitigated IPCC Special Report on Emission Scenarios (SRES: B1, A1B and A2) scenarios (including only anthropogenic forcing) run by the AOGCMs (warming averaged for 2011 to 2030 compared to 1980 to 1999 is between +0.64°C and +0.69°C, with a range of only 0.05°C). Thus, this warming rate is affected little by different scenario assumptions or different model sensitivities, and is consistent with that observed for the past few decades (see Chapter 3). Possible future variations in natural forcings (e.g., a large volcanic eruption) could change those values somewhat, but about half of the early 21st-century warming is committed in the sense that it would occur even if atmospheric concentrations were held fixed at year 2000 values. By mid-century (2046–2065), the choice of scenario becomes more important for the magnitude of multi-model globally averaged SAT warming, with values of +1.3°C, +1.8°C and +1.7°C from the AOGCMs for B1, A1B and A2, respectively. About a third of that warming is projected to be due to climate change that is already committed. By late century (2090–2099), differences between scenarios are large, and only about 20% of that warming arises from climate change that is already committed.

An assessment based on AOGCM projections, probabilistic methods, EMICs, a simple model tuned to the AOGCM responses, as well as coupled climate carbon cycle models, suggests that for non-mitigation scenarios, the future increase in global mean SAT is *likely* to fall within –40 to +60% of the multi-model AOGCM mean warming simulated for a given scenario. The greater uncertainty at higher values results in part from uncertainties in the carbon cycle feedbacks. The multi-model mean SAT warming and associated uncertainty ranges for 2090 to 2099 relative to 1980 to 1999 are B1: +1.8°C (1.1°C to 2.9°C), B2: +2.4°C (1.4°C to 3.8°C), A1B: +2.8°C (1.7°C to 4.4°C), A1T: 2.4°C (1.4°C to 3.8°C), A2: +3.4°C (2.0°C to 5.4°C) and A1FI: +4.0°C (2.4°C to 6.4°C). It is not appropriate to compare the lowest and highest values across these ranges against the single range given in the TAR, because the TAR range resulted only from projections using an SCM and covered all SRES scenarios, whereas here a number of different and independent modelling approaches are combined to estimate ranges for the six illustrative scenarios separately. Additionally, in contrast to the TAR, carbon cycle uncertainties are now included in these ranges. These uncertainty ranges include only anthropogenically forced changes.

Geographical patterns of projected SAT warming show greatest temperature increases over land (roughly twice the global average temperature increase) and at high northern latitudes, and less warming over the southern oceans and North Atlantic, consistent with observations during the latter part of the 20th century (see Chapter 3). The pattern of zonal mean warming in the atmosphere, with a maximum in the upper tropical troposphere and cooling throughout the stratosphere, is notable already early in the 21st century, while zonal mean warming in the ocean progresses from near the surface and in the northern mid-latitudes early in the 21st century, to gradual penetration downward during the course of the 21st century.

An expert assessment based on the combination of available constraints from observations (assessed in Chapter 9) and the strength of known feedbacks simulated in the models used to produce the climate change projections in this chapter indicates that the equilibrium global mean SAT warming for a doubling of atmospheric carbon dioxide (CO_2), or 'equilibrium climate sensitivity', is *likely* to lie in the range 2°C to 4.5°C, with a most likely value of about 3°C. Equilibrium climate sensitivity is *very likely* larger than 1.5°C. For fundamental physical reasons, as well as data limitations, values substantially higher than 4.5°C still cannot be excluded, but agreement with observations and proxy data is generally worse for those high values than for values in the 2°C to 4.5°C range. The 'transient climate response' (TCR, defined as the globally averaged SAT change at the time of CO_2 doubling in the 1% yr^{-1} transient CO_2 increase experiment) is better constrained than equilibrium climate sensitivity. The TCR is *very likely* larger than 1°C and *very unlikely* greater than 3°C based on climate models, in agreement with constraints from the observed surface warming.

Temperature Extremes

It is *very likely* that heat waves will be more intense, more frequent and longer lasting in a future warmer climate. Cold episodes are projected to decrease significantly in a future warmer climate. Almost everywhere, daily minimum temperatures are projected to increase faster than daily maximum temperatures, leading to a decrease in diurnal temperature range. Decreases in frost days are projected to occur almost everywhere in the middle and high latitudes, with a comparable increase in growing season length.

Mean Precipitation

For a future warmer climate, the current generation of models indicates that precipitation generally increases in the areas of regional tropical precipitation maxima (such as the monsoon regimes) and over the tropical Pacific in particular, with general decreases in the subtropics, and increases at high latitudes as a consequence of a general intensification of the global hydrological cycle. Globally averaged mean water vapour, evaporation and precipitation are projected to increase.

Precipitation Extremes and Droughts

Intensity of precipitation events is projected to increase, particularly in tropical and high latitude areas that experience increases in mean precipitation. Even in areas where mean precipitation decreases (most subtropical and mid-latitude regions), precipitation intensity is projected to increase but there would be longer periods between rainfall events. There is a tendency for drying of the mid-continental areas during summer, indicating a greater risk of droughts in those regions. Precipitation extremes increase more than does the mean in most tropical and mid- and high-latitude areas.

Snow and Ice

As the climate warms, snow cover and sea ice extent decrease; glaciers and ice caps lose mass owing to a dominance of summer melting over winter precipitation increases. This contributes to sea level rise as documented for the previous generation of models in the TAR. There is a projected reduction of sea ice in the 21st century in both the Arctic and Antarctic with a rather large range of model responses. The projected reduction is accelerated in the Arctic, where some models project summer sea ice cover to disappear entirely in the high-emission A2 scenario in the latter part of the 21st century. Widespread increases in thaw depth over much of the permafrost regions are projected to occur in response to warming over the next century.

Carbon Cycle

There is unanimous agreement among the coupled climate-carbon cycle models driven by emission scenarios run so far that future climate change would reduce the efficiency of the Earth system (land and ocean) to absorb anthropogenic CO_2. As a result, an increasingly large fraction of anthropogenic CO_2 would stay airborne in the atmosphere under a warmer climate. For the A2 emission scenario, this positive feedback leads to additional atmospheric CO_2 concentration varying between 20 and 220 ppm among the models by 2100. Atmospheric CO_2 concentrations simulated by these coupled climate-carbon cycle models range between 730 and 1,020 ppm by 2100. Comparing these values with the standard value of 836 ppm (calculated beforehand by the Bern carbon cycle-climate model without an interactive carbon cycle) provides an indication of the uncertainty in global warming due to future changes in the carbon cycle. In the context of atmospheric CO_2 concentration stabilisation scenarios, the positive climate-carbon cycle feedback reduces the land and ocean uptake of CO_2, implying that it leads to a reduction of the compatible emissions required to achieve a given atmospheric CO_2 stabilisation. The higher the stabilisation scenario, the larger the climate change, the larger the impact on the carbon cycle, and hence the larger the required emission reduction.

Ocean Acidification

Increasing atmospheric CO_2 concentrations lead directly to increasing acidification of the surface ocean. Multi-model projections based on SRES scenarios give reductions in pH of between 0.14 and 0.35 units in the 21st century, adding to the present decrease of 0.1 units from pre-industrial times. Southern Ocean surface waters are projected to exhibit undersaturation with regard to calcium carbonate for CO_2 concentrations higher than 600 ppm, a level exceeded during the second half of the century in most of the SRES scenarios. Low-latitude regions and the deep ocean will be affected as well. Ocean acidification would lead to dissolution of shallow-water carbonate sediments and could affect marine calcifying organisms. However, the net effect on the biological cycling of carbon in the oceans is not well understood.

Sea Level

Sea level is projected to rise between the present (1980–1999) and the end of this century (2090–2099) under the SRES B1 scenario by 0.18 to 0.38 m, B2 by 0.20 to 0.43 m, A1B by 0.21 to 0.48 m, A1T by 0.20 to 0.45 m, A2 by 0.23 to 0.51 m, and A1FI by 0.26 to 0.59 m. These are 5 to 95% ranges based on the spread of AOGCM results, not including uncertainty in carbon cycle feedbacks. For each scenario, the midpoint of the range is within 10% of the TAR model average for 2090-2099. The ranges are narrower than in the TAR mainly because of improved information about some uncertainties in the projected contributions. In all scenarios, the average rate of rise during the 21st century *very likely* exceeds the 1961 to 2003 average rate (1.8 ± 0.5 mm yr^{-1}). During 2090 to 2099 under A1B, the central estimate of the rate of rise is 3.8 mm yr^{-1}. For an average model, the scenario spread in sea level rise is only 0.02 m by the middle of the century, and by the end of the century it is 0.15 m.

Thermal expansion is the largest component, contributing 70 to 75% of the central estimate in these projections for all scenarios. Glaciers, ice caps and the Greenland Ice Sheet are also projected to contribute positively to sea level. General Circulation Models indicate that the Antarctic Ice Sheet will receive increased snowfall without experiencing substantial surface melting, thus gaining mass and contributing negatively to sea level. Further accelerations in ice flow of the kind recently observed in some Greenland outlet glaciers and West Antarctic ice streams could substantially increase the contribution from the ice sheets. For example, if ice discharge from these processes were to scale up in future in proportion to global average surface temperature change (taken as a measure of global climate change), it would add 0.1 to 0.2 m to the upper bound of sea level rise by 2090 to 2099. In this example, during 2090 to 2099 the rate of scaled-up Antarctic discharge would roughly balance the expected increased rate of Antarctic accumulation, being under A1B a factor of 5 to 10 greater than in recent years. Understanding of these effects is too limited to assess their likelihood or to give a best estimate.

Sea level rise during the 21st century is projected to have substantial geographical variability. The model median spatial standard deviation is 0.08 m under A1B. The patterns from different models are not generally similar in detail, but have some common features, including smaller than average sea level rise in the Southern Ocean, larger than average in the Arctic, and a narrow band of pronounced sea level rise stretching across the southern Atlantic and Indian Oceans.

Mean Tropical Pacific Climate Change

Multi-model averages show a weak shift towards average background conditions which may be described as 'El Niño-like', with sea surface temperatures in the central and east equatorial Pacific warming more than those in the west, weakened tropical circulations and an eastward shift in mean precipitation.

El Niño

All models show continued El Niño-Southern Oscillation (ENSO) interannual variability in the future no matter what the change in average background conditions, but changes in ENSO interannual variability differ from model to model. Based on various assessments of the current multi-model data set, in which present-day El Niño events are now much better simulated than in the TAR, there is no consistent indication at this time of discernible changes in projected ENSO amplitude or frequency in the 21st century.

Monsoons

An increase in precipitation is projected in the Asian monsoon (along with an increase in interannual season-averaged precipitation variability) and the southern part of the west African monsoon with some decrease in the Sahel in northern summer, as well as an increase in the Australian monsoon in southern summer in a warmer climate. The monsoonal precipitation in Mexico and Central America is projected to decrease in association with increasing precipitation over the eastern equatorial Pacific through Walker Circulation and local Hadley Circulation changes. However, the uncertain role of aerosols in general, and carbon aerosols in particular, complicates the nature of future projections of monsoon precipitation, particularly in the Asian monsoon.

Sea Level Pressure

Sea level pressure is projected to increase over the subtropics and mid-latitudes, and decrease over high latitudes (order several millibars by the end of the 21st century) associated with a poleward expansion and weakening of the Hadley Circulation and a poleward shift of the storm tracks of several degrees latitude with a consequent increase in cyclonic circulation patterns over the high-latitude arctic and antarctic regions. Thus, there is a projected positive trend of the Northern Annular Mode (NAM) and the closely related North Atlantic Oscillation (NAO) as well as the Southern Annular Mode (SAM). There is considerable spread among the models for the NAO, but the magnitude of the increase for the SAM is generally more consistent across models.

Tropical Cyclones (Hurricanes and Typhoons)

Results from embedded high-resolution models and global models, ranging in grid spacing from 100 km to 9 km, project a *likely* increase of peak wind intensities and notably, where analysed, increased near-storm precipitation in future tropical cyclones. Most recent published modelling studies investigating tropical storm frequency simulate a decrease in the overall number of storms, though there is less confidence in these projections and in the projected decrease of relatively weak storms in most basins, with an increase in the numbers of the most intense tropical cyclones.

Mid-latitude Storms

Model projections show fewer mid-latitude storms averaged over each hemisphere, associated with the poleward shift of the storm tracks that is particularly notable in the Southern Hemisphere, with lower central pressures for these poleward-shifted storms. The increased wind speeds result in more extreme wave heights in those regions.

Atlantic Ocean Meridional Overturning Circulation

Based on current simulations, it is *very likely* that the Atlantic Ocean Meridional Overturning Circulation (MOC) will slow down during the course of the 21st century. A multi-model ensemble shows an average reduction of 25% with a broad range from virtually no change to a reduction of over 50% averaged over 2080 to 2099. In spite of a slowdown of the MOC in most models, there is still warming of surface temperatures around the North Atlantic Ocean and Europe due to the much larger radiative effects of the increase in greenhouse gases. Although the MOC weakens in most model runs for the three SRES scenarios, none shows a collapse of the MOC by the year 2100 for the scenarios considered. No coupled model simulation of the Atlantic MOC shows a mean increase in the MOC in response to global warming by 2100. It is *very unlikely* that the MOC will undergo a large abrupt transition during the course of the 21st century. At this stage, it is too early to assess the likelihood of a large abrupt change of the MOC beyond the end of the 21st century. In experiments with the low (B1) and medium (A1B) scenarios, and for which the atmospheric greenhouse gas concentrations are stabilised beyond 2100, the MOC recovers from initial weakening within one to several centuries after 2100 in some of the models. In other models the reduction persists.

Radiative Forcing

The radiative forcings by long-lived greenhouse gases computed with the radiative transfer codes in twenty of the AOGCMs used in the Fourth Assessment Report have been compared against results from benchmark line-by-line (LBL) models. The mean AOGCM forcing over the period 1860 to 2000 agrees with the mean LBL value to within 0.1 W m^{-2} at the tropopause. However, there is a range of 25% in longwave forcing due to doubling atmospheric CO_2 from its concentration in 1860 across the ensemble of AOGCM codes. There is a 47% relative range in longwave forcing in 2100 contributed by all greenhouse gases in the A1B scenario across the ensemble of AOGCM simulations. These results imply that the ranges in climate sensitivity and climate response from models discussed in this chapter may be due in part to differences in the formulation and treatment of radiative processes among the AOGCMs.

Climate Change Commitment (Temperature and Sea Level)

Results from the AOGCM multi-model climate change commitment experiments (concentrations stabilised for 100 years at year 2000 for 20th-century commitment, and at 2100 values for B1 and A1B commitment) indicate that if greenhouse gases were stabilised, then a further warming of 0.5°C would occur. This should not be confused with 'unavoidable climate change' over the next half century, which would be greater because forcing cannot be instantly stabilised. In the very long term, it is plausible that climate change could be less than in a commitment run since forcing could be reduced below current levels. Most of this warming occurs in the first several decades after stabilisation; afterwards the rate of increase steadily declines. The globally averaged precipitation commitment 100 years after stabilising greenhouse gas concentrations amounts to roughly an additional increase of 1 to 2% compared to the precipitation values at the time of stabilisation.

If concentrations were stabilised at A1B levels in 2100, sea level rise due to thermal expansion in the 22nd century would be similar to that in the 21st, and would amount to 0.3 to 0.8 m (relative to 1980 to 1999) above present by 2300. The ranges of thermal expansion overlap substantially for stabilisation at different levels, since model uncertainty is dominant; A1B is given here because most model results are available for that scenario. Thermal expansion would continue over many centuries at a gradually decreasing rate, reaching an eventual level of 0.2 to 0.6 m per °C of global warming relative to present. Under sustained elevated temperatures, some glacier volume may persist at high altitudes, but most could disappear over centuries.

If greenhouse gas concentrations could be reduced, global temperatures would begin to decrease within a decade, although sea level would continue to rise due to thermal expansion for at least another century. Earth System Models of Intermediate Complexity with coupled carbon cycle model components show that for a reduction to zero emissions at year 2100 the climate would take of the order of 1 kyr to stabilise. At year 3000, the model range for temperature increase is 1.1°C to 3.7°C and for sea level rise due to thermal expansion is 0.23 to 1.05 m. Hence, they are projected to remain well above their pre-industrial values.

The Greenland Ice Sheet is projected to contribute to sea level after 2100, initially at a rate of 0.03 to 0.21 m per century for stabilisation in 2100 at A1B concentrations. The contribution would be greater if dynamical processes omitted from current models increased the rate of ice flow, as has been observed in recent years. Except for remnant glaciers in the mountains, the Greenland Ice Sheet would largely be eliminated, raising sea level by about 7 m, if a sufficiently warm climate were maintained for millennia; it would happen more rapidly if ice flow accelerated. Models suggest that the global warming required lies in the range 1.9°C to 4.6°C relative to the pre-industrial temperature. Even if temperatures were to decrease later, it is possible that the reduction of the ice sheet to a much smaller extent would be irreversible.

The Antarctic Ice Sheet is projected to remain too cold for widespread surface melting, and to receive increased snowfall, leading to a gain of ice. Loss of ice from the ice sheet could occur through increased ice discharge into the ocean following weakening of ice shelves by melting at the base or on the surface. In current models, the net projected contribution to sea level rise is negative for coming centuries, but it is possible that acceleration of ice discharge could become dominant, causing a net positive contribution. Owing to limited understanding of the relevant ice flow processes, there is presently no consensus on the long-term future of the ice sheet or its contribution to sea level rise.

10.1   Introduction

Since the Third Assessment Report (TAR), the scientific community has undertaken the largest coordinated global coupled climate model experiment ever attempted in order to provide the most comprehensive multi-model perspective on climate change of any IPCC assessment, the World Climate Research Programme (WCRP) Coupled Model Intercomparison Project phase three (CMIP3), also referred to generically throughout this report as the 'multi-model data set' (MMD) archived at the Program for Climate Model Diagnosis and Intercomparison (PCMDI). This open process involves experiments with idealised climate change scenarios (i.e., 1% yr^{-1} carbon dioxide (CO_2) increase, also included in the earlier WCRP model intercomparison projects CMIP2 and CMIP2+ (e.g., Covey et al., 2003; Meehl et al., 2005b), equilibrium $2 \times CO_2$ experiments with atmospheric models coupled to non-dynamic slab oceans, and idealised stabilised climate change experiments at $2 \times CO_2$ and $4 \times$ atmospheric CO_2 levels in the 1% yr^{-1} CO_2 increase simulations).

In the idealised 1% yr^{-1} CO_2 increase experiments, there is no actual real year time line. Thus, the rate of climate change is not the issue in these experiments, but what is studied are the types of climate changes that occur at the time of doubling or quadrupling of atmospheric CO_2 and the range of, and difference in, model responses. Simulations of 20th-century climate have been completed that include temporally evolving natural and anthropogenic forcings. For projected climate change in the 21st century, a subset of three IPCC Special Report on Emission Scenarios (SRES; Nakićenović and Swart, 2000) scenario simulations have been selected from the six commonly used marker scenarios. With respect to emissions, this subset (B1, A1B and A2) consists of a 'low', 'medium' and 'high' scenario

among the marker scenarios, and this choice is solely made by the constraints of available computer resources that did not allow for the calculation of all six scenarios. This choice, therefore, does not imply a qualification of, or preference over, the six marker scenarios. In addition, it is not within the scope of the Working Group I contribution to the Fourth Assessment Report (AR4) to assess the plausibility or likelihood of emission scenarios.

In addition to these non-mitigation scenarios, a series of idealised model projections is presented, each of which implies some form and level of intervention: (i) stabilisation scenarios in which greenhouse gas concentrations are stabilised at various levels, (ii) constant composition commitment scenarios in which greenhouse gas concentrations are fixed at year 2000 levels, (iii) zero emission commitment scenarios in which emissions are set to zero in the year 2100 and (iv) overshoot scenarios in which greenhouse gas concentrations are reduced after year 2150.

The simulations with the subset A1B, B1 and A2 were performed to the year 2100. Three different stabilisation scenarios were run, the first with all atmospheric constituents fixed at year 2000 values and the models run for an additional 100 years, and the second and third with constituents fixed at year 2100 values for A1B and B1, respectively, for another 100 to 200 years. Consequently, the concept of climate change commitment (for details and definitions see Section 10.7) is addressed in much wider scope and greater detail than in any previous IPCC assessment. Results based on this Atmosphere-Ocean General Circulation Model (AOGCM) multi-model data set are featured in Section 10.3.

Uncertainty in climate change projections has always been a subject of previous IPCC assessments, and a substantial amount of new work is assessed in this chapter. Uncertainty arises in various steps towards a climate projection (Figure 10.1). For

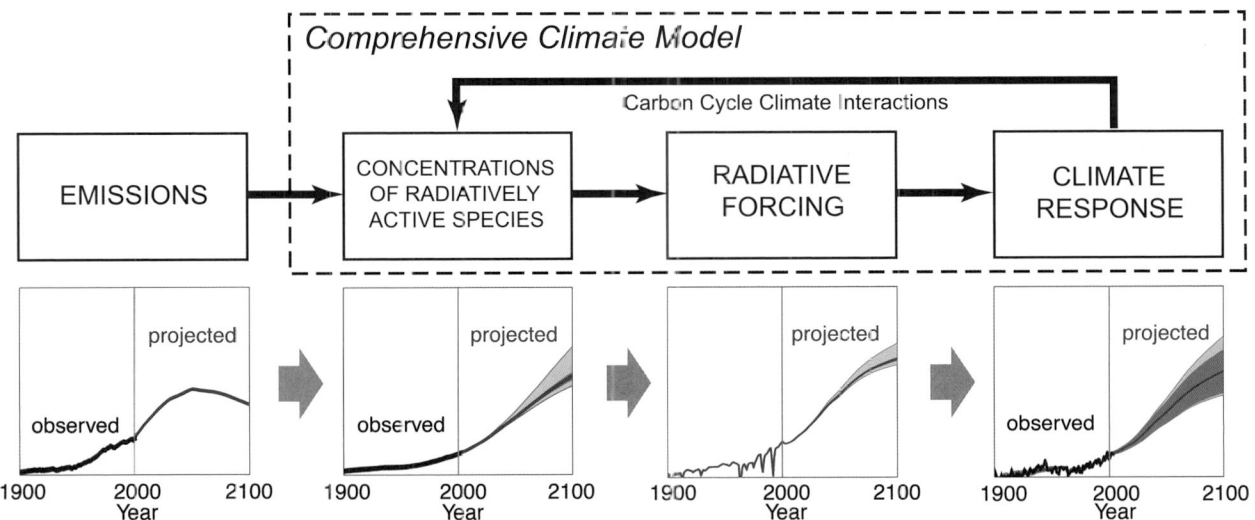

Figure 10.1. *Several steps from emissions to climate response contribute to the overall uncertainty of a climate model projection. These uncertainties can be quantified through a combined effort of observation, process understanding, a hierarchy of climate models, and ensemble simulations. In a comprehensive climate model, physical and chemical representations of processes permit a consistent quantification of uncertainty. Note that the uncertainty associated with the future emission path is of an entirely different nature and not addressed in Chapter 10. Bottom row adapted from Figure 10.26, A1B scenario, for illustration only.*

a given emissions scenario, various biogeochemical models are used to calculate concentrations of constituents in the atmosphere. Various radiation schemes and parametrizations are required to convert these concentrations to radiative forcing. Finally, the response of the different climate system components (atmosphere, ocean, sea ice, land surface, chemical status of atmosphere and ocean, etc.) is calculated in a comprehensive climate model. In addition, the formulation of, and interaction with, the carbon cycle in climate models introduces important feedbacks which produce additional uncertainties. In a comprehensive climate model, physical and chemical representations of processes permit a consistent quantification of uncertainty. Note that the uncertainties associated with the future emission path are of an entirely different nature and not considered in this chapter.

Many of the figures in Chapter 10 are based on the mean and spread of the multi-model ensemble of comprehensive AOGCMs. The reason to focus on the multi-model mean is that averages across structurally different models empirically show better large-scale agreement with observations, because individual model biases tend to cancel (see Chapter 8). The expanded use of multi-model ensembles of projections of future climate change therefore provides higher quality and more quantitative climate change information compared to the TAR. Even though the ability to simulate present-day mean climate and variability, as well as observed trends, differs across models, no weighting of individual models is applied in calculating the mean. Since the ensemble is strictly an 'ensemble of opportunity', without sampling protocol, the spread of models does not necessarily span the full possible range of uncertainty, and a statistical interpretation of the model spread is therefore problematic. However, attempts are made to quantify uncertainty throughout the chapter based on various other lines of evidence, including perturbed physics ensembles specifically designed to study uncertainty within one model framework, and Bayesian methods using observational constraints.

In addition to this coordinated international multi-model experiment, a number of entirely new types of experiments have been performed since the TAR to quantify uncertainty regarding climate model response to external forcings. The extent to which uncertainties in parametrizations translate into the uncertainty in climate change projections is addressed in much greater detail. New calculations of future climate change from the larger suite of SRES scenarios with simple models and Earth System Models of Intermediate Complexity (EMICs) provide additional information regarding uncertainty related to the choice of scenario. Such models also provide estimates of long-term evolution of global mean temperature, ocean heat uptake and sea level rise due to thermal expansion beyond the 21st century, and thus allow climate change commitments to be better constrained.

Climate sensitivity has always been a focus in the IPCC assessments, and this chapter assesses more quantitative estimates of equilibrium climate sensitivity and transient climate response (TCR) in terms of not only ranges but also probabilities within these ranges. Some of these probabilities are now derived from ensemble simulations subject to various observational constraints, and no longer rely solely on expert judgement. This permits a much more complete assessment of model response uncertainties from these sources than ever before. These are now standard benchmark calculations with the global coupled climate models, and are useful to assess model response in the subsequent time-evolving climate change scenario experiments.

With regard to these time-evolving experiments simulating 21st-century climate, since the TAR increased computing capabilities now allow routine performance of multi-member ensembles in climate change scenario experiments with global coupled climate models. This provides the capability to analyse more multi-model results and multi-member ensembles, and yields more probabilistic estimates of time-evolving climate change in the 21st century.

Finally, while future changes in some weather and climate extremes (e.g., heat waves) were addressed in the TAR, there were relatively few studies on this topic available for assessment at that time. Since then, more analyses have been performed regarding possible future changes in a variety of extremes. It is now possible to assess, for the first time, multi-model ensemble results for certain types of extreme events (e.g., heat waves, frost days, etc.). These new studies provide a more complete range of results for assessment regarding possible future changes in these important phenomena with their notable impacts on human societies and ecosystems. A synthesis of results from studies of extremes from observations and model is provided in Chapter 11.

The use of multi-model ensembles has been shown in other modelling applications to produce simulated climate features that are improved over single models alone (see discussion in Chapters 8 and 9). In addition, a hierarchy of models ranging from simple to intermediate to complex allows better quantification of the consequences of various parametrizations and formulations. Very large ensembles (order hundreds) with single models provide the means to quantify parametrization uncertainty. Finally, observed climate characteristics are now being used to better constrain future climate model projections.

10.2 Projected Changes in Emissions, Concentrations and Radiative Forcing

The global projections discussed in this chapter are extensions of the simulations of the observational record discussed in Chapter 9. The simulations of the 19th and 20th centuries are based upon changes in long-lived greenhouse gases (LLGHGs) that are reasonably constrained by the observational record. Therefore, the models have qualitatively similar temporal evolutions of their radiative forcing time histories for LLGHGs (e.g., see Figure 2.23). However, estimates of future concentrations of LLGHGs and other radiatively active species are clearly subject to significant uncertainties. The evolution of these species is governed by a variety of factors that are difficult to predict, including changes in population, energy use, energy sources and emissions. For these reasons, a range of projections of future climate change has been conducted using coupled AOGCMs. The future concentrations of LLGHGs and the anthropogenic emissions of sulphur dioxide (SO_2), a chemical precursor of sulphate aerosol, are obtained from several scenarios considered representative of low, medium and high emission trajectories. These basic scenarios and other forcing agents incorporated in the AOGCM projections, including several types of natural and anthropogenic aerosols, are discussed in Section 10.2.1. Developments in projecting radiatively active species and radiative forcing for the early 21st century are considered in Section 10.2.2.

10.2.1 Emissions Scenarios and Radiative Forcing in the Multi-Model Climate Projections

The temporal evolution of the LLGHGs, aerosols and other forcing agents are described in Sections 10.2.1.1 and 10.2.1.2. Typically, the future projections are based upon initial conditions extracted from the end of the simulations of the 20th century. Therefore, the radiative forcing at the beginning of the model projections should be approximately equal to the radiative forcing for present-day concentrations relative to pre-industrial conditions. The relationship between the modelled radiative forcing for the year 2000 and the estimates derived in Chapter 2 is evaluated in Section 10.2.1.3. Estimates of the radiative forcing in the multi-model integrations for one of the standard scenarios are also presented in this section. Possible explanations for the range of radiative forcings projected for 2100 are discussed in Section 10.2.1.4, including evidence for systematic errors in the formulations of radiative transfer used in AOGCMs. Possible implications of these findings for the range of global temperature change and other climate responses are summarised in Section 10.2.1.5.

10.2.1.1 The Special Report on Emission Scenarios and Constant-Concentration Commitment Scenarios

The future projections discussed in this chapter are based upon the standard A2, A1B and B2 SRES scenarios (Nakićenović and Swart, 2000). The emissions of CO_2, methane (CH_4) and SO_2, the concentrations of CO_2, CH_4 and nitrous oxide (N_2O) and the total radiative forcing for the SRES scenarios are illustrated in Figure 10.26 and summarised for the A1B scenario in Figure 10.1. The models have been integrated to year 2100 using the projected concentrations of LLGHGs and emissions of SO_2 specified by the A1B, B1 and A2 emissions scenarios. Some of the AOGCMs do not include sulphur chemistry, and the simulations from these models are based upon concentrations of sulphate aerosols from Boucher and Pham (2002; see Section 10.2.1.2). The simulations for the three scenarios were continued for another 100 to 200 years with all anthropogenic forcing agents held fixed at values applicable to the year 2100. There is also a new constant-concentration commitment scenario that assumes concentrations are held fixed at year 2000 levels (Section 10.7.1). In this idealised scenario, models are initialised from the end of the simulations for the 20th century, the concentrations of radiatively active species are held constant at year 2000 values from these simulations, and the models are integrated to 2100.

For comparison with this constant composition case, it is useful to note that constant emissions would lead to much larger radiative forcing. For example, constant CO_2 emissions at year 2000 values would lead to concentrations reaching about 520 ppm by 2100, close to the B1 case (Friedlingstein and Solomon, 2005; Hare and Munschausen, 2006; see also FAQ 10.3).

10.2.1.2 Forcing by Additional Species and Mechanisms

The forcing agents applied to each AOGCM used to make climate projections are summarised in Table 10.1. The radiatively active species specified by the SRES scenarios are CO_2, CH_4, N_2O, chlorofluorocarbons (CFCs) and SO_2, which is listed in its aerosol form as sulphate (SO_4) in the table. The inclusion, magnitude and temporal evolution of the remaining forcing agents listed in Table 10.1 were left to the discretion of the individual modelling groups. These agents include tropospheric and stratospheric ozone, all of the non-sulphate aerosols, the indirect effects of aerosols on cloud albedo and lifetime, the effects of land use and solar variability.

The scope of the treatments of aerosol effects in AOGCMs has increased markedly since the TAR. Seven of the AOGCMs include the first indirect effects and five include the second indirect effects of aerosols on cloud properties (Section 2.4.5). Under the more emissions-intensive scenarios considered in this chapter, the magnitude of the first indirect (Twomey) effect can saturate. Johns et al. (2003) parametrize the first indirect effect of anthropogenic sulphur (S) emissions as perturbations to the effective radii of cloud drops in simulations of the B1, B2, A2 and A1FI scenarios using UKMO-HadCM3. At 2100, the first indirect forcing ranges from –0.50 to

Table 10.1. Radiative forcing agents in the multi-model global climate projections. See Table 8.1 for descriptions of the models. Entries mean Y: forcing agent is included; C: forcing agent varies with time during the 20th Century Climate in Coupled Models (20C3M) simulations and is set to constant or annually cyclic distribution for scenario integrations; E: forcing agent represented using equivalent CO_2; and n.a.: forcing agent is not specified in either the 20th-century or scenario integrations. Numeric codes indicate that the forcing agent is included using data described at 1: http://www.cnrm.meteo.fr/ensembles/public/results/results.html; 2: Boucher and Pham (2002); 3: Yukimoto et al. (2006); 4: Meehl, et al., 2006b; 5: http://aom.giss.nasa.gov/IN/GHGA1B.LP; and 6: http://sres.ciesin.org/final_data.html.

Model	Greenhouse Gases						Aerosols										Other	
	CO₂	CH₄	N₂O	Stratospheric Ozone	Tropospheric Ozone	CFCs	SO₄	Urban	Black carbon	Organic carbon	Nitrate	1st Indirect	2nd Indirect	Dust	Volcanic	Sea Salt	Land Use	Solar
BCC-CM1	Y	Y	Y	Y	C	4	4	n.a.	n.a.	n.a.	n.a.	n.a.	n.a.	n.a.	C	n.a.	C	C
BCCR-BCM2.0	1	1	1	C	C	1	2	C	n.a.	n.a.	n.a.	n.a.	n.a.	n.a.	n.a.	n.a.	C	C
CCSM3	4	4	4	4	4	4	4	n.a.	4	4	n.a.	n.a.	n.a.	Y	C	Y	n.a.	C
CGCM3.1(T47)	Y	Y	Y	C	C	Y	2	n.a.	n.a.	n.a.	n.a.	n.a.	n.a.	Y	C	C	C	C
CGCM3.1(T63)	Y	Y	Y	C	C	Y	2	C	n.a.	n.a.	n.a.	n.a.	n.a.	C	C	C	C	C
CNRM-CM3	1	1	1	Y	Y	1	2	C	n.a.	n.a.	n.a.	n.a.	n.a.	C	n.a.	C	C	C
CSIRO-MK3.0	Y	E	E	Y	Y	E	Y	n.a.	n.a.	n.a.	n.a.	Y	n.a.	n.a.	n.a.	n.a.	n.a.	n.a.
ECHAM5/MPI-OM	1	1	1	Y	C	1	2	n.a.	n.a.	n.a.	n.a.	Y	n.a.	n.a.	C	n.a.	n.a.	n.a.
ECHO-G	1	1	1	C	Y	1	6	n.a.	n.a.	n.a.	n.a.	n.a.	n.a.	n.a.	C	n.a.	C	C
FGOALS-g1.0	4	4	4	C	C	4	4	n.a.	n.a.	n.a.	n.a.	n.a.	n.a.	C	C	C	C	C
GFDL-CM2.0	Y	Y	Y	Y	Y	Y	Y	n.a.	Y	Y	n.a.	n.a.	n.a.	C	C	C	C	C
GFDL-CM2.1	Y	Y	Y	Y	Y	Y	Y	n.a.	Y	Y	n.a.	n.a.	n.a.	n.a.	C	C	C	n.a.
GISS-AOM	5	5	5	C	C	5	2	n.a.	n.a.	n.a.	n.a.	n.a.	n.a.	C	n.a.	Y	n.a.	Y
GISS-EH	Y	Y	Y	Y	Y	Y	Y	n.a.	Y	Y	Y	Y	Y	Y	Y	C	Y	Y
GISS-ER	Y	Y	Y	Y	Y	Y	Y	n.a.	Y	Y	Y	Y	Y	C	Y	C	Y	Y
INM-CM3.0	4	4	4	C	C	n.a.	4	n.a.	n.a.	n.a.	n.a.	n.a.	n.a.	n.a.	C	n.a.	n.a.	n.a.
IPSL-CM4	1	1	1	n.a.	n.a.	1	2	n.a.	Y	Y	n.a.	Y	Y	Y	n.a.	Y	C	C
MIROC3.2(H)	Y	Y	Y	Y	Y	Y	Y	n.a.	Y	Y	n.a.	Y	Y	Y	C	Y	C	C
MIROC3.2(M)	Y	Y	Y	Y	Y	Y	Y	n.a.	Y	Y	n.a.	Y	Y	Y	C	Y	C	C
MRI-CGCM2.3.2	3	3	3	C	C	3	3	n.a.	n.a.	n.a.	n.a.	n.a.	n.a.	n.a.	C	n.a.	n.a.	C
PCM	Y	Y	Y	Y	Y	Y	Y	n.a.	n.a.	n.a.	n.a.	n.a.	n.a.	n.a.	C	n.a.	n.a.	C
UKMO-HadCM3	Y	Y	Y	Y	Y	Y	Y	n.a.	n.a.	n.a.	n.a.	Y	n.a.	n.a.	C	n.a.	C	C
UKMO-HadGEM1	Y	Y	Y	Y	Y	Y	Y	n.a.	Y	Y	Y	Y	Y	n.a.	C	Y	Y	C

–0.79 W m^{-2}. The normalised indirect forcing (the ratio of the forcing (W m^{-2}) to the mass burden of a species (mg m^{-2}), leaving units of W mg^{-1}) decreases by a factor of four, from approximately –7 W mgS^{-1} in 1860 to between –1 and –2 W mgS^{-1} by the year 2100. Boucher and Pham (2002) and Pham et al. (2005) find a comparable projected decrease in forcing efficiency of the indirect effect, from –9.6 W mgS^{-1} in 1860 to between –2.1 and –4.4 W mgS^{-1} in 2100. Johns et al. (2003) and Pham et al. (2005) attribute the projected decline to the decreased sensitivity of clouds to greater sulphate concentrations at sufficiently large aerosol burdens.

10.2.1.3 Comparison of Modelled Forcings to Estimates in Chapter 2

The forcings used to generate climate projections for the standard SRES scenarios are not necessarily uniform across the multi-model ensemble. Differences among models may be caused by different projections for radiatively active species (see Section 10.2.1.2) and by differences in the formulation of radiative transfer (see Section 10.2.1.4). The AOGCMs in the ensemble include many species that are not specified or constrained by the SRES scenarios, including ozone, tropospheric non-sulphate aerosols, and stratospheric volcanic aerosols. Other types of forcing that vary across the ensemble include solar variability, the indirect effects of aerosols on clouds and the effects of land use change on land surface albedo and other land surface properties (Table 10.1). While the time series of LLGHGs for the future scenarios are mostly identical across the ensemble, the concentrations of these gases in the 19th and early 20th centuries were left to the discretion of individual modelling groups. The differences in radiatively active species and the formulation of radiative transfer affect both the 19th- and 20th-century simulations and the scenario integrations initiated from these historical simulations. The resulting differences in the forcing complicate the separation of forcing and response across the multi-model ensemble. These differences can be quantified by comparing the range of shortwave and longwave forcings across the multi-model ensemble against standard estimates of radiative forcing over the historical record. Shortwave and longwave forcing refer to modifications of the solar and infrared atmospheric radiation fluxes, respectively, that are caused by external changes to the climate system (Section 2.2).

The longwave radiative forcings for the SRES A1B scenario from climate model simulations are compared against estimates using the TAR formulae (see Chapter 2) in Figure 10.2a. The graph shows the longwave forcings from the TAR and 20 AOGCMs in the multi-model ensemble from 2000 to 2100. The forcings from the models are diagnosed from changes in top-of-atmosphere fluxes and the forcing for doubled atmospheric CO$_2$ (Forster and Taylor, 2006). The TAR and median model estimates of the longwave forcing are in very good agreement over the 21st century, with differences ranging from –0.37 to +0.06 W m^{-2}. For the year 2000, the global mean values from the TAR and median model differ by only –0.13 W m^{-2}. However,

the 5th to 95th percentile range of the models for the period 2080 to 2099 is approximately 3.1 W m^{-2}, or approximately 47% of the median longwave forcing for that time period.

The corresponding time series of shortwave forcings for the SRES A1B scenario are plotted in Figure 10.2b. It is evident that the relative differences among the models and between the models and the TAR estimates are larger for the shortwave band. The TAR value is larger than the median model forcing by 0.2 to 0.3 W m^{-2} for individual 20-year segments of the integrations. For the year 2000, the TAR estimate is larger by 0.42 W m^{-2}. In addition, the range of modelled forcings is sufficiently large that it includes positive and negative values

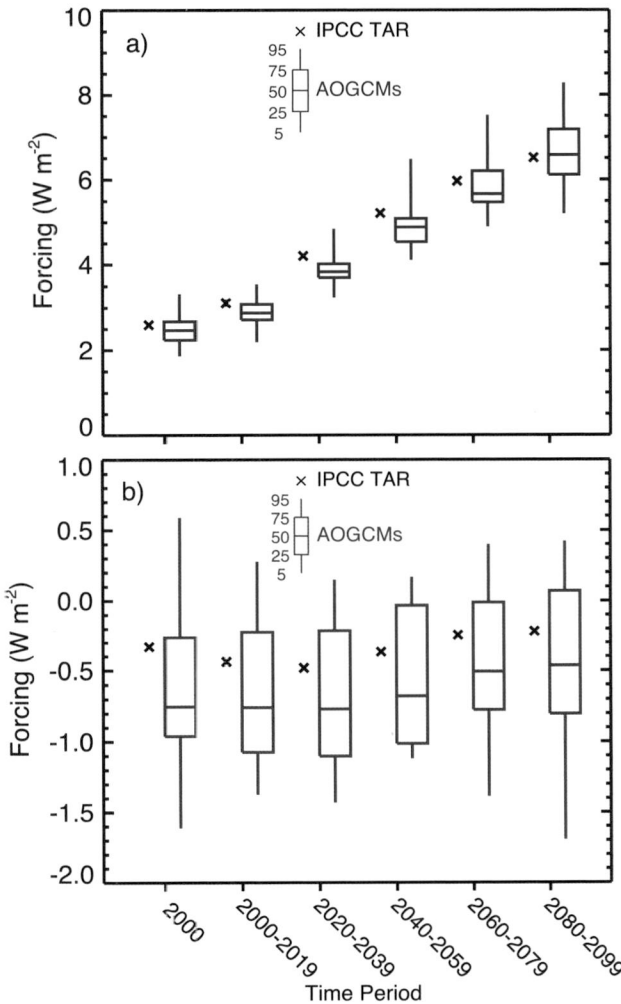

Figure 10.2. *Radiative forcings for the period 2000 to 2100 for the SRES A1B scenario diagnosed from AOGCMs and from the TAR (IPCC, 2001) forcing formulas (Forster and Taylor, 2006). (a) Longwave forcing; (b) shortwave forcing. The AOGCM results are plotted with box-and-whisker diagrams representing percentiles of forcings computed from 20 models in the AR4 multi-model ensemble. The central line within each box represents the median value of the model ensemble. The top and bottom of each box shows the 75th and 25th percentiles, and the top and bottom of each whisker displays the 95th and 5th percentile values in the ensemble, respectively. The models included are CCSM3, CGCM3.1 (T47 and T63), CNRM-CM3, CSIRO-MK3, ECHAM5/MPI-OM, ECHO-G, FGOALS-g1.0, GFDL-CM2.0, GFDL-CM2.1, GISS-EH, GISS-ER, INM-CM3.0, IPSL-CM4, MIROC3.2 (medium and high resolution), MPI-CGCM2.3.2, PCM1, UKMO-HadCM3 and UKMO-HadGEM1 (see Table 8.1 for model details).*

Table 10.2. *All-sky radiative forcing for doubled atmospheric CO_2. See Table 8.1 for model details.*

Model[Source]	Longwave (W m^{-2})	Shortwave (W m^{-2})
CGCM 3.1 (T47/T63)[a]	3.39	−0.07
CSIRO-MK3.0[b]	3.42	0.05
GISS-EH/ER[a]	4.21	−0.15
GFDL-CM2.0/2.1[b]	3.62	−0.12
IPSL-CM4[c]	3.50	−0.02
MIROC 3.2-hires[d]	3.06	0.08
MIROC 3.2-medres[d]	2.99	0.10
ECHAM5/MPI-OM[a]	3.98	0.03
MRI-CGCM2.3.2[b]	3.75	−0.28
CCSM3[a]	4.23	−0.28
UKMO-HadCM3[a]	4.03	−0.22
UKMO-HadGEM1[a]	4.02	−0.24
Mean ± standard deviation[e]	3.80 ± 0.33	−0.13 ± 0.11

Notes:

[a] Forster and Taylor (2006) based upon forcing data from PCMDI for 200 hPa. Longwave forcing accounts for stratospheric adjustment; shortwave forcing does not.

[b] Forcings derived by individual modelling groups using the method of Gregory et al. (2004b).

[c] Based upon forcing data from PCMDI for 200 hPa. Longwave and shortwave forcing account for stratospheric adjustment.

[d] Forcings at diagnosed tropopause.

[e] Mean and standard deviation are calculated just using forcings at 200 hPa, with each model and model version counted once.

for every 20-year period. For the year 2100, the shortwave forcing from individual AOGCMs ranges from approximately −1.7 W m^{-2} to +0.4 W m^{-2} (5th to 95th percentile). The reasons for this large range include the variety of the aerosol treatments and parametrizations for the indirect effects of aerosols in the multi-model ensemble.

Since the large range in both longwave and shortwave forcings may be caused by a variety of factors, it is useful to determine the range caused just by differences in model formulation for a given (identical) change in radiatively active species. A standard metric is the global mean, annually averaged all-sky forcing at the tropopause for doubled atmospheric CO_2. Estimates of

this forcing for 15 of the models in the ensemble are given in Table 10.2. The shortwave forcing is caused by absorption in the near-infrared bands of CO_2. The range in the longwave forcing at 200 mb is 0.84 W m^{-2}, and the coefficient of variation, or ratio of the standard deviation to mean forcing, is 0.09. These results suggest that up to 35% of the range in longwave forcing in the ensemble for the period 2080 to 2099 is due to the spread in forcing estimates for the specified increase in CO_2. The findings also imply that it is not appropriate to use a single best value of the forcing from doubled atmospheric CO_2 to relate forcing and response (e.g., climate sensitivity) across a multi-model ensemble. The relationships for a given model should be derived using the radiative forcing produced by the radiative parametrizations in that model. Although the shortwave forcing has a coefficient of variation close to one, the range across the ensemble explains less than 17% of the range in shortwave forcing at the end of the 21st-century simulations. This suggests that species and forcing agents other than CO_2 cause the large variation among modelled shortwave forcings.

10.2.1.4 Results from the Radiative-Transfer Model Intercomparison Project: Implications for Fidelity of Forcing Projections

Differences in radiative forcing across the multi-model ensemble illustrated in Table 10.2 have been quantified in the Radiative-Transfer Model Intercomparison Project (RTMIP, W.D. Collins et al., 2006). The basis of RTMIP is an evaluation of the forcings computed by 20 AOGCMs using five benchmark line-by-line (LBL) radiative transfer codes. The comparison is focused on the instantaneous clear-sky radiative forcing by the LLGHGs CO_2, CH_4, N_2O, CFC-11, CFC-12 and the increased water vapour expected in warmer climates. The results of this intercomparison are not directly comparable to the estimates of forcing at the tropopause (Chapter 2), since the latter include the effects of stratospheric adjustment. The effects of adjustment on forcing are approximately −2% for CH_4, −4% for N_2O, +5% for CFC-11, +8% for CFC-12 and −13% for CO_2 (IPCC, 1995; Hansen et al., 1997). The total (longwave plus shortwave) radiative forcings at 200 mb, a surrogate for the tropopause, are shown in Table 10.3 for climatological mid-latitude summer conditions.

Table 10.3. *Total instantaneous forcing at 200 hPa (W m^{-2}) from AOGCMs and LBL codes in RTMIP (W.D. Collins et al., 2006). Calculations are for cloud-free climatological mid-latitude summer conditions.*

Radiative Species	CO_2	CO_2	N_2O + CFCs	CH_4 + CFCs	All LLGHGs	Water Vapour
Forcing[a]	2000–1860	2x–1x	2000–1860	2000–1860	2000–1860	1.2x–1x
AOGCM mean	1.56	4.28	0.47	0.95	2.68	4.82
AOGCM std. dev.	0.23	0.66	0.15	0.30	0.30	0.34
LBL mean	1.69	4.75	0.38	0.73	2.58	5.08
LBL std. dev.	0.02	0.04	0.12	0.12	0.11	0.16

Notes:

[a] 2000–1860 is the forcing due to an increase in the concentrations of radiative species between 1860 and 2000. 2x–1x^{x} and 1.2x–1x are forcings from increases in radiative species by 100% and 20% relative to 1860 concentrations.

Total forcings calculated from the AOGCM and LBL codes due to the increase in LLGHGs from 1860 to 2000 differ by less than 0.04, 0.49 and 0.10 W m^{-2} at the top of model, surface and pseudo-tropopause at 200mb, respectively (Table 10.3). Based upon the Student t-test, none of the differences in mean forcings shown in Table 10.3 is statistically significant at the 0.01 level. This indicates that the ensemble mean forcings are in reasonable agreement with the LBL codes. However, the forcings from individual models, for example from doubled atmospheric CO_2, span a range at least 10 times larger than that exhibited by the LBL models.

The forcings from doubling atmospheric CO_2 from its concentration at 1860 AD are shown in Figure 10.3a at the top of the model (TOM), 200 hPa (Table 10.3), and the surface. The AOGCMs tend to underestimate the longwave forcing at these three levels. The relative differences in the mean forcings are less than 8% for the pseudo-tropopause at 200 hPa but increase to approximately 13% at the TOM and to 33% at the surface. In general, the mean shortwave forcings from the LBL and AOGCM codes are in good agreement at all three surfaces. However, the range in shortwave forcing at the surface from individual AOGCMs is quite large. The coefficient of variation (the ratio of the standard deviation to the mean) for the surface shortwave forcing from AOGCMs is 0.95. In response to a doubling in atmospheric CO_2, the specific humidity increases by approximately 20% through much of the troposphere. The changes in shortwave and longwave fluxes due to a 20% increase in water vapour are illustrated in Figure 10.3b. The mean longwave forcing from increasing water vapour is quite well simulated with the AOGCM codes. In the shortwave, the only significant difference between the AOGCM and LBL calculations occurs at the surface, where the AOGCMs tend to underestimate the magnitude of the reduction in insolation. In general, the biases in the AOGCM forcings are largest at the surface level.

10.2.1.5   Implications for Range in Climate Response

The results from RTMIP imply that the spread in climate response discussed in this chapter is due in part to the diverse representations of radiative transfer among the members of the multi-model ensemble. Even if the concentrations of LLGHGs were identical across the ensemble, differences in radiative transfer parametrizations among the ensemble members would lead to different estimates of radiative forcing by these species. Many of the climate responses (e.g., global mean temperature) scale linearly with the radiative forcing to first approximation. Therefore, systematic errors in the calculations of radiative forcing should produce a corresponding range in climate responses. Assuming that the RTMIP results (Table 10.3) are globally applicable, the range of forcings for 1860 to 2000 in the AOGCMs should introduce a ±18% relative range (the 5 to 95% confidence interval) for 2000 in the responses that scale with forcing. The corresponding relative range for doubled atmospheric CO_2, which is comparable to the change in CO_2 in the B1 scenario by 2100, is ± 25%.

10.2.2   Recent Developments in Projections of Radiative Species and Forcing for the 21st Century

Estimation of ozone forcing for the 21st century is complicated by the short chemical lifetime of ozone compared to atmospheric transport time scales and by the sensitivity of the radiative forcing to the vertical distribution of ozone. Gauss et al. (2003) calculate the forcing by anthropogenic increases

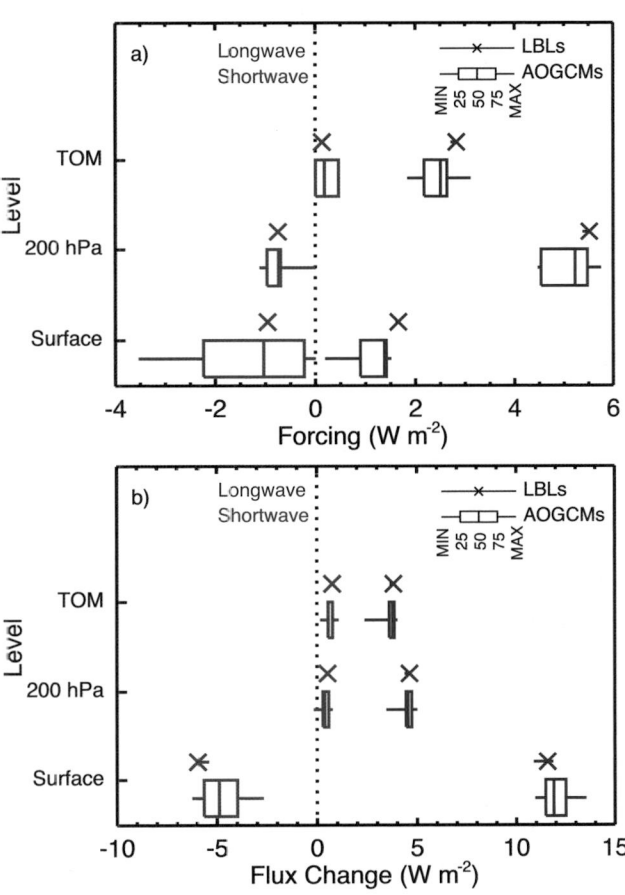

Figure 10.3. *Comparison of shortwave and longwave instantaneous radiative forcings and flux changes computed from AOGCMs and line-by-line (LBL) radiative transfer codes (W.D. Collins et al., 2006). (a) Instantaneous forcing from doubling atmospheric CO_2 from its concentration in 1860; b) changes in radiative fluxes caused by the 20% increase in water vapour expected in the climate produced from doubling atmospheric CO_2. The forcings and flux changes are computed for clear-sky conditions in mid-latitude summer and do not include effects of stratospheric adjustment. No other well-mixed greenhouse gases are included. The minimum-to-maximum range and median are plotted for five representative LBL codes. The AOGCM results are plotted with box-and-whisker diagrams (see caption for Figure 10.2) representing percentiles of forcings from 20 models in the AR4 multi-model ensemble. The AOGCMs included are BCCR-BCM2.0, CCSM3, CGCM3.1(T47 and T63), CNRM-CM3, ECHAM5/MPI-OM, ECHO-G, FGOALS-g1.0, GFDL-CM2.0, GFDL-CM2.1, GISS-EH, GISS-ER, INM-CM3.0, IPSL-CM4, MIROC3.2 (medium and high resolution), MRI-CGCM2.3.2, PCM, UKMO-HadCM3, and UKMO-HadGEM1 (see Table 8.1 for model details). The LBL codes are the Geophysical Fluid Dynamics Laboratory (GFDL) LBL, the Goddard Institute for Space Studies (GISS) LBL3, the National Center for Atmospheric Research (NCAR)/Imperial College of Science, Technology and Medicine (ICSTM) general LBL GENLN2, the National Aeronautics and Space Administration (NASA) Langley Research Center MRTA and the University of Reading Reference Forward Model (RFM).*

of tropospheric ozone through 2100 from 11 different chemical transport models integrated with the SRES A2p scenario. The A2p scenario is the preliminary version of the marker A2 scenario and has nearly identical time series of LLGHGs and forcing. Since the emissions of CH_4, carbon monoxide (CO), reactive nitrogen oxides (NO_x) and volatile organic compounds (VOCs), which strongly affect the formation of ozone, are maximised in the A2p scenario, the modelled forcings should represent an upper bound for the forcing produced under more constrained emissions scenarios. The 11 models simulate an increase in tropospheric ozone of 11.4 to 20.5 Dobson units (DU) by 2100, corresponding to a range of radiative forcing from 0.40 to 0.78 W m^{-2}. Under this scenario, stratospheric ozone increases by between 7.5 and 9.3 DU, which raises the radiative forcing by an additional 0.15 to 0.17 W m^{-2}.

One aspect of future direct aerosol radiative forcing omitted from all but 2 (the GISS-EH and GISS-ER models) of the 23 AOGCMS analysed in AR4 (see Table 8.1 for list) is the role of nitrate aerosols. Rapid increases in NO_x emissions could produce enough nitrate aerosol to offset the expected decline in sulphate forcing by 2100. Adams et al. (2001) compute the radiative forcing by sulphate and nitrate accounting for the interactions among sulphate, nitrate and ammonia. For 2000, the sulphate and nitrate forcing are –0.95 and –0.19 W m^{-2}, respectively. Under the SRES A2 scenario, by 2100 declining SO_2 emissions cause the sulphate forcing to drop to –0.85 W m^{-2}, while the nitrate forcing rises to –1.28 W m^{-2}. Hence, the total sulphate-nitrate forcing increases in magnitude from –1.14 W m^{-2} to –2.13 W m^{-2} rather than declining as models that omit nitrates would suggest. This projection is consistent with the large increase in coal burning forecast as part of the A2 scenario.

Recent field programs focused on Asian aerosols have demonstrated the importance of black carbon (BC) and organic carbon (OC) for regional climate, including potentially significant perturbations of the surface energy budget and hydrological cycle (Ramanathan et al., 2001). Modelling groups have developed a multiplicity of projections for the concentrations of these aerosol species. For example, Takemura et al. (2001) use data sets for BC released by fossil fuel and biomass burning (Cooke and Wilson, 1996) under current conditions and scale them by the ratio of future to present-day CO_2. The emissions of OC are derived using OC:BC ratios estimated for each source and fuel type. Koch (2001) models the future radiative forcing of BC by scaling a different set of present-day emission inventories by the ratio of future to present-day CO_2 emissions. There are still large uncertainties associated with current inventories of BC and OC (Bond et al., 2004), the ad hoc scaling methods used to produce future emissions, and considerable variation among estimates of the optical properties of carbonaceous aerosols (Kinne et al., 2006). Given these uncertainties, future projections of forcing by BC and OC should be quite model dependent.

Recent evidence suggests that there are detectable anthropogenic increases in stratospheric sulphate (e.g., Myhre et al., 2004), water vapour (e.g., Forster and Shine, 2002), and

condensed water in the form of aircraft contrails. However, recent modelling studies suggest that these forcings are relatively minor compared to the major LLGHGs and aerosol species. Marquart et al. (2003) estimate that the radiative forcing by contrails will increase from 0.035 W m^{-2} in 1992 to 0.094 W m^{-2} in 2015 and to 0.148 W m^{-2} in 2050. The rise in forcing is due to an increase in subsonic aircraft traffic following estimates of future fuel consumption (Penner et al., 1999). These estimates are still subject to considerable uncertainties related to poor constraints on the microphysical properties, optical depths and diurnal cycle of contrails (Myhre and Stordal, 2001, 2002; Marquart et al., 2003). Pitari et al. (2002) examine the effect of future emissions under the A2 scenario on stratospheric concentrations of sulphate aerosol and ozone. By 2030, the mass of stratospheric sulphate increases by approximately 33%, with the majority of the increase contributed by enhanced upward fluxes of anthropogenic SO_2 through the tropopause. The increase in direct shortwave forcing by stratospheric aerosols in the A2 scenario during 2000 to 2030 is –0.06 W m^{-2}.

Some recent studies have suggested that the global atmospheric burden of soil dust aerosols could decrease by between 20 and 60% due to reductions in desert areas associated with climate change (Mahowald and Luo, 2003). Tegen et al. (2004a,b) compared simulations by the European Centre for Medium Range Weather Forecasts/Max Planck Institute for Meteorology Atmospheric GCM (ECHAM4) and UKMO-HadCM3 that included the effects of climate-induced changes in atmospheric conditions and vegetation cover and the effects of increased CO_2 concentrations on vegetation density. These simulations are forced with identical (IS92a) time series for LLGHGs. Their findings suggest that future projections of changes in dust loading are quite model dependent, since the net changes in global atmospheric dust loading produced by the two models have opposite signs. They also conclude that dust from agriculturally disturbed soils is less than 10% of the current burden, and that climate-induced changes in dust concentrations would dominate land use changes under both minimum and maximum estimates of increased agricultural area by 2050.

10.3   Projected Changes in the Physical Climate System

The context for the climate change results presented here is set in Chapter 8 (evaluation of simulation skill of the control runs and inherent natural variability of the global coupled climate models), and in Chapter 9 (evaluation of the simulations of 20th-century climate using the global coupled climate models). Table 8.1 describes the characteristics of the models, and Table 10.4 summarises the climate change experiments that have been performed with the AOGCMs and other models that are assessed in this chapter.

Table 10.4. *Summary of climate change model experiments produced with AOGCMs. Numbers in each scenario column indicate how many ensemble members were produced for each model. Coloured fields indicate that some but not necessarily all variables of the specific data type (separated by climate system component and time interval) were available for download at the PCMDI to be used in this report; ISCCP is the International Satellite Cloud Climatology Project. Additional data has been submitted for some models and may subsequently become available. Where different colour shadings are given in the legend, the colour indicates whether data from a single or from multiple ensemble members is available. Details on the scenarios, variables and models can be found at the PCMDI webpage (http://www-pcmdi.llnl.gov/ipcc/about_ipcc.php). Model IDs are the same as in Table 8.1, which provides details of the models.*

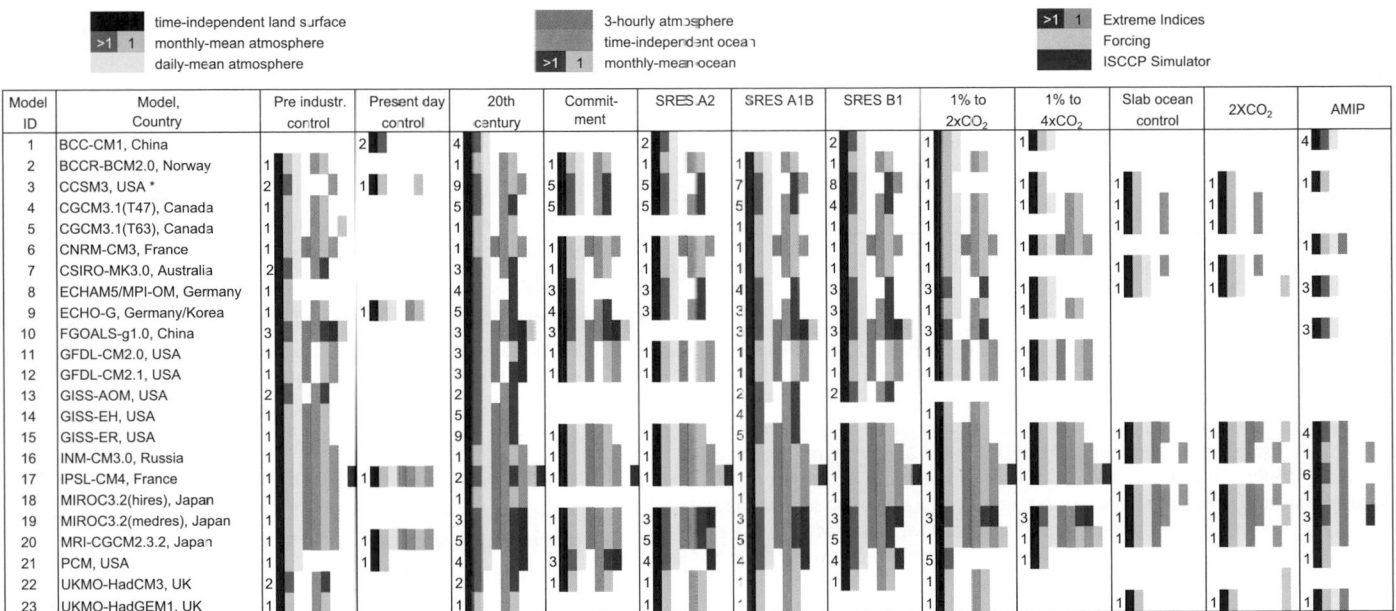

Legend:
- time-independent land surface
- monthly-mean atmosphere
- daily-mean atmosphere
- 3-hourly atmosphere
- time-independent ocean
- monthly-mean ocean
- Extreme Indices
- Forcing
- ISCCP Simulator

Model ID	Model, Country	Pre industr. control	Present day control	20th century	Commitment	SRES A2	SRES A1B	SRES B1	1% to 2xCO2	1% to 4xCO2	Slab ocean control	2XCO2	AMIP
1	BCC-CM1, China		2	4		2		2	1				4
2	BCCR-BCM2.0, Norway	1	1	1	1	1	1	1	1				1
3	CCSM3, USA *	2	1	9	5	5	7	8	1	1	1	1	1
4	CGCM3.1(T47), Canada	1		5	5	5	5	1	1	1	1	1	
5	CGCM3.1(T63), Canada	1		1		1	1	1	1	1	1	1	
6	CNRM-CM3, France	1		1		1	1	1	1				1
7	CSIRO-MK3.0, Australia	2		3		1	1	1	1		1	1	
8	ECHAM5/MPI-OM, Germany	1		4		3	4	3	1	1			3
9	ECHO-G, Germany/Korea	1	1	5	4	3	3	3	1				1
10	FGOALS-g1.0, China	3		3		3	3	3	3				3
11	GFDL-CM2.0, USA	1		3		1	1	1	1				
12	GFDL-CM2.1, USA	1		3		1	1	1	1				
13	GISS-AOM, USA	2		2		2	2	2					
14	GISS-EH, USA	1		5		4	5		1				
15	GISS-ER, USA	1		9		5	5	1	1		1	1	4
16	INM-CM3.0, Russia	1		1		1	1	1	1				1
17	IPSL-CM4, France	1	1	2		1	1	1	1				6
18	MIROC3.2(hires), Japan	1		1			1	1	1				1
19	MIROC3.2(medres), Japan	1		3		3	3	3	3	3	1	1	3
20	MRI-CGCM2.3.2, Japan	1	1	5		5	5	5	1		1	1	1
21	PCM, USA	1	1	4		3	4	4	5				1
22	UKMO-HadCM3, UK	2		2		1	1	1					1
23	UKMO-HadGEM1, UK	1		1		1	1		1		1	1	1

* Some of the ensemble members using the CCSM3 were run on the Earth Simulator in Japan in collaboration with the Central Research Institute of Electric Power Industry (CRIEPI).

The TAR showed multi-model results for future changes in climate from simple 1% yr^{-1} CO_2 increase experiments, and from several scenarios including the older IS92a, and, new to the TAR, two SRES scenarios (A2 and B2). For the latter, results from nine models were shown for globally averaged temperature change and regional changes. As noted in Section 10.1, since the TAR, an unprecedented internationally coordinated climate change experiment has been performed by 23 models from around the world, listed in Table 10.4 along with the results submitted. This larger number of models running the same experiments allows better quantification of the multi-model signal as well as uncertainty regarding spread across the models (in this section), and also points the way to probabilistic estimates of future climate change (Section 10.5). The emission scenarios considered here include one of the SRES scenarios from the TAR, scenario A2, along with two additional scenarios, A1B and B1 (see Section 10.2 for details regarding the scenarios). This is a subset of the SRES marker scenarios used in the TAR, and they represent 'low' (B1), 'medium' (A1B) and 'high' (A2) scenarios with respect to the prescribed concentrations and the resulting radiative forcing, relative to the SRES range. This choice was made solely due to the limited computational resources for multi-model simulations using comprehensive AOGCMs and does not imply any preference or qualification of these three scenarios over the others. Qualitative conclusions derived from those three scenarios are in most cases also valid for other SRES scenarios.

Additionally, three climate change commitment experiments were performed, one where concentrations of greenhouse gases were held fixed at year 2000 values (constant composition commitment) and the models were run to 2100 (termed 20th-century stabilisation here), and two where concentrations were held fixed at year 2100 values for A1B and B1, and the models were run for an additional 100 to 200 years (see Section 10.7). The span of the experiments is shown in Figure 10.4.

This section considers the basic changes in climate over the next hundred years simulated by current climate models under non-mitigation anthropogenic forcing scenarios. While we assess all studies in this field, the focus is on results derived by the authors from the new data set for the three SRES scenarios. Following the TAR, means across the multi-model ensemble are used to illustrate representative changes. Means are able to simulate the contemporary climate more accurately than individual models, due to biases tending to compensate each other (Phillips and Gleckler, 2006). It is anticipated that this holds for changes in climate also (Chapter 9). The mean temperature trends from the 20th-century simulations are included in Figure 10.4. While the range of model results is indicated here, the consideration of uncertainty resulting from this range is addressed more completely in Section 10.5. The use of means has the additional advantage of reducing the 'noise' associated with internal or unforced variability in the simulations. Models are equally weighted here, but other options are noted in Section 10.5. Lists of the models used in the results are provided in the Supplementary Material for this Chapter.

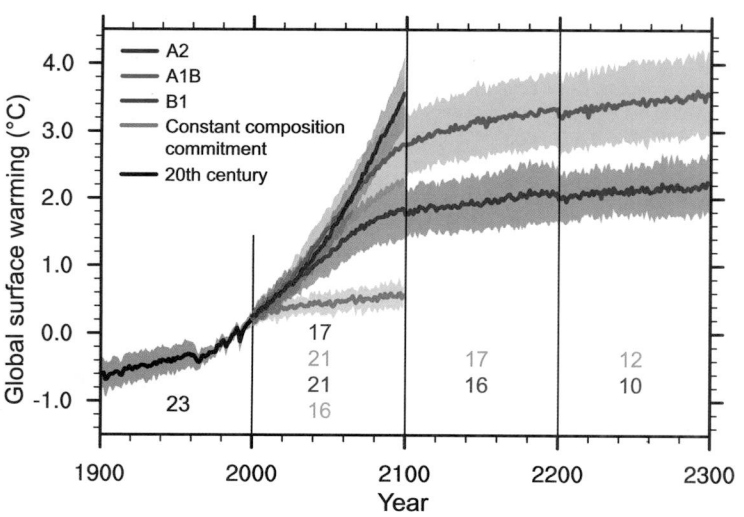

Figure 10.4. *Multi-model means of surface warming (relative to 1980–1999) for the scenarios A2, A1B and B1, shown as continuations of the 20th-century simulation. Values beyond 2100 are for the stabilisation scenarios (see Section 10.7). Linear trends from the corresponding control runs have been removed from these time series. Lines show the multi-model means, shading denotes the ±1 standard deviation range of individual model annual means. Discontinuities between different periods have no physical meaning and are caused by the fact that the number of models that have run a given scenario is different for each period and scenario, as indicated by the coloured numbers given for each period and scenario at the bottom of the panel. For the same reason, uncertainty across scenarios should not be interpreted from this figure (see Section 10.5.4.6 for uncertainty estimates).*

Standard metrics for response of global coupled models are the equilibrium climate sensitivity, defined as the equilibrium globally averaged surface air temperature change for a doubling of CO_2 for the atmosphere coupled to a non-dynamic slab ocean, and the TCR, defined as the globally averaged surface air temperature change at the time of CO_2 doubling in the 1% yr^{-1} transient CO_2 increase experiment. The TAR showed results for these 1% simulations, and Section 10.5.2 discusses equilibrium climate sensitivity, TCR and other aspects of response. Chapter 8 includes processes and feedbacks involved with these metrics.

10.3.1 Time-Evolving Global Change

The globally averaged surface warming time series from each model in the MMD is shown in Figure 10.5, either as a single member (if that was all that was available) or a multi-member ensemble mean, for each scenario in turn. The multi-model ensemble mean warming is also plotted for each case. The surface air temperature is used, averaged over each year, shown as an anomaly relative to the 1980 to 1999 period and offset by any drift in the corresponding control runs in order to extract the forced response. The base period was chosen to match the contemporary climate simulation that is the focus of previous chapters. Similar results have been shown in studies of these models (e.g., Xu et al., 2005; Meehl et al., 2006b; Yukimoto et al., 2006). Interannual variability is evident in each single-model series, but little remains in the ensemble mean because most of this is unforced and is a result of internal variability, as was presented in detail in Section 9.2.2 of TAR. Clearly, there is a range of model results for each year, but over time this

range due to internal variability becomes smaller as a fraction of the mean warming. The range is somewhat smaller than the range of warming at the end of the 21st century for the A2 scenario in the comparable Figure 9.6 of the TAR, despite the larger number of models here (the ensemble mean warming is comparable, +3.0°C in the TAR for 2071 to 2100 relative to 1961 to 1990, and +3.13°C here for 2080 to 2099 relative to 1980 to 1999, Table 10.5). Consistent with the range of forcing presented in Section 10.2, the warming by 2100 is largest in the high greenhouse gas growth scenario A2, intermediate in the moderate growth A1B, and lowest in the low growth B1. Naturally, models with high sensitivity tend to simulate above-average warming in each scenario. The trends of the multi-model mean temperature vary somewhat over the century because of the varying forcings, including that of aerosols (see Section 10.2). This is illustrated in Figure 10.4, which shows the mean for A1B exceeding that for A2 around 2040. The time series beyond 2100 are derived from the extensions of the simulations (those available) under the idealised constant composition commitment experiments (Section 10.7.1).

Internal variability in the model response is reduced by averaging over 20-year time periods. This span is shorter than the traditional 30-year climatological period, in recognition of the transient nature of the simulations, and of the larger size of the ensemble. This analysis focuses on three periods over the coming century: an early-century period 2011 to 2030, a mid-century period 2046 to 2065 and the late-century period 2080 to 2099, all relative to the 1980 to 1999 means. The multi-model ensemble mean warmings for the three future periods in the different experiments are given in Table 10.5, among other results. The close agreement of warming for the early century, with a range of only 0.05°C among the SRES cases, shows that no matter which of these non-mitigation scenarios is followed, the warming is similar on the time scale of the next decade or two. Note that the precision given here is only relevant for comparison between these means. As evident in Figure 10.4 and discussed in Section 10.5, uncertainties in the projections are larger. It is also worth noting that half of the early-century climate change arises from warming that is already committed to under constant composition (0.37°C for the early century). By mid-century, the choice of scenario becomes more important for the magnitude of warming, with a range of 0.46°C, and with about one-third of that warming due to climate change that is already committed to. By the late century, there are clear consequences for which scenario is followed, with a range of 1.3°C in these results, with as little as 18% of that warming coming from climate change that is already committed to.

Global mean precipitation increases in all scenarios (Figure 10.5, right column), indicating an intensification of the hydrological cycle. Douville et al. (2002) show that this is associated with increased water-holding capacity of the atmosphere in addition to other processes. The multi-model

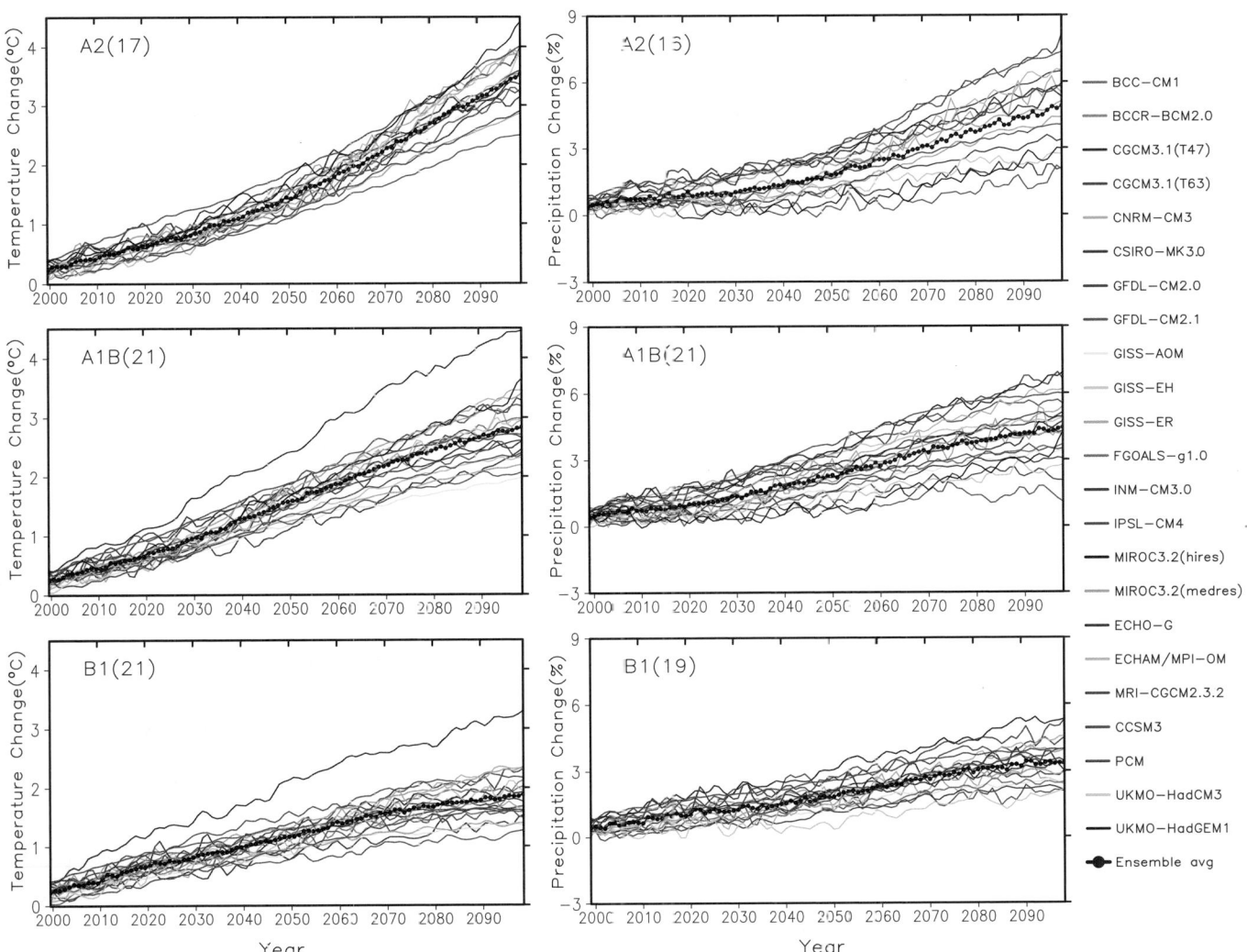

Figure 10.5. *Time series of globally averaged (left) surface warming (surface air temperature change, °C) and (right) precipitation change (%) from the various global coupled models for the scenarios A2 (top), A1B (middle) and B1 (bottom). Numbers in parentheses following the scenario name represent the number of simulations shown. Values are annual means, relative to the 1980 to 1999 average from the corresponding 20th-century simulations, with any linear trends in the corresponding control run simulations removed. A three-point smoothing was applied. Multi-model (ensemble) mean series are marked with black dots. See Table 8.1 for model details.*

Table 10.5. *Global mean warming (annual mean surface air temperature change) from the multi-model ensemble mean for four time periods relative to 1980 to 1999 for each of the available scenarios. (The mean for the base period is 13.6°C). Also given are two measures of agreement of the geographic scaled patterns of warming (the fields in Figure 10.8 normalised by the global mean), relative to the A1B 2080 to 2099 case. First the non-dimensional M value (see Section 10.3.2.1) and second (in italics) the global mean absolute error (mae, or difference, in °C/°C) between the fields, both multiplied by 100 for brevity. Here $M = (2/\pi) \arcsin[1 - mse / (V_X + V_Y + (G_X - G_Y)^2)]$, with mse the mean square error between the two fields X and Y, and V and G are variance and global mean of the fields (as subscripted). Values of 1 for M and 0 for mae indicate perfect agreement with the standard pattern. 'Commit' refers to the constant composition commitment experiment. Note that warming values for the end of the 21st century, given here as the average of years 2080 to 2099, are for a somewhat different averaging period than used in Figure 10.29 (2090–2099); the longer averaging period here is consistent with the comparable averaging period for the geographic plots in this section and is intended to smooth spatial noise.*

	Global mean warming (°C)				Measures of agreement (M × 100, *mae* × 100)			
	2011–2030	2046–2065	2080–2099	2180–2199	2011–2030	2046–2065	2080–2099	2180–2199
A2	0.64	1.65	3.13		83, *8*	91, *4*	93, *3*	
A1B	0.69	1.75	2.65	3.36	88, *5*	94, *4*	100, *0*	90, *5*
B1	0.66	1.29	1.79	2.10	86, *6*	89, *4*	92, *3*	86, *6*
Commit[a]	0.37	0.47	0.56		74, *11*	66, *13*	68, *13*	

Notes:

[a] Committed warming values are given relative to the 1980 to 1999 base period, whereas the commitment experiments started with stabilisation at year 2000. The committed warming trend is about 0.1°C per decade over the next two decades with a reduced rate after that (see Figure 10.4).

mean varies approximately in proportion to the mean warming, though uncertainties in future hydrological cycle behaviour arise due in part to the different responses of tropical precipitation across models (Douville et al., 2005). Expressed as a percentage of the mean simulated change for 1980 to 1999 (2.83 mm day^{-1}), the rate varies from about 1.4% °C^{-1} in A2 to 2.3% °C^{-1} in the constant composition commitment experiment (for a table corresponding to Table 10.5 but for precipitation, see the Supplementary Material, Table S10.1). These increases are less than increases in extreme precipitation events, consistent with energetic constraints (see Sections 9.5.4.2 and 10.3.6.1)

10.3.2 Patterns of Change in the 21st Century

10.3.2.1 Warming

The TAR noted that much of the regional variation of the annual mean warming in the multi-model means is associated with high- to low-latitude contrast. This can be better quantified from the new multi-model mean in terms of zonal averages. A further contrast is provided by partitioning the land and ocean values based on model data interpolated to a standard grid. Figure 10.6 illustrates the late-century A2 case, with all values shown both in absolute terms and relative to the global mean warming. Warming over land is greater than the mean except in the southern mid-latitudes, where the warming over ocean is a

minimum. Warming over ocean is smaller than the mean except at high latitudes, where sea ice changes have an influence. This pattern of change illustrated by the ratios is quite similar across the scenarios. The commitment case (shown), discussed in Section 10.7.1, has relatively smaller warming of land, except in the far south, which warms closer to the global rate. At nearly all latitudes, the A1B and B1 warming ratios lie between A2 and commitment, with A1B particularly close to the A2 results. Aside from the commitment case, the ratios for the other time periods are also quite similar to those for A2. Regional patterns and precipitation contrasts are discussed in Section 10.3.2.3.

Figure 10.7 shows the zonal mean warming for the A1B scenario at each latitude from the bottom of the ocean to the top of the atmosphere for the three 21st-century periods used in Table 10.5. To produce this ensemble mean, the model data were first interpolated to standard ocean depths and atmospheric pressures. Consistent with the global transfer of excess heat from the atmosphere to the ocean, and the difference between warming over land and ocean, there is some discontinuity between the plotted means of the lower atmosphere and the upper ocean. The relatively uniform warming of the troposphere and cooling of the stratosphere in this multi-model mean are consistent with the changes shown in Figure 9.8 of the TAR, but now its evolution during the 21st century under this scenario can also be seen. Upper-tropospheric warming reaches a maximum in the tropics and is seen even in the early-century

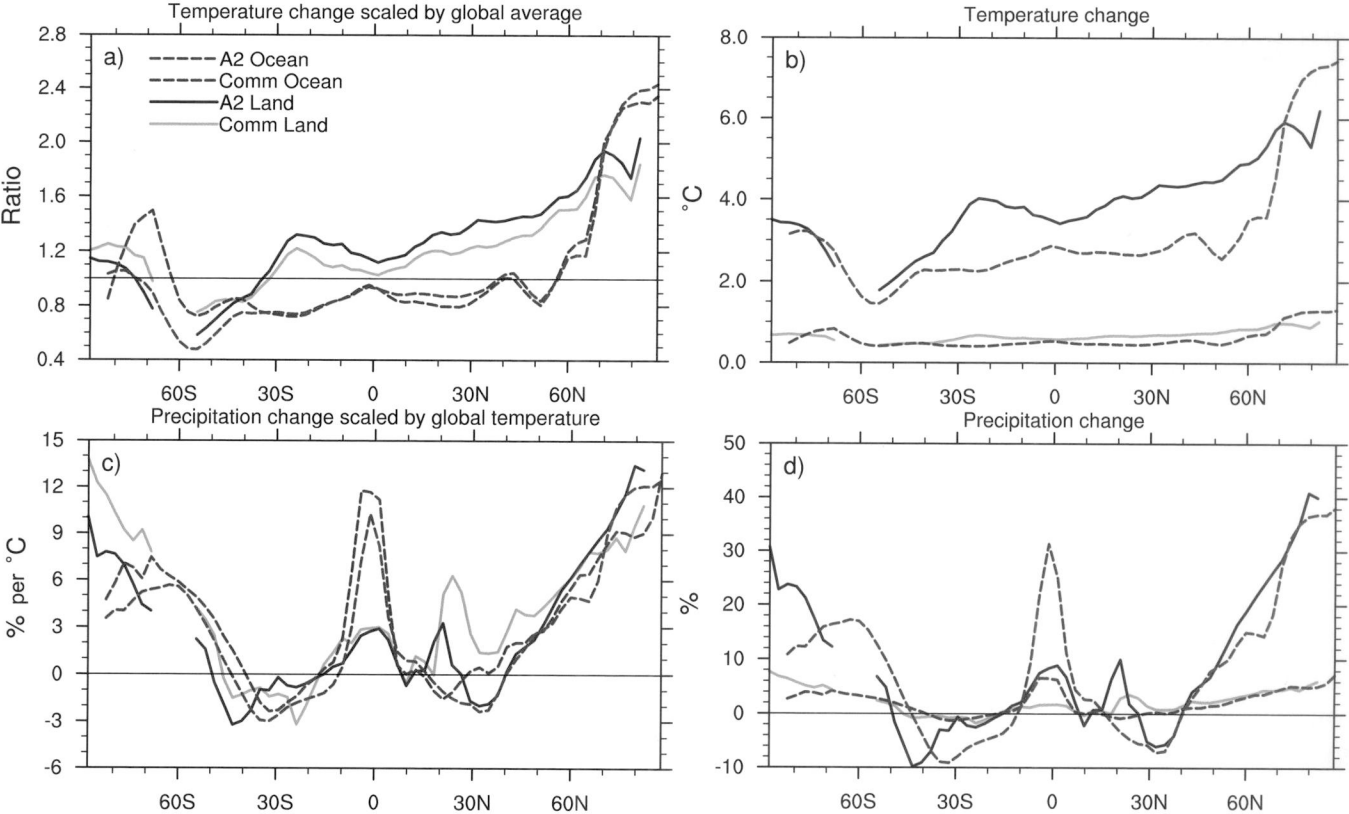

Figure 10.6. Zonal means over land and ocean separately, for annual mean surface warming (a, b) and precipitation (c, d), shown as ratios scaled with the global mean warming (a, c) and not scaled (b, d). Multi-model mean results are shown for two scenarios, A2 and Commitment (see Section 10.7), for the period 2080 to 2099 relative to the zonal means for 1980 to 1999. Results for individual models can be seen in the Supplementary Material for this chapter.

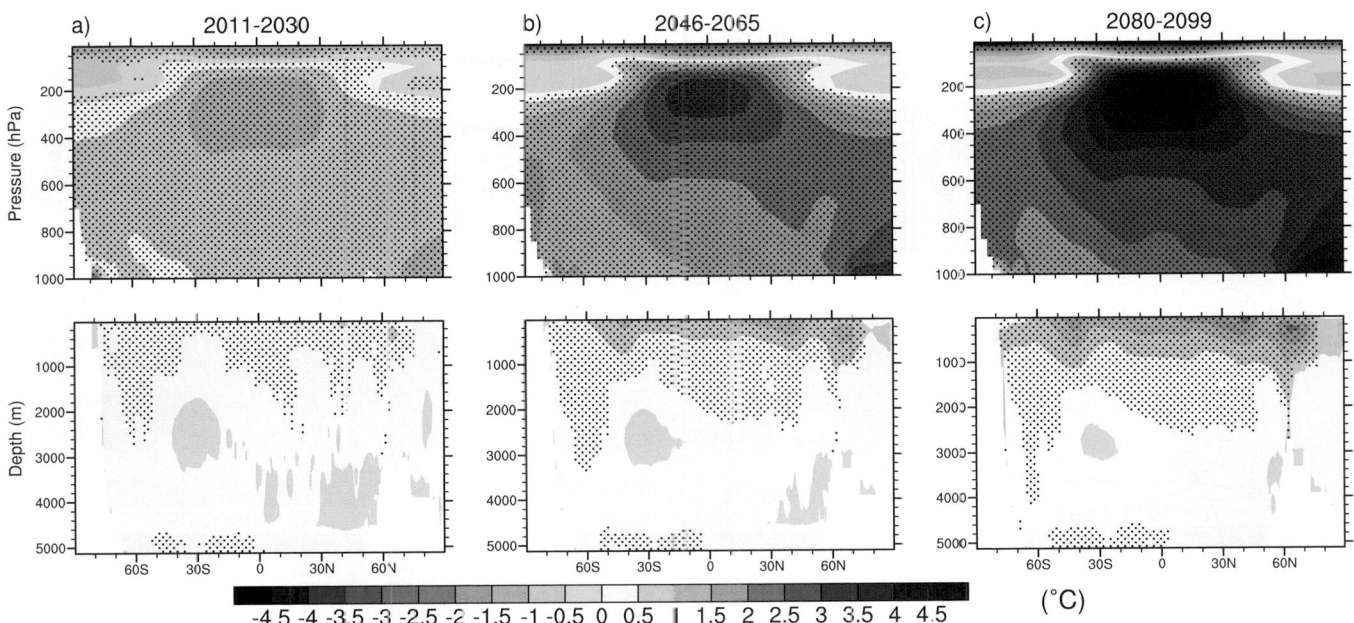

Figure 10.7. *Zonal means of change in atmospheric (top) and oceanic (bottom) temperatures (°C), shown as cross sections. Values are the multi-model means for the A1B scenario for three periods (a–c). Stippling denotes regions where the multi-model ensemble mean divided by the multi-model standard deviation exceeds 1.0 (in magnitude). Anomalies are relative to the average of the period 1980 to 1999. Results for individual models can be seen in the Supplementary Material for this chapter.*

time period. The pattern is very similar over the three periods, consistent with the rapid adjustment of the atmosphere to the forcing. These changes are simulated with good consistency among the models. The larger values of both signs are stippled, indicating that the ensemble mean is larger in magnitude than the inter-model standard deviation. The ratio of mean to standard deviation can be related to formal tests of statistical significance and confidence intervals, if the individual model results were to be considered a sample.

The ocean warming evolves more slowly. There is initially little warming below the mixed layer, except at some high latitudes. Even as a ratio with mean surface warming, later in the century the temperature increases more rapidly in the deep ocean, consistent with results from individual models (e.g., Watterson, 2003; Stouffer, 2004). This rapid warming of the atmosphere and the slow penetration of the warming into the ocean has implications for the time scales of climate change commitment (Section 10.7). It has been noted in a five-member multi-model ensemble analysis that, associated with the changes in temperature of the upper ocean in Figure 10.7, the tropical Pacific Ocean heat transport remains nearly constant with increasing greenhouse gases due to the compensation of the subtropical cells and the horizontal gyre variations, even as the subtropical cells change in response to changes in the trade winds (Hazeleger, 2005). Additionally, a southward shift of the Antarctic Circumpolar Current is projected to occur in a 15-member multi-model ensemble, due to changes in surface winds in a future warmer climate (Fyfe and Saenko, 2005). This is associated with a poleward shift of the westerlies at the surface (see Section 10.3.6) and in the upper troposphere particularly notable in the Southern Hemisphere (SH) (Stone and Fyfe, 2005), and increased relative angular momentum from stronger

westerlies (Räisänen, 2003) and westerly momentum flux in the lower stratosphere particularly in the tropics and southern mid-latitudes (Watanabe et al., 2005). The surface wind changes are associated with corresponding changes in wind stress curl and horizontal mass transport in the ocean (Saenko et al., 2005).

Global-scale patterns for each of the three scenarios and time periods are given in Figure 10.8. In each case, greater warming over most land areas is evident (e.g., Kunkel and Liang, 2005). Over the ocean, warming is relatively large in the Arctic and along the equator in the eastern Pacific (see Sections 10.3.5.2 and 10.3.5.3), with less warming over the North Atlantic and the Southern Ocean (e.g., Xu et al., 2005). Enhanced oceanic warming along the equator is also evident in the zonal means of Figure 10.6, and can be associated with oceanic heat flux changes (Watterson, 2003) and forced by the atmosphere (Liu et al., 2005).

Fields of temperature change have a similar structure, with the linear correlation coefficient as high as 0.994 between the late-century A2 and A1B cases. As for the zonal means, the fields normalised by the mean warming are very similar. The strict agreement between the A1B field, as a standard, and the others is quantified in Table 10.5, by the absolute measure M (Watterson, 1996; a transformation of a measure of Mielke, 1991), with unity meaning identical fields and zero meaning no similarity (the expected value under random rearrangement of the data on the grid of the measure prior to the arcsin transformation). Values of M become progressively larger later in the 21st century, with values of 0.9 or larger for the late 21st century, thus confirming the closeness of the scaled patterns in the late-century cases. The deviation from unity is approximately proportional to the mean absolute difference. The earlier warming patterns are also similar to the standard case,

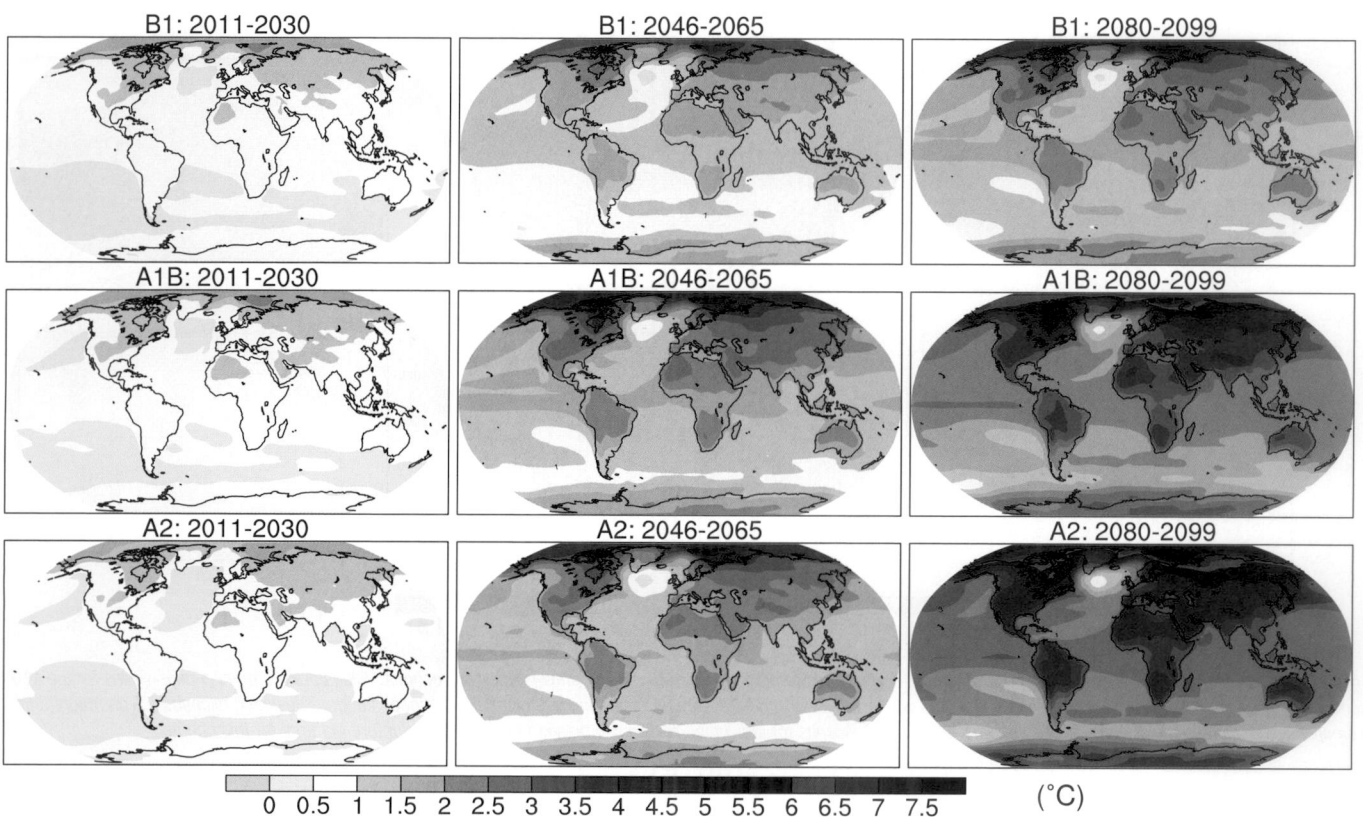

Figure 10.8. *Multi-model mean of annual mean surface warming (surface air temperature change, °C) for the scenarios B1 (top), A1B (middle) and A2 (bottom), and three time periods, 2011 to 2030 (left), 2046 to 2065 (middle) and 2080 to 2099 (right). Stippling is omitted for clarity (see text). Anomalies are relative to the average of the period 1980 to 1999. Results for individual models can be seen in the Supplementary Material for this chapter.*

particularly for the same scenario A1B. Furthermore, the zonal means over land and ocean considered above are representative of much of the small differences in warming ratio. While there is some influence of differences in forcing patterns among the scenarios, and of effects of oceanic uptake and heat transport in modifying the patterns over time, there is also support for the role of atmospheric heat transport in offsetting such influences (e.g., Boer and Yu, 2003b; Watterson and Dix, 2005). Dufresne et al. (2005) show that aerosol contributes a modest cooling of the Northern Hemisphere (NH) up to the mid-21st century in the A2 scenario.

Such similarities in patterns of change have been described by Mitchell (2003) and Harvey (2004). They aid the efficient presentation of the broad scale multi-model results, as patterns depicted for the standard A1B 2080 to 2099 case are usually typical of other cases. This largely applies to other seasons and also other variables under consideration here. Where there is similarity of normalised changes, values for other cases can be estimated by scaling by the appropriate ratio of global means from Table 10.5. Note that for some quantities like variability and extremes, such scaling is unlikely to work. The use of such scaled results in combination with global warmings from simple models is discussed in Section 11.10.1.

As for the zonal means (aside from the Arctic Ocean), consistency in local warmings among the models is high (stippling is omitted in Figure 10.8 for clarity). Only in the

central North Atlantic and the far south Pacific in 2011 to 2030 is the mean change less than the standard deviation, in part a result of ocean model limitations there (Section 8.3.2). Some regions of high-latitude surface cooling occur in individual models.

The surface warming fields for the extratropical winter and summer seasons, December to February (DJF) and June to August (JJA), are shown for scenario A1B in Figure 10.9. The high-latitude warming is rather seasonal, being larger in winter as a result of sea ice and snow, as noted in Chapter 9 of the TAR. However, the relatively small warming in southern South America is more extensive in southern winter. Similar patterns of change in earlier model simulations are described by Giorgi et al. (2001).

10.3.2.2   *Cloud and Diurnal Cycle*

In addition to being an important link to humidity and precipitation, cloud cover plays an important role for the sensitivity of the general circulation models (GCMs; e.g., Soden and Held, 2006) and for the diurnal temperature range (DTR) over land (e.g., Dai and Trenberth, 2004 and references therein) so this section considers the projection of these variables now made possible by multi-model ensembles. Cloud radiative feedbacks to greenhouse gas forcing are sensitive to the elevation, latitude and hence temperature of the clouds, in addition to their optical

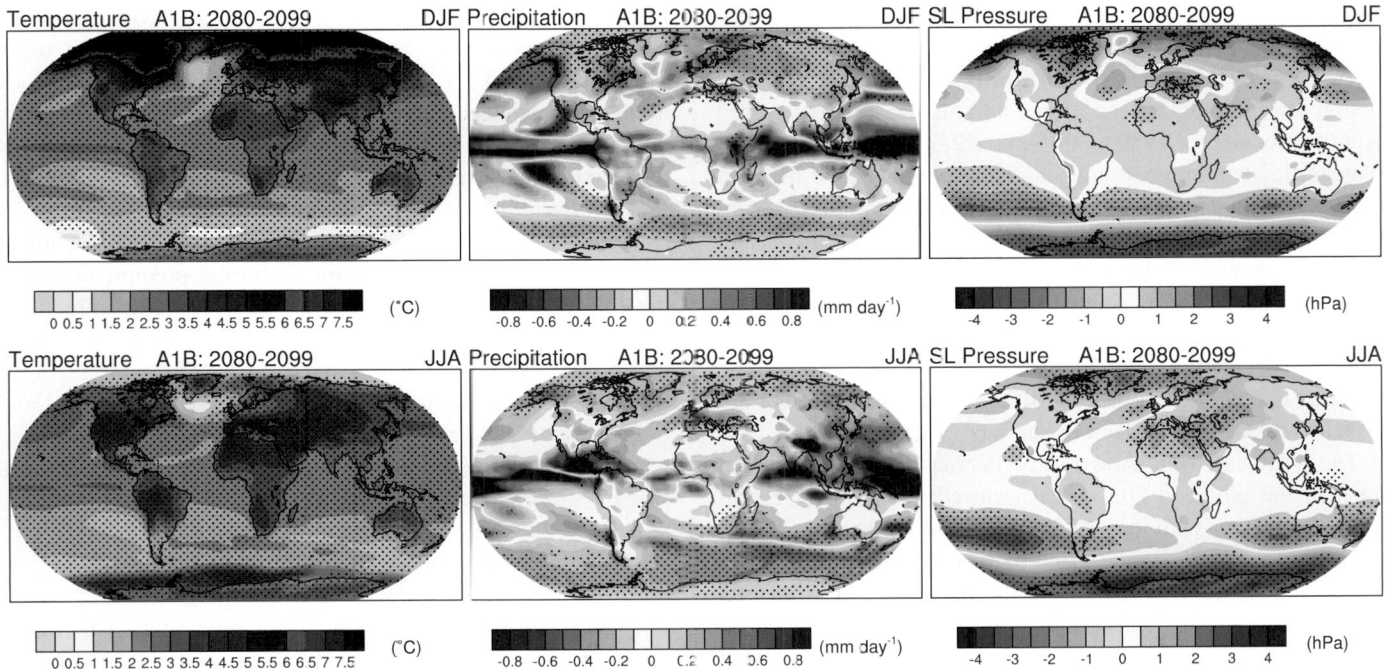

Figure 10.9. *Multi-model mean changes in surface air temperature (°C, left), precipitation (mm day⁻¹, middle) and sea level pressure (hPa, right) for boreal winter (DJF, top) and summer (JJA, bottom). Changes are given for the SRES A1B scenario, for the period 2080 to 2099 relative to 1980 to 1999. Stippling denotes areas where the magnitude of the multi-model ensemble mean exceeds the inter-model standard deviation. Results for individual models can be seen in the Supplementary Material for this chapter.*

depth and their atmospheric environment (see Section 8.6.3.2). Current GCMs simulate clouds through various complex parametrizations (see Section 8.2.1.3) to produce cloud cover quantified by an area fraction within each grid square and each atmospheric layer. Taking multi-model ensemble zonal means of this quantity interpolated to standard pressure levels and latitudes shows increases in cloud cover at all latitudes in the vicinity of the tropopause, and mostly decreases below, indicating an increase in the altitude of clouds overall (Figure 10.10a). This shift occurs consistently across models. Outside the tropics the increases aloft are rather consistent, as indicated by the stippling in the figure. Near-surface amounts increase at some latitudes. The mid-level mid-latitude decreases are very consistent, amounting to as much as one-fifth of the average cloud fraction simulated for 1980 to 1999.

The total cloud area fraction from an individual model represents the net coverage over all the layers, after allowance for the overlap of clouds, and is an output included in the data set. The change in the ensemble mean of this field is shown in Figure 10.10b. Much of the low and middle latitudes experience a decrease in cloud cover, simulated with some consistency. There are a few low-latitude regions of increase, as well as substantial increases at high latitudes. The larger changes relate well to changes in precipitation discussed in Section 10.3.2.3. While clouds need not be precipitating, moderate spatial correlation between cloud cover and precipitation holds for seasonal means of both the present climate and future changes.

The radiative effect of clouds is represented by the cloud radiative forcing diagnostic (see Section 8.6.3.2). This can be

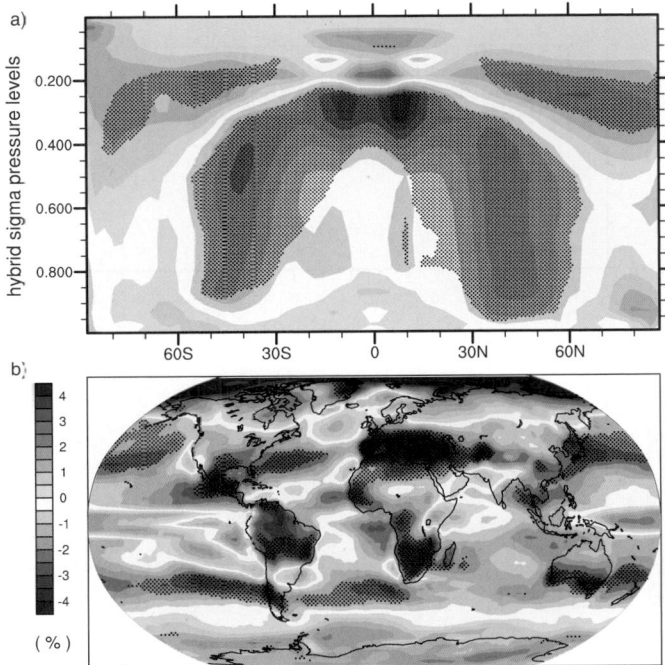

Figure 10.10. *Multi-model mean changes in (a) zonal mean cloud fraction (%), shown as a cross section though the atmosphere, and (b) total cloud area fraction (percent cover from all models). Changes are given as annual means for the SRES A1B scenario for the period 2080 to 2099 relative to 1980 to 1999. Stippling denotes areas where the magnitude of the multi-model ensemble mean exceeds the inter-model standard deviation. Results for individual models can be seen in the Supplementary Material for this chapter.*

767

evaluated from radiative fluxes at the top of the atmosphere calculated with or without the presence of clouds that are output by the GCMs. In the multi-model mean (not shown) values vary in sign over the globe. The global and annual mean averaged over the models, for 1980 to 1999, is –22.3 W m^{-2}. The change in mean cloud radiative forcing has been shown to have different signs in a limited number of previous modelling studies (Meehl et al., 2004b; Tsushima et al., 2006). Figure 10.11a shows globally averaged cloud radiative forcing changes for 2080 to 2099 under the A1B scenario for individual models of the data set, which have a variety of different magnitudes and even signs. The ensemble mean change is –0.6 W m^{-2}. This range indicates that cloud feedback is still an uncertain feature of the global coupled models (see Section 8.6.3.2.2).

The DTR has been shown to be decreasing in several land areas of the globe in 20th-century observations (see Section 3.2.2.7), together with increasing cloud cover (see also Section 9.4.2.3). In the multi-model mean of present climate, DTR over land is indeed closely spatially anti-correlated with the total cloud cover field. This is true also of the 21st-century changes in the fields under the A1B scenario, as can be seen by comparing

the change in DTR shown in Figure 10.11b with the cloud area fraction shown in Figure 10.10b. Changes in DTR reach a magnitude of 0.5°C in some regions, with some consistency among the models. Smaller widespread decreases are likely due to the radiative effect of the enhanced greenhouse gases including water vapour (see also Stone and Weaver, 2002). Further discussion of DTR is provided in Section 10.3.6.2.

In addition to the DTR, Kitoh and Arakawa (2005) document changes in the regional patterns of diurnal precipitation over the Indonesian region, and show that over ocean, nighttime precipitation decreases and daytime precipitation increases, while over land the opposite is the case, thus producing a decrease in the diurnal precipitation amplitude over land and ocean. They attribute these changes to a larger nighttime temperature increase over land due to increased greenhouse gases.

10.3.2.3   Precipitation and Surface Water

Models simulate that global mean precipitation increases with global warming. However, there are substantial spatial and seasonal variations in this field even in the multi-model means depicted in Figure 10.9. There are fewer areas stippled for precipitation than for the warming, indicating more variation in the magnitude of change among the ensemble of models. Increases in precipitation at high latitudes in both seasons are very consistent across models. The increases in precipitation over the tropical oceans and in some of the monsoon regimes (e.g., South Asian monsoon in JJA, Australian monsoon in DJF) are notable, and while not as consistent locally, considerable agreement is found at the broader scale in the tropics (Neelin et al., 2006). There are widespread decreases in mid-latitude summer precipitation, except for increases in eastern Asia. Decreases in precipitation over many subtropical areas are evident in the multi-model ensemble mean, and consistency in the sign of change among the models is often high (Wang, 2005), particularly in some regions like the tropical Central American-Caribbean (Neelin et al., 2006). Further discussion of regional changes is presented in Chapter 11.

The global map of the A1B 2080 to 2099 change in annual mean precipitation is shown in Figure 10.12, along with other hydrological quantities from the multi-model ensemble. Emori and Brown (2005) show percentage changes of annual precipitation from the ensemble. Increases of over 20% occur at most high latitudes, as well as in eastern Africa, central Asia and the equatorial Pacific Ocean. The change over the ocean between 10°S and 10°N accounts for about half the increase in the global mean (Figure 10.5). Substantial decreases, reaching 20%, occur in the Mediterranean region (Rowell and Jones, 2006), the Caribbean region (Neelin et al., 2006) and the subtropical western coasts of each continent. Overall, precipitation over land increases by about 5%, while precipitation over ocean increases 4%, but with regional changes of both signs. The net change over land accounts for 24% of the global mean increase in precipitation, a little less than the areal proportion of land (29%). In Figure 10.12, stippling indicates that the sign of the

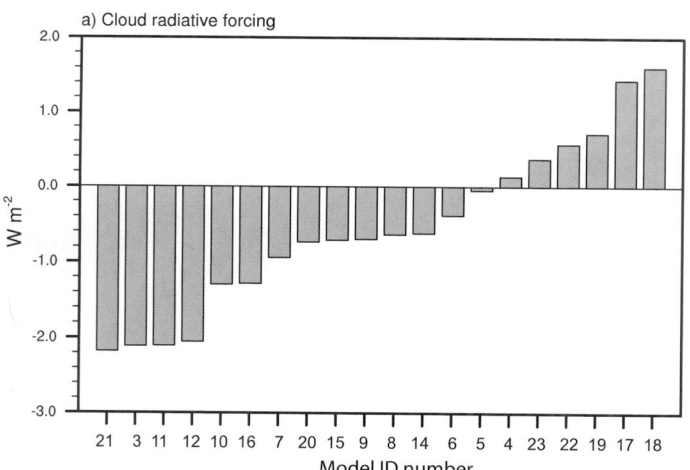

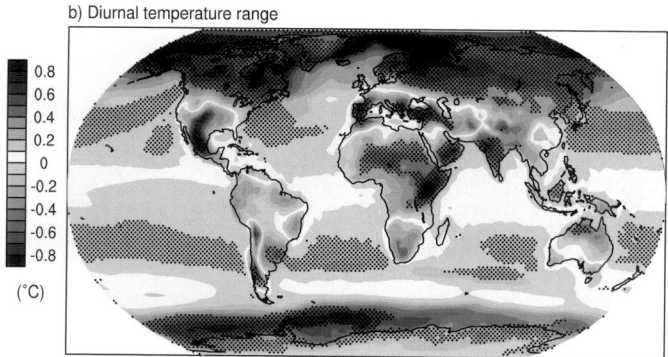

Figure 10.11. *Changes in (a) global mean cloud radiative forcing (W m^{-2}) from individual models (see Table 10.4 for the list of models) and (b) multi-model mean diurnal temperature range (°C). Changes are annual means for the SRES A1B scenario for the period 2080 to 2099 relative to 1980 to 1999. Stippling denotes areas where the magnitude of the multi-model ensemble mean exceeds the inter-model standard deviation. Results for individual models can be seen in the Supplementary Material for this chapter.*

a) Precipitation
b) Soil moisture
c) Runoff
d) Evaporation

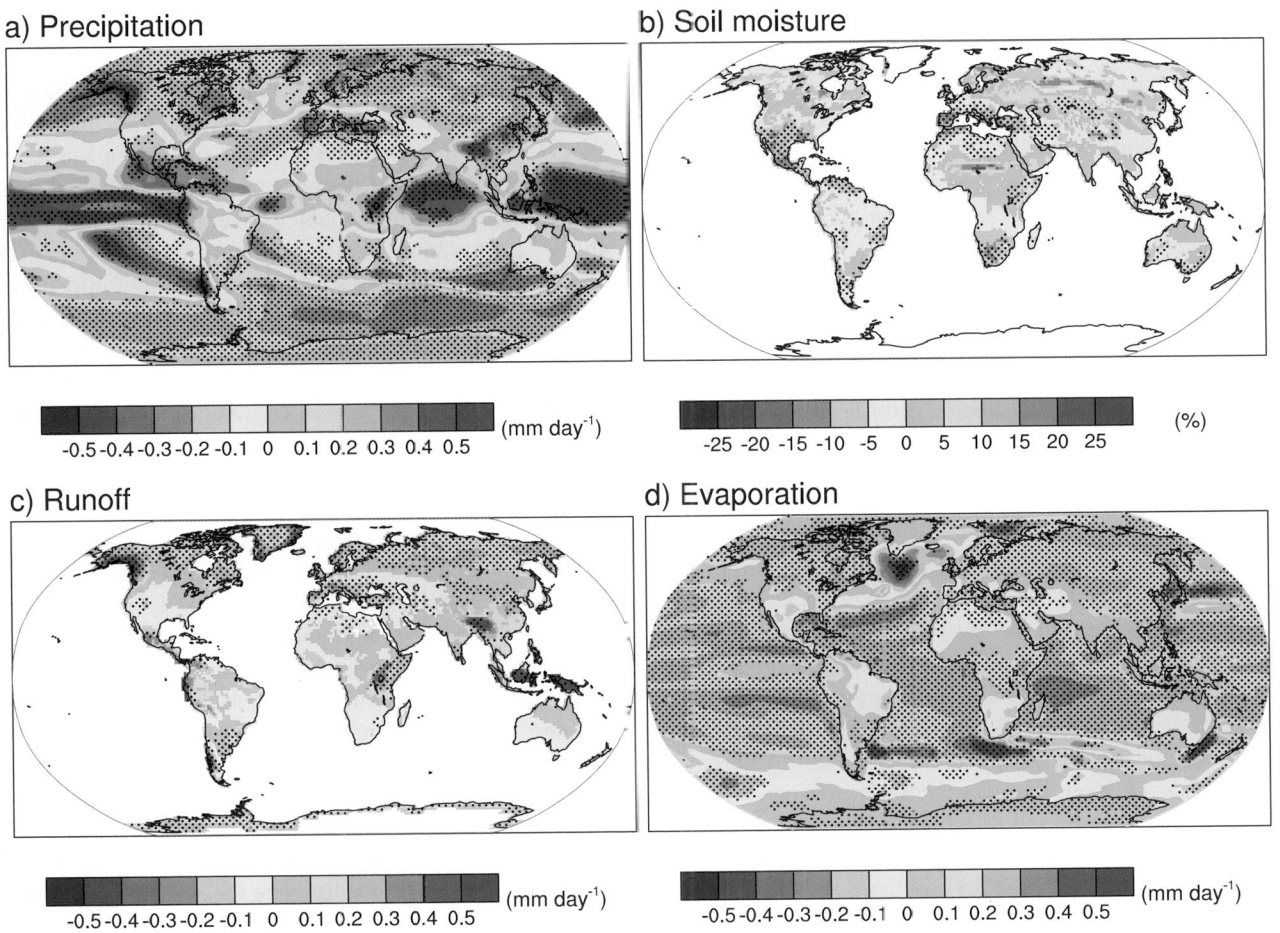

Figure 10.12. *Multi-model mean changes in (a) precipitation (mm day⁻¹), (b) soil moisture content (%), (c) runoff (mm day⁻¹) and (d) evaporation (mm day⁻¹). To indicate consistency in the sign of change, regions are stippled where at least 80% of models agree on the sign of the mean change. Changes are annual means for the SRES A1B scenario for the period 2080 to 2099 relative to 1980 to 1999. Soil moisture and runoff changes are shown at land points with valid data from at least 10 models. Details of the method and results for individual models can be found in the Supplementary Material for this chapter.*

local change is common to at least 80% of the models (with the alternative test shown in the Supplementary Material). This simpler test for consistency is of particular interest for quantities where the magnitudes for the base climate vary across models.

These patterns of change occur in the other scenarios, although with agreement (by the metric M) a little lower than for the warming. The predominance of increases near the equator and at high latitudes, for both land and ocean, is clear from the zonal mean changes of precipitation included in Figure 10.6. The results for change scaled by global mean warming are rather similar across the four scenarios, an exception being a relatively large increase over the equatorial ocean for the commitment case. As with surface temperature, the A1B and B1 scaled values are always close to the A2 results. The zonal means of the percentage change map (shown in Figure 10.6) feature substantial decreases in the subtropics and lower mid-latitudes of both hemispheres in the A2 case, even if increases occur over some regions.

Wetherald and Manabe (2002) provide a good description of the mechanism of hydrological change simulated by GCMs. In GCMs, the global mean evaporation changes closely balance the precipitation change, but not locally because of changes in the atmospheric transport of water vapour. Annual average evaporation (Figure 10.12) increases over much of the ocean, with spatial variations tending to relate to those in the surface warming (Figure 10.8). As found by Kutzbach et al. (2005) and Bosilovich et al. (2005), atmospheric moisture convergence increases over the equatorial oceans and over high latitudes. Over land, rainfall changes tend to be balanced by both evaporation and runoff. Runoff (Figure 10.12) is notably reduced in southern Europe and increased in Southeast Asia and at high latitudes, where there is consistency among models in the sign of change (although less consistency in the magnitude of change). The larger changes reach 20% or more of the simulated 1980 to 1999 values, which range from 1 to 5 mm day⁻¹ in wetter regions to below 0.2 mm day⁻¹ in deserts. Runoff from the melting of ice sheets (Section 10.3.3) is not included here. Nohara et al. (2006) and Milly et al. (2005) assess the impacts of these changes in terms of river flow, and find that discharges from high-latitude rivers increase, while those from major rivers in the Middle East, Europe and Central America tend to decrease.

Models simulate the moisture in the upper few metres of the land surface in varying ways, and evaluation of the soil moisture content is still difficult (See Section 8.2.3.2; Wang, 2005; Gao and Dirmeyer, 2006 for multi-model analyses). The average of the total soil moisture content quantity submitted to the data set is presented here to indicate typical trends. In the annual mean (Figure 10.12), decreases are common in the subtropics and the Mediterranean region. There are increases in east Africa, central Asia, and some other regions with increased precipitation. Decreases also occur at high latitudes, where snow cover diminishes (Section 10.3.3). While the magnitudes of change are quite uncertain, there is good consistency in the signs of change in many of these regions. Similar patterns of change occur in seasonal results (Wang, 2005). Regional hydrological changes are considered in Chapter 11 and in the IPCC Working Group II report.

10.3.2.4 Sea Level Pressure and Atmospheric Circulation

As a basic component of the mean atmospheric circulations and weather patterns, projections of the mean sea level pressure for the medium scenario A1B are considered. Seasonal mean changes for DJF and JJA are shown in Figure 10.9 (matching results in Wang and Swail, 2006b). Sea level pressure differences show decreases at high latitudes in both seasons in both hemispheres. The compensating increases are predominantly over the mid-latitude and subtropical ocean regions, extending across South America, Australia and southern Asia in JJA, and the Mediterranean in DJF. Many of these increases are consistent across the models. This pattern of change, discussed further in Section 10.3.5.3, has been linked to an expansion of the Hadley Circulation and a poleward shift of the mid-latitude storm tracks (Yin, 2005). This helps explain, in part, the increases in precipitation at high latitudes and decreases in the subtropics and parts of the mid-latitudes. Further analysis of the regional details of these changes is given in Chapter 11. The pattern of pressure change implies increased westerly flows across the western parts of the continents. These contribute to increases in mean precipitation (Figure 10.9) and increased precipitation intensity (Meehl et al., 2005a).

10.3.3 Changes in Ocean/Ice and High-Latitude Climate

10.3.3.1 Changes in Sea Ice Cover

Models of the 21st century project that future warming is amplified at high latitudes resulting from positive feedbacks involving snow and sea ice, and other processes (Section 8.6.3.3). The warming is particularly large in autumn and early winter (Manabe and Stouffer, 1980; Holland and Bitz, 2003) when sea ice is thinnest and the snow depth is insufficient to blur the relationship between surface air temperature and sea ice thickness (Maykut and Untersteiner, 1971). As shown by Zhang and Walsh (2006), the coupled models show a range of responses in NH sea ice areal extent ranging from very little

change to a strong and accelerating reduction over the 21st century (Figure 10.13a,b).

An important characteristic of the projected change is for summer ice area to decline far more rapidly than winter ice area (Gordon and O'Farrell, 1997), and hence sea ice rapidly approaches a seasonal ice cover in both hemispheres (Figures 10.13b and 10.14). Seasonal ice cover is, however, rather robust and persists to some extent throughout the 21st century in most (if not all) models. Bitz and Roe (2004) note that future projections show that arctic sea ice thins fastest where it is initially thickest, a characteristic that future climate projections share with sea ice thinning observed in the late 20th century (Rothrock et al., 1999). Consistent with these results, a projection by Gregory et al. (2002b) shows that arctic sea ice volume decreases more quickly than sea ice area (because trends in winter ice area are low) in the 21st century.

In 20th- and 21st-century simulations, antarctic sea ice cover is projected to decrease more slowly than in the Arctic (Figures 10.13c,d and 10.14), particularly in the vicinity of the Ross Sea where most models predict a local minimum in surface warming. This is commensurate with the region with the greatest reduction in ocean heat loss, which results from reduced vertical mixing in the ocean (Gregory, 2000). The ocean stores much of its increased heat below 1 km depth in the Southern Ocean. In contrast, horizontal heat transport poleward of about 60°N increases in many models (Holland and Bitz, 2003), but much of this heat remains in the upper 1 km of the northern subpolar seas and Arctic Ocean (Gregory, 2000; Bitz et al., 2006). Bitz et al. (2006) argue that these differences in the depth where heat is accumulating in the high-latitude oceans have consequences for the relative rates of sea ice decay in the Arctic and Antarctic.

While most climate models share these common characteristics (peak surface warming in autumn and early winter, sea ice rapidly becomes seasonal, arctic ice decays faster than antarctic ice, and northward ocean heat transport increases into the northern high latitudes), models have poor agreement on the amount of thinning of sea ice (Flato and Participating CMIP Modeling Groups, 2004; Arzel et al., 2006) and the overall climate change in the polar regions (IPCC, 2001; Holland and Bitz, 2003). Flato (2004) shows that the basic state of the sea ice and the reduction in thickness and/or extent have little to do with sea ice model physics among CMIP2 models. Holland and Bitz (2003) and Arzel et al. (2006) find serious biases in the basic state of simulated sea ice thickness and extent. Further, Rind et al. (1995), Holland and Bitz (2003) and Flato (2004) show that the basic state of the sea ice thickness and extent have a significant influence on the projected change in sea ice thickness in the Arctic and extent in the Antarctic.

10.3.3.2 Changes in Snow Cover and Frozen Ground

Snow cover is an integrated response to both temperature and precipitation and exhibits strong negative correlation with air temperature in most areas with a seasonal snow cover (see Section 8.6.3.3 for an evaluation of model-simulated

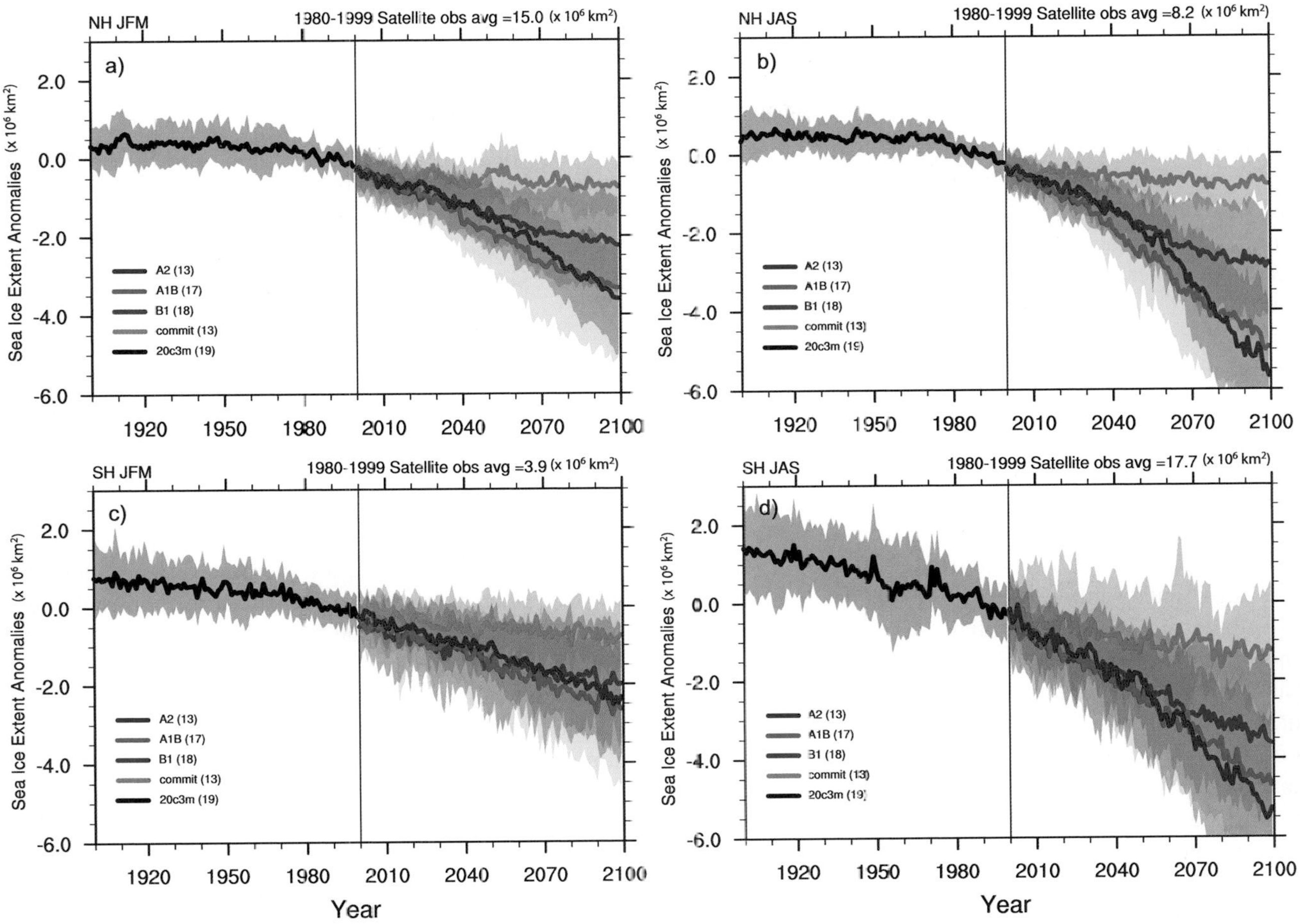

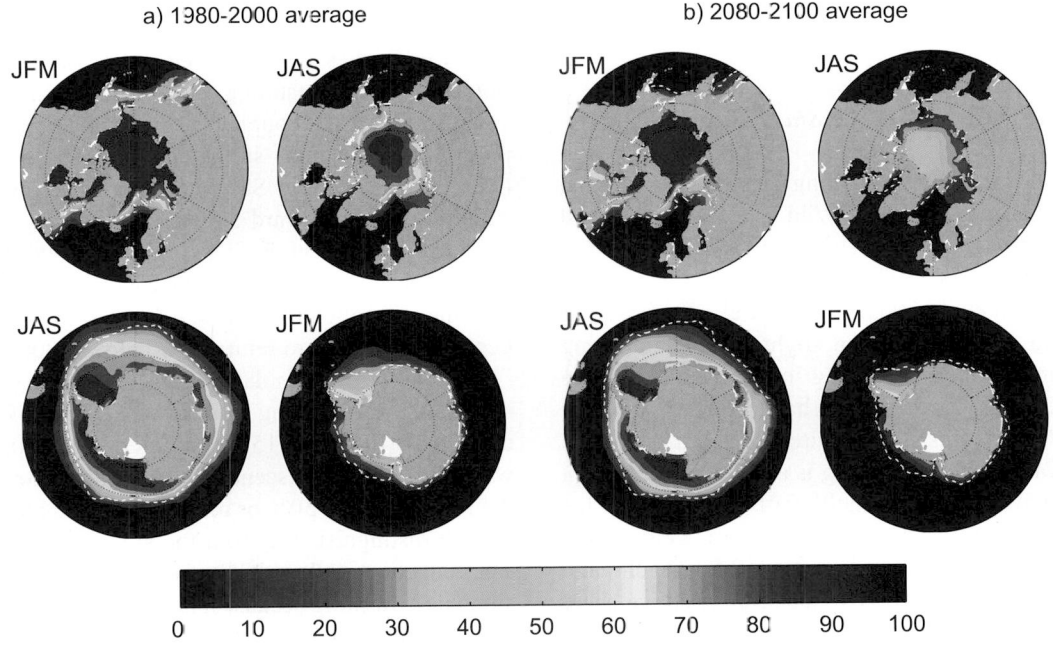

Figure 10.13. *Multi-model simulated anomalies in sea ice extent for the 20th century (20c3m) and 21st century using the SRES A2, A1B and B1 as well as the commitment scenario for (a) Northern Hemisphere January to March (JFM), (b) Northern Hemisphere July to September (JAS). Panels (c) and (d) are as for (a) and (b) but for the Southern Hemisphere. The solid lines show the multi-model mean, shaded areas denote ±1 standard deviation. Sea ice extent is defined as the total area where sea ice concentration exceeds 15%. Anomalies are relative to the period 1980 to 2000. The number of models is given in the legend and is different for each scenario.*

a) 1980-2000 average   b) 2080-2100 average

Figure 10.14. *Multi-model mean sea ice concentration (%) for January to March (JFM) and June to September (JAS), in the Arctic (top) and Antarctic (bottom) for the periods (a) 1980 to 2000 and b) 2080 to 2100 for the SRES A1B scenario. The dashed white line indicates the present-day 15% average sea ice concentration limit. Modified from Flato et al. (2004).*

present-day snow cover). Because of this temperature association, the simulations project widespread reductions in snow cover over the 21st century (Supplementary Material, Figure S10.1). For the Arctic Climate Impact Assessment (ACIA) model mean, at the end of the 21st century the projected reduction in the annual mean NH snow cover is 13% under the B2 scenario (ACIA, 2004). The individual model projections range from reductions of 9 to 17%. The actual reductions are greatest in spring and late autumn/early winter, indicating a shortened snow cover season (ACIA, 2004). The beginning of the snow accumulation season (the end of the snowmelt season) is projected to be later (earlier), and the fractional snow coverage is projected to decrease during the snow season (Hosaka et al., 2005).

Warming at high northern latitudes in climate model simulations is also associated with large increases in simulated thaw depth over much of the permafrost regions (Lawrence and Slater, 2005; Yamaguchi et al., 2005; Kitabata et al., 2006). Yamaguchi et al. (2005) show that initially soil moisture increases during the summer. In the late 21st century when the thaw depth has increased substantially, a reduction in summer soil moisture eventually occurs (Kitabata et al., 2006). Stendel and Christensen (2002) show poleward movement of permafrost extent, and a 30 to 40% increase in active layer thickness for most of the permafrost area in the NH, with the largest relative increases concentrated in the northernmost locations.

Regionally, the changes are a response to both increased temperature and increased precipitation (changes in circulation patterns) and are complicated by the competing effects of warming and increased snowfall in those regions that remain below freezing (see Section 4.2 for a further discussion of processes that affect snow cover). In general, snow amount and snow coverage decreases in the NH (Supplementary Material, Figure S10.1). However, in a few regions (e.g., Siberia), snow amount is projected to increase. This is attributed to the increase in precipitation (snowfall) from autumn to winter (Meleshko et al., 2004; Hosaka et al., 2005).

10.3.3.3 *Changes in Greenland Ice Sheet Mass Balance*

As noted in Section 10.6, modelling studies (e.g., Hanna et al., 2002; Kiilsholm et al., 2003; Wild et al., 2003) as well as satellite observations, airborne altimeter surveys and other studies (Abdalati et al., 2001; Thomas et al., 2001; Krabill et al., 2004; Johannessen et al., 2005; Zwally et al., 2005; Rignot and Kanagaratnam, 2006) suggest a slight inland thickening and strong marginal thinning resulting in an overall negative Greenland Ice Sheet mass balance which has accelerated recently (see Section 4.6.2.2.). A consistent feature of all climate models is that projected 21st-century warming is amplified in northern latitudes. This suggests continued melting of the Greenland Ice Sheet, since increased summer melting dominates over increased winter precipitation in model projections of future climate. Ridley et al. (2005) coupled UKMO-HadCM3 to an ice sheet model to explore the melting of the Greenland Ice Sheet under elevated (four times pre-industrial) levels of atmospheric CO_2 (see Section 10.7.4.3, Figure 10.38). While the entire Greenland

Ice Sheet eventually completely ablated (after 3 kyr), the peak rate of melting was 0.06 Sv (1 Sv = 10^6 m³ s⁻¹) corresponding to about 5.5 mm yr⁻¹ global sea level rise (see Sections 10.3.4 and 10.6.6). Toniazzo et al. (2004) further show that in UKMO-HadCM3, the complete melting of the Greenland Ice sheet is an irreversible process even if pre-industrial levels of atmospheric CO_2 are re-established after it melts.

10.3.4 Changes in the Atlantic Meridional Overturning Circulation

A feature common to all climate model projections is the increase in high-latitude temperature as well as an increase in high-latitude precipitation. This was reported in the TAR and is confirmed by the projections using the latest versions of comprehensive climate models (see Section 10.3.2). Both of these effects tend to make the high-latitude surface waters less dense and hence increase their stability, thereby inhibiting convective processes. As more coupled models have become available since the TAR, the evolution of the Atlantic Meridional Overturning Circulation (MOC) can be more thoroughly assessed. Figure 10.15 shows simulations from 19 coupled models integrated from 1850 to 2100 under SRES A1B atmospheric CO_2 and aerosol scenarios up to year 2100, and constant concentrations thereafter (see Figure 10.5). All of the models, except CGCM3.1, INM-CM3.0 and MRI-CGCM2.3.2, were run without flux adjustments (see Table 8.1). The MOC is influenced by the density structure of the Atlantic Ocean, small-scale mixing and the surface momentum and buoyancy fluxes. Some models simulate a MOC strength that is inconsistent with the range of present-day estimates (Smethie and Fine, 2001; Ganachaud, 2003; Lumpkin and Speer, 2003; Talley, 2003). The MOC for these models is shown for completeness but is not used in assessing potential future changes in the MOC in response to various emissions scenarios.

Fewer studies have focused on projected changes in the Southern Ocean resulting from future climate warming. A common feature of coupled model simulations is the projected poleward shift and strengthening of the SH westerlies (Yin, 2005; Fyfe and Saenko, 2006). This in turn leads to a strengthening, poleward shift and narrowing of the Antarctic Circumpolar Current. Fyfe and Saenko (2006) further note that the enhanced equatorward surface Ekman transport, associated with the intensified westerlies, is balanced by an enhanced deep geostrophic poleward return flow below 2,000 m.

Generally, the simulated late-20th century Atlantic MOC shows a spread ranging from a weak MOC of about 12 Sv to over 20 Sv (Figure 10.15; Schmittner et al., 2005). When forced with the SRES A1B scenario, the models show a reduction in the MOC of up to 50% or more, but in one model, the changes are not distinguishable from the simulated natural variability. The reduction in the MOC proceeds on the time scale of the simulated warming because it is a direct response to the increase in buoyancy at the ocean surface. A positive North Atlantic Oscillation (NAO) trend might delay this response by a few decades but not prevent it (Delworth and Dixon, 2000). Such

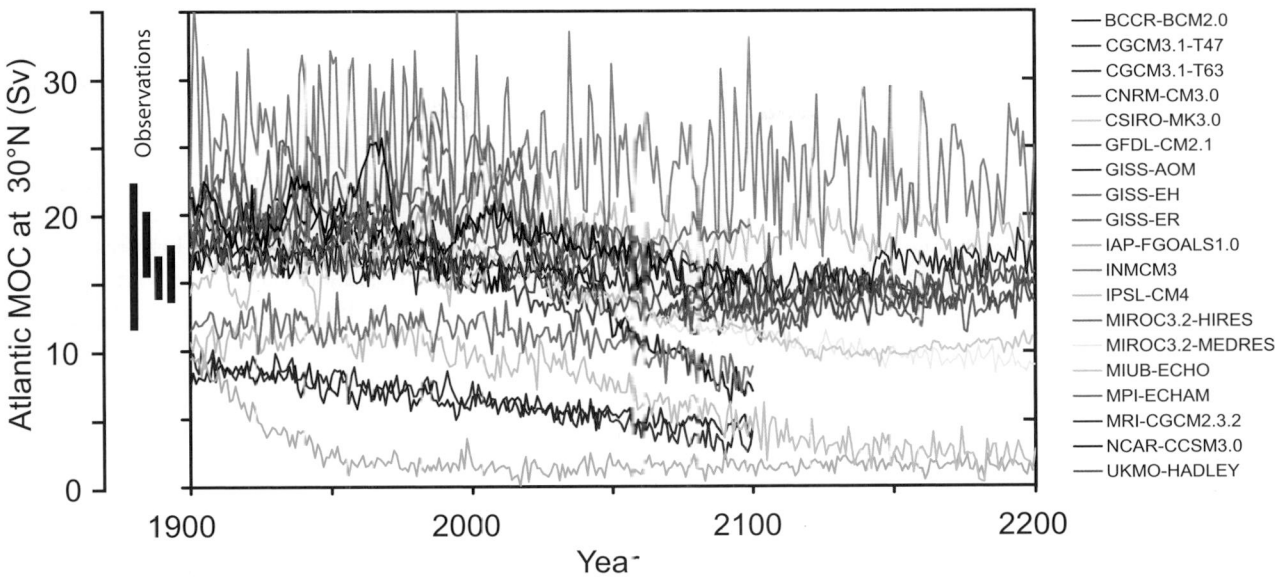

Figure 10.15. *Evolution of the Atlantic meridional overturning circulation (MOC) at 30°N in simulations with the suite of comprehensive coupled climate models (see Table 8.1 for model details) from 1850 to 2100 using 20th Century Climate in Coupled Models (20C3M) simulations for 1850 to 1999 and the SRES A1B emissions scenario for 1999 to 2100. Some of the models continue the integration to year 2200 with the forcing held constant at the values of year 2100. Observationally based estimates of late-20th century MOC are shown as vertical bars on the left. Three simulations show a steady or rapid slow down of the MOC that is unrelated to the forcing; a few others have late-20th century simulated values that are inconsistent with observational estimates. Of the model simulations consistent with the late-20th century observational estimates, no simulation shows an increase in the MOC during the 21st century; reductions range from indistinguishable within the simulated natural variability to over 50% relative to the 1960 to 1990 mean; and none of the models projects an abrupt transition to an off state of the MOC. Adapted from Schmittner et al. (2005) with additions.*

a weakening of the MOC in future climate causes reduced sea surface temperature (SST) and salinity in the region of the Gulf Stream and North Atlantic Current (Dai et al., 2005). This can produce a decrease in northward heat transport south of 60°N, but increased northward heat transport north of 60°N (A. Hu et al., 2004). No model shows an increase in the MOC in response to the increase in greenhouse gases, and no model simulates an abrupt shut-down of the MOC within the 21st century. One study suggests that inherent low-frequency variability in the Atlantic region, the Atlantic Multidecadal Oscillation, may produce a natural weakening of the MOC over the next few decades that could further accentuate the decrease due to anthropogenic climate change (Knight et al., 2005; see Section 8.4.6).

In some of the older models (e.g., Dixon et al., 1999), increased high-latitude precipitation dominates over increased high-latitude warming in causing the weakening, while in others (e.g., Mikolajewicz and Voss, 2000), the opposite is found. In a recent model intercomparison, Gregory et al. (2005) find that for all 11 models analysed, the MOC reduction is caused more by changes in surface heat flux than changes in surface freshwater flux. In addition, simulations using models of varying complexity (Stocker et al., 1992b; Saenko et al., 2003; Weaver et al., 2003) show that freshening or warming in the Southern Ocean acts to increase or stabilise the Atlantic MOC. This is likely a consequence of the complex coupling of Southern Ocean processes with North Atlantic Deep Water production.

A few simulations using coupled models are available that permit the assessment of the long-term stability of the MOC (Stouffer and Manabe, 1999; Voss and Mikolajewicz, 2001;

Stouffer and Manabe, 2003; Wood et al., 2003; Yoshida et al., 2005; Bryan et al., 2006). Most of these simulations assume an idealised increase in atmospheric CO_2 by 1% yr^{-1} to various levels ranging from two to four times pre-industrial levels. One study also considers slower increases (Stouffer and Manabe, 1999), or a reduction in CO_2 (Stouffer and Manabe, 2003). The more recent models are not flux adjusted and have higher resolution (about 1.0°) (Yoshida et al., 2005; Bryan et al., 2006). A common feature of all simulations is a reduction in the MOC in response to the warming and a stabilisation or recovery of the MOC when the concentration is kept constant after achieving a level of two to four times the pre-industrial atmospheric CO_2 concentration. None of these models shows a shutdown of the MOC that continues after the forcing is kept constant. But such a long-term shutdown cannot be excluded if the amount of warming and its rate exceed certain thresholds as shown using an EMIC (Stocker and Schmittner, 1997). Complete shut-downs, although not permanent, were also simulated by a flux-adjusted coupled model (Manabe and Stouffer, 1994; Stouffer and Manabe, 2003; see also Chan and Motoi, 2005). In none of these AOGCM simulations were the thresholds, as determined by the EMIC, passed (Stocker and Schmittner, 1997). As such, the long-term stability of the MOC found in the present AOGCM simulations is consistent with the results from the simpler models.

The reduction in MOC strength associated with increasing greenhouse gases represents a negative feedback for the warming in and around the North Atlantic. That is, through reducing the transport of heat from low to high latitudes, SSTs are cooler than they would otherwise be if the MOC was unchanged. As such, warming is reduced over and downstream of the North Atlantic. It is important to note that in models where the MOC weakens, warming still occurs downstream over Europe due to the overall dominant role of the radiative forcing associated with increasing greenhouse gases (Gregory et al., 2005). Many future projections show that once the radiative forcing is held fixed, re-establishment of the MOC occurs to a state similar to that of the present day. The partial or complete re-establishment of the MOC is slow and causes additional warming in and around the North Atlantic. While the oceanic meridional heat flux at low latitudes is reduced upon a slowdown of the MOC, many simulations show increasing meridional heat flux into the Arctic which contributes to accelerated warming and sea ice melting there. This is due to both the advection of warmer water and an intensification of the influx of North Atlantic water into the Arctic (A. Hu et al., 2004).

Climate models that simulated a complete shutdown of the MOC in response to sustained warming were flux-adjusted coupled GCMs or EMICs. A robust result from such simulations is that the shutdown of the MOC takes several centuries after the forcing is kept fixed (e.g., at 4 × atmospheric CO_2 concentration). Besides the forcing amplitude and rate (Stocker and Schmittner, 1997), the amount of mixing in the ocean also appears to determine the stability of the MOC: increased vertical and horizontal mixing tends to stabilise the MOC and to eliminate the possibility of a second equilibrium state (Manabe and Stouffer, 1999; Knutti and Stocker, 2000; Longworth et al., 2005). Random internal variability or noise, often not present in simpler models, may also be important in determining the effective MOC stability (Knutti and Stocker, 2002; Monahan, 2002).

The MOC is not necessarily a comprehensive indicator of ocean circulation changes in response to global warming. In a transient 2 × atmospheric CO_2 experiment using a coupled AOGCM, the MOC changes were small, but convection in the Labrador Sea stopped due to warmer and hence less dense waters that inflow from the Greenland-Iceland-Norwegian Sea (GIN Sea) (Wood et al., 1999; Stouffer et al., 2006a). Similar results were found by A. Hu et al. (2004), who also report an increase in convection in the GIN Sea due to the influx of more saline waters from the North Atlantic. Various simulations using coupled models of different complexity find significant reductions in convection in the GIN Sea in response to warming (Schaeffer et al., 2004; Bryan et al., 2006). Presumably, a delicate balance exists in the GIN Sea between the circum-arctic river runoff, sea ice production and advection of saline waters from the North Atlantic, and on a longer time scale, the inflow

of freshwater through Bering Strait. The projected increases in circum-arctic river runoff (Wu et al., 2005) may enhance the tendency towards a reduction in GIN Sea convection (Stocker and Raible, 2005; Wu et al., 2005). Cessation of convection in the Labrador Sea in the next few decades is also simulated in a high-resolution model of the Atlantic Ocean driven by surface fluxes from two AOGCMs (Schweckendiek and Willebrand, 2005). The large-scale responses of the high-resolution ocean model (e.g., MOC, Labrador Seas) agree with those from the AOGCMs. The grid resolution of the ocean components in the coupled AOGCMs has significantly increased since the TAR, and some consistent patterns of changes in convection and water mass properties in the Atlantic Ocean emerge in response to the warming, but models still show a variety of responses in the details.

The best estimate of sea level from 1993 to 2003 (see Section 5.5.5.2) associated with the slight net negative mass balance from Greenland is 0.1 to 0.3 mm yr^{-1} over the total ocean surface. This converts to only about 0.002 to 0.003 Sv of freshwater forcing. Such an amount, even when added directly and exclusively to the North Atlantic, has been suggested to be too small to affect the North Atlantic MOC (see Weaver and Hillaire-Marcel, 2004a). While one model exhibits a MOC weakening in the later part of the 21st century due to Greenland Ice Sheet melting (Fichefet et al., 2003), this same model had a very large downward drift of its overturning in the control climate, making it difficult to actually attribute the model MOC changes to the ice sheet melting. As noted in Section 10.3.3.3, Ridley et al. (2005) find the peak rate of Greenland Ice Sheet melting is about 0.1 Sv when they instantaneously elevate greenhouse gas levels in UKMO-HadCM3. They further note that this has little effect on the North Atlantic meridional overturning, although 0.1 Sv is sufficiently large to cause more dramatic transient changes in the strength of the MOC in other models (Stouffer et al., 2006b).

Taken together, it is very likely that the MOC, based on currently available simulations, will decrease, perhaps associated with a significant reduction in Labrador Sea Water formation, but very unlikely that the MOC will undergo an abrupt transition during the course of the 21st century. At this stage, it is too early to assess the likelihood of an abrupt change of the MOC beyond the end of the 21st century, but the possibility cannot be excluded (see Box 10.1). The few available simulations with models of different complexity instead suggest a centennial slowdown. Recovery of the MOC is simulated in some models if the radiative forcing is stabilised but would take several centuries; in other models, the reduction persists.

Box 10.1: Future Abrupt Climate Change, 'Climate Surprises', and Irreversible Changes

Theory, models and palaeoclimatic reconstructions (see Chapter 6) have established the fact that changes in the climate system can be abrupt and widespread. A working definition of 'abrupt climate change' is given in Alley et al. (2002): 'Technically, an abrupt climate change occurs when the climate system is forced to cross some threshold, triggering a transition to a new state at a rate determined by the climate system itself and faster than the cause'. More generally, a gradual change in some determining quantity of the climate system (e.g., radiation balance, land surface properties, sea ice, etc.) can cause a variety of structurally different responses (Box 10.1, Figure 1). The response of a purely linear system scales with the forcing, and at stabilisation of the forcing, a new equilibrium is achieved which is structurally similar, but not necessarily close to the original state. However, if the system contains more than one equilibrium state, transitions to structurally different states are possible. Upon the crossing of a tipping point (bifurcation point), the evolution of the system is no longer controlled by the time scale of the forcing, but rather determined by its internal dynamics, which can either be much faster than the forcing, or significantly slower. Only the former case would be termed 'abrupt climate change', but the latter case is of equal importance. For the long-term evolution of a climate variable one must distinguish between reversible and irreversible changes. The notion of 'climate surprises' usually refers to abrupt transitions and temporary or permanent transitions to a different state in parts of the climate system such as, for example the 8.2 kyr event (see Section 6.5.2.1).

Atlantic Meridional Overturning Circulation and other ocean circulation changes:

The best-documented type of abrupt climate change in the palaeoclimatic archives is that associated with changes in the ocean circulation (Stocker, 2000). Since the TAR, many new results from climate models of different complexity have provided a more detailed

view on the anticipated changes in the Atlantic MOC in response to global warming. Most models agree that the MOC weakens over the next 100 years and that this reduction ranges from indistinguishable from natural variability to over 50% by 2100 (Figure 10.15). None of the AOGCM simulations shows an abrupt change when forced with the SRES emissions scenarios until 2100, but some long-term model simulations suggest that a complete cessation can result for large forcings (Stouffer and Manabe, 2003). Models of intermediate complexity indicate that thresholds in the MOC may be present but that they depend on the amount and rate of warming for a given model (Stocker and Schmittner, 1997). The few long-term simulations from AOGCMs indicate that even complete shutdowns of the MOC may be reversible (Stouffer and Manabe, 2003; Yoshida et al., 2005; Stouffer et al., 2006b). However, until millennial simulations with AOGCMs are available, the important question of potential irreversibility of an MOC shutdown remains unanswered. Both simplified models and AOGCMs agree,

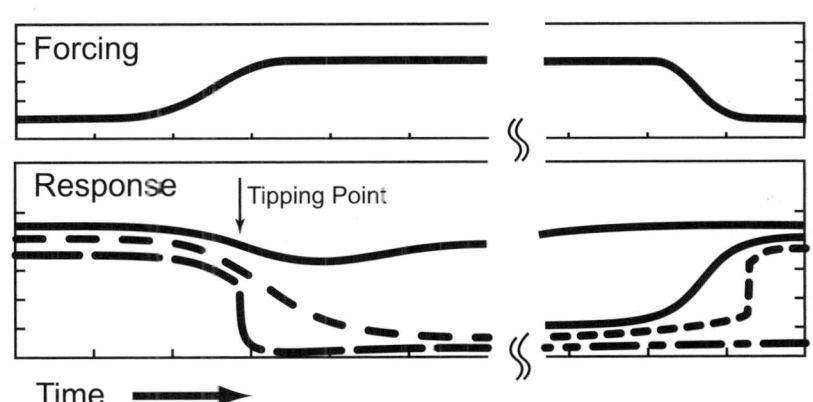

Box 10.1, Figure 1. *Schematic illustration of various responses of a climate variable to forcing. The forcing (top panels) reaches a new stable level (left part of figure), and later approaches the original level on very long time scales (right part of the figure). The response of the climate variable (bottom panels) can be smooth (solid line) or cross a tipping point inducing a transition to a structurally different state (dashed lines). That transition can be rapid (abrupt change, long-dashed), or gradual (short-dashed), but is usually dictated by the internal dynamics of the climate system rather than the forcing. The long-term behaviour (right part) also exhibits different possibilities. Changes can be irreversible (dash-dotted) with the system settling at a different stable state, or reversible (solid, dotted) when the forcing is set back to its original value. In the latter case, the transition again can be gradual or abrupt. An example for illustration, but not the only one, is the response of the Atlantic meridional overturning circulation to a gradual change in radiative forcing.*

however, that a potentially complete shut-down of the MOC, induced by global warming, would take many decades to more than a century. There is no direct model evidence that the MOC could collapse within a few decades in response to global warming. However, a few studies do show the potential for rapid changes in the MOC (Manabe and Stouffer, 1999), and the processes concerned are poorly understood (see Section 8.7). This is not inconsistent with the palaeoclimate records. The cooling events during the last ice ages registered in the Greenland ice cores developed over a couple of centuries to millennia. In contrast, there were also a number of very rapid warmings, the so-called Dansgaard-Oeschger events (NorthGRIP Members, 2004), or rapid cooling (LeGrande et al., 2006), which evolved over decades or less, most probably associated with rapid latitudinal shifts in ocean convection sites and changes in strength of the MOC (see Section 6.3.2).   *(continued)*

Recent simulations with models with ocean components that resolve topography in sufficient detail obtain a consistent pattern of a strong to complete reduction of convection in the Labrador Sea (Wood et al., 1999; Schweckendiek and Willebrand, 2005). Such changes in the convection, with implications for the atmospheric circulation, can develop within a few years (Schaeffer et al., 2002). The long-term and regional-to-hemispheric scale effects of such changes in water mass properties have not yet been investigated.

With a reduction in the MOC, the meridional heat flux also decreases in the subtropical and mid-latitudes with large-scale effects on the atmospheric circulation. In consequence, the warming of the North Atlantic surface proceeds more slowly. Even for strong reductions in MOC towards the end of the 21st century, no cooling is observed in the regions around the North Atlantic because it is overcompensated by the radiative forcing that caused the ocean response in the first place.

At high latitudes, an increase in the oceanic meridional heat flux is simulated by these models. This increase is due to both an increase in the overturning circulation in the Arctic and the advection of warmer waters from lower latitudes and thus contributes significantly to continuing sea ice reduction in the Atlantic sector of the Arctic (A. Hu et al., 2004). Few simulations have also addressed the changes in overturning in the South Atlantic and Southern Ocean. In addition to water mass modifications, this also has an effect on the transport by the Antarctic Circumpolar Current, but results are not yet conclusive.

Current understanding of the processes responsible for the initiation of an ice age indicate that a reduction or collapse of the MOC in response to global warming could not start an ice age (Berger and Loutre, 2002; Crucifix and Loutre, 2002; Yoshimori et al., 2002; Weaver and Hillaire-Marcel, 2004b).

Arctic sea ice:

Arctic sea ice is responding sensitively to global warming. While changes in winter sea ice cover are moderate, late summer sea ice is projected to disappear almost completely towards the end of the 21st century. A number of positive feedbacks in the climate system accelerate the melt back of sea ice. The ice-albedo feedback allows open water to receive more heat from the Sun during summer, and the increase in ocean heat transport to the Arctic through the advection of warmer waters and stronger circulation further reduces ice cover. Minimum arctic sea ice cover is observed in September. Model simulations indicate that the September sea ice cover decreases substantially in response to global warming, generally evolving on the time scale of the warming. With sustained warming, the late summer disappearance of a major fraction of arctic sea ice is permanent.

Glaciers and ice caps:

Glaciers and ice caps are sensitive to changes in temperature and precipitation. Observations point to a reduction in volume over the last 20 years (see Section 4.5.2), with a rate during 1993 to 2003 corresponding to 0.77 ± 0.22 mm yr^{-1} sea level equivalent, with a larger mean central estimate than that for 1961 to 1998 (corresponding to 0.50 ± 0.18 mm yr^{-1} sea level equivalent). Rapid changes are therefore already underway and enhanced by positive feedbacks associated with the surface energy balance of shrinking glaciers and newly exposed land surface in periglacial areas. Acceleration of glacier loss over the next few decades is likely (see Section 10.6.3). Based on simulations of 11 glaciers in various regions, a volume loss of 60% of these glaciers is projected by the year 2050 (Schneeberger et al., 2003). Glaciated areas in the Americas are also affected. A comparative study including seven GCM simulations at $2 \times$ atmospheric CO_2 conditions inferred that many glaciers may disappear completely due to an increase in the equilibrium line altitude (Bradley et al., 2004). The disappearance of these ice bodies is much faster than a potential re-glaciation several centuries hence, and may in some areas be irreversible.

Greenland and West Antarctic Ice Sheets:

Satellite and *in situ* measurement networks have demonstrated increasing melting and accelerated ice flow around the periphery of the Greenland Ice Sheet (GIS) over the past 25 years (see Section 4.6.2). The few simulations of long-term ice sheet simulations suggest that the GIS will significantly decrease in volume and area over the coming centuries if a warmer climate is maintained (Gregory et al., 2004a; Huybrechts et al., 2004; Ridley et al., 2005). A threshold of annual mean warming of 1.9°C to 4.6°C in Greenland has been estimated for elimination of the GIS (Gregory and Huybrechts, 2006; see section 10.7.3.3), a process which would take many centuries to complete. Even if temperatures were to decrease later, the reduction of the GIS to a much smaller extent might be irreversible, because the climate of an ice-free Greenland could be too warm for accumulation; however, this result is model dependent (see Section 10.7.3.3). The positive feedbacks involved here are that once the ice sheet gets thinner, temperatures in the accumulation region are higher, increasing the melting and causing more precipitation to fall as rain rather than snow; that the lower albedo of the exposed ice-free land causes a local climatic warming; and that surface melt water might accelerate ice flow (see Section 10.6.4.2).

A collapse of the West Antarctic Ice Sheet (WAIS) has been discussed as a potential response to global warming for many years (Bindschadler, 1998; Oppenheimer, 1998; Vaughan, 2007). A complete collapse would cause a global sea level rise of about 5 m. The observed acceleration of ice streams in the Amundsen Sea sector of the WAIS, the rapidity of propagation of this signal upstream and the acceleration of glaciers that fed the Larsen B Ice Shelf after its collapse have renewed these concerns (see Section 10.6.4.2).

(continued)

It is possible that the presence of ice shelves tends to stabilise the ice sheet, at least regionally. Therefore, a weakening or collapse of ice shelves, caused by melting on the surface or by melting at the bottom by a warmer ocean, might contribute to a potential destabilisation of the WAIS, which could proceed through the positive feedback of grounding-line retreat. Present understanding is insufficient for prediction of the possible speed or extent of such a collapse (see Box 4.1 and Section 10.7.3.4).

Vegetation cover:

Irreversible and relatively rapid changes in vegetation cover and composition have occurred frequently in the past. The most prominent example is the desertification of the Sahara region about 4 to 6 ka (Claussen et al., 1999). The reason for this behaviour is believed to lie in the limits of plant communities with respect to temperature and precipitation. Once critical levels are crossed, certain species can no longer compete within their ecosystem. Areas close to vegetation boundaries will experience particularly large and rapid changes due to the slow migration of these boundaries induced by global warming. A climate model simulation into the future shows that drying and warming in South America leads to a continuous reduction in the forest of Amazonia (Cox et al., 2000, 2004). While evolving continuously over the 21st century, such a change and ultimate disappearance could be irreversible, although this result could be model dependent since an analysis of 11 AOGCMs shows a wide range of future possible rainfall changes over the Amazon (Li et al., 2006).

One of the possible 'climate surprises' concerns the role of the soil in the global carbon cycle. As the concentration of CO_2 is increasing, the soil is acting, in the global mean, as a carbon sink by assimilating carbon due to accelerated growth of the terrestrial biosphere (see also Section 7.3.3.1.1). However, by about 2050, a model simulation suggests that the soil changes to a source of carbon by releasing previously accumulated carbon due to increased respiration (Cox et al., 2000) induced by increasing temperature and precipitation. This represents a positive feedback to the increase in atmospheric CO_2. While different models agree regarding the sign of the feedback, large uncertainties exist regarding the strength (Cox et al., 2000; Dufresne et al., 2002; Friedlingstein et al., 2006). However, the respiration increase is caused by a warmer and wetter climate. The switch from moderate sink to strong source of atmospheric carbon is rather rapid and occurs within two decades (Cox et al. 2004), but the timing of the onset is uncertain (Huntingford et al., 2004). A model intercomparison reveals that once set in motion, the increase in respiration continues even after the CO_2 levels are held constant (Cramer et al., 2001). Although considerable uncertainties still exist, it is clear that feedback mechanisms between the terrestrial biosphere and the physical climate system exist which can qualitatively and quantitatively alter the response to an increase in radiative forcing.

Atmospheric and ocean-atmosphere regimes:

Changes in weather patterns and regimes can be abrupt processes that might occur spontaneously due to dynamical interactions in the atmosphere-ice-ocean system, or manifest as the crossing of a threshold in the system due to slow external forcing. Such shifts have been reported in SST in the tropical Pacific, leading to a more positive ENSO phase (Trenberth, 1990), in the stratospheric polar vortex (Christiansen, 2003), in a shut-down of deep convection in the Greenland Sea (Bönisch et al., 1997; Ronski and Budeus, 2005) and in an abrupt freshening of the Labrador Sea (Dickson et al., 2002). In the latter, the freshening evolved throughout the entire depth but the shift in salinity was particularly rapid: the 34.87 psu isohaline plunged from seasonally surface to 1,600 metres within 2 years with no return since 1973.

In a long, unforced model simulation, a period of a few decades with anomalously cold temperatures (up to 10 standard deviations below average) in the region south of Greenland was found (Hall and Stouffer, 2001). It was caused by persistent winds that changed the stratification of the ocean and inhibited convection, thereby reducing heat transfer from the ocean to the atmosphere. Similar results were found in a different model in which the major convection site in the North Atlantic spontaneously switched to a more southerly location for several decades to centuries (Goosse et al., 2002). Other simulations show that the slowly increasing radiative forcing is able to cause transitions in the convective activity in the Greenland-Iceland-Norwegian Sea that have an influence on the atmospheric circulation over Greenland and Western Europe (Schaeffer et al., 2002). The changes unfold within a few years and indicate that the system has crossed a threshold.

A multi-model analysis of regimes of polar variability (NAO, Arctic and Antarctic Oscillations) reveals that the simulated trends in the 21st century influence the Arctic and Antarctic Oscillations and point towards more zonal circulation (Rauthe et al., 2004). Temperature changes associated with changes in atmospheric circulation regimes such as the NAO can exceed in certain regions (e.g., Northern Europe) the long-term global warming that causes such inter-decadal regime shifts (Dorn et al., 2003).

10.3.5 Changes in Properties of Modes of Variability

10.3.5.1 Interannual Variability in Surface Air Temperature and Precipitation

Future changes in anthropogenic forcing will result not only in changes in the mean climate state but also in the variability of climate. Addressing the interannual variability in monthly mean surface air temperature and precipitation of 19 AOGCMs in CMIP2, Räisänen (2002) finds a decrease in temperature variability during the cold season in the extratropical NH and a slight increase in temperature variability in low latitudes and in warm season northern mid-latitudes. The former is likely due to the decrease of sea ice and snow with increasing temperature. The summer decrease in soil moisture over the mid-latitude land surfaces contributes to the latter. Räisänen (2002) also finds an increase in monthly mean precipitation variability in most areas, both in absolute value (standard deviation) and in relative value (coefficient of variation). However, the significance level of these variability changes is markedly lower than that for time mean climate change. Similar results were obtained from 18 AOGCM simulations under the SRES A2 scenario (Giorgi and Bi, 2005).

10.3.5.2 Monsoons

In the tropics, an increase in precipitation is projected by the end of the 21st century in the Asian monsoon and the southern part of the West African monsoon with some decreases in the Sahel in northern summer (Cook and Vizy, 2006), as well as increases in the Australian monsoon in southern summer in a warmer climate (Figure 10.9). The monsoonal precipitation in Mexico and Central America is projected to decrease in association with increasing precipitation over the eastern equatorial Pacific that affects Walker Circulation and local Hadley Circulation changes (Figure 10.9). A more detailed assessment of regional monsoon changes is provided in Chapter 11.

As a projected global warming will be more rapid over land than over the oceans, the continental-scale land-sea thermal contrast will become larger in summer and smaller in winter. Based on this, a simple idea is that the summer monsoon will be stronger and the winter monsoon will be weaker in the future than the present. However, model results are not as straightforward as this simple consideration. Tanaka et al. (2005) define the intensities of Hadley, Walker and monsoon circulations using the velocity potential fields at 200 hPa. Using 15 AOGCMs, they show a weakening of these tropical circulations by 9%, 8% and 14%, respectively, by the late 21st century compared to the late 20th century. Using eight AOGCMs, Ueda et al. (2006) demonstrate that pronounced warming over the tropics results in a weakening of the Asian summer monsoon circulations in relation to a reduction in the meridional thermal gradients between the Asian continent and adjacent oceans.

Despite weakening of the dynamical monsoon circulation, atmospheric moisture buildup due to increased greenhouse gases and consequent temperature increase results in a larger moisture flux and more precipitation for the Indian monsoon (Douville et al., 2000; IPCC, 2001; Ashrit et al., 2003; Meehl and Arblaster, 2003; May, 2004; Ashrit et al., 2005). For the South Asian summer monsoon, models suggest a northward shift of lower-tropospheric monsoon wind systems with a weakening of the westerly flow over the northern Indian Ocean (Ashrit et al., 2003, 2005). Over Africa in northern summer, multi-model analysis projects an increase in rainfall in East and Central Africa, a decrease in the Sahel, and increases along the Gulf of Guinea coast (Figure 10.9). However, some individual models project an increase of rainfall in more extensive areas of West Africa related to a projected northward movement of the Sahara and the Sahel (Liu et al., 2002; Haarsma et al., 2005). Whether the Sahel will be more or less wet in the future is thus uncertain, although a multi-model assessment of the West African monsoon indicates that the Sahel could become marginally more dry (Cook and Vizy, 2006). This inconsistency of the rainfall projections may be related to AOGCM biases, or an unclear relationship between Gulf of Guinea and Indian Ocean warming, land use change and the West African monsoon. Nonlinear feedbacks that may exist within the West African climate system should also be considered (Jenkins et al., 2005).

Most model results project increased interannual variability in season-averaged Asian monsoon precipitation associated with an increase in its long-term mean value (e.g., Hu et al., 2000b; Räisänen, 2002; Meehl and Arblaster, 2003). Hu et al. (2000a) relate this to increased variability in the tropical Pacific SST (El Niño variability) in their model. Meehl and Arblaster (2003) relate the increased monsoon precipitation variability to increased variability in evaporation and precipitation in the Pacific due to increased SSTs. Thus, the South Asian monsoon variability is affected through the Walker Circulation such that the role of the Pacific Ocean dominates and that of the Indian Ocean is secondary.

Atmospheric aerosol loading affects regional climate and its future changes (see Chapter 7). If the direct effect of the aerosol increase is considered, surface temperatures will not get as warm because the aerosols reflect solar radiation. For this reason, land-sea temperature contrast becomes smaller than in the case without the direct aerosol effect, and the summer monsoon becomes weaker. Model simulations of the Asian monsoon project that the sulphate aerosols' direct effect reduces the magnitude of precipitation change compared with the case of only greenhouse gas increases (Emori et al., 1999; Roeckner et al., 1999; Lal and Singh, 2001). However, the relative cooling effect of sulphate aerosols is dominated by the effects of increasing greenhouse gases by the end of the 21st century in the SRES marker scenarios (Figure 10.26), leading to the increased monsoon precipitation at the end of the 21st century in these scenarios (see Section 10.3.2.3). Furthermore, it is suggested that aerosols with high absorptivity such as black carbon absorb solar radiation in the lower atmosphere, cool the surface, stabilise the atmosphere and reduce precipitation (Ramanathan et al., 2001). The solar radiation reaching the surface decreases as much as 50% locally, which could reduce the surface warming by greenhouse gases (Ramanathan et al., 2005). These atmospheric brown clouds could cause precipitation to increase over the Indian Ocean in winter and decrease in the surrounding Indonesia region and the western Pacific Ocean (Chung et al., 2002), and could reduce the summer monsoon precipitation in South and East Asia (Menon et al., 2002; Ramanathan et al., 2005). However, the total influence on monsoon precipitation of temporally varying direct and indirect effects of various aerosol species is still not resolved and the subject of active research.

10.3.5.3   Mean Tropical Pacific Climate Change

This subsection assesses changes in mean tropical Pacific climate. Enhanced greenhouse gas concentrations result in a general increase in SST, which will not be spatially uniform in association with a general reduction in tropical circulations in a warmer climate (see Section 10.3.5.2). Figures 10.8 and 10.9 indicate that SST increases more over the eastern tropical Pacific than over the western tropical Pacific, together with a decrease in the sea level pressure (SLP) gradient along the equator and an eastward shift of the tropical Pacific rainfall distribution. These background tropical Pacific changes can be called an El Niño-like mean state change (upon which individual El Niño-Southern Oscillation (ENSO) events occur). Although individual models show a large scatter of 'ENSO-ness' (Collins and The CMIP Modelling Groups, 2005; Yamaguchi and Noda, 2006), an ENSO-like global warming pattern with positive polarity (i.e., El Niño-like mean state change) is simulated based on the spatial anomaly patterns of SST, SLP and precipitation (Figure 10.16; Yamaguchi and Noda, 2006). The El Niño-like change may be attributable to the general reduction in tropical circulations resulting from the increased dry static stability in the tropics in a warmer climate (Knutson and Manabe, 1995; Sugi et al., 2002; Figure 10.7). An eastward displacement of precipitation in the tropical Pacific accompanies an intensified and south-westward displaced subtropical anticyclone in the western Pacific, which can be effective in transporting moisture from the low latitudes to the Meiyu/Baiu region, thus generating more precipitation in the East Asian summer monsoon (Kitoh and Uchiyama, 2006).

In summary, the multi-model mean projects a weak shift towards conditions which may be described as 'El Niño-like', with SSTs in the central and eastern equatorial Pacific warming more than those in the west, and with an eastward shift in mean precipitation, associated with weaker tropical circulations.

10.3.5.4   El Niño

This subsection addresses the projected change in the amplitude, frequency and spatial pattern of El Niño. Guilyardi (2006) assessed mean state, coupling strength and modes (SST mode resulting from local SST-wind interaction or thermocline mode resulting from remote wind-thermocline feedbacks), using the pre-industrial control and stabilised 2 × and 4 × atmospheric

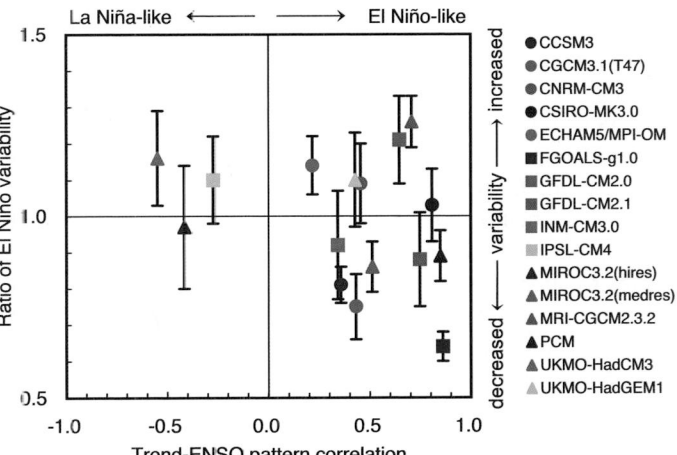

Figure 10.16. *Base state change in average tropical Pacific SSTs and change in El Niño variability simulated by AOGCMs (see Table 8.1 for model details). The base state change (horizontal axis) is denoted by the spatial anomaly pattern correlation coefficient between the linear trend of SST in the 1% yr⁻¹ CO_2 increase climate change experiment and the first Empirical Orthogonal Function (EOF) of SST in the control experiment over the area 10°S to 10°N, 120°E to 80°W (reproduced from Yamaguchi and Noda, 2006). Positive correlation values indicate that the mean climate change has an El Niño-like pattern, and negative values are La Niña-like. The change in El Niño variability (vertical axis) is denoted by the ratio of the standard deviation of the first EOF of sea level pressure (SLP) between the current climate and the last 50 years of the SRES A2 experiments (2051–2100), except for FGOALS-g1.0 and MIROC3.2(hires), for which the SRES A1B was used, and UKMO-HadGEM1 for which the 1% yr⁻¹ CO_2 increase climate change experiment was used, in the region 30°S to 30°N, 30°E to 60°W with a five-month running mean (reproduced from Oldenborgh et al., 2005). Error bars indicate the 95% confidence interval. Note that tropical Pacific base state climate changes with either El Niño-like or La Niña-like patterns are not permanent El Niño or La Niña events, and all still have ENSO inter-annual variability superimposed on that new average climate state in a future warmer climate.*

CO_2 simulations in a multi-model ensemble. The models that exhibit the largest El Niño amplitude change in scenario experiments are those that shift towards a thermocline mode. The observed 1976 climate shift in the tropical Pacific actually involved such a mode shift (Fedorov and Philander, 2001). The mean state change, through change in the sensitivity of SST variability to surface wind stress, plays a key role in determining the ENSO variance characteristics (Z. Hu et al., 2004; Zelle et al., 2005). For example, a more stable ENSO system is less sensitive to changes in the background state than one that is closer to instability (Zelle et al., 2005). Thus, GCMs with an improper simulation of present-day climate mean state and air-sea coupling strength are not suitable for ENSO amplitude projections. Van Oldenborgh et al. (2005) calculate the change in ENSO variability by the ratio of the standard deviation of the first Empirical Orthogonal Function (EOF) of SLP between the current climate and in the future (Figure 10.16), which shows that changes in ENSO interannual variability differ from model to model. They categorised 19 models based on their skill in the present-day ENSO simulations. Using the most realistic 6 out of 19 models, they find no statistically significant changes in the amplitude of ENSO variability in the future. Large uncertainty in the skewness of the variability limits the assessment of the future relative strength of El Niño and La Niña events.

Merryfield (2006) also analysed a multi-model ensemble and finds a wide range of behaviour for future El Niño amplitude, ranging from little change to larger El Niño events to smaller El Niño events, although several models that simulated some observed aspects of present-day El Niño events showed future increases in El Niño amplitude. However, significant multi-decadal fluctuations in El Niño amplitude in observations and in long coupled model control runs add another complicating factor to attempting to discern whether any future changes in El Niño amplitude are due to external forcing or are simply a manifestation of internal multi-decadal variability (Meehl et al., 2006a). Even with the larger warming scenario under 4 × atmospheric CO_2 climate, Yeh and Kirtman (2005) find that despite the large changes in the tropical Pacific mean state, the changes in ENSO amplitude are highly model dependent. Therefore, there are no clear indications at this time regarding future changes in El Niño amplitude in a warmer climate. However, as first noted in the TAR, ENSO teleconnections over North America appear to weaken due at least in part to the mean change of base state mid-latitude atmospheric circulation (Meehl et al., 2006a).

In summary, all models show continued ENSO interannual variability in the future no matter what the change in average background conditions, but changes in ENSO interannual variability differ from model to model. Based on various assessments of the current multi-model archive, in which present-day El Niño events are now much better simulated than in the TAR, there is no consistent indication at this time of discernible future changes in ENSO amplitude or frequency.

10.3.5.5   ENSO-Monsoon Relationship

The El Niño-Southern Oscillation affects interannual variability throughout the tropics through changes in the Walker Circulation. Analysis of observational data finds a significant correlation between ENSO and tropical circulation and precipitation such that there is a tendency for less Indian summer monsoon rainfall in El Niño years and above normal rainfall in La Niña years. Recent analyses have revealed that the correlation between ENSO and the Indian summer monsoon has decreased recently, and many hypotheses have been put forward (see Chapter 3). With respect to global warming, one hypothesis is that the Walker Circulation (accompanying ENSO) shifted south-eastward, reducing downward motion in the Indian monsoon region, which originally suppressed precipitation in that region at the time of El Niño, but now produces normal precipitation as a result (Krishna Kumar et al., 1999). Another explanation is that as the ground temperature of the Eurasian continent has risen in the winter-spring season, the temperature difference between the continent and the ocean has increased, thereby causing more precipitation, and the Indian monsoon is normal in spite of the occurrence of El Niño (Ashrit et al., 2001).

An earlier version of an AOGCM developed at the Max Planck Institute (MPI) (Ashrit et al., 2001) and the Action de Recherche Petite Echelle Grande Echelle/Océan Parallélisé (ARPEGE/OPA) model (Ashrit et al., 2003) simulated no global-warming related change in the ENSO-monsoon relationship, although a decadal-scale fluctuation is seen, suggesting that a weakening of the relationship might be part of the natural variability. However, Ashrit et al. (2001) show that while the impact of La Niña does not change, the influence of El Niño on the monsoon becomes small, suggesting the possibility of asymmetric behaviour of the changes in the ENSO-monsoon relationship. On the other hand, the MRI-CGCM2 (see Table 8.1 for model details) indicates a weakening of the correlation into the 21st century, particularly after 2050 (Ashrit et al., 2005). The MRI-CGCM2 model results support the above hypothesis that the Walker Circulation shifts eastward and no longer influences India at the time of El Niño in a warmer climate. Camberlin et al. (2004) and van Oldenborgh and Burgers (2005) find decadal fluctuations in the effect of ENSO on regional precipitation. In most cases, these fluctuations may reflect natural variability in the ENSO teleconnection, and long-term correlation trends may be comparatively weaker.

The Tropospheric Biennial Oscillation (TBO) has been suggested as a fundamental set of coupled interactions in the Indo-Pacific region that encompasses ENSO and the Asian-Australian monsoon, and the TBO has been shown to be simulated by current AOGCMs (see Chapter 8). Nanjundiah et al. (2005) analyse a multi-model data set to show that, for models that successfully simulate the TBO for present-day climate, the TBO becomes more prominent in a future warmer climate due to changes in the base state climate, although, as with ENSO, there is considerable inherent decadal variability in the relative dominance of TBO and ENSO.

In summary, the ENSO-monsoon relationship can vary due to natural variability. Model projections suggest that a future weakening of the ENSO-monsoon relationship could occur in a future warmer climate.

10.3.5.6   Annular Modes and Mid-Latitude Circulation Changes

Many simulations project some decrease in the arctic surface pressure in the 21st century, as seen in the multi-model average (see Figure 10.9). This contributes to an increase in indices of the Northern Annular Mode (NAM) or the Arctic Oscillation (AO), as well as the NAO, which is closely related to the NAM in the Atlantic sector (see Chapter 8). In the recent multi-model analyses, more than half of the models exhibit a positive trend in the NAM (Rauthe et al., 2004; Miller et al., 2006) and/or NAO (Osborn, 2004; Kuzmina et al., 2005). Although the magnitude of the trends shows a large variation among different models, Miller et al. (2006) find that none of the 14 models exhibits a trend towards a lower NAM index and higher arctic SLP. In another multi-model analysis, Stephenson et al. (2006) show that of the 15 models able to simulate the NAO pressure dipole, 13 predict a positive increase in the NAO index with increasing CO_2 concentrations, although the magnitude of the response is generally small and model dependent. However, the multi-model average from the larger number (21) of models shown in

Figure 10.9 indicates that it is likely that the NAM index would not notably decrease in a future warmer climate. The average of IPCC-AR4 simulations from 13 models suggests the increase of the NAM index becomes statistically significant early in the 21st century (Figure 10.17a, Miller et al., 2006).

The spatial patterns of the simulated SLP trends vary among different models, in spite of close correlations of the models' leading patterns of interannual (or internal) variability with the observations (Osborn, 2004; Miller et al., 2006). However, at the hemispheric scale of SLP change, the reduction in the Arctic is seen in the multi-model mean (Figure 10.9), although the change is smaller than the inter-model standard deviation. Besides the decrease in the arctic region, increases over the North Pacific and the Mediterranean Sea exceed the inter-model standard deviation; the latter suggests an association with a north-eastward shift of the NAO's centre of action (Hu and Wu, 2004). The diversity of the patterns seems to reflect different responses in the Aleutian Low (Rauthe et al., 2004) in the North Pacific. Yamaguchi and Noda (2006) discuss the modelled response of ENSO versus AO, and find that many models project a positive AO-like change. In the North Pacific at high latitudes, however, the SLP anomalies are incompatible between the El Niño-like change and the positive AO-like change, because models that project an El Niño-like change over the Pacific simulate a non-AO-like pattern in the polar region. As a result, the present models cannot fully determine the relative importance of the mechanisms inducing the positive AO-like change and those inducing the ENSO-like change, leading to scatter in global warming patterns at regional scales over the North Pacific. Rauthe et al. (2004) suggest that the effects of sulphate aerosols contribute to a deepening of the Aleutian Low resulting in a slower or smaller increase in the AO index.

Analyses of results from various models indicate that the NAM can respond to increasing greenhouse gas concentrations through tropospheric processes (Fyfe et al., 1999; Gillett et al., 2003; Miller et al., 2006). Greenhouse gases can also drive a positive NAM trend through changes in the stratospheric circulation, similar to the mechanism by which volcanic aerosols in the stratosphere force positive annular changes (Shindell et al., 2001). Models with their upper boundaries extending farther into the stratosphere exhibit, on average, a relatively larger increase in the NAM index and respond consistently to the observed volcanic forcing (Figure 10.17a, Miller et al., 2006), implying the importance of the connection between the troposphere and the stratosphere.

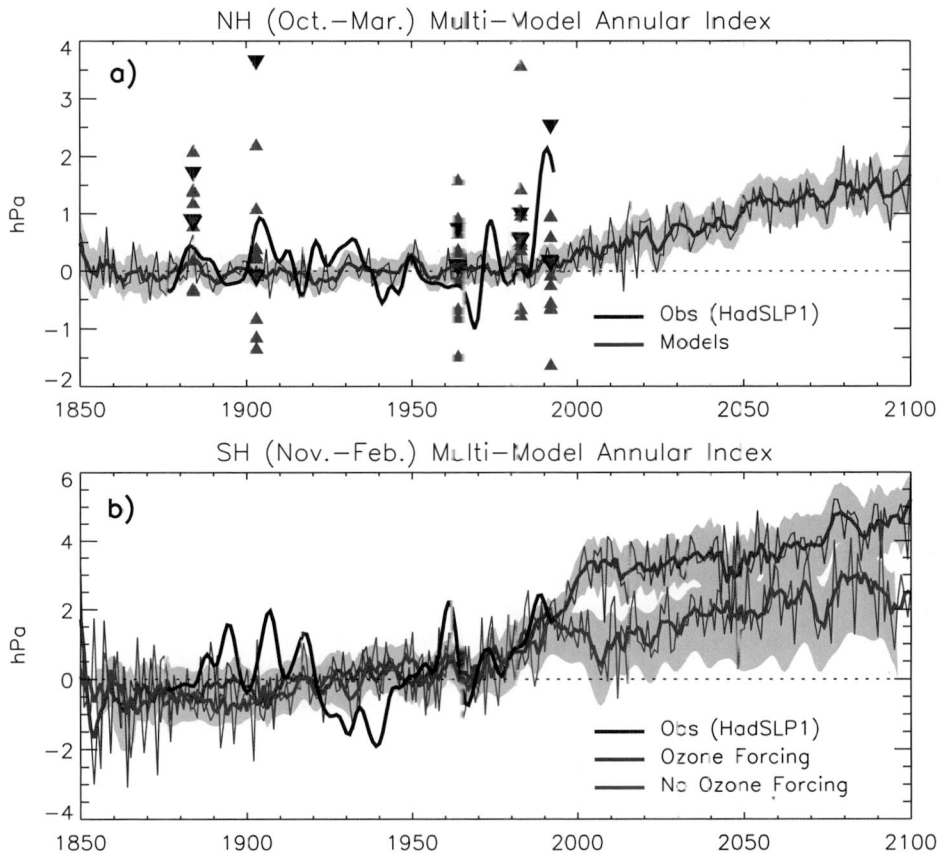

Figure 10.17. *(a) Multi-model mean of the regression of the leading EOF of ensemble mean Northern Hemisphere sea level pressure (NH SLP, thin red line). The time series of regression coefficients has zero mean between year 1900 and 1970. The thick red line is a 10-year low-pass filtered version of the mean. The grey shading represents the inter-model spread at the 95% confidence level and is filtered. A filtered version of the observed SLP from the Hadley Centre (HadSLP1) is shown in black. The regression coefficient for the winter following a major tropical eruption is marked by red, blue and black triangles for the multi-model mean, the individual model mean and observations, respectively. (b) As in (a) for Southern Hemisphere SLP for models with (red) and without (blue) ozone forcing. Adapted from Miller et al. (2006).*

A plausible explanation for the cause of the upward NAM trend simulated by the models is an intensification of the polar vortex resulting from both tropospheric warming and stratospheric cooling mainly due to the increase in greenhouse gases (Shindell et al., 2001; Sigmond et al., 2004; Rind et al., 2005a). The response may not be linear with the magnitude of radiative forcing (Gillett et al., 2002) since the polar vortex response is attributable to an equatorward refraction of planetary waves (Eichelberger and Holton, 2002) rather than radiative forcing itself. Since the long-term variation in the NAO is closely related to SST variations (Rodwell et al., 1999), it is considered essential that the projection of the changes in the tropical SST (Hoerling et al., 2004; Hurrell et al., 2004) and/or meridional gradient of the SST change (Rind et al., 2005b) is reliable.

The future trend in the Southern Annular Mode (SAM) or the Antarctic Oscillation (AAO) has been projected in a number of model simulations (Gillett and Thompson, 2003; Shindell and Schmidt, 2004; Arblaster and Meehl, 2006; Miller et al., 2006). According to the latest multi-model analysis (Miller et al., 2006), most models indicate a positive trend in the SAM index, and a declining trend in the antarctic SLP (as seen in Figure 10.9), with a higher likelihood than for the future NAM trend. On average, a larger positive trend is projected during the late 20th century by models that include stratospheric ozone changes than those that do not (Figure 10.17b), although during the 21st century, when ozone changes are smaller, the SAM trends of models with and without ozone are similar. The cause of the positive SAM trend in the second half of the 20th century is mainly attributed to stratospheric ozone depletion, evidenced by the fact that the signal is largest in the lower stratosphere in austral spring through summer (Thompson and Solomon, 2002; Arblaster and Meehl, 2006). However, increases in greenhouse gases are also important factors (Shindell and Schmidt, 2004; Arblaster and Meehl, 2006) for the year-round positive SAM trend induced by meridional temperature gradient changes (Brandefelt and Källén, 2004). During the 21st century, although the ozone amount is expected to stabilise or recover, the polar vortex intensification is likely to continue due to the increases in greenhouse gases (Arblaster and Meehl, 2006).

It is implied that the future change in the annular modes leads to modifications of the future change in various fields such as surface temperatures, precipitation and sea ice with regional features similar to those for the modes of natural variability (e.g., Hurrell et al., 2003). For instance, the surface warming in winter would be intensified in northern Eurasia and most of North America while weakened in the western North Atlantic, and winter precipitation would increase in northern Europe while decreasing in southern Europe. The atmospheric circulation change would also affect the ocean circulations. Sakamoto et al. (2005) simulate an intensification of the Kuroshio Current but no shift in the Kuroshio Extension in response to an AO-like circulation change for the 21st century. However, Sato et al. (2006) simulate a northward shift of the Kuroshio Extension, which leads to a strong warming off the eastern coast of Japan.

In summary, the future changes in the extratropical circulation variability are likely to be characterised by increases in positive phases of both the NAM and the SAM. The response in the NAM to anthropogenic forcing might not be distinct from the larger multi-decadal internal variability in the first half of the 21st century. The change in the SAM would appear earlier than in the NAM since stratospheric ozone depletion acts as an additional forcing. The positive trends in annular modes would influence the regional changes in temperature, precipitation and other fields, similar to those that accompany the NAM and the SAM in the present climate, but would be superimposed on the global-scale changes in a future warmer climate.

10.3.6 Future Changes in Weather and Climate Extremes

Projections of future changes in extremes rely on an increasingly sophisticated set of models and statistical techniques. Studies assessed in this section rely on multi-member ensembles (three to five members) from single models, analyses of multi-model ensembles ranging from 8 to 15 or more AOGCMs, and a perturbed physics ensemble with a single mixed-layer model with over 50 members. The discussion here is intended to identify general characteristics of changes in extremes in a global context. Chapter 3 provides a definition of weather and climate extremes, and Chapter 11 addresses changes in extremes for specific regions.

10.3.6.1 Precipitation Extremes

A long-standing result from global coupled models noted in the TAR is a projected increase in the chance of summer drying in the mid-latitudes in a future warmer climate with associated increased risk of drought. This is shown in Figure 10.12, and has been documented in the more recent generation of models (Burke et al., 2006; Meehl et al., 2006b; Rowell and Jones, 2006). For example, Wang (2005) analyse 15 recent AOGCMs and show that in a future warmer climate, the models simulate summer dryness in most parts of the northern subtropics and mid-latitudes, but with a large range in the amplitude of summer dryness across models. Droughts associated with this summer drying could result in regional vegetation die-offs (Breshears et al., 2005) and contribute to an increase in the percentage of land area experiencing drought at any one time, for example, extreme drought increasing from 1% of present-day land area to 30% by the end of the century in the A2 scenario (Burke et al., 2006). Drier soil conditions can also contribute to more severe heat waves as discussed in Section 10.3.6.2 (Brabson et al., 2005).

Associated with the risk of drying is a projected increase in the chance of intense precipitation and flooding. Although somewhat counter-intuitive, this is because precipitation is projected to be concentrated into more intense events, with longer periods of little precipitation in between. Therefore, intense and heavy episodic rainfall events with high runoff amounts are interspersed with longer relatively dry periods with increased evapotranspiration, particularly in the subtropics

Frequently Asked Question 10.1

Are Extreme Events, Like Heat Waves, Droughts or Floods, Expected to Change as the Earth's Climate Changes?

Yes; the type, frequency and intensity of extreme events are expected to change as Earth's climate changes, and these changes could occur even with relatively small mean climate changes. Changes in some types of extreme events have already been observed, for example, increases in the frequency and intensity of heat waves and heavy precipitation events (see FAQ 3.3).

In a warmer future climate, there will be an increased risk of more intense, more frequent and longer-lasting heat waves. The European heat wave of 2003 is an example of the type of extreme heat event lasting from several days to over a week that is likely to become more common in a warmer future climate. A related aspect of temperature extremes is that there is likely to be a decrease in the daily (diurnal) temperature range in most regions. It is also likely that a warmer future climate would have fewer frost days (i.e., nights where the temperature dips below freezing). Growing season length is related to number of frost days, and has been projected to increase as climate warms. There is likely to be a decline in the frequency of cold air outbreaks (i.e., periods of extreme cold lasting from several days to over a week) in NH winter in most areas. Exceptions could occur in areas with the smallest reductions of extreme cold in western North America, the North Atlantic and southern Europe and Asia due to atmospheric circulation changes.

In a warmer future climate, most Atmosphere-Ocean General Circulation Models project increased summer dryness and winter wetness in most parts of the northern middle and high latitudes. Summer dryness indicates a greater risk of drought. Along with the risk of drying, there is an increased chance of intense precipitation and flooding due to the greater water-holding capacity of a warmer atmosphere. This has already been observed and is projected to continue because in a warmer world, precipitation tends to be concentrated into more intense events, with longer periods of little precipitation in between. Therefore, intense and heavy downpours would be interspersed with longer relatively dry periods. Another aspect of these projected changes is that wet extremes are projected to become more severe in many areas where mean precipitation is expected to increase, and dry extremes are projected to become more severe in areas where mean precipitation is projected to decrease.

In concert with the results for increased extremes of intense precipitation, even if the wind strength of storms in a future climate did not change, there would be an increase in extreme rainfall intensity. In particular, over NH land, an increase in the likelihood of very wet winters is projected over much of central and northern Europe due to the increase in intense precipitation during storm events, suggesting an increased chance of flooding over Europe and other mid-latitude regions due to more intense rainfall and snowfall events producing more runoff. Similar results apply for summer precipitation, with implications for more flooding in the Asian monsoon region and other tropical areas. The increased risk of floods in a number of major river basins in a future warmer climate has been related to an increase in river discharge with an increased risk of future intense storm-related precipitation events and flooding. Some of these changes would be extensions of trends already underway.

There is evidence from modelling studies that future tropical cyclones could become more severe, with greater wind speeds and more intense precipitation. Studies suggest that such changes may already be underway; there are indications that the average number of Category 4 and 5 hurricanes per year has increased over the past 30 years. Some modelling studies have projected a decrease in the number of tropical cyclones globally due to the increased stability of the tropical troposphere in a warmer climate, characterised by fewer weak storms and greater numbers of intense storms. A number of modelling studies have also projected a general tendency for more intense but fewer storms outside the tropics, with a tendency towards more extreme wind events and higher ocean waves in several regions in association with those deepened cyclones. Models also project a poleward shift of storm tracks in both hemispheres by several degrees of latitude.

as discussed in Section 10.3.6.2 in relation to Figure 10.19 (Frei et al., 1998; Allen and Ingram, 2002; Palmer and Räisänen, 2002; Christensen and Christensen, 2003; Beniston, 2004; Christensen and Christensen, 2004; Pal et al., 2004; Meehl et al., 2005a). However, increases in the frequency of dry days do not necessarily mean a decrease in the frequency of extreme high rainfall events depending on the threshold used to define such events (Barnett et al., 2006). Another aspect of these changes has been related to the mean changes in precipitation, with wet extremes becoming more severe in many areas where mean precipitation increases, and dry extremes where the mean precipitation decreases (Kharin and Zwiers, 2005; Meehl et al., 2005a; Räisänen, 2005a; Barnett et al., 2006). However, analysis of the 53-member perturbed physics ensemble indicates that the change in the frequency of extreme precipitation at an individual location can be difficult to estimate definitively due to model parametrization uncertainty (Barnett et al., 2006). Some specific regional aspects of these changes in precipitation extremes are discussed further in Chapter 11.

Climate models continue to confirm the earlier results that in a future climate warmed by increasing greenhouse gases, precipitation intensity (e.g., proportionately more precipitation per precipitation event) is projected to increase over most regions (Wilby and Wigley, 2002; Kharin and Zwiers, 2005; Meehl et al., 2005a; Barnett et al., 2006), and the increase in precipitation extremes is greater than changes in mean precipitation (Kharin and Zwiers, 2005). As discussed in Chapter 9, this is related to the fact that the energy budget of the atmosphere constrains increases in large-scale mean precipitation, but extreme precipitation relates to increases in moisture content and thus the nonlinearities involved with the Clausius-Clapeyron relationship such that, for a given increase in temperature, increases in extreme precipitation can be more than the mean precipitation increase (e.g., Allen and Ingram, 2002). Additionally, time scale can play a role whereby increases in the frequency of seasonal mean rainfall extremes can be greater than the increases in the frequency of daily extremes (Barnett et al., 2006). The increase in mean and extreme precipitation in various regions has been attributed to contributions from both dynamic and thermodynamic processes associated with global warming (Emori and Brown, 2005). The greater increase in extreme precipitation compared to the mean is attributed to the greater thermodynamic effect on the extremes due to increases in water vapour, mainly over subtropical areas. The thermodynamic effect is important nearly everywhere, but changes in circulation also contribute to the pattern of precipitation intensity changes at middle and high latitudes (Meehl et al., 2005a). Kharin and Zwiers (2005) show that changes in both the location and scale of the extreme value distribution produce increases in precipitation extremes substantially greater than increases in annual mean precipitation. An increase in the scale parameter from the gamma distribution represents an increase in precipitation intensity, and various regions such as the NH land areas in winter showed particularly high values of increased scale parameter (Semenov and Bengtsson, 2002; Watterson and Dix, 2003). Time-slice

simulations with a higher-resolution model (~1°) show similar results using changes in the gamma distribution, namely increased extremes in the hydrological cycle (Voss et al., 2002). However, some regional decreases are also projected such as over the subtropical oceans (Semenov and Bengtsson, 2002).

A number of studies have noted the connection between increased rainfall intensity and an implied increase in flooding. McCabe et al. (2001) and Watterson (2005) show a projected increase in extreme rainfall intensity with the extra-tropical surface lows, particularly over NH land, with an implied increase in flooding. In a multi-model analysis of the CMIP models, Palmer and Räisänen (2002) show an increased likelihood of very wet winters over much of central and northern Europe due to an increase in intense precipitation associated with mid-latitude storms, suggesting more floods across Europe (see also Chapter 11). They found similar results for summer precipitation with implications for greater flooding in the Asian monsoon region in a future warmer climate. Similarly, Milly et al. (2002), Arora and Boer (2001) and Voss et al. (2002) relate the increased risk of floods in a number of major river basins in a future warmer climate to an increase in spring river discharge related to increased winter snow depth in some regions. Christensen and Christensen (2003) conclude that there could be an increased risk of summer flooding in Europe.

Globally averaged time series of the Frich et al. (2002) indices in the multi-model analysis of Tebaldi et al. (2006) show simulated increases in precipitation intensity during the 20th century continuing through the 21st century (Figure 10.18a,b), along with a somewhat weaker and less consistent trend of increasing dry periods between rainfall events for all scenarios (Figure 10.18c,d). Part of the reason for these results is shown in the geographic maps for these quantities, where precipitation intensity increases almost everywhere, but particularly at middle and high latitudes where mean precipitation also increases (Meehl et al., 2005a; compare Figure 10.18b to Figure 10.9). However, in Figure 10.18d, there are regions of increased runs of dry days between precipitation events in the subtropics and lower mid-latitudes, but decreased runs of dry days at higher mid-latitudes and high latitudes where mean precipitation increases (compare Figure 10.9 with Figure 10.18d). Since there are areas of both increases and decreases in consecutive dry days between precipitation events in the multi-model average (Figure 10.9), the global mean trends are smaller and less consistent across models as shown in Figure 10.18. Consistency of response in a perturbed physics ensemble with one model shows only limited areas of increased frequency of wet days in July, and a larger range of changes in precipitation extremes relative to the control ensemble mean in contrast to the more consistent response of temperature extremes (Section 10.6.3.2), indicating a less consistent response for precipitation extremes in general compared to temperature extremes (Barnett et al., 2006). Analysis of the Frich et al. (2002) precipitation indices in a 20-km resolution global model shows similar results to those in Figure 10.18, with particularly large increases in precipitation intensity in South Asia and West Africa (Kamiguchi et al., 2005).

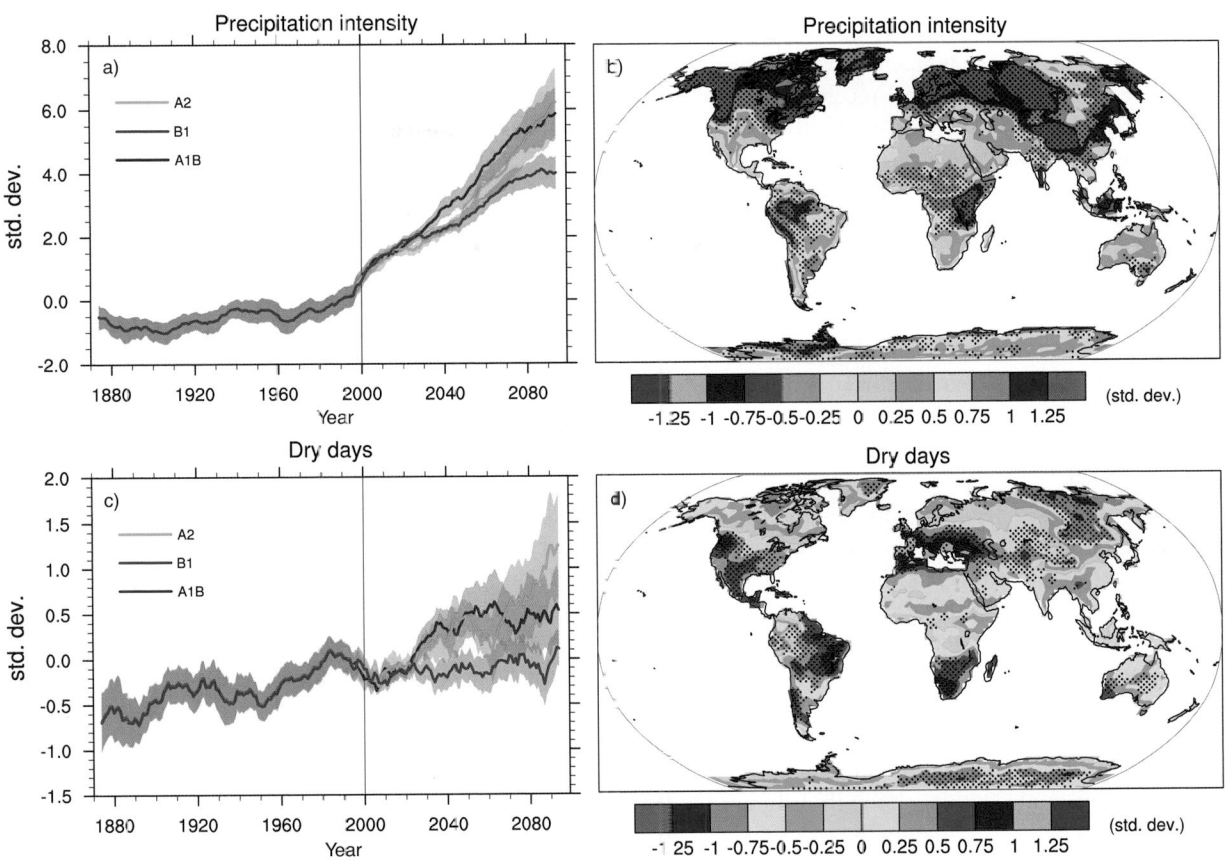

Figure 10.18. *Changes in extremes based on multi-model simulations from nine global coupled climate models, adapted from Tebaldi et al. (2006). (a) Globally averaged changes in precipitation intensity (defined as the annual total precipitation divided by the number of wet days) for a low (SRES B1), middle (SRES A1B) and high (SRES A2) scenario. (b) Changes in spatial patterns of simulated precipitation intensity between two 20-year means (2080–2099 minus 1980–1999) for the A1B scenario. (c) Globally averaged changes in dry days (defined as the annual maximum number of consecutive dry days). (d) Changes in spatial patterns of simulated dry days between two 20-year means (2080–2099 minus 1980–1999) for the A1B scenario. Solid lines in (a) and (c) are the 10-year smoothed multi-model ensemble means; the envelope indicates the ensemble mean standard deviation. Stippling in (b) and (d) denotes areas where at least five of the nine models concur in determining that the change is statistically significant. Extreme indices are calculated only over land following Frich et al. (2002). Each model's time series was centred on its 1980 to 1999 average and normalised (rescaled) by its standard deviation computed (after de-trending) over the period 1960 to 2099. The models were then aggregated into an ensemble average, both at the global and at the grid-box level. Thus, changes are given in units of standard deviations.*

10.3.6.2 Temperature Extremes

The TAR concluded that there was a very likely risk of increased high temperature extremes (and reduced risk of low temperature extremes) with more extreme heat episodes in a future climate. The latter result has been confirmed in subsequent studies (Yonetani and Gordon, 2001). Kharin and Zwiers (2005) show in a single model that future increases in temperature extremes follow increases in mean temperature over most of the world except where surface properties change (melting snow, drying soil). Furthermore, they show that in most instances warm extremes correspond to increases in daily maximum temperature, but cold extremes warm up faster than daily minimum temperatures, although this result is less consistent when model parameters are varied in a perturbed physics ensemble where there are increased daily temperature maxima for nearly the entire land surface. However, the range in magnitude of increases was substantial indicating a sensitivity to model formulations (Clark et al., 2006).

Weisheimer and Palmer (2005) examine changes in extreme seasonal (DJF and JJA) temperatures in 14 models for three scenarios. They show that by the end of 21st century, the probability of such extreme warm seasons is projected to rise in many areas. This result is consistent with the perturbed physics ensemble where, for nearly all land areas, extreme JJA temperatures were at least 20 times and in some areas 100 times more frequent compared to the control ensemble mean, making these changes greater than the ensemble spread.

Since the TAR, possible future cold air outbreaks have been studied. Vavrus et al. (2006) analyse seven AOGCMs run with the A1B scenario, and define a cold air outbreak as two or more consecutive days when the daily temperatures are at least two standard deviations below the present-day winter mean. For a future warmer climate, they document a 50 to 100% decline in the frequency of cold air outbreaks in NH winter in most areas compared to the present, with the smallest reductions occurring in western North America, the North Atlantic and southern Europe and Asia due to atmospheric circulation changes associated with the increase in greenhouse gases.

No studies at the time of the TAR specifically documented changes in heat waves (very high temperatures over a sustained period of days, see Chapter 3). Several recent studies address possible future changes in heat waves explicitly, and find an increased risk of more intense, longer-lasting and more frequent heat waves in a future climate (Meehl and Tebaldi, 2004; Schär et al., 2004; Clark et al., 2006). Meehl and Tebaldi (2004) show that the pattern of future changes in heat waves, with greatest intensity increases over western Europe, the Mediterranean and the southeast and western USA, is related in part to base state circulation changes due to the increase in greenhouse gases. An additional factor leading to extreme heat is drier soils in a future warmer climate (Brabson et al., 2005; Clark et al., 2006). Schär et al. (2004), Stott et al. (2004) and Beniston (2004) use the European 2003 heat wave as an example of the types of heat waves that are likely to become more common in a future warmer climate. Schär et al. (2004) note that the increase in the frequency of extreme warm conditions is also associated with a change in interannual variability, such that the statistical distribution of mean summer temperatures is not merely shifted towards warmer conditions but also becomes wider. A multi-model ensemble shows that heat waves are simulated to have been increasing over the latter part of the 20th century, and are projected to increase globally and over most regions (Figure 10.19; Tebaldi et al., 2006), although different model parameters can contribute to the range in the magnitude of this response (Clark et al., 2006).

A decrease in DTR in most regions in a future warmer climate was reported in the TAR, and is substantiated by more recent studies (e.g., Stone and Weaver, 2002; also discussed in relation to Figure 10.11b and in Chapter 11). For a quantity related to the DTR, the TAR concluded that it would be likely that a future warmer climate would also be characterised by a decrease in the number of frost days, although there were no studies at that time from global coupled climate models that addressed this issue explicitly. It has since been shown that there would indeed be decreases in frost days in a future warmer climate in the extratropics (Meehl et al., 2004a), with the pattern of the decreases dictated by the changes in atmospheric circulation due to the increase in greenhouse gases (Meehl et al., 2004a). Results from a nine-member multi-model ensemble show simulated decreases in frost days for the 20th century continuing into the 21st century globally and in most regions (Figure 10.19). A quantity related to frost days in many mid- and high-latitude areas, particularly in the NH, is growing season length as defined by Frich et al. (2002), and this has been projected to increase in future climate (Tebaldi et al., 2006). This result is also shown in a nine-member multi-model ensemble where the simulated increase in growing season length in the 20th century continues into the 21st century globally and in most regions (Figure 10.19). The globally averaged extremes indices in Figures 10.18 and 10.19 have non-uniform changes across the scenarios compared to the more consistent relative increases in Figure 10.5 for globally averaged temperature. This indicates that patterns that scale well by radiative forcing for temperature (e.g., Figure 10.8) would not scale for extremes.

10.3.6.3 Tropical Cyclones (Hurricanes)

Earlier studies assessed in the TAR showed that future tropical cyclones would likely become more severe with greater wind speeds and more intense precipitation. More recent modelling experiments have addressed possible changes in tropical cyclones in a warmer climate and generally confirmed those earlier results. These studies fall into two categories: those with model grid resolutions that only roughly represent some aspects of individual tropical cyclones, and those with model grids of sufficient resolution to reasonably simulate individual tropical cyclones.

In the first category, a number of climate change experiments with global models have started to simulate some characteristics of individual tropical cyclones, although classes of models with 50 to 100 km resolution or lower cannot accurately simulate observed tropical cyclone intensities due to the limitations of the relatively coarse grid spacing (e.g., Yoshimura et al., 2006). A study with roughly 100-km grid spacing shows a decrease in tropical cyclone frequency globally and in the North Pacific but a regional increase over the North Atlantic and no significant changes in maximum intensity (Sugi et al., 2002). Yoshimura et al. (2006) use the same model but different SST patterns and two different convection schemes, and show a decrease in the global frequency of relatively weak tropical cyclones but no significant change in the frequency of intense storms. They also show that the regional changes are dependent on the SST pattern, and precipitation near the storm centres could increase in the future. Another study using a 50 km resolution model confirms this dependence on SST pattern, and also shows a consistent increase in precipitation intensity in future tropical cyclones (Chauvin et al., 2006). Another global modelling study with roughly a 100-km grid spacing finds a 6% decrease in tropical storms globally and a slight increase in intensity, with both increases and decreases regionally related to the El Niño-like base state response in the tropical Pacific to increased greenhouse gases (McDonald et al., 2005). Another study with the same resolution model indicates decreases in tropical cyclone frequency and intensity but more mean and extreme precipitation from the tropical cyclones simulated in the future in the western north Pacific (Hasegawa and Emori, 2005). An AOGCM analysis with a coarser-resolution atmospheric model (T63, or about 200-km grid spacing) shows little change in overall numbers of tropical storms in that model, but a slight decrease in medium-intensity storms in a warmer climate (Bengtsson et al., 2006). In a global warming simulation with a coarse-resolution atmospheric model (T42, or about 300-km grid spacing), the frequency of global tropical cyclone occurrence did not change significantly, but the mean intensity of the global tropical cyclones increased significantly (Tsutsui, 2002). Thus, from this category of coarser-grid models that can only represent rudimentary aspects of tropical cyclones, there is no consistent evidence for large changes in either frequency or intensity of these models' representation of tropical cyclones, but there is a consistent response of more intense precipitation from future storms in a warmer climate. Also note that the

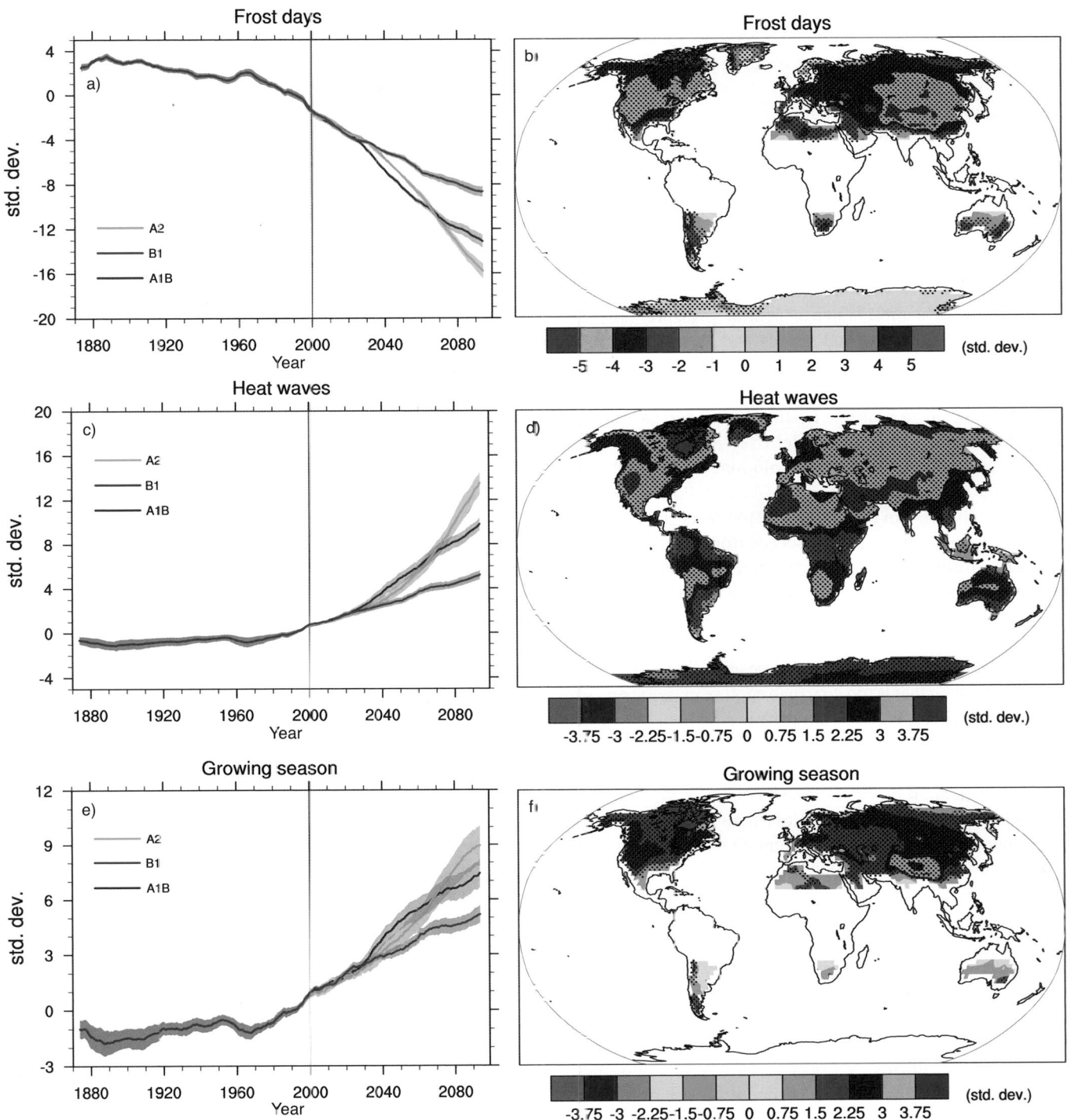

Figure 10.19. *Changes in extremes based on multi-model simulations from nine global coupled climate models, adapted from Tebaldi et al. (2006). (a) Globally averaged changes in the frost day index (defined as the total number of days in a year with absolute minimum temperature below 0°C) for a low (SRES B1), middle (SRES A1B) and high (SRES A2) scenario. (b) Changes in spatial patterns of simulated frost days between two 20-year means (2080–2099 minus 1980–1999) for the A1B scenario. (c) Globally averaged changes in heat waves (defined as the longest period in the year of at least five consecutive days with maximum temperature at least 5°C higher than the climatology of the same calendar day). (d) Changes in spatial patterns of simulated heat waves between two 20-year means (2080–2099 minus 1980–1999) for the A1B scenario. (e) Globally averaged changes in growing season length (defined as the length of the period between the first spell of five consecutive days with mean temperature above 5°C and the last such spell of the year). (f) Changes in spatial patterns of simulated growing season length between two 20-year means (2080–2099 minus 1980–1999) for the A1B scenario. Solid lines in (a), (c) and (e) show the 10-year smoothed multi-model ensemble means; the envelope indicates the ensemble mean standard deviation. Stippling in (b), (d) and (f) denotes areas where at least five of the nine models concur in determining that the change is statistically significant. Extreme indices are calculated only over land. Frost days and growing season are only calculated in the extratropics. Extremes indices are calculated following Frich et al. (2002). Each model's time series was centred around its 1980 to 1999 average and normalised (rescaled) by its standard deviation computed (after de-trending) over the period 1960 to 2099. The models were then aggregated into an ensemble average, both at the global and at the grid-box level. Thus, changes are given in units of standard deviations.*

decreasing tropical precipitation in future climate in Yoshimura et al. (2006) is for SSTs held fixed as atmospheric CO_2 is increased, a situation that does not occur in any global coupled model.

In the second category, studies have been performed with models that have been able to credibly simulate many aspects of tropical cyclones. For example, Knutson and Tuleya (2004) use a high-resolution (down to 9 km) mesoscale hurricane model to simulate hurricanes with intensities reaching about 60 to 70 m s^{-1}, depending on the treatment of moist convection in the model. They use mean tropical conditions from nine global climate models with increased CO_2 to simulate tropical cyclones with 14% more intense central pressure falls, 6% higher maximum surface wind speeds and about 20% greater near-storm rainfall after an idealised 80-year buildup of CO_2 at 1% yr^{-1} compounded (warming given by TCR shown for models in Chapter 8). Using a multiple nesting technique, an AOGCM was used to force a regional model over Australasia and the western Pacific with 125-km grid resolution, with an embedded 30-km resolution model over the south-western Pacific (Walsh et al., 2004). At that 30-km resolution, the model is able to closely simulate the climatology of the observed tropical cyclone lower wind speed threshold of 17 m s^{-1}. Tropical cyclone occurrence (in terms of days of tropical cyclone activity) is slightly greater than observed, and the somewhat weaker than observed pressure gradients near the storm centres are associated with lower than observed maximum wind speeds, likely due to the 30-km grid spacing that is too coarse to capture extreme pressure gradients and winds. For 3 × atmospheric CO_2 in that model configuration, the simulated tropical cyclones experienced a 56% increase in the number of storms with maximum wind speed greater than 30 m s^{-1} and a 26% increase in the number of storms with central pressures less than 970 hPa, with no large changes in frequency and movement of tropical cyclones for that southwest Pacific region. It should also be noted that ENSO fluctuations have a strong impact on patterns of tropical cyclone occurrence in the southern Pacific (Nguyen and Walsh, 2001), and that uncertainty with respect future ENSO behaviour (Section 10.3.5.1) contributes to uncertainty with respect to tropical cyclones (Walsh, 2004).

In another experiment with a high resolution global model that is able to generate tropical cyclones that begin to approximate real storms, a global 20-km grid atmospheric model was run in time slice experiments for a present-day 10-year period and a 10-year period at the end of the 21st century for the A1B scenario to examine changes in tropical cyclones. Observed climatological SSTs were used to force the atmospheric model for the 10-year period at the end of the 20th century, time-mean SST anomalies from an AOGCM simulation for the future climate were added to the observed SSTs and atmospheric composition was changed in the model to be consistent with the A1B scenario. At that resolution, tropical cyclone characteristics, numbers and tracks were relatively well simulated for present-day climate, although simulated wind speed intensities were somewhat weaker than observed intensities (Oouchi et al., 2006). In that study, tropical

cyclone frequency decreased 30% globally (but increased about 34% in the North Atlantic). The strongest tropical cyclones with extreme surface winds increased in number while weaker storms decreased. The tracks were not appreciably altered, and maximum peak wind speeds in future simulated tropical cyclones increased by about 14% in that model, although statistically significant increases were not found in all basins. As noted above, the competing effects of greater stabilisation of the tropical troposphere (less storms) and greater SSTs (the storms that form are more intense) likely contribute to these changes except for the tropical North Atlantic where there are greater SST increases than in the other basins in that model. Therefore, the SST warming has a greater effect than the vertical stabilisation in the Atlantic and produces not only more storms but also more intense storms there. However, these regional changes are largely dependent on the spatial pattern of future simulated SST changes (Yoshimura et al., 2006).

Sugi et al. (2002) show that the global-scale reduction in tropical cyclone frequency is closely related to weakening of tropospheric circulation in the tropics in terms of vertical mass flux. They note that a significant increase in dry static stability in the tropical troposphere and little increase in tropical precipitation (or convective heating) are the main factors contributing to the weakening of the tropospheric circulation. Sugi and Yoshimura (2004) investigate a mechanism of this tropical precipitation change. They show that the effect of CO_2 enhancement (without changing SST conditions, which is not realistic as noted above) is a decrease in mean precipitation (Sugi and Yoshimura, 2004) and a decrease in the number of tropical cyclones as simulated in an atmospheric model with about 100 km resolution (Yoshimura and Sugi, 2005). Future changes in the large-scale steering flow as a mechanism to deduce possible changes in tropical cyclone tracks in the western North Pacific (Wu and Wang, 2004) were analysed to show different shifts at different times in future climate change experiments along with a dependence of such shifts on the degree of El Niño-like mean climate change in the Pacific (see Section 10.3.5).

A synthesis of the model results to date indicates that, for a future warmer climate, coarse-resolution models show few consistent changes in tropical cyclones, with results dependent on the model, although those models do show a consistent increase in precipitation intensity in future storms. Higher-resolution models that more credibly simulate tropical cyclones project some consistent increase in peak wind intensities, but a more consistent projected increase in mean and peak precipitation intensities in future tropical cyclones. There is also a less certain possibility of a decrease in the number of relatively weak tropical cyclones, increased numbers of intense tropical cyclones and a global decrease in total numbers of tropical cyclones.

10.3.6.4   Extratropical Storms and Ocean Wave Height

The TAR noted that there could be a future tendency for more intense extratropical storms, although the number of storms could be less. A more consistent result that has emerged more recently, in agreement with earlier results (e.g., Schubert et al., 1998), is a tendency for a poleward shift of several degrees latitude in mid-latitude storm tracks in both hemispheres (Geng and Sugi, 2003; Fischer-Bruns et al., 2005; Yin, 2005; Bengtsson et al., 2006). Consistent with these shifts in storm track activity, Cassano et al. (2006), using a 10-member multi-model ensemble, show a future change to a more cyclonically dominated circulation pattern in winter and summer over the Arctic, and increasing cyclonicity and stronger westerlies in the same multi-model ensemble for the Antarctic (Lynch et al., 2006).

Some studies have shown little change in extratropical cyclone characteristics (Kharin and Zwiers, 2005; Watterson, 2005). But a regional study showed a tendency towards more intense systems, particularly in the A2 scenario in another global coupled climate model analysis (Leckebusch and Ulbrich, 2004), with more extreme wind events in association with those deepened cyclones for several regions of Western Europe, with similar changes in the B2 simulation although less pronounced in amplitude. Geng and Sugi (2003) use a higher-resolution (about 100 km resolution) atmospheric GCM (AGCM) with time-slice experiments and find a decrease in cyclone density (number of cyclones in a 4.5° by 4.5° area per season) in the mid-latitudes of both hemispheres in a warmer climate in both the DJF and JJA seasons, associated with the changes in the baroclinicity in the lower troposphere, in general agreement with earlier results and coarser GCM results (e.g., Dai et al., 2001b). They also find that the density of strong cyclones increases while the density of weak and medium-strength cyclones decreases. Several studies have shown a possible reduction in mid-latitude storms in the NH but a decrease in central pressures in these storms (Lambert and Fyfe, 2006, for a 15-member multi-model ensemble) and in the SH (Fyfe, 2003, with a possible 30% reduction in sub-antarctic cyclones). The latter two studies did not definitively identify a poleward shift of storm tracks, but their methodologies used a relatively coarse grid that may not have been able to detect shifts of several degrees latitude and they used only identification of central pressures which could imply an identification of semi-permanent features like the sub-antarctic trough. More regional aspects of these changes were addressed for the NH in a single model study by Inatsu and Kimoto (2005), who show a more active storm track in the western Pacific in the future but weaker elsewhere. Fischer-Bruns et al. (2005) document storm activity increasing over the North Atlantic and Southern Ocean and decreasing over the Pacific Ocean.

By analysing stratosphere-troposphere exchanges using time-slice experiments with the middle atmosphere version of ECHAM4, Land and Feichter (2003) suggest that cyclonic and blocking activity becomes weaker poleward of 30°N in a warmer climate at least in part due to decreased baroclinicity below 400 hPa, while cyclonic activity becomes stronger in the SH associated with increased baroclinicity above 400 hPa. The atmospheric circulation variability on inter-decadal time scales may also change due to increasing greenhouse gases and aerosols. One model result (Hu et al., 2001) showed that inter-decadal variability of the SLP and 500 hPa height fields increased over the tropics and decreased at high latitudes due to global warming.

In summary, the most consistent results from the majority of the current generation of models show, for a future warmer climate, a poleward shift of storm tracks in both hemispheres that is particularly evident in the SH, with greater storm activity at higher latitudes.

A new feature that has been studied related to extreme conditions over the oceans is wave height. Studies by Wang et al. (2004), Wang and Swail (2006a,b) and Caires et al. (2006) have shown that for many regions of the mid-latitude oceans, an increase in extreme wave height is likely to occur in a future warmer climate. This is related to increased wind speed associated with mid-latitude storms, resulting in higher waves produced by these storms, and is consistent with the studies noted above that showed decreased numbers of mid-latitude storms but more intense storms.

10.4 Changes Associated with Biogeochemical Feedbacks and Ocean Acidification

10.4.1 Carbon Cycle/Vegetation Feedbacks

As a parallel activity to the standard IPCC AR4 climate projection simulations described in this chapter, the Coupled Climate-Carbon Cycle Model Intercomparison Project (C[4]MIP) supported by WCRP and the International Geosphere-Biosphere Programme (IGBP) was initiated. Eleven climate models with a representation of the land and ocean carbon cycle (see Chapter 7) performed simulations where the model was driven by an anthropogenic CO_2 emissions scenario for the 1860 to 2100 time period (instead of an atmospheric CO_2 concentration scenario as in the standard IPCC AR4 simulations). Each C[4]MIP model performed two simulations, a 'coupled' simulation where the growth of atmospheric CO_2 induces a climate change which affects the carbon cycle, and an 'uncoupled' simulation, where atmospheric CO_2 radiative forcing is held fixed at pre-industrial levels, in order to estimate the atmospheric CO_2 growth rate that would occur if the carbon cycle was unperturbed by the climate. Emissions were taken from the observations for the historical period (Houghton and Hackler, 2000; Marland et al., 2005) and from the SRES A2 scenario for the future (Leemans et al., 1998).

Chapter 7 describes the major results of the C[4]MIP models in terms of climate impact on the carbon cycle. This section starts from these impacts to infer the feedback effect on atmospheric CO_2 and therefore on the climate system. There is unanimous agreement among the models that future climate change will reduce the efficiency of the land and ocean carbon cycle to absorb anthropogenic CO_2, essentially owing to a reduction in land carbon uptake. The latter is driven by a combination of reduced net primary productivity and increased soil respiration of CO_2 under a warmer climate. As a result, a larger fraction of anthropogenic CO_2 will stay airborne if climate change controls the carbon cycle. By the end of the 21st century, this additional CO_2 varies between 20 and 220 ppm for the two extreme models, with most of the models lying between 50 and 100 ppm (Friedlingstein et al., 2006). This additional CO_2 leads to an additional radiative forcing of between 0.1 and 1.3 W m^{-2} and hence an additional warming of between 0.1°C and 1.5°C.

All of the C^{4}MIP models simulate a higher atmospheric CO_2 growth rate in the coupled runs than in the uncoupled runs. For the A2 emission scenario, this positive feedback leads to a greater atmospheric CO_2 concentration (Friedlingstein et al., 2006) as noted above, which is in addition to the concentrations in the standard coupled models assessed in the AR4 (e.g., Meehl et al., 2005b). By 2100, atmospheric CO_2 varies between 730 and 1,020 ppm for the C^{4}MIP models, compared with 836 ppm for the standard SRES A2 concentration in the multi-model data set (e.g., Meehl et al., 2005b). This uncertainty due to future changes in the carbon cycle is illustrated in Figure 10.20a where the CO_2 concentration envelope of the C^{4}MIP uncoupled simulations is centred on the standard SRES A2 concentration value. The range reflects the uncertainty in the carbon cycle. It should be noted that the standard SRES A2 concentration value of 836 ppm was calculated in the TAR with the Bern carbon cycle-climate model (BERN-CC; Joos et al., 2001) that accounted for the climate-carbon cycle feedback. Parameter sensitivity studies were performed with the BERN-CC model at that time and gave a range of 735 ppm to 1,080 ppm, comparable to the range of the C^{4}MIP study. The effects of climate feedback uncertainties on the carbon cycle have also been considered probabilistically by Wigley and Raper (2001). A later paper (Wigley, 2004) considers individual emissions scenarios, accounting for carbon cycle feedbacks in the same way as Wigley and Raper (2001). The results of these studies are consistent with the more recent C^{4}MIP results. For the A2 scenario considered in C^{4}MIP, the CO_2 concentration range in 2100 using the Wigley and Raper model is 769 to 1,088 ppm, compared with 730 to 1,020 ppm in the C^{4}MIP study (which ignored the additional warming effect due to non-CO_2 gases). Similarly, using neural networks, Knutti et al. (2003) show that the climate-carbon cycle feedback leads to an increase of about 0.6°C over the central estimate for the SRES A2 scenario and an increase of about 1.5°C for the upper bound of the uncertainty range.

Further uncertainties regarding carbon uptake were addressed with a 14-member multi-model ensemble using the CMIP2 models to quantify contributions to uncertainty from inter-model variability as opposed to internal variability (Berthelot et al., 2002). They found that the AOGCMs with the largest climate sensitivity also had the largest drying of soils in the tropics and thus the largest reduction in carbon uptake.

The C^{4}MIP protocol did not account for the evolution of non-CO_2 greenhouse gases and aerosols. In order to compare the C^{4}MIP simulated warming with the IPCC AR4 climate models, the SRES A2 radiative forcings of CO_2 alone and total forcing (CO_2 plus non-CO_2 greenhouse gases and aerosols) as given

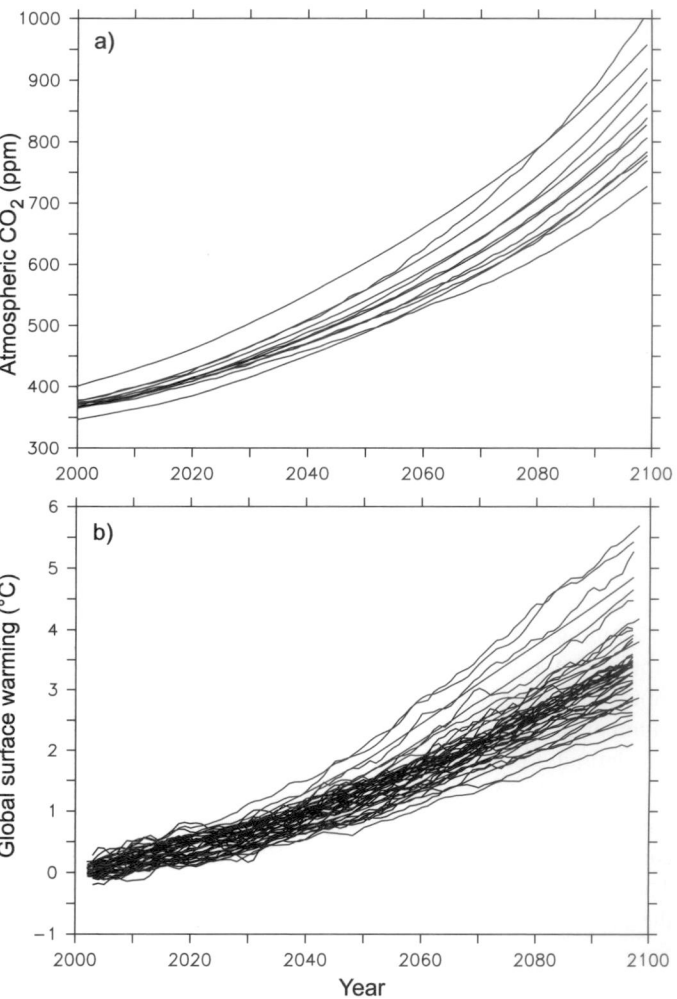

Figure 10.20. *(a) 21st-century atmospheric CO_2 concentration as simulated by the 11 C^{4}MIP models for the SRES A2 emission scenario (red) compared with the standard atmospheric CO_2 concentration used as a forcing for many IPCC AR4 climate models (black). The standard CO_2 concentration values were calculated by the BERN-CC model and are identical to those used in the TAR. For some IPCC-AR4 models, different carbon cycle models were used to convert carbon emissions to atmospheric concentrations. (b) Globally averaged surface temperature change (relative to 2000) simulated by the C^{4}MIP models forced by CO_2 emissions (red) compared to global warming simulated by the IPCC AR4 models forced by CO_2 concentration (black). The C^{4}MIP global temperature change has been corrected to account for the non-CO_2 radiative forcing used by the standard IPCC AR4 climate models.*

in Appendix II of the TAR were used. Using these numbers and knowing the climate sensitivity of each C^{4}MIP model, the warming that would have been simulated by the C^{4}MIP models if they had included the non-CO_2 greenhouse gases and aerosols can be estimated. For the SRES A2 scenario, these estimates show that the C^{4}MIP range of global temperature increase by the end of the 21st century would be 2.4°C to 5.6°C, compared with 2.6°C to 4.1°C for standard IPCC-AR4 climate models (Figure 10.20b). As a result of a much larger CO_2 concentration by 2100 in most of the C^{4}MIP models, the upper estimate of the global warming by 2100 is up to 1.5°C higher than for standard SRES A2 simulations.

The C^{4}MIP results highlight the importance of coupling the climate system and the carbon cycle in order to simulate, for a given scenario of CO_2 emissions, a climate change that takes into account the dynamic evolution of the Earth's capacity to absorb the CO_2 perturbation.

Conversely, the climate-carbon cycle feedback will have an impact on the estimate of the projected CO_2 emissions leading to stabilisation of atmospheric CO_2 at a given level. The TAR showed the range of future emissions for the Wigley, Richels and Edmonds (WRE; Wigley et al., 1996) stabilisation concentration scenarios, using different model parametrizations (including the climate-carbon feedback, Joos et al., 2001; Kheshgi and Jain,

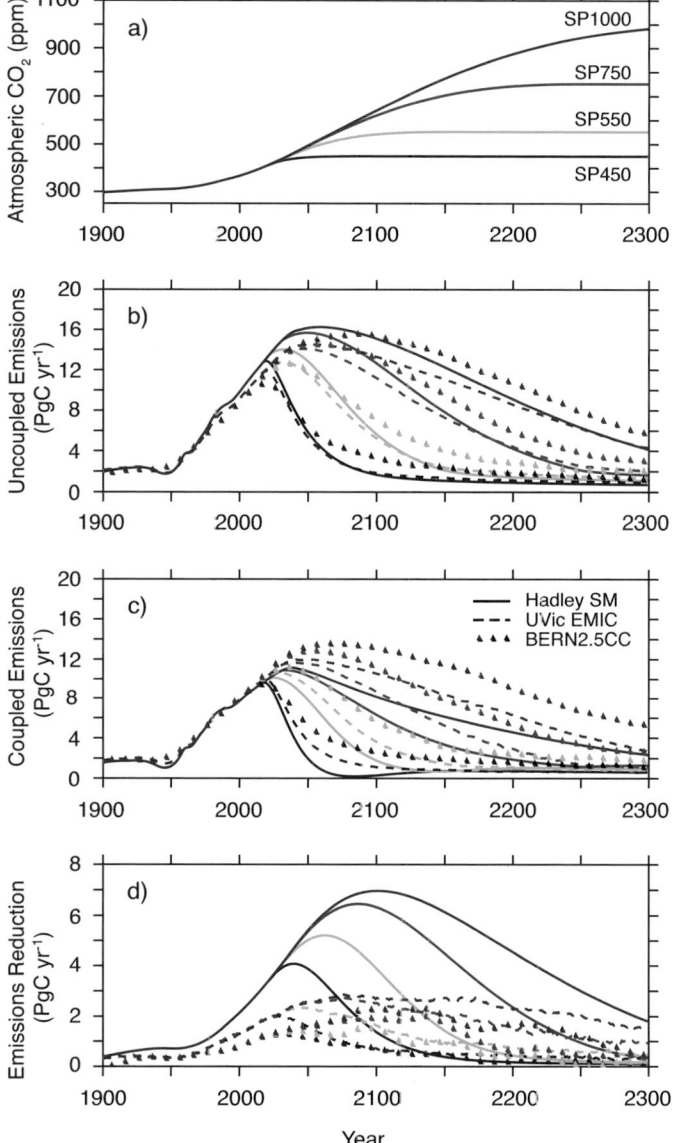

Figure 10.21. (a) Atmospheric CO_2 stabilisation scenarios SP1000 (red), SP750 (blue), SP550 (green) and SP450 (black). (b) Compatible annual emissions calculated by three models, the Hadley simple model (Jones et al., 2006; solid), the UVic EMIC (Matthews, 2005; dashed) and the BERN2.5CC EMIC (Joos et al., 2001; Plattner et al., 2001; triangles) for the three stabilisation scenarios without accounting for the impact of climate on the carbon cycle (see Table 8.3 for details of the latter two models). (c) As for (b) but with the climate impact on the carbon cycle accounted for. (d) The difference between (b) and (c) showing the impact of the climate-carbon cycle feedback on the calculation of compatible emissions.

2003). However, the emission reduction due to this feedback was not quantified. Similar to the C4MIP protocol, coupled and uncoupled simulations have been recently performed in order to specifically evaluate the impact of climate change on the future CO_2 emissions required to achieve stabilisation (Matthews, 2005; Jones et al., 2006). Figure 10.21 shows the emissions required to achieve CO_2 stabilisation for the stabilisation profiles SP450, SP550, SP750 and SP1000 (SP450 refers to stabilisation at a CO_2 concentration of 450 ppm, etc.) as simulated by three climate-carbon cycle models. As detailed above, the climate-carbon cycle feedback reduces the land and ocean uptake of CO_2, leading to a reduction in the emissions compatible with a given atmospheric CO_2 stabilisation pathway. The higher the stabilisation scenario, the larger the climate change, the larger the impact on the carbon cycle, and hence the larger the emission reduction relative to the case without climate-carbon cycle feedback. For example, stabilising atmospheric CO_2 at 450 ppm, which will likely result in a global equilibrium warming of 1.4°C to 3.1°C, with a best guess of about 2.1°C, would require a reduction of current annual greenhouse gas emissions by 52 to 90% by 2100. Positive carbon cycle feedbacks (i.e., reduced ocean and terrestrial carbon uptake caused by the warming) reduce the total (cumulative) emissions over the 21st century compatible with a stabilisation of CO_2 concentration at 450 ppm by 105 to 300 GtC relative to a hypothetical case where the carbon cycle does not respond to temperature. The uncertainty regarding the strength of the climate-carbon cycle feedback highlighted in the C4MIP analysis is also evident in Figure 10.21. For higher stabilisation scenarios such as SP550, SP750 and SP1000, the larger warming (2.9°C, 4.3°C and 5.5°C, respectively) requires an increasingly larger reduction (130 to 425 GtC, 160 to 500 GtC and 165 to 510 GtC, respectively) in the cumulated compatible emissions.

The current uncertainty involving processes driving the land and ocean carbon uptake will translate into an uncertainty in the future emissions of CO_2 required to achieve stabilisation. In Figure 10.22, the carbon-cycle related uncertainty is addressed using the BERN2.5CC carbon cycle EMIC (Joos et al., 2001; Plattner et al., 2001; see Table 8.3 for model details) and the series of S450 to SP1000 CO_2 stabilisation scenarios. The range of emission uncertainty was derived using identical assumptions as made in the TAR, varying ocean transport parameters and parametrizations describing the cycling of carbon through the terrestrial biosphere. Results are thus very closely comparable, and the small differences can be largely explained by the different CO_2 trajectories and the use of a dynamic ocean model here compared to the TAR.

The model results confirm that for stabilisation of atmospheric CO_2, emissions need to be reduced well below year 2000 values in all scenarios. This is true for the full range of simulations covering carbon cycle uncertainty, even including the upper bound, which is based on rather extreme assumptions of terrestrial carbon cycle processes.

Cumulative emissions for the period from 2000 to 2100 (to 2300) range between 596 GtC (933 GtC) for SP450, and 1,236 GtC (3,052 GtC) for SP1000. The emission uncertainty varies

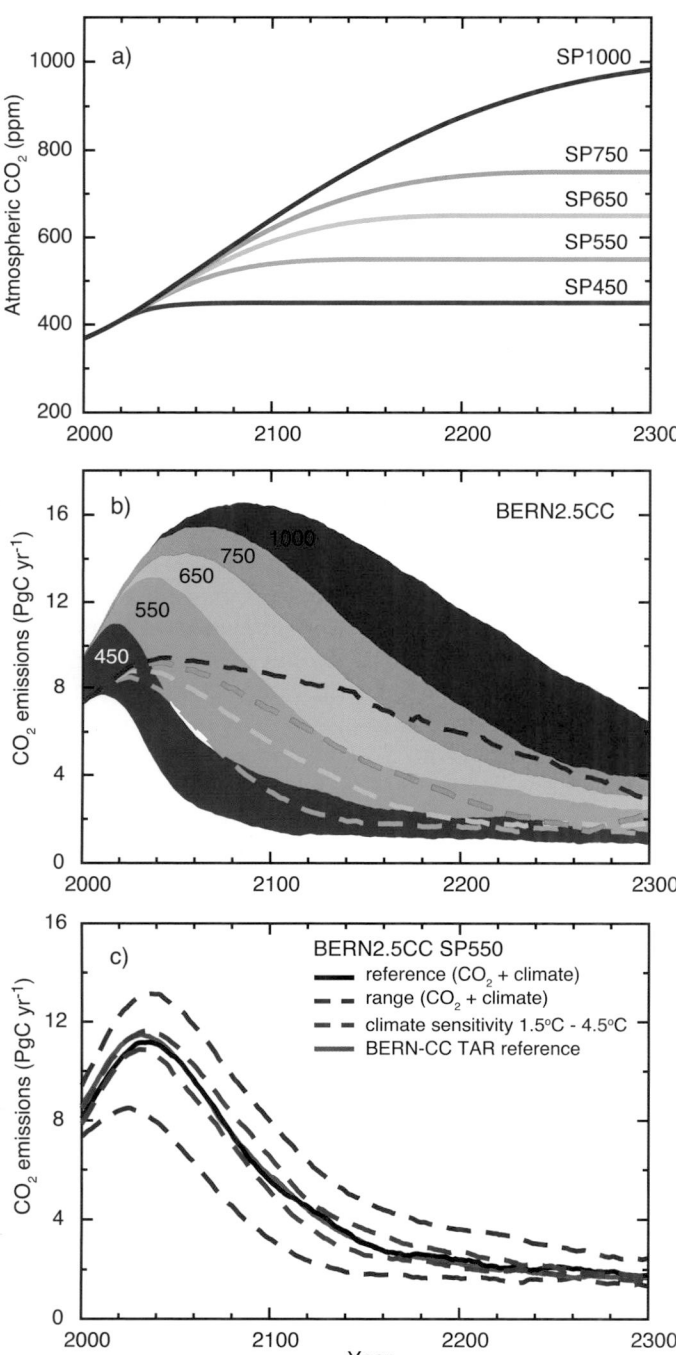

Figure 10.22. *Projected CO$_2$ emissions leading to stabilisation of atmospheric CO$_2$ concentrations at different levels and the effect of uncertainty in carbon cycle processes on calculated emissions. Panel (a) shows the assumed trajectories of CO$_2$ concentration (SP scenarios)(Knutti et al., 2005); (b) and (c) show the implied CO$_2$ emissions, as projected with the Bern2.5CC EMIC (Joos et al., 2001; Plattner et al., 2001). The ranges given in (b) for each of the SP scenarios represent effects of different model parametrizations and assumptions illustrated for scenario SP550 in panel (c) (range for 'CO$_2$ + climate'). The upper and lower bounds in (b) are indicated by the top and bottom of the shaded areas. Alternatively, the lower bound (where hidden) is indicated by a dashed line. Panel (c) illustrates emission ranges and sensitivities for scenario SP550.*

between –26 and +28% about the reference cases in year 2100 and between –26 and +34% in year 2300, increasing with time. The range of uncertainty thus depends on the magnitude of the CO$_2$ stabilisation level and the induced climate change. The additional uncertainty in projected emissions due to uncertainty in climate sensitivity is illustrated by two additional simulations with 1.5°C and 4.5°C climate sensitivities (see Box 10.2). The resulting emissions for this range of climate sensitivities lie within the range covered by the uncertainty in processes driving the carbon cycle.

Both the standard IPCC-AR4 and the C^{4}MIP models ignore the effect of land cover change in future projections. However, as described in Chapters 2 and 7, past and future changes in land cover may affect the climate through several processes. First, they may change surface characteristics such as albedo. Second, they may affect the ratio of latent to sensible heat and therefore affect surface temperature. Third, they may induce additional CO$_2$ emissions from the land. Fourth, they can affect the capacity of the land to take up atmospheric CO$_2$. So far, no comprehensive coupled AOGCM has addressed these four components all together. Using AGCMs, DeFries et al. (2004) studied the impact of future land cover change on the climate, while Maynard and Royer (2004) performed a similar experiment on Africa only. DeFries et al. (2002) forced the Colorado State University GCM (Randall et al., 1996) with Atmospheric Model Intercomparison Project (AMIP) climatological sea surface temperatures and with either the present-day vegetation cover or a 2050 vegetation map adapted from a low-growth scenario of the Integrated Model to Assess the Global Environment (IMAGE-2; Leemans et al., 1998). The study finds that in the tropics and subtropics, replacement of forests by grassland or cropland leads to a reduction in carbon assimilation, and therefore in latent heat flux. The latter reduction leads to a surface warming of up to 1.5°C in deforested tropical regions. Using the ARPEGE-Climat AGCM (Déqué et al., 1994) with a higher resolution over Africa, Maynard et al. (2002) performed two experiments, one simulation with 2 × atmospheric CO$_2$ SSTs taken from a previous ARPEGE transient SRES B2 simulation and present-day vegetation, and one with the same SSTs but the vegetation taken from a SRES B2 simulation of the IMAGE-2 model (Leemans et al., 1998). Similar to DeFries et al. (2002), they find that future deforestation in tropical Africa leads to a redistribution of latent and sensible heat that leads to a warming of the surface. However, this warming is relatively small (0.4°C) and represents about 20% of the warming due to the atmospheric CO$_2$ doubling.

Two recent studies further investigated the relative roles of future changes in greenhouse gases compared with future changes in land cover. Using a similar model design as Maynard and Royer (2004), Voldoire (2006) compared the climate change simulated under a 2050 SRES B2 greenhouse gases scenario to the one under a 2050 SRES B2 land cover change scenario. They show that the relative impact of vegetation change compared to greenhouse gas concentration increase is of the order of 10%, and can reach 30% over localised tropical regions. In a more comprehensive study, Feddema et al. (2005) applied the same methodology for the SRES A2 and B1 scenario over the 2000 to 2100 period. Similarly, they find no significant effect at the global scale, but a potentially large effect at the regional scale, such as a warming of 2°C by 2100 over the Amazon for the A2 land cover change scenario, associated with a reduction in the DTR. The general finding of these studies is that the climate change due to land cover changes may be important relative to greenhouse gases at the regional level, where intense land cover change occurs. Globally, the impact of greenhouse gas concentrations dominates over the impact of land cover change.

10.4.2 Ocean Acidification Due to Increasing Atmospheric Carbon Dioxide

Increasing atmospheric CO_2 concentrations lower oceanic pH and carbonate ion concentrations, thereby decreasing the saturation state with respect to calcium carbonate (Feely et al., 2004). The main driver of these changes is the direct geochemical effect due to the addition of anthropogenic CO_2 to the surface ocean (see Box 7.3). Surface ocean pH today is already 0.1 unit lower than pre-industrial values (Section 5.4.2.3). In the multi-model median shown in Figure 10.23, pH is projected to decrease by another 0.3 to 0.4 units under the IS92a scenario by 2100. This translates into a 100 to 150% increase in the concentration of H^+ ions (Orr et al., 2005). Simultaneously, carbonate ion concentrations will decrease. When water is undersaturated with respect to calcium carbonate, marine organisms can no longer form calcium carbonate shells (Raven et al., 2005).

Under scenario IS92a, the multi-model projection shows large decreases in pH and carbonate ion concentrations throughout the world oceans (Orr et al., 2005; Figures 10.23 and 10.24). The decrease in surface carbonate ion concentrations is found to be largest at low and mid-latitudes, although undersaturation is projected to occur at high southern latitudes first (Figure 10.24). The present-day surface saturation state is strongly influenced by temperature and is lowest at high latitudes, with minima in the Southern Ocean. The model simulations project that undersaturation will be reached in a few decades. Therefore, conditions detrimental to high-latitude ecosystems could develop within decades, not centuries as suggested previously (Orr et al., 2005).

While the projected changes are largest at the ocean surface, the penetration of anthropogenic CO_2 into the ocean interior will alter the chemical composition over the 21st century down to several thousand metres, albeit with substantial regional differences (Figure 10.23). The total volume of water in the ocean that is undersaturated with regard to calcite (not shown) or aragonite, a meta-stable form of calcium carbonate, increases substantially as atmospheric CO_2 concentrations continue to rise (Figure 10.23). In the multi-model projections, the aragonite saturation horizon (i.e., the 100% line separating over- and undersaturated regions) reaches the surface in the Southern Ocean by about 2050 and substantially shoals by 2100 in the South Pacific (by >1,000 m) and throughout the Atlantic (between 800 m and 2,200 m).

Ocean acidification could thus conceivably lead to undersaturation and dissolution of calcium carbonate in parts of the surface ocean during the 21st century, depending on the evolution of atmospheric CO_2 (Orr et al., 2005). Southern Ocean surface water is projected to become undersaturated with respect to aragonite at a CO_2 concentration of approximately 600 ppm. This concentration threshold is largely independent of emission scenarios.

Uncertainty in these projections due to potential future climate change effects on the ocean carbon cycle (mainly through changes in temperature, ocean stratification and marine biological production and re-mineralization; see Box 7.3) are small compared to the direct effect of rising atmospheric CO_2 from anthropogenic emissions. Orr et al. (2005) estimate that 21st century climate change could possibly counteract less than 10% of the projected direct geochemical changes. By far the largest uncertainty in the future evolution of these ocean interior changes is thus associated with the future pathway of atmospheric CO_2.

10.4.3 Simulations of Future Evolution of Methane, Ozone and Oxidants

Simulations using coupled chemistry-climate models indicate that the trend in upper-stratospheric ozone changes sign sometime between 2000 and 2005 due to the gradual reduction in halocarbons. While ozone concentrations in the upper stratosphere decreased at a rate of 400 ppb (–6%) per decade during 1980 to 2000, they are projected to increase at a rate of 100 ppb (1 to 2%) per decade from 2000 to 2020 (Austin and Butchart, 2003). On longer time scales, simulations show significant changes in ozone and CH_4 relative to current concentrations. The changes are related to a variety of factors, including increased emissions of chemical precursors, changes in gas-phase and heterogeneous chemistry, altered climate conditions due to global warming and greater transport and mixing across the tropopause. The impacts on CH_4 and ozone from increased emissions are a direct effect of anthropogenic activity, while the impacts of different climate conditions and stratosphere-troposphere exchange represent indirect effects of these emissions (Grewe et al., 2001).

The projections for ozone based upon scenarios with high emissions (IS92a; Leggett et al., 1992) and SRES A2 (Nakićenović and Swart, 2000) indicate that concentrations of tropospheric ozone might increase throughout the 21st century, primarily as a result of these emissions. Simulations for the period 2015 through 2050 project increases in ozone of 20 to 25% (Grewe et al., 2001; Hauglustaine and Brasseur, 2001), and simulations through 2100 indicate that ozone below 250 mb may grow by 40 to 60% (Stevenson et al., 2000; Grenfell et al., 2003; Zeng and Pyle, 2003; Hauglustaine et al., 2005; Yoshimura et al., 2006). The primary species contributing to the increase in tropospheric ozone are anthropogenic emissions of NO_x, CH_4, CO and compounds from fossil fuel combustion. The photochemical reactions that produce smog are accelerated by increases of 2.6 times the present flux of NO_x, 2.5 times the

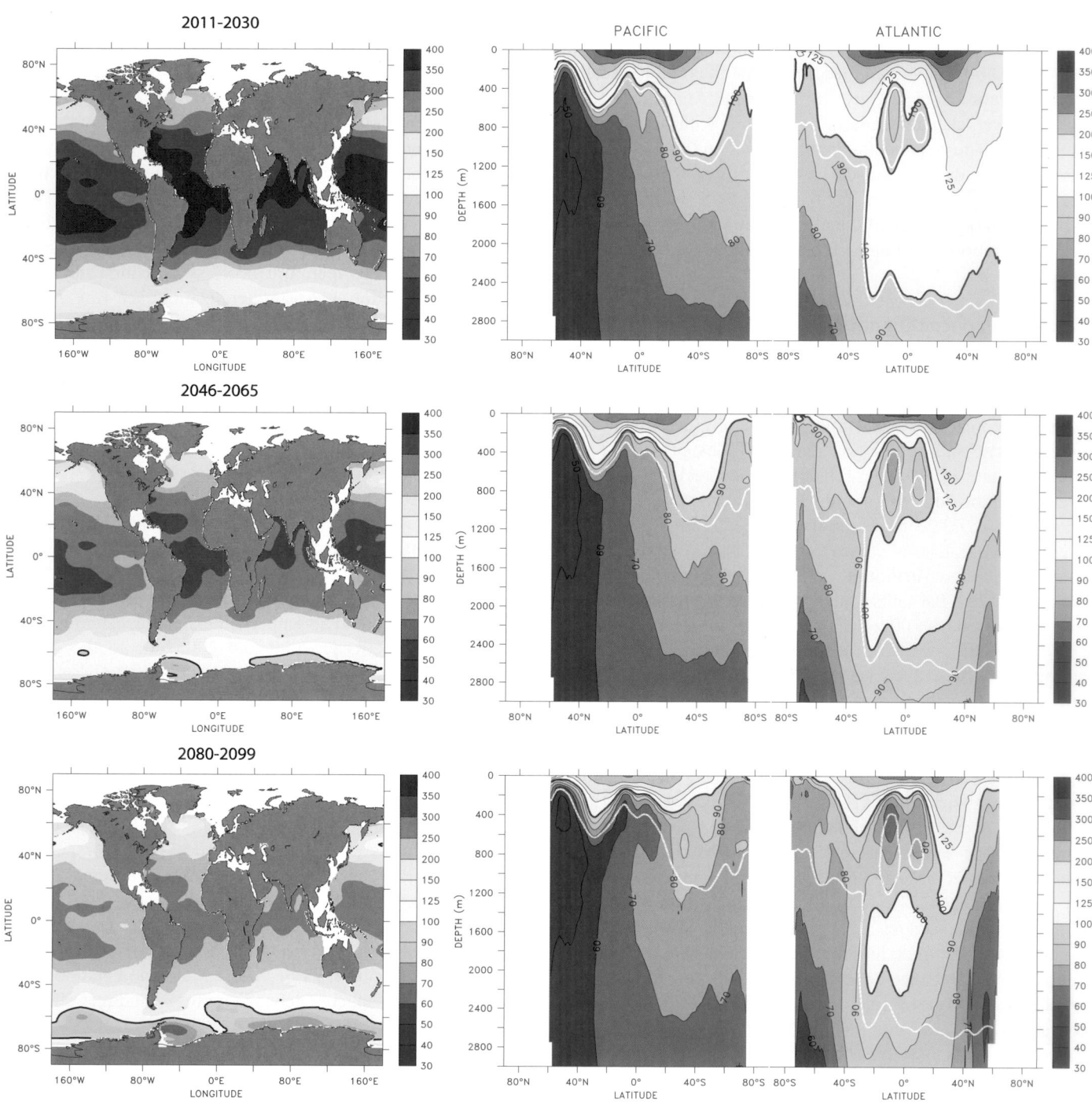

Figure 10.23. *Multi-model median for projected levels of saturation (%) with respect to aragonite, a meta-stable form of calcium carbonate, over the 21st century from the Ocean Carbon-Cycle Model Intercomparison Project (OCMIP-2) models (adapted from Orr et al., 2005). Calcium carbonate dissolves at levels below 100%. Surface maps (left) and combined Pacific/Atlantic zonal mean sections (right) are given for scenario IS92a as averages over three time periods: 2011 to 2030 (top), 2045 to 2065 (middle) and 2080 to 2099 (bottom). Atmospheric CO_2 concentrations for these three periods average 440, 570 and 730 ppm, respectively. Latitude-depth sections start in the North Pacific (at the left border), extend to the Southern Ocean Pacific section and return through the Southern Ocean Atlantic section to the North Atlantic (right border). At 100%, waters are saturated (solid black line - the aragonite saturation horizon); values larger than 100% indicate super-saturation; values lower than 100% indicate undersaturation. The observation-based (Global Ocean Data Analysis Project; GLODAP) 1994 saturation horizon (solid white line) is also shown to illustrate the projected changes in the saturation horizon compared to the present.*

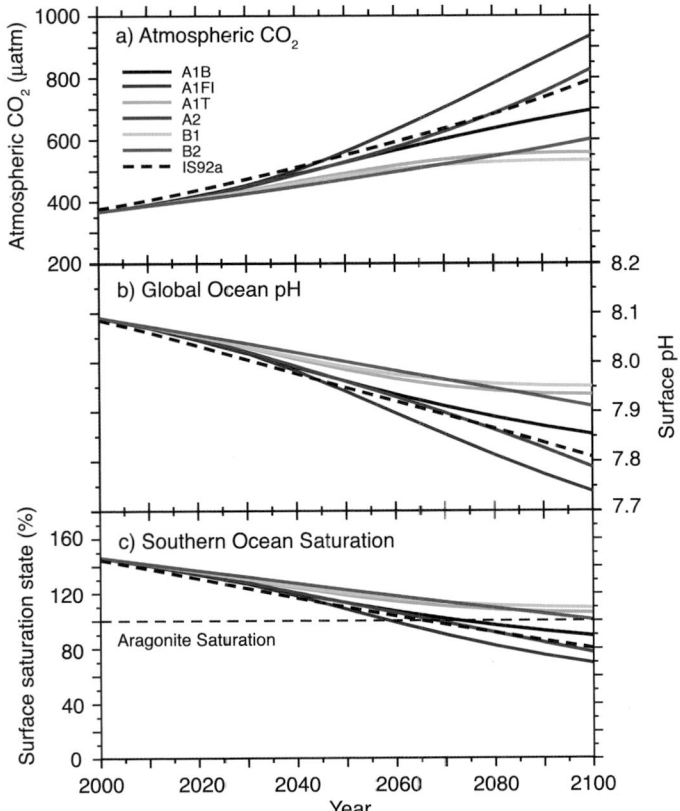

Figure 10.24. *Changes in global average surface pH and saturation state with respect to aragonite in the Southern Ocean under various SRES scenarios. Time series of (a) atmospheric CO_2 for the six illustrative SRES scenarios, (b) projected global average surface pH and (c) projected average saturation state in the Southern Ocean from the BERN2.5D EMIC (Plattner et al., 2001). The results for the SRES scenarios A1T and A2 are similar to those for the non-SRES scenarios S650 and IS92a, respectively. Modified from Orr et al. (2005).*

present flux of CH_4 and 1.8 times the present flux of CO in the A2 scenario. Between 91 and 92% of the higher concentrations in ozone are related to direct effects of these emissions, with the remainder of the increase attributable to secondary effects of climate change (Zeng and Pyle, 2003) combined with biogenic precursor emissions (Hauglustaine et al., 2005). These emissions may also lead to higher concentrations of oxidants including the hydroxyl radical (OH), possibly leading to an 8% reduction in the lifetime of tropospheric CH_4 (Grewe et al., 2001).

Since the projected growth in emissions occurs primarily in low latitudes, the ozone increases are largest in the tropics and subtropics (Grenfell et al., 2003). In particular, the concentrations in Southeast Asia, India and Central America increase by 60 to 80% by 2050 under the A2 scenario. However, the effects of tropical emissions are not highly localised, since the ozone spreads throughout the lower atmosphere in plumes emanating from these regions. As a result, the ozone in remote marine regions in the SH may grow by 10 to 20% over present-day levels by 2050. The ozone may also be distributed through vertical transport in tropical convection followed by lateral transport on isentropic surfaces. Ozone concentrations can also be increased by emissions of biogenic hydrocarbons (e.g., Hauglustaine et

al., 2005), in particular isoprene emitted by broadleaf forests. Under the A2 scenario, biogenic hydrocarbons are projected to increase by between 27% (Sanderson et al., 2003) and 59% (Hauglustaine et al., 2005) contributing to a 30 to 50% increase in ozone formation over northern continental regions.

Developing countries have begun reducing emissions from mobile sources through stricter standards. New projections of the evolution of ozone precursors that account for these reductions have been developed with the Regional Air Pollution Information and Simulation (RAINS) model (Amann et al., 2004). One set of projections is consistent with source strengths permitted under the Current Legislation (CLE) scenario. A second set of projections is consistent with lower emissions under a Maximum Feasible Reduction (MFR) scenario. The concentrations of ozone and CH_4 have been simulated for the MFR, CLE and A2 scenarios for the period 2000 through 2030 using an ensemble of 26 chemical transport models (Dentener et al., 2006; Stevenson et al., 2006). The changes in NO_x emissions for these three scenarios are –27%, +12% and +55%, respectively, relative to year 2000. The corresponding changes in ensemble-mean burdens in tropospheric ozone are –5%, +6% and +18% for the MFR, CLE and A2 scenarios, respectively. There are substantial inter-model differences of order ±25% in these results. The ozone decreases throughout the troposphere in the MFR scenario, but the zonal annual mean concentrations increase by up to 6 ppb in the CLE scenario and by typically 6 to 10 ppb in the A2 scenario (Supplementary Material, Figure S10.2).

The radiative forcing by the combination of ozone and CH_4 changes by –0.05, 0.18, and 0.30 W m^{-2} for the MFR, CLE and A2 scenarios, respectively. These projections indicate that the growth in tropospheric ozone between 2000 and 2030 could be reduced or reversed depending on emission controls.

The major issues in the fidelity of these simulations for future tropospheric ozone are the sensitivities to the representation of the stratospheric production, destruction and transport of ozone and the exchange of species between the stratosphere and troposphere. Few of the models include the effects of non-methane hydrocarbons (NMHCs), and the sign of the effects of NMHCs on ozone are not consistent among the models that do (Hauglustaine and Brasseur, 2001; Grenfell et al., 2003).

The effect of more stratosphere-troposphere exchange (STE) in response to climate change is projected to increase the concentrations of ozone in the upper troposphere due to the much greater concentrations of ozone in the lower stratosphere than in the upper troposphere. While the sign of the effect is consistent in recent simulations, the magnitude of the change in STE and its effects on ozone are very model dependent. In a simulation forced by the SRES A1FI scenario, Collins et al. (2003) project that the downward flux of ozone increases by 37% from the 1990s to the 2090s. As a result, the concentration of ozone in the upper troposphere at mid-latitudes increases by 5 to 15%. For the A2 scenarios, projections of the increase in ozone by 2100 due to STE range from 35% (Hauglustaine et al., 2005) to 80% (Sudo et al., 2003; Zeng and Pyle, 2003). The increase in STE is driven by increases in the descending branches of the Brewer-Dobson Circulation at mid-latitudes and is caused by changes in meridional temperature gradients in the upper troposphere and lower stratosphere (Rind et al., 2001). The effects of the enhanced STE are sensitive to the simulation of processes in the stratosphere, including the effects of lower temperatures and the evolution of chlorine, bromine and NO_x concentrations. Since the greenhouse effect of ozone is largest in the upper troposphere, the treatment of STE remains a significant source of uncertainty in the calculation of the total greenhouse effect of tropospheric ozone.

The effects of climate change, in particular increased tropospheric temperatures and water vapour, tend to offset some of the increase in ozone driven by emissions. The higher water vapour is projected to offset the increase in ozone by between 10% (Hauglustaine et al., 2005) and 17% (Stevenson et al., 2000). The water vapour both decelerates the chemical production and accelerates the chemical destruction of ozone. The photochemical production depends on the concentrations of NO_y (reactive odd nitrogen), and the additional water vapour causes a larger fraction of NO_y to be converted to nitric acid, which can be efficiently removed from the atmosphere in precipitation (Grewe et al., 2001). The water vapour also increases the concentrations of OH through reaction with the oxygen radical in the 1D excited state ($O(^1D)$), and the removal of $O(^1D)$ from the atmosphere slows the formation of ozone. The increased concentrations of OH and the increased rates of CH_4 oxidation with higher temperature further reduce the lifetime of tropospheric CH_4 by 12% by 2100 (Stevenson et al., 2000; Johnson et al., 2001). Decreases in CH_4 concentrations also tend to reduce tropospheric ozone (Stevenson et al., 2000).

Recent measurements show that CH_4 growth rates have declined and were negative for several years in the early 21st century (see Section 2.3.2). The observed rate of increase of 0.8 ppb yr^{-1} for the period 1999 to 2004 is considerably less than the rate of 6 ppb yr^{-1} assumed in all the SRES scenarios for the period 1990 to 2000 (Nakićenović and Swart, 2000; TAR Appendix II). Recent studies (Dentener et al., 2005) have considered lower emission scenarios (see above) that take account of new pollution control techniques adopted in major developing countries. In the CLE scenario, emissions of CH_4 are comparable to the B2 scenario and increase from 340 Tg yr^{-1} in 2000 to 450 Tg yr^{-1} in 2030. The CH_4 concentrations increase from 1,750 ppb in 2000 to between 2,090 and 2,200 ppb in 2030 under this scenario. In the MFR scenario, the emissions are sufficiently low that the concentrations in 2030 are unchanged at 1,750 ppb. Under these conditions, the changes in radiative forcing due to CH_4 between the 1990s and 2020s are less than 0.01 W m^{-2}.

Current understanding of the magnitude and variation of CH_4 sources and sinks is covered in Section 7.4, where it is noted that there are substantial uncertainties although the modelling has progressed. There is some evidence for a coupling between climate and wetland emissions. For example, calculations using atmospheric concentrations and small-scale emission measurements as input differ by 60% (Shindell and Schmidt, 2004). Concurrent changes in natural sources of CH_4 are

now being estimated to first order using simple models of the biosphere coupled to AOGCMs. Simulations of the response of wetlands to climate change from doubling atmospheric CO_2 show that wetland emissions increase by 78% (Shindell and Schmidt, 2004). Most of this effect is caused by growth in the flux of CH_4 from existing tropical wetlands. The increase would be equivalent to approximately 20% of current inventories and would contribute an additional 430 ppb to atmospheric concentrations. Global radiative forcing would increase by approximately 4 to 5% from the effects of wetland emissions by 2100 (Gedney et al., 2004).

10.4.4 Simulations of Future Evolution of Major Aerosol Species

The time-dependent evolution of major aerosol species and the interaction of these species with climate represent some of the major sources of uncertainty in projections of climate change. An increasing number of AOGCMs have included multiple types of tropospheric aerosols including sulphates, nitrates, black and organic carbon, sea salt and soil dust. Of the 23 models represented in the multi-model ensemble of climate-change simulations for IPCC AR4, 13 include other tropospheric species besides sulphates. Of these, seven have the non-sulphate species represented with parametrizations that interact with the remainder of the model physics. Nitrates are treated in just two of the models in the ensemble. Recent projections of nitrate and sulphate loading under the SRES A2 scenario suggest that forcing by nitrates may exceed forcing by sulphates by the end of the 21st century (Adams et al., 2001). This result is of course strongly dependent upon the evolution of precursor emissions for these aerosol species.

The black and organic carbon aerosols in the atmosphere include a very complex system of primary organic aerosols (POA) and secondary organic aerosols (SOA), which are formed by oxidation of biogenic VOCs. The models used for climate projections typically use highly simplified bulk parametrizations for POA and SOA. More detailed parametrizations for the formation of SOA that trace oxidation pathways have only recently been developed and used to estimate the direct radiative forcing by SOA for present-day conditions (Chung and Seinfeld, 2002). The forcing by SOA is an emerging issue for simulations of present-day and future climate since the rate of chemical formation of SOA may be 60% or more of the emissions rate for primary carbonaceous aerosols (Kanakidou et al., 2005). In addition, two-way coupling between reactive chemistry and tropospheric aerosols has not been explored comprehensively in climate change simulations. Unified models that treat tropospheric ozone-NO_x-hydrocarbon chemistry, aerosol formation, heterogeneous processes in clouds and on aerosols, and gas-phase photolysis have been developed and applied to the current climate (Liao et al., 2003). However, these unified models have not yet been used extensively to study the evolution of the chemical state of the atmosphere under future scenarios.

The interaction of soil dust with climate is under active investigation. Whether emissions of soil dust aerosols increase or decrease in response to changes in atmospheric state and circulation is still unresolved (Tegen et al., 2004a). Several recent studies have suggested that the total surface area where dust can be mobilised will decrease in a warmer climate with higher concentrations of CO_2 (e.g., Harrison et al., 2001). The net effects of reductions in dust emissions from natural sources combined with land use change could potentially be significant but have not been systematically modelled as part of climate change assessment.

Uncertainty regarding the scenario simulations is compounded by inherently unpredictable natural forcings from future volcanic eruptions and solar variability. The eruptions that produce climatologically significant forcing represent just the extremes of global volcanic activity (Naveau and Ammann, 2005). Global simulations can account for the effects of future natural forcings using stochastic representations based upon prior eruptions and variations in solar luminosity. The relative contribution of these forcings to the projections of global mean temperature anomalies are largest in the period up to 2030 (Stott and Kettleborough, 2002).

10.5 Quantifying the Range of Climate Change Projections

10.5.1 Sources of Uncertainty and Hierarchy of Models

Uncertainty in predictions of anthropogenic climate change arises at all stages of the modelling process described in Section 10.1. The specification of future emissions of greenhouse gases, aerosols and their precursors is uncertain (e.g., Nakićenović and Swart, 2000). It is then necessary to convert these emissions into concentrations of radiatively active species, calculate the associated forcing and predict the response of climate system variables such as surface temperature and precipitation (Figure 10.1). At each step, uncertainty in the true signal of climate change is introduced both by errors in the representation of Earth system processes in models (e.g., Palmer et al., 2005) and by internal climate variability (e.g., Selten et al., 2004). The effects of internal variability can be quantified by running models many times from different initial conditions, provided that simulated variability is consistent with observations. The effects of uncertainty in the knowledge of Earth system processes can be partially quantified by constructing ensembles of models that sample different parametrizations of these processes. However, some processes may be missing from the set of available models, and alternative parametrizations of other processes may share common systematic biases. Such limitations imply that distributions of future climate responses from ensemble simulations are themselves subject to uncertainty (Smith, 2002), and would be wider were uncertainty

due to structural model errors accounted for. These distributions may be modified to reflect observational constraints expressed through metrics of the agreement between the observed historical climate and the simulations of individual ensemble members, for example through Bayesian methods (see Chapter 9 Supplementary Material, Appendix 9.B). In this case, the choice of observations and their associated errors introduce further sources of uncertainty. In addition, some sources of future radiative forcing are yet to be accounted for in the ensemble projections, including those from land use change, variations in solar and volcanic activity (Kettleborough et al., 2007), and CH_4 release from permafrost or ocean hydrates (see Section 8.7).

A spectrum or hierarchy of models of varying complexity has been developed (Claussen et al., 2002; Stocker and Knutti, 2003) to assess the range of future changes consistent with the understanding of known uncertainties. Simple climate models (SCMs) typically represent the ocean-atmosphere system as a set of global or hemispheric boxes, predicting global surface temperature using an energy balance equation, a prescribed value of climate sensitivity and a basic representation of ocean heat uptake (see Section 8.8.2). Their role is to perform comprehensive analyses of the interactions between global variables, based on prior estimates of uncertainty in their controlling parameters obtained from observations, expert judgement and from tuning to complex models. By coupling SCMs to simple models of biogeochemical cycles they can be used to extrapolate the results of AOGCM simulations to a wide range of alternative forcing scenarios (e.g., Wigley and Raper, 2001; see Section 10.5.3).

Compared to SCMs, EMICs include more of the processes simulated in AOGCMs, but in a less detailed, more highly parametrized form (see Section 8.8.3), and at coarser resolution. Consequently, EMICs are not suitable for quantifying uncertainties in regional climate change or extreme events, however they can be used to investigate the large-scale effects of coupling between multiple Earth system components in large ensembles or long simulations (e.g., Forest et al., 2002; Knutti et al., 2002), which is not yet possible with AOGCMs due to their greater computational expense. Some EMICs therefore include modules such as vegetation dynamics, the terrestrial and ocean carbon cycles and atmospheric chemistry (Plattner et al., 2001; Claussen et al., 2002), filling a gap in the spectrum of models between AOGCMs and SCMs. Thorough sampling of parameter space is computationally feasible for some EMICs (e.g., Stocker and Schmittner, 1997; Forest et al., 2002; Knutti et al., 2002), as for SCMs (Wigley and Raper, 2001), and is used to obtain probabilistic projections (see Section 10.5.4.5). In some EMICs, climate sensitivity is an adjustable parameter, as in SCMs. In other EMICs, climate sensitivity is dependent on multiple model parameters, as in AOGCMs. Probabilistic estimates of climate sensitivity and TCR from SCMs and EMICs are assessed in Section 9.6 and compared with estimates from AOGCMs in Box 10.2.

The high resolution and detailed parametrizations in AOGCMs enable them to simulate more comprehensively the

Box 10.2: Equilibrium Climate Sensitivity

The likely range[1] for equilibrium climate sensitivity was estimated in the TAR (Technical Summary, Section F.3; Cubasch et al., 2001) to be 1.5°C to 4.5°C. The range was the same as in an early report of the National Research Council (Charney, 1979), and the two previous IPCC assessment reports (Mitchell et al., 1990; Kattenberg et al., 1996). These estimates were expert assessments largely based on equilibrium climate sensitivities simulated by atmospheric GCMs coupled to non-dynamic slab oceans. The mean ±1 standard deviation values from these models were 3.8°C ± 0.78°C in the SAR (17 models), 3.5°C ± 0.92°C in the TAR (15 models) and in this assessment 3.26°C ± 0.69°C (18 models).

Considerable work has been done since the TAR (IPCC, 2001) to estimate climate sensitivity and to provide a better quantification of relative probabilities, including a most likely value, rather than just a subjective range of uncertainty. Since climate sensitivity of the real climate system cannot be measured directly, new methods have been used since the TAR to establish a relationship between sensitivity and some observable quantity (either directly or through a model), and to estimate a range or probability density function (PDF) of climate sensitivity consistent with observations. These methods are summarised separately in Chapters 9 and 10, and here we synthesize that information into an assessment. The information comes from two main categories: constraints from past climate change on various time scales, and the spread of results for climate sensitivity from ensembles of models.

The first category of methods (see Section 9.6) uses the historical transient evolution of surface temperature, upper air temperature, ocean temperature, estimates of the radiative forcing, satellite data, proxy data over the last millennium, or a subset thereof to calculate ranges or PDFs for sensitivity (e.g., Wigley et al., 1997b; Tol and De Vos, 1998; Andronova and Schlesinger, 2001; Forest et al., 2002; Gregory et al., 2002a; Harvey and Kaufmann, 2002; Knutti et al., 2002, 2003; Frame et al., 2005; Forest et al., 2006; Forster and Gregory, 2006; Hegerl et al., 2006). A summary of all PDFs of climate sensitivity from those methods is shown

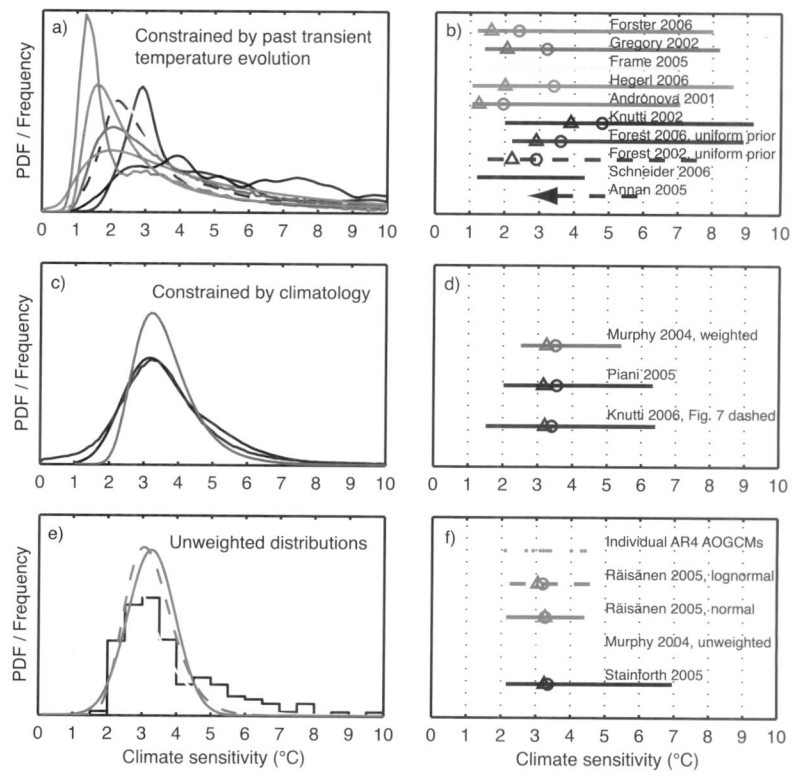

Box 10.2, Figure 1. *(a) PDFs or frequency distributions constrained by the transient evolution of the atmospheric temperature, radiative forcing and ocean heat uptake, (b) as in (a) and (b) but 5 to 95% ranges, medians (circles) and maximum probabilities (triangles), (c) and (d) as in (a) but using constraints from present-day climatology, and (e) and (f) unweighted or fitted distributions from different models or from perturbing parameters in a single model. Distributions in (e) and (f) should not be strictly interpreted as PDFs. See Chapter 9 text, Figure 9.20 and Table 9.3 for details. Note that Annan et al. (2005b) only provide an upper but no lower bound. All PDFs are truncated at 10°C for consistency, some are shown for different prior distributions than in the original studies, and ranges may differ from numbers reported in individual studies.*

in Figure 9.20 and in Box 10.2, Figure 1a. Median values, most likely values (modes) and 5 to 95% uncertainty ranges are shown in Box 10.2, Figure 1b for each PDF. Most of the results confirm that climate sensitivity is very unlikely below 1.5°C. The upper bound is more difficult to constrain because of a nonlinear relationship between climate sensitivity and the observed transient response, and is further hampered by the limited length of the observational record and uncertainties in the observations, which are particularly large for ocean heat uptake and for the magnitude of the aerosol radiative forcing. Studies that take all the important known uncertainties in observed historical trends into account cannot rule out the possibility that the climate sensitivity exceeds 4.5°C, although such high values are consistently found to be less likely than values of around 2.0°C to 3.5°C. Observations of transient climate change provide better constraints for the TCR (see Section 9.6.1.3).

Two recent studies use a modelled relation between climate sensitivity and tropical SSTs in the Last Glacial Maximum (LGM) and proxy records of the latter to estimate ranges of climate sensitivity (Annan et al., 2005b; Schneider von Deimling et al., 2006; see

(continued)

---

[1] Though the TAR Technical Summary attached 'likely' to the 1.5°C - 4.5°C range, the word 'likely' was used there in a general sense rather than in a specific calibrated sense. No calibrated confidence assessment was given in either the Summary for Policymakers or in Chapter 9 of the TAR, and no probabilistic studies on climate sensitivity were cited in Chapter 9 where the range was assessed.

Section 9.6). While both of these estimates overlap with results from the instrumental period and results from other AOGCMS, the results differ substantially due to different forcings and the different relationships between LGM SSTs and sensitivity in the models used. Therefore, LGM proxy data provide support for the range of climate sensitivity based on other lines of evidence.

Studies comparing the observed transient response of surface temperature after large volcanic eruptions with results obtained from models with different climate sensitivities (see Section. 9.6) do not provide PDFs, but find best agreement with sensitivities around 3°C, and reasonable agreement within the 1.5°C to 4.5°C range (Wigley et al., 2005). They are not able to exclude sensitivities above 4.5°C.

The second category of methods examines climate sensitivity in GCMs. Climate sensitivity is not a single tuneable parameter in these models, but depends on many processes and feedbacks. Three PDFs of climate sensitivity were obtained by comparing different variables of the simulated present-day climatology and variability against observations in a perturbed physics ensemble (Murphy et al., 2004; Piani et al., 2005; Knutti et al., 2006, Box 10.2, Figure 1c,d; see Section 10.5.4.2). Equilibrium climate sensitivity is found to be most likely around 3.2°C, and very unlikely to be below about 2°C. The upper bound is sensitive to how model parameters are sampled and to the method used to compare with observations.

Box 10.2, Figure 1e,f show the frequency distributions obtained by different methods when perturbing parameters in the Hadley Centre Atmospheric Model (HadAM3) but before weighting with observations (Section 10.5.4). Murphy et al. (2004; unweighted) sampled 29 parameters and assumed individual effects to combine linearly.

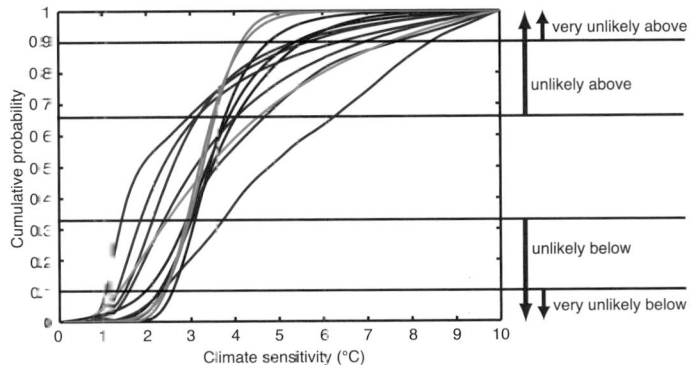

Box 10.2, Figure 2. Individual cumulative distributions of climate sensitivity from the observed 20th-century warming (red), model climatology (blue) and proxy evidence (cyan, taken from Box 10.2, Figure 1a, c (except LGM studies and Forest et al. (2002), which is superseded by Forest et al. (2006)) and cumulative distributions fitted to the AOGCMs' climate sensitivities (green) from Box 10.2, Figure 1e. Horizontal lines and arrows mark the edges of the likelihood estimates according to IPCC guidelines.

Stainforth et al. (2005) found nonlinearities when simulating multiple combinations of a subset of key parameters. The most frequently occurring climate sensitivity values are grouped around 3°C, but this could reflect the sensitivity of the unperturbed model. Some, but not all, of the simulations by high-sensitivity models have been found to agree poorly with observations and are therefore unlikely, hence even very high values are not excluded. This inability to rule out very high values is common to many methods, since for well-understood physical reasons, the rate of change (against sensitivity) of most quantities that can be observed tends to zero as the sensitivity increases (Hansen et al., 1985; Knutti et al., 2005; Allen et al., 2006b).

There is no well-established formal way of estimating a single PDF from the individual results, taking account of the different assumptions in each study. Most studies do not account for structural uncertainty, and thus probably tend to underestimate the uncertainty. On the other hand, since several largely independent lines of evidence indicate similar most likely values and ranges, climate sensitivity values are likely to be better constrained than those found by methods based on single data sets (Annan and Hargreaves, 2006; Hegerl et al., 2006).

The equilibrium climate sensitivity values for the AR4 AOGCMs coupled to non-dynamic slab ocean models are given for comparison (Box 10.2, Figure 1e,f; see also Table 8.2). These estimates come from models that represent the current best efforts from the international global climate modelling community at simulating climate. A normal fit yields a 5 to 95% range of about 2.1°C to 4.4°C with a mean value of equilibrium climate sensitivity of about 3.3°C (2.2°C to 4.6°C for a lognormal distribution, median 3.2°C) (Räisänen, 2005b). A probabilistic interpretation of the results is problematic, because each model is assumed to be equally credible and the results depend upon the assumed shape of the fitted distribution. Although the AOGCMs used in IPCC reports are an 'ensemble of opportunity' not designed to sample modelling uncertainties systematically or randomly, the range of sensitivities covered has been rather stable over many years. This occurs in spite of substantial model developments, considerable progress in simulating many aspects of the large-scale climate, and evaluation of those models against observations. Progress has been made since the TAR in diagnosing and understanding inter-model differences in climate feedbacks and equilibrium climate sensitivity. Confidence has increased in the strength of water vapour-lapse rate feedbacks, whereas cloud feedbacks (particularly from low-level clouds) have been confirmed as the primary source of climate sensitivity differences (see Section 8.6).

Since the TAR, the levels of scientific understanding and confidence in quantitative estimates of equilibrium climate sensitivity have increased substantially. Basing our assessment on a combination of several independent lines of evidence, as summarised in Box 10.2 Figures 1 and 2, including observed climate change and the strength of known feedbacks simulated in GCMs, we conclude that the global mean equilibrium warming for doubling CO_2, or 'equilibrium climate sensitivity', is likely to lie in the range 2°C to 4.5°C, with a most likely value of about 3°C. Equilibrium climate sensitivity is very likely larger than 1.5°C.

For fundamental physical reasons as well as data limitations, values substantially higher than 4.5°C still cannot be excluded, but agreement with observations and proxy data is generally worse for those high values than for values in the 2°C to 4.5°C range.

processes giving rise to internal variability (see Section 8.4), extreme events (see Section 8.5) and climate change feedbacks, particularly at the regional scale (Boer and Yu, 2003a; Bony and Dufresne, 2005; Bony et al., 2006; Soden and Held, 2006). Given that ocean dynamics influence regional feedbacks (Boer and Yu, 2003b), quantification of regional uncertainties in time-dependent climate change requires multi-model ensemble simulations with AOGCMs containing a full, three-dimensional dynamic ocean component. However, downscaling methods (see Chapter 11) are required to obtain credible information at spatial scales near or below the AOGCM grid scale (125 to 400 km in the AR4 AOGCMs, see Table 8.1).

10.5.2 Range of Responses from Different Models

10.5.2.1 Comprehensive AOGCMs

The way a climate model responds to changes in external forcing, such as an increase in anthropogenic greenhouse gases, is characterised by two standard measures: (1) 'equilibrium climate sensitivity' (the equilibrium change in global surface temperature following a doubling of the atmospheric equivalent CO_2 concentration; see Glossary), and (2) 'transient climate response' (the change in global surface temperature in a global coupled climate model in a 1% yr^{-1} CO_2 increase experiment at the time of atmospheric CO_2 doubling; see Glossary). The first measure provides an indication of feedbacks mainly residing in the atmospheric model but also in the land surface and sea ice components, and the latter quantifies the response of the fully coupled climate system including aspects of transient ocean heat uptake (e.g., Sokolov et al., 2003). These two measures have become standard for quantifying how an AOGCM will react to more complicated forcings in scenario simulations.

Historically, the equilibrium climate sensitivity has been given in the range from 1.5°C to 4.5°C. This range was reported in the TAR with no indication of a probability distribution within this range. However, considerable recent work has addressed the range of equilibrium climate sensitivity, and attempted to assign probabilities to climate sensitivity.

Equilibrium climate sensitivity and TCR are not independent (Figure 10.25a). For a given AOGCM, the TCR is smaller than the equilibrium climate sensitivity because ocean heat uptake delays the atmospheric warming. A large ensemble of the BERN2.5D EMIC has been used to explore the relationship of TCR and equilibrium sensitivity over a wide range of ocean heat uptake parametrizations (Knutti et al., 2005). Good agreement with the available results from AOGCMs is found, and the BERN2.5D EMIC covers almost the entire range of structurally different models. The percent change in precipitation is closely related to the equilibrium climate sensitivity for the current generation of AOGCMs (Figure 10.25b), with values from the current models falling within the range of the models from the TAR. Figure 10.25c shows the percent change in globally averaged precipitation as a function of TCR at the time of atmospheric CO_2 doubling, as simulated by 1% yr^{-1} transient CO_2 increase experiments with AOGCMs. The figure suggests

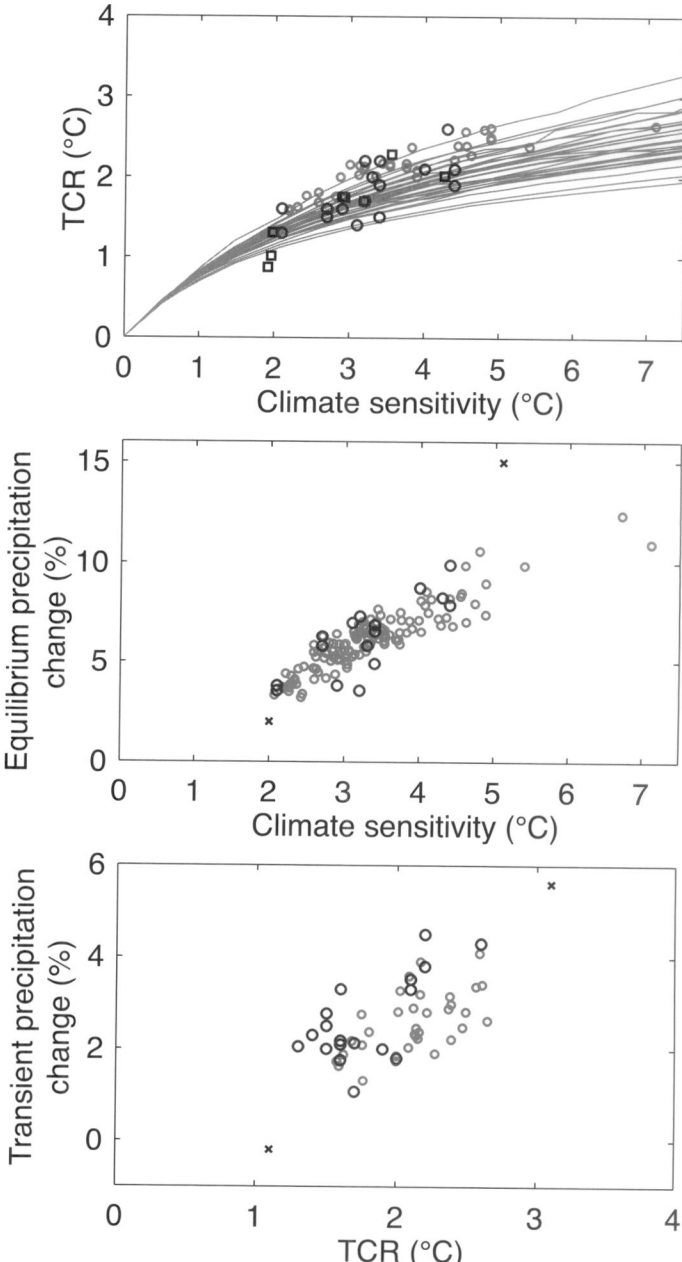

Figure 10.25. (a) TCR versus equilibrium climate sensitivity for all AOGCMs (red), EMICs (blue), a perturbed physics ensemble of the UKMO-HadCM3 AOGCM (green; an updated ensemble based on M. Collins et al., 2006) and from a large ensemble of the Bern2.5D EMIC (Knutti et al., 2005) using different ocean vertical diffusivities and mixing parametrizations (grey lines). (b) Global mean precipitation change (%) as a function of global mean temperature change at equilibrium for doubled CO_2 in atmospheric GCMs coupled to a non-dynamic slab ocean (red all AOGCMS, green from a perturbed physics ensemble of the atmosphere-slab ocean version of UKMO-HadCM3 (Webb et al., 2006)). (c) Global mean precipitation change (%) as a function of global mean temperature change (TCR) at the time of CO_2 doubling in a transient 1% yr^{-1} CO_2 increase scenario, simulated by coupled AOGCMs (red) and the UKMO-HadCM3 perturbed physics ensemble (green). Black crosses in (b) and (c) mark ranges covered by the TAR AOGCMs (IPCC, 2001) for each quantity.

a broadly positive correlation between these two quantities similar to that for equilibrium climate sensitivity, with these values from the new models also falling within the range of the previous generation of AOGCMs assessed in the TAR. Note that the apparent relationships may not hold for other forcings or at smaller scales. Values for an ensemble with perturbations made to parameters in the atmospheric component of UKMO-HadCM3 (M. Collins et al., 2006) cover similar ranges and are shown in Figure 10.25 for comparison.

Fitting normal distributions to the results, the 5 to 95% uncertainty range for equilibrium climate sensitivity from the AOGCMs is approximately 2.1°C to 4.4°C and that for TCR is 1.2°C to 2.4°C (using the method of Räisänen, 2005b). The mean for climate sensitivity is 3.26°C and that for TCR is 1.76°C. These numbers are practically the same for both the normal and the lognormal distribution (see Box 10.2). The assumption of a (log) normal fit is not well supported by the limited sample of AOGCM data. In addition, the AOGCMs represent an 'ensemble of opportunity' and are by design not sampled in a random way. However, most studies aiming to constrain climate sensitivity with observations do indeed indicate a similar to lognormal probability distribution of climate sensitivity and an approximately normal distribution of the uncertainty in future warming and thus TCR (see Box 10.2). Those studies also suggest that the current AOGCMs may not cover the full range of uncertainty for climate sensitivity. An assessment of all the evidence on equilibrium climate sensitivity is provided in Box 10.2. The spread of the AOGCM climate sensitivities is discussed in Section 8.6 and the AOGCM values for climate sensitivity and TCR are listed in Table 8.2.

The nonlinear relationship between TCR and equilibrium climate sensitivity shown in Figure 10.25a also indicates that on time scales well short of equilibrium, the model's TCR is not particularly sensitive to the model's climate sensitivity. The implication is that transient climate change is better constrained than the equilibrium climate sensitivity, that is, models with different sensitivity might still show good agreement for projections on decadal time scales. Therefore, in the absence of unusual solar or volcanic activity, climate change is well constrained for the coming few decades, because differences in some feedbacks will only become important on long time scales (see also Section 10.5.4.5) and because over the next few decades, about half of the projected warming would occur as a result of radiative forcing being held constant at year 2000 levels (constant composition commitment, see Section 10.7).

Comparing observed thermal expansion with those AR4 20th-century simulations that have natural forcings indicates that ocean heat uptake in the models may be 25% larger than observed, although both could be consistent within their uncertainties. This difference is possibly due to a combination of overestimated ocean heat uptake in the models, observational uncertainties and limited data coverage in the deep ocean (see Sections 9.5.1.1, 9.5.2, and 9.6.2.1). Assigning this difference solely to overestimated ocean heat uptake, the TCR estimates could increase by 0.6°C at most. This is in line with evidence for a relatively weak dependence of TCR on ocean mixing based

on SCMs and EMICS (Allen et al., 2000; Knutti et al., 2005). The range of TCR covered by an ensemble with perturbations made to parameters in the atmospheric component of UKMO-HadCM3 is 1.5 to 2.6°C (M. Collins et al., 2006), similar to the AR4 AOGCM range. Therefore, based on the range covered by AOGCMs, and taking into account structural uncertainties and possible biases in transient heat uptake, TCR is assessed as very likely larger than 1°C and very likely greater than 3°C (i.e., 1.0°C to 3.0°C is a 10 to 90% range). Because the dependence of TCR on sensitivity becomes small as sensitivity increases, uncertainties in the upper bound on sensitivity only weakly affect the range of TCR (see Figure 10.25; Chapter 9; Knutti et al., 2005; Allen et al., 2006b). Observational constraints based on detection and attribution studies provide further support for this TCR range (see Section 9.6.2.3).

10.5.2.2 Earth System Models of Intermediate Complexity

Over the last few years, a range of climate models has been developed that are dynamically simpler and of lower resolution than comprehensive AOGCMs, although they might well be more 'complete' in terms of climate system components that are included. The class of such models, usually referred to as EMICs (Claussen et al., 2002), is very heterogeneous, ranging from zonally averaged ocean models coupled to energy balance models (Stocker et al., 1992a) or to statistical-dynamical models of the atmosphere (Petoukhov et al., 2000), to low resolution three-dimensional ocean models, coupled to energy balance or simple dynamical models of the atmosphere (Opsteegh et al., 1998; Edwards and Marsh, 2005; Müller et al., 2006). Some EMICs have a radiation code and prescribe greenhouse gases, while others use simplified equations to project radiative forcing from projected concentrations and abundances (Joos et al., 2001; see Chapter 2 and the TAR, Appendix II, Table II.3.11). Compared to comprehensive models, EMICs have hardly any computational constraints, and therefore many simulations can be performed. This allows for the creation of large ensembles, or the systematic exploration of long-term changes many centuries hence. However, because of the reduced resolution, only results at the largest scales (continental to global) are to be interpreted (Stocker and Knutti, 2003). Table 8.3 lists all EMICs used in this section, including their components and resolution.

A set of simulations is used to compare EMICs with AOGCMs for the SRES A1B scenario with stable atmospheric concentrations after year 2100 (see Section 10.7.2). For global mean temperature and sea level, the EMICs generally reproduce the AOGCM behaviour quite well. Two of the EMICs have values for climate sensitivity and transient response below the AOGCM range. However, climate sensitivity is a tuneable parameter in some EMICs, and no attempt was made here to match the range of response of the AOGCMs. The transient reduction of the MOC in most EMICs is also similar to the AOGCMs (see also Sections 10.3.4 and 10.7.2 and Figure 10.34), providing support that this class of models can be used for both long-term commitment projections (see Section 10.7) and probabilistic projections involving hundreds to thousands of simulations (see Section 10.5.4.5). If the forcing is strong enough, and lasts long enough (e.g., 4 × CO_2), a complete and irreversible collapse of the MOC can be induced in a few models. This is in line with earlier results using EMICs (Stocker and Schmittner, 1997; Rahmstorf and Ganopolski, 1999) or a coupled model (Stouffer and Manabe, 1999).

10.5.3 Global Mean Responses from Different Scenarios

The TAR projections with an SCM presented a range of warming over the 21st century for 35 SRES scenarios. The SRES emission scenarios assume that no climate policies are implemented (Nakićenović and Swart, 2000). The construction of Figure 9.14 of the TAR was pragmatic. It used a simple model tuned to AOGCMs that had a climate sensitivity within the long-standing range of 1.5°C to 4.5°C (e.g., Charney, 1979; and stated in earlier IPCC Assessment Reports). Models with climate sensitivity outside that range were discussed in the text and allowed the statement that the presented range was not the extreme range indicated by AOGCMs. The figure was based on a single anthropogenic-forcing estimate for 1750 to 2000, which is well within the range of values recommended by TAR Chapter 6, and is also consistent with that deduced from model simulations and the observed temperature record (TAR Chapter 12.). To be consistent with TAR Chapter 3, climate feedbacks on the carbon cycle were included. The resulting range of global mean temperature change from 1990 to 2100 given by the full set of SRES scenarios was 1.4°C to 5.8°C.

Since the TAR, several studies have examined the TAR projections and attempted probabilistic assessments. Allen et al. (2000) show that the forcing and simple climate model tunings used in the TAR give projections that are in agreement with the observationally constrained probabilistic forecast, reported in TAR Chapter 12.

As noted by Moss and Schneider (2000), giving only a range of warming results is potentially misleading unless some guidance is given as to what the range means in probabilistic terms. Wigley and Raper (2001) interpret the warming range in probabilistic terms, accounting for uncertainties in emissions, the climate sensitivity, the carbon cycle, ocean mixing and aerosol forcing. They give a 90% probability interval for 1990 to 2100 warming of 1.7°C to 4°C. As pointed out by Wigley and Raper (2001), such results are only as realistic as the assumptions upon which they are based. Key assumptions in this study were that each SRES scenario was equally likely, that 1.5°C to 4.5°C corresponds to the 90% confidence interval for the climate sensitivity, and that carbon cycle feedback uncertainties can be characterised by the full uncertainty range of abundance in 2100 of 490 to 1,260 ppm given in the TAR. The aerosol probability density function (PDF) was based on the uncertainty estimates given in the TAR together with constraints based on fitting the SCM to observed global and hemispheric mean temperatures.

The most controversial assumption in the Wigley and Raper (2001) probabilistic assessment was the assumption that each SRES scenario was equally likely. The *Special Report on Emissions Scenarios* (Nakićenović and Swart, 2000) states that 'No judgment is offered in this report as to the preference for any of the scenarios and they are not assigned probabilities of occurrence, neither must they be interpreted as policy recommendations.'

Webster et al. (2003) use the probabilistic emissions projections of Webster et al. (2002), which consider present uncertainty in SO_2 emissions, and allow the possibility of continuing increases in SO_2 emissions over the 21st century, as well as the declining emissions consistent with SRES scenarios. Since their climate model parameter PDFs were constrained by observations and are mutually dependent, the effect of the lower present-day aerosol forcing on the projections is not easy to separate, but there is no doubt that their projections tend to be lower where they admit higher and increasing SO_2 emissions.

Irrespective of the question of whether it is possible to assign probabilities to specific emissions scenarios, it is important to distinguish different sources of uncertainties in temperature projections up to 2100. Different emission scenarios arise because future greenhouse gas emissions are largely dependent on key socioeconomic drivers, technological development and political decisions. Clearly, one factor leading to different temperature projections is the choice of scenario. On the other hand, the 'response uncertainty' is defined as the range in projections for a particular emission scenario and arises from the limited knowledge of how the climate system will react to the anthropogenic perturbations. In the following, all given uncertainty ranges reflect the response uncertainty of the climate system and should therefore be seen as conditional on a specific emission scenario.

The following paragraphs describe the construction of the AR4 temperature projections for the six illustrative SRES scenarios, using the SCM tuned to 19 models from the MMD (see Section 8.8). These 19 tuned simple model versions have effective climate sensitivities in the range 1.9°C to 5.9°C. The simple model sensitivities are derived from the fully coupled 2 × and 4 × CO_2 1% yr^{-1} CO_2 increase AOGCM simulations and in some cases differ from the equilibrium slab ocean model sensitivities given in Table 8.2.

The SRES emission scenarios used here were designed to represent plausible futures assuming that no climate policies will be implemented. This chapter does not analyse any scenarios with explicit climate change mitigation policies. Still, there is a wide variation across these SRES scenarios in terms of anthropogenic emissions, such as those of fossil CO_2, CH_4 and SO_2 (Nakićenović and Swart, 2000) as shown in the top three panels of Figure 10.26. As a direct consequence of the different emissions, the projected concentrations vary widely for the six illustrative SRES scenarios (see panel rows four to six in Figure 10.26 for the concentrations of the main greenhouse gases, CO_2, CH_4 and N_2O). These results incorporate the effect of carbon cycle uncertainties (see Section 10.4.1), which were not explored with the SCM in the TAR. Projected CH_4 concentrations are influenced by the temperature-dependent water vapour feedback on the lifetime of CH_4.

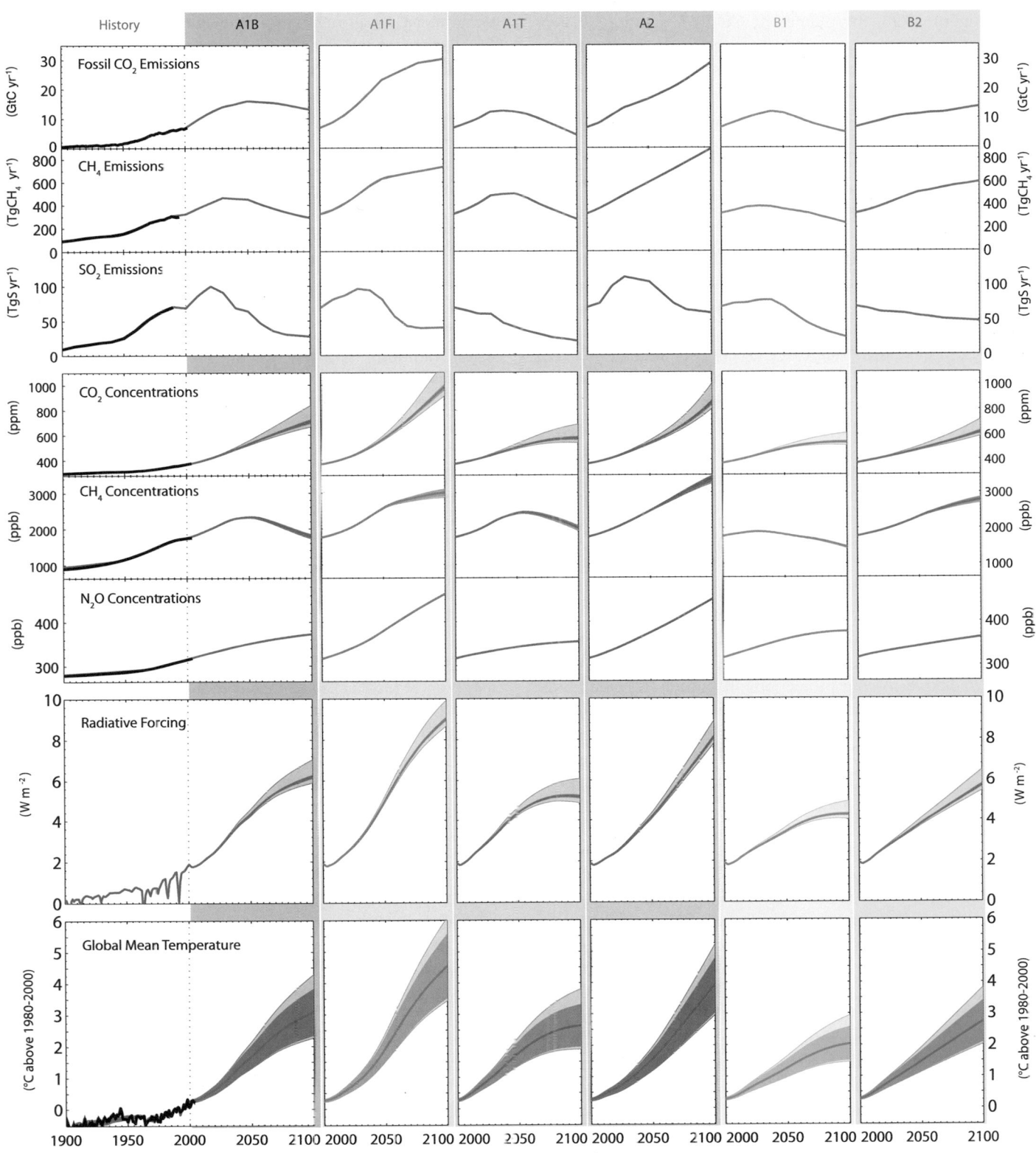

Figure 10.26. *Fossil CO_2, CH_4 and SO_2 emissions for six illustrative SRES non-mitigation emission scenarios, their corresponding CO_2, CH_4 and N_2O concentrations, radiative forcing and global mean temperature projections based on an SCM tuned to 19 AOGCMs. The dark shaded areas in the bottom temperature panel represent the mean ±1 standard deviation for the 19 model tunings. The lighter shaded areas depict the change in this uncertainty range, if carbon cycle feedbacks are assumed to be lower or higher than in the medium setting. Mean projections for mid-range carbon cycle assumptions for the six illustrative SRES scenarios are shown as thick coloured lines. Historical emissions (black lines) are shown for fossil and industrial CO_2 (Marland et al., 2005), for SO_2 (van Aardenne et al., 2001) and for CH_4 (van Aardenne et al., 2001, adjusted to Olivier and Berdowski, 2001). Observed CO_2, CH_4 and N_2O concentrations (black lines) are as presented in Chapter 6. Global mean temperature results from the SCM for anthropogenic and natural forcing compare favourably with 20th-century observations (black line) as shown in the lower left panel (Folland et al., 2001; Jones et al., 2001; Jones and Moberg, 2003).*

In Figure 10.26, the plumes of CO_2 concentration reflect high and low carbon cycle feedback settings of the applied SCM. Their derivation is described as follows. The carbon cycle model in the SCM used here (Model for the Assessment of Greenhouse-gas Induced Climate Change: MAGICC) includes a number of climate-related carbon cycle feedbacks driven by global mean temperature. The parametrization of the overall effect of carbon cycle feedbacks is tuned to the more complex and physically realistic carbon cycle models of the C[4]MIP (Friedlingstein et al, 2006; see also Section 10.4) and the results are comparable to the BERN-CC model results across the six illustrative scenarios. This allows the SCM to produce projections of future CO_2 concentration change that are consistent with state-of-the-art carbon cycle model results. Specifically, the C[4]MIP range of CO_2 concentrations for the A2 emission scenario in 2100 is 730 to 1,020 ppm, while the SCM results presented here show an uncertainty range of 806 ppm to 1,008 ppm. The lower bound of this SCM uncertainty range is the mean minus one standard deviation for low carbon cycle feedback settings and the 19 AOGCM tunings, while the upper bound represents the mean plus one standard deviation for high carbon cycle settings. For comparison, the 90% confidence interval from Wigley and Raper (2001) is 770 to 1,090 ppm. The simple model CO_2 concentration projections can be slightly higher than under the C[4]MIP because the SCM's carbon cycle is driven by the full temperature changes in the A2 scenario, while the C[4]MIP values are driven by the component of A2 climate change due to CO_2 alone.

The radiative forcing projections in Figure 10.26 combine anthropogenic and natural (solar and volcanic) forcing. The forcing plumes reflect primarily the sensitivity of the forcing to carbon cycle uncertainties. Results are based on a forcing of 3.71 W m^{-2} for a doubling of the atmospheric CO_2 concentration. The anthropogenic forcing is based on Table 2.12 but uses a value of -0.8 W m^{-2} for the present-day indirect aerosol forcing. Solar forcing for the historical period is prescribed according to Lean et al. (1995) and volcanic forcing according to Ammann et al. (2003). The historical solar forcing series is extended into the future using its average over the most recent 22 years. The volcanic forcing is adjusted to have a zero mean over the past 100 years and the anomaly is assumed to be zero for the future. In the TAR, the anthropogenic forcing was used alone even though the projections started in 1765. There are several advantages of using both natural and anthropogenic forcing for the past. First, this was done by most of the AOGCMs the simple models are emulating. Second, it allows the simulations to be compared with observations. Third, the warming commitments accrued over the instrumental period are reflected in the projections. The disadvantage of including natural forcing is that the warming projections in 2100 are dependent to a few tenths of a degree on the necessary assumptions made about the natural forcing (Bertrand et al., 2002). These assumptions include how the natural forcing is projected into the future and whether to reference the volcanic forcing to a past reference

period mean value. In addition, the choice of data set for both solar and volcanic forcing affects the results (see Section 2.7 for discussion about uncertainty in natural forcings).

The temperature projections for the six illustrative scenarios are shown in the bottom panel of Figure 10.26. Model results are shown as anomalies from the mean of observations (Folland et al., 2001; Jones et al., 2001; Jones and Moberg, 2003) over the 1980 to 2000 period and the corresponding observed temperature anomalies are shown for comparison. The inner (darker) plumes show the ±1 standard deviation uncertainty due to the 19 model tunings and the outer (lighter) plumes show results for the corresponding high and low carbon cycle settings. Note that the asymmetry in the carbon cycle uncertainty causes global mean temperature projections to be skewed towards higher warming.

Considering only the mean of the SCM results with mid-range carbon cycle settings, the projected global mean temperature rise above 1980 to 2000 levels for the lower-emission SRES scenario B1 is 2.0°C in 2100. For a higher-emission scenario, for example, the SRES A2 scenario, the global mean temperature is projected to rise by 3.9°C above 1980 to 2000 levels in 2100. This clear difference in projected mean warming highlights the importance of assessing different emission scenarios separately. As mentioned above, the 'response uncertainty' is defined as the range in projections for a particular emission scenario. For the A2 emission scenario, the temperature change projections with the SCM span a ±1 standard deviation range of about 1.8°C, from 3.0°C to 4.8°C above 1980 to 2000 levels in 2100. If carbon cycle feedbacks are considered to be low, the lower end of this range decreases only slightly and is unchanged to one decimal place. For the higher carbon cycle feedback settings, the upper bound of the ±1 standard deviation range increases to 5.2°C. For lower-emission scenarios, this uncertainty range is smaller. For example, the B1 scenario projections span a range of about 1.4°C, from 1.5°C to 2.9°C, including carbon cycle uncertainties. The corresponding results for the medium-emission scenario A1B are 2.3°C to 4.3°C, and for the higher-emission scenario A1FI, they are 3.4°C to 6.1°C. Note that these uncertainty ranges are not the minimum to maximum bounds of the projected warming across all SCM runs, which are higher, namely 2.7°C to 7.1°C for the A2 scenario and 1.3°C to 4.2°C for the B1 scenario (not shown).

The SCM results presented here are a sensitivity study with different model tunings and carbon cycle feedback parameters. Note that forcing uncertainties have not been assessed and that the AOGCM model results available for SCM tuning may not span the full range of possible climate response. For example, studies that constrain forecasts based on model fits to historic or present-day observations generally allow for a somewhat wider 'response uncertainty' (see Section 10.5.4). The concatenation of all such uncertainties would require a probabilistic approach because the extreme ranges have low probability. A synthesis of the uncertainty in global temperature increase by the year 2100 is provided in Section 10.5.4.6.

10.5.4 Sampling Uncertainty and Estimating Probabilities

Uncertainty in the response of an AOGCM arises from the effects of internal variability, which can be sampled in isolation by creating ensembles of simulations of a single model using alternative initial conditions, and from modelling uncertainties, which arise from errors introduced by the discretization of the equations of motion on a finite resolution grid, and the parametrization of sub-grid scale processes (radiative transfer, cloud formation, convection, etc). Modelling uncertainties are manifested in alternative structural choices (for example, choices of resolution and the basic physical assumptions on which parametrizations are based), and in the values of poorly constrained parameters within parametrization schemes. Ensemble approaches are used to quantify the effects of uncertainties arising from variations in model structure and parameter settings. These are assessed in Sections 10.5.4.1 to 10.5.4.3, followed by a discussion of observational constraints in Section 10.5.4.4 and methods used to obtain probabilistic predictions in Sections 10.5.4.5 to 10.5.4.7.

While ensemble projections carried out to date give a wide range of responses, they do not sample all possible sources of modelling uncertainty. For example, the AR4 multi-model ensemble relies on specified concentrations of CO_2, thus neglecting uncertainties in carbon cycle feedbacks (see Section 10.4.1), although this can be partially addressed by using less detailed models to extrapolate the AOGCM results (see Section 10.5.3). More generally, the set of available models may share fundamental inadequacies, the effects of which cannot be quantified (Kennedy and O'Hagan, 2001). For example, climate models currently implement a restricted approach to the parametrization of sub-grid scale processes, using deterministic bulk formulae coupled to the resolved flow exclusively at the grid scale. Palmer et al. (2005) argue that the outputs of parametrization schemes should be sampled from statistical distributions consistent with a range of possible sub-grid scale states, following a stochastic approach that has been tried in numerical weather forecasting (e.g., Buizza et al., 1999; Palmer, 2001). The potential for missing or inadequately parametrized processes to broaden the simulated range of future changes is not clear, however, this is an important caveat for the results discussed below.

10.5.4.1 The Multi-Model Ensemble Approach

The use of ensembles of AOGCMs developed at different modelling centres has become established in climate prediction/projection on both seasonal-to-interannual and centennial time scales. To the extent that simulation errors in different AOGCMs are independent, the mean of the ensemble can be expected to outperform individual ensemble members, thus providing an improved 'best estimate' forecast. Results show this to be the case, both in verification of seasonal forecasts (Palmer et al., 2004; Hagedorn et al., 2005) and of the present-day climate from long term simulations (Lambert and Boer, 2001). By

sampling modelling uncertainties, ensembles of AOGCMs should provide an improved basis for probabilistic projections compared with ensembles of a single model sampling only uncertainty in the initial state (Palmer et al., 2005). However, members of a multi-model ensemble share common systematic errors (Lambert and Boer, 2001), and cannot span the full range of possible model configurations due to resource constraints. Verification of future climate change projections is not possible, however, Räisänen and Palmer (2001) used a 'perfect model approach' (treating one member of an ensemble as truth and predicting its response using the other members) to show that the hypothetical economic costs associated with climate events can be reduced by calculating the probability of the event across the ensemble, rather than using a deterministic prediction from an individual ensemble member.

An additional strength of multi-model ensembles is that each member is subjected to careful testing in order to obtain a plausible and stable control simulation, although the process of tuning model parameters to achieve this (Section 8.1.3.1) involves subjective judgement, and is not guaranteed to identify the optimum location in the model parameter space.

10.5.4.2 Perturbed Physics Ensembles

The AOGCMs featured in Section 10.5.2 are built by selecting components from a pool of alternative parametrizations, each based on a given set of physical assumptions and including a number of uncertain parameters. In principle, the range of predictions consistent with these components could be quantified by constructing very large ensembles with systematic sampling of multiple options for parametrization schemes and parameter values, while avoiding combinations likely to double-count the effect of perturbing a given physical process. Such an approach has been taken using simple climate models and EMICs (Wigley and Raper, 2001; Knutti et al., 2002), and Murphy et al. (2004) and Stainforth et al. (2005) describe the first steps in this direction using AOGCMs, constructing large ensembles by perturbing poorly constrained parameters in the atmospheric component of UKMO-HadCM3 coupled to a mixed layer ocean. These experiments quantify the range of equilibrium responses to doubled atmospheric CO_2 consistent with uncertain parameters in a single GCM. Murphy et al. (2004) perturbed 29 parameters one at a time, assuming that effects of individual parameters were additive but making a simple allowance for additional uncertainty introduced by nonlinear interactions. They find a probability distribution for climate sensitivity with a 5 to 95% range of 2.4°C to 5.4°C when weighting the models with a broadly based metric of the agreement between simulated and observed climatology, compared to 1.9°C to 5.3°C when all model versions are assumed equally reliable (Box 10.2, Figure 1c).

Stainforth et al. (2005) deployed a distributed computing approach (Allen, 1999) to run a very large ensemble of 2,578 simulations sampling combinations of high, intermediate and low values of six parameters known to affect climate sensitivity. They find climate sensitivities ranging from 2°C to

11°C, with 4.2% of model versions exceeding 8°C, and show that the high-sensitivity models cannot be ruled out, based on a comparison with surface annual mean climatology. By utilising multivariate linear relationships between climate sensitivity and spatial fields of several present-day observables, the 5 to 95% range of climate sensitivity is estimated at 2.2°C to 6.8°C from the same data set (Piani et al., 2005; Box 10.2 Figure 1c). In this ensemble, Knutti et al. (2006) find a strong relationship between climate sensitivity and the amplitude of the seasonal cycle in surface temperature in the present-day simulations. Most of the simulations with high sensitivities overestimate the observed amplitude. Based on this relationship, the 5 to 95% range of climate sensitivity is estimated at 1.5°C to 6.4°C (Box 10.2, Figure 1c). The differences between the PDFs in Box 10.2, Figure 1c, which are all based on the same climate model, reflect uncertainties in methodology arising from choices of uncertain parameters, their expert-specified prior distributions and alternative applications of observational constraints. They do not account for uncertainties associated with changes in ocean circulation, and do not account for structural model errors (Smith, 2002; Goldstein and Rougier, 2004)

Annan et al. (2005a) use an ensemble Kalman Filter technique to obtain uncertainty ranges for model parameters in an EMIC subject to the constraint of minimising simulation errors with respect to a set of climatological observations. Using this method, Hargreaves and Annan (2006) find that the risk of a collapse in the Atlantic MOC (in response to increasing CO_2) depends on the set of observations to which the EMIC parameters are tuned. Section 9.6.3 assesses perturbed physics studies of the link between climate sensitivity and cooling during the Last Glacial Maximum (Annan et al., 2005b; Schneider von Deimling et al., 2006).

10.5.4.3 Diagnosing Drivers of Uncertainty from Ensemble Results

Figure 10.27a shows the agreement between annual changes simulated by members of the AR4 multi-model ensemble for 2080 to 2099 relative to 1980 to 1999 for the A1B scenario, calculated as in Räisänen (2001). For precipitation, the agreement increases with spatial scale. For surface temperature, the agreement is high even at local scales, indicating the robustness of the simulated warming (see also Figure 10.8, discussed in Section 10.3.2.1). Differences in model formulation are the dominant contributor to ensemble spread, though the role of internal variability increases at smaller scales (Figure 10.27b). The agreement between AR4 ensemble members is slightly higher compared with the earlier CMIP2 ensemble of Räisänen (2001) (also reported in the TAR), and internal variability explains a smaller fraction of the ensemble spread. This is expected, given the larger forcing and responses in the A1B scenario for 2080 to 2099 compared to the transient response to doubled CO_2 considered by Räisänen (2001), although the use of an updated set of models may also contribute. For seasonal changes, internal variability is found to be comparable with model differences as a source of

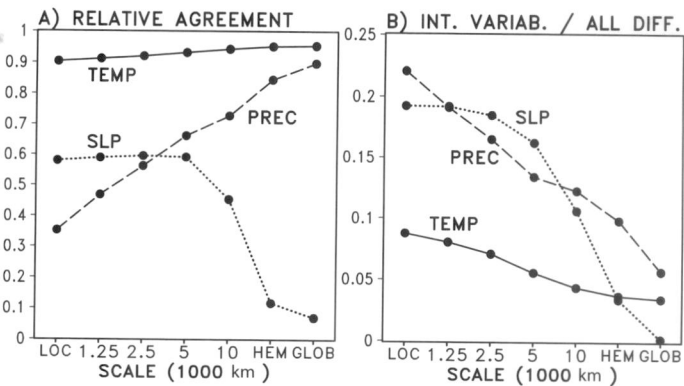

Figure 10.27. *Statistics of annual mean responses to the SRES A1B scenario, for 2080 to 2099 relative to 1980 to 1999, calculated from the 21-member AR4 multi-model ensemble using the methodology of Räisänen (2001). Results are expressed as a function of horizontal scale on the x axis ('Loc': grid box scale; 'Hem': hemispheric scale; 'Glob': global mean) plotted against the y axis showing (a) the relative agreement between ensemble members, a dimensionless quantity defined as the square of the ensemble-mean response (corrected to avoid sampling bias) divided by the mean squared response of individual ensemble members, and (b) the dimensionless fraction of internal variability relative to the ensemble variance of responses. Values are shown for surface air temperature, precipitation and sea level pressure. The low agreement of SLP changes at hemispheric and global scales reflects problems with the conservation of total atmospheric mass in some of the models, however, this has no practical significance because SLP changes at these scales are extremely small.*

uncertainty in local precipitation and SLP changes (although not for surface temperature) in both multi-model and perturbed physics ensembles (Räisänen, 2001; Murphy et al., 2004). Consequently the local seasonal changes for precipitation and SLP are not consistent in the AR4 ensemble over large areas of the globe (i.e., the multi-model mean change does not exceed the ensemble standard deviation; see Figure 10.9), whereas surface temperature changes are consistent almost everywhere, as discussed in Section 10.3.2.1.

Wang and Swail (2006b) examine the relative importance of internal variability, differences in radiative forcing and model differences in explaining the transient response of ocean wave height using three AOGCMs each run for three plausible forcing scenarios, and find model differences to be the largest source of uncertainty in the simulated changes.

Selten et al. (2004) report a 62-member initial condition ensemble of simulations of 1940 to 2080 including natural and anthropogenic forcings. They find an individual member that reproduces the observed trend in the NAO over the past few decades, but no trend in the ensemble mean, and suggest that the observed change can be explained through internal variability associated with a mode driven by increases in precipitation over the tropical Indian Ocean. Terray et al. (2004) find that the ARPEGE coupled ocean-atmosphere model shows small increases in the residence frequency of the positive phase of the NAO in response to SRES A2 and B2 forcing, whereas larger increases are found when SST changes prescribed from the coupled experiments are used to drive a version of the atmosphere model with enhanced resolution over the North Atlantic and Europe (Gibelin and Déqué, 2003).

Figure 10.25 compares global mean transient and equilibrium changes simulated by the AR4 multi-model ensembles against perturbed physics ensembles (M. Collins et al., 2006; Webb et al., 2006) designed to produce credible present-day simulations while sampling a wide range of multiple parameter perturbations and climate sensitivities. The AR4 ensembles partially sample structural variations in model components, whereas the perturbed physics ensembles sample atmospheric parameter uncertainties for a fixed choice of model structure. The results show similar relationships between TCR, climate sensitivity and precipitation change in both types of ensemble. The perturbed physics ensembles contain several members with sensitivities higher than the multi-model range, while some of the multi-model transient simulations give TCR values slightly below the range found in the perturbed physics ensemble (Figure 10.25a,b).

Soden and Held (2006) find that differences in cloud feedback are the dominant source of uncertainty in the transient response of surface temperature in the AR4 ensemble (see also Section 8.6.3.2), as in previous IPCC assessments. Webb et al. (2006) compare equilibrium radiative feedbacks in a 9-member multi-model ensemble against those simulated in a 128-member perturbed physics ensemble with multiple parameter perturbations. They find that the ranges of climate sensitivity in both ensembles are explained mainly by differences in the response of shortwave cloud forcing in areas where changes in low-level clouds predominate. Bony and Dufresne (2005) find that marine boundary layer clouds in areas of large-scale subsidence provide the largest source of spread in tropical cloud feedbacks in the AR4 ensemble. Narrowing the uncertainty in cloud feedback may require both improved parametrizations of cloud microphysical properties (e.g., Tsushima et al., 2006) and improved representations of cloud macrophysical properties, through improved parametrizations of other physical processes (e.g., Williams et al., 2001) and/or increases in resolution (Palmer, 2005).

10.5.4.4 Observational Constraints

A range of observables has been used since the TAR to explore methods for constraining uncertainties in future climate change in studies using simple climate models, EMICs and AOGCMs. Probabilistic estimates of global climate sensitivity have been obtained from the historical transient evolution of surface temperature, upper-air temperature, ocean temperature, estimates of the radiative forcing, satellite data, proxy data over the last millennium, or a subset thereof (Wigley et al., 1997a; Tol and De Vos, 1998; Andronova and Schlesinger, 2001; Forest et al., 2002; Gregory et al., 2002a; Knutti et al., 2002, 2003; Frame et al., 2005; Forest et al., 2006; Forster and Gregory, 2006; Hegerl et al., 2006; see Section 9.6). Some of these studies also constrain the transient response to projected future emissions (see section 10.5.4.5). For climate sensitivity, further probabilistic estimates have been obtained using statistical measures of the correspondence between simulated and observed fields of present-day climate (Murphy et al.,

2004; Piani et al., 2005), the climatological seasonal cycle of surface temperature (Knutti et al., 2006) and the response to palaeoclimatic forcings (Annan et al., 2005b; Schneider von Deimling et al., 2006). For the purpose of constraining regional climate projections, spatial averages or fields of time-averaged regional climate have been used (Giorgi and Mearns, 2003; Tebaldi et al., 2004, 2005; Laurent and Cai, 2007), as have past regional- or continental-scale trends in surface temperature (Greene et al., 2006; Stott et al., 2006a).

Further observables have been suggested as potential constraints on future changes, but are not yet used in formal probabilistic estimates. These include measures of climate variability related to cloud feedbacks (Bony et al., 2004; Bony and Dufresne, 2005; Williams et al., 2005), radiative damping of the seasonal cycle (Tsushima et al., 2005), the relative entropy of simulated and observed surface temperature variations (Shukla et al., 2006), major volcanic eruptions (Wigley et al., 2005; Yokohata et al., 2005; see Section 9.6) and trends in multiple variables derived from reanalysis data sets (Lucarini and Russell, 2002).

Additional constraints could also be found, for example, from evaluation of ensemble climate prediction systems on shorter time scales for which verification data exist. These could include assessment of the reliability of seasonal to interannual probabilistic forecasts (Palmer et al., 2004; Hagedorn et al., 2005) and the evaluation of model parametrizations in short-range weather predictions (Phillips et al., 2004; Palmer, 2005). Annan and Hargreaves (2006) point out the potential for narrowing uncertainty by combining multiple lines of evidence. This will require objective quantification of the impact of different constraints and their degree of independence, estimation of the effects of structural modelling errors and the development of comprehensive probabilistic frameworks in which to combine these elements (e.g., Rougier, 2007).

10.5.4.5 Probabilistic Projections - Global Mean

A number of methods for providing probabilistic climate change projections, both for global means (discussed in this section) and geographical depictions (discussed in the following section) have emerged since the TAR.

Methods of constraining climate sensitivity using observations of present-day climate are discussed in Section 10.5.4.2. Results from both the AR4 multi-model ensemble and from perturbed physics ensembles suggest a very low probability for a climate sensitivity below 2°C, despite exploring the effects of a wide range of alternative modelling assumptions on the global radiative feedbacks arising from lapse rate, water vapour, surface albedo and cloud (Bony et al., 2006; Soden and Held, 2006; Webb et al., 2006; Box 10.2). However, exclusive reliance on AOGCM ensembles can be questioned on the basis that models share components, and therefore errors, and may not sample the full range of possible outcomes (e.g., Allen and Ingram, 2002).

Observationally constrained probability distributions for climate sensitivity have also been derived from physical relationships based on energy balance considerations, and from instrumental observations of historical changes during the past 50 to 150 years or proxy reconstructions of surface temperature during the past millennium (Section 9.6). The results vary according to the choice of verifying observations, the forcings considered and their specified uncertainties, however, all these studies report a high upper limit for climate sensitivity, with the 95th percentile of the distributions invariably exceeding 6°C (Box 10.2). Frame et al. (2005) demonstrate that uncertainty ranges for sensitivity are dependent on the choices made about prior distributions of uncertain quantities before the observations are applied. Frame et al. (2005) and Piani et al. (2005) show that many observable variables are likely to scale inversely with climate sensitivity, implying that projections of quantities that are inversely related to sensitivity will be more strongly constrained by observations than climate sensitivity itself, particularly with respect to the estimated upper limit (Allen et al., 2006b).

In the case of transient climate change, optimal detection techniques have been used to determine factors by which hindcasts of global surface temperature from AOGCMs can be scaled up or down while remaining consistent with past changes, accounting for uncertainty due to internal variability (Section 9.4.1.6). Uncertainty is propagated forward in time by assuming that the fractional error found in model hindcasts of global mean temperature change will remain constant in projections of future changes. Using this approach, Stott and Kettleborough (2002) find that probabilistic projections of global mean temperature derived from UKMO-HadCM3 simulations were insensitive to differences between four representative SRES emissions scenarios over the first few decades of the 21st century, but that much larger differences emerged between the response to different SRES scenarios by the end of the 21st century (see also Section 10.5.3 and Figure 10.28). Stott et al. (2006b) show that scaling the responses of three models with different sensitivities brings their projections into better agreement. Stott et al. (2006a) extend their approach to obtain probabilistic projections of future warming averaged over continental-scale regions under the SRES A2 scenario. Fractional errors in the past continental warming simulated by UKMO-HadCM3 are used to scale future changes, yielding wide uncertainty ranges, notably for North America and Europe where the 5 to 95% ranges for warming during the 21st century are 2°C to 12°C and 2°C to 11°C respectively. These estimates do not account for potential constraints arising from regionally differentiated warming rates. Tighter ranges of 4°C to 8°C for North America and 4°C to 7°C for Europe are obtained if fractional errors in past global mean temperature are used to scale the future continental changes, although this neglects uncertainty in the relationship between global and regional temperature changes.

Allen and Ingram (2002) suggest that probabilistic projections for some variables may be made by searching for 'emergent constraints'. These are relationships between variables that can be directly constrained by observations, such as global surface temperature, and variables that may be indirectly constrained by establishing a consistent, physically

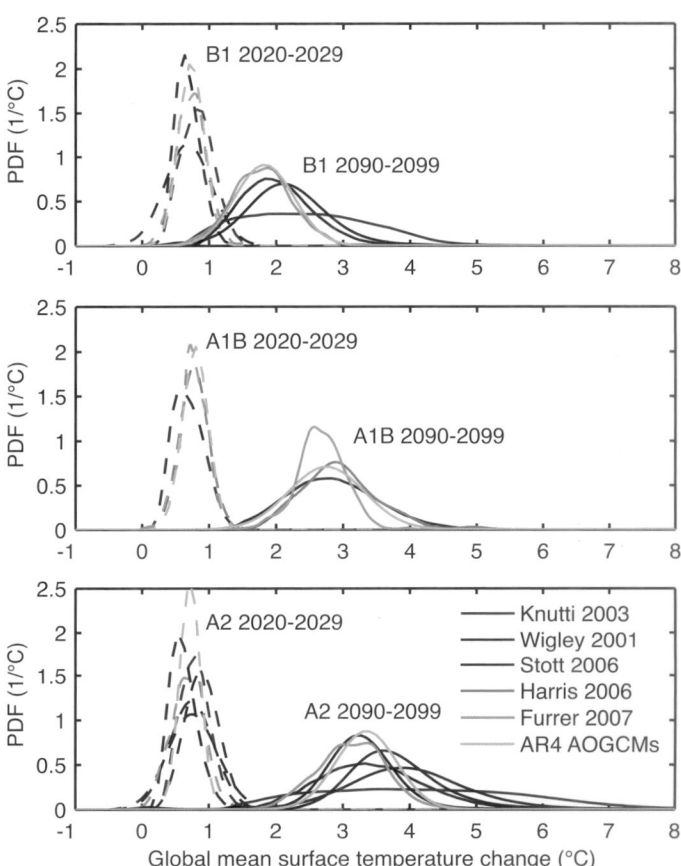

Figure 10.28. *Probability density functions from different studies for global mean temperature change for the SRES scenarios B1, A1B and A2 and for the decades 2020 to 2029 and 2090 to 2099 relative to the 1980 to 1999 average (Wigley and Raper, 2001; Knutti et al., 2002; Furrer et al., 2007; Harris et al., 2006; Stott et al., 2006b). A normal distribution fitted to the multi-model ensemble is shown for comparison.*

based relationship which holds across a wide range of models. They present an example in which future changes in global mean precipitation are constrained using a probability distribution for global temperature obtained from a large EMIC ensemble (Forest et al., 2002) and a relationship between precipitation and temperature obtained from multi-model ensembles of the response to doubled atmospheric CO_2. These methods are designed to produce distributions constrained by observations, and are relatively model independent (Allen and Stainforth, 2002; Allen et al., 2006a). This can be achieved provided the inter-variable relationships are robust to alternative modelling assumptions Piani et al. (2005) and Knutti et al. (2006) (described in Section 10.5.4.2) follow this approach, noting that in these cases the inter-variable relationships are derived from perturbed versions of a single model, and need to be confirmed using other models.

A synthesis of published probabilistic global mean projections for the SRES scenarios B1, A1B and A2 is given in Figure 10.28. Probability density functions are given for short-term projections (2020–2030) and the end of the century (2090–2100). For comparison, normal distributions fitted to results from AOGCMs in the multi-model archive (see Section

10.3.1) are also given, although these curve fits should not be regarded as PDFs. The five methods of producing PDFs are all based on different models and/or techniques, described in Section 10.5. In short, Wigley and Raper (2001) use a large ensemble of a simple model with expert prior distributions for climate sensitivity, ocean heat uptake, sulphate forcing and the carbon cycle, without applying constraints. Knutti et al. (2002, 2003) use a large ensemble of EMIC simulations with non-informative prior distributions, consider uncertainties in climate sensitivity, ocean heat uptake, radiative forcing and the carbon cycle, and apply observational constraints. Neither method considers natural variability explicitly. Stott et al. (2006b) apply the fingerprint scaling method to AOGCM simulations to obtain PDFs which implicitly account for uncertainties in forcing, climate sensitivity and internal unforced as well as forced natural variability. For the A2 scenario, results obtained from three different AOGCMs are shown, illustrating the extent to which the Stott et al. PDFs depend on the model used. Harris et al. (2006) obtain PDFs by boosting a 17-member perturbed physics ensemble of the UKMO-HadCM3 model using scaled equilibrium responses from a larger ensemble of simulations. Furrer et al. (2007) use a Bayesian method described in Section 10.5.4.7 to calculate PDFs from the AR4 multi-model ensemble. The Stott et al. (2006b), Harris et al. (2006) and Furrer et al. (2007) methods neglect carbon cycle uncertainties.

Two key points emerge from Figure 10.28. For the projected short-term warming (i) there is more agreement among models and methods (narrow width of the PDFs) compared to later in

the century (wider PDFs), and (ii) the warming is similar across different scenarios, compared to later in the century where the choice of scenario significantly affects the projections. These conclusions are consistent with the results obtained with SCMs (Section 10.5.3).

Additionally, projection uncertainties increase close to linearly with temperature in most studies. The different methods show relatively good agreement in the shape and width of the PDFs, but with some offsets due to different methodological choices. Only Stott et al. (2006b) account for variations in future natural forcing, and hence project a small probability of cooling over the next few decades not seen in the other PDFs. The results of Knutti et al. (2003) show wider PDFs for the end of the century because they sample uniformly in climate sensitivity (see Section 9.6.2 and Box 10.2). Resampling uniformly in observables (Frame et al., 2005) would bring their PDFs closer to the others. In sum, probabilistic estimates of uncertainties for the next few decades seem robust across a variety of models and methods, while results for the end of the century depend on the assumptions made.

10.5.4.6 Synthesis of Projected Global Temperature at Year 2100

All available estimates for projected warming by the end of the 21st century are summarised in Figure 10.29 for the six SRES non-intervention marker scenarios. Among the various techniques, the AR4 AOGCM ensemble provides the most

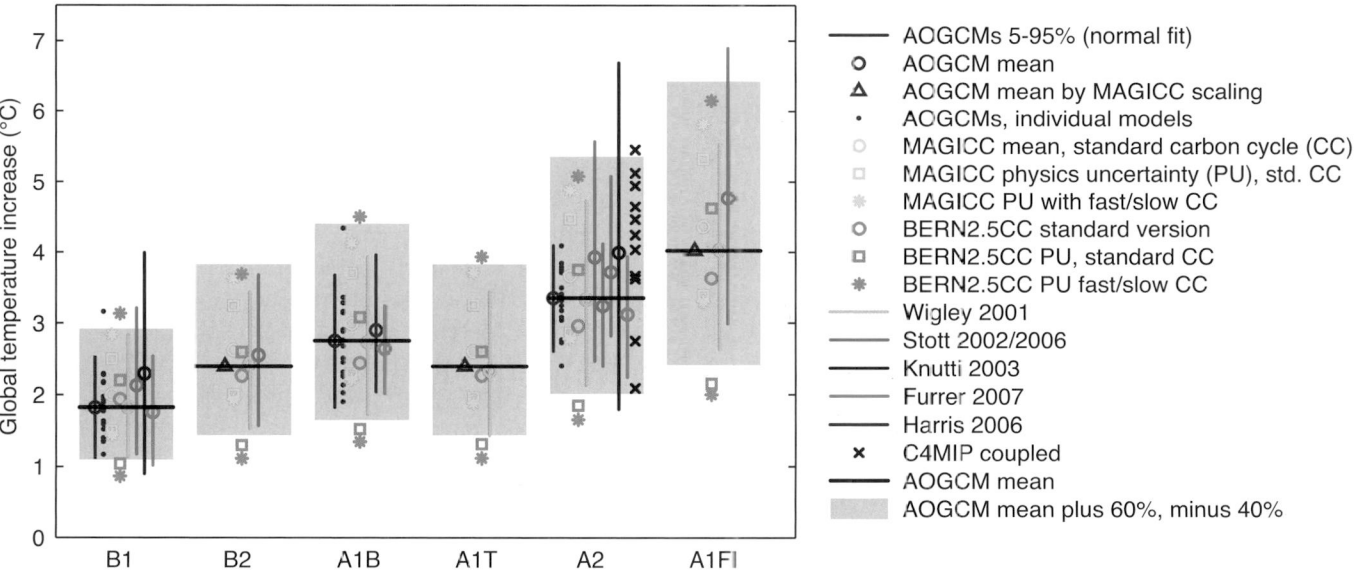

Figure 10.29. Projections and uncertainties for global mean temperature increase in 2090 to 2099 (relative to the 1980 to 1999 average) for the six SRES marker scenarios. The AOGCM means and the uncertainty ranges of the mean –40% to +60% are shown as black horizontal solid lines and grey bars, respectively. For comparison, results are shown for the individual models (red dots) of the multi-model AOGCM ensemble for B1, A1B and A2, with a mean and 5 to 95% range (red line and circle) from a fitted normal distribution. The AOGCM mean estimates for B2, A1T and A1FI (red triangles) are obtained by scaling the A1B AOGCM mean with ratios obtained from the SCM (see text). The mean (light green circle) and one standard deviation (light green square) of the MAGICC SCM tuned to all AOGCMs (representing the physics uncertainty) are shown for standard carbon cycle settings, as well as for a slow and fast carbon cycle assumption (light green stars). Similarly, results from the BERN2.5CC EMIC are shown for standard carbon cycle settings and for climate sensitivities of 3.2°C (AOGCM average, dark green circle), 1.5°C and 4.5°C (dark green squares). High climate sensitivity/low carbon cycle and low climate sensitivity/high carbon cycle combinations are shown as dark green stars. The 5 to 95% ranges (vertical lines) and medians (circles) are shown from probabilistic methods (Wigley and Raper, 2001; Stott and Kettleborough, 2002; Knutti et al., 2003; Furrer et al., 2007; Harris et al., 2006; Stott et al., 2006b). Individual model results are shown for the C4MIP models (blue crosses, see Figure 10.20).

sophisticated set of models in terms of the range of processes included and consequent realism of the simulations compared to observations (see Chapters 8 and 9). On average, this ensemble projects an increase in global mean surface air temperature of 1.8°C, 2.8°C and 3.4°C in the B1, A1B and A2 scenarios, respectively, by 2090 to 2099 relative to 1980 to 1999 (note that in Table 10.5, the years 2080 to 2099 were used for those globally averaged values to be consistent with the comparable averaging period for the geographic plots in Section 10.3; this longer averaging period smoothes spatial noise in the geographic plots). A scaling method is used to estimate AOGCM mean results for the three missing scenarios B2, A1T and A1FI. The ratio of the AOGCM mean values for B1 relative to A1B and A2 relative to A1B are almost identical to the ratios obtained with the MAGICC SCM, although the absolute values for the SCM are higher. Thus, the AOGCM mean response for the scenarios B2, A1T and A1FI can be estimated as 2.4°C, 2.4°C and 4.0°C by multiplying the AOGCM A1B mean by the SCM-derived ratios B2/A1B, A1T/A1B and A1FI/A1B, respectively (for details see Appendix 10.A.1).

The AOGCMs cannot sample the full range of possible warming, in particular because they do not include uncertainties in the carbon cycle. In addition to the range derived directly from the AR4 multi-model ensemble, Figure 10.29 depicts additional uncertainty estimates obtained from published probabilistic methods using different types of models and observational constraints: the MAGICC SCM and the BERN2.5CC coupled climate-carbon cycle EMIC tuned to different climate sensitivities and carbon cycle settings, and the C⁴MIP coupled climate-carbon cycle models. Based on these results, the future increase in global mean temperature is likely to fall within –40 to +60% of the multi-model AOGCM mean warming simulated for each scenario. This range results from an expert judgement of the multiple lines of evidence presented in Figure 10.29, and assumes that the models approximately capture the range of uncertainties in the carbon cycle. The range is well constrained at the lower bound since climate sensitivity is better constrained at the low end (see Box 10.2), and carbon cycle uncertainty only weakly affects the lower bound. The upper bound is less certain as there is more variation across the different models and methods, partly because carbon cycle feedback uncertainties are greater with larger warming. The uncertainty ranges derived from the above percentages for the warming by 2090 to 2099 relative to 1980 to 1999 are 1.1°C to 2.9°C, 1.4°C to 3.8°C, 1.7°C to 4.4°C, 1.4°C to 3.8°C, 2.0°C to 5.4°C and 2.4°C to 6.4°C for the scenarios B1, B2, A1B, A1T, A2 and A1FI, respectively. It is not appropriate to compare the lowest and highest values across these ranges against the single range given in the TAR, because the TAR range resulted only from projections using an SCM and covered all SRES scenarios, whereas here a number of different and independent modelling approaches are combined to estimate ranges for the six illustrative scenarios separately. Additionally, in contrast to the TAR, carbon cycle uncertainties are now included in these ranges. These uncertainty ranges include only anthropogenically forced changes.

10.5.4.7   *Probabilistic Projections - Geographical Depictions*

Tebaldi et al. (2005) present a Bayesian approach to regional climate prediction, developed from the ideas of Giorgi and Mearns (2002, 2003). Non-informative prior distributions for regional temperature and precipitation are updated using observations and results from AOGCM ensembles to produce probability distributions of future changes. Key assumptions are that each model and the observations differ randomly and independently from the true climate, and that the weight given to a model prediction should depend on the bias in its present-day simulation and its degree of convergence with the weighted ensemble mean of the predicted future change. Lopez et al. (2006) apply the Tebaldi et al. (2005) method to a 15-member multi-model ensemble to predict future changes in global surface temperature under a 1% yr⁻¹ increase in atmospheric CO_2. They compare it with the method developed by Allen et al. (2000) and Stott and Kettleborough (2002) (ASK), which aims to provide relatively model independent probabilities consistent with observed changes (see Section 10.5.4.5). The Bayesian method predicts a much narrower uncertainty range than ASK. However its results depend on choices made in its design, particularly the convergence criterion for up-weighting models close to the ensemble mean, relaxation of which substantially reduces the discrepancy with ASK.

Another method by Furrer et al. (2007) employs a hierarchical Bayesian model to construct PDFs of temperature change at each grid point from a multi-model ensemble. The main assumptions are that the true climate change signal is a common large-scale structure represented to some degree in each of the model simulations, and that the signal unexplained by climate change is AOGCM-specific in terms of small-scale structure, but can be regarded as noise when averaged over all AOGCMs. In this method, spatial fields of future minus present temperature difference from each ensemble member are regressed upon basis functions. One of the basis functions is a map of differences of observed temperatures from late-minus mid-20th century, and others are spherical harmonics. The statistical model then estimates the regression coefficients and their associated errors, which account for the deviation in each AOGCM from the (assumed) true pattern of change. By recombining the coefficients with the basis functions, an estimate is derived of the true climate change field and its associated uncertainty, thus providing joint probabilities for climate change at all grid points around the globe.

Estimates of uncertainty derived from multi-model ensembles of 10 to 20 members are potentially sensitive to outliers (Räisänen, 2001). Harris et al. (2006) therefore augment a 17-member ensemble of AOGCM transient simulations by scaling the equilibrium response patterns of a large perturbed physics ensemble. Transient responses are emulated by scaling equilibrium response patterns according to global temperature (predicted from an energy balance model tuned to the relevant climate sensitivities). For surface temperature, the scaled equilibrium patterns correspond well to the transient response patterns, while scaling errors for precipitation vary more widely with location. A correction field is added to account for ensemble-mean differences between the equilibrium and transient patterns, and uncertainty is allowed for in the emulated result. The correction field and emulation errors are determined by comparing the responses of model versions for which both transient and equilibrium simulations exist. Results are used to obtain frequency distributions of transient regional changes in surface temperature and precipitation in response to increasing atmospheric CO_2, arising from the combined effects of atmospheric parameter perturbations and internal variability in UKMO-HadCM3.

Figure 10.30 shows probabilities of a temperature change larger than 2°C by the end of the 21st century under the A1B scenario, comparing values estimated from the 21-member AR4 multi-model ensemble (Furrer et al., 2007) against values estimated by combining transient and equilibrium perturbed physics ensembles of 17 and 128 members, respectively (Harris et al., 2006). Although the methods use different ensembles and different statistical approaches, the large-scale patterns are similar in many respects. Both methods show larger probabilities (typically 80% or more) over land, and at high latitudes in the winter hemisphere, with relatively low values (typically less than 50%) over the southern oceans. However, the plots also reveal some substantial differences at a regional level, notably over the North Atlantic Ocean, the sub-tropical Atlantic and Pacific Oceans in the SH, and at high northern latitudes during June to August.

10.5.4.8 Summary

Significant progress has been made since the TAR in exploring ensemble approaches to provide uncertainty ranges and probabilities for global and regional climate change. Different methods show consistency in some aspects of their results, but differ significantly in others (see Box 10.2; Figures 10.28 and 10.30), because they depend to varying degrees on the nature and use of observational constraints, the nature and design of model ensembles and the specification of prior distributions for uncertain inputs (see, e.g., Table 11.3). A preferred method cannot yet be recommended, but the assumptions and limitations underlying the various approaches, and the sensitivity of the results to them, should be communicated to users. A good example concerns the treatment of model error in Bayesian methods, the uncertainty in which affects the calculation of the likelihood of different model versions, but is difficult to specify (Rougier, 2007). Awareness of this issue is growing in the field of climate prediction (Annan et al., 2005b; Knutti et al., 2006), however, it is yet to be thoroughly addressed. Probabilistic depictions, particularly at the regional level, are new to climate change science and are being facilitated by the recently available multi-model ensembles. These are discussed further in Section 11.10.2.

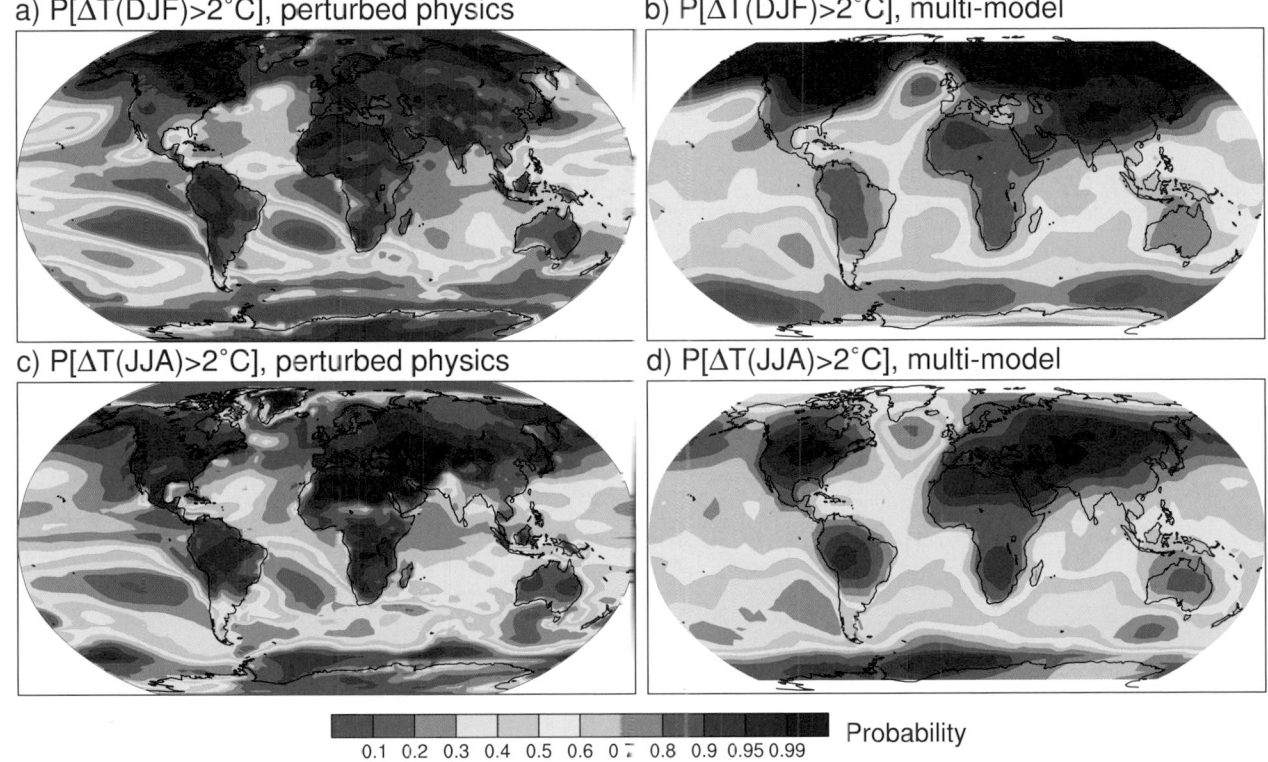

Figure 10.30. *Estimated probabilities for a mean surface temperature change exceeding 2°C in 2080 to 2099 relative to 1980 to 1999 under the SRES A1B scenario. Results obtained from a perturbed physics ensemble of a single model (a, c), based on Harris et al. (2006), are compared with results from the AR4 multi-model ensemble (b, d), based on Furrer et al. (2007), for December to February (DJF, a, b) and June to August (JJA, c, d).*

10.6 Sea Level Change in the 21st Century

10.6.1 Global Average Sea Level Rise Due to Thermal Expansion

As seawater warms up, it expands, increasing the volume of the global ocean and producing thermosteric sea level rise (see Section 5.5.3). Global average thermal expansion can be calculated directly from simulated changes in ocean temperature. Results are available from 17 AOGCMs for the 21st century for SRES scenarios A1B, A2 and B1 (Figure 10.31), continuing from simulations of the 20th century. One ensemble member was used for each model and scenario. The time series are rather smooth compared with global average temperature time series, because thermal expansion reflects heat storage in the entire ocean, being approximately proportional to the time integral of temperature change (Gregory et al., 2001).

During 2000 to 2020 under scenario SRES A1B in the ensemble of AOGCMs, the rate of thermal expansion is 1.3 ± 0.7 mm yr^{-1}, and is not significantly different under A2 or B1. This rate is more than twice the observationally derived rate of 0.42 ± 0.12 mm yr^{-1} during 1961 to 2003. It is similar to the rate of 1.6 ± 0.5 mm yr^{-1} during 1993 to 2003 (see Section 5.5.3), which may be larger than that of previous decades partly because of natural forcing and internal variability (see Sections 5.5.2.4, 5.5.3 and 9.5.2). In particular, many of the AOGCM experiments do not include the influence of Mt. Pinatubo, the omission of which may reduce the projected rate of thermal expansion during the early 21st century.

During 2080 to 2100, the rate of thermal expansion is projected to be 1.9 ± 1.0, 2.9 ± 1.4 and 3.8 ± 1.3 mm yr^{-1} under scenarios SRES B1, A1B and A2 respectively in the AOGCM ensemble (the width of the range is affected by the different numbers of models under each scenario). The acceleration is caused by the increased climatic warming. Results are shown for all SRES marker scenarios in Table 10.7 (see Appendix 10.A for methods). In the AOGCM ensemble, under any given SRES scenario, there is some correlation of the global average temperature change across models with thermal expansion and its rate of change, suggesting that the spread in thermal expansion for that scenario is caused both by the spread in surface warming and by model-dependent ocean heat uptake efficiency (Raper et al., 2002; Table 8.2) and the distribution of added heat within the ocean (Russell et al., 2000).

10.6.2 Local Sea Level Change Due to Change in Ocean Density and Dynamics

The geographical pattern of mean sea level relative to the geoid (the dynamic topography) is an aspect of the dynamical balance relating the ocean's density structure and its circulation, which are maintained by air-sea fluxes of heat, freshwater and momentum. Over much of the ocean on multi-annual time scales, a good approximation to the pattern of dynamic topography change is given by the steric sea level change, which can be calculated straightforwardly from local temperature and salinity change (Gregory et al., 2001; Lowe and Gregory, 2006). In much of the world, salinity changes are as important as temperature changes in determining the pattern of dynamic topography change in the future, and their contributions can be opposed (Landerer et al., 2007; and as in the past, Section 5.5.4.1). Lowe and Gregory (2006) show that in the UKMO-HadCM3 AOGCM, changes in heat fluxes are the cause of many of the large-scale features of sea level change, but freshwater

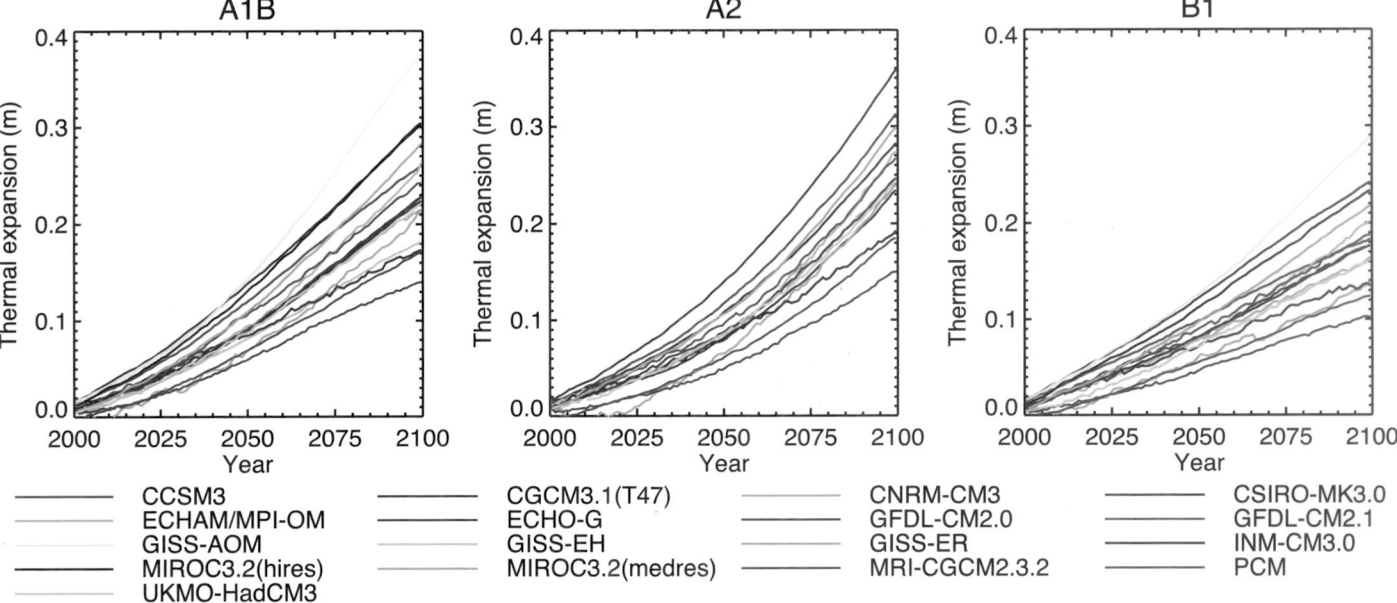

Figure 10.31. *Projected global average sea level rise (m) due to thermal expansion during the 21st century relative to 1980 to 1999 under SRES scenarios A1B, A2 and B1. See Table 8.1 for model descriptions.*

flux change dominates the North Atlantic and momentum flux change has a signature in the north and low-latitude Pacific and the Southern Ocean.

Results are available for local sea level change due to ocean density and circulation change from AOGCMs in the multi-model ensemble for the 20th century and the 21st century. There is substantial spatial variability in all models (i.e., sea level change is not uniform), and as the geographical pattern of climate change intensifies, the spatial standard deviation of local sea level change increases (Church et al., 2001; Gregory et al., 2001). Suzuki et al. (2005) show that, in their high-resolution model, enhanced eddy activity contributes to this increase, but across models there is no significant correlation of the spatial standard deviation with model spatial resolution. This section evaluates sea level change between 1980 to 1999 and 2080 to 2099 projected by 16 models forced with SRES scenario A1B. (Other scenarios are qualitatively similar, but fewer models are available.) The ratio of spatial standard deviation to global average thermal expansion varies among models, but is mostly within the range 0.3 to 0.4. The model median spatial standard deviation of thermal expansion is 0.08 m, which is about 25% of the central estimate of global average sea level rise during the 21st century under A1B (Table 10.7).

The geographical patterns of sea level change from different models are not generally similar in detail, although they have more similarity than those analysed in the TAR by Church et al.

(2001). The largest spatial correlation coefficient between any pair is 0.75, but only 25% of correlation coefficients exceed 0.5. To identify common features, an ensemble mean (Figure 10.32) is examined. There are only limited areas where the model ensemble mean change exceeds the inter-model standard deviation, unlike for surface air temperature change (Section 10.3.2.1).

Like Church et al. (2001) and Gregory et al. (2001), Figure 10.32 shows smaller than average sea level rise in the Southern Ocean and larger than average in the Arctic, the former possibly due to wind stress change (Landerer et al., 2007) or low thermal expansivity (Lowe and Gregory, 2006) and the latter due to freshening. Another obvious feature is a narrow band of pronounced sea level rise stretching across the southern Atlantic and Indian Oceans and discernible in the southern Pacific. This could be associated with a southward shift in the circumpolar front (Suzuki et al., 2005) or subduction of warm anomalies in the region of formation of sub antarctic mode water (Banks et al., 2002). In the zonal mean, there are maxima of sea level rise in 30°S to 45°S and 30°N to 45°N. Similar indications are present in the altimetric and thermosteric patterns of sea level change for 1993 to 2003 (Figure 5.15). The model projections do not share other aspects of the observed pattern of sea level rise, such as in the western Pacific, which could be related to interannual variability.

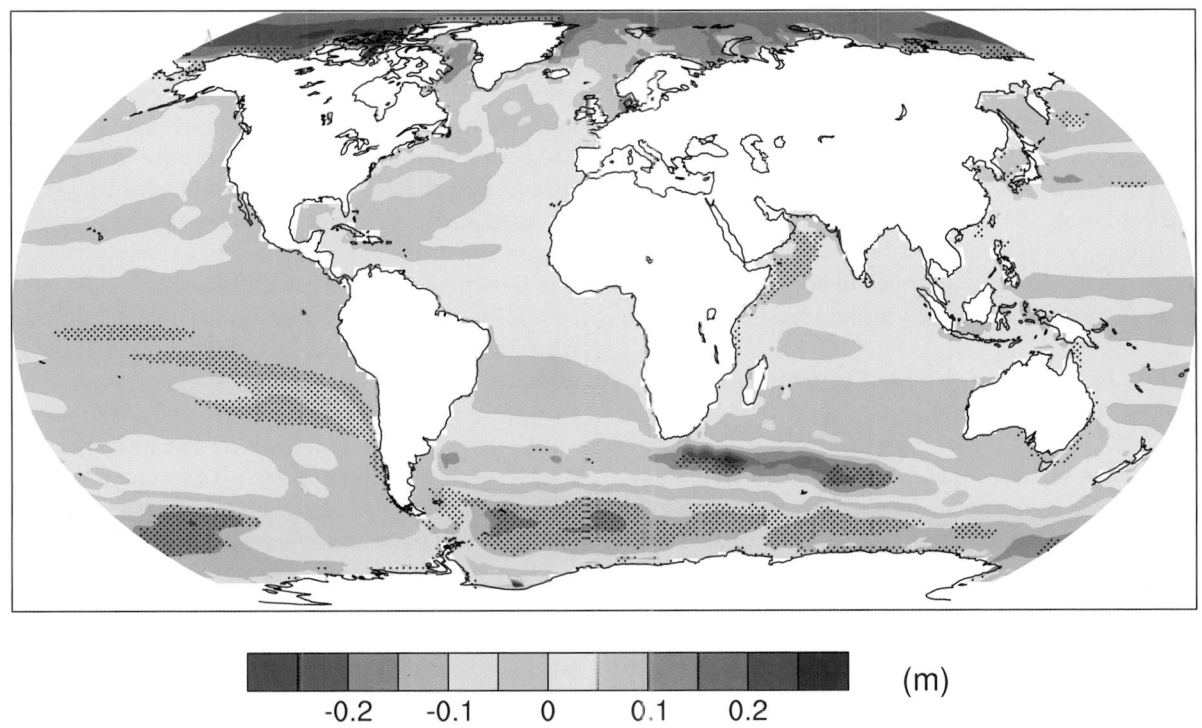

Figure 10.32. *Local sea level change (m) due to ocean density and circulation change relative to the global average (i.e., positive values indicate greater local sea level change than global) during the 21st century, calculated as the difference between averages for 2080 to 2099 and 1980 to 1999, as an ensemble mean over 16 AOGCMs forced with the SRES A1B scenario. Stippling denotes regions where the magnitude of the multi-model ensemble mean divided by the multi-model standard deviation exceeds 1.0.*

The North Atlantic dipole pattern noted by Church et al. (2001), that is, reduced rise to the south of the Gulf Stream extension, enhanced to the north, consistent with a weakening of the circulation, is present in some models; a more complex feature is described by Landerer et al. (2007). The reverse is apparent in the north Pacific, which Suzuki et al. (2005) associate with a wind-driven intensification of the Kuroshio Current. Using simplified models, Hsieh and Bryan (1996) and Johnson and Marshall (2002) show how upper-ocean velocities and sea level would be affected in North Atlantic coastal regions within months of a cessation of sinking in the North Atlantic as a result of propagation by coastal and equatorial Kelvin waves, but would take decades to adjust in the central regions and the south Atlantic. Levermann et al. (2005) show that a sea level rise of several tenths of a metre could be realised in coastal regions of the North Atlantic within a few decades (i.e., tens of millimetres per year) of a collapse of the MOC. Such changes to dynamic topography would be much more rapid than global average sea level change. However, it should be emphasized that these studies are sensitivity tests, not projections; the Atlantic MOC does not collapse in the SRES scenario runs evaluated here (see Section 10.3.4).

The geographical pattern of sea level change is affected also by changes in atmospheric surface pressure, but this is a relatively small effect given the projected pressure changes (Figure 10.9; a pressure increase of 1 hPa causes a drop in local sea level of 0.01 m; see Section 5.5.4.3). Land movements and changes in the gravitational field resulting from the changing loading of the crust by water and ice also have effects which are small over most of the ocean (see Section 5.5.4.4).

10.6.3 Glaciers and Ice Caps

Glaciers and ice caps (G&IC, see also Section 4.5.1) comprise all land ice except for the ice sheets of Greenland and Antarctica (see Sections 4.6.1 and 10.6.4). The mass of G&IC can change because of changes in surface mass balance (Section 10.6.3.1). Changes in mass balance cause changes in area and thickness (Section 10.6.3.2), with feedbacks on surface mass balance.

10.6.3.1 Mass Balance Sensitivity to Temperature and Precipitation

Since G&IC mass balance depends strongly on their altitude and aspect, use of data from climate models to make projections requires a method of downscaling, because individual G&IC are much smaller than typical AOGCM grid boxes. Statistical relations for meteorological quantities can be developed between the GCM and local scales (Reichert et al., 2002), but they may not continue to hold in future climates. Hence, for projections the approach usually adopted is to use GCM simulations of changes in climate parameters to perturb the observed climatology or mass balance (Gregory and Oerlemans, 1998; Schneeberger et al., 2003).

Change in ablation (mostly melting) of a glacier or ice cap is modelled using b_T (in m yr^{-1} °C^{-1}), the sensitivity of the mean specific surface mass balance to temperature (refer to Section 4.5 for a discussion of the relation of mass balance to climate). One approach determines b_T by energy balance modelling, including evolution of albedo and refreezing of melt water within the firn (Zuo and Oerlemans, 1997). Oerlemans and Reichert (2000), Oerlemans (2001) and Oerlemans et al. (2006) refine this approach to include dependence on monthly temperature and precipitation changes. Another approach uses a degree-day method, in which ablation is proportional to the integral of mean daily temperature above the freezing point (Braithwaite et al., 2003). Braithwaite and Raper (2002) show that there is excellent consistency between the two approaches, which indicates a similar relationship between b_T and climatological precipitation. Schneeberger et al. (2000, 2003) use a degree-day method for ablation modified to include incident solar radiation, again obtaining similar results. De Woul and Hock (2006) find somewhat larger sensitivities for arctic G&IC from the degree-day method than the energy balance method. Calculations of b_T are estimated to have an uncertainty of ±15% (standard deviation) (Gregory and Oerlemans, 1998; Raper and Braithwaite, 2006).

The global average sensitivity of G&IC surface mass balance to temperature is estimated by weighting the local sensitivities by land ice area in various regions. For a geographically and seasonally uniform rise in global temperature, Oerlemans and Fortuin (1992) derive a global average G&IC surface mass balance sensitivity of –0.40 m yr^{-1} °C^{-1}, Dyurgerov and Meier (2000) –0.37 m yr^{-1} °C^{-1} (from observations), Braithwaite and Raper (2002) –0.41 m yr^{-1} °C^{-1} and Raper and Braithwaite (2005) –0.35 m yr^{-1} °C^{-1}. Applying the scheme of Oerlemans (2001) and Oerlemans et al. (2006) worldwide gives a smaller value of –0.32 m yr^{-1} °C^{-1}, the reduction being due to the modified treatment of albedo by Oerlemans (2001).

These global average sensitivities for uniform temperature change are given only for scenario-independent comparison of the various methods; they cannot be used for projections, which require regional and seasonal temperature changes (Gregory and Oerlemans, 1998; van de Wal and Wild, 2001). Using monthly temperature changes simulated in G&IC regions by 17 AR4 AOGCMs for scenarios A1B, A2 and B1, the global total surface mass balance sensitivity to global average temperature change for all G&IC outside Greenland and Antarctica is 0.61 ± 0.12 mm yr^{-1} °C^{-1} (sea level equivalent) with the b_T of Zuo and Oerlemans (1997) or 0.49 ± 0.13 mm yr^{-1} °C^{-1} with those of Oerlemans (2001) and Oerlemans et al. (2006), subject to uncertainty in G&IC area (see Section 4.5.2 and Table 4.4).

Hansen and Nazarenko (2004) collate measurements of soot (fossil fuel black carbon) in snow and estimate consequent reductions in snow and ice albedo of between 0.001 for the pristine conditions of Antarctica and over 0.10 for polluted NH land areas. They argue that glacial ablation would be increased by this effect. While it is true that soot has not been explicitly considered in existing sensitivity estimates, it may already be included because the albedo and degree-day parametrizations have been empirically derived from data collected in affected regions.

For seasonally uniform temperature rise, Oerlemans et al. (1998) find that an increase in precipitation of 20 to 50% °C^{-1} is required to balance increased ablation, while Braithwaite et al. (2003) report a required precipitation increase of 29 to 41% °C^{-1}, in both cases for a sample of G&IC representing a variety of climatic regimes. Oerlemans et al. (2006) require a precipitation increase of 20 to 43% °C^{-1} to balance ablation increase, and de Woul and Hock (2006) approximately 20% °C^{-1} for Arctic G&IC. Although AOGCMs generally project larger than average precipitation change in northern mid- and high-latitude regions, the global average is 1 to 2% °C^{-1} (Section 10.3.1), so ablation increases would be expected to dominate worldwide. However, precipitation changes may sometimes dominate locally (see Section 4.5.3).

Regressing observed global total mass balance changes of all G&IC outside Greenland and Antarctica against global average surface temperature change gives a global total mass balance sensitivity which is greater than model results (see Appendix 10.A). The current state of knowledge does not permit a satisfactory explanation of the difference. Giving more weight to the observational record but enlarging the uncertainty to allow for systematic error, a value of 0.80 ± 0.33 mm yr^{-1} °C^{-1} (5 to 95% range) is adopted for projections. The regression indicates that the climate of 1865 to 1895 was 0.13°C warmer globally than the climate that gives a steady state for G&IC (cf., Zuo and Oerlemans, 1997; Gregory et al., 2006). Model results for the 20th century are sensitive to this value, but the projected temperature change in the 21st century is large by comparison, making the effect relatively less important for projections (see Appendix 10.A).

10.6.3.2   Dynamic Response and Feedback on Mass Balance

As glacier volume is lost, glacier area declines so the ablation decreases. Oerlemans et al. (1998) calculate that omitting this effect leads to overestimates of ablation of about 25% by 2100. Church et al. (2001), following Bahr et al. (1997) and Van de Wal and Wild (2001), make some allowance for it by diminishing the area A of a glacier of volume V according to $V \propto A^{1.375}$. This is a scaling relation derived for glaciers in a steady state, which may hold only approximately during retreat. For example, thinning in the ablation zone will steepen the surface slope and tend to increase the flow. Comparison with a simple flow model suggests the deviations do not exceed 20% (van de Wal and Wild, 2001). Schneeberger et al. (2003) find that the scaling relation produced a mixture of over- and underestimates of volume loss for their sample of glaciers compared with more detailed dynamic modelling. In some regions where G&IC flow into the sea or lakes there is accelerated dynamic discharge (Rignot et al., 2003) that is not included in currently available glacier models, leading to an underestimate of G&IC mass loss.

The mean specific surface mass balance of the glacier or ice cap will change as volume is lost: lowering the ice surface as the ice thins will tend to make it more negative, but the predominant loss of area at lower altitude in the ablation zone

will tend to make it less negative (Braithwaite and Raper, 2002). For rapid thinning rates in the ablation zone, of several metres per year, lowering the surface will give enhanced local warmings comparable to the rate of projected climatic warming. However, those areas of the ablation zone of valley glaciers that thin most rapidly will soon be removed altogether, resulting in retreat of the glacier. The enhancement of ablation by surface lowering can only be sustained in glaciers with a relatively large, thick and flat ablation area. On multi-decadal time scales, for the majority of G&IC, the loss of area is more important than lowering of the surface (Schneeberger et al., 2003).

The dynamical approach (Oerlemans et al., 1998; Schneeberger et al., 2003) cannot be applied to all the world's glaciers individually as the required data are unknown for the vast majority of them. Instead, it might be applied to a representative ensemble derived from statistics of size distributions of G&IC. Raper et al. (2000) developed a geometrical approach, in which the width, thickness and length of a glacier are reduced as its volume and area declines. When applied statistically to the world population of glaciers and individually to ice caps, this approach shows that the reduction of area of glaciers strongly reduces the ablation during the 21st century (Raper and Braithwaite, 2006), by about 45% under scenario SRES A1B for the GFDL-CM2.0 and PCM AOGCMs (see Table 8.1 for model details). For the same cases, using the mass-balance sensitivities to temperature of Oerlemans (2001) and Oerlemans et al. (2006), G&IC mass loss is reduced by about 35% following the area scaling of Van de Wal and Wild (2001), suggesting that the area scaling and the geometrical model have a similar effect in reducing estimated ablation for the 21st century. The effect is greater when using the observationally derived mass balance sensitivity (Section 10.6.3.1), which is larger, implying faster mass loss for fixed area. The uncertainty in present-day glacier volume (Table 4.4) introduces a 5 to 10% uncertainty into the results of area scaling. For projections, the area scaling of Van de Wal and Wild (2001) is applied, using three estimates of world glacier volume (see Table 4.4 and Appendix 10.A). The scaling reduces the projections of the G&IC contribution up to the mid-21st century by 25% and over the whole century by 40 to 50% with respect to fixed G&IC area.

10.6.3.3   Glaciers and Ice Caps on Greenland and Antarctica

The G&IC on Greenland and Antarctica (apart from the ice sheets) have been less studied and projections for them are consequently more uncertain. A model estimate for the G&IC on Greenland indicates an addition of about 6% to the G&IC sea level contribution in the 21st century (van de Wal and Wild, 2001). Using a degree-day scheme, Vaughan (2006) estimates that ablation of glaciers in the Antarctic Peninsula presently amounts to 0.008 to 0.055 mm yr^{-1} of sea level, 1 to 9% of the contribution from G&IC outside Greenland and Antarctica (Table 4.4). Morris and Mulvaney (2004) find that accumulation increases on the Antarctic Peninsula were larger than ablation increases during 1972 to 1998, giving a small net *negative* sea level contribution from the region. However, because ablation increases nonlinearly with temperature, they estimate that for future warming the contribution would become positive, with a sensitivity of 0.07 ± 0.03 mm yr^{-1} °C^{-1} to uniform temperature change in Antarctica, that is, about 10% of the global sensitivity of G&IC outside Greenland and Antarctica (Section 10.6.3.1).

These results suggest that the Antarctic and Greenland G&IC will together give 10 to 20% of the sea level contribution of other G&IC in future decades. In recent decades, the G&IC on Greenland and Antarctica have together made a contribution of about 20% of the total of other G&IC (see Section 4.5.2). On these grounds, the global G&IC sea level contribution is increased by a factor of 1.2 to include those in Greenland and Antarctica in projections for the 21st century (see Section 10.6.5 and Table 10.7). Dynamical acceleration of glaciers in Greenland and Antarctica following removal of ice shelves, as has recently happened on the Antarctic Peninsula (Sections 4.6.2.2 and 10.6.4.2), would add further to this, and is included in projections of that effect (Section 10.6.4.3).

10.6.4 Ice Sheets

The mass of ice grounded on land in the Greenland and Antarctic Ice Sheets (see also Section 4.6.1) can change as a result of changes in surface mass balance (the sum of accumulation and ablation; Section 10.6.4.1) or in the flux of ice crossing the grounding line, which is determined by the dynamics of the ice sheet (Section 10.6.4.2). Surface mass balance and dynamics together both determine and are affected by the change in surface topography.

10.6.4.1 Surface Mass Balance

Surface mass balance (SMB) is immediately influenced by climate change. A good simulation of the ice sheet SMB requires a resolution exceeding that of AGCMs used for long climate experiments, because of the steep slopes at the margins of the ice sheet, where the majority of the precipitation and all of the ablation occur. Precipitation over ice sheets is typically overestimated by AGCMs, because their smooth topography does not present a sufficient barrier to inland penetration (Ohmura et al., 1996; Glover, 1999; Murphy et al., 2002). Ablation also tends to be overestimated because the area at low altitude around the margins of the ice sheet, where melting preferentially occurs, is exaggerated (Glover, 1999; Wild et al., 2003). In addition, AGCMs do not generally have a representation of the refreezing of surface melt water within the snowpack and may not include albedo variations dependent on snow ageing and its conversion to ice.

To address these issues, several groups have computed SMB at resolutions of tens of kilometres or less, with results that compare acceptably well with observations (e.g., van Lipzig et al., 2002; Wild et al., 2003). Ablation is calculated either by schemes based on temperature (degree-day or other temperature index methods) or by energy balance modelling. In the studies listed in Table 10.6, changes in SMB have been calculated

from climate change simulations with high-resolution AGCMs or by perturbing a high-resolution observational climatology with climate model output, rather than by direct use of low-resolution GCM results. The models used for projected SMB changes are similar in kind to those used to study recent SMB changes (Section 4.6.3.1).

All the models show an increase in accumulation, but there is considerable uncertainty in its size (Table 10.6; van de Wal et al., 2001; Huybrechts et al., 2004). Precipitation increase could be determined by atmospheric radiative balance, increase in saturation specific humidity with temperature, circulation changes, retreat of sea ice permitting greater evaporation or a combination of these (van Lipzig et al., 2002). Accumulation also depends on change in local temperature, which strongly affects whether precipitation is solid or liquid (Janssens and Huybrechts, 2000), tending to make the accumulation increase smaller than the precipitation increase for a given temperature rise. For Antarctica, accumulation increases by 6 to 9% °C^{-1} in the high-resolution AGCMs. Precipitation increases somewhat less in AR4 AOGCMs (typically of lower resolution), by 3 to 8% °C^{-1}. For Greenland, accumulation derived from the high-resolution AGCMs increases by 5 to 9% °C^{-1}. Precipitation increases by 4 to 7% °C^{-1} in the AR4 AOGCMs.

Kapsner et al. (1995) do not find a relationship between precipitation and temperature variability inferred from Greenland ice cores for the Holocene, although both show large changes from the Last Glacial Maximum (LGM) to the Holocene. In the UKMO-HadCM3 AOGCM, the relationship is strong for climate change forced by greenhouse gases and the glacial-interglacial transition, but weaker for naturally forced variability (Gregory et al., 2006). Increasing precipitation in conjunction with warming has been observed in recent years in Greenland (Section 4.6.3.1).

All studies for the 21st century project that antarctic SMB changes will contribute negatively to sea level, owing to increasing accumulation exceeding any ablation increase (see Table 10.6). This tendency has not been observed in the average over Antarctica in reanalysis products for the last two decades (see Section 4.6.3.1), but during this period Antarctica as a whole has not warmed; on the other hand, precipitation has increased on the Antarctic Peninsula, where there has been strong warming.

In projections for Greenland, ablation increase is important but uncertain, being particularly sensitive to temperature change around the margins. Climate models project less warming in these low-altitude regions than the Greenland average, and less warming in summer (when ablation occurs) than the annual average, but greater warming in Greenland than the global average (Church et al., 2001; Huybrechts et al., 2004; Chylek and Lohmann, 2005; Gregory and Huybrechts, 2006). In most studies, Greenland SMB changes represent a net positive contribution to sea level in the 21st century (Table 10.6; Kiilsholm et al., 2003) because the ablation increase is larger than the precipitation increase. Only Wild et al. (2003) find the opposite, so that the net SMB change contributes negatively to sea level in the 21st century. Wild et al. (2003) attribute this

Table 10.6. *Comparison of ice sheet (grounded ice area) SMB changes calculated from high-resolution climate models. ΔP/ΔT is the change in accumulation divided by change in temperature over the ice sheet, expressed as sea level equivalent (positive for falling sea level), and ΔR/ΔT the corresponding quantity for ablation (positive for rising sea level). Note that ablation increases more rapidly than linearly with ΔT (van de Wal et al., 2001; Gregory and Huybrechts, 2006). To convert from mm yr⁻¹ °C⁻¹ to kg yr⁻¹ °C⁻¹, multiply by 3.6 × 10¹⁴ m². To convert mm yr⁻¹ °C⁻¹ of sea level equivalent to mm yr⁻¹ °C⁻¹ averaged over the ice sheet, multiply by −206 for Greenland and −26 for Antarctica. ΔP/(PΔT) is the fractional change in accumulation divided by the change in temperature.*

Study	Climate model[a]	Model resolution and SMB source[b]	Greenland ΔP/ΔT (mm yr⁻¹ °C⁻¹)	Greenland ΔP/(PΔT) (% °C⁻¹)	Greenland ΔR/ΔT (mm yr⁻¹ °C⁻¹)	Antarctica ΔP/ΔT (mm yr⁻¹ °C⁻¹)	Antarctica ΔP/(PΔT) (% °C⁻¹)
Van de Wal et al. (2001)	ECHAM4	20 km EB	0.14	8.5	0.16	n.a.	n.a.
Wild and Ohmura (2000)	ECHAM4	T106 ≈ 1.1° EB	0.13	8.2	0.22	0.47	7.4
Wild et al. (2003)	ECHAM4	2 km TI	0.13	8.2	0.04	0.47	7.4
Bugnion and Stone (2002)	ECHAM4	20 km EB	0.10	6.4	0.13	n.a.	n.a.
Huybrechts et al. (2004)	ECHAM4	20 km TI	0.13[c]	7.6[c]	0.14	0.49[c]	7.3[c]
Huybrechts et al. (2004)	HadAM3H	20 km TI	0.09[c]	4.7[c]	0.23	0.37[c]	5.5[c]
Van Lipzig et al. (2002)	RACMO	55 km EB	n.a.	n.a.	n.a.	0.53	9.0
Krinner et al. (2007)	LMDZ4	60 km EB	n.a.	n.a.	n.a.	0.49	8.4

Notes:

[a] ECHAM4: Max Planck Institute for Meteorology AGCM; HadAM3: high-resolution Met Office Hadley Centre AGCM; RACMO: Regional Atmospheric Climate Model (for Antarctica); LMDZ4: Laboratoire de Météorologie Dynamique AGCM (with high resolution over Antarctica).

[b] EB: SMB calculated from energy balance; TI: SMB calculated from temperature index.

[c] In these cases *P* is precipitation rather than accumulation.

difference to the reduced ablation area in their higher-resolution grid. A positive SMB change is not consistent with analyses of recent changes in Greenland SMB (see Section 4.6.3.1).

For an average temperature change of 3°C over each ice sheet, a combination of four high-resolution AGCM simulations and 18 AR4 AOGCMs (Huybrechts et al., 2004; Gregory and Huybrechts, 2006) gives SMB changes of 0.3 ± 0.3 mm yr⁻¹ for Greenland and −0.9 ± 0.5 mm yr⁻¹ for Antarctica (sea level equivalent), that is, sensitivities of 0.11 ± 0.09 mm yr⁻¹ °C⁻¹ for Greenland and −0.29 ± 0.18 mm yr⁻¹ °C⁻¹ for Antarctica. These results generally cover the range shown in Table 10.6, but tend to give more positive (Greenland) or less negative (Antarctica) sea level rise because of the smaller precipitation increases projected by the AOGCMs than by the high-resolution AGCMs. The uncertainties are from the spatial and seasonal patterns of precipitation and temperature change over the ice sheets, and from the ablation calculation. Projections under SRES scenarios for the 21st century are shown in Table 10.7.

10.6.4.2 Dynamics

Ice sheet flow reacts to changes in topography produced by SMB change. Projections for the 21st century are given in Section 10.6.5 and Table 10.7, based on the discussion in this section. In Antarctica, topographic change tends to increase ice flow and discharge. In Greenland, lowering of the surface tends to increase the ablation, while a steepening slope in the ablation zone opposes the lowering, and thinning of outlet glaciers reduces discharge. Topographic and dynamic changes simulated by ice flow models (Huybrechts and De Wolde, 1999; van de Wal et al., 2001; Huybrechts et al., 2002, 2004; Gregory and Huybrechts, 2006) can be roughly represented as modifying the sea level changes due to SMB change with fixed topography by −5% ± 5% from Antarctica, and 0% ±10% from Greenland (± one standard deviation) during the 21st century.

The TAR concluded that accelerated sea level rise caused by rapid dynamic response of the ice sheets to climate change is very unlikely during the 21st century (Church et al., 2001). However, new evidence of recent rapid changes in the Antarctic Peninsula, West Antarctica and Greenland (see Section 4.6.3.3) has again raised the possibility of larger dynamical changes in the future than are projected by state-of-the-art continental models, such as cited above, because these models do not incorporate all the processes responsible for the rapid marginal thinning currently taking place (Box 4.1; Alley et al., 2005a; Vaughan, 2007).

The main uncertainty is the degree to which the presence of ice shelves affects the flow of inland ice across the grounding

Frequently Asked Question 10.2

How Likely are Major or Abrupt Climate Changes, such as Loss of Ice Sheets or Changes in Global Ocean Circulation?

Abrupt climate changes, such as the collapse of the West Antarctic Ice Sheet, the rapid loss of the Greenland Ice Sheet or large-scale changes of ocean circulation systems, are not considered likely to occur in the 21st century, based on currently available model results. However, the occurrence of such changes becomes increasingly more likely as the perturbation of the climate system progresses.

Physical, chemical and biological analyses from Greenland ice cores, marine sediments from the North Atlantic and elsewhere and many other archives of past climate have demonstrated that local temperatures, wind regimes and water cycles can change rapidly within just a few years. The comparison of results from records in different locations of the world shows that in the past major changes of hemispheric to global extent occurred. This has led to the notion of an unstable past climate that underwent phases of abrupt change. Therefore, an important concern is that the continued growth of greenhouse gas concentrations in the atmosphere may constitute a perturbation sufficiently strong to trigger abrupt changes in the climate system. Such interference with the climate system could be considered dangerous, because it would have major global consequences.

Before discussing a few examples of such changes, it is useful to define the terms 'abrupt' and 'major'. 'Abrupt' conveys the meaning that the changes occur much faster than the perturbation inducing the change; in other words, the response is nonlinear. A 'major' climate change is one that involves changes that exceed the range of current natural variability and have a spatial extent ranging from several thousand kilometres to global. At local to regional scales, abrupt changes are a common characteristic of natural climate variability. Here, isolated, short-lived events that are more appropriately referred to as 'extreme events' are not considered, but rather large-scale changes that evolve rapidly and persist for several years to decades. For instance, the mid-1970s shift in sea surface temperatures in the Eastern Pacific, or the salinity reduction in the upper 1,000 m of the Labrador Sea since the mid-1980s, are examples of abrupt events with local to regional consequences, as opposed to the larger-scale, longer-term events that are the focus here.

One example is the potential collapse, or shut-down of the Gulf Stream, which has received broad public attention. The Gulf Stream is a primarily horizontal current in the north-western Atlantic Ocean driven by winds. Although a stable feature of the general circulation of the ocean, its northern extension, which feeds deep-water formation in the Greenland-Norwegian-Iceland Seas and thereby delivers substantial amounts of heat to these seas and nearby land areas, is influenced strongly by changes in the density of the surface waters in these areas. This current

constitutes the northern end of a basin-scale meridional overturning circulation (MOC) that is established along the western boundary of the Atlantic basin. A consistent result from climate model simulations is that if the density of the surface waters in the North Atlantic decreases due to warming or a reduction in salinity, the strength of the MOC is decreased, and with it, the delivery of heat into these areas. Strong sustained reductions in salinity could induce even more substantial reduction, or complete shut-down of the MOC in all climate model projections. Such changes have indeed happened in the distant past.

The issue now is whether the increasing human influence on the atmosphere constitutes a strong enough perturbation to the MOC that such a change might be induced. The increase in greenhouse gases in the atmosphere leads to warming and an intensification of the hydrological cycle, with the latter making the surface waters in the North Atlantic less salty as increased rain leads to more freshwater runoff to the ocean from the region's rivers. Warming also causes land ice to melt, adding more freshwater and further reducing the salinity of ocean surface waters. Both effects would reduce the density of the surface waters (which must be dense and heavy enough to sink in order to drive the MOC), leading to a reduction in the MOC in the 21st century. This reduction is predicted to proceed in lockstep with the warming: none of the current models simulates an abrupt (nonlinear) reduction or a complete shut-down in this century. There is still a large spread among the models' simulated reduction in the MOC, ranging from virtually no response to a reduction of over 50% by the end of the 21st century. This cross-model variation is due to differences in the strengths of atmosphere and ocean feedbacks simulated in these models.

Uncertainty also exists about the long-term fate of the MOC. Many models show a recovery of the MOC once climate is stabilised. But some models have thresholds for the MOC, and they are passed when the forcing is strong enough and lasts long enough. Such simulations then show a gradual reduction of the MOC that continues even after climate is stabilised. A quantification of the likelihood of this occurring is not possible at this stage. Nevertheless, even if this were to occur, Europe would still experience warming, since the radiative forcing caused by increasing greenhouse gases would overwhelm the cooling associated with the MOC reduction. Catastrophic scenarios suggesting the beginning of an ice age triggered by a shutdown of the MOC are thus mere speculations, and no climate model has produced such an outcome. In fact, the processes leading to an ice age are sufficiently well understood and so completely different from those discussed here, that we can confidently exclude this scenario.

(continued)

Irrespective of the long-term evolution of the MOC, model simulations agree that the warming and resulting decline in salinity will significantly reduce deep and intermediate water formation in the Labrador Sea during the next few decades. This will alter the characteristics of the intermediate water masses in the North Atlantic and eventually affect the deep ocean. The long-term effects of such a change are unknown.

Other widely discussed examples of abrupt climate changes are the rapid disintegration of the Greenland Ice Sheet, or the sudden collapse of the West Antarctic Ice Sheet. Model simulations and observations indicate that warming in the high latitudes of the Northern Hemisphere is accelerating the melting of the Greenland Ice Sheet, and that increased snowfall due to the intensified hydrological cycle is unable to compensate for this melting. As a consequence, the Greenland Ice Sheet may shrink substantially in the coming centuries. Moreover, results suggest that there is a critical temperature threshold beyond which the Greenland Ice Sheet would be committed to disappearing completely, and that threshold could be crossed in this century. However, the total melting of the Greenland Ice Sheet, which

would raise global sea level by about seven metres, is a slow process that would take many hundreds of years to complete.

Recent satellite and *in situ* observations of ice streams behind disintegrating ice shelves highlight some rapid reactions of ice sheet systems. This raises new concern about the overall stability of the West Antarctic Ice Sheet, the collapse of which would trigger another five to six metres of sea level rise. While these streams appear buttressed by the shelves in front of them, it is currently unknown whether a reduction or failure of this buttressing of relatively limited areas of the ice sheet could actually trigger a widespread discharge of many ice streams and hence a destabilisation of the entire West Antarctic Ice Sheet. Ice sheet models are only beginning to capture such small-scale dynamical processes that involve complicated interactions with the glacier bed and the ocean at the perimeter of the ice sheet. Therefore, no quantitative information is available from the current generation of ice sheet models as to the likelihood or timing of such an event.

line. A strong argument for enhanced flow when the ice shelf is removed is yielded by the acceleration of Jakobshavn Glacier (Greenland) following the loss of its floating tongue, and of the glaciers supplying the Larsen B Ice Shelf (Antarctic Peninsula) after it collapsed (see Section 4.6.3.3). The onset of disintegration of the Larsen B Ice Shelf has been attributed to enhanced fracturing by crevasses promoted by surface melt water (Scambos et al., 2000). Large portions of the Ross and Filchner-Ronne Ice Shelves (West Antarctica) currently have mean summer surface temperatures of around –5°C (Comiso, 2000, updated). Four high-resolution GCMs (Gregory and Huybrechts, 2006) project summer surface warming in these major ice shelf regions of between 0.2 and 1.3 times the antarctic annual average warming, which in turn will be a factor 1.1 ± 0.3 greater than global average warming according to AOGCM simulations using SRES scenarios. These figures indicate that a local mean summer warming of 5°C is unlikely for a global warming of less than 5°C (see Appendix 10.A). This suggests that ice shelf collapse due to surface melting is unlikely under most SRES scenarios during the 21st century, but we have low confidence in the inference because there is evidently large systematic uncertainty in the regional climate projections, and it is not known whether episodic surface melting might initiate disintegration in a warmer climate while mean summer temperatures remain below freezing.

In the Amundsen Sea sector of West Antarctica, ice shelves are not so extensive and the cause of ice shelf thinning is not surface melting, but bottom melting at the grounding line (Rignot and Jacobs, 2002). Shepherd et al. (2004) find an average ice-

shelf thinning rate of 1.5 ± 0.5 m yr^{-1}. At the same time as the basal melting, accelerated inland flow has been observed for Pine Island, Thwaites and other glaciers in the sector (Rignot, 1998, 2001; Thomas et al., 2004). The synchronicity of these changes strongly implies that their cause lies in oceanographic change in the Amundsen Sea, but this has not been attributed to anthropogenic climate change and could be connected with variability in the SAM.

Because the acceleration took place in only a few years (Rignot et al., 2002; Joughin et al., 2003) but appears up to about 150 km inland, it implies that the dynamical response to changes in the ice shelf can propagate rapidly up the ice stream. This conclusion is supported by modelling studies of Pine Island Glacier by Payne et al. (2004) and Dupont and Alley (2005), in which a single and instantaneous reduction of the basal or lateral drag at the ice front is imposed in idealised ways, such as a step retreat of the grounding line. The simulated acceleration and inland thinning are rapid but transient; the rate of contribution to sea level declines as a new steady state is reached over a few decades. In the study of Payne et al. (2004) the imposed perturbations were designed to resemble loss of drag in the 'ice plain', a partially grounded region near the ice front, and produced a velocity increase of about 1 km yr^{-1} there. Thomas et al. (2005) suggest the ice plain will become ungrounded during the next decade and obtain a similar velocity increase using a simplified approach.

Most of inland ice of West Antarctica is grounded below sea level and so it could float if it thinned sufficiently; discharge therefore promotes inland retreat of the grounding line, which represents a positive feedback by further reducing basal traction. Unlike the one-time change in the idealised studies, this would represent a sustained dynamical forcing that would prolong the contribution to sea level rise. Grounding line retreat of the ice streams has been observed recently at rates of up to about 1 km yr^{-1} (Rignot, 1998, 2001; Shepherd et al., 2002), but a numerical model formulation is difficult to construct (Vieli and Payne, 2005).

The majority of West Antarctic ice discharge is through the ice streams that feed the Ross and Ronne-Filchner ice shelves, but in these regions no accelerated flow causing thinning is currently observed; on the contrary, they are thickening or near balance (Zwally et al., 2005). Excluding these regions, and likewise those parts of the East Antarctic Ice Sheet that drain into the large Amery ice shelf, the total area of ice streams (areas flowing faster than 100 m yr^{-1}) discharging directly into the sea or via a small ice shelf is 270,000 km^2. If all these areas thinned at 2 m yr^{-1}, the order of magnitude of the larger rates observed in fast-flowing areas of the Amundsen Sea sector (Shepherd et al., 2001, 2002), the contribution to sea level rise would be about 1.5 mm yr^{-1}. This would require sustained retreat simultaneously on many fronts, and should be taken as an indicative upper limit for the 21st century (see also Section 10.6.5).

The observation in west-central Greenland of seasonal variation in ice flow rate and of a correlation with summer temperature variation (Zwally et al., 2002) suggest that surface melt water may join a sub-glacially routed drainage system lubricating the ice flow (although this implies that it penetrates more than 1,200 m of subfreezing ice). By this mechanism, increased surface melting during the 21st century could cause

acceleration of ice flow and discharge; a sensitivity study (Parizek and Alley, 2004) indicated that this might increase the sea level contribution from the Greenland Ice Sheet during the 21st century by up to 0.2 m, depending on the warming and other assumptions. However, other studies (Echelmeyer and Harrison, 1990; Joughin et al., 2004) found no evidence of seasonal fluctuations in the flow rate of nearby Jakobshavn Glacier despite a substantial supply of surface melt water.

10.6.5 Projections of Global Average Sea Level Change for the 21st Century

Table 10.7 and Figure 10.33 show projected changes in global average sea level under the SRES marker scenarios for the 21st century due to thermal expansion and land ice changes based on AR4 AOGCM results (see Sections 10.6.1, 10.6.3 and 10.6.4 for discussion). The ranges given are 5 to 95% intervals characterising the spread of model results, but we are not able to assess their likelihood in the way we have done for temperature change (Section 10.5.4.6), for two main reasons. First, the observational constraint on sea level rise projections is weaker, because records are shorter and subject to more uncertainty. Second, current scientific understanding leaves poorly known uncertainties in the methods used to make projections for land ice (Sections 10.6.3 and 10.6.4). Since the AOGCMs are integrated with scenarios of CO_2 concentration, uncertainties in carbon cycle feedbacks are not included in the results. The carbon cycle uncertainty in projections of temperature change cannot be translated into sea level rise because thermal expansion is a major contributor and its relation to temperature change is uncertain (Section 10.6.1).

Table 10.7. *Projected global average sea level rise during the 21st century and its components under SRES marker scenarios. The upper row in each pair gives the 5 to 95% range (m) of the rise in sea level between 1980 to 1999 and 2090 to 2099. The lower row in each pair gives the range of the rate of sea level rise (mm yr^{-1}) during 2090 to 2099. The land ice sum comprises G&IC and ice sheets, including dynamics, but excludes the scaled-up ice sheet discharge (see text). The sea level rise comprises thermal expansion and the land ice sum. Note that for each scenario the lower/upper bound for sea level rise is larger/smaller than the total of the lower/upper bounds of the contributions, since the uncertainties of the contributions are largely independent. See Appendix 10.A for methods.*

		B1		B2		A1B		A1T		A2		A1FI	
Thermal expansion	m	0.10	0.24	0.12	0.28	0.13	0.32	0.12	0.30	0.14	0.35	0.17	0.41
	mm yr^{-1}	1.1	2.6	1.6	4.0	1.7	4.2	1.3	3.2	2.6	6.3	2.8	6.8
G&IC	m	0.07	0.14	0.07	0.15	0.08	0.15	0.08	0.15	0.08	0.16	0.08	0.17
	mm yr^{-1}	0.5	1.3	0.5	1.5	0.6	1.6	0.5	1.4	0.6	1.9	0.7	2.0
Greenland Ice Sheet SMB	m	0.01	0.05	0.01	0.06	0.01	0.08	0.01	0.07	0.01	0.08	0.02	0.12
	mm yr^{-1}	0.2	1.0	0.2	1.5	0.3	1.9	0.2	1.5	0.3	2.8	0.4	3.9
Antarctic Ice Sheet SMB	m	-0.10	-0.02	-0.11	-0.02	-0.12	-0.02	-0.12	-0.02	-0.12	-0.03	-0.14	-0.03
	mm yr^{-1}	-1.4	-0.3	-1.7	-0.3	-1.9	-0.4	-1.7	-0.3	-2.3	-0.4	-2.7	-0.5
Land ice sum	m	0.04	0.18	0.04	0.19	0.04	0.20	0.04	0.20	0.04	0.20	0.04	0.23
	mm yr^{-1}	0.0	1.8	-0.1	2.2	-0.2	2.5	-0.1	2.1	-0.4	3.2	-0.8	4.0
Sea level rise	m	0.18	0.38	0.20	0.43	0.21	0.48	0.20	0.45	0.23	0.51	0.26	0.59
	mm yr^{-1}	1.5	3.9	2.1	5.6	2.1	6.0	1.7	4.7	3.0	8.5	3.0	9.7
Scaled-up ice sheet discharge	m	0.00	0.09	0.00	0.11	-0.01	0.13	-0.01	0.13	-0.01	0.13	-0.01	0.17
	mm yr^{-1}	0.0	1.7	0.0	2.3	0.0	2.6	0.0	2.3	-0.1	3.2	-0.1	3.9

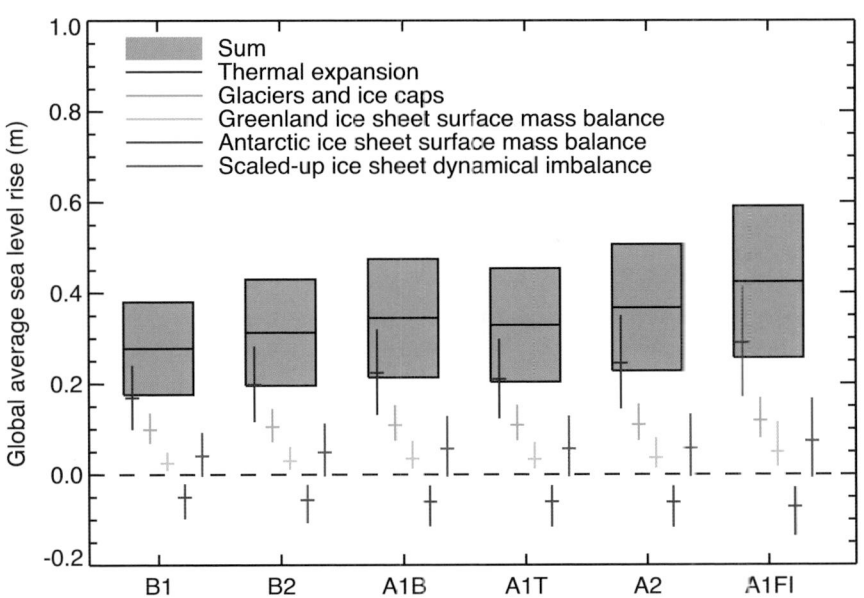

Figure 10.33. *Projections and uncertainties (5 to 95% ranges) of global average sea level rise and its components in 2090 to 2099 (relative to 1980 to 1999) for the six SRES marker scenarios. The projected sea level rise assumes that the part of the present-day ice sheet mass imbalance that is due to recent ice flow acceleration will persist unchanged. It does not include the contribution shown from scaled-up ice sheet discharge, which is an alternative possibility. It is also possible that the present imbalance might be transient, in which case the projected sea level rise is reduced by 0.02 m. It must be emphasized that we cannot assess the likelihood of any of these three alternatives, which are presented as illustrative. The state of understanding prevents a best estimate from being made.*

In all scenarios, the average rate of rise during the 21st century is very likely to exceed the 1961 to 2003 average rate of 1.8 ± 0.5 mm yr^{-1} (see Section 5.5.2.1). The central estimate of the rate of sea level rise during 2090 to 2099 is 3.8 mm yr^{-1} under A1B, which exceeds the central estimate of 3.1 mm yr^{-1} for 1993 to 2003 (see Section 5.5.2.2). The 1993 to 2003 rate may have a contribution of about 1 mm yr^{-1} from internally generated or naturally forced decadal variability (see Sections 5.5.2.4 and 9.5.2). These sources of variability are not predictable and not included in the projections; the actual rate during any future decade might therefore be more or less than the projected rate by a similar amount. Although simulated and observed sea level rise agree reasonably well for 1993 to 2003, the observed rise for 1961 to 2003 is not satisfactorily explained (Section 9.5.2), as the sum of observationally estimated components is 0.7 ± 0.7 mm yr^{-1} less than the observed rate of rise (Section 5.5.6). This indicates a deficiency in current scientific understanding of sea level change and may imply an underestimate in projections.

For an average model (the central estimate for each scenario), the scenario spread (from B1 to A1FI) in sea level rise is only 0.02 m by the middle of the century. This is small because of the time-integrating effect of sea level rise, on which the divergence among the scenarios has had little effect by then. By 2090 to 2099 it is 0.15 m.

In all scenarios, the central estimate for thermal expansion by the end of the century is 70 to 75% of the central estimate for the sea level rise. In all scenarios, the average rate of expansion

during the 21st century is larger than central estimate of 1.6 mm yr^{-1} for 1993 to 2003 (Section 5.5.3). Likewise, in all scenarios the average rate of mass loss by G&IC during the 21st century is greater than the central estimate of 0.77 mm yr^{-1} for 1993 to 2003 (Section 4.5.2). By the end of the century, a large fraction of the present global G&IC mass is projected to have been lost (see, e.g., Table 4.3). The G&IC projections are rather insensitive to the scenario because the main uncertainties come from the G&IC model.

Further accelerations in ice flow of the kind recently observed in some Greenland outlet glaciers and West Antarctic ice streams could increase the ice sheet contributions substantially, but quantitative projections cannot be made with confidence (see Section 10.6.4.2). The land ice sum in Table 10.7 includes the effect of dynamical changes in the ice sheets that can be simulated with a continental ice sheet model (Section 10.6.4.2). It also includes a scenario-independent term of 0.32 ± 0.35 mm yr^{-1} (0.035 ± 0.039 m in 110 years). This is the central estimate for 1993 to 2003 of the sea level contribution from the Antarctic Ice Sheet, plus half of that from Greenland (Sections 4.6.2.2 and 5.5.5.2). We take this as an estimate of the part of the present ice sheet mass imbalance that is due to recent ice flow acceleration (Section 4.6.3.2), and assume that this contribution will persist unchanged.

We also evaluate the contribution of rapid dynamical changes under two alternative assumptions (see, e.g., Alley et al., 2005b). First, the present imbalance might be a rapid short-term adjustment, which will diminish during coming decades. We take an e-folding time of 100 years, on the basis of an idealised model study (Payne et al., 2004). This assumption reduces the sea level rise in Table 10.7 by 0.02 m. Second, the present imbalance might be a response to recent climate change, perhaps through oceanic or surface warming (Section 10.6.4.2). No models are available for such a link, so we assume that the imbalance might scale up with global average surface temperature change, which we take as a measure of the magnitude of climate change (see Appendix 10.A). This assumption adds 0.1 to 0.2 m to the estimated upper bound for sea level rise depending on the scenario (Table 10.7). During 2090 to 2099, the rate of scaled-up antarctic discharge roughly balances the increased rate of antarctic accumulation (SMB). The central estimate for the increased antarctic discharge under the SRES scenario A1FI is about 1.3 mm yr^{-1}, a factor of 5 to 10 greater than in recent years, and similar to the order-of-magnitude upper limit of Section 10.6.4.2. It must be emphasized that we cannot assess the likelihood of any of these three alternatives, which are presented as illustrative. The state of understanding prevents a best estimate from being made.

The central estimates for sea level rise in Table 10.7 are smaller than the TAR model means (Church et al., 2001) by 0.03 to 0.07 m, depending on scenario, for two reasons. First, these projections are for 2090-2099, whereas the TAR projections were for 2100. Second, the TAR included some small constant additional contributions to sea level rise which are omitted here (see below regarding permafrost). If the TAR model means are adjusted for this, they are within 10% of the central estimates from Table 10.7. (See Appendix 10.A for further information.) For each scenario, the upper bound of sea level rise in Table 10.7 is smaller than in the TAR, and the lower bound is larger than in the TAR. This is because the uncertainty on the sea level projection has been reduced, for a combination of reasons (see Appendix 10.A for details). The TAR would have had similar ranges to those shown here if it had treated the uncertainties in the same way.

Thawing of permafrost is projected to contribute about 5 mm during the 21st century under the SRES scenario A2 (calculated from Lawrence and Slater, 2005). The mass of the ocean will also be changed by climatically driven alteration in other water storage, in the forms of atmospheric water vapour, seasonal snow cover, soil moisture, groundwater, lakes and rivers. All of these are expected to be relatively small terms, but there may be substantial contributions from anthropogenic change in terrestrial water storage, through extraction from aquifers and impounding in reservoirs (see Sections 5.5.5.3 and 5.5.5.4).

10.7   Long Term Climate Change and Commitment

10.7.1   Climate Change Commitment to Year 2300 Based on AOGCMs

Building on Wigley (2005), we use three specific definitions of climate change commitment: (i) the 'constant composition commitment', which denotes the further change of temperature ('constant composition temperature commitment' or 'committed warming'), sea level ('constant composition sea level commitment') or any other quantity in the climate system, since the time the composition of the atmosphere, and hence the radiative forcing, has been held at a constant value; (ii) the 'constant emission commitment', which denotes the further change of, for example, temperature ('constant emission temperature commitment') since the time the greenhouse gas emissions have been held at a constant value; and (iii) the 'zero emission commitment', which denotes the further change of, for example, temperature ('zero emission temperature commitment') since the time the greenhouse gas emissions have been set to zero.

The concept that the climate system exhibits commitment when radiative forcing has changed is mainly due to the thermal inertia of the oceans, and was discussed independently by Wigley (1984), Hansen et al. (1984) and Siegenthaler and Oeschger

(1984). The term 'commitment' in this regard was introduced by Ramanathan (1988). In the TAR, this was illustrated in idealised scenarios of doubling and quadrupling atmospheric CO_2, and stabilisation at 2050 and 2100 after an IS92a forcing scenario. Various temperature commitment values were reported (about 0.3°C per century with much model dependency), and EMIC simulations were used to illustrate the long-term influence of the ocean owing to long mixing times and the MOC. Subsequent studies have confirmed this behaviour of the climate system and ascribed it to the inherent property of the climate system that the thermal inertia of the ocean introduces a lag to the warming of the climate system after concentrations of greenhouse gases are stabilised (Mitchell et al., 2000; Wetherald et al., 2001; Wigley and Raper, 2003; Hansen et al., 2005b; Meehl et al., 2005c; Wigley, 2005). Climate change commitment as discussed here should not be confused with 'unavoidable climate change' over the next half century, which would surely be greater because forcing cannot be instantly stabilised. Furthermore, in the very long term it is plausible that climate change could be less than in a commitment run since forcing could plausibly be reduced below current levels as illustrated in the overshoot simulations and zero emission commitment simulations discussed below.

Three constant composition commitment experiments have recently been performed by the global coupled climate modelling community: (1) stabilising concentrations of greenhouse gases at year 2000 values after a 20th-century climate simulation, and running the model for an additional 100 years; (2) stabilising concentrations of greenhouse gases at year 2100 values after a 21st-century B1 experiment (e.g., CO_2 near 550 ppm) and running the model for an additional 100 years (with some models run to 200 years); and (3) stabilising concentrations of greenhouse gases at year 2100 values after a 21st-century A1B experiment (e.g., CO_2 near 700 ppm), and running the model for an additional 100 years (and some models to 200 years). Multi-model mean warming in these experiments is depicted in Figure 10.4. Time series of the globally averaged surface temperature and percent precipitation change after stabilisation are shown for all the models in the Supplementary Material, Figure S10.3.

The multi-model average warming for all radiative forcing agents held constant at year 2000 (reported earlier for several of the models by Meehl et al., 2005c), is about 0.6°C for the period 2090 to 2099 relative to the 1980 to 1999 reference period. This is roughly the magnitude of warming simulated in the 20th century. Applying the same uncertainty assessment as for the SRES scenarios in Fig. 10.29 (–40 to +60%), the likely uncertainty range is 0.3°C to 0.9°C. Hansen et al. (2005a) calculate the current energy imbalance of the Earth to be 0.85 W m^{-2}, implying that the unrealised global warming is about 0.6°C without any further increase in radiative forcing. The committed warming trend values show a rate of warming averaged over the first two decades of the 21st century of about 0.1°C per decade, due mainly to the slow response of the oceans. About twice as much warming (0.2°C per decade) would be expected if emissions are within the range of the SRES scenarios.

For the B1 constant composition commitment run, the additional warming after 100 years is also about 0.5°C, and roughly the same for the A1B constant composition commitment (Supplementary Material, Figure S10.3). These new results quantify what was postulated in the TAR in that the warming commitment after stabilising concentrations is about 0.5°C for the first century, and considerably smaller after that, with most of the warming commitment occurring in the first several decades of the 22nd century.

Constant composition precipitation commitment for the multi-model ensemble average is about 1.1% by 2100 for the 20th-century constant composition commitment experiment, and for the B1 constant composition commitment experiment it is 0.8% by 2200 and 1.5% by 2300, while for the A1B constant composition commitment experiment it is 1.5% by 2200 and 2% by 2300.

The patterns of change in temperature in the B1 and A1B experiments, relative to the pre-industrial period, do not change greatly after stabilisation (Table 10.5). Even the 20th-century stabilisation case warms with some similarity to the A1B pattern (Table 10.5). However, there is some contrast in the land and ocean warming rates, as seen from Figure 10.6. Mid- and low-latitude land warms at rates closer to the global mean of that of A1B, while high-latitude ocean warming is larger.

10.7.2 Climate Change Commitment to Year 3000 and Beyond to Equilibrium

Earth System Models of Intermediate Complexity are used to extend the projections for a scenario that follows A1B to 2100 and then keeps atmospheric composition, and hence radiative forcing, constant to the year 3000 (see Figure 10.34). By 2100, the projected warming is between 1.2°C and 4.1°C, similar to the range projected by AOGCMs. A large constant composition temperature and sea level commitment is evident in the simulations and is slowly realised over coming centuries. By the year 3000, the warming range is 1.9°C to 5.6°C. While surface temperatures approach equilibrium relatively quickly, sea level continues to rise for many centuries.

Five of these EMICs include interactive representations of the marine and terrestrial carbon cycle and, therefore, can be used to assess carbon cycle-climate feedbacks and effects of

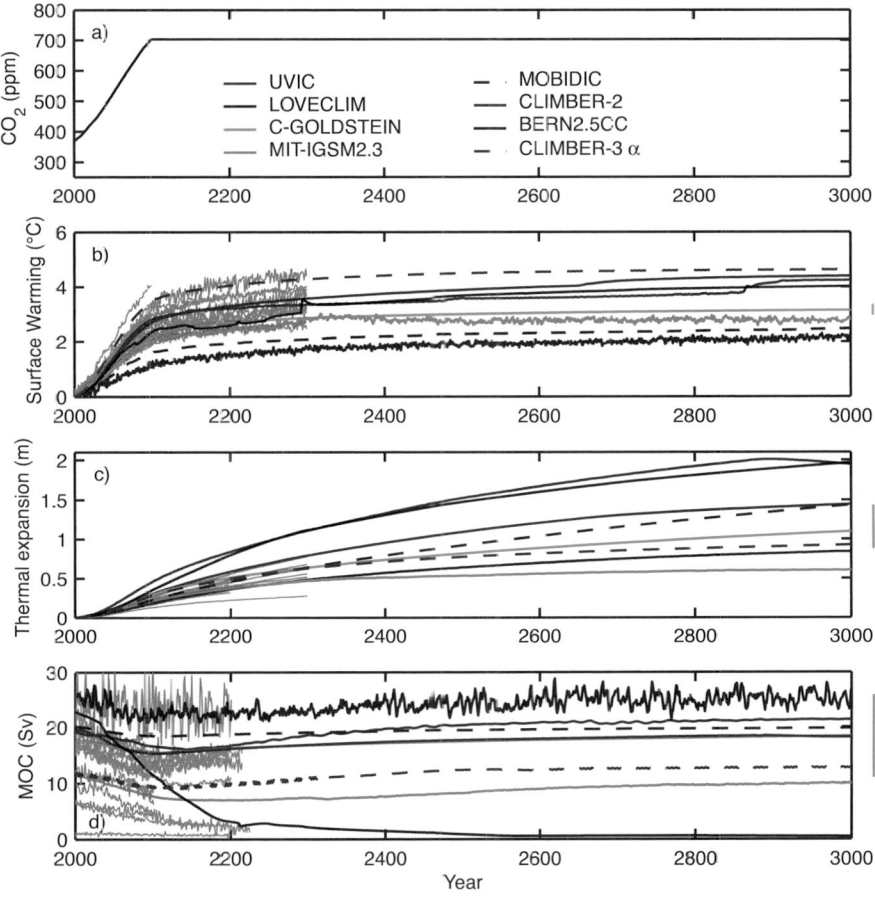

Figure 10.34. *(a) Atmospheric CO₂, (b) global mean surface warming, (c) sea level rise from thermal expansion and (d) Atlantic meridional overturning circulation (MOC) calculated by eight EMICs for the SRES A1B scenario and stable radiative forcing after 2100, showing long-term commitment after stabilisation. Coloured lines are results from EMICs, grey lines indicate AOGCM results where available for comparison. Anomalies in (b) and (c) are given relative to the year 2000. Vertical bars indicate ±2 standard deviation uncertainties due to ocean parameter perturbations in the C-GOLDSTEIN model. The MOC shuts down in the BERN2.5CC model, leading to an additional contribution to sea level rise. Individual EMICs (see Table 8.3 for model details) treat the effect from non-CO₂ greenhouse gases and the direct and indirect aerosol effects on radiative forcing differently. Despite similar atmospheric CO₂ concentrations, radiative forcing among EMICs can thus differ within the uncertainty ranges currently available for present-day radiative forcing (see Chapter 2).*

Frequently Asked Question 10.3

If Emissions of Greenhouse Gases are Reduced, How Quickly do Their Concentrations in the Atmosphere Decrease?

The adjustment of greenhouse gas concentrations in the atmosphere to reductions in emissions depends on the chemical and physical processes that remove each gas from the atmosphere. Concentrations of some greenhouse gases decrease almost immediately in response to emission reduction, while others can actually continue to increase for centuries even with reduced emissions.

The concentration of a greenhouse gas in the atmosphere depends on the competition between the rates of emission of the gas into the atmosphere and the rates of processes that remove it from the atmosphere. For example, carbon dioxide (CO_2) is exchanged between the atmosphere, the ocean and the land through processes such as atmosphere-ocean gas transfer and chemical (e.g., weathering) and biological (e.g., photosynthesis) processes. While more than half of the CO_2 emitted is currently removed from the atmosphere within a century, some fraction (about 20%) of emitted CO_2 remains in the atmosphere for many millennia. Because of slow removal processes, atmospheric CO_2 will continue to increase in the long term even if its emission is substantially reduced from present levels. Methane (CH_4) is removed by chemical processes in the atmosphere, while nitrous oxide (N_2O) and some halocarbons are destroyed in the upper atmosphere by solar radiation. These processes each operate at different time scales ranging from years to millennia. A measure for this is the lifetime of a gas in the atmosphere, defined as the time it takes for a perturbation to be reduced to 37% of its initial amount. While for CH_4, N_2O, and other trace gases such as hydrochlorofluorocarbon-22 (HCFC-22), a refrigerant fluid, such lifetimes can be reasonably determined (for CH_4 it is about 12 yr, for N_2O about 110 yr and for HCFC-22 about 12 yr), a lifetime for CO_2 cannot be defined.

The change in concentration of any trace gas depends in part on how its emissions evolve over time. If emissions increase with time, the atmospheric concentration will also increase with time, regardless of the atmospheric lifetime of the gas. However, if actions are taken to reduce the emissions, the fate of the trace gas concentration will depend on the relative changes not only of emissions but also of its removal processes. Here we show how the lifetimes and removal processes of different gases dictate the evolution of concentrations when emissions are reduced.

As examples, FAQ 10.3, Figure 1 shows test cases illustrating how the future concentration of three trace gases would respond to illustrative changes in emissions (represented here as a response to an imposed pulse change in emission). We consider CO_2, which has no specific lifetime, as well as a trace gas with a well-defined long lifetime on the order of a century (e.g., N_2O), and a trace gas with a well-defined short lifetime on the order of decade (such as CH_4, HCFC-22 or other halocarbons). For each gas, five illustrative cases of future emissions are presented: stabilisation of emissions at present-day levels, and immediate emission reduction by 10%, 30%, 50% and 100%.

The behaviour of CO_2 (Figure 1a) is completely different from the trace gases with well-defined lifetimes. Stabilisation of CO_2 emissions at current levels would result in a continuous increase of atmospheric CO_2 over the 21st century and beyond, whereas for a gas with a lifetime on the order of a century (Figure 1b) or a decade (Figure 1c), stabilisation of emissions at current levels would lead to a stabilisation of its concentration at a level higher than today within a couple of centuries, or decades, respectively. In fact, only in the case of essentially complete elimination of

(continued)

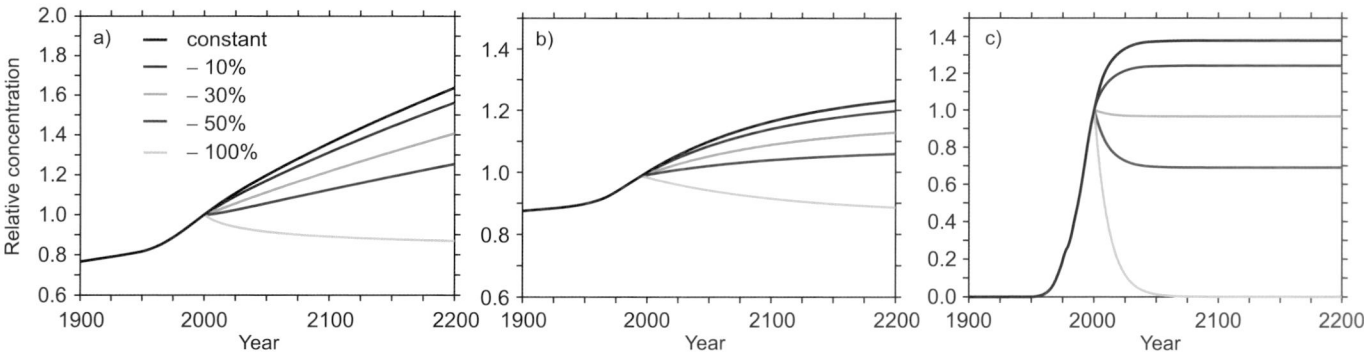

FAQ 10.3, Figure 1. *(a) Simulated changes in atmospheric CO_2 concentration relative to the present-day for emissions stabilised at the current level (black), or at 10% (red), 30% (green), 50% (dark blue) and 100% (light blue) lower than the current level; (b) as in (a) for a trace gas with a lifetime of 120 years, driven by natural and anthropogenic fluxes; and (c) as in (a) for a trace gas with a lifetime of 12 years, driven by only anthropogenic fluxes.*

emissions can the atmospheric concentration of CO_2 ultimately be stabilised at a constant level. All other cases of moderate CO_2 emission reductions show increasing concentrations because of the characteristic exchange processes associated with the cycling of carbon in the climate system.

More specifically, the rate of emission of CO_2 currently greatly exceeds its rate of removal, and the slow and incomplete removal implies that small to moderate reductions in its emissions would not result in stabilisation of CO_2 concentrations, but rather would only reduce the rate of its growth in coming decades. A 10% reduction in CO_2 emissions would be expected to reduce the growth rate by 10%, while a 30% reduction in emissions would similarly reduce the growth rate of atmospheric CO_2 concentrations by 30%. A 50% reduction would stabilise atmospheric CO_2, but only for less than a decade. After that, atmospheric CO_2 would be expected to rise again as the land and ocean sinks decline owing to well-known chemical and biological adjustments. Complete elimination of CO_2 emissions is estimated to lead to a slow decrease in atmospheric CO_2 of about 40 ppm over the 21st century.

The situation is completely different for the trace gases with a well-defined lifetime. For the illustrative trace gas with a lifetime of the order of a century (e.g., N_2O), emission reduction of more than 50% is required to stabilise the concentrations close to present-day values (Figure 1b). Constant emission leads to a stabilisation of the concentration within a few centuries.

In the case of the illustrative gas with the short lifetime, the present-day loss is around 70% of the emissions. A reduction in emissions of less than 30% would still produce a short-term increase in concentration in this case, but, in contrast to CO_2, would lead to stabilisation of its concentration within a couple of decades (Figure 1c). The decrease in the level at which the concentration of such a gas would stabilise is directly proportional to the emission reduction. Thus, in this illustrative example, a reduction in emissions of this trace gas larger than 30% would be required to stabilise concentrations at levels significantly below those at present. A complete cut-off of the emissions would lead to a return to pre-industrial concentrations within less than a century for a trace gas with a lifetime of the order of a decade.

carbon emission reductions on atmospheric CO_2 and climate. Although carbon cycle processes in these models are simplified, global-scale quantities are in good agreement with more complex models (Doney et al., 2004).

Results for one carbon emission scenario are shown in Figure 10.35, where anthropogenic emissions follow a path towards stabilisation of atmospheric CO_2 at 750 ppm but at year 2100 are reduced to zero. This permits the determination of the zero emission climate change commitment. The prescribed emissions were calculated from the SP750 profile (Knutti et al., 2005) using the BERN-CC model (Joos et al., 2001). Although unrealistic, such a scenario permits the calculation of zero emission commitment, i.e., climate change due to 21st-century emissions. Even though emissions are instantly reduced to zero at year 2100, it takes about 100 to 400 years in the different models for the atmospheric CO_2 concentration to drop from the maximum (ranges between 650 to 700 ppm) to below the level of doubled pre-industrial CO_2 (~560 ppm) owing to a continuous transfer of carbon from the atmosphere into the terrestrial and oceanic reservoirs. Emissions during the 21st century continue to have an impact even at year 3000 when both surface temperature and sea level rise due to thermal expansion are still substantially higher than pre-industrial. Also shown are atmospheric CO_2 concentrations and ocean/terrestrial carbon inventories at year 3000 versus total emitted carbon for similar emission pathways targeting (but not actually reaching) 450, 550, 750 and 1,000 ppm atmospheric CO_2 and with carbon emissions reduced to zero at year 2100. Atmospheric CO_2 at year 3000 is approximately linearly related to the total amount of carbon emitted in each model, but with a substantial spread among the models in both slope and absolute values, because the redistribution of carbon between the different reservoirs is

model dependent. In summary, the model results show that 21st-century emissions represent a minimum commitment of climate change for several centuries, irrespective of later emissions. A reduction of this 'minimum' commitment is possible only if, in addition to avoiding CO_2 emissions after 2100, CO_2 were actively removed from the atmosphere.

Using a similar approach, Friedlingstein and Solomon (2005) show that even if emissions were immediately cut to zero, the system would continue to warm for several more decades before starting to cool. It is important also to note that ocean heat content and changes in the cryosphere evolve on time scales extending over centuries.

On very long time scales (order several thousand years as estimated by AOGCM experiments, Bi et al., 2001; Stouffer, 2004), equilibrium climate sensitivity is a useful concept to characterise the ultimate response of climate models to different future levels of greenhouse gas radiative forcing. This concept can be applied to climate models irrespective of their complexity. Based on a global energy balance argument, equilibrium climate sensitivity S and global mean surface temperature increase ΔT at equilibrium relative to pre-industrial for an equivalent stable CO_2 concentration are linearly related according to $\Delta T = S \times \log(CO_2 / 280 \text{ ppm}) / \log(2)$, which follows from the definition of climate sensitivity and simplified expressions for the radiative forcing of CO_2 (Section 6.3.5 of the TAR). Because the combination of various lines of modelling results and expert judgement yields a quantified range of climate sensitivity S (see Box 10.2), this can be carried over to equilibrium temperature increase. Most likely values, and the likely range, as well as a very likely lower bound for the warming, all consistent with the quantified range of S, are given in Table 10.8.

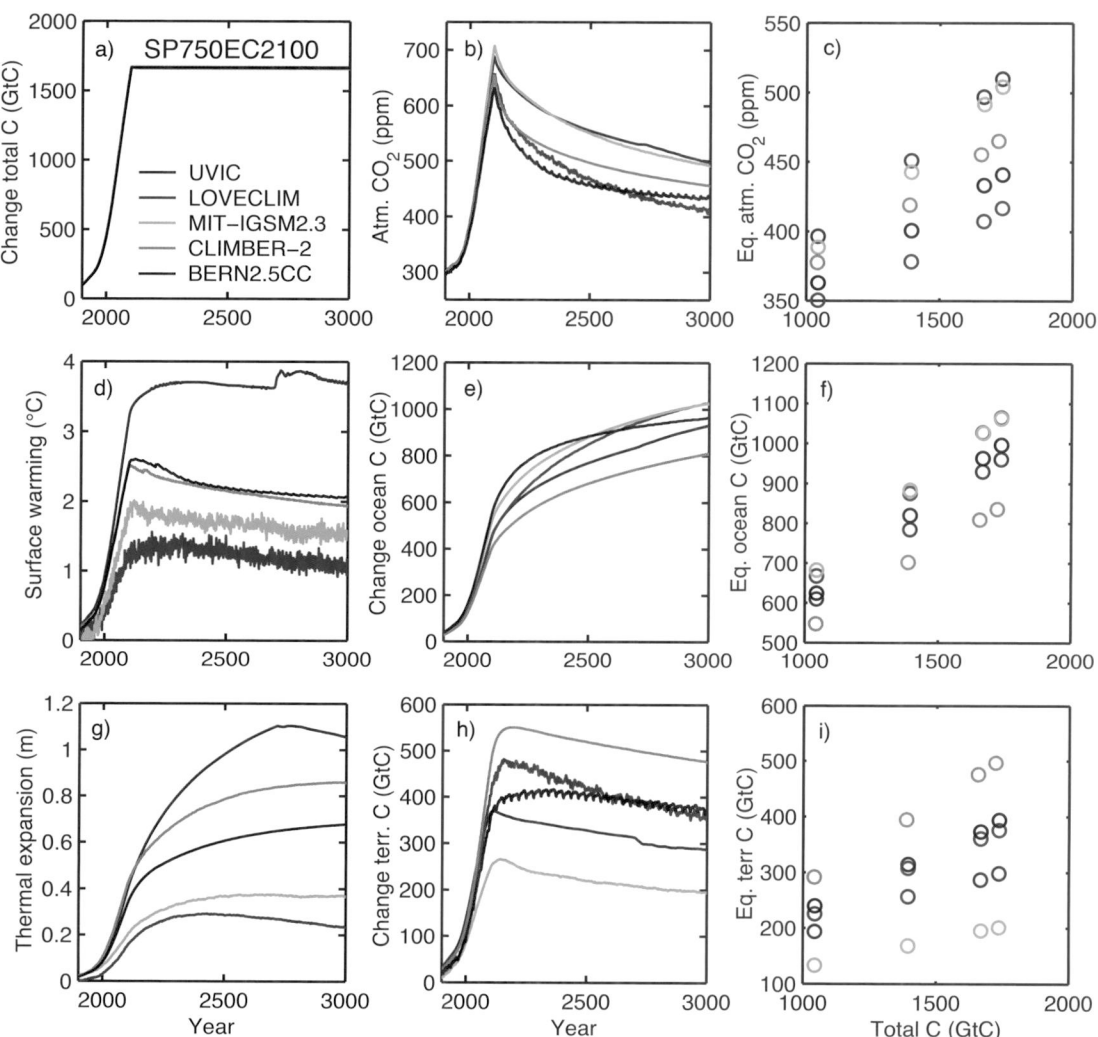

Figure 10.35. *Changes in carbon inventories and climate response relative to the pre-industrial period simulated by five different intermediate complexity models (see Table 8.3 for model descriptions) for a scenario where emissions follow a pathway leading to stabilisation of atmospheric CO₂ at 750 ppm, but before reaching this target, emissions are reduced to zero instantly at year 2100. (a) Change in total carbon, (b) atmospheric CO₂, (d) change in surface temperature, (e) change in ocean carbon, (g) sea level rise from thermal expansion and (h) change in terrestrial carbon. Right column: (c) atmospheric CO₂ and the change in (f) oceanic and (i) terrestrial carbon inventories at year 3000 relative to the pre-industrial period for several emission scenarios of similar shape but with different total carbon emissions.*

Table 10.8. *Best guess (i.e. most likely), likely and very likely bounds/ranges of global mean equilibrium surface temperature increase ΔT(°C) above pre-industrial temperatures for different levels of CO₂ equivalent concentrations (ppm), based on the assessment of climate sensitivity given in Box 10.2.*

Equivalent CO₂	Best Guess	Very Likely Above	Likely in the Range
350	1.0	0.5	0.6–1.4
450	2.1	1.0	1.4–3.1
550	2.9	1.5	1.9–4.4
650	3.6	1.8	2.4–5.5
750	4.3	2.1	2.8–6.4
1,000	5.5	2.8	3.7–8.3
1,200	6.3	3.1	4.2–9.4

It is emphasized that this table does not contain more information than the best knowledge of *S* and that the numbers are not the result of any climate model simulation. Rather it is assumed that the above relationship between temperature increase and CO₂ holds true for the entire range of equivalent CO₂ concentrations. There are limitations to the concept of radiative forcing and climate sensitivity (Senior and Mitchell, 2000; Joshi et al., 2003; Shine et al., 2003; Hansen et al., 2005b). Only a few AOGCMs have been run to equilibrium under elevated CO₂ concentrations, and some results show that nonlinearities in the feedbacks (e.g., clouds, sea ice and snow cover) may cause a time dependence of the effective climate sensitivity and substantial deviations from the linear relation assumed above (Manabe and Stouffer, 1994; Senior and Mitchell, 2000; Voss and Mikolajewicz, 2001; Gregory et al., 2004b), with effective climate sensitivity tending to grow with time in some of the AR4 AOGCMs. Some studies suggest that climate sensitivities larger than the likely estimate given below (which would suggest greater warming) cannot be ruled out (see Box 10.2 on climate sensitivity).

Another way to address eventual equilibrium temperature for different CO_2 concentrations is to use the projections from the AOGCMs in Figure 10.4, and an idealised 1% yr^{-1} CO_2 increase to 4 × CO_2. The equivalent CO_2 concentrations in the AOGCMs can be estimated from the forcings given in Table 6.14 in the TAR. The actual CO_2 concentrations for A1B and B1 are roughly 715 ppm and 550 ppm (depending on which model is used to convert emissions to concentrations), and equivalent CO_2 concentrations are estimated to be about 835 ppm and 590 ppm, respectively. Using the equation above for an equilibrium climate sensitivity of 3.0°C, eventual equilibrium warming in these experiments would be 4.8°C and 3.3°C, respectively. The multi-model average warming in the AOGCMs at the end of the 21st century (relative to pre-industrial temperature) is 3.1°C and 2.3°C, or about 65 to 70% of the eventual estimated equilibrium warming. Given rates of CO_2 increase of between 0.5 and 1.0% yr^{-1} in these two scenarios, this can be compared to the calculated fraction of eventual warming of around 50% in AOGCM experiments with those CO_2 increase rates (Stouffer and Manabe, 1999). The Stouffer and Manabe (1999) model has somewhat higher equilibrium climate sensitivity, and was actually run to equilibrium in a 4-kyr integration to enable comparison of transient and equilibrium warming. Therefore, the AOGCM results combined with the estimated equilibrium warming seem roughly consistent with earlier AOGCM experiments of transient warming rates. Additionally, similar numbers for the 4 × CO_2 stabilisation experiments performed with the AOGCMs can be computed. In that case, the actual and equivalent CO_2 concentrations are the same, since there are no other radiatively active species changing in the models, and the multi-model CO_2 concentration at quadrupling would produce an eventual equilibrium warming of 6°C, where the multi-model average warming at the time of quadrupling is about 4.0°C or 66% of eventual equilibrium. This is consistent with the numbers for the A1B and B1 scenario integrations with the AOGCMs.

It can be estimated how much closer to equilibrium the climate system is 100 years after stabilisation in these AOGCM experiments. After 100 years of stabilised concentrations, the warming relative to pre-industrial temperature is 3.8°C in A1B and 2.6°C in B1, or about 80% of the estimated equilibrium warming. For the stabilised 4 × CO_2 experiment, after 100 years of stabilised CO_2 concentrations the warming is 4.7°C, or 78% of the estimated equilibrium warming. Therefore, about an additional 10 to 15% of the eventual equilibrium warming is achieved after 100 years of stabilised concentrations (Stouffer, 2004). This emphasizes that the approach to equilibrium takes a long time, and even after 100 years of stabilised atmospheric concentrations, only about 80% of the eventual equilibrium warming is realised.

10.7.3 Long-Term Integrations: Idealised Overshoot Experiments

The concept of mitigation related to overshoot scenarios has implications for IPCC Working Groups II and III and was addressed in the Second Assessment Report. A new suite of mitigation scenarios is currently being assessed for the AR4. Working Group I does not have the expertise to assess such scenarios, so this section assesses the processes and response of the physical climate system in a very idealised overshoot experiment. Plausible new mitigation and overshoot scenarios will be run subsequently by modelling groups and assessed in the next IPCC report.

An idealised overshoot scenario has been run in an AOGCM where the CO_2 concentration decreases from the A1B stabilised level to the B1 stabilised level between 2150 and 2250, followed by 200 years of integration with that constant B1 level (Figure 10.36a). This reduction in CO_2 concentration would require large reductions in emissions, but such an idealised experiment illustrates the processes involved in how the climate system would respond to such a large change in emissions and concentrations. Yoshida et al. (2005) and Tsutsui et al. (2007) show that there is a relatively fast response in the surface and upper ocean, which start to recover to temperatures at the B1 level after several decades, but a much more sluggish response with more commitment in the deep ocean. As shown in Figure 10.36b and c, the overshoot scenario temperatures only slowly decrease to approach the lower temperatures of the B1 experiment, and continue a slow convergence that has still not cooled to the B1 level at the year 2350, or 100 years after the CO_2 concentration in the overshoot experiment was reduced to equal the concentration in the B1 experiment. However, Dai et al. (2001a) show that reducing emissions to achieve a stabilised CO_2 concentration in the 21st century reduces warming moderately (less than 0.5°C) by the end of the 21st century in comparison to a business-as-usual scenario, but the warming reduction is about 1.5°C by the end of the 22nd century in that experiment. Other climate system responses include the North Atlantic MOC and sea ice volume that almost recover to the B1 level in the overshoot scenario experiment, except for a significant hysteresis effect that is shown in the sea level change due to thermal expansion (Yoshida et al., 2005; Nakashiki et al., 2006).

Such stabilisation and overshoot scenarios have implications for risk assessment as suggested by Yoshida et al. (2005) and others. For example, in a probabilistic study using an SCM and multi-gas scenarios, Meinshausen (2006) estimated that the probability of exceeding a 2°C warming is between 68 and 99% for a stabilisation of equivalent CO_2 at 550 ppm. They also considered scenarios with peaking CO_2 and subsequent stabilisation at lower levels as an alternative pathway and found that if the risk of exceeding a warming of 2°C is not to be greater than 30%, it is necessary to peak equivalent CO_2 concentrations around 475 ppm before returning to lower concentrations of about 400 ppm. These overshoot and targeted climate change estimations take into account the climate change commitment

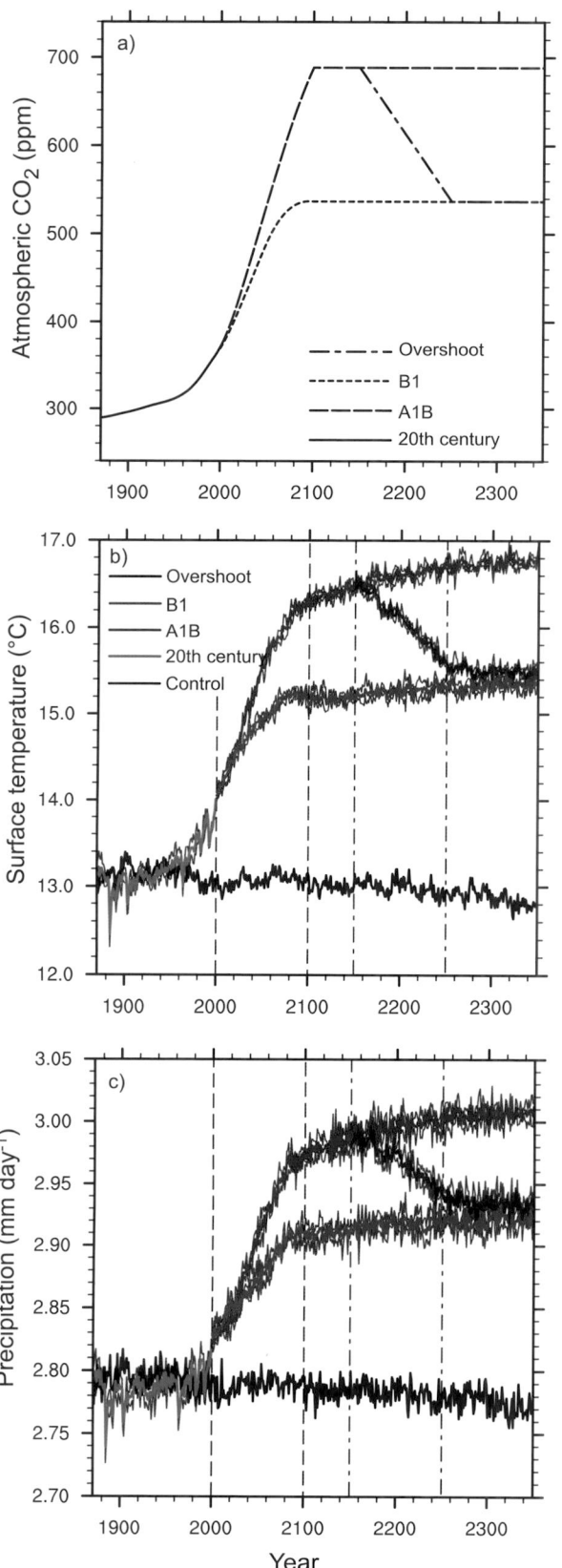

Figure 10.36. *(a) Atmospheric CO_2 concentrations for several experiments simulated with an AOGCM; (b) globally averaged surface air temperatures for the overshoot scenario and the A1B and B1 experiments; (c) same as in (b) but for globally averaged precipitation rate. Modified from Yoshida et al. (2005).*

in the system that must be overcome on the time scale of any overshoot or emissions target calculation. The probabilistic studies also show that when certain thresholds of climate change are to be avoided, emission pathways depend on the certainty requested of not exceeding the threshold.

Earth System Models of Intermediate Complexity have been used to calculate the long-term climate response to stabilisation of atmospheric CO_2, although EMICs have not been adjusted to take into account the full range of AOGCM sensitivities. The newly developed stabilisation profiles were constructed following Enting et al. (1994) and Wigley et al. (1996) using the most recent atmospheric CO_2 observations, CO_2 projections with the BERN-CC model (Joos et al., 2001) for the A1T scenario over the next few decades, and a ratio of two polynomials (Enting et al., 1994) leading to stabilisation at levels of 450, 550, 650, 750 and 1,000 ppm atmospheric CO_2 equivalent. Other forcings are not considered. Supplementary Material, Figure S10.4a shows the equilibrium surface warming for seven different EMICs and six stabilisation levels. Model differences arise mainly from the models having different climate sensitivities.

Knutti et al. (2005) explore this further with an EMIC using several published PDFs of climate sensitivity and different ocean heat uptake parametrizations and calculate probabilities of not overshooting a certain temperature threshold given an equivalent CO_2 stabilisation level (Supplementary Material, Figure S10.4b). This plot illustrates, for example, that for low values of stabilised CO_2, the range of response of possible warming is smaller than for high values of stabilised CO_2. This is because with greater CO_2 forcing, there is a greater spread of outcomes as illustrated in Figure 10.26. Figure S10.4b also shows that for any given temperature threshold, the smaller the desired probability of exceeding the target is, the lower the stabilisation level that must be chosen. Stabilisation of atmospheric greenhouse gases below about 400 ppm CO_2 equivalent is required to keep the global temperature increase likely less than 2°C above pre-industrial temperature (Knutti et al., 2005).

10.7.4 Commitment to Sea Level Rise

10.7.4.1 Thermal Expansion

The sea level rise commitment due to thermal expansion has much longer time scales than the surface warming commitment, owing to the slow processes that mix heat into the deep ocean (Church et al., 2001). If atmospheric composition were stabilised at A1B levels in 2100, thermal expansion in the 22nd century would be similar to in the 21st (see, e.g., Section 10.6.1; Meehl et al., 2005c), reaching 0.3 to 0.8 m by 2300 (Figure 10.37). The ranges of thermal expansion overlap substantially for stabilisation at different levels, since model uncertainty is dominant; A1B is given here because results are available from more models for this scenario than for other scenarios. Thermal expansion would continue over many centuries at a gradually decreasing rate (Figure 10.34). There is a wide spread among

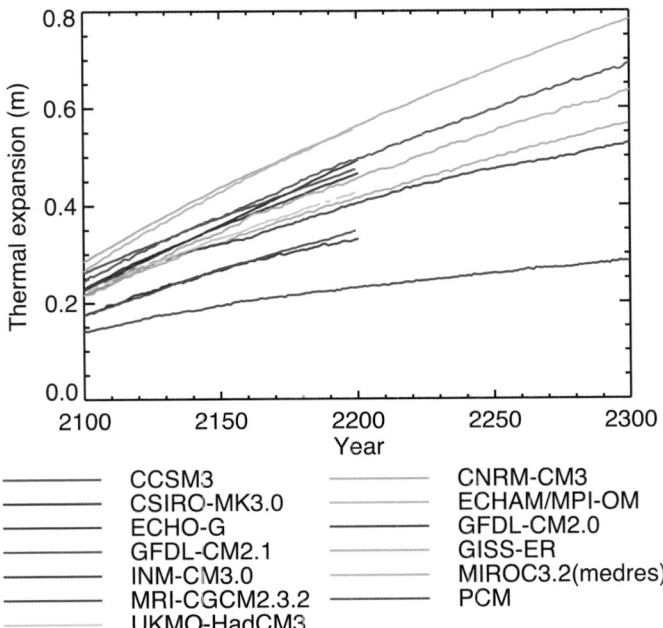

CCSM3   CNRM-CM3
CSIRO-MK3.0   ECHAM/MPI-OM
ECHO-G   GFDL-CM2.0
GFDL-CM2.1   GISS-ER
INM-CM3.0   MIROC3.2(medres)
MRI-CGCM2.3.2   PCM
UKMO-HadCM3

Figure 10.37. *Globally averaged sea level rise from thermal expansion relative to the period 1980 to 1999 for the A1B commitment experiment calculated from AOGCMs. See Table 8.1 for model details.*

the models for the thermal expansion commitment at constant composition due partly to climate sensitivity, and partly to differences in the parametrization of vertical mixing affecting ocean heat uptake (e.g., Weaver and Wiebe, 1999). If there is deep-water formation in the final steady state as in the present day, the ocean will eventually warm up fairly uniformly by the amount of the global average surface temperature change (Stouffer and Manabe, 2003), which would result in about 0.5 m of thermal expansion per degree celsius of warming, calculated from observed climatology; the EMICs in Figure 10.34 indicate 0.2 to 0.6 m °C^{-1} for their final steady state (year 3000) relative to 2000. If deep-water formation is weakened or suppressed, the deep ocean will warm up more (Knutti and Stocker, 2000). For instance, in the 3 × CO$_2$ experiment of Bi et al. (2001) with the CSIRO AOGCM, both North Atlantic Deep Water and Antarctic Bottom Water formation cease, and the steady-state thermal expansion is 4.5 m. Although these commitments to sea level rise are large compared with 21st-century changes, the eventual contributions from the ice sheets could be larger still.

10.7.4.2 Glaciers and Ice Caps

Steady-state projections for G&IC require a model that evolves their area-altitude distribution (see, e.g., Section 10.6.3.3). Little information is available on this. A comparative study including seven GCM simulations at 2 × CO$_2$ conditions inferred that many glaciers may disappear completely due to an increase of the equilibrium line altitude (Bradley et al., 2004), but even in a warmer climate, some glacier volume may persist at high altitude. With a geographically uniform warming relative to 1900 of 4°C maintained after 2100, about 60% of G&IC volume would vanish by 2200 and practically all by 3000

(Raper and Braithwaite, 2006). Nonetheless, this commitment to sea level rise is relatively small (<1 m; Table 4.4) compared with those from thermal expansion and ice sheets.

10.7.4.3 Greenland Ice Sheet

The present SMB of Greenland is a net accumulation estimated as 0.6 mm yr^{-1} of sea level equivalent from a compilation of studies (Church et al., 2001) and 0.47 mm yr^{-1} for 1988 to 2004 (Box et al., 2006). In a steady state, the net accumulation would be balanced by calving of icebergs. General Circulation Models suggest that ablation increases more rapidly than accumulation with temperature (van de Wal et al., 2001; Gregory and Huybrechts, 2006), so warming will tend to reduce the SMB, as has been observed in recent years (see Section 4.6.3), and is projected for the 21st century (Section 10.6.4.1). Sufficient warming will reduce the SMB to zero. This gives a threshold for the long-term viability of the ice sheet because negative SMB means that the ice sheet must contract even if ice discharge has ceased owing to retreat from the coast. If a warmer climate is maintained, the ice sheet will eventually be eliminated, except perhaps for remnant glaciers in the mountains, raising sea level by about 7 m (see Table 4.1). Huybrechts et al. (1991) evaluated the threshold as 2.7°C of seasonally and geographically uniform warming over Greenland relative to a steady state (i.e. pre-industrial temperature). Gregory et al. (2004a) examine the probability of this threshold being reached under various CO$_2$ stabilisation scenarios for 450 to 1000 ppm using TAR projections, and find that it was exceeded in 34 out of 35 combinations of AOGCM and CO$_2$ concentration considering seasonally uniform warming, and 24 out of 35 considering summer warming and using an upper bound on the threshold.

Assuming the warming to be uniform underestimates the threshold, because warming is projected by GCMs to be weaker in the ablation area and in summer, when ablation occurs. Using geographical and seasonal patterns of simulated temperature change derived from a combination of four high-resolution AGCM simulations and 18 AR4 AOGCMs raises the threshold to 3.2°C to 6.2°C in annual- and area-average warming in Greenland, and 1.9°C to 4.6°C in the global average (Gregory and Huybrechts, 2006), relative to pre-industrial temperatures. This is likely to be reached by 2100 under the SRES A1B scenario, for instance (Figure 10.29). These results are supported by evidence from the last interglacial, when the temperature in Greenland was 3°C to 5°C warmer than today and the ice sheet survived, but may have been smaller by 2 to 4 m in sea level equivalent (including contributions from arctic ice caps, see Section 6.4.3). However, a lower threshold of 1°C (Hansen, 2005) in global warming above present-day temperatures has also been suggested, on the basis that global mean (rather than Greenland) temperatures during previous interglacials exceeded today's temperatures by no more than that.

For stabilisation in 2100 with SRES A1B atmospheric composition, Greenland would initially contribute 0.3 to

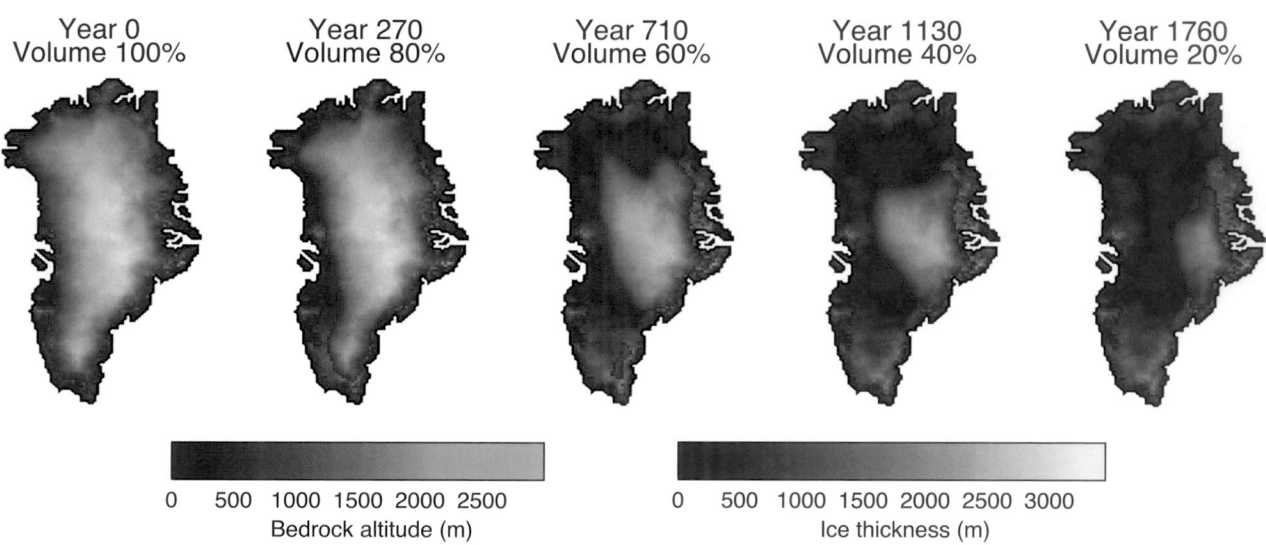

| Year 0 Volume 100% | Year 270 Volume 80% | Year 710 Volume 60% | Year 1130 Volume 40% | Year 1760 Volume 20% |

0   500 1000 1500 2000 2500
Bedrock altitude (m)

0   500 1000 1500 2000 2500 3000
Ice thickness (m)

Figure 10.38. *Evolution of Greenland surface elevation and ice sheet volume versus time in the experiment of Ridley et al. (2005) with the UKMO-HadCM3 AOGCM coupled to the Greenland Ice Sheet model of Huybrechts and De Wolde (1999) under a climate of constant quadrupled pre-industrial atmospheric CO_2.*

2.1 mm yr^{-1} to sea level (Table 10.7). The greater the warming, the faster the loss of mass. Ablation would be further enhanced by the lowering of the surface, which is not included in the calculations in Table 10.7. To include this and other climate feedbacks in calculating long-term rates of sea level rise requires coupling an ice sheet model to a climate model. Ridley et al. (2005) couple the Greenland Ice Sheet model of Huybrechts and De Wolde (1999) to the UKMO-HadCM3 AOGCM. Under constant $4 \times CO_2$, the sea level contribution is 5.5 mm yr^{-1} over the first 300 years and declines as the ice sheet contracts; after 1 kyr only about 40% of the original volume remains and after 3 kyr only 4% (Figure 10.38). The rate of deglaciation would increase if ice flow accelerated, as in recent years (Section 4.6.3.3). Basal lubrication due to surface melt water might cause such an effect (see Section 10.6.4.2). The best estimate of Parizek and Alley (2004) is that this could add an extra 0.15 to 0.40 m to sea level by 2500, compared with 0.4 to 3.2 m calculated by Huybrechts and De Wolde (1999) without this effect. The processes whereby melt water might penetrate through subfreezing ice to the bed are unclear and only conceptual models exist at present (Alley et al., 2005b).

Under pre-industrial or present-day atmospheric CO_2 concentrations, the climate of Greenland would be much warmer without the ice sheet, because of lower surface altitude and albedo, so it is possible that Greenland deglaciation and the resulting sea level rise would be irreversible. Toniazzo et al. (2004) find that snow does not accumulate anywhere on an ice-free Greenland with pre-industrial atmospheric CO_2, whereas Lunt et al. (2004) obtain a substantial regenerated ice sheet in east and central Greenland using a higher-resolution model.

10.7.4.4 Antarctic Ice Sheet

With rising global temperature, GCMs indicate increasingly positive SMB for the Antarctic Ice Sheet as a whole because

of greater accumulation (Section 10.6.4.1). For stabilisation in 2100 with SRES A1B atmospheric composition, antarctic SMB would contribute 0.4 to 2.0 mm yr^{-1} of sea level fall (Table 10.7). Continental ice sheet models indicate that this would be offset by tens of percent by increased ice discharge (Section 10.6.4.2), but still give a negative contribution to sea level, of –0.8 m by 3000 in one simulation with antarctic warming of about 4.5°C (Huybrechts and De Wolde, 1999).

However, discharge could increase substantially if buttressing due to the major West Antarctic ice shelves were reduced (see Sections 4.6.3.3 and 10.6.4.2), and could outweigh the accumulation increase, leading to a net positive antarctic sea level contribution in the long term. If the Amundsen Sea sector were eventually deglaciated, it would add about 1.5 m to sea level, while the entire West Antarctic Ice Sheet (WAIS) would account for about 5 m (Vaughan, 2007). Contributions could also come in this manner from the limited marine-based portions of East Antarctica that discharge into large ice shelves.

Weakening or collapse of the ice shelves could be caused either by surface melting or by thinning due to basal melting. In equilibrium experiments with mixed-layer ocean models, the ratio of antarctic to global annual warming is 1.4 ± 0.3. Following reasoning in Section 10.6.4.2 and Appendix 10.A, it appears that mean summer temperatures over the major West Antarctic ice shelves are about as likely as not to pass the melting point if global warming exceeds 5°C, and disintegration might be initiated earlier by surface melting. Observational and modelling studies indicate that basal melt rates depend on water temperature near to the base, with a constant of proportionality of about 10 m yr^{-1} °C^{-1} indicated for the Amundsen Sea ice shelves (Rignot and Jacobs, 2002; Shepherd et al., 2004) and 0.5 to 10 m yr^{-1} °C^{-1} for the Amery ice shelf (Williams et al., 2002). If this order of magnitude applies to future changes, a warming of about 1°C under the major ice shelves would eliminate them within centuries. We are not able to relate this quantitatively to global warming with any confidence, because the issue has so far received little attention, and current models may be inadequate to treat it because of limited resolution and poorly understood processes. Nonetheless, it is reasonable to suppose that sustained global warming would eventually lead to warming in the seawater circulating beneath the ice shelves.

Because the available models do not include all relevant processes, there is much uncertainty and no consensus about what dynamical changes could occur in the Antarctic Ice Sheet (see, e.g., Vaughan and Spouge, 2002; Alley et al., 2005a). One line of argument is to consider an analogy with palaeoclimate (see Box 4.1). Palaeoclimatic evidence that sea level was 4 to 6 m above present during the last interglacial may not all be explained by reduction in the Greenland Ice Sheet, implying a contribution from the Antarctic Ice Sheet (see Section 6.4.3). On this basis, using the limited available evidence, sustained global warming of 2°C (Oppenheimer and Alley, 2005) above present-day temperatures has been suggested as a threshold beyond which there will be a commitment to a large sea level contribution from the WAIS. The maximum rates of sea level rise during previous glacial terminations were of the order of 10 mm yr^{-1} (Church et al., 2001). We can be confident that future accelerated discharge from WAIS will not exceed this size, which is roughly an order of magnitude increase in present-day WAIS discharge, since no observed recent acceleration has exceeded a factor of ten.

Another line of argument is that there is insufficient evidence that rates of dynamical discharge of this magnitude could be sustained over long periods. The WAIS is 20 times smaller than the LGM NH ice sheets that contributed most of the melt water during the last deglaciation at rates that can be explained by surface melting alone (Zweck and Huybrechts, 2005). In the study of Huybrechts and De Wolde (1999), the largest simulated rate of sea level rise from the Antarctic Ice Sheet over the next 1 kyr is 2.5 mm yr^{-1}. This is dominated by dynamical discharge associated with grounding line retreat. The model did not simulate ice streams, for which widespread acceleration would give larger rates. However, the maximum loss of ice possible from rapid discharge of existing ice streams is the volume in excess of flotation in the regions occupied by these ice streams (defined as regions of flow exceeding 100 m yr^{-1}; see Section 10.6.4.2). This volume (in both West and East Antarctica) is 230,000 km^3, equivalent to about 0.6 m of sea level, or about 1% of the mass of the Antarctic Ice Sheet, most of which does not flow in ice streams. Loss of ice affecting larger portions of the ice sheet could be sustained at rapid rates only if new ice streams developed in currently slow-moving ice. The possible extent and rate of such changes cannot presently be estimated, since there is only very limited understanding of controls on the development and variability of ice streams. In this argument, rapid discharge may be transient and the long-term sign of the antarctic contribution to sea level depends on whether increased accumulation is more important than large-scale retreat of the grounding line.

References

Abdalati, W., et al., 2001: Outlet glacier and margin elevation changes: Near-coastal thinning of the Greenland ice sheet. *J. Geophys. Res.*, **106**, 33729–33741.

ACIA, 2004: *Arctic Climate Impact Assessment (ACIA): Impacts of a Warming Arctic.* Cambridge University Press, New York, NY, 140 pp.

Adams, P.J., et al., 2001: General circulation model assessment of direct radiative forcing by the sulfate-nitrate-ammonium-water inorganic aerosol system. *J. Geophys. Res.*, **106**, 1097–1111.

Allen, M.R., 1999: Do-it-yourself climate prediction. *Nature*, **401**, 627.

Allen, M.R., and W.J. Ingram, 2002: Constraints on future changes in climate and the hydrologic cycle. *Nature*, **419**, 224–232.

Allen, M.R., and D.A. Stainforth, 2002: Towards objective probabilistic climate forecasting. *Nature*, **419**, 228.

Allen, M.R., D.J. Frame, J.A. Kettleborough, and D.A. Stainforth, 2006a: Model error in weather and climate forecasting. In: *Predictability of Weather and Climate* [Palmer, T., and R. Hagedorn (eds.)]. Cambridge University Press, New York, NY, pp. 391–427.

Allen, M.R., et al., 2000: Quantifying the uncertainty in forecasts of anthropogenic climate change. *Nature*, **407**, 617–620.

Allen, M.R., et al., 2006b: Observational constraints on climate sensitivity. In: *Avoiding Dangerous Climate Change* [Schellnhuber, H.J., et al. (eds.)]. Cambridge University Press, New York, NY, pp. 281–289.

Alley, R.B., P.U. Clark, P. Huybrechts, and I. Joughin, 2005a: Ice-sheet and sea-level changes. *Science*, **310**, 456–460.

Alley, R.B., T.K. Dupont, B.R. Parizek, and S. Anandakrishnan, 2005b: Access of surface meltwater to beds of sub-freezing glaciers: Preliminary insights. *Ann. Glaciol.*, **40**, 8–14.

Alley, R.B., et al., 2002: *Abrupt Climate Change: Inevitable Surprises.* US National Research Council Report, National Academy Press, Washington, DC, 230 pp.

Amann, M., et al., 2004: *The RAINS Model. Documentation of the Model Approach Prepared for the RAINS Peer Review 2004.* International Institute for Applied Systems Analysis, Laxenburg, Austria, 156 pp.

Ammann, C.M., G.A. Meehl, W.M. Washington, and C.S. Zender, 2003: A monthly and latitudinally varying volcanic forcing dataset in simulations of 20th century climate. *Geophys. Res. Lett.*, **30**, 1657.

Andronova, N.G., and M.E. Schlesinger, 2001: Objective estimation of the probability density function for climate sensitivity. *J. Geophys. Res.*, **106**, 22605–22612.

Annan, J.D., and J.C. Hargreaves, 2006: Using multiple observationally-based constraints to estimate climate sensitivity. *Geophys. Res. Lett.*, **33**, L06704, doi:10.1029/2005GL025259.

Annan, J.D., J.C. Hargreaves, N.R. Edwards, and R. Marsh, 2005a: Parameter estimation in an intermediate complexity earth system model using an ensemble Kalman filter. *Ocean Modelling*, **8**, 135–154.

Annan, J.D., et al., 2005b: Efficiently constraining climate sensitivity with ensembles of paleoclimate simulations. *Scientific Online Letters on the Atmosphere*, **1**, 181–184.

Arblaster, J.M., and G.A. Meehl, 2006: Contributions of external forcings to Southern Annular Mode trends. *J. Clim.*, **19**, 2896–2905.

Arora, V.K., and G.J. Boer, 2001: Effects of simulated climate change on the hydrology of major river basins. *J. Geophys. Res.*, **106**, 3335–3348.

Arzel, O., T. Fichefet, and H. Goosse, 2006: Sea ice evolution over the 20th and 21st centuries as simulated by current AOGCMs. *Ocean Modelling*, **12**, 401–415.

Ashrit, R.G., K. Rupa Kumar, and K. Krishna Kumar, 2001: ENSO-monsoon relationships in a greenhouse warming scenario. *Geophys. Res. Lett.*, **28**, 1727–1730.

Ashrit, R.G., H. Douville, and K. Rupa Kumar, 2003: Response of the Indian monsoon and ENSO-monsoon teleconnection to enhanced greenhouse effect in the CNRM coupled model. *J. Meteorol. Soc. Japan*, **81**, 779–803.

Ashrit, R.G., A. Kitoh, and S. Yukimoto, 2005: Transient response of ENSO-monsoon teleconnection in MRI.CGCM2 climate change simulations. *J. Meteorol. Soc. Japan*, **83**, 273–291.

Austin, J., and N. Butchart, 2003: Coupled chemistry-climate model simulations for the period 1980 to 2020: Ozone depletion and the start of ozone recovery. *Q. J. R. Meteorol. Soc.*, **129**, 3225–3249.

Bahr, D.B., M.F. Meier, and S.D. Peckham, 1997: The physical basis of glacier volume-area scaling. *J. Geophys. Res.*, **102**, 20355–20362.

Banks, H., R.A. Wood, and J.M. Gregory, 2002: Changes to Indian Ocean subantarctic mode water in a coupled climate model as CO_2 forcing increases. *J. Phys. Oceanogr.*, **32**, 2816–2827.

Barnett, D.N., et al., 2006: Quantifying uncertainty in changes in extreme event frequency in response to doubled CO_2 using a large ensemble of GCM simulations. *Clim. Dyn.*, **26**, 489–511.

Bengtsson, L., K.I. Hodges, and E. Roeckner, 2006: Storm tracks and climate change. *J. Clim.*, **19**, 3518–3543.

Beniston, M., 2004: The 2003 heat wave in Europe: A shape of things to come? An analysis based on Swiss climatological data and model simulations. *Geophys. Res. Lett.*, **31**, L02202.

Berger, A., and M.F. Loutre, 2002: An exceptionally long interglacial ahead? *Science*, **297**, 1287–1288.

Berthelot, M., et al., 2002: Global response of the terrestrial biosphere to CO_2 and climate change using a coupled climate-carbon cycle model. *Global Biogeochem. Cycles*, **16**, 1084.

Bertrand, C., J.P. Van Ypersele, and A. Berger, 2002: Are natural climate forcings able to counteract the projected global warming. *Clim. Change*, **55**, 413–427.

Bi, D.H., W.F. Budd, A.C. Hirst, and X.R. Wu, 2001: Collapse and reorganisation of the Southern Ocean overturning under global warming in a coupled model. *Geophys. Res. Lett.*, **28**, 3927–3930.

Bindschadler, R., 1998: Future of the West Antarctic ice sheet. *Science*, **282**, 428–429.

Bitz, C.M., and G.H. Roe, 2004: A mechanism for the high rate of sea-ice thinning in the Arctic Ocean. *J. Clim.*, **18**, 3622–3631.

Bitz, C.M., et al., 2006: The influence of sea ice on ocean heat uptake in response to increasing CO_2. *J. Clim.*, **19**, 2437–2450.

Boer, G.J., and B. Yu, 2003a: Climate sensitivity and response. *Clim. Dyn.*, **20**, 415–429.

Boer, G.J., and B. Yu, 2003b: Dynamical aspects of climate sensitivity. *Geophys. Res. Lett.*, **30**, 1135.

Bond, T.C., et al., 2004: A technology-based global inventory of black and organic carbon emissions from combustion. *J. Geophys. Res.*, **109**, D14203, doi:10.1029/2003JD003697.

Bönisch, G., et al., 1997: Long-term trends of temperature, salinity, density, and transient tracers in the central Greenland Sea. *J. Geophys. Res.*, **102**, 18553–18571.

Bony, S., and J.-L. Dufresne, 2005: Marine boundary layer clouds at the heart of cloud feedback uncertainties in climate models. *Geophys. Res. Lett.*, **32**, L20806, doi:10.1029/2005GL023851.

Bony, S., et al., 2004: On dynamic and thermodynamic components of cloud changes. *Clim. Dyn.*, **22**, 71–86.

Bony, S., et al., 2006: How well do we understand and evaluate climate change feedback processes? *J. Clim.*, **19**, 3445–3482.

Bosilovich, M.G., S.D. Schubert, and G.K. Walker, 2005: Global changes of the water cycle intensity. *J. Clim.*, **18**, 1591–1608.

Boucher, O., and M. Pham, 2002: History of sulfate aerosol radiative forcings. *Geophys. Res. Lett.*, **29**, L1308, doi:10.1029/2001GL014048.

Box, J.E., et al., 2006: Greenland ice sheet surface mass balance variability (1988-2004) from calibrated Polar MM5 output. *J. Clim.*, **19**, 2783–2800.

Brabson, B.B., D.H. Lister, P.D. Jones, and J.P. Palutikof, 2005: Soil moisture and predicted spells of extreme temperatures in Britain. *J. Geophys. Res.*, **110**, D05104, doi:10.1029/2004JD005156.

Bradley, R.S., F.T. Keimig, and H.F. Diaz, 2004: Projected temperature changes along the American cordillera and the planned GCOS network. *Geophys. Res. Lett.*, **31**, L16210, doi:10.1029/2004GL020229.

Braithwaite, R.J., and S.C.B. Raper, 2002: Glaciers and their contribution to sea level change. *Phys. Chem. Earth*, **27**, 1445–1454.

Braithwaite, R.J., Y. Zhang, and S.C.B. Raper, 2003: Temperature sensitivity of the mass balance of mountain glaciers and ice caps as a climatological characteristic. *Z. Gletscherk. Glazialgeol.*, **38**, 35–61.

Brandefelt, J., and E. Källén, 2004: The response of the Southern Hemisphere atmospheric circulation to an enhanced greenhouse gas forcing. *J. Clim.*, **17**, 4425–4442.

Breshears, D.D., et al., 2005: Regional vegetation die-off in response to global-change-type drought. *Proc. Natl. Acad. Sci. U.S.A.*, **102**, 15144–15148.

Bryan, F.O., et al., 2006: Response of the North Atlantic thermohaline circulation and ventilation to increasing carbon dioxide in CCSM3. *J. Clim.*, **19**, 2382–2397.

Bugnion, V., and P.H. Stone, 2002: Snowpack model estimates of the mass balance of the Greenland ice sheet and its changes over the twenty-first century. *Clim. Dyn.*, **20**, 87–106.

Buizza, R., M. Miller, and T.N. Palmer, 1999: Stochastic representation of model uncertainties in the ECMWF ensemble prediction system. *Q. J. R. Meteorol. Soc.*, **125**, 2887–2908.

Burke, E.J., S.J. Brown, and N. Christidis, 2006: Modelling the recent evolution of global drought and projections for the twenty-first century with the Hadley Centre climate model. *J. Hydrometeorol.*, **7**, 1113–1125.

Caires, S., V.R. Swail, and X.L. Wang, 2006: Projection and analysis of extreme wave climate. *J. Clim.*, **19**, 5581–5605.

Camberlin, P., F. Chauvin, H. Douville, and Y. Zhao, 2004: Simulated ENSO-tropical rainfall teleconnections in present-day and under enhanced greenhouse gases conditions. *Clim. Dyn.*, **23**, 641–657.

Cassano, J.J., P. Uotila, and A. Lynch, 2006: Changes in synoptic weather patterns in the polar regions in the twentieth and twenty-first centuries, Part 1: Arctic. *Int. J. Climatol.*, **26**, 1027–1049.

Chan, W.L., and T. Motoi, 2005: Response of thermohaline circulation and thermal structure to removal of ice sheets and high atmospheric CO_2 concentration. *Geophys. Res. Lett.*, **32**, L07601, doi:10.1029/2004GL021951.

Charney, J.G., 1979: *Carbon Dioxide and Climate: A Scientific Assessment.* National Academy of Science, Washington, DC, 22 pp.

Chauvin, F., J.-F. Royer, and M. Déqué, 2006: Response of hurricane-type vortices to global warming as simulated by ARPEGE-Climat at high resolution. *Clim. Dyn.*, **27**, 377–399.

Christensen, J.H., and O.B. Christensen, 2003: Severe summertime flooding in Europe. *Nature*, **421**, 805–806.

Christensen, O.B., and J.H. Christensen, 2004: Intensification of extreme European summer precipitation in a warmer climate. *Global Planet. Change*, **44**, 107–117.

Christiansen, B., 2003: Evidence for nonlinear climate change: Two stratospheric regimes and a regime shift. *J. Clim.*, **16**, 3681–3690.

Chung, C.E., V. Ramanathan, and J.T. Kiehl, 2002: Effects of the South Asian absorbing haze on the northeast monsoon and surface-air heat exchange. *J. Clim.*, **15**, 2462–2476.

Chung, S.H., and J.H. Seinfeld, 2002: Global distribution and climate forcing of carbonaceous aerosols. *J. Geophys. Res.*, **107**, 4407.

Church, J.A., et al., 2001: Changes in sea level. In: *Climate Change 2001: The Scientific Basis. Contribution of Working Group I to the Third Assessment Report of the Intergovernmental Panel on Climate Change* [Houghton, J.T., et al. (eds.)]. Cambridge University Press, Cambridge, United Kingdom and New York, NY, USA, pp. 639–693.

Chylek, P., and U. Lohmann, 2005: Ratio of the Greenland to global temperature change: Comparison of observations and climate modeling results. *Geophys. Res. Lett.*, **32**, L14705, doi:10.1029/2005GL023552.

Clark, R., S. Brown, and J. Murphy, 2006: Modelling northern hemisphere summer heat extreme changes and their uncertainties using a physics ensemble of climate sensitivity experiments. *J. Clim.*, **19**, 4418–4435.

Claussen, M., et al., 1999: Simulation of an abrupt change in Saharan vegetation in the mid-Holocene. *Geophys. Res. Lett.*, **26**, 2037–2040.

Claussen, M., et al., 2002: Earth system models of intermediate complexity: closing the gap in the spectrum of climate system models. *Clim. Dyn.*, **18**, 579–586.

Collins, M., and The CMIP Modelling Groups, 2005: El Niño- or La Niña-like climate change? *Clim. Dyn.*, **24**, 89–104.

Collins, M., et al., 2006: Towards quantifying uncertainty in transient climate change. *Clim. Dyn.*, **27**, 127–147.

Collins, W.D., et al., 2006: Radiative forcing by well-mixed greenhouse gases: Estimates from climate models in the Intergovernmental Panel on Climate Change (IPCC) Fourth Assessment Report (AR4). *J. Geophys. Res.*, **111**, D14317, doi:10.1029/2005JD006713.

Collins, W.J., et al., 2003: Effect of stratosphere-troposphere exchange on the future tropospheric ozone trend. *J. Geophys. Res.*, **108**, 8528.

Comiso, J.C., 2000: Variability and trends in Antarctic surface temperatures from in situ and satellite infrared measurements. *J. Clim.*, **13**, 1675–1696.

Cook, K.H., and E.K. Vizy, 2006: Coupled model simulations of the West African Monsoon system: Twentieth- and twenty-first-century simulations. *J. Clim.*, **19**, 3681–3703.

Cooke, W.F., and J.J.N. Wilson, 1996: A global black carbon aerosol model. *J. Geophys. Res.*, **101**, 19395–19409.

Covey, C., et al., 2003: An overview of results from the Coupled Model Intercomparison Project. *Global Planet. Change*, **37**, 103–133.

Cox, P.M., et al., 2000: Acceleration of global warming due to carbon-cycle feedbacks in a coupled climate model. *Nature*, **408**, 184–187.

Cox, P.M., et al., 2004: Amazonian forest dieback under climate-carbon cycle projections for the 21st century. *Theor. Appl. Climatol.*, **78**, 137–156.

Cramer, W., et al., 2001: Global response of terrestrial ecosystem structure and function to CO_2 and climate change: results from six dynamic global vegetation models. *Global Change Biol.*, **7**, 357–373.

Crucifix, M., and M.F. Loutre, 2002: Transient simulations over the last interglacial period (126-115 kyr BP): feedback and forcing analysis. *Clim. Dyn.*, **19**, 417–433.

Cubasch, U., et al., 2001: Projections of future climate change. In: *Climate Change 2001: The Scientific Basis. Contribution of Working Group I to the Third Assessment Report of the Intergovernmental Panel on Climate Change* [Houghton, J.T., et al. (eds.)]. Cambridge University Press, Cambridge, United Kingdom and New York, NY, USA, pp. 525–582.

Dai, A.G., and K.E. Trenberth, 2004: The diurnal cycle and its depiction in the Community Climate System Model. *J. Clim.*, **17**, 930–951.

Dai, A.G., T.M.L. Wigley, G.A. Meehl, and W.M. Washington, 2001a: Effects of stabilizing atmospheric CO_2 on global climate in the next two centuries. *Geophys. Res. Lett.*, **28**, 4511–4514.

Dai, A., et al., 2001b: Climates of the twentieth and twenty-first centuries simulated by the NCAR Climate System Model. *J. Clim.*, **14**, 485–519.

Dai, A., et al., 2005: Atlantic thermohaline circulation in a coupled general circulation model: Unforced variations versus forced changes. *J. Clim.*, **18**, 3270–3293.

de Woul, M., and R. Hock, 2006: Static mass balance sensitivity of Arctic glaciers and ice caps using a degree-day approach. *Ann. Glaciol.*, **42**, 217–224.

DeFries, R.S., L. Bounoua, and G.J. Collatz, 2002: Human modification of the landscape and surface climate in the next fifty years. *Global Change Biol.*, **8**, 438–458.

DeFries, R.S., J.A. Foley, and G.P. Asner, 2004: Land-use choices: balancing human needs and ecosystem function. *Frontiers Ecol. Environ.*, **2**, 249–257.

Delworth, T.L., and K.W. Dixon, 2000: Implications of the recent trend in the Arctic/North Atlantic Oscillation for the North Atlantic thermohaline circulation. *J. Clim.*, **13**, 3721–3727.

Dentener, F., et al., 2005: The impact of air pollutant and methane emission controls on tropospheric ozone and radiative forcing: CTM calculations for the period 1990-2030. *Atmos. Chem. Phys.*, **5**, 1731–1755.

Dentener, F., et al., 2006: The global atmospheric environment for the next generation. *Environ. Sci. Technol.*, **40**, 3586–3594.

Déqué, M., C. Dreveton, A. Braun, and D. Cariolle, 1994: The ARPEGE/ IFS atmosphere model: a contribution to the French community climate modelling. *Clim. Dyn.*, **10**, 249–266.

Dickson, B., et al., 2002: Rapid freshening of the deep North Atlantic Ocean over the past four decades. *Nature*, **416**, 832–837.

Dixon, K.W., T.L. Delworth, M.J. Spelman, and R.J. Stouffer, 1999: The influence of transient surface fluxes on North Atlantic overturning in a coupled GCM climate change experiment. *Geophys. Res. Lett.*, **26**, 2749–2752.

Doney, S.C., et al., 2004: Evaluating global ocean carbon models: The importance of realistic physics. *Global Biogeochem. Cycles*, **18**, GB3017, doi:10.1029/2003GB002150.

Dorn, W., K. Dethloff, A. Rinke, and E. Roeckner, 2003: Competition of NAO regime changes and increasing greenhouse gases and aerosols with respect to Arctic climate projections. *Clim. Dyn.*, **21**, 447–458.

Douville, H., D. Salas-Mélia, and S. Tyteca, 2005: On the tropical origin of uncertainties in the global land precipitation response to global warming. *Clim. Dyn.*, **26**, 367–385.

Douville, H., et al., 2000: Impact of CO_2 doubling on the Asian summer monsoon: Robust versus model-dependent responses. *J. Meteorol. Soc. Japan*, **78**, 421–439.

Douville, H., et al., 2002: Sensitivity of the hydrological cycle to increasing amounts of greenhouse gases and aerosols. *Clim. Dyn.*, **20**, 45–68.

Dufresne, J.L., et al., 2002: On the magnitude of positive feedback between future climate change and the carbon cycle. *Geophys. Res. Lett.*, **29**, 1405.

Dufresne, J.L., et al., 2005: Contrasts in the effects on climate of anthropogenic sulfate aerosols between the 20th and the 21st century. *Geophys. Res. Lett.*, **32**, L21703, doi:10.1029/2005GL023619.

Dupont, T.K., and R.B. Alley, 2005: Assessment of the importance of ice-shelf buttressing to ice-sheet flow. *Geophys. Res. Lett.*, **32**, L04503, doi:10.1029/2004GL022024.

Dyurgerov, M.B., and M.F. Meier, 2000: Twentieth century climate change: evidence from small glaciers. *Proc. Natl. Acad. Sci. U.S.A.*, **97**, 1406–1411.

Echelmeyer, K., and W.D. Harrison, 1990: Jakobshavn Isbræ, West Greenland: Seasonal variations in velocity - or lack thereof. *J. Glaciol.*, **36**, 82–88.

Edwards, N., and R. Marsh, 2005: Uncertainties due to transport-parameter sensitivity in an efficient 3-D ocean-climate model. *Clim. Dyn.*, **24**, 415–433.

Eichelberger, S.J., and J.R. Holton, 2002: A mechanistic model of the northern annular mode. *J. Geophys. Res.*, **107**, 4388.

Emori, S., and S.J. Brown, 2005: Dynamic and thermodynamic changes in mean and extreme precipitation under changed climate. *Geophys. Res. Lett.*, **32**, L17706, doi:10.1029/2005GL023272.

Emori, S., et al., 1999: Coupled ocean-atmosphere model experiments of future climate with an explicit representation of sulphate aerosol scattering. *J. Meteorol. Soc. Japan*, **77**, 1299–1307.

Enting, I.G., T.M.L. Wigley, and M. Heimann, 1994: *Future Emissions and Concentrations of Carbon Dioxide: Key Ocean/ Atmosphere/ Land Analyses.* Technical Report 31, Commonwealth Scientific and Industrial Research Organisation, Div. of Atmospheric Research, Melbourne, Australia.

Feddema, J.J., et al., 2005: The importance of land-cover change in simulating future climates. *Science*, **310**, 1674–1678.

Fedorov, A.V., and S.G. Philander, 2001: A stability analysis of tropical ocean-atmosphere interactions: Bridging measurements and theory for El Niño. *J. Clim.*, **14**, 3086–3101.

Feely, R.A., et al., 2004: Impact of anthropogenic CO_2 on the $CaCO_3$ system in the oceans. *Science*, **305**, 362–366.

Fichefet, T., et al., 2003: Implications of changes in freshwater flux from the Greenland ice sheet for the climate of the 21st century. *Geophys. Res. Lett.*, **30**, 1911.

Fischer-Bruns, I., H. Von Storch, J.F. Gonzalez-Rouco, and E. Zorita, 2005: Modelling the variability of midlatitude storm activity on decadal to century time scales. *Clim. Dyn.*, **25**, 461–476.

Flato, G.M., and Participating CMIP Modeling Groups, 2004: Sea-ice and its response to CO_2 forcing as simulated by global climate change studies. *Clim. Dyn.*, **23**, 220–241.

Folland, C.K., et al., 2001: Global temperature change and its uncertainties since 1861. *Geophys. Res. Lett.*, **28**, 2621–2624.

Forest, C.E., P.H. Stone, and A.P. Sokolov, 2006: Estimated PDFs of climate system properties including natural and anthropogenic forcings. *Geophys. Res. Lett.*, **33**, L01705, doi:10.1029/2005GL023977.

Forest, C.E., et al., 2002: Quantifying uncertainties in climate system properties with the use of recent climate observations. *Science*, **295**, 113–117.

Forster, P.M.D., and K.P. Shine, 2002: Assessing the climate impact of trends in stratospheric water vapor. *Geophys. Res. Lett.*, **29**, 1086.

Forster, P.M.D., and J.M. Gregory, 2006: The climate sensitivity and its components diagnosed from Earth radiation budget data. *J. Clim.*, **19**, 39–52.

Forster, P.M.D., and K.E. Taylor, 2006: Climate forcings and climate sensitivities diagnosed from coupled climate model integrations. *J. Clim.*, **19**, 6181–6194.

Frame, D.J., et al., 2005: Constraining climate forecasts: The role of prior assumptions. *Geophys. Res. Lett.*, **32**, L09702, doi:10.1029/2004GL022241.

Frei, C., C. Schär, D. Lüthi, and H.C. Davies, 1998: Heavy precipitation processes in a warmer climate. *Geophys. Res. Lett.*, **25**, 1431–1434.

Frich, P., et al., 2002: Observed coherent changes in climatic extremes during the second half of the twentieth century. *Clim. Res.*, **19**, 193–212.

Friedlingstein, P., and S. Solomon, 2005: Contributions of past and present human generations to committed warming caused by carbon dioxide. *Proc. Natl. Acad. Sci. U.S.A.*, **102**, 10832–10836.

Friedlingstein, P., et al., 2006: Climate-carbon cycle feedback analysis: Results from the C⁴MIP model intercomparison. *J. Clim.*, **19**, 3337–3353.

Furrer, R., S.R. Sain, D. Nychka, and G.A. Meehl, 2007: Multivariate Bayesian analysis of atmosphere-ocean general circulation models. *Environ. Ecol. Stat.*, in press.

Fyfe, J.C., 2003: Extratropical southern hemisphere cyclones: Harbingers of climate change? *J. Clim.*, **16**, 2802–2805.

Fyfe, J.C., and O.A. Saenko, 2005: Human-induced change in the Antarctic Circumpolar Current. *J. Clim.*, **18**, 3068–3073.

Fyfe, J.C., and O.A. Saenko, 2006: Simulated changes in the extratropical Southern Hemisphere winds and currents. *Geophys. Res. Lett.*, **33**, L06701, doi:10.1029/2005GL025332.

Fyfe, J.C., G.J. Boer, and G.M. Flato, 1999: The Arctic and Antarctic oscillations and their projected changes under global warming. *Geophys. Res. Lett.*, **26**, 1601–1604.

Ganachaud, A., 2003: Large-scale mass transports, water mass formation, and diffusivities estimated from World Ocean Circulation Experiment (WOCE) hydrographic data. *J. Geophys. Res.*, **108**, 3213.

Gao, X., and P.A. Dirmeyer, 2006: A multimodel analysis, validation and transferability study of global soil wetness products. *J. Hydrometeorol.*, **7**, 1218–1236.

Gauss, M., et al., 2003: Radiative forcing in the 21st century due to ozone changes in the troposphere and the lower stratosphere. *J. Geophys. Res.*, **108**, 4292.

Gedney, N., P.M. Cox, and C. Huntingford, 2004: Climate feedback from wetland methane emissions. *Geophys. Res. Lett.*, **31**, L20503, doi:10.1029/2004GL020919.

Geng, Q.Z., and M. Sugi, 2003: Possible change of extratropical cyclone activity due to enhanced greenhouse gases and sulfate aerosols - Study with a high-resolution AGCM. *J. Clim.*, **16**, 2262–2274.

Gibelin, A.-L., and M. Déqué, 2003: Anthropogenic climate change over the Mediterranean region simulated by a global variable resolution model. *Clim. Dyn.*, **20**, 327–339.

Gillett, N.P., and D.W.J. Thompson, 2003: Simulation of recent Southern Hemisphere climate change. *Science*, **302**, 273–275.

Gillett, N.P., M.R. Allen, and K.D. Williams, 2003: Modeling the atmospheric response to doubled CO$_2$ and depleted stratospheric ozone using a stratosphere-resolving coupled GCM. *Q. J. R. Meteorol. Soc.*, **129**, 947–966.

Gillett, N.P., et al., 2002: How linear is the Arctic Oscillation response to greenhouse gases? *J. Geophys. Res.*, **107**, 4022.

Giorgi, F., and L.O. Mearns, 2002: Calculation of average, uncertainty range and reliability of regional climate changes from AOGCM simulations via the reliability ensemble averaging (REA) method. *J. Clim.*, **15**, 1141–1158.

Giorgi, F., and L.O. Mearns, 2003: Probability of regional climate change based on the Reliability Ensemble Averaging (REA) method. *Geophys. Res. Lett.*, **30**, 1629.

Giorgi, F., and X. Bi, 2005: Regional changes in surface climate interannual variability for the 21st century from ensembles of global model simulations. *Geophys. Res. Lett.*, **32**, L13701, doi:10.1029/2005GL023002.

Giorgi, F., et al., 2001: Emerging patterns of simulated regional climatic changes for the 21st century due to anthropogenic forcings. *Geophys. Res. Lett.*, **28**, 3317–3320.

Glover, R.W., 1999: Influence of spatial resolution and treatment of orography on GCM estimates of the surface mass balance of the Greenland ice sheet. *J. Clim.*, **12**, 551–563.

Goldstein, M., and J.C. Rougier, 2004: Probabilistic formulations for transferring inferences from mathematical models to physical systems. *SIAM J. Sci. Computing*, **26**, 467–487.

Goosse, H., et al., 2002: Potential causes of abrupt climate events: A numerical study with a three-dimensional climate model. *Geophys. Res. Lett.*, **29**, 1860.

Gordon, H.B., and S.P. O'Farrell, 1997: Transient climate change in the CSIRO coupled model with dynamic sea ice. *Mon. Weather Rev.*, **125**, 875–907.

Greene, A.M., L. Goddard, and U. Lall, 2006: Probabilistic multimodel regional temperature change projections. *J. Clim.*, **19**, 4326–4346.

Gregory, J.M., 2000: Vertical heat transports in the ocean and their effect on time-dependent climate change. *Clim. Dyn.*, **16**, 501–515.

Gregory, J.M., and J. Oerlemans, 1998: Simulated future sea-level rise due to glacier melt based on regionally and seasonally resolved temperature changes. *Nature*, **391**, 474–476.

Gregory, J.M., and P. Huybrechts, 2006: Ice-sheet contributions to future sea-level change. *Philos. Trans. R. Soc. London Ser. A*, **364**, 1709–1731.

Gregory, J.M., P. Huybrechts, and S.C.B. Raper, 2004a: Threatened loss of the Greenland ice-sheet. *Nature*, **428**, 616.

Gregory, J.M., J.A. Lowe, and S.F.B. Tett, 2006: Simulated global-mean sea-level changes over the last half-millennium. *J. Clim.*, **19**, 4576–4591.

Gregory, J.M., et al., 2001: Comparison of results from several AOGCMs for global and regional sea-level change 1900-2100. *Clim. Dyn.*, **18**, 241–253.

Gregory, J.M., et al., 2002a: An observationally based estimate of the climate sensitivity. *J. Clim.*, **15**, 3117–3121.

Gregory, J.M., et al., 2002b: Recent and future changes in Arctic sea ice simulated by the HadCM3 AOGCM. *Geophys. Res. Lett.*, **29**, 2175.

Gregory, J.M., et al., 2004b: A new method for diagnosing radiative forcing and climate sensitivity. *Geophys. Res. Lett.*, **31**, L03205, doi:10.1029/2003GL018747.

Gregory, J.M., et al., 2005: A model intercomparison of changes in the Atlantic thermohaline circulation in response to increasing atmospheric CO$_2$ concentration. *Geophys. Res. Lett.*, **32**, L12703, doi:10.1029/2005GL023209.

Grenfell, J.L., D.T. Shindell, and V. Grewe, 2003: Sensitivity studies of oxidative changes in the troposphere in 2100 using the GISS GCM. *Atmos. Chem. Phys.*, **3**, 1267–1283.

Grewe, V., et al., 2001: Future changes of the atmospheric composition and the impact of climate change. *Tellus*, **53B**, 103–121.

Guilyardi, E., 2006: El Niño-mean state-seasonal cycle interactions in a multi-model ensemble. *Clim. Dyn.*, **26**, 329–348.

Haarsma, R.J., F.M. Selten, S.L. Weber, and M. Kliphuis, 2005: Sahel rainfall variability and response to greenhouse warming. *Geophys. Res. Lett.*, **32**, L17702, doi:10.1029/2005GL023232.

Hagedorn, R., F.J. Doblas-Reyes, and T.N. Palmer, 2005: The rationale behind the success of multi-model ensembles in seasonal forecasting. Part I: Basic concept. *Tellus*, **57A**, 219–233.

Hall, A., and R.J. Stouffer, 2001: An abrupt climate event in a coupled ocean-atmosphere simulation without external forcing. *Nature*, **409**, 171–174.

Hanna, E., P. Huybrechts, and T.L. Mote, 2002: Surface mass balance of the Greenland ice sheet from climate-analysis data and accumulation/runoff models. *Ann. Glaciol.*, **35**, 67–72.

Hansen, J., 2005: A slippery slope: How much global warming constitutes dangerous anthropogenic interference? *Clim. Change*, **68**, 269–279.

Hansen, J., and L. Nazarenko, 2004: Soot climate forcing via snow and ice albedos. *Proc. Natl. Acad. Sci. U.S.A.*, **101**, 423–428.

Hansen, J.E., M. Sato, and R. Ruedy, 1997: Radiative forcing and climate response. *J. Geophys. Res.*, **102**, 6831–6864.

Hansen, J., et al., 1984: Climate sensitivity: Analysis of feedback mechanisms. In: *Climate Processes and Climate Sensitivity* [Hansen, J., and T. Takahashi (eds.)]. Geophysical Monograph Vol. 29, American Geophysical Union, Washington, DC, pp. 130–163.

Hansen, J., et al., 1985: Climate response-times - Dependence on climate sensitivity and ocean mixing. *Science*, **229**, 857–859.

Hansen, J., et al., 2005a: Earth's energy imbalance: Confirmation and implications. *Science*, **308**, 1431–1435.

Hansen, J., et al., 2005b: Efficacy of climate forcings. *J. Geophys. Res.*, **110**, D18104, doi:10.1029/2005JD005776.

Hare, W.L., and M. Meinshausen, 2006: How much warming are we committed to and how much can be avoided? *Clim. Change*, **75**, 111–149.

Hargreaves, J.C., and J.D. Annan, 2006: Using ensemble prediction methods to examine regional climate variation under global warming scenarios. *Ocean Modelling*, **11**, 174–192.

Harris, G., et al., 2006: Frequency distributions of transient regional climate change from perturbed physics ensembles of general circulation model simulations. *Clim. Dyn.*, **27**, 357–375.

Harrison, S.P., K.E. Kohfeld, C. Roelandt, and T. Claquin, 2001: The role of dust in climate changes today, at the last glacial maximum, and in the future. *Earth Sci. Rev.*, **54**, 43–80.

Harvey, L.D.D., 2004: Characterizing the annual-mean climatic effect of anthropogenic CO$_2$ and aerosol emissions in eight coupled atmosphere-ocean GCMs. *Clim. Dyn.*, **23**, 569–599.

Harvey, L.D.D., and R.K. Kaufmann, 2002: Simultaneously constraining climate sensitivity and aerosol radiative forcing. *J. Clim.*, **15**, 2837–2861.

Hasegawa, A., and S. Emori, 2005: Tropical cyclones and associated precipitation over the western North Pacific: T106 atmospheric GCM simulation for present-day and doubled CO$_2$ climates. *Scientific Online Letters on the Atmosphere*, **1**, 145–148.

Hauglustaine, D.A., and G.P. Brasseur, 2001: Evolution of tropospheric ozone under anthropogenic activities and associated radiative forcing of climate. *J. Geophys. Res.*, **106**, 32337–32360.

Hauglustaine, D.A., J. Lathiere, S. Szopa, and G.A. Folberth, 2005: Future tropospheric ozone simulated with a climate-chemistry-biosphere model. *Geophys. Res. Lett.*, **32**, L24807, doi:10.1029/2005GL024031.

Hazeleger, W., 2005: Can global warming affect tropical ocean heat transport? *Geophys. Res. Lett.*, **32**, L22701, doi:10.1029/2005GL023450.

Hegerl, G.C., T.J. Crowley, W.T. Hyde, and D.J. Frame, 2006: Climate sensitivity constrained by temperature reconstructions over the past seven centuries. *Nature*, **440**, 1029–1032.

Hoerling, M.P., et al., 2004: Twentieth century North Atlantic climate change. Part II: Understanding the effect of Indian Ocean warming. *Clim. Dyn.*, **23**, 391–405.

Holland, M.M., and C.M. Bitz, 2003: Polar amplification of climate change in the Coupled Model Intercomparison Project. *Clim. Dyn.*, **21**, 221–232.

Hosaka, M., D. Nohara, and A. Kitoh, 2005: Changes in snow coverage and snow water equivalent due to global warming simulated by a 20km-mesh global atmospheric model. *Scientific Online Letters on the Atmosphere*, **1**, 93–96.

Houghton, R.A., and J.L. Hackler, 2000: Changes in terrestrial carbon storage in the United States. 1: The roles of agriculture and forestry. *Global Ecol. Biogeogr.*, **9**, 125–144.

Hsieh, W.W., and K. Bryan, 1996: Redistribution of sea level rise associated with enhanced greenhouse warming: A simple model study. *Clim. Dyn.*, **12**, 535–544.

Hu, A., G.A. Meehl, W.M. Washington, and A. Dai, 2004: Response of the Atlantic thermohaline circulation to increased atmospheric CO_2 in a coupled model. *J. Clim.*, **17**, 4267–4279.

Hu, Z.-Z., and A. Wu, 2004: The intensification and shift of the annual North Atlantic Oscillation in a global warming scenario simulation. *Tellus*, **56A**, 112–124.

Hu, Z.-Z., L. Bengtsson, and K. Arpe, 2000a: Impact of the global warming on the Asian winter monsoon in a coupled GCM. *J. Geophys. Res.*, **105**, 4607–4624.

Hu, Z.-Z., M. Latif, E. Roeckner, and L. Bengtsson, 2000b: Intensified Asian summer monsoon and its variability in a coupled model forced by increasing greenhouse gas concentrations. *Geophys. Res. Lett.*, **27**, 2681–2684.

Hu, Z.-Z., E.K. Schneider, U.S. Bhatt, and B.P. Kirtman, 2004: Potential mechanism for response of El Niño-Southern Oscillation variability to change in land surface energy budget. *J. Geophys. Res.*, **109**, D21113, doi:10.1029/2004JD004771.

Hu, Z.-Z., et al., 2001: Impact of global warming on the interannual and interdecadal climate modes in a coupled GCM. *Clim. Dyn.*, **17**, 361–374.

Huntingford, C., et al., 2004: Using a GCM analogue model to investigate the potential for Amazonian forest dieback. *Theor. Appl. Climatol.*, **78**, 177–185.

Hurrell, J.W., Y. Kushnir, G. Ottersen, and M. Visbeck, 2003: An overview of the North Atlantic Oscillation. In: *The North Atlantic Oscillation: Climatic Significance and Environmental Impact* [Hurrell, J.W., et al. (eds.)]. Geophysical Monograph Vol. 134, American Geophysical Union, Washington, DC, pp. 1–35.

Hurrell, J.W., M.P. Hoerling, A. Phillips, and T. Xu, 2004: Twentieth century North Atlantic climate change. Part I: Assessing determinism. *Clim. Dyn.*, **23**, 371–389.

Huybrechts, P., and J. De Wolde, 1999: The dynamic response of the Greenland and Antarctic ice sheets to multiple-century climatic warming. *J. Clim.*, **12**, 2169–2188.

Huybrechts, P., A. Letréguilly, and N. Reeh, 1991: The Greenland ice sheet and greenhouse warming. *Palaeogeogr. Palaeoclimatol. Palaeoecol.*, **89**, 399–412.

Huybrechts, P., I. Janssens, C. Poncin, and T. Fichefet, 2002: The response of the Greenland ice sheet to climate changes in the 21st century by interactive coupling of an AOGCM with a thermomechanical ice-sheet model. *Ann. Glaciol.*, **35**, 409–415.

Huybrechts, P., J. Gregory, I. Janssens, and M. Wild, 2004: Modelling Antarctic and Greenland volume changes during the 20th and 21st centuries forced by GCM time slice integrations. *Global Planet. Change*, **42**, 83–105.

Inatsu, M., and M. Kimoto, 2005: Two types of interannual variability of the mid-winter storm-track and their relationship to global warming. *Scientific Online Letters on the Atmosphere*, **1**, 61–64.

IPCC, 1995: *Climate Change 1994: Radiative Forcing of Climate Change* [Houghton, J.T., et al. (eds.)]. Intergovernmental Panel on Climate Change, Cambridge University Press, Cambridge, United Kingdom and New York, NY, USA, 339 pp.

IPCC, 2001: *Climate Change 2001: The Scientific Basis. Contribution of Working Group I to the Third Assessment Report of the Intergovernmental Panel on Climate Change* [Houghton, J.T., et al. (eds.)]. Cambridge University Press, Cambridge, United Kingdom and New York, NY, USA, 881 pp.

Janssens, I., and P. Huybrechts, 2000: The treatment of meltwater retention in mass-balance parameterizations of the Greenland ice sheet. *Ann. Glaciol.*, **31**, 133–140.

Jenkins, G.S., A.T. Gaye, and B. Sylla, 2005: Late 20th century attribution of drying trends in the Sahel from the Regional Climate Model (RegCM3). *Geophys. Res. Lett.*, **32**, L22705, doi:10.1029/2005GL024225.

Johannessen, O.M., K. Khvorostovsky, M.W. Miles, and L.P. Bobylev, 2005: Recent ice-sheet growth in the interior of Greenland. *Science*, **310**, 1013–1016.

Johns, T.C., et al., 2003: Anthropogenic climate change for 1860 to 2100 simulated with the HadCM3 model under updated emissions scenarios. *Clim. Dyn.*, **20**, 583–612.

Johnson, C.E., D.S. Stevenson, W.J. Collins, and R.G. Derwent, 2001: Role of climate feedback on methane and ozone studied with a coupled ocean-atmosphere-chemistry model. *Geophys. Res. Lett.*, **28**, 1723–1726.

Johnson, H.L., and D.P. Marshall, 2002: A theory for surface Atlantic response to thermohaline variability. *J. Phys. Oceanogr.*, **32**, 1121–1132.

Jones, C.D., P.M. Cox, and C. Huntingford, 2006: Impact of climate-carbon cycle feedbacks on emission scenarios to achieve stabilisation. In: *Avoiding Dangerous Climate Change* [Schellnhuber, H.J., et al. (eds.)]. Cambridge University Press, New York, NY, pp. 323–331.

Jones, P.D., and A. Moberg, 2003: Hemispheric and large-scale surface air temperature variations: An extensive revision and an update to 2001. *J. Clim.*, **16**, 206–223.

Jones, P.D., et al., 2001: Adjusting for sampling density in grid box land and ocean surface temperature time series. *J. Geophys. Res.*, **106**, 3371–3380.

Joos, F., et al., 2001: Global warming feedbacks on terrestrial carbon uptake under the Intergovernmental Panel on Climate Change (IPCC) emission scenarios. *Global Biogeochem. Cycles*, **15**, 891–908.

Joshi, M., et al., 2003: A comparison of climate response to different radiative forcings in three general circulation models: towards an improved metric of climate change. *Clim. Dyn.*, **20**, 843–854.

Joughin, I., W. Abdalati, and M. Fahnestock, 2004: Large fluctuations in speed on Greenland's Jakobshavn Isbræ. *Nature*, **432**, 608–610.

Joughin, I., et al., 2003: Timing of recent accelerations of Pine Island Glacier, Antarctica. *Geophys. Res. Lett.*, **30**, 1706.

Kamiguchi, K., et al., 2005: Changes in precipitation-based extremes indices due to global warming projected by a global 20-km-mesh atmospheric model. *Scientific Online Letters on the Atmosphere*, **2**, 64–77.

Kanakidou, M., et al., 2005: Organic aerosol and global climate modelling: a review. *Atmos. Chem. Phys.*, **5**, 1053–1123.

Kapsner, W.R., et al., 1995: Dominant influence of atmospheric circulation on snow accumulation in Greenland over the past 18,000 years. *Nature*, **373**, 52–54.

Kattenberg, A., et al., 1996: Climate models – Projections of future climate. In: *Climate Change 1995: The Science of Climate Change. Contribution of Working Group I to the Second Assessment Report of the Intergovernmental Panel on Climate Change* [Houghton, J.T., et al. (eds.)]. Cambridge University Press, Cambridge United Kingdom and New York, NY, USA, pp. 285–357.

Kennedy, M., and A. O'Hagan, 2001: Bayesian calibration of computer models. *J. Roy. Stat. Soc.*, **63B**, 425–464.

Kettleborough, J.A., B.B.B. Booth, P.A. Stott, and M.R. Allen, 2007: Estimates of uncertainty in predictions of global mean surface temperature. *J. Clim.*, **20**, 843–855.

Kharin, V.V., and F.W. Zwiers, 2005: Estimating extremes in transient climate change simulations. *J. Clim.*, **18**, 1156–1173.

Kheshgi, H.S., and A.K. Jain, 2003: Projecting future climate change: Implications of carbon cycle model intercomparisons. *Global Biogeochem. Cycles*, **17**, GB001842, doi:10.1029/2001GB001842.

Kiilsholm, S., J.H. Christensen, K. Dethloff, and A. Rinke, 2003: Net accumulation of the Greenland ice sheet: High resolution modeling of climate changes. *Geophys. Res. Lett.*, **30**, 1485.

Kinne, S., et al., 2006: An AeroCom initial assessment - optical properties in aerosol component modules of global models. *Atmos. Chem. Phys.*, **6**, 1815–1834.

Kitabata, H., K. Nishizawa, Y. Yoshida, and K. Maruyama, 2006: Permafrost thawing in circum-Arctic and highland under climatic change scenarios projected by CCSM3. *Scientific Online Letters on the Atmosphere*, **2**, 53–56.

Kitoh, A., and O. Arakawa, 2005: Reduction in tropical rainfall diurnal variation by global warming simulated by a 20-km mesh climate model. *Geophys. Res. Lett.*, **32**, L18709, doi:10.1029/2005GL023350.

Kitoh, A., and T. Uchiyama, 2006: Changes in onset and withdrawal of the East Asian summer rainy season by multi-model global warming experiments. *J. Meteorol. Soc. Japan*, **84**, 247–258.

Knight, J.R., et al., 2005: A signature of persistent natural thermohaline circulation cycles in observed climate. *Geophys. Res. Lett.*, **32**, L20708, doi:10.1029/2005GL024233.

Knutson, T.R., and S. Manabe, 1995: Time-mean response over the tropical Pacific to increased CO_2 in a coupled ocean-atmosphere model. *J. Clim.*, **8**, 2181–2199.

Knutson, T.R., and R.E. Tuleya, 2004: Impact of CO_2-induced warming on simulated hurricane intensity and precipitation: Sensitivity to the choice of climate model and convective parameterization. *J. Clim.*, **17**, 3477–3495.

Knutti, R., and T.F. Stocker, 2000: Influence of the thermohaline circulation on projected sea level rise. *J. Clim.*, **13**, 1997–2001.

Knutti, R., and T.F. Stocker, 2002: Limited predictability of the future thermohaline circulation close to an instability threshold. *J. Clim.*, **15**, 179–186.

Knutti, R., T.F. Stocker, F. Joos, and G.-K. Plattner, 2002: Constraints on radiative forcing and future climate change from observations and climate model ensembles. *Nature*, **416**, 719–723.

Knutti, R., T.F. Stocker, F. Joos, and G.-K. Plattner, 2003: Probabilistic climate change projections using neural networks. *Clim. Dyn.*, **21**, 257–272.

Knutti, R., G.A. Meehl, M.R. Allen, and D.A. Stainforth, 2006: Constraining climate sensitivity from the seasonal cycle in surface temperature. *J. Clim.*, **19**, 4224–4233.

Knutti, R., et al., 2005: Probabilistic climate change projections for CO_2 stabilization profiles. *Geophys. Res. Lett.*, **32**, L20707, doi:10.1029/2005GL023294.

Koch, D., 2001: Transport and direct radiative forcing of carbonaceous and sulfate aerosols in the GISS GCM. *J. Geophys. Res.*, **106**, 20311–20332.

Krabill, W., et al., 2004: Greenland Ice Sheet: Increased coastal thinning. *Geophys. Res. Lett.*, **31**, L24402, doi:10.1029/2004GL021533.

Krinner, G., et al., 2007: Simulated Antarctic precipitation and surface mass balance at the end of the twentieth and twenty-first centuries. *Clim. Dyn.*, **28**, 215–230.

Krishna Kumar, K., B. Rajagopalan, and M.A. Cane, 1999: On the weakening relationship between the Indian monsoon and ENSO. *Science*, **284**, 2156–2159.

Kunkel, K.E., and X.Z. Liang, 2005: GCM simulations of the climate in the central United States. *J. Clim.*, **18**, 1016–1031.

Kutzbach, J.E., J.W. Williams, and S.J. Vavrus, 2005: Simulated 21st century changes in regional water balance of the Great Lakes region and links to changes in global temperature and poleward moisture transport. *Geophys. Res. Lett.*, **32**, L17707, doi:10.1029/2005GL023506.

Kuzmina, S.I., et al., 2005: The North Atlantic Oscillation and greenhouse-gas forcing. *Geophys. Res. Lett.*, **32**, L04703, doi:10.1029/2004GL021064.

Lal, M., and S.K. Singh, 2001: Global warming and monsoon climate *Mausam*, **52**, 245–262.

Lambert, S.J., and G.J. Boer, 2001: CMIP1 evaluation and intercomparison of coupled climate models. *Clim. Dyn.*, **17**, 83–106.

Lambert, S.J., and J.C. Fyfe, 2006: Changes in winter cyclone frequencies and strengths simulated in enhanced greenhouse warming experiments: results from the models participating in the IPCC diagnostic exercise. *Clim. Dyn.*, **26**, 713–728.

Land, C., and J. Feichter, 2003: Stratosphere-troposphere exchange in a changing climate simulated with the general circulation model MAECHAM4. *J. Geophys. Res.*, **108**, 8523.

Landerer, F.W., J.H. Jungclaus, and J. Marotzke, 2007: Regional dynamic and steric sea level change in response to the IPCC-A1B scenario. *J. Phys. Oceanogr.*, in press.

Laurent, R., and X. Cai, 2007: A maximum entropy method for combining AOGCMs for regional intra-year climate change assessment. *Clim. Change*, **82**, 411–435.

Lawrence, D.M., and A.G. Slater, 2005: A projection of severe near-surface permafrost degradation during the 21st century. *Geophys. Res. Lett.*, **32**, L24401, doi:10.1029/2005GL025080.

Lean, J., J. Beer, and R.S. Bradley, 1995: Reconstruction of solar irradiance since 1610: Implications for climate change. *Geophys. Res. Lett.*, **22**, 3195–3198.

Leckebusch, G.C., and U. Ulbrich, 2004: On the relationship between cyclones and extreme windstorm events over Europe under climate change. *Global Planet. Change*, **44**, 181–193.

Leemans, R., et al., 1998: *The IMAGE User Support System: Global Change Scenarios from IMAGE 2.1.* CD-ROM 4815006, National Institute of Public Health and the Environment (RIVM), Bilthoven, The Netherlands.

Leggett, J., W.J. Pepper, and R. Swart, 1992: Emissions scenarios for IPCC: An update. In: *Climate Change 1992. The Supplementary Report to the IPCC Scientific Assessment* [Houghton, J.T., B.A. Calalnder, and S.K. Varney (eds.)]. Cambridge University Press, Cambridge, United Kingdom and New York, NY, USA, pp. 69–95.

LeGrande, A.N., et al., 2006: Consistent simulations of multiple proxy responses to an abrupt climate change event. *Proc. Natl. Acad. Sci. U.S.A.*, **103**, 837–842.

Levermann, A., et al., 2005: Dynamic sea level changes following changes in the thermohaline circulation. *Clim. Dyn.*, **24**, 347–354.

Li, W., R. Fu, and R.E. Dickinson, 2006: Rainfall and its seasonality over the Amazon in the 21st century as assessed by the coupled models for the IPCC AR4. *J. Geophys. Res.*, **111**, D02111, doi:10.1029/2005JD006355.

Liao, H., et al., 2003: Interactions between tropospheric chemistry and aerosols in a unified general circulation model. *J. Geophys. Res.*, **108**, 4001.

Liu, P., G.A. Meehl, and G. Wu, 2002: Multi-model trends in the Sahara induced by increasing CO_2. *Geophys. Res. Lett.*, **29**, 1881.

Liu, Z., et al., 2005: Rethinking tropical ocean response to global warming: The enhanced equatorial warming. *J. Clim.*, **18**, 4684–4700.

Longworth, H., J. Marotzke, and T.F. Stocker, 2005: Ocean gyres and abrupt change in the thermohaline circulation: A conceptual analysis. *J. Clim.*, **18**, 2403–2416.

Lopez, A., et al., 2006: Two approaches to quantifying uncertainty in global temperature changes. *J. Clim.*, **19**, 4785–4796.

Lowe, J.A., and J.M. Gregory, 2006: Understanding projections of sea level rise in a Hadley Centre coupled climate model. *J. Geophys. Res.*, **111**, C11014, doi:10.1029/2005JC003421.

Lucarini, V., and G.L. Russell, 2002: Comparison of mean climate trends in the northern hemisphere between National Centers for Environmental Prediction and two atmosphere-ocean model forced runs. *J. Geophys. Res.*, **107**, 4269.

Lumpkin, R., and K. Speer, 2003: Large-scale vertical and horizontal circulation in the North Atlantic Ocean. *J. Phys. Oceanogr.*, **33**, 1902–1920.

Lunt, D.J., N. de Noblet-Ducoudré, and S. Charbit, 2004: Effects of a melted Greenland ice sheet on climate, vegetation, and the cryosphere. *Clim. Dyn.*, **23**, 679–694.

Lynch, A., P. Uotila, and J.J. Cassano, 2006: Changes in synoptic weather patterns in the polar regions in the twentieth and twenty-first centuries, Part 2: Antarctic. *Int. J. Climatol.*, **26**, 1181–1199.

Mahowald, N.M., and C. Luo, 2003: A less dusty future? *Geophys. Res. Lett.*, **30**, 1903.

Manabe, S., and R.J. Stouffer, 1980: Sensitivity of a global climate model to an increase of CO_2 concentration in the atmosphere. *J. Geophys. Res.*, **85**, 5529–5554.

Manabe, S., and R.J. Stouffer, 1994: Multiple-century response of a coupled ocean-atmosphere model to an increase of atmospheric carbon dioxide. *J. Clim.*, **7**, 5–23.

Manabe, S., and R.J. Stouffer, 1999: Are two modes of the thermohaline circulation stable? *Tellus*, **51A**, 400–411.

Marland, G., T.A. Boden, and R.J. Andres, 2005: Global, regional and national fossil fuel CO_2 emissions. In: *Trends: A Compendium of Data on Global Change*. Carbon Dioxide Information Analysis Center, Oak Ridge National Laboratory, U.S. Department of Energy, Oak Ridge, TN.

Marquart, S., M. Ponater, F. Mager, and R. Sausen, 2003: Future development of contrail cover, optical depth, and radiative forcing: Impacts of increasing air traffic and climate change. *J. Clim.*, **16**, 2890–2904.

Matthews, H.D., 2005: Decrease of emissions required to stabilize atmospheric CO_2 due to positive carbon cycle-climate feedbacks. *Geophys. Res. Lett.*, **32**, L21707, doi:10.1029/2005GL023435.

May, W., 2004: Potential future changes in the Indian summer monsoon due to greenhouse warming: analysis of mechanisms in a global time-slice experiment. *Clim. Dyn.*, **22**, 389–414.

Maykut, G.A., and N. Untersteiner, 1971: Some results from a time-dependent thermodynamic model of sea ice. *J. Geophys. Res.*, **76**, 1550–1575.

Maynard, K., and J.-F. Royer, 2004: Effects of "realistic" land-cover change on a greenhouse-warmed African climate. *Clim. Dyn.*, **22**, 343–358.

Maynard, K., J.-F. Royer, and F. Chauvin, 2002: Impact of greenhouse warming on the West African summer monsoon. *Clim. Dyn.*, **19**, 499–514.

McCabe, G.J., M.P. Clark, and M.C. Serreze, 2001: Trends in Northern Hemisphere surface cyclone frequency and intensity. *J. Clim.*, **14**, 2763–2768.

McDonald, R.E., et al., 2005: Tropical storms: Representation and diagnosis in climate models and the impacts of climate change. *Clim. Dyn.*, **25**, 19–36.

Meehl, G.A., and J.M. Arblaster, 2003: Mechanisms for projected future changes in south Asian monsoon precipitation. *Clim. Dyn.*, **21**, 659–675.

Meehl, G.A., and C. Tebaldi, 2004: More intense, more frequent, and longer lasting heat waves in the 21st century. *Science*, **305**, 994–997.

Meehl, G.A., C. Tebaldi, and D. Nychka, 2004a: Changes in frost days in simulations of twenty-first century climate. *Clim. Dyn.*, **23**, 495–511.

Meehl, G.A., J.M. Arblaster, and C. Tebaldi, 2005a: Understanding future patterns of precipitation extremes in climate model simulations. *Geophys. Res. Lett.*, **32**, L18719, doi:10.1029/2005GL023680.

Meehl, G.A., H. Teng, and G.W. Branstator, 2006a: Future changes of El Niño in two global coupled climate models. *Clim. Dyn.*, **26**, 549–566.

Meehl, G.A., W.M. Washington, J.M. Arblaster, and A. Hu, 2004b: Factors affecting climate sensitivity in global coupled models. *J. Clim.*, **17**, 1584–1596.

Meehl, G.A., et al., 2005b: Overview of the coupled model intercomparison project. *Bull. Am. Meteorol. Soc.*, **86**, 89–93.

Meehl, G.A., et al., 2005c: How much more global warming and sea level rise? *Science*, **307**, 1769–1772.

Meehl, G.A., et al., 2006b: Climate change projections for the twenty-first century and climate change commitment in the CCSM3. *J. Clim.*, **19**, 2597–2616.

Meinshausen, M., 2006: What does a 2°C target mean for greenhouse gas concentrations? A brief analysis based on multi-gas emission pathways and several climate sensitivity uncertainty estimates. In: *Avoiding Dangerous Climate Change* [Schellnhuber, H.J., et al. (eds.)]. Cambridge University Press, New York, NY, pp. 265–279.

Meleshko, V.P., et al., 2004: Anthropogenic climate change in 21st century over Northern Eurasia. *Meteorol. Hydrol.*, **7**, 5–26.

Menon, S., J. Hansen, L. Nazarenko, and Y. Luo, 2002: Climate effects of black carbon aerosols in China and India. *Science*, **297**, 2250–2253.

Merryfield, W., 2006: Changes to ENSO under CO_2 doubling in a multi-model ensemble. *J. Clim.*, **19**, 4009–4027.

Mielke, P.W., 1991: The application of multivariate permutation methods based on distance functions in the earth sciences. *Earth Sci. Rev.*, **31**, 55–71.

Mikolajewicz, U., and R. Voss, 2000: The role of the individual air-sea flux components in CO_2-induced changes of the ocean's circulation and climate. *Clim. Dyn.*, **16**, 627–642.

Miller, R.L., G.A. Schmidt, and D.T. Shindell, 2006: Forced annular variations in the 20th century IPCC AR4 simulations. *J. Geophys. Res.*, **111**, D18101, doi:10.1029/2005JD006323.

Milly, P.C.D., K.A. Dunne, and A.V. Vecchia, 2005: Global pattern of trends in streamflow and water availability in a changing climate. *Nature*, **438**, 347–350.

Milly, P.C.D., R.T. Wetherald, K.A. Dunne, and T.L. Delworth, 2002: Increasing risk of great floods in a changing climate. *Nature*, **415**, 514–517.

Mitchell, J.F.B., S. Manabe, V. Meleshko, and T. Tokioka, 1990: Equilibrium climate change – and its implications for the future. In: *Climate Change. The IPCC Scientific Assessment. Contribution of Working Group 1 to the First Assessment Report of the Intergovernmental Panel on Climate Change* [Houghton, J.L., G.J. Jenkins, and J.J. Ephraums (eds.)]. Cambridge University Press, Cambridge, United Kingdom and New York, NY, USA, pp. 137–164.

Mitchell, J.F.B., T.C. Johns, W.J. Ingram, and J.A. Lowe, 2000: The effect of stabilising atmospheric carbon dioxide concentrations on global and regional climate change. *Geophys. Res. Lett.*, **27**, 2977–2980.

Mitchell, T.D., 2003: Pattern scaling - An examination of the accuracy of the technique for describing future climates. *Clim. Change*, **60**, 217–242.

Monahan, A.H., 2002: Stabilisation of climate regimes by noise in a simple model of the thermohaline circulation. *J. Phys. Oceanogr.*, **32**, 2072–2085.

Morris, E.M., and R. Mulvaney, 2004: Recent variations in surface mass balance of the Antarctic Peninsula ice sheet. *J. Glaciol.*, **50**, 257–267.

Moss, R.H., and S.H. Schneider, 2000: Uncertainties in the IPCC TAR: Recommendations to Lead Authors for more consistent assessment and reporting. In: *Guidance Papers on the Cross-Cutting Issues of the Third Assessment Report of the IPCC* [Pachauri, R., et al. (eds.)]. Intergovernmental Panel on Climate Change, Geneva, pp. 33–51.

Müller, S.A., F. Joos, N.R. Edwards, and T.F. Stocker, 2006: Water mass distribution and ventilation time scales in a cost-efficient, 3-dimensional ocean model. *J. Clim.*, **19**, 5479–5499.

Murphy, B.F., I. Marsiat, and P. Valdes, 2002: Atmospheric contributions to the surface mass balance of Greenland in the HadAM3 atmospheric model. *J. Geophys. Res.*, **107**, 4556.

Murphy, J.M., et al., 2004: Quantification of modelling uncertainties in a large ensemble of climate change simulations. *Nature*, **429**, 768–772.

Myhre, G., and F. Stordal, 2001: On the tradeoff of the solar and thermal infrared radiative impact of contrails. *Geophys. Res. Lett.*, **28**, 3119–3122.

Myhre, G., T.F. Berglen, C.E.L. Myhre, and I.S.A. Isaksen, 2004: The radiative effect of the anthropogenic influence on the stratospheric sulfate aerosol layer. *Tellus*, **56B**, 294–299.

Nakashiki, N., et al., 2006: Recovery of thermohaline circulation under CO_2 stabilization and overshoot scenarios. *Ocean Modelling*, **15**, 200–217.

Nakićenović, N., and R. Swart (eds.), 2000: *Special Report on Emissions Scenarios. A Special Report of Working Group III of the Intergovernmental Panel on Climate Change*. Cambridge University Press, Cambridge, United Kingdom and New York, NY, USA, 599 pp.

Nanjundiah, R.S., V. Vidyunmala, and J. Srinivasan, 2005: The impact of increase in CO_2 on the simulation of tropical biennial oscillations (TBO) in 12 coupled general circulation models. *Atmos. Sci. Lett.*, **6**, 183–191.

Naveau, P., and C.M. Ammann, 2005: Statistical distributions of ice core sulphate from climatically relevant volcanic eruptions. *Geophys. Res. Lett.*, **32**, L05711, doi:10.1029/2004GL021732.

Neelin, J.D., et al., 2006: Tropical drying trends in global warming models and observations. *Proc. Natl. Acad. Sci. U.S.A.*, **103**, 6110–6115.

Nguyen, K.C., and K.J.E. Walsh, 2001: Interannual, decadal, and transient greenhouse simulation of tropical cyclone-like vortices in a regional climate model of the South Pacific. *J. Clim.*, **14**, 3043–3054.

Nohara, D., A. Kitoh, M. Hosaka, and T. Oki, 2006: Impact of climate change on river discharge projected by multi-model ensemble. *J. Hydrometeorol.*, **7**, 1076–1089.

NorthGRIP Members, 2004: High-resolution climate record of the northern hemisphere back into the last interglacial period. *Nature*, **431**, 147–151.

Oerlemans, J., 2001: *Glaciers and Climate Change*. A. A. Balkema, Lisse, The Netherlands, 148 pp.

Oerlemans, J., and J.P.F. Fortuin, 1992: Sensitivity of glaciers and small ice caps to greenhouse warming. *Science*, **258**, 115–117.

Oerlemans, J., and B.K. Reichert, 2000: Relating glacier mass balance to meteorological data by using a seasonal sensitivity characteristic. *J. Glaciol.*, **46**, 1–6.

Oerlemans, J., et al., 1998: Modeling the response of glaciers to climate warming. *Clim. Dyn.*, **14**, 267–274.

Oerlemans, J., et al., 2006: Estimating the contribution from Arctic glaciers to sea-level change in the next hundred years. *Ann. Glaciol.*, **42**, 230–236.

Ohmura, A., M. Wild, and L. Bengtsson, 1996: Present and future mass balance of the ice sheets simulated with GCM. *Ann. Glaciol.*, **23**, 187–193.

Olivier, J.G.J., and J.J.M. Berdowski, 2001: Global emissions sources and sinks. In: *The Climate System* [Berdowski, J., R. Guicherit, and B.J. Heij (eds.)]. A. A. Balkema, Lisse, The Netherlands, pp. 33–78.

Oouchi, K., et al., 2006: Tropical cyclone climatology in a global-warming climate as simulated in a 20km-mesh global atmospheric model: Frequency and wind intensity analyses. *J. Meteorol. Soc. Japan*, **84**, 259–276.

Oppenheimer, M., 1998: Global warming and the stability of the West Antarctic Ice Sheet. *Nature*, **393**, 325–332.

Oppenheimer, M., and R.B. Alley, 2005: Ice sheets, global warming, and Article 2 of the UNFCCC. *Clim. Change*, **68**, 257–267.

Opsteegh, J.D., R.J. Haarsma, F.M. Selten, and A. Kattenberg, 1998: ECBILT: A dynamic alternative to mixed boundary conditions in ocean models. *Tellus*, **50A**, 348–367.

Orr, J.C., et al., 2005: Anthropogenic ocean acidification over the twenty-first century and its impact on calcifying organisms. *Nature*, **437**, 681–686.

Osborn, T.J., 2004: Simulating the winter North Atlantic Oscillation: The roles of internal variability and greenhouse forcing. *Clim. Dyn.*, **22**, 605–623.

Pal, J.S., F. Giorgi, and X. Bi, 2004: Consistency of recent European summer precipitation trends and extremes with future regional climate projections. *Geophys. Res. Lett.*, **31**, L13202, doi:10.1029/2004GL019836.

Palmer, T.N., 2001: A nonlinear dynamical perspective on model error: A proposal for non-local stochastic-dynamic parametrization in weather and climate prediction models. *Q. J. R. Meteorol. Soc.*, **127**, 279–303.

Palmer, T.N., 2005: Global warming in a nonlinear climate - Can we be sure? *Europhys. News*, **36**, 42–46.

Palmer, T.N., and J. Räisänen, 2002: Quantifying the risk of extreme seasonal precipitation events in a changing climate. *Nature*, **415**, 514–517.

Palmer, T.N., et al., 2004: Development of a European multimodel ensemble system for seasonal-to-interannual prediction (DEMETER). *Bull. Am. Meteorol. Soc.*, **85**, 853–872.

Palmer, T.N., et al., 2005: Representing model uncertainty in weather and climate prediction. *Annu. Rev. Earth Planet. Sci.*, **33**, 4.1–4.31.

Parizek, B.R., and R.B. Alley, 2004: Implications of increased Greenland surface melt under global-warming scenarios: Ice-sheet simulations. *Quat. Sci. Rev.*, **23**, 1013–1027.

Payne, A.J., et al., 2004: Recent dramatic thinning of largest West Antarctic ice stream triggered by oceans. *Geophys. Res. Lett.*, **31**, L23401, doi:10.1029/2004GL021284.

Penner, J.E., et al. (eds.), 1999: *Aviation and the Global Atmosphere*. Cambridge University Press, Cambridge, United Kingdom and New York, NY, USA, 373 pp.

Petoukhov, V., et al., 2000: CLIMBER-2: a climate system model of intermediate complexity. Part I: model description and performance for present climate. *Clim. Dyn.*, **16**, 1–17.

Pham, M., O. Boucher, and D. Hauglustaine, 2005: Changes in atmospheric sulfur burdens and concentrations and resulting radiative forcings under IPCC SRES emission scenarios for 1990-2100. *J. Geophys. Res.*, **110**, D06112, doi:10.1029/2004JD005125.

Phillips, T.J., and P.J. Gleckler, 2006: Evaluation of continental precipitation in 20th century climate simulations: The utility of multi-model statistics. *Water Resour. Res.*, **42**, W03202, doi:10.1029/2005WR004313.

Phillips, T.J., et al., 2004: Evaluating parameterizations in general circulation models. *Bull. Am. Meteorol. Soc.*, **85**, 1903–1915.

Piani, C., D.J. Frame, D.A. Stainforth, and M.R. Allen, 2005: Constraints on climate change from a multi-thousand member ensemble of simulations. *Geophys. Res. Lett.*, **32**, L23825, doi:10.1029/2005GL024452.

Pitari, G., E. Mancini, V. Rizi, and D.T. Shindell, 2002: Impact of future climate and emission changes on stratospheric aerosols and ozone. *J. Atmos. Sci.*, **59**, 414–440.

Plattner, G.-K., F. Joos, T.F. Stocker, and O. Marchal, 2001: Feedback mechanisms and sensitivities of ocean carbon uptake under global warming. *Tellus*, **53B**, 564–592.

Rahmstorf, S., and A. Ganopolski, 1999: Long-term global warming scenarios computed with an efficient coupled climate model. *Clim. Change*, **43**, 353–367.

Räisänen, J., 2001: CO_2-induced climate change in CMIP2 experiments: Quantification of agreement and role of internal variability. *J. Clim.*, **14**, 2088–2104.

Räisänen, J., 2002: CO_2-induced changes in interannual temperature and precipitation variability in 19 CMIP2 experiments. *J. Clim.*, **15**, 2395–2411.

Räisänen, J., 2003: CO_2-induced changes in angular momentum in CMIP2 experiments. *J. Clim.*, **16**, 132–143.

Räisänen, J., 2005a: Impact of increasing CO_2 on monthly-to-annual precipitation extremes: Analysis of the CMIP2 experiments. *Clim. Dyn.*, **24**, 309–323.

Räisänen, J., 2005b: Probability distributions of CO_2-induced global warming as inferred directly from multimodel ensemble simulations. *Geophysica*, **41**, 19–30.

Räisänen, J., and T.N. Palmer, 2001: A probability and decision-model analysis of a multimodel ensemble of climate change simulations. *J. Clim.*, **14**, 3212–3226.

Ramanathan, V., 1988: The greenhouse theory of climate change: A test by an inadvertent global experiment. *Science*, **240**, 293–299.

Ramanathan, V., P.J. Crutzen, J.T. Kiehl, and D. Rosenfeld, 2001: Aerosols, climate, and the hydrologic cycle. *Science*, **294**, 2119–2124.

Ramanathan, V., et al., 2005: Atmospheric brown clouds: Impacts on South Asian climate and hydrological cycle. *Proc. Natl. Acad. Sci. U.S.A.*, **102**, 5326–5333.

Randall, D.A., et al., 1996: A revised land surface parameterization (SiB2) for GCMs. Part III: The greening of the Colorado State University General Circulation Model. *J. Clim.*, **9**, 738–763.

Raper, S.C.B., and R.J. Braithwaite, 2005: The potential for sea level rise: New estimates from glacier and ice cap area and volume distributions. *Geophys. Res. Lett.*, **32**, L05502, doi:10.1029/2004GL021981.

Raper, S.C.B., and R.J. Braithwaite, 2006: Low sea level rise projections from mountain glaciers and icecaps under global warming. *Nature*, **439**, 311–313.

Raper, S.C.B., O. Brown, and R.J. Braithwaite, 2000: A geometric glacier model for sea-level change calculations. *J. Glaciol.*, **46**, 357–368.

Raper, S.C.B., J.M. Gregory, and R.J. Stouffer, 2002: The role of climate sensitivity and ocean heat uptake on AOGCM transient temperature response. *J. Clim.*, **15**, 124–130.

Rauthe, M., A. Hense, and H. Paeth, 2004: A model intercomparison study of climate change-signals in extratropical circulation. *Int. J. Climatol.*, **24**, 643–662.

Raven, J., et al., 2005: *Ocean Acidification Due to Increasing Atmospheric Carbon Dioxide*. The Royal Society, London, 60 pp.

Reichert, B.K., L. Bengtsson, and J. Oerlemans, 2002: Recent glacier retreat exceeds internal variability. *J. Clim.*, **15**, 3069–3081.

Ridley, J.K., P. Huybrechts, J.M. Gregory, and J.A. Lowe, 2005: Elimination of the Greenland ice sheet in a high CO_2 climate. *J. Clim.*, **17**, 3409–3427.

Rignot, E., 1998: Fast recession of a West Antarctic glacier. *Science*, **281**, 549–551.

Rignot, E., 2001: Evidence for a rapid retreat and mass loss of Thwaites Glacier, West Antarctica. *J. Glaciol.*, **47**, 213–222.

Rignot, E., and S.S. Jacobs, 2002: Rapid bottom melting widespread near Antarctic ice sheet grounding lines. *Science*, **296**, 2020–2023.

Rignot, E., and P. Kanagaratnam, 2006: Changes in the velocity structure of the Greenland ice sheet. *Science*, **311**, 986–990.

Rignot, E., A. Rivera, and G. Casassa, 2003: Contribution of the Patagonia ice fields of South America to sea level rise. *Science*, **302**, 434–437.

Rignot, E., et al., 2002: Acceleration of Pine Island and Thwaites Glaciers, West Antarctica. *Ann. Glaciol.*, **34**, 189–194.

Rind, D., J. Lerner, and C. McLinden, 2001: Changes of tracer distributions in the doubled CO_2 climate. *J. Geophys. Res.*, **106**, 28061–28079.

Rind, D., J. Perlwitz, and P. Lonergan, 2005a: AO/NAO response to climate change: 1. Respective influences of stratospheric and tropospheric climate changes. *J. Geophys. Res.*, **110**, D12107, doi:10.1029/2004JD005103.

Rind, D., R.J. Healy, C. Parkinson, and D. Martinson, 1995: The role of sea ice in $2{\times}CO_2$ climate model sensitivity. 1. The total influence of sea ice thickness and extent. *J. Clim.*, **8**, 449–463.

Rind, D., J. Perlwitz, P. Lonergan, and J. Lerner, 2005b: AO/NAO response to climate change: 2. Relative importance of low- and high-latitude temperature changes. *J. Geophys. Res.*, **110**, D12108, doi:10.1029/2004JD005686.

Rodwell, M.J., D.P. Rowell, and C.K. Folland, 1999: Oceanic forcing of the wintertime North Atlantic oscillation and European climate. *Nature*, **398**, 320–323.

Roeckner, E., L. Bengtsson, and J. Feichter, 1999: Transient climate change simulations with a coupled atmosphere-ocean GCM including the tropospheric sulfur cycle. *J. Clim.*, **12**, 3004–3032.

Ronski, S., and G. Budeus, 2005: Time series of winter convection in the Greenland Sea. *J. Geophys. Res.*, **110**, C04015, doi:10.1029/2004JC002318.

Rothrock, D.A., Y. Yu, and G.A. Maykut, 1999: Thinning of the Arctic sea-ice cover. *Geophys. Res. Lett.*, **26**, 3469–3472.

Rougier, J.C., 2007: Probabilistic inference for future climate using an ensemble of climate model evaluations. *Clim. Change*, **81**, 247–264.

Rowell, D.P., and R.G. Jones, 2006: Causes and uncertainty of future summer drying over Europe. *Clim. Dyn.*, **27**, 281–299.

Russell, G.L., V. Gornitz, and J.R. Miller, 2000: Regional sea-level changes projected by the NASA/GISS atmosphere-ocean model. *Clim. Dyn.*, **16**, 789–797.

Saenko, O.A., A.J. Weaver, and J.M. Gregory, 2003: On the link between the two modes of the ocean thermohaline circulation and the formation of global-scale water masses. *J. Clim.*, **16**, 2797–2801.

Saenko, O.A., J.C. Fyfe, and M.H. England, 2005: On the response of the oceanic wind-driven circulation to atmospheric CO_2 increase. *Clim. Dyn.*, **25**, 415–426.

Sakamoto, T.T., et al., 2005: Responses of the Kuroshio and the Kuroshio Extension to global warming in a high-resolution climate model. *Geophys. Res. Lett.*, **32**, L14617, doi:10.1029/2005GL023384.

Sanderson, M.G., et al., 2003: Effect of climate change on isoprene emissions and surface ozone levels. *Geophys. Res. Lett.*, **30**, 1936.

Sato, Y., et al., 2006: Response of North Pacific ocean circulation in a Kuroshio-resolving ocean model to an Arctic Oscillation (AO)-like change in Northern Hemisphere atmospheric circulation due to greenhouse-gas forcing. *J. Meteorol. Soc. Japan*, **84**, 295–309.

Scambos, T.A., C. Hulbe, M.A. Fahnestock, and J. Bohlander, 2000: The link between climate warming and break-up of ice shelves in the Antarctic Peninsula. *J. Glaciol.*, **46**, 516–530.

Schaeffer, M., F.M. Selten, J.D. Opsteegh, and H. Goosse, 2002: Intrinsic limits to predictability of abrupt regional climate change in IPCC SRES scenarios. *Geophys. Res. Lett.*, **29**, 1767.

Schaeffer, M., F.M. Selten, J.D. Opsteegh, and H. Goosse, 2004: The influence of ocean convection patterns on high-latitude climate projections. *J. Clim.*, **17**, 4316–4329.

Schär, C., et al., 2004: The role of increasing temperature variability in European summer heat waves. *Nature*, **427**, 332–336.

Schmittner, A., M. Latif, and B. Schneider, 2005: Model projections of the North Atlantic thermohaline circulation for the 21st century assessed by observations. *Geophys. Res. Lett.*, **32**, L23710, doi:10.1029/2005GL024368.

Schneeberger, C., H. Blatter, A. Abe-Ouchi, and M. Wild, 2003: Modelling changes in the mass balance of glaciers of the northern hemisphere for a transient $2{\times}CO_2$ scenario. *J. Hydrol.*, **282**, 145–163.

Schneeberger, C., et al., 2000: Størglacieren in doubling CO_2 climate. *Clim. Dyn.*, **17**, 825–834.

Schneider von Deimling, T., H. Held, A. Ganopolski, and S. Rahmstorf, 2006: Climate sensitivity estimated from ensemble simulations of glacial climate. *Clim. Dyn.*, **27**, 149–163.

Schubert, M., et al., 1998: North Atlantic cyclones in CO_2-induced warm climate simulations: frequency, intensity, and tracks. *Clim. Dyn.*, **14**, 827–838.

Schweckendiek, U., and J. Willebrand, 2005: Mechanisms for the overturning response in global warming simulations. *J. Clim.*, **18**, 4925–4936.

Selten, F.M., G.W. Branstator, M. Kliphuis, and H.A. Dijkstra, 2004: Tropical origins for recent and future northern hemisphere climate change. *Geophys. Res. Lett.*, **31**, L21205, doi:10.1029/2004GL020739.

Semenov, V.A., and L. Bengtsson, 2002: Secular trends in daily precipitation characteristics: Greenhouse gas simulation with a coupled AOGCM. *Clim. Dyn.*, **19**, 123–140.

Senior, C.A., and J.F.B. Mitchell, 2000: The time-dependence of climate sensitivity. *Geophys. Res. Lett.*, **27**, 2685–2688.

Shepherd, A., D.J. Wingham, and J.A.D. Mansley, 2002: Inland thinning of the Amundsen Sea sector, West Antarctica. *Geophys. Res. Lett.*, **29**, 1364.

Shepherd, A., D. Wingham, and E. Rignot, 2004: Warm ocean is eroding West Antarctic ice sheet. *Geophys. Res. Lett.*, **31**, L23402, doi:10.1029/2004GL021106.

Shepherd, A., D.J. Wingham, J.A.D. Mansley, and H.F.J. Corr, 2001: Inland thinning of Pine Island Glacier, West Antarctica. *Science*, **291**, 862–864.

Shindell, D.T., and G.A. Schmidt, 2004: Southern hemisphere climate response to ozone changes and greenhouse gas increases. *Geophys. Res. Lett.*, **31**, L18209, doi:10.1029/2004GL020724.

Shindell, D.T., G.A. Schmidt, R.L. Miller, and D. Rind, 2001: Northern hemisphere winter climate response to greenhouse gas, ozone, and volcanic forcing. *J. Geophys. Res.*, **106**, 7193–7210.

Shine, K.P., J. Cook, E.J. Highwood, and M.M. Joshi, 2003: An alternative to radiative forcing for estimating the relative importance of climate change mechanisms. *Geophys. Res. Lett.*, **30**, 2047.

Shukla, J., et al., 2006: Climate model fidelity and projections of climate change. *Geophys. Res. Lett.*, **33**, L07702, doi:10.1029/2005GL025579.

Siegenthaler, U., and H. Oeschger, 1984: Transient temperature changes due to increasing CO_2 using simple models. *Ann. Glaciol.*, **5**, 153–159.

Sigmond, M., P.C. Siegmund, E. Manzini, and H. Kelder, 2004: A simulation of the separate climate effects of middle-atmosphere and tropospheric CO_2 doubling. *J. Clim.*, **17**, 2352–2367.

Smethie, W.M., and R.A. Fine, 2001: Rates of North Atlantic Deep Water formation calculated from chlorofluorocarbon inventories. *Deep-Sea Res. I*, **48**, 189–215.

Smith, L.A., 2002: What might we learn from climate forecasts? *Proc. Natl. Acad. Sci. U.S.A.*, **99**, 2487–2492.

Soden, B.J., and I.M. Held, 2006: An assessment of climate feedbacks in coupled ocean-atmosphere models. *J. Clim.*, **19**, 3354–3360.

Sokolov, A., C.E. Forest, and P.H. Stone, 2003: Comparing oceanic heat uptake in AOGCM transient climate change experiments. *J. Clim.*, **16**, 1573–1582.

Stainforth, D.A., et al., 2005: Uncertainty in predictions of the climate response to rising levels of greenhouse gases. *Nature*, **433**, 403–406.

Stendel, M., and J.H. Christensen, 2002: Impact of global warming on permafrost conditions in a coupled GCM. *Geophys. Res. Lett.*, **29**, 1632.

Stephenson, D.B., et al., 2006: North Atlantic Oscillation response to transient greenhouse gas forcing and the impact on European winter climate: A CMIP2 multi-model assessment. *Clim. Dyn.*, **27**, 401–420.

Stevenson, D.S., et al., 2000: Future estimates of tropospheric ozone radiative forcing and methane turnover - the impact of climate change. *Geophys. Res. Lett.*, **27**, 2073–2076.

Stevenson, D.S., et al., 2006: Multi-model ensemble simulations of present-day and near-future tropospheric ozone. *J. Geophys. Res.*, **111**, D08301, doi:10.1029/2005JD006338.

Stocker, T.F., 2000: Past and future reorganisations in the climate system. *Quat. Sci. Rev.*, **19**, 301–319.

Stocker, T.F., and A. Schmittner, 1997: Influence of CO_2 emission rates on the stability of the thermohaline circulation. *Nature*, **388**, 862–865.

Stocker, T.F., and R. Knutti, 2003: Do simplified climate models have any useful skill? *CLIVAR Exchanges*, **8**, 7–10.

Stocker, T.F., and C.C. Raible, 2005: Climate change - Water cycle shifts gear. *Nature*, **434**, 830–833.

Stocker, T.F., D.G. Wright, and L.A. Mysak, 1992a: A zonally averaged, coupled ocean-atmosphere model for paleoclimate studies. *J. Clim.*, **5**, 773–797.

Stocker, T.F., D.G. Wright, and W.S. Broecker, 1992b: The influence of high-latitude surface forcing on the global thermohaline circulation. *Paleoceanogr.*, **7**, 529–541.

Stone, D.A., and A.J. Weaver, 2002: Daily maximum and minimum temperature trends in a climate model. *Geophys. Res. Lett.*, **29**, 1356.

Stone, D.A., and J.C. Fyfe, 2005: The effect of ocean mixing parametrisation on the enhanced CO_2 response of the Southern Hemisphere midlatitude jet. *Geophys. Res. Lett.*, **32**, L06811, doi:10.1029/2004GL022007.

Stott, P.A., and J.A. Kettleborough, 2002: Origins and estimates of uncertainty in predictions of twenty-first century temperature rise. *Nature*, **416**, 723–726.

Stott, P.A., D.A. Stone, and M.R. Allen, 2004: Human contribution to the European heatwave of 2003. *Nature*, **432**, 610–613.

Stott, P.A., J.A. Kettleborough, and M.R. Allen, 2006a: Uncertainty in continental-scale temperature predictions. *Geophys. Res. Lett.*, **33**, L02708, doi:10.1029/2005GL024423.

Stott, P.A., et al., 2006b: Observational constraints on past attributable warming and predictions of future global warming. *J. Clim.*, **19**, 3055–3069.

Stouffer, R.J., 2004: Time scales of climate response. *J. Clim.*, **17**, 209–217.

Stouffer, R.J., and S. Manabe, 1999: Response of a coupled ocean-atmosphere model to increasing atmospheric carbon dioxide: sensitivity to the rate of increase. *J. Clim.*, **12**, 2224–2237.

Stouffer, R.J., and S. Manabe, 2003: Equilibrium response of thermohaline circulation to large changes in atmospheric CO_2 concentration. *Clim. Dyn.*, **20**, 759–773.

Stouffer, R.J., et al., 2006a: GFDL's CM2 global coupled climate models. Part IV: Idealized climate response. *J. Clim.*, **19**, 723–740.

Stouffer, R.J., et al., 2006b: Investigating the causes of the response of the thermohaline circulation to past and future climate changes. *J. Clim.*, **19**, 1365–1387.

Sudo, K., M. Takahashi, and H. Akimoto, 2003: Future changes in stratosphere-troposphere exchange and their impacts on future tropospheric ozone simulations. *Geophys. Res. Lett.*, **30**, 2256.

Sugi, M., and J. Yoshimura, 2004: A mechanism of tropical precipitation change due to CO_2 increase. *J. Clim.*, **17**, 238–243.

Sugi, M., A. Noda, and N. Sato, 2002: Influence of the global warming on tropical cyclone climatology: An experiment with the JMA global model. *J. Meteoroi. Soc. Japan*, **80**, 249–272.

Suzuki, T., et al., 2005: Projection of future sea level and its variability in a high-resolution climate model: Ocean processes and Greenland and Antarctic ice-melt contributions. *Geophys. Res. Lett.*, **32**, L19706, doi:10.1029/2005GL023677.

Takemura, T., T. Nakajima, T. Nozawa, and K. Aoki, 2001: Simulation of future aerosol distribution, radiative forcing, and long-range transport in East Asia. *J. Meteorol. Soc. Japan*, **79**, 1139–1155.

Talley, L.D., 2003: Shallow, intermediate, and deep overturning components of the global heat budget. *J. Phys. Oceanogr.*, **33**, 530–560.

Tanaka, H.L., N. Ishizaki, and D. Nohara, 2005: Intercomparison of the intensities and trends of Hadley, Walker and monsoon circulations in the global warming projections. *Scientific Online Letters on the Atmosphere*, **1**, 77–80.

Tebaldi, C., L.O. Mearns, D. Nychka, and R.L. Smith, 2004: Regional probabilities of precipitation change: A Bayesian analysis of multimodel simulations. *Geophys. Res. Lett.*, **31**, L24213, doi:10.1029/2004GL021276.

Tebaldi, C., R.W. Smith, D. Nychka, and L.O. Mearns, 2005: Quantifying uncertainty in projections of regional climate change: A Bayesian approach to the analysis of multi-model ensembles. *J. Clim.*, **18**, 1524–1540.

Tebaldi, C., K. Hayhoe, J.M. Arblaster, and G.A. Meehl, 2006: Going to the extremes: An intercomparison of model-simulated historical and future changes in extreme events. *Clim. Change*, **79**, 185-211

Tegen, I., M. Werner, S.P. Harrison, and K.E. Kohfeld, 2004a: Reply to comment by N. M. Mahowald et al. on "Relative importance of climate and land use in determining present and future global soil dust emission". *Geophys. Res. Lett.*, **31**, L24106, doi:10.1029/2004GL021560.

Tegen, I., M. Werner, S.P. Harrison, and K.E. Kohfeld, 2004b: Relative importance of climate and land use in determining present and future global soil dust emission'. *Geophys. Res. Lett.*, **31**, L05105, doi:10.1029/2003GL019216.

Terray, L., et al., 2004: Simulation of late twenty-first century changes in wintertime atmospheric circulation over Europe de to anthropogenic causes. *J. Clim.*, **17**, 4630–4635.

Thomas, R., et al., 2001: Mass balance of higher-elevation parts of the Greenland ice sheet. *J. Geophys. Res.*, **106**, 33707–33716.

Thomas, R., et al., 2004: Accelerated sea level rise from West Antarctica. *Science*, **306**, 255–258.

Thomas, R., et al., 2005: Force-perturbation analysis of Pine Island Glacier, Antarctica, suggests cause for recent acceleration. *Ann. Glaciol.*, **39**, 133–138.

Thompson, D.W., and S. Solomon, 2002: Interpretation of recent Southern Hemisphere climate change. *Science*, **296**, 895–899.

Tol, R.S.J., and A.F. De Vos, 1998: A Bayesian statistical analysis of the enhanced greenhouse effect. *Clim. Change*, **38**, 87–112.

Toniazzo, T., J.M. Gregory, and P. Huybrechts, 2004: Climatic impact of Greenland deglaciation and its possible irreversibility. *J. Clim.*, **17**, 21–33.

Trenberth, K.E., 1990: Recent observed interdecadal climate changes in the Northern Hemisphere. *Bull. Am. Meteorol. Soc.*, **71**, 988–993.

Tsushima, Y., A. Abe-Ouchi, and S. Manabe, 2005: Radiative damping of annual variation in global mean surface temperature: Comparison between observed and simulated feedback. *Clim. Dyn.*, **24**, 591–597.

Tsushima, Y., et al., 2006: Importance of the mixed-phase cloud distribution in the control climate for assessing the response of clouds to carbon dioxide increase: a multi-model study. *Clim. Dyn.*, **27**, 113–126.

Tsutsui, J., 2002: Implications of anthropogenic climate change for tropical cyclone activity: A case study with the NCAR CCM2. *J. Meteorol. Soc. Japan*, **80**, 45–65.

Tsutsui, J., et al., 2007: Long-term climate response to stabilized and overshoot anthropogenic forcings beyond the 21st century. *Clim. Dyn.*, **28**, 199–214.

Ueda, H., A. Iwai, K. Kuwako, and M.E. Hori, 2006: Impact of anthropogenic forcing on the Asian summer monsoon as simulated by eight GCMs. *Geophys. Res. Lett.*, **33**, L06703, doi:10.1029/2005GL025336.

van Aardenne, J.A., et al., 2001: A 1×1 degree resolution dataset of historical anthropogenic trace gas emissions for the period 1890-1990. *Global Biogeochem. Cycles*, **15**, 909–928.

van de Wal, R.S.W., and M. Wild, 2001: Modelling the response of glaciers to climate change by applying volume-area scaling in combination with a high resolution GCM. *Clim. Dyn.*, **18**, 359–366.

van de Wal, R.S.W., M. Wild, and J. de Wolde, 2001: Short-term volume change of the Greenland ice sheet in response to doubled CO_2 conditions. *Tellus*, **53B**, 94–102.

van der Veen, C.J., 2002: Polar ice sheets and global sea level: how well can we predict the future? *Global Planet. Change*, **32**, 165–194.

van Lipzig, N.M., E. van Meijgaard, and J. Oerlemans, 2002: Temperature sensitivity of the Antarctic surface mass balance in a regional atmospheric climate model. *J. Clim.*, **15**, 2758–2774.

van Oldenborgh, G.J., and G. Burgers, 2005: Searching for decadal variations in ENSO precipitation teleconnections. *Geophys. Res. Lett.*, **32**, L15701, doi:10.1029/2005GL023110.

van Oldenborgh, G.J., S.Y. Philip, and M. Collins, 2005: El Niño in a changing climate: a multi-model study. *Ocean Sci.*, **1**, 81–95.

Vaughan, D.G., 2006: Recent trends in melting conditions on the Antarctic Peninsula and their implications for ice-sheet mass balance and sea level. *Arctic, Antarctic, and Alpine Res.*, **38**, 147–152.

Vaughan, D.G., 2007: West Antarctic Ice Sheet collapse – the fall and rise of a paradigm. *Clim. Change*, in press.

Vaughan, D.G., and J.R. Spouge, 2002: Risk estimation of collapse of the West Antarctic Ice Sheet. *Clim. Change*, **52**, 65–91.

Vavrus, S.J., J.E. Walsh, W.L. Chapman, and D. Portis, 2006: The behavior of extreme cold air outbreaks under greenhouse warming. *Int. J. Climatol.*, **26**, 1133–1147.

Vieli, A., and A.J. Payne, 2005: Assessing the ability of numerical ice sheet models to simulate grounding line migration. *J. Geophys. Res.*, **110**, F01003, doi:10.1029/2004JF000202.

Voldoire, A., 2006: Quantifying the impact of future land-use changes against increases in GHG concentrations. *Geophys. Res. Lett.*, **33**, L04701, doi:10.1029/2005GL024354.

Voss, R., and U. Mikolajewicz, 2001: Long-term climate changes due to increased CO_2 concentration in the coupled atmosphere-ocean general circulation model ECHAM3/LSG. *Clim. Dyn.*, **17**, 45–60.

Voss, R., W. May, and E. Roeckner, 2002: Enhanced resolution modeling study on anthropogenic climate change: changes in the extremes of the hydrological cycle. *Int. J. Climatol.*, **22**, 755–777.

Walsh, K., 2004: Tropical cyclones and climate change: Unresolved issues. *Clim. Res.*, **27**, 77–84.

Walsh, K.J.E., K.C. Nguyen, and J.L. McGregor, 2004: Fine-resolution regional climate model simulations of the impact of climate change on tropical cyclones near Australia. *Clim. Dyn.*, **22**, 47–56.

Wang, G., 2005: Agricultural drought in a future climate: Results from 15 global climate models participating in the IPCC 4th Assessment. *Clim. Dyn.*, **25**, 739–753.

Wang, X.L., and V.R. Swail, 2006a: Historical and possible future changes of wave heights in northern hemisphere ocean. In: *Atmosphere-Ocean Interactions* [Perrie, W. (ed.)]. Vol. 2, Wessex Institute of Technology Press, Southampton, pp. 240.

Wang, X.L., and V.R. Swail, 2006b: Climate change signal and uncertainty in projections of ocean wave heights. *Clim. Dyn.*, **26**, 109–126.

Wang, X.L., F.W. Zwiers, and V.R. Swail, 2004: North Atlantic Ocean wave climate change scenarios for the twenty-first century. *J. Clim.*, **17**, 2368–2383.

Watanabe, S., T. Nagashima, and S. Emori, 2005: Impact of global warming on gravity wave momentum flux in the lower stratosphere. *Scientific Online Letters on the Atmosphere*, **1**, 189–192.

Watterson, I.G., 1996: Non-dimensional measures of climate model performance. *Int. J. Climatol.*, **16**, 379–391.

Watterson, I.G., 2003: Effects of a dynamic ocean on simulated climate sensitivity to greenhouse gases. *Clim. Dyn.*, **21**, 197–209.

Watterson, I.G., 2005: Simulated changes due to global warming in the variability of precipitation, and their interpretation using a gamma-distributed stochastic model. *Adv. Water Res.*, **28**, 1368–1381.

Watterson, I.G., and M.R. Dix, 2003: Simulated changes due to global warming in daily precipitation means and extremes and their interpretation using the gamma distribution. *J. Geophys. Res.*, **108**, 4379.

Watterson, I.G., and M.R. Dix, 2005: Effective sensitivity and heat capacity in the response of climate models to greenhouse gas and aerosol forcings. *Q. J. R. Meteorol. Soc.*, **131**, 259–279.

Weaver, A.J., and E.C. Wiebe, 1999: On the sensitivity of projected oceanic thermal expansion to the parameterisation of sub-grid scale ocean mixing. *Geophys. Res. Lett.*, **26**, 3461–3464.

Weaver, A.J., and C. Hillaire-Marcel, 2004a: Ice growth in the greenhouse: A seductive paradox but unrealistic scenario. *Geoscience Canada*, **31**, 77–85.

Weaver, A.J., and C. Hillaire-Marcel, 2004b: Global warming and the next ice age. *Science*, **304**, 400–402.

Weaver, A.J., O.A. Saenko, P.U. Clark, and J.X. Mitrovica, 2003: Meltwater pulse 1A from Antarctica as a trigger of the Bølling-Allerød warm interval. *Science*, **299**, 1709–1713.

Webb, M.J., et al., 2006: On the contribution of local feedback mechanisms to the range of climate sensitivity in two GCM ensembles. *Clim. Dyn.*, **27**, 17–38.

Webster, M.D., et al., 2002: Uncertainty in emissions projections for climate models. *Atmos. Environ.*, **36**, 3659–3670.

Webster, M., et al., 2003: Uncertainty analysis of climate change and policy response. *Clim. Change*, **61**, 295–320.

Weisheimer, A., and T.N. Palmer, 2005: Changing frequency of occurrence of extreme seasonal-mean temperatures under global warming. *Geophys. Res. Lett.*, **32**, L20721, doi:10.1029/2005GL023365.

Wetherald, R.T., and S. Manabe, 2002: Simulation of hydrologic changes associated with global warming. *J. Geophys. Res.*, **107**, 4379.

Wetherald, R.T., R.J. Stouffer, and K.W. Dixon, 2001: Committed warming and its implications for climate change. *Geophys. Res. Lett.*, **28**, 1535–1538.

Wigley, T.M.L., 1984: Carbon dioxide, trace gases and global warming. *Climate Monitor*, **13**, 133–148.

Wigley, T.M.L., 2004: Modeling climate change under no-policy and policy emissions pathways. In: *The Benefits of Climate Change Policies: Analytical and Framework Issues*. OECD Publications, Paris, pp. 221–248.

Wigley, T.M.L., 2005: The climate change commitment. *Science*, **307**, 1766–1769.

Wigley, T.M.L., and S.C.B. Raper, 2001: Interpretation of high projections for global-mean warming. *Science*, **293**, 451–454.

Wigley, T.M.L., and S.C.B. Raper, 2003: Future changes in global-mean temperature and sea level. In: *Climate and Sea Level Change: Observations, Projections and Implications* [Warrick, R.A., et al. (eds.)]. Cambridge University Press, Cambridge, pp. 111–133.

Wigley, T.M.L., and S.C.B. Raper, 2005: Extended scenarios for glacier melt due to anthropogenic forcing. *Geophys. Res. Lett.*, **32**, L05704, doi:10.1029/2004GL021238.

Wigley, T.M.L., R. Richels, and J.A. Edmonds, 1996: Economic and environmental choices in the stabilization of atmospheric CO_2 concentrations. *Nature*, **379**, 240–243.

Wigley, T.M.L., P.D. Jones, and S.C.B. Raper, 1997a: The observed global warming record: What does it tell us? *Proc. Natl. Acad. Sci. U.S.A.*, **94**, 8314–8320.

Wigley, T.M.L., C.M. Ammann, B.D. Santer, and S.C.B. Raper, 2005: Effect of climate sensitivity on the response to volcanic forcing. *J. Geophys. Res.*, **110**, D09107, doi:10.1029/2004JD005557.

Wigley, T.M.L., et al., 1997b: *Implications of Proposed CO_2 Emissions Limitations*. IPCC Technical Paper IV, Intergovernmental Panel on Climate Change, Geneva, 51 pp.

Wilby, R.L., and T.M.L. Wigley, 2002: Future changes in the distribution of daily precipitation totals across North America. *Geophys. Res. Lett.*, **29**, 1135.

Wild, M., and A. Ohmura, 2000: Changes in mass balance of the polar ice sheets and sea level from high resolution GCM simulations of global warming. *Ann. Glaciol.*, **30**, 197–203.

Wild, M., P. Calanca, S. Scherrer, and A. Ohmura, 2003: Effects of polar ice sheets on global sea level in high resolution greenhouse scenarios. *J. Geophys. Res.*, **108**, 4165.

Williams, K.D., C.A. Senior, and J.F.B. Mitchell, 2001: Transient climate change in the Hadley Centre models: The role of physical processes. *J. Clim.*, **14**, 2659–2674.

Williams, K.D., et al., 2005: Evaluation of a component of the cloud response to climate change in an intercomparison of climate models. *Clim. Dyn.*, **26**, 145–165.

Williams, M.J.M., R.C. Warner, and W.F. Budd, 2002: Sensitivity of the Amery ice shelf, Antarctica, to changes in the climate of the Southern Ocean. *J. Clim.*, **15**, 2740–2757.

Wood, R.A., M. Vellinga, and R. Thorpe, 2003: Global warming and thermohaline circulation stability. *Philos. Trans. R. Soc. London Ser. A*, **361**, 1961–1975.

Wood, R.A., A.B. Keen, J.F.B. Mitchell, and J.M. Gregory, 1999: Changing spatial structure of the thermohaline circulation in response to atmospheric CO_2 forcing in a climate model. *Nature*, **399**, 572–575.

Wu, L., and B. Wang, 2004: Assessing impacts of global warming on tropical cyclone tracks. *J. Clim.*, **17**, 1686–1698.

Wu, P., R. Wood, and P. Stott, 2005: Human influence on increasing Arctic river discharge. *Geophys. Res. Lett.*, **32**, L02703, doi:10.1029/2004GL021570.

Xu, Y., Z.-C. Zhao, Y. Luo, and X. Gao, 2005: Climate change projections for the 21st century by the NCC/IAP T63 with SRES scenarios. *Acta Meteorol. Sin.*, **19**, 407–417.

Yamaguchi, K., and A. Noda, 2006: Global warming patterns over the North Pacific: ENSO versus AO. *J. Meteorol. Soc. Japan*, **84**, 221–241.

Yamaguchi, K., A. Noda, and A. Kitoh, 2005: The changes in permafrost induced by greenhouse warming: A numerical study applying multiple-layer ground model. *J. Meteorol. Soc. Japan*, **83**, 799–815.

Yeh, S.-W., and B.P. Kirtman, 2005: Pacific decadal variability and decadal ENSO amplitude modulation. *Geophys. Res. Lett.*, **32**, L05703, doi:10.1029/2004GL021731.

Yin, J.H., 2005: A consistent poleward shift of the storm tracks in simulations of 21st century climate. *Geophys. Res. Lett.*, **32**, L18701, doi:10.1029/2005GL023684.

Yokohata, T., et al., 2005: Climate response to volcanic forcing: Validation of climate sensitivity of a coupled atmosphere-ocean general circulation model. *Geophys. Res. Lett.*, **32**, L21710, doi:10.1029/2005GL023542.

Yonetani, T., and H.B. Gordon, 2001: Simulated changes in the frequency of extremes and regional features of seasonal/annual temperature and precipitation when atmospheric CO_2 is doubled. *J. Clim.*, **14**, 1765–1779.

Yoshida, Y., et al., 2005: Multi-century ensemble global warming projections using the Community Climate System Model (CCSM3). *J. Earth Simulator*, **3**, 2–10.

Yoshimori, M., M.C. Reader, A.J. Weaver, and N.A. McFarlane, 2002: On the causes of glacial inception at 116 kaBP. *Clim. Dyn.*, **18**, 383–402.

Yoshimura, J., and M. Sugi, 2005: Tropical cyclone climatology in a high-resolution AGCM: Impacts of SST warming and CO_2 increase. *Scientific Online Letters on the Atmosphere*, **1**, 133–136.

Yoshimura, J., M. Sugi, and A. Noda, 2006: Influence of greenhouse warming on tropical cyclone frequency. *J. Meteorol. Soc. Japan*, **84**, 405–428.

Yukimoto, S., et al., 2006: Climate change of the twentieth through the twenty-first centuries simulated by the MRI-CGCM2.3. *Pap. Meteorol. Geophys.*, **56**, 9–24.

Zelle, H., G.J. van Oldenborgh, G. Burgers, and H.A. Dijkstra, 2005: El Niño and greenhouse warming: Results from ensemble simulations with the NCAR CCSM. *J. Clim.*, **18**, 4669–4683.

Zeng, G., and J.A. Pyle, 2003: Changes in tropospheric ozone between 2000 and 2100 modeled in a chemistry-climate model. *Geophys. Res. Lett.*, **30**, 1392.

Zhang, X.D., and J.E. Walsh, 2006: Toward a seasonally ice-covered Arctic Ocean: Scenarios from the IPCC AR4 model simulations. *J. Clim.*, **19**, 1730–1747.

Zuo, Z., and J. Oerlemans, 1997: Contribution of glacial melt to sea level rise since AD 1865: A regionally differentiated calculation. *Clim. Dyn.*, **13**, 835–845.

Zwally, H.J., et al., 2002: Surface melt-induced acceleration of Greenland ice-sheet flow. *Science*, **297**, 218-222.

Zwally, H.J., et al., 2005: Mass changes of the Greenland and Antarctic ice sheets and shelves and contributions to sea level rise: 1992-2002. *J. Glaciol.*, **175**, 509–527.

Zweck, C., and P. Huybrechts, 2005: Modeling of the northern hemisphere ice sheets during the last glacial cycle and glaciological sensitivity. *J. Geophys. Res.*, **110**, D07103, doi:10.1029/2004JD005489.

Appendix 10.A: Methods for Sea Level Projections for the 21st Century

10.A.1   Scaling MAGICC Results

The MAGICC SCM was tuned to emulate global average surface air temperature change and radiative flux at the top of the atmosphere (assumed equal to ocean heat uptake on decadal time scales; Section 5.2.2.3 and Figure 5.4) simulated by each of 19 AOGCMs in scenarios with CO_2 increasing at 1% yr^{-1} (Section 10.5.3). Under SRES scenarios for which AOGCMs have been run (B1, A1B and A2), the ensemble average of the tuned versions of MAGICC gives about 10% greater temperature rise and 25% more thermal expansion over the 21st century (2090 to 2099 minus 1980 to 1999) than the average of the corresponding AOGCMs. The MAGICC radiative forcing is close to that of the AOGCMs (as estimated for A1B by Forster and Taylor, 2006), so the mismatch suggests there may be structural limitations on the accurate emulation of AOGCMs by the SCM. We therefore do not use the tuned SCM results directly to make projections, unlike in the TAR. The TAR model means for thermal expansion were 0.06–0.10 m larger than the central estimates in Table 10.7, probably because the simple climate model used in the TAR overestimated the TAR AOGCM results.

The SCM may nonetheless be used to estimate results for scenarios that have not been run in AOGCMs, by calculating time-dependent ratios between pairs of scenarios (Section 10.5.4.6). This procedure is supported by the close match between the ratios derived from the AOGCM and MAGICC ensemble averages under the scenarios for which AOGCMs are available. Applying the MAGICC ratios to the A1B AOGCM results yields estimates of temperature rise and thermal expansion for B1 and A2 differing by less than 5% from the AOGCM ensemble averages. We have high confidence that the procedure will yield similarly accurate estimates for the results that the AOGCMs would give under scenarios B2, A1T and A1FI.

The spread of MAGICC models is much narrower than the AOGCM ensemble because the AOGCMs have internally generated climate variability and a wider range of forcings. We assume inter-model standard deviations of 20% of the model average for temperature rise and 25% for thermal expansion, since these proportions are found to be fairly time and scenario independent in the AOGCM ensemble.

10.A.2   Mass Balance Sensitivity of Glaciers and Ice Caps

A linear relationship $r_g = b_g \times (T - T_0)$ is found for the period 1961 to 2003 between the observational time series of the contribution r_g to the rate of sea level rise from the world's glaciers and ice caps (G&IC, excluding those on Antarctica and Greenland; Section 4.5.2, Figure 4.14) and global average surface air temperature T (Hadley Centre/Climatic Research Unit gridded surface temperature dataset HadCRUT3; Section 3.2.2.4, Figure 3.6), where b_g is the global total G&IC mass balance sensitivity and T_0 is the global average temperature of the climate in which G&IC are in a steady state, T and T_0 being expressed relative to the average of 1865 to 1894. The correlation coefficient is 0.88. Weighted least-squares regression gives a slope $b_g = 0.84 \pm 0.15$ (one standard deviation) mm yr^{-1} °C^{-1}, with $T_0 = -0.13$°C. Surface mass balance models driven with climate change scenarios from AOGCMs (Section 10.6.3.1) also indicate such a linear relationship, but the model results give a somewhat lower b_g of around 0.5 to 0.6 mm yr^{-1} °C^{-1} (Section 10.6.3.1). To cover both observations and models, we adopt a value of $b_g = 0.8 \pm 0.2$ (one standard deviation) mm yr^{-1} °C^{-1}. This uncertainty of ±25% is smaller than that of ±40% used in the TAR because of the improved observational constraint now available. To make projections, we choose a set of values of b_g randomly from a normal distribution. We use $T_0 = T - r_g/b_g$, where $T = 0.40$ °C and $r_g = 0.45$ mm yr^{-1}, are the averages over the period 1961 to 2003. This choice of T_0 minimises the root mean square difference of the predicted r_g from the observed, and gives T_0 in the range −0.5°C to 0.0°C (5 to 95%). Note that a constant b_g is not expected to be a good approximation if glacier area changes substantially (see Section 10.A.3).

10.A.3   Area Scaling of Glaciers and Ice Caps

Model results using area-volume scaling of G&IC (Section 10.6.3.2) are approximately described by the relations $b_g / b_1 = (A_g / A_1)^{1.96}$ and $A_g / A_1 = (V_g / V_1)^{0.84}$, where A_g and V_g are the global G&IC area and volume (excluding those on Greenland and Antarctica) and variable X_1 is the initial value of X_g. The first relation describes how total SMB sensitivity declines as the most sensitive areas are ablated most rapidly. The second relation follows Wigley and Raper (2005) in its form, and describes how area declines as volume is lost, with $dV_g / dt = -r_g$ (expressing V as sea level equivalent, i.e., the liquid-water-equivalent volume of ice divided by the surface area of the world ocean). Projections are made starting from 1990 using T from Section 10.A.1 with initial values of the present-day b_g from Section 10.A.2 and the three recent estimates $V_g = 0.15$, 0.24 and 0.37 m from Table 4.4, which are assumed equally likely. We use $T = 0.48$°C at 1990 relative to 1865 to 1894, and choose T_0 as in Section 10.A.2. An uncertainty of 10% (one standard deviation) is assumed because of the scaling relations. The results are multiplied by 1.2 (Section 10.6.3.3) to include contributions from G&IC on Greenland and Antarctica (apart from the ice sheets). These scaling relations are expected to give a decreasingly adequate approximation as greater area and volume is lost, because they do not model hypsometry explicitly; they predict that V will tend eventually to zero in any steady-state warmer climate, for instance, although this is not necessarily the case. A similar scaling procedure was used in the TAR. Current estimates of present-day G&IC mass are smaller than those used in the TAR, leading to more rapid wastage of area. Hence, the central estimates for the G&IC contribution to sea level rise in Table 10.7 are similar to those in the TAR, despite our use of a larger mass balance sensitivity (Section 10.A.2).

10.A.4 Changes in Ice Sheet Surface Mass Balance

Quadratic fits are made to the results of Gregory and Huybrechts (2006) (Section 10.6.4.1) for the SMB change of each ice sheet as a function of global average temperature change relative to a steady state, which is taken to be the late 19th century (1865–1894). The spread of results for the various models used by Gregory and Huybrechts represents uncertainty in the patterns of temperature and precipitation change. The Greenland contribution has a further uncertainty of 20% (one standard deviation) from the ablation calculation. The Antarctic SMB projections are similar to those of the TAR, while the Greenland SMB projections are larger by 0.01–0.04 m because of the use of a quadratic fit to temperature change rather than the constant sensitivity of the TAR, which gave an underestimate for larger warming.

10.A.5 Changes in Ice Sheet Dynamics

Topographic and dynamic changes that can be simulated by currently available ice flow models are roughly represented as modifying the sea level changes due to SMB change by –5% ± 5% from Antarctica, and 0% ± 10% from Greenland (± one standard deviation) (Section 10.6.4.2).

The contribution from scaled-up ice sheet discharge, given as an illustration of the effect of accelerated ice flow (Section 10.6.5), is calculated as $r_1 \times T / T_1$, with T and T_1 expressed relative to the 1865 to 1894 average, where $r_1 = 0.32$ mm yr^{-1} is an estimate of the contribution during 1993 to 2003 due to recent acceleration and $T_1 = 0.63°C$ is the global average temperature during that period.

10.A.6 Combination of Uncertainties

For each scenario, time series of temperature rise and the consequent land ice contributions to sea level are generated using a Monte Carlo simulation (van der Veen, 2002). Temperature rise and thermal expansion have some correlation for a given scenario in AOGCM results (Section 10.6.1). In the Monte Carlo simulation, we assume them to be perfectly correlated; by correlating the uncertainties in the thermal expansion and land ice contributions, this increases the resulting uncertainty in the sea level rise projections. However, the uncertainty in the projections of the land ice contributions is dominated by the various uncertainties in the land ice models themselves (Sections 10.A.2–4) rather than in the temperature projections. We assume the uncertainties in land ice models and temperature projections to be uncorrelated. The procedure used in the TAR, however, effectively assumed the land ice model uncertainty

to be correlated with the temperature and expansion projection uncertainty. This is the main reason why the TAR ranges for sea level rise under each of the scenarios are wider than those of Table 10.7. Also, the TAR gave uncertainty ranges of ±2 standard deviations, whereas the present report gives ±1.65 standard deviations (5 to 95%).

10.A.7 Change in Surface Air Temperature Over the Major West Antarctic Ice Shelves

The mean surface air temperature change over the area of the Ross and Filchner-Ronne ice shelves in December and January, divided by the mean annual antarctic surface air temperature change, is $F_1 = 0.62 \pm 0.48$ (one standard deviation) on the basis of the climate change simulations from the four high-resolution GCMs used by Gregory and Huybrechts (2006). From AR4 AOGCMs, the ratio of mean annual antarctic temperature change to global mean temperature change is $F_2 = 1.1 \pm 0.2$ (one standard deviation) under SRES scenarios with stabilisation beyond 2100 (Gregory and Huybrechts, 2006), while from AR4 AGCMs coupled to mixed-layer ocean models it is $F_2 = 1.4 = 0.2$ (one standard deviation) at equilibrium under doubled CO_2. To evaluate the probability of ice shelf mean summer temperature increase exceeding a particular value, given the global temperature rise, a Monte Carlo distribution of $F_1 \times F_2$ is used, generated by assuming the two factors to be normal and independent random variables. Since this procedure is based on a small number of models, and given other caveats noted in Sections 10.6.4.2 and 10.7.4.4, we have low confidence in these probabilities.

11

Regional Climate Projections

Coordinating Lead Authors:
Jens Hesselbjerg Christensen (Denmark), Bruce Hewitson (South Africa)

Lead Authors:
Aristita Busuioc (Romania), Anthony Chen (Jamaica), Xuejie Gao (China), Isaac Held (USA), Richard Jones (UK), Rupa Kumar Kolli (India), Won-Tae Kwon (Republic of Korea), René Laprise (Canada), Victor Magaña Rueda (Mexico), Linda Mearns (USA), Claudio Guillermo Menéndez (Argentina), Jouni Räisänen (Finland), Annette Rinke (Germany), Abdoulaye Sarr (Senegal), Penny Whetton (Australia)

Contributing Authors:
R. Arritt (USA), R. Benestad (Norway), M. Beniston (Switzerland), D. Bromwich (USA), D. Caya (Canada), J. Comiso (USA), R. de Elía (Canada, Argentina), K. Dethloff (Germany), S. Emori (Japan), J. Feddema (USA), R. Gerdes (Germany), J.F. González-Rouco (Spain), W. Gutowski (USA), I. Hanssen-Bauer (Norway), C. Jones (Canada), R. Katz (USA), A. Kitoh (Japan), R. Knutti (Switzerland), R. Leung (USA), J. Lowe (UK), A.H. Lynch (Australia), C. Matulla (Canada, Austria), K. McInnes (Australia), A.V. Mescherskaya (Russian Federation), A.B. Mullan (New Zealand), M. New (UK), M.H. Nokhandan (Iran), J.S. Pal (USA, Italy), D. Plummer (Canada), M. Rummukainen (Sweden, Finland), C. Schär (Switzerland), S. Somot (France), D.A. Stone (UK, Canada), R. Suppiah (Australia), M. Tadross (South Africa), C. Tebaldi (USA), W. Tennant (South Africa), M. Widmann (Germany, UK), R. Wilby (UK), B.L. Wyman (USA)

Review Editors:
Congbin Fu (China), Filippo Giorgi (Italy)

This chapter should be cited as:
Christensen, J.H., B. Hewitson, A. Busuioc, A. Chen, X. Gao, I. Held, R. Jones, R.K. Kolli, W.-T. Kwon, R. Laprise, V. Magaña Rueda, L. Mearns, C.G. Menéndez, J. Räisänen, A. Rinke, A. Sarr and P. Whetton, 2007: Regional Climate Projections. In: *Climate Change 2007: The Physical Science Basis. Contribution of Working Group I to the Fourth Assessment Report of the Intergovernmental Panel on Climate Change* [Solomon, S., D. Qin, M. Manning, Z. Chen, M. Marquis, K.B. Averyt, M. Tignor and H.L. Miller (eds.)]. Cambridge University Press, Cambridge, United Kingdom and New York, NY, USA.

Table of Contents

Supplementary Material

The following supplementary material is available on CD-ROM and in on-line versions of this report.

Supplementary Figures S11.1–S11.37

Supplementary Tables S1.1 and S1.2

Supplementary References

Executive Summary

Increasingly reliable regional climate change projections are now available for many regions of the world due to advances in modelling and understanding of the physical processes of the climate system. A number of important themes have emerged:

- Warming over many land areas is greater than global annual mean warming due to less water availability for evaporative cooling and a smaller thermal inertia as compared to the oceans.

- Warming generally increases the spatial variability of precipitation, contributing to a reduction of rainfall in the subtropics and an increase at higher latitudes and in parts of the tropics. The precise location of boundaries between regions of robust increase and decrease remains uncertain and this is commonly where Atmosphere-Ocean General Circulation Model (AOGCM) projections disagree.

- The poleward expansion of the subtropical highs, combined with the general tendency towards reductions in subtropical precipitation, creates especially robust projections of a reduction in precipitation at the poleward edges of the subtropics. Most of the regional projections of reductions in precipitation in the 21st century are associated with areas adjacent to these subtropical highs.

- There is a tendency for monsoonal circulations to result in increased precipitation due to enhanced moisture convergence, despite a tendency towards weakening of the monsoonal flows themselves. However, many aspects of tropical climatic responses remain uncertain.

Atmosphere-Ocean General Circulation Models remain the primary source of regional information on the range of possible future climates. A clearer picture of the robust aspects of regional climate change is emerging due to improvement in model resolution, the simulation of processes of importance for regional change and the expanding set of available simulations. Advances have been made in developing probabilistic information at regional scales from the AOGCM simulations, but these methods remain in the exploratory phase. There has been less development extending this to downscaled regional information. However, downscaling methods have matured since the Third Assessment Report (TAR; IPCC, 2001) and have been more widely applied, although only in some regions has large-scale coordination of multi-model downscaling of climate change simulations been achieved.

Regional climate change projections presented here are assessed drawing on information from four potential sources: AOGCM simulations; downscaling of AOGCM-simulated data using techniques to enhance regional detail; physical understanding of the processes governing regional responses; and recent historical climate change.

Previous chapters describe observed climate change on regional scales (Chapter 3) and compare global model simulations with these changes (Chapter 9). Comparisons of model simulations of temperature change with observations can be used to help constrain future regional temperature projections. Regional assessments of precipitation change rely primarily on convergence in both global and downscaling models along with physical insights. Where there is near unanimity among models with good supporting physical arguments, as is more typical for middle and higher latitudes, these factors encourage stronger statements as to the likelihood of a regional climate change. In some circumstances, physical insights alone clearly indicate the direction of future change.

The summary likelihood statements on projected regional climate are as follows:

- *Temperature projections:* These are comparable in magnitude to those of the TAR and confidence in the regional projections is now higher due to a larger number and variety of simulations, improved models, a better understanding of the role of model deficiencies and more detailed analyses of the results. Warming, often greater than the global mean, is *very likely* over all landmasses.

- *Precipitation projections:* Overall patterns of change are comparable to those of TAR, with greater confidence in the projections for some regions. Model agreement is seen over more and larger regions. For some regions, there are grounds for stating that the projected precipitation changes are *likely* or *very likely*. For other regions, confidence in the projected change remains weak.

- *Extremes:* There has been a large increase in the available analyses of changes in extremes. This allows for a more comprehensive assessment for most regions. The general findings are in line with the assessment made in TAR and now have a higher level of confidence derived from multiple sources of information. The most notable improvements in confidence relate to the regional statements concerning heat waves, heavy precipitation and droughts. Despite these advances, specific analyses of models are not available for some regions, which is reflected in the robust statements on extremes. In particular, projections concerning extreme events in the tropics remain uncertain. The difficulty in projecting the distribution of tropical cyclones adds to this uncertainty. Changes in extra-tropical cyclones are dependent on details of regional atmospheric circulation response, some of which remain uncertain.

The following summarises the robust findings of the projected regional change over the 21st century. Supporting narratives are provided in Sections 11.2 to 11.9. These changes are assessed as *likely* to *very likely* taking into account the uncertainties in climate sensitivity and emission trajectories (in the Special Report on Emission Scenarios (SRES) B1/A1B/B2 scenario range) discussed in earlier chapters.

All land regions:

It is *very likely* that all land regions will warm in the 21st century.

Africa:

Warming is *very likely* to be larger than the global annual mean warming throughout the continent and in all seasons, with drier subtropical regions warming more than the moister tropics. Annual rainfall is *likely* to decrease in much of Mediterranean Africa and the northern Sahara, with a greater likelihood of decreasing rainfall as the Mediterranean coast is approached. Rainfall in southern Africa is *likely* to decrease in much of the winter rainfall region and western margins. There is *likely* to be an increase in annual mean rainfall in East Africa. It is unclear how rainfall in the Sahel, the Guinean Coast and the southern Sahara will evolve.

Mediterranean and Europe:

Annual mean temperatures in Europe are *likely* to increase more than the global mean. Seasonally, the largest warming is *likely* to be in northern Europe in winter and in the Mediterranean area in summer. Minimum winter temperatures are *likely* to increase more than the average in northern Europe. Maximum summer temperatures are *likely* to increase more than the average in southern and central Europe. Annual precipitation is *very likely* to increase in most of northern Europe and decrease in most of the Mediterranean area. In central Europe, precipitation is *likely* to increase in winter but decrease in summer. Extremes of daily precipitation are *very likely* to increase in northern Europe. The annual number of precipitation days is *very likely* to decrease in the Mediterranean area. Risk of summer drought is *likely* to increase in central Europe and in the Mediterranean area. The duration of the snow season is *very likely* to shorten, and snow depth is *likely* to decrease in most of Europe.

Asia:

Warming is *likely* to be well above the global mean in central Asia, the Tibetan Plateau and northern Asia, above the global mean in eastern Asia and South Asia, and similar to the global mean in Southeast Asia. Precipitation in boreal winter is *very likely* to increase in northern Asia and the Tibetan Plateau, and *likely* to increase in eastern Asia and the southern parts of Southeast Asia. Precipitation in summer is *likely* to increase in northern Asia, East Asia, South Asia and most of Southeast Asia, but is *likely* to decrease in central Asia. It is *very likely* that heat waves/hot spells in summer will be of longer duration, more intense and more frequent in East Asia. Fewer very cold days are *very likely* in East Asia and South Asia. There is *very likely* to be an increase in the frequency of intense precipitation events in parts of South Asia, and in East Asia. Extreme rainfall and winds associated with tropical cyclones are *likely* to increase in East Asia, Southeast Asia and South Asia.

North America:

The annual mean warming is *likely* to exceed the global mean warming in most areas. Seasonally, warming is *likely* to be largest in winter in northern regions and in summer in the southwest. Minimum winter temperatures are *likely* to increase more than the average in northern North America. Maximum summer temperatures are *likely* to increase more than the average in the southwest. Annual mean precipitation is *very likely* to increase in Canada and the northeast USA, and *likely* to decrease in the southwest. In southern Canada, precipitation is *likely* to increase in winter and spring but decrease in summer. Snow season length and snow depth are *very likely* to decrease in most of North America except in the northernmost part of Canada where maximum snow depth is *likely* to increase.

Central and South America:

The annual mean warming is *likely* to be similar to the global mean warming in southern South America but larger than the global mean warming in the rest of the area. Annual precipitation is *likely* to decrease in most of Central America and in the southern Andes, although changes in atmospheric circulation may induce large local variability in precipitation response in mountainous areas. Winter precipitation in Tierra del Fuego and summer precipitation in south-eastern South America is *likely* to increase. It is uncertain how annual and seasonal mean rainfall will change over northern South America, including the Amazon forest. However, there is qualitative consistency among the simulations in some areas (rainfall increasing in Ecuador and northern Peru, and decreasing at the northern tip of the continent and in southern northeast Brazil).

Australia and New Zealand:

Warming is *likely* to be larger than that of the surrounding oceans, but comparable to the global mean. The warming is less in the south, especially in winter, with the warming in the South Island of New Zealand *likely* to remain less than the global mean. Precipitation is *likely* to decrease in southern Australia in winter and spring. Precipitation is *very likely* to decrease in south-western Australia in winter. Precipitation is *likely* to increase in the west of the South Island of New Zealand. Changes in rainfall in northern and central Australia are uncertain. Increased mean wind speed is *likely* across the South Island of New Zealand, particularly in winter. Increased frequency of extreme high daily temperatures in Australia and New Zealand, and a decrease in the frequency of cold extremes is *very likely*. Extremes of daily precipitation are *very likely* to increase, except possibly in areas of significant decrease in mean rainfall (southern Australia in winter and spring). Increased risk of drought in southern areas of Australia is *likely*.

Polar regions:

The Arctic is *very likely* to warm during this century more than the global mean. Warming is projected to be largest in winter and smallest in summer. Annual arctic precipitation is *very likely* to increase. It is *very likely* that the relative precipitation increase will be largest in winter and smallest in summer. Arctic sea ice is *very likely* to decrease in its extent and thickness. It is uncertain how the Arctic Ocean circulation will change. The Antarctic is *likely* to warm and the precipitation is *likely* to increase over the continent. It is uncertain to what extent the frequency of extreme temperature and precipitation events will change in the polar regions.

Small Islands:

Sea levels are *likely* to rise on average during the century around the small islands of the Caribbean Sea, Indian Ocean and northern and southern Pacific Oceans. The rise will *likely* not be geographically uniform but large deviations among models make regional estimates across the Caribbean, Indian and Pacific Oceans uncertain. All Caribbean, Indian Ocean and North and South Pacific islands are *very likely* to warm during this century. The warming is *likely* to be somewhat smaller than the global annual mean. Summer rainfall in the Caribbean is *likely* to decrease in the vicinity of the Greater Antilles but changes elsewhere and in winter are uncertain. Annual rainfall is *likely* to increase in the northern Indian Ocean with increases *likely* in the vicinity of the Seychelles in December, January and February, and in the vicinity of the Maldives in June, July and August, while decreases are *likely* in the vicinity of Mauritius in June, July and August. Annual rainfall is *likely* to increase in the equatorial Pacific, while decreases are projected by most models for just east of French Polynesia in December, January and February.

11.1   Introduction

Increasingly reliable regional climate change projections are now available for many regions of the world due to advances in modelling and understanding of the physical processes of the climate system. Atmosphere-Ocean General Circulation Models (AOGCMs) remain the foundation for projections while downscaling techniques now provide valuable additional detail. Atmosphere-Ocean General Circulation Models cannot provide information at scales finer than their computational grid (typically of the order of 200 km) and processes at the unresolved scales are important. Providing information at finer scales can be achieved through using high resolution in dynamical models or empirical statistical downscaling. Development of downscaling methodologies remains an important focus. Downscaled climate change projections tailored to specific needs are only now starting to become available.

11.1.1   Summary of the Third Assessment Report

The assessment of regional climate projections in the Third Assessment Report (TAR; Chapter 10 of IPCC, 2001) was largely restricted to General Circulation Model (GCM)-derived temperature with limited precipitation statements. The major assessment of temperature change was that it is very likely all land areas will warm more than the global average (with the exception of Southeast Asia and South America in June, July and August; JJA), with amplification at high latitudes. The changes in precipitation assessed to be likely were: an increase over northern mid-latitude regions in winter and over high-latitude regions in both winter and summer; in December, January and February (DJF), an increase in tropical Africa, little change in Southeast Asia, and a decrease in Central America; an increase or little change in JJA over South Asia and a decrease over Australia and the Mediterranean region. These projections were almost entirely based on analysis of nine coarse-resolution AOGCMs that had performed transient experiments for the 20th century with the specifications for the A2 and B2 emission scenarios. Chapter 10 of the TAR noted that studies with regional models indicate that changes at finer scales may be substantially different in magnitude from these large sub-continental findings.

Information available for assessment regarding climate variability and extremes at the regional scale was too sparse for it to be meaningfully drawn together in a systematic manner. However, some statements of a more generic nature were made. It was assessed that the variability of daily to interannual temperatures is likely to decrease in winter and increase in summer for mid-latitude Northern Hemisphere (NH) land areas, daily high temperature extremes are likely to increase and future increases in mean precipitation are very likely to lead to an increase in variability. In some specifically analysed regions, it was assessed that extreme precipitation may increase and there were indications that droughts or dry spells may increase in occurrence in Europe, North America and Australia.

11.1.2   Introduction to Regional Projections

Assessments of climate change projections are provided here on a region-by-region basis. The discussion is organised according to the same continental-scale regions used by Working Group II (WGII) in the Fourth Assessment Report (AR4) and in earlier assessments: Africa, Europe and Mediterranean, Asia, North America, Central and South America, Australia-New Zealand, Polar Regions and Small Islands. While the topics covered vary somewhat from region to region, each section includes a discussion of key processes of importance for climate change in that region, relevant aspects of model skill in simulating current climate, and projections of future regional climate change based on global models and downscaling techniques.

Each of these continental-scale regions encompasses a broad range of climates and is too large to be used as a basis for conveying quantitative regional climate change information. Therefore, each is subdivided into a number of sub-continental or oceanic regions. The sub-continental regions as defined in Table 11.1 are the framework for developing specific regional or sub-continental robust statements of projected change.

Area-averaged temperature and precipitation changes are presented from the coordinated set of climate model simulations archived at the Program for Climate Model Diagnosis and Intercomparison (PCMDI; subsequently called the multi-model data set or MMD). The regions are very close to those initially devised by Giorgi and Francesco (2000) with some minor modifications similar to those of Ruosteenoja et al. (2003). They have simple shapes and are no smaller than the horizontal scales on which current AOGCMs are useful for climate simulations (typically judged to be roughly 1,000 km).

These regional averages have some deficiencies for discussion of the AOGCM projections. In several instances, the simple definition of these boxes results in spatial averaging over regions in which precipitation is projected to increase and decrease. There are also sub-regions where the case can be made for a robust and physically plausible hydrological response, information about which is lost in the regional averages. Partially to help in discussing these features, this chapter also uses maps of temperature and precipitation responses, interpolated to a grid with 128 longitudes by 64 latitudes which is typical of many of the lower-resolution atmospheric models in the MMD.

In the regional discussion to follow, the starting points are temperature and precipitation. Changes in temperature are introduced in each continental section by plotting for each of the regions the evolution of the range of projected decadal mean change for the A1B scenario through the 21st century (simulations hereafter referred to as MMD-A1B). These are put into the context of observed changes in the 20th century by plotting the observed changes and how well the models reproduce these. This summary information is displayed for continental regions in Box 11.1, which also contains details of how the figures were constructed. The equivalent figures for the individual regions of each continental-scale region are displayed in the following sections. These are constructed in the same way as Box 11.1, Figure 1. The 20th-century parts of these figures are also displayed in Section 9.4, where more details on their construction are provided. The discussion on precipitation provides a limited view of hydrological changes. Supplementary Material Figure S11.1 expands on this issue by comparing the annual mean responses in precipitation and in precipitation minus evaporation over the 21st century in the MMD-A1B projections. Over North America and Europe, for example, the region of drying in the sense of precipitation minus evaporation is shifted poleward compared to the region of reduced precipitation. A summary of the more significant hydrological cycle changes from the regional discussions is presented in Box 11.1.

Table 11.1 provides detailed information for each region generated from the MMD-A1B models focusing on the change in climate between the 1980 to 1999 period in the 20th-century integrations and the 2080 to 2099 period. The distribution of the annual and seasonal mean surface air temperature response and percentage change in precipitation are described by the median, the 25 and 75% values (half of the models lie between these two values) and the maximum and minimum values in the model ensemble. Information on model biases in these regional averages for the 1980 to 1999 simulations is provided in Supplementary Material Table S11.1 in a similar format. Maps of biases are referred to in some of the following and are included in the Supplementary Material as well. Data sources used in these comparisons are listed in the table and figure captions where these biases are displayed.

Most of the discussion focuses on the A1B scenario. The global mean near-surface temperature responses (between the period 1980 to 1999 of the 20th-century integrations and the period 2080 to 2099) in the ensemble mean of the MMD models are in the ratio 0.69:1:1.17 for the B1:A1B:A2 scenarios. The local temperature responses in nearly all regions closely follow the same ratio, as discussed in Chapter 10 and as illustrated in Supplementary Material Figures S11.2 to S11.4. Therefore, little is gained by repeating the discussion of the A1B scenario for the other scenarios. The ensemble mean local precipitation responses also approximately scale with the global mean temperature response, although not as precisely as the temperature itself. Given the substantial uncertainties in hydrological responses, the generally smaller signal/noise ratio and the similarities in the basic structure of the AOGCM precipitation responses in the different scenarios, a focus on A1B seems justified for the precipitation as well. The overall regional assessments, however, do rely on all available scenario information.

Given the dominantly linear response of the models, the 2080 to 2099 period allows the greatest clarity of the background climate change underlying the interannual and decadal variability. In the ensemble mean AOGCM projections there is no indication of abrupt climate change, nor does the literature on individual models provide any strong suggestions of robust nonlinearities. Some local temporal nonlinearities are to be expected, for example as the sea ice boundary retreats from a particular location in the Arctic. While the possibility

exists that changes of more abrupt character could happen, such as major ocean circulation or land surface/vegetation change, there is little basis to judge the plausibility of these factors (see Chapter 10). Therefore, this discussion is based on this linear picture.

Table 11.1 also provides some simple estimates of the signal-to-noise ratio. The signal is the change in 20-year means of seasonal or annual mean temperature or precipitation. The noise is an estimate of the internal variability of 20-year means of seasonal or annual mean temperature or precipitation, as generated by the models. The signal-to-noise ratio is converted into the time interval that is required before the signal is clearly discernible, assuming that the signal grows linearly over the century at the average rate in the ensemble mean A1B projection. 'Clearly discernible' is defined in this context as distinguishable with 95% confidence. As an example, the annual mean precipitation increase in northern Europe (NEU) (Table 11.1) is clearly discernible in these models after 45 years, meaning that the 20-year average from 2025 to 2044 will be greater than the 20-year mean over 1980 to 1999 with 95% confidence, accounting only for the internal variability in the models and no other sources of uncertainty. In contrast, the annual temperature response in Southeast Asia (SEA) rises above the noise by this measure after only 10 years, implying that the average temperature over the period 1990 to 2009 is clearly discernible in the models from the average over the control period 1980 to 1999. This measure is likely an overestimate of the time of emergence of the signal as compared to that obtained with more refined detection strategies (of the kind discussed in Chapter 9). This noise estimate is solely based on the models and must be treated with caution, but it would be wrong to assume that models always underestimate this internal variability. Some models overestimate and some underestimate the amplitude of the El Niño-Southern Oscillation (ENSO), for example, thereby over- or underestimating the most important source of interannual variability in the tropics. On the other hand, few models capture the range of decadal variability of rainfall in West Africa, for example (Hoerling, et al., 2006; Section 8.4).

Also included in Table 11.1 is an estimate of the probability of extremely warm, extremely wet and extremely dry seasons, for the A1B scenario and for the time period 2080 to 2099. An 'extremely warm' summer is defined as follows. Examining all of the summers simulated in a particular realisation of a model in the 1980 to 1999 control period, the warmest of these 20 summers can be computed as an estimate of the temperature of the warmest 5% of all summers in the control climate. The period 2080 to 2099 is then examined, and the fraction of the summers exceeding this warmth determined. This is referred to as the probability of extremely warm summers. The results are tabulated after averaging over models, and similarly for both extremely low and extremely high seasonal precipitation amounts. Values smaller (larger) than 5% indicate a decrease (increase) in the frequency of extremes. This follows the approach in Weisheimer and Palmer (2005) except that this chapter compares each model's future with its own 20th century to help avoid distortions due to differing biases in the different

Table 11.1. *Regional averages of temperature and precipitation projections from a set of 21 global models in the MMD for the A1B scenario. The mean temperature and precipitation responses are first averaged for each model over all available realisations of the 1980 to 1999 period from the 20th Century Climate in Coupled Models (20C3M) simulations and the 2080 to 2099 period of A1B. Computing the difference between these two periods, the table shows the minimum, maximum, median (50%), and 25 and 75% quartile values among the 21 models, for temperature (°C) and precipitation (%) change. Regions in which the middle half (25–75%) of this distribution is all of the same sign in the precipitation response are coloured light brown for decreasing and light blue for increasing precipitation. Signal-to-noise ratios for these 20-year mean responses is indicated by first computing a consensus standard deviation of 20-year means, using those models that have at least three realisations of the 20C3M simulations and using all 20-year periods in the 20th century. The signal is assumed to increase linearly in time, and the time required for the median signal to reach 2.83 (2 × √2) times the standard deviation is displayed as an estimate of when this signal is significant at the 95% level. These estimates of the times for emergence of a clearly discernible signal are only shown for precipitation when the models are in general agreement on the sign of the response, as indicated by the colouring. The frequency (%) of extremely warm, wet and dry seasons, averaged over the models, is also presented, as described in Section 11.2.1. Values are only shown when at least 14 out of the 21 models agree on an increase (bold) or a decrease in the extremes. A value of 5% indicates no change, as this is the nominal value for the control period by construction. The regions are defined by rectangular latitude/longitude boxes and the coordinates of the bottom left-hand and top right-hand corners of these are given in degrees in the first column under the region acronym (see table notes for full names of regions). Information is provided for land areas contained in the boxes except for the Small Islands regions where sea areas are used and for Antarctica where both land and sea areas are used.*

Region[a]	Season	Temperature Response (°C)						Precipitation Response (%)						Extreme Seasons (%)		
		Min	25	50	75	Max	T yrs	Min	25	50	75	Max	T yrs	Warm	Wet	Dry
AFRICA																
WAF	DJF	2.3	2.7	3.0	3.5	4.6	10	-16	-2	6	13	23		100	21	4
	MAM	1.7	2.8	3.5	3.6	4.8	10	-11	-7	-3	5	11		100		
12S,20W	JJA	1.5	2.7	3.2	3.7	4.7	10	-18	-2	2	7	16		100	19	
to	SON	1.9	2.5	3.3	3.7	4.7	10	-12	0	1	10	15		100	15	
22N,18E	Annual	1.8	2.7	3.3	3.6	4.7	10	-9	-2	2	7	13		100	22	
EAF	DJF	2.0	2.6	3.1	3.4	4.2	10	-3	6	13	16	33	55	100	25	1
	MAM	1.7	2.7	3.2	3.5	4.5	10	-9	2	6	9	20	>100	100	15	4
12S,22E	JJA	1.6	2.7	3.4	3.6	4.7	10	-18	-2	4	7	16		100		
to	SON	1.9	2.6	3.1	3.6	4.3	10	-10	3	7	13	38	95	100	21	3
18N,52E	Annual	1.8	2.5	3.2	3.4	4.3	10	-3	2	7	11	25	60	100	30	1
SAF	DJF	1.8	2.7	3.1	3.4	4.7	10	-6	-3	0	5	10		100	11	
	MAM	1.7	2.9	3.1	3.8	4.7	10	-25	-8	0	4	12		98		
35S,10E	JJA	1.9	3.0	3.4	3.6	4.8	10	-43	-27	-23	-7	-3	70	100	1	23
to	SON	2.1	3.0	3.7	4.0	5.0	10	-43	-20	-13	-8	3	90	100	1	20
12S,52E	Annual	1.9	2.9	3.4	3.7	4.8	10	-12	-9	-4	2	6		100	4	13
SAH	DJF	2.4	2.9	3.2	3.5	5.0	15	-47	-31	-18	-12	31	>100	97		12
	MAM	2.3	3.3	3.6	3.8	5.2	10	-42	-37	-18	-10	13	>100	100	2	21
18N,20E	JJA	2.6	3.6	4.1	4.4	5.8	10	-53	-28	-4	16	74		100		
to	SON	2.8	3.4	3.7	4.3	5.4	10	-52	-15	6	23	64		100		
30N,65E	Annual	2.6	3.2	3.6	4.0	5.4	10	-44	-24	-6	3	57		100		
EUROPE																
NEU	DJF	2.6	3.6	4.3	5.5	8.2	40	9	13	15	22	25	50	82	43	0
	MAM	2.1	2.4	3.1	4.3	5.3	35	0	8	12	15	21	60	79	28	2
48N,10W	JJA	1.4	1.9	2.7	3.3	5.0	25	-21	-5	2	7	16		88	11	
to	SON	1.9	2.6	2.9	4.2	5.4	30	-5	4	8	11	13	80	87	20	2
75N,40E	Annual	2.3	2.7	3.2	4.5	5.3	25	0	6	9	11	16	45	96	48	2
SEM	DJF	1.7	2.5	2.6	3.3	4.6	25	-16	-10	-6	-1	6	>100	93	3	12
	MAM	2.0	3.0	3.2	3.5	4.5	20	-24	-17	-16	-8	-2	60	98	1	31
30N,10W	JJA	2.7	3.7	4.1	5.0	6.5	15	-53	-35	-24	-14	-3	55	100	1	42
to	SON	2.3	2.8	3.3	4.0	5.2	15	-29	-15	-12	-9	-2	90	100	1	21
48N,40E	Annual	2.2	3.0	3.5	4.0	5.1	15	-27	-16	-12	-9	-4	45	100	0	46

Table 11.1 (continued)

Region[a]	Season	Temperature Response (°C)						Precipitation Response (%)						Extreme Seasons (%)		
		Min	25	50	75	Max	T yrs	Min	25	50	75	Max	T yrs	Warm	Wet	Dry
ASIA																
NAS	DJF	2.9	4.8	6.0	6.6	8.7	20	12	20	26	37	55	30	93	68	0
	MAM	2.0	2.9	3.7	5.0	6.8	25	2	16	18	24	26	30	89	66	1
50N,40E	JJA	2.0	2.7	3.0	4.9	5.6	15	-1	6	9	12	16	40	100	51	2
to	SON	2.8	3.6	4.8	5.8	6.9	15	7	15	17	19	29	30	99	65	0
70N,180E	Annual	2.7	3.4	4.3	5.3	6.4	15	10	12	15	19	25	20	100	92	0
CAS	DJF	2.2	2.6	3.2	3.9	5.2	25	-11	0	4	9	22		84	8	
	MAM	2.3	3.1	3.9	4.5	4.9	20	-26	-14	-9	-4	3	>100	94		16
30N,40E	JJA	2.7	3.7	4.1	4.9	5.7	10	-58	-28	-13	-4	21	>100	100	3	20
to	SON	2.5	3.2	3.8	4.1	4.9	15	-18	-4	3	9	24		99		
50N,75E	Annual	2.6	3.2	3.7	4.4	5.2	10	-18	-6	-3	2	6		100		12
TIB	DJF	2.8	3.7	4.1	4.9	6.9	20	1	12	19	26	36	45	95	40	0
	MAM	2.5	2.9	3.6	4.3	6.3	15	-3	4	10	14	34	70	96	34	2
30N,50E	JJA	2.7	3.2	4.0	4.7	5.4	10	-11	0	4	10	28		100	24	
to	SON	2.7	3.3	3.8	4.6	6.2	15	-8	-4	8	14	21		100	20	
75N,100E	Annual	2.8	3.2	3.8	4.5	6.1	10	-1	2	10	13	28	45	100	46	1
EAS	DJF	2.1	3.1	3.6	4.4	5.4	20	-4	6	10	17	42	>100	96	18	2
	MAM	2.1	2.6	3.3	3.8	4.6	15	0	7	11	14	20	55	98	35	2
20N,100E	JJA	1.9	2.5	3.0	3.9	5.0	10	-2	5	9	11	17	45	100	32	1
to	SON	2.2	2.7	3.3	4.2	5.0	15	-13	-1	9	15	29		100	20	3
50N,145E	Annual	2.3	2.8	3.3	4.1	4.9	10	2	4	9	14	20	40	100	47	1
SAS	DJF	2.7	3.2	3.6	3.9	4.8	10	-35	-9	-5	1	15		99		
	MAM	2.1	3.0	3.5	3.8	5.3	10	-30	-2	9	18	26		100	14	
5N,64E	JJA	1.2	2.2	2.7	3.2	4.4	15	-3	4	11	16	23	45	96	32	1
to	SON	2.0	2.5	3.1	3.5	4.4	10	-12	8	15	20	26	50	100	29	3
50N,100E	Annual	2.0	2.7	3.3	3.6	4.7	10	-15	4	11	15	20	40	100	39	3
SEA	DJF	1.6	2.1	2.5	2.9	3.6	10	-4	3	6	10	12	80	99	23	2
	MAM	1.5	2.2	2.7	3.1	3.9	10	-4	2	7	9	17	75	100	27	1
11S,95E	JJA	1.5	2.2	2.4	2.9	3.8	10	-3	3	7	9	17	70	100	24	2
to	SON	1.6	2.2	2.4	2.9	3.6	10	-2	2	6	10	21	85	99	26	3
20N,115E	Annual	1.5	2.2	2.5	3.0	3.7	10	-2	3	7	8	15	40	100	44	1
NORTH AMERICA																
ALA	DJF	4.4	5.6	6.3	7.5	11.0	30	6	20	28	34	56	40	80	39	0
	MAM	2.3	3.2	3.5	4.7	7.7	35	2	13	17	23	38	40	69	45	0
60N,170W	JJA	1.3	1.8	2.4	3.8	5.7	25	1	8	14	20	30	45	86	51	1
to	SON	2.3	3.6	4.5	5.3	7.4	25	6	14	19	31	36	40	86	51	0
72N,103W	Annual	3.0	3.7	4.5	5.2	7.4	20	6	13	21	24	32	25	97	80	0
CGI	DJF	3.3	5.2	5.9	7.2	8.5	20	6	15	26	32	42	30	95	58	0
	MAM	2.4	3.2	3.8	4.6	7.2	20	4	13	17	20	34	35	94	49	1
50N,103W	JJA	1.5	2.1	2.8	3.7	5.6	15	0	8	11	12	19	35	99	46	1
to	SON	2.7	3.4	4.0	5.7	7.3	20	7	14	16	22	37	35	99	62	0
85N,10W	Annual	2.8	3.5	4.3	5.0	7.1	15	8	12	15	20	31	25	100	90	0

Table 11.1 (continued)

Region[a]	Season	Temperature Response (°C)						Precipitation Response (%)						Extreme Seasons (%)		
		Min	25	50	75	Max	T yrs	Min	25	50	75	Max	T yrs	Warm	Wet	Dry
NORTH AMERICA (continued)																
WNA	DJF	1.6	3.1	3.6	4.4	5.8	25	-4	2	7	11	36	>100	80	18	3
	MAM	1.5	2.4	3.1	3.4	6.0	20	-7	2	5	8	14	>100	87	14	
30N,50E	JJA	2.3	3.2	3.8	4.7	5.7	10	-18	-10	-1	2	10		100	3	
to	SON	2.0	2.8	3.1	4.5	5.3	20	-3	3	6	12	18	>100	95	17	2
75N,100E	Annual	2.1	2.9	3.4	4.1	5.7	15	-3	0	5	9	14	70	100	21	2
CNA	DJF	2.0	2.9	3.5	4.2	6.1	30	-18	0	5	8	14		71	7	
	MAM	1.9	2.8	3.3	3.9	5.7	25	-17	2	7	12	17	>100	81	19	4
30N,103W	JJA	2.4	3.1	4.1	5.1	6.4	20	-31	-15	-3	4	20	>100	93		15
to	SON	2.4	3.0	3.5	4.6	5.8	20	-17	-4	4	11	24		91	11	
50N,85W	Annual	2.3	3.0	3.5	4.4	5.8	15	-16	-3	3	7	15		98		
ENA	DJF	2.1	3.1	3.8	4.6	6.0	25	2	9	11	19	28	85	78	24	
	MAM	2.3	2.7	3.5	3.9	5.9	20	-4	7	12	16	23	60	86	23	2
25N,85W	JJA	2.1	2.6	3.3	4.3	5.4	15	-17	-3	1	6	13		98		
to	SON	2.2	2.8	3.5	4.4	5.7	20	-7	4	7	11	17	>100	97	19	
50N,50W	Annual	2.3	2.8	3.6	4.3	5.6	15	-3	5	7	10	15	55	100	29	1
CENTRAL AND SOUTH AMERICA																
CAM	DJF	1.4	2.2	2.6	3.5	4.6	15	-57	-18	-14	-9	0	>100	96	2	25
	MAM	1.9	2.7	3.6	3.8	5.2	10	-46	-25	-16	-10	15	75	100	2	18
10N,116W	JJA	1.8	2.7	3.4	3.6	5.5	10	-44	-25	-9	-4	12	90	100		24
to	SON	2.0	2.7	3.2	3.7	4.6	10	-45	-10	-4	7	24		100		15
30N,83W	Annual	1.8	2.6	3.2	3.6	5.0	10	-48	-16	-9	-5	9	65	100	2	33
AMZ	DJF	1.7	2.4	3.0	3.7	4.6	10	-13	0	4	11	17	>100	93	27	4
	MAM	1.7	2.5	3.0	3.7	4.6	10	-13	-1	1	4	14		100	18	
20S,82W	JJA	2.0	2.7	3.5	3.9	5.6	10	-38	-10	-3	2	13		100		
to	SON	1.8	2.8	3.5	4.1	5.4	10	-35	-12	-2	8	21		100		
12N,34W	Annual	1.8	2.6	3.3	3.7	5.1	10	-21	-3	0	6	14		100		
SSA	DJF	1.5	2.5	2.7	3.3	4.3	10	-16	-2	1	7	10		100		
	MAM	1.8	2.3	2.6	3.0	4.2	15	-11	-2	1	5	7		98	8	
56S,76W	JJA	1.7	2.1	2.4	2.8	3.6	15	-20	-7	0	3	17		95		
to	SON	1.8	2.2	2.7	3.2	4.0	15	-20	-12	1	6	11		99		
20S,40W	Annual	1.7	2.3	2.5	3.1	3.9	10	-12	-1	3	5	7		100		
AUSTRALIA AND NEW ZEALAND																
NAU	DJF	2.2	2.6	3.1	3.7	4.6	20	-20	-8	1	8	27		89		
	MAM	2.1	2.7	3.1	3.3	4.3	20	-24	-12	1	15	40		92		3
30S,110E	JJA	2.0	2.7	3.0	3.3	4.3	25	-54	-20	-14	3	26		94	3	
to	SON	2.5	3.0	3.2	3.8	5.0	20	-58	-32	-12	2	20		98		
11S,155E	Annual	2.2	2.8	3.0	3.5	4.5	15	-25	-8	-4	8	23		99		
SAU	DJF	2.0	2.4	2.7	3.2	4.2	20	-23	-12	-2	12	30		95		
	MAM	2.0	2.2	2.5	2.8	3.9	20	-31	-9	-5	13	32		90		6
45S,110E	JJA	1.7	2.0	2.3	2.5	3.5	15	-37	-20	-11	-4	9	>100	95		17
to	SON	2.0	2.6	2.8	3.0	4.1	20	-42	-27	-14	-5	4	>100	95		15
30S,155E	Annual	1.9	2.4	2.6	2.8	3.9	15	-27	-13	-4	3	12		100		

Table 11.1 (continued)

Region[a]	Season	Temperature Response (°C)						Precipitation Response (%)						Extreme Seasons (%)		
		Min	25	50	75	Max	T yrs	Min	25	50	75	Max	T yrs	Warm	Wet	Dry
POLAR REGIONS																
ARC[b]	DJF	4.3	6.0	6.9	8.4	11.4	15	11	19	26	29	39	25	100	90	0
	MAM	2.4	3.7	4.4	4.9	7.3	15	9	14	16	21	32	25	100	79	0
60N,180E	JJA	1.2	1.6	2.1	3.0	5.3	15	4	10	14	17	20	25	100	85	0
to	SON	2.9	4.8	6.0	7.2	8.9	15	9	17	21	26	35	20	100	96	0
90N,180W	Annual	2.8	4.0	4.9	5.6	7.8	15	10	15	18	22	28	20	100	100	0
ANT[c]	DJF	0.8	2.2	2.6	2.8	4.6	20	-11	5	9	14	31	50	85	34	3
	MAM	1.3	2.2	2.6	3.3	5.3	20	1	8	12	19	40	40	88	54	0
90S,180E	JJA	1.4	2.3	2.8	3.3	5.2	25	5	14	19	24	41	30	83	59	0
to	SON	1.3	2.1	2.3	3.2	4.8	25	-2	9	12	18	36	45	79	42	1
60S,180W	Annual	1.4	2.3	2.6	3.0	5.0	15	-2	9	14	17	35	25	99	81	1
SMALL ISLANDS																
CAR	DJF	1.4	1.8	2.1	2.4	3.2	10	-21	-11	-6	0	10		100	2	
	MAM	1.3	1.8	2.2	2.4	3.2	10	-28	-20	-13	-6	6	>100	100	3	18
10N,85W	JJA	1.3	1.8	2.0	2.4	3.2	10	-57	-35	-20	-6	8	60	100	2	40
to	SON	1.6	1.9	2.0	2.5	3.4	10	-38	-18	-6	1	19		100		22
25N,60W	Annual	1.4	1.8	2.0	2.4	3.2	10	-39	-19	-12	-3	11	60	100	3	39
IND	DJF	1.4	2.0	2.1	2.4	3.8	10	-4	2	4	9	20	>100	100	19	1
	MAM	1.5	2.0	2.2	2.5	3.8	10	0	3	5	6	20	80	100	22	1
35S,50E	JJA	1.4	1.9	2.1	2.4	3.7	10	-3	-1	3	5	20		100	17	
to	SON	1.4	1.9	2.0	2.3	3.6	10	-5	2	4	7	21	>100	100	17	2
17.5N,100E	Annual	1.4	1.9	2.1	2.4	3.7	10	-2	3	4	5	20	65	100	30	2
MED	DJF	1.5	2.0	2.3	2.7	4.2	25	-25	-16	-14	-10	-2	85	96	1	18
	MAM	1.5	2.1	2.4	2.7	3.7	20	-32	-23	-19	-16	-6	65	99	0	32
30N,5W	JJA	2.0	2.6	3.1	3.7	4.7	15	-64	-34	-29	-20	-3	60	100	1	36
to	SON	1.9	2.3	2.7	3.2	4.4	20	-33	-16	-10	-5	9	>100	99	2	21
45N,35E	Annual	1.7	2.2	2.7	3.0	4.2	15	-30	-16	-15	-10	-6	45	100	0	50
TNE	DJF	1.4	1.9	2.1	2.3	3.3	10	-35	-8	-6	3	10	>100	100		
	MAM	1.5	1.9	2.0	2.2	3.1	15	-16	-7	-2	6	39	>100	100		
0,30W	JJA	1.4	1.9	2.1	2.4	3.6	15	-8	-2	2	7	13	>100	100		
to	SON	1.5	2.0	2.2	2.6	3.7	15	-16	-5	-1	3	9	>100	100		
40N,10W	Annual	1.4	1.9	2.1	2.4	3.5	15	-7	-3	1	3	7	>100	100		
NPA	DJF	1.5	1.9	2.4	2.5	3.6	10	-5	1	3	6	17	>100	100	20	2
	MAM	1.4	1.9	2.3	2.5	3.5	10	-17	-1	1	3	17		100	14	
0,150E	JJA	1.4	1.9	2.3	2.7	3.9	10	1	5	8	14	25	55	100	43	1
to	SON	1.6	1.9	2.4	2.9	3.9	10	1	5	6	13	22	50	100	31	1
40N,120W	Annual	1.5	1.9	2.3	2.6	3.7	10	0	3	5	10	19	60	100	35	1
SPA	DJF	1.4	1.7	1.8	2.1	3.2	10	-6	1	4	7	15	80	100	19	4
	MAM	1.4	1.8	1.9	2.1	3.2	10	-3	3	6	8	17	35	100	35	1
55S,150E	JJA	1.4	1.7	1.8	2.0	3.1	10	-2	1	3	5	12	70	100	27	3
to	SON	1.4	1.6	1.8	2.0	3.0	10	-8	-2	2	4	5		100		
0,80W	Annual	1.4	1.7	1.8	2.0	3.1	10	-4	3	3	6	11	40	100	40	3

Notes: [a] Regions are: West Africa (WAF), East Africa (EAF), South Africa (SAF), Sahara (SAH), Northern Europe (NEU), Southern Europe and Mediterranean (SEM), Northern Asia (NAS), Central Asia (CAS), Tibetan Plateau (TIB), East Asia (EAS), South Asia (SAS), Southeast Asia (SEA), Alaska (ALA), East Canada, Greenland and Iceland (CGI), Western North America (WNA), Central North America (CNA), Eastern North America (ENA), Central America (CAM), Amazonia (AMZ), Southern South America (SSA), North Australia (NAU), South Australia (SAU), Arctic (ARC), Antarctic (ANT), Caribbean (CAR), Indian Ocean (IND), Mediterranean Basin (MED), Tropical Northeast Atlantic (TNE), North Pacific Ocean (NPA), and South Pacific Ocean (SPA).

       [b] land and ocean

       [c] land only

Box 11.1: Summary of Regional Responses

As an introduction to the more detailed regional analysis presented in this chapter, Box 11.1, Figure 1 illustrates how continental-scale warming is projected to evolve in the 21st century using the MMD models. This warming is also put into the context of the observed warming during the 20th century by comparing results from that subset of the models incorporating a representation of all known forcings with the observed evolution (see Section 9.4 for more details). Thus for the six continental regions, the figure displays: 1) the observed time series of the evolution of decadally averaged surface air temperature from 1906 to 2005 as an anomaly from the 1901 to 1950 average; 2) the range of the equivalent anomalies derived from 20th-century simulations by the MMD models that contain a full set of historical forcings; 3) the evolution of the range of this anomaly in MMD-A1B projections between 2000 and 2100; and 4) the range of the projected anomaly for the last decade of the 21st century for the B1, A1B, and A2 scenarios. For the observed part of these graphs, the decadal averages are centred on the decade boundaries (i.e., the last point is for 1996 to 2005), whereas for the future period they are centred on the decade mid-points (i.e., the first point is for 2001 to 2010). The width of the shading and the bars represents the 5 to 95% range of the model results. To construct the ranges, all simulations from the set of models involved were considered independent realisations of the possible evolution of the climate given the forcings applied. This involved 58 simulations from 14 models for the observed period and 47 simulations from 18 models for the future. Important in this representation is that the models' estimate of natural climate variability is included and thus the ranges include both the potential mitigating and amplifying effects of variability on the underlying signal. In contrast, the bars representing the range of projected change at the end of the century are constructed from ensemble mean changes from the models and thus provide a measure of the forced response. These bars were constructed from decadal mean anomalies from 21 models using A1B scenario forcings, from the 20 of these models that used the B1 forcings and the 17 that used the A2 forcings. The bars for the B1 and A2 scenarios were scaled to approximate ranges for the full set of models. The scaling factor for B1 was derived from the ratio between its range and the A1B range of the corresponding 20 models. The same procedure was used to obtain the A2 scaling factor. Only 18 models were used to display the ranges of projected temperature evolution as the control simulations for the other 3 had a drift of >0.2°C per century, which precludes clearly defining the decadal anomalies from these models. However, anomalies from all 21 models were included in calculating the bars in order to provide the fullest possible representation of projected changes in the MMD. Comparison of these different representations shows that the main messages from the MMD about projected continental temperature change are insensitive to the choices made. Finally, results are not shown here for Antarctica because the observational record is not long enough to provide the relevant information for the first part of the 20th century. Results of a similar nature to those shown here using the observations that are available are presented in Section 11.8.   *(continued)*

Box 11.1, Figure 1. *Temperature anomalies with respect to 1901 to 1950 for six continental-scale regions for 1906 to 2005 (black line) and as simulated (red envelope) by MMD models incorporating known forcings; and as projected for 2001 to 2100 by MMD models for the A1B scenario (orange envelope). The bars at the end of the orange envelope represent the range of projected changes for 2091 to 2100 for the B1 scenario (blue), the A1B scenario (orange) and the A2 scenario (red). The black line is dashed where observations are present for less than 50% of the area in the decade concerned. More details on the construction of these figures are given in Section 11.1.2.*

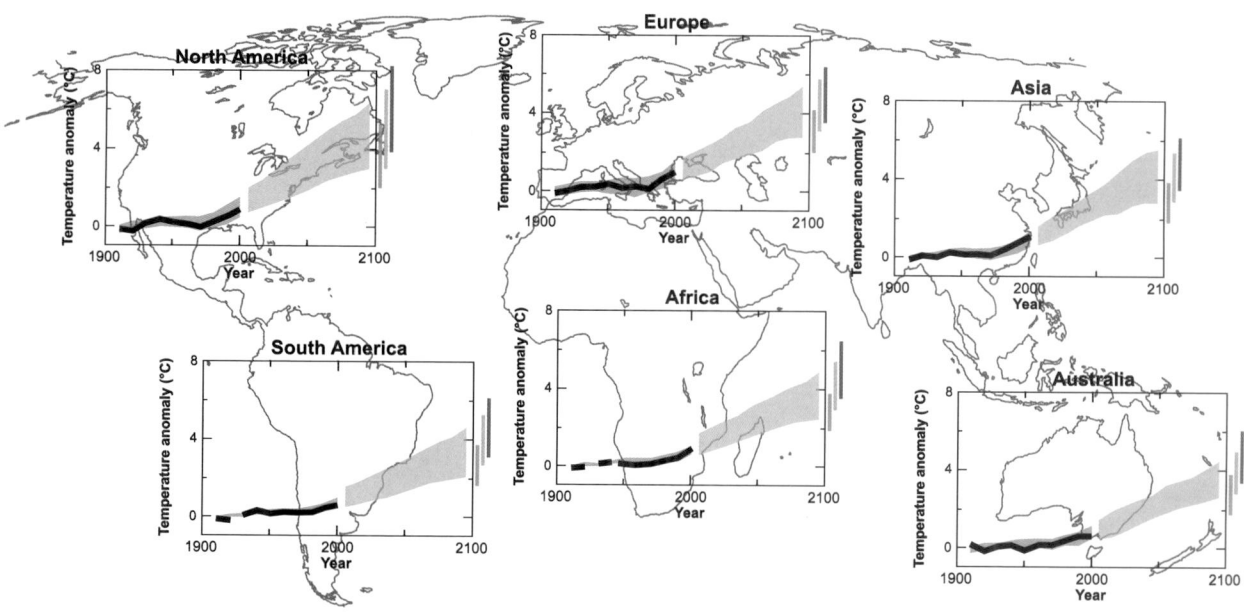

Box 11.1, Figure 2 serves to illustrate some of the more significant hydrological changes, with the two panels corresponding to DJF and JJA. The backdrop to these figures is the fraction of the AOGCMs (out of the 21 considered for this purpose) that predict an increase in mean precipitation in that grid cell (using the A1B scenario and comparing the period 2080 to 2099 with the control period 1980 to 1999). Aspects of this pattern are examined more closely in the separate regional discussions.

Robust findings on regional climate change for mean and extreme precipitation, drought and snow are highlighted in the figure with further detail in the accompanying notes.

(continued)

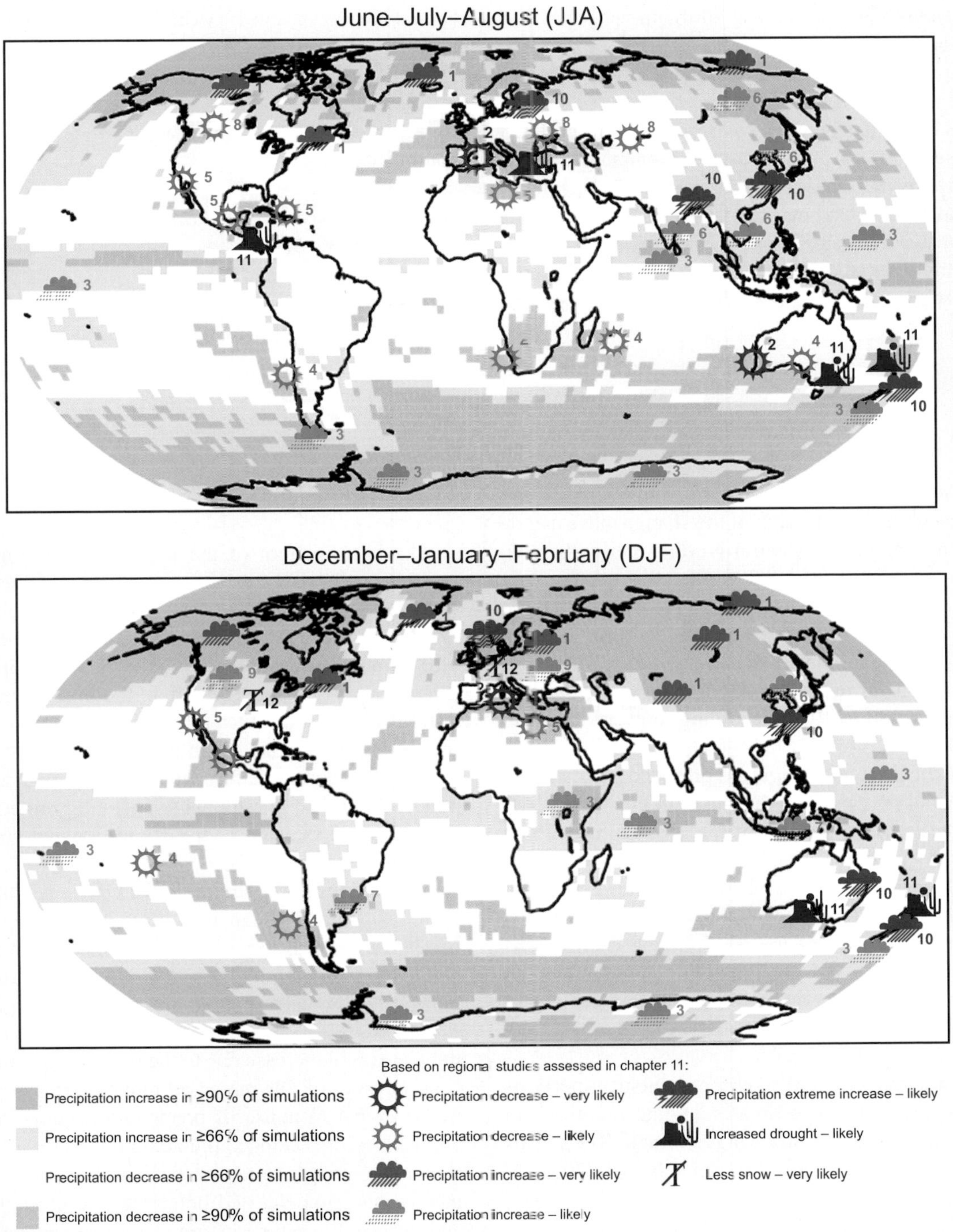

June–July–August (JJA)

December–January–February (DJF)

Based on regional studies assessed in chapter 11:

Precipitation increase in ≥90% of simulations	Precipitation decrease – very likely	Precipitation extreme increase – likely
Precipitation increase in ≥66% of simulations	Precipitation decrease – likely	Increased drought – likely
Precipitation decrease in ≥66% of simulations	Precipitation increase – very likely	Less snow – very likely
Precipitation decrease in ≥90% of simulations	Precipitation increase – likely	

Box 11.1, Figure 2. *Robust findings on regional climate change for mean and extreme precipitation, drought, and snow. This regional assessment is based upon AOGCM based studies, Regional Climate Models, statistical downscaling and process understanding. More detail on these findings may be found in the notes below, and their full description, including sources is given in the text. The background map indicates the degree of consistency between AR4 AOGCM simulations (21 simulations used) in the direction of simulated precipitation change.*

(1)   Very likely annual mean increase in most of northern Europe and the Arctic (largest in cold season), Canada, and the North-East USA; and winter (DJF) mean increase in Northern Asia and the Tibetan Plateau.

(2)   Very likely annual mean decrease in most of the Mediterranean area, and winter (JJA) decrease in southwestern Australia.

(3)   Likely annual mean increase in tropical and East Africa, Northern Pacific, the northern Indian Ocean, the South Pacific (slight, mainly equatorial regions), the west of the South Island of New Zealand, Antarctica and winter (JJA) increase in Tierra del Fuego.

(4)   Likely annual mean decrease in and along the southern Andes, summer (DJF) decrease in eastern French Polynesia, winter (JJA) decrease for Southern Africa and in the vicinity of Mauritius, and winter and spring decrease in southern Australia.

(5)   Likely annual mean decrease in North Africa, northern Sahara, Central America (and in the vicinity of the Greater Antilles in JJA) and in South-West USA.

(6)   Likely summer (JJA) mean increase in Northern Asia, East Asia, South Asia and most of Southeast Asia, and likely winter (DJF) increase in East Asia.

(7)   Likely summer (DJF) mean increase in southern Southeast Asia and southeastern South America

(8)   Likely summer (JJA) mean decrease in Central Asia, Central Europe and Southern Canada.

(9)   Likely winter (DJF) mean increase in central Europe, and southern Canada

(10)   Likely increase in extremes of daily precipitation in northern Europe, South Asia, East Asia, Australia and New Zealand.

(11)   Likely increase in risk of drought in Australia and eastern New Zealand; the Mediterranean, central Europe (summer drought); in Central America (boreal spring and dry periods of the annual cycle).

(12)   Very likely decrease in snow season length and likely to very likely decrease in snow depth in most of Europe and North America.

models. The results are shown in Table 11.1 only when 14 out of the 21 models agree as to the sign of the change in frequency of extremes. For example, in Central North America (CNA), 15% of the summers in 2080 to 2099 in the A1B scenario are projected to be extremely dry, corresponding to a factor of three increase in the frequency of these events. In contrast, in many regions and seasons, the frequency of extreme warmth is 100%, implying that all seasons in 2080 to 2099 are warmer than the warmest season in 1980 to 1999, according to every model in this ensemble.

In each continental section, a figure is provided summarising the temperature and precipitation responses in the MMD-A1B projection for the last two decades of the 21st century. These figures portray a multi-model mean comprising individual models or model ensemble means where ensembles exist. Also shown is the simple statistic of the number of these models that show agreement in the sign of the precipitation change. The annual mean temperature and precipitation responses in each of the 21 separate AOGCMs are provided in Supplementary Material Figures S11.5 to 11.12 and S11.13 to 11.20, respectively.

Recent explorations of multi-model ensemble projections seek to develop probabilistic estimates of uncertainties and are provided in the Supplementary Material Table S11.2. This information is based on the approach of Tebaldi et al. (2004a,b; see also section 11.10.2).

11.1.3 Some Unifying Themes

The basic pattern of the projected warming as described in Chapter 10 is little changed from previous assessments. Examining the spread across the MMD models, temperature projections in many regions are strongly correlated with the global mean projections, with the most sensitive models in global mean temperature often the most sensitive locally. Differing treatments of regional processes and the dynamical interactions between a given region and the rest of the climate system are responsible for some spread. However, a substantial part of the spread in regional temperature projections is due to differences in the sum of the feedbacks that control transient climate sensitivity (see also Chapter 10).

The response of the hydrological cycle is controlled in part by fundamental consequences of warmer temperatures and the increase in water vapour in the atmosphere (Chapter 3). Water is transported horizontally by the atmosphere from regions of moisture divergence (particularly in the subtropics) to regions of convergence. Even if the circulation does not change, these transports will increase due to the increase in water vapour. The consequences of this increased moisture transport can be seen in the global response of precipitation, described in Chapter 10, where, on average, precipitation increases in the inter-tropical convergence zones, decreases in the subtropics, and increases in subpolar and polar regions. Over North America and Europe, the pattern of subpolar moistening and subtropical drying dominates the 21st-century projections. This pattern is also described in Section 9.5.4, which assesses the extent to which this pattern is visible over land during the 20th century in precipitation observations and model simulations. Regions of large uncertainty often lie near the boundaries between these robust moistening and drying regions, with boundaries placed differently by each model.

High-resolution model results indicate that in regions with strong orographic forcing, some of these large-scale findings can be considerably altered locally. In some cases, this may result in changes in the opposite direction to the more general large-scale behaviour. In addition, large-area and grid-box average projections for precipitation are often very different from local changes within the area (Good and Lowe, 2006). These issues demonstrate the inadequacy of inferring the behaviour at fine scales from that of large-area averages.

Another important theme in the 21st-century projections is the poleward expansion of the subtropical highs, and the poleward displacement of the mid-latitude westerlies and associated storm tracks. This circulation response is often referred to as an enhanced positive phase of the Northern or Southern Annular Mode, or when focusing on the North Atlantic, the positive phase of the North Atlantic Oscillation (NAO). In regions without strong orographic forcing, superposition of the tendency towards subtropical drying and poleward expansion of the subtropical highs creates especially robust drying responses at the poleward boundaries of the five subtropical oceanic high centres in the South Indian, South Atlantic, South Pacific, North Atlantic and, less robustly, the North Pacific (where a tendency towards El-Niño like conditions in the Pacific in the models tends to counteract this expansion). Most of the regional projections of strong drying tendencies over land in the 21st century are immediately downstream of these centres (south-western Australia, the Western Cape Provinces of South Africa, the southern Andes, the Mediterranean and Mexico). The robustness of this large-scale circulation signal is discussed in Chapter 10, while Chapters 3, 8 and 9 describe the observed poleward shifts in the late 20th century and the ability of models to simulate these shifts.

The retreats of snow and ice cover are important for local climates. The difficulty of quantifying these effects in regions of substantial topographic relief is a significant limitation of global models (see Section 11.4.3.2, Box 11.3) and is improved with dynamical and statistical downscaling. The drying effect of an earlier spring snowmelt and, more generally, the earlier reduction in soil moisture (Manabe and Wetherald, 1987) is a continuing theme in discussion of summer continental climates.

The strong interactions between sea surface temperature gradients and tropical rainfall variability provides an important unifying theme for tropical climates. Models can differ in their projections of small changes in tropical ocean temperature gradients and in the simulation of the potentially large shifts in rainfall that are related to these oceanic changes. Chou and Neelin (2004) provide a guide to some of the complexity

involved in diagnosing and evaluating hydrological responses in the tropics. With a few exceptions, the spread in projections of hydrological changes is still too large to make strong statements about the future of tropical climates at regional scales (see also Section 10.3). Many AOGCMs project large tropical precipitation changes, so uncertainty as to the regional pattern of these changes should not be taken as evidence that these changes are likely to be small.

Assessments of the regional and sub-regional climate change projections have primarily been based on the AOGCM projections summarised in Table 11.1 and an analysis of the biases in the AOGCM simulations, regional downscaling studies available for some regions with either physical or statistical models or both, and reference to plausible physical mechanisms.

To assist the reader in placing the various regional assessments in a global context, Box 11.1 displays many of the detailed assessments documented in the following regional sections. Likewise, an overview of projected changes in various types of extreme weather statistics is summarised in Table 11.2, which contains information from the assessments within this chapter and from Chapter 10. Thus, the details of the assessment that lead to each individual statement can all be found in either Chapter 10, or the respective regional sections, and links for each statement are identifiable from Table 11.2.

Table 11.2. *Projected changes in climate extremes. This table summarises key phenomena for which there is confidence in the direction of projected change based on the current scientific evidence. The included phenomena are those where confidence ranges between medium and very likely, and are listed with the notation of VL (very likely), L (likely), and M (medium confidence). maxTmax refers to the highest maximum temperature, maxTmin to the highest minimum temperature, minTmax to the lowest maximum temperature, and minTmin to the lowest minimum temperature. In addition to changes listed in the table, there are two phenomena of note for which there is little confidence. The issue of drying and associated risk of drought in the Sahel remains uncertain as discussed in Section 11.2.4.2. The change in mean duration of tropical cyclones cannot be assessed with confidence at this stage due to insufficient studies.*

Temperature-Related Phenomena	
Change in phenomenon	**Projected changes**
Higher monthly absolute maximum of daily maximum temperatures (maxTmax) more hot / warm summer days	**VL (consistent across model projections)** maxTmax increases at same rate as the mean or median[1] over northern Europe,[2] Australia and New Zealand[3] **L (fairly consistent across models, but sensitivity to land surface treatment)** maxTmax increases more than the median over southern and central Europe,[4] and southwest USA[5] **L (consistent with projected large increase in mean temperature)** Large increase in probability of extreme warm seasons over most parts of the world[6]
Longer duration, more intense, more frequent heat waves / hot spells in summer	**VL (consistent across model projections)** Over almost all continents[7], but particularly central Europe,[8] western USA,[9] East Asia[10] and Korea[11]
Higher monthly absolute maximum of daily minimum temperatures (maxTmin); more warm and fewer cold nights	**VL (consistent with higher mean temperatures)** Over most continents[12]
Higher monthly absolute minimum of daily minimum temperatures (minTmin)	**VL (consistent across model projections)** minTmin increases more than the mean in many mid- and high-latitude locations,[13] particularly in winter over most of Europe except the southwest[14]
Higher monthly absolute minimum of daily maximum temperatures (minTmax), fewer cold days	**L (consistent with warmer mean temperatures)** minTmin increases more than the mean in some areas[15]
Fewer frost days	**VL (consistent across model projections)** Decrease in number of days with below-freezing temperatures everywhere[16]
Fewer cold outbreaks; fewer, shorter, less intense cold spells / cold extremes in winter	**VL (consistent across model projections)** Northern Europe, South Asia, East Asia[17] **L (consistent with warmer mean temperatures)** Most other regions[18]
Reduced diurnal temperature range	**L (consistent across model projections)** Over most continental regions, night temperatures increase faster than the day temperatures[19]
Temperature variability on interannual and daily time scales	**L (general consensus across model projections)** Reduced in winter over most of Europe[20] Increase in central Europe in summer[21]

1   Kharin and Zwiers (2005)
2   §11.3.3.3, Supplementary Material Figure S11.23, PRUDENCE, Kjellström et al. (2007)
3   §11.7.3.5, CSIRO (2001)
4   §11.3.3.3, PRUDENCE, Kjellström et al. (2007)
5   §11.5.3.3, Bell et al. (2004),
6   Table 11.1
7   §11.3.3.3, Tebaldi et al. (2006), Meehl and Tebaldi (2004)
8   §11.5.3.3, Barnett et al. (2006), Clark et al. (2006), Tebaldi et al. (2006), Gregory and Mitchell (1995), Zwiers and Kharin (1998), Hegerl et al. (2004), Meehl and Tebaldi (2004)
9   §11.5.3.3, Bell et al. (2004), Leung et al. (2004)
10 §11.4.3.2, Gao et al. (2002)
11 §11.4.3.2, Kwon et al. (2005), Boo et al. (2006)

12 §11.3.3.2, §11.4.3.1
13 Kharin and Zwiers (2005)
14 §11.3.3.2, Fig. 11.3.3.3, PRUDENCE
15 §11.7.3.5, Whetton et al. (2002)
16 Tebaldi et al. (2006), Meehl and Tebaldi (2004), §11.3.3.2, PRUDENCE, §11.7.3.1, CSIRO (2001), Mullan et al. (2001b)
17 §11.3.3.2, PRUDENCE, Kjellström et al. (2007), §11.4.3.2, Gao et al. (2002), Rupa Kumar et al. (2006)
18 §11.1.3
19 §11.5.3.3, Bell et al. (2004), Leung et al. (2004), §11.4.3.2, Rupa Kumar et al. (2006), Mizuta et al. (2005)
20 §11.3.3.2, Räisänen (2001), Räisänen and Alexandersson (2003), Giorgi and Bi (2005), Zwiers and Kharin (1998), Hegerl et al. (2004), Kjellström et al. (2007)
21 §11.3.3.2, PRUDENCE, Schär et al. (2004), Vidale et al. (2007)

Table 11.2. (continued)

Moisture-Related Phenomena	
Phenomenon	**Projected changes**
Intense precipitation events	**VL (consistent across model projections; empirical evidence, generally higher precipitation extremes in warmer climates)** Much larger increase in the frequency than in the magnitude of precipitation extremes over most land areas in middle latitudes,[22] particularly over northern Europe,[23] Australia and New Zealand[24] Large increase during the Indian summer monsoon season over Arabian Sea, tropical Indian Ocean, South Asia[25] Increase in summer over south China, Korea and Japan[26]
Intense precipitation events	**L (some inconsistencies across model projections)** Increase over central Europe in winter[27] Increase associated with tropical cyclones over Southeast Asia, Japan[28] **Uncertain** Changes in summer over Mediterranean and central Europe[29] **L decrease (consistent across model projections)** Iberian Peninsula[30]
Wet days	**L (consistent across model projections)** Increase in number of days at high latitudes in winter, and over northwest China[31] Increase over the Inter-Tropical Convergence Zone[32] Decrease in South Asia[33] and the Mediterranean area[34]
Dry spells (periods of consecutive dry days)	**VL (consistent across model projections)** Increase in length and frequency over the Mediterranean area[35], southern areas of Australia, New Zealand[36] **L (consistent across model projections)** Increase in most subtropical areas[37] Little change over northern Europe[38]
Continental drying and associated risk of drought	**L (consistent across model projections; consistent change in precipitation minus evaporation, but sensitivity to formulation of land surface processes)** Increased in summer over many mid-latitude continental interiors, e.g., central[39] and southern Europe, Mediterranean area,[40] in boreal spring and dry periods of the annual cycle over Central America[41]

[22] §11.3.3.4, Groisman et al. (2005), Kharin and Zwiers (2005), Hegerl et al. (2004), Semenov and Bengtsson (2002), Meehl et al. (2006)

[23] §11.3.3.4, Räisänen (2002), Giorgi and Bi (2005), Räisänen (2005)

[24] §11.1.3, §11.7.3.2, §11.3.3.4, Huntingford et al. (2003), Barnett et al. (2006), Frei et al. (2006), Hennessy et al. (1997), Whetton et al. (2002), Watterson and Dix (2003), Suppiah et al. (2004), McInnes et al. (2003), Hennessy et al. (2004b), Abbs (2004), Semenov and Bengtsson (2002)

[25] §11.4.3.2, May (2004a), Rupa Kumar et al. (2006)

[26] §11.4.3.2, Gao et al. (2002), Boo et al. (2006), Kimoto et al. (2005), Kitoh et al. (2005), Mizuta et al. (2005)

[27] §11.3.3.4, PRUDENCE, Frei et al. (2006), Christensen and Christensen (2003, 2004)

[28] §11.1.3, §11.4.3.2, Kimoto et al. (2005), Mizuta et al. (2005), Hasegawa and Emori (2005), Kanada et al. (2005)

[29] §11.3.3.4, PRUDENCE, Frei et al. (2006), Christensen and Christensen (2004), Tebaldi et al. (2006)

[30] §11.3.3.4, PRUDENCE, Frei et al. (2006)

[31] §11.4.3.2, Gao et al. (2002), Hasegawa and Emori (2005)

[32] Semenov and Bengtsson (2002)

[33] §11.4.3.2 Krishna Kumar et al. (2003)

[34] §11.3.3.4, Semenov and Bengtsson (2002), Voss et al. (2002); Räisänen et al. (2004); Frei et al. (2006)

[35] §11.3.3.4, Semenov and Bengtsson, 2002; Voss et al., 2002; Hegerl et al., 2004; Wehner, 2004; Kharin and Zwiers, 2005; Tebaldi et al., 2006

[36] §11.1.3, §11.7.3.2, §11.7.3.4, Whetton and Suppiah (2003), McInnes et al. (2003), Walsh et al. (2002), Hennessy et al. (2004c), Mullan et al. (2005)

[37] §11.1.3

[38] §11.3.3.4, Beniston et al. (2007), Tebaldi et al. (2006), Voss et al. (2002)

[39] §11.3.3.2, Rowell and Jones (2006)

[40] §11.1.3, §11.3.3.4, Voss et al. (2002)

[4] §11.1.3

Table 11.2. (continued)

Tropical Cyclones (typhoons and hurricanes)	
Change in phenomenon	**Projected changes**
Increase in peak wind intensities	**L (high-resolution Atmospheric GCM (AGCM) and embedded hurricane model projections)** Over most tropical cyclone areas[42]
Increase in mean and peak precipitation intensities	**L (high-resolution AGCM projections and embedded hurricane model projections)** Over most tropical cyclone areas,[43] South,[44] East[45] and southeast Asia[46]
Changes in frequency of occurrence	**M (some high-resolution AGCM projections)** Decrease in number of weak storms, increase in number of strong storms[47] **M (several climate model projections)** Globally averaged decrease in number, but specific regional changes dependent on sea surface temperature change[48] Possible increase over the North Atlantic[49]

Extratropical Cyclones	
Change in phenomenon	**Projected changes**
Changes in frequency and position	**L (consistent in AOGCM projections)** Decrease in the total number of extratropical cyclones[50] Slight poleward shift of storm track and associated precipitation, particularly in winter[51]
Change in storm intensity and winds	**L (consistent in most AOGCM projections, but not explicitly analysed for all models)** Increased number of intense cyclones[52] and associated strong winds, particularly in winter over the North Atlantic,[53] central Europe[54] and Southern Island of New Zealand[55] **More likely than not** Increased windiness in northern Europe and reduced windiness in Mediterranean Europe[56]
Increased wave height	**L (based on projected changes in extratropical storms)** Increased occurrence of high waves in most mid-latitude areas analysed, particularly the North Sea[57]

[42] Knutson and Tuleya (2004)
[43] Knutson and Tuleya (2004)
[44] §11.4.3.2, Unnikrishnan et al. (2006)
[45] §11.3.4, Hasegawa and Emori (2005)
[46] §11.3.4, Hasegawa and Emori (2005), Knutson and Tuleya (2004)
[47] Oouchi et al. (2006)
[48] Hasegawa and Emori (2005)
[49] Sugi et al. (2002), Oouchi et al. (2006)
[50] §11.3.3.6, Yin (2005), Lambert and Fyfe (2006), §11.3.3.5, Lionello et al. (2002), Leckebusch et al. (2006), Vérant (2004), Somot (2005)

[51] §11.1.3, Yin (2005), Lambert and Fyfe (2006)
[52] §11.1.2, §11.3.3.5, Yin (2005), Lambert and Fyfe (2006)
[53] §11.3.3.5, Leckebusch and Ulbrich (2004)
[54] §11.3.3.5, Zwiers and Kharin (1998), Knippertz et al. (2000), Leckebusch and Ulbrich (2004), Pryor et al. (2005a), Lionello et al. (2002), Leckebusch et al. (2006), Vérant (2004), Somot (2005)
[55] §11.1.3, §11.7.3.7
[56] §11.3.3.5, Lionello et al. (2002), Leckebusch et al. (2006), Vérant (2004), Somot (2005)
[57] X.L. Wang et al. (2004)

Frequently Asked Question 11.1
Do Projected Changes in Climate Vary from Region to Region?

Climate varies from region to region. This variation is driven by the uneven distribution of solar heating, the individual responses of the atmosphere, oceans and land surface, the interactions between these, and the physical characteristics of the regions. The perturbations of the atmospheric constituents that lead to global changes affect certain aspects of these complex interactions. Some human-induced factors that affect climate ('forcings') are global in nature, while others differ from one region to another. For example, carbon dioxide, which causes warming, is distributed evenly around the globe, regardless of where the emissions originate, whereas sulphate aerosols (small particles) that offset some of the warming tend to be regional in their distribution. Furthermore, the response to forcings is partly governed by feedback processes that may operate in different regions from those in which the forcing is greatest. Thus, the projected changes in climate will also vary from region to region.

Latitude is a good starting point for considering how changes in climate will affect a region. For example, while warming is expected everywhere on Earth, the amount of projected warming generally increases from the tropics to the poles in the Northern Hemisphere. Precipitation is more complex, but also has some latitude-dependent features. At latitudes adjacent to the polar regions, precipitation is projected to increase, while decreases are projected in many regions adjacent to the tropics (see Figure 1). Increases in tropical precipitation are projected during rainy seasons (e.g., monsoons), and over the tropical Pacific in particular.

Location with respect to oceans and mountain ranges is also an important factor. Generally, the interiors of continents are projected to warm more than the coastal areas. Precipitation responses are especially sensitive not only to the continental geometry, but to the shape of nearby mountain ranges and wind flow direction. Monsoons, extratropical cyclones and hurricanes/typhoons are all influenced in different ways by these region-specific features.

Some of the most difficult aspects of understanding and projecting changes in regional climate relate to possible changes in the circulation of the atmosphere and oceans, and their patterns of variability. Although general statements covering a variety of regions with qualitatively similar climates can be made in some cases, nearly every region is idiosyncratic in some ways. This is true whether it is the coastal zones surrounding the subtropical Mediterranean Sea, the extreme weather in the North American interior that depends on moisture transport from the Gulf of Mexico, or the interactions between vegetation distribution, oceanic temperatures and atmospheric circulation that help control the southern limit of the Sahara Desert.

While developing an understanding of the correct balance of global and regional factors remains a challenge, the understanding of these factors is steadily growing, increasing our confidence in regional projections.

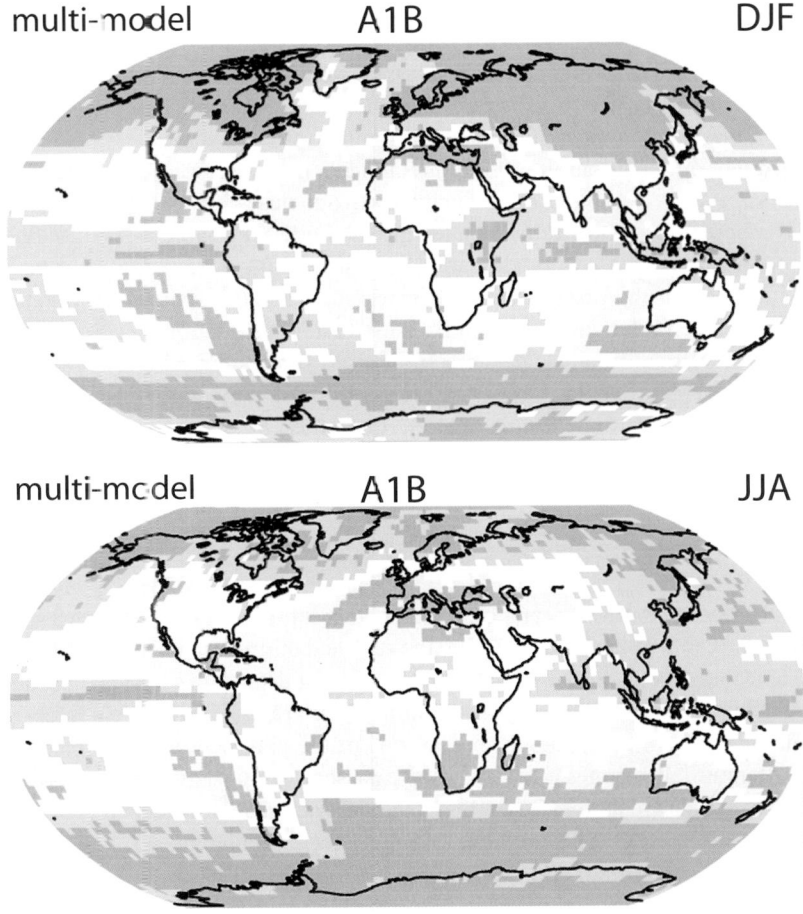

FAQ 11.1, Figure 1. *Blue and green areas on the map are by the end of the century projected to experience increases in precipitation, while areas in yellow and pink are projected to have decreases. The top panel shows projections for the period covering December, January and February, while the bottom panel shows projections for the period covering June, July and August.*

11.2 Africa

Assessment of projected climate changes for Africa:

All of Africa is very likely to warm during this century. The warming is very likely to be larger than the global, annual mean warming throughout the continent and in all seasons, with drier subtropical regions warming more than the moister tropics.

Annual rainfall is likely to decrease in much of Mediterranean Africa and northern Sahara, with the likelihood of a decrease in rainfall increasing as the Mediterranean coast is approached. Rainfall in southern Africa is likely to decrease in much of the winter rainfall region and on western margins. There is likely to be an increase in annual mean rainfall in East Africa. It is uncertain how rainfall in the Sahel, the Guinean Coast and the southern Sahara will evolve in this century.

The MMD models have significant systematic errors in and around Africa (excessive rainfall in the south, southward displacement of the Atlantic Inter-Tropical Convergence Zone (ITCZ), insufficient upwelling off the West Coast) making it difficult to assess the consequences for climate projections. The absence of realistic variability in the Sahel in most 20th-century simulations casts some doubt on the reliability of coupled models in this region. Vegetation feedbacks and feedbacks from dust aerosol production are not included in the global models. Possible future land surface modification is also not taken into account in the projections. The extent to which current regional models can successfully downscale precipitation over Africa is unclear, and limitations of empirical downscaling results for Africa are not fully understood. There is insufficient information on which to assess possible changes in the spatial distribution and frequency of tropical cyclones affecting Africa.

11.2.1 Key Processes

The bulk of the African continent is tropical or subtropical with the central phenomenon being the seasonal migration of the tropical rain belts. Small shifts in the position of these rain belts result in large local changes in rainfall. There are also regions on the northern and southern boundaries of the continent with winter rainfall regimes governed by the passage of mid-latitude fronts, which are therefore sensitive to a poleward displacement of the storm tracks. This is evident from the correlation between South African rainfall and the Southern Annular Mode (Reason and Rouault, 2005) and between North African rainfall and the NAO (Lamb and Peppler, 1987). Troughs penetrating into the tropics from mid-latitudes also influence warm season rainfall, especially in southern Africa, and can contribute to a sensitivity of warm season rains to a displacement of the circulation (Todd and Washington, 1999). Any change in tropical cyclone

distribution and intensity will affect the southeast coastal regions, including Madagascar (Reason and Keibel, 2004).

The factors that determine the southern boundary of the Sahara and rainfall in the Sahel have attracted special interest because of the extended drought experienced by this region in the 1970s and 1980s. The field has moved steadily away from explanations for rainfall variations in this region as primarily due to land use changes and towards explanations based on changes in sea surface temperatures (SSTs). The early SST perturbation Atmospheric GCM (AGCM) experiments (Palmer, 1986; Rowell, et al., 1995) are reinforced by the results from the most recent models (Giannini et al., 2003; Lu and Delworth, 2005; Hoerling et al., 2006). The north-south inter-hemispheric gradient, with colder NH oceans conducive to an equatorward shift and/or a reduction in Sahel rainfall, is important. This has created interest in the possibility that aerosol cooling localised in the NH could dry the Sahel (Rotstayn and Lohmann, 2002; see also Section 9.5.4.3.1). However, temperatures over other oceanic regions, including the Mediterranean (Rowell, 2003), are also important.

In southern Africa, changing SSTs are also thought to be more important than changing land use patterns in controlling warm season rainfall variability and trends. Evidence has been presented for strong links with Indian Ocean temperatures (Hoerling et al., 2006). The warming of the troposphere over South Africa, possibly a consequence of warming of the Indo-Pacific, has been linked with the increase in days with stable inversion layers over southern Africa (Freiman and Tyson, 2000; Tadross et al., 2005a, 2006) in the late 20th century.

In addition to the importance of ocean temperatures, vegetation patterns help shape the climatic zones throughout much of Africa (e.g., Wang and Eltahir, 2000; Maynard and Royer, 2004a; Paeth and Henre, 2004; see also Section 11.7, Box 11.4). In the past, land surface changes have primarily acted as feedbacks generated by the underlying response to SST anomalies, and vegetation changes are thought to provide a positive feedback to climate change. The plausibility of this positive feedback is enhanced by recent work suggesting that land surface feedbacks may also play an important role in both intra-seasonal variability and rainy season onset in southern Africa (New et al., 2003; Anyah and Semazzi, 2004; Tadross et al., 2005a,b).

The MMD models prescribe vegetation cover; they would likely respond more strongly to large-scale forcing if they predicted vegetation, especially in semi-arid areas. The possibility of multiple stable modes of African climate due to vegetation-climate interactions has been raised, especially in the context of discussions of the very wet Sahara during the mid-Holocene 6 to 8 ka (Claussen et al., 1999; Foley et al., 2003). One implication is that centennial time-scale feedbacks associated with vegetation patterns may have the potential to make climate changes over Africa less reversible.

11.2.2 Skill of Models in Simulating Present and Past Climates

There are biases in the simulations of African climate that are systematic across the MMD models, with 90% of models overestimating precipitation in southern Africa, by more than 20% on average (and in some cases by as much as 80%) over a wide area often extending into equatorial Africa. The temperature biases over land are not considered large enough to directly affect the credibility of the model projections (see Supplementary Material Figure S11.21 and Table S11.1).

The ITCZ in the Atlantic is displaced equatorward in nearly all of these AOGCM simulations. Ocean temperatures are too warm by an average of 1°C to 2°C in the Gulf of Guinea and typically by 3°C off the southwest coast in the region of intense upwelling, which is clearly too weak in many models. In several of the models there is no West African monsoon as the summer rains fail to move from the Gulf onto land, but most of the models do have a monsoonal climate albeit with some distortion. Moderately realistic interannual variability of SSTs in the Gulf of Guinea and the associated dipolar rainfall variations in the Sahel and the Guinean Coast are, by the criteria of Cook and Vizy (2006), only present in 4 of the 18 models examined. Tennant (2003) describes biases in several AGCMs, such as the equatorward displacement of the mid-latitude jet in austral summer, a deficiency that persists in the most recent simulations (Chapter 8).

Despite these deficiencies, AGCMs can simulate the basic pattern of rainfall trends in the second half of the 20th century if given the observed SST evolution as boundary conditions, as described in the multi-model analysis of Hoerling et al. (2006) and the growing literature on the interannual variability and trends in individual models (e.g., Rowell et al., 1995; Bader and Latif, 2003; Giannini et al., 2003; Haarsma et al., 2005; Kamga et al., 2005; Lu and Delworth, 2005). However, there is less confidence in the ability of AOGCMs to generate interannual variability in the SSTs of the type known to affect African rainfall, as evidenced by the fact that very few AOGCMs produce droughts comparable in magnitude to the Sahel drought of the 1970s and 1980s (Hoerling et al., 2006). There are exceptions, but what distinguishes these from the bulk of the models is not understood.

The very wet Sahara 6 to 8 ka is thought to have been a response to the increased summer insolation due to changes in the Earth's orbital configuration. Modelling studies of this response provide background information on the quality of a model's African monsoon, but the processes controlling the response to changing seasonal insolation may be different from those controlling the response to increasing greenhouse gases. The fact that GCMs have difficulty in simulating the full magnitude of the mid-Holocene wet period, especially in the absence of vegetation feedbacks, may indicate a lack of sensitivity to other kinds of forcing (Jolly et al., 1996; Kutzbach et al., 1996).

Regional climate modelling has mostly focused on southern Africa, where the models generally improve on the climate simulated by global models but also share some of the biases in the global models. For example, Engelbrecht et al. (2002) and Arnell et al. (2003) both simulate excessive rainfall in parts of southern Africa, reminiscent of the bias in the MMD. Hewitson et al. (2004) and Tadross et al. (2006) note strong sensitivity to the choice of convective parametrization, and to changes in soil moisture and vegetative cover (New et al., 2003; Tadross et al., 2005a), reinforcing the view (Rowell et al., 1995) that land surface feedbacks enhance regional climate sensitivity over Africa's semi-arid regions. Over West Africa, the number of Regional Climate Model (RCM) investigations is even more limited (Jenkins et al., 2002; Vizy and Cook, 2002). The quality of the 25-year simulation undertaken by Paeth et al. (2005) is encouraging, emphasizing the role of regional SSTs and changes in the land surface in forcing West African rainfall anomalies. Several recent AGCM time-slice simulations focusing on tropical Africa show good simulation of the rainy season (Coppola and Giorgi, 2005; Caminade et al., 2006; Oouchi et al., 2006).

Hewitson and Crane (2005) developed empirical downscaling for point-scale precipitation at sites spanning the continent, as well as a 0.1° resolution grid over South Africa. The downscaled precipitation forced by reanalysis data provides a close match to the historical climate record, including regions such as the eastern escarpment of the sub-continent that have proven difficult for RCMs.

11.2.3 Climate Projections

11.2.3.1 Mean Temperature

The differences in near-surface temperature between the years 2080 to 2099 and the years 1980 to 1999 in the MMD-A1B projections, averaged over the West African (WAF), East African (EAF), South African (SAF) and Saharan (SAH) sub-regions, are provided in Table 11.1, with the temporal evolution displayed in Figure 11.1. The Mediterranean coast is discussed together with southern Europe in Section 11.3. In all four regions and in all seasons, the median temperature increase lies between 3°C and 4°C, roughly 1.5 times the global mean response. Half of the models project warming within about 0.5°C of these median values. The distributions estimated by Tebaldi et al. (2004a,b; see also Supplementary Material Table S11.2) have a very similar half-width, but reduce the likelihood of the extreme high limit as compared to the raw quartiles in Table 11.1. There is a strong correlation across these AOGCMs between the global mean temperature response and the response in Africa. The signal-to-noise ratio is very large for these 20-year mean temperatures and 10 years is typically adequate to obtain a clearly discernible signal, as defined in Section 11.1.2. Regionally averaged temperatures averaged over the period 1990 to 2009 are clearly discernible from the 1980 to 1999 averages.

The upper panels in Figure 11.2 show the geographical structure of the ensemble-mean projected warming for the A1B scenario in more detail. Smaller values of projected warming, near 3°C, are found in equatorial and coastal areas and larger values, above 4°C, in the western Sahara. The largest temperature responses in North Africa are projected to occur in JJA, while the largest responses in southern Africa occur in September, October and November (SON). But the seasonal structure in the temperature response over Africa is modest as compared to extratropical regions. The basic pattern of projected warming has been robust to changes in models since the TAR, as indicated by comparison with Hulme et al. (2001).

To date there is insufficient evidence from RCMs to modify the large-scale temperature projections from GCMs, although Tadross et al. (2005a) project changes in the A2 scenario for southern Africa that are lower than those in the forcing GCM and near the low end of the spread in the MMD models, likely due to a weaker drying tendency than in most of the global models.

11.2.3.2 Mean Precipitation

Figure 11.2 and Table 11.1 illustrate some of the robust aspects of the precipitation response over Africa in the MMD-A1B projections. The fractional changes in annual mean precipitation in each of the 21 models are provided in Supplementary Material Figure S11.13. With respect to the most robust features (drying in the Mediterranean and much of southern Africa, and increases in rainfall in East Africa), there is qualitative agreement with the results in Hulme et al. (2001) and Ruosteenoja et al. (2003), which summarise results from the models available at the time of the TAR.

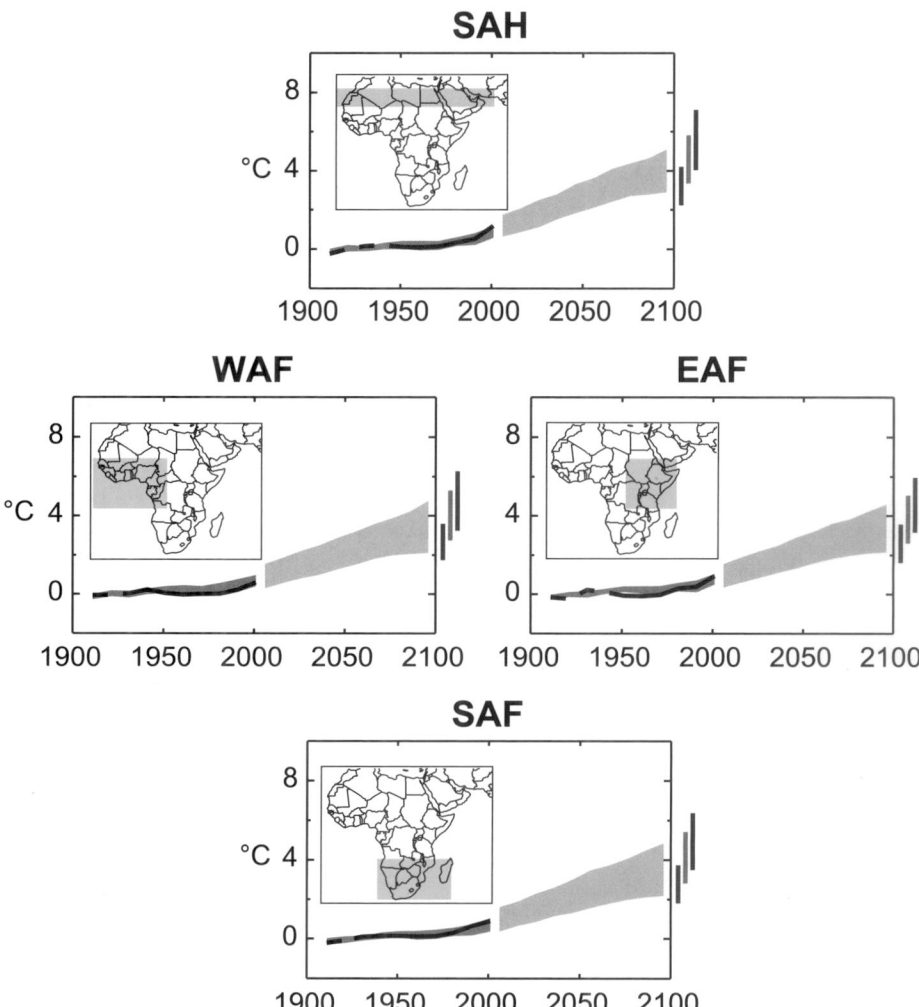

Figure 11.1. *Temperature anomalies with respect to 1901 to 1950 for four African land regions for 1906 to 2005 (black line) and as simulated (red envelope) by MMD models incorporating known forcings; and as projected for 2001 to 2100 by MMD models for the A1B scenario (orange envelope). The bars at the end of the orange envelope represent the range of projected changes for 2091 to 2100 for the B1 scenario (blue), the A1B scenario (orange) and the A2 scenario (red). The black line is dashed where observations are present for less than 50% of the area in the decade concerned. More details on the construction of these figures are given in Box 11.1 and Section 11.1.2.*

The large-scale picture is one of drying in much of the subtropics and an increase (or little change) in precipitation in the tropics, increasing the rainfall gradients. This is a plausible hydrological response to a warmer atmosphere, a consequence of the increase in water vapour and the resulting increase in vapour transport in the atmosphere from regions of moisture divergence to regions of moisture convergence (see Chapter 9 and Section 11.2.1).

The drying along Africa's Mediterranean coast is a component of a larger-scale drying pattern surrounding the Mediterranean and is discussed further in the section on Europe (Section 11.3). A 20% drying in the annual mean is typical along the African Mediterranean coast in A1B by the end of the 21st century. Drying is seen throughout the year and is generated by nearly every MMD model. The drying signal in this composite

extends into the northern Sahara, and down the West Coast as far as 15°N. The processes involved include increased moisture divergence and a systematic poleward shift of the storm tracks affecting the winter rains, with positive feedback from decreasing soil moisture in summer (see Section 11.3).

In southern Africa, a similar set of processes produces drying that is especially robust in the extreme southwest in winter, a manifestation of a much broader-scale poleward shift in the circulation across the South Atlantic and Indian Oceans. However, the drying is subject to the caveat that strong orographic forcing may result in locally different changes (as discussed in Section 11.4.3.2, Box 11.3). With the exception of the winter rainfall region in the southwest, the robust drying in winter corresponds to the dry season over most of the sub-continent and does not contribute to the bulk of the annual mean

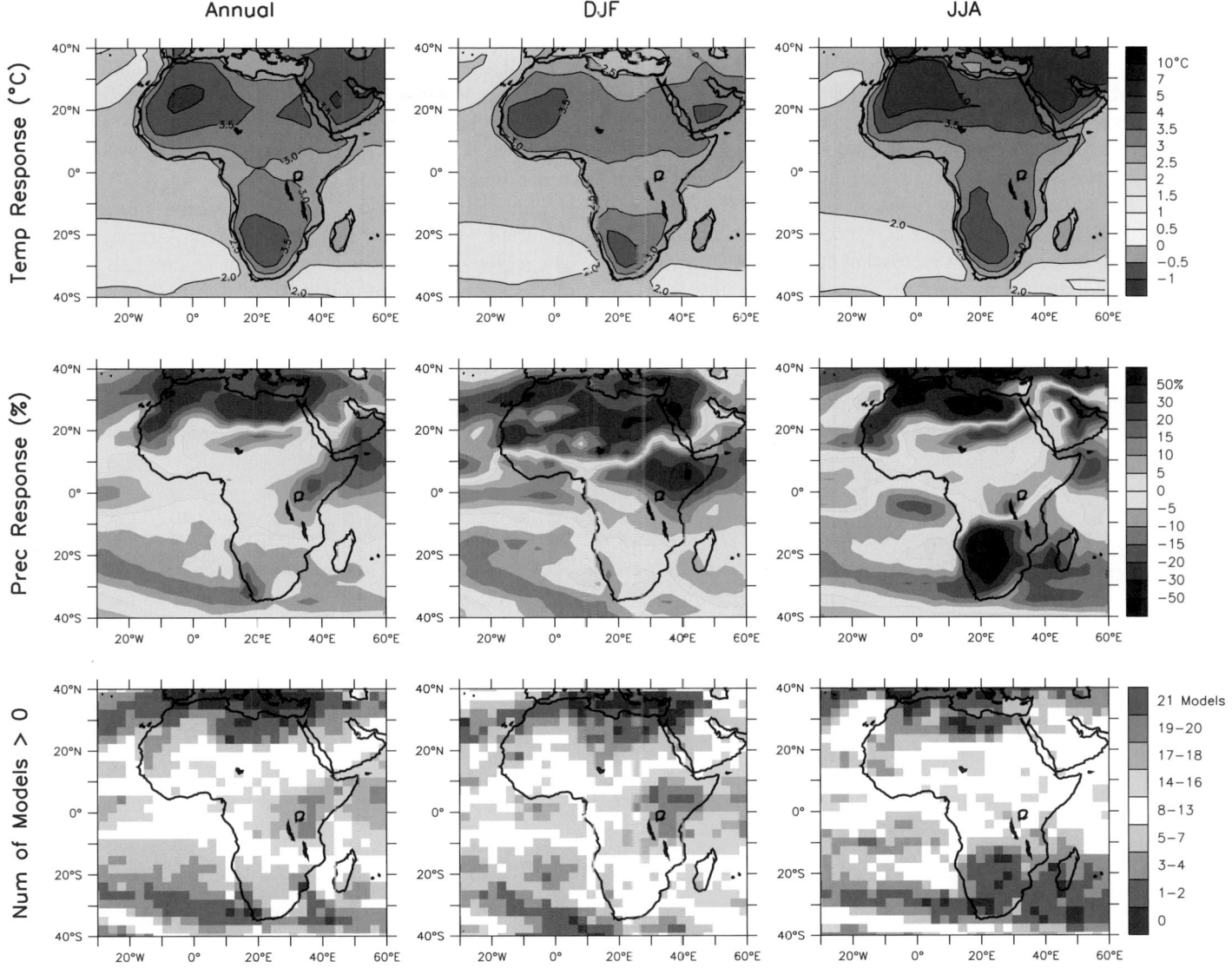

Figure 11.2. *Temperature and precipitation changes over Africa from the MMD-A1B simulations. Top row: Annual mean, DJF and JJA temperature change between 1980 to 1999 and 2080 to 2099, averaged over 21 models. Middle row: same as top, but for fractional change in precipitation. Bottom row: number of models out of 21 that project increases in precipitation.*

drying. More than half of the annual mean reduction occurs in the spring and is mirrored in some RCM simulations for this region (see below). To an extent, this can be thought of as a delay in the onset of the rainy season. This spring drying suppresses evaporation, contributing to the spring maximum in the temperature response.

The increase in rainfall in East Africa, extending into the Horn of Africa, is also robust across the ensemble of models, with 18 of 21 models projecting an increase in the core of this region, east of the Great Lakes. This East African increase is also evident in Hulme et al. (2001) and Ruosteenoja et al. (2003). The Guinean coastal rain belts and the Sahel do not show as robust a response. A straight average across the ensemble results in modest moistening in the Sahel with little change on the Guinean coast. The composite MMD simulations have a weak drying trend in the Sahel in the 20th century that does

not continue in the future projections (Biasutti and Giannini, 2006; Hoerling, et al., 2006), implying that the weak 20th-century drying trend in the composite 20th-century simulations is unlikely to be forced by greenhouse gases, but is more likely forced by aerosols, as in Rotstayn and Lohmann (2002), or a result of low-frequency internal variability of the climate.

Individual models generate large, but disparate, responses in the Sahel. Two outliers are GFDL/CM2.1, which projects very strong drying in the Sahel and throughout the Sahara, and MIROC3.2_midres, which shows a very strong trend towards increased rainfall in the same region (see Supplementary Figure S11.13; and see Table 8.1 for model descriptions). Cook and Vizy (2006) find moderately realistic interannual variability in the Gulf of Guinea and Sahel in both models. While the drying in the GFDL model is extreme within the ensemble, it generates a plausible simulation of 20th-century Sahel rainfall trends (Held et al., 2005; Hoerling et al., 2006) and an empirical downscaling from AOGCMs (Hewitson and Crane, 2006) shows a similar response (see below). More research is needed to understand the variety of modelled precipitation responses in the Sahel and elsewhere in the tropics. Progress is being made in developing new methodologies for this purpose (e.g., Chou and Neelin, 2004; Lintner and Chiang, 2005; Chou et al., 2007), leading to better appreciation of the sources of model differences. Haarsma et al. (2005) describe a plausible mechanism associated with increasing land-ocean temperature contrast and decreasing surface pressures over the Sahara, which contributes to the increase in Sahel precipitation with warming in some models.

It has been argued (e.g., Paethe and Hense, 2004) that the partial amelioration of the Sahel drought since the 1990s may be a sign of a greenhouse-gas driven increase in rainfall,

providing support for those models that moisten the Sahel into the 21st century (e.g., Maynard et al., 2002; Haarsma et al., 2005; Kamga et al., 2005). However, it is premature to take this partial amelioration as evidence of a global warming signature, given the likely influence of internal variability on the inter-hemispheric SST gradients that influence Sahel rainfall, as well as the influence of aerosol variations.

Table 11.1 provides information on the spread of model-projected precipitation change in the four African sub-regions. The regions and seasons for which the central half (25 to 75%) of the projections are uniformly of one sign are: EAF where there is an increase in DJF, March, April and May (MAM), SON and in the annual mean; SAF where there is a decrease in austral winter and spring; and SAH where there is a decrease in boreal winter and spring. The Tebaldi et al. (2004a,b) Bayesian estimates (Supplementary Material Table S11.2)

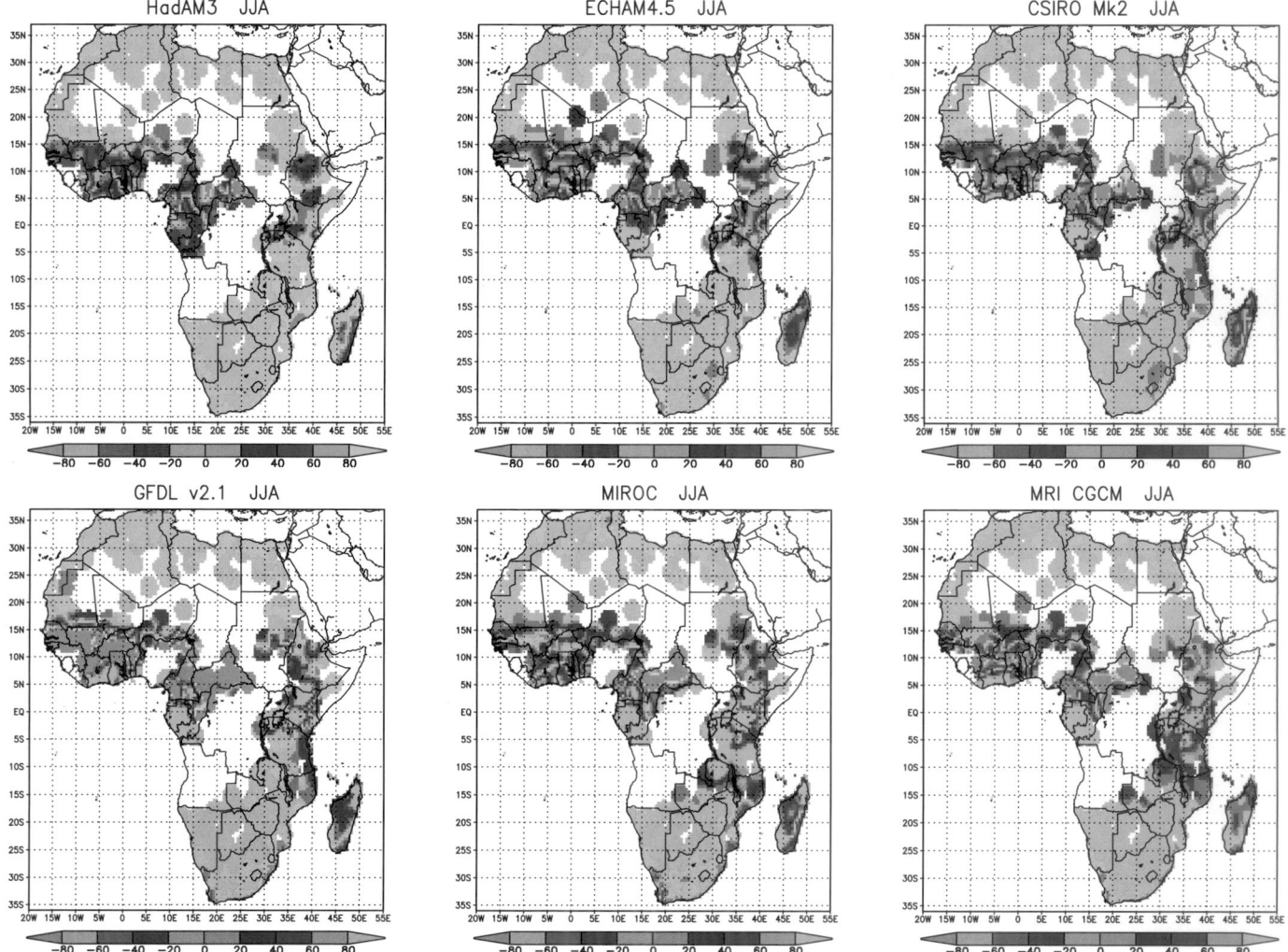

Figure 11.3. *Anomaly of mean monthly precipitation (mm) using daily data empirically downscaled from six GCMs (ECHAM4.5, Hadley Centre Atmospheric Model (HadAM3), CSIRO Mk2, GFDL 2.1, MRI, MIROC; see Table 8.1 for descriptions of most of these models) to 858 station locations. The GCMs were forced by the SRES A2 scenario. Anomalies are for the future period (2070 to 2099 for the first three models, and 2080 to 2099 for the latter three models) minus a control 30-year period (from Hewitson and Crane, 2006).*

do not change this distinction between robust and non-robust regions and seasons. The time required for emergence of a clearly discernible signal in these robust regions and seasons is typically 50 to 100 years, except in the Sahara where even longer times are required.

Land use change is a potential contributor to climate change in the 21st century (see also Section 11.7, Box 11.4). C.M. Taylor et al. (2002) project drying over the Sahel of 4% from 1996 to 2015 due to changing land use, but suggest that the potential exists for this contribution to grow substantially further into the century. Maynard and Royer (2004a) suggest that estimated land use change scenarios for the mid-21st century would have only a modest compensating effect on the greenhouse-gas induced moistening in their model. Neither of these studies includes a dynamic vegetation model.

Several climate change projections based on RCM simulations are available for southern Africa but are much scarcer for other regions. Tadross et al. (2005a) examine two RCMs, Providing Regional Impacts for Climate Studies (PRECIS) and Mesoscale Model version 5 (MM5), nested for southern Africa in a time-slice AGCM based in turn on lower-resolution Hadley Centre Coupled Model (HadCM3) coupled simulations for the Special Report on Emission Scenarios (SRES) A2 scenario. During the early summer season, October to December, both models predict drying over the tropical western side of the continent, responding to the increase in high-pressure systems entering from the west, with MM5 indicating that the drying extends further south and PRECIS further east. The drying in the west continues into late summer, but there are increases in total rainfall towards the east in January and February, a feature barely present in the ensemble mean of the MMD models. Results obtained by downscaling one global model must be assessed in the context of the variety of responses in southern Africa among the MMD models (Supplementary Material Figure S11.13).

Hewitson and Crane (2006) use empirical downscaling to provide projections for daily precipitation as a function of six GCM simulations. The degree of convergence in the downscaled results for the SRES A2 scenario near the end of the 21st century suggests more commonality in GCM-projected changes in daily circulation, on which the downscaling is based, than in the GCM precipitation responses. Figure 11.3 shows the response of mean JJA monthly total precipitation for station locations across Africa. The ensemble mean of these downscaling results shows increased precipitation in east Africa extending into southern Africa, especially in JJA, strong drying in the core Sahel in JJA with some coastal wetting, and moderate wetting in DJF. There is also drying along the Mediterranean coast, and, in most models, drying in the western portion of southern Africa. The downscaling also shows marked local-scale variation in the projected changes, for example, the contrasting changes in the west and east of Madagascar, and on the coastal and inland borders of the Sahel.

While this result is generally consistent with the underlying GCMs and the composite MMD projections, there is a tendency for greater Sahel drying than in the underlying GCMs, providing further rationale (alongside the large spread in model responses and poor coupled model performance in simulating droughts of the magnitude observed in the 20th century) for viewing with caution the projection for a modest increase in Sahel rainfall in the ensemble mean of the MMD models.

11.2.3.3 Extremes

Research on changes in extremes specific to Africa, in either models or observations, is limited. A general increase in the intensity of high-rainfall events, associated in part with the increase in atmospheric water vapour, is expected in Africa, as in other regions. Regional modelling and downscaling results (Tadross et al., 2005a) both support an increase in the rainfall intensity in southern Africa. In regions of mean drying, there is generally a proportionally larger decrease in the number of rain days, indicating compensation between intensity and frequency of rain. In the downscaling results of Hewitson and Crane (2006) and Tadross et al. (2005a), changes in the median precipitation event magnitude at the station scale do not always mirror the projected changes in seasonal totals.

There is little modelling guidance on possible changes in tropical cyclones affecting the southeast coast of Africa. Thermodynamic arguments for increases in precipitation rates and intensity of tropical storms (see Chapter 10) are applicable to these Indian Ocean storms as for other regions, but changes in frequency and spatial distribution remain uncertain. In a time-slice simulation with a 20-km resolution AGCM, Oouchi et al. (2006) obtain a significant reduction in the frequency of tropical storms in the Indian Ocean.

Using the definition of 'extreme seasons' given in Section 11.1.2, the probability of extremely warm, wet and dry seasons, as estimated by the MMD models, is provided in Table 11.1. As in most tropical regions, all seasons are extremely warm by the end of the 21st century, with very high confidence under the A1B scenario. Although the mean precipitation response in West Africa is less robust than in East Africa, the increase in the number of extremely wet seasons is comparable in both, increasing to roughly 20% (i.e., 1 in 5 of the seasons are extremely wet, as compared to 1 in 20 in the control period in the late 20th century). In southern Africa, the frequency of extremely dry austral winters and springs increases to roughly 20%, while the frequency of extremely wet austral summers doubles in this ensemble of models.

11.3 Europe and the Mediterranean

Assessments of projected climate change for Europe:

Annual mean temperatures in Europe are likely to increase more than the global mean. The warming in northern Europe is likely to be largest in winter and that in the Mediterranean area largest in summer. The lowest winter temperatures are likely to increase more than average winter temperature in northern Europe, and the highest summer temperatures are likely to increase more than average summer temperature in southern and central Europe.

Annual precipitation is very likely to increase in most of northern Europe and decrease in most of the Mediterranean area. In central Europe, precipitation is likely to increase in winter but decrease in summer. Extremes of daily precipitation are very likely to increase in northern Europe. The annual number of precipitation days is very likely to decrease in the Mediterranean area. The risk of summer drought is likely to increase in central Europe and in the Mediterranean area.

Confidence in future changes in windiness is relatively low, but it seems more likely than not that there will be an increase in average and extreme wind speeds in northern Europe.

The duration of the snow season is very likely to shorten in all of Europe, and snow depth is likely to decrease in at least most of Europe.

Although many features of the simulated climate change in Europe and the Mediterranean area are qualitatively consistent among models and qualitatively well understood in physical terms, substantial uncertainties remain. Simulated seasonal-mean temperature changes vary even at the sub-continental scale by a factor of two to three among the current generation of AOGCMs. Similarly, while agreeing on a large-scale increase in winter half-year precipitation in the northern parts of the area and a decrease in summer half-year precipitation in the southern parts of the area, models disagree on the magnitude and geographical details of precipitation change. These uncertainties reflect the sensitivity of the European climate change to the magnitude of the global warming and the changes in the atmospheric circulation and the Atlantic Meridional Overturning Circulation (MOC). Deficiencies in modelling the processes that regulate the local water and energy cycles in Europe also introduce uncertainty, for both the changes in mean conditions and extremes. Finally, the substantial natural variability of European climate is a major uncertainty, particularly for short-term climate projections in the area (e.g., Hulme et al., 1999).

11.3.1 Key Processes

In addition to global warming and its direct thermodynamic consequences such as increased water vapour transport from low to high latitudes (Section 11.1.3), several other factors may shape future climate changes in Europe and the Mediterranean area. Variations in the atmospheric circulation influence the European climate both on interannual and longer time scales. Recent examples include the central European heat wave in the summer 2003, characterised by a long period of anticyclonic weather (see Box 3.5), the severe cyclone-induced flooding in central Europe in August 2002 (see Box 3.6), and the strong warming of winters in northern Europe from the 1960s to 1990s that was affected by a trend toward a more positive phase of the NAO (Hurrell and van Loon, 1997; Räisänen and Alexandersson, 2003; Scaife et al., 2005). At fine geographical scales, the effects of atmospheric circulation are modified by topography, particularly in areas of complex terrain (Fernandez et al., 2003; Bojariu and Giorgi, 2005).

Europe, particularly its north-western parts, owes its relatively mild climate partly to the northward heat transport by the Atlantic MOC (e.g., Stouffer et al., 2006). Most models suggest increased greenhouse gas concentrations will lead to a weakening of the MOC (see Section 10.3), which will act to reduce the warming in Europe. However, in the light of present understanding, it is very unlikely to reverse the warming to cooling (see Section 11.3.3.1).

Local thermodynamic factors also affect the European climate and are potentially important for its future changes. In those parts of Europe that are presently snow-covered in winter, a decrease in snow cover is likely to induce a positive feedback, further amplifying the warming. In the Mediterranean region and at times in central Europe, feedbacks associated with the drying of the soil in summer are important even in the present climate. For example, they acted to exacerbate the heat wave of 2003 (Black et al., 2004; Fink et al., 2004).

11.3.2 Skill of Models in Simulating Present Climate

Atmosphere-Ocean General Circulation Models show a range of performance in simulating the climate in Europe and the Mediterranean area. Simulated temperatures in the MMD models vary on both sides of the observational estimates in summer but are mostly lower than observed in the winter half-year, particularly in northern Europe (Supplementary Material Table S11.1). Excluding one model that simulates extremely cold winters in northern Europe, the seasonal area mean temperature biases in the northern Europe region (NEU) vary from –5°C to 3°C and those in the southern Europe and Mediterranean region (SEM) from –5°C to 6°C, depending on model and season. The cold bias in northern Europe tends to increase towards the northeast, reaching –7°C in the ensemble mean in the northeast of European Russia in winter. This cold bias coincides with a north-south gradient in the winter mean sea level pressure that is weaker than observed, which implies

Box 11.2: The PRUDENCE Project

The Prediction of Regional scenarios and Uncertainties for Defining European Climate change risks and Effects (PRUDENCE) project involved more than 20 European research groups. The main objectives of the project were to provide dynamically downscaled high-resolution climate change scenarios for Europe at the end of the 21st century, and to explore the uncertainty in these projections. Four sources of uncertainty were studied: (i) sampling uncertainty due to the fact that model climate is estimated as an average over a finite number (30) of years, (ii) regional model uncertainty due to the fact that RCMs use different techniques to discretize the equations and to represent sub-grid effects, (iii) emission uncertainty due to choice of IPCC SRES emission scenario, and (iv) Boundary uncertainty due to the different boundary conditions obtained from different global climate models.

Each PRUDENCE experiment consisted of a control simulation representing the period 1961 to 1990 and a future scenario simulation representing 2071 to 2100. A large fraction of the simulations used the same boundary data (from the Hadley Centre Atmospheric Model (HadAM3H) for the A2 scenario) to provide a detailed understanding of the regional model uncertainty. Some simulations were also made for the B2 scenario, and by using driving data from two other GCMs and from different ensemble members from the same GCM. More details are provided in, for example, Christensen et al. (2007), Déqué et al. (2005) and http://prudence.dmi.dk.

weaker than observed westerly flow from the Atlantic Ocean to northern Europe in most models (Supplementary Material Figure S11.22).

Biases in simulated precipitation vary substantially with season and location. The average simulated precipitation in NEU exceeds that observed from autumn to spring (Supplementary Material Table S11.1), but the interpretation of the difference is complicated by the observational uncertainty associated with the undercatch of, in particular, solid precipitation (e.g., Adam and Lettenmaier, 2003). In summer, most models simulate too little precipitation, particularly in the eastern parts of the area. In SEM, the area and ensemble mean precipitation is close to observations.

Regional Climate Models capture the geographical variation of temperature and precipitation in Europe better than global models but tend to simulate conditions that are too dry and warm in southeastern Europe in summer, both when driven by analysed boundary conditions (Hagemann et al., 2004) and when driven by GCM data (e.g., Jacob et al., 2007). Most but not all RCMs also overestimate the interannual variability of summer temperatures in southern and central Europe (Jacob et al., 2007; Lenderink et al., 2007; Vidale et al., 2007). The excessive temperature variability coincides with excessive interannual variability in either shortwave radiation or evaporation, or

both (Lenderink et al., 2007). A need for improvement in the modelling of soil, boundary layer and cloud processes is implied. One of the key model parameters may be the depth of the hydrological soil reservoir, which appears to be too small in many RCMs (van den Hurk et al., 2005).

The ability of RCMs to simulate climate extremes in Europe has been addressed in several studies. In the Prediction of Regional scenarios and Uncertainties for Defining European Climate change risks and Effects (PRUDENCE) simulations (Box 11.2), the biases in the tails of the temperature distribution varied substantially between models but were generally larger than the biases in average temperatures (Kjellström et al., 2007). Inspection of the individual models showed similarity between the biases in daily and interannual variability, suggesting that similar mechanisms may be affecting both.

The magnitude of precipitation extremes in RCMs is model-dependent. In a comparison of the PRUDENCE RCMs, Frei et al. (2006) find that the area-mean five-year return values of one-day precipitation in the vicinity of the European Alps vary by up to a factor of two between the models. However, except for too-low extremes in the southern parts of the area in summer, the set of models as a whole showed no systematic tendency to over- or underestimate the magnitude of the extremes when compared with gridded observations. A similar level of skill has been found in other model verification studies made for European regions (e.g., Booij, 2002; Semmler and Jacob, 2004; Fowler et al., 2005; see also Frei et al., 2003).

Evidence of model skill in simulation of wind extremes is mixed. Weisse et al. (2005) find that an RCM simulated a very realistic wind climate over the North Sea, including the number and intensity of storms, when driven by analysed boundary conditions. However, most PRUDENCE RCMs, while quite realistic over sea, severely underestimate the occurrence of very high wind speeds over land and coastal areas (Rockel and Woth, 2007). Realistic frequencies of high wind speeds were only found in the two models that used a gust parametrization to mimic the large local and temporal variability of near-surface winds over land.

11.3.3 Climate Projections

11.3.3.1 Mean Temperature

The observed evolution of European temperatures in the 20th century, characterised by a warming trend modulated by multi-decadal variability, was well within the envelope of the MMD simulations (Figure 11.4).

In this century, the warming is projected to continue at a rate somewhat greater than its global mean, with the increase in 20-year mean temperatures (from its values in 1980 to 1999) becoming clearly discernible (as defined in Section 11.1.2) within a few decades. Under the A1B scenario, the simulated area and annual mean warming from 1980 to 1999 to 2080 to 2099 varies from 2.3°C to 5.3°C in NEU and from 2.2°C to 5.1°C in SEM. The warming in northern Europe is likely to be largest in winter and that in the Mediterranean area largest in summer (Figure 11.5). Seasonal mean temperature changes typically vary by a factor of three among the MMD models (Table 11.1); however, the upper end of the range in NEU in DJF is reduced from 8.1°C to 6.7°C when one model with an extreme cold bias in present-day winter climate is excluded. Further details are given in Table 11.1 and Supplementary Material Figures S11.2 to S11.4.

Although changes in atmospheric circulation have a significant potential to affect temperature in Europe (e.g., Dorn et al., 2003), they are not the main cause of the projected warming (e.g., Rauthe and Paeth, 2004; Stephenson et al., 2006; van Ulden et al., 2007). A regression-based study using five of the MMD models (van Ulden and van Oldenborgh, 2006) indicated that in a region comprising mainly Germany, circulation changes enhanced the warming in most models in winter (due to an increase in westerly flow) and late summer (due to a decrease in westerly flow), but reduced the warming slightly in May and June. However, the circulation contribution to the simulated temperature changes (typically –1°C to 1.5°C depending on model and month) was generally much smaller than the total simulated warming in the late 21st century.

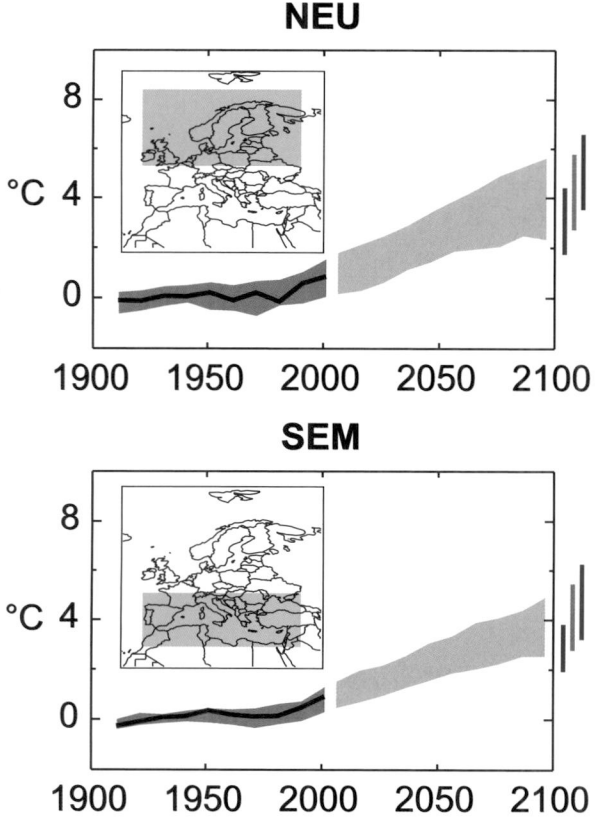

Figure 11.4. *Temperature anomalies with respect to 1901 to 1950 for two Europe land regions for 1906 to 2005 (black line) and as simulated (red envelope) by MMD models incorporating known forcings; and as projected for 2001 to 2100 by MMD models for the A1B scenario (orange envelope). The bars at the end of the orange envelope represent the range of projected changes for 2091 to 2100 for the B1 scenario (blue), the A1B scenario (orange) and the A2 scenario (red). More details on the construction of these figures are given in Box 11.1 and Section 11.1.2.*

Despite a decrease in the North Atlantic MOC in most models (see Section 10.3), all the MMD simulations show warming in Great Britain and continental Europe, as other climatic effects of increased greenhouse gases dominate over the changes in ocean circulation. The same holds for earlier simulations with increased greenhouse gas concentrations, except for a very few (Russell and Rind, 1999; Schaeffer et al., 2004) with slight cooling along the north-western coastlines of Europe but warming over the rest of the continent. The impact of MOC changes depends on the regional details of the change, being largest if ocean convection is suppressed at high latitudes where the sea ice feedback may amplify atmospheric cooling (Schaeffer et al., 2004). Sensitivity studies using AOGCMs with an artificial shutdown of the MOC and no changes in greenhouse gas concentrations typically show a 2°C to 4°C annual mean cooling in most of Europe, with larger cooling in the extreme north-western parts (e.g., Stouffer et al., 2006).

Statistical downscaling (SD) studies tend to show a large-scale warming similar to that of dynamical models but with finer-scale regional details affected by factors such as distance from the coast and altitude (e.g., Benestad, 2005; Hanssen-Bauer et al., 2005). Comparing RCM and SD projections for Norway downscaled from the same GCM, Hanssen-Bauer et al. (2003) found the largest differences between the two approaches in winter and/or spring at locations with frequent temperature inversions in the present climate. A larger warming at these locations in the SD projections was found, consistent with increased winter wind speed in the driving GCM and reduced snow cover, both of which suppress formation of ground inversions.

11.3.3.2   Mean Precipitation

A south-north contrast in precipitation changes across Europe is indicated by AOGCMs, with increases in the north and decreases in the south (Figure 11.5). The annual area-mean change from 1980 to 1999 to 2080 to 2099 in the MMD-A1B projections varies from 0 to 16% in NEU and from –4 to –27% in SEM (Table 11.1). The largest increases in northern and central Europe are simulated in winter. In summer, the NEU area mean changes vary in sign between models, although most models simulate increased (decreased) precipitation north (south) of about 55°N. In SEM, the most consistent and, in percentage terms, largest decreases, occur in summer, but the area mean precipitation in the other seasons also decreases in most or all models. More detailed statistics are given in Table 11.1. Increasing evaporation makes the simulated decreases in annual precipitation minus evaporation extend a few hundred kilometres further north in central Europe than the decreases in precipitation (Supplementary Material Figure S11.1).

Both circulation changes and thermodynamic factors appear to affect the simulated seasonal cycle of precipitation changes in Europe. Applying a regression method to five of the MMD simulations, van Ulden and van Oldenborgh (2006) found that in a region comprising mainly Germany, circulation changes played a major role in all seasons. In most models, increases in winter precipitation were enhanced by increased westerly winds, with decreases in summer precipitation largely due to more easterly and anticyclonic flow. However, differences in the simulated circulation changes among the individual models were accompanied by large differences in precipitation change, particularly in summer. The residual precipitation change varied less with season and among models, being generally positive as expected from the increased moisture transport capacity of a warmer atmosphere. In a more detailed study of one model, HadAM3P, Rowell and Jones (2006) showed that decreases in summer precipitation in continental and southeastern Europe were mainly associated with thermodynamic factors. These included reduced relative humidity resulting from larger continental warming compared to surrounding sea areas and reduced soil moisture due mainly to spring warming causing earlier snowmelt. Given the confidence in the warming patterns

driving these changes, the reliability of the simulated drying was assessed as being high.

Changes in precipitation may vary substantially on relatively small horizontal scales, particularly in areas of complex topography. Details of these variations are sensitive to changes in the atmospheric circulation, as illustrated in Figure 11.6 for two PRUDENCE simulations that only differ with respect to the driving global model. In one, an increase in westerly flow from the Atlantic Ocean (caused by a large increase in the north-south pressure gradient) is accompanied by increases of up to 70% in annual precipitation over the Scandinavian mountains. In the other, with little change in the average pressure pattern, the increase is in the range of 0 to 20%. When compared with circulation changes in the more recent MMD simulations, these two cases fall in the opposite ends of the range. Most MMD models suggest an increased north-south pressure gradient

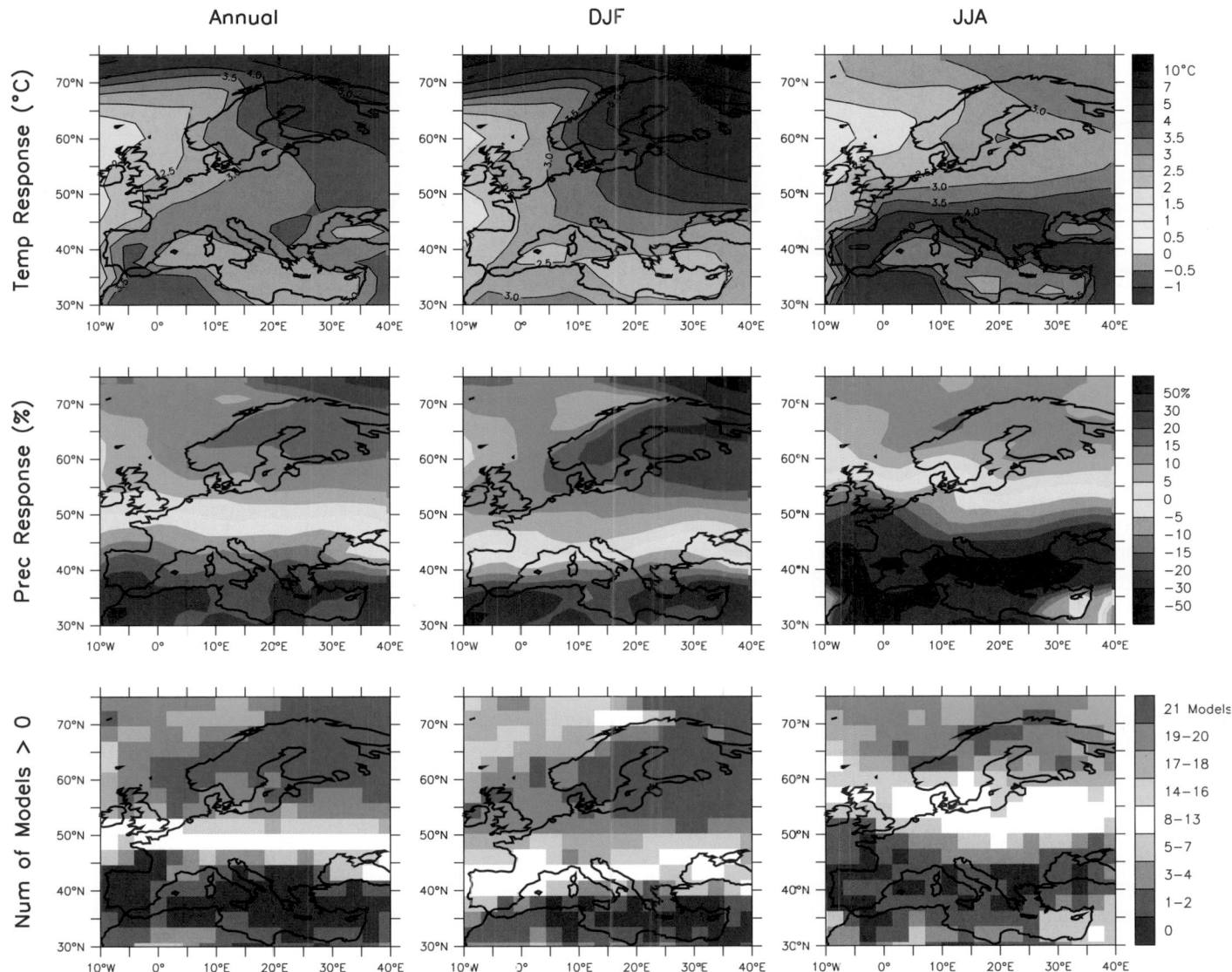

Figure 11.5. *Temperature and precipitation changes over Europe from the MMD-A1B simulations. Top row: Annual mean, DJF and JJA temperature change between 1980 to 1999 and 2080 to 2099, averaged over 21 models. Middle row: same as top, but for fractional change in precipitation. Bottom row: number of models out of 21 that project increases in precipitation.*

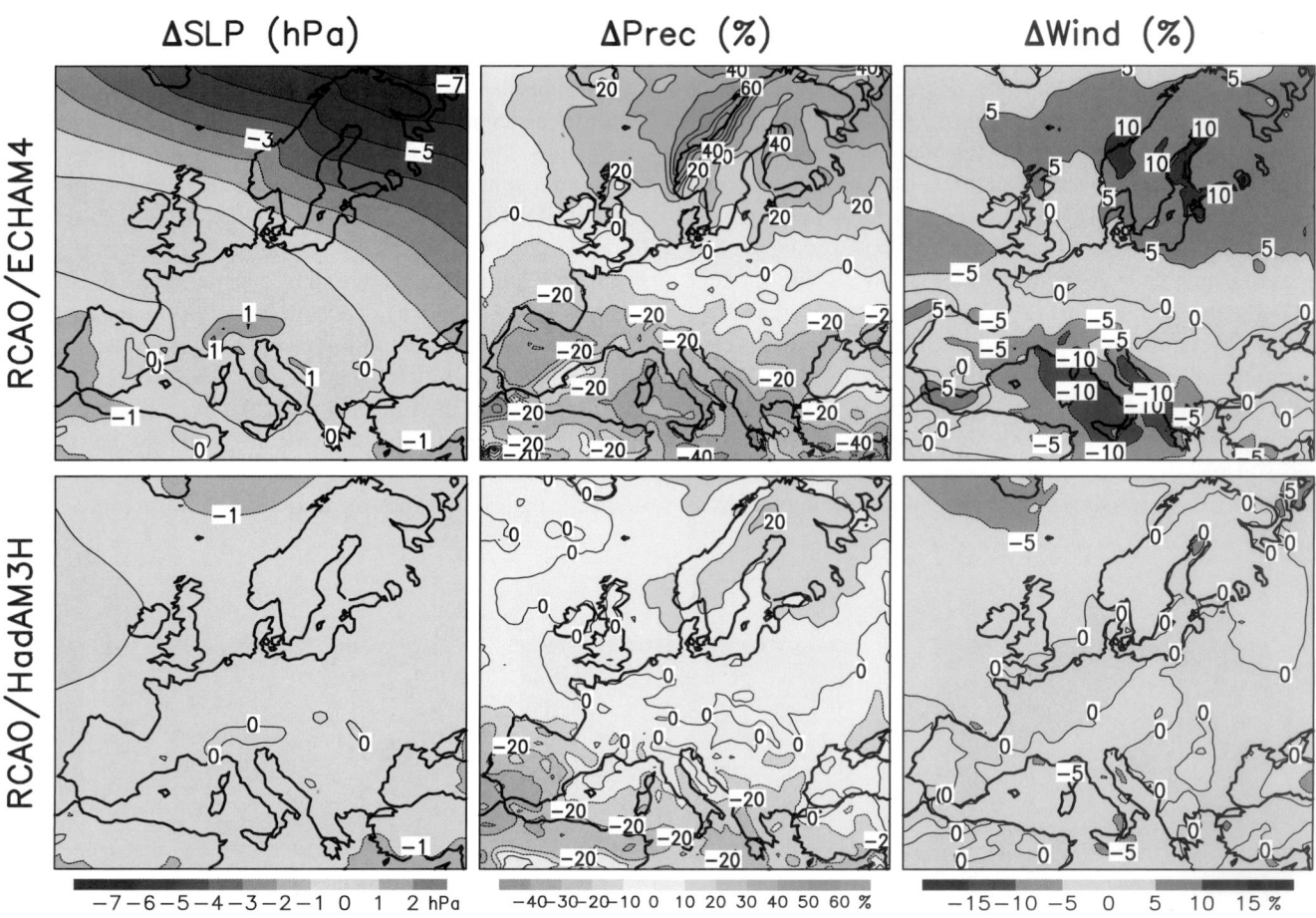

Figure 11.6. *Simulated changes in annual mean sea level pressure (ΔSLP), precipitation (ΔPrec) and mean 10-m level wind speed (ΔWind) from the years 1961 to 1990 to the years 2071 to 2100. The results are based on the SRES A2 scenario and were produced by the same RCM (Rossby Centre regional Atmosphere-Ocean model; RCAO) using boundary data from two global models: ECHAM4/OPYC3 (top) and HadAM3H (bottom) (redrawn from Rummukainen et al., 2004).*

across northern Europe, but the change is generally smaller than in the top row of Figure 11.6.

Projections of precipitation change in Europe based on SD tend to support the large-scale picture from dynamical models (e.g., Busuioc et al., 2001; Beckmann and Buishand, 2002; Hanssen-Bauer et al., 2003, 2005; Benestad, 2005; Busuioc et al., 2006), although variations between SD methods and the dependence on the GCM data sets used (see Section 11.10.1.3) make it difficult to draw quantitative conclusions. However, some SD studies have suggested a larger small-scale variability of precipitation changes than indicated by GCM and RCM results, particularly in areas of complex topography (Hellström et al., 2001).

The decrease in precipitation together with enhanced evaporation in spring and early summer is very likely to lead to reduced summer soil moisture in the Mediterranean region and parts of central Europe (Douville et al., 2002; Wang, 2005). In northern Europe, where increased precipitation competes with earlier snowmelt and increased evaporation, the MMD models disagree on whether summer soil moisture will increase or decrease (Wang, 2005).

11.3.3.3 Temperature Variability and Extremes

Based on both GCM (Giorgi and Bi, 2005; Rowell, 2005; Clark et al., 2006) and RCM simulations (Schär et al., 2004; Vidale et al., 2007), interannual temperature variability is likely to increase in summer in most areas. However, the magnitude of change is uncertain, even in central Europe where the evidence for increased variability is strongest. In some PRUDENCE simulations, interannual summer temperature variability in central Europe doubled between 1961 to 1990 and 2071 to 2100 under the A2 scenario, while other simulations showed almost no change (Vidale et al., 2007). Possible reasons for the increase in temperature variability are reduced soil moisture, which reduces the capability of evaporation to damp temperature variations, and increased land-sea contrast in average summer temperature (Rowell, 2005; Lenderink et al., 2007).

Simulated increases in summer temperature variability also extend to daily time scales. Kjellström et al. (2007) analyse the PRUDENCE simulations and find that the inter-model differences in the simulated temperature change increase towards the extreme ends of the distribution. However, a general increase in summer daily temperature variability is evident, especially in southern and central parts of Europe, with the highest maximum temperatures increasing more than the median daily maximum temperature (Supplementary Material Figure S11.23). Similarly, Shkolnik et al. (2006) report a simulated increase in summer daily time-scale temperature variability in mid-latitude western Russia. These RCM results are supported by GCM studies of Hegerl et al. (2004), Meehl and Tebaldi (2004) and Clark et al. (2006).

In contrast with summer, models project reduced temperature variability in most of Europe in winter, both on interannual (Räisänen, 2001; Räisänen et al., 2003; Giorgi et al., 2004; Giorgi and Bi, 2005; Rowell, 2005) and daily time scales (Hegerl et al., 2004; Kjellström et al., 2007). In the PRUDENCE simulations, the lowest winter minimum temperatures increased more than the median minimum temperature especially in eastern, central and northern Europe, although the magnitude of this change was strongly model-dependent (Supplementary Material Figure S11.23). The geographical patterns of the change indicate a feedback from reduced snow cover, with a large warming of the cold extremes where snow retreats but a more moderate warming in the mostly snow-free south-western Europe (Rowell, 2005; Kjellström et al., 2007).

Along with the overall warming and changes in variability, heat waves are very likely to increase in frequency, intensity and duration (Barnett et al., 2006; Clark et al., 2006; Tebaldi et al., 2006). Conversely, the number of frost days is very likely to decrease (Tebaldi et al., 2006).

11.3.3.4   Precipitation Variability and Extremes

In northern Europe and in central Europe in winter, where time mean precipitation is simulated to increase, high extremes of precipitation are very likely to increase in magnitude and frequency. In the Mediterranean area and in central Europe in summer, where reduced mean precipitation is projected, extreme short-term precipitation may either increase (due to the increased water vapour content of a warmer atmosphere) or decrease (due to a decreased number of precipitation days, which if acting alone would also make heavy precipitation less common). These conclusions are based on several GCM (e.g., Semenov and Bengtsson, 2002; Voss et al., 2002; Hegerl et al., 2004; Wehner, 2004; Kharin and Zwiers, 2005; Tebaldi et al., 2006) and RCM (e.g., Jones and Reid, 2001; Räisänen and Joelsson, 2001; Booij, 2002; Christensen and Christensen, 2003, 2004; Pal et al., 2004; Räisänen et al., 2004; Sánchez et al., 2004; Ekström et al., 2005; Frei et al., 2006; Gao et al., 2006a; Shkolnik et al., 2006; Beniston et al., 2007) studies. However, there is still a lot of quantitative uncertainty in the changes in both mean and extreme precipitation.

Time scale also matters. Although there are some indications of increased interannual variability, particularly in summer precipitation (Räisänen, 2002; Giorgi and Bi, 2005; Rowell, 2005), changes in the magnitude of long-term (monthly to annual) extremes are expected to follow the changes in mean precipitation more closely than are those in short-term extremes (Räisänen, 2005). On the other hand, changes in the frequency

of extremes tend to increase with increasing time scale even when this is not the case for the changes in the magnitude of extremes (Barnett et al., 2006).

Figure 11.7 illustrates the possible characteristics of precipitation change. The eight models in this PRUDENCE study (Frei et al., 2006) projected an increase in mean precipitation in winter in both southern Scandinavia and central Europe, due to both increased wet day frequency and increased mean precipitation for the wet days. In summer, a decrease in the number of wet days led to a decrease in mean precipitation, particularly in central Europe. Changes in extreme short-term precipitation were broadly similar to the change in average wet-day precipitation in winter. In summer, extreme daily precipitation increased in most models despite the decrease in mean precipitation, although the magnitude of the change was highly model-dependent. However, this study only covered the uncertainties associated with the choice of the RCM, not those associated with the driving GCM and the emissions scenario.

Much larger changes are expected in the recurrence frequency of precipitation extremes than in the magnitude of extremes (Huntingford et al., 2003; Barnett et al., 2006; Frei et al., 2006). For example, Frei et al. (2006) estimate that, in Scandinavia under the A2 scenario, the highest five-day winter precipitation totals occurring once in 5 years in 2071 to 2100 would be similar to those presently occurring once in 8 to 18 years (the range reflects variation between the PRUDENCE models). In the MMD simulations, large increases occur in the frequencies of both high winter precipitation in northern Europe and low summer precipitation in southern Europe and the Mediterranean area (Table 11.1).

The risk of drought is likely to increase in southern and central Europe. Several model studies have indicated a decrease in the number of precipitation days (e.g., Semenov and Bengtsson, 2002; Voss et al., 2002; Räisänen et al., 2003, 2004; Frei et al., 2006) and an increase in the length of the longest dry spells in this area (Voss et al., 2002; Pal et al., 2004; Beniston et al., 2007; Gao et al., 2006a; Tebaldi et al., 2006). By contrast, the same studies do not suggest major changes in dry-spell length in northern Europe.

11.3.3.5   Wind Speed

Confidence in future changes in windiness in Europe remains relatively low. Several model studies (e.g., Zwiers and Kharin, 1998; Knippertz et al., 2000; Leckebusch and Ulbrich, 2004; Pryor et al., 2005a; van den Hurk et al., 2006) have suggested increased average and/or extreme wind speeds in northern and/or central Europe, but some studies point in the opposite direction (e.g., Pryor et al., 2005b). The changes in both average and extreme wind speeds may be seasonally variable, but the details of this variation appear to be model-dependent (e.g., Räisänen et al., 2004; Rockel and Woth, 2007).

A key factor is the change in the large-scale atmospheric circulation (Räisänen et al., 2004; Leckebusch et al., 2006). Simulations with an increased north-south pressure gradient across northern Europe (e.g., top row of Figure 11.6) tend to

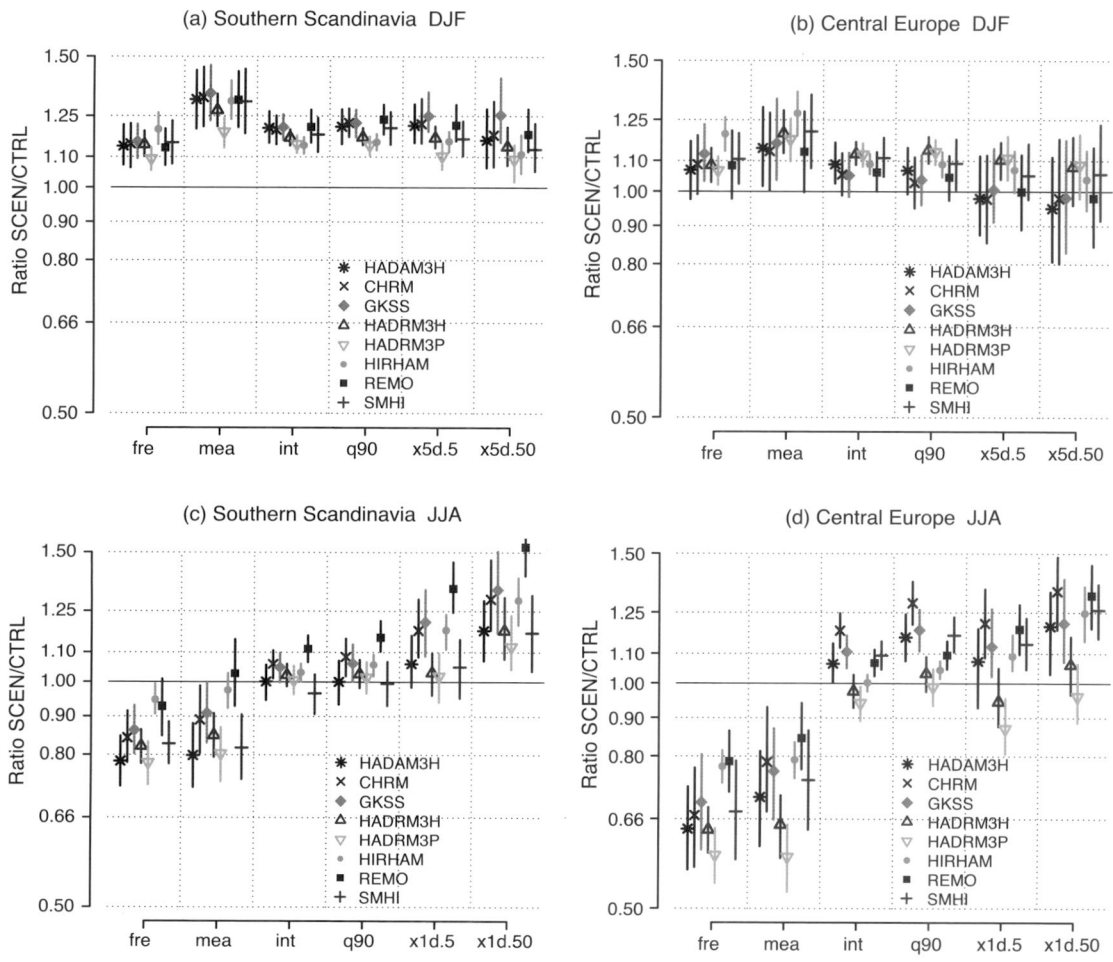

Figure 11.7. *Changes (ratio 2071–2100 / 1961–1990 for the A2 scenario) in domain-mean precipitation diagnostics in the PRUDENCE simulations in southern Scandinavia (5°E–20°E, 55°N–62°N) and central Europe (5°E–15°E, 48°N–54°N) in winter (top) and in summer (bottom). fre = wet-day frequency; mea = mean seasonal precipitation; int = mean wet-day precipitation; q90 = 90th percentile of wet-day precipitation; x1d.5 and x1d.50 = 5- and 50-year return values of one-day precipitation; x5d.5 and x5d.50 = 5- and 50-year return values of five-day precipitation. For each of the eight models, the vertical bar gives the 95% confidence interval associated with sampling uncertainty (redrawn from Frei et al., 2006). Models are the Hadley Centre Atmospheric Model (HadAM3H), the Climate High Resolution Model (CHRM), the climate version of the 'Lokalmodell' (CLM), the Hadley Centre Regional Model (HadRM3H and HadRM3P), the combination of the High-Resolution Limited Area Model (HIRLAM) and the European Centre Hamburg (ECHAM4) GCM (HIRHAM), the regional climate model REMO, and the Rossby Centre regional Atmosphere-Ocean model (RCAO).*

indicate stronger winds in northern Europe, because of both the larger time-averaged pressure gradient and a northward shift in cyclone activity. Conversely, the northward shift in cyclone activity tends to reduce windiness in the Mediterranean area. On the other hand, simulations with little change in the pressure pattern tend to show only small changes in the mean wind speed (bottom row of Figure 11.6). Most of the MMD-projected pressure changes fall between the two PRUDENCE simulations shown in Figure 11.6, which suggests that the most likely outcome for windiness might be between these two cases.

Extreme wind speeds in Europe are mostly associated with strong winter cyclones (e.g., Leckebush and Ullbrich, 2004), the occurrence of which is only indirectly related to the time-mean circulation. Nevertheless, models suggest a general similarity between the changes in average and extreme wind speeds (Knippertz et al., 2000; Räisänen et al., 2004). A caveat to this conclusion is that, even in most RCMs, the simulated

extremes of wind speed over land tend to be too low (see Section 11.3.2).

11.3.3.6 Mediterranean Cyclones

Several studies have suggested a decrease in the total number of cyclones in the Mediterranean Sea (Lionello et al., 2002; Vérant, 2004; Somot 2005; Leckebusch et al., 2006; Pinto et al., 2006; Ulbrich et al., 2006), but there is no agreement on whether the number of intense cyclones will increase or decrease (Lionello et al., 2002; Pinto et al., 2006).

11.3.3.7 Snow and Sea Ice

The overall warming is very likely to shorten the snow season in all of Europe. Snow depth is also likely to be reduced, at least in most areas, although increases in total winter precipitation may counteract the increased melting and decreased fraction of solid precipitation associated with the warming. The changes may be large, including potentially a one-to-three month shortening of the snow season in northern Europe (Räisänen et al., 2003) and a 50 to 100% decrease in snow depth in most of Europe (Räisänen et al., 2003; Rowell, 2005) by the late 21st century. However, snow conditions in the coldest parts of Europe, such as northern Scandinavia and north-western Russia (Räisänen et al., 2003; Shkolnik et al., 2006) and the highest peaks of the Alps (Beniston et al., 2003) appear to be less sensitive to the temperature and precipitation changes projected for this century than those at lower latitudes and altitudes (see also Section 11.4.3.2, Box 11.3).

The Baltic Sea is likely to lose a large part of its seasonal ice cover during this century. Using a regional atmosphere-Baltic Sea model (Meier et al., 2004), the average winter maximum ice extent decreased by about 70% (60%) between 1961 to 1990 and 2071 to 2100 under the A2 (B2) scenario. The length of the ice season was projected to decrease by one to two months in northern parts and two to three months in the central parts. Comparable decreases in Baltic Sea ice cover were projected by earlier studies (Haapala et al., 2001; Meier, 2002).

11.4   Asia

Assessment of projected climate change for Asia:

All of Asia is very likely to warm during this century; the warming is likely to be well above the global mean in central Asia, the Tibetan Plateau and northern Asia, above the global mean in East and South Asia, and similar to the global mean in Southeast Asia. It is very likely that summer heat waves/hot spells in East Asia will be of longer duration, more intense, and more frequent. It is very likely that there will be fewer very cold days in East Asia and South Asia.

Boreal winter precipitation is very likely to increase in northern Asia and the Tibetan Plateau, and likely to increase in eastern Asia and the southern parts of Southeast Asia. Summer precipitation is likely to increase in northern Asia, East and South Asia and most of Southeast Asia, but it is likely to decrease in central Asia. An increase in the frequency of intense precipitation events in parts of South Asia, and in East Asia, is very likely.

Extreme rainfall and winds associated with tropical cyclones are likely to increase in East, Southeast and South Asia. Monsoonal flows and the tropical large-scale circulation are likely to be weakened.

While broad aspects of Asian climate change show consistency among AOGCM simulations, a number of sources of uncertainty remain. A lack of observational data in some areas limits model assessment. There has been little assessment of the projected changes in regional climatic means and extremes.

There are substantial inter-model differences in representing monsoon processes, and a lack of clarity over changes in ENSO further contributes to uncertainty about future regional monsoon and tropical cyclone behaviour. Consequently, quantitative estimates of projected precipitation change are difficult to obtain. It is likely that some local climate changes will vary significantly from regional trends due to the region's very complex topography and marine influences.

11.4.1   Key Processes

As monsoons are the dominant phenomena over much of Asia, the factors that influence the monsoonal flow and precipitation are of central importance for understanding climate change in this region. Precipitation is affected both by the strength of the monsoonal flows and the amount of water vapour transported. Monsoonal flows and the tropical large-scale circulation often weaken in global warming simulations (e.g., Knutson and Manabe, 1995). This arises out of an increase in dry static stability associated with the tropical warming in these models, and the reduction in adiabatic warming/cooling needed to balance a given amount of radiative cooling/condensational heating (e.g., Betts, 1998). But there is an emerging consensus that the effect of enhanced moisture convergence in a warmer, moister atmosphere dominates over any such weakening of the circulation, resulting in increased monsoonal precipitation (e.g., Douville et al., 2000; Giorgi et al., 2001a,b; Stephenson et al., 2001; Dairaku and Emori, 2006; Ueda et al., 2006).

There is an association of the phase of ENSO with the strength of the summer monsoons (Pant and Rupa Kumar, 1997), so changes in ENSO will have an impact on these monsoons. However, such an impact can be compounded by a change in the ENSO-South Asian monsoon connection under greenhouse gas warming (Krishna Kumar et al., 1999; Ashrit et al., 2001; Sarkar et al., 2004; see Section 3.7). Moreover, there is a link between Eurasian snow cover and the strength of the monsoon (see also Section 3.7), with the monsoon strengthening if snow cover retreats. Aerosols, particularly absorbing aerosols, further modify monsoonal precipitation (e.g., Ramanathan et al., 2005 for South Asia), as do modifications of vegetation cover (e.g., Chen et al., 2004 for East Asia). However, most emission scenarios suggest that future changes in regional climate are still likely to be dominated by increasing greenhouse gas forcing rather than changes in sulphate and absorbing aerosols, at least over the South Asian region.

For South Asia, the monsoon depressions and tropical cyclones generated over the Indian seas modulate the monsoon anomalies. For East Asia, the monsoonal circulations are strengthened by extratropical cyclones energised in the lee of the Tibetan Plateau and by the strong temperature gradient along the East Coast. The influence of ENSO on the position and strength of the subtropical high in the North Pacific influences both typhoons and other damaging heavy rainfall events, and has been implicated in observed inter-decadal variations in typhoon tracks (Ho et al., 2004). This suggests that the spatial structure of warming in the Pacific will be relevant for changes in these features. The dynamics of the Meiyu-Changma-Baiu rains in the early summer, which derive from baroclinic disturbances strongly modified by latent heat release, remain poorly understood. While an increase in rainfall in the absence of circulation shifts is expected, relatively modest shifts or changes in timing can significantly affect East Chinese, Korean and Japanese climates.

Over central and Southeast Asia and the maritime continent, interannual rainfall variability is significantly affected by ENSO (e.g., McBride et al., 2003), particularly the June to November rainfall in southern and eastern parts of the Indonesian Archipelago, which is reduced in El Niño years (Aldrian and Susanto, 2003). Consequently, the pattern of ocean temperature change across the Pacific is of central importance to climate change in this region.

In central Asia, including the Tibetan Plateau, the temperature response is strongly influenced by changes in winter and spring snow cover, the isolation from maritime influences, and the spread of the larger winter arctic warming into the region. With regard to precipitation, a key issue is related to the moisture transport from the northwest by westerlies and polar fronts. How far the projected drying of the neighbouring Mediterranean penetrates into these regions is likely to be strongly dependent on accurate simulation of these moisture transport processes. The dynamics of climate change in the Tibetan Plateau are further complicated by the high altitude of this region and its complex topography with large elevation differences.

11.4.2 Skill of Models in Simulating Present Climate

Regional mean temperature and precipitation in the MMD models show biases when compared with observed climate (Supplementary Material Table S11.1). The multi-model mean shows a cold and wet bias in all regions and in most seasons, and the bias of the annual average temperature ranges from –2.5°C over the Tibetan Plateau (TIB) to –1.4°C over South Asia (SAS). For most regions, there is a 6°C to 7°C range in the biases from individual models with a reduced bias range in Southeast Asia (SEA) of 3.6°C. The median bias in precipitation is small (less than 10%) in Southeast Asia, South Asia, and Central Asia (CAS), larger in northern Asia and East Asia (NAS and EAS, around +23%), and very large in the Tibetan Plateau (+110%). Annual biases in individual models are in the range of –50 to +60% across all regions except the Tibetan Plateau, where some models simulate annual precipitation 2.5 times that observed and even larger seasonal biases occur in winter and spring. These global models clearly have significant problems over Tibet, due to the difficulty in simulating the effects of the dramatic topographic relief, as well as the distorted albedo feedbacks due to extensive snow cover. However, with only limited observations available, predominantly in valleys, large errors in temperature and significant underestimates of precipitation are likely.

South Asia

Over South Asia, the summer is dominated by the southwest monsoon, which spans the four months from June to September and dominates the seasonal cycles of the climatic parameters. While most models simulate the general migration of seasonal tropical rain, the observed maximum rainfall during the monsoon season along the west coast of India, the north Bay of Bengal and adjoining northeast India is poorly simulated by many models (Lal and Harasawa, 2001; Rupa Kumar and Ashrit, 2001; Rupa Kumar et al., 2002, 2003). This is likely linked to the coarse resolution of the models, as the heavy rainfall over these regions is generally associated with the steep orography. However, the simulated annual cycles in South Asian mean precipitation and surface air temperature are reasonably close to the observed (Supplementary Material Figure S11.24). The MMD models capture the general regional features of the monsoon, such as the low rainfall amounts coupled with high variability over northwest India. However, there has not yet been sufficient analysis of whether finer details of regional significance are simulated more adequately in the MMD models.

Recent work indicates that time-slice experiments using an AGCM with prescribed SSTs, as opposed to a fully coupled system, are not able to accurately capture the South Asian monsoon response (Douville, 2005). Thus, neglecting the short-term SST feedback and variability seems to have a significant impact on the projected monsoon response to global warming, complicating the regional downscaling problem. However, May (2004a) notes that the high-resolution (about 1.5 degrees) European Centre-Hamburg (ECHAM4) GCM simulates the variability and extremes of daily rainfall (intensity as well as frequency of wet days) in good agreement with the observations (Global Precipitation Climatology Project, Huffman et al., 2001).

Three-member ensembles of baseline simulations (1961–1990) from an RCM (PRECIS) at 50 km resolution have confirmed that significant improvements in the representation of regional processes over South Asia can be achieved (Rupa Kumar et al., 2006). For example, the steep gradients in monsoon precipitation with a maximum along the western coast of India are well represented in PRECIS.

East Asia

Simulated temperatures in most MMD models are too low in all seasons over East Asia; the mean cold bias is largest in winter and smallest in summer. Zhou and Yu (2006) show that over China, the models perform reasonably in simulating the dominant variations of the mean temperature over China, but not the spatial distributions. The annual precipitation over East Asia exceeds the observed estimates in almost all models and the rain band in the mid-latitudes is shifted northward in seasons other than summer. This bias in the placement of the rains in central China also occurred in earlier models (e.g., Zhou and Li, 2002; Gao et al., 2004). In winter, the area-mean precipitation is overestimated by more than 50% on average due to strengthening of the rain band associated with extratropical systems over South China. The bias and inter-model differences in precipitation are smallest in summer but the northward shift of this rain band results in large discrepancies in summer rainfall distribution over Korea, Japan and adjacent seas.

Kusunoki et al. (2006) find that the simulation of the Meiyu-Changma-Baiu rains in the East Asian monsoon is improved substantially with increasing horizontal resolution. Confirming the importance of resolution, RCMs simulate more realistic climatic characteristics over East Asia than AOGCMs, whether driven by re-analyses or by AOGCMs (e.g., Ding et al., 2003; Oh et al., 2004; Fu et al., 2005; Zhang et al., 2005a, Ding et al., 2006; Sasaki et al., 2006b). Several studies reproduce the fine-scale climatology of small areas using a multiply nested RCM (Im et al., 2006) and a very-high resolution (5 km) RCM (Yasunaga et al., 2006). Gao et al. (2006b) report that simulated East Asia large-scale precipitation patterns are significantly affected by resolution, particularly during the mid- to late-monsoon months, when smaller-scale convective processes dominate.

Southeast Asia

The broad-scale spatial distribution of temperature and precipitation in DJF and JJA averaged across the MMD models compares well with observations. Rajendran et al. (2004) examine the simulation of current climate in the MRI coupled model (see Table 8.1 for model details). Large-scale features were well simulated, but errors in the timing of peak rainfall over Indochina were considered a major shortcoming. Collier et al. (2004) assess the performance of the CCSM3 model (see Table 8.1 for model details) in simulating tropical precipitation forced by observed SST. Simulation was good over the Maritime continent compared to the simulation for other tropical regions. B. Wang et al. (2004) assess the ability of 11 AGCMs in the Asian-Australian monsoon region simulation forced with observed SST variations. They found that the models' ability to simulate observed interannual rainfall variations was poorest in the Southeast Asian portion of the domain. Since current AOGCMs continue to have some significant shortcomings in representing ENSO variability (see Section 8.4), the difficulty of projecting changes in ENSO-related rainfall in this region is compounded.

Rainfall simulation across the region at finer scales has been examined in some studies. The Commonwealth Scientific and Industrial Research Organisation (CSIRO) stretched-grid Conformal-Cubic Atmospheric Model (CCAM) at 80-km resolution shows reasonable precipitation simulation in JJA, although Indochina tended to be drier than in the observations (McGregor and Nguyen, 2003). Aldrian et al. (2004a) conducted a number of simulations with the Max-Planck Institute (MPI) regional model for an Indonesian domain, forced by reanalyses and by the ECHAM4 GCM. The model was able to represent the spatial pattern of seasonal rainfall. It was found that a resolution of at least 50 km was required to simulate rainfall seasonality correctly over Sulawesi. The formulation of a coupled regional model improves regional rainfall simulation over the oceans (Aldrian et al., 2004b). Arakawa and Kitoh (2005) demonstrate an accurate simulation of the diurnal cycle of rainfall over Indonesia with an AGCM of 20-km horizontal resolution.

Central Asia and Tibet

Due to the complex topography and the associated mesoscale weather systems of the high-altitude and arid areas, GCMs typically perform poorly over the region. Importantly, the GCMs, and to a lesser extent RCMs, tend to overestimate the precipitation over arid and semi-arid areas in the north (e.g., Small et al., 1999; Gao et al., 2001; Elguindi and Giorgi, 2006).

Over Tibet, the few available RCM simulations generally exhibit improved performance in the simulation of present-day climate compared to GCMs (e.g., Gao et al., 2003a,b; Zhang et al., 2005b). For example, the GCM simulation of Gao et al. (2003a) overestimated the precipitation over the north-western Tibetan Plateau by a factor of five to six, while in an RCM nested in this model, the overestimate was less than a factor of two.

11.4.3 Climate Projections

11.4.3.1 Temperature

The temperature projections for the 21st century based on the MMD-A1B models (Figure 11.8 and Table 11.1) represent a strong warming over the 21st century. Warming is similar to the global mean warming in Southeast Asia (mean warming between 1980 to 1999 and 2080 to 2099 of 2.5°C). Warming greater than the global mean is projected for South Asia (3.3°C) and East Asia (3.3°C), and much more than the global mean in the continental interior of Asia (3.7°C in central Asia, 3.8°C in Tibet and 4.3°C in northern Asia). In four out of the six regions, the largest warming occurs in DJF, but in central Asia, the maximum occurs in JJA. In Southeast Asia, the warming is nearly the same throughout the year. Model-to-model variation in projected warming is typically about three-quarters of the mean warming (e.g., 2.0°C to 4.7°C for annual mean warming in South Asia). The 5 to 95% ranges based on Tebaldi et al. (2004a) suggest a slightly smaller uncertainty than the full range of the model results (Supplementary Material Table S 1.2). Because the projected warming is large compared to the interannual temperature variability, a large majority of individual years and seasons in the late 21st century are likely to be extremely warm by present standards (Table 11.1). The projections of changes in mean temperature and, where available, temperature extremes, are discussed below in more detail for individual Asian regions.

South Asia

For the A1B scenario, the MMD-A1B models show a median increase of 3.3°C (see Table 11.1) in annual mean temperature by the end of the 21st century. The median warming varies seasonally from 2.7°C in JJA to 3.6°C in DJF, and is likely to increase northward in the area, particularly in winter, and from sea to land (Figure 11.9). Studies based on earlier AOGCM simulations (Douville et al., 2000; Lal and Harasawa, 2001; Lal et al., 2001; Rupa Kumar and Ashrit, 2001; Rupa Kumar et al., 2002, 2003; Ashrit et al., 2003; May, 2004b) support this picture. The tendency of the warming to be more pronounced in

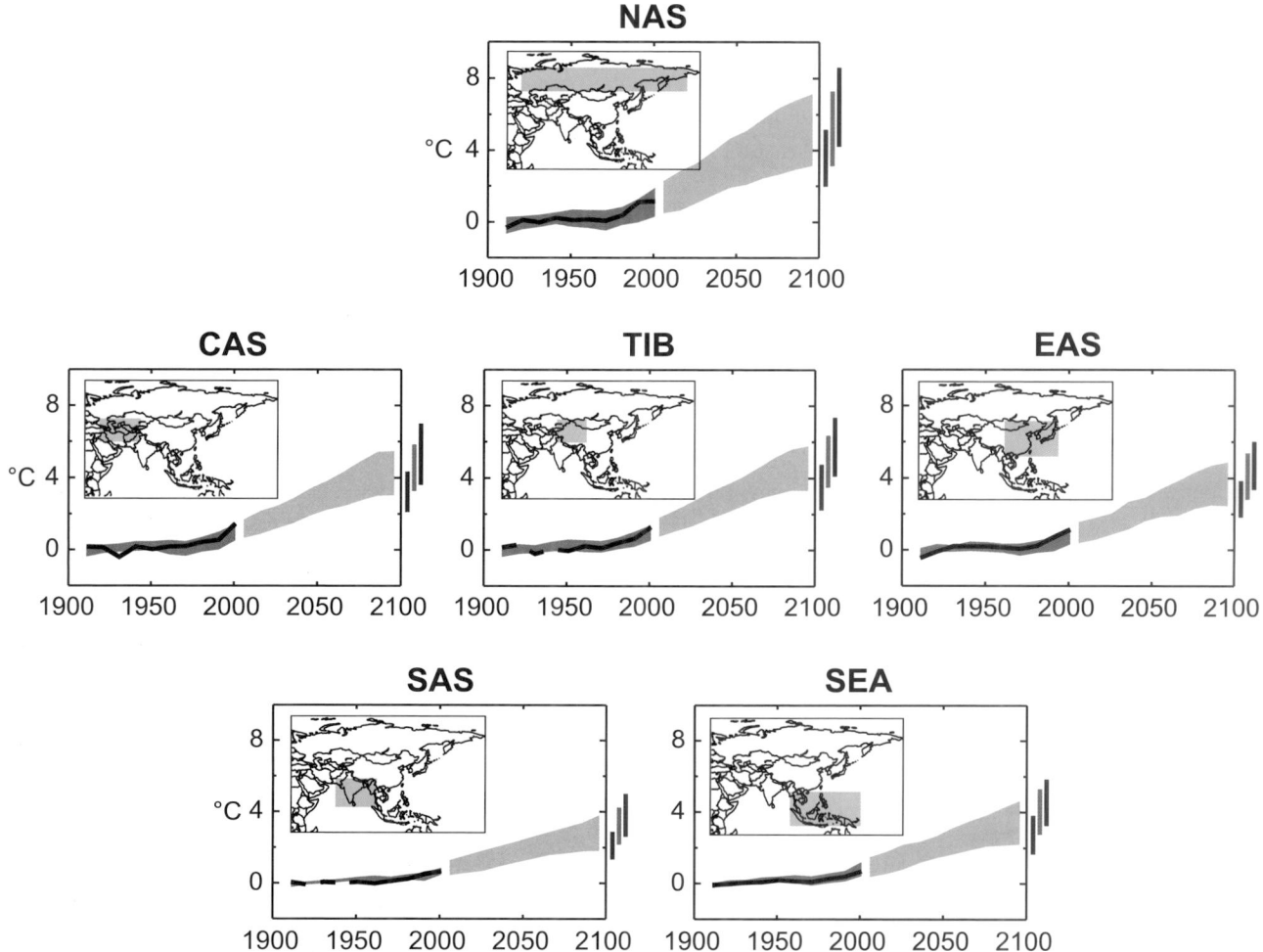

Figure 11.8. *Temperature anomalies with respect to 1901 to 1950 for six Asian land regions for 1906 to 2005 (black line) and as simulated (red envelope) by MMD models incorporating known forcings; and as projected for 2001 to 2100 by MMD models for the A1B scenario (orange envelope). The bars at the end of the orange envelope represent the range of projected changes for 2091 to 2100 for the B1 scenario (blue), the A1B scenario (orange) and the A2 scenario (red). The black line is dashed where observations are present for less than 50% of the area in the decade concerned. More details on the construction of these figures are given in Box 11.1 and Section 11.1.2.*

winter is also a conspicuous feature of the observed temperature trends over India (Rupa Kumar et al., 2002, 2003).

Downscaled projections using the Hadley Centre Regional Model (HadRM2) indicate future increases in extreme daily maximum and minimum temperatures throughout South Asia due to the increase in greenhouse gas concentrations. This projected increase is of the order of 2°C to 4°C in the mid-21st century under the IPCC Scenario IS92a in both minimum and maximum temperatures (Krishna Kumar et al., 2003). Results from a more recent RCM, PRECIS, indicate that the night temperatures increase faster than the day temperatures, with the implication that cold extremes are very likely to be less severe in the future (Rupa Kumar et al., 2006).

East Asia

The MMD-A1B models project a median warming of 3.3°C (Table 11.1) by the end of the 21st century, which varies seasonally from 3.0°C in JJA to 3.6°C in DJF. The warming tends to be largest in winter, especially in the northern inland area (Figure 11.9), but the area-mean difference from the other

seasons is not large. There is no obvious relationship between model bias and the magnitude of the warming. The spatial pattern of larger warming over northwest EAS (Figure 11.9) is very similar to the ensemble mean of pre-MMD models. Regional Climate Model simulations show mean temperature increases similar to those simulated by AOGCMs (Gao et al., 2001, 2002; Kwon et al., 2003; Jiang, 2005; Kurihara et al., 2005; Y.L. Xu et al., 2005).

Daily maximum and minimum temperatures are very likely to increase in East Asia, resulting in more severe warm but less severe cold extremes (Gao et al., 2002; Mizuta et al., 2005; Y.L. Xu et al., 2005; Boo et al., 2006). Mizuta et al. (2005) analyse temperature-based extreme indices over Japan with a 20-km mesh AGCM and find the changes in the indices to be basically those expected from the mean temperature increase, with changes in the distribution around the mean not playing a large role. Boo et al. (2005) report similar results for Korea. Gao et al. (2002) and Y.L. Xu et al. (2005) find a reduced diurnal temperature range in China and larger increases in daily minimum than maximum temperatures.

Southeast Asia

In the MMD-A1B simulations, the median warming for the region is 2.5°C by the end of the 21st century, with little seasonal variation (Table 11.1). Simulations by the CSIRO Division of Atmospheric Research Limited Area Model (DARLAM; McGregor et al., 1998) and more recently by the CSIRO stretched-grid model (McGregor and Dix, 2001) centred on the Indochina Peninsula (AIACC, 2004) at a resolution of 14 km have demonstrated the potential for significant local variation in warming, particularly the tendency for warming to be significantly stronger over the interior of the landmasses than over the surrounding coastal regions. A tendency for the warming to be stronger over Indochina and the larger landmasses of the archipelago is also visible in the MMD models (Figures 10.8 and 11.9). As in other regions, the magnitude of the warming depends on the forcing scenario.

Central Asia and Tibet

In the MMD-A1B simulations, central Asia warms by a median of 3.7°C, and Tibet by 3.8°C (Table 11.1) by the end of the 21st century. The seasonal variation in the simulated warming is modest. Findings from earlier multi-model studies (Zhao et al., 2002; Xu et al., 2003a,b; Meleshko et al., 2004; Y. Xu et al., 2005) are consistent with the MMD models' results.

An RCM study by Gao et al. (2003b) indicates greater warming over the Plateau compared to surrounding areas, with the largest warming at highest altitudes, for example, over the Himalayas (see also Box 11.3). The higher temperature increase over high-altitude areas can be explained by the decrease in surface albedo associated with the melting of snow and ice (Giorgi et al., 1997). This phenomenon is found to different extents in some, although not all, of the MMD models, and it is visible in the multi-model mean changes, particularly in the winter (Figure 11.9).

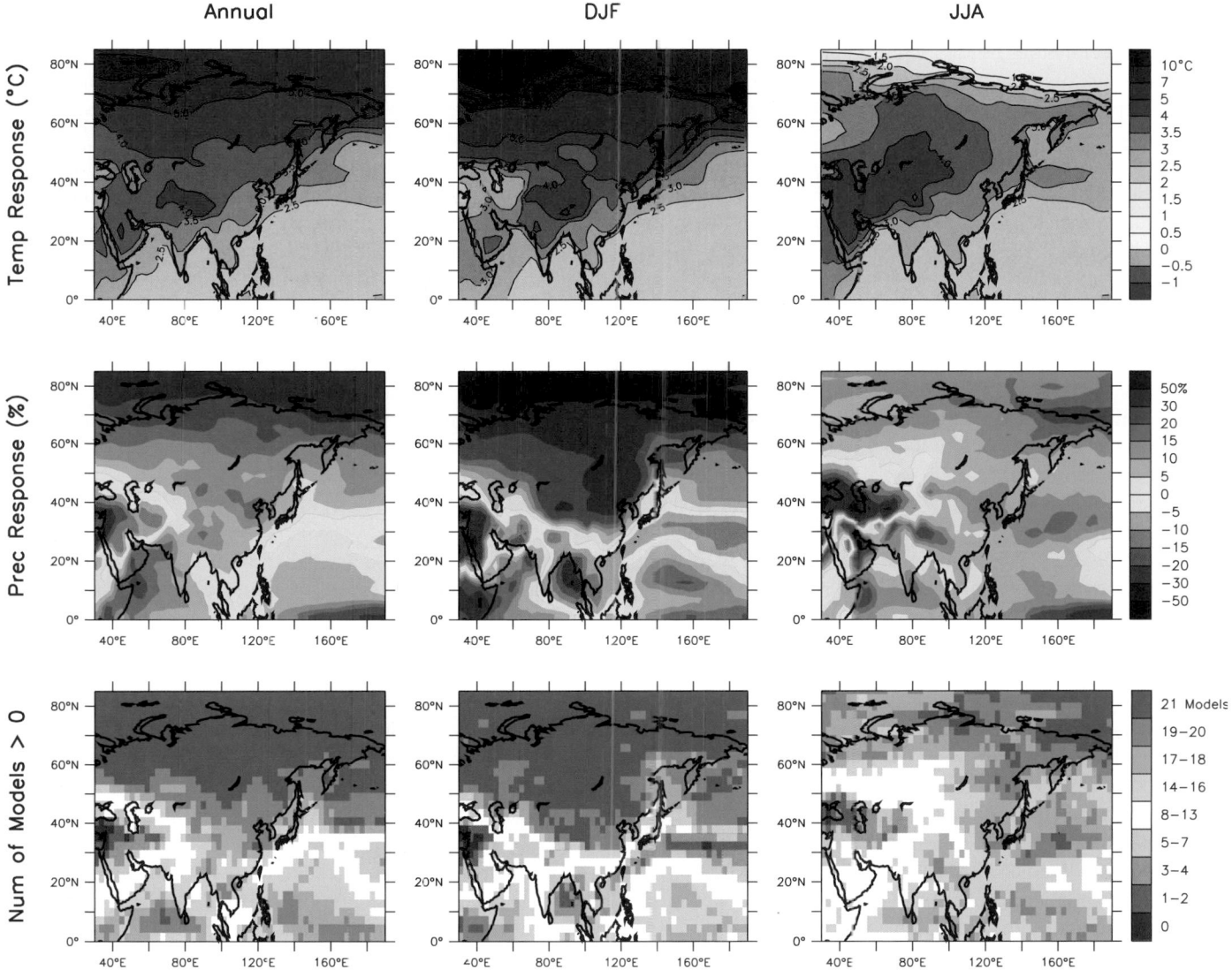

Figure 11.9. *Temperature and precipitation changes over Asia from the MMD-A1B simulations. Top row: Annual mean, DJF and JJA temperature change between 1980 to 1999 and 2080 to 2099, averaged over 21 models. Middle row: same as top, but for fractional change in precipitation. Bottom row: number of models out of 21 that project increases in precipitation.*

11.4.3.2   Precipitation and Associated Circulation Systems

The MMD models indicate an increase in annual precipitation in most of Asia during this century, the percentage increase being largest and most consistent among models in North and East Asia (Figure 11.9 and Table 11.1). The main exception is central Asia, particularly its western parts, where most models simulate reduced precipitation in the summer. Based on these simulations, sub-continental boreal winter precipitation is very likely to increase in northern Asia and the Tibetan Plateau, and likely to increase in eastern Asia. It is also likely to increase in the southern parts of Southeast Asia. Summer precipitation is likely to increase in North, South, Southeast and East Asia, but decrease in central Asia. Probability estimates from Tebaldi et al. (2004a; see Supplementary Material Table S11.2) support these judgments.

The projected decrease in mean precipitation in central Asia is accompanied by an increase in the frequency of very dry spring, summer and autumn seasons; conversely, in winter, where models project increases in the mean precipitation, very high precipitation becomes more common (Table 11.1). The projections of changes in mean precipitation and, where available, precipitation extremes, are discussed in more detail below for individual Asian regions. Where appropriate, the connection to changes in circulation systems that bring precipitation is also discussed. Smaller (slightly larger) changes are generally projected for the B1 (A2) scenario, but the inter-scenario differences are small compared with the inter-model differences.

South Asia

Most of the MMD-A1B models project a decrease in precipitation in DJF (the dry season), and an increase during the rest of the year. The median change is 11% by the end of the 21st century, and seasonally is –5% in DJF and 11% in JJA, with a large inter-model spread (Table 11.1). The probabilistic method of Tebaldi et al. (2004a) similarly shows a large spread, although only 3 of the 21 models project a decrease in annual precipitation. This qualitative agreement on increasing precipitation for most of the year is also supported by earlier AOGCM simulations (Lal and Harasawa, 2001; Lal et al., 2001; Rupa Kumar and Ashrit, 2001; Rupa Kumar et al., 2002, 2003; Ashrit et al., 2003; May, 2004b).

In a study with four GCMs, Douville et al. (2000) find a significant spread in the summer monsoon precipitation anomalies despite a general weakening of the monsoon circulation (see also May, 2004b). They conclude that the changes in atmospheric water content, precipitation and land surface hydrology under greenhouse forcing could be more important than the increase in the land-sea thermal gradient for the future evolution of monsoon precipitation. Stephenson et al. (2001) propose that the consequences of climate change could manifest in different ways in the physical and dynamical components of monsoon circulation. Douville et al. (2000) also argue that the weakening of the ENSO-monsoon correlation could be explained by a possible increase in precipitable water as a result of global warming, rather than by an increased land-sea thermal gradient. However, model diagnostics using ECHAM4 to investigate this aspect indicate that both the above mechanisms can play a role in monsoon changes in a greenhouse-gas warming scenario. Ashrit et al. (2001) show that the monsoon deficiency due to El Niño might not be as severe, while the favourable impact of La Niña seems to remain unchanged. In a later study using the Centre National de Recherches Météorologiques (CNRM) GCM, Ashrit et al. (2003) find that the simulated ENSO-monsoon teleconnection shows a strong modulation on multi-decadal time scales, but no systematic change with increasing amounts of greenhouse gases.

Time-slice experiments with ECHAM4 indicate a general increase in the intensity of heavy rainfall events in the future, with large increases over the Arabian Sea and the tropical Indian Ocean, in northern Pakistan and northwest India, as well as in northeast India, Bangladesh and Myanmar (May, 2004a). The HadRM2 RCM shows an overall decrease by up to 15 days in the annual number of rainy days over a large part of South Asia, under the IS92a scenario in the 2050s, but with an increase in the precipitation intensity as well as extreme precipitation (Krishna Kumar et al., 2003). Simulations with the PRECIS RCM also project substantial increases in extreme precipitation over a large area, particularly over the west coast of India and west central India (Rupa Kumar et al., 2006). Dairaku and Emori (2006) show from a high-resolution AGCM simulation (about 1.5 degrees) that the increased extreme precipitation over land in South Asia would arise mainly from dynamic effects, that is, enhanced upward motion due to the northward shift of monsoon circulation.

Based on regional HadRM2 simulations, Unnikrishnan et al. (2006) report increases in the frequency as well as intensities of tropical cyclones in the 2050s under the IS92a scenario in the Bay of Bengal, which will cause more heavy precipitation in the surrounding coastal regions of South Asia, during both southwest and northeast monsoon seasons.

East Asia

The MMD-A1B models project an increase in precipitation in East Asia in all seasons. The median change at the end of the 21st century is +9% in the annual mean with little seasonal difference, and a large model spread in DJF (Table 11.1). In winter, this increase contrasts with a decrease in precipitation over the ocean to the southeast, where reduced precipitation corresponds well with increased mean sea level pressure. While the projections have good qualitative agreement, there remain large quantitative differences among the models, which is consistent with previous studies (e.g., Giorgi et al., 2001a; Hu et al., 2003; Min et al., 2004).

Based on the MMD models, Kimoto (2005) projects increased Meiyu-Changma-Baiu activity associated with the strengthening of anticyclonic cells to its south and north, and Kwon et al. (2005) show increased East Asia summer precipitation due to an enhanced monsoon circulation in the decaying phase of El Niño. A 20-km mesh AGCM simulation shows that Meiyu-Changma-Baiu rainfall increases over the Yangtze River valley,

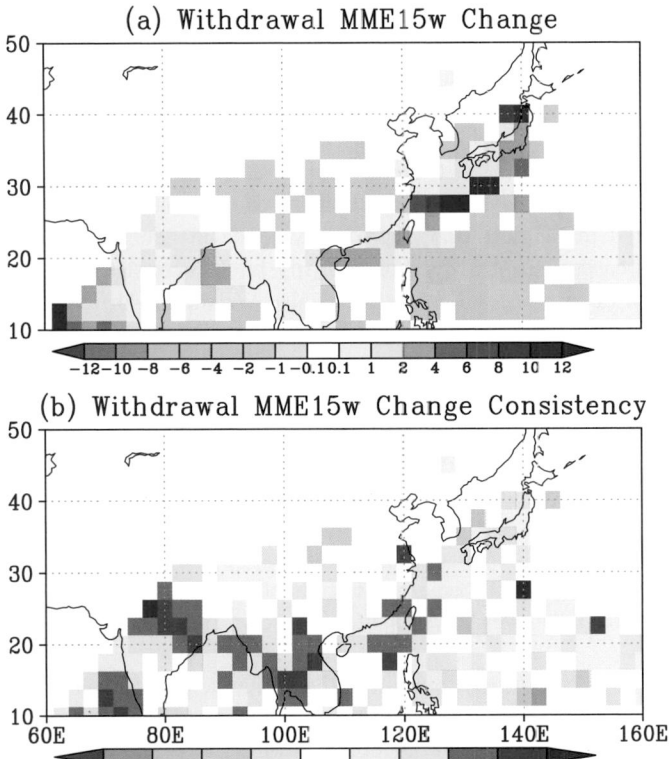

Figure 11.10. *(a) The ensemble mean change in withdrawal date of the summer rainy season between the MMD-A1B projections in 2081 to 2100 as compared with the 1981 to 2000 period in the 20C3M simulations. A positive value indicates a later withdrawal date in the A1B scenario. Units are 5 days. (b) Fraction of the models projecting a positive difference in withdrawal date. (Kitoh and Uchiyama, 2006).*

the East China Sea and western Japan, while rainfall decreases to the north of these areas mostly due to the lengthening of the Meiyu-Changma-Baiu (Kusunoki et al., 2006). Simulations by RCMs support the results from AOGCMs. For example, Kurihara et al. (2005) show an increase in precipitation over western Japan in summer.

Kitoh and Uchiyama (2006) investigated the onset and withdrawal times of the Asian summer rainfall season in 15 MMD simulations (Figure 11.10). They find a delay in early summer rain withdrawal over the region extending from Taiwan to the Ryukyu Islands to the south of Japan, but an earlier withdrawal over the Yangtze Basin, although the latter is not significant due to large inter-model variation. Changes in onset dates are smaller.

Yasunaga et al. (2006) used a 5-km mesh cloud-resolving RCM to investigate summer rainfall in Japan. They find no changes in June rainfall but increased July rainfall in a warmer climate. The increase in July can be attributed to the more frequent large-precipitation systems.

Intense precipitation events are very likely to increase in East Asia, consistent with the historical trend in this region (Fujibé et al., 2005; Zhai et al., 2005). Kanada et al. (2005) show, using a 5-km resolution RCM, that the confluence of disturbances from the Chinese continent and from the East China Sea would

often cause extremely heavy precipitation over Japan's Kyushu Island in July in a warmer climate. An increase in the frequency and intensity of heavy precipitation events also occurs in Korea in the long RCM simulation of Boo et al. (2006). Similarly based on RCM simulations, Y.L. Xu et al. (2005) report more extreme precipitation events over China. Gao et al. (2002) find a simulated increase in the number of rainy days in northwest China, and a decrease in rain days but an increase in days with heavy rain over South China. Kitoh et al. (2005) report similar results in South China from an AOGCM simulation.

Kimoto et al. (2005) suggest that the frequencies of non-precipitating and heavy (≥ 30 mm day^{-1}) rainfall days would increase significantly at the expense of relatively weak (1–20 mm day^{-1}) rainfall days in Japan. Mizuta et al. (2005) find significantly more days with heavy precipitation and stronger average precipitation intensity in western Japan and Hokkaido Island. Hasegawa and Emori (2005) show that daily precipitation associated with tropical cyclones over the western North Pacific would increase.

The previously noted weakening of the East Asian winter monsoon (e.g., Hu et al., 2000) is further confirmed by recent studies (e.g., Kimoto, 2005; Hori and Ueda, 2005).

Southeast Asia

Area-mean precipitation over Southeast Asia increases in most MMD model simulations, with a median change of about 7% in all seasons (Table 11.1), but the projected seasonal changes vary strongly within the region. The seasonal confidence intervals based on the methods of Tebaldi et al. (2004a,b) are similar for DJF and JJA (roughly –4% to 17%). The strongest and most consistent increases broadly follow the ITCZ, lying over northern Indonesia and Indochina in JJA, and over southern Indonesia and Papua New Guinea in DJF (Figure 11.9). Away from the ITCZ, precipitation decreases are often simulated. The pattern is broadly one of wet season rainfall increase and dry season decrease.

Earlier studies of precipitation change in the area in some cases have suggested a worse inter-model agreement than found for the MMD models. Both Giorgi et al. (2001a) and Ruosteenoja et al. (2003) find inconsistency in the simulated direction of precipitation change in the region, but a relatively narrow range of possible changes; similar results were found over an Indonesian domain by Boer and Faqih (2004). Compositing the projections from a range of earlier simulations forced by the IS92a scenario, Hulme and Sheard (1999a,b) find a pattern of rainfall increase across northern Indonesia and the Philippines, and decrease over the southern Indonesian archipelago. More recently, Boer and Faqih (2004) compared patterns of change across Indonesia from five AOGCMs and obtained highly contrasting results. They conclude that 'no generalisation could be made on the impact of global warming on rainfall' in the region.

The regional high-resolution simulations of McGregor et al. (1998), McGregor and Dix (2001) and AIACC (2004) have demonstrated the potential for significant local variation in projected precipitation change. The simulations showed considerable regional detail in the simulated patterns of change, but little consistency across the three simulations. The authors related this result to significant deficiencies in the current-climate simulations of the models for this region.

Rainfall variability will be affected by changes in ENSO and its effect on monsoon variability, but this is not well understood (see Section 10.3). However, as Boer and Faqih (2004) note, those parts of Indonesia that experience a mean rainfall decrease are likely to also experience increases in drought risk. The region is also likely to share the general tendency for daily

extreme precipitation to become more intense under enhanced greenhouse conditions, particularly where the mean precipitation is projected to increase. This has been demonstrated in a range of global and regional studies (see Section 10.3), but needs explicit study for the Southeast Asian region.

The northern part of the Southeast Asian region will be affected by any change in tropical cyclone characteristics. As noted in Section 10.3, there is evidence in general of likely increases in tropical cyclone intensity, but less consistency about how occurrence will change (see also Walsh, 2004). The likely

Box 11.3: Climatic Change in Mountain Regions

Although mountains differ considerably from one region to another, one common feature is the complexity of their topography. Related characteristics include rapid and systematic changes in climatic parameters, in particular temperature and precipitation, over very short distances (Becker and Bugmann, 1997); greatly enhanced direct runoff and erosion; systematic variation of other climatic (e.g., radiation) and environmental (e.g., soil types) factors. In some mountain regions, it has been shown that temperature trends and anomalies have an elevation dependence (Giorgi et al., 1997), a feature that is not, however, systematically observed in all upland areas (e.g., Vuille and Bradley, 2000, for the Andes).

Few model simulations have attempted to directly address issues related specifically to future climatic change in mountain regions, primarily because the current spatial resolution of GCMs and even RCMs is generally too crude to adequately represent the topographic detail of most mountain regions and other climate-relevant features such as land cover that are important determinants in modulating climate in the mountains (Beniston et al., 2003). High-resolution RCM simulations (5-km and 1-km grid scales) are used for specific investigations of processes such as surface runoff, infiltration, evaporation and extreme events such as precipitation (Weisman et al., 1997; Walser and Schär, 2004; Kanada et al., 2005; Yasunaga et al., 2006) and damaging wind storms (Goyette et al., 2003), but these simulations are too costly to operate in a 'climate mode'. Because of the highly complex terrain, empirical and statistical downscaling techniques have often been seen as a very valuable tool to generate climate change information for mountainous regions (e.g., Benestad, 2005; Hanssen-Bauer et al., 2005).

Projections of changes in precipitation patterns in mountains are unreliable in most GCMs because the controls of topography on precipitation are not adequately represented. In addition, it is now recognised that the superimposed effects of natural modes of climatic variability such as ENSO or the NAO can perturb mean precipitation patterns on time scales ranging from seasons to decades (Beniston and Jungo, 2001). Even though there has been progress in reproducing some of these mechanisms in coupled ocean-atmosphere models (Osborn et al., 1999), deficiencies remain and prevent a good simulation of these large-scale modes of variability (see also Section 8.4). However, several studies indicate that the higher resolution of RCMs and GCMs can represent observed mesoscale patterns of the precipitation climate that are not resolved in coarse-resolution GCMs (Frei et al., 2003; Kanada et al., 2005; Schmidli et al., 2006; Yasunaga et al., 2006).

Snow and ice are, for many mountain ranges, a key component of the hydrological cycle, and the seasonal character and amount of runoff is closely linked to cryospheric processes. In temperate mountain regions, the snowpack is often close to its melting point, so that it may respond rapidly to minor changes in temperature. As warming increases in the future, regions where snowfall is the current norm will increasingly experience precipitation in the form of rain (e.g., Leung et al., 2004). For every degree celsius increase in temperature, the snow line will on average rise by about 150 m. Although the snow line is difficult to determine in the field, it is established that at lower elevations the snow line is very likely to rise by more than this simple average estimate (e.g., Martin et al., 1994; Vincent, 2002; Gerbaux et al., 2005; see also Section 4.2). Beniston et al. (2003) show that for a 4°C shift in mean winter temperatures in the European Alps, as projected by recent RCM simulations for climatic change in Europe under the A2 emissions scenario, snow duration is likely to be reduced by 50% at altitudes near 2,000 m and by 95% at levels below 1,000 m. Where some models predict an increase in winter precipitation, this increase does not compensate for the effect of changing temperature. Similar reductions in snow cover that will affect other mountain regions of the world will have a number of implications, in particular for early seasonal runoff (e.g., Beniston, 2003), and the triggering of the annual cycle of mountain vegetation (Cayan et al., 2001; Keller et al., 2005).

Because mountains are the source region for over 50% of the globe's rivers, the impacts of climatic change on mountain hydrology not only affect the mountains themselves but also populated lowland regions that depend on mountain water resources for domestic, agricultural, energy and industrial supply. Water resources for populated lowland regions are influenced by mountain climates and vegetation; shifts in intra-annual precipitation regimes could lead to critical water amounts resulting in greater flood or drought episodes (e.g., Barnett et al., 2005; Graham et al., 2007).

increase in intensity (precipitation and winds) is supported for the northwest Pacific (and other regions) by the recent modelling study of Knutson and Tuleya (2004). The high-resolution time-slice modelling experiment of Hasegawa and Emori (2005) also demonstrates an increase in tropical cyclone precipitation in the western North Pacific, but not an increase in tropical cyclone intensity. Wu and Wang (2004) examined possible changes in tracks in the northwest Pacific due to changes in steering flow in two Geophysical Fluid Dynamics Laboratory (GFDL) enhanced greenhouse gas experiments. Tracks moved more north-easterly, possibly reducing tropical cyclone frequency in the Southeast Asian region. Since most of the tropical cyclones form along the monsoon trough and are also influenced by ENSO, changes in the occurrence, intensity and characteristics of tropical cyclones and their interannual variability will be affected by changes in ENSO (see Section 10.3).

Central Asia and Tibet

Precipitation over central Asia increases in most MMD-A1B projections for DJF but decreases in the other seasons. The median change by the end of the 21st century is –3% in the annual mean, with +4% in DJF and –13% in JJA (the dry season) (Table 11.1). This seasonal variation in the changes is broadly consistent with the earlier multi-model study of Meleshko et al. (2004), although they find an increase in summer precipitation in the northern part of the area.

Over the Tibetan Plateau, all MMD-A1B models project increased precipitation in DJF (median 19%). Most but not all models also simulate increased precipitation in the other seasons (Table 11.1). Earlier studies both by AOGCMs and RCMs are consistent with these findings (Gao et al., 2003b; Y. Xu et al., 2003a,b, 2005).

11.5   North America

Assessment of projected climate change for North America:

All of North America is very likely to warm during this century, and the annual mean warming is likely to exceed the global mean warming in most areas. In northern regions, warming is likely to be largest in winter, and in the southwest USA largest in summer. The lowest winter temperatures are likely to increase more than the average winter temperature in northern North America, and the highest summer temperatures are likely to increase more than the average summer temperature in the southwest USA.

Annual mean precipitation is very likely to increase in Canada and the northeast USA, and likely to decrease in the southwest USA. In southern Canada, precipitation is likely to increase in winter and spring, but decrease in summer.

Snow season length and snow depth are very likely to decrease in most of North America, except in the northernmost part of Canada where maximum snow depth is likely to increase.

The uncertainties in regional climate changes over North America are strongly linked to the ability of AOGCMs to reproduce the dynamical features affecting the region (Chapter 10). Atmosphere-Ocean General Circulation Models exhibit large model-to-model differences in ENSO and NAO/Arctic Oscillation (AO) responses to climate changes. Changes in the Atlantic MOC are uncertain, and thus so is the magnitude of consequent reduced warming in the extreme north-eastern part of North America; cooling here cannot be totally excluded. The Hudson Bay and Canadian Archipelago are poorly resolved by AOGCMs, contributing to uncertainty in ocean circulation and sea ice changes and their influence on the climate of northern regions. Tropical cyclones are not resolved by the MMD models and inferred changes in the frequency, intensity and tracks of disturbances making landfall in southeast regions remain uncertain. At the coarse horizontal resolution of the MMD models, high-altitude terrain is poorly resolved, which likely results in an underestimation of warming associated with snow-albedo feedback at high elevations in western regions. Little is known about the dynamical consequences of the larger warming over land than over ocean, which may affect the northward displacement and intensification of the subtropical anticyclone off the West Coast. This could affect the subtropical North Pacific eastern boundary current, the offshore Ekman transport, the upwelling and its cooling effect on SST, the persistent marine stratus clouds and thus precipitation in the southwest USA.

The uncertainty associated with RCM projections of climate change over North America remains large despite the investments made in increasing horizontal resolution. All reported RCM projections were driven by earlier AOGCMs that exhibited larger biases than the MMD models. Coordinated ensemble RCM projections over North America are not yet available, making it difficult to compare results.

11.5.1   Key Processes

Central and northern regions of North America are under the influence of mid-latitude cyclones. Projections by AOGCMs (Chapter 10) generally indicate a slight poleward shift in storm tracks, an increase in the number of strong cyclones but a reduction in medium-strength cyclones over Canada and poleward of 70°N. Consequent with the projected warming, the atmospheric moisture transport and convergence is projected to increase, resulting in a widespread increase in annual precipitation over most of the continent except the south and south-western part of the USA and over Mexico.

The southwest region is very arid, under the general influence of a subtropical ridge of high pressure associated with the thermal contrast between land and adjacent ocean. The North American Monsoon System develops in early July (e.g., Higgins and Mo, 1997); the prevailing winds over the Gulf of California undergo a seasonal reversal, from northerly in winter to southerly in summer, bringing a pronounced increase in rainfall over the southwest USA and ending the late spring wet period in the Great Plains (e.g., Bordoni et al., 2004). The projection of smaller warming over the Pacific Ocean than over the continent, and amplification and northward displacement of the subtropical anticyclone, is likely to induce a decrease in annual precipitation in the south-western USA and northern Mexico.

The Great Plains Low-Level Jet (LLJ) is a dynamical feature that transports considerable moisture from the Gulf of Mexico into the central USA, playing a critical role in the summer precipitation there. Several factors, including the land-sea thermal contrast, contribute to the strength of the moisture convergence during the night and early morning, resulting in prominent nocturnal maximum precipitation in the plains of the USA (such as Nebraska and Iowa; e.g., Augustine and Caracena, 1994). The projections of climate changes indicate an increased land-sea thermal contrast in summer, with anticipated repercussions on the LLJ.

Interannual variability over North America is connected to two large-scale oscillation patterns (see Chapter 3), ENSO and the NAO/AO. The MMD model projections indicate an intensification of the polar vortex and many models project a decrease in the arctic surface pressure, which contributes to an increase in the AO/NAO index; the uncertainty is large, however, due to the diverse responses of AOGCMs in simulating the Aleutian Low (Chapter 10). The MMD model projections indicate a shift towards mean El-Niño like conditions, with the eastern Pacific warming more than the western Pacific; there is a wide range of behaviour among the current models, with no clear indication of possible changes in the amplitude or period of El Niño (Chapter 10).

11.5.2 Skill of Models in Simulating Present Climate

Individual AOGCMs in the MMD vary in their ability to reproduce the observed patterns of pressure, surface air temperature and precipitation over North America (Chapter 8). The ensemble mean of MMD models reproduces very well the annual-mean mean sea level pressure distribution (Section 8.4). The maximum error is of the order of ±2 hPa, with the simulated Aleutian Low pressure extending somewhat too far north, probably due to the inability of coarse-resolution models to adequately resolve the high topography of the Rocky Mountains that blocks incoming cyclones in the Gulf of Alaska. Conversely, the pressure trough over the Labrador Sea is not deep enough. The depth of the thermal low pressure over the southwest region in summer is somewhat excessive.

The MMD models simulate successfully the overall pattern of surface air temperature over North America, with reduced biases compared to those reported in the TAR. Ensemble-mean regional mean bias ranges from –4.5°C to 1.9°C for the 25th to 75th percentile range, and medians vary from –2.4°C to +0.4°C depending on region and season (Supplementary

Material Table S11.1). The ensemble mean of MMD models reproduces the overall distribution of annual mean precipitation (Supplementary Material Table S11.1), but almost all models overestimate precipitation for western and northern regions. The ensemble-mean regional mean precipitation bias medians vary from –16% to +93% depending on region and season. The ensemble-mean precipitation is excessive on the windward side of major mountain ranges, with the excess reaching 1 to 2 mm day^{-1} over high terrain in the west of the continent.

Regional Climate Models are quite successful in reproducing the overall climate of North America when driven by reanalyses. Over a 10° × 10° Southern Plains region, an ensemble of six RCMs in the North American Regional Climate Change Assessment Program (NARCCAP; Mearns et al., 2005) had 76% of all monthly temperature biases within ±2°C and 82% of all monthly precipitation biases within ±50%, based on preliminary results for a single year. RCM simulations over North America exhibit rather high sensitivity to parameters such as domain size (e.g., Juang and Hong, 2001; Pan et al., 2001; Vannitsem and Chomé, 2005) and the intensity of large-scale nudging (providing large-scale information to the interior of the model domain, see e.g., von Storch et al., 2000; Miguez-Macho et al., 2004) if used. In general, RCMs are more skilful at reproducing cold-season temperature and precipitation (e.g., Pan et al., 2001; Han and Roads, 2004; Plummer et al., 2006) because the warm-season climate is more controlled by mesoscale and convective-scale precipitation events, which are harder to simulate (Giorgi et al., 2001a; Leung et al., 2003; Liang et al., 2004; Jiao and Caya, 2006). On the other hand, Gutowski et al. (2004) find that spatial patterns of monthly precipitation for the USA, when viewed as a whole rather than broken into individual regions, are better simulated in summer than winter. Several studies point to the large sensitivity of RCMs to parametrization of moist convection, including the vertical transport of moisture from the boundary layer (Chaboureau et al., 2004; Jiao and Caya, 2006) and entrainment mixing between convective plumes and the local environment (Derbyshire et al., 2004). In a study of the simulation of the 1993 summer flood in the central USA by 13 RCMs, Anderson et al. (2003) find that all models produced a precipitation maximum that represented the flood, but most underestimated it to some degree, and 10 out of 13 of the models succeeded in reproducing the observed nocturnal maxima of precipitation. Leung et al. (2003) examined the 95th percentile of daily precipitation and find generally good agreement across many areas of the western USA.

A survey of recently published RCM current-climate simulations driven with AOGCMs data reveals that biases in surface air temperature and precipitation are two to three times larger than the simulations driven with reanalyses. The sensitivity of simulated surface air temperature to changing lateral boundary conditions from reanalyses to AOGCMs appears to be high in winter and low in summer (Han and Roads, 2004; Plummer et al., 2006). Most RCM simulations to date for North America have been made for time slices that are too short to properly sample natural variability. Some

RCMs have employed less than optimal formulations, such as outdated parametrizations (e.g., bucket land surface scheme), too few levels in the vertical (e.g., 14) or a too-low uppermost computational level (e.g., 100 hPa).

11.5.3 Climate Projections

11.5.3.1 Surface Air Temperature

The ensemble mean of the MMD models projects a generalised warming for the entire continent with the magnitude projected to increase almost linearly with time (Figure 11.11). On an annual-mean basis, projected surface air temperature warming varies from 2°C to 3°C along the western, southern and eastern continental edges (where at least 16 out of the 21 models project a warming in excess of 2°C) up to more than 5°C in the northern region (where 16 out of the 21 AOGCMs project a warming in excess of 4°C). This warming exceeds the spread among models by a factor of three to four over most of the continent. The warming in the USA is projected to exceed 2°C by nearly all the models, and to exceed 4°C by more than 5 AOGCMs out of 21. More regional and seasonal detail on ranges of projected warming is provided in Table 11.1 and Supplementary Table S11.2.

The largest warming is projected to occur in winter over northern parts of Alaska and Canada, reaching 10°C in the northernmost parts, due to the positive feedback from a reduced period of snow cover. The ensemble-mean northern warming varies from more than 7°C in winter (nearly all AOGCMs project a warming exceeding 4°C) to as little as 2°C in summer.

In summer, ensemble-mean projected warming ranges between 3°C and 5°C over most of the continent, with smaller values near the coasts. In western, central and eastern regions, the projected warming has less seasonal variation and is more modest, especially near the coast, consistent with less warming over the oceans. The warming could be larger in winter over elevated areas as a result of snow-albedo feedback, an effect that is poorly modelled by AOGCMs due to insufficient horizontal resolution (see also Box 11.3). In winter, the northern part of the eastern region is projected to warm most while coastal areas are projected to warm by only 2°C to 3°C.

The climate change response of RCMs is sometimes different from that of the driving AOGCM. This appears to be the result of a combination of factors, including the use of different parametrizations (convection and land surface processes are particularly important over North America in summer) and resolution (different resolution may lead to differing behaviour of the same parametrization). For example, Chen et al. (2003) find that two RCMs project larger temperature changes in summer than their driving AOGCM. In contrast, the projected warming of an RCM compared to its driving AOGCM was found to be 1.5°C less in the central USA (Pan et al., 2004; Liang et al., 2006), a region where observations have shown a cooling trend in recent decades. This resulted in an area of little warming that may have been due to a changing pattern of the LLJ frequency and associated moisture convergence. It is argued that the improved simulation of the LLJ in the RCM is made possible owing to its increased horizontal and vertical resolution. However, other RCMs with similar resolution do not produce the same response.

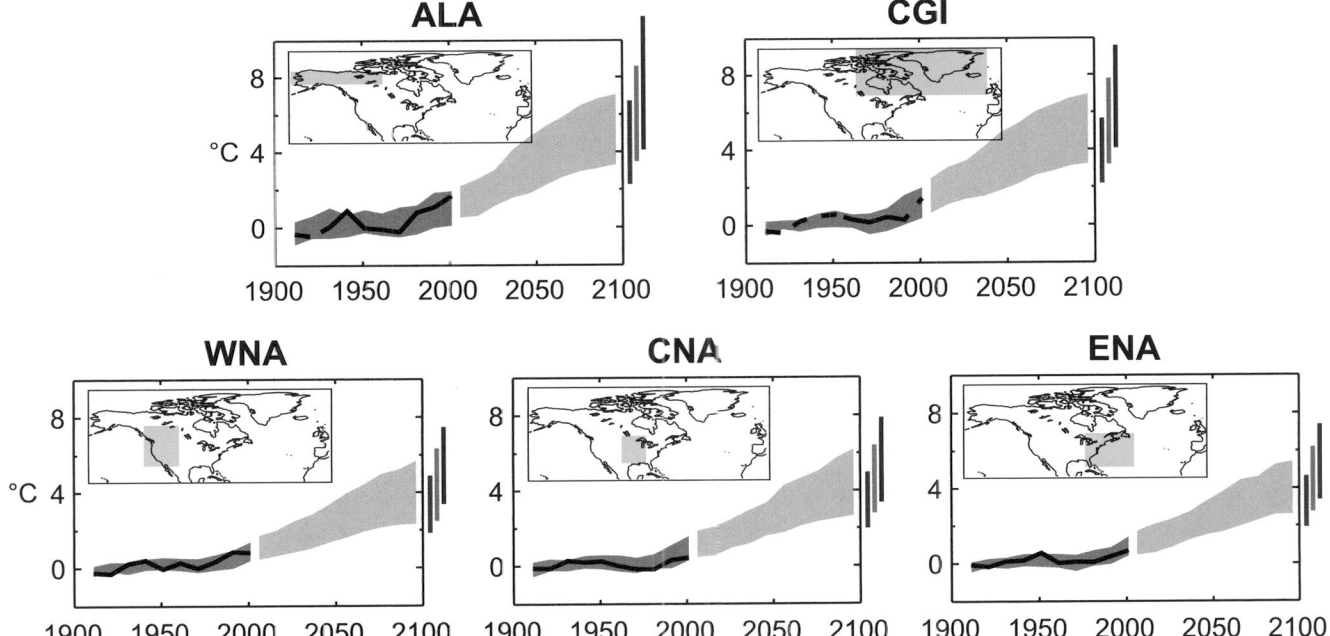

Figure 11.11. *Temperature anomalies with respect to 1901 to 1950 for five North American land regions for 1906 to 2005 (black line) and as simulated (red envelope) by MMD models incorporating known forcings; and as projected for 2001 to 2100 by MMD models for the A1B scenario (orange envelope). The bars at the end of the orange envelope represent the range of projected changes for 2091 to 2100 for the B1 scenario (blue), the A1B scenario (orange) and the A2 scenario (red). The black line is dashed where observations are present for less than 50% of the area in the decade concerned. More details on the construction of these figures are given in Box 11.1 and Section 11.1.2.*

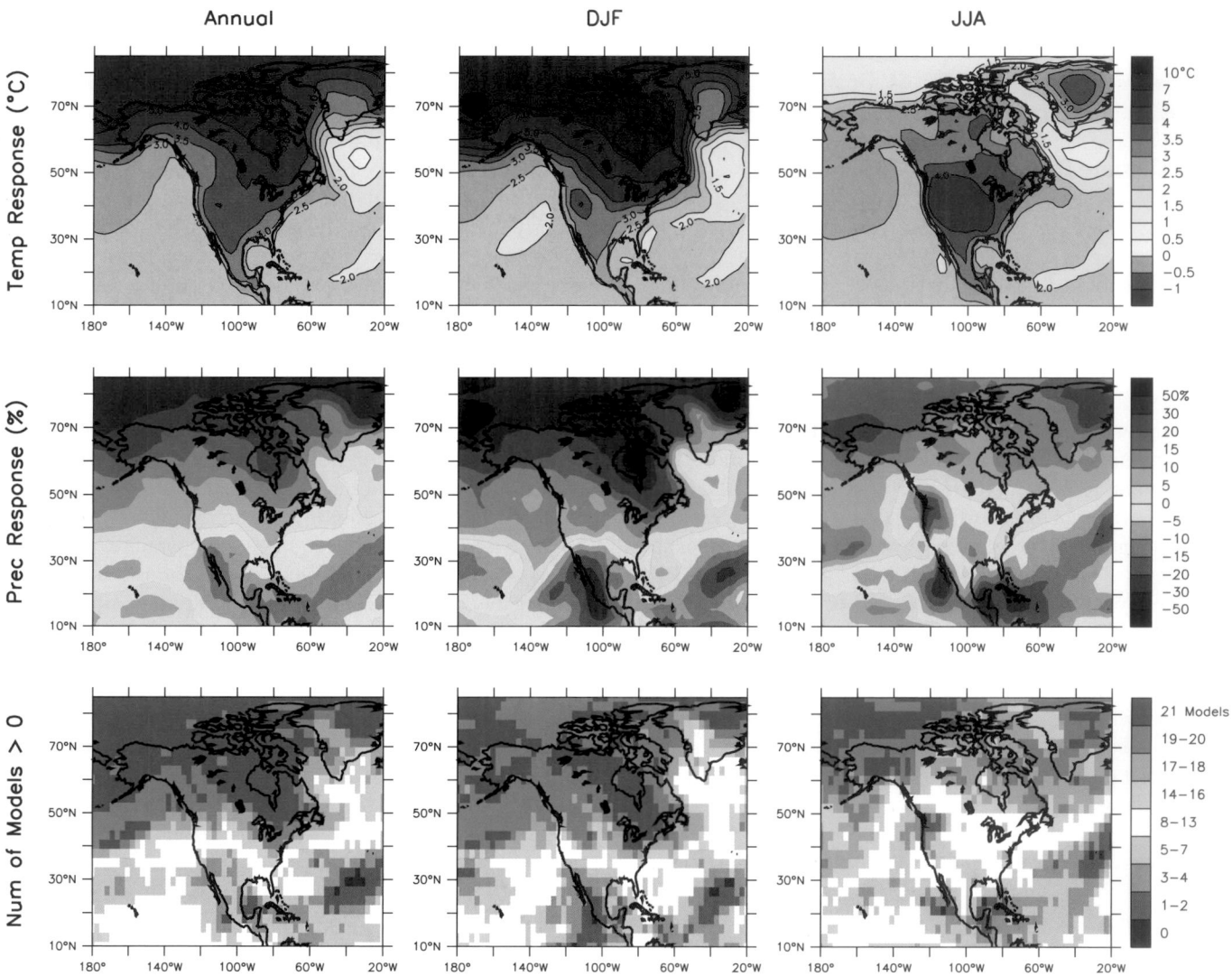

Figure 11.12. *Temperature and precipitation changes over North America from the MMD-A1B simulations. Top row: Annual mean, DJF and JJA temperature change between 1980 to 1999 and 2080 to 2099, averaged over 21 models. Middle row: same as top, but for fractional change in precipitation. Bottom row: number of models out of 21 that project increases in precipitation.*

11.5.3.2 Precipitation

As a consequence of the temperature dependence of the saturation vapour pressure in the atmosphere, the projected warming is expected to be accompanied by an increase in atmospheric moisture flux and its convergence/divergence intensity. This results in a general increase in precipitation over most of the continent except the most south-westerly part (Figure 11.12). The ensemble mean of MMD models projects an increase in annual mean precipitation in the north reaching +20%, which is twice the inter-model spread, so likely significant; the projected increase reaches as much as +30% in winter. Because the increased saturation vapour pressure can also yield greater evaporation, projected increases in annual precipitation are partially offset by increases in evaporation; regions in central North America may experience net surface drying as a result (see Supplementary Material Figure S11.1). See Table 11.1 and Supplementary Table S11.2 for more

regional and seasonal details, noting that regional averaging hides important north-south differences.

In keeping with the projected northward displacement of the westerlies and the intensification of the Aleutian Low (Section 11.5.3.3), northern region precipitation is projected to increase, by the largest amount in autumn and by the largest fraction in winter. Due to the increased precipitable water, the increase in precipitation amount is likely to be larger on the windward slopes of the mountains in the west with orographic precipitation. In western regions, modest changes in annual mean precipitation are projected, but the majority of AOGCMs indicate an increase in winter and a decrease in summer. Models show greater consensus on winter increases (ensemble mean maximum of 15%) to the north and on summer decreases (ensemble mean maximum of –20%) to the south. These decreases are consistent with enhanced subsidence and flow of drier air masses in the southwest USA and northern Mexico resulting from an amplification of the subtropical anticyclone off the West Coast due to the land-sea contrast in warming (e.g., Mote and Mantua, 2002). However, this reduction is close to the inter-model spread so it contains large uncertainty, an assessment that is reinforced by the fact that some AOGCMs project an increase in precipitation.

In central and eastern regions, projections from the MMD models show the same characteristics as in the west, with greater consensus for winter increases to the north and summer decreases to the south. The line of zero change is oriented more or less west-to-east and moves north from winter to summer. The line of zero change is also projected to lie further to the north under SRES scenarios with larger greenhouse gas amounts. However, uncertainty around the projected changes is large and the changes do not scale well across different SRES scenarios.

Govindasamy (2003) finds that, averaged over the USA, the few existing time-slice simulations with high-resolution AGCM results do not significantly differ from those obtained with AOGCMs. Available RCM simulations provide little extra information on average changes. Some RCMs project precipitation changes of different sign, either locally (Chen et al., 2003) or over the entire continental USA (Han and Roads, 2004, where in summer the AOGCM generally produced a small increase and the RCM a substantial decrease). In contrast, Plummer et al. (2006) find only small differences in precipitation responses using two sets of physical parametrizations in their RCM, despite the fact that one corrected significant summer precipitation excess present in the other.

11.5.3.3 Temperature and Precipitation Extremes

Several RCM studies focused particularly on changes in extreme temperature events. Bell et al. (2004) examine changes in temperature extremes in their simulations centred on California. They find increases in extreme temperature events, both as distribution percentiles and threshold events, prolonged hot spells and increased diurnal temperature range. Leung et al. (2004) examine changes in extremes in their RCM simulations of the western USA; in general, they find increases in diurnal temperature range in six sub-regions of their domain in summer. Diffenbaugh et al. (2005) find that the frequency and magnitude of extreme temperature events changes dramatically under SRES A2, with increases in extreme hot events and a decrease in extreme cold events.

In a study of precipitation extremes over California, Bell et al. (2004) find that changes in precipitation exceeding the 95th percentile followed changes in mean precipitation, with decreases in heavy precipitation in most areas. Leung et al. (2004) find that extremes in precipitation during the cold season increase in the northern Rockies, the Cascades, the Sierra Nevada and British Columbia by up to 10% for 2040 to 2060, although mean precipitation was mostly reduced, in accord with earlier studies (Giorgi et al., 2001a). In a large river basin in the Pacific Northwest, increases in rainfall over snowfall and rain-on-snow events increased extreme runoff by 11%, which would contribute to more severe flooding. In their 25-km RCM

simulations covering the entire USA, Diffenbaugh et al. (2005) find widespread increases in extreme precipitation events under SRES A2, which they determine to be significant.

11.5.3.4 Atmospheric Circulation

In general, the projected climate changes over North America follow the overall features of those over the NH (Chapter 10). The MMD models project a northward displacement and strengthening of the mid-latitude westerly flow, most pronounced in autumn and winter. Surface pressure is projected to decrease in the north, with a northward displacement of the Aleutian low-pressure centre and a north-westward displacement of the Labrador Sea trough, and to decrease slightly in the south. The reductions in surface pressure in the north are projected to be strongest in winter, reaching –1.5 to –3 hPa, in part as a result of the warming of the continental arctic air mass. On an annual basis, the pressure decrease in the north exceeds the spread among models by a factor of 3 on an annual-mean basis and a factor of 1.5 in summer, so it is significant. The East Pacific subtropical anticyclone is projected to intensify in summer, particularly off the coast of California and Baja California, resulting in an increased air mass subsidence and drier airflow over south-western North America. The pressure increase (less than 0.5 hPa) is small compared to the spread among models, so this projection is rather uncertain.

11.5.3.5 Snowpack, Snowmelt and River Flow

The ensemble mean of the MMD models projects a general decrease in snow depth (Chapter 10) as a result of delayed autumn snowfall and earlier spring snowmelt. In some regions where winter precipitation is projected to increase, the increased snowfall can more than make up for the shorter snow season and yield increased snow accumulation. Snow depth increases are projected by some GCMs over some land around the Arctic Ocean (Figure S10.1) and by some RCMs in the northernmost part of the Northwest Territories (Figure 11.13). In principle a similar situation could arise at lower latitudes at high elevations in the Rocky Mountains, although most models project a widespread decrease of snow depth there (Kim et al., 2002; Snyder et al., 2003; Leung et al., 2004; see also Box 11.3).

Much SD research activity has focused on resolving future water resources in the complex terrain of the western USA. Studies typically point to a decline in winter snowpack and hastening of the onset of snowmelt caused by regional warming (Hayhoe et al., 2004; Salathé, 2005). Comparable trends towards increased annual mean river flows and earlier spring peak flows have also been projected by two SD techniques for the Saguenay watershed in northern Québec, Canada (Dibike and Coulibaly, 2005). Such changes in the flow regime also favour increased risk of winter flooding and lower summer soil moisture and river flows. However, differences in snowpack behaviour derived from AOGCMs depend critically on the realism of downscaled winter temperature variability and its

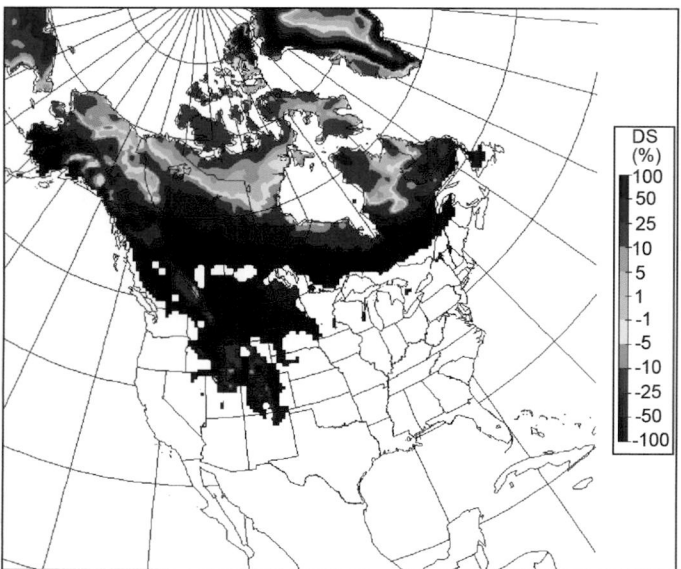

Figure 11.13. *Percent snow depth changes in March (only calculated where climatological snow amounts exceed 5 mm of water equivalent), as projected by the Canadian Regional Climate Model (CRCM; Plummer et al., 2006), driven by the Canadian General Circulation Model (CGCM), for 2041 to 2070 under SRES A2 compared to 1961 to 1990.*

interplay with precipitation and snowpack accumulation and melt (Salathé, 2005). Hayhoe et al. (2004) produced a standard set of statistically downscaled temperature and precipitation scenarios for California; under both the A1F1 and B1 scenarios, they find overall declines in snowpack.

11.6   Central and South America

Assessment of projected climate change for Central and South America:

All of Central and South America is very likely to warm during this century. The annual mean warming is likely to be similar to the global mean warming in southern South America but larger than the global mean warming in the rest of the area.

Annual precipitation is likely to decrease in most of Central America, with the relatively dry boreal spring becoming drier. Annual precipitation is likely to decrease in the southern Andes, with relative precipitation changes being largest in summer. A caveat at the local scale is that changes in atmospheric circulation may induce large local variability in precipitation changes in mountainous areas. Precipitation is likely to increase in Tierra del Fuego during winter and in south-eastern South America during summer.

It is uncertain how annual and seasonal mean rainfall will change over northern South America, including

the Amazon forest. In some regions, there is qualitative consistency among the simulations (rainfall increasing in Ecuador and northern Peru, and decreasing at the northern tip of the continent and in southern northeast Brazil).

The systematic errors in simulating current mean tropical climate and its variability (Section 8.6) and the large inter-model differences in future changes in El Niño amplitude (Section 10.3) preclude a conclusive assessment of the regional changes over large areas of Central and South America. Most MMD models are poor at reproducing the regional precipitation patterns in their control experiments and have a small signal-to-noise ratio, in particular over most of Amazonia (AMZ). The high and sharp Andes Mountains are unresolved in low-resolution models, affecting the assessment over much of the continent. As with all landmasses, the feedbacks from land use and land cover change are not well accommodated, and lend some degree of uncertainty. The potential for abrupt changes in biogeochemical systems in AMZ remains as a source of uncertainty (see Box 10.1). Large differences in the projected climate sensitivities in the climate models incorporating these processes and a lack of understanding of processes have been identified (Friedlingstein et al., 2003). Over Central America, tropical cyclones may become an additional source of uncertainty for regional scenarios of climate change, since the summer precipitation over this region may be affected by systematic changes in hurricane tracks and intensity.

11.6.1   Key Processes

Over much of Central and South America, changes in the intensity and location of tropical convection are the fundamental concern, but extratropical disturbances also play a role in Mexico's winter climate and throughout the year in southern South America. A continental barrier over Central America and along the Pacific coast in South America and the world's largest rainforest are unique geographical features that shape the climate in the area.

Climate over most of Mexico and Central America is characterised by a relatively dry winter and a well-defined rainy season from May through October (Magaña et al., 1999). The seasonal evolution of the rainy season is largely the result of air-sea interactions over the Americas' warm pools and the effects of topography over a dominant easterly flow, as well as the temporal evolution of the ITCZ. During the boreal winter, the atmospheric circulation over the Gulf of Mexico and the Caribbean Sea is dominated by the seasonal fluctuation of the Subtropical North Atlantic Anticyclone, with invasions of extratropical systems that affect mainly Mexico and the western portion of the Great Antilles.

A warm season precipitation maximum, associated with the South American Monsoon System (Vera et al., 2006), dominates the mean seasonal cycle of precipitation in tropical and subtropical latitudes over South America. Amazonia has had increasing rainfall over the last 40 years, despite deforestation, due to global-scale water vapour convergence (Chen et al., 2001; see also Section 3.3). The future of the rainforest is not only of vital ecological importance, but also central to the future evolution of the global carbon cycle, and as a driver of regional climate change. The monsoon system is strongly influenced by ENSO (e.g., Lau and Zhou, 2003), and thus future changes in ENSO will induce complementary changes in the region. Displacements of the South Atlantic Convergence Zone have important regional impacts such as the large positive precipitation trend over the recent decades centred over southern Brazil (Liebmann et al., 2004). There are well-defined teleconnection patterns (the Pacific-South American modes, Mo and Nogués-Paegle, 2001) whose preferential excitation could help shape regional changes. The Mediterranean climate of much of Chile makes it sensitive to drying as a consequence of poleward expansion of the South Pacific subtropical high, in close analogy to other regions downstream of oceanic subtropical highs in the Southern Hemisphere (SH). South-eastern South America would experience an increase in precipitation from the same poleward storm track displacement.

11.6.2 Skill of Models in Simulating Present Climate

In the Central America (CAM) and AMZ regions, most models in the MMD have a cold bias of 0°C to 3°C, except in AMZ in SON (Supplementary Material Table S11.1). In southern South America (SSA) average biases are close to zero. The biases are unevenly geographically distributed (Supplementary Material Figure S11.25). The MMD mean climate shows a warm bias around 30°S (particularly in summer) and in parts of central South America (especially in SON). Over the rest of South America (central and northern Andes, eastern Brazil, Patagonia) the biases tend to be predominantly negative. The SST biases along the western coasts of South America are likely related to weakness in oceanic upwelling.

For the CAM region, the multi-model scatter in precipitation is substantial, but half of the models lie in the range of –15 to 25% in the annual mean. The largest biases occur during the boreal winter and spring seasons, when precipitation is meagre (Supplementary Material Table S11.1). For both AMZ and SSA, the ensemble annual mean climate exhibits drier than observed conditions, with about 60% of the models having a negative bias. Unfortunately, this choice of regions for averaging is particularly misleading for South America since it does not clearly bring out critical regional biases such as those related to rainfall underestimation in the Amazon and La Plata Basins (Supplementary Material Figure S11.26). Simulation of the regional climate is seriously affected by model deficiencies at low latitudes. In particular, the MMD ensemble tends to depict a relatively weak ITCZ, which extends southward of its observed position. The simulations have a systematic bias towards underestimated rainfall over the Amazon Basin. The simulated subtropical climate is typically also adversely affected by a dry bias over most of south-eastern South America and in the South Atlantic Convergence Zone, especially during the rainy season.

In contrast, rainfall along the Andes and in northeast Brazil is excessive in the ensemble mean.

Some aspects of the simulation of tropical climate with ACGCMs have improved. However, in general, the largest errors are found where the annual cycle is weakest, such as over tropical South America (see, e.g., Section 8.3). Atmospheric GCMs approximate the spatial distribution of precipitation over the tropical Americas, but they do not correctly reproduce the temporal evolution of the annual cycle in precipitation, specifically the mid-summer drought (Magaña and Caetano, 2005). Tropical cyclones are important contributors to precipitation in the region. If close to the continent, they will produce large amounts of precipitation over land, and if far from the coast, moisture divergence over the continental region enhances drier conditions.

Zhou and Lau (2002) analyse the precipitation and circulation biases in a set of six AGCMs provided by the Climate Variability and Predictability Programme (CLIVAR) Asian-Australian Monsoon AGCM Intercomparison Project (Kang et al., 2002). This model ensemble captures some large-scale features of the South American monsoon system reasonably well, including the seasonal migration of monsoon rainfall and the rainfall associated with the South America Convergence Zone. However, the South Atlantic subtropical high and the Amazonia low are too strong, whereas low-level flow tends to be too strong during austral summer and too weak during austral winter. The model ensemble captures the Pacific-South American pattern quite well, but its amplitude is generally underestimated.

Regional models are still being tested and developed for this region. Relatively few studies using RCMs for Central and South America exist, and those that do are constrained by short simulation length. Some studies (Chou et al., 2000; Nobre et al., 2001; Druyan et al., 2002) examine the skill of experimental dynamic downscaling of seasonal predictions over Brazil. Results suggest that both more realistic GCM forcing and improvements in the RCMs are needed. Seth and Rojas (2003) performed seasonal integrations driven by reanalyses, with emphasis on tropical South America. The model was able to simulate the different rainfall anomalies and large-scale circulations but, as a result of weak low-level moisture transport from the Atlantic, rainfall over the western Amazon was underestimated. Vernekar et al. (2003) follow a similar approach to study the low-level jets and report that the RCM produces better regional circulation details than does the reanalysis. However, an ensemble of four RCMs did not provide a noticeable improvement in precipitation over the driving large-scale reanalyses (Roads et al., 2003).

Other studies (Misra et al., 2003; Rojas and Seth, 2003) analyse seasonal RCM simulations driven by AGCM simulations. Relative to the AGCMs, regional models generally improve the rainfall simulation and the tropospheric circulation over both tropical and subtropical South America. However, AGCM-driven RCMs degrade compared with the reanalyses-driven integrations and they could even exacerbate the dry bias over sectors of AMZ and perpetuate the erroneous ITCZ over the neighbouring ocean basins from the AGCMs. Menéndez et al. (2001) used a RCM driven by a stretched-grid AGCM with higher resolution over the southern mid-latitudes to simulate the winter climatology of SSA. They find that both the AGCM and the regional model have similar systematic errors but the biases are reduced in the RCM. Analogously, other RCM simulations for SSA give too little precipitation over the subtropical plains and too much over elevated terrain (e.g., Nicolini et al., 2002; Menéndez et al., 2004).

11.6.3 Climate Projections

11.6.3.1 Temperature

The warming as simulated by the MMD-A1B projections increases approximately linearly with time during this century, but the magnitude of the change and the inter-model range are greater over CAM and AMZ than over SSA (Figure 11.14). The annual mean warming under the A1B scenario between 1980 to 1999 and 2080 to 2099 varies in the CAM region from 1.8°C to 5.0°C, with half of the models within 2.6°C to 3.6°C and a median of 3.2°C. The corresponding numbers for AMZ are 1.8°C to 5.1°C, 2.6°C to 3.7°C and 3.3°C, and those for SSA 1.7°C to 3.9°C, 2.3°C to 3.1°C and 2.5°C (Table 11.1). The median warming is close to the global ensemble mean in SSA but about 30% above the global mean in the other two regions. As in the rest of the tropics, the signal-to-noise ratio is large for temperature, and it requires only 10 years for a 20-year mean temperature, growing at the rate of the median A1B response, to be clearly discernible above the models' internal variability.

The simulated warming is generally largest in the most continental regions, such as inner Amazonia and northern Mexico (Figure 11.15). Seasonal variation in the regional area mean warming is relatively modest, except in CAM where there is a difference of 1°C in median values between DJF and MAM (Table 11.1). The warming in central Amazonia tends to be larger in JJA than in DJF, while the reverse is true over the Altiplano where, in other words, the seasonal cycle of temperature is projected to increase (Figure 11.15). Similar results were found by Boulanger et al. (2006), who studied the regional thermal response over South America by applying a statistical method based on neural networks and Bayesian statistics to find optimal weights for a linear combination of MMD models.

For the variation of seasonal warming between the individual models, see Table 11.1. As an alternative approach to estimating uncertainty in the magnitude of the warming, the 5th and 95th percentiles for temperature change at the end of the 21st century, assessed using the method of Tebaldi et al. (2004a), are typically within ±1°C of the median value in all three of these regions (Supplementary Material Table S11.2).

11.6.3.2 Precipitation

The MMD models suggest a general decrease in precipitation over most of Central America, consistent with Neelin et al. (2006), where the median annual change by the end of the 21st

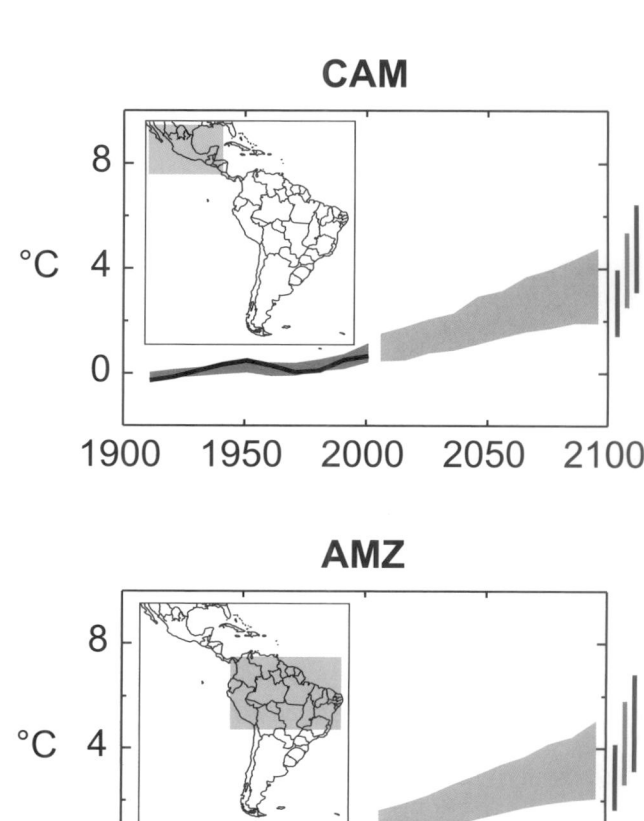

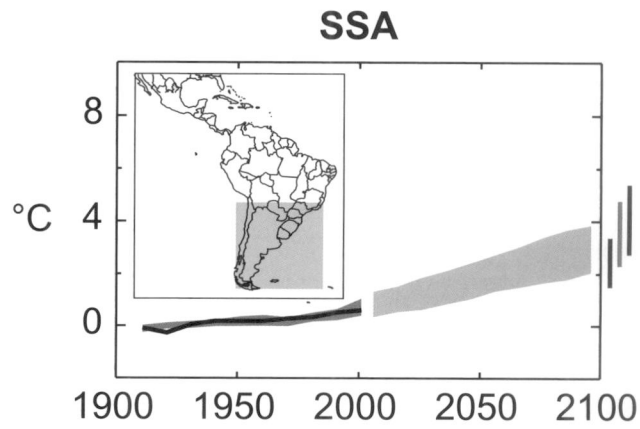

Figure 11.14. *Temperature anomalies with respect to 1901 to 1950 for three Central and South American land regions for 1906 to 2005 (black line) and as simulated (red envelope) by MMD models incorporating known forcings; and as projected for 2001 to 2100 by MMD models for the A1B scenario (orange envelope). The bars at the end of the orange envelope represent the range of projected changes for 2091 to 2100 for the B1 scenario (blue), the A1B scenario (orange) and the A2 scenario (red). The black line is dashed where observations are present for less than 50% of the area in the decade concerned. More details on the construction of these figures are given in Box 11.1 and Section 11.1.2.*

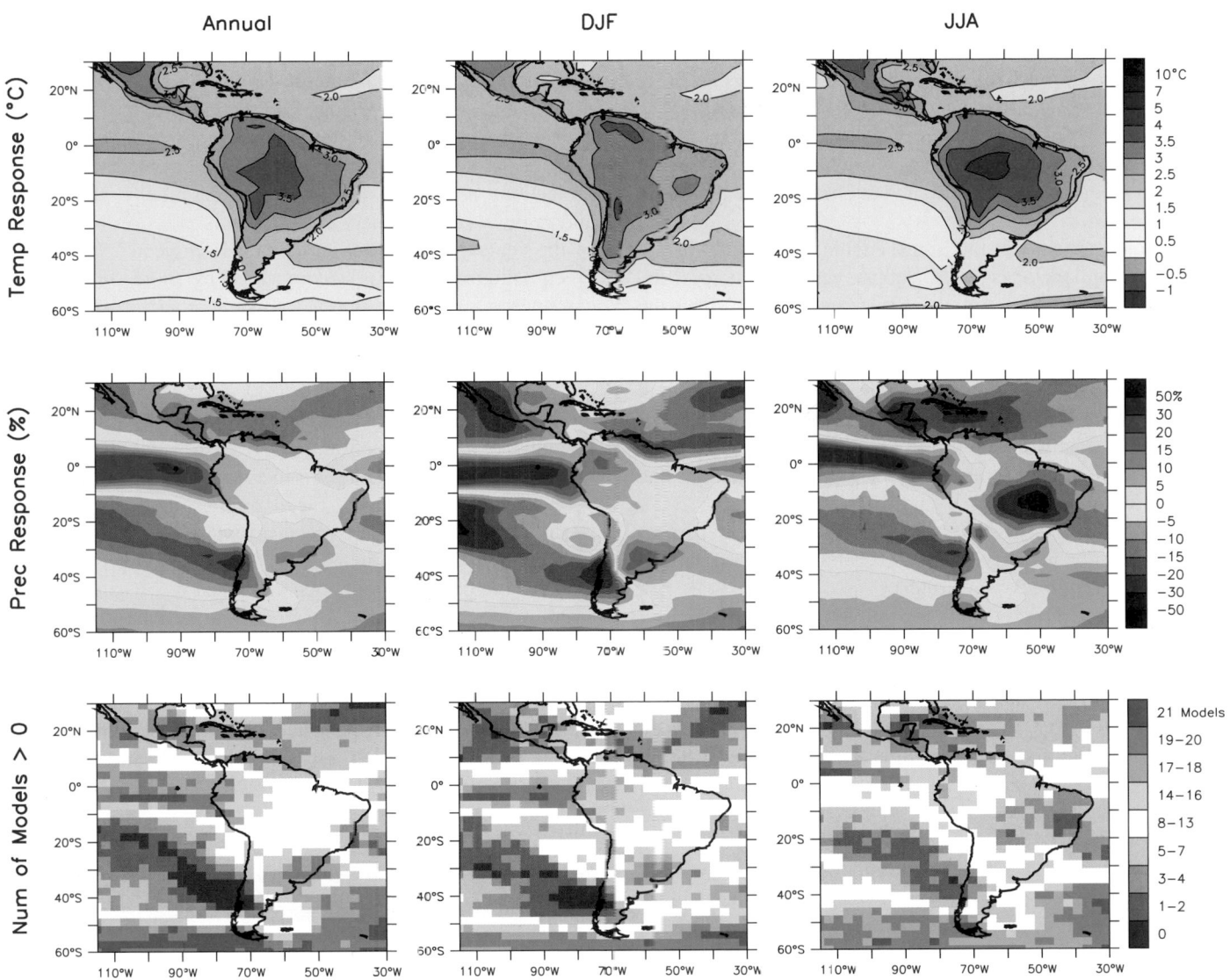

Figure 11.15. *Temperature and precipitation changes over Central and South America from the MMD-A1B simulations. Top row: Annual mean, DJF and JJA temperature change between 1980 to 1999 and 2080 to 2099, averaged over 21 models. Middle row: same as top, but for fractional change in precipitation. Bottom row: number of models out of 21 that project increases in precipitation.*

century is –9% under the A1B scenario, and half of the models project area mean changes from –16 to –5%, although the full range of the projections extends from –48 to 9%. Median changes in area mean precipitation in Amazonia and southern South America are small and the variation between the models is also more modest than in Central America, but the area means hide marked regional differences (Table 11.1, Figure 11.15).

Area mean precipitation in Central America decreases in most models in all seasons. It is only in some parts of north-eastern Mexico and over the eastern Pacific, where the ITCZ forms during JJA, that increases in summer precipitation are projected (Figure 11.15). However, since tropical storms can contribute a significant fraction of the rainfall in the hurricane season in this region, these conclusions might be modified by the possibility of increased rainfall in storms not well captured by these global models. In particular, if the number of storms

does not change, Knutson and Tuleya (2004) estimate nearly a 20% increase in average precipitation rate within 100 km of the storm centre at the time of atmospheric carbon dioxide (CO_2) doubling.

For South America, the multi-model mean precipitation response (Figure 11.15) indicates marked regional variations. The annual mean precipitation is projected to decrease over northern South America near the Caribbean coasts, as well as over large parts of northern Brazil, Chile and Patagonia, while it is projected to increase in Colombia, Ecuador and Peru, around the equator and in south-eastern South America. The seasonal cycle modulates this mean change, especially over the Amazon Basin where monsoon precipitation increases in DJF and decreases in JJA. In other regions (e.g., Pacific coasts of northern South America, a region centered over Uruguay, Patagonia) the sign of the response is preserved throughout the seasonal cycle.

As seen in the bottom panels of Figure 11.15, most models project a wetter climate near the Rio de la Plata and drier conditions along much of the southern Andes, especially in DJF. However, when estimating the likelihood of this response, the qualitative consensus within this set of models should be weighed against the fact that most models show considerable biases in regional precipitation patterns in their control simulations.

The poleward shift of the South Pacific and South Atlantic subtropical anticyclones is a robust response across the models. Parts of Chile and Patagonia are influenced by the polar boundary of the subtropical anticyclone in the South Pacific and experience particularly strong drying because of the combination of the poleward shift in circulation and increase in moisture divergence. The strength and position of the subtropical anticyclone in the South Atlantic is known to influence the climate of south-eastern South America and the South Atlantic Convergence Zone (Robertson et al., 2003; Liebmann et al., 2004). The increase in rainfall in south-eastern South America is related to a corresponding poleward shift in the Atlantic storm track (Yin, 2005).

Some projected changes in precipitation (such as the drying over east-central Amazonia and northeast Brazil and the wetter conditions over south-eastern South America) could be a partial consequence of the El-Niño like response projected by the models (Section 10.3). The accompanying shift and alterations in the Walker Circulation would directly affect tropical South America (Cazes Boezio et al., 2003) and affect southern South America through extratropical teleconnections (Mo and Nogués-Paegle, 2001).

Although feedbacks from carbon cycle and dynamic vegetation are not included in MMD models, a number of coupled carbon cycle-climate projections have been performed since the TAR (see Sections 7.2 and 10.4.1). The initial carbon-climate simulations suggest that drying of the Amazon potentially contributes to acceleration of the rate of anthropogenic global warming by increasing atmospheric CO_2 (Cox et al., 2000; Friedlingstein et al., 2001; Dufresne et al., 2002; Jones et al., 2003). These models display large uncertainty in climate projections and differ in the timing and sharpness of the changes (Friedlingstein et al., 2003). Changes in CO_2 are related to precipitation changes in regions such as the northern Amazon (Zeng et al., 2004). In a version of the HadCM3 model with dynamic vegetation and an interactive global carbon cycle (Betts et al., 2004), a tendency to a more El Niño-like state contributes to reduced rainfall and vegetation dieback in the Amazon (Cox et al., 2004). But the version of HadCM3 participating in the MMD projects by far the largest reduction in annual rainfall over AMZ (–21% for the A1B scenario). This stresses the necessity of being very cautious in interpreting carbon cycle impacts on the regional climate and ecosystem change until there is more convergence among models on rainfall projections for the Amazon with fixed vegetation. Box 11.4 summarises some of the major issues related to regional land use/land changes in the context of climate change.

11.6.4 Extremes

Little research is available on extremes of temperature and precipitation for this region. Table 11.1 provides estimates on how frequently the seasonal temperature and precipitation extremes as simulated in 1980 to 1999 are exceeded in model projections using the A1B scenario. Essentially all seasons and regions are extremely warm by this criterion by the end of the century. In Central America, the projected time mean precipitation decrease is accompanied by more frequent dry extremes in all seasons. In AMZ, models project extremely wet seasons in about 27% (18%) of all DJF (MAM) seasons in the period 2080 to 2099. Significant changes are not projected in the frequency of extremely wet or dry seasons over SSA.

On the daily time scale, Hegerl et al. (2004) analyse an ensemble of simulations from two AOGCMs and find that both models simulate a temperature increase in the warmest night of the year larger than the mean response over the Amazon Basin but smaller than the mean response over parts of SSA. Concerning extreme precipitation, both models project more intense wet days per year over large parts of south-eastern South America and central Amazonia and weaker precipitation extremes over the coasts of northeast Brazil. Intensification of the rainfall amounts are consistent, given the agreement between the MMD model simulations over parts of south-eastern South America and most of AMZ but with longer periods between rainfall events, except in north-western South America, where the models project that it will rain more frequently (Meehl et al., 2005; Tebaldi et al., 2006).

11.7 Australia – New Zealand

Assessment of projected climate change for Australia and New Zealand:

All of Australia and New Zealand are very likely to warm during this century, with amplitude somewhat larger than that of the surrounding oceans, but comparable overall to the global mean warming. The warming is smaller in the south, especially in winter, with the warming in the South Island of New Zealand likely to remain smaller than the global mean. Increased frequency of extreme high daily temperatures in Australia and New Zealand, and decrease in the frequency of cold extremes is very likely.

Precipitation is likely to decrease in southern Australia in winter and spring. Precipitation is very likely to decrease in south-western Australia in winter. Precipitation is likely to increase in the west of the South Island of New Zealand. Changes in rainfall in northern and central Australia are uncertain. Extremes of daily precipitation are very likely to increase. The effect may be offset or reversed in areas of significant decrease in

Box 11.4: Land Use and Land Cover Change Experiments Related to Climate Change

Land use and land cover change significantly affect climate at the regional and local scales (e.g., Hansen et al., 1998; Bonan, 2001; Kabat et al., 2002; Foley et al., 2005). Recent modelling studies also show that in some instances these effects can extend beyond the areas where the land cover changes occur, through climate teleconnection processes (e.g., Gaertner et al., 2001; Pielke et al., 2002; Marland et al., 2003). Changes in vegetation result in alteration of surface properties, such as albedo and roughness length, and alter the efficiency of ecosystem exchange of water, energy and CO_2 with the atmosphere (for more details see Section 7.2). The effects differ widely based on the type and location of the altered ecosystem. The effects of land use and land cover change on climate can also be divided into biogeochemical and biophysical effects (Brovkin et al., 1999; see Sections 7.2 and 2.5 for discussion of these effects).

The net effect of human land cover activities increases the concentration of greenhouse gases in the atmosphere, thus increasing warming (see Sections 7.2 and 10.4 for further discussion); it has been suggested that these land cover emissions have been underestimated in the future climate projections used in the SRES scenarios (Sitch, 2005). Climate models assessed in this report incorporate various aspects of the effects of land cover change including representation of the biogeochemical flux, inclusion of dynamic land use where natural vegetation shifts as climate changes, and explicit human land cover forcing. In all cases, these efforts should be considered at early stages of development (see Chapters 2 and 7, and Table 10.1 for more details on many of these aspects).

One important impact of land cover conversion, generally not simulated in GCMs, is urbanisation. Although small in aerial extent, conversion to urban land cover creates urban heat islands associated with considerable warming (Arnfield, 2003). Since much of the world population lives in urban environments (and this proportion may increase, thus expanding urban areas), many people will be exposed to climates that combine expanded urban heat island effects and increased temperature from greenhouse gas forcing (see Box 7.2 for more details on urban land use effects).

One major shift in land use, relevant historically and in the future, is conversion of forest to agriculture and agriculture back to forest. Most areas well suited to large-scale agriculture have already been converted to this land use/cover type. Yet land cover conversion to agriculture may continue in the future, especially in parts of western North America, tropical areas of south and central America and arable regions in Africa and south and central Asia (IPCC, 2001; RIVM, 2002). In the future, mid-latitude agricultural areal expansion (especially into forested areas) could possibly result in cooling that would offset a portion of the expected warming due to greenhouse gas effects alone. In contrast, reforestation may occur in eastern North America and the eastern portion of Europe. In these areas, climate effects may include local warming associated with reforestation due to decreased albedo values (Feddema et al., 2005).

Tropical land cover change results in a very different climate response compared to mid-latitude areas. Changes in plant cover and the reduced ability of the vegetation to transpire water to the atmosphere lead to temperatures that are warmer by as much as 2°C in regions of deforestation (Costa and Foley, 2000; Gedney and Valdes, 2000; De Fries et al., 2002). The decrease in transpiration acts to reduce precipitation, but this effect may be modified by changes in atmospheric moisture convergence. Most model simulations of Amazonian deforestation suggest reduced moisture convergence, which would amplify the decrease in precipitation (e.g., McGuffie et al., 1995; Costa and Foley, 2000; Avissar and Worth, 2005). However, increased precipitation and moisture convergence in Amazonia during the last few decades contrast with this expectation, suggesting that deforestation has not been the dominant driver of the observed changes (see Section 11.6).

Tropical regions also have the potential to affect climates beyond their immediate areal extent (Chase et al., 2000; Delire et al., 2001; Voldoire and Royer, 2004; Avissar and Werth, 2005; Feddema et al., 2005; Snyder, 2006). For example, changes in convection patterns can affect the Hadley Circulation and thus propagate climate perturbations into the mid-latitudes. In addition, tropical deforestation in the Amazon has been found to affect SSTs in nearby ocean locations, further amplifying teleconnections (Avissar and Werth, 2005; Feddema et al., 2005; Neelin and Su, 2005; Voldoire and Royer, 2005). However, studies also indicate that there are significantly different responses to similar land use changes in other tropical regions and that responses are typically linked to dry season conditions (Voldoire and Royer, 2004a; Feddema et al., 2005). However, tropical land cover change in Africa and southeast Asia appears to have weaker local impacts largely due to influences of the Asian and African monsoon circulation systems (Mabuchi et al., 2005a,b; Voldoire and Royer, 2005).

Several land cover change studies have explicitly assessed the potential impacts (limited to biophysical effects) associated with specific future SRES land cover change scenarios, and the interaction between land cover change and greenhouse gas forcings (De Fries et al., 2002; Maynard and Royer, 2004a; Feddema et al., 2005; Sitch et al., 2005; Voldoire, 2006). In the A2 scenario, large-scale Amazon deforestation could double the expected warming in the region (De Fries et al., 2002; Feddema et al., 2005). Lesser local impacts are expected in tropical Africa and south Asia, in part because of the difference in regional circulation patterns (Delire et al., 2001; Maynard and Royer, 2004a,b; Feddema et al., 2005; Mabuchi et al., 2005a,b). In mid-latitude regions, land-cover induced cooling could offset some of the greenhouse-gas induced warming. Feddema et al. (2005) suggest that in the B1 scenario (where reforestation occurs in many areas and there are other low-impact tropical land cover changes) there are few local tropical climate or teleconnection effects. However, in this scenario, mid-latitude reforestation could lead to additional local warming compared to greenhouse-gas forcing scenarios alone.   *(continued)*

These simulations suggest that the effects of future land cover change will be a complex interaction of local land cover change impacts combined with teleconnection effects due to land cover change elsewhere, in particular the Amazon, and areas surrounding the Indian Ocean. However, projecting the potential outcomes of future climate effects due to land cover change is difficult for two reasons. First, there is considerable uncertainty regarding how land cover will change in the future. In this context, the past may not be a good indicator of the types of land transformation that may occur in the future. For example, if land cover change becomes a part of climate change mitigation (e.g., carbon trading) then a number of additional factors that include carbon sequestration in soils and additional land cover change processes will need to be incorporated in scenario development schemes. Second, current land process models cannot simulate all the potential impacts of human land cover transformation. Such processes as adequate simulation of urban systems, agricultural systems, ecosystem disturbance regimes (e.g., fire) and soil impacts are not yet well represented.

mean rainfall (southern Australian in winter and spring). An increase in potential evaporation is likely. Increased risk of drought in southern areas of Australia is likely.

Increased mean wind speed across the Southern Island of New Zealand, particularly in winter, is likely.

Significant factors contribute to uncertainty in projected climate change for the region. The El Niño-Southern Oscillation significantly influences rainfall, drought and tropical cyclone behaviour in the region and it is uncertain how ENSO will change in the future. Monsoon rainfall simulations and projections vary substantially from model to model, thus we have little confidence in model precipitation projections for northern Australia. More broadly, across the continent summer rainfall projections vary substantially from model to model, reducing confidence in their reliability. In addition, no detailed assessment of MMD model performance over Australia or New Zealand is available, which hinders efforts to establish the reliability of projections from these models. Finally, downscaling of MMD model projections are not yet available for New Zealand but are much needed because of the strong topographical control of New Zealand rainfall.

11.7.1 Key Processes

Key climate processes affecting the Australian region include the Australian monsoon (the SH counterpart of the Asian monsoon), the Southeast trade wind circulation, the subtropical high-pressure belt and the mid-latitude westerly wind circulation with its embedded disturbances. The latter two systems also predominate over New Zealand. Climatic variability in Australia and New Zealand is also strongly affected by the ENSO system (McBride and Nicholls, 1983; Mullan, 1995) modulated by the Inter-decadal Pacific Oscillation (IPO; Power et al., 1999; Salinger et al., 2001). Tropical cyclones occur in the region, and are a major source of extreme rainfall and wind events in northern coastal Australia, and, more rarely, on the North Island of New Zealand (Sinclair, 2002). Rainfall patterns in New Zealand are also strongly influenced by the interaction of the predominantly westerly circulation with its very mountainous topography.

Apart from the general increase in temperature that the region will share with most other parts of the globe, details of anthropogenic climate change in the Australia-New Zealand

region will depend on the response of the Australian monsoon, tropical cyclones, the strength and latitude of the mid-latitude westerlies, and ENSO.

11.7.2 Skill of Models in Simulating Present Climate

There are relatively few studies of the quality of the MMD global model simulations in the Australia-New Zealand area. The ensemble mean of the MMD model simulations has a systematic low-pressure bias near 50°S at all longitudes in the SH, including the Australia-New Zealand sector, corresponding to an equatorward displacement of the mid-latitude westerlies (see Chapter 8). On average, mid-latitude storm track eddies are displaced equatorward (Yin, 2005) and deep winter troughs over southwest Western Australia are over-represented (Hope 2006a,b). How this bias might affect climate change simulations is unclear. It can be hypothesised that by spreading the effects of mid-latitude depressions too far inland, the consequences of a poleward displacement of the westerlies and the storm track might be exaggerated, but the studies needed to test this hypothesis are not yet available.

The simulated surface temperatures in the surrounding oceans are typically warmer than observed, but at most by 1°C in the composite. Despite this slight warm bias, the ensemble mean temperatures are biased cold over land, especially in winter in the southeast and southwest of the Australian continent, where the cold bias is larger than 2°C. At large scales, the precipitation also has some systematic biases (see Supplementary Material Table S11.1). Averaged across northern Australia, the median model error is 20% more precipitation than observed, but the range of biases in individual models is large (–71 to +131%). This is discouraging with regard to confidence in many of the individual models. Consistent with this, Moise et al. (2005) identify simulation of Australian monsoon rainfall as a major deficiency of many of the AOGCM simulations included in Phase 2 of the Coupled Model Intercomparison Project (CMIP2). The median annual bias in the southern Australian region is –6%, and the range of biases –59 to +36%. In most models, the northwest is too wet and the northeast and east coast too dry, and the central arid zone is insufficiently arid.

The Australasian simulations in the AOGCMs utilised in the TAR have recently been scrutinised more closely, in part as a component of a series of national and state-based climate change projection studies (e.g., Whetton et al., 2001; Cai et al., 2003b; McInnes et al., 2003; Hennessy et al., 2004a,b; McInnes et al., 2004). Some high-resolution regional simulations were also considered in this process. The general conclusion is that large-scale features of Australian climate were quite well simulated. In winter, temperature patterns were more poorly simulated in the south where topographic variations have a stronger influence, although this was alleviated in the higher-resolution simulations. A set of the TAR AOGCM simulations was also assessed for the New Zealand region by Mullan et al. (2001a) with similar conclusions. The models were able to represent ENSO-related variability in the Pacific and the temperature and rainfall teleconnection patterns at the Pacific-wide scale, but there was considerable variation in model performance at finer scales (such as over the New Zealand region).

Decadal-scale variability patterns in the Australian region as simulated by the CSIRO AOGCM were considered by Walland et al. (2000) and found 'broadly consistent' with the observational studies of Power et al. (1998). At smaller scales, Suppiah et al. (2004) directly assessed rainfall-producing processes by comparing the simulated correlation between rainfall anomalies and pressure anomalies in Victoria against observations. They find that this link was simulated well by most models in winter and autumn, but less well in spring and summer. As a result of this, they warn that the spring and summer projected rainfall changes should be viewed as less reliable.

Pitman and McAvaney (2004) examine the sensitivity of GCM simulations of Australian climate to methods of representation of the surface energy balance. They find that the quality of the simulation of variability is strongly affected by the land surface model, but that simulation of climate means, and the changes in those means in global warming simulations, is less sensitive to the scheme employed.

Statistical downscaling methods have been employed in the Australian region and have demonstrated good performance at representing means, variability and extremes of station temperature and rainfall (Timbal and McAvaney, 2001; Charles et al., 2004; Timbal, 2004) based on broad-scale observational or climate model predictor fields. The method of Charles et al. (2004) is able to represent spatial coherence at the daily time scale in station rainfall, thus enhancing its relevance to hydrological applications.

11.7.3 Climate Projections

In addition to the MMD models, numerous studies have been conducted with earlier models. Recent regional average projections are provided in Giorgi et al. (2001b) and Ruosteenoja et al. (2003). The most recent national climate change projections of CSIRO (2001) were based on the results of eight AOGCMs and one higher-resolution regional simulation. The methodology (and simulations) used in these projections is described in Whetton et al. (2005) and follows closely that described for earlier projections in Whetton et al (1996). More detailed projections for individual states and other regions have also been prepared in recent years (Whetton

et al., 2001; Cai et al., 2003b; McInnes et al., 2003, 2004; Hennessy et al., 2004a,b; IOCI, 2005). This work has focused on temperature and precipitation, with additional variables such as potential evaporation and winds being included in the more recent assessments.

A range of dynamically downscaled projections have been undertaken for Australia using the DARLAM regional model (Whetton et al., 2001) and the CCAM stretched grid model (McGregor and Dix, 2001) at resolutions of 60 km across Australia and down to 14 km for Tasmania (McGregor, 2004). These projections use forcing from recent CSIRO AOGCM projections. Downscaled projected climate change using statistical methods has also been recently undertaken for parts of Australia (e.g., Timbal and McAvaney, 2001; Charles et al., 2004; Timbal, 2004) and New Zealand (Mullan et al., 2001a; Ministry for the Environment, 2004).

11.7.3.1 Mean Temperature

In both the southern and northern Australia regions, the projected MMD-A1B warming in the 21st century represents a significant acceleration of warming over that observed in the 20th century (Figure 11.16). The warming is larger than over the surrounding oceans, but only comparable to, or slightly larger than the global mean warming. Averaging over the region south of 30°S (SAU), the median 2100 warming among all of the models is 2.6°C (with an inter-quartile range of 2.4°C to 2.9°C) whereas the median warming averaged over the region north of 30°S (NAU) is 3.0°C (range of 2.8°C to 3.5°C). The seasonal cycle in the warming is weak, but with larger values (and larger spread among model projections) in summer (DJF). Across the MMD models, the warming is well correlated with the global mean warming, with a correlation coefficient of 0.79, so that more than half of the variance among models is controlled by global rather than local factors, as in many other regions. The range of responses is comparable but slightly smaller than the range in global mean temperature responses, and warming over equivalent time periods under the B1, A1B, and A2 scenarios is close to the ratios of the global mean responses The warming varies sub-regionally, with less warming in coastal regions, Tasmania and the South Island of New Zealand, and greater warming in central and northwest Australia (see Figure 10.8).

These results are broadly (and in many details) similar to those described in earlier studies, so other aspects of these earlier studies can be assumed to remain relevant. For the CSIRO (2001) projections, pattern-scaling methods were used to provide patterns of change rescaled by the range of global warming given by IPCC (2001) for 2030 and 2070 based on the SRES scenarios. By 2030, the warming is 0.4°C to 2°C over most of Australia, with slightly less warming in some coastal areas and Tasmania, and slightly more warming in the northwest. By 2070, annual average temperatures increase by 1°C to 6°C over most of Australia with spatial variations similar to those for 2030. Dynamically downscaled mean temperature change typically does not differ very significantly from the picture based on AOGCMs (e.g., see Whetton et al., 2002).

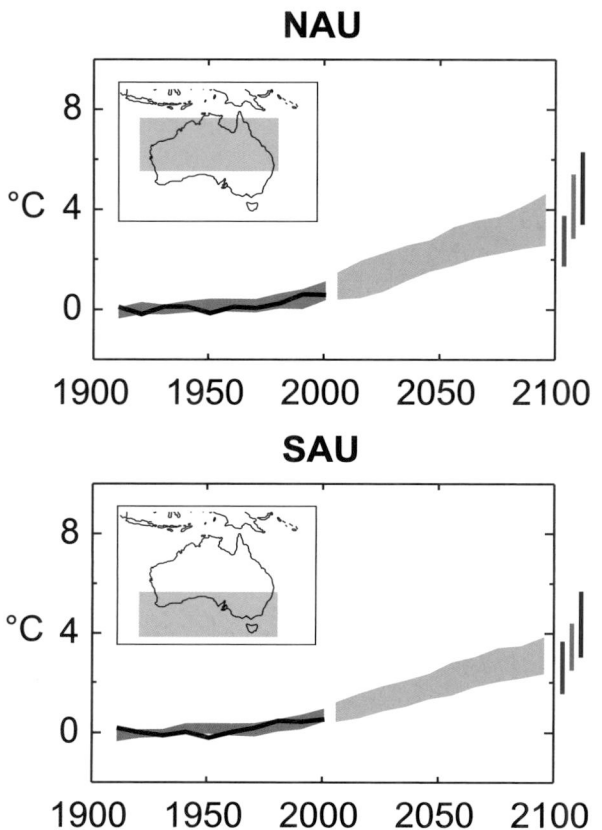

Figure 11.16. *Temperature anomalies with respect to 1901 to 1950 for two Australian land regions for 1906 to 2005 (black line) and as simulated (red envelope) by MMD models incorporating known forcings; and as projected for 2001 to 2100 by MMD models for the A1B scenario (orange envelope). The bars at the end of the orange envelope represent the range of projected changes for 2091 to 2100 for the B1 scenario (blue), the A1B scenario (orange) and the A2 scenario (red). More details on the construction of these figures are given in Box 11.1 and Section 11.1.2.*

Projected warming over New Zealand (allowing for the IPCC (2001) range of global warming and differences in the regional results of six GCMs used for downscaling) is 0.2°C to 1.3°C by the 2030s and 0.5°C to 3.5°C by the 2080s (Ministry for the Environment, 2004).

11.7.3.2   Mean Precipitation

A summary of projected precipitation changes from the MMD models is presented in Figure 11.17 and Table 11.1. The most robust feature is the reduction in rainfall along the south coast in JJA (not including Tasmania) and in the annual mean, and a decrease is also strongly evident in SON. The percentage JJA change in 2100 under the A1B scenario for southern Australia has an inter-quartile range of –26 to –7% and by comparison the same range using the probabilistic method of Tebaldi et al. (2004b) is –13 to –6% (Supplemental Material Table S11.2). There are large reductions to the south of the continent in all seasons, due to the poleward movement of the westerlies and embedded depressions (Cai et al., 2003a; Yin, 2005; Chapter 10), but this reduction extends over land during

winter when the storm track is placed furthest equatorward. Due to poleward drift of the storm track as it crosses Australian longitudes, the strongest effect is in the southwest, where the ensemble mean drying is in the 15 to 20% range. Hope (2006a,b) shows a southward or longitudinal shift in storms away from south-western Australia in the MMD simulations. To the east of Australia and over New Zealand, the primary storm track is more equatorward, and the north/south drying/moistening pattern associated with the poleward displacement is shifted equatorward as well. The result is a robust projection of increased rainfall on the South Island (especially its southern half), possibly accompanied by a decrease in the north part of the North Island. The South Island increase is likely to be modulated by the strong topography (see Box 11.3) and to appear mainly upwind of the main mountain range.

Other aspects of simulated precipitation change appear less robust. On the east coast of Australia, there is a tendency in the models for an increase in rain in the summer and a decrease in winter, with a slight annual decrease. However, consistency among the models on these features is weak.

These results are broadly consistent with results based on earlier GCM simulations. In the CSIRO (2001) projections based on a range of nine simulations, projected ranges of annual average rainfall change tend towards a decrease in the southwest and south but show more mixed results elsewhere (Whetton et al., 2005). Seasonal results showed that rainfall tended to decrease in southern and eastern Australia in winter and spring, increase inland in autumn and increase along the east coast in summer. Moise et al. (2005) also find a tendency for winter rainfall decreases across southern Australia and a slight tendency for rainfall increases in eastern Australia in 18 CMIP2 simulations under a 1% yr^{-1} atmospheric CO_2 increase.

Whetton et al. (2001) demonstrate that inclusion of high-resolution topography could reverse the simulated direction of rainfall change in parts of Victoria (see Box 11.3). In a region of strong rainfall decrease as simulated directly by the GCMs, two different downscaling methods (Charles et al., 2004; Timbal, 2004) have been applied to obtain the characteristics of rainfall change at stations (IOCI, 2002, 2005; Timbal, 2004). The downscaled results continued to show the simulated decrease, although the magnitude of the changes was moderated relative to the GCM in the Timbal (2004) study. Downscaled rainfall projections for New Zealand (incorporating differing results from some six GCMs) showed a strong variation across the islands (Ministry for the Environment, 2004). The picture that emerges is that the pattern of precipitation changes described above in the global simulations is still present, but with the precipitation changes focused on the upwind sides of the islands, with the increase in rainfall in the south concentrated in the west, and the decrease in the north concentrated in the east.

11.7.3.3   Snow Cover

The likelihood that precipitation will fall as snow will decrease as temperature rises. Hennessy et al. (2003) modelled snowfall and snow cover in the Australian Alps under the

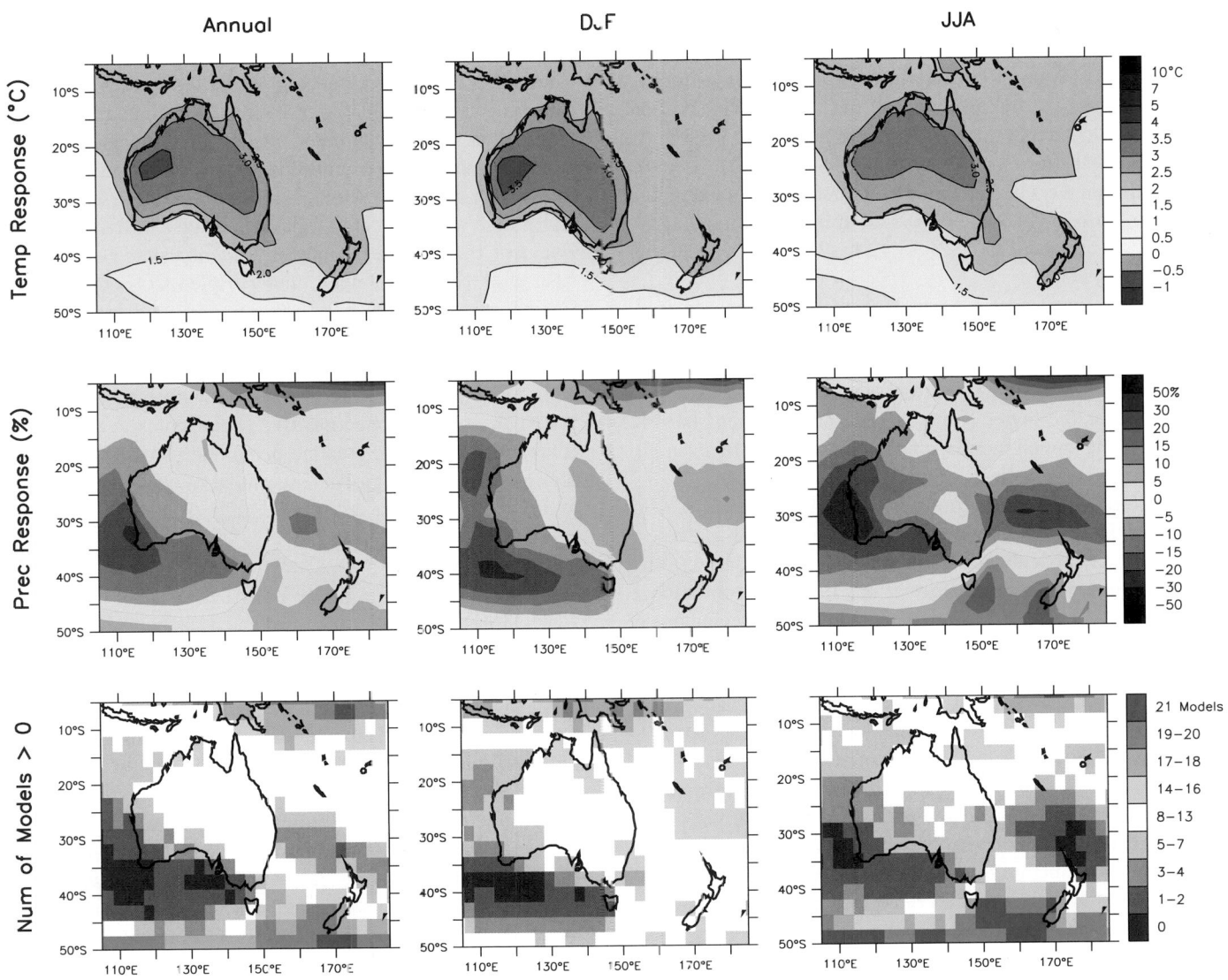

Figure 11.17. *Temperature and precipitation changes over Australia and New Zealand from the MMD-A1B simulations. Top row: Annual mean, DJF and JJA temperature change between 1980 to 1999 and 2080 to 2099, averaged over 21 models. Middle row: same as top, but for fractional change in precipitation. Bottom row: number of models out of 21 that project increases in precipitation.*

CSIRO (2001) projected temperature and precipitation changes and obtained very marked reductions in snow. The total alpine area with at least 30 days of snow cover decreases 14 to 54% by 2020, and 30 to 93% by 2050. Because of projected increased winter precipitation over the southern Alps, it is less clear that mountain snow will be reduced in New Zealand (Ministry for the Environment, 2004; see also Box 11.3). However, marked decreases in average snow water over New Zealand (60% by 2040 under the A1B scenario) have been simulated by Ghan and Shippert (2006) using a high-resolution sub-grid scale orography in a global model that simulates little change in precipitation.

11.7.3.4   Potential Evaporation

Using the method of Walsh et al. (1999), changes in potential evaporation in the Australian region have been calculated for

a range of enhanced greenhouse climate model simulations (Whetton et al., 2002; Cai et al., 2003b; McInnes et al., 2003, 2004; Hennessy et al., 2004a,b). In all cases, increases in potential evaporation were simulated, and in almost all cases, the moisture balance deficit became larger. This has provided a strong indication of the Australian environment becoming drier under enhanced greenhouse conditions.

11.7.3.5   Temperature and Precipitation Extremes

Where the analysis has been done for Australia (e.g., Whetton et al., 2002), the effect on changes in extreme temperature due to simulated changes in variability is small relative to the effect of the change in the mean. Therefore, most regional assessments of changes in extreme temperatures have been based on adding a projected mean temperature change to each day of a station-observed data set. Based on the CSIRO (2001) projected mean temperature change scenarios, the average number of days over 35°C each summer in Melbourne would increase from 8 at present to 9 to 12 by 2030 and 10 to 20 by 2070 (CSIRO, 2001). In Perth, such hot days would rise from 15 at present to 16 to 22 by 2030 and 18 to 39 by 2070 (CSIRO, 2001). On the other hand, cold days become much less frequent. For example, Canberra's current 44 winter days of minimum temperature below 0°C is projected to be 30 to 42 by 2030 and 6 to 38 by 2070 (CSIRO, 2001).

Changes in extremes in New Zealand have been assessed using a similar methodology and simulations (Mullan et al., 2001b). Decreases in the annual frequency of days below 0°C of 5 to 30 days by 2100 are projected for New Zealand, particularly for the lower North Island and the South Island. Increases in the annual number of days above 25°C of 10 to 50 days by 2100 are projected.

A range of GCM and regional modelling studies in recent years have identified a tendency for daily rainfall extremes to increase under enhanced greenhouse conditions in the Australian region (e.g., Hennessy et al., 1997; Whetton et al., 2002; McInnes et al., 2003; Watterson and Dix, 2003; Hennessy et al., 2004b; Suppiah et al., 2004; Kharin and Zwiers, 2005). Commonly, return periods of extreme rainfall events halve in late 21st-century simulations. This tendency can apply even when average rainfall is simulated to decrease, but not necessarily when this decrease is marked (see Timbal, 2004). Recently, Abbs (2004) dynamically downscaled to a resolution of 7 km current and enhanced greenhouse cases of extreme daily rainfall occurrence in northern New South Wales and southern Queensland as simulated by the CSIRO GCM. The downscaled extreme events for a range of return periods compared well with observations and the enhanced greenhouse simulations for 2040 showed increases of around 30% in magnitude, with the 1-in-40 year event becoming the 1-in-15 year event. Less work has been done on projected changes in rainfall extremes in New Zealand, although the recent analysis of Ministry for the Environment (2004) based on Semenov and Bengtsson (2002) indicates the potential for extreme winter rainfall (95th percentile) to change by between –6% and +40%.

Where GCMs simulate a decrease in average rainfall, it may be expected that there would be an increase in the frequency of dry extremes (droughts). Whetton and Suppiah (2003) examine simulated monthly frequencies of serious rainfall deficiency for Victoria, which show strong average rainfall decreases in most simulations considered. There is a marked increase in the frequency of rainfall deficiencies in most simulations, with doubling in some cases by 2050. Using a slightly different approach, likely increases in the frequency of drought have also been established for the states of South Australia, New South Wales and Queensland (Walsh et al., 2002; McInnes et al., 2003; Hennessy et al., 2004c). Mullan et al. (2005) show that by the 2080s in New Zealand there may be significant increases in drought frequency in the east of both islands.

11.7.3.6 Tropical Cyclones

A number of recent regional model-based studies of changes in tropical cyclone behaviour in the Australian region have examined aspects of number, tracks and intensities under enhanced greenhouse conditions (e.g., Walsh and Katzfey, 2000; Walsh and Ryan, 2000; Walsh et al., 2004). There is no clear picture with respect to regional changes in frequency and movement, but increases in intensity are indicated. For example, Walsh et al. (2004) obtained, under tripled CO_2 conditions, a 56% increase in storms with a maximum wind speed greater than 30 m s^{-1}. It should also be noted that ENSO fluctuations have a strong impact on patterns of tropical cyclone occurrence in the region, and therefore uncertainty with respect to future ENSO behaviour (see Section 10.3) contributes to uncertainty with respect to tropical cyclone behaviour (Walsh, 2004). See Section 10.3.6.3 for a global assessment of changes in tropical cyclone characteristics.

11.7.3.7 Winds

The MMD ensemble mean projected change in winter sea level pressure is shown in Figure 10.9. Much of Australia lies to the north of the centre of the high-pressure anomaly. With the mean latitude of maximum pressure near 30°S at this season, this corresponds to a modest strengthening of the mean wind over inland and northern areas and a slight weakening of the mean westerlies on the southern coast, consistent with Hennessy et al. (2004b). Studies of daily extreme winds in the region using high-resolution model output (McInnes et al., 2003) indicate increases of up to 10% across much of the northern half of Australia and the adjacent oceans during summer by 2030. In winter, the pressure gradient is projected to increase over the South Island of New Zealand (see Figure 10.9), implying increased windiness. This increase is present in all of the MMD-A1B projections.

11.8   Polar Regions

Assessment of projected climate change for the polar regions:

The Arctic is very likely to warm during this century in most areas, and the annual mean warming is very likely to exceed the global mean warming. Warming is projected to be largest in winter and smallest in summer.

Annual arctic precipitation is very likely to increase. It is very likely that the relative precipitation increase will be largest in the winter and smallest in summer.

Arctic sea ice is very likely to decrease in extent and thickness. It is uncertain how the Arctic Ocean circulation will change.

It is likely that the Antarctic will be warmer and that precipitation will increase over the continent.

It is uncertain to what extent the frequency of extreme temperature and precipitation events will change in the polar regions.

Polar climate involves large natural variability on interannual, decadal and longer time scales, which is an important source of uncertainty. The projections of the trends in the underlying teleconnections, such as the Northern Annular Mode (NAM) or ENSO, contain substantial uncertainty (see Chapter 10). Further, understanding of the polar climate system is still incomplete due to its complex atmosphere-land-cryosphere-ocean-ecosystem interactions involving a variety of distinctive feedbacks. Processes that are not particularly well represented in the models are clouds, planetary boundary layer processes and sea ice. Additionally, the resolution of global models is still not adequate to resolve important processes in the polar seas. All this contributes to a rather large range of present-day and future simulations, which may reduce confidence in the future projections. A serious problem is the lack of observations against which to assess models, and for developing process knowledge, particularly over Antarctica.

11.8.1 Arctic

11.8.1.1 Key Processes

Arctic climate is characterised by a distinctive complexity due to numerous nonlinear interactions between and within the atmosphere, cryosphere, ocean, land and ecosystems. Sea ice plays a crucial role in the arctic climate, particularly through its albedo. Reduction of ice extent leads to warming due to increased absorption of solar radiation at the surface. Substantial low-frequency variability is evident in various atmosphere and ice parameters (Polyakov et al., 2003a,b), complicating the detection and attribution of arctic changes. Natural multi-decadal variability has been suggested as partly responsible for the large warming in the 1920s to 1940s (Bengtsson et al., 2004; Johannessen et al., 2004) followed by cooling until the 1960s. In both models and observations, the interannual variability of monthly temperatures is at a maximum at high latitudes (Räisänen, 2002). Natural atmospheric patterns of variability on annual and decadal time scales play an important role in the arctic climate. Such patterns include the NAM, the NAO, the Pacific-North American (PNA) pattern and the Pacific Decadal Oscillation (PDO), which are associated with prominent arctic regional precipitation and temperature anomalies (see Box 3.4 and Section 3.6). For instance, the positive NAM/NAO phase is associated with warmer, wetter winters in Siberia and colder, drier winters in western Greenland and north-eastern Canada. The NAM/NAO showed a trend towards its positive phase over the last three to four decades, although it returned to near its long-term mean state in the last five years (see Section 3.6). In the future, global models project a positive trend in the NAO/NAM during the 21st century (see Section 10.3). There was

substantial decadal-to-inter-decadal atmospheric variability in the North Pacific over the 20th century, associated with fluctuations in the strength of the winter Aleutian Low that co-vary with North Pacific SST in the PDO (see Section 3.6). A deeper and eastward-shifted Aleutian Low advects warmer and moister air into Alaska. While some studies have suggested that the Brooks Range effectively isolates arctic Alaska from much of the variability associated with North Pacific teleconnection patterns (e.g., L'Heureux et al., 2004), other studies find relationships between the Alaskan and Beaufort-Chukchi region's climate and North Pacific variability (Stone, 1997; Curtis et al., 1998; Lynch et al., 2004). Patterns of variability in the North Pacific, and their implications for climate change, are especially difficult to sort out due to the presence of several patterns (NAM, PDO, PNA) with potentially different underlying mechanisms (see Chapter 3).

11.8.1.2 Skill of Models in Simulating Present Climate

Many processes are still poorly understood and thus continue to pose a challenge for climate models (ACIA, 2005). In addition, evaluating simulations of the Arctic is difficult because of the uncertainty in the observations. The few available observations are sparsely distributed in space and time and different data sets often differ considerably (Serreze and Hurst, 2000; ACIA, 2005; Liu et al., 2005; Wyser and Jones, 2005). This holds especially for precipitation measurements, which are problematic in cold environments (Goodison et al., 1998; Bogdanova et al., 2002).

Few pan-arctic atmospheric RCMs are in use. When driven by analysed lateral and sea ice boundary conditions, RCMs tend to show smaller temperature and precipitation biases in the Arctic compared to GCMs, indicating that sea ice simulation biases and biases originating from lower latitudes contribute substantially to the contamination of GCM results in the Arctic (e.g., Dethloff et al., 2001; Wei et al., 2002; Lynch et al., 2003; Semmler et al., 2005). However, even under a very constrained experimental design, there can be considerable across-model scatter in RCM simulations (Tjernström et al., 2005; Rinke et al., 2006). The construction of coupled atmosphere-ice-ocean RCMs for the Arctic is a recent development (Maslanik et al., 2000; Debernard et al., 2003; Rinke et al., 2003; Mikolajewicz et al., 2005).

Temperature

The simulated spatial patterns of the MMD ensemble mean temperatures agree closely with those of the observations throughout the annual cycle. Generally, the simulations are 1°C to 2°C colder than the European Centre for Medium-Range Weather Forecasts 40-year (ERA40) reanalyses with the exception of a cold bias maximum of 6°C to 8°C in the Barents Sea (particularly in winter/spring) caused by overestimated sea ice in this region (Chapman and Walsh, 2007; see also Section 8.3). Compared with earlier model versions, the annual temperature simulations improved in the Barents and Norwegian Seas and Sea of Okhotsk, but some deterioration is noted in the central Arctic Ocean and the high terrain areas of Alaska and northwest Canada (Chapman and Walsh, 2007). The mean model ensemble bias is relatively small compared to the across-model scatter of temperatures. The annual mean root-mean-squared error in the individual MMD models ranges from 2°C to 7°C (Chapman and Walsh, 2007). Compared with previous models, the MMD-simulated temperatures are more consistent across the models in winter, but somewhat less so in summer. There is considerable agreement between the modelled and observed interannual variability both in magnitude and spatial pattern.

Precipitation

The AOGCM-simulated monthly precipitation varies substantially among the models throughout the year but the MMD ensemble mean monthly means are within the range of different observational data sets. This is an improvement compared to earlier simulations (Walsh et al., 2002; ACIA, 2005), particularly from autumn to spring (Kattsov et al., 2007). The ensemble mean bias varies with season and remains greatest in spring and smallest in summer. The annual bias pattern (positive over most parts of the Arctic) can be partly attributed to coarse orography and to biased atmospheric storm tracks and sea ice cover (see Chapter 8). The MMD models capture the observed increase in the annual precipitation through the 20th century (see Section 3.3).

Sea Ice and Ocean

Arctic sea ice biases in present-day MMD simulations are discussed in Section 8.3. Arctic ocean-sea ice RCMs under realistic atmospheric forcing are increasingly capable of reproducing the known features of the Arctic Ocean circulation and observed sea ice drift patterns. The inflow of the two branches of Atlantic origin via the Fram Strait and the Barents Sea and their subsequent passage at mid-depths in several cyclonic circulation cells are present in most recent simulations (Karcher et al., 2003; Maslowski et al., 2004; Steiner et al., 2004). Most of the models are biased towards overly salty values in the Beaufort Gyre and thus too little freshwater storage in the arctic halocline. Several potential causes have been identified, among them a biased simulation of arctic shelf processes and wind forcing. Most hindcast simulations with these RCMs show a reduction in the arctic ice volume over recent decades (Holloway and Sou, 2002).

11.8.1.3   Climate Projections

Temperature

A northern high-latitude maximum in the warming ('polar amplification') is consistently found in all AOGCM simulations (see Section 10.3). The simulated annual mean arctic warming exceeds the global mean warming by roughly a factor of two in the MMD models, while the winter warming in the central arctic is a factor of four larger than the global annual mean when averaged over the models. These magnitudes are comparable to those obtained in previous studies (Holland and Bitz, 2003;

ACIA, 2005). The consistency between observations and the ensemble mean 20th-century simulations (Figure 11.18), combined with the fact that the near-future projections (2010–2029) continue the late 20th-century trends in temperature, ice extent and thickness with little modification (Serreze and Francis, 2006), increases confidence in this basic polar-amplified warming pattern, despite the inter-model differences in the amount of polar amplification.

At the end of the 21st century, the projected annual warming in the Arctic is 5°C, estimated by the MMD-A1B ensemble mean projection (Section 11.8.2.3, Figure 11.21). There is a considerable across-model range of 2.8°C to 7.8°C (Table 11.1). Larger (smaller) mean warming is found for the A2 (B1) scenario of 5.9°C (3.4°C), with a proportional across-model range. The across-model and across-scenario variability in the projected temperatures are both considerable and of comparable amplitude (Chapman and Walsh, 2007).

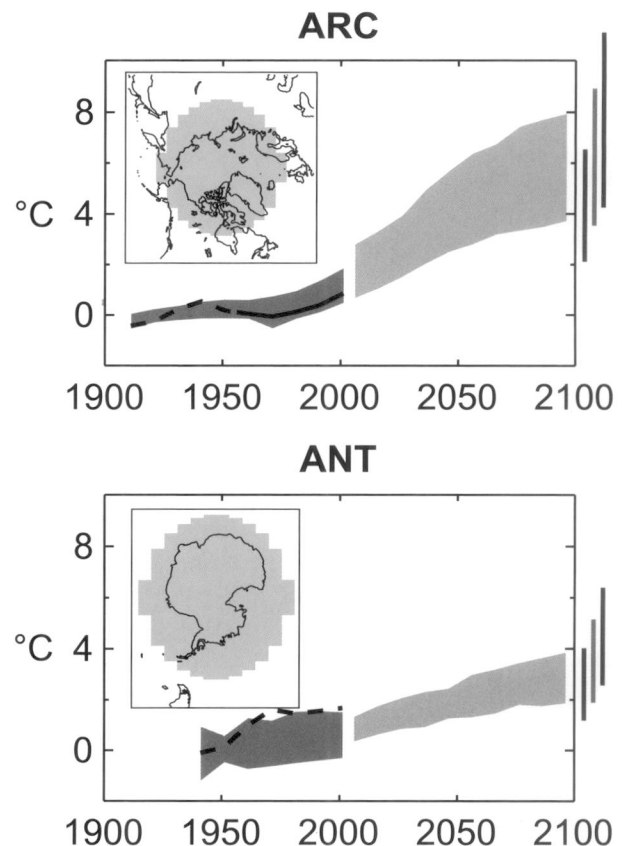

Figure 11.18. *Top panels: Temperature anomalies with respect to 1901 to 1950 for the whole Arctic for 1906 to 2005 (black line) as simulated (red envelope) by MMD models incorporating known forcings; and as projected for 2001 to 2100 by MMD models for the A1B scenario (orange envelope). The bars at the end of the orange envelope represent the range of projected changes for 2091 to 2100 for the B1 scenario (blue), the A1B scenario (orange) and the A2 scenario (red). The black line is dashed where observations are present for less than 50% of the area in the decade concerned. Bottom panels: The same for Antarctic land, but with observations for 1936 to 2005 and anomalies calculated with respect to 1951 to 2000. More details on the construction of these figures are given in Box 11.1 and Section 11.1.2.*

Over both ocean and land, the largest (smallest) warming is projected in winter (summer) (Table 11.1, Figure 11.19). But the seasonal amplitude of the temperature change is much larger over ocean than over land due the presence of melting sea ice in summer keeping the temperatures close to the freezing point. The surface air temperature over the Arctic Ocean region is generally warmed more than over arctic land areas (except in summer). The range between the individual simulated changes remains large (Figure 11.19, Table 11.1). By the end of the century, the mean warming ranges from 4.3°C to 11.4°C in winter, and from 1.2°C to 5.3°C in summer under the A1B scenario. The corresponding 5 to 95% confidence intervals are given in Supplementary Material Table S11.2. In addition to the overall differences in global warming, difficulties in simulating sea ice, partly related to biases in the surface wind fields, as well as deficiencies in cloud schemes, are likely responsible for much of the inter-model scatter. Internal variability plays a secondary role when examining these late-21st century responses.

The annual mean temperature response pattern at the end of the 21st century under the A1B scenario (Supplementary Material Figures S11.27 and S11.11) is characterised by a robust and large warming over the central Arctic Ocean (5°C to 7°C), dominated by the warming in winter/autumn associated with the reduced sea ice. The maximum warming is projected over the Barents Sea, although this could result from an overestimated albedo feedback caused by removal of the present-day simulations' excessive sea ice cover. A region of reduced warming (<2°C, even slight cooling in several models) is projected over the northern North Atlantic, which is consistent among the models. This is due to weakening of the MOC (see Section 10.3).

While the natural variability in arctic temperatures is large compared to other regions, the signals are still large enough

to emerge quickly from the noise (Table 11.1). Looking more locally, as described by Chapman and Walsh (2007), Alaska is perhaps the land region with the smallest signal-to-noise ratio, and is the only arctic region in which the 20-year mean 2010 to 2029 temperature is not clearly discernible from the 1981 to 2000 mean in the MMD models. But even here the signal is clear by mid-century in all three scenarios.

The regional temperature responses are modified by changes in circulation patterns (Chapter 10). In winter, shifts in NAO phase can induce inter-decadal temperature variations of up to 5°C in the eastern Arctic (Dorn et al., 2003). The MMD models project winter circulation changes consistent with an increasingly positive NAM/NAO (see Section 10.3), which acts to enhance the warming in Eurasia and western North America. In summer, circulation changes are projected to favour warm anomalies north of Scandinavia and extending into the eastern Arctic, with cold anomalies over much of Alaska (Cassano et al., 2006). However, deficiencies in the arctic summer synoptic activity in these models reduce confidence in the detailed spatial structure. In addition, these circulation-induced temperature changes are not large enough to change the relatively uniform summer warming seen in the MMD models

The patterns of temperature changes simulated by RCMs are quite similar to those simulated by GCMs. However, they show an increased warming along the sea ice margin possibly due to a better description of the mesoscale weather systems and air-sea fluxes associated with the ice edge (ACIA, 2005). The warming over most of the central Arctic and Siberia, particularly in summer, tends to be lower in RCM simulations (by up to 2°C) probably due to more realistic present-day snowpack simulations (ACIA, 2005). The warming is modulated by the topographical height, snow cover and associated albedo feedback as shown for the region of northern Canada and Alaska (see Section 11.5.3).

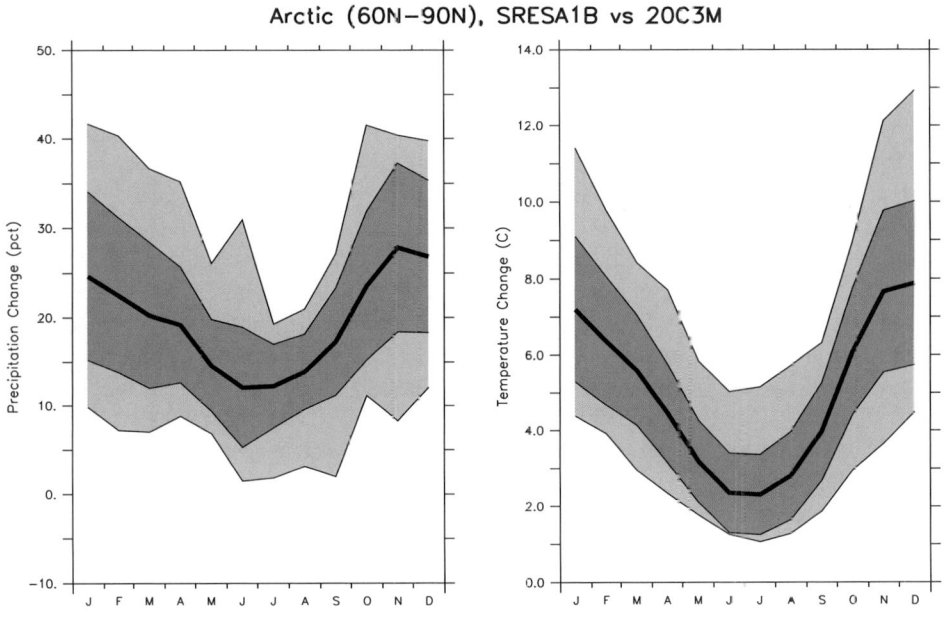

Arctic (60N–90N), SRESA1B vs 20C3M

Figure 11.19. *Annual cycle of arctic area mean temperature and percentage precipitation changes (averaged over the area north of 60°N) for 2080 to 2099 minus 1980 to 1999, under the A1B scenario. Thick lines represent the ensemble median of the 21 MMD models. The dark grey area represents the 25 and 75% quartile values among the 21 models, while the light grey area shows the total range of the models.*

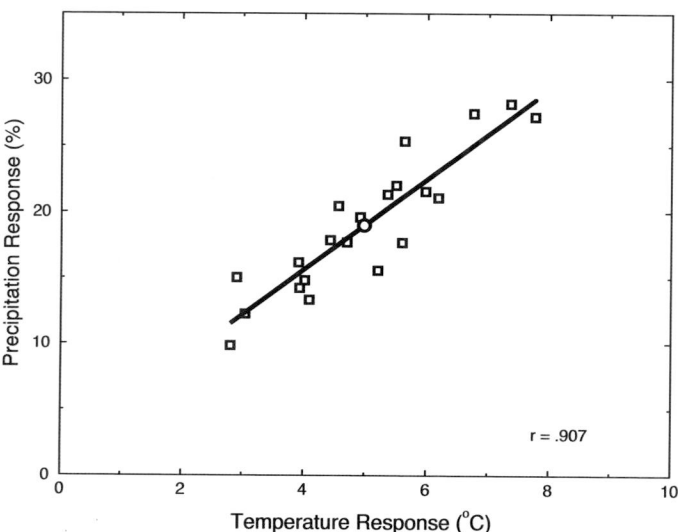

Figure 11.20. *Relationship between the change in annual precipitation (%) and temperature (°C) (2080–2099 minus 1980–1999) in the Arctic (averaged over the area north of 60°N) in the MMD-A1B projections. Each point represents one model. The model ensemble mean response is indicated by the circle.*

Precipitation

The MMD simulations show a general increase in precipitation over the Arctic at the end of the 21st century (Table 11.1; Supplementary Material Figure S11.28). The precipitation increase is robust among the models (Table 11.1; Supplementary Material Figure S11.19) and qualitatively well understood, attributed to the projected warming and related increased moisture convergence (Section 10.3). The very strong correlation between the temperature and precipitation changes (approximately a 5% precipitation increase per degree celsius warming) across the model ensemble is worth noting (Figure 11.20). Thus, both the sign and the magnitude (per degree warming) of the percentage precipitation change are robust among the models.

The spatial pattern of the projected change (Supplementary Material Figure S11.28) shows the greatest percentage increase over the Arctic Ocean (30 to 40%) and smallest (and even slight decrease) over the northern North Atlantic (<5%). By the end of the 21st century, the projected change in the annual mean arctic precipitation varies from 10 to 28%, with an MMD-A1B ensemble median of 18% (Table 11.1). Larger (smaller) mean precipitation increases are found for the A2 (B1) scenario with 22% (13%). The percentage precipitation increase is largest in winter and smallest in summer, consistent with the projected warming (Figure 11.19; Table 11.1). The across-model scatter of the precipitation projections is substantial (Figure 11.19; Table 11.1). The Tebaldi et al. (2004a) 5th to 95th percentile confidence interval of percentage precipitation change in winter is 13 to 36% and in summer 5 to 19% (Supplementary Material Table S11.2).

Differences between the projections for different scenarios are small in the first half of the 21st century but increase later.

Differences among the models increase rapidly as the spatial domain becomes smaller (ACIA, 2005). The geographical variation of precipitation changes is determined largely by changes in the synoptic circulation patterns. During winter, the MMD models project a decreased (increased) frequency of strong Arctic high (Icelandic low) pressure patterns that favour precipitation increases along the Canadian west coast, southeast Alaska and North Atlantic extending into Scandinavia (Cassano et al., 2006). Projections with RCMs support the broad-scale messages while adding expected local and regional detail (ACIA, 2005).

By the end of the 21st century, the MMD-A1B ensemble-projected precipitation increase is significant (Table 11.1), particularly the annual and cold season (winter/autumn) precipitation. However, local precipitation changes in some regions and seasons (particularly in the Atlantic sector and generally in summer) remain difficult to discern from natural variability (ACIA, 2005).

Extremes of Temperature and Precipitation

Very little work has been done in analysing future changes in extreme events in the Arctic. However, the MMD simulations indicate that the increase in mean temperature and precipitation will be combined with an increase in the frequency of very warm and wet winters and summers. Using the definition of extreme season in Section 11.1.2, every DJF and JJA season, in all model projections, is 'extremely' warm in the period 2080 to 2099 (Table 11.1). The corresponding numbers for extremely wet seasons are 90 and 85% for DJF and JJA. For the other scenarios, the frequency of extremes is very similar, except that for the wet seasons under B1, which is smaller (~63%).

Cryosphere

Northern Hemisphere sea ice, snow and permafrost projections are discussed in Section 10.3; projected changes in the surface mass balance of arctic glaciers and of the Greenland Ice Sheet are discussed in Sections 10.3 and 10.7.

Arctic Ocean

A systematic analysis of future projections for the Arctic Ocean circulation is still lacking. Coarse resolution in global models prevents the proper representation of local processes that are of global importance (such as the convection in the Greenland Sea that affects the deep waters in the Arctic Ocean and the intermediate waters that form overflow waters). The MMD models project a reduction in the MOC in the Atlantic Ocean (see Section 10.3). Correspondingly, the northward oceanic heat transport decreases south of 60°N in the Atlantic. However, the CMIP2 model assessment showed a projected increase in the oceanic heat transport at higher latitudes, associated with a stronger subarctic gyre circulation in the models (Holland and Bitz, 2003). The Atlantic Ocean north of 60°N freshens during the 21st century, in pronounced contrast to the observed development in the late 20th century (Wu et al., 2003).

11.8.2 Antarctic

11.8.2.1 Key Processes

Over Antarctica, there is special interest in changes in snow accumulation expected to accompany global climate change as well as the pattern of temperature change, particularly any differences in warming over the peninsula and the interior of the ice sheet. As in the Arctic, warming of the troposphere is expected to increase precipitation. However, circulation changes in both ocean and atmosphere can alter the pattern of air masses, which would modify both precipitation and temperature patterns substantially over the region.

The dominant patterns controlling the atmospheric seasonal-to-interannual variability of the SH extratropics are the Southern Annular Mode (SAM) and ENSO (see Section 3.6). Signatures of these patterns in the Antarctic have been revealed in many studies (reviews by Carleton, 2003 and Turner, 2004). The positive phase of the SAM is associated with cold anomalies over most of Antarctica and warm anomalies over the Antarctic Peninsula (Kwok and Comiso, 2002a). Over recent decades, a drift towards the positive phase in the SAM is evident (see Section 3.6). Observational studies have presented evidence of pronounced warming over the Antarctic Peninsula, but little change over the rest of the continent during the last half of the 20th century (see Sections 3.6 and 4.6). The response of the SAM in transient warming simulations is a robust positive trend, but the response to the ozone hole in the late 20th century, which is also a positive perturbation to the SAM, makes any simple extrapolation of current trends into the future uncertain (see Section 10.3).

Compared to the SAM, the Southern Oscillation (SO) shows weaker association with surface temperature over Antarctica but the correlation with SST and sea ice variability in the Pacific sector of the Southern Ocean is significant (e.g., Kwok and Comiso, 2002b; Renwick, 2002; Bertler et al., 2004; Yuan, 2004). Correlation between the SO index and antarctic precipitation and accumulation has also been studied but the persistence of the signal is not clear (Bromwich et al., 2000, 2004a; Genthon and Cosme, 2003; Guo et al., 2004; Genthon et al., 2005). Recent work suggests that this intermittence is due to nonlinear interactions between ENSO and SAM that vary on decadal time scales (Fogt and Bromwich, 2006; L'Heureux and Thompson, 2006). The SO index has a negative trend over recent decades (corresponding to a tendency towards more El Niño-like conditions in the equatorial Pacific; see Section 3.6) associated with sea ice cover anomalies in the Pacific sector, namely negative (positive) anomalies in the Ross and Amundsen Seas (Bellingshausen and Weddell Seas) (Kwok and Comiso, 2002a). However, a definitive assessment of ENSO amplitude and frequency changes in the 21st century cannot be made (see Chapter 10).

11.8.2.2 Skill of Models in Simulating Present Climate

Evaluating temperature and precipitation simulations over Antarctica is difficult due to sparse observations and often relies on numerical weather prediction (re)analyses. However, significant differences between those have been found, and comparisons with station observations show that the surface temperature can be subject to considerable biases (Connolley and Harangozo, 2001; Bromwich and Fogt, 2004). Marked improvement in the bias is seen after the satellite era (~1978) (Simmons et al., 2004), and parts of the bias are explained by the reanalyses' smoothing of the sharp changes in the terrain near coastal stations. Satellite-derived monthly surface temperatures agree with antarctic station data with an accuracy of 3°C (Comiso, 2000). Precipitation evaluation is even more challenging and the different (re)analyses differ significantly (Connolley and Harangozo, 2001; Zou et al., 2004). Very few direct precipitation gauge and detailed snow accumulation data are available, and these are uncertain to varying degrees (see Section 4.6).

Major challenges face the simulation of the atmospheric conditions and precipitation patterns of the polar desert in the high interior of East Antarctica (Guo et al., 2003; Bromwich et al., 2004a; Pavolonis et al., 2004, Van de Berg et al., 2005). Driven by analysed boundary conditions, RCMs tend to show smaller temperature and precipitation biases in the Antarctic compared to the GCMs (Bailey and Lynch, 2000; Van Lipzig et al., 2002a,b; Van den Broeke and Van Lipzig, 2003; Bromwich et al., 2004b; Monaghan et al., 2006). Krinner et al. (1997) show the value of a stretched model grid with higher horizontal resolution over the Antarctic as compared to standard GCM formulations. Despite these promising developments, since the TAR there has been no coordinated comparison of the performance of GCMs, RCMs and other alternatives to global GCMs over Antarctica.

Temperature

Compared to National Centers for Environmental Prediction (NCEP) reanalyses, the MMD ensemble annual surface temperatures are in general slightly warmer in the Southern Ocean to the north of the sea ice region. The mean bias is predominantly less than 2°C (Carril et al., 2005), which may indicate a slight improvement compared to previous models due to better simulation of the position and depth of the Antarctic trough (Carril et al., 2005; Raphael and Holland, 2006). The temperature bias over sea ice is larger. Biases over the continent are several degrees where the model topography is erroneous (Turner et al., 2006). However, as emphasized above, the biases have to be viewed in the context of the uncertainty in the observations. Changes in cloud and radiation parametrizations have been shown to change the temperature simulation significantly (Hines et al., 2004). A lateral nudging of a stretched-grid GCM (imposing the correct synoptic cyclones from 60°S and lower latitudes) brings the model in better agreement with observations but significant biases remain (Genthon et al., 2002).

The spread in the individual MMD-simulated patterns of surface temperature trends over the past 50 years is very large, but in contrast to previous models, the multi-model composite of the MMD models qualitatively captures the observed enhanced warming trend over the Antarctic Peninsula (Chapman and Walsh, 2006). The general improvements in resolution, sea ice models and cloud-radiation packages have evidently contributed to improved simulations. The ensemble-mean temperature trends show similarity to the observed spatial pattern of the warming, for both annual and seasonal trends. For the annual trend, this includes the warming of the peninsula and near-coastal Antarctica and neutral or slight cooling over the sea-ice covered regions of the Southern Ocean. While the large spread among the models is not encouraging, this level of agreement suggests that some confidence in the ensemble mean 21st-century projection is appropriate.

Precipitation

The MMD models simulate the position of the storm tracks reasonably well but nearly all show some deficiency in the distribution and level of cyclone activity compared to reanalyses (see Section 8.3). Regional Climate Models generally capture the cyclonic events affecting the coast and the associated synoptic variability of precipitation with more fidelity (Adams, 2004; Bromwich et al., 2004a). Over the 20th century, the MMD models simulate changes in storm track position that are generally consistent with observed changes (i.e., poleward displacement of the storm tracks; see Sections 9.5 and 10.3).

The precipitation simulations by both GCMs and RCMs contain uncertainty, on all time scales (Covey et al., 2003; Bromwich et al., 2004a,b; Van de Berg et al., 2005), as a result of model physics limitations. All atmospheric models, including the models underlying the reanalyses, have incomplete parametrizations of polar cloud microphysics and ice crystal precipitation. The simulated precipitation depends, among other things, on the simulated sea ice concentrations, and is strongly affected by biases in the sea ice simulations (Weatherly, 2004). Recent RCM simulations driven by observed sea ice conditions demonstrate good precipitation skill (Van de Berg et al., 2005; Monaghan et al., 2006). However, as emphasized above, the observational uncertainty contributes to uncertainty in the differences between observations and simulations.

Sea Ice

The performance biases of SH sea ice conditions in present-day MMD simulations are discussed in Section 8.3.

11.8.2.3   Climate Projections

Very little effort has been spent to model the future climate of Antarctica at a spatial scale finer than that of GCMs.

Temperature

At the end of the 21st century, the annual warming over the Antarctic continent is moderate but significant (Figure 11.21;

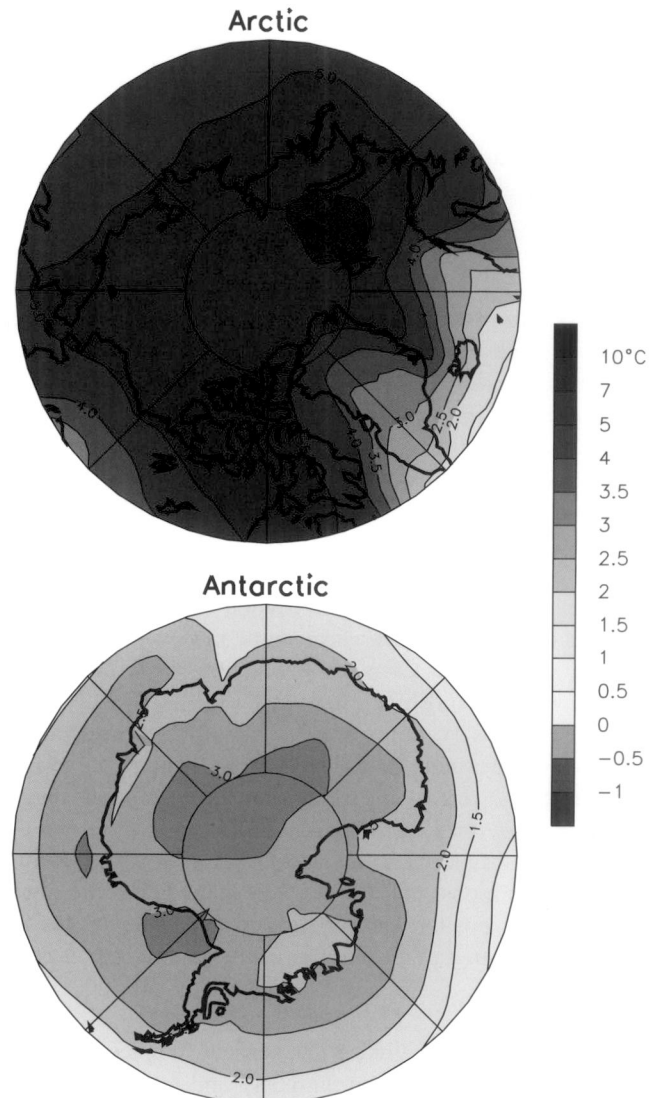

Figure 11.21. *Annual surface temperature change between 1980 to 1999 and 2080 to 2099 in the Arctic and Antarctic from the MMD-A1B projections.*

Table 11.1; Chapman and Walsh, 2006). It is estimated to be 2.6°C by the median of the MMD-A1B models with a range from 1.4°C to 5.0°C across the models (Table 11.1). Larger (smaller) warming is found for the A2 (B1) scenario with mean value of 3.1°C (1.8°C). These warming magnitudes are similar to previous estimates (Covey et al., 2003). The annual mean MMD model projections show a relatively uniform warming over the entire continent (with a maximum in the Weddell Sea) (Figure 11.21; Carril et al., 2005; Chapman and Walsh, 2006). They do not show a local maximum warming over the Antarctic Peninsula. This is a robust feature among the individual models (Supplementary Material Figure S11.12). Thus, the pattern of observed temperature trends in the last half of the 20th century (warming over the Antarctic Peninsula, little change over the rest of the continent) is not projected to continue throughout the 21st century, despite a projected positive SAM trend (see

Section 10.3). It has been argued that two distinct factors have contributed to the observed SAM trend: greenhouse gas forcing and the ozone hole formation (Stone et al., 2001; Shindell and Schmidt, 2004). Their relative importance for the peninsular warming is not readily understood (see Chapter 10).

The mean antarctic temperature change does not show a strong seasonal dependency; the MMD-A1B ensemble mean winter (summer) warming is 2.8°C (2.6°C) (Table 11.1; Supplementary Material Figure S11.29; Chapman and Walsh, 2006). This is also illustrated by how close the Tebaldi et al. (2004a) 5 to 95% confidence interval for the two seasons is: 0.1°C to 5.7°C in summer and 1.0°C to 4.8°C in winter (Supplementary Material Table S11.2). However, over the Southern Oceans, the temperature change is larger in winter/autumn than in summer/spring, which can primarily be attributed to the sea ice retreat (see Section 10.3).

Precipitation

Almost all MMD models simulate a robust precipitation increase in the 21st century (Supplementary Material Figures S11.29 and S11.30; Table 11.1). However, the scatter among the individual models is considerable. By the end of the 21st century, the projected change in the annual precipitation over the Antarctic continent varies from –2% to 35%, with a MMD-A1B ensemble median of 14% (Table 11.1). Similar (smaller) mean precipitation increase is found for the A2 (B1) scenario, with values of 15% (10%). The spatial pattern of the annual change is rather uniform (Supplementary Material Figure S11.30). The projected relative precipitation change shows a seasonal dependency, and is larger in winter than in summer (Supplementary Material Figure S11.29). The Tebaldi et al. (2004a) 5 to 95% confidence interval for winter is –1 to 34% and in summer –6 to 22% (Supplementary Material Table S11.2). The projected increase in precipitation over Antarctica and thus greater accumulation of snow, without substantial surface melting, will contribute negatively to sea level rise relative to the present day (see Section 10.6). It is notable that the most recent model studies of antarctic precipitation show no significant contemporary trends (Van de Berg et al., 2005; Monaghan et al., 2006; Van den Broeke et al., 2006; see Section 4.6).

The moisture transport to the continent by synoptic activity represents a large fraction of net precipitation (Noone and Simmonds, 2002; Massom et al., 2004). During summer and winter, a systematic shift towards strong cyclonic events is projected by the MMD models (see Section 10.3). In particular, the frequency of occurrence of deep cyclones in the Ross Sea to Bellingshausen Sea sector is projected to increase by 20 to 40% (63%) in summer (winter) by the middle of the 21st century (Lynch et al., 2006). Related to this, the precipitation over the sub-antarctic seas and Antarctic Peninsula is projected to increase.

Extremes of Temperature and Precipitation

Very little work has been done in analysing future changes in extreme events in the Antarctic. However, the MMD simulations indicate that the increase in mean temperature and precipitation

will be combined with an increase in the frequency of very warm and wet winters and summers. Using the definition of 'extreme' seasons provided in Section 11.1.2, the MMD models predict extremely warm seasons in about 85% of all DJF and 83% of all JJA seasons in the period 2080 to 2099, as averaged over all models (Table 11.1). The corresponding numbers for extremely wet seasons are 34% and 59%. For the B1 scenario, the frequency of extremes is smaller, with little difference between A1B and A2.

Sea Ice and Antarctic Ice Sheet

Southern Hemisphere sea ice projections are discussed in Section 10.3. The projections of the Antarctic Ice Sheet surface mass balance are discussed in Section 10.6.

11.9   Small Islands

Assessment of projected climate change for Small Islands regions:

Sea levels are likely to continue to rise on average during the century around the small islands of the Caribbean Sea, Indian Ocean and Northern and Southern Pacific Oceans. Models indicate that the rise will not be geographically uniform but large deviations among models make regional estimates across the Caribbean, Indian and Pacific Oceans uncertain.

All Caribbean, Indian Ocean and North and South Pacific islands are very likely to warm during this century. The warming is likely to be somewhat smaller than the global annual mean warming in all seasons.

Summer rainfall in the Caribbean is likely to decrease in the vicinity of the Greater Antilles but changes elsewhere and in winter are uncertain. Annual rainfall is likely to increase in the northern Indian Ocean with increases likely in the vicinity of the Seychelles in DJF, and in the vicinity of the Maldives in JJA while decreases are likely in the vicinity of Mauritius in JJA. Annual rainfall is likely to increase in the equatorial Pacific, while most models project decreases just east of French Polynesia in DJF.

Since AOGCMs do not have sufficiently fine resolution to see the islands, the projections are given over ocean surfaces rather than over land and very little work has been done in downscaling these projections to individual islands. Assessments are also difficult because some climatic processes are still not well understood, such as the midsummer drought in the Caribbean and the ocean-atmosphere interaction in the Indian Ocean. Furthermore, there is insufficient information on future SST changes to determine the regional distribution of cyclone changes. Large deviations among models make the regional distribution of sea level rise uncertain and the number of models addressing storm surges is very limited.

11.9.1 Key Processes

Climate change scenarios for small islands of the Caribbean Sea, Indian Ocean and Pacific Ocean are included in the AR4 for a number of reasons. Ocean-atmosphere interactions play a major role in determining the climate of the islands and including their climate in the projections for neighbours with larger landmasses would miss features peculiar to the islands themselves. Many small islands are sufficiently removed from large landmasses so that atmospheric circulation may be different over the smaller islands compared to their larger neighbours (e.g., in the Pacific Ocean). For the Caribbean, which is close to large landmasses in Central America and northern South America, some islands partly share climate features of one, while others partly share features of the other. At the same time, the Caribbean islands share many common features that are more important than are those shared with the larger landmasses, such as the strong relationship of their climate to SST.

11.9.1.1 Caribbean

The Caribbean region spans roughly the area between 10°N to 25°N and 85°W to 60°W. Its climate can be broadly characterised as dry winter/wet summer with orography and elevation being significant modifiers at the sub-regional scale (Taylor and Alfero, 2005). The dominant synoptic influence is the North Atlantic subtropical high (NAH). During the winter, the NAH is southernmost and the region is generally at its driest. With the onset of the spring, the NAH moves northward, the trade wind intensity decreases and the equatorial flank of the NAH becomes convergent. Concurrently, easterly waves traverse the Atlantic from the coast of Africa into the Caribbean. These waves frequently mature into storms and hurricanes under warm SSTs and low vertical wind shear, generally within a 10°N to 20°N latitudinal band. They represent the primary rainfall source and their onset in June and demise in November roughly coincide with the mean Caribbean rainy season. In the coastal zones of Venezuela and Columbia, the wet season occurs later, from October to January (Martis et al., 2002). Interannual variability of the rainfall is influenced mainly by ENSO events through their effect on SSTs in the Atlantic and Caribbean Basins. The late rainfall season tends to be drier in El Niño years and wetter in La Niña years (Giannini et al., 2000, Martis et al., 2002, M. Taylor et al., 2002) and tropical cyclone activity diminishes over the Caribbean during El Niño summers (Gray, 1984). However, the early rainfall season in the central and southern Caribbean tends to be wetter in the year after an El Niño and drier in a La Niña year (Chen and Taylor, 2002). The phase of the NAO modulates the behaviour of warm ENSO events (Giannini et al., 2001). A positive NAO phase implies a stronger than normal NAH and amplifies the drying during a warm ENSO. On the other hand, a negative NAO phase amplifies the precipitation in the early rainfall season in the year after an El Niño.

11.9.1.2 Indian Ocean

The Indian Ocean region refers to the area between 35°S to 17.5°N and 50°E to 100°E. The climate of the region is influenced primarily by the Asian monsoons (see Section 11.4.1 for processes influencing monsoons). During January, the ITCZ is located primarily in the SH. The region north of the ITCZ then experiences north-easterly trade winds (northeast monsoons) and the region to the south, the south-easterly trades. During northern summer, the ITCZ is located in the north and virtually covers the entire Bay of Bengal, the surrounding lands and the eastern Arabian Sea. The winds in the north turn into strong south-westerlies (southwest monsoons), while the south-easterlies persist in the south. Precipitation and wind stress bring about a response that is distinctly different in the northern and southern parts of the Indian Ocean (International CLIVAR Project Office, 2006). The wet (dry) season in the Maldives occurs during the southwest (northeast) monsoons. From May to October, the southeast trades dominate in the Seychelles and the climate is relatively cool and dry, and December to March is the principal wet season with winds mainly from west to northwest.

While the monsoons recur each year, their irregularity at a range of time scales from weeks to years depends on feedback from the ocean in ways that are not fully understood. Intraseasonal variability is associated with the Monsoon Intra-Seasonal Oscillation (MISO) and the Madden-Julian Oscillation (MJO), which are long-lasting weather patterns that evolve in a systematic way for periods of four to eight weeks. On an interannual and decadal scale, statistical methods have shown that while there are periods of high correlation between ENSO and monsoon variation, there are decades where there appears to be little or no association (International CLIVAR Project Office, 2006; see also Section 10.3.5.4). A modulating factor is the Indian Ocean Dipole or Indian Ocean Zonal Mode (IOZM), a large interannual variation in zonal SST gradient (see Section 3.6). The magnitude of the secondary rainfall maximum from October to December in East Africa is strongly correlated with IOZM events, and the positive phase of IOZM, with higher SSTs in the west, counters the drying effect that ENSO has on monsoon rainfall (Ashok et al., 2001).

11.9.1.3 Pacific

The Pacific region refers to equatorial, tropical and subtropical region of the Pacific in which there is a high density of inhabited small islands. Broadly, it is the region between 20°N and 30°S and 120°E to 120°W. The major climatic processes that play a key role in the climate of this region are the easterly trade winds (both north and south of the equator), the SH high pressure belt, the ITCZ and the South Pacific Convergence Zone (SPCZ; see Vincent, 1994), which extends from the ICTZ near the equator due north of New Zealand south-eastward to at least 21°S, 130°W. The region has a warm, highly maritime climate and rainfall is abundant. The highest rainfall follows the seasonal migration of the ITCZ and SPCZ.

Year-to-year climatic variability in the region is very strongly affected by ENSO events. During El Niño conditions, rainfall increases in the zone northeast of the SPCZ (Vincent, 1994). Tropical cyclones are also a feature of the climate of the region, except within 10 degrees of the equator, and are associated with extreme rainfall, strong winds and storm surge. Many islands in the region are very low lying, but there are also many with strong topographical variations. In the case of the latter, orographic effects on rainfall amount and seasonal distribution can be strong.

11.9.2 Skill of Models in Simulating Present Climate

The ability of the MMD models to simulate present climate in the Caribbean, Indian Ocean and North and South Pacific Ocean is summarised in Supplementary Material Table S11.1. In general, the biases in about half of the temperature simulations are less than 1°C in all seasons, so that the model performances are, on the whole, satisfactory. There are, however, large spreads in precipitation simulations. During the last decade, steady progress has been made in simulating and predicting ENSO using coupled GCMs. However, serious systematic errors in both the simulated mean climate and the natural variability persist (see Section 8.4.7)

11.9.2.1   Caribbean

Simulations of the annual Caribbean temperature in the 20th century (1980–1999) by the MMD models give an average that agrees closely with climatology, differing by less than 0.1°C. The inter-quartile range difference between individual models and climatology ranged from –0.3°C to +0.3°C. Thus, the models have good skill in simulating annual temperature. The average of the MMD simulations of precipitation, however, underestimates the observed precipitation by approximately 30%. The deviations in individual models range from –64 to +20%, a much greater range than the deviations in temperature simulations. Recently the Parallel Climate Model (at T42 resolution – about 3.75 degrees), a fully coupled global climate model, was found to be capable of simulating the main climate features over the Caribbean region (Angeles et al., 2007), but it also underestimated the area average precipitation across the Caribbean. Martinez-Castro et al. (2006), in a sensitivity experiment, conclude that the Regional Climate Model (RegCM3), using the Anthes-Kuo cumulus parametrization scheme, can be used for long-term area-averaged climatology.

11.9.2.2   Indian Ocean

For annual temperature in the Indian Ocean in the 20th century (1980–1999), the mean value of the MMD outputs overestimated the climatology by 0.6°C, with 50% of deviations ranging from 0.2°C to 1.0°C. For rainfall, the multi-model ensemble average was only slightly below the mean precipitation by 3%, and the model deviations ranged from –22 to +20%. There are, however, problems with the simulation of

year-to-year variation. Many of the important climatic effects of the MJO, including its impacts on rainfall variability in the monsoons, are still poorly simulated by contemporary climate models (see Section 8.4).

11.9.2.3 Pacific

Climate model simulations of current-climate means of temperature and precipitation were investigated by Jones et al. (2000, 2002) and Lal et al. (2002) for the South Pacific. The AOGCMs available at the time of these studies simulated well the broad-scale patterns of temperature and precipitation across the region, with the precipitation patterns more variable than for temperature in the models considered, and with some significantly underestimating or overestimating of the intensity of rainfall in the high-rainfall zones. All models simulated a broad rainfall maximum stretching across the SPCZ and ITCZ, but not all models resolved a rainfall minimum between these two regions. A problem of simulating the spatial structure of the MJO resulting in tendencies for the convective anomaly to split into double ITCZs in the Pacific is also discussed in Section 8.4.8.

Analysis of the MMD simulations shows that the average model value overestimated the annual mean temperature from 1980 to 1999 by 0.9°C over a southern Pacific region, with 50% of the deviations varying from 0.6°C to 1.2°C. Over the North Pacific, the simulated ensemble average temperature for the same period was only 0.7°C above the climatology, with half of the model deviations from the climatology ranging from 0.2°C to 1.0°C. Average precipitation was overestimated by 10%, but individual model values varied from –7 to 31% in the southern Pacific region, whereas in the northern Pacific the mean model output for precipitation almost agreed with the climatology, while the individual models deviated from –13 to 13%. Thus, the models are better at simulating rainfall in the northern Pacific than in the southern Pacific and the quality of the simulations, both north and south, were not much different from those for the Indian Ocean.

11.9.3 Temperature and Precipitation Projections

Scenarios of temperature change and percentage precipitation change between 1980 to 1999 and 2080 to 2099 are summarised in Table 11.1 (described in Section 11.1.3). A small value of T implies a large signal-to-noise ratio and it can be seen that, in general, the signal-to-noise ratio is greater for temperature than for precipitation change. The probability of extreme warm seasons is 100% in all cases for the small islands and the scenarios of warming are all very significant by the end of the century. Approximate results for the A2 and B1 scenarios and for other future times in this century can obtained by scaling the A1B values, as described in Section 11.1.3.

The temporal evolution of temperature as simulated by the MMD models for the 20th and 21st centuries is also shown in Figure 11.22 for oceanic regions including the Caribbean (CAR), Indian Ocean (IND), North Pacific Ocean (NPA) and South

Pacific Ocean (SPA). In general, it can be seen, by comparison with Box 11.1, Figure 1, that the temperature increases for the small islands are less than for the continental regions. The almost linear nature of the evolution is also apparent in the figure. Temperature and precipitation projections for the small island regions are discussed below in the context of Table 11.1.

11.9.3.1   Caribbean

The MMD-simulated annual temperature increases at the end of the 21st century range from 1.4°C to 3.2°C with a median of 2.0°C, somewhat below the global average. Fifty percent of the models give values differing from the median by only ±0.4°C. Statistical downscaling of HadCM3 results using the A2 and B2 emission scenarios gives around a 2°C rise in temperature by the 2080s, approximately the same as the HadCM3 model. The agreement between the AOGCMs and the downscaling analysis gives a high level of confidence in the temperature simulations. The downscaling was performed with the use of the Statistical DownScaling Model (SDSM) developed by Wilby et al. (2002) as part of an Assessments of Impacts and Adaptations to Climate Change in Multiple Regions and Sectors (AIACC) Small Island States SIS06 project (http://www.aiaccproject.org). Angeles et al. (2007) also simulate an approximately 1°C rise in SST up to the 2050s using the IS92a scenario. There were no noticeable differences in monthly changes (see Supplementary Material Figure S11.31). Observations suggest that warming is ongoing (Peterson et al., 2002).

According to Table 11.1, most models project decreases in annual precipitation and a few increases, varying from –39 to +11%, with a median of –12%. Figure 11.23 shows that the annual mean decrease is spread across the entire region. In DJF,

some areas of increases are noted and in JJA, the region-wide decrease is enhanced, especially in the region of the Greater Antilles, where the model consensus is also strong. Monthly changes in the Caribbean are shown in Supplementary Material Figure S11.32, which also shows that the simulations for the Caribbean have a greater spread compared to the other oceanic regions (IND, NPA and SPA in S11.32). Results from HadCM3 downscaled for the A2 and B2 emission scenarios using the SDSM also show a near-linear decrease in summer precipitation to the 2080s for a station in Jamaica. Downscaled results from the SDSM for stations in Barbados and Trinidad, however, show increases rather than decreases. Thus, there is consensus between the MMD results and the downscaled results for the Greater Antilles in JJA but not for the other islands, and also not on an annual basis. Angeles et al. (2007) also simulate decreases up to the middle of the century in the vicinity of the Greater Antilles but not in the other islands in the late rainfall season. Table 11.1 shows that the decrease in JJA has the largest signal-to-noise ratio. The decrease is in agreement with the expected drying in the subtropics discussed in Sections 9.5 and 11.1. In the multi-model analysis, most models show shift to a more positive phase of the NAO (see Section 10.3), and consensus on temperature changes in the Pacific indicates an El Niño-like pattern with higher temperatures in the eastern Pacific (see Section 10.3). These conditions are associated with drying in the Caribbean. Observed trends in precipitation are unclear. While Peterson et al. (2002) find no statistically significant trends in mean precipitation amounts from the 1950s to 2000, Neelin et al. (2006) note a modest but statistically significant summer drying trend over recent decades in the Caribbean in several observational data sets.

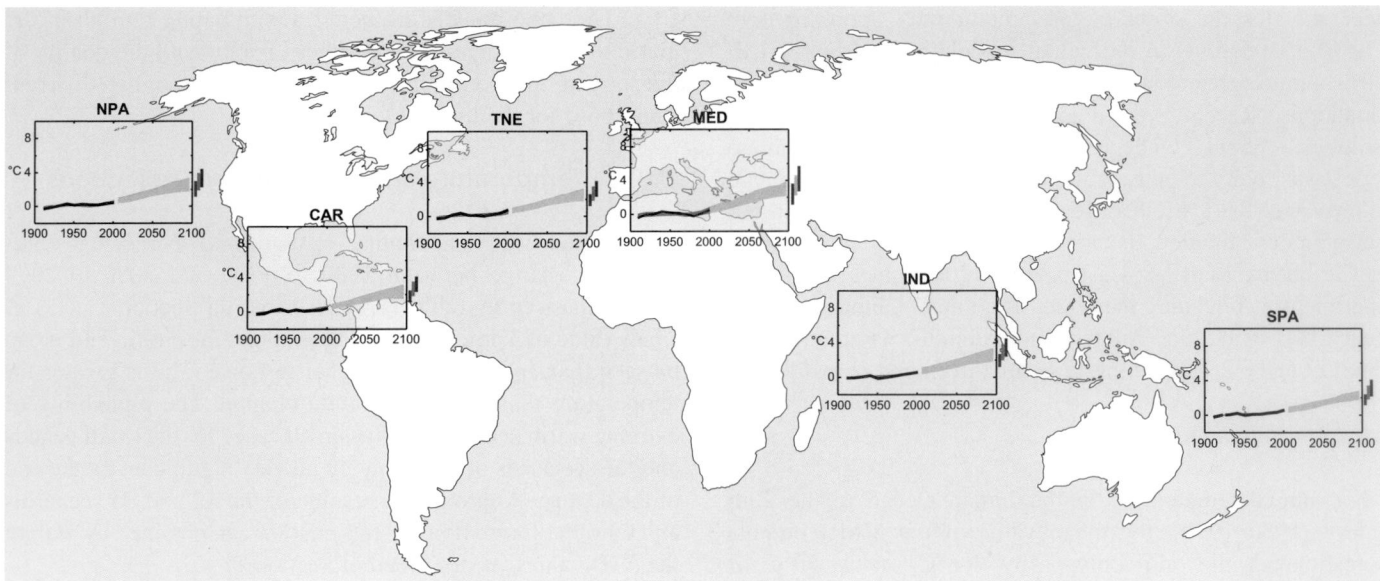

Figure 11.22. *Temperature anomalies with respect to 1901 to 1950 for six oceanic regions for 1906 to 2005 (black line) and as simulated (red envelope) by MMD models incorporating known forcings; and as projected for 2001 to 2100 by MMD models for the A1B scenario (orange envelope). The bars at the end of the orange envelope represent the range of projected changes for 2091 to 2100 for the B1 scenario (blue), the A1B scenario (orange) and the A2 scenario (red). The black line is dashed where observations are present for less than 50% of the area in the decade concerned. More details on the construction of these figures are given in Box 11.1 and Section 11.1.2.*

11.9.3.2   Indian Ocean

Based on the MMD ensemble mean the annual temperature is projected to increase by about 2.1°C, somewhat below the global average, with individual models ranging from 1.4°C to 3.7°C and at least half of the models giving values quite close to the mean. All models show temperature increases in all months with no significant seasonal variation (Supplementary Material Figure S11.31). Evidence of temperature increases from 1961 to 1990 in the Seychelles is provided by Easterling et al. (2003), who find that the percentage of time when the minimum temperature is below the 10th percentile is decreasing, and the percentage of time where the minimum temperature exceeds the 90th percentile is increasing. Similar results were obtained for the maximum temperatures. This is consistent with general patterns of warming elsewhere (see Chapter 3).

The annual precipitation changes projected by individual MMD models varied from –2 to 20% with a median of 4% and 50% of the models projecting changes between 3 and 5%. Thus, there is some level of confidence in the precipitation results although not as high as for temperature. Figure 11.24 shows that the annual increase is restricted mainly to the north Indian Ocean, where the model consensus is greatest, especially in the vicinity of the Maldives. In DJF, some increases are noted in the south. Model agreement on increases is greatest for the Seychelles in DJF and for the Maldives in JJA. There is also strong agreement on decreases in the vicinity of Mauritius in

JJA. Sections 10.3.5 and 11.4 discuss changes in monsoon behaviour in a warmer climate. There is an emerging consensus that the effect of enhanced moisture convergence in a warmer atmosphere will dominate over possible weaker monsoonal flows and tropical large-scale circulation in global warming simulations, resulting in increased monsoonal precipitation. Easterling et al. (2003) find evidence that extreme rainfall tended to increase from 1961 to 1990 (see also Section 11.4.3, South Asia projections).

11.9.3.3 Pacific

Projected regional temperature changes in the South Pacific from a range of AOGCMs have been prepared by Lal et al. (2002), Lal (2004) and Ruosteenoja et al. (2003). Jones et al. (2000, 2002) and Whetton and Suppiah (2003) also consider patterns of change. Broadly, simulated warming in the South Pacific closely follows the global average warming rate. However, there is a tendency in many models for the warming to be a little stronger in the central equatorial Pacific (North Polynesia) and a little weaker to the south (South Polynesia).

The MMD-A1B projections for the period 2080 to 2099 show an increase in annual temperature of 1.8°C, somewhat below the global average over the South Pacific (Table 11.1). The individual model values vary from 1.4°C to 3.1°C and at least half of the models project values very close to the mean. All models show increases, slightly less in the second half of

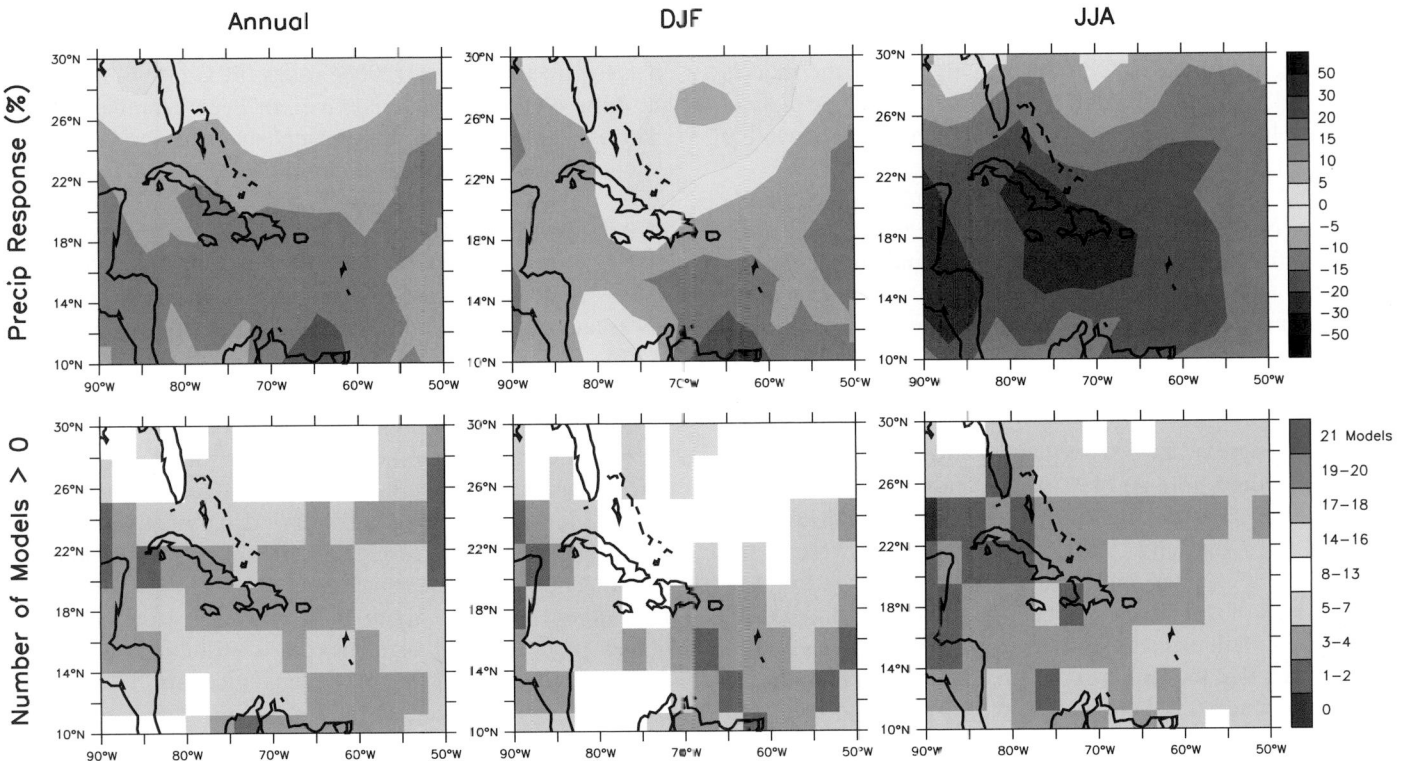

Figure 11.23. *Precipitation changes over the Caribbean from the MMD-A1B simulations. Top row: Annual mean, DJF and JJA fractional precipitation change between 1980 to 1999 and 2080 to 2099, averaged over 21 models. Bottom row: number of models out of 21 that project increases in precipitation.*

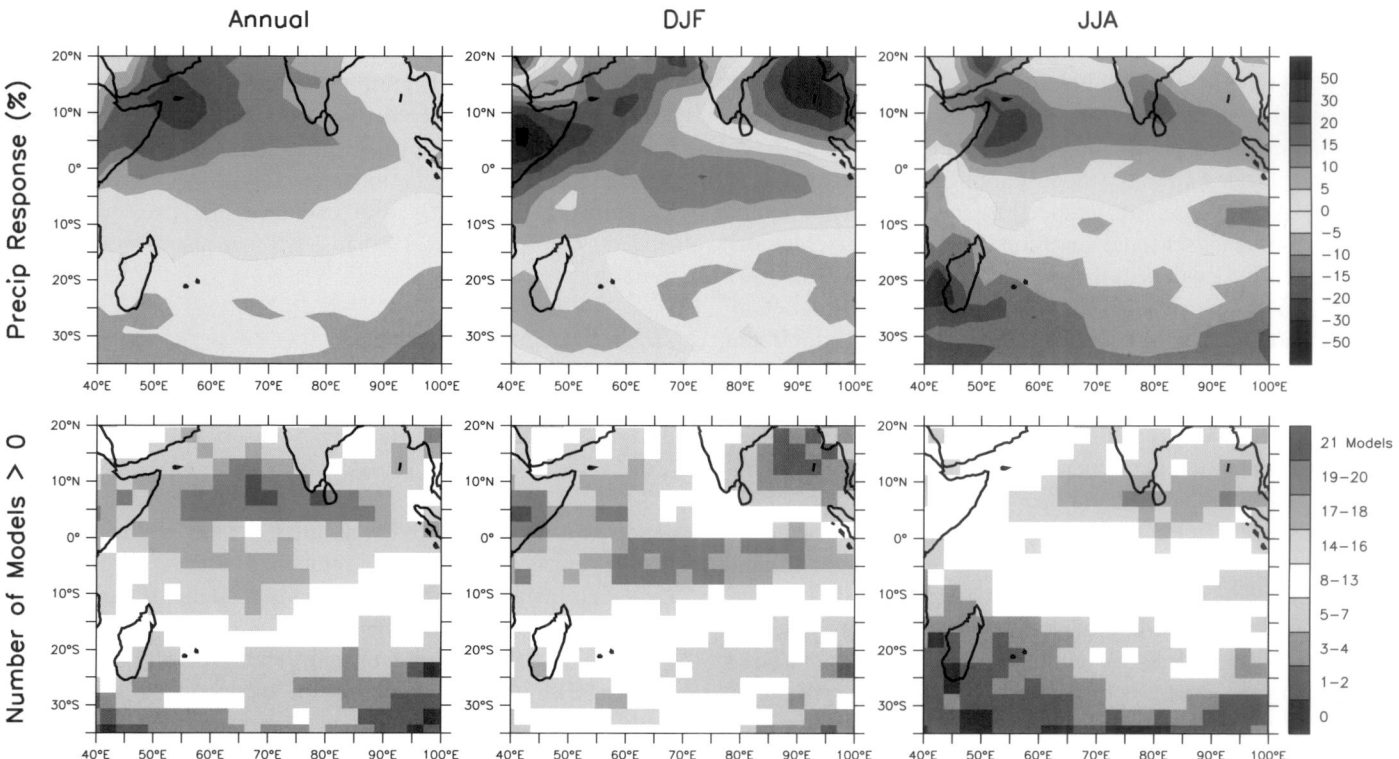

Figure 11.24. *As for Figure 11.23 but for the Indian Ocean.*

the year compared to the first (Supplementary Material Figure S11.31). Over the North Pacific, the models simulate an increase in temperature of 2.3°C, slightly below the global average, with values ranging from 1.5°C to 3.7°C and 50% of the models within ±0.4°C of the mean. All models show increases, more in the second half of the year compared to the first (Supplementary Material Figure S11.31).

For the same period, 2080 to 2099, annual precipitation increases over the southern Pacific when averaged over all MMD models are close to 3%, with individual models projecting values from –4 to +11% and 50% of the models showing increases between 3 and 6%. The time to reach a discernible signal is relatively short. (Table 11.1). Most of these increases were in the first half of the year (Supplementary Material Figure S11.32). For precipitation in the northern Pacific, an increase of 5% is found, with individual models projecting values from 0 to 19% and at least half of the models within –2 to +5% of the median. The time to reach a discernible signal is relatively long. Most of these increases were in the latter half of the year (Supplementary Material Figure S11.32). Figure 11.25 illustrates the spatial distribution of annual, DJF and JJA rainfall changes and inter-model consistency. The figure shows that the tendency for precipitation increase in the Pacific is strongest in the region of the ITCZ due to increased moisture transport described in Section 11.1.3.1. Griffiths et al. (2003) find an increasing trend from 1961 to 2000 in mean rainfall in and northeast of the SPCZ in the southern Pacific. As for the Indian Ocean, there is some level of confidence in the precipitation results for the Pacific, but not as high as for the temperature results.

Changes in rainfall variability in the South Pacific were analysed by Jones et al. (2000) using IPCC (1996) scenarios, but more recent simulations have not been examined. These changes will be strongly driven by changes in ENSO, and this is not well understood (see Section 10.3).

11.9.4 Sea Level Rise

Sea level is projected to rise between the present (1980–1999) and the end of this century (2090–2099) by 0.35 m (0.23 to 0.47 m) for the A1B scenario (see Section 10.6). Due to ocean density and circulation changes, the distribution will not be uniform and Figure 10.32 shows a distribution of local sea level change based on ensemble mean of 14 AOGCMs. A lower-than-average rise in the Southern Ocean can be seen, possibly due to increased wind stress. Also obvious is a narrow band of pronounced sea level rise stretching across the southern Atlantic and Indian Oceans at about 40°S. This is also seen in the southern Pacific at about 30°S. However, large deviations among models make estimates of distribution across the Caribbean, Indian and Pacific Oceans uncertain. Extreme sea level changes, including storm surges, are discussed in Box 11.5 in a broader context. The range of uncertainty cannot be reliably quantified due to the limited set of models addressing the problem.

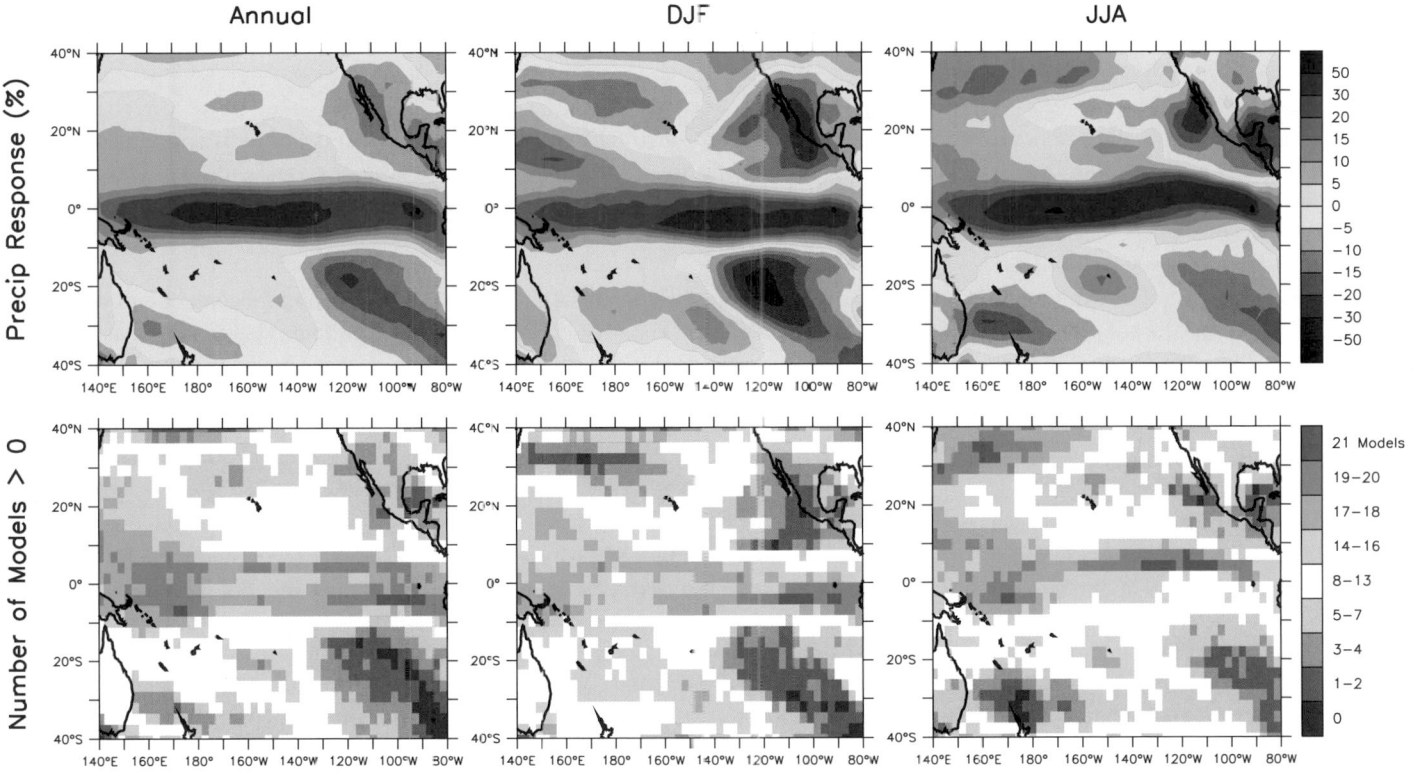

Figure 11.25. *As for Figure 11.23 but for the northern and southern Pacific Ocean.*

Global sea level rise over the 20th century is discussed in Section 5.5; the best estimate is 0.17 ± 0.05 m. From estimates of observed sea level rise from 1950 to 2000 by Church et al. (2004), the rise in the Caribbean appeared to be near the global mean. Church et al. (2006) also estimate the average rise in the region of the Indian and Pacific Ocean to be close to the global average. There have been large observed variations in sea level rise in the Pacific Ocean mainly due to ocean circulation changes associated with ENSO events. From 1993 to 2001, all the data show large rates of sea level rise over the western Pacific and eastern Indian Ocean and sea level falls in the eastern Pacific and western Indian Ocean (Church et al., 2006). Observed sea level rise in the Pacific and Indian Oceans is discussed in Chapter 5.

11.9.5 Tropical Cyclones

Fewer models have simulated tropical cyclones in the context of climate change than those simulating temperature and precipitation changes and sea level rise, mainly because of the computational burden associated with the high resolution needed to capture the characteristics of tropical cyclones. Accordingly, there is less certainty about the changes in frequency and intensity of tropical cyclones on a regional basis than for temperature and precipitation changes. An assessment of results for projected changes in tropical cyclones is presented in Section 10.3.6.3, and a synthesis is given at the end of the section. Regional model-based studies of changes in tropical cyclone behaviour in the southwest Pacific include works by Nguyen and Walsh (2001) and Walsh (2004). Walsh concludes that in general there is no clear picture with respect to regional changes in frequency and movement, but increases in intensity are indicated. It should also be noted that ENSO fluctuations have a strong impact on patterns of tropical cyclone occurrence in the southern Pacific, and that therefore uncertainty with respect to future ENSO behaviour (see Section 10.3) contributes to uncertainty with respect to tropical cyclone behaviour (Walsh, 2004).

Box 11.5: Coastal Zone Climate Change

Introduction

Climate change has the potential to interact with the coastal zone in a number of ways including inundation, erosion and salt water intrusion into the water table. Inundation and intrusion will clearly be affected by the relatively slow increases in mean sea level over the next century and beyond. Mean sea level is addressed in Chapter 10; this box concentrates on changes in extreme sea level that have the potential to significantly affect the coastal zone. There is insufficient information on changes in waves or near-coastal currents to provide an assessment of the effects of climate change on erosion.

The characteristics of extreme sea level events are dependent on the atmospheric storm intensity and movement and coastal geometry. In many locations, the risk of extreme sea levels is poorly defined under current climate conditions because of sparse tide gauge networks and relatively short temporal records. This gives a poor baseline for assessing future changes and detecting changes in observed records. Using results from 141 sites worldwide for the last four decades, Woodworth and Blackman (2004) find that at some locations extreme sea levels have increased and that the relative contribution from changes in mean sea level and atmospheric storminess depends on location.

Methods of simulating extreme sea levels

Climate-driven changes in extreme sea level will come about because of the increases in mean sea level and changes in the track, frequency or intensity of atmospheric storms. (From the perspective of coastal flooding, the vertical movement of land, for instance due to post glacial rebound, is also important when considering the contribution from mean sea level change.) To provide the large-scale context for these changes, global climate models are required, although their resolution (typically 150 to 300 km horizontally) is too coarse to represent the details of tropical cyclones or even the extreme winds associated with mid-latitude cyclones. However, some studies have used global climate model forcing to drive storm surge models in order to provide estimates of changes in extreme sea level (e.g., Flather and Williams, 2000). To obtain more realistic simulations from the large-scale drivers, three approaches are used: dynamical and statistical downscaling and a stochastic method (see Section 11.10 for general details).

As few RCMs currently have an ocean component, these are used to provide high-resolution (typically 25 to 50 km horizontally) surface winds and pressure to drive a storm surge model (e.g., Lowe et al., 2001). This sequence of one-way coupled models is usually carried out for a present-day (Debenard et al., 2003) or historical baseline (e.g., Flather et al., 1998) and a period in the future (e.g., Lowe et al., 2001; Debenard et al., 2003). In the statistical approach, relationships between large-scale synoptic conditions and local extreme sea levels are constructed. These relationships can be developed either by using analyses from weather prediction models and observed extreme sea levels, or by using global climate models and present-day simulations of extreme water level generated using the dynamic methods described above. Simulations of future extreme sea level are then derived from applying the statistical relationships to the future large-scale atmospheric synoptic conditions simulated by a global climate model (e.g., von Storch and Reichardt, 1997). The statistical and dynamical approach can be combined, using a statistical model to produce the high-resolution wind fields forcing the wave and storm surge dynamical models (Lionello et al., 2003). Similarly, the stochastic sampling method identifies the key characteristics of synoptic weather events responsible for extreme sea levels (intensity and movement) and represents these by frequency distributions. For each event, simple models are used to generate the surface wind and pressure fields and these are applied to the storm surge model (e.g., Hubbert and McInnes, 1999). Modifications to the frequency distributions of the weather events to represent changes under enhanced greenhouse conditions are derived from global climate models and then used to infer a future storm surge climatology.

Extreme sea level changes – sample projections from three regions

1. Australia

In a study of storm surge impacts in northern Australia, a region with only a few short sea level records, McInnes et al. (2005) used stochastic sampling and dynamical modelling to investigate the implications of climate change on extreme storm surges and inundation. Cyclones occurring in the Cairns region from 1907 to 1997 were used to develop probability distribution functions governing the cyclone characteristics of speed and direction with an extreme value distribution fitted to the cyclone intensity. Cyclone intensity distribution was then modified for enhanced greenhouse conditions based on Walsh and Ryan (2000), in which cyclones off northeast Australia were found to increase in intensity by about 10%. No changes were imposed upon cyclone frequency or direction since no reliable information is available on the future behaviour of the main influences on these, respectively ENSO or mid-level winds. Analysis of the surges resulting from 1,000 randomly selected cyclones with current and future intensities shows that the increased intensity leads to an increase in the height of the 1-in-100 year event from 2.6 m to 2.9 m with the 1-in-100 year event becoming the 1-in-70 year event. This also results in the areal extent of inundation more than doubling (from approximately 32 to 71 km^2). Similar increases for Cairns and other coastal locations were found by Hardy et al. (2004). *(continued)*

2. Europe

Several dynamically downscaled projections of climate-driven changes in extreme water levels in the European shelf region have been carried out. Woth (2005) explored the effect of two different GCMs and their projected climates changes due to two different emissions scenarios (SRES A2 and B2) on storm surges along the North Sea coast. She used data from one RCM downscaling the four GCM simulations (Woth et al., 2006) (using data from four RCMs driven by one GCM produced indistinguishable results) and demonstrates significant increases in the top 1% of events (10 to 20 cm above average sea level change) over the continental European North Sea coast. The changes projected by the different experiments were statistically indistinguishable, although those from the models incorporating the A2 emissions scenario were consistently larger. When including the effects of global mean sea level rise and vertical land movements, Lowe and Gregory (2005) find that increases in extreme sea level are projected for the entire UK coastline, using a storm surge model driven by one of the RCMs analysed by Woth et al. (2006) (Box 11.5, Figure 1). Using a Baltic Sea model driven by data from four RCM simulations, Meier (2006) finds that the changes in storm surges vary strongly between the simulations but with some tendency for larger increases in the 100-year surges than in the mean sea level.

Lionello et al. (2003) estimate the effect of atmospheric CO_2 doubling on the frequency and intensity of high wind waves and storm surge events in the Adriatic Sea. The regional surface wind fields were derived from the sea level pressure field in a 30-year long ECHAM4 high-resolution (about 1.5 degrees) time slice experiment by statistical downscaling and then used to force a wave and an ocean model. They find no statistically significant changes in the extreme surge level and a decrease in the extreme wave height with increased atmospheric CO_2. An underestimation of the observed wave heights and surge levels calls for caution in the interpretation of these results. Using AOGCM projections, X.L. Wang et al. (2004) infer an increase in winter and autumn seasonal mean and extreme wave heights in the northeast and southwest North Atlantic, but a decrease in the mid-latitudes of the North Atlantic. Not all changes were significant and in some regions (e.g., the North Sea), their sign was found to depend on the emissions scenario.

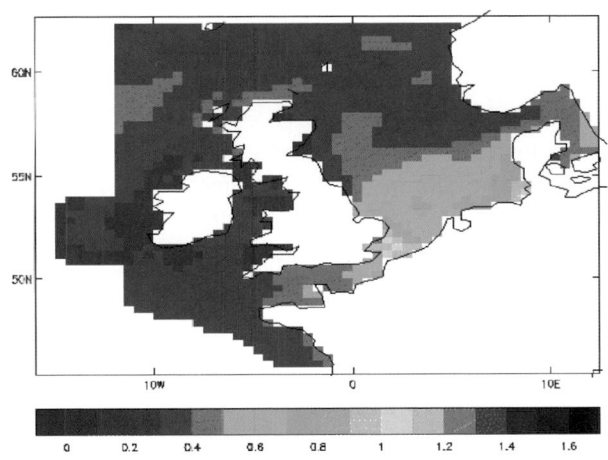

Box 11.5, Figure 1. *The change with respect to 1961-1990 in the 50-year return period extreme water level (m) in the North Sea due to changes in atmospheric storminess, mean sea level and vertical land movements for the period 2071 to 2100 under the A2 scenario (from Lowe and Gregory, 2005).*

3. Bay of Bengal

Several dynamic simulations of storm surges have been carried out for the region but these have often involved using results from a small set of historical storms with simple adjustments (such as adding on a mean sea level or increasing wind speeds by 10%) to account for future climate change (e.g., Flather and Khandker, 1993). This technique has the disadvantage that by taking a relatively small and potentially biased set of storms it may lead to a biased distribution of water levels with an unrealistic count of extreme events. In one study using dynamical models driven by RCM simulations of current and future climates, Unnikrishnan et al. (2006) show that despite no significant change in the frequency of cyclones there are large increases in the frequency of the highest storm surges.

Uncertainty

Changes in storm surges and wave heights have been addressed for only a limited set of models. Thus, we cannot reliably quantify the range of uncertainty in estimates of future coastal flooding and can only make crude estimates of the minimum values (Lowe and Gregory, 2005). There is some evidence that the dynamical downscaling step in providing data for storm surge modelling is robust (i.e., does not add to the uncertainty). However, the general low level of confidence in projected circulation changes from AOGCMs implies a substantial uncertainty in these projections.

11.10 Assessment of Regional Climate Projection Methods

The assessment of methods recognises the challenges posed by the complex interactions that occur at many spatial and temporal scales, involving the general circulation, cross-scale feedbacks and regional-scale forcing.

11.10.1 Methods for Generating Regional Climate Information

Atmosphere-Ocean General Circulation Models constitute the primary tool for capturing the global climate system behaviour. They are used to investigate the processes responsible for maintaining the general circulation and its natural and forced variability (Chapter 8), to assess the role of various forcing factors in observed climate change (Chapter 9) and to provide projections of the response of the system to scenarios of future external forcing (Chapter 10). As AOGCMs seek to represent the whole climate system, clearly they provide information on regional climate and climate change and relevant processes directly. For example, the skill in simulating the climate of the last century when accounting for all known forcings demonstrates the causes of recent climate change (Chapter 9) and this information can be used to constrain the likelihood of future regional climate change (Stott et al., 2006; see also Section 11.10.2). AOGCM projections provide plausible future regional climate scenarios, although methods to establish the reliability of the regional AOGCM scales have yet to mature. The spread within an ensemble of AOGCMs is often used to characterise the uncertainty in projected future climate changes. Some regional responses are consistent across AOGCM simulations, although for other regions the spread remains large (see Sections 11.2 to 11.9).

Because of their significant complexity and the need to provide multi-century integrations, horizontal resolutions of the atmospheric components of the AOGCMs in the MMD range from 400 to 125 km. Generating information below the grid scale of AOGCMs is referred to as downscaling. There are two main approaches, dynamical and statistical. Dynamical downscaling uses high-resolution climate models to represent global or regional sub-domains, and uses either observed or lower-resolution AOGCM data as their boundary conditions. Dynamical downscaling has the potential for capturing mesoscale nonlinear effects and providing coherent information among multiple climate variables. These models are formulated using physical principles and they can credibly reproduce a broad range of climates around the world, which increases confidence in their ability to downscale realistically future climates. The main drawbacks of dynamical models are their computational cost and that in future climates the parametrization schemes they use to represent sub-grid scale processes may be operating outside the range for which they were designed.

Empirical SD methods use cross-scale relationships that have been derived from observed data, and apply these to climate model data. Statistical downscaling methods have the advantage of being computationally inexpensive, able to access finer scales than dynamical methods and applicable to parameters that cannot be directly obtained from the RCM outputs. They require observational data at the desired scale for a long enough period to allow the method to be well trained and validated. The main drawbacks of SD methods are that they assume that the derived cross-scale relationships remain stable when the climate is perturbed, they cannot effectively accommodate regional feedbacks and, in some methods, can lack coherency among multiple climate variables.

11.10.1.1 High-Resolution Atmosphere-Only GCMs

Atmosphere-only GCMs (AGCMs) include interactive land surface schemes as in an AOGCM but require information on SST and sea ice as a lower boundary condition. Given the short time scales associated with the atmosphere and land surface components compared to those in the ocean, relatively short time slices (a few decades) can be run at high resolution. The SST and sea ice information required can be derived from observations or AOGCMs. The use of observations can improve simulations of current climate but combining these with AOGCM-derived changes for the future climate (e.g., Rowell, 2005) increases the risk of inconsistency in the projected climate. The absence of two-way feedback between the atmosphere and ocean in AGCMs can cause a significant distortion of the climatic variability (Bretherton and Battisti, 2000), as documented over regions such as the Indian Ocean and the South Asian monsoon (Douville, 2005; Inatsu and Kimoto, 2005). The large-scale climate responses of AGCMs and AOGCMs appear to be similar in many regions; when and where they differ, the consistency of the oceanic surface boundary condition may be questioned (May and Roeckner, 2001; Govindasamy et al., 2003). Further research is required to determine if the similarity is sufficient for the time-slice approach with AGCMs to be considered a robust downscaling technique.

Model grids of 100 km and finer have become feasible and 50 km will likely be the norm in the near future (Bengtsson, 1996; May and Roeckner, 2001; Déqué and Gibelin, 2002; Govindaswamy, 2003). High-performance computer systems now allow global computations at 20 km (e.g., May, 2004a; Mizuta et al., 2006), although for short time slices only. Evaluated on the scale typical of current AOGCMs, nearly all quantities simulated by high-resolution AGCMs agree better with observations, but the improvements vary significantly for different regions (Duffy et al., 2003) and specific variables, and extensive recalibration of parametrizations is often required. Notable improvements occur in orographic precipitation and dynamics of mid-latitude weather systems (see Chapter 10). The highest resolution offers the prospect of credible simulations of the climatology of tropical cyclones (e.g., May, 2004a; Mizuta et al., 2006). Coordinated multi-model experiments are needed, however, to optimise the value of these high-resolution studies for general assessment.

An alternative to uniform high-resolution AGCMs is Variable-Resolution AGCMs (VRGCMs; e.g., Déqué and Piedelievre, 1995; Krinner et al., 1997; Fox-Rabinovitz et al., 2001; McGregor et al., 2002; Gibelin and Déqué, 2003). The VRGCM approach is attractive as it permits, within a unified modelling framework, a regional increase in resolution while retaining the interaction of all regions of the atmosphere. Numerical artefacts due to stretching have been shown to be small when using modest stretching factors (e.g., Lorant and Royer, 2001). The results from VRGCMs capture, over the high-resolution region, finer-scale details than uniform-resolution models while retaining global skill similar to uniform-resolution simulations with the same number of grid points.

11.10.1.2   Nested Regional Climate Models

The principle behind nested modelling is that, consistent with the large-scale atmospheric circulation, realistic regional climate information can be generated by integrating an RCM if the following premises are satisfied: time-varying large-scale atmospheric fields (winds, temperature and moisture) are supplied as lateral boundary conditions (LBCs) and SST and sea ice as lower boundary conditions; the control from the LBCs keeps the interior solution of the RCM consistent with the driving atmospheric circulation; and sub-grid scale physical processes are suitably parametrized, including fine-scale surface forcing such as orography, land-sea contrast and land use.

A typical RCM grid for climate change projections is around 50 km, although some climate simulations have been performed using grids of 15 or 20 km (e.g., Leung et al., 2003, 2004; Christensen and Christensen, 2004; Kleinn et al., 2005). Recently, projections of climate changes for East Asia were completed with a 5-km non-hydrostatic RCM (Kanada et al., 2005; Yoshizaki et al., 2005; Yasunaga et al., 2006), but only for short simulations. Following the trend in global modelling, RCMs are increasingly coupled interactively with other components of the climate system, such as regional ocean and sea ice (e.g., Bailey and Lynch 2000; Döscher et al., 2002; Rinke et al., 2003; Bailey et al., 2004; Meier et al., 2004; Sasaki et al., 2006a), hydrology, and with interactive vegetation (Gao and Yu, 1998; Xue et al., 2000).

Multi-decadal RCM experiments are becoming standard (e.g., Whetton et al., 2000; Kwon et al., 2003; Leung et al., 2004; Kjellström et al., 2007; Plummer et al., 2006), including the use of ensembles (Christensen et al., 2002), enabling a more thorough validation and exploration of projected changes. In multi-year ensemble simulations driven by reanalyses of atmospheric observations, Vidale et al. (2003) show that RCMs have skill in reproducing interannual variability in precipitation and surface air temperature. The use of ensemble simulations has enabled quantitative estimates regarding the sources of uncertainty in projections of regional climate changes (Rowell, 2005; Déqué et al., 2005, 2007; Beniston et al., 2007; Frei et al.,

2006; Graham et al., 2007). Combining information from four RCM simulations, Christensen et al. (2001) and Rummukainen et al. (2003) demonstrate that it is feasible to explore not only uncertainties related to projections in the mean climate state, but also for higher-order statistics.

The difficulties associated with the implementation of LBCs in nested models are well documented (e.g., Davies, 1976; Warner et al., 1997). As time progresses in a climate simulation, the RCM solution gradually turns from an initial-value problem more into a boundary value problem. The mathematical interpretation is that nested models represent a fundamentally ill-posed boundary value problem (Staniforth, 1997; Laprise, 2003). The control exerted by LBCs on the internal solution generated by RCMs appears to vary with the size of the computational domain (e.g., Rinke and Dethloff, 2000), as well as location and season (e.g., Caya and Biner, 2004). In some applications, the flow developing within the RCM domain may become inconsistent with the driving LBC. This may (Jones et al., 1997) or may not (Caya and Biner, 2004) affect climate statistics. Normally, RCMs are only driven by LBCs with high time resolution to capture the temporal variations of large-scale flow. Some RCMs also use nudging or relaxation of large scales in the interior of the domain (e.g., Kida et al., 1991; Biner et al., 2000; von Storch et al., 2000). This has proved useful to minimise the distortion of the large scales in RCMs (von Storch et al., 2000; Mabuchi et al., 2002; Miguez-Macho et al., 2004), although it can also hide model biases. One-way RCM-GCM coupling is mostly used, although recently a two-way nested RCM has been developed (Lorenz and Jacob, 2005) thus achieving interaction with the global atmosphere as with variable-resolution AGCMs.

The ability of RCMs to simulate the regional climate depends strongly on the realism of the large-scale circulation that is provided by the LBCs (e.g., Pan et al., 2001). Latif et al. (2001) and Davey et al. (2002) show that strong biases in the tropical climatology of AOGCMs can negatively affect downscaling studies for several regions of the world. Nonetheless, the reliability of nested models, that is, their ability to generate meaningful fine-scale structures that are absent in the LBCs, is clear. A number of studies have shown that the climate statistics of atmospheric small scales can be re-created with the right amplitude and spatial distribution, even if these small scales are absent in the LBCs (Denis et al., 2002, 2003; Antic et al., 2005; Dimitrijevic and Laprise, 2005). This implies that RCMs can add value at small scales to climate statistics when driven by AOGCMs with accurate large scales. Overall, the skill at simulating current climate has improved with the MMD AOGCMs (Chapter 8), which will lead to higher quality LBCs for RCMs.

11.10.1.3   Empirical and Statistical Downscaling Methods

A complementary technique to RCMs is to use derived relationships linking large-scale atmospheric variables (predictors) and local/regional climate variables (predictands).

The local/regional-scale climate change information is then obtained by applying the relationships to equivalent predictors from AOGCM simulations. The guidance document (Wilby et al., 2004) from the IPCC Task Group on Data and Scenario Support for Impact and Climate Analysis (TGICA) provides a comprehensive background on this approach and covers important issues in using SD applications. Statistical downscaling methods cover regression-type models including both linear and nonlinear relationships, unconditional or conditional weather generators for generating synthetic sequences of local variables, techniques based on weather classification that draw on the more skilful attributes of models to simulate circulation patterns, and analogue methods that seek equivalent weather states from the historical record; a combination of these techniques possibly being most appropriate. An extension to SD is the statistical-dynamical downscaling technique (e.g., Fuentes and Heimann, 2000), which combines weather classification with RCM simulations. A further development is the application of SD to high-resolution climate model output (Lionello et al., 2003; Imbert and Benestad, 2005).

Research on SD has shown an extensive growth in application, and includes an increased availability of generic tools for the impact community (e.g., SDSM, Wilby et al., 2002; the clim.pact package, Benestad, 2004b; the pyclimate package, Fernández and Sáenz, 2003); applications in new regions (e.g., Asia, Chen and Chen, 2003); the use of techniques to address exotic variables such as phenological series (Matulla et al., 2003), extreme heat-related mortality (Hayhoe et al., 2004), ski season (Scott et al., 2003), land use (Solecki and Oliveri, 2004), streamflow or aquatic ecosystems (Cannon and Whitfield, 2002; Blenckner and Chen, 2003); the treatment of climate extremes (e.g., Katz et al., 2002; Seem, 2004; X.L. Wang et al., 2004; Caires et al., 2006); intercomparison studies evaluating methods (e.g., STAtistical and Regional dynamical Downscaling of EXtremes for European regions (STARDEX), Haylock et al., 2006; Schmidli et al., 2006); application to multi-model and multi-ensemble simulations in order to express model uncertainty alongside other key uncertainties (e.g., Benestad, 2002a,b; Hewitson and Crane, 2006; Wang and Swail, 2006b); assessing non-stationarity in climate relationships (Hewitson and Crane, 2006); and spatial interpolation using geographical dependencies (Benestad, 2005). In some cases SD methods have been used to project statistical attributes instead of raw values of the predictand, for example, the probability of rainfall occurrence, precipitation, wind or wave height distribution parameters and extreme event frequency (e.g., Beckmann and Buishand, 2002; Buishand et al., 2004; Busuioc and von Storch, 2003; Abaurrea and Asin, 2005; Diaz-Nieto and Wilby, 2005; Pryor et al., 2005a,b; Wang and Swail, 2006a,b).

Evaluation of SD is done most commonly through cross-validation with observational data for a period that represents an independent or different 'climate regime' (e.g., Busuioc et al. 2001; Trigo and Palutikof, 2001; Bartman et al., 2003; Hanssen-Bauer et al., 2003). Stationarity, that is, whether the statistical relationships are valid under future climate regimes, remains a

concern with SD methods. This is only weakly assessed through cross-validation tests because future changes in climate are likely to be substantially larger than observed historical changes. This issue was assessed in Hewitson and Crane (2006) where, within the SD method used, the non-stationarity was shown to result in an underestimation of the magnitude of the change. In general, the most effective SD methods are those that combine elements of deterministic transfer functions and stochastic components (e.g., Hansen and Mavromatis, 2001; Palutikof et al., 2002; Beersma and Buishand, 2003; Busuioc and von Storch, 2003; Katz et al., 2003; Lionello et al., 2003; Wilby et al., 2003; X.L. Wang et al., 2004; Hewitson and Crane, 2006). Regarding the predictors, the best choice appears to combine dynamical and moisture variables, especially in cases where precipitation is the predictand (e.g., Wilby et al., 2003).

Pattern scaling is a simple statistical method for projecting regional climate change, which involves normalising AOGCM response patterns according to the global mean temperature. These normalised patterns are then rescaled using global mean temperature responses estimated for different emissions scenarios from a simple climate model (see Chapter 10). Some developments were made using various versions of scaling techniques (e.g., Christensen et al., 2001; Mitchell, 2003; Salathé, 2005; Ruosteenoja et al., 2007). For example, Ruosteenoja et al. (2007) developed a pattern-scaling method using linear regression to represent the relationship between the local AOGCM-simulated temperature and precipitation response and the global mean temperature change. Another simple statistical technique is to use the GCM output for the variable of interest (i.e., the predictand) as the predictor and then apply a simple local change factor/scaling procedure (e.g., Chapter 13 of IPCC, 2001; Hanssen-Bauer et al., 2003; Widmann et al., 2003; Diaz-Nieto and Wilby, 2005).

Many studies have been performed since the TAR comparing various SD methods. In general, conclusions about one method compared to another are dependent on region and the criteria used for comparison, and on the inherent attributes of each method. For example, Diaz-Nieto and Wilby (2005) downscale river flow and find that while two methods give comparable results, they differ in responses as a function of how the methods treat multi-decadal variability.

When comparing the merits of SD methods based on daily and monthly downscaling models, in terms of their ability to predict monthly means, daily models are better (e.g., Buishand et al., 2004). In terms of nonlinearity in downscaling relationships, Trigo and Palutikof (2001) note that complex nonlinear models may not be better than simpler linear/slightly nonlinear approaches for some applications. However, Haylock et al. (2006) find that SD methods based on nonlinear artificial neural networks are best at modelling the interannual variability of heavy precipitation but they underestimate extremes. Much downscaling work remains unreported, as SD activities are often implemented pragmatically for serving specific project needs, rather than for use by a broader scientific community; this is especially the case in developing nations. In some cases, this work is only found within the "grey" literature, for example, the AIACC project (http://www.aiaccproject.org/), which supports impact studies in developing nations.

11.10.1.4   *Intercomparison of Downscaling Methods*

At the time of the TAR, SD methods were viewed as a complementary technique to RCMs for downscaling regional climate, each approach having distinctive strengths and weaknesses. The conclusion of the TAR that SD methods and RCMs are comparable for simulating current climate still holds.

Since the TAR, a few additional studies have systematically compared the SD and RCM approaches (e.g., Huth et al., 2001; Hanssen-Bauer et al., 2003, 2005; Wood et al., 2004, Busuioc et al., 2006; Haylock et al., 2006; Schmidli et al., 2006). These related mainly to the similarity of the climate change signal (e.g., Hanssen-Bauer et al., 2003). A more complex study considered additional information about the RCM skill in simulating the current regional climate features and reproducing the connection between large- and regional-scale patterns used for fitting the SD method (Busuioc et al., 2006). Other studies following the STARDEX project (e.g., Haylock et al., 2006; Schmidli et al., 2006) compared the two approaches in terms of their skill in reproducing current climate features, as well as in terms of the climate change signal derived from their outputs, focusing on climate extremes and complex topography processes over Europe.

11.10.2   Quantifying Uncertainties

11.10.2.1 *Sources of Regional Uncertainty*

Most sources of uncertainty at regional scales are similar to those at the global scale (Section 10.5 and Box 10.2), but there are both changes in emphasis and new issues that arise in the regional context. Spatial inhomogeneity of both land use and land cover change (De Fries et al., 2002; Chapter 2, Section 7.2 and Box 11.4) and aerosol forcing adds to regional uncertainty. When analysing studies involving models to add local detail, the full cascade of uncertainty through the chain of models has to be considered. The degree to which these uncertainties influence the regional projections of different climate variables is not uniform. An indication of this is, for example, that models agree more readily on the sign and magnitude of temperature changes than of precipitation changes.

The regional impact of these uncertainties in climate projections has been illustrated by several authors. For example, incorporating a model of the carbon cycle into a coupled AOGCM gave a dramatically enhanced response to climate change over the Amazon Basin (Cox et al., 2000; Jones et al., 2003) and Borneo (Kumagi et al., 2004). Further, the scale of the resolved processes in a climate model can significantly affect its simulation of climate over large regional scales (Pope and Stratton, 2002; Lorenz and Jacob, 2005). Frei et al. (2003) show that models with the same representation of resolved

processes but different representations of sub-grid scale processes can represent the climate differently. The regional impact of changes in the representation of the land surface feedback is demonstrated by, for example, Oleson et al. (2004) and Feddema et al. (2005) (see also Box 11.4).

Evaluation of uncertainties at regional and local scales is complicated by the smaller ratio of the signal to the internal variability, especially for precipitation, which makes the detection of a response more difficult. In addition, the climate may itself be poorly known on regional scales in many data-sparse regions. Thus, evaluation of model performance as a component of an analysis of uncertainty can itself be problematic.

11.10.2.2   *Characterising and Quantifying Regional Uncertainty*

11.10.2.2.1 *Review of regional uncertainty portrayed in the TAR*

In the TAR, uncertainties in regional climate projections were discussed, but methods for quantifying them were relatively primitive. For example, in the TAR chapter on regional projections (Giorgi et al., 2001a), uncertainties in regional projections of climate change (e.g., large or small increases/decreases in precipitation) from different GCMs were qualitatively portrayed based only on simple agreement heuristics (e.g., seven of the nine models showed increases). Early examples of quantitative estimates of regional uncertainty include portraying the median and inter-model range of a variable (e.g., temperature) across a series of model projections and attaching probabilities to a group of scenarios on a regional scale (Jones, 2000; New and Hulme, 2000).

11.10.2.2.2   *Using multi-model ensembles*

A number of studies have taken advantage of multi-model ensembles formed by GCMs that have been driven by the same forcing scenarios to generate quantitative measures of uncertainty, particularly probabilistic information at a regional scale. Table 11.3 summarises aspects of the methods reviewed in this section and in Section 11.10.2.2.3. The results highlighted in Section 10.5 and Box 10.2 on climate sensitivity demonstrate that multi-model ensembles explore only a limited range of the uncertainty. In addition, the distribution of GCM sensitivities is not by construction a representative sample from those probability distributions and thus the regional probabilities generated using multi-model ensembles will not represent the full spread of possible regional changes.

Räisänen and Palmer (2001) used 17 GCMs forced with an idealised annual increase in atmospheric CO_2 of 1% to calculate the probability of exceedance of thresholds of temperature increase (e.g., >1°C) and precipitation change (e.g., <–10%). These were used to demonstrate that a probabilistic approach has advantages over conventional deterministic estimates by demonstrating the economic value of a probabilistic assessment of future climate change. Giorgi and Mearns (2002) developed measures of uncertainty for regional temperature

Table 11.3. *Methods for generating probabilistic information from future climate simulations at continental and sub-continental scales, SRES-scenario specific. Results from the methods of Greene et al. (2006) and Tebaldi et al. (2004a,b) are displayed in Figure 11.26.*

Reference	Experiment	Input Type — Spatial Scale	Time Resolution	Synthesis Method and Results	Methodological Assumptions	Model Performance Evaluation
Furrer et al. (2007)	Multi-model Ensemble	Grid points (after interpolation to common grid)	Seasonal multi-decadal averages	Bayesian approach. AOGCMs are assumed independent. Large-scale patterns projected on basis functions, small-scale modelled as an isotropic Gaussian process. Spatial dependence fully accounted for by spatial model. PDFs at grid point level, jointly derived accounting for spatial dependence	AOGCMs are assumed independent.	Model performance not explicitly brought to bear.
Giorgi and Mearns (2003)	Multi-model Ensemble	Regional averages (Giorgi and Francisco, 2000)	Seasonal multi-decadal averages	Cumulative Distribution Functions (CDFs) derived by counting threshold exceedances among members, and weighing the counts by the REA method. Stepwise CDFs at the regional levels		Model performance (bias and convergence) explicitly quantified in each AOGCMs' weight. Observable at same spatial scale and time resolution, for period 1961 to 1990.
Greene et al. (2006)	Multi-model Ensemble	Regional averages (Giorgi and Francisco, 2000)	Seasonal and annual averages	Bayesian approach. AOGCMs dependence is modelled. Linear regression of observed values on model's values (similar to Model-Output-Statistics approach used in weather forecasting and seasonal forecasting) with coefficients estimates applied to future simulations. PDFs at regional level	AOGCMs dependence is modelled.	Model performance measured on 1902 to 1998 historical trend reproduction at same spatial scale and time resolution.
Harris et al. (2006)	Perturbed Physics Ensemble (PPE)	Grid points	Seasonal multi-annual averages	Scaled equilibrium response patterns from a large slab-model Perturbed Physics Ensemble (PPE), using transient responses of an Energy Balance Model driven by PPE climate feedbacks. Quantifying scaling error, against a smaller PPE of transient simulations, to include in PDFs. PDFs at arbitrary level of aggregation		All model versions assumed equally likely.
Stott et al. (2006a)	Single Model (HadCM3)	Continental averages	Annual decadal averages	Linear scaling factor estimated through optimal fingerprinting approach at continental scales or at global scale and applied to future projections, with estimated uncertainty. Natural variability estimated from control run added as additional uncertainty component. PDFs at the continental-scale level		Not applicable
Tebaldi et al. (2004a,b)	Multi-model Ensemble	Regional averages (Giorgi and Francisco, 2000)	Seasonal multi-decadal averages	Bayesian approach. AOGCMs are assumed independent. Normal likelihood for their projections, with AOGCM-specific variability. PDFs at the regional level	AOGCMs are assumed independent.	Model performance (bias and convergence) implicitly brought to bear through likelihood assumptions. Observable at same spatial scale and time resolution, for period 1961 to 1990 in original papers, for period 1980 to 1999 for results displayed in this report.

and precipitation change by weighting model results according to biases in their simulation of present-day climate and convergence of their projections to the ensemble's mean. Their Reliability Ensemble Average (REA) method was applied to the nine GCMs assessed in the TAR to provide uncertainty estimates separately for the SRES A2 and B2 emission scenarios for 22 large sub-continental regions.

Tebaldi et al. (2004a,b) used a Bayesian approach to define a formal statistical model for deriving probabilities from an ensemble of projections forced by a given SRES scenario. Using the Giorgi and Mearns (2002, 2003) approach, model bias and convergence criteria determine the shape and width of the posterior probability density functions (PDFs) of temperature and precipitation change signals. Expert judgement can be incorporated in the form of prior distributions that have the effect of assigning different relative weights to the two criteria (Tebaldi et al., 2004b; Lopez et al., 2006). The method developed by Furrer et al. (2007) to combine GCM output at the grid point scale into probabilistic projections is described in detail in Chapter 10. By straightforward area averaging, PDFs of climate change at the regional scale can be obtained. When this is done for the Giorgi and Francisco (2000) regions, the regional PDFs from Furrer et al. (2007) agree overall with the

empirical histogram of the ensemble projections and the Tebaldi et al. (2004b) PDFs, with relatively small differences in spread and generally no clear difference in location.

Greene et al. (2006) used a Bayesian framework to model an ensemble of GCM projections under individual SRES scenarios by an extension of methods used for seasonal ensemble forecasting. The set of GCM simulations of the observed period 1902 to 1998 are individually aggregated in area-averaged annual or seasonal time series and jointly calibrated through a linear model to the corresponding observed regional trend. The calibration coefficients and their uncertainty are estimated and then applied to the future projections to provide probabilistic forecasts of future trends. Two critical assumptions are responsible for this method's results being so different from the ensemble projections or the PDFs produced by Tebaldi et al. (2004a,b) (see Figure 11.26 and Supplementary Material Figures S11.33 to S11.35). Firstly, the method attributes large uncertainty to models that are unable to reproduce historical trends despite the uncertainty in the relatively weak forcings in the historical period and the large natural variability at regional scales. Second, a strong stationarity assumption is required to extrapolate the relationship derived over the historical record to future trends, which involve a different combination of and

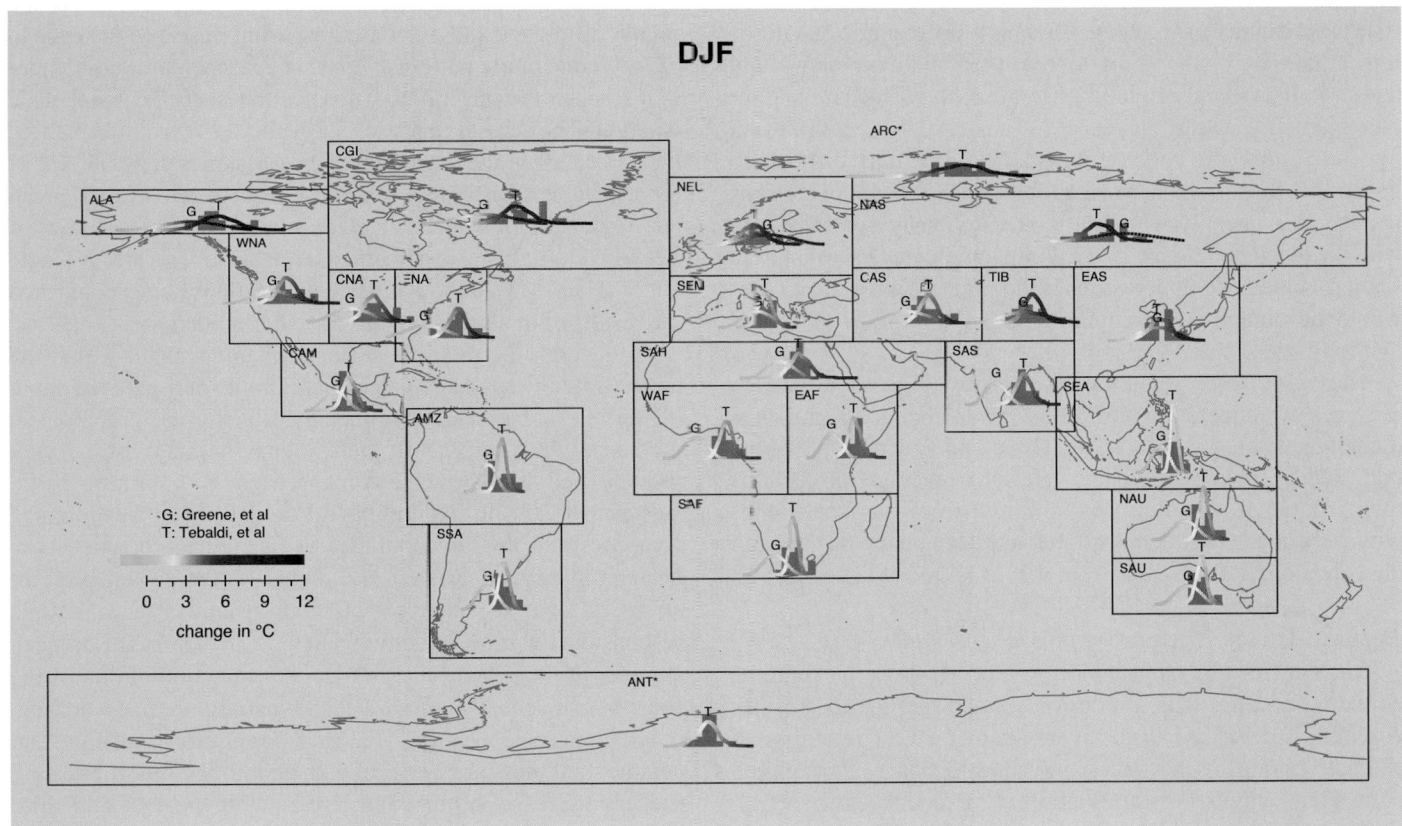

Figure 11.26. *Map comparing PDFs of change in temperature (2080 to 2099 compared to 1980 to 1999) from Tebaldi et al. (2004a,b) and Greene et al. (2006) as well as the raw model projections (represented by shaded histograms) for the Giorgi and Francisco (2000) regions. Areas under the curves and areas covered by the histograms have been scaled to equal unity. The scenario is SRES A1B and the season is NH winter (DJF). Asterisks adjacent to ARC and ANT regions indicate that only the Tebaldi et al. results were available.*

some significantly stronger forcings. The significantly smaller warming and the large width of the PDFs (at times including negative values) labelled by a 'G' in Figure 11.26 are then interpretable as a result of this stationarity constraint and the large uncertainty in the fitting of the trends. They contrast starkly with the larger warming represented in the histograms of model projections and their synthesis in the Tebaldi et al. (2004a,b) and in the Furrer et al. (2007, not shown) PDFs. This is particularly so in the lower-latitude regions of Africa, South Asia and the SH, possibly as a consequence of particularly weak trends in the observations and/or relatively worse performance of the GCMs.

Dessai et al. (2005) apply the idea of simple pattern scaling (Santer et al., 1990) to a multi-model ensemble of AOGCMs. They 'modulate' the normalised regional patterns of change by the global mean temperature changes generated under many SRES scenarios and climate sensitivities through the Model for the Assessment of Greenhouse-Gas Induced Climate Change (MAGICC), a simple probabilistic energy balance model (Wigley and Raper, 2001). Their work focuses on measuring the changes in PDFs as a function of different sources of uncertainty. In this analysis, the impact of the SRES scenarios turns out to be the most relevant for temperature changes, particularly in the upper tail of the distributions, while the GCM weighting does not produce substantial differences. This result is probably dependent on the long horizon of the projections considered (late 21st century). Arguably, the emission scenario would be less important in the short to mid-term. Climate sensitivity has an impact mainly in the lower tail of the distributions. For precipitation changes, all sources of uncertainty seem relevant but the results are very region-specific and thus difficult to generalise. More work to test the robustness of these conclusions is needed, especially when these are obviously not consistent with the results in Figure 10.29. For example, the use of pattern scaling is likely to underestimate the range of projections that would be obtained by running a larger ensemble of GCMs (Murphy et al., 2004).

The work described above has involved either large-area averages of temperature and precipitation change or statistical modelling at the grid box scale. Good and Lowe (2006) show that trends in large-area and grid-box average projections of precipitation are often very different from the local trends within the area. This demonstrates the inadequacy of inferring the behaviour at fine scales from that of large-area averages.

11.10.2.2.3   Using perturbed physics ensembles

Another method for exploring uncertainties in regional climate projections is the use of large perturbed physics ensembles (described in detail in Chapter 10). These allow a characterisation of the uncertainty due to poorly constrained parameters within the formulation of a model. Harris et al. (2006) combined the results from a 17-member ensemble (Collins et al., 2006) with a larger perturbed physics ensemble, investigating the equilibrium climate response to a doubling of atmospheric CO_2 (Webb et al., 2006). They developed a

bridge between spatial patterns of the transient and equilibrium climate response by way of simple pattern scaling (Santer et al., 1990), allowing results from the large ensemble to be translated into PDFs of time-dependent regional changes. Uncertainties in surface temperature and precipitation changes are derived (Supplementary Material Figures S11.36 and S11.37), which arise from the poorly constrained atmospheric model parameters, internal variability and pattern scaling errors. The latter are quantified by comparing the scaled equilibrium response with the transient response for 17 model versions with identical parameter settings. Errors introduced by the pattern-scaling technique are largest when the transient response varies nonlinearly with global temperature, as is the case for precipitation in certain regions.

11.10.2.2.4   Other approaches to quantifying regional uncertainty

As described in Chapter 10, Stott and Kettleborough (2002) provide PDFs of future change in climate by making use of the robust observational constraints on a climate model's response to greenhouse gas and sulphate aerosol forcings that underpin the attribution of recent climate change to anthropogenic sources. The study by Stott et al. (2006a) is the first to adapt this method for continental scales. It considers two methods of constraining future continental temperature projections, one based on using observed historical changes only over the region of interest and one based on using observed changes in global temperature patterns. The first approach produces wider PDFs, since the uncertainty of detection at the regional scale is larger. The second approach incorporates more information, hence reducing the uncertainty, but assumes that the GCM represents correctly the relationship between global mean and regional temperature change. In contrast to the studies of Section 11.10.2.2.2, this work uses projections from a single GCM (HadCM3), although Stott et al. (2006b) have confirmed the results of this methodology for other models.

In general, the regional sections of this chapter assess the uncertainty in regional changes based on expert understanding of the relevant processes, rather than by formal probabilistic methods, which are still in their infancy and currently do not provide definitive results. An approach to a process-based assessment of the reliability of modelled climate change responses and thus uncertainties in its future projections has been proposed by Rowell and Jones (2006). They perform an assessment of the physical and dynamical mechanisms responsible for a specific future outcome, in their case European summer drying. Their analysis isolates the contribution of the four major mechanisms analysed: the spatial pattern of warming, other large-scale changes, reduced spring soil moisture and summer soil moisture feedbacks. In certain regions, the second process makes a minor contribution with the first and third dominating. This leads to the conclusion that the sign of the change is robust as confidence in the processes underlying these mechanisms is high.

11.10.2.2.5   Combined uncertainties: General Circulation Models, emissions and downscaling techniques

It is important to quantify the relative importance of the uncertainty arising from the downscaling step (from the RCM formulation or the assumptions underlying an empirical SD method) against the other sources of uncertainty. For example, in the application of SD methods to probabilistic scenarios, Benestad (2002b, 2004a) used a multi-model ensemble coupled to SD to derive tentative probabilistic scenarios at a regional scale for northern Europe.

The PRUDENCE project (Box 11.2) provided the first opportunity to weigh these various sources of uncertainty for simulations over Europe. Rowell (2005) evaluated a four-dimensional matrix of climate modelling experiments that included two different emissions scenarios, four different GCM experiments, multiple ensemble members within the latter to assess internal variability, and nine different RCMs, for the area of the British Isles. He found that the dynamical downscaling added a small amount of uncertainty compared to the other sources for temperature evaluated as monthly/seasonal averages. For precipitation, the relative contributions of the four sources of uncertainty are more balanced. Déqué et al. (2005, 2007) show similar results for the whole of Europe, as do Ruosteenoja et al. (2007) for subsections of Europe. Kjellström et al. (2007) find that the differences among different RCMs driven by the same GCM become comparable to those among the same RCM driven by different GCMs when evaluating daily maximum and minimum temperatures. However, mean responses in the PRUDENCE RCMs were often quite different from that of the driving GCM. This suggests that some of the spread in RCM responses may be unrealistic due to model inconsistency (Jones et al., 1997). However, it should be noted that only a few of the RCMs in PRUDENCE were driven by more than one GCM, which adds further uncertainty regarding these conclusions. Other programs similar to PRUDENCE have begun for other regions of the world, such as NARCCAP over North America (Mearns et al., 2005), Regional Climate Change Scenarios for South America (CREAS; Marengo and Ambrizzi, 2006), and the Europe-South America Network for Climate Change Assessment and Impact Studies (CLARIS; http://www.claris-eu.org) over South America.

References

Abaurrea, J., and J. Asin, 2005: Forecasting local daily precipitation patterns in a climate change scenario. *Clim. Res.*, **28**, 183–197.

Abbs, D.J., 2004: A high resolution modelling study of the effect of climate change on the intensity of extreme rainfall events. In: *Staying Afloat: Floodplain Management Authorities of NSW 44th Annual Conference: Conference Proceedings, Coffs Harbour, NSW*. Floodplain Management Authorities of New South Wales, Tamworth, pp. 17–24.

ACIA, 2005: *Arctic Climate Impact Assessment*. Cambridge University Press, New York, 1042 pp.

Adam, J.C., and D.P. Lettenmeier, 2003: Adjustment of global gridded precipitation for systematic bias. *J. Geophys. Res.*, **108**, 4257–4272.

Adams, N., 2004: A numerical modelling study of the weather in East Antarctica and the surrounding Southern Ocean. *Weather Forecasting*, **19**, 653–672.

AIACC (Assessments of Impacts and Adaptations to Climate Change in Multiple Regions and Sectors), 2004: *AIACC Regional Study AS07: Southeast Asia Regional Vulnerability to Changing Water Resource and Extreme Hydrological Events due to Climate Change. Progress Report: Period Year-end 2003*. 8 pp., http://sedac.ciesin.columbia.edu/aiacc/progress/AS07_Jan04.pdf.

Aldrian, E., and R. Dwi Susanto, 2003: Identification of three dominant rainfall regions within Indonesia and their relationship to sea surface temperature. *Int. J. Climatol.*, **23**(12), 1435–1452.

Aldrian, E., et al., 2004a: Long term simulation of the Indonesian rainfall with the MPI Regional Model. *Clim. Dyn.*, **22**(8), 794–814, doi:10.1007/s00382-004-0418-9.

Aldrian, E., et al., 2004b: Modelling Indonesian rainfall with a coupled regional model. *Clim. Dyn.*, **25**(1), 1–17, doi:10.1007/s00382-004-0483-0.

Anderson, C.J., et al., 2003: Hydrological processes in regional climate model simulations of the Central United States flood of June-July 1993. *J. Hydrometeor.*, **4**, 584–598.

Angeles, M.E., J.E. Gonzalez, D.J. Erickson, and J.L. Hernández, 2007: Predictions of future climate change in the Caribbean region using global general circulation models. *Int. J. Climatol.*, **27**, 555-569, doi:10.1002/joc.1416.

Antic, S., R. Laprise, B. Denis, and R. de Elia, 2005: Testing the downscaling ability of a one-way nested regional climate model in regions of complex topography. *Clim. Dyn.*, **23**, 473–493.

Anyah, R., and F. Semazzi, 2004: Simulation of the sensitivity of Lake Victoria basin climate to lake surface temperatures. *Theor. Appl. Climatol.*, **79**(1–2), 55–69.

Arakawa, O., and A. Kitoh, 2005: Rainfall diurnal variation over the Indonesian Maritime Continent simulated by 20km-mesh GCM. *Scientific Online Letters on the Atmosphere*, **1**, 109–112.

Arnell, N., D. Hudson, and R. Jones, 2003: Climate change scenarios from a regional climate model: Estimating change in runoff in southern Africa. *J. Geophys. Res.*, **108**(D16), 4519, doi:10.1029/2002JD002782.

Arnfield, A.J., 2003: Two decades of urban climate research: a review of turbulence, exchanges of energy and water, and the urban heat island. *Int. J. Climatol.*, **23**, 1–26.

Ashok, K., Z.Y. Guan, and T. Yamagata, 2001: Impact of the Indian Ocean Dipole on the relationship between the Indian monsoon rainfall and ENSO. *Geophys. Res. Lett.*, **28**, 4499–4502.

Ashrit, R.G., K. Rupa Kumar, and K. Krishna Kumar, 2001: ENSO-monsoon relationships in a greenhouse warming scenario. *Geophys. Res. Lett.*, **29**, 1727–1730.

Ashrit, R.G., H. Douville, and K. Rupa Kumar, 2003: Response of the Indian monsoon and ENSO-monsoon teleconnection to enhanced greenhouse effect in the CNRM coupled model. *J. Meteorol. Soc. Japan*, **81**, 779–803.

Augustine, J.A., and F. Caracena, 1994: Lower-tropospheric precursors to nocturnal MCS development over central United States. *Weather Forecasting*, **9**, 116–135.

Avissar, R., and D. Werth, 2005: Global hydroclimatological teleconnections resulting from tropical deforestation. *J. Hydrometeor.*, **6**, 134–145.

Bader, J., and M. Latif, 2003: The impact of decadal-scale Indian Ocean sea surface temperature anomalies on Sahelian rainfall and the North Atlantic Oscillation. *Geophys. Res. Lett.*, **30**(22), 2166–2169, doi.10.1029/2003GL018426.

Bailey, D.A., and A.H. Lynch, 2000: Development of an Antarctic regional climate system model: Part 2. Station validation and surface energy balance. *J. Clim.*, **13**, 1351–1361.

Bailey, D.A., A.H. Lynch, and T.E. Arbetter, 2004: The relationship between synoptic forcing and polynya formation in the Cosmonaut Sea, II: Polynya simulation. *J. Geophys. Res*, **109**, doi:10.1029/2003JC001838.

Barnett, D.N., et al., 2006: Quantifying uncertainty in changes in extreme event frequency in response to doubled CO_2 using a large ensemble of GCM simulations. *Clim. Dyn.*, **26**, 489–511.

Barnett, T.P., J.C. Adam, and D.P. Lettenmeier, 2005: Potential impacts of a warming climate on water availability in snow-dominated regions. *Nature*, **438**, 303–309, doi:10.1038/nature04141.82511-825179.

Bartman, A.G., W.A. Landman, and C.J. de W. Ratenbach, 2003: Recalibration of general circulation model output to Austral summer rainfall over Southern Africa. *Int. J. Climatol.*, **23**, 1407–1419.

Becker, A., and H. Bugmann (eds.), 1997: *Predicting Global Change Impacts on Mountain Hydrology and Ecology: Integrated Catchment Hydrology/Altitudinal Gradient Studies*. IGBP Report 43, International Geosphere-Biosphere Programme, Stockholm.

Beckmann, B.R., and T.A. Buishand, 2002: Statistical downscaling relationship for precipitation in the Netherlands and North Germany. *Int. J. Climatol.*, **22**, 15–32.

Beersma, J.J., and T.A. Buishand, 2003: Multi-site simulation of daily precipitation and temperature conditional on atmospheric circulation. *Clim. Res.*, **25**, 121–133.

Bell, J.L., L.C. Sloan, and M.A. Snyder, 2004: Changes in extreme climatic events: A future climate scenario. *J. Clim.*, **17**(1), 81–87.

Benestad, R.E., 2002a: Empirically downscaled temperature scenarios for Northern Europe based on a multi-model ensemble. *Clim. Res.*, **21**(2), 105–125.

Benestad, R.E., 2002b: Empirically downscaled multimodel ensemble temperature and precipitation scenarios for Norway. *J. Clim.*, **15**, 3008–3027.

Benestad, R.E., 2004a: Tentative probabilistic temperature scenarios for Northern Europe. *Tellus*, **56A**(2), 89–101.

Benestad, R.E., 2004b: Empirical-statistical downscaling in climate modeling. *Eos*, **85**(42), 417.

Benestad, R.E., 2005: Climate change scenarios for northern Europe from multi-model IPCC AR4 climate simulations. *Geophys. Res. Lett.*, **32**, L17704, doi:10.1029/2005GL023401.

Bengtsson, L., 1996: The climate response to the changing greenhouse gas concentration in the atmosphere. In: *Decadal Climate Variability, Dynamics And Variability* [Anderson, D.L.T., and J. Willebrand (eds.)]. NATO ASI Series 44, Springer, Berlin, 493 pp.

Bengtsson, L., V.A. Semenov, and O.M. Johannessen, 2004: The early twentieth-century warming in the Arctic - a possible mechanism. *J. Clim.*, **17**, 4045–4057.

Beniston, M., and P. Jungo, 2001: Shifts in the distributions of pressure, temperature and moisture in the alpine region in response to the behavior of the North Atlantic Oscillation. *Theor. Appl. Climatol.*, **71**, 29–42.

Beniston, M., F. Keller, B. Koffi, and S. Goyette, 2003: Estimates of snow accumulation and volume in the Swiss Alps under changing climatic conditions. *Theor. Appl. Climatol.*, **76**, 125–140.

Beniston, M., et al., 2007: Future extreme events in European climate: An exploration of regional climate model projections. *Clim. Change*, doi:10.1007/s10584-006-9226-z.

Bertler, N.A.N., P.J. Barrett, P.A. Mayewski, and R.L. Fogt, 2004: El Niño suppresses Antarctic warming. *Geophys. Res. Lett.*, **31**, L15207, doi:10.1029/2004GL020749.

Betts, A.K., 1998: Climate-convection feedbacks: Some further issues. *Clim. Dyn.*, **39**(1), 35–38.

Betts, R.A., et al., 2004: The role of ecosystem-atmosphere interactions in simulated Amazonian precipitation decrease and forest dieback under global climate warming. *Theor. Appl. Climatol.*, **78**, 157–175.

Biasutti, M., and A. Giannini, 2006: Robust Sahel drying in response to late 20th century forcings. *Geophys. Res. Lett.*, **33**, L11706, doi:10.1029/2006GRL026067.

Biner, S., D. Caya, R. Laprise and L. Spacek, 2000: Nesting of RCMs by imposing large scales. In: *Research Activities in Atmospheric and Oceanic Modelling*. WMO/TD No. 987, Report No. 30, World Meteorological Organization, Geneva, pp. 7.3–7.4.

Black, E., et al., 2004: Factors contributing to the summer 2003 European heatwave. *Weather*, **59**, 217–223.

Blenckner, T. and D. Chen, 2003: Comparison of the impact of regional and North-Atlantic atmospheric circulation on an aquatic ecosystem. *Clim. Res.*, **23**, 131–136.

Boer, R., and A. Faqih, 2004: *Current and Future Rainfall Variability in Indonesia*. AIACC Technical Report 021, http://sedac.ciesin.columbia.edu/aiacc/progress/AS21_Jan04.pdf.

Bogdanova, E.G, B.M. Ilyin, and I.V. Dragomilova, 2002: Application of a comprehensive bias correction model to application of a comprehensive bias correction model to stations. *J. Hydrometeorol.*, **3**, 700–713.

Bojariu, R., and F. Giorgi, 2005: The North Atlantic Oscillation signal in a regional climate simulation for the European region. *Tellus*, **57A**(4), 641–653.

Bonan, G.B., 2001: Observational evidence for reduction of daily maximum temperature by croplands in the Midwest United States. *J. Clim.*, **14**, 2430–2442.

Boo, K.-O., W.-T. Kwon, and J.-K. Kim, 2005: Vegetation changes in the regional surface climate over East Asia due to global warming using BIOME4. *Il Nuovo Cimento*, **27**(4), 317–327.

Boo, K.-O., W.-T. Kwon, and H.-J. Baek, 2006: Change of extreme events of temperature and precipitation over Korea using regional projection of future climate change. *Geophys. Res. Lett.*, **33**(1), L01701, doi:10.1029/2005GL023378.

Booij, M.J., 2002: Extreme daily precipitation in western Europe with climate change at appropriate spatial scales. *Int. J. Climatol.*, **22**, 69–85.

Bordoni, S., et al., 2004: The low-level circulation of the North American Monsoon as revealed by QuikSCAT. *Geophys. Res. Lett.*, **31**, L10109, doi:10.1029/2004GL020009.

Boulanger, J.P., F. Martinez, and E.C. Segura, 2006: Projection of future climate change conditions using IPCC simulations, neural networks and Bayesian statistics. Part 1: Temperature mean state and seasonal cycle in South America. *Clim. Dyn.*, **27**, 233–259.

Bretherton, C.S., and D.S. Battisti, 2000: An interpretation of the results from atmospheric general circulation models forced by the time history of the observed sea surface temperature distribution. *Geophys. Res. Lett.*, **27**, 767–770.

Bromwich, D.H., and R.L. Fogt, 2004: Strong trends in the skill of the ERA-40 and NCEP/NCAR Reanalyses in the high and middle latitudes of the Southern Hemisphere, 1958-2001. *J. Clim.*, **17**, 4603–4619.

Bromwich, D.H., A.J. Monaghan, and Z. Guo, 2004a: Modeling the ENSO modulation of Antarctic climate in the late 1990s with the Polar MM5. *J. Clim.*, **17**, 109–132.

Bromwich, D.H., A.J. Monaghan, K.W. Manning, and J.G. Powers, 2004b: Real-time forecasting for the Antarctic: An evaluation of the Antarctic Mesoscale Prediction System (AMPS). *Mon. Weather Rev.*, **133**, 579–603.

Bromwich, D.H., et al., 2000: ECMWF analyses and reanalyses depiction of ENSO signal in Antarctic precipitation. *J. Clim.*, **13**, 1406–1420.

Brovkin, V., et al., 1999: Modelling climate response to historical land cover change. *Global Ecol. Biogeogr.*, **8**, 509–517.

Buishand, T.A., M.V. Shabalova, and T. Brandsma, 2004: On the choice of the temporal aggregation level for statistical downscaling of precipitation. *J. Clim.*, **17**, 1816–1827.

Busuioc, A., and H. von Storch, 2003: Conditional stochastic model for generating daily precipitation time series. *Clim. Res.*, **24**, 181–195.

Busuioc, A., D. Chen, and C. Hellström, 2001: Performance of statistical downscaling models in GCM validation and regional climate change estimates: application for Swedish precipitation. *Int. J. Climatol.*, **21**(5), 557–578.

Busuioc, A., F. Giorgi, X. Bi, and M. Ionita, 2006: Comparison of regional climate model and statistical downscaling simulations of different winter precipitation change scenarios over Romania. *Theor. Appl. Climatol.*, **86**, 101–120.

Cai, W., P.H. Whetton, and D.J. Karoly, 2003a: The response of the Antarctic Oscillation to increasing and stabilized atmospheric CO_2. *J. Clim.*, **16**, 1525–1538.

Cai, W., et al., 2003b: *Climate Change in Queensland under Enhanced Greenhouse Conditions. Annual Report, 2003*. CSIRO Atmospheric Research, Aspendale, Vic., 74 pp, http://www.longpaddock.qld.gov.au/ClimateChanges/pub/CSIRO2003.html#end.

Caires, S., V.R. Swail, and X.L. Wang, 2006: Projection and analysis of extreme wave climate. *J. Clim.*, **19**, 5581–5605.

Caminade, C., L. Teray, and E. Maisonnave, 2006: West African monsoon system response to greenhouse gas and sulphate aerosol forcing under two emission scenarios. *Clim. Dyn.*, **26**, 531–547.

Cannon, A., and P. Whitfield, 2002: Downscaling recent streamflow conditions in British Columbia, Canada using ensemble neural network models. *J. Hydrol.*, **259**(1–4), 136–151.

Carleton, A.M., 2003: Atmospheric teleconnections involving the Southern Ocean. *J. Geophys. Res.*, **108**, 8080, doi:10.1029/2000JC000379.

Carril, A.F., C.G. Menéndez, and A. Navarra, 2005: Climate response associated with the Southern Annular Mode in the surroundings of Antarctic Peninsula: a multi-model ensemble analysis. *Geophys. Res. Lett.*, **32**, L16713, doi:10.1029/2005GL023581.

Cassano, J.J., P. Uotila, and A. Lynch, 2006: Changes in synoptic weather patterns in the polar regions in the 20th and 21st centuries. Part 1: Arctic. *Int. J. Climatol.*, **26**, 1027–1049, doi:10.1002/joc.1306.

Caya, D., and S. Biner, 2004: Internal variability of RCM simulations over an annual cycle. *Clim. Dyn.*, **22**, 33–46.

Cayan, D.R., et al., 2001: Changes in the onset of spring in the western United States. *Bull. Am. Meteorol. Soc.*, **82**, 399–415.

Cazes Boezio, G., A.W. Robertson, and C.R. Mechoso, 2003: Seasonal dependence of ENSO teleconnections over South America and relationships with precipitation in Uruguay. *J. Clim.*, **16**(8), 1159–1176.

Chaboureau, J.P., F. Guichard, J.L. Redelsperger, and J.P. Lafore, 2004: The role of stability and moisture in the diurnal cycle of convection over land. *Q. J. R. Meteorol. Soc.*, **130**, 3105–3117.

Chapman, W.L., and J.E. Walsh, 2006: A synthesis of Antarctic temperatures. *J. Clim.*, **26**, 1181–2119, doi:10.1002/joc.1305.

Chapman, W.L., and J.E. Walsh, 2007: Simulations of Arctic temperature and pressure by global coupled models. *J. Clim.*, **20**, 609-632, doi:10.1175/JCLI4026.1.

Charles, S.P., B.C. Bates, I.N. Smith, and J.P. Hughes, 2004: Statistical downscaling of daily precipitation from observed and modelled atmospheric fields. *Hydrolog. Process.*, **18**(8), 1373–1394.

Chase, T.N., et al., 2000: Simulated impacts of historical land cover changes on global climate in northern winter. *Clim. Dyn.*, **16**, 93–105.

Chen, A.A., and M.A. Taylor, 2002: Investigating the link between early season Caribbean rainfall and the El Niño +1 year. *Int. J. Climatol.*, **22**, 87–106.

Chen, D.L., and Y.M. Chen, 2003: Association between winter temperature in China and upper air circulation over East Asia revealed by canonical correlation analysis. *Global Planet. Change*, **37**, 315–325.

Chen, M., D. Pollard, and E.J. Barron, 2003: Comparison of future climate change over North America simulated by two regional climate models. *J. Geophys. Res.*, **108**(D12), 4348, doi:10.1029/2002JD002738.

Chen, T.-C., J.-H. Yoon, K.J. St. Croix, and E.S. Takle, 2001: Suppressing impacts of the Amazonian deforestation by the global circulation change. *Bull. Am. Meteorol. Soc.*, **82**, 2209–2216.

Chen, T.-C., S.-Y. Wang, W.-R. Huang, and M.-C. Yen, 2004: Variation of the East Asian summer monsoon rainfall. *J. Clim.*, **17**, 744–762.

Chou, C., and J.D. Neelin, 2004: Mechanisms of global warming impacts of regional tropical precipitation. *J. Clim.*, **17**, 2688–2701.

Chou, C., J.D. Neelin, J.-Y. Tu, and C.-T. Chen, 2007: Regional tropical precipitation change mechanisms in ECHAM4/OPYC3 under global warming. *J. Clim.* **19**, 4207-4223..

Chou, S.C., A.M.B. Nunes, and I.F.A. Cavalcanti, 2000: Extended range forecasts over South America using the regional eta model. *J. Geophys. Res.*, **105**, 10147–10160.

Christensen, J.H., and O.B. Christensen, 2003. Severe summertime flooding in Europe. *Nature*, **421**, 805–806.

Christensen, J.H., T. Carter, and F. Giorgi, 2002: PRUDENCE employs new methods to assess European climate change. *Eos*, **83**, 147.

Christensen, J.H., T.R. Carter, M. Rummukainen, and G. Amanatidis, 2007: Evaluating the performance and utility of regional climate models: the PRUDENCE project. *Clim. Change*, doi:10.1007/s10584-006-9211-6.

Christensen, J.H., et al., 2001: A synthesis of regional climate change simulations – A Scandinavian perspective. *Geophys. Res. Lett.*, **28**(6), 1003–1006.

Christensen, O.B., and J.H. Christensen, 2004: Intensification of extreme European summer precipitation in a warmer climate. *Global Planet. Change*, **44**, 107–117.

Church, J.A., N.J. White, and J.R. Hunter, 2006: Sea level rise at tropical Pacific and Indian Ocean islands. *Global Planet. Change*, **53**(3), 155–168.

Church, J.A., et al., 2004: Estimates of regional distribution of sea level rise over the 1950-2000 period. *J. Clim.*, **17**, 2609–2625.

Clark, R., S. Brown, and J. Murphy, 2006: Modelling Northern Hemisphere summer heat extreme changes and their uncertainties using a physics ensemble of climate sensitivity experiments. *J. Clim.*, **19**, 4418–4435.

Claussen, M., C. Kutzbaki, V. Brovkin, and A. Ganapolski, 1999: Simulation of an abrupt change in Saharan vegetation in the mid-Holocene. *Geophys. Res. Lett.*, **26**, 2037–2040.

Collier, J.C., K.P. Bowman, and G.R. North, 2004: A comparison of tropical precipitation simulated by the community climate model with that measured by the tropical rainfall measuring mission satellite. *J. Clim.*, **17**, 3319–3333.

Collins, M., et al., 2006: Towards quantifying uncertainty in transient climate change. *Clim. Dyn.*, **27**, 127–147.

Comiso, J.C., 2000: Variability and trends in Antarctic surface temperatures from in situ and satellite infrared measurements. *J. Clim.*, **13**, 1674–1696.

Connolley, W.M., and S.A. Harangozo, 2001: A comparison of five numerical weather prediction analysis climatologies in southern high latitudes. *J. Clim.*, **14**, 30–44.

Cook, K.H., and E.K. Vizy, 2006: Coupled model simulations of the West African monsoon system: twentieth-century simulations and twenty-first-century predictions. *J. Clim.*, **19**, 3681–3703.

Coppola, E., and F. Giorgi, 2005: Climate change in tropical regions from high-resolution time-slice AGCM experiments. *Q. J. R. Meteorol. Soc.*, **131**(612), 3123–3145.

Costa, M.H., and J.A. Foley, 2000: Combined effects of deforestation and doubled atmospheric CO_2 concentrations on the climate of Amazonia. *J. Clim.*, **13**, 35–58.

Covey, C., et al., 2003: An overview of results from the Coupled Model Intercomparison Project (CMIP). *Global Planet. Change*, **37**, 103–133, doi:10.1016/S0921-8181(02)00193-5.

Cox, P.M., et al., 2000: Acceleration of global warming due to carbon-cycle feedbacks in a coupled climate model. *Nature*, **408**, 184–187.

Cox, P.M., et al., 2004: Amazonian forest dieback under climate-carbon cycle projections for the 21st century. *Theor. Appl. Clim.*, **78**, 137–156.

CSIRO (Commonwealth Scientific and Industrial Research Organisation), 2001: *Climate Projections for Australia*. CSIRO Atmospheric Research, Melbourne, 8 pp., http://www.dar.csiro.au/publications/projections2001.pdf.

Curtis, J., G. Wendler, R. Stone, and E. Dutton, 1998: Precipitation decrease in the western Arctic, with special emphasis on Barrow and Barter Island, Alaska. *Int. J. Climatol.*, **18**, 1687–1707.

Dairaku, K., and S. Emori, 2006: Dynamic and thermodynamic influences on intensified daily rainfall during the Asian summer monsoon under doubled atmospheric CO_2 conditions. *Geophys. Res. Lett.*, **33**, L01704, doi:10.1029/2005GL024754.

Davey, M.K., et al., 2002: STOIC: A study of coupled model climatology and variability in tropical ocean regions. *Clim. Dyn.*, **18**, 403–420.

Davies, H.C., 1976: A lateral boundary formulation for multi-levels prediction models. *Q. J. R. Meteorol. Soc.*, **102**, 405–418.

Debernard, J., M.Ø. Køltzow, J.E. Haugen, and L.P. Røed, 2003: Improvements in the sea-ice module of the regional coupled atmosphere-ice-ocean model and the strategy for the coupling of the three spheres. In: *RegClim General Technical Report No. 7* [Iversen, T., and M. Lystad (eds)]. Norwegian Meteorological Institute, Oslo, pp. 59–69.

DeFries, R.S., L. Bounoua, and G.J. Collatz, 2002: Human modification of the landscape and surface climate in the next fifty years. *Global Change Biol.*, **8**, 438–458.

Delire, C., et al., 2001: Simulated response of the atmosphere-ocean system to deforestation in the Indonesian Archipelago. *Geophys Res Lett*, **28**(10), 2081–2084.

Denis, B., R. Laprise, and D. Caya, 2003: Sensitivity of a regional climate model to the spatial resolution and temporal updating frequency of the lateral boundary conditions. *Clim. Dyn.*, **20**, 107–126.

Denis, B., R. Laprise, D. Caya, and J. Côté, 2002: Downscaling ability of one-way-nested regional climate models: The big-brother experiment. *Clim. Dyn.*, **18**, 627–646.

Déqué, M., and J.P. Piedelievre, 1995: High resolution climate simulation over Europe. *Clim. Dyn.*, **11**, 321–339.

Déqué, M., and A.L. Gibelin, 2002: High versus variable resolution in climate modelling. In: *Research Activities in Atmospheric and Oceanic Modelling* [Ritchie, H. (ed.)]. WMO/TD No. 1105, Report No. 32, World Meteorological Organization, Geneva, pp. 74–75.

Déqué, M., et al., 2005: Global high resolution versus Limited Area Model climate change scenarios over Europe: results from the PRUDENCE project. *Clim. Dyn.*, **25**, 653–670, 10.1007/s00382-005-0052-1.

Déqué, M., et al., 2007: An intercomparison of regional climate simulations for Europe: assessing uncertainties in model projections. *Clim. Change*, doi:10.1007/s10584-006-9228-x.

Derbyshire, S.H., et al., 2004: Sensitivity of moist convection to environmental humidity. *Q. J. R. Meteorol. Soc.*, **130**, 3055–3079.

Dessai, S., X. Lu, and M. Hulme, 2005: Limited sensitivity analysis of regional climate change probabilities for the 21st century. *J. Geophys. Res.*, **110**, D19108, doi:10.1029/2005JD005919.

Dethloff, K., et al., 2001: Sensitivity of Arctic climate simulations to different boundary layer parameterizations in a regional climate model. *Tellus*, **53**, 1–26.

Diaz-Nieto, J., and R.L. Wilby, 2005: A comparison of statistical downscaling and climate change factor methods: impacts on low flows in the River Thames, United Kingdom. *Clim. Change*, **69**, 245–268.

Dibike, Y.B., and P. Coulibaly, 2005: Hydrologic impact of climate change in the Saguenay watershed: Comparison of downscaling methods and hydrologic models. *J. Hydrol.*, **307**, 145–163.

Diffenbaugh, N.S., J.S. Pal, R.J. Trapp, and F. Giorgi, 2005: Fine-scale processes regulate the response of extreme events to global climate change. *Proc. Natl. Acad. Sci. U.S.A.*, **102**(44), 15774–15778, doi:10.1073/pnas.0506042102.

Dimitrijevic, M., and R. Laprise, 2005: Validation of the nesting technique in a regional climate model through sensitivity tests to spatial resolution and the time interval of lateral boundary conditions during summer. *Clim. Dyn.*, **25**, 555–580.

Ding, Y.H., Y.M. Liu, X.L. Shi, and Q.Q. Li, 2003: The experimental use of the regional climate model in the seasonal prediction in China National Climate Center. In: *Proceedings of the 2nd Workshop on Regional Climate Model, March 3-6, 2003, Yokohama, Japan.* GAME Publication No. 39, pp. 9–14.

Ding, Y.H., et al., 2006: Multi-year simulations and experimental seasonal predictions for rainy seasons in China by using a nested regional climate model (RegCM_NCC). Part I: Sensitivity study. *Adv. Atmos. Sci.*, **23**(3), 323–341.

Dorn, W., K. Dethloff, and A. Rinke, 2003: Competition of NAO regime changes and increasing greenhouse gases and aerosols with respect to Arctic climate estimate. *Clim. Dyn.*, **21**(5–6), 447–458, doi:10.1007/s00382-003-0344-2.

Döscher, R., et al., 2002: The development of the coupled ocean-atmosphere model RCAO. *Boreal Environ. Res.*, **7**, 183–192.

Douville, H., 2005: Limitations of time-slice experiments for predicting regional climate change over South Asia. *Clim. Dyn.*, **24**(4), 373–391.

Douville, H., et al., 2000: Impact of CO_2 doubling on the Asian summer monsoon: Robust versus model dependent responses. *J. Meteorol. Soc. Japan*, **78**, 1–19.

Douville, H., et al., 2002: Sensitivity of the hydrological cycle to increasing amounts of greenhouse gases and aerosols. *Clim. Dyn.*, **20**, 45–68.

Druyan, L.M., M. Fulakeza, and P. Lonergan, 2002: Dynamic downscaling of seasonal climate predictions over Brazil. *J. Clim.*, **15**, 3411–3426.

Duffy, P.B., et al., 2003: High-resolution simulations of global climate, part 1: Present climate. *Clim. Dyn.*, **21**, 371–390.

Dufresne, J.-L., et al., 2002: On the magnitude of positive feedback between future climate change and the carbon cycle, *Geophys. Res. Lett.*, **29**(10), 1405, doi:10.1029/2001GL013777.

Easterling, D.R., L.V. Alexander, A. Mokssit, and V. Detemmerman, 2003: CCl/Clivar workshop to develop priority climate indices. *Bull. Am. Meteorol. Soc.*, **84**, 1403–1407.

Ekström, M., H.J. Fowler, C.G. Kilsby, and P.D. Jones, 2005: New estimates of future changes in extreme rainfall across the UK using regional climate model integrations. 2. Future estimates and use in impact studies. *J. Hydrol.*, **300**, 234–251.

Elguindi, N., and F. Giorgi, 2006: Simulating multi-decadal variability of Caspian Sea level changes using regional climate model outputs. *Clim. Dyn.*, **26**: 167–181.

Engelbrecht, F., C. Rautenbach, J. McGregor, and J. Katzfey, 2002: January and July climate simulations over the SADC region using the limited-area model DARLAM. *Water SA*, **28**(4), 361–374.

Feddema, J.J., et al., 2005: A comparison of a GCM response to historical anthropogenic land cover change and model sensitivity to uncertainty in present-day land cover representations. *Clim. Dyn.*, **25**, 581–609.

Fernández, J., and J. Sáenz, 2003: Improved field reconstruction with the analog method: searching the CCA space. *Clim. Res.*, **24**, 199–213.

Fernandez, J., J. Sáenz, and E. Zorita, 2003: Analysis of wintertime atmospheric moisture transport and its variability over southern Europe in the NCEP reanalyses. *Clim. Res.*, **23**, 195–215.

Fink, A.H., et al., 2004: The 2003 European summer heatwaves and drought – synoptic diagnostics and impacts. *Weather*, **59**, 209–216.

Flather, R.A., and H. Khandker, 1993: The storm surge problem and possible effects of sea level changes on coastal flooding in the Bay of Bengal. In: *Climate and Sea Level Change* [Warrick, R.A., E.M. Barrow, and T. Wigley (eds)]. Cambridge University Press, Cambridge, UK, pp 229–245.

Flather, R.A., and J.A. Williams, 2000: Climate change effects on storm surges: methodologies and results. In: *Climate Scenarios for Water-Related and Coastal Impacts* [Beersma, J., M. Agnew, D. Viner, and M. Hulme (eds)]. ECLAT-2 Workshop Report No. 3, KNMI, The Netherlands, pp. 66–78.

Flather, R.A., et al., 1998: Direct estimates of extreme storm surge elevations from a 40-year numerical model simulation and from observations. *Global Atmos. Ocean System*, **6**, 165–176.

Fogt, R.L., and D.H. Bromwich, 2006: Decadal variability of the ENSO teleconnection to the high latitude South Pacific governed by coupling with the Southern Annular Mode. *J. Clim.*, **19**, 979–997.

Foley, J.A., M.T. Coe, M. Scheffer, and G. Wang, 2003: Regime shifts in the Sahara and Sahel: Interactions between ecological systems in Southern Africa. *Ecosystems*, **6**, 524–539.

Foley, J.A., et al., 2005: Global consequences of land use. *Science*, **309**, 570–574.

Fowler, H.J., M. Ekström, C.G. Kilsby, and P.D. Jones, 2005: New estimates of future changes in extreme rainfall across the UK using regional climate model integrations. 1. Assessment of control climate. *J. Hydrol.*, **300**, 212–233.

Fox-Rabinovitz, M.S., L.L. Takacs, R.C. Govindaraju, and M.J. Suarez, 2001: A variable-resolution stretched-grid general circulation model: Regional climate simulation. *Mon. Weather Rev.*, **129**(3), 453–469.

Frei, C., et al., 2003: Daily precipitation statistics in regional climate models: Evaluation and intercomparison for the European Alps. *J. Geophys. Res.*, **108**(D3), 4124, doi:10.1029/2002JD002287.

Frei, C., et al., 2006: Future change of precipitation extremes in Europe: Intercomparison of scenarios from regional climate models. *J. Geophys. Res.*, **111**, D06105, doi:10.1029/2005JD005965.

Freiman, M., and P. Tyson, 2000: The thermodynamic structure of the atmosphere over South Africa: Implications for water vapour transport. *Water SA*, **26**(2), 153–158.

Friedlingstein, P., J.-L. Dufresne, P.M. Cox, and P. Rayner, 2003: How positive is the feedback between climate change and the carbon cycle? *Tellus*, **55B**, 692–700.

Friedlingstein, P., et al., 2001: Positive feedback between future climate change and the carbon cycle. *Geophys. Res. Lett.*, **28**, 1543–1546.

Fu, C.B., et al., 2005: Regional Climate Model Intercomparison project for Asia. *Bull. Am. Meteorol. Soc.*, **86**(2), 257–266, doi:10.11/BAMS-86-2-257.

Fuentes, U., and D. Heimann, 2000: An improved statistical-dynamical downscaling scheme and its application to the alpine precipitation climatology. *Theor. Appl. Climatol.*, **65**, 119–135.

Fujibe, F., N. Yamazaki, M. Katsuyama, and K. Kobayashi, 2005: The increasing trend of intense precipitation in Japan based on four-hourly data for a hundred years. *Scientific Online Letters on the Atmosphere*, **1**, 41–44.

Furrer, R., S.R. Sain, D.W. Nychka, and G.A. Meehl, 2007: Multivariate Bayesian analysis of atmosphere-ocean general circulation models. *Environ. Ecol. Stat.*, in press.

Gaertner, M.A., et al., 2001: The impact of deforestation on the hydrological cycle in the western Mediterranean: an ensemble study with two regional climate models. *Clim. Dyn.*, **17**, 857-873

Gao, Q., and M. Yu, 1998: A model of regional vegetation dynamics and its application to the study of Northeast China Transect (NECT) responses to global change. *Global Biogeochem. Cycles*, **12**(2), 329–344.

Gao, X.J., Z.C. Zhao, and F. Giorgi, 2002: Changes of extreme events in regional climate simulations over East Asia. *Adv. Atmos. Sci.*, **19**, 927–942.

Gao, X.J., Z.C. Zhao, and Y.H. Ding, 2003a: Climate change due to greenhouse effects in Northwest China as simulated by a regional climate model. *J. Glaciol. Geocryol.*, **25**(2), 165–169.

Gao, X.J., J.S. Pal, and F. Giorgi, 2006a: Projected changes in mean and extreme precipitation over the Mediterranean region from a high resolution double nested RCM simulation. *Geophys. Res. Lett.*, **33**, L03706, doi:10.1029/2005GL024954.

Gao, X.J., D.L. Li, Z.C. Zhao, and F. Giorgi, 2003b: Climate change due to greenhouse effects in Qinghai-Xizang Plateau and along the Qinghai-Tibet Railway. *Plateau Meteorol.*, **22**(5), 458–463.

Gao, X.J., W.T. Lin, Z.C. Zhao, and F. Kucharsky, 2004: Simulation of climate and short-term climate prediction in China by CCM3 driven by observed SST. *Chin. J. Atmos. Sci.*, **28**, 63–76.

Gao, X.J., et al., 2001: Climate change due to greenhouse effects in China as simulated by a regional climate model. *Adv. Atmos. Sci.*, **18**, 1224–1230.

Gao, X.J., et al., 2006b: Impacts of horizontal resolution and topography on the numerical simulation of East Asia precipitation. *Chin. J. Atmos. Sci.*, **30**, 185–192.

Gedney, N., and P.J. Valdes, 2000: The effect of Amazonian deforestation on the Northern Hemisphere circulation and climate. *Geophys. Res. Lett.*, **27**(19), 3053–3056.

Genthon, C., and E. Cosme, 2003: Intermittent signature of ENSO in west-Antarctic precipitation. *Geophys. Res. Lett.*, **30**, 2081, doi:10.1029/2003GL018280.

Genthon, C., G. Krinner, and E. Cosme, 2002: Free and laterally-nudged Antarctic climate of an atmospheric general circulation model. *Mon. Weather Rev.*, **130**, 1601–1616.

Genthon, C., S. Kapari, and P.A. Mayewski, 2005: Interannual variability of the surface mass balance of West Antarctica from ITASE cores and ERA40 reanalyses. *Clim. Dyn.*, **24**, 759–770.

Gerbaux, M., et al., 2005: Surface mass balance of glaciers in the French Alps: distributed modeling and sensitivity to climate change. *J. Glaciol.*, **51**(175), 561–572.

Ghan, S.J., and T. Shippert, 2006: Physically-based global downscaling climate change projections for a full century, *J. Clim.*, **19**, 1589–1604.

Giannini, A., Y. Kushnir, and M.A. Cane, 2000: Interannual variability of Caribbean rainfall, ENSO and the Atlantic Ocean. *J. Clim.*, **13**, 297–311.

Giannini, A., M.A. Cane, and Y. Kushnir, 2001: Interdecadal changes in the ENSO teleconnection to the Caribbean region and North Atlantic Oscillation. *J. Clim.*, **14**, 2867–2879.

Giannini, A., R. Saravanan, and P. Chang, 2003: Oceanic forcing of Sahel rainfall on interannual to interdecadal time scales. *Science*, **302**, 1027–1030.

Gibelin, A.L., and Déqué, M., 2003: Anthropogenic climate change over the Mediterranean region simulated by a global variable resolution model. *Clim. Dyn.*, **20**, 327–339.

Giorgi, F., and R. Francesco, 2000: Evaluating uncertainties in the prediction of regional climate change. *Geophys. Res. Lett.*, **27**, 1295–1298.

Giorgi, F., and L.O. Mearns, 2002: Calculation of average, uncertainty range, and reliability of regional climate changes from AOGCM simulations via the reliability ensemble averaging (REA) method. *J. Clim.*, **15**, 1141–1158.

Giorgi, F., and L.O. Mearns, 2003: Probability of regional climate change based on the Reliability Ensemble Averaging (REA) method. *Geophys. Res. Lett.*, **30**(12), 1629, doi:10.1029/2003GL017130.

Giorgi, F., and X. Bi, 2005: Regional changes in surface climate interannual variability for the 21st century from ensembles of global model simulations. *Geophys. Res. Lett.*, **32**, L13701, doi:10.1029/2005GL023002.

Giorgi, F., X. Bi, and J.S. Pal, 2004: Mean, interannual variability and trends in a regional climate change experiment over Europe. II: climate change scenarios (2071-2100). *Clim. Dyn.*, **23**, 839–858.

Giorgi, F., J.W. Hurrell, M.R. Marinucci, and M. Beniston, 1997: Elevation signal in surface climate change: A model study. *J. Clim.*, **10**, 288–296.

Giorgi, F., et al., 2001a: Regional climate information – Evaluation and projections. In: *Climate Change 2001: The Scientific Basis. Contribution of Working Group 1 to the Third Assessment Report of the Intergovernmental Panel on Climate Change* [Houghton, J.T., et al. (eds.)]. Cambridge University Press, Cambridge, United Kingdom and New York, NY, USA, pp. 583–638.

Giorgi, F., et al., 2001b: Emerging patterns of simulated regional climatic changes for the 21st century due to anthropogenic forcings. *Geophys. Res. Lett.*, **28**(17), 3317–3320.

Good, P., and J. Lowe, 2006: Emergent behavior and uncertainty in multi-model climate projections of precipitation trends at small spatial scales. *J. Clim.*, **27**(4), 357–375.

Goodison, B.E., P.Y.T. Louie, and D. Yang, 1998: *WMO Solid Precipitation Measurement Intercomparison, Final Report*. WMO/TD No.872, World Meteorological Organization, Geneva, 212 pp.

Govindasamy, B., P.B. Duffy, and J. Coquard, 2003: High resolution simulations of global climate, part 2: Effects of increased greenhouse gases. *Clim. Dyn.*, **21**, 391–404.

Goyette, S., O. Brasseur, and M. Beniston, 2003: Application of a new wind gust parameterisation: multi-scale case studies performed with the Canadian RCM. *J. Geophys. Res.*, **108**, 4371–4389.

Graham, L.P., S. Hagemann, S. Jaun, and M. Beniston, 2007: On interpreting hydrological change from regional climate models. *Clim. Change*, doi:10.1007/s10584-006-9217-0.

Gray, W.M., 1984: Atlantic seasonal hurricane frequency. Part I: El Niño and 30 mb quasi-biennial oscillation influences. *Mon. Weather Rev.*, **112**, 1649–1668.

Greene, A.M., L. Goddard, and U. Lall, 2006: Probabilistic multimodel regional temperature change projections. *J. Clim.*, **19**, 4326–4343.

Gregory, J.M., and J.F.B. Mitchell, 1995: Simulation of daily variability of surface temperature and precipitation over Europe in the current and $2 \times CO_2$ climate using the UKMO climate model. *Q. J. R. Meteorol. Soc.*, **121**, 1451–1476.

Griffiths, G.M., M.J. Salinger, and I. Leleu, 2003: Trends in extreme daily rainfall across the South Pacific and relationship to the South Pacific convergence zone. *Int. J. Climatol.*, **23**, 847- 869.

Groisman, P.Y., et al., 2005: Trends in intense precipitation in the climate record. *J. Clim.*, **18**, 1326–1350.

Guo, Z., D.H. Bromwich, and J.J. Cassano, 2003: Evaluation of Polar MM5 simulations of Antarctic atmospheric circulation. *Mon. Weather Rev.*, **131**, 384–411.

Guo, Z., D.H. Bromwich, and K.M. Hines, 2004: Modeled Antarctic precipitation. Part II: ENSO modulation over West Antarctica. *J. Clim.*, **17**, 448–465.

Gutowski, W.J., et al., 2004: Diagnosis and attribution of a seasonal precipitation deficit in a U.S. regional climate simulation. *J. Hydrometeorol.*, **5**(1), 230–242.

Haapala, J., H.E.M. Meier, and J. Rinne, 2001: Numerical investigations of future ice conditions in the Baltic Sea. *Ambio*, **30**, 237–244.

Haarsma, R.J., F. Selten, S. Weber, and M. Kliphuis, 2005: Sahel rainfall variability and response to greenhouse warming. *Geophys. Res. Lett.*, **32**, L17702, doi:10.1029/2005GL023232.

Hagemann, S., et al., 2004: Evaluation of water and energy budgets in regional climate models applied over Europe. *Clim. Dyn.*, **23**, 547–607.

Han, J., and J. Roads, 2004: US climate sensitivity simulated with the NCEP Regional Spectral Model. *Clim. Change*, **62**, 115–154, doi:10.1023/B:CLIM.0000013675.66917.15.

Hansen, J.E., et al., 1998: Climate forcings in the industrial era. *Proc. Natl. Acad. Sci. U.S.A.*, **95**, 12753–12758.

Hansen, J.W., and T. Mavromatis, 2001: Correcting low-frequency variability bias in stochastic weather generators. *Agr. For. Meteorol.*, **109**, 297–310.

Hanssen-Bauer, I., E.J. Førland, J.E. Haugen, and O.E. Tveito, 2003: Temperature and precipitation scenarios for Norway: comparison of results from dynamical and empirical downscaling. *Clim. Res.*, **25**(1), 15–27.

Hanssen-Bauer, I., et al., 2005: Statistical downscaling of climate scenarios over Scandinavia: A review. *Clim. Res.*, **29**, 255–268.

Hardy, T., L. Mason, A. Astorquia, and B. Harper 2004: *Queensland Climate Change and Community Vulnerability to Tropical Cyclones: Ocean Hazards Assessment*. Report to Queensland Government, 45 pp. +7 appendices, http://www.longpaddock.qld.gov.au/ClimateChanges/pub/OceanHazards/Stage2LowRes.pdf.

Harris, G.R., et al., 2006: Frequency distributions of transient regional climate change from perturbed physics ensembles of general circulation model simulations. *Clim. Dyn.*, **27**, 357–375.

Hasegawa, A., and S. Emori, 2005: Tropical cyclones and associated precipitation over the Western North Pacific: T106 atmospheric GCM simulation for present-day and doubled CO_2 climates. *Scientific Online Letters on the Atmosphere*, **1**, 145–148.

Hayhoe, K., et al., 2004: Emissions pathways, climate change, and impacts on California. *Proc. Natl. Acad. Sci. U.S.A.*, **101**, 12422–12427.

Haylock, M.R., et al., 2006: Downscaling heavy precipitation over the UK: A comparison of dynamical and statistical methods and their future scenarios. *Int. J. Climatol.*, **26**(10), 1397–1415, doi:10.1002/joc.1318.

Hegerl, G.C., F.W. Zwiers, P.A. Stott, and V.V. Kharin, 2004: Detectability of anthropogenic changes in annual temperature and precipitation extremes. *J. Clim.*, **17**, 3683–3700.

Held, I.M., et al., 2005: Simulation of Sahel drought in the 20th and 21st centuries. *Proc. Natl. Acad. Sci. U.S.A.*, **102**(50), 17891–17896.

Hellström, C., D. Chen, C. Achberger, and J. Räisänen, 2001: A comparison of climate change scenarios for Sweden based on statistical and dynamical downscaling of monthly precipitation. *Clim. Res.*, **19**, 45–55.

Hennessy, K.J., J.M. Gregory, and J.F.B. Mitchell, 1997: Changes in daily precipitation under enhanced greenhouse conditions. *Clim. Dyn.*, **13**, 667–680.

Hennessy, K.J., et al., 2003: *The Impact of Climate Change on Snow Conditions in Mainland Australia.* CSIRO Atmospheric Research, Aspendale, 47 pp., http://www.cmar.csiro.au/e-print/open/hennessy_2003a.pdf.

Hennessy, K.J., et al., 2004a: *Climate Change in the Northern Territory.* Consultancy report for the Northern Territory Department of Infrastructure, Planning and Environment by CSIRO Atmospheric Research Climate Impact Group and Melbourne University School of Earth Sciences, Northern Territory Government, Darwin, 65 pp.

Hennessy, K.J., et al., 2004b: *Climate Change in New South Wales – Part 1: Past Climate Variability and Projected Changes in Average Climate.* Consultancy report for the New South Wales Greenhouse Office by CSIRO Atmospheric Research and Australian Government Bureau of Meteorology, 46 pp., http://www.dar.csiro.au/publications/hennessy_2004b.pdf.

Hennessy, K.J., et al., 2004c: *Climate Change in New South Wales– Part 2, Projected Changes in Climate Extremes.* Consultancy report for the New South Wales Greenhouse Office. CSIRO Atmospheric Research, Aspendale, 79 pp.

Hewitson, B.C., and R.G. Crane, 2005: Gridded area-averaged daily precipitation via conditional interpolation. *J. Clim.*, **18**, 41–51.

Hewitson, B.C., and R.G. Crane, 2006: Consensus between GCM climate change projections with empirical downscaling: precipitation downscaling over South Africa. *Int. J. Climatol.*, **26**, 1315–1337

Hewitson, B.C., et al., 2004: *Dynamical Modelling of the Present and Future Climate System.* Technical Report to the Water Research Commission, Report No. 1154/1/04 , Pretoria, South Africa.

Higgins, R.W., and K.C. Mo, 1997: Persistent North Pacific circulation anomalies and the tropical intraseasonal oscillation. *J. Clim.*, **10**, 223–244.

Hines, K.M., D.H. Bromwich, P.J. Rasch, and M.J. Iacono, 2004: Antarctic clouds and radiation within the NCAR climate models. *J. Clim.*, **17**, 1198–1212.

Ho, C.-H., J.-J. Baik, J.-H. Kim, and D.Y. Gong, 2004: Interdecadal changes in summertime typhoon tracks. *J. Clim.*, **17**, 1767–1776.

Hoerling, M.P., J.W. Hurrell, and J. Eischeid, 2006: Detection and attribution of 20th century Northern and Southern African monsoon change. *J. Clim.* **19**(16), 3989–4008.

Holland, M.M., and C.M. Bitz, 2003: Polar amplification of climate change in the coupled model intercomparison project. *Clim. Dyn.*, **21**, 221–232.

Holloway, G., and T. Sou, 2002: Has Arctic sea ice rapidly thinned? *J. Clim.*, **15**, 1691–1701.

Hope, P.K., 2006a: Shifts in synoptic systems influencing southwest Western Australia. *Clim. Dyn.*, **26**, 751–764.

Hope, P.K., 2006b: Future changes in synoptic systems influencing southwest Western Australia. *Clim. Dyn.*, **26**, 765–780.

Hori, M.E., and H. Ueda, 2006: Impact of global warming on the East Asian winter monsoon as revealed by nine coupled atmosphere-ocean GCMs. *Geophys. Res. Lett.*, **33**, L03713, doi:10.1029/2005GL024961.

Hu, Z.Z., L. Bengtsson, and K. Arpe, 2000: Impact of global warming on the Asian winter monsoon in a coupled GCM. *J. Geophys. Res.*, **105**(D4), 4607–4624.

Hu, Z.Z., S. Yang, and R. Wu, 2003: Long-term climate variations in China and global warming signals. *J. Geophys. Res.*, **108**(D19), 4614, doi:10.1029/2003JD003651.

Hubbert, G.D., and K.L. McInnes, 1999: A storm surge inundation model for coastal planning and impact studies. *J. Coastal Res.*, **15**, 168–185.

Huffman, G.J., et al., 2001: Global precipitation at one-degree daily resolution from multisatellite observations. *J. Hydrometeorol.*, **2**, 36–50.

Hulme, M., and N. Sheard, 1999a: *Climate Change Scenarios for Indonesia.* Climatic Research Unit, Norwich, UK, 6 pp.

Hulme, M., and N. Sheard, 1999b: *Climate Change Scenarios for the Philippines.* Climatic Research Unit, Norwich, UK, 6 pp.

Hulme, M., R. Doherty, and T. Ngara, 2001: African climate change: 1900-2100. *Clim. Res.*, **17**, 145–168.

Hulme, M., et al., 1999: Relative impacts of human-induced climate change and natural variability. *Nature*, **397**, 688–691.

Huntingford, C., et al., 2003: Regional climate-model predictions of extreme rainfall for a changing climate. *Q. J. R. Meteorol. Soc.*, **129**, 1607–1621.

Hurrell, J.W., and H. van Loon, 1997: Decadal variations in climate associated with the North Atlantic Oscillation. *Clim. Change*, **36**, 301–326.

Huth, R., J. Kysely, and M. Dubrovsky, 2001: Time structure of observed, GCM-simulated, downscaled, and stochastically generated daily temperature series. *J. Clim.*, **14**, 4047–4061.

Im E.S., E-H. Park, W.-T. Kwon, and F. Giorgi, 2006: Present climate simulation over Korea with a regional climate model using a one-way double-nested system. *Theor. Appl. Climatol.* **86**, 183–196.

Imbert, A., and R.E. Benestad, 2005: An improvement of analog model strategy for more reliable local climate change scenarios. *Theor. Appl. Climatol.*, **82**, 245–255.

Inatsu, M., and M. Kimoto, 2005: Difference of boreal summer climate between coupled and atmosphere-only GCMs. *Scientific Online Letters on the Atmosphere*, **1**, 105–108.

International CLIVAR Project Office, 2006: *Understanding the Role of the Indian Ocean in the Climate System — Implementation Plan for Sustained Observations.* CLIVAR Publication Series No.100, International CLIVAR Project Office, Southampton, UK, 76 pp.

IOCI, 2002: *Climate Variability and Change in South West Western Australia.* Technical Report, Indian Ocean Climate Initiative Panel, Perth, Australia, 34 pp.

ICCI, 2005: *Indian Ocean Climate Initiative Stage 2: Report of Phase 1 Activity.* Indian Ocean Climate Initiative Panel, Perth, Australia, 42 pp., http://www.ioci.org.au/publications/pdf/2005202-IOCI%20reportvis2.pdf.

IPCC, 1996: Technical summary. In: *Climate Change 1995: The Science of Climate Change. Contribution of Working Group I to the Second Assessment Report of the Intergovernmental Panel on Climate Change* [Houghton, J.T., et al. (eds.)]. Cambridge University Press, Cambridge, United Kingdom and New York, NY, USA, pp. 9–49.

IPCC, 2001: *Climate Change 2001: The Scientific Basis. Contribution of Working Group I to the Third Assessment Report of the Intergovernmental Panel on Climate Change* [Houghton, J.T., et al. (eds.)]. Cambridge University Press, Cambridge, United Kingdom and New York, NY, USA, 881 pp.

Jacob, D., et al., 2007: An intercomparison of regional climate models for Europe: design of the experiments and model performance. *Clim. Change*, doi: 10.1007/s10584-006-9213-4.

Jenkins, G.S., G. Adamou, and S. Fongang, 2002: The challenges of modeling climate variability and change in West Africa. *Clim. Change*, **52**, 263–286.

Jiang, Y.D., 2005: *The Northward Shift of Climatic Belts in China during the Last 50 Years, and the Possible Future Changes*. PhD Thesis, Institute of Atmospheric Physics, China Academy of Science, Beijing, 137 pp.

Jiao, Y., and D. Caya, 2006: An investigation of summer precipitation simulated by the Canadian regional climate model. *Mon. Weather Rev.*, **134**, 919–932.

Johannessen, O.M., et al., 2004: Arctic climate change: observed and modelled temperature and sea-ice variability. *Tellus*, **56A**(4), 328.

Jolly, D., S.P. Harrison, B. Damnati, and E. Bonnefille, 1996: Simulated climate and biomes of Africa during the late Quaternary: comparison with pollen and lake status data. *Quat. Sci. Rev.*, **17**, 629–657.

Jones, C.D., et al., 2003: Strong carbon cycle feedbacks in a climate model with interactive CO_2 and sulphate aerosols. *Geophys. Res. Lett.*, **30**(9), 1479, doi:10.1029/2003GL016867.

Jones, P.D., and P.A. Reid, 2001: Assessing future changes in extreme precipitation over Britain using regional climate model integrations. *Int. J. Climatol.*, **21**, 1337–1356.

Jones, R.G., J.M. Murphy, M. Noguer, and A.B. Keen, 1997: Simulation of climate change over Europe using a nested regional climate model. II: Comparison of driving and regional model responses to a doubling of carbon dioxide. *Q. J. R. Meteorol. Soc.*, **123**, 265–292.

Jones, R.N., 2000: Managing uncertainty in climate change projections – issues for impact assessment. *Clim. Change*, **45**, 403–419.

Jones, R.N., et al., 2000: *An Analysis of the Effects of The Kyoto Protocol on Pacific Island Countries, Part Two: Regional Climate Change Scenarios and Risk Assessment Methods*. South Pacific Regional Environment Programme, Apia, Samoa, 68 pp., available from sprep@sprep.org.ws.

Jones, R.N., et al., 2002: Scenarios and projected ranges of change for mean climate and climate variability for the South Pacific. *Asia Pac. J. Environ. Dev.*, **9**(1–2), 1–42.

Juang, H.M.H., and S.Y. Hong, 2001: Sensitivity of the NCEP regional spectral model to domain size and nesting strategy. *Mon. Weather Rev.*, **129**, 2904–2922.

Kabat, P., et al., 2002: *Vegetation, Water, Humans and the Climate Change: A New Perspective on an Interactive System*. Springer, Heidelberg, Germany, 566 pp.

Kamga, A.F., et al., 2005: Evaluating the National Center for Atmospheric Research climate system model over West Africa: Present-day and the 21st century A1 scenario. *J. Geophys. Res.*, **110**(D03106), doi:10.1029/2004JD004689.

Kanada, S., et al., 2005: Structure of mesoscale convective systems during the late Baiu season in the global warming climate simulated by a non-hydrostatic regional model. *Scientific Online Letters on the Atmosphere*, **1**, 117–120.

Kang I.-S., et al., 2002: Intercomparison of atmospheric GCM simulated anomalies associated with the 1997/98 El Niño. *J. Clim.*, **15**, 2791–2805.

Karcher, M.J., R. Gerdes, F. Kauker, and C. Köberle, 2003: Arctic warming: Evolution and spreading of the 1990s warm event in the Nordic seas and the Arctic Ocean. *J. Geophys. Res.*, **108**(C2), 3034, doi:10.1029/2001JC001265.

Kattsov, V.M., et al., 2007: Simulation and projection of Arctic freshwater budget components by the IPCC AR4 global climate models. *J. Hydrometeorol.*, **8**, in press.

Katz, R.W., M.B. Parlange, and P. Naveau, 2002: Statistics of extremes in hydrology. *Adv. Water Resour.*, **25**, 1287–1304.

Katz, R.W., M.B. Parlange, and C. Tebaldi, 2003: Stochastic modelling of the effects of large-scale circulation on daily weather in the southeastern US. *Clim. Change*, **60**, 189–216.

Keller, F., S. Goyette, and M. Beniston, 2005: Sensitivity analysis of snow cover to climate change scenarios and their impact on plant habitats in alpine terrain. *Clim. Change*, **72**, 299–319.

Kharin, V.V., and F.W. Zwiers, 2005: Estimating extremes in transient climate change simulations. *J. Clim.*, **18**, 1156–1173.

Kida, H., T. Koide, H. Sasaki, and M. Chiba, 1991: A new approach for coupling a limited area model to a GCM for regional climate simulations. *J. Meteorol. Soc. Japan.*, **69**, 723–728.

Kim, J., T.-K. Kim, R.W. Arritt, and N.L. Miller, 2002: Impacts of increased atmospheric CO_2 on the hydroclimate of the Western United States. *J. Clim.*, **15**(14), 1926–1942.

Kimoto, M., 2005: Simulated change of the east Asian circulation under global warming scenario. *Geophys. Res. Lett.*, **32**, L16701, doi:10.1029/2005GRL023383.

Kimoto, M., N. Yasutomi, C. Yokoyama, and S. Emori, 2005: Projected changes in precipitation characteristics around Japan under the global warming. *Scientific Online Letters on the Atmosphere*, **1**, 85–88.

Kitoh, A., and T. Uchiyama, 2006: Changes in onset and withdrawal of the East Asian summer rainy season by multi-model global warming experiments. *J. Meteorol. Soc. Japan*, **84**, 247–258.

Kitoh, A., M. Hosaka, Y. Adachi and K. Kamiguchi, 2005: Future projections of precipitation characteristics in East Asia simulated by the MRI CGCM2. *Adv. Atmos. Sci.*, **22**(4), 467–478.

Kjellström, E., et al., 2007: Variability in daily maximum and minimum temperatures: recent and future changes over Europe. *Clim. Change*, doi: 10.1007/s10584-006-9220-5.

Kleinn, J., et al., 2005: Hydrological simulations in the Rhine basin, driven by a regional climate model. *J. Geophys. Res.*, **110**, D04102, doi:10.1029/2004JD005143.

Knippertz, P., U. Ulbrich, and P. Speth, 2000: Changing cyclones and surface wind speeds over the North-Atlantic and Europe in a transient GHG experiment. *Clim. Res.*, **15**, 109–122.

Knutson, T.R., and S. Manabe, 1995: Time-mean response over the tropical Pacific to increased CO_2 in a coupled ocean-atmosphere model. *J. Clim.*, **8**, 2181–2199.

Knutson, T.R., and R.E. Tuleya, 2004: Impacts of CO_2-induced warming on simulated hurricane intensities and precipitation: sensitivity to the choice of climate model and convective parameterization. *J. Clim.*, **17**, 3477–3495.

Krinner, G., C. Genthon, Z. Li, and P.L. Van, 1997: Studies of the Antarctic climate with a stretched-grid general circulation model. *J. Geophys. Res.*, **102**, 13731–13745.

Krishna Kumar, K., B. Rajagopalan, and M.A. Cane, 1999: On the weakening relationship between the Indian monsoon and ENSO. *Science*, **284**, 2156–2159.

Krishna Kumar, K., et al., 2003: Future scenarios of extreme rainfall and temperature over India. In: *Proceedings of the Workshop on Scenarios and Future Emissions, Indian Institute of Management (IIM), Ahmedabad, July 22, 2003*. NATCOM Project Management Cell, Ministry of Environment and Forests, Government of India, New Delhi, pp. 56–68.

Kumagi, T., G.G. Katul, and A. Porporato, 2004: Carbon and water cycling in a Bornean tropical rainforest under current and future climate scenarios. *Adv. Water Resour.*, **27**, 1135–1150.

Kurihara, K., et al., 2005: Projection of climatic change over Japan due to global warming by high-resolution regional climate model in MRI. *Scientific Online Letters on the Atmosphere*, **1**, 97–100.

Kusunoki, S., et al., 2006: Change of Baiu rain band in global warming projection by an atmospheric general circulation model with a 20-km grid size. *J. Meteorol. Soc. Japan*, **84**(4), 581–611

Kutzbach, J.E., G. Bonan, J. Foley, and S. Harrison, 1996: Vegetation and soil feedbacks on the response of the African monsoon to forcing in the early to middle Holocene. *Nature*, **384**, 623–626.

Kwok, R., and J.C. Comiso, 2002a: Spatial patterns of variability in Antarctic surface temperature: Connections to the Southern Hemisphere annular mode and the Southern Oscillation. *Geophys. Res. Lett.*, **29**(14), 1705, doi:10.1029/2002GL015415.

Kwok, R., and J.C. Comiso, 2002b: Southern Ocean climate and sea ice anomalies associated with the Southern Oscillation. *J. Clim.*, **15**, 487–501.

Kwon, W.-T., et al., 2003: *The Development of Regional Climate Change Scenario for the National Climate Change Report (II)*. METRI Technical Report MR030CR09, Meteorological Research Institute, Seoul, Korea, 502 pp (in Korean).

Kwon, W.-T., et al., 2005: *The Application of Regional Climate Change Scenario of the National Climate Change Report (I)*. METRI Technical Report MR050C03, Meteorological Research Institute, Seoul, Korea, 408 pp (in Korean).

Lal, M., 2004: Climate change and small island developing countries of the South Pacific. *Fijian Studies*, **2**(1), 1–15.

Lal, M., and H. Harasawa, 2001: Future climate change scenarios for Asia as inferred from selected coupled atmosphere-ocean global climate models. *J. Meteorol. Soc. Japan*, **79**, 219–227.

Lal, M., H. Harasawa, and K. Takahashi, 2002: Future climate change and it's impacts over small island states. *Clim. Res.*, **19**, 179–192.

Lal, M., et al., 2001: Future climate change: Implications for Indian summer monsoon and its variability. *Curr. Sci.*, **81**, 1196–1207.

Lamb, P.J., and R.A. Peppler, 1987: North Atlantic Oscillation: Concept and an application. *Bull. Am. Meteorol. Soc.*, **68**(10), 1218–1225.

Lambert, S.J., and J.C. Fyfe, 2006: Changes in winter cyclone frequencies and strengths simulated in enhanced greenhouse warming experiments: results from the models participating in the IPCC diagnostic exercise. *J. Clim. Dyn.*, **26**, 713–728.

Laprise, R., 2003: Resolved scales and nonlinear interactions in limited-area models. *J. Atmos. Sci.*, **60**(5), 768–779.

Latif, M., et al., 2001: ENSIP: The El Niño Simulation Intercomparison Project. *Clim. Dyn*, **18**, 255–276.

Lau, K.M., and J. Zhou, 2003: Responses of the South American Summer Monsoon climate system to ENSO during 1997–99. *Int. J. Climatol.*, **23**, 529–539.

Leckebusch, G.C., and U. Ulbrich, 2004: On the relationship between cyclones and extreme windstorm events over Europe under climate change. *Global Planet. Change*, **44**, 181–193.

Leckebusch, G.C., et al., 2007: Analysis of frequency and intensity of winter storm events in Europe on synoptic and regional scales from a multi-model perspective. *Clim. Res.* **31**, 59–74.

Lenderink, G., A. van Ulden, B. van den Hurk, and E. van Meijgaard, 2007: Summertime inter-annual temperature variability in an ensemble of regional model simulations: analysis of the surface energy budget. *Clim. Change*, doi: 10.1007/s10584-006-9229-9

Leung, L.R., L.O. Mearns, F. Giorgi, and R.L. Wilby, 2003: Regional climate research: needs and opportunities. *Bull. Am. Meteorol. Soc.*, **84**, 89–95.

Leung, L.R., et al., 2004: Mid-century ensemble regional climate change scenarios for the western United States. *Clim. Change*, **62**, 75–113.

L'Heureux, M.L., and D.W.J. Thompson, 2006: Observed relationships between the El-Niño/Southern Oscillation and the extratropical zonal mean circulation. *J. Clim.*, **19**, 276–287.

L'Heureux, M.L., et al., 2004: Atmospheric circulation influences on seasonal precipitation patterns in Alaska during the latter 20th century. *J. Geophys. Res.*, **109**(6), D06106, doi:10.1029/2003JD003845.

Liang, X.Z., et al., 2004: Regional climate model simulation of U.S. precipitation during 1982–2002. Part I: Annual cycle. *J. Clim.*, **17**(18), 3510–3529.

Liang, X.Z., et al., 2006: Regional climate model downscaling of the U.S. summer climate and future change. *J. Geophys. Res.*, **111**, D10108, doi:10.1029/2005JD006685.

Liebmann, B., et al., 2004: An observed trend in Central South American precipitation. *J. Clim.*, **17**, 4357–4367.

Lintner, B.R., and J.C.H. Chiang, 2005: Reorganization of tropical climate during El Niño: a weak temperature gradient approach. *J. Clim.*, **18**(24), 5312–5329.

Lionello, P., F. Dalan, and E. Elvini, 2002: Cyclones in the Mediterranean region: the present and the doubled CO$_2$ climate scenarios. *Clim. Res.*, **22**, 147–159.

Lionello, P., E. Elvini, and A. Nizzero, 2003: Ocean waves and storm surges in the Adriatic Sea: intercomparison between the present and the doubled CO$_2$ climate scenarios. *Clim. Res.*, **23**, 217–231.

Liu J., et al., 2005: Comparison of surface radiative flux data sets over the Arctic Ocean. *J. Geophys. Res.*, **110**, C02015, doi:10.1029/2004JC002381.

Lopez, A., et al., 2006: Two approaches to quantifying uncertainty in global temperature changes. *J. Clim.*, **19**, 4785–4796.

Lorant, V., and J-F. Royer, 2001: Sensitivity of equatorial convection to horizontal resolution in aquaplanet simulations with a variable-resolution GCM. *Mon. Weather Rev.*, **129**(11), 2730–2745.

Lorenz, P., and D. Jacob, 2005: Influence of regional scale information on the global circulation: a two-way nested climate simulation. *Geophys. Res. Lett.*, **32**, L18706, doi:10.1029/2005GL023351.

Lowe, J.A., and J.M. Gregory, 2005: The effects of climate change on storm surges around the United Kingdom. *Philos. Trans. R. Soc. London Ser. A*, **363**, 1313–1328.

Lowe, J.A., J.M. Gregory, and R.A. Flather, 2001: Changes in the occurrence of storm surges around the United Kingdom under a future climate scenario using a dynamic storm surge model driven by the Hadley Centre climate models. *Clim. Dyn.*, **18**(3–4), 179–188.

Lu J., and T.L. Delworth, 2005: Oceanic forcing of late 20th century Sahel drought. *Geophys. Res. Lett.*, **32**, L22706, doi:10.1029/2005GL023316.

Lynch, A., P. Uotila, and J.J. Cassano, 2006: Changes in synoptic weather patterns in the polar regions in the 20th and 21st centuries, Part 2: Antarctic. *Int. J. Climatol.*, **26**, 1181–2119.

Lynch, A.H., E.N. Cassano, J.J. Cassano, and L. Lestak, 2003: Case studies of high wind events in Barrow, Alaska: Climatological context and development processes. *Mon. Weather Rev.*, **131**, 719–732.

Lynch, A.H., J.A. Curry, R.D. Brunner, and J.A. Maslanik, 2004: Toward an integrated assessment of the impacts of extreme wind events on Barrow, Alaska. *Bull. Am. Meteorol. Soc.*, **85**, 209–221.

Mabuchi, K., Y. Sato, and H. Kida, 2002: Verification of the climatic features of a regional climate model with BAIM. *J. Meteorol. Soc. Japan*, **80**(4), 621–644.

Mabuchi, K., Y. Sato, and H. Kida, 2005a: Climatic impact of vegetation change in the Asian tropical region. Part I: Case of the Northern Hemisphere summer. *J. Clim.*, **18**(3), 410–428.

Mabuchi, K., Y. Sato, and H. Kida, 2005b: Climatic impact of vegetation change in the Asian tropical region. Part II: Case of the Northern Hemisphere winter and impact on the extratropical circulation. *J. Clim.*, **18**(3), 429–446.

Magaña, V., and E. Caetano, 2005: Temporal evolution of summer convective activity over the Americas warm pools. *Geophys. Res. Lett.*, **32**, L02803, doi:10.1029/2004GL021033.

Magaña, V., J.A. Amador, and S. Medina, 1999: The mid-summer drought over Mexico and Central America. *J. Clim.*, **12**, 1577–1588.

Manabe, S., and R.T. Wetherald, 1987: Large-scale changes of soil wetness induced by an increase in atmospheric carbon dioxide. *J. Atmos. Sci.*, **44**, 1211–1235.

Marengo, J.A., and T. Ambrizzi, 2006: Use of regional climate models in impacts assessments and adaptation studies from continental to regional and local scales. In: *Proceedings of the 8th International Conference on Southern Hemisphere Meteorology and Oceanography (ICSHMO), Foz do Iguaçu, Brazil, 24-28 April 2006*. Brazilian Institute for Space Research (INPE), São José dos Campos, pp. 291–296.

Marland, G., et al., 2003: The climatic impacts of land surface change and carbon management, and the implications for climate-change mitigation policy. *Clim. Policy*. **3**, 149–157.

Martin E., E. Brun, and Y. Durand, 1994: Sensitivity of the French Alps snow cover to the variation of climatic variables. *Ann. Geophys.*, **12**, 469–477.

Martinez-Castro, D., et al., 2006: Sensitivity studies of the RegCM-3 simulation of summer precipitation, temperature and local wind field in the Caribbean region. *Theor. Appl. Climatol.*, **86**, 5–22.

Martis, A., G.J. van Oldenborgh, and G. Burgers, 2002: Predicting rainfall in the Dutch Caribbean – More than El Niño? *Int. J. Climatol.*, **22**, 1219–1234.

Maslanik, J.A., A.H. Lynch, M.C. Serreze, and W. Wu, 2000: A case study of regional climate anomalies in the Arctic: performance requirements for a coupled model. *J. Clim.*, **13**, 383–401.

Maslowski, W., et al., 2004: On climatological mass, heat, and salt transports through the Barents Sea and Fram Strait from a pan-Arctic coupled ice-ocean model simulation. *J. Geophys. Res.*, **109**, C03032, doi:10.1029/2001JC001039.

Massom, R.A., et al., 2004: Precipitation over the interior East Antarctic ice sheet related to midlatitude blocking-high activity. *J. Clim.*, **17**, 1914–1928.

Matulla C., H. Scheifinger, A. Menzel, and E. Koch, 2003: Exploring two methods for statistical downscaling of Central European phenological time series. *Int. J. Biometeorol.*, **48**, 56–64.

May, W., 2004a: Simulation of the variability and extremes of daily rainfall during the Indian summer monsoon for present and future times in a global time-slice experiment. *Clim. Dyn.*, **22**, 183–204.

May, W., 2004b: Potential of future changes in the Indian summer monsoon due to greenhouse warming: analysis of mechanisms in a global time-slice experiment. *Clim. Dyn.*, **22**, 389–414.

May, W., and E. Roeckner, 2001: A time-slice experiment with the ECHAM4 AGCM at high resolution: The impact of horizontal resolution on annual mean climate change. *Clim. Dyn.*, **17**, 407–420.

Maynard, K., and J.-F. Royer, 2004a: Effects of realistic land-cover change on a greenhouse-warmed African climate. *Clim. Dyn.*, **22**(4), 343–358.

Maynard, K., and J.-F. Royer, 2004b: Sensitivity of a general circulation model to land surface parameters in African tropical deforestation experiments. *Clim. Dyn.*, **22**(6/7), 555–572.

Maynard, K., J.-F. Royer, and F. Chauvin, 2002: Impact of greenhouse warming on the West African summer monsoon. *Clim. Dyn.*, **19**, 499–514.

McBride, J.L., and N. Nicholls, 1983: Seasonal relationships between Australian rainfall and the Southern Oscillation. *Mon. Weather Rev.*, **111**, 1998–2004.

McBride, J.L., M.R. Haylock, and N. Nicholls, 2003: Relationships between the Maritime Continent heat source and the El Niño-Southern Oscillation phenomenon. *J. Clim.*, **16**, 2905–2914.

McGregor, J.L., 2004: Regional climate modelling activities at CSIRO. In: *Symposium on Water Resource and its Variability in Asia in the 21st Century, 1-2 March 2004, Epochal Tsukuba (International Congress Center), Tsukuba, Ibaraki, Japan.* Meteorological Research Institute, Japan Meteorological Agency, Tsukuba, Japan, pp. 68–71.

McGregor, J.L., and M.R. Dix, 2001: The CSIRO conformal-cubic atmospheric GCM. In: *IUTAM Symposium on Advances in Mathematical Modelling of Atmosphere and Ocean Dynamics* [Hodnett, P.F. (ed.)]. Kluwer Academic, Dordrecht, pp. 307–315.

McGregor, J.L., and K.C. Nguyen, 2003: Simulations of the East Asian and Australian monsoons using a variable-resolution model. In: *Proceedings of the 2nd Workshop on Regional Climate Modeling for Monsoon System, Yokohama, Japan.* GAME Publication No. 39, FRSGC and GAME International Science Panel, Yokohama, pp.117–120.

McGregor, J.L., J.J. Katzfey, and K.C. Nguyen, 1998: *Fine Resolution Simulations of Climate Change for Southeast Asia.* Final report for a research project commissioned by Southeast Asian Regional Committee for START (SARCS), CSIRO Atmospheric Research, Aspendale, Vic., 35 pp. + 3 CD-ROMs.

McGregor, J.L., K.C. Nguyen, and J.J. Katzfey, 2002: Regional climate simulations using a stretched-grid global model. In: *Research Activities in Atmospheric and Oceanic Modelling* [Ritchie, H. (ed.)]. Report No. 32, WMO/TD-No. 1105, World Meteorological Organisation, Geneva, pp. 15–16.

McGuffie, K., et al., 1995: Global climate sensitivity to tropical deforestation. *Global Planet. Change*, **10**, 97–128.

McInnes, K.L., et al, 2003: *Assessment of climate change, impacts and possible adaptation strategies relevant to South Australia.* Consultancy report undertaken for the South Australian Government by the Climate Impact Group, CSIRO Atmospheric Research, Aspendale, VIC, Australia, 61pp. http://www.cmar.csiro.au/e-print/open/mcinnes_2003a.pdf

McInnes, K. L., et al. (2004). *Climate change in Tasmania.* A report undertaken for Hydro Tasmania by the Climate Impact GroupC/0919. CSIRO Atmospheric Research, Aspendale, Vic, 49 pp., http://www.cmar.csiro.au/e-print/open/mcinnesskl_2004a.pdf

McInnes, K. L., et al. (2005). *Climate change in Eastern Victoria*: *Stage 2 report*: *the effect of climate change on storm surges.* A project undertaken for the Gippsland Coastal Board. CSIRO Marine and Atmospheric Research, Aspendale, Vic.: 37 pp. http://www.cmar.csiro.au/e-print/open/mcinnes_2005b.pdf

Mearns, L.O., et al., 2005: NARCCAP, North American Regional Climate Change Assessment Program, A multiple AOGCM and RCM climate scenario project over North America. *Preprints of the American Meteorological Society 16th Conference on Climate Variations and Change. 9-13 January, 2005.* Paper J6.10, American Meteorological Society, Washington, DC, pp. 235–238.

Meehl, G.A., and C. Tebaldi, 2004: More intense, more frequent, and longer lasting heat waves in the 21st century. *Science*, **305**, 994–997.

Meehl, G.A., J.M. Arblaster, and C. Tebaldi, 2005: Understanding future patterns of increased precipitation intensity in climate model simulations. *Geophys. Res. Lett.*, **32**, L18719, doi:10.1029/2005GL023680.

Meehl, G.A., et al., 2006: Climate change in the 20th and 21st centuries and climate change commitment in the CCSM3. *J. Clim.*, **19**, 2597–2616.

Meier, H.E.M, 2002: Regional ocean climate simulations with a 3D ice-ocean model for the Baltic Sea. Part 2: results for sea ice. *Clim. Dyn.*, **19**, 255–266.

Meier, H.E.M, 2006: Baltic Sea climate in the late twenty-first century: a dynamical downscaling approach using two global models and two emission scenarios. *Clim. Dyn.*, **27**(1), 39–68, doi:10.1007/s00382-006-0124-x.

Meier, H.E.M., R. Döscher, and A. Halkka, 2004: Simulated distributions of Baltic sea-ice in warming climate and consequences for the winter habitat of the Baltic Sea ringed seal. *Ambio*, **33**, 249–256.

Meleshko, V.P., et al., 2004: Anthropogenic climate changes in Northern Eurasia in the 21st century. *Russ. Meteorol. Hydrol.*, **7**, 5–26.

Menéndez, C.G., A.C. Saulo, and Z.-X. Li, 2001: Simulation of South American wintertime climate with a nesting system. *Clim. Dyn.*, **17**, 219–231.

Menéndez, C.G., M.F. Cabré, and M.N. Nuñez, 2004: Interannual and diurnal variability of January precipitation over subtropical South America simulated by a regional climate model. *CLIVAR Exchanges*, **29**, 1–3.

Miguez-Macho, G., G.L. Stenchikov, and A. Robock, 2004: Spectral nudging to eliminate the effects of domain position and geometry in regional climate model simulations. *J. Geophys. Res.*, **109**, D13104, doi:10.1029/2003JD004495.

Mikolajewicz, U., et al., 2005: Simulating Arctic sea ice variability with a coupled regional atmosphere-ocean-sea ice model. *Meteorol. Z.*, **14**, 793–800.

Min, S.K., E.H. Park, and W.T. Kwon, 2004: Future projections of East Asian climate change from Multi-AOGCM ensembles of IPCC SRES scenario simulations. *J. Meteorol. Soc. Japan*, **82**(4), 1187–1211.

Ministry for the Environment, 2004: *Climate Change Effects and Impacts Assessment: A Guidance Manual for Local Government in New Zealand.* New Zealand Climate Change Office, Ministry for the Environment, Wellington, http://www.climatechange.govt.nz/resources/local-govt/effects-impacts-may04/index.html.

Misra, V., P.A. Dirmeyer, and B.P. Kirtman, 2003: Dynamic downscaling of seasonal simulations over South America. *J. Clim.*, **16**, 103–117.

Mitchell, T.D., 2003: Pattern scaling: an examination of the accuracy of the technique for describing future climates. *Clim. Change*, **60**(3), 217–242.

Mizuta, R., et al., 2005: Changes in extremes indices over Japan due to global warming projected by a global 20-km-mesh atmospheric model. *Scientific Online Letters on the Atmosphere*, **1** 153–156.

Mizuta, R., et al., 2006: 20km-mesh global climate simulations using JMA-GSM model. Mean climate states. *J. Meteorol. Soc. Japan.*, **84**, 165–185.

Mo, K.C., and J. Nogués-Paegle, 2001: The Pacific-South American modes and their downstream effects. *Int. J. Climatol.*, **21**, 1211–1229.

Moise, A., R. Colman, and H. Zhang, 2005: Coupled model simulations of current Australian surface climate and its changes under greenhouse warming: An analysis of 18 CMIP2 models. *Aust. Meteorol. Mag.*, **54**, 291–307.

Monaghan, A.J., D.H. Bromwich, and S.-H. Wang, 2006: Recent trends in Antarctic snow accumulation from Polar MM5 simulations. *Philos. Trans. R. Soc. London Ser. A.*, **364**, 1683–1708.

Mote, P.W., and N.J. Mantua, 2002: Coastal upwelling in a warmer future. *Geophys. Res. Lett.*, **29**(23), 2138, doi:10.1029/2002GL016086.

Mullan, A.B., 1995: On the linearity and stability of Southern Oscillation - climate relationships for New Zealand. *Int. J. Climatol.*, **15**, 1365–1386.

Mullan, A.B., D.S. Wratt, and J.A. Renwick, 2001a: Transient model scenarios of climate changes for New Zealand. *Weather and Climate*, **21**, 3–34.

Mullan, A.B., M.J. Salinger, C.S. Thompson, and A.S. Porteous, 2001b: The New Zealand climate: present and future. In: *Effects of Climate Change and Variation in New Zealand: An Assessment using the CLIMPACTS System* [Warrick, R.A., G.J. Kenny, and J.J. Harman, (eds.)]. International Global Change Institute, University of Waikato, pp. 11–31.

Mullan, B., A. Porteous, D. Wratt, and M. Hollis, 2005: *Changes in Drought Risk with Climate Change*. NIWA Client Report WLG2005-23, National Institute for Water and Atmosphere Research, Wellington, New Zealand, 68 pp.

Murphy, J.M., et al., 2004: Quantification of modelling uncertainties in a large ensemble of climate change simulations. *Nature*, **430**, 768–772.

Neelin, J.D., and H. Su, 2005: Moist teleconnection mechanisms for the tropical South American and Atlantic sector. *J. Clim.*, **18**(18), 3928–3950.

Neelin, J.D., et al., 2006: Tropical drying trends in global warming models and observations. *Proc. Natl. Acad. Sci. U.S.A.*, **103**, 6110–6115.

New, M., and M. Hulme, 2000: Representing uncertainty in climate change scenarios: a Monte Carlo approach. *Integr. Assess. J.*, **1**, 203–213.

New, M., B.C. Hewitson, C. Jack, and R. Washington, 2003: Sensitivity of southern African rainfall to soil moisture. *CLIVAR Exchanges*, **27**, 45–47.

Nguyen, K.C., and K.J.E. Walsh, 2001: Interannual, decadal, and transient greenhouse simulation of tropical cyclone-like vortices in a regional climate model of the South Pacific. *J. Clim.*, **14**(13), 3043–3054.

Nicolini, M., et al., 2002: January and July regional climate simulation over South America. *J. Geophys. Res.*, **107**(D22), 4637, doi:10.1029/2001JD000736.

Nobre, P., A. Moura, and L. Sun, 2001: Dynamic downscaling of seasonal climate prediction over Nordeste Brazil with ECHAM3 and NCEP's regional spectral models at IRI. *Bull. Am. Meteorol. Soc.*, **82**, 2787–2796.

Noone, D., and I. Simmonds, 2002: Annular variations in moisture transport mechanisms and the abundance of delta O-18 in Antarctic snow. *J. Geophys. Res.*, **107**, 4742, doi:10.1029/2002JD002262.

Oh, J-H et al., 2004: Regional climate simulation for Korea using dynamic downscaling and statistical adjustment. *J. Meteorol. Soc. Japan*, **82**(6), 1629–1643.

Oleson, K.W., G.B. Bonan, S. Levis, and M. Vertenstein, 2004: Effects of land use change on U.S. climate: Impact of surface datasets and model biogeophysics. *Clim. Dyn.*, **23**, 117–132.

Oouchi, K., et al., 2006: Tropical cyclone climatology in a global-warming climate as simulated in a 20 km-mesh global atmospheric model: Frequency and wind intensity analyses. *J. Meteorol. Soc. Japan*, **84**, 259–276.

Osborn, T.J., et al., 1999: Evaluation of the North Atlantic Oscillation as simulated by a coupled climate model. *Clim. Dyn.*, **15**, 685–702.

Paeth, H., and A. Hense. 2004: SST versus climate change signals in West African rainfall: 20th-century variations and future projections. *Clim. Change*, **65**(1–2): 179–208.

Paeth, H., K. Born, D. Jacob, and R. Podzun, 2005: Regional dynamic downscaling over West Africa: model validation and comparison of wet and dry years. *Meteorol. Z.*, **14**(3), 349–367.

Pal, J.S., F. Giorgi, and X. Bi, 2004: Consistency of recent European summer precipitation trends and extremes with future regional climate projections. *Geophys. Res. Lett.*, **31**, L13202, doi:10.1029/2004GL019836.

Palmer, T.N., 1986: Influence of the Atlantic, Pacific and Indian Oceans on Sahel rainfall, *Nature*, **322**, 251–253.

Palutikof, J.P., C.M. Goodess, S.J. Watkins, and T. Holt, 2002: Generating rainfall and temperature scenarios at multiple sites: examples from the Mediterranean. *J. Clim.*, **15**, 3529–3548.

Pan, Z., E.S. Takle, and F. Otieno., 2001: Evaluation of uncertainties in regional climate change simulations. *J. Geophys. Res.*, **106**(D16), 17735–17752.

Pan, Z., M. Segal, R.W. Arritt, and E.S. Takle, 2004: On the potential change in solar radiation over the US due to increases of atmospheric greenhouse gases. *Renew. Energy*, **29**(11), 1923–1928.

Pant, G.B., and K. Rupa Kumar, 1997: *Climates of South Asia*. John Wiley & Sons, Chichester, 320 pp.

Pavolonis, M.J., J.R. Key, and J.J. Cassano, 2004: A study of the Antarctic surface energy budget using a polar regional atmospheric model forced with satellite-derived cloud properties. *Mon. Weather Rev.*, **132**, 654–661.

Peterson, T.C., et al., 2002: Recent changes in climate extremes in the Caribbean region. *J. Geophys. Res.*, **107**(D21), 4601, doi:10.1029/2002JD002251.

Pielke, R.A. Sr., et al., 2002: The influence of land-use change and landscape dynamics on the climate system – Relevance to climate change policy beyond the radiative effect of greenhouse gases. *Philos. Trans. R. Soc. London Ser. A*, **360**, 1705–1719.

Pinto, J.G., T. Spangehl, U. Ulbrich, and P. Speth, 2006: Assessment of winter cyclone activity in a transient ECHAM4-OPYC3 GHG experiment. *Meteorol. Z.*, **15**, 279–291.

Pitman, A.J., and B.J. McAvaney, 2004: Impact of varying the complexity of the land surface energy balance on the sensitivity of the Australian climate to increasing carbon dioxide. *Clim. Res.*, **25**(3), 191–203.

Plummer, D.A., et al., 2006: Climate and climate change over North America as simulated by the Canadian Regional Climate Model. *J. Clim.*, **19**, 3112–3132.

Polyakov, I.V., et al., 2003a: Long-term ice variability in Arctic marginal seas. *J. Clim.*, **16**, 2078–2085.

Polyakov, I.V., et al., 2003b: Variability and trends of air temperature and pressure in the maritime Arctic, 1875-2000. *J. Clim.*, **16**, 2067–2077.

Pope, V.D., and R.A. Stratton, 2002: The processes governing resolution sensitivity in a climate model. *Clim. Dyn.*, **19**, 211–236.

Power, S., C. Folland, A. Colman, and V. Mehta, 1999: Inter-decadal modulation of the impact of ENSO on Australia. *Clim. Dyn.*, **15**, 319–324.

Power, S., et al., 1998: Australian temperature, Australian rainfall, and the Southern Oscillation, 1910-1996: Coherent variability and recent changes. *Aust. Meteorol. Mag.*, **47**, 85–101.

Pryor, S.C., R.J. Barthelmie, and E. Kjellström, 2005a: Potential climate change impact on wind energy resources in northern Europe: Analyses using a regional climate model. *Clim. Dyn.*, **25**, 815–835.

Pryor, S.C., J.T. School, and R.J. Barthelmie, 2005b: Potential climate change impacts on wind speeds and wind energy density in northern Europe: Results from empirical downscaling of multiple AOGCMs. *Clim. Res.*, **29**, 183–198.

Räisänen, J., 2001: Hiilidioksidin lisääntymisen vaikutus Pohjois-Euroopan ilmastoon globaaleissa ilmastomalleissa (The impact of increasing carbon dioxide on the climate of northern Europe in global climate models). *Terra*, **113**, 139–151.

Räisänen, J., 2002: CO_2-induced changes in interannual temperature and precipitation variability in 19 CMIP2 experiments. *J. Clim.*, **15**, 2395–2411.

Räisänen, J., 2005: CO_2-induced impact of increasing CO_2 on monthly-to-annual precipitation extremes: analysis of the CMIP2 experiments. *Clim. Dyn.*, **24**, 309–323.

Räisänen, J., and R. Joelsson, 2001: Changes in average and extreme precipitation in two regional climate model experiments. *Tellus*, **53A**, 547–566.

Räisänen, J., and T.N. Palmer, 2001: A probability and decision-model analysis of a multi-model ensemble of climate change simulations. *J. Clim.*, **14**, 3212–3226.

Räisänen, J., and H. Alexandersson, 2003: A probabilistic view on recent and near future climate change in Sweden. *Tellus*, **55A**, 113–125.

Räisänen, J., et al., 2003: *GCM Driven Simulations of Recent and Future Climate with the Rossby Centre Coupled Atmosphere – Baltic Sea Regional Climate Model RCAO.* Reports Meteorology and Climatology 101, Swedish Meteorological and Hydrological Institute, Norrköping, Sweden, 61 pp.

Räisänen, J., et al., 2004: European climate in the late 21st century: regional simulations with two driving global models and two forcing scenarios. *Clim. Dyn.*, **22**, 13–31.

Rajendran, K., A. Kitoh, and S. Yukimoto, 2004: South and East Asian summer monsoon climate and variation in the MRI coupled model (MRI-CGM2). *J. Clim.*, **17**, 763–782.

Ramanathan, V., et al., 2005: Atmospheric brown clouds: Impacts on South Asian climate and hydrological cycle. *Proc. Natl. Acad. Sci. U.S.A.*, **102**(15), 5326–5333.

Raphael, M.N., and M.M. Holland, 2006: Twentieth century simulation of the Southern hemisphere climate in coupled models. Part 1: Large-scale circulation variability. *Clim. Dyn.*, **26**, 217–228, doi:10.1007/s00382-005-0082-8.

Rauthe, M., and H. Paeth, 2004: Relative importance of Northern Hemisphere circulation modes in predicting regional climate change. *J. Clim.*, **17**, 4180–4189.

Reason, C., and A. Keibel, 2004: Tropical Cyclone Eline and its unusual penetration and impacts over the southern African mainland. *Weather Forecasting*, **19**(5), 789–805.

Reason, C.J.C., and M. Rouault, 2005: Links between the Antarctic Oscillation and winter rainfall over western South Africa. *Geophys. Res. Lett.*, **32**, L07705, doi:10.1029/2005GL022419.

Renwick, J.A., 2002: Southern Hemisphere circulation and relations with sea ice and sea surface temperature. *J. Clim.*, **15**, 3058–3068.

Rinke, A., and K. Dethloff, 2000: On the sensitivity of a regional Arctic climate model to initial and boundary conditions. *Clim. Res.*, **14**(2), 101–113.

Rinke, A., et al., 2003: A case study of the anomalous Arctic sea ice conditions during 1990: Insights from coupled and uncoupled regional climate model simulations. *J. Geophys. Res.*, **108**, 4275, doi:10.1029/2002JD003146.

Rinke, A., et al., 2006: Evaluation of an ensemble of Arctic regional climate models: Spatial patterns and height profiles. *Clim. Dyn.*, **26**(5), 459–472, doi:10.1007/s00382-005-0095-3.

RIVM (Rijks Instituut voor Volksgezondheid en Milieu), 2002: *The IMAGE 2.2 Implementation of the SRES Scenarios: A Comprehensive Analysis of Emissions, Climate Change and Impacts in the 21st Century.* CD-ROM, http://www.rivm.nl/bibliotheek/rapporten/481508018.html.

Roads, J., et al., 2003: International Research Institute/Applied Research Centers (IRI/ARCs) regional model intercomparison over South America. *J. Geophys. Res.*, **108**(D14), 4425, doi:10.1029/2002jd003201.

Robertson, A.W., J.D. Farrara, and C.R. Mechoso, 2003: Simulations of the atmospheric response to South Atlantic sea surface temperature anomalies. *J. Clim.*, **16**, 2540–2551.

Rockel, B., and K. Woth, 2007: Future changes in near surface wind speed extremes over Europe from an ensemble of RCM simulations. *Clim. Change*, doi: 10.1007/s10584-006-9227-y.

Rojas, M., and A. Seth, 2003: Simulation and sensitivity in a nested modeling system for South America. Part II: GCM boundary forcing. *J. Clim.*, **16**, 2454–2471.

Rotstayn, L.D., and U. Lohmann, 2002: Tropical rainfall trends and the indirect aerosol effect. *J. Clim.*, **15** (15), 2103–2116.

Rowell, D.P., 2003: The impact of Mediterranean SSTs on the Sahelian rainfall season. *J. Clim.*, **16** (5), 849–862.

Rowell, D.P., 2005: A scenario of European climate change for the late 21st century: seasonal means and interannual variability. *Clim. Dyn.*, **25**, 837–849.

Rowell, D.P., and R.G. Jones, 2006: Causes and uncertainty of future summer drying over Europe. *Clim. Dyn.*, **27**, 281–299.

Rowell, D.P., C.K. Folland, K. Maskell, and N.M. Ward, 1995: Variability of summer rainfall over Tropical North Africa (1906-92): Observations and modelling. *Q. J. R. Meteorol. Soc.*, **121**, 669–704.

Rummukainen, M., et al., 2003: Regional climate scenarios for use in Nordic water resources studies. *Nord Hydrol.*, **34**(5), 399–412.

Rummukainen, M., et al., 2004: The Swedish Regional Climate Modelling Programme, SWECLIM: a review. *Ambio*, **33**, 176–182.

Ruosteenoja, K., H. Tuomenvirta, and K. Jylhä, 2007: GCM-based regional temperature and precipitation change estimates for Europe under four SRES scenarios applying a super-ensemble pattern-scaling method. *Clim. Change*, doi: 10.1007/s10584-006-9222-3.

Ruosteenoja, K, T.R. Carter, K. Jylhä, and H. Tuomenvirta, 2003: *Future Climate in World Regions: And Intercomparison of Model-Based Projections for the New IPCC Emissions Scenarios.* Finnish Environment Institute, Helsinki, 83 pp.

Rupa Kumar, K., and R.G. Ashrit, 2001: Regional aspects of global climate change simulations: Validation and assessment of climate response over Indian monsoon region to transient increase of greenhouse gases and sulfate aerosols. *Mausam, Special Issue on Climate Change*, **52**, 229–244.

Rupa Kumar, K., et al., 2002: Climate change in India: Observations and model projections, In: *Climate Change and India: Issues, Concerns and Opportunities* [Shukla, P.R., et al., (eds.)]. Tata McGraw-Hill Publishing Co. Ltd., New Delhi, pp. 24–75.

Rupa Kumar, K., et al., 2003: Future climate scenarios. In: *Climate Change and India: Vulnerability Assessment and Adaptation* [Shukla, P.R., et al. (eds.)]. Universities Press, Hyderabad, pp. 69–127.

Rupa Kumar, K., et al., 2006: High-resolution climate change scenarios for India for the 21st century. *Curr. Sci. India.*, **90**, 334–345.

Russell, G.L., and D. Rind, 1999: Response to CO_2 transient increase in the GISS model: regional coolings in a warming climate. *J. Clim.*, **12**, 531–539.

Salathé, E.P., 2005: Downscaling simulations of future global climate with application to hydrologic modelling. *Int. J. Climatol.*, **25**, 419–436.

Salinger, M.J., J.A. Renwick, and A.B. Mullan, 2001: Interdecadal Pacific Oscillation and South Pacific climate. *Int. J. Climatol.*, **21**, 1705–1721.

Sánchez, E., C. Gallardo, M.A. Gaertner, A. Arribas and M. Castro, 2004: Future climate extreme events in the Mediterranean simulated by a regional climate model: first approach. *Global Planet. Change*, **44**, 163-180.

Santer, B.D., T.W.L. Wigley, M.E. Schlesinger, and J.F.B. Mitchell, 1990: *Developing Climate Scenarios from Equilibrium GCM Results.* Report No. 47, Max-Plank Institute for Meteorology, Hamburg, 29 pp.

Sarkar, S., R.P. Singh, and M. Kafatos, 2004: Further evidences for the weakening relationship of Indian rainfall and ENSO over India. *Geophys. Res. Lett.*, **31**, L13209, doi:10.1029/2004GL020259.

Sasaki, H., K. Kurihara, and I. Takayabu, 2006a: Comparison of climate reproducibilities between a super-high-resolution atmosphere general circulation model and a Meteorological Institute regional climate model. *Scientific Online Letters on the Atmosphere*, **1**, 81–84.

Sasaki, H., et al., 2006b: Preliminary results from the coupled atmosphere-ocean regional climate model developed at Meteorological Research Institute. *J. Meteorol. Soc. Japan*, **84**, 389–403.

Scaife, A., J.R. Knight, G.K. Vallis, and C.K. Folland, 2005: A stratospheric influence on the winter NAO and North Atlantic surface climate. *Geophys. Res. Lett.*, **32**, L18715, doi:10.1029/2005GL023226.

Schaeffer, M., F.M. Selten, J.D. Opsteegh, and H. Goosse, 2004: The influence of ocean convection patterns on high-latitude climate projections. *J. Clim.*, **17**, 4316–4329.

Schär, C., et al., 2004: The role of increasing temperature variability in European summer heatwaves. *Nature*, **427**, 332–336.

Schmidli, J., C. Frei, and P.L. Vidale, 2006: Downscaling from GCM precipitation: A benchmark for dynamical and statistical downscaling. *Int. J. Climatol.*, **26**, 679–689.

Scott, D., G. McBoyle, and B. Mills, 2003: Climate change and the skiing industry in southern Ontario (Canada): exploring the importance of snowmaking as a technical adaptation. *Clim. Res.*, **23**, 171–181.

Seem, R., 2004: Forecasting plant disease in a changing climate: a question of scale. *Can. J. Plant Pathol.*, **26**(3), 274–283.

Semenov, V.A., and L. Bengtsson, 2002: Secular trends in daily precipitation characteristics: greenhouse gas simulation with a coupled AOGCM. *Clim. Dyn.*, **19**, 123–140.

Semmler, T., and D. Jacob, 2004: Modeling extreme precipitation events – a climate change simulation for Europe. *Global Planet. Change*, **44**, 119–127.

Semmler, T., D. Jacob, K.H. Schluenzen, and R. Podzun, 2005: The water and energy budget of the Arctic atmosphere. *J. Clim.*, **18**, 2515–2530, doi:10.1175/JCLI3414.1.

Serreze, M.C., and C.M. Hurst, 2000: Representation of mean Arctic precipitation from NCEP-NCAR and ERA reanalyses. *J. Clim.*, **13**, 182–201.

Serreze, M.C., and J. Francis, 2006: The Arctic amplification debate. *Clim. Change*, **76**, 241–264, doi:10.1007/s10584-005-9017-y.

Seth, A., and M. Rojas, 2003: Simulation and sensitivity in a nested modeling system for South America. Part I: Reanalyses boundary forcing. *J. Clim.*, **16**, 2437–2453.

Shindell, D.T., and G.A. Schmidt, 2004: Southern hemisphere climate response to ozone changes and greenhouse gas increases. *Geophys. Res. Lett.*, **31**, L18209, doi:10.1029/2004GL020724.

Shkolnik, I.M., V.P. Meleshko, and V.M. Kattsov, 2006: Climate change in the 21st century over the Western Russia: a simulation with the MGO Regional Climate Model. *Russ. Meteorol. Hydrol.*, **3**, 5–17.

Simmons, A.J., et al., 2004: Comparison of trends and low-frequency variability in CRU, ERA-40, and NCEP/ NCAR analyses of surface air temperature. *J. Geophys. Res.*, **109**, D24115, doi:10.1029/2004JD005306.

Sinclair, M.R., 2002: Extratropical transition of southwest Pacific tropical cyclones. Part I: Climatology and mean structure changes. *Mon. Weather Rev.*, **130**, 590–609.

Sitch, S., et al., 2005: Impacts of future land cover changes on atmospheric CO_2 and climate. *Global Biogeochem. Cycles*, **19**(2), GB2013, doi:10.1029/2004GB002311.

Small, E., F. Giorgi, and L.C. Sloan, 1999: Regional climate model simulation of precipitation in central Asia: Mean and interannual variability. *J. Geophys. Res.*, **104**, 6563–6582.

Snyder, M.A., L.C. Sloan, N.S. Diffenbaugh, and J.L. Bell, 2003: Future climate change and upwelling in the California Current. *Geophys. Res. Lett.*, **30**(15), 1823, doi:10.1029/2003GL017647.

Solecki, W.D., and C. Oliveri, 2004: Downscaling climate change scenarios in an urban land use change model. *J. Environ. Manage.*, **72**, 105–115.

Somot, S., 2005: *Modélisation Climatique du Bassin Méditerranéen: Variabilité et Scénarios de Changement Climatique*. PhD Thesis, Université Paul Sabatier, Toulouse, France, 333 pp.

Staniforth, A., 1997: Regional modeling: A theoretical discussion. *Meteorol. Atmos. Phys.*, **63**, 15–29.

Steiner, N., et al., 2004: Comparing modelled streamfunction, heat and freshwater content in the Arctic Ocean. *Ocean Model.*, **6**(3–4), 265–284.

Stephenson, D.B., H. Douville, and K. Rupa Kumar, 2001: Searching for a fingerprint of global warming in the Asian summer monsoon. *Mausam*, **52**, 213–220.

Stephenson, D.B., et al., 2006: North Atlantic Oscillation response to transient greenhouse gas forcing and the impact on European winter climate: A CMIP2 multi-model assessment. *Clim. Dyn.*, **27**, 401–420.

Stone, D.A., A.J. Weaver, and R.J. Stouffer, 2001: Projection of climate change onto modes of atmospheric variability. *J. Clim.*, **14**, 3551–3565.

Stone, R.S., 1997: Variations in western Arctic temperatures in response to cloud radiative and synoptic-scale influences. *J. Geophys. Res.*, **102**, 21769–21776.

Stott, P.A., and J.A. Kettleborough, 2002: Origins and estimates of uncertainty in predictions of twenty-first century temperature rise. *Nature*, **416**, 723–726.

Stott, P.A., J.A. Kettleborough, and M.R. Allen, 2006a: Uncertainty in continental-scale temperature predictions. *Geophys. Res. Lett.*, **33**, L02708, doi:10.1029/2005GL024423.

Stott, P.A., et al., 2006b: Observational constraints on past attributable warming and predictions of future global warming. *J. Clim.*, **19**, 3055–3069.

Stouffer, R.J., et al., 2006: Investigating the causes of the response of the thermohaline circulation to past and future climate changes. *J. Clim.*, **19**(8), 1365–1387.

Sugi, M., A. Noda, and N. Sato, 2002: Influence of the global warming on tropical cyclone climatology: An experiment with the JMA Global Model. *J. Meteorol. Soc. Japan*, **80**, 249–272.

Suppiah, R., P.H. Whetton, and I.G. Watterson, 2004: *Climate Change in Victoria: Assessment of Climate Change for Victoria: 2001-2002*. Undertaken for Victorian Department of Sustainability and Environment. CSIRO Atmospheric Research, Aspendale, Vic., 33 pp.

Tadross, M.A., B.C. Hewitson, and M.T. Usman, 2005a: The interannual variability of the onset of the maize growing season over South Africa and Zimbabwe. *J. Clim.*, **18**(16), 3356–3372.

Tadross M.A., C. Jack, and B. Hewitson, 2005b: On RCM-based projections of change in southern African summer climate. *Geophys. Res. Lett.*, **32**, L23713, doi:10.1029/2005GL024460.

Tadross, M.A., et al., 2006: MM5 simulations of interannual change and the diurnal cycle of southern African regional climate. *Theor. Appl. Climatol.*, **86**, 63–80.

Taylor, C.M., et al., 2002: The influence of land use change on climate in the Sahel. *J. Clim.*, **15**, 3615–3629.

Taylor, M., and E. Alfero, 2005: Climate of Central America and the Caribbean. In: *The Encyclopedia of World Climatology* [Oliver, J. (ed.)]. Encyclopedia of Earth Sciences Series, Springer Press, 854 pp.

Taylor, M., D. Enfield, and A. Chen, 2002: The influence of the tropical Atlantic vs. the tropical Pacific on Caribbean rainfall. *J. Geophys. Res.*, **107** (C9), 3127, doi:10.1029/2001JC001097.

Tebaldi, C., L.O. Mearns, D. Nychka, and R. Smith, 2004a: Regional probabilities of precipitation change: A Bayesian analysis of multi-model simulations. *Geophys. Res. Lett.*, **31**, L24213, doi:10.1029/2004GL021276.

Tebaldi, C., R. Smith, D. Nychka, and L.O. Mearns, 2004b: Quantifying uncertainty in projections of regional climate change: A Bayesian Approach. *J. Clim.*, **18**(10), 1524–1540.

Tebaldi, C., K. Hayhoe, J.M. Arblaster, and G.E. Meehl, 2006: Going to the extremes: an intercomparison of model-simulated historical and future changes in extreme events. *Clim. Change*, **79**, 185–211.

Tennant, W., 2003: An assessment of intraseasonal variability from 13-yr GCM simulations. *Mon. Weather Rev.*, **131**(9), 1975–1991.

Timbal, B., 2004: Southwest Australia past and future rainfall trends. *Clim. Res.*, **26**, 233–249.

Timbal, B., and B.J. McAvaney, 2001: An analogue-based method to downscale surface air temperature: Application for Australia. *Clim. Dyn.*, **17**(12), 947–963.

Tjernström, M., et al., 2005: Modeling the Arctic boundary layer: An evaluation of six ARCMIP regional-scale models with data from the SHEBA project. *Bound.-Lay. Meteorol.*, **117**, 337–381.

Todd, M., and R. Washington, 1999: Circulation anomalies associated with tropical-temperate troughs in southern Africa and the south west Indian Ocean. *Clim. Dyn.*, **15**(12), 937–951.

Trigo, R.M., and J.P. Palutikof, 2001: Precipitation scenarios over Iberia: a comparison between direct GCM output and different downscaling techniques. *J. Clim.*, **14**, 4422–4446.

Turner, J., 2004: The El Niño-Southern Oscillation and Antarctica. *Int. J. Climatol.*, **24**, 1–31.

Turner, J., W.M. Connolley, T.A. Lachlan-Cope, and G.J. Marshall, 2006: The performance of the Hadley Centre climate model (HadCM3) in high southern latitudes. *Int. J. Climatol.*, **26**, 91–112.

Ueda, H., A. Iwai, K. Kuwako, and M.E. Hori, 2006: Impact of anthropogenic forcing on the Asian summer monsoon as simulated by eight GCMs. *Geophys. Res. Lett.*, **33**, L06703, doi:10.1029/2005GL025336.

Ulbrich, U., et al., 2006: The Mediterranean climate change under global warming. In: *Mediterranean Climate Variability* [Lionello, P., P. Malanotte, and R. Boscolo (eds.)]. Elsevier B.V, pp. 399–415.

Unnikrishnan, A.S., et al., 2006: Sea level changes along the Indian coast: Observations and projections. *Curr. Sci. India*, **90**, 362–368.

van de Berg, W.J., M.R. van den Broeke, C.H. Reijmer, and E. van Meijgaard, 2005: Characteristics of the Antarctic surface mass balance (1958-2002) using a regional atmospheric climate model. *Ann. Glaciol.*, **41**, 97–104.

van den Broeke, M.R., and N.P.M. van Lipzig, 2003: Factors controlling the near-surface wind field in Antarctica. *Mon. Weather Rev.*, **131**, 733–743.

van den Broeke, M., W.J. van de Berg, and E. van Meijgaard, 2006: Snowfall in coastal West Antarctica much greater than previously assumed. *Geophys. Res. Lett.*, **33**, L02505, doi:10.1029/2005GL025239.

van den Hurk, B., et al., 2005: Soil control on runoff response to climate change in regional climate model simulations. *J. Clim.*, **18**, 3536–3551.

van den Hurk, B., et al., 2006: *KNMI Climate Change Scenarios 2006 for the Netherlands.* KNMI WR-2006-01, KNMI, The Netherlands, 82 pp.

van Lipzig, N.P.M., E.W. van Meijgaard, and J. Oerlemans, 2002a: The spatial and temporal variability of the surface mass balance in Antarctica: results from a regional atmospheric climate model. *Int. J. Climatol.*, **22**, 1197–1217.

van Lipzig, N.P.M., E.W. van Meijgaard, and J. Oerlemans, 2002b: Temperature sensitivity of the Antarctic surface mass balance in a regional atmospheric climate model. *J. Clim.*, **15**, 2758–2774.

van Ulden, A.P., and G.J. van Oldenborgh, 2006: Large-scale atmospheric circulation biases and changes in global climate model simulations and their importance for climate change in Central Europe. *Atmos. Chem. Phys.*, **6**, 863–881.

van Ulden, A., G. Lenderink, B. van den Hurk, and E. van Meijgaard, 2007: Circulation statistics and climate change in Central Europe: Prudence simulations and observations. *Clim. Change*, doi: 10.1007/s10584-006-9212-5.

Vannitsem, S., and F. Chomé, 2005: One-way nested regional climate simulations and domain size. *J. Clim.*, **18**, 229–233.

Vera, C., et al., 2006: Towards a unified view of the American monsoon systems. *J. Clim.*, **19**, 4977–5000.

Vérant, S., 2004: *Etude des Dépressions sur l'Europe de l'Ouest : Climat Actuel et Changement Climatique.* PhD thesis, Université Paris VI, Paris, France, 204 pp.

Vernekar, A.D., B.P. Kirtman, and M.J. Fennessy, 2003: Low-level jets and their effects on the South American summer climate as simulated by the NCEP Eta model. *J. Clim.*, **16**, 297–311.

Vidale, P.L., D. Lüthi, R. Wegmann, and C. Schär, 2007: European climate variability in a heterogeneous multi-model ensemble. *Clim. Change*, doi: 10.1007/s10584-006-9218-z.

Vidale, P.L., et al., 2003: Predictability and uncertainty in a regional climate model. *J. Geophys. Res.*, **108**(D18), 4586, doi:10.1029/2002JD002810.

Vincent, C., 2002: Influence of climate change over the 20th century on 4 French glacier mass balances. *J. Geophys. Res.*, **107**, 4375, doi:10.1029/2001JD000832.

Vincent, D.G., 1994: The South Pacific Convergence Zone (SPCZ): A review. *Mon. Weather Rev.*, **122**, 1949–1970.

Vizy, E.K., and K.H. Cook, 2002: Development and application of a mesoscale climate model for the tropics: Influence of sea surface temperature anomalies on the West African monsoon. *J. Geophys. Res.*, **107**(D3), 4023, doi:10.1029/2001JD000686.

Voldoire, A., 2006: Quantifying the impact of future land-use changes against increases in GHG concentrations. *Geophys. Res. Lett.*, **33**, L04701, doi:10.1029/2005GL024354.

Voldoire, A., and J.F. Royer, 2004: Tropical deforestation and climate variability. *Clim. Dyn.*, **22**, 857–874.

Voldoire, A., and J.F. Royer, 2005: Climate sensitivity to tropical land surface changes with coupled versus prescribed SSTs, *Clim. Dyn.*, **24**, 843–862.

von Storch, H., and H. Reichardt, 1997: A scenario of storm surge statistics for the German Bight at the expected time of doubled atmospheric carbon dioxide concentration. *J. Clim.*, **10**, 2653–2662.

von Storch, H., H. Langenberg, and F. Feser, 2000: A spectral nudging technique for dynamical downscaling purposes. *Mon. Weather Rev.*, **128**, 3664–3673.

Voss, R., W. May, and E. Roeckner, 2002: Enhanced resolution modelling study on anthropogenic climate change: Changes in extremes of the hydrological cycle. *Int. J. Climatol.*, **22**, 755–777.

Vuille, M., and Bradley, R.S., 2000: Mean annual temperature trends and their vertical structure in the tropical Andes. *Geophys. Res. Lett.*, **27**, 3885–3888.

Walland, D.J., S.B. Power, and A.C. Hirst, 2000: Decadal climate variability simulated in a coupled general circulation model. *Clim. Dyn.*, **16**(2–3), 201–211.

Walser, A., and C. Schär, 2004: Convection-resolving precipitation forecasting and its predictability in Alpine river catchments. *J. Hydrol.*, **288**, 57–73.

Walsh, J.E., et al., 2002: Comparison of Arctic climate simulations by uncoupled and coupled global models. *J. Clim.*, **15**, 1429–1446.

Walsh, K.J., 2004: Tropical cyclones and climate change: unresolved issues. *Clim. Res.*, **27**, 77–84.

Walsh, K.J.E., and J.J. Katzfey, 2000: The impact of climate change on the poleward movement of tropical cyclone-like vortices in a regional climate model. *J. Clim.*, **13**(6), 1116–1132.

Walsh, K.J.E., and B.F. Ryan, 2000: Tropical cyclone intensity increase near Australia as a result of climate change. *J. Clim.*, **13**(16), 3029–3036.

Walsh, K.J., K.C. Nguyen, and J.L. McGregor, 2004: Fine-resolution regional climate model simulations of the impact of climate change on tropical cyclones near Australia. *Clim. Dyn.*, **22**(1), 47–56.

Walsh, K., et al., 1999: *Climate Change in Queensland Under Enhanced Greenhouse Conditions: First Annual Report, 1997-1998.* CSIRO Atmospheric Research, Melbourne, 84 pp.

Wang, B., I.S. Kang, and J.Y. Lee, 2004: Ensemble simulations of Asian-Australian monsoon variability by 11 GCMs. *J. Clim.*, **17**, 803–818.

Wang, G., 2005: Agricultural drought in a future climate: results from 15 global climate models participating in the IPCC 4th assessment. *Clim. Dyn.*, **25**, 739–753.

Wang, G., and E.A.B. Eltahir, 2000: Role of vegetation dynamics in enhancing the low-frequency variability of the Sahel rainfall. *Water Resour. Res.*, **36**(4), 1013–1021.

Wang, X.L., and V.R. Swail, 2006a: Climate change signal and uncertainty in projections of ocean wave heights. *Clim. Dyn.*, **26**, 106–126, doi:10.1007/s00382-005-0080-x.

Wang, X.L., and V.R. Swail, 2006b: Historical and possible future changes of wave heights in Northern Hemisphere oceans. In: *Atmosphere-Ocean Interactions – Vol. 2* [Perrie, W. (ed.)]. Advances in Fluid Mechanics Series, Vol 39. Wessex Institute of Technology Press, Southampton, UK, 240 pp.

Wang, X.L., F. Zwiers, and V. Swail, 2004: North Atlantic ocean wave climate change scenarios for the twenty-first century. *J. Clim.*, **17**, 2368–2383.

Warner, T.T., R.A. Peterson, and R.E. Treadon, 1997: A tutorial on lateral conditions as a basic and potentially serious limitation to regional numerical weather prediction. *Bull. Am. Meteorol. Soc.*, **78**(11), 2599–2617.

Watterson, I.G., and M.R. Dix, 2003: Simulated changes due to global warming in daily precipitation means and extremes and their interpretation using the gamma distribution. *J. Geophys. Res.*, **108**(D13), 4379, doi:10.1029/2002JD002928.

Weatherly, J., 2004: Sensitivity of Antarctic precipitation to sea ice concentrations in a general circulation model. *J. Clim.*, **17**, 3214–3223.

Webb, M., et al., 2006: On the contribution of local feedback mechanisms to the range of climate sensitivity in two GCM ensembles. *Clim. Dyn.*, **27**, 17–38.

Wehner, M.F., 2004: Predicted twenty-first-century changes in seasonal extreme precipitation events in the Parallel Climate Model. *J. Clim.*, **17**, 4281–4290.

Wei, H., W.J. Gutowski, C.J. Vorosmarty, and B.M. Fekete, 2002: Calibration and validation of a regional climate model for pan-Arctic hydrologic simulation. *J. Clim.*, **15**, 3222–3236.

Weisheimer, A., and T. Palmer, 2005: Changing frequency of occurrence of extreme seasonal temperatures under global warming. *Geophys. Res. Lett.*, **32**, L20721, doi:10.1029/2005GL023365.

Weisman, M.L., W.S. Skamarock, and J.B. Klemp, 1997: The resolution dependence of explicitly modeled convective systems. *Mon. Weather Rev.*, **125**, 527–548.

Weisse, R., H. von Storch, and F. Feser, 2005: Northeast Atlantic and North Sea storminess as simulated by a regional climate model 1958-2001 and comparison with observations. *J. Clim.*, **18**, 465–479.

Whetton, P.H., and R. Suppiah, 2003: Climate change projections and drought. In: *Science for Drought. Proceedings of the National Drought Forum, Carlton Crest Hotel, Brisbane* [Stone, R., and I. Partridge (eds.)]. Queensland Department of Primary Industries, Brisbane, Qld., pp. 130–136.

Whetton, P.H., A.B. Mullan, and A.B. Pittock, 1996: Climate-change scenarios for Australia and New Zealand. In: *Greenhouse: Coping with Climate Change* [Bouma, W.J., G.I. Pearman, and M.R. Manning (eds.)]. CSIRO, Collingwood, Vic., pp. 145–168.

Whetton, P.H., et al., 2000: *Climate Averages Based on a Doubled CO$_2$ Simulation*. Victorian Dept. of Natural Resources and Environment, Melbourne, 43 pp.

Whetton, P.H., et al., 2001: Developing scenarios of climate change for Southeastern Australia: An example using regional climate model output. *Clim. Res.*, **16**(3), 181–201.

Whetton, P. H., et al. 2002. *Climate change in Victoria : high resolution regional assessment of climate change impacts*. Undertaken for the Victorian Department of Natural Resources and Environment. Dept. of Natural Resources and Environment, East Melbourne, VIC, 44 pp., http://www.greenhouse.vic.gov.au/climatechange.pdf

Whetton, P.H., et al., 2005: *Australian Climate Change Projections for Impact Assessment and Policy Application: A Review.* CSIRO Marine and Atmospheric Research Paper 001, CSIRO Marine and Atmospheric Research, Aspendale, Vic., 34 pp.

Widmann, M., C.S. Bretherton, and E.P. Salathé Jr., 2003: Statistical precipitation downscaling over the Northwestern United States using numerically simulated precipitation as a predictor. *J. Clim.*, **16**, 799–816.

Wigley, T.M.L., and S.C.B. Raper, 2001: Interpretation of high projections for global-mean warming. *Science*, **293**, 451–454.

Wilby, R.L., C.W. Dawson, and E.M. Barrow, 2002: SDSM – A decision support tool for the assessment of regional climate change impacts. *Environ. Model. Software*, **17**, 147–159.

Wilby, R.L., O.J. Tomlinson, and C.W. Dawson, 2003: Multi-site simulation of precipitation by conditional resampling. *Clim. Res.*, **23**, 183–194.

Wilby, R.L., et al., 2004: *Guidelines for Use of Climate Scenarios Developed from Statistical Downscaling Methods.* IPCC Task Group on Data and Scenario Support for Impact and Climate Analysis (TGICA), http://ipcc-ddc.cru.uea.ac.uk/guidelines/StatDown_Guide.pdf.

Wood, A.W., L.R. Leung, V. Sridhar, and D.P. Lettenmaier, 2004: Hydrologic implications of dynamical and statistical approaches to downscaling climate model outputs. *Clim. Change*, **62**, 189–216.

Woodworth, R., and D.L. Blackman, 2004: Evidence for systematic changes in extreme high waters since the mid-1970s. *J. Clim.*, **17**(6), 1190–1197.

Woth, K., 2005: North Sea storm surge statistics based on projections in a warmer climate: How important are the driving GCM and the chosen emission scenario? *Geophys. Res. Lett.*, **32**, L22708, doi:10.1029/2005GL023762.

Woth, K., R. Weisse, and H. von Storch, 2006: Climate change and North Sea storm surge extremes: An ensemble study of storm surge extremes expected in a changed climate projected by four different Regional Climate Models. *Ocean Dyn.*, **56**, 3–15, doi:10.1007/s10236-005-0024-3.

Wu, B., and B. Wang, 2004: Assessing impacts of global warming on tropical cyclone tracks. *J. Clim.*, **17**(8), 1686–1698.

Wu, P., R. Wood, and P. Scott, 2003: Does the recent freshening trend in the North Atlantic indicate a weakening thermohaline circulation? *Geophys. Res. Lett.*, **31**, doi:10.1029/2003GL018584.

Wyser, K., and C.G. Jones, 2005: Modeled and observed clouds during Surface Heat Budget of the Arctic Ocean (SHEBA). *J. Geophys. Res.*, **110**, D09207, doi:10.1029/2004JD004751.

Xu, Y., Y.H. Ding., and L.D. Li, 2003a: Climate change projection in Qinghai-Xizang Plateau in the future 100 years. *Plateau Meteorol.*, **22**(5), 451–457.

Xu, Y., Y.H. Ding, and Z.C. Zhao, 2003b: Scenario of temperature and precipitation changes in Northwest China due to effects of human activities in 21st century. *J. Glaciol.*, **25**(3), 327–330.

Xu, Y., Z.C. Zhao, and L.D. Li, 2005: The simulated result analyses on climate changes over Qinghai-Xizang Plateau and along the Railway in the coming 50 Years, *Plateau Meteorol.*, **24**(5), 700–707.

Xu, Y.L., et al., 2005: Statistical analyses of climate change scenario over China in the 21st century. *Adv. Clim. Change Res.*, **1**, 80–83.

Xue, M., K.K. Droegemeier, and V. Wong, 2000: The Advanced Regional Prediction System (ARPS) - A multi-scale nonhydrostatic atmospheric simulation and prediction model. Part I: Model dynamics and verification. *Meteorol. Atmos. Phys.*, **75**(3 – 4), 161–193.

Yasunaga, K., et al., 2006: Changes in the Baiu frontal activity in the future climate simulated by super-high-resolution global and cloud-resolving regional climate models. *J. Meteorol. Soc. Japan*, **84**, 199–220.

Yin, J.H., 2005: A consistent poleward shift of the storm tracks in simulations of 21st century climate. *Geophys. Res. Lett.*, **32**, L18701, doi:10.1029/2005GL023684.

Yoshizaki, M., et al., 2005: Changes of Baiu (Mei-yu) frontal activity in the global warming climate simulated by a non-hydrostatic regional model. *Scientific Online Letters on the Atmosphere*, **1**, 25–28.

Yuan, X.J., 2004: ENSO-related impacts on Antarctic sea ice: a synthesis of phenomenon and mechanisms. *Antarct. Sci.*, **16**, 415–425.

Zeng, N., H. Qian, E. Munoz, and R. Iacono, 2004: How strong is carbon cycle-climate feedback under global warming? *Geophys. Res. Lett.*, **31**, L20203, doi:10.1029/2004GL020904.

Zhai, P.M., X.B. Zhang, H. Wan, and X.H. Pan, 2005: Trends in total precipitation and frequency of daily precipitation extremes over China. *J. Clim.*, **18**, 1096–1108.

Zhang, D.F., X.J. Gao, and Z.C. Zhao, 2005a: Simulation of climate in China by RegCM3. *Adv. Clim. Change Res.*, **1**(3), 119–121.

Zhang, D.F., X.J. Gao, H.Z. Bai, and D.L. Li, 2005b: Simulation of climate over Qinghai-Xizang Plateau utilizing RegCM3. *Plateau Meteorol.*, **24**, 714–720.

Zhao, Z.C., X.J. Gao, M.C. Tang, and Y. Xu, 2002: Climate change projections. In: *Assessment for Evolution of Environment in Western China, Vol. 2* [Qin, D.H., and Y.D. Ding (eds.)]. Science Press, Beijing, pp.16–46.

Zhou, J., and K.-M. Lau, 2002: Intercomparison of model simulations of the impact of 1997/98 El Niño on South American Summer Monsoon. *Meteorologica*, **27**(1–2), 99–116.

Zhou, T.J., and Z.X. Li, 2002: Simulation of the East Asian summer monsoon by using a variable resolution atmospheric GCM. *Clim. Dyn.*, **19**, 167–180.

Zhou, T.J., and R.C. Yu, 2006: 20th century surface air temperature over China and the globe simulated by coupled climate models. *J. Clim.*, **19**, 5843–5858.

Zou, C.-Z., M.L. van Woert, C. Xu, and K. Syed, 2004: Assessment of the NCEP–DOE Reanalysis-2 and TOVS pathfinder a moisture fields and their use in Antarctic net precipitation estimates. *Mon. Weather Rev.*, **132**, 2463–2476.

Zwiers, F.W., and V.V. Kharin, 1998: Changes in the extremes of the climate simulated by CCC GCM2 under CO_2 doubling. *J. Clim.*, **11**, 2200–2222.

Annex I

Glossary

Editor: A.P.M. Baede (Netherlands)

Notes: This glossary defines some specific terms as the lead authors intend them to be interpreted in the context of this report. Red, italicised words indicate that the term is defined in the Glossary

8.2ka event Following the last post-glacial warming, a rapid *climate* oscillation with a cooling lasting about 400 years occurred about 8.2 ka. This event is also referred to as the *8.2kyr event*.

Abrupt climate change The *nonlinearity* of the *climate system* may lead to abrupt climate change, sometimes called *rapid climate change, abrupt events or even surprises*. The term *abrupt* often refers to time scales faster than the typical time scale of the responsible forcing. However, not all abrupt climate changes need be *externally forced*. Some possible abrupt events that have been proposed include a dramatic reorganisation of the *thermohaline circulation*, rapid deglaciation and massive melting of *permafrost* or increases in soil *respiration* leading to fast changes in the *carbon cycle*. Others may be truly unexpected, resulting from a strong, rapidly changing forcing of a nonlinear system.

Active layer The layer of ground that is subject to annual thawing and freezing in areas underlain by *permafrost* (Van Everdingen, 1998).

Adiabatic process An adiabatic process is a process in which no external heat is gained or lost by the system. The opposite is called a *diabatic process*.

Adjustment time See *Lifetime*; see also *Response time*.

Advection Transport of water or air along with its properties (e.g., temperature, chemical tracers) by the motion of the fluid. Regarding the general distinction between advection and *convection*, the former describes the predominantly horizontal, large-scale motions of the *atmosphere* or ocean, while convection describes the predominantly vertical, locally induced motions.

Aerosols A collection of airborne solid or liquid particles, with a typical size between 0.01 and 10 μm that reside in the *atmosphere* for at least several hours. Aerosols may be of either natural or *anthropogenic* origin. Aerosols may influence *climate* in several ways: directly through scattering and absorbing radiation, and indirectly by acting as *cloud condensation nuclei* or modifying the optical properties and lifetime of clouds (see *Indirect aerosol effect*).

Afforestation Planting of new forests on lands that historically have not contained forests. For a discussion of the term *forest* and related terms such as afforestation, *reforestation* and *deforestation*, see the IPCC Special Report on Land Use, Land-Use Change and Forestry (IPCC, 2000). See also the report on Definitions and Methodological Options to Inventory Emissions from Direct Human-induced Degradation of Forests and Devegetation of Other Vegetation Types (IPCC, 2003).

Air mass A widespread body of air, the approximately homogeneous properties of which (1) have been established while that air was situated over a particular *region* of the Earth's surface, and (2) undergo specific modifications while in transit away from the source region (AMS, 2000).

Albedo The fraction of *solar radiation* reflected by a surface or object, often expressed as a percentage. Snow-covered surfaces have a high albedo, the surface albedo of soils ranges from high to low, and vegetation-covered surfaces and oceans have a low albedo. The Earth's planetary albedo varies mainly through varying cloudiness, snow, ice, leaf area and land cover changes.

Albedo feedback A *climate feedback* involving changes in the Earth's *albedo*. It usually refers to changes in the *cryosphere*, which has an albedo much larger (~0.8) than the average planetary albedo (~0.3). In a warming *climate*, it is anticipated that the cryosphere would shrink, the Earth's overall albedo would decrease and more *solar radiation* would be absorbed to warm the Earth still further.

Alkalinity A measure of the capacity of a solution to neutralize acids.

Altimetry A technique for measuring the height of the sea, lake or river, land or ice surface with respect to the centre of the Earth within a defined terrestrial reference frame. More conventionally, the height is with respect to a standard *reference ellipsoid* approximating the Earth's oblateness, and can be measured from space by using radar or laser with centimetric precision at present. Altimetry has the advantages of being a geocentric measurement, rather than a measurement relative to the Earth's crust as for a *tide gauge*, and of affording quasi-global coverage.

Annular modes Preferred patterns of change in atmospheric circulation corresponding to changes in the zonally averaged mid-latitude westerlies. The *Northern Annular Mode* has a bias to the North Atlantic and has a large correlation with the *North Atlantic Oscillation*. The *Southern Annular Mode* occurs in the Southern Hemisphere. The variability of the mid-latitude westerlies has also been known as *zonal flow* (or *wind*) vacillation, and defined through a *zonal index*. For the corresponding circulation indices, see Box 3.4.

Anthropogenic Resulting from or produced by human beings.

Atlantic Multi-decadal Oscillation (AMO) A multi-decadal (65 to 75 year) fluctuation in the North Atlantic, in which *sea surface temperatures* showed warm phases during roughly 1860 to 1880 and 1930 to 1960 and cool phases during 1905 to 1925 and 1970 to 1990 with a range of order 0.4°C.

Atmosphere The gaseous envelope surrounding the Earth. The dry atmosphere consists almost entirely of nitrogen (78.1% *volume mixing ratio*) and oxygen (20.9% volume mixing ratio), together with a number of trace gases, such as argon (0.93% volume mixing ratio), helium and radiatively active *greenhouse gases* such as *carbon dioxide* (0.035% volume mixing ratio) and *ozone*. In addition, the atmosphere contains the greenhouse gas water vapour, whose amounts are highly variable but typically around 1% volume mixing ratio. The atmosphere also contains clouds and *aerosols*.

Atmospheric boundary layer The atmospheric layer adjacent to the Earth's surface that is affected by friction against that boundary surface, and possibly by transport of heat and other variables across that surface (AMS, 2000). The lowest 10 metres or so of the boundary layer, where mechanical generation of turbulence is dominant, is called the *surface boundary layer* or *surface layer*.

Atmospheric lifetime See *Lifetime*.

Attribution See *Detection and attribution*.

Autotrophic respiration *Respiration* by *photosynthetic* organisms (plants).

Bayesian method A Bayesian method is a method by which a statistical analysis of an unknown or uncertain quantity is carried out in two steps. First, a prior probability distribution is formulated on the basis of existing knowledge (either by eliciting expert opinion or by using existing data and studies). At this first stage, an element of subjectivity may influence the choice, but in many cases, the prior probability distribution is chosen as neutrally as possible, in order not to influence the final outcome of the analysis. In the second step, newly acquired data are introduced, using a theorem formulated by and named after the British mathematician Bayes (1702–1761), to update the prior distribution into a posterior distribution.

Biomass The total mass of living organisms in a given area or volume; dead plant material can be included as dead biomass.

Biome A biome is a major and distinct regional element of the *biosphere*, typically consisting of several *ecosystems* (e.g. *forests*, rivers, ponds, swamps within a *region*). Biomes are characterised by typical communities of plants and animals.

Biosphere (terrestrial and marine) The part of the Earth system comprising all *ecosystems* and living organisms, in the *atmosphere*, on land (*terrestrial biosphere*) or in the oceans (*marine biosphere*), including derived dead organic matter, such as litter, soil organic matter and oceanic detritus.

Black carbon (BC) Operationally defined *aerosol* species based on measurement of light absorption and chemical reactivity and/or thermal stability; consists of *soot*, *charcoal* and/or possible light-absorbing refractory organic matter (Charlson and Heintzenberg, 1995, p. 401).

Blocking anticyclone An anticyclone that remains nearly stationary for a week or more at middle to high latitudes, so that it blocks the normal eastward progression of high- and low-pressure systems.

Bowen ratio The ratio of *sensible* to *latent heat fluxes* from the Earth's surface up into the *atmosphere*. Values are low (order 0.1) for wet surfaces like the ocean, and greater than 2 for deserts and *drought* regions.

Burden The total mass of a gaseous substance of concern in the *atmosphere*.

^{13}C Stable isotope of carbon having an atomic weight of approximately 13. Measurements of the ratio of ^{13}C/^{12}C in *carbon dioxide* molecules are used to infer the importance of different *carbon cycle* and *climate* processes and the size of the terrestrial carbon *reservoir*.

^{14}C Unstable isotope of carbon having an atomic weight of approximately 14, and a half-life of about 5,700 years. It is often used for dating purposes going back some 40 kyr. Its variation in time is affected by the magnetic fields of the Sun and Earth, which influence its production from cosmic rays (see *Cosmogenic isotopes*).

C3 plants Plants that produce a three-carbon compound during *photosynthesis*, including most trees and agricultural crops such as rice, wheat, soybeans, potatoes and vegetables.

C4 plants Plants that produce a four-carbon compound during *photosynthesis*, mainly of tropical origin, including grasses and the agriculturally important crops maize, sugar cane, millet and sorghum.

Carbonaceous aerosol *Aerosol* consisting predominantly of organic substances and various forms of *black carbon* (Charlson and Heintzenberg, 1995, p. 401).

Carbon cycle The term used to describe the flow of carbon (in various forms, e.g., as *carbon dioxide*) through the *atmosphere*, ocean, terrestrial *biosphere* and *lithosphere*.

Carbon dioxide (CO$_2$) A naturally occurring gas, also a by-product of burning fossil fuels from fossil carbon deposits, such as oil, gas and coal, of burning *biomass* and of *land use* changes and other industrial processes. It is the principal *anthropogenic greenhouse gas* that affects the Earth's radiative balance. It is the reference gas against which other greenhouse gases are measured and therefore has a *Global Warming Potential* of 1.

Carbon dioxide (CO$_2$) fertilization The enhancement of the growth of plants as a result of increased atmospheric *carbon dioxide* (CO$_2$) concentration. Depending on their mechanism of *photosynthesis*, certain types of plants are more sensitive to changes in atmospheric CO$_2$ concentration. In particular, *C3 plants* generally show a larger response to CO$_2$ than *C4 plants*.

CFC See *Halocarbons*.

Chaos A *dynamical system* such as the *climate system*, governed by nonlinear deterministic equations (see *Nonlinearity*), may exhibit erratic or chaotic behaviour in the sense that very small changes in the initial state of the system in time lead to large and apparently unpredictable changes in its temporal evolution. Such chaotic behaviour may limit the *predictability* of nonlinear dynamical systems.

Charcoal Material resulting from charring of *biomass*, usually retaining some of the microscopic texture typical of plant tissues; chemically it consists mainly of carbon with a disturbed graphitic structure, with lesser amounts of oxygen and hydrogen (Charlson and Heintzenberg, 1995, p. 402). See *Black carbon*; *Soot*.

Chronology Arrangement of events according to dates or times of occurrence.

Clathrate (methane) A partly frozen slushy mix of methane gas and ice, usually found in sediments.

Climate Climate in a narrow sense is usually defined as the average weather, or more rigorously, as the statistical description in terms of the mean and variability of relevant quantities over a period of time ranging from months to thousands or millions of years. The classical period for averaging these variables is 30 years, as defined by the World Meteorological Organization. The relevant quantities are most often surface variables such as temperature, precipitation and wind. Climate in a wider sense is the state, including a statistical description, of the *climate system*. In various chapters in this report different averaging periods, such as a period of 20 years, are also used.

Climate change Climate change refers to a change in the state of the *climate* that can be identified (e.g., by using statistical tests) by changes in the mean and/or the variability of its properties, and that persists for an extended period, typically decades or longer. Climate change may be due to natural internal processes or *external forcings*, or to persistent *anthropogenic* changes in the composition of the *atmosphere* or in *land use*. Note that the *Framework Convention on Climate Change* (UNFCCC), in its Article 1, defines *climate change* as: 'a change of climate which is attributed directly or indirectly to human activity that alters the composition of the global atmosphere and which is in addition to natural climate variability observed over comparable time periods'. The UNFCCC thus makes a distinction between climate change attributable to human activities altering the atmospheric composition, and *climate variability* attributable to natural causes. See also *Climate variability; Detection and Attribution*.

Climate change commitment Due to the thermal inertia of the ocean and slow processes in the *biosphere*, the *cryosphere* and land surfaces, the *climate* would continue to change even if the atmospheric composition were held fixed at today's values. Past change in atmospheric composition leads to a *committed climate change*, which continues for as long as a radiative imbalance persists and until all components of the *climate system* have adjusted to a new state. The further change in temperature after the composition of the *atmosphere* is held constant is referred to as the *constant composition temperature commitment* or simply *committed warming* or *warming commitment*. Climate change commitment includes other future changes, for example in the hydrological cycle, in *extreme weather* and climate events, and in *sea level change*.

Climate feedback An interaction mechanism between processes in the *climate system* is called a climate feedback when the result of an initial process triggers changes in a second process that in turn influences the initial one. A positive feedback intensifies the original process, and a negative feedback reduces it.

Climate Feedback Parameter A way to quantify the radiative response of the *climate system* to a *global surface temperature* change induced by a *radiative forcing* (units: W m^{-2} °C^{-1}). It varies as the inverse of the effective *climate sensitivity*. Formally, the Climate Feedback Parameter (Λ) is defined as: $\Lambda = (\Delta Q - \Delta F) / \Delta T$, where Q is the global mean radiative forcing, T is the global mean air surface temperature, F is the heat flux into the ocean and Δ represents a change with respect to an unperturbed *climate*.

Climate model (spectrum or hierarchy) A numerical representation of the *climate system* based on the physical, chemical and biological properties of its components, their interactions and *feedback* processes, and accounting for all or some of its known properties. The climate system can be represented by models of varying complexity, that is, for any one component or combination of components a *spectrum* or *hierarchy* of models can be identified, differing in such aspects as the number of spatial dimensions, the extent to which physical, chemical or biological processes are explicitly represented, or the level at which empirical *parametrizations* are involved. Coupled Atmosphere-Ocean General Circulation Models (AOGCMs) provide a representation of the climate system that is near the most comprehensive end of the spectrum currently available. There is an evolution towards more complex models with interactive chemistry and biology (see Chapter 8). Climate models are applied as a research tool to study and simulate the climate, and for operational purposes, including monthly, seasonal and interannual *climate predictions*.

Climate prediction A climate prediction or *climate forecast* is the result of an attempt to produce an estimate of the actual evolution of the *climate* in the future, for example, at seasonal, interannual or long-term time scales. Since the future evolution of the *climate system* may be highly sensitive to initial conditions, such predictions are usually probabilistic in nature. See also *Climate projection; Climate scenario; Predictability*.

Climate projection A *projection* of the response of the *climate system* to *emission or concentration scenarios* of *greenhouse gases* and *aerosols*, or *radiative forcing* scenarios, often based upon simulations by *climate models*. Climate projections are distinguished from *climate predictions* in order to emphasize that climate projections depend upon the emission/concentration/radiative forcing scenario used, which are based on assumptions concerning, for example, future socioeconomic and technological developments that may or may not be realised and are therefore subject to substantial *uncertainty*.

Climate response
See *Climate sensitivity*.

Climate scenario A plausible and often simplified representation of the future *climate*, based on an internally consistent set of climatological relationships that has been constructed for explicit use in investigating the potential consequences of *anthropogenic climate change*, often serving as input to impact models. *Climate projections* often serve as the raw material for constructing climate scenarios, but climate scenarios usually require additional information such as about the observed current climate. A *climate change scenario* is the difference between a climate scenario and the current climate.

Climate sensitivity In IPCC reports, *equilibrium climate sensitivity* refers to the equilibrium change in the annual mean *global surface temperature* following a doubling of the atmospheric *equivalent carbon dioxide concentration*. Due to computational constraints, the equilibrium climate sensitivity in a *climate model* is usually estimated by running an atmospheric general circulation model coupled to a mixed-layer ocean model, because equilibrium climate sensitivity is largely determined by atmospheric processes. Efficient models can be run to equilibrium with a dynamic ocean.

The *effective climate sensitivity* is a related measure that circumvents the requirement of equilibrium. It is evaluated from model output for evolving non-equilibrium conditions. It is a measure of the strengths of the *climate feedbacks* at a particular time and may vary with forcing history and *climate* state. The climate sensitivity parameter (units: °C (W m^{-2})$^{-1}$) refers to the equilibrium change in the annual mean *global surface temperature* following a unit change in *radiative forcing*.

The *transient climate response* is the change in the global surface temperature, averaged over a 20-year period, centred at the time of atmospheric carbon dioxide doubling, that is, at year 70 in a 1% yr^{-1} compound carbon dioxide increase experiment with a global coupled climate model. It is a measure of the strength and rapidity of the surface temperature response to *greenhouse gas* forcing.

Climate shift or climate regime shift An abrupt shift or jump in mean values signalling a change in *regime*. Most widely used in conjunction with the 1976/1977 climate shift that seems to correspond to a change in *El Niño-Southern Oscillation* behavior.

Climate system The climate system is the highly complex system consisting of five major components: the *atmosphere*, the *hydrosphere*, the *cryosphere*, the land surface and the *biosphere*, and the interactions between them. The climate system evolves in time under the influence of its own internal dynamics and because of *external forcings* such as volcanic eruptions, solar variations and *anthropogenic* forcings such as the changing composition of the atmosphere and *land use change*.

Climate variability Climate variability refers to variations in the mean state and other statistics (such as standard deviations, the occurrence of extremes, etc.) of the *climate* on all *spatial and temporal scales* beyond that of individual weather events. Variability may be due to natural internal processes within the *climate system* (*internal variability*), or to variations in natural or *anthropogenic external forcing* (*external variability*). See also *Climate change*.

Cloud condensation nuclei (CCN) Airborne particles that serve as an initial site for the condensation of liquid water, which can lead to the formation of cloud droplets. See also *Aerosols*.

Cloud feedback A *climate feedback* involving changes in any of the properties of clouds as a response to other atmospheric changes. Understanding cloud feedbacks and determining their magnitude and sign require an understanding of how a change in *climate* may affect the spectrum of cloud types, the cloud fraction and height, and the radiative properties of clouds, and an estimate of the impact of these changes on the Earth's radiation budget. At present, cloud feedbacks remain the largest source of *uncertainty* in *climate sensitivity* estimates. See also *Cloud radiative forcing*; *Radiative forcing*.

Cloud radiative forcing Cloud radiative forcing is the difference between the all-sky Earth's radiation budget and the clear-sky Earth's radiation budget (units: W m^{-2}).

CO$_2$-equivalent See *Equivalent carbon dioxide*.

Confidence The *level of confidence* in the correctness of a result is expressed in this report, using a standard terminology defined in Box 1.1. See also *Likelihood*; *Uncertainty*.

Convection Vertical motion driven by buoyancy forces arising from static instability, usually caused by near-surface cooling or increases in salinity in the case of the ocean and near-surface warming in the case of the *atmosphere*. At the location of convection, the horizontal scale is approximately the same as the vertical scale, as opposed to the large contrast between these scales in the *general circulation*. The net vertical mass transport is usually much smaller than the upward and downward exchange.

Cosmogenic isotopes Rare isotopes that are created when a high-energy cosmic ray interacts with the nucleus of an *in situ* atom. They are often used as indications of solar magnetic activity (which can shield cosmic rays) or as tracers of atmospheric transport, and are also called *cosmogenic nuclides*.

Cryosphere The component of the *climate system* consisting of all snow, ice and *frozen ground* (including *permafrost*) on and beneath the surface of the Earth and ocean. See also *Glacier*; *Ice sheet*.

Dansgaard-Oeschger events Abrupt warming events followed by gradual cooling. The abrupt warming and gradual cooling is primarily seen in Greenland *ice cores* and in *palaeoclimate* records from the nearby North Atlantic, while a more general warming followed by a gradual cooling has been observed in other areas as well, at intervals of 1.5 to 7 kyr during glacial times.

Deforestation Conversion of forest to non-forest. For a discussion of the term *forest* and related terms such as *afforestation*, *reforestation*, and deforestation see the IPCC Special Report on Land Use, Land-Use Change and Forestry (IPCC, 2000). See also the report on Definitions and Methodological Options to Inventory Emissions from Direct Human-induced Degradation of Forests and Devegetation of Other Vegetation Types (IPCC, 2003).

Desertification Land degradation in arid, semi-arid, and dry sub-humid areas resulting from various factors, including climatic variations and human activities. The United Nations Convention to Combat Desertification defines land degradation as a reduction or loss in arid, semi-arid, and dry sub-humid areas, of the biological or economic productivity and complexity of rain-fed cropland, irrigated cropland, or range, pasture, *forest*, and woodlands resulting from *land uses* or from a process or combination of processes, including processes arising from human activities and habitation patterns, such as (i) soil erosion caused by wind and/or water; (ii) deterioration of the physical, chemical and biological or economic properties of soil; and (iii) long-term loss of natural vegetation.

Detection and attribution *Climate* varies continually on all time scales. *Detection* of *climate change* is the process of demonstrating that climate has changed in some defined statistical sense, without providing a reason for that change. *Attribution* of causes of climate change is the process of establishing the most likely causes for the detected change with some defined level of *confidence*.

Diatoms Silt-sized algae that live in surface waters of lakes, rivers and oceans and form shells of opal. Their species distribution in ocean cores is often related to past *sea surface temperatures*.

Diurnal temperature range The difference between the maximum and minimum temperature during a 24-hour period.

Dobson unit (DU) A unit to measure the total amount of *ozone* in a vertical column above the Earth's surface (*total column ozone*). The number of Dobson units is the thickness in units of 10^{-5} m that the ozone column would occupy if compressed into a layer of uniform density at a pressure of 1,013 hPa and a temperature of 0°C. One DU corresponds to a column of ozone containing $2.69 \times 1,020$ molecules per square metre. A typical value for the amount of ozone in a column of the Earth's atmosphere, although very variable, is 300 DU.

Downscaling Downscaling is a method that derives local- to regional-scale (10 to 100 km) information from larger-scale models or data analyses. Two main methods are distinguished: *dynamical downscaling* and *empirical/statistical downscaling*. The dynamical method uses the output of regional *climate models*, global models with variable spatial resolution or high-resolution global models. The empirical/statistical methods develop statistical relationships that link the large-scale atmospheric variables with local/regional climate variables. In all cases, the quality of the downscaled product depends on the quality of the driving model.

Drought In general terms, drought is a 'prolonged absence or marked deficiency of precipitation', a 'deficiency that results in water shortage for some activity or for some group', or a 'period of abnormally dry weather sufficiently prolonged for the lack of precipitation to cause a serious hydrological imbalance' (Heim, 2002). Drought has been defined in a number of ways. *Agricultural drought* relates to moisture deficits in the topmost 1 metre or so of soil (the root zone) that affect crops, *meteorological drought* is mainly a prolonged deficit of precipitation, and *hydrologic drought* is related to below-normal streamflow, lake and groundwater levels. A *megadrought* is a long-drawn out and pervasive drought, lasting much longer than normal, usually a decade or more. For further information, see Box 3.1.

Dynamical system A process or set of processes whose evolution in time is governed by a set of deterministic physical laws. The *climate system* is a dynamical system. See *Abrupt climate change*; *Chaos*; *Nonlinearity*; *Predictability*.

Ecosystem A system of living organisms interacting with each other and their physical environment. The boundaries of what could be called an ecosystem are somewhat arbitrary, depending on the focus of interest or study. Thus, the extent of an ecosystem may range from very small spatial scales to. ultimately, the entire Earth.

Efficacy A measure of how effective a *radiative forcing* from a given *anthropogenic* or natural mechanism is at changing the equilibrium *global surface temperature* compared to an equivalent *radiative forcing* from *carbon dioxide*. A carbon dioxide increase by definition has an efficacy of 1.0.

Ekman pumping Frictional stress at the surface between two fluids (*atmosphere* and ocean) or between a fluid and the adjacent solid surface (Earth's surface) forces a circulation. When the resulting mass transport is converging, mass conservation requires a vertical flow away from the surface. This is called Ekman pumping. The opposite effect, in case of divergence, is called *Ekman suction*. The effect is important in both the *atmosphere* and the ocean.

Ekman transport The total transport resulting from a balance between the Coriolis force and the frictional stress due to the action of the wind on the ocean surface. See also *Ekman pumping*.

El Niño-Southern Oscillation (ENSO) The term *El Niño* was initially used to describe a warm-water current that periodically flows along the coast of Ecuador and Perú, disrupting the local fishery. It has since become identified with a basin-wide warming of the tropical Pacific Ocean east of the dateline. This oceanic event is associated with a fluctuation of a global-scale tropical and subtropical surface pressure pattern called the Southern Oscillation. This coupled *atmosphere*-ocean phenomenon, with preferred time scales of two to about seven years, is collectively known as the El Niño-Southern Oscillation (ENSO). It is often measured by the surface pressure anomaly difference between Darwin and Tahiti and the *sea surface temperatures* in the central and eastern equatorial Pacific. During an ENSO event, the prevailing trade winds weaken, reducing upwelling and altering ocean currents such that the sea surface temperatures warm, further weakening the trade winds. This event has a great impact on the wind, sea surface temperature and precipitation patterns in the tropical Pacific. It has climatic effects throughout the Pacific *region* and in many other parts of the world, through global *teleconnections*. The cold phase of ENSO is called *La Niña*.

Emission scenario A plausible representation of the future development of emissions of substances that are potentially radiatively active (e.g., *greenhouse gases, aerosols*), based on a coherent and internally consistent set of assumptions about driving forces (such as demographic and socioeconomic development, technological change) and their key relationships. *Concentration scenarios*, derived from emission scenarios, are used as input to a *climate model* to compute *climate projections*. In IPCC (1992) a set of emission scenarios was presented which were used as a basis for the climate projections in IPCC (1996). These emission scenarios are referred to as the IS92 scenarios. In the IPCC Special Report on Emission Scenarios (Nakićenović and Swart, 2000) new emission scenarios, the so-called SRES scenarios, were published, some of which were used, among others, as a basis for the climate projections presented in Chapters 9 to 11 of IPCC (2001) and Chapters 10 and 11 of this report. For the meaning of some terms related to these scenarios, see *SRES scenarios*.

Energy balance The difference between the total incoming and total outgoing energy. If this balance is positive, warming occurs; if it is negative, cooling occurs. Averaged over the globe and over long time periods, this balance must be zero. Because the *climate system* derives virtually all its energy from the Sun, zero balance implies that, globally, the amount of incoming *solar radiation* on average must be equal to the sum of the outgoing reflected solar radiation and the outgoing *thermal infrared radiation* emitted by the climate system. A perturbation of this global radiation balance, be it *anthropogenic* or natural, is called *radiative forcing*.

Ensemble A group of parallel model simulations used for *climate projections*. Variation of the results across the ensemble members gives an estimate of *uncertainty*. Ensembles made with the same model but different initial conditions only characterise the uncertainty associated with internal *climate variability*, whereas multi-model ensembles including simulations by several models also include the impact of model differences. Perturbed-parameter ensembles, in which model parameters are varied in a systematic manner, aim to produce a more objective estimate of modelling uncertainty than is possible with traditional multi-model ensembles.

Equilibrium and transient climate experiment An *equilibrium climate experiment* is an experiment in which a *climate model* is allowed to fully adjust to a change in *radiative forcing*. Such experiments provide information on the difference between the initial and final states of the model, but not on the time-dependent response. If the forcing is allowed to evolve gradually according to a prescribed *emission scenario*, the time-dependent response of a climate model may be analysed. Such an experiment is called a *transient climate experiment*. See *Climate projection*.

Equilibrium line The boundary between the region on a *glacier* where there is a net annual loss of ice mass (ablation area) and that where there is a net annual gain (accumulation area). The altitude of this boundary is referred to as *equilibrium line altitude*.

Equivalent carbon dioxide (CO$_2$) concentration
The concentration of *carbon dioxide* that would cause the same amount of *radiative forcing* as a given mixture of carbon dioxide and other *greenhouse gases*.

Equivalent carbon dioxide (CO$_2$) emission The amount of *carbon dioxide* emission that would cause the same integrated *radiative forcing*, over a given time horizon, as an emitted amount of a well mixed *greenhouse gas* or a mixture of well mixed greenhouse gases. The equivalent carbon dioxide emission is obtained by multiplying the emission of a well mixed greenhouse gas by its *Global Warming Potential* for the given time horizon. For a mix of greenhouse gases it is obtained by summing the equivalent carbon dioxide emissions of each gas. Equivalent carbon dioxide emission is a standard and useful *metric* for comparing emissions of different greenhouse gases but does not imply exact equivalence of the corresponding *climate change* responses (see Section 2.10).

Evapotranspiration The combined process of evaporation from the Earth's surface and transpiration from vegetation.

External forcing External forcing refers to a forcing agent outside the *climate system* causing a change in the climate system. Volcanic eruptions, solar variations and *anthropogenic* changes in the composition of the *atmosphere* and *land use change* are external forcings.

Extreme weather event An extreme weather event is an event that is rare at a particular place and time of year. Definitions of *rare* vary, but an extreme weather event would normally be as rare as or rarer than the 10th or 90th *percentile* of the observed *probability density function*. By definition, the characteristics of what is called *extreme weather* may vary from place to place in an absolute sense. Single extreme events cannot be simply and directly attributed to *anthropogenic* climate change, as there is always a finite chance the event in question might have occurred naturally. When a pattern of extreme weather persists for some time, such as a season, it may be classed as an *extreme climate event*, especially if it yields an average or total that is itself extreme (e.g., *drought* or heavy rainfall over a season).

Faculae   Bright patches on the Sun. The area covered by faculae is greater during periods of high *solar activity*.

Feedback   See *Climate feedback*.

Fingerprint   The *climate* response pattern in space and/or time to a specific forcing is commonly referred to as a fingerprint. Fingerprints are used to detect the presence of this response in observations and are typically estimated using forced *climate model* simulations.

Flux adjustment   To avoid the problem of coupled *Atmosphere*-Ocean General Circulation Models (AOGCMs) drifting into some unrealistic *climate* state, adjustment terms can be applied to the atmosphere-ocean fluxes of heat and moisture (and sometimes the surface stresses resulting from the effect of the wind on the ocean surface) before these fluxes are imposed on the model ocean and atmosphere. Because these adjustments are pre-computed and therefore independent of the coupled model integration, they are uncorrelated with the anomalies that develop during the integration. Chapter 8 of this report concludes that most models used in this report (Fourth Assessment Report AOGCMs) do not use flux adjustments, and that in general, fewer models use them.

Forest   A vegetation type dominated by trees. Many definitions of the term *forest* are in use throughout the world, reflecting wide differences in biogeophysical conditions, social structure and economics. For a discussion of the term *forest* and related terms such as *afforestation, reforestation* and *deforestation* see the IPCC Report on Land Use, Land-Use Change and Forestry (IPCC, 2000). See also the Report on Definitions and Methodological Options to Inventory Emissions from Direct Human-induced Degradation of Forests and Devegetation of Other Vegetation Types (IPCC, 2003).

Fossil fuel emissions   Emissions of *greenhouse gases* (in particular *carbon dioxide*) resulting from the combustion of fuels from fossil carbon deposits such as oil, gas and coal.

Framework Convention on Climate Change   See *United Nations Framework Convention on Climate Change* (UNFCCC).

Free atmosphere
The atmospheric layer that is negligibly affected by friction against the Earth's surface, and which is above the *atmospheric boundary layer*.

Frozen ground   Soil or rock in which part or all of the *pore water* is frozen (Van Everdingen, 1998). Frozen ground includes *permafrost*. Ground that freezes and thaws annually is called *seasonally frozen ground*.

General circulation   The large-scale motions of the *atmosphere* and the ocean as a consequence of differential heating on a rotating Earth, which tend to restore the *energy balance* of the system through transport of heat and momentum.

General Circulation Model (GCM)   See *Climate model*.

Geoid   The equipotential surface (i.e., having the same gravity potential at each point) that best fits the mean sea level (see *relative sea level*) in the absence of astronomical tides; ocean circulations; hydrological, cryospheric and atmospheric effects; Earth rotation variations and polar motion; nutation and precession; tectonics and other effects such as *post-glacial rebound*. The geoid is global and extends over continents, oceans and *ice sheets*, and at present includes the effect of the permanent tides (zero-frequency gravitational effect from the Sun and the Moon). It is the surface of reference for astronomical observations, geodetic levelling, and for ocean, hydrological, glaciological and climate modelling. In practice, there exist various operational definitions of the geoid, depending on the way the time-variable effects mentioned above are modelled.

Geostrophic winds or currents   A wind or current that is in balance with the horizontal pressure gradient and the Coriolis force, and thus is outside of the influence of friction. Thus, the wind or current is directly parallel to isobars and its speed is inversely proportional to the spacing of the isobaric contours.

Glacial isostatic adjustment   See *Post-glacial rebound*.

Glacier   A mass of land ice that flows downhill under gravity (through internal deformation and/or sliding at the base) and is constrained by internal stress and friction at the base and sides. A glacier is maintained by accumulation of snow at high altitudes, balanced by melting at low altitudes or discharge into the sea. See *Equilibrium line*; *Mass balance*.

Global dimming   Global dimming refers to perceived widespread reduction of *solar radiation* received at the surface of the Earth from about the year 1961 to around 1990.

Global surface temperature   The global surface temperature is an estimate of the global mean surface air temperature. However, for changes over time, only anomalies, as departures from a climatology, are used, most commonly based on the area-weighted global average of the *sea surface temperature* anomaly and *land surface air temperature* anomaly.

Global Warming Potential (GWP)   An index, based upon radiative properties of well-mixed *greenhouse gases*, measuring the *radiative forcing* of a unit mass of a given well-mixed greenhouse gas in the present-day *atmosphere* integrated over a chosen time horizon, relative to that of *carbon dioxide*. The GWP represents the combined effect of the differing times these gases remain in the atmosphere and their relative effectiveness in absorbing outgoing *thermal infrared radiation*. The *Kyoto Protocol* is based on GWPs from pulse emissions over a 100-year time frame.

Greenhouse effect   *Greenhouse gases* effectively absorb *thermal infrared radiation*, emitted by the Earth's surface, by the *atmosphere* itself due to the same gases, and by clouds. Atmospheric radiation is emitted to all sides, including downward to the Earth's surface. Thus, greenhouse gases trap heat within the surface-*troposphere* system. This is called the *greenhouse effect*. Thermal infrared radiation in the troposphere is strongly coupled to the temperature of the atmosphere at the altitude at which it is emitted. In the troposphere, the temperature generally decreases with height. Effectively, infrared radiation emitted to space originates from an altitude with a temperature of, on average, $-19°C$, in balance with the net incoming *solar radiation*, whereas the Earth's surface is kept at a much higher temperature of, on average, $+14°C$. An increase in the concentration of greenhouse gases leads to an increased infrared opacity of the atmosphere, and therefore to an effective radiation into space from a higher altitude at a lower temperature. This causes a *radiative forcing* that leads to an enhancement of the greenhouse effect, the so-called *enhanced greenhouse effect*.

Greenhouse gas (GHG) Greenhouse gases are those gaseous constituents of the *atmosphere*, both natural and *anthropogenic*, that absorb and emit radiation at specific wavelengths within the spectrum of *thermal infrared radiation* emitted by the Earth's surface, the atmosphere itself, and by clouds. This property causes the *greenhouse effect*. Water vapour (H_2O), *carbon dioxide* (CO_2), nitrous oxide (N_2O), methane (CH_4) and *ozone* (O_3) are the primary greenhouse gases in the Earth's atmosphere. Moreover, there are a number of entirely human-made greenhouse gases in the atmosphere, such as the *halocarbons* and other chlorine- and bromine-containing substances, dealt with under the *Montreal Protocol*. Beside CO_2, N_2O and CH_4, the *Kyoto Protocol* deals with the greenhouse gases sulphur hexafluoride (SF6), hydrofluorocarbons (HFCs) and perfluorocarbons (PFCs).

Gross Primary Production (GPP) The amount of energy fixed from the *atmosphere* through *photosynthesis*.

Ground ice A general term referring to all types of ice contained in freezing and seasonally *frozen ground* and *permafrost* (Van Everdingen, 1998).

Ground temperature The temperature of the ground near the surface (often within the first 10 cm). It is often called *soil temperature*.

Grounding line/zone The junction between a *glacier* or *ice sheet* and *ice shelf*; the place where ice starts to float.

Gyre Basin-scale ocean horizontal circulation pattern with slow flow circulating around the ocean basin, closed by a strong and narrow (100–200 km wide) boundary current on the western side. The subtropical gyres in each ocean are associated with high pressure in the centre of the gyres; the subpolar gyres are associated with low pressure.

Hadley Circulation A direct, thermally driven overturning cell in the *atmosphere* consisting of poleward flow in the upper *troposphere*, subsiding air into the subtropical anticyclones, return flow as part of the trade winds near the surface, and with rising air near the equator in the so-called *Inter-Tropical Convergence Zone*.

Halocarbons A collective term for the group of partially halogenated organic species, including the chlorofluorocarbons (CFCs), hydrochlorofluorocarbons (HCFCs), hydrofluorocarbons (HFCs), halons, methyl chloride, methyl bromide, etc. Many of the halocarbons have large *Global Warming Potentials*. The chlorine- and bromine-containing halocarbons are also involved in the depletion of the *ozone layer*.

Halosteric See *Sea level change*.

HCFC See *Halocarbons*.

HFC See *Halocarbons*.

Heterotrophic respiration The conversion of organic matter to *carbon dioxide* by organisms other than plants.

Holocene The Holocene geological epoch is the latter of two *Quaternary* epochs, extending from about 11.6 ka to and including the present.

Hydrosphere The component of the *climate system* comprising liquid surface and subterranean water, such as oceans, seas, rivers, fresh water lakes, underground water, etc.

Ice age An ice age or *glacial period* is characterised by a long-term reduction in the temperature of the Earth's *climate*, resulting in growth of continental *ice sheets* and mountain *glaciers* (*glaciation*).

Ice cap A dome shaped ice mass, usually covering a highland area, which is considerably smaller in extent than an *ice sheet*.

Ice core A cylinder of ice drilled out of a *glacier* or *ice sheet*.

Ice sheet A mass of land ice that is sufficiently deep to cover most of the underlying bedrock topography, so that its shape is mainly determined by its dynamics (the flow of the ice as it deforms internally and/or slides at its base). An ice sheet flows outward from a high central ice plateau with a small average surface slope. The margins usually slope more steeply, and most ice is discharged through fast-flowing *ice streams* or outlet *glaciers*, in some cases into the sea or into *ice shelves* floating on the sea. There are only three large ice sheets in the modern world, one on Greenland and two on Antarctica, the East and West Antarctic Ice Sheets, divided by the Transantarctic Mountains. During glacial periods there were others.

Ice shelf A floating slab of ice of considerable thickness extending from the coast (usually of great horizontal extent with a level or gently sloping surface), often filling embayments in the coastline of the *ice sheets*. Nearly all ice shelves are in Antarctica, where most of the ice discharged seaward flows into ice shelves.

Ice stream A stream of ice flowing faster than the surrounding *ice sheet*. It can be thought of as a *glacier* flowing between walls of slower-moving ice instead of rock.

Indirect aerosol effect *Aerosols* may lead to an indirect *radiative forcing* of the *climate system* through acting as *cloud condensation nuclei* or modifying the optical properties and lifetime of clouds. Two indirect effects are distinguished:

> **Cloud albedo effect** A radiative forcing induced by an increase in *anthropogenic* aerosols that cause an initial increase in droplet concentration and a decrease in droplet size for fixed liquid water content, leading to an increase in cloud *albedo*. This effect is also known as the *first indirect effect* or *Twomey effect*.

> **Cloud lifetime effect** A forcing induced by an increase in anthropogenic aerosols that cause a decrease in droplet size, reducing the precipitation efficiency, thereby modifying the liquid water content, cloud thickness and cloud life time. This effect is also known as the *second indirect effect* or *Albrecht effect*.
>
> Apart from these indirect effects, aerosols may have a *semi-direct effect*. This refers to the absorption of *solar radiation* by absorbing aerosol, which heats the air and tends to increase the static stability relative to the surface. It may also cause evaporation of cloud droplets.

Industrial revolution A period of rapid industrial growth with far-reaching social and economic consequences, beginning in Britain during the second half of the eighteenth century and spreading to Europe and later to other countries including the United States. The invention of the steam engine was an important trigger of this development. The industrial revolution marks the beginning of a strong increase in the use of fossil fuels and emission of, in particular, fossil *carbon dioxide*. In this report the terms *pre-industrial* and *industrial* refer, somewhat arbitrarily, to the periods before and after 1750, respectively.

Infrared radiation See *Thermal infrared radiation*.

Insolation The amount of *solar radiation* reaching the Earth by latitude and by season. Usually *insolation* refers to the radiation arriving at the top of the *atmosphere*. Sometimes it is specified as referring to the radiation arriving at the Earth's surface. See also: *Total Solar Irradiance*.

Interglacials The warm periods between *ice age* glaciations. The previous interglacial, dated approximately from 129 to 116 ka, is referred to as the *Last Interglacial* (AMS, 2000)

Internal variability See *Climate variability*.

Inter-Tropical Convergence Zone (ITCZ) The Inter-Tropical Convergence Zone is an equatorial zonal belt of low pressure near the equator where the northeast trade winds meet the southeast trade winds. As these winds converge, moist air is forced upward, resulting in a band of heavy precipitation. This band moves seasonally.

Isostatic or Isostasy Isostasy refers to the way in which the *lithosphere* and mantle respond visco-elastically to changes in surface loads. When the loading of the lithosphere and/or the mantle is changed by alterations in land ice mass, ocean mass, sedimentation, erosion or mountain building, vertical isostatic adjustment results, in order to balance the new load.

Kyoto Protocol The Kyoto Protocol to the *United Nations Framework Convention on Climate Change* (UNFCCC) was adopted in 1997 in Kyoto, Japan, at the Third Session of the Conference of the Parties (COP) to the UNFCCC. It contains legally binding commitments, in addition to those included in the UNFCCC. Countries included in Annex B of the Protocol (most Organisation for Economic Cooperation and Development countries and countries with economies in transition) agreed to reduce their *anthropogenic greenhouse gas* emissions (*carbon dioxide*, methane, nitrous oxide, hydrofluorocarbons, perfluorocarbons, and sulphur hexafluoride) by at least 5% below 1990 levels in the commitment period 2008 to 2012. The Kyoto Protocol entered into force on 16 February 2005.

Land use and Land use change *Land use* refers to the total of arrangements, activities and inputs undertaken in a certain land cover type (a set of human actions). The term land use is also used in the sense of the social and economic purposes for which land is managed (e.g., grazing, timber extraction and conservation). *Land use change* refers to a change in the use or management of land by humans, which may lead to a change in land cover. Land cover and land use change may have an impact on the surface *albedo*, *evapotranspiration*, *sources* and *sinks* of *greenhouse gases*, or other properties of the *climate system* and may thus have a *radiative forcing* and/or other impacts on *climate*, locally or globally. See also the IPCC Report on Land Use, Land-Use Change, and Forestry (IPCC, 2000).

La Niña See *El Niño-Southern Oscillation*.

Land surface air temperature The surface air temperature as measured in well-ventilated screens over land at 1.5 m above the ground.

Lapse rate The rate of change of an atmospheric variable, usually temperature, with height. The lapse rate is considered positive when the variable decreases with height.

Last Glacial Maximum (LGM) The Last Glacial Maximum refers to the time of maximum extent of the *ice sheets* during the last glaciation, approximately 21 ka. This period has been widely studied because the *radiative forcings* and boundary conditions are relatively well known and because the global cooling during that period is comparable with the projected warming over the 21st century.

Last Interglacial (LIG) See *Interglacial*.

Latent heat flux The flux of heat from the Earth's surface to the *atmosphere* that is associated with evaporation or condensation of water vapour at the surface; a component of the surface energy budget.

Level of Scientific Understanding (LOSU) This is an index on a 5-step scale (high, medium, medium-low, low and very low) designed to characterise the degree of scientific understanding of the *radiative forcing* agents that affect *climate change*. For each agent, the index represents a subjective judgement about the evidence for the physical/chemical mechanisms determining the forcing and the consensus surrounding the quantitative estimate and its *uncertainty*.

Lifetime Lifetime is a general term used for various time scales characterising the rate of processes affecting the concentration of trace gases. The following lifetimes may be distinguished:

Turnover time (T) (also called *global atmospheric lifetime*) is the ratio of the mass M of a *reservoir* (e.g., a gaseous compound in the *atmosphere*) and the total rate of removal S from the reservoir: $T = M / S$. For each removal process, separate turnover times can be defined. In soil carbon biology, this is referred to as *Mean Residence Time*.

Adjustment time or *response time* (T_a) is the time scale characterising the decay of an instantaneous pulse input into the reservoir. The term adjustment time is also used to characterise the adjustment of the mass of a reservoir following a step change in the *source* strength. *Half-life* or *decay constant* is used to quantify a first-order exponential decay process. See *response time* for a different definition pertinent to *climate* variations.

The term *lifetime* is sometimes used, for simplicity, as a surrogate for *adjustment time*.

In simple cases, where the global removal of the compound is directly proportional to the total mass of the reservoir, the adjustment time equals the turnover time: $T = T_a$. An example is *CFC*-11, which is removed from the *atmosphere* only by photochemical processes in the *stratosphere*. In more complicated cases, where several reservoirs are involved or where the removal is not proportional to the total mass, the equality $T = T_a$ no longer holds. *Carbon dioxide* (CO_2) is an extreme example. Its turnover time is only about four years because of the rapid exchange between the atmosphere and the ocean and terrestrial biota. However, a large part of that CO_2 is returned to the atmosphere within a few years. Thus, the adjustment time of CO_2 in the atmosphere is actually determined by the rate of removal of carbon from the surface layer of the oceans into its deeper layers. Although an approximate value of 100 years may be given for the adjustment time of CO_2 in the atmosphere, the actual adjustment is faster initially and slower later on. In the case of methane (CH_4), the adjustment time is different from the turnover time because the removal is mainly through a chemical reaction with the hydroxyl radical OH, the concentration of which itself depends on the CH_4 concentration. Therefore, the CH_4 removal rate S is not proportional to its total mass M.

Likelihood The likelihood of an occurrence, an outcome or a result, where this can be estimated probabilistically, is expressed in this report using a standard terminology, defined in Box 1.1. See also *Uncertainty*; *Confidence*.

Lithosphere The upper layer of the solid Earth, both continental and oceanic, which comprises all crustal rocks and the cold, mainly elastic part of the uppermost mantle. Volcanic activity, although part of the lithosphere, is not considered as part of the *climate system*, but acts as an *external forcing* factor. See *Isostatic*.

Little Ice Age (LIA) An interval between approximately AD 1400 and 1900 when temperatures in the Northern Hemisphere were generally colder than today's, especially in Europe.

Mass balance (of glaciers, ice caps or ice sheets) The balance between the mass input to the ice body (accumulation) and the mass loss (ablation, iceberg calving). Mass balance terms include the following:

Specific mass balance: net mass loss or gain over a hydrological cycle at a point on the surface of a *glacier*.

Total mass balance (of the glacier): The specific mass balance spatially integrated over the entire glacier area; the total mass a glacier gains or loses over a hydrological cycle.

Mean specific mass balance: The total mass balance per unit area of the glacier. If *surface* is specified (*specific surface mass balance*, etc.) then ice flow contributions are not considered; otherwise, mass balance includes contributions from ice flow and iceberg calving. The specific surface mass balance is positive in the accumulation area and negative in the ablation area.

Mean sea level See *Relative sea level*.

Medieval Warm Period (MWP) An interval between AD 1000 and 1300 in which some Northern Hemisphere *regions* were warmer than during the *Little Ice Age* that followed.

Meridional Overturning Circulation (MOC) Meridional (north-south) overturning circulation in the ocean quantified by zonal (east-west) sums of mass transports in depth or density layers. In the North Atlantic, away from the subpolar *regions*, the MOC (which is in principle an observable quantity) is often identified with the *Thermohaline Circulation* (THC), which is a conceptual interpretation. However, it must be borne in mind that the MOC can also include shallower, wind-driven overturning cells such as occur in the upper ocean in the tropics and subtropics, in which warm (light) waters moving poleward are transformed to slightly denser waters and *subducted* equatorward at deeper levels.

Metadata Information about meteorological and climatological data concerning how and when they were measured, their quality, known problems and other characteristics.

Metric A consistent measurement of a characteristic of an object or activity that is otherwise difficult to quantify.

Mitigation A human intervention to reduce the *sources* or enhance the *sinks* of greenhouse gases.

Mixing ratio See *Mole fraction*.

Model hierarchy See *Climate model* (spectrum or hierarchy).

Modes of climate variability Natural variability of the *climate system*, in particular on seasonal and longer time scales, predominantly occurs with preferred spatial patterns and time scales, through the dynamical characteristics of the atmospheric circulation and through interactions with the land and ocean surfaces. Such patterns are often called *regimes*, *modes* or *teleconnections*. Examples are the *North Atlantic Oscillation* (NAO), the *Pacific-North American pattern* (PNA), the *El Niño-Southern Oscillation* (ENSO), the *Northern Annular Mode* (NAM; previously called Arctic Oscillation, AO) and the *Southern Annular Mode* (SAM; previously called the Antarctic Oscillation, AAO). Many of the prominent modes of climate variability are discussed in section 3.6. See also *Patterns of climate variability*.

Mole fraction Mole fraction, or *mixing ratio*, is the ratio of the number of moles of a constituent in a given volume to the total number of moles of all constituents in that volume. It is usually reported for dry air. Typical values for long-lived *greenhouse gases* are in the order of μmol mol⁻¹ (parts per million: *ppm*), nmol mol⁻¹ (parts per billion: *ppb*), and fmol mol⁻¹ (parts per trillion: *ppt*). Mole fraction differs from *volume mixing ratio*, often expressed in ppmv etc., by the corrections for non-ideality of gases. This correction is significant relative to measurement precision for many greenhouse gases. (Schwartz and Warneck, 1995).

Monsoon A monsoon is a tropical and subtropical seasonal reversal in both the surface winds and associated precipitation, caused by differential heating between a continental-scale land mass and the adjacent ocean. Monsoon rains occur mainly over land in summer.

Montreal Protocol The Montreal Protocol on Substances that Deplete the Ozone Layer was adopted in Montreal in 1987, and subsequently adjusted and amended in London (1990), Copenhagen (1992), Vienna (1995), Montreal (1997) and Beijing (1999). It controls the consumption and production of chlorine- and bromine-containing chemicals that destroy stratospheric *ozone*, such as chlorofluorocarbons, methyl chloroform, carbon tetrachloride and many others.

Microwave Sounding Unit (MSU) A satellite-borne microwave sounder that estimates the temperature of thick layers of the *atmosphere* by measuring the thermal emission of oxygen molecules from a complex of emission lines near 60 GHz. A series of nine MSUs began making this kind of measurement in late 1978. Beginning in mid 1998, a follow-on series of instruments, the Advanced Microwave Sounding Units (AMSUs), began operation.

MSU See *Microwave Sounding Unit*.

Nonlinearity A process is called *nonlinear* when there is no simple proportional relation between cause and effect. The *climate system* contains many such nonlinear processes, resulting in a system with a potentially very complex behaviour. Such complexity may lead to *abrupt climate change*. See also *Chaos*; *Predictability*.

North Atlantic Oscillation (NAO) The North Atlantic Oscillation consists of opposing variations of barometric pressure near Iceland and near the Azores. It therefore corresponds to fluctuations in the strength of the main westerly winds across the Atlantic into Europe, and thus to fluctuations in the embedded cyclones with their associated frontal systems. See NAO Index, Box 3.4.

Northern Annular Mode (NAM) A winter fluctuation in the amplitude of a pattern characterised by low surface pressure in the Arctic and strong mid-latitude westerlies. The NAM has links with the northern polar vortex into the *stratosphere*. Its pattern has a bias to the North Atlantic and has a large correlation with the *North Atlantic Oscillation*. See NAM Index, Box 3.4.

Ocean acidification A decrease in the *pH* of sea water due to the *uptake* of *anthropogenic carbon dioxide*.

Ocean heat uptake efficiency This is a measure (W m⁻² °C⁻¹) of the rate at which heat storage by the global ocean increases as global surface temperature rises. It is a useful parameter for *climate change* experiments in which the *radiative forcing* is changing monotonically, when it can be compared with the climate sensitivity parameter to gauge the relative importance of climate response and ocean heat uptake in determining the rate of climate change. It can be estimated from a 1% yr⁻¹ atmospheric *carbon dioxide* increase experiment as the ratio of the global average top-of-*atmosphere* net downward radiative flux to the *transient climate response* (see *climate sensitivity*).

Organic aerosol *Aerosol* particles consisting predominantly of organic compounds, mainly carbon, hydrogen, oxygen and lesser amounts of other elements. (Charlson and Heintzenberg, 1995, p. 405). See *Carbonaceous aerosol*.

Ozone Ozone, the triatomic form of oxygen (O_3), is a gaseous atmospheric constituent. In the *troposphere*, it is created both naturally and by photochemical reactions involving gases resulting from human activities (*smog*). Tropospheric ozone acts as a *greenhouse gas*. In the *stratosphere*, it is created by the interaction between solar ultraviolet radiation and molecular oxygen (O_2). Stratospheric ozone plays a dominant role in the stratospheric radiative balance. Its concentration is highest in the *ozone layer*.

Ozone hole See *Ozone layer*.

Ozone layer The *stratosphere* contains a layer in which the concentration of *ozone* is greatest, the so-called ozone layer. The layer extends from about 12 to 40 km above the Earth's surface. The ozone concentration reaches a maximum between about 20 and 25 km. This layer is being depleted by human emissions of chlorine and bromine compounds. Every year, during the Southern Hemisphere spring, a very strong depletion of the ozone layer takes place over the antarctic *region*, caused by *anthropogenic* chlorine and bromine compounds in combination with the specific meteorological conditions of that region. This phenomenon is called the *ozone hole*. See *Montreal Protocol*.

Pacific decadal variability Coupled decadal-to-inter-decadal variability of the atmospheric circulation and underlying ocean in the Pacific Basin. It is most prominent in the North Pacific, where fluctuations in the strength of the winter Aleutian Low pressure system co-vary with North Pacific *sea surface temperatures*, and are linked to decadal variations in atmospheric circulation, sea surface temperatures and ocean circulation throughout the whole Pacific Basin. Such fluctuations have the effect of modulating the *El Niño-Southern Oscillation* cycle. Key measures of Pacific decadal variability are the *North Pacific Index (NPI)*, the *Pacific Decadal Oscillation (PDO)* index and the *Inter-decadal Pacific Oscillation (IPO)* index, all defined in Box 3.4.

Pacific-North American (PNA) pattern An atmospheric large-scale wave pattern featuring a sequence of tropospheric high- and low-pressure anomalies stretching from the subtropical west Pacific to the east coast of North America. See PNA pattern index, Box 3.4.

Palaeoclimate *Climate* during periods prior to the development of measuring instruments, including historic and geologic time, for which only *proxy* climate records are available.

Parametrization In *climate models*, this term refers to the technique of representing processes that cannot be explicitly resolved at the spatial or temporal resolution of the model (sub-grid scale processes) by relationships between model-resolved larger-scale flow and the area- or time-averaged effect of such sub-grid scale processes.

Patterns of climate variability See *Modes of climate variability*.

Percentile A percentile is a value on a scale of one hundred that indicates the percentage of the data set values that is equal to or below it. The percentile is often used to estimate the extremes of a distribution. For example, the 90th (10th) percentile may be used to refer to the threshold for the upper (lower) extremes.

Permafrost Ground (soil or rock and included ice and organic material) that remains at or below 0°C for at least two consecutive years (Van Everdingen, 1998).

pH pH is a dimensionless measure of the acidity of water (or any solution) given by its concentration of hydrogen ions (H^+). pH is measured on a logarithmic scale where

pH $= -\log_{10}(H^+)$. Thus, a pH <u>decrease</u> of 1 unit corresponds to a 10-fold <u>increase</u> in the concentration of H^+, or acidity.

Photosynthesis The process by which plants take carbon dioxide from the air (or bicarbonate in water) to build carbohydrates, releasing oxygen in the process. There are several pathways of photosynthesis with different responses to atmospheric carbon dioxide concentrations. See Carbon dioxide fertilization; C3 plants; C4 plants.

Plankton Microorganisms living in the upper layers of aquatic systems. A distinction is made between *phytoplankton*, which depend on *photosynthesis* for their energy supply, and *zooplankton*, which feed on phytoplankton.

Pleistocene The earlier of two *Quaternary* epochs, extending from the end of the Pliocene, about 1.8 Ma, until the beginning of the *Holocene* about 11.6 ka.

Pollen analysis A technique of both relative dating and environmental *reconstruction*, consisting of the identification and counting of pollen types preserved in peat, lake sediments and other deposits. See *Proxy*.

Post-glacial rebound The vertical movement of the land and sea floor following the reduction of the load of an ice mass, for example, since the *Last Glacial Maximum* (21 ka). The rebound is an *isostatic* land movement.

Precipitable water The total amount of atmospheric water vapour in a vertical column of unit cross-sectional area. It is commonly expressed in terms of the height of the water if completely condensed and collected in a vessel of the same unit cross section.

Precursors Atmospheric compounds that are not *greenhouse gases* or *aerosols*, but that have an effect on greenhouse gas or aerosol concentrations by taking part in physical or chemical processes regulating their production or destruction rates.

Predictability The extent to which future states of a system may be predicted based on knowledge of current and past states of the system.

Since knowledge of the *climate system*'s past and current states is generally imperfect, as are the models that utilise this knowledge to produce a *climate prediction*, and since the climate system is inherently *nonlinear* and *chaotic*, predictability of the climate system is inherently limited. Even with arbitrarily accurate models and observations, there may still be limits to the predictability of such a nonlinear system (AMS, 2000)

Pre-industrial See *Industrial revolution*.

Probability Density Function (PDF) A probability density function is a function that indicates the relative chances of occurrence of different outcomes of a variable. The function integrates to unity over the domain for which it is defined and has the property that the integral over a sub-domain equals the probability that the outcome of the variable lies within that sub-domain. For example, the probability that a temperature anomaly defined in a particular way is greater than zero is obtained from its PDF by integrating the PDF over all possible temperature anomalies greater than zero. Probability density functions that describe two or more variables simultaneously are similarly defined.

Projection A projection is a potential future evolution of a quantity or set of quantities, often computed with the aid of a model.

Projections are distinguished from *predictions* in order to emphasize that projections involve assumptions concerning, for example, future socioeconomic and technological developments that may or may not be realised, and are therefore subject to substantial *uncertainty*. See also *Climate projection*; *Climate prediction*.

Proxy A proxy *climate* indicator is a local record that is interpreted, using physical and biophysical principles, to represent some combination of climate-related variations back in time. Climate-related data derived in this way are referred to as proxy data. Examples of proxies include *pollen analysis*, *tree ring* records, characteristics of corals and various data derived from *ice cores*.

Quaternary The period of geological time following the *Tertiary* (65 Ma to 1.8 Ma). Following the current definition (which is under revision at present) the Quaternary extends from 1.8 Ma until the present. It is formed of two epochs, the *Pleistocene* and the *Holocene*.

Radiative forcing Radiative forcing is the change in the net, downward minus upward, irradiance (expressed in W m^{-2}) at the *tropopause* due to a change in an external driver of *climate change*, such as, for example, a change in the concentration of *carbon dioxide* or the output of the Sun. Radiative forcing is computed with all tropospheric properties held fixed at their unperturbed values, and after allowing for stratospheric temperatures, if perturbed, to readjust to radiative-dynamical equilibrium. Radiative forcing is called *instantaneous* if no change in stratospheric temperature is accounted for. For the purposes of this report, radiative forcing is further defined as the change relative to the year 1750 and, unless otherwise noted, refers to a global and annual average value. Radiative forcing is not to be confused with *cloud radiative forcing*, a similar terminology for describing an unrelated measure of the impact of clouds on the irradiance at the top of the *atmosphere*.

Radiative forcing scenario A plausible representation of the future development of *radiative forcing* associated, for example, with changes in atmospheric composition or *land use change*, or with external factors such as variations in *solar activity*. Radiative forcing scenarios can be used as input into simplified *climate models* to compute *climate projections*.

Rapid climate change See *Abrupt climate change*.

Reanalysis Reanalyses are atmospheric and oceanic analyses of temperature, wind, current, and other meteorological and oceanographic quantities, created by processing past meteorological and oceanographic data using fixed state-of-the-art weather forecasting models and data assimilation techniques. Using fixed data assimilation avoids effects from the changing analysis system that occurs in operational analyses. Although continuity is improved, global reanalyses still suffer from changing coverage and biases in the observing systems.

Reconstruction The use of *climate* indicators to help determine (generally past) climates.

Reforestation Planting of forests on lands that have previously contained forests but that have been converted to some other use. For a discussion of the term *forest* and related terms such as *afforestation*, reforestation and *deforestation*, see the IPCC Report on Land Use, Land-Use Change and Forestry (IPCC, 2000). See also the Report on Definitions and Methodological Options to Inventory Emissions from Direct Human-induced Degradation of Forests and Devegetation of Other Vegetation Types (IPCC, 2003)

Regime A regime is preferred states of the *climate system*, often representing one phase of dominant patterns or *modes of climate variability*.

Region A region is a territory characterised by specific geographical and climatological features. The *climate* of a region is affected by regional and local scale forcings like topography, *land use* characteristics, lakes, etc., as well as remote influences from other regions. See *Teleconnection*.

Relative sea level Sea level measured by a *tide gauge* with respect to the land upon which it is situated. *Mean sea level* is normally defined as the average relative sea level over a period, such as a month or a year, long enough to average out transients such as waves and tides. See *Sea level change*.

Reservoir A component of the *climate system*, other than the *atmosphere*, which has the capacity to store, accumulate or release a substance of concern, for example, carbon, a *greenhouse gas* or a *precursor*. Oceans, soils and *forests* are examples of reservoirs of carbon. *Pool* is an equivalent term (note that the definition of pool often includes the atmosphere). The absolute quantity of the substance of concern held within a reservoir at a specified time is called the *stock*.

Respiration The process whereby living organisms convert organic matter to *carbon dioxide*, releasing energy and consuming molecular oxygen.

Response time The response time or *adjustment time* is the time needed for the *climate system* or its components to re-equilibrate to a new state, following a forcing resulting from external and internal processes or *feedbacks*. It is very different for various components of the climate system. The response time of the *troposphere* is relatively short, from days to weeks, whereas the *stratosphere* reaches equilibrium on a time scale of typically a few months. Due to their large heat capacity, the oceans have a much longer response time: typically decades, but up to centuries or millennia. The response time of the strongly coupled surface-troposphere system is, therefore, slow compared to that of the stratosphere, and mainly determined by the oceans. The *biosphere* may respond quickly (e.g., to *droughts*), but also very slowly to imposed changes. See *lifetime* for a different definition of response time pertinent to the rate of processes affecting the concentration of trace gases.

Return period The average time between occurrences of a defined event (AMS, 2000).

Return value The highest (or, alternatively, lowest) value of a given variable, on average occurring once in a given period of time (e.g., in 10 years).

Scenario A plausible and often simplified description of how the future may develop, based on a coherent and internally consistent set of assumptions about driving forces and key relationships. Scenarios may be derived from *projections*, but are often based on additional information from other sources, sometimes combined with a *narrative storyline*. See also *SRES scenarios*; *Climate scenario*; *Emission scenario*.

Sea ice Any form of ice found at sea that has originated from the freezing of seawater. Sea ice may be discontinuous pieces (ice floes) moved on the ocean surface by wind and currents (pack ice), or a motionless sheet attached to the coast (land-fast ice). Sea ice less than one year old is called *first-year ice*. *Multi-year ice* is sea ice that has survived at least one summer melt season.

Sea level change   Sea level can change, both globally and locally, due to (i) changes in the shape of the ocean basins, (ii) changes in the total mass of water and (iii) changes in water density. Sea level changes induced by changes in water density are called *steric*. Density changes induced by temperature changes only are called *thermosteric*, while density changes induced by salinity changes are called *halosteric*. See also *Relative Sea Level*; *Thermal expansion*.

Sea level equivalent (SLE)   The change in global average sea level that would occur if a given amount of water or ice were added to or removed from the oceans.

Seasonally frozen ground   See *Frozen ground*.

Sea surface temperature (SST)   The sea surface temperature is the temperature of the subsurface bulk temperature in the top few metres of the ocean, measured by ships, buoys and drifters. From ships, measurements of water samples in buckets were mostly switched in the 1940s to samples from engine intake water. Satellite measurements of *skin temperature* (uppermost layer; a fraction of a millimetre thick) in the infrared or the top centimetre or so in the microwave are also used, but must be adjusted to be compatible with the bulk temperature.

Sensible heat flux   The flux of heat from the Earth's surface to the *atmosphere* that is not associated with phase changes of water; a component of the surface energy budget.

Sequestration   See *Uptake*.

Significant wave height   The average height of the highest one-third of the wave heights (sea and swell) occurring in a particular time period.

Sink   Any process, activity or mechanism that removes a *greenhouse gas*, an *aerosol* or a *precursor* of a greenhouse gas or aerosol from the *atmosphere*.

Slab-ocean model   A simplified presentation in a *climate model* of the ocean as a motionless layer of water with a depth of 50 to 100 m. Climate models with a slab ocean can only be used for estimating the equilibrium response of climate to a given forcing, not the transient evolution of climate. See *Equilibrium and transient climate experiment*.

Snow line   The lower limit of permanent snow cover, below which snow does not accumulate.

Soil moisture   Water stored in or at the land surface and available for evaporation.

Soil temperature   See *Ground temperature*.

Solar activity   The Sun exhibits periods of high activity observed in numbers of *sunspots*, as well as radiative output, magnetic activity and emission of high-energy particles. These variations take place on a range of time scales from millions of years to minutes. See *Solar cycle*.

Solar ('11 year') cycle   A quasi-regular modulation of *solar activity* with varying amplitude and a period of between 9 and 13 years.

Solar radiation   Electromagnetic radiation emitted by the Sun. It is also referred to as *shortwave radiation*. Solar radiation has a distinctive range of wavelengths (spectrum) determined by the temperature of the Sun, peaking in visible wavelengths. See also: *Thermal infrared radiation, Insolation*.

Soot   Particles formed during the quenching of gases at the outer edge of flames of organic vapours, consisting predominantly of carbon, with lesser amounts of oxygen and hydrogen present as carboxyl and phenolic groups and exhibiting an imperfect graphitic structure. See *Black carbon*; *Charcoal* (Charlson and Heintzenberg, 1995, p. 406).

Source   Any process, activity or mechanism that releases a *greenhouse gas*, an *aerosol* or a *precursor* of a greenhouse gas or aerosol into the *atmosphere*.

Southern Annular Mode (SAM)   The fluctuation of a pattern like the *Northern Annular Mode*, but in the Southern Hemisphere. See SAM Index, Box 3.4.

Southern Oscillation   See *El Niño-Southern Oscillation* (ENSO).

Spatial and temporal scales   *Climate* may vary on a large range of spatial and temporal scales. Spatial scales may range from local (less than 100,000 km^2), through regional (100,000 to 10 million km^2) to continental (10 to 100 million km^2). Temporal scales may range from seasonal to geological (up to hundreds of millions of years).

SRES scenarios   SRES scenarios are *emission scenarios* developed by Nakićenović and Swart (2000) and used, among others, as a basis for some of the *climate projections* shown in Chapter 10 of this report. The following terms are relevant for a better understanding of the structure and use of the set of SRES scenarios:

> **Scenario family**   Scenarios that have a similar demographic, societal, economic and technical change storyline. Four scenario families comprise the SRES scenario set: A1, A2, B1 and B2.

> **Illustrative Scenario**   A scenario that is illustrative for each of the six scenario groups reflected in the Summary for Policymakers of Nakićenović and Swart (2000). They include four revised *scenario markers* for the scenario groups A1B, A2, B1, B2, and two additional scenarios for the A1FI and A1T groups. All scenario groups are equally sound.

> **Marker Scenario**   A scenario that was originally posted in draft form on the SRES website to represent a given scenario family. The choice of markers was based on which of the initial quantifications best reflected the storyline, and the features of specific models. Markers are no more likely than other scenarios, but are considered by the SRES writing team as illustrative of a particular storyline. They are included in revised form in Nakićenović and Swart (2000). These scenarios received the closest scrutiny of the entire writing team and via the SRES open process. Scenarios were also selected to illustrate the other two scenario groups.

> **Storyline**   A narrative description of a scenario (or family of scenarios), highlighting the main scenario characteristics, relationships between key driving forces and the dynamics of their evolution.

Steric   See *Sea level change*.

Stock   See *Reservoir*.

Storm surge   The temporary increase, at a particular locality, in the height of the sea due to extreme meteorological conditions (low atmospheric pressure and/or strong winds). The storm surge is defined as being the excess above the level expected from the tidal variation alone at that time and place.

Storm tracks   Originally, a term referring to the tracks of individual cyclonic weather systems, but now often generalised to refer to the *regions* where the main tracks of extratropical disturbances occur as sequences of low (cyclonic) and high (anticyclonic) pressure systems.

Stratosphere   The highly stratified region of the *atmosphere* above the *troposphere* extending from about 10 km (ranging from 9 km at high latitudes to 16 km in the tropics on average) to about 50 km altitude.

Subduction   Ocean process in which surface waters enter the ocean interior from the surface mixed layer through *Ekman pumping* and lateral *advection*. The latter occurs when surface waters are advected to a region where the local surface layer is less dense and therefore must slide below the surface layer, usually with no change in density.

Sunspots   Small dark areas on the Sun. The number of sunspots is higher during periods of high *solar activity*, and varies in particular with the *solar cycle*.

Surface layer   See *Atmospheric boundary layer*.

Surface temperature   See *Global surface temperature*; *Ground temperature*; *Land surface air temperature*; *Sea surface temperature*.

Teleconnection   A connection between *climate variations* over widely separated parts of the world. In physical terms, teleconnections are often a consequence of large-scale wave motions, whereby energy is transferred from source regions along preferred paths in the *atmosphere*.

Thermal expansion   In connection with sea level, this refers to the increase in volume (and decrease in density) that results from warming water. A warming of the ocean leads to an expansion of the ocean volume and hence an increase in sea level. See *Sea level change*.

Thermal infrared radiation   Radiation emitted by the Earth's surface, the *atmosphere* and the clouds. It is also known as *terrestrial* or *longwave radiation*, and is to be distinguished from the near-infrared radiation that is part of the solar spectrum. Infrared radiation, in general, has a distinctive range of wavelengths (*spectrum*) longer than the wavelength of the red colour in the visible part of the spectrum. The spectrum of thermal infrared radiation is practically distinct from that of shortwave or *solar radiation* because of the difference in temperature between the Sun and the Earth-atmosphere system.

Thermocline   The layer of maximum vertical temperature gradient in the ocean, lying between the surface ocean and the abyssal ocean. In subtropical regions, its source waters are typically surface waters at higher latitudes that have *subducted* and moved equatorward. At high latitudes, it is sometimes absent, replaced by a *halocline*, which is a layer of maximum vertical salinity gradient.

Thermohaline circulation (THC)   Large-scale circulation in the ocean that transforms low-density upper ocean waters to higher-density intermediate and deep waters and returns those waters back to the upper ocean. The circulation is asymmetric, with conversion to dense waters in restricted regions at high latitudes and the return to the surface involving slow upwelling and diffusive processes over much larger geographic regions. The THC is driven by high densities at or near the surface, caused by cold temperatures and/or high salinities, but despite its suggestive though common name, is also driven by mechanical forces such as wind and tides. Frequently,

the name THC has been used synonymously with *Meridional Overturning Circulation*.

Thermokarst   The process by which characteristic landforms result from the thawing of ice-rich *permafrost* or the melting of massive ground ice (Van Everdingen, 1998).

Thermosteric   See *Sea level change*.

Tide gauge   A device at a coastal location (and some deep-sea locations) that continuously measures the level of the sea with respect to the adjacent land. Time averaging of the sea level so recorded gives the observed secular changes of the *relative sea level*.

Total solar irradiance (TSI)   The amount of *solar radiation* received outside the Earth's *atmosphere* on a surface normal to the incident radiation, and at the Earth's mean distance from the Sun.

Reliable measurements of solar radiation can only be made from space and the precise record extends back only to 1978. The generally accepted value is 1,368 W m^{-2} with an accuracy of about 0.2%. Variations of a few tenths of a percent are common, usually associated with the passage of *sunspots* across the solar disk. The *solar cycle* variation of TSI is of the order of 0.1% (AMS, 2000). See also *Insolation*.

Transient climate response   See *Climate sensitivity*.

Tree rings   Concentric rings of secondary wood evident in a cross-section of the stem of a woody plant. The difference between the dense, small-celled late wood of one season and the wide-celled early wood of the following spring enables the age of a tree to be estimated, and the ring widths or density can be related to *climate* parameters such as temperature and precipitation. See *Proxy*.

Trend   In this report, the word *trend* designates a change, generally monotonic in time, in the value of a variable.

Tropopause   The boundary between the *troposphere* and the *stratosphere*.

Troposphere   The lowest part of the *atmosphere*, from the surface to about 10 km in altitude at mid-latitudes (ranging from 9 km at high latitudes to 16 km in the tropics on average), where clouds and weather phenomena occur. In the troposphere, temperatures generally decrease with height.

Turnover time   See *Lifetime*.

Uncertainty   An expression of the degree to which a value (e.g., the future state of the *climate system*) is unknown. Uncertainty can result from lack of information or from disagreement about what is known or even knowable. It may have many types of sources, from quantifiable errors in the data to ambiguously defined concepts or terminology, or uncertain *projections* of human behaviour. Uncertainty can therefore be represented by quantitative measures, for example, a range of values calculated by various models, or by qualitative statements, for example, reflecting the judgement of a team of experts (see Moss and Schneider, 2000; Manning et al., 2004). See also *Likelihood*; *Confidence*.

United Nations Framework Convention on Climate Change (UNFCCC)
The Convention was adopted on 9 May 1992 in New York and signed at the 1992 Earth Summit in Rio de Janeiro by more than 150 countries and the European Community. Its ultimate objective is the 'stabilisation of greenhouse gas concentrations in the atmosphere at a level that would prevent dangerous anthropogenic interference with the climate system'. It contains commitments for all Parties. Under the Convention, Parties included in Annex I (all OECD

countries and countries with economies in transition) aim to return *greenhouse gas* emissions not controlled by the *Montreal Protocol* to 1990 levels by the year 2000. The convention entered in force in March 1994. See *Kyoto Protocol*.

Uptake   The addition of a substance of concern to a *reservoir*. The uptake of carbon containing substances, in particular *carbon dioxide*, is often called (carbon) *sequestration*.

Urban heat island (UHI)   The relative warmth of a city compared with surrounding rural areas, associated with changes in runoff, the *concrete jungle* effects on heat retention, changes in surface *albedo*, changes in pollution and *aerosols*, and so on.

Ventilation   The exchange of ocean properties with the *atmospheric surface layer* such that property concentrations are brought closer to equilibrium values with the *atmosphere* (AMS, 2000).

Volume mixing ratio   See *Mole fraction*.

Walker Circulation   Direct thermally driven zonal overturning circulation in the *atmosphere* over the tropical Pacific Ocean, with rising air in the western and sinking air in the eastern Pacific.

Water mass   A volume of ocean water with identifiable properties (temperature, salinity, density, chemical tracers) resulting from its unique formation process. Water masses are often identified through a vertical or horizontal extremum of a property such as salinity.

Younger Dryas   A period 12.9 to 11.6 kya, during the deglaciation, characterised by a temporary return to colder conditions in many locations, especially around the North Atlantic.

REFERENCES

AMS, 2000: *AMS Glossary of Meteorology*, 2nd Ed. American Meteorological Society, Boston, MA,http://amsglossary. allenpress.com/glossary/browse.

Charlson, R.J., and J. Heintzenberg (eds.), 1995: *Aerosol Forcing of Climate*. John Wiley and Sons Limited, pp. 91–108. Copyright 1995 ©John Wiley and Sons Limited. Reproduced with permission.

Heim, R.R., 2002: *A Review of Twentieth-Century Drought Indices Used in the United States. Bull. Am. Meteorol. Soc.*, **83**, 1149–1165

IPCC, 1992: *Climate Change 1992: The Supplementary Report to the IPCC Scientific Assessment* [Houghton, J.T., B.A. Callander, and S.K. Varney (eds.)]. Cambridge University Press, Cambridge, United Kingdom and New York, NY, USA, 116 pp.

IPCC, 1996: *Climate Change 1995: The Science of Climate Change. Contribution of Working Group I to the Second Assessment Report of the Intergovernmental Panel on Climate Change* [Houghton., J.T., et al. (eds.)]. Cambridge University Press, Cambridge, United Kingdom and New York, NY, USA, 572 pp.

IPCC, 2000: *Land Use, Land-Use Change, and Forestry. Special Report of the Intergovernmental Panel on Climate Change* [Watson, R.T., et al. (eds.)]. Cambridge University Press, Cambridge, United Kingdom and New York, NY, USA, 377 pp.

IPCC, 2001: *Climate Change 2001: The Scientific Basis. Contribution of Working Group I to the Third Assessment Report of the Intergovernmental Panel on Climate Change* [Houghton, J.T., et al. (eds.)]. Cambridge University Press, Cambridge, United Kingdom and New York, NY, USA, 881 pp.

IPCC, 2003: *Definitions and Methodological Options to Inventory Emissions from Direct Human-Induced Degradation of Forests and Devegetation of Other Vegetation Types* [Penman, J., et al. (eds.)]. The Institute for Global Environmental Strategies (IGES), Japan , 32 pp.

Manning, M., et al., 2004: *IPCC Workshop on Describing Scientific Uncertainties in Climate Change to Support Analysis of Risk of Options*. Workshop Report. Intergovernmental Panel on Climate Change, Geneva.

Moss, R., and S. Schneider, 2000: *Uncertainties in the IPCC TAR: Recommendations to Lead Authors for More Consistent Assessment and Reporting*. In: IPCC Supporting Material: Guidance Papers on Cross Cutting Issues in the Third Assessment Report of the IPCC. [Pachauri, R., T. Taniguchi, and K. Tanaka (eds.)]. Intergovernmental Panel on Climate Change, Geneva, pp. 33–51.

Nakićenović, N., and R. Swart (eds.), 2000: *Special Report on Emissions Scenarios. A Special Report of Working Group III of the Intergovernmental Panel on Climate Change*. Cambridge University Press, Cambridge, United Kingdom and New York, NY, USA, 599 pp.

Schwartz, S.E., and P. Warneck, 1995: Units for use in atmospheric chemistry. *Pure Appl. Chem.*, 67, 1377–1406.

Van Everdingen, R. (ed.): 1998. *Multi-Language Glossary of Permafrost and Related Ground-Ice Terms*, revised May 2005. National Snow and Ice Data Center/World Data Center for Glaciology, Boulder, CO, http://nsidc.org/fgdc/glossary/.

Annex II

Contributors to the IPCC WGI Fourth Assessment Report

ACHUTARAO, Krishna
Lawrence Livermore National Laboratory
USA

ADLER, Robert
National Aeronautics and
Space Administration
USA

ALEXANDER, Lisa
Hadley Centre for Climate Prediction
and Research, Met Office
UK, Australia, Ireland

ALEXANDERSSON, Hans
Swedish Meteorological and
Hydrological Institute
Sweden

ALLAN, Richard
Environmental Systems Science
Centre, University of Reading
UK

ALLEN, Myles
Climate Dynamics Group, Atmospheric,
Oceanic and Planetary Physics, Department
of Physics, University of Oxford
UK

ALLEY, Richard B.
Department of Geosciences,
Pennsylvania State University
USA

ALLISON, Ian
Australian Antarctic Division and
Antarctic Climate and Ecosystems
Cooperative Research Centre
Australia

AMBENJE, Peter
Kenya Meteorological Department
Kenya

AMMANN, Caspar
Climate and Global Dynamics Division,
National Center for Atmospheric Research
USA

ANDRONOVA, Natalia
University of Michigan
USA

ANNAN, James
Frontier Research Center for Global
Change, Japan Agency for Marine-
Earth Science and Technology
Japan, UK

ANTONOV, John
National Oceanic and
Atmospheric Administration
USA, Russian Federation

ARBLASTER, Julie
National Center for Atmospheric Research
and Bureau of Meteorology Research Center
USA, Australia

ARCHER, David
University of Chicago
USA

ARORA, Vivek
Canadian Centre for Climate Modelling
and Analysis, Environment Canada
Canada

ARRITT, Raymond
Iowa State University
USA

ARTALE, Vincenzo
Italian National Agency for
New Technologies, Energy and
the Environment (ENEA)
Italy

ARTAXO, Paulo
Instituto de Fisica, Universidade
de Sao Paulo
Brazil

AUER, Ingeborg
Central Institute for Meteorology
and Geodynamics
Austria

AUSTIN, John
National Oceanic and Atmospheric
Administration, Geophysical
Fluid Dynamics Laboratory
USA

BAEDE, Alphonsus
Royal Netherlands Meteorological
Institute (KNMI) and Ministry of Housing,
Spatial Planning and the Environment
Netherlands

BAKER, David
National Center for Atmospheric Research
USA

BALDWIN, Mark P.
Northwest Research Associates
USA

BARNOLA, Jean-Marc
Laboratoire de Glaciologie et
Géophysique de l'Environnement
France

BARRY, Roger
National Snow and Ice Data
Center, University of Colorado
USA

BATES, Nicholas Robert
Bermuda Institute of Ocean Sciences
Bermuda

BAUER, Eva
Potsdam Institute for Climate
Impact Research
Germany

BENESTAD, Rasmus
Norwegian Meteorological Institute
Norway

BENISTON, Martin
University of Geneva
Switzerland

BERGER, André
Université catholique de Louvain,
Institut d'Astronomie et de
Géophysique G. Lemaitre
Belgium

BERNTSEN, Terje
Centre for International Climate and
Environmantal Research (CICERO)
Norway

BERRY, Joseph A.
Carnegie Institute of Washington,
Department of Global Ecology
USA

BETTS, Richard A.
Hadley Centre for Climate Prediction
and Research, Met Office
UK

BIERCAMP, Joachim
Deutsches Klimarechenzentrum GmbH
Germany

BINDOFF, Nathaniel L.
Antarctic Climate and Ecosystems
Cooperative Research Centre and CSIRO
Marine and Atmospheric Research
Australia

BITZ, Cecilia
University of Washington
USA

BLATTER, Heinz
Institute for Atmospheric and
Climate Science, ETH Zurich
Switzerland

BODEKER, Greg
National Institute of Water and
Atmospheric Research
New Zealand

BOJARIU, Roxana
National Institute of Meteorology
and Hydrology (NIMH)
Romania

BONAN, Gordon
National Center for Atmospheric Research
USA

Coordinating lead authors, lead authors, and contributing authors are listed alphabetically by surname.

BONFILS, Cèline
School of Natural Sciences,
Univerity of California, Merced
USA, France

BONY, Sandrine
Laboratoire de Météorologie Dynamique,
Institut Pierre Simon Laplace
France

BOONE, Aaron
CNRS CNRM at Meteo France
France, USA

BOONPRAGOB, Kansri
Department of Biology, Faculty of
Science, Ramkhamhaeng University
Thailand

BOUCHER, Olivier
Hadley Centre for Climate Prediction
and Research, Met Office
UK, France

BOUSQUET, Philippe
Institut Pierre Simon Laplace,
Laboratoire des Sciences du
Climat et de l'Environnement
France

BOX, Jason
Ohio State University
USA

BOYER, Tim
National Oceanic and
Atmospheric Administration
USA

BRACONNOT, Pascale
Pascale Braconnot Institu Pierre Simon
Laplace, Laboratoire des Sciences
du Climat et de l'Environnement
France

BRADY, Esther
National Center for Atmospheric Research
USA

BRASSEUR, Guy
Earth and Sun Systems Laboratory,
National Center for Atmospheric Research
USA, Germany

BRETHERTON, Christopher
Department of Atmospheric Sciences,
University of Washington
USA

BRIFFA, Keith R.
Climatic Research Unit, School
of Environmental Sciences,
University of East Anglia
UK

BROCCOLI, Anthony J.
Rutgers University
USA

BROCKMANN, Patrick
Laboratoire des Sciences du
Climat et de l'Environnement
France

BROMWICH, David
Byrd Polar Research Center,
The Ohio State University
USA

BROVKIN, Victor
Potsdam Institute for Climate
Impact Research
Germany, Russian Federation

BROWN, Ross
Environment Canada
Canada

BUJA, Lawrence
National Center for Atmospheric Research
USA

BUSUIOC, Aristita
National Meteorological Administration
Romania

CADULE, Patricia
Institut Pierre Simon Laplace
France

CAI, Wenju
CSIRO Marine and Atmospheric Research
Australia

CAMILLONI, Inés
Universidad de Buenos Aires, Cwentro de
Investigaciones del Mar y la Atmósfera
Argentina

CANADELL, Josep
Global Carbon Project, CSIRO
Australia

CARRASCO, Jorge
Direccion Meteorologica de Chile
and Centro de Estudios Cientificos
Chile

CASSOU, Christophe
Centre National de Recherche Scientifique,
Centre Europeen de Recherche et de
Formation Avancee en Calcul Scientifique
France

CAYA, Daniel
Consortium Ouranos
Canada

CAYAN, Daniel R.
Scripps Institution of Oceanography,
University of California, San Diego
USA

CAZENAVE, Anny
Laboratoire d'Etudes en Géophysique et
Océanographie Spatiale (LEGOS), CNES
France

CHAMBERS, Don
Center for Space Research, The
University of Texas at Austin
USA

CHANDLER, Mark
Columbia University and NASA
Goddard Institute for Space Studies
USA

CHANG, Edmund K.M.
Stony Brook University, State
University of New York
USA

CHAO, Ben
NASA Goddard Institute for Space Studies
USA

CHEN, Anthony
Department of Physics, University
of the West Indies
Jamaica

CHEN, Zhenlin
Dept of International Cooperation,
China Meteorological Administration
China

CHIDTHAISONG, Amnat
The Joint Graduate School of Energy
and Environment, King Mongkut's
University of Technology Thonburi
Thailand

CHRISTENSEN, Jens Hesselbjerg
Danish Meteorological Institute
Denmark

CHRISTIAN, James
Fisheries and Oceans, canada, Candian
Centre for Climate Modelling and Analysis
Canada

CHRISTY, John
University of Alabama in Huntsville
USA

CHURCH, John
CSIRO Marine and Atmospheric
Research and Ecosystems
Cooperative Research Centre
Australia

CIAIS, Philippe
Laboratoire des Sciences du
Climat et de l'Environnement
France

CLARK, Deborah A.
University of Missouri, St. Louis
USA

CLARKE, Garry
Earth and Ocean Sciences,
University of British Columbia
Canada

CLAUSSEN, Martin
Potsdam Institute for Climate
Impact Research
Germany

CLEMENT, Amy
University of Miami, Rosenstiel School
of Marine and Atmospheric Science
USA

COGLEY, J. Graham
Department of Geography, Trent University
Canada

COLE, Julia
University of Arizona
USA

COLLIER, Mark
CSIRO Marine and Atmospheric Research
Australia

COLLINS, Matthew
Hadley Centre for Climate Prediction
and Research, Met Office
UK

COLLINS, William D.
Climate and Global Dynamics Division,
National Center for Atmospheric Research
USA

COLMAN, Robert
Bureau of Meteorology Research Centre
Australia

COMISO, Josefino
National Aeronautics and Space
Administration, Goddard
Space Flight Center
USA

CONWAY, Thomas J.
National Oceanic and Atmospheric
Administration, Earth System
Research Laboratory
USA

COOK, Edward
Lamont-Doherty Earth Observatory
USA

CORTIJO, Elsa
Laboratoire des Sciences du Climat et de
l'Environnement, CNRS-CEA-UVSQ
France

COVEY, Curt
Lawrence Livermore National Laboratory
USA

COX, Peter M.
School of Engineering, Computer Science
and Mathematics, University of Exeter
UK

CROOKS, Simon
University of Oxford
UK

CUBASCH, Ulrich
Institut für Meteorologie,
Freie Universität Berlin
Germany

CURRY, Ruth
Woods Hole Oceanographic Institution
USA

DAI, Aiguo
National Center for Atmospheric Research
USA

DAMERIS, Martin
German Aerospace Center
Germany

DE ELÍA, Ramón
Ouranos Consortium
Canada, Argentina

DELWORTH, Thomas L.
Geophysical Fluid Dynamics
Laboratory, National Oceanic and
Atmospheric Administration
USA

DENMAN, Kenneth L.
Canadian Centre for Climate Modelling
and Analysis, Environment Canada and
Department of Fisheries and Oceans
Canada

DENTENER, Frank
European Commission Joint Research
Centre; Institute of Environment and
Sustainability Climate Change Unit
EU

DESER, Clara
National Center for Atmospheric Research
USA

DETHLOFF, Klaus
Alfred Wegener Institute for Polar and
Marine Research, Research Unit Potsdam
Germany

DIANSKY, Nikolay A.
Institute of Numerical Mathematics,
Russian Academy of Sciences
Russian Federation

DICKINSON, Robert E.
School of Earth and Atmospheric Sciences,
Georgia Institute of Technology
USA

DING, Yihui
National Climate Centre, China
Meteorological Administration
China

DIRMEYER, Paul
Center for Ocean-Land-Atmosphere Studies
USA

DIX, Martin
CSIRO
Australia

DIXON, Keith
National Oceanic and
Atmospheric Administration
USA

DLUGOKENCKY, Ed
National Oceanic and Atmospheric
Administration, Earth System
Research Laboratory
USA

DOKKEN, Trond
Bjerknes Centre for Climate Research
Norway

DOTZEK, Nikolai
Deutsches Zentrum für Luft und Raumfahrt,
Institut für Physik der Atmosphäre
Germany

DOUTRIAUX, Charles
Program for Climate Model
Diagnosis and Intercomparison
USA, France

DRANGE, Helge
Nansen Environmental and
Remote Sensing Center, Bjerknes
Centre for Climate Research
Norway

DRIESSCHAERT, Emmanuelle
Université catholique de Louvain,
Institut d'Astronomie et de
Géophysique G. Lemaitre
Belgium

DUFRESNE, Jean-Louis
Laboratoire de Météorologie Dynamique,
Institut Pierre Simon Laplace
France

DUPLESSY, Jean-Claude
Centre National dela Recerche
Scientifique, Laboratoire des Sciences
du Climat et de l'Environnement
France

DYURGEROV, Mark
Institute of Arctic and Alpine Research,
University of Colorado at Boulder
& Department of Geography and
Quaternary Geology at Stockholm
Sweden, USA

EASTERLING, David
National Oceanic and Atmospheric
Administration, Earth System
Research Laboratory
USA

EBY, Michael
University of Victoria
Canada

EDWARDS, Neil R.
The Open University
UK

ELKINS, James W.
National Oceanic and Atmospheric
Administration, Earth System
Research Laboratory
USA

EMERSON, Steven
School of Oceanography,
University of Washington
USA

EMORI, Seita
National Institute for Environmental
Studies and Frontier Research Center
for Global Change, Japan Agency for
Marine-Earth Science and Technology
Japan

ETHERIDGE, David
CSIRO Marine and Atmospheric Research
Australia

EYRING, Veronika
Deutsches Zentrum für Luft und Raumfahrt,
Institut für Physik der Atmosphäre
Germany

FAHEY, David W.
National Oceanic and Atmospheric
Administration, Earth System
Research Laboratory
USA

FASULLO, John
National Center for Atmospheric Research
USA

FEDDEMA, Johannes
University of Kansas
USA

FEELY, Richard
National Oceanic and Atmospheric
Administration, Pacific Marine
Environmental Laboratory
USA

FEICHTER, Johann
Max Planck Institute for Meteorology
Germany

FICHEFET, Thierry
Université catholique de Louvain,
Institut d'Astronomie et de
Géophysique G. Lemaitre
Belgium

FITZHARRIS, Blair
Department of Geography,
University of Otago
New Zealand

FLATO, Gregory
Canadian Centre for Climate Modelling
and Analysis, Environment Canada
Canada

FLEITMANN, Dominik
Institute of Geological Sciences,
Uniersity of Bern
Switzerland, Germany

FLEMING, James Rodger
Colby College
USA

FOGT, Ryan
Polar Meteorology Group, Byrd Polar
Research Center and Atmospheric
Sciences Program, Department of
geography, The Ohio State University
USA

FOLLAND, Christopher
Hadley Centre for Climate Prediction
and Research, Met Office
UK

FOREST, Chris
Massachusetts Institute of Technology
USA

FORSTER, Piers
School of Earth and Environment,
University of Leeds
UK

FOUKAL, Peter
Heliophysics, Inc.
USA

FRASER, Paul
CSIRO Marine and Atmospheric Research
Australia

FRAUENFELD, Oliver
National Snow and Ice Data Center,
University of Colorado at Boulder
USA, Austria

FREE, Melissa
Air Resources Laboratory, National
Oceanic and Atmospheric Administration
USA

FREI, Allan
Hunter College, City
University of New York
USA

FREI, Christoph
Federal Office of Meteorology
and Climatology MeteoSwiss
Switzerland

FRICKER, Helen
Scripps Institution of Oceanography,
University of California, San Diego
USA

FRIEDLINGSTEIN, Pierre
Institut Pierre Simon Laplace,
Laboratoire des Sciences du
Climat et de l'Environnement
France, Belgium

FU, Congbin
Start Regional Center for Temperate
East Asia, Institute of Atmospheric
Physics, Chinese Academy of Science
China

FUJII, Yoshiyuki
Arctic Environment Research Center,
National Institute of Polar Research
Japan

FUNG, Inez
University of California, Berkeley
USA

FURRER, Reinhard
Colorado School of Mines
USA, Switzerland

FUZZI, Sandro
National Research Council, Institute of
Atmospheric Sciences and Climate
Italy

FYFE, John
Canadian Centre for Climate Modelling
and Analysis, Environment Canada
Canada

GANOPOLSKI, Andrey
Potsdam Institute for Climate
Impact Research
Germany

GAO, Xuejie
Laboratory for Climate Change,
National Climate Centre, China
Meteorological Administration
China

GARCIA, Hernan
National Oceanic and Atmospheric
Administration, National
Oceanographic Data Center
USA

GARCÍA-HERRERA, Ricardo
Universidad Complutense de Madrid
Spain

GAYE, Amadou Thierno
Laboratory of Atmospheric Physics,
ESP/CAD, Dakar University
Senegal

GELLER, Marvin
Stony Brook University
USA

GENT, Peter
National Center for Atmospheric Research
USA

GERDES, Rüdiger
Alfred-Wegener-Institute für
Polar und Meeresforschung
Germany

GILLETT, Nathan P.
Climatic Research Unit, School
of Environmental Sciences,
University of East Anglia
UK

GIORGI, Filippo
Abdus Salam International Centre
for Theoretical Physics
Italy

GLEASON, Byron
National Climatic Data Center, National
Oceanic and Atmospheric Administration
USA

GLECKLER, Peter
Lawrence Livermore National Laboratory
USA

GONG, Sunling
Air Quality Researcch Division, Science &
Technology Branch, Environment Canada
Canada

GONZÁLEZ-DAVÍLA, Melchor
University of Las Palmas de Gran Canaria
Spain

GONZÁLEZ-ROUCO, Jesus Fidel
Universidad Complutense de Madrid
Spain

GOOSSE, Hugues
Université catholique de Louvain
Belgium

GRAHAM, Richard
Hadley Centre, Met Office
UK

GREGORY, Jonathan M.
Department of Meteorology, University of
Reading and Hadley Centre for Climate
Prediction and Research, Met Office
UK

GRIESER, Jürgen
Deutscher Wetterdienst, Global
Precipitatioin Climatology Centre
Germany

GRIGGS, David
Hadley Centre for Climate Prediction
and Research, Met Office
UK

GROISMAN, Pavel
University Corporation for Atmospheric
Research at the National Climatic
Data Center, National Oceanic and
Atmospheric Administration
USA, Russian Federation

GRUBER, Nicolas
Institute of Geophysics and Planetary
Physics, University of California,
Los Angeles and Department of
Environmental Sciences, ETH Zurich
USA, Switzerland

GUDGEL, Richard
National Oceanic and
Atmospheric Administration
USA

GUDMUNDSSON, G. Hilmar
British Antarctic Survey
UK, Iceland

GUENTHER, Alex
National Center for Atmospheric Research
USA

GULEV, Sergey
P. P. Shirshov Institute of Oceanography
Russian Federation

GURNEY, Kevin
Department of Earth and Atmospheric
Science, Purdue University
USA

GUTOWSKI, William
Iowa State University
USA

HAAS, Christian
Alfred Wegener Institute
Germany

HABIBI NOKHANDAN, Majid
National Center for Climatology
Iran

HAGEN, Jon Ove
University of Oslo
Norway

HAIGH, Joanna
Imperial College London
UK

HALL, Alex
Department of Atmospheric and
Oceanic Sciences, University
of California, Los Angeles
USA

HALLEGATTE, Stéphane
Centre International de Recherche sur
l'Environnement et le Developpement,
Ecole Nationale des Ponts-et-Chaussées
and Centre National de Recherches
Meteorologique, Meteo-France
USA, France

HANAWA, Kimio
Physical Oceanography Laboratry,
Department of Geophysics, Graduate
School of Science, Tohoku University
Japan

HANSEN, James
Goddard Institute for Space Studies
USA

HANSSEN-BAUER, Inger
Norwegian Meteorological Institute
Norway

HARRIS, Charles
School of Earth, Ocean and Planetary
Science, Cardiff University
UK

HARRIS, Glen
Hadley Centre for Climate Prediction
and Research, Met Office
UK, New Zealand

HARVEY, Danny
University of Toronto
Canada

HASUMI, Hiroyasu
Center for Climate System
Research, University of Tokyo
Japan

HAUGLUSTAINE, Didier
Institut Pierre Simon Laplace,
Laboratoire des Sciences du Climat et de
l'Environnement, CEA-CNRS-UVSQ
France

HAYWOOD, James
Hadley Centre for Climate Prediction
and Research, Met Office
UK

HEGERL, Gabriele C.
Division of Earth and Ocean Sciences,
Nicholas School for the Environment
and Earth Sciences, Duke University
USA, Germany

HEIMANN, Martin
Max-Planck-Institut für Biogeochemie
Germany, Switzerland

HEINZE, Christoph
University of Bergen, Geophysical Institute
and Bjerknes Centre for Climate Research
Norway, Germany

HELD, Isaac
National Oceanic and Atmospheric
Administration, Geophysical
Fluid Dynamics Laboratory
USA

HENDERSON-SELLERS, Ann
World Meteorological Organization
Switzerland

HENDON, Henry
Bureau of Meteorology Research Centre
Australia

HEWITSON, Bruce
Department of Environmental
and Geographical Sciences,
University of Cape Town
South Africa

HINZMAN, Larry
University of Alaska, Fairbanks
USA

HOCK, Regine
Stockholm University
Sweden

HODGES, Kevin
Environmental Systems Science Centre
UK

HOELZLE, Martin
University of Zürich,
Department of Geography
Switzerland

HOLLAND, Elisabeth
Atmospheric Chemistry Division, National
Center for Atmospheric Research (NCAR)
USA

HOLLAND, Marika
National Center for Atmospheric Research
USA

HOLTSLAG, Albert A. M.
Wageningen University
Netherlands

HOSKINS, Brian J.
Department of Meteorology,
University of Reading
UK

HOUSE, Joanna
Quantifying and Understanding the Earth
System Programme, University of Bristol
UK

HU, Aixue
National Center for Atmospheric Research
USA, China

HUNKE, Elizabeth
Los Alamos National Laboratory
USA

HURRELL, James
National Center for Atmospheric Research
USA

HUYBRECHTS, Philippe
Departement Geografie, Vrije
Universiteir Brussel
Belgium

INGRAM, William
Hadley Centre for Climate Prediction
and Research, Met Office
UK

ISAKSEN, Ketil
Norwegian Meteorological Institute
Norway

ISHII, Masayoshi
Fronteir Research Center for Global
Change, Japan Agency for Marine-
Earth Science and Technology
Japan

JACOB, Daniel
Department of Earth and Planetary
Sciences, Harvard University
USA, France

JALLOW, Bubu
Department of Water Resources
The Gambia

JANSEN, Eystein
University of Bergen, Department
of Earth Sciences and Bjerknes
Centre for Climate Research
Norway

JANSSON, Peter
Department of Physical Geography and
Quaternary Geology, Stockholm University
Sweden

JENKINS, Adrian
British Antarctic Survey, Natural
Environment Research Council
UK

JONES, Andy
Hadley Centre for Climate Prediction
and Research, Met Office
UK

JONES, Christopher
Hadley Centre for Climate Prediction
and Research, Met Office
UK

JONES, Colin
Universite du Quebec a Montreal, Canadian
Regional Climate Modelling Network
Canada

JONES, Gareth S.
Hadley Centre for Climate Prediction
and Research, Met Office
UK

JONES, Julie
GKSS Research Centre
Germany, UK

JONES, Philip D.
Climatic Research Unit, School
of Environmental Sciences,
University of East Anglia
UK

JONES, Richard
Hadley Centre for Climate Prediction
and Research, Met Office
UK

JOOS, Fortunat
Climate and Environmental Physics,
Physics Institute, University of Bern
Switzerland

JOSEY, Simon
National Oceanography Centre,
University of Southampton
UK

JOUGHIN, Ian
Applied Physics Laboratory,
University of Washington
USA

JOUZEL, Jean
Institut Pierre Simon Laplace,
Laboratoire des Sciences du Climat et de
l'Environnement, CEA-CNRS-UVSQ
France

JOYCE, Terrence
Woods Hole Oceanographic Institution
USA

JUNGCLAUS, Johann H.
Max Planck Institute for Meteorology
Germany

KAGEYAMA, Masa
Laboratoire des Sciences du
Climat et de l'Environnement
France

KÅLLBERG, Per
European Centre for Medium-
Range Weather Forecasts
ECMWF

KÄRCHER, Bernd
Deutsches Zentrum für Luft und Raumfahrt,
Institut für Physik der Atmosphäre
Germany

KARL, Thomas R.
National Oceanic and Atmospheric
Administration, National
Climatic Data Center
USA

KAROLY, David J.
University of Oklahoma
USA, Australia

KASER, Georg
Institut für Geographie,
University of Innsbruck
Austria, Italy

KATTSOV, Vladimir
Voeikov Main Geophysical Observatory
Russian Federation

KATZ, Robert
National Center for Atmospheric Research
USA

KAWAMIYA, Michio
Frontier Research Center for Global
Change, Japan Agency for Marine-
Earth Science and Technology
Japan

KEELING, C. David
Scripps Institution of Oceanography
USA

KEELING, Ralph
Scripps Institution of Oceanography
USA

KENNEDY, John
Hadley Centre, Met Office
UK

KENYON, Jesse
Duke University
USA

KETTLEBOROUGH, Jamie
British Atmospheric Data Centre,
Space Science and Technology
Department, Council for the Central
Laboratory of the Research Councils
UK

KHARIN, Viatcheslar
Canadian Centre for Climate Modelling
and Analysis, Environment Canada
Canada

KHODRI, Myriam
Institut de Recherche Pour
le Developpement
France

KILADIS, George
National Oceanic and
Atmospheric Administration
USA

KIM, Kuh
Seoul National University
Republic of Korea

KIMOTO, Masahide
Center for Climate System
Research, University of Tokyo
Japan

KING, Brian
National Oceanography
Centre, Southampton
UK

KINNE, Stefan
Max-Planck Institute for Meteorology
Germany

KIRTMAN, Ben
Center for Ocean-Land-Atmosphere
Studies, George Mason University
USA

KITOH, Akio
First Research Laboratory, Climate Research
Department, Meteorological Research
Institute, Japan Meteorological Agency
Japan

KLEIN, Stephen A.
Lawrence Livermore National Laboratory
USA

KLEIN TANK, Albert
Royal Netherlands Meteorological
Institute (KNMI)
Netherlands

KNUTSON, Thomas
Geophysical Fluid Dynamics
Laboratory, National Oceanic and
Atmospheric Administration
USA

KNUTTI, Reto
Climate and Global Dynamics Division,
National Center for Atmospheric Research
Switzerland

KOERTZINGER, Arne
Leibniz Institut für Meereswissenschaften
an der Universitat Kiel und Institut
fur Ostseeforschung Warnemunde
Germany

KOIKE, Toshio
Department of Civil Engineering,
University of Tokyo
Japan

KOLLI, Rupa Kumar
Climatology and Hydrometeorology
Division, Indian Institute of
Tropical Meteorology
India

KOSTER, Randal
National Aeronautics and
Space Administration
USA

KOTTMEIER, Christoph
Institut für Meteorologie, und
Klimaforschung, Universitat Karlsruhe/
Forschungszentrum Karlsruhe
Germany

KRIPALANI, Ramesh
Indian Institute of Tropical Meteorology
India

KRYNYTZKY, Marta
University of Washington
USA

KUNKEL, Kenneth
Illinois State Water Survey
USA

KUSHNER, Paul J.
Department of Physics,
University of Toronto
Canada

KWOK, Ron
Jet Propulsion Laboratory, California
Institute of Technology
USA

KWON, Won-Tae
Climate Research Laboratory,
Meteorological Research Institute (METRI),
Korean Meteorological Administration
Republic of Korea

LABEYRIE, Laurent
Laboratoire des Sciences du
Climat et de l'Environnement
France

LAINE, Alexandre
Laboratoire des Sciences du
Climat et de l'Environnement
France

LAM, Chiu-Ying
Hong Kong Observatory
China

LAMBECK, Kurt
Australia National University
Australia

LAMBERT, F. Hugo
Atmospheric, Oceanic and Planetary
Physics, University of Oxford
UK

LANZANTE, John
National Oceanic and
Atmospheric Administration
USA

LAPRISE, René
Deprtement des Sciences de la Terra
et de l'Atmosphere, University
of Quebec at Montreal
Canada

LASSEY, Keith
National Institute of Water and
Atmospheric Research
New Zealand

LATIF, Mojib
Leibniz Institut für Meereswissenschaften,
IFM-GEOMAR
Germany

LAU, Ngar-Cheung
Geophysical Fluid Dynamics
Laboratory, National Oceanic and
Atmospheric Administration
USA

LAVAL, Katia
Laboratoire de Météorologie
Dynamique du CNRS
France

LAVINE, Michael
Duke University
USA

LAWRENCE, David
National Center for Atmospheric Research
USA

LAWRIMORE, Jay
National Oceanic and Atmospheric
Administration, National
Climatic Data Center
USA

LAXON, Seymour
Centre for Polar Observation and
Modelling, University College London
UK

LE BROCQ, Anne
Centre for Polar Observation and
Modelling, University of Bristol
UK

LE QUÉRÉ, Corrine
University of East Anglia and
British Antarctic Survey
UK, France, Canada

LE TREUT, Hervé
Laboratoire de Météorologie
Dynamique du CNRS
France

LEAN, Judith
Naval Research Laboratory
USA

LECK, Caroline
Department of Metorology,
Stockholm University
Sweden

LEE, Terry C.K.
University of Victoria
Canada

LEE-TAYLOR, Julia
National Center for Atmospheric Research
USA, UK

LEFEVRE, Nathalie
Institut de Recherche Pour le
Developpement, Laboratoire
d'Oceanographie et de Climatologie
France

LEMKE, Peter
Alfred Wegener Institute for
Polar and Marine Research
Germany

LEULIETTE, Eric
University of Colorado, Boulder
USA

LEUNG, Ruby
Pacific Northwest National
Laboratory, National Oceanic and
Atmospheric Administration
USA

LEVERMANN, Anders
Potsdam Institute for Climate
Impact Research
Germany

LEVINSON, David
National Oceanic and Atmospheric
Administration, National
Climatic Data Center
USA

LEVITUS, Sydney
National Oceanic and
Atmospheric Administration
USA

LIE, Øyvind
Bjerknes Centre for Climate Research
Norway

LIEPERT, Beate
Lamont-Doherty Earth Observatory,
Columbia University
USA

LIU, Shiyin
Cold and Arid Regions Environmental
and Engineering Research Institute,
Chinese Academy of Sciences
China

LOHMANN, Ulrike
ETH Zürich, Institute for Atmospheric
and Climate Science
Switzerland

LOUTRE, Marie-France
Université catholique de Louvain,
Institut d'Astronomie et de
Géophysique G. Lemaitre
Belgium

LOWE, David C.
National Institute of Water and
Atmospheric Research
New Zealand

LOWE, Jason
Hadley Centre for Climate Prediction
and Research, Met Office
UK

LUO, Yong
Laboratory for Climate Change,
National Climate Centre, China
Meteorological Administration
China

LUTERBACHER, Jürg
Institute of Geography, Climatology
and Meteorology, and National
Centre of Competence in Research
on Climate, University of Bern
Switzerland

LYNCH, Amanda H.
School of Geography and Environmental
Science, Monash University
Australia

MACAYEAL, Douglas
University of Chicago
USA

MACCRACKEN, Michael
Climate Institute
USA

MAGAÑA RUEDA, Victor
Centro de Ciencias de la Atmósfera,
Ciudad Universitaria, Universidad
Nacional Autonomia de Mexico
Mexico

MALHI, Yadvinder
University of Oxford
UK

MALANOTTE-RIZZOLI, Paola
Massachusetts Institute of Technology
USA, Italy

MANNING, Andrew C.
University of East Anglia
UK, New Zealand

MANNING, Martin
IPCC WGI TSU, National Oceanic
and Atmospheric Administration,
Earth System Research Laboratory
USA, New Zealand

MANZINI, Elisa
National Institute for Geophysics
and Volcanology
Italy

MARENGO ORSINI, Jose Antonio
CPTEC/INPE
Brazil, Peru

MARSH, Robert
National Oceanography Centre,
University of Southampton
UK

MARSHALL, Gareth
British Antarctic Survey
UK

MARTELO, Maria
Ministerio del Ambiente y los Rcursos
Naturales, Dir. de Hidrologia y Meteorologia
Venezuela

MASARIE , Ken
National Oceanic and Atmospheric
Administration, Earth System Research
Laboratory, Global Monitoring Division
USA

MASSON-DELMOTTE, Valérie
Laboratoire des Sciences du
Climat et de l'Environnement
France

MATSUMOTO, Katsumi
University of Minnesota, Twin Cities
USA

MATSUNO, Taroh
Frontier Research Center for Global
Change, Japan Agency for Marine-
Earth Science and Technology
Japan

MATTHEWS, H. Damon
University of Calgary and
Concordia University
Canada

MATULLA, Christoph
Environment Canada
Canada, Austria

MAURITZEN, Cecilie
Norwegian Meteorological Institute
Norway

MCAVANEY, Bryant
Bureau of Meteorology Research Centre
Australia

MCFIGGANS, Gordon
University of Manchester
UK

MCINNES, Kathleen
CSIRO, Marine and Atmospheric
Chemistry Research
Australia

MCPHADEN, Michael
National Oceanic and
Atmospheric Administration
USA

MEARNS, Linda
National Center for Atmospheric Research
USA

MEARS, Carl
Remote Sensing Systems
USA

MEEHL, Gerald A.
Climate and Global Dynamics Division,
National Center for Atmospheric Research
USA

MEINSHAUSEN, Malte
Potsdam Institute for Climate
Impact Research
Germany

MELLING, Humphrey
Fisheries and Oceans Canada
Canada

MENÉNDEZ, Claudio Guillermo
Centro de Investigaciones del Mar y
de la Atmósfera, (CONICET-UBA)
Argentina

MENON, Surabi
Lawrence Berkeley National Laboratory
USA

MESCHERSKAYA, Anna V.
Russian Federation

MILLER, John B.
National Oceanic and
Atmospheric Administration
USA

MILLOT, Claude
Centre National dela Recherche Scientifique
France

MILLY, Chris
United States Geological Survey
USA

MITCHELL, John
Hadley Centre for Climate Prediction
and Research, Met Office
UK

MOKSSIT, Abdalah
Direction de la météorologie Nationale
Morocco

MOLINA, Mario
Scripps Institution of Oceanography,
Dept. of Chemistry and Biochemistry,
University of California, San Diego
USA, Mexico

MOLINARI, Robert
National Oceanic and Atmospheric
Administration, Atlantic Oceanographic
and Meteorological Laboratory
USA

MONAHAN, Adam H.
School of Earth and Ocean
Sciences, University of Victoria
Canada

MONNIN, Eric
Climate and Environmental Physics,
Physics Institute, University of Bern
Switzerland

MONTZKA, Steve
National Oceanic and
Atmospheric Administration
USA

MOSLEY-THOMPSON, Ellen
Ohio State University
USA

MOTE, Philip
Climate Impacts Group, Joint Institute for
the Study of the Atmosphere and Oceans
(JIASO), University of Washington
USA

MUHS, Daniel
United States Geological Survey
USA

MULLAN, A. Brett
National Institute of Water and
Atmospheric Research
New Zealand

MÜLLER, Simon A.
Climate and Environmental Physics,
Physics Institute, University of Bern
Switzerland

MURPHY, James M.
Hadley Centre for Climate Prediction
and Research, Met Office
UK

MUSCHELER, Raimund
Goddard Earth Sciences and Technology
Center, University of Maryland &
NASA/Goddard Space Flight Center,
Climate & Radiation Branch
USA

MYHRE, Gunnar
Department of Geosciences,
University of Oslo
Norway

NAKAJIMA, Teruyuki
Center for Climate System
Research, University of Tokyo
Japan

NAKAMURA, Hisashi
Department of Earth, Planetary
Science, University of Tokyo
Japan

NAWRATH, Susanne
Potsdam Institute for Climate
Impact Research
Germany

NEREM, R. Steven
University of Colorado at Boulder
USA

NEW, Mark
Centre for the Environment,
University of Oxford
UK

NGANGA, John
University of Nairobi
Kenya

NICHOLLS, Neville
Monash University
Australia

NODA, Akira
Meteorological Research Institute,
Japan Meteorological Agency
Japan

NOJIRI, Yukihiro
Secretariat of Council for Science and
Technology Policy, Cabinet Office
Japan

NOKHANDAN, Majid Habibi
Iranian Meteorological Organization
Iran

NORRIS, Joel
Scripps Institution of Oceanography
USA

NOZAWA, Toru
National Institute for Environmental Studies
Japan

OERLEMANS, Johannes
Institute for Marine and Atmospheric
Research, Utrecht University
Netherlands

OGALLO, Laban
IGAD Climate Prediction and
Application Centre
Kenya

OHMURA, Atsumu
Swiss Federal Institute of Technology
Switzerland

OKI, Taikan
Institute of Industrial Science
The University of Tokyo
Japan

OLAGO, Daniel
Department of Geology,
University of Nairobi
Kenya

ONO, Tsuneo
Hokkaido National Fisheries Research
Institute, Fisheries Research Agency
Japan

OPPENHEIMER, Michae
Princeton University
USA

ORAM, David
University of East Anglia
UK

ORR, James C.
Marine Environment Laboratories,
International Atomic Energy Agency
Monaco, USA

OSBORN, Tim
University of East Anglia
UK

O'SHAUGHNESSY, Kath
National Institute of Water and
Atmospheric Research
New Zealand

OTTO-BLIESNER, Bette
Climate and Global Dynamics Division,
National Center for Atmospheric Research
USA

OVERPECK, Jonathan
Institute for the Study of Planet
Earth, University of Arizona
USA

PAASCHE, Øyvind
Bjerknes Centre for Climate Research
Norway

PAHLOW, Markus
Dalhousie University, Bedford
Institute of Oceanography
Canada

PAL, Jeremy S.
Loyola Marymount University,
The Abdus Salam International
Centre for Theoretical Physics
USA, Italy

PALMER, Timothy
European Centre for Medium-
Range Weather Forecasting
ECMWF, UK

PANT, Govind Ballabh
Indian Institute of Tropical Meteorology
India

PARKER, David
Hadley Centre for Climate Prediction
and Research, Met Office
UK

PARRENIN, Frédéric
Laboratoire de Glaciologie et
Géophysique de l'Environnement
France

PAVLOVA, Tatyana
Voeikov Main Geophysical Observatory
Russian Federation

PAYNE, Antony
University of Bristol
UK

PELTIER, W. Richard
Department of Physics,
University of Toronto
Canada

PENG, Tsung-Hung
Atlantic Oceanographic and Meteorological
Laboratory, National Oceanic and
Atmospheric Administration
USA

PENNER, Joyce E.
Department of Atmospheric, Oceanic, and
Space Sciences, University of Michigan
USA

PETERSON, Thomas
National Oceanic and Atmospheric
Administration, National
Climatic Data Center
USA

PETOUKHOV, Vladimir
Potsdam Institute for Climate
Impact Research
Germany

PEYLIN, Philippe
Laboratoire des Modélisation du
Climat et de l'Environnement
France

PFISTER, Christian
University of Bern
Switzerland

PHILLIPS, Thomas
Program for Climate Model Diagnosis
and Intercomparison, Lawrence
Livermore National Laboratory
USA

PIERCE, David
Scripps Institution of Oceanography
USA

PIPER, Stephen
Scripps Institution of Oceanography
USA

PITMAN, Andrew
Department of Physical Geography,
Macquarie University
Australia

PLANTON, Serge
Météo-France
France

PLATTNER, Gian-Kasper
Climate and Environmental Physics,
Physics Institute, University of Bern
Switzerland

PLUMMER, David
Environment Canada
Canada

POLLACK, Henry
University of Michigan
USA

PONATER, Michael
Deutsches Zentrum für Luft und Raumfahrt,
Institut für Physik der Atmosphäre
Germany

POWER, Scott
Bureau of Meteorology Research Centre
Australia

PRATHER, Michael
Earth System Science Department,
University of California at Irvine
USA

PRINN, Ronald
Department of Earth, Atmospheric
and Planetary Sciences, Massachusetts
Institute of Technology
USA, New Zealand

PROSHUTINSKY, Andrey
Woods Hole Oceanographic Institution
USA

PROWSE, Terry
Environment Canada, University of Victoria
Canada

QIN, Dahe
Co-Chair, IPCC WGI, China
Meteorological Administration
China

QIU, Bo
University of Hawaii
USA

QUAAS, Johannes
Max Planck Institute for Meteorology
Germany

QUADFASEL, Detlef
Institut für Meereskunde, Centre for Marine
and Atmospheric Sciences Hamburg
Germany

RAGA, Graciela
Centro de Ciencias de la Atmósfera,
Universidad Nacional Autonoma de Mexico
Mexico, Argentina

RAHIMZADEH, Fatemeh
Atmospheric Science & Meteorological
Research Center (ASMERC), I.R. of Iran
Meteorological Organization (IRIMO)
Iran

RAHMSTORF, Stefan
Potsdam Institute for Climate
Impact Research
Germany

RÄISÄNEN, Jouni
Department of Physical Sciences,
University of Helsinki
Finland

RAMACHANDRAN, Srikanthan
Space & Atmospheric Sciences Division,
Physical Research Laboratory
India

RAMANATHAN, Veerabhadran
Scripps Institution of Oceanography
USA

RAMANKUTTY, Navin
University of Wisconsin, Madison
USA, India

RAMASWAMY, Venkatachalam
National Oceanic and Atmospheric
Administration, Geophysical
Fluid Dynamics Laboratory
USA

RAMESH, Rengaswamy
Physical Research Laboratory
India

RANDALL, David A.
Department of Atmospheric Science,
Colorado State University
USA

RAPER, Sarah C.B.
Manchester Metropolitan University
UK

RAUP, Bruce H.
National Snow and Ice Data
Center, University of Colorado
USA

RAUPACH, Michael
CSIRO
Australia

RAYMOND, Charles
University of Washington, Department
of Earth and Space Sciences
USA

RAYNAUD, Dominique
Laboratoire de Glaciologie et
Géophysique de l'Environnement
France

RAYNER, Peter
Institut Pierre Simon Laplace,
Laboratoire des Sciences du
Climat et de l'Environnement
France

REHDER, Gregor
Leibniz Institut für Meereswissenschaften
an der Universitat Kiel and Institut
fur Ostseeforschung Warnemunde
Germany

REID, George
National Oceanic and
Atmospheric Administration
USA

REN, Jiawen
Cold and Arid Regions Environmental
and Engineering Research Institute,
Chinese Academy of Sciences
China

RENSSEN, Hans
Faculty of Earth and Life Sciences,
Vrije Universiteit Amsterdam
Netherlands

RENWICK, James A.
National Institute of Water and
Atmospheric Research
New Zealand

RIEBESELL, Ulf
Leibniz Institute for Marine
Sciences, IFM-GEOMAR
Germany

RIGNOT, Eric
Jet Propulsion Laboratory
USA

RIGOR, Ignatius
Polar Science Center, Applied Physics
Laboratory, University of Washington
USA

RIND, David
National Aeronautics and Space
Administration, Goddard
Institute for Space Studies
USA

RINKE, Annette
Alfred Wegener Institute for
Polar and Marine Research
Germany

RINTOUL, Stephen
CSIRO, Marine and Atmospheric
Research and Antarctic Climate and
Ecosystems Cooperative Research Centre
Australia

RIXEN, Michel
University of Liege and NATO
Undersea Research Center
NATO, Belgium

RIZZOLI, Paola
Massachusetts Institute of Technology
USA, Italy

ROBERTS, Malcolm
Hadley Centre for Climate Prediction
and Research, Met Office
UK

ROBERTSON, Franklin R.
National Aeronautics and
Space Administration
USA

ROBINSON, David
Rutgers University
USA

RÖDENBECK, Christian
Max Planck Institute for
Biogeochemistry Jena
Germany

ROECKNER, Erich
Max Planck Institute for Meteorology
Germany

ROSATI, Anthony
National Oceanic and
Atmospheric Administration
USA

ROSENLOF, Karen
National Oceanic and
Atmospheric Administration
USA

ROTHROCK, David
University of Washington
USA

ROTSTAYN, Leon
CSIRO Marine and Atmospheric Research
Australia

ROULET, Nigel
McGill University
Canada

RUMMUKAINEN, Markku
Rossby Centre, Swedish Meteorological
and Hydrological Institute
Sweden, Finland

RUSSELL, Gary L.
National Aeronautics and Space
Administration, Goddard
Institute for Space Studies
USA

RUSTICUCCI, Matilde
Departamento de Ciencias de la
Atmósfera y los Océanos, FCEN,
Universidad de Buenos Aires
Argentina

SABINE, Christopher
National Oceanic and Atmospheric
Administration, Pacific Marine
Environmental Laboratory
USA

SAHAGIAN, Dork
Lehigh University
USA

SALAS Y MÉLIA, David
Météo-France, Centre National de
Recherches Météorologiques
France

SANTER, Ben D.
Program for Climate Model Diagnosis
and Intercomparison, Lawrence
Livermore National Laboratory
USA

SARR, Abdoulaye
Service Météorologique, DMN Sénégal
Senegal

SAUSEN, Robert
Deutsches Zentrum für Luft und Raumfahrt,
Institut für Physik der Atmosphäre
Germany

SCHÄR, Christoph
ETH Zürich, Institute for Atmospheric
and Climate Science
Switzerland

SCHERRER, Simon Christian
Federal Office of Meteorology
and Climatology MeteoSwiss
Switzerland

SCHMIDT, Gavin
National Aeronautics and Space
Administration, Goddard
Institute for Space Studies
USA, UK

SCHMITTNER, Andreas
College of Oceanic and Atmospheric
Sciences, Oregon State University
USA, Germany

SCHNEIDER, Birgit
Leibniz Institut für Meereswissenschaften
Germany

SCHOTT, Friedrich
Leibniz Institut für Meereswissenschaften,
IFM-GEOMAR
Germany

SCHULTZ , Martin G.
Max Planck Institute for Meteorology
Germany

SCHULZ, Michael
Institut Pierre Simon Laplace,
Laboratoire des Sciences du Climat et de
l'Environnement, CEA-CNRS-UVSQ
France, Germany

SCHWARTZ, Stephen E
Brookhaven National Laboratory
USA

SCHWARZKOPF, Dan
National Oceanic and
Atmospheric Administration
USA

SCINOCCA, John
Canadian Centre for Climate Modelling
and Analysis, Environment Canada
Canada

SEIDOV, Dan
Pennsylvania State University
USA

SEMAZZI, Fred H.
North Carolina State University
USA

SENIOR, Catherine
Hadley Centre for Climate Prediction
and Research, Met Office
UK

SEXTON, David
Hadley Centre for Climate Prediction
and Research, Met Office
UK

SHEA, Dennis
National Center for Atmospheric Research
USA

SHEPHERD, Andrew
School of Geosciences, The
University of Ediburgh
UK

SHEPHERD, J. Marshall
University of Georgia,
Department of Geography
USA

SHEPHERD, Theodore G.
University of Toronto
Canada

SHERWOOD, Steven
Yale University
USA

SHUKLA, Jagadish
Center for Ocean-Land-Atmosphere
Studies, George Mason University
USA

SHUM, C.K.
Geodetic Science, School of Earth
Sciences, The Ohio State University
USA

SIEGMUND, Peter
Royal Netherlands Meteorological
Institute (KNMI)
Netherlands

SILVA DIAS, Pedro Leite da
Universidade de Sao Paulo
Brazil

SIMMONDS, Ian
University of Melbourne
Australia

SIMMONS, Adrian
European Centre for Medium-
Range Weather Forecasts
ECMWF, UK

SIROCKO, Frank
University of Mainz
Germany

SLATER, Andrew G.
Cooperative Institute for Research
in Environmental Sciences,
University of Colorado, Boulder
USA, Australia

SLINGO, Julia
National Centre for Atmospheric
Science, University of Reading
UK

SMITH, Doug
Hadley Centre for Climate Prediction
and Research, Met Office
UK

SMITH, Sharon
Geological Survey of Canada,
Natural Resources Canada
Canada

SODEN, Brian
University of Miami, Rosentiel School
for Marine and Atmospheric Science
USA

SOKOLOV, Andrei
Massachusetts Institute of Technology
USA, Russian Federation

SOLANKI, Sami K.
Max Planck Institute for
Solar System Research
Germany, Switzerland

SOLOMINA, Olga
Institute of Geography RAS
Russian Federation

SOLOMON, Susan
Co-Chair, IPCC WGI, National Oceanic
and Atmospheric Administration,
Earth System Research Laboratory
USA

SOMERVILLE, Richard
Scripps Institution of Oceanography,
University of California, San Diego
USA

SOMOT, Samuel
Météo-France, Centre National de
Recherches Météorologiques
France

SONG, Yuhe
Jet Propulsion Laboratory
USA

SPAHNI, Renato
Climate and Environmental Physics,
Physics Institute, University of Bern
Switzerland

SRINIVASAN, Jayaraman
Centre for Atmospheric and Oceanic
Sciences, Indian Institute of Science
India

STAINFORTH, David
Atmospheric, Oceanic and
Planetary Physics, Department of
Physics, University of Oxford
UK

STAMMER, Detlef
Institut fuer Meereskunde Zentrum
fuer Meeres und Klimaforschung
Universitaet Hamburg
Germany

STANIFORTH, Andrew
Hadley Centre for Climate Prediction
and Research, Met Office
UK

STARK, Sheila
Hadley Centre for Climate Prediction
and Research, Met Office
UK

STEFFEN, Will
Australian National University
Australia

STENCHIKOV, Georgiy
Rutgers, The State University of New Jersey
USA

STERN, William
National Oceanic and
Atmospheric Administration
USA

STEVENSON, David
University of Edinburgh
UK

STOCKER, Thomas F.
Climate and Environmental Physics,
Physics Institute, University of Bern
Switzerland

STONE, Daíthí A.
University of Oxford
UK, Canada

STOTT, Lowell D.
Department of Earth Sciences,
University of Southern California
USA

STOTT, Peter A.
Hadley Centre for Climate Prediction
and Research, Met Office
UK

STOUFFER, Ronald J.
National Oceanic and Atmospheric
Administration, Geophysical
Fluid Dynamics Laboratory
USA

STUBER, Nicola
Department of Meteorology,
University of Reading
UK, Germany

SUDO, Kengo
Nagoya University
Japan

SUGA, Toshio
Tohoku University
Japan

SUMI, Akimasa
Center for Climate System
Research, University of Tokyo
Japan

SUPPIAH, Ramasamy
CSIRO
Australia

SWEENEY, Colm
Princeton University
USA

TADROSS, Mark
Climate Systems Analysis Group,
University of Cape Town
South Africa

TAKEMURA, Toshihiko
Research Institute for Applied
Mechanics, Kyushu University
Japan

TALLEY, Lynne D.
Scripps Institution of Oceanography,
University of California, San Diego
USA

TAMISIEA, Mark
Harvard-Smithsonian Center
for Astrophysics
USA

TAYLOR, Karl E.
Program for Climate Model Diagnosis
and Intercomparison, Lawrence
Livermore National Laboratory
USA

TEBALDI, Claudia
National Center for Atmospheric Research
USA

TENG, Haiyan
National Center for Atmospheric Research
USA, China

TENNANT, Warren
South African Weather Service
South Africa

TERRAY, Laurent
Eoropean Centre for Research and Advanced
Training in Scientific Computation
France

TETT, Simon
Hadley Centre for Climate Prediction
and Research, Met Office
UK

TEXTOR, Christiane
Laboratoire des Sciences du
Climat et de l'Environnement
France, Germany

THOMAS, Robert H.
EG&G Technical Services, Inc. and
Centro de Estudios Cientificos (CECS)
USA, Chile

THOMPSON, Lonnie
Ohio State University
USA

THORNCROFT, Chris
Department of Earth and Atmospheric
Science, University at Albany, SUNY
USA, UK

THORNE, Peter
Hadley Centre for Climate Prediction
and Research, Met Office
UK

TIAN, Yuhong
Georgia Institute of Technology
USA, China

TRENBERTH, Kevin E.
Climate Analysis Section, National
Center for Atmospheric Research
USA

TSELIOUDIS, George
National Aeronautics and Space
Administration, Goddard Institute for
Space Studies, Columbia University
USA, Greece

TSIMPLIS, Michael
National Oceanography Centre,
University of Southampton
UK, Greece

UNNIKRISHNAN, Alakkat S.
National Institute of Oceanography
India

UPPALA, Sakari
European Centre for Medium-
Range Weather Forecasts
ECMWF

VAN DE WAL, Roderik Sylvester Willo
Institute for Marine and Atmospheric
Research, Utrecht University
Netherlands

VAN DORLAND, Robert
Royal Netherlands Meteorological
Institute (KNMI)
Netherlands

VAN NOIJE, Twan
Royal Netherlands Meteorological
Institute (KNMI)
Netherlands

VAUGHAN, David
British Antarctic Survey
UK

VILLALBA, Ricardo
Departmento de Dendrocronología e
Historia Ambiental, Instituto Argentino
de Novologia, Glaciologia y Ciencias
Ambientales (IANIGLA - CRICYT)
Argentina

VOLODIN, Evgeny M.
Institute of Numerical Mathematics
of Russian Academy of Sciences
Russian Federation

VOSE, Russell
National Oceanic and Atmospheric
Administration, National
Climatic Data Center
USA

WAELBROECK, Claire
Institut Pierre Simon Laplace,
Laboratoire des Sciences du Climat
et de l'Environnement, CNRS
France

WALSH, John
University of Alaska
USA

WANG, Bin
National Key Laboratory of Numerical
Modeling for Atmospheric Sciences
and Geophysical Fluid Dynamics,
institute of Atmospheric Physics,
Chinese Academy of Sciences
China

WANG, Bin
University of Hawaii
USA

WANG, Minghuai
Department of Atmospheric, Oceanic, and
Space Sciences, University of Michigan
USA

WANG, Ray
Georgia Institute of Technology
USA

WANNINKHOF, Rik
Atlantic Oceanographic and Meteorological
Laboratory, National Oceanic and
Atmospheric Administration
USA

WARREN, Stephen
University of Washington
USA

WASHINGTON, Richard
UK, South Africa

WATTERSON, Ian G.
CSIRO Marine and Atmospheric Research
Australia

WEAVER, Andrew J.
School of Earth and Ocean
Sciences, University of Victoria
Canada

WEBB, Mark
Hadley Centre for Climate Prediction
and Research, Met Office
UK

WEISHEIMER, Antje
European Centre for Medium-
Range Weather Forecasting and
Free University, Berlin
ECMWF, Germany

WEISS, Ray
Scripps Institution of Oceanography,
University of California, San Diego
USA

WHEELER, Matthew
Bureau of Meteorology Research Centre
Australia

WHETTON, Penny
CSIRO Marine and Atmospheric Research
Australia

WHORF, Tim
Scripps Institution of Oceanography,
University of California, San Diego
USA

WIDMANN, Martin
GKSS Research Centre, Geesthacht
and School of Geography, Earth
and Envrionmental Sciences,
University of Birmingham
Germany, UK

WIELICKI, Bruce
National Aeronautics and Space
Administration, Langley Research Center
USA

WIGLEY, Tom M.L.
National Center for Atmospheric Research
USA

WILBY, Rob
Environment Agency of England and Wales
UK

WILD, Martin
ETH Zürich, Institute for Atmospheric
and Climate Sciencce
Switzerland

WILD, Oliver
Frontier Research Center for Global
Change, Japan Agency for Marine-
Earth Science and Technology
Japan, UK

WILES, Gregory
The College of Wooster
USA

WILLEBRAND, Jürgen
Leibniz Institut für Meereswissenschaften
an der Universität Kiel
Germany

WILLIS, Josh
Jet Propulsion Laboratory
USA

WOFSY, Steven C.
Division of Engineering and Applied
Science, Harvard University
USA

WONG, A.P.S.
School of Oceanography,
University of Washington
USA, Australia

WONG, Takmeng
National Aeronautics and Space
Administration, Langley Research Center
USA

WOOD, Richard A.
Hadley Centre for Climate Prediction
and Research, Met Office
UK

WOODWORTH, Philip
Proudman Oceanographic Laboratory
UK

WORBY, Anthony
Australian Antarctic Division and
Antarctic Climate and Ecosystems
Cooperative Research Centre
Australia

WRATT, David
National Climate Centre, National Institute
of Water and Atmospheric Research
New Zealand

WUERTZ, David
National Oceanic and Atmospheric
Administration, National
Climatic Data Center
USA

WYMAN, Bruce L.
Geophysical Fluid Dynamics
Laboratory, National Oceanic and
Atmospheric Administration
USA

XU, Li
Department of Atmospheric, Oceanic, and
Space Sciences, University of Michigan
USA, China

YAMADA, Tomomi
Japanese Society of Snow and Ice
Japan

YASHAYAEV, Igor
Maritimes Region of the Department
of Fisheries and Oceans
Canada

YASUDA, Ichiro
University of Tokyo
Japan

YOSHIMURA, Jun
Meteorological Research Institute
Japan

YU, Rucong
China Meteorological Administration
China

YUKIMOTO, Seiji
Meteorological Research Institute
Japan

ZACHOS, James
University of California, Santa Cruz
USA

ZHAI, Panmao
National Climate Center, China
Meteorological Administration
China

ZHANG, De'er
National Climate Center, China
Meteorological Administration
China

ZHANG, Tingjun
National Snow and Ice Data Center, CIRES,
University of Colorado at Boulder
USA, China

ZHANG, Xiaoye
Chinese Academy of Meteorological
Sciences, Centre for Atmophere
Watch & Services
China

ZHANG, Xuebin
Climate Research Division,
Environment Canada
Canada

ZHAO, Lin
Cold and Arid Regions Environmental
and Engineering Research Institute,
Chinese Academy of Science
China

ZHAO, Zong-Ci
National Climate Center, China
Meteorological Administration
China

ZHENGTENG, Guo
Institute of Geology and Geophysics,
Chinese Academy of Science
China

ZHOU, Liming
Georgia Institute of Technology
USA, China

ZORITA, Eduardo
Helmholtz Zentrum Geesthacht
Germany, Spain

ZWIERS, Francis
Canadian Centre for Climate Modelling
and Analysis, Environment Canada
Canada

Annex III

Reviewers of the IPCC WGI Fourth Assessment Report

Algeria

AMAR, Matari
IHFR, Oran

MATARI, Amar
IHFR, Oran

Australia

CAI, Wenju
CSIRO Marine and Atmospheric Research

CHURCH, John
CSIRO Marine and Atmospheric
Research and Ecosystems
Cooperative Research Centre

COLMAN, Robert
Bureau of Meteorology Research Centre

ENTING, Ian
University of Melbourne

GIFFORD, Roger
CSIRO Plant Industry

HIRST, Anthony
CSIRO Marine and Atmospheric Research

HOBBINS, Michael
Australian National University

HOWARD, William
Antarctic Climate and Ecosystems
Cooperative Research Centre

HUNTER, John
Antarctic Climate and Ecosystems
Cooperative Research Centre

JONES, Roger
CSIRO Marine and Atmospheric Research

KININMONTH, William

LYNCH, Amanda H.
School of Geography and Environmental
Science, Monash University

MANTON, Michael
Bureau of Meteorology Research Centre

MCAVANEY, Bryant
Bureau of Meteorology Research Centre

MCDOUGALL, Trevor
CSIRO Marine and Atmospheric Research

MCGREGOR, John
CSIRO Marine and Atmospheric Research

MCNEIL, Ben
University of New South Wales

MOISE, Aurel
Bureau of Meteorology Research Centre

NICHOLLS, Neville
Monash University

PITMAN, Andrew
Department of Physical Geography,
Macquarie University

RAUPACH, Michael
CSIRO

RINTOUL, Stephen
CSIRO, Marine and Atmospheric
Research and Antarctic Climate and
Ecosystems Cooperative Research Centre

RODERICK, Michael
Australian National University

ROTSTAYN, Leon
CSIRO Marine and Atmospheric Research

SIEMS, Steven
Monash University

SIMMONDS, Ian
University of Melbourne

TREWIN, Blair
National Climate Centre,
Bureau of Meteorology

VAN OMMEN, Tas
Australian Antarctic Division

WALSH, Kevin
School of Earth Sciences,
University of Melbourne

WATKINS, Andrew
National Cliamte Centre,
Bureau of Meteorology

WHEELER, Matthew
Bureau of Meteorology Research Centre

WHITE, Neil
CSIRO Marine and Atmospheric Research

Austria

BÖHM, Reinhard
Central Institute for Meteorology
and Geodynamics

KIRCHENGAST, Gottfried
University of Graz

O'NEILL, Brian
IIASA and Brown University

RADUNSKY, Klaus
Umweltbundesamt

Belgium

BERGER, André
Université catholique de Louvain,
Institut d'Astronomie et de
Géophysique G. Lemaitre

DE BACKER, Hugo
Royal Meteorological Institute

GOOSSE, Hugues
Université catholique de Louvain

JANSSENS, Ivan A.
University of Antwerp

LOUTRE, Marie-France
Université catholique de Louvain,
Institut d'Astronomie et de
Géophysique G. Lemaitre

VAN LIPZIG, Nicole
Katholieke Universiteit Leuven

Benin

BOKO, Michel
Universite de Bourgogne

GUENDEHOU, G. H. Sabin
Benin Centre for Scientific
and Technical Review

VISSIN, Expédit Wilfrid
LECREDE/DGAT/FLASH/
Université d'Abomey-Calavi

YABI, Ibouraïma
Laboratoire de Climatologie/DGAT/UAC

Brazil

CARDIA SIMÕES, Jefferson
Departamento de Geografia, Instituto
de Geociências, Universidade
Federal do Rio Grande do Sul

GOMES, Marcos S.P.
Department of Mechanical
Research, Pontifical Catholic
University of Rio de Janeiro

MARENGO ORSINI, Jose Antonio
CPTEC/INPE

Canada

BELTRAMI, Hugo
St. Francis Xavier University

BROWN, Ross
Environment Canada

Expert reviewers are listed by country. Experts from international organizations are listed at the end.

CAYA, Daniel
Consortium Ouranos

CHYLEK, Petr
Dalhousie University, Departments
of Physics and Oceanography

CLARKE, Garry
Earth and Ocean Sciences,
University of British Columbia

CLARKE, R. Allyn
Bedford Institute of Oceanography

CULLEN, John
Dalhousie University

DERKSEN, Chris
Climate Research Branch,
Meteorological Service of Canada

FERNANDES, Richard
Canada Centre for Remote Sensing,
Natural Resources Canada

FORBES, Donald L.
Natural Resources Canada,
Geological Survey of Canada

FREELAND, Howard
Department of Fisheries and Oceans

GARRETT, Chris
University of Victoria

HARVEY, Danny
University of Toronto

ISAAC, George
Environment Canada

JAMES, Thomas
Geological Survey of Canada,
Natural Resources Canada

LEWIS, C.F. Michael
Geological Survey of Canada,
Natural Resources Canada

MACDONALD, Robie
Department of Fisheries and Oceans

MATTHEWS, H. Damon
University of Calgary and
Concordia University

MCINTYRE, Stephen
University of Toronto

MCKITRICK, Ross
University of Guelph

PELTIER, Wm. Richard
Department of Physics,
University of Toronto

SAVARD, Martine M.
Geological Survey of Canada,
Natural Resources Canada

SMITH, Sharon
Geological Survey of Canada,
Natural Resources Canada

TRISHCHENKO, Alexander P.
Canada Centre for Remote Sensing,
Natural Resources Canada

WANG, Shusen
Canada Centre for Remote Sensing,
Natural Resources Canada

WANG, Xiaolan L.
Climate Research Branch,
Meteorological Service of Canada

ZWIERS, Francis
Canadian Centre for Climate Modelling
and Analysis, Environment Canada

Chile

ACEITUNO, Patricio
Department Geophysics,
Universidad de Chile

China

CAI, Zucong
Institute of Soil Science, Chinese
Academy of Sciences

CHAN, Johnny
City University of Hong Kong

DONG, Zhaoqian
Polar Research Institute of China

GONG, Dao-Yi
College of Resources Science and
Technology, Beijing Normal University

GUO, Xueliang
Institute of Atmospheric Physics,
Chinese Academy of Sciences

LAM, Chiu-Ying
Hong Kong Observatory

REN, Guoyu
National Climate Center, China
Meteorological Administration

SHI, Guang-yu
Institute of Atmospheric Physics,
Chinese Academy of Sciences

SU, Jilan
Lab of Ocean Dynamic Processes and
Satellite Oceanography,Second Institute of
Oceanography, State Oceanic Administration

SUN, Junying
Centre for Atmosphere Watch and
Services, Chinese Academy of
Meteorological Sciences, CMA

WANG, Dongxiao
South China Sea Institute of Oceanology,
Chinese Academy of Sciences

WANG, Mingxing
Institute of Atmospheric Physics,
Chinese Academy of Sciences

XIE, Zhenghui
Institute of Atmospheric Physics,
Chinese Academy of Sciences

XU, Xiaobin
Chinese Academy of
Meteorological Sciences

YU, Rucong
China Meteorological Administration

ZHAO, Zong-Ci
National Climate Center, China
Meteorological Administration

ZHOU, Tianjun
Institute of Atmospheric Physics,
Chinese Academy of Sciences

Denmark

GLEISNER, Hans
Atmosphere Space Research
Division, Danish Met. Institute

STENDEL, Martin
Danish Meteorological Institute

Egypt

EL-SHAHAWY, Mohamed
Cairo University, Egyptian
Environmemntal Affairs Agency

Estonia

JAAGUS, Jaak
University of Tartu

Fiji

LAL, Murari
University of the South Pacific

Finland

CARTER, Timothy
Finnish Environment Institute

KORTELAINEN , Pirkko
Finnish Environment Institute

KULMALA, Markku
University of Helsinki

LAAKSONEN, Ari
University of Kuopio

MÄKIPÄÄ, Raisa
Finnish Forest Research Institute

RÄISÄNEN, Jouni
Department of Physical Sciences,
University of Helsinki

SAVOLAINEN, Ilkka
Technical Research Centre of Finland

France

BONY, Sandrine
Laboratoire de Météorologie Dynamique,
Institut Pierre Simon Laplace

BOUSQUET, Philippe
Institut Pierre Simon Laplace,
Laboratoire des Sciences du
Climat et de l'Environnement

BRACONNOT, Pascale
Pascale Braconnot Institu Pierre Simon
Laplace, Laboratoire des Sciences
du Climat et de l'Environnement

CAZENAVE, Anny
Laboratoire d'Etudes en Géophysique et
Océanographie Spatiale (LEGOS), CNES

CLERBAUX, Cathy
Centre National de Recherche Scientifique

CORTIJO, Elsa
Laboratoire des Sciences du Climat et de
l'Environnement, CNRS-CEA-UVSQ

DELECLUSE, Pascale
CEA, CNRS

DÉQUÉ, Michel
Météo-France

DUFRESNE, Jean-Louis
Laboratoire de Météorologie Dynamique,
Institut Pierre Simon Laplace

FRIEDLINGSTEIN, Pierre
Institut Pierre Simon Laplace,
Laboratoire des Sciences du
Climat et de l'Environnement

GENTHON, Christophe
Centre National de Recherche
Scientifique, Laboratoire de Glaciologie
et Géophysique de l'Environnement

GUILYARDI, Eric
Laboratoire des Sciences du
Climat et de l'Environnement

GUIOT, Joel
CEREGE, Centre National de
Recherche Scientifique

HAUGLUSTAINE, Didier
Institut Pierre Simon Laplace,
Laboratoire des Sciences du Climat et de
l'Environnement, CEA-CNRS-UVSQ

JOUSSAUME, Sylvie
Centre National de Recherche Scientifique

KANDEL, Robert
Laboratoire de Météorologie
Dynamique, Ecole Polytechnique

KHODRI, Myriam
Institut de Recherche Pour
le Developpement

LABEYRIE, Laurent
Laboratoire des Sciences du
Climat et de l'Environnement

MARTIN, Eric
Météo-France

MOISSELIN, Jean-Marc
Météo-France

PAILLARD, Didier
Laboratoire des Sciences du
Climat et de l'Environnement

PETIT, Michel
CGTI

PLANTON, Serge
Météo-France

RAMSTEIN, Gilles
Laboratoire des Sciences du
Climat et de l'Environnement

SCHULZ, Michael
Institut Pierre Simon Laplace,
Laboratoire des Sciences du Climat et de
l'Environnement, CEA-CNRS-UVSQ

SEGUIN, Bernard
INRA

TEXTOR, Christiane
Laboratoire des Sciences du
Climat et de l'Environnement

WAELBROECK, Claire
Institut Pierre Simon Laplace,
Laboratoire des Sciences du Climat
et de l'Environnement, CNRS

Germany

BANGE, Hermann W.
Leibniz Institut für Meereswissenschaften,
IFM-GEOMAR

BAUER, Eva
Potsdam Institute for Climate
Impact Research

BECK, Christoph
Global Precipitation Climatology Centre

BROVKIN, Victor
Potsdam Institute for Climate
Impact Research

CHURKINA, Galina
Max Planck Institute for Biogeochemistry

COTRIM DA CUNHA, Leticia
Max-Planck-Institut für Biogeochemie

DOTZEK, Nikolai
Deutsches Zentrum für Luft und Raumfahrt,
Institut für Physik der Atmosphäre

FEICHTER, Johann
Max Planck Institute for Meteorology

GANOPOLSKI, Andrey
Potsdam Institute for Climate
Impact Research

GIORGETTA, Marco A.
Max Planck Institute for Meteorology

GRASSL, Hartmut
Max Planck Institute for Meteorology

GREWE, Volker
Deutsches Zentrum für Luft und Raumfahrt,
Institut für Physik der Atmosphäre

GRIESER, Jürgen
Deutscher Wetterdienst, Global
Precipitatioin Climatology Centre

HARE, William
Potsdam Institute for Climate
Impact Research

HELD, Hermann
Potsdam Institute for Climate
Impact Research

HOFZUMAHAUS, Andreas
Forschungszentrum Jülich, Institut
für Chemie und Dynamik der
Geosphäre II: Troposphäre

KOPPMANN, Ralf
Institut für Chemie und Dynamik
der Geosphaere, Institut II:
Troposphaere, Forschungszentrum
Juelich, Juelich, Germany

LATIF, Mojib
Leibniz Institut für Meereswissenschaften,
IFM-GEOMAR

LAWRENCE, Mark
Max Planck Institute for Chemistry

LELIEVELD, Jos
Max Planck Institute for Chemistry

LEVERMANN, Anders
Potsdam Institute for Climate
Impact Research

LINGNER, Stephan
Europäische Akademie Bad
Neuenahr-Ahrweiler GmbH

LUCHT, Wolfgang
Potsdam Institute for Climate
Impact Research

MAROTZKE, Jochem
Max Planck Institute for Meteorology

MATA, Louis Jose
Center for Development Research,
University of Bonn

MEINSHAUSEN, Malte
Potsdam Institute for Climate
Impact Research

MICHAELOWA, Axel
Hamburg Institute of
International Economics

MÜLLER, Rolf
Research Centre Jülich

RAHMSTORF, Stefan
Potsdam Institute for Climate
Impact Research

RHEIN, Monika
Institute for Environmental
Physics, University Bremen

SAUSEN, Robert
Deutsches Zentrum für Luft und Raumfahrt,
Institut für Physik der Atmosphäre

SCHOENWIESE, Christian-D.
University Frankfurt a.M., Institute
for Atmosphere and Environment

SCHOTT, Friedrich
Leibniz Institut für Meereswissenschaften,
IFM-GEOMAR

SCHULZ, Michael
University of Bremen

SCHÜTZENMEISTER, Falk
Technische Universität Dresden,
Institut für Soziologie

STAMMER, Detlef
Institut fuer Meereskunde Zentrum
fuer Meeres und Klimaforschung
Universitaet Hamburg

TEGEN, Ina
Institute for Tropospheric Research

VÖLKER, Christoph
Alfred Wegener Institute for
Polar and Marine Research

WEFER, Gerold
University of Bremen, Research
Center Ocean Margins

WURZLER, Sabine
North Rhine-Westphalia State
Environment Agency

ZENK, Walter
Leibniz Institut für Meereswissenschaften,
IFM-GEOMAR

ZOLINA, Olga
Meteorologisches Institut
der Universität Bonn

ZORITA, Eduardo
Helmholtz Zentrum Geesthacht

Hungary

ZAGONI, Miklos
Budapest University

India

SRIKANTHAN, Ramachandran
Physical Research Laboratory

TULKENS, Philippe
The Energy and Research Institute (TERI)

Iran

RAHIMZADEH, Fatemeh
Atmospheric Science & Meteorological
Research Center (ASMERC), I.R. of Iran
Meteorological Organization (IRIMO)

Ireland

FEALY, Rowan
National University of Ireland, Maynooth

SWEENEY, John
National University of Ireland, Maynooth

Italy

ARTALE, Vincenzo
Italian National Agency for
New Technologies, Energy and
the Environment (ENEA)

BALDI, Marina
Consiglio Nazionale delle Ricerche
(CNR), Inst of Biometeorology

BERGAMASCHI, Peter
European Commission, Joint
Research Centre, Institute for
Environment and Sustainability

BRUNETTI, Michele
Istituto di Scienze dell'atmosfera
e del Clima (ISAC) Consiglio
Nazionale delle Ricerche (CNR)

CAMPOSTRINI, Pierpaolo
CORILA

COLOMBO, Tiziano
Italian Met Service

CORTI, Susanna
Istituto di Scienze dell'atmosfera
e del Clima (ISAC) Consiglio
Nazionale delle Ricerche (CNR)

DESIATO, Franco
Agenzia per la protezione dell'ambiente
e per i servizi tecnici (APAT)

DI SARRA, Alcide
Italian National Agency for
New Technologies, Energy and
the Environment (ENEA)

DRAGONI, Walter
Perugia University

ETIOPE, Giuseppe
Istituto Nazionale di Geofisica e
Vulcanologia

FACCHINI, Maria Cristina
Consiglio Nazionale delle Ricerche (CNR)

GIORGI, Filippo
Abdus Salam International Centre
for Theoretical Physics

LIONELLO, Piero
Univ. of Lecce, Dept."Scienza dei materiali"

MARIOTTI, Annarita
Italian National Agency for New
Technologies, Energy and the Environment
(ENEA) and Earth System Science
Interdisciplinary Center (ESSIC-USA)

MOSETTI, Renzo
OGS

NANNI, Teresa
Istituto di Scienze dell'atmosfera e
del Clima (ISAC) Consiglio
Nazionale delle Ricerche (CNR)

RUTI, Paolo Michele
Italian National Agency for New
Technologies, Energy and the Environment

SANTINELLI, Chiara
Consiglio Nazionale delle Ricerche (CNR)

VAN DINGENEN, Rita
European Commission, Joint
Research Centre, Institute for
Environment and Sustainability

VIGNUDELLI, Stefano
Consiglio Nazionale delle Ricerche
(CNR), Istituto di Biofisica

Japan

ALEXANDROV, Georgii
National Institute for Environmental Studies

ANNAN, James
Frontier Research Center for Global
Change, Japan Agency for Marine-
Earth Science and Technology

AOKI, Teruo
Meteorological Research Institute,
Japan Meteorological Agency

AWAJI, Toshiyuki
Kyoto University

EMORI, Seita
National Institute for Environmental
Studies and Frontier Research Center
for Global Change, Japan Agency for
Marine-Earth Science and Technology

HARGREAVES, Julia
Frontier Research Center for Global
Change, Japan Agency for Marine-
Earth Science and Technology

HAYASAKA, Tadahiro
Research Institute for Humanity and Nature

IKEDA, Motoyoshi
Hokkaido University

ITOH, Kiminori
Yokohama National University

KAWAMIYA, Michio
Frontier Research Center for Global
Change, Japan Agency for Marine-
Earth Science and Technology

KIMOTO, Masahide
Center for Climate System
Research, University of Tokyo

KITOH, Akio
First Research Laboratory, Climate Research
Department, Meteorological Research
Institute, Japan Meteorological Agency

KOBAYASHI, Shigeki
TRDL

KONDO, Hiroki
Frontier Research Center for Global
Change, Japan Agency for Marine-
Earth Science and Technology

MAKI, Takashi
Meteorological Research Institute,
Japan Meteorological Agency

MAKSYUTOV, Shamil
National Institute for Environmental
Studies

MARUYAMA, Koki
CRIEPI

MATSUNO, Taroh
Frontier Research Center for Global
Change, Japan Agency for Marine-
Earth Science and Technology

MIKAMI, Masao
Meteorological Research Institute,
Japan Meteorological Agency

MIKAMI, Takehiko
Tokyo Metropolitan University

NAKAJIMA, Teruyuki
Center for Climate System
Research, University of Tokyo

NAKAWO, Masayoshi
Research Institute for Humanity and
Nature

NODA, Akira
Meteorological Research Institute,
Japan Meteorological Agency

OHATO, Tetsuo
JAMSTEC

ONO, Tsuneo
Hokkaido National Fisheries Research
Institute, Fisheries Research Agency

SASAKI, Hidetaka
Meteorological Research Institute,
Japan Meteorological Agency

SATO, Yasuo
Meteorological Research Institute,
Japan Meteorological Agency

SEKIYA, Akira
National Institute of Advanced Industrial
Science and Technology (AIST)

SHINODA, Masato
Tottori University, Arid Land
Research Center

SUGA, Toshio
Tohoku University

SUGI, Masato
Meteorological Research Institute,
Japan Meteorological Agency

TOKIOKA, Tatsushi
Frontier Research Center for Global
Change, Japan Agency for Marine-
Earth Science and Technology

TOKUHASHI, Kazuaki
National Institute of Advanced Industrial
Science and Technology (AIST)

TSUSHIMA, Yoko
Japan Agency for Marine-Earth
Science and Technology

UCHIYAMA, Akihiro
Meteorological Research Institute,
Japan Meteorological Agency

YAMAMOTO, Susumu
Graduate School of Environmental
Science, Okayama University

YAMANOUCHI, Takashi
National Institute of Polar Research

YAMASAKI, Masanori
Japan Agency for Marine-Earth
Science and Technology

YAMAZAKI, Koji
Graduate School of Environmental
Science, Hokkaido University

YOKOYAMA, Yusuke
Department of Earth and Planetary
Sciences, University of Tokyo

TSUTSUMI, Yukitomo
Meteorological Research Institute,
Japan Meteorological Agency

Republic of Korea

KIM, Kyung-Ryul
Seoul National University, School of
Earth and Environmental Services

Mexico

LLUCH-BELDA, Daniel
Centro Interdisciplinario de
Ciencias Marinas del IPN

Mozambique

QUEFACE, Antonio Joaquim
Physics Department, Eduardo
Mondlane University

Netherlands, Antilles and Aruba

MARTIS, Albert
Climate Research Center, Meteorological
Service Netherlands, Antilles & Aruba

Netherlands

BAEDE, Alphonsus
Royal Netherlands Meteorological
Institute (KNMI) and Ministry of Housing,
Spatial Planning and the Environment

BURGERS, Gerrit
Royal Netherlands Meteorological
Institute (KNMI)

DE BRUIN, Henk
Meteorology and Air Quality
Group, Wageningen University

DE WIT, Florens

DILLINGH, Douwe
National Institute for Coastal and
Marine Management / RIKZ

HAARSMA, Reindert
Royal Netherlands Meteorological
Institute (KNMI)

HAZELEGER, Wilco
Royal Netherlands Meteorological
Institute (KNMI)

HOLTSLAG, Albert A. M.
Wageningen University

KROON, Dick
Vrije Universiteit, Amsterdam

SIEGMUND, Peter
Royal Netherlands Meteorological
Institute (KNMI)

STERL, Andreas
Royal Netherlands Meteorological
Institute (KNMI)

VAN AKEN, Hendrik M.
Royal Netherlands Institute for
Sea Research (NIOZ)

VAN DE WAL, Roderik Sylvester Willo
Institute for Marine and Atmospheric
Research, Utrecht University

VAN DEN HURK, Bart
Royal Netherlands Meteorological
Institute (KNMI)

VAN NOIJE, Twan
Royal Netherlands Meteorological
Institute (KNMI)

VAN VELTHOVEN, Peter
Royal Netherlands Meteorological
Institute (KNMI)

VANDENBERGHE, Jef
Vrije Universiteit, Inst. of Earth Sciences

VEEFKIND, Pepijn
Royal Netherlands Meteorological
Institute (KNMI)

VELDERS, Guus J.M.
Netherlands Environmental
Assessment Agency (MNP)

New Zealand

ALLOWAY, Brent
Institute of Geological and Nuclear Sciences

BARRETT, Peter
Antarctic Research Centre, Victoria
University of Wellington

BODEKER, Greg
National Institute of Water and
Atmospheric Research

BOWEN, Melissa
National Institute of Water and
Atmospheric Research

CRAMPTON, James
Institute of Geological and Nuclear Sciences

GRAY, Vincent
Climate Consultant

LASSEY, Keith
National Institute of Water and
Atmospheric Research

LAW, Cliff
National Institute of Water and
Atmospheric Research

MACLAREN, Piers
NZ Forest Research Institute

MULLAN, A. Brett
National Institute of Water and
Atmospheric Research

NODDER, Scott
National Institute of Water and
Atmospheric Research

RENWICK, James A.
National Institute of Water and
Atmospheric Research

SALINGER, M. James
National Institute of Water and
Atmospheric Research

SHULMEISTER, James
University of Canterbury

WILLIAMS, Paul W.
Auckland University

WRATT, David
National Climate Centre, National Institute
of Water and Atmospheric Research

Norway

BENESTAD, Rasmus
Norwegian Meteorological Institute

FUGLESTVEDT, Jan
Centre for International Climate and
Environmantal Research (CICERO)

GODAL, Odd
Department of Economics,
University of Bergen

HANSSEN-BAUER, Inger
Norwegian Meteorological Institute

ISAKSEN, Ketil
Norwegian Meteorological Institute

JOHANNESSEN, Ola M.
Nansen Environmental and
Remote Sensing Center

KRISTJÁNSSON , Jón Egill
University of Oslo

NESJE, Atle
Department of Earth Science,
University of Bergen

PAASCHE, Øyvind
Bjerknes Centre for Climate Research

Peru

GAMBOA, Nadia
Pontificia Universidad Carolica del Pero

Romania

BOJARIU, Roxana
National Institute of Meteorology
and Hydrology (NIMH)

BORONEANT, Constanta-Emilia
National Meteorological Administration

BUSUIOC, Aristita
National Meteorological Administration

MARES, Constantin
Romanian Academy, Geodynamics Institute

MARES, Ileana
Romanian Academy of Technical Studies

Russian Federation

MELESHKO, Valentin
Voeykov Main Geophysical Observatory

Slovakia

LAPIN, Milan
Slovak National Climate Program

Spain

AGUILAR, Enric
Climate Change Research Group,
Universitat Rovira i Virgili de Tarragona

BLADÉ, Ileana
Department of Astronomy and
Meteorology. University of Barcelona

BRUNET, Manola
University Rovira i Virgili

CALVO COSTA , Eva
Institut de Ciències del Mar

GARCÍA-HERRERA, Ricardo
Universidad Complutense de Madrid

GONZÁLEZ-ROUCO, Jesus Fidel
Universidad Complutense de Madrid

LAVIN, Alicia M.
Instituto Espanol de Oceanografia

MARTIN-VIDE, Javier
Physical Geography of the
University of Barcelona

MONTOYA, Marisa
Dpto. Astrofisica y Fisica de la
Atmosfera, Facultad de Ciencias Fisicas,
Universidad Complutense de Madrid

PELEJERO, Carles
Institut de Ciències del Mar, CMIMA-CSIC

RIBERA, Pedro
Universidad Pablo de Olavide

Sweden

HOLMLUND, Per
Stockholm University

KJELLSTRÖM, Erik
Swedish Meteorological and
Hydrological Institute

LECK, Caroline
Department of Metorology,
Stockholm University

RUMM AINEN, Markku
Rossby Centre, Swedish Meteorological
and Hydrological Institute

Switzerland

APPENZELLER, Christof
Federal Office of Meteorology
and Climatology MeteoSwiss

BLUNIER, Thomas
Climate and Environmental
Physics, University of Bern

BRÖNNIMANN, Stefan
ETH Zürich

CASTY, Carlo
Climate and Environmental Physics

CHERUBINI, Paolo
Swiss Federal Research Institute WSL

ESPER, Jan
Swiss Federal Research Institute WSL

FREI, Christoph
Federal Office of Meteorology
and Climatology MeteoSwiss

GHOSH, Sucharita
Swiss Federal Research Institute WSL

HAEBERLI, Wilfried
Geography Department,
University of Zürich

JOOS, Fortunat
Climate and Environmental Physics,
Physics Institute, University of Bern

KNUTTI, Reto
Climate and Global Dynamics Division,
National Center for Atmospheric Research

LUTERBACHER, Jürg
Institute of Geography, Climatology
and Meteorology, and National
Centre of Competence in Research
on Climate, University of Bern

MARCOLLI, Claudia
ETH Zürich, Institute for
Atmosphere and Climate

PETER, Thomas
ETH Zürich

PHILIPONA, Rolf
Observatory Davos

PLATTNER, Gian-Kasper
Climate and Environmental Physics,
Physics Institute, University of Bern

RAIBLE, C. Christoph
Climate and Environmental
Physics, University of Bern

REBETEZ, Martine
Swiss Federal Research Institute WSL

ROSSI, Michel J.
Ecole Polytechnique Fédérale de
Lausanne, Laboratoire de Pollution
Atmosphérique et Sol

ROZANOV, Eugene
IAC ETHZ and PMOD/WRC

SCHÄR, Christoph
ETH Zürich, Institute for Atmospheric
and Climate Science

SIDDALL, Mark
Climate and Environmental
Physics, University of Bern

SPAHNI, Renato
Climate and Environmental Physics,
Physics Institute, University of Bern

STAEHELIN, Johannes
ETH Zürich

STOCKER, Thomas F.
Climate and Environmental Physics,
Physics Institute, University of Bern

WANNER, Heinz
National Centre of Competence in
Research on Climate, University of Bern

WILD, Martin
ETH Zürich, Institute for Atmospheric
and Climate Sciencce

Thailand

GARIVAIT, Savitri
The Joint Graduate School of Energy
and Environment, King Mongkut's
University of Technology Thonburi

LIMMEECHOKCHAI, Bundit
Sirindhorn International Institute of
Technology, Thammasat Univ.

Togo

AJAVON, Ayite-Lo N.
Atmospheric Chemistry Laboratory

UK

ALEXANDER, Lisa
Hadley Centre for Climate Prediction
and Research, Met Office

ALLAN, Richard
Environmental Systems Science
Centre, University of Reading

BANKS, Helene
Hadley Centre for Climate Prediction
and Research, Met Office

BETTS, Richard A.
Hadley Centre for Climate Prediction
and Research, Met Office

BODAS-SALCEDO, Alejandro
Hadley Centre for Climate Prediction
and Research, Met Office

BOUCHER, Olivier
Hadley Centre for Climate Prediction
and Research, Met Office

BROWN, Simon
Hadley Centre for Climate Prediction
and Research, Met Office

BRYDEN, Harry
University of Southampton

CAESAR, John
Hadley Centre for Climate Prediction
and Research, Met Office

CARSLAW, Kenneth
University of Leeds

COLLINS, Matthew
Hadley Centre for Climate Prediction
and Research, Met Office

COLLINS, William
Hadley Centre for Climate Prediction
and Research, Met Office

CONNOLLEY, William
British Antarctic Survey

COURTNEY, Richard S.
European Science and Environment Forum

CRUCIFIX, Michel
Hadley Centre for Climate Prediction
and Research, Met Office

FALLOON, Pete
Hadley Centre for Climate Prediction
and Research, Met Office

FOLLAND, Christopher
Hadley Centre for Climate Prediction
and Research, Met Office

FORSTER, Piers
School of Earth and Environment,
University of Leeds

FOWLER, Hayley
Newcastle University

GEDNEY, Nicola
Hadley Centre for Climate Prediction
and Research, Met Office

GILLETT, Nathan P.
Climatic Research Unit, School
of Environmental Sciences,
University of East Anglia

GRAY, Lesley
Reading University

GREGORY, Jonathan M.
Department of Meteorology, University of
Reading and Hadley Centre for Climate
Prediction and Research, Met Office

GRIGGS, David
Hadley Centre for Climate Prediction
and Research, Met Office

HAIGH, Joanna
Imperial College London

HARANGOZO, Steve
British Antarctic Survey

HAWKINS, Stephen J.
The Marine Biological
Association of the UK

HIGHWOOD, Eleanor
University of Reading

HINDMARSH, Richard
British Antarctic Survey

HOSKINS, Brian J.
Department of Meteorology,
University of Reading

HOUSE, Joanna
Quantifying and Understanding the Earth
System Programme, University of Bristol

INGRAM, William
Hadley Centre for Climate Prediction
and Research, Met Office

JOHNS, Timothy
Hadley Centre for Climate Prediction
and Research, Met Office

JONES, Christopher
Hadley Centre for Climate Prediction
and Research, Met Office

JONES, Gareth S.
Hadley Centre for Climate Prediction
and Research, Met Office

JONES, Philip D.
Climatic Research Unit, School
of Environmental Sciences,
University of East Anglia

JOSEY, Simon
National Oceanography Centre,
University of Southampton

KING, John
British Antarctic Survey

LE QUÉRÉ, Corrine
University of East Anglia and
British Antarctic Survey

LEE, David
Manchester Metropolitan University

LOWE, Jason
Hadley Centre for Climate Prediction
and Research, Met Office

MARSH, Robert
National Oceanography Centre,
University of Southampton

MARTIN, Gill
Hadley Centre for Climate Prediction
and Research, Met Office

MCCARTHY, Mark
Hadley Centre for Climate Prediction
and Research, Met Office

MCDONALD, Ruth
Hadley Centre for Climate Prediction
and Research, Met Office

MITCHELL, John
Hadley Centre for Climate Prediction
and Research, Met Office

MURPHY, James
Hadley Centre for Climate Prediction
and Research, Met Office

NICHOLLS, Robert
School of Civil Engineering and the
Environment, University of Southampton

PARKER, David
Hadley Centre for Climate Prediction
and Research, Met Office

PRENTICE, Iain Colin
Quantifying and Understanding the Earth
System Programme, Department of
Earth Sciences, University of Bristol

RAPER, Sarah
Manchester Metropolitan University

RAYNER, Nick
Hadley Centre for Climate Prediction
and Research, Met Office

REISINGER, Andy
IPCC Synthesis Report TSU

RIDLEY, Jeff
Hadley Centre for Climate Prediction
and Research, Met Office

ROBERTS, C. Neil
University of Plymouth,
School of Geography

RODGER, Alan
British Antarctic Survey

ROSCOE, Howard
British Antarctic Survey

ROUGIER, Jonathan
Durham University

ROWELL, Dave
Hadley Centre for Climate Prediction
and Research, Met Office

SENIOR, Catherine
Hadley Centre for Climate Prediction
and Research, Met Office

SEXTON, David
Hadley Centre for Climate Prediction
and Research, Met Office

SHINE, Keith
University of Reading

SLINGO, Julia
National Centre for Atmospheric
Science, University of Reading

SMITH, Leonard A.
London School of Economics

SROKOSZ, Meric
National Oceanography Centre

STARK, Sheila
Hadley Centre for Climate Prediction
and Research, Met Office

STEPHENSON, David
Department of Meteorology,
University of Reading
STONE, Daíthí A.
University of Oxford

STOTT, Peter A.
Hadley Centre for Climate Prediction
and Research, Met Office

THORNE, Peter
Hadley Centre for Climate Prediction
and Research, Met Office

TSIMPLIS, Michael
National Oceanography Centre,
University of Southampton

TURNER, John
British Antarctic Survey

VAUGHAN, David
British Antarctic Survey

VELLINGA, Michael
Hadley Centre for Climate Prediction
and Research, Met Office

WASDELL, David
Meridian Programme

WILLIAMS, Keith
Hadley Centre for Climate Prediction
and Research, Met Office

WOLFF, Eric
British Antarctic Survey

WOOD, Richard A.
Hadley Centre for Climate Prediction
and Research, Met Office

WOODWORTH, Philip
Proudman Oceanographic Laboratory

WU, Peili
Hadley Centre for Climate Prediction
and Research, Met Office

Uruguay

BIDEGAIN, Mario
Universidad de la Republica

USA

ALEXANDER, Becky
University of Washington

ALEXANDER, Michael
National Oceanic and Atmospheric
Administration, Climate Diagnostics
Brach, Pysical Science Division,
Earth System Research Lab

ALLEY, Richard B.
Department of Geosciences,
Pennsylvania State University

ANDERSON, David M.
National Center for Atmospheric
Research, Paleoclimatology

ANDERSON, Theodore
University of Washington

ANDERSON, Wilmer
University of Wisconsin, Madison,
Physics Department

ANTHES, Richard
University Corporation for
Atmospheric Research

ARRITT, Raymond
Iowa State University

AVERYT, Kristen
IPCC WGI TSU, National Oceanic
and Atmospheric Administration,
Earth System Research Laboratory

BAER, Paul
Stanford University, Center for
Environmental Science and Policy

BAKER, Marcia
University of Washington

BARRY, Roger
National Snow and Ice Data
Center, University of Colorado

BATES, Timothy
National Oceanic and
Atmospheric Administration

BAUGHCUM, Steven
Boeing Company

BENTLEY, Charles R.
University of Wisconsin, Madison

BERNSTEIN, Lenny
International Petroleum Industry
Envirional Conservation Association
& L.S. Bernstein & Associates, LLC

BOND, Tami
University of Illinois at Urbana-Champaign

BROCCOLI, Anthony J.
Rutgers University

BROMWICH, David
Byrd Polar Research Center,
The Ohio State University

BROOKS, Harold
National Oceanic and Atmospheric
Administration, National Severe
Storms Laboratory

BRYAN, Frank
National Center for Atmospheric Research

CAMERON-SMITH, Philip
Lawrence Livermore National Laboratory

CHIN, Mian
National Aeronautics and Space
Administration, Goddard
Space Flight Center

CHRISTY, John
University of Alabama in Huntsville

CLEMENS, Steven
Brown University

COFFEY, Michael
National Center for Atmospheric Research

COLLINS, William D.
Climate and Global Dynamics Division,
National Center for Atmospheric Research

CROWLEY, Thomas
Duke University

CUNNOLD, Derek
School of Earth and Atmospheric Sciences,
Georgia Institute of Technology

DAI, Aiguo
National Center for Atmospheric Research

DANIEL, John S.
National Oceanic and Atmospheric
Administration, Earth System
Research Laboratory

DANILIN, Mikhail
The Boeing Company

D'ARRIGO, Rosanne
Lamont Doherty Earth Observatory

DAVIES, Roger
Jet Propulsion Laboratory, California
Institute of Technology

DEL GENIO, Anthony
National Aeronautics and Space
Administration, Goddard
Institute for Space Studies

DIAZ, Henry
National Oceanic and Atmospheric
Administration, Climate Diagnostics
Brach, Pysical Science Division,
Earth System Research Lab

DICKINSON, Robert E.
School of Earth and Atmospheric Sciences,
Georgia Institute of Technology

DIXON, Keith
National Oceanic and
Atmospheric Administration

DONNER, Leo
Geophysical Fluid Dynamics
Laboratory, National Oceanic and
Atmospheric Administration

DOUGLAS, Bruce
International Hurricane Research Center

DOUGLASS, Anne
National Aeronautics and Space
Administration, Goddard
Space Flight Center

DUTTON, Ellsworth
National Oceanic and Atmospheric
Administration, Earth System Research
Laboratory, Global Monitoring Division

EASTERLING, David
National Oceanic and Atmospheric
Administration, Earth System
Research Laboratory

EMANUEL, Kerry A.
Massachusetts Institute of Technology

EVANS, Wayne F.J.
North West Research Associates

FAHEY, David W.
National Oceanic and Atmospheric
Administration, Earth System
Research Laboratory

FEELY, Richard
National Oceanic and Atmospheric
Administration, Pacific Marine
Environmental Laboratory

FEINGOLD, Graham
National Oceanic and
Atmospheric Administration

FELDMAN, Howard
American Petroleum Institute

FEYNMAN, Joan
Jet Propulsion Laboratory, California
Institute of Technology

FITZPATRICK, Melanie
University of Washington

FOGT, Ryan
Polar Meteorology Group, Byrd Polar
Research Center and Atmospheric
Sciences Program, Department of
geography, The Ohio State University

FREE, Melissa
Air Resources Laboratory, National
Oceanic and Atmospheric Administration

FU, Qiang
Department of Atmospheric Sciences,
University of Washington

GALLO, Kevin
National Oceanic and Atmospheric
Administration, NESDIS

GARCIA, Hernan
National Oceanic and Atmospheric
Administration, National
Oceanographic Data Center

GASSÓ, Santiago
University of Maryland, Baltimore
County and NASA

GENT, Peter
National Center for Atmospheric Research

GERHARD, Lee C.
Thomasson Partner Associates

GHAN, Steven
Pacific Northwest National Laboratory

GNANADESIKAN, Anand
National Oceanic and Atmospheric
Administration, Geophysical
Fluid Dynamics Laboratory

GORNITZ, Vivien
National Aeronautics and Space
Administration, Goddard Institute for
Space Studies, Columbia University

GROISMAN, Pavel
University Corporation for Atmospheric
Research at the National Climatic
Data Center, National Oceanic and
Atmospheric Administration

GRUBER, Nicolas
Institute of Geophysics and Planetary
Physics, University of California,
Los Angeles and Department of
Environmental Sciences, ETH Zurich

GURWICK, Noel
Carnegie Institution of Washington,
Department of Global Ecology

HAKKARINEN, Chuck
Electric Power Research Institute, retired

HALLEGATTE, Stéphane
Centre International de Recherche sur
l'Environnement et le Developpement,
Ecole Nationale des Ponts-et-Chaussées
and Centre National de Recherches
Meteorologique, Meteo-France

HALLETT, John
Desert Research Institute

HAMILL, Patrick
San Jose State University

HARTMANN, Dennis
University of Washington

HAYHOE, Katharine
Texas Tech University

HEGERL, Gabriele
Division of Earth and Ocean Sciences,
Nicholas School for the Environment
and Earth Sciences, Duke University

HELD, Isaac
National Oceanic and Atmospheric
Administration, Geophysical
Fluid Dynamics Laboratory

HEMMING, Sidney
Lamont Doherty Earth Observatory,
Columbia University

HOULTON, Benjamin
Stanford Unviersity, Dept. of Biological
Sciences; Carnegie Institution of
Washington, Dept. of Global Ecology

HU, Aixue
National Center for Atmospheric Research

HUGHES, Dan
Hughes and Associates

ICHOKU, Charles
Science Systems & Applications,
Inc. (SSAI), NASA-GSFC

JACOB, Daniel
Department of Earth and Planetary
Sciences, Harvard University

JACOBSON, Mark
Stanford University

JIN, Menglin
Department of Atmospheric and
Oceanic Sciences, University of
Maryland, College Park

JOYCE, Terrence
Woods Hole Oceanographic Institution

KARL, Thomas R.
National Oceanic and Atmospheric
Administration, National
Climatic Data Center

KAROLY, David J.
University of Oklahoma

KAUFMAN, Yoram
National Aeronautics and Space
Administration, Goddard
Space Flight Center

KELLER, Klaus
Pennsylvania State University

KHESHGI, Haroon
ExxonMobil Research and
Engineering Company

KNUTSON, Thomas
Geophysical Fluid Dynamics
Laboratory, National Oceanic and
Atmospheric Administration

KO, Malcolm
National Aeronautics and Space
Administration, Langley Research Center

KOUTNIK, Michelle
University of Washington

KUETER, Jeffrey
Marshall Institue

LACIS, Andrew
National Aeronautics and Space
Administration, Goddard
Institute for Space Studies

LASZLO, Istvan
National Oceanic and
Atmospheric Administration

LEULIETTE, Eric
University of Colorado, Boulder

LEVY, Robert
Science Systems & Applications,
Inc. (SSAI), NASA-GSFC

LEWITT, Martin

LI, Zhanqing
University of Maryland,
Department of Atmospheric and
Oceanic Science and ESSIC

LIU, Yangang
Brookhaven National Laboratory

LOVEJOY, Edward R.
National Oceanic and
Atmospheric Administration

LUNCH, Claire
Stanford University, Carnegie
Institution of Washington

LUPO, Anthony
University of Missouri, Columbia

MACCRACKEN, Michael
Climate Institute

MAGI, Brian
University of Washington

MAHLMAN, Jerry
National Center for Atmospheric Research

MAHOWALD, Natalie
National Center for Atmospheric Research

MANN, Michael
Pennsylvania State University

MANNING, Martin
IPCC WGI TSU, National Oceanic
and Atmospheric Administration,
Earth System Research Laboratory

MARQUIS, Melinda
IPCC WGI TSU, National Oceanic
and Atmospheric Administration,
Earth System Research Laboratory

MARTIN, Scot
Harvard University

MASSIE, Steven
National Center for Atmospheric Research

MASTRANDREA, Michael
Stanford University

MATSUMOTO, Katsumi
University of Minnesota, Twin Cities

MATSUOKA, Kenichi
University of Washington

MAURICE, Lourdes
Federal Aviation Administration

MICHAELS, Patrick
University of Virginia

MILLER, Charles
Jet Propulsion Laboratory, California
Institute of Technology

MILLER, Laury
National Oceanic and Atmospheric
Administration, Lab for Satellite Altimetry

MILLER, Ron
National Aeronautics and Space
Administration, Goddard
Institute for Space Studies

MILLET, Dylan
Harvard University

MILLY, Chris
United States Geological Survey

MINNIS, Patrick
National Aeronautics and Space
Administration, Langley Research Center

MOLINARI, Robert
National Oceanic and Atmospheric
Administration, Atlantic Oceanographic
and Meteorological Laboratory

MOTE, Philip
Climate Impacts Group, Joint Institute for
the Study of the Atmosphere and Oceans
(JIASO), University of Washington

MURPHY, Daniel
National Oceanic and Atmospheric
Administration, Earth System
Research Laboratory

MUSCHELER, Raimund
Goddard Earth Sciences and Technology
Center, University of Maryland &
NASA/Goddard Space Flight Center,
Climate & Radiation Branch

NEELIN, J. David
University of California, Los Angeles

NELSON, Frederick
Department of Geography,
University of Delaware

NEREM, R. Steven
University of Colorado at Boulder

NOLIN, Anne
Oregon State University

NORRIS, Joel
Scripps Institution of Oceanography

OPPENHEIMER, Michael
Princeton University

OTTO-BLIESNER, Bette
Climate and Global Dynamics Division,
National Center for Atmospheric Research

OVERPECK, Jonathan
Institute for the Study of Planet
Earth, University of Arizona

OWENS, John
3M

PATT, Anthony
Boston University

PENNER, Joyce E.
Department of Atmospheric, Oceanic, and
Space Sciences, University of Michigan

PETERS, Halton
Carnegie Institution of Washington,
Department of Global Ecology

PRINN, Ronald
Department of Earth, Atmospheric
and Planetary Sciences, Massachusetts
Institute of Technology

PROFETA, Timothy H.
Nicholas Institute of Environmental
Policy Solutions, D e University

RAMANATHAN, Veerabhadran
Scripps Institution of Oceanography

RAMASWAMY, Venkatachalam
National Oceanic and Atmospheric
Administration, Geophysical
Fluid Dynamics Laboratory

RANDERSON, James
University of California, Irvine

RAVISHANKARA, A. R.
National Oceanic and
Atmospheric Administration

RIGNOT, Eric
Jet Propulsion Laboratory

RIND, David
National Aeronautics and Space
Administration, Goddard
Institute for Space Studies

RITSON, David
Stanford University

ROBOCK, Alan
Rutgers University

RUSSO, Felicita
UMBC/JCET

SABINE, Christopher
National Oceanic and Atmospheric
Administration, Pacific Marine
Environmental Laboratory

SCHIMEL, David
National Center for Atmospheric Research

SCHMIDT, Gavin
National Aeronautics and Space
Administration, Goddard
Institute for Space Studies

SCHWARTZ, Stephen E.
Brookhaven National Laboratory

SCHWING, Franklin
National Oceanic and Atmospheric
Administration Fisheries
Service, SWFSC/ERD

SEIDEL, Dian
National Oceanic and Atmospheric
Administration, Air Resources Laboratory

SEINFELD, John
California Institute of Technology

SETH, Anji
University of Connecticut,
Department of Geography

SEVERINGHAUS, Jeffrey
Scripps Institution of Oceanography,
University of California, San Diego

SHERWOOD, Steven
Yale University

SHINDELL, Drew
National Aeronautics and Space
Administration, Goddard
Institute for Space Studies

SHUKLA, Jagadish
Center for Ocean-Land-Atmosphere
Studies, George Mason University

SIEVERING, Herman
University of Colorado
- Boulder and Denver

SODEN, Brian
University of Miami, Rosentiel School
for Marine and Atmospheric Science

SOLOMON, Susan
Co-Chair, IPCC WGI, National Oceanic
and Atmospheric Administration,
Earth System Research Laboratory

SOULEN, Richard

STEFFAN, Konrad
University of Colorado

STEIG, Eric
University of Washington

STEVENS, Bjorn
UCLA Department of Atmospheric
& Oceanic Sciences

STONE, Peter
Massachusetts Institute of Technology

STOUFFER, Ronald J.
National Oceanic and Atmospheric
Administration, Geophysical
Fluid Dynamics Laboratory

TAKLE, Eugene
Iowa State University

TAMISIEA, Mark
Harvard-Smithsonian Center
for Astrophysics

TERRY, Joyce
Woods Hole Oceanographic Institution

THOMPSON, Anne
Pennsylvania State University,
Department of Meteorology

THOMPSON, David
Department of Atmospheric Science,
Colorado State University

THOMPSON, LuAnne
University of Washington

THOMPSON, Robert
United States Geological Survey

TRENBERTH, Kevin E
Climate Analysis Section, National
Center for Atmospheric Research

VINNIKOV, Konstantin
University of Maryland

VONDER HAAR, Thomas
Colorado State University

WAITZ, Ian
Massachusetts Institute of Technology

WANG, James S.
Environmental Defense

WEBB, Robert
National Oceanic and Atmospheric
Administration, Earth System
Research Laboratory

WEISS, Ray
Scripps Institution of Oceanography,
University of California, San Diego

WELTON, Ellsworth
National Aeronautics and Space
Administration, Goddard
Space Flight Center

WIELICKI, Bruce
National Aeronautics and Space
Administration, Langley Research Center

WILES, Gregory
The College of Wooster

WINTON, Michael
Geophysical Fluid Dynamics
Laboratory, National Oceanic and
Atmospheric Administration

WOODHOUSE, Connie
National Climatic Data Center

YU, Hongbin
National Aeronautics and Space
Administration, Goddard
Space Flight Center

YU, Jin-Yi
University of California, Irvine

ZENDER, Charles
University of California, Irvine

ZHAO, Xuepeng
ESSIC/UMCP & National Oceanic
and Atmospheric Administration

International Organizations

PALMER, Timothy
European Centre for Medium-
Range Weather Forecasting

RIXEN, Michel
University of Liege and NATO
Undersea Research Center

SIMMONS, Adrian
European Centre for Medium-
Range Weather Forecasts

Annex IV

Acronyms & Regional Abbreviations

Acronyms

μmol	micromole
20C3M	20th Century Climate in Coupled Models
AABW	Antarctic Bottom Water
AAIW	Antarctic Intermediate Water
AAO	Antarctic Oscillation
AATSR	Advanced Along Track Scanning Radiometer
ACC	Antarctic Circumpolar Current
ACCENT	Atmospheric Composition Change: a European Network
ACE	Accumulated Cyclone Energy or Aerosol Characterization Experiment
ACRIM	Active Cavity Radiometer Irradiance Monitor
ACRIMSAT	Active Cavity Radiometer Irradiance Monitor Satellite
ACW	Antarctic circumpolar wave
ADEC	Aeolian Dust Experiment on Climate
ADNET	Asian Dust Network
AeroCom	Aerosol Model Intercomparison
AERONET	Aerosol RObotic NETwork
AGAGE	Advanced Global Atmospheric Gases Experiment
AGCM	Atmospheric General Circulation Model
AGWP	Absolute Global Warming Potential
AIACC	Assessments of Impacts and Adaptations to Climate Change in Multiple Regions and Sectors
AIC	aviation-induced cloudiness
ALAS	Autonomous LAgrangian Current Explorer
ALE	Atmospheric Lifetime Experiment
AMIP	Atmospheric Model Intercomparison Project
AMO	Atlantic Multi-decadal Oscillation
AMSU	Advanced Microwave Sounding Unit
AO	Arctic Oscillation
AOGCM	Atmosphere-Ocean General Circulation Model
APEX	Atmospheric Particulate Environment Change Studies
AR4	Fourth Assessment Report
ARM	Atmospheric Radiation Measurement

ASOS	Automated Surface Observation Systems
ASTEX	Atlantic Stratocumulus Transition Experiment
ATCM	Atmospheric Transport and Chemical Model
ATSR	Along Track Scanning Radiometer
AVHRR	Advanced Very High Resolution Radiometer
BATS	Bermuda Atlantic Time-series Study
BC	black carbon
BCC	Beijing Climate Center
BCCR	Bjerknes Centre for Climate Research
BIOME 6000	Global Palaeovegetation Mapping project
BMRC	Bureau of Meteorology Research Centre
C^{4}MIP	Coupled Carbon Cycle Climate Model Intercomparison Project
CaCO$_3$	calcium carbonate
CAMS	Climate Anomaly Monitoring System (NOAA)
CAPE	Convective Available Potential Energy
CCl$_4$	carbon tetrachloride
CCM	Chemistry-Climate Model
CCCma	Canadian Centre for Climate Modelling and Analysis
CCN	cloud condensation nuclei
CCSR	Centre for Climate System Research
CDIAC	Carbon Dioxide Information Analysis Center
CDW	Circumpolar Deep Water
CERES	Clouds and the Earth's Radiant Energy System
CERFACS	Centre Europeen de Recherche et de Formation Avancee en Calcul Scientific
CF$_4$	perfluoromethane
CFC	chlorofluorocarbon
CFCl$_3$	CFC-11
CH$_2$I$_2$	di-iodomethane (methylene iodide)
CH$_2$O	formaldehyde
CH$_3$CCl$_3$	methyl chloroform
CH$_3$COOH	acetic acid
CH$_4$	methane

CLAMS	Chesapeake Lighthouse and Aircraft Measurements for Satellites		**DTR**	diurnal temperature range
CLARIS	Europe-South America Network for Climate Change Assessment and Impact Studies		**DU**	Dobson unit
CLIMAP	Climate: Long-range Investigation, Mapping, and Prediction		**EARLINET**	European Aerosol Research Lidar Network
CLIVAR	Climate Variability and Predictability Programme		**EBM**	Energy Balance Model
CMAP	CPC Merged Analysis of Precipitation		**ECMWF**	European Centre for Medium Range Weather Forecasts
CMDL	Climate Monitoring and Diagnostics Laboratory (NOAA)		**ECS**	equilibrium climate sensitivity
CMIP	Coupled Model Intercomparison Project		**EDGAR**	Emission Database for Global Atmospheric Research
CNRM	Centre National de Recherches Météorologiques		**EMIC**	Earth System Model of Intermediate Complexity
CO	carbon monoxide		**ENSO**	El Niño-Southern Oscillation
CO$_2$	carbon dioxide		**EOF**	Empirical Orthogonal Function
CO$_3^{2-}$	carbonate		**EOS**	Earth Observing System
COADS	Comprehensive Ocean-Atmosphere Data Set		**EPICA**	European Programme for Ice Coring in Antarctica
COARE	Coupled Ocean-Atmosphere Response Experiment		**ERA-15**	ECMWF 15-year reanalysis
COBE-SST	Centennial in-situ Observation-Based Estimates of SSTs		**ERA-40**	ECMWF 40-year reanalysis
COWL	Cold Ocean-Warm Land		**ERBE**	Earth Radiation Budget Experiment
CPC	Climate Prediction Center (NOAA)		**ERBS**	Earth Radiation Budget Satellite
CREAS	Regional Climate Change Scenarios for South America		**ERS**	European Remote Sensing satellite
CRIEPI	Central Research Institute of Electric Power Industry		**ESRL**	Earth System Research Library (NOAA)
CRUTEM2v	CRU/Hadley Centre gridded land-surface air temperature version 2v		**ESTOC**	European Station for Time-series in the Ocean
CRUTEM3	CRU/Hadley Centre gridded land-surface air temperature version 3		**EUROCS**	EUROpean Cloud Systems
CSIRO	Commonwealth Scientific and Industrial Research Organization		**FACE**	Free Air CO$_2$ Enrichment
CTM	Chemical Transport Model		**FAO**	Food and Agriculture Organization (UN)
DEMETER	Development of a European Multimodel Ensemble System for Seasonal to Interannual Prediction		**FAR**	First Assessment Report
			FRCGC	Frontier Research Center for Global Change
DIC	dissolved inorganic carbon		**FRSGC**	Frontier Research System for Global Change
DJF	December, January, February		**GAGE**	Global Atmospheric Gases Experiment
DLR	Deutsches Zentrum für Luft- und Raumfahrt		**GARP**	Global Atmospheric Research Program
DMS	dimethyl sulphide		**GATE**	GARP Atlantic Tropical Experiment
D-O	Dansgaard-Oeschger		**GAW**	Global Atmosphere Watch
DOC	dissolved organic carbon		**GCM**	General Circulation Model
DORIS	Determination d'Orbite et Radiopositionnement Intégrés par Satellite		**GCOS**	Global Climate Observing System
			GCSS	GEWEX Cloud System Study
DSOW	Denmark Strait Overflow Water		**GEIA**	Global Emissions Inventory Activity
			GEOS	Goddard Earth Observing System
DSP	Dynamical Seasonal Prediction		**GEWEX**	Global Energy and Water Cycle Experiment
			GFDL	Geophysical Fluid Dynamics Laboratory

GHCN	Global Historical Climatology Network	HCO_3^-	bicarbonate
GHG	greenhouse gas	HFC	hydrofluorocarbon
GIA	glacial isostatic adjustment	HIRS	High Resolution Infrared Radiation Sounder
GIN Sea	Greenland-Iceland-Norwegian Sea	HLM	High Latitude Mode
GISP2	Greenland Ice Sheet Project 2	HNO_3	nitric acid
GISS	Goddard Institute for Space Studies	HO_2	hydroperoxyl radical
GLACE	Global Land Atmosphere Coupling Experiment	HONO	nitrous acid
GLAMAP	Glacial Ocean Mapping	HOT	Hawaii Ocean Time-Series
GLAS	Geoscience Laser Altimeter System	hPa	hectopascal
GLODAP	Global Ocean Data Analysis Project	HYDE	HistorY Database of the Environment
GLOSS	Global Sea Level Observing System	IABP	International Arctic Buoy Programme
GMD	Global Monitoring Division (NOAA)	ICESat	Ice, Cloud and land Elevation Satellite
GOME	Global Ozone Monitoring Experiment	ICOADS	International Comprehensive Ocean-Atmosphere Data Set
GPCC	Global Precipitation Climatology Centre	ICSTM	Imperial College of Science, Technology and Medicine
GPCP	Global Precipitation Climatology Project	IGBP	International Geosphere-Biosphere Programme
GPS	Global Positioning System		
GRACE	Gravity Recovery and Climate Experiment	IGBP-DIS	IGBP Data and Information System
GRIP	Greenland Ice Core Project	IGRA	Integrated Global Radiosonde Archive
GSA	Great Salinity Anomaly	IMO	International Meteorological Organization
Gt	gigatonne (10^9 tonnes)	INDOEX	Indian Ocean Experiment
GWE	Global Weather Experiment	InSAR	Interferometric Synthetic Aperture Radar
GWP	Global Warming Potential	IO	iodine monoxide
H_2	molecular hydrogen	IOCI	Indian Ocean Climate Initiative
HadAT	Hadley Centre Atmospheric Temperature data set	IOD	Indian Ocean Dipole
HadAT2	Hadley Centre Atmospheric Temperature data set Version 2	IOZM	Indian Ocean Zonal Mode
		IPAB	International Programme for Antarctic Buoys
HadCRUT2v	Hadley Centre/CRU gridded surface temperature data set version 2v	IPO	Inter-decadal Pacific Oscillation
HadCRUT3	Hadley Centre/CRU gridded surface temperature data set version 3	IPSL	Institut Pierre Simon Laplace
		IS92	IPCC Scenarios 1992
HadISST	Hadley Centre Sea Ice and Sea Surface Temperature data set	ISCCP	International Satellite Cloud Climatology Project
HadMAT	Hadley Centre Marine Air Temperature data set	ITCZ	Inter-Tropical Convergence Zone
HadRT	Hadley Centre Radiosonde Temperature data set	JAMSTEC	Japan Marine Science and Technology Center
		JJA	June, July, August
HadRT2	Hadley Centre Radiosonde Temperature data set	JMA	Japan Meteorological Agency
HadSLP2	Hadley Centre MSLP data set version 2	ka	thousand years ago
HadSST2	Hadley Centre SST data set version 2	KMA	Korea Meteorological Administration
HALOE	Halogen Occultation Experiment	KNMI	Royal Netherlands Meteorological Institute
HCFC	hydrochlorofluorocarbon	kyr	thousand years

LASG	National Key Laboratory of Numerical Modeling for Atmospheric Sciences and Geophysical Fluid Dynamics		**MODIS**	Moderate Resolution Imaging Spectrometer
LBA	Large-Scale Biosphere-Atmosphere Experiment in Amazonia		**mol**	mole
LBC	lateral boundary condition		**MONEX**	Monsoon Experiment
LBL	line-by-line		**MOPITT**	Measurements of Pollution in the Troposphere
LGM	Last Glacial Maximum		**MOZAIC**	Measurement of Ozone by Airbus In-service Aircraft
LIG	Last Interglacial		**MPI**	Max Planck Institute
LKS	Lanzante-Klein-Seidel		**MPIC**	Max Planck Institute for Chemistry
LLGHG	long-lived greenhouse gas		**MPLNET**	Micro-Pulse Lidar Network
LLJ	Low-Level Jet		**MRI**	Meteorological Research Institute of JMA
LLNL	Lawrence Livermore National Laboratory		**MSLP**	mean sea level pressure
LMD	Laboratoire de Météorologie Dynamique		**MSU**	Microwave Sounding Unit
LOA	Laboratoire d'Optique Atmospherique		**Myr**	million years
LOSU	level of scientific understanding		**N_2**	molecular nitrogen
LSCE	Laboratoire des Sciences du Climat et de l'Environnement		**N_2O**	nitrous oxide
LSM	land surface model		**N_2O_5**	dinitrogen pentoxide
LSW	Labrador Sea Water		**NADW**	North Atlantic Deep Water
LW	longwave		**NAH**	North Atlantic subtropical high
LWP	liquid water path		**NAM**	Northern Annular Mode
Ma	million years ago		**NAMS**	North American Monsoon System
MAM	March, April, May		**NAO**	North Atlantic Oscillation
MARGO	Multiproxy Approach for the Reconstruction of the Glacial Ocean surface		**NARCCAP**	North American Regional Climate Change Assessment Program
mb	millibar		**NASA**	National Aeronautics and Space Administration
MDI	Michelson Doppler Imager		**NCAR**	National Center for Atmospheric Research
Meteosat	European geostationary meteorological satellite		**NCDC**	National Climatic Data Center
MFR	Maximum Feasible Reduction		**NCEP**	National Centers for Environmental Prediction
MHT	meridional heat transport		**NEAQS**	New England Air Quality Study
MINOS	Mediterranean Intensive Oxidants Study		**NEP**	net ecosystem production
MIP	Model Intercomparison Project		**NESDIS**	National Environmental Satellite, Data and Information Service
MIRAGE	Megacity Impacts on Regional and Global Environments		**NGRIP**	North Greenland Ice Core Project
MISO	Monsoon Intra-Seasonal Oscillation		**NH**	Northern Hemisphere
MISR	Multi-angle Imaging Spectro-Radiometer		**NH_3**	ammonia
MJO	Madden-Julian Oscillation		**NH_4^+**	ammonium ion
MLS	Microwave Limb Sounder		**NIES**	National Institute for Environmental Studies
MMD	Multi-Model Data set (at PCMDI)		**NIWA**	National Institute of Water and Atmospheric Research
MOC	Meridional Overturning Circulation		**NMAT**	Nighttime Marine Air Temperature

NMHC	non-methane hydrocarbon	**PMOD**	Physikalisch-Meteorologisches Observatorium Davos
NMVOC	non-methane volatile organic compound	**PNA**	Pacific-North American pattern
NO	nitric oxide	**PNNL**	Pacific Northwest National Laboratory
NO$_2$	nitrogen dioxide	**PNV**	potential natural vegetation
NO$_3$	nitrate radical	**POA**	primary organic aerosol
NOAA	National Oceanic and Atmospheric Administration	**POC**	particulate organic carbon
NO$_x$	reactive nitrogen oxides (the sum of NO and NO$_2$)	**POLDER**	Polarization and Directionality of the Earth's Reflectance
NPI	North Pacific Index	**POM**	particulate organic matter
NPIW	North Pacific Intermediate Water	**ppb**	parts per billion
NPP	net primary productivity	**ppm**	parts per million
NRA	NCEP/NCAR reanalysis	**PR**	Precipitation Radar
NVAP	NASA Water Vapor Project	**PREC/L**	Precipitation Reconstruction over Land (PREC/L)
O(^{1}D)	oxygen radical in the 1D excited state	**PROVOST**	Prediction of Climate Variations on Seasonal to Interannual Time Scales
O$_2$	molecular oxygen	**PRP**	Partial Radiative Perturbation
O$_3$	ozone	**PSA**	Pacific-South American pattern
OASIS	Ocean Atmosphere Sea Ice Soil	**PSC**	polar stratospheric cloud
OCTS	Ocean Colour and Temperature Scanner	**PSMSL**	Permanent Service for Mean Sea Level
ODS	ozone-depleting substances	**PSU**	Pennsylvania State University
OECD	Organisation for Economic Co-operation and Development	**psu**	Practical Salinity Unit
OGCM	Ocean General Circulation Model	**QBO**	Quasi-Biennial Oscillation
OH	hydroxyl radical	**RATPAC**	Radiosonde Atmospheric Temperature Products for Assessing Climate
OIO	iodine dioxide	**RCM**	Regional Climate Model
OLR	outgoing longwave radiation	**REA**	Reliability Ensemble Average
OMI	Ozone Monitoring Instrument	**REML**	restricted maximum likelihood
OPAC	Optical Parameters of Aerosols and Clouds	**RF**	radiative forcing
PCMDI	Program for Climate Model Diagnosis and Intercomparison	**RFI**	Radiative Forcing Index
pCO$_2$	partial pressure of CO$_2$	**RH**	relative humidity
PDF	probability density function	**RMS**	root-mean square
PDI	Power Dissipation Index	**RSL**	relative sea level
PDO	Pacific Decadal Oscillation	**RSS**	Remote Sensing Systems
PDSI	Palmer Drought Severity Index	**RTMIP**	Radiative-Transfer Model Intercomparison Project
PET	potential evapotranspiration	**SACZ**	South Atlantic Convergence Zone
PETM	Palaeocene-Eocene Thermal Maximum	**SAFARI**	Southern African Regional Science Initiative
PFC	perfluorocarbon	**SAGE**	Stratospheric Aerosol and Gas Experiment or Centre for Sustainability and the Global Environment
Pg	petagram (10^{15} grams)		
PMIP	Paleoclimate Modelling Intercomparison Project	**SAM**	Southern Annular Mode or Stratospheric Aerosol Measurement

SAMS	South American Monsoon System
SAMW	Subantarctic Mode Water
SAR	Second Assessment Report or Synthetic Aperture Radar
SARB	Surface and Atmosphere Radiation Budget
SARR	Space Absolute Radiometric Reference
SAT	surface air temperature
SCA	snow-covered area
SCIAMACHY	SCanning Imaging Absorption SpectroMeter for Atmospheric CHartographY
SCM	Simple Climate Model
SeaWiFs	Sea-Viewing Wide Field-of-View Sensor
SF$_6$	sulphur hexafluoride
SH	Southern Hemisphere
SIO	Scripps Institution of Oceanography
SIS	Small Island States
SLE	sea level equivalent
SLP	sea level pressure
SMB	surface mass balance
SMM	Solar Maximum Mission
SMMR	Scanning Multichannel Microwave Radiometer
SO	Southern Oscillation
SO$_2$	sulphur dioxide
SO$_4$	sulphate
SOA	secondary organic aerosol
SOHO	Solar Heliospheric Observatory
SOI	Southern Oscillation Index
SOM	soil organic matter
SON	September, October, November
SORCE	Solar Radiation and Climate Experiment
SPARC	Stratospheric Processes and their Role in Climate
SPCZ	South Pacific Convergence Zone
SPM	Summary for Policymakers
SRALT	Satellite radar altimetry
SRES	Special Report on Emission Scenarios
SSM/I	Special Sensor Microwave/Imager
SST	sea surface temperature

STARDEX	STAtistical and Regional dynamical Downscaling of EXtremes for European regions
STE	stratosphere-troposphere exchange
STMW	Subtropical Mode Water
SUNY	State University of New York
Sv	Sverdrup (10^6 m^3 s^{-1})
SW	shortwave
SWE	snow water equivalent
SWH	significant wave height
T/P	TOPEX/Poseidon
T12	HIRS channel 12
T2	MSU channel 2
T2$_{LT}$	MSU lower-troposphere channel
T3	MSU channel 3
T4	MSU channel 4
TAR	Third Assessment Report
TARFOX	Tropospheric Aerosol Radiative Forcing Experiment
TBO	Tropospheric Biennial Oscillation
TCR	transient climate response
TEAP	Technology and Economic Assessment Panel
TGBM	Tide Gauge Bench Mark
TGICA	Task Group on Data and Scenario Support for Impact and Climate Analysis (IPCC)
THC	Thermohaline Circulation
THIR	Temperature Humidity Infrared Radiometer
TIM	Total Solar Irradiance Monitor
TIROS	Television InfraRed Observation Satellite
TMI	TRMM microwave imager
TOA	top of the atmosphere
TOGA	Tropical Ocean Global Atmosphere
TOM	top of the model
TOMS	Total Ozone Mapping Spectrometer
TOPEX	TOPography EXperiment
TOVS	TIROS Operational Vertical Sounder
TransCom 3	Atmospheric Tracer Transport Model Intercomparison Project
TRMM	Tropical Rainfall Measuring Mission
TSI	total solar irradiance
UAH	University of Alabama in Huntsville

UARS	Upper Atmosphere Research Satellite	
UCDW	Upper Circumpolar Deep Water	
UCI	University of California at Irvine	
UEA	University of East Anglia	
UHI	Urban Heat Island	
UIO	University of Oslo	
UKMO	United Kingdom Meteorological Office	
ULAQ	University of L'Aquila	
UMD	University of Maryland	
UMI	University of Michigan	
UNEP	United Nations Environment Programme	
UNFCCC	United Nations Framework Convention on Climate Change	
USHCN	US Historical Climatology Network	
UTC	Coordinated Universal Time	
UTRH	upper-tropospheric relative humidity	
UV	ultraviolet	
UVic	University of Victoria	
VIRGO	Variability of Irradiance and Gravity Oscillations	
VIRS	Visible Infrared Scanner	
VOC	volatile organic compound	
VOS	Voluntary Observing Ships	
VRGCM	Variable-Resolution General Circulation Model	
W	watt	
WAIS	West Antarctic Ice Sheet	
WCRP	World Climate Research Programme	
WDCGG	World Data Centre for Greenhouse Gases	
WGI	IPCC Working Group I	
WGII	IPCC Working Group II	
WGIII	IPCC Working Group III	
WGMS	World Glacier Monitoring Service	
WMDW	Western Mediterranean Deep Water	
WMO	World Meteorological Organization	
WOCE	World Ocean Circulation Experiment	
WRE	Wigley, Richels and Edmonds (1996)	
WWR	World Weather Records	
ZIA	0°C isotherm altitude	
τ_{aer}	aerosol optical depth	

Regional Abbreviations used in Chapter 11

ALA	Alaska
AMZ	Amazonia
ANT	Antarctic
ARC	Arctic
CAM	Central America
CAR	Caribbean
CAS	Central Asia
CGI	East Canada, Greenland and Iceland
CNA	Central North America
EAF	East Africa
EAS	East Asia
ENA	Eastern North America
IND	Indian Ocean
MED	Mediterrranean Basin
NAS	Northern Asia
NAU	North Australia
NEU	Northern Europe
NPA	North Pacific Ocean
SAF	South Africa
SAH	Sahara
SAS	South Asia
SAU	South Australia
SEA	Southeast Asia
SEM	Southern Europe and Mediterranean
SPA	South Pacific Ocean
SSA	Southern South America
TIB	Tibetan Plateau
TNE	Tropical Northeast Atlantic
WAF	West Africa
WNA	Western North America

Index

Note: * indicates the term also appears in the Glossary (Annex I). Page numbers in italics denote tables, figures and boxed material; page numbers for boxed material are followed by B. Page numbers in bold indicate page spans for entire chapters.

Winning Ways

Winning Ways

for your mathematical plays

VOLUME 2: GAMES IN PARTICULAR

Elwyn R. Berlekamp
John H. Conway
Richard K. Guy

1982

ACADEMIC PRESS

London   New York
Paris   San Diego   San Francisco
São Paulo   Sydney   Tokyo   Toronto

A Subsidiary of Harcourt Brace Jovanovich, Publishers

ACADEMIC PRESS INC. (LONDON) LTD
24/28 Oval Road,
London NW1 7DX

United States Edition published by
ACADEMIC PRESS INC.
111 Fifth Avenue,
New York, New York 10003

British Library Cataloguing in Publication Data

Berlekamp, E. R.
  Winning ways, Vol. 2
  1. Mathematical recreations
  I. Title   II. Conway, J. H.   III. Guy, R. K.
  793.7'4   QA95   LCCCN 81-66678
ISBN 01-12-091102-7

Text set in 10/12pt Times, printed
in Great Britain by Page Bros (Norwich) Ltd.
Mile Cross Lane, Norwich

Elwyn Berlekamp was born in Dover, Ohio, on September 6, 1940. After spending two years as Assistant Professor at the University of California, Berkeley, and five years at the Bell Telephone Laboratories, in 1971 he became Professor of Mathematics and Electrical Engineering–Computer Science at Berkeley.

His book *Algebraic Coding Theory* received the best research paper award of the IEEE Information Theory Group. Eta Kappa Nu named him the "Outstanding Young Electrical Engineer" of 1971 in the US, and he has been President of the IEEE Information Theory Society. In 1977 he was elected to membership of the US National Academy of Engineering.

John Conway was born in Liverpool, England, on December 26, 1937. He is a Fellow of Gonville and Caius College and a former Fellow of Sidney Sussex College, Cambridge, and is Reader in Pure Mathematics at the University of Cambridge. He has held visiting professorships at several universities and has made original contributions to many branches of mathematics, notably transfinite arithmetic, the theory of knots, many-dimensional geometry and the theory of symmetry (group theory).

He has published two previous books, *Regular Algebra and Finite Machines* and *On Numbers and Games*. He has recently been made a Fellow of the Royal Society.

Richard Guy was born in Nuneaton, England, on September 30, 1916. He has taught mathematics at many levels and in many places—England, Singapore, India, Canada. Since 1965 he has been Professor of Mathematics at the University of Calgary and he is a member of the Board of Governors of the Mathematical Association of America.

He edits the Unsolved Problems section of American Mathematical Monthly; he wrote the volume on Number Theory for the series *Unsolved Problems in Intuitive Mathematics* and is preparing another on Combinatorics, Graph Theory and Game Theory. He is a keen member of the Alpine Club of Canada.

Preface

Does a book need a Preface? What more, after fifteen years of toil, do three talented authors have to add.

We can reassure the bookstore browser, "Yes, this is just the book you want!"

We can direct you, if you want to know quickly what's in the book, to the last page of this preliminary material. This in turn directs you to pages 1, 255, 427 and 695.

We can supply the reviewer, faced with the task of ploughing through nearly a thousand information-packed pages, with some pithy criticisms by indicating the horns of the polylemma the book finds itself on. It is not an encyclopedia. It is encyclopedic, but there are still too many games missing for it to claim to be complete. It is not a book on recreational mathematics because there's too much serious mathematics in it. On the other hand, for us, as for our predecessors Rouse Ball, Dudeney, Martin Gardner, Kraitchik, Sam Loyd, Lucas, Tom O'Beirne and Fred. Schuh, mathematics itself is a recreation. It is not an undergraduate text, since the exercises are not set out in an orderly fashion, with the easy ones at the beginning. They are there though, and with the hundred and sixty-three mistakes we've left in, provide plenty of opportunity for reader participation. So don't just stand back and admire it, work of art though it is. It is not a graduate text, since it's too expensive and contains far more than any graduate student can be expected to learn. But it does carry you to the frontiers of research in combinatorial game theory and the many unsolved problems will stimulate further discoveries.

We thank Patrick Browne for our title. This exercised us for quite a time. One morning, while walking to the university, John and Richard came up with "Whose game?" but realized they couldn't spell it (there are three tooze in English) so it became a one-line joke on line one of the text. There isn't room to explain all the jokes, not even the fifty-nine private ones (each of our birthdays appears more than once in the book).

Omar started as a joke, but soon materialized as Kimberley King. Louise Guy also helped with proof-reading, but her greater contribution was the hospitality which enabled the three of us to work together on several occasions. Louise also did technical typing after many drafts had been made by Karen McDermid and Betty Teare.

Our thanks for many contributions to content may be measured by the number of names in the index. To do real justice would take too much space. Here's an abridged list of helpers: Richard Austin, Clive Bach, John Beasley, Aviezri Fraenkel, David Fremlin, Solomon Golomb, Steve Grantham, Mike Guy, Dean Hickerson, Hendrick Lenstra, Richard Nowakowski, Anne Scott, David Seal, John Selfridge, Cedric Smith and Steve Tschantz.

No small part of the reason for the assured success of the book is owed to the well-informed and sympathetic guidance of Len Cegielka and the willingness of the staff of Academic Press and of Page Bros. to adapt to the idiosyncrasies of the authors, who grasped every opportunity to modify grammar, strain semantics, pervert punctuation, alter orthography, tamper with traditional typography and commit outrageous puns and inside jokes.

Thanks also the the Isaak Walton Killam Foundation for Richard's Resident Fellowship at The University of Calgary during the compilation of a critical draft, and to the National (Science & Engineering) Research Council of Canada for a grant which enabled Elwyn and John to visit him more frequently than our widely scattered habitats would normally allow.

And thank you, Simon!

University of California, Berkeley, CA 94720   *Elwyn Berlekamp*
University of Cambridge, England, CB2 1SB   *John Conway*
University of Calgary, Canada, T2N 1N4   *Richard Guy*

Contents

♣

GAMES IN CLUBS!

Chapter 15 **Chips and Strips** **457**

Chapter 16 **Dots-and-Boxes** **507**

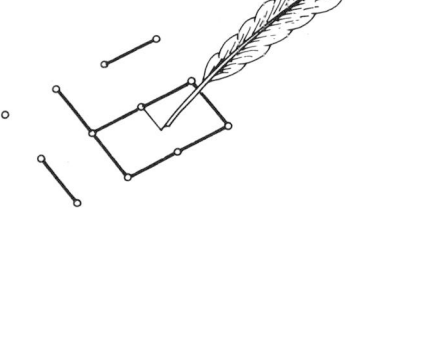

Chapter 17 Spots and Sprouts 551

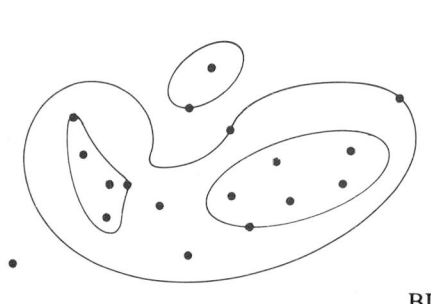

Chapter 18   **The Emperor and His Money**   575

{16}, {18}, {24}, {27}, {32}, {36}, . . . ?

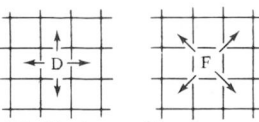

Chapter 19   **The King and the Consumer**   607

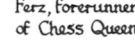

Sol Golomb's   Ferz, forerunner
Duke   of Chess Queen

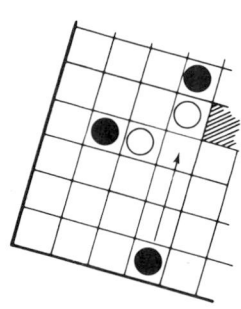

Chapter 20   Fox and Geese   635

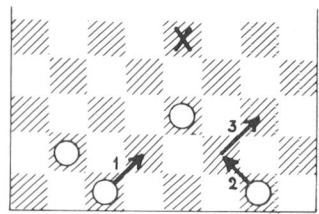

Chapter 21   Hare and Hounds   647

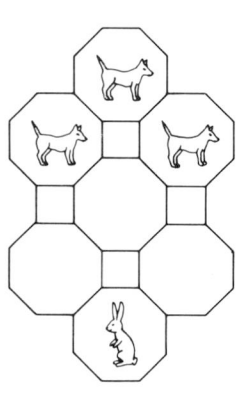

Chapter 22   Lines and Squares   667

SOLITAIRE DIAMONDS!

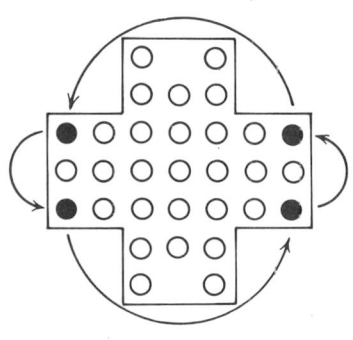

Chapter 24   Pursuing Puzzles Purposefully   735

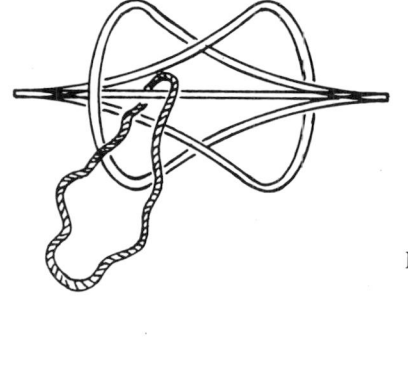

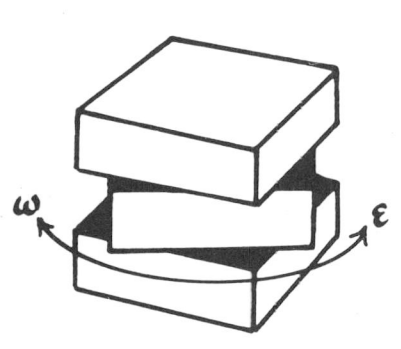

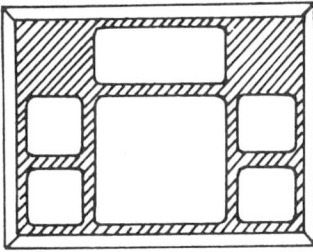

Chapter 25   **What is Life?**   **817**

 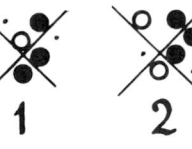 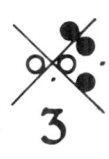

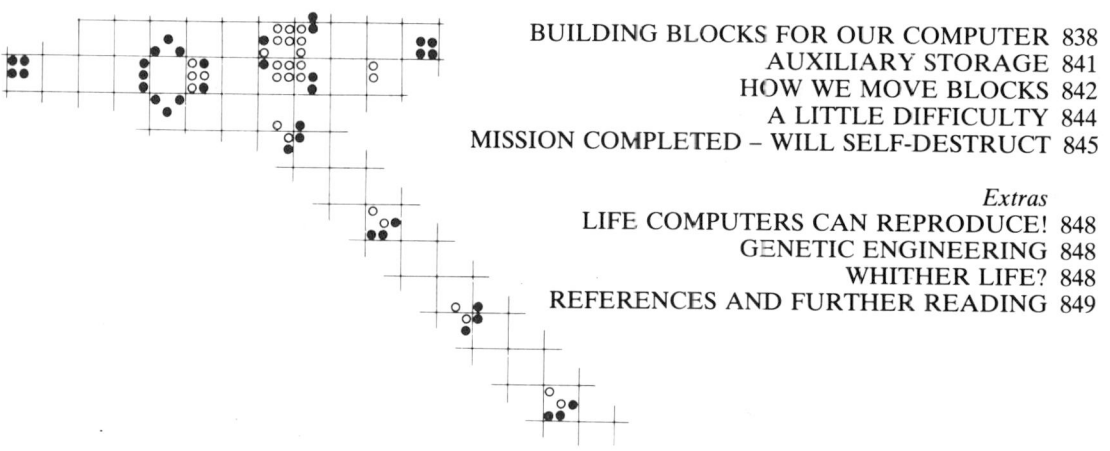

Contents of Volume 1

♠

SPADE WORK!

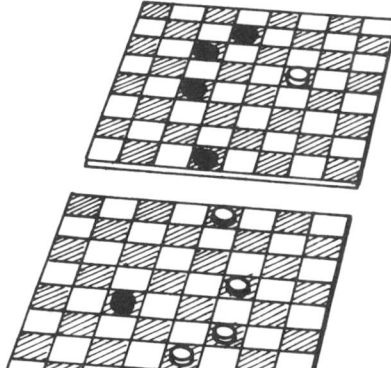

Chapter 3   Some Hard Games and How to Make Them Easier   55

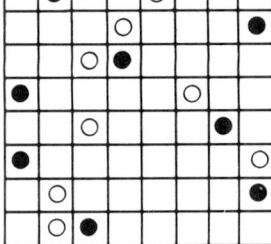

Chapter 4 **Taking and Breaking** 81

0	1	2	3
0	1	2	3
2	4	1	2	3
10	9	8	7	6

Chapter 5 **Numbers, Nimbers and Numberless Wonders** 117

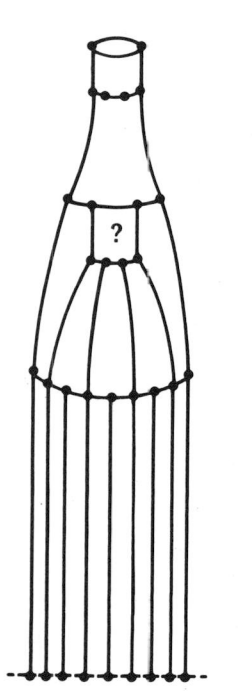

Chapter 8   It's a Small, Small, Small, Small World

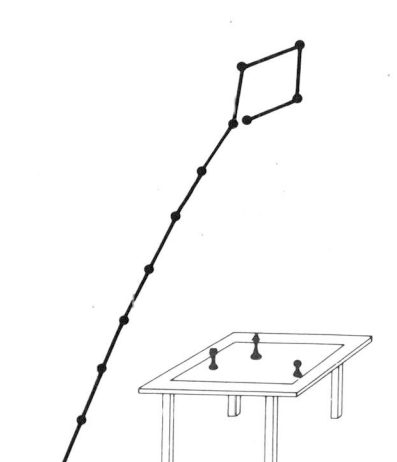

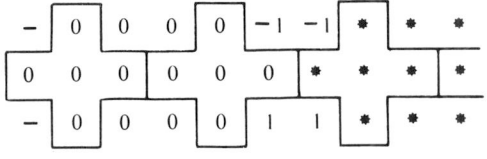

CHANGE OF HEART!

Chapter 11   Games Infinite and Indefinite   307

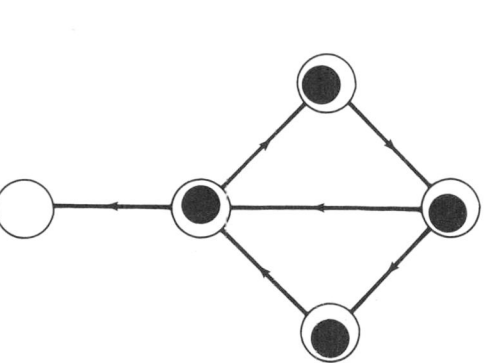

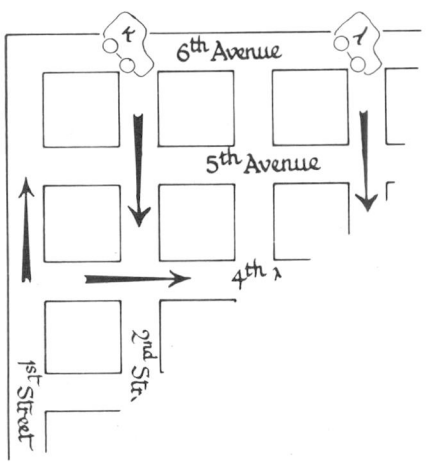

Chapter 12 **Games Eternal—Games Entailed**   **359**

Chapter 13 **Survival in the Lost World** **393**

You are
now here

If you want to know roughly what's elsewhere,
turn to the little notes about our four main themes:

There are a number of other connexions between various chapters of the book:

However, you should be able to pick any chapter and read almost all of it
without reference to anything earlier, except perhaps the basic ideas at the start of the book.

GAMES IN CLUBS!

To be an Englishman is to belong to the most exclusive
club there is.
Ogden Nash, *England Expects*.

There are lots of games for which the theories we've now developed are useful, and even more for which they're not, and we've grouped them into clubs according to how you play them.

First some games you can play with coins, either by turning them over (Chapter 14) or moving them along strips or about in heaps (Chapter 15).

Then games for which you'll need pencil and paper, perhaps to draw straight lines (Chapter 16), or curved ones (Chapter 17) or merely to do the calculations in Chapter 18.

And for board games we have three case studies in which one player wins by trapping his opponent (Chapters 19, 20, 21) and finally many more which are usually won by the first player to establish some kind of winning configuration (Chapter 22).

Chapter 14

Turn and Turn About

Because I do not hope to turn again
Because I do not hope
Because I do not hope to turn.
　　　T.S. Eliot, *Ash Wednesday*, I.

Open not thine heart to every man, lest he requite thee
with a shrewd turn.
　　　Ecclesiasticus, 8:19.

These games, based on an idea of H.W. Lenstra, are similar in that they all involve turning things over, but we shall see that they call for a variety of strategies.

TURNING TURTLES

Figure 1. Playing Turning Turtles.

429

In Fig. 1 the Walrus and the Carpenter are playing a rather cruel game. At each move a player must put one turtle on its back and may also turn over any single turtle to the left of it. This second turtle, unlike the first, may be turned either onto its feet or onto its back. The player wins who turns the last turtle upside-down. Which turtles should the Walrus (*l.*) turn?

Like most readers of this book, he wearily suspects another disguise for Nim. Here only turtles 3, 4, 6, 8 and 10 are on their feet, and since the nim-sum of 3, 4 and 6 is 1, he may turn 10 onto its back and 9 onto its feet, producing 3, 4, 6, 8, 9, a $\mathscr{P}$-position since $8 \not{+} 9 = 1$. The Carpenter (*r.*) responds by turning 8 and 5 producing the position 3, 4, 5, 6, 9 as in Fig. 2. In Nim

Figure 2. After the Carpenter's Reply.

there is only one good move from this position—reduce 9 to 4, so as to produce 3, 4, 4, 5, 6, which, since two equal Nim heaps may be cancelled, is much the same as 3, 5, 6, which the Walrus reaches by turning both 9 and 4 on their backs (Fig. 3).

Figure 3. How The Walrus Won.

Nim moves become turtle turns as follows. We reduce a heap to a size not already present by turning one turtle on its back and putting another on its feet, as in the Walrus's opening move. If a heap of the reduced size is already present we turn two turtles on their backs as in the Walrus's response to the Carpenter's move (cancelling two equal heaps). To eliminate a heap entirely we merely turn the appropriate turtle. So since 4, 6, 8, 10 is a $\mathscr{P}$-position, the Walrus could have won from Fig. 1 by just turning turtle 3.

Since all our turning games are impartial they are solved by computing the nim-values, and often may be thought of as heap games in disguise; but many games with interesting theories are more naturally suggested by the turning version.

MOCK TURTLES

Figure 4. The Mock Turtle Joins in.

Let the players turn up to three turtles subject only to the condition that the rightmost of these must be turned from his feet onto his back. We may think of this as a game with numbers in which any number may be replaced by 0, 1 or 2 smaller ones. So $\mathscr{G}(n)$ is the least number not of any form

$$0, \quad \mathscr{G}(a), \quad \mathscr{G}(a) \overset{*}{+} \mathscr{G}(b),$$

in which a and b are any numbers less than n.

If we number the positions from 0 we find the nim-values shown in Table 1.

$n =$	0	1	2	3	4	5	6	7	8	9	10	11	12	13	14	15	16	17	18	...
$\mathscr{G}(n) =$	1	2	4	7	8	11	13	14	16	19	21	22	25	26	28	31	32	35	37	...

Table 1. Nim-values for Mock Turtles.

We see that $\mathscr{G}(n)$ is always $2n$ or $2n+1$, so that its binary expansion is obtained by adjoining a digit 0 or 1 to that of n. Which shall it be?

$n =$	0	1	10	11	100	101	110	111	1000	1001	1010	...
$\mathscr{G}(n) =$	1	10	100	111	1000	1011	1101	1110	10000	10011	10101	...

Table 2. The Odious Numbers Revealed.

Table 2 suggests we choose whichever makes the total number of 1-digits *odd*.

ODIOUS AND EVIL NUMBERS

Every number is **odious** or **evil** according to the number of 1's in its binary expansion (odious for odd, evil for even). These behave under Nim addition like odd and even numbers under ordinary addition:

$$\text{EVIL} \overset{*}{+} \text{EVIL} \quad = \quad \text{EVIL} \quad = \text{ODIOUS} \overset{*}{+} \text{ODIOUS},$$

$$\text{EVIL} \overset{*}{+} \text{ODIOUS} = \text{ODIOUS} = \text{ODIOUS} \overset{*}{+} \text{EVIL}.$$

When we compute $\mathscr{G}(n)$ in Mock Turtles, the next odious number is *never* excluded, because the nim-sum of two odious numbers is evil, but smaller evil numbers always *are* excluded.

If $a_1, a_2, \ldots, a_n$ is a $\mathscr{P}$-position in Nim, so that

$$a_1 + a_2 + \ldots + a_n = 0,$$

then for the corresponding odious numbers $\mathscr{G}(a_i)$ in Mock Turtles we shall have

$$\mathscr{G}(a_1) + \mathscr{G}(a_2) + \ldots + \mathscr{G}(a_n) = 0 \text{ or } 1.$$

But if n is even, this nim-sum is evil, and so 0; while if n is odd it is odious, and so 1. The $\mathscr{P}$-positions in Mock Turtles are therefore just those $\mathscr{P}$-positions in Nim for which n is even.

Note that in Mock Turtles we number the turtles from 0. The turtle numbered 0, called the Mock Turtle, must take his turn with the rest and cannot be neglected in the conversion to Nim. To obtain a $\mathscr{P}$-position in Mock Turtles from the Turning Turtles position of Fig. 3, the Mock Turtle must be brought into the game with his four feet on the ground. In Mock Turtles, 3, 5, 6, is *not* a $\mathscr{P}$-position, but 0, 3, 5, 6 is (Fig. 4).

MOEBIUS, MOGUL AND GOLD MOIDORES

Table 3 shows the nim-values, kindly checked for us on the computer by M.J.T. Guy, for similar games in which we may turn over up to t objects for $t = 1, 2, \ldots$. Because the numbers get much larger than the other nim-values in this book, we have written them in base 8 (octal) notation. Nim-sums of octal numbers may be computed digit by digit thus:

$$
\begin{array}{c}
1\ 2\ 3\ 4\ 5\ 6\ 7\ 0 \\
1\ 3\ 5\ 7\ 0\ 2\ 4\ 6 \\
\hline
1\ 6\ 3\ 5\ 4\ 3\ 6.
\end{array}
$$

In the table we have only named the most interesting cases: $t = 3, 5, 7$ and 9. Note that C, E, G and I are the 3rd, 5th, 7th and 9th letters of the alphabet. For convenience, and to avoid cruelty to turtles, the reader may play these games with coins. The coins will show heads or tails according as the turtle is on his feet or on his back, and the rightmost coin that is turned must change from heads to tails.

THE MOCK TURTLE THEOREM

Take a $\mathscr{P}$-position in the game for an even value of t, $t = 2m$, and place an extra coin (the Mock Turtle) at the left, whichever way up will ensure an even number of heads. Positions obtained in this way will be called "good" positions for the next odd value of t, $t = 2m + 1$. We assert that the good positions are precisely the $\mathscr{P}$-positions for the game $t = 2m + 1$.

n	$t=1$	2	MOCK TURTLES 3	4	MOEBIUS 5	6	MOGUL 7	8	MOIDORES 9
THE MOCK TURTLE	1		1		1		1		1
1	1	1	2	1	2	1	2	1	2
2	1	2	4	2	4	2	4	2	4
3	1	3	7	4	10	4	10	4	10
4	1	4	10	10	20	10	20	10	20
5	1	5	13	17	37	20	40	20	40
6	1	6	15	20	40	40	100	40	100
7	1	7	16	40	100	77	177	100	200
8	1	10	20	63	147	100	200	200	400
9	1	11	23	100	200	200	400	377	777
10	1	12	25	125	253	400	1000	400	1000
11	1	13	26	152	325	707	1617	1000	2000
12	1	14	31	200	400	1000	2000	2000	4000
13	1	15	32	226	455	1331	2663	4000	10000
14	1	16	34	253	526	1552	3325	7417	17037
15	1	17	37	333	667	1664	3551	10000	20000
16	1	20	40	355	733	2000	4000	20000	40000
17	1	21	43	367	756	2353	4726	31463	63147
18	1	22	45	400	1000	2561	5343	40000	100000
19	1	23	46	427	1056	2635	5472	52525	125253
20	1	24	51	451	1123	3174	6370	65252	152525
21	1	25	52	707	1617	3216	6435	100000	200000
22	1	26	54	1000	2000	3447	7116	113152	226325
23	1	27	57	1031	2063	3722	7644	200000	400000
24	1	30	61	1055	2132	4000	10000	213630	427461
25	1	31	62	1122	2245	10000	20000	263723	547646
26	1	32	64	1203	2407	20000	40000	306136	614274
27	1	33	67	1443	3106	34007	70017	400000	1000000
28	1	34	70	1537	3277	40000	100000	416246	1034515
29	1	35	73	1746	3714	54031	130063	521055	1242133
30	1	36	75	2000	4000	64052	150125	724616	1651435
31	1	37	76	2033	4066	70064	160151	1000000	2000000
32	1	40	100	2056	4134	100000	200000	1023305	2046613
33	1	41	103	2130	4261	114053	230126	1347214	2716431
34	1	42	105	2221	4443	124061	250143	2000000	4000000
35	1	43	106	2465	5153	130035	260072	2027151	4056322
36	1	44	111	2501	5203	144074	310170	2457261	5136542
37	1	45	112	3124	6250	150016	320035	3166444	6355111
38	1	46	114	3512	7225	160047	340116	4000000	10000000
39	1	47	117	4000	10000	174022	370044	4055666	10133554
40	1	50	121	4034	10071	200000	400000	4632577	11465377
41	1	51	122	4045	10113	214301	430603	5251417	12523036
42	1	52	124	4211	10423	224502	451205	7514712	17231625
43	1	53	127	4504	11211	230604	461411	10000000	20000000

Table 3. These Nim-values are in Octal (base 8), *not* Decimal.

We show first that there is no way of changing from one good position to another by turning at most $2m+1$ coins. If there were, the number of coins turned would necessarily be even, since the good positions have evenly many heads, and so would actually be at most $2m$. But this would entail a move between two $\mathscr{P}$-positions in the $2m$ game.

It remains to show that from any bad position in the $2m+1$ game there is a move to some good position. If the position is bad because it corresponds to an $\mathscr{N}$-position in the $2m$ game, there is a move in that game to some $\mathscr{P}$-position, and, by turning the Mock Turtle if necessary, we obtain a move to a good position in the $2m+1$ game. The other bad positions correspond to $\mathscr{P}$-positions in the $2m$ game, but have an odd number of heads. In this case, by turning over the rightmost head, we obtain a position that gives an $\mathscr{N}$-position in the $2m$ game. We can now turn over at most $2m$ further coins to make this a $\mathscr{P}$-position and then, if necessary to obtain a good position, also turn the Mock Turtle. We have turned at most $2m+2$ coins in all, but since we started with an odd number of heads and finished with an even number, we have in fact turned over at most $2m+1$ coins, and so have made a legal move in the $2m+1$ game.

This result is equivalent to the statement:

> Every nim-value for the $2m+1$ game
> is an odious number,
> and the corresponding value for the $2m$ game
> is obtained by dropping the final binary digit.

THE MOCK TURTLE THEOREM

WHY MOEBIUS?

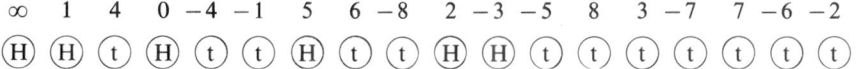

Figure 5. Moebius Labels Make $\mathscr{P}$-positions Easy to Find.

When restricted to 18 coins, the $\mathscr{P}$-positions of the game with $t=5$ possess a remarkable symmetry. To see this, name the heads of a position by the numbers shown in Fig. 5. For example, the $\mathscr{P}$-position with heads in just the first 6 places is $\infty, 0, \pm 1, \pm 4$. In this notation $\mathscr{P}$-positions remain $\mathscr{P}$-positions when their numbers are increased by any fixed amount, modulo 17, leaving ∞ unchanged. Adding 1 to the numbers $\infty, 0, \pm 1, \pm 4$ we find $\infty, 1, 2, 0, 5, -3$, so that the position displayed in Fig. 5 is another $\mathscr{P}$-position. The 15 positions shown in Table 4 yield a total of $15 \times 17 = 255$ $\mathscr{P}$-positions in this way. It is also true that a $\mathscr{P}$-position remains a $\mathscr{P}$-position if we interchange heads and tails in every place. The positions with all tails or all heads are therefore both $\mathscr{P}$-positions, giving $2 \times 255 + 2 = 512$ $\mathscr{P}$-positions in all, distributed as follows:

Number of heads	0	6	8	10	12	18
Number of $\mathscr{P}$-positions	1	102	153	153	102	1

6 heads

8 heads

$\infty, 0,$	$\pm 1,$	± 4
$\infty, 0,$	$\pm 2,$	± 8
$\pm 1,$	$\pm 3,$	± 6
$\pm 2,$	$\pm 5,$	± 6
$\pm 4,$	$\pm 5,$	∓ 7
$\pm 3,$	$\pm 7,$	± 8

$\infty, 0,$	$\pm 1,$	$\pm 5,$	± 7
$\infty, 0,$	$\pm 2,$	$\pm 3,$	± 7
$\infty, 0,$	$\pm 3,$	$\pm 4,$	± 6
$\infty, 0,$	$\pm 5,$	$\pm 6,$	± 8
$\pm 1,$	$\pm 2,$	$\pm 4,$	± 8
$\pm 1,$	$\pm 2,$	$\pm 3,$	± 5
$\pm 2,$	$\pm 4,$	$\pm 6,$	± 7
$\pm 3,$	$\pm 4,$	$\pm 5,$	± 8
$\pm 1,$	$\pm 6,$	$\pm 7,$	± 8

Table 4. The $\mathscr{P}$-positions for Moebius.

Dropping the Mock Turtle (at ∞) we find that the $\mathscr{P}$-positions for the game $t = 4$ on 17 coins are distributed:

Number of heads	0	5	6	7	8	9	10	11	12	17
Number of $\mathscr{P}$-positions	1	34	68	68	85	85	68	68	34	1.

We can also double the numbers (modulo 17) of any $\mathscr{P}$-position to give another. Thus ∞, $0, 1, 2, -3, 5$ of Fig. 5 becomes $\infty, 0, 2, 4, -6, -7$. We can invert them modulo 17; since $1/2 = -8$, $1/3 = 6$ and $1/5 = 7$, Fig. 5 inverts into $0, \infty, 1, -8, -6, 7$. In fact we can make any transformation (modulo 17)

$$ x \to \frac{ax + b}{cx + d}, \qquad ad - bc = 1. $$

Since these are known as the Möbius transformations, we have named our game after that distinguished mathematician.

MOGUL

On 24 coins the game for $t = 7$ displays even more symmetries. The $\mathscr{P}$-positions among the first 24 places are distributed as follows:

Number of heads	0	8	12	16	24
Number of $\mathscr{P}$-positions	1	759	2576	759	1.

Figure 6 enables us to find the 759 $\mathscr{P}$-positions with just 8 heads, or equally those with 8 tails. In either case the set of 8 places involved is called an **octad**. In Fig. 6 there are 35 **pictures** and each picture shows the 24 places colored in six sets of four (the 6 colors used are black, white, star, circle, plus and dot). Any two sets of 4 (any two colors) in the same picture make an octad: in particular this gives every octad with just 4 places in the last pair of (black and white) rows,

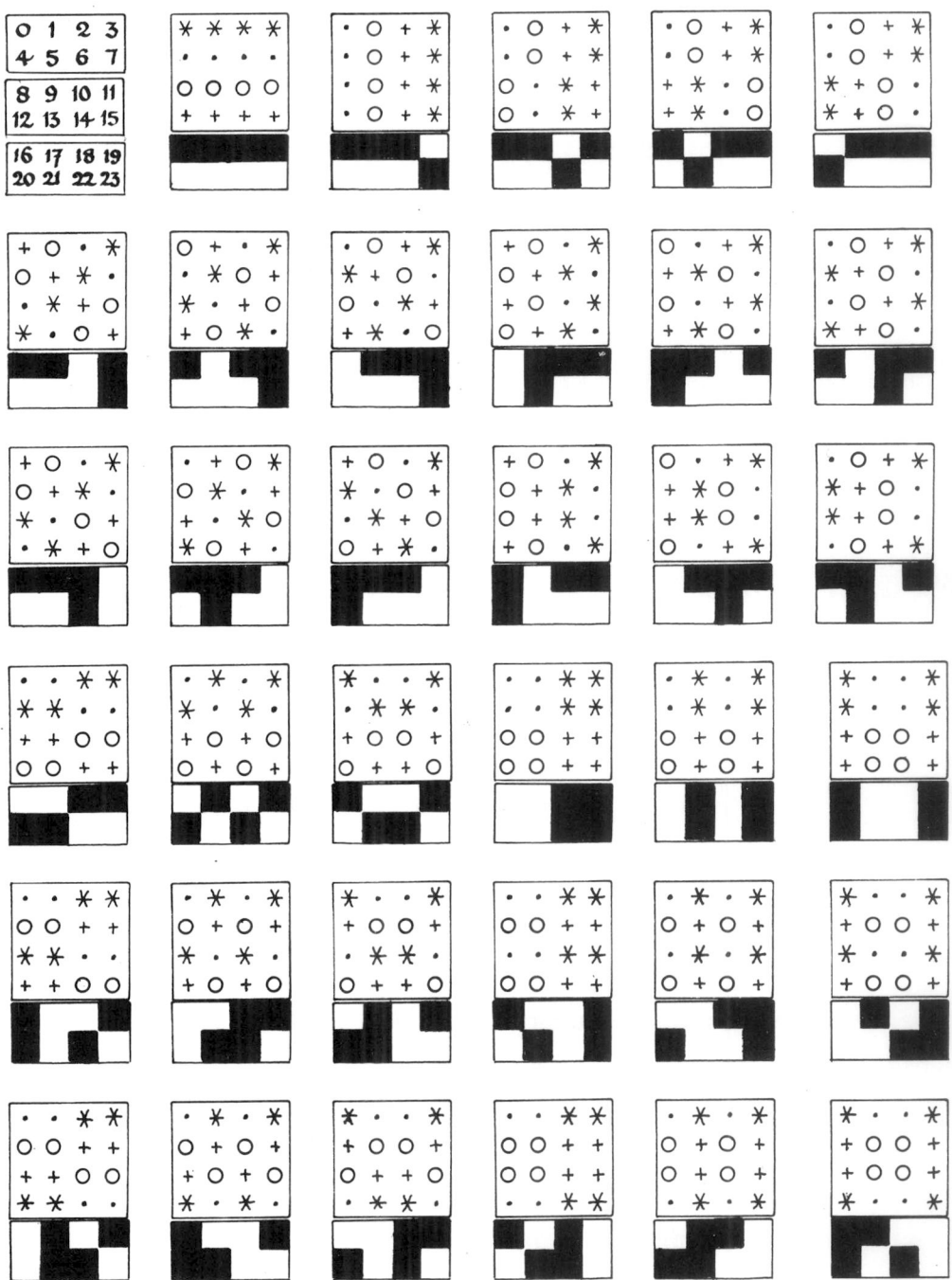

Figure 6. Curtis's Miracle Octad Generator.

and this pair of rows themselves form an octad. By interchanging this last pair of rows with the first pair, or the middle pair, of the same picture, we can now find all the octads, since it can be shown that these pairs of rows form octads and that every other octad meets at least one of them in just 4 places.

This Miracle Octad Generator, or MOG, is due to R.T. Curtis, but we have modified it slightly for the Mogul player's convenience. Various regular features of its arrangement make it easy for the practised user to locate the unique octad containing any five given places. It seems to be the case that the winner in 24-place Mogul need never play into a 12-head $\mathscr{P}$-position.

MOTLEY

This is the game in which any number of coins may be turned. When well played it lasts at most one move, since we can turn all the heads to tails instantly! The nim-values are the powers of 2:

$$1, 2, 4, 8, 16, 32, 64, 128, 256, 512, \ldots$$

so, when played with several rows, Motley is yet another disguise for Nim; the heads in a row are binary digits 1 in the number of beans in the corresponding Nim-heap.

TWINS, TRIPLETS, ETC.

We can also play the game **Twins**, in which we must turn *exactly* two coins, or **Triplets**, in which we turn exactly three, etc. The nim-value sequence for the game in which we turn exactly t coins consists of $t-1$ zeros followed by the nim-value sequence for the game in which we turn *at most* t coins. Thus the nim-values for Triplets are

$$0, 0, 1, 2, 4, 7, 8, 11, 13, 14, 16, 19, 21, 22, 25, \ldots .$$

We may think of the first $t-1$ coins as $t-1$ Mock Turtles which may be used to fill out our move to its proper complement of turns.

THE RULER GAME

If the coins we turn must be *consecutive* but are otherwise unrestricted (except that the rightmost coin must be turned from heads to tails), then the nim-values are computed by the rule:

$$\mathscr{G}(n) = \operatorname{mex} \begin{cases} 0 \\ \mathscr{G}(n-1) \\ \mathscr{G}(n-1) \overset{*}{+} \mathscr{G}(n-2) \\ \mathscr{G}(n-1) \overset{*}{+} \mathscr{G}(n-2) \overset{*}{+} \mathscr{G}(n-3) \\ \ldots\ldots\ldots\ldots\ldots\ldots\ldots\ldots \end{cases},$$

and are found to be reminiscent of Dividing Rulers (Fig. 7 of Chapter 13).

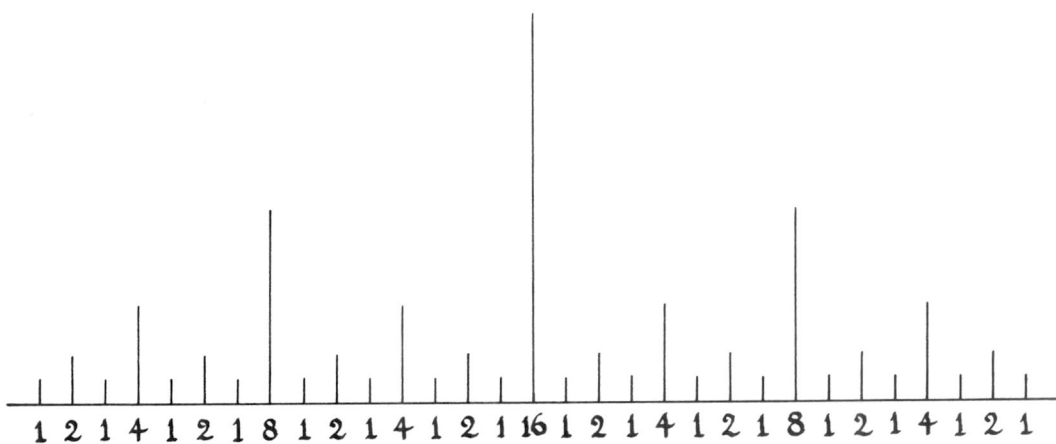

1 2 1 4 1 2 1 8 1 2 1 4 1 2 1 16 1 2 1 4 1 2 1 8 1 2 1 4 1 2 1

Figure 7. Nim-values for the Ruler Game.

If the coins are numbered starting from 1, $\mathcal{G}(n)$ is just the highest power of 2 dividing n.

CIRCUMSCRIBED GAMES

We can play any of these games under the additional restriction that the coins to be turned may not be too far apart. Thus in **Mock Turtle Fives** we may turn *up to three* of five consecutive coins. In **Triplet Fives** we turn *exactly three* out of five consecutive coins. In **Ruler Fives** we may turn 1, 2, 3, 4 or 5 consecutive coins. The nim-values for these three games are:

Mock Turtle Fives: 1 2 4 7 8 1 2 4 7 8 1 2 4 7 8 1 2 4 7 8 ...

Triplet Fives: 0 0 1 2 4 0 0 1 2 4 0 0 1 2 4 0 0 1 2 4 ...

Ruler Fives: 1 2 1 4 1 2 1 4 1 2 1 4 1 2 1 4 1 2 1 4 ....

These are parts of general patterns. Thus, **Moebius Nineteens**, for example, would have the first 19 values of Moebius repeated indefinitely. This happens for all the above games except the Ruler game; Ruler Fours, Sixes and Sevens have the same values as Ruler Fives, while Ruler Eights to Fifteens all have nim-values:

1 2 1 4 1 2 1 8 1 2 1 4 1 2 1 8 1 2 1 4 1 2 1 8 1 2 1 4 1 2 1 8 1 ....

TURNIPS (or TERNUPS)

This game has a richer theory, but it is a great pity that the full theory is only needed by people wealthy enough to play with a very large number of coins. The move is to turn over any three equally spaced coins, the rightmost going from heads to tails as usual. Numbering from 0 we find that the nim-values for 0 to 100 are:

0– 8	0	0	1	0	0	1	2	2	1		
9–17	0	0	1	0	0	1	2	2	1		
18–26	4	4	1	4	4	1	2	2	1		
27–35	0	0	1	0	0	1	2	2	1		
36–44	0	0	1	0	0	1	2	2	1		
45–53	4	4	1	4	4	1	2	2	1		
54–62	7	7	1	7	7	1	2	2	1		
63–71	7	7	1	7	7	1	2	2	1		
72–80	4	4	1	4	4	1	2	2	1		
81–89	0	0	1	0	0	1	2	2	1		
90–100	0	0	1	0	0	1	2	2	1	4	4 …

Table 5. The Nim-values for Turnips.

To find $\mathscr{G}(n)$ in general, we expand n in base 3:

	n in ternary	$\mathscr{G}(n)$	
$\phi = 0$ or 1	$\ldots\ \phi\ \phi\ \phi\ \phi\ \phi\ \phi\ \phi$	0	
$? = 0, 1$ or 2	$\ldots\ ?\ ?\ ?\ ?\ ?\ ?\ 2$	1	the
	$\ldots\ ?\ ?\ ?\ ?\ ?\ 2\ \phi$	2	odious
	$\ldots\ ?\ ?\ ?\ ?\ 2\ \phi\ \phi$	4	numbers
	$\ldots\ ?\ ?\ ?\ 2\ \phi\ \phi\ \phi$	7	in
	$\ldots\ ?\ ?\ 2\ \phi\ \phi\ \phi\ \phi$	8	order
	$\ldots\ ?\ 2\ \phi\ \phi\ \phi\ \phi\ \phi$	11	
	$\ldots\ldots\ldots\ldots\ldots\ldots$	…	…

In words, $\mathscr{G}(n) = 0$ if the ternary expansion of n has no 2-digit, but is the kth odious number if the last 2-digit is in the kth place from the right, when we call n a k-**number**. The numbers n whose ternary expansions have no 2-digit will be called **empty numbers**.

To see all this, note that $\mathscr{G}(n)$ is the mex of all the numbers

$$\mathscr{G}(n-\delta)\overset{*}{+}\mathscr{G}(n-2\delta) \qquad \text{for } \delta = 1,2,\ldots\ .$$

We show first that the putative value for $\mathscr{G}(n)$ is not one of these numbers, or equivalently that

$$\mathscr{G}(n)\overset{*}{+}\mathscr{G}(n-\delta)\overset{*}{+}\mathscr{G}(n-2\delta) \neq 0.$$

Since the nim-sum of three odious numbers is odious, this will be true unless one of

$$\mathscr{G}(n),\ \ \mathscr{G}(n-\delta),\ \ \mathscr{G}(n-2\delta)$$

is zero and the other two coincide. But if the last non-zero ternary digit ($x = 1$ or 2) of δ is in the kth place, the expansions of n, $n-\delta$, $n-2\delta$ look like:

```
              k         j                              k
δ:        ? ? ? x 0 0 0 0 0 0              δ:       ? ? ? x 0 0 0 0 0
─────────────────────────────                ─────────────────────────────
 n    ⎧ ? ? ? 0 ? ? ? ? 2 φ φ                 n    ⎧ ? ? ? 0 φ φ φ φ φ
n-δ  :⎨ ? ? ? 1 ? ? ? ? 2 φ φ      or        n-δ  :⎨ ? ? ? 1 φ φ φ φ φ
n-2δ  ⎩ ? ? ? 2 ? ? ? ? 2 φ φ                 n-2δ  ⎩ ? ? ? 2 φ φ φ φ φ
```

according as n has or has not a 2-digit in some j place, $j < k$. In the first case the three putative nim-values are all the jth odious number, and in the second exactly one of them is the kth odious number, so they cannot have zero nim-sum.

Now we know from our analysis of Mock Turtles that each odious number is the first number not the nim-sum of two or fewer earlier ones. It suffices to show that if n is a k-number, we can choose δ so as to make $n-\delta$ and $n-2\delta$ i- and j-numbers or empty, for any i and j less than k. The subtraction sums in Table 6 show how to do this.

```
            k   j   i                        δ     1 0 0 0 1 0 1 0 0 1 0 0      k
δ  :           2 0 0 1 0 0                    n  :  φ 2 φ φ 2 φ 2 φ φ 2 φ φ
n  : ... ? ? 2 φ φ 1 φ φ 1 φ φ               n-δ :  φ 1 φ φ 1 φ 1 φ φ 1 φ φ
n-δ: ... ? ? ? ? ? 2 φ φ 0 φ φ               n-2δ:  φ 0 φ φ 0 φ 0 φ φ 0 φ φ
n-2δ:... ? ? ? ? ? ? ? ? 2 φ φ

δ  :           1 0 0 1 0 0                                               k   i
n  : ... ? ? 2 φ φ 0 φ φ 1 φ φ               n  :   φ 2 φ φ 2 φ 2 φ φ 1 φ φ
n-δ: ... ? ? ? ? ? 2 φ φ 0 φ φ               n-δ :  ? ? ? ? ? ? ? ? ? ? 2 φ φ
n-2δ:... ? ? ? ? ? ? ? ? 2 φ φ               n-2δ:  0 0 0 0 0 0 0 φ 0 0 0 φ φ

δ  :           1 2 2 2 0 0
n  : ... ? ? 2 φ φ 1 φ φ 0 φ φ               n  :   φ 2 φ φ 2 φ 2 φ φ 0 φ φ
n-δ: ... ? ? ? ? ? 2 φ φ 1 φ φ               n-δ :  ? ? ? ? ? ? ? ? ? ? 2 φ φ
n-2δ:... ? ? ? ? ? ? ? ? 2 φ φ               n-2δ:  0 0 0 0 0 0 0 φ 0 0 1 φ φ

δ  :           0 2 2 2 0 0
n  : ... ? ? 2 φ φ 0 φ φ 0 φ φ               In the last two cases above, the first φ of
n-δ: ... ? ? ? ? ? 2 φ φ 1 φ φ               the last line is whichever of 0 or 1 makes
n-2δ:... ? ? ? ? ? ? ? ? 2 φ φ               n-2δ have the same parity as n. Then δ
                                             can be found from n and n-2δ.
```

Table 6. How to Make $n-\delta$, $n-2\delta$ into i- and j-numbers, or Empty.

GRUNT

In a move of this game one must turn over four symmetrically arranged coins of which the first must be the leftmost coin of the game and the last must be turned from heads to tails. Numbering from 0 the restriction is that we turn numbers

$$0, a, n-a \text{ and } n, \qquad 0 < a < \tfrac{1}{2}n,$$

and we find the nim-values:

```
n    0 1 2 3 4 5 6 7 8 9 10 11 12 13 14 15 16 17 18 19 20 21 22 ...
𝒢(n) 0 0 0 1 0 2 1 0 2 1  0  2  1  3  2  1  3  2  4  3  0  4  3 ....
```

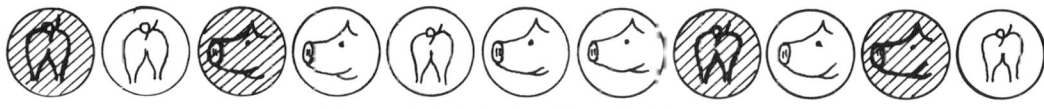

Figure 8. A Winning Move in Grunt.

Since $\mathscr{G}(0)=0$, $\mathscr{G}(n)$ can more easily be computed as the mex of all numbers of the form

$$\mathscr{G}(a)\overset{*}{+}\mathscr{G}(n-a), \qquad 0 < a < \tfrac{1}{2}n,$$

and so the game is a disguise for Grundy's Game (see Chapter 4) in which any heap may be split into two smaller heaps of different sizes.

SYM

As an example where the nim-values display no recognizable pattern, let us turn over any symmetrically arranged set of coins, not necessarily including the leftmost coin, number 0. We find

$$n = 0\ 1\ 2\ 3\ 4\ 5\ 6\ \ 7\ \ 8\ \ 9\ 10\ 11\ 12\ 13\ \ 14\ 15\ \ 16\ \ldots$$
$$\mathscr{G}(n) = 1\ 2\ 4\ 3\ 6\ 7\ 8\ 16\ 18\ 25\ 32\ 11\ 64\ 31\ 128\ 10\ 256\ \ldots .$$

The reader can also try to solve the game **Sympler** in which the leftmost coin *is* to be included in the symmetrical set of coins turned.

TWO-DIMENSIONAL TURNING GAMES

All our one-dimensional games were played with the restriction that the rightmost coin to be turned was to be changed from heads to tails. In the two-dimensional games the corresponding requirement is that the most "south-easterly" coin which is turned must go from heads to tails. In such games we'll write $\mathscr{G}(a,b)$ for the value of a coin in row a and column b.

ACROSTIC TWINS

We start with a very simple game. The move is to turn two coins which must either be in the same row or in the same column. The typical entry in the nim-value table is therefore the least number not appearing earlier in the same row or column, and we find Table 7. So we see that Acrostic Twins defines nim-addition:

$$\mathscr{G}(a,b) = a\overset{*}{+}b.$$

TURNING CORNERS

This is a much more interesting game. The move is to turn over the four corners of any rectangle with horizontal and vertical sides. The nim-values can be computed using

$$\mathscr{G}(a,b) = \text{mex}\{\mathscr{G}(a',b)\overset{*}{+}\mathscr{G}(a,b')\overset{*}{+}\mathscr{G}(a',b')\},$$

where a' and b' are any numbers respectively less than a and b (see Fig. 9). Table 8 gives values for a and b less than 16.

0	1	2	3	4	5	6	7	8	9	10	11	12	13	14	15
1	0	3	2	5	4	7	6	9	8	11	10	13	12	15	14
2	3	0	1	6	7	4	5	10	11	8	9	14	15	12	13
3	2	1	0	7	6	5	4	11	10	9	8	15	14	13	12
4	5	6	7	0	1	2	3	12	13	14	15	8	9	10	11
5	4	7	6	1	0	3	2	13	12	15	14	9	8	11	10
6	7	4	5	2	3	0	1	14	15	12	13	10	11	8	9
7	6	5	4	3	2	1	0	15	14	13	12	11	10	9	8
8	9	10	11	12	13	14	15	0	1	2	3	4	5	6	7
9	8	11	10	13	12	15	14	1	0	3	2	5	4	7	6
10	11	8	9	14	15	12	13	2	3	0	1	6	7	4	5
11	10	9	8	15	14	13	12	3	2	1	0	7	6	5	4
12	13	14	15	8	9	10	11	4	5	6	7	0	1	2	3
13	12	15	14	9	8	11	10	5	4	7	6	1	0	3	2
14	15	12	13	10	11	8	9	6	7	4	5	2	3	0	1
15	14	13	12	11	10	9	8	7	6	5	4	3	2	1	0

Table 7. How to Play Acrostic Twins.

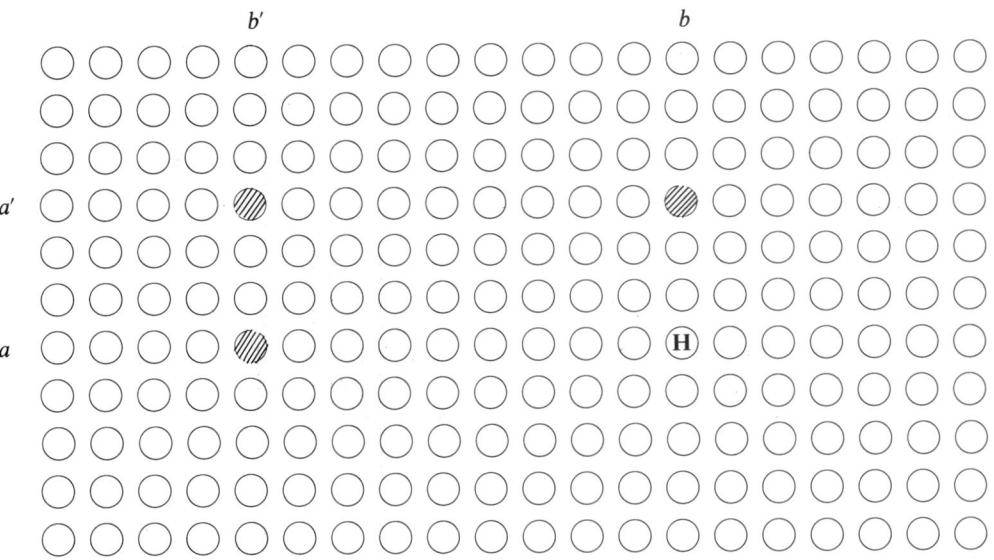

Figure 9. A Typical Move in Turning Corners.

0	0	0	0	0	0	0	0	0	0	0	0	0	0	0	0
0	1	2	3	4	5	6	7	8	9	10	11	12	13	14	15
0	2	3	1	8	10	11	9	12	14	15	13	4	6	7	5
0	3	1	2	12	15	13	14	4	7	5	6	8	11	9	10
0	4	8	12	6	2	14	10	11	15	3	7	13	9	5	1
0	5	10	15	2	7	8	13	3	6	9	12	1	4	11	14
0	6	11	13	14	8	5	3	7	1	12	10	9	15	2	4
0	7	9	14	10	13	3	4	15	8	6	1	5	2	12	11
0	8	12	4	11	3	7	15	13	5	1	9	6	14	10	2
0	9	14	7	15	6	1	8	5	12	11	2	10	3	4	13
0	10	15	5	3	9	12	6	1	11	14	4	2	8	13	7
0	11	13	6	7	12	10	1	9	2	4	15	14	5	3	8
0	12	4	8	13	1	9	5	6	10	2	14	11	7	15	3
0	13	6	11	9	4	15	2	14	3	8	5	7	10	1	12
0	14	7	9	5	11	2	12	10	4	13	3	15	1	8	6
0	15	5	10	1	14	4	11	2	13	7	8	3	12	6	9

Table 8. Have You Learnt Your Tims Table?

NIM-MULTIPLICATION

Observing that the nim-value $\mathscr{G}(0,n)=0$, while $\mathscr{G}(1,n)=n$, we guess that this might be a kind of multiplication, so we shall write

$$a \overset{*}{\times} b$$

(and you will read "a **tims** b") for the nim-value $\mathscr{G}(a,b)$ of the general coin in Turning Corners. We shall call this the **nim-product** of a and b.

It is shown in ONAG (Chapter 6) that this remarkable operation has all the usual algebraic properties of multiplication, and in particular obeys the distributive law

$$a \overset{*}{\times} (b \overset{*}{+} c) = a \overset{*}{\times} b \overset{*}{+} a \overset{*}{\times} c$$

with nim-addition. For example

$$7 \overset{*}{\times} (5 \overset{*}{+} 6) = 7 \overset{*}{\times} 3 = 14,$$

$$7 \overset{*}{\times} 5 \overset{*}{+} 7 \overset{*}{\times} 6 = 13 \overset{*}{+} 3 = 14.$$

But note, for example, that the nim-sum of 6 and 6 is not two sixes but no sixes, since $1 \overset{*}{+} 1$ is not 2, but 0.

In computing nim-products of larger numbers, the **Fermat powers** of 2,

$$2, \ 4, \ 16, \ 256, \ 65536, \ 4294967296, \ \ldots, \ 2^{2^n}, \ \ldots$$

play a role similar to that played by *all* the powers of 2 in nim-addition. Recall that in nim-addition if N is any power of 2 we have:

$$
\begin{aligned}
N \overset{*}{+} n &= N + n \quad \text{for } n < N, \\
N \overset{*}{+} N &= 0.
\end{aligned}
$$

For nim-multiplication, if N is any *Fermat* power of 2 we have:

$$
\begin{aligned}
N \overset{*}{\times} n &= N \times n \quad \text{for } n < N, \\
N \overset{*}{\times} N &= \tfrac{3}{2} N.
\end{aligned}
$$

For example, $16 \overset{*}{\times} 5 = 80$, as usual, but $16 \overset{*}{\times} 16 = 24$. Table 9 gives products of powers of 2.

1	2	4	8	16	32	64	128	256 ...
2	3	8	12	32	48	128	192	512 ...
4	8	6	11	64	128	96	176	1024 ...
8	12	11	13	128	192	176	208	2048 ...
16	32	64	128	24	44	75	141	4096 ...
32	48	128	192	44	52	141	198	8192 ...
64	128	96	176	75	141	103	185	16384 ...
128	192	176	208	141	198	185	222	32768 ...
256	512	1024	2048	4096	8192	16384	32768	384 ...

Table 9. Nim-products of Powers of 2.

SWIRLING TARTANS

Figure 10 indicates the coins which may be turned in a typical move of this game. The boxed places form what we call a **tartan**. In general we select a certain number of rows and a certain number of columns and the places of the tartan are where our chosen rows meet our chosen columns. In **Swirling Tartans** we may turn the coins of *any* tartan, but in other games there will be restrictions on the rows and columns we may choose. Table 9 is actually a table of nim-values for Swirling Tartans. This is a particular case of the following theory.

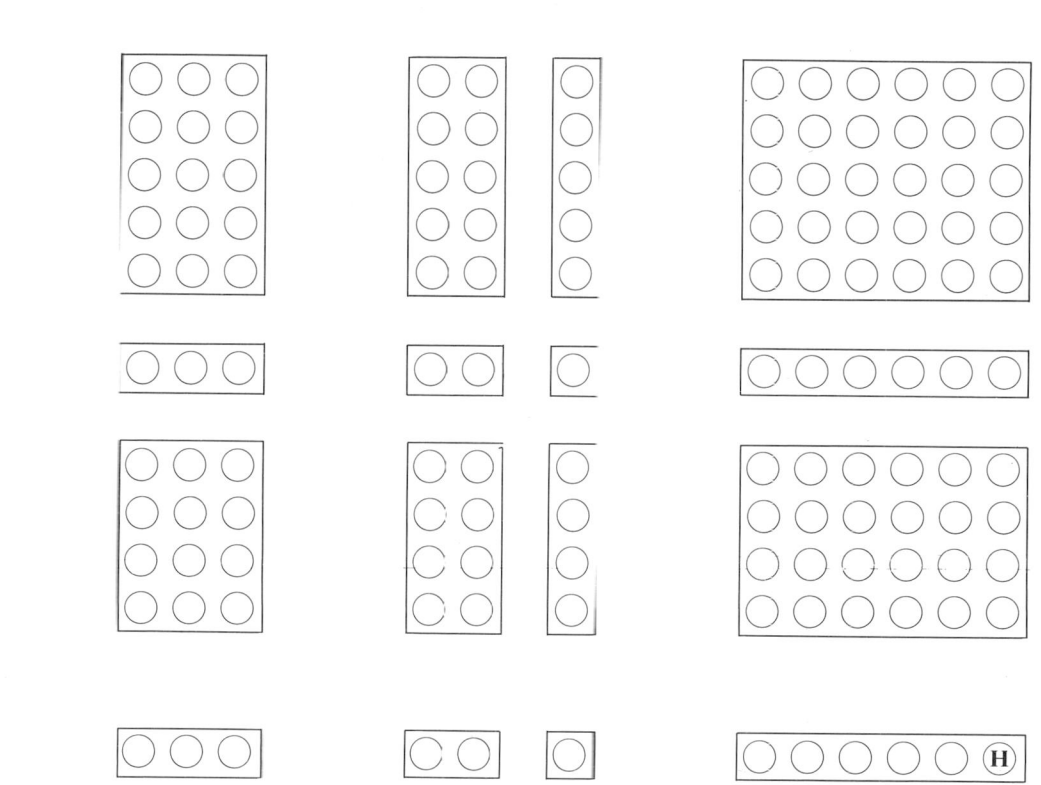

Figure 10. A Tartan.

THE TARTAN THEOREM

We can build a **tartan game**, $A \times B$, from two one-dimensional turning games, A and B, by specifying that the rows of the tartan shall correspond to the coins which may be turned in a move of game A and the columns to the coins which may be turned in a move of game B. Taking both A and B to be the game of Motley, in which *any* sets of coins may be turned, we see that

$$\text{MOTLEY} \times \text{MOTLEY} = \text{SWIRLING TARTANS}.$$

It follows from the Tartan Theorem that the nim-values for Swirling Tartans are the nim-products of those for two games of Motley—since the latter nim-values are just the powers of 2; this justifies our assertion about Table 9. More generally:

> the nim-values for the tartan game $A \times B$
> are the nim-products of those for A and B:
> $$\mathscr{G}_{A \times B}(a,b) = \mathscr{G}_A(a) \overset{*}{\times} \mathscr{G}_B(b).$$

THE TARTAN THEOREM

The proof, which we do not give, depends on the following characterizing property of the nim-product $a \overset{*}{\times} b$.

> If $x_1, x_2, \ldots$ are numbers for which
> $$a = \mathrm{mex}(a \overset{*}{+} x_i)$$
> and $y_1, y_2, \ldots$ are numbers for which
> $$b = \mathrm{mex}(b \overset{*}{+} y_j)$$
> then we have
> $$a \overset{*}{\times} b = \mathrm{mex}(a \overset{*}{\times} b \ \overset{*}{+} \ x_i \overset{*}{\times} y_j).$$

This can be deduced from a result on p. 55 of ONAG.

RUGS, CARPETS, WINDOWS AND DOORS

In Turning Corners we turned over the corners of a rectangle, so that

$$\text{TURNING CORNERS} = \text{TWINS} \times \text{TWINS}.$$

In **Rugs** we turn over *all* the coins in some solid rectangle, in other words the tartan must be defined by a block of consecutive rows and a block of consecutive columns. Since in the Ruler Game a move was to turn over a block of consecutive coins, we have

$$\text{RULER} \times \text{RULER} = \text{RUGS},$$

and so the nim-values are those of Table 10.

A **carpet** is a tartan in which both rows and columns form symmetrical sets, as in Fig. 11, and the corresponding game, **Carpets**, has therefore the nim-values of Table 11:

$$\text{CARPETS} = \text{SYM} \times \text{SYM}.$$

1	2	1	4	1	2	1	8	1	2	1	4	1	2	1	16
2	3	2	8	2	3	2	12	2	3	2	8	2	3	2	32
1	2	1	4	1	2	1	8	1	2	1	4	1	2	1	16
4	8	4	6	4	8	4	11	4	8	4	6	4	8	4	64
1	2	1	4	1	2	1	8	1	2	1	4	1	2	1	16
2	3	2	8	2	3	2	12	2	3	2	8	2	3	2	32
1	2	1	4	1	2	1	8	1	2	1	4	1	2	1	16
8	12	8	11	8	12	8	13	8	12	8	11	8	12	8	128

Table 10. A Rug with a Table on it.

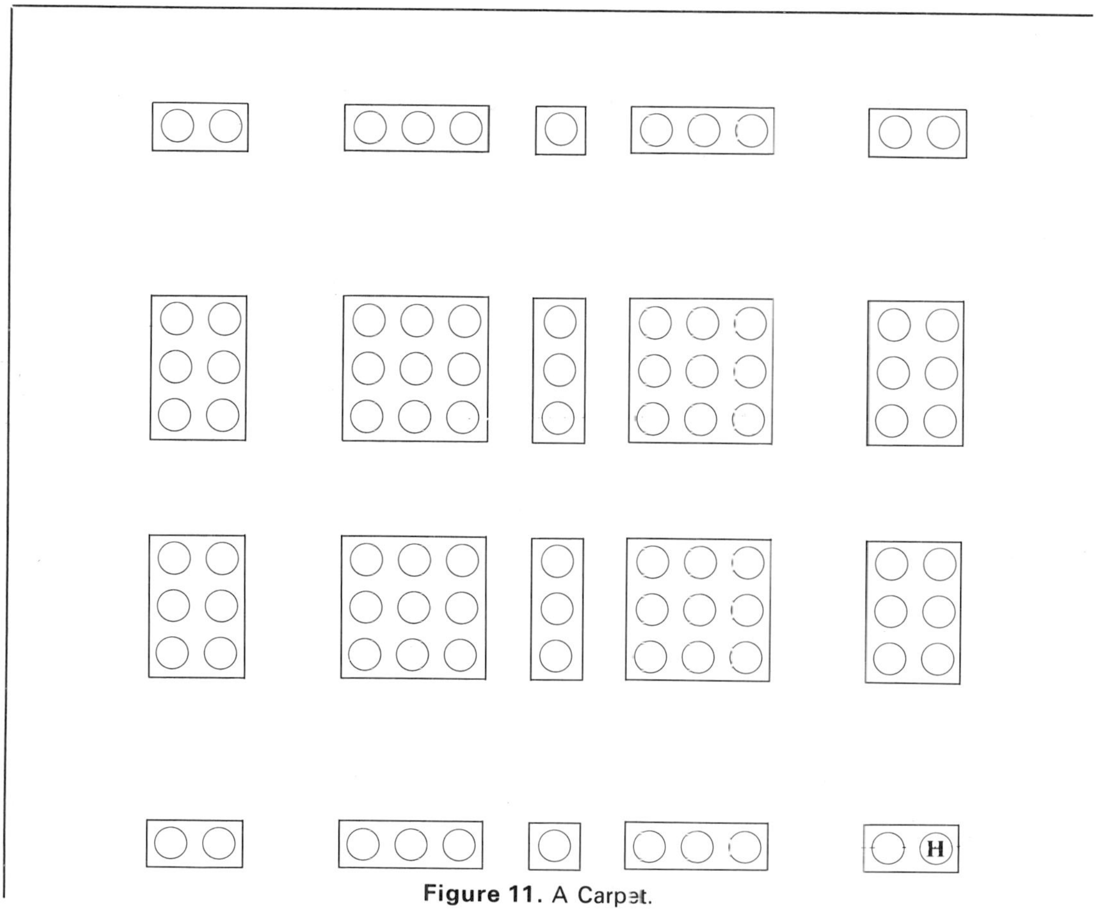

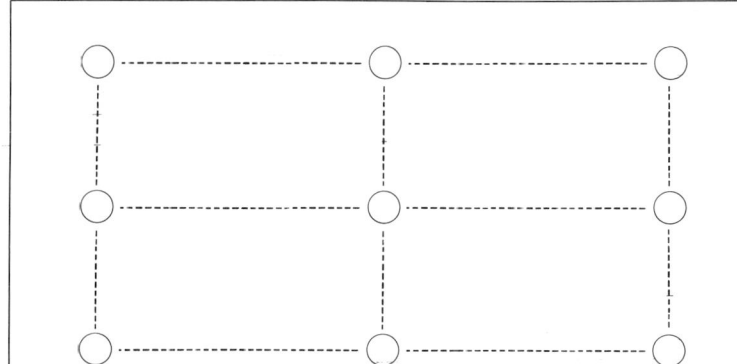

Figure 11. A Carpet.

In **Fitted Carpets** one is only allowed to turn carpets which fit snugly into the corner of the room, so

$$\text{FITTED CARPETS} = \text{SYMPLER} \times \text{SYMPLER}$$

whose analysis is left to the reader.

Figure 12. A Move in Windows.

1	2	4	3	6	7	8	16	18	25	32	11	64	31	128	10	256
2	3	8	1	11	9	12	32	35	46	48	13	128	37	192	15	512
4	8	6	12	14	10	11	64	72	79	128	7	96	65	176	3	1024
3	1	12	2	13	14	4	48	49	55	16	6	192	58	64	5	768
6	11	14	13	5	3	7	96	107	97	176	10	224	100	112	12	1536
7	9	10	14	3	4	15	112	121	120	144	1	160	123	240	6	1792
8	12	11	4	7	15	13	128	140	133	192	9	176	130	208	1	2048
16	32	64	48	96	112	128	24	56	136	44	176	75	232	141	160	4096
18	35	72	49	107	121	140	56	27	166	28	189	203	205	77	175	4608
25	46	79	55	97	120	133	136	166	20	204	178	187	117	221	171	6400
32	48	128	16	176	144	192	44	28	204	52	208	141	124	198	240	8192
11	13	7	6	10	1	9	176	189	178	208	15	112	184	144	4	2816
64	128	96	192	224	160	176	75	203	187	141	112	103	91	185	48	16384
31	37	65	58	100	123	130	232	205	117	124	184	91	17	173	167	7936
128	192	176	64	112	240	208	141	77	221	198	144	185	173	222	16	32768
10	15	3	5	12	6	1	160	175	171	240	4	48	167	16	14	2560
256	512	1024	768	1536	1792	2048	4096	4608	6400	8192	2816	16384	7936	32768	2560	384

Table 11. A Greatly-Valued Carpet.

In the game **Windows** we turn the nine coins where three equally spaced rows meet three equally spaced columns, as in Fig. 12, so that

$$\text{WINDOWS} = \text{TURNIPS} \times \text{TURNIPS}.$$

The nim-values form the most complex system we have yet discovered. To calculate the outcome of a given position we must perform no fewer than four successive operations:

1. Expand the two coordinates of a head in base 3 and find the last 2-digit (if any) in each.
2. Replace the coordinates by the corresponding odious numbers (or zero). This involves a further expansion, in base 2.
3. For each head find the *nim-product* of the two numbers so obtained.
4. Find the *nim-sum* of the numbers so found for all the heads.

In all these games there has been the condition that the coin most to the South-East in any move be turned from heads to tails. So that our next game deserves its name we will play it "upside-down" and impose the condition on the most North-East coin. The move in **Doors** is to turn over the twelve coins where any three equally spaced columns meet four symmetrically arranged rows, which must include the bottom row, as in Fig. 13. This shows that

$$\text{DOORS} = \text{TURNIPS} \times \text{GRUNT}$$

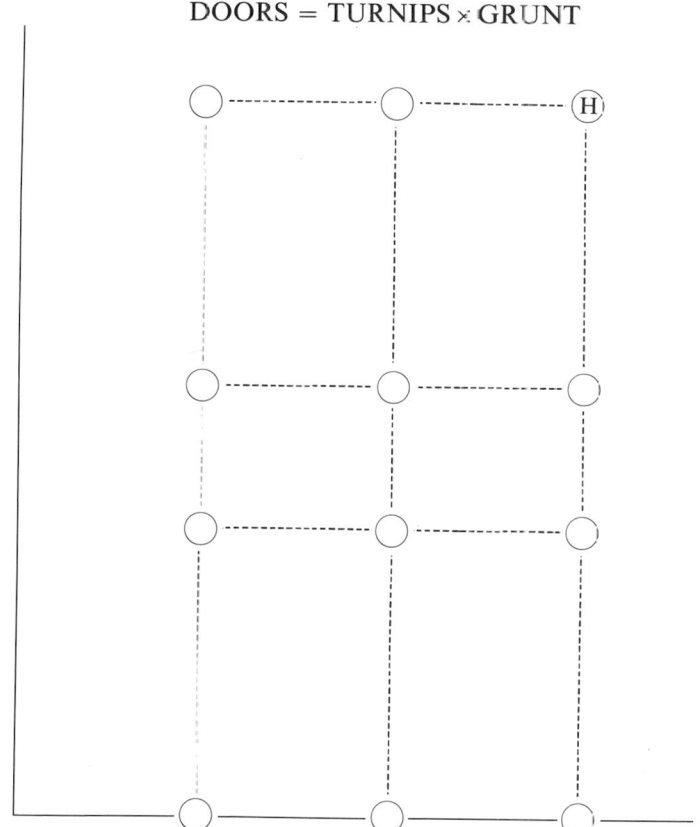

Figure 13. A Typical Move in Doors.

so you should soon be able to find the nim-value of the Doors position in which there is a single head in the 100th row and 100th column. (Beware: the first row is row number 0.)

ACROSTIC GAMES

There is another way to build a two-dimensional game out of two one-dimensional turning games, A and B. In the **acrostic product** $A \cup B$, the coins we turn must either all be in the same

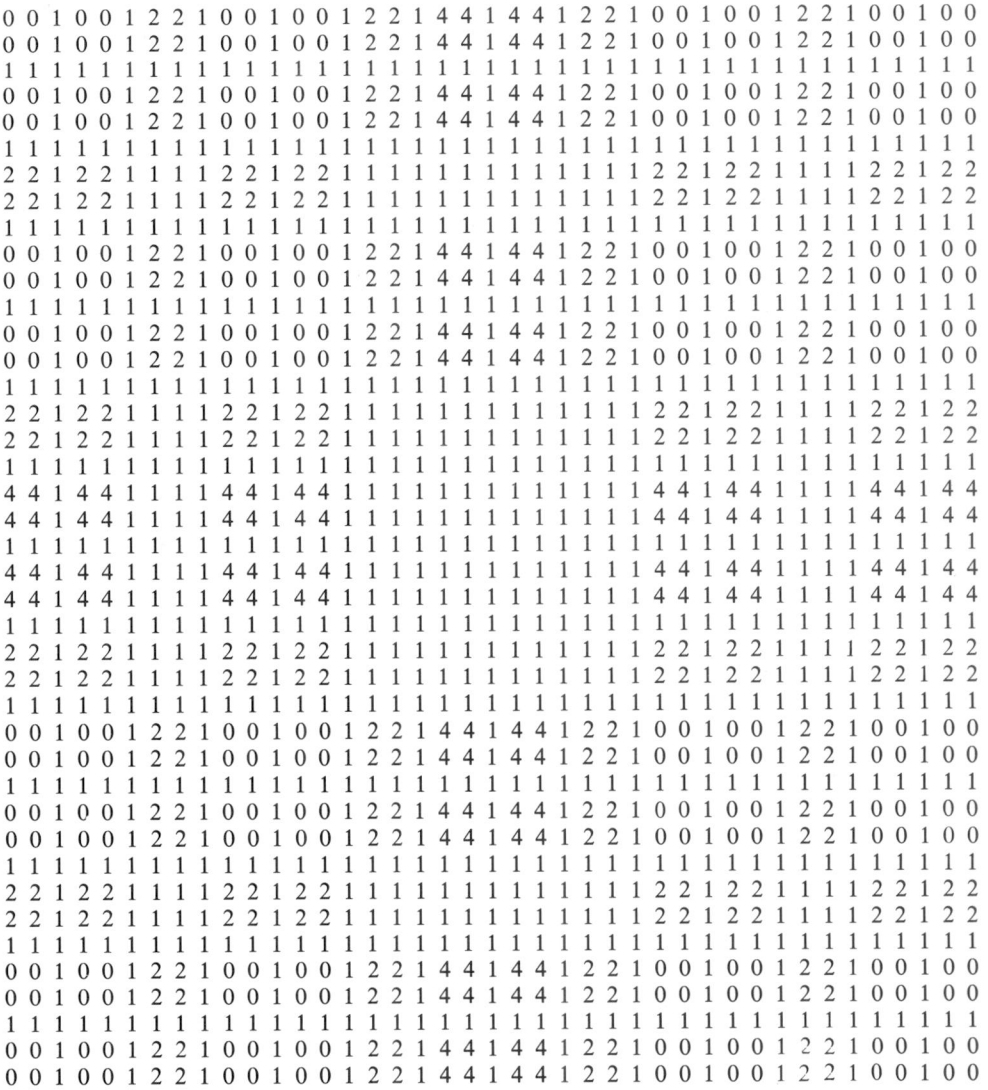

Table 12. A Field of Turnips.

column or all be in the same row. If the coins are all in a column, they must correspond to a move of game A; if all in a row, to a move of game B. We have already met one game of this type:

$$\text{ACROSTIC TWINS} = \text{TWINS} \cup \text{TWINS}.$$

In **Acrostic Turnips** we must, of course, upturn three turnips which are equally spaced and either in the same row or in the same column, the furthest from the corner of the field being turned from top to tail.

$$\text{ACROSTIC TURNIPS} = \text{TURNIPS} \cup \text{TURNIPS}.$$

The first 1681 nim-values are displayed in Table 12. It is not hard to prove that a row or column that begins with zero repeats the nim-sequence for Turnips itself, while the values not in such a row or column are 1.

STRIPPING AND STREAKING

We have no idea how to play the general acrostic product $A \cup B$, even when the one-dimensional games A and B are fully understood. But if it just should happen, as sometimes it does, that every nim-value of a place in each of your games is either 7 or a power of 2, we can offer you some help. We first discuss two easy games of this kind.

In **Streaking** we turn over any collection (**streak**) of coins all in the same row or column, so that

$$\text{STREAKING} = \text{MOTLEY} \cup \text{MOTLEY}.$$

Since the nim-values for Motley are exactly the powers of 2 and each nim-value for Streaking is the mex of all numbers that are sums of earlier nim-values from the same row or earlier nim-values from the same column, we find Table 13.

In **Stripping**, the coins we turn must be consecutive (form a **strip**) in either a row or a column. The entries in Table 14, of nim-values for

$$\text{STRIPPING} = \text{RULER} \cup \text{RULER},$$

can apparently be obtained from entries in Table 13. How do we explain this?

UGLIFICATION AND DERISION

We introduce an ambitious distraction. We shall call an entry in Table 13 (or 14) the **ugly product** of the two powers of 2 that head its row and column, and write

$$4 \overset{*}{\circlearrowup} 8 = 10 \quad (\text{"four \textbf{uggles} eight is ten"})$$

for example. Ugly products of other numbers can then be found using the distributive law:

$$4 \overset{*}{\circlearrowup} 11 = 4 \overset{*}{\circlearrowup} (8 \overset{*}{+} 2 \overset{*}{+} 1) = 10 \overset{*}{+} 5 \overset{*}{+} 4 = 11,$$
$$5 \overset{*}{\circlearrowup} 11 = (4 \overset{*}{+} 1) \overset{*}{\circlearrowup} 11 = 11 \overset{*}{+} 11 = 0.$$

The latter equation shows that 5 and 11 are **deriders of zero**. An uglification table up to 16 is given in Table 15.

1	2	4	8	16	32	64	128	256	...
2	1	5	9	17	33	65	129	257	...
4	5	2	10	18	34	66	130	258	...
8	9	10	4	20	36	68	132	260	...
16	17	18	20	8	40	72	136	264	...
32	33	34	36	40	16	80	144	272	...
64	65	66	68	72	80	32	160	288	...
128	129	130	132	136	144	160	64	320	...
256	257	258	260	264	272	288	320	128	...

....

Table 13. Streaking Values.

1	2	1	4	1	2	1	8	1	2	1	4	1	2	1	16
2	1	2	5	2	1	2	9	2	1	2	5	2	1	2	17
1	2	1	4	1	2	1	8	1	2	1	4	1	2	1	16
4	5	4	2	4	5	4	10	4	5	4	2	4	5	4	18
1	2	1	4	1	2	1	8	1	2	1	4	1	2	1	16
2	1	2	5	2	1	2	9	2	1	2	5	2	1	2	17
1	2	1	4	1	2	1	8	1	2	1	4	1	2	1	16
8	9	8	10	8	9	8	4	8	9	8	10	8	9	8	20
1	2	1	4	1	2	1	8	1	2	1	4	1	2	1	16
2	1	2	5	2	1	2	9	2	1	2	5	2	1	2	17
1	2	1	4	1	2	1	8	1	2	1	4	1	2	1	16
4	5	4	2	4	5	4	10	4	5	4	2	4	5	4	18
1	2	1	4	1	2	1	8	1	2	1	4	1	2	1	16
2	1	2	5	2	1	2	9	2	1	2	5	2	1	2	17
1	2	1	4	1	2	1	8	1	2	1	4	1	2	1	16
16	17	16	18	16	17	16	20	16	17	16	18	16	17	16	8

Table 14. Stripping Values.

0	0	0	0	0	0	0	0	0	0	0	0	0	0	0	0	0
0	1	2	3	4	5	6	7	8	9	10	11	12	13	14	15	16
0	2	1	3	5	7	4	6	9	11	8	10	12	14	13	15	17
0	3	3	0	1	2	2	1	1	2	2	1	0	3	3	0	1
0	4	5	1	2	6	7	3	10	14	15	11	8	12	13	9	18
0	5	7	2	6	3	1	4	2	7	5	0	4	1	3	6	2
0	6	4	2	7	1	3	5	3	5	7	0	4	2	0	6	3
0	7	6	1	3	4	5	2	11	12	13	10	8	15	14	9	19
0	8	9	1	10	2	3	11	4	12	13	5	14	6	7	15	20
0	9	11	2	14	7	5	12	12	5	7	14	2	11	9	0	4
0	10	8	2	15	5	7	13	13	7	5	15	2	8	10	0	5
0	11	10	1	11	0	1	10	5	14	15	4	14	5	4	15	21
0	12	12	0	8	4	4	8	14	2	2	14	6	10	10	6	6
0	13	14	3	12	1	1	15	6	11	8	5	10	7	4	9	22
0	14	13	3	13	3	0	14	7	9	10	4	10	4	7	9	23
0	15	15	0	9	6	6	9	15	0	0	15	6	9	9	6	7
0	16	17	1	18	2	3	19	20	4	5	21	6	22	23	7	8

Table 15. The Uglification Table up to 16s.

For larger numbers we may use the following rules:

> If N is any power of 2 and $n < N$, we have
> $$N \overset{*}{\cup} n = \begin{cases} N + \lfloor \tfrac{1}{2}n \rfloor & \text{if } n \text{ is odious} \\ \lfloor \tfrac{1}{2}n \rfloor & \text{if } n \text{ is evil,} \end{cases}$$
> $$N \overset{*}{\cup} N = \lceil \tfrac{1}{2}N \rceil.$$

The rows and columns of Table 15 which correspond to 7 and powers of 2 have been printed in bold type and their intersections have been boxed. We shall use the following properties of these rows.

> 1. The entries in any **bold** row are distinct. In symbols, if a is 7 or a power of 2, and $b \neq \bar{b}$, then
>
> $$a \,\overset{*}{\circleddash}\, b \neq a \,\overset{*}{\circleddash}\, \bar{b}.$$
>
> 2. Each boxed entry is the mex of all previous entries in its row and column. That is, if both a and b are 7 or powers of 2, then $a \,\overset{*}{\circleddash}\, b$ is the mex of all numbers of the form
>
> $$a' \,\overset{*}{\circleddash}\, b \text{ or } a \,\overset{*}{\circleddash}\, b', \qquad a' < a, \quad b' < b.$$

These are used in proving the following theorem.

> If every nim-value of a place in each of A and B is 7 or a power of 2, then the nim-values of the acrostic game $A \cup B$ are obtained by uglification of those for A and B:
>
> $$\mathcal{G}_{A \cup B}(a,b) = \mathcal{G}_A(a) \,\overset{*}{\circleddash}\, \mathcal{G}_B(b).$$

THE UGLIFICATION THEOREM

To see this, let the typical move in game A or B turn the coins in places

$$a_1 < a_2 < \ldots < a$$

or

$$b_1 < b_2 < \ldots < b$$

respectively. We will denote the nim-values of these places by

$$\alpha_1, \alpha_2, \ldots, \alpha$$

and

$$\beta_1, \beta_2, \ldots, \beta$$

respectively. Then the nim-value $\mathcal{G}(a,b)$ in the acrostic product $A \cup B$ is the mex of all numbers of the form

$$\alpha_1 \,\overset{*}{\circleddash}\, \beta \,\overset{*}{\mp}\, \alpha_2 \,\overset{*}{\circleddash}\, \beta \,\overset{*}{\mp}\, \ldots$$

or

$$\alpha \,\overset{*}{\circleddash}\, \beta_1 \,\overset{*}{\mp}\, \alpha \,\overset{*}{\circleddash}\, \beta_2 \,\overset{*}{\mp}\, \ldots,$$

that is to say the mex of all numbers of the form

$$\bar{\alpha} \,\overset{*}{\circleddash}\, \beta \quad \text{or} \quad \alpha \,\overset{*}{\circleddash}\, \bar{\beta},$$

where

$$\bar{\alpha} = \alpha_1 \,\overset{*}{\mp}\, \alpha_2 \,\overset{*}{\mp}\, \ldots$$
$$\bar{\beta} = \beta_1 \,\overset{*}{\mp}\, \beta_2 \,\overset{*}{\mp}\, \ldots.$$

Now α is the mex of all numbers $\bar{\alpha}$ which arise in this way, and β is the mex of all numbers $\bar{\beta}$, so that every $\alpha' < \alpha$ is one of the numbers $\bar{\alpha}$ and every $\beta' < \beta$ is a $\bar{\beta}$. But each of α and β is either 7 or a power of 2 by assumption, so that certainly $\alpha \overset{*}{\cup} \beta$ is the mex of all numbers of the form

$$\alpha' \overset{*}{\cup} \beta \quad \text{or} \quad \alpha \overset{*}{\cup} \beta'.$$

Also, α is distinct from all the numbers $\bar{\alpha}$, and so $\alpha \overset{*}{\cup} \beta$ is distinct from all the numbers $\bar{\alpha} \overset{*}{\cup} \beta$, and similarly from $\alpha \overset{*}{\cup} \bar{\beta}$, and so $\alpha \overset{*}{\cup} \beta$ is *their* mex also.

This explains the values we found for Stripping and enables us to discuss a few similar games. For instance, in **Strip and Streak**, where we may turn coins in a horizontal strip or a vertical streak, the first few nim-values are as in Table 16.

1	2	1	4	1	2	1	8	1	2	1	4	1	2	1	16	1	2	...
2	1	2	5	2	1	2	9	2	1	2	5	2	1	2	17	2	1	...
4	5	4	2	4	5	4	10	4	5	4	2	4	5	4	18	4	5	...
8	9	8	10	8	9	8	4	8	9	8	10	8	9	8	20	8	9	...
16	17	16	18	16	17	16	20	16	17	16	18	16	17	16	8	16	17	...

Table 16. Strip and Streak.

Table 17 gives the nim-values for **Acrostic Mock Turtle Fives** in which we turn *up to three* coins provided that these are all contained in some horizontal or vertical strip of *five*.

1	2	4	7	8	1	2	4	7	8	1	2	4	7	8	...
2	1	5	6	9	2	1	5	6	9	2	1	5	6	9	...
4	5	2	3	10	4	5	2	3	10	4	5	2	3	10	...
7	6	3	2	11	7	6	3	2	11	7	6	3	2	11	...
8	9	10	11	4	8	9	10	11	4	8	9	10	11	4	...
1	2	4	7	8	1	2	4	7	8	1	2	4	7	8	...
2	1	5	6	9	2	1	5	6	9	2	1	5	6	9	...

Table 17. Acrostic Mock Turtle Fives.

EXTRAS

UNLOCKING DOORS

For Turnips, Table 5 shows that $\mathcal{G}(99)=4$, while for Grunt, the discussion of Grundy's Game in Chapter 4 shows that $\mathcal{G}(99)=5$; so a coin in the 100th row and 100th column of Doors has value

$$4 \text{ tims } 5 = 2.$$

SPARRING, BOXING AND FENCING

Turning games can be played in any number of dimensions. We mention just three 3-dimensional games. In **Sparring** we turn over any two coins in the same row, column or vertical, that is to say the two ends of a "spar". The typical entry in the nim-value table is the mex of previous entries in its row, column or vertical, so we have

$$\mathcal{G}(a,b,c) = a \overset{*}{+} b \overset{*}{+} c.$$

In **Boxing** we turn the eight corners of a rectangular "box". This is the 3-dimensional version of Turning Corners and its nim-values are three-term nim-products

$$\mathcal{G}(a,b,c) = a \overset{*}{\times} b \overset{*}{\times} c.$$

In **Fencing** we turn the four corners of a rectangular "fence" whose edges are parallel to any two of the three coordinate axes. It can be shown that

$$\mathcal{G}(a,b,c) = b \overset{*}{\times} c \overset{*}{+} c \overset{*}{\times} a \overset{*}{+} a \overset{*}{\times} b.$$

In each case the furthest turned coin from the origin must go from heads to tails.

"COINS" (OR HEAPS) WITH INFINITELY MANY (OR 2^{2^N}) "SIDES"

may be used to give lots of new "turning" games whose theory also involves Nim-multiplication. Thus if the move is to alter at most two of the heaps $H_{-1}, H_0, H_1, \ldots$ of which the rightmost one must be reduced, the $\mathcal{P}$-positions are those with

$$H_0 \overset{*}{+} H_1 \overset{*}{+} H_2 \overset{*}{+} \ldots = 0 \quad \text{and} \quad 0 \overset{*}{\times} H_0 \overset{*}{+} 1 \overset{*}{\times} H_1 \overset{*}{+} 2 \overset{*}{\times} H_2 \overset{*}{+} \ldots = H_{-1}$$

REFERENCES AND FURTHER READING

J.H. Conway, "On Numbers and Games", Academic Press, London and New York, 1976, Chapter 6.

Richard K. Guy, She loves me, she loves me not; relatives of two games of Lenstra, Een Pak met een Korte Broek, Papers presented to H.W. Lenstra, 77:05:18, Mathematisch Centrum, Amsterdam.

H.W. Lenstra, Nim multiplication, Séminaire de Théorie des Nombres, 1977–78 exposé No. 11, Université de Bordeaux.

Chapter 15

Chips and Strips

There is some ill a-brewing towards my rest
For I did dream of money-bags tonight.
William Shakespeare, *The Merchant of Venice*, II, v, 17.

Many of the games in this chapter are derived in some way from Nim. Although Nim is usually played with heaps of chips it can also be played with coins on a strip, the move being to shift any coin leftwards any number of squares. Figure 1 shows the same Nim position in both versions. Moving a coin leftwards corresponds to reducing a heap.

We obtain many generalizations of Nim by varying the conditions under which heaps can be reduced, or coins moved.

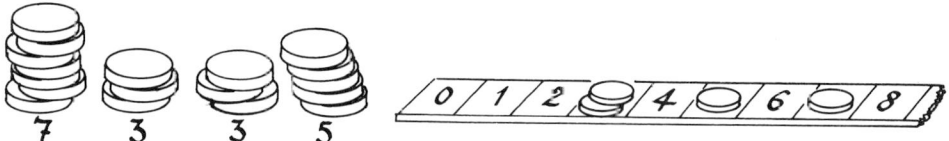

Figure 1. Two Forms of Nim.

THE SILVER DOLLAR GAME

In our first variant we allow at most one coin per square and do *not* allow one coin to jump over another. It can take quite a long time to discover that this is a cunning disguise for Nim, related to the game of Poker-Nim in Chapter 3. The sizes of the Nim-heaps are the lengths of *alternate* gaps between the coins starting from the rightmost coin (Fig. 2).

Observe that any *decrease* of one of these numbers is possible (by moving the coin at the *right* end of the gap) and that some *increases* are also possible (by moving the coin at the *left* end of a gap). We've indicated sample moves of both types in the figure. But just as in the theory of Poker-Nim, the increasing moves are mere reversible delaying moves and the winner wins by playing Nim.

457

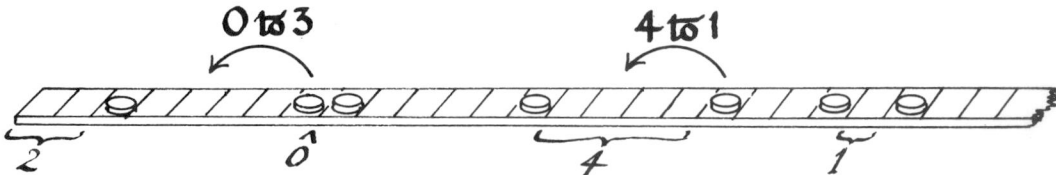

Figure 2. The Silver Dollar Game Without the Dollar.

N.G. de Bruijn has made the game more interesting by turning the leftmost square into a moneybag capable of holding any number of coins and making one of the coins a Silver Dollar, more valuable than all the others put together. Now the leftmost coin not already in the moneybag, may be put into the moneybag, as a move. The person who bags the dollar loses the game, because we also allow another move—pocket the moneybag!

Figure 3. De Bruijn's Silver Dollar Game.

In this version the moneybag counts as a *full* square when the first coin to the right of it is the Silver Dollar; otherwise as an *empty* one (it's because we don't want to put the dollar into the bag that we think of it as full when the dollar is the nearest coin to it!) If we win the Nim game we won't be forced to put the dollar into the bag.

If we allow whoever bags the dollar to pocket the bag all in one move, we count the bag as *full* only when there's just one other coin between it and the dollar. (We don't want to put *this* coin in the bag because our opponent will make sure it is immediately followed by the dollar!)

Find the winning moves in Fig. 3 in both versions.

PROFIT FROM GAMING TABLES

What do you do when you meet a game that's *not* analyzed in *Winning Ways*? You might be very lucky and get the hang of it after your first few games, but if you can't quite see what's going on and our theories don't seem to provide much of a clue, the best thing to do is to compile a **gaming table**. To do this profitably can take some skill in organizing the information. There'll be some varied examples in this chapter.

ANTONIM

Antipathetic Nim is Nim in which no two heaps are allowed to have the same number of chips. Of course, we don't notice empty heaps, so if you want to play it with coins on a strip the condition is that no two coins may be on the same square unless this square is the moneybag (square 0).

We can analyze 3-coin Antonim in a single table (Table 1). The headings are the sizes of two of the heaps and the entry is the unique size of heap that completes these to a $\mathscr{P}$-position. The typical entry is filled in as the least number not coinciding with any earlier entry in the same row or column, nor coinciding with either the row or column heading. An X denotes an illegal position. There is an obvious pattern showing that:

> (a, b, c) is a $\mathscr{P}$-position in Antonim just when
> $(a+1, b+1, c+1)$ is a $\mathscr{P}$-position in Nim.

	0	1	2	3	4	5	6	7	8	9	10	11	12	13	14
0	0	2	1	4	3	6	5	8	7	10	9	12	11	14	13
1	2	X	0	5	6	3	4	9	10	7	8	13	14	11	12
2	1	0	X	6	5	4	3	10	9	8	7	14	13	12	11
3	4	5	6	X	0	1	2	11	12	13	14	7	8	9	10
4	3	6	5	0	X	2	1	12	11	14	13	8	7	10	9
5	6	3	4	1	2	X	0	13	14	11	12	9	10	7	8
6	5	4	3	2	1	0	X	14	13	12	11	10	9	8	7
7	8	9	10	11	12	13	14	X	0	1	2	3	4	5	6
8	7	10	9	12	11	14	13	0	X	2	1	4	3	6	5
9	10	7	8	13	14	11	12	1	2	X	0	5	6	3	4
10	9	8	7	14	13	12	11	2	1	0	X	6	5	4	3
11	12	13	14	7	8	9	10	3	4	5	6	X	0	1	2
12	11	14	13	8	7	10	9	4	3	6	5	0	X	2	1
13	14	11	12	9	10	7	8	5	6	3	4	1	2	X	0
14	13	12	11	10	9	8	7	6	5	4	3	2	1	0	X

Table 1. $\mathscr{P}$-positions for Three-Coin Antonim.

For Antonim with 4 coins we need a 3-D table, which we can build in layers. The next layer (Table 2) suggests that there is unlikely to be a simple rule, even for positions

$$(1, a, b, c),$$

so we have cut short the further layers (Table 3).

0	X	1	2	3	4	5	6	7	8	9	10	11	12	13	14
X	2	X	0	5	6	3	4	9	10	7	8	13	14	11	12
1	X	X	X	X	X	X	X	X	X	X	X	X	X	X	X
2	0	X	X	4	3	6	5	8	7	10	9	12	11	14	13
3	5	X	4	X	2	0	7	6	9	8	11	10	13	12	15
4	6	X	3	2	X	7	0	5	12	11	14	9	8	15	10
5	3	X	6	0	7	X	2	4	11	12	13	8	9	10	16
6	4	X	5	7	0	2	X	3	14	13	12	15	10	9	8
7	9	X	8	6	5	4	3	X	2	0	15	14	16	17	11
8	10	X	7	9	12	11	14	2	X	3	0	5	4	16	6
9	7	X	10	8	11	12	13	0	3	X	2	4	5	6	17
10	8	X	9	11	14	13	12	15	0	2	X	3	6	5	4
11	13	X	12	10	9	8	15	14	5	4	3	X	2	0	7
12	14	X	11	13	8	9	10	16	4	5	6	2	X	3	0
13	11	X	14	12	15	10	9	17	16	6	5	0	3	X	2
14	12	X	13	15	10	16	8	11	6	17	4	7	0	2	X

Table 2. $\mathscr{P}$-positions $(1,a,b,c)$ for Antonim.

It's not hard to show that the $\mathscr{P}$-positions with numbers $\leqslant 7$ are just

$$(0)12, \quad (0)34, \quad (0)56, \quad 135, \quad 146, \quad 236, \quad 245,$$
$$1234, \quad 1256, \quad 1367, \quad 1457, \quad 2357, \quad 2467, \quad 3456$$
$$(0)123456.$$

The 3-heap $\mathscr{P}$-positions are the lines of Fig. 4, counting the top node as 0; the 4-heap ones are the complements of lines, counting it as 7.

SYNONIM

In **Sympathetic Nim** all heaps of the same size must be treated alike—if you reduce one heap of a given size you must reduce all heaps of that size and by the same amount (no move may affect heaps of different sizes). In the strip version the move is to take *all* the coins from some square and put them on to any earlier square.

0	X	1	2	3	4	5	6	7
X	1	0	X	6	5	4	3	10
1	0	X	X	4	3	6	5	8
2	X	X	X	X	X	X	X	X
3	6	4	X	X	1	7	0	5
4	5	3	X	1	X	0	7	6
5	4	6	X	7	0	X	1	3
6	3	5	X	0	7	1	X	4
7	10	8	X	5	6	3	4	X

0	X	1	2	3	4	5	6	7
X	4	5	6	X	0	1	2	11
1	5	X	4	X	2	0	7	6
2	6	4	X	X	1	7	0	5
3	X	X	X	X	X	X	X	X
4	0	2	1	X	X	6	5	8
5	1	0	7	X	6	X	4	2
6	2	7	0	X	5	4	X	1
7	11	6	5	X	8	2	1	X

0	X	1	2	3	4	5	6	7
X	3	6	5	0	X	2	1	12
1	6	X	3	2	X	7	0	5
2	5	3	X	1	X	0	7	6
3	0	2	1	X	X	6	5	8
4	X	X	X	X	X	X	X	X
5	2	7	0	6	X	X	3	1
6	1	0	7	5	X	3	X	2
7	12	5	6	8	X	1	2	X

0	X	1	2	3	4	5	6	7
X	6	3	4	1	2	X	0	13
1	3	X	6	0	7	X	2	4
2	4	6	X	7	0	X	1	3
3	1	0	7	X	6	X	4	2
4	2	7	0	6	X	X	3	1
5	X	X	X	X	X	X	X	X
6	0	2	1	4	3	X	X	8
7	13	4	3	2	1	X	8	X

Table 3. $\mathcal{P}$-positions (k,a,b,c) for Antonim, $2 \leqslant k \leqslant 5$.

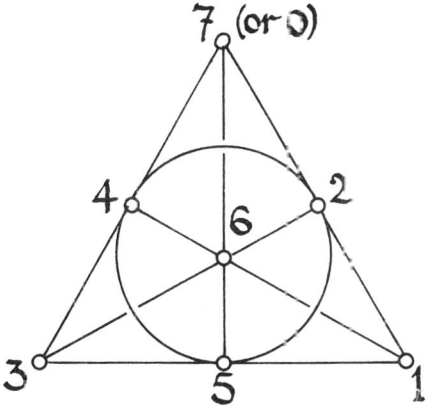

Figure 4. Fano's Fancy Antonim Finder.

This game need not detain us long. Since all the heaps of a given size must be treated in the same way they may be regarded as a single heap. A move reducing this heap to the size of an already existing heap has the same effect as removing the heap entirely. We might as well say that the heaps must always be of different sizes, so

> SYNONIM is
> just a synonym
> for ANTONIM!

SIMONIM

SIimilar MOve NIM, or **Simonim**, was rediscovered by Simon Norton. It is just Nim with the additional feature that a player may make any number of moves provided that these are all exactly similar, i.e. that they all reduce some number a to another number b. It differs from Synonim in that we are *not* required to reduce *all* the heaps of a given size. If we play it with coins on a strip the rule becomes that any number of coins may be moved from any square to any earlier square, occupied or not. Table 4 is a bit harder to compute.

	0	1	2	3	4	5	6	7	8	9	10	11	12	13
0	→↓0	2	1	4	3	6	5	8	7	10	9	12	11	14
1	2	3	0	↓1	→4	7	8	5	6	11	12	9	10	15
2	1	0	4	→3	↓2	8	7	6	5	12	11	10	9	16
3	4	→1	↓3	2	0	9	10	11	12	5	6	7	8	17
4	3	↓4	→2	0	1	10	9	12	11	6	5	8	7	18
5	6	7	8	9	10	11	0	1	2	3	4	↓5	→12	19
6	5	8	7	10	9	0	12	2	1	4	3	→11	↓6	20
7	8	5	6	11	12	1	2	9	0	↓7	→10	3	4	21
8	7	6	5	12	11	2	1	0	10	→9	↓8	4	3	22
9	10	11	12	5	6	3	4	→7	↓9	8	0	1	2	23
10	9	12	11	6	5	4	3	↓10	→8	0	7	2	1	24
11	12	9	10	7	8	→5	↓11	3	4	1	2	6	0	25
12	11	10	9	8	7	↓12	→6	4	3	2	1	0	5	26
13	14	15	16	17	18	19	20	21	22	23	24	25	26	27

Table 4. $\mathscr{P}$-positions for Three-Coin Simonim.

As usual we try to fill in the least number not seen earlier in the row or column (or diagonal, if the entry is on the diagonal). But at most one entry n (written $\vec{n}$) in a row may equal its column label, at most one entry m (written $\downarrow m$) in a column may equal its row label and only the diagonal entry 0 may coincide with its row *and* column label.

When you've stared at Table 4 for an hour or two you'll notice various patterns which make the structure crystal clear. The solid dividing lines mark the closing up of the leading

$$1 \times 1, \, 5 \times 5, \, 13 \times 13,$$

and, in general

$$2^n - 3 \quad \text{by} \quad 2^n - 3$$

portions, which form Latin squares. The arrowed entries fall into 2×2 boxes. Various portions of the table resemble the nim-addition table.

After we'd extended the table into three dimensions and enlarged it in two, we were able to work out a general rule for 4-heap Simonim. With Simon's help we were even able to prove it!

Partition the positive integers into **ranges**

$$1, \quad 2\text{–}3, \quad 4\text{–}7, \quad 8\text{–}15, \quad 16\text{–}31, \quad \ldots.$$

Then transform the Simonim position as follows:
Replace the first occurrence of a number n by n',
a second occurrence by n'', and a third by n''', where

$$n' = \begin{cases} n+3, & \text{if this is in the largest range} \\ & \text{that is represented in the} \\ & \text{transformed position,} \\ n+1, & \text{if this is in the next largest range} \\ & \text{that is represented in the} \\ & \text{transformed position,} \\ n, & \text{otherwise.} \end{cases}$$

$$n'' = \begin{cases} \text{the largest number in the range before } n', \text{ or} \\ \text{the next-to-largest number in this range,} \\ \quad \text{if } n \text{ is the next-to-largest number} \\ \quad \text{in the original position.} \end{cases}$$

$$n''' = \quad \text{the largest number from the range before } n''$$

The original position will be a $\mathscr{P}$-position in SIMONIM just if the transformed position is a $\mathscr{P}$-position in NIM.

RULE FOR 4-HEAP SIMONIM

In applying the rule it's best to write the numbers in descending order. What should we do from

$$n \quad 16 \quad 9 \quad 4 \quad 1?$$

We find

$$n' \quad 19 \quad 10 \quad 4 \quad 1$$

whose nim-sum involves 16, so it can't be a $\mathscr{P}$-position. We must therefore decrease 16 to some value x for which $x+3$ won't be in the 16–31 range. Then

$$n \quad x \quad 9 \quad 4 \quad 1$$

$$n' \quad ? \quad 12 \quad 5 \quad 1$$

so ? must be $12 \overset{*}{+} 5 \overset{*}{+} 1 = 8$. Since this is in the largest range to appear in the transformed position,

$$x \text{ must be } 8 - 3 = 5.$$

Let's do

$$n \quad 9 \quad 9 \quad 7 \quad 2$$

Here we find $\begin{cases} n' & 12 & & 10 & 2 \\ n'' & & 7 & \end{cases}$ nim-sum 3.

We can change the nim-sum to 0 by changing 2 into 1, 10 into 9 or 7 into 4, yielding the positions

n	9	9	7	1		9	9	6	2		9	3	7	2
n'	12		10	1		12		9	2		12	4	10	2
n''		7						7						

A really tricky example is

$$n \quad 44 \quad 33 \quad 22 \quad 11$$

$$n' \quad 47 \quad 36 \quad 23 \quad 11 \quad \text{nim-sum 23}$$

There's no hope of changing any of the nim-values 47, 36, or 11 by 23, while if we *remove* the 22 heap:

$$n \quad 44 \quad 33 \quad 0 \quad 11$$

$$n' \quad 47 \quad 36 \quad 0 \quad 12 \quad \text{nim-sum 7}$$

we don't arrive at a $\mathscr{P}$-position. The trick is to equalize the two small heaps:

$$n \quad 44 \quad 33 \quad 11 \quad 11$$

$$\begin{matrix} n' & 47 & 36 & 12 & \\ n'' & & & & 7 \end{matrix} \Big\} \text{ nim-sum 0.}$$

STAIRCASE FIVES

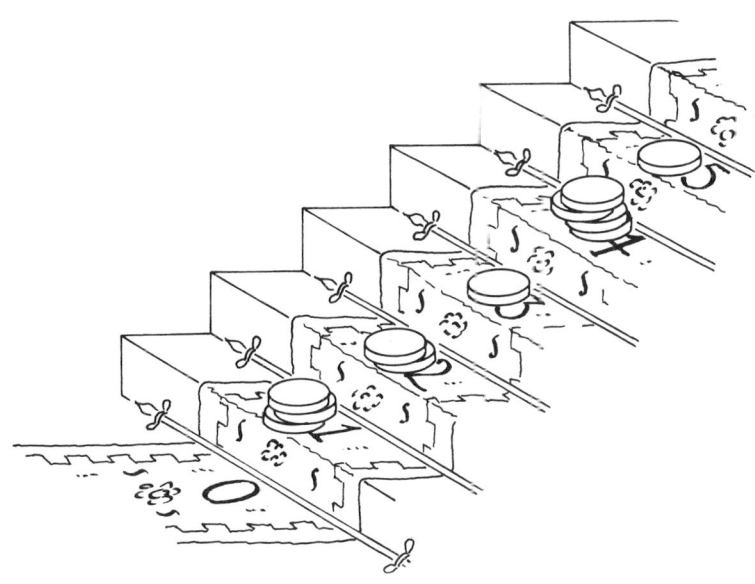

Figure 5. Stacks on Stairs.

You play this with coins on a staircase (Fig. 5). The move is to take any number, *less than five*, of coins from one step and put them on any lower step, *less than five* stairs away. The winner is the one who puts the last coin on the bottommost step.

If there are only 4 coins and 5 steps, the "five" restrictions don't matter and the game reduces to Antonim. A study of the upper 5×5 portion of Tables 1–3 provides an unexpectedly simple rule:

> Mentally interchange the coins on steps 2 and 4.
> Then the position is a $\mathscr{P}$-position just if the sum
> of the heights of all the coins is a multiple of 5.

Thus you should arrange that after your move, if there are

$$a \text{ coins on } 0, \quad b \text{ on } 1, \quad c \text{ on } 2, \quad d \text{ on } 3 \text{ and } e \text{ on } 4,$$

then

$$0 \cdot a + 1 \cdot b + 4 \cdot c + 3 \cdot d + 2 \cdot e$$

is divisible by 5.

The rule continues to apply with more coins and more steps provided we interchange steps $5n+2$ and $5n+4$.

TWOPINS (pronounced "Tuppins")

is a bowling game which generalizes Kayles ($\cdot$77) and Dawson's Kayles ($\cdot$07) in Chapter 4. This time the pins are set up in columns of 1 or 2 and the condition is that, as in Kayles, a legal shot must remove just 1 or 2 adjacent columns. But there is the additional rule that it is illegal to remove just a single pin.

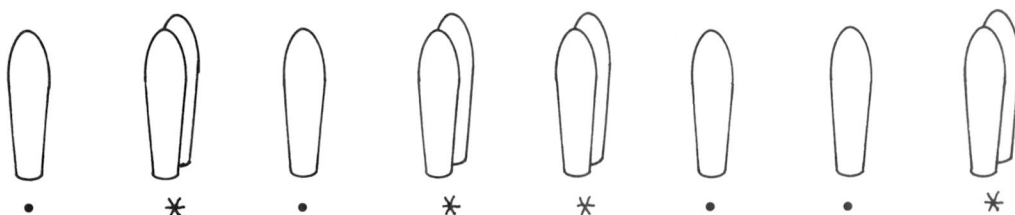

Figure 6. A Game of Twopins.

In discussing Twopins configurations we'll use

$\cdot$ for a column of one,

$*$ for a column of two,

so that the 2^n possible configurations of n non-empty Twopins columns are represented by the 2^n sequences of n $*$'s and $\cdot$'s. For instance we find

$$\cdot = 0, \quad \cdot\,\cdot = \cdot\,\cdot\,\cdot = \cdot\,* \ = \cdot\,*\cdot = *, \quad *\cdot* = *+* = 0,$$

and happily $*$, $**$, $***$ have values $*$, $*2$, $*3$; however, $**** = *$. Fortunately we don't need to list all possible sequences separately because there are several useful equivalences. For example it's easy to see that

$$\sim\!\sim\!\sim *\cdot\,, \quad \sim\!\sim\!\sim \cdot\,\cdot \quad \text{and} \quad \sim\!\sim\!\sim *$$

all behave the same in play, while

$$\sim\!\sim\!\sim *\cdot\,\cdot\,* \sim\!\sim\!\sim \quad \text{behaves like} \quad \sim\!\sim\!\sim *\,*\,* \sim\!\sim\!\sim$$

There is also a useful **Twopins Decomposition Theorem**:

$$\sim\!\sim\!\sim *\cdot\,* \sim\!\sim\!\sim \ = \ \sim\!\sim\!\sim *\ +\ * \sim\!\sim\!\sim$$

After these theorems you can suppose that all strings have stars at each end and that dots come in internal blocks of three or more. Also the sequence $(*)^n$ of n stars behaves like the Kayles position K_n, while the sequences $(\cdot)^n$ and $*(\cdot)^{n-4}*$ behave like D_n in Dawson's Kayles, so you can read off their values from Chapter 4. Our Twopins-Wheel (Fig. 7) gives the nim-values of all other Twopins sequences of length 9 or less, except

$$*\cdot\,\cdot\,\cdot\,*\,\cdot\,\cdot\,\cdot\,*, \quad \text{nim-value 1}$$

Figure 7. A Twopins-Wheel.

All our equivalences remain valid in misère play, but the entries in the Twopins-Wheel should be replaced according to the scheme

For	read	genus	For	read	genus
0	0	0^1	**0**	$2+2$	0^0
1	1	1^0	**1**	$3+2$	1^1
2	2	2^2	**2**	$k_1 k 3_2 2_2 30$	2^2
3	3	3^3	**3**	$2_2 21 = d$	3^{1431}
4	$2_2 321 = k$	4^{146}	**4**	$3_2 320$	4^{046}
5	$k+1$	5^{057}	**5**	$kd3_2 210$	5^{3146}

Twopins has applications to Dots-and-Boxes (Chapter 16) and to

CRAM

which is Martin Gardner's name for impartial Domineering. It has also been called Plugg and Dots-and-Pairs, and is associated with the names of Geoffrey Mott-Smith, Sol Golomb and John Conway. You play it just like ordinary Domineering (Chapter 5; Martin Gardner called it Crosscram) except that *either* player may place his dominoes in *either* direction. You just cram them in however you can.

If you start with a rectangle with an even number of squares, then there's a simple symmetry strategy for the second player if the aim is to be the last to move, so it's a good idea to declare that the last one is the *loser*, i.e. to play Misère Cram.

It helps to see what's going on if you replace the available regions by graphs with nodes for squares, joined by edges when they're adjacent, as we did for Col (Chapter 2) and Snort (Chapter 6).

In this form the move is to delete two adjacent nodes and all edges running up to them, and you can play the game on arbitrary graphs. Only the abstract structure of the graph matters, so that many differently shaped regions can have the same graph, e.g.

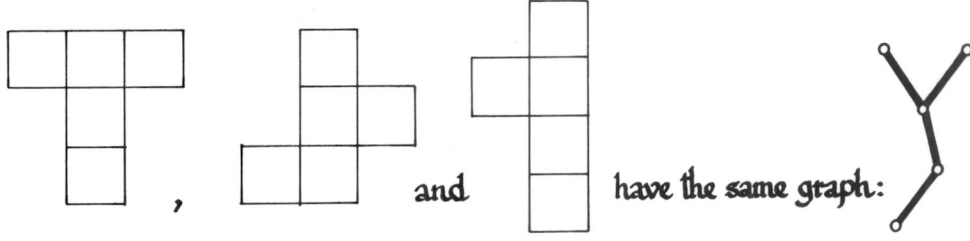

This graph, like many others, is a caterpillar. Formally, a **caterpillar** is a graph whose **body**

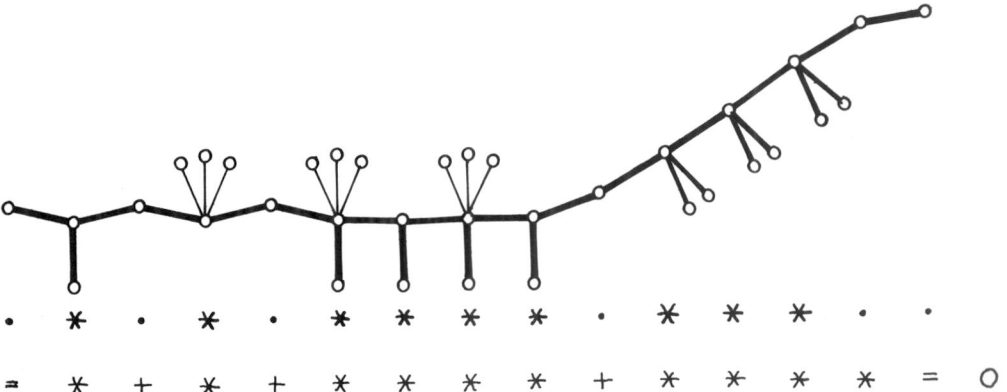

Figure 8. Even a Complicated Cram Caterpillar is a Twopins Position.

consists of a chain of nodes, some of which may have **tufts** (or legs) i.e. 1 or more edges leading to otherwise isolated nodes. Luckily, even the most complicated caterpillar (Fig. 8) is equivalent to a Twopins configuration, by letting

- · replace each untufted body node, and
- * represent any tufted one.

In this notation our Twopins equivalences become

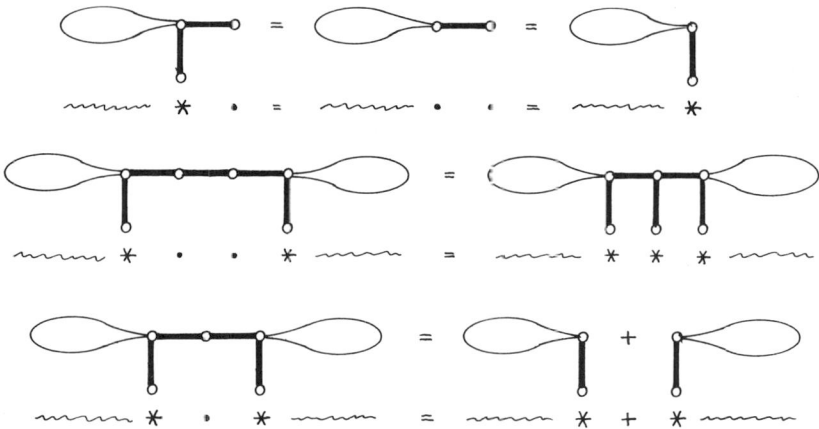

the last of which we might indicate in a single picture, using thin lines for edges which can all be omitted without affecting its value.

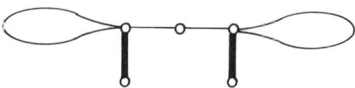

The balloons in our pictures need not be caterpillar-shaped and can even be allowed to meet, so our last identity becomes

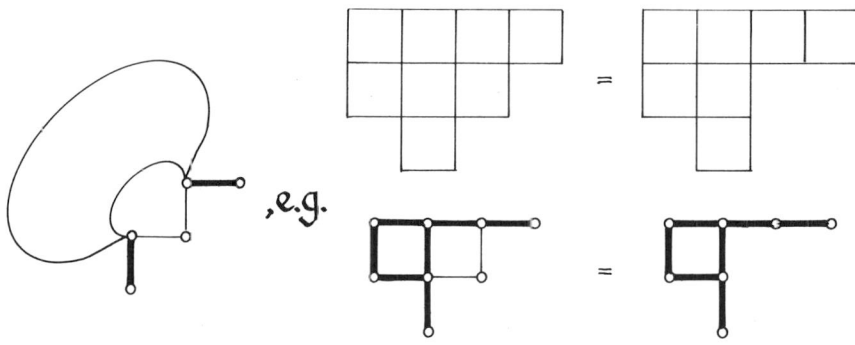

The diagrams in Fig. 9 have similar properties for example, the deletion of *all* thin edges in a diagram will not affect its value.

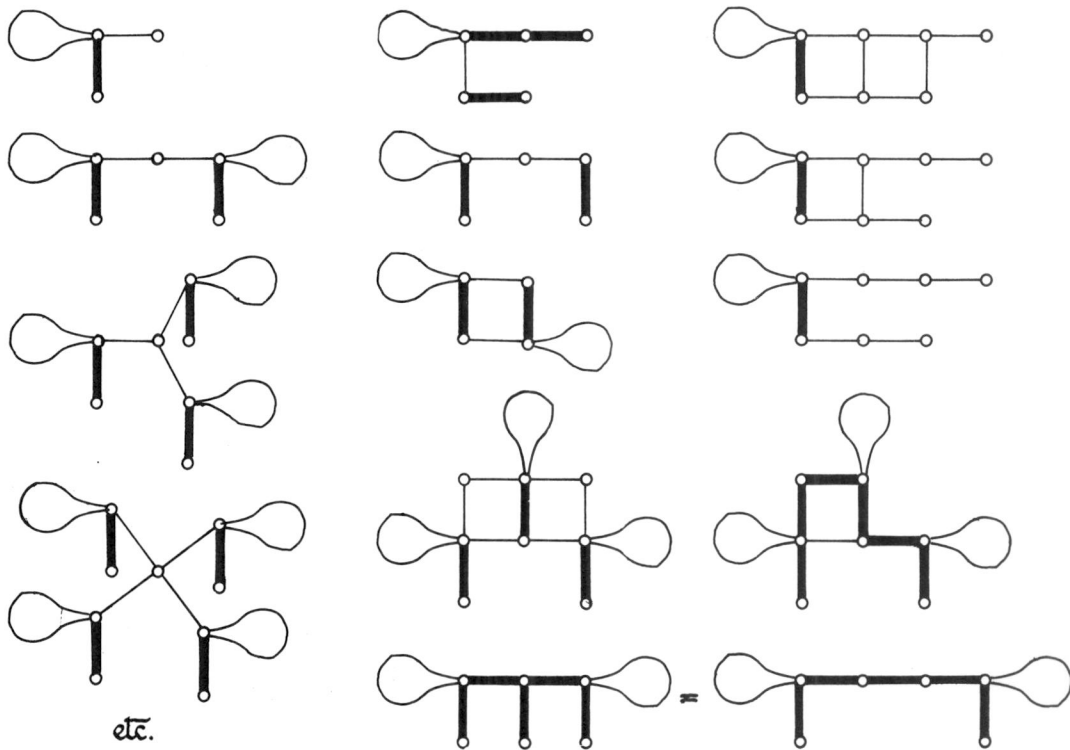

Figure 9. A Packet of Cram Crackers.

Table 5 gives values of a number of Cram positions. The dotted line indicates a chain of n edges where n is at the head of the table. We've given the full genus so that you can play Misère Cram. We use the letters

k for 2_2321 (position K_5 in Kayles),
d for 2_121 (position D_{10} in Dawson's Kayles),
e for 2_231 (arises in the *Ex*-Officer's Game, ·**06**), and
f for 2_21 (arises in Flanigan's Game, ·**34**)

in the last column, to list the games which are not Nim-heaps.

Some other values appear in Fig. 10. The ladder values in particular are easy to remember and the remark about tufts makes them extremely useful.

number of edges in the dotted line	0	1	2	3	4	5	6	non-Nim-heaps
	1	1	2	0	3	1	1	–
	2	2	3	3	1	2	4^{146}	k
	3	3	1	2	4^{146}	3	3	k
	1	1	4^{146}	0	3	5^{057}	2^2	$k, k+1, k_1k3_22_230$
	3	3	2	2	0	3	5^{057}	$kd3_2320$
	1	1	2	0	3	1	2^{0520}	$k3_230$
	0	0	1	1	2	2^{1420}	3	f
	0	3	1	2	2^{1420}	3	5^{3146}	$e, ked3_210$
	3	1	2	0	3^{1431}	3	2^{0520}	$d, d+1$
	1	1	2	0	3	1	2^{0520}	$kd3_230$
		3	1	0^0	4^{146}	1^1	3	$2_2, k, 3_2$
		1	2	2	3	3^{32}	5^{057}	$kf3_2210, k+1$
		2	2^{1420}	0	1	1	2	f

Table 5. The Genus of Various Cram Positions.

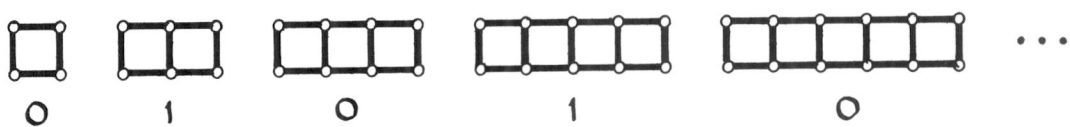

(The values of these ladders alternate and are unaffected by the addition of up to two tufts.)

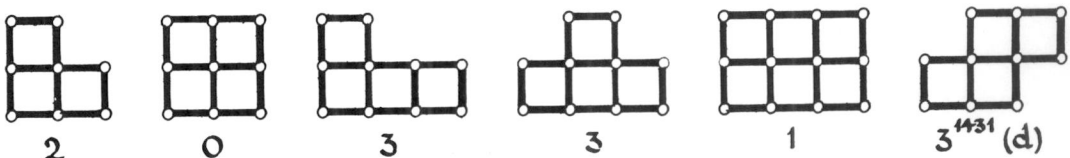

   2   0   3   3   1 3^{1431} (d)

Figure 10. A Few More Cram Values.

WELTER'S GAME

This is the coin-on-strip game in which at most one coin may be on a square, and any leftwards move of a coin onto an empty square is permitted, even if it passes over other coins. Although the simplest cases were investigated by Roland Sprague, C.P. Welter discovered many remarkable properties of the general case. A simplified version of the theory is given in ONAG. Here we'll just tell you the answers and describe some new discoveries.

We'll write

$$[a|b|c|\ldots]_k \qquad (\text{"}a \textbf{ welt } b \text{ welt } c \text{ welt } \ldots\text{"})$$

for the nim-value of the Welter's Game position with coins on the k different squares

$$a, b, c, \ldots,$$

and will often omit k when the number of terms is clear. The easiest way to compute this **Welter function** is the **Mating Method**.

Mate those two of the k numbers that are congruent modulo the highest power of 2. Then select a pair of mates from the remaining $k-2$ numbers by the same rule, and so on. Eventually we have mated all except possibly one of the numbers (the **spinster**, s), say as

$$(a, b), (c, d), \ldots, \text{and possibly } s.$$

Then

$$[a|b|c|\ldots]_k = [a|b] \overset{*}{+} [c|d] \overset{*}{+} \ldots \qquad (\overset{*}{+} s \text{ if } k \text{ is odd}).$$

The two-term Welter function can be evaluated using the formula

$$[x|y] = (x \overset{*}{+} y) - 1.$$

For example,

$$[2|3|5|7|11|13|17|19]_8$$
$$= [3|19] \overset{*}{+} [5|13] \overset{*}{+} [7|11] \overset{*}{+} [2|17]$$
$$= \quad 15 \quad \overset{*}{+} \quad 7 \quad \overset{*}{+} \quad 11 \quad \overset{*}{+} \quad 18 \quad = 17,$$

there being no spinster in this case, while in

$$[0|1|4|9|16|25|36]$$
$$= [4|36] \overset{*}{+} [0|16] \overset{*}{+} [9|25] \overset{*}{+} 1$$
$$= \quad 31 \quad \overset{*}{+} \quad 15 \quad \overset{*}{+} \quad 15 \quad \overset{*}{+} 1 = 30$$

we see that 1 is the spinster. In this example there were two equally well mated pairs, (0, 16), (9, 25). In such cases it doesn't matter which pair we mate first.

FOUR-COIN WELTER IS JUST NIM

When you play a few games you'll soon notice, like many other people, that a Nim-like strategy suffices for Welter's Game with four coins, so that

$$[a|b|c|d] = 0 \quad \text{just if} \quad a \overset{*}{+} b \overset{*}{+} c \overset{*}{+} d = 0.$$

Welter's theory explains this by noting that if a, b and c, d are the mates, these equalities reduce to

$$[a|b] = [c|d] \quad \text{and} \quad a \overset{*}{+} b = c \overset{*}{+} d$$

which are equivalent since $[x|y] = (x \overset{*}{+} y) - 1$.

AND SO'S THREE-COIN WELTER!

If one of your four coins is on 0, you're really just playing three-coin Welter with the others, but shifted one place. In symbols

$$[0|a|b|c] = [a-1 \mid b-1 \mid c-1]$$

or

$$[a|b|c] = [0 \mid a+1 \mid b+1 \mid c+1].$$

Thus the Welter position with coins on 2, 5, 7 is equivalent to the Welter or Nim position with coins on 0,3,6,8, which is cured by moving 8 to 5, so in the three-coin position we should move 7 to 4.

THE CONGRUENCE MODULO 16

Although the Mating Method makes it very easy to work out nim-values, it's not so easy to find which move you should make to restore the nim-value to 0. But if the number of coins is a multiple of 4, there is a remarkable connexion with Nim:

$$[a|b|c|\ldots]_{4k} \equiv 0, \text{ mod } 16$$
$$\text{exactly when}$$
$$a \overset{*}{+} b \overset{*}{+} c \overset{*}{+} \ldots \equiv 0, \text{ mod } 16.$$

This ensures in particular that when the $4k$ coins are among the first 16 places the Welter's game $\mathcal{P}$-positions are exactly the $\mathcal{P}$-positions in Nim that have distinct numbers. What are the good moves from

$$(0,1,2,3,5,7,11,13)?$$

The numbers	0	1	2	3	5	7	11	13	nim-add to 4,
and we get	4	5	6	7	1	3	15	9	by nim-adding 4 to them.
But the marks	×	×	×	×	×	×	×	√	show that only the last of these is legal (the rest

involve increases or moves to occupied squares) so the only good move is from 13 to 9.

Now let's look at

	2	3	5	7	11	13	17	19,	nim-sum 7,
giving	5	4	2	0	12	10	22	20,	on nim-adding 7,
which reduce to	5	4	2	0	12	10	6	4,	modulo 16.
	×	×	×		×				

So the only hopeful moves are

$$7 \text{ to } 0, \quad 13 \text{ to } 10, \quad 17 \text{ to } 6 \quad \text{and} \quad 19 \text{ to } 4.$$

But of the Welter functions

$$[2|3|5|0|11|13|17|19], \quad [2|3|5|7|11|10|17|19], \quad [2|3|5|7|11|13|6|19], \quad [2|3|5|7|11|13|17|4]$$

only the third can be zero (glance at the mate of 17 to see that the binary expansion of the others must have a 16-digit). So the unique good move is from 17 to 6.

What happens when the number of coins *isn't* a multiple of 4? If there are 6 coins, say, on positions

$$1, 2, 3, 5, 8, 13$$

you can pretend that there are really 8 coins on places

$$-2, -1, 1, 2, 3, 5, 8, 13$$

on a strip you've perversely numbered starting from -2. Renumbering from 0 we see the position

	0	1	3	4	5	7	10	15	nim-sum 1
yielding	1	0	2	5	4	6	11	14	
	×	×		×	×		×		

so this time there are three good moves, from

$$3 \text{ to } 2, \quad 7 \text{ to } 6, \quad 15 \text{ to } 14, \quad \text{in the new notation,}$$
$$\text{or} \quad 1 \text{ to } 0, \quad 5 \text{ to } 4, \quad 13 \text{ to } 12, \quad \text{in the old.}$$

If there are 5 coins, say on

			2	3	5	7	11	
we increase by 3:	0	1	2	5	6	8	10	14	nim-sum 12
yielding			9	10	4	6	2	
			×	×		×	×	
decreasing again:					1			

showing that the only good move is from 5 to 1.

FRIEZE PATTERNS

Patterns of numbers such as

```
1    1    1    1    1    1    1    1    1    1    1    1    1    ...
   1    2    2    3    1    2    4    1    2    2    3    1    ...
      1    3    5    2    1    7    3    1    3    5    2    ...
         1    7    3    1    3    5    2    1    7    3    ...
            1    2    4    1    2    2    3    1    2    4    ...
         1    1    1    1    1    1    1    1    1    1    ...
```

(in which each diamond of numbers

$$
\begin{matrix} & b & \\ a & & d \\ & c & \end{matrix}
\quad \text{satisfies} \quad ad = bc + 1 \quad \text{so that} \quad d = \frac{bc + 1}{a}),
$$

have many wonderful properties. For example, if you start with two horizontal rows of 1's connected by any zigzag of intermediate 1's, say

```
1    1    1    1    1    1    1    1    1    1    1
   1    ?    ?    ?    ?    !    ?    ?    !    .    .
      1    ?    ?    ?    !    ?    ?    ?    !    .    .
   1    ?    ?    ?    !    ?    ?    ?    !    .    .
      1    ?    ?    ?    !    ?    ?    ?    !    .
         1    ?    ?    !    ?    ?    ?    ?    !    .
            1    1    1    1    1    1    1    1    1
```

you'll find that all the entries are whole numbers and that each ! is 1 so that the pattern repeats itself alternately one way up then the other. These self-checking properties mean that your children can have fun while practising their arithmetic. If you want to check your own arithmetic on the above example, see the Extras.

G.C. Shephard has observed that we can replace multiplication by addition, making each diamond

$$
\begin{matrix} & b & \\ a & & d \\ & c & \end{matrix}
\quad \text{satisfy} \quad (a+d) = (b+c) + 1 \quad \text{so that} \quad d = b + c + 1 - a.
$$

If we replace the starting 1's by 0's the resulting pattern

```
0    0    0    0    0    0    0    0    0    0    0    0    0   ...
  0    1    2    4    3    0    1    2    4    3    0    1      ...
     0    2    5    6    2    0    2    5    6    2    0    2   ...
        0    4    6    4    1    0    4    6    4    1    0     ...
     0    1    4    3    2    0    1    4    3    2    0    1   ...
  0    0    0    0    0    0    0    0    0    0    0    0      ...
```

has similar properties, but this time it repeats itself the same way up.

We thought it might be a good idea to take the basic operation as nim-addition rather than ordinary multiplication or addition, and to our great surprise found that we had discovered a new way of calculating the Welter function!

You start with a row of zeros above the Welter position you want to evaluate, and work downwards making each diamond

$$b$$
$$a \qquad d \qquad \text{satisfy} \quad (a \overset{*}{+} d) = (b \overset{*}{+} c) + 1 \quad \text{so that} \quad c = ((a \overset{*}{+} d) - 1) \overset{*}{+} b,$$
$$c$$

and, in the unlikely event that you make no mistakes, you'll find the Welter function at the bottom of the triangle; e.g.

```
        0     0     0     0     0     0     0     0     0
           2     3     5     7    11    13    17    19
              0     5     1    11     5    27     1
                 7     6    14     6    16     8
                    5     6    12    16    12
                       4     7    29    11
                          4    21     5
                            23    18
                               17
```

yields the same answer as before, so *we* probably haven't made any mistakes! This rule is equivalent to the identity

$$[a|b|...|y|z]_{k+1} = [[a|b|...|y]_k \mid [b|...|y|z]_k] \overset{*}{+} [b|...|y]_{k-1}$$

which you can find in ONAG (p. 159).

Although by hand this calculation seems much longer, it's quite a good technique to use if you want to teach your computer to play Welter's Game.

INVERTING THE WELTER FUNCTION

Suppose that you've evaluated

$$[a|b|c|\ldots] = n$$

and have in mind a number $n' \neq n$. Then there are unique numbers

$$a' \neq a, \quad b' \neq b, \quad c' \neq c, \quad \ldots$$

for which

$$[a'|b|c|\ldots] = n',$$
$$[a|b'|c|\ldots] = n',$$
$$[a|b|c'|\ldots] = n',$$
$$\ldots\ldots\ldots\ldots\ldots$$

Moreover it can be shown that the equation

$$[a|b|c|\ldots] = n$$

remains true if any *even* number of the letters $a, b, c, \ldots, n$ are replaced by the corresponding primed letters. We express this happy state of affairs by the single "equation"

$$\begin{bmatrix} a & b & c & \ldots \\ a' & b' & c' & \ldots \end{bmatrix} = \frac{n}{n'}$$

Using this Even Alteration Theorem and the properties of frieze patterns your computer can invert the Welter function.

For example, if you have five numbers with Welter function

$$[a|b|c|d|e] = n$$

and want to find the numbers

$$a', b', c', d', e'$$

for which

$$\begin{bmatrix} a & b & c & d & e \\ a' & b' & c' & d' & e' \end{bmatrix} = \frac{n}{n'}$$

you should complete the frieze pattern

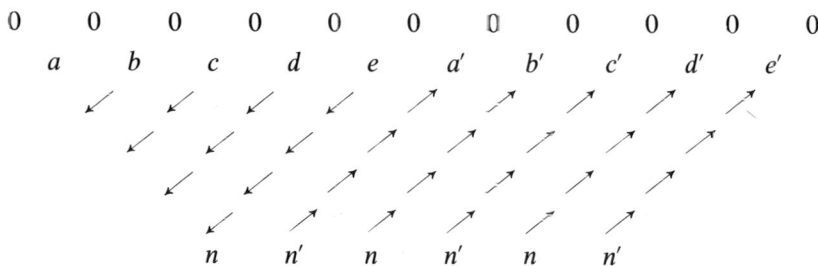

in which n and n' alternate along the bottom row, by working in the directions shown by the arrows.

Thus to find the good moves from

$$1 \quad 4 \quad 9 \quad 16 \quad 25$$

we must change one number to make the Welter function 0. The calculation

0	0	0	0	0	0	0	0	0	0	0	0	0
1	4	9	16	25	36	33	12	13	28	1	4	
4	12	24	8	60	4	44	0	16	28	4		
3	26	31	42	19	6	39	2	23	22			
20	28	60	4	16	12	36	4	28				
29	0	29	0	29	0	29	0					

(in which the rightmost two diagonals are only for checking) shows that

$$\begin{bmatrix} 1 & 4 & 9 & 16 & 25 \\ 36 & 33 & 12 & 13 & 28 \end{bmatrix} = \frac{29}{0}$$

and so the only good move is from 16 to 13.

THE ABACUS POSITIONS

One day we idly wrote down the infinite frieze pattern

...	0	0	0	0	0	0	0	0	**0**	0	0	0	0	0	0	0	0	...
...	14	12	10	8	6	4	2	0	1	3	5	7	9	11	13	15		...
...	1	5	1	13	1	5	1	**0**	1	5	1	13	1	5	1			...
...	15	9	3	13	7	1	0	1	0	6	12	2	8	14				...
...	0	8	0	8	0	1	**0**	1	0	8	0	8	0					...
...	14	4	10	0	1	0	1	0	1	11	5	15						...
...	1	13	1	0	1	**0**	1	0	1	13	1							...
...	15	1	0	1	**0**	1	0	1	0	14								...
...	0	1	0	1	**0**	1	0	1	0									...

which suggested to us the sequence of equations

$$\begin{bmatrix} 0 \\ 1 \end{bmatrix} = \frac{0}{1} \qquad \begin{bmatrix} 2 & 0 \\ 1 & 3 \end{bmatrix} = \frac{1}{0} \qquad \begin{bmatrix} 4 & 2 & 0 \\ 1 & 3 & 5 \end{bmatrix} = \frac{1}{0} \qquad \begin{bmatrix} 6 & 4 & 2 & 0 \\ 1 & 3 & 5 & 7 \end{bmatrix} = \frac{0}{1} \cdots$$

$$\cdots \begin{bmatrix} 14 & 12 & 10 & 8 & 6 & 4 & 2 & 0 \\ 1 & 3 & 5 & 7 & 9 & 11 & 13 & 15 \end{bmatrix} = \frac{0}{1} \cdots$$

Since we can interchange any even number of the pairs

$$(a,a'),(b,b'),(c,c'),\ldots,(n,n'),$$

we can reorder these equations to say

$$\begin{bmatrix} 0 \\ 1 \end{bmatrix} = \begin{bmatrix} 0 & 1 \\ 3 & 2 \end{bmatrix} = \begin{bmatrix} 0 & 1 & 2 \\ 5 & 4 & 3 \end{bmatrix} = \begin{bmatrix} 0 & 1 & 2 & 3 \\ 7 & 6 & 5 & 4 \end{bmatrix} = \begin{bmatrix} 0 & 1 & 2 & 3 & 4 \\ 9 & 8 & 7 & 6 & 5 \end{bmatrix} = \cdots = \frac{0}{1}$$

For some reason the particular equation

$$\begin{bmatrix} 0 & 1 & 2 & 3 & 4 \\ 9 & 8 & 7 & 6 & 5 \end{bmatrix} = \begin{matrix} 0 \\ 1 \end{matrix}$$

made us think of our abacus (Fig. 11) so we call the positions evaluated in the equation

$$\begin{bmatrix} 0 & 1 & 2 & \ldots & k-3 & k-2 & k-1 \\ 2k-1 & 2k-2 & 2k-3 & \ldots & k+2 & k+1 & k \end{bmatrix}_k = \begin{matrix} 0 \\ 1 \end{matrix}$$

the k-coin **Abacus Positions**. Thus the equation

$$[9|1|2|6|5]_5 = 1$$

shows a 5-coin Abacus Position with its Welter function (or nim-value) 1.

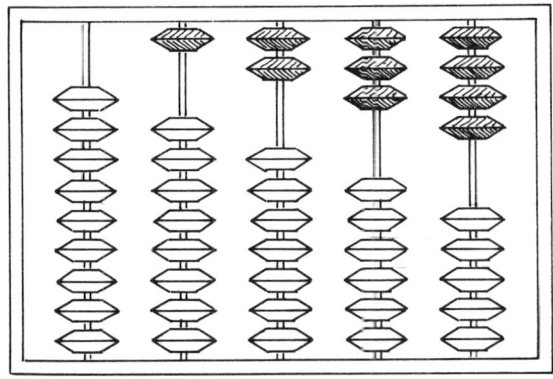

Figure 11. Swanpan.

THE ABACUS STRATEGY

We can give an explicit strategy for the Abacus Positions. Let the putative equation

$$[a|b|c|\ldots]_k = 0$$

represent one of the Abacus Positions which we believe has Welter function 0. Define

$$a' = 2k - 1 - a, \qquad b' = 2k - 1 - b, \qquad c' = 2k - 1 - c, \qquad \ldots$$

and note that we have an even number of

$$a > a', \quad b > b', \quad c > c', \quad \ldots .$$

Now suppose your opponent makes the move which replaces a by x. Then since every number $\leqslant 2k - 1$ appears in the list

$$a, a', b, b', c, c', \ldots ,$$

x must be one of

$$a', b', c', \ldots ,$$

say b' or a'. If $x = b'$ we must have

$$a > b', \quad \text{and so} \quad b > a'$$

and we can respond with the move from b to a' since

$$[b'|a'|c\ldots]_k = 0$$

represents a simpler Abacus Position. If $x = a'$, then we have $a > a'$ and therefore an *odd* number of

$$b > b', \quad c > c', \quad \ldots$$

so that we can respond with one of the moves

$$b \text{ to } b', \quad c \text{ to } c', \quad \ldots.$$

A similar strategy shows that if

$$[a|b|c|\ldots]_k = 1$$

represents one of the Abacus Positions asserted to have Welter function 1, then for every move our opponent makes except the very last move of the game, we can reply with a move to another such position.

THE MISÈRE FORM OF WELTER'S GAME

The remarks we've just made show not only that the Abacus Positions really do have the asserted nim-values 0 and 1, but actually that they are equivalent to nim-heaps of sizes 0 and 1 even in the misère form of Welter's Game, for it is easy to see that there is a move from any non-terminal Abacus Position to an Abacus Position of the other value. In the language of Chapter 13,

> every Abacus Position is *fickle*,

because nim-heaps of sizes 0 and 1 swap outcomes when we change from normal to misère play. On the other hand,

> every *non*-Abacus Position is *firm*,

a result which establishes that

> Welter's Game
> is really tame!

It suffices to show that if we can move from some *non*-Abacus Position

$$(x, b, c, \ldots) \ldots$$

to some Abacus Position

$$(a, b, c, \ldots)$$

then we could also have moved to an Abacus Position of the opposite value. But in our previous notation, if $x > a'$ we can move to

$$(a', b, c, \ldots).$$

Otherwise we must have $x < a'$, since

$$(a', b, c, \ldots)$$

is an Abacus Position. Since all numbers $\leqslant 2k - 1$ appear among

$$a, a', b, b', c, c', \ldots$$

we can suppose that $x = b'$, say, whence

$$b' < a' \quad \text{and so} \quad a < b$$

so that we could have moved to

$$(b', a, c, \ldots).$$

> If you intend to *lose* Welter's Game,
> play as if you meant to win, until
> this would make you move into an
> Abacus Position, and then move instead to
> an Abacus Position of the opposite kind.

T.H. O'Beirne considered the misère form of Welter's Game. However our complete analysis, which independently reaches the same conclusions as that of Yamasaki, shows that his simple rule only works for very small numbers.

KOTZIG'S NIM

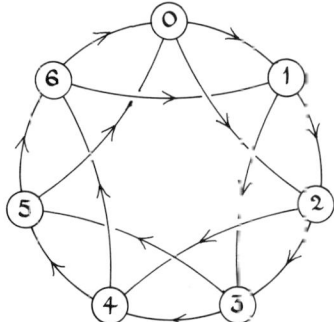

Figure 12. Kotzig's Nim on a 7-Place Strip with Move Set {1,2}.

You play this by placing coins on a circular strip. Start by placing a coin on any square—after that each player in turn puts a coin just *m* places further round the strip in a clockwise direction from the last coin placed. You must choose *m* from a previously decided **move set**. You lose if all the places where you might put coins are already occupied—you're only allowed to put one coin in any one place. Figure 12 is the directed graph (that's the way Anton Kotzig originally described his game) showing the successive places which may be occupied when the move set is {1, 2} and you play on a 7-place strip. What happens?

By symmetry we can assume that the first player plays on 0, and then he can win as follows:

2nd	1st	2nd	1st	2nd	1st
1?	3!	∼	6!		
2?	4!	5?	6	1	3.
		6?	1	3	5.

The sign ∼ means "any legal move". Where there's a choice of moves, we've put ! or ? to indicate winning or losing; other moves are forced.

If the move set contains only one move, *m*, the game is just She-Loves-Me, She-Loves-Me-Not. For if there are *n* places on the strip, there will be just *n*/*d* moves made, where *d* is the g.c.d. of *m* and *n*. If *n*/*d* is even, the second player wins; if odd, the first.

If the move set is {1, 2}, all values of *n* are $\mathscr{P}$-positions, except for *n* = 1, 3 and 7. We've already seen that *n* = 7 is an $\mathscr{N}$-position, and it's easy to check that *n* = 1 and 3 are, too. Here's a strategy for the second player in all other cases:

	1st	2nd	1st	2nd	1st	2nd	2nd	1st	2nd	1st	2nd
$n = 3k+2 \ (k \geqslant 0)$	0?	1!	∼	4!	∼	7!	∼ ... ∼ $(3k+1)!$				
$n = 3k \ (k \geqslant 2)$	0?	2!	3?	4!	∼	7!	∼ ... ∼ $(3k-2)!$	$3k-1$	1		
			4?	5!	∼	8!	∼ ... ∼ $(3k-1)!$	1	3		
$n = 3k+1 \ (k=1, k \geqslant 3)$	0?	2!	3?	5!	∼	8!	∼ ... ∼ $(3k-1)!$	$3k$	1		
			4?	6!	7?	8!	∼ ... ∼ $(3k-1)!$	$3k$	1	3	5
					8?	9!	∼ ... ∼ $(3k)!$	1	3	5	7

(in the last case, if *k* = 1, the move 4 is illegal, so play goes 0? 2! 3 1).

We've got quite used to games which behave regularly after a while, with a few exceptions near the beginning. If the move set is {1, 3} the game is *exactly* periodic with period 6. The $\mathscr{N}$-positions are just those with $n \equiv 1$ or 3, mod 6. Here's Richard Nowakowski's explanation.

If *n* is even the first player always plays on even places, so the second player wins, since the move *m* = 1 is always available to him.

If *n* is odd, then on the first tour of the strip, if a player A responds to a coin placed on *p*, say, with a coin on *p* + 1, then the other player wins by putting a coin on *p* + 2 (so long as *p* + 2 < *n*) and A will find himself blocked the next time round. So each player uses the move *n* = 3 as long as he can.

So if $n \equiv 3$, mod 6, the first player will arrive on $n-3$, forcing the second player to $n-2$, leaving $n-1$ to the first player, who wins next time round.

If $n \equiv 1$, mod 6, the first player arrives on $n-1$. The second time round both players are forced to play on places $p \equiv 2$, mod 3, and the last time round on places $p \equiv 1$, mod 3. Since n is odd, the first player wins.

If $n \equiv 5$, mod 6, the second player wins since first time round he arrives at $n-2$. The first player now plays on $n-1$ or on 1, and the corresponding winning replies are 2 and 4.

For the move set $\{2, 3\}$ we leave the reader to verify that

$$n \equiv 0, 1 \text{ and } 4, \quad \text{mod } 5 \text{ are } \mathscr{P}\text{-positions, } except \text{ } for \text{ } n = 1, 5 \text{ and } 11. \text{ and}$$
$$n \equiv 2 \text{ and } 3, \quad \text{mod } 5 \text{ are } \mathscr{N}\text{-positions, } except \text{ } for \text{ } n = 2.$$

Omar will also confirm that if the move set is $\{1, 2, 3\}$, then

the $\mathscr{P}$-positions are $n \equiv 0, 1, 2$, mod 4, except for $n = 1$ and 5, and
the $\mathscr{N}$-positions are $n \equiv 3$, mod 4, except for $n = 7$,

and will go on to examine more complicated move sets.

FIBONACCI NIM

Suppose you play with just one heap of chips, and let the first player take away any number he likes, but not the whole heap. After that each player may take at most twice as many as the previous player took. Who wins?

The $\mathscr{P}$-positions turn out to be heaps with a Fibonacci number

$$u_1 = u_2 = 1, \quad u_3 = 2, \quad u_4 = 3, \quad u_5 = 5, \quad 8, \quad 13, \quad 21, \quad 24, \quad 55, \quad 89, \quad \dots$$

of chips. Zeckendorf has a remarkable theorem which says that any whole number has a *unique* expression as the sum of *non-neighboring* Fibonacci numbers, for instance

$$54 = 34 + 13 + 5 + 2.$$

If the heap has a *non*-Fibonacci number of chips, the next player can win by taking any number of small terms from such an expansion, provided their total is *less than half* the next largest term. E.g. from a heap of 54 take 2, but not $2 + 5 = 7$ in case your opponent then takes 13.

MORE GENERALLY BOUNDED NIM

Suppose the rules are changed very slightly to read

"may take less than twice as many as"

instead of

"may take at most twice as many as";

does it make much difference? Curiously enough we get the same result as if we had changed the rules to read

"may take no more than".

If the number of chips is a power of 2 it's a $\mathscr{P}$-position in either case; otherwise the next player can win by taking the highest power of 2 which divides the number of chips.

Of course, if the rules say

"may take less than",

then, provided there's more than one chip, you win immediately by taking one, since your opponent then has no legal move. It's a disguise for She-Loves-Me-Constantly.

There are two whole series of such games, in which the rules read

"may take less than k times as many as"

or

"may take at most k times as many as".

For the "less than" games the sequence of $\mathscr{P}$-positions $\{a_n\}$ satisfies the recurrence relation

$$a_{n+1} = a_n + a_{n-1} \qquad \text{for } n \geqslant n_l,$$

and for the "at most" games the relation is

$$a_{n+1} = a_n + a_{n-m} \qquad \text{for } n \geqslant n_m,$$

where l, m, n_l and n_m are given in the table:

$k =$	1	2	3	4	5	6	7	...
$l =$	–	0	2	5	7	10	13	...
$m =$	0	1	3	5	7	10	13	...
$n_l =$	–	2	5	13	14	23	28	...
$n_m =$	2	3	6	9	11	19	24	...

To be consistent with the usual labelling of the Fibonacci numbers we start each sequence with $a_2 = 1$. The "less than" sequences continue with

$$a_i = i - 1 \quad (2 \leqslant i \leqslant k+1), \qquad a_i = 2i - k - 2 \ (k+1 \leqslant i \leqslant (3k+2)/2), \qquad \ldots$$

and the "at most" ones with

$$a_i = i - 1 \quad (2 \leqslant i \leqslant k+2), \qquad a_i = 2i - k - 3 \ \ (k+2 \leqslant i \leqslant (3k+5)/2), \qquad \ldots$$

but as k increases, it takes longer and longer before the sequences settle down. Omar, having stayed with us so far, will doubtless find the exact way in which these sequences and the above table continue.

EPSTEIN'S PUT-OR-TAKE-A-SQUARE GAME

This is also played with just one heap of chips. At each turn there are just two options: to add or take away the largest perfect square number of chips that there is in the heap. For example, if the number in the heap *is* a perfect square other than 0, the next player can win by taking the whole heap.

This is a loopy game! If we start from a heap of 2, the legal moves are to add or subtract 1. The first player won't take 1, leaving a perfect square, so he adds 1 to make 3. For the same reason his opponent doesn't *add* 1, so he takes 1 and the game is drawn.

But 5 is a $\mathscr{P}$-position since 5 ± 4 are both squares! And $4 \times 5 = 20$, $9 \times 5 = 45$, $16 \times 5 = 80$ are also $\mathscr{P}$-positions; why not 125? A slightly more interesting $\mathscr{P}$-position is 29. The next player won't *subtract* 25, but when he adds it to make 54, his opponent can go to 5 and win.

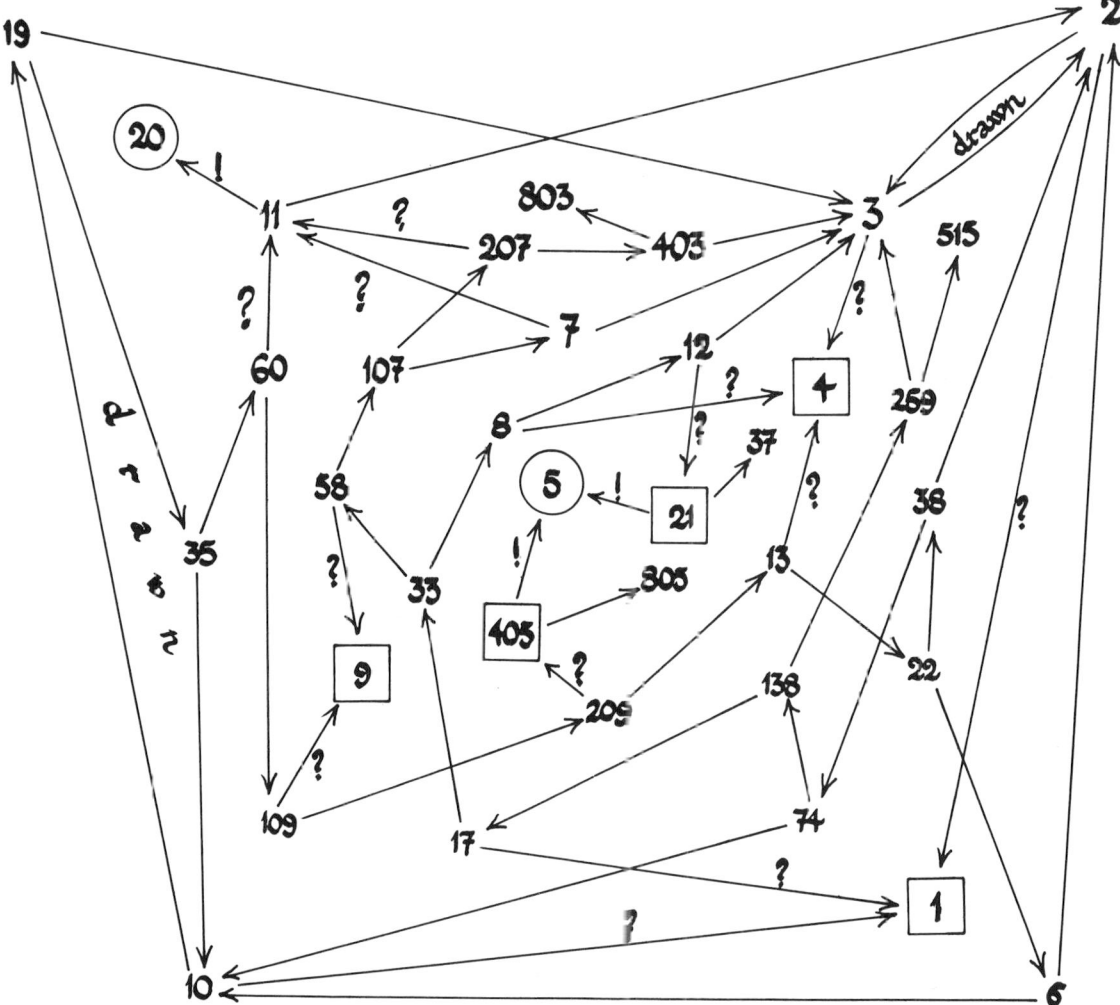

Figure 13. Partial Analysis of a Heap of 10.

Figure 13 shows part of the analysis of games starting from 10. If you continue the figure you'll soon realize why we don't give a complete analysis of Epstein's game. Squares and other $\mathscr{N}$-positions are in square boxes, $\mathscr{P}$-positions are circled.

Here is a list which includes all $\mathscr{P}$-positions of remoteness $\leqslant 14$ below 5000, and a few of the more interesting $\mathscr{N}$-positions.

$\mathcal{P}$-positions	$\mathcal{N}$-positions
Remoteness 0 : 0	
	Remoteness 1 : all squares
Remoteness 2 : 5, 20, 45, 80, 145, 580, 949, 1305, 1649, 2320, 3625, 4901, 5220, …	
	Remoteness 3 : 11, 14, 21, 30, 41, 44, 54, 69, 86, 105, 120, 126, 141, 149, 164, 174, 189, 216, 291, …
Remoteness 4 : 29, 101, 116, 135, 165, 236, 404, 445, 540, 565, 585, 845, 885, 909, 944, 954, 975, 1125, 1310, 1350, 1380, 1445, 1616, 1654, 1669, 2325, 2340, 2405, 2541, 2586, 2705, 3079, 3150, 3185, 3365, 3380, 3405, 3601, 3630, 3705, 4239, 4921, 4981, 5225, 5265, …	
	Remoteness 5: 52, 71, 84, 208, 254, 284, 296, 444, …
Remoteness 6 : 257, 397, 629, 836, 1177, 1440, 1818, 1833, 1901, 1937, 1988, 2210, 2263, 2280, 2501, 2516, 2612, 2845, 2861, 3039, 3188, 3389, 3621, 3654, 3860, 4053, 4105, 4541, 4693, 4708, 4813, 4930, …	
	Remoteness 7 : 136, 436, 1291, …
Remoteness 8 : 477, 666, 5036, …	*Remoteness* 9 : 252, 342, …
Remoteness 10 : 173, …	*Remoteness* 11 : 92, …
Remoteness 12 : 3341, 3573, 3898, 4177, 4229, 4581, …	*Remoteness* 13 : 1809, 1962, …
Remoteness 14 : 1918, …	

If you want to know how to win from a heap of 92, look in the Extras.

The misère form of the game is uninteresting, because the increasing move is always available.

TRIBULATIONS AND FIBULATIONS

What happens if we use another system of numbers instead of squares? An easy case is $2^k - 1$, but more interesting ones have been suggested to us, namely the triangular numbers 1, 3, 6, 10, 15, 21, … (Simon Norton) and the Fibonacci numbers plus one, 1, 2, 3, 4, 6, 9, 14, 22, 35, … (Mike Guy). See the Extras.

THIRD ONE LUCKY

Ordinary Nim ends when a player takes the last stick. Misère Nim may be thought of as over when only one stick remains. What happens if we say the game's over when exactly two sticks remain, the winner being the player who takes the third last stick? Even the three-heap version of this game is quite hard.

If there are m sticks in the first heap, and n in the second, there is a unique size for the third heap to make a $\mathscr{P}$-position. For fixed m, this size is eventually arithmetico-periodic in n. The periods for

$$m = 1\ 2\ 3\ 4\quad 5\quad 6\quad 7\ 8\ 9\ 10\ 11\ 12\ 13\ 14\ 15\ 16\ 17\ 18\quad 19\ \dots$$
$$\text{are}\quad 1\ 2\ 4\ 2\ 12\ 12\ 12\ 8\ 8\ 10\ 60\ 60\ 84\ 84\ 84\ 16\ 18\ 180\ 20\ \dots$$

HICKORY, DICKORY, DOCK

Dean Hickerson suggested this game in which a move replaces a heap of n by *three* heaps of sizes

$$k, n-k, n-2k \qquad \text{where} \quad 1 \leqslant k \leqslant \tfrac{1}{2}n.$$

It looks rather like Turnips in the last chapter, but in fact the nim-values for $n = 1, 2, 3, \dots$ are the exponents, $0, 1, 0, 2, 0, 1, 0, 3, 0, \dots$ of the nim-values (Fig. 7 of Chapter 14) for the Ruler Game.

D.U.D.E.N.E.Y

is a game,

Deductions Unfailing, Disallowing Echoes, Not Exceeding Y;

a particular case of which was described by Dudeney as "The 37 Puzzle Game".

From a single heap either player may subtract a number from 1 to Y:

"Not Exceeding Y"

except that the immediately previous deduction may not be repeated:

"Disallowing Echoes"

and you win if you can always move:

"Deductions Unfailing",

but at some stage your opponent cannot.

If echoes were *not* disallowed, the $\mathscr{P}$-positions would be the multiples of $Y+1$ and the winner would always follow a deduction of X by one of $Y+1-X$. This strategy still works for D.U.D.E.N.E.Y when Y is even because it is impossible for $Y+1-X$ to equal X. So we'll suppose from now on that Y is odd.

Here are the good moves from N when $Y = 3$:

from N =	0	1	2	3	4	5	6	7	8	9	10	11	...
deduct	?	1	1, 2	3	?	1	1, 2, 3	3	?	1	1, 2, 3	3	...

and here they are for Y = 5:

from N =	0	1	2	3	4	5	6	7	8	9	10	11	12
deduct		?	1	1, 2	3	4	5	3	?	1, 4	2, 3	3, 5	4	5

from N =	13	14	15	16	17	18	19	20	21	22	23	24	25	...
deduct		?	1	1, 2, 4	3	4, 5	5	3	?	1	1, 2, 3	3, 5	4	5	...

For Y = 3 there is an ultimate period of 4; for Y = 5 a period of 13. The easiest way to win is to move to one of those **pearls** among numbers, which have a ? entry, indicating that the next player has no good move at all. Pearls are $\mathcal{P}$-positions no matter what the previous move, but there are other $\mathcal{P}$-positions in which the only winning move is disallowed by the echo rule.

In general the pearls are spaced at intervals of either

$$E = Y + 1, \quad \text{the next even number after Y, or}$$
$$D = Y + 2, \quad \text{the next odd one.}$$

For, if P is a pearl then after most moves from $P+E$ or $P+D$ we can go immediately to P. The only exceptions are the moves from

$$P+E \text{ to } P+\tfrac{1}{2}E \quad \text{and} \quad P+D \text{ to } P+E.$$

If the first of these is a bad move, $P + E$ is a pearl, and if it's a good one, $P + D$ is a pearl.

From a given position there's usually only one move which will prevent your opponent from reducing to an earlier pearl, though sometimes there can be two. So it's fairly easy to determine the status of the critical move

$$P+E \text{ to } P+\tfrac{1}{2}E$$

by searching back along one or two alleys. In the case Y = 5, the critical moves from

$$13 \text{ to } 10 \quad \text{and} \quad 26 \text{ to } 23$$

are bad, because they can be answered by

$$10 \text{ to } 5 \quad \text{and} \quad 23 \text{ to } 18$$

but those from

$$6 \text{ to } 3 \quad \text{and} \quad 19 \text{ to } 16$$

are good, and are indicated by bold 3's.

STRINGS OF PEARLS

Knowing the pearls helps you to win the game: move directly to a pearl if you can, and otherwise prevent your opponent from doing so. So all we need tell you is the sequence of D's and E's separating the pearls. These are (ultimately—see entries 55 and 95) periodic. In Table 6 the periods are in parentheses, and $r \geqslant 0$. Much of the work was done by John Selfridge and Roger Eggleton.

In his chapter on subtraction games, Schuh discusses this game, together with its misère form, and the two variants in which the outcome is changed if play terminates at 1.

Y	Pearl-string	Y	Pearl-string
3 or $8r+3$	(E)EEE ...	41	(DDDEDE) ...
5 or $8r+5$	(DE)DEDE ...	55	DD(EDE)EDE ...
7	(DEE)DEE ...	63 or $128r+53$	(E)EE ...
9	(DDE)DDE ...	65 or $128r+55$	(DDDE)DDDE ...
15 or $32r+15$	(E)EE ...	71 or $64r+71$	(DE)DEDE ...
17 or $32r+17$	(DDE)DDE ...	73 or $128r+73$	(DDEDE)DDEDE ...
23	(DDEDDDEE) ...	87 or $128r+87$	(DDE)DDE ...
25 or $32r+25$	(DE)DE ...	95	DDEE(DDE)DDE ...
31 or $128r+31$	(DEE)DEE ...	97	(DDEDDDE) ...
33	(DDDEDDE) ...	103	(DE)DEDE ...
39 or $128r+39$	(DEE)DEE ...	105 or $128r-105$	(DE)DEDE ...

Table 6. Strings of Pearls for D.U.D.E.N.E.Y for $\frac{53}{64}$ of the odd Values of Y.

SCHUHSTRINGS

Prof. Schuh also discusses the variation in which 0 is a permissible deduction, but the person who first gets to 0 wins.

Starting from any positive number n you'll always have at least one good move, because if no positive deduction wins for you, you can deduct 0 and present your opponent with a similar situation, except that 0 is now illegal. How can a positive deduction

$$n \text{ to } n-g$$

possibly be a good move? The only move it prohibits from $n-g$ is

$$n-g \text{ to } n-2g$$

and so this must be the unique good move from $n-g$. But then similarly

$$n-2g \text{ to } n-3g,$$
$$n-3g \text{ to } n-4g,$$
$$\dots\dots\dots\dots .$$

must be good moves, the last of which must be from

$$g \text{ to } 0.$$

A positive deduction g can therefore only be good at a **string**,

$$g, 2g, 3g, \dots, kg, \dots,$$

of multiples of g. It will be good from $(k+1)g$ if and only if it was the *unique* good move from kg. The first multiple of g from which there's another good move will terminate the g-string.

Thus, when the permissible deductions are 0, 1, 2, 3, 4, 5, we find that the good moves are:

from $n=1$ 2   3 4 5 6 7 8 9 10   11 12 13 14 15 16 17 18 19 20 ...
deduct   1 1,2] 3 4 5 3 0 4 3 5   0 3,4] 0 0 5 0 0 0 0 5 ...

The 1- and 2-strings terminate at 2 and the 3- and 4-strings at 12, but the 5-string continues indefinitely.

In general, if there are two or more numbers

$$a, b, \ldots$$

whose strings have not yet terminated, then the first number to occur in two or more strings will terminate those strings. At most one string continues forever. In Table 7 an entry (a, b) means that the a- and b-strings terminate at their l.c.m., while an entry $g\infty$ corresponds to an infinite g-string, and is only relevant when the largest deduction is odd. It is not known whether there is any Schuhstring game in which three or more strings terminate simultaneously.

2 or 3	4 or 5	6 or 7	8 or 9	10 or 11	12 or 13	14 or 15
(1,2)	(1,2)	(1,2)	(1,2)	(1,2)	(1,2)	(1,2)
3∞	(3,4)	(3,6)	(3,6)	(3,6)	(3,6)	(3,6)
	5∞	(4,5)	(4,8)	(4,8)	(4,8)	(4,8)
(1,2)		7∞	(5,7)	(5,10)	(5,10)	(5,10)
(3,6)		(1,2)	9∞	(7,9)	(9,12)	(7,14)
(4,8)		(3,6)	(1,2)	11∞	(7,11)	(9,12)
(5,10)		(4,8)	(3,6)	(1,2)	13∞	(11,13)
(7,14)		(5,10)	(4,8)	(3,6)	(1,2)	15∞
(9,18)		(7,14)	(5,10)	(4,8)	(3,6)	(1,2)
(11,22)		(9,18)	(7,14)	(5,10)	(4,8)	(3,6)
(12,24)		(11,22)	(9,18)	(7,14)	(5,10)	(4,8)
(13,26)		(12,24)	(11,22)	(9,18)	(7,14)	(5,10)
(15,20)		(15,20)	(12,16)	(12,16)	(9,18)	(7,14)
(21,27)	(16,17)	(13,16)	(15,20)	(15,20)	(12,16)	(9,12)
(16,17)	(19,21)	(17,19)	(13,17)	(11,13)	(11,13)	(11,13)
(19,23)	(23,25)	(21,23)	(19,21)	(17,19)	(15,17)	(15,16)
25∞	25∞	25∞	23∞	21∞	19∞	17∞
27	26	25 or 24	23 or 22	21 or 20	19 or 18	17 or 16

Table 7. Schuhstrings Corresponding to Various Maximum Deductions.

THE PRINCESS AND THE ROSES

When we originally planned *Winning Ways*, the Princess was to have had a chapter all to herself, but as with other beautiful and intriguing women, we probably dallied too long in her company and it now seems more discreet to limit our memoirs to a brief résumé of our rencontres.

Perhaps our narrative will steer a course between the original bare account of Prof. Schuh and the later flights of fancy of Monsieur Filet de Carteblanche.

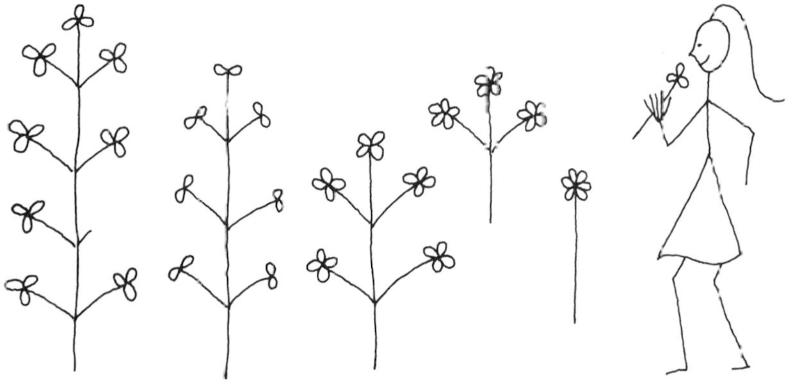

Figure 14. Princess Romantica Smells Charming Charles's Rose.

The Princess Romantica is known to have had two princely suitors, Handsome Hans and Charming Charles. Each suitor went in turn to the rose-garden and would bring back a rose, or two roses from different bushes. In Fig. 14 you can see the Princess smelling the beautiful rose that Charming Charles has just brought from the largest bush. Eventually one suitor, finding himself unable to bring her a rose because none was left in the garden, crept despondently away, and left the other to claim her hand as in Fig. 15. Which was the lucky Prince?

Figure 15. Who Won the Hand of the Princess?

Of course, you can play the game as a heap game in which the legal move is to take any one chip, or any two, one from each of two distinct heaps. Prof. Schuh showed that a worldly prince in a 5-bush garden should always arrange that when the numbers are put in descending order, they form one of the patterns

$$\text{even–even–even–even–even,}$$
$$\text{even–odd –odd –odd –odd,}$$
$$\text{odd –even–even–odd –odd,}$$
$$\text{odd –odd –odd –even–even.}$$

Obviously Charles knew what he was doing and Prof. Schuh's researches leave little doubt that he must be the man depicted in Fig. 15.

You can see from Prof. Schuh's rule that when there's only a small number of bushes (which may contain a large number of roses).

> parity considerations are paramount.

However, when there are many few-rose bushes then

> it is triality that triumphs

because the $\mathscr{P}$-positions ultimately are just those in which the total number of roses is a multiple of three.

This reveals itself by the final subscript 3's in Tables 8, 9 and 10 which respectively list all $\mathscr{P}$-positions of the forms

$$3^x 2^y 1^z \qquad \text{or} \qquad a.2^y 1^z \qquad \text{or} \qquad a.b.1^z,$$

i.e.

$$
\left.
\begin{array}{l}
x \text{ 3-rose bushes,} \\
y \text{ 2-rose bushes,} \\
z \text{ 1-rose bushes}
\end{array}
\right\}
\text{ or }
\left.
\begin{array}{l}
1 \text{ } a\text{-rose bush,} \\
y \text{ 2-rose bushes,} \\
z \text{ 1-rose bushes}
\end{array}
\right\}
\text{ or }
\begin{array}{l}
1 \text{ } a\text{-rose bush,} \\
1 \text{ } b\text{-rose bush,} \\
z \text{ 1-rose bushes.}
\end{array}
$$

In these tables an entry n_3 represents all numbers of the infinite arithmetic progression

$$n, n+3, n+6, n+9, \ldots$$

while $m_d n$ represents the finite progression

$$m, m+d, m+2d, \ldots, n,$$

and so on; for example, the entry $_6 7_5 17_3$ represents 1, 7, 12, 17, 20, 23, 26, 29, …

x \ y	0	1	2	3	4	5	6	7	8	9	10	11
0	0_3	${}_4 4_3$	${}_5 5_3$	0_3	${}_4 4_3$	${}_5 5_3$	0_3	${}_4 4_3$	${}_5 5_3$	0_3	${}_4 4_3$	${}_5 5_3$
1	${}_4 6_3$	4_3	2_3	3_3	4_3	2_3	3_3	4_3	2_3	3_3	4_3	2_3
2	${}_5 6_3$	4_3	2_3	3_3	1_3	2_3	3_3	1_3	2_3	3_3	1_3	2_3
3	0_3	${}_4 4_3$	${}_5 5_2$	0_3	1_3	2_3	0_3	1_3	2_3	0_3	1_3	2_3
4	3_3	4_3	2_2	0_3	1_3	2_3	0_3	1_3	2_3	0_3	1_3	2_3
5	${}_5 6_3$	4_3	2_3	0_3	1_3	2_3	0_3	1_3	2_3	0_2	1_3	2_3
6	0_3	1_3	2_3	0_3	1_3	2_3	0_3	1_3	2_3	0_2	1_3	2_3
7	0_3	1_3	2_3	0_3	1_3	2_3	0_3	1_3	2_3	0_2	1_3	2_3

Table 8. $\mathscr{P}$-positions of Type $3^x 2^y 1^z$. Entries Are Sets of Values of z.

y \ a	0	1	2	3	4	5	6	7	8	9	10	11
0	0_3	2_3	${}_4 4_3$	${}_4 6_3$	${}_4 8_2$	${}_4 10_3$	${}_4 12_3$	${}_4 14_3$	${}_4 16_3$	${}_4 18_3$	${}_4 20_3$	${}_4 22_3$
1	${}_4 4_3$	3_3	${}_5 5_3$	4_3	${}_6 6_3$	$4_4 8_3$	$6_{64}10_3$	$4_4 12_3$	$6_{64}14_3$	$4_4 16_3$	$6_{64}18_3$	$4_4 20_3$
2	${}_5 5_3$	4_3	0_3	2_3	${}_4 4_3$	${}_4 6_3$	${}_4 8_3$	${}_4 10_3$	${}_4 12_3$	${}_4 14_3$	${}_4 16_3$	${}_4 18_3$
3	0_3	2_3	${}_4 4_3$	3_3	${}_5 5_3$	4_3	${}_6 6_3$	$4_4 8_3$	$6_{64}10_3$	$4_4 12_3$	$6_{64}14_3$	$4_4 16_3$
4	${}_4 4_3$	3_3	${}_5 5_3$	4_3	0_3	2_3	${}_4 4_3$	${}_4 6_3$	${}_4 8_3$	${}_4 10_3$	${}_4 12_3$	${}_4 14_3$
5	${}_5 5_3$	4_3	0_3	2_3	${}_4 4_3$	3_3	${}_5 5_3$	4_3	${}_6 6_3$	$4_4 8_3$	$6_{64}10_3$	$4_4 12_3$
6	0_3	2_3	${}_4 4_3$	3_3	${}_5 5_3$	4_3	0_3	2_3	${}_4 4_3$	${}_4 6_3$	${}_4 8_3$	${}_4 10_3$
7	${}_4 4_3$	3_3	${}_5 5_3$	4_3	0_3	2_3	${}_4 4_3$	3_3	${}_5 5_3$	4_3	${}_6 6_3$	$4_4 8_3$

Table 9. $\mathscr{P}$-positions of Type $a \cdot 2^y 1^z$. Entries Are Sets of Values of z.

b \ a	0	1	2	3	4	5	6	7	8	9	10	11
0	0_3	2_3	$_4 4_3$	$_4 6_3$	$_4 8_3$	$_4 10_3$	$_4 12_3$	$_4 14_3$	$_4 16_3$	$_4 18_3$	$_4 20_3$	$_4 22_3$
1	2_3	1_3	3_3	$_4 5_3$	$_4 7_3$	$_4 9_3$	$_4 11_3$	$_4 13_3$	$_4 15_3$	$_4 17_3$	$_4 19_3$	$_4 21_3$
2	$_4 4_3$	3_3	$_5 5_3$	4_3	6_3	$_4 8_3$	$_6 10_3$	$_4 12_3$	$_6 14_3$	$_4 16_3$	$_6 18_3$	$_4 20_3$
3	$_4 6_3$	$_4 5_3$	4_3	$_5 6_3$	$_5 8_3$	$_6 7_3$	$_6 9_3$	$_{67} 11_3$	$_{69} 13_3$	$_{67} 15_3$	$_{69} 17_3$	$_{67} 19_3$
4	$_4 8_3$	$_4 7_3$	6_3	$_5 8_3$	$_5 10_3$	$_5 9_3$	$_6 6_5 11_3$	$_6 10_3$	$_6 12_3$	$_6 10_4 14_3$	$_6 12_4 16_3$	$_6 10_4 18_3$
5	$_4 10_3$	$_4 9_3$	$4_4 8_3$	$_6 7_3$	$_5 9_3$	$_5 11_3$	$_5 13_3$	$_6 7_5 12_3$	$_6 9_5 14_3$	$_6 13_3$	$_6 15_3$	$_6 13_4 17_3$
6	$_4 12_3$	$_4 11_3$	$6_4 10_3$	$_6 9_3$	$_6 6_5 11_3$	$_5 13_3$	$_5 15_3$	$_5 14_3$	$_6 6_5 16_3$	$_6 10_5 15_3$	$_6 12_5 17_3$	$_6 16_3$
7	$_4 14_3$	$_4 13_3$	$4_4 12_3$	$_6 7_4 11_3$	$_6 10_3$	$_6 7_5 12_3$	$_5 14_3$	$_5 16_3$	$3_5 18_3$	$_6 7_5 17_3$	$_6 9_5 19_3$	$_6 13_5 18_3$
8.	$_4 16_3$	$_4 15_3$	$6_4 14_3$	$_6 9_4 13_3$	$_6 12_3$	$_6 9_5 14_3$	$_6 6_5 16_3$	$3_5 18_3$	$_5 20_3$	$_5 19_3$	$_6 6_5 21_3$	$_6 10_5 20_3$
9	$_4 18_3$	$_4 17_3$	$4_4 16_3$	$_6 7_4 15_3$	$_6 10_4 14_3$	$_6 13_3$	$_6 10_5 15_3$	$_6 7_5 17_3$	$_5 19_3$	$_5 21_3$	$_5 23_3$	$_6 7_5 22_3$
10	$_4 20_3$	$_4 19_3$	$6_4 18_3$	$_6 9_4 17_3$	$_6 12_4 16_3$	$_6 15_3$	$_6 12_5 17_3$	$_6 9_5 19_3$	$_6 6_5 21_3$	$_5 23_3$	$_5 25_3$	$_5 24_3$
11	$_4 22_3$	$_4 21_3$	$4_4 20_3$	$_6 7_4 19_3$	$_6 10_4 18_3$	$_6 13_4 17_3$	$_6 16_3$	$_6 13_5 18_3$	$_6 10_5 20_3$	$_6 7_5 22_3$	$_5 24_3$	$_5 26_3$

Table 10. $\mathscr{P}$-positions of Type $a \cdot b \cdot 1^z$. Entries Are Sets of Values of z.

The tables also illustrate that there are places between the parity and triality regions in which the outcome depends on considerations mod 4 and 5, so that

> quaternity's a quality,

> quinticity can be quintessential,

and there are even hints that

> sex may be significant;

but we have explored other regions in which it sadly seems that

> randomness reigns.

    In his first paper on this subject *M.* de Carteblanche asks for a code of behavior for princes in a 6-bush garden wherein there's a bush with only 1 rose. You'll find one in the Extras. In a second paper he further describes how the princes, after their weddings to Romantica and her even more beautiful younger sister Belladonna, transformed the rose game into a different one with chocolates and discovered some more interesting games to play.

ONE-STEP, TWO-STEP

This is the strip game in which arbitrarily many coins are allowed on a square, and the legal move is to make either one step or two steps, a **step** being to move a single coin just one space leftwards. The coins moved in the two steps of a 2-step move may be the same or different.

Letting a_n be the number of coins on square n, we can ask when

$$a_0 a_1 a_2 \ldots$$

represents a $\mathscr{P}$-position. The answer certainly won't depend on a_0 since coins on square 0 will never be moved again.

There's a surprising connexion between this game and our previous one. In fact the position described above behaves exactly like

$$a_1 + a_2 + a_3 + \ldots + a_n, \qquad a_2 + a_3 + \ldots + a_n, \qquad a_3 + \ldots + a_n, \qquad \ldots, a_{n-1} + a_n, \qquad a_n$$

in the Princess-and-Roses game! We leave it to Omar to work out why.

So when your coins are all in the first 6 places, you can translate Prof. Schuh's rules to give the $\mathscr{P}$-positions, which are

$$?eeeee, \qquad ?deeed, \qquad ?deded, \qquad ?eedee,$$

where e means even, d means odd and $?$ means anything.

MORE ON SUBTRACTION GAMES

Since $\mathscr{G}(n)$ for a heap of n beans in the subtraction game (see Chapter 4)

$$S(s_1, s_2, \ldots, s_k)$$

depends only on k earlier values, namely

$$\mathscr{G}(n-s_1), \mathscr{G}(n-s_2), \ldots, \mathscr{G}(n-s_k),$$

we see that $\mathscr{G}(n) \leqslant k$. Moreover this sequence of k values must eventually repeat so the nim-sequences of all subtraction games are (ultimately) periodic. But the bound on the length of the period given by this argument seems astronomical when compared with the facts. Can you find something nearer the truth?

We have seen that if the g.c.d. $(s_1, s_2, \ldots, s_k)$ is $d > 1$, then the game is just the d-plicate of a simpler game. Thus $S(s_1)$ is the s_1-plicate of $S(1)$, She-Loves-Me, She-Loves-Me-Not, and so has period $2s_1$ and nim-sequence $\dot{0}.00\ldots0111\ldots\dot{1}$.

We can also analyse $S(s_1, s_2)$ and $S(s_1, s_2, s_1+s_2)$ completely. Write

$$s_1 = a, \qquad s_2 = b = 2ha \pm r \qquad \text{for } 0 \leqslant r \leqslant a$$

and suitable h. After the g.c.d. remark we needn't consider $r=0$ or a unless $a=1$.

$S(1, 2h)$ has period $2h+1$ and nim-sequence $\dot{0}.10101\ldots01\dot{2}$, and $S(1, 2h+1) = S(1)$. (In fact $s_1 = 1$ and all s_i odd gives She-Loves-Me, She-Loves-Me-Not.)

For $a > 1$ the period of $S(a, b)$ contains $a + b$ digits, alternating blocks of a 0's and a 1's, except that the last $a - r$ 0's are replaced by 2's, where r is as above. For example: $a = 3$, $r = 1$; the nim-sequences for $S(3, 11)$ and $S(3, 13)$ are

$$\dot{0}.001110001112\dot{2} \quad \text{and} \quad \dot{0}.001110001110022\dot{1}.$$

Here is a general method for analyzing $S(s_1, s_2, \ldots, s_k)$. Write the numbers in $k + 1$ columns. The first row is

$$0, \quad s_1, \quad s_2, \quad \ldots \quad s_k.$$

Each later row is of the form

$$l, \quad l + s_1, \quad l + s_2, \quad \ldots \quad l + s_k,$$

where l is the least whole number which hasn't appeared in earlier rows. The table will eventually become periodic in that a block of c consecutive rows can be obtained from the preceding block of c by adding p to all the entries, for a suitable c and p.

The first column contains all numbers n for which $\mathscr{G}(n) = 0$, and the second, by Ferguson's pairing property, just those for which $\mathscr{G}(n) = 1$. Later columns contain numbers for which $\mathscr{G}(n) \geqslant 2$, apart from repetitions of entries in the second column. We illustrate with $S(1, b, b + 1)$. If b is even (Fig. 16(a)) there are no such repetitions, the period is $2b$ and the nim-sequence is

$$\dot{0}.101\ldots012323\ldots2\dot{3}.$$

$\mathscr{G}(n) =$	0	1	2	3
$n =$	0	1	10	11
	2	3	12	13
	4	5	14	15
	6	7	16	17
	8	9	18	19
	20	21	30	31
	22	23	32	33
	24	25	34	35
	26	27	36	37
	28	29	38	39
	40	41	50	51
	42	...		

(a)

$\mathscr{G}(n) =$	0	1	3	2	except that
$n =$	0	1	9	10	$\mathscr{G}(9) = 1$
	2	3	11	12	
	4	5	13	14	
	6	7	15	16	
	8	9	17	18	
	19	20	28	29	$\mathscr{G}(28) = 1$
	21	22	30	31	
	23	24	32	33	
	25	26	34	35	
	27	28	36	37	
	38	39	47	48	$\mathscr{G}(47) = 1$
	40	...			

(b)

Figure 16. The Subtraction Games $S(1,10,11)$ and $S(1,9,10)$.

If b is odd (Fig. 16b) there is one repetition $(9, 28, 47, \ldots)$ in each period, whose length is $2b + 1$. The nim-sequence is as before, but with the final 3 omitted:

$$\dot{0}.101\ldots012323\ldots\dot{2}.$$

To complete the analysis of $S(a, b, a + b)$ note that the case $a > 1$, $b = 2ha - r$, $0 < r < a$ is fairly straightforward. It is illustrated by a period of ha rows in Fig. 17; there are r repetitions (boxed in the figure) so the period is $4ha - r = 2b + r$. The period comprises h blocks of a 0's and a 1's followed by $h - 1$ blocks of a 2's and a 3's, then a 2's and $a - r$ 3's.

$G(n) =$ 0 1

$n =$ (0)	(1)	(0)	(1)
0	a	$2ha-r$ ▢	$(2h+1)a-r$
1	$a+1$	$2ha-r+1$	$(2h+1)a-r+1$
2	$a+2$	$2ha-r+2$	$(2h+1)a-r+2$
$\cdots$	$\cdots$	$\cdots$	$\cdots$
$r-1$	$a+r-1$	$2ha-1$ ▢	$(2h+1)a-1$
r	$a+r$	$2ha$	$(2h+1)a$
$\cdots$	$\cdots$	$\cdots$	$\cdots$
$a-1$	$2a-1$	$(2h+1)a-r-1$	$(2h+2)a-r-1$
$2a$	$3a$	$(2h+2)a-r$	$(2h+3)a-r$
$2a+1$	$3a+1$	$(2h+2)a-r+1$	$(2h+3)a-r+1$
$\cdots$	$\cdots$	$\cdots$	$\cdots$
$3a-1$	$4a-1$	$(2h+3)a-r-1$	$(2h+4)a-r-1$
$4a$	$5a$	$(2h+4)a-r$	$(2h+5)a-r$
$4a+1$	$5a+1$	$(2h+4)a-r+1$	$(2h+5)a-r+1$
$\cdots$	$\cdots$	$\cdots$	$\cdots$
$5a-1$	$6a-1$	$(2h+5)a-r-1$	$(2h+6)a-r-1$
$\cdots$	$\cdots$	$\cdots$	$\cdots$
$(2h-2)a$	$(2h-1)a$	$(4h-2)a-r$	$(4h-1)a-r$
$(2h-2)a+1$	$(2h-1)a+1$	$(4h-2)a-r+1$	$(4h-1)a-r+1$
$\cdots$	$\cdots$	$\cdots$	$\cdots$
$(2h-1)a-r-1$	$2ha-r-1$	$(4h-1)a-2r-1$	$4ha-2r-1$
$(2h-1)a-r$	$2ha-r$ ▢	$(4h-1)a-2r$	$4ha-2r$
$\cdots$	$\cdots$	$\cdots$	$\cdots$
$(2h-1)a-1$	$2ha-1$ ▢	$(4h-1)a-r-1$	$4ha-r-1$

Figure 17. Analysis of $S(a,b,a+b)$, $b = 2ha-r$. $0 < r < a$, $(a,b) = 1$.

The case $b = 2ha+r$, $0 < r < a$, is more complicated. The period is a times as long, $(2b+r)a$. We illustrate it with the particular case $a = 5$, $b = 43$, $h = 4$, $r = 3$ in Fig. 18. The $ar\,(=15)$ repetitions are shown boxed.

In either of the cases $b = 2ha \pm r$, the ith value of n for which $\mathscr{G}(n) = 0$ is

$$n_i = i + \left\lfloor \frac{i}{a} \right\rfloor a + \left\lfloor \frac{2i}{b+r} \right\rfloor b$$

and the ith value for which $\mathscr{G}(n) = 1$ is $a + n_i$ by Ferguson's pairing property.

```
 0   5  43  48     93  98 136 141                        269 274 312 317
 1   6  44  49     94  99 137 142                        270 275 313 318
 2   7 [45] 50     95 100 [138]143                       271 276 [314]319
 3   8 [46] 51     96 101 139 144                        272 277 [315]320
 4   9 [47] 52     97 102 140 145                        273 278 316 321

10  15  53  58    103 108 146 151   186 191 229 234      279 284 322 327    362 367 405 410
11  16  54  59    104 109 147 152   187 192 230 235      280 285 323 328    363 368 406 411
12  17  55  60    105 110 148 153   188 193 231 236      281 286 324 329    364 369 407 412
13  18  56  61    106 111 149 154   189 194 232 237      282 287 325 330    365 370 408 413
14  19  57  62    107 112 150 155   190 195 233 238      283 288 326 331    366 371 409 414

20  25  63  68    113 118 156 161   196 201 239 244      289 294 332 337    372 377 415 420
21  26  64  69    114 119 157 162   197 202 240 245      290 295 333 338    373 378 416 421
22  27  65  70    115 120 158 163   198 203 241 246      291 296 334 339    374 379 417 422
23  28  66  71    116 121 159 164   199 204 242 247      292 297 335 340    375 380 418 423
24  29  67  72    117 122 160 165   200 205 243 248      293 298 336 341    376 381 419 424

30  35  73  78    123 128 166 171   206 211 249 254      299 304 342 347    382 387 425 430
31  36  74  79    124[129]167 172   207 212 250 255      300 305 343 348    383 388 426 431
32  37  75  80    125[130]168 173   208 213 251 256      301[306]344 349    384 389 427 432
33  38  76  81    126 131 169 174   209 214 252 257      302 307 345 350    385 390 428 433
34  39  77  82    127 132 170 175   210 215 253 258      303 308 346 351    386 391 429 434

40 [45] 83  88    133[138]176 181   216[221]259 264      309[314]352 357    392[397]435 440
41 [46] 84  89                      217[222]260 265      310[315]353 358    393[398]436 441
42 [47] 85  90    177 182 220 225   218[223]261 266                         394[399]437 442
                  178 183[221]226   219 224 262 267      354 359[397]402    395 400 438 443
86  91[129]134    179 184 222 227                        355 360 398 403    396 401 439 444
87  92[130]135    180 185[223]228   263 268[306]311      356 361[399]404
```

Figure 18. Analysis of $S(5,43,48)$.

MOORE'S NIM$_k$

E.H. Moore suggested the heap game in which the legal move is to reduce the size of any positive number, up to k, of heaps. Thus Nim$_1$ is ordinary Nim, and Nim$_2$ is the game in which you can reduce just *one* or *two* heaps. The theory rather surprisingly involves calculations in base two and in base $k+1$. You

> *Expand* the numbers in base 2, and
> *Add* them in base $k+1$, without carrying.

You should move to positions in which this "sum" is zero.

For example, if $k=2$, and you are confronted with

5 =	101		5 =	101	
6 =	110		6 =	110	
9 =	1001		3 =	11	
10 =	1010		7 =	11:	
	————			———	
	2222			000	

you must reduce the 9 and 10 heaps, replacing them by 3 and 7.

Smith's analysis of Subselective Compounds (Chapter 12) is similar. Nim-values for Moore's game have been found by Jenkyns and Mayberry. Yamasaki has shown that it is tame, and that a position is fickle only if its nonzero heaps are all of size 1, and the number of them is 0 or 1 modulo $k+1$.

THE MORE THE MERRIER

Bob Li has suggested that ordinary Nim can be adapted for n players. They take turns in a fixed cyclic order and there are different grades of winner.

> *First prize* goes to the player who makes the last move,
> *Second prize* to the immediately previous player, and so on, until the
> *Booby prize*, which goes to the player who was first unable to move.

Sharing of prizes is not permitted: as soon as the game ends each player must take his prize and set off for his home town without any under-the-table payoffs for help he might have received from some of the other players.

Li was surprised to find that his game was very similar to Moore's. Take the position you're faced with and add the binary expansions of your numbers in base n without carrying. You're the Booby only when the sum is 0.

MOORE AND MORE

If we allow each of the n players to reduce any number up to k of the heaps at his move, the theory, also due to Li, is similar. The Booby this time is the player who sees 0 when he adds the binary expansions of the numbers modulo $k(n-1)+1$, without carrying.

We have not discussed games for more than 2 players elsewhere in this book because the stipulations to prevent coalitions are somewhat artificial, and lead to paradoxes of the "surprise exam" type. See the article by Paul Hudson in the References.

There are so many generalizations of Nim with interesting theories that we certainly haven't said the last word on the subject, and so

> This is the way the chapter ends,
> This is the way the chapter ends,
> This is the way the chapter ends,

NOT WITH A BANG BUT A WHIM

Perhaps you might like to play Nim, but on just one occasion one of the two players is allowed, instead of his usual Nim move, to exercise his **Whim** to decide whether the outcome will be decided by normal or by misère play. If we use

> 0-nim to mean normal Nim
> 1-nim to mean misère Nim, and
> 2-nim to mean Whim,

then we can continue the sequence with

> 3-nim = **Trim**,
> 4-nim = **Quam**, etc.

The move in d-nim ("**Denim**"), $d \geqslant 2$, is

> *either*   to move as in Nim
> *or*   to reduce d (but not both).

This is easy to analyze if you introduce a **quiddity heap**, to keep account of just which game you're playing. Then each of these games becomes like Nim with an extra heap. If $2^k \leqslant d < 2^{k+1}$, the quiddity heap behaves like a heap of size d when all other heaps are of size less than 2^{k+1}, but like a heap of size $d-1$ when they're not.

EXTRAS

DID YOU WIN THE SILVER DOLLAR?

You did if you moved the coin behind the $ just 2 squares or 1, leaving (3, 2, 1) or (2, 3, 1), depending on which version you're playing. Notice that, in the latter case, there's only one coin between the $ and the bag.

HOW WAS YOUR ARITHMETIC?

When you filled in the frieze pattern it should have looked like

```
1    1    1    1    1    1    1    1    1    1    1    1    ...
   1    2    3    2    2    1    4    3    1    2    3    ...
      1    5    5    3    1    3   11    2    1    5    5    ...
   1    2    8    7    1    2    8    7    1    2    8    ...
      1    3   11    2    1    5    5    3    1    3   11    ...
   1    4    3    1    2    3    2    2    1    4    3    ...
      1    1    1    1    1    1    1    1    1    1    1    ...
```

IN PUT-OR-TAKE-A-SQUARE, 92 IS AN $\mathcal{N}$-POSITION

In Fig. 19 the $\mathcal{P}$-positions are in the centre column, squares on the right and other $\mathcal{N}$-positions on the left.

TRIBULATIONS AND FIBULATIONS

Norton conjectures that in his game of **Tribulations** no position is drawn and $\mathcal{N}$-positions are more numerous than $\mathcal{P}$-positions in golden ratio. Richard Parker has verified these assertions for numbers < 5000 (for which the calculations sometimes run into the millions). The remoteness and suspense numbers (which are probably always finite) are shown in Table 11. Play from 51, 52 and 56 is especially interesting; draw diagrams like Fig. 19.

For Mike Guy's game of **Fibulations** we have proved the corresponding assertions and can in fact give a complete analysis. It is well known that any number can be economically expressed as the sum of Fibonacci numbers by the **Zeckendorf algorithm**: always subtract the largest Fibonacci number you can. Less economically we can use the **Secondoff algorithm**: always take off the *second* largest Fibonacci number you can, e.g.

$$100 = 89 + 8 + 3 \text{ (Zeckendorf)} \quad \text{or} \quad 55 + 21 + 13 - 5 + 3 + 2 + 1 \text{ (Secondoff)}$$

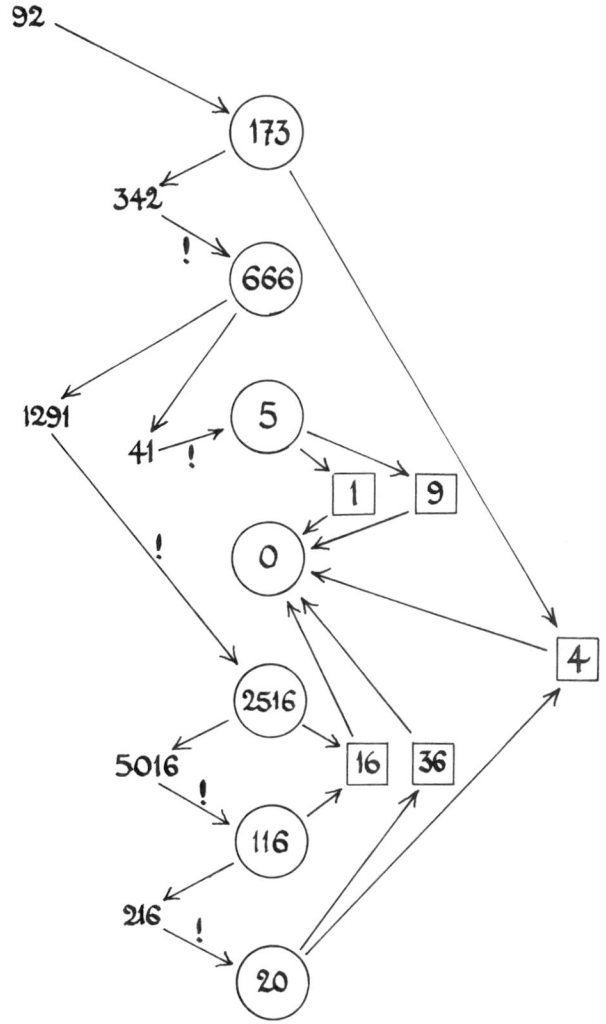

Figure 19. How to Win Epstein's Game starting from 92.

n	1	2	3	4	5	6	7	8	9	10	11	12	13	14	15	16	17	18	19	20	21	22	23	24	25
R	1	2	1	6	3	1	5	3	2	1	2	3	4	3	1	9	3	6	7	8	1	10	3	2	3
S	3	2	1	4	3	1	3	3	2	5	4	5	2	5	1	3	3	2	5	4	3	4	5	2	5

n	26	27	28	29	30	31	32	33	34	35	36	37	38	39	40	41	42	43	44	45	46	47	48	49	50
R	4	5	1	4	3	8	7	5	9	7	1	14	3	4	7	4	2	9	4	1	2	3	4	7	8
S	4	3	5	4	3	2	5	3	5	5	3	4	5	2	5	4	2	5	4	7	2	5	2	5	4

n	51	52	53	54	55	56	57	58	59	60	61	62	63	64	65	66	67	68	69	70	71	72	73	74	75
R	12	16	9	3	1	12	3	14	7	6	4	8	6	3	2	1	6	3	5	7	11	4	13	8	3
S	2	4	5	3	5	4	5	2	5	4	2	4	4	7	2	1	4	3	3	5	5	2	3	4	5

Table 11. emoteness and Suspense Numbers for Tribulations.

A number is a $\mathscr{P}$-position in Fibulations if and only if

> *either* its Secondoff expansion ends $3 + 1 - 1$ or $5 + 2 + 1$
> *or*   it is 3 more than a Fibonacci number which is at least 8.

The numbers $0, 11, 5, 8, 13, 21, 34, 55, 89, 144, \ldots$

have remotenesses $0, \; 8, 2, 2, \; 2, \; 4, \; 6, \; 8, 10, \; 12, \ldots$

The remoteness of any other $\mathscr{P}$-position is found by adding twice the number of Secondoff steps you take to get to one of these and the remoteness of an $\mathscr{N}$-position is one more than the smallest remoteness of its $\mathscr{P}$-options. E.g. from 1000 we get to 34 (remoteness 6) after 4 subtractions (of 610, 233, 89 and 34) so 1000 has remoteness $6 + (4 \times 2) = 14$. On the other hand, 1001 has remoteness 3 (move to 13). We believe, and Omar might confirm, that the suspense numbers (which are all finite) and nim-values (which are all 0, 1, 2 or ∞_0) have similar patterns.

OUR CODE OF BEHAVIOR FOR PRINCES

in that 6-bush rose-garden with a 1-rose bush is best described by translating it into the One-Step, Two-Step game. It can be checked that all the resulting positions except

?edede1	?eddde1	?ddede1	?dddde1
?ededd1	?eddd1	?ddedd1	?ddddd1
?0eeee1	?0edee1	?ed0ed1	?dd0ee1

can be moved to one of Schuh's $\mathscr{P}$-positions, and that these 12 classes of position, when joined by possible moves, form the graph of Fig. 20. In the figure a boxed position is $\mathscr{P}$, an unboxed one is $\mathscr{N}$, and

> e means any even number, including 0,
> d means any odd number,
> E means any even number $\geqslant 2$,
> D means any odd number $\geqslant 3$, and
> ?   means any number,

and the dotted arrow indicates that moves can only be made in that direction. The positions of form ?00dee1 and ?00eee1 have been omitted from the figure because they cannot be reached from the other ones. To complete the figure, adjoin

?00dee1	?00eee1
except	except
?00dEE1	?00EEE1
?001021	?000E01
?001041	?002021.

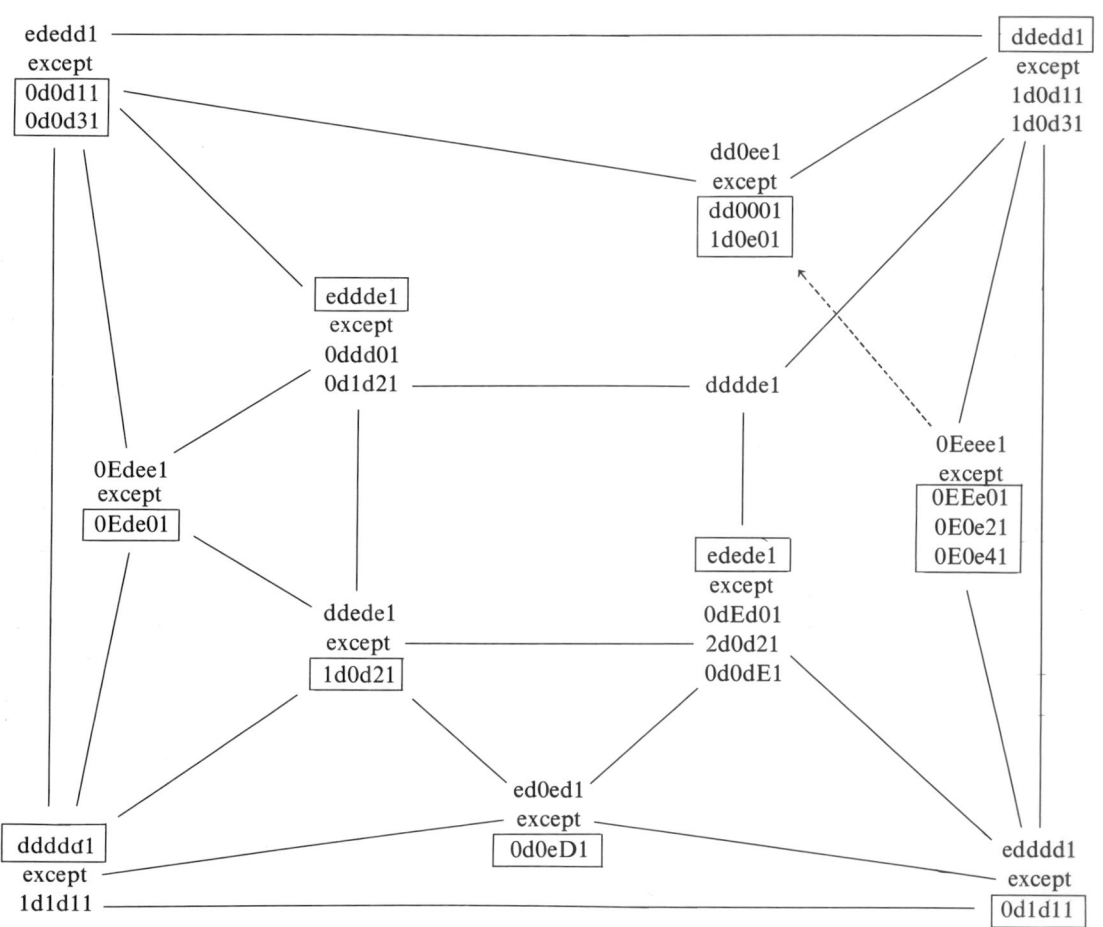

Figure 20. Our Code of Behavior for Princes in 6-bush Rose-gardens with a 1-rose Bush.

REFERENCES AND FURTHER READING

Richard Austin, Impartial and Partisan Games, M.Sc. thesis, University of Calgary, 1976.

W.W. Rouse Ball and H.S.M. Coxeter, "Mathematical Recreations & Essays", 12th edn., University of Toronto Press, 1974 (esp. pp. 38–39).

E.R. Berlekamp, Unsolved problem #4, in W.T. Tutte (ed.) "Recent Progress in Combinatorics", Academic Press, New York and London, 1969, pp. 342–343.

E.R. Berlekamp, Some recent results on the combinatorial game called Welter's Nim, Proc. 6th Conf. Information Sci. and Systems, Princeton, 1972, 203–204.

F. de Carteblanche, The princess and the roses, J. Recreational Math. **3** (1970) 238–239.

F. de Carte Blanche, The roses and the princes, ibid. **7** (1974) 295–298.

J.H. Conway, "On Numbers and Games", Academic Press, London and New York, 1976, Chapters 11 and 13, and p. 181.

J.H. Conway and H.S.M. Coxeter, Triangulated polygons and frieze patterns, Math. Gaz. **57** (1973) 87–94, 175–183; MR **57** #1254–5.

H.E. Dudeney, "536 Puzzles and Curious Problems" (ed. Martin Gardner) Chas. Scribner's Sons, N.Y., 1969; #475 The 37 Puzzle Game, 186–187, 392–393.

Robert J. Epp and Thomas S. Ferguson, Remarks on take-away games and Dawson's Game, Abstract 742-90-3, Notices Amer. Math. Soc. **24** (1977) A-179.

Jim Flanigan, Generalized two-pile Fibonacci Nim, Fibonacci Quart, **16** (1978) 459–469.

Richard K. Guy, Anyone for Twopins?, in David Klarner (ed.) "The Mathematical Gardner", Prindle Weber and Schmidt, 1980.

Paul D.C. Hudson, The logic of social conflict: a game-theoretic approach in one lesson, Bull. Inst. Math. Appl. **14** (1978) 54–66.

Thomas A. Jenkyns and John P. Mayberry, The skeleton of an impartial game and the nim-function of Moore's Nim$_k$, Internat. J. Game Theory **9** (1980) 51–63.

S.-Y.R. Li, N-person Nim and N-person Moore's games, Internat. J. Game Theory, **7** (1978) 31–36; MR **58** #4367.

Eliakim H. Moore, A generalization of the game called Nim Ann. of Math. Princeton (2) **11** (1910) 93–94.

J. von Neumann and O. Morganstern, "Theory of Games and Economic Behavior", Princeton, 1944.

T.H. O'Beirne, "Puzzles and Paradoxes", Oxford University Press, London, 1965, Chapter 9.

I.C. Pond and D.F. Howells, More on Fibonacci Nim, Fibonacci Quart. **3** (1965) 61.

Fred. Schuh, "The Master Book of Mathematical Recreations", (transl. F. Göbel, from "Wonderlijke Problemen; Leerzaam Tijdverdrijf Door Puzzle en Spel", W.J. Thieme, Zutphen, 1943; ed. T. H. O'Beirne) Dover, London, 1968. Chapter VI, 131–154; Chapter XII, 263–230.

Allen J. Schwenk, Take-away games, Fibonacci Quart. **8** (1970) 225–234, 241; MR **44** #1446.

G.C. Shephard, Additive frieze patterns and multiplication tables, Math. Gaz. **60** (1976) 178–184; MR **58** #16353.

Roland Sprague, "Recreations in Mathematics" (trans. T.H. O'Beirne) Blackie, 1963; #14: Pieces to be moved, pp. 12–14, 41–42.

R. Sprague, Bemerkungen über eine spezielle Abelsche Gruppe, Math. Z. **51** (1947) 82–84; MR **9**, 330–331.

C.P. Welter, The advancing operation in a special abelian group, Nederl. Akad. Wetensch. Proc. Ser. A **55** = Indagationes Math. **14** (1952) 304–314; MR **14**, 132.

C.P. Welter, The theory of a class of games on a sequence of squares, in terms of the advancing operation in a special group, ibid. **57** = **16** (1954) 194–200; MR **15**, 682; **17**, 1436.

Michael J. Whinihan, Fibonacci Nim, Fibonacci Quart. **1** (1963) 9–13.

Yōhei Yamasaki, On misère Nim-type games, J. Math. Soc. Japan **32** (1980) 461–475.

Chapter 16

Dots-and-Boxes

Come, children, let us shut up the box.
William Makepeace Thackeray, *Vanity Fair*, Ch. 67.

I could never make out what those darned dots meant.
Lord Randolph Churchill.

Dots-and-Boxes is a familiar paper and pencil game for two players and has other names in various parts of the world. Two players start from a rectangular array of dots and take turns to join two horizontally or vertically adjacent dots. If a player completes the fourth side of a unit square (**box**) he initials that box and must then draw another line (so that completing a box is a complimenting move). When all the boxes have been completed the game ends and whoever has initialled more boxes is declared the winner.

A player who *can* complete a box is not obliged to do so if he has something else he prefers to do. Play would become significantly simpler were this obligation imposed; see the article by Holladay mentioned in the references.

Figure 1 shows Arthur's and Bertha's first game, in which Arthur started. Nothing was given away in the fairly typical opening until Arthur was forced to make the unlucky thirteenth move, releasing 2 boxes for Bertha. Her last bonus move enabled Arthur to take the bottom 3 boxes, but he then had to surrender the last 4.

This is how most children play, but Bertha is brighter than most. She started the return match with the opening that Arthur had used. He was happy to copy Bertha's replies from that game, and was delighted to see her follow it even as far as that unlucky thirteenth move, which had proved his undoing (Fig. 2). He grabbed those 2 boxes and happily surrendered the bottom 3, expecting 4 in return. But Bertha astounded him by giving him back 2. He pounced on these, but when he came to make his bonus move, realized he was doubled-crossed!

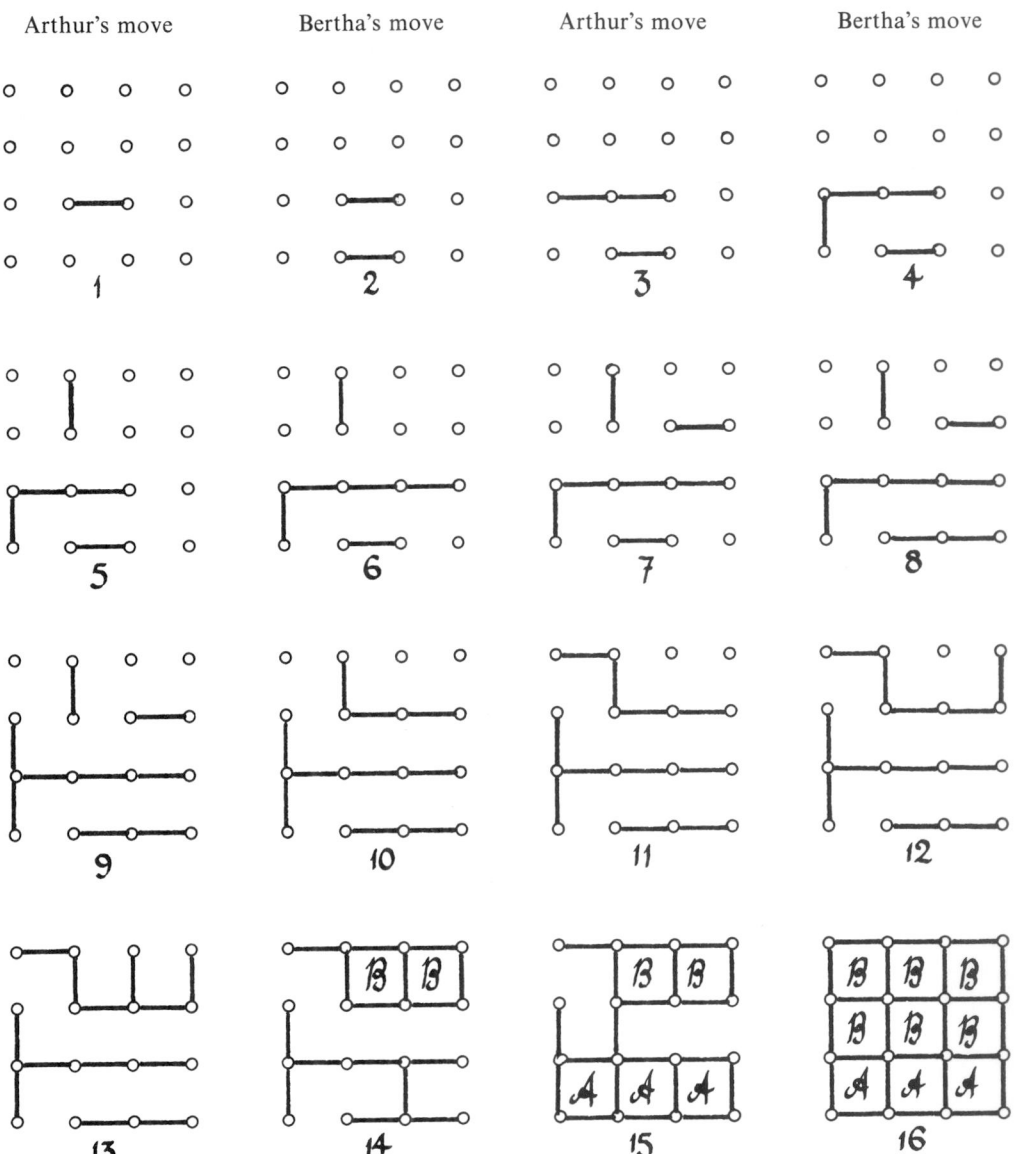

Figure 1. Arthur's and Bertha's First Game.

Bertha beats all her friends in this double-dealing way. Most children play at random unless they've looked quite hard and found that every move opens up some chain of boxes. Then they give the shortest chain away and get back the next shortest in return, and so on.

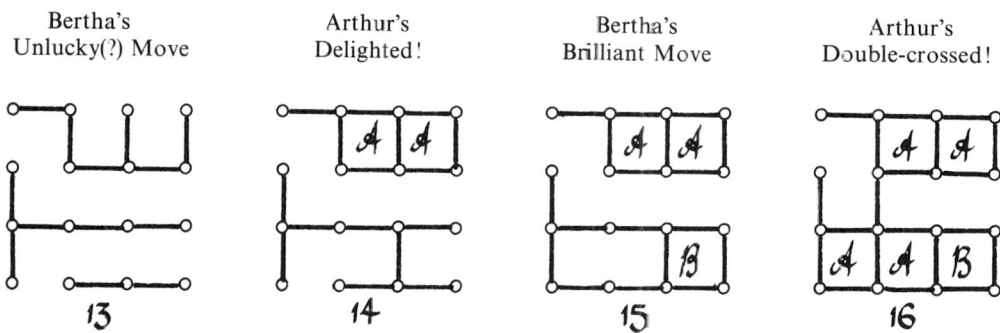

Figure 2. Bertha's Brilliance Astounds Arthur.

But when you open a long chain for Bertha, she may close it off with a double-dealing move which gives you the last 2 boxes but forces you to open the next chain for her (Fig. 3). In this way *she* keeps control right to the end of the game.

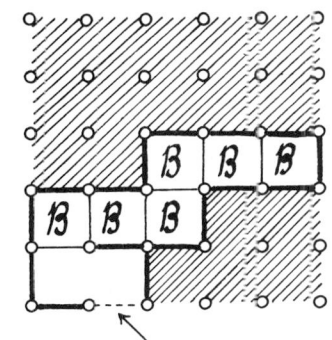

Figure 3. Bertha's Double-Dealing Move.

You can see in Fig. 4 just how effective this strategy can be. By politely rejecting two cakes on every plate but the last you offer her, Bertha helps herself to a resounding 19 to 6 victory. In the same position you'd have defeated the ordinary child 14 to 11.

DOUBLE-DEALING LEADS TO DOUBLE-CROSSES

Each double-dealing move is followed, usually immediately, by a move in which two boxes are completed with a single stroke of the pen (Fig. 5). These moves are very important in the theory. We'll call them **doublecrossed** moves, because whoever makes them usually has been!

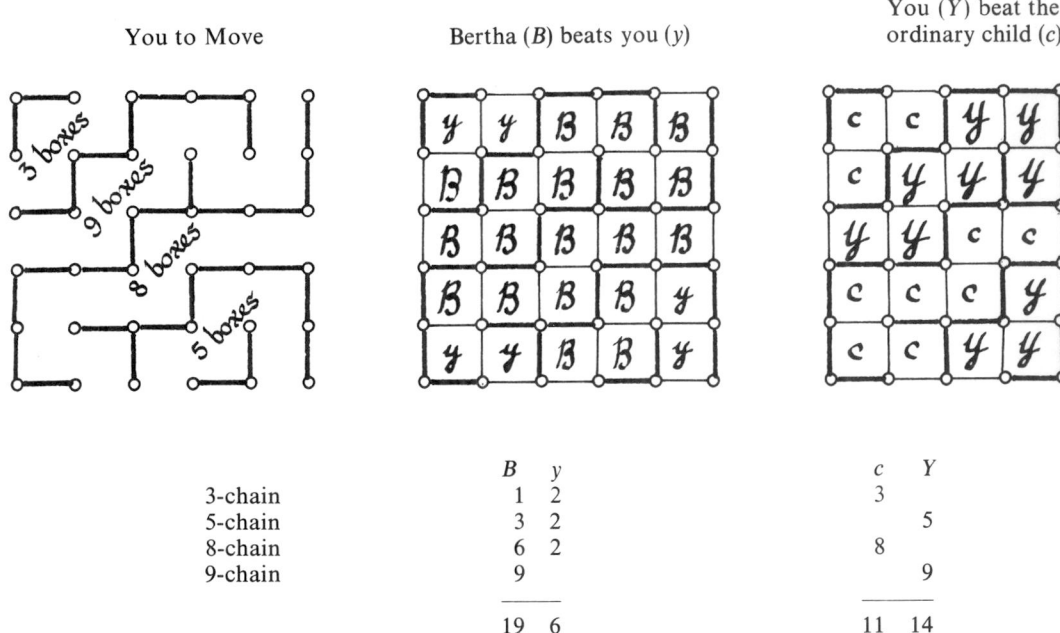

You to Move   Bertha (*B*) beats you (*y*)   You (*Y*) beat the ordinary child (*c*)

	B	*y*		*c*	*Y*
3-chain	1	2		3	
5-chain	3	2			5
8-chain	6	2		8	
9-chain	9				9
	19	6		11	14

Figure 4. Double-Dealing Pays Off!

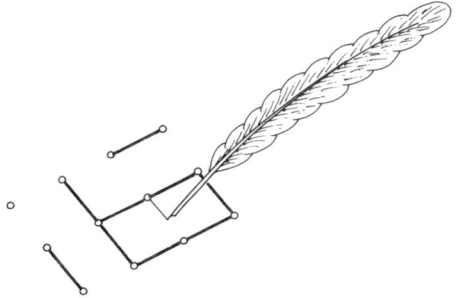

Figure 5. A Doublecross—Two Boxes at a Single Stroke.

Now Bertha's strategy suggests the following policy:

> Make sure there are long chains about
> and try to force your opponent to be
> the first to open one.

Try To Get Control ...

We'll say that whoever can force her opponent to open a long chain has **control** of the game. Then:

> When you have control, make sure you
> keep it by politely declining 2 boxes
> of every long chain except the last.

... And Then Keep It.

The player who has control usually wins decisively when there are several long chains.

So the fight is really about control. How can you make sure of acquiring this valuable commodity? It depends on whether you're playing the odd- or even-numbered turns ... .

Figure 6. Which is Dodie and Which is Evie?

Arthur and Bertha live next to the Parr family, in which there are two little sisters called Dodie and Evie (Bertha often teases them by calling them the Parrotty Girls!). You can see them playing the 4-box game in Fig. 6. Dodie's a year younger than Evie and so always has first turn in any game they play. They've got so used to playing like this that even when they're playing somebody else, Dodie always insists on taking the odd-numbered moves while Evie will only take the even-numbered ones:

> Dodie Parr: odd parity,
> Evie Parr: even parity.

The rule that helps them take control is:

> *Dodie*
> tries to make the number of
> initial dots + doublecrossed moves
> *odd.*
> *Evie*
> tries to make this number
> *even.*

Be SELFish about Dots + Doublecrosses!

In simple games the number of doublecrosses will be one less than the number of long chains and this rule becomes:

THE LONG CHAIN RULE

> Try to make the number of
> initial dots + eventual long chains
> *even* if your opponent is *Evie*,
> *odd* if your opponent is *Dodie*.

The OPPOsite for Dots + Long Chains!

The reason for these rules is that whatever shape board you have on your paper, you'll find that:

Number of dots you start with

+ Number of doublecrosses

= Total number of turns in the game.

We'll show this in the Extras.

HOW LONG IS "LONG"?

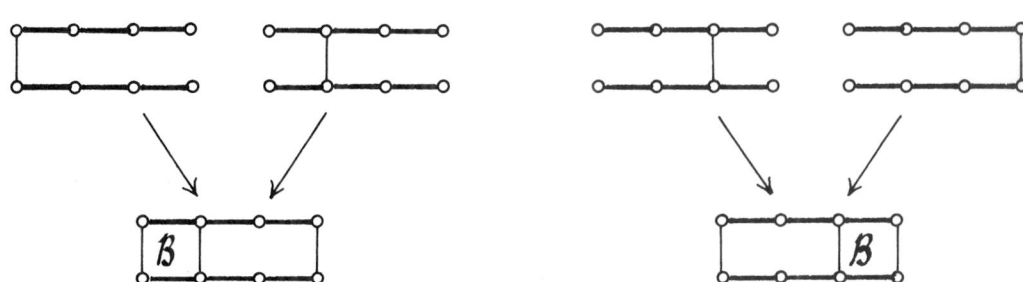

Figure 7. Bertha's Endgame Technique.

We can find the proper definition of long by thinking about Bertha's endgame technique. A **long chain** is one which contains 3 or more squares. This is because whichever edge Arthur draws in such a chain, Bertha can take all but 2 of the boxes in it, and complete her turn by drawing an edge which does not complete a box. Figure 7 shows this for the 3-square chain. A chain of 2 squares is *short* because our opponent might insert the *middle* edge, leaving us with no way of finishing our turn in the same chain. This is called (Fig. 8(a)) the **hard-hearted handout**.

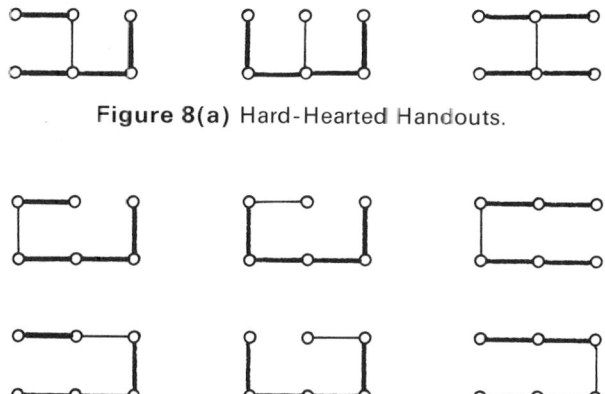

Figure 8(a) Hard-Hearted Handouts.

Figure 8(b). Half-Hearted Handouts.

When you think you are winning, but are forced to give away a pair of boxes, you should always make a hard-hearted handout, so that your opponent has no option but to accept. If you use a **half-hearted** one (Fig. 8(b)) he might reply with a double-dealing move and regain control. But if you're losing, you might try a half-hearted handout on the Enough Rope Principle (Chapter 1 Extras). Officially this is a bad move, since your opponent, if he has any sense, will grab both squares. But boys by billions, being bemused by Bertha's brilliance, blindly blunder both boxes back.

THE 4-BOX GAME

When Dodie was *very* young, the girls often played the 4-box game and offset Dodie's first move advantage by calling it a win for Evie (the second player) when they each got 2 boxes:

> TWO TWOS IS A WIN TO TWO

At first Dodie would never give away a box if she could see something else to do, and Evie, who you can see is a very symmetrical player, would always win by copying Dodie's moves on the opposite side of the board. But after watching Bertha playing Evie, Dodie found how to counter this strategy by making a Greek gift on her 7th move. Evie can still win if Dodie dares to stray from the Path of Righteousness but must resist her temptation to make *every* move a symmetrical one (Fig. 9).

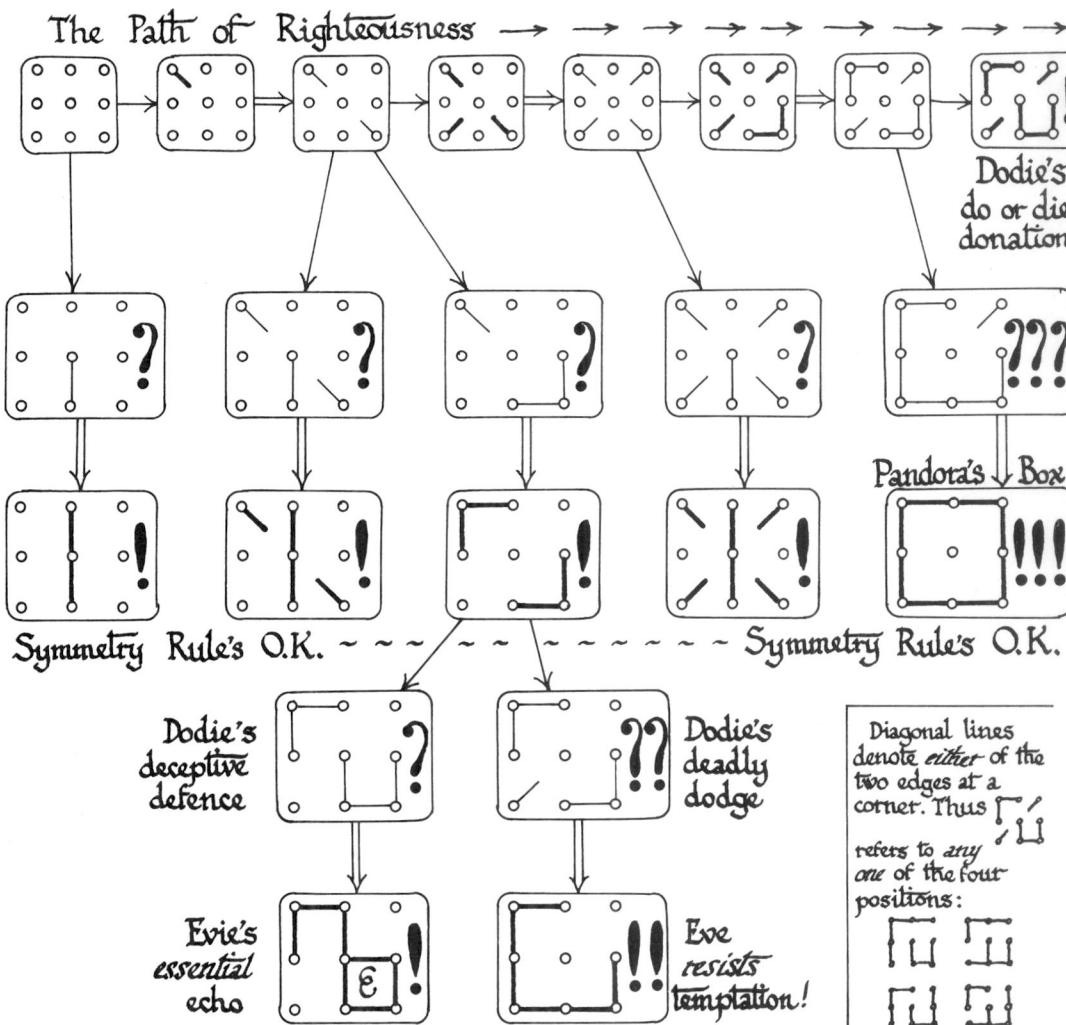

Figure 9. Evie Envisaging Every Eventuality.

Even though Dodie has the win, it's much harder to write out in full her best plays against sufficiently cunning opponents. In Fig. 34 of the Extras we give an adequate strategy for Dodie and in Fig. 35 a complete list of $\mathscr{P}$-positions for both players. This little game is full of traps for the unwary, and those of you who have written to us for advice on becoming Professional Boxers will find these tables very useful in the bruising preliminary contests on the 4-box board.

If the chain lengths are

		loop of 4
4		2 + 2
	or	
3 + 1		1 + 1 + 1 + 1

the winner will usually be

| Dodie | or | Evie |

in agreement with the Long Chain Rule, but on this small board, Dodie should often defy the rule and win by splitting the chains as $2+1+1$.

THE 9-BOX GAME

Surprisingly, the Long Chain Rule makes the 9-box game seem easier than the 4-box one. This time Evie wins, and her basic strategy is to draw 4 spokes as in Fig. 10, forcing every long chain to go through the centre. Against most children this wins for Evie by at least 6–3, but Dodie can hold her down to 5–4, perhaps by sacrificing the centre square, after which Evie should abandon her spoke strategy. Of course, Evie's real aim is to arrange that there's just one long chain, and she often improves her score by forming this chain in some other way.

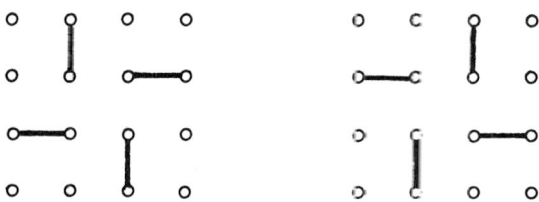

Figure 10. Lucky Charms Ward Off More Than One Long Chain; Evie Puts Spokes in Dodie's Wheel.

Evie usually prefers to put her spokes in squares where another side is already drawn, and she's careful to draw spokes in only *one* of the two swastika patterns of Fig. 10. There usually aren't any double-crossed moves, so that Evie wins at the $(16+0=)$ 16th turn.

Dodie tries to arrange *her* moves so that some spoke can only be inserted as a sacrifice, and *either* cuts up the chains as much as possible (maybe with a centre sacrifice) *or* forms *two* long chains when Evie isn't thinking. Every now and then a half-hearted handout has saved the game for her just when she thought that all was lost.

THE 16-BOX GAME

We don't know who wins on the 4×4 box board, which makes a very interesting game to play. Evie tries to make the number of long chains 2, while Dodie tries to cut it down to 1 or force

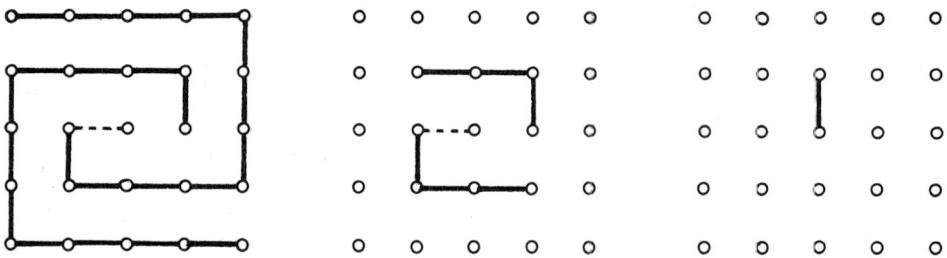

Figure 11. "Come into my symmetrical parlor!"

it up to 3. Evie beats many children with her symmetry strategy, but Dodie remembers her trick from the 4-box game. If she thinks her opponent will mimic her every move, she can lure him into the spider's web of Fig. 11(a), but when he's less predictable she finds it safer just to use the middle of the web (Fig. 11(b)). Dodie doesn't usually open with Fig. 11(c), because she finds the symmetry strategy very hard to beat.

OTHER SHAPES OF BOARD

To beat all your friends on larger square and rectangular boards you'll really need the Long Chain Rule. Remember to count a closed loop of 4 or more cells as *two* long chains and that each doublecross, no matter who makes it, changes the number of long chains you want. (Think of a doublecross as a long chain that's already been filled in.) It's good tactics to make the long chains as long as possible and avoid closed loops when you can, because you forfeit *four* boxes when declining a loop. These rules work for all large boards and even for triangular Dots-and-Boxes boards, like that in Fig. 12.

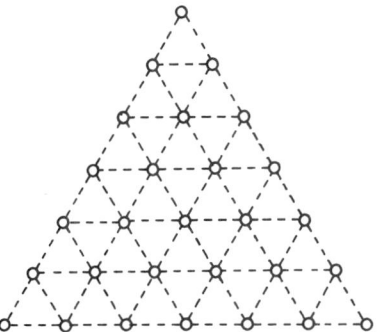

Figure 12. A Board with 28 Dots and 36 Triangular Cells.

Of course, if your opponent is also using the Long Chain Rule, the fight for control might be quite hard. The game of Nimstring, discussed in the rest of this chapter, is what control is all about. There's a piece in the Extras that describes some of the rare occasions when you might find it wise to lose control.

DOTS-AND-BOXES AND STRINGS-AND-COINS

You can play a dual form of Dots-and-Boxes, called **Strings-and-Coins**, with strings, coins and scissors. The ends of each piece of string are glued to two different coins or to a coin and the ground (each string has at most one end glued to the ground) and each player in turn cuts a new string. If your cut completely detaches a coin, you pocket it and must then cut another string (if there's one still uncut). The game ends when all coins are detached and the player who pockets the greater number is the winner.

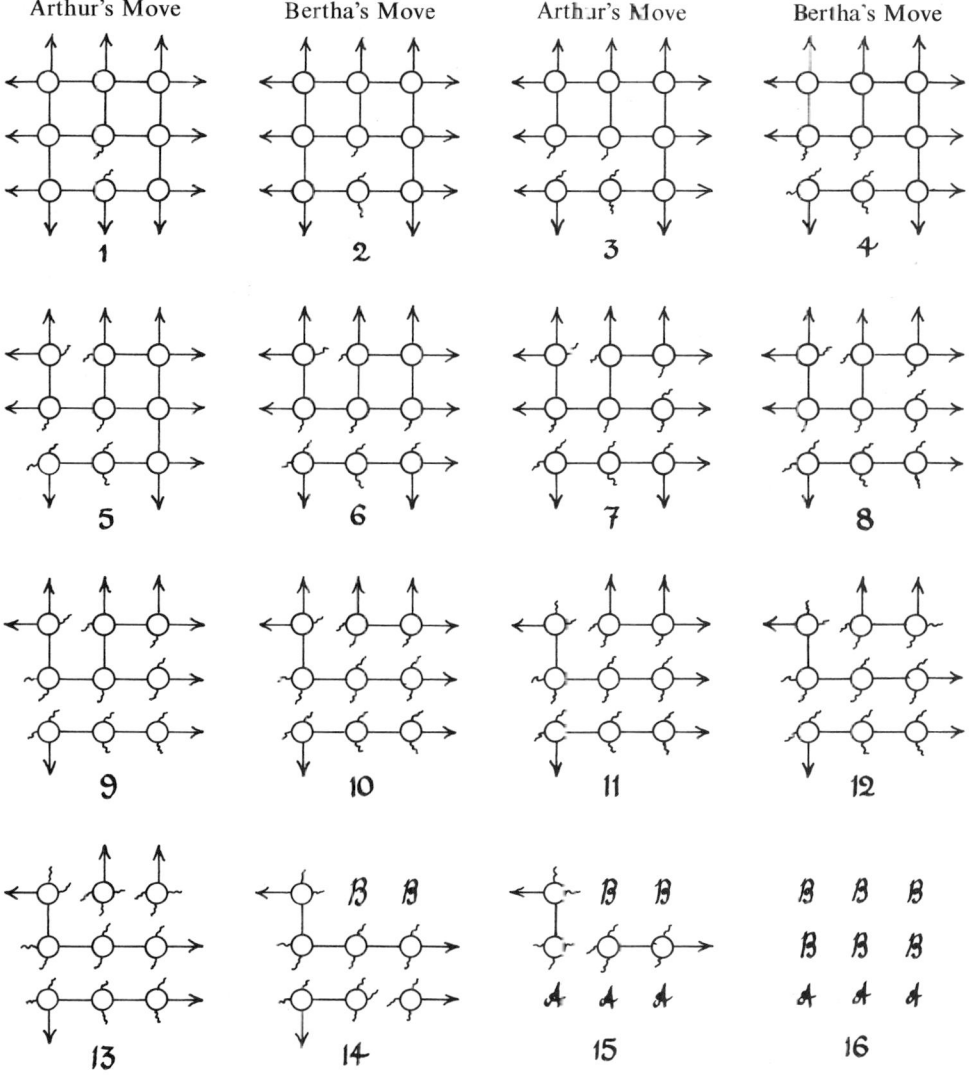

Figure 13. A Strings-and-Coins Game—The Dual of Figure 1.

Figure 13 shows the dual of Arthur's and Bertha's first game (compare it with Fig. 1). It started with 9 coins connected by 24 strings, 12 of them between coins and coins, the other 12 between coins and the ground. We use little arrows for strings that run to the ground. The coins and strings form the nodes and edges of a **graph**. It's easy to make a graph to correspond to any Dots-and-Boxes position. However, there are lots of graphs which *don't* correspond to such positions; for example the graph may have cycles of odd length or nodes with more than 4 edges, or the graph may be *non-planar*. In fact Strings-and-Coins is a generalization of Dots-and-Boxes.

NIMSTRING

The game of **Nimstring** is played on exactly the same kind of graphs as Strings-and-Coins, and you make exactly the same move by cutting a string (which is a *complimenting* move whenever you detach a coin). In Strings-and-Coins the winner is the player who detaches the larger number of coins, but Nimstring is played instead according to the Normal Play Rule. So, for ordinary Nimstring positions you *lose* when you detach the last coin, for then the rules require you to make a further move when it is impossible to do so. (But a Nimstring graph *may* have a string joining the ground to itself, and if the last move cuts *this* it doesn't detach a coin, and so *wins*.)

Nimstring looks quite different from Strings-and-Coins, but closer investigation shows that Nimstring is in fact a special case of Strings-and-Coins.

> You can't know all about Strings-and-Coins
> unless you know all about Nimstring!

Figure 14(a). A Hard Nimstring Problem.

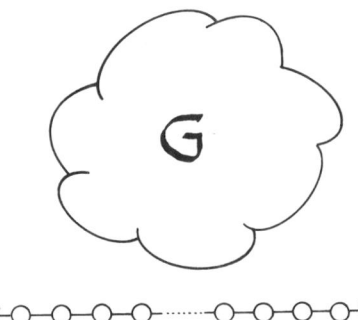

Figure 14(b). This Strings-and-Coins Problem is Just as Hard.

Figure 14 shows the construction which proves this. If *G* represents an arbitrary Nimstring problem, we add a long chain to it and consider the resulting Strings-and-Coins game—the long chain should have more coins than *G*. Because the chain is so long and whoever first cuts a string of it allows his opponent to capture all the coins of the chain on his next turn, both players will try to avoid cutting any string of the chain. Neither player can force his opponent to move on the chain until all the strings of *G* have been cut. In other words, the only way to win the Strings-and-Coins game of Fig. 14(b) is to play a winning game of Nimstring on the graph *G*.

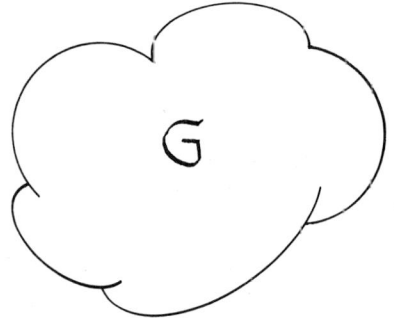

Figure 15(a). Another Nimstring Game.

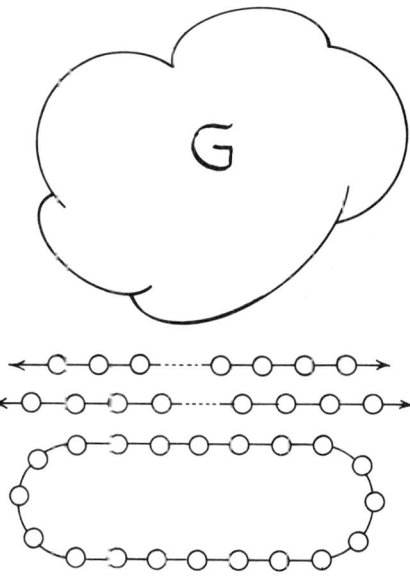

Figure 15(b). A Corresponding Strings-and-Coins Game.

Figure 15 shows another construction. This time we get the Strings-and-Coins game by adding several long chains and cycles to the Nimstring game G. If these are long enough the winning strategy for the Strings-and-Coins game is then:

> *If your opponent moves in G,*
> reply in G with a move from
> the winning Nimstring strategy.
> *If he moves in a long chain,*
> take *all but two* coins of that
> chain, leaving just the string
> which joins them.
> *If he moves in a long cycle,*
> take *all but four* coins of the
> cycle, leaving them as *two pairs*
> each joined by a string.

This strategy gives you all but 2 coins of each long chain and all but 4 of each long cycle, so it will win for you if the total number of nodes in the added chains and cycles exceeds

$$\text{(the number of nodes in } G\text{)}$$
$$+ \; 4 \times \text{(the number of added long chains)}$$
$$+ \; 8 \times \text{(the number of added long cycles)}.$$

In practice the Nimstring position will often contain (potential) long chains of its own, so that the strategy is of wider application. Recall that the "all but 2" principle was used by Bertha in her second game with Arthur (Fig. 2). Well-played games of Dots-and-Boxes are usually played like the corresponding Nimstring games, except at the very end. The last long chain in a Nimstring game is treated like any other; the winner takes all but the last 2 coins, which he gives to the loser by a hard-hearted handout. For the last chain in Dots-and-Boxes, of course, winner takes all!

WHY LONG IS LONG

The argument explains why "long" must be defined precisely as follows. We should call a chain **long** if it contains 3 or more coins, because no matter which string of such a chain our opponent might cut, we may take all but 2 of its coins and finish by cutting another string of the chain. We must call a chain of 2 coins **short**, because he might cut the middle string and prevent us from declining those 2 vital coins (the hard-hearted handout). For a similar reason a closed loop of 2 or 3 coins would be called **short** (short loops don't arise in rectangular Dots-and-Boxes). However, a loop with at least 4 coins is called **long**, because we can politely decline the last 4 coins no matter which string our opponent cuts. Figure 16(a) shows how to do this on a 6-loop. When your opponent has cut the first string as shown, you only take 2 of the coins and then cut the string in the middle of the remaining 4. Figure 16(b) shows how this corresponds with Bertha's way of playing Dots-and-Boxes.

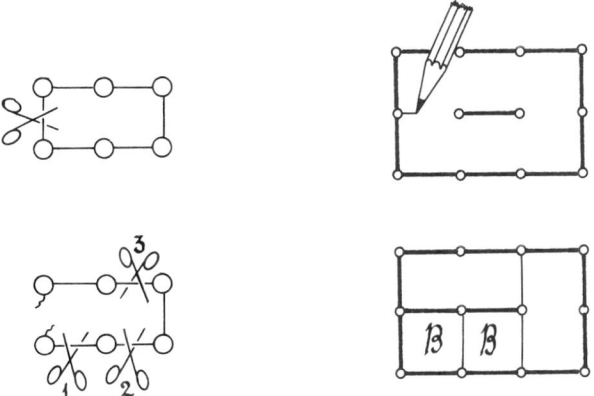

Figure 16. Bertha Politely Declines a Long Loop.

Well-played games of Dots-and-Boxes frequently lead to the duals of positions like those in Fig. 15(b). Most of the coins are in the long chains and loops, and the winner is whoever can force his opponent to cut the first string in one of these. It seems to be almost always the case that the winning strategy for Nimstring also gives the winning strategy for Strings-and-Coins. There are many other graphs than those satisfying the conditions of Fig. 15(b) for which this can be proved to happen. To win a game of Dots-and-Boxes or Strings-and-Coins, you should try to win the corresponding game of Nimstring and at the same time arrange that there are some fairly long chains about. In the rest of this chapter we'll teach you how to become an expert at Nimstring.

TO TAKE OR NOT TO TAKE A COIN IN NIMSTRING

A coin which has only a single string attached is **capturable**. Whenever there's a capturable coin the next player has the option of removing the corresponding branch, thereby detaching the coin and getting another (complimentary) move. For some graphs this is the best move; for others, including one of those encountered by Bertha in the game of Fig. 2, the winning strategy is to refuse to detach the coin. As you might guess, the decision as to whether it's better to take a coin or decline it often depends on the entire graph. However a great deal can be deduced by examining only local properties of the graph near the capturable coin.

Any capturable coin must look like one of the six possibilities in Fig. 17. The string from the capturable coin goes either to the ground (Fig. 17(a)) or to another coin. If to another coin, the

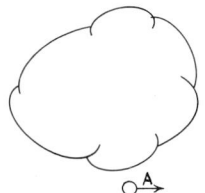

Figure 17(a). TAKE!
A free coin.

Figure 17(b). TAKE!
Two free coins and a
doublecross.

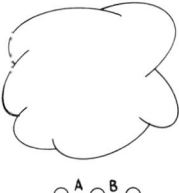

Figure 17(c). TAKE!
Three free coins and a
doublecross.

Figure 17(d). TAKE!
A free coin.

Figure 17(e). WIN!

Figure 17(f). WIN!

Half-hearted handouts.

number of strings there is either one (Fig. 17(b)), two (Figs. 17(c), (e) and (f)) or three or more (Fig. 17(d)). If there are two strings, the second goes either to another capturable coin (Fig. 17(c)), or to the ground (Fig. 17(e)), or to a coin with two more or strings (Fig. 17(f)). In each of the six cases the cloud contains all the coins and strings not regarded as near enough to the capturable coin. The dotted lines in Figs. 17(d) and (f) are possible additional strings which may or may not be present.

We claim that in the first four cases (Figs. 17(a)–(d)) the player to move might as well cut string *A* and capture the coin, and in Fig. 17(c), he might as well continue by cutting string *B*, taking two more coins. For suppose you have a winning strategy starting from one of these graphs. If this tells you to complete your first turn by cutting only certain *un*lettered strings, then your opponent has the option of beginning *his* turn by cutting the lettered ones. But the same position will be reached if instead you first cut all the lettered strings and then cut the same unlettered ones as before. If there's any winning strategy at all, starting from these four cases, there's one which begins by cutting the lettered strings. So there's no loss in generality in supposing that a good player will TAKE a capturable coin of one of the four types in Figs. 17(a)–(d).

The other two positions (Figs. 17(e) and (f) are much more interesting. If it's your turn to move in one of these two cases, you can either *detach* the capturable coin by cutting string *A*, or *decline* to take it by cutting string *B*. No matter what the rest of the graph might be, *one or other* of these two moves will WIN. But you might need to look at the whole graph to decide whether your winning strategy begins by cutting string *A* or string *B*!

This somewhat surprising result is proved by a cunning use of Strategy Stealing (Fig. 18).

We ask, for the games of Figs. 17(e) and (f):

> *who wins the smaller game G consisting of just the unlettered strings (Figs.* 18(*e*) *and* 18(*f*))?

This is either the player who has to move from *G* or the player who doesn't. Whoever this fortunate player is, you should arrange to steal his strategy. If the player to move from *G* can win, then when playing from Fig. 17(e) or (f) you should start by cutting string *A* (which detaches a coin, so you continue), then cut string *B* (detaching another coin, so you continue again) and then begin the game on *G*, which of course you will play according to the winning strategy for the first player. On the other hand, if there's no winning move for the first player from *G* then, starting from Fig. 17(e) or (f), you should finish your turn immediately by cutting string *B* and so force your opponent to start the game *G* (he might as well start by cutting string *A*; if he doesn't, you will later).

The fact that the declining move forfeits 2 coins to your opponent makes no difference in Nimstring, where the winner is determined by the last move. In Strings-and-Coins (and Dots-and-Boxes) it *might* matter, but is unlikely to when there are long chains about.

SPRAGUE–GRUNDY THEORY FOR NIMSTRING GRAPHS

We now try to define values for arbitrary Nimstring graphs. We'd like these values to be nimbers so that we can use the ordinary Mex and Nim Addition Rules. The only trouble is that there are positions like that shown in Fig. 19.

Figure 17 (e) Figure 17 (f)

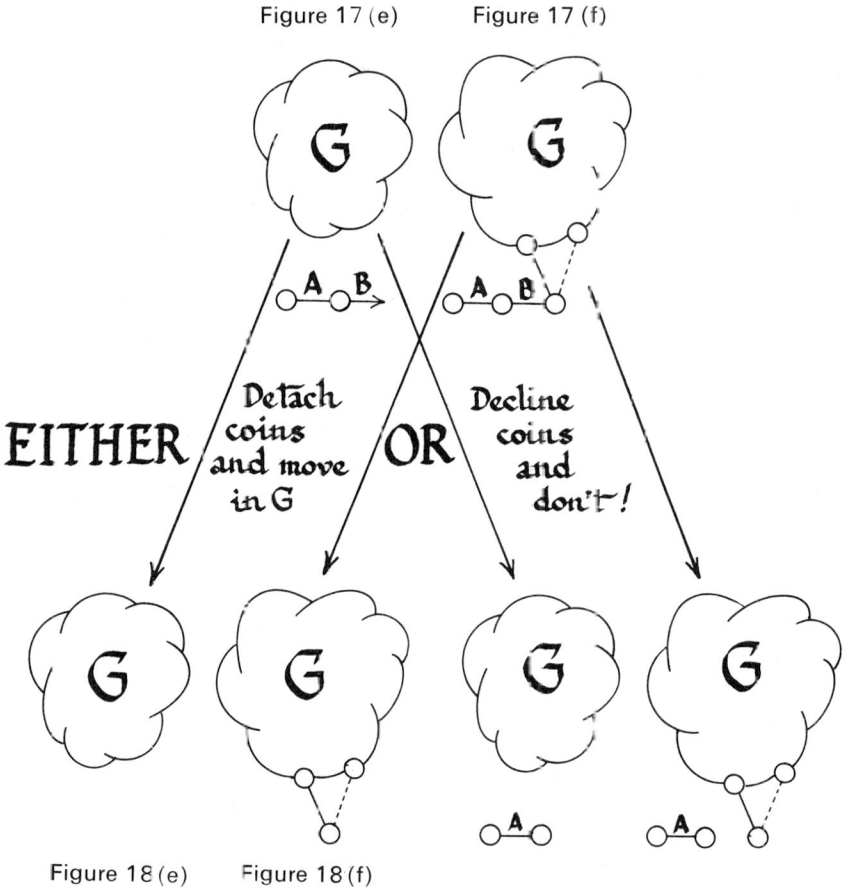

Figure 18 (e) Figure 18 (f)

Figure 18. Strategy Stealing after a Half-Hearted Handout.

Figure 19. A Loony Nimstring (or Dots-and-Boxes) Position.

Our discussion of Fig. 17(e) shows that no matter what graph G is added to this, the result is a win for the first player. The supposed value, $*x$, for Fig. 19 must therefore have the property that

$$*x + *y \neq 0$$

for every nimber $*y$, including even $*x$ itself, so that in particular

$$*x + *x \neq 0.$$

Those of you who have read Chapter 12 will see at once how to resolve this paradox. Figure 19 is what we call a **loony** position, whose value is 𝒟. The theory of complimenting and complimentary moves in that chapter applies to Nimstring (where the complimenting moves are those that capture coins) and shows that every position has either an ordinary nimber value or the special value 𝒟. But don't reread Chapter 12 now, because we can easily summarize the rules for finding values at Nimstring:

The value of a graph without strings is 0.

The value of a graph with a capturable coin of one of the four types in Figs. 17(a)–(d) is equal to that of the subgraph obtained by removing the capturable coin(s) and its string(s).

The value of a graph with a capturable coin of one of the two types in Figs. 17(e) and (f) is 𝒟.

The value of a graph with no capturable coins is found from the values of the graphs left after cutting single strings by using the Mex Rule (Chapter 4).

VALUES FOR NIMSTRING

When adding these values, remember that

$$\mathcal{D} + 0 = \mathcal{D} + {*}1 = \mathcal{D} + {*}2 = \ldots = \mathcal{D} + \mathcal{D} = \mathcal{D},$$

as well as the ordinary nim-addition rules.

We show the calculation for some graphs in Fig. 20. When there are no capturable coins we write against each string the nim-value of the sub-graph obtained by cutting that string. Thus the last picture has options of nim-values 0, 1, 3, i.e. values $*0$, $*1$, $*3$, and so its own value is $*2$, because $2 = \text{mex}(0, 1, 3)$. Strings marked 𝒟 are loony options for the first player—if he cuts such a string he will LOSE against proper play even if some other graphs are added to the position. The nim-value of each graph is found from the mex of the numbers against its strings—in this you should ignore the 𝒟 values, which correspond to suiciding moves.

Although Dodie wins the 4-box game of Dots-and-Boxes, we can deduce from Fig. 20 that Evie wins the corresponding Nimstring position:

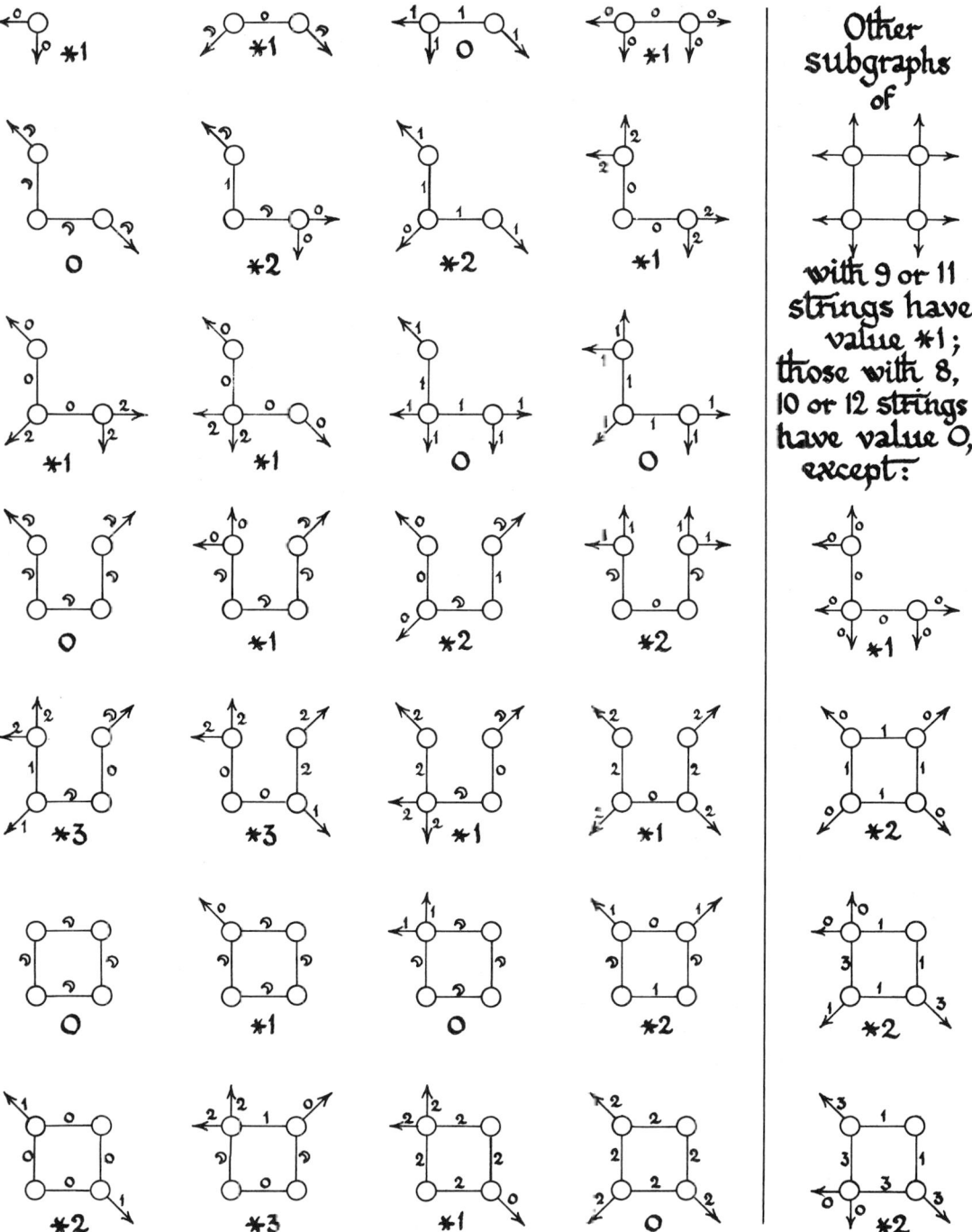

Figure 20. Working Out Values for Nimstring Graphs.

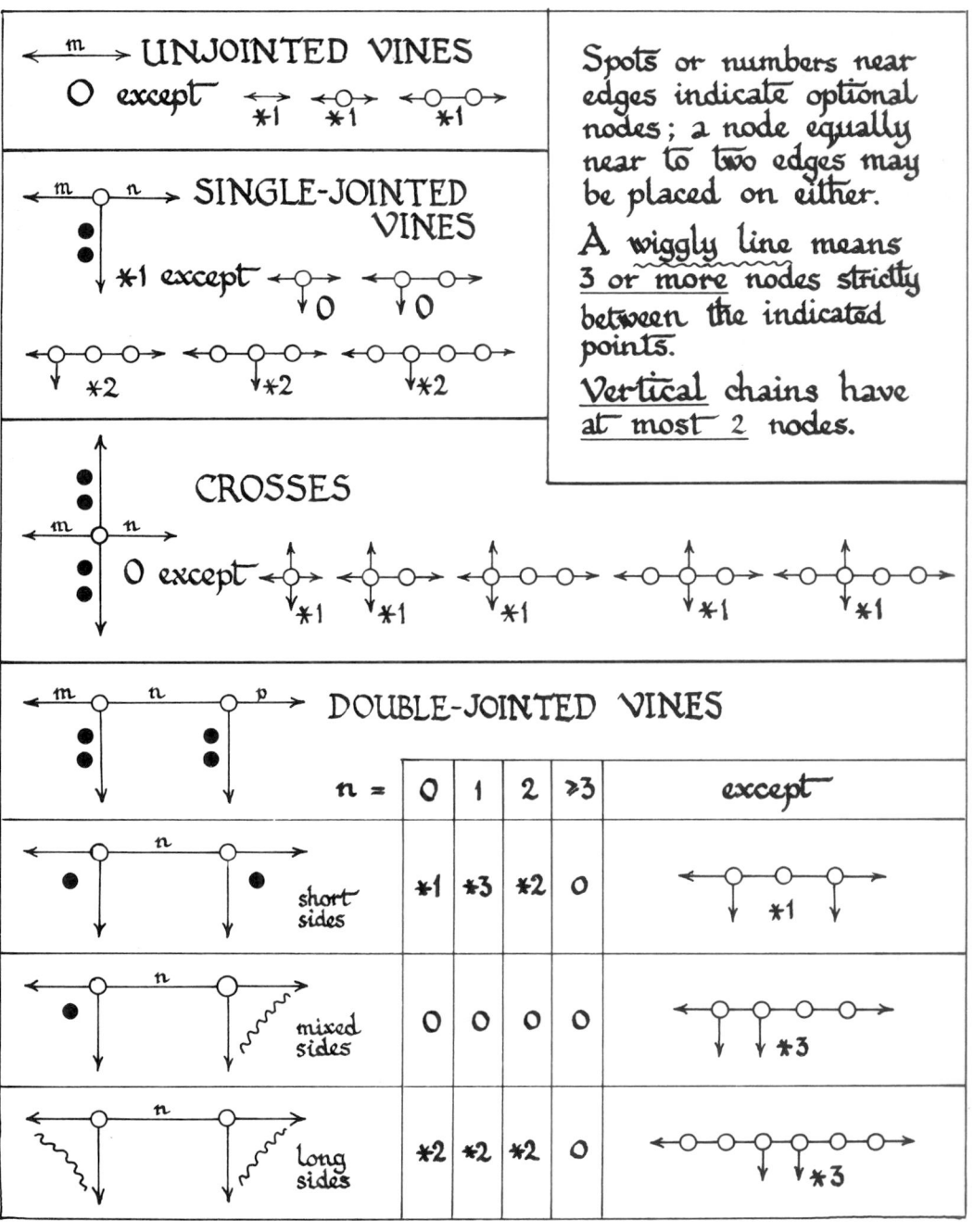

Figure 21. Noteworthy Nimstring Nimbers.

This means that even in Dots-and-Boxes she should win

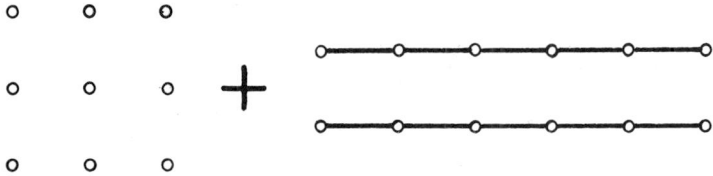

Figure 21 assembles answers for all graphs with at most 4 ends and no internal circuit. You'll find an extended table in the Extras that covers tree-like graphs with 5 ends.

ALL LONG CHAINS ARE THE SAME

Look at the various positions of Fig. 22, in which the clouds all conceal exactly the same thing, and the necklaces that hang from them all have at least three beads. The graphs all behave the same way in Nimstring because all the visible edges will always be loony moves.

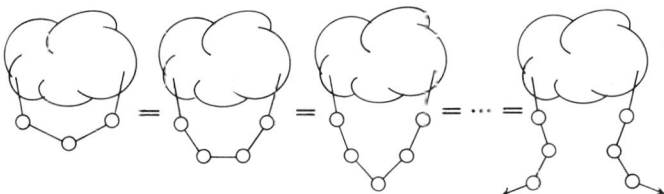

Figure 22. Three or More's a Crowd.

> Provided a chain has
> 3 or more nodes along it,
> the exact number
> doesn't make any difference
> to the value.

This makes it handy to have a special notation for long chains:

$$—\!\bigwedge\!\!\bigwedge\!—$$

means

—○—○—○— or —○—○—○—○— or —○—○—○—○—○— or ···

WHICH MUTATIONS ARE HARMLESS?

More generally we can put in or take out some beads on any Nimstring graph G to obtain **mutations** of the graph (a **bead**, of course, is a node with just 2 edges). Figure 23 shows a graph G and two mutations, H and K.

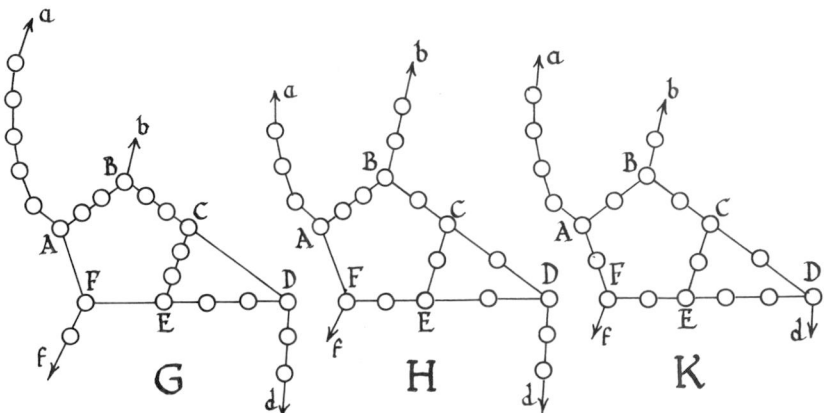

Figure 23. A Graph, a Harmless Mutation and a Killing One.

We'll use the word **stop** to mean *either* an arrowhead where the graph goes to ground (an **end**) *or* any of the nodes which have 3 or more edges (the **joints**). A path between two stops is **long** if it passes through 3 or more intermediate nodes, **short** otherwise. Mutation usually affects the value, but there are a lot of **harmless** mutations that don't:

> A mutation between two graphs
> will certainly be harmless if
> every short path between stops
> in either graph corresponds to a
> short path in the other.

THE HARMLESS MUTATION THEOREM

In Fig. 23, H is a harmless mutation of G, since the only short paths are AE, Af, Ef, and the ones other than Aa that don't pass through a stop. But AE is long in K, and Cd is short, so this mutation is not covered by our theorem. In fact G and H have value $*2$, while K has value 0.

When G and H are related by a harmless mutation you just play H like G. A non-loony move must cut some string of a short chain between two points A and B that were stops at least until the move was made. A and B must have been stops in the original graph and we can find a similar non-loony move in the mutated graph because the distance between A and B will be short. (Fig. 24).

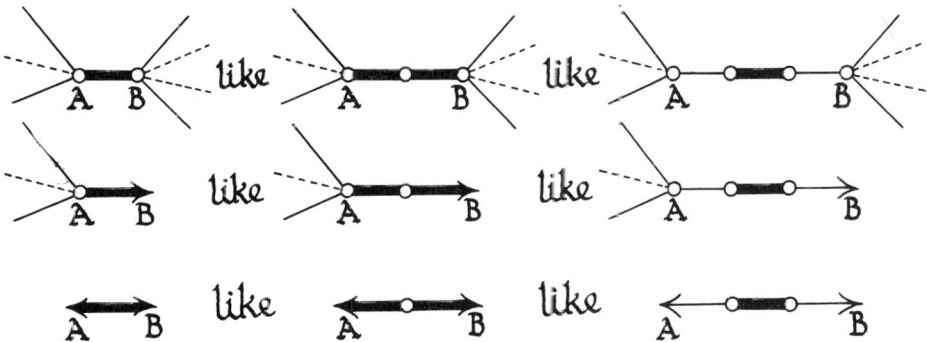

Figure 24. Like Moves in Harmless Mutations.

We can strengthen our Mutation Theorem a little:

> If the path between two stops
> passes through one end of a
> long chain, you needn't worry
> about the length of the path.

(For in a graph like Fig. 25—in which A or B might have been ends—AB won't become a chain unless someone makes a loony move cutting the long chain ending at C.)

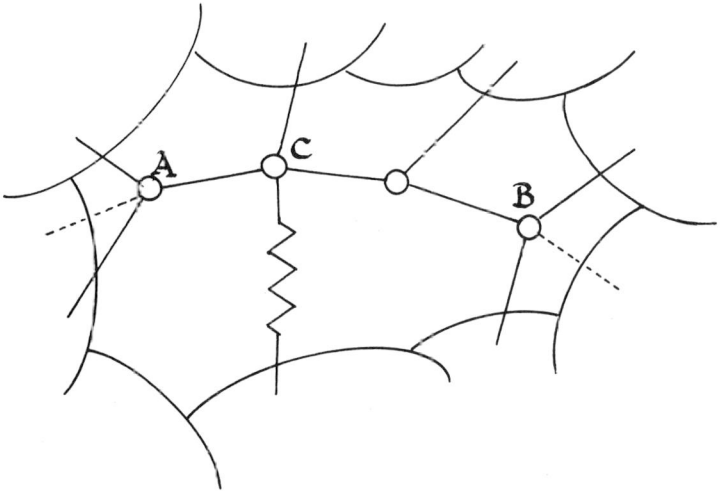

Figure 25. The Path *AB* Passes a Long Chain at *C*.

CHOPPING AND CHANGING

There are lots of more drastic changes we can make to Nimstring graphs without affecting their values; for instance:

> Long chains snap!

This was hinted at in Fig. 22, and Fig. 26(a) shows how it's written in our long chain notation.

The remaining equivalences of Fig. 26 are more interesting. The middle equivalence of Fig. 26(b) is particularly useful (the left equivalence is a long chain that snaps). It asserts that when an edge runs to a node from which two long chains emanate, then this edge may be replaced by an edge running directly to the ground. More generally, when two long chains are attached to a node, all other edges ending at this node may be replaced by edges running to the ground (Fig. 26(c)).

The idea of the proof is that a node at the end of two long chains can't be captured until after someone concedes the game by making a loony move. We can apply the equality between the first and last parts of Fig. 26(b) to every branch that runs to the ground, so as to eliminate ground branches from every graph. But usually it's more convenient to use it in the other direction, eliminating many branches and nodes by introducing new ends. Sometimes, as in Fig. 26(d), this gives rise to a branch joining the ground to itself (a 0 by 1 game of Dots-and-Boxes!); such a branch contributes *1 to the value.

The equality between the first three parts of Fig. 26(e) follows from the Harmless Mutation Theorem, but that between these and the last three doesn't, because some short chains have become long. The letters label corresponding moves, and the 𝒟's show moves which we should ignore. Figures 26(b), (d), (f) and (g) show that we can sometimes eliminate circuits from our graphs—the last diagram of Fig. 26(f) is our shorthand notation for any of the previous three, which are harmless mutations of each other. Figure 26(h) has many variants, abbreviated in Fig. 26(i) (using the notation of Fig. 21).

VINES

A **vine** is a Nimstring graph without circuits or capturable nodes in which all the joints lie on a single long path (the **stem**) and each joint belongs to just 3 edges. The chain joining an end to its nearest joint is called a **tendril**, so a single-jointed vine has 3 tendrils (Fig. 21). Vines with more joints have 2 tendrils at their endmost joints and just 1 at intermediate ones. If the distance between two neighboring joints is long, the vine decomposes into two smaller ones because long chains snap, so we can suppose such distances *short*, if we like.

A **Twopins-vine** is one whose every distance between *non*-neighboring stops (which may be either ends or joints) is *long*. It is a remarkable fact that the value of any Twopins-vine is equal to that of a corresponding configuration in the game of Twopins (Chapter 15). Each joint with a *short* tendril becomes a column of *two* pins (even if it has also a long tendril); each other joint becomes a column of *one*; and two neighboring joints a long distance apart correspond to an *empty* Twopins column (Fig. 27). A bowling shot which removes a *single* column at Twopins corresponds to a *tendril* move at Nimstring; one which removes a *pair* of columns corresponds to a move on the *stem* of the vine.

(a) Snap every Long Chain!

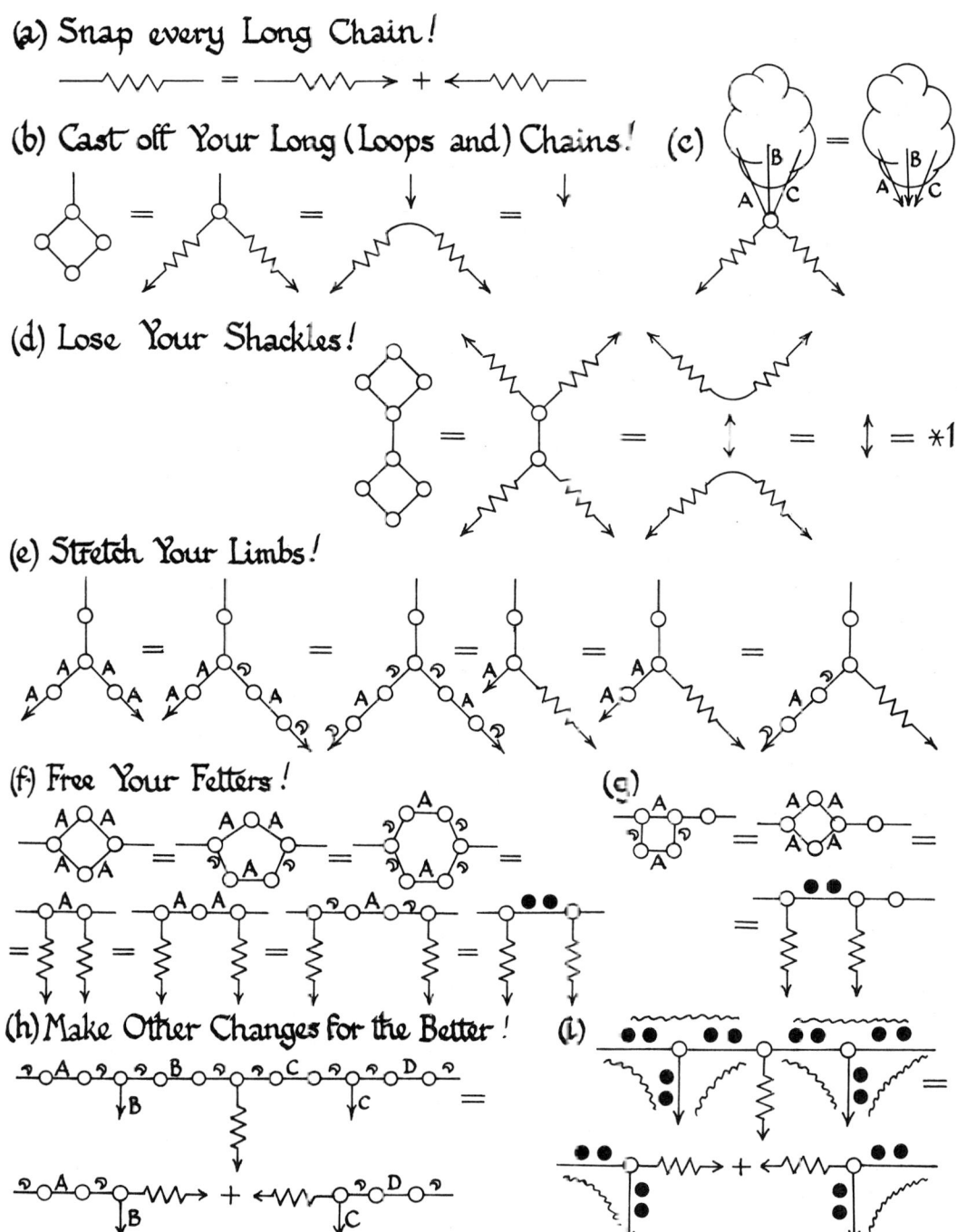

(b) Cast off Your Long (Loops and) Chains!   (c)

(d) Lose Your Shackles!

= *1

(e) Stretch Your Limbs!

(f) Free Your Fetters!   (g)

(h) Make Other Changes for the Better!   (i)

Figure 26. Some Useful Nimstring Equivalences.

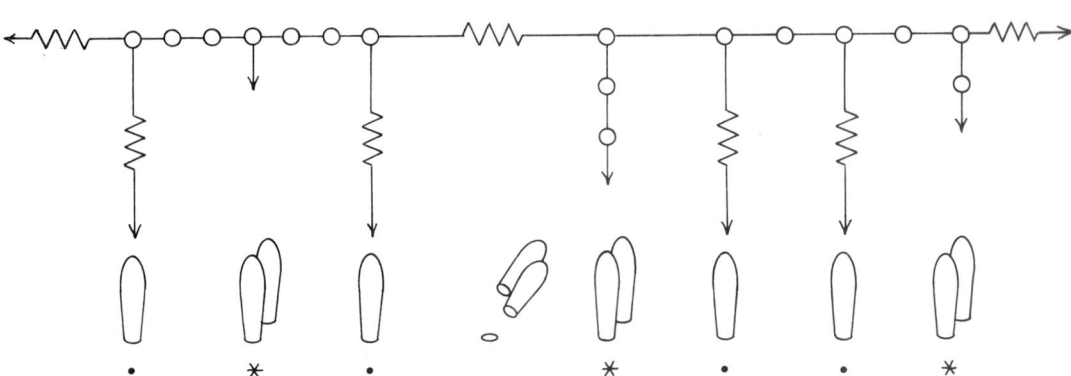

Figure 27. A Twopins-vine and a Game of Twopins.

Our remarks about vines show that:

> You can't know all about Nimstring
> without knowing all about Twopins!

If you've read Chapter 15 you'll know that Kayles and Dawson's Kayles are just special cases of Twopins, so, combining several slogans of this chapter:

> You can't know all about
> Dots-and-Boxes
> unless you know all about
> Kayles and Dawson's Kayles!

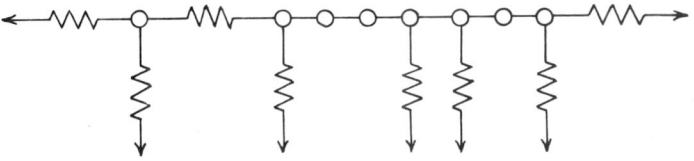

Figure 28. A Snappable Dawson's-vine, $D_1 + D_4$, value $0 + *2 = *2$.

A **Dawson's-vine** (*Parthenocissus dawsonia*) is a Twopins-vine all of whose tendrils are long. Of course if any distance between neighboring joints is *long* the Dawson's-vine will snap, like the one in Fig. 28. If all these distances are *short*, the nim-values, D_n, of n-jointed Dawson's-vines are (Chapter 4):

n	0	1	2	3	4	5	6	7	8	9	11	13	15	17	19	21	23	25	27	29	31	33
D_n	0	0	1	1	2	0	3	1	1	0 3 3 2 2 4	0 5 2	2 3 3	0 1 1	3 0	2 1	1 0	4 5	2	7			
D_{n+34}	4	0	1	1	2	0	3	1	1	0 3 3 2 2 4	4 5 5	2 3 3	3 0 1 1	3 0	2 1	1 0	4 5	3	7			
D_{n+68}	4	8	1	1	2	0	3	1	1	0 3 3 2 2 4	4 5 5	9 3 3	3 0 1 1	3 0	2 ...							

A *Kayles* position corresponds to a Twopins-vine with a *short* tendril at every joint. However, we can extend this class by observing that we don't need to worry about some of the distances between joints and ends:

> A vine is a **Kayles-vine** if
> (i) every joint has a *short* tendril, and
> (ii) every distance between two ends or two non-neighboring joints is *long*.

Again, if any distance between neighboring joints of your Kayles-vine is *long*, it snaps (Fig. 29). From Chapter 4, the nim-values, K_n, of unsnappable n-jointed Kayles-vines are:

n	0	1	2	3	4	5	6	7	8	9	10	11	12	13	14	15	16	17	18	19	20	21	22	23
K_n	0	1	2	3	1	4	3	2	1	4	2	6	4	1	2	7	1	4	3	2	1	4	6	7
K_{n+24}	4	1	2	8	5	4	7	2	1	8	6	7	4	1	2	3	1	4	7	2	1	8	2	7
K_{n+48}	4	1	2	8	1	4	7	2	1	4	2	7	4	1	2	8	1	4	7	2	1	8	6	7
K_{n+72}	4	1	2	8	1	4	7	2	1	8	2	7	4	1	2	8	1	4	7	2	1	8	2	7

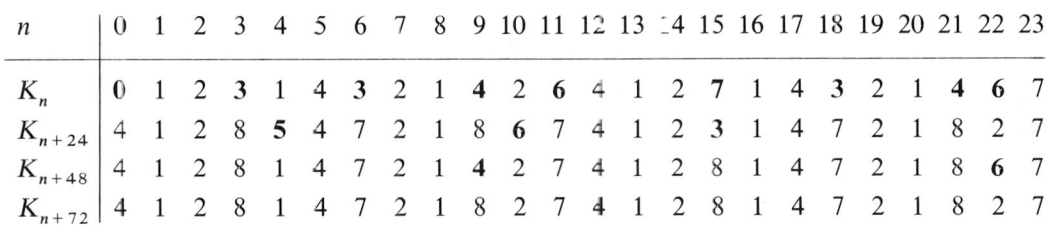

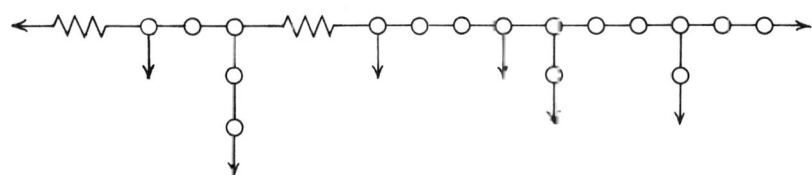

Figure 29. A Snappable Kayles-vine, $K_2 + K_4$ value $*2 + *1 = *3$.

The correspondence between the Twopins-vines and the game of Twopins enables us to interpret the Decomposition Theorem of Figs. 26(h) and (i) as a generalization of the Decomposition Theorem for Twopins. There are Nimstring generalizations for all the Twopins equivalences:

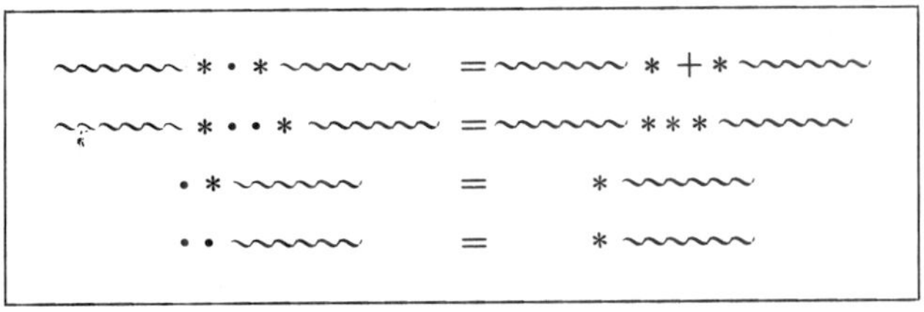

$$\begin{array}{ccc}
\sim\!\!\sim * \bullet * \sim\!\!\sim & = & \sim\!\!\sim * + * \sim\!\!\sim \\
\sim\!\!\sim * \bullet \bullet * \sim\!\!\sim & = & \sim\!\!\sim * * * \sim\!\!\sim \\
\bullet * \sim\!\!\sim & = & * \sim\!\!\sim \\
\bullet \bullet \sim\!\!\sim & = & * \sim\!\!\sim
\end{array}$$

The last two enable us to suppose that the endmost joints of a Twopins-vine have short tendrils (they correspond to uses of Fig. 26(b)). Collectively the Twopins equivalences allow us to suppose that the joints with no short tendrils come in strictly internal blocks of 3 or more and so all the simplest Twopins-vines reduce to compounds of Kayles-vines. Figure 30 is a small Twopins dictionary culled from Chapters 4 and 15. The equivalences of Fig. 26 enable us to show that many graphs that don't look like vines are really equivalent to them; for instance Fig. 31(a) is equivalent to D_8.

Kayles-vines, K_n		n	Dawson's-vines, D_n				Other Twopins-vines	
*	$= *1$	1	•	=		0	* • * • • • *	
* *	$= *2$	2	• •	=	*	$= *1$	* * * * • • • *	$= *3$
* * *	$= *3$	3	• • •	=	* •	$= *1$	* * • • • * * *	$= *1$
* * * *	$= *1$	4	• • • •	=	* *	$= *2$	* * • • • • • *	$= *4$
* * * * *	$= *4$	5	• • • • •	=	* + *	$= 0$	* * * * * • • • *	$= *5$
* * * * * *	$= *3$	6	• • • • • •	=	* * *	$= *3$	* • * • • • * *	$= *4$
* * * * * * *	$= *2$	7	• • • • • • •	=	* • • • • *	$= *1$	* * * • • • • • *	$= *3$
* * * * * * * *	$= *1$	8	• • • • • • • •	=	* • • • • • *	$= *1$	* * • • • • • * *	$= *3$
* * * * * * * * *	$= *4$	9	• • • • • • • • •	=	* • • • • • • *	$= 0$	* * • • • • • • *	$= *3$
* * * * * * * * * *	$= *2$	10	• • • • • • • • • •	=	* • • • • • • • *	$= *3$	* • • • * • • • * *	$= *1$

Figure 30. Various Vines Values.

Twopins-vines are **decomposing** in the sense that when any branch of the vine is removed the new vine decomposes—often by snapping a chain—into two smaller ones. Some other vines, including that of Fig. 31(b), are decomposing in the same sense. It is rather straightforward to compute the value of a decomposing vine from the values of those of its subvines which include all of a consecutive sequence of the original tendrils. Since the number of such subvines is proportional only to the square of the number of tendrils this idea is feasible for quite long decomposing vines, and can easily be implemented on a computer.

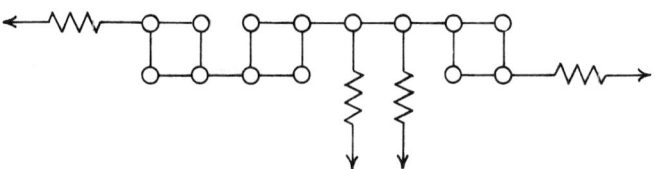

(a) A graph equivalent to the Dawson's-vine D_8.

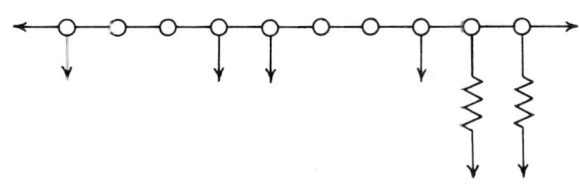

(b) A decomposing vine.

Figure 31. Two Uses of Figure 26.

Dots-and-Boxes is, like other good games, remarkable in that it can be played on several different levels of sophistication.

First, there's Arthur's way: you just don't open up any boxes unless you have to and then you open as few as possible. This seems to be the only level that many players reach.

Then there's Bertha's double-dealing endgame technique which gets the winner a lot of boxes at the finish and makes it seem likely that Nimstring will be useful.

Next comes the Parrotty girls' parity rule for long chains.

Then we realize that to get the right parity is an exercise in Sprague–Grundy theory, so we need tables of nim-values.

The unwieldiness of these tables forces us to use equivalence theorems whenever we can, and to look for interesting classes of analyzable graphs.

We can use Twopins theory to reduce many games to positions in the well-known games of Kayles and Dawson's Kayles.

Finally, experts will need to know something about the rare occasions when the Nimstring theory does not give the correct Dots-and-Boxes winner.

What moves would you recommend for the positions in Fig. 32? We give our own recommendations in the Extras.

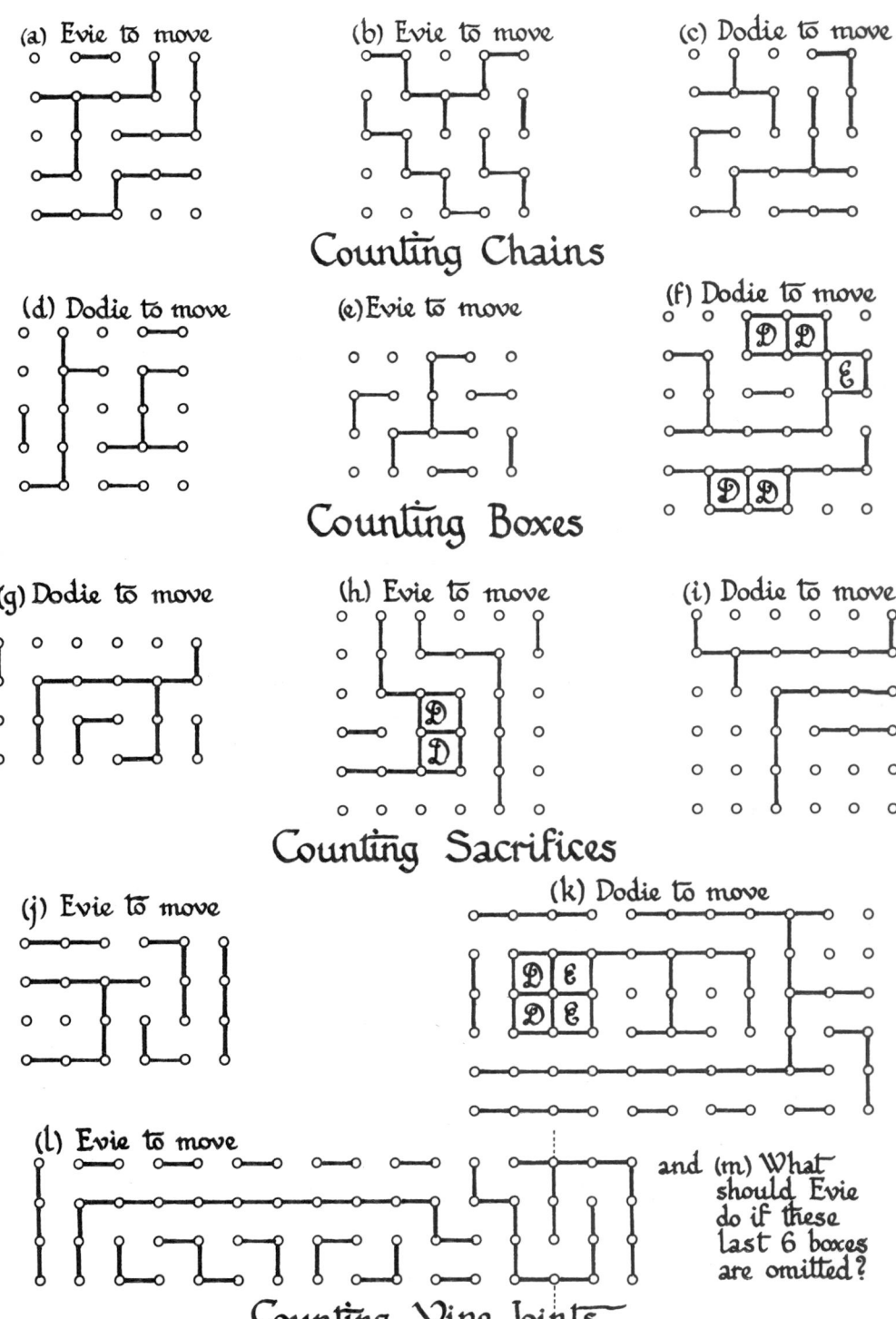

Figure 32. Try These Dots-and-Boxes Problems.

EXTRAS

DOTS + DOUBLECROSSES = TURNS

Suppose we play a Dots-and-Boxes game, starting with D dots, that takes T turns to draw L lines and finish with B boxes. Then if there are no doublecrosses, each line except the last either creates just one box or hands the turn to the next player, so

$$L = B + T - 1.$$

However Fig. 33 shows that

$$L = B + D - 1,$$

so a game with no doublecrosses lasts for exactly the same number of turns as the initial number of dots. But each doublecross creates 2 boxes instead of 1, so in general the number of turns will be the number of dots we started with plus the number of doublecrosses.

When we've broken
B edges to flood
the B boxes ...

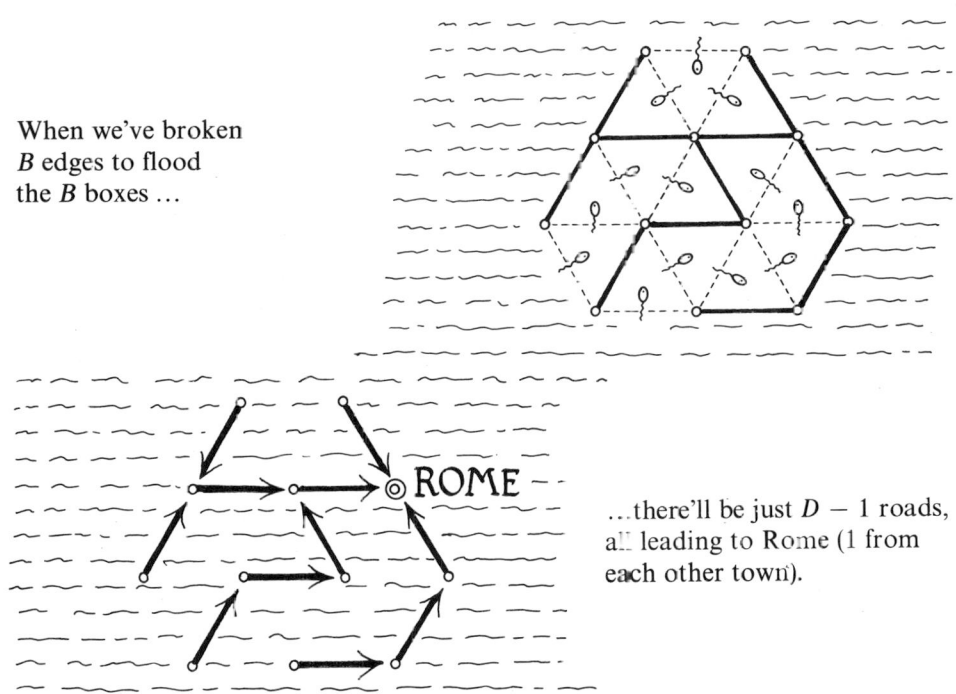

...there'll be just $D - 1$ roads,
all leading to Rome (1 from
each other town).

Figure 33. Euler via Rademacher and Toeplitz.

537

HOW DODIE CAN WIN THE 4-BOX GAME

Figure 34 shows a sufficient set of $\mathscr{P}$-positions to enable Dodie to win the 4-box game. Figure 35 shows *all* $\mathscr{P}$-positions except those in which a player has already signed enough boxes to win, classified according to the number of moves made. Figure 36 shows the three $\mathscr{N}$-positions in which a sacrifice wins but a non-sacrifice *loses*. In these figures broken lines indicate boxes with three sides already drawn.

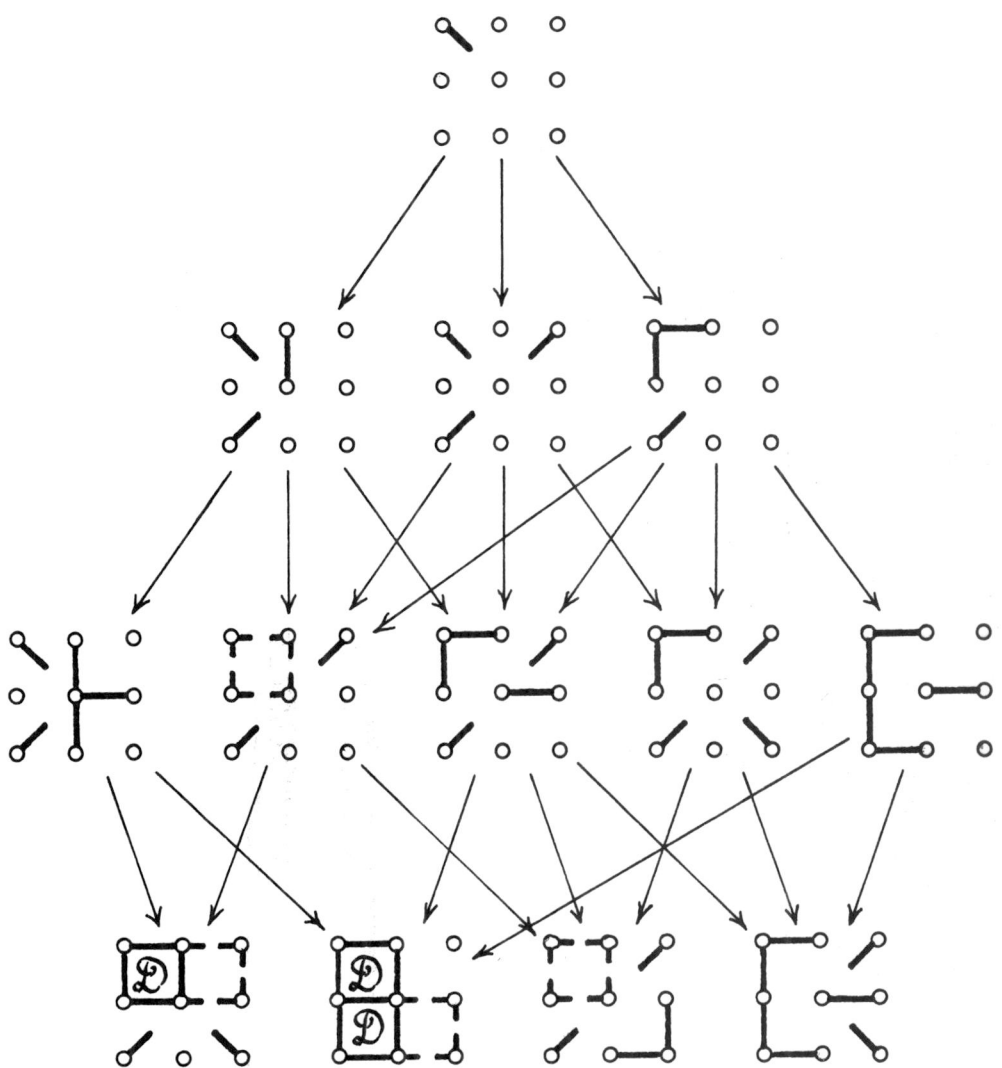

Figure 34. A Winning Strategy for Dodie.

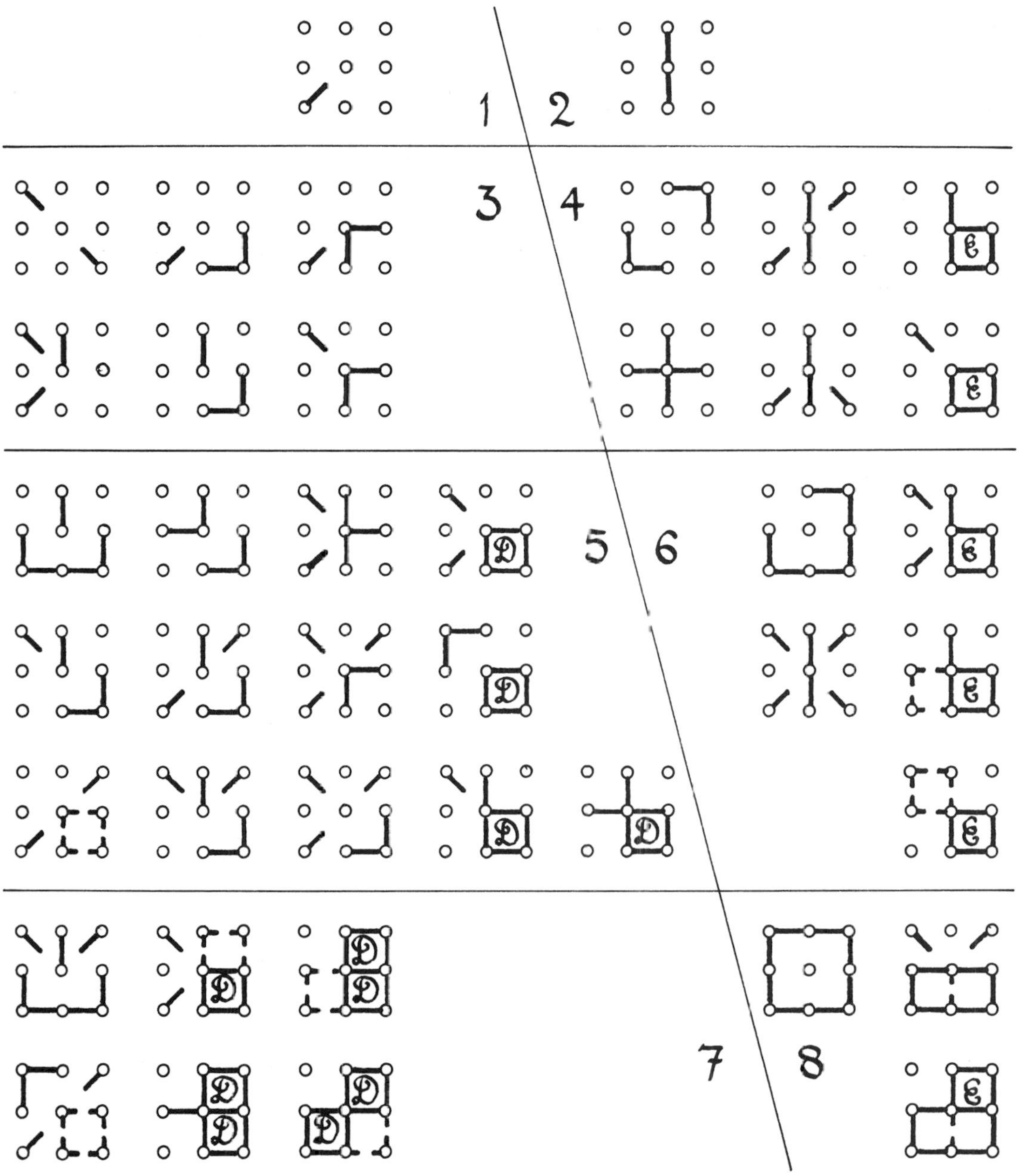

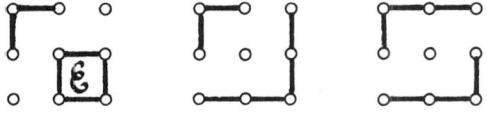

Figure 35. All $\mathscr{P}$-positions in the 4-box Game.

Figure 36. Three Tricky $\mathscr{N}$-positions.

WHEN IS IT BEST TO LOSE CONTROL?

It is clearly not *always* a good idea to keep control. For suppose all that's left of a game is 1001 chains, all of length 3, and your opponent has just opened the first of these. If you slavishly insist on keeping control to the very end, you'll have to give away 2 boxes in each chain but the last and so you'll only get 1003 to your opponent's 2000.

What the question really boils down to is this: when your opponent has just opened a long chain or made a half-hearted handout, should you, like Bertha, decline the last 2 boxes, or, like Arthur, grab them all and be forced to move elsewhere? Suppose for a moment that you use Bertha's tactic and give up 2 boxes in order to force your opponent to move first in the rest of the position, in which, by playing perfectly, you get D more boxes than your opponent. Then Arthur's strategy would take those 2 boxes and give you D less of the rest than your opponent. Comparing:

Bertha's technique	Arthur's technique
$-2 + D$	$2 - D$

shows you should

> adopt Bertha's technique unless
> D is less than 2.
> (Either will do, if D is just 2.)

That's all very well, but you still won't know which is best if you don't know the value of D. We can't say much about this in general, but we have a rule which gives D when the position is made up entirely of long chains of lengths

$$a, b, c, \ldots .$$

For such a position

$$D = (a-4) + (b-4) + (c-4) + \ldots + 4$$

provided the right side is positive; otherwise $D = 1$ or 2.

Using this rule we can answer our question for positions made entirely of chains:

> You should keep or gain control, *unless*
> there are *evenly* many short chains, *and*
> *either* there are no long chains
> *or* the long chains can be partitioned
> into two sets, each of average
> length strictly less than 4.

The A.B.C. of Control when Long Chains are Short.

Of course, for such positions, keeping or gaining control involves making Bertha's move if this would leave an even number of unopened short chains, and Arthur's move otherwise.

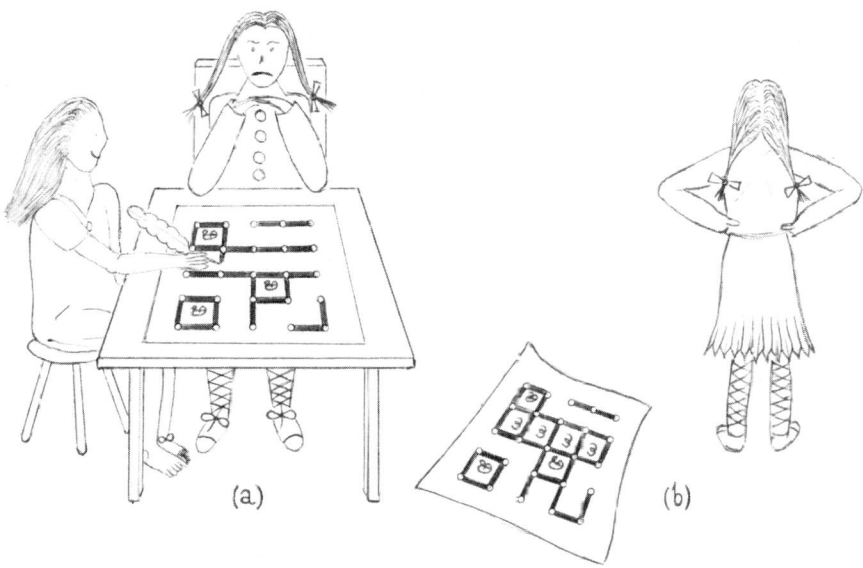

Figure 37. Dodie's Drawing the Game—Evie Loses Control.

Figure 37(a) shows a position where Evie, who's not feeling very well, has only managed to keep control so far by sacrificing 3 boxes. Dodie's now foolishly opening a chain of length 4 with the move shown. Since the remaining long chains satisfy the exceptional condition in our A.B.C., Evie's best move (Fig. 37(b)) is to *lose* control and take all 4 squares of the chain as Arthur would. The boxes are then divided:

		4-chain	three 3-chains				
to Evie		4	2	2		an $\frac{8}{8}$ tie,
to Dodie	3 already +		1	1	3	

where Bertha's response would give

to Evie		2		1	1	3	a $\frac{7}{9}$ loss.
to Dodie	3 already +		2	2	2	

COMPUTING THE VALUES OF VINES

On most graphs made of Twopins-vines and long chains a winning Nimstring strategy really does win at Dots-and-Boxes. For let V be the number of separate vines, counting long chains as unjointed vines. *Don't* use Fig. 26(a) to decompose vines with long stems—*instead* let I denote the number of these internal long stems. Let J be the total number of joints and L the Nimstring loser's present score. We study the quantities

$$f = L + 2J + 2V, \qquad g = I + 2L + 3J + 4V$$

during any double turn, consisting of the loser's move and the winner's reply, assigning any boxes given away by a move to that move rather than to the next one. First for the Nimstring winner's move:

| | Change in | | | | | |
Move on	I	L	J	V	f	g
Stem	0	$\leqslant 2$	-2	1	$\leqslant 0$	$\leqslant 2$
Inner tendril	$\leqslant 1$	$\leqslant 2$	-1	0	$\leqslant 0$	$\leqslant 2$
End tendril	$\leqslant 0$	$\leqslant 2$	-1	0	$\leqslant 0$	$\leqslant 1$

And now for the Nimstring loser's move:

| | Change in | | | | | | and including winner's reply | |
Move on	I	L	J	V	f	g	f	g
Stem	0	0	-2	1	-2	-2		
Inner tendril	$\leqslant 1$	0	-1	0	-2	$\leqslant -2$	$\leqslant -2$	$\leqslant 0$
End tendril	$\leqslant 0$	0	-1	0	-2	$\leqslant -3$		
Loony stem move that winner must accept	$\leqslant 0$	0	-2	1	-2	$\leqslant -2$	$\leqslant -2$	$\leqslant 0$
Loony tendril move that winner must accept	$\leqslant 1$	0	-1	0	-2	$\leqslant -2$	$\leqslant -2$	$\leqslant 0$
Loony chain move (declined by winner)	0	2	0	-1	0	0	0	0
Loony stem move (declined by winner)	$\leqslant 0$	2	-2	1	0	$\leqslant 2$	0	$\leqslant 2$[†]
Loony tendril move (declined by winner)	$\leqslant 1$	2	-1	0	0	$\leqslant 2$	0	$\leqslant 2$[†]

(No loony chain move makes the winner *accept*, because by *declining* he leaves the value unchanged.) The next to last column shows that f *never* increases and so:

> If the number of nodes, N, in the game exceeds
>
> $$4(J + V)$$
>
> then the Nimstring winner wins the Dots-and-Boxes game.

(For the loser's score at the end of the game will be less than $N/2$.)

Since all Dawson's-vines and many Twopins-vines have more than four nodes per joint, they satisfy this condition. *If g* never increased, we could similarly assert that the Nimstring strategy works for Twopins-vines with

$$N > I + 3J + 4V,$$

but since the daggered entries can be positive a skilful Nimstring loser might win the occasional Dots-and-Boxes game.

Such cases are very rare. The Nimstring loser can increase *g* only by choosing a loony stem or tendril move that the Nimstring winner must decline (*most* loony moves can be accepted). Usually the winner has other opportunities to decrease *g* by playing on an end tendril or making a move conceding fewer than two boxes.

Even though we have been able to construct some examples (Fig. 32(m)). the difficulties of composing them and the closeness of their scores reinforce our opinion that:

> Your best chances at
> Dots-and-Boxes
> are likely to be found
> by the Nimstring strategy

LOONY ENDGAMES ARE NP-HARD

If you're faced with a position in which all the edges are on long chains you'll lose at Nimstring because only loony moves are possible. But if you've already got lots of boxes you might still manage to win the Dots-and-Boxes game. How do you find *which* loony move to make to stop your opponent from catching up?

To simplify the argument we'll suppose that the last move will take place on a chain (between ground and ground) that's long enough to ensure that your opponent's best strategy for the remaining boxes is the Nimstring strategy, which requires that he conclude each turn, except the last, with a double-dealing move. Any of the *m* moves you make on isolated cycles will give you 4 boxes, while any of the other *n* moves on chains (except the last) will give you 2 boxes each, so your score will be

$$4m + 2n - 2.$$

Suppose the graph has *j* joints with total valence *v*, counting grounded ends as having valence 1 each. A move on an isolated cycle doesn't change the valence, but a move on a chain decreases the valence by 1 at each end, except that whenever the valence of a joint changes from 3 to 2 that joint disappears. This happens just once for each joint, so

$$v = 2n + 2j$$

and your score will be

$$4m + v - 2j - 2.$$

Since *v* and *j* are fixed, we want to make as many moves on isolated cycles as possible. These isolated cycles are disjoint, and any disjoint set of cycles can be isolated just by playing all chain moves first.

> You can't know all about (possibly
> generalized) Dots-and-Boxes unless
> you know all about how to find
> a largest set of node-disjoint cycles
> in an arbitrary (possibly non-planar)
> graph.

Finding a largest set of node-disjoint cycles in arbitrary graphs is known to be NP-hard (cf. the Extras to Chapter 7).

SOLUTIONS TO DOTS-AND-BOXES PROBLEMS

Here are *our* answers to the problems in Fig. 32.

(a) Evie wants an even number of long chains. She immediately establishes just two by drawing either edge at the top left-hand corner.

(b) This time Evie establishes two long chains by sacrificing the box whose lower left corner is the central dot. An additional sacrifice in the lower left-hand corner will be needed if Dodie tries to make a third long chain there.

(c) Dodie wants an odd number of long chains. She should sacrifice two boxes by a hard-hearted handout drawn rightwards from the central dot, and will win 9–7.

(d) Neither player can afford to sacrifice four boxes of the central loop, so the long chain theory doesn't really apply. Either player can force a tie (making no sacrifices). A well-played endgame will have chains of length 3 at top and bottom and a loop of 4 in the centre. The left side may be a single chain of length 4 or a pair of chains of lengths 1 and 3; either position is a tie.

(e) A little trap for Nimstring players! The *dotted* move is the only good Nimstring move, but involves too much sacrifice and will lose the Dots-and-Boxes game 5–7 against an extremely skilful opponent. The *dashed* move loses at Nimstring, but only if the opponent sacrifices two boxes on the next turn, and with the board then broken into many small pieces we get a tie.

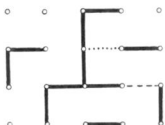

(f) The Nimstring game is essentially over, but at Dots-and-Boxes what matters is whether the top left-hand corner becomes a chain of length 2 or two chains of length 1. Dodie should force the former by drawing the second edge on the top row and will win 13–12 rather than losing 12–13.

(g) Dodie forces a treble sacrifice! She wants an even number of long chains but can see only one. Her dotted opening move threatens to create a second long chain at EFG on the next move. Since Evie can be prevented from making a third long chain, she must cut between F and G, sacrificing two boxes. Accepting these, Dodie repeats her threat by the dashed move, which threatens a long chain at CDE, forcing Evie to sacrifice D and E. Dodie accepts these and repeats the threat yet again, drawing the left edge of B. Although Evie wins the Nimstring game by sacrificing B and C, Dodie will have 8 boxes: $F, G; D, E; B, C; N, O$; enough to win the Dots-and-Boxes game.

(h) Evie does likewise! Starting with the next to last of the top edges she repeatedly threatens to construct a third long chain at the right. Dodie can stop this only by three 2-box sacrifices. Evie then makes a loony move conceding the chain of length 3 to the left of the 2 captured boxes, acquiring 2 boxes after the resulting doublecross. Since she has now changed sides, she threatens to build another long chain in the top left-hand corner, forcing a further sacrifice from Dodie. She then stops further growth of the length 7 chain and awaits her last 3 boxes to win 13–12.

(i) A very complicated position! Dodie *must* prevent a third long chain from forming in the top row. She first sacrifices one of the top corner boxes (and will probably need more sacrifices) and strives vigilantly to chew up as much of the empty space as she can by extending her long chains. If Evie sacrifices either long chain too soon, Dodie accepts.

(j) An easy one! There is a treble-jointed Kayles-vine, value $*3$, and four boxes at the lower left, value $*2$ from Fig. 20. Evie wins by drawing the middle top edge or rightmost bottom one, which reduces K_3 to K_2. (There are other moves that do this, but they sacrifice too much.)

(k) There's a 4-jointed Kayles-vine, value $*1$, at the bottom, and four boxes at top right, worth $*3$ from Fig. 20. The rest of the figure is a 5-jointed Kayles-vine, value $*4$, under a disguise you can strip off by looking at Figs. 26(a) and (b). Dodie's Nimstring move must therefore replace K_5 by $K_3 + K_1$, which she can do only by drawing the vertical edge at the top left-hand corner or by isolating the loop just to the right of the captured boxes. In this problem it's only if Dodie plays carefully that her Nimstring strategy will also win at Dots-and-Boxes. When she has a choice of several Nimstring moves, she should select whichever scores more boxes. On the Kayles-vines any stem move (by either player) leads ultimately to another long chain, which will give two more boxes to Evie. So Dodie prefers to make tendril moves whenever possible, while Evie selects stem moves which make Dodie respond with more stem moves.

(l) A unique winning move! Cut the 12-jointed Kayles-vine into two 5-jointed ones by separating N from O. The rest is easy!

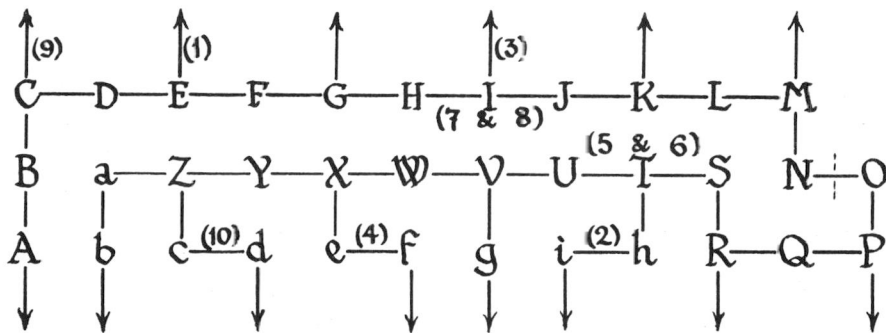

(m) In the modified problem, if Evie really respected Dodie's skill, she would resign! If she opened as before by sacrificing N, O, Dodie could unground E. Evie is then faced with $K_3 + K_1 + K_5$, and the Nimstring replies give Dodie two more boxes, e, f or h, i, say the latter (2). Dodie then (3) ungrounds I, leaving $4K_1 + K_3$, and Evie can't do better than (4) giving Dodie e and f (say) too. Evie now has $6K_1$ and has won the Nimstring game, but Dodie makes a loony move (5) on STU, which Evie's sadly forced to decline (6), conceding two more boxes. Dodie can make another loony move (7) on HIJ and collect two of those as well (8). Finally (9), Dodie can reduce $2K_1$ to K_1 by ungrounding C, which makes Evie (10) sacrifice c, d (or a, b). The resulting position

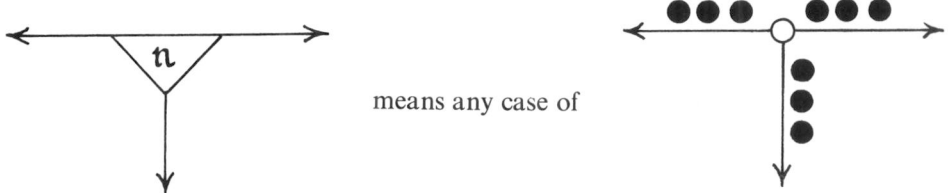

has five chains, lengths 3, 3, 4, 7, 8, but Evie has only 2 boxes to Dodie's 12. Although of the remaining chains Evie collects 17 boxes to Dodie's 8, Dodie wins 20 to 19! Rather than risk this disgrace, we recommend to Evie a timid opening such as ungrounding E; it's just conceivable that Dodie doesn't remember the first nine Kayles values!

SOME MORE NIMSTRING VALUES

As in Fig. 21, dots near edges are optional additional nodes and a dot equally near two edges may be placed on either. A wiggly line means 3 or more nodes strictly between the indicated points. The symbol

means any case of

that has value $*n$ (see Fig. 21).

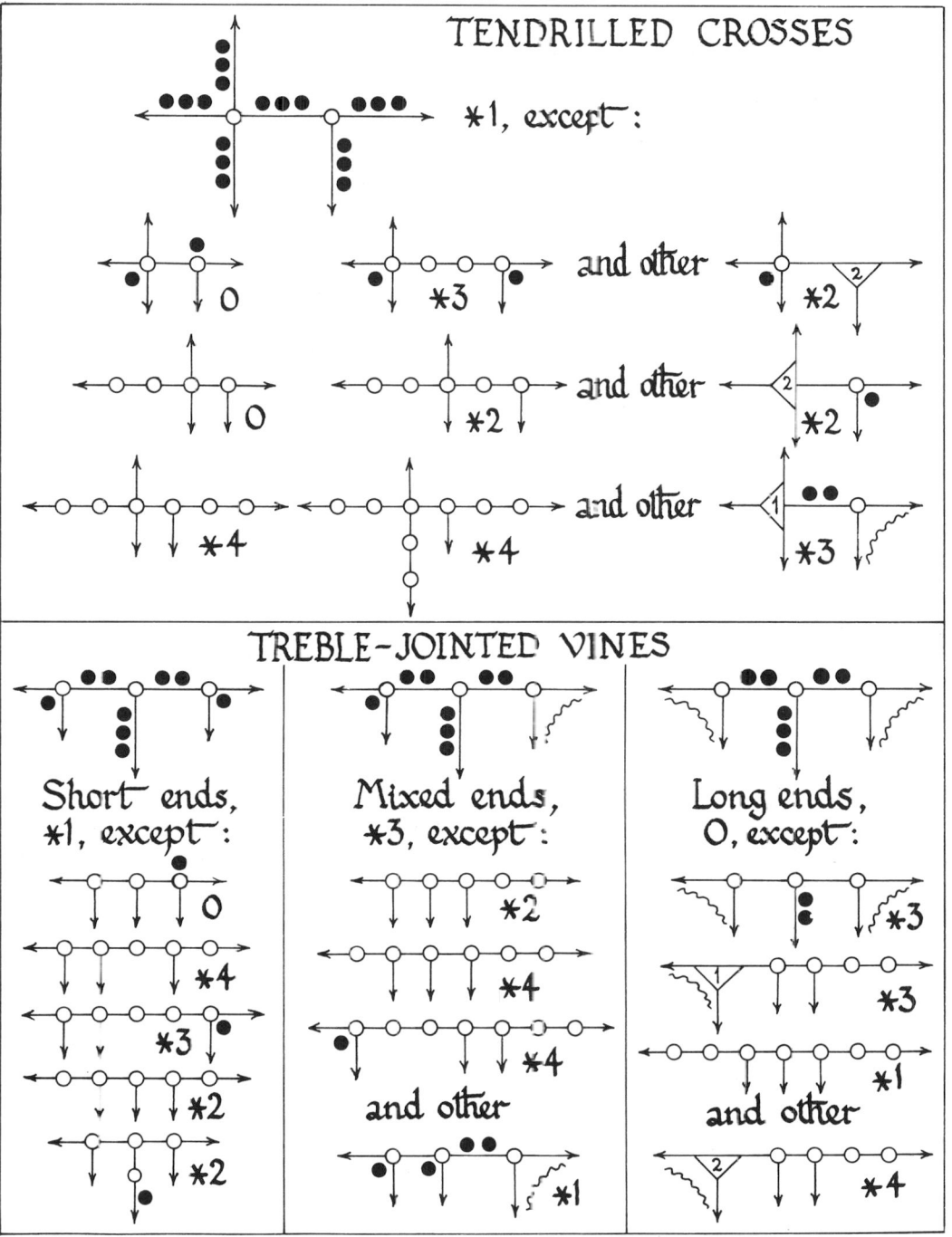

Figure 38. The Next Volume of the Nimstring Dictionary.

NIMBERS FOR NIMSTRING ARRAYS

To show the sides on which rectangular arrays are grounded we give their dimensions with primes (or double primes), for example:

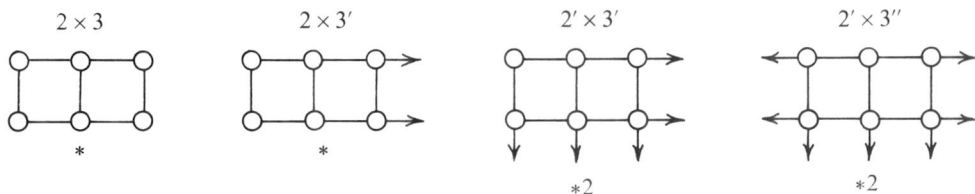

Tables 1 and 2 give values for such rectangular arrays:

	1′	2	2′	3	3′	4	4′	5	5′	6	6′	7	7′	8
2	*	0	*2	*	*	0	*2	*	*3	0	*2	*	*3	0
3	*2	*	*2	*	*2	*	*2	*						

Table 1. Nimstring Values for Ungrounded Rectangular Arrays or Arrays Grounded along One Edge.

n	2	3	4	5	6	7	8	9	10	11
$1′ \times n$	*	*2	*	*2	*3	0	*3	0	*	*2
$1′ \times n′$	0	*	0	*3	*2	*3	*2	*5	*4	*5
$1′ \times n″$	*	0	*	0	*	*2	*3	*	*3	
$2′ \times n$	*2	*2	*	*	*					
$2′ \times n′$	*2	*2	*	*						
$2′ \times n″$	0	*2	*5	*						

Table 2. Nimstring Values for Arrays Grounded at 1, 2 or 3 Edges.

Figure 39 shows Nimstring values for some less regular arrays.

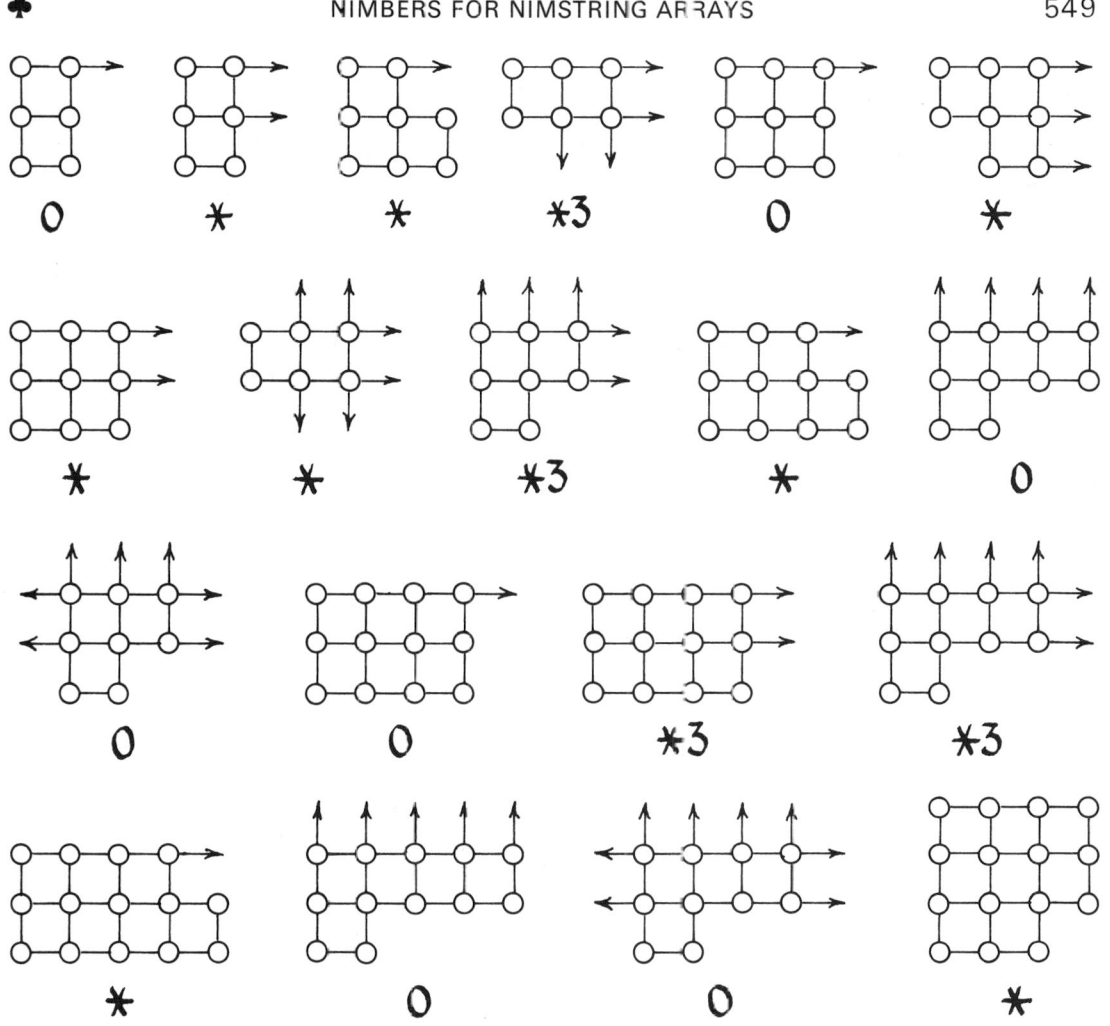

Figure 39. Nimbers for Various Arrays.

We have seen that we can play Nimstring on any graphs; here are the nim-values of small complete graphs, K_n and complete bipartite graphs, $K_{m,n}$:

n	2	3	4	5	6	7	8	9	10
K_n	0	*	0	*	*				
$K_{2,n}$	0	*	0	*	0	*	0	*	0
$K_{3,n}$	*	*	*	*	*	*			
$K_{4,n}$	0	*	0	*					

Does the value of $K_{m,n}$ depend only on the parity of $(m-1)(n-1)$?

REFERENCES AND FURTHER READING

John C. Holladay, A note on the game of dots, Amer. Math. Monthly, **73** (1966) 717–720; M.R.

Hans Rademacher and Otto Toeplitz, "The Enjoyment of Mathematics", Princeton University Press, 1957. Pages 75–76 give the proof of Euler's theorem.

Chapter 17

Spots and Sprouts

He shall not live, with a spot I damn him.
William Shakespeare, Julius Caesar IV, i, 6.

The games we treat here are played with spots (or crosses) on a piece of paper, the move being to join two spots by a curve satisfying various conditions specified in the rules of the game. *We shall always demand that no curve crosses itself or another curve*. We have just devoted a whole chapter to such a game, but here we shall consider games whose theories, while not all trivial (or even all complete) will occupy only a few pages each. We had to make an exception for Lucasta, with whom we fell in love.

RIMS

Here the move is simply to draw a loop passing through at least one and arbitrarily many of the spots. The only further condition is that no two loops may cross. A typical Rims position is shown in Fig. 1. What should be our next move, supposing normal play?

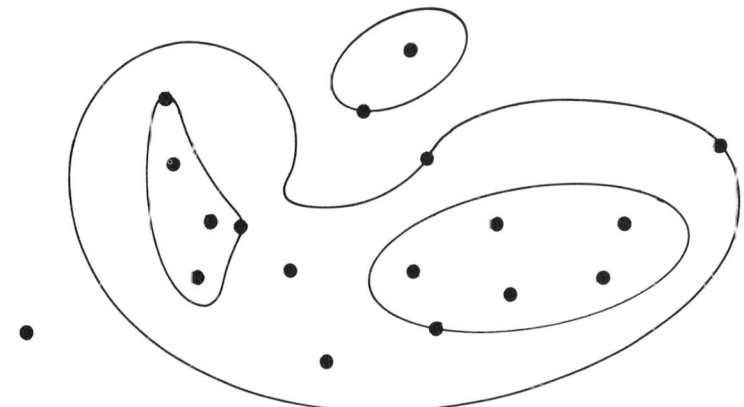

Figure 1. A Game of Rims. (or Rails).

On examining the position, we see that the loops divide the plane into regions containing respectively 5, 2, 3, 1, 1 spots (and sometimes other regions with internal spots). When we make a move in a region with n spots, we automatically divide it into two regions with a and b spots, where $a+b$ is less than n, but a and b are otherwise arbitrary. It follows that Rims is merely a disguised form of Nim with the additional possibility of dividing the heap we've just reduced into two smaller ones. This is the game $0\cdot\dot{7}$ in the octal notation of Chapter 4, where we saw that the extra possibility doesn't affect the strategy, so the only good move in Fig. 1 is to draw a loop through just 4 of the spots in the 5-spot region. The theory of Misère Nim tells us that exactly the same move should be made in Misère Rims. In general we move so that the nim-sum of the spot counts is zero, except that in the misère form we must make the nim-sum 1 if every spot count is 0 or 1.

RAILS

Let us require that the loop must pass through just *one* or *two* spots, the rules otherwise being as in Rims. What should now be our move in Fig. 1? The legitimate moves in an n-spot region produce regions of a and b spots, where we require that $a+b = n-1$ or $n-2$. Since these moves correspond exactly to the legal moves in Kayles, our moves can be deduced from the theory of that game. For the Rails position of Fig. 1, this tells us to draw a loop through just one of the spots in the 5-spot region, in either normal or misère play.

Many other octal games can be reformulated very nicely as spot and loop games, and we find by observation that more people can be persuaded to play them this way. Often the geometrical form suggests particular rules very naturally, and sometimes the rules suggested do not quite correspond to natural games with heaps. Here are two further examples.

LOOPS-AND-BRANCHES

The move is to join two spots together, *or* join a single spot to itself so as to form a loop. No spot may be involved in two different moves. The game is isomorphic to the octal game $\cdot\mathbf{73}$, for which we computed the nim-values in Chapter 4 (Table 6) and the reduced forms in Chapter 13 (Extras, Table 5, Notes A and T, Adders). The patterns in Table 1 continue indefinitely.

n	0	1	2	3	4	5	6	7	8	9	...
nim-value	0	1	2	3	0	1	2	3	0	1	...
reduced form	0	1	2	3	2+2	3+2	2+2+2	3+2+2	2+2+2+2	3+2+2+2	...

Table 1. Nim-values and Reduced Forms for Loops-and-Branches.

So we have complete strategies in both normal and misère play. In both cases we move so that the nim-sum of the nim-values is zero, *except* that in misère play we must make the nim-sum one if every region has at most one spot.

CONTOURS

This game is rather more interesting. The move is to draw a closed loop (or *contour*) through just one spot, with the side condition that every loop must have at least one spot strictly inside (possibly internal to some further contours). In other words, when we view the position as a system of contours drawn on a map, every hill must have its peak marked (and every valley its bottom).

In this game, we must distinguish a region containing n spots and nothing else (type n) from one which, in addition to n free spots, contains a contour or contours with their internal spots (type $\hat{n}$). But the number or structure of the contours within a region of type $\hat{n}$ is immaterial, and the spots inside them do not count in computing n. So Fig. 2 has five regions, of types $\hat{5}$, 5, $\hat{3}$, 3, 2. What should be our move here?

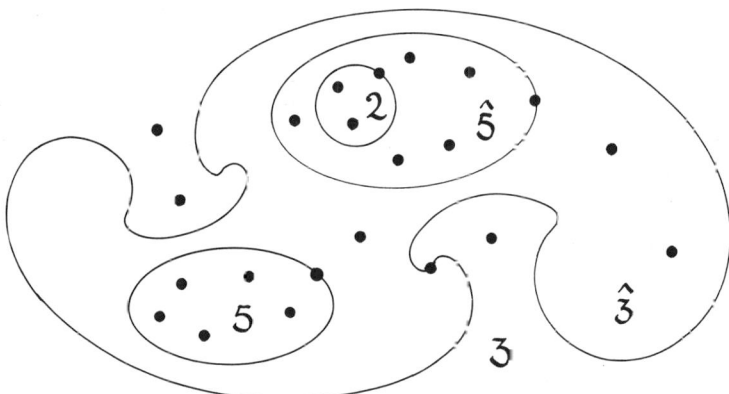

Figure 2. A Game of Contours.

The moves available from the general position are:

$$n \text{ or } \hat{n} \text{ to } a + \hat{b} \quad (a > 0)$$
$$\hat{n} \text{ to } \hat{a} + \hat{b}$$

where in each case $a + b = n - 1$. So we can draw up a table of nim-values, as in Table 2.

n:	0	1	2	3	4	5	6	7	8	9	10	11	12	13	14	15	16	17	18	19	20	
$\mathscr{G}(n)$:		0	1	0	1	0	3	2	0	5	2	0	1	4	3	2	0	5	2	3	1	...
$\mathscr{G}(\hat{n})$:		0	1	2	3	1	4	3	2	0	5	2	3	1	4	3	2	0	5	2	3	1	...

Table 2. Nim-values for Contours.

We see that for $n \geqslant 12$ the two nim-sequences coincide and have period 8. So the starting position with n spots is a $\mathscr{P}$-position in normal play only if n is 1, 3, 5, 11, or a multiple of 8. We have not found a complete misère analysis, but Table 3 gives the start of a genus analysis (Chapter 13).

n	0	1	2	3	4	5	6	7	8	9	10	11	12	13	14	15	16
genus of n		0	1	0	1	0	3	2	0	$5^{057}_{A_1}$	2^2_C	0^3	1^0	4^{146}	3^3	2^{20}	0^1
genus of $\hat{n}$	0	1	2	3	1	4^{146}_A	3	2	0^1_B	$5^{057}_{A_1}$	2^2_D	3^3	1^0	4^{146}	3^3	2^{20}	0^1

Table 3. The Genus of Contours Positions.

$A = 2_2 321$, $B = A_2 A_1 321$. For every position with all numbers $\leqslant 10$, the genus is correctly computed by pretending that $A + A = B = 0$, $C = D = 2$.

In the position of Fig. 2, the nim-values are $4, 0, 3, 0, 1$, and so we must make a move changing nim-value 4 to 2. This can only be done by converting the region of type $\hat{5}$ into two of types $\hat{3}$ and $\hat{1}$, so we must draw a loop surrounding the inner contour of this region and just 1 or 3 more points, at least in normal play. It turns out that we have exactly the same good moves in misère play also.

LUCASTA

This is an old game first described by Lucas, and since it does not seem previously to have had a proper name, we have named it for him. It is quite remarkable that we can give complete strategies for both normal and misère play from the starting position, although the general theory is very complicated. Our strategy for normal play is easily proved when once found, but the misère play strategy is very tricky indeed.

The move is to draw a curve having as endpoints two distinct spots. These may *not* be the two endpoints of a single previously drawn curve (though they *may* be linked together by a chain of curves through intermediate spots). No two curves may cross, and no spot may be an endpoint of more than two curves, so that the curves can only build up into chains or into closed loops which must go through three or more spots.

The loops separate the plane into connected regions, as in the previous games, but now the situation within one of these regions needs a triple (a, b, c) of three numbers to describe it adequately. Here a is the number of *atoms*, or isolated spots, b the number of *branches* joining two otherwise isolated spots, and c is the number of *chains* consisting of three or more spots joined by a sequence of edges. It turns out that the number of spots in a chain is immaterial, except that chains (3 or more spots) must be distinguished from atoms and branches (1 and 2 spots).

The possible moves are classified as follows:

> aa: join two atoms to form a branch,
> ab: join an atom to a branch, making a chain,
> bb: connect two branches, forming a chain,
> ac: lengthen a chain by adjoining some atom,
> bc: extend a chain by attaching a branch,
> $c!$: *pull* a chain by joining its ends together.

Since pulling a chain divides a region into two, the result may depend on how it separates the remaining atoms, branches and chains from each other. We denote this by (for example) $c!(a^3)$ or $c!(ab)$, which mean that we separate 3 atoms or an atom and a branch into a region of their own. It is also possible to make a move cc: joining two chains to form a longer one: but the same effect could be achieved by *sharply* ($c!!$) pulling one of the chains, that is, with no separation.

A CHILD'S GUIDE TO NORMAL LUCASTA

We were fortunate that the nim-values we computed for Lucasta suggested a pattern for the outcomes of all positions with at most one chain. This pattern is displayed in Table 4 in which the entry (a, b) is

> P if $(a, b, 0)$ is a $\mathscr{P}$-position, so $(a, b, 1)$ is an $\mathscr{N}$-position
> + if $(a, b, 1)$ is a $\mathscr{P}$-position, so $(a, b, 0)$ is an $\mathscr{N}$-position
> − if $(a, b, 0)$ and $(a, b, 1)$ are both $\mathscr{N}$-positions.

Notice that the columns repeat with period 4, after the first four, while the rows alternate.

$a = 0$	1	2	3	4	5	6	7	8	9	10	11	12	13	14	15
$b = 0$ P	P	−	+	P	P	P	−	P	P	P	−	P	P	P	−
1 P	P	−	−	−	P	−	−	−	P	−	−	−	P	−	−
2 P	P	−	+	P	P	P	−	P	P	P	−	P	P	P	−
3 P	P	−	−	−	P	−	−	−	P	−	−	−	P	−	−
4 P	P	−	+	P	P	P	−	P	P	P	−	P	P	P	−
5 P	P	−	−	−	P	−	−	−	P	−	−	−	P	−	−
6 P	P	−	+	P	P	P	−	P	P	P	−	P	P	P	−
7 P	P	−	−	−	P	−	−	−	P	−	−	−	P	−	−

Table 4. Lucasta With At Most One Chain.

A complete analysis may be difficult, though a machine attack will probably show that the nim-values have period 2 in b and c. However we can give a strategy which enables you to win all the positions you deserve to, when the number of chains is small. This strategy also proves that the pattern of Table 4 persists indefinitely. It uses the *special $\mathscr{P}$-positions*

$$(0, b, 0), \quad (1+4k, b, 0), \quad (3, 2m, 1), \quad (4+2k, 2m, 0) \quad \text{and} \quad (0, 2m, 2), \quad b, k, m \geq 0.$$

It is almost always bad to leave chains in a position because they can be pulled in a number of different ways.

Our opponent can move from one of the special positions so as to leave two or more chains in only a few ways. If he joins two branches to form a chain we pull it smartly; if he joins a branch to a chain we join another. In either case the total effect is to remove two branches. The only other case is the ab move from $(3, 2m, 1)$ to $(2, 2m-1, 2)$ after which we join the two atoms to get the position $(0, 2m, 2)$. Our responses to positions with at most one chain are given in Table 5. Observe that we have completely justified Table 4. The nim-values on which we based our strategy are shown in Table 6. The entry (a, b) gives the sequence of nim-values for $c = 0, 1, 2, \ldots$; the last pair of values always repeats indefinitely, so that 13145 abbreviates 131454545.... Each unprinted row in the first five columns has the same entries as that with b decreased by 2. We have shown that all $\mathscr{N}$-positions $(a, b, 0)$ except $(2, 2m+2, 0)$ and $(6, 1, 0)$, have nim-value 1 and all $\mathscr{N}$-positions $(a, b, 1)$, except $(0, 2m, 1), (1, 2m+1, 1)$ and $(5, 0, 1)$ have nim-value at least 2.

	$a = 0, 1, 5, \ldots, 1+4k$	$a = 2$	$a = 3$	$a = 4, 6, \ldots, 4+2k$	$a = 7, 11, \ldots, 7+4k$
$c = 0$	$\mathscr{P}$-positions; bad luck! Hope for a blunder	aa gives $(0, b+1, 0)$	aa gives $(1, b+1, 0)$	b even: bad luck! b odd: ab gives $(3, 2m, 1)$ if $k=0$. aa gives $(2+2k, 2m+2, 0)$ otherwise	aa gives $(5+4k, b+1, 0)$
$c = 1$	$c!!$ smartly gives $(0, b, 0)$ or $(1+4k, b, 0)$	$c!(a)$ round a solitary atom: $(1, 0, 0) + (1, b, 0)$	b even: bad luck! b odd: bc gives $(3, 2m, 1)$	$c!$ separating atoms from branches, $(0, b, 0) + (4+2k, 0, 0)$	$c!$ separating all but one of the atoms from the branches: $(1, b, 0) + (6+4k, 0, 0)$

Table 5. How to Win at Lucasta.

$a = 0$	1	2	3	4	5	6	7
$b = 0$　01	023	13145	10201	0351732	01023245	0245713101	13169498
1　023	01	124567	13132	1464601	02518189	230645	154578Xx
2　01	023	2356745	10401	0258589	046262Tt	06798	1316XTFf
3　023	01	15478967	13132	1567Xx	020101tFf		
4　01	023	2376945	10401	0278549t98	046292TfTt		
5　023	01	15498X67	13132	15696x6xX	020101fF		
6　01	023	2376X45	10401	027854Tt98	0462X2tSs		
7　023	01	15498x67	13132	15696T6xSxX			
8　01	023	2376X45	10401	027854F89			
9　023	01	15498x67	13132	15696T6xSxX			

$$X = 10$$
$$x = 11$$
$$T = 12$$
$$t = 13$$
$$F = 14$$
$$f = 15$$
$$S = 16$$
$$s = 17$$

Table 6. Nim-values for Positions (a, b, c) in Lucasta.

THE MISÈRE FORM OF LUCASTA

It is remarkable that we can still give a strategy for misère Lucasta from any starting position $(a, 0, 0)$. This is largely because the player who wins can do so without allowing the creation of too many chains, for of course positions with many chains are very difficult to analyze. For fairly small values of a, b, c we can of course compute the genus, as in Table 9, given later, which show that the complete theory is very complicated. In fact, Table 9 was first used in constructing our other tables and figures, and it then suggested our general strategy. This strategy is described in Table 7, Fig. 3, Table 8 and the explanatory notes to these. In Table 7 the notation is as in Table 4, and the patterns continue.

a =	0	1	2	3	4	5	6	7	8	9	10	11	12	13	14	15	16	17	18	19
b = 0	+	−	P	P	−	P	−	P	−	P	−	−	P	P	−	−	−	P	−	−
1	−	+	P	−	P	−	P	−	P	P	−	+	P	P	P	−	P	P	P	−
2	+	P	−	P	−	P	−	−	P	P	−	−	−	P	−	−	−	P	−	−
3	P	−	P	−	P	P	−	+	P	P	P	−	P	P	P	−	P	P	P	−
4	−	P	−	−	P	P	−	−	−	P	−	−	−	P	−	−	−	P	−	−
5	P	P	−	+	P	P	P	−	P	P	P	−	P	P	P	−	P	P	P	−
6	P	P	−	−	−	P	−	−	−	P	−	−	−	P	−	−	−	P	−	−
7	P	P	−	+	P	P	P	−	P	P	P	−	P	P	P	−	P	P	P	−
8	P	P	−	−	−	P	−	−	−	P	−	−	−	P	−	−	−	P	−	−
9	P	P	−	+	P	P	P	−	P	P	P	−	P	P	P	−	P	P	P	−

Table 7. Outcomes of Some Misère Lucasta Positions.

Table 7 gives the outcome of positions of form $(a, b, 0)$ or $(a, b, 1)$, and is the skeleton of our strategy. Most of the rest of our discussion is concerned only with the justification of the + entries. First we show how the remaining entries can be deduced from these. We use three principles.

(1) The entry (a, b) is P if and only if it is non-terminal and there is no entry of form

$$\begin{array}{llllll}
\text{P} & \text{in} & (a-2, b+1) & \text{the only} & aa & \text{to} & (a-2, b+1, 0) \\
+ & \text{in} & (a, b-2) & \text{moves from} & bb & \text{to} & (a, b-2, 1) \\
\text{or } + & \text{in} & (a-1, b-1) & (a, b, 0) \text{ being} & ab & \text{to} & (a-1, b-1, 1).
\end{array}$$

(2) The entry (a, b) cannot be + if there is any entry of form

$$\begin{array}{lll}
\text{P} & \text{in} & (a, b-2) \\
\text{P} & \text{in} & (a-1, b-1) \\
\text{P or } + & \text{in} & (a-1, b) \\
\text{P or } + & \text{in} & (a, b-1)
\end{array}
\quad \begin{array}{c} \text{because} \\ \text{there are} \\ \text{moves} \end{array}
\quad \begin{array}{lll}
c!(bb) & \text{from } (a, b, 1) & (a, b-2, 0) + (0, 2, 0) \\
c!(ab) & \text{to each} & (a-1, b-1, 0) + (1, 1, 0) \\
c!(a) \text{ or } ac & \text{of the} & (a-1, b, 0) + (1, 0, 0) \text{ or } (a-1, b, 1) \\
c!(b) \text{ or } bc & \text{positions} & (a, b-1, 0) + (0, 1, 0) \text{ or } (a, b-1, 1)
\end{array}$$

and the positions $(0, 2, 0)$, $(1, 1, 0)$, $(1, 0, 0)$, $(0, 1, 0)$ can be neglected, since they necessarily last for exactly 0 or 2 moves.

(3) The entry $(a, 0)$ cannot be + if there is a P entry in $(a-4, 1)$ or $(a-6, 0)$. (For from the position $(a, 0, 1)$ we can move to $(a-2, 0, 0) + (2, 0, 0)$, and whatever our opponent does to this, we can move to the sum $(a-4, 1, 0) + (0, 1, 0)$ on our next move, in which $(0, 1, 0)$ can be neglected. We can also move to $(a-6, 0, 0) + (6, 0, 0)$ and we shall show later that $(6, 0, 0)$ can be neglected, being equivalent to 0.)

The reader should now check that all the entries in Table 7 follow from the + entries using only these three principles, and the obvious fact that each entry is P or + or −, since $(a, b, 0)$ and $(a, b, 1)$ cannot both be P.

It is not such a routine matter to justify the + entries themselves, the main difficulty being that our opponent might try to create two or more chains, and we cannot allow this to persist, or the position will become too complicated for words (or pictures). The backbone of our strategy (supporting the skeleton of Table 7) is illustrated in Fig. 3, which illustrates winning strategies for the second player from each of the positions

$$(0, 0, 1), (1, 1, 1) = (0, 2, 1), (3, 5, 1), (3, 7, 1), (3, 9, 1), \ldots .$$

which are all but two of the $\mathscr{P}$-positions corresponding to the + entries in Table 7. We write $(1, 1, 1) = (0, 2, 1)$ because a single atom has exactly the same effect on the game as another branch. For the same reason, we have systematically replaced any position $(1, b, c)$ which should appear in Fig. 3 by the equivalent position $(0, b+1, c)$.

Some further remarks need to be made about Fig. 3. The positions surrounded by double boxes represent $\mathscr{P}$-positions which will be dealt with shortly. All other $\mathscr{P}$-positions are surrounded by single boxes, and all their options also appear on the Figure. Every unboxed position on the diagram represents an $\mathscr{N}$-position, for which a $\mathscr{P}$-option is always given. The symbol $abcD$ denotes the sum of the position (a, b, c) with another position (like $(0, 0, 1)$) which necessarily lasts an *odd* number of moves (usually *one* move), however played, while $abcE$ denotes the sum of (a, b, c) with a position (like $(0, 2, 0)$ or $(1, 1, 0)$) which necessarily lasts an *even* number of moves (usually *two*). In any later analysis, we have always supposed that these odd and even numbers were *one* and *zero*, respectively. Finally, $*abc$ denotes the sum of any two positions (x, y, z) and $(a-x, b-y, c-z)$. To continue the figure downwards increase b by 2.

The two $\mathscr{P}$-positions $(7. 3, 1)$ and $(11, 1, 1)$ corresponding to the only + entries in Table 7 not yet verified, are discussed in Table 8.

It remains to discuss the double-boxed positions of Fig. 3.

Theorem. The sum of any number of positions of the form $(0, b, 0)$ together with a game which necessarily lasts for exactly n moves, is a $\mathscr{P}$-position if and only if:

> *either* n is odd, and all the numbers b are 0, 1, 2, or 4
> *or* n is even, and at least one of the numbers b is *not* 0, 1, 2, or 4.

Proof. The positions $(0, 0, 0)$ and $(0, 1, 0)$ are ended, while $(0, 2, 0)$ lasts exactly two moves, so all of these positions can be neglected. In fact positions $(0, 4, 0)$ can be neglected as well, since we can always arrange that they last an even number of moves. The only line of play from $(0, 4, 0)$ lasting an odd number of moves is

$$(0, 4, 0) \text{ to } (0, 2, 1) \text{ to } (0, 1, 1) \text{ to } (0, 1, 0).$$

We need never make the move from $(0, 2, 1)$ to $(0, 1, 1)$, and if our opponent does so, we can immediately reply with a move from $(0, 1, 1)$ to $(0, 0, 1)$ which makes the game last an extra move.

Neglecting $(0, 4, 0)$ and positions which always last an even number of moves, the only real assertion is that a sum of positions $(0, b, 0)$, with each b either $=3$ or $\geqslant 5$, is a $\mathscr{P}$-position. The only move from $(0, b, 0)$ is to $(0, b-2, 1)$ from which we can move to any position $(0, x, 0) + (0, y, 0)$ with $x+y = b-2$. However our opponent moves, we can use this to restore the position to another one covered by the theorem, *unless* it is just the single position $(0, 3, 0)$, from which our opponent can only move to $(0, 1, 1)$, and we then move to $(0, 0, 1)$, leaving him to make the last (losing) move.

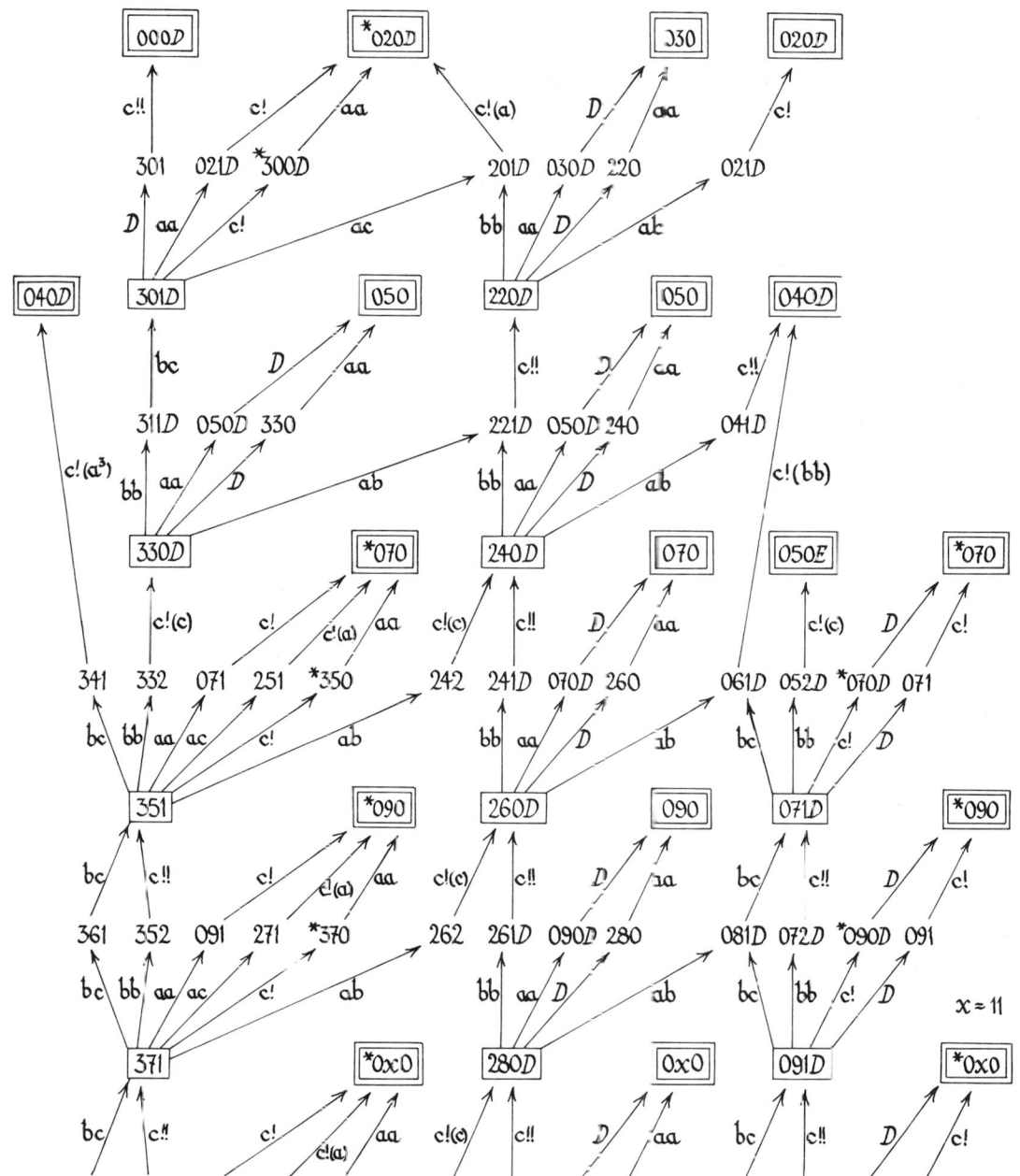

Figure 3. Strategy for Misère Lucasta.

THE POSITIONS (7, 3, 1) AND (11, 1, 1)

In Table 8, to each option of either of these positions, we give a response, which is in every case expressed as the sum of a $\mathscr{P}$-position from Table 7 and some position which can be verified to be equivalent to 0 from Fig. 4. In fact the + entries in Table 7 corresponding to (7, 3, 1) and (11, 1, 1) are not needed to verify any other entry, and so are not used in our strategy from any initial position, so that Table 8 is not really necessary for the strategy.

The options of (7, 3, 1)	have good replies	The options of (11, 1, 1)	have good replies
aa (5, 4, 1)	(5, 2, 0) + (0, 2, 0)	(9, 2, 1)	(9, 0, 0) + (0, 2, 0)
ab (6, 2, 2)	(3, 2, 0) + (3, 0, 1)	(10, 0, 2)	(7, 0, 0) + (3, 0, 1)
bb (7, 1, 2)	(1, 1, 1) + (6, 0, 0)		
ac (6, 3, 1)	(5, 3, 0) + (1, 0, 0)	(10, 1, 1)	(9, 1, 0) + (1, 0, 0)
bc (7, 2, 1)	(7, 0, 0) + (0, 2, 0)	(11, 0, 1)	(7, 0, 0) + (4, 0, 0)

The other options are all sums of two positions $(a, b, 0)$, and in every case we move in the region which has just 2, 3, 6, 7, 10, or 11 atoms, and join two of these atoms together.

Table 8. (7, 3, 1) and (11, 1, 1) are $\mathscr{P}$-positions.

It is interesting to note that the positions

$$000, \; 010 = 100, \; 020 = 110, \; 040 = 130,$$
$$400, \; 420, \; 510, \; 600, \; 800,$$
$$301, \; 022 = 112, \; 002, \; 004, \; 006, \ldots$$

are equivalent to 0 in the misère sense. (This remark can be useful in play from more complicated positions than those which need arise if our strategy is followed.) To prove that a position is misère-equivalent to 0 it is necessary and sufficient to show, first, that it is an $\mathscr{N}$-position and second, that each of its options has itself an option misère-equivalent to 0. This is done for the above positions in Fig. 4, in which the subscript

P denotes a $\mathscr{P}$-position
N denotes an $\mathscr{N}$-position *not* misère-equivalent to 0
O denotes an $\mathscr{N}$-position misère-equivalent to 0.

In a strategically fought game of Misère Lucasta, we find three phases. In the first phase, both players join pairs of atoms together to form branches. If either player dares to form a chain, his opponent can certainly win by closing the chain around some small number of atoms and branches (which can be neglected), and converting the rest of the position to a $\mathscr{P}$-position. When the number of atoms is reduced to just above three, the winner is the player able to convert the position to $(3, 2n + 1, 1)$, and the game enters its second phase, in which play follows the lines of Fig. 3. The third phase is reached when the position becomes a sum of positions $(0, b, 0)$ in which only branches (with isolated atoms) remain, together possibly with some rather trivial game. From then on, the winner always restores the position to a similar form, except that near the end of the game he is careful to restore the position $(0, 3, 0)$ to a single chain $(0, 0, 1)$.

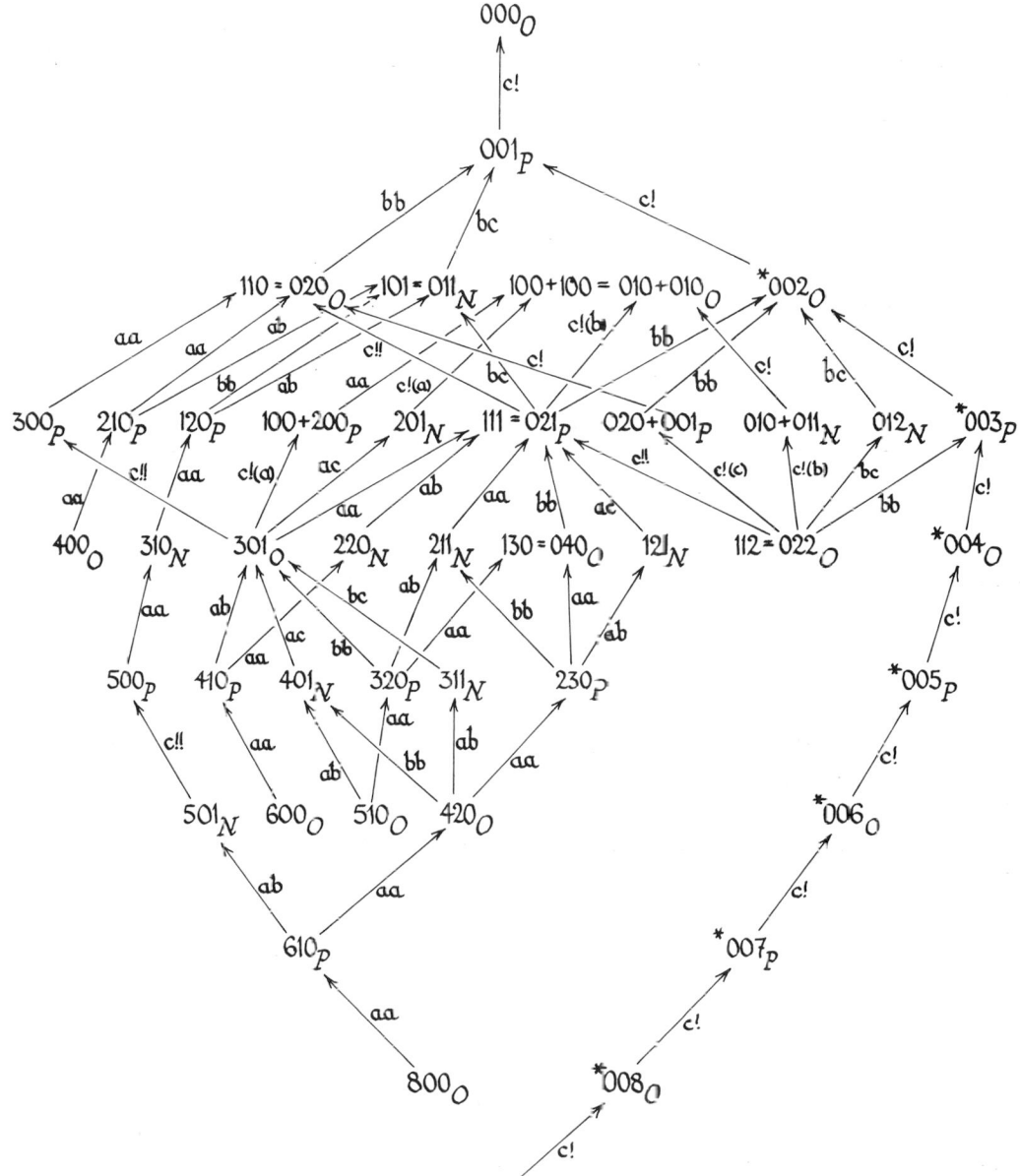

Figure 4. Proof of Misère-equivalence to Zero.

	a = 0									a = 1									a = 2						
c =	0	1	2	3	4	5	6	7	8	0	1	2	3	4	5	6	7	8	0	1	2	3	4	5	6
b = 0	0	1	0	1	0	1	0	1	0	0	2	3	2	3	2	3	2	3	1	3	1	p	p_1	p	p_1
1	0	2	3	2	3	2	3	2	3	0	1	0	a	a_1	a	a_1	a	a_1	1	2	4	q	r	s	s_1
2	0	1	0	a	a_1	a	a_1	a	a_1	2_+	2	3	b	c	d	d_1	d	d_1	t	u	v	6^{686}			
3	2_+	2	3	b	c	d	d_1	d	d_1	0	e	f	g	h	i	j	k	k_1	1	w	4^{04}				
4	0	e	f	g	h	i	j	k	k_1	2_+	l	3^{04}							K	3^3	5^{16}				
5	2_+	l	3^{04}							e_+	1^3	0^{52}							1^1	5^4					
6	e_+	1^3	0^{52}							l_+	2^1								2^1	3^{203}					
7	l_+	2^1								0^0	1^2								1^2						
8	0^0	1^2								0^0									2^1						
9	0^0									0^0															
10	0^0																								

Table 9. The Genus of Lucasta Positions.

genus	name	structure
1^{4313}	a	2_2320
2^{2020}	b	$a2_+30$
3^{0431}	c	$ba_1a2_{+1}2$
2^{1520}	d	$cb_1a_2a_1a2_+3$
1^{3131}	e	2_+20
0^{1202}	f	$ea2_{+2}321$
1^{4313}	g	$fe_1ba_12_{+3}2_230$
0^{5202}	h	$gf_1ecba2_{+2}3_21$
1^{4313}	i	$hg_1fe_1dcb_2a_12_{+3}2_20$
0^{5202}	j	$ih_1gf_1ed_1dc_2b_3a2_{+2}3_21$
1^{4313}	k	$ji_1hg_1fe_1d_2d_1dc_3b_2a_aa_12_{+3}2_20$
0^{0202}	a_a	$a_{22}a_3a_2a$
2^{2020}	l	$e2_+30$
4^{1464}	p	2_2321
5^{5757}	q	$pa3_243210$
6^{6846}	r	$qp_1pa_1a2_25320$
7^{7957}	s	$rq_1p_2p_1pa_2a_1a3_24321$
2^{1420}	t	2_+31
3^{3131}	u	$t2_+210$
5^{5757}	v	$ut_1pa2_{+2}43210$
5^{2057}	w	$ute2_{+1}2_+4310$

genus	name	structure
0^{3131}	A	$pa3_221$
1^{2020}	B	$Ap_1pa_12_230$
0^{3131}	C	$BA_1p_2p_1pa3_221$
1^{5313}	D	$p2_+4320$
3^{6464}	E	$DAqp_1b2_{+1}2_2421$
1^{1313}	F	$2_+3 = 2_{+1}$
1^{2020}	H	F_+30
2^{0313}	I	$H0$
1^{1313}	J	$u2_+3$
2^{1313}	K	$ue2_+$
0^{0202}		$2_+, e_+, l_+, F_+$

	$a = 3$	$a = 4$	$a = 5$	$a = 6$	7	8	9
$c =$	0 1 2 3 4 5 6	0 1	0 1	0 1	0	0	0
$b = 0$	1 0 2 A B C C_1	0 3	F_+ H	0 2^3	1	0	0^0
1	F 3 D E	1 4^4	0	I	1^1		
2	1	0	0^0				
3	J						

Note: If $c \geqslant 2b + 2a$, then
(a, b, c) has value x_1, where $(a, b, c - 1)$ has value x.

To save space in Table 9 the abbreviating conventions are *not* the same as those of Chapter 13. In the table

$$g^{a \cdots x} \quad \text{means} \quad g^{a \cdots xyxy \cdots} \quad \text{where} \quad y = x \ast 2$$

and no assertions about tameness or restiveness are intended. In the notes opposite, the genus is given to four superscripts even when the period starts earlier, and now the last two superscripts repeat indefinitely.

CABBAGES; OR BUGS, CATERPILLARS AND COCOONS

If we modify Lucasta by allowing the move which completes a closed loop passing through only two spots and consisting of two curves joining them, we get a simpler game. Here, we call the isolated spots *bugs*, chains of two or more spots *caterpillars*, and closed loops *cocoons*. The cocoons separate the plane into regions, so that the general position is a sum of positions (b, c), where these numbers specify the numbers of bugs and caterpillars per region.

It turns out that the position (b, c) in this game behaves just like the position $(0, b, c)$ in Lucasta, so we have the analysis already. (Using our nim-value table for Lucasta we can in fact analyze arbitrary positions in normal play.) In particular, we have:

The initial position $(n, 0)$ is a $\mathscr{P}$-position in normal play for all n,
and   in misère play for all n except 0, 1, 2, 4.

JOCASTA

We obtain an even simpler game by allowing in addition the move which joins an isolated spot to itself to form a closed loop passing only through that spot.

SPROUTS

This game (introduced by M.S. Paterson and J.H. Conway some time ago) has a novel feature which complicates the analysis to such an extent that the normal outcome of the 7-spot game is still unknown. Even the 2-spot game is remarkably complicated.

The move in Sprouts is to join two spots, or a single spot to itself (Fig. 5) by a curve which does not meet any previously drawn curve or spot. But when this curve is drawn, a new spot must be placed upon it. No spot may have more than three parts of curves ending at it.

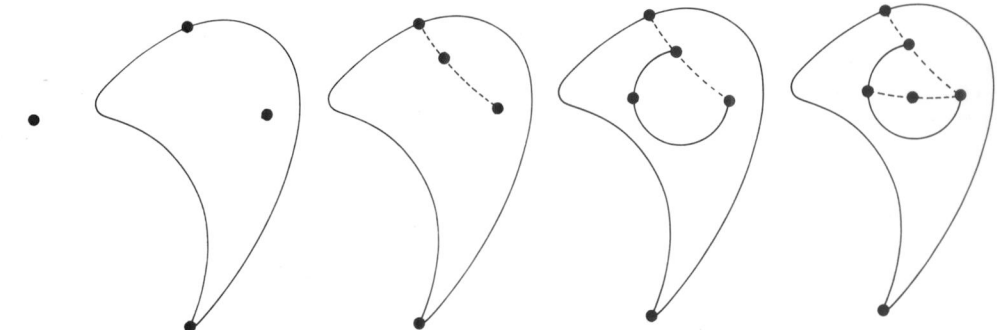

Figure 5. A Short Game of Sprouts.

A typical game is shown in Fig. 5, with the second player's moves drawn as dotted lines. Since the new spots can still be used in later moves, a Sprouts game will last longer than a Cabbages game from the same initial position, and it is perhaps not even obvious that it need ever end. But there is a simple argument which shows that in fact a Sprouts game starting from n spots can last at most $3n - 1$ moves. We take the 3-spot game as an example. Each spot has potentially 3 ends of curves available to it, which we shall call its three *lives*, so initially the 3-spot game has 9 lives. But each move takes one life away from the two spots it joins (or two lives away from a spot joined to itself), and adds a new spot which has just one life. Therefore each move reduces the total number of lives by one. Since the very last spot to be created is still alive at the end of the game, the total number of moves is at most $9 - 1 = 8$. But Fig. 6 shows just how complicated even the 2-spot game really is.

One of the most interesting theorems about Sprouts (due to D. Mollison and J.H. Conway) is the Fundamental Theorem of Zeroth Order Moribundity (FTOZOM). We shall not prove it here, but will at least state it. The FTOZOM asserts that the n-spot Sprouts game must last at least $2n$ moves, and that if it lasts exactly this amount, the final configuration is made up of the insects shown in Fig. 7.

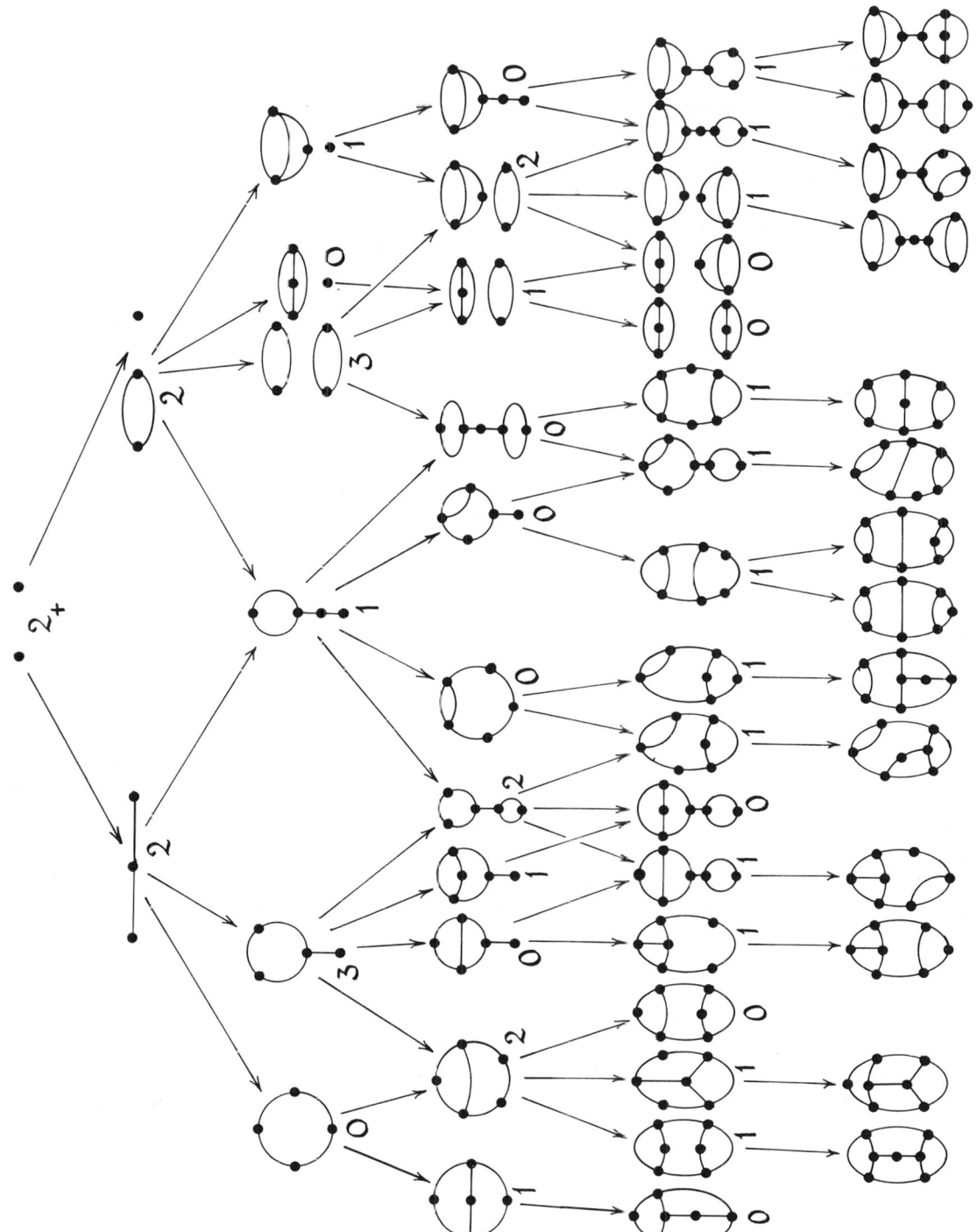

Figure 6. Two-spot Sprouts, with Reduced Forms.

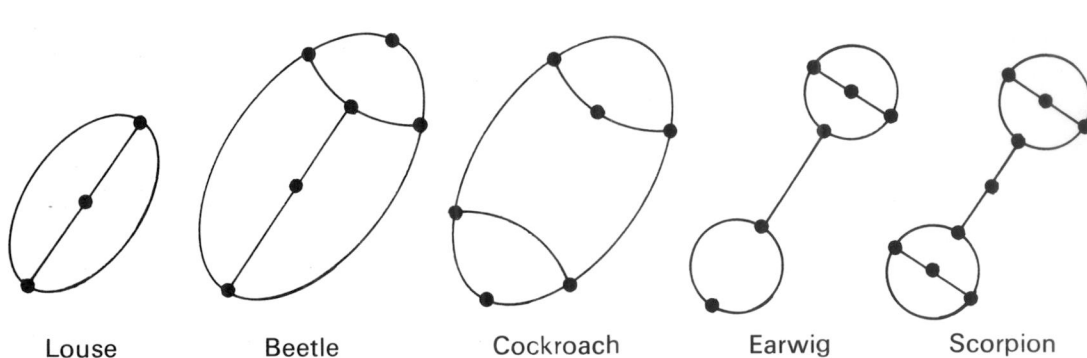

Figure 7. The Five Fundamental Insects.

To be more precise, the final configuration must consist of just one of these insects (which might perhaps be turned inside out in some way) infected by an arbitrarily large number of lice (some of which might infect others). One of the possible configurations is shown as Fig. 8—it consists of an inside-out scorpion inside an inside-out louse, liberally infected with other lice!

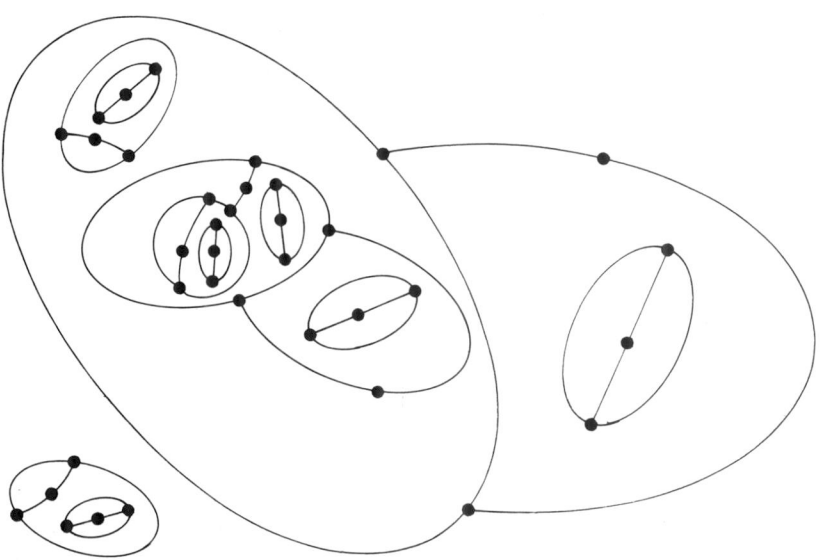

Figure 8. The Lousy End of a Short Sprouts Game.

How should we play if we wish to win a Sprouts game? It is clear that whether the play is normal or misère, the outcome only depends on whether the total number of moves in the game is odd or even, so in some sense winning is controlling the number of moves. Now the 3-spot game necessarily lasts for 6, 7, or 8 moves, and it is very difficult to make it last 8 moves, so that really the fight is between 6 and 7 moves. Apparently the same thing happens in larger games— essentially one player tries to make the game last m moves, while the other tries to drag it out to $m+1$, all other numbers being very unlikely.

To see how to control the number of moves, we examine the situation at the end of the game, which we suppose to have started with n spots and lasted for m moves. The final number of spots is $n+m$, and the total life at the end of the game is $l = 3n-m$, since we started with $3n$ lives, and subtracted one per move. Each of the live spots at the end of the game has two dead spots as its two nearest neighbors, and the remaining dead spots are called *Pharisees*. (The concept of neighbor is quite subtle—in Fig. 9 we show the two different ways in which two dead spots can be neighbors of a live one.)

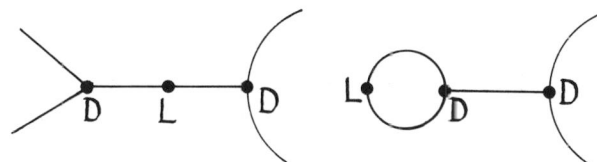

Figure 9. Two Live Spots (L) and their Dead Neighbors (D).

Now no dead spot can be a neighbor of two different live spots, for otherwise we could join these two spots and continue the game. So the number ϕ of Pharisees is given by the equation

$$\phi = (n+m) - (l+2l) = (n+m) - 3(3n-m) = 4m - 8n$$

and we have the *Moribundity Equation*:

$$m = 2n + \tfrac{1}{4}\phi.$$

From this equation we can deduce several things:

(i) The number of moves is at least $2n$.
(ii) The number of Pharisees is a multiple of 4.
(iii) If at any time in the game we can ensure that the final position has at least P Pharisees, then the game will last at least $2n+\tfrac{1}{4}P$ moves.

There is a corresponding result to (iii) in the opposite direction:

(iv) If at any time in the game we can ensure that the final position has at least l live spots, then the game will last at most $3n-l$ moves.

So, according to our previous ideas, one player will try to lengthen the game by producing Pharisees, while his opponent tries to shorten it by producing spots which must remain alive.

There is a useful way to estimate the number of live spots there will be at the end of the game. *If any region defined by curves of the game has a live spot strictly inside, then there will be a live spot inside that region at all later times.* So in Fig. 10 we can regard, if we like, the plane as divided

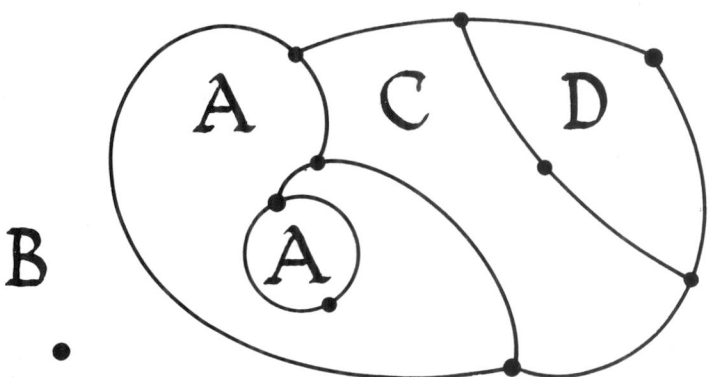

Figure 10. A Sprouts Position with One Pharisee.

into four regions A, B, C, D, and the regions A and B each have live spots strictly inside. Any move made in either of these regions creates a new live spot, and so each of A and B will contain a live spot at the end of the game. We cannot say the same of C and D, whose only live spots lie on their borders, but if we regard C and D together as forming a single region, then this new region has just one spot strictly inside. So we can see that the game will have at least 3 live spots in its final position. It also has presently one Pharisee P, and so (since it developed from an initial position with $n=4$ spots) we can see that it will last *at most* $3n-3 = 9$ moves, and *at least* $2n+\frac{1}{4} = 8\frac{1}{4}$ moves. Since it is difficult to see how the game could last for *exactly* $8\frac{1}{4}$ moves, we conclude that the total length of the game will be 9 moves, however it is played from now on! (Actually, 6 moves have already been made, so just 3 more moves are to follow.) Accordingly, this is either a normal play game about to be won by the first player, or a misère play one being won by the second player.

Using these ideas, it is fairly easy to give analyses of games with small numbers of spots. We give the results we have obtained in Table 10, which shows the number of moves that the winner can arrange for the game to last.

no. of spots:	0	1	2	3	4	5	6
normal play:	0P	2P	4P	7N	9N	11N	14P
misère play:	0N	2N	5P	7P	9P		

Table 10. Outcomes of the Smallest Sprouts Games.

The fact that 6-spot normal Sprouts is a $\mathscr{P}$-position was first proved (to win a bet) by Denis Mollison, whose analysis of the game ran to 47 pages! Using the ideas above, we can shorten this considerably, but we have not yet been able to analyze 5-spot Sprouts with misère play.

BRUSSELS SPROUTS

Here is another game, which should be more interesting than Sprouts. We start with a number of crosses, instead of spots. The move is to continue one arm of a cross by some curve which ends at another arm of the same or a different cross, and then to add a new cross-bar at some point along this curve. A 2-cross game of Brussels Sprouts is shown as Fig. 11. After playing a few games of Brussels Sprouts, the skilful reader will be able to suggest a good starting strategy.

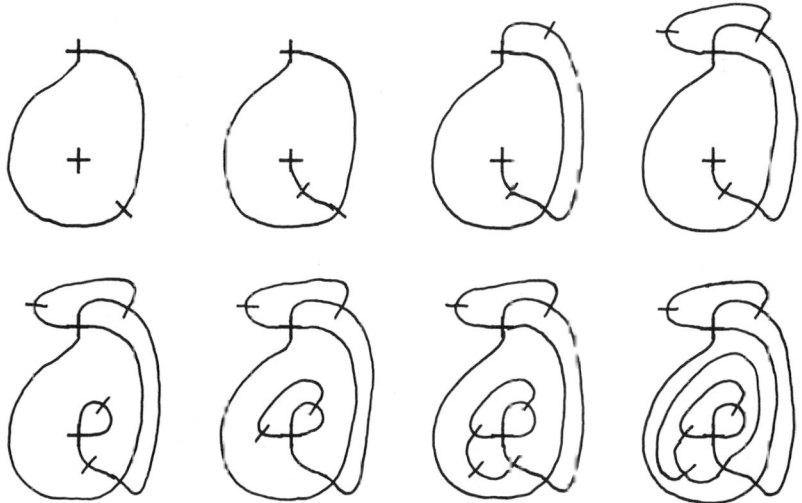

Figure 11. A 2-cross Game of Brussels Sprouts.

STARS-AND-STRIPES

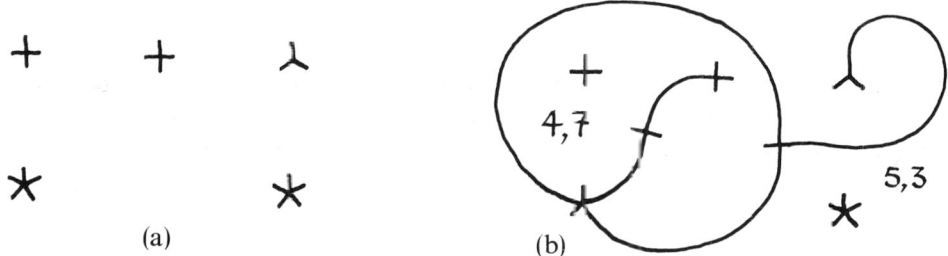

(a)   (b)

Figure 12. A Game of Stars-and-Stripes.

Suppose we make addition of the cross-bar optional in Brussels Sprouts. It is natural at the same time to allow "stars" with any number of arms instead of just crosses with exactly 4 arms, and to call the cross-bar a *stripe*. An initial position $(5, 5, 4, 4, 3)$ is shown in Fig. 12(a), along with a position 3 moves later (Fig. 12(b)). In the analysis, the game becomes a disjunctive sum of regions, and we can pretend that each region contains only stars. In general, a connected portion of the picture which has just n arms sticking into the region concerned counts as an n-arm star inside the region. (Even the boundary of the region counts as a star.) In Fig. 12(b) we have therefore labelled each region with numbers showing the sizes of the stars in that region.

The one-star game is isomorphic to the octal game **4·07**, since a move with cross-bar essentially splits an n-arm star into two stars of sizes a and b, with $a+b = n$, $a, b \neq 0$, and the move without crossbar splits it into stars of sizes a and b with $a+b = n-2$. The nim-values for this game (Chapter 4, Tables 7(a), 6(b)) are $0.\dot{0}12\dot{3}$ and the genus appears in Table 11.

n	0	1	2	3	4	5	6	7	8	9	10	11
genus of n	0	0	1	2	3	0	1_a^{431}	2	3_b^{31}	$0_{a_1}^{520}$	1_c^{431}	2_d^{0420}

$$a = 2_2 320 \qquad b = a_1 a 2_2 20 \qquad c = b_1 b a_3 a_1 2_2 20 \qquad d = c b_2 b a_2 a_1 a 3_2 3$$

Table 11. The Genus of Stars-and-Stripes Positions.

BUSHENHACK

is another pencil and paper game. It's played with a number of rooted trees, but now when you chop an edge, all edges connecting it to the ground disappear, leaving a number of floating bits of tree to be rerooted as in Fig. 13. Its theory involves yet another property of Nim.

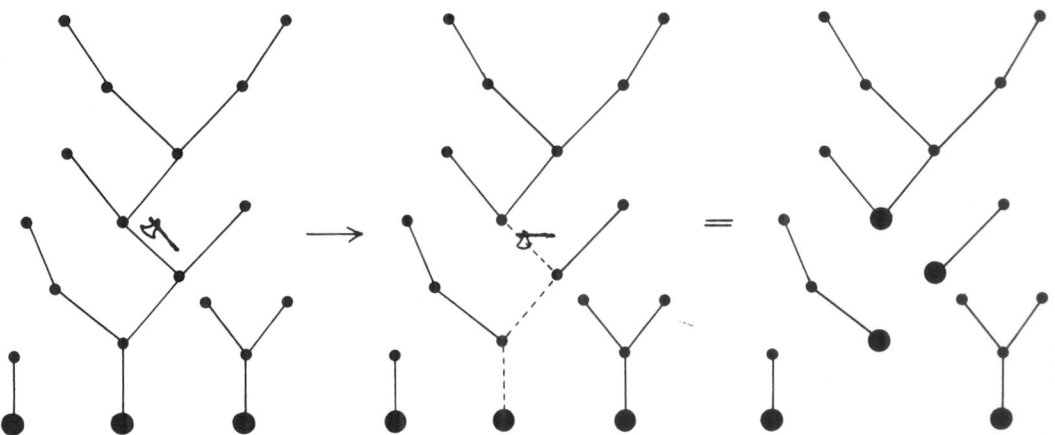

Figure 13. A Bushenhack Move.

GENETIC CODES FOR NIM

If you tell me that you're in a Nim-position of some nim-value (e.g. 9) and can move to positions having exactly so many (e.g. 13) other nim-values, I can tell you exactly what (in this case 0, 1, 2, 3, 4, 5, 6, 7, 8, 12, 13, 14, 15) those values are!

To see why, we enlarge upon a notation from the Extras to Chapter 7, in which the single Nim-heaps are

$$0_{\{\ \}},\ 1_{\{1\}},\ 2_{\{2,\,3\}},\ 3_{\{1,\,2,\,3\}},\ 4_{\{4,\,5,\,6,\,7\}},\ 5_{\{1,\,4,\,5,\,6,\,7\}},\ \ldots,$$

and in general $n_{[n]}$, where $[n]$, the *variation set*, is the set of changes in nim-value that are possible in one move. The variation set $[n]$ for an arbitrary Nim-heap consists of all numbers whose leftmost binary digit 1 is present in the binary expansion of n, and so it can be found as the union of the appropriate selection of

$$[1] = \{1\}, \quad [2] = \{2, 3\}, \quad [4] = \{4, 5, 6, 7\}, \quad [8] = \{8, 9, \ldots, 15\}, \quad \ldots$$

E.g., since $13 = 1+4+8$, $[13] = \{1, 4, 5, 6, 7, 8, 9, \ldots, 15\}$.

We'll say that a position has **genetic code** A if it has the same variation set as the Nim-heap of size A. Arbitrary Nim-positions have genetic codes, because when you *add* positions you *unite* their variation sets; e.g., $5+12$ has genetic code 13, because

$$5 + 12 = 5_{\{1,\,4,\,5,\,6,\,7\}} + 12_{\{4,\,5,\,6,\,7,\,8,\,9,\,\ldots,\,15\}} = 9_{\{1,\,4,\,5,\,6,\,7,\,8,\,9,\,\ldots,\,15\}} = 9_{[13]},$$

and the options of $9_{[13]}$ can be found by nim-adding 9 to the members of the variation set $[13]$.

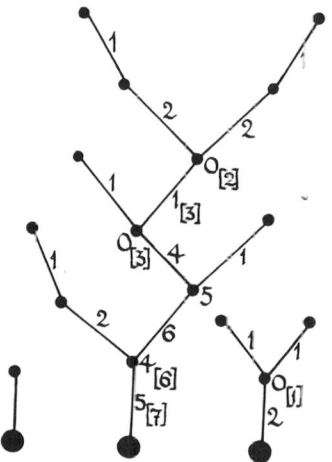

Figure 14. What's the Winning Move? (see the Extras).

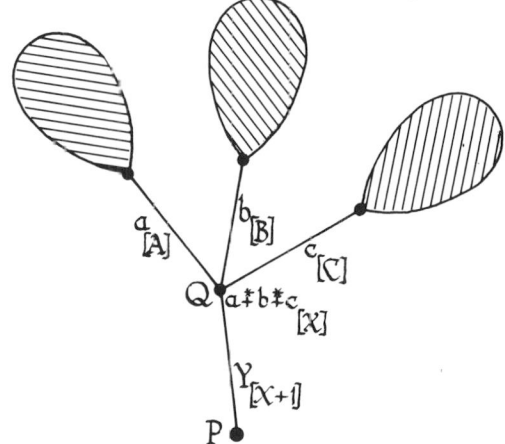

Figure 15. Calculating the Value and the Code.

BUSHENHACK POSITIONS HAVE GENETIC CODES!

In Fig. 14 the symbol $a_{[A]}$ against any edge gives the value and genetic code for the subtree whose trunk is that edge, while at every node where there are several branches we've given this information for the sum of the corresponding subtrees. (An isolated digit a means $a_{[a]}$.) The numbers are calculated as in Fig. 15 where X is the number whose binary expansion has a 1 wherever there is a 1 in *any* of A, B, C, so that

$$[X] = [A] \cup [B] \cup [C]$$

and Y is the smallest number greater than $a \overset{*}{+} b \overset{*}{+} c$ that is divisible by exactly that power of 2 which divides $X + 1$.

Suppose that X has binary expansion

$$\dots \; ? \; ? \; ? \; 0 \; 1 \; 1 \; 1 \; \dots \; 1 \quad \text{(ending in } k \text{ 1's)}.$$

Then that of $a \overset{*}{+} b \overset{*}{+} c$ will have the form

$$\dots \; p \quad q \quad r \quad 0 \quad t \quad u \quad v \; \dots \; z$$

We already know why these numbers are the code and value for the sum of the three subtrees above Q in Fig. 15. So we need only show that

$$X + 1 = \dots \; ? \; ? \; ? \; 1 \; 0 \; 0 \; 0 \; \dots \; 0$$

and

$$Y = \dots \; p \quad q \quad r \quad 1 \; 0 \; 0 \; 0 \; \dots \; 0$$

are the code and value for the subtree with trunk PQ. The options for this tree have nim-values

$$a \overset{*}{+} b \overset{*}{+} c \quad (\text{chop } PQ) \quad \text{and} \quad a \overset{*}{+} b \overset{*}{+} c \overset{*}{+} \Delta$$

for any number Δ whose first digit 1 is present in X. In particular, we can move to all nim-values that are $\leqslant a \overset{*}{+} b \overset{*}{+} c$ or differ from it only in the last k places. So we can move to all numbers smaller than Y, which differs from $a \overset{*}{+} b \overset{*}{+} c$ in the $(k+1)$st place from the right, corresponding to the rightmost zero in X. The nim-values of the options are exactly those binary numbers whose leftmost difference from Y corresponds to a digit 1 in $X + 1$, which is therefore the genetic code.

VON NEUMANN HACKENBUSH

when played on trees, is an exactly equivalent game in which the move is to delete a *node* together with all nodes on the path connecting it to the ground and all edges meeting these. To convert to Bushenhack, just add a new trunk to every tree. Von Neumann proved, by a strategy-stealing argument, that a single tree was always an $\mathcal{N}$-position, and Úlehla gave an explicit strategy for trees, which prompted our own discussion.

Bushenhack is really just the theory of $A + B$ and $A:*$, whereas ordinary Hackenbush is concerned with $A + B$ and $*:B$. The most general version of von Neumann Hackenbush is played on any directed graph (remove a node and all nodes it points to). Its analysis involves the properties of $A + B$ and $A:B$ for arbitrary variation sets (Extras to Chapter 7).

EXTRAS

THE JOKE IN JOCASTA

is that the n-spot game always last for n moves, because each spot has two lives and each move uses two. The game is therefore just another form of She-Loves-Me, She-Loves-Me-Not.

THE WORM IN BRUSSELS SPROUTS

is similar but more subtle. The n-cross game always lasts for just $5n-2$ moves, but Brussels Sprouts is definitely more interesting on surfaces of higher genus, e.g. the torus.

BUSHENHACK

The winning move in Fig. 14 is that shown in Fig. 13.

REFERENCES AND FURTHER READING

Piers Anthony, "Macroscope", Avon, 1972.

Hugo D'Alarcao and Thomas E. Moore, Euler's formula and a game of Conway, J. Recreational Math. **9** (1977) 249–251; Zbl. 355.05021.

Martin Gardner, Mathematical Games: Of sprouts and Brussels sprouts; games with a topological flavor, Sci. Amer. **217** #1 (July 1967) 112–115.

Martin Gardner, "Mathematical Carnival", Alfred A. Knopf, New York 1975, Chapter 1.

Emmanuel Lasker, Brettspiele der Völker,

E. Lucas, "Récréations Mathématiques", Gauthiers-Villars, 1882–94; Blanchard, Paris, 1960.

Gordon Pritchett, The game of Sprouts, Two-Year Coll. Math. J. **7** #4 (Dec. 1976) 21–25.

J.M.S. Simões-Pereira and Isabel Maria S.N. Zuzarte, Some remarks on a game with graphs, J. Recreational Math. 6 (1973) 54–60; Zbl. 339.05129.

J. Úlehla, A complete analysis of von Neumann's Hackendot, Internat. J. Game Theory, **9** (1980) 107–115.

Chapter 18

The Emperor and His Money

For good ye are and bad, and like to coirs,
Some true, some light, but every one of you
Stamp'd with the image of the King.
> Alfred, Lord Tennyson, *The Idylls of the King, The Holy Grail*, l.25.

Figure 1. The Emperor's Declaration.

"...Emperor Nu took power by overthrowing the divisive My-Nus dynasty. The Nu régime introduced many positive reforms, and in particular abolished the old (An-Tsient) irrational currency, which had his predecessor's head on it, and introduced the Nu system. The masters of the Imperial Mint, Hi and Lo, were alternately to decide the value of each new denomination, and after each decision, sufficiently many coins of this value were to be struck. All went well until Hi ordered the striking of a coin of value one, so throwing the Workers of the Mint into unemployment. They rose in a body, and threw the unfortunate Hi from the tower at the quiet end of the capital, which has been known as the Hi Tower ever since."

My-Nus—Some Divisive Times.

575

SYLVER COINAGE

Had Hi and Lo read this book, they would have realized they were only playing a game, the game of **Sylver Coinage**. In this the players alternately name different numbers, but are not allowed to name *any* number that is a sum of previously named ones. So, if 3 and 5 have been named, for example, neither of the players is allowed to play any of the numbers

$$3, \quad 5, \quad 6 = 3+3, \quad 8 = 3+5, \quad 9 = 3+3+3, \quad 10 = 5+5, \quad 11 = 3+3+5, \quad \ldots$$

When will this game end? If neither player has played 1, 1 will still be playable. But, of course, as soon as 1 has been played, every number

$$1, \quad 2 = 1+1, \quad 3 = 1+1+1, \quad 4 = 1+1+1+1, \quad 5 = 1+1+1+1+1, \quad \ldots$$

is illegal, and so the game ends. Because the player who names 1 is declared the *loser*, Sylver Coinage is a *misère* game. (Skilful players won't spend much time on the normal play version!).

We had better point out that because the old currency had been rather irrational (with coins of value $\sqrt{2}$, e and π) the Emperor declared that there was to be a new monetary unit, the **You-Nit**, and the value of each coin was to be an integral number of You-Nits. (You can see the Emperor making this declaration in Fig. 1!).

And recalling how people were nonplussed by the great financial scandal of the My-Nus dynasty when they had to take away Teh Kah-Weh for issuing currency of negative value, Emperor Nu decided that each coin's value must be a *positive* number of You-Nits.

HOW LONG WILL IT LAST?

It might take quite a long time. To see that it can last for a thousand moves, we need only consider the game

$$1000, 999, 998, \ldots, 4, 3, 2, 1.$$

And of course a thousand can be replaced by any other number, so that the game is **unbounded**. Many other games have this property, for example Green Hackenbush (Chapter 2) played with an infinite snake, but are *boundedly* unbounded because after some fixed number of moves the end will be in sight. Thus after the first move in the Hackenbush game only a finite amount of snake is left.

But Sylver Coinage is not like that! No matter what number you choose, Hi and Lo can find a way to play that number of moves so that what's left of the game will still be unbounded. Their first thousand moves might be

$$2^{1000}, 2^{999}, 2^{998}, \ldots, 2^4, 2^3, 2^2, 2^1$$

and the rest of the game can still last as long as you like:

$$1000001, 999999, 999997, \ldots, 7, 5, 3, 1.$$

In other words Sylver Coinage is *unboundedly unbounded*. And this isn't all. It's *unboundedly unboundedly unbounded* and unboundedly like that, and so (unboundedly) on!

Nevertheless, it can't go on for ever; in the language of Chapter 11 it's an *ender*. It is because the little theorem which proves this is due to the famous mathematician J.J. Sylvester that we have called the game *Sylver* Coinage.

For, at any time after the first move, let g be the greatest common divisor (g.c.d.) of the moves made. Then it's not hard to see that only finitely many multiples of g are *not* expressible as sums of numbers already played. So after at most this known number of moves the g.c.d. must be reduced. Eventually we must arrive at a position with $g = 1$ and can bound the number of moves yet to be made. So although we may not be able to bound the game after any given number of moves, we *can* bound the number of moves it will take to reduce the g.c.d.

SOME OPENINGS ARE BAD

The proof we gave in the Extras to Chapter 2 shows that from any position in Sylver Coinage there *is* a winning strategy for one of the two players *but* because of the infinite nature of the game we cannot work through all positions and guarantee to find winning strategies when they exist. In fact we do not know of (and there may not exist) any way of working out in a finite time who wins from an arbitrarily given position. But we do know the answers for some easy positions.

If at any time you name 1, you lose by definition.

If you name 2, my reply will be 3 if it's still available, and then all larger numbers

$$4 = 2+2, \quad 5 = 2+3, \quad 6 = 2+2+2, \quad 7 = 2+2+3, \quad 8 = 2+2+2+2, \quad \ldots$$

are excluded and you will be forced to name 1.

If you name 3, then for the same reason, 2 is a good reply.

So whoever first names any of 1, 2 and 3 will lose. In particular the first three numbers are bad opening moves. What will you reply if I open with 4? Maybe 5? If so the g.c.d. becomes 1 and there will be only finitely many numbers left. We can find out which by arranging the numbers as in Fig. 2. The circled numbers are excluded because they're multiples of 5 and these exclude the lower numbers by adding 4's. So only 1, 2, 3, 6, 7, 11 remain.

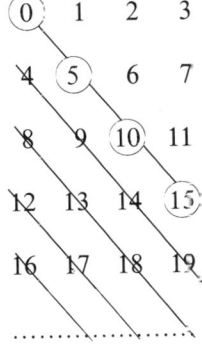

Figure 2. What's Left After {4, 5}.

I won't take 1, 2 or 3. If *I* say 6 or 7, you'll say the other, since these dismiss 11 and leave only 1, 2, 3 for me. So I'll say 11 and make *you* say 6 or 7 instead.

> {4,5,11} is a $\mathcal{P}$-position.

Here's what happens after 4 and 6.

$$\begin{array}{cccc} \textcircled{0} & 1 & 2 & 3 \\ \cancel{4} & 5 & \textcircled{6} & 7 \\ \cancel{8} & 9 & \cancel{10} & 11 \\ \cancel{12} & 13 & \cancel{14} & 15 \\ \cancel{16} & 17 & \cancel{18} & 19 \end{array}$$

. . . .

Since 5 and 7 exclude all large numbers, they kill each other. Similarly for 9 and 11, and for 13 and 15, and so on.

> After $\{4,6\}$ the pairs
> $(2,3), (5,7), (9,11), ..., (4k+1, 4k+3)$,
> for $k \geqslant 1$, are mates.

So if you open with 4, I shall respond with 6; if you open with 6, I shall respond with 4. A few similar strategies are known.

> After $\{8,12\}$ the pairs
> $(2,3), (5,7), (9,11), ..., (4k+1, 4k+3)$
> and
> $(4,6), (10,14), (18,22), ..., (8k+2, 8k+6)$,
> for $k \geqslant 1$, are mates.

There is a slightly more complicated strategy showing that another good reply to 6 is 9.

> After $\{6,9\}$ mate the pairs
> $(4,11), (5,8), (7,10)$ and $(3k+1, 3k+2)$ for $k \geqslant 4$,
> but *then*
> after 4,11 mate 5 with 7
> after 5,8 mate 4 with 7
> after 7,10 mate 4 with 5. 8 with 11.

We have proved that

$$
\begin{array}{c}
\{2,3\}\ \{4,6\}\ \{6,9\}\ \{8,12\} \\
\text{are all } \mathscr{P}\text{-positions,}
\end{array}
$$

and so

$$
\begin{array}{c}
\{1\}\ \{2\}\ \{3\}\ \{4\}\ \{6\}\ \{8\}\ \{9\}\ \{12\} \\
\text{are all } \mathscr{N}\text{-positions.}
\end{array}
$$

The numbers 1,2,3,4,6,8,9 and 12 are the only first moves for which explicit strategies have been found. You might expect that pairs (2,3), $(4k+1, 4k+3)$, (4,6), $(8k+2, 8k+6)$, (8,12), $(16k+4, 16k+12)$ provide a strategy after {16,24} but unfortunately 12 is *not* a legal move from the position {16,24,5,7,8}. On the other hand, for the strategies given above, both members of a pair are legal whenever one is. In fact 8 is a good reply to {16,24,5,7} because it makes 16 and 24 irrelevant and we shall soon see that

$$
\{5,7,8\}\ \text{is a } \mathscr{P}\text{-position.}
$$

We don't know whether 24 is a good reply to 16, nor even whether 16 *has* any good reply.

ARE ALL OPENINGS BAD?

If on observing the fate of 1, 2 and 3 you thought maybe that all openings were bad, then probably our discussions of 4, 6, 8, 9 and 12 have tended to confirm your suspicions. In this section we'll try to analyze 5 and 7. The discussion of possible replies is made a lot easier by the **clique technique**.

You've already seen some cliques: The number 1 forms a rather special clique all by itself; 2 and 3 form another because they exclude all larger numbers. In our discussion of {4,5}, 6 and 7 formed a clique since they excluded 11. Cliques have the property that any reply to a clique member must also be a clique member and these two numbers must together exclude all numbers outside the clique.

We illustrate the clique technique by discussing {6, 7} (Fig. 3).

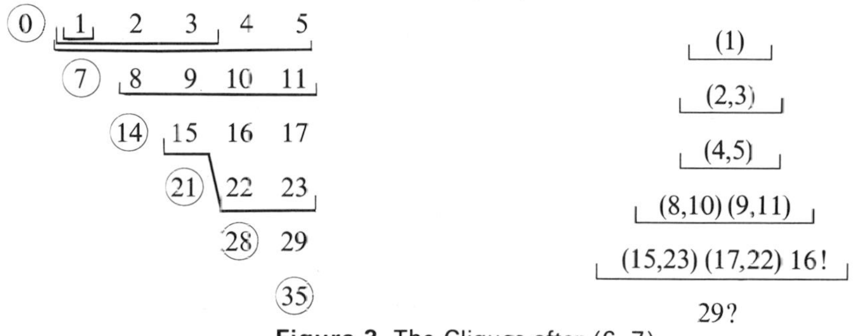

0	1	2	3	4	5
7	8	9	10	11	
14	15	16	17		
21	22	23			
28	29				
35					

(1)
(2,3)
(4,5)
(8,10) (9,11)
(15,23) (17,22) 16!
29?

Figure 3. The Cliques after (6, 7).

As usual, we can disregard 1, 2 and 3 which form the innermost cliques in *every* position. Now 4 and 5 *together* exclude all larger numbers and so form a third clique. *No matter what larger numbers have been named,* 4 will answer 5 and 5 will answer 4. We can therefore afford to neglect them in discussing larger numbers.

Now we assert that 8, 9, 10 and 11 form the next clique, because 8 and 10 together exclude all but 9 and 11, and these together exclude all but 8 and 10. Even when some larger numbers have already been named, 8 will answer 10, 9 will answer 11, and vice versa, and we can dismiss all four from the subsequent discussion.

We now know that any good reply to any of the remaining numbers

$$15 \quad 16 \quad 17$$

$$22 \quad 23$$

$$29$$

must be another of these. We see that 15 answers 23 and vice versa since these leave only 16 and 17. Similarly 17 and 22 are mates. But since 16 excludes *both* 22 and 23, leaving only 15 and 17, it's a good move by itself. These five numbers form a clique, since 29 is always excluded.

> 16 is the unique
> good reply to {6,7}.

Table 6 in the Extras exhibits complete strategies in a similar way for all the positions

$$\{4,5\}, \quad \{4,7\}, \quad \{4,9\},$$
$$\{5,6\}, \quad \{5,7\}, \quad \{5,8\}, \quad \{5,9\},$$
$$\{6,7\}, \quad \{7,8\}, \quad \{7,9\}.$$

In particular it shows that

> $\{4,5,11\}, \quad \{4,7,13\}, \quad \{4,9,19\},$
> $\{5,6,19\}, \quad \{5,7,8\}, \quad \{5,9,31\},$
> $\{6,7,16\}, \quad \{7,9,19\}, \quad \{7,9,24\},$
> are $\mathscr{P}$-positions.

We deduce that any good reply to 5 or 7 must be at least a two-digit number. The smallest two-digit number, 10, isn't a legal answer to 5; is it a good answer to 7? No!

> $\{7,10,12\}$ is a $\mathscr{P}$-position.

This is proved by Fig. 4. Since the clique technique isn't as helpful as it might have been, we've added extra notes for three of the pairs.

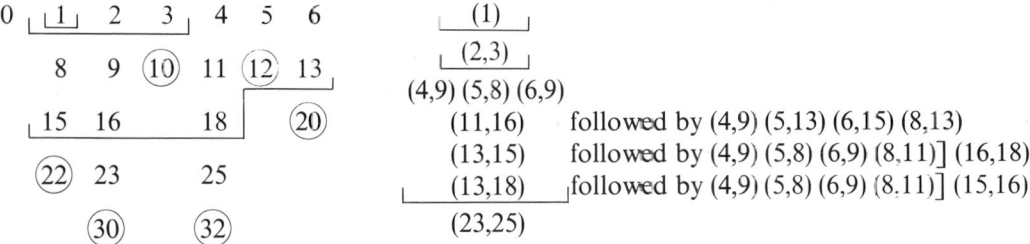

$$
\begin{array}{llllllll}
0 & \lfloor 1 \rfloor & 2 & 3 & 4 & 5 & 6 \\
& 8 & 9 & ⑩ & 11 & ⑫ & 13 \\
& 15 & 16 & & 18 & & ⑳ \\
& ㉒ & 23 & & 25 \\
& & ㉚ & & ㉜
\end{array}
$$

(1)
(2,3)
(4,9) (5,8) (6,9)
(11,16)   followed by (4,9) (5,13) (6,15) (8,13)
(13,15)   followed by (4,9) (5,8) (6,9) (8,11)] (16,18)
(13,18)   followed by (4,9) (5,8) (6,9) (8,11)] (15,16)
(23,25)

Figure 4. The Position {7, 10, 12}.

NOT ALL OPENINGS ARE BAD

R.L. Hutchings has proved that there can't be any good replies to 5 or 7! His main theorem is

> If a and b are coprime $(g = 1)$
> and $\{a,b\} \neq \{2,3\}$, then $\{a,b\}$
> is an $\mathcal{N}$-position.

From this he deduces his **p-theorem**:

> If $p \geqslant 5$ is a prime number,
> $\{p\}$ is a $\mathcal{P}$-position,

p-positions are $\mathcal{P}$-positions.

(For any legal reply produces a position with a g.c.d. of 1.) And from the p-theorem he deduces in turn his **n-theorem**:

> If n is a composite number
> not of the form $2^a 3^b$, then
> $\{n\}$ is an $\mathcal{N}$-position,

n-positions are $\mathcal{N}$-positions.

(Since n has a prime divisor $p \geqslant 5$, which is a good reply.) Together these account for the first few missing numbers:

> $\{5\}, \{7\}, \{11\}, \{13\}, \{17\}, \ldots$ are $\mathcal{P}$-positions.
> $\{10\}, \{14\}, \{15\}, \{20\}, \{21\}, \ldots$ are $\mathcal{N}$-positions.

Our explicit strategies accounted for the eight smallest numbers $2^a 3^b$:

> $\{1\}, \{2\}, \{3\}, \{4\}, \{6\}, \{8\}, \{9\}, \{12\}$
> are $\mathcal{N}$-positions.

But

> Nobody knows about
> $\{16\}, \{18\}, \{24\}, \{27\}, \{32\}, \{36\}, \ldots!$

(We'd be glad to be proved wrong.)

STRATEGY STEALING

Hutchings proves his main theorem by a fine piece of strategy stealing. He considers the topmost number, t, that is not excluded by $\{a,b\}$ and proves that if t is *not* a good reply, then some other number *is*!

We shall call $\{a,b\}$ an **end-position** because, as we'll see in a moment, the topmost number is excluded by every other legal move.

Now let's ask:

> Is t a good reply to $\{a,b\}$?

If the answer is "yes", then $\{a,b\}$ is an $\mathcal{N}$-position.

If the answer is "no", then either the game is over or there is a good reply s to $\{a, b, t\}$. But since a, b and s exclude t, s is itself a good reply to $\{a,b\}$. We can say that the player to move from $\{a,b\}$ finds his strategy by stealing the second player's strategy, if he has one, for $\{a,b,t\}$.

In some cases, e.g. $\{5,9\}$, t (here 31) is a good reply. But in others, e.g. $\{5,7\}$ (where $t = 23$) it *isn't*. The strategy stealing argument only tells us that good moves exist, not what they are. Theft is no substitute for honest toil!

In general,

> An end position with $t > 1$
> is an $\mathcal{N}$-position,

end-positions are $\mathcal{N}$-positions.

But the end-position $\{2,3\}$ is *not* an $\mathcal{N}$-position. This is because $t = 1$ and the only legal move ends the whole game.

Why is $\{a,b\}$ an end-position if its g.c.d. is 1? In Fig. 5 we illustrate with $\{9,11\}$ for which the authors know no good reply.

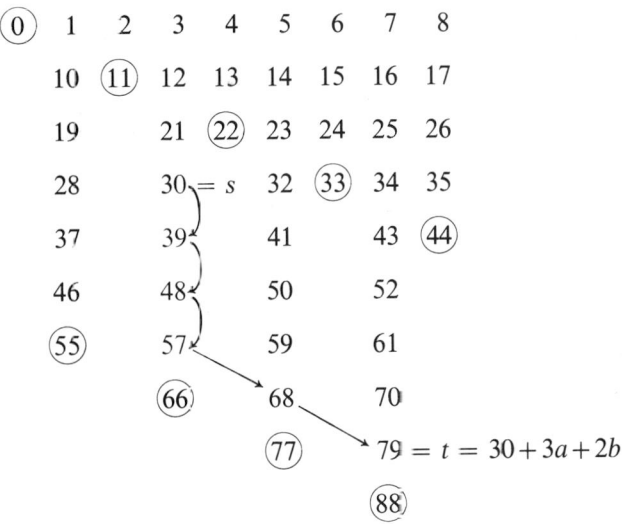

Figure 5. Hutchings's Theorem for $a = 9$, $b = 11$.

Writing the numbers in a columns, as is our wont, we see that in each column the *first excluded* (circled) number is a multiple of b, so the *last included* numbers must differ by multiples of b. Now from any legal move s we can get to the last legal number in its column by adding a's and from this we can get to t by adding b's showing that s excludes t (e.g. $s = 30$ in Fig. 5). The argument also provides a proof of Sylvester's well-known formula

$$t = (a-1)b - a = ab - (a+b).$$

QUIET ENDS

Suppose Hi and Lo have named two coprime numbers a and b and Hi is considering making the move s. Then we know that the topmost number t will be obtainable using sufficiently many coins of values s, a and b. But our argument proved that only *one* copy of the new coin will be needed:

$$t = s + ma + nb.$$

More generally from a position $\{a,b,c,\ldots\}$ we shall say that s **quietly** excludes t if t can be made up using any numbers of $a,b,c,\ldots$ together with *just one copy* of s:

$$t = ma + nb + \ldots + s.$$

A **quiet end-position** is one in which the topmost legal move is *quietly* excluded by every number not already excluded.

> If a is coprime with each of
> b and b_1, then
> $$S = \{a, bc, bd, be,...\}$$
> is a quiet end-position if
> and only if
> $$S_1 = \{a, b_1c, b_1d, b_1e,...\}$$
> is.

THE QUIET END THEOREM

Thus

$$\{7, 1 \times 3, 1 \times 4\},$$

which is really the same position as $\{3,4\}$, is a quiet end-position, so that

$$\{7,9,12\} = \{7, 3 \times 3, 3 \times 4\}$$

and

$$\{7,15,20\} = \{7, 5 \times 3, 5 \times 4\}$$

are. In particular, these are end-positions and so are $\mathcal{N}$-positions by the strategy stealing argument. As usual we aren't told what the good replies are.

We shall use $\{7,9,12\}$ and $\{7,15,20\}$ to illustrate our proof of the quiet end theorem. Once again we write out the numbers in a (here 7) columns and circle the first excluded number in every column (Fig. 6). We assert that these numbers for the positions S and S_1 are in the proportion $b:b_1$ (3:5 in the example; see Fig. 7).

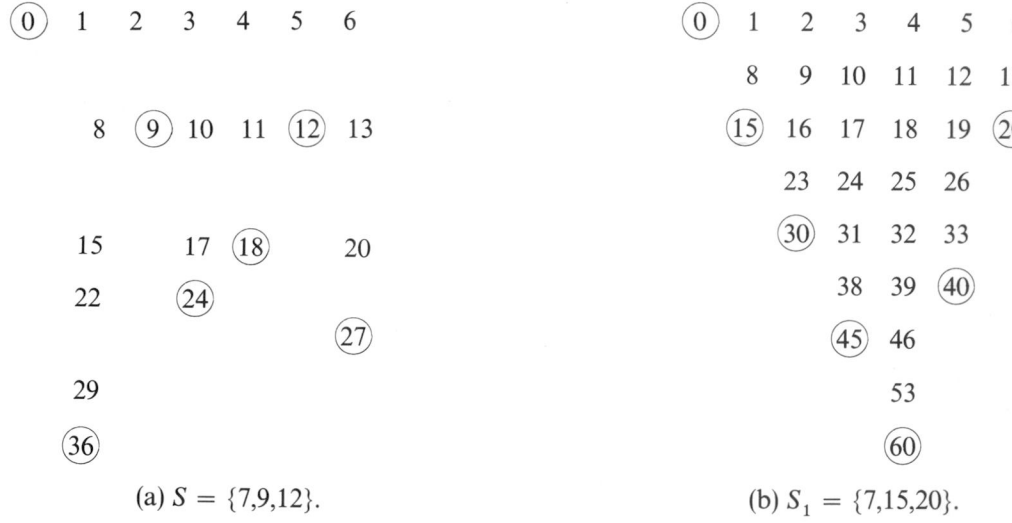

(a) $S = \{7,9,12\}$.   (b) $S_1 = \{7,15,20\}$.

Figure 6. Circled Numbers in Proportion.

$$\begin{array}{ccccccc} & & & & & & \text{in the} \\ 29 & 2 & 17 & 11 & 5 & 20 & \text{proportion} \end{array}$$

$S:$ ⓪ ㊱ ⑨ ㉔ ⑱ ⑫ ㉗   3

to

$$\begin{array}{cccccc} 53 & 8 & 33 & 23 & 13 & 38 \end{array}$$

$S_1:$ ⓪ 60 15 40 30 20 45   5

Figure 7. The Circled Numbers Sorted.

We first see that the circled numbers for S really are multiples of b. Recall that we **circle** n for S if n is excluded by S, but $n-a$ is not. Since n is excluded, it has the form

$$n = ak + bm$$

where m would be excluded by $\{c,d,e,\dots\}$. But if k were positive,

$$n - a = a(k-1) - bm$$

would also be excluded by S, so $k=0$ and we have simply

$$n = bm.$$

Now our assertion is that bm is circled for S only if $b_1 m$ is circled for S_1. Now $b_1 m$ is certainly *excluded* by S_1 and so is circled unless

$$b_1 m - a$$

is also excluded. But then we must have

$$b_1 m - a = ak + b_1 m'$$

for some m' excluded by $\{c,d,e,\dots\}$, and

$$b_1 m = a(k+1) + b_1 m'.$$

showing that b_1 divides $k+1$ since it is coprime with a. We can now divide by b_1 and multiply by b to obtain

$$bm = ak' + bm'$$

for some positive number k', showing that

$$bm - a = a(k'-1) + bm'$$

was excluded, and bm was *not* circled for S.

In its modest way, the quiet end theorem is quite powerful. It often gives the quietus to infinitely many replies with a single blow.

> No odd number is a
> good reply to $\{16,24\}$.

For 1 clearly isn't and if a is any other odd number then $\{a,2,3\}$ is really the same as the quiet end position $\{2,3\}$. By the quiet end theorem $\{a,16,24\}$ is a quiet end-position and so an $\mathcal{N}$-position.

In a similar way it proves that $\{4,6\}$ and $\{6,9\}$ are $\mathcal{P}$-positions without bothering to provide a detailed strategy. Let's use it to discuss the position $\{8,10\}$. After $\{4,5\}$ we found that the only remaining moves were

$$1, \quad 2, \quad 3, \quad 6, \quad 7, \quad 11,$$

so after $\{8,10\}$ the only remaining even numbers will be twice these,

$$2, \quad 4, \quad 6, \quad 12, \quad 14, \quad 22.$$

The quiet end theorem enables us to say that any good reply to $\{8,10\}$ must be in one of these two sets, for otherwise it is an odd number a excluded by $\{4,5\}$ so $\{a,4,5\}$ and therefore $\{a,8,10\}$ will be quiet end-positions. Now,

1	loses instantly,
(2,3)	are mated as usual,
(4,6)	eliminate 8,10 and will mate, as will
(7,11)	(see $\{6,7\}$ in Table 6 in the Extras) and
(12,14)	by our strategy for $\{8,12\}$.

So 22 is the only hope for a good reply to $\{8,10\}$. We shall see later that

$$\boxed{\{8,10,22\} \text{ is a } \mathcal{P}\text{-position.}}$$

DOUBLING AND TRIPLING?

Note that the $\mathcal{P}$-position $\{8,10,22\}$ is the *double* of $\{4,5,11\}$. Our $\{8,12\}$ strategy shows that all $\mathcal{P}$-positions arising in the $\{4,6\}$ strategy have doubles that are also $\mathcal{P}$-positions. Maybe every $\mathcal{P}$-position doubles to another? No! For $\{5,6,19\}$ is $\mathcal{P}$, but $\{10,12,38\}$ is answered by 7 since $\{10,12,38,7\}$ is really the same as $\{7,10,12\}$.

Maybe the *triple* of every $\mathcal{P}$-position is another? No! This time $\{4,5,11\}$ is $\mathcal{P}$, but $\{12,15,33\}$ is answered by 5 since $\{5,12,33\}$ is a $\mathcal{P}$-position, as we'll soon see.

HALVING AND THIRDING?

Nevertheless there are many $\mathcal{P}$-positions whose doubles and triples are still $\mathcal{P}$. We conjecture:

$$\boxed{\text{¿If } \{2a,2b,2c,\ldots\} \text{ is } \mathcal{P} \text{ so is } \{a,b,c,\ldots\}?}$$
and
$$\boxed{\text{¿If } \{3a,3b,3c,\ldots\} \text{ is } \mathcal{P} \text{ so is } \{a,b,c,\ldots\}?}$$

FINDING THE RIGHT COMBINATIONS

How should you start a game of Sylver Coinage? Now that you know so much you will perhaps name 5 for your first move. You now have a strategy for every move I might make and probably feel a little safe. But those stolen strategies are firmly locked inside that little safe you're feeling and more than sensitive fingers are needed to find the right combinations.

You know the first few: 1 needs no reply and you should make the pairs (2,3), (4,11), (6,19), (7,8) and (9,31). Is there any general rule? In trying to answer this question for you we went to a lot of trouble and eventually found a fairly efficient way of breaking open the safe. But the winning combinations it reveals (Fig. 8) suggest that there is no simple answer.

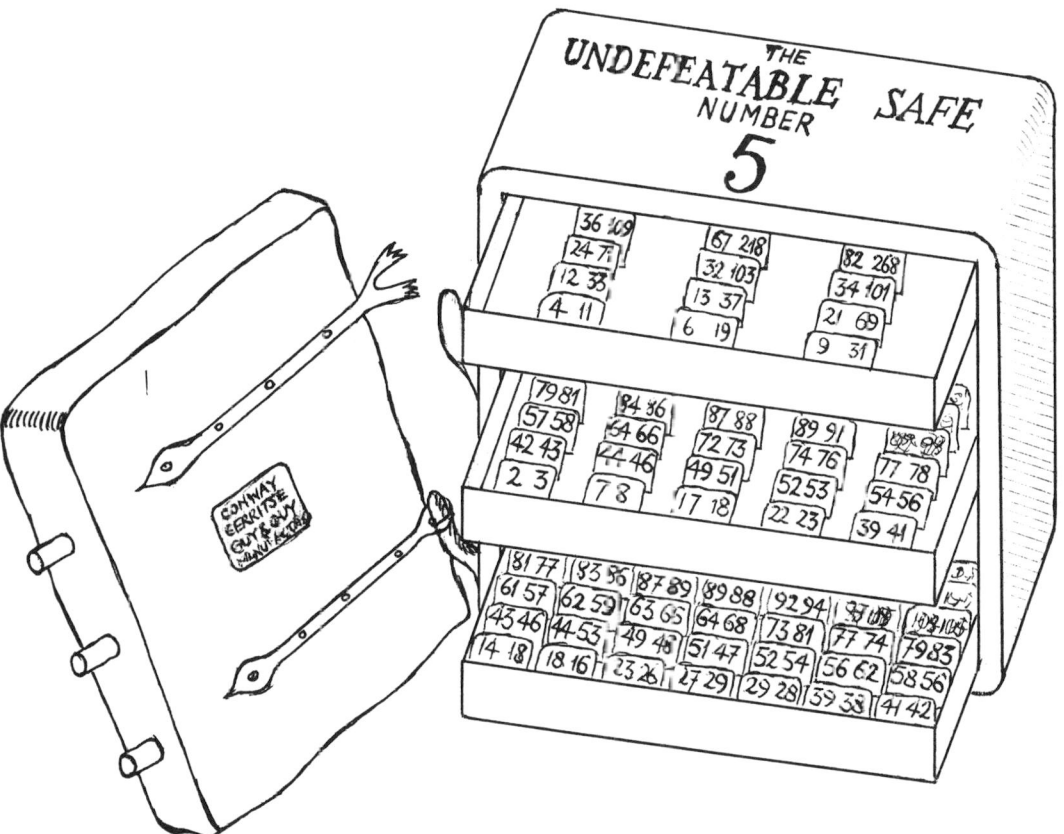

Figure 8. The Stolen Secrets of Safe Number 5.

Let's take a closer look at a position in which 5 and some other numbers have been named. If we were to write the numbers in five columns as usual we would circle 0 and just four other numbers a,b,c,d in the 1-,2-,3-,4-columns respectively, as in Fig. 9. We now make a three-dimensional table of 𝒫-positions using just three of these numbers as headings and the fourth as an entry.

Table 1(a) shows the case in which a is the entry and b,c,d the row, column and layer headings. Tables 1(b,c,d) have b,c,d as entries.

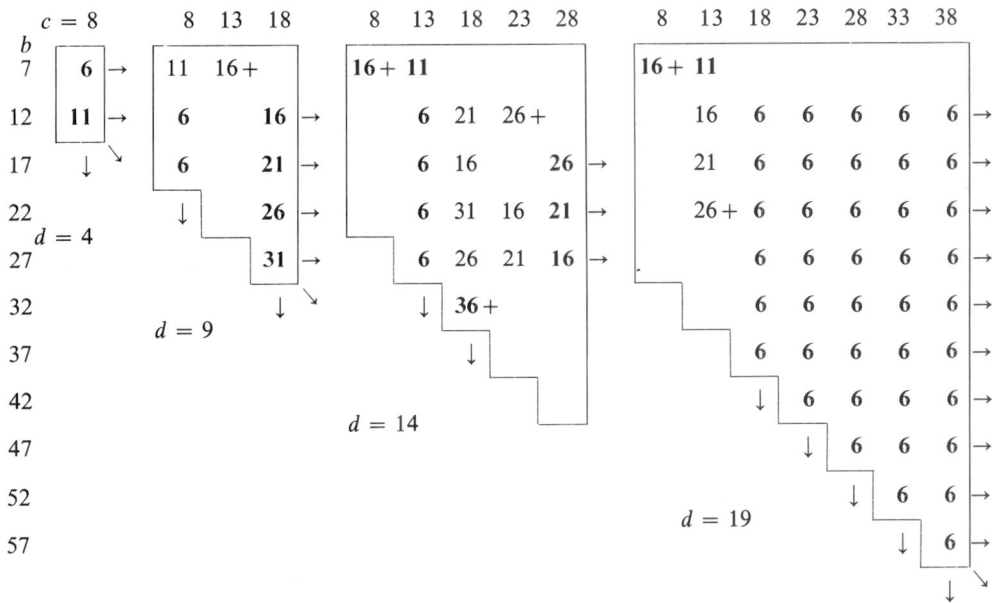

Table 1(a) Entries a for $\mathscr{P}$-positions $\{5,a,b,c,d\}$.

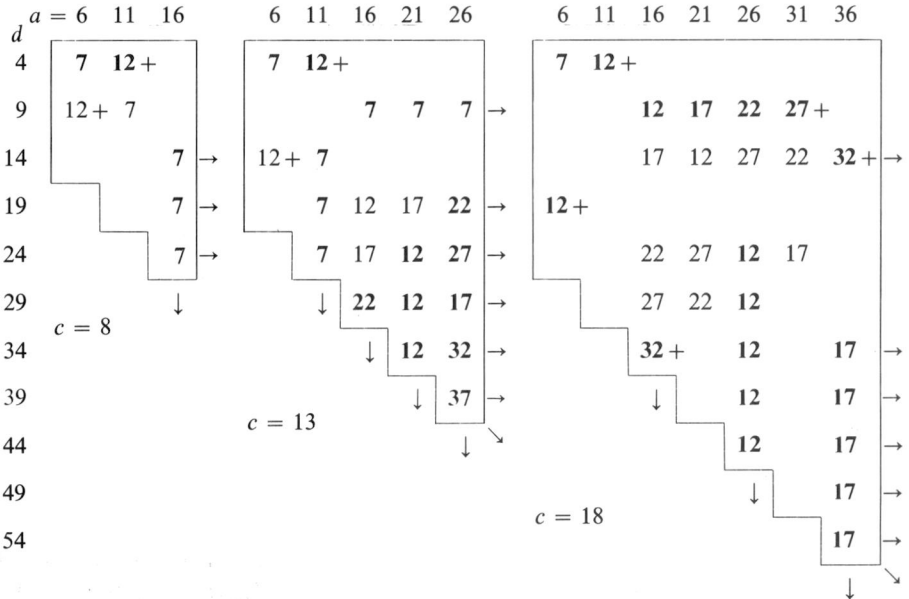

Table 1(b). Entries b for $\mathscr{P}$-positions $\{5,a,b,c,d\}$.

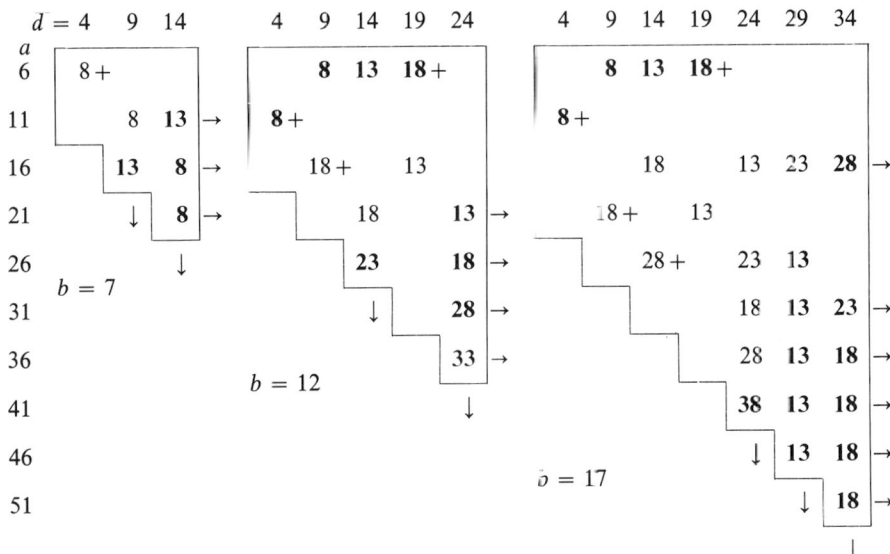

Table 1(c). Entries c for $\mathscr{P}$-positions $\{5,a,b,c,d\}$

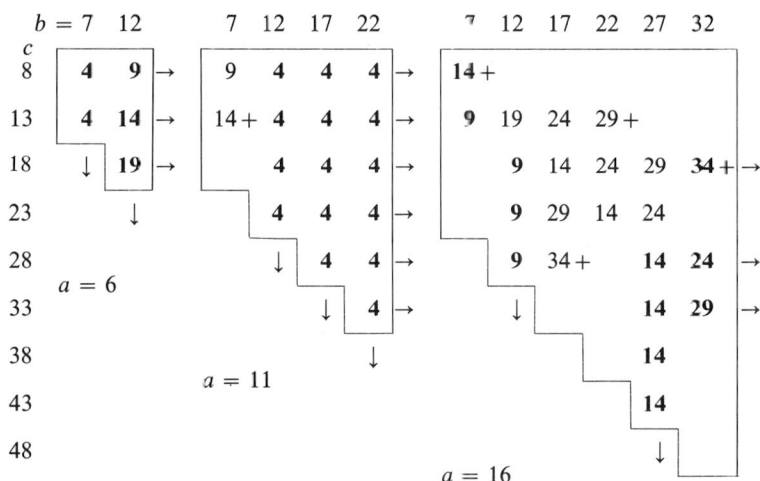

Table 1(d). Entries d for $\mathscr{P}$-positions $\{5,a,b,c,d\}$.

Some positions will appear repeatedly because a heading is redundant. These are indicated by bold figures. For example

$$\{5,6,12,13,14\}, \quad \{5,6,17,13,14\}, \quad \{5,6,22,13,14\}, \quad \ldots$$

are really the same position because $12 = 6+6$ is redundant and so we have a column of 6's in layer 14 of Table 1(a). In $\{5,6,12,18,19\}$ both 12 and 18 are redundant, so the 19 layer of that table is almost entirely made up of 6's.

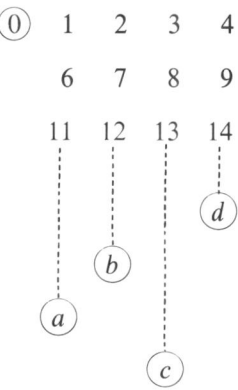

Figure 9. The General Position {5,*x,y,*. . .}.

In {5,16,7,13,9} it is the *entry* 16 = 7+9 that is redundant, so 16 can be replaced by any of

$$21, 26, 31, 36, 41, \ldots$$

and we have written 16+ to indicate this. Really an entry $n+$ is short for infinitely many entries

$$n, n+5, n+10, n+15, n+20, \ldots .$$

The entries in Table 1(a) were computed in lexicographic order, by making due allowance for these repetitions and otherwise entering the least number $5k+1$ not appearing earlier in the same row, column or file.

You'll probably find the method easier to follow in Table 2 which deals with positions {4,*a,b,c*} in a similar way. This time each entry is the smallest number $b = 4k+2$ which has not appeared earlier in its row or column and an entry $b+$, shorthand for

$$b, b+4, b+8, b+12, b+16, \ldots$$

is made when $b=2a$ or $2c$. It can be deduced from the quiet end theorem that other kinds of repetition will not appear.

Table 3 gives pairs $x < y$ for which {4,*x,y*} is already a $\mathscr{P}$-position, extracted from an extended version of Table 2, kindly calculated for us by Richard Gerritse. It seems that the ratio y/x approaches 2·56... .

As soon as 4 or 5 arises in your game you should refer to the appropriate one of these tables. If 6 turns up first, see the corresponding Table 7 in the Extras.

$c =$	7	11	15	19	23	27	31	35	39	43	47	51	55	59	63	67	71	75	79	83	87	91
a																						
5		6	10+																			
9	10	6	14	18+																		
13	14+		6	10																		
17			10	6	14	18	22	26	30	34+												
21			18	14	6	10	26	22	34	30	38	42−										
25			22		10	6	14	18	26		30	34	38	42	46	50+						
29			26		18	14	6	10	22		34	30	42	38	50	46	54	58+				
33			30+		22	26	10	6	14	18												
37					26	22	18	14	6	10	42	38	30	34	54		46	50	58	62	66	70
41					30	34	38	42	10	6	14	18	22	26	58		50	46	54	66	62	74
45					34	30	42	38	18	14	6	10	26	22	62		58	54	46	50	70	66
49					38	42	30	34	46	22	10	6	14	18	26		62		50	54	58	78
53					42	38	34	30	50	26	18	14	6	10	22		66		62	46	54	58
57					46+				38		22	26	10	6	14	18	30	34	42			
61						46	50	54	42		26	22	18	14	6	10	34	30	38	58	74	62
65						50	46	58	54		62		34	30	10	6	14	18	22	26	38	42
69						54+		46			50				18	14	6	10	26	22	30	34
73							54	50	58		46		62	66	30	22	10	6	14	18	26	38
77							58	62	66		54		46	50	34	26	18	14	6	10	22	30
81								62+			58		50	46	38	30	22	26	10	6	14	18
85									66	62	70		54	58	42	34	26	22	18	14	6	10
89									70+		66		58	54		38	42		30	34	10	6
93											70		74	66	62	78	42	38	34	30	18	14
97											74		78	70	82	66	86	38	90	42	34	22
101										78+			74	70				42	66	38	46	26
105											82		78	74	70		90		86	94	42	46
109											86		82	78	74		70		94	90	50	54
113											90		86	94	82		74		70	78	98	50
117											94+		90	86			78		74	70	82	

Table 2. Values of b for which $\{4,a,b,c\}$ is a $\mathscr{P}$-position.

x	y	x	y	x	y	x	y	x	y	x	y	x	y
5	11	107	269	205	531	303	777	405	1043	501	1291	603	1549
7	13	109	279	207	529	305	783	407	1045	503	1289	605	1555
9	19	111	277	211	541	311	797	409	1051	505	1299	607	1557
15	33	113	287	213	547	313	807	411	1053	511	1309	609	1567
17	43	119	301	219	557	315	805	413	1063	513	1319	615	1577
21	51	121	307	221	567	317	819	415	1065	519	1329	617	1583
23	57	123	309	223	569	321	823	419	1077	521	1339	621	1595
25	67	125	319	227	585	323	829	421	1079	523	1341	623	1593
27	69	129	331	229	583	325	839	425	1095	525	1347	625	1603
29	75	131	333	231	593	327	841	427	1097	527	1349	627	1609
31	81	133	343	233	595	329	851	429	1103	533	1371	633	1627
35	89	135	345	237	611	335	857	431	1105	535	1373	635	1625
37	95	141	363	239	613	337	871	433	1115	537	1379	637	1635
39	101	143	365	241	619	339	869	439	1125	539	1381	639	1637
41	103	147	373	245	631	341	879	441	1135	543	1393	643	1649
45	115	149	379	247	629	347	885	447	1145	545	1395	645	1655
47	117	151	385	251	641	349	899	449	1151	549	1411	651	1665
49	127	153	391	253	647	351	901	451	1157	551	1413	653	1679
53	139	155	397	255	649	353	911	455	1165	553	1423	655	1681
55	137	157	399	257	659	355	909	457	1171	555	1425	657	1687
59	145	163	417	259	665	357	919	459	1177	559	1433	661	1699
61	159	165	423	261	671	359	917	461	1183	561	1443	663	1697
63	161	169	435	265	683	361	927	463	1189	563	1445	667	1709
65	167	171	437	267	685	367	941	465	1195	565	1451	669	1719
71	177	173	443	271	697	369	951	469	1207	571	1465	673	1731
73	183	175	445	273	699	371	953	471	1209	573	1471	675	1729
77	195	179	453	275	705	375	961	473	1215	575	1477	677	1739
79	193	181	467	281	723	377	971	479	1225	577	1483	679	1741
83	209	185	475	283	725	381	983	481	1235	579	1485	681	1751
85	215	187	477	285	731	383	981	485	1247	581	1495	687	1757
87	217	189	483	289	743	387	993	487	1245	587	1505	689	1771
91	225	191	489	291	745	389	1003	491	1257	589	1511	691	1769
93	235	197	507	293	755	393	1011	493	1267	591	1517	693	1783
97	243	199	509	295	757	395	1013	495	1269	597	1531	695	1781
99	249	201	515	297	767	401	1031	497	1279	599	1537	701	1803
105	263	203	517	299	765	403	1029	499	1277	601	1543	703	1801
												707	1813

Table 3. Pairs x,y for which $\{4,x,y\}$ is a $\mathcal{P}$-position.

WHAT SHALL I DO WHEN g IS TWO?

Example: $\{8,10,22\}$

Apparently we have to examine infinitely many possible replies. Fortunately there is a way of doing this in a finite time. A similar method will work for *any* position with $g = 2$.

Let's see how the position will look after some play from $\{8,10,22\}$. If 1, 2 or 3 has been played, we know what to do. Otherwise the only even numbers that can have been played are

4, 6, 12, 14

and the even part of the position must look like one of

$$\{4,6\} \; \{4,10\} \; \{6,8,10\}$$
$$\{8,10,12,14\} \; \{8,10,12\} \; \{8,10,14\} \; \{8,10,22\}$$

    What odd numbers have been played? If the least of them is n, then since $\{8,10,22\}$ excludes all of

$$16, 18, 20, 22, 24, \ldots$$

we can suppose that the only relevant odd numbers are among

$$n, n+2, n+4, n+6, n+12, n+14.$$

And if any even moves have been made they will restrict the possibilities still further. For instance if 6 has been played we can suppose that the odd numbers are one of the sets

$$n, n+2, n+4 \qquad n, n+2 \qquad n, n+4 \qquad n$$

Table 4 shows the status of the positions classified in this way. Since the last four columns repeat indefinitely, this finite table contains the information for every odd n. How was it computed and why is it periodic?

    Let's take a typical entry:

$$\{8,10,14,n,n+6,n+12\}.$$

From this position there are three kinds of option:

  (a) the *even* numbers 4, 6 or 12,
  (b) the *small odd* numbers $m \leqslant n-14$,
  (c) the *large odd* numbers $n-12, n-10, n-8, n-6, n-4, n-2, n+2, n+4$.

    Case (a) leads to a position (in an earlier segment of the table) with even part

$$\{4,10\}, \; \{6,8,10\} \text{ or } \{8,10,12,14\}$$

and we can suppose that these have already been analyzed and found to be ultimately periodic in n.

    A case (b) move leads to

$$\{8,10,14,m\}$$

since m excludes n and all larger odd numbers. If there is any odd m for which this is a $\mathscr{P}$-position, then $\{8, 10, 14, n, n+6, n-12\}$ will be an $\mathscr{N}$-position for all $n \geqslant m+14$. If not, we can reject moves in case (b).

    Finally, case (c) moves either leave n unchanged or decrease it by at most 12. We conclude that the outcome of every position in the table is computed in a fixed way from

> ultimately periodic information (case (a)),
> ultimately constant information (case (b)), and
> information in the last few columns (case (c));

it must therefore be ultimately periodic in n.

Table values use columns headed by odd numbers 5–51. "P" = P-position, "–" = blank.

Set	Row	5	7	9	11	13	15	17	19	21	23	25	27	29	31	33	35	37	39	41	43	45	47	49	51
{4,6} (P)	$n, n+2$	P	–	P	–	P	–	P	–	P	–	P	–	P	–	P	–	P	–	P	–	P	–	P	–
	n	–	–	–	–	–	–	–	–	–	–	–	–	–	–	–	–	–	–	–	–	–	–	–	–
{4,10} (N)	$n, n+2$	–	P	–	–	–	P	–	–	–	P	–	–	–	P	–	–	–	P	–	–	–	P	–	–
	$n, n+6$	P	–	–	–	P	–	–	–	P	–	–	–	P	–	–	–	P	–	–	–	P	–	–	–
	n	–	–	–	–	–	–	–	–	–	–	–	–	–	–	–	–	–	–	–	–	–	–	–	–
{6,8,10} (N)	$n, n+2, n+4$	–	P	–	P	–	–	–	–	–	–	–	–	–	–	–	–	–	–	–	–	–	–	–	–
	$n, n+2$	–	–	–	–	–	–	–	–	–	–	–	–	–	–	–	–	–	–	–	–	–	–	–	–
	$n, n+4$	P	–	–	–	–	–	–	–	–	–	–	–	–	–	–	–	–	–	–	–	–	–	–	–
	n	–	P	–	P	–	–	–	–	–	–	–	–	–	–	–	–	–	–	–	–	–	–	–	–
{8,10,12,14} (P)	$n, n+2, n+4, n+6$	P	–	P	–	P	–	P	–	P	–	P	–	P	–	P	–	P	–	P	–	P	–	P	–
	$n, n+2, n+4$	–	–	–	–	–	–	–	–	–	–	–	–	–	–	–	–	–	–	–	–	–	–	–	–
	$n, n+2, n+6$	–	–	–	–	–	–	–	P	–	–	–	P	–	–	–	P	–	–	–	P	–	–	–	P
	$n, n+2$	P	–	P	–	P	–	P	–	P	–	P	–	P	–	P	–	P	–	P	–	P	–	P	–
	$n, n+4, n+6$	–	P	–	P	–	–	–	P	–	–	–	P	–	–	–	P	–	–	–	P	–	–	–	P
	$n, n+4$	–	–	P	–	–	–	P	–	–	–	P	–	–	–	P	–	–	–	P	–	–	–	P	–
	$n, n+6$	–	–	P	–	–	–	P	–	–	–	P	–	–	–	P	–	–	–	P	–	–	–	P	–
	n	–	–	–	–	–	–	–	–	–	–	–	–	–	–	–	–	–	–	–	–	–	–	–	–
{8,10,12} (N)	$n, n+2, n+4, n+6$	P	–	–	–	–	–	–	–	–	–	–	–	–	–	–	–	P	–	–	P	–	–	–	P
	$n, n+2, n+4$	–	–	P	–	–	–	P	–	–	–	–	–	–	–	–	–	–	–	–	–	–	–	–	–
	$n, n+2, n+6$	–	–	P	–	–	–	–	–	–	–	–	–	–	–	–	–	P	–	–	P	–	–	–	–
	$n, n+2$	P	–	–	–	–	–	–	–	–	–	–	–	–	–	–	–	P	–	–	P	–	–	–	P
	$n, n+4, n+6$	–	P	P	–	–	–	P	–	–	–	–	–	–	–	–	–	–	–	–	–	–	–	–	–
	$n, n+4$	–	–	–	P	–	–	P	–	–	P	–	–	–	P	P	–	–	P	P	–	–	–	–	P
	$n, n+6$	–	–	–	–	–	–	–	P	–	–	P	–	–	–	P	–	–	–	P	–	–	–	P	–
	$n, n+14$	–	–	P	–	–	–	P	–	–	–	–	–	–	–	–	–	–	–	–	–	–	–	–	–
	n	–	–	–	–	–	–	–	–	–	–	–	–	–	–	–	–	–	–	–	–	–	–	–	–
{8,10,14} (N)	$n, n+2, n+4, n+6$	P	–	–	–	–	–	–	–	–	–	–	–	–	–	–	–	P	–	–	–	–	–	–	–
	$n, n+2, n+4$	–	–	P	–	–	–	–	–	–	–	–	–	P	–	P	–	–	–	–	–	–	–	–	–
	$n, n+2, n+6$	–	–	P	–	–	–	–	–	–	–	–	–	P	–	P	–	–	–	–	–	–	–	–	–
	$n, n+2$	–	–	–	–	–	–	–	–	–	–	–	–	–	–	–	–	–	–	–	–	–	–	–	–
	$n, n+4, n+6$	–	–	P	–	P	–	–	–	–	–	–	–	–	–	–	–	P	–	P	–	–	–	–	–
	$n, n+4$	–	P	–	–	–	–	–	P	–	–	–	–	–	–	–	–	–	–	–	–	P	–	–	–
	$n, n+6, n+12$	–	–	–	–	–	–	–	P	–	–	–	–	–	P	–	P	–	–	–	–	–	–	–	–
	$n, n+6$	–	–	–	–	–	–	–	–	–	–	–	–	–	–	–	–	–	–	–	–	–	–	–	–
	$n, n+12$	P	–	P	–	P	P	–	–	–	–	–	–	–	P	–	P	P	–	–	–	–	–	–	–
	n	–	–	–	–	–	–	–	–	–	–	–	–	–	P	–	P	–	–	–	–	–	–	–	–
{8,10,22} (P)	$n, n+2, n+4, n+6$	P	–	P	–	P	–	P	–	P	–	P	–	P	–	P	–	P	–	P	–	P	–	P	–
	$n, n+2, n+4$	–	–	–	–	–	–	–	–	–	–	–	–	–	–	–	–	–	–	–	–	–	–	–	–
	$n, n+2, n+6$	–	–	–	P	–	–	–	–	–	–	–	P	–	–	–	P	–	–	–	P	–	–	–	P
	$n, n+2, n+14$	–	–	P	–	P	–	P	–	P	–	P	–	P	–	P	–	P	–	P	–	P	–	P	–
	$n, n+2$	P	–	–	–	–	–	–	–	–	–	–	–	–	–	–	–	–	–	–	–	–	–	–	–
	$n, n+4, n+6$	–	–	–	–	–	–	–	–	–	–	–	–	–	–	–	–	–	–	–	–	–	–	–	–
	$n, n+4$	–	P	P	–	–	P	P	–	–	P	P	–	–	P	P	–	–	P	P	–	–	P	P	–
	$n, n+6, n+12$	–	–	P	–	–	–	–	–	–	–	P	–	–	–	P	–	–	–	P	–	–	–	P	–
	$n, n+6$	–	–	–	–	–	–	–	–	–	–	–	–	–	–	–	–	–	–	–	–	–	–	–	–
	$n, n+12, n+14$	–	–	–	–	–	–	–	–	–	–	–	–	–	–	–	–	–	–	–	–	–	–	–	–
	$n, n+12$	–	–	–	–	–	–	–	–	–	–	–	–	–	–	–	–	–	–	–	–	–	–	–	–
	$n, n+14$	P	–	–	–	–	–	–	–	–	–	–	–	–	–	–	–	–	–	–	–	–	–	–	–
	n	–	–	–	–	–	–	–	–	–	–	–	–	–	–	–	–	–	–	–	–	–	–	–	–

Table 4. The Position {8,10,22}.

Every position with $g = 2$ can be handled in this way. When we have computed enough to verify the period, we can decide in particular whether there is any good reply. For $\{8,10,22\}$ there isn't one, so it is a $\mathscr{P}$-position.

THE GREAT UNKNOWN

We can best describe our knowledge in terms of the number g. When

$$g = 1$$

the position is bounded so you can find what to do by working through all positions. Of course this might take a long time even if one of our theorems already tells you the outcome. We know that there must a be good reply to $\{31,37\}$ but don't know any method which guarantees to find one in the next millenium. When

$$g = 2$$

the method we have just described will compute the outcome in a finite but probably even longer time. If

$$g \text{ is divisible by a prime } p \geqslant 5$$

then p is a good reply when it hasn't already been named, when of course there isn't any.

The authors have only been able to examine a few particular positions with other values of g. Table 8 in the Extras contains a complete discussion of $\{6,9\}$. Although this is a two-dimensional table, a periodicity develops which enables us to analyze the position to infinity. Maybe a similar thing happens for some other positions with $g = 3$. We computed a much larger three-dimensional table for $\{8, 12\}$ ($g = 4$), but could detect no structure outside the range covered by our explicit strategy.

16 is the first opening move whose status is in doubt. We don't know whether $\{16\}$ has a good reply nor even any way of finding out in any finite time. You might consider working upwards testing each possible reply in turn and hoping to detect some structure, but even this is impossible. We don't know any way to test the reply 24, say, in any finite time. We don't even know how to test 100, say, as a possible reply to $\{16,24\}$!

The quiet end theorem often eliminates infinitely many replies, for example all odd replies to $\{16\}$ or to $\{16,24\}$, but it never eliminates any reply that would be infinitely hard to analyze.

	{7,9,11}	{7,9}	{7,11}	{7}	{9,11}	{9}	{11}	{}
{6,8,10}	[]	[11]	[11]	[]	[4,5,7]	[5]	[]	[4,7,11]
{6,8}	[10]	[]	[]	[9,10,11]	[4,5]	[5,7]	[7,10]	[4]
{6,10}	[8]	[]	[15]	[8,9]	[4]	[7]	[8,13]	[4]
{6}	[]	[8,10,11]	[8,9]	[16]	[4,7]	[]	[26]	[4,9]
{8,10,12}	[4,5,6]	[4]	[13]	[5,6]	[13,14,15]	[23]	[6]	[14]
{8,12}	[5]	[6]	[6]	[5]	[]	[11,15]	[9,13]	[]
{10,12}	[4]	[4,6]	[16]	[]	[]	[11,13,14]	[9	[7
{12}	[6]	[15]	[27]	[10	[8,10]	[6]	[	[8
{8,10}	[4,5,6]	[4]	[]	[5,6,11]	[23]	[13,15]	[6,7]	[22]
{8}	[5]	[6]	[6,10]	[5]	[12]	[21	[	[12
{10}	[4]	[4,6]	[8]	[12]	[12]	[	[	[5
{}	[6]	[19,24]	[24,34]	[]	[	[6]	[]	[5,7,11,13,...

Table 5. Status of Subsets of {6,7,8,9,10,11,12} and Known Good Replies.

Even members of set at Left; Odd members at head. Bracket is closed when *all* good replies are known, so that [] indicates a $\mathscr{P}$-position, but [always indicates an $\mathscr{N}$-position for which no good reply is known. The last entry contains all primes greater than 3; and *may* contain some entries 2^a3^b.

Table 5 tells you the outcome and all the good replies we know to every position made from the numbers

$$6, 7, 8, 9, 10, 11, 12$$

(if 4 or 5 is involved, Tables 2, 3, 1 and Fig. 8 go much further). If you can add any more to this table or decide whether any number 2^a3^b is a good opening move we would like to hear from you.

ARE OUTCOMES COMPUTABLE?

We can prove that there *must be* a way of programming a computer to find the outcome of {n} even though we don't know what that way *is*! The reason is:

> There can only be finitely many good opening moves 2^a3^b.

For no one of these can divide any other, so that no two can have the same value of a or the same value of b. So if $2^{a_0}3^{b_0}$ is such a number with a_0 as small as possible, and 2^a3^b is any other, then we must have $b < b_0$ and so there are at most $b_0 + 1$ such numbers, say $n_1, n_2, \ldots, n_k$. We suspect there are none!

If you only knew what these numbers were, then you could program your machine with PORN (Fig. 10) and work out the outcome of any {n}. This argument shows that in the purely technical sense this is a computable function of n, even though we don't know what function it is.

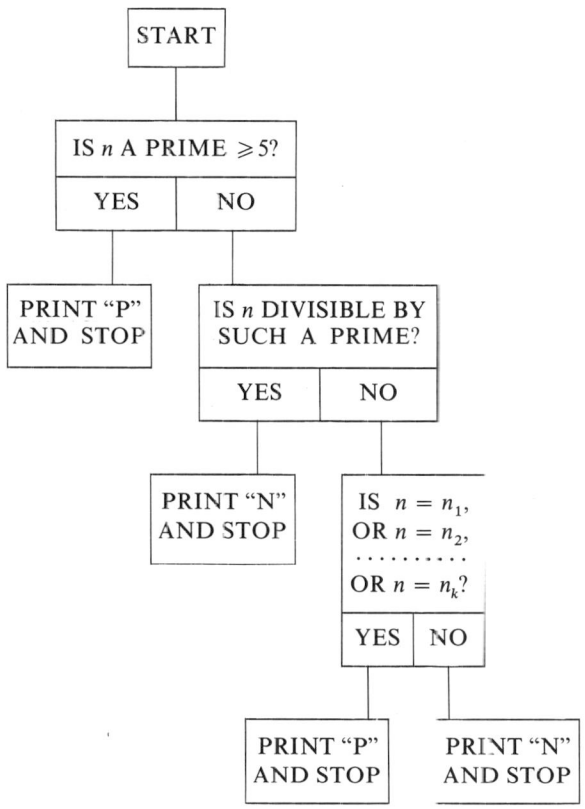

Figure 10. PORN, A Program Which Decides if $\{n\}$ is $\mathscr{P}$ or $\mathscr{N}$.

THE ETIQUETTE OF SYLVER COINAGE

Few Western readers can understand the subtleties of etiquette in the oriental country from which our game comes. But at least we can save you from the more obvious gaffes by pointing out that in Sylver Coinage it is customary for a player who knows he is winning to resign by naming 1, 2 or 3. This quaint custom is said to originate in the tradition that Hi, who could see much further than Lo, nobly took upon himself the fate that was about to befall his beloved brother.

When it's plain to all the world that you have a win, any move but 1 will insult your opponent, but in other cases we advise you to name 3 (2 is possible, but may be misunderstood). If your opponent concurs in your analysis, he will respond with 2, but you have allowed him to express another opinion by naming 1. (Replies to 3 other than 1 or 2 may also be available but their nuances are harder to interpret.)

Of course, one of the greatest insults you can offer is to name 1, 2 or 3 at the very start of the game, for this is the philosopher Hu Tchings' prerogative, at least until someone finds a new way to win.

EXTRAS

CHOMP

Here is a game with similar rules to Sylver Coinage. For some fixed number N, the players alternately name divisors of N which may not be multiples of previously named numbers. Whoever names 1 loses. If $N=432=2^43^3$, for example, a move is essentially to eat a square (e.g. 36) from the chocolate bar in Fig. 11, together with all squares below and/or to the right of it. Square number 1 is poisoned!

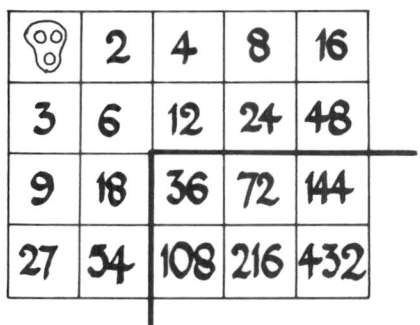

Figure 11. Chomping at a Chocolate Bar.

The first few $\mathscr{P}$-positions are shown in Fig. 12. Strategy stealing shows that rectangles larger than 1×1 are $\mathscr{N}$-positions; the replies are unique if either side is at most 3, but Ken Thompson found that 4×5 and 5×2 bites both answer 8×10.

The arithmetic form of the game is due to Fred. Schuh, the geometric one to David Gale.

ZIG-ZAG

Two players alternately name distinct numbers (which are allowed to be fractional or negative) and the game ends as soon as the resulting sequence contains either an increasing subsequence (**zig**) of length a or a decreasing one (**zag**) of length b. The normal play $a+1$, $b+1$ game is really the same as the misère a, b game, and so we consider only the latter.

Zig-Zag, which was suggested to us by S. Fajtlowicz, sounds difficult to analyze, but fortunately there is a rather clever transformation into a geometrical game like Chomp. We regard square (r, s) in Fig. 13 as eaten if the number sequence so far contains a rising zig of length r and a sagging zag of length s that end with the same number. Then the moves are as in Chomp except that the first move may eat square (1,1) only, and the innermost square eaten on any subsequent move must be adjacent to a previously eaten square. The squares $(a,1)$ and $(1,b)$ are poisoned, so play really goes on inside the outlined $a-1$ by $b-1$ **chocolate bar** of Fig. 13.

598

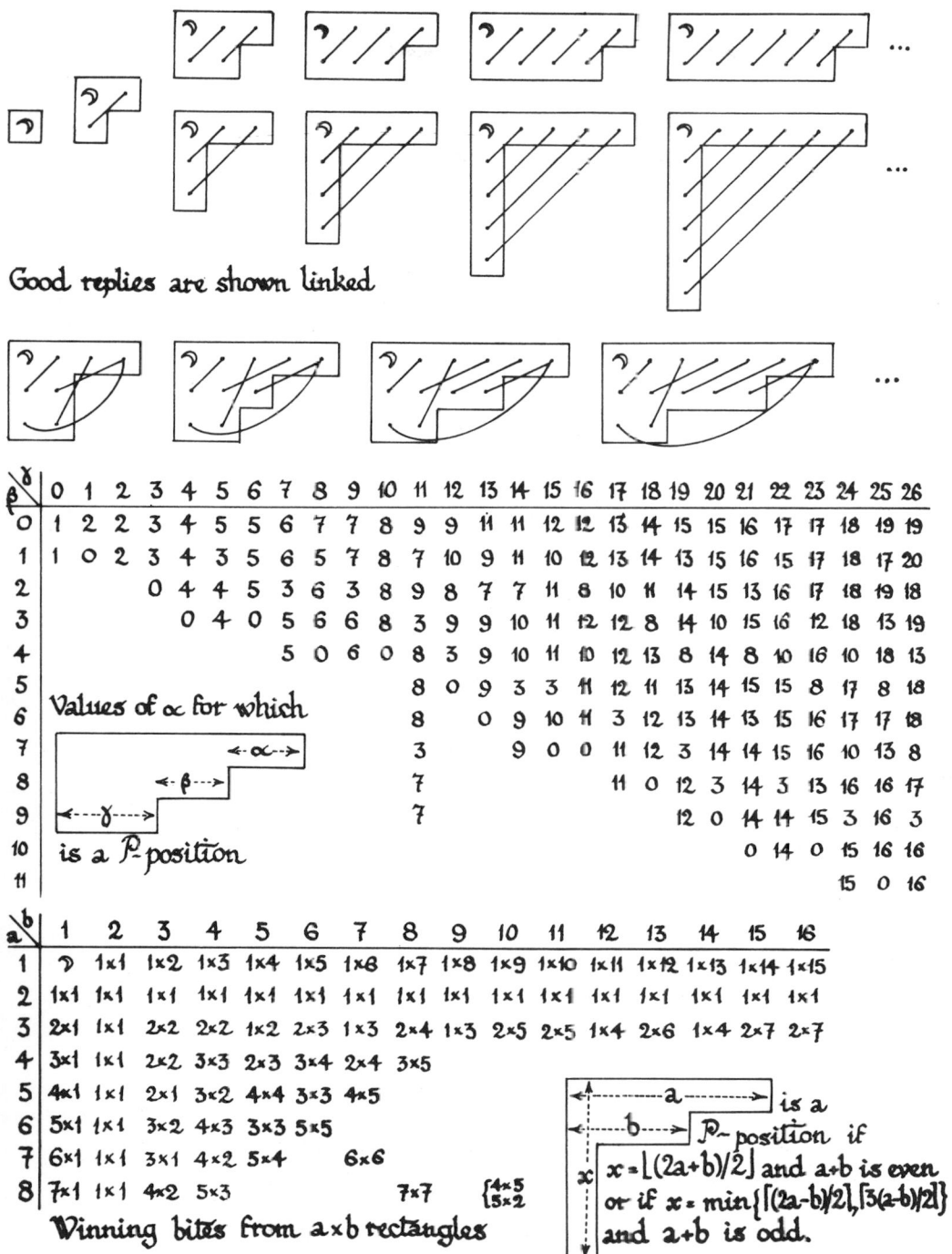

Good replies are shown linked

γ β	0	1	2	3	4	5	6	7	8	9	10	11	12	13	14	15	16	17	18	19	20	21	22	23	24	25	26
0	1	2	2	3	4	5	5	6	7	7	8	9	9	11	11	12	12	13	14	15	15	16	17	17	18	19	19
1	1	0	2	3	4	3	5	6	5	7	8	7	10	9	11	10	12	13	14	13	15	16	15	17	18	17	20
2			0	4	4	5	3	6	3	8	9	8	7	7	11	8	10	11	14	15	13	16	17	18	19	18	
3				0	4	0	5	6	6	8	3	9	9	10	11	12	12	8	14	10	15	16	12	18	13	19	
4					5	0	6	0	8	3	9	10	11	10	12	13	8	14	8	10	16	10	18	13			
5						8	0	9	3	3	11	12	11	13	14	15	15	8	17	8	18						
6							8	0	9	10	11	3	12	13	14	13	15	16	17	17	18						
7							3		9	0	0	11	12	3	14	14	15	16	10	13	8						
8							7				11	0	12	3	14	3	13	16	16	17							
9							7					12	0	14	14	15	3	16	3								
10													0	14	0	15	16	16									
11														15	0	16											

Values of α for which ⌐ is a P-position

b a	1	2	3	4	5	6	7	8	9	10	11	12	13	14	15	16
1	⌐	1×1	1×2	1×3	1×4	1×5	1×6	1×7	1×8	1×9	1×10	1×11	1×12	1×13	1×14	1×15
2	1×1	1×1	1×1	1×1	1×1	1×1	1×1	1×1	1×1	1×1	1×1	1×1	1×1	1×1	1×1	1×1
3	2×1	1×1	2×2	2×2	1×2	2×3	1×3	2×4	1×3	2×5	2×5	1×4	2×6	1×4	2×7	2×7
4	3×1	1×1	2×2	3×3	2×3	3×4	2×4	3×5								
5	4×1	1×1	2×1	3×2	4×4	3×3	4×5									
6	5×1	1×1	3×2	4×3	3×3	5×5										
7	6×1	1×1	3×1	4×2	5×4		6×6									
8	7×1	1×1	4×2	5×3				7×7	{4×5, 5×2}							

Winning bites from a×b rectangles

⌐ is a P-position if $x = \lfloor (2a+b)/2 \rfloor$ and $a+b$ is even or if $x = \min\{\lceil (2a-b)/2 \rceil, \lceil 3(a-b)/2 \rceil\}$ and $a+b$ is odd.

Figure 12. 𝒫-positions in Chomp.

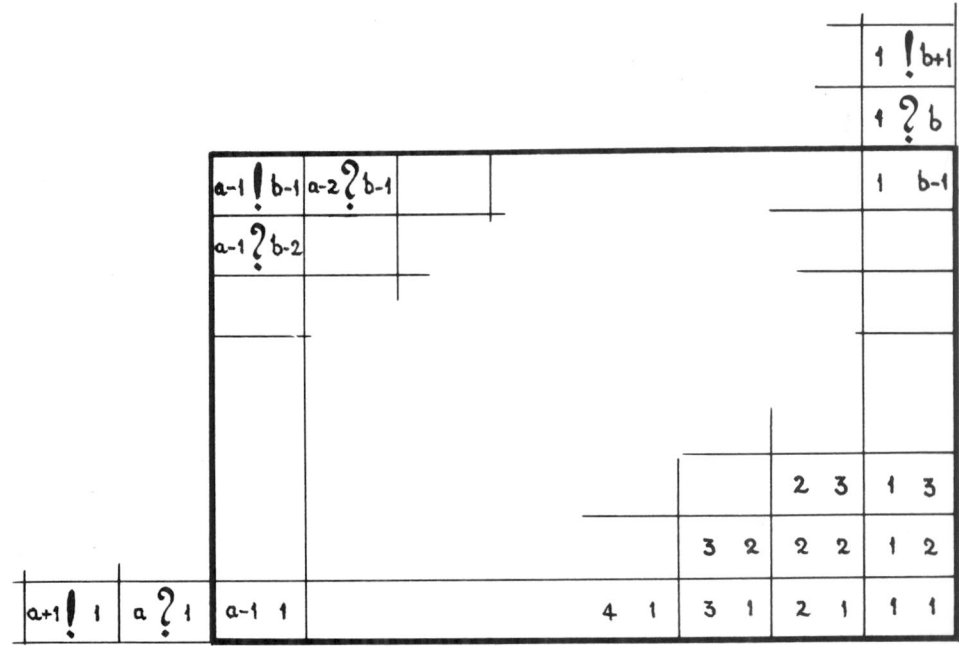

Figure 13. The Chocolate Bar Form of Zig-Zag.

If $a \geqslant 3$, $b \geqslant 3$ and $a+b \leqslant 17$, the first player wins the misère a-Zig, b-Zag game, because David Seal's calculations show that the corresponding $a-1$ by $b-1$ chocolate bars are $\mathcal{N}$-positions.

By assigning Heads to Horizontal edges and Tails to verTical ones we get an equivalent game with coin sequences, involving moves of a head rightwards over tails or a tail leftwards over heads, and Seal used this idea to compute Fig. 14 showing all $\mathcal{P}$-positions for which the uneaten part of the chocolate bar fits inside a 5×5 square.

To find $\mathcal{P}$-positions in both Chomp and Zig-Zag, we used the tabular technique of Chapter 15, and the Clique Technique of this one.

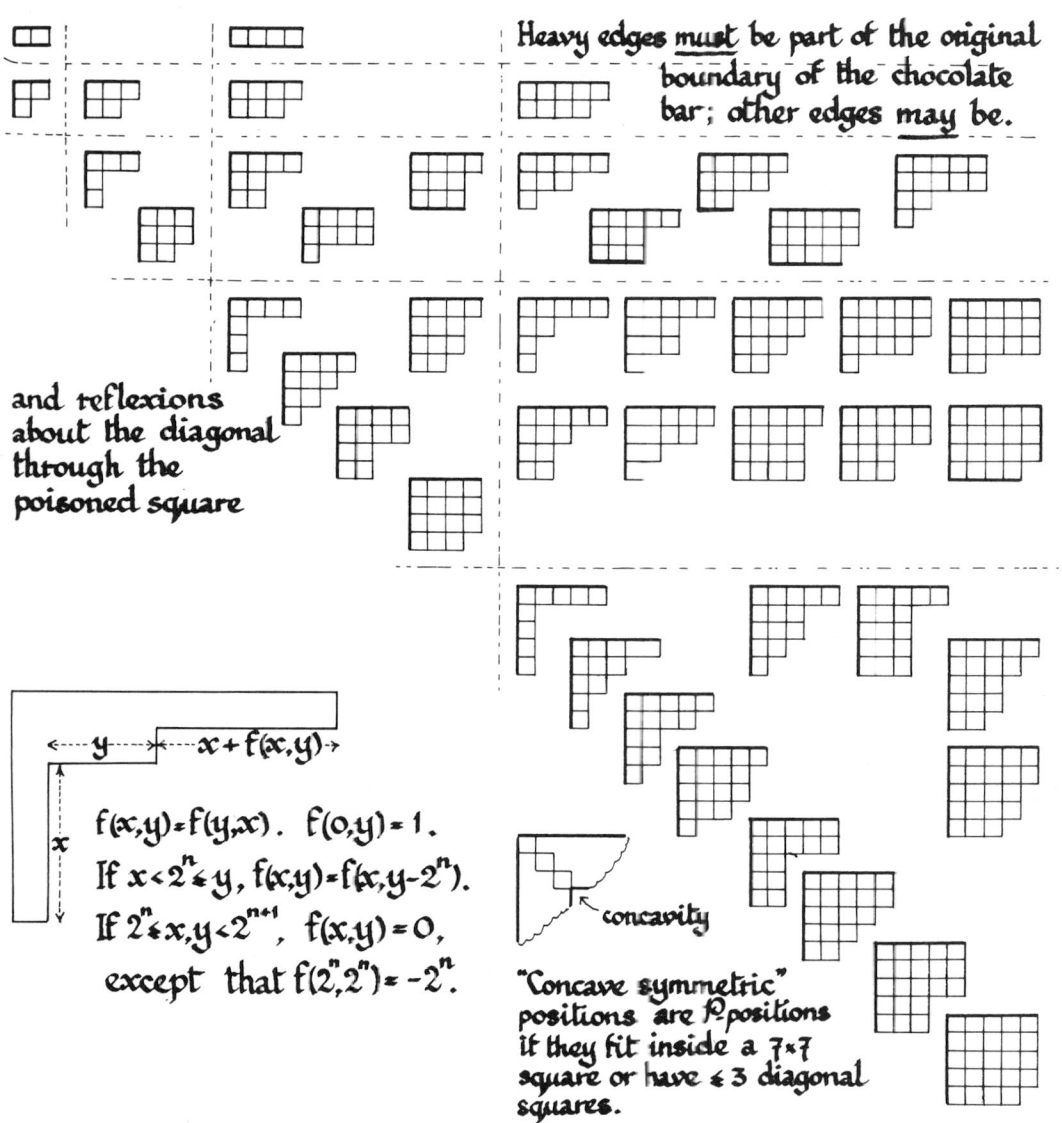

Heavy edges <u>must</u> be part of the original boundary of the chocolate bar; other edges <u>may</u> be.

and reflexions about the diagonal through the poisoned square

$f(x,y) = f(y,x)$. $f(0,y) = 1$.
If $x < 2^n \leq y$, $f(x,y) = f(x, y - 2^n)$.
If $2^n \leq x, y < 2^{n+1}$, $f(x,y) = 0$,
except that $f(2^n, 2^n) = -2^n$.

concavity

"Concave symmetric" positions are $\mathscr{P}$-positions if they fit inside a 7×7 square or have ≤ 3 diagonal squares.

Figure 14. $\mathscr{P}$-positions for Zig-Zag.

MORE CLIQUES FOR SYLVER COINAGE

To follow the cliques in Table 6, we advise you to set out the remaining numbers as we did in Figs. 2, 3, 4 for the cases {4,5}, {6,7} and {7,10,12}. Numbers not mentioned are excluded by a good reply.

position	replies	strategy, with cliques indicated by]
{4,5}	11!	1?](2,3)](6,7)]11!
{4,7}	13!	1?](2,3)](5,6)](9,10)]13!
{4,9}	19!	1?](2,3)](5,11)(6,11)(7,10)](14,15)]19!
{5,6}	19!	1?](2,3)](4,7)](8,9)](13,14)]19!
{5,7}	8!	1?](2,3)](4,6)(9,13)(11,13)8!
{5,8}	7!	1?](2,3)](4,11)(6,9)7!
{5,9}	31!	1?](2,3)](4,11)(6,8)(7,13)](12,16)](17,21)](22,26)]31!
{6,7}	16!	1?](2,3)](4,5)](8,9)(8,10)(8,11)(9,10)(9,11)](15,23)(17,22)16!
{7,8}	5!	1?](2,3)](4,13)(6,9)(6,10)(6,11)5!
{7,9}	19!24!	1?](2,3)](4,10)(5,13)(6,8)(6,10)(6,11)](12,15)(17,20)(22,26)(29,33)19!24!

after {7,9,22,26} (12,17)(19,24)(15,20)

{7,9,29,33} (12,15)(17,20)(19,31)(22,26)(24,26) …

{7,9,19}   (12,15)(15,17)(15,20)](22,24)(29,31)

{7,9,24}   (12,15)(17,20)(19,22)](26,29)

Table 6. Some Complete Strategies for Sylver Coinage.

5-PAIRS

The safe combinations {5,x,y} are of three types. In the top drawer in Fig. 8 are those with y so much larger than x that the coordinates {a,b,c,d} are {x,2x,3x,y}. For these it seems that y/x tends to 3. But are there infinitely many numbers in the top drawer?

The middle drawer contains the remaining ones for which $x+y$ is a multiple of 5. It seems that for these, x and y always differ by 1 or 2.

In the bottom drawer we have arranged the pairs with coordinates {x, y, x + y, 2y} where x and y are in the order given. It seems that here, as in the second drawer, y/x tends to 1.

POSITIONS CONTAINING 6

As in our other analyses, we write the numbers in six columns and circle 0 and five other numbers a,b,c,d,e, one in each of the 1-, 2-, 3-, 4- and 5-columns respectively. We tabulate $\mathscr{P}$-positions by entries c in a 4-dimensional table (Table 7) whose coordinates, a,b,d,e are congruent to 1,2,4,5, modulo 6.

Entries outside the areas enclosed by full lines are found by repeating entries according to the arrows, where appropriate. The tables for b=8, d=4 and b=8, d=16 can be extended indefinitely by repeating the portions between the pecked lines and increasing all entries by 12 or 60 respectively. The two tables with d=10, b=8 or 14 contain no further entries. All further entries in that for b=14, d=16 are **15**.

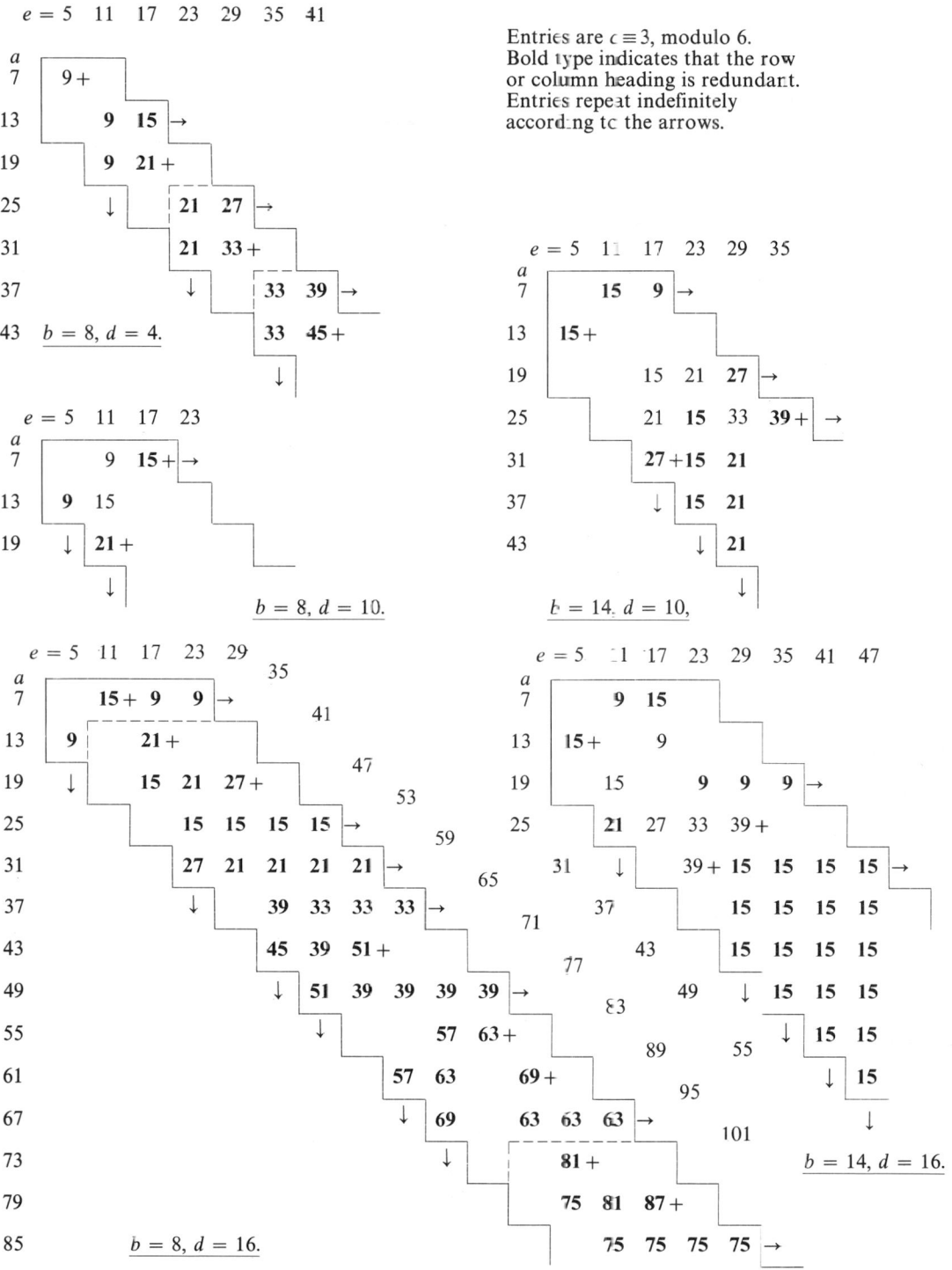

Entries are $c \equiv 3$, modulo 6.
Bold type indicates that the row
or column heading is redundant.
Entries repeat indefinitely
according to the arrows.

Table 7. $\mathscr{P}$-Positions Containing 6,

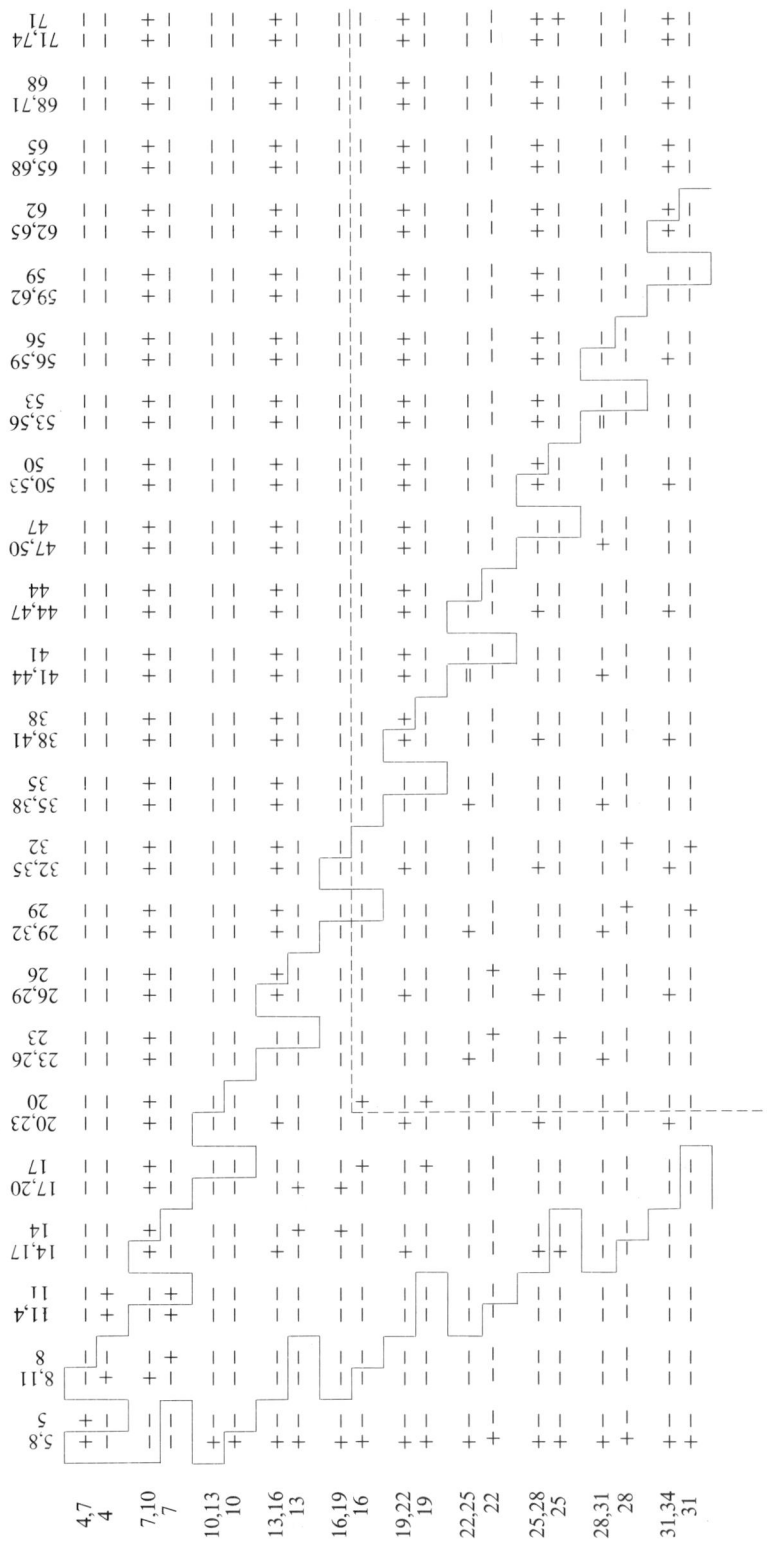

Table 8. A Complete Discussion of {6,9}.

The table represents a complete discussion of positions containing {6,9}. **Reduced positions contain one, or possibly two neighboring, numbers of form $3k+1$, and of form $3k+2$; pairs of rows and columns refer to the latter and former possibilities.** Positions represented by cells outside the crenellated line are not reduced. The pattern within the rectangular quadrant continues indefinitely. The minus signs denote $\mathcal{N}$-positions, the plus signs $\mathcal{P}$-positions and the "$=$" signs $\mathcal{N}$-positions which, at a casual glance at the pattern, might be mistaken for $\mathcal{P}$-positions.

SYLVER COINAGE HAS INFINITE NIM-VALUES

If we make naming 1 an *illegal* rather than a *stupid* move, Sylver Coinage becomes a normal play rather than a misère play game and we could consider adding it to other games using the Sprague–Grundy theory. However since some positions have infinitely many options, we can expect infinite nim-values and indeed they happen!

For example, $\mathcal{G}(2,2n+3)=n$ $(n \geqslant 0)$, so $\mathcal{G}(2)=\omega$. On the other hand, $\mathcal{G}(3,3n+1,3n+2)=1$ $(n \geqslant 1)$, so $\mathcal{G}(3)=1$. Here are some other nim-values:

k	= 4	5	6	7	8	9	10	11	13	14	15	16	17	19	20	22	23	25	26	28	29	31	32
$\mathcal{G}(3,k)$ =	2	3	1	4	6	1		7		8		9	11		1	12	14	15	15	17	19	20	21	23	24	26	27
$\mathcal{G}(4,k)$ =	?	3	0	5	?	8		1		9	14		4	15		?	16	19		?

$$\mathcal{G}(4,6,4n-1,4n+1)=1 \qquad \mathcal{G}(4,6,4n+1,4n+3)=0 \qquad (n \geqslant 1)$$
$$\mathcal{G}(5,6)=7, \ \mathcal{G}(5,7)=8, \ \mathcal{G}(5,8)=10, \qquad \mathcal{G}(6,7)=9, \qquad \mathcal{G}(3,3n-1,3n+1)=5 \ (n \geqslant 6),$$
$$\mathcal{G}(3,9n-8,9n-4)=\mathcal{G}(3,9n+2,9n+7)=\mathcal{G}(3,9n+8,9n+13)=10 \qquad (n \geqslant 3).$$

A FEW FINAL QUESTIONS

Is there any effective technique for computing the outcome and all good replies for the *general position*?

If the game is played "between intelligent players", is the first person to make the game bounded the loser?

Is there a winning strategy of bounded length?

Is there an $\mathcal{N}$-position with $g>1$ for which *all* good replies lead to positions with $g=1$?

Is $\mathcal{G}(4)=\omega+1$? or is $\mathcal{G}(4)=6$, say?

REFERENCES AND FURTHER READING

Morton Davis, Infinite games of perfect information, Ann. of Math. Studies, Princeton, **52** (1963) 85–101.

David Gale and F.M. Stewart, Infinite games with perfect information, Ann. of Math. Studies, Princeton, **28** (1953) 245–266.

Martin Gardner, Mathematical Games, Sci. Amer. **228** 1, 2, 5 (Jan, Feb, May 1973)

Richard K. Guy, Twenty questions concerning Conway's Sylver Coinage, Amer. Math. Monthly, **83** (1976) 634–637.

Michael O. Rabin, Effective computability of winning strategies, Ann. of Math. Studies, Princeton, **39** (1957) 147–157: M.R. 20 # 263.

Fred. Schuh, The game of divisions, Nieuw Tijdschrift voor Wiskunde, **39** (1952) 299–304.

J.J. Sylvester, Math. Quest. Educ. Times, **41** (1884) 21.

Chapter 19

The King and the Consumer

For fools rush in where angels fear to tread.
Alexander Pope, *Essay on Criticism.*

... because your adversary the devil, as a roaring lion,
walketh about, seeking whom he may devour.
1 Peter 5:8

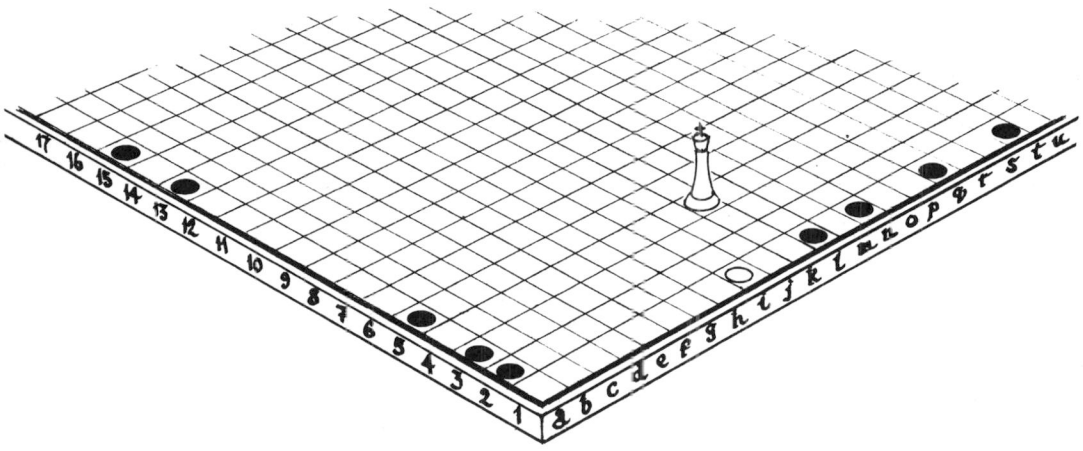

Figure 1. Chas. Plays Geo.

CHESSGO, KINGGO AND DUKEGO

These games are played on some *i* by *j* board. One player, Chas., plays Chess with a lone chess piece which might be a King, or a Knight, or a Duke, or a Ferz, whose moves are shown in Fig. 2. The variants of Chessgo are named for various real and Fairy Chess pieces, Kinggo for a King, etc. Only Kinggo and Dukego will be considered in any detail here.

607

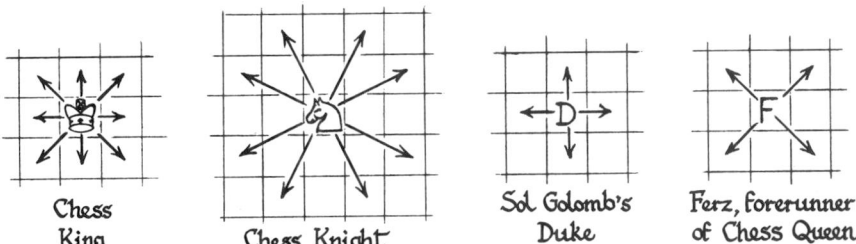

Figure 2. Various Chesspersons.

Chas.'s opponent, Geo., has a number of black (blocking) Go stones and a number of white (wandering) ones. The game starts with the chess piece on a specified square of the otherwise empty board. At each turn the chess piece moves to any legitimate empty square and Geo. then does one of the following:

(a) puts a new Go stone (of either color) on any empty square,
(b) moves a wandering (white) stone already on the board to any other empty square,
(c) passes.

If the chess piece reaches any square on the edge of the board, Chas. wins. If Geo. succeeds in surrounding the chess piece so that it has no legal moves, he wins. A game that continues for ever is declared a draw.

QUADRAPHAGE

This is the special case, invented by R. Epstein, where there are no wandering stones and enough blocking ones to cover the whole board. The title of this chapter refers to the case of Quadraphage in which the chessperson is the King. In Epstein's language, Geo. is a square-eater (graeco-latin *tesseravore*, latino-greek *quadraphage*). Because Geo. eats a square at every turn, this game ends after at most $ij-1$ turns on an i by j board. The starting position for the chessperson is conventionally the middle of the board, or as near as possible if i or j is even.

Since having the first move is never a disadvantage, a strategy-copying argument shows that there are only three possible outcomes for a well-played Quadraphage game from a given starting position on a finite board. Either Geo. wins (even if Chas. moves first) or Chas. wins (even if Geo. moves first) or the first player to move wins. A **fair position** is one in which the first player to move can win.

We'll show that the fair starting positions for the Duke on a quarter-infinite board are all of the squares on the third rank or file, *except* those that are also on the first or second file or rank. We'll also show that the fair starting positions for the King on this board are all of the squares on the ninth rank or file, except those which are also on a lower file or rank. Finally we'll show that the square board which is fair (from the conventional starting position) for a Duke is the ordinary 8 by 8 chessboard, and we assert that the only fair and square boards for a King are 33 by 33 and 34 by 34. On boards smaller than these, Chas. should win even if Geo. starts first and the reverse should happen on larger boards.

THE ANGEL AND THE SQUARE-EATER

The game of Chessgo is not well understood and it's very difficult to exhibit explicit winning strategies for Chas. even on modest sized boards. For example it seems very likely indeed that the Knight can draw on an infinite board although this seems extremely difficult to prove.

Indeed it's never been shown that there is *any* generalized chess piece that can draw on the infinite board. This suggests the following problem. An **angel** (of power 1000) is a chessperson who can fly in one move to any empty square which could be reached by a thousand King moves. Angels, of course, have wings, so it won't matter if some of the intervening squares have been eaten.

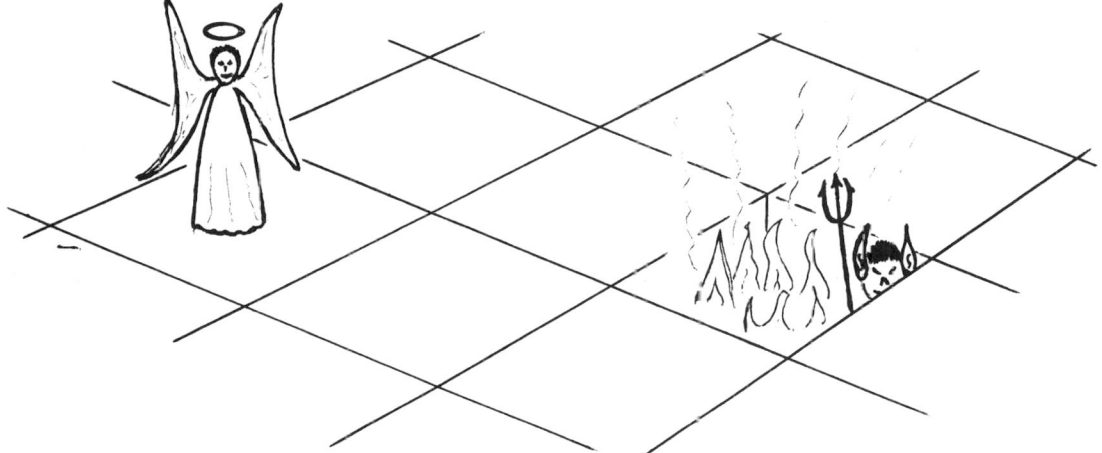

Figure 3. The Angel and the Square-Eater.

We'll say the angel wins by continuing forever (i.e. drawing the game of Quadraphage) against a square-eating **devil** (who can devour *any* square of the board, no matter how far away it is from his previous moves). The devil, of course, wins if he can surround the angel with a sulphurous moat, a thousand squares wide, of eaten squares. Can you give an explicit strategy that's *guaranteed* to win for the angel?

If the devil adopts certain cunning tactics worked out for him by Andreas Blass and John Conway, then infinitely often the angel will find itself decreasing its distance from the centre by arbitrarily large amounts. Although the angel never seems to be in any real danger, its path must also contain arbitrarily convoluted spirals.

STRATEGY AND TACTICS

In both Dukego and Kinggo it's possible to distinguish between strategic moves and tactical ones. In either game Geo. wins, on large enough boards, by first playing a few *strategic* stones on squares far away from the chess piece. When the chess piece gets closer to the edge of the board, Geo. switches to *tactical* moves fairly close to him. Whenever the chess piece is driven away from the edge towards the centre of the board, Geo. reverts to strategic moves.

DUKEGO

Dukego is much simpler than Kinggo and so we consider it first. You might like to try playing it yourself before reading this section. The optimal strategies we present here were first discovered by Solomon Golomb. We consider various infinite boards first.

On an infinite half-plane the Duke can win only if he can get to the edge at his first move. In any other situation Geo. can draw by playing directly between the Duke and the edge. In fact Geo. needs only one white (wandering) stone.

On an infinite strip of width i with $i \leq 4$ the Duke, moving first, can win immediately. If $i \geq 4$ and Geo. moves first he can draw by playing between the Duke and the nearest edge and again needs only one stone, if it's a wandering one.

On an infinite quarter-plane the Duke, moving first, can win if he starts within a three squares wide border. His initial move attacks the edge and Geo. has no choice but to move directly between Duke and edge. The Duke then charges towards the corner. At each move Geo. is forced to play between the Duke and the edge and eventually the Duke wins by reaching one of the two squares next to the corner.

If Geo. moves first against the Duke on the third rank or file of an infinite quarter plane, he can draw using just one blocking stone and one wandering one. He first puts his blocking stone at the strategic position diagonally next to the corner (Fig. 4). This blocks the only square from which the Duke might attack two boundary squares at once. Whenever the Duke moves onto a lower case letter, Geo. puts his wandering stone on the corresponding capital letter.

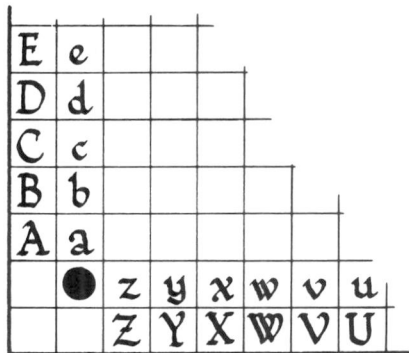

Figure 4. Geo. Beats the Duke on a Quarter-Infinite Board.

On an 8 by j board for any value of *j*, the Duke can win if he moves first no matter how many stones Geo. has. The Duke first advances towards the nearest edge, getting onto the third rank or file. If Geo. blocks his advance, the Duke charges along the edge to win in one of the corners. If Geo. does not immediately block his advance, the Duke's second move places him next to an edge square and Geo. puts a second stone on the board. One of Geo.'s stones must be on the edge square next to the Duke; if the other is to the left of this, the Duke charges to the right, and if to the right, he charges leftward. In either case the Duke eventually wins in the other corner.

On a 7 by j board the Duke can win even if Geo. starts, by attacking whichever half of the board does not contain Geo.'s opening stone.

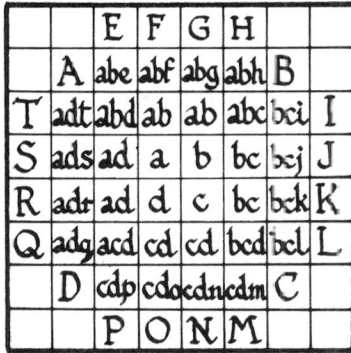

Figure 5. Geo. Beats the Duke on an Ordinary Chessboard.

On the 8 by 8 board Geo. can draw using only three wandering stones (Fig. 5). He always arranges to have his stones on the capital letters corresponding to the small letters covered by the Duke. Since the combinations of small letters on any two contiguous squares never differ by more than one letter, this is always possible. Geo. can draw with just two wandering stones and two blocking ones, by placing the blocking stones on *A* and *C*. This works because every square with three letters includes just one of *a* and *c*. He can also draw with just one wandering stone and four blocking ones placed at *A, B, C, D*. It's much harder to find how many blocking stones Geo. needs when he has no wandering ones. Table 1 summarizes the fair starting positions in Dukego.

Size of Board	Starting Position	Least Number of Stones Giving Geo. at least a Draw, Moving First
$4 \times \infty$	centre.	1 wandering.
quarter-infinite.	3rd rank or file, excluding 1st or 2nd file or rank.	1 wandering, 1 blocking.
$8 \times j, j \geqslant 8$	centre.	3 wandering, *or* 2 wandering, 2 blocking, *or* 1 wandering, 4 blocking, *or* ? blocking.

Table 1. Fair Boards for Dukego.

THE GAME OF KINGGO

The remaining sections of this chapter are devoted to Kinggo.

THE EDGE ATTACK

Figure 6 shows how the King can force his way to a nearby edge of the board if this is in-adequately defended. The solid line indicates the edge of the board and the dot shows the present

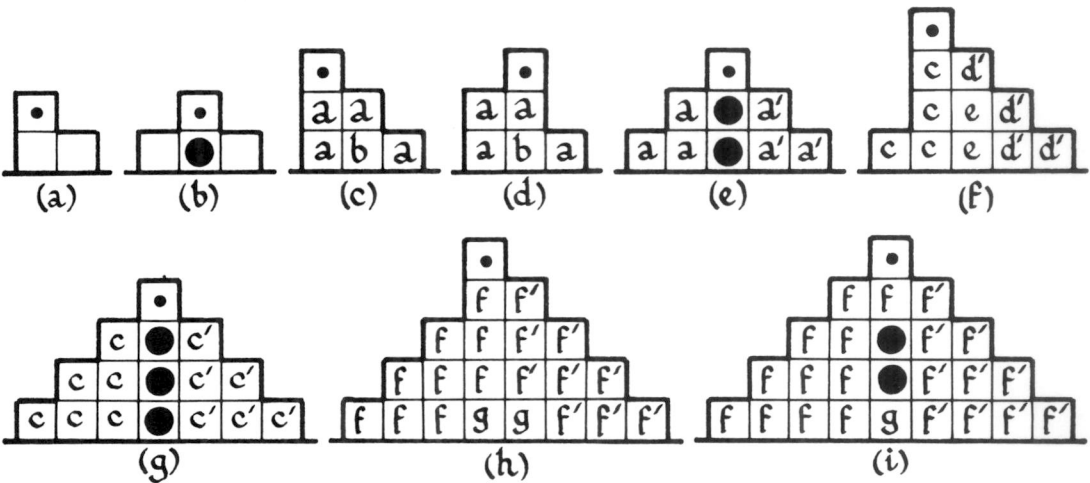

Figure 6. How the King Gets to the Edge.

position of the King. We suppose that the lettered squares are empty; Go stones may occupy any or all of the other squares. In each case Geo. is to move.

If Geo. moves outside the region shown, the King simply advances towards the edge, while if Geo. moves on to a letter x or x' ($x = a, b, \ldots, g$) the King makes a move which results in a case of Fig. 6(x) or its reflexion. If, for example, Geo. puts a stone in the lower right corner (labelled d') of Fig. 6(f), the King moves downwards and achieves a reflexion of Fig. 6(d),

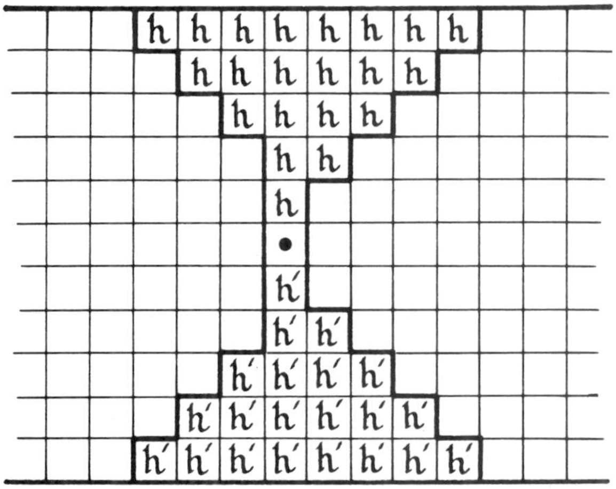

Figure 7. How the King Wins on an Infinite Strip of Width Eleven.

Now a glance at Fig. 7 and Fig. 6(h) shows that:

> the King can win
> on an infinite strip
> of width at most 11,
> even if Geo. goes first.

THE EDGE DEFENCE

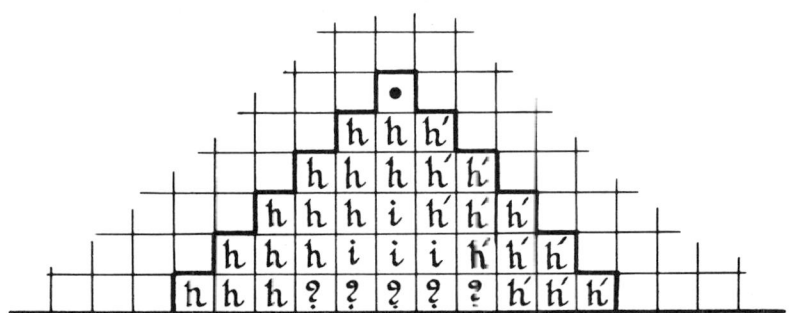

Figure 8. How Geo. Guards the Edge Against the King.

Figure 8, which again refers back to Fig. 6, shows that there are only five possible moves (?, ?, ?, ?, ?) which give Geo. any chance of stopping the King approaching from the sixth rank of an empty board. Figure 9(k) shows how Geo. can successfully defend the edge with any of these five moves. The King may move from any of the shaded squares. If he remains on such a square Geo. passes. When the King moves on to a letter x or x' ($x=j, k, l, ..., q$) Geo. can move into a case of Fig. 9(x) or its reflexion (as he did in Fig. 6). Note that the proof of each of Figs. 9(j) to 9(q) depends on the others, because the King can nip from one of these positions to another in ingenious ways.

Since none of these positions has more than three stones. Geo. can defend the edge with only three wandering stones.

A MEMORYLESS EDGE DEFENCE

Figure 10, which uses the same conventions as Fig. 5, shows another way that Geo. can stop the King approaching from the sixth rank of an empty board using just three wandering stones. Unlike Fig. 9, this is memoryless in the sense that the positions of these stones depend only on the position of the King and not on how he got there.

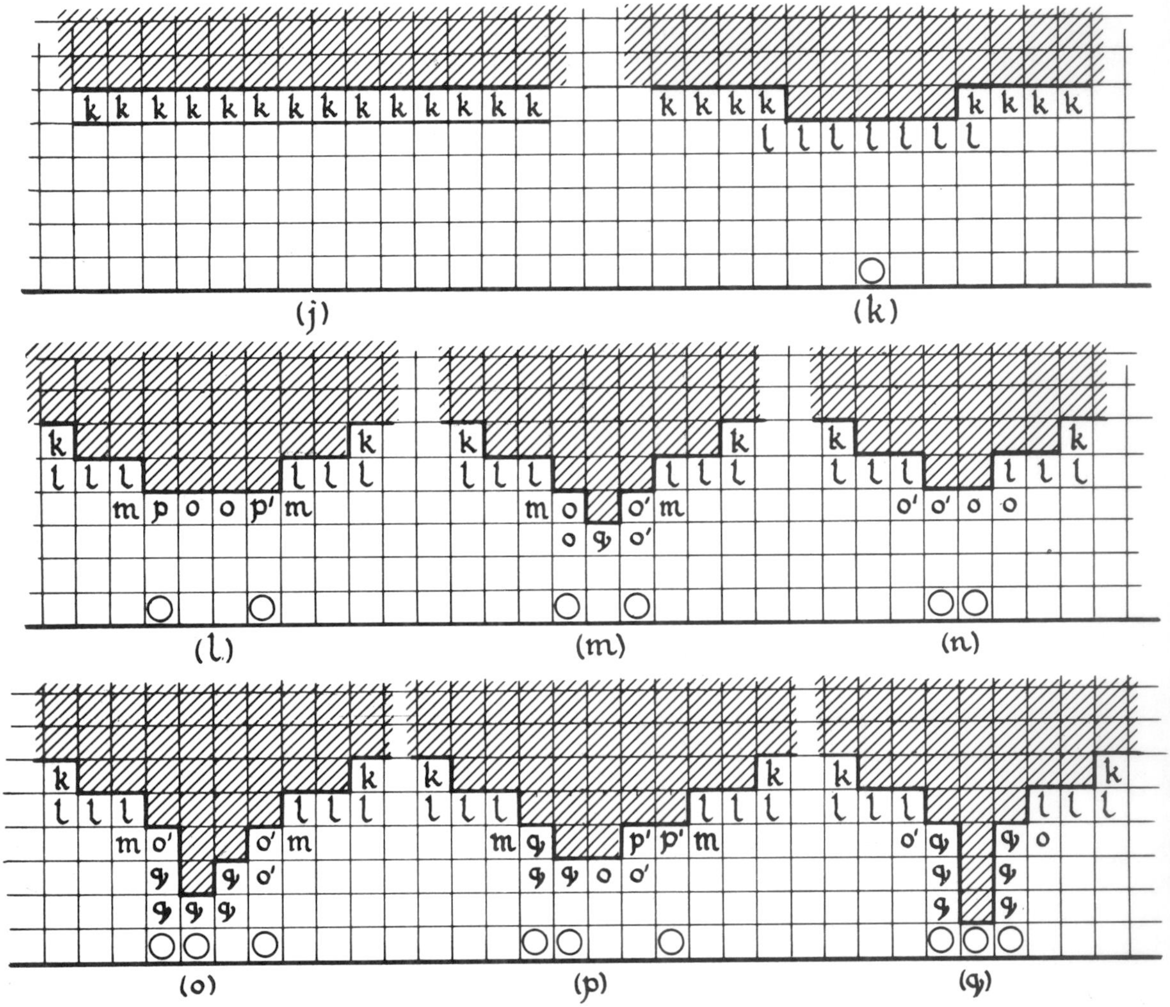

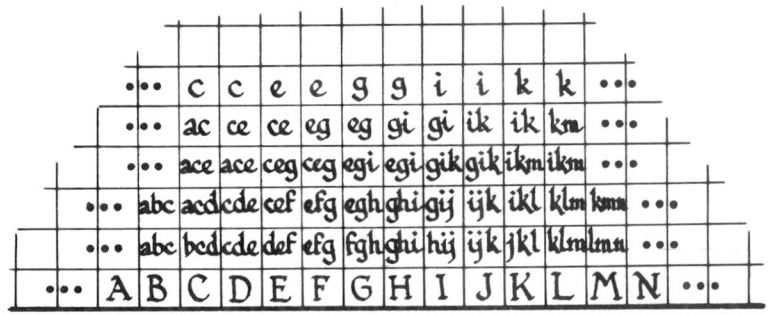

Figure 9. Three Wandering Stones Ward Off the King.

Figure 10. A Memoryless Kinggo Edge Defence.

In later sections of this chapter Geo. will want to patch together several copies of this defence (Figs. 11 and 12). Figure 11 shows how it may be joined to its left-right mirror image, and Fig. 12 shows how to change its phase by one square. Many other memoryless edge defences can be obtained by joining various combinations of Figs. 10, 11, 12 and their translates and reflexions.

Figure 11. Wedding Figure 10 With Its Reflexion.

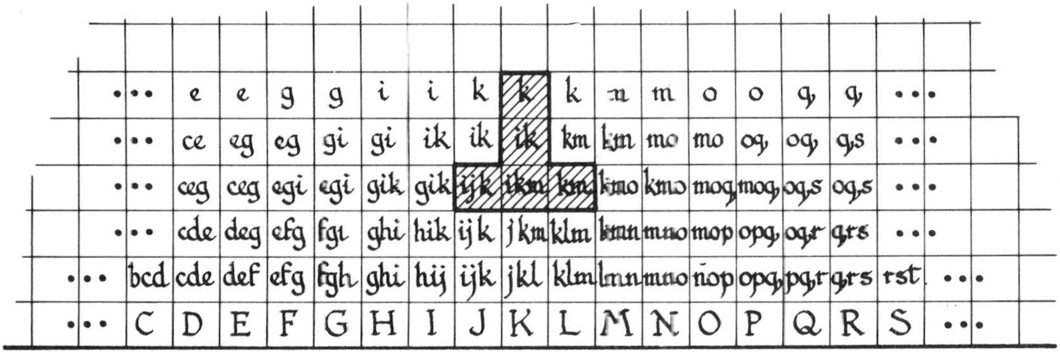

Figure 12. Getting Figure 10 One Square Out of Step with Itself.

Some results for infinite strips follow immediately from Figs. 9, 10, 11, 12:

> On an infinite strip
> of width at least 12,
> Geo., moving first, can draw
> with just 3 wandering stones.
>
> ———
>
> On an infinite strip
> of width at least 13,
> he can draw even if
> the King moves first.

If the King advances towards the edge he will be stopped at Fig. 9(q) and if the King refuses to attack the edge, Geo. can still obtain 3 consecutive stones as in that figure.

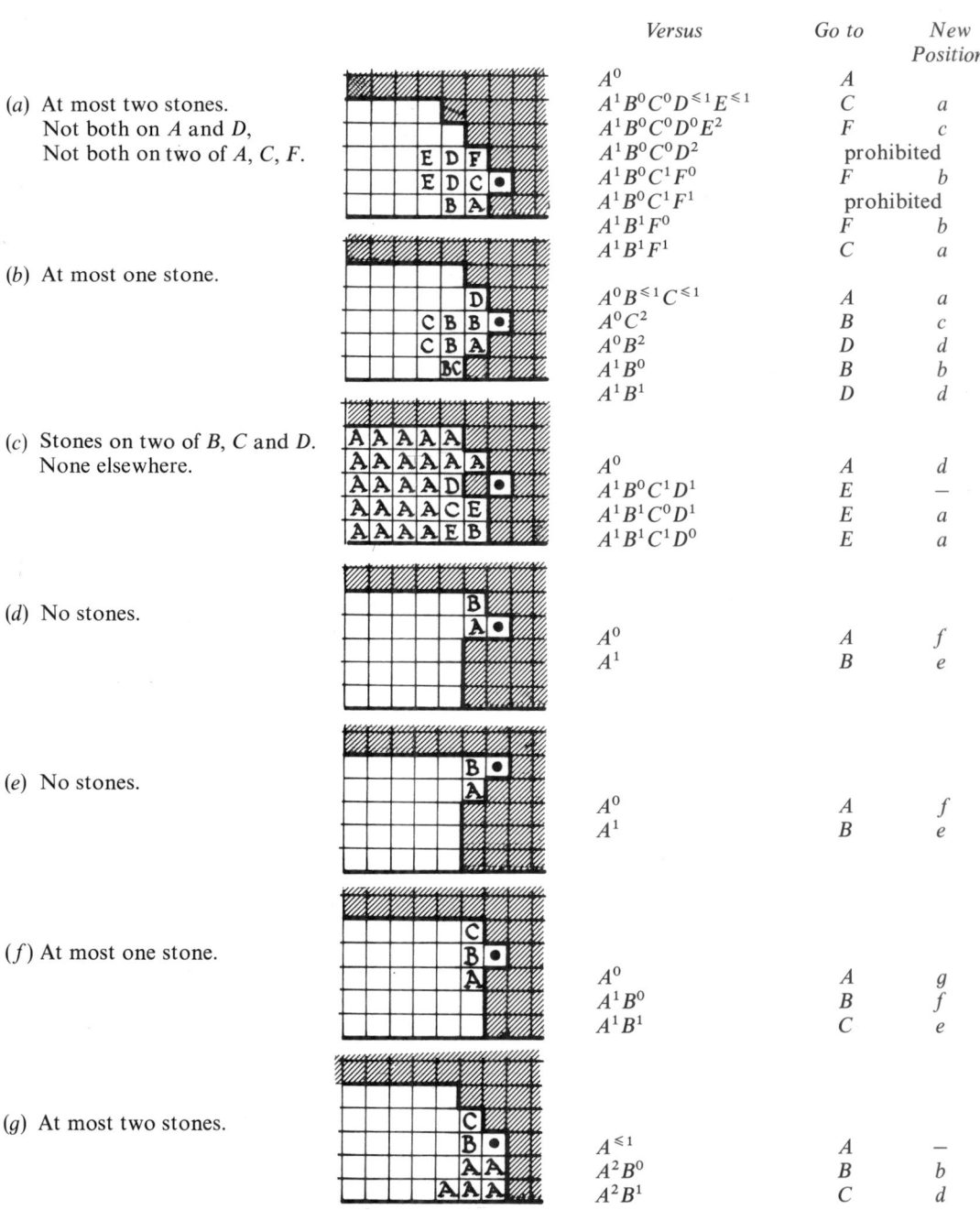

		Versus	Go to	New Position
(a) At most two stones.		A^0	A	
Not both on A and D,		$A^1B^0C^0D^{\leqslant 1}E^{\leqslant 1}$	C	a
Not both on two of A, C, F.		$A^1B^0C^0D^0E^2$	F	c
		$A^1B^0C^0D^2$	prohibited	
		$A^1B^0C^1F^0$	F	b
		$A^1B^0C^1F^1$	prohibited	
		$A^1B^1F^0$	F	b
		$A^1B^1F^1$	C	a
(b) At most one stone.				
		$A^0B^{\leqslant 1}C^{\leqslant 1}$	A	a
		A^0C^2	B	c
		A^0B^2	D	d
		A^1B^0	B	b
		A^1B^1	D	d
(c) Stones on two of B, C and D.				
None elsewhere.		A^0	A	d
		$A^1B^0C^1D^1$	E	$-$
		$A^1B^1C^0D^1$	E	a
		$A^1B^1C^1D^0$	E	a
(d) No stones.				
		A^0	A	f
		A^1	B	e
(e) No stones.				
		A^0	A	f
		A^1	B	e
(f) At most one stone.				
		A^0	A	g
		A^1B^0	B	f
		A^1B^1	C	e
(g) At most two stones.				
		$A^{\leqslant 1}$	A	$-$
		A^2B^0	B	b
		A^2B^1	C	d

Figure 13. The Edge-Corner Attack.

THE EDGE-CORNER ATTACK

On an infinite strip of width 13, the King can't *win*, but *can* force his way to the second rank, as in Fig. 9(q). He may then charge along this rank in either direction, forcing Geo. to accompany him. Even if Geo. has a large supply of stones he can do no more than build up a solid wall along the first rank and on a finite board the edge-charging King will eventually reach a corner.

We now claim that for an adequate defence, Geo. must have at least three strategic stones stationed somewhere between the edge-charging King and the corner. All of these three stones must be positioned somewhere in the first five ranks. The proof of this follows from Fig. 13, which lists the appropriate moves for the King against all positions not satisfying these conditions. There are squares with one of the first seven capital letters and infinitely many squares with no letters at all. At Geo.'s move he has at most 3 stones in the figure. The line

$$\text{"Versus } A^0, \text{ go to } A, \text{ position } -\text{"}$$

means that if none (superscript 0) of the 3 stones are on A, then the King moves to A and wins at once. The line

$$\text{"Versus } A^1 B^0 C^0 D^{\leqslant 1} E^{\leqslant 1}, \text{ go to } C, \text{ position } a\text{"}$$

means that if Geo. has one stone on A, none on B or C, and at most one (superscript $\leqslant 1$) on each of D and E, the King should move to C and obtain a translate of the position shown in Fig. 13(a).

Since in every case the King counters Geo.'s moves to any of Figs. 13(a) to 13(g) by a move resulting in another of these figures, Geo. can never force the King above the fifth rank or prevent him from continuing the edge-corner attack, although he *can* keep him moving to and fro among these seven figures.

On the other hand, almost all combinations of three strategic stones along the first rank of the board *will* suffice for Geo. to stop the edge-corner attack. Figure 14 shows the only exceptions.

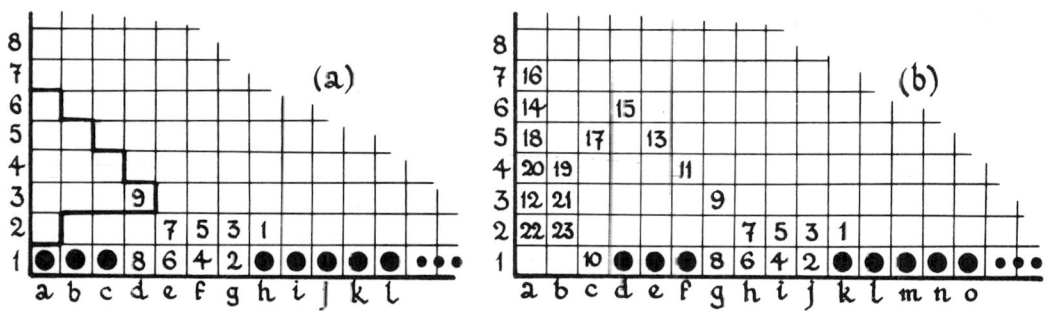

Figure 14. Triplets Which Fail to Stop the Edge-Charging King.

In Fig. 14(a) Geo. has three strategic stones at a1, b1, c1, as well as his tactical stones, h1, i1, ..., defending the edge near the King. The King moves to 1; Geo. is forced to put a stone at 2; the King moves to 3; and so on. The King's move to 9 guarantees him a win by Fig. 6(f) (reflected).

Figure 14(b) uses a similar notation to show how the King wins if Geo.'s strategic stones are at d1, e1, f1.

STRATEGIC AND TACTICAL STONES

Since Geo. can stop an edge corner attack with just three extra stones, and most combinations of three stones suffice, it's convenient to call his three most distant stones in the first five ranks along any edge of the board **strategic** stones; his other stones are **tactical** ones. The tactical stones try to stop the King winning along the side and the strategic ones then prevent him from winning the ensuing edge-corner attack.

Let's consider for instance the game on an infinite strip of width 23 (Figure 15). The King starts at 1; Geo. puts a stone at 2; the King moves to 3; Geo. puts a stone at 4; and so on. The crucial position arises when Geo. puts a stone at 16. Where should the King move now? Although various moves look plausible, only one succeeds!

You must distinguish between strategy and tactics if you're to find the right move. The stones 4, 6 and 8 defend the right flank, so 10, 12 and 16 are needed to defend the left one. Since the stone at 16 is required for *strategic* purposes it is *tactically* worthless.

So the King pretends that 16 is empty and moves to 17, which would give him a tactical victory via Fig. 6(g)! Any other King move would lose to a defence at α or β.

Of course, since 16 *isn't* empty, the game *won't* end on the lower edge, for Geo. can stop the edge attack, but only by using the stone at 16. Eventually the King would have an opportunity to move to 16 if it were vacant. Instead of doing this he embarks on an unstoppable edge-corner attack, running along the second rank towards the left. Geo. can eventually use his stones 10 and 12 to divert the King into various positions of Fig. 13 but can't halt the edge-corner assault.

This sort of argument shows that:

> if Geo. is required to place his first 10 stones on the top and bottom ranks, the King can draw on an infinite strip of width 23, even when Geo. moves first

We believe that this remains true when we remove the constraints on Geo.'s initial moves, since it seems very unlikely that he gains any advantage by putting his stones nearer to the middle. Although such moves seem futile, we haven't managed to exhibit a precise strategy by which the King can refute them.

CORNER TACTICS

Figure 16 shows how Geo. defends the corner against an attack from either edge using three consecutive blocking stones and three wandering ones. The edges can be continued using Fig. 10 to give a strategy for Geo. on a quarter-infinite board.

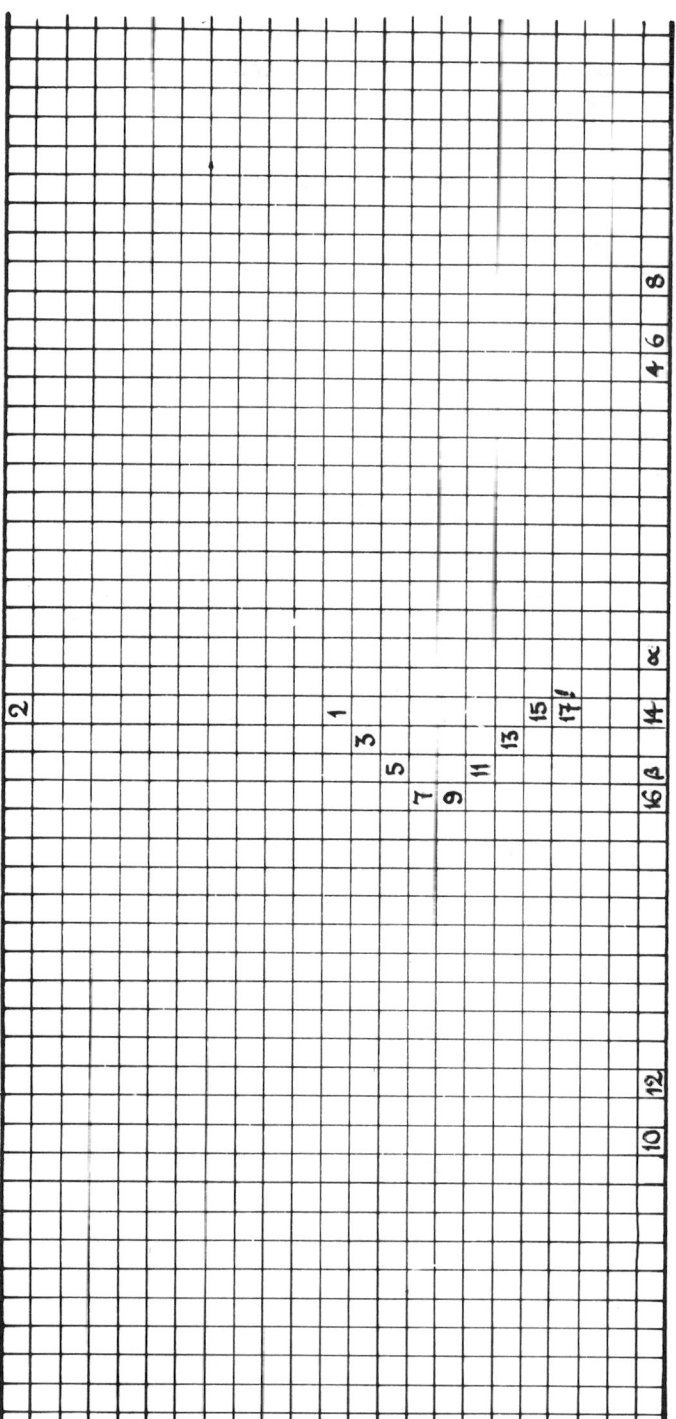

Figure 15. The King Draws a Typical Game on an Infinite Strip of Width 23.

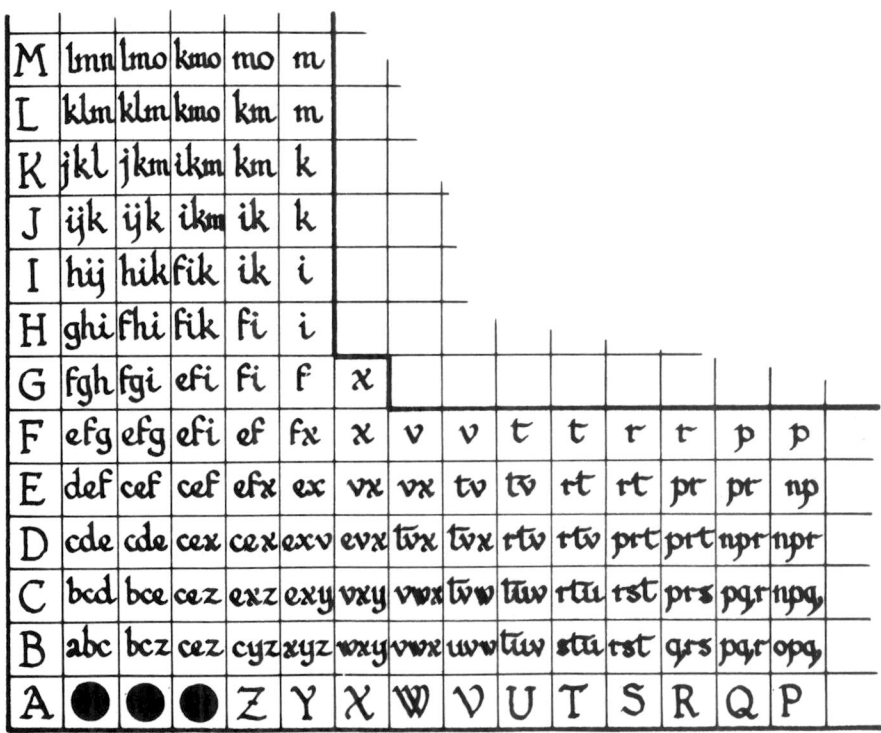

Figure 16. Three Blocking and Three Wandering Stones Defend the Corner.

Although it defends the corner from attack along either edge, it provides only a weak defence against a direct attack towards the corner along the diagonal. It defends against a King on the tenth rank and tenth file of an empty board *only* when Geo. moves first. Since Geo. must first enter his three strategic stones, if the King moves first he will arrive at the sixth rank and file before Geo. has put any wandering stone on the board, and Fig. 16 now requires Geo. to put wandering stones on both F and X. In fact Fig. 17 (which should be used in conjunction with Fig. 6) shows that the King can now win against *any* strategy for Geo., even if there are stones on all the indicated squares of Fig. 17(a). This figure depends on Figs. 17(b) to 17(e) whose proofs are left to the reader.

Figure 17 shows that Geo.'s only hope of defending the corner against a diagonally attacking King, starting from the tenth rank and file, requires that his first three stones be placed elsewhere. One promising possibility uses squares a2, a3, a5 along one edge, when Geo.'s major problem is to find an appropriate continuation when the King arrives on the sixth rank and file. Figure 18(a) shows that there is only one possibility (indicated by ?). The proof of Fig. 18 depends on Fig. 17 and Fig. 6, but we again leave some of these proofs to the reader.

So there's only one move with which Geo. can successfully defend Fig. 18(a)! His complete strategy appears in Fig. 19 with the edges extended by Fig. 10. With the three blocking stones positioned as shown, he defends the corner against attacks along edges or diagonal.

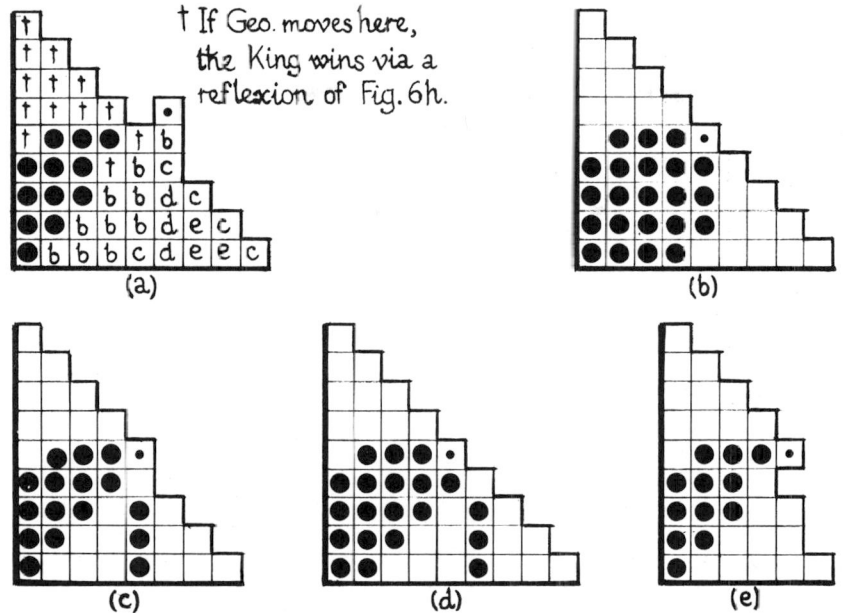

Figure 17. Three Consecutive Blocking Stones Won't Defend the Corner Against the King on the Sixth Rank and File.

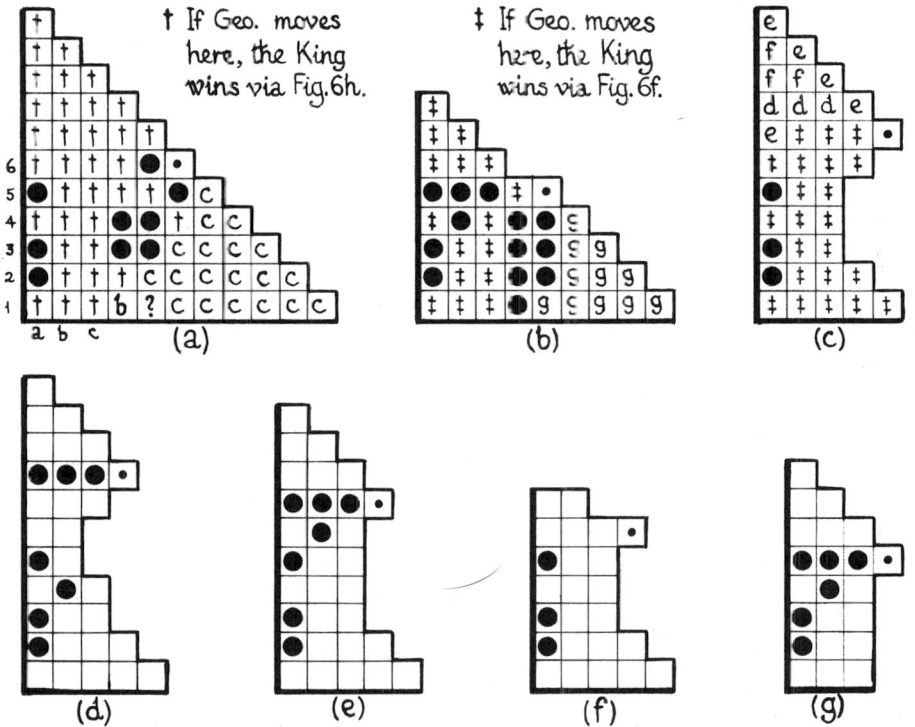

Figure 18. With Three Well Placed Stones and the Initial Move, Geo. Defends the Corner Against the Diagonal Attack.

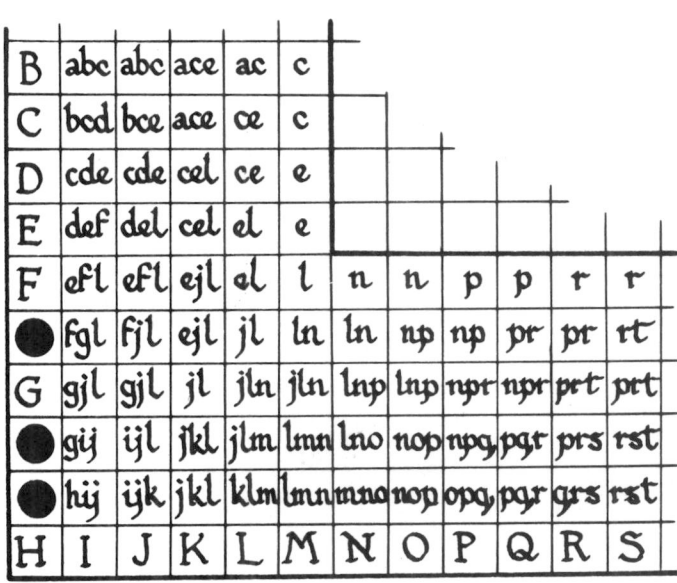

Figure 19. Memoryless Corner Defence.

Combining Figs 13 and 19 we have:

> the fair starting positions
> on the quarter-infinite board
> are those on the ninth rank or file,
> excluding lesser files or ranks.

Geo., moving first, can defend any such position with just three blocking stones and three wandering ones according to Fig. 19. But if the King moves first, he attacks the nearest edge, ignoring the three further stones between him and the corner as in the sample game of Fig. 15. Since Fig. 13 shows that Geo. needs three strategic stones to defend the corner, the King can *either* win on the edge *or* divert to a winning edge-corner assault as in Fig. 13.

DEFENCE ON LARGE SQUARE BOARDS

We've seen that Geo. can defend a corner with three blocking stones and three wandering ones, so he can defend a large enough square board with twelve blocking stones (three in each corner) and three wandering ones. He first puts the twelve blocking stones in their permanent places. If the board is 35 × 35 or bigger, the King begins at least 18 squares from any edge, so he's still at least 6 squares away from the edge after Geo. has placed his 12-strategic stones. So,

Geo., moving first,
can win on a square board
of size 35 × 35 or larger.

THE 33 × 33 BOARD

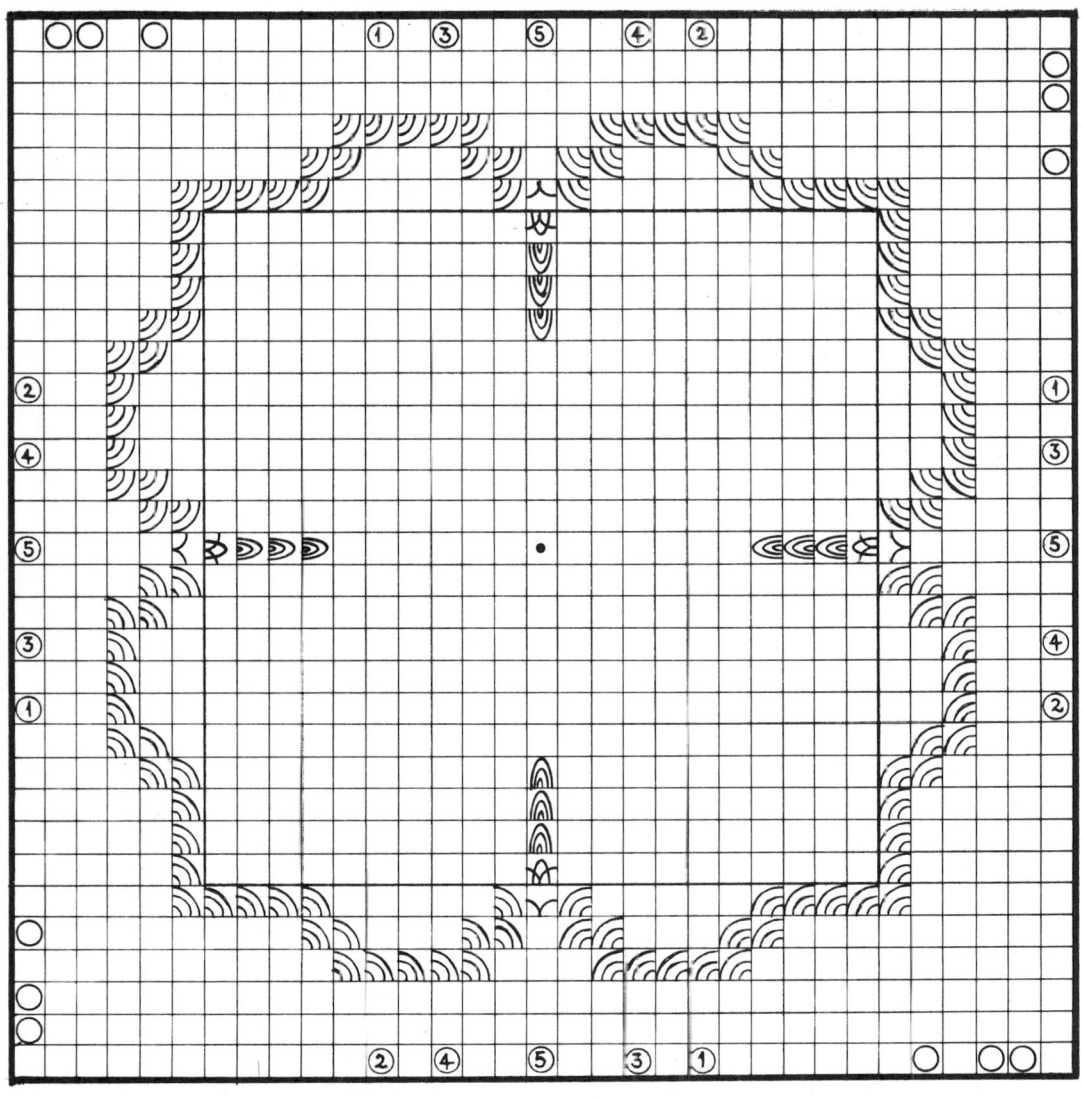

Figure 20. The Centred King on a 33×33 Board.

We'll now show you a more intricate defence which allows Geo., moving first, to survive on a 33×33 board with just 12 (wandering) stones. The details are in Figs. 20 to 26.

THE CENTRED KING

So long as the King stays in the central region of Fig. 20, Geo. puts stones on certain strategic squares, marked with circles on the perimeter of the board. There are 32 of these, 3 near each corner and 5 on each edge. Geo. puts the first four stones one on each edge, and the distribution of his stones after the King has made four or more moves is shown in Fig. 21, a close-up of part of Fig. 20 (the four quarters of the board are congruent). Most of the squares in the central region are divided into nine subsquares, the central one of which is always empty. The other eight subsquares tell Geo. how many stones he should have in each corresponding area. For example, if the King moves to a square marked

3		
1	4	

then Geo. moves so that he has three stones on the left edge, one near the bottom left corner and four on the bottom edge. The order in which Geo. puts his stones in the three squares near the corner doesn't matter, but of the five strategic squares on each edge, it's the middle one that must be occupied last. A reasonable order is indicated by the numbers 1,2,3,4,5 in the circles in Figs. 20 and 21.

A few squares on the main diagonals of Fig. 21 contain arrows:

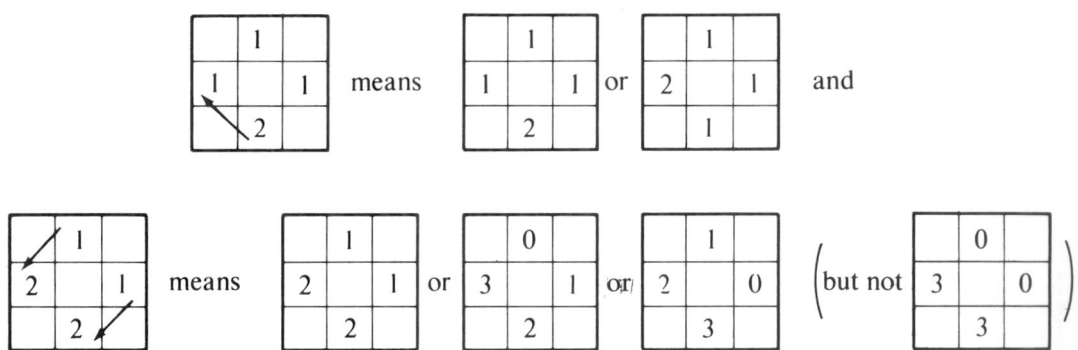

Geo. can use any of these alternatives as a satisfactory defence.

LEAVING THE CENTRAL REGION

If the King leaves the central region of Fig. 20 via a square marked as in Fig. 22(a), we'll say that he's **cornered** in the lower left of the board, and then Geo. will keep him inside the region

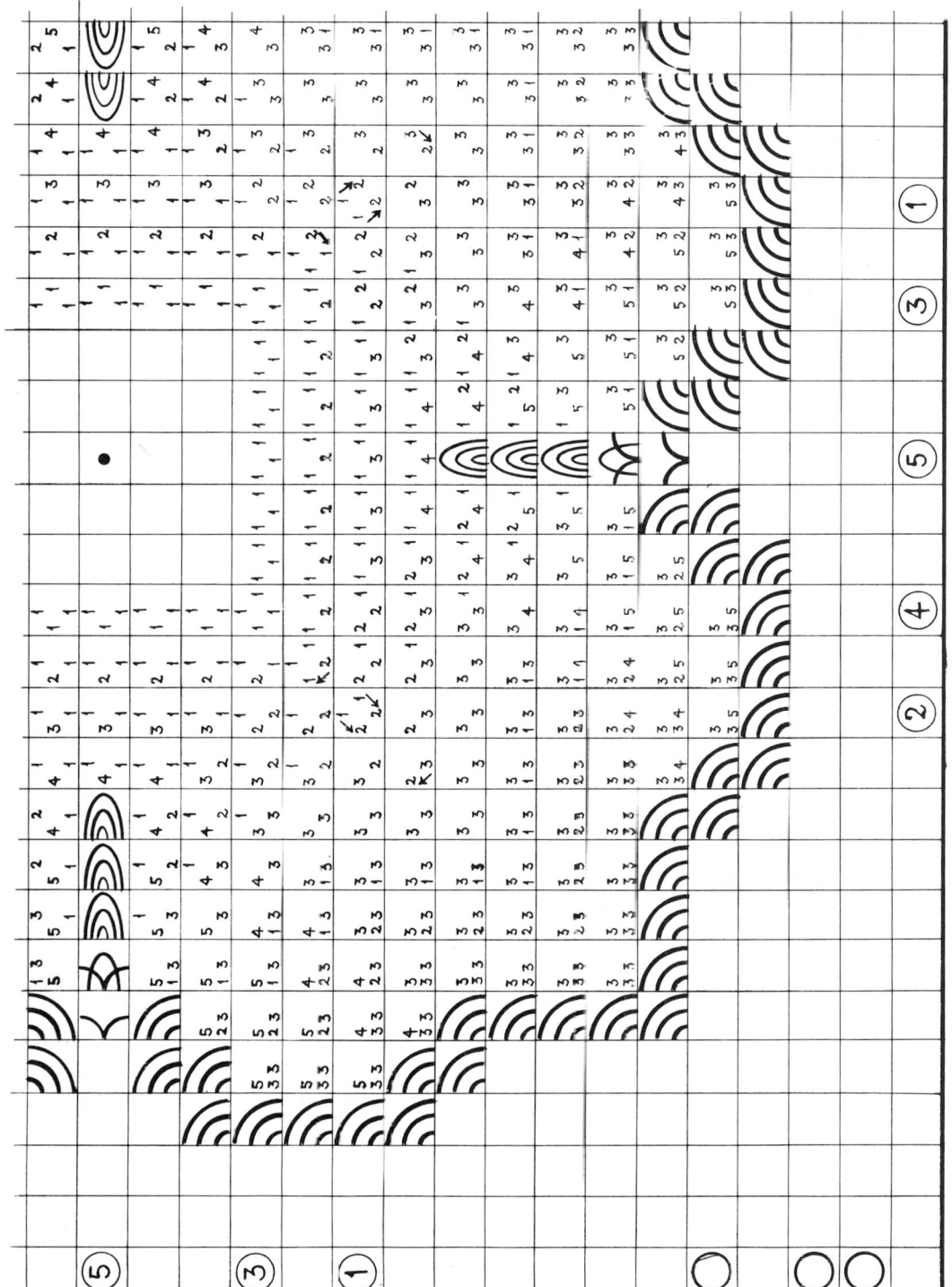

Figure 21. Close-up of Figure 20.

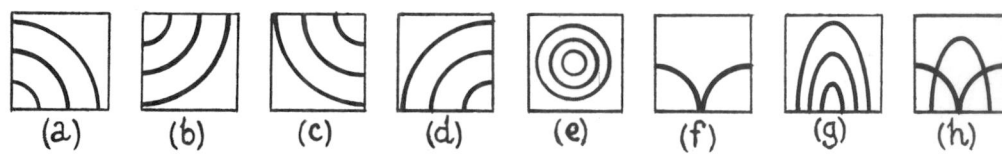

Figure 22. Key to Markings in Figures 20, 21, 23, 24, 25 (see text).

shown in Fig. 23 by making tactical moves which prevent the King from reaching a shaded square, so that he can only "re-centre" himself by moving to a square marked as in Fig. 22(e).

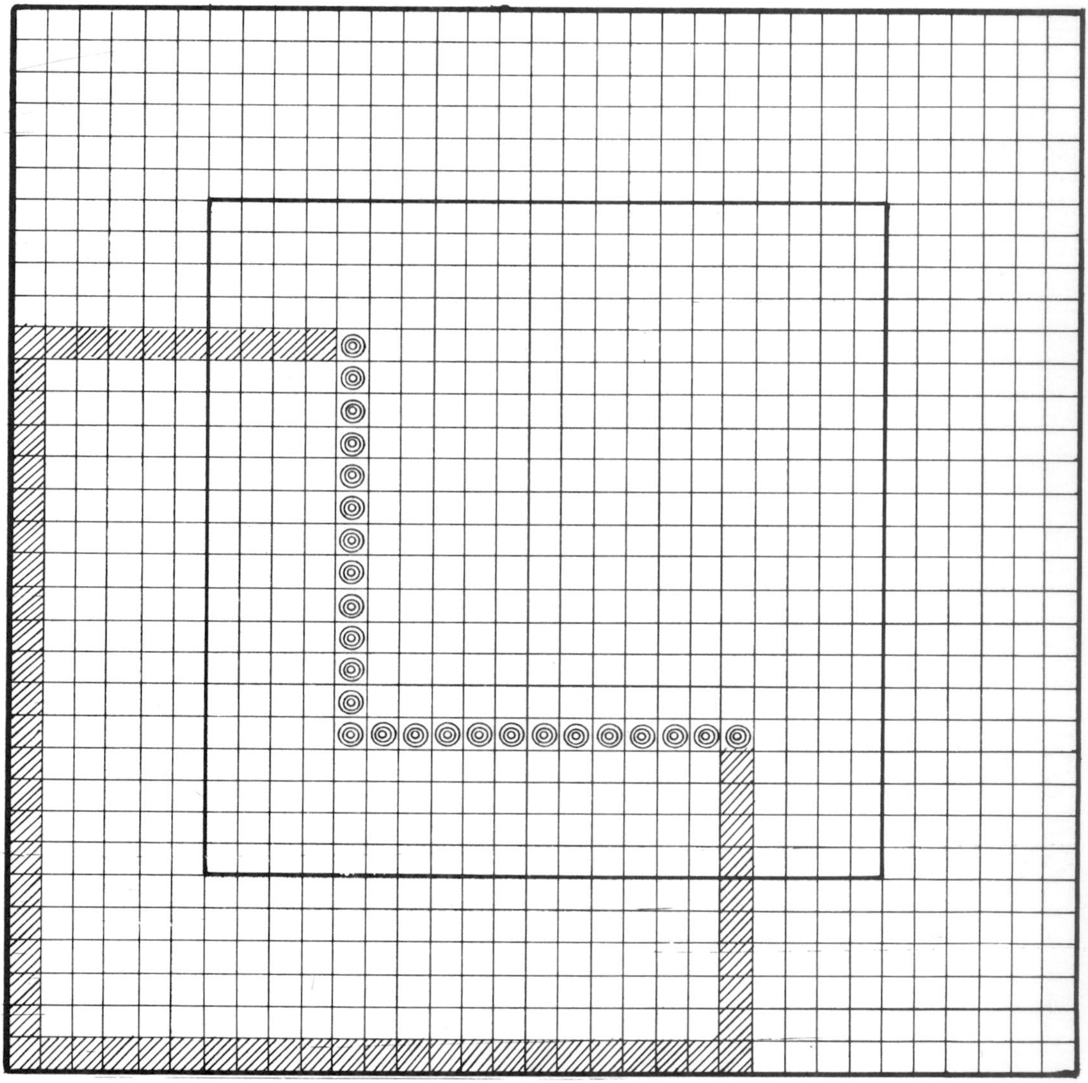

Figure 23. The Cornered King.

If the King moves to squares marked as in Fig. 22(b), (c) or (d) he is correspondingly cornered in the upper left, upper right or lower right of the board. If the King moves to a square labelled as in Fig. 22(f) he is cornered in the lower left or lower right of the board, depending on the direction he came from. If he moves to a label like Fig. 22(g) he is **sidelined** (see later) and when he moves to one like Fig. 22(h) he is either sidelined or cornered, again depending on which way he came; if diagonally, he's sidelined; if horizontally then he'll be pushed back to the corner whence he came.

THE CORNERED KING

Figure 24, a close-up of Fig. 23, reveals the tactical details that Geo. uses to keep the King cornered with just three wandering stones and nine **static** ones (semi-stationary, both strategic and tactical). Of course, when the King first becomes cornered by moving to a square marked as in Fig. 22(a), Geo. may not have his nine static stones in the exact places shown in Fig. 24, but he will have three stones between the King and the lower left corner, and three on the bottom edge and three on the left edge. Geo. uses the stones already on the boundary as substitutes for any stones missing from Fig. 24. When the tactics call for placing a stone on a square already occupied, Geo. places a stone on an unoccupied circle in Fig. 24.

Suppose, for example, that the King leaves the central area of Fig. 20 by moving to square $k4$ (see Fig. 1). He must have come from $l5$, marked

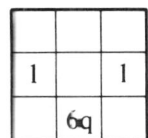

so there are already three stones on the left-edge, three as indicated near the lower left corner and five in the squares 2, 4, 5 and those next to Z and A and between them ("3" and "1"). The King is now on a square labelled "$s24$" so Geo. puts his last stone (the white one in Fig. 1) on S and continues to follow Fig. 24 with the stone on "5" substituting for the missing one between Z and A.

The right arrow in certain squares near the top right of Fig. 24 means that the third genuinely wandering (non-static) stone belongs on a strategic square on the right edge.

THE SIDELINED KING

If the King leaves the central area of Fig. 20 by moving onto a square marked as in Fig. 22(g) he is sidelined as in Fig. 25. Geo. tactically keeps the King off the shaded squares and the King can only re-centre himself by moving onto a square marked as in Fig. 22(e).

The notation in Fig. 26 (a close-up of Fig. 25) is as in Figs. 21 and 24, but we now have some squares

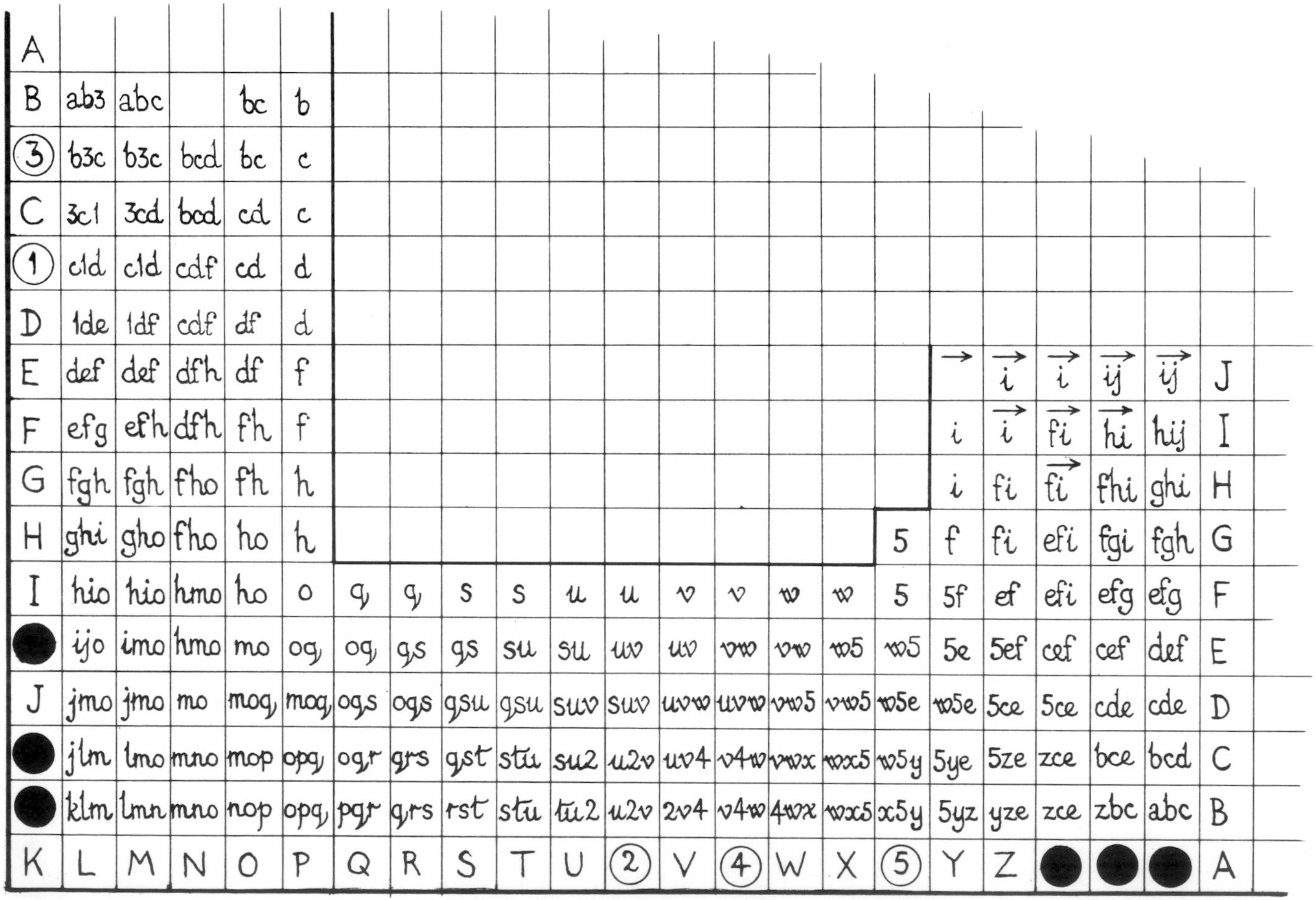

Figure 24. Close-up of Figure 23.

These advise Geo. to have one stone on each of the left and right edges (shown in Fig. 25 in their lowest and highest positions) and *six* on the bottom edge, not only the usual five but one on *J* or *Q* as well. [Assume in Fig. 26 that Geo. has 6 static stones on 1, 2, 3, 4, *J*, *Q*, and that one of the last two is substituting for 5.]

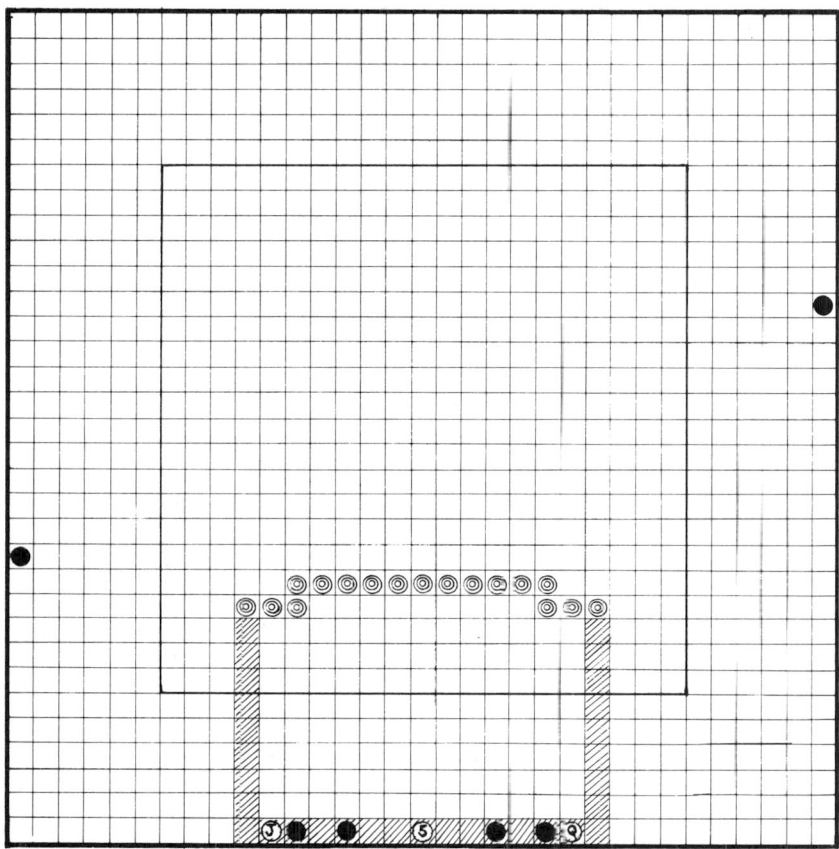

Figure 25. The Sidelined King.

HOW CHAS. CAN WIN ON A 34 × 34 BOARD

Geo., going first, can survive in Kinggo with 12 wandering stones on a 33 × 33 board, so he can certainly survive on a 35 × 35 or larger board, even if the King goes first.

However it seems that the King can win on a 34 × 34 board if he moves first. Here's how he does it. His first three moves diagonally attack the nearest corner. He then turns left or right and attacks the adjacent corner in the half of the board where Geo. has at most one stone. After 9 more moves he is at *l6*, say, and Geo. has been unable to get 9 useful stones on the board. If the corner is adequately defended (with 3 stones), then one flank or the other is weak and a carefully executed edge-corner attack eventually leads to victory.

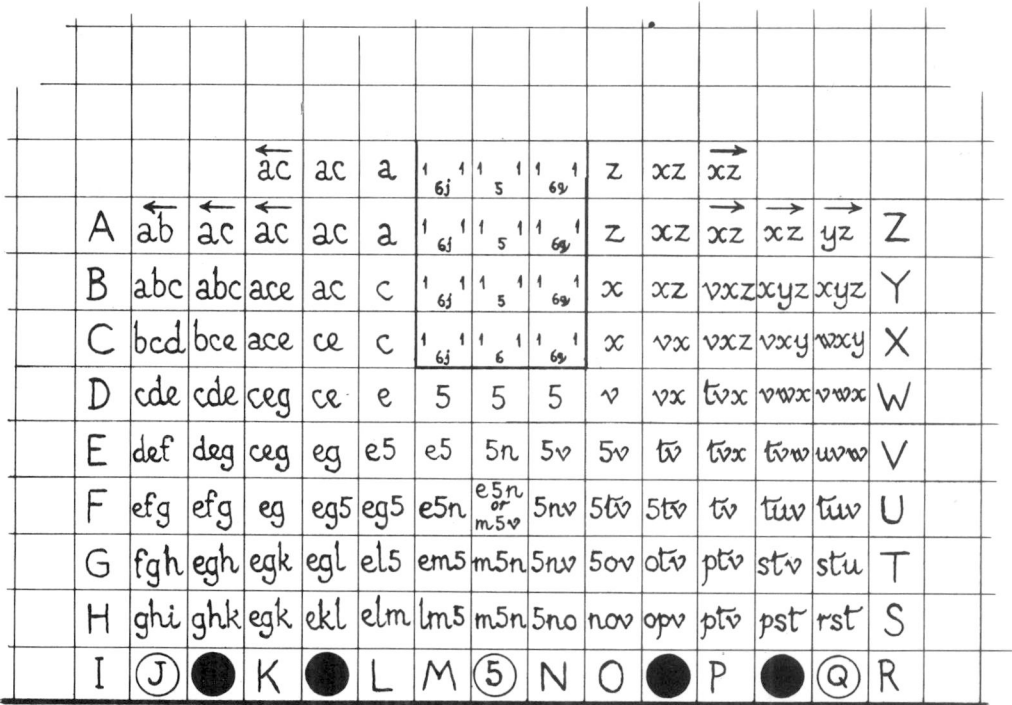

Figure 26. Close-up of Figure 25.

Unfortunately we haven't been able to formalize these remarks into a strategy for Chas. that's even as explicit as Geo.'s 33×33 one.

RECTANGULAR BOARDS

Geo. can't beat the King on the infinite strip of width 23, even if he moves first and has an unlimited supply of stones. However, if he moves first on a 24 by n board, he can win for sufficiently large n. The minimum value of n seems to be about 63. The King is immediately sidelined along whichever long edge he's nearest to. The King can circumvent the pseudocorners (I and R in Fig. 26) of Geo.'s sideline defence, but only by moving back to squares about midway between the two long edges. Geo. can then defend a second pseudocorner between the real corner and the first one. By the time the King reaches the corner, Geo. has prepared defences of both corners along a short edge of the board and a pair of opposite pesudocorners somewhere between the King and the unattacked short edge.

For each value of i, $24 \leqslant i \leqslant 37$, there appears to be a range of values of j for which the i by j board is a fair battleground for a Quadraphage game against the Chess King. We believe that the 32×33 board is fair and that Geo., moving first, wins with a strategy similar to that we gave for the 33×33 board. We leave the problem of determining the dimensions of all fair Quadraphage boards as a challenge for Omar.

EXTRAS

MANY-DIMENSIONAL ANGELS

can escape from the corresponding hypercube-eaters. This has been proved by Tom Körner who thinks that his proof could conceivably be adapted to the two-dimensional game. Don't write to us with *your* solution to the Angel problem unless you've taken account of the remarks on p. 609!

GAMES OF ENCIRCLEMENT

The games of this Chapter, and of the next two, are ones cf *encirclement* or *escape*. There are many games, going a long way back in history, in which this idea is combined with varying kinds of *capture*. Here are a few examples.

WOLVES-AND-SHEEP

There are several games played on Solitaire-like boards (Chapter 23). In **Wolves-and-Sheep** (Fig. 27(a)) the shepherd has 20 sheep, which have first move. They move one place forward or sideways only, onto unoccupied places. The two wolves can move similarly but on any of the indicated lines and can capture in these directions by jumping as in checkers (draughts), including multiple captures. A wolf failing to make a possible capture may be removed by the shepherd, so the sheep may be used as decoys. The shepherd wins if he gets nine sheep into the *fold* (top 9 positions of board).

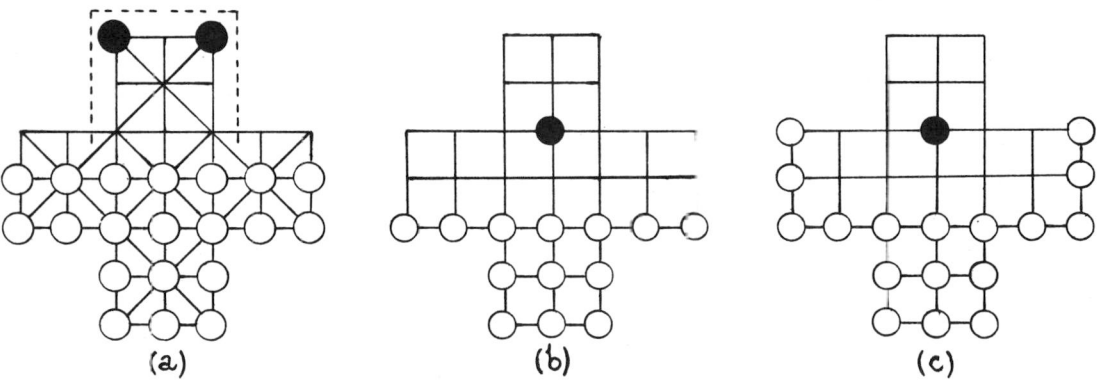

Figure 27. Wolves, Sheep and Other Animals.

The games shown in Figs. 27(b) and (c) are called Fox and Geese, although we use this name for a different game in the next chapter. They are similar to Wolves-and-Sheep, but there are no diagonal moves. The fox starts in any unoccupied position, and the geese try to crowd the fox into a corner. In Fig. 27(b) the 13 geese can move in any of the four orthogonal directions, but the 17 geese of Fig. 27(c) can't move backwards; they move like the sheep in Wolves and Sheep.

Hala-tafl (the Fox Game), and **Freystafl** are mentioned in the later Icelandic sagas. As in the Chapter 20 version of Fox and Geese, the more numerous animals win with correct play, but it's very easy to make mistakes!

TABLUT

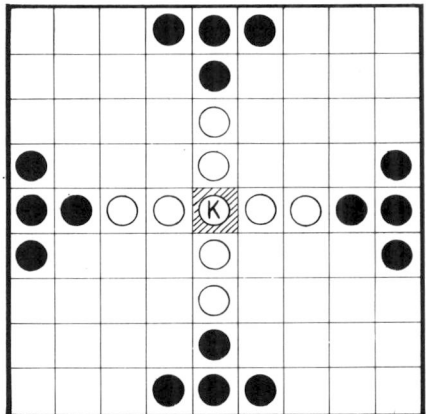

Figure 28. The Start of a Game of Tablut.

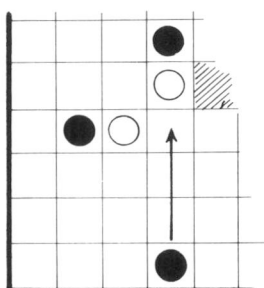

Figure 29. A Muscovite Captures Two Swedes.

Linnaeus, on his 1732 visit to Lapland, recorded a game played on a 9 × 9 board (Fig. 28) whose centre square, the **Konakis** or throne, may only be occupied by the Swedish King. He is protected by 8 blond Swedes and confronted by the 16 swarthy Muscovites. All the pieces move like the rook in Chess, any distance orthogonally. Capture of the King is by surrounding him, N, S, E and W by four Muscovites or by three Muscovites with the Konakis as the fourth square. Any other piece is removed by **custodian** capture, i.e. by placing two opposing pieces to the immediate N and S, or E and W of it. Figure 29 shows a Muscovite capturing two Swedes. A piece may move "into custody" without being captured. The aim of the Swedes is to get their King to the edge of the board.

SAXON HNEFATAFL

Only a fragment of a board has been found; it is probable that the game was played using the 19 × 19 positions of a modern Go board. See R.C. Bell's excellent little book for a possible reconstruction from a tenth century English manuscript. The game was evidently like Tablut apart from the size of the board and the number and position of the pieces.

We finish this chapter with two Chess problems which also involve escape or encirclement.

KING AND ROOK VERSUS KING

Most beginning Chess players soon learn how to win this ending, so it's a surprise to find a couple of non-trivial problems which use just this material, albeit on a quarter-infinite board.

In Fig. 30, can White win? If so, in how few moves? Simon Norton says it's better to ask, "what is the smallest board (if any) that White can win on if Black is given a win if he walks off the North or East edges of the board?" Can Omar prove that it's 9×11?

Figure 31 shows Leo Moser's problem: can White win if he's allowed to make *only one move with the Rook*? If you find yourself frustrated by this, partition the squares in the first three columns into the four sets

$$a1,a3,a5,\ldots,c2,c4,c6,\ldots$$
$$b1,b3,b5,\ldots$$
$$a2,a4,a6,\ldots,c1,c3,c5,\ldots$$
$$b2,b4,b6,\ldots.$$

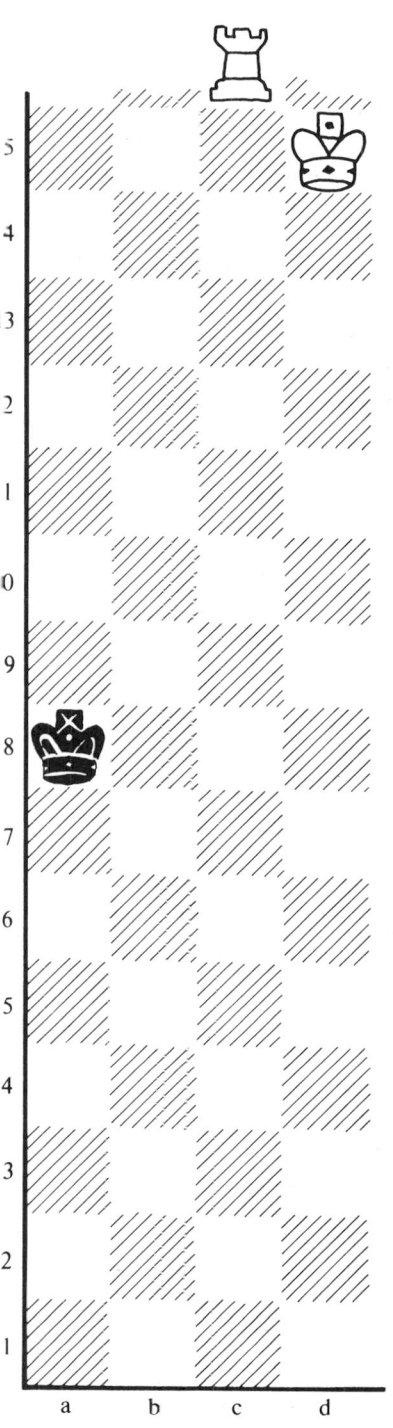

Figure 30. Simon Norton's Problem.

Figure 31. Leo Moser's Problem.

REFERENCES AND FURTHER READING

Robert Charles Bell, "Board and Table Games from Many Civilizations", Oxford University Press, London, 1969.

Richard A. Epstein, "Theory of Gambling and Statistical Logic", Academic Press, New York and London, 1967, p. 406.

Martin Gardner, Mathematical games: Cram, crosscram, and quadraphage: new games having elusive winning strategies, Sci. Amer. **230** #2 (Feb. 1974) 106–108.

C. Linnaeus, "Lachesis Lapponica", London, 1811, ii, 55.

H.J.R. Murray, "A History of Board Games other than Chess", Clarendon Press, Oxford, 1952.

David L. Silverman, "Your Move", McGraw–Hill, 1971, p. 186.

Chapter 20

Fox and Geese

While the one eludes, must the other pursue.
Robert Browning, *Life in a Love*.

The twelve good rules, the royal game of goose.
Oliver Goldsmith, *The Deserted Village*, l.232.

Figure 1. Playing a Game of Fox and Geese.

635

The game of Fox and Geese is played on an ordinary checkerboard between the *Fox*, who has just one black or red piece, and the *Geese*, who have four white ones. The players use squares of only one color (as in Checkers), and the Geese are initially placed in the squares marked O in Fig. 2. The Fox is usually placed at **X** in Fig. 2, but since the Geese seem to have the better chances, it is perhaps wiser to allow the Fox to choose his own starting square (provided this has the correct color), and then let the Geese have first move.

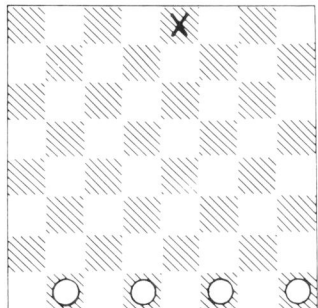

Figure 2. The Usual Starting Position.

The Geese move diagonally one place forward—like ordinary checkers they may not retreat. The Fox also moves diagonally one place, but like a King in Checkers, he may move in any one of the four diagonal directions. There is no taking or jumping. The Geese aim to trap the Fox so that he has no legal move, while conversely the Fox tries to break through the barrier of Geese so that he can stay alive indefinitely. We can therefore say simply that the first player unable to move is the loser, the usual normal play convention.

It is the general opinion that between expert players the Geese should win, but even against most moderately competent players a wily Fox can usually win a game every now and then, and if we let him choose his starting position, he should be able to defeat most novices for quite a long time. Perhaps those of our readers who have not met the game should take some time off to play a few games before reading further.

The question we shall ask and answer in this chapter is just how much of an advantage do the Geese have in this game? Perhaps we should first of all prove that the Geese really do have a winning strategy, even when the Fox is allowed the extra dispensation we suggest. In Fig. 3 we show the five types of position that our own favorite strategy relies on. The O's indicate the positions of the Geese, and the **X**'s indicate particularly critical positions for the Fox. When the Fox is in one of these places, we shall say that the Geese are *in danger*.

If the Geese follow our advice, they will play the game as follows. Most of the time they should play with their eyes closed, so that they will not be alarmed unnecessarily by the Fox's manoeuvres. We can offer them a guarantee that whenever they open their eyes the position will be like one of *A,B,C,D,E*, possibly left-right reflected, and that all they need do before closing their eyes again is see whether or not they are in danger. If not, they should make the moves indicated by the digits before the solidus (/) in Fig. 3, and if in danger, those indicated by the digits after the solidus. We also show by letters before and after the solidus which type *A,B,C,D,E* of position will be seen next. The indicated moves can be made with eyes closed, since we can also guarantee that the Fox will never be in the way, and so the Geese need only open eyes again when the sequence has been completed and the position is once again one of *A,B,C,D,E.*

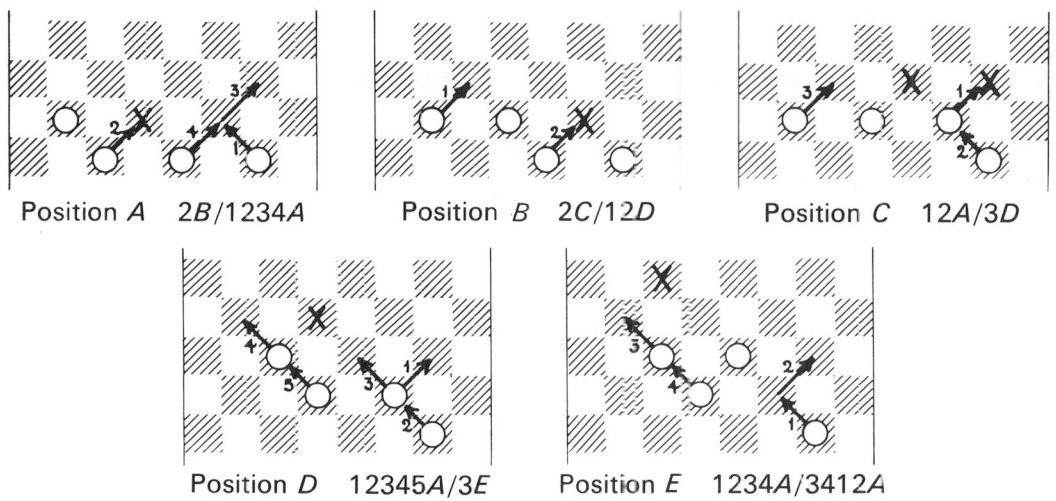

Figure 3. The Most Concise Strategy.

It is very easy to prove that the strategy works, when once the Geese have got into position A. As an example, we consider the position D. If the Geese have been behaving as we suggest, they can only have arrived at a position D from a position B or C in which they were in danger, and so the Fox can only be in one of a limited number of places. In fact he will be in one of **X,Y,Z,T** of Fig. 4_0.

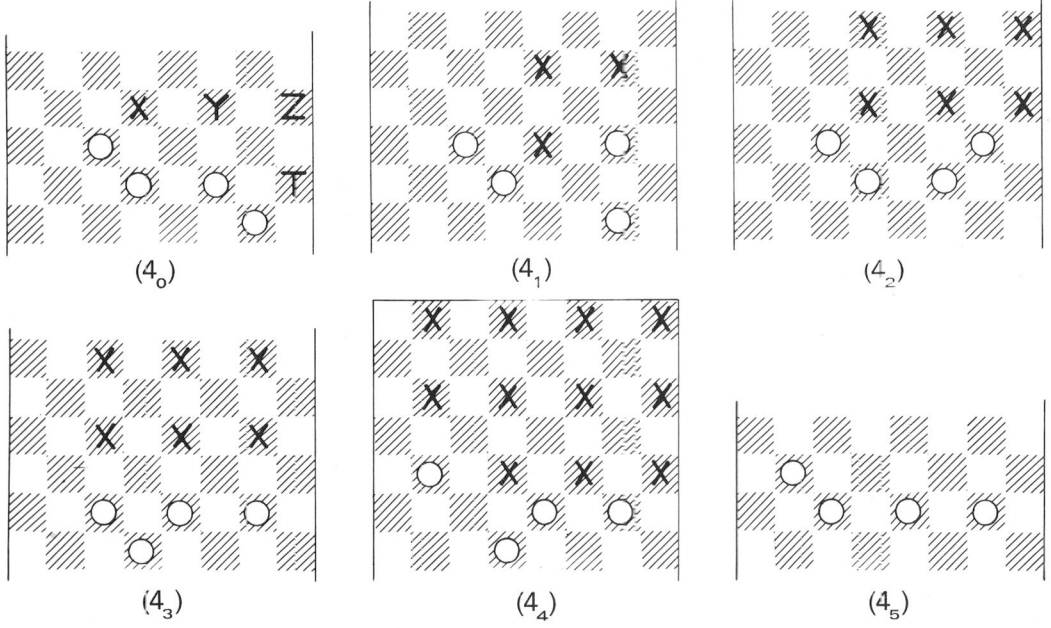

Figure 4. Analysis of Position D

Now if the Geese are in danger (i.e., the Fox is at **X**), they make a single move and arrive immediately at position *E*. So we can suppose that the Fox is at **Y**, **Z**, or **T**, and the Geese are told to make moves 1,2,3,4,5. Figures 4_1, 4_2, 4_3, 4_4, 4_5 show the position after each of these, and show why the next move in the sequence is legal, for in Figs 4_1 to 4_4 we have marked **X** in every possible place the Fox can be in just before the next move takes place. Since the position shown in Fig. 4_5 is of type A, this verifies that the strategy works from positions of shape *D*. Note that if the Fox were at **T** in Fig. 4_0 then he lost instantly after move 1, being trapped at the edge of the board.

We leave to the reader the corresponding discussion for positions of shapes *A,B,C,E*, noting merely that since *E* can be reached only from a position *D* with the Geese in danger, we need consider just two places for the Fox in position *E*, namely that marked **X** in Fig. 3, and the place two squares to the right of it. But for positions *A,B,C* the Fox may be on any square of the right color that is above the line of Geese.

How do the Geese start the strategy? The answer is that they can move into a position *A* on their first move, unless the Fox chooses the starting position *F* of Fig. 5. In this case, we advise them to make moves 1,2,3 and then open their eyes again to behold another anomalous position, *G*, from which we can put them back on track by giving them another sequence of three moves to be performed with eyes closed.

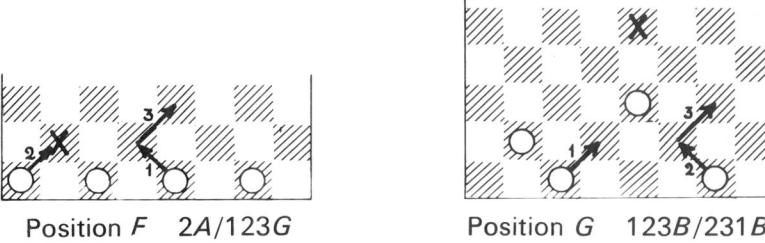

Position *F*   2A/123G   Position *G*   123B/231B

Figure 5. The Anomalous Starting Position.

SOME PROPERTIES OF OUR STRATEGY

From the standard starting position, and indeed from any position except the anomalous position *F* of Fig. 5, our strategy leads only to positions in which the Geese never occupy places in either the leftmost or the rightmost column. This is interesting because most reasonably competent players like to move the Geese so that they straddle a horizontal row whenever possible. But if the Geese do this, a cunning Fox can force them into a wider variety of positions than occurs in our strategy, so that the Geese then need to know a lot more about the game to be certain of winning. In fact a competent Fox can force the Geese into positions of all the types *A,B,C,D,E*, no matter what winning strategy they adopt. However, only if they adopt our strategy is he unable to force them into any other position in which they need open their eyes because their action cannot be automatic. Since our strategy is the only one with as few as five positions in which the Geese need to take a decision, it is the unique *minimal* winning strategy.

Many people who think they know winning strategies can be caught out every now and then by a clever Fox who seduces them into an unfamiliar part of the game. In fact it takes considerable skill to play the Fox against ordinary players so as to exploit to the full any deficiencies in their knowledge. We can give few hints here beyond remarking that the Fox should stay near to the Geese and try to bring them into the middle of the board around him before stepping sideways to slip through any gap they may leave at either side. The best starting position is from the square near to the Geese and directly below the **X** of Fig. 2 (and so two squares right of **X** in position F), but position F itself is also useful.

THE SIZE OF THE GEESE'S ADVANTAGE

Since the loser in Fox and Geese is the first player unable to move, we should be able to apply the general theory of the rest of the book and evaluate the size of the advantage for the Geese, at least approximately. Is it perhaps $\frac{1}{2}$ of a move? Or maybe $\uparrow$? More probably it is some very complicated value about which we can say little of any use.

The answer is something of a surprise, in that we can evaluate the advantage very precisely indeed, and it turns out to have a rather unusual nature. Let us show first that the advantage is strictly greater than 1 move. To see this, of course, we play the game

Fox and Geese $-$ 1

(with the Geese being Left), and show that the Geese can still win. The new game is just Fox and Geese in which the Fox is allowed to pass just once in the game (i.e., when he moves in the component -1). In fact it seems that this allowance is not of very much use to the Fox, and in almost all positions it will only help the Geese if the Fox passes.

However, the analysis illustrated in Fig. 4 shows that our minimal strategy no longer works, because it often makes heavy use of parity to keep the Fox out of the way. In Fig. 4_1, for instance, the Fox might be in the position nearest the Geese and then stay there in Fig. 4_2, making the move to Fig. 4_3 illegal. But such antics are not of much use, and in every case the Geese can survive, although they need to know about a larger number of positions. It even appears that the Geese can allow the Fox to pass arbitrarily many times, provided that he does not pass on two successive moves. In any case, these remarks are enough to show that

Fox and Geese > 1.

We now argue that on the other hand we have

Fox and Geese < 2.

The idea is that the Fox waits until the Geese are very near the top of the board, and then passes twice in succession. The Geese are then forced to advance past him, and the Fox can slip between them and wait until they lose by being jammed at the top of the board. It would be very difficult to give a formal proof of this, because of the very great number of possible moves for the Geese. However, if the Fox hugs the barrier formed by the Geese and seizes any opportunity to get past, all but a few moves of the Geese from any given position are plainly disastrous, and the Geese will necessarily advance in a fairly straight formation. It seems that the Fox can then arrange to move into an advantageous position, perhaps H or I of Fig. 6. In position H he should pass twice, as soon as the Geese have moved, and when they have moved from position I, the

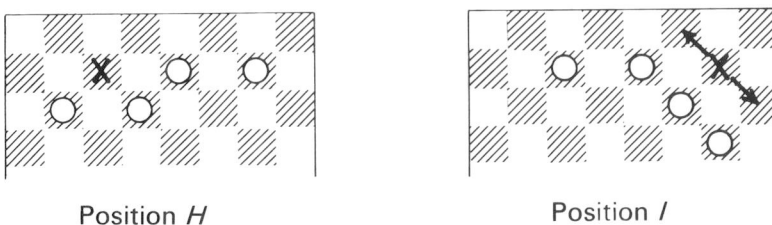

Position *H*   Position *I*

Figure 6. When Should the Fox Pass?

Fox makes one of the indicated moves and then passes twice. Of course this does not amount to a formal proof, but the interested reader should play a few games for himself to see how very easy it is to force such a position, if the Geese haven't in fact lost much earlier. A complete proof should not be difficult, but would certainly be very long, since there are many tedious cases to consider. We prefer to suppose that the above argument will convince the reader.

Obviously, the next value to examine is $1\frac{1}{2}$. The tree for $-1\frac{1}{2}$ appears as Fig. 7, and shows that the game

$$\text{Fox and Geese} \; - \; 1\tfrac{1}{2}$$

is Fox and Geese in which the Fox (Right) is allowed to pass twice, and the Geese may also pass, but only once, and only provided that the Fox has not yet passed.

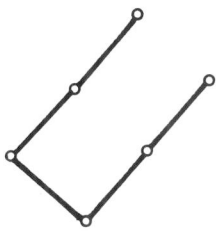

Figure 7. Minus One and a Half.

We need not examine this any further. The Geese obviously can't afford to waste time by passing just when they need all the moves they can get to stop the Fox slipping through, so they will not be amused at our suggestion that they might like to do so. It might have been some use to allow them to pass *after* the Fox's first pass, because then their problem is that they are forced to move past him, but of course this is exactly what we don't intend to permit. So $1\frac{1}{2}$ is too big.

Exactly the same argument shows that $1\frac{1}{4}$, $1\frac{1}{8}$, etc., are also too big. The Geese will only be insulted by the suggestion that they be allowed to pass up to three times (say) provided they do so before the first of the Fox's allowance of two passes. But Fig. 8 shows that this is equivalent to subtracting $1\frac{1}{8}$ from the game. It is plain that we can now assert that

$$1 < \text{Fox and Geese} < 1 + \frac{1}{2^n}$$

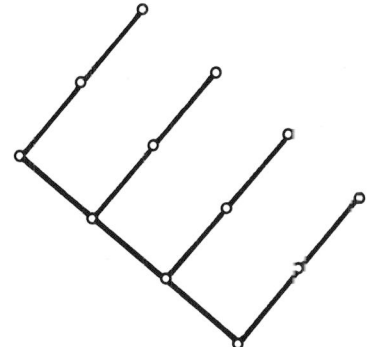

Figure 8. Minus One and One Eighth.

for all $n=0,1,2,3,\ldots$. In other words, Fox and Geese is only greater than 1 by an infinitesimal amount.

It's amusing that we can say such a thing about a real-life children's game, but of course we need not stop with this assertion. We know many games that are positive but smaller than every positive number, the game $\uparrow$ being the most notorious. How does Fox and Geese compare with $1+\uparrow$?

The difference

$$\text{Fox and Geese} - (1+\uparrow)$$

is best thought of as the game

$$\text{Fox and Geese} + \downarrow$$

with an additional allowance of one pass for the Fox. What should the Geese do? They plainly want to eliminate the component $\downarrow$, for we already know that a single pass for the Fox will not prevent the Geese from winning.

So before they make any move in the game of Fox and Geese proper, the Geese should move in $\downarrow$, leaving $*$, and then if the Fox doesn't do so, at the next move the Geese should replace $*$ by 0. If the Fox moves in $\downarrow$ or $*$, so much the better. So in fact

$$\text{Fox and Geese} > 1 + \uparrow,$$

and an exactly similar argument shows that we have

$$\text{Fox and Geese} > 1 + n.\hat{\ }$$

for $n=1,2,3,\ldots$. The point is that $n.\uparrow$ is an all small game, since whenever there is still a legal move for one player, so is there for the other. The Geese can therefore gobble up all the all small component before starting the game proper, and it doesn't really matter what the Fox does meanwhile—he can either idle away this time on the checkerboard or chew up some of the all small game himself, since it would be rather foolish of him to waste his free pass—but whatever he does, his retribution will come when the real game starts.

THE PARADOX

At this point, we seem to have found a contradiction. No matter who starts, a game of Fox and Geese can last at most 57 moves, since each of the four Goose pieces can advance at most seven places, and alternating with these 28 moves of the Geese there can be at most 29 moves for the Fox. But one of our theorems asserts that every finite game that is less than every positive number is necessarily less than some finite multiple of ↑, and we seem to have shown that

$$n \cdot \uparrow < \text{Fox and Geese} - 1 < \frac{1}{2^n}$$

for all n. What is the explanation?

Perhaps the trouble lies with the admittedly rather fluid argument which purported to make it plain that the Fox could win with two passes? Or maybe he can win with two, but the extension suggesting that $1 + \frac{1}{2^n}$ was still enough is not quite so insultingly trivial after all? The arguments about 1, and $1 + n \cdot \uparrow$, seems pretty safe, but maybe we overlooked something even there?

It's easy for you to make such suggestions, but rather harder to substantiate them. I'm absolutely certain you can't beat me if I play the Geese in

$$\text{Fox and Geese} - (1 + n \cdot \uparrow),$$

and would be prepared to bet quite a large sum that on the other hand I can win playing the Fox in

$$\text{Fox and Geese} - \left(1 + \frac{1}{2^n}\right),$$

no matter what number n you choose.

Oh, of course! I see it now! Fox and Geese isn't really a finite game at all! It's true that it can last at most 57 moves if played by itself, but we are no longer playing a single game but a sum, and when we do that the Fox may make several consecutive moves in the Fox and Geese game while the Geese are moving elsewhere. So Fox and Geese is really an infinite game, and we know from Chapter 11 that there are infinite games (even some numbers) greater than all multiples of ↑, yet still less than $\frac{1}{2^n}$ for every finite number n, for example

$$\frac{1}{\omega} = \{0 \mid 1, \tfrac{1}{2}, \tfrac{1}{4}, \ldots\}.$$

We've just got to examine Fox and Geese a little more closely, that's all. Let's start by comparing it with $1 + \frac{1}{\omega}$. What does that amount to? We allow the Fox to pass just twice, as before, but this time the Geese can pass any finite number of times provided they do so before the Fox's first pass, and provided they announce when making their first pass just how many passes they intend to allow themselves. In other words, at their first pass they can choose what allowance to give the Fox from then on from the numbers $-2, -1\frac{1}{2}, -1\frac{1}{4}$, etc., in conformity with the equation

$$-1 - \frac{1}{\omega} = \{-2, -1\tfrac{1}{2}, -1\tfrac{1}{4}, \ldots \mid -1\}.$$

Just how offensive can you get? The Geese weren't amused when we offered them one or two passes under these insulting restrictions, and to offer them an unbounded number only adds injury to insult. Obviously

$$\text{Fox and Geese} < 1 + \frac{1}{\omega}.$$

Let's *see* just how offensive we *can* get, if we really try. The equation

$$-1 - \frac{1}{2\omega} = \left\{ -1 - \frac{1}{\omega} \,\middle|\, -1 \right\}$$

gives us a hint. For every ordinal number β, we have the equation

$$-1 - \frac{1}{2^{\beta+1}} = \left\{ -1 - \frac{1}{2^\beta} \,\middle|\, -1 \right\},$$

and more generally

$$-1 - \frac{1}{2^\alpha} = \left\{ -1 - \frac{1}{2^\beta} \text{ (for all } \beta < \alpha) \,\middle|\, -1 \right\}$$

for every ordinal number α, with β ranging over all smaller ones. We can make the following superb offer to the Geese. They can choose any ordinal number α they like, no matter how large say $\omega, \omega^2, \omega^\omega, \omega^{\omega^\omega}, \ldots$. Then whenever they pass they must replace the number by a strictly smaller one, and at their next pass by a smaller one again, and so on. There's just one small restriction which we hope they won't mind, after all this generosity. The Fox will be allowed to pass just twice, even though the Geese may have passed a million times, but would the Geese kindly not pass *after* the Fox has done so even once. The Geese's probable reaction suggests very strongly that

$$\text{Fox and Geese} < 1 + \frac{1}{2^\alpha}$$

for every ordinal number α, no matter how infinite it may be.

On the other hand, we can compare Fox and Geese with 1 plus infinite multiples of ↑. Such games exist, but even the infinite multiples of ↑ are all small games, and the argument we gave for the finite multiples of ↑ still applies. The only difference is that we can no longer give a bound for the length of time it will take the Geese to gobble up the ups. So we can strengthen our previous result to

$$1 + any \text{ multiple of } ↑ < \text{Fox and Geese} < 1 + any \text{ positive number.}$$

Even for infinite ending games such a thing can't happen. However in Fox and Geese the Fox can move infinitely many times even if the Geese don't. So Fox and Geese *isn't* an ender, even though it *is* a stopper (see Chapter 11).

PUNCHING THE CLOCK

We can make Fox and Geese into a genuine ending game by giving the Fox a kind of time clock, as in Fig. 1. At any time, the Fox's side of the clock bears a number, and whenever he moves, he must *punch the clock*—that is, replace this number by a smaller one. We could, say, restrict him to at most 100 moves by starting the clock with 100 showing. After our recent discussion, this seems much too restrictive, however, and we should allow him to choose any ordinal number, even an infinite one, to start with. Of course the Geese's clock need never show any number larger than 28, and could even be dispensed with, because a clock for the Geese is built in to the very nature of the game.

Let's call the modified game, with the Fox's clock started at α, (Fox and Geese)$_\alpha$. We've made things worse for the Fox, and so should expect

$$(\text{Fox and Geese})_\alpha > \text{Fox and Geese}.$$

In fact we can show that

$$(\text{Fox and Geese})_{\alpha+30} < 1 + \frac{1}{2^\alpha} < (\text{Fox and Geese})_\alpha.$$

For, the Geese can win

$$(\text{Fox and Geese})_\alpha - 1 - \frac{1}{2^\alpha}$$

by making use of those extremely insulting passes we offered them. Whenever the Fox punches the clock, replacing β by γ, say, the Geese make use of their pass move which replaces $-1 - \frac{1}{2^\beta}$ by $-1 - \frac{1}{2^\gamma}$. The Geese make no move in the Fox and Geese game proper, for the Fox will lose on time before they need to!

On the other hand, the Fox can win

$$(\text{Fox and Geese})_{\alpha+30} - 1 - \frac{1}{2^\alpha}$$

in a similarly underhand way. He can keep his clock showing at least 30 more moves than the number of passes left to the Geese, and when the Geese have run out of pass moves, can win the game making use of his own two passes. This takes at most 30 moves.

So how big *is* Fox and Geese? It seems natural to regard Fox and Geese as the limit of the truncated version (Fox and Geese)$_\alpha$ as α increases indefinitely through the Class **on** of all **ordinal** numbers. Or in symbols

$$\text{Fox and Geese} = (\text{Fox and Geese})_{\textbf{on}},$$

since the name **on** is also used for the limit of all the proper ordinal numbers. The concept **on** is not really an ordinal number itself (just as ∞ is not really an integer, but the limit of the integers) but it does no harm to think of **on** as if it were the largest ordinal number (which doesn't really exist but can be thought of as the absolutely largest infinite number).

Since $\alpha + 30$ tends to **on** as α does, this makes it sensible to write

$$\text{Fox and Geese} = 1 + \frac{1}{\mathbf{on}}$$

in which the left-hand side isn't a genuine game, and the right-hand side isn't a genuine number! Although our argument doesn't amount to a formal proof, it seems very likely that this equation holds exactly in the loopy game sense of Chapter 11.

There is a famous mathematical argument called the Burali–Forti paradox, according to which there should really be a largest ordinal number. We can say that the amount by which Fox and Geese fails to be a genuine game is exactly the same as the extent to which the Burali–Forti argument is genuinely paradoxical!

EXTRAS

MAHARAJAH AND SEPOYS

As we said in the Extras to Chapter 19, there are very many games involving encirclement, often mixed with various forms of capture. The name Fox and Geese, for example, has also been used for various games played on the English Solitaire board (Chapter 23) a couple of which are described in the Extras to Chapter 19. Most of these games are typified by a considerable numerical imbalance between the opposing forces. This is compensated by much greater mobility of the numerically inferior pieces. An extreme example is Maharajah and Sepoys which is played like ordinary Chess. White has a standard set of 16 pieces, starting from their usual positions, while Black has a single piece, the Maharajah, who starts on any unoccupied square and can move like a Chess Queen *or* a Chess Knight. The object of both sides is checkmate, of either the White King or the Maharajah. As in most of these games the Lord is on the side of the big battalions, and White wins with correct play.

Other games, where the forces are more equal in number and character are mentioned in Chapter 22.

REFERENCES AND FURTHER READING

Robert Charles Bell, Board and Table Games from Many Civilizations", Oxford University Press, London, 1969.

Maurice Kraitchik, "Mathematical Recreations", George Allen and Unwin, London, 1943.

Fred. Schuh, "The Master Book of Mathematical Recreations", transl. F. Göbel, ed. T. H. O'Beirne, Dover, N.Y., 1968. Chapter X: Some Games of Encirclement, pp. 214–244.

Chapter 21

Hare and Hounds

I like the hunting of the hare
Better than that of the fox.
　　　Wilfred Scawen Blunt, *The Old Squire*.

THE FRENCH MILITARY HUNT

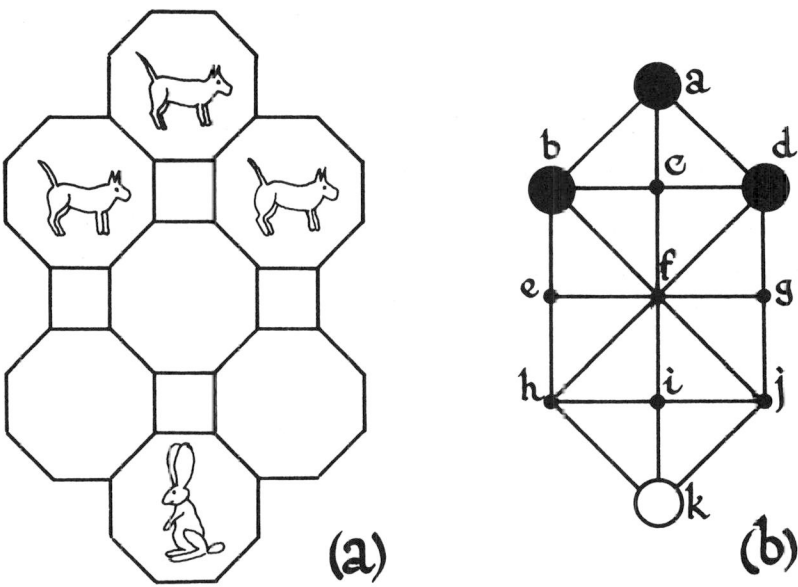

Figure 1. The French Military Hunt. Hare and Hounds on the Small Board.

This little game is very like Fox and Geese. It features a hunter whose three hounds (*dogs*) try to trap a hare (*rabbit*) on the board shown in Fig. 1(a). If you can't persuade enough animals to make the right manoeuvres, you can play with four coins on the nodes of the equivalent board shown in Fig. 1(b). It becomes more interesting on the larger board of Fig. 2. At each turn the hunter moves any one hound to a neighboring empty place, and the hare makes a similar move. However the hounds, starting from the top, may not retreat, although a hound may go back and forth horizontally as between *e* and *f* in Fig. 1(b). The hare is completely free to advance or retreat or move horizontally. The hounds win by trapping the hare so that he cannot move at his turn. If the hounds fail to advance in ten consecutive moves, the game is usually declared a win for the hare.

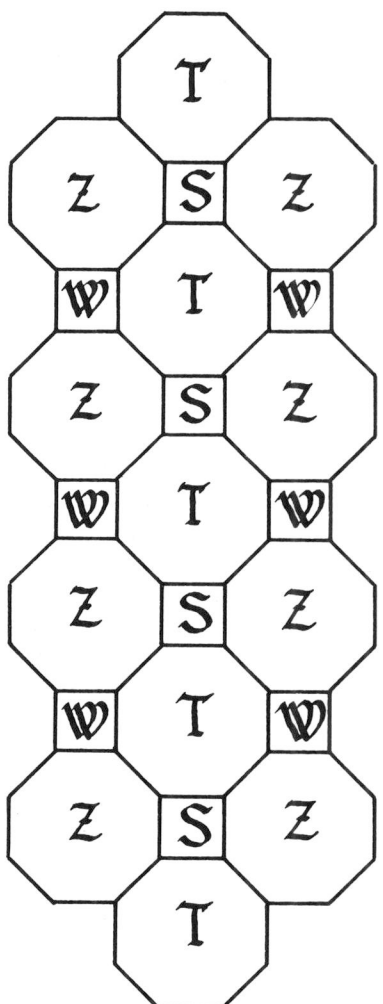

Figure 2. The Larger Board, With Four Types of Place.

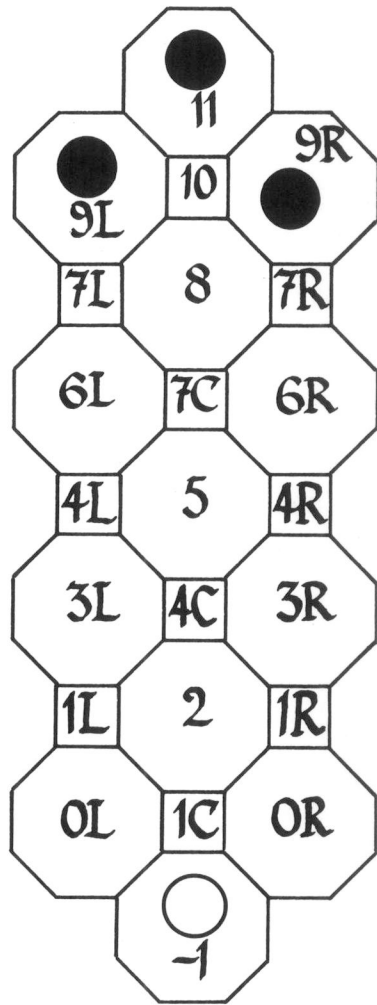

Figure 3. The Larger Board, Numbered for the Trace.

TWO TRIAL GAMES

If you want to see how the game goes, first set up the board and watch an expert hunter against a novice hare:

hounds: *abd cbd fbd fed fhd fhg fhj ihj* (wins)

hare: *k*   *i*   *j*   *g*   *j*   *i*   *k*

First game.

The chase looks so easy that the novice decides to direct the hounds in pursuit of an expert hare:

hounds: *abd cbd fbd fed feg fhg fig eig fig fij*

hare: *k*   *j*   *i*   *h*   *k*   *j*   *k*   *j*   *k*   *h*

Second game.

and now the hare will escape by *e* or *f*.

If expert hounds chase an expert hare on Fig. 1, who wins? And what if the hare makes the first move? Or starts from a different place? (See the Extras.) And (when you've become more expert) what about Fig. 3?

HISTORY

According to Lucas the game (on Fig. 1) was popular among French military officers in the nineteenth century. Some say it was invented by Louis Dyen; others attribute it to Constant Roy. It was solved by Lucas (1893) and Schuh (1943) and popularized (again) by Martin Gardner (1963). Schuh's analysis was based on a list of 18 classes of winning positions for the hounds (reproduced in the Extras) and he recognized that "the opposition" plays a key role, but he had no exact definition for it. In a later section we'll give a definition which simplifies the game on Fig. 1 and also allows us to solve that on Fig. 3.

THE DIFFERENT KINDS OF PLACE

Let's look at the board more closely. There are really two types of octagon: central ones (T in Fig. 2) and side ones (Z). There are also two types of square: central squares (S) and side squares (W). Except near the very top or bottom of the figure, each T or Z is next to at least one place of every other type, but each W or S is only next to octagons, T and Z. Since W and S are never adjacent, it's sometimes convenient to lump them together into a single class, N. Of the three types T, Z and N, every place, even the ones at the top and bottom, is next to at least one place of each other type, but to none of its own type. The letters correspond to remainders after division by 3 of the numbers from Fig. 3:

Remainder Zero : Z
Remainder oNe : N = Weak or Strong
Remainder Two : T

In Fig. 3 the difference of two numbers in adjacent places is always 1 or 2.

The sum of the numbers occupied by the four animals is an important property of the position; we call it the **trace**. Every move changes the trace by 1 or 2. If the hounds succeed in trapping the hare at the bottom of the board, then the hounds are at 0, 1, 0 against the trapped hare at −1 and the trace is 0. If instead the hounds trap the hare on the side of the board, say at 1L, then the hounds end on 3, 2, 0 against the hare on 1, and the trace is 6. It can easily be checked that

> No matter where
> You trap the hare,
> The trace you'll see
> Divides by *three*

TRIALITY TRAPS!

THE OPPOSITION

The best way for the hounds to make their trap is to move so that they leave the trace a multiple of 3 at every turn. We call this "keeping the opposition". If they do this, the hare's move must be to a non-multiple of 3, because it changes the trace by 1 or 2. But whenever the trace is not divisible by 3 the hunter usually has a choice of several hound moves which restore it to a multiple of 3, and among these he should find one which restores a winning position.

> Threefold traces
> Win most chases.

KEEPING THE OPPOSITION

If you check the traces for our first game, with the board numbered as in Fig. 4, you'll see that the hounds always kept the opposition:

hare:	*k*	*i*	*j*	*g*	*j*	*i*	*k*
hounds:	*cbd*	*fbd*	*fed*	*fhd*	*fhg*	*fhj*	*ihj*
trace:	9	9	6	6	3	3	0

Since the hare doesn't want to be trapped, he doesn't want the hunter to move to positions whose trace is divisible by 3. The best way to prevent this is for the hare to grab the opposition by moving to such a position himself. Then any hound move will change the trace to a non-multiple of 3 and the hare is likely to be able to regrab the opposition. This is the way the hare won our second game. The hounds blundered on their second move by playing from 4 to 2, giving a trace of 8, and from then on the hare managed to retain the opposition at every turn:

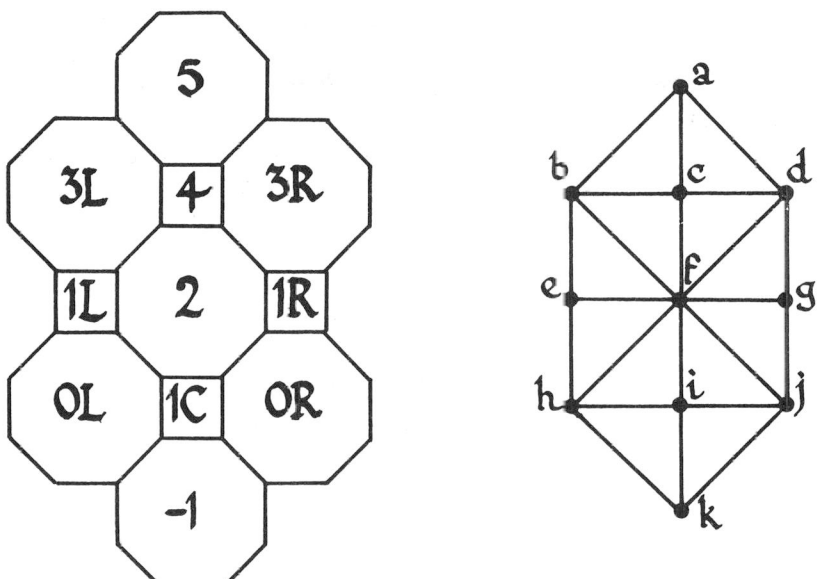

Figure 4. The Board Numbered for Determining the Opposition.

hounds:	*abd*	*cbd*	*fbd?*	*fed*	*feg*	*fhg*	*fig*	*eig*	*fig*	*fij*
hare:	*k*	*j*	*i!*	*h*	*k*	*j*	*k*	*j*	*k*	*h*
trace:	10	10	**9**	6	3	3	3	3	3	3

So whoever can move to a position whose trace divides by 3 is said to have the opposition. The opposition is certainly a valuable commodity which both players desire. But it's not all there is to the game, because sometimes the *hounds* may have the opposition but be unable to keep it without letting the hare escape behind them. In other cases the *hare* may have the opposition for several moves, but then lose it because the hounds block his only moves to places which would restore it. However, such positions are rather rare, and the average player who combines the principle with a little commonsense will usually trap a novice hare on the small board. An annotated example appears on p. 652.

WHEN HAS THE HARE ESCAPED?

He has **escaped** if he has passed or is passing two hounds, unless he is on a *square* place (W or S) and the hounds can immediately occupy the neighboring octagons (Z or T) aside or ahead of him.

Although he may have not escaped, the hare is **free** in some other positions in which the hounds can never force him to retreat. This certainly happens if he's strictly passed a hound and is not on a Weak (W) square, or, if he's on a central octagon (T) and is past or passing at least one hound.

Third Game

Hounds	Hare	Trace	Comments
3L, 5, 3R	−1	10	
3L, 4, 3R		9	Taking the opposition
	0R	10	
2, 4, 3R		9	A novice hunter might have moved 4 to 2, giving a "solid" position, but losing the opposition.
	1C	10	
2, 3L, 3R		9	The other "reasonable" move, 3R to 1R, changes the trace by the wrong amount. Since the move from 2 to 1 would allow the hare to escape, there's really only one choice.
	−1	7	
1C, 3L, 3R(!)		6	Because the hounds can't retreat, they can never increase the trace by 2, so to gain the opposition they must decrease 7 to 6 by moving a hound from 2 to 1. A move to 1R or 1L won't lose, but wastes time, since the hare can force the hounds back to the present position position by going to 1C.
	0R	7	
1C, 2, 3R		6	The other two moves (3R to 2, 1C to 0L) that restore the trace to 6 would let the hare escape.
	−1	5	
1C, 2, 4(!)		6	Once again, the other moves (3 to 1, 2 to 0) keeping the opposition would let the hare escape, leaving only this unlikely looking move.
	0R	7	
0L, 2, 4(!)		6	4 to 3R repeats; 2 to 1 allows escape; only 1C to 0L makes progress
	1R	7	
0L, 2, 3R		6	
	0R	5	Obvious
0L, 2, 1R		3	
	−1	2	Hare's last gasp.
0L, 0R, 1R		0	The novice hunter might now lose by playing from 1R to 0R.
	1C	2	
0L, 0R, 2		3	The only time the hounds reach a trace larger than their previous one.
	−1	1	
0L, 0R, 1C		0	Wins.

LOSING THE OPPOSITION

To analyze the exceptional positions, when someone wins in spite of not having the opposition, it's best to consider the types of place the animals occupy. For example, all the positions where the hounds have just won are of type Z^2NT, meaning that 2 animals are on Z places, 1 on N and 1 on T.

Some of the exceptional cases arise from the difference between the Strong and the Weak types of N places. Each Strong (central) N square is next to *four* other places, while each Weak (side) square is next to only three. Other things being equal, an animal should prefer a Strong place to a Weak one, since both make the same contribution to the opposition; but the Strong place is likely to offer him more choices later. For example, one exceptional case arises when the hounds move to Fig. 5. Despite the fact that the hounds have the opposition (trace 3), the

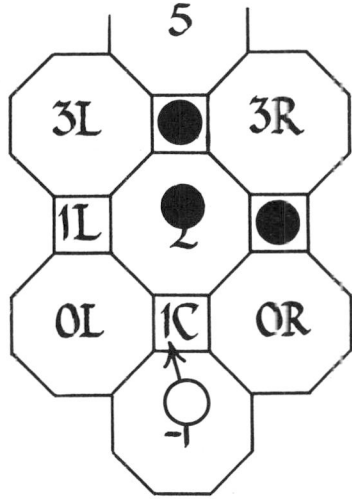

Figure 5. An Exceptional Hare and Hounds Position.

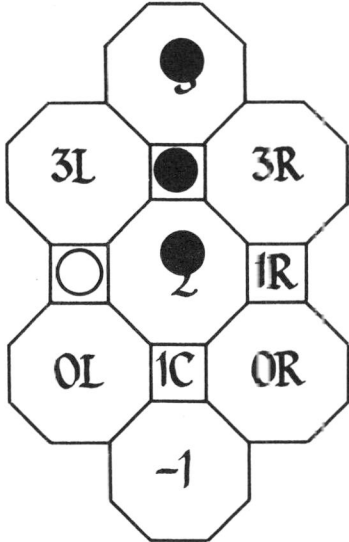

Figure 6. Another Exception to the Opposition Principle.

hare wins by playing to 1C, because now the only hound moves which keep the opposition let the hare escape. In some sense this N^2T^2 position loses because the hound at 1R is on a Weak square. On the other hand, we saw in our third game that a hare on -1 has no defence against hounds on 4C, 2, 1C (another N^2T^2 position). Unless the hare has passed one or more hounds, S^2T^2 wins for the hounds, but SWT^2 often loses.

As another example, suppose the *hare* has just moved to the position of Fig. 6. He has the opposition, but after the hound on 4C moves to 3L, the hare must retreat to 0L, losing the opposition and the game. But a hare in place 1C against these hounds would have both the opposition and a winning position. Once again, the difference between a Strong and a Weak square means the difference between winning and losing, this time for the hare.

A STRATEGY FOR THE HARE

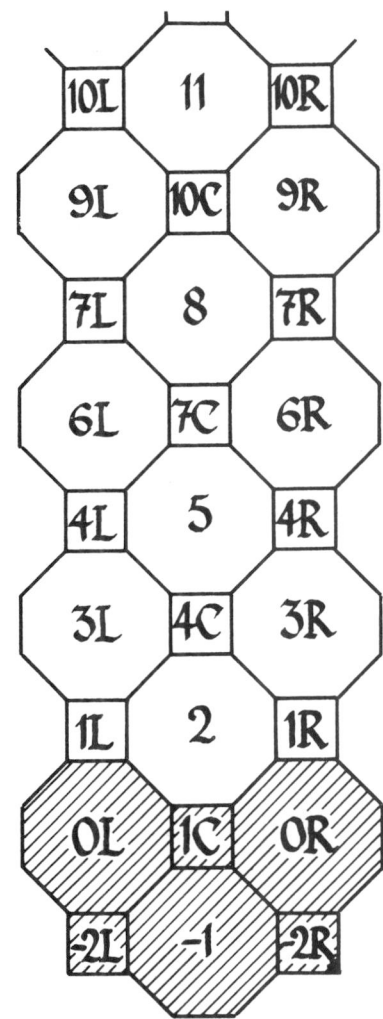

Figure 7. Keeping the Opposition on a Semi-infinite Board.

We'll show that an expert hare that has the opposition on the semi-infinite board of Fig. 7 can either keep it indefinitely or escape, unless he has to start from the **Scare'm Hare'm** position (Fig. 8). In fact the hare will always stay on the six shaded places numbered 1C, 0L, 0R, −1, −2L and −2R, unless the hounds let him out. His basic strategy is to keep the opposition.

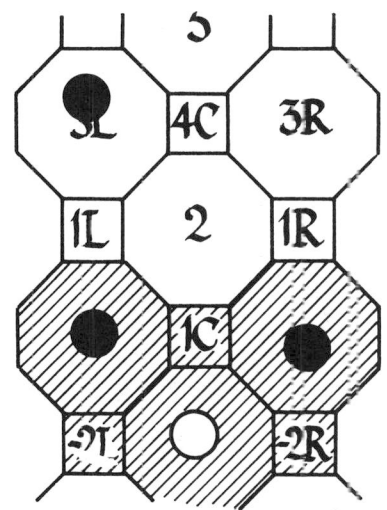

Figure 8. The Scare'm Hare'm Position.

If possible, escape or gain your freedom!
Otherwise, keep on the six shaded places, and
if you can keep the opposition by a move to a
non-Weak place, do so.
If a move to S(1C) is blocked, then
(A) against hounds on T^2S, move to W(−2L or −2R),
(B) against hounds on ZN^2, advance to Z (losing the opposition) on the other side of the board from the hound-occupied Z.
If a move to Z (0) is blocked,
(C) go to T (−1) (losing the opposition).

THE HARE'S STRATEGY

If these rules allow two or more moves, choose any one. If they allow none, resign (or hope for a mistake)!

First we show that if the hounds reach Fig. 8, a recent hare's move must have been of type (A), (B) or (C). For if the hounds came from a position in which they *had* the opposition, then the hare, after his last move, *didn't*, and the present position must have been reached by (B) or (C). Otherwise the hounds have come from a position whose trace was congruent to 1, mod 3, and hence from Z^2N, since they are on Z^3 in the figure. At the hare's last move 2 was vacant and either 0L or 0R was occupied by a hound. But if the hare came from 0L, 1C or 0R he could have escaped by moving to 2 and so he must have come from the Weak square -2, which he can only have reached by a move of type (A).

Suppose you've just made a move of this strategy which was not of type (A), (B) *or* (C). *Then you have* the opposition and you're *not* on a Weak square and the table below shows that the Hare's Strategy always gives you another move, unless you're faced with the Scare'm Hare'm Position.

To From	Z	S	T
Z	—	(A) or gain freedom by advance to T.	already free
S	escape by advance to T.	—	already free.
T	escape, since *not* Fig. 8.	(A) or (B)	—

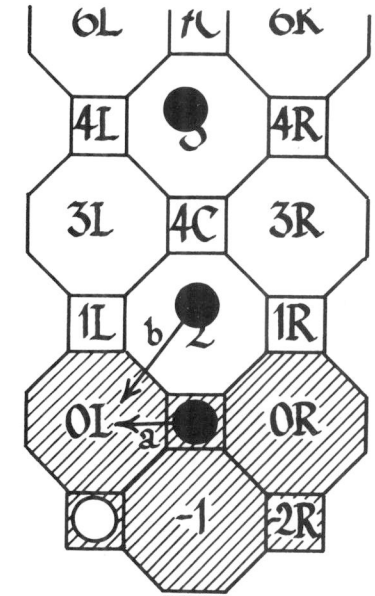

Figure 9. Position After a Move of Type (A).

Next suppose you've just made a move of type (A). Then in the next few moves you can either escape or regain the opposition by a move *not* of type (A), (B) or (C), and from which the hounds can't immediately move to Fig. 8. This is because when (A) is applied, the Strong square 1C must be occupied and also two central octagons (not including −1 because the hare has not escaped); see Fig. 9. Now the only way the Hare's Strategy can lose the opposition from an N square is by a move of type (C) after a hound moves to 0L. But after move (a) in Fig. 9 the hare regains the opposition, while after move (b) he soon escapes. The hounds can't reach the Scare'm Hare'm Position in time.

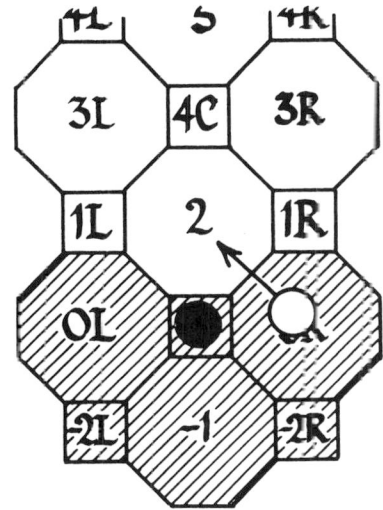

Figure 10. Position After a Move of Type (B).

Now suppose you've just made a move of type (B) (Fig. 10). Then you threaten to escape by moving to the empty T place ahead of you. If the hounds fill this from N, you escape by advancing to W, and if a hound from Z fills it you can reacquire the opposition by retreating to T. The hounds can't straight away reach the Scare'm Hare'm Position.

Finally, *if you've just made a move of type* (C), and were on a Strong square, *both* adjacent Z's must be occupied and you could have escaped. So you were on a Weak square and we have already discussed the situation following your previous move, which must have been of type (A).

ON THE SMALL BOARD

The Hare's Strategy shows that if they don't have the opposition the hounds can only win on the small board by keeping a hound on 5 until they can grab the opposition by moving him to 4 or 3. If they move first from 3L, 5, 3R the hounds can beat a hare starting anywhere except 4. Here is a sample game.

Hounds	Hare	Remarks
3L, 5, 3R	1C	(Or the hare could start on 1L or 1R.)
3L, 5, 2		
	−1	If instead to 0, the hounds take the opposition by moving from 5 to 4.
1L, 5, 2		
	1C	If instead to 0, the hounds take the opposition by moving from 5 to 3.
0L, 5, 2		
	−1	If instead to 0, the hounds take the opposition by moving from 5 to 4.
1C, 5, 2		
		Now, since there is no place −3 on this board, the hare is forced to give the hounds the opposition and the game.

ON THE MEDIUM AND LARGER BOARDS

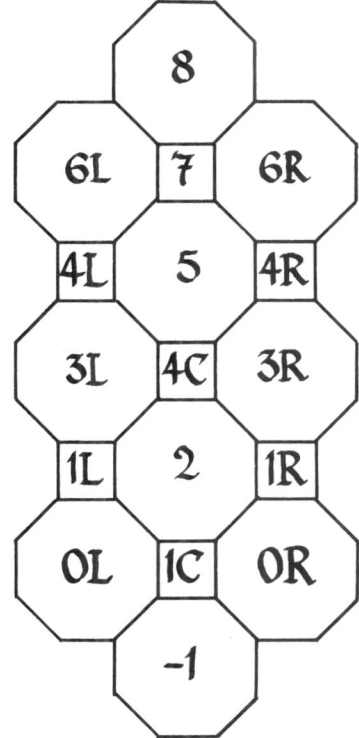

Figure 11. The Medium Board

By a slight extension of this argument, the hounds, moving from 6L, 8, 6R on the Medium Board (Fig. 11) can trap a hare starting on −1, 0, 2, 3 or 5. Since they have the opposition they can certainly win on the Small Board got by dropping numbers −1, 0 and 1 (Fig. 1). The hare on 2 may reach one of the positions of Fig. 12, forcing the hounds to give him the opposition in return for his retreat, but it is too late, since the hounds can play to 3L, 5, 3R, which wins for them, even without the opposition, because places numbered −2 are not on the board. What if the hare now goes to 0L? See the Extras.

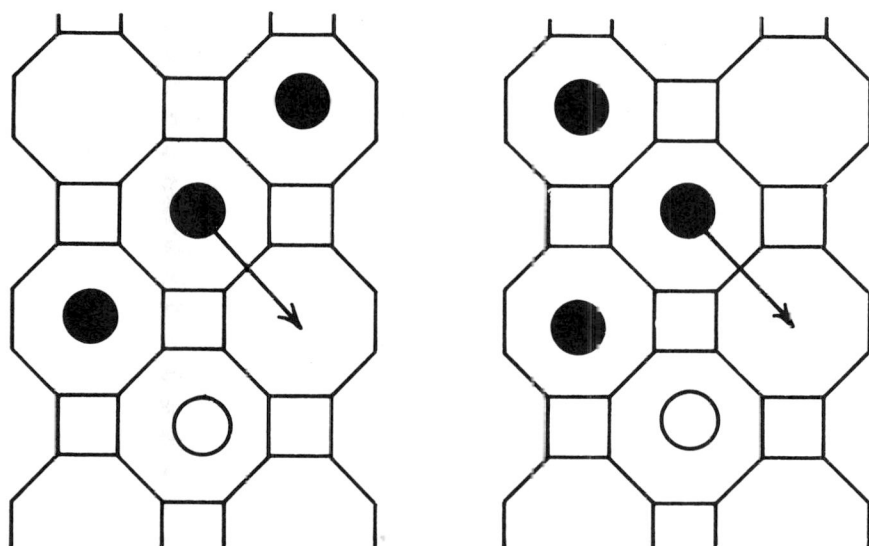

Figure 12. A Sound Bound for a Hound?

It is interesting that the hounds win if the configuration of Fig. 12(a) occurs at 3L, 5, 6R against 2, but not if it is higher (on the Larger Board of Fig. 3) at 6L, 8, 9R against 5. After the hound moves from 8 to 6, the Hare snatches the opposition by retreating to 3 and then follows his Strategy, but using the next set of six squares (4C, 3L, 3R, 2, 1L and 1R) up the board.

It should now be clear that the Hare's Strategy can be improved. If the Hare doesn't have the opposition, he should try to reach a position like 5 against hounds on 6L, 8, 9R (all such positions have trace 28). The way to force the hounds to move into such a position is to move to one whose trace is larger than the desired one by a small multiple of 3. In fact we can prove that

> on the Larger Board (Fig. 3) the
> hounds can win from a position
> of trace 31 only if the Hare is on a
> Weak square or the position is
> 6, 10, 11 *versus* 4C (Fig. 13).

THE THIRTY-ONE THEOREM

The proof is sketched in the Extras.

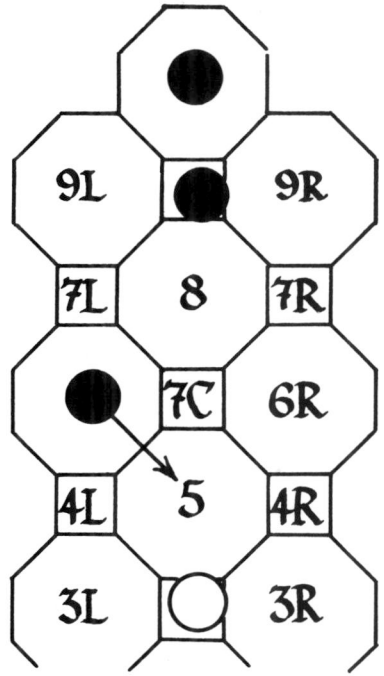

Figure 13. The Hound-Dog Position.

EXTRAS

ANSWERS TO QUESTIONS

Against the hounds placed as in Fig. 1, the hare can win only if he starts at c and requires the hounds to start first.

With the hounds on 9L, 11, 9R in Fig. 3 the hare can win from any position provided he has first move. The hardest case is when he starts on 1C, so that the hounds have the opposition. He wins by playing to 2 and using the Thirty-one Theorem. Of course if he *starts* on 2 he plays to the Strong square 4C *and* gains the opposition.

If the hounds move first they can win *only* if the hare starts on -1. They must play with great care, not only maintaining the opposition but also preventing the hare from escaping or achieving the trace 31. Surprisingly, even though it gains the opposition, the opening move from 11 to 10 loses! The difficulty is that if the hare advances via 0, 1 and 3 to 5, the hounds must then be able to reach 6L, 10, 6R. The defence 6L, 7, 6R is unattainable against a hare who is determined to keep the trace at least as high as 27. The defence 6L, 7R, 9R fails when the hare moves from 5 to 7C, forcing the 7R hound to occupy 8, and then retreats to 5 again and wins as in Fig. 12. If the hounds try to prevent the hare from reaching 5 by occupying 5, 9, 10, say, when the hare is on 3, then he escapes via a weak 4. But how can the hounds reach 6L, 10, 6R if the hare plays via 0, 1 and 3 to 5? They must have come from 6, 8, 10 if they had the opposition with the hare on 3; but where were they before that with the hare on 1? There is no position leading to 6, 8, 10 in which they had the opposition!

A SOUND BOUND FOR A HOUND?

If the hare is 0L and the hounds are on 3L, 5, 3R, how do they win? Answer 3R to 2. If hare takes the opposition by going to -1 then 2 to 0R, and if hare to 1C, then 3L to 2. If hare to -1 again, 0R to 1C wins; the trick is to hold back the hound on 5 until they're ready for the kill.

ALL IS FOUND FOR THE SMALL BOARD HOUND

In this chapter we've normally taken the point of view of the hare. To redress the balance, Figs. 14 and 15, which are adapted from Figs. 92 and 93 on pp. 241, 243 of Fred. Schuh's Master Book of Mathematical Recreations, show all the winning positions for the hounds on the Small Board. Figure 14 is a minimal set of 24 $\mathscr{P}$-positions (hounds win if hare has to move) which will ensure victory in all the positions the hounds deserve to win. Figure 15 shows 13 other $\mathscr{P}$-positions for the hounds which they can use for variety in seeking to hide their strategy from inquisitive hares. In each of the 37 positions the hare's place is indicated by a number, the remoteness function for the position.

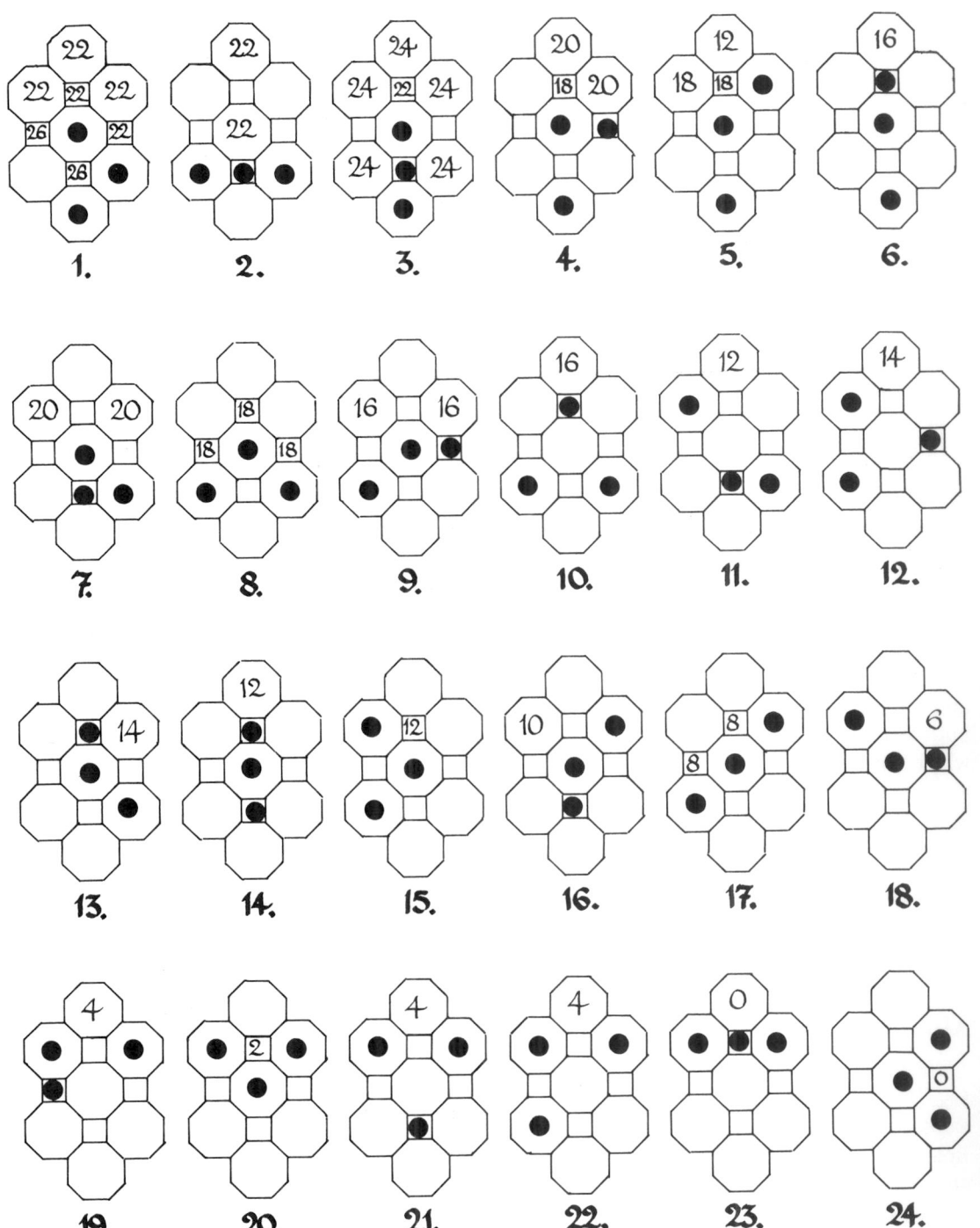

Figure 14. Twenty-four $\mathscr{P}$-positions for the Hounds which Provide a Minimal Winning Strategy.

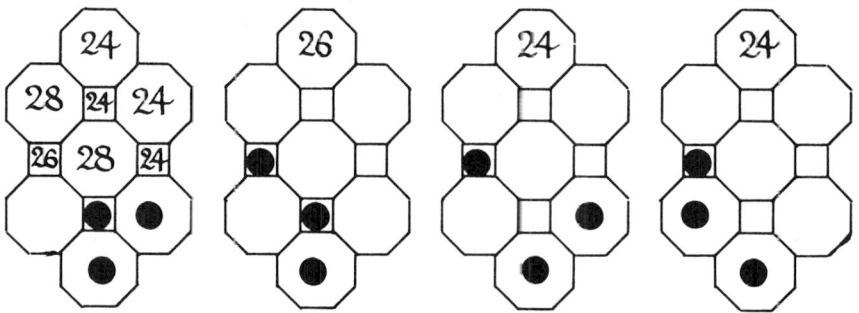

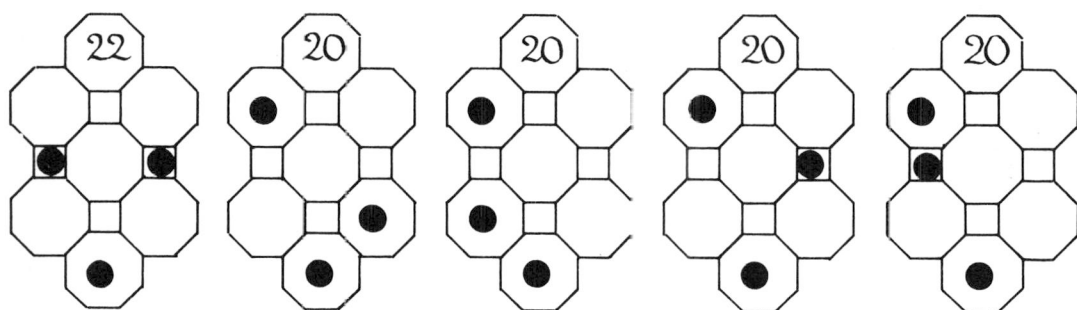

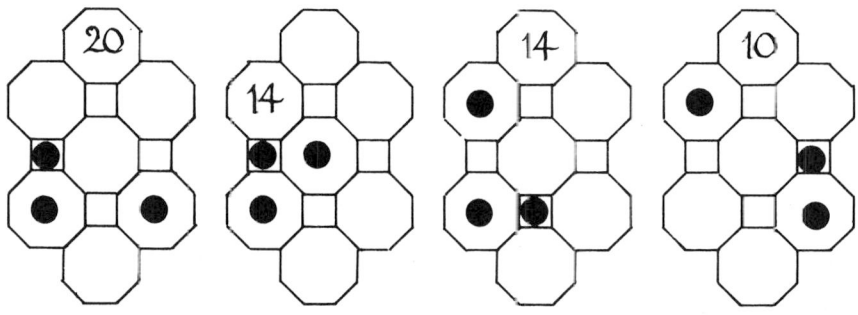

Figure 15. Thirteen More $\mathscr{P}$-positions for the Hounds.

Table 1 gives a winning strategy for the hounds, based on the 20 $\mathscr{P}$-positions of Fig. 14. The remarks are listed here (L and R are *hare's*; in Figs. 14, 15 left and right are the *hounds'*):

(a) the hounds do not have the opposition, but the hare is now forced to 1R, whereupon the hounds go to 3R, 5, 2 (position 1, reflected), still without the opposition, but the hare is forced again. After he goes to 0R, the hounds to go 3R, 4, 2 (position 7, reflected).

(b) also without the opposition, but see position 4.

(c) still without the opposition, but see position 5.

(d) even now the hounds don't have the opposition, but (position 6) the hare is forced to a zero place and the hounds go to position 13 or its reflexion.

(e) if now or later the hare goes to -1, play as from position 18: the hound on 2 goes to 0, (position 21 or 22), and then the rear hound comes to 2 (position 20).

From:	if hare plays on:	hounds reply by moving to:	arriving at:	with remoteness:	and trace:	Remarks
initial	4,1R,1C,0,-1	3L 5 2	1.	26,26,22,22,22	14,11,11,10,9	
position	2	3 4 3	2.	22	12	
1.	3R	4 5 2	3.	24	14	(a)
	-1	1L 5 2	4.	20	7	(b)
1. or 2.	0	3L 4 2	7.	20	9	
	1	3 2 3	8.	18	9	
4.	1C!	0L 5 2	5.	18	8	(c)
	0	1L 3R 2	9.	16	6	
5.	-1!	1C 5 2	6.	16	7	(d)
	0R	0 4 2	16.	10	6	
6.	0L	1C 3L 2	13.	14	6	
7.	1!	3 3 2	8.	18	9	
	-1	3L 4 0R	11.	12	6	
8.	0	1L 2 3R	9.	16	6	
	-1	3 1C 3	10.	16	6	
9.	-1!	1L 3R 0R!	12.	14	3	
	1	0L 2 3R	17.	8	6	
10.	0L	3L 1C 2	13.	14	6	
11.	0L	2 4 0R	16. refl.	10	6	
	1C	3L 2 0R	17. refl.	8	6	
12.	1C	2 3R 0R	15.	12	6	
	0L	1L 2 0R	18.	6	3	
13.	-1	4 1C 2	14.	12	6	
	1L?	3L 0L 2	trapped!	0	6	
14. or 15.	0L	4 0R 2	16. refl.	10	6	(e)
16.	1	3R 0L 2	17.	8	6	(e)
17.	0R	1R 0L 2	18. refl.	6	3	(e)
18.	-1	1L 0 0	19.	4	0	
	1C	0 0 2	20.	2	3	
19.	1C	2 0 0	20.	2	3	
20.	-1	1C 0 0	trapped!	0	0	

Table 1. A Winning Strategy for the Hounds Using Just the Positions of Figure 14.

PROOF OF THE THIRTY-ONE THEOREM

The hounds can only keep the opposition from a position of trace 31 by moving to 30 (a move to 33 would involve an illegal retreat). If they go to 30, the hare will move to 31 if he can; the hounds will move back to 30 and the hare will win by repetition. The hounds can only win by getting the hare on to a Weak square, or by preventing him from moving to 31. How might they do that? If the hare is on r, the hounds must be blocking any strong neighboring place $r+1$. Suppose the other hounds are on x and y where $x+y \leqslant 11+10$.

$$r + (r + 1) + x + y = 30,$$

$2r+1 \geqslant 9$, $r \geqslant 4$. If $r \geqslant 8$, $x+y \leqslant 13$ and the hare has escaped.

If $r=7$, a hound must be blocking 8, $x+y=15$, and, unless the hare has escaped, $x=6$, $y=9$. The hare moves to 5 and reaches Fig. 12.

If $r=6$, a Z place, the hounds must be blocking 7C and also $x=8$, to prevent escape, so $y=9$. The hare plays to 5. If the hounds restore the trace to 30 the hare returns to 6 and wins by repetition.

If $r=5$, a T place, the hounds must be blocking 6L, 6R and $y=13$, off the board.

If $r=4$ and not a Weak square, the hare is on 4C. A hound must be blocking 5, so $x+y=21$, $x=10$, $y=11$. This is the exceptional Hound-Dog Position (Fig. 13) which the hare can't win. If he goes to 3, the hound on 10 moves to 8. If the hare then moves to 4R, a hound moves from 8 to 6R forcing hare to retreat to 3R, after which the hound moves from 11 to 10C and regains control.

REFERENCES AND FURTHER READING

Martin Gardner, Mathematical Games: About two new and two old mathematical board games, Sci. Amer. **209** #4 (Oct. 1963) 124–130.

Martin Gardner, "Sixth Book of Mathematical Games from Scientific American", W.H. Freeman, San Francisco, 1971, Ch. 5.

Édouard Lucas, Récréations Mathématiques, Blanchard, Paris, Vol. III, 1882, 1960, 105–116.

Sydney Sackson, "A Gamut of Games", Random House, 1969.

Frederick Schuh, Wonderlijke Problemen; Leerzam Tijdverdrijf Door Puzzle en Spel, W.J. Thieme & Cie, Zutphen, 1943, 189–192.

Frederick Schuh, "The Master Book of Mathematical Recreations" (transl. F. Göbel, ed. T.H. O'Beirne) Dover Publications, New York, 1968, 239–244.

Chapter 22

Lines and Squares

And I say to them, "Bears,
Just look how I'm walking in all of the squares!"
And the little bears growl to each other, "He's mine,
As soon as he's silly and steps on a line."
A.A. Milne, *When We Were Very Young.*

On the square, to the left, was elegantly engraved in capital
letters this sentence: ALL THINGS MOVE TO THEIR END.
François Rabelais, *Pantagruel*, V, 37.

If you find you're bored to pieces with our other games, you should find your board and pieces to play these ones. The chapter contains several old friends and some new ones, but we'll avoid the really grown-up games like Chess and Go.

TIT-TAT-TOE, MY FIRST GO,
THREE JOLLY BUTCHER BOYS ALL IN A ROW

Oxford Book of Mother Goose Rhymes, 1951, p. 406.

The game is more usually known as Tic-Tac-Toe. or Noughts-and-Crosses, depending on which side of the Atlantic you are. Whoever moves first puts a cross (**X**) in one of the nine spaces in the board of Fig. 1. His opponent then puts a nought (**O**) into any other space and then they alternate **X**'s and **O**'s in the remaining empty spaces until one player wins by getting three of his own kind on one of the eight lines of Fig. 2. If teacher isn't listening he then shouts a suitably

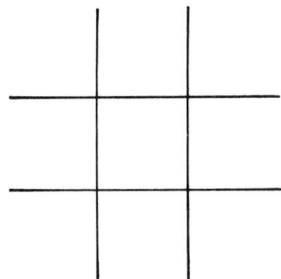

Figure 1. Tic-Tac-Toe Board.

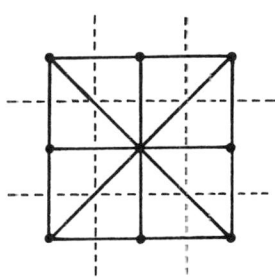

Figure 2. Its Eight Lines.

667

triumphant phrase, which in some parts of America is

"Tic-Tac-Toe, three in a row",

and, in Holland, according to Fred. Schuh, is

"Boter, melk en kaas, ik ben de baas".

When neither player is able to make a line we have a tied game. We have no doubt that most of our readers were bright enough as children to discover that this always happens when the game is properly played, and only the authors of books like *Winning Ways* retain sufficient interest to study the game in any detail.

But have you ever tried a complete analysis? If so, you've probably found that it took more space than you first thought it should. Later on we'll give a more concise analysis than most, though we admit that our rough work took more than one sheet of paper. But first let's look at three non-board games.

MAGIC FIFTEEN

In this game the players alternately select numbers from 1 to 9 and no digit may be used twice. You win by getting three numbers whose sum is 15. This game was suggested by E. Pericoloso Sporgersi.

SPIT NOT SO, FAT FOP, AS IF IN PAN!

is a sentence for which we are indebted to Anne Duncan. It suggests the following game. Write the nine words on nine separate cards and have the two players alternately select cards, a player winning if he can collect all the cards which contain a given letter. This game was suggested by Leo Moser's game of **Hot** in which the nine words were HOT, FORM, WOES, TANK, HEAR, WASP, TIED, BRIM, SHIP, and the winner must collect *three* words with a common letter.

JAM

John A. Michon's game of **Jam** is played on Fig. 3. The players alternately select roads (straight lines) and whoever manages to take all the roads through a town wins.

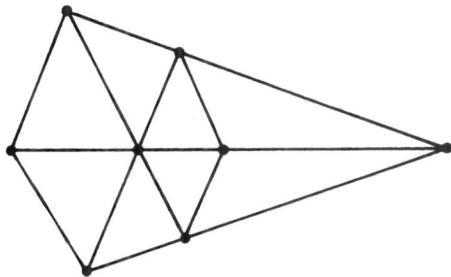

Figure 3. A Jam Board.

HOW LONG CAN YOU FOOL YOUR FRIENDS?

We'll bet you can fool most of them for quite a long time, playing any one of the above games. But they're all Tic-Tac-Toe in disguise, so you should be able to make the right moves while they are floundering! You can see why these games are all the same by arranging the numbers for Magic Fifteen as a magic square (Fig. 4(a)); the words for Spit, Etc. as in Fig. 4(b); and naming the towns or numbering the roads for Jam as in Fig. 4(c). For Hot you can prove the same thing by writing the words on Fig. 1 in the order we gave them. Can you find a better sentence than Anne's, possibly using redundant letters as in Hot? It would be nice if the words of your sentence could be written across the board in order!

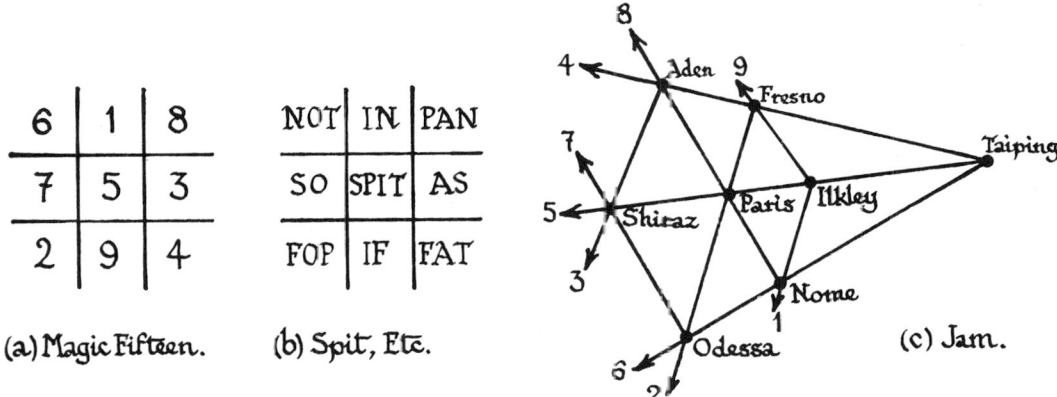

Figure 4. The Game's the Same By Any Name.

ANALYSIS OF TIC-TAC-TOE

For convenience we number the board as in Magic Fifteen and suppose by symmetry that the first move (**X**) is in 5 (Fig. 5), 6 (Fig. 6) or 7 (Fig. 7). We'll also suppose that each player is sensible enough to

(a) complete a line of his kind if he can, and
(b) prevent his opponent from doing so on his next move.

In the analysis,

  bold numbers represent such **forced** moves,
  ! denotes a move that's better than some others,
  ? denotes a move that's worse than some others,
  X denotes a win for Cross,
  O denotes a win for Nought,
  ⊗ denotes a tied game,
  ~ denotes an arbitrary move, and
  v. is a cross-reference to another column in the analysis.

The plays are given in numerical order, apart from the convention about the initial digit.

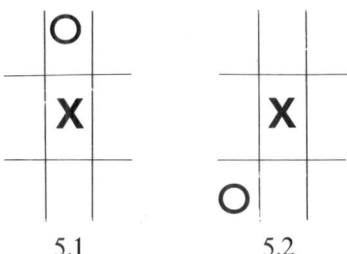

5.1   5.2

Figure 5. Starting in the Centre.

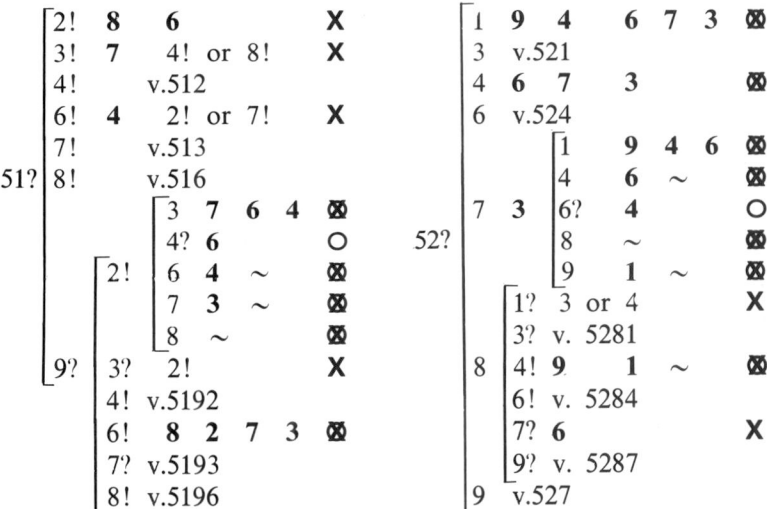

```
        ⎡2!  8    6                 X        ⎡1   9   4     6  7  3   ⊗
         3!  7    4! or 8!          X         3   v.521
         4!       v.512                        4   6   7     3          ⊗
         6!  4    2! or 7!          X         6   v.524
         7!       v.513                                   ⎡1    9   4   6   ⊗
   51?  ⎢8!       v.516                                    ⎢4    6   ~       ⊗
                      ⎡3  7  6  4  ⊗    52?   7   3   ⎢6?   4           O
                       4? 6        O                       ⎢8    ~           ⊗
                 ⎡2!  ⎢6  4  ~     ⊗                       ⎣9    1   ~       ⊗
                      ⎢7  3  ~     ⊗                   ⎡1?  3 or 4           X
                      ⎣8  ~        ⊗                    3?  v. 5281
         9?  ⎢3?  2!               X         8   ⎢4!  9    1   ~       ⊗
              4!  v.5192                          6!  v. 5284
              6!  8  2  7  3       ⊗              7?  6                 X
              7?  v.5193                          9?  v. 5287
              8!  v.5196                      9   v.527
```

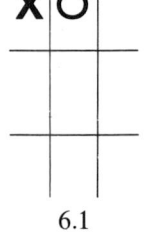   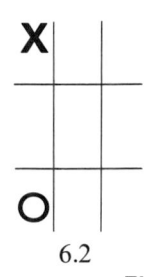   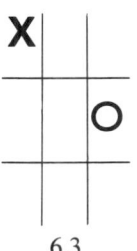   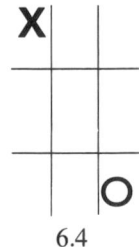   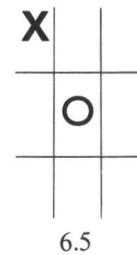

6.1   6.2   6.3   6.4   6.5

Figure 6. Starting in a Corner.

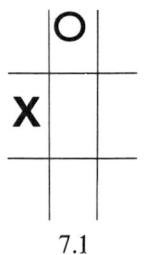   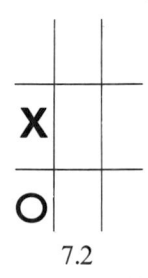   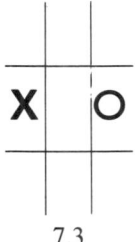   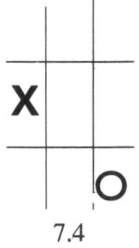   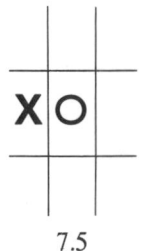

7.1   7.2   7.3   7.4   7.5

Figure 7. Starting on a Side.

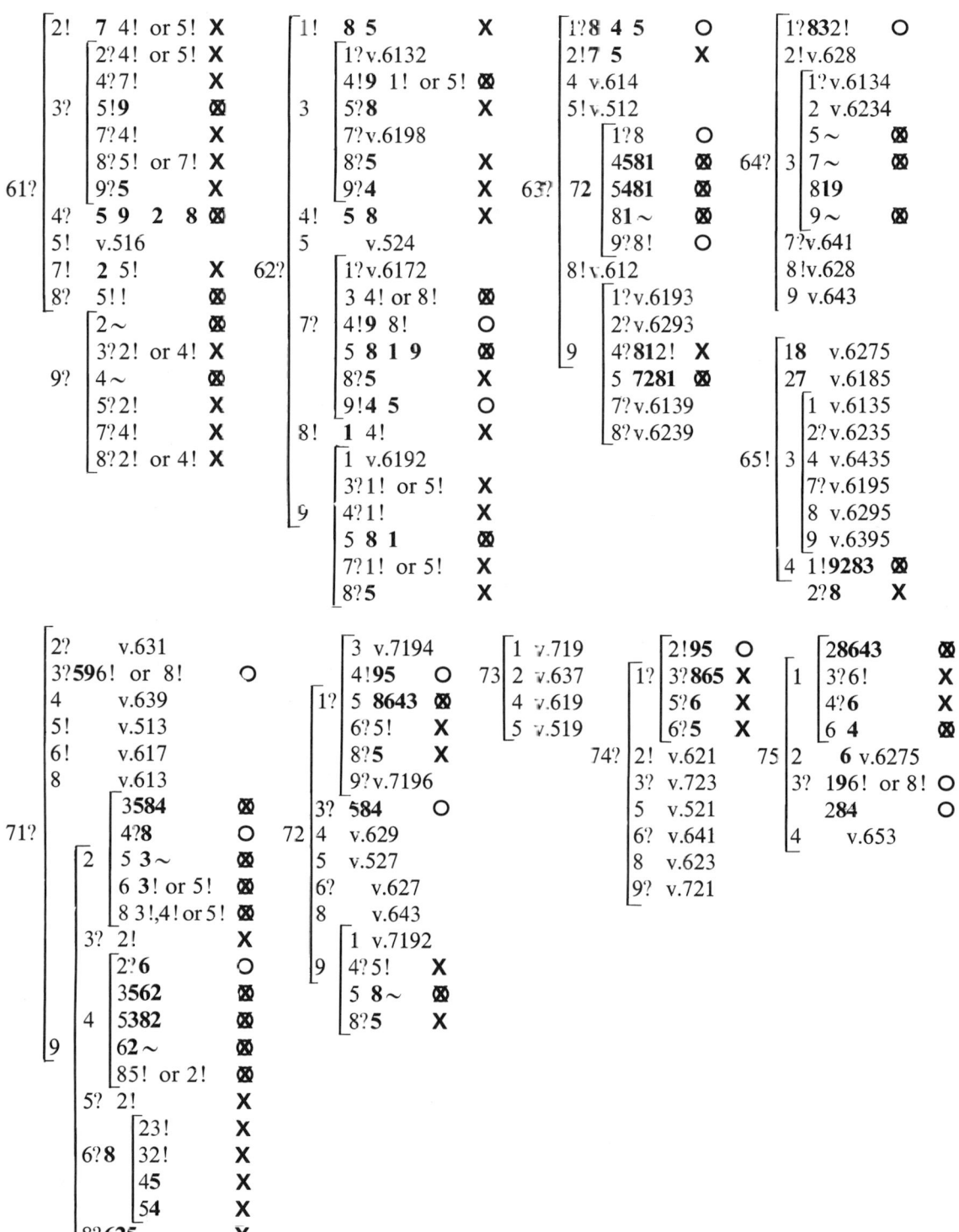

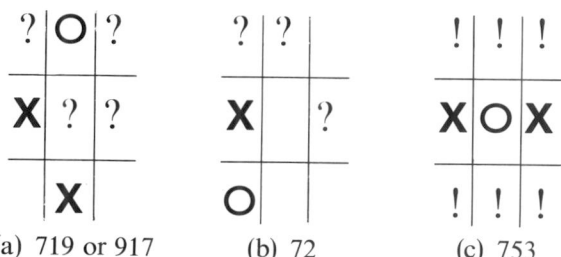

(a) 719 or 917   (b) 72   (c) 753

Figure 8. Lesser Known Byways of Tic-Tac-Toe.

In "The Scientific American Book of Mathematical Puzzles and Diversions" Martin Gardner remarks (and we agree) that many players have the mistaken impression that because they are unbeatable they have nothing more to learn. He gives three examples (Fig. 8) showing how a master player can take the best possible advantage of a bad play. In Fig. 8(a) **X**'s last move was chosen so as to give **O** four losing chances out of six (the Enough Rope Principle). Against **X**'s opening of 7, Gardner recommends **O** to reply with 2 since this offers **X** three losing chances (Fig. 8(b)). In Fig. 8(c) **O** can let **X** choose his move for him, since it is impossible for **O** to play without setting a winning trap!

OVID'S GAME, HOPSCOTCH, LES PENDUS

In his *Ars Amatoris*, Ovid advises young women to learn certain games to amuse their lovers. He mentions in particular a certain *ludus terni lapilli* played on a *tabella* which is conjectured to be a moving form of tic-tac-toe, played with 3 black pebbles and 3 white ones. Several such games are known to have been popular in ancient China, Greece and Rome and in medieval England and France.

In the version nowadays known as **Ovid's Game**, the players take turns placing their pebbles on the board until all 6 are down. If neither player has won by getting the 3 of his kind in a row they continue playing by moving on each turn a single one of their pebbles to any orthogonally adjacent square. The first player has a sure win by playing in the centre:

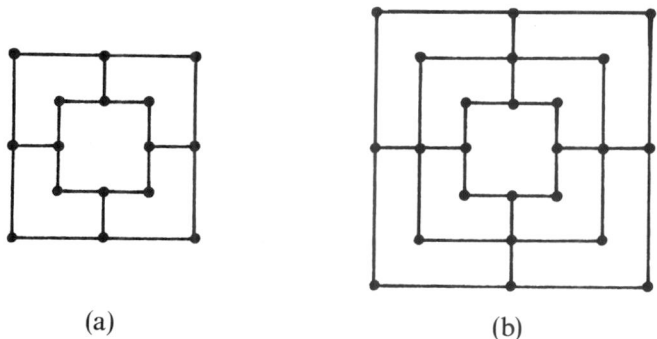

(a)   (b)

Figure 9. Six and Nine Men's Morris Boards.

$$5! \begin{bmatrix} 1 & 4 & 6 & 8 & 3, & 4 \text{ to } 9, \text{ any. } 9 \text{ to } 2, \text{ or} \\ 2 & 1 & 9 & 4 & 6, & 1 \text{ to } 8, \text{ any, } 5 \text{ to } 3, \end{bmatrix}$$

so the central opening is usually forbidden, making the game a draw. However, the loopiness of the game allows many variations to occur in a single play and the game teems with traps.

We will use the name **Three Men's Morris** for the version in which, in the moving part of the game, the players are allowed any chess king move along the 8 lines of Fig. 2. An American Indian version, which has been called **Hopscotch**, allows any king move, whether on the 8 lines or not. It is a draw, even when the central opening move is allowed, as is the French version, **Les Pendus**, in which a pebble can be moved to *any* empty space.

SIX MEN'S MORRIS

is played on the board of Fig. 9(a). Each player has 6 counters and the game has two phases as in Ovid's Game. First the counters are placed alternately by the two players. Then the counters are moved from one of the 16 nodes to an adjacent one along a line of the board. If a player gets three in a row he removes an opposing counter. A player wins when he reduces the opposing force to two counters.

NINE MEN'S MORRIS

is played similarly with 9 counters for each player on a square or rectangular board designed as in Fig. 9(b). When a player forms a **mill** (gets three in a row) he again removes an opposing counter, but is not allowed to take one from an opposing mill. There are a number of variations and many names (Merrilees, Morelles, Mill, Mühle); see the books of R.C. Bell or H.J.R. Murray for details.

THREE UP

This is a vertical three-in-a-row game. Each player starts with six checkers of his own color. They play alternately by putting a checker onto the table or onto a previous stack, and each tries to complete a stack three high of his own color. When all the checkers are placed, the players continue by alternately transferring single checkers of their own color from the top of one stack to the top of another, or possibly onto the table. At no time may any stack be more than three high.

It's very easy for a skilled player to beat a novice at this game, which has many cunning features. But Vasek Chvátal has shown that if you never try to win (by putting two of your pieces in a stack) then you can't lose! For if your opponent has set up t (≥ 1) threats (stacks beginning with two of his pieces) then he can cover at most $6 - 2t$ of your checkers, so you have at least $2t$ uncovered ones—more than enough to deal with his threats.

FOUR-IN-A-ROW

It's clear that the first player can get 1-in-a-row on a 1×1 board, and 2-in-a-row on a 2×2 board, and we have seen that he can't get 3-in-a-row on a 3×3 board, but it's not hard to show

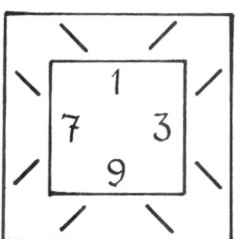

Figure 10. Four-in-a-Row is a Second Player Tie on a 5×5 Board.

that he can get 3-in-a-row on any bigger board, even with just one extra square. How big a board is needed to get 4-in-a-row? C.Y. Lee observes that the second player can tie on a 5 × 5 board. His strategy is to play as in Tic-Tac-Toe whenever the first player plays in the central 3 × 3 square. You won't have too much difficulty if you remember this and note that when you play in the squares marked with a diagonal line in Fig. 10 you sabotage your opponent's chance of getting 4-in-a-row on the border, and also on a diagonal involving two of the squares 1, 3, 9 and 7.

Lustenberger has used a computer to show that 4-in-a-row is a win for the first player on a 4 × 30 board.

By far the most interesting and popular version is the 3-dimensional one, played on a 4 × 4 × 4 cube. Oren Patashnik has shown that the first player can always win at 4 × 4 × 4 **tic-toc-tac-toe**. Patashnik's solution now includes a computerized dictionary of several thousand openings. This dictionary was obtained by patient and skilful interaction between Patashnik and a computer over a period of many months. It is too large to be accessible other than by computer. Several skeptical computer scientists have recently examined Patashnik's dictionary and it is now accepted as complete and correct.

FIVE-IN-A-ROW

It's quite a good game just to try to get 5 in a row orthogonally or diagonally on any reasonably large board. Mathematicians will prefer to play **Five-in-a-Row** on an infinite board.

In this kind of game there are several well defined degrees of threat and when playing with children and good friends it's nice to announce these by suitable cries. We recommend

SHOT! for a threat to win next move, e.g.

SHOTS! for two or more SHOTs at once, e.g.

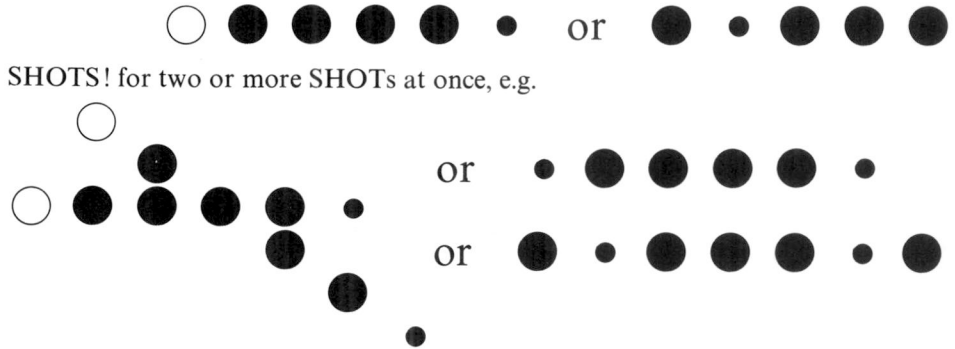

POT! if you can guarantee a SHOT next move, e.g.

● · ● ● ● · · or ● ● · ● ● · ● ·

POTSHOT! for a POT and a SHOT at the same time, e.g.

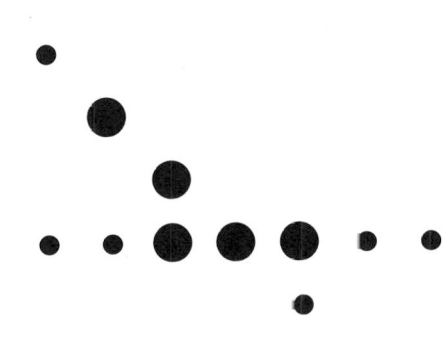

, and

POTS! for two or more POTs at once, e.g.

These can be of great help in understanding the effects of forced moves, for example:

A SHOT, typically a line of 4 open at one end, must be blocked instantly. So a pair of SHOTS wins next move.

A POT, typically a line of 3 open at both ends, must either be blocked immediately or staved off with a SHOT. So a POTSHOT wins unless possibly the move that blocks the SHOT is a countering SHOT. Against a pair of POTS you can only hope to defend by making a sequence of SHOTS until one of them happens to block one of the POTS.

These terms can be applied to many games of this type, for instance in 4-in-a-row,

○ ● ● ● ·

is a SHOT and

· · ● ● · ·

is a POT. Similar cries are used in Phutball (see later in this chapter); there are obvious connexions with the notion of *remoteness* in Chapter 9.

Five-in-a-row has been called Go-Bang in England for at least a hundred years and has more recently been called Pegotty or Pegity (Parker Bros., U.S.A.).

GO-MOKU

In Japan there are several perfect players who can always claim their win in their version, **Go-Moku**, of 5-in-a-row on a Go board, size 19 × 19, even though the first player is handicapped by not being allowed to make the **fork threat** of a pair of open lines of 3 (we'd cry POTS! for this) and *six* in a row is *not* counted as a win.

SIX, SEVEN, EIGHT, NINE, ..., IN A ROW

A.W. Hales and R.I. Jewett have produced an ingenious pairing strategy which shows that many games of this type are tied or drawn. For instance here is a quick proof that 5-in-a-row is tied on a 5 × 5 board. All you have to do is to make sure that for every move of your opponent in a marked square in Fig. 11 you take the similarly marked square in the direction indicated by the mark. So you could give her the centre square *and* let her make the first move as well. If the position you're presented with already satisfies the condition, make a random move. At the end of the game there will be at least one of your counters in every conceivable winning line.

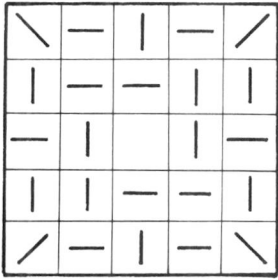

Figure 11. A Hales-Jewett Pairing.

You can see that 9-in-a-row is a draw on an infinite board with the Hales–Jewett pairing of Fig. 12. When your opponent takes the cell at one end of a line in the figure, you take the one at the other. The result was first proved in 1954 by Henry Oliver Pollak and Claude Elwood Shannon, using the following strategy. Tile the board with H-shaped heptominoes: the second player plays ordinary tic-tac-toe in each of these regions, concentrating on preventing a line of 3 in either a diagonal, or the horizontal, or the right vertical. John Lewis Selfridge also gave a Hales–Jewett pairing on an 8 × 8 board, which could be used to tile an infinite one and give the same result.

T.G.L. Zetters (nom de guerre of some Amsterdam combinatorists) recently showed that the second player can even draw 8-in-a-row. Their proof uses a parallelogram-shaped tile of 12 cells, and goes some way towards showing that 7-in-a-row is also a draw.

Figure 12. Nine-in-a-Row is a Draw on an Infinite Board.

S.W. Golomb has found a Hales–Jewett pairing for 8-in-a-row on an $8 \times 8 \times 8$ cube. It will be easier to explain if we first describe the analogous two-dimensional solution for 6-in-a-row on a 6×6 square. Figure 13(a) is like Fig. 11 except that you may reply to a move on a diagonal with *any* other move on the same diagonal. Note that the figure has the mirror symmetries indicated by the two thick lines, so it would suffice to indicate only one quadrant as in Fig. 13(b).

Figure 14 indicates one octant of Golomb's $8 \times 8 \times 8$ pairing in a similar way. The symbols —, | and \ are in the horizontal layer shown, while ● is all you can see of a vertical line. The arrows pierce the layers and represent lines of obvious directions in various diagonal planes. The three mid-planes of the $8 \times 8 \times 8$ cube (represented by the thick lines in Fig. 14) are reflecting planes and, as in the 6×6 pairing of Fig. 13, you may respond to *any* move on a body diagonal with another on the same diagonal. In fact Golomb can give you *any* six cells on *each* of the four body diagonals *and* allow you to have first move and still tie the game.

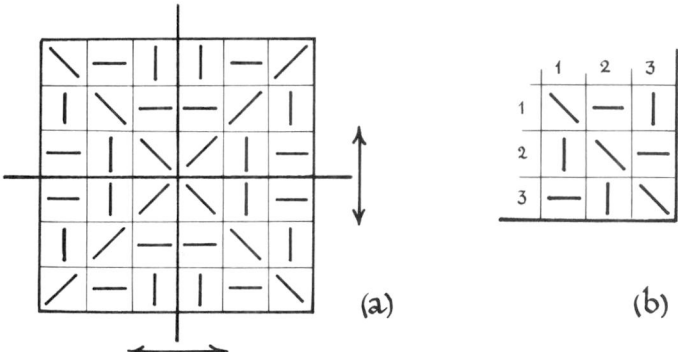

Figure 13. A Pairing for Six-in-a-Row on a 6×6 Board.

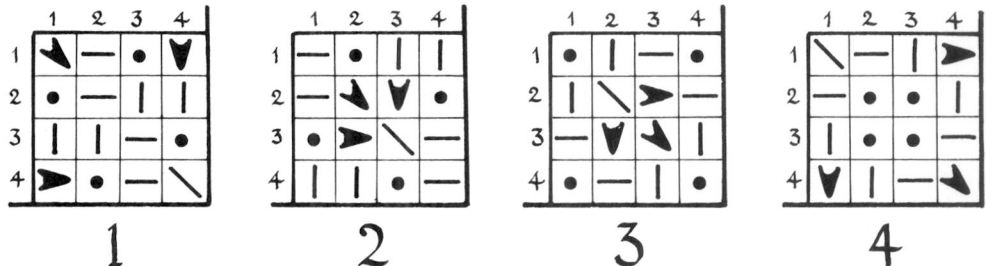

Figure 14. Golomb's Pairing for Eight-in-a-Row on an 8×8×8 Cube.

n-DIMENSIONAL *k*-IN-A-ROW

Hales and Jewett consider the game of *k*-in-a-row on the *n*-dimensional

$$k \times k \times k \times \ldots \times k$$

board. They prove that if *k* is sufficiently large, namely

$$k \geqslant 3^n - 1 \quad (k \text{ odd}) \text{ or}$$
$$k \geqslant 2^{n+1} - 2 \quad (k \text{ even})$$

the game is tied by a suitable pairing strategy, and on the other hand that if *n* is sufficiently large compared to *k*, it is a first player win by the strategy stealing argument described below. They conjecture that the game is tied if there are at least twice as many cells as lines.

How many lines are there? Leo Moser has remarked that each line is determined by either one of the two cells which extend it into the surrounding

$$(k+2) \times (k+2) \times (k+2) \times \ldots \times (k+2)$$

cube, so that the total number of lines is exactly

$$\tfrac{1}{2}\{(k+2)^n - k^n\}.$$

The Hales–Jewett conjecture is therefore that the game is tied whenever

$$k^n \geqslant (k+2)^n - k^r,$$

i.e.

$$2k^n \geqslant (k+2)^n.$$

So it should be true if $k \geqslant 3n$, for example; Leo Moser has proved that it's true if $k > cn \log n$ for some constant c.

STRATEGY STEALING IN TIC-TAC-TOE GAMES

For almost all forms of tic-tac-toe game there is a strategy stealing argument which shows that the second player cannot have a winning strategy. Though earlier authors probably knew it, this was formally proved by Hales and Jewett. We suppose that each player has an indefinite supply of his own kind of piece, that the pieces don't move after they're once put down, and that each player's aim is to produce a winning configuration with some of his pieces.

The assertion is that all such games in which the winning configurations for the two players are similar, are either wins for the first player or are tied under best play. For if the second player had a winning strategy, then the first player could steal it as follows. After a random first move he could pretend to *be* the second player, ignoring his opening move and making a random move whenever the stolen strategy would otherwise repeat a move already made. We conclude that if the second player had a winning strategy, so would the first since an additional piece on the board can never harm him! Obviously *both* players can't win at once, so the supposed winning strategy for the second player cannot exist.

The argument applies to n-in-a-row on any shape of board, provided no special restrictions, like those of Go-Moku, are added. In this case the winning configurations are just the appropriate lines of n, and are exactly the same for each player.

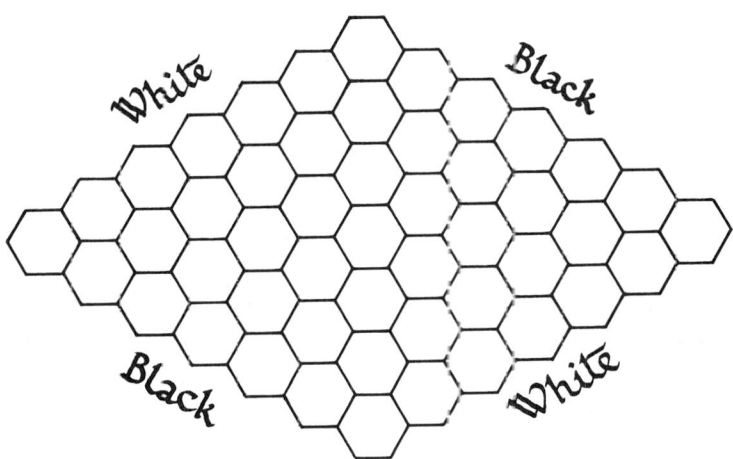

Figure 15. A 7×7 Hex Board.

However, in the most notorious cases of strategy theft the winning configurations are not identical but related by a symmetry of the board (and so are still *similar*). These are, in chronological order,

HEX,

played on a rhombus of hexagons like that of Fig. 15. Black wins if his pieces connect one pair of opposite sides of the board and White if his connect the other pair.

Hex was invented by Piet Hein and the strategy stealing argument found by Nash.

BRIDGIT

(or Gale) is played on two interlaced n by $n+1$ lattices. Left joins two adjacent (horizontal or vertical) spots of the black lattice and Right makes similar moves in the white one. No two moves may cross. In Fig. 16 Left has just won since he has formed a chain connecting a topmost spot to a bottommost one. Bridgit was invented by David Gale and its strategy stealing argument by Tarjan.

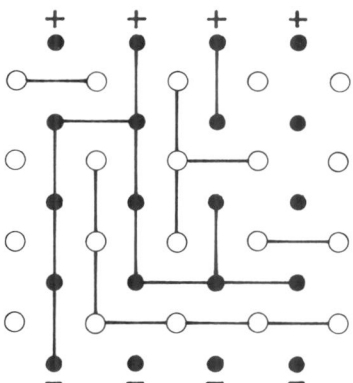

Figure 16. Left Forms a Black Chain in Bridgit.

HOW DOES THE FIRST PLAYER WIN?

In these cases, as in the examples considered by Hales and Jewett, it is impossible for a completed game to be tied, so that the argument actually proves that the first player can win, but does not give much help in finding an explicit winning strategy for him. No explicit strategy for Hex is known, and Tarjan and Even have shown that, in the technical sense, generalized Hex is hard. But for Bridgit an explicit pairing strategy was found by Oliver Gross, and many other strategies can be deduced from Alfred Lehman's subsequent theory of

THE SHANNON SWITCHING GAME

which generalizes Bridgit. It is played on a graph representing an electrical network in which certain nodes are labelled + and some others are labelled −. Each edge (begin the game with them drawn in *pencil*) represents a permissible connexion between the nodes at its ends. *Mr. Shortt*, at his move may *establish* one of these connexions permanently (*ink over* a pencilled edge) and attempts to form a chain between some + node and a − one. His opponent, *Mr. Cutt* may permanently *prevent* a possible connexion (*erase* a pencilled edge) and tries to separate + from − forever. Figure 17(a) shows a Shannon game equivalent to our Bridgit one. You can always suppose that there's only one positive node and one negative one by making identifications as in Fig. 17(b).

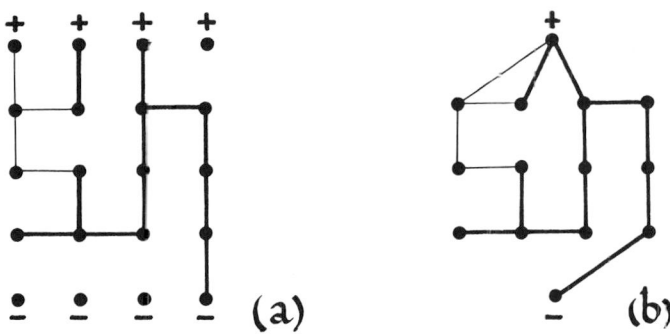

Figure 17. Bridgit Played as a Shannon Switching Game.

Supposing this, Lehman has proved that Mr. Shortt can win as second player if and only if he can find two edge-disjoint trees which each contain all the nodes of some subgraph containing + and −. The "only if" part is hard, but there's an easy strategy which proves "if": whenever Mr. Cutt's move separates one of the trees into two parts, *A* and *B*, Mr. Shortt makes a move on the other tree joining a vertex of *A* to one of *B*.

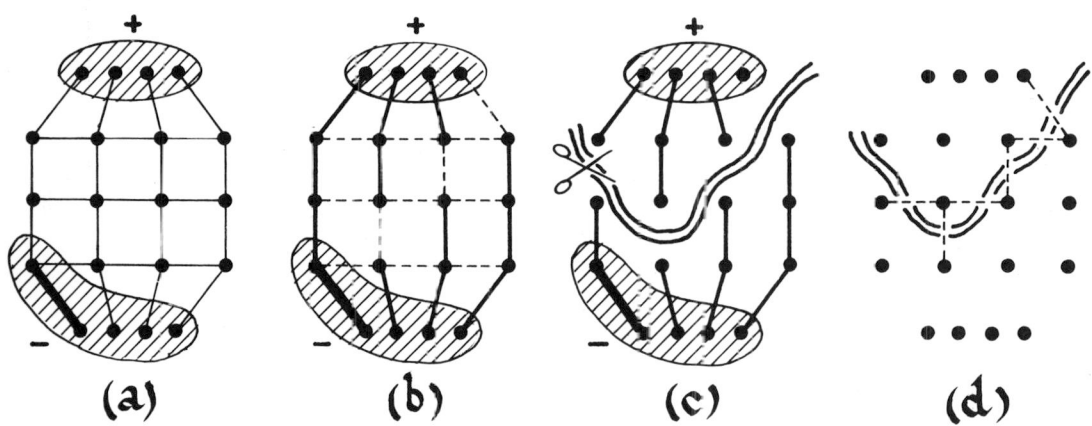

Figure 18. How Mr. Shortt Wins a Game of Bridgit.

Let's use Lehman's theory to show how the first player, who should regard himself as Mr. Shortt, can win in Bridgit. After his first move (Fig. 18(a)) Mr. Shortt (who's now *second* in line to move) can see the two edge-disjoint trees indicated by the thick and pecked lines of Fig. 18(b) (remember to regard each of the + and − sets as connected, including the node that has been Shortted to −). If now Mr. Cutt disconnects one of the trees, for example by erasing the scissored edge in Fig. 18(c), then Mr. Shortt should secure one of the six bridges (Fig. 18(d)) across the imaginary river which now separates the two parts of the tree severed by Mr. Cutt.

The game can be generalized to make the winning configurations for Mr. Shortt just those which contain a specified family P of sets of edges. (In the original game P was the family of paths from + to − .) Lehman proves the "only if" part of his theorem by taking P to be the family of all trees containing every vertex (spanning trees).

If Mr. Shortt, as second player, has a win in the modified game, it's *very* easy to see that there must be two edge-disjoint spanning trees. For since an extra move is no disadvantage, both players can play Mr. Shortt's strategy! If they do this, *two* spanning trees will be established, using disjoint sets of edges. Conversely, if two such trees exist, our previous strategy for Mr. Shortt actually wins for him as second player, even in the modified game.

The more detailed part of Lehman's argument establishes that, in a suitable sense, the modified game reduces to the original one.

THE BLACK PATH GAME

This elegant little game was invented by Larry Black in 1960. You can play it on a rectangular piece of paper ruled into squares as in Fig. 19. At any time the squares that have been used will each contain one of the three patterns shown in Fig. 19(b) and will include a path like the black path in Fig. 19(a) which begins at the starting arrow. The player to move must continue the black path by drawing one of the permissible patterns in the next square. You lose if your move makes the black path run into the edge of the board. The numbers 1 to 8 show the order of the first eight moves in our sample game, and the next player must now move in the square marked 9. You'll see that pattern 1. loses instantly, 2. wins quickly and 3. loses slowly.

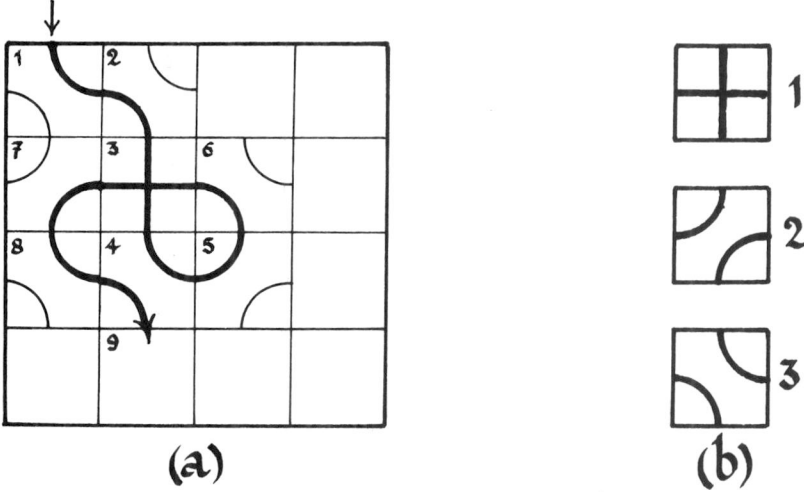

(a)   **(b)**

Figure 19. Forming a Black Path.

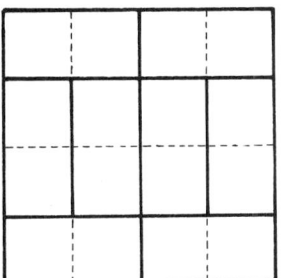

Figure 20. Black Path Game Board Divided into Dominoes.

We have a pairing strategy by which the first player can win on any rectangular board with an even number of squares. He imagines the board divided into 2×1 dominoes in any way he likes, for instance Fig. 20, and then plays so as to leave the end of the path in the middle of a domino (which can never be the edge of the board!). On an odd by odd board it is the second player who can win, by dividing all of the board except the opening square into dominoes.

LEWTHWAITE'S GAME

Domino pairing (e.g. Fig. 21(b)) also enables the second player to win a game invented by G.W. Lewthwaite in which 12 white and 12 black squares are slid alternately in a 5×5 box from the starting position of Fig. 21(a), and a player who, at his turn, cannot move any piece of his color, loses. What happens if a player is also allowed to slide a row or column of 2, 3 or 4 squares, provided both end squares are of his color?

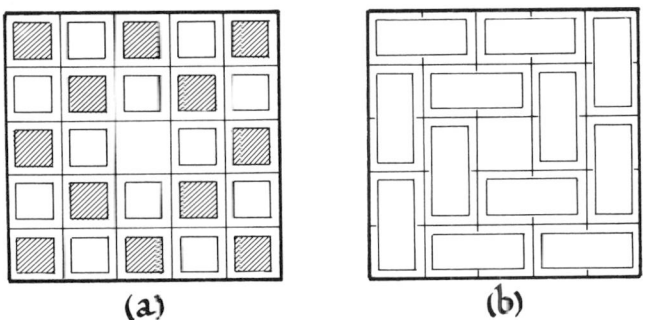

(a)   (b)

Figure 21. Pairing Gives a Second Player Win in Lewthwaite's Game.

MEANDER

was also invented by Lewthwaite and is also played with 24 tiles in a 5×5 box, but the tiles are now patterned as in Fig. 22(a), and are slid by either player. Figure 22(b) shows the starting position. The winner is the first player to produce a continuous curve connecting the boundary to itself and involving at least three tiles, as in Fig. 22(c). There are two versions of the game. In the first, players alternately slide just one tile; in the other, a row or column of 1, 2, 3 or 4 tiles may be slid as a single move.

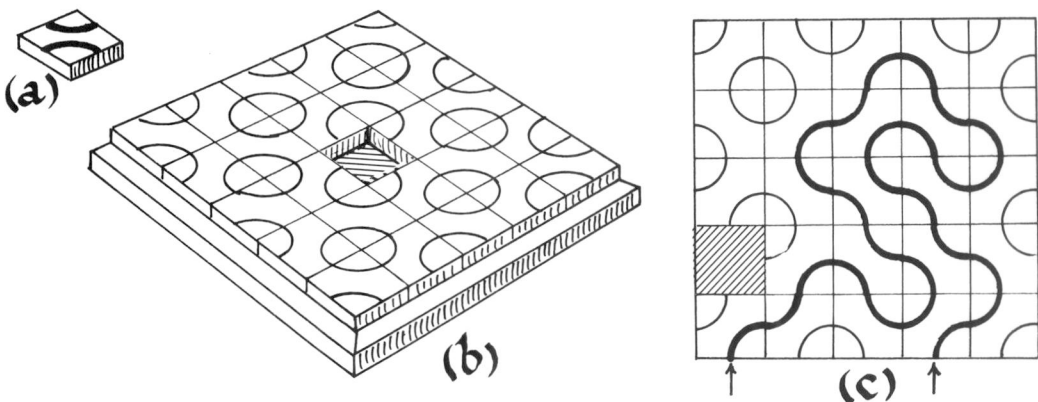

Figure 22. Meander.

WINNERS AND LOSERS

Frank Harary has proposed a family of games, one for each polyomino, P, all played on an infinite board. On alternate turns, Left makes a square black, and Right makes one white, and Left's aim is to produce a black copy of P, while Right tries to foil him. Harary calls P a **winner** if Left has a winning strategy—otherwise a **loser**.

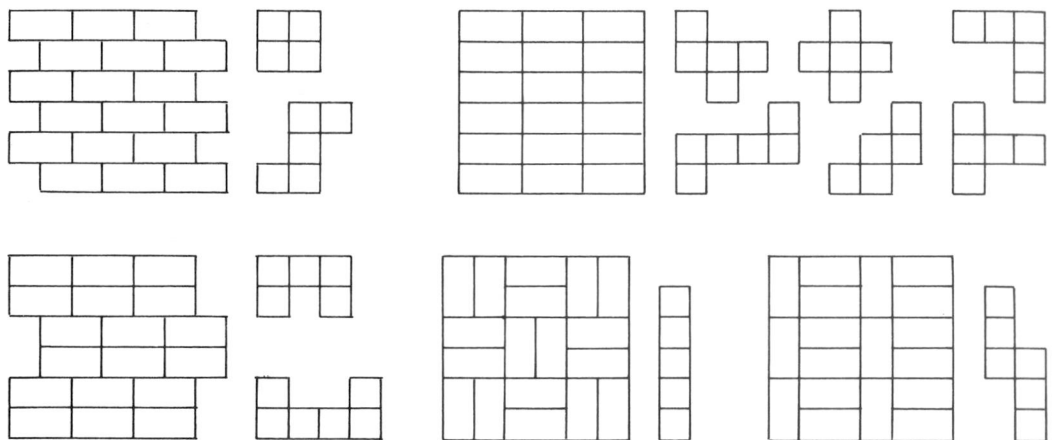

Figure 23. Hales–Jewett Pairings Make Twelve Polyomino Losers.

The twelve polyominoes of Fig. 23 can be proved to be losers using the indicated Hales–Jewett pairings, mostly found by Andreas Blass. If P contains one of these it is therefore a loser. The only polyominoes *not* containing one of these twelve are the twelve shown in Fig. 24. Eleven of these are known to be winners, with known strategies which win in m moves on a $b \times b$ board (see the figure). The last, called "snaky" by Harary, is also conjectured to be a winner.

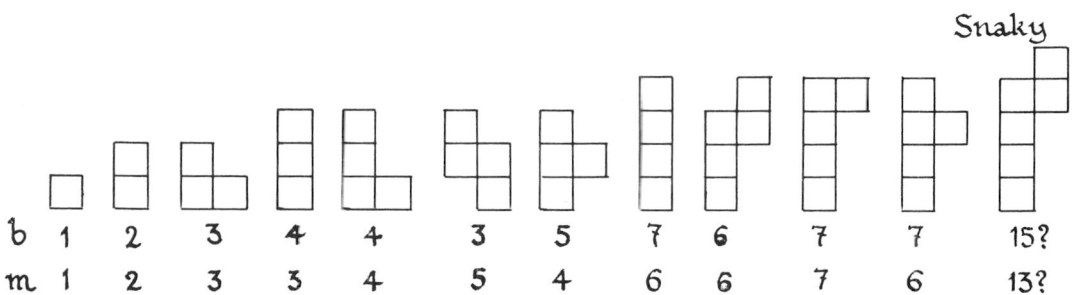

Snaky

	b	1	2	3	4	4	3	5	7	6	7	7	15?
	m	1	2	3	3	4	5	4	6	6	7	6	13?

Figure 24. Twelve Polyomino Winners with Board Sizes and Numbers of Moves
(but Snaky's is a bit shaky).

DODGEM

Colin Vout invented this excellent little game played with two black cars and two white ones on a 3 × 3 board, starting as in Fig. 25(a). The players alternately move one of their cars one square in one of the three permitted directions (E, N or S for Black; N, E or W for White) and the first player to get *both* of his cars off the board wins. Black's cars may only leave the board across its right-hand edge and White's cars only leave across the top edge. Only one car is permitted on a square, and you lose if you prevent your opponent from moving.

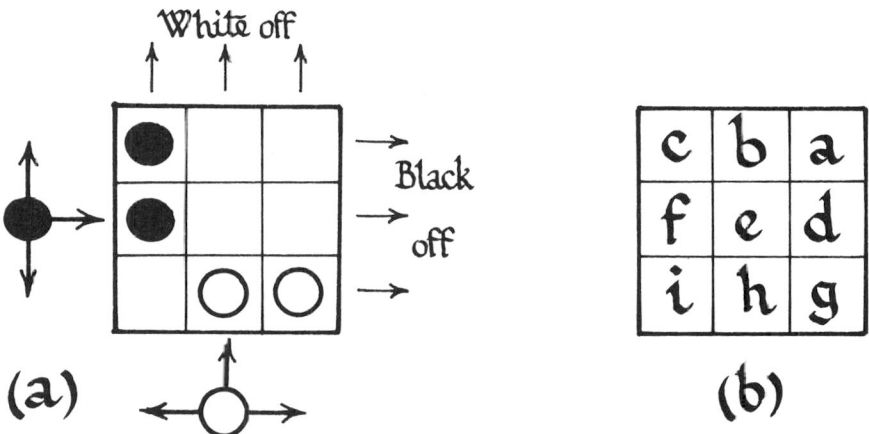

Figure 25. Colin Vout's Game of Dodgem.

Although the board is the same size as that for Tic-Tac-Toe, this game is much more interesting to play. Table 1 contains the outcome of every position; the column gives the positions of the black cars, the row those of the white, labelled by pairs of letters from Fig. 25(b). A blank entry represents an illegal position, since only one car is allowed on each square.

> \+ is a win for Black (Left),
> − is a win for Right (White),
> ○ is a win for the second player,
> * is a win for the first player.

Left's Position

```
f c c h h h e e e b b b e b b g g g d d d a a a   i f c g g g d d d a a a   h e b   d a a   g d a
i i f i f c i f c i f c h h e i f c i f c i f c       h e b h e b h e b         g g d
```

	Right's Position
	a
	b
	c
	ab
	ac
	bc
	d
	e
	f
	ad
	ae
	af
	bd
	be
	bf
	cd
	ce
	cf
	g
	h
	i
	ag
	ah
	ai
	bg
	bh
	bi
	cg
	ch
	ci
	de
	df
	ef
	dg
	dh
	di
	eg
	eh
	ei
	fg
	fh
	fi
	gh
	gi
	hi

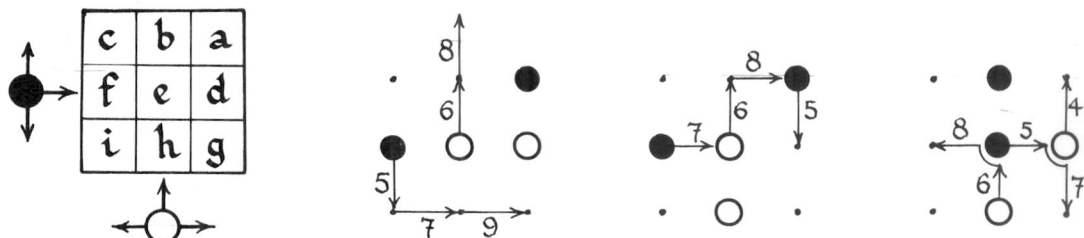

Table 1. Outcomes of Positions in Dodgem and Some Good Moves.

If you haven't got our table drawn on the back of your hand, you'll find this little game hard to play against an expert, who'll spring all sorts of little traps for you. It's often *not* a good idea to push your car off as soon as you can since it may be more useful blocking your opponent. In many situations it's a good idea to aim for the top right hand corner.

When you are expert, you can try playing Dodgem with $n-1$ cars of each color, on an $n \times n$ board, starting in the first column and row, with the SW corner empty.

DODGERYDOO

This game is played with two Dodgem cars on a quarter-infinite board. Now either player may move either car any distance North or West in a single move, provided it does not jump on to or over the other car. If you can't move you lose.

It's not hard to see that

> any position in which
> the two cars are
> on neighboring squares
> is a $\mathscr{P}$-position,

because whatever the next player does you can continue to shadow him. As a consequence,

> any other position with
> the cars in the
> same row or column,
> or in adjacent ones,
> is an $\mathscr{N}$-position,

because the next player can immediately creep one car up to the other. So in analyzing later positions we might as well make it illegal to have both cars in the same row or column, or in adjacent ones.

Let (x_1, y_1) and (x_2, y_2) be the positions of the two cars in this restricted game. Then on these numbers we are playing a nim-like game with four heaps in which we can reduce any one of the four numbers x_1, x_2, y_1, y_2 provided we ensure that neither $x_1 - x_2$ nor $y_1 - y_2$ is 0 or ± 1. Since the x's and the y's don't interact, we can regard this as the sum of two games, one played on the x's, the other on the y's. Table 2 gives the nim-values for either of these games—apart from the positions described in the boxes above. An X denotes an illegal position in the restricted game.

> The position (x_1, y_1), (x_2, y_2)
> is a Dodgerydoo $\mathscr{P}$-position
> just if $f(x_1, x_2) = f(y_1, y_2)$,

(since their nim-sum is then 0), where $f(x_1, x_2)$ is the function given in Table 2.

x_1 \ x_2	0	1	2	3	4	5	6	7	8	9	10	11	12	13	14	15	16
0	X	X	0	1	2	3	4	5	6	7	8	9	10	11	12	13	14
1	X	X	X	0	1	2	3	4	5	6	7	8	9	10	11	12	13
2	0	X	X	X	3	1	2	6	4	5	9	7	8	12	10	11	15
3	1	0	X	X	X	4	5	2	3	8	6	10	7	9	13	14	11
4	2	1	3	X	X	X	0	7	8	4	5	6	11	13	9	10	12
5	3	2	1	4	X	X	X	0	7	9	10	5	6	8	14	15	16
6	4	3	2	5	0	X	X	X	1	10	11	12	13	6	7	8	9
7	5	4	6	2	7	0	X	X	X	1	3	11	12	14	8	9	10
8	6	5	4	3	8	7	1	X	X	X	0	2	14	15	16	17	18
9	7	6	5	8	4	9	10	1	X	X	X	0	2	3	15	16	17
10	8	7	9	6	5	10	11	3	0	X	X	X	1	2	4	18	19
11	9	8	7	10	6	5	12	11	2	0	X	X	X	1	3	4	20
12	10	9	8	7	11	6	13	12	14	2	1	X	X	X	0	3	4
13	11	10	12	9	13	8	6	14	15	3	2	1	X	X	X	0	5

Table 2. Dodgerydoo Values, $f(x_1, x_2)$.

There doesn't seem to be much pattern in the table, once we get away from the edge, but at least the first few rows (and columns) are arithmetico-periodic. The (ultimate) periods and salt-uses in the first five rows are 1, 1, 3, 9, 36, and are valid outside the heavy line. The same table solves two-car Dodgerydoo in three dimensions, for which the $\mathscr{P}$-position condition becomes

$$f(x_1,x_2) \overset{*}{+} f(y_1,y_2) \overset{*}{+} f(z_1,z_2) = 0.$$

PHILOSOPHER'S FOOTBALL

or PHUTBALL (registered J.H. Conway) for short, is a very playable game that you can read about for the first time in this book. It is usually played on the 15×19 intersections of the board shown in Fig. 26, or on a 19×19 Go board, using one black stone (the **ball**) and a large supply of white ones (**men**). All pieces are common to both players and indeed both players have the same legal moves although their aims are different.

Start with the pitch empty except for the ball which starts at the central spot. Then each player, when he moves, must

       either place a new man at any unoccupied intersection
       or   **jump** the ball, removing the men jumped over.

(He may not do both.)

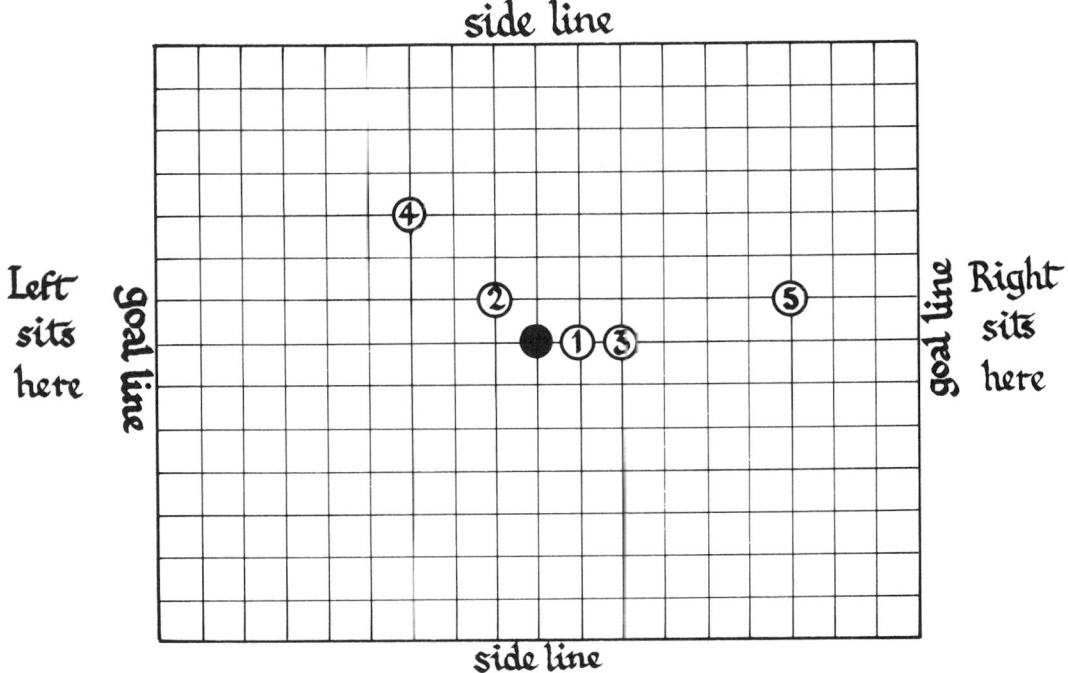

Figure 26. The Phutball Pitch and the First Five Moves of a Game.

A single jump of the ball may be in any of the eight standard compass directions, N, NE, E, SE, S, SW, W, NW on to the first empty point in that direction provided at least one man is jumped over. All the men jumped over are removed instantly. A player may take several consecutive such jumps in various of the eight directions as a single move. But because the men are removed instantly, the same man cannot be jumped over more than once in a move, and no man can be placed on the board in a jumping move.

It is legal for the ball to land on any of the goal lines or side lines. It is also legal for the ball to leave the board, but only by jumping over a man on the goal line, and only as the last move of the game. In fact Left's aim is to arrange that at the *end* of a move the ball is either *on or over* Right's goal line, while Right's is to get it on or over Left's. However a defender can sometimes successfully use his own goal line by jumping the ball onto and off it during a single move.

In the standard opening,

Left,   Right,   Left,   Right,   Left,

will place the stones

1,   2,   3,   4,   5,

of Fig. 26, building chains towards their opponents' goals. Right is now frightened by Left's threat to make a long jump over 1 and 3 and later establish a chain through 5. He therefore makes two short jumps himself over 2 and 4.

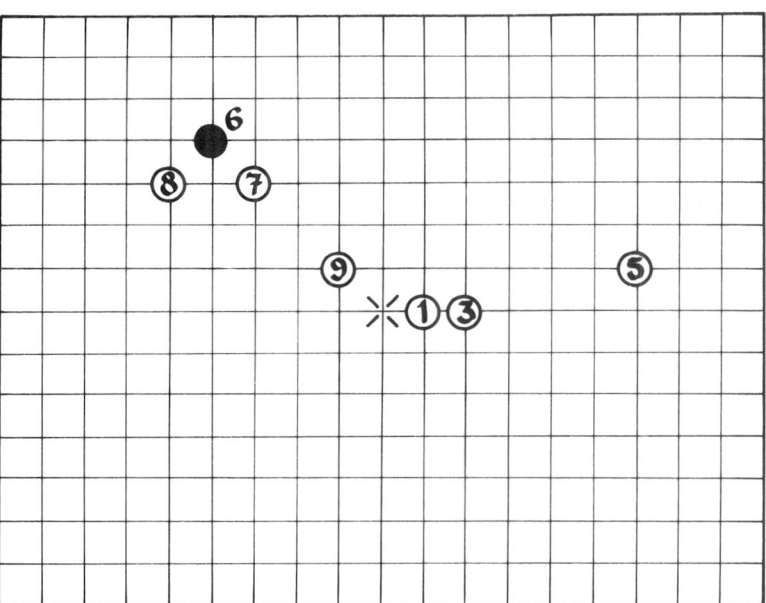

Figure 27. The Next Four Moves.

In the subsequent moves

7,   8,   9,

of

Left, Right, Left,

Left tries to reëstablish his chain while Right prepares the way for a sideways jump to defend against this. If it were Left's turn to move in Fig. 27 he could in one move make two jumps over 7 and 9 and a longer jump over 1 and 3 (it would probably be better for him not to make this last jump; as in Chess, a threat is often more powerful than its execution). However, it's Right's turn so he jumps over 8, and the next few moves are shown in Fig. 28.

These are all rather subtle. Left's move 11 is much better than reïnstating 8, which Right could too easily **tackle** by placing a man where the ball was in Fig. 27 (after a jump of these two stones, Left would find it very difficult to reëstablish a useful connexion with the rest of his chain). Right's move 12 is even more subtle! A direct threat to win at this point would make Left jump over 11 and 7, and arrive at a commanding position. Move 12 provides a way back after this jump and also prepares the way for a move at 14, followed by a roundabout triple jump over 11, 12 and 14, which both gets Right near to the Left goal line and removes some pieces useful to his opponent. The move 12 has even more hidden secrets: if Left places 13, *Right* can make the jump over 11 and 7, and then any Left threat to connect with his old chain equally helps Right to connect with 13 and 12.

Almost all these moves have become standard, but from now on experts differ. The game has many subtle tactics (tackling, poisoning one's opponent's threats, devastating U-turns, ...), and we'll only offer a few hints.

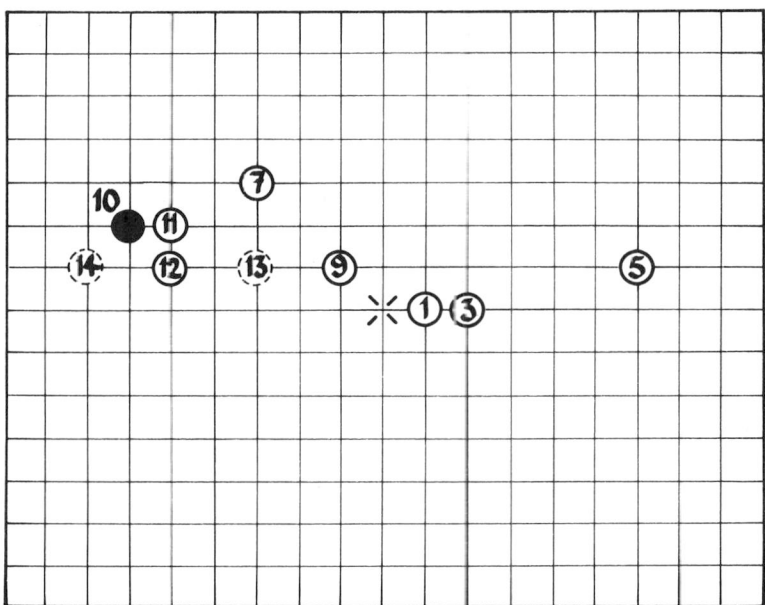

Figure 28. The Game Continues.

Try not to jump until you really have to, and then only as far as you really must. If you will have a stone within three of the place your opponent will jump to, but *not* a knight's move away, you can probably use it to get back and needn't be too frightened by his jump (which he probably shouldn't be making!). Remember that a stone a knight's move away from the ball is almost always useless. Such stones are called **poultry** (a corruption of paltry and parity). A threatened chain becomes much more useful if it can be jumped along in several different ways. Don't forget that the stone you place may be useful to your opponent—possibly in a devastating U-turn.

A pleasing feature of the game is that an expert can still enjoy a game against a novice provided they start with the ball much nearer the expert's goal-line.

Like Chess and Go, and unlike most of the games in this book, Phutball is *not* the kind of game for which one can expect a complete analysis.

EXTRAS

REFERENCES AND FURTHER READING

William N. Anderson, Maximum matching and the game of Slither, J. Combinatorial Theory Ser. B, **17** (1974) 234–239.

Charles Babbage, Passages from the Life of a Philosopher, Longman, Green, Longman, Roberts and Green, London, 1864; reprinted Augustus M. Kelley, New York, 1969, pp. 467–471.

J.P.V.D. Balsdon, Life and Leisure in Ancient Rome, McGraw–Hill, New York. 1969, pp. 156 ff.

A.G. Bell, Kalah on Atlas, in D. Michie (ed.) "Machine Intelligence, 3", Oliver & Boyd, London, 1968, 181–193.

A.G. Bell, "Games Playing with Computers", George Allen & Unwin. London, 1972, pp. 27–33.

Robert Charles Bell, "Board and Table Games from Many Civilizations", Oxford University Press, London, 1960, 1969.

Richard A. Brualdi, Networks and the Shannon Switching Game, Delta, **4** (1974) 1–23.

Gottfried Bruckner, Verallgemeinerung eines Satzes über arithmetische Progressionen, Math. Nachr. **56** (1973) 179–188; M.R. **49** #10562.

L. Csmiraz, On a combinatorial game with an application to Go-Moku, Discrete Math. **29** (1980) 19–23.

D.W. Davies, A theory of Chess and Noughts and Crosses, Sci. News, **16** (1950) 40–64.

H.E. Dudeney, "The Canterbury Puzzles and other Curious Problems", Thomas Nelson and Sons, London, 1907; Dover, New York, 1958.

H.E. Dudeney, "536 Puzzles and Curious Problems", ed. Martin Gardner, Chas. Scribner's Sons, New York, 1967.

J. Edmonds, Lehman's Switching Game and a theorem of Tutte and Nash-Williams, J. Res. Nat. Bur. Standards, **69B** (1965) 73–77.

P. Erdös and J.L. Selfridge, On a combinatorial game, J. Combinatorial Theory Ser. B, **14** (1973) 298–301.

Ronald J. Evans, A winning opening in Reverse Hex, J. Recreational Math. **7** (1974) 189–192.

Ronald J. Evans, Some variants of Hex, J. Recreational Math. **8** (1975–76) 120–122.

Edward Falkener, "Games Ancient and Oriental and How to Play Them", Longmans Green, London, 1892; Dover, New York, 1961.

G.E. Felton and R.H. Macmillan, Noughts and Crosses, Eureka. **11** (1949) 5–9.

William Funkenbusch and Edwin Eagle, Hyperspacial Tit-Tat-Toe or Tit-Tat-Toe in four dimensions, Nat. Math. Mag. **19** #3 (Dec. 1944) 119–122.

David Gale, The game of Hex and the Brouwer fixed-point theorem, Amer. Math. Monthly, **86** (1979) 818–827.

Martin Gardner, "The Scientific American Book of Mathematical Puzzles and Diversions", Simon & Schuster, New York, 1959.

Martin Gardner, Mathematical Games, Scientific Amer., each issue, but especially **196** #3 (Mar. 1957) 160–166; **209** #4 (Oct. 1963) 124–130; **209** #5 (Nov. 1967) 144–154; **216** #2 (Feb. 1967) 116–120; **225** #2 (Aug. 1971) 102–105; **232** #6 (June 1975) 106–111; **233** #6 (Dec. 1975) 116–119; **240** #4 (Apr. 1979) 18–28.

Martin Gardner, "Sixth Book of Mathematical Games from Scientific American", W.H. Freeman, San Francisco, 1971; 39–47.

Martin Gardner, Mathematical Carnival, W.H. Freeman, San Francisco 1975, chap. 16.

Richard K. Guy and J.L. Selfridge, Problem S. 10, Amer. Math. Monthly, **86** (1979) 306; solution T.G.L. Zetters **87** (1980) 575–576.

A.W. Hales and R. I. Jewett, Regularity and positional games. Trans. Amer. Math. Soc. **106** (1963) 222–229; M.R. **26** #1265.

Professor Hoffman (Angelo Lewis), "The Book of Table Games", Geo. Routledge & Sons, London, 1894, pp. 599–603.

Isidor, Bishop of Saville, "Origines", Book 18, Chap. 64.

Edward Lasker, "Go and Go-Moku", Alfred A. Knopf, New York, 1934; 2nd revised edition, Dover, New York, 1960.

Alfred Lehman, A solution of the Shannon switching game. SIAM J. **12** (1964) 687–725.

E. Lucas, "Récréations Mathématiques", Gauthier-Villars, 1882–1894, Blanchard, Paris, 1960.

Carlyle Lustenberger, M.S. thesis, Pennsylvania State University, 1967.

Leo Moser, Solution to problem E773[1947,281], Amer. Math. Monthly. **55** (1948) 99.

Geoffrey Mott-Smith, "Mathematical Puzzles", Dover. New York, 1954; ch. 13 Board Games.

H.J.R. Murray, "A History of Board Games other than Chess", Oxford University Press, 1952; Hacker Art Books, New York, 1978; chap. 3, Games of alignment and configuration.

T.H. O'Beirne, New boards for old games, New Scientist, **269** (62:01:11).

T.H. O'Beirne, "Puzzles and Paradoxes", Oxford University Press, 1965.

Ovid, "Ars Amatoria", ii, 208, iii, 358.

Jerome L. Paul, The q-regularity of lattice point paths in R^n, Bull. Amer. Math. Soc. **81** (1975) 492; Addendum, ibid. 1136.

Jerome L. Paul, Tic-Tac-Toe in n dimensions, Math. Mag. **51** (1978) 45–49.

Jerome L. Paul, Partitioning the lattice points in R^n, J. Combin. Theory Ser. A, **26** (1979) 238–248.

Harry D. Ruderman, The games of Tick-Tack-Toe, Math. Teacher, **44** (1951) 344–346.

Sidney Sackson, "A Gamut of Games", Random House, New York, 1969.

John Scarne. Scarne's Encyclopedia of Games. Harper and Row. New York, 1973.

Fred. Schuh, "The Master Book of Mathematical Recreations", trans. F. Göbel, ed. T.H. O'Beirne, Dover, New York, 1968; ch. 3, The game of Noughts and Crosses.

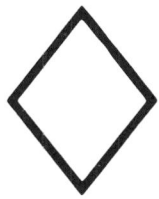

SOLITAIRE DIAMONDS!

Twinkle, twinkle, little star,
How I wonder what you are!
Up above the world so high,
Like a diamond in the sky!
      Jane Taylor, *The Star*.

We are all in the dumps, For diamonds are trumps;
The kittens are gone to St. Paul's.
The babies are bit, The Moon's in a fit,
And the houses are built without walls.
      Nursery Rhyme.

If you've followed everything in *Winning Ways* so far, you're probably finding it hard to get people to play with you, so will need something to do on your own. Here are our favorite solitaire diamonds:

The classical games of Peg Solitaire, treated by old and new methods in Chapter 23.

A host of puzzles, pastimes and other party tricks in Chapter 24.

And finally, every automaton will enjoy playing the notorious game of Life (Chapter 25).

Chapter 23

Purging Pegs Properly

We can merely mention bean-bags, peg-boards, size and form boards,
as some of the apparatus found useful for the purpose of amusing
and instructing the weak-minded.
Allbutt's Systematic Medicine, 1899, VIII, 246.

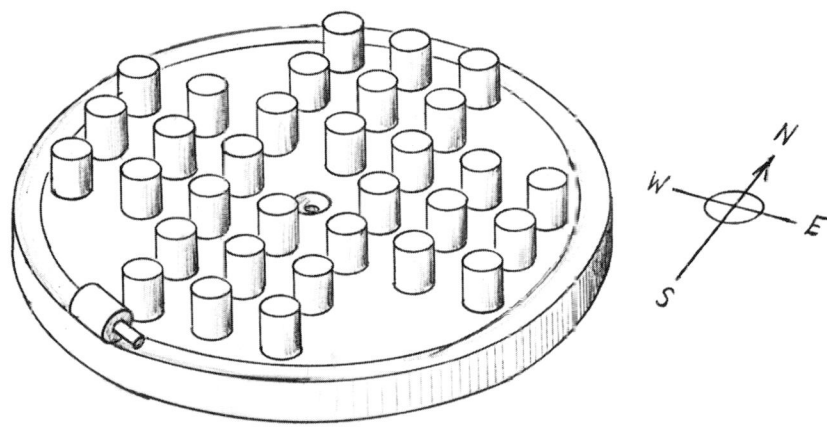

Figure 1. The English Solitaire Board.

Figure 1 shows the English Board on which the game of Peg Solitaire is usually played. It's easier to refill the board if you use marbles, but pegs are steadier when it comes to analysis.

The game is played (by one person of course) as shown in Fig. 2. If in some row or column two adjacent pegs are next to an empty space as in Fig. 2(a), then we may jump the peg *p* over *r* into the space *s* (Fig. 2(b)). The peg *r* that has been jumped over is then removed (Fig. 2(c)). Jumps are like captures in Draughts or Checkers, but they *never* take place diagonally, but only in the East, South, West or North directions.

697

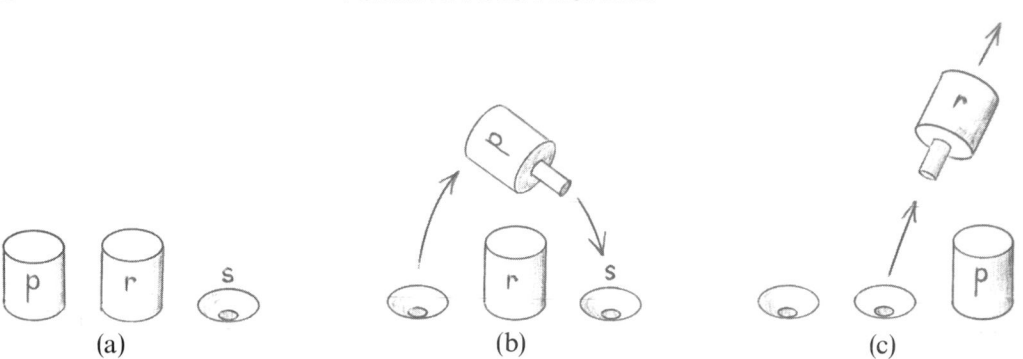

Figure 2. Making a Solitaire Jump.

CENTRAL SOLITAIRE

The standard problem is to start as in Fig. 1, with a peg in every hole except the centre, and then aim, by making a series of these jumping moves, to reduce the situation to a single peg in the central hole (Fig. 3).

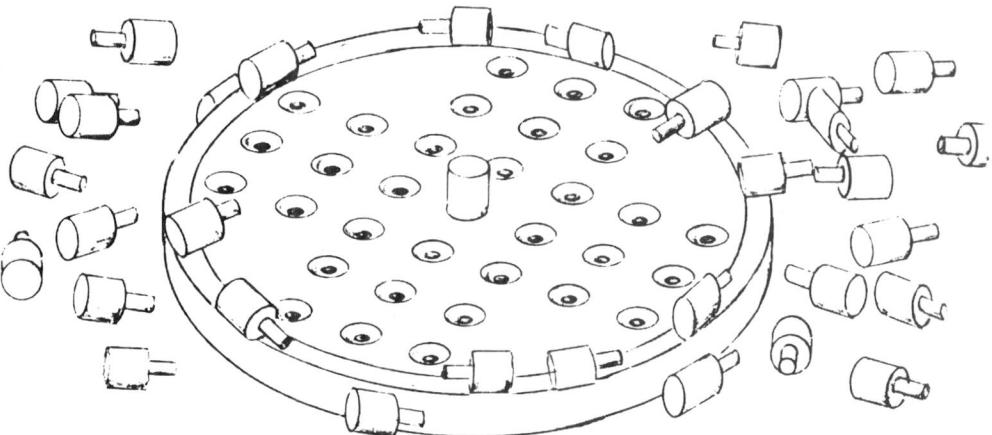

Figure 3. Success!

Like many card solitaire ("Patience") games, Solitaire is probably called a *game* rather than a puzzle because one often feels one is playing against an invisible opponent. Many people not normally interested in puzzles will recall some period of their lives when they have struggled with this opponent for days at a time; yet it seems that most of those who can readily solve simple Solitaire problems have been taught the trick by someone else as a child. It is rare indeed to find someone who has acquired the knack single-handed, and surely Peg Solitaire (nowadays selling in many parts of the world under the trade name of Hi-Q) must be the hardest game of its kind to have gained substantial popularity. It is an ideal game to while away hours of enforced idleness during illness or long journeys, and perhaps we should believe those old books which tell us that the game was invented by a French nobleman who first played it on the stone tiles of his prison cell.

If you haven't played this game before, put down this book, go out right now, buy a board, and try to solve the Central Solitaire Game. Those of you who are left will have plenty of time to read the chapter before the novices come back in a week or so—why not learn a particularly elegant solution to impress them all?

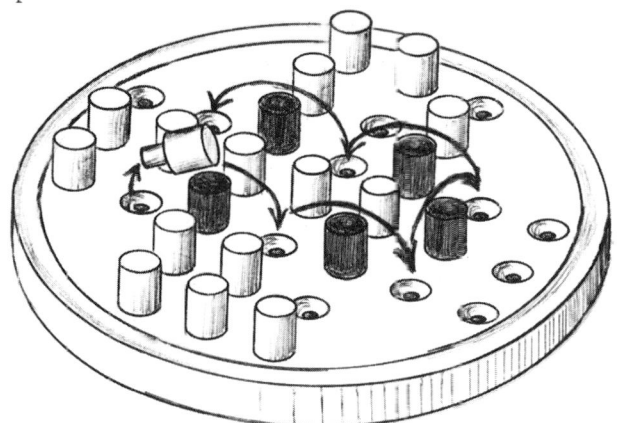

Figure 4. A Move of Five Jumps.

DUDENEY, BERGHOLT AND BEASLEY

Since you must already know how to solve the problem, you'll want to do it quickly, so let's agree to count any number of consecutive jumps made with a single peg as just one **move**. Figure 4 shows such a move—the five shaded jumped-over pegs are to be taken off as part of the move.

		a	b	c		
	y	d	e	f	z	
g	h	i	j	k	l	m
n	o	p	x	P	O	N
M	L	K	J	I	H	G
	Z	F	E	D	Y	
		C	B	A		

Figure 5. Labelling the Places.

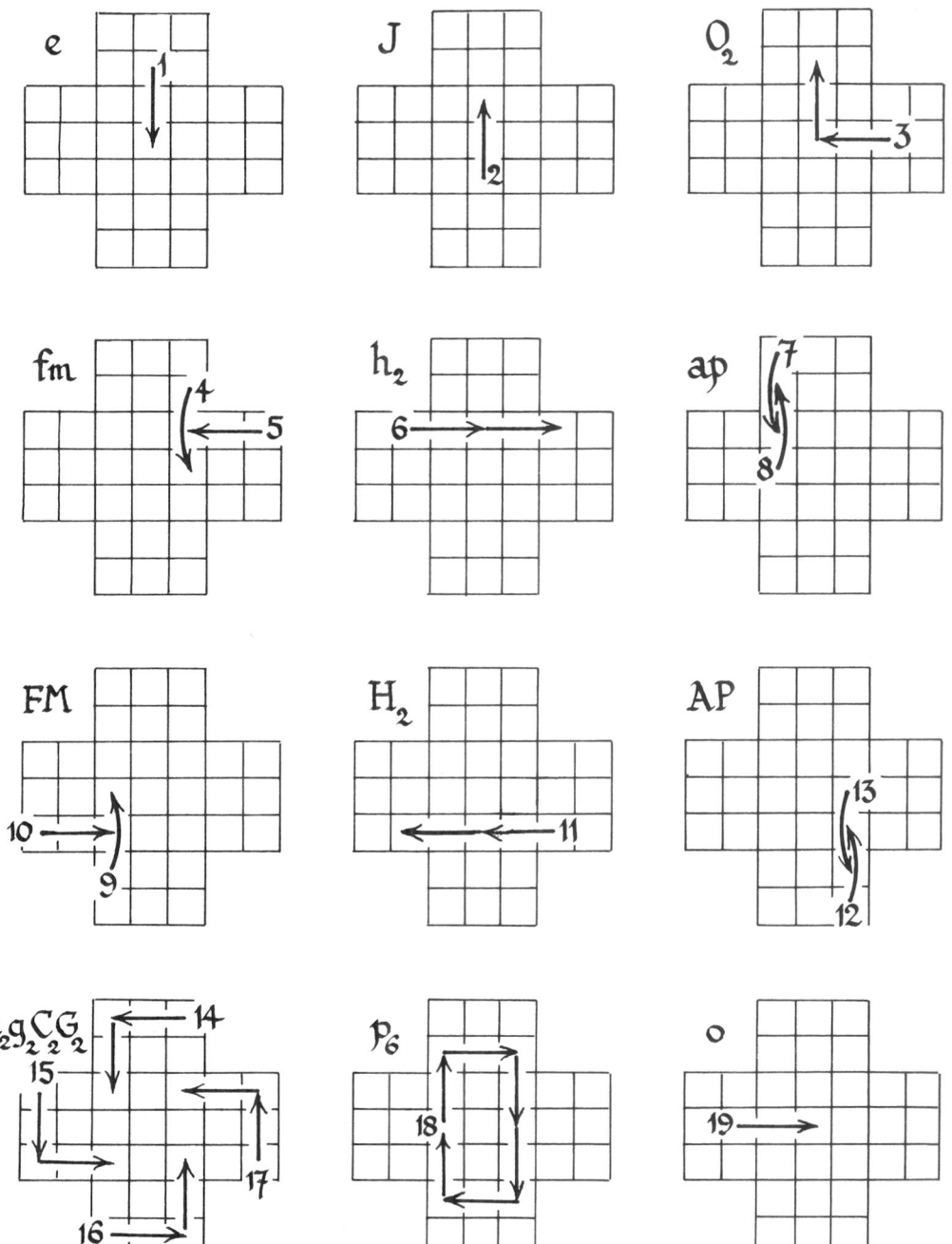

Figure 6. Dudeney's 19-move Solution for Central Solitaire.

In order to describe a solution concisely, we label the places as in Fig. 5, and write S_t for a jump from S to t and shorten this to S when we don't need to indicate the direction. The 5-jump move of Fig. 4 is L_{JHljh} which we will abbreviate to L_5 when it is unambiguous (we can't do that here since L_5 could also mean L_{hjJHl} and various other things). In this notation, Dudeney's elegant 19-move solution of his Central Solitaire problem is

$$eJO_2 \; fmh_2ap \;\; FMH_2AP \;\; c_2g_2C_2G_2 \;\; p_6o,$$

and this is set out in Fig. 6.

Dudeney thought that the number 19 could not be improved, but, in *The Queen* four years later, Ernest Bergholt gave an 18-move solution, unfortunately not quite as symmetrical as Dudeney's:

$$elcPDGJm_2igL_5CpA_2M_2a_3d_5o.$$

Here the notation L_5 is ambiguous, but the intended 5-jump move is the one depicted in Fig. 4. The move d_5 is also ambiguous, but either interpretation leads to the same result.

The whole truth emerged only 52 years later, in 1964, when John Beasley used the methods described in this chapter to prove that a solution in fewer than 18 moves is impossible. With Beasley's kind permission we publish his proof for the first time in the Extras to this chapter. It is very condensed, so the reader who wishes to follow it should first study the chapter diligently!

PACKAGES AND PURGES

It's nice to be able to know the effect of a whole collection of moves before you make them, so let us sell you some of our instant **packages**. When a package is used to clear all the pegs from a region, we call it a **purge**.

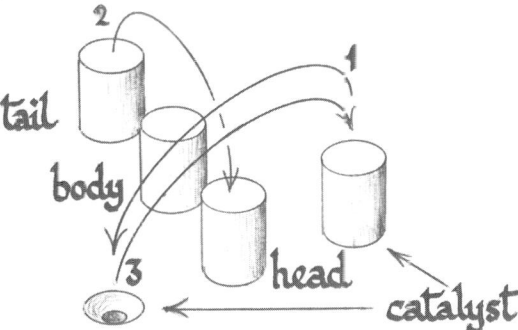

Figure 7. Purging Three Pegs.

Figure 7 shows the handy little **3-purge**, our most popular package. When three pegs—the *tail*, the *body* and the *head*—are adjacent in line, this will remove them all, provided the head has an additional peg on one side of it, and an empty space on the other, as in the figure. Move 1 of the package jumps the additional peg *over* the head; move 2 jumps tail over body *into* head; and move 3 jumps *over* the head, back to its original position. Since the peg and the space on either side of the head are essential to the package, but are restored to their original state, we call them the **catalyst**.

In Figs. 8(a) to 8(h), ● indicates a peg to be purged, ○ a space to be filled, and **XX** indicate catalyst places of which one must be full and the other empty. In most of the purges there are two catalyst moves in opposite directions over the same position (which may be a peg or an empty space) and the remaining moves form one or two packages which deliver pegs to that place. For the 3-purge (8(a)), one peg was already in place and the second is delivered by a single jump which we might call the "2-package" (8(b)). The 6-purge is usually accomplished (8(c)) using a 2-package to deliver the first peg and a 4-package (8(d)) for the second.

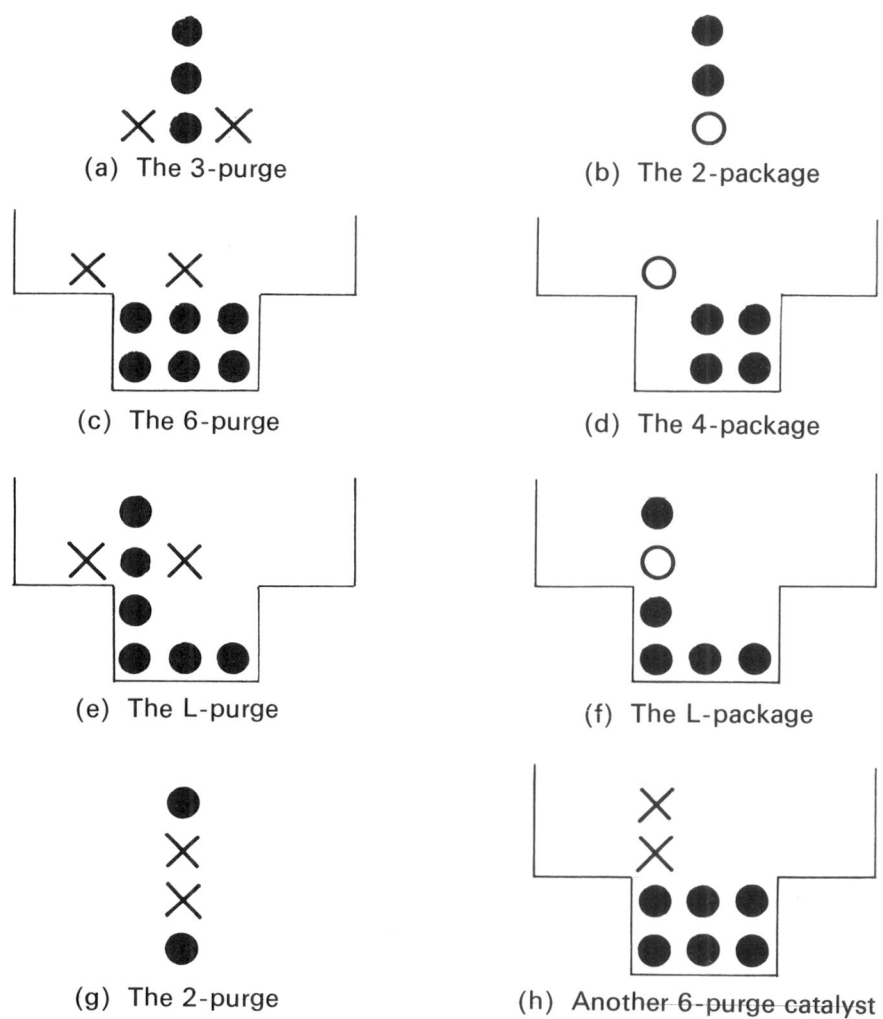

(a) The 3-purge

(b) The 2-package

(c) The 6-purge

(d) The 4-package

(e) The L-purge

(f) The L-package

(g) The 2-purge

(h) Another 6-purge catalyst

Figure 8. A Parcel of Packages.

The L-purge (8(e)) and L-package (8(f)) are very useful indeed. The first peg for the L-purge is already in place and an L-package supplies the second. The first two moves of the L-package form a 2-purge (8(g)) which can also be used in other situations. The catalyst for the 2-purge is restored in a rather unorthodox way, as is the alternative catalyst for the 6-purge shown in Fig. 8(h).

PACKAGES PROVIDE PERFECT PANACEA

Plenty of problems are performed with panache by people who purchase our packages.

In Fig. 9 we can see at a glance a solution for Central Solitaire, consisting of two 3-purges (1 and 2) followed by three 6-purges (3, 4 and 5) and an L-purge, leaving only the final jump to be made. You should check that every purge has the catalyst it needs.

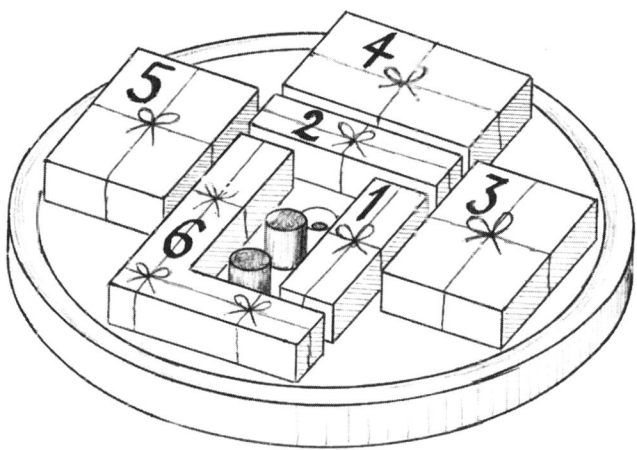

Figure 9. Central Solitaire Painlessly Packaged.

Instead of Central Solitaire we can consider other one-peg reversal problems: start with only one empty space and finish with only one peg in the same place. Figures 9 and 10 show that most such problems can be solved by purely purgatory methods, but in Fig. 10(e) we start with a 4-package indicated by the arrow (1), and the notorious problem (b) needs more complicated methods.

To clarify our notation we explain our solution for (e) in detail. For the first jump we have no choice but to jump from the place marked 1 in the figure. Our second jump, from the place marked 2, clears a space which enables us to make the L-*package*, indicated by the bent arrow (3). We now have a catalyst for the L-purge (4) which is followed by a single jump from the place marked 5. We are now on the home run with purges 6, 7 and 8 followed by a single jump from place 9. If the reader plays this through she will find that we have set up a spectacular 5-jump move from the place marked 10_5.

The reader might like to try her hand at some *two*-peg reversal problems—start with just two spaces on the board and end with just two pegs in those places.

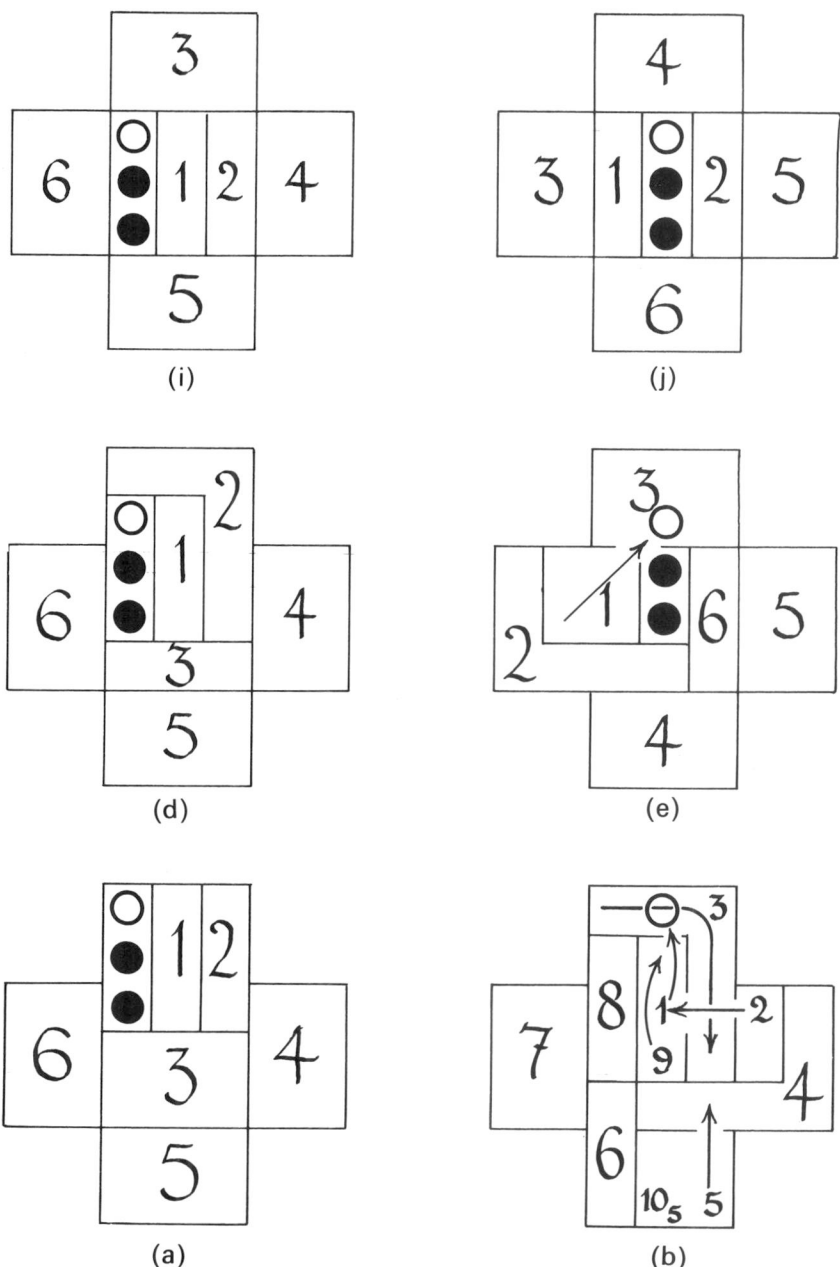

Figure 10. The Other Six One-Peg Reversals.

THE RULE OF TWO AND THE RULE OF THREE

Here is another type of problem (Fig. 11). We start with just one empty space and declare that some particular peg is to be the **finalist** (last on the board). In the example the initial hole is at position d and we want the finalist to be the peg that starts at b. Where must it end?

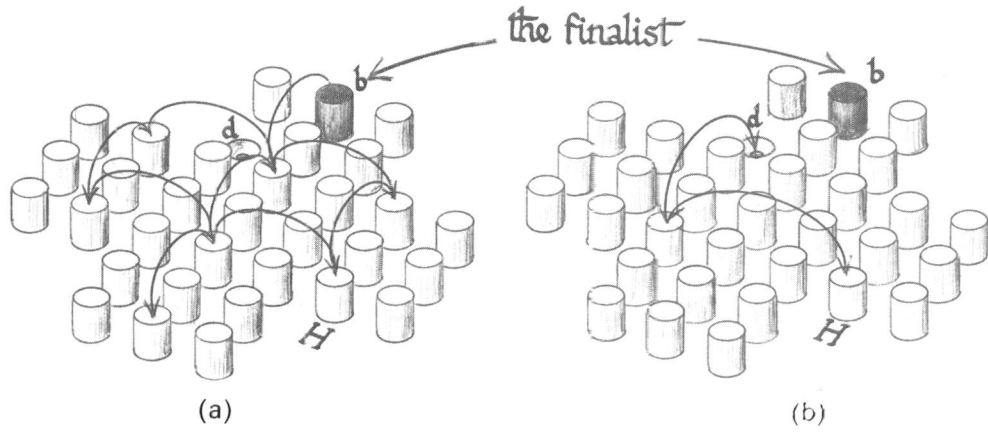

Figure 11. Find Where the Finalist Finishes!

There is an obvious **Rule of Two**—the peg can only jump an even number of places in either direction, as indicated by the arrows in Fig. 11(a). But there is a much more interesting **Rule of Three**. One of the consequences of this is that if we start with a single space on the English board and end with a single peg, then we can move in steps of three from the initial space to that of the finalist, as in Fig. 11(b).

The Rule of Two and the Rule of Three, taken together, can lead to surprises. See how they point to the unique finishing place H in Fig. 11(a) and (b). Now that we know that H is the only place permitted by both the Rule of Two and the Rule of Three, the problem is a lot easier than it might have been. Figure 12 shows a neatly packaged solution; how did we find it?

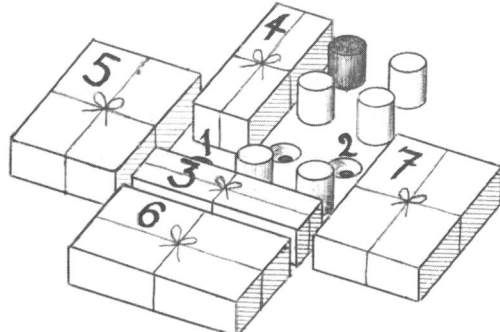

Figure 12(a). The Position After the First Two Moves.

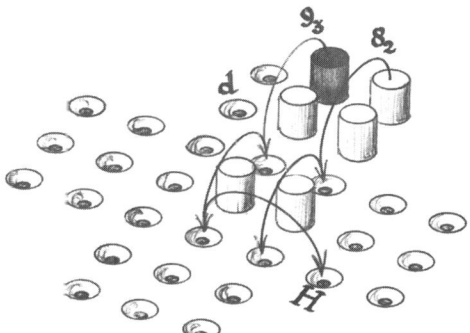

Figure 12(b). The Position Before the Last Two Moves.

What we did was plan the 3-jump move 9_3 which puts the finalist in his place, and our second jump was to clear a space for this. But after we made this second jump most of the pegs parcelled themselves up naturally. The one apparent exception was the peg starting just right of the finalist, and the best way of clearing this seemed to be to use it as in move 8_2 to provide the final jump.

For other problems, gentle reader, we recommend a similar procedure. Plan the last few moves of your solution and let the first few be used to smooth the way for these and leave the remaining pegs in tidy packages. Remember that the catalyst for the very last purge must be among the pegs in your planned finale.

Here's a nice finalist problem for you. Let the initial hole be in position B and the finalist be the peg which starts at J. Can you end with only this peg?

SOME PEGS ARE MORE EQUAL THAN OTHERS

How do we explain the Rule of Three? The best way is to introduce "multiplication" for Solitaire positions. In Fig. 13(a) the two adjacent pegs s and t can obviously be replaced by a single peg at r, so we write

$$st = r,$$

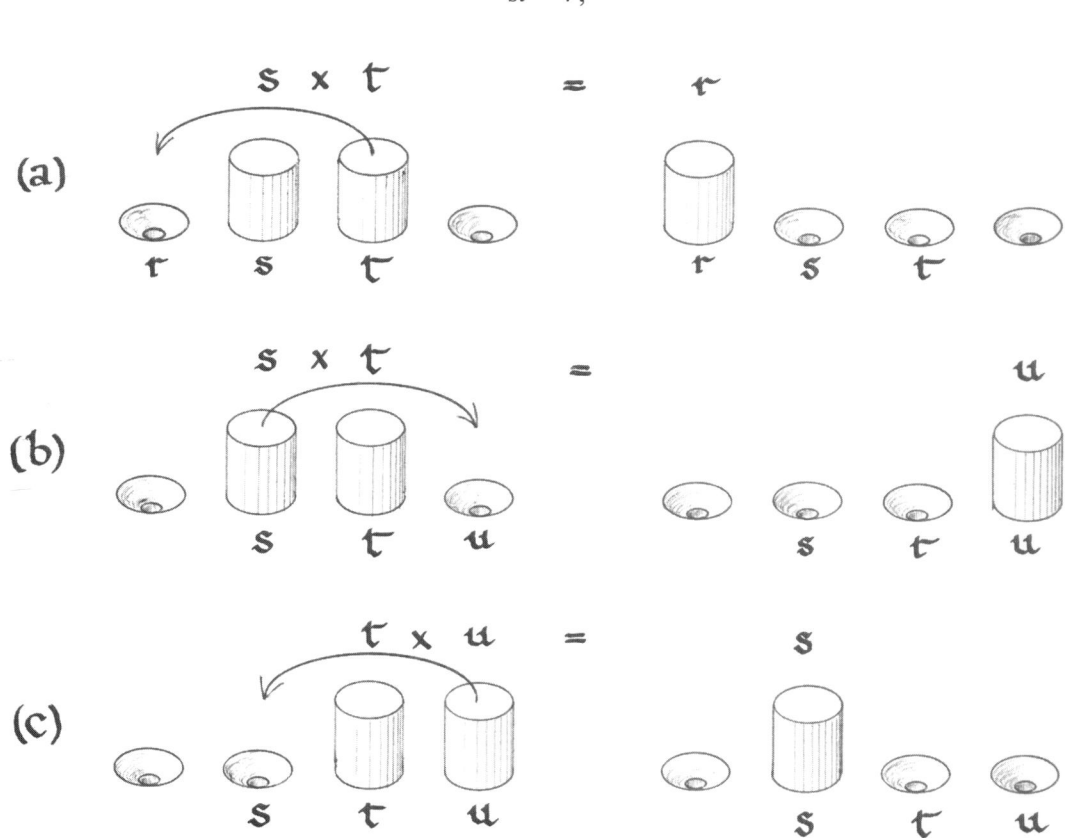

Figure 13. Multiplying Pegs.

but Fig. 13(b) shows that we can also write

$$st = u.$$

Now Euclid tells us that things that are equal to the same thing are equal, so we must agree that $r = u$.

> Places three apart
> in any line are
> considered equal.

Let's see what other rules of algebra tell us. Combining Figs. 13(b) and 13(c), we have

$$st = u, \qquad tu = s$$
$$st^2u = us,$$

or, cancelling,

$$t^2 = 1,$$

which seems to tell us that

> two pegs in the
> same place cancel.

Remember how catalysts do precisely this—they remove two pegs which are delivered to the same place by the other moves of a purge. In fact it follows from our algebra that

> any set of pegs
> that can be
> purged cancel.

For example, in Fig. 13(c), $tu = s$, so

$$stu = ss = 1.$$

> Three adjacent pegs
> in line cancel,   (3-purge)

and since $r = u$,

$$ru = uu = 1.$$

┌─────────────────────────────┐
│   Two pegs at   │
│   distance three cancel.   │   (2-purge)
└─────────────────────────────┘

But in the algebra there are less obvious equalities: for since $s^2 = 1 = rst$, we find

$$s = rt.$$

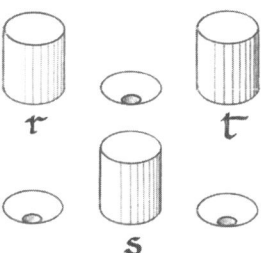

Figure 14. $s = rt$.

REISS'S 16 SOLITAIRE POSITION CLASSES

We've now said enough to see how our algebra cuts the Solitaire board down to size, for since places three apart are algebraically equal, every place is equal to one of the nine in the middle of the board (Fig. 15); for example $a = p$. Now we can use our most recent rule to express each of these nine in terms of the four corner ones, i, k, I, K:

$$j = ik \qquad P = Ik$$
$$p = iK \qquad J = IK$$
$$x = jJ = ikIK.$$

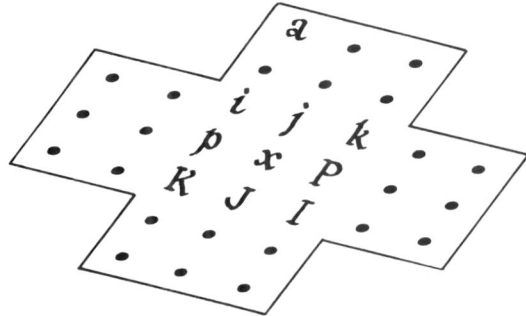

Figure 15. Stripping Down to Essentials.

Since equal pegs cancel,

> every Solitaire position
> is algebraically equal to
> one of the 16 combinations
> of the places i, k, I, K.

THE SIXTEEN REISS CLASSES

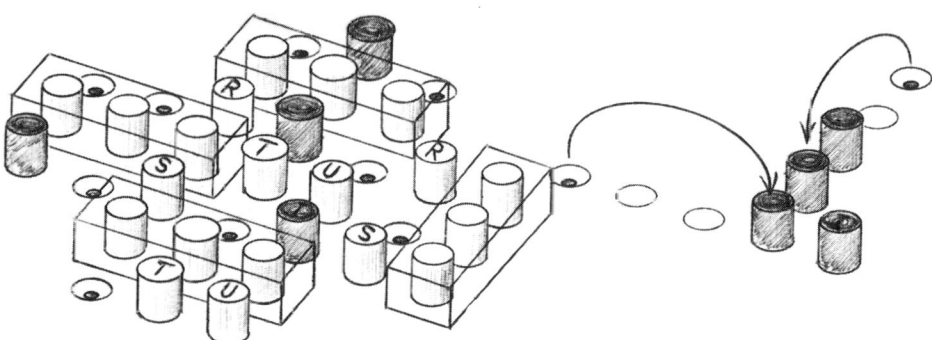

Figure 16(a). Found near Split? **Figure 16(b).** Reduced to Size.

The position of Fig. 16(a) was found unattended in a Jugoslav railway train. Those filmy packages and letters weren't there—but just came into our mind's eye when we pondered the possibility of reducing the position to a single peg. Where must this single peg be?

Our rules allow us to cancel those four packages of three and then the four pairs RR, SS, TT, UU, so that the position is algebraically equal to the four shaded pegs. We then can move two of these three spaces and cancel another 3-package as in Fig. 16(b) to see that the position equals a single peg at I. So the Rule of Three says that the finalist must be at I, L or f. For which of these places can you find solutions?

How do we know that Reiss's sixteen classes are really different? Might not our algebraic rules imply perhaps that $i = kK$? No! For consider the numbers ± 1 shown in the places of Fig. 17(i). Whenever three of these numbers

$$r, s, t$$

are adjacent in line, we really do have

$$rs = t,$$

and from this we can see that all our algebraic rules hold for these numbers. But in this system we have

$$i = -1, \qquad k = K = +1,$$

so we can't prove $i = kK$! In fact Figs. 17(i, k, I, K) show that all 16 combinations of the pegs i, k, I, K are algebraically distinct: for example the value on Fig. 17(i) is -1 just if i is involved in the combination. Making a Solitaire move or applying any of our algebraic rules will never change the value in any of the four Figures.

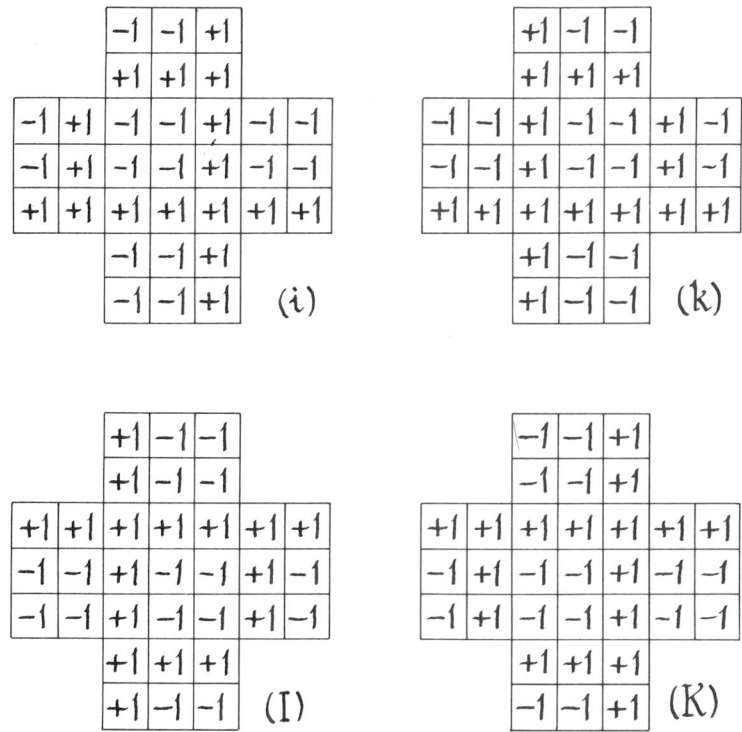

Figure 17. "Answers" to the Algebra.

In algebraic language, the first thing we told you about the Rule of Three may be restated— a position with just one empty space is algebraically equal to the **complementary** position in which *only* that place is full. More generally,

> any position on the English board
> is algebraically equal to the
> complementary position which has
> empty spaces replacing pegs
> and pegs replacing empty spaces.

For our rules allow us to complement any line of three adjacent places and the whole board can be parcelled into such threes.

This property fails for

THE CONTINENTAL BOARD

which has the four extra holes at *y, z, Y, Z* in Fig. 5. So no reversal problems are possible on this board. Which of the problems which start with a single hole and end with a single peg are solvable on this board? See the Extras.

PLAYING BACKWARDS AND FORWARDS

"The game called Solitaire pleases me much. I take it in reverse order. That is to say that instead of making a configuration according to the rules of the game, which is to jump to an empty place and remove the piece over which one has jumped, I thought it was better to reconstruct what had been demolished, by filling an empty hole over which one has leaped."

Leibniz.

The famous philosopher plainly thought that playing Solitaire backwards was different from playing it forwards, but really it's exactly the same game! For let's see what happens when he makes one of his backward moves from Figs. 18(a) to 18(c). Leibniz regards this as jumping *piece t* into *hole r* and *filling* the empty *hole s* over which he has leaped, but Fig. 18(b) shows that we can regard him as jumping the *hole* at *r* over the *hole* at *s* into the *piece* at *t* and removing the *hole* over which he has jumped. (Of course to *remove* a hole he *inserts* a piece!)

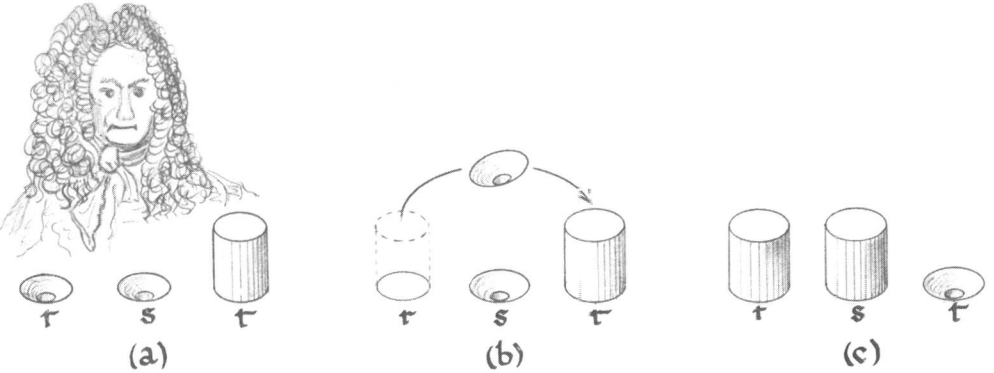

Figure 18. The Philosophy of Leibniz.

> Backwards Solitaire is just forwards Solitaire with the notions "empty" and "full" interchanged.

TIME-REVERSAL = ANTI-MATTER?

This can be useful as well as interesting. A quite spectacular Solitaire finale happens in Beasley's remarkable 16-move solution of the *i*-reversal problem:

$$apc_2F_2gdM_2IAP_\downarrow f_\downarrow C_3Gm\ldots.$$

After 14 of the 16 moves the board still seems quite full (Fig. 19(a)) but can be cleared to a single peg in just 2 moves. (Can you find them?)

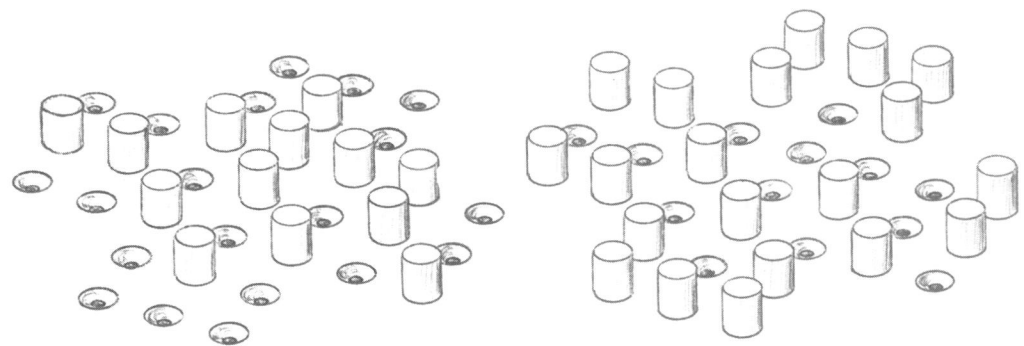

Figure 19. Two moves to Go!   . . .   And How to Get Back?

How would *you* find the moves leading up to this position? The time-reversal trick should make it easy. Instead of reducing a position with only one space, at *i*, to Fig. 19(a), try to reduce the complementary position (Fig. 19(b)) to just one peg at *i*. If you've been doing your homework and practising diligently, you won't find this too hard. You too can astonish your friends with grand finales to other Solitaire problems set up by the time-reversal trick.

PAGODA FUNCTIONS

Reiss's algebraic theory (known to many!) applies even when we allow you to make moves backward in time (like Leibniz) as well as the ordinary forward ones. Of course this lets you take back any of your bad moves, but you may also "undo" moves you haven't even made! If two positions are in different Reiss classes, then we can never get from one to the other by normal moves, by Leibniz's backward moves nor by any mixture of the two.

Unfortunately this means, of course, that the Reiss theory can never tell you when you've made a bad move, because the Reiss class never changes. You need something like the **Pagoda Functions** (known to few!) we are about to show you, that can *change* when you make a move, albeit in a restricted way. Mike Boardman was one of those who helped us to develop these.

Those friends of yours should now be back from the store with their Solitaire boards, so why not present them with a couple of innocent-looking problems? Since these are *reversal* problems, your friends won't be able to prove them impossible even if they've got as far as the last section.

The two problems are shown in Figs. 21(a) and (b) where circles show the only places which are initially empty and which must also be the only places which are finally full. Figures 21(c) and (d) show two pagoda functions which prove the problems impossible. In general, if pag is any such function and *X* any Solitaire position, we shall write

$$\text{pag } X$$

for the *sum* of the numbers that pag assigns to the pegs which are present in X. If X is partitioned into smaller positions Y and Z, then, in our algebraic notation we have

$$X = YZ,$$

and

$$\text{pag } X = \text{pag } Y + \text{pag } Z.$$

so that pagoda functions behave like logarithms.

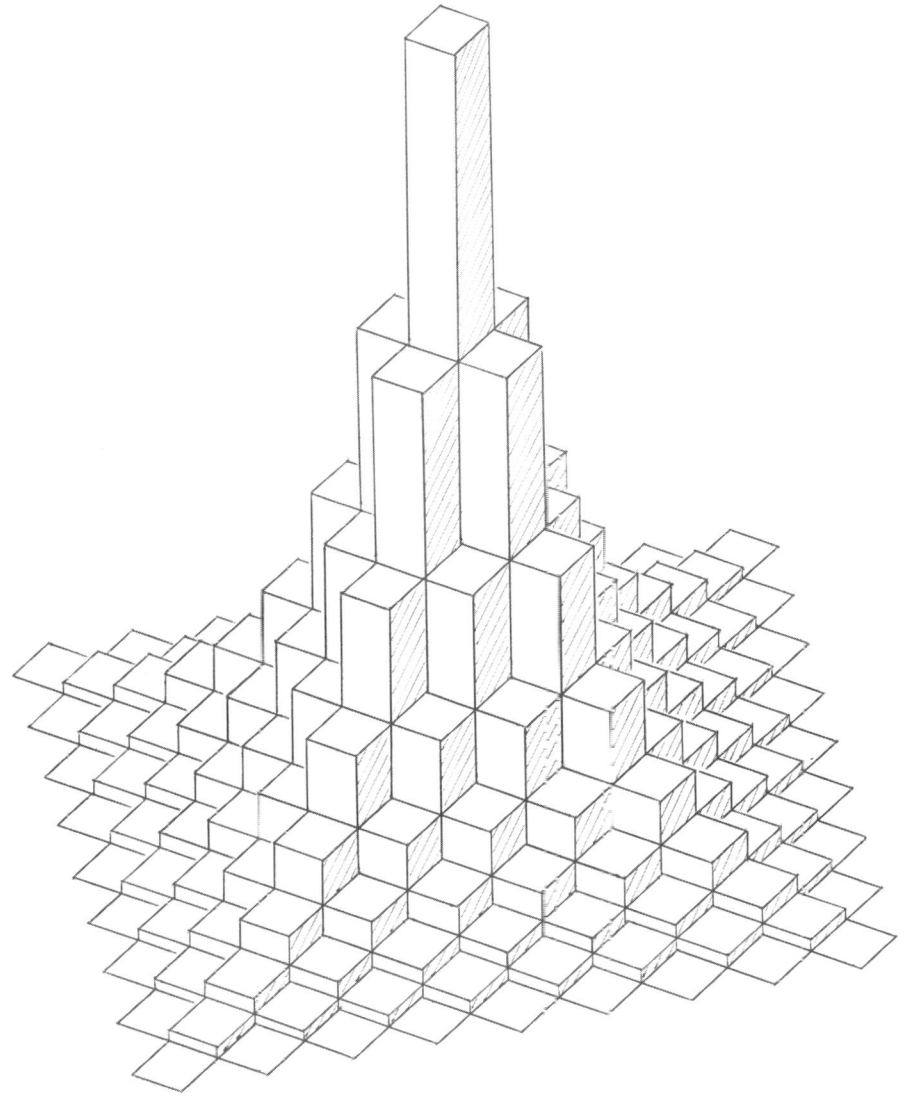

Figure 20. The Golden Pagoda.

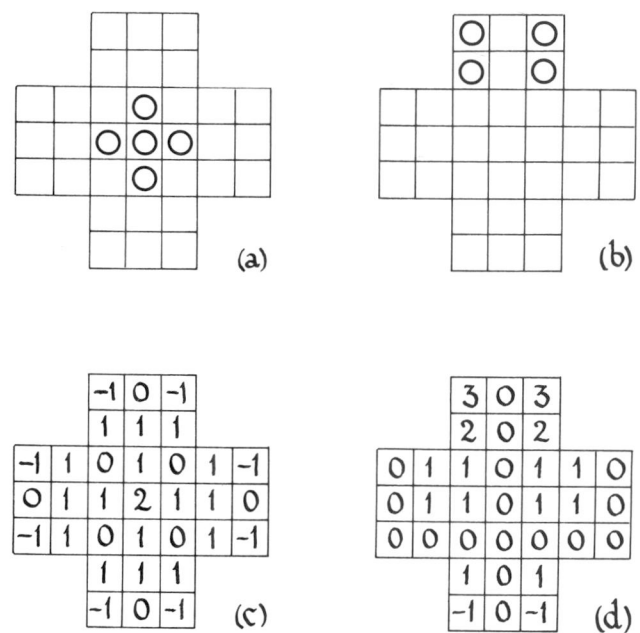

Figure 21. Two Impossible Reversal Problems.

The essential property which defines pagoda functions is that *no move may increase the value.* To check this condition you must make sure that

THE PAGODA FUNCTION CONDITION

$$\mathrm{pag}\,r + \mathrm{pag}\,s \geqslant \mathrm{pag}\,t$$

holds for every conceivable Solitaire jump r over s into t.

CHECK THIS CONDITION NOW IN FIGURES 21(c) AND (d)!

When you've done that you will see the impossibility of our two problems, since the pag (Fig. 21(c)) of the initial position in 21(a), namely 4, can't be increased to 6, the pag of the final position; nor can 8 be increased to 10 (Figs. 21(b) and (c)).

In Fig. 22 we show the pagoda functions you're most likely to find useful; so you'd better check the Pagoda Function Condition for each of them! The values in the blank spaces are zero and you can make any of the indicated swaps. Figures 22(c, d, h, and v) are obvious pagoda functions since they just indicate all the places that a given peg can go to. The 12 places in 22(c) are called **corners** and the 5 in 22(d) are the **dodos**, because one of the easiest mistakes you can make is to let your dodos become extinct when you need one in your final position. Those extra minus ones often make 22(a) and (b) more useful than (h) and (v).

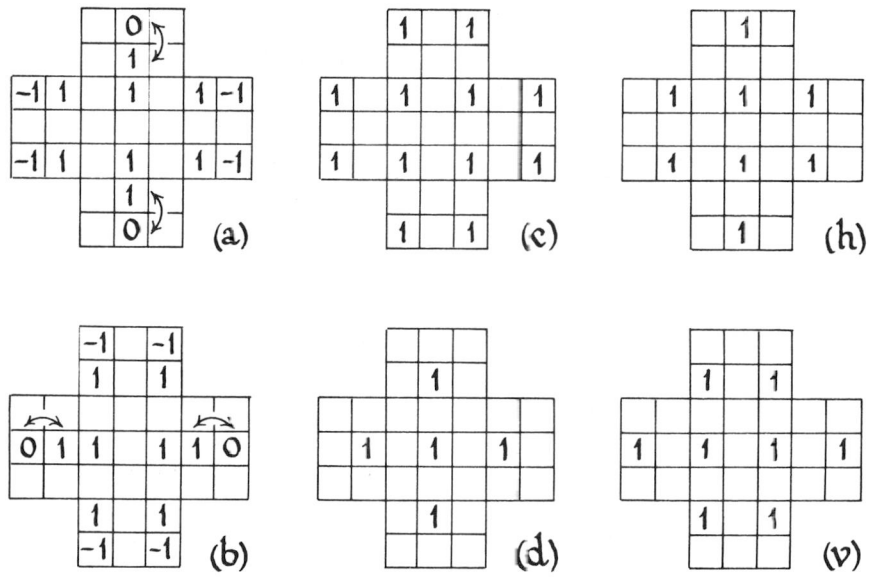

Figure 22. Some Useful Pagoda Functions.

THE SOLITAIRE ARMY

A number of Solitaire men stand initially on one side of a straight line beyond which is an infinite empty desert (Fig. 23). How many men do we need to send a scout just 0, 1, 2, 3, 4, or 5 paces out into the desert?

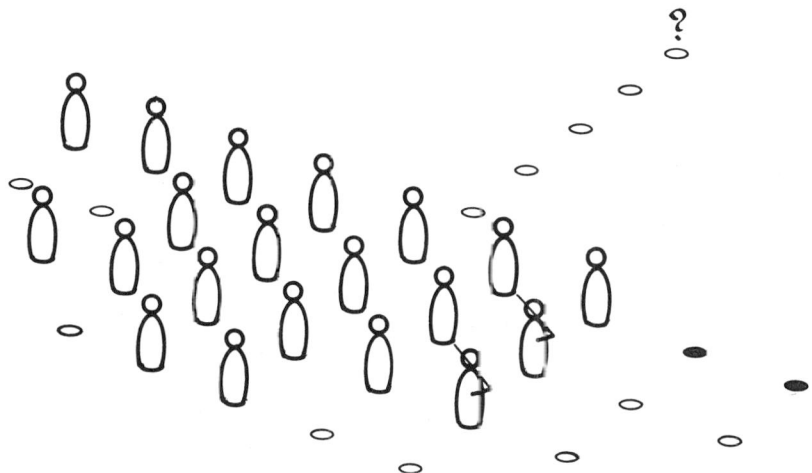

Figure 23. How About Sending a Scout Out?

It's not hard to see that the answers for 0, 1, 2 and 3 paces are 1, 2, 4 and 8 men, so you might guess that the next two answers are 16 and 32. But in fact no less than 20 men are needed to get 4 paces out. Can you find the two possible configurations of 20 men? (See the Extras.)

For 5 paces the answer is even more surprising—it is *impossible* to send a scout five paces into the desert, no matter how large an army we hire! The pagoda function which proves this is shown in Fig. 24. It was the shape of the graph of this function (Fig. 20) which first suggested the name "pagoda". The number σ is determined by the golden ratio:

$$\sigma = \tfrac{1}{2}(\sqrt{5} - 1) = 0{\cdot}618\ldots,$$

$$\sigma^2 + \sigma = 1.$$

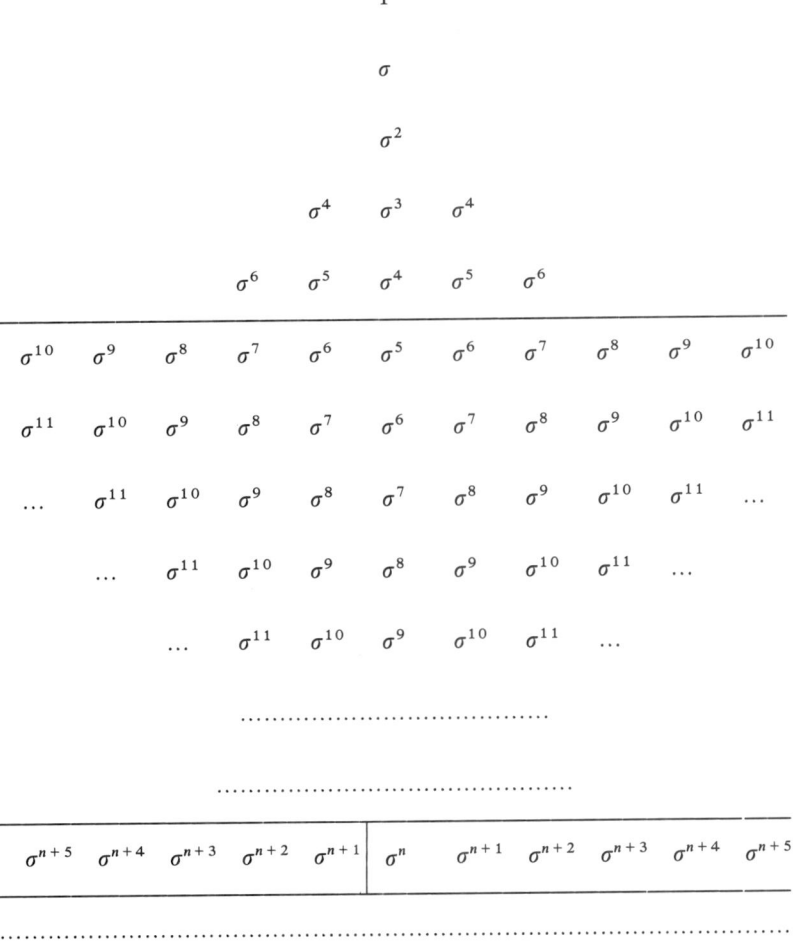

Figure 24. Pagoda Function for the Solitaire Army.

By some easy mathematics we have

$$\sigma^n + \sigma^{n+1} + \sigma^{n+2} + \dots = \frac{\sigma^n}{1-\sigma} = \sigma^{n-2},$$

so that the total score of the line whose middle element is σ^n is

$$\sigma^{n-2} + \sigma^{(n+1)-2} = \sigma^{n-3},$$

and the total score of this line and all lower lines is

$$\frac{\sigma^{n-3}}{1-\sigma} = \sigma^{n-5}.$$

In particular, the sum of *all* the men on or below the σ^5 line is *exactly* 1, so no finite number of these men will suffice to send a scout to the place whose score is 1. But infinitely many men are *almost* enough, because we once showed that if any man of our army is allowed to carry a comrade on his shoulders at the start, then no matter how far away the extra man is, the problem can now be solved.

MANAGING YOUR RESOURCES

Your score on a pagoda function is in some sense a measure of your resources, which you should not consume too rapidly. But mere worldly goods are not enough: they must be capably managed to preserve a balance between your commitments in various directions.

The **Balance Sheet** of Fig. 25 has been cunningly devised to do just this. The subtlety of the English board is that you are often forced to consume assets in order to maintain the **balance**,

		b^{-1}	β	$b^{-1}\beta$		
		b	$\alpha\beta$	$b\beta$		
a^{-1}	a	1	a	1	a	a^{-1}
α	$b\alpha$	b	$c\alpha\beta$	$b\beta$	$b\alpha\beta$	α
$a^{-1}\alpha$	$a\alpha$	1	$a\alpha$	1	$a\alpha$	$a^{-1}\alpha$
		b	$a\alpha\beta$	$b\beta$		
		b^{-1}	β	$b^{-1}\beta$		

$\beta^2 = 1$

$\alpha^2 = 1$

Figure 25. The Balance Sheet.

as measured by the greek letters α and β, of your position in the North-South and East-West directions. The latin letters a, b and c measure the **assets** on a number of pagoda functions simultaneously (a and b for Figs. 22(a) and 22(b) and abc^2 for Fig. 21(c)).

To estimate the overall capacity of a position, find the product of the resources of all its pegs in Fig. 25, using the relations

$$\alpha^2 = \beta^2 = 1.$$

A problem has two such products, the **raw product** (for its initial position), which must be taken to the **finished product** (for the final position) while consuming the **available resources**:

$$\frac{\text{raw product}}{\text{finished product}} = \text{available resources}.$$

In Fig. 26, all the jumps that change the product are shown to do so in **units** of sizes

$$
\begin{array}{cccccc}
a & a\alpha & c\alpha & a^2c^{-1}\alpha = A & a^2 \\
b & b\beta & c\beta & b^2c^{-1}\beta = B & b^2
\end{array}
$$

so your available resources will only be **productive** if they can be made up of such units.

Central Solitaire, for example, has raw product a^4b^4 and finished product $c\alpha\beta$ so that its available resources are

$$\frac{a^4b^4}{c\alpha\beta} = a^4b^4c^{-1}\alpha\beta.$$

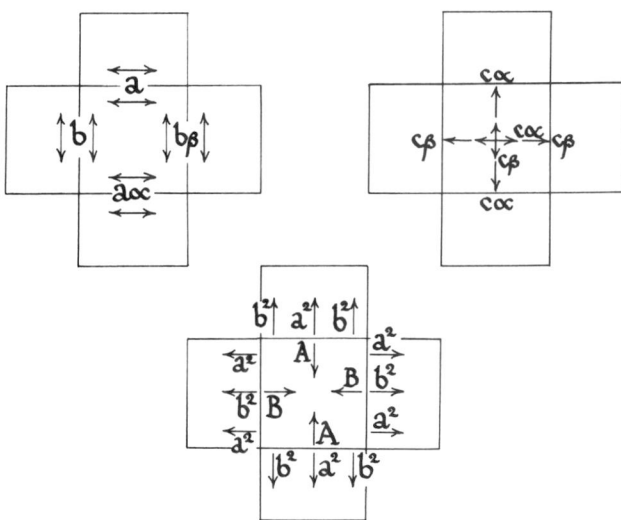

Figure 26. Using Resources in Various Units.

In Dudeney's solution only the opening and closing few moves actually use any of these:

move	e	J	O_2	$fmh_2apFMH_2APc_2G_2C_2G_2$	P_6	o
resources	A	$c\beta$	$B.c\alpha$	$\longleftarrow$ free moves $\longrightarrow$ $1.a.1.1.a\alpha.1$		B

UNPRODUCTIVITY AND THE PRODIGAL SON

Many problems are impossible for the simple reason that

$$b^2\alpha \text{ and } a^2\beta$$
$$\text{are unproductive!}$$

Why is this? In the case $b^2\alpha$, for example, we are hamstrung for lack of a's, so the α forces us to make a jump $c\alpha$, leaving only b^2c^{-1} for the remaining moves, in which c^{-1} demands a move $b^2c^{-1}\beta = B$, and we then have no assets with which to adjust the remaining β.

THE PRODIGAL SON'S OPENING

Jump into centre; jump over centre;
jump into centre; jump back over centre;

is the only way Central Solitaire can go wrong in as few as four moves. What's so bad about these moves? The prodigality lies in the second and fourth moves which both use $c\alpha$ or both use $c\beta$ and therefore leave only

$$a^4b^4c^{-1}\alpha\beta/c^2 = a^4b^4c^{-3}\alpha\beta$$

for the remaining moves. But

$$a^4b^4c^{-3}\alpha\beta$$
$$\text{is unproductive!}$$

For the only way to cope with c^{-3} without overspending either a or b is to use the units

$$A,A,B \text{ or } A,B,B$$

which leave only the unproductive products

$$b^2\alpha \text{ or } a^2\beta.$$

Of course the same argument shows that *no* two moves in *any* solution of Central Solitaire can have product c^2.

Can you find the only way (**Fool's Solitaire**) of getting absolutely pegbound (unable to move) in six jumps? And can you **Succour the Sucker** by solving the position reached after five of these moves? And can you flag yourself down to *another* pegbound position in as few as ten jumps from the start?

DEFICIT ACCOUNTING AND THE G.N.P.

The **deficit** of a problem is the amount by which its initial position lacks the resources of the entire board, combined with the total resources of the final position and the costs of any moves you intend to make. Since the resources of the entire board are

$$a^4b^4c\alpha\beta \text{ (the (English) Gross National Product)}$$

we have

$$\text{remaining resources} = \frac{a^4b^4c\alpha\beta}{\text{deficit}}.$$

The deficit is found very easily by multiplying the initial hole values by the final peg values. For Central Solitaire, the basic deficit is

$$c\alpha\beta \cdot c\alpha\beta = c^2,$$

which the Prodigal Son's bad moves extravagantly enlarged to c^4. He clearly didn't know the **Deficit Rule**:

> If deficit$/c^4$ *IS* productive,
> your remaining resources *AREN'T*!

This is because (G.N.P.)$/c^4$ is our unproductive product $\alpha^4\beta^4c^{-3}\alpha\beta$.

ACCOUNTING FOR TWO-PEG REVERSAL PROBLEMS

We know that all the one-peg reversal problems are possible, but there are just four different impossible two-peg reversal problems. The first of these is **Hamlet's Memorable Problem** (to be or not to be):

Get to only b, e present   (to be)
from only b, e absent   (not be).

Deficit account for Hamlet's Problem

To:	
Initial holes @ b & e :	$\beta \cdot \alpha\beta = a$
Final pegs @ b & e :	$\beta \cdot \alpha\beta = a$
First & last jumps into e :	$c\alpha \cdot c\alpha = c^2$
Jump into b :	a^2
Deficit :	a^4c^2

Since

$$\frac{a^4c^2}{c^4} = a^4c^{-2} = A^2 \ldots$$

is productive, Hamlet's Problem succumbs to the Deficit Rule. The other three impossible two-peg reversals are the **Dodo Problems**, for which the two places are two of the five dodo pegs (Fig. 22(d)). Deficit accounts for the typical problems *eo*, *ex* and *eE* are:

Dodo Problem	*eo*	*ex*	*eE*
Initial holes and } final pegs }	$(a\beta \cdot bd)^2$	$(a\beta \cdot ca\ell)^2$	$(a\beta \cdot a\alpha\beta)^2$
Required moves	$c\alpha \cdot c\beta$	$c\alpha$	$c\alpha \cdot c\alpha$
Deficit	$a^2b^2c^2\alpha\beta$	$x^2c^3\alpha$	a^4c^2
Deficit$/c^4$	$A \cdot B$	A	$A \cdot A$

The reader who has been paying attention will have no difficulty in finding solutions to any other two-peg reversal problem.

John Conway, Mike Guy and Bob Hutchings have shown that the only impossible *three*-peg reversals are typified by

(1) The **Bumble-bee Problems** (*b,e* and any third place other than *g,m,M,G*),
(2) The **Deader Dodo Problems** (two dodos and any third place other than an outside corner *acgmMGCA*),
(3) The **Three B'ars Problems** (any three of the unlucky 13 places in the three rows *def*, *nopxPON*, *FED*).

These can be shown to be impossible by deficit accounting. In fact in any reversal problem, an additional peg other than an outside corner merely aggravates the deficit.

FORGETTING THE ORDER CAN BE USEFUL

If you allow yourself to have 2 or more, or -1 or less, pegs in a hole, you can make your moves in any order! It's a good idea to alter a hard problem in this way, and when you've solved the altered problem, go back and find a sensible order for the original one.

We'll do out loud for you the tricky 3-peg reversal:

start with 0 pegs in *b*, *N*, *n*;　1 peg everywhere else;
end　with 1 peg in *b*, *N*, *n*;　0 pegs anywhere else.

In the altered problem it's easier to

start with -1 peg in *b*, *N*, *n*;　1 peg everywhere else;
end　with 0 pegs anywhere.

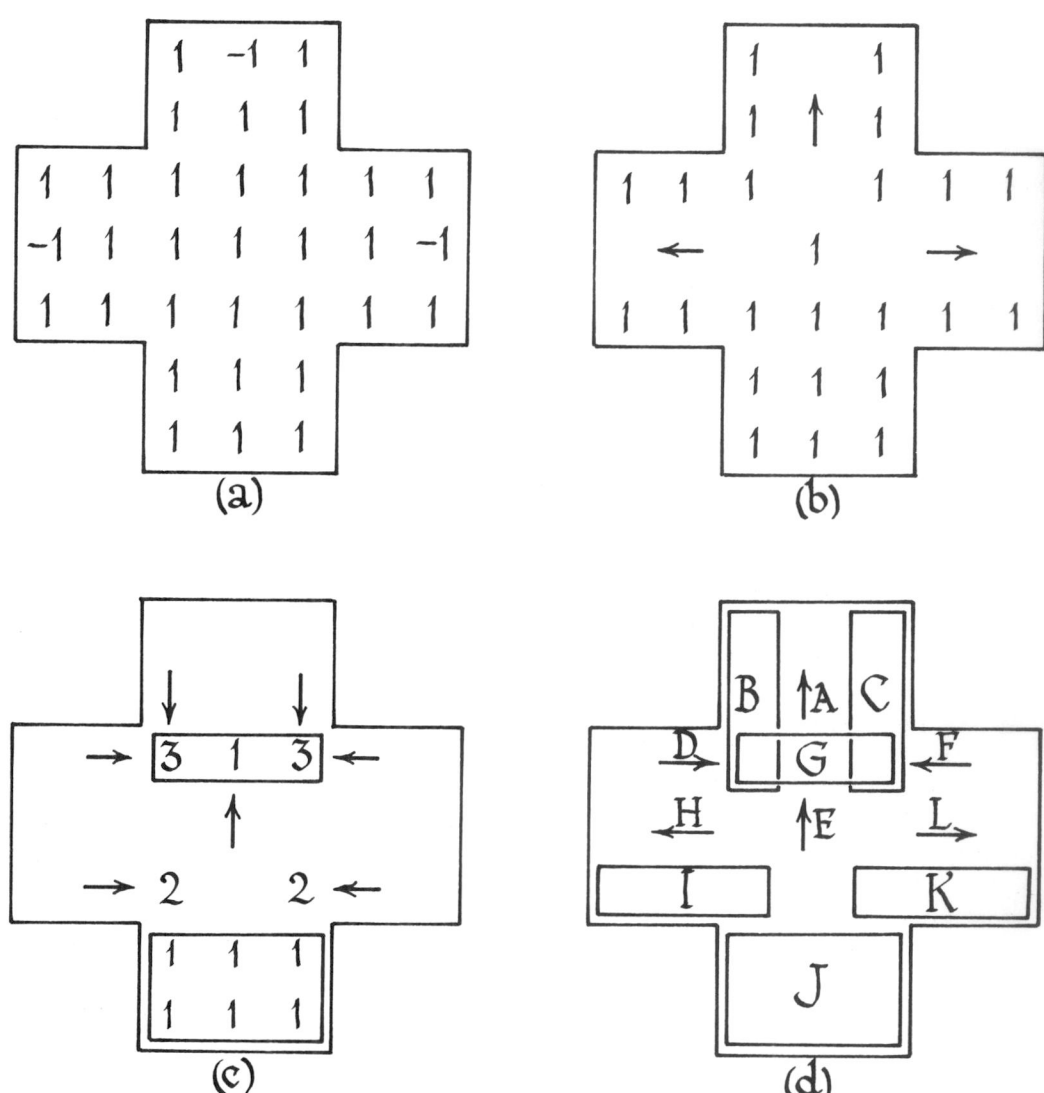

Figure 27. Solving a Tricky 3-peg Reversal Problem.

From the starting position, Fig. 27(a), we'll need, at some time, to make three jumps to fill those −1's, so we make these three jumps *now*, reaching Fig. 27(b). The only plausible way to deal with the six isolated corner pegs in this is to jump them inwards, and after the indicated upward jump over the centre we reach Fig. 27(c). The remaining pegs in this can be cleared by a 3-purge, a 6-purge and four double jumps over the inner corners.

If you follow Fig. 27(d) in the order A to L you'll find yourself making all the above moves in a legal way. The double jumps have been incorporated into the 3-purges B, C, I and K.

BEASLEY'S EXIT THEOREMS

Sometimes you can work out exactly what moves to make in a problem, but find it hard to get them into the right order. The following remarks can help you get your moves in order, or prove that it can't be done.

> A region of at least three squares
> that starts full *or* ends empty
> needs at least one *exit move*.
> A region of at least three squares
> that starts full *and* ends empty
> needs at least two exit moves.

BEASLEY'S FIRST AND SECOND EXIT THEOREMS

An **exit move** for a region is a jump that empties some square in the region and fills some square outside the region. To justify Beasley's Second Exit Theorem, note that the first and last moves affecting a region must both be exits. We'll illustrate with

A STOLID SURVIVOR PROBLEM

Suppose we want to do an a-reversal, with the added condition that peg K is the **stolid survivor**, i.e. that the first move of K is also the final move from K to a. Can the grand finale be a 6-chain?

The ideas of the first part of our discussion are often useful in long chain problems. Then we'll try to put the moves we've found into order, using Beasley's Exit Theorems.

How do we use the 16 side pegs $h^8 v^8$ of Figs. 22(h) and (v)? Each of the outer corner pegs must at some time be jumped into the central 3×3 square, and those at C and M must sidestep first to avoid the stolid survivor at K. So the jumps mentioned use up side pegs as follows:

$$
\begin{array}{ccccccc}
c & m & G & A & C & M & g \\
v & h & h & v & hv & vh & h
\end{array}
$$

leaving $h^3 v^4$ for the remaining jumps. Since the first move uses one side peg and the final chain six, we have accounted for all the side pegs and no *other* move can destroy one.

This forces us to make the first move c_a, since the alternative, i_a, would move an inside corner peg to the outside and make us use another side peg to bring it back later. Next k_c would use another side peg, so the second move is j_b, and this peg must stay at b until the grand finale because a move refilling e would use yet another side peg. We now know that the final 6-chain uses $h^2 v^4$ and involves a horizontal jump over b, so it must be as in Fig. 28(a).

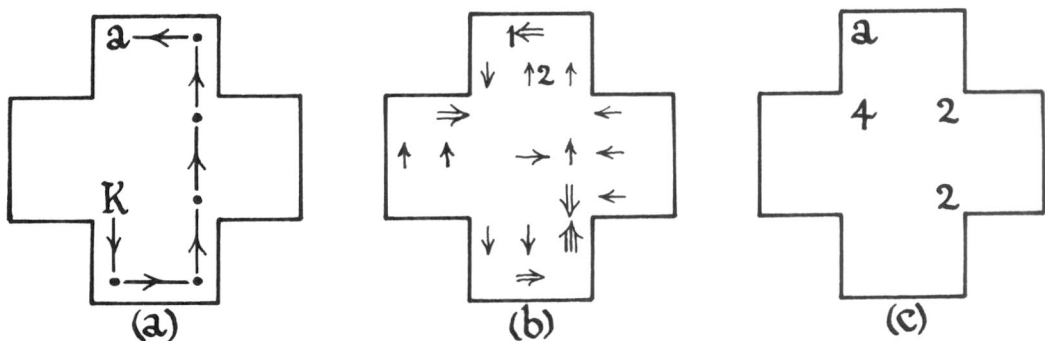

Figure 28. Can the Stolid Survivor Make a Grand 6-chain Finale?

Since K doesn't move till the end, L can't be jumped over and can only be cleared by the upward jump L_h. We need to make two jumps over B, once to get corner peg C out, and once in the finale, so we must deliver an extra peg there by a downward jump J_B. For similar reasons *two* extra pegs are needed at D, so we must make two downward jumps P_D. For the second of these, and for the finale, we need two more pegs delivered at P; these must come from N and p. We've now found 23 (Fig. 28(b)) of the 31 jumps. If we make these we arrive at Fig. 28(c). The two pegs on each of I and k must be cleared by pairs of vertical or horizontal to and fro jumps, and the four on i by two such pairs.

To find the right order in which to make these moves we use Beasley's Second Exit Theorem. Consider the region of Fig. 29(a). The moves we've copied from Fig. 28(b) incorporate just one exit from the region; the vertical jump across P. To make sure there's another we must remove the two pegs on k by a *vertical* pair of to and fro jumps as in Fig. 29(b). But the region of *that* figure can now have only one exit, the vertical jump across f. So our problem's impossible!

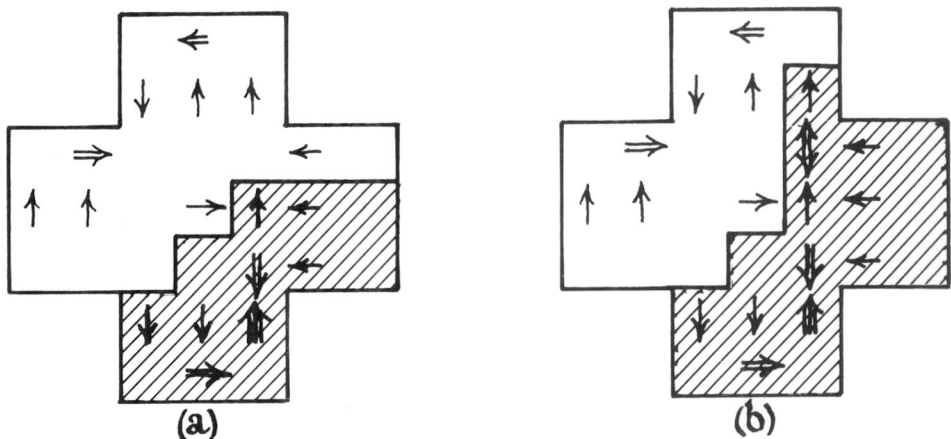

Figure 29. Using Beasley's Second Exit Theorem.

ANOTHER HARD PROBLEM

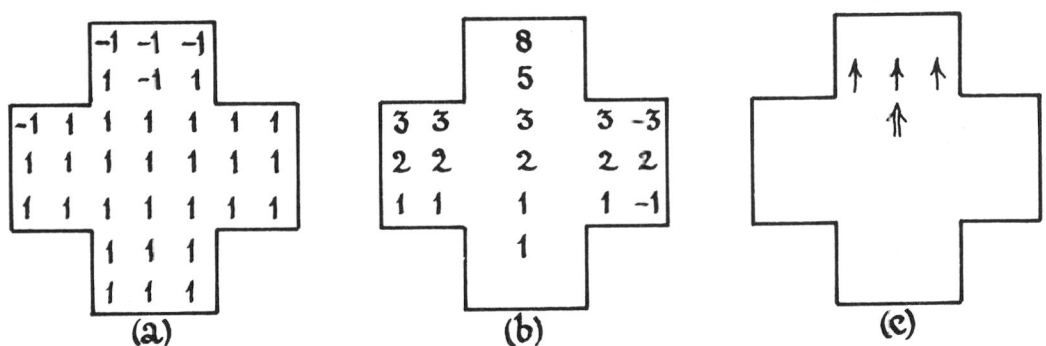

Figure 30. The Reversal Problem *abceg*.

We'll try the 5-peg reversal problem *abceg*, i.e. start with the board full except for spaces at *a*, *b*, *c*, *e*, *g*, and finish with pegs in just those places. An equivalent problem is to clear the board of Fig. 30(a), which has negpegs (or **negs**) in each of the places *a*, *b*, *c*, *e*, *g* and pegs in the other 28 places.

For the original problem the pagoda function of Fig. 30(b) changes from 20 to 16, or, in the form of Fig. 30(a), from 4 to 0. This pag kills any possible move across *b*, which would lose 8, so the jumps i_a, k_c shown in Fig. 30(c) are forced, as is the jump j_b to fill *b*. In order to make this last jump a peg must be delivered to *e*, and *e* must also be full by the end, so two jumps x_e are also needed. If we make these five jumps, using negs where needed, we arrive at Fig. 31(a), whose

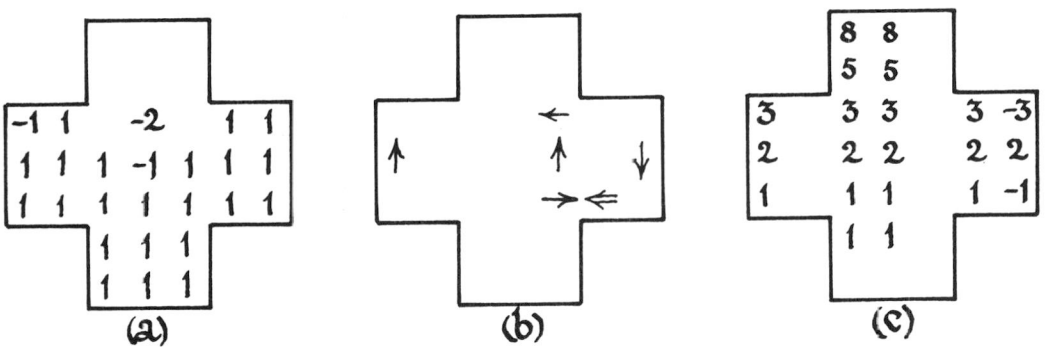

Figure 31. We Make Some Progress.

resources are $a^2b^4c^{-1}\alpha\beta$, showing a deficit of a^2c^2. The Deficit Rule tells us we can't make another move of value a^2, since $a^4c^2/c^4 = A^2$, so the jump M_g (Fig. 31(b)) is forced. Now use the pagoda function of Fig. 31(c). Its value for Fig. 31(a) is 2, so the peg at *N* can't jump inwards, nor can we jump over it upwards, since these moves lose 4 on this pag. So the jump m_G (Fig. 31(b)) is forced. The two pegs at *G* must now both jump to *I*, and a peg must be delivered to *H* for the second of these. This can't come from l, as this loses 4 on the 1st pag, so the jump J_H is also forced. Moreover, as l can't jump downwards, and can't be jumped over (this would lose a^2) it must make the jump l_j over k. This needs delivery of a peg at k, which can't come from i (loses 6 on the last pag) so the jump I_k is forced. If we make all these moves, which have been collected in Fig. 31(b), we arrive at Fig. 32(a), for which the resources are now

$$a^2 b^3 c^{-1} \alpha = Ab^3 \quad \text{or} \quad A.B.b.c\beta \quad \text{or} \quad B.a.a\alpha.b\beta.$$

so there is no jump $c\alpha$.

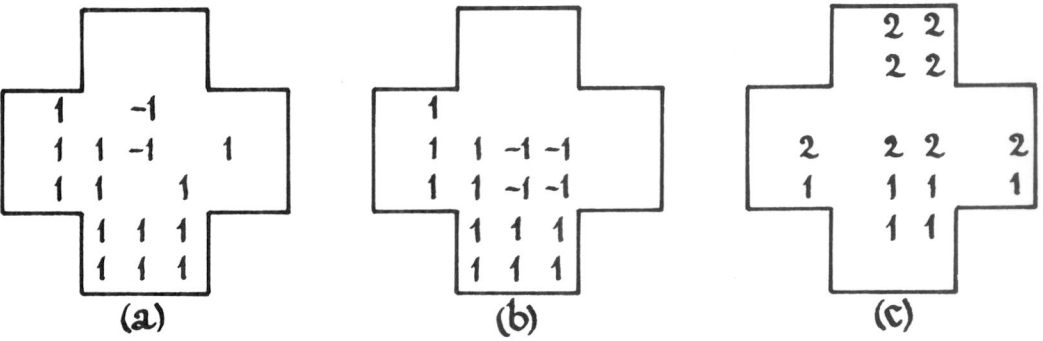

Figure 32. A Cul-de-Sac!

There are two ways in which we might remove the peg at O: by j_{lHJ} or by O_x. The former (after the necessary delivery I_k) leads to Fig. 32(b) which is impossible to clear, as the pag of Fig. 32(c) shows. So O_x is forced (Fig. 33(a)) and this requires the delivery D_P (horizontal delivery is prohibited by the pag of Fig. 31(c)). These two jumps lead to a position whose resources $a^2\alpha\beta$ are uniquely productive: $(a^2 c^{-1}\alpha)(c\beta)$ and the jump E_x is forced. The L-package of Fig. 33(a) will deliver a second peg to K and the board is cleared by L_J and h_{LJj}.

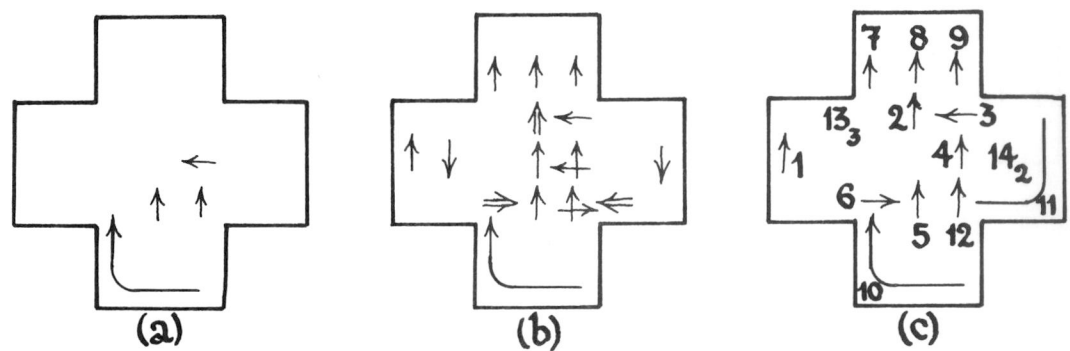

Figure 33. The Problem Solved.

The 23 jumps are shown in Fig. 33(b). How do we do them in practice? In what order? The answer isn't unique, but one possibility is given in Fig. 33(c). It involves two L-packages, 10 and 11, and two chain moves, 13_3 and 14_2.

THE SPINNER

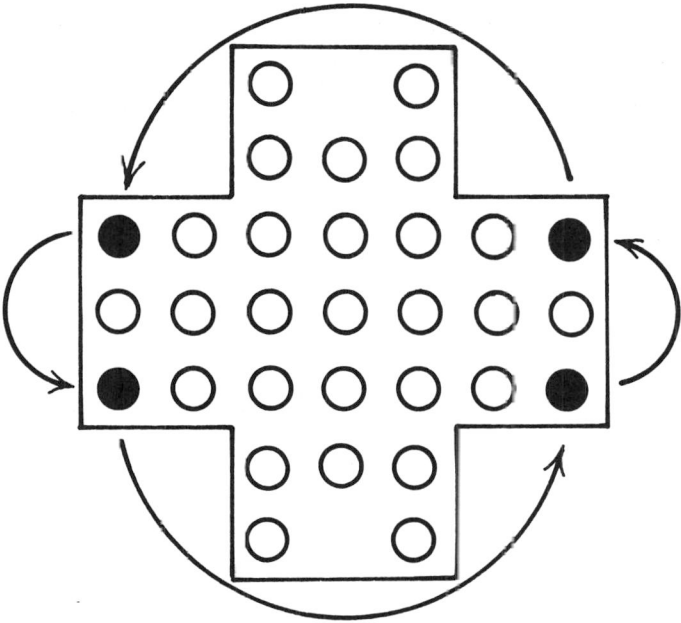

Figure 34. The Spinner.

If you start with empty spaces at b, B and marked pegs at g, M, G, m, can you finish with just the four marked pegs on the board in the respective positions M, G, m, g?

EXTRAS

OUR FINE FINALIST

The Rule of Two and the Rule of Three together tells us that if the initial hole is at B, then a finalist that starts at J must end in either B or b. Here's a solution for b:

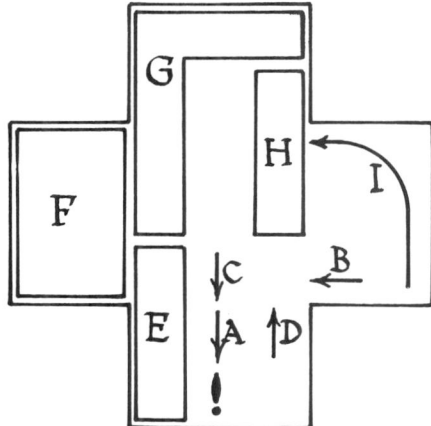

The letters indicate the order of the successive moves, except for the finalizing flourish. The bent arrow we've used for move I is our notation for an L-*package*, as distinct from an L-*purge*.

However, it's impossible for the finalist to finish at B. This is because there are forced moves

$$J_B \qquad x_E \qquad x_E \qquad J_B$$

which consume

$$a^2 \qquad c\alpha \qquad c\alpha \qquad a^2$$

on the Balance Sheet, giving a deficit of a^4c^2. Since

$$\text{deficit}/c^4 = a^4c^{-2}$$

is productive, your remaining resources *aren't*.

DOING THE SPLITS

If you start from Fig. 16(a) and make the moves A to I indicated in Fig. 35, where the pairs of circles C, F, G indicate 2-purges, you'll reach a 5-peg configuration which can easily be reduced to I, L or f. (We found this solution by the ordering process after subtracting this 5-peg configuration from the starting position.)

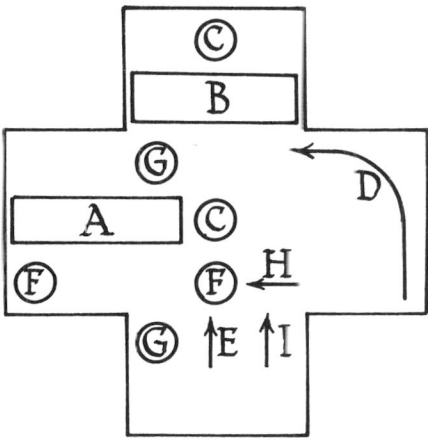

Figure 35. The Train was going to IvanicGrad, Ljubljana or foća.

ALL SOLUBLE ONE-PEG PROBLEMS ON THE CONTINENTAL BOARD

were found by Reiss using his theory. In our language, the Continental board is algebraically equal to its centre, and so for a one-peg problem to be soluble we must have

(initial hole) × (final peg) = centre,

in our algebraic sense. You can easily check that the initial hole and final peg are at opposite ends of an arrow in

$$(a\,p\,C\,O) \leftrightarrow (A\,P\,c\,o) \qquad (e\,G\,J\,M) \leftrightarrow (E\,g\,j\,m).$$

There is a 41-hole board for which Lucas gives all the soluble problems; but see the appendix to his *second* edition, because he first conjectured that most of the problems were insoluble!

THE LAST TWO MOVES

in Fig. 19(a) are $n_c\,G_3$.

A 20-MAN SOLITAIRE ARMY

can get a scout 4 places out by arranging itself as shown in Fig. 23. The two men with guns can be moved to the shaded places so as to obtain the only other arrangement.

FOOL'S SOLITAIRE, ETC.

If each of your moves is confined to the middle row or column you'll reach a position like Fig. 36(a) after six jumps. The next pegbound position is the Hammer and Sickle position of Fig. 36(b), reached after ten jumps.

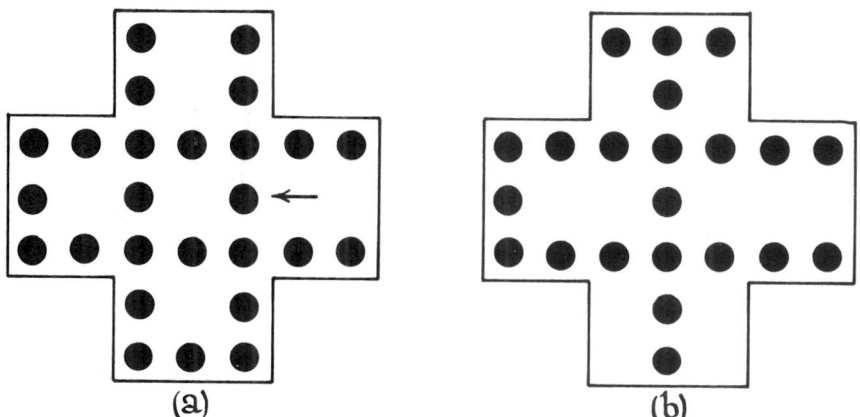

Figure 36. (a) Sickle and Sickle. (b) Hammer and Sickle.

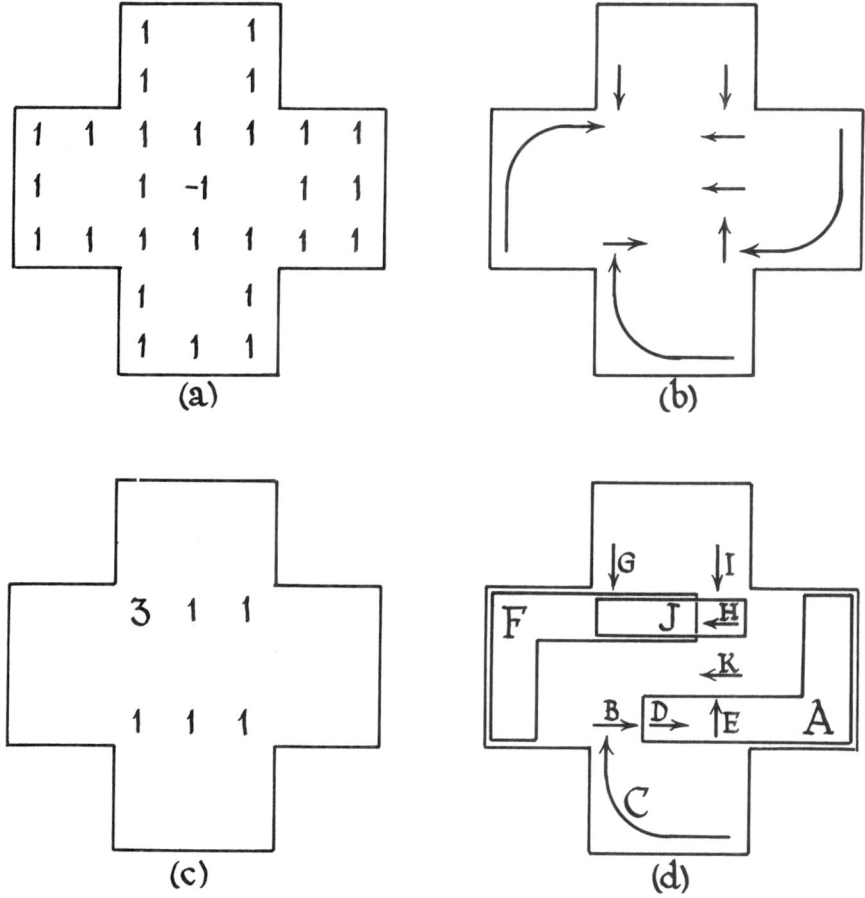

Figure 37. Succouring the Sucker.

To succour the sucker who's made five of the six moves leading to Fig. 36(a) it's best to try to clear Fig. 37(a) to zero. If you set this up on the board (use an upside down peg for the -1!) you should see how the moves of Fig. 37(b) suggest themselves in order, leading to the easily cleared position, Fig. 37(c). The L-moves in Fig. 37(b) are L-*packages*, not L-purges. You then have the tricky little problem of arranging the moves in order, one solution of which is given in Fig. 37(d), in which A and F are L-*purges*, but C is an L-*package*.

BEASLEY PROVES BERGHOLT IS BEST

Suppose there were a 17-move solution to Central Solitaire. Then Beasley first uses the scoring function of Fig. 38(a) ("score" refers to this function—which is *not* a pagoda function—throughout the proof) and his First Exit Theorem, to show that no move begins or ends on b, n, B or N. The initial and final scores are 20 and 0. Moves which begin or end on b, n, B or N *increase* the score by at least 1. Others decrease it by at most 2 (the careful reader will make a table of score changes for each type of move).

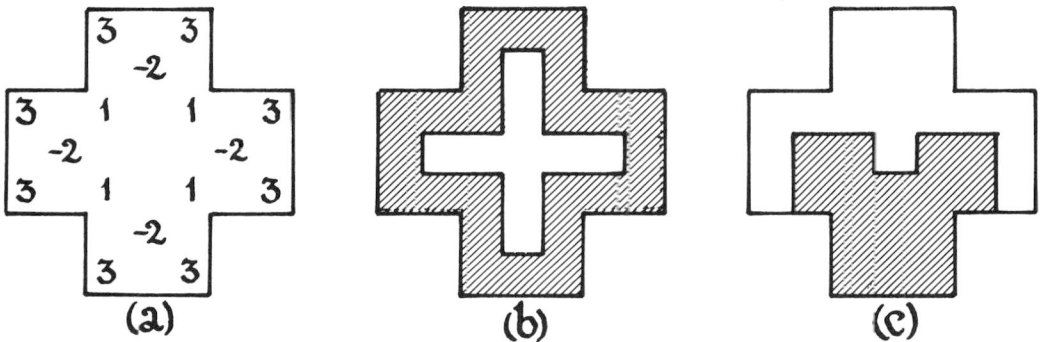

Figure 38. Scoring Function and Regions Used in Beasley's Proof.

Any solution to Central Solitaire contains 11 **reserved** moves:
    the first, which we'll take to be e_x,
    the last, a single jump into the centre,
    the penultimate one, taking a peg to j, p, J or P, and
    eight moves bringing the outside corner pegs to inside corner
        squares so they may be captured.

The first and last moves each increase the score by 2, the penultimate one decreases it by at most 1 and each of the other eight decreases it by at most 2. So the other six (**loose**) moves must decrease the score by at least 7.

The second move is a loose move, either J_j or h_j, say. The move J_j doesn't change the score and it leaves the region of Fig. 38(b) full. The first exit from this region is a loose move, either of type b_j or of type (ending with) h_j. The former increases the score by 2 and the other four loose moves would have to decrease it by at least 9, which is impossible. The latter decreases the score by 1, and our four loose moves have to reduce it further by at least 6. If any of these increased the score, the others could not then reduce it to zero, so moves starting or ending on b, n, B or N

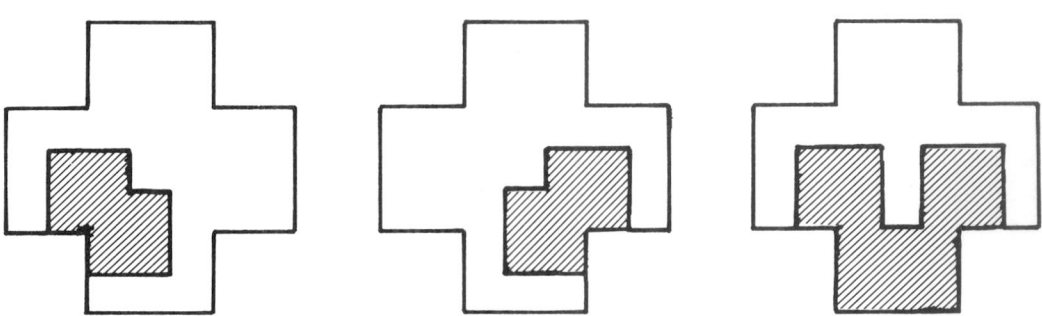

Figure 39. What is the First Exit?

are again impossible. Such a move *might* occur as the penultimate one, but the six loose moves would then have to reduce the score by 10, and the same argument shows this is impossible.

The second move h_j reduces the score by 1 *and* is a first exit from the region of Fig. 38(b). The other five loose moves must reduce the score by at least 6. What is the first exit from the region of Fig. 38(c)? There are several possibilities, all of them loose moves, which we'll leave the reader to pursue. In some cases he'll want to ask a further question about one of the regions of Fig. 39, whichever is still full. From now on we'll assume that no move begins or ends on b, n, B or N.

How do we clear a, b and c? We've proved that b can't jump out, so there must be a jump over it, say c_a. The two pegs at a now force two jumps a_i and a jump into d, which we shall call a **side delivery**. The four jumps

$$c_a \quad a_i \quad a_i \quad ?_d$$

are parts of at least three moves

$$a_i \ldots \quad ?_{\ldots d} \quad c_{ai \ldots} \quad \text{(the normal case), or}$$

$$a_i \ldots \quad ?_{\ldots d} \quad ?_{\ldots kcai} \quad \text{(a U-turn).}$$

However a U-turn demands a previous clearance of c and an extra side delivery to f.

Since the same argument applies at n, B and N, we shall need at least four side deliveries, none of which can be among our 11 reserved moves, and none of which can be the first exit from Fig. 38(b). This accounts for 16 moves; call the other the **spare**. Moreover, if a U-turn is involved we have a further side delivery, and so no spare. Note that after *eha*, p is a side delivery, but j doesn't count as one while g is still occupied, because we'll still need one to clear gnM.

The final stage of Beasley's proof just enumerates all the variations. In the list below the spare move is in **bold**. In the first two variations the first exit from Fig. 38(b) is L, and in all the others it's h. Each variation ends with ‡, § or a colon and a number.

> ‡   means that the next move can't be a corner move
>   or a side delivery and the spare has already been used,
> §   means that there aren't enough moves left to reduce
>   the score to zero, and
> :9 refers to variation number 9, for example.

This list of variations covers the cases where no U-turns are used. If there is a U-turn there is *no* spare move so we have only the variation $ehapc_2$‡ (cf. 56).

1 $eJLCpA_2\ddagger$	17 $ehKMJg_2\ddagger$	33 $enajgpL\S$	48 $ehapFc_3\ddagger$
2 $D\S$	18 $CD\S$	34 $c_2L\S$	49 $M_2\ddagger$
	19 $P_2g_2\ddagger$	35 $J_2M_2\ddagger$	50 $c_2MJg_2\ddagger$
3 $ehxaf\ddagger$	20 $A_2\ddagger$	36 $l_2M_2\ddagger$	51 $Mc_2{:}50$
4 $pc_2\ddagger$	21 $d_2g_2\ddagger$	37 $L\S$	52 $Jc_3\ddagger$
5 $Lap\S$	22 A_2	38 $J_2M_2p_dc_2\ddagger$	53 $g_3\ddagger$
6 $gj\S$	23 $j_2\S$	39 $l_2M_2p_dc_2\ddagger$	54 $c_2{:}50$
7 $kcPa_2\ddagger$	24 $CD\S$		55 $g_2c_2\ddagger$
8 $mH\S$	25 $P_2A_2\ddagger$	40 $ehapc{:}30$	56 $c_2P_2\ddagger$
9 $J_2a_2\ddagger$	26 $MJ{:}19$	41 $k_2\ddagger$	57 $F_2MJg_2\ddagger$
10 $G_2\ddagger$	27 $j_2\S$	42 $x{:}4$	58 $x{:}4$
11 $L_3\S$	28 $d_2A_2\ddagger$	43 $L{:}5$	59 $L\S$
12 $mH\S$	29 $MJ{:}21$	44 $kmH\S$	60 $jgL\S$
13 $J_lG_2\ddagger$		45 $J_lG_2\ddagger$	61 $J_2M_2\ddagger$
14 $cP{:}9$	30 $ehacpa\ddagger$	46 $L_3\S$	62 $l_2M_2\ddagger$
15 $L_3G_2ap\S$	31 $x{:}3$	47 $Pc_2\ddagger$	63 $P\ddagger$
16 $c_2\S$	32 $L{:}5$		64 $FMj_2\S$
			65 $Jg_2\ddagger$

THE CLASSICAL PROBLEMS

These are: start with one empty space, finish with a single peg. They include the reversals, for which Bergholt's results were:

$$
\begin{array}{cccccccc}
 & a\text{-} & b\text{-} & d\text{-} & e\text{-} & i\text{-} & j\text{-} & x\text{-} \quad \text{reversal} \\
\text{in} & 16 & 18 & 16 & 19 & 16 & 16 & 18 \quad \text{moves.}
\end{array}
$$

We've just seen that his x-reversal is best possible, but Harry O. Davis has given a 15-move solution of the i-reversal:

$$kmh_2cPKCD_{PF}A_3MG_2H_4a_2a_5g_3.$$

And here are his solutions, which equal Bergholt's, for the b- and j-reversals:

$$jhapc_2xl_hIf_PA_2GJm_2gL_{Jh}M_2CB_5,$$

$$hKCd_2MJkmH_{Jl}G_3cA_2D_{Fd}g_2ab_7.$$

Hermary identified the 21 distinct problems, one place empty to one place full (see Lucas) and Davis has made a table of best known solutions (see Martin Gardner, "The Unexpected Hanging and Other Mathematical Diversions"). The numbers of moves are:

aa	ap	aO	aC	bb	bn	bx	bB	dd	dK	dH	ee	eM	eJ	ii	il	jj	jg	jE	xx	xb
16	16	17	16	18	17	18	18	16	15	16	19	17	17	15	16	16	16	17	18	17

For this information we thank Wade E. Philpott, who has copies of the solutions. Omar will want to find better ones, or prove them best possible.

REFERENCES AND FURTHER READING

J.D. Beasley, Some notes on Solitaire, Eureka, **25** (1962) 13–18.

E. Bergholt, The Queen, May 11, 1912; and "The Game of Solitaire", Routledge, London, 1920.

N.G. de Bruijn, A Solitaire game and its relation to a finite field, J. Recreational Math. **5** (1972) 133–137.

Busschop, "Recherches sur le jeu de Solitaire" Bruges, 1879.

M. Charosh, The Math. Student J., U.S.A, March 1961.

Donald C. Cross, Square Solitaire and variations, J. Recreational Math. **1** (1968) 121–123.

Harry O. Davis, 33-solitaire, new limits, small and large, Math. Gaz. **51** (1967) 91–100.

H.E. Dudeney, The Strand Magazine, April, 1908; and see "Amusements in Mathematics", problems 227, 359, 360, Nelson, London 1917. pp. 63–64, 107–108, 195, 234.

M. Gardner, Sci. Amer. **206** #6 (June 1962) 156–166; **214** #2 (Feb. 1966) 112–113; **214** #5 (May 1966) 127.

M. Gardner, "The Unexpected Hanging and other Mathematical Diversions", Simon and Schuster, 1969, p. 126.

Heinz Haber, Das Solitaire-Spiel, in "Das Mathematische Kabinett", Vol. 2, Bild der Wissenschaft, D.V-A., Stuttgart 1970, pp. 53–57.

Irvin Roy Hentzel, Triangular puzzle peg, J. Recreational Math. **6** (1973) 280–283.

Hermary, Sur le jeu du Solitaire, Assoc. franc. pour l'avancement des sci., Congrès de Montpellier, 1879.

Ross Honsberger, "Mathematical Gems II", Mathematical Association America, 1976, chap. 3, 23–28.

M. Kraitchik, "Mathematical Recreations", George Allen & Unwin, London, 1943, pp. 297–298 quotes letter 1716:1:17 Leibniz to Monmort.

E. Lucas, "Récréations Mathématiques", Blanchard, Paris 1882, Vol. 1, part 5, 89–141 is mainly concerned with the continental board (37 places) but pp. 132–138 refer to the English board and much of the whole is applicable. He attributes (pp. 114–115) the 3-*purge* to Hermary.

M. Reiss, Beitrage zur Theorie der Solitär-Spiels, Crelle's J., **54** (1857) 344–379.

Ruchonet, Théorie du solitaire, par feu le docteur Reiss, librement traduit de l'allemand, Nouv. Corr. math., t. III p. 231, Bruxelles 1877.

B.M. Stewart, "Theory of Numbers," Macmillan, New York, 1952, 1964, pp. 20–26. Analyzes Solitaire on a 7×5 rectangular board. He colors the diagonals with 3 colors in either direction and obtains the Rule of Three as an example of congruences (mod 3). Exercise 4.5 on p. 24, due to F. Gozreh, asks you to start from an even length row of pegs with the second peg missing and finish with just the second peg from the other end. A pattern of 2-purges does the trick. If you start with the fifth peg missing, then a 2-purge and a double jump by the 1st peg reduces the problem to the earlier one. Can you clear the row with other pegs missing? Which missing pegs enable you to clear an *odd* length row?

Chapter 24

Pursuing Puzzles Purposefully

The chapter of accidents is the longest chapter in the book.
John Wilkes

I shall proceed to such Recreations as adorn the Mind; of
which those of the Mathematicks are inferior to none.
William Leybourne; *Pleasure with Profit.*

We know you want to use your winning ways mostly when playing with other people, but there are quite a lot of puzzles that are so interesting that you really feel you're playing a game against some invisible opponent—perhaps the puzzle's designer—maybe a malevolent deity. In this chapter we'll discuss a few cases where some kind of strategic thinking simplifies the problem. But because we don't want to spoil your fun we'll try to arrange not always to give the *whole* game away.

SOMA

This elegant little puzzle was devised by Piet Hein. Figure 1 shows the seven non-convex shapes that can be made by sticking 4 or fewer $1 \times 1 \times 1$ cubes together. Piet Hein's puzzle is to assemble these as a $3 \times 3 \times 3$ cube.

1 = W	2 = Y	3 = G	4 = O	5 = L	6 = R	7 = B

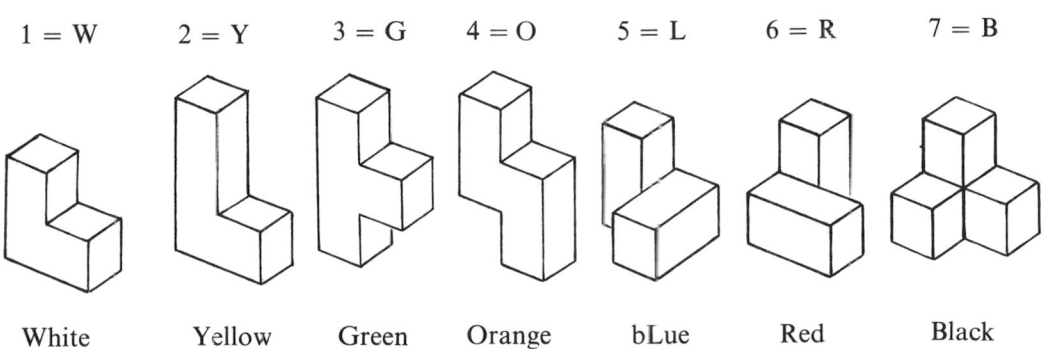

White · Yellow · Green · Orange · bLue · Red · Black

Figure 1. The Seven Pieces of Soma.

735

We advise you to use seven different colors for your pieces as in the figure. Many people solve this puzzle in under ten minutes, so it can't be terribly hard. But we've got a distinct feeling that it's much harder than it ought to be. Is this just because the pieces have such awkwardly wriggly shapes?

BLOCKS-IN-A-BOX

Here is another puzzle invented by one of us some years ago, in which all the pieces are rectangular cuboids but it still seems undeservedly hard to fit them together. We are asked to pack one $2 \times 2 \times 2$ *cube*, one $2 \times 2 \times 1$ *square*, three $3 \times 1 \times 1$ *rods* and thirteen $4 \times 2 \times 1$ *planks* into a $5 \times 5 \times 5$ box (Fig. 2). It's quite easy to get all but one of the blocks into the box, but somehow one piece always seems to stick out somewhere. A friend of ours once spent many evenings without ever finding a solution. Why is it so much harder than it seems to be?

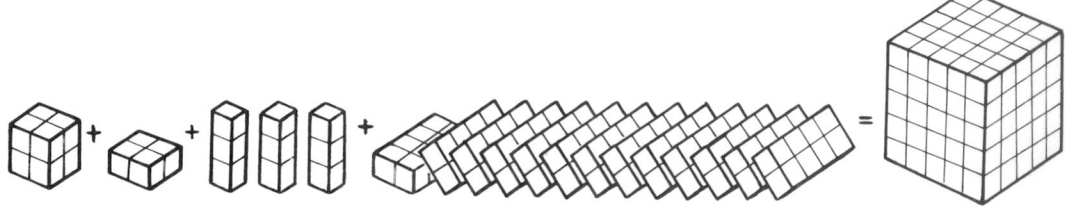

Figure 2. The Eighteen Pieces for Blocks-in-a-Box.

HIDDEN SECRETS

In our view the good puzzles are those with simple pieces but difficult solutions. Anyone can make a hard puzzle with lots of complicated pieces but how can you possibly make a hard puzzle out of a few easy pieces?

When a seemingly simple puzzle is unexpectedly difficult, it's usually because, as well as the obvious problem, there are some hidden ones to be attended to. Both Soma and Blocks-in-a-Box have such hidden secrets, but let's look at a much simpler puzzle, to fit six $2 \times 2 \times 1$ squares into a $3 \times 3 \times 3$ box, leaving three of the $1 \times 1 \times 1$ cells empty—the *holes* (Fig. 3). This now seems fairly trivial, but even so there's a hidden secret which sometimes makes people take more than 5 minutes over it. This hidden problem comes from the fact that the square pieces can only occupy

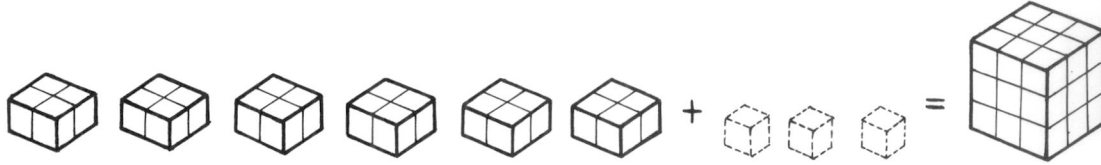

Figure 3. A Much Simpler Puzzle.

an even number of the cells in each horizontal layer. So since 9 is odd each horizontal layer must have a hole and there are only just enough holes to go round. Of course these holes must also manage to meet each of the three layers in each of two vertical directions—you can't afford to have two holes in any layer, because some other layers would have to go without.

The problem wasn't really to fit the *pieces* in but rather the *holes*. Only when you've realized this do you see why the unique solution (Fig. 4) has to be so awkward looking, with the holes strung out in a line between opposite corners rather than neatly arranged at the top of the box.

Perhaps you'd like to try the big Blocks-in-a-Box problem now, before looking at the extra hints in the Extras.

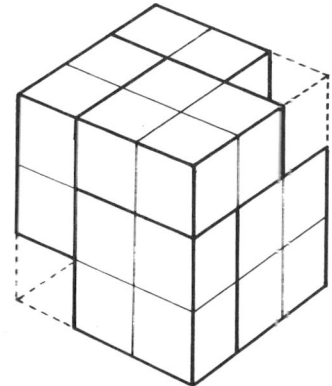

Figure 4. Six Squares in a 3×3×3 Box.

THE HIDDEN SECRETS OF SOMA

It's because the Soma puzzle pieces have to satisfy some hidden constraints as well as the obvious ones, that it causes most people more trouble than it should. Let's see why.

The $3 \times 3 \times 3$ cube has 8 *vertex* cells, 12 *edge* cells, 6 *face* cells and 1 *central* cell as in Fig. 5.

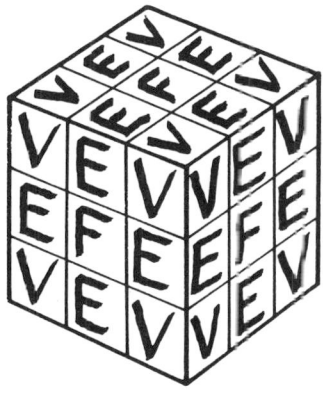

Figure 5. The Vertex, Edge, Face and Central (invisible) Cells.

Now the respective pieces can occupy at most

$$
\begin{array}{ccccccc}
\text{W} & \text{Y} & \text{G} & \text{O} & \text{L} & \text{R} & \text{B} \\
1, & 2, & 2, & 1, & 1, & 1, & 1
\end{array}
$$

of the vertex cells, so just one piece, the **deficient** one, must occupy just one less vertex-cell than it might. The green piece can't be deficient without being doubly so, and therefore:

> the Green piece has
> its spine along an
> edge of the cube.

Now let's color the 27 cells of the cube in two alternating colors,

> Flame for the 14 FaVored cells, F and V,
> Emerald for the 13 ExCeeded ones, E and C.

Then in *one* solution that we know, the respective pieces occupy

$$
\begin{array}{ccccccc}
\text{W} & \text{Y} & \text{G} & \text{O} & \text{L} & \text{R} & \text{B} \\
\end{array}
$$
$$2+2+3+2+2+2+1 = 14 \text{ F, V cells.}$$
$$1+2+1+2+2+2+3 = 13 \text{ E, C cells,}$$

but the Yellow, Orange, bLue and Red pieces, and we now know also the Green piece, *must* occupy these numbers in *every* solution, and therefore so must the White and the Black, since an interchange of colors in either or both of these would alter the totals.

> The White piece occupies
> 2 FV cells, 1 EC cell.

> The Black piece occupies
> 1 FV cell and 3 EC ones.

For the placing of a single piece within the box, these considerations leave only the positions of Fig. 6 (which all arise). You'll see that up to rotations of the cube, the placement of any single piece is determined by whether or not it is deficient and whether or not it occupies the central cell.

The hidden secrets of Soma make it quite likely that one of the first few pieces you put in may already be wrong, when of course you'll spend a lot of time assembling more pieces before such a mistake shows its effect. This would happen for instance if you started by putting the corner of

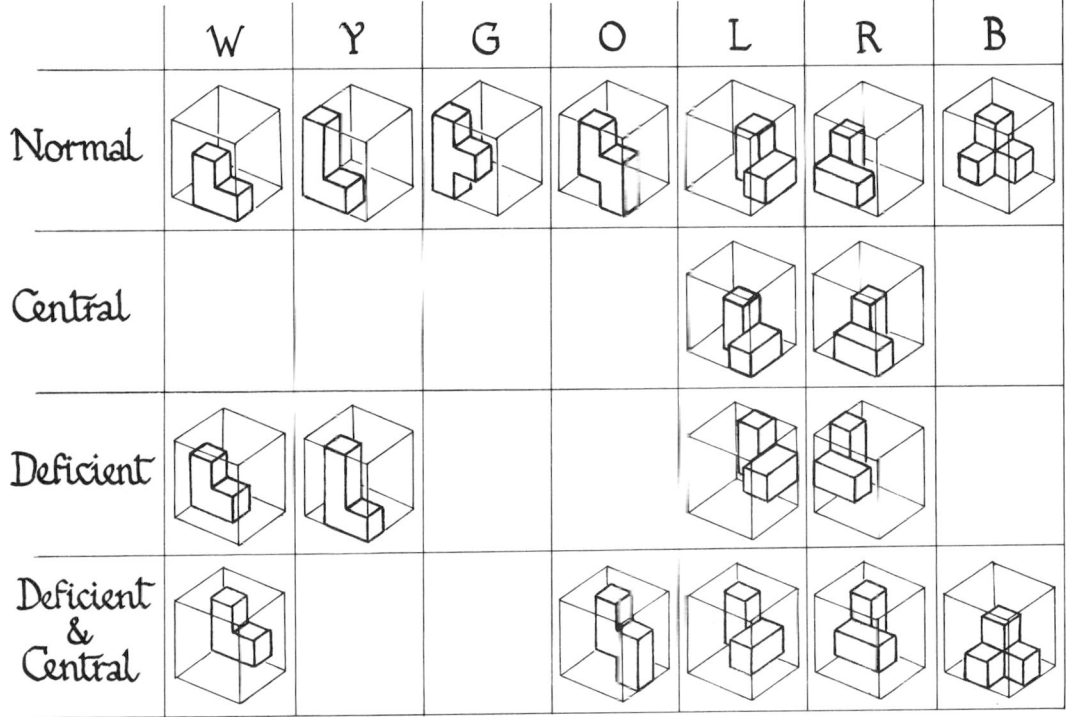

Figure 6. All Possible Positions for the Seven Soma Pieces.

the White piece into a corner of the cube. But if you only put the pieces into the allowed positions, you'll find a solution almost as soon as you start. The complete list of 240 Soma solutions was made by hand by J.H. Conway and M.J.T. Guy one particularly rainy afternoon in 1961. The SOMAP in the Extras enables you to get to 239 of them, when you've found one—*and* located it on the map!

HOFFMAN'S ARITHMETICO-GEOMETRIC PUZZLE

A well-known mathematical theorem is the inequality between the arithmetic and geometric means:

$$\sqrt{ab} \leqslant \frac{a+b}{2}.$$

Figure 7 provides a neat proof of this in the form
$$4ab \leqslant (a+b)^2$$
and the three variable version
$$27abc \leqslant (a+b-c)^3$$

has prompted Dean Hoffman to enquire whether 27 $a \times b \times c$ blocks can always be fitted into a cube of side $a+b+c$. This turns out to be quite a hard puzzle if a, b, c are fairly close together

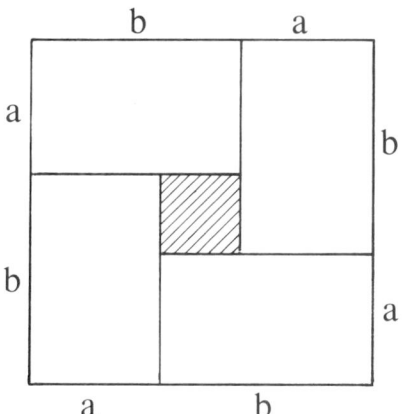

Figure 7. Proof of the Arithmetico-Geometric Inequality.

but not equal. A good practical problem is to fit

27     $4 \times 5 \times 6$ blocks into a $15 \times 15 \times 15$ box.

With these choices, as for any others with

$$\tfrac{1}{4}(a + b + c) < a < b < c,$$

it can be shown that each vertical stack of three blocks must contain just one of each height a,b,c, while there must be just three of each height in each horizontal layer. There must be the same unused area on each face (just 3 square units in the $4 \times 5 \times 6$ case).

It's almost impossible to solve the puzzle if you don't keep these hidden secrets constantly in mind because you'll make irretrievable mistakes like making a stack of three height 5 blocks, or leaving a 2×2 empty hole on some face. When you *do* keep them in mind, the puzzle becomes much easier, being only extremely difficult! You'll find some information about solutions to Hoffman's puzzle in the Extras.

COLORING THREE-BY-THREE-BY-THREE BY THREE, BAR THREE

In Hoffman's $3 \times 3 \times 3$ puzzle, the three lengths along any line of three had to be different. Can you color the cells of a $3 \times 3 \times 3$ tic-tac-toe board with

three different colors,

using all

three colors the same

number (9) of times, in such a way that *none* of the $\tfrac{1}{2}(5^3 - 3^3) = 49$ tic-tac-toe lines uses

three different colors,

nor has all its

three colors the same?

WIRE AND STRING PUZZLES

Figure 8 shows a number of topological puzzles which can be made with wire and string. It's a pity that manufacturers don't seem to know about all of these.

You wouldn't expect to be able to say much about such varied looking objects, but in fact there's a quite general principle which helps you to solve a lot of them.

THE MAGIC MIRROR METHOD

We'll just take the one-knot version of the puzzle shown in Fig. 8(c) which has been commercially sold as The Loony Loop (Trolbourne Ltd., London). You're to take the string off the rigid wire frame in Fig. 9(a).

If only that rigid wire were a bit stretchable, the puzzle would be quite easy. After squashing the string up (Fig. 9(b)) so as not to get in the way, we could stretch the loops over the ends (Fig. 9(c)) and shrink them again (Fig. 9(d)). After this we can take the string right off (Fig. 9(e)) and then put the loops back as they were (Fig. 9(f)) so as not to upset the owner.

Now the change from Fig. 9(b) to Fig. 9(d) could be accomplished by continuously distorting space. Think of embedding the puzzle in a flexible jelly, if Mother has one made up. Now old-fashioned fairgrounds had special mirrors which seemed to distort space in very funny ways. Now let's imagine a magic mirror with the wonderful property that the distortion is just what's required to make Fig. 9(b) look like Fig. 9(d). Hold the wire frame absolutely still before the magic mirror (Fig. 10(a)) and bunch the string up until it's almost a single point on the axis. Because the space distortion was continuous, its image will also be almost a single point on the image axis.

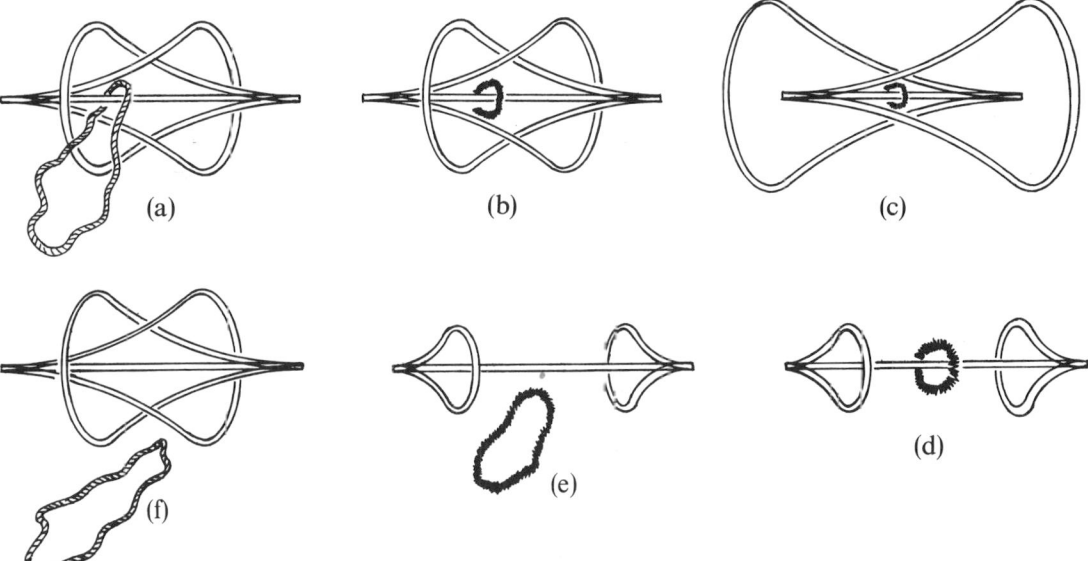

Figure 9. Solving The Loony Loop.

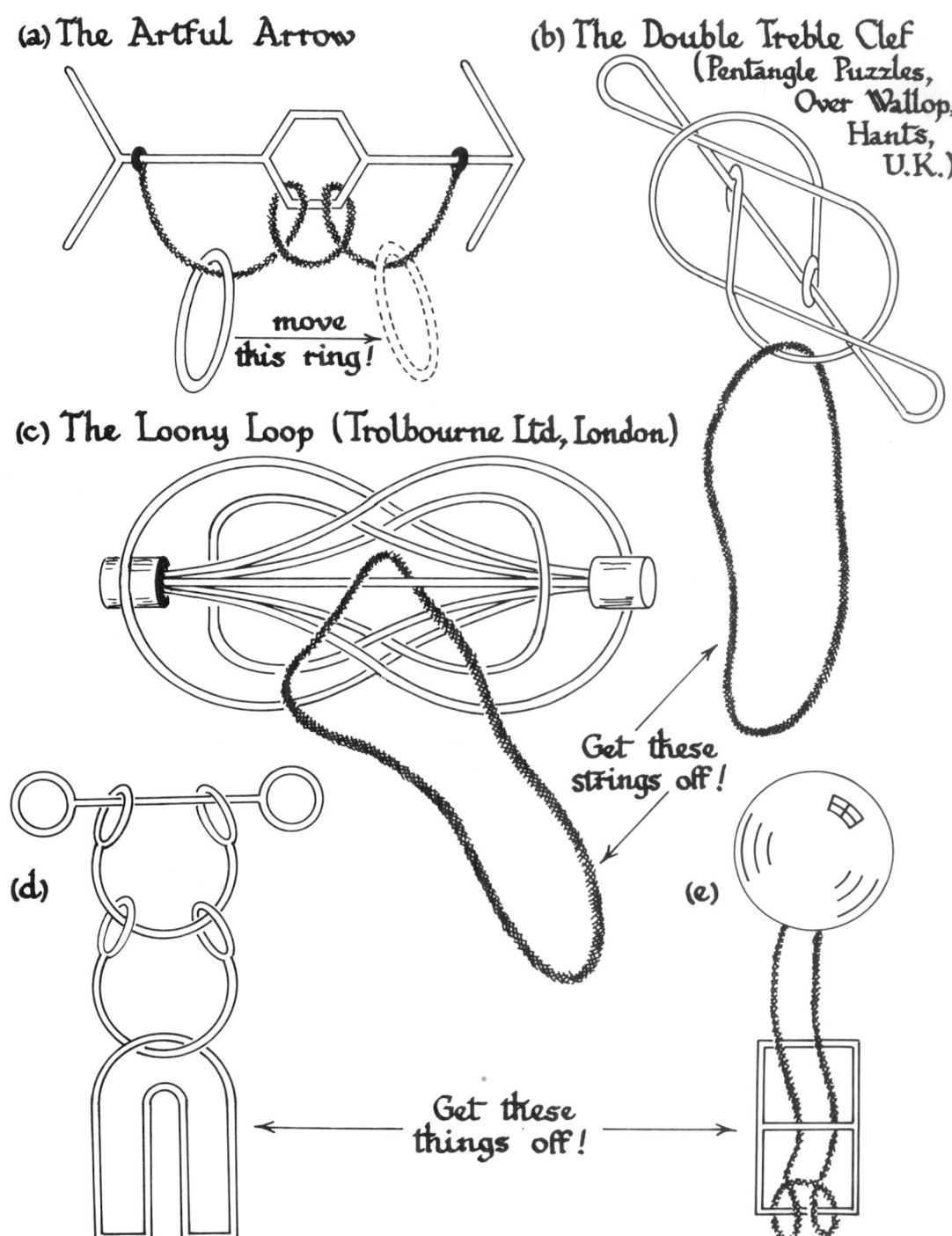

(a) The Artful Arrow

move this ring!

(b) The Double Treble Clef (Pentangle Puzzles, Over Wallop, Hants, U.K.)

(c) The Loony Loop (Trolbourne Ltd, London)

Get these strings off!

(d)

(e)

Get these things off!

Figure 8. Shifting Rings, Strings . . . . . .

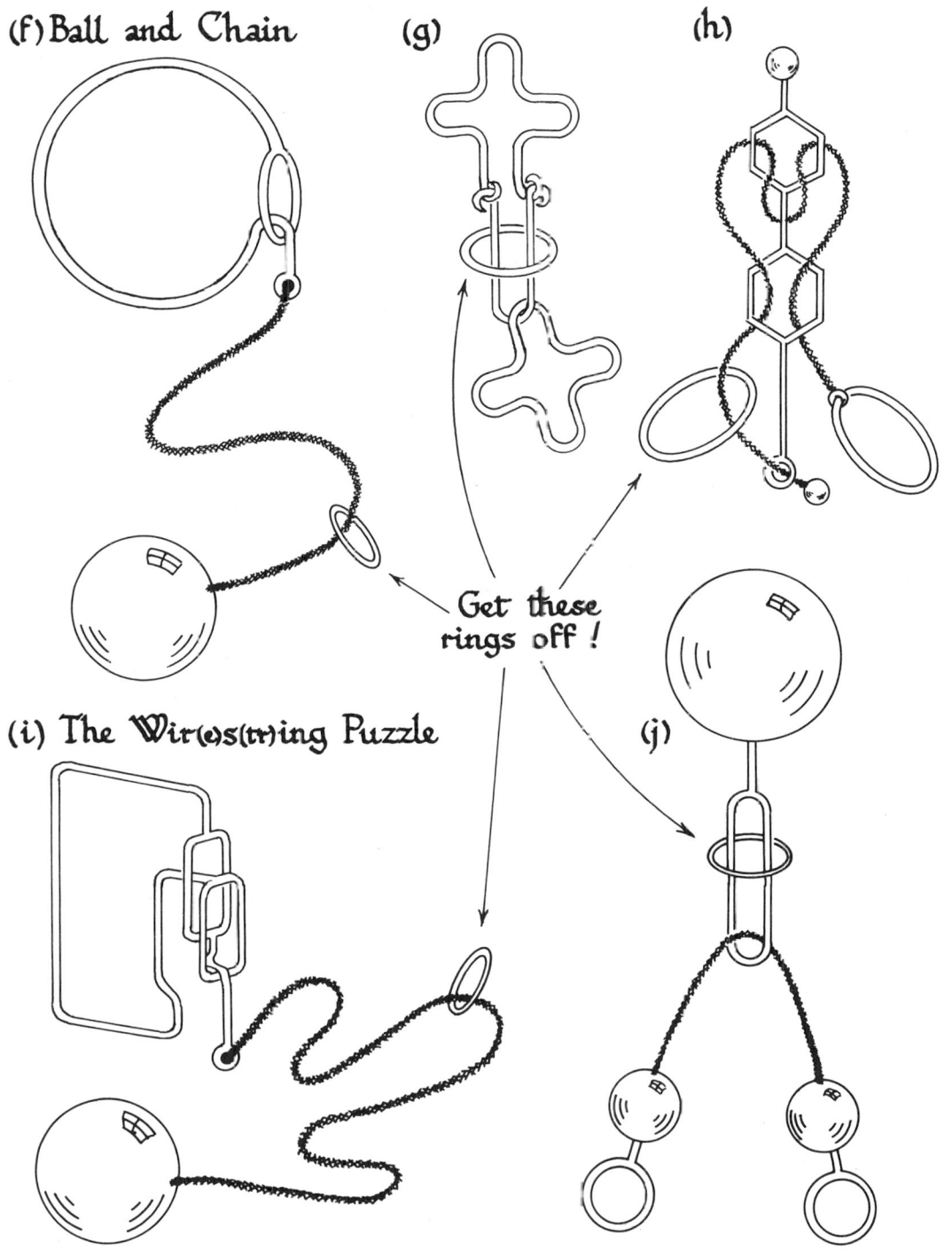

(f) Ball and Chain

(g)

(h)

Get these rings off !

(i) The Wir(e)s(tr)ing Puzzle

(j)

. . . . . . and Other Things.

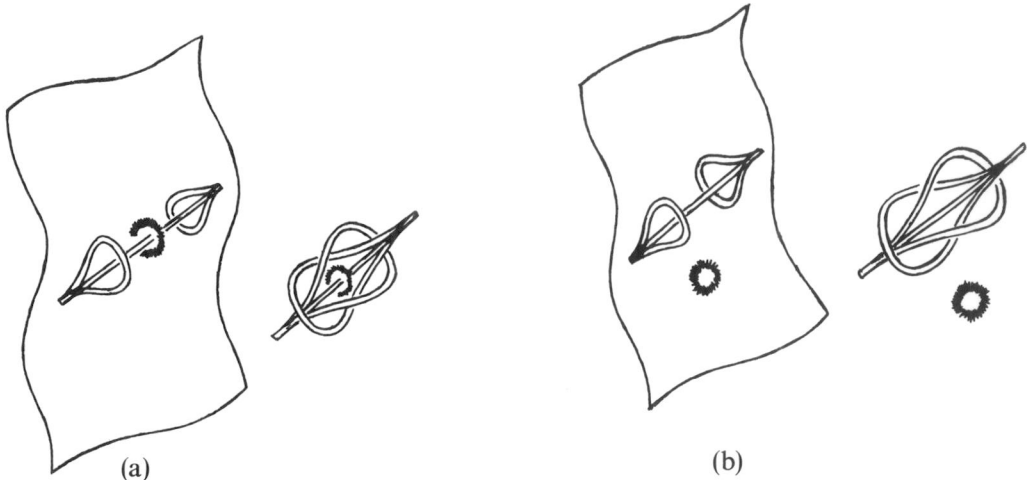

(a)   (b)

Figure 10. The Magic Mirror.

Now, very carefully, move the string in just such a way that its *image* in the magic mirror moves completely away from the wire and shrinks to a small point at some little distance from it. Once again, because the distortion was continuous, the real string must now be almost a point, some distance from the wire, and you've solved the puzzle. Easy, wasn't it?

In such cases it often helps to imagine an intermediate distortion. In Fig. 11 we show two stages in an intermediately distorted one-knot Loony Loop. Perhaps you're ready for the two knot version (Fig. 8(c))? Or the Double Treble Clef (Fig. 8(b)) (Pentangle Puzzles, Over Wallop, Hants, U.K.)?

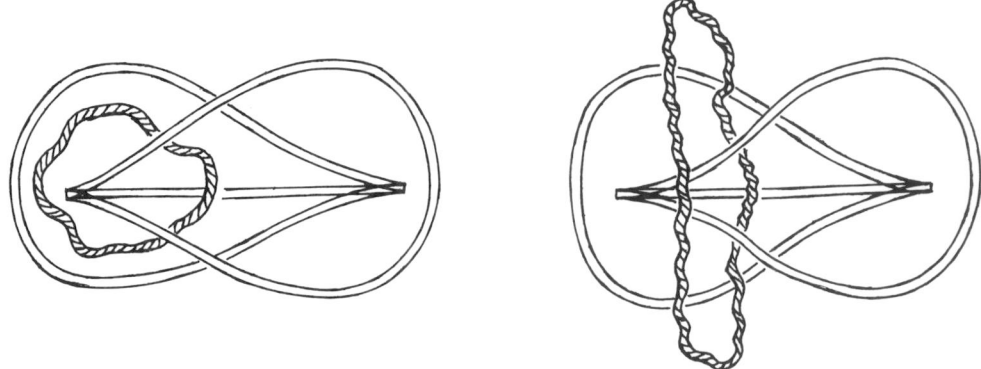

Figure 11. A Less Distorting Mirror.

If a puzzle has got just one completely rigid piece and a number of completely flexible pieces then you can often use the magic mirror method to pretend that the rigid piece is also flexible. For instance, although it may seem impossible to make the braided piece of paper in Fig. 12

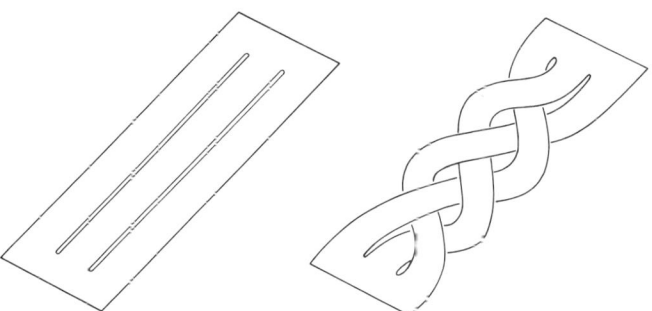

Figure 12. Can You Braid this Strip of Paper?

without glue, it can be undone quite easily. This principle is quite familiar to craftsmen in leather. (To make it you should start braiding at one end and undo the tangle which forms at the other.) The

BARMY BRAID

problem appears for the first time in this book. It's to take the string off the rigid wire frame in Fig. 13(a). You know you can do it, because in a suitable magic mirror it looks like Fig. 13(b).

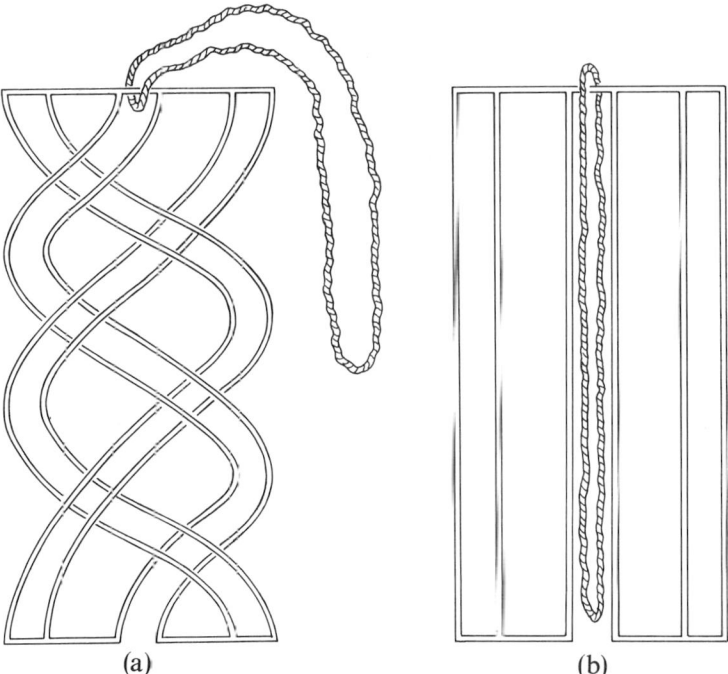

(a)   (b)

Figure 13. Barmy Braid Meets Magic Mirror.

THE ARTFUL ARROW

Figure 8(a) is our version of a puzzle that appears in many different forms. The basic framework is often a bar of wood with a drill hole in place of our hexagon. We have even seen a version in which the ends of our arrow are a giant's arms and the central hole his nostrils, but the solution is always the same! You can solve this puzzle, and some similar ones, by a modification of the Magic Mirror Method which we call

THE MAGIC MOVIE METHOD

If the Artful Arrow had a much smaller ring, there'd be no difficulty about solving it; we'd just slide the ring along the string from the tail of the arrow to its head. Let's suppose we have a kinematic friend who takes a movie of this, but that through some accident with his filters, the string doesn't show up too well, so that what the movie shows is the rigid arrow framework and a little ring that wanders about in space. In fact the ring moves downward through the hexagon (1 to 2 to 3 in Fig. 14(a)), sweeps around (3 to 4 to 5) and then comes safely back up again (5 to 6 to 7).

What we want to do is to watch this movie in a sort of hyperspace magic mirror which

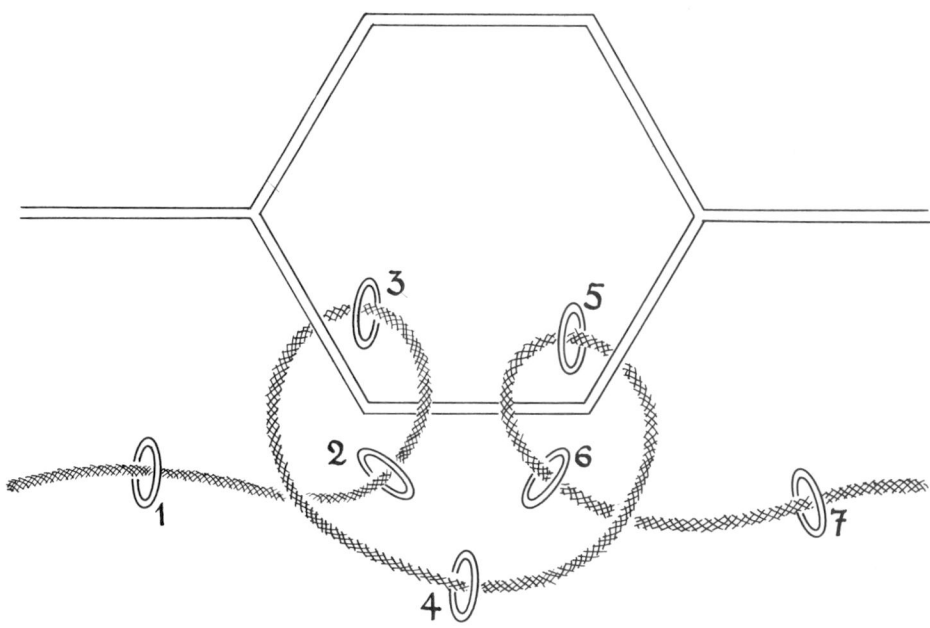

Figure 14(a). The Magic Movie M_o.

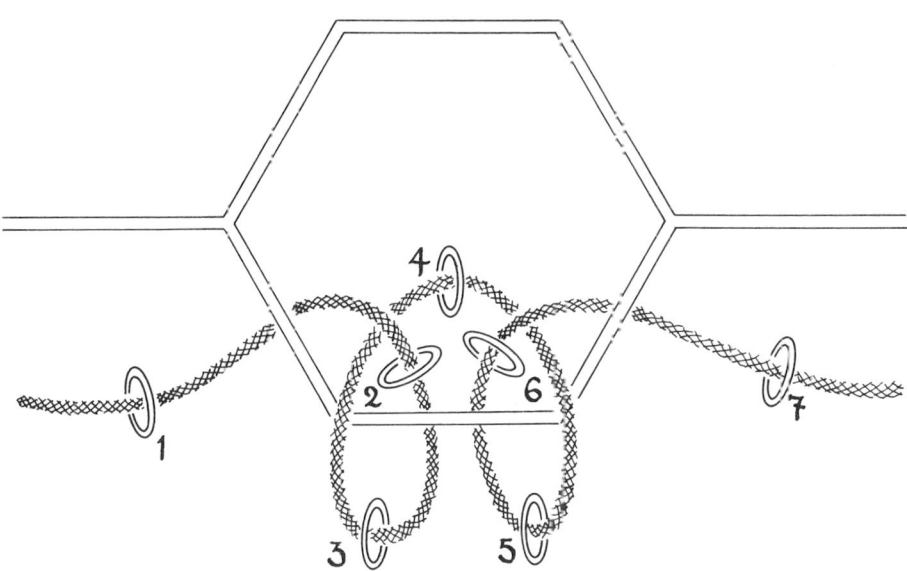

Figure 14(b). An Intermediate Half-Magic Movie.

distorts both space and time. Our friend can arrange this for us by taking the movie M_0 to the animation department where they can change the whole movie bit by bit, first to M_1, in which the ring goes down through the hexagon and wanders about a bit less before it comes back up again, then to M_2, in which it wanders hardly at all before coming up, then to M_3 in which it only takes a timid dip through the hexagon, then in M_4 not at all, while in M_5, M_6, ... the size of the ring gradually increases until it is too big to go through the hexagon.

The trouble with all these movies is that we can't see the string! But since we intend the sequence of movies to realize a continuous distortion of space-time, we can ask the animation department to work overtime and fill in the position of the string as well. The final movie, M_{10} say, should satisfy the producer as representing a solution to the puzzle.

As usual, it helps if the whole process is only half-magic. What must actually happen in this sequence of movies is that the excursion of the ring through the hexagon is gradually replaced by a pulling up of the central loop of string (Mahomet coming to the mountain). In Fig. 14(b) we show an intermediate movie in which you can hardly tell whether this loop, as it passes through the ring in position 4, is above or below the hexagon. You can therefore solve this puzzle by passing the ring from 1 to 2 to 3 while the loop is *below* the hexagon, then lifting the loop a bit while you slide the ring from 3 to 4 to 5 and drop it again so that you can go from 5 to 6 to 7. Since all these movies can be made with a full-sized ring, this will solve the puzzle.

This argument allows us to extend the idea we noted when introducing the Barmy Braid. Suppose that a puzzle has *any number* of rigid pieces (like our arrow and ring) and some arbitrarily flexible ones (e.g., our string) and you could find a solution if the rigid pieces were made flexible. Then, if the motion of the rigid pieces in your solution can be continuously distorted into a rigidly permissible motion, you can use the Magic Movie Method to solve the original puzzle. In topologists' technical language we are using the *Isotopy Extension Principle*.

PARTY TRICKS AND CHINESE RINGS

Figure 15. Girl Meets Boy.

You must have met the party trick where the boy and the girl have to separate themselves without untying the knots in the string. Usually they have lots of fun stepping through one another's arms without effect before they find the real answer.

Let's look at one of those fists more closely (Fig. 16(a)). With a really magic mirror this looks like (Fig. 16(b)) and the solution is obvious, but as usual it's slightly easier to see what to do if your mirror is only half magic (Fig. 16(c)).

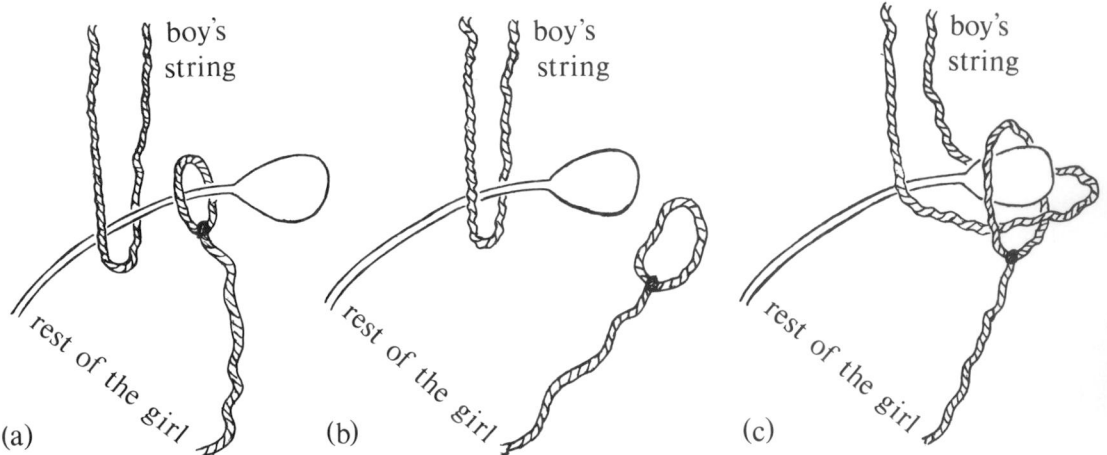

Figure 16. Boy Leaves Girl.

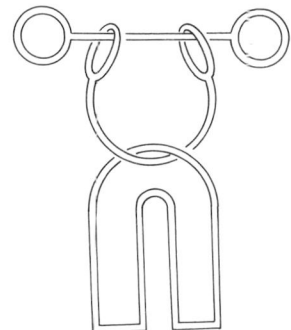

Figure 17(a). Pajamas or Hanger.

One of our wire and string puzzles is very like this. The pajama-shaped frame at the bottom of Fig. 17(a) is made of wire rather than string, but it happens to be just about the shape that a piece of string would need to get to while being taken off. In Fig. 8(d) you'll see there's a similar puzzle, but with an extra piece.

The magic mirror in Fig. 17(b) shows that this puzzle can certainly be solved if the wire pajama shape is replaced by a completely flexible string—once again this funny shape is sufficient to overcome its lack of flexibility.

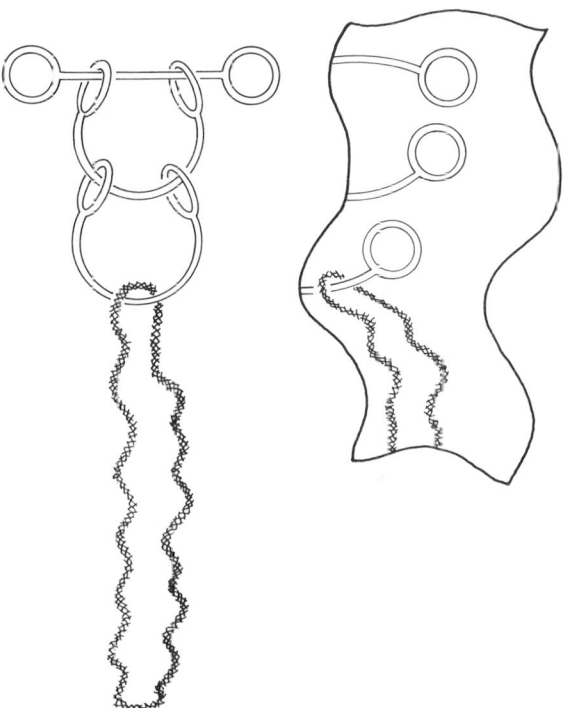

Figure 17(b). Another Look in the Magic Mirror.

The Chinese rings are an indefinite extension of this principle. The magic mirror method shows that the string in Fig. 18(a) can be taken right off. In the course of doing so it reaches a position like that of the wire loop in Fig. 18(b), and removal of this is the usual Chinese Rings puzzle.

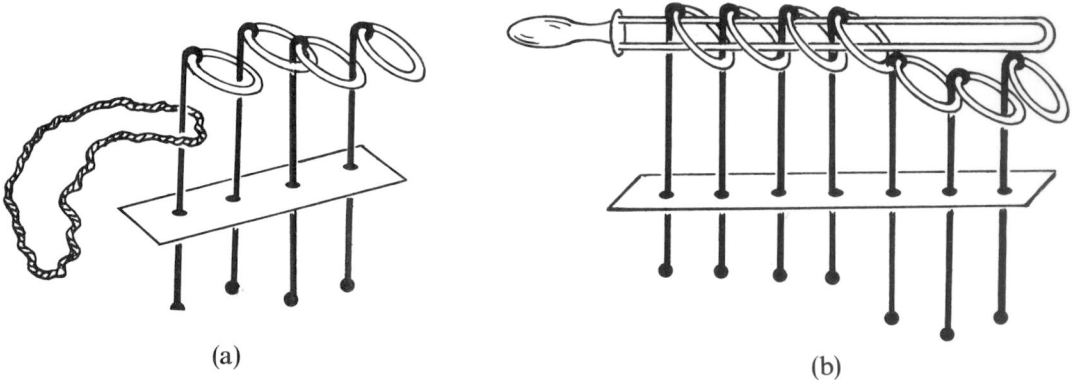

(a)   (b)

Figure 18. The Chinese (st)ring Puzzle.

CHINESE RINGS AND THE GRAY CODE

Figure 19(a) shows a certain position of a 7-ring Chinese Rings puzzle. We call this position

1   0   1   1   1   0   0

because the rings we've numbered

64   32   16   8   4   2   1

are respectively

on   off   on   on   on   off   off

the loop. ("On" means that the ring's retaining wire passes through the loop.) Which positions neighbor this?

You hardly need a magic mirror to see how the state of the rightmost ring, number 1, can always be changed (Fig. 19(b)), showing that our position neighbors

1   0   1   1   1   0   1.

But it also neighbors

1   0   1   0   1   0   0

as well!

To see this, slip ring number 8 up over the end of the loop as suggested by the dotted arrow in Fig. 19(a) and then drop it down through the loop as hinted in Fig. 19(c).

In general the rightmost ring, number 1, can always be slipped on or off the loop, so that

... ? ? ? 0 neighbors ... ? ? ? 1.

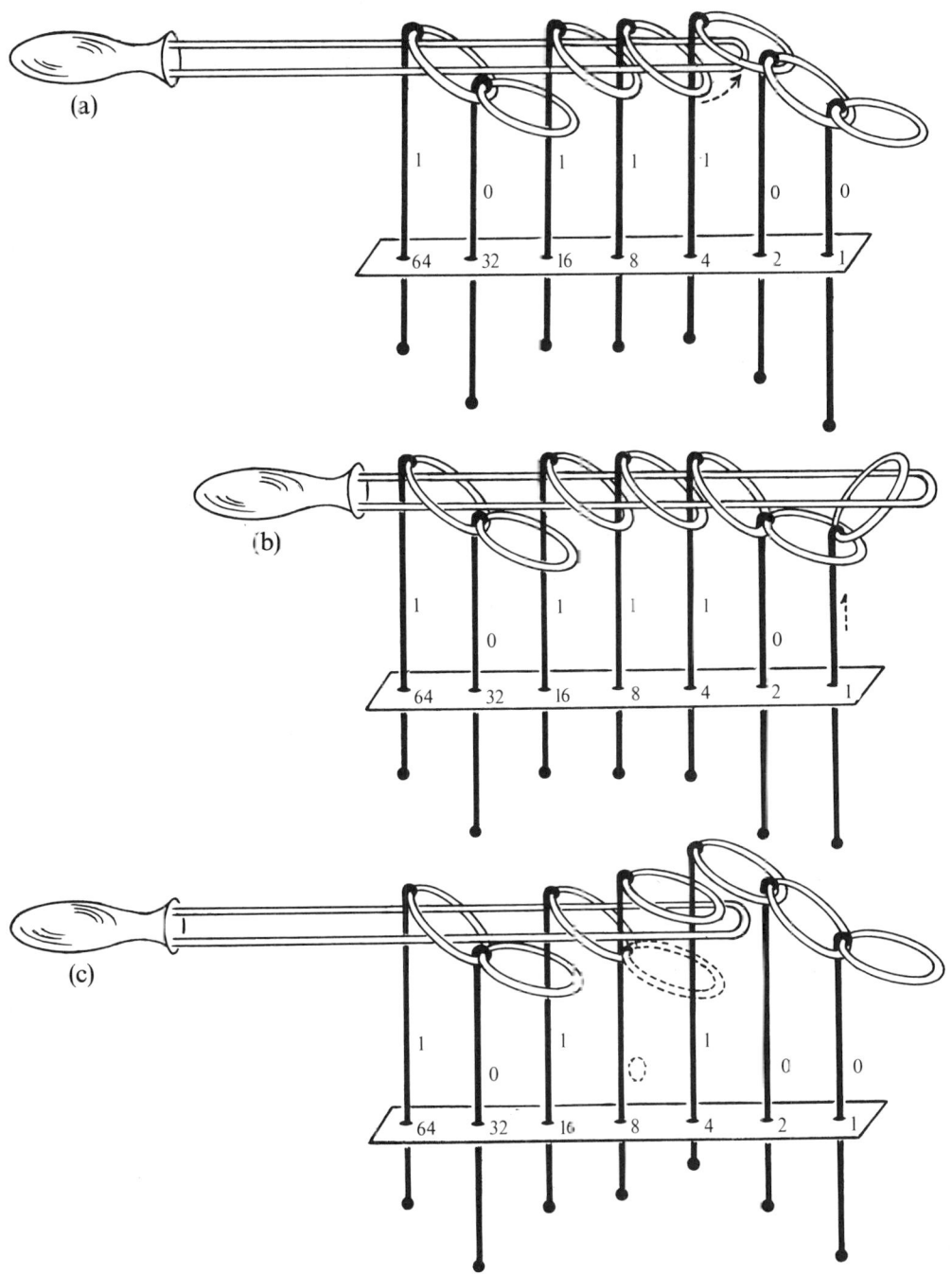

Figure 19. Gray Code and Chinese Rings.

But also a ring can be slipped on or off provided that the ring just right of it is *on* and all rings right of *that* are *off*, so that

... ? 1 1 0 0 0 neighbors ... ? 0 1 0 0 0.

With these neighboring rules the entire set of 2^n positions in the n-ring puzzle form one continuous sequence, which for $n=4$ is:

ring #	8	4	2	1	
state # 8, i.e.	1	0	0	0	is 15 moves from being off,
state # 9, i.e.	1	0	0	1	is 14 moves from being off,
state # 11, i.e.	1	0	1	1	is 13 moves from being off,
state # 10, i.e.	1	0	1	0	is 12 moves from being off,
state # 14, i.e.	1	1	1	0	is 11 moves from being off,
state # 15, i.e.	1	1	1	1	is 10 moves from being off,
state # 13, i.e.	1	1	0	1	is 9 moves from being off,
state # 12, i.e.	1	1	0	0	is 8 moves from being off,
state # 4, i.e.	0	1	0	0	is 7 moves from being off,
state # 5, i.e.	0	1	0	1	is 6 moves from being off,
state # 7, i.e.	0	1	1	1	is 5 moves from being off,
state # 6, i.e.	0	1	1	0	is 4 moves from being kff,
state # 2, i.e.	0	0	1	0	is 3 moves from being off,
state # 3, i.e.	0	0	1	1	is 2 moves from being off,
state # 1, i.e.	0	0	0	1	is 1 moves from being off,
and state # 0, i.e.	0	0	0	0	is OFF!

How do we tell how many moves it takes to get all the rings off if we're given only the state number, i.e. the sum of the numbers of the rings that are on? The answer displays a remarkable connexion with nim-addition! When you're in state number n, it will take you exactly

$$n \overset{*}{+} \lfloor n/2 \rfloor \overset{*}{+} \lfloor n/4 \rfloor \overset{*}{+} \lfloor n/8 \rfloor \overset{*}{+} \ldots = m$$

moves to get off. For example in state 13 you're just

$$13 \overset{*}{+} 6 \overset{*}{+} 3 \overset{*}{+} 1 = 9$$

moves away. And if you're given a number m, then state number

$$m \overset{*}{+} \lfloor m/2 \rfloor = n$$

is the one that's just m moves from off. For example

$$9 \overset{*}{+} 4 = 13.$$

Let's find the position that's 99 moves from off in the 7-ring puzzle. Because the binary expansions of 99 and $\lfloor 99/2 \rfloor$ are

$$1 \quad 1 \quad 0 \quad 0 \quad 0 \quad 1 \quad 1$$

and $\quad 1 \quad 1 \quad 0 \quad 0 \quad 0 \quad 1$

the answer is $\quad 1 \quad 0 \quad 1 \quad 0 \quad 0 \quad 1 \quad 0.$

How many moves is state

$$1 \quad 1 \quad 0 \quad 1 \quad 1 \quad 1 \quad 1$$

from off? The answer is found by the 7-term nim-sum

$$
\begin{array}{ccccccc}
1 & 1 & 0 & 1 & 1 & 1 & 1 \\
 & 1 & 1 & 0 & 1 & 1 & 1 \\
 & & 1 & 1 & 0 & 1 & 1 \\
 & & & 1 & 1 & 0 & 1 \\
 & & & & 1 & 1 & 0 \\
 & & & & & 1 & 1 \\
 & & & & & & 1 \\
\hline
= 1 & 0 & 0 & 1 & 0 & 1 & 0,
\end{array}
$$

which is the binary expansion of 74.

In various kinds of control device it's important to code numbers in such a way that the codes from adjacent numbers differ in only one place and the code that appears above, known to engineers as the Gray code, has this useful property. It has also been used in transmitting television signals. However, its connexion with the Chinese Rings puzzle was known to Monsieur L. Gros, more than a century ago. Incidentally, the multiknot Loony Loop is connected with a ternary version of the Gray code.

The Chinese Rings, despite their name, appear to be a medieval Scandinavian discovery, having originally been used as a sort of combination lock. In recent years several mechanical and electronic puzzles, completely different in appearance, but employing the same mathematical structure, have appeared on the market.

THE TOWER OF HANOÏ

In happier times, Hanoï was mainly known to puzzlers as the fabled site of that temple where monks were ceaselessly engaged in transferring 64 gold discs from the first to the last of three pegs according to the conditions that

> only one disc may be moved at a time, and
> no disc may be placed above a smaller one.

Figure 20(a) shows the initial position in a smaller version of the puzzle and Fig. 20(b) shows the position 13 moves later.

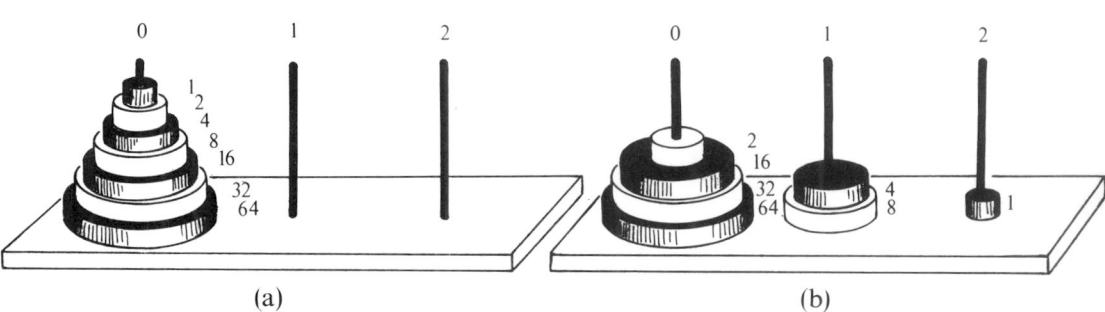

Figure 20. The Tower of Hanoï.

In this puzzle it's possible to make mistakes, unlike in the Chinese Rings where the only mistake you can make is to start travelling in the wrong direction. However, you won't make too many mistakes if you use discs that are alternately gold and silver and

> never place a disc immediately
> above another of the same metal.

To find out where you should be after m moves, expand m in binary, and then, according as the total number of discs is

	even	or	odd,
replace a 1 digit by the ternary number	1	or	2,
replace a 2 digit by the ternary number	21	or	12,
replace a 4 digit by the ternary number	122	or	211,
replace an 8 digit by the ternary number	2111	or	1222,
replace a 16 digit by the ternary number	12222	or	21111,
replace a 32 digit by the ternary number	211111	or	122222,
replace a 64 digit by the ternary number	1222222	or	2111111,

... .

These ternary numbers, when added mod 3 without carrying, show you what peg each disc should be on. For 13 moves and a 7-disc tower, since 7 is odd and

$$\left. \begin{array}{r} 1 \\ +4 \\ +8 \end{array} \right\} \quad \text{we find the ternary numbers} \quad \left\{ \begin{array}{l} 2 \\ 211 \\ 1222 \end{array} \right.$$

$$= 13. \qquad\qquad\qquad\qquad\qquad \overline{0001102,}$$

showing that disc 1 should be on peg 2, discs 4 and 8 on peg 1, and the rest on peg 0 as in Fig. 20(b).

The Tower of Hanoï puzzle and the fable which usually accompanies it were invented by Monsieurs Claus (Édouard Lucas) and De Parville in 1883 and 1884.

A SOLITAIRE-LIKE PUZZLE AND SOME COIN-SLIDING PROBLEMS

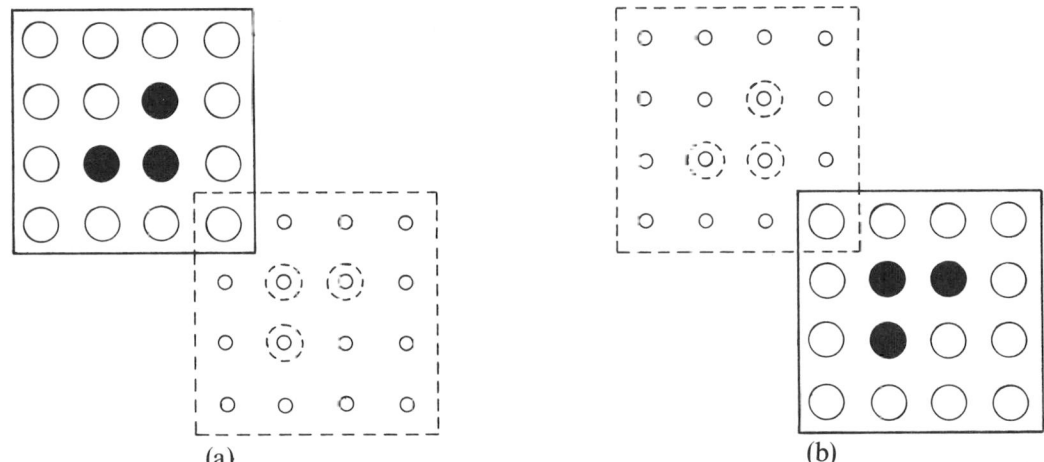

Figure 21. A Solitaire-Like Puzzle.

A little puzzle we came across recently is played in a way very similar to the game of Peg Solitaire, except that the pegs are not removed after jumping. Starting from the position of Fig. 21(a), go to the "opposite" position of Fig. 21(b) jumping only in the N–S and W–E directions. The three special pegs are to move to the three special places.

H	H	H		T	T	T

Figure 22. Swap the Hares and Tortoises.

This is rather like various two-dimensional forms of the familiar Hares and Tortoises (or sheep and goats) puzzle (Fig. 22) in which the animals (you can use coins) have to change places and the permitted moves are as in the game of Toads and Frogs in Chapter 1. Other problems with the same coins are:

(i) get from Fig. 23(a) to Fig. 23(b) with just 3 moves of 2 contiguous coins (the coins to be slid on the table, remaining in the same orientation and touching throughout);
(ii) the same, but reversing the orientation of each pair of coins as it is moved;
(iii) similar problems, but with more coins;
(iv) form the six coins of Fig. 24(a) into a ring (Fig. 24(b)) with just three moves. At each move one coin must be slid on the table, without disturbing any of the others, and positioned by touching it against just two coins. For example, you might try Fig. 24(c) for your first move, but then you wouldn't be able to slide the middle one out.

Figure 23. Make Three Moves of Two Contiguous Coins.

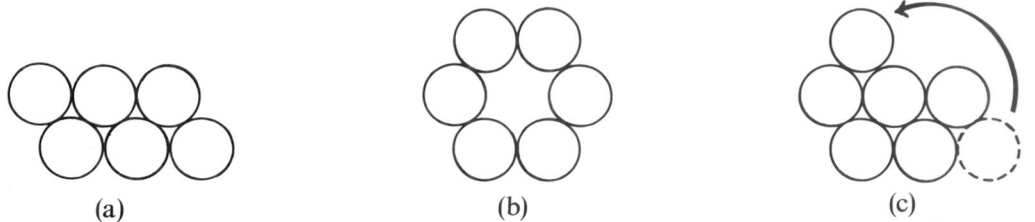

(a)   (b)   (c)

Figure 24. Ringing the Changes.

THE FIFTEEN PUZZLE AND THE LUCKY SEVEN PUZZLE

Figure 25. Sam Loyd's Fifteen Puzzle.

The most famous sliding puzzle is Sam Loyd's *Fifteen Puzzle* in which the home position is Fig. 25 and the move is to slide one square at a time into the empty space. You are required to get home from the random position you usually find the puzzle in. Nowadays the puzzle is usually sold with pieces so designed that it is impossible to remove them from the base.

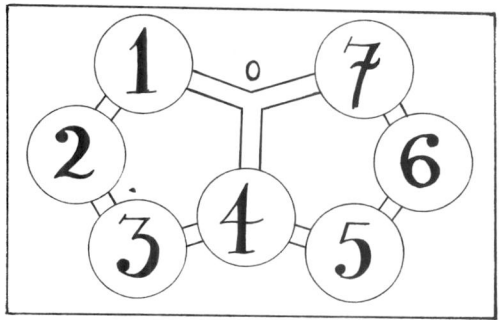

Figure 26. The Lucky Seven Puzzle.

A more interesting puzzle is the **Lucky Seven Puzzle** for which the home state is displayed in Fig. 26 and similar rules apply.

In such puzzles there are certain basic permutations of the pieces that bring the empty space back to its standard position. For the Seven Puzzle you can either move the four discs in the left pentagon in the order 1, 2, 3, 4, 1, leading to the position of Fig. 27(a) or treat the right pentagon similarly, moving 7, 6, 5, 4, 7, leading to the position of Fig. 27(b).

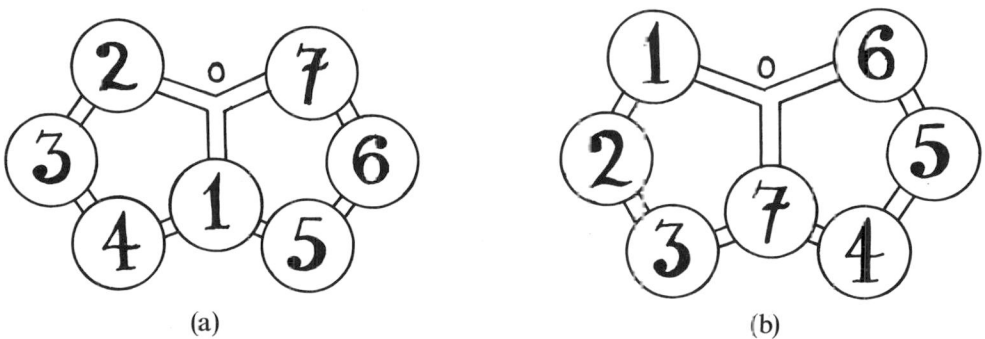

(a)   (b)

Figure 27. After a Few Moves.

In the first case we have effected the permutation α in which

$$\begin{array}{ccccccc} \text{disc} & 1 & 2 & 3 & 4 & 5 & 6 & 7 \\ \text{goes to place} & 4 & 1 & 2 & 3 & 5 & 6 & 7 \end{array}\Big\} \text{or, for short,} \ (1432)\,(5)\,(6)\,(7),$$

and in the second case the permutation β in which

$$\begin{array}{ccccccc} \text{disc} & 1 & 2 & 3 & 4 & 5 & 6 & 7 \\ \text{goes to place} & 1 & 2 & 3 & 5 & 6 & 7 & 4 \end{array}\Big\} \text{or } (1)\,(2)\,(3)\,(4567).$$

We can obviously combine these basic permutations to any extent. For instance, by performing the sequence

$$1 \xrightarrow{\alpha} 4 \xrightarrow{\beta} 5 \xrightarrow{\alpha} 5 \xrightarrow{\alpha} 5 \xrightarrow{\beta} 6$$
$$2 \to 1 \to 1 \to 4 \to 3 \to 3$$
$$3 \to 2 \to 2 \to 1 \to 4 \to 5$$
$$4 \to 3 \to 3 \to 2 \to 1 \to 1$$
$$5 \to 5 \to 6 \to 6 \to 6 \to 7$$
$$6 \to 6 \to 7 \to 7 \to 7 \to 4$$
$$7 \to 7 \to 4 \to 3 \to 2 \to 2$$

$$\begin{array}{ccccccc} \text{disc} & 1 & 2 & 3 & 4 & 5 & 6 & 7 \\ \text{gets to place} & 6 & 3 & 5 & 1 & 7 & 4 & 2 \end{array}\Big\} \text{or } (164)\,(2357).$$

By combining any given permutations in all possible ways we get what mathematicians call a **group** of permutations. Is there an easy way to see which permutations belong to the group of the Lucky Seven Puzzle? Yes! The trick, as always in such cases, is to find some permutations which keep most of the objects fixed. In the case of the Seven Puzzle it seems best to regard the outer edges as forming a complete circle across which there is a single bridge between places 0 and 4 (Fig. 28). In this form the seven discs can be freely cycled round the outer circle (which

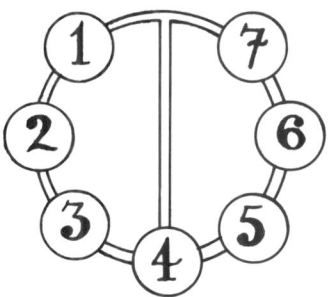

Figure 28. Crossing Bridges.

we hardly count as a move) or else a single disc may be slid across the bridge (remember that in the actual form of the puzzle the bridge is too short for several discs to traverse it at once). It doesn't really matter whether the disc we slide across the bridge goes upwards or downwards, since this has the same effect on the cyclic order, so we'll always slide our discs *downward*.

If we think of the puzzle in this way and, starting from the home position, slide discs 2, 4, 2, 4, 2 down the bridge, we reach the position of Fig. 29 in which discs 2 and 4 have been interchanged

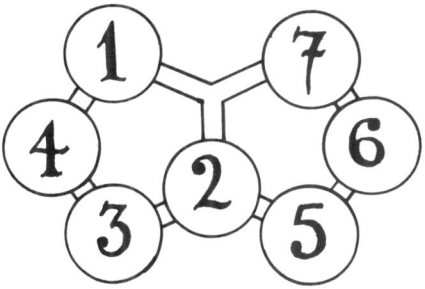

Figure 29. Swapping Two and Four.

and all the others are in their original places. Obviously we can interchange any pair of discs which are two places apart round the circle in this way. It's not hard to see how *any* desired rearrangement can be reached by a succession of such interchanges. For instance if we wanted to get

$$\begin{array}{llllllll} \text{discs} & 1 & 2 & 3 & 4 & 5 & 6 & 7 \\ \text{to places} & 7 & 6 & 5 & 4 & 3 & 2 & 1 \end{array}$$

we might perform the interchanges of the following scheme

$$
\begin{array}{ccccccc}
1 & 2 & 3 & 4 & 5 & 6 & 7 \\
3 & 2 & 1 & 4 & 5 & 6 & 7 \\
3 & 2 & 5 & 4 & 1 & 6 & 7 \\
3 & 2 & 5 & 4 & 7 & 6 & 1 \\
3 & 4 & 5 & 2 & 7 & 6 & 1 \\
3 & 4 & 5 & 6 & 7 & 2 & 1 \\
5 & 4 & 3 & 6 & 7 & 2 & 1 \\
5 & 4 & 7 & 6 & 3 & 2 & 1 \\
5 & 6 & 7 & 4 & 3 & 2 & 1 \\
7 & 6 & 5 & 4 & 3 & 2 & 1
\end{array}
$$

Get 1 in position first,

then 2,

then 3,

then 4,

then 5 (5 and 7).

leading to a solution in which 45 discs have crossed the bridge. This method is not very efficient but it has the great advantage of providing an almost mechanical technique by which you can obtain any position. Can you find a shorter solution to the above problem?

ALL OTHER COURSES FOR POINT-TO-POINT

The history of the Fifteen Puzzle has been given too many times to bear further repetition here. Exactly half of the

$$
15! = 1 \times 2 \times 3 \times \ldots \times 15 = 1\,307\,674\,368\,000
$$

permutations (the so called *even* permutations) can be obtained. In technical language, the available permutations form the **alternating group**, A_{15}, whereas for the Lucky Seven Puzzle we have the full **symmetric group**, S_7, of $7! = 7 \times 6 \times 5 \times 4 \times 3 \times 2 \times 1 = 5040$ permutations.

You can make a puzzle of this type by putting counters on all but one of the nodes of any connected graph and then sliding them, point to point, always along an edge into the currently empty node. We can afford to ignore the *degenerate* cases, when your graph is a cycle, or is made by putting two smaller graphs together at a single node, because then the puzzle is trivial, or degenerates into the two smaller puzzles corresponding to the two smaller graphs.

Rick Wilson has proved the remarkable theorem that for every non-degenerate case but one we get either the full symmetric group (if some circuit is odd) or the alternating group (otherwise). The single exception is the graph of the **Tricky Six Puzzle** (Fig. 30) for which the group consists of all possible Möbius transformations

$$
x \rightarrow \frac{ax + b}{cx + d} \pmod 5
$$

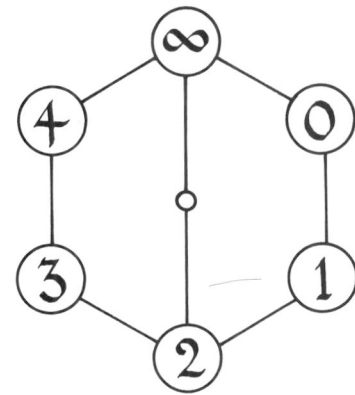

Figure 30. Rick's Tricky Six Puzzle.

THE HUNGARIAN CUBE—BŰVÖS KOCKA

The Hungarian words actually mean "magic cube". If you're crazy enough to get one of these you'll see that when it comes to you from the manufacturer it has just one color on each face (Fig. 31(a)) but your Hungarian cube is unlikely to stay in this beautiful state because you can rotate the nine little **cubelets** that make up any face (Fig. 31(b)) and so disturb the color scheme. For example, if you complete the turn started in Fig. 31(b), and then turn the top face clockwise you'll arrive at Fig. 31(c). After three more turns the colors are all over the place (Fig. 31(d)) and you'll find it very hard to recover the original arrangement; in other words to get each of the cubelets back into its own **cubicle**, *and* the right way round.

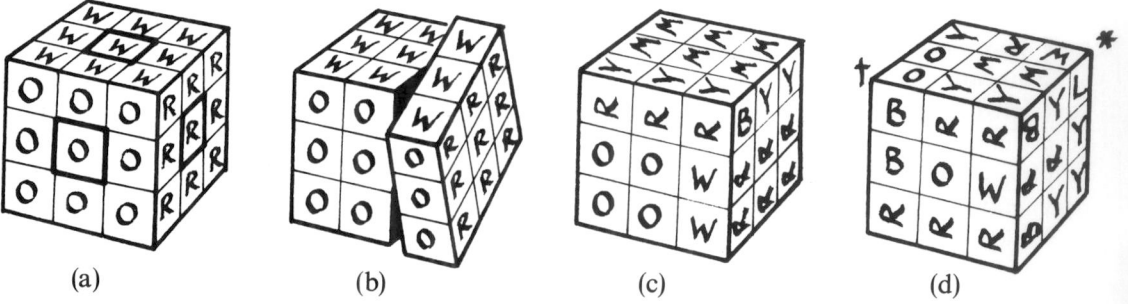

(a)   (b)   (c)   (d)

Figure 31. The Hungarian Magic Cube.

There are really two problems about this elegant little puzzle. The first is how its brilliant designer, Ernö Rubik, can possibly have managed to make all those motions feasible without all the cubelets falling apart. We'll leave that one to you! The other is, of course, to provide a method by which we can guarantee to get home from any position our friends have muddled the cube into.

JUST HOW CHAOTIC CAN THE CUBE GET?

At least there are six permanent landmarks: the cubelets at the centres of the faces always stay in their own cubicles although they may be rotated. We call these the **face cubelets** and have framed them in Fig. 31(a). No matter how confused your cube looks, you can tell what the final color of each face should be, just by looking at the face cubelet at its centre. So, for instance, in Fig. 31(d) we call the top face **white** even though only one third of it really is.

So you can work out the home cubicle for any cubelet by just looking at its colors and thinking which faces these belong to. For instance the LWO cubelet ∗ in Fig. 31(d) should end up at † (in our cube the colors opposite R, W, O are L, B, Y). We recommend the nervous novice always to hold the cube with its white face uppermost and then to take a careful note of the color of the bottom face, which we call the **ground** color.

Since the other 20 visible cubes are of two types,

> 8 **corner cubes**, which have 3 possible orientations in their cubicles,
> and 12 **edge cubes**, which have 2,

there are at most

$$3^8 \times 2^{12} \times 8! \times 12! = 519\,024\,039\,293\,878\,272\,000$$

conceivable arrangements. However, Anne Scott proved that only one-twelfth of this number, namely

$$43\,252\,003\,274\,489\,856\,000$$

are attainable.

CHIEF COLORS AND CHIEF FACES

These notions help us keep track of the orientations of cubelets, even when they're not in their home cubicles. We'll call the **chief face** of a *cubicle* the one in the top or bottom surface of the cube, if there is one, and otherwise the one in the right or left wall. The **chief color** of a *cubelet* is the color that should be in the chief place when the cubelet gets home. In other words White or the Ground color if possible, and otherwise the color that should end up in the left or right wall of the cube.

If a cubelet, no matter where it is, has its chief color in the chief face of its current cubicle we'll call it **sane** and otherwise **flipped** (if it's an edge cubelet) or **twisted** (if it's a corner one). There's only one way to make an edge-flip (e), but a corner may be twisted anticlockwise (a) or clockwise (c).

Now, as shown in Fig. 32, turning the top (or bottom) preserves the chiefness of every cubelet. Turning the front (or back) changes the chiefness at four corners and turning the left (or right) changes it at four corners and four edges. Since each turn flips an even number of edges, you can see that for attainable positions

> the total number of edge-flips
> will always be even.

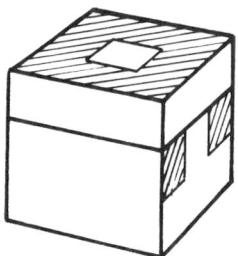 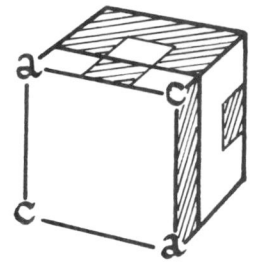 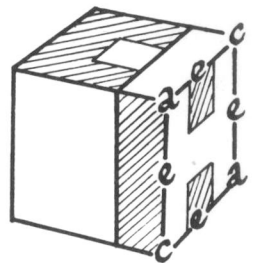

Figure 32. Changes in Chiefness.

And since each turn produces equal numbers of clockwise and anticlockwise twists

> the total corner twisting
> will always be zero, mod 3.

In computing corner twists we count $+1$ for clockwise and -1 for anticlockwise — of course three clockwise twists of a cubelet produce no effect. Finally, for reasons as in the Fifteen Puzzle

> the total permutation of all the
> 20 movable cubelets must be *even*.

An **even permutation** is one we might imagine making by an even number of interchanges.

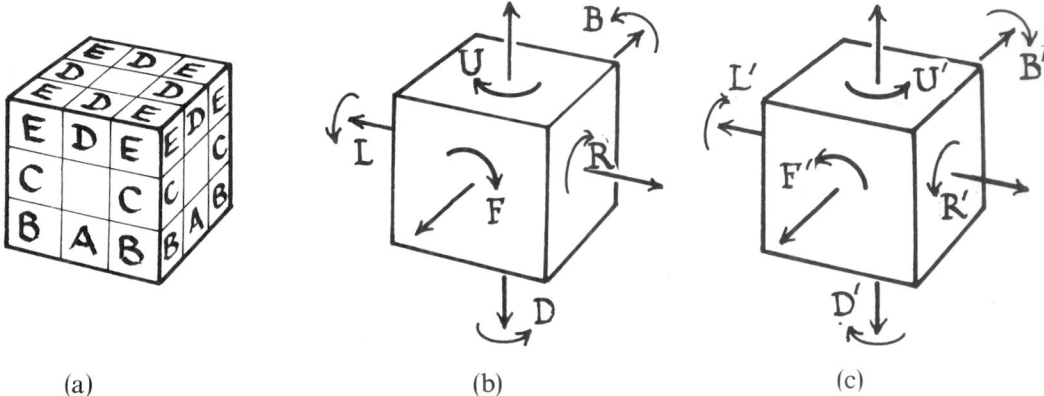

(a)   (b)   (c)

Figure 33. The Six Stages and Our Notation for the Moves.

CURING THE CUBE

Benson, Conway and Seal have simplified Anne Scott's proof that you really can get home from any position for which

   (i) the total edge flipping is zero, mod 2,
   (ii) the total corner twisting is zero, mod 3, and
   (iii) the total permutation of all 20 movable cubes s even.

We have adapted our names for the moves (Fig. 33) so as to agree with David Singmaster's in the hope that a single notation will rapidly become universal. Note that the unprimed letters L,R,F,B,U,D, refer to *clockwise* turns, and the primed letters L′,R′,F′,B′,U′,D′ to *anticlockwise* ones. Our notation for the **slice moves** is illustrated in Fig. 34. Note that in these moves only the *middle* layer of the cube is turned. We shall also use the common notation in which, for example, X^2 means "do X twice" and X^{-1} means "undo X".

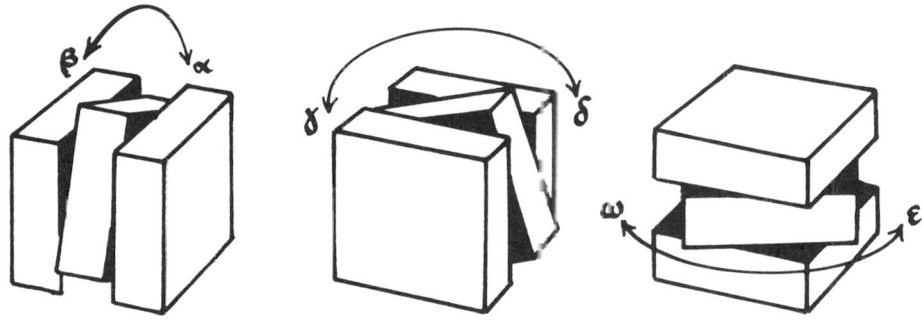

Figure 34. Slice Moves.

Our method has six stages which correspond roughly to the letters in Fig. 33(a).

A: Aloft, Around (Adjust) and About.
B: Bottom Layer Corner Cubelets.
C: Central Layer Edge Cubelets.

D: Domiciling the Top Edge Cubelets.
E: Exchanging Pairs of Top Corners.
F: Finishing Flips and Fiddles.

We've collected the figures for these stages in Fig. 35 for easy reference, so keep a finger on page 765.

Warning: Be very careful when applying this algorithm. Think of "tightening" or "loosening" a screw-cap, so that you never mistake a clockwise turn for an anticlockwise one, even from behind. Be aware at all times which way up you are holding the cube, and don't stop to think in the middle of a sequence of moves. Remember that if you make a tiny mistake you'll probably have to go all the way back to Stage A.

> THE CUBE SELDOM FORGIVES!

A: ALOFT, AROUND (ADJUST) AND ABOUT

Our first stage (Fig. 35A) gets the bottom edge cubelets (A in Fig. 33(a)) into their correct cubicles, the right way round. You bring the ground (=chief) color of such a cubelet into the topmost surface (Aloft) then turn the top layer Around to put this cubelet into the correct side wall which can be turned About to home the cubelet. Sometimes this disturbs a bottom edge cubelet that's already home, but this can be Adjusted by turning the appropriate side wall just before the About step.

B: BOTTOM LAYER CORNER CUBELETS

Now, without disturbing the bottom layer edge cubelets, you must get the bottom layer corner cubelets home.

If the cubelet that's to stand on the shaded square of Fig. 35B is in the top layer, turn the top layer until this cubelet's ground color is in one of the three numbered positions. Then do the appropriate one of

$$B1: F'U'F \qquad B2: RUR' \qquad B3: F'UF.RU^2R'$$

If the cubelet is already *in* the bottom layer, but wrongly placed, use one of these to put any corner cubelet from the top layer into its current position, thereby evicting it into the top layer. Then work as above to put it into the proper place. Repeat this procedure for the other three bottom layer corner cubelets.

C: CENTRAL LAYER EDGE CUBELETS

This stage corrects the central layer edge cubelets without affecting the bottom layer.

If the cubelet destined for the shaded cubicle of Fig. 35C is in the top layer, turn the top layer until you want to move this cubelet in one of the two ways of Fig. 35C (its side face will then be just above the face cubelet of the same color). Then do the appropriate one of

$$C1: URU'R'.U'F'UF \qquad C2: U'F'UF.URU'R'$$

If the cubelet is already *in* the central layer, but wrongly placed, use one of these to evict it into the top layer. Then work as above. Repeat the procedure for the other three central layer edge cubelets.

D: DOMICILING THE TOP EDGE CUBELETS

i.e. putting the top layer edge cubelets into their own home cubicles without as yet worrying about their orientations.

You can do this by a sequence of swaps of adjacent edge cubelets as in Fig. 35D for which the moves are

$$UF.RUR'U'.F'$$

Of course you can first turn the top layer to reduce the number of swaps needed.

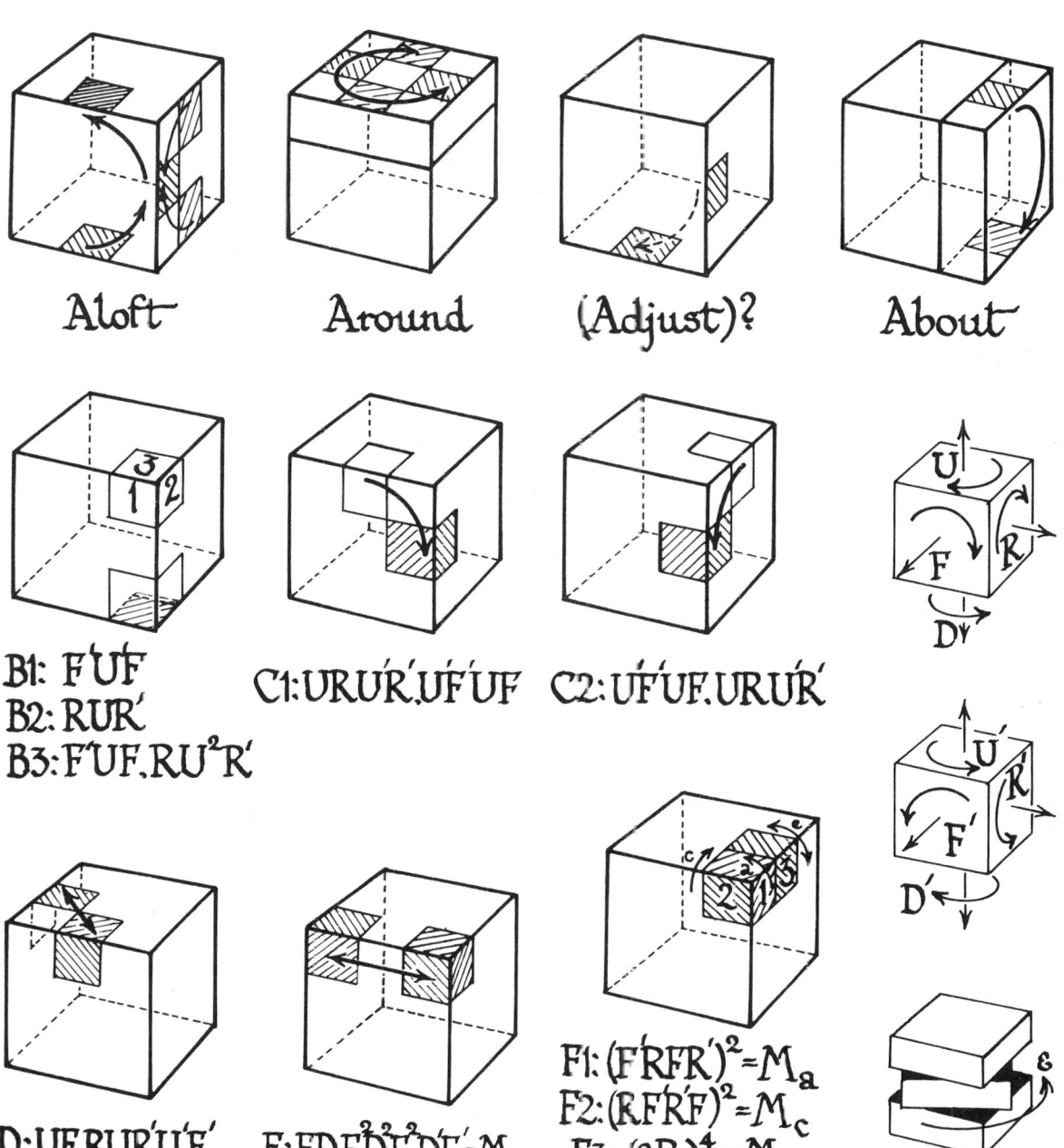

Aloft Around (Adjust)? About

B1: F'UF
B2: RUR'
B3: F'UF.RU²R'

C1: URUR'.UF'UF

C2: UF'UF.URUR'

D: UF.RUR'U.F'

E: FD.F²D²F.DF'=M_s

F1: (F'RFR')²=M_a
F2: (RFR'F)²=M_c
F3: (ƐR)⁴=M_e

Figure 35. Six Simple Stages Cure Chaotic Cubes.

E: EXCHANGING PAIRS OF TOP CORNERS

Now you must get the top layer corner cubelets into their own cubicles by moves that, when they are finally completed, won't have affected the bottom two layers or moved the top layer edge cubelets. Usually you can do this in just two swaps of adjacent corners, but sometimes four will be needed.

Correct performance requires some care. Work out a pair of successive swaps of adjacent corner cubelets that will improve things. Then turn the cube until the first required swap is as in Fig. 35E and do our

$$\text{monoswap, } M_S = FD.F^2D^2F^2.D'F'$$

Then turn **THE TOP LAYER** *ONLY* to bring the second desired swap into the position of Fig. 35E, do another monoswap, and then return the top layer to its original position.

Since the bottom two layers are disordered by a single monoswap, but restored by a second one, it's important not to move these layers (by turning the cube, say) between the two monoswaps of each pair.

F: FINISHING FLIPS AND FIDDLES

Since every cubelet should now be in its own cubicle, the only remaining problems can be solved by edge-flips and corner twists in the top layer. To tackle any particular top layer cubelet, turn **THE TOP LAYER** *ONLY* to bring that cubelet into one of the two shaded cubicles of Fig. 35F and then, according as its white face is in position

1,	2	or	3

do our

anticlockwise monotwist, $M_a = (F'RFR')^2$	**clockwise monotwist** $M_c = (RF'R'F)^2$	or	**edge monoflip** $M_e = (\varepsilon R)^4$

where ε is a slice move (Fig. 34).

Once again it's important not to move the bottom two layers by turning the cube between operations, since individual monotwists and monoflips affect these layers. However, the entire set of operations needed to correct the top layer will automatically correct the bottom two layers as well.

EXPLANATIONS

Stage E works because our monoswap operation M_S leaves the top layer unchanged except for the desired swap of the two near corner cubelets, while two copies of the monoswap cancel ($M_S^2 = 1$).

So a sequence such as

monoswap, turn top clockwise, monoswap, turn top back

doesn't really disturb the bottom two layers, which "feel" only the two cancelling monoswaps. The top layer, however, effectively undergoes a swap of the two near corners followed by a swap of two right corners, which are brought into position by the first top turn and returned by the second.

Stage F works similarly because M_a, M_c and M_c have exactly the desired effects on the top layer, and enjoy the properties

$$M_e^2 = M_c^3 = 1, \qquad M_c M_e = M_e M_c, \qquad M_a = M_c^{-1}.$$

So Anne Scott's laws ensure that the bottom two layers feel a cancelling combination of operations, while the top layer undergoes the desired flips and twists.

IMPROVEMENTS

Our method is easy to explain, perform and remember, but usually takes more moves than an expert would. If you're prepared to take more trouble and have a rather larger memory, you can often shorten it considerably. For instance, the original monoflips and monotwists (due to David Seal and David Goto) are shorter:

$$m_e = R\varepsilon R^2\varepsilon^2 R \qquad m_e^{-1} = R'\omega^2 R^2 \omega R' \qquad m_c = R'DRFD\bar{\partial}' \qquad m_a = m_c^{-1} = FD'F'R'D'R$$

but with these you must always be careful to follow a mono-operation by the corresponding inverse one.

Explore the effects of the following moves, which many people have found useful. The first few only affect the top layer. Here and elsewhere we've credited moves to those who first told them to us. We expect that many facts about the cube were found by clever Hungarians long before we learnt of them. For the Greek letter slice moves, see Fig. 33.

David Benson's "special" $RUR^2.FRF^2.UFU^2$
David Singmaster's "Sigma" $FURU'R'F'$
Margaret Bumby's top edge-tricycle $\beta U^{\pm 1}\alpha . U^2.\beta U^{\pm 1}\alpha$
Two more top edge-tricycles $U^2F.\alpha U\beta.U^2.\alpha U\beta.FL^2$; $\quad FUF'UFU^2F'U^2$
Top corner tricycle $RU'L'UR'U'LU$
Clive Bach's cross-swap $(\alpha^2 U^2 \alpha^2 U)^2$
Kati Fried's edge-tricycle $\beta F^2 \alpha F^2$
Tamas Varga's corner tricycle $((FR'F'R)^3 U^2)^2$
Two double edge-swaps $(R^2 U^2)^3$; $\quad (\alpha^2 U^2)^2$
Andrew Taylor's Stage C moves $F^2(RF)^2(R'F')^3$; $\quad (FR)^3(F'R')^2F^2$
Other Stage C moves $FUFUF.U'F'U'F'U'$; $\quad R'U'R'U'R'.URURU$

In the Extras you'll find lists of the shortest known words (improvements welcome!) to achieve any rearrangement, or any reorientation of the top layer. These are quoted from an algorithm due to Benson, Conway and Seal which guarantees to cure the cube in at most 85 moves (a half turn still counts as one move, but a slice counts as two). Morwen Thistlethwaite has recently constructed an impressive algorithm which never takes more than 52 moves.

Because there are 18 choices for the first move, but only 15 (non-cancelling) choices for subsequent ones, the number of positions after 16 moves is at most

$$18 \times 15^{15} = 7\,882\,090\,026\,855\,468\,750 < 43\,252\,003\,274\,489\,856\,000$$

proving that there are many positions that need 17 or more moves to cure. We can improve this to 18 moves by using the estimates $u_1 = 18, u_2 = 27 + 12u_1 = 243, u_{n+2} \leqslant 18u_n + 12u_{n+1}$, which take into account relations like $LR = RL$.

ELENA'S ELEMENTS

Elena Conway likes making her cube into pretty patterns. Here are some ways to do this, most of which she discovered herself.

"4 Windows"	"6 Windows"	"Chequers"	"Harlequin"
$\alpha\gamma^2\beta\delta^2$	$\alpha\gamma\beta\delta$	$\alpha^2\gamma^2\varepsilon^2$	$\alpha\gamma\beta\delta\alpha^2\gamma^2\varepsilon^2$

"Stripey"	"Zigzag"	"4 Crosses"	"6 Crosses"
$(L^2F^2R^2)^2 \cdot LR'$	$(LRFB)^3$	$(LRFB)^3(FBLR)^3$ or $(\gamma^2L'\gamma^2R)^3$	$(\gamma^2L'\gamma^2R)^3(\alpha^2B'\alpha^2F)^3$

And try following "6 Crosses" with any of the earlier ones.

ARE YOU PARTIAL TO PARTIAL PUZZLES?

It's interesting to see what you can do using only *some* of the available moves. You might restrict yourself to just a specified selection of faces, to half-turns, to slice moves, or to the **helislice moves** like LR. Mathematically these correspond to subgroups we call the 2-, 3-, 4- and 5-**face groups**, the **square group**, the **slice group** and the **helislice group**.

Beginners are recommended to stay in the slice group because they cannot get lost. From any position you can cure the edge-cubes in 3 slices, getting to "4 Windows" or "6 Windows" and so home in 4 more slices. Frank O'Hara has shown that in fact at most 5 slices are needed in all. The slice group has order $4^3.4!/2 = 768$ and the helislice group has order $2^{11}.3 = 6144$.

The 2-face group has been intensively studied by Morwen Thistlethwaite. It's interesting to notice that it involves both the lucky Seven Puzzle (on the edge cubelets that move) and Rick Wilson's Tricky Six Puzzle (on the corners).

Roger Penrose first proved that everything can be done using just 5 faces. David Benson has a simple proof:

$$RL'F^2B^2RL' \cdot U \cdot RL'F^2B^2RL' = D.$$

OTHER "HUNGARIAN" OBJECTS

A $2 \times 2 \times 2$ cube and $2 \times 3 \times 3$ "domino" have also been manufactured. Their design seems even more mysterious, although as puzzles they're much easier. One can *imagine* Hungarian tetrahedra, octahedra, dodecahedra, icosahedra, etc. Although, as far as we know, these have neither been manufactured nor completely solved, Andrew Taylor has found a neat proof that (for any choice of chief faces and colors)

the total permutation on edges and corners is even,
the number of edge-flips is even, and
the total corner twisting is zero, modulo the corner valence.

Despondent Domino dabblers should need but three little words (with effects):

X = EhEhEh	Y = EcEhNcE	Z = cYcYc
(28)	(13)(26)(1'3')(2'6')	(13)(26)

(c is a clockwise $\frac{1}{4}$-turn of the top; h,E,N $\frac{1}{2}$-turns of top, East, North).

A TRIO OF SLIDING BLOCK PUZZLES

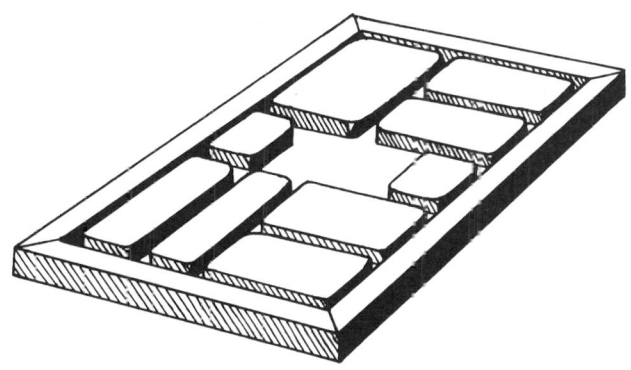

Figure 36. Dad's Puzzler.

Dad's Puzzler (Fig. 36) is unfortunately almost the only sliding block puzzle that's generally available from toy stores. although it goes under many different names. The problem is to slide the pieces without lifting any out of the tray, until the 2×2 square arrives in the lower left hand corner. Fifty years ago the puzzle represented Dad's furniture-removing difficulties, and the 2×2 block was the piano; at other times it has been depicted as a pennant, a car, a mountain, or space capsule but the puzzle has remained unchanged, probably for a hundred years. Some more enterprising manufacturer should sell a set containing one 2×2, four 1×1 and six 2×1 pieces which can be used either for Dad's Puzzler or for the following more interesting puzzles.

In the **Donkey** puzzle the initial arrangement is as in Fig. 37(a) and the problem is to move the 2×2 square to the middle of the bottom row. The name arises from the picture of a red donkey which adorned the 2×2 square in the original French version (L'Âne Rouge, which probably goes back to the last century) but we think that our choice of starting position already looks quite like a donkey's face.

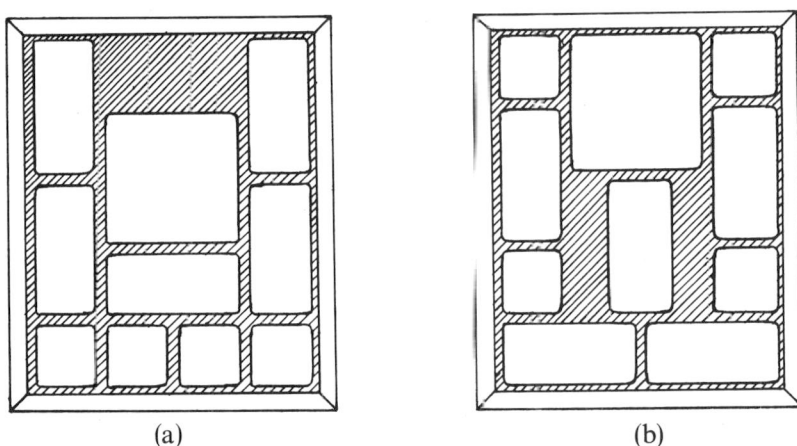

(a)   (b)

Figure 37. The Donkey and The Century (and a Half).

The **Century Puzzle**, published for the first time in *Winning Ways*, was discovered by one of us several years ago as a result of a systematic search for the hardest puzzle of this size. Start from Fig. 37(b) and, as in the Donkey, end with the 2×2 block in the middle of the bottom row. Or, if you're a real expert, you might try the **Century-and-a-Half Puzzle** in which you're to end in the position got by turning Fig. 37(b) upside-down.

TACTICS FOR SOLVING SUCH PUZZLES

As in our previous sliding puzzles the basic idea is to see what can be done while quite a lot of the pieces are kept fixed. In all three of these examples one occasionally sees one of the configurations of Fig. 38 somewhere, and any of these can be exchanged for any other, moving only the

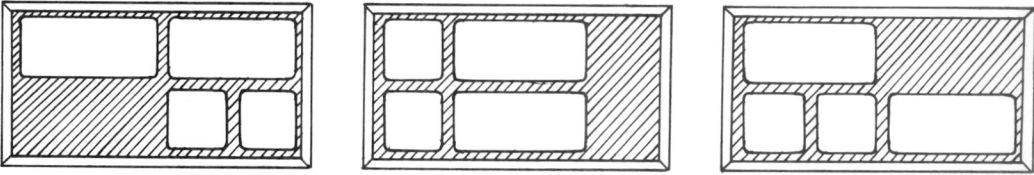

Figure 38. A Micropuzzle.

pieces in the area shown. They form a kind of micro-puzzle within the larger one. Figure 39 is a complete "map" of Dad's Puzzler showing how it consists of a dozen of these micro-puzzles joined by various paths of moves that are more or less forced. Using this map, you'll find it easy to get from anywhere to anywhere else.

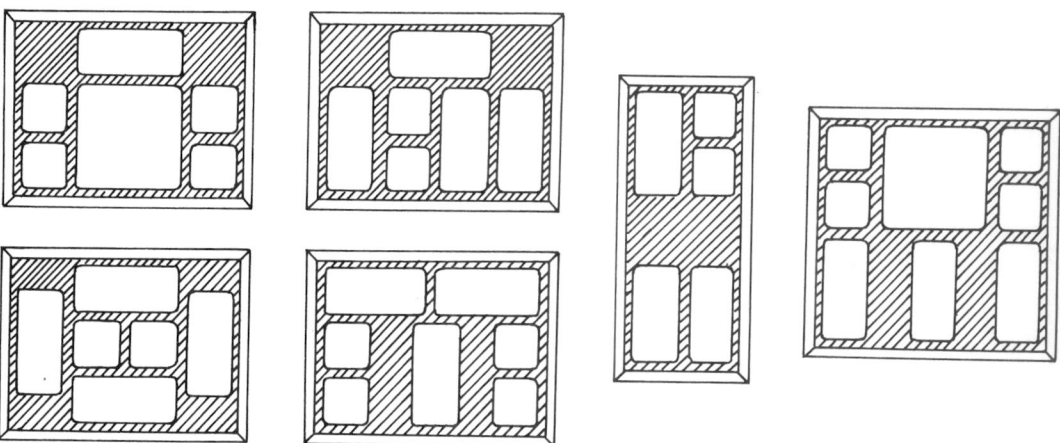

Figure 40. Micro- and Mini-puzzles Found in Donkey and Century.

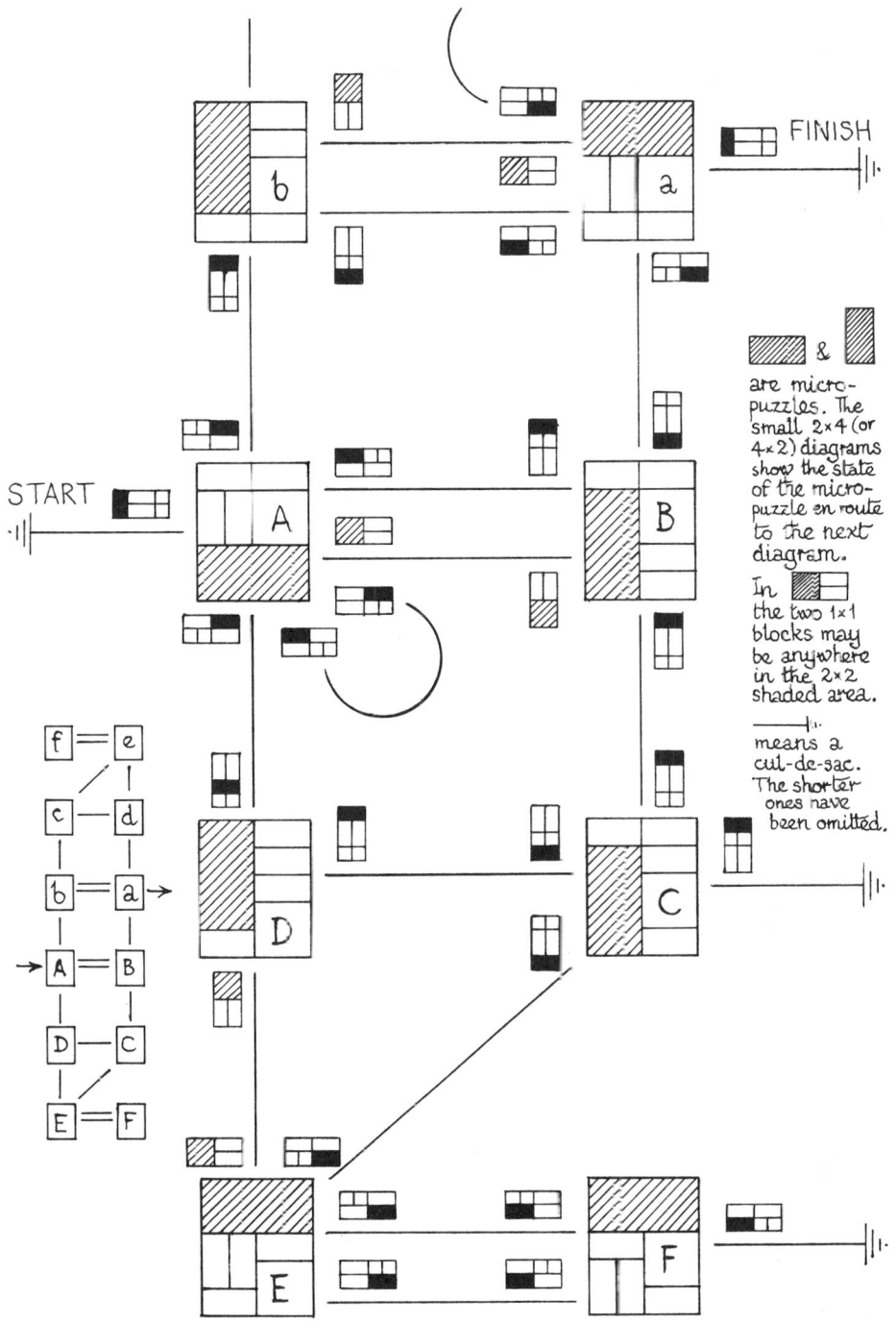

Figure 39. Map of Dad's Puzzler.

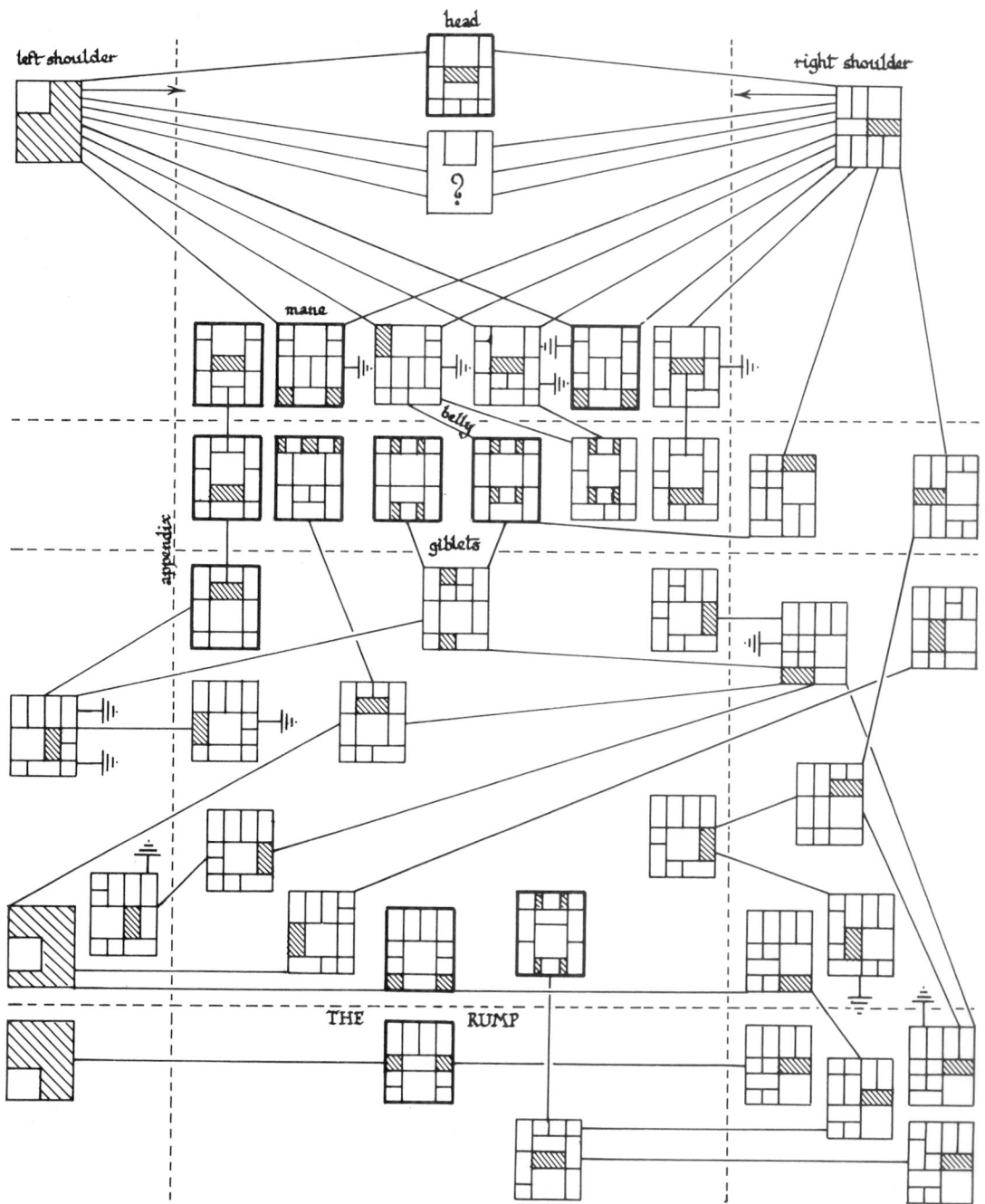

Figure 41. Map of the Donkey.

In the Donkey and Century puzzles there are several micro- and mini-puzzles: see what moves you can make inside the regions shown in Fig. 40. The Century and Donkey puzzles will never become easy but it will help if you become an adept at these minipuzzles. Figure 41 is our map of the Donkey. The positions are classified according to the location of the 2×2 square and in most cases we have only drawn one of a left-right mirror-image pair. Some unimportant culs-de-sac will be found in the directions indicated by the signs ⊣‖▮ and the rectangle containing (?) represents many positions connected to the left and right shoulders. The arrows indicate other connexions to the shoulders. Left-right symmetric positions are boldly bordered.

The Century puzzle is very much larger, and we need more abbreviations to draw its map within a reasonable compass. The positions are best classified by the position of the large square together with information about which of the two horizontal pieces should be counted as "above" or "below" the square. We remark that in Fig. 42 both horizontal pieces should be counted as *below* the square despite their appearance, because the only way to move these pieces takes the horizontals *down* and the square *up*.

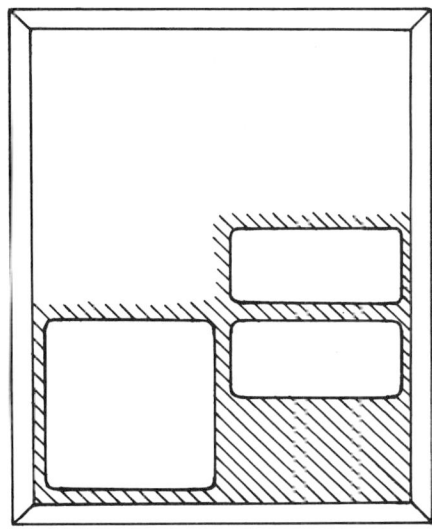

Figure 42. The Two Horizontal Pieces are Below the Square!

The key to the puzzle is to find one of the two possible **narrow bridges** in the map at which the first horizontal piece changes from *below* the square to *above* it. In fact it's best to think out the possible configurations in which this can happen and then work the puzzle backwards and forwards from one of these. Very few people have ever solved the puzzle by starting at the initial configuration and moving steadily towards its end. A much abbreviated map appears as Fig. 43.

Our maps were prepared with much help from some computer calculations made by David Fremlin at the University of Essex, who found incidentally that the Donkey pieces may be placed in the tray in 65880 positions and the Century pieces in 109260 ways. Although the Century puzzle can be inverted (this is our Century-and-a-Half problem) Fremlin's computer found that the Donkey cannot. It would be nice to have a more perspicuous proof of this.

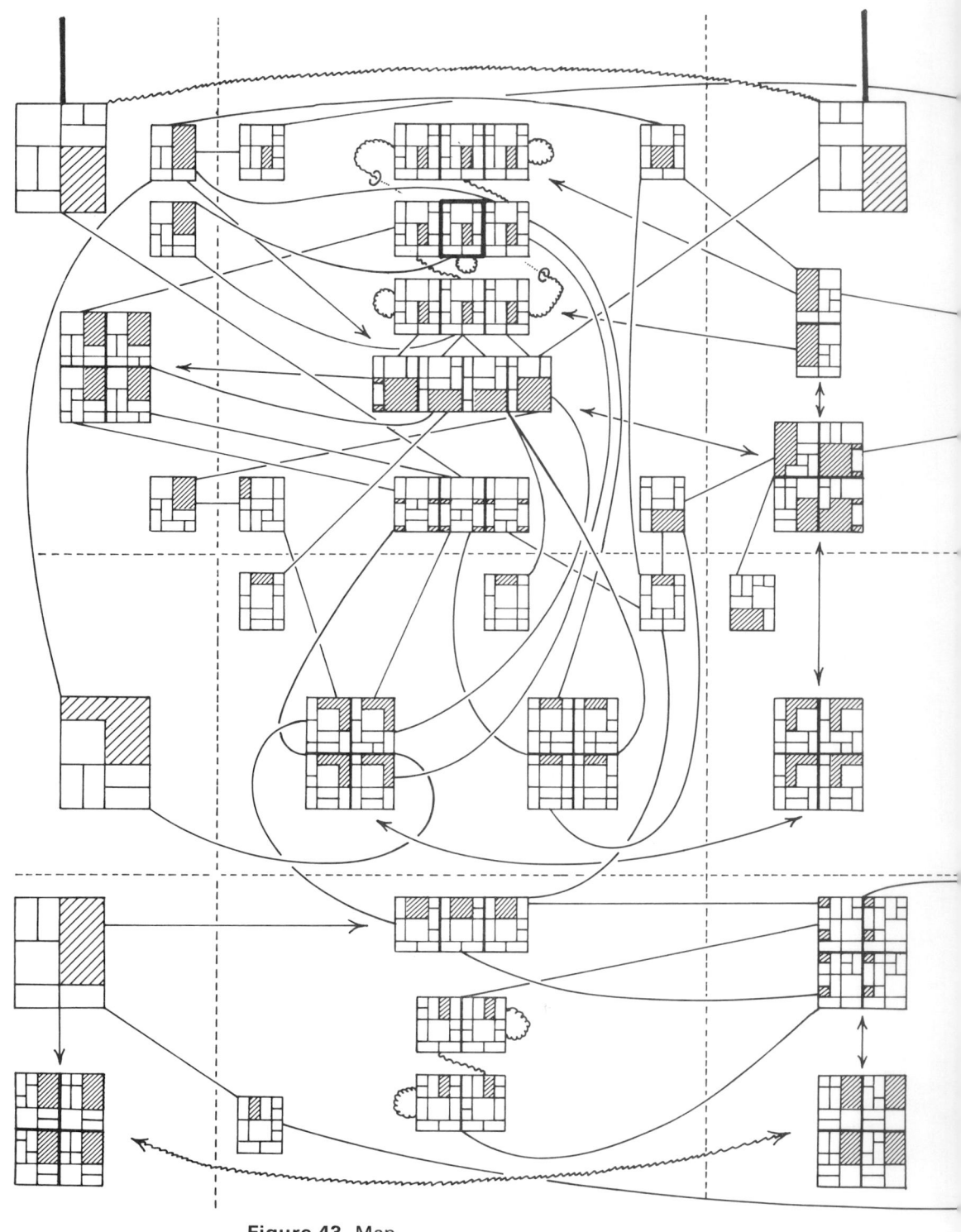

Figure 43. Map

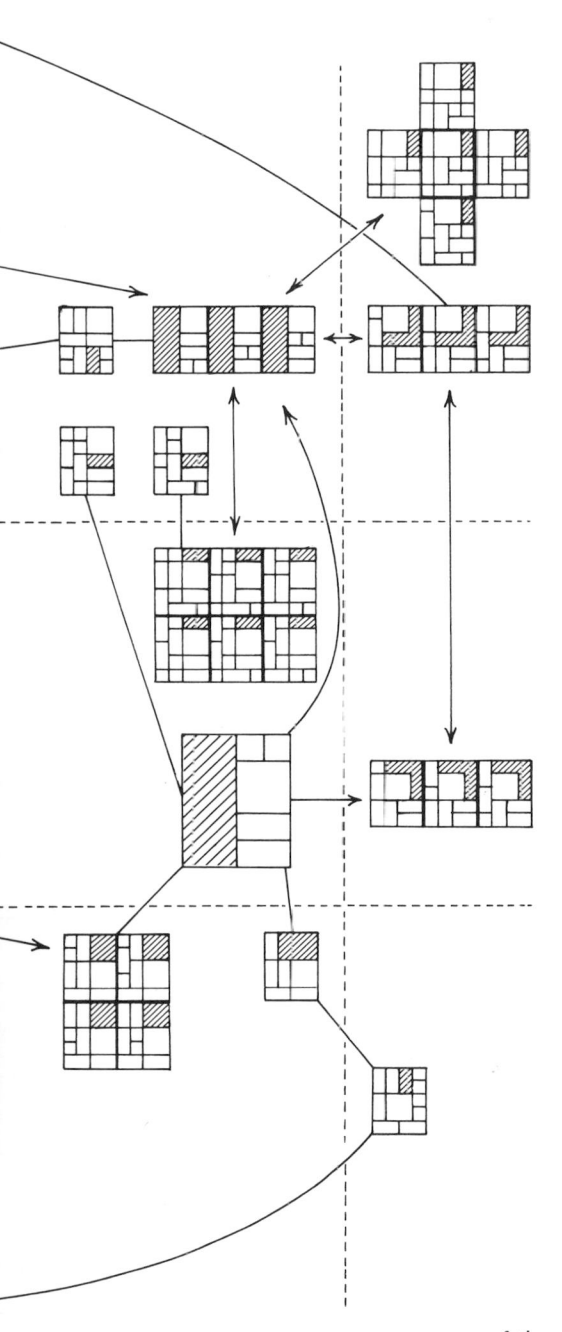

The starting position is heavily outlined; see centre column of opposite page, near top

Positions are classified thus:

overleaf: Square "between" horizontals.
opposite: Two verticals left, one right.
this page: Three verticals left.

They are further classified according to the location of the square:

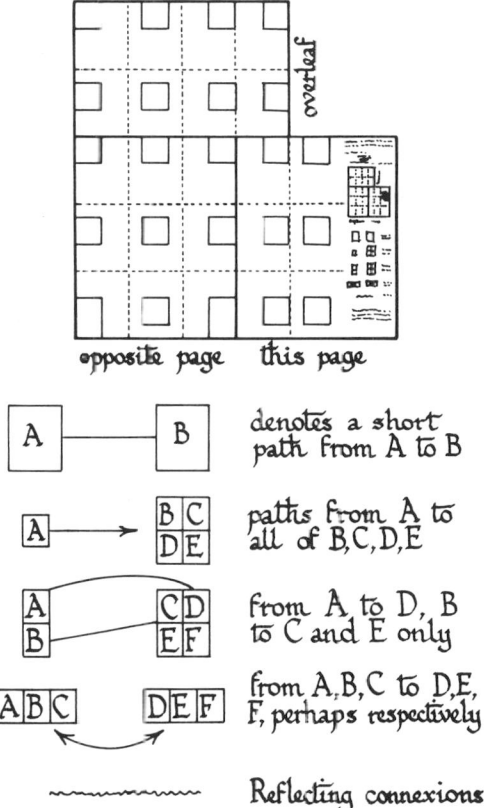

opposite page   this page

A ——— B   denotes a short path from A to B

A ——→ B C / D E   paths from A to all of B, C, D, E

A / B ——→ C D / E F   from A to D, B to C and E only

A B C ⟷ D E F   from A, B, C to D, E, F, perhaps respectively

～～～～ Reflecting connexions

The "narrow bridges" are the two thick connexions between the top of the opposite page and the left of overleaf.

. . . . of the . . . .

Hint: You're still near the start
if the puzzle looks like:

"Freedom Square"

with one vertical removed, two fused to make a big square, and
two subdivided into four little squares.

. . .   .   Century.

COUNTING YOUR MOVES

It's customary to follow Martin Gardner and declare that any kind of motion involving just one piece counts as a single move. It takes 58 moves to solve Dad's Puzzler and 83 to solve the Donkey. How many do you need to solve the Century puzzle? And how many for the Century-and-a-Half?

PARADOXICAL PENNIES

You tell me your favorite sequence of three Heads or Tails and then I'll tell you mine. We then spin a penny until the first time either of our sequences appears as the result of three consecutive throws. I bet you 2 to 1 it's mine!

The graph

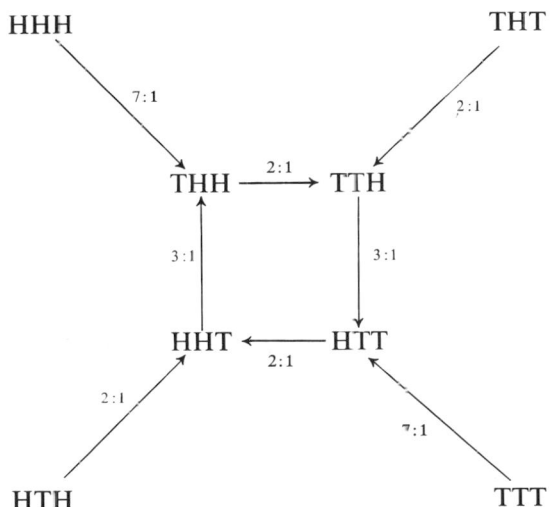

shows the sequence I'll choose for each possible sequence of yours, together with the odds that I win. You'll see that it's always at least 2 to 1 in my favor.

Here's a rule for computing the odds. Given two Head-Tail sequences a and b of the same length, n, we compute the **leading number**, aLb, by scoring 2^{k-1} for every positive k for which the last k letters of a coincide with the first k of b. Then we can show that the odds, that b beats a in Paradoxical Pennies, are exactly

$$aLa - aLb \quad \text{to} \quad bLb - bLa.$$

Leo Guibas and Andy Odlyzko have proved that, given a, the best choice for b is one of the two sequences obtained by dropping the last digit of a and prefixing a new first digit. Notice the paradoxical fact that in the length 3 game:

THH beats HHT beats HTT beats TTH beats THH.

PARADOXICAL DICE

You can make three dice, A, B, C, with a similar paradoxical property, using the magic square:

	D	E	F
A	6	1	8
B	7	5	3
C	2	9	4

Each die has the numbers of one row of the square on its faces (opposite faces bearing the same number). For these dice

$$A \text{ beats } B \text{ beats } C \text{ beats } A,$$

all by 5 to 4 odds! Similarly for the three dice, D, E, F, obtained from the columns. The only other paradoxical triples of dice using the same numbers are those obtained from A, B, C by interchanging 3 with 4 and/or 6 with 7. These interchanges improve the odds.

It's possible to put positive integers on the faces of two dice in a unique non-standard way that gives the same probability for each total as the standard one. Algebraically, the problem reduces to factorizing

$$x^2 + 2x^3 + 3x^4 + 4x^5 + 5x^6 + 6x^7 + 5x^8 + 4x^9 + 3x^{10} + 2x^{11} + x^{12}$$

into the form $f(x)g(x)$ with $f(0)=g(0)=0$ and $f(1)=g(1)=6$. The two factorizations are

$$(x + x^2 + x^3 + x^4 + x^5 + x^6)^2 \qquad \text{and}$$

$$(x + 2x^2 + 2x^3 + x^4)(x + x^3 + x^4 + x^5 + x^6 + x^8),$$

so the new pair of dice have the numbers

$$1, 2, 2, 3, 3, 4 \qquad \text{and} \qquad 1, 3, 4, 5, 6, 8.$$

MORE ON MAGIC SQUARES

It's an old puzzle to arrange the numbers from 1 to n^2 in an array so that all the rows and columns and both the diagonals have the same sum, which turns out to be $\frac{1}{2}n(n^2+1)$. The only 3×3 magic square (see the last section), often called the Lo-Shu, was discovered several dynasties ago by the Chinese. We also used it in Chapter 22. In 1693 Frenicle de Bessy had worked out the 880 magic squares of order 4. In this section we'll show you how to find all these.

It's handy to subtract 1 from all the numbers, because the numbers 0 to 15 are closed under nim-addition. With this convention the magic sum is 30. We shall call a square **perfect** if we can nim-add *any* number from 0 to 15 to its entries and still obtain a magic square; if only $\frac{1}{2}$ of these additions are possible we'll call it $\frac{1}{2}$-**perfect**, and so on. Since nim-adding 15 is the same as complementing in 15, it always preserves the magic property, showing that *every* square is at least $\frac{1}{8}$-perfect. We shall also classify the squares by the disposition of complementary pairs as in Fig. 44.

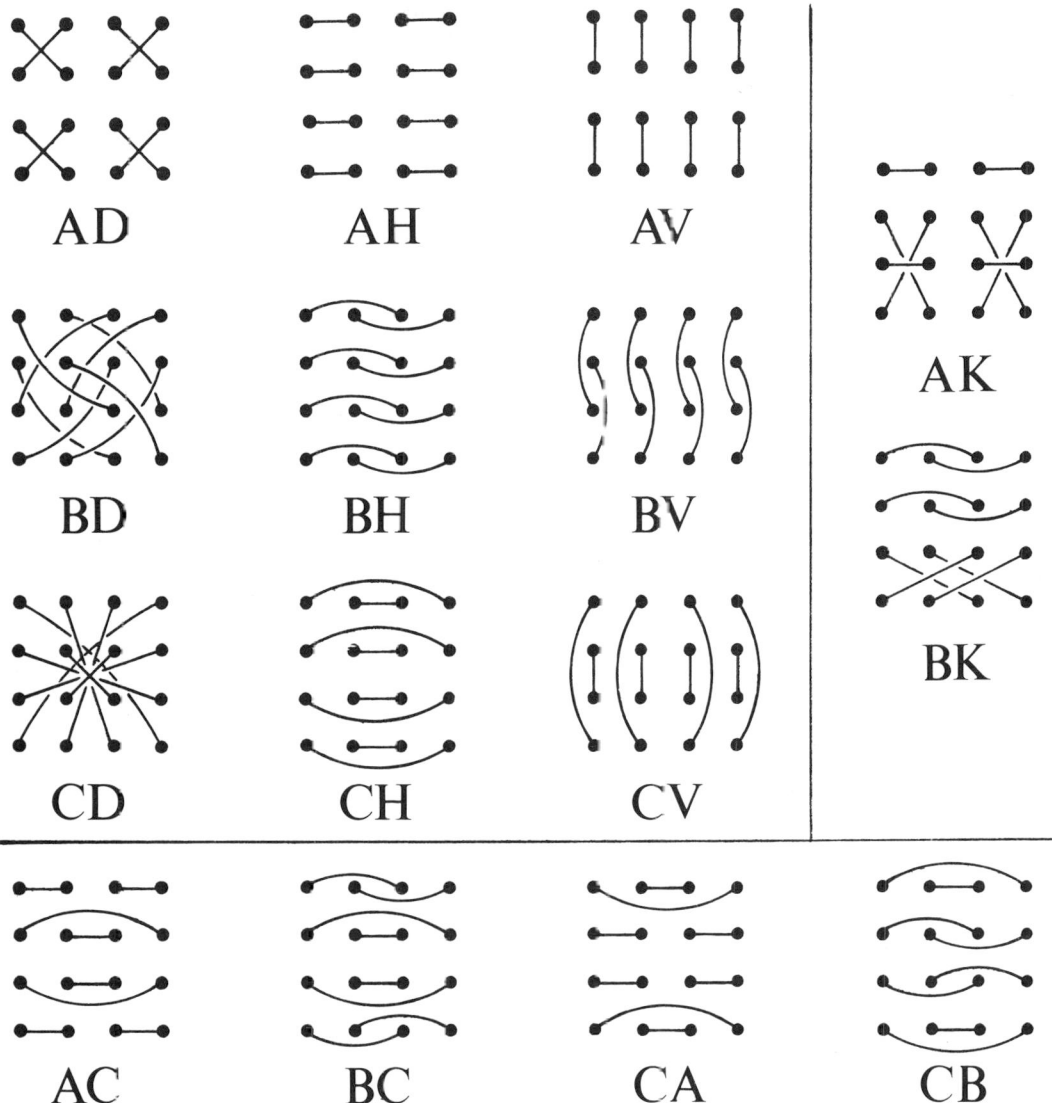

Figure 44. Classifying Squares by Complementing Pairs.

There are essentially just three ways to write the numbers from 0 to 15 as an addition table:

0	1	2	3
4	5	6	7
8	9	10	11
12	13	14	15

0	1	4	5
2	3	6	7
8	9	12	13
10	11	14	15

0	2	4	6
1	3	5	7
8	10	12	14
9	11	13	15

but you can then freely permute the rows and columns in any of these. Take any table obtained in this way, say

$$
\begin{array}{cccc}
15 & 11 & 14 & 10 \\
13 & 9 & 12 & 8 \\
7 & 3 & 6 & 2 \\
5 & 1 & 4 & 0
\end{array}
$$

Apply the interchanges indicated by our **Quaquaversal Quadrimagifier**

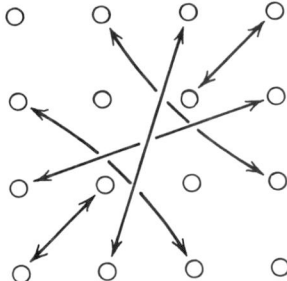

and you get a magic square:

$$
\begin{array}{cccc}
15 & 2 & 1 & 12 \\
4 & 9 & 10 & 7 \\
8 & 5 & 6 & 11 \\
3 & 14 & 13 & 0
\end{array}
\qquad
\begin{array}{cccc}
16 & 3 & 2 & 13 \\
5 & 10 & 11 & 8 \\
9 & 6 & 7 & 12 \\
4 & \boxed{15 \quad 14} & & 1
\end{array}
$$

Adding 1 to this particular example we obtain the right hand square which features in Albrecht Dürer's famous self-portrait, *Melencolia I*, in which the boxed figures indicate the date of the work. In this case complementary numbers appear according to the scheme called *central* in Fig. 44, and so this square is called central.

By applying the Quaquaversal Quadrimagifier to the other forms of addition table we can get 432 essentially different perfect magic squares. The complementary pairs enable us to classify these as:

48 Adjacent Diagonal (AD),
48 Broken Diagonal (BD),
48 Central Diagonal (CD),
96 Adjacent Horizontal (AH) or Adjacent Vertical (AV),
96 Broken Horizontal (BH) or Broken Vertical (BV), and
96 Central Horizontal (CH) or Central Vertical (CV).

Because we don't count squares different when they are related merely by a reflexion or a rotation of the diagram, we must regard adjacent-horizontal and adjacent-vertical squares as the same type. You can find out what type your square will be by looking at the position occupied by the complement of the addition table's leading entry before Quadrimagification:

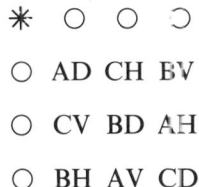

```
 *   ○   ○   ⊃
 ○  AD  CH  BV
 ○  CV  BD  AH
 ○  BH  AV  CD
```

Now take the above 96 central-horizontal squares and apply the flip operation

```
 ○   ○   ○   ○
 ○   ○   ○   ○
 ↕               ↕
 ○   ○   ○   ○
 ○   ○   ○   ○
```

and you'll get 96 more central-horizontal squares. All squares so far found are perfect.

There are 112 more central-horizontal squares that are only $\tfrac{1}{4}$- or $\tfrac{1}{8}$-perfect. They can be found by taking any of the seven squares:

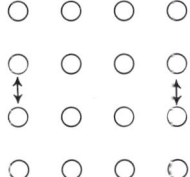

6	10	5	9
13	12	3	2
0	7	8	15
11	1	14	4

$a_8 \leftrightarrow$

14	2	13	1
5	4	11	10
8	15	0	7
3	9	6	12

0	13	2	15
11	8	7	4
14	3	12	1
5	6	9	10

$c \nearrow$

10	1	14	5
13	8	7	2
4	15	0	11
3	6	9	12

$\downarrow d$ $\downarrow d$ $\downarrow c$

$c \searrow$

12	5	10	3
11	9	6	4
0	14	1	15
7	2	13	8

$a_1 \leftrightarrow$

13	4	11	2
10	8	7	5
1	15	0	14
6	3	12	9

12	11	4	3
1	8	7	14
2	5	10	13
15	6	9	0

$\tfrac{1}{4}$-perfect $\tfrac{1}{8}$-perfect

and applying any combination of the four operations:

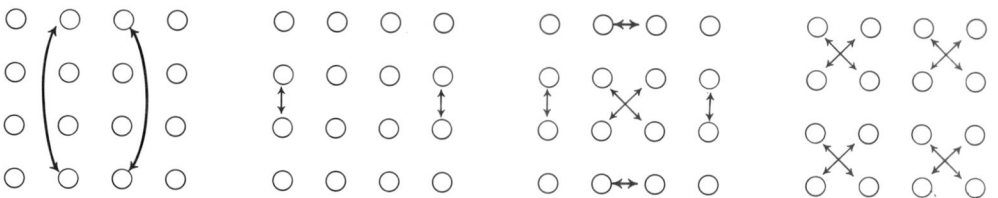

Now take the 14 squares of Fig. 45 and apply any combination of complementation and the last three of our operations and you'll get a total of 224 squares, 56 of each of the types

> Adjacent Central (AC),
> Broken Central (BC),
> Central Adjacent (CA), and
> Central Broken (CB).

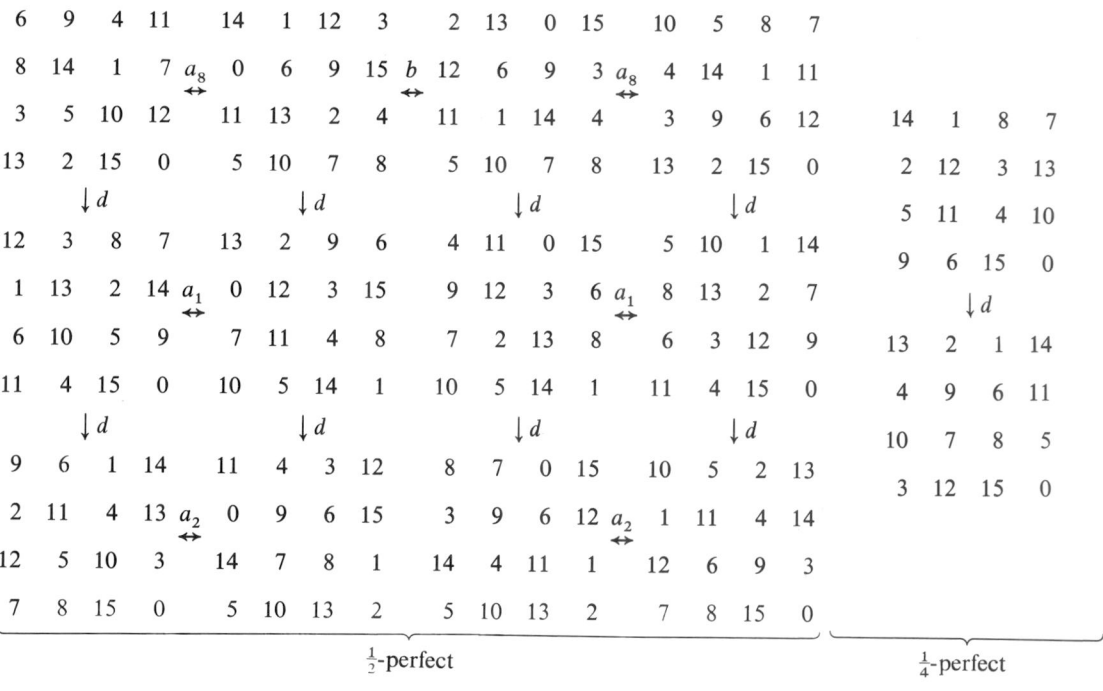

```
 6  9  4 11     14  1 12  3      2 13  0 15     10  5  8  7
 8 14  1  7 a₈   0  6  9 15 b    12  6  9  3 a₈   4 14  1 11
 3  5 10 12     11 13  2  4     11  1 14  4      3  9  6 12          14  1  8  7
13  2 15  0      5 10  7  8      5 10  7  8     13  2 15  0           2 12  3 13
   ↓d              ↓d              ↓d              ↓d                 5 11  4 10
12  3  8  7     13  2  9  6      4 11  0 15      5 10  1 14           9  6 15  0
 1 13  2 14 a₁   0 12  3 15      9 12  3  6 a₁   8 13  2  7              ↓d
 6 10  5  9      7 11  4  8      7  2 13  8      6  3 12  9          13  2  1 14
11  4 15  0     10  5 14  1     10  5 14  1     11  4 15  0           4  9  6 11
   ↓d              ↓d              ↓d              ↓d                10  7  8  5
 9  6  1 14     11  4  3 12      8  7  0 15     10  5  2 13           3 12 15  0
 2 11  4 13 a₂   0  9  6 15      3  9  6 12 a₂   1 11  4 14
12  5 10  3     14  7  8  1     14  4 11  1     12  6  9  3
 7  8 15  0      5 10 13  2      5 10 13  2      7  8 15  0
```

$\frac{1}{2}$-perfect $\frac{1}{4}$-perfect

Figure 45. Adjacent and Broken Central and Central Adjacent and Broken Squares.

There remain only 16, rather irregular, squares to be found. You can get them by applying any combination of complementation and the last *two* of our operations to the two $\frac{1}{2}$-perfect squares

$$
\begin{array}{cccc}
1 & 14 & 9 & 6 \\
10 & 3 & 4 & 13 \\
7 & 8 & 15 & 0 \\
12 & 5 & 2 & 11
\end{array}
\qquad \xrightarrow{\;d\;} \qquad
\begin{array}{cccc}
2 & 13 & 3 & 12 \\
5 & 6 & 8 & 11 \\
14 & 1 & 15 & 0 \\
9 & 10 & 4 & 7
\end{array}
$$

and they're 8 each of the types

Adjacent Knighted (AK),
Broken Knighted (BK).

There are various permutations of the 16 numbers that occasionally lead from one magic square to another, namely

a_n: nim-*add n*, for example $a_6 = (0\ 6)(1\ 7)(2\ 4)(3\ 5)(8\ 14)(9\ 15)(10\ 12)(11\ 13)$
b: the *big* swap $(0\ 12)(1\ 13)(14\ 2)(15\ 3)$
c: *circle* $(0\ 10\ 12)(1\ 11\ 13)(14\ 4\ 2)(15\ 5\ 3)$
d: *double*, mod 15 $(1\ 2\ 4\ 8)(3\ 6\ 12\ 9)(5\ 10)(7\ 14\ 13\ 11)$

and we've indicated some of these in the figures.

THE MAGIC TESSERACT

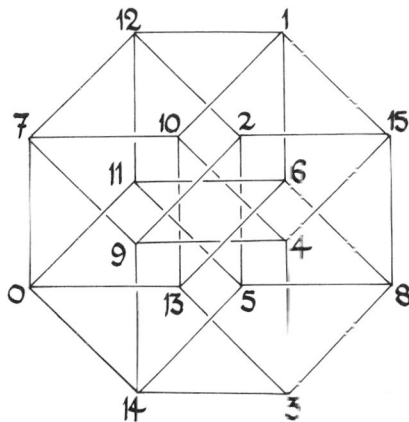

We'll leave it to you to rediscover the many remarkable relations between the 48 BD squares, sometimes called **pandiagonal** or **Nasik** squares, and our **Magic Tesseract** in which the vertices of every square add to 30. By projecting this along three different directions, you can find three magic cubes in which each face adds to 14. These are the duals of the three octahedral dice found by Andreas and Coxeter. Alternate vertices in the magic tesseract are the odious and evil numbers, and if you replace each odious number by its opposite (nim-sum with 15) you'll see how the tesseract was made.

ADAMS'S AMAZING MAGIC HEXAGON

Starting from the pattern

```
              1
        2           3
    4           5           6
        7           8
    9           10          11
        12          13
    14          15          16
        17          18
              19
```

can you reorder the numbers from 1 to 19, taking less than 47 years, so that all five rows in each of the three directions have the same sum?

THE GREAT TANTALIZER

This is a tantalizing puzzle which surfaces every now and then with a new alias. We've chosen one of the older names. An early American version was the Katzenjammer puzzle, but most recently it has emerged under yet another name, Instant Insanity. The manufacturers seem to be very good at selecting new names, but they never change the underlying puzzle. The problem

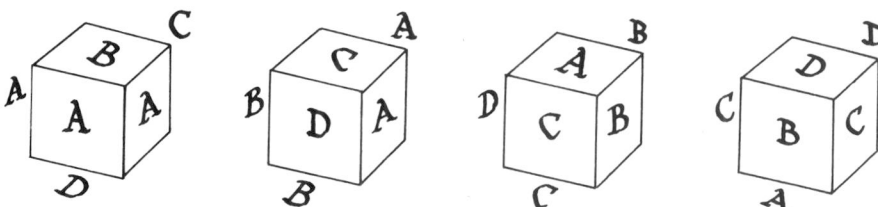

Figure 46. Pieces for The Great Tantalizer.

is to assemble the four cubes of Fig. 46 (in which the outer letters refer to the hidden faces) into a vertical $1 \times 1 \times 4$ tower in which each wall displays all four "colors", A, B, C, D. If you don't go instantly insane on playing with the cubes, you'll probably be greatly tantalized by them.

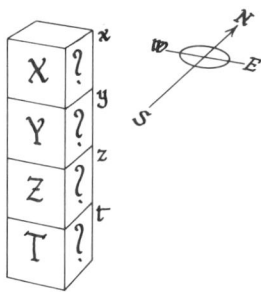

Figure 47. The Tantalizer Solved?

T.H. O'Beirne seems to have been the first to publish a general way of solving such problems and we think his solution is still the best. Let's imagine the problem solved and concentrate on the North and South walls of the tower (Fig. 47). Then X, Y, Z, T will be A, B, C, D in some order, as will x, y, z, t. Write the four letters A, B, C, D on a piece of paper and join

$$\text{X to x,} \quad \text{Y to y,} \quad \text{Z to z} \quad \text{and} \quad \text{T to t.}$$

What you'll get will probably be a way of joining ABCD into a circuit, but it might perhaps be several circuits which together include each letter just once. For example if

$$\text{X Y Z T x y z t}$$

are

$$\text{A B C D D C A B}$$

we get the single circuit

while if they were

$$\text{A B C D D A C B}$$

you'd get two circuits of different lengths

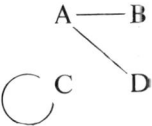

There will be a similar circuit, or system of circuits, for the East–West walls. Each of the two systems will contain every vertex just once and have one edge for each cube.

It's now easy to solve the puzzle by drawing the following graph (Fig. 48). The vertices of the graph are the colors A, B, C, D and the ith cube yields three edges labelled i joining pairs of vertices corresponding to its pairs of opposite faces. All you have to do is to select from this graph the two separate systems of circuits which each use all four numbers and all four vertices just once.

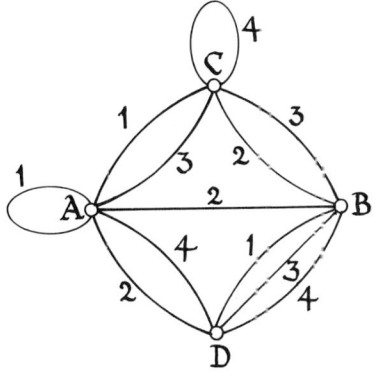

Figure 48. Solving the Tantalizer.

What are the possibilities for such circuit systems in the example? By considering each possibility

$$1\ 1\ 1\ 1, \quad 2\ 1\ 1, \quad 2\ 2, \quad 3\ 1, \quad 4$$

for the circuit lengths, you'll rapidly conclude that both systems must consist of a single 4-circuit which can only use the letters in the cyclic order ACBD. There is only one way of selecting two such systems without using any edge twice:

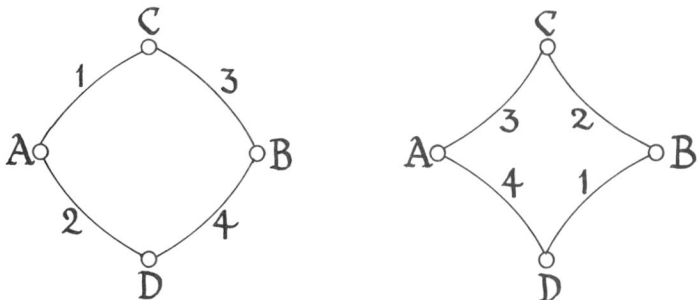

So the Great Tantalizer has a unique solution (up to reordering the cubes and rotating or inverting the whole tower). You can get it by pushing the cubes of Fig. 46 together left to right and tipping the result on end.

O'Beirne takes as his basic example a five cube puzzle of this type which dates from the first World War (Fig. 49) and uses the flags of the allies Belgium, France, Japan, Russia and the United Kingdom. You might like to check his assertion that this has just two essentially different solutions.

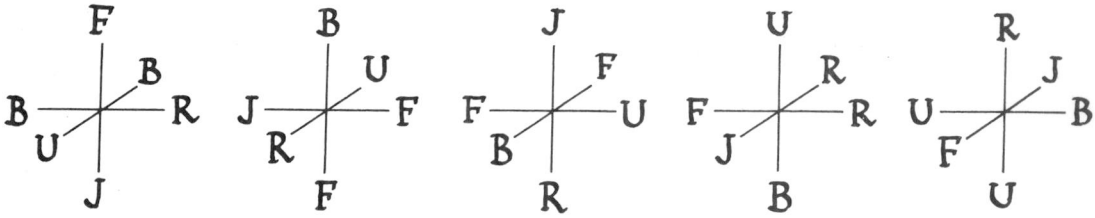

Figure 49. The "Flags of the Allies" Puzzle.

POLYOMINOES, POLYIAMONDS AND SEARCHING POLICY

A domino is made of two squares stuck together, so S.W. Golomb has suggested the words tromino, tetromino, etc. for the figures that can be made by sticking 3, 4, or more equal squares together. He has registered the particular names pentomino (5 squares) and polyomino (*n* squares) as trade-marks. Unfortunately few of the puzzles that have been proposed have hidden secrets, so they yield to nothing better than trial and error (or systematic search). As Rouse Ball says about Tangrams in early editions of *Mathematical Recreations and Essays*, "the recreation is not mathematical and I reluctantly content myself with a bare mention of it".

Here is the type of puzzle that arises. Up to rotations and reflexions there are just 12 pentominoes, for which you'll find our naming system in Chapter 25, with a total area of 60 square units. Which of the candidate rectangles

$$3 \times 20 \quad 4 \times 15 \quad 5 \times 12 \quad 6 \times 10$$

can be packed with them? Figure 50 shows a way of solving two of these problems at once, and also, if the pieces are regarded as made of five cubes each, of packing the $2 \times 5 \times 6$ box (they will also pack a $3 \times 4 \times 5$ box). Such problems are peculiarly susceptible to idle computers and the 6×10 pentomino rectangle was one of the first to be tackled in this way when C.B. Haselgrove found its 2339 solutions in 1960.

 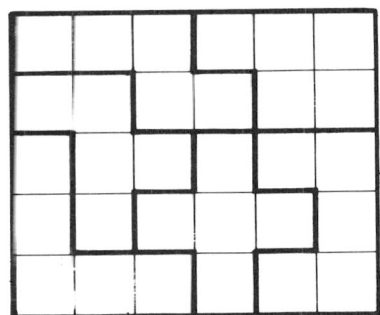

Figure 50. Packing Pentominoes.

Noting that two equilateral triangles can form a diamond, T.H. O'Beirne has proposed the terms triamond, etc., for figures made from three of more. Counting reflexions as distinct this time we find there are 19 hexiamonds, named in Fig. 51, which will pack into the shape of Fig. 52 in many thousands of different ways. The packing shown in the figure is probably the most symmetric about the North-South line. We'd like to see a similarly symmetric one for the East–West line.

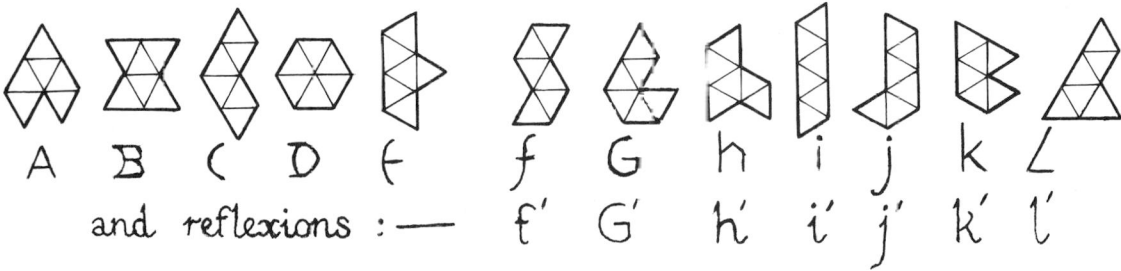

Figure 51. The Nineteen Hexiamonds.

This prompts a few remarks about sensible search procedures when solving puzzles or finding strategies for games that may be too large for complete discussion. Even when you have a large computer it's wise to have some idea where to look. Symmetry is usually a valuable consideration. For instance the (nearly) left-right symmetric solutions of the hexiamond puzzle admittedly form

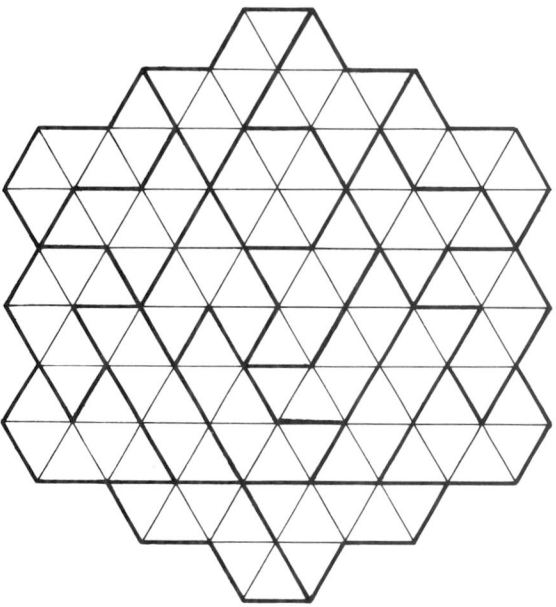

Figure 52. The Most Symmetric Hexiamond Solution?

only a small corner of the space to be searched, but this one is likely to be a profitable one because the constraints on opposite sides of the board are satisfied simultaneously. However, symmetry is not the only consideration. In analyzing a game it's wise to try to find out what the players are really fighting for (the game's hidden secrets). For example the French Military Hunt game on the Small Board is small enough that you can give an exhaustive analysis without needing to understand what's really going on. But when you've discovered that the players are really just fighting over the opposition you can extend the analysis to much larger boards for which a complete analysis would be prohibitive, even by computer.

Many of the analyses in *Winning Ways* were found in this way. Only when we realized that Dots and Boxes was really more concerned with parity than with box counting were we able to make any headway. And it's impossibly complicated to evaluate a reasonably sized position in Hackenbush Hotchpotch exactly, but we got a head start when we realized that often the atomic weight was the only thing that really mattered. In Peg Solitaire the hidden secret turned out to be the notion of balance represented by α and β in Chapter 23.

Even though polyomino type problems may have no hidden secrets, some people are much better at them than others because they subconsciously search in more likely places. Experienced polyominists don't undo their good work by repeatedly starting from scratch but keep most of the puzzle in place while fiddling with just a few pieces at any time. When they've found one solution, they can usually transform it into others by similar manipulation. For example, from Fig. 50 you can obtain another solution by repacking pentominoes R and S, and in Fig. 52 we can interchange the two (f, h) pairs or rotate the central (A,D,E, j, j') hexagon.

Exercise for Experts: For what values of n can you pack n^2 copies of hexiamond A into a replica of A on n times the scale?

ALAN SCHOEN'S CYCLOTOME

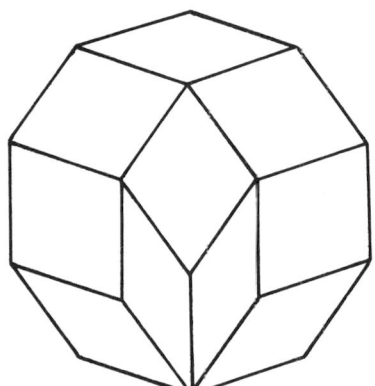 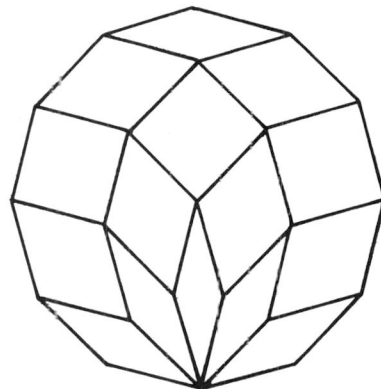

Figure 53. Dissections of $2n$-gons into Rhombs, $n = 5$ and 6.

Alan Schoen is patenting the interesting sequence of puzzles he derived from the well-known dissections of $2n$-gons into $\binom{n}{2}$ rhombs of angles $\pi k/n$ $1 \leqslant k \leqslant n-1$ (Fig. 53). He takes one of each of the $\lfloor n/2 \rfloor$ shapes of rhomb and one of each of the shapes you can make by joining two rhombs in every possible way to form a hexagon. The hexagon must not contain a straight angle, since he observes that no packing of rhombs in the $2n$-gon contains a pair of parallel edges, except those which form the rungs of the "ladders" which run between each pair of opposite sides in every packing. This non-convexity condition is similar to that imposed by Piet Hein in designing the Soma pieces, but here it arises naturally. Reflexions are not counted as different. This set of rhombs and hexagons (cyclotominoes?) will pack into the original $2n$-gon. In fact for

$$n = \quad 2 \quad 3 \quad 4 \quad 5 \quad \text{and} \quad 6$$
$$\text{there are} \quad 1 \quad 1 \quad 3 \quad 14 \quad \text{and} \quad \text{more than } 150$$

essentially different packings. Schoen gave one of us a set of pieces for $n=8$ and we were able to assemble them as in Fig. 54. We've numbered the pieces with the values of k, where $\pi k/n$ is the smaller angle of the rhomb. Where two shapes of piece are made from the same pair of rhombs, the one with the straighter reflex angle has its digits in natural order.

Solutions can be obtained from one another much as in O'Beirne's Hexiamond, or as on our Somap. In Fig. 54 the pieces 4 and 22 may be rotated or exchanged with 2 and 24, which in turn can be rotated or reflected. After this exchange, with 2 touching 11 and 34, we have a rotatable decagon, 1,2,34,11,32,3 & 4 of which the last four pieces form a rotatable octagon. As 3 & 4 are contiguous, they will exchange with 34, after which 4 & 1 and 2 & 3 are contiguous, and will swap with 14 and 23. After the original exchange, 2 may instead have two sides in common with 13 and these two will rotate, after which 21 and 12 may be interchanged if 1 & 2 are moved as well. Or again, 2 may touch 23 & 24, so that after the 34 exchange, 2 & 3 will swap with 32, and then 2 & 11 form a symmetric hexagon. And so on and on, yielding well over a hundred solutions.

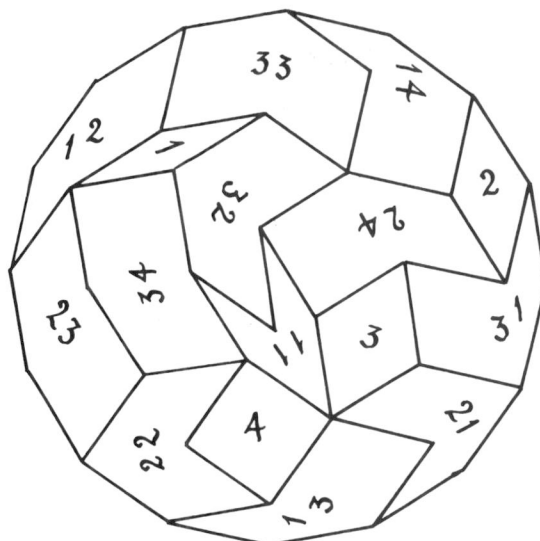

Figure 54. Schoen's 16-piece 16-gon. A Century or So of Solutions.

How many pieces are there in a set of cyclotominoes? According as $n = 2m$ or $2m+1$, there are $m^2 - m$ or m^2 hexagons, and m rhombs in either case, so there are m^2 or $m^2 + m$ altogether. You can use sets for a variety of games and puzzles, ranging from Tangram-like pictures (Fig. 55) to quite sophisticated packing problems. It's early to say if these last contain any hidden secrets (though Alan Schoen has noted the one about parallel edges); there's perhaps a better chance since there is more structure in the shapes than there was in polyominoes and polyiamonds.

Many pleasing patterns can be produced: for example, take r^2 sets of pieces and pack them in nesting $2n$-gons of side lengths $1,2,\ldots, r$.

The exponential difficulty of this sequence of puzzles prompts us to add another remark about searching. A typical combinatorial puzzle or search of "size" n takes something like $n!$ trials to complete, and this is much more like n^n than c^n, no matter how big you take c. On the other hand the number of solutions may only be c^n, and while this goes up fast, your chance of finding one of them is only $(c/n)^n$ and this gets very small very fast as soon as n is bigger than c.

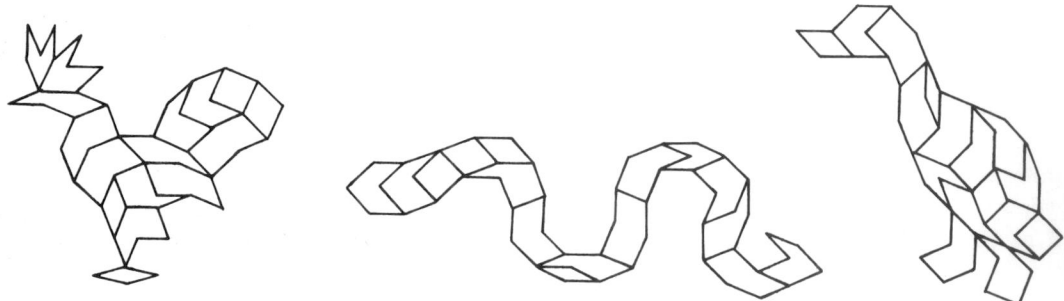

Figure 55. Schoen-Shapes Made with a Sixteen Set: Rooster, Serpent and Gosling.

MACMAHON'S SUPERDOMINOES

In his *New Mathematical Pastimes*, MacMahon proposed a different kind of generalized domino, got by dividing a regular polygon into colored triangles. We'll discuss just two examples. If we use just four colors, there are exactly 24 ways of coloring a triangular superdomino, and the standard problem is to pack these into a regular hexagon with an all black perimeter and adjacent colors alike, as in Fig. 56.

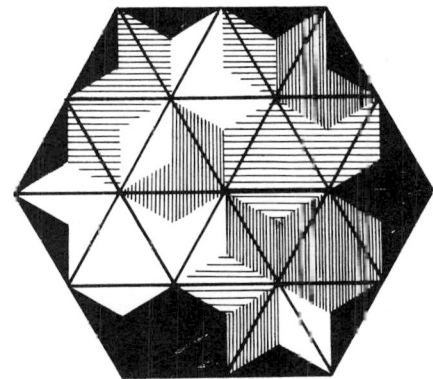

Figure 56. MacMahon's Four-Colored Triangular Superdominoes.

In this case it's hard to keep the secret hidden for very long. There are barely enough black edges to go round, and once you've found a suitable arrangement for them the rest is fairly easy.

When we consider the 24 three-colored square superdominoes, with which the usual problem is to make a 4×6 rectangle under similar conditions, the black edge problem is much more subtle. It can be shown that every solution to this problem has a column of four squares in which every horizontal edge is black (the **ladder**). In Fig. 57(a) the ladder occupies the second column and in Fig. 57(b) it occupies the third. In the Extras you'll find every possible configuration for the black edges.

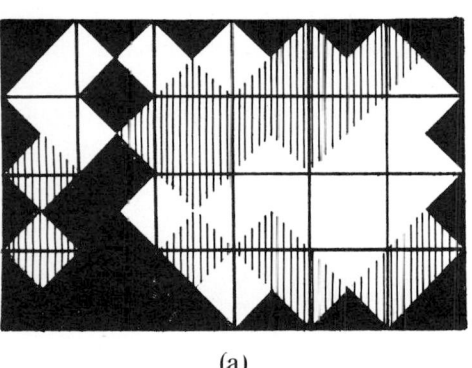

(a)

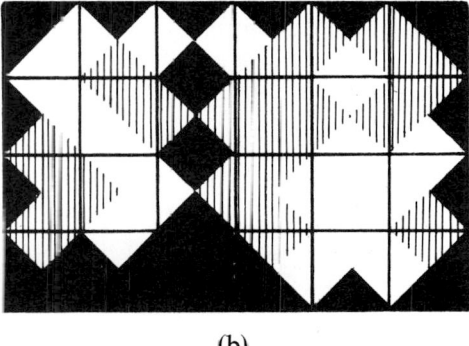

(b)

Figure 57. Three-Colored Square Superdomino Solutions Showing the Ladder.

MacMahon's superdomino problems can be made into jigsaw puzzles by using differently shaped edges in place of colors. Thus for the three colors in MacMahon's square problem one can use either the three edge shapes of Fig. 58(a) or those of Fig. 58(b) (which alter the matching condition).

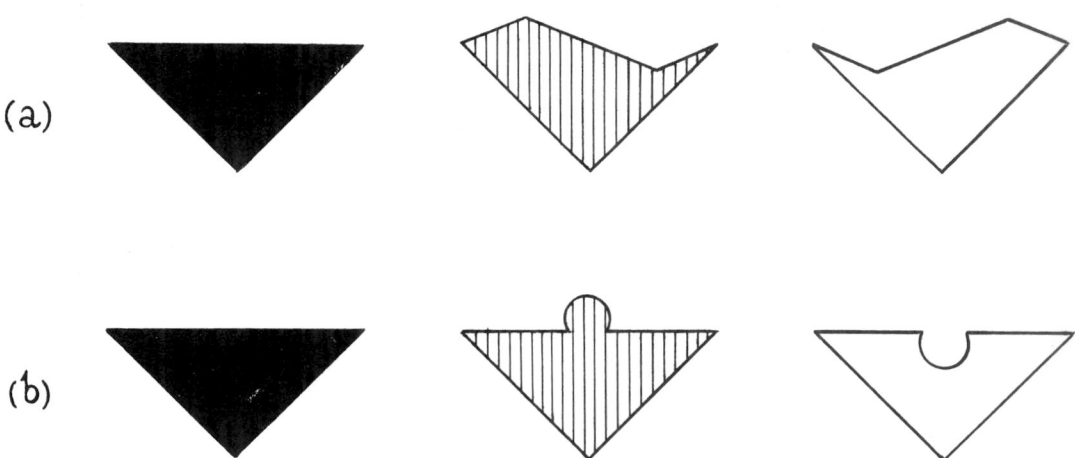

Figure 58. Two Ways of Making a MacMahon Jigsaw Puzzle.

Some years ago one of us sent out a Christmas card (Fig. 59) in the form of a jigsaw puzzle based on Fig. 58(b). The assembly in Fig. 59 is *not* a solution because it contains heads connected directly to hands and necks connected directly to arms. Can you turn it into an anatomically correct solution? Figure 60 is M.S. Paterson's modification of this idea, using another shape system. You must rearrange the pieces so that each wrestler has a properly connected body consisting of one head, one torso, one pair of shorts, two arms and two legs!

QUINTOMINAL DODECAHEDRA

The MacMahon superdominoes with five or more sides have not received much attention, but here's a nice little problem. There are 12 different **quintominoes** if we use five different colors once each and allow turning over. Can you fit them, colors matching, onto the 12 faces of a regular dodecahedron?

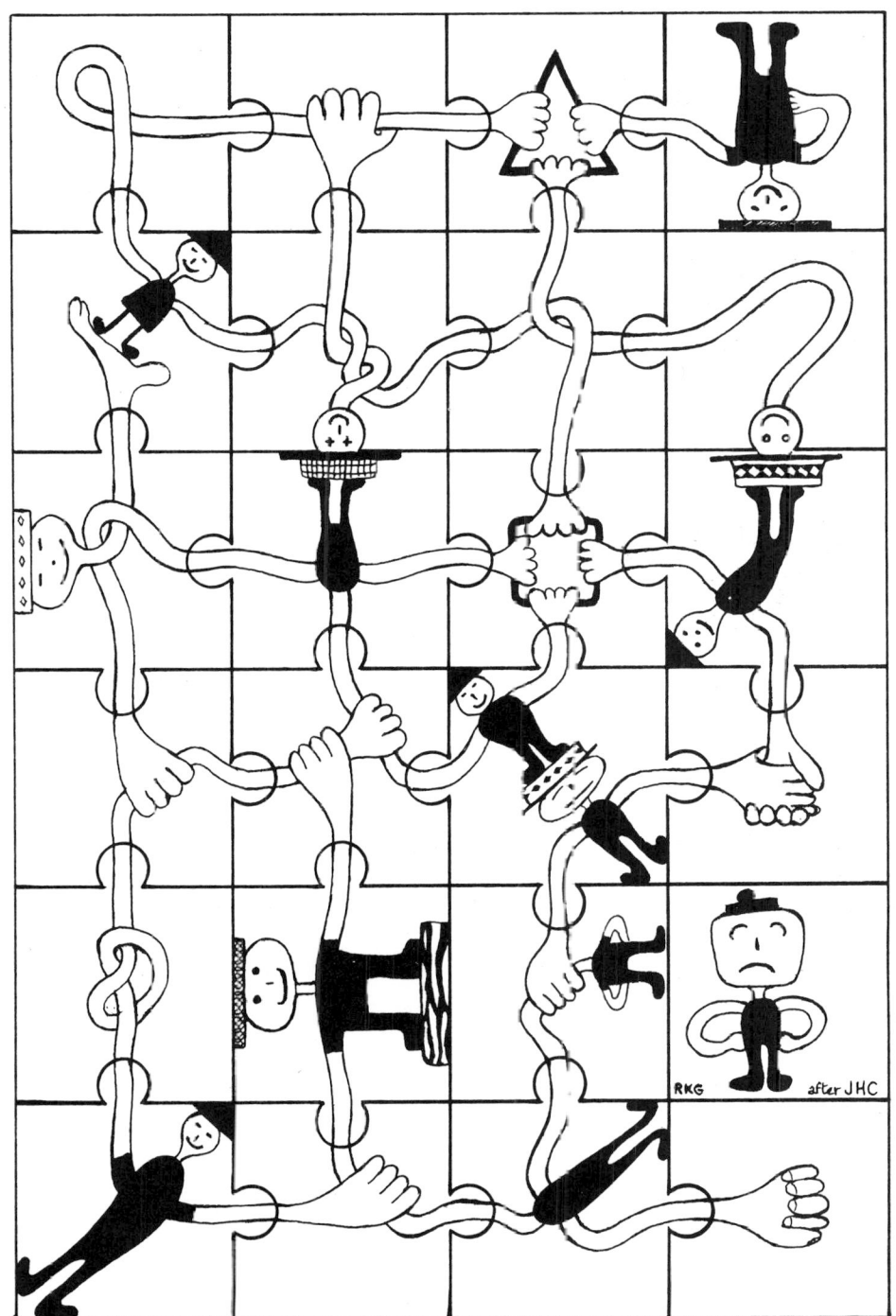

Figure 59. Conway's Christmas Card, 1968.

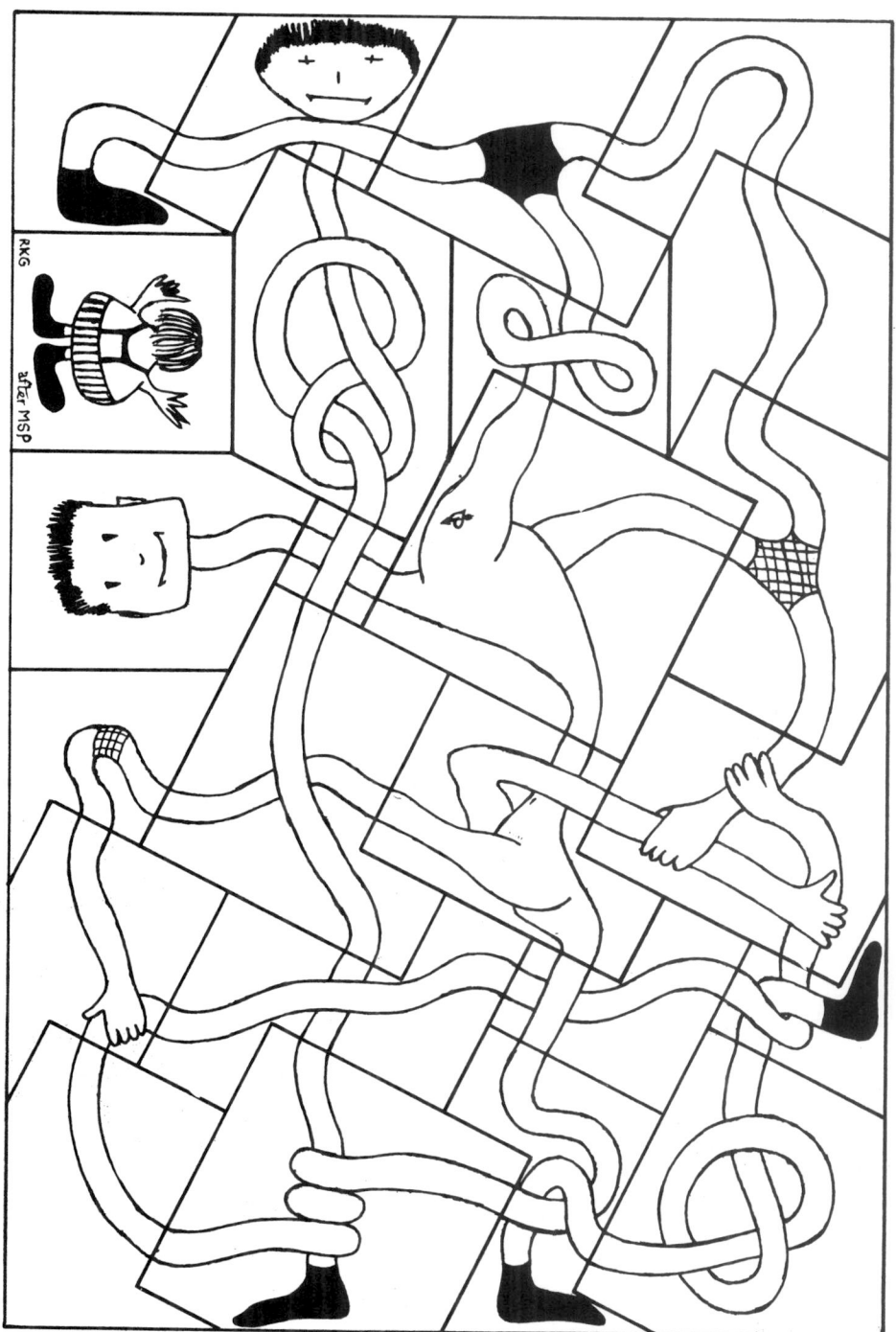

Figure 60. Paterson's Wrestling Match.

THE DOOMSDAY RULE

Here's an easy way to find the day of the week for an arbitrary date in an arbitrary year. The day of the week on which the last day of February falls in any given year will be called the **doomsday** for that year. For instance, in year 1000, doomsday (Feb. 29) was a Thursday (THOUSday). Then the following dates in *any* year are all doomsdays:

$$\text{Feb } 28/29 \qquad \text{Jan } 31/32$$

(the second alternative in leap years), otherwise for even months,

$$\text{Apr } 4 \qquad \text{Jun } 6 \qquad \text{Aug } 8 \qquad \text{Oct } 10 \qquad \text{Dec } 12$$

(the number of the month in the year), and for odd ones,

$$\text{Mar } 3+4 \qquad \text{May } 5+4 \qquad \text{Jul } 7+4 \qquad \text{Sep } 9-4 \qquad \text{Nov } 11-4$$

(add 4 for the 31-day, **long**, months; subtract 4 for 30-day, **short**, ones). Here's a summary with memos.

Jan	Feb	Mar	Apr	May	Jun	Jul	Aug	Sep	Oct	Nov	Dec
31/32	28/29	7	4	9	6	11	8	5	10	7	12
"last"	last	long 3	even 4	long 5	even 6	long 7	even 8	short 9	even 10	short 11	even 12

You should get used to finding other doomsdays in each month by changing the given one by weeks or fortnights; for example, since

$$\text{Jul 11 is a doomsday, so is Jul 4 (Independence Day),}$$

and since

$$\text{Dec 12 is a doomsday, so is Dec 26 (Boxing Day).}$$

On what day of the week was May-Day in the year 1000? May 9, and so May 2, were doomsdays (Thursdays in year 1000), so May 1 was a Wednesday.

It's easy to go wrong when adding numbers to days, so we suggest you use our mnemonics

NUN-day	ONE-day	TWOS-day	TREBLES-day	FOURS-day	FIVE-day	SIXER-day	SE'EN-day
Sunday	Monday	Tuesday	Wednesday	Thursday	Friday	Saturday	Sunday

Let's suppose we want Michaelmas Day (Sep 29) in the year 1000: we say

Sep 5 (short 9) and so Sep 26 are doomsdays (Thursdays—FOURS-days) so
Sep 29 is 3 on FOURS-day = SE'EN-day (Sunday).

To find doomsday for any year in a given century, you should add to the doomsday for the century year,

the number of *dozens* after that year,
the *remainder* after this, and
the number of *fours* in the remainder.

For example, for the year 1066 we say

THOUS⎫
Thurs ⎬day,   5 dozen,   6   and   1,   and since
FOURS⎭   (60)   (remainder)   (4's in 6)
   4   +5   +6   +   1   ≡ 2, mod 7,

doomsday in 1066 was a TWOS-day, and so the Battle of Hastings (Oct 14) was fought on a

4 on TWOS-day = SIXER-day (Saturday).

Let's do some years in our own century, given that 1900 = Wednesday = TREBLES-day.

Aug 4,   19——————14
4 off   TREBLES-day, 1 dozen, 2 (and 0) = TWOS-day (Tuesday),

Nov 11,   19——————18
4 on   TREBLES-day, 1 dozen, 6 and 1 = 15-day = ONE-day (Monday).

Of course, whole weeks can be cancelled, so the parentheses in

(4 on TREBLES)   1,   (6 and 1)

can be forgotten, making the answer immediate.

In the Julian calendar (as instituted by Julius Caesar) each century was one day earlier than the last, and so

0	100	200	300	400	500	600
700	800	900	1000	1100	1200	1300
1400	1500	1600	1700	...		

were

Sunday   Saturday   Friday   Thursday   Wednesday   Tuesday   Monday.

But in the modern, Gregorian, calendar (as reformed by Pope Gregory XIII)

		...	1500
1600	1700	1800	1900
2000	2100	2200	...

are

Tuesday   Sunday   Friday   Wednesday

because each century year that is *not* a multiple of 400 drops its leap day, and so is *two* days earlier than the previous one. In practice, remember that 1900 was a Wednesday, and that each step *backwards* to 1800, 1700, 1600 *adds* two days.

Thus, since Jul 4 is a doomsday,

Jul 4,   17————76

was

exactly   Sunday, 6 dozen, 4 and 1 = Thursday.

Various countries adopted the Gregorian reform by omitting various days; for example,
in Italy, France and Spain,   Oct 5–14, 1582.
in Britain and the American colonies,   Sep 3–13, 1752,
in Sweden,   leap days, 1700–1740,
elsewhere, various dates between 1583 (Poland) and 1923 (Greece).

You should also remember that the start of the year has not always been Jan 1. For some time before 1066 it was Christmas Day of the previous year, and for several centuries it was Mar 25 (so called Old Style dating, which was abolished in 1752). Such things are ignored in the Doomsday Rule, but, along with varying national conventions, must be accounted for in subtle examples:

Apr 23, 1616 (England) = 2 off Friday, 1 dozen, 4 and 1 = Tuesday (Shakespeare's deathday).
Apr 23, 1616 (Spain) = 2 off Tuesday, 1 dozen, 4 and 1 = Saturday (Cervantes' deathday),
Feb 29, 1603 (England) = exactly Friday, 0 dozen, 4 and 1 = Wednesday (Whitgift's deathday).

This "1603" must obviously be 1604 (New Style). Archbishop Whitgift was Queen Elizabeth's "worthy prelate" and first chairman of the commission which eventually produced the Authorized Version of the Bible.

The ambiguous days from Jan 1 through Mar 24 in years between about 1300 and 1752 were usually written in the "double dating" convention; e.g. Queen Elizabeth's deathday was Mar. 24, 1602/3 for which we find "3 on Fri + 3" = Thursday.

When calculating a B.C. date, it's best to add a big enough multiple of 28 (or 700) years to make it into an A.D. one, remembering that there was no year 0 (1 B.C. was immediately followed by 1 A.D.). Thus, in the Julian system we add 4200 to

$$\text{Oct 23, 4004 B.C.,} \quad \text{getting} \quad \text{Oct 23, 197 A.D.} \ (not \ 196),$$

and giving

$$1 \text{ off SIXER-day, 8 dozen, 1 (and 0)} = \text{SE'EN-day} = \text{Sunday}$$

for the day of Creation, according to Archbishop Ussher.

Problems   1. A man was nearly 48 years old on celebrating his first birthday. Where, when and what day of the week was it?
2. On what weekday is the 13th of the month most likely to fall in the Gregorian calendar?

... AND EASTER EASILY

A number of sources give more or less complicated rules for determining Easter. These usually apply only over limited ranges and are sometimes incorrect, even in reputable works, because they neglect the exceptions in the simple rule below.

Easter Day is defined to be the first Sunday strictly later than the **Paschal full moon**, which is a kind of arithmetical approximation to the astronomical one. The Paschal full moon is given by the formula

$$(\text{Apr 19} = \text{Mar 50}) - (11G + C)_{\text{mod } 30}$$

except that when the formula gives

Apr 19   you should take   Apr 18

and when it gives

Apr 18 *and* $G \geqslant 12$,   you should take   Apr 17.

In the formula,

$$G(\text{the } \textbf{Golden number}) = \text{Year}_{\text{mod } 19} + 1 \text{ (never forget to add the 1!)}$$
$$C(\text{the Century term}) \quad = +3 \text{ for all Julian years}$$
$$\left. \begin{array}{l} -4 \text{ for 15xx, 16xx} \\ -5 \text{ for 17xx, 18xx} \\ -6 \text{ for } \textbf{19xx}, \text{ 20xx, 21xx} \end{array} \right\} \text{Gregorian}$$

The general formula for C in a Gregorian year Hxx is

$$-H + \lfloor H/4 \rfloor + \lfloor 8(H+11)/25 \rfloor.$$

The next Sunday is then easily found by the Doomsday rule. Example

$$1945 \equiv 7, \text{ mod 19 so } G = 8 \text{ and we find for the Paschal full moon:}$$

$$\text{Mar } 50 - (88-6)_{\text{mod } 30} = \text{Mar } 50 - 22 = \text{Mar } 28.$$

Because this is a Doomsday, it's very easy to work out that it is

$$\text{"exactly Wed } (+3+9+2)\text{"}.$$

Easter Day, 1945, was therefore Mar 32, April Fool's Day.
    For 1981 ($\equiv 5$, mod 19) the formula gives

$$\text{Apr } 19 - (66-6)_{\text{mod } 30} = \text{Apr } 19,$$

so the Paschal full moon is

$$\text{Apr } 18 = \text{Doomsday}, 1981 = \text{Saturday},$$

so Easter Sunday, in 1981, is Apr 19.
    Here is an example in the Julian system:

$$1573: \text{P.F.M.} = \text{Mar } 50 - (176+3)_{\text{mod } 30} = \text{Mar } 50 - 29 = \text{Mar } 21 = \text{Saturday},$$

so Easter Day, 1573 was Mar 22. Since this date is still in the Old Style 1572, we can say that that year contained two Easters!
    You should use the Julian system even today if you want to know when the Orthodox churches celebrate Easter. Example:

$$\text{Julian P.F.M. } 1984 = \text{Apr } 19 - (99+3)_{\text{mod } 30} = \text{Apr } 7.$$

The next Doomsday is Apr 11, which is, still in the Julian system,

$$\text{Tuesday}, \qquad 7 \text{ dozen} = \text{Tuesday},$$
$$(\text{Julian } 1900)$$

so that Orthodox Easter Day, 1984 is the Julian date, Apr 9. Since the Julian calendar is now 13 days out of date, this is Apr 22 in the Gregorian system.
    Differences between Julian and Gregorian dates:

15xx,	16xx,	17xx,	18xx,	19xx,	20xx,	21xx,	...
10 days,	10 days,	11 days,	12 days,	13 days,	13 days,	14 days,	....

HOW OLD IS THE MOON?

If you stand on the earth and watch the sun and moon going round you, you'll see that they take about $365\frac{1}{4}$ [365·242199] and 30 [29·530588 or $29\frac{5}{9}$] days to do so, on average [brackets like these contain better approximations to various numbers].

From these facts you can deduce that the number of days that have passed since the last new moon is approximately:

(day number) + (month number) + (year number) + (century number),

all reduced mod 30 [$29\frac{5}{9}$].

The **day number** is the number of the day in the month.

The **month number**

for	Jan	Feb	Mar	Apr	May	Jun	Jul	Aug	Sep	Oct	Nov	Dec
is	3	4	3	4	5	6	7	8	9	10	11	12
[	$2\frac{2}{3}$	4	$2\frac{1}{3}$	$3\frac{8}{9}$	$4\frac{4}{9}$	6	$6\frac{5}{9}$	8	$9\frac{5}{9}$	$10\frac{1}{9}$	$11\frac{5}{9}$	$11\frac{8}{9}$]

(or just remember that the rule is about $\frac{1}{2}$ a day late/early in the long/short odd months).

The **year number** for a year whose last two digits are congruent, modulo 19,

to 0	$+1$	$+2$	$+3$	$+4$	$+5$	$+6$	$+7$	$+8$	$+9$
is 0	$+11$	$+22$	$+03$	$+14$	$+25$	$+06$	$+17$	$+28$	$+09$
[0	$+10\frac{8}{9}$	$+21\frac{7}{9}$	$+3\frac{1}{9}$	$+14$	$+24\frac{8}{9}$	$+6\frac{2}{5}$	$+17\frac{1}{9}$	$+28$	$+9\frac{1}{3}$]

[with an additional

	$\frac{1}{2}$		$\frac{1}{4}$	0	$-\frac{1}{4}$		$-\frac{1}{2}$
in years	$4n$ (after leap day)		$4n+1$	$4n+2$	$4n+3$		$4n+4$ (before leap day)].

The **century number** for the Gregorian centuries

	15xx	16xx	17xx	18xx	19xx	20xx	21xx	22xx	23xx	24xx
is	$16\frac{1}{3}$	12	$6\frac{2}{3}$	$1\frac{1}{3}$	-4	$-8\frac{1}{3}$	$-13\frac{2}{3}$	-19	$-24\frac{1}{3}$	$-28\frac{2}{3}$

and, for the Julian centuries

	8xx	9xx	10xx	11xx	12xx	13xx	14xx	15xx	16xx	17xx
	27	$22\frac{2}{3}$	$18\frac{2}{3}$	14	$9\frac{2}{3}$	$5\frac{1}{3}$	$+1$	$-3\frac{1}{3}$	$-7\frac{2}{3}$	-12

To remember these,

the day number is easy,
the month number also, except for Jan = 3, Feb = 4.
the year number's tens digit is its units digit reduced, modulo 3,
the centuries 14xx and 19xx are $+1$ and -4; and a short century (36524 days)
　drops back $5\frac{1}{3}$ days, while a long century (36525 days) drops back $4\frac{1}{3}$ days
　(because 1273 lunations take $36529\frac{1}{3}$ [36529·337] days).

Thus (using only the rough numbers) on Christmas Day 1984, the moon will be

$$25 + 12 + (+28) - 4 \ (\text{mod } 30) = 1 \text{ day old,}$$

since $84 \equiv +8$, mod 19 and $8 \equiv 2$, mod 3. But, applying the formula to New Year's Day 1985 we find

$$1 + 3(!) + (+09) - 4 = 9 \text{ days old}$$

despite the interval of exactly 7 days. The true motion of the moon is very complicated, and such a simple rule can only hope to give answers to within a day or so. If you're watching the moon late at night, for instance, remember that 11:00 p.m. is nearer tomorrow than today because the rule is attuned to the start of the day.

Of course a moon's age of about

$$0, \qquad 7\tfrac{1}{2}, \qquad 15, \qquad 22\tfrac{1}{2},$$

days corresponds to

New Moon,   First Quarter,   Full Moon,   Last Quarter.

Those who like to keep mental track of the moon throughout a year should remember the total number for that year, e.g.

in 1982, day number + month number +2,
in 1983, day number + month number +13 etc.

1984,	1985,	1986,	1987,	1988,	1989,	1990,	1991,	1992,	1993,	1994
−6,	+5,	−13,	−2,	+9,	−10,	+1,	+12,	−7,	+4,	−15

JEWISH NEW YEAR (ROSH HASHANA)

The Jewish New Year's Day in a Gregorian year Y A.D. (i.e. the first day of the Jewish year Y + 3761) happens on September N, where

$$\{\lfloor Y/100 \rfloor - \lfloor Y/400 \rfloor - 2\} + \frac{765433}{492480}(12G)_{\bmod 19} + \tfrac{1}{4}(Y)_{\bmod 4} - \frac{313Y + 89091}{98496} = N + \text{fraction}$$

and G is the Golden number, except that it must be postponed from any

			TUE	MON
SUN	WED	FRI	if fraction $\geqslant \dfrac{1367}{2160}$	if fraction $\geqslant \dfrac{23269}{25920}$
			and if $(12G)_{\bmod 19} > 6$	and if $(12G)_{\bmod 19} > 11$

to the following

MON	THU	SAT	THU (*not* WED)	TUE

(omit the terms { } for Julian years).

EXTRAS

BLOCKS-IN-A-BOX

The key to this puzzle is that every piece except the three $3 \times 1 \times 1$ rods occupies as many "black" cells as "white" in every layer. The rods must therefore be arranged so as to correct the color compositions in all fifteen layers simultaneously. It turns out that there is a unique arrangement which does this. Figure 61 also shows the only three dispositions for the $2 \times 2 \times 2$ cube and $2 \times 2 \times 1$ square. With these five pieces in place, the puzzle becomes easy.

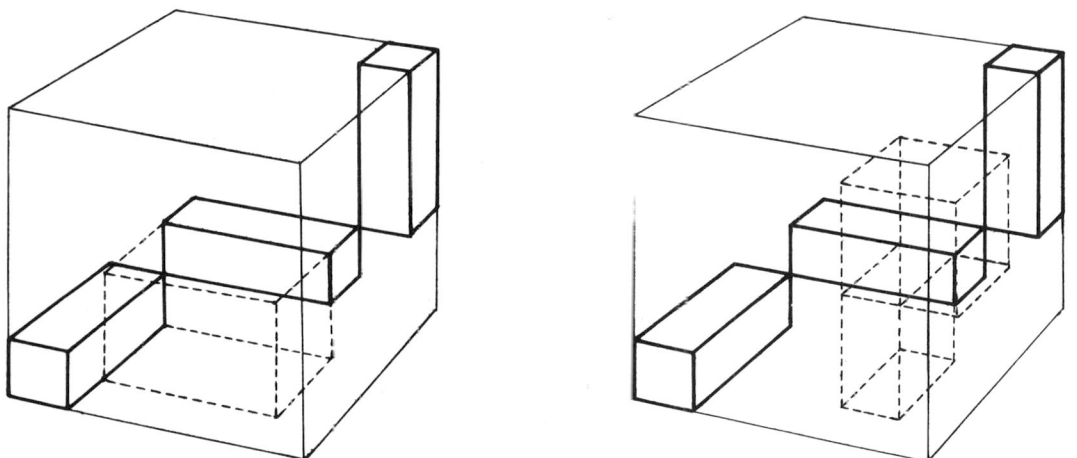

Figure 61. Were You Able to Fit the Blocks-in-a-Box?

A much harder puzzle is to pack 41 $1 \times 2 \times 4$ planks (together with 15 $1 \times 1 \times 1$ holes) into a $7 \times 7 \times 7$ box (see reference to Foregger, and to Mather, who proves that 42 planks can't be packed.)

THE SOMAP

The Soma pieces $1 = W$, $2 = Y$ and $4 = O$, while themselves symmetrical, may appear on the surface of the cube in either the *dexter* fashion

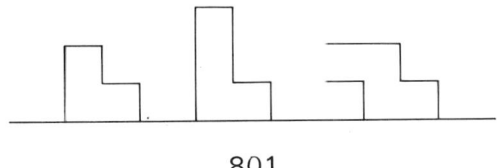

801

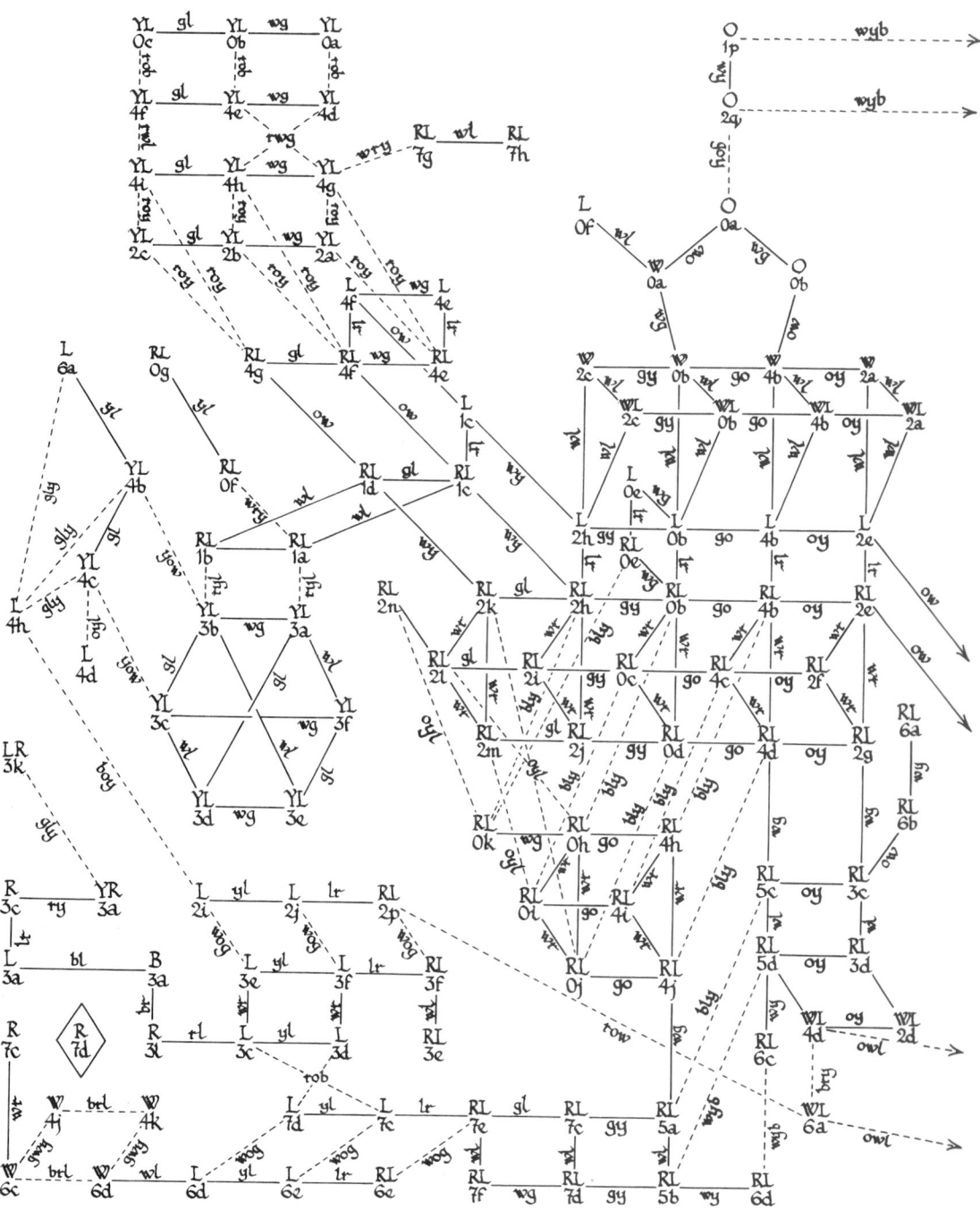

The diamond's gory secrets are seven seas away!

Figure 62. The . . . .

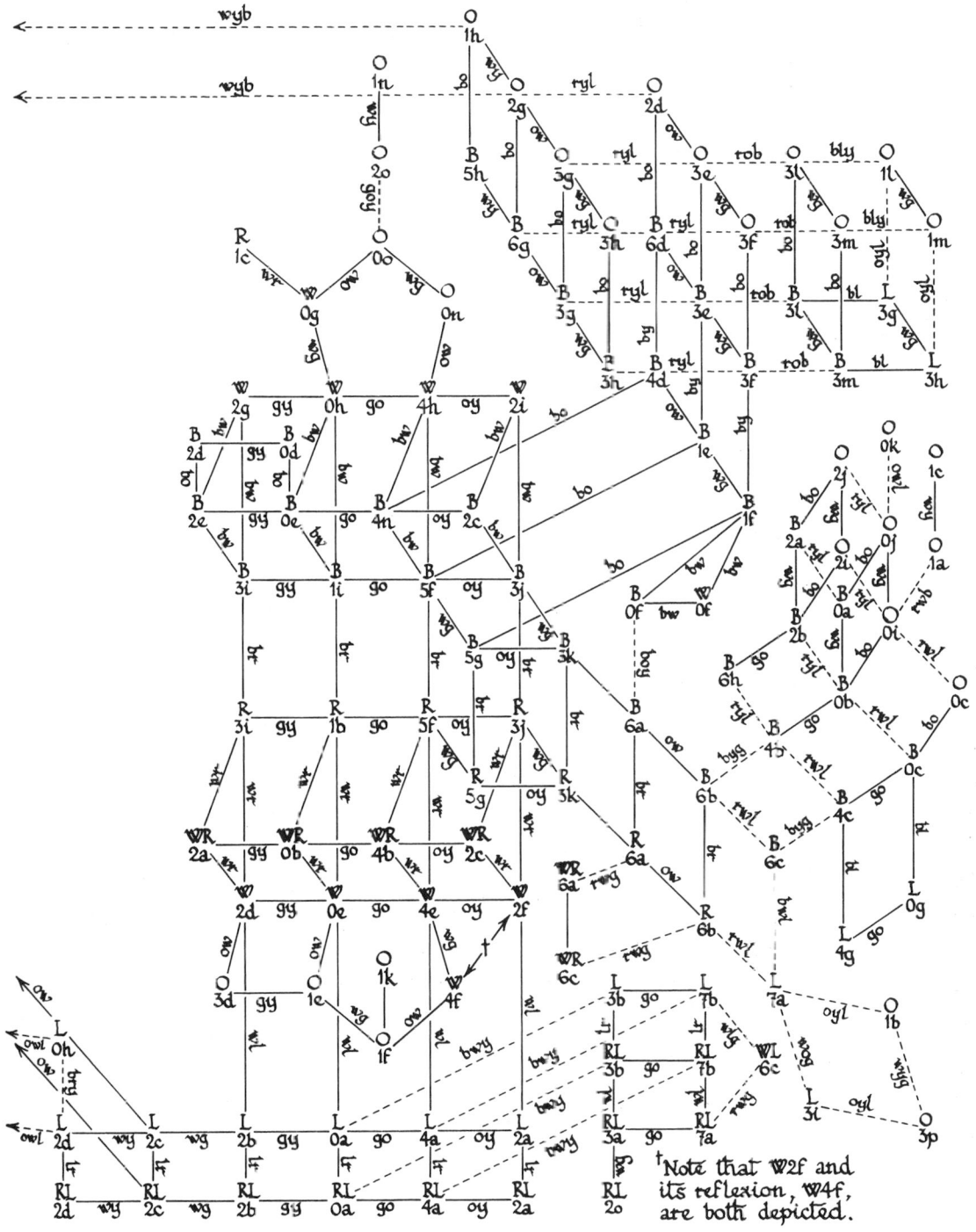

Somap.

or the *sinister* one

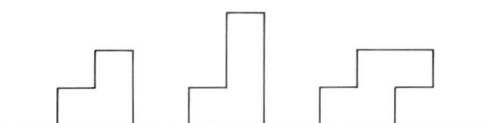

so you can tell which of these pieces are dexter by giving the sum of their numbers, which we call the **dexterity** of the solution. The symbols

$$\begin{array}{ccc} \text{DC} & \text{DC} & \text{DC} \\ na & nb & nc \end{array}$$

refer to different solutions having deficient piece D, central piece C and dexterity n, a single capital letter indicating that the same piece is both deficient and central. Thus

$$\begin{array}{cccc} \text{RL} & \text{RL} & \text{RL} & \text{RL} \\ 5a & 5b & 5c & 5d \end{array}$$

are four solutions in which Red is deficient, bLue is central and pieces 1 and 4 are dexter ($1+4=5$), while

$$\begin{array}{ccc} \text{B} & \text{B} & \text{B} \\ 6a & 6b & 6c \end{array} \cdots$$

are solutions in which Black is deficient *and* central while 2 and 4 are dexter.

Along with the solutions in Fig. 62, there are their reflexions whose names are found by interchanging R and L and replacing n by

$$3-n, \qquad 6-n, \qquad 7-n,$$

in the cases

$$\text{O central}, \qquad \text{W central}, \qquad \text{otherwise}.$$

When two solutions are related by changing just two pieces, P and Q, this is indicated by a solid line PQ. Some three-piece changes are indicated by dashed lines in a similar way. So all that's left for you to do is to find a suitable solution which you can locate on the Somap which will then lead you to all the others except R7d.

SOLUTIONS TO THE ARITHMETICO-GEOMETRIC PUZZLE

Figure 63 shows how we indicate layers in this puzzle by using a or α, according to orientation, for an a-high block, etc. The 21 solutions to Hoffman's puzzle are exhibited in Table 1 in this notation. When, as usual, only the middle layer is shown, another layer is separated from it by a letter S, and the remaining one is the special layer of Fig. 63. The meanings of the other letters in Table 1 are:

R: reflect the special layer across the dotted diagonal,
S: swap the two non-special layers,
S′: swap two adjacent layers in a different direction,
T: tamper with a $2 \times 2 \times 2$ corner, not involving the special layer,
T′: tamper with a $2 \times 2 \times 2$ corner, which does involve the special layer.

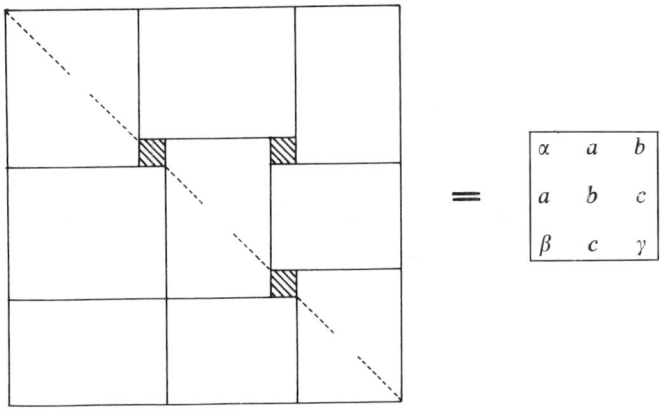

Figure 63. The Special Layer.

We'll leave it to you to work out why this gives just 21 solutions, and to verify that of these, exactly 17 have **duals**, obtained by replacing the dimensions a, b, c by c, b, a. Just one of the solutions (which?) is self-dual. This solution has the remarkable property that it can be repeatedly transformed (into rotations of itself!) by transporting either of two special faces to the opposite side.

Raphael Robinson and David Seal have found ways of combining solutions to the Arithmetico-Geometric puzzle in various dimensions to produce higher-dimensional ones. For example, if

$$a = a_1 + a_2 + a_3 \quad \text{and} \quad b = b_1 + b_2 + b_3$$

we know how to pack 27

$$a_1 \times a_2 \times a_3 \quad \text{or} \quad b_1 \times b_2 \times b_3$$

blocks into an

$$a \times a \times a \quad \text{or} \quad b \times b \times b$$

cube. The Cartesian product of these gives us a way of packing $27^2 = 729$

$$a_1 \times a_2 \times a_3 \times b_1 \times b_2 \times b_3$$

6-dimensional hyperblocks into a single

$$a \times a \times a \times b \times b \times b$$

hyperblock. But now the Cartesian product of three copies of Fig. 7 gives us a way to pack $4^3 = 64$ of these

$$a \quad \times \quad b \quad \times \quad a \quad \times \quad b \quad \times \quad a \quad \times \quad b$$

hyperblocks into an

$$(a+b) \times (a+b) \times (a+b) \times (a+b) \times (a+b) \times (a+b)$$

hypercube.

In general the method combines m-dimensional and n-dimensional solutions to give an mn-dimensional one. We hope Omar will tell us how to deal with dimensions 5, 7, 11 and so on.

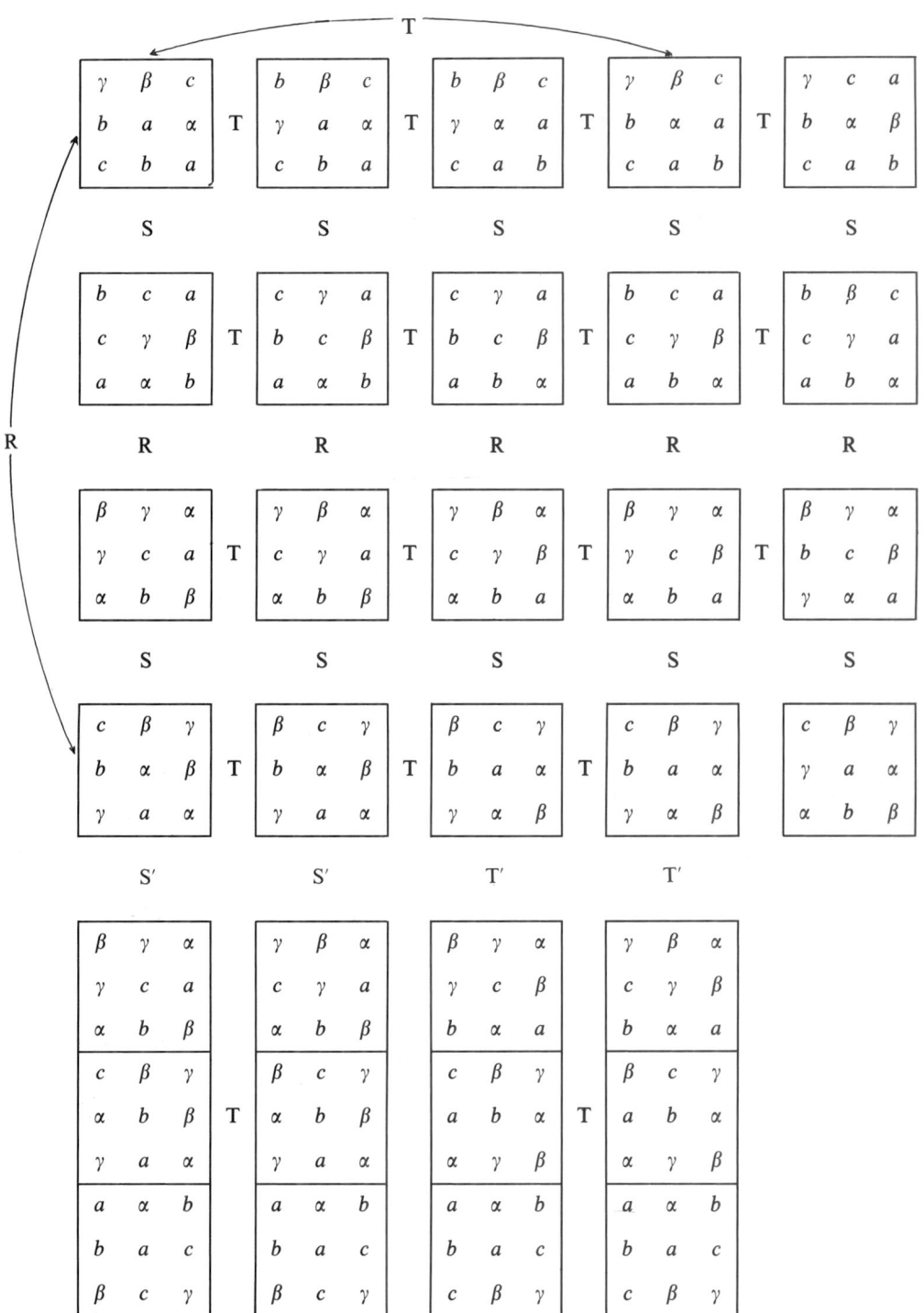

Table 1. The 21 Solutions to Hoffman's Puzzle.

... AND ONE FOR "THREE" TOO!

0	0	2		1	1	0		1	0	0
0	0	2		1	2	1		1	2	2
2	2	1		0	1	1		0	2	2

There's only one other solution. Hint: add $x + y + z$.

HARES AND TORTOISES

Make the moves in this order (jumps are bold):

H, **T**, **T**, **H**, **H**, **H**, **T**, **T**, **T**, **H**, **H**, **H**, T, T, H.

If you move only one kind of animal for as long as you can before moving the other kind, you'll soon see how to swap 57 Hares with 57 Tortoises.

Solutions to the other coin problems (heads are **bold**) are:

Start from 012**345**; move 01 to 67, **56** to **89** and 23 to 56;

or 01 to 76, 23 to 98 and 56 to 65.

Start from 01234**567**; move 12 to 89, **45** to **12**, 78 to 45 and 01 to 78;

or **67** to **98**, 01 to 76, 34 to 43 and 78 to 87.

M. Delannoy has shown that the first problem with n pairs of coins can always be solved in just n moves. However the second problem, due to Tait, requires $n + 1$ moves if $n > 4$. For some reason which we don't understand, we have always found these little problems confusing and can never remember their solutions!

The last little coin puzzle is one of the simplest examples we know of a psychological block. You notice that four coins are already in position (Fig. 64(a)), so you're reluctant either to move one of them (Fig. 64(b)) or to waste time by replacing it (Fig. 64(c)), but that's the only way to get to Fig. 64(d) in three moves. There's a four-move version in which you start with a triangle.

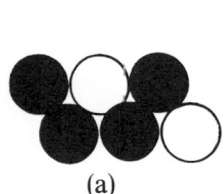 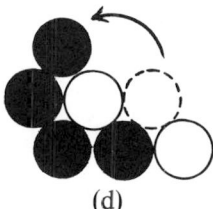 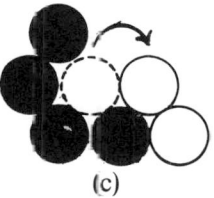 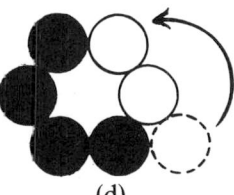

(a)   (d)   (c)   (d)

Figure 64. How to Infuriate Your Friends.

THE LUCKY SEVEN PUZZLE

has a solution in which just seven discs are slid down the bridge, alternately from the left and right sides:

1, 7, 2, 6, 3, 5, 4.

TOP FACE ALTERATIONS FOR THE HUNGARIAN CUBE

We give the shortest known sequences for all permutations (Table 2) and for all combinations of flips and twists (Table 3) in the top layer. The numbers are the numbers of moves, but not counting any final top turns (U^k) which can all be saved to the end. David Seal has proved that most of these are best possible.

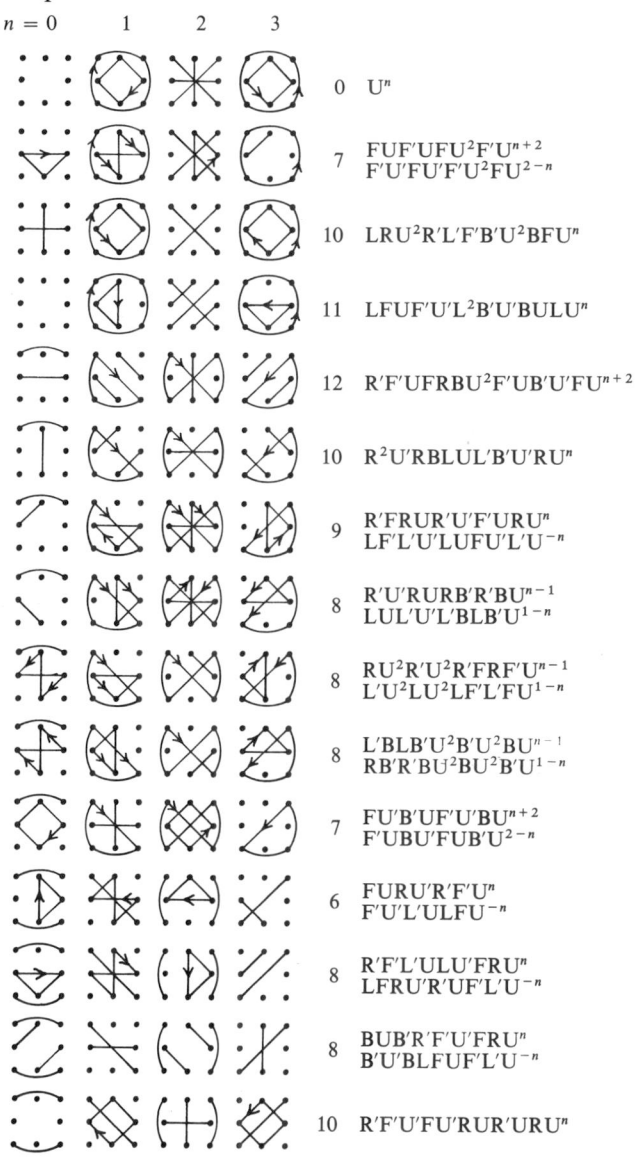

$n = 0$	1	2	3	0	U^n
				7	$FUF'UFU^2F'U^{n+2}$
					$F'U'FU'F'U^2FU^{2-n}$
				10	$LRU^2R'L'F'B'U^2BFU^n$
				11	$LFUF'U'L^2B'U'BULU^n$
				12	$R'F'UFRBU^2F'UB'U'FU^{n+2}$
				10	$R^2U'RBLUL'B'U'RU^n$
				9	$R'FRUR'U'F'URU^n$
					$LF'L'U'LUFU'L'U^{-n}$
				8	$R'U'RURB'R'BU^{n-1}$
					$LUL'U'L'BLB'U^{1-n}$
				8	$RU^2R'U^2R'FRF'U^{n-1}$
					$L'U^2LU^2LF'L'FU^{1-n}$
				8	$L'BLB'U^2B'U^2BU^{n-1}$
					$RB'R'BU^2BU^2B'U^{1-n}$
				7	$FU'B'UF'U'BU^{n+2}$
					$F'UBU'FUB'U'^{2-n}$
				6	$FURU'R'F'U^n$
					$F'U'L'ULFU^{-n}$
				8	$R'F'L'ULU'FRU^n$
					$LFRU'R'UF'L'U^{-n}$
				8	$BUB'R'F'U'FRU^n$
					$B'U'BLFUF'L'U^{-n}$
				10	$R'F'U'FU'RUR'URU^n$

Table 2. Top Layer Permutation Sequences.
(the lower sequence of a pair refers to the left-right reflected picture)

State	Moves	Sequence
(no dots)	0	no moves required
	12	F'U'F²DRUR'D'U'F²U²FU'
	13	LF'UL'FB'UR'FU'RF'BU'
	13	L²F²L²U²R'LFL'RU²L²F²L²U
	12	R'U'LU²R'F²RF²U'RU²L'U² / RU²L'U'B²L'B²LU²R'ULU²
	13	RU'LU²R²F'U'FUR²U²R'L'U / BFU²F²U'L'ULF²U²B'UF'U'
	13	R'UL'U²R²BUB'U'R²U²RLU' / F'B'U'B²ULU'L'B²U²FU'BU
	14	F²DF'UFD'FL'U'LUF²U²F'U / LU²L²U'B'UBL DL'U'LD'L²U'
	14	B²D'BU'B'DB'LUL'U'B²U²BU' / L'U²L²UFU'F'LD'LUL'DL²U
	14	RUR²F²D'R²BL'B'R²DF'RF'U² / BLBD'L²FRF²L²DB²L²U'L'U²
	14	LF'D'L'BL'B²U²L'BDF'L²F²U² / B²R²BD'F'RU²F²RF'RDBR'U²
	15	BU²BR²FD²FLFL'F²D²F'R²B²U / R²F²LD²L²B²L'B'L'D²L'F²R'U²R'U'
	11	R'BD²B'RU²R'BD²B'RU²
	14	B'UFU'BU²F²L'U'LUF²U²F' / LU'R'UL'U²R²BUB'U'R²U²R
	13	L'UBL'D'BD²R²D'B²L²UF²U² / B²U'R²F²DL²D²F'DRF'U'RU²
	13	R'B²F'L'DF'L²FD'LFB²RU²
	13	BR²LFD'LF²L'DF'L'R²B'U²
	16	BUB²RBR²URL'U'L²F'L'F²U'F'
	12	RU'L'UDB²D'R²U²LU'RU' / B'UF'U²B²DL²U'D'FUB'U'
	11	LR'U'R²B'R'B²U'B'U'L' / LU²BUB²RBR²URL'
	11	F'U²R'U'R²B'R'B²U'B'F / F'BUB²RBR²URU²F
	14	F²R²FU'F²R'F'R²U'R'U²F²R²F² / F²R²F²U²RUR²FRF²UF'R²F²
	8	B'U'B²L'B'L²U'L'U² / RUR²FRF²UFU²
	12	BLU²B'U'B²L'B'L²U'L²B' / BL²UL²BLB²UBU²L'B'
	12	R'F²U'F²R'F'R²U'R'U²FR / R'F'U²RUR²FRF²UF²R
	12	L'U'B'UBLBLUL'U'B'U² / FURU'R'F'R'F'U'FURU²
	14	R²B²R²U'RL'DL²U²LD'L²U²R'U
	12	LUFU'F'L'R'U'F'UFR / R'F'U'FURLFUF'U'L'
	12	RBUB'U'R'L'B'U'BUL / L'U'B'UBLRUBU'B'R'
	14	F'U'F²UL'U'RLU'R'UF²U'F'U²
	14	R'U'F'L'R'U'F'UFRLUFR
	13	LU²F'U'F²LF²L'F²UFU²L'U²
	15	R'U²RU²R²B'D'R'FR²F'DBU'R'U'
	15	R'F²U'D'L'FLFDULFL'FR / R'F'LF'L'U'D'F'L'FLDUF²R
	14	R'UB'R²U'RUR²BU²R²B'R'BU² / F'LFL²U²F'L²U'L'UL²FU'LU²
	14	B'D²FU²RF'D²BLU²FU²F'LU

Table 3. Top Layer Flip and Twist Sequences.
(the lower sequence of a pair refers to the picture with a and c swapped)

THE CENTURY PUZZLE

is so called because it takes exactly 100 moves, and the Century-and-a-Half takes 151 according to the official rules, but since the first and last are only half-moves, we can obviously count it as 150 whole moves. You can see solutions to both puzzles in Fig. 65, and by turning the book upside-down you'll see the only other 100-move solution to the Century.

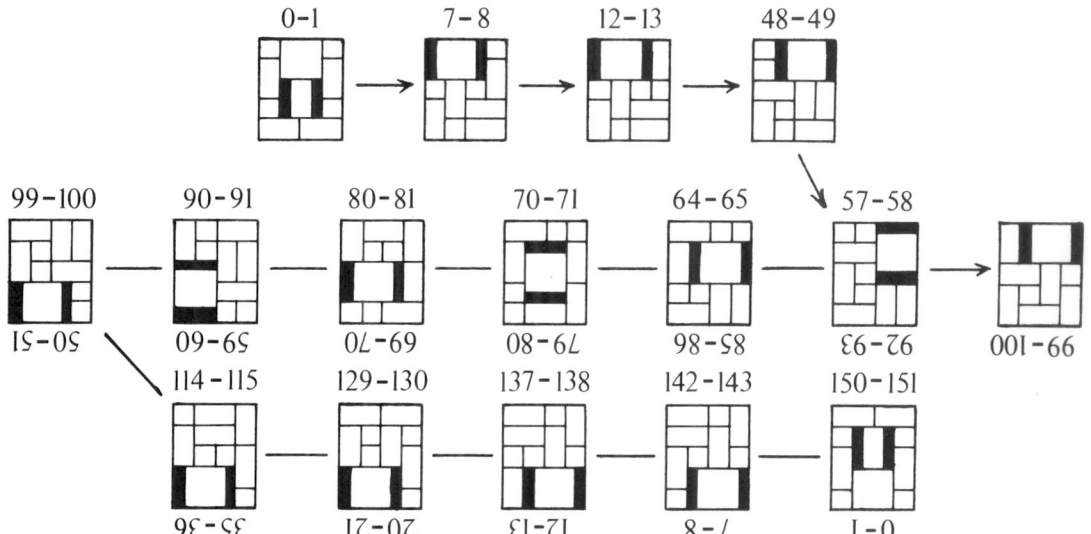

Figure 65. Solutions to the Century and Century-and-a-Half Puzzles.

ADAMS'S AMAZING MAGIC HEXAGON

```
           15
        14     13
     9      8      10
        6      4
    11      5      12
        1      2
    18      7      16
       17     19
           3
```

In Martin Gardner's *Sixth Book of Mathematical Games* you can read the remarkable history of Clifford W. Adams's discovery and of Charles W. Trigg's uniqueness proof. It's easy to see that a diameter d magic hexagon uses the numbers from 1 to $(3d^2 + 1)/4$, which add to

$$\frac{1}{2}\left(\frac{3d^2 + 1}{4}\right)\left(\frac{3d^2 + 5}{4}\right) = \frac{1}{32}(9d^4 + 18d^2 + 5),$$

so that each of its d columns must add to

$$\frac{1}{32}\left(9d^3 + 18d + \frac{5}{d}\right)$$

which can only be an integer if d divides 5.

FLAGS OF THE ALLIES SOLUTION

If you use the O'Beirne method you will find the two pairs of 5-circuits

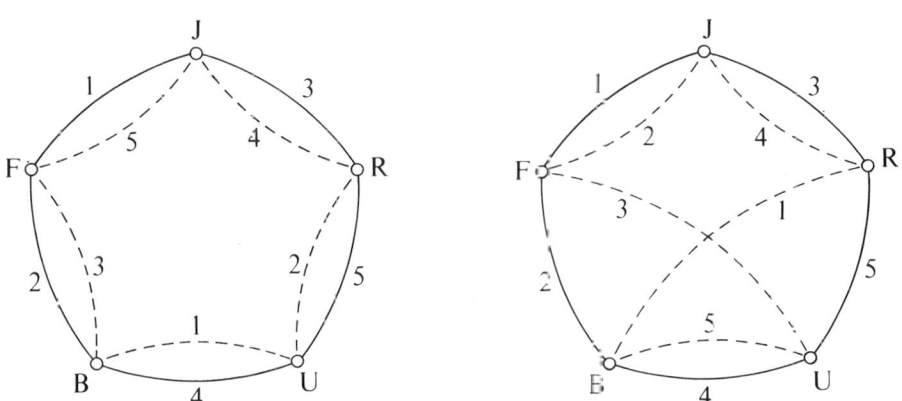

which lead to the solutions shown in Fig. 49 and Fig. 66.

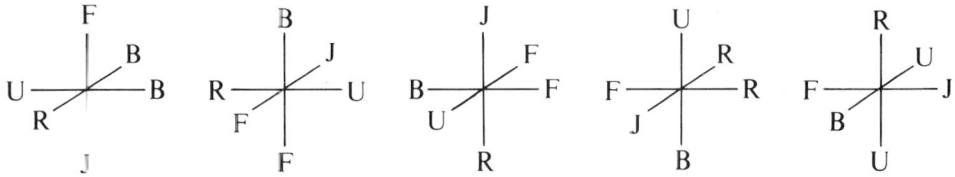

Figure 66. The Other Solution to the Flags of the Allies Problems.

ANSWER TO EXERCISE FOR EXPERTS

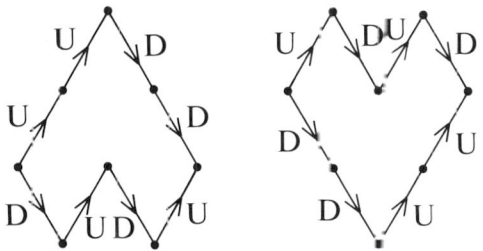

We have a rather complicated proof that n^2 copies of hexiamond A can be used to replicate A on a larger scale only if $n \equiv 0$ or ± 1 mod 6. Our proof establishes that these are the only values of n for which the relations (look at the foot of the previous page)

$$U^2D^2 = DUDU \quad \text{and} \quad D^2U^2 = UDUD$$

imply

$$U^{2n}D^{2n} = D^nU^nD^nU^n.$$

We've also shown that none of the usual kinds of coloring argument excludes other values of n.

WHERE DO THE BLACK EDGES OF MACMAHON SQUARES GO?

Round the outside, of course, but there are six more inside. These can be arranged in 20 different ways. In the first two the ladder is in the *third* column, otherwise it's in the second. The last row of Fig. 67 contains $6+6+2$ arrangements: the dotted lines are alternative positions for the sixth black edge.

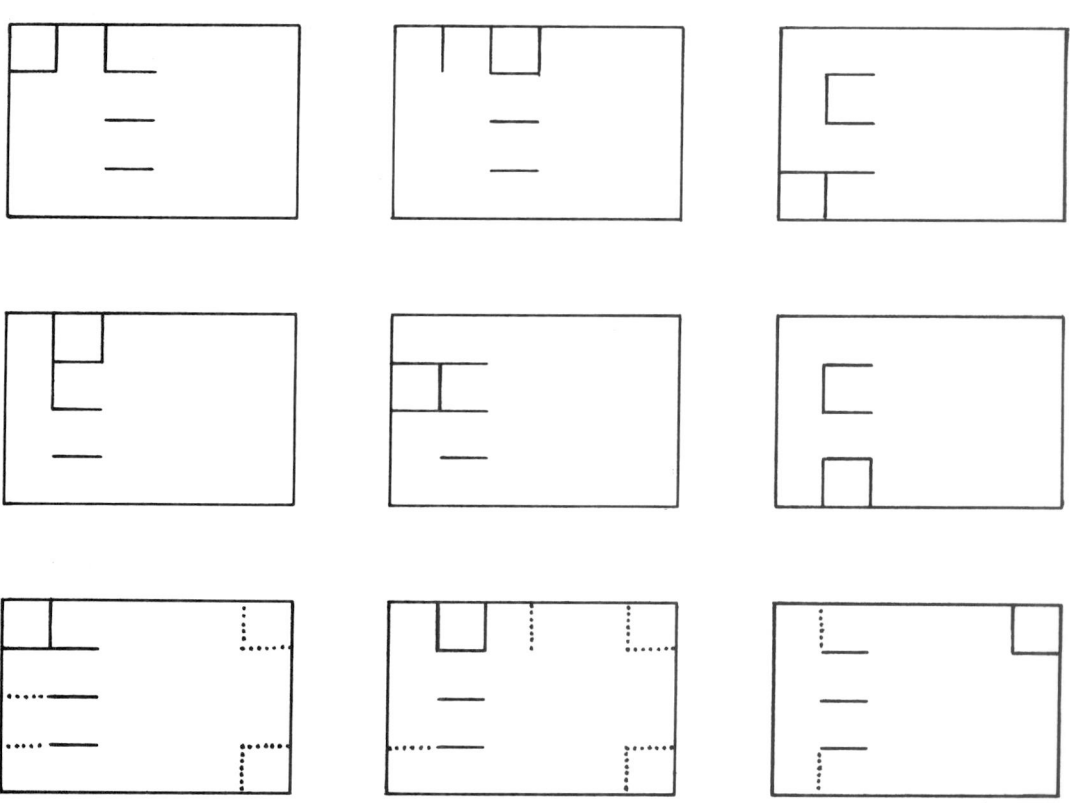

Figure 67. The Twenty Black Edge Arrangements for MacMahon Squares.

THE THREE QUINTOMINAL DODECAHEDRA

should be recoverable from

| 12345 = A | 12354 = B | 12435 = C | 12453 = D | 12534 = E | 12543 = F |
| 13245 = G | 13254 = H | 13425 = J | 13524 = K | 14235 = L | 14325 = M. |

DOOMSDAY ANSWERS

1. In Sweden there were no leap-days between Saturday 29 Feb 1696 and Saturday 29 Feb 1744, so the answer is Sweden, 29 Feb 1744, Saturday. At the time he was 11 days short of being 48 years old. This riot-avoiding scheme for losing the controversial 11 days was mooted in 1645 by John Greaves, an Oxford professor of astronomy.

2. A tedious enumeration shows that in the 400 years of the Gregorian cycle Doomsday is

	Sun	Mon	Tue	Wed	Thu	Fri	Sat	
for	43	43	43	43	44	43	44	ordinary years
and	13	15	13	15	13	14	14	leap years

From this you can work out that the 13th day falls on

	Sun	Mon	Tue	Wed	Thu	Fri	Sat	
in	687	685	685	687	684	688	584	months,

verifying B.H. Brown's assertion that the 13th of a month is just a little bit more likely to be a Friday than any other day of the week!

REFERENCES AND FURTHER READING

W.S. Andrews, "Magic Squares and Cubes", Open Court, 1917, reprinted Dover, 1960.

A.K. Austin, The 14–15 puzzle, Note 63.5, Math. Gaz. **63** (1979) 45–46.

W.W. Rouse Ball and H.S.M. Coxeter, "Mathematical Recreations and Essays", 12th edn. University of Toronto Press, 1974, pp. 26–27 (calendar problems), pp. 116–118 (shunting problems), p. 121 (sliding coins), pp. 193–221 (magic squares), pp. 312–322 (Fifteen Puzzle, Tower of Hanoï, Chinese rings). See early editions for Tangrams. Pages 141–144 on equilateral zonohedra and the references there, are related to Schoen's Cyclotome puzzles.

C.J. Bouwkamp, Catalogue of solutions of the rectangular $2 \times 5 \times 6$ solid pentomino problem, Nederl. Akad. Wet. Proc. Ser. A, **81** (1978) 177–186; Zbl. 384.42011.

Bro. Alfred Brousseau, Tower of Hanoï with more pegs, J. Recreational Math., **8** (1975–76) 169–176.

B.H. Brown, Problem E36, Amer. Math. Monthly, **40** (1933) 295 (calendar).

T.A. Brown, A note on "Instant Insanity", Math. Mag. **41** (1968) 68.

Cardan, "De Subtilitate", book xv, para 2; ed. Sponius vol. III, p. 587 (Chinese rings).

F. de Carteblanche, The coloured cubes problem, Eureka, **9** (1947) 9–11 (Tantalizer).

T.R. Dawson and W.E. Lester, A notation for dissection problems, Fairy Chess Review, **3** (Apr 1937) 5, 46–47 (polyominoes).

M. Delannoy, La Nature, June 1887, p. 10 (sliding coins).

A.P. Domoryad, "Mathematical Games and Pastimes", Pergamon, 1963, pp. 71–74 (Chinese rings); pp. 75–76 (Tower of Hanoï); pp. 79–85 (Fifteen puzzle); pp. 97–104 (magic squares); pp. 127–128 (sliding coins); pp. 142–144 (cf. Schoen's puzzle).

Henry Ernest Dudeney, "The Canterbury Puzzles", Nelson, London, 1907, 1919 reprinted Dover, 1958, No. 74 The Broken Chessboard, pp. 119–121, 220–221 (pentominoes).

Henry Ernest Dudeney, "536 Puzzles and Curious Problems", ed. Martin Gardner, Chas. Scribner's Sons, New York 1967, No. 383 The six pennies, pp. 138, 343; No. 377 Black and white, pp. 135, 340; No. 516 A calendar puzzle, pp. 212, 409–410; No. 528 A leap year puzzle, pp. 217, 413.

T.H. Foregger, Problem E2524, Amer Math. Monthly, **82** (1975) 300; solution Michael Mather, **83** (1976) 741–742.

Martin Gardner, Mathematical Games, Sci. Amer., **196** #5 (May 1957) (Tower of Hanoï); **197** #6 (Dec. 1957) **203** #5 (Nov 1960) 186–194, **207** #5 (Nov 1962) 151–159 (Polyominoes); **199** #3 (Sept 1958) 182–188 (Soma); **210** #2 (Feb 1964) 122–126, **222** #2 (Feb 1970) (Fifteen puzzle. Sliding block puzzles); **210** #3 (Mar 1964) 126–127 (Magic squares); **219** #4 (Oct 1968) 120–125 (MacMahon triangles; Conway's "three" puzzle); **223** #6 (Dec 1970) 110–114 (non-transitive dice; quintominal dodecahedra).

Martin Gardner, "The Scientific American Book of Mathematical Puzzles and Diversions", Simon and Schuster, New York 1959, pp. 15–22 (magic squares); pp. 55–62 (Tower of Hanoï); pp. 88 (Fifteen puzzle); pp. 124–140 (polyominoes).

Martin Gardner, "The 2nd Scientific American Book of Mathematical Puzzles and Diversions", Simon and Schuster, New York, 1961, pp. 55–56, 59 (sliding pennies); pp. 65–77 (Soma); pp. 130–140 (Magic squares); pp. 214–215, 218–219 (another Solitaire-like puzzle).

Martin Gardner, "Sixth Book of Mathematical Games from Scientific American", Chas. Scribner's Sons, New York, 1971, pp. 23–24 (magic hexagon); pp. 64–70 (sliding block puzzles); pp. 173–182 (polyiamonds).

Martin Gardner, "Mathematical Puzzles of Sam Loyd", Dover, New York 1959, No. 73 pp. 70, 146–147.

S.W. Golomb, Checkerboards and polyominoes, Amer. Math. Monthly, **61** (1954) 675–682.

S.W. Golomb, The general theory of polyominoes, Recreational Math. Mag. **4** (Aug 1961) 3–12; **5** (Oct 1961) 3–12; **6** (Dec 1961) 3–20; **8** (Apr 1962) 7–16.

S.W. Golomb, "Polyominoes", Chas. Scribner's Sons, New York, 1965.

A.P. Grecos and R.W. Gibberd, A diagrammatic solution to "Instant Insanity" problem, Math. Mag. **44** (1971) 71.

L. Gros, "Théorie du Baguenodier", Lyons, 1872 (Chinese rings).

L.J. Guibas and A.M. Odlyzko, Periods in strings, J. Combin. Theory Ser. A **30** (1981) 19–42.

L.J. Guibas and A.M. Odlyzko, String overlaps, pattern matching and non-transitive games, J. Combin. Theory Ser. A **30** (1981) 183–208.

Béla Hajtman, On coverings of generalized checkerboards, I, Magyar Tud. Akad. Math. Kutato Int. Köz, **7** (1962) 53–71.

Sir Paul Harvey, "The Oxford Companion to English Literature", 4th ed., Oxford, 1967, Appendix III, The Calendar.

C.B. and Jenifer Haselgrove, A computer program for pentominoes, Eureka, **23** (1960) 16–18.

Kersten Meier, Restoring the Rubik's Cube. A manual for beginners, an improved translation of "Puzzles-pass mit dem Rubik's Cube" 4c Hulme, Escondido Village. Stanford CA 94305 USA or Henning-Storm-Str. 5, 221 Itzehoe, W. Germany, 1981:01:20.

J.A. Hunter and Joseph S. Madachy, "Mathematical Diversions", Van Nostrand, New York, 1963, Chapter 8, Fun with Shapes, pp. 77–89.

Maurice Kraitchik, "Mathematical Recreations", George Allen and Unwin, 1943, pp. 89–93 (Chinese rings, Tower of Hanoï); pp. 109–116 (calendar); pp. 142–192 (magic squares); pp. 222–226 (shunting puzzles); pp. 302–308 (Fifteen puzzle).

Kay P. Litchfield, A 2 × 2 × 1 solution to "Instant Insanity", Pi Mu Epsilon J. **5** (1972) 334–337.

E. Lucas, "Récréations Mathematiques", Gauthier-Villars 1882–94; Blanchard. Paris, 1960.

Major P.A. MacMahon, "New Mathematical Pastimes", Cambridge University Press, 1921.

Douglas R. Hofstadter, Metamagical Themas: The Magic Cube's cubies are twiddled by cubists and solved by cubemeisters, Sci. Amer. **244** #3 (Mar. 1981) 20–39.

J.C.P. Miller, Pentominoes, Eureka, **23** (1960) 13–16.

T.H. O'Beirne, "Puzzles and Paradoxes", Oxford University Press, London, 1965, pp. 112–129 (Tantalizer); pp. 168–184 (Easter).

T.H. O'Beirne, Puzzles and Paradoxes, in *New Scientist*, **258** (61:10:26) 260–261; **259** (61:11:02) 316–317; **260** (61:11:9) 379–380; **266** (61:12:21) 751–752; **270** (62:01:18) 158–159.

Ozanam, "Recreations", 1723, vol. IV 439, (Chinese rings).

De Parville, La Nature, Paris, 1884, part i, 285–286 (Tower of Hanoï).

B.D. Price, Pyramid Patience, Eureka, **8** (1944) 5–7.

R.C. Read, Contributions to the cell growth problem, Canad. J. Math. **14** (1962) 1–20 (polyominoes).

J.E. Reeve and J.A. Tyrrell, Maestro puzzles, Math. Gaz. **45** (1961) 97–99 (polyominoes).

Raphael Robinson, Solution to problem E36, Amer. Math. Monthly, **40** (1933) 607.

Barkley Rosser and R.J. Walker, On the transformation group for diabolic magic squares of order four, Bull. Amer. Math. Soc. **44** (1938) 416–420.

Barkley Rosser and R.J. Walker, The algebraic theory of diabolic magic squares, Duke Math. J. **5** (1939) 705–728.

T. Roth, The Tower of Bramah revisited, J. Recreational Math. **7** (1974) 116–119.

Wolfgang Alexander Schocken, "The Calculated Confusion of Calendars", Vantage Press, New York, 1976.

Leslie E. Shader, Cleopatra's pyramid. Math. Mag. **51** (1978) 57–60 (Tantalizer variant).

David Singmaster, Notes on the 'magic cube', Dept. Math. Sci. & Comput. Polytech. of S. Bank, London SE AA, England, 1979, 5th edition, 1980, £2·00 or $5·00.

W. Stead, Dissection, Fairy Chess Review, **9** (Dec 1954) 2–4 (polyominoes).

James Ussher, "Annales Veteris Testamenti", Vol. 8, Dublin ed. 1864, p. 13 ("beginning of night leading into Oct. 23").

Joan Vandeventer, Instant Insanity, in "The Many Facets of Graph Theory", Springer Lecture Notes **110** (1969) 283–286.

Wallis, "Algebra", latin edition 1693. Opera, Vol. II, Chap. cxi, pp. 472–478 (Chinese rings).

Harold Watkins, "Time counts; the story of the calendar", Neville Spearman, London, 1954.

Richard M. Wilson, Graph puzzles, homotopy and the alternating group, J. Combin. Theory Ser. B **16** (1974) 86–96.

Chapter 25

What is Life?

Life's not always as simple as mathematics, Abraham!
Mrs. Abraham Fraenkel.

Life's too important a matter to be taken seriously.
Oscar Wilde.

... in real life mistakes are likely to be irrevocable.
Computer simulation, however, makes it economically practical
to make mistakes on purpose. If you are astute, therefore,
you can learn much more than they cost. Furthermore, if you
are at all discreet, no one but you need ever know you made
a mistake.
John McLeod and John Osborn, *Natural Automata*
and *Useful Simulations*, Macmillan, 1966.

Most of this book has been about two-player games, and our last two chapters were about
one player games. Now we're going to talk about a no-player game, the **Game of Life**! Our
younger readers won't have learned much about Life, so we'd better tell you some of the facts.

Life is a "game" played on an infinite squared board. At any time some of the cells will be
live and others **dead**. Which cells are live at time 0 is up to you! But then you've nothing else to
do, because the state at any later time follows inexorably from the previous one by the rules of
the game:

BIRTH. A cell that's *dead* at time t becomes *live* at $t+1$ only if *exactly three* of its eight neighbors
were live at t.

DEATH by overcrowding. A cell that's live at t and has four or more of its eight neighbors live
at t will be dead by time $t+1$.

DEATH by exposure. A live cell that has only one live neighbor, or none at all, at time t, will
also be dead at $t+1$.

These are the only causes of death, so we can take a more positive viewpoint and describe
instead the rule for

SURVIVAL. A cell that was live at time t will remain live at $t+1$ if and only if it had just 2 or
3 live neighbors at time t.

817

> Just 3 for BIRTH
> 2 or 3 for SURVIVAL

A fairly typical Life History is shown in Fig. 1. We chose a simple line of five live cells for our generation 0. In the figures a circle denotes a live cell.

Which of these will survive to generation 1? The two end cells have just one neighbor each and so will die of exposure. But the three inner ones have two living neighbors and so will survive. That's why we've filled in those circles.

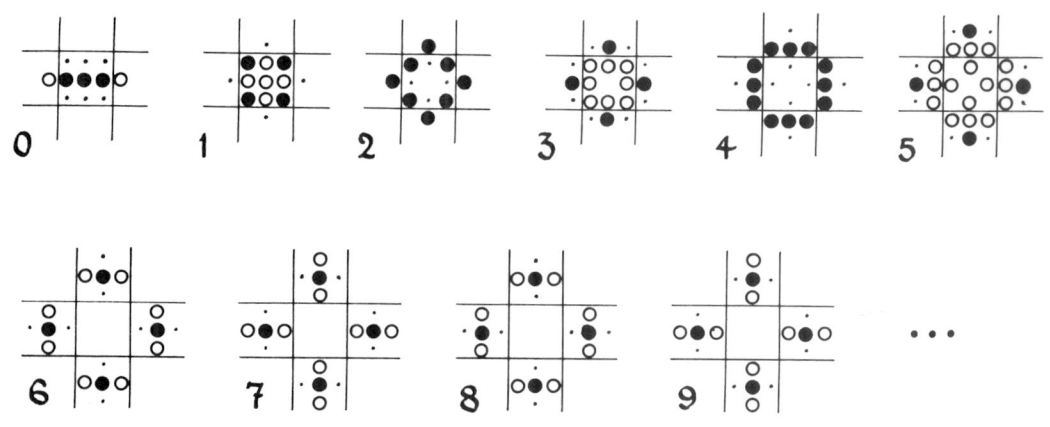

Figure 1. A Line of Five Becomes Traffic Lights.

What about births at time 1? There are three cells on either side of the line that are dead at time 0, but have exactly three live neighbors, so will come to life at time 1. We've shown these prospective births by dots in the figure.

So at time 1 the configuration will be a solid 3 × 3 square. Let's briefly follow its later progress.

Time 1–2: The corners will survive, having 3 neighbors each, but everything else will die of overcrowding. There will be 4 births, one by the middle of each side.

2–3: We see a ring in which each live cell has 2 neighbors so everything survives; there are 4 births inside.

3–4: Massive overcrowding kills off all except the 4 outer cells, but neighbors of these are born to form:

4–5: another survival ring with 8 happy events about to take place.

5–6: More overcrowding again leaves just 4 survivors. This time the neighboring births form:

6–7: four separated lines of 3, called **Blinkers**, which will never interact again.

7–8–9–10–... At each generation the tips of the Blinkers die of exposure but the births on each side reform the line in a perpendicular direction.

The configuration will therefore oscillate with period two forever. The final pair of configurations is sufficiently common to deserve a name. We call them **Traffic Lights**.

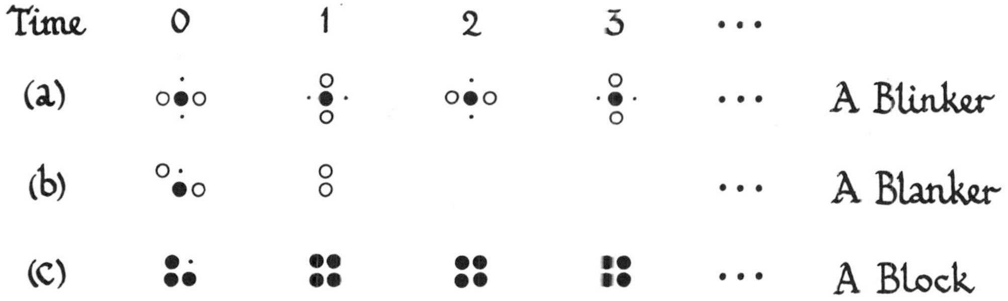

Time	0	1	2	3	...	
(a)	o●o	●.	o●o	●.	...	A Blinker
(b)	○.●o	○○			...	A Blanker
(c)	●●.	●●	●●	●●	...	A Block

Figure 2. If Three Survive, They'll Make a Blinker or a Block.

The Blinker is also quite common on its own (Fig 2a). Most other starting configurations of three live cells will blank out completely in two moves (Fig. 2(b)). But if you start with three of the four cells of a 2×2 block, the fourth cell will be born and then the **Block** will be stable (Fig. 2(c)) because each cell is neighbored by the three others.

STILL LIFE

It's easy to find other stable configurations. The commonest such **Still Life** can be seen in Fig. 3 along with their traditional names. The simple cases are usually loops in which each live cell has two or three neighbors according to local curvature, but the exact shape of the loop is important for effective birth control.

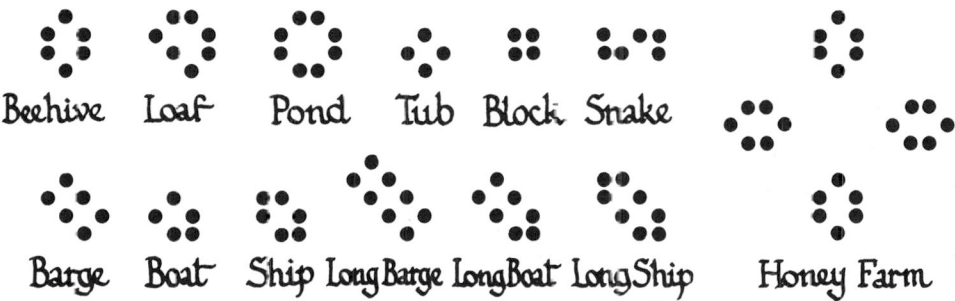

Beehive Loaf Pond Tub Block Snake

Barge Boat Ship LongBarge LongBoat LongShip Honey Farm

Figure 3. Some of the Commoner Forms of Still Life.

LIFE CYCLES

The blinker is the simplest example of a configuration whose life history repeats itself with period >1. Lifenthusiasts (a word due to Robert T. Wainwright) have found many other such configurations, a number of which are shown in Figs. 4 to 8.

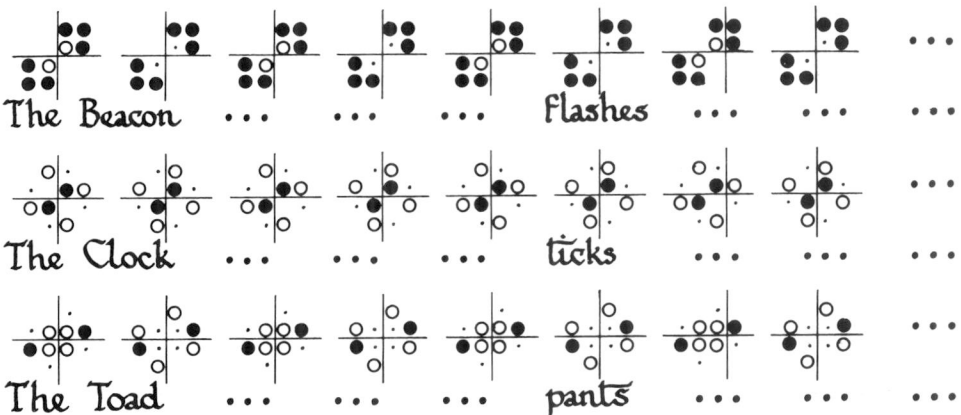

Figure 4. Three Life Cycles with Period Two.

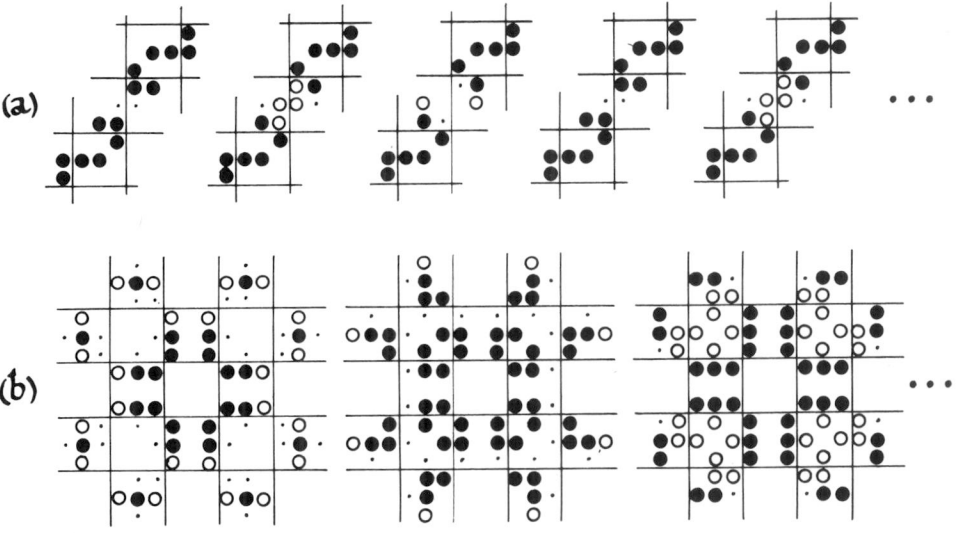

Figure 5. Two Life Cycles with Period Three. (a) Two Eaters Gnash at Each Other. (b) The Cambridge Pulsar CP 48-56-72.

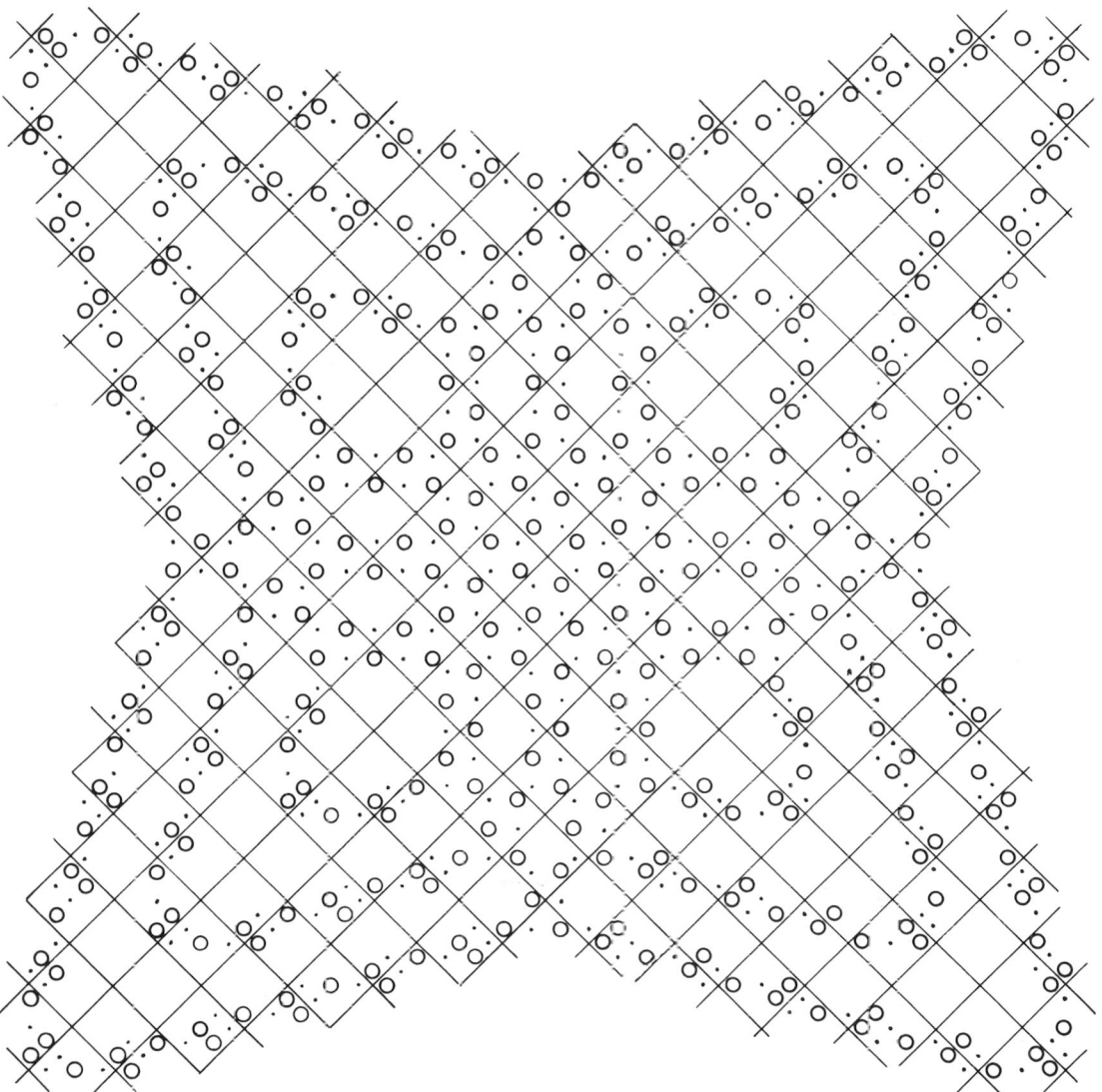

Figure 6. A Flip-Flop by the Gosper Group.

THE GLIDER AND OTHER SPACE SHIPS

When we first tracked the *r*-pentomino (you'll hear about that soon) some guy suddenly said, "Come over here, there's a piece that's walking!" We came over and found Fig. 9.

You'll see that generation 4 is just like generation 0 but moved one diagonal place, so that the configuration will steadily move across the plane. Because the arrangements at times 2, 6, 10, ... are related to those at times 0, 4, 8, 12, ... by the symmetry that geometers call a *glide reflexion*, we call this creature the **glider**. But when you see Life played at the right speed by a computer

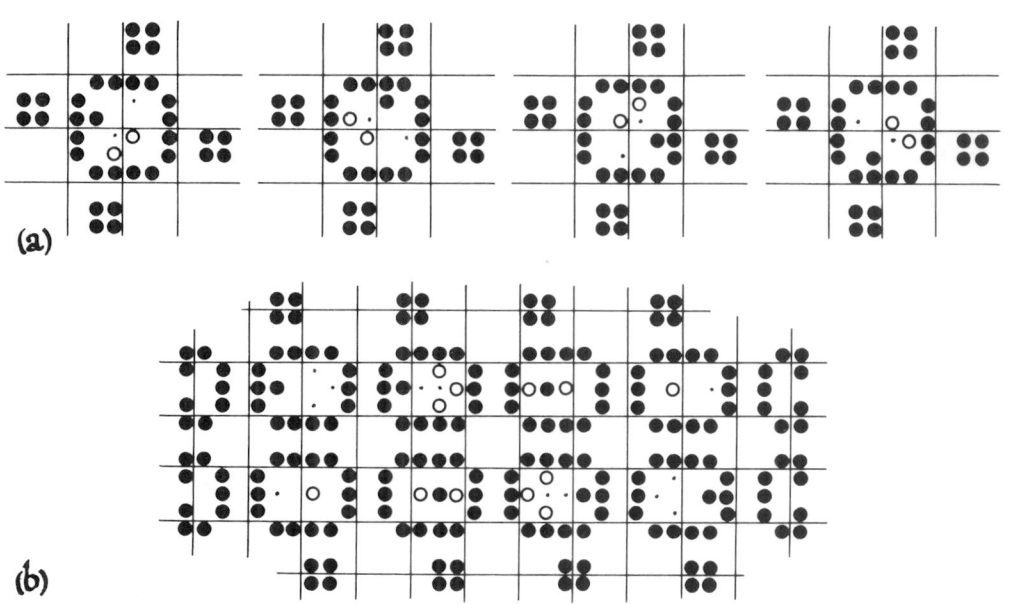

Figure 7. (a) Catherine Wheel. (b) Hertz Oscillator. Still Life Induction Coils Keep Field Stable.

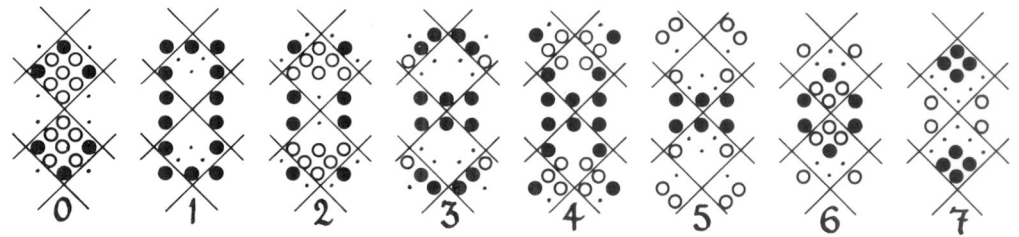

Figure 8. Figure 8.

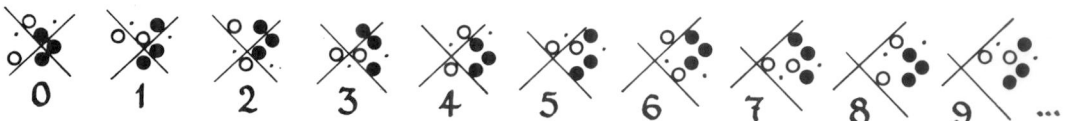

Figure 9. The Glider Moves One Square Diagonally Each Four Generations.

on a visual display, you'll see that the glider walks quite seductively, wagging its tail behind it. We'll see quite a lot of the glider in this chapter.

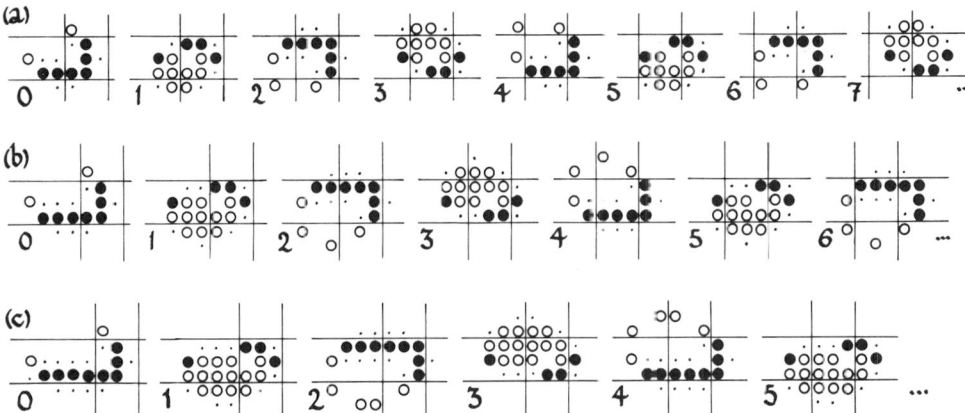

Figure 10. (a) Lightweight, (b) Middleweight, (c) Heavyweight Spaceships.

It was at just such a visual computer display that one of us first noticed the **spaceship** of Fig. 10(a) (and was very lucky to be able to stop the machine just before it would have crashed into another configuration). This **lightweight spaceship** immediately generalizes to the **middleweight** and **heavyweight** ones (Figs. 10(b) and (c)) but longer versions turn out to be unstable. It was later discovered, however, that arbitrarily long spaceships can still travel provided they are suitably escorted by small ones (Fig. 11). All the spaceships, as drawn, move Eastwards.

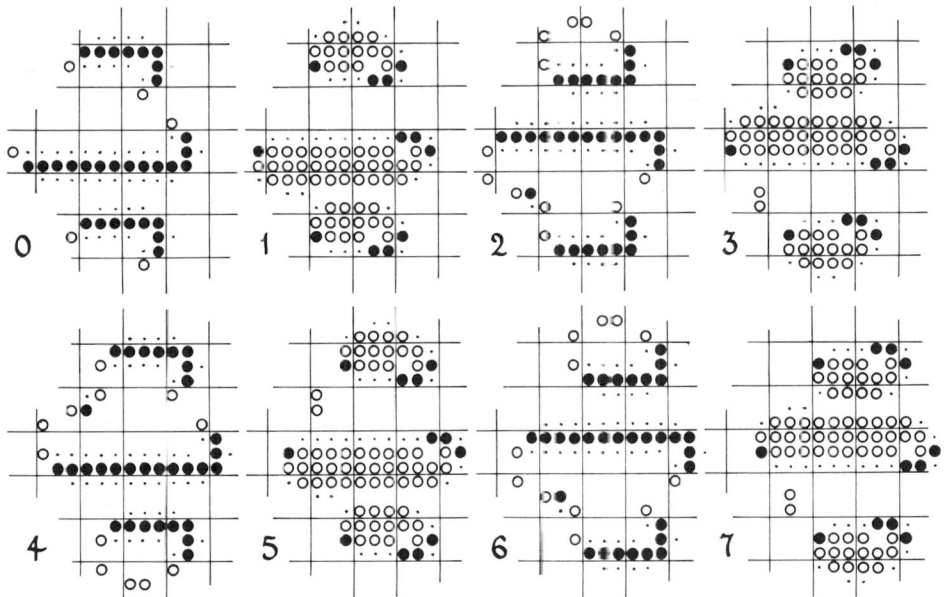

Figure 11. An Overweight Spaceship Escorted by Two Heavyweight Ones.

THE UNPREDICTABILITY OF LIFE

Is there some way to foretell the destiny of a Life pattern? Will it eventually fade away completely? Or become static? Oscillate? Travel across the plane, or maybe expand indefinitely?

Let's look at what *should* be a very simple starting configuration—a straight line of n live cells.

$n = 1$ or 2 fades immediately,

$n = 3$ is the Blinker;

$n = 4$ becomes a Beehive at time 2,

$n = 5$ gave Traffic Lights (Fig. 1) at time 6,

$n = 6$ fades at $t = 12$,

$n = 7$ makes a beautifully symmetric display before terminating in
             the **Honey Farm** (Fig. 3) at $t = 14$;

$n = 8$ gives 4 blocks and 4 beehives,

$n = 9$ makes two sets of Traffic Lights,

$n = 10$ turns into the **pentadecathlon**, with a life cycle of 15,

$n = 11$ becomes two blinkers,

$n = 12$ makes two beehives,

$n = 13$ turns into two blinkers,

$n = 14$
and   } vanish completely,
$n = 15$

$n = 16$ makes a big set of Traffic Lights with 8 blinkers,

$n = 17$ becomes 4 blocks

$n = 18$
and   } fade away entirely,
$n = 19$

$n = 20$ makes just 2 blocks,

and so on.

What's the general pattern? Even when we follow the configurations which start with a very small number of cells, it's not easy to see what goes on. There are 12 edge-connected regions with 5 cells (S.W. Golomb calls them pentominoes). Here are their histories:

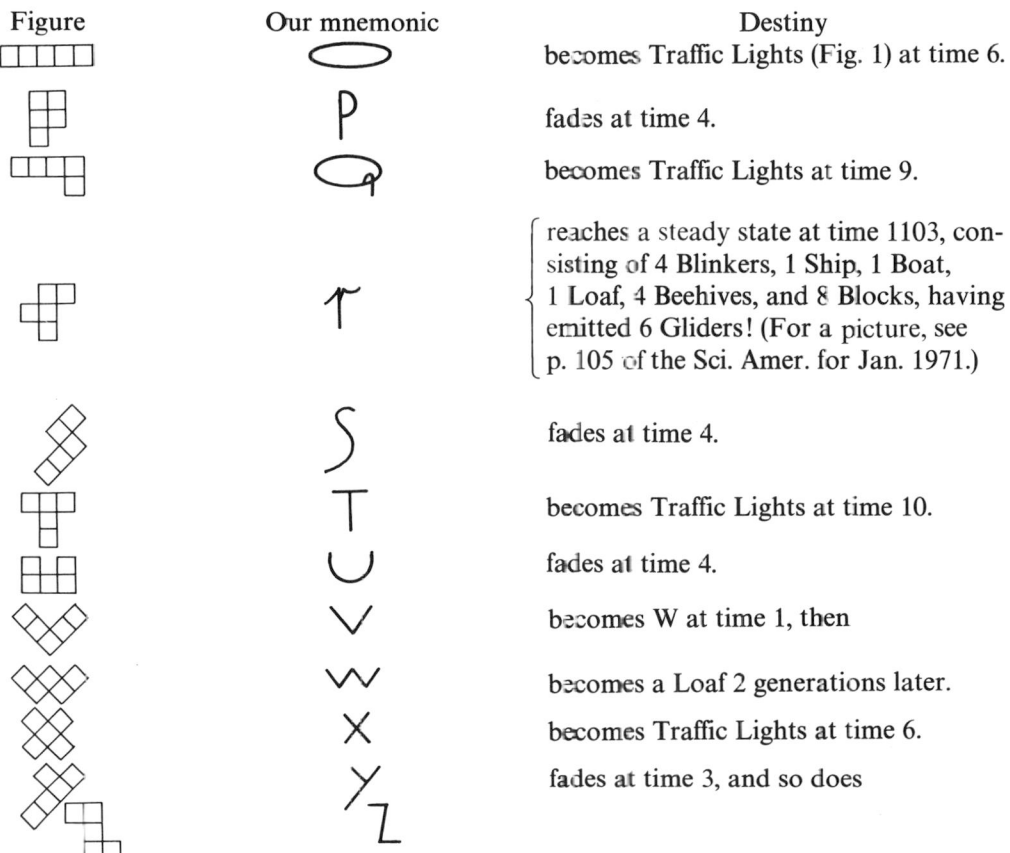

Figure	Our mnemonic	Destiny
	⬭	becomes Traffic Lights (Fig. 1) at time 6.
	P	fades at time 4.
	Q	becomes Traffic Lights at time 9.
	r	reaches a steady state at time 1103, consisting of 4 Blinkers, 1 Ship, 1 Boat, 1 Loaf, 4 Beehives, and 8 Blocks, having emitted 6 Gliders! (For a picture, see p. 105 of the Sci. Amer. for Jan. 1971.)
	S	fades at time 4.
	T	becomes Traffic Lights at time 10.
	U	fades at time 4.
	V	becomes W at time 1, then
	W	becomes a Loaf 2 generations later.
	X	becomes Traffic Lights at time 6.
	Y	fades at time 3, and so does
	Z	

Once again, it doesn't seem easy to detect any general rule.

Here, in Figs. 12 and 13 are some other configurations with specially interesting Life Histories, for you to try your skill with.

Can the population of a Life configuration grow without limit? Yes! The $50.00 prize that one of us offered for settling this question was won in November 1970 by a group at M.I.T. headed by R.W. Gosper. Gosper's ingenious **glider gun** (Fig. 14) emits a new glider every 30 generations. Fortunately it was just what we wanted to complete our proof that

Life is really unpredictable!

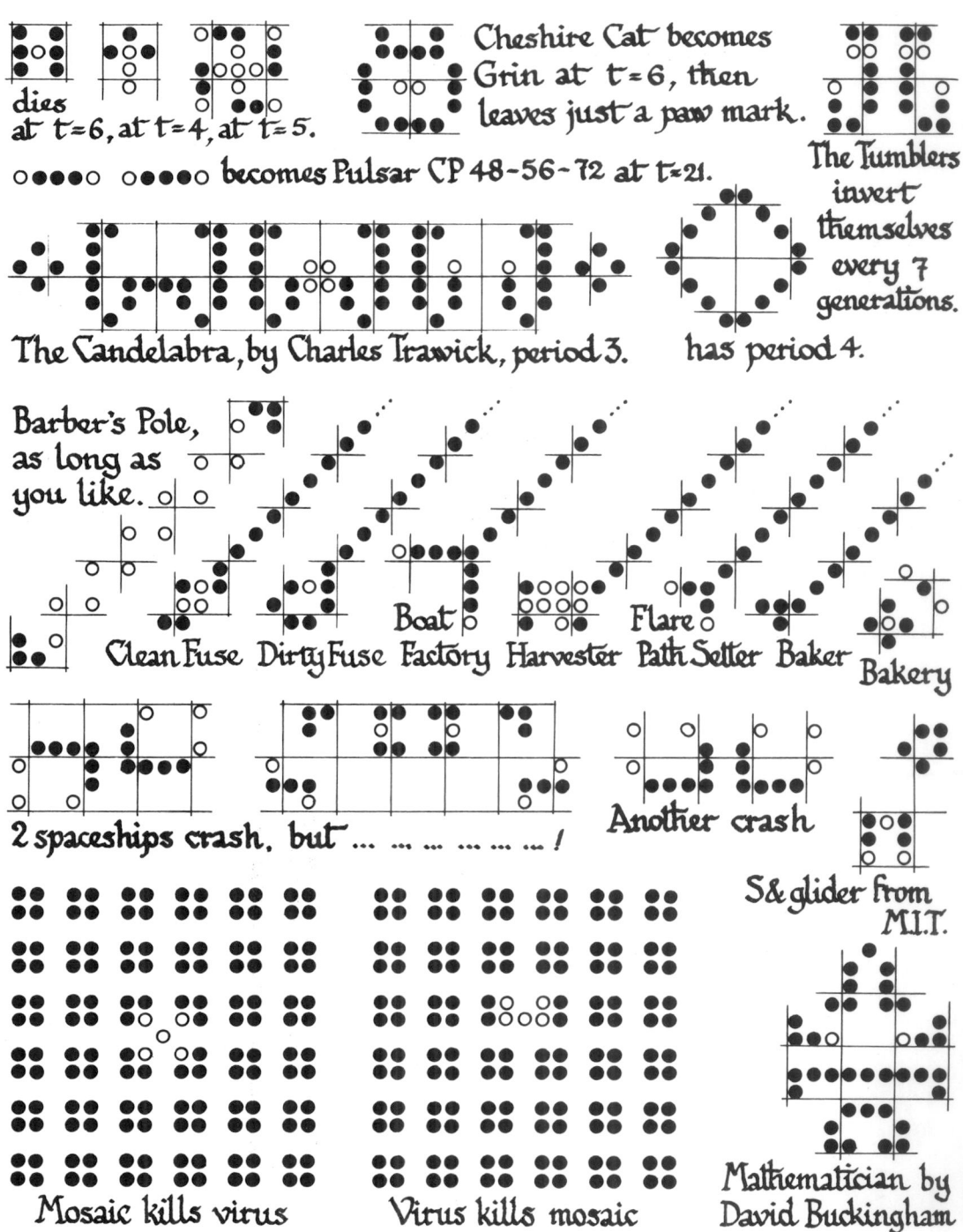

dies
at t≈6, at t≈4, at t≈5.

Cheshire Cat becomes
Grin at t=6, then
leaves just a paw mark.

The Tumblers
invert
themselves
every 7
generations.

○●●●○ ○●●●○ becomes Pulsar CP 48-56-72 at t≈21.

The Candelabra, by Charles Trawick, period 3.

has period 4.

Barber's Pole,
as long as
you like.

Clean Fuse   Dirty Fuse   Factory   Harvester   Path Setter   Baker

Boat

Flare

Bakery

2 spaceships crash, but ... ...!

Another crash

S & glider from
M.I.T.

Mosaic kills virus

Virus kills mosaic

Mathematician by
David Buckingham

Figure 12. Exercises for the Reader.

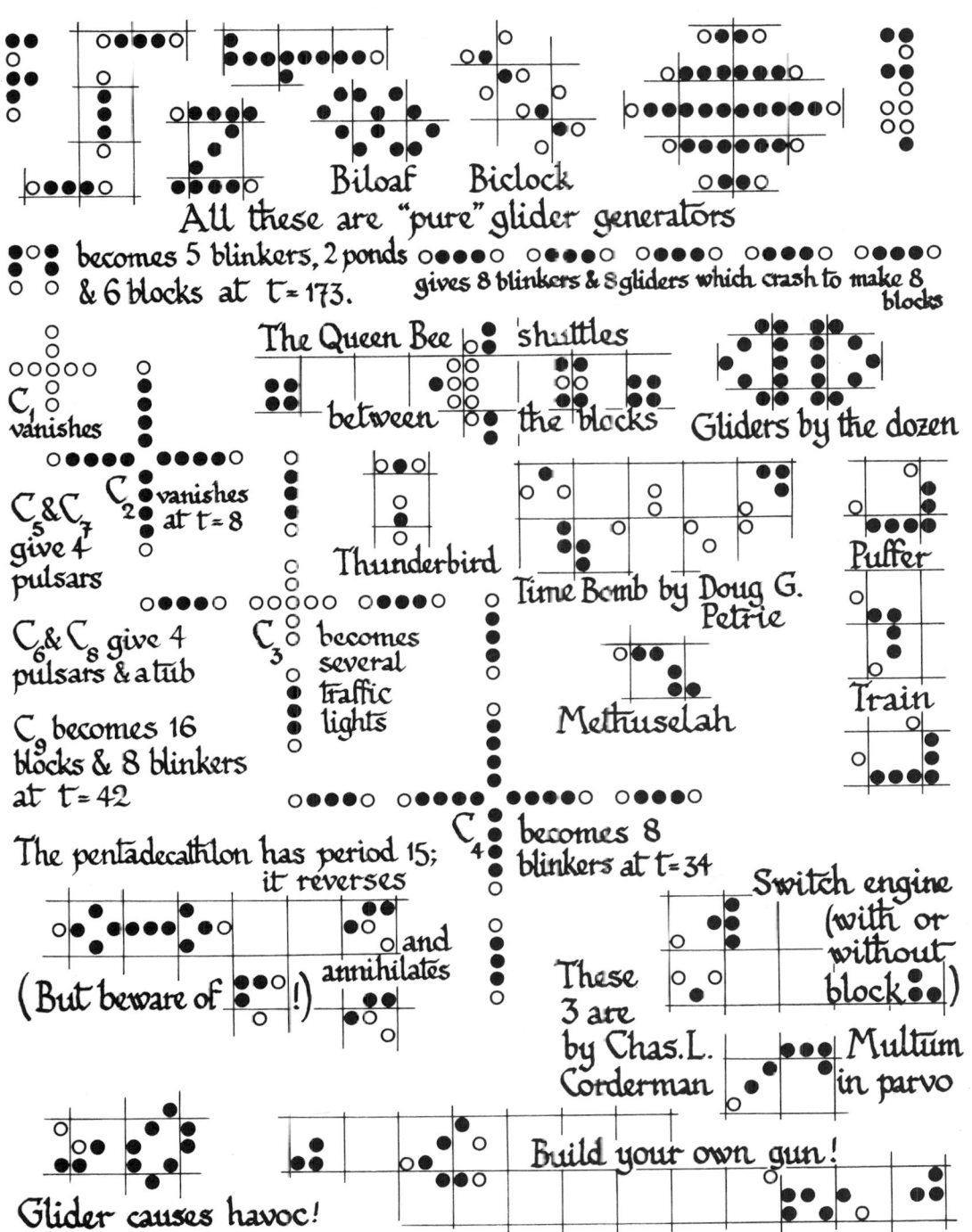

Biloaf Biclock

All these are "pure" glider generators

becomes 5 blinkers, 2 ponds & 6 blocks at $t = 173$.

gives 8 blinkers & 8 gliders which crash to make 8 blocks

The Queen Bee shuttles

C_1 vanishes

between the blocks

Gliders by the dozen

C_5 & C_7 give 4 pulsars

C_2 vanishes at $t = 8$

Thunderbird

Time Bomb by Doug G. Petrie

Puffer

C_6 & C_8 give 4 pulsars & a tub

C_3 becomes several traffic lights

Methuselah

Train

C_9 becomes 16 blocks & 8 blinkers at $t = 42$

The pentadecathlon has period 15; it reverses

C_4 becomes 8 blinkers at $t = 34$

Switch engine (with or without block)

(But beware of !) and annihilates

These 3 are by Chas. L. Corderman

Multum in parvo

Glider causes havoc!

Build your own gun!

Figure 13. Mainly for Computer Buffs.

827

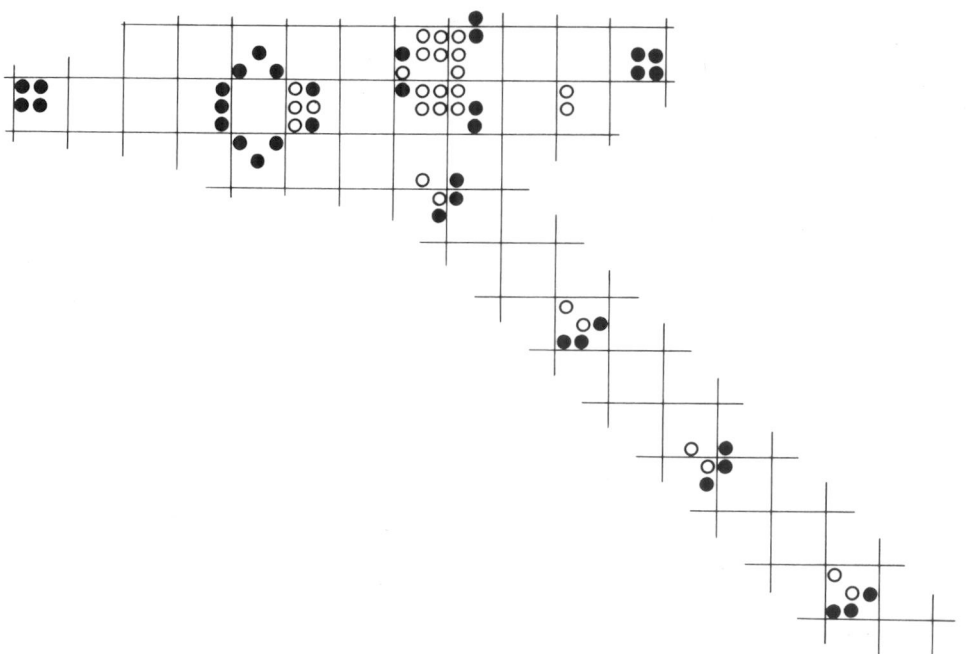

Figure 14. Gosper's Glider Gun.

GARDENS OF EDEN

There are Life configurations that can only arise as the initial state, because they have no ancestors!

We'll prove that if n is sufficiently large, there is some configuration within a $5n-2$ by $5n-2$ square that has no parent. It will suffice to examine that part of a prospective parent that lies in the surrounding $5n \times 5n$ square (Fig. 15). If any one of the component 5×5 squares is empty, it can be replaced as in Fig. 15(b) without affecting subsequent generations. So we need consider only

$$(2^{25} - 1)^{n^2} = 2^{24 \cdot 999999957004337\ldots n^2} \text{ of the } 2^{25n^2}$$

configurations in the $5n$ by $5n$ square. However, there are exactly

$$2^{(5n-2)^2} = 2^{25n^2 - 20n + 4}$$

possible configurations in the $5n-2$ by $5n-2$ square, so that if

$$24 \cdot 999999957004337\ldots n^2 < 25n^2 - 20n + 4,$$

one of these will have no parent! We calculate that this happens for $n = 465163200$, so that there is a Garden of Eden configuration that will fit comfortably inside a

2325816000 by 2325816000 square!

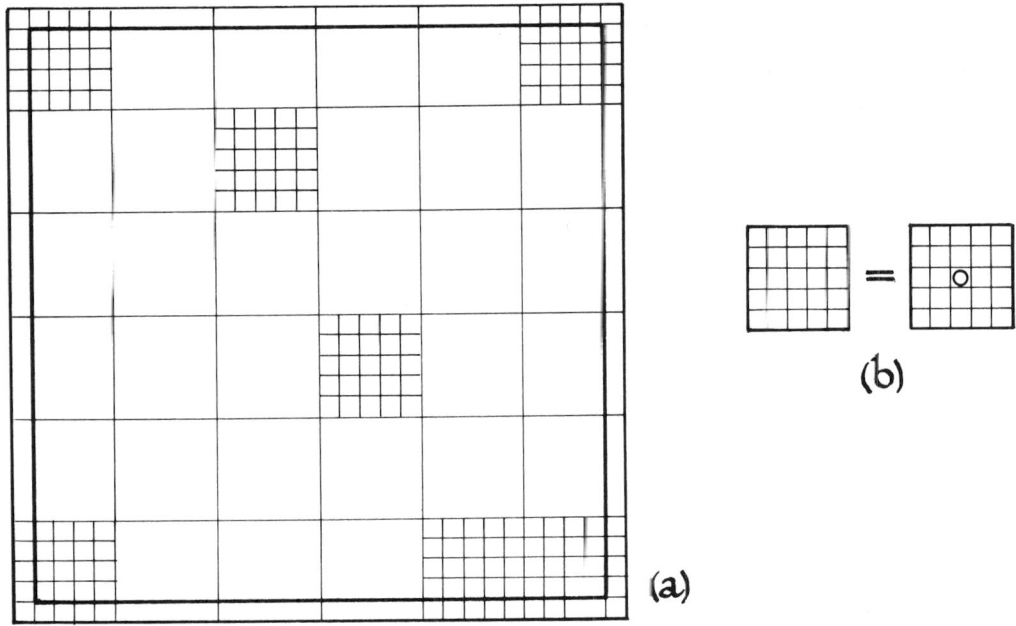

Figure 15. Location of the Garden of Eden.

This type of argument was first used by E.F. Moore in a more general context. More careful counting in the Life case has brought the size down to 1400 by 1400. However, using completely different ideas and many hours of computer time the M.I.T. group managed to produce an explicit example (Fig. 16).

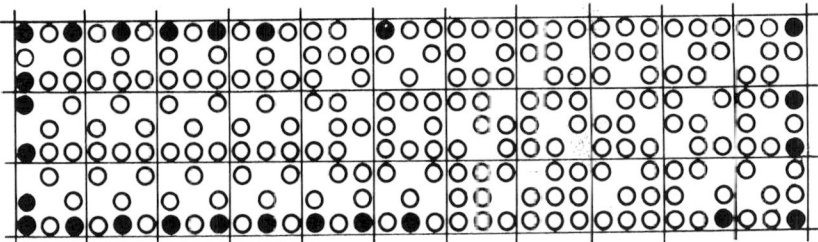

Figure 16. An Orphan Found by Roger Banks, Mike Beeler, Rick Schroeppel, Steve Ward, *et al.*

LIFE'S PROBLEMS ARE HARD!

The questions we posed about the ultimate destiny of Life configurations may not seem very mathematical. After all, Life's but a game! Surely there aren't any difficult mathematical problems there?

Well, yes there are! Indeed we can prove the astonishing fact that *every* sufficiently well-stated mathematical problem can be reduced to a question about Life! Those apparently trivial problems about Life histories can be arbitrarily difficult!

Here, for instance, is a tricky little problem that's kept mathematicians busy ever since Pierre de Fermat proposed it over 300 years ago. Is it possible for a perfect nth power to be the sum of two smaller ones for any n larger than 2? Despite many learned investigations by many learned mathematicians we still don't know! But if you had an infallible way to foretell the destiny of a given Life configuration, you'd be able to answer this question!

The reason is that we can design for you a finite starting pattern P_0 which will fade away completely if and only if there is a way of breaking an nth power into two smaller ones. If you had a mechanical method which would accept as input an arbitrary finite Life pattern P, and is guaranteed to respond with

> FADE, if the rules of Life will eventually cause P to disappear completely, and
> STAY, if not,

then you could apply it to P_0 and settle Fermat's question.

Even better, we could design a pattern P_1 which will tell you what those perfect powers are. If

$$a^n + b^n = c^n$$

 is the first solution of Fermat's problem in a certain dictionary order, then eventually P_1 will lead to a configuration in which there are

> a gliders, travelling North–West,
> b gliders, travelling North–East,
> c gliders, travelling South–West,
> n gliders, travelling South–East,

and nothing else at all! We can do the same sort of thing for other mathematical problems.

MAKING A LIFE COMPUTER

Many computers have been programmed to play the game of Life. We shall now return the compliment by showing how to define Life patterns that can imitate computers. Many remarkable consequences will follow from this idea.

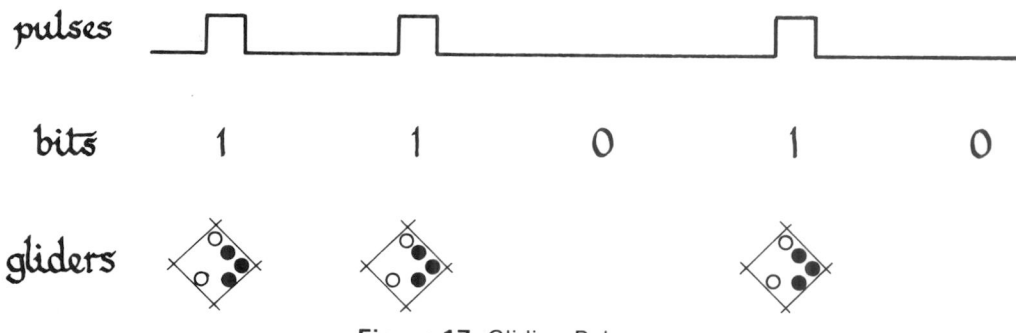

Figure 17. Gliding Pulses.

Good old fashioned computers are made from pieces of wire along which pulses of electricity go. Our basic idea is to mimic these by certain lines in the plane along which gliders travel (Fig. 17). (Because gliders travel diagonally, from now on we'll turn the plane through 45°, so they move across, or up and down, the page.) Somewhere in the machine there is a part called the **clock** which generates pulses at regular intervals and most of the working parts of the machine are made up of logical **gates**, like those drawn in Fig. 18. Obviously we can use Glider Guns as pulse generators. What should we do about the logical gates? Let's study the possible interactions of two gliders which crash at right angles.

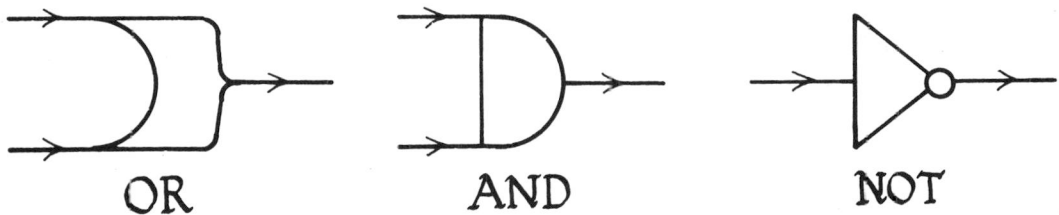

OR   AND   NOT

Figure 18. The Three Logical Gates.

WHEN GLIDER MEETS GLIDER

There are lots of different ways in which two gliders can meet, because there are lots of different possibilities for their exact arrangement and timing. Figure 19 shows them crashing (a) to form a blinker, (b) to form a block, (c) to form a pond, or (d) in one of several ways in which they can annihilate themselves completely. This last may seem rather unconstructive, but these **vanishing reactions** turn out to be surprisingly useful!

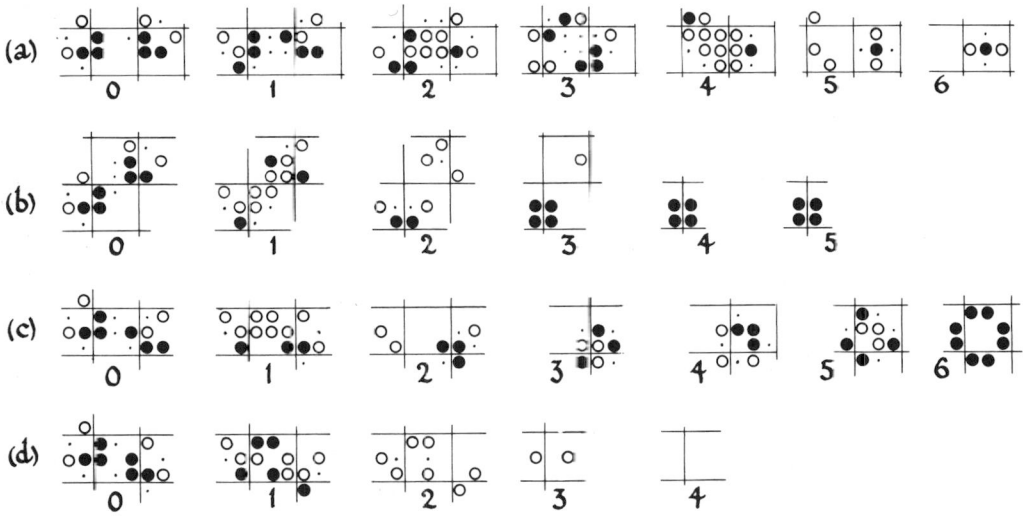

Figure 19. Gliders Crashin' in Diverse Fashion.

HOW TO MAKE A NOT GATE

We can use a vanishing reaction, together with a Glider Gun, to create a NOT gate (Fig. 20). The input stream enters at the left of the figure, and the Glider Gun is positioned and timed so that every space in the input stream allows just one glider to escape from the gun, while a glider in the stream necessarily crashes with one from the gun in a vanishing reaction (indicated by *). Figure 20 shows the periodic stream

$$1\ 1\ 0\ 1\ 1\ 0\ 1\ 1\ 0\ \dots$$

being complemented to

$$0\ 0\ 1\ 0\ 0\ 1\ 0\ 0\ 1\ \dots.$$

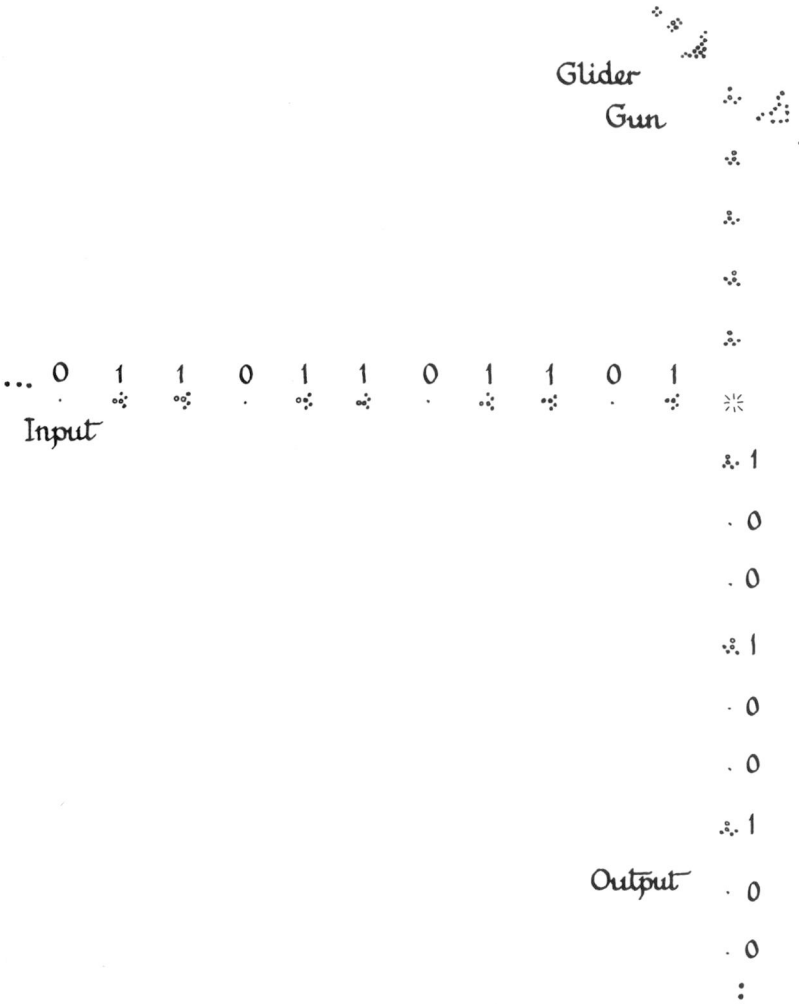

Figure 20. A Glider Gun and a Vanishing Reaction Make a NOT Gate.

Fortunately there are several vanishing reactions with different positions and timings in which the decay is so fast that later gliders from the same gun stream will not be affected (Fig. 21). This means that we can reposition a glider stream arbitrarily by turning it through sufficiently many corners (Fig. 22).

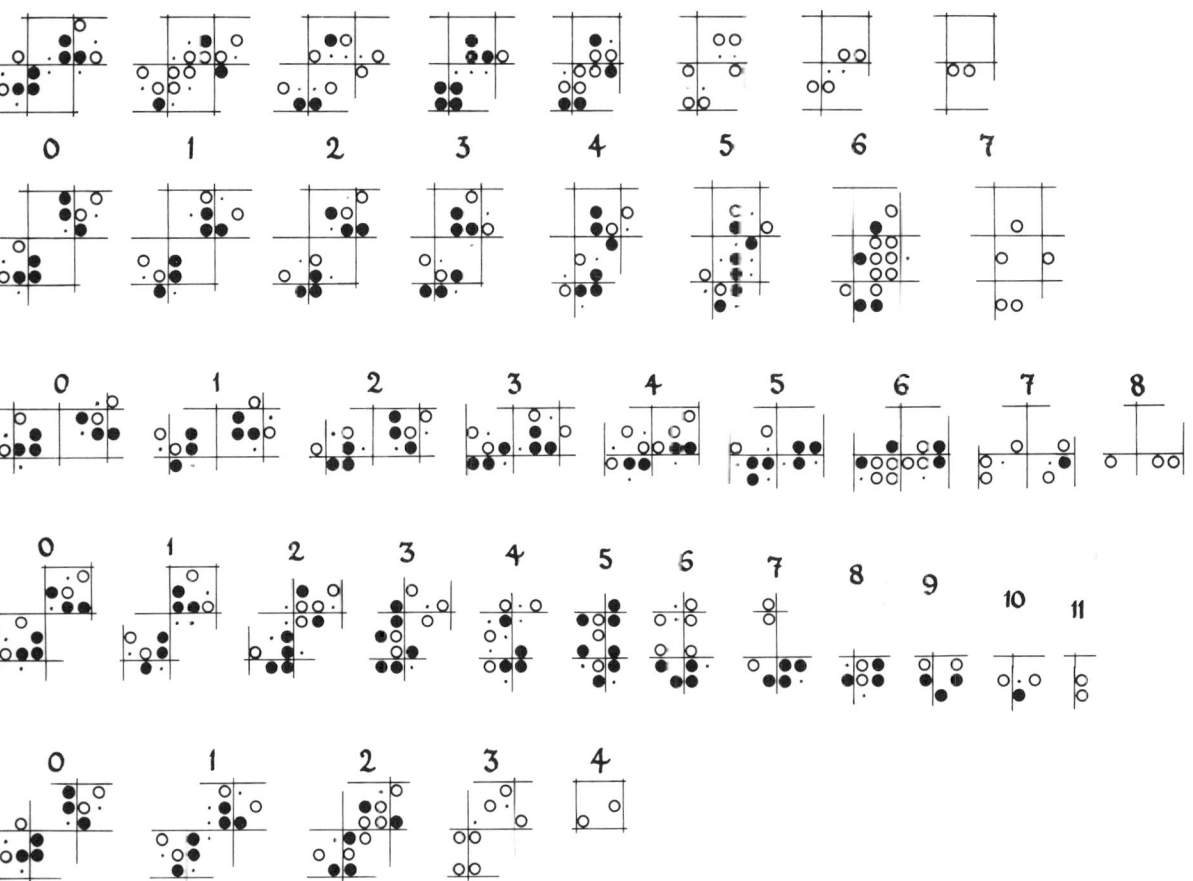

Figure 21. A Variety of Vanishing Reactions Between Crashing Gliders.

THE EATER

What else can happen when glider meets glider? Lots of things! One of them is to make an **eater** (Fig. 23) and an eater can eat lots of things without suffering any indisposition. The eater, which was discovered by Gosper, will be very useful to us; in Fig. 24 you can see it enjoying a varied diet of (a) a blinker, (b) a pre-beehive, (c) a lightweight spaceship, (d) a middleweight spaceship, and (e) a glider. If it attempts a heavyweight spaceship it gets indigestion and leaves a loaf behind; if it tries a blinker in the wrong orientation it leaves a baker's shop!

Sometimes glider streams are embarrassing to have around, so it's especially useful then— it just sits there and eats up the whole stream!

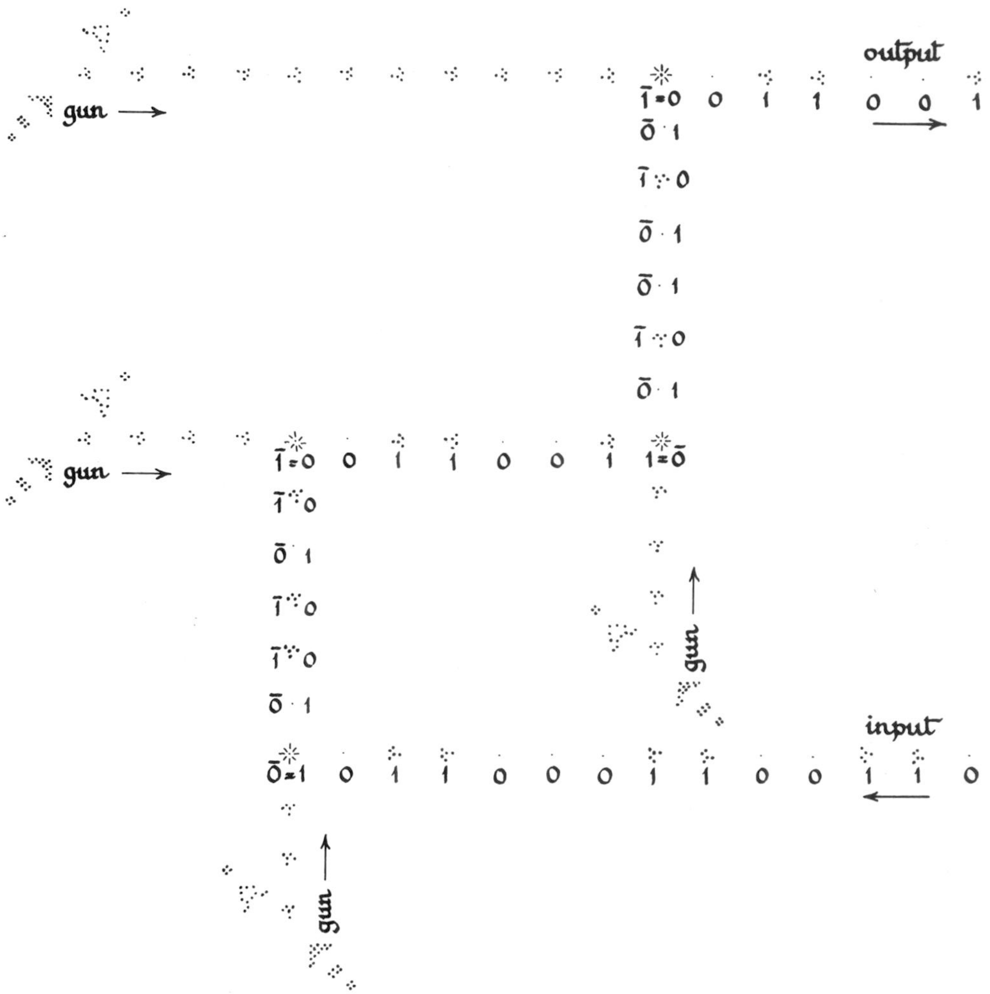

gun → output

$\bar{1}=0$ 0 1 1 0 0 1 →

$\bar{0}\cdot 1$

$\bar{1}\cdot 0$

$\bar{0}\cdot 1$

$\bar{0}\cdot 1$

$\bar{1}\cdot 0$

$\bar{0}\cdot 1$

gun → $\bar{1}=0$ 0 1 1 0 0 1 $1=\bar{0}$

$\bar{1}\cdot 0$

$\bar{0}\cdot 1$

$\bar{1}\cdot 0$

$\bar{1}\cdot 0$

$\bar{0}\cdot 1$

gun ↑

$\bar{0}=1$ 0 1 1 0 0 0 1 1 0 0 1 1 0   input ←

gun ↑

Figure 22. Repositioning and Delaying a Glider Stream.

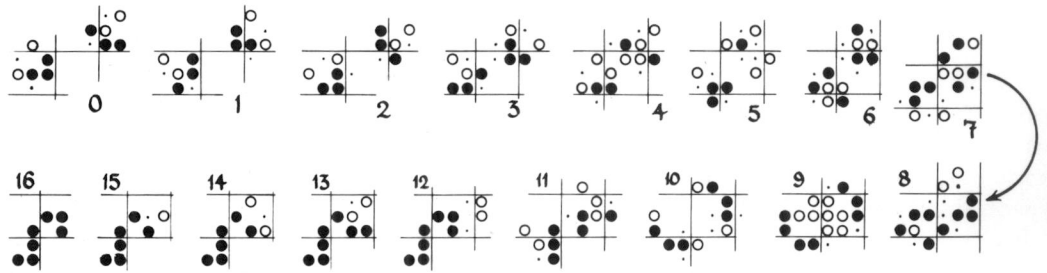

Figure 23. Two Gliders Crash to Form an Eater.

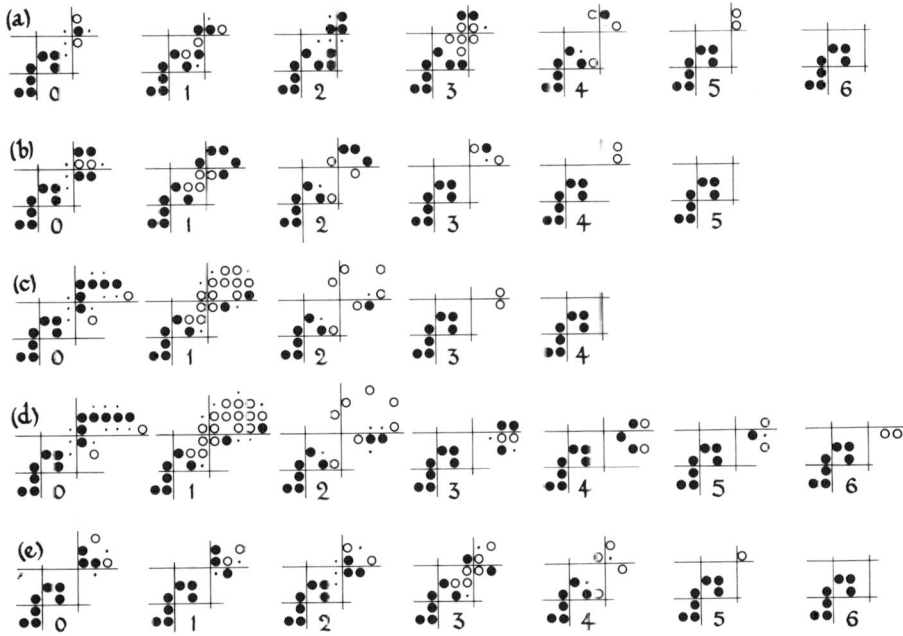

Figure 24. The Voracious Eater Devours a Varied Meal.

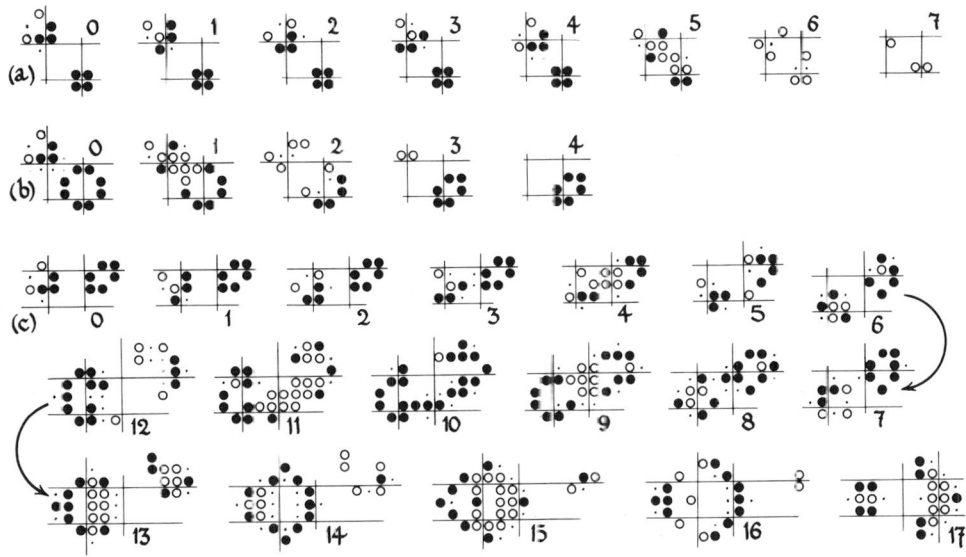

Figure 25. (a) Blockbusting Glider. (b) Glider Dives into Pond and Comes Up With Ship. (c) Glider Crashes into Ship and Makes Part of Glider Gun.

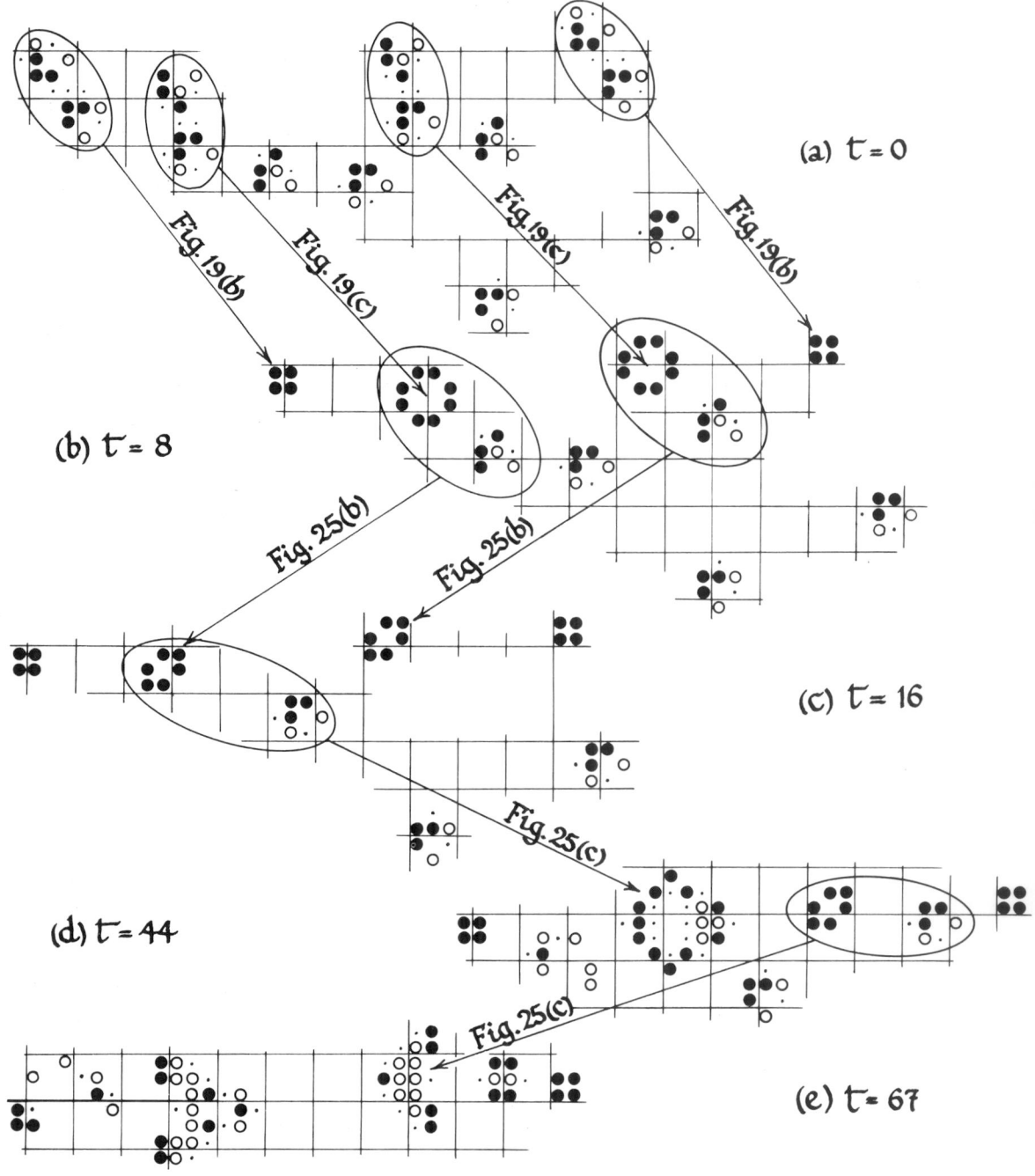

Figure 26. Thirteen Gliders Build Their Own Gun.

GLIDERS CAN BUILD THEIR OWN GUNS!

What happens when a glider meets other things? We have seen it get eaten by an eater. It can also annihilate a block (and itself! Fig. 25(a)). But more constructively it can turn a pond into a ship (Fig. 25(b)) and a ship into a part of the glider gun (Fig. 25(c)). And since gliders can crash to make blocks (Fig. 19(b)) and ponds (Fig. 19(c)), they can make a whole glider gun! The 13 gliders in Fig. 26(a) do this in 67 generations. Figures 26(b,c,d) show the positions after 8, 16 and 44 generations. The extra glider then slips in to deal with an incipient beehive, and by 67 generations (Fig. 26(e)) the gun is in full working order and launches its first glider 25 generations later.

THE KICKBACK REACTION

Yet another very useful reaction between gliders is the **kickback** (Fig. 27(a)) in which the decay product is a glider travelling along a line closely parallel to one of the original ones, but in the opposite direction. We think of this glider as having been *kicked back* by the other one. Figure 27(b) shows our notation for the kickback.

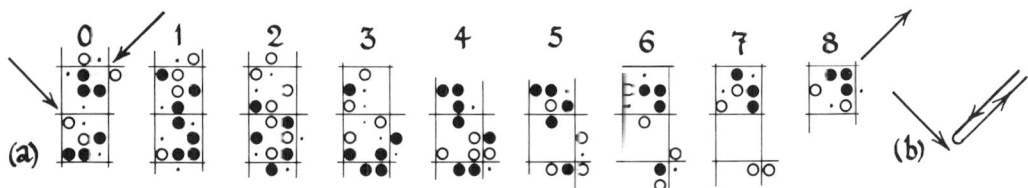

Figure 27. The Kickback.

All the working parts of our computer will be moving glider streams, meeting in vanishing and kickback reactions. The only static parts will be glider guns and eaters (indicated by G and E in the figures).

THINNING A GLIDER STREAM

The glider streams that emerge from normal guns are so dense that they cannot interpenetrate without interfering. If we try to build a computer using streams of this density we couldn't allow any two wires of this kind to cross each other, so we'd better find some way to reduce the pulse rate.

In Fig. 28 the guns G_1 and G_2 produce normal glider streams in parallel but opposite directions. But there is a glider g which will travel West until at A it is kicked East by a glider from the G_1 stream. The timing and phasing are such that at B it will be kicked back towards A again, so that it repeatedly "loops the loop", removing one glider from each of the two streams per cycle. After this every Nth glider is missing from each of these streams. We don't want the G_1 stream, so we feed it into an eater, but we feed the G_2 stream into a vanishing reaction with a stream from a third gun G_3. Every glider from G_2 now dies, but every Nth one from G_3 escapes through a hole in the G_2 stream! So the whole pattern acts as a **thin gun**, producing just one Nth as many gliders as the normal gun. To get the phasing right, N must be divisible by 4, but it can be arbitrarily

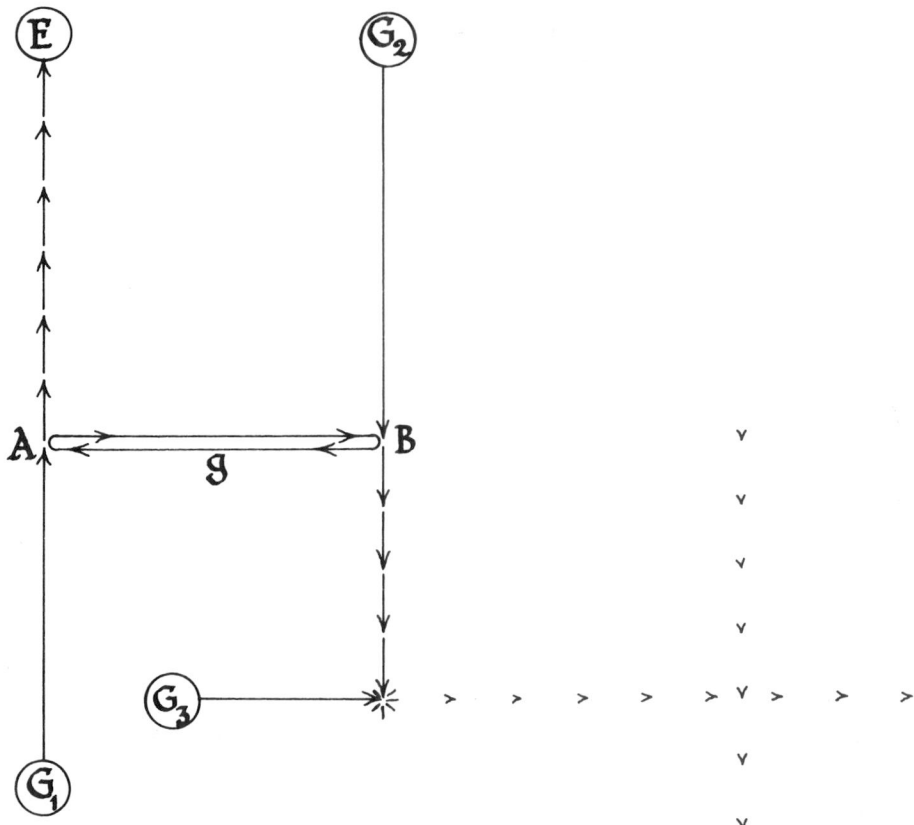

Figure 28. Thinning a Glider Stream.

large and so we can make an arbitrarily thin stream. Now two such streams can cross without interacting as in the right hand part of Fig. 28, provided things are properly timed. So from now on we can use the word **gun** to mean an arbitrarily thin gun. Perhaps a thinning factor of 1000 will make all our constructions work.

BUILDING BLOCKS FOR OUR COMPUTER

In Fig. 29 we see how to build logical gates using only vanishing reactions (we've already seen the NOT gate in more detail in Fig. 20). But there's a problem! The output streams from the AND and OR gates are *parallel* to the input, but the output stream from the NOT gate is at *right angles* to the input. We need a way to turn streams round corners without complementing them, or of complementing them without turning them round corners. Fortunately the solution to our *next* problem automatically solves *this* one.

The new problem is to provide several copies of the information from a given glider stream, and we found it a hard problem to solve. To get some clues, let's see what happens when we use one glider to kick back a glider from a gun stream.

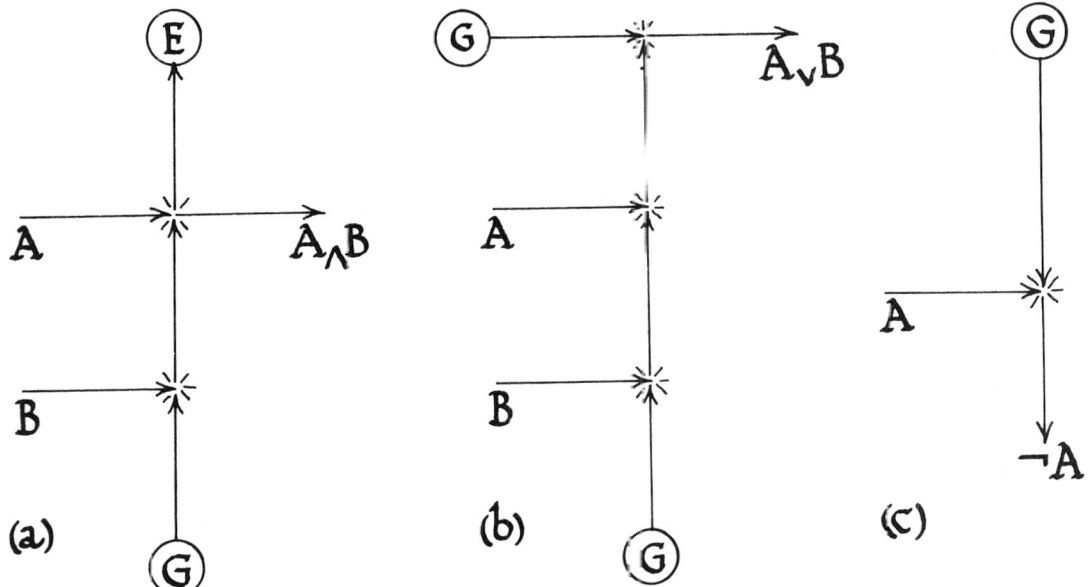

Figure 29. (a) An AND Gate. (b) An OR Gate. (c) A NOT Gate.

We suppose that the gun stream, the **full stream**, produces a glider every 120 generations (a quarter of the original gun density; $N = 4$ in the previous section). Then it turns out that when we kick back the first glider, the effect is to remove just three gliders from the stream! This happens as follows:

(i) The first glider is kicked back (Fig. 27) along the full stream.
(ii) The second glider crashes into the first, forming a block (somewhat as in Fig. 19(b)).
(iii) The third glider annihilates the block (Fig. 25(a)).
(iv) All subsequent gliders from the full stream escape unharmed.

We can use this curious behavior as follows. Suppose that our information-carrying stream operates at one tenth, say, of the density of the full stream, so that the last 9 of every 10 places on it will be empty, while the first place might or might not be full. If we use 0 for a hole and block the places in tens, our stream looks like

$$\ldots\; 000000000D\; 000000000C\; 000000000B\; 000000000A \rightarrow$$

We first feed it into an OR gate with a stream of type

$$\ldots\; 00000000g0\; 00000000g0\; 00000000g0\; 00000000g0 \rightarrow$$

the g's denoting gliders that are definitely present. The result is a stream

$$\ldots\; 00000000gD\; 00000000gC\; 00000000gB\; 00000000gA \rightarrow$$

in which every information-carrying place is definitely followed by a glider g. This stream is used to kick back a full stream whose gliders are numbered:

$$\ldots\; \ldots\ldots\ldots\ldots\; \ldots\ldots\ldots\; X987654321\; X987654321 \rightarrow$$

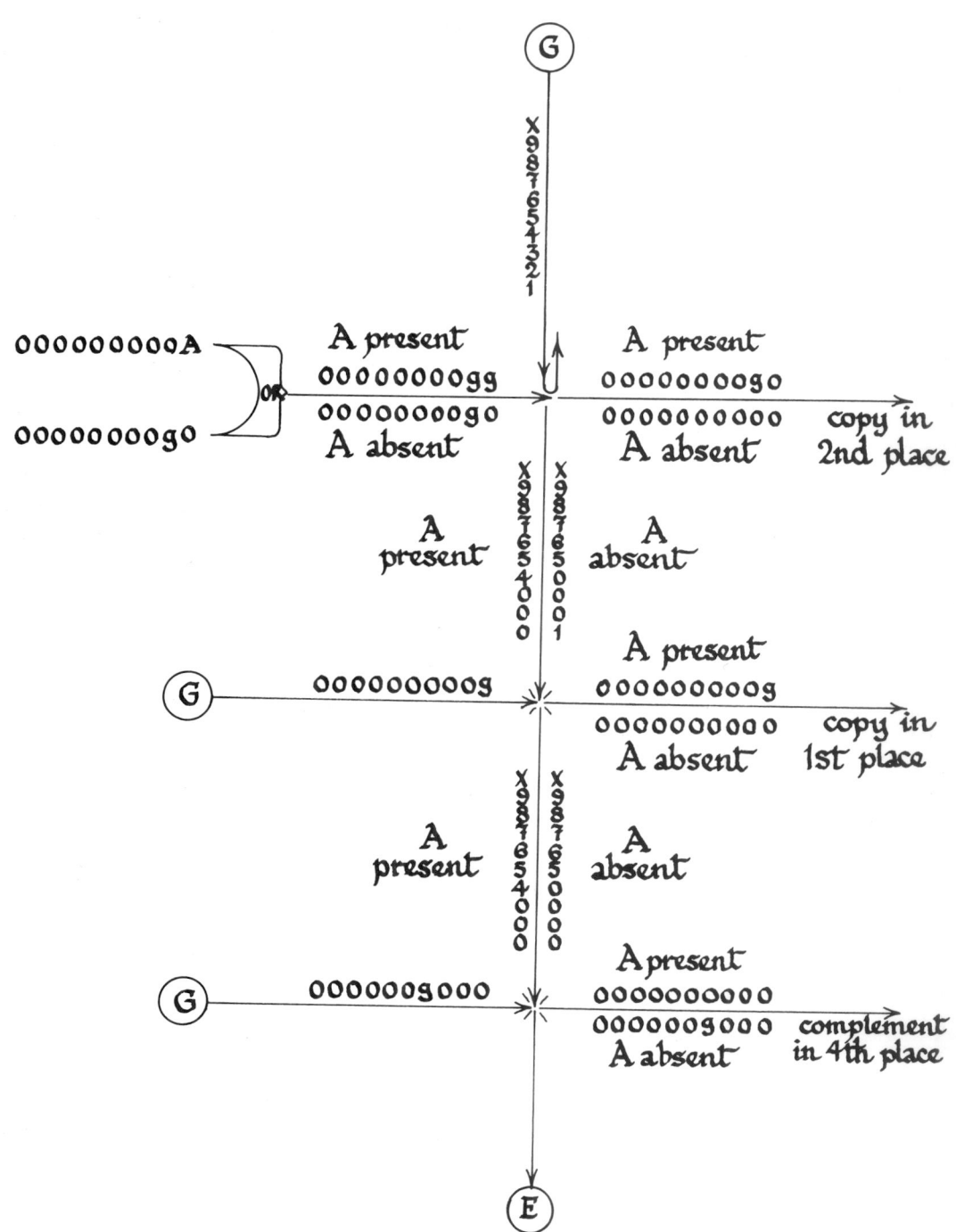

Figure 30. Copying a Glider Stream.

If glider *A* is *present*, it will obliterate gliders 1, 2 and 3 of the full stream and the following glider *g* can escape in the confusion. But if *A* is *absent*, then full stream glider 1 escapes and gliders 2, 3, 4 are removed instead by the following glider *g*. So the stream which emerges is definitely empty except for the second of every ten places and these places carry a copy of the input stream. The original full stream now manages to carry the information *twice*, in the first and fourth digits of each block, the first digit carrying the *complemented* version (which has *not* been turned through a right angle). By feeding this stream into vanishing reactions with suitably thin streams we can recover the original stream either complemented or not, and freed from undesirable accompanying gliders! Figure 30 shows these techniques in action.

From here on it's just an engineering problem to construct an arbitrarily large finite (and very slow!) computer. Our engineer has been given the tools—let him finish the job! We know that such computers can be programmed to do many things. The most important ones that we will want it to do involve emitting sequences of gliders at precisely controlled positions and times.

AUXILIARY STORAGE

Of course the engineer will probably have designed an internal memory for our computer using circulating delay lines of glider streams. Unfortunately this won't be enough for the kind of problem we have in mind, and we'll have to find some way of adjoining an *external* memory, capable of holding arbitrarily large numbers. To build this memory, we'll need an additional static piece (the block).

For instance we might ask the computer to compute

$$a^n + b^n \text{ and } c^n$$

for *all* quadruples (a,b,c,n) in turn and stop when it finds a quadruple for which

$$a^n + b^n = c^n.$$

We don't know how big a, b, c and n might get, and they'll almost certainly get too large even to be written in the internal memory.

So we're going to adjoin some auxiliary storage registers, each of which will store an arbitrarily large number. Figure 31 shows the general plan. Each register contains a block, whose distance from the computer (on a certain scale) indicates the number it contains. In the figure, register *A* contains 3, *B* contains 7, *C* contains 0 and *D* contains 2. When the contents of a register is 0, the block is just inside the computer. All we have to do is to provide a way for the computer to

> *increase* the contents of a register by 1,
> *decrease* the contents of a register by 1, and
> *test*   whether the contents are 0.

Fortunately each of these can be accomplished by a suitable fleet of gliders. One such fleet is off to increase register *B* by one! And another glider is about to discover that register *C* contains 0.

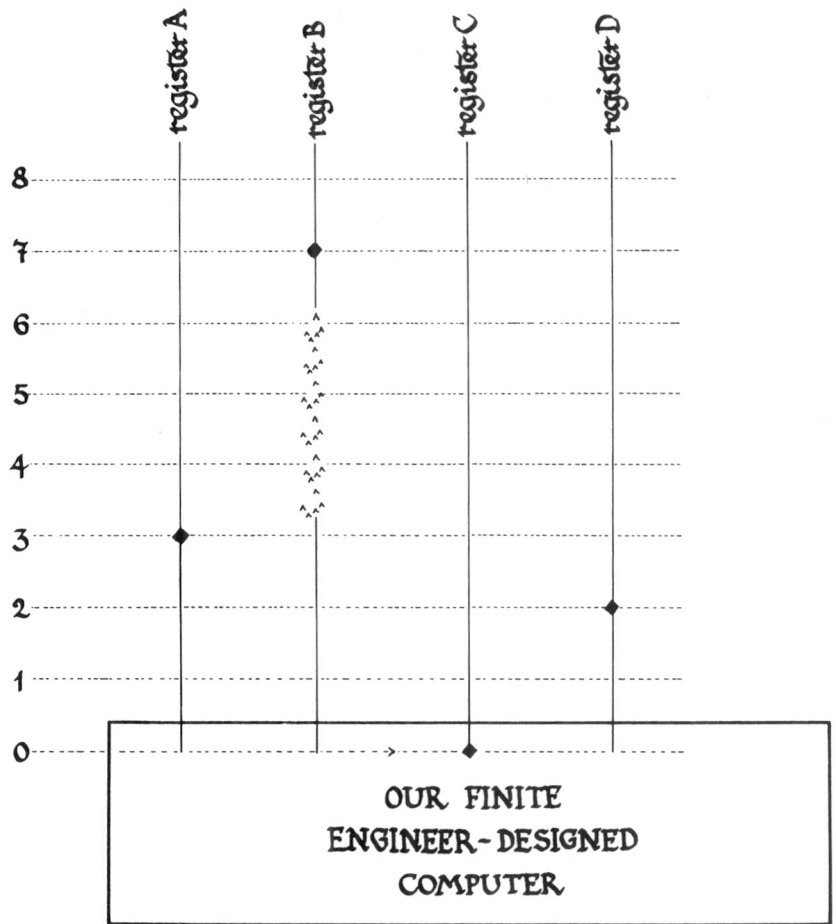

Figure 31. Auxiliary Storage.

HOW WE MOVE BLOCKS

To find these fleets we studied the six possible glider–block crashes. One of them does indeed bring the block in a bit, but unfortunately by a knight's move. However the block can be brought back onto the proper diagonal by repeating the process with a reflected glider on a parallel course. The combined effect of this pair of gliders is to pull the block back three diagonal places (Fig. 32).

Unfortunately there is no single glider-block crash which moves the block further away, but there is a crash which produces the arrangement of 4 beehives we call a honey farm, and two of these four are slightly further away, and so we can send in second, third and fourth gliders to annihilate three of the beehives, and then a fifth glider which converts the remaining beehive back into a block. The total effect again pushes the block off the proper diagonal, but a second team of five gliders will restore this, resulting in a block just one diagonal place further out! (Fig. 33).

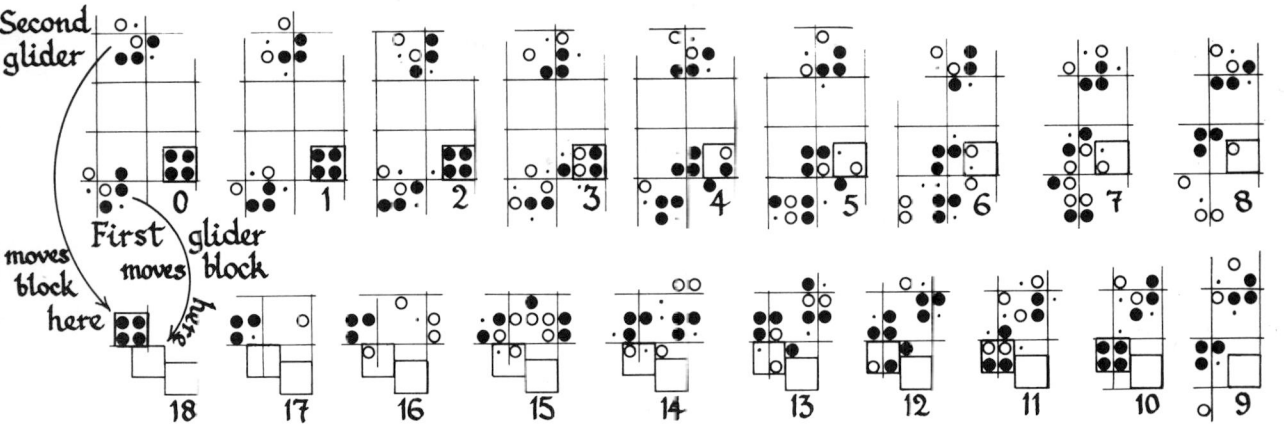

Figure 32. Two Gliders Pull a Block Back Three Diagonal Places.

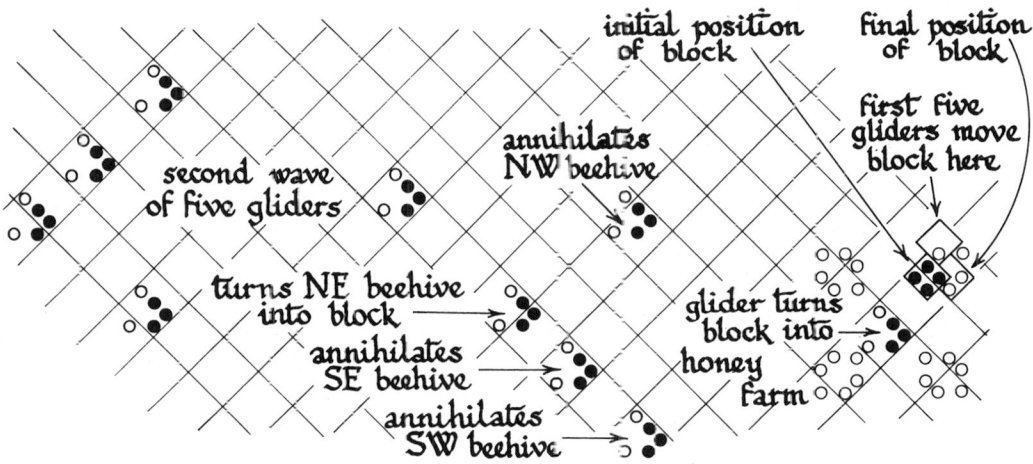

Figure 33. Ten Gliders Move a Block Just One Diagonal Place.

We therefore choose a diagonal distance of 3 to represent a change of 1 in a register and can decrease the contents of a register using a pair of gliders, or increase it using 3 flotillas of 10.

Apart from the difficulty discussed in the next section we have now finished the work, for Minsky has shown that a finite computer, equipped with memory registers like the ones in Fig. 30, can be programmed to attack arbitrarily complicated mathematical problems.

A LITTLE DIFFICULTY

But now comes the problem. Every glider in our finite computer has at some time been produced by a glider gun, so how could we arrange to send those gliders along closely parallel, but distinct paths? Surely one gun would have to fire right through another (Fig. 34)? Our technique of **side-tracking** uses three computer controlled guns G_1, G_2, G_3 as in Fig. 35. These are programmed to emit gliders exactly when we want them to.

Figure 34. How Can One Gun Fire Through Another?

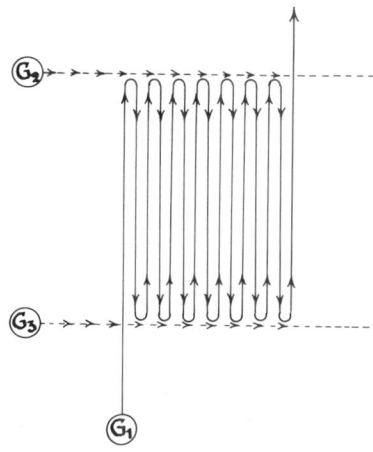

Figure 35. Side-tracking.

Firstly, G_1 emits a glider g travelling upwards,
Secondly, G_2 emits a glider at just the right time to kick g back downwards,
Thirdly, G_3 kicks g back up again,

and so on, alternately, until at a suitable time G_2 fails to fire and g is released. By controlling the number of times G_2 and G_3 fire, the *same* guns can be used to send a succession of gliders along distinct parallel paths.

MISSION COMPLETED—WILL SELF-DESTRUCT

Side-tracking can be used for a much more spectacular juggling act! We can actually program our computer to throw a glider into the air *and* bring it back down again. In Fig. 36, G_1, G_2, G_3 behave as before and can be programmed to arrange that a glider g ends up travelling Eastwards arbitrarily far above the ground. But G_4 has been arranged to emit a glider which will be kicked back down by g. We could even arrange to kick it back up again, then down again, then up again,... as suggested by the dotted lines in Fig. 36.

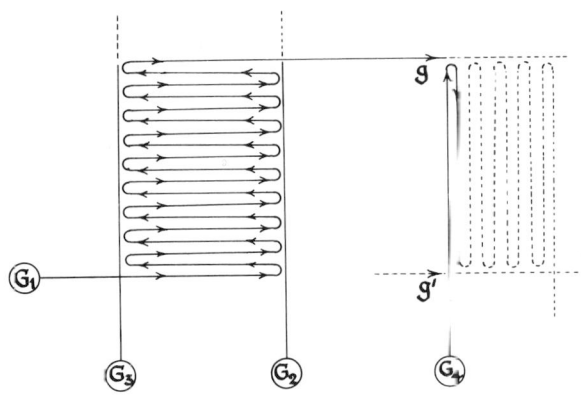

Figure 36. Double Side-tracking.

Using such techniques we can design a program for our computer which will send large numbers of gliders far out into space and then turn them round so that they head back towards the computer along precisely defined tracks (Fig. 37).

Now comes the clever part. Figures 38(a), 38(b) and 25(a) show that the eaters, the guns' moving parts, and blocks, can all be destroyed by aiming suitably positioned gliders from behind their backs. If the computer is cleverly designed we can even destroy it completely by an appropriate configuration of gliders!

Here's the idea. We design the computer so that every glider emitted by a gun or circulating in a loop would, if not deflected by meeting other gliders, be eventually consumed by an appropriately placed eater. Then we design our attacking force of gliders to shoot the computer down, guns first. After each gun is destroyed we wait until any gliders it has already emitted have percolated through the system and either been destroyed by other gliders, or swallowed by eaters, before attacking the next gun. When all the guns are destroyed we shoot down the eaters and blocks.

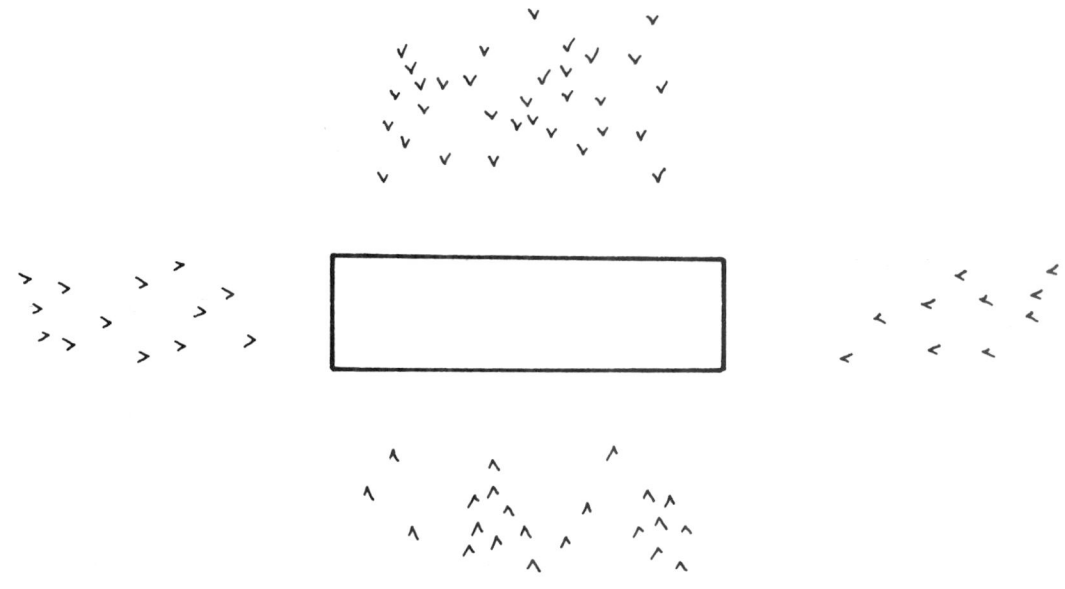

Figure 37. Self-attack from All Directions!

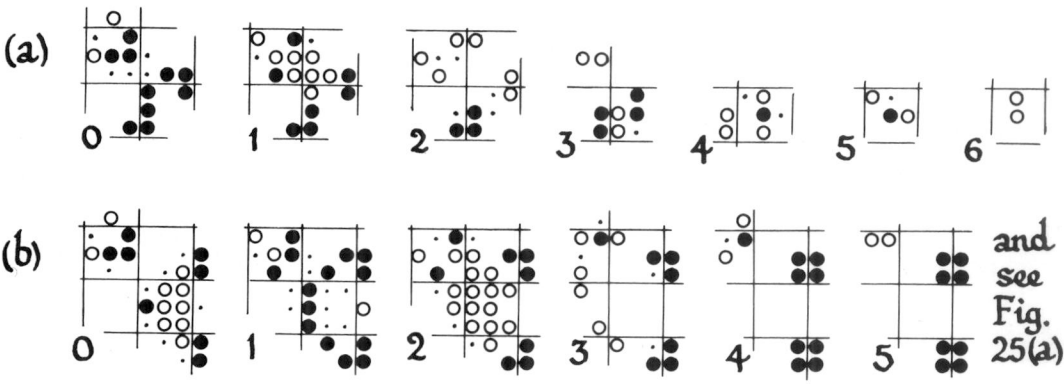

Figure 38. (a) The Eater Eaten! (b) The Gun Gunned Down!

The whole process requires some care. Each gun G_i must have a matching eater E_i, and G_i and E_i lie in a strip of the plane which contains no other static parts of the computer (Fig. 39). The gliders $g_1, g_2, g_3, \ldots$ with which we shoot down a given gun can be arbitrarily widely spaced in time provided they come in along the right tracks. Moreover we can arrange to shoot down the successive guns, eaters and blocks after increasingly long intervals of time.

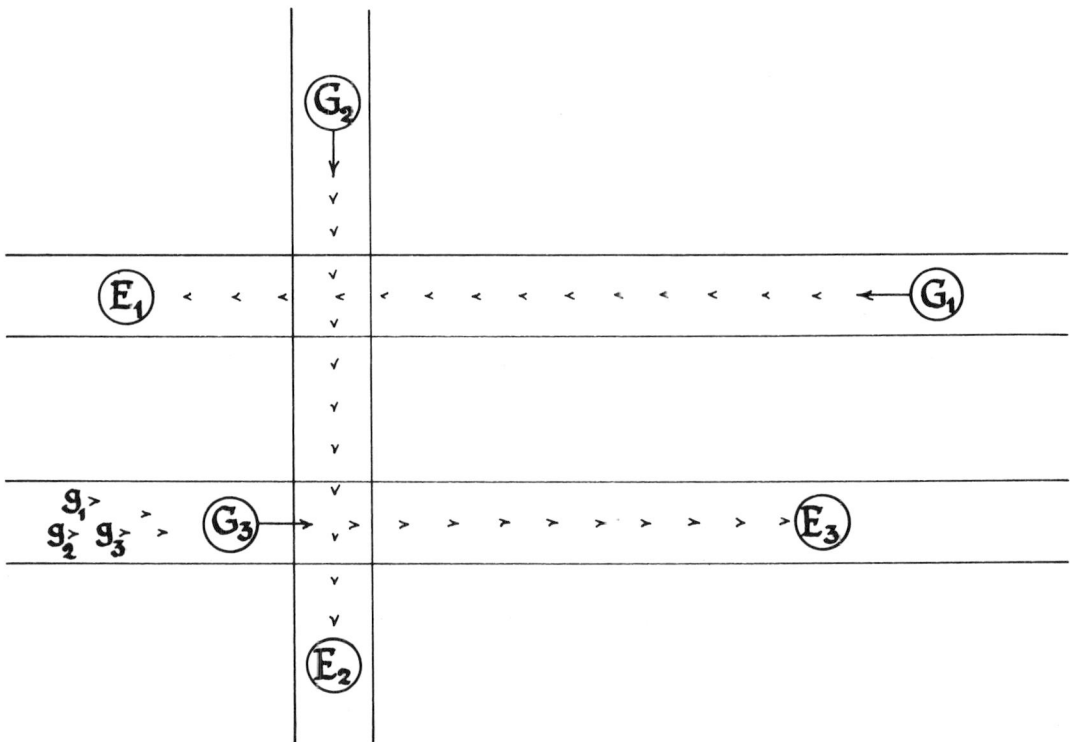

Figure 39. Arranging Destroyable Guns.

However, it *can* be done! We intend to use it like this. Program the computer to look for a solution of an arbitrarily hard problem, such as Fermat's. If it never finds a solution it will just go on forever. However, if it *does* find a solution we instruct it to throw into the air a precisely arranged army of gliders, then reduce all its storage numbers to zero (this brings all the blocks inside the computer), switch off, and await its fate. The attacking glider army, of course, is exactly what's needed to obliterate the computer, leaving no trace. It's important to realize that a *fixed* computer can be programmed to produce many different patterns of gliders and in particular the one required to kill itself. The information about this glider pattern can be held by the numbers in the memory of the computer and not in the computer's design.

Since mathematical logicians have proved that there's no technique which guarantees to tell when arbitrary arithmetical problems have solutions, there's no technique that's guaranteed to tell even when a Life configuration will fade away completely. The kind of computer we have simulated is technically known as a *universal machine* because it can be programmed to perform any desired calculation. We can summarize our result in this answer to our chapter heading:

> LIFE IS UNIVERSAL!

LIFE COMPUTERS CAN REPRODUCE!

Eaters and guns can be made by crashing suitable fleets of gliders, so it's possible to build a computer simply by crashing some enormously large initial pattern of gliders. Moreover, we can design a computer whose sole aim in Life is to throw just such a pattern of gliders into the air. In this way one computer can give birth to another, which can, if we like, be an exact copy of the first. Alternatively, we could arrange that the first computer eliminates itself after giving birth; then we would regard the second as a reincarnation of the first.

> There are Life patterns which behave
> like self-replicating animals.

> There are Life patterns which move
> steadily in any desired rational
> direction, recovering their initial
> form exactly after some fixed
> number of generations.

GENETIC ENGINEERING

We've now shown that among finite Life patterns there is a very small proportion behaving like self-replicating animals. Moreover, it is presumably possible to design such patterns which will survive inside the typical Life environment (a sort of primordial broth made of blocks, blinkers, gliders, ...). It might for instance do this by shooting out masses of gliders to detect nearby objects and then take appropriate action to eliminate them. So one of these "animals" could be more or less adjusted to its environment than another. If both were self-replicating and shared a common territory, presumably more copies of the better adapted one would survive and replicate.

WHITHER LIFE?

From here on is a familiar story. Inside any sufficiently large random broth, we expect, *just by chance*, that there will be some of these self-replicating creatures! Any particularly well adapted ones will gradually come to populate their territory. Sometimes one of the creatures will be accidentally modified by some unusual object which it was not programmed to avoid. Most of these modifications, or **mutations**, are likely to be harmful and will adversely affect the animal's chances of survival, but very occasionally, there will be some *beneficial* mutations. In these cases the modified animals will slowly come to predominate in their territory, and so on. There seems to be no limit to this process of evolution.

> It's probable, given a large enough
> Life space, initially in a random state,
> that after a long time, intelligent
> self-reproducing animals will emerge and
> populate some parts of the space.

This is more than mere speculation, since the earlier parts are based on precisely proved theorems. Of course, "sufficiently large" means very large indeed, and we can't prove that "living" animals of any kind are likely to emerge in any Life space we can construct in practice.

It's remarkable how such a simple system of genetic rules can lead to such far-reaching results. It may be argued that the small configurations so far looked at correspond roughly to the molecular level in the real world. If a two-state cellular automaton can produce such varied and esoteric phenomena from these simple laws, how much more so in our own universe?

Analogies with real life processes are impossible to resist. If a primordial broth of amino-acids is large enough, and there is sufficient time, self-replicating moving automata may result from transition rules built into the structure of matter and the laws of nature. There is even the possibility that space-time itself is granular, composed of discrete units, and that the universe, as Edward Fredkin of M.I.T. and others have suggested, is a cellular automaton run by an enormous computer. If so, what we call motion may be only simulated motion. A moving particle in the ultimate microlevel may be essentially the same as one of our gliders, appearing to move on the macrolevel, whereas actually there is only an alteration of states of basic space-time cells in obedience to transition rules that have yet to be discovered.

REFERENCES AND FURTHER READING

Clark C. Abt, "Serious Games: The Art and Science of Games that Simulate Life", The Viking Press, 1970.

Michael A. Arbib, Simple self-reproducing universal automata, Information and Control, **9** (1966) 177–189.

E.R. Banks, Information Processing and Transmission in Cellular Automata, Ph.D. thesis, M.I.T., 71:01:15.

E.F. Codd, "Cellular Automata", Academic Press, New York and London, 1968.

Martin Gardner, Mathematical Games, Sci. Amer. **223** #4 (Oct. 1970) 120–123; **223** #5 (Nov. 1970) 118; **223** #6 (Dec. 1970) 114; **224** #1 (Jan. 1971) 108; **224** #2 (Feb. 1971) 112–117; **224** #3 (Mar. 1971) 108–109; **224** #4 (Apr. 1971) 116–117; **225** #5 (Nov. 1971) 120–121; **226** #1 (Jan. 1972) 107; **233** #6 (Dec. 1975).

M.J.E. Golay, Hexagonal parallel pattern transformations, IEEE Trans. Computers **C18** (1969) 733–740.

Chester Lee, Synthesis of a cellular universal machine using the 29-state model of von Neumann, Automata Theory Notes, Univ. of Michigan Engg. Summer Conf., 1964.

Marvin L. Minsky, "Computation: Finite and Infinite Machines", Prentice-Hall, Englewood Cliffs, N.J., 1967.

Edward F. Moore, Mathematics in the biological sciences, Sci. Amer. **211** #3 (Sep. 1964) 148–164.

Edward F. Moore, Machine models of self-reproduction, Proc. Symp. Appl. Math. **14**, Amer. Math. Soc. 1962, 17–34.

Edward F. Moore, John Myhill, in Arthur W. Burks (ed.) "Essays in Cellular Automata", University of Illinois Press, 1970.

C.E. Shannon, A universal Turing machine with two internal states, in C.E. Shannon and J. McCarthy (eds.) "Automata Studies", Princeton University Press, 1956.

Alvy Ray Smith, Cellular automata theory, Tech. Report No. 2, Digital Systems Lab., Stanford Electronics Labs., Stanford Univ., 1969.

J.W. Thatcher, Universality in the von Neumann cellular model, Tech. Report 03105-30-T, ORA, Univ. of Michigan, 1964.

A.M. Turing, Computing machinery and intelligence, Mind, **59** (1950) 433–460.

Robert T. Wainwright (editor) Lifeline; a quarterly newsletter for enthusiasts of John Conway's game of Life, **1–11**, Mar., Jun., Sep., Dec. 1971, Sep., Oct., Nov., Dec. 1972, Mar., Jun., Sep. 1973.

Index

GLOSSARY OF SYMBOLS

See also the Appendix (pp. 225–228) to ONAG (reference on p. 24).

A = **ace** = $\{0 \mid \textbf{tiny}\}$, 337

$\overline{A}$ = $-$**ace** = $\{\textbf{miny} \mid 0\}$, 339

$A-$ = $\{\textbf{on} \mid A \parallel 0\}$, 339

$\overline{A}+$ = $\{0 \parallel \overline{A} \mid \textbf{off}\}$, 339

$\aleph_0$, aleph-zero, 309

$\lceil\ \rceil$, "ceiling", least integer not less than. 453

$(x)_n$, $\{x \mid -y\}_x$, Childish Hackenbush values, 230

$\overline{1}\clubsuit = \{\clubsuit \mid 0\}$
$\clubsuit = 0\ \clubsuit = \{1\clubsuit \mid 0\}$, clubs, 339
$1\clubsuit = \{\textbf{deuce} \mid 0\}$

$a\langle b\rangle$, class a and variety b, 343

●, ◯, ☢, Col and Snort positions, 49, 142

$\gamma°$, degree of loopiness, 341

$2\clubsuit = \{0 \mid \textbf{ace}\} = \textbf{ace} + \textbf{ace} = \textbf{deuce}$, 337

$\overline{1}\diamond = \{\overline{J} \mid 0\}$
$\diamond = 0\ \diamond = \{\textbf{ace} \mid \overline{1}\diamond\}$, diamonds, 339
$1\diamond = \{0 \mid \diamond\}$

$\Downarrow = \{\downarrow* \mid 0\} = \downarrow+\downarrow$, double-down, 70, 71, 73

$\Uparrow = \{0 \mid \uparrow*\} = \uparrow+\uparrow$, double-up, 70, 71, 73

$\Uparrow* = \{0 \mid \uparrow\} = \uparrow+\uparrow+*$, double-up-star, 73

$\downarrow = \{* \mid 0\}$, down, 66

$\downarrow_2 = \{\uparrow* \mid 0\}$, down-second, 227, 228

$\downarrow_3 = \{\uparrow+\uparrow^2 +* \mid 0\}$, down-third, 227, 228

$\downarrow_{abc...}$, $\overline{\downarrow}_{abc...}$, 252

$\curlyvee$, downsum, 316, 337

dud = $\{\textbf{dud} \mid \textbf{dud}\}$, **d**eathless **u**niversal **d**raw, 317

ε, epsilon, small positive number, 308

$\doteqdot$, equally uppity, 233, 238, 240

$\lfloor\ \rfloor$, "floor", greatest integer not greater than, 53

$\parallel 0$, fuzzy, 31, 34, 35

G, general game, 30–33

$G \parallel 0$, G fuzzy, 2nd player wins
$G <0$, G negative, R wins
$G >0$, G positive, L wins $\Big\}$, 31
$G =0$, G zero, 1st player wins

$G + H$, sum of games
G^L, (set of) L option(s)
G^R, (set of) R option(s)

$\mathscr{G}(n)$, nim-value, 82

$G.\uparrow = \{G^L.\uparrow + \Uparrow* \mid G^R.\uparrow + \Downarrow*\}$, 238, 246, 249

$>$, greater than, 34

$\geqslant$, greater than or equal, 34

$\rhd$, greater than or incomparable, 34, 37

$\gtrdot$, at least as uppity, 233, 238

$\frac{1}{2} = \{0 \mid 1\}$, half, 9, 22

$\overline{1}\heartsuit = \{\heartsuit \mid G\}$
$\heartsuit = 0\heartsuit = \{1\heartsuit \mid \overline{A}\}$, hearts, 339
$1\heartsuit = \{0 \mid \textbf{joker}\}$

hi = $\{\textbf{on} \parallel 0 \mid \textbf{off}\}$, 335

hot = $\{\textbf{on} \mid \textbf{off}\}$, 335

$\parallel$, incomparable, 37

$\infty = \mathbb{Z} \parallel \mathbb{Z} \mid \mathbb{Z}$, infinity, 309, 314, 371

$\pm \infty = \infty \mid -\infty = \mathbb{Z} \mid \mathbb{Z} = \int^{\mathbb{Z}} *$, 314

$\infty \pm \infty = \infty \mid 0 = \mathbb{Z} \mid 0$, 314

$\infty + \infty = 2.\infty = \mathbb{Z} \parallel \mathbb{Z} \mid 0$, double infinity, 314

$\infty_{abc...}$, 367–375

$\infty_{\beta\gamma\delta...}$, 313

$\int$, integral, 163–176, 314, 346, 347

$J = \{0 \mid \overline{A}+\} = \textbf{ace} \curlywedge (-\textbf{ace}) = \textbf{joker}$, 338

$\overline{J} = \{A- \mid 0\} = \textbf{ace} \curlyvee (-\textbf{ace}) = -\textbf{joker}$, 339

L, Left, 4

LnL, LnR, RnR, positions in Seating games, 46, 130, 251

$<$, less than, 34

$\leqslant$, less than or equal, 34

$\lhd$, less than or incomparable, 34, 37

lo = $\{\textbf{on} \mid 0 \parallel \textbf{off}\}$, 335

⅃, loony, 377–387

$\gamma, \gamma^*, \gamma^+, \gamma^-$, loopy games, 315

$s\&t$ loopy game, 316

$-1 = \{\ \mid 0\}$, minus one, 21

$-\textbf{on} = \{\textbf{on} \mid 0 \parallel 0\}$, **miny**, 333

$-_{\frac{1}{4}} = \{\frac{1}{4} \mid 0 \parallel 0\}$, miny-a-quarter, 133

$-_x = \{x \mid 0 \parallel 0\}$, miny-$x$, 124

$\overset{*}{\times}$, nim-product, 443

$\overset{*}{+}$, nim-sum, 60, 75, 82, 108

off = $\{\ \mid \textbf{off}\}$, 316–320, 337

$\omega = \{0, 1, 2, ... \mid \ \}$, omega
$\omega+1 = \{\omega \mid \ \}$, omega plus one
$\omega \times 2 = \{\omega, \omega+1, \omega+2, ... \mid \ \} = \omega+\omega$ $\Big\}$, 309–313
$\omega^2 = \{\omega, \omega \times 2, \omega \times 3, ... \mid \ \} = \omega \times \omega$

on = $\{\textbf{on} \mid \ \}$, 316–321

1

Mosby's
Review Questions
for NCLEX-RN

Mosby's
Review Questions for
NCLEX-RN

FOURTH EDITION

Editor

Dolores F. Saxton, R.N., B.S.Ed., M.A., M.P.S., Ed.D.

Associate Editors

Phyllis K. Pelikan, R.N., A.A.S., B.S., M.A.

Patricia M. Nugent, R.N., A.A.S., B.S., M.S., Ed.M., Ed.D.

Content Editors

Phyllis Portnoy Cohen, R.N.C., B.S.N., M.S.

JoAnn Schmidt-Festa, R.N.C., A.A.S., B.S., M.S., Ph.D.

Colleen Glavinspiehs, R.N., B.S., M.S.N., D.N.Sc., F.N.P.

Christina Algiere Kasprisin, R.N., M.S.

Anita Throwe, R.N., B.S.N., M.S., C.S.

A Harcourt Health Sciences Company

St. Louis London Philadelphia Sydney Toronto

A Harcourt Health Sciences Company

Vice President and Publishing Director, Nursing: Sally Schrefer
Senior Editor: Loren Wilson
Senior Developmental Editor: Nancy L. O'Brien
Project Manager: Peggy Fagen
Designer: Judi Lang

FOURTH EDITION
Copyright © 2001 by Mosby, Inc.

NOTICE

Pharmacology is an ever-changing field. Standard safety precautions must be followed, but as new research and clinical experience broaden our knowledge, changes in treatment and drug therapy may become necessary or appropriate. Readers are advised to check the most current product information provided by the manufacturer of each drug to be administered to verify the recommended dose, the method and duration of administration, and contraindications. It is the responsibility of the licensed health care provider, relying on experience and knowledge of the patient, to determine dosages and the best treatment for each individual patient. Neither the publisher nor the editor assumes any liability for any injury and/or damage to persons or property arising from this publication.

Mosby, Inc.
A Harcourt Health Sciences Company
11830 Westline Industrial Drive
St. Louis, Missouri 63146

Printed in the United States of America

International Standard Book Number 0-323-01273-6

01 02 03 04 05 CL/KPT 9 8 7 6 5 4 3 2 1

PREFACE

This text was developed to meet the requests of students for "still more questions—with answers and rationales." We believe that, along with our other publications, *Mosby's Comprehensive Review of Nursing* and *Mosby's AssessTest,* we have provided the necessary tools for students to base their study and review for both coursework and the NCLEX-RN. This fourth edition has been increased to more than 3500 single-item questions reflecting the NCLEX-RN format. There are integrated quizzes after each clinical chapter and two comprehensive examinations that parallel the NCLEX-RN test plan. To enhance learning and preparation for NCLEX-RN, a dual platform CD-ROM is included, with 500, single-item questions that are not duplicated in the text. For those who wish to sharpen their computer-based testing skills, an interactive *NCLEX-RN Review of Nursing* on a dual platform CD-ROM and *Mosby's Online Computer Adaptive Test (CAT) for NCLEX-RN* are available.

To meet the needs of students who have different study styles and learning needs, the questions are presented in four distinct formats:

- The questions in the clinical chapters are grouped according to categories of concern for a specific clinical area. The categories of concern reflect specific content areas within a broad clinical area from which the material in the question has been drawn. We have presented these questions in the traditional clinical groupings because we believe that, even when preparing for an integrated examination like the NCLEX-RN, most students will need to study all the distinct parts before attempting to put them together.

- Four 50-question quizzes are included at the end of the medical-surgical nursing review (Chapter 5) and two 50-question quizzes are included at the end of the pediatric nursing review (Chapter 2), the childbearing and women's health nursing review (Chapter 3), and the mental health nursing review (Chapter 4). The quizzes integrate the content from the various categories of concern within a specific clinical area. These quizzes provide a bridge for moving from the clinical chapters to the comprehensive examinations.

- Two comprehensive examinations, consisting of 265 questions each, are provided to approximate the NCLEX-RN test plan. To parallel the NCLEX-RN, the first 75 questions reflect the minimal testing experience for students taking the NCLEX-RN CAT. The total test of 265 questions reflects the maximum number of questions that a student will be asked on NCLEX-RN. Although NCLEX-RN is computerized, the substance of the test remains constant. It is our belief that if students study the material and develop a strong knowledge base, the method of testing should not have a major influence on their performance.

- A dual-based CD-ROM containing 500 test items, which can be used in both a study and a test format, is included with the textbook. The questions on the CD-ROM are not repeated in the text. To reinforce learned information and build confidence, we suggest that students practice answering questions on a computer to simulate the NCLEX-RN CAT.

For each question in the textbook and on the CD-ROM, the reason why the correct answer is correct, as well as why each of the other options is incorrect, is included. In addition, each question has been analyzed as to the area of client needs, the step in the nursing process, the clinical area, and the category of concern (specific area of content). Each question in the textbook has also been analyzed as to the level of difficulty. Questions incorporate material from the basic sciences, nutrition, and pharmacology, as well as current information relative to topics such as gerontology, rehabilitation, the DSM-IV, and the delivery of health care.

All the questions in this textbook were developed by outstanding and experienced nursing educators and practitioners. The editorial panel reviewed all questions initially submitted, selecting and editing the most pertinent for inclusion in a mass field-testing project. Students graduating from baccalaureate, associate degree, and diploma nursing programs in various locations in the United States and Canada provided a diverse testing group. The results were statistically analyzed, and this analysis was used to select only the highest quality questions for inclusion in this text and to determine each question's level of difficulty.

We would like to express our sincere appreciation to our many colleagues for their contributions: to Loren Wilson, our editor, and Nancy O'Brien, our developmental editor, for their assistance and support; to our content editors for their careful reading and thoughtful comments and critique of our manuscript; and to our families and friends for their love, understanding, and encouragement.

Dolores F. Saxton
Phyllis K. Pelikan
Patricia M. Nugent

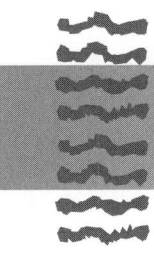

CONTENTS

five
Medical-Surgical Nursing

six
Comprehensive Examination 1

seven
Comprehensive Examination 2

Mosby's
Review Questions
for NCLEX-RN

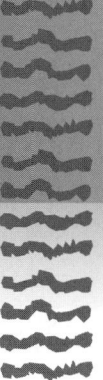

CHAPTER 1

Preparing for the Licensure Examination

INTRODUCTION

Licensure examinations in the United States and Canada have been integrated and comprehensive for many years. Nursing candidates in both countries are required to answer questions that necessitate a recognition and understanding of the physiologic, biologic, and social sciences as well as the specific nursing skills and abilities involved in any given client situation.

Both the U.S. and Canadian tests contain objective multiple-choice questions based on the steps of the nursing process and recognition of client needs. To answer the questions appropriately, a candidate needs to understand and correlate certain aspects of anatomy and physiology, the behavioral sciences, basic nursing, the effects of medications administered, the client's attitude toward illness, and other pertinent factors (e.g., legal responsibilities). Most questions are based on nursing situations similar to those with which candidates have had experience, because both the United States and Canada emphasize the nursing care of clients with representative common national health problems. Some questions, however, require candidates to apply basic principles and techniques to clinical situations with which they have had little if any actual experience.

To adequately prepare for an integrated comprehensive examination, it is necessary to understand the discrete parts that compose the universe under consideration. This is one of the major principles of learning on which our review and study materials have been developed.

Using this concept, this text first presents questions for each major clinical area that test the student's knowledge of principles and theories underlying nursing care in a variety of situations (acute, critical, and long term); in a variety of settings (acute-care hospitals, nursing homes, and the community); and with a variety of nursing approaches to promote health and prevent illness (including primary, secondary, and tertiary care). The text concludes with two integrated comprehensive tests reflecting the licensure examinations. In other words, the questions in the integrated comprehensive tests require the student to cross clinical disciplines and respond to individual and specific needs associated with given health problems.

Answers to all of the questions and rationales supporting the correct answers are provided. Explanations also are presented to document why each of the other choices is inappropriate. Reviewing the rationales enables the student to verify information and reinforce knowledge.

HOW TO USE THIS BOOK IN STUDYING

A. Start in one clinical area. Answer all of the questions in the area. Do not worry if you select the same numbered answer repeatedly; there usually is no pattern to the answers.
B. As you answer each question, write a few words about why you think that answer was correct; in other words, justify why you selected the answer.
C. If you guess at an answer in this text, you should make a special mark to identify it as a guess. This will permit you to recognize areas that need further review. It also will help you to see how correct your guessing can be.
D. You may find it easier to tear out the sheets with answers and rationales for the area you are reviewing and compare your answers with those provided. If you answered the item correctly, check your reason for selecting the answer with the rationale presented. If you answered the item incorrectly, read the ration-

ale to determine why the answer you selected was incorrect. In addition, you should review the correct answer and rationale for each item answered incorrectly. If you still do not understand your mistake, look up the theory pertaining to these questions. You should carefully review all questions and rationales for items you identified as guesses since you did not have mastery of the material being questioned.

E. After the rationale for the correct answer you will find a number—**1, 2,** or **3**—in parentheses. These numbers indicate the level of difficulty of the question and reflect the percentage of tested students answering the question correctly. These can serve as a guide in your studying. (See the sample questions on p. 4.)

1. The number 1 signifies that 75% or more—but less than 89%—of the students in the testing group answered the question correctly. Sample question 3 is a level 1 question.

2. The number 2 signifies that 50% or more—but less than 75%—of the students in the testing group answered the question correctly. Sample questions 1 and 2 are level 2 questions.

3. The number 3 signifies that 25% or more—but less than 50%—of the students in the testing group answered the question correctly. Sample questions 4 and 5 are level 3 questions.

All the questions in this text were answered correctly by 25% to 89% of the students in the testing group. Any questions falling outside these parameters were not included.

F. In addition you will find a grouping of letters that classify the question according to various categories.

In the **Clinical Chapters** the questions are all from the specific clinical area and are grouped in the chapters by their category of concern (this reflects the specific content area within the broad clinical area from which the material in the question has been drawn). Therefore, two letters following the correct answer classify the question by Client Need and Phase of the Nursing Process.

In the **Quizzes** following the clinical chapters, the questions are all from the specific clinical area but they are not grouped by category of concern. Therefore, the letters following the correct answer classify each question by Client Need, Phase of the Nursing Process, and Category of Concern.

In the **Comprehensive Examinations** in Chapters 6 and 7, the questions are not grouped in any manner. Therefore, the letters following the correct answer classify each question by Client Need, Phase of the Nursing Process, Clinical Area, and Category of Concern.

The series of letters always appear in the same order for all questions.

G. The following descriptions and the five sample questions on p. 4 are presented to assist the reader in understanding and reviewing these classifications.

CLIENT NEEDS are those health care needs of the client that the nurse must address. Ten areas of client needs are grouped under four categories:

1. **A safe and therapeutic environment.** Addressing this need includes but is not limited to:

    a. **Coordination of Care (CC):** focuses on coordination of health staff; continuity and delegation of goal-directed care; health teaching and evaluation of client's response; legal responsibility; advocacy, including client rights, confidentiality, organ donation, informed consent, referrals, and ethical practices.

    b. **Promotion of Safety and Prevention of Infection (SI):** focuses on provision of safe and effective care; prevention of errors in care; prevention of accidents; safe handling of contaminated or hazardous material; consistent use of standard or specific precautions; management of potential disaster situations; use of medical and surgical asepsis.

2. **Physiologic and anatomic equilibrium.** Addressing this need includes but is not limited to:

    a. **Provision of Care and Comfort (PC):** focuses on provision of nutrition and oral fluids; reduction of risks that interfere with psychologic, physiologic, or anatomic integrity; promotion of hygiene, comfort, elimination, mobility, rest, and sleep; safe use of assistive devices.

    b. **Management of Pharmacologic and Parenteral Therapy (PT):** focuses on peripheral or central devices to administer parenteral fluids, blood and blood products, total parenteral nutrition, chemotherapy, and pharmacologic agents; medication therapy including purpose, route, range of dose, interactions, and expected, side, and untoward effects.

    c. **Prevention of Complications and Iatrogenic Problems (CI):** focuses on reducing the risk involved in health care provision, diagnostic testing, invasive and noninvasive procedures; modifying, limiting, or assisting with psychologic, physiologic, and anatomic adaptations; evaluating laboratory values, diagnostic test results, and physical states.

    d. **Direction and Assistance with Adaptations to Physiologic Problems (DA):** focuses on acute, chronic, or life-threatening conditions; severe untoward response to therapies; imbalances in fluids, electrolytes, and hemodynamics; infectious diseases; emergencies; pathologic responses; respiratory care; radiation therapy; alterations in body systems.

3. **Health promotion throughout the life span.** Addressing this need includes but is not limited to:

    a. **Support of Developmental Factors Throughout the Life Cycle (DF):** focuses on supporting the client's optimal growth and development to provide for the achievement of the highest possible level of functioning; encouraging the use of support systems and self-care; promoting recognition and acceptance of physiologic and psychologic changes, body image alterations, and aging associated with the developmental stages; acknowledging the role of the family system and the function of human sexuality, including reproduction and family planning.

    b. **Health Promotion and the Prevention of Illness (HI):** focuses on promoting the prevention, recognition, and early treatment of disease throughout the life cycle by use of assessment, education, screening, and immunization; supporting programs designed to promote health and life-style choices.

4. **Psychosocial and emotional equilibrium.** Addressing this need includes but is not limited to:

    a. **Support in Coping with Stress (SC):** focuses on supporting individual coping and adaptation to promote stress management and reduction; recognizing sociocultural, religious, spiritual, and emotional influences on health and use of support systems; dealing with real or imagined threats to self-esteem; adapting to grief and loss; accepting body image and situational role changes; employing counseling techniques and mental health concepts.

    b. **Assistance in Adapting to Emotional and Psychosocial Problems (EP):** focuses on limiting or modifying those responses to crises that produce psychophysiologic or psychopathologic consequences, such as chemical dependency, child neglect and abuse, domestic violence, elder neglect and abuse, and sexual abuse; creating an environment that supports, fosters, and promotes optimal emotional health; using behavioral interventions to limit psychopathology.

NURSING BEHAVIOR (PHASE OF THE NURSING PROCESS). Test questions are also classified by phases of the nursing process. These five components of the nursing process represent the various types of nursing behavior.

1. **Assessment (AS).** The assessment phase involves gathering subjective and objective data about the client's health status from meaningful sources, grouping the data into categories, and communicating the information to others. The database for making nursing decisions is determined through the assessment phase.

2. **Analysis (AN).** In the analysis phases, the nurse interprets the data obtained during the assessment phase to identify the client's actual or potential health care needs and to formulate nursing diagnoses.

3. **Planning (PL).** During the planning phase, the nurse designs strategies to correct, minimize, or prevent problems identified during the assessment and analysis phase; sets priorities for the problems diagnosed; develops both short- and long-term goals with the client and client's family; establishes outcome criteria for nursing interventions; and writes the nursing care plan.

4. **Implementation (IM).** In the implementation phase, the nurse initiates and completes the plan of care. The nurse may perform the care or assist, teach, counsel, or supervise the client, the client's significant others, or other health team members to perform specific interventions based on the client's identified needs, diagnoses, priorities, and goals.

5. **Evaluation (EV).** Through the evaluation component, the nurse determines the effectiveness of nursing intervention. In doing so, the nurse compares the actual outcomes with the expected outcomes to determine client compliance with and response to the intervention or therapy. The nurse uses the evaluation phase to identify whether the health care need still exists, which would require modification of the plan, or whether new health care needs have developed, which would require new interventions.

CATEGORY OF CONCERN reflects the specific content area within the broad clinical area from which the material in the question has been drawn.

1. **Medicine, Surgery, and Pediatrics**
   Blood and immunity (BI)
   Cardiovascular (CV)
   Drug-related responses (DR)
   Emotional needs related to health problems (EH)
   Endocrine (EN)
   Fluid and electrolyte (FE)
   Gastrointestinal (GI)
   Growth and development (GD)
   Integumentary (IT)
   Neuromuscular (NM)
   Reproductive and genitourinary (RG)
   Respiratory (RE)
   Skeletal (SK)

2. **Childbearing and Women's Health**
   Drug-related responses (DR)
   Emotional needs related to childbearing and women's health (EC)
   Healthy childbearing (HC)
   High-risk maternal-fetal conditions affecting childbearing (HP)
   High-risk neonate (HN)

Normal neonate (NN)
Reproductive choices (RC)
Reproductive problems (RP)
Women's health (WH)

3. **Mental Health**
Anxiety, somatoform, factitious, and dissociative disorders (AX)
Crisis situations (CS)
Dementia, delirium, and other cognitive disorders (DD)
Disorders first evident before adulthood (BA)
Disorders of mood (MO)
Disorders of personality (PR)
Drug-related responses (DR)
Eating and sleep disorders (ES)
Emotional disorders related to physical health and childbearing (ED)
Personality development (PD)
Schizophrenic disorders (SD)
Substance abuse (SA)
Therapeutic relationships (TR)

The **Clinical Area** reflects the specialized area of nursing knowledge.

1. **Medicine-Surgery (MS).** These questions include the care of adult clients who have health problems that may or may not require surgical intervention or invasive techniques. Sample questions 1 and 5 are medical-surgical nursing questions.
2. **Childbearing and Women's Health (CW).** These questions include the care of clients preparing for or experiencing childbirth, those involved in family planning, and health problems associated with women. Sample question 2 is a childbearing and women's health nursing question.
3. **Pediatric (PE).** These questions include health problems and the care of clients from birth to young adulthood. Sample question 4 is a pediatric nursing question.
4. **Mental Health (MH).** These questions include the care of clients experiencing emotional stress with or without overt psychiatric behavior in all settings. Sample question 3 is a mental health nursing question.

SAMPLE QUESTIONS

1. During a routine physical examination a client is found to have a blood pressure of 150/96 mm Hg, and hypertension is suspected. In obtaining the health history, an early sign of hypertension that the nurse should expect the client to complain of is:
   1. Swollen ankles
   2. Recent weight loss
   3. Palpitations of the heart
   4. Early morning headaches
      [Correct answer is 4. (2) (CI; AS; CV; MS)]

2. For a woman, identification with the parenting role begins:
   1. Early in life
   2. During adolescence
   3. After the baby has been born
   4. When pregnancy is confirmed
      [Correct answer is 1. (2) (DF; AN; EC; CW)]

3. A young male has a history of an antisocial personality disorder. His parents tell the nurse that their son is very manipulative and causes havoc in their home. The nurse should include in the teaching plan ways that they can cope with their son by using an approach that is:
   1. Rigid
   2. Flexible
   3. Accepting
   4. Consistent
      [Correct answer is 4. (1) (EP; PL; PR; MH)]

4. A 3-year-old girl is admitted for surgery. When her mother leaves, she begins to sob. The nurse should:
   1. Tell her to be a big girl, her mother will be right back
   2. Put up the side rails on the crib and let her calm down by herself
   3. Distract her with her teddy, expecting her to forget her mother has gone
   4. Hold her and explain that her mother had to go but will return in 2 hours
      [Correct answer is 3. (3) (PC; IM; GD; PE)]

5. Following a mastectomy, tamoxifen (Nolvadex) is prescribed for a client. The nurse knows that teaching about this drug was understood when the client states, "I will:
   1. Drink 4 glasses of milk every day while I am taking this drug."
   2. Expect pain at the site of the tumor when I am taking this drug."
   3. Take a stool softener every day while I am taking this medication."
   4. Rise from a sitting position slowly when I am taking this medication."
      [Correct answer is 2. (3) (PT; EV; DR; MS)]

H. A few days later, review the area again. If you miss the same question a second time, you need further study of the material.

I. After you have completed the clinical area questions, begin taking the comprehensive tests because they will assist you in applying knowledge and principles from the specific clinical area to any nursing situation. The following steps are suggested:
   1. Arrange for a quiet, uninterrupted time span for each of the comprehensive tests
   2. Pace yourself during the testing period; allow about 1 minute per question
   3. Do not rush
   4. Answer every question

J. Since most examinations have specified time limits, you will need to pace yourself during the practice testing period, working as quickly and accurately as possible. It is helpful to estimate the time that can be spent on each item and still complete the examination in the allotted time. You can obtain this figure by dividing the testing time by the number of items on the test. For example, with a 1-hour (60-minute) testing period and approximately 60 items, an average of 1 minute per item will be the appropriate pace.

K. To help analyze your mistakes on the comprehensive examinations and to provide a database for making future study plans, two types of worksheets are included in this text. One is designed to aid you in identifying and recording errors in the way you process information. The other is to help you identify and record gaps in knowledge. These worksheets follow the Answers and Rationales for each test in the Comprehensive Test Section. Instructions for their use appear on each worksheet.

L. After completing your worksheets, do the following:
   1. Use Worksheet 1 to identify the frequency with which you made particular errors. As you review material in class notes or study material such as *Mosby's Comprehensive Review of Nursing,* pay special attention to correcting your most common problems.
   2. Use Worksheet 2 to identify the topics you want to review. It might be helpful to set priorities; review the most difficult topics first so that you will have time to review them more than once.

M. Use this opportunity to learn from your mistakes.
   1. Because you receive immediate feedback on your performance, you have an excellent opportunity to learn from your mistakes. Answer every question. Do not leave any questions unanswered; use educated or pure guesses.
   2. The mistakes you make on the questions in this text will be as valuable to you as the confident feeling you get from answering correctly.

READINESS FOR THE LICENSURE EXAMINATION

A few individuals can improve their scores significantly by a highly concentrated period of study immediately before taking an examination. Most, however, profit by spreading their review over a much longer period. Cramming usually does not help. Identification of your own specific strengths and weaknesses should eliminate much of the anxiety of deciding what material to study by giving you a sense of direction and a means of setting priorities.

Reduce Stress

Stress is a part of life. While there is no way to prevent it, it is possible to reduce it by diffusing your emotional responses before stress gets the better of you. Controlling stress allows you to use it instead of being abused by it. The following are tips to reduce and control stress:

1. Talk it out, but try to talk it out with someone who is not as stressed as you are. This relieves the burden of coping alone and helps put things in perspective. Try talking with people who have had the same experience and understand what you are going through.
2. Obtain as much information as you can. STUDY!!!
3. Keep fit. Good nutrition, regular exercise, and ample sleep help.
4. Try relaxation exercises. Relaxation is essential to reduce stress.
5. Sort out the important things. Take stock of your strengths. Set realistic deadlines. Drop the nonessentials.
6. Spend time on yourself and your needs outside of nursing.
7. Be greedy and put yourself first. Be flexible with yourself. Do not set rigid, unmanageable goals.
8. Discover your positive defenses and use them.
9. Become familiar with reading questions on a computer screen. Familiarity reduces anxiety and decreases errors.

Manage Test-Taking Time

The computerized NCLEX-RN is not a timed test per se, although there is a maximum 5-hour time period. The computerized test reportedly has been designed to measure the individual's level of knowledge, skills, and abilities to determine that the competency level is achieved. The test length will vary depending on each candidate's performance but will be somewhere between 75 and 265 questions.

Although certain questions will be more difficult than others and will require more time, spending too much time on these difficult items may compromise your overall performance.

Do not be pressured into finishing early. Do not rush! Students who achieve higher scores on examinations are typically those who use all the time available.

Build Test-Taking Confidence

You should feel confident and competent if you have studied and reviewed the content to be tested and you are armed with methods for reading and answering questions. Your emotional state is vitally important when thinking about, preparing for, and taking any test. Think positively.

While you are taking the test, you may have problems with a question. On a written examination it is often best to move on to another question that you can answer and come back to the more difficult question. However, on the computerized NCLEX-RN you must answer the question before you can go on to the next question, so remain calm. Anxiety can block the recall of familiar information required to answer questions, so control it early. Do not stop to think about gaps in your preparation or waste time and emotional energy building anxiety. Focus on the positive. You have the ability to make sound "educated guesses." Now is the time to use it. Questions that seem complicated at first glance often can be answered by just such guesses. Remain calm and confident.

You will find that practice test-taking experiences will give you confidence for the actual examination. After you have completed studying in this book, you may wish to take a simulated examination such as *Mosby's AssessTest* before you take the licensure examination. The *AssessTest* is a computer-scored, multiple-choice examination designed to test nursing knowledge and evaluate your ability to apply that knowledge in clinical situations. The extensive computer analysis of your performance, which is the most outstanding feature of this test, will help you design effective and efficient plans for further study and review.

TAKING THE LICENSURE EXAMINATION

On the NCLEX-RN each of the five steps of the nursing process and each of the clinical areas are represented. There appears to be a deemphasis in the areas of maternity (obstetric) nursing and severe mental illness while there seems to be a greater emphasis on medical-surgical principles and interpersonal skills, especially communication. The category of client health care needs reflects the need for support and promotion of physiologic and anatomic equilibrium; the need for a therapeutic environment; the need for education and health promotion; the need for support and promotion of psychosocial and emotional equilibrium. The score on the examination is reported as pass or fail.

The most crucial requisite for doing well on the licensure examination is a sound understanding of the subject and a high level of reading comprehension. Determination to do well and a degree of confidence will further enhance the well-prepared individual's chance of passing and achieving the recognition deserved.

At least three other requirements must be met if an individual's performance is to accurately reflect professional competence. First, the candidate must follow explicitly the directions given by the examiner and those appearing at the beginning of the test. Second, the candidate must read each question carefully before deciding how to answer it. Third, the candidate must record the answers in the manner specified.

The computerized NCLEX-RN is an individualized testing experience in which the computer chooses your next question based on the ability and competency you have demonstrated on previous questions. The minimum number of questions will be 75 and the maximum

265. You must answer each question before the computer will present the next question, and you cannot go back to any previously answered questions. There is no deduction for incorrect answers so you are not penalized for guessing. You cannot leave an answer blank, and since you have a 1 in 4 (25%) chance of guessing the correct answer, go for it. Remember: you do not have to get all the questions correct to pass.

You also should keep in mind that if you practice and learn the material, the method of testing (oral, written, or computer) should not significantly influence your performance.

TEST-TAKING SKILLS

Test-taking skills and techniques are not a substitute for good study habits or an adequate grasp of the content and abilities measured in an examination. Memorization is of little help because few questions require simple recall and most require the use of higher, more complex thought processes. If you have a thorough understanding of the knowledge content measured by an examination, however, good test-taking skills will enhance your overall performance.

The question in its entirety is called a test item. The portion of the test item that poses the question or problem is called the stem. Potential answers to the question or problem posed are called options. In well-constructed multiple-choice items there is only one correct answer among the options supplied; the incorrect options are called distractors. Remember, test questions are meant to measure your nursing knowledge. The items may be easy to read, but the answers to questions are not intended to be readily apparent. The questions draw on your ability to apply nursing knowledge from a variety of sources.

■ *Read Questions Carefully*

Scores on tests are greatly affected by reading ability. In answering a test item, you should begin by carefully reading the stem and then asking yourself the following questions:
- What is the question really asking?
- Are there any key words?
- What information relevant to answering this question is included in the stem?
- How would I ask this question in my own words?
- How would I answer this question in my own words?

After you have answered these questions, carefully read the options and then ask yourself the following questions:
- Is there an option that is similar to my answer?
- Is this option the best, most complete answer to the question?

Deal with the question as it is stated, without reading anything into it, or making assumptions about it. Answer the question asked, not the one you would like to answer. For simple recall items the self-questioning process usually will be completed quickly. For more complex items the self-questioning process may take longer, but it should assist you in clarifying the item and selecting the best response.

■ *Identify Key Words*

Certain key words in the stem, the options, or both should alert you to the need for caution in choosing your answer. Because few things are absolute without exception, avoid selecting answers that include words such as *always, never, all, every, only, must, no, except,* and *none.* Answers containing these key word are rarely correct because they place special limitations and qualifications on potentially correct answers. For example:

All of the following are services of the National Kidney Foundation except:
1. Public education programs
2. Research about kidney disease
3. Fund-raising affairs for research activities
4. Identification of potential transplant recipients

This stem contains two key words: *all* and *except.* They limit the correct answer choice to the one option that does not represent a service of the National Kidney Foundation. When *except, not,* or a phrase such as *all but one of the following* appears in the stem, the inappropriate option is the correct answer—in this instance, option **4.**

If the options in an item do not seem to make sense because more than one option is correct, reread the question; you may have missed one of the key words in the stem. Also be on guard when you see one of the key words in an option; it may limit the context in which such an option would be correct.

■ *Pay Attention to Specific Details*

The well-written multiple-choice question is precisely stated, providing you with only the information needed to make the question or problem clear and specific. Careful reading of details in the stem can provide important clues to the correct option. For example:

A male client is told that he will no longer be able to ingest alcohol if he wants to live. To effect a change in his behavior while he is in the hospital, the nurse should attempt to:
1. Help the client set short-term dietary goals
2. Discuss his hopes and dreams for the future
3. Discuss the pathophysiology of the liver with him
4. Withhold approval until he agrees to stop drinking

The specific clause *to effect a change in his behavior while he is in the hospital* is critical. Option **2** is not really related to his alcoholism. Option **3** may be part of educating the alcoholic, but you would not expect a behav-

ioral change observable in the hospital to emerge from this discussion. Option **4** rejects the client as well as his behavior instead of only his behavior. Option **1,** the correct answer, could result in an observable behavioral change while the client is hospitalized; for example, he could define ways to achieve short-term goals relating to diet and alcohol while in the hospital.

■ *Eliminate Clearly Wrong or Incorrect Answers*

Eliminate clearly incorrect, inappropriate, and unlikely answers to the question asked in the stem. By systematically eliminating distractors that are unlikely in the context of a given question, you increase the probability of selecting the correct answer. Eliminating obvious distractors also allows you more time to focus on the options that appear to be potentially sound answers to the question. For example:

The four levels of cognitive ability are:
1. Assessing, analyzing, applying, evaluating
2. Knowledge, analysis, assessing, comprehension
3. Knowledge, comprehension, application, analysis
4. Medical-surgical nursing, obstetric nursing, psychiatric nursing

Option **1** contains both cognitive levels and nursing behaviors, thus eliminating it from consideration. Option **4** is clearly inappropriate since the choices are all clinical areas. Both options **2** and **3** contain levels of cognitive ability; however, option **2** includes assessing, which is a nursing behavior. Therefore option **3** is correct. By reducing the plausible options, you reduce the material to consider and increase the probability of selecting the correct option.

■ *Identify Similar Options*

When an item contains two or more options that are similar in meaning, the successful test taker knows that all are correct, in which case it is a poor question, or that none is correct, which is more likely to be the case. The correct option usually will either include all the similar options or exclude them entirely. For example:

When teaching newly diagnosed diabetic clients about their condition, it is important for the nurse to focus on:
1. Dietary modifications
2. Use of sugar substitutes
3. Their present understanding of diabetes
4. Use of diabetic nutritional exchange lists

Options **1, 2,** and **4** deal only with the diabetic diet, involving no other aspect of diabetic teaching; it is impossible to select the most correct option because each represents equally plausible, though limited, answers to the question. Option **3** is the best choice because it is most complete and allows the other three options to be excluded.

As another example:

A child's intelligence is influenced by:
1. A variety of factors
2. Socioeconomic factors
3. Heredity and environment
4. Environment and experience

The most correct answer is option **1**. It includes the material covered by the other options, eliminating the need for an impossible choice, since each of the other options is only partially correct.

■ *Identify Answer (Option) Components*

When an answer contains two or more parts, you can reduce the number of potentially correct answers by identifying one part as incorrect. For example:

After a cholecystectomy the postoperative diet is usually:
1. High fat, low calorie
2. High fat, low protein
3. Low fat, high calorie
4. Low fat, high protein

If you know, for instance, that the diet after a cholecystectomy is usually low or moderate in fat, you can eliminate options **1** and **2** from consideration. If you know that the cholecystectomy client usually is overweight, you can eliminate option **3** from consideration. Therefore option **4** is correct.

■ *Identify Specific Determiners*

When the options of a test item contain words that are identical or similar to words in the stem, the alert test taker recognizes the similarities as clues about the likely answer to the question. The stem word that clues you to a similar word in the option or that limits potential options is known as a specific determiner. For example:

The government agency responsible for administering the nursing practice act in each state is the:
1. Board of regents
2. Board of nursing
3. State nurses' association
4. State hospital association

Options **2** and **3** contain the closely related words nurse and nursing. The word *nursing*, used both in the stem and in option **2**, is a clue to the correct answer.

■ *Identify Words in the Options That Are Closely Associated With Words in the Stem*

Be alert to words in the options that may be closely associated with but not identical to a word or words in the stem. For example:

When a person develops symptoms of physical illness for which psychogenic factors act as causative agents, the resulting illness is classified as:
1. Dissociative
2. Compensatory
3. Psychophysiologic
4. Reaction formation

Option **3** should strike you as a likely answer since it combines physical and psychologic factors, like those referred to in the stem.

■ *Watch for Grammatical Inconsistencies*

If one or more of the options are not grammatically consistent with the stem, the alert test taker usually can eliminate these distractors. The correct option must be consistent with the form of the question. If the question demands a response in the singular, plural options usually can be safely eliminated. When the stem is in the form of an incomplete sentence, each option should complete the sentence in a grammatically correct way. For example:

Communicating with a male client who is deaf will be facilitated by:
1. Use gestures
2. Speaking loudly
3. Find out if he has a hearing aid
4. Facing the client while speaking

Options **1** and **3** do not complete the sentence in a grammatically correct way and can therefore be eliminated. Option **2** would be of no assistance with a deaf client, so option **4** is the correct answer.

■ *Be Alert to Relevant Information From Earlier Questions*

Occasionally, remembering information from one question may provide you with a clue for answering a later question. For example:

A client has an intestinal tube inserted for treatment of intestinal obstruction. Intestinal suction can result in excessive loss of:
1. Protein enzymes
2. Energy carbohydrates
3. Water and electrolytes
4. Vitamins and minerals

If you determined that the correct answer to this question was option **3**, it may help you to answer a later question. For example:

Critical assessment of a client with intestinal suction should include observation for:
1. Edema
2. Nausea
3. Belching
4. Dehydration

The correct answer is option **4.** If you knew that excessive loss of water and electrolytes may lead to dehydration, you could have used the clue provided in the earlier question to assist you in answering the later question.

■ *Make Educated Guesses*

When you are unsure about the correct answer to a question, it is better to make an educated guess than not to answer the question. You generally can eliminate one or more of the distractors by using partial knowledge and the methods just listed. The elimination process increases your chances of selecting the correct option from those remaining. Elimination of two distractors on a four-option multiple-choice item increases your probability of selecting the correct answer from 25% to 50%.

GENERAL STRATEGIES

1. Develop a plan for study and stick to it. A good plan is to allow 1 week per clinical area.
2. As you study, identify your problem areas that need attention.
3. Avoid planning things that will add stress to your life between now and the time you take the NCLEX-RN examination. Enough things will happen spontaneously; do not plan to add to them.
4. Do not change your pattern of study. It has obviously contributed to your being here, so it worked. If you have studied alone, continue to study alone. If you have studied in a group, form a study group.
5. Practice timed tests and stay within their suggested time frames. You usually will have about 1 minute per question on most examinations.

6. Pace yourself during the testing period and work as accurately as possible. Do not rush. Excessive pressure on yourself early in the examination can result in early fade out.
7. Although certain questions are more difficult and require more time, do not spend too much time on one question because it can compromise your overall performance.
8. If you find you tend to reread test answers and change the right ones to wrong ones, stop going back, even if you are taking tests that permit you to go back. If you find that going back helps you to correct wrong answers, by all means go back and review your answers, if the test permits you to go back. Remember: your first answer is usually correct and should not be changed without reason. You cannot go back on the NCLEX-RN.
9. Do not read information into questions, and avoid speculating. Reading into questions creates errors in judgment.
10. Make certain that the answer you select is reasonable and obtainable under ordinary circumstances and that the action can be carried out in the given situation.
11. Avoid selecting answers that state hospital rules or regulations as a reason or rationale for action.
12. Look for answers that focus on the client or that are directed toward the client's feelings.
13. If the question asks for an immediate action or response, all the answers may be correct, so base your selection on identified priorities for action.
14. Do not select answers that contain exceptions to the general rule, controversial material, or responses that appear to be degrading.
15. Do not be pressured into finishing early. Use all the time necessary without pressuring yourself.

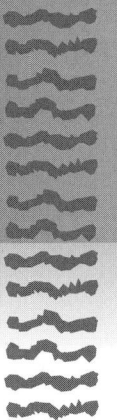

CHAPTER 2

Pediatric Nursing

REVIEW QUESTIONS

GROWTH AND DEVELOPMENT (GD)

1. To bring about effective communication with any child, the nurse must first take into consideration the child's:
   1. State of health
   2. Developmental level
   3. Ability at self-expression
   4. Fear of authoritarian figures

2. The nurse understands that the first activity of daily living that should be taught to a developmentally disabled child is:
   1. Toileting
   2. Dressing
   3. Self-feeding
   4. Combing hair

3. An appropriate toy for a 3-month-old infant would be a:
   1. Push-pull toy
   2. Metallic mirror
   3. Stuffed animal
   4. Large plastic ball

4. The nurse is aware that the play of a 5-month-old infant is most likely to consist of:
   1. Picking up a rattle or toy and putting it into the mouth
   2. Exploratory searching when a cuddly toy is hidden from view
   3. Simultaneously kicking the legs and batting the hands in the air
   4. Waving and clenching fists and dropping toys placed in the hands

5. The mother of a 7-month-old infant, who is to be catheterized to obtain a sterile urine specimen, expresses fear that this procedure may traumatize the child psychologically. The nurse reassures the mother that:
   1. Her fear is justified and the nurse will obtain a "clean catch" specimen
   2. She has every right to refuse catheterization since her concerns are realistic
   3. Her concern is appropriate but the need for a sterile specimen is a higher priority
   4. The procedure, though slightly uncomfortable, should not have any damaging effect

6. The nurse counsels a mother of an 8-month-old child to make sure the floors are free of small objects when her child is crawling on the floor. The major rationale for this instruction is that:
   1. An 8-month-old infant can easily pick up small objects
   2. Sharp objects can injure the fragile skin of an 8-month-old
   3. It is a health hazard for babies to pick things up off the floor
   4. The infant could hide small objects, making them difficult to locate

7. The nurse notes that a 22-month-old uses two- or three-word phrases (telegraphic speech), has a vocabulary of about 200 words, and often uses the word "me." The nurse would interpret the child's language development as being:
   1. A developmental lag
   2. Slow for the child's age
   3. Normal for the child's age
   4. Advanced for the child's age

11

8. Based on an understanding of normal preschool behavior, during hospitalization the nurse is aware that a 4-year-old will probably:
   1. Refuse to cooperate with nurses during the parents' absence
   2. Demonstrate despair if parents do not visit at least once a week
   3. Cry when the parents leave and return but not during their absence ✓
   4. Be unable to relate to peers in the playroom if there are parents present

9. A 2½-year-old male child who has fallen from a tree tells his parents, "Bad, bad tree." The nurse recognizes that the child is within the cognitive developmental norm of Piaget's:
   1. Concrete operations
   2. Concept of reversibility
   3. Preconceptual operations ✓
   4. Sensorimotor development

10. The nurse is aware that the most reliable indicator of pain in a 4-year-old child is:
    1. Crying and sobbing
    2. Changes in behavior ✓
    3. Decreased heart rate
    4. Verbal reports of pain ✓

11. To properly visualize the auditory canal of a 4-year-old during an otoscopic examination, the nurse should pull the pinna of the ear:
    1. Up and back ✓
    2. Up and forward
    3. Down and back
    4. Down and forward

12. The nurse is aware that the toy that would be most appropriate for a 3-year-old would be a:
    1. Fuzzy stuffed animal
    2. Seven-piece jigsaw puzzle
    3. Lunch box filled with plastic figures ✓
    4. Blunt scissors and pictures to cut out

13. The nurse is aware that corrective surgery for a newborn's hypospadias will be done:
    1. Shortly after birth
    2. Within a few months after birth
    3. At approximately 5½ years of age
    4. When the child is between 3 and 4 years old ✓

14. An 8-year-old child, admitted to the hospital for intrathecal methotrexate chemotherapy, is prescribed allopurinol (Zyloprim) and asks the nurse why this medication has to be taken. The nurse's best response would be:
    1. "Because this pill helps the other medicines get rid of the things that are making you sick." ✓
    2. "To protect your body from developing other problems after your treatment has been stopped."
    3. "To stop your sick white cells from going to other parts of your body where they can cause problems."
    4. "Because your doctor ordered it. Your doctor would not order anything for you unless it was very important."

15. The primary nurse, assigned to a 5½-month-old girl being admitted to the hospital, understands that the infant's emotional development should make the infant:
    1. Cry when the nurse approaches her for the first time
    2. Welcome the attention that the primary nurse gives her ✓
    3. Smile socially in recognition of the primary nurse's face
    4. Cling furiously to her mother when the nurse tries to take her away

16. After teaching a mother about the appropriate play for an 8-month-old infant, the nurse is aware that the mother needs additional teaching when the mother states that she will buy a:
    1. Stuffed animal
    2. Play telephone
    3. Hanging mobile
    4. Book with textures

17. A 2½-year-old girl whose older sibling has recently died has started hitting her mother and refusing to go to bed at night. The nurse in the pediatric well-child clinic tells the mother that the toddler is probably:
    1. Fearful of dying in her sleep
    2. Trying to get more of her mother's attention
    3. Just going through the "terrible twos" developmental stage
    4. Reacting appropriately to anxiety generated by the family upheaval

18. When talking with a 4-year-old, the nurse observes that the child is shy and stutters. The nurse is aware that stuttering in a 4-year-old child is considered:
    1. A sign of a delay in neural development
    2. A normal characteristic for a preschooler
    3. The result of a serious emotional problem
    4. An indication of a serious permanent impairment

19. The most appropriate toys for a 6-month-old infant would be:
    1. Push-pull toys
    2. Wooden blocks
    3. Soft stuffed animals
    4. Shape-matching toys

20. A mother indicates to the nurse in the pediatric clinic that she is concerned that her 20-month-old baby's bedtime thumb-sucking will cause the teeth to protrude. The nurse's most appropriate response would be:
1. "You should seek counseling; the thumb-sucking may indicate an emotional problem."
2. "You should switch the baby to a pacifier in the next 2 months to prevent protrusion of the teeth."
3. "You need to restrain the baby from sucking the thumb because it prematurely loosens the first teeth."
4. "You need not be concerned about the teeth protruding unless it persists after permanent teeth appear."

21. The nurse knows that an appropriate toy for a 6-year-old in a spica cast would be a:
1. Ball and jacks
2. Game of checkers
3. Set of building blocks
4. Coloring book and crayons

22. A developmental assessment of a 9-month-old would be expected to reveal:
1. A two- to three-word vocabulary
2. An ability to feed self with a spoon
3. The ability to sit steadily without support
4. Closure of both anterior and posterior fontanels

23. The nurse would assess a 4-year-old child's abdominal pain by:
1. Asking the child to point to where it hurts
2. Auscultating the abdomen for bowel sounds
3. Asking the parents about the child's bowel movements
4. Observing the position and behavior while the child is moving

24. The nurse is aware that a 10-month-old infant on a regular diet could be fed:
1. Applesauce, carrots, chicken, and formula
2. Pears, green beans, turkey, and whole milk
3. Bananas, sweet potatoes, ham, and formula
4. Peaches, corn, cottage cheese, and whole milk

25. When assessing a 4-year-old, the nurse would expect the child to:
1. Ask the definitions of new words
2. Have a vocabulary of 1500 words
3. Name two or three different colors
4. Use just three- or four-word sentences

26. When teaching parents to instill eardrops in an 18-month-old child, the nurse shows them how to:
1. Cleanse the ear canal by pulling the pinna up and down
2. Apply medicated ear wicks tightly before instilling the eardrops
3. Straighten the auditory canal by pulling the earlobe up and back
4. Straighten the auditory canal by pulling the pinna down and back

27. The nurse is reinforcing previous learning about cystic fibrosis and its treatment with a 9-year-old with the disease. The most suitable information for this child's stage of development would be:
1. "The postural drainage will help you feel better."
2. "The dietitian says this meal schedule is best for you."
3. "Your medication is scheduled at this time because your doctor has prescribed it this way."
4. "Your mucus is thick because cystic fibrosis interferes with how your mucous glands work."

28. While caring for a 6-month-old infant, it is likely that the nurse will observe the presence of the reflex called:
1. Startle
2. Babinski
3. Extrusion
4. Tonic neck

29. Developmentally, 2-year-old children are at risk for lead poisoning primarily because:
1. Lead is easily available to them
2. Their vascular system is very fragile
3. They have a high level of oral activity
4. Motor vehicle use and pollution have increased

30. Specific preoperative teaching before an orchiopexy in a 4-year-old child should include a:
1. Doll with an intravenous tube in the arm
2. Demonstration of the use of the abdominal binder
3. Picture of a boy with a bandage on his lower abdomen
4. Doll with a rubber band stretched from perineum to thigh

31. The nurse knows that a child is performing normal developmental tasks for a 5-year-old when the child:
1. Is ritualistic when playing
2. Makes up rules for a new game
3. Asks for a pacifier when uncomfortable
4. Plays near others quietly, but not with them

32. The nurse recognizes that behaviors typical of an 8-month-old include:
1. Drinking from a cup, using the words "mama" and "dada," and standing alone
2. Smiling spontaneously, clasping hands, and keeping the head steady when sitting
3. Removing some clothing, building a tower of two cubes, and stooping to pick up toys
4. Being shy with strangers, playing peek-a-boo, and standing by holding onto furniture

33. Preparation for surgery on a 4-year-old must include consideration of the child's age-related fear of:
    1. Strangers
    2. Intrusive procedures
    3. Disruption of routines
    4. Separation from parents

34. The nurse is aware that the most therapeutic play for a 4-year-old child on bed rest would be:
    1. Finger-painting on blank sheets of paper
    2. Using crayons to color in a coloring book
    3. Engaging in a checker game with the father
    4. Playing dominos with an 8-year-old roommate

35. Once the crisis has passed, when listing all the problems of a teenage client who has sickle cell anemia, the nurse recognizes that priority must be given to the client's:
    1. Restriction of movement during periods of arthralgia
    2. Altered body image resulting from skeletal deformities
    3. Separation from family during periods of hospitalization
    4. Interruption of learning as a result of multiple hospitalizations

36. A 1-year-old visits the playroom. The toy selected and used that would indicate an appropriate growth and developmental level would be a:
    1. Rocking horse
    2. Stuffed animal
    3. Four-piece puzzle
    4. Plastic toy that squeaks

37. In planning self-care that would foster independence, the nurse would expect a 4-year-old child to be able to:
    1. Button a shirt
    2. Tie shoelaces
    3. Part and comb hair
    4. Cut the meat at dinner

38. The nurse should understand that to a preschooler, death is thought of as:
    1. An end to life
    2. A reversible separation
    3. Something that happens to old people
    4. A persona who takes one away from one's family

39. When caring for a 15-year-old client receiving chemotherapy for leukemia, the nurse should keep in mind that an adolescent of this age will:
    1. Feel dependent and enjoy the "sick role"
    2. Be most bothered by having to limit activities
    3. Feel different because of an altered body image
    4. Be preoccupied by concerns about missed schoolwork

40. When evaluating growth and development of a 6-month-old infant, the nurse would expect the infant to be able to:
    1. Sit alone, display pincer grasp, wave bye-bye
    2. Crawl, transfer toy from one hand to the other, display fear of strangers
    3. Pull self to a standing position, release a toy by choice, play peek-a-boo
    4. Turn completely over, sit momentarily without support, reach to be picked up

41. The parents of an 18-month-old child are anxious to know why their child has experienced several episodes of suppurative otitis. The nurse should explain the:
    1. Immunologic difference between the young child and the adult
    2. Structural difference between the eustachian tube of younger and older children
    3. Difference between the size of the middle ear cavity in infants and older children
    4. Functional difference between an infant's eustachian tube and that of an older child

42. Following surgery, a 5-year-old is experiencing intense pain and an analgesic is prescribed. When administering the analgesic, the nurse should consider that:
    1. Even though children do not like medicines, analgesics will make them more comfortable
    2. Pain is not as strongly felt by children as by adults, so analgesics are not needed frequently
    3. Children should rarely receive analgesics because this may result in addiction or respiratory depression
    4. Children do not need analgesics because they are easily distracted and quickly return to playing or sleeping

43. An 11-year-old is diagnosed with acute lymphocytic leukemia (ALL), and the physician discusses the diagnosis and treatment with the family. The assessment data that indicate age-appropriate behavior for an 11-year-old regarding a diagnosis implying death and dying is the child:
    1. Is rude, impolite, and insolent
    2. Says that an uncle died and went to heaven
    3. Is afraid to go to sleep for fear of not awakening
    4. Tells the nurse that death is punishment for not being good

44. An assessment of a 6-month-old infant's growth and developmental level should reveal that the infant can:
    1. Say "mama"
    2. Crawl forward

3. Turn pages in a book
4. Hold a bottle without help

45. At the well-baby clinic, the nurse discusses the food and feeding needs of a toddler with the mother of a 2-year-old. Considering a toddler's food and feeding needs, the nurse should teach the mother that:
1. Growth rate is increased at the age of 2 years, so the child needs more protein per unit of body size
2. A child's energy requirements during the toddler stage are so high that more calories are needed to meet them
3. A child's normal struggle for independence at this age often involves refusal of food, but children will eat the amount they need
4. Because the child often refuses food, the mother should prepare only the food the child likes and should avoid snacks between meals

46. The nurse is aware that 5-year-old children engage in play that is known as:
1. Parallel
2. Ritualistic
3. Aggressive
4. Cooperative

47. The social development of a 9-month-old is best promoted by having the infant:
1 Manipulate soft clay
2. Pound on a pegboard
3. Play peek-a-boo and bye-bye
4. Play with a large ball with a bell

48. The nurse is aware that 18-month-old children with normal hearing have usually acquired a vocabulary sufficient to enable them to communicate by:
1. Pointing and grunting
2. Using at least six words
3. Making babbling sounds
4. Using complete sentences

49. A 15-year-old girl is grounded for 2 weeks by her parents for smoking in school. The adolescent tells the school nurse, "It's not fair that I get punished when my friends get away with doing the same thing." The nurse's most appropriate response would be:
1. "The others will pay someday for lying to school authorities."
2. "I intend to report your friends to the principal so they can also be punished."
3. "It is difficult enough to get teenagers to tell the truth. When parents don't act, it reinforces deceptive behavior."
4. "When errors in judgment are made, people must be prepared to take the consequences of their actions."

50. The nurse is aware that the kind of play 2-year-old children engage in is called:
1. Group play
2. Parallel play
3. Dramatic play
4. Cooperative play

51. The play activity that would be appropriate for a 6-year-old whose energy level has improved following an acute episode of Hirschsprung's disease would be:
1. Using a set of building blocks
2. Finger-painting on a large paper surface
3. Taking apart and putting together a truck
4. Drawing or writing with a pencil or marker

52. The mother of a 15-year-old female who is being treated for allergies privately tells the nurse that she thinks her daughter is becoming a hypochondriac. The nurse can be most therapeutic by:
1. Discussing the underlying causes of hypochondriasis
2. Discussing the developmental behavior of adolescents
3. Explaining the potentially serious complications of allergies
4. Explaining that the mother may be transferring her own fears to her daughter

53. A 15-year-old type 1 diabetic has a history of noncompliance with therapy. The nurse is aware that the noncompliance is developmentally related to:
1. The need for attention
2. A denial of the diabetes
3. The struggle for identity
4. A regression associated with illness

EMOTIONAL NEEDS RELATED TO HEALTH PROBLEMS (EH)

54. Three days after being admitted to the hospital with meningococcal meningitis, a 12-year-old client is afebrile and asymptomatic but appears very sad and cries frequently. To assist the child to verbalize thoughts and feelings, the nurse should:
1. Encourage the parents to speak with their child
2. Ask the child directly what seems to be the trouble
3. Show the child some photos of hospitalized children and have the child tell stories about them
4. Have the child watch videotapes about sick children and answer any questions that the child may have

55. To be most therapeutic when giving a 3-year-old toddler an intramuscular injection, the nurse should approach the child and say:
    1. "Act like a big child and we can be done quickly."
    2. "You are afraid of having a shot because of the pain."
    3. "I know this might hurt, but it's important that you hold still."
    4. "I brought another nurse along to help me give your medicine."

56. A 9-year-old male child, who has been newly diagnosed with diabetes mellitus, is being discharged. The nurse suspects that there may be a problem with family dynamics when the child's mother states:
    1. "We want to encourage our son to do as much as he can for himself."
    2. "We know our child is special, and we'll have to go easy on the discipline for him."
    3. "We know our child and the rest of the family are in for a lot of ups and downs over the years."
    4. "We really hope our son can still be in the Boy Scouts and participate with his Little League baseball team."

57. A 16-month-old male infant has been in the hospital for 3 weeks and has become increasingly withdrawn and mute. It would be most appropriate for the nurse to:
    1. Move him in with other children
    2. Provide him with distracting toys
    3. Encourage the parents to stay with him as much as possible
    4. Assign different nurses to be with him to provide sensory stimuli

58. A 5-year-old, newly arrived from Latin America, attends a nursery school where everyone speaks English. The child's mother tells the nurse in the well-child clinic that her child is no longer outgoing and is very passive in the classroom. The nurse suspects that the child:
    1. May be experiencing discrimination
    2. Lacks adequate motivation for school
    3. Is not mature enough for kindergarten
    4. Is undergoing a state of cultural shock

59. As the nurse plans to teach a 9-year-old male with a learning disability about his diabetes, the parents intervene and state, "That won't be necessary. With our son's disability we recognize that he is unable to care for himself." The best response by the nurse would be:
    1. "Then I will just teach you what he needs to have done."

2. "This material is not difficult; even a slow child can learn it."
3. "Your son cannot always depend on you for his health needs."
4. "Your son seems bright enough to me. I think he can learn this."

60. The nursing plan for an 8-year-old boy with celiac disease should include helping the child to:
    1. Express his feelings, while focusing on ways in which he can still be normal like his friends
    2. Select meals from those high-residue, high-carbohydrate foods that are gluten-free and permitted
    3. Understand the relationship of diet to disease so that he will be more willing to adhere to his diet and refrain from eating snack foods
    4. Learn which snack foods can be substituted for the hot dogs and hamburgers that have wheat fillers, because occasional noncompliance is permitted

61. In the management of a child newly diagnosed with chronic celiac disease, the primary nursing goal is to:
    1. Prevent celiac crisis and resulting complications
    2. Prevent complications from respiratory involvement
    3. Teach the parents to control the diet to promote normal growth
    4. Help the parents and child adjust to the lifelong dietary restrictions

62. Before a 4-year-old child with a new colostomy is discharged, the nurse prepares a teaching plan for the parents that includes telling them that:
    1. An enterostomal therapist is available to assist with home care
    2. They should try correcting the child's poor eating habits at mealtime
    3. Fluids should be limited between meals, although permitted at meals
    4. The child should not take part in physical education when attending school

63. The parents of a sick child constantly blame each other for their child's illness. The response by the parents that would indicate that the nurse's attempts to point out reality had been successful would be:
    1. The father bringing the child many expensive gifts
    2. The parents promising the child a trip to Disney World
    3. The parents making an appointment with a family counselor
    4. The mother assuming the blame for not paying attention to the child's complaints

64. A child who has barely survived a near-drowning episode is in critical condition in an intensive care unit. At one point the child opens the eyes and smiles, prompting the mother to say to the nurse, "Look, I think my child will get better now." The nurse's best response would be:
1. "Yes, you are right; this is a very good sign."
2. "See if you can get your child to hold your hand, too."
3. "God must have certainly been watching over your child."
4. "We are doing everything we can to help your child to recover."

65. Following treatment for Lyme disease, a child expresses fear of going camping again because of the ticks. The nurse's best response would be:
1. "Tell me more about your fears about camping."
2. "You are afraid to go camping just because of a tick?"
3. "Frequent checks for ticks are a defense against infection."
4. "Oh, camping is fun. Just think of what you will be missing."

66. An infant is scheduled for emergency surgery. The nurse notes that the baby's mother is 13 years old and the father is 16 years old. The baby's father and the paternal grandmother, who cares for the baby, are at the bedside. The nurse should obtain the informed consent from the:
1. Paternal grandmother
2. Hospital administrator
3. Sixteen-year-old father
4. Thirteen-year-old mother

67. When caring for children in a family that is economically deprived, the nurse should understand that the characteristic most common to those living in poverty is:
1. Long-term feelings of powerlessness
2. A willingness to postpone gratification
3. Open and direct expressions of anger
4. Compliance with health recommendations

68. When working with a family as the unit of service, the public health nurse should consider that:
1. Separating health problems from other aspects of this family's life is essential to help them
2. Certain members of the family may be capable of providing more support than the nurse can
3. Assessing each member of this family is not necessary to plan the care for the family as a whole
4. Values, beliefs, and attitudes held by the family have limited influence on how they will perceive assistance

69. When planning teaching for the parents of a child with tetralogy of Fallot, who are both employed full time, the nurse should:
1. Schedule a whole evening for teaching
2. Insist both parents attend the teaching sessions
3. Provide written and oral information in short sessions
4. Point things out to them when they are visiting their child

70. A 16-year-old, her 1-month-old baby, and the baby's grandmother come to the emergency room saying that the infant accidentally fell down the stairs. Legally, consent for the baby's medical care:
1. Should be obtained from the grandmother, who must sign the consent
2. Must be decided by family court because the baby's mother is a minor
3. Is not necessary because this is an emergency and no consent is needed
4. Is the responsibility of the baby's mother, and she should sign the consent

71. The nursing diagnosis that would apply to families of all children with cerebral palsy is high risk for alteration in parenting related to:
1. Lack of social support
2. An unrealistic expectation of self
3. Having a mentally retarded child
4. Loss of the expected normal child

72. A 6-year-old boy is receiving chemotherapy for a neuroblastoma, stage IV. He had his first chemotherapy session last week and has arrived with his mother for this week's session. The nurse should approach the child and his mother by asking:
1. "Did your son vomit after the last dose?"
2. "How did you feel after your last medicine?"
3. "Aren't you happy that two sessions will be finished?"
4. "How are you feeling this week? Ready for another dose?"

73. A 16-year-old with full-thickness burns of the entire right arm states, "I'll never be able to use my arm again. I'll be scarred forever." The nurse's best initial response would be:
1. "The staff will take steps to minimize scarring."
2. "Think about how lucky you are. You are alive."
3. "I know you are worried, but it is still too early to tell."
4. "Try not to worry; concentrate on doing your range-of-motion exercises."

74. A child who had been severely beaten was found wandering in the streets and was admitted to the hospital. The nursing diagnosis that will most profoundly influence the planning of care for this child would be:
    1. Altered parenting
    2. Impaired skin integrity
    3. Post-traumatic response
    4. Sensory/perceptual alteration: visual

75. An abused child, after being hospitalized for severe injuries, is being placed in temporary foster care. The foster family is coming into the hospital to meet the child. The nurse should plan to facilitate this meeting by:
    1. Decorating the child's room with "welcome" signs
    2. Providing a private room for the foster family and the child
    3. Encouraging the child to draw a picture for the new mother
    4. Answering all the child's questions about the foster family ahead of time

76. When the adolescent mother of an infant admitted to the hospital with multiple trauma sees her infant in the intensive care unit for the first time, she cries out, "I didn't mean to hurt her." The nurse should:
    1. Encourage the young mother's family to come and comfort her
    2. Notify the Child Abuse Hotline of this probable instance of abuse
    3. Respond by saying, "You caused your baby's injury and feel guilty."
    4. Put an arm around the young mother and say, "This must be difficult for you."

77. The nurse is aware that abusing parents:
    1. Are mature, independent individuals
    2. Have few available personal resources
    3. Are aware of the abilities of young children
    4. Often have been raised with little discipline

78. When assessing the family dynamics of a suspected abusing family, the nurse would be surprised to observe that the:
    1. Parents provide little emotional support to the child
    2. Parents offer consistent, detailed stories about the injuries
    3. Child cringes and appears unduly afraid when approached
    4. Child has many unexplained old injuries, scars, and bruises

79. A female client who has been abusing her son is undergoing treatment to control her behavior. A statement by the client that indicates the development of some insight into her behavior as a parent would be:
    1. "I promise that I won't get so angry when my son causes trouble again."
    2. "Once my son gets straightened out, we should not have these problems."
    3. "I think the root of the problem is when my husband comes home after drinking."
    4. "If I feel angry at my son again, I'm going to go into the bedroom and punch a pillow."

80. A mother of three children who was abandoned by her husband shortly after the birth of her youngest daughter brings the child, now 9 months old, into the hospital with a diagnosis of failure to thrive. As the mother leaves, the nurse is not surprised to see the daughter react by:
    1. Clinging to the mother and expressing fear of the nurse
    2. Crying at first but then letting the nurse hold and comfort her
    3. Sustaining eye contact with the mother and refusing the nurse's arms
    4. Readily allowing the nurse to take her but remaining stiff while being held

FLUIDS AND ELECTROLYTES (FE)

81. In children with nephrotic syndrome, the best indicator of fluid balance is the daily measurement of:
    1. Body weight
    2. Urinary output
    3. Abdominal girth
    4. Urine osmolality

82. A 4-year-old child with nephrotic syndrome has been restricted to 600 ml of fluid for 24 hours. The nursing intervention that would be most appropriate in assisting the child to cope with such a limitation is:
    1. Dividing fluid intake equally among each shift (200 ml each shift)
    2. Allowing the child to drink fluids as desired until the 600 ml limit is reached
    3. Withholding fluids from 7 pm to 7 am and giving the entire 600 ml from 7 am to 7 pm
    4. Providing the child a minimum of 1 ounce of fluid in small, 1-ounce cups each waking hour

83. Children with acute glomerulonephritis are placed on diets low in:
    1. Fat
    2. KCl

3. Protein

4. Glucose

84. A 5½-month-old infant is admitted to the hospital with a fever and a history of vomiting for 48 hours. In view of this infant's responses, the assessment by the nurse that would initially influence the child's care is:
   1. Inspecting the baby's skin for poor turgor
   2. Determining the baby's vital signs and weight
   3. Checking the baby's neurologic status and urinary output
   4. Asking the mother whether the baby is breast-fed or bottle-fed

85. To best ascertain the magnitude of fluid loss in an infant with gastroenteritis and diarrhea, the nurse should:
   1. Evaluate the infant's skin turgor carefully
   2. Note the elevation of the infant's hematocrit value
   3. Assess the moistness of the infant's mucous membranes
   4. Compare the infant's preillness weight with the current weight

86. An initial nursing assessment of an infant with severe dehydration will most likely reveal:
   1. Stools that are frothy
   2. A weak, decreased pulse
   3. Bulging of the occipital fontanel
   4. An elevated urine specific gravity

87. A 3-month-old infant with gastroenteritis and dehydration is admitted to the hospital and placed on contact precautions. Nursing assessment of this child will probably reveal:
   1. A bulging fontanel
   2. Resilient skin turgor
   3. Marked restlessness
   4. Decreased urinary output

88. When assessing a 4-month-old infant with gastroenteritis and dehydration, the nurse would expect to find a:
   1. Specific gravity of 1.014
   2. Urinary output of 50 ml/hr
   3. Depressed anterior fontanel
   4. History of allergies to various foods

89. A 5-month-old with a history of frequent bouts of diarrhea is to be discharged. A priority concern that the nurse should include in the teaching plan for the mother is the:
   1. Effects of antibiotics on viral gastroenteritis
   2. Potential hazards of fluid loss in young children

3. Importance of a well-balanced diet for the infant

4. Need for cleanliness of foods and feeding utensils

90. The nurse is aware that normal arterial blood gas values in the child would be most accurately reflected by readings of:
   1. pH 7.40, Po_2 85 mm Hg, Pco_2 40 mm Hg, base excess 0
   2. pH 7.50, Po_2 85 mm Hg, Pco_2 35 mm Hg, base excess 0
   3. pH 7.25, Po_2 60 mm Hg, Pco_2 50 mm Hg, base excess −4
   4. pH 7.45, Po_2 70 mm Hg, Pco_2 25 mm Hg, base excess +4

91. The nurse should recognize that the sequence of events that occurs in the respiratory response to acidosis is:
   1. Hypoventilation; increased CO_2 elimination; decreased blood H ions; increased pH
   2. Hypoventilation; decreased blood H ions; increased CO_2 elimination; decreased pH
   3. Hyperventilation; increased CO_2 elimination; decreased blood H ions; increased pH
   4. Hyperventilation; decreased CO_2 elimination; decreased blood H ions; decreased pH

92. A critically ill child develops Cheyne-Stokes respirations, and the nurse suspects an increasing acid-base imbalance related to:
   1. Respiratory alkalosis from overbreathing and excess carbon dioxide output
   2. Respiratory acidosis from impeded breathing and the retention of carbon dioxide
   3. Metabolic acidosis from the concentration of cations in body fluids, which displace bicarbonate
   4. Metabolic alkalosis from an increase in base bicarbonate resulting from the primary health problem

93. The physiologic compensatory mechanism that is activated to counteract the effects of acid-base imbalance in a child with severe dehydration is:
   1. Profuse diaphoresis
   2. Renal retention of H+
   3. Elevated temperature
   4. Increased respirations

94. The blood gas report that would most likely reflect the acid-base balance found in a child admitted to the hospital with severe dehydration is:
   1. A pH of 7.50 and a Pco_2 of 34 mm Hg
   2. A pH of 7.20 and a Pco_2 of 20 mm Hg
   3. A pH of 7.23 and a Pco_2 of 70 mm Hg
   4. A pH of 7.56 and a Pco_2 of 20 mm Hg

95. Following surgery for the repair of a meningomyelocele, an infant develops diarrhea and metabolic acidosis with a decreased urinary output. Because of the infant's status, the nurse anticipates that the physician will order:
    1. Plasmanate
    2. Isotonic saline
    3. Sodium lactate
    4. Potassium chloride

96. With persistent diarrhea an infant is subject to significant fluid and electrolyte alterations. Because of this the nurse should be aware that an infant with diarrhea may develop:
    1. Hypovolemia, hypercalcemia
    2. Hypernatremia, hypervolemia
    3. Metabolic acidosis, hypovolemia
    4. Decreased hematocrit, hyponatremia

97. A 9-year-old with insulin-dependent diabetes mellitus is admitted to the hospital with deep, rapid respirations; flushed, dry cheeks; abdominal pain with nausea; and increased thirst. Laboratory tests would be expected to show:
    1. A blood pH of 7.25 with a blood glucose level of 60 mg/dl
    2. A blood pH of 7.50 with a blood glucose level of 60 mg/dl
    3. A blood pH of 7.50 with a blood glucose level of 460 mg/dl
    4. A blood pH of 7.25 with a blood glucose level of 460 mg/dl

98. When a 3-month-old infant is receiving IV fluids via a scalp vein, the nurse should:
    1. Check the baby's pupils for reaction every hour
    2. Observe behind the baby's ear and occiput for infiltration
    3. Restrain the baby's arms and legs when nobody is present
    4. Explain to the parents why they can't hold the baby during IV therapy

99. The nurse notes that an infant with a diagnosis of failure to thrive, who has been on tube feedings for 3 days, has very dry skin and mucous membranes. The nurse verifies that all feedings have been retained, but the urinary output is consistently 250 ml and the infant has lost weight. The nurse should:
    1. Increase the intravenous flow of half-normal saline and call physician
    2. Realize this is probably normal for babies and infants with failure to thrive
    3. Recognize undernutrition and call the physician to increase the caloric intake
    4. Recognize underhydration and call the physician to increase the infant's fluid intake

100. The best method for assessing an infant's response to rehydration therapy is for the nurse to:
    1. Measure the infant's abdominal girth
    2. Assess the color of the infant's stools
    3. Weigh the infant at the same time daily
    4. Monitor the infant's skin turgor frequently

101. After abdominal surgery, a priority nursing intervention for a young infant with an IV is:
    1. Administering oral fluids
    2. Limiting handling by the parents
    3. Weighing the diapers after voiding
    4. Placing elbow restraints on both arms

102. An IV of D5W is infusing when a child returns to the pediatric unit from surgery. The postoperative orders do not indicate the desired rate of infusion. The most appropriate action for the nurse to take is to:
    1. Adjust flow to the rate child was receiving prior to surgery
    2. Reduce the flow rate to keep the vein open and call the physician
    3. Regulate the flow rate to 25 ml per hour until the physician makes rounds
    4. Maintain the present flow rate and call the operating room to verify the correct rate

103. An 8-month-old child who weighs 18 lb, 12 oz is receiving 8 oz of full-strength formula every 4 hours. Based on the recommended caloric intake for an 8-month-old infant (108 kcal/kg), this child's caloric intake:
    1. Is difficult to evaluate without further information
    2. Meets the recommended requirements for growth
    3. Exceeds the recommended requirements for growth
    4. Is less than the recommended requirements for growth

104. The physician orders multielectrolyte solution (MES) 150 ml per kg of body weight per 24 hours for a child weighing 13 pounds. The nurse is aware that the intake of MES for this child should be:
    1. 500 ml/24 hr
    2. 750 ml/24 hr
    3. 885 ml/24 hr
    4. 965 ml/24 hr

105. The physician has ordered 500 ml of a balanced electrolyte solution to run over 18 hours for a 5-

year-old. The IV set delivers 60 gtt/ml. To administer the 500 ml over 18 hours, the nurse should set the IV drip rate at:
1. 0.5 gtt/min
2. 3.0 gtt/min
3. 28.0 gtt/min
4. 36.0 gtt/min

106. The physician orders 700 ml of IV fluid over 24 hours. The IV tubing has a drop factor of 60 gtt/ml. The nurse should set the flow to provide:
1. 15 gtt/min
2. 20 gtt/min
3. 29 gtt/min
4. 34 gtt/min

107. A child is to receive 500 ml of D5NS and 250 ml of RL over 24 hours. The tubing delivers 60 gtt/ml. The nurse should set the flow to provide:
1. 11 gtt/min
2. 21 gtt/min
3. 31 gtt/min
4. 41 gtt/min

BLOOD AND IMMUNITY (BI)

108. The nurse is aware that children with AIDS are even more prone to infection than adults with AIDS because:
1. Even the immune system of a healthy child is incapable of producing antibodies
2. The AIDS virus attacks children's immune systems through different mechanisms
3. Children with AIDS are exposed to many more pathogens than are adults with AIDS
4. Children have fewer circulating antibodies resulting from a lack of previous exposure to pathogens

109. The nurse is aware that immunization of infants does not begin until 2 months of age because:
1. The neonatal spleen is unable to produce efficient antibodies
2. Infants younger than 2 months are rarely exposed to infectious diseases
3. Maternal antibodies interfere with development of active antibodies by the infant
4. The immunization would attack the immature infant's body and produce the disease

110. Before giving a 2-month-old child the first DTP immunization, the nurse discusses with the mother the possible reactions that may occur because these reactions are:
1. Often serious and may require hospitalization
2. Quite common and may be either local or systemic

3. Often responsible for permanent neurologic damage
4. Sometimes responsible for deep ulceration at the site of injection

111. An infant receives the first DTP immunization at 2 months. The nurse instructs the parent to:
1. Apply heat to the injection site for the first day; afterward apply ice if the arm is sore
2. Apply ice to the injection site if soreness is present; call the physician if a fever develops
3. Give the baby aspirin for pain; if swelling at the injection site develops, call the physician
4. Give acetaminophen for fever; call the physician if marked drowsiness or convulsions occur

112. If a 5½-month-old infant's immunizations are on schedule, the nurse can assume that the baby has already received:
1. Measles, mumps, and rubella vaccine
2. A booster dose of trivalent oral polio vaccine
3. Two doses of diphtheria, tetanus, and pertussis vaccine
4. The first booster dose of diphtheria, tetanus, and pertussis vaccine

113. If a 9-month-old's immunization schedule is up to date, the next immunization that the infant should receive at 15 months of age is:
1. Polio vaccine
2. Tetanus toxoid
3. Measles, mumps, and rubella vaccine
4. Diphtheria, tetanus, and pertussis vaccine

114. A newborn's immunization schedule is started. When discussing the immunization schedule for the first 6 months with the parents, the nurse should inform them that infants are not usually immunized against:
1. Polio
2. Tetanus
3. Measles
4. Hepatitis B

115. A child with leukemia is to continue taking prednisone at home after discharge. The nurse discovers that the child's sibling is home with chickenpox. The nurse plans the discharge teaching based on the knowledge that:
1. Chickenpox can be fatal to clients with leukemia
2. The child must be immunized before going home
3. Clients receiving prednisone are immune to chickenpox
4. If direct contact between the two children is prevented, the client can go home

116. The nurse should suggest to the mother of an 18-month-old infant with anemia that the child be fed:
1. Pieces of pumpkin pie
2. Slices of a whole apple
3. A cup of seedless grapes
4. Bread pudding with raisins

117. A mother tells the nurse that her 8-month-old will eat only mashed potatoes and drink only milk, and she is concerned about the baby's diet, even though the baby is receiving Poly-Vi-Sol daily. The nurse recognizes that the child's diet could lead to:
1. A potassium deficit
2. A vitamin deficiency
3. An amino acid deficiency
4. An iron-deficiency anemia

118. The nurse, preparing a 12-year-old child for a bone marrow aspiration, would know that the child does not understand the teaching about the procedure when the child states:
1. "I can get out of bed after the doctor is finished."
2. "I will have a tight dressing to put pressure on the area."
3. "The doctor is going to inject a needle into the center of one of my hip bones."
4. "The only pain I should feel is when the doctor puts in the shot so it won't hurt."

119. The mother of a child who has been recently diagnosed as having hemophilia is pregnant with her second child. She asks the nurse what the chances are that this baby will also have hemophilia. The nurse's best response would be:
1. "There is no chance the baby will be affected."
2. "There is a 25% chance the baby will be affected."
3. "There is a 50% chance the baby will be affected."
4. "There is a 75% chance the baby will be affected."

120. The most common area for bleeding to develop in a child with hemophilia is the:
1. Brain
2. Joints
3. Abdomen
4. Pericardium

121. In addition to the relief of pain, the nurse should direct the care for a client with sickle cell crisis toward:
1. Antibiotics and narcotic regulation
2. Oxygenation and adequate hydration
3. Hydration and psychologic counseling
4. Oxygenation and factor VIII replacement

122. A 10-year-old is admitted to the hospital in thrombocytic sickle cell crisis. When assigning a room, it is most appropriate for the nurse to place the child with a roommate who has:
1. Pneumonia
2. Thalassemia
3. Osteomyelitis
4. Acute pharyngitis

123. The nurse recognizes that the teaching about sickle cell anemia was understood when a 16-year-old with the disease states, "I know that symptoms will appear when I have:
1. A low iron intake."
2. A decreased amount of fluids."
3. A breakdown of thrombocytes."
4. An increased WBC production."

124. The mother of a 13-year-old who has sickle cell anemia tells the nurse that the family is going camping this summer. She asks what activities would be appropriate for the child. The nurse suggests:
1. Softball games with the family
2. Collecting logs for the campfire
3. Motorboat rides around the lake
4. Walking along the mountain trails

125. When obtaining a health history from the parents of a toddler who is admitted to the hospital with acute lymphocytic leukemia (ALL), the nurse would be surprised if the parents report that the first sign they observed was:
1. A loss of appetite
2. Sores in the mouth
3. A paleness of the skin
4. Purplish spots on the skin

126. When giving nursing care to a child with leukemia, the nurse notes blood on the pillow case and several bloody tissues. The nurse should check the child's laboratory report for the:
1. Platelet count
2. Uric acid level
3. Prothrombin time
4. Red blood cell count

127. When discharging a 5-year-old girl who is to continue chemotherapy at home, the nurse would know that the parents understood the discharge instructions when they say, "We should:
1. Isolate her from other children her age."
2. Allow her to eat her food at her own pace."
3. Have her rinse her mouth with mouthwash."
4. Provide her with structured activities each day."

128. A child with leukemia is to be sent home on a protocol that includes several antineoplastics after an intrathecal administration of methotrexate. Before discharge the nurse instructs the child's parents to:
    1. Limit contact with peers because they tend to have communicable diseases
    2. Return weekly for bone marrow aspiration to determine effectiveness of therapy
    3. Schedule routine laboratory appointments for evaluation of response to medication
    4. Withhold medications when nausea occurs to prevent additional episodes of vomiting

129. The nurse must assess the child who had a bone marrow transplant for infection, as evidenced by:
    1. Fever and lethargy
    2. Blood antibody titers
    3. Delayed bone growth
    4. The presence of leukopenia

130. Caring for an infant who is HIV positive can best be accomplished in:
    1. The pediatric unit
    2. A critical care unit
    3. The home environment
    4. An extended care facility

131. When a child with AIDS is awaiting adoption, to protect the staff, the nurse should immediately institute:
    1. Droplet precautions
    2. Contact precautions
    3. Airborne precautions
    4. Standard precautions

132. The nursing diagnosis with the highest priority for a child with AIDS would be:
    1. High risk for injury
    2. High risk for infection
    3. Alteration in growth and development
    4. Alteration in nutrition: less than body requirements

INTEGUMENTARY (IT)

133. A 3-month-old with severe dehydration has an excoriated diaper area. The infant's mother is quite upset when she finds the nurse has left her infant without a diaper. The nurse should explain that:
    1. Air drying of the skin prevents the diaper from sticking to it
    2. Increasing the exposed areas helps to reduce the infant's fever
    3. Cleansing of the skin followed by air drying reduces excoriation

4. This action allows the nurse to observe more quickly when the infant stools

134. The nurse learns that a 4-month-old child recently had a fever, runny nose, cough, and white spots in the mouth for 3 days. The child then developed a rash that started on the face and spread to the whole body. The nurse should suspect that the child had:
    1. Rubella
    2. Rubeola
    3. Varicella
    4. Scarlet fever

135. The nurse is aware that eczema in infants is a nonspecific ailment that is:
    1. Easily treated
    2. Highly contagious
    3. Predominantly found in infants
    4. Associated with chronic respiratory infections

136. The most important nursing care for infants with eczema is:
    1. Promotion of physical growth
    2. Provision of sufficient hydration
    3. Identification of causative factors
    4. Prevention of secondary infections

137. Allergic reactions in eczematous infants and young children are most often caused by:
    1. Fruit, eggs, and wheat
    2. Milk, eggs, and peanuts
    3. Woolens, meat, and milk
    4. Woolens, house dust, and dog hairs

138. The nurse evaluates that the mother of a 6-month-old with eczema needs more teaching regarding her baby's care when the mother states:
    1. "I will be careful not to cut the baby's nails short."
    2. "I have given all of the baby's woolen blankets to my nephew."
    3. "I will make sure not to give the baby any whole milk products."
    4. "I am going to buy my baby a whole new wardrobe of cotton clothing."

139. The symptoms that would most probably lead the nurse to suspect that a 3-year-old child has rubella are:
    1. Conjunctivitis and sensitivity to light
    2. Severe headache and nuchal rigidity
    3. Koplik's spots on the soft palate and buccal mucosa
    4. Enlargement of the posterior cervical and postauricular nodes

140. A child with rubella should be isolated from an unimmunized:
1. 20-year-old brother living at home
2. 3-year-old girl friend who lives next door
3. 12-year-old sister who had rubeola as a child
4. 18-year-old female cousin who has recently married

141. When extensive eschar formation is present on the arms of a child admitted to the hospital with severe burns, the priority nursing action would be to:
1. Remove blisters
2. Check radial pulses
3. Perform range of motion
4. Enforce respiratory isolation

142. A 12-year-old incurred partial-thickness burns of the entire right arm, upper left arm, and anterior chest when trying to start a campfire. According to the "rule of nines," the nurse estimates burn injury for this child would be:
1. 15%-24%
2. 25%-34%
3. 35%-44%
4. 45%-54%

143. When doing a dressing change, the nurse understands that a basic principle of surgical asepsis is:
1. The entire sterile field is considered sterile
2. Sterile items held below the waist are considered sterile
3. Sterile objects in contact with clean objects are considered clean
4. Wounds with exudates are contaminated, and dry wounds are sterile

144. When teaching a mother about communicable diseases, the nurse informs her that chickenpox is:
1. Communicable until all the vesicles have dried
2. Still communicable even when just dry scabs remain
3. No longer communicable after a high fever has subsided
4. Not communicable as long as the vesicles are intact and surrounded by a red areola

145. Understanding that there is a need to protect susceptible persons from exposure to chickenpox during the acute phase, the nurse should question the mother of a child with chickenpox about relatives or friends who are receiving:
1. Long-term anticonvulsant therapy
2. Prolonged topical antibiotic therapy
3. High doses of systemic steroid therapy
4. Therapeutic doses of vitamins and minerals

146. To hasten the drying of the lesions and relieve the itch in a child with chickenpox, the nurse suggests that the mother should try:
1. Rubbing bacitracin ointment into the open lesions
2. Using wet to dry saline dressings over the oozing vesicles
3. Having the child wear mittens and cutting the fingernails short
4. Patting the lesions gently with a paste of baking soda and warm water

147. As a mother leaves the hospital with her infant, the nurse notes that the mother covers the infant with a blanket. The nurse understands that the mother is preventing heat loss by the principle of:
1. Radiation
2. Conduction
3. Active transport
4. Fluid vaporization

148. The nurse teaches a teenager about the need for special mouth care because of the potential for lesions from the chemotherapy being administered. The nurse evaluates that the instructions were understood when the teenager says, "I should:
1. Brush my teeth with a toothbrush."
2. Rinse my mouth with hydrogen peroxide."
3. Rinse my mouth with undiluted mouthwash."
4. Brush my teeth with a foam-tipped applicator."

149. The nurse recognizes that when fever is suspected in preschool children with leukemia who are receiving chemotherapy that:
1. Rectal temperatures are too upsetting for this age group
2. Oral temperatures alone are inaccurate in children with leukemia
3. Tympanic temperatures are not accurate when a fever is suspected
4. Rectal temperatures are avoided to reduce the risk of rectal trauma

150. The nurse is aware that rubeola is commonly known as:
1. Measles
2. Chickenpox
3. Whooping cough
4. German measles

151. The nurse is aware that rubeola often causes children to have:
1. A macular rash
2. A paroxysmal cough
3. Enlarged parotid glands
4. Generalized vesicular lesions

152. The nurse is aware that the most common secondary infection to head lice (pediculosis capitis) is:
1. Eczema
2. Cellulitis
3. Impetigo
4. Tinea capitis

153. A child, recently returned from a camping trip, complains of a rash, chills, fever, and a headache and is taken to the clinic by the parents. The nurse in the clinic recognizes that this child's history and physical assessment should include:
1. A history of allergies and duration of symptoms
2. A developmental screening and history of exposure to chickenpox
3. Sports played on the trip and when the child has to return to school
4. The date the child received a flu vaccination and a history of any sunburn

154. When a 12-year-old boy who received several tick bites on a camping trip becomes ill, he is told that he may have Lyme disease. He asks the nurse, "What is Lyme disease?" The nurse's best response would be:
1. "It's a spirochetal infection that penicillin will treat."
2. "You sound upset, but we have medicine that will make you better."
3. "You are concerned. Why don't you ask me what you want to know."
4. "The insect bites gave you an infection, but there is medication that will stop it."

155. A client asks the nurse what is the best way to remove a tick from the skin. The nurse should reply:
1. "Touch the tick with a lighted cigarette."
2. "Remove the tick carefully with tweezers."
3. "Pour ammonia over the tick and it will shrivel up."
4. "Spray the tick with insect repellent and it will fall off."

SKELETAL (SK)

156. When doing an assessment of a child who has just had a cast applied to the right arm, the nurse finds that the fingers of the child's hand are cool. The nurse should first:
1. Elevate the right arm to reduce the swelling
2. Clip the edge of the cast to reduce pressure

3. Compare the temperature of the right and left hands
4. Call the physician to report the circulatory impairment

157. The nurse, when teaching parents how to care for their child's plaster cast, tells them that to keep the cast clean they should:
1. Cover the cast with a piece of plastic
2. Remove surface dirt with a damp cloth
3. Rub the dirty area with a diluted bleach solution
4. Scrub the cast with a soft brush and mild abrasive

158. The nurse is aware that the finding that would differentiate talipes equinovarus as a true clubfoot from a pseudo-clubbed foot is that the affected foot:
1. Exhibits little movement
2. Cannot be rotated past the midline
3. Is cooler to the touch than the other foot
4. Rotates past the midline but returns quickly when released

159. After a boot cast is applied to an infant with a clubfoot, a nursing diagnosis of "Altered peripheral tissue perfusion related to casting" is considered. The vascular assessment that would not contribute to evaluating the outcome while the cast is in place would be:
1. Color
2. Warmth
3. Blanching
4. Pedal pulse

160. A child sustains a fractured femur in a bicycle accident. However, the admission x-ray films reveal evidence of fractures of other long bones in various stages of healing. The nurse determines that this child should be assessed for:
1. Child abuse
2. Vitamin D deficiency
3. Osteogenesis imperfecta
4. Inadequate calcium intake

161. A newborn with talipes equinovarus has a unilateral boot cast applied to the involved foot. When moving the infant with a newly applied cast, the nurse should:
1. Touch the cast with just the fingertips
2. Turn the infant without touching the cast
3. Handle the cast with the palms of the hands
4. Move the infant's body and let the cast slide on the bed

162. Teaching for parents whose baby is undergoing frequent casting to correct a foot deformity should include information on plaster cast care, such as:
    1. Covering damp cast edges with adhesive petals
    2. Applying lotion to the skin at cast edges to keep it soft
    3. Checking the skin at the edges of the cast daily for redness
    4. Immersing the cast briefly during the tub bath and wiping it lightly

163. During a newborn assessment, a positive Ortolani sign would be indicated by:
    1. A unilateral droop of the hip
    2. A broadening of the perineum
    3. An apparent shortening of one leg
    4. An audible click on hip manipulation

164. When assessing a child suspected of having congenital hip dysplasia, the nurse would expect that an assessment of the child's orthopedic status would reveal:
    1. An apparent shortening of one leg
    2. A limited ability to adduct the affected leg
    3. A narrowing of the perineum with an anal stricture
    4. An inability to palpate movement of the femoral head

165. The statement by the mother of a 6-week-old female infant that would lead the nurse to assess the child for the presence of an abnormality is:
    1. "She wants to sleep curled up on her stomach all the time."
    2. "Her feet look very flat when I put both of her booties on her."
    3. "I seem to have a hard time getting her diaper between her legs."
    4. "I can't get her to straighten out her legs when I try to stand her up."

166. Six weeks after birth, an infant is diagnosed as having congenital hip dysplasia. The infant is admitted to the hospital for immediate corrective measures because:
    1. Infants are easier to manage in spica casts than toddlers
    2. Mobility will be delayed if correction is postponed until later
    3. The infant's hip joint is still cartilaginous and molding of the acetabulum is possible
    4. Bryant's traction cannot be used effectively after a child reaches the age of 2 years

167. Before discharging a newborn infant with congenital hip dysplasia, the nurse teaches the parents that hip dislocation can be minimized if the infant is:
    1. Carried straddling the hip
    2. Tightly swaddled in blankets
    3. Periodically strapped to a cradleboard
    4. Placed in an infant seat on a set schedule

168. When planning home care for a 6-month-old infant who has just been placed in a hip spica cast, the nurse should emphasize to the parents that:
    1. No special precautions will be necessary when diapering
    2. The entire cast should be wrapped in plastic wrap to prevent soiling
    3. Baby oil and powder should be used liberally around the diaper area
    4. The edges of the cast in the perineal area should be covered with plastic wrap

169. The nurse is aware that a community screening program for scoliosis is most effective in identifying the problem when the children are:
    1. 3 to 5 years of age
    2. 6 to 8 years of age
    3. 9 to 12 years of age
    4. 13 to 16 years of age

170. The nurse understands that a food that should be restricted for young orthopedic clients on prolonged bed rest, even though they may be on a regular diet, is:
    1. Beef
    2. Liver
    3. Broth
    4. Cheese

171. Three days following the application of a spica cast, a child has a temperature of 101.4° F. Suspecting an infection, the nurse should first assess the child for:
    1. Rapid irregular respirations
    2. A foul smell coming from the cast
    3. Any complaints of tingling in the toes
    4. The presence of itching around the top of the cast

172. After giving a teenager discharge instructions regarding cast care, the nurse evaluates that the instructions were understood when the teenager says, "If I am itchy around the cast I will:
    1. Gently rub the itchy area."
    2. Pat the area with an alcohol swab."
    3. Ask the physician for an antihistamine."
    4. Sprinkle a layer of powder around the itchy spots."

173. The nurse understands that the correct way to turn a 10-year-old in a spica cast is to:
1. Logroll the body as one unit
2. Use the crossbar between the legs
3. Have the child assist by using the overhead trapeze
4. Teach the child to sit up and help when changing position

NEUROMUSCULAR (NM)

174. The parents of an infant with cerebral palsy should be taught to:
1. Focus on cognitive rather than motor skills
2. Preserve muscle tone to prevent contractures
3. Maintain prolonged immobility of limbs with splints
4. Encourage strenuous exercise to build the infant's muscle tone

175. A 3-year-old male has recently been diagnosed with X-linked Duchenne's muscular dystrophy. Neither parent has muscular dystrophy. The statement by the parents that indicates an understanding of the disease's transmission is:
1. "Our daughters could be carriers."
2. "Our sons or daughters could have the disease."
3. "We each contributed a gene that caused our son to have the disease."
4. "By mendelian law, our son's having muscular dystrophy limits its occurrence in other children."

176. The nurse is aware that discharge planning related to care of a child with Duchenne's muscular dystrophy should include teaching the parents about:
1. Range-of-motion exercises
2. Maintaining a high-calorie diet
3. Instituting seizure precautions
4. Restricting the use of larger muscles

177. The nurse teaches the parents that when a child with Duchenne's muscular dystrophy reaches adolescence additional problems will probably develop with the:
1. Nutritional system
2. Neurologic system
3. Musculoskeletal system
4. Cardiopulmonary system

178. The nurse understands that the genetic etiologic factor of Down syndrome is an:
1. Extra chromosome
2. Intrauterine infection
3. X-linked chromosome
4. Autosomal recessive gene

179. The nurse should recognize that the most common serious anomaly associated with Down syndrome is:
1. Renal disease
2. Hepatic defects
3. Congenital heart disease
4. Endocrine gland malfunction

180. The nurse suspects that the concept that a developmentally disabled child could probably learn the fastest is:
1. Love vs hate
2. Life vs death
3. Large vs small
4. Right vs wrong

181. When parents ask the nurse what to do about their preschooler's stuttering, the nurse should suggest that they:
1. Identify situations that increase stuttering and avoid or ignore the hesitancy
2. Avoid looking at the child when the child has difficulty forming or expressing words
3. Help the child by supplying the correct word when the child is experiencing a block
4. Stop the conversation and tell the child to speak slowly and think before starting again

182. Shortly following birth, a newborn is diagnosed as having Erb's palsy. The nurse is aware that this problem is caused by:
1. A disease acquired in utero
2. An X-linked inheritance pattern
3. A tumor arising from muscle tissue
4. An injury to the brachial plexus during birth

183. The nurse recognizes that a couple who have a newborn with Erb's palsy have an accurate understanding of their infant's prognosis when they state:
1. "This is a progressive disease with no cure."
2. "A year of physical therapy will be necessary."
3. "Correction can be achieved only through surgery."
4. "Complete recovery should occur in about 3 months."

184. The nurse is aware that a high level of lead in the blood of children with lead poisoning (plumbism) can lead to:
1. Liver damage
2. Marked anemia
3. Increased urination
4. Severe malnutrition

185. The nursing diagnosis that is most appropriate for children with lead poisoning is:
1. Risk for injury
2. Chronic pain
3. Altered nutrition
4. Unilateral neglect

186. After a craniotomy for the removal of a hematoma sustained in a fall, a child is returned to the recovery room. The nurse places the child in a semi-Fowler's position to:
1. Increase cranial drainage and prevent accumulation of fluid
2. Reduce subdural pressure and promote reaction from the anesthetic
3. Decrease pressure on diaphragm and increase thoracic expansion
4. Decrease the cardiac workload and facilitate oxygenation after surgery

187. The day after brain surgery, a 9-year-old child with diabetes mellitus develops a temperature of 103° F. The nurse understands that:
1. A slight fever is to be expected following any surgery
2. Anyone with diabetes will develop an infection following surgery
3. Edema following the surgery often causes pressure on the hypothalamus
4. An excess of viscid secretions has caused inadequate respiratory ventilation

188. The nurse is aware that infants with a myelomeningocele, not just a meningocele, usually have:
1. Hydrocephalus
2. Lower extremity paralysis
3. A sac over the lumbar area
4. Infections of the spinal fluid

189. The abnormal finding that a nurse would expect to observe during an assessment of a 1-month-old infant admitted to the hospital with hydrocephalus would be that:
1. The infant's anterior fontanel is tense on palpation
2. The infant demonstrates poor eye muscle coordination
3. The infant is unable to support the head and shoulders while prone
4. The infant's head circumference is larger than the chest circumference

190. An infant with a myelomeningocele is scheduled for surgery to close the defect. The nursing action that would best facilitate parent-child relationships in the preoperative period is:

1. Allowing the parents to cuddle the child in their arms
2. Demonstrating feeding techniques in the prone position
3. Encouraging the parents to stroke and comfort the child
4. Referring the parents to the Spina Bifida Association of America

191. A 3-month-old infant has a ventriculoperitoneal shunt inserted. The nurse plans to:
1. Keep the infant in the prone position
2. Apply sterile moist dressings to the incision
3. Observe for signs of leakage of cerebrospinal fluid
4. Teach the parents the signs of increased intracranial pressure

192. A 1-month-old infant with hydrocephalus is scheduled for surgery for the insertion of a ventriculoperitoneal (VP) shunt. A short-term preoperative goal for the infant would be to:
1. Keep the infant as comfortable as possible to limit crying
2. Use a thick head bandage to protect the infant's head from injury
3. Establish and maintain a strict fixed feeding schedule to ensure hydration
4. Provide a wide variety of play objects to maintain age-appropriate stimulation

193. Preoperatively, the parents of a child undergoing an insertion of a right ventriculoperitoneal (VP) shunt for hydrocephalus are taught about postoperative positioning. The nurse can evaluate their understanding of the teaching when they state, "We will avoid putting pressure on the valve site by positioning our baby:
1. In the position that provides the most comfort."
2. On the back with a small support beneath the neck."
3. On the abdomen with a small support against the left side of the head."
4. Flat and side-lying with a small support against the right side of the head and back."

194. Following the repair of a myelomeningocele in a newborn, the nurse should:
1. Keep the child in the supine position
2. Apply sterile moist dressings to the incision
3. Observe for signs of leakage of cerebrospinal fluid
4. Teach the parents intermittent clean catheterization

195. On the day after surgery for insertion of a ventriculoperitoneal shunt for hydrocephalus, an infant's temperature rises to 103° F. The nurse should first notify the physician and then:
1. Sponge the infant with tepid alcohol
2. Recheck the temperature in 2 hours
3. Remove any excess clothing from the infant
4. Record the temperature on the infant's chart

196. Following the insertion of a shunt to treat hydrocephalus, the nurse evaluates the function of the shunt by:
1. Noting the frequency of voiding
2. Assessing for periorbital edema
3. Palpating the child's anterior fontanel
4. Observing for symmetric Moro reflexes

197. A 2½-year-old undergoes a shunt revision. Before discharge, the nurse, recognizing normal behavior for this age group, tells the parents to call the physician if the child:
1. Tries to copy all of the father's mannerisms
2. Talks incessantly regardless of the presence of others
3. Becomes fussy when frustrated and displays a shortened attention span
4. Frequently starts arguments with playmates by claiming all toys are "mine"

198. The nurse suspects that a 7-month-old, who is brought to the well-baby clinic for the first time, may have a hearing deficit when:
1. The mother says the infant is unable to learn the word "Mama"
2. The infant does not always turn the head when the name is called
3. The infant fails to demonstrate a Moro reflex in response to hand clapping
4. The mother says the infant stopped making verbal sounds about a month ago

199. In addition to systemic chemotherapy, the nurse is aware that cranial radiation is done on children with leukemia to:
1. Improve the quality of the child's life
2. Reduce the risk of systemic infection
3. Avoid metastasis to the lymphatic system
4. Prevent central nervous system involvement

200. While in the playroom, a 7-year-old child has a twitching of the right arm and leg that almost immediately progresses to a generalized tonic-clonic seizure with clenched jaws. The nurse's best initial action would be to:
1. Take the other children to their rooms
2. Put a plastic airway into the child's mouth

3. Place a large pillow under the child's head
4. Move the toys and furniture away from the child

201. During a generalized clonic-tonic seizure a child becomes cyanotic. The nurse should:
1. Insert an oral airway
2. Administer oxygen by mask
3. Continue to observe the seizure
4. Notify the physician immediately

202. A 4½-year-old admitted to the hospital with a diagnosis of cerebellar astrocytoma has surgery, and a large tumor is excised. When the nurse is preparing for the child's admission to the intensive care unit, it would be inappropriate for the nurse to:
1. Place a hypothermic blanket on the bed
2. Raise the foot of the bed on shock blocks
3. Secure an IV pump to closely monitor fluids
4. Obtain a cardiorespiratory monitor and sphygmomanometer

203. The nurse knows that among infants and children, otitis media is considered the most common:
1. Viral infection
2. Fungal infection
3. Bacterial infection
4. Rickettsial infection

204. When explaining a myringotomy procedure, the nurse should emphasize that the incision:
1. Takes several days to heal and often leaves a scar
2. Provides immediate relief of pressure in the middle ear
3. Widens the perforation in the eardrum to allow for drainage
4. Often results in permanent perforation of the tympanic membrane

205. The most important nursing responsibility during a myringotomy procedure on an 18-month-old child is to:
1. Collect the aspirated drainage in a culture tube
2. Maintain the continuous flow of local anesthetic
3. Have the mother stay and hold the child in her arms
4. Keep the child restrained and completely immobilized

206. To help the parents promote the effectiveness of their child's myringotomy procedure, the nurse suggests that they should:
1. Maintain the child in the supine position
2. Position the child with the affected ear down
3. Keep the child with the affected ear uppermost
4. Observe the child for bleeding from the operative site

207. Before discharge after a myringotomy, a potential complication the nurse teaches the child's parents to report is:
1. Bleeding and diminished pain
2. Mild or moderate hearing loss
3. Lack of drainage and increased pain
4. Low-grade temperature and headache

208. The purpose of isolation for a child with an infectious disease is to:
1. Separate the infected child from noninfected persons
2. Interrupt the infectious process as quickly as possible
3. Protect the child with a decreased resistance to infection
4. Prevent nosocomial infection during the hospitalized period

209. The parents of an 18-month-old who has developed signs of tetanus are concerned about how the disease will affect their child's intellectual ability to function in the future. The nurse can best respond that the child's intellectual functioning:
1. May be damaged
2. Should remain intact
3. May be temporarily retarded
4. Depends on the severity of complications

210. A child is admitted to the hospital with a diagnosis of meningococcal meningitis. The nurse is aware that isolation:
1. Will be required for 7 days
2. Of any kind is not required
3. Must be maintained during the incubation period
4. Is required for 24 to 72 hours after onset of antibiotic therapy

211. The nurse, caring for a child with symptoms of tetanus, responds to the child's parents' request to assist with their daughter's care by stating:
1. "You may talk but don't touch her since it could cause uncontrolled seizures."
2. "I can understand you wanting to help, but we must avoid all unnecessary stimuli at this time."
3. "Tell me more about your concerns; we realize you must be terribly upset and worried about her."
4. "We encourage you to speak to her and touch her even though she is unable to respond right now."

212. The early symptoms that usually bring the child with Reye's syndrome to the hospital are:
1. Diarrhea and a rash
2. Jaundice and oliguria
3. Low-grade fever and petechiae
4. Intractable vomiting and confusion

213. When reviewing nursing care of the child with seizures, the nurse charts a clonic episode as:
1. Generalized rigidity
2. A loss of consciousness
3. Tremors of the upper extremities
4. A spasmodic jerking of the entire body

214. When caring for a child with Reye's syndrome the nurse must be alert for critical manifestations such as:
1. Bladder distention and overflow
2. A macular rash on face and trunk
3. Marked periorbital edema from renal shutdown
4. Bleeding and ecchymosis from liver involvement

215. Following oral surgery, the physician writes an order for pain medication for an 18-month-old. The nurse recognizes that it would be unlikely that the physician would order:
1. Aspirin
2. Codeine
3. Percodan
4. Acetaminophen

ENDOCRINE (EN)

216. The regulation of diabetes in a newly diagnosed juvenile is best accomplished by:
1. Insulin, dietary control, and exercise
2. Dietary control, exercise, and urine testing
3. Dietary control, exercise, and blood glucose monitoring
4. Oral hypoglycemic agents, dietary control, and exercise

217. After assessing what a newly diagnosed juvenile knows about diabetes, the nurse should:
1. Develop a rapport with the client
2. Set goals with the client and the client's family
3. Teach the client how to do blood glucose testing
4. Teach the client how to administer the insulin injections required daily

218. Since prevention of infection is of utmost importance in children with diabetes, the nurse should emphasize to the child and parents the importance of:
1. Inspecting both feet frequently and carefully
2. Soaking the feet at least once daily for 30 minutes in hot water
3. Drying the feet thoroughly after a bath by rubbing vigorously with a towel
4. Treating minor cuts on the feet immediately with a strong antiseptic such as iodine

219. The statement that reflects why the nurse teaches the mother of a young child with diabetes how to test the child's urine at home even though blood glucose testing is being done 4 times a day is:
    1. The urine should be tested for acetone during illness and when the blood glucose level is more than 250
    2. Blood glucose testing before meals and at bedtime may be stopped once the child is stabilized on insulin
    3. Urine testing remains the most accurate way to check for high glucose levels if double-voided specimens are used
    4. It is now thought that voided urine specimens reflect short-term glucose levels more accurately than blood glucose, especially in children

220. When reviewing the plan of care for a child with diabetes, the nurse is aware that the most accurate method to evaluate the effectiveness of diet and insulin therapy over time is the laboratory test that measures:
    1. Urine ketones
    2. Serum protein levels
    3. Serum glucose levels
    4. Glycosylated hemoglobin

221. If surgery is to be performed on a child with diabetes, the nurse should be cognizant that:
    1. Urine test results provide the best gauge of diabetic control after the surgery
    2. The greatest danger during the surgical procedure is from diabetic ketoacidosis
    3. The stress of surgery causes a rise in blood glucose levels during the postoperative period
    4. If insulin was not required before surgery, it generally will not be required in the postoperative period

222. An 8-year-old child is receiving 45 units of Humulin N (NPH) insulin at 7 am and 7 pm. The most appropriate information for the nurse to give the parents concerning a bedtime snack would be:
    1. Provide a snack at bedtime to prevent hypoglycemia during the night
    2. Give the child a snack at bedtime if any signs of hyperglycemia are displayed
    3. Keep the snack at the bedside in case the child becomes hungry during the night
    4. Bedtime snacks are not recommended for diabetic children treated with long-acting insulin

223. A young girl with diabetes has just joined the school's soccer team, and her mother is unsure whether to tell the coach of her child's condition. The mother asks the nurse in the pediatric clinic for guidance. The nurse's response to the mother's concern should be based on the fact that:
    1. The coach might discuss the child's condition with other faculty members
    2. Hyperglycemia can be treated by the school nurse if symptoms are recognized early
    3. The child would be dropped from the team if school authorities learn she has a chronic disease
    4. Episodes of hypoglycemia are associated with children with diabetes who participate in sports activities

224. A 9-year-old child with diabetes is hospitalized for dosage regulation of insulin. The child appears to be very manipulative and has been observed sneaking food and trying to talk the mother into providing sweets. Based on this behavior, when the child complains of hypoglycemia, the most appropriate nursing action would be to:
    1. Test the urine for glucose
    2. Obtain a blood glucose level
    3. Administer orange juice with sugar
    4. Ask the child the last time food was eaten

225. A child with diabetes who is also learning disabled has trouble correctly measuring the required insulin dose. The child frequently draws up 42 units of insulin instead of the prescribed 24 units. The most appropriate intervention to ensure dosage safety would be to:
    1. Teach the child to use a magnifying glass to read the numbers on the syringe
    2. Exchange the insulin syringe the child has been using for a tuberculin syringe
    3. Provide the child with preset syringe guides that were developed for the blind
    4. Allow the child to have the number written down on paper when filling the syringe

226. An adolescent, newly diagnosed as having type 1 diabetes, asks what will happen when the blood glucose declines. Before answering, the nurse recalls that a hormone is secreted by the islets of Langerhans and causes liver glycogenolysis. This hormone is:
    1. ACTH
    2. Insulin
    3. Glucagon
    4. Epinephrine

227. A nurse suspects that a child with diabetes might be hypoglycemic when the child manifests:
    1. Redness of the face and deep, rapid breathing
    2. A change in behavior, hunger, and diaphoresis
    3. Increased thirst, sleepiness, and some vomiting
    4. A decreased level of consciousness and dry mouth

228. When teaching about insulin self-administration to a 10-year-old child newly diagnosed with diabetes, the nurse should teach the child to:
    1. Always wash the hands prior to preparing the insulin injection
    2. Shake the bottle of insulin thoroughly before drawing the dose
    3. Briskly rub the injection site for a minute after giving the injection
    4. Give the insulin injections primarily in the opposite arm or either leg

229. The nurse plans to teach a child with diabetes, who is receiving both Humulin N and Humulin R insulin daily, how to self-administer the insulin before discharge. When learning to give the injections, the child should:
    1. Alternate the sites until the best one to use is found
    2. Learn to use the needle and syringe by practicing on an orange first
    3. Administer the injections immediately after being taught the technique
    4. Draw up the Humulin N insulin first and then draw up the Humulin R insulin

230. The nurse is caring for an 11-year-old child with type 1 diabetes. Two hours after breakfast, the child becomes pale, diaphoretic, and shaky. The nurse should:
    1. Notify the physician
    2. Administer the supplemental order of insulin
    3. Compare these findings with those in the child's chart
    4. Give the child 4 ounces of orange juice and a slice of bread

231. A 13-year-old type 1 diabetic adolescent with a history of poor adherence to therapy is admitted to the hospital with a blood glucose level of 700 mg/dl. A continuous insulin infusion is begun. When developing a plan of care for this adolescent, the nurse should be alert for possible:
    1. Hypovolemia
    2. Hypokalemia
    3. Hypernatremia
    4. Hypercalcemia

232. A 14-year-old with diabetes wants to go for a pizza after a volleyball game. The teenager asks the school nurse whether this would be permissible on the insulin-diet-exercise regimen prescribed. The most appropriate response by the nurse would be:
    1. "Fast foods are unhealthy, especially for teenagers with diabetes."
    2. "It would be best if you ate at home where your diet can be controlled."
    3. "Go with your friends but make an effort to eat something other than pizza."
    4. "I will teach you how to use a fast food exchange list for individuals with diabetes."

233. A 16-year-old type 1 diabetic adolescent is brought to the emergency room unconscious. The adolescent's blood glucose level is 742 mg/dl. During the initial assessment the nurse would expect to note:
    1. Pyrexia
    2. Hyperpnea
    3. Bradycardia
    4. Hypertension

234. A primary long-term goal for a 16-year-old admitted to the hospital for the control of type 1 diabetes is to:
    1. Keep free of glucosuria
    2. Adhere to a routine exercise program
    3. Develop a life-style that will be nonstressful
    4. Maintain normoglycemia with few episodes of hypoglycemia or hyperglycemia

RESPIRATORY (RE)

235. An adolescent has an order for placement of a pulse oximeter. To ensure accuracy of the pulse oximeter reading, the nurse should:
    1. Place the probe on a finger or earlobe
    2. Calibrate the oximeter at least every 8 hours
    3. Place the probe on the abdomen or upper leg
    4. After application wait 30 minutes before obtaining a reading

236. The nurse understands that in the child, as in the adult, respiratory patterns are controlled by the:
    1. Medulla
    2. Cerebellum
    3. Hypothalamus
    4. Cerebral cortex

237. When examining the throat of a 5-year-old, the nurse should position a tongue blade to the side of the child's tongue primarily to avoid:
1. Eliciting the gag reflex
2. Obstructing the airway
3. Hurting any of the teeth
4. Interfering with the visual examination

238. Immediately after being placed flat, a child experiences shortness of breath and must sit up to breathe. The nurse knows that the term that best describes this phenomenon is:
1. Apnea
2. Dyspnea
3. Orthopnea
4. Hyperpnea

239. The nurse is aware that when administering routine oxygen therapy to a child the oxygen:
1. Should be labeled as flammable
2. Is warmed before administration
3. Must be humidified before administration
4. May be administered without a prescription

240. A 15-year-old high school student with hay fever has been taking a prescribed long-acting antihistamine/decongestant q8h for the past 3 days. The adolescent tells the nurse, "This medication is making me sleepy. Can you change it to something else?" The nurse's best response would be:
1. "Take only half a tablet before school."
2. "I think you should omit the early morning dose."
3. "The drowsiness will usually diminish after a few days."
4. "I'll ask the physician to change you to a medication containing ephedrine."

241. A 14-year-old develops sinusitis and is placed on a broad-spectrum oral antibiotic to be taken 4 times a day. To maintain the blood level, the nurse should recommend that the medication be taken at:
1. 8 am, 12 pm, 4 pm, and 8 pm
2. 8 am, 4 pm, 12 am, and 4 am
3. 6 am, 12 pm, 6 pm, and 12 am
4. 10 am, 2 pm, 10 pm, and 2 am

242. A 10-year-old who is mentally challenged and blind must be fed all meals. The child has problems swallowing and frequently chokes and coughs during the feeding. When feeding this child, the nurse should:
1. Liquefy the feedings and use a soft-tip bulb syringe
2. Place the child in the supine position with the head turned to the right

3. Prop the child in a semi-sitting position, chop up the food, and place it into the child's mouth with plastic tableware
4. Seat the child in the wheelchair and give small bites of food with metal tableware, encouraging the child's participation

243. When making an assessment of a 6-month-old infant with bronchiolitis, the nurse would expect:
1. A decreased heart rate
2. Increased breath sounds
3. A prolonged expiratory phase
4. Intercostal and subcostal retractions

244. Based on the problems associated with bronchiolitis, the treatment of choice for a child with this illness should consist of:
1. Humidified air and adequate hydration
2. Postural drainage and corticosteroids
3. Adequate hydration and bronchodilators
4. Croupette and broad-spectrum antibiotics

245. The nurse organizes care for an infant with bronchiolitis to allow for uninterrupted periods of rest. This plan is:
1. Inappropriate because constant care is necessary in the acute stage
2. Appropriate because the cool mist helps to maintain hydration status
3. Appropriate because this action promotes decreased oxygen demands
4. Inappropriate because frequent assessment by auscultation is required

246. An important nursing measure for a 6-month-old infant with bronchiolitis is:
1. Promoting stimulating activities that meet the infant's developmental needs
2. Making frequent observations of the infant's skin color, anterior fontanel, and vital signs
3. Discouraging visits from the parents during the acute phase to conserve the infant's energy
4. Keeping the infant on airborne precautions and using a gown, cap, mask, and gloves when giving care

247. A 3-year-old is admitted to the pediatric unit with a diagnosis of acute asthma. The child is short of breath, the respirations are 56, the pulse is 102, and there is a nonproductive cough. The nurse would expect the child's blood gas values to indicate a:
1. pH of 7.32
2. Po_2 of 95 mm Hg
3. Pco_2 of 40 mm Hg
4. HCO_3 of 26 mEq/L

248. An 8-year-old child with asthma is being assessed by the nurse. An assessment that requires immediate intervention would be:
1. A round face
2. Persistent wheezing
3. Regular use of inhalers
4. A respiratory rate of 30 per minute

249. An 8-year-old has a tonsillectomy. During the immediate postoperative period it is most important for the nurse to ensure that the child maintains:
1. Hydration by providing cool liquids frequently
2. Aeration by assisting with coughing and deep breathing
3. Airway patency by placing the child in a side-lying position
4. Consciousness by encouraging the mother to interact with the child

250. In the immediate postoperative period following a tonsillectomy, the mother of a 9-year-old should be encouraged to give her child:
1. Ice cream
2. Cold soda
3. Tepid milk
4. Orange juice

251. A male client with cystic fibrosis (CF) becomes romantically involved with a female with the same disease. He asks the nurse about the chances of having an affected child like himself. The most appropriate response by the nurse would be:
1. "Use condoms for protection from pregnancy."
2. "Young women with cystic fibrosis are not fertile."
3. "All of your children would be carriers of cystic fibrosis."
4. "You are probably not able to father children because of your cystic fibrosis."

252. The nurse understands that a male teenager who has a sibling with cystic fibrosis (CF) understands genetic counseling when he states, "To determine whether I am a carrier, I will have to undergo:
1. A chest x-ray."
2. Enzyme assays."
3. DNA probe testing."
4. Sweat chloride tests."

253. The respiratory assessment of an 8-year-old admitted with viral pneumonia demonstrates bronchial breath sounds over areas of consolidation, mild substernal retractions, profuse mucus production, pallor, and temperature of 102° F. The nursing action with the highest priority would be to:
1. Contact the respiratory therapist to set up oxygen

2. Start an intravenous line to provide fluids and electrolytes
3. Notify the physician of the fever so an antipyretic can be given
4. Place the child in a semi-Fowler's position and suction the nasopharynx

254. The nurse understands that the pathophysiologic factor primarily responsible for respiratory manifestations of cystic fibrosis is that:
1. There is acute inflammation of lung parenchyma
2. Increased irritability of airways causes obstruction
3. Endocrine glands secrete abnormal levels of hormones
4. Abnormally thick mucus leads to obstruction of the airways

255. The nurse provides clapping, percussion, and postural drainage every 4 hours for a 3-month-old infant with cystic fibrosis. The nurse is aware that the best time for scheduling this chest physiotherapy is:
1. After every feeding
2. Before every feeding
3. During every feeding
4. Midway between every feeding

256. A 20-year-old female is known to be heterozygous for the cystic fibrosis (CF) gene. Her husband's genotype is unknown at present and the couple is expecting their first child. The chance that their baby will have cystic fibrosis is:
1. 25% or less
2. 50% or more
3. Extremely rare
4. Unknown at this time

257. A 2-year-old, is admitted with croup and ¼ humidified oxygen via cannula is started because it:
1. Congeals the mucous secretions and relieves the dyspnea
2. Liquefies the mucous secretions making them easier to expectorate
3. Triggers the cough reflex and facilitates the expectoration of mucus
4. Dilates the blood vessels in the bronchi and alleviates the congestion

258. When assessing a newborn who has been diagnosed as having a diaphragmatic hernia, the nurse would expect to note the presence of:
1. Blood in the stool
2. A barrel shaped chest
3. Breath sounds over the abdomen
4. An increased abdominal circumference

259. A nursing diagnosis of impaired gas exchange is made for an infant with a diaphragmatic hernia. The causative factor for this diagnosis is the:
    1. Diaphragmatic hernia
    2. Decrease in oxygen intake
    3. Presence of excessive secretions
    4. Increase in the basal metabolic rate

260. In addition to raising the head of the bed of an infant who has had a surgical repair of a diaphragmatic hernia, the nurse should place the infant in the:
    1. Contour position in an infant seat
    2. Supine position with the knees flexed
    3. Prone position with the head to the side
    4. Side-lying position on the operative side

261. After the repair of a diaphragmatic hernia, the nurse would assess that the infant's respiratory condition is improving when:
    1. The infant stops crying
    2. The blood pH decreases to 7.31
    3. Breath sounds are heard bilaterally
    4. 1 oz of formula is ingested and retained

262. A child who was found face down in a water-filled ditch is brought to the emergency room. The child, who has a pulse of 50 beats per minute but no spontaneous respirations, is intubated and bagged with 100% oxygen. The most important nursing measure at this time is to:
    1. Start an intravenous line to provide fluid and electrolytes
    2. Assist the physician in delivering intracardiac medications
    3. Suction the endotracheal tube, mouth, and nasal passages
    4. Call the pediatric ICU to inform them of the child's admission

263. A child survives a near-drowning episode in a cold pond but still has many problems to overcome. The nurse is aware that the ultimate prognosis will depend mainly on the extent of damage resulting from the:
    1. Hypoxia
    2. Hyperthermia
    3. Emotional trauma
    4. Aspiration pneumonia

264. A young baby has an open repair of a fractured sternum and has a chest tube. The nurse explains to the baby's mother that the chest tube:
    1. Will be removed once the baby is feeding well and is afebrile
    2. Does not cause discomfort and is put in place for emergency use
    3. Is left in to drain the air from the chest cavity that entered during surgery
    4. Drains the extra air in the baby's chest that accumulated following the punctured lung

CARDIOVASCULAR (CV)

265. An infant born at 39 weeks gestation is sent to the intensive care nursery. The nurse suspects a possible cardiac anomaly when the admission assessment reveals:
    1. Projectile vomiting
    2. An irregular respiratory rhythm
    3. Hyperreflexia of the extremities
    4. Unequal peripheral blood pressures

266. A newborn with a cardiac defect is fed in the semi-Fowler's position. After the nurse feeds and burps the infant and changes the infant's position, the infant has a bowel movement and almost immediately becomes cyanotic, diaphoretic, and limp. These symptoms are most likely caused by the:
    1. Burping
    2. Formula
    3. Position change
    4. Bowel movement

267. When examining the laboratory work of a child with the diagnosis of rheumatic fever, the nurse would expect the findings to demonstrate:
    1. A negative C-reactive protein
    2. A positive antistreptolysin titer
    3. An elevated reticulocyte count
    4. A decreased erythrocyte sedimentation rate

268. A cardiac catheterization is scheduled for a 5-year-old with a ventricular septal defect to:
    1. Identify the degree of cardiomegaly present
    2. Demonstrate the exact location of the defect
    3. Confirm the presence of a pansystolic murmur
    4. Establish the presence of ventricular hypertrophy

269. A child returns to the unit following a cardiac catheterization. The statement on the child's progress made during the change-of-shift report 2 hours after the catheterization that should be questioned by the oncoming nurse would be that the child:
    1. Is on bed rest with bathroom privileges
    2. Has a pressure dressing over the entry site
    3. Has voided only 100 ml since the procedure
    4. Has to have the blood pressure checked every 2 hours

270. Discharge instructions for a child following a cardiac catheterization should include:
    1. Giving a sponge bath for the first 3 days at home
    2. Using ice compresses to relieve swelling at the entry site
    3. Limiting fluid intake for the next 3 days to prevent nausea
    4. Returning to the clinic in 5 days for removal of the pressure dressing

271. The physician orders a complete blood workup for a 5-month-old infant with tetralogy of Fallot. Because of the infant's heart disease, the nurse would expect the report to show:
    1. Polycythemia
    2. Agranulocytosis
    3. Thrombocytopenia
    4. A decreased hematocrit level

272. When caring for a 4-month-old infant with tetralogy of Fallot and congestive heart failure, the nurse should:
    1. Force nutritional fluids
    2. Provide small, frequent feedings
    3. Measure the head circumference daily
    4. Position the infant flat on the abdomen

273. The nurse is aware that the aim of palliative surgery for children with tetralogy of Fallot is to directly increase the blood flow to the:
    1. Brain
    2. Lungs
    3. Myocardium
    4. Right ventricle

274. A newborn is diagnosed with coarctation of the aorta. The baby is discharged with a prescription for digoxin (Lanoxin) 0.01 mg po q12h. The bottle of digoxin is labeled 0.01 mg in ½ teaspoon. The nurse should teach the mother to administer the medication by using:
    1. A nipple
    2. A plastic baby spoon
    3. The calibrated dropper in the bottle
    4. The small size baby bottle with 1 oz of water

275. The nurse is aware that in infants with congestive heart failure (CHF):
    1. The illness is an acquired congenital anomaly
    2. The treatment differs vastly from adult treatment
    3. Treatment is experimental because infants rarely develop congestive heart failure
    4. Digoxin (Lanoxin) and furosemide (Lasix) are the most commonly used medications

276. A 4-month-old who has a congenital heart defect develops congestive heart failure and is exhibiting marked dyspnea at rest. This finding is attributed to:
    1. Anemia
    2. Hypovolemia
    3. Pulmonary edema
    4. Metabolic acidosis

277. The mother of a 5-month-old infant with congestive heart failure questions the necessity of weighing the infant every morning. The nurse's response should be based on the fact that this daily information is important in determining:
    1. Renal failure
    2. Fluid retention
    3. Nutritional status
    4. Medication dosage

278. When attempting to identify the presence of tetralogy of Fallot in an infant, the nurse should understand that:
    1. In the absence of cyanosis, poor sucking is insignificant
    2. Many infants retain mucus that may interfere with feeding
    3. Feeding problems are fairly common in infants during the first year
    4. Poor sucking and swallowing may be early indications of heart defects

279. The nurse is aware that a common adaptation of children with tetralogy of Fallot is:
    1. Clubbing of fingers
    2. Slow, irregular respirations
    3. Subcutaneous hemorrhages
    4. Decreased red blood cell count

280. An infant with tetralogy of Fallot becomes cyanotic and dyspneic after a crying episode. To relieve the cyanosis and dyspnea, the nurse should place the infant in the:
    1. Orthopneic position
    2. Knee-chest position
    3. Lateral Sims' position
    4. Semi-Fowler's position

281. A 3½-year-old child returns to the room after a cardiac catheterization. Post-procedure nursing care for the child should include:
    1. Encouraging early ambulation
    2. Monitoring the insertion site for bleeding
    3. Restricting fluids until blood pressure is stabilized
    4. Comparing blood pressure in affected and unaffected extremities

282. A 5-year-old returns following cardiac surgery. The child has a left chest tube attached to water-seal drainage, an IV of D5 ½NS at 40 ml/hr, and a nasogastric tube to gravity. The child is attached to a cardiac monitor and has a left chest dressing. Upon admission to the unit, the nurse should first:
 1. Take the vital signs
 2. Check the identification bracelet
 3. Measure the chest and gastric drainage
 4. Check the suction pressure of the water-seal drainage

283. A parent asks the nurse, "The doctor said my baby has pulmonic stenosis. What does that mean?" The nurse's best response would be:
 1. "What else did your doctor say?"
 2. "Your baby has a heart problem."
 3. "Are you concerned about your baby?"
 4. "I'll page your physician so that you can discuss this again."

REPRODUCTIVE AND GENITOURINARY (RG)

284. The nurse understands that surgery is needed for a 4-year-old child with undescended testes because:
 1. Future malignancy may be prevented
 2. Maturation of the testes starts at about age 7
 3. The puboscrotal ring is more elastic at this age
 4. Early surgery produces less psychologic damage

285. The nurse is aware that uncorrected bilateral cryptorchidism can cause:
 1. Sterility
 2. Hydrocele
 3. Varicocele
 4. Epididymitis

286. The nurse explains to parents whose infant son has hypospadias that this defect occurred in the:
 1. First trimester
 2. Third trimester
 3. Second trimester
 4. Implantation phase

287. The care plan for a newborn with hypospadias should include:
 1. Keeping the penis wrapped with petrolatum gauze
 2. Preparing the infant for the insertion of a cystostomy tube
 3. Explaining to the parents why a circumcision will not be done
 4. Carefully explaining the genetic basis for the defect to the parents

288. After a child has a surgical correction for hypospadias, it is important for the nurse to:
 1. Ensure that the child's privacy is maintained
 2. Maintain the surgically implanted tension device
 3. Keep the child properly immobilized with restraints
 4. Gradually increase the time that the catheter is clamped

289. A nursing diagnosis that should have priority for an infant born with exstrophy of the bladder is:
 1. Urinary retention
 2. Risk for infection
 3. Sexual dysfunction
 4. Fluid volume deficit

290. An infant has exstrophy of the bladder. To protect the actual exposed bladder area, the nurse should expect the physician to order:
 1. Antibacterial ointments
 2. Pediatric urine collectors
 3. Warm moist compresses
 4. Sterile nonadherent dressings

291. After surgical repair of a urinary tract malformation, a child is to be discharged with an indwelling catheter. The parents should be taught that if no urine appears in the urinary drainage bag for a period of an hour or longer they should first:
 1. Call the physician
 2. Give the child extra fluids to drink
 3. Check for blockage of the drainage tubing
 4. Place firm pressure on the abdominal wall just above the bladder

292. The mother of a 6-year-old brings the child to the pediatric clinic and complains that the child has malaise, weakness, lethargy, anorexia, headaches, and smoky urine. When taking the nursing history, the nurse asks the mother whether the child has had a:
 1. Pain in the shoulders and knees
 2. Recent weight loss of at least 2 pounds
 3. Streptococcal infection within the last 2 weeks
 4. Rash on the palms and feet within the last 3 weeks

293. The nurse is aware that in order to confirm a diagnosis of acute glomerulonephritis in a 6-year-old, the tests that the physician will order will include:
 1. A routine urinalysis, a chest x-ray film, blood glucose levels, and an IVP
 2. An electrocardiogram, a heterophile antibody test, a routine urinalysis, and a chest x-ray
 3. A routine urinalysis, a complete blood chemistry, a nasopharyngeal culture, and an ASO titer
 4. An upper GI series, a 24-hour urine specimen, a complete blood chemistry, and a nasopharyngeal culture

294. A child is admitted to the pediatric unit with a diagnosis of acute glomerulonephritis. The nursing action that has priority is:
1. Assessing for dysuria
2. Observing for jaundice
3. Monitoring blood pressure
4. Testing vomitus for occult blood

295. A mother whose child has glomerulonephritis is fearful that her other child may get the disease. To allay the fears of the mother, the nurse should tell her that:
1. "The cause of acute glomerulonephritis is unknown, so it is difficult to know how to prevent it."
2. "Acute glomerulonephritis is inherited by an autosomal recessive trait but usually occurs only in males."
3. "Acute glomerulonephritis is caused by clot formation in the small renal tubules secondary to systemic infection."
4. "Acute glomerulonephritis is caused by an antigen-antibody response secondary to group A beta-hemolytic streptococcus."

296. When assessing a child with glomerulonephritis, the nurse should expect to find:
1. A decrease in joint mobility
2. An increase in urine volume
3. The presence of periorbital edema
4. The occurrence of an intermittent fever

297. The nurse encourages a child with glomerulonephritis to choose combinations of foods that include:
1. Corn on the cob, baked chicken breast, rice, applesauce, milk
2. Hot dog on a bun, potato chips, dill pickle slices, brownie, milk
3. Baked potato, ground beef, canned carrots, banana, buttermilk
4. Canned green beans, baked ham, bread and butter, peach, milk

298. When testing the urine and assessing the condition of a child with acute glomerulonephritis, the nurse would not be surprised to note a:
1. Normal blood pressure, anorexia, 1+ proteinuria, and 3+ glycosuria
2. Decreased blood pressure, anorexia, hematuria, and 1+ proteinuria
3. Decreased blood pressure, periorbital edema, 1+ proteinuria, and a specific gravity of 1.001
4. Moderately elevated blood pressure, periorbital edema, 4+ proteinuria, and a specific gravity of 1.030

299. When caring for a child with acute glomerulonephritis the nurse plans to:
1. Maintain bed rest, use isolation techniques, encourage fluids, and provide meticulous skin care
2. Maintain bed rest, provide a low-sodium diet, monitor blood pressure every hour, and monitor IV therapy
3. Prevent chilling, monitor vital signs every 2 hours, provide a no-sodium diet, and get the child up in a chair
4. Promote rest, monitor child's intake and output, weigh the child daily, and provide a regular diet with no added salt

300. Before discharging a child who has been treated for acute glomerulonephritis, the nurse should plan to provide the parents with:
1. Suggestions about activities that will keep the child active for long periods of time
2. The nurse's phone number so that the parents can call if they have any questions
3. Instructions as to when the child should return for a workup for a kidney transplant
4. A sample of a sodium-restricted diet because the child will continue on this diet at home

301. Close monitoring of the urine of a child with nephrotic syndrome who has been admitted to the hospital with massive edema and decreased urinary output would be expected to reveal:
1. High protein levels
2. Crystalline particles
3. Normal specific gravity
4. Numerous red blood cells

302. When admitting a 4-year-old child with nephrotic syndrome to the hospital, the nurse should assess for:
1. Severe lethargy
2. Chronic hypertension
3. Dark, frothy urine output
4. Flushed, ruddy complexion

303. The nurse assigns a 4-year-old boy admitted to the hospital with nephrotic syndrome to a room with a:
1. 2-year-old boy with croup
2. 3-year-old boy with impetigo
3. 4-year-old girl with conjunctivitis
4. 5-year-old girl with a fractured femur

304. When planning nursing care for a child with nephrotic syndrome, the nurse includes:
1. Provision of meticulous skin care
2. A diet low in carbohydrates and protein
3. Restriction of fluids to 500 ml each shift
4. A laboratory test for blood type and crossmatch

305. The adaptation that indicates that a child may have nephrotic syndrome rather than glomerulonephritis is the presence of:
1. Edema
2. Lethargy
3. Protein in the urine
4. A slightly decreased blood pressure

306. The nurse realizes that the parents of a child with nephrotic syndrome need further discharge instructions when they state, "We will:
1. Ignore any weight gain since it's normal."
2. Look at our child's eyelids every morning."
3. Need to test our child's urine for specific gravity."
4. Give our child the prednisone with meals or milk."

307. The urinary output of a child with acute glomerulonephritis decreases to less than 100 ml/24 hours and the creatinine clearance is 60 ml/min. The physician determines that the child has acute renal failure. The nurse notes that the child has an irregular apical pulse. The most significant laboratory finding would be:
1. Serum sodium 126 mEq/L
2. Serum creatinine 1.3 mg/dl
3. Serum potassium 6.1 mEq/L
4. Blood urea nitrogen 40 mg/dl

308. A child in renal failure, who has had the creation of an arteriovenous fistula access, begins hemodialysis three times a week. The nurse would know the child's mother needs further teaching when the mother states, "I will:
1. Call the doctor if my child develops vomiting or diarrhea."
2. Check the pulse at the wrist of the arm with the fistula daily."
3. Take a blood pressure in the arm with the fistula once a day."
4. Ensure that my child drinks the appropriate amount of fluid in warm weather."

309. The nurse understands that dialysis will be necessary when a child with chronic kidney disease exhibits:
1. Hypotension
2. Hypokalemia
3. Hypervolemia
4. Hypercalcemia

310. A child with chronic kidney disease is receiving peritoneal dialysis. To monitor for complications associated with this procedure, the nurse should assess the:
1. Abdomen for a bruit
2. Blood glucose levels
3. Clarity of return dialysate solution
4. Skin and mucous membrane for petechiae

311. A 7-year-old child must remain quietly in bed while having peritoneal dialysis. The most appropriate activity the nurse should plan for this child would be:
1. Learning to play chess with one parent
2. Gluing together a model airplane without help
3. Working a 100-piece puzzle with another child
4. Using a sponge ball to play catch with a roommate

312. A 4-year-old child has a nephrectomy because of a Wilms' tumor. Following a nephrectomy it is essential that the parents:
1. Maintain fluid restrictions
2. Prevent urinary tract infections
3. Restrict the child's intake of sodium
4. Prepare the child for a kidney transplant

313. When a child with an uncorrected hypospadias with chordee becomes an adult, he will be at increased risk for:
1. Renal failure
2. Testicular torsion
3. Testicular cancer
4. Reproductive dysfunction

314. The statement by a teenage female with cystic fibrosis that best reflects her understanding of healthy sexuality would be:
1. "I can never become pregnant."
2. "Having sex is not possible for me."
3. "A diaphragm is my best protection."
4. "I will not have sex without condoms."

315. To confirm a suspected diagnosis of gonorrhea in a 16-year-old male who has come to the clinic with a complaint of a thick urethral discharge, the nurse should:
1. Get a sexual history
2. Draw blood for a VDRL
3. Obtain a urethral culture
4. Collect a urine specimen

316. The nurse in an adolescent clinic is aware that an early diagnosis of syphilis is important and its presence is often determined by:
1. Evidence of a rash
2. A lesion on the genitals
3. Multiple gummatous lesions
4. A discharge from the genitals

GASTROINTESTINAL (GI)

317. Immediate nursing care for a neonate born with a cleft lip is directed primarily toward:
    1. Modifying feeding methods
    2. Keeping the baby from crying
    3. Minimizing handling by parents
    4. Preventing the occurrence of infection

318. A newborn with a cleft lip is fed with a special nipple. To minimize regurgitation of the feedings the nurse instructs the mother to:
    1. Give the baby the thickened formula as ordered
    2. Hold and burp the baby frequently while feeding
    3. Lay the baby on the side with the bottle firmly propped
    4. Feed the baby while sitting the baby up in an infant seat

319. A mother asks why her 1½-year-old toddler's cleft palate was not repaired at the time the cleft lip was repaired at 3 months of age. The nurse's best response would be:
    1. "Waiting leaves time for other birth defects to be detected and corrected."
    2. "The cleft lip was so disfiguring that plastic surgery was done as quickly as possible."
    3. "Your surgeon prefers to separate the operations to minimize and prevent complications."
    4. "The palate is corrected after teething and before your child talks so that correct speech may be learned."

320. After the repair of a cleft lip, the nurse will provide nutrition for the baby via:
    1. A plastic teaspoon
    2. Intravenous feedings
    3. A rubber-tipped syringe
    4. Nasogastric tube feedings

321. The first action following each feeding of a newborn with a fresh surgical repair of a cleft lip should be to:
    1. Burp the infant several times
    2. Place the infant on the abdomen
    3. Cuddle the infant for a few minutes
    4. Clean and rinse the suture line of the lip

322. A priority nursing measure for an infant during the immediate postoperative period following a surgical repair of a cleft lip is to:
    1. Minimize the infant's crying
    2. Restrain the infant at all times
    3. Oxygenate the infant frequently
    4. Handle the infant as little as possible

323. The physician orders arm restraints for a 1½-year-old who just had surgery for a cleft palate. The reason for the restraints is to prevent the child from:
    1. Playing with unsterile toys
    2. Rolling to a supine position
    3. Putting fingers into the mouth
    4. Pulling out the nasogastric tube

324. When a toddler with a cleft palate repair is able to tolerate fluids, the nurse should administer the fluids with a:
    1. Small cup
    2. Bulb syringe
    3. Lamb's nipple
    4. Teflon-coated spoon

325. When assessing an infant with a suspected diagnosis of hypertrophic pyloric stenosis, the nurse would expect to find:
    1. Visible peristaltic waves across the lower abdomen
    2. A palpable mass in the epigastrium to the right of the umbilicus
    3. Tenderness over the epigastric region not relieved by heat application
    4. Lower abdominal distention with vomiting of bile-stained gastric contents

326. A 25-day-old infant is admitted to the hospital after 3 days of vomiting, and pyloric stenosis is diagnosed. The most important nursing assessment at the time of admission is the:
    1. Character, amount, and times when the baby vomited
    2. Time of last feeding, type of formula, and amount taken
    3. Presence of an olive-shaped mass in the lower abdomen
    4. Amount and color of last voiding, skin turgor, and respiratory status

327. The nurse explains to the parents of an infant with pyloric stenosis that the type of surgery scheduled for this problem has a high success rate when:
    1. The fluid and electrolyte imbalances are corrected preoperatively
    2. Gastric decompression is monitored for amount and type of drainage
    3. It is performed before the infant's vomiting becomes severe and projectile
    4. The infant receives small, frequent feedings of thickened formula preoperatively

328. The mother of an infant with pyloric stenosis asks the nurse many questions about the problem. When

answering these questions the nurse should convey the idea that:

1. It is unlikely that surgery will be necessary
2. This is a condition with an excellent prognosis
3. This condition results from an inborn error of metabolism
4. Special feedings and handling will be needed for a few months

329. A 10-week-old is diagnosed as having pyloric stenosis and is scheduled for surgery. Oral feedings are usually initiated a few hours after surgery. The nurse expects that initially the baby will receive:

1. Clear liquids
2. Half-strength formula
3. Thickened formula with cereal
4. Oral electrolyte feedings (Pedialyte)

330. An infant is to be discharged following surgery for pyloric stenosis. The mother should be instructed to:

1. Give the baby creamy cereal at each feeding followed by the regular formula
2. Continue the regular formula, hold the baby during all feedings, feed the baby slowly, and bubble frequently
3. Give the baby about 1 ounce of regular formula per hour for the next 2 weeks; progressing slowly, as tolerated, to larger amounts
4. Feed the regular formula while the baby is in the crib positioned on the right side; handle the baby as little as possible for 2 hours after each feeding

331. A 6-week-old infant is admitted to the pediatric unit with a diagnosis of gastroesophageal reflux. When discussing care with this child's mother, the nurse plans to include:

1. Ways of handling the child after surgery to repair the esophageal defect
2. Discontinuing breast-feeding and placing the child in an infant seat after feedings
3. Feeding cereal with a spoon and administering a drug to help the stomach to empty
4. Giving formula thickened with cereal and placing the child prone with the head elevated 30 degrees

332. Because of the relaxation of the lower esophageal sphincter and the frequent return of stomach contents into the upper GI tract, an infant with gastroesophageal reflux needs to be carefully monitored for:

1. Abdominal distention
2. Increased hematocrit
3. Respiratory symptoms
4. Lower bowel obstruction

333. A mother whose 20-month-old infant has just developed diarrhea calls the pediatric clinic and asks what she should do. The nurse practitioner tells the mother to:

1. Limit the child's activities, hold all oral feedings, observe carefully, and call back in 4 hours
2. Allow the child to continue normal activities, hold all feeding for 24 hours, and call back tomorrow
3. Wrap the child snugly, give child sugar water, and bring the child in to see the physician immediately
4. Continue to feed the child as usual, make an appointment with the receptionist, and bring the child to the clinic tomorrow

334. An infant with persistent diarrhea is placed on npo. The nurse understands that this is done to:

1. Allow the intestinal tract to rest
2. Correct the electrolyte imbalance
3. Determine the cause of the diarrhea
4. Prevent irritation of the perineum from diarrhea

335. The parents of an infant admitted to the hospital with gastroenteritis and dehydration want to be involved with the baby's care. The nurse recognizes that they understand the teaching about the maintenance of standard precautions when they state, "We should:

1. Wear a mask when we are holding the baby."
2. Weigh the diaper each time we change the baby."
3. Wear gloves each time we change the baby's diaper."
4. Keep the door to the baby's room closed most of the time."

336. After receiving and tolerating a water and electrolyte formula (Pedialyte) because of dehydration from diarrhea, a child, 20 months old, improves and is advanced to soft foods. The nurse understands that a food that would be contraindicated is:

1. Creamed soup
2. Strained carrots
3. Animal crackers
4. Mashed bananas

337. To help a child retain tube feedings and avoid aspiration, the nurse should place the child in the:

1. Prone position
2. Left side-lying position
3. Semi-Fowler's position
4. Supine position with head turned

338. The food choice that would ensure maintenance of nitrogen balance after surgery in a 5-year-old child would be:
    1. Chicken soup
    2. A bacon sandwich
    3. Cut-up orange slices
    4. A hamburger on a bun

339. The nurse is aware that thiamine, which helps to produce more energy for both children and adults, is found in foods such as:
    1. Eggs
    2. Fruits
    3. Green vegetables
    4. Whole or enriched grains

340. The nurse recognizes that the diagnosis of celiac disease can be confirmed when a jejunal biopsy reveals:
    1. Small areas of fatty plaques
    2. Atrophic changes in the mucosal wall
    3. Irregular areas of superficial ulcerations
    4. Diffuse degenerative fibrosis of the acini

341. When taking the health history and assessing a 6-year-old child with celiac disease, the nurse would expect to find:
    1. Diarrhea, malnutrition, rickets, anemia, steatorrhea
    2. Constipation, abdominal distention, flatulence, rickets
    3. Diarrhea, muscle wasting, anemia, osteomalacia, steatorrhea
    4. Constipation, abdominal distention, peripheral edema, decreased clotting time

342. The nurse explains to the mother of a child with celiac disease who has been placed on a low-gluten diet that this type of diet will mainly restrict:
    1. Milk and dairy products
    2. The grains of corn and rice
    3. Saturated and unsaturated fats
    4. The grains of wheat, rye, oat, and barley

343. The nurse goes over dietary instructions with the mother of a 9-month-old who has a diagnosis of gluten-induced enteropathy. The nurse feels that the teaching has been understood when the mother states, "In planning my child's diet I will avoid:
    1. Beef, pork, chicken."
    2. Corn, spinach, cheese."
    3. Eggs, milk, Rice Krispies."
    4. Chocolate milk, whole wheat toast, fruit."

344. The effectiveness of a gluten-restricted diet in a child with celiac disease can be assessed on the second day by having the nurse and mother evaluate the child for:
    1. Decreased irritability
    2. Maintenance of weight
    3. Normal bowel movements
    4. Disappearance of steatorrhea

345. The nurse recognizes that anemia in a child with celiac disease is caused by:
    1. The poor absorption of iron and folic acid
    2. An inadequate amount of the intrinsic factor
    3. The small amount of iron included in the diet
    4. A low food intake and the child's minimal appetite

346. After being on a dietary regimen for celiac disease for 6 months, the child's compliance to the diet can be evaluated by assessing the:
    1. Physical and emotional progress
    2. Ability to handle stressful situations
    3. Understanding of the disease process
    4. Knowledge of foods allowed on the diet

347. The mother of a 3-year-old tells the nurse that her child frequently has constipation. The nurse asks how the mother handles the child's toileting. The nurse plans further teaching when the mother says:
    1. "I give my child a lot of those high-fiber foods at meals and as snacks."
    2. "I encourage my child to drink a lot of fluids, including prune juice, every day."
    3. "I notice that my child always has just one bowel movement a day except when ill."
    4. "I schedule my child's toileting time first thing in the morning so we are done before breakfast."

348. A 5-year-old child is admitted to the hospital complaining of colicky abdominal pain with guarding, nausea, anorexia, and a low-grade fever. Palpation of the RLQ elicits pain. The nurse prepares to implement care associated with:
    1. Constipation
    2. An irritated bowel
    3. A parasitic infestation
    4. An inflamed appendix

349. After an emergency appendectomy, the nurse should place the child in a semi-Fowler's or right Sims' position because:
    1. The lungs can aerate fully in both of these positions
    2. Drainage is facilitated, preventing subdiaphragmatic abscesses
    3. Movement is easier, thus reducing complications from immobility

4. Splinting of the wound is accomplished by pressure on the operative site

350. The nurse's charting for a child who has had an appendectomy should include, in addition to coughing and deep breathing, documentation of:
   1. Intake and output and bowel sounds
   2. Mouth care and frequency of dressing changes
   3. Bowel sounds and teaching about the low-residue diet
   4. Teaching to prevent dumping syndrome and early ambulation

351. The mother of a 3-month-old tells the nurse that her baby has occasional bouts of diarrhea. The nurse explains that:
   1. This is the way all infants' bodies respond to a developing infection
   2. It is important to have a consultation with a pediatric gastroenterologist
   3. This may be an allergic response to the ground meat that has been added to the diet
   4. The immunologic properties of the intestinal mucosal lining are immature in young babies

352. The nurse is aware that the parents of a 4-year-old with Hirschsprung's disease understand what their child will require when they tell the nurse that they know that care at home will most likely include:
   1. A low-protein diet
   2. A high-calorie diet
   3. Soapsuds enemas
   4. Nasogastric feedings

353. The nurse recognizes that the parents of a child with Hirschsprung's disease need further teaching about their child's diet when they indicate that they are going to allow the child to have:
   1. Apples
   2. Spaghetti
   3. Ice cream
   4. Ripe bananas

354. The nurse teaches a mother how to obtain a specimen to detect pinworms from her 5-year-old child by instructing the mother to:
   1. Give the child a tap-water enema and save all returns for testing
   2. Tape a 4 × 4 gauze bandage tightly over the child's anus during the night
   3. Make an anal impression on cellophane tape before the child uses the bathroom in the morning
   4. Insert a cotton-tipped swab into the child's rectum to collect a small amount of stool after a bowel movement

355. The nurse is aware that enterobiasis (pinworm infestation) is most commonly diagnosed by:
   1. Anal itching
   2. Scaly skin patches
   3. Maculopapular rash
   4. A bald spot on the head

356. A 1-day-old was born with an imperforate anus and undergoes a pull-through procedure with an anoplasty. The nurse knows that it is most appropriate postoperatively to place the infant:
   1. In Buck's traction
   2. In Trendelenburg's position
   3. Prone with the head of the crib elevated
   4. Supine with the legs suspended at a 90-degree angle to the trunk

357. The nurse working with infants who are preterm should be alert to the fact that these infants may later develop intestinal obstruction because of:
   1. Meconium ileus
   2. Imperforate anus
   3. Duodenal atresia
   4. Necrotizing enterocolitis

358. After several episodes of abdominal pain and vomiting, a 5-month-old child is taken to the hospital. A diagnosis of intussusception is made. To assist in confirming the diagnosis, the priority assessment should be:
   1. Noting frequency of crying
   2. Listening for bowel sounds
   3. Measuring fluid intake and output
   4. Observing characteristics of stools

DRUG-RELATED RESPONSES (DR)

359. Ferrous fumarate (Feostat) 30 mg is ordered for an infant. The Feostat solution contains 45 mg/0.6 ml. The nurse should administer:
   1. 0.2 ml
   2. 0.4 ml
   3. 0.6 ml
   4. 0.9 ml

360. Cough syrup ½ tsp is ordered for a 4-year-old. Each teaspoonful contains dextromethorphan hydrobromide 7.5 mg. When administering the cough syrup, the nurse should provide:
   1. 0.5 ml
   2. 2.5 ml
   3. 7.5 ml
   4. 3.75 ml

361. A child with cystic fibrosis is ready for discharge and the nurse is reviewing the discharge instructions with the mother. The statement by the mother that would indicate that she knows how to administer the pancreatic enzyme replacement would be:
1. "I will give the medication with feedings."
2. "I will give the pills in applesauce every morning."
3. "I must dissolve the enteric-coated pills in the formula."
4. "I must give the medication every 6 hours including during the night."

362. After taking levothyroxine (Synthroid) for 3 months for congenital hypothyroidism, an infant returns to the clinic for a checkup. The nurse evaluates that the drug is effective when the mother states that the infant's:
1. Activity level has decreased
2. Fine tremors have decreased
3. Skin is cool and dry to the touch
4. Bowel movements have increased to two soft stools daily

363. A 10-year-old is newly diagnosed with diabetes mellitus and is started on insulin therapy. The child and the family are taught how to give the injections. The child dislikes the injections and asks the nurse why the insulin cannot be taken by mouth. The nurse explains that insulin:
1. Is a protein and would be inactivated by digestion
2. Would irritate the stomach lining and lose its potency
3. Has a carbohydrate portion that would add to the blood glucose
4. Is alkaline and would be neutralized by gastric hydrochloric acid

364. Screening for hearing loss should be planned for the child who is receiving:
1. Penicillin
2. Tetracycline
3. Streptomycin
4. Chloramphenicol

365. The nurse would know that the teaching about administration of tetracycline was effective when a teenage client says that the drug should be taken:
1. Just before meals
2. With meals or milk
3. At least 1 hour before meals
4. Approximately 30 minutes after meals

366. The nurse explains to a teenager who is receiving penicillin G and probenecid for syphilis that the rationale for both drugs being used is:
1. Each drug attacks the organism during different stages of cell multiplication
2. The penicillin treats the syphilis whereas the probenecid relieves the severe urethritis
3. Probenecid delays excretion of penicillin by the kidneys to maintain effective blood levels for longer periods
4. Probenecid decreases the potential for an allergic reaction developing to the penicillin, which treats the syphilis

367. A 12-year-old with cystic fibrosis is to receive three Pancrease capsules 5 times a day. The nurse is aware that this medication is given to:
1. Promote excretion of fats
2. Prevent iron-deficiency anemia
3. Facilitate utilization of nutrients
4. Promote adequate oxygenation

368. The nurse would determine that the teaching about the side effects of tetracycline was understood when the client says that the medication could cause:
1. Vertigo
2. Tinnitus
3. Diarrhea
4. Constipation

369. When caring for a child receiving prednisone, it is important for the nurse to know that adrenocorticosteroid therapy:
1. May produce hyperkalemia
2. Accelerates wound healing
3. Increases production of antibodies by the blood
4. Suppresses the inflammatory symptoms of infection

370. The alkylating agent cyclophosphamide (Cytoxan) is ordered for a child with cancer. When the child is receiving this drug the nurse should assess for:
1. Extent of hydration
2. Increased irritability
3. Unexpected nausea
4. Hyperplasia of gums

371. An 8-year-old girl in the hospital who is receiving methotrexate and cranial radiation is very weak, and her mother asks the nurse whether her daughter could receive some vitamin therapy to give her strength. The nurse's best response would be:
1. "That is an excellent idea; I'll ask the physician to order some for her."
2. "Some vitamins contain folic acid, which interferes with methotrexate."

3. "Unfortunately, vitamin supplements won't make her feel any better now."

4. "Your daughter will benefit from vitamins and will be receiving them soon."

372. When assessing a child with leukemia for the possible side effects of vincristine (Oncovin), the nurse should be aware that a sign of toxicity is:
1. Diarrhea
2. Alopecia
3. Hemorrhagic cystitis
4. Peripheral neuropathy

373. When assessing the status of a child with leukemia who is receiving vincristine (Oncovin), the nurse would know that the fluid intake should be increased when the child's:
1. Temperature is 99.8° F
2. Uric acid level is elevated
3. Urine's specific gravity is 1.026
4. Output for the last 24 hours totaled 1700 ml

374. A child is receiving vincristine (Oncovin) as a part of the chemotherapy protocol. The nurse should monitor the child for untoward reactions affecting the:
1. Emotional status
2. Neurologic status
3. Respiratory status
4. Cardiovascular status

375. An adolescent who is receiving prednisone and vincristine for leukemia complains of constipation. The nurse is aware that the constipation is most probably caused by:
1. A side effect of the vincristine
2. A toxic effect of the prednisone
3. An enlarged spleen compressing the bowel
4. An obstruction of the bowel by a leukemic mass

376. An adolescent with acute lymphocytic leukemia (ALL) completes parenteral chemotherapy and is discharged home with a prescription for mercaptopurine (6-MP) one tablet daily by mouth. The statement by the adolescent that indicates an understanding of the reason for this therapy is:
1. "These pills prepare me for additional IV drugs."
2. "Taking pills will be better than having brain radiation."
3. "This medication should help prevent another relapse of my disease."
4. "This medication should prevent the cancer from spreading to my stomach."

377. When considering the side effects of dactinomycin (Cosmegen) and doxorubicin (Adriamycin) therapy, the nurse can suggest to the parents of a child receiving these drugs that the child:
1. Avoid dairy products
2. Wear a baseball cap
3. Dress in light clothing
4. Eat three large meals daily

378. A 1-year-old is in the pediatric unit for management of AIDS. The child is receiving zidovudine (AZT) every 6 hours. The nurse evaluates that the child is in life-threatening AZT toxicity when the child manifests:
1. Fatigue and lethargy
2. A progressive weight loss
3. An increased urine output
4. Multiple bruises on the limbs and trunk

379. Priority nursing care for children on chelation therapy for lead poisoning should include:
1. Scrupulous care of the skin
2. Providing a high-protein diet
3. Careful monitoring of intake and output
4. Drawing blood daily for liver function tests

380. In addition to removing lead from the blood, chelation therapy with calcium disodium edetate (Ca EDTA) predisposes the child to:
1. Anemia
2. Hyperkalemia
3. Hypocalcemia
4. Hypoglycemia

381. The physician orders calcium disodium edetate (Ca EDTA), a chelating agent for a child with lead poisoning. The nurse recognizes that chelating agents cause:
1. Lead to be removed from the bone
2. Free lead to be secreted in the feces
3. Lead to be combined with hemoglobin
4. Lead to bind and be excreted in the urine

382. The nurse knows that the chelating agent calcium disodium edetate (Ca EDTA) is generally given to children by:
1. Mouth with milk
2. Mouth with water
3. Subcutaneous injection
4. Deep intramuscular injection

383. The nurse teaches the parents of a child on long-term phenytoin (Dilantin) therapy about care pertinent to this medication. The nurse recognizes that the teaching is effective when the parents say, "We should:
    1. Give our child the medication 2 hours after breakfast and dinner."
    2. Supplement the diet with high-calorie foods and encourage fluids."
    3. Provide oral hygiene, especially gum massage and flossing of teeth."
    4. Observe our child's urine for the complication of a reddish-brown discoloration."

384. The nurse is aware that the most common reason for status epilepticus in children is that the prescribed dosage of phenytoin (Dilantin) is:
    1. Toxic to the child
    2. At the therapeutic level
    3. Probably not taken consistently
    4. Insufficient to cover the child's activities

385. When administering Pancrease capsules to a child with cystic fibrosis, the nurse knows they should be given:
    1. With meals and snacks
    2. Every 3 hours while awake
    3. On awakening, following meals, and at bedtime
    4. After each bowel movement and after postural drainage

386. The nurse explains to the parents of a child who is receiving mannitol (Osmitrol) after a craniotomy that the medication is being given to:
    1. Increase the filtration rate of the bladder
    2. Decrease the peripheral retention of fluid
    3. Reduce the amount of glucose in the urine
    4. Relieve cerebral pressure following surgery

387. The nurse tells a 13-year-old with hay fever that the ordered phenylephrine nasal spray must be used exactly as directed to prevent the development of:
    1. Nasal polyps
    2. Bleeding tendencies
    3. Tinnitus and diplopia
    4. Increased nasal congestion

388. The nurse recognizes that the drug of choice for a child with status asthmaticus is:
    1. Prednisone
    2. Epinephrine
    3. Cefaclor (Ceclor)
    4. Clemastine (Tavist)

389. When reviewing medication instructions with the parents of an infant who is receiving digoxin (Lanoxin) and spironolactone (Aldactone), the nurse would know that the parents understood the instructions when they indicate:
    1. Their infant must have orange juice daily
    2. Any vomiting should be reported to the physician
    3. Their infant's activity should be carefully restricted
    4. Aspirin should be avoided while the infant is taking Aldactone

390. The pediatric nurse is aware that digoxin (Lanoxin) toxicity in children is most commonly manifested by:
    1. Oliguria
    2. Tachypnea
    3. Bradycardia
    4. Splenomegaly

391. A child is having a cardiac arrest. The physician orders epinephrine as a cardiac stimulant. The nurse is aware that the one factor that would still permit administration of an epinephrine solution that is on hand would be that the reconstituted solution:
    1. Is slightly discolored
    2. Has been exposed to light
    3. Is no more than 72 hours old
    4. Contains only slight sediment

392. When assessing a child after the administration of epinephrine, a side effect the nurse should be aware of is:
    1. Tachycardia
    2. Hypoglycemia
    3. Constricted pupils
    4. Decreased blood pressure

393. When a 5-year-old child is receiving dactinomycin (Cosmegen) and doxorubicin (Adriamycin) therapy following a nephrectomy for a Wilms' tumor, nursing care should include:
    1. Administering aspirin for pain
    2. Serving citrus juices with meals
    3. Using an anesthetic mouthwash
    4. Providing age-appropriate books

394. A nurse is planning discharge teaching for a 10-year-old male who is taking prednisone for asthma. A critical medication instruction for the child and his parents would be:
    1. "Observe for a moon-shaped face."
    2. "Do not stop taking prednisone abruptly."
    3. "This drug will protect you from infection."
    4. "Prednisone may cause an earlier growth spurt."

395. An important nursing intervention for a child who has been receiving long-term prednisone therapy for nephrotic syndrome would be:
1. Daily checking of pulse for irregularities
2. Frequent checking of stools for occult blood
3. Regular checking of urine for mucous threads
4. Daily checking of the oral mucous membrane for ulcers

396. When a child with acute glomerulonephritis is found to be hypertensive, the nurse would expect the physician to order:
1. Digoxin (Lanoxin)
2. Diazepam (Valium)
3. Phenytoin (Dilantin)
4. Hydralazine (Apresoline)

397. After the severe effects of dehydration are under control in a 3-month-old infant, the physician orders lactobacilli granules (Lactinex) to:
1. Diminish inflammatory mucosal edema
2. Relieve the pain caused by gastric hyperacidity
3. Relieve the pain of gas in the gastrointestinal tract
4. Recolonize the normal flora of the gastrointestinal tract

398. Mebendazole (Vermox) is prescribed for a child with pinworms. When teaching about the medication, the nurse tells the mother and child that:
1. The drug may precipitate transient diarrhea
2. Rectal itching will be relieved rapidly once the drug is started

3. Only the child and no other family member will need treatment
4. A single course of treatment is all that is needed to control the problem

399. The nurse evaluates that pancreatic enzyme replacement being taken by a child with cystic fibrosis is inadequate when the child complains of:
1. Anorexia
2. Constipation
3. Sudden weight gain
4. Abdominal cramping

400. A 14-year-old female has a spinal fusion. On the first postoperative day, her face is red and she is rigid and crying that she hurts. She has orders for both meperidine and an acetaminophen-codeine compound. When considering analgesia for this client, the nurse should remember that:
1. One dose of meperidine can be given, but she should be restricted thereafter because meperidine is highly addictive
2. Adolescents often tend to exaggerate their discomfort, particularly when they are hospitalized and immobilized by surgery or injury
3. Spinal fusion results in considerable pain during the first few postoperative days, and meperidine would be most effective analgesic medication
4. It would be better to give her the acetaminophen-codeine compound, because meperidine can cause respiratory depression, addiction, and respiratory arrest

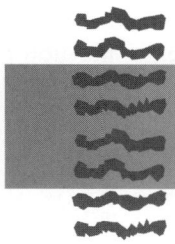

PEDIATRIC NURSING ANSWERS AND RATIONALES

GROWTH AND DEVELOPMENT (GD)

1. ② With each age group, there are different means of communication; the approach used with a school-age child should differ from that used with a toddler or a teenager. (1) (DF; AS)
1 This might modify the approach, but knowing the child's developmental level is the most important factor.
3 This would be related to the child's developmental level, which should be assessed first.
4 Although children may fear authoritarian figures, this is only one aspect included in the assessment of developmental level, a more inclusive assessment.

2. ③ This follows the normal course of growth and development skills and is no different with a child who is mentally challenged. (2) (DF; PL)
1 This would not be taught before self-feeding.
2 Same as answer 1.
4 Same as answer 1.

3. ② The 3-month-old infant is interested in self-recognition and playing with the baby in the mirror. (3) (DF; IM)
1 This is not appropriate for a 3-month-old.
3 Same as answer 1.
4 Same as answer 1.

4. ① During the oral stage, infants tend to complete the exploration of all objects by putting the objects in the mouth as a final step. (2) (DF; AS)
2 Infants 9 to 10 months old play this way as they learn that objects continue to exist even though they are not visible.
3 These are the random reflexive movements of 1- to 2-month-olds, whose voluntary control of distal extremities has not developed.
4 This is the momentary grasp reflex of neonates before the development of eye-hand-mouth coordination.

5. ④ A 7-month-old infant is used to having the perineal area exposed and cared for and is not in a developmental stage where fears related to sexuality are present. (1) (DF; IM)
1 A "clean catch" at this age is often contaminated; the physician ordered a catheterization.
2 The mother does have the right to refuse, but these concerns are not realistic for an infant at this age.
3 Although the mother may be concerned, a catheterization is not a problem for an infant and the mother needs to be educated.

6. ① Eight-month-old infants have the ability to use their fingers and thumbs in opposition (pincer grasp); this enables them to pick up small objects and put them in their mouths and aspirate them. (1) (SI; AN)
2 Although this statement is true, the major concern is preventing infants from putting foreign objects in their mouths where they can be aspirated.
3 It is not a health hazard if the floor is clean and the object is large enough so that there is no danger of the child aspirating it and obstructing the airway.
4 The danger is not that the items would be hidden but that they would be put into the mouth and aspirated.

7. ③ Brief messages, with only essential words included (telegraphic speech), are a normal pattern for a child 18 months to 2½ years of age. (2) (DF; AN)
1 A child with a severe developmental lag would have no obvious recognizable speech pattern and would only make a few sounds.
2 A child slow for this age would have a smaller vocabulary and would use only single words to identify familiar objects.
4 A child advanced for this age would have a larger vocabulary and would use three- to four-word sentences rather than telegraphic speech.

8. ③ Preschoolers can tolerate brief periods of separation from their parents; however, emotions associated with separation and perhaps anger at being left are difficult to hide when parents arrive or leave. (2) (DF; IM)

1 Preschoolers usually are quite docile and cooperative because they are afraid of being totally abandoned.

2 The child would demonstrate despair long before the week was over.

4 The presence of other children's parents would be unrelated to their relationship with peers.

9. ③ In the toddler, two- and three-word phrases are used with an increased vocabulary; attributing lifelike qualities to inanimate objects (animism) is also associated with preconceptual thought. (3) (DF; AN)

1 This is related to school-age children.

2 This is a phase of concrete operations seen in school-age children.

4 This is related to infants.

10. ② Although none of the choices is always indicative of pain, a change in behavior is the indicator that occurs most often in children. (2) (DF; AS)

1 Many things can cause crying, including pain, fear, separation, and unhappiness; crying does not always indicate pain.

3 Vital signs are often normal in children, even in the presence of pain.

4 Children often hide their pain; they may perceive it as punishment, or they may fear the treatment that would be given to relieve the pain.

11. ① The external auditory canal curves downward and forward in a child older than 3 years and is approximately 1 inch long; to adequately view the tympanic membrane in a child this age, the pinna must be pulled up and back; in a child younger than 3 years, the pinna should be pulled down and back. (3) (HI; AS)

2 This positioning would impede visualization of the tympanic membrane; the pinna should be pulled backward, not forward.

3 This is how the pinna of a child younger than 3 years should be positioned for otoscopic examination.

4 This positioning would impede visualization of the tympanic membrane; it is exactly the opposite of what should be done in a child older than 3 years.

12. ③ A child this age loves to collect and manipulate; this meets the need to develop fine motor skills. (2) (DF; IM)

1 This is more appropriate for an older infant.

2 This would be frustrating for this child's developmental level.

4 The child is too young for scissors and fragile toys.

13. ④ At this age the phallus is large enough for surgical repair and the child has not reached the age at which fear of mutilation develops. (3) (DF; IM)

1 The phallus is not developed enough for surgery to be done at this age.

2 Same as answer 1.

3 The child is in the oedipal stage of development, which is accompanied by fear of mutilation; surgery is inadvisable.

14. ① This is the most accurate and age-appropriate response to the question. (3) (PT; IM)

2 This is inaccurate; not being truthful interferes with the development of trust.

3 This is inaccurate and may instill more fear.

4 This response is insensitive to the question and does not provide any explanation.

15. ② The infant has not yet recognized boundaries between herself and her mother and is not particular about who meets and resolves needs. (1) (SC; EV)

1 The infant does not yet differentiate familiar faces from those of strangers.

3 This behavior is that of a younger infant and does not indicate recognition of a specific person but only a human face.

4 Because the concept of self-boundaries has not yet developed, the infant does not really know or fear separation from the mother.

16. ② This is inappropriate for an 8-month-old; this is appropriate for a toddler to promote imitative play. (3) (DF; EV)

1 A stuffed animal is appropriate; it promotes manipulative play.

3 A hanging mobile is appropriate; it promotes visual stimulation.

4 A textured book is appropriate; it promotes tactile stimulation and touch discrimination.

17. ④ Changes in the daily routines in the home and anxiety expressed by family members lead to anxiety in toddlers. (2) (SC; IM)

1 This is incorrect because the toddler has no reality-based concept of death.

2 This may be true, but the primary motivation for the behavior is a response to the upheaval in the family.

3 This is false reassurance.

18. (2) Stuttering occurs because the child's advancing mental ability and level of comprehension exceed the vocabulary acquisitions in the preschool years. (2) (DF; AN)

1 This is not true; stuttering is common in the preschool years.

3 Same as answer 1.

4 Same as answer 1.

19. (3) A stuffed animal is the most appropriate toy for the 6-month-old because it is safe and cuddly and requires only gross motor movement. (1) (DF; PL)

1 A push-pull toy is appropriate for the older infant (9 to 12 months) and the toddler because it encourages walking.

2 These are inappropriate; a child at this age puts toys in the mouth; playing with blocks requires motor development beyond this age.

4 Shape-matching toys require intellectual and motor development beyond that of this age group.

20. (4) Lips and teeth closed around the finger create suction and can move permanent teeth forward, causing malocclusions. (1) (DF; IM)

1 If thumb-sucking is practiced only in relation to sleep, no treatment is necessary because it involves only a short period of time.

2 The pacifier will also cause malocclusions when permanent teeth appear.

3 There is no indication that the first teeth are loosened by thumb-sucking.

21. (4) This is appropriate for the child's age and suited to the child's limited motion. (1) (DF; AN)

1 The child will not have enough mobility to engage in this type of play and is too young for this activity.

2 The child is too young for this activity; children of 7 years or older are able to play checkers.

3 Because of the spica cast, the child will not have enough mobility to play with this type of toy.

22. (3) This usually occurs by age 8 months. (1) (DF; AS)

1 A two- to three-word vocabulary is an expectation of a 12-month-old child.

2 This is accomplished by the 2-year-old, not the 9-month-old.

4 Whereas the posterior fontanel is closed by age 2 months, the anterior fontanel closes between ages 18 to 24 months.

23. (4) The child with abdominal pain may assume the side-lying position with the knees flexed to the abdomen and/or may self-splint when moving. (1) (HI; AS)

1 A 4-year-old may be unable to define the exact location of the pain; in addition, the pain may be generalized rather than localized.

2 This might be included in the physical assessment, but it is not specific to the assessment of pain.

3 This might be included in the health history, but it is not specific to the assessment of pain.

24. (1) These easily digested foods have usually been introduced by 10 months of age; breast milk or formula is recommended for the first year of life. (2) (PC; PL)

2 Whole milk makes this incorrect; breast milk or formula, rather than cow's milk, is recommended for the first year of life.

3 Ham makes this incorrect; it is too high in fat content for a 6-month-old.

4 Corn is too difficult for a child this age to digest; formula is recommended for the first year of life.

25. (2) Because of expanded experiences and developing cognitive ability, the 4-year-old should have a vocabulary of approximately 1500 words. (2) (DF; AS)

1 At 5 years of age, children ask the definitions of new words.

3 At 2½ to 3 years of age, children can name colors.

4 At 3 years of age, children use three- or four-word sentences.

26. (4) The canal curves upward in children, and this straightens the canal so that medication will reach the inflamed eardrum. (3) (DF; IM)

1 This is an incorrect technique; the auditory canal must be pulled down and back to straighten it.

2 This is not advised; it can only add more pressure within the ear and prevent the drops from reaching the middle ear.

3 This is an incorrect technique for children younger than 3 years because it will not straighten the ear canal.

27. ④ This explanation illustrates that the child can understand cause-effect relationships and offers information to increase the child's understanding of the illness. (2) (DF; IM)
   1 This is too general and does not explain why the child will feel better.
   2 This is too authoritarian; the child needs information that will increase understanding and compliance with the regimen.
   3 Same as answer 2.

28. ② The Babinski reflex, present at birth, should remain positive throughout the first 12 months of life. (2) (DF; AS)
   1 This reflex, present at birth, disappears by 4 months of age.
   3 Same as answer 1.
   4 Same as answer 1.

29. ③ Young children have an increased propensity for putting things in their mouths; this age group uses this as a means of exploring the environment. (1) (HI; AN)
   1 Although this may be true in older homes or in the inner city, it is the activity of putting things into the mouth that is the primary cause.
   2 This is untrue; toddlers do not have a fragile vascular system; children with a fragile vascular system are severely compromised.
   4 This is not true; although gas fumes in areas of heavy traffic have increased pollution, most gasoline used today does not contain lead.

30. ④ A visual display, which simulates a suture holding the testes to the thigh, aids in explanation and understanding; it may help the child express feelings about surgery. (3) (SC; PL)
   1 This is used to explain IVs, but it is not specific to this surgery.
   2 A binder will not be used after this surgery.
   3 This is used to explain abdominal surgery such as an appendectomy; it would not apply to this situation.

31. ② A 5-year-old is able to negotiate and use make believe to play. (3) (DF; AS)
   1 Children in the middle childhood years need conformity and rituals, whether they play games or amass collections; rules to games are fixed, unvarying, and rigid; knowing the rules means belonging.
   3 The use of a pacifier for oral satisfaction is normal for infants.
   4 Parallel play occurs in children ages 2 to 3 years.

32. ④ These are typical behaviors of an 8-month-old. (2) (DF; AS)
   1 These are typical behaviors of a 12-month-old.
   2 These are typical behaviors of a 3-month-old.
   3 These are typical behaviors of an 18-month-old.

33. ② Intrusive procedures threaten the developing body image of the preschooler. (2) (SC; PL)
   1 The preschooler is more tolerant of strangers than is a younger child.
   3 Routines are still important to the preschooler, but some deviations in structure of activities can be tolerated.
   4 The preschooler can tolerate short periods of separation from parents.

34. ① This is appropriate for this age child; it would give the child the opportunity for free expression and its free-form nature can give the child a sense of mobility. (3) (DF; PL)
   2 This is less than optimal because coloring within lines of pictures in a coloring book requires more skill than most 4-year-olds possess; also this does not allow freedom of expression or movement.
   3 Checkers is a game with too many rules for a 4-year-old to comprehend.
   4 Playing dominos requires the ability to count and conserve numbers, which most 4-year-olds do not possess.

35. ② The teenage child is concerned with body image and fears change or mutilation of body parts; in sickle cell anemia, bone weakened because of hyperplasia and congestion of the marrow can cause lordosis and kyphosis. (1) (DF; AN)
   1 Restriction of movement is not a major problem because when the pain is relieved and the crisis is over, activity can be resumed; for the teenager, the change in body image produces greater anxiety.
   3 Teenagers can easily tolerate extended periods of separation from the family.
   4 Although this could be a concern at this time for a teenager, altered body image is a more fearful threat.

36. ④ A plastic toy that squeaks provides auditory, tactile, and visual stimulation. (3) (DF; EV)
   1 The potential for injury is too great for a 1-year-old on a rocking horse.

2 A stuffed animal would not be kept in a play-room because it could not be washed between use by different children.

3 A 1-year-old child is too young for a puzzle.

37. (1) A 4-year-old can manage large buttons on a shirt. (1) (DF; AS)

2 A child of 4 years can put on shoes but is usually unable to tie them until age 5.

3 A child of 4 years will be able to comb but not part the hair.

4 A child of 4 years can handle a fork and spoon but cannot hold the meat with the fork to cut it with the knife; the child is usually 7 years old before this can be managed.

38. (2) Death is viewed as a separation and preschoolers believe they will return to life and former activity; this is part of the fantasy world of the child. (2) (DF; AN)

1 At about 9 or 10 years of age, the child develops an adult concept of death and views it as inevitable, irreversible, and universal.

3 This is true for all age groups.

4 This is true of the 6- to 7-year-old child.

39. (3) The 15-year-old is normally preoccupied with appearance; the side effects of the antineoplastics and prednisone will result in the client feeling different and may cause a poor body image. (1) (DF; AN)

1 A normal 15-year-old enjoys and strives for independence; the sick role would force the client to be dependent.

2 This may be a possible concern but is not likely to be the outstanding concern or feeling.

4 Same as answer 2.

40. (4) These abilities are age-appropriate for the 6-month-old. (2) (DF; PL)

1 These abilities should be developed by 10 months of age.

2 Same as answer 1.

3 Same as answer 1.

41. (2) The eustachian tube in young children is shorter, lacks tone, and opens inappropriately, allowing a reflux of nasopharyngeal secretions. (3) (DF; IM)

1 Immunologic differences are not a factor in the development of otitis media.

3 The size of the middle ear does not play a role in the common occurrence of otitis media in very young children.

4 There is no difference in the functioning of the eustachian tube among age groups.

42. (1) Children feel pain and should receive analgesics when needed. (3) (PC; AN)

2 This is a myth; it may be difficult for children to communicate pain.

3 This is a common, but unsound, belief; addiction and respiratory depression are rare.

4 Some sources suggest this may be a child's way of coping with unrelieved pain; however, it is no reason to withhold medication.

43. (1) This is appropriate for an 11-year-old, who sees dying as loss of control over every aspect of living; the child may convey this meaning by physically attempting to run away or by pushing others away by rude behavior; it is a plea for some self-control and power. (3) (SC; AS)

2 This is characteristic of the toddler, who is egocentric and has a vague separation of fact and fantasy, which makes it impossible to understand the absence of life.

3 This is characteristic of the preschooler, who does not have logical thinking.

4 This is more typical of the adolescent, who sees deviation from accepted behavior as the reason for becoming ill.

44. (4) Six-month-old infants are capable of holding their bottles. (2) (PC; AS)

1 This is a skill of older infants.

2 Same as answer 1.

3 Same as answer 1.

45. (3) A toddler's increasing mobility and growing independence in behavior, including food behavior, are normal aspects of psychologic development; slowed physical growth at this age requires relatively less caloric intake. (3) (DF; IM)

1 A toddler's growth rate and energy requirements decrease in comparison with the first year of life.

2 Same as answer 1.

4 Nutritious snacks between meals should be encouraged if the child is not eating adequate meals.

46. (4) School-aged children begin to see the need for rules and are able to play with each other, not just beside each other. (2) (DF; AN)

1 This type of play is characteristic of a 2-year-old.

2 This is a type of behavior, not a type of play.

3 Same as answer 2.

47. ③ These age-appropriate games help the infant's social development by fostering a sense of object constancy and object permanency. (2) (DF; PL)

  1 This is age-appropriate play for the toddler; it promotes gross and fine motor development, not social development.

  2 This is age-appropriate play for preschoolers; it helps develop motor, not social, skills.

  4 This is age-appropriate play for a child older than 9 months; it promotes psychomotor, not social, development.

48. ② A vocabulary consisting of a minimum of six words with telegraphic-type speech is normal for a child this age. (2) (DF; AS)

  1 The child with a hearing impairment communicates in this way because the child has not acquired the rudiments of language.

  3 Babbling is normal communication for an 8-month-old infant, even one with a moderate hearing loss.

  4 This language skill is seen in the 5-year-old child.

49. ④ As part of the maturation process, adolescents need to be made to accept the consequences of their actions. (1) (SC; IM)

  1 This is false reassurance; there is no way to predict what will be the outcome of her friends' behavior in the future.

  2 The focus should be on pointing out that the girl should be accountable for her own behavior, not that her friends should also be punished.

  3 The focus should be on the girl's actions and not those of her friends' parents.

50. ② Toddlers play independently but beside other children; they are unable to follow rules or negotiate. (1) (DF; AN)

  1 This kind of play is characteristic of older children.

  3 Dramatic play or acting is characteristic of older children; they assume and act out roles.

  4 Same as answer 1.

51. ④ This provides a 6-year-old, who is of school age, an appropriate way to express feelings, either by writing or by drawing pictures. (3) (DF; PL)

  1 This would be appropriate for preschoolers, whose imaginations are unlimited.

  2 This would be appropriate for preschoolers, who enjoy experimenting with different textures.

  3 This would be appropriate for preschoolers, who like repetition.

52. ② Adolescents are very aware of their changing bodies and become especially concerned with any alteration resulting from illness or injury. (3) (DF; IM)

  1 This does not educate the mother about concepts concerning the developing adolescent; a discussion about hypochondriasis may reinforce the mother's concern.

  3 This does not address concepts related to growth and development of the adolescent and could cause unnecessary concern about the daughter's physical condition.

  4 This could reinforce the mother's concern as well as promote feelings of guilt; it does not include concepts about growth and development of the adolescent.

53. ③ Striving to attain identity and independence is a task of the adolescent, and rebellion against established norms may be exhibited. (3) (DF; AN)

  1 This behavior is not a bid for attention, rather it is an attempt to establish an identity, which is a normal developmental task of the adolescent.

  2 Although the adolescent may be using denial, denial is not developmentally related to adolescence.

  4 This behavior is not regression; it is an attempt to attain identity by rebellion against established norms.

EMOTIONAL NEEDS RELATED TO HEALTH PROBLEMS (EH)

54. ② The child is old enough to respond when asked a direct question. (2) (SC; IM)

  1 The parents are too emotionally involved with the child and may not be trained in principles of mental health and therapeutic communication.

  3 A younger child, about 8 to 10 years old, would benefit from this projective technique.

  4 This may be productive with a younger child.

55. ③ This is a truthful statement; the nurse recognizes the fact that this might hurt and requests expected behavior. (2) (PC; IM)

  1 This puts unrealistic expectations on the child.

  2 This puts a thought in the mind of the child.

  4 This would be too threatening for the child.

56. ② Children with diabetes mellitus need to be treated normally; they need discipline and should have limits set for their behavior. (1) (SC; EV)

1 This is correct; parents should foster independence in the child with diabetes mellitus.

3 This statement is correct; it is realistic to think that the family will have ups and downs.

4 This is correct; the child with diabetes mellitus should be encouraged to maintain normal interests and activities.

57. ③ These behaviors are associated with separation anxiety; parental contact should be encouraged. (1) (SC; PL)

1 Separation anxiety can be minimized by increasing contact with parents, not peers.

2 Separation anxiety can be minimized by increasing contact with parents, not by distraction with toys.

4 This would increase feelings of anxiety; the same nurse should care for the child to promote consistency, continuity, and the development of trust.

58. ④ The child learned to think and solve problems in a very different culture and used a different language and may feel helpless in the new classroom. (1) (SC; AN)

1 There are not enough data to substantiate this.

2 This is untrue; 5-year-olds are inquisitive and adapt well to school.

3 Most 5-year-olds adapt well to kindergarten.

59. ③ The parents need to recognize that their child must be taught responsibility for self-care. (2) (PC; IM)

1 This supports the parents' need to keep the child totally dependent.

2 This demeans the child and inhibits the parents from expressing additional feelings.

4 This denigrates the parents and does not allow for further expression of feelings.

60. ① This child needs help adjusting; focusing on feelings and abilities promotes effective coping and raises self-esteem. (2) (SC; PL)

2 In general, the diet is limited to simple carbohydrates; bowel inflammation necessitates avoidance of high-roughage foods.

3 Teaching the relationship of diet to the disease process does not ensure compliance with the diet.

4 Occasional noncompliance is not permitted; it eventually causes a relapse.

61. ④ Lifelong adherence to dietary restrictions prevents complications and celiac crisis. (2) (PC; PL)

1 Celiac crisis usually develops as a result of nonadherence to the diet, so adherence would be a primary goal.

2 Respiratory involvement is not a primary problem in celiac disease.

3 Regardless of adherence to the diet, there is an interference with normal growth.

62. ① Colostomy care may seem overwhelming to the parents, and it may reassure them to know that a therapist is available. (1) (CC; PL)

2 Mealtime should be a pleasant time; also, this assumes that eating habits are poor.

3 Increased fluids are often needed to compensate for fecal fluid loss.

4 This is untrue; physical activity will probably not be limited.

63. ③ The parents need assistance in exploring their feelings and their family relationships with a professional. (1) (CC; EV)

1 The gifts are attempts to relieve guilt feelings; the parent still feels responsible.

2 This is a gift to the child that helps the parents relieve their guilt feelings.

4 The parent is assuming the martyr role and accepting responsibility for the child's illness.

64. ④ The nurse must emphasize that everything possible is being done because the outcome cannot be predicted. (3) (SC; IM)

1 The outcome is still in doubt; encouraging the parent's positive interpretation of the child's reflexive behavior raises false hope.

2 Same as answer 1.

3 The outcome is still in doubt; this response by the nurse may raise false hope; the parent's statement did not ask for the nurse's religious viewpoint.

65. ③ This response identifies concern and presents an appropriate protective intervention; regular and prompt removal of ticks decreases the chances of the spread of Lyme disease to humans. (2) (SI; IM)

1 The response centers on camping, not on the fear of ticks.

2 This response belittles the child's feelings.

4 This is an inappropriate response because it focuses on the wrong fear.

66. ③ Regardless of age, parenthood confers the rights of an adult on the teenager. (2) (CC; PL)
1 It is unnecessary and not legal for the grandmother to sign the consent; the father is present.
2 This would be done only if neither parent was available to give consent in an emergency.
4 The mother has a legal right to give consent but is not available.

67. ① People living in poverty have long-term feelings of powerlessness because they do not have buying power or social status to influence change. (1) (SC; AN)
2 The opposite is true; they are focused on the present, not the future.
3 Their anger is covert and not direct; in addition, the anger rarely resolves their situation, resulting in feelings of powerlessness and hopelessness.
4 People in poverty tend to focus on today; health recommendations may not be delivered under optimal circumstances or may be misunderstood, confusing, or of little value.

68. ② Family strengths must be identified and utilized by the nurse. (3) (SC; AN)
1 The family members and their problems must be viewed as a whole.
3 The opposite is true.
4 This is untrue; values, beliefs, and attitudes greatly influence perceptions.

69. ③ The parents will probably be anxious and will benefit most from short teaching sessions and written material to review at their leisure. (1) (HI; PL)
1 This would be overwhelming, and the parents would not be able to retain everything presented.
2 The nurse could recommend, but not insist, that both parents attend the teaching sessions.
4 The most effective teaching and learning sessions occur in an area with minimal distractions; being in the room with their child at this time would present a major distraction to the parents.

70. ④ In most states, the age of majority is 18 years; however, mothers younger than 18 years are considered emancipated minors and can sign consents for themselves and their children. (2) (CC; AN)
1 The grandmother has no legal right to give consent; the 16-year-old mother is present and can legally give consent.

2 This is unnecessary; the client is an emancipated minor, and this confers adult status.
3 Consent is always needed; the 16-year-old mother is present and can legally give consent.

71. ④ All parents initially grieve over the loss of health in their children. (1) (SC; AN)
1 Many parents have excellent support systems.
2 This may be true of some, not all, parents.
3 At least one third of the children with cerebral palsy are not mentally retarded.

72. ② This allows the client to volunteer information first and thus feel in control; the nurse can ask validating questions later. (1) (CI; AS)
1 This focuses the assessment on vomiting, thus predisposing the client to vomiting during this treatment.
3 This is an unfeeling response; it reminds the child and mother of the many sessions remaining and brings little consolation to the child for the discomfort or to the mother for her worry about the prognosis.
4 This response is flippant.

73. ③ This is a truthful answer that offers some hope without false reassurance. (2) (SC; IM)
1 Although true, this response shuts off communication and discourages further ventilation of feelings.
2 This response produces guilt and denies the adolescent's realistic fears.
4 This response denies the adolescent's feelings.

74. ③ This child experienced trauma; the child's reactions to this experience will influence every aspect of the rest of the care plan. (1) (EP; PL)
1 This is a nursing diagnosis related to the parents of the child, not the child.
2 Although this is an appropriate nursing diagnosis for this child, the child's post-traumatic response is primary and will most profoundly permeate all aspects of care.
4 There is no evidence in the data given that the child's vision is impaired.

75. ② A private room will provide a secure environment for the child and the family to get to know one another. (2) (SC; PL)
1 This is not therapeutic because it may make the child feel guilty about leaving the biologic family.
3 Same as answer 1.
4 Although some information may be given, too much information about the family may promote preconceived ideas that may be inaccurate.

76. (4) This response is accepting of the individual and communicates concern. (3) (SC; IM)
    1 This is the nurse's responsibility and should not be transferred immediately to the client's family.
    2 There is no indication that the injuries were deliberate abuse.
    3 This response interprets the client's statement as guilt, which may or may not be the true interpretation.

77. (2) Abusers lack personal strengths and adequate support systems, which could help them handle stress and frustration. (1) (SC; AN)
    1 Abusers tend to be young, immature, and dependent.
    3 Abusers have an incorrect concept of what the small child can do; their expectations are unrealistic.
    4 Most abusers were abused as children; physical discipline was probably excessive.

78. (2) Because parents are trying to hide the fact of abuse, the explanations are fabricated and vague. (2) (EP; AS)
    1 This is expected; the parents are unable to provide emotional support.
    3 The child behaves in this manner because in past experiences adults have inflicted pain rather than provided comfort.
    4 This is no surprise; parents do not discuss them because this would be an admission of child abuse.

79. (4) This plan for behavior shows the potential for increased impulse control, which is important for the prevention of further abuse. (2) (EP; EV)
    1 This is unrealistic because all parents become angry with their children at some time or another.
    2 This places the blame on the child, rather than on the parent's own behavior.
    3 This places the blame on the spouse, rather than the self.

80. (4) Going to a stranger without protest usually indicates the lack of a meaningful relationship with the mother. (1) (EP; AS)
    1 This is a healthy, normal reaction to strangers that is uncommon in children with nonorganic failure to thrive syndrome.
    2 Same as answer 1.
    3 Children who fail to thrive avoid eye contact with their mothers and do not prefer them over others.

FLUIDS AND ELECTROLYTES (FE)

81. (1) In nephrotic syndrome a large proportion of the child's body weight is composed of retained fluid; the loss of fluid would be readily reflected by a loss of weight. (3) (DA; AS)
    2 It is very difficult to get an accurate recording of output in a young child, especially if vomiting and diarrhea also occur.
    3 With nephrotic syndrome it would be difficult to evaluate return to fluid balance in this way because the edema is generalized, not concentrated in the abdomen.
    4 Osmolality reflects kidney activity, not the reduction in edema.

82. (4) This allows the child to get a full cup (1 oz medicine cup) without long waits; a full cup, even if it is a small cup, creates the illusion of receiving more. (3) (PC; PL)
    1 When fluid is limited, a smaller amount should be apportioned to sleeping hours.
    2 If the child were allowed to drink as much as desired until the limit is reached, 15 to 20 hours might elapse before any fluid would be permitted again.
    3 Although fluids can be limited more easily during sleeping hours, 12 hours is too long for a young child to tolerate.

83. (2) KCl is always restricted in the presence of oliguria to prevent cardiac dysrhythmias associated with hyperkalemia. (3) (PC; PL)
    1 Glucose and fat are not restricted; they are usually prime sources of calories.
    3 Protein restriction is used only when severe azotemia with prolonged oliguria is present.
    4 Same as answer 1.

84. (2) The degree of dehydration is correlated with weight loss; continued fever aggravates fluid losses through evaporation. (2) (HI; AS)
    1 Poor skin turgor may not occur after only 48 hours of vomiting.
    3 This is not relevant; the neurologic status is not altered; the urinary output may show signs of decreasing.
    4 This is not relevant because the child has been vomiting and will most likely be npo to rest the gastrointestinal tract.

85. (4) Loss of weight is the best way to evaluate the magnitude of fluid loss in the infant; 1 liter of fluid weighs 2.2 pounds. (2) (DA; AS)
    1 This is a subjective assessment; measurement of weight is an objective assessment.

2   Although this would indicate dehydration, it is not an effective monitoring method for assessing fluid loss.

3   This is a subjective and inaccurate assessment.

86. (4) This is a normal adaptation to a state of dehydration; the urine will be concentrated. (2) (DA; AS)

1   There is no indication of celiac disease.

2   The initial response to decreased circulating fluids would be an increased pulse rate.

3   One of the signs of dehydration in an infant is a sunken, not a bulging, fontanel.

87. (4) A decreased urinary output is expected with dehydration in a young infant because of decreased circulating fluid volume. (2) (DA; AS)

1   This is associated with increased intracranial pressure, not dehydration; the fontanel would be depressed with dehydration.

2   This is associated with an adequate fluid balance.

3   Because of loss of fluid and electrolytes, the infant would be lethargic, not restless.

88. (3) This is a classic sign of fluid volume deficit in infants. (1) (DA; AS)

1   This is within the normal limits of 1.010 to 1.030.

2   This indicates adequate hydration; the urinary output would be decreased in dehydration.

4   Dehydration is unrelated to allergies.

89. (2) Infants have a higher fluid-to-body mass ratio than do older children or adults; severe dehydration can occur more rapidly and the correction of fluid loss problems is more difficult; early, immediate medical intervention is necessary. (2) (HI; PL)

1   This is unrelated since data do not indicate any administration of antibiotics; antibiotics are not given for viral infections.

3   This is important at all times, but the priority at this time is to make the mother understand how serious diarrhea can be.

4   This is related but not the priority; diarrhea can be contracted despite cleanliness.

90. (1) Normal arterial blood gas values are pH 7.35 to 7.45; Po_2 83 to 108 mm Hg; Pco_2 35 to 45 mm Hg; base excess -3 to $+3$. (2) (CI; AS)

2   The pH is alkalotic.

3   The pH is acidotic, the Po_2 is low, the Pco_2 is high, and the base excess is outside the normal range.

4   The Po_2 is low (hypoxic); the Pco_2 is low (hypocapnic); the base excess is high.

91. (3) Respiratory compensation to acidosis involves increased CO_2 elimination through hyperventilation, with a resulting increase in pH to normal limits. (2) (HI; AN)

1   Hypoventilation would not increase expiration of CO_2 with the ultimate increase in pH.

2   If the client is hypoventilating, blood H ions would increase, not decrease, because of CO_2 retention, not elimination; pH would decrease.

4   With hyperventilation, there would be an increase in CO_2 elimination, not a decrease; the pH would increase, not decrease.

92. (3) Metabolic acidosis results from an excess concentration of hydrogen cations; potassium increases; the kidneys cannot convert ammonium (NH_3) to ammonia (NH_4); there is inadequate base bicarbonate to maintain an appropriate acid-base balance. (3) (DA; AN)

1   This child will have an excess of hydrogen ions, resulting in metabolic acidosis; carbonic acid blown off as CO_2 results in respiratory alkalosis.

2   This child has an excess of hydrogen ions from a metabolic problem rather than an excess of carbonic acid resulting from retained CO_2.

4   This child will have an excess of hydrogen ions, the opposite of an excess of base bicarbonate.

93. (4) In metabolic acidosis the lungs try to compensate by blowing off excess carbonic acid in the form of carbon dioxide. (3) (DA; AN)

1   This is a compensatory mechanism to reduce fever by evaporation.

2   This indicates renal compensation for respiratory alkalosis.

3   This is not an adaptation to metabolic acidosis; fever with dehydration results from inadequate fluid for perspiring and cooling.

94. (2) A low blood pH and bicarbonate level indicate metabolic acidosis. (3) (DA; AS)

1   These findings indicate metabolic alkalosis.

3   These findings indicate respiratory acidosis.

4   Same as answer 1.

95. (3) Sodium lactate is converted to sodium bicarbonate; it helps correct the sodium deficit and the metabolic acidosis. (3) (PT; PL)

1 Plasmanate is a colloid used as a substitute when plasma is needed; it is not used in the treatment of metabolic acidosis.

2 Saline results in the chloride combining with the hydrogen ion, intensifying the acidosis.

4 Potassium is not administered until urinary function is restored.

96. ③ Loss of bicarbonate and sodium in the stools causes metabolic acidosis; fluid loss in liquid stools causes hypovolemia. (2) (DA; AN)

1 Fluid loss does cause hypovolemia, but hypercalcemia does not occur.

2 Hypovolemia, not hypervolemia, would result from diarrhea; there may or may not be an increase in sodium.

4 The hematocrit would be elevated because of fluid loss (hemoconcentration); sodium may be increased, decreased, or normal.

97. ④ The symptoms indicate ketoacidosis so both these values would be expected; the pH indicates acidosis (metabolic or ketoacidosis) and the blood glucose level, elevated more than the normal range of 70 to 105 mg/dl, indicates severe hyperglycemia. (3) (CI; AS)

1 Although the blood pH indicates acidosis, the blood glucose level is below the normal range of 70 to 105 mg/dl; with ketoacidosis, the client would be hyperglycemic.

2 Both values would be unexpected in ketoacidosis; in ketoacidosis, the pH would be lowered and blood glucose level elevated.

3 Although the blood glucose level would be elevated in ketoacidosis, the pH would be lowered not elevated; a pH of 7.50 indicates alkalosis.

98. ③ The extremities need to be restrained because the child will use all extremities in an attempt to dislodge the needle. (2) (SI; IM)

1 Pupillary responses are unrelated to dehydration and fluid replacement.

2 Scalp veins used for IVs are not located in these areas.

4 The parents can be taught how to hold a child with an IV infusing via a scalp vein.

99. ④ These are classic signs of dehydration and hyponatremia; a physician's order to increase fluids is needed. (2) (PT; EV)

1 The nurse must have a physician's order for this; also, dehydration can be corrected with fluids via IV or feeding tube.

2 It is not common for the condition of these infants to continue to deteriorate once therapy is implemented.

3 The symptoms indicate dehydration, not undernutrition.

100. ③ One liter of fluid weighs 2.2 pounds; this is the most objective and accurate way to assess fluid loss or gain; weights measured at the same time each day provide for daily comparisons. (2) (HI; EV)

1 This would be appropriate for assessing the progression of ascites, not for assessing rehydration.

2 Color of stools is unrelated to fluid balance; noting consistency would be more important, although subjective.

4 Although this would be done, it is subjective and inaccurate.

101. ④ Safety is a priority; the infant may inadvertently dislodge the circulatory access. (2) (SI; IM)

1 This is contraindicated; oral fluids will not be administered until peristalsis returns.

2 Parent-infant contact should be encouraged.

3 This is unnecessary; the number of voidings should be assessed.

102. ② To prevent fluid overload, the IV infusion should be maintained at the slowest rate possible to keep the circulatory access patent until the physician can be reached to verify the correct rate. (2) (DA; IM)

1 This is unsafe; after surgery all previous orders are canceled and new orders must be written by the physician.

3 This is unsafe; the administration of intravenous fluids requires a physician's order.

4 Same as answer 3.

103. ③ The present caloric intake for a 24-hour period is 8 oz × 20 calories × 6 feedings = 960 calories; the recommended intake is 108 calories × 8.52 kg = 920.16 calories. (3) (PC; EV)

1 The data indicated can be used to calculate the child's daily caloric intake (oz × calories × number of feedings), which can be compared with the recommended caloric intake (108 calories × kg of body weight).

2 The present caloric intake exceeds the daily recommended requirements by 40 calories.

4 Same as answer 2.

104. ③ 2.2 lb = 1 kg; 13 lb = 5.9 kg; 150 × 5.9 = 885 ml. (2) (PT; AN)
 1 This is inaccurate; this is less than the ordered amount of fluid.
 2 Same as answer 1.
 4 This is inaccurate; this exceeds the ordered amount of fluid.

105. ③ Use ratio and proportion: multiply the amount to be infused (500 ml) by the drop factor (60), and divide the result by the amount of time in minutes (18 hours × 60 minutes). (2) (PT; AN)
 1 It is impossible to deliver a half of a drop.
 2 This rate is too slow; less than the ordered amount would infuse.
 4 This is too fast; the fluid would be delivered in a shorter period of time than ordered.

106. ③ This is the correct flow rate; multiply the amount to be infused (700 ml) by the drop factor (60), and divide the result by the amount of time in minutes (24 hr × 60 min). (2) (PT; AN)
 1 This rate is too slow; less than the ordered amount would be infused.
 2 Same as answer 1.
 4 This is too fast; the fluid would be administered in a shorter period of time than ordered.

107. ③ This is the correct flow rate; multiply the sum of all IV fluid to be infused for the period (750 ml) by the drop factor (60); divide the results by the amount of time in minutes (24 hours × 60 minutes); thus 45,000 ÷ 1440 = 31 gtt/min. (2) (PT; AN)
 1 This is too slow to deliver the required amount of fluid in 24 hours.
 2 Same as answer 1.
 4 This rate of flow is too fast; the fluid will be infused before 24 hours.

BLOOD AND IMMUNITY (BI)

108. ④ Previously formed antibodies, acquired through active immunity, offer some resistance to infection; adults have higher levels of antibodies than children because over time they have been exposed to more pathogens. (1) (HI; AN)
 1 The immune systems of children are as capable of producing antibodies as are those of adults.
 2 The pathophysiologic mechanism of AIDS is the same in both children and adults.

 3 Exposure to pathogens in the environment does not differ significantly between children and adults.

109. ③ The passive antibodies received from the mother would be diminished by age 8 weeks and would not interfere with the development of active immunity after this time. (2) (HI; IM)
 1 The spleen does not produce antibodies.
 2 This is untrue; infants are often exposed to infectious diseases; passive immunity from the mother offers some protection.
 4 This is untrue; these immunizations are attenuated; they may cause irritability and fever, but they will not cause the related disease.

110. ② Mild reactions are redness and induration at the injection site, slight fever, and irritability. (2) (HI; EV)
 1 Serious reactions are not common.
 3 Occasionally a DTP injection may precipitate a febrile seizure, but it does not cause permanent brain damage.
 4 Induration at the site of injection may occur, but deep ulceration does not.

111. ④ Fever is a common reaction to the immunizations; acetaminophen helps to reduce fever; both loss of consciousness and convulsions are rare, but serious, complications of the pertussis vaccine. (2) (HI; IM)
 1 This would cause an extension of the inflammatory response and should be avoided.
 2 Infants do not respond well to the application of ice; fever is expected and requires no intervention other than administration of acetaminophen.
 3 Aspirin should not be given to children because it is associated with Reye's syndrome.

112. ③ The schedule for active immunization is three doses of DTP at 2-month intervals beginning at 2 months of age. (2) (HI; AN)
 1 Measles vaccine is not given until 12 to 15 months because maternal antibodies block the formation of the infant's antibodies.
 2 TOPV booster is due at 18 months; it is given at the same time as the DTP booster dose.
 4 This is given at 18 months, or approximately 1 year after the third dose that is given at 6 months of age.

113. ③ The American Academy of Pediatrics recommends that infants be given the MMR combination vaccine at 12 to 15 months of age. (1) (HI; PL)

1 The infant should have received this immunization at 2 months, 4 months, and 6 months of age; the next booster will be at 18 months.
2 Same as answer 1.
4 Same as answer 1.

114. ③ MMR vaccine for measles, mumps, and rubella is not given until 12 months. (1) (HI; PL)
1 In the first 6 months, the polio vaccine is given to infants at 2 and 4 months of age.
2 This is part of the DTP vaccine; in the first 6 months, it is given at 2, 4, and 6 months of age.
4 Hepatitis B immune globulin is recommended to be given routinely to all newborns.

115. ① Children with leukemia are immunosuppressed; the chickenpox virus can cause death in the individual without an intact immune system. (2) (SI; PL)
2 The vaccine against chickenpox is a live virus and should not be given to an immunosuppressed child.
3 Prednisone does not confer immunity to the chickenpox virus.
4 This would be unsafe; chickenpox can be spread by airborne droplets in the prodromal stage and by fomites that have come in contact with pox that are oozing.

116. ④ This supplies iron and some protein; it can be eaten with a spoon, encouraging mastery of fine motor muscles. (1) (PC; PL)
1 This provides some protein and iron but has a spicy taste that is not generally a favorite of this age group.
2 This is low in protein and iron.
3 Same as answer 2.

117. ④ Potatoes and whole milk are not adequate sources of iron; at age 8 months, fetal iron stores are depleted. (2) (PC; AN)
1 Potatoes are a rich source of potassium.
2 The infant is receiving Poly-Vi-Sol, which contains vitamins but no iron; Poly-Vi-Sol with iron exists, but the data indicate that plain Poly Vi-Sol is being used.
3 There are some amino acids in the foods that are being eaten.

118. ④ The physician will probably use a local anesthetic, which can hurt; however, there will also be pain as the aspiration needle is inserted through the bone to the periosteum as well as pain and pressure when the bone marrow is withdrawn. (3) (PC; EV)
1 This is true; although the site may be sore and the child may prefer to remain quiet, activity is not usually restricted.
2 This is true; this is done to prevent bleeding from the puncture site.
3 This is true; the anterior or posterior iliac crest is the site most often used for a bone marrow aspiration in children.

119. ② Before the sex of the unborn child is known, the odds are 25%; 50% of pregnancies will result in boys, and each has a 50% chance of having hemophilia. (3) (DF; AN)
1 Because the disease is genetically transmitted, this is not likely.
3 This is too high; there is only a 25% chance that the baby will be affected.
4 Same as answer 3.

120. ② Joints are the most commonly involved areas, probably because of weight bearing and constant movement of joints. (3) (HI; AS)
1 This is not the most common site; however, bleeding can occur here.
3 Same as answer 1.
4 Same as answer 1.

121. ② During sickle cell crisis, the RBCs are sickled and clumped and the hemoglobin is ineffective in providing oxygen; therefore fluids to liquefy the clumping cells and additional oxygen are necessary. (3) (DA; PL)
1 Neither one of these is required during sickle cell crisis.
3 Hydration is important at this time; even if counseling were needed, it would not be done until the client was out of crisis.
4 Oxygenation is needed; factor VIII is used in hemophilia.

122. ② Thalassemia is a hemolytic anemia that is not communicable; roommates with infectious diseases should be avoided because a child with sickle cell anemia is susceptible to infections. (2) (SI; PL)
1 The child with sickle cell anemia is susceptible to infection; pneumonia is an infection of the lung.
3 The child with sickle cell anemia is susceptible to infection; osteomyelitis is an infection of the bone.
4 The child with sickle cell anemia is susceptible to infection; pharyngitis is an upper respiratory infection.

123. (2) Dehydration is a major causative factor for a sickle cell episode. (2) (HI; AN)

1 It is not the iron intake but ineffective hemoglobin that causes the anemia.

3 A breakdown of platelets (thrombocytes) is unrelated to sickle cell episodes.

4 The WBCs are not increased unless infection is present.

124. (3) This is a relatively passive, innocuous activity that would not increase oxygen demands of the body, which could precipitate sickling. (2) (CI; IM)

1 This may lead to increased cellular metabolism and increased tissue hypoxia, which could precipitate sickling.

2 Same as answer 1.

4 High altitudes should be avoided because of the lower oxygen concentration of the air, which might trigger a crisis.

125. (2) Sores in the mouth are not a presenting sign but often result from chemotherapy. (3) (CI; IM)

1 Anorexia is a presenting symptom of leukemia and may be the result of enlarged lymph nodes and areas of inflammation in the intestinal tract.

3 Pallor is a presenting sign of leukemia and reflects the anemia present because of decreased erythrocytes.

4 Decreased platelet production with petechiae and bleeding is a presenting sign of leukemia.

126. (1) The platelet count is reduced as a result of the bone marrow depression associated with leukemia. (3) (CI; IM)

2 The uric acid level reflects urinary function, not blood clotting.

3 Prothrombin time is influenced by vitamin K factors, not lack of platelets.

4 The red blood cell count will indicate the hematocrit and hemoglobin levels, which would neither provide the reason for nor cause the bleeding.

127. (2) Good nutrition is extremely important to the child's overall health; best results are attained when a child receiving chemotherapy is allowed to eat as desired. (2) (PC; EV)

1 The child should be isolated only from other children with known or possible infections.

3 Mouthwashes can irritate the fragile mucosa, so saline should be used; nutrition is the priority.

4 Although activities are important to the child's health, these should be provided according to the child's interest and should not always be structured.

128. (3) Blood tests indicate response to therapy; if the WBC count drops severely, therapy may be temporarily halted. (2) (PT; PL)

1 These children receive therapy for extended periods, and prolonged isolation from their peers may lead to destructive social isolation.

2 This is a very painful procedure and is not done weekly.

4 Nausea commonly occurs with this therapy; although antiemetic measures are instituted, the drug is not withdrawn.

129. (1) A fever occurs with an infection because pyrogens affect the temperature-regulating center in the hypothalamus. Lethargy occurs with an infection because of the related increased basal metabolic rate. (1) (SI; EV)

2 Antibody titers only indicate exposure to the microorganisms, not the presence of disease.

3 This is not an indication of infection.

4 Infection usually causes an increase, not a decrease, in white blood cells.

130. (3) Unless the child has an episode of acute illness, the home is the best place for the child; this prevents nosocomial infection and promotes family interaction. (2) (CC; PL)

1 This is required only for episodes of acute illness that cannot be handled at home.

2 This is not required unless the illness exacerbates.

4 This should be used only if a home environment is not available.

131. (4) The Centers for Disease Control and Prevention recommends standard precautions when caring for those who have HIV infection and/or AIDS without opportunistic infections. (1) (SI; IM)

1 Droplet precautions are not necessary because HIV is not transmitted by large particle respiratory droplets.

2 Contact precautions would not be necessary unless the HIV or AIDS were complicated by the presence of disease or infection, necessitating the addition of these precautions to standard precautions.

3 Airborne precautions are unnecessary because HIV is not spread by airborne droplet nuclei; these precautions would be used in addition to standard precautions if an opportunistic infection such as *Mycobacterium tuberculosis* were present.

132. (2) Children with AIDS have a dysfunction of the immune system (depressed or ineffective T cells, B cells, and immunoglobulins) and are susceptible to opportunistic infections. (1) (SI; AN)
　1　All children have a high risk for injury because of their curiosity, inexperience, and lack of judgment.
　3　Although children with AIDS are most likely small for their ages, altered growth and development are not as life threatening as an infection.
　4　Although this can occur in children with AIDS, the prevention of infection is the priority.

INTEGUMENTARY (IT)

133. (3) Air drying promotes healing; moisture macerates the skin and provides a medium for the growth of microorganisms. (1) (PC; IM)
　1　Although this statement is true, it is not the reason for leaving the baby without a diaper.
　2　This is not the reason for leaving the baby exposed; body heat effectively leaves the body through the head; also, the situation does not convey that the baby has a fever.
　4　Same as answer 1.

134. (2) White spots (Koplik's spots) and the rash with coryza are very indicative of measles (rubeola). (3) (DA; AN)
　1　Rubella (German measles) does not cause Koplik's spots.
　3　Varicella (chickenpox) has skin lesions rather than a rash and lesions in the mouth.
　4　Scarlet fever does not cause Koplik's spots but a strawberry-red tongue.

135. (3) It is most often found in infants at about 2 to 4 months of age. (1) (HI; AN)
　1　The age of the child, the elusive causative factor, and the multifaceted modalities used for therapy make it very difficult to treat.
　2　It is not contagious.
　4　It is not associated with any respiratory infections.

136. (4) The skin integrity of these children is highly compromised because of their constant scratching; they are prone to streptococcal and staphylococcal infections. (1) (SI; PL)
　1　This is always important for infants, but the priority is prevention of secondary infections.
　2　Same as answer 1.
　3　This is the physician's, not the nurse's, responsibility.

137. (2) All of these contain protein to which the eczematous child is allergic. (2) (HI; AN)
　1　Fruit does not contain protein, the food element to which the child is allergic.
　3　Woolens provoke itching but do not cause the child to break out.
　4　Environmental inhalants cause eczema in the older child; in infants, protein is the offender.

138. (1) The baby's nails should be cut very short to minimize injury from scratching. (1) (SI; EV)
　2　This statement is correct; woolens tend to further irritate the eczematous rash.
　3　This is a correct statement; infants with eczema should avoid milk.
　4　This is a correct statement; this kind of clothing seems to be less irritating.

139. (4) Lymphadenopathy and the development of a rash after a day of fever, sneezing, and coughing characterize rubella. (2) (DA; AS)
　1　These are symptoms of rubeola, not rubella.
　2　These are symptoms associated with meningitis and encephalitis, not rubella.
　3　The spots in the mucous membrane of the soft palate in rubella are called Forschheimer spots; Koplik's spots are present with measles.

140. (4) An unimmunized woman who is exposed to the rubella virus may contract the disease and transmit it to the fetus; there is a potential for pregnancy in this cousin. (1) (SI; PL)
　1　There is less risk if a young male adult contracts rubella than if a young adult female, who may be pregnant, does.
　2　If the playmate should contract the rubella from the child, the disease would probably be mild and confer immunity.
　3　Rubeola has no relationship to rubella; the sister would be at risk only if she were pregnant.

141. (2) Eschar is rigid and may restrict circulation and lead to loss of limb perfusion. (3) (DA; AS)
　1　This is not the role of the nurse; blisters are a protective adaptation.
　3　Although this would be done, adequate arterial perfusion is the priority.
　4　This is unnecessary.

142. (2) The child's burn injury totals 31.5% burned (9% right arm, 4.5% left arm, 18% anterior chest); the "rule of nines" adult chart is applicable for a 12-year-old. (1) (DA; AS)
　1　This is too little (9% right arm, 4.5% left arm, 18% anterior chest = 31.5% burned); the "rule of nines" adult chart is applicable for a 12-year-old.

3 This is too much (9% right arm, 4.5% left arm, 18% anterior chest = 31.5% burned); the "rule of nines" adult chart is applicable for a 12-year-old.

4 Same as answer 3.

143. ③ Once a sterile object comes into contact with any object that is not sterile, it is no longer considered sterile. (1) (SI; AN)

1 This is untrue; a 1-inch border around the sterile field is considered contaminated.

2 This is untrue; the object is considered contaminated.

4 This is untrue; dry wounds are considered clean.

144. ① When all vesicles are dried, chickenpox is no longer transmissible; dried vesicles do not harbor the varicella virus. (2) (HI; IM)

2 This is not true; dry scabs do not transmit the virus.

3 Chickenpox is not associated with a high fever unless a bacterial complication such as pneumonia is present.

4 These vesicles are mature vesicles that occur in successive crops; these vesicles contain the varicella virus.

145. ③ Individuals taking steroids have lowered resistance and may become fatally ill if exposed to the varicella virus. (1) (HI; IM)

1 This does not lower body resistance; therefore it does not increase susceptibility.

2 This does not affect body resistance because topical antibiotics do not have a systemic effect.

4 This may increase resistance rather than decrease it.

146. ④ Patting the lesions will not disturb them, and baking soda is an effective drying agent. (1) (PC; IM)

1 This may prevent secondary infection but has no drying effect.

2 This may tear off the vesicles and lead to scar formation.

3 This may minimize scratching but will not relieve pruritus.

147. ① Radiation, or the transferring of heat from a warm object to the atmosphere, is prevented by reducing the surface and covering the child with a blanket. (2) (PC; AN)

2 Conduction is the transfer of heat from one molecule to another with contact between the two; very little body heat is lost by conduction.

3 Active transport is not related to loss of heat; this is a process that moves ions or molecules across a cell membrane against a concentration gradient.

4 Vaporization is the conversion of liquid or solid into a vapor; it would occur if a person were perspiring.

148. ④ Foam is soft, so it will not damage the oral mucosa. (1) (CI; EV)

1 A toothbrush will injure the oral mucosa.

2 Hydrogen peroxide will irritate the mucosa and has an offensive taste.

3 Mouthwash may irritate the oral mucosa and should always be diluted.

149. ④ Chemotherapy causes severe alteration in mucous membranes; a rectal thermometer may damage delicate rectal tissue. (2) (SI; AN)

1 Although this may be true, it is not the primary reason to avoid taking rectal temperatures in children receiving chemotherapy for leukemia.

2 Oral temperatures are accurate, provided the child can hold the thermometer in the mouth correctly.

3 This is incorrect; tympanic temperatures are accurate.

150. ① Measles is another name for rubeola. (1) (HI; AN)

2 This is known as varicella.

3 This is pertussis.

4 This is rubella.

151. ① Measles starts with a discrete maculopapular rash on the face and spreads downward, eventually becoming confluent. (2) (HI; AS)

2 This occurs with whooping cough.

3 This occurs with mumps.

4 This occurs with chickenpox.

152. ③ Impetigo may develop as a secondary bacterial infection because of breaks in the skin from scratching. (3) (SI; AN)

1 Eczema is an allergic response, not an infection.

2 This is an extended inflammation that is not commonly found in children with pediculosis.

4 This is a fungal infection of the scalp; it usually occurs by itself, not as a secondary infection to pediculosis.

153. ① The nurse needs to gather information regarding the symptoms because they can be related to many factors. (2) (HI; AS)

2 A developmental screening is not necessary in an acute situation.

3 This is unnecessary; this information is not related to the situation.

4 A child in good health would not be in a high-risk group and receive an influenza vaccination; a rash is unusual after a sunburn.

154. ④ This is a straightforward, truthful answer at a level that a 12-year-old would comprehend. (2) (DF; IM)

1 The answer is full of medical jargon that the child might not understand.

2 This identifies a feeling but avoids answering the question.

3 This identifies a feeling but disregards the fact that the child has already asked a question.

155. ② The tick must be carefully removed with tweezers or forceps so that the tick does not further inoculate the individual. (1) (SI; IM)

1 This is an unsafe method of removing a tick; the tick may further inoculate the individual, and the method may hurt the child.

3 Same as answer 1.

4 Same as answer 1.

SKELETAL (SK)

156. ③ Cool fingers are a sign of circulatory impairment caused by the pressure of the cast; however, if both hands feel cool, it indicates some factor other than circulatory impairment is responsible. (1) (CI; EV)

1 Further assessment to determine the cause of temperature change is indicated before taking immediate remedial action.

2 This should not be done without a physician's order.

4 Further assessment should be done before informing the physician.

157. ② A damp cloth will remove surface soil but will not damage the plaster cast. (1) (SI; IM)

1 Plastic will cause condensation and wetness, which would soften the cast.

3 Excessive water would soften and damage the cast; cleansing agents could injure the skin.

4 Same as answer 3.

158. ② This finding indicates that tendons, ligaments, and bone changes are involved. (2) (DF; AS)

1 Club foot is not associated with neurologic damage; there is no paralysis or lessening of movement involved.

3 This finding could indicate vascular damage; it is not related to club foot.

4 This finding indicates a positional (pseudo) club foot.

159. ④ This would not be measurable under a boot cast. (2) (CI; EV)

1 This is an appropriate neurovascular check and it is measurable.

2 Same as answer 1.

3 Same as answer 1.

160. ① Injuries in various stages of healing are the classic sign of child abuse. (1) (EP; AS)

2 This can be evaluated after child abuse has been ruled out.

3 Same as answer 2.

4 Same as answer 2.

161. ③ The palm of the hand provides a wide base of support for the extremity with the cast and prevents indentation of the cast; this preserves the integrity of the cast and prevents neurovascular injury to the lower extremity. (1) (CI; IM)

1 This could cause indentations in the cast that would put pressure on the lower extremity, which could compromise the skin and/or neurovascular functioning.

2 This could injure the infant and could compromise the integrity of the cast; the lower extremity and cast must be supported.

4 Same as answer 2.

162. ③ Rough cast edges can cause skin irritation and breakdown; the skin at the edges of the cast should be assessed for edema, signs of pressure, and evidence of skin breakdown. (1) (CI; EV)

1 Adhesive petals will not adhere to a damp cast.

2 Lotions applied to the skin at the edges of a cast can promote skin breakdown.

4 This is contraindicated; plaster casts should be kept dry to maintain their shape; the skin under the cast may become macerated from inadequate drying after water immersion.

163. ④ With specific manipulation, an audible click may be heard or felt as the femoral head slips into the acetabulum. (1) (HI; AS)

1 This is Trendelenburg's sign; it is associated with weight bearing.

2 This is not the Ortolani sign; this is associated with bilateral dislocation.

3 This is Allis' sign.

164. ① The affected leg appears to be shorter because the femoral head is displaced upward. (2) (HI; AS)

2   There is a limited ability to abduct, not adduct, the affected leg.

3   This does not occur with congenital hip dysplasia.

4   When the femoral head slips out of the acetabulum, it is easily palpable.

165. (3) Difficulty with abduction may indicate a congenital hip problem. (1) (HI; AS)

1   Flexion of extremities is the normal position for a young infant.

2   This is a normal finding in a young infant.

4   Same as answer 2.

166. (3) This is the basis for abduction devices and spica casts when the infant is very young. (1) (DF; AN)

1   This may be true, but the easy moldability of the bones at this age favors corrective devices.

2   Congenital hip dysplasia usually is not painful and does not limit ambulation for the young child.

4   This traction is no longer used; other traction can be used if necessary.

167. (1) This position promotes hip abduction and flexion. (1) (PC; PL)

2   This practice limits hip abduction and puts stress on the hip joint; if mild dysplasia is present, it will be aggravated.

3   Same as answer 2.

4   This allows free movement in the flexed position but does not promote abduction.

168. (4) This is the preferred method of protecting the cast from soiling by excreta. (2) (SI; PL)

1   Special precautions are definitely required to keep the cast clean.

2   Plaster needs to "breathe" and should not be completely covered with occlusive material.

3   They should not be used together because clumping of powder, with resultant irritation, may occur.

169. (4) Adolescence, a period of rapid growth, is the time when scoliosis is most likely to become evident. (2) (DF; AN)

1   Although scoliosis may occur at any time, idiopathic scoliosis, the most common type, tends to become evident in the teenage years.

2   Same as answer 1.

3   Same as answer 1.

170. (4) Cheese contains calcium, which is excreted by the kidneys and favors formation of kidney stones; this adds to client's risk because immobility causes bone decalcification. (2) (PC; PL)

1   Beef contains protein, which promotes wound healing.

2   Liver is high in iron, which promotes RBC production; this may be beneficial because many orthopedic injuries tend to ooze and bleed.

3   Broth contains some protein, which facilitates wound healing, and fluid, which prevents dehydration and urinary calculi.

171. (2) This may be indicative of an infection under the cast and would probably cause a fever. (1) (CI; EV)

1   Respirations may increase but do not become irregular with a fever.

3   This would not cause a fever; it may indicate neurovascular impairment.

4   This would not cause a fever.

172. (1) Gentle rubbing may soothe the skin; stimulation of sensory neurons by rubbing may decrease the itching sensation. (2) (PC; EV)

2   Alcohol is a drying agent and should not be used.

3   Antihistamines are generally not given unless all other measures fail.

4   Powder may become caked, slip under the cast, and cause additional discomfort.

173. (1) The child should be rolled as one unit, with shoulders and hips turned at the same time to prevent injury. (1) (CI; IM)

2   The crossbar is not used to turn because it may dislodge and weaken the cast.

3   This would be used for lifting, not turning.

4   The child will not be able to sit up because the cast immobilizes the hips.

NEUROMUSCULAR (NM)

174. (2) Children with cerebral palsy are especially prone to muscle tone disorders, including spasticity, which can lead to contractures. (1) (CC; PL)

1   The therapy program must be balanced to promote progress in all areas of growth and development.

3   This is contraindicated because prolonged immobility promotes the development of contractures.

4   In a therapeutic regimen there must be a balance between exercise and rest.

175. (1) Duchenne's muscular dystrophy follows an X-linked recessive inheritance pattern; when the father is unaffected and the mother is a carrier, there is a 50% chance that a son will be affected and there is a 25% chance that a daughter will be affected. (3) (DF; EV)
  2 Sex-linked transmission rarely results in females with the condition; males are predominantly affected, and females tend to be carriers.
  3 This sex-linked condition is transmitted from the recessive gene carried by the mother only.
  4 When the father is unaffected and the mother is a carrier, each son has a 50% chance of being affected.

176. (1) Range-of-motion exercises are essential to help achieve the primary objectives of maintaining optimal muscle function as long as possible and preventing the development of contractures. (1) (HI; IM)
  2 A high-calorie diet may result in obesity, which may accelerate the time when a wheelchair will be necessary.
  3 Seizures are not usually associated with Duchenne's muscular dystrophy.
  4 Restricting the use of large muscles may result in disuse atrophy and contractures.

177. (4) Muscular degeneration is advanced in the adolescent; the disease process involves the diaphragm, auxiliary muscles of respiration, and the heart, resulting in life-threatening respiratory infections and heart failure. (3) (DA; IM)
  1 Nutritional problems are less of a priority when compared to cardiopulmonary problems.
  2 Central nervous system functioning is not affected by Duchenne's muscular dystrophy.
  3 Although the musculoskeletal system will exhibit marked degeneration, it is second in priority to the cardiopulmonary changes.

178. (1) Down syndrome (trisomy 21) results from an extra chromosome 21. (1) (DF; AN)
  2 Down syndrome is not infectious in origin.
  3 Down syndrome is not related to an X-linked or Y-linked gene.
  4 This is not a cause of Down syndrome, although translocation of chromosomes 15 and 21 or 22 is a genetic aberration found in 4% to 6% of the children with Down syndrome.

179. (3) Forty percent of children with Down syndrome have cardiac anomalies. (2) (DF; AN)

  1 This is not a characteristic finding in children with Down syndrome.
  2 Same as answer 1.
  4 Same as answer 1.

180. (3) Children who are mentally challenged can learn concrete concepts faster than they can learn abstract concepts. (1) (DF; AN)
  1 This is an abstract concept that a child begins to learn between the ages of 7 and 11 years.
  2 Same as answer 1.
  4 Same as answer 1.

181. (1) This prevents placing undue emphasis on the speech pattern, thus preventing inadvertent reinforcement of the pattern. (3) (CC; IM)
  2 This is demeaning; it may decrease self-esteem and increase stuttering.
  3 Same as answer 2.
  4 Same as answer 2.

182. (4) Erb's palsy results from forces that alter the normal position and relationship of the arm, shoulder, and neck; stretching or pulling away of the shoulder from the head during delivery damages the brachial plexus. (3) (HI; AN)
  1 Erb's palsy is a birth injury that is acquired during delivery, not in utero.
  2 Erb's palsy is a birth injury; it is not an X-linked inherited disease.
  3 Erb's palsy is a birth injury involving the brachial plexus, not a tumor.

183. (4) The nerves that have been stretched normally take about 3 months to recover from the trauma sustained during delivery. (3) (CI; EV)
  1 The paralysis is not progressive and the prognosis is usually excellent.
  2 This is unnecessary; passive range of motion and intermittent splinting performed by a family member are all that are necessary; recovery usually occurs in 3 months.
  3 Intermittent splinting and passive range of motion are all that are required; only in rare instances, when avulsion of the nerves results in permanent damage, is orthopedic or surgical intervention necessary.

184. (2) Lead blocks formation of normal RBCs because it is very toxic to the biosynthesis of heme; this leads to anemia, an initial sign of the disease. (2) (DA; AN)
  1 This does not occur; the child usually has anemia and some CNS disturbances.
  3 Same as answer 1.
  4 Same as answer 1.

185. ① This is related to lead toxicity or buildup that causes fluid shifts into brain tissue, producing cell ischemia and destruction; this ultimately results in convulsions, mental retardation, and death. (2) (CC; AN)

2 Although abdominal cramps and headache may be symptoms of chronic lead poisoning, these are not the primary problems as lead poisoning progresses and central nervous system symptoms appear.

3 The child is not necessarily malnourished; the child usually eats the paint or plaster chips containing lead (pica) in addition to the diet.

4 This is not related to this disorder; it is related to hemianopia or hemiparesis.

186. ① This position utilizes gravity to drain fluid from the head and prevent fluid accumulation. (1) (CI; IM)

2 The semi-Fowler's position has no effect on chemicals, but it does reduce subdural pressure.

3 This is true, but it is primarily done to prevent cerebral edema.

4 Cardiac workload is not reduced in the semi-Fowler's position, but it is in the supine position; the semi-Fowler's position does facilitate oxygenation.

187. ③ Pressure on the hypothalamus, the temperature-regulating mechanism of the brain, causes temperature imbalances. (2) (CI; EV)

1 After an operation, a temperature from the inflammatory response rarely exceeds 101° F; this temperature is not slight.

2 This is not true when aseptic technique is observed.

4 These would not cause such a high temperature unless infection were present.

188. ② A defective development of the spinal cord resulting in lower extremity paralysis is found only in myelomeningocele. (3) (DA; AN)

1 Hydrocephalus results from associated ventricular abnormalities and can occur with either defect.

3 A saclike cyst containing meninges and spinal fluid may be present in either defect.

4 Infection is possible with either defect because of the exposure of the meninges.

189. ① This sign is indicative of increased intracranial pressure, which is caused by the fluid accumulation associated with hydrocephalus. (1) (DA; AS)

2 This is a normal finding; conjugate gaze does not occur until 3 to 4 months of age when eye muscles are mature.

3 This is a normal finding; a baby does not do this before 1 to 1½ months of age.

4 This is a normal finding; the head is the largest part of the body at this age.

190. ③ Because the infant cannot be readily held, tactile stimulation meets the infant's needs and fosters bonding with the parents. (1) (SC; IM)

1 An infant with an unrepaired myelomeningocele cannot be held in the arms.

2 Demonstration of feeding techniques is helpful but may not improve parent-child relationships.

4 This intervention will be more beneficial at a later time.

191. ④ The parents must be taught to identify increased intracranial pressure, since this can develop if shunt malfunction occurs. (3) (CC; IM)

1 The prone position places too much pressure on the shunt; the infant should be flat and turned on the unoperative side.

2 Dry, sterile dressings are applied postoperatively to prevent infection.

3 Cerebrospinal fluid would not drain from the incision.

192. ① This will avoid sudden increases in intracranial pressure. (2) (SI; PL)

2 This is inappropriate; it may be frightening for the parents.

3 Young infants, especially with this health problem, tolerate a demand schedule better, and it may diminish the possibility of vomiting.

4 This is inappropriate for a 1-month-old infant.

193. ④ The side-lying position and use of supports prevent pressure on the valve, which is on the right side; the flat position prevents too rapid drainage of cerebrospinal fluid. (3) (CC; EV)

1 This is inappropriate in the immediate postoperative period; the infant should be kept flat.

2 Neck supports should not be used with infants; they may cause airway occlusion.

3 The support could push the head to the right and exert pressure on the valve, which is on the right side, possibly causing it to close; this would heighten the risk of increased intracranial pressure.

194. ③ Leakage of cerebrospinal fluid indicates incomplete closure of the defect. (1) (CI; EV)
    1. The supine position places too much pressure on the operative site.
    2. Moist dressings are applied preoperatively to prevent drying out of the sac.
    4. Teaching clean catheterization is not appropriate at this time.

195. ③ This may help to reduce the infant's temperature; chilling should be avoided. (2) (DA; IM)
    1. Alcohol should never be used with infants or children; it causes severe chilling, which can lead to increased metabolic activity and a higher temperature.
    2. This fever requires more frequent readings than every 2 hours.
    4. This is not a priority; temperature reduction should be done first; recording can be done later.

196. ③ A bulging fontanel is the most significant sign of increased intracranial pressure in an infant. (3) (CI; EV)
    1. This is not a significant indicator of increased intracranial pressure.
    2. Same as answer 1.
    4. Same as answer 1.

197. ③ Shortened attention span and fussy behavior may indicate a change in intracranial pressure and/or shunt malfunction. (2) (DF; EV)
    1. This is normal behavior for a 2½-year-old.
    2. Same as answer 1.
    4. Same as answer 1.

198. ④ Deaf infants commonly babble until they are 6 months old; if they cannot hear, their vocalizations are not reinforced and often stop at this time. (2) (DF; AS)
    1. This skill is learned by the infant at about 11 to 12 months of age.
    2. Infants with no hearing impairment do not always respond to a name all of the time.
    3. Inappropriate test at 7 months; the Moro reflex usually disappears when the infant is 3 to 4 months old.

199. ④ Cranial radiation destroys leukemic cells in the brain because chemotherapeutic agents are poorly absorbed through the blood-brain barrier. (3) (DA; AN)
    1. This is not the primary reason for the treatment; it is a curative measure.
    2. This is not the reason for cranial radiation.
    3. This is inaccurate; leukemia is an abnormality of the bone marrow and lymphatic system.

200. ④ Safety is the priority during the seizure. (2) (SI; IM)
    1. It would be unsafe to leave the child having the seizure.
    2. Attempting to open clenched jaws could result in injury to the child's teeth and jaw.
    3. This may cause airway occlusion by forcing the neck onto the chin; a small, flat blanket is more effective.

201. ③ The child's status and the progression of the seizure should be monitored; the child will not breathe until the seizure is over, and cyanosis should subside at that time. (1) (HI; IM)
    1. Attempting to open clenched jaws could result in injury to the child.
    2. Oxygen will be useless until the child breathes when the seizure is over.
    4. The physician can be notified later; provision of safety and observation are the priorities.

202. ② Raising the foot of the bed would increase blood flow to the brain and increase intracranial pressure. (2) (DA; PL)
    1. Temperature elevations are expected after a craniotomy because of stimulation of the hypothalamus.
    3. Fluids should be monitored closely to reduce the possibility of cerebral edema.
    4. Vital signs are a major component of the neurologic check.

203. ③ This is one of the most prevalent diseases in young children; approximately 70% have otitis media at least once, and one third of children younger than age 3 have had at least three episodes. (1) (SI; AN)
    1. This is not the causative agent.
    2. Same as answer 1.
    4. Same as answer 1.

204. ② The incision allows for drainage, which produces relief of pressure and results in immediate relief of pain. (2) (PC; IM)
    1. This incision does not leave any scar because healing by primary intention occurs within 24 hours.
    3. A myringotomy is performed to prevent the trauma of perforation.
    4. This incision is very small and heals spontaneously within 24 hours.

205. ④ Movement by the child will impede the procedure and may cause additional injuries to the surrounding structures. (2) (SI; IM)
    1. This is not essential to the accomplishment of the procedure.

2 The child should have had the local anesthetic applied before the procedure.

3 This will not guarantee that the child will keep still during the procedure.

206. (2) This position facilitates drainage by gravity. (1) (CI; IM)

1 This position will not allow for proper drainage.

3 This position will promote pooling of drainage in the operative site and may lead to reinfection.

4 This is rare and should not occur from this tiny incision.

207. (3) These may indicate the need for a repeat myringotomy because of ineffective drainage. (2) (CC; EV)

1 Bleeding is not seen in otitis media or after a myringotomy.

2 This is characteristic of otitis media and does not indicate an infection.

4 These are not expected complications of a myringotomy.

208. (1) This precaution reduces the transmission of infection from client to client (cross-infection). (1) (SI; AN)

2 The act of isolation has no effect on the infectious process.

3 This is protective (reverse) isolation; it is not used for infectious clients but rather to protect clients with lowered resistance, for example, clients receiving chemotherapy.

4 Thorough hand-washing and careful aseptic technique limit nosocomial infections.

209. (2) The higher brain centers are not invaded by the exotoxin. (3) (HI; IM)

1 Interference with intellectual functions occurs as a result of inadequate management of oxygen needs during episodes of respiratory difficulty, not the disease process.

3 Same as answer 1.

4 Interference with intellectual functioning results from inadequate management, not complications associated with the disease process.

210. (4) The meningococcal organism is rendered inactive after 24 to 72 hours of antibiotic therapy; therefore isolation is required at least for this time. (2) (CC; PL)

1 Treatment with antibiotic therapy for 24 to 72 hours will render the microorganism inactive; after that, isolation is usually not required.

2 Meningococcal meningitis is a serious contagious disease; isolation is required for at least 24 to 72 hours after beginning antibiotics.

3 The disease is not evident in the incubation period; because the disease is undiagnosed, isolation would not have been instituted.

211. (2) This gives the parents empathy along with an honest appraisal of the child's needs at this time. (3) (CI; IM)

1 Unsafe; any stimuli may precipitate muscular spasm; this statement could increase parents' anxiety.

3 This recognizes parental concern but does not answer the question.

4 Unsafe; unnecessary auditory, tactile, and visual stimuli are likely to set off muscular spasms.

212. (4) These are the symptoms that bring the child to the hospital; they reflect central nervous system involvement. (2) (HI; AS)

1 Vomiting rather than diarrhea occurs; the presence of a rash.

2 These are not early symptoms; these may occur with multiple organ failure.

3 The fever is usually high; petechiae may or may not be present.

213. (4) Clonus is the rapid rhythmic extension and relaxation of muscle groups during seizure. (2) (HI; AN)

1 This occurs during a tonic seizure.

2 This is not synonymous with clonus; however, a client can lose consciousness during a clonic seizure.

3 Clonus generally occurs throughout entire body; the movements during a clonic seizure are more marked than the movements of a tremor.

214. (4) Reye's syndrome affects the liver, causing problems with blood coagulation because the liver-dependent clotting factors such as prothrombin are diminished. (3) (DA; AS)

1 Liver and CNS, not bladder, functions are impaired.

2 This would not be critical; a rash may or may not be present.

3 Cerebral, not periorbital, edema would be significant at this time.

215. (1) Aspirin interferes with platelet ability to form clots; the high vascularity of the oropharynx requires intact clotting mechanisms to prevent bleeding. Also aspirin should not be given to

infants and small children because it is associated with Reye's syndrome. (2) (SI; IM)

2 This drug is not contraindicated for a client who has had oral surgery.

3 Same as answer 2.

4 Same as answer 2.

ENDOCRINE (EN)

216. ① Most juveniles are type 1 diabetics and have little or no endogenous insulin; diet control and exercise reduce the amount of exogenous insulin needed. (2) (PC; PL)

2 Those having type 1 diabetes have little or no endogenous insulin and need exogenous insulin; this regimen is for type 2 diabetics; blood glucose monitoring is usually used.

3 Those having type 1 diabetes need insulin for control because of a lack of endogenous insulin.

4 Oral hypoglycemics are ineffective in stimulating insulin secretion in clients having type 1 diabetes because they have no endogenous insulin.

217. ② Negotiation of goals precedes and is essential to successful learning; mutual goal-setting provides a focus for learning. (3) (CC; AN)

1 A rapport should already be developed before beginning the teaching-learning process.

3 If the client does not identify this need or set this as a goal, there will probably be little motivation to learn this task.

4 Same as answer 3.

218. ① Because paresthesias may be present and circulation may be compromised, adequate inspection of the feet is necessary and is the quickest and easiest measure to identify pressure sites and prevent infection. (1) (CI; IM)

2 Hot water should never be used because it can cause injury by burning the skin.

3 The feet should be patted dry, not rubbed vigorously; rubbing can cause abrasion and injure the skin.

4 Strong antiseptics are too harsh and should not be used because they can cause injury to the skin.

219. ① Urine testing is primarily helpful in detecting ketones, which are most likely to be present during illness and hyperglycemia. (3) (DA; AN)

2 Because of the complexity of the medical regimen and the variety of factors that influence serum glucose levels, such as food ingested, exercise, medications, and the stresses of growth and development, serum glucose levels in children can fluctuate and should therefore be measured by monitoring the serum glucose levels before meals and at bedtime.

3 Blood, not urine, is the best specimen to be tested for determining glucose levels.

4 The opposite is true.

220. ④ The glycosylated hemoglobin (GHb) test provides an accurate long-term index of the client's average blood glucose level for the 100- to 120-day period before the test; the more glucose the RBC was exposed to, the greater the GHb percentage; the test result is not affected by short-term variations. (3) (CI; AS)

1 The presence of ketones in the urine may indicate only short-term variations and extreme hyperglycemia.

2 Serum protein levels do not reflect the effectiveness of glucose management in diabetes mellitus.

3 Serum glucose levels reflect short-term (hours) variations.

221. ③ The stress of surgery causes the release of epinephrine and glucocorticoids, which raise blood glucose levels. (2) (DA; AN)

1 Urine test results are affected by many variables, such as renal threshold, so they are not accurate enough when control is precarious; serum glucose levels are preferred.

2 Hypoglycemia can result because the client has taken nothing by mouth and body fluids are being lost.

4 Most clients with diabetes who are diet-controlled require insulin for a short period after surgery, especially when receiving IV glucose; most children have type 1 diabetes.

222. ① Humulin N peaks in 8 to 12 hours; a bedtime snack will prevent hypoglycemia during the night. (1) (PT; PL)

2 This is unsafe because it would intensify the hyperglycemia; if hyperglycemia is present, the child needs insulin.

3 When hypoglycemia develops the child will be asleep; the snack should be eaten before going to bed.

4 Humulin N is an intermediate, not a long-acting, insulin for which bedtime snacks are recommended.

223. (4) Frequent episodes of hypoglycemia result from difficulty in balancing food, exercise, and insulin in active children. (1) (CC; AN)
1   The people associated with the school who are interacting with the child need to know about the child's condition but should be asked to keep the information confidential.
2   With increased activity, the child will experience episodes of hypoglycemia, not hyperglycemia.
3   This would be a form of discrimination; the child with diabetes mellitus should be allowed to engage in activities as long as the diabetes remains under control.

224. (2) A quick check of the blood glucose level will confirm whether the client is hypoglycemic. (2) (CI; AS)
1   This is inaccurate and does not reflect the present status.
3   Although this might be appropriate to counter hypoglycemia, it does not determine whether the client is hypoglycemic or is being manipulative.
4   Same as answer·3.

225. (3) The client's trouble stems from perceptual difficulties; the preset syringe removes the need to differentiate between 24 and 42 units. (2) (CC; PL)
1   This would not solve the transposition of the numbers; the problem is not caused by the inability to see the numbers but by the child's perception of them.
2   This would not solve the transposition of the numbers.
4   Same as answer 2.

226. (3) Glucagon promotes liver glycogenolysis, resulting in the release of glucose into the blood. (3) (CC; AN)
1   ACTH is not directly related to glycogenolysis; it is released from the anterior pituitary.
2   The level of insulin will decrease as the glucose level declines; it is not directly related to glycogenesis.
4   Epinephrine is not directly related to glycogenolysis; it is released from the adrenal medulla and sympathetic nerve endings.

227. (2) These are the most common signs of hypoglycemia in children. (2) (DA; AS)
1   These are signs of hyperglycemia.
3   Same as answer 1.
4   Same as answer 1.

228. (1) Proper hand-washing is the best infection-prevention technique and should always precede preparation of an injection. (1) (SI; PL)
2   Shaking causes air bubbles, which can interfere with preparing the dosage accurately; the bottle should be gently rotated.
3   The injection site should not be rubbed because this affects absorption of the insulin and also causes reactions at the site.
4   The abdomen, not the upper extremities, is the preferred site for self-administration of insulin; sites should be rotated.

229. (2) The child's confidence, readiness, and skill for giving self-injections is essential for long-term management of diabetes. (3) (PT; PL)
1   The sites must be rotated at all times.
3   Learning responsibility for injections should be a gradual process with continuous support and guidance.
4   The recommended procedure is to draw up the Humulin R first and then the Humulin N to prevent contamination of the multidose vial of Humulin R by the intermediate-acting Humulin N.

230. (4) The client is demonstrating signs and symptoms of hypoglycemia and needs glucose. The orange juice is a simple carbohydrate, and the bread is a complex carbohydrate. (1) (DA; IM)
1   This may be done after the child receives glucose.
2   Administering insulin will exacerbate the hypoglycemia and endanger the client.
3   Same as answer 1.

231. (2) Insulin causes potassium to move into the cells along with glucose, thus lowering the serum potassium level. (3) (DA; PL)
1   Insulin does not lead to reduced blood volume.
3   Insulin does not directly alter sodium levels.
4   Insulin does not affect calcium mobilization.

232. (4) A fast food exchange list allows participation in postgame activities without feeling different from peers, which is important to the adolescent. (1) (PC; IM)
1   The nutritional benefits of fast foods are not the issue.
2   The adolescent needs to learn how to select appropriate foods when away from the home environment. This limits social interaction with peers.

3 This would make the adolescent feel different from peers.

233. (2) This is an attempt by the respiratory system to eliminate the excess carbon dioxide; it is a characteristic compensatory mechanism of metabolic acidosis. (1) (DA; AS)
1 An increased temperature will only occur if an infection is present.
3 Tachycardia, not bradycardia, results from the hypovolemia of dehydration.
4 Hypotension, not hypertension, may result from decreased vascular volume.

234. (4) Normoglycemia decreases the chances of developing long-term complications such as neuropathy, retinopathy, and atherosclerosis. (2) (CI; AN)
1 Blood glucose levels are more accurate than urine glucose levels.
2 This is only one part of the regimen; this would be both a short- and a long-term goal.
3 Stress is difficult to control, particularly in the adolescent; a nonstressful life-style would be utopia.

RESPIRATORY (RE)

235. (1) Capillary beds are closest to the surface in a finger or earlobe; this proximity allows for more accurate measurement of the arterial oxygen saturation. (1) (SI; IM)
2 The pulse oximeter requires no routine calibration.
3 Capillary beds are closest to the surface in a finger, toe, or earlobe and not on the abdomen or upper thigh.
4 An almost instantaneous accurate readout can be obtained with the pulse oximeter.

236. (1) The medulla oblongata contains the respiratory center, and the neurons that supply the respiratory muscles originate here; they produce the rhythmic pattern of inspiration and expiration. (2) (DF; AN)
2 This is concerned with the control of skeletal muscles.
3 This links the nervous system to the endocrine system and functions as a relay station between the cerebral cortex and the lower autonomic centers.
4 This is unrelated to respirations; this is the thin surface layer of the cerebrum.

237. (1) The gag reflex is elicited by pressing on the posterior pharynx, resulting in glossopharyn-geal stimulation; inserting the tongue blade on the side of the mouth limits this stimulation. (1) (PC; IM)
2 Although this is important, it is not the reason for inserting the tongue blade on the side of the tongue.
3 Same as answer 2.
4 Same as answer 2.

238. (3) Orthopnea is shortness of breath in any position but the erect sitting or standing position. (2) (DA; AS)
1 This is a temporary cessation of breathing.
2 This is labored or difficult breathing regardless of the position.
4 This is an increased respiratory rate, not shortness of breath.

239. (3) Because of the drying nature of oxygen, most oxygen is humidified before it is administered. (2) (CI; PL)
1 Oxygen is combustible and supports fire; it does not ignite; it is not flammable.
2 Oxygen is not usually warmed before administration; it is cool on routine administration.
4 Oxygen is considered a drug and therefore must be prescribed.

240. (3) This reply addresses the client's concern; CNS depressant effects may diminish or spontaneously disappear after several days. (2) (PT; IM)
1 Administration of medication is a dependent function of the nurse, and the nurse has no legal authority to tell the client to alter the dose.
2 Administration of medication is a dependent function of the nurse, and the nurse has no legal authority to tell the client to omit a dose.
4 This is unnecessary because the side effect of drowsiness should diminish in several days; the client needs teaching about the drug.

241. (3) Antibiotics should be administered with doses equally spaced over 24 hours so that a constant blood level of the drug is maintained. (2) (PT; IM)
1 The 12 hours between the 8 pm and 8 am doses is too long; the blood level of the drug will drop and the therapy will not be effective.
2 Doses are not equally spaced over 24 hours, and the blood level of the drug will not remain constant.
4 Same as answer 2.

242. (4) An upright position helps prevent aspiration; metal tableware is safer than plastic because it is unbreakable; encouraging participation attempts to socialize and treat the client with dignity. (3) (SI; IM)
    1 Using a syringe is essentially a form of forced feedings; in addition, the client should be encouraged to eat solids.
    2 Feeding the client in the supine position puts the client at risk for aspiration and choking.
    3 A mentally challenged client can easily bite down on plastic tableware and break it.

243. (3) The infectious and mechanical changes narrow the bronchial passages and make it difficult for air to leave the lungs. (2) (DA; AS)
    1 As a result of increased respiratory effort and decreased oxygen exchange, tachycardia may develop.
    2 Breath sounds may be diminished because of the swelling of the bronchiolar mucosa and filling of the lumina with mucus and exudate.
    4 Intercostal retractions are unlikely because of the overinflation of the chest with air and the shallow, rapid breathing.

244. (1) Adequate hydration and high humidity are essential to loosen tenacious secretions and minimize fluid loss. (1) (DA; PL)
    2 Corticosteroids are not used because they have not proved effective.
    3 Bronchodilators are not used because the bronchial tree is not in spasm.
    4 Antibiotics are ineffectual because the etiologic agent is viral.

245. (3) The infant is having difficulty with breathing; disturbing the infant frequently causes an excessive expenditure of energy and increases oxygen demands. (2) (PC; PL)
    1 Constant observation, not constant physical disturbance, is important; the infant needs to rest to minimize oxygen demands.
    2 Cool mist helps to liquefy secretions and keeps the temperature down; cool mist does not maintain hydration.
    4 Too frequent auscultation will disturb the infant's rest, causing an excessive expenditure of energy and increased oxygen demands.

246. (2) These observations are vital to assess the child's hydration status. (2) (DA; AS)
    1 The child is too ill to be involved in stimulating activities; energy should be conserved and oxygen demands kept at a minimum.
    3 The child needs the parents to limit separation anxiety.

4 Contact isolation is recommended for bronchiolitis; airborne precautions are recommended for diseases such as measles, varicella, and tuberculosis.

247. (1) This is less than the normal range of 7.35 to 7.45; hypoxia causes hypercapnia, resulting in a fall in the pH. (3) (DA; AS)
    2 This is within the normal range of 80 to 100 mm Hg.
    3 This is within the normal range of 35 to 45 mm Hg.
    4 This is within the normal range of 21 to 28 mEq/L.

248. (2) Wheezing is indicative of bronchopulmonary constriction or obstruction. (3) (DA; AS)
    1 Clients with asthma often receive corticosteroids, which could cause a round face; however, this does not require nursing intervention.
    3 The use of inhalant medications is expected.
    4 This is the normal respiratory rate for an 8-year-old child.

249. (3) Positioning on the side permits flow of oral secretions that could block the child's airway; a patent airway takes precedence; if the client is not able to take air in, all other measures are futile. (1) (DA; AN)
    1 This becomes important after a patent airway is established; the client can receive tepid or cool liquids.
    2 After airway patency is established, deep breathing can be encouraged; coughing is contraindicated because it could dislodge a clot.
    4 Airway patency is of primary importance.

250. (2) Cold liquid promotes vasoconstriction, decreases bleeding, and limits pain. (1) (CI; IM)
    1 Ice cream has a milk component, which increases the viscosity of mucus.
    3 Milk increases the viscosity of mucus.
    4 Orange juice will irritate inflamed tissue.

251. (4) With few exceptions males are sterile; failure of normal development of the wolffian duct structures (vas deferens, epididymis, and seminal vesicles) and blockage of the vas deferens by abnormal secretions result in decreased or absent sperm production. (3) (DF; IM)
    1 This does not answer the client's question.
    2 Females with CF generally have normal ovaries and fallopian tubes and are fertile; however, fertility can be inhibited by highly viscous cervical secretions.

3   Theoretically, all offspring of couples who are homozygous for a recessive gene will have the disease; however, with cystic fibrosis, affected men are usually sterile.

252. (3) This test establishes genotype with the lowest margin of error. (3) (DF; EV)
1   The results of this test will not determine whether the individual is a carrier of CF; this may be part of the testing that is conducted when a client is suspected of having CF.
2   The results of these tests will not determine whether the individual is a carrier of CF; these may be part of the testing that is conducted when a client is suspected of having CF.
4   Same as answer 2.

253. (4) Maintenance of a patent airway is always the priority. (2) (DA; PL)
1   An order is needed to begin routine oxygen.
2   An IV will probably be started but this is not essential immediately; establishing a patent airway takes priority.
3   This delays attention to the immediate problem of respiratory distress.

254. (4) Dysfunction of the exocrine glands leads to an abnormal accumulation of thick mucus, a slower flow rate of mucus, and incomplete expectoration of mucus, all of which contribute to airway obstruction. (1) (DA; AN)
1   This is associated with pneumonia.
2   This is associated with asthma.
3   The endocrine glands are not affected in cystic fibrosis.

255. (4) Chest physiotherapy is done midway between feedings to lessen vomiting and increase drainage for suctioning. (3) (PC; IM)
1   This is inadvisable; the infant may vomit and nutritional intake will be impaired.
2   Doing chest physiotherapy at this time will tire the infant and possibly lead to an impaired nutritional intake.
3   Same as answer 1.

256. (1) Males with cystic fibrosis usually are sterile; therefore, the father does not have cystic fibrosis, but he could be a carrier; if both parents are heterozygous carriers, the chance of having a child with CF is 25%; when one parent is a heterozygous carrier and the other has two normal genes, the chance of having a child that has CF is 0% but the chance of having a child that is a carrier is 50%. (3) (DF; AS)

2   If both parents are heterozygous carriers or if one parent is a heterozygous carrier and one parent has two normal genes, the chance of having a child that is a carrier, not a child that is affected, is 50%.
3   This is an inaccurate conclusion when the father's genotype is unknown.
4   With the data provided, the mother is a heterozygous carrier and the father is either a heterozygous carrier or has two normal genes; under these circumstances, it is safe to conclude that the chance that their baby will have CF is 25% or less.

257. (2) Respiratory distress will be reduced when excess mucus is mobilized and removed from the respiratory system. (3) (DA; AN)
1   Congealed mucus would obstruct air passageways and increase respiratory distress.
3   Cool moist air will not stimulate this reflex.
4   Cool moist air constricts the blood vessels in the bronchi; if dilation of the blood vessels occurred, airway diameters would be narrowed and respiratory distress increased.

258. (2) The chest is barrel shaped because of the protrusion of abdominal viscera through the defect into the thoracic cavity. (3) (DA; AS)
1   This is not associated with a diaphragmatic hernia; assessments related to the bowel include colicky pain and constipation.
3   There are no breath sounds over the abdomen, and there are diminished or absent breath sounds on the affected side of the thorax; bowel sounds may be auscultated over the affected thorax.
4   The abdomen is markedly scaphoid (sunken).

259. (2) The presence of abdominal viscera in the thoracic cavity impinges on the lung, limiting the amount of air that can enter. (3) (DA; AN)
1   A medical diagnosis cannot be used as part of a nursing diagnosis.
3   There are no excessive secretions with a diaphragmatic hernia.
4   The basal metabolic rate is not increased with a diaphragmatic hernia.

260. (4) Placing the infant on the operative side promotes gas exchange in the unimpaired lung. (3) (CI; IM)
1   This would not maximally promote aeration of the unaffected lung.
2   Same as answer 1.

3 This would not maximally promote aeration of the unaffected lung; the prone position increases the effort of breathing because respiratory excursion is impeded by the weight of the body.

261. ③ Bilateral lung sounds indicate that the hypoplastic lung has begun functioning. (3) (CI; EV)
    1 This is not a reliable indicator that the respiratory status is improving; it could actually indicate that the infant is hypoxic and too fatigued to cry.
    2 A normal pH is 7.35 to 7.45; a decreasing pH indicates respiratory acidosis, which can be attributed to decreased gas exchange.
    4 Retention of formula is unrelated to gas exchange.

262. ③ Maintenance of a patent airway is always the priority. (2) (DA; IM)
    1 An IV can be started later; suctioning to ensure airway patency is the priority.
    2 The primary focus now is to establish breathing; the child has a pulse of 50, which can be addressed later.
    4 The ICU can be called once the child's vital signs are stabilized and a patent airway is ensured.

263. ① The degree of the hypoxia and asphyxia the child had will determine the extent of the neurologic, liver, and renal damage. (3) (DA; AN)
    2 The child is hypothermic, not hyperthermic.
    3 Although emotional trauma can be all-encompassing, it usually does not influence the ultimate physical prognosis as does hypoxia.
    4 Although initially severe, aspiration pneumonia does not result in long-term sequelae as does hypoxia.

264. ③ The chest was opened during surgery for the sternal repair, and air was allowed into the thorax; the air must be removed for the lungs to expand properly. (2) (CI; IM)
    1 The chest tube is unrelated to the baby's ability to retain feedings.
    2 Chest tubes are uncomfortable; also, this response discounts the importance of the chest tube to the baby's respiratory status.
    4 The baby did not have a punctured lung.

CARDIOVASCULAR (CV)

265. ④ A discrepancy in blood pressures from the arms to the legs indicates vascular stenosis. (2) (DA; AS)

1 Projectile vomiting results from a gastrointestinal problem and is usually not manifested immediately after birth.
2 An irregular respiratory rhythm is common and expected in the newborn.
3 Hyperreflexia of the extremities may be indicative of a neurologic problem.

266. ④ During a bowel movement the Valsalva maneuver can occasionally initiate a hypercyanotic spell (tet spell, blue spell); the Valsalva maneuver causes increased intrathoracic pressure, slowing of the pulse, decreased return of blood to the heart, and increased venous pressure. (3) (DA; AN)
    1 This would not influence cardiovascular functioning.
    2 Same as answer 1.
    3 Same as answer 1.

267. ② A positive antistreptolysin titer is present with rheumatic fever because of previous infection with streptococci. (3) (CI; AS)
    1 A positive, not a negative, C-reactive protein would be present; this is indicative of an inflammatory process.
    3 This is usually related to a decrease in mature RBCs caused by hemorrhage or other blood diseases; it is unrelated to an infectious or inflammatory process.
    4 The ESR would be elevated, not decreased, indicating the presence of an inflammatory process.

268. ② A cardiac catheterization will identify the exact location of the ventricular septal defect as well as assess pulmonary pressures. (1) (CI; AN)
    1 This is demonstrated by electrocardiographic and echocardiographic examinations.
    3 Murmurs can be heard with a stethoscope placed at the left lower sternal border.
    4 Same as answer 1.

269. ① Children are kept on complete bed rest after a cardiac catheterization to reduce the risk of bleeding or trauma to the insertion site; the order for bathroom privileges should be questioned. (3) (CC; EV)
    2 This is an expected part of postcatheterization care; a pressure dressing is placed over the insertion site to reduce the possibility of bleeding.
    3 This urinary output is within acceptable limits; the child was kept npo before and during the procedure; after the procedure oral fluids need to be encouraged to promote hydration and voiding.

4 Frequent blood pressure checks are part of routine postcatheterization care.

270. ① The catheter insertion site should not be submerged in water; sponge baths limit trauma and infection at the insertion site. (1) (SI; IM)
   2 This is contraindicated; ice will cause vasoconstriction and could compromise circulation.
   3 Fluids should be encouraged to enhance excretion of the contrast media used during the procedure.
   4 The child is not sent home with a pressure dressing.

271. ① The body responds to the chronic hypoxia caused by the heart defect by increasing the production of red blood cells in an attempt to increase the oxygen-carrying capacity of the blood. (3) (CI; AS)
   2 This does not result from hypoxia; it occurs in disease processes in which the WBCs drop to very low levels and neutropenia becomes pronounced.
   3 Thrombocytopenia (low platelet production) does not result from hypoxia; it occurs in disease processes in which platelet production is suppressed, as in leukemia.
   4 The hematocrit level would be elevated because the body increases the production of red blood cells in an attempt to make more cells available to carry oxygen.

272. ② Small, frequent feedings with adequate rest periods in between may improve the child's intake at each feeding; these children become extremely fatigued while sucking. (2) (PC; PL)
   1 To reduce cardiac workload, fluids are usually not forced and may sometimes be restricted in clients with congestive heart failure.
   3 The head circumference is not a parameter in congenital heart disease; head measurement would be done when infants have hydrocephaly.
   4 Positioning with the head elevated facilitates ease in respiration; placing the infant flat on the stomach restricts lung expansion.

273. ② This defect causes blood to bypass the lungs; surgery increases blood flow to the lungs. (3) (CI; AN)
   1 This would not improve the oxygen content of the blood.
   3 Same as answer 1.
   4 Same as answer 1.

274. ③ A calibrated dropper is the most accurate way to measure the medication. (1) (PT; IM)
   1 This is not an accurate way to measure medication.
   2 Same as answer 1.
   4 If the dose of medication is diluted in 1 oz of water and the infant does not drink the entire ounce, the resulting dose will be insufficient.

275. ④ Because the mechanism of CHF is the same in all children, the same basic treatment of a cardiac glycoside and a loop diuretic (Lasix) is used, although the dosage may vary. (1) (PT; AN)
   1 CHF in infants is not in itself a congenital defect but results from a congenital defect of the heart.
   2 This is untrue; the treatment of CHF is basically the same whether the client is an infant or a senior citizen.
   3 Children can develop CHF just as adults can; if there is cardiac decompensation, the treatment is the well-established combination of Lanoxin and Lasix.

276. ③ The increased blood volume and pressure in the lungs resulting from left ventricular failure cause pulmonary edema; dyspnea, an early sign of failure, is probably caused by the decreased distensibility of the lungs. (3) (CI; AN)
   1 Anemia is fairly well tolerated in infants and does not cause dyspnea.
   2 Dyspnea, not hypovolemia, is an early sign of pulmonary edema; hypovolemia would cause signs of shock.
   4 Respiratory, not metabolic, acidosis could develop; this occurs because of the pulmonary insufficiency resulting in retention of carbon dioxide.

277. ② Fluid retention is reflected by an excessive weight gain in a short period of time in a child with CHF; inadequate cardiac output decreases blood flow to the kidneys, thus leading to increased intracellular fluid and hypervolemia. (1) (DA; IM)
   1 This would be an appropriate answer if a renal condition or hypovolemia were present; however, other assessments such as hourly output and BUN values would then also be indicated.
   3 Weight gain resulting from nutritional intake is gradual and would not vary greatly on a day-to-day basis.
   4 Weight is helpful in determining drug dosages but the drug dosage would not need to be recalculated daily according to weight changes.

278. (4) Compromised heart function and inadequate oxygen reserves in the infant often result in feeding problems such as cyanosis and fatigue while sucking and swallowing. (2) (CI; AN)
1   Poor sucking is always significant.
2   This may be true in the first days of life but is not true as the infant grows, unless there is a major health problem.
3   Same as answer 2.

279. (1) Hypoxia leads to poor peripheral circulation; clubbing occurs as a result of additional capillary development and tissue hypertrophy of the fingertips. (2) (CI; AS)
2   Respirations will be increased.
3   The child's problems are related to decreased oxygenation, not to a clotting deficit.
4   The body attempts to compensate for the hypoxemia by increased erythropoiesis.

280. (2) This position has the same effect as squatting, which decreases venous return from the legs; the blood returning to the heart and lungs has a higher O_2 content. (1) (DA; IM)
1   Although this would reduce pressure of abdominal organs on the diaphragm, it does not put enough pressure on the femoral veins and vena cava to sufficiently reduce venous return to the heart.
3   This position does not reduce venous return to the heart.
4   Same as answer 1.

281. (2) Postprocedure hemorrhage is a major life-threatening complication following cardiac catheterization because arterial blood is under pressure and the catheter has entered an artery. (2) (CI; PL)
1   The child has an oxygen deficit; rest should be encouraged; flexion of the insertion site should be avoided to prevent disturbance of the clot.
3   The blood pressure should not be unstable unless a problem developed; fluids should be administered as ordered.
4   This is unnecessary; the distal pulses would be monitored.

282. (1) The vital signs must be taken first to determine the child's postoperative status and to compare them with the previously recorded vital signs. (3) (CI; AS)
2   Although this is important, obtaining the vital signs is the priority.
3   Same as answer 2.
4   Same as answer 2.

283. (1) The nurse needs to know how much information the parent has before responding. Pulmonic stenosis is a narrowing at the entrance to the pulmonary artery and may vary in severity; treatment may vary from balloon angioplasty to valvotomy. (3) (HI; IM)
2   This response is too blunt. The nurse needs to know how much the parent knows and understands first.
3   The parent is concerned or the question would not have been asked.
4   This response abdicates the nurse's role.

REPRODUCTIVE AND GENITOURINARY (RG)

284. (4) Surgery before school age reduces concerns about body image in relation to peers. (2) (DF; AN)
1   Malignancy may develop with or without surgical correction.
2   Maturation of testes starts about age 5; surgery should be done before maturation to prevent sterility.
3   The puboscrotal ring has nothing to do with the outcome of this surgical procedure.

285. (1) In cryptorchidism, sperm-producing abilities of the testes are destroyed, resulting in sterility. (1) (DF; AN)
2   This is enlargement of the scrotum with fluid; it is not related to cryptorchidism.
3   This is an abnormal dilation and tortuosity of the scrotal veins; it is not caused by undescended testicles.
4   Inflammation of the epididymis may occur whether or not cryptorchidism is corrected.

286. (1) This is the critical period of organogenesis, during which fetal development is most likely to be adversely affected. (1) (DF; IM)
2   The fetus is less vulnerable during this period because development is almost complete.
3   The fetus is less vulnerable to major anomalies during this period because all major organ systems are already formed.
4   At the time of implantation, cellular differentiation has not occurred; the genital bud appears in the seventh week.

287. (3) Circumcisions are never done because the foreskin may be needed for repair and reconstruction of the penis. (2) (DA; PL)
1   The penis does not need to be wrapped in petrolatum gauze because no surgery has been done.

2 A cystotomy tube is not inserted because there is no interference with voiding.

4 Hypospadias is not a genetic disorder, although there appears to be some evidence that it may be familial.

288. ③ Arm and leg restraints are necessary to maintain the position of the urethral stent to ensure optimum healing of the newly formed urethra. (3) (SI; PL)

1 Although this is important, the site must be assessed frequently and safety is the priority.

2 There is no tension device.

4 The indwelling catheter is never clamped because backup pressure may disturb the suture line.

289. ② The constant seepage of urine from the exposed ureteral orifices makes the area very susceptible to infection; infection must be prevented or controlled because an infection could ultimately lead to renal failure. (1) (SI; AN)

1 This will not occur because of the constant seepage of urine.

3 Although this could be a problem when the child reaches puberty or later, it is not the primary nursing diagnosis at this time.

4 Although this could occur if the infant is not well hydrated, risk for infection is the priority nursing diagnosis for the infant at this time.

290. ④ These help prevent infection and ulceration of the surrounding skin, as well as prevent the diaper from adhering to the mucosa. (3) (SI; PL)

1 Seepage of urine would prevent ointments from remaining on the exposed mucosa for more than a few moments; also, ointments may irritate the mucosa and result in bleeding.

2 Pediatric urine collectors would not adhere because of the moist environment; also, the adhesive backing would be irritating to the skin.

3 These are contraindicated because they would increase the moisture and temperature in the area, which would enhance the growth of microorganisms and the potential for infection.

291. ③ Kinking or twisting of the tubing, which can result in an obstruction of urine flow, can be easily resolved by the parents. (1) (CI; IM)

1 This is not the first action; the parents should not call the physician until they have attempted to resolve the problem.

2 Although it is important to keep the child adequately hydrated, the patency of the tubing should be assessed first.

4 Although this eventually may be done to assess for distention, the first action should be to determine patency of the tubing.

292. ③ In view of the smoky urine and the other symptoms, the nurse may suspect glomerulonephritis, which usually occurs after a recent streptococcal infection. (2) (HI; AS)

1 This kind of pain is found in rheumatic fever, which never results in smoky-colored urine.

2 Weight loss generally occurs with children who have developed diabetes mellitus, not glomerulonephritis.

4 This rash would be related to scarlet fever, with which there is no smoky-colored urine.

293. ③ These tests would confirm glomerulonephritis mainly because they identify the causative organism and its pathologic effects; also, the ASO titer confirms a past presence of streptococci. (2) (CI; AN)

1 A urinalysis would be done, but a chest x-ray film, a blood glucose level test, and an IVP would not confirm a diagnosis of glomerulonephritis.

2 A urinalysis would be done, but a chest x-ray film, an electrocardiogram, and a heterophile antibody test are not specific for the symptoms described.

4 A blood chemistry and nasopharyngeal culture may be done; an upper GI series and a 24-hour urinalysis are not specific for the symptoms described.

294. ③ Acute hypertension, which may occur in these children, must be anticipated and identified early to prevent any unpredictable complications. (2) (DA; PL)

1 This should be noted, but identifying hypertension is the priority.

2 This does not occur with glomerulonephritis.

4 Same as answer 2.

295. ④ The beta-hemolytic streptococcal immune complex becomes trapped in the glomerular capillary loop, causing glomerulonephritis. (2) (HI; AN)

1 The cause is known; prevention depends on treating infected individuals with antibiotics to eliminate the organism.

2 This is not an inherited but an acquired disease; incidence in males outnumbers that in females by 2:1.

3 The precipitating streptococcal infection is usually a localized pharyngitis, and clots do not form in the small renal tubules.

296. (3) Because of glomerular dysfunction, there is decreased filtration of plasma, leading to excessive water accumulation and sodium retention; this leads to congestion and edema. (3) (CI; AS)
1 This does not occur with glomerulonephritis.
2 There is usually a decrease in urine volume.
4 Same as answer 1.

297. (1) All these foods are permitted on low-sodium, low-potassium diets. (1) (PC; PL)
2 All are fairly high in sodium and/or potassium.
3 Carrots are high in sodium, a banana is high in potassium, and buttermilk is high in both sodium and potassium.
4 All but the peach are high in sodium; all but the butter are fairly high in potassium.

298. (4) The glomerular filtration rate is reduced; this results in sodium retention, protein loss, and fluid accumulation, producing these symptoms. (2) (CI; AS)
1 The blood pressure would be elevated; proteinuria would be greater than 1+; glycosuria is unrelated; anorexia would be present.
2 The blood pressure would be elevated; proteinuria would be greater than 1+; anorexia and hematuria would be present.
3 The blood pressure and specific gravity are elevated; proteinuria would be greater than 1+.

299. (4) Bed rest conserves energy; a diet with no added salt is permitted; the other care monitors the response to therapy. (2) (DA; PL)
1 Isolation is unnecessary because the disease is not communicable; fluids would not be encouraged but limited or permitted as desired.
2 The blood pressure is not monitored every hour, and IV therapy is not used unless there is no oral intake or convulsions occur.
3 Vital signs do not need to be monitored every 2 hours, a no-sodium diet is not possible, and bed rest is maintained to conserve energy.

300. (4) Sodium is usually limited to control or prevent edema and/or hypertension until the child is asymptomatic. (2) (CC; PL)
1 The child should not be kept active for long periods because rest is needed; the child will not usually need a long convalescence.

2 The nurse would not give a home phone number; the mother should contact the physician for follow-up care.
3 Glomerulonephritis does not cause such severe kidney damage that a kidney transplant would be necessary.

301. (1) Protein (albumin) is present in the urine in nephrotic syndrome and is evidence of kidney injury or disease; if the urine is gently shaken, it will foam if protein is present. (2) (CI; AS)
2 Crystals are not found in the urine of clients with nephrotic syndrome.
3 A large amount of protein in the urine would result in a high specific gravity.
4 Only rarely do RBCs or RBC casts get through the glomerular basement membrane in nephrotic syndrome.

302. (3) This is characteristic of a child in nephrotic syndrome; large amounts of protein in the urine cause it to have a dark, frothy appearance. (2) (CI; AS)
1 The child may be somewhat lethargic but usually not severely.
2 Blood pressure is normal or decreased; hypertension is associated with glomerulonephritis.
4 These children are usually pale.

303. (4) In children with nephrotic syndrome, infection is always a threat because of lowered resistance; the child with a fractured femur is noninfectious and therefore is appropriate as a roommate; in addition, the closeness of age will provide for preschool socialization. (3) (CC; IM)
1 This disorder is caused by a pathogen; it exposes the client to infection.
2 Same as answer 1.
3 Same as answer 1.

304. (1) The massive edema predisposes to skin breakdown. (3) (PC; PL)
2 Carbohydrates are not restricted; proteins may be limited.
3 This is far too much fluid; the damaged kidneys would not be able to handle this amount.
4 These children usually do not receive blood transfusions.

305. (4) With nephrotic syndrome, the child's blood pressure will be normal or slightly decreased; with glomerulonephritis, the child's blood pressure will be elevated. (3) (CI; AS)
1 This occurs in both nephrotic syndrome and glomerulonephritis.

2 Same as answer 1.

3 Same as answer 1.

306. ① This is incorrect; weight gain must be monitored carefully because it could be indicative of an accumulation of fluid. (1) (CC; EV)

1 This is a correct statement; the child should be monitored for edema.

3 This is a correct statement; this is to determine whether kidney functioning is impaired.

4 This is a correct statement; steroids are given with milk or food to prevent gastric irritation.

307. ③ High potassium levels can cause cardiac dysrhythmias; the normal range for serum potassium in the child is 3.4 to 4.7 mEq/L. (1) (CI; AS)

1 Hyponatremia is expected with acute renal failure.

2 An increase is expected with acute renal failure.

4 Same as answer 2.

308. ③ This is contraindicated because the pressure of the inflated cuff could disrupt the integrity of the arteriovenous fistula. (1) (SI; EV)

1 This would be desirable because vomiting or diarrhea could lead to dehydration and an acid-base imbalance.

2 Not only should this be done to assess vascular functioning distal to the arteriovenous fistula, but this assessment should be done bilaterally and the results compared.

4 This would be desirable because an inadequate fluid intake could result in dehydration and an acid-base imbalance.

309. ③ This will result when kidneys have failed and are no longer putting out urine, the blood pressure is high, and cardiac overload is imminent. (2) (DA; AN)

1 Hypertension is present when there is kidney failure.

2 Hyperkalemia occurs in kidney failure and is relieved by dialysis.

4 Hypocalcemia is present when kidney failure occurs.

310. ③ A return of a cloudy dialysate solution is indicative of infection. (2) (SI; EV)

1 There is no danger of an abdominal bruit developing during dialysis.

2 Dialysis does not affect blood glucose level.

4 Petechiae do not occur during dialysis treatments.

311. ③ This provides quiet activity that will not jeopardize placement of the peritoneal catheter. Also, it is appropriate for the child's cognitive level and allows for social interaction with a peer. (2) (SI; PL)

1 Chess requires cognitive abilities beyond the scope of a 7-year-old.

2 Although this is a good quiet activity, it would probably be too difficult for a 7-year-old to do without help from an adult.

4 This activity could result in displacement of the peritoneal catheter.

312. ② Because the child has only one kidney, efforts to prevent urinary tract infections, which can compromise kidney function, must be ongoing. (3) (SI; AN)

1 Fluids are not restricted; adequate fluid is encouraged to prevent urinary tract infections.

3 After surgery sodium is usually not restricted.

4 With a unilateral tumor, a kidney transplant is not necessary because the child still has one kidney.

313. ④ The presence of a chordee can affect the child's future reproductive capabilities related to the inability to inseminate directly. (3) (DF; AN)

1 Kidney function is not affected.

2 The risk of testicular torsion is not increased.

3 The incidence of testicular cancer is not increased.

314. ④ This response indicates that the teenager understands that she can have sexual intercourse even though she has cystic fibrosis; also, other than abstinence, condoms offer the best protection from sexually transmitted diseases. (3) (DF; EV)

1 Although fertility can be inhibited by highly viscous cervical secretions, which act as a plug blocking sperm entry, pregnancy is possible; in vitro fertilization also makes pregnancy possible.

2 A woman with cystic fibrosis can have sex.

3 A diaphragm provides protection against pregnancy, but it does not protect the individual from sexually transmitted diseases.

315. ③ The *Gonococcus* organism is present in the genitourinary tract of males and is easy to identify with a culture. (1) (HI; IM)

1 This does not confirm the diagnosis; it may identify sexual activity and partners.

2 The *Gonococcus* organism is in the genitourinary tract, not the blood; VDRL is a test for syphilis, not for gonorrhea.

4 Although urine may contain *Gonococcus* organisms, the urine would dilute the concentration; the organisms are more concentrated in the urethral discharge.

316. (2) A chancre is the earliest symptom of syphilis; a dark-field examination of the scraping will reveal the *Treponema* organism. (3) (HI; AS)

1 A rash is found in the secondary stage of syphilis; if a rash is found, it is too late for early diagnosis.

3 These are late manifestations of syphilis.

4 This is associated with gonorrhea.

GASTROINTESTINAL (GI)

317. (1) Because of the anomalous structure of the upper lip, the neonate may have difficulty sucking on a nipple; duckbill nipples and other modified devices may have to be used. (1) (PC; PL)

2 This is not an immediate concern; it is necessary after surgery to prevent tension on the suture line.

3 The infant should be cuddled like any newborn.

4 The cleft lip does not predispose the neonate to infection; difficulty in feeding is the main problem.

318. (2) Because of the cleft (opening) in the lip, the infant tends to suck in more air than usual; burping will prevent frequent regurgitation of formula. (1) (PC; PL)

1 Thickened formula is given to an infant with reflux problems, such as vomiting after each feeding.

3 The infant's bottle should never be propped; the infant can aspirate.

4 The baby should be held while being fed.

319. (4) This surgical repair is done after the teeth appear to prevent damage to the tooth buds and before the child talks, so that the child can speak with the proper anatomic structures in place in the mouth. (2) (CC; IM)

1 While both cleft lip and palate may occur with other birth defects, it is not always so; most birth defects are diagnosed early, not at 1½ years of age.

2 Focusing on the disfigurement may raise anxiety and precipitate feelings of guilt, which can delay acceptance.

3 This response may raise anxiety; the statement implies that the surgical decision is based on the whim of the surgeon.

320. (3) Feeding the infant this way does not put any stress or pressure on the suture line. (1) (CI; IM)

1 A plastic spoon may be too hard against the lip; also, it is inflexible and it may be difficult to get formula into the side of the infant's mouth.

2 This is unnecessary and nonnutritious; the infant can be fed formula with a rubber-tipped syringe as soon as it can be tolerated.

4 A nasogastric tube may be uncomfortable and cause the infant to cry; the infant can be fed with a rubber-tipped syringe.

321. (4) Meticulous care of the suture line is necessary because inflammation and sloughing of tissue disrupt healing. (1) (SI; IM)

1 This would be done throughout the feeding; the priority care after a feeding is cleansing the suture line.

2 This is contraindicated; the infant may rub the face on the sheet and irritate the suture line.

3 The infant can be cuddled at all times; priority care after feeding is cleansing the suture line.

322. (1) This is important; crying would put tension on the suture line. (2) (SI; PL)

2 The infant should be out of restraints periodically.

3 This is unnecessary; the infant should have no respiratory difficulty.

4 The infant needs to be cuddled frequently; parents are encouraged to pick the baby up as much as possible.

323. (3) The suture lines in the mouth must be protected from accidental harm; because an infant explores the world through the mouth, elbow restraints must be used to prevent the placement of fingers or objects in the mouth. (2) (SI; PL)

1 The child should have time to play with toys, but with supervision to prevent mouthing activities that would disrupt the suture line.

2 This is an acceptable position; the prone position is contraindicated because friction from the bed linen could harm the suture line.

4 A nasogastric tube is not used.

324. (1) Feeding with a small cup is best because liquids can be given slowly without stress on the suture lines; also, using a cup is age-appropriate for a toddler. (2) (SI; IM)

2 Feeding with a syringe increases the chance of aspiration.

3 Sucking on a nipple may cause pressure on the suture line; also, a cup is more age-appropriate.

4 Feeding with a spoon increases the risk of damage to the suture line.

325. ② Hypertrophy of the circular muscle of the pylorus forms this palpable mass. (1) (HI; AS)

1 There are visible peristaltic waves that move from left to right across the epigastric area, not the lower abdomen.

3 Discomfort is felt because of hunger, but there is no pain or tenderness.

4 Abdominal distention is not seen; food never reaches the intestines because of the gastric obstruction.

326. ④ When a baby has scanty dark urine, poor skin turgor, and increased depth of respirations, it is likely that dehydration and metabolic alkalosis are present; these occur because of the fluid and hydrochloric acid loss and the potassium depletion; immediate intervention is necessary. (2) (DA; AS)

1 This would be difficult to determine accurately; it is far more important to assess the infant for signs of dehydration and metabolic alkalosis.

2 These are not indicators of immediate needs; assessment for signs of dehydration and metabolic alkalosis are more important.

3 The olive-shaped mass is in the epigastric area, not the lower abdomen.

327. ① This corrects the metabolic alkalosis; fluid and electrolytes should be in balance when the child is undergoing the stress of anesthesia and surgery. (2) (DA; IM)

2 This does not include the necessary fluid replacement.

3 Conservative treatment rather than surgery would be used if vomiting were not severe and projectile.

4 This may help restore protein, but it will not balance fluids and electrolytes.

328. ② In the absence of severe dehydration and malnutrition the mortality rate is very low; immediate fluid and electrolyte replacement followed by surgery usually results in full recovery of the infant with pyloric stenosis. (1) (SC; IM)

1 The success rate with surgery is extremely high and produces a rapid recovery; surgery is usually necessary.

3 Pyloric stenosis is a structural defect; hypertrophy of the circular muscle of the pylorus causes obstruction at the pyloric sphincter; it is not caused by an inborn error of metabolism.

4 The infant will be feeding normally within a week following surgery.

329. ① Initial feedings postoperatively consist of clear fluids until tolerance for feeding is determined. (2) (CI; PL)

2 An increase in feeding osmolarity is attempted after the tolerance for clear liquids is assessed.

3 Thickened formula with cereal is used when an infant experiences gastroesophageal reflux.

4 Fluid and electrolyte status is stabilized preoperatively; if necessary MES would be given IV.

330. ② When there are no complications, the infant is usually feeding normally within a few days of the surgery. (2) (CC; IM)

1 Within a few days of the surgery, the infant will be feeding normally and needs no special dietary modification.

3 Same as answer 1.

4 Holding the infant should be encouraged because it is an important part of the parent-child relationship at this age.

331. ④ This is the most currently accepted form of therapy; the thickened formula and gravity limit reflux. (2) (PC; PL)

1 Surgery would be done only if complications, such as respiratory distress, esophagitis, or esophageal stricture, occur.

2 Breast milk can be placed in a bottle and cereal can be added to thicken it; this method of positioning has been replaced by a prone position with the head of the mattress elevated.

3 A 6-week-old baby cannot take food from a spoon and swallow it.

332. ③ Reflux of gastric contents to the pharynx predisposes the infant to aspiration and the development of respiratory problems. (2) (CI; AS)

1 This does not occur because gastric contents are forcefully vomited.

2 This would occur with excessive vomiting that produces dehydration, not just reflux.

4 There is no obstruction in the lower bowel; the problem is a relaxed or incompetent lower esophageal sphincter.

333. ① This reduces activities, allows the intestines to rest, and provides for early follow-up. (1) (SI; IM)

2   This provides neither for rest nor for more immediate contact with a health professional.

3   Wrapping elevates temperature; sugar water does not include electrolytes and may cause further gastric irritation; there is no emergency.

4   Food may increase the diarrhea; the immediate intervention does not support the mother's attempt to manage the situation at home; there is no emergency.

334. ① Withholding food decreases intestinal activity, thus minimizing the diarrhea and loss of fluid. (1) (CI; AN)

2   An electrolyte imbalance is corrected by IV therapy.

3   Stool cultures are used to determine the cause of the diarrhea.

4   Irritation of the perineum is prevented by meticulous skin care, not by withholding food.

335. ③ The organisms causing gastroenteritis are eliminated along with the feces; gloves provide a protective barrier and are used for medical asepsis. (2) (SI; EV)

1   This is required with airborne, not standard, precautions.

2   This is not necessary with standard precautions; this is necessary for an accurate measure of intake and output.

4   Same as answer 1.

336. ① This contains milk, which irritates the gastrointestinal tract; milk products are usually withheld for at least 1 week. (2) (PC; PL)

2   These are an appropriate soft food; they also replace the sodium lost in diarrhea.

3   These are an appropriate bland food; they are not irritating to the GI tract.

4   This is an appropriate soft food; it also replaces the potassium lost in diarrhea.

337. ③ This position limits the potential for aspiration; the infant will be partially upright, and fluid is held in the stomach by gravity. (2) (SI; IM)

1   This position allows gastric reflux and may lead to aspiration.

2   Same as answer 1.

4   Same as answer 1.

338. ④ This is the best source of the complete protein that is needed to maintain nitrogen balance. (2) (PC; PL)

1   This is a source of protein, but not the best source.

2   Same as answer 1.

3   This is high in vitamin C, which is needed for healing but not for nitrogen balance.

339. ④ Whole grains, legumes, and meat are excellent sources of thiamine, an essential coenzyme factor in energy metabolism, with an RDA standard of 0.5 mg/1000 kcal intake. (3) (PC; IM)

1   Eggs are only a fair source of thiamine.

2   Fruits do not contain thiamine.

3   Vegetables are only a fair source of thiamine.

340. ② Celiac disease is a primary defect in which the intestinal mucosal transport system is impaired; the inability to digest gliadin results in an accumulation of glutamine, which is toxic to mucosal cells and causes atrophy of the villi. (3) (CI; AS)

1   This does not occur in celiac disease.

3   Same as answer 1.

4   The pancreatic acini degenerate in cystic fibrosis.

341. ③ These are classic signs of celiac disease caused by pathologic mucous cells and atrophy of intestinal villi, which result in poor absorption of nutrients and increased stools. (2) (CI; AS)

1   Rickets is not seen because bone growth is arrested in celiac disease; the remaining symptoms do occur.

2   Constipation leading to fecal impaction is not expected, although this may occur because of decreased peristalsis; rickets is not seen.

4   There is an inability of the villi to absorb vitamin K, with a tendency toward inadequate or prolonged blood coagulation.

342. ④ These grains, mainly wheat, are major dietary sources of gluten, the gliadin fraction of which causes celiac syndrome. (1) (PC; IM)

1   There is no gluten in milk and dairy products.

2   Corn and rice are good substitute grain foods because they contain no gluten.

3   There is no gluten in saturated or unsaturated fats.

343. ④ Primary sources of gluten are wheat, rye, barley, and oats; chocolate and whole wheat bread contain wheat flour or wheat by-products; these foods must be avoided. (2) (PC; EV)

1   Meats, in small to average portions, are allowed on gluten-free diets because their protein is digestible; these foods need not be avoided.

2 Spinach and cheese are gluten-free; corn is a grain that can be tolerated; these foods need not be avoided.

3 Eggs and milk are gluten-free; rice is a grain that can be tolerated; these foods need not be avoided.

344. (1) After the removal of gluten, most children demonstrate a favorable personality change within 48 hours. (2) (CI; EV)

2 Weight is not affected for several days or weeks.

3 Abnormal bowel movements usually persist for several days or weeks.

4 Steatorrhea usually persists for several days or weeks.

345. (1) Because mucosal lesions limit nutrient absorption, there are inadequate nutrients for hemoglobin synthesis (iron and folic acid) causing anemia. (1) (HI; AN)

2 This causes pernicious anemia.

3 This child's anemia is caused by poor absorption rather than the quantity consumed.

4 Same as answer 3.

346. (1) Weight gain, improved appetite, improved behavior, and the disappearance of steatorrhea should have occurred. (1) (CI; EV)

2 This is not a stress-related disease; it is caused by a basic defect in metabolism or an immunologic response.

3 Even when the child understands the disease process, adherence to the diet may be relaxed; in time, symptoms may recur.

4 It is important to assess this, but it does not guarantee the child will select the foods on the diet.

347. (4) Evacuation time should be unhurried and without time limits; evenings are often less hurried. (2) (PC; EV)

1 High-fiber foods aid treatment of constipation; no additional teaching is needed.

2 Increasing fluid intake may help to relieve or allay constipation; no additional teaching is needed.

3 One bowel movement per day makes scheduling easier; no additional intervention is needed.

348. (4) These are the classic signs and symptoms related to acute appendicitis. They are caused by inflammation and altered gastrointestinal functioning. (1) (CI; PL)

1 The client adaptations do not indicate constipation. Constipation is associated with dry,

hard, infrequent stools; rectal pressure; and a feeling of fullness.

2 The client would be exhibiting diarrhea if there were an irritated bowel.

3 These adaptations are not related to a parasitic infestation.

349. (2) Drainage is promoted by the principle of gravity, preventing fluid accumulation and possible subdiaphragmatic abscess formation. (2) (CI; PL)

1 The lungs aerate well in any position if a subdiaphragmatic abscess does not form; with an abscess, the child will splint breathing.

3 Deep breathing and coughing, leg exercises, and ambulation must be employed to prevent problems with immobility; maintaining a constant position does not provide mobility.

4 Splinting of an abdominal wound is best accomplished by direct external pressure to the site, not by positional changes.

350. (1) Assessment of fluid balance and possible paralytic ileus are important aspects of care after surgery. (1) (CI; PL)

2 Mouth care is not specific, and following an appendectomy there is little drainage.

3 A low-residue diet is considered to be one of the factors leading to appendicitis because it tends to decrease peristalsis.

4 Dumping syndrome is a phenomenon associated with a gastrectomy.

351. (4) This fact makes young infants, not children, more vulnerable to common infections. (2) (DF; IM)

1 Many, but not all, infants respond this way.

2 Occasional bouts of diarrhea do not require medical intervention; the nurse can explain about the immaturity of the immunologic properties in the intestinal mucosa.

3 It is unlikely that the infant is eating ground meat; an unlikely response.

352. (2) A high-calorie, low-residue, high-protein diet is important to improve the child's nutritional status. (1) (PC; EV)

1 A high-protein diet is important to promote nutritional status.

3 Isotonic, not soapsuds, enemas are given.

4 Because the obstruction is in the large intestine, there is no indication for nasogastric feedings.

353. (1) Most raw fruits are high-residue foods; the colon cannot handle this type of food. (1) (PC; EV)

2 This is a low-residue food.

3 Same as answer 2.

4 Ripe bananas are raw fruit but are not classified as high residue.

354. ③ The impression is made on the sticky side of the tape; the specimen is collected before toileting so the ova deposited in the perianal area during the night have not been removed. (2) (HI; IM)

1 This will wash away the eggs; they must be directly collected from the perianal area.

2 This will prevent the female worm from migrating to the perianal area and laying her eggs.

4 The eggs are deposited in the perianal area, not in the stool.

355. ① With enterobiasis, the adult worm lays her eggs around the anal opening, producing an irritation and thus an itch. (2) (SI; AS)

2 This is commonly seen with eczema or dermatitis.

3 This may be seen with hookworm infestations but is more commonly associated with a localized allergen

4 This is produced by a ringworm fungus, *Tinea capitis*.

356. ④ This is one of the preferred positions to prevent pressure on the perineal sutures following this surgery, which is done to correct intermediate anorectal malformations. (3) (CI; IM)

1 Buck's traction is usually applied in the supine position, and this would put pressure on the perineal sutures; it would also contribute to prolonged contamination of the operative site by feces.

2 This would not promote healing of the anal area and could impede respiratory excursion.

3 This would increase pressure in the perineal area; the child may be placed in the side-lying prone (Sims') position with the hips elevated, not with the head of the bed elevated.

357. ④ Necrotizing enterocolitis (NEC) is an inflammatory disease of the gastrointestinal mucosa related to several factors including prematurity, hypoxemia, and high-solute feedings; it includes shunting of blood from the GI tract, decreased secretion of mucus, greater permeability of the mucosa, and increased gas-forming bacteria, eventually causing obstruction. (3) (DF; AS)

1 Meconium ileus is not related to the development of NEC; it is a common complication of cystic fibrosis.

2 Imperforate anus is the failure of fetal tissue to connect early in gestation and is present at birth.

3 Duodenal atresia may be a genetic or environmental defect that occurs early in gestation and is present at birth.

358. ④ Because intussusception creates intestinal obstruction where the intestine "telescopes" and becomes trapped, passage of intestinal contents is lessened; stools are red and currant jelly–like from the mixing of stool with blood and mucus. (2) (CI; AS)

1 This is not as important to assessment as is observable behavior associated with crying.

2 Bowel sounds would not be significantly affected.

3 Accurate fluid intake and output records are important, but they are not essential to confirming this diagnosis.

DRUG-RELATED RESPONSES (DR)

359. ② This is the correct amount; no conversion is necessary because desired and available dosage are both in milligrams. (2) (PT; AN)

1 This is too little.

3 This is too much.

4 Same as answer 3.

360. ② 5 ml = 1 teaspoon; therefore 2.5 ml = ½ (0.5) teaspoon. (2) (PT; AN)

1 This is too little; 0.5 ml = 0.10 teaspoon.

3 This is too much; 7.5 ml = 1½ teaspoons.

4 This is too much; 3.75 ml = 0.75 teaspoon.

361. ① Pancreatic enzyme replacements are given just before or with every meal to help aid digestion. (2) (PT; EV)

2 Inappropriate; the medication must be given just before or with every meal to aid digestion.

3 Breaking up and dissolving the medication will enhance its degradation by gastric secretions and interfere with its efficiency.

4 Same as answer 2.

362. ④ Because Synthroid speeds up the basal metabolic rate, an absence of constipation is a therapeutic response to the medication. (3) (PT; EV)

1 This is a clinical sign of hypothyroidism and is related to a slow basal metabolic rate.

2 Fine hand tremors are related to hyperthyroidism and would not have been present in an infant with hypothyroidism.

3 Same as answer 1.

363. (1) Insulin is a hormone and protein; when taken orally it is destroyed by the digestive enzymes, particularly pepsin. (2) (PT; IM)
2 The potency of insulin is not just reduced, it is totally inactivated.
3 This is untrue; insulin does not contain a carbohydrate that would raise the blood glucose.
4 Insulin is not neutralized, it is inactivated by digestive enzymes.

364. (3) Streptomycin is potentially ototoxic and nephrotoxic. (3) (PT; PL)
1 Penicillin reactions are usually allergic reactions.
2 Tetracycline causes discoloration of forming teeth.
4 Chloramphenicol may cause blood dyscrasias.

365. (3) Absorption of tetracycline is enhanced when the stomach is empty. (3) (PT; EV)
1 Food interferes with absorption.
2 Same as answer 1.
4 Same as answer 1.

366. (3) Probenecid results in better utilization of the penicillin by delaying the excretion of the penicillin through the kidneys. (3) (PT; IM)
1 This is untrue; the penicillin destroys the *Treponema* during all stages of its development; the probenecid delays the excretion of penicillin.
2 The probenecid does not treat the urethritis; it delays excretion of the penicillin.
4 Probenecid does not prevent allergic reactions.

367. (3) Pancreatic enzyme replacement (Pancrease) is needed because pancreatic enzymes are absent in children with cystic fibrosis. (1) (PT; AN)
1 Pancrease promotes the body's ability to metabolize and absorb fat rather than excrete it.
2 Not the effect of Pancrease; Pancrease contains enzymes to breakdown fats, proteins, and carbohydrates.
4 Same as answer 2.

368. (3) Diarrhea is initially related to GI irritation; later, overgrowth of drug-resistant microbes can result in superinfection, which also causes diarrhea. (2) (PT; EV)
1 Vertigo is unrelated to tetracycline.
2 Tinnitus is unrelated to tetracycline; this is associated with ASA toxicity.
4 The opposite occurs because GI irritation increases motility.

369. (4) Because of the suppression of the inflammatory symptoms of infection, such as an increase in body temperature, the nurse must be alert to the subtle signs of infection such as changes in appetite, sleep patterns, and behavior. (1) (PT; AS)
1 Adrenocorticosteroid therapy may cause hypokalemia, not hyperkalemia, because of the retention of sodium and fluid.
2 Adrenocorticosteroid therapy delays, not accelerates, wound healing.
3 Adrenocorticosteroid therapy decreases, not increases, the production of antibodies.

370. (1) Cystitis is a potentially serious adverse reaction to Cytoxan, which can sometimes be prevented by increased hydration because the fluid flushes the bladder. (3) (PT; PL)
2 Irritability may be present but is not necessarily a result of Cytoxan administration.
3 This is an expected but not serious side effect of Cytoxan.
4 This is unrelated to Cytoxan administration; it occurs with Dilantin therapy.

371. (2) Many vitamins contain folic acid, which is contraindicated with methotrexate, a folic acid antagonist. (1) (PT; IM)
1 Vitamins would be contraindicated because folic acid interferes with the action of methotrexate.
3 This is true but does not answer the question; it permits vitamin use in the near future, which long-term chemotherapy contraindicates.
4 This is inaccurate; vitamin use is contraindicated.

372. (4) Neurotoxicity is a specific common side effect to this drug; the client can become numb and ataxic. (3) (PT; AS)
1 Vincristine causes adynamic ileus, resulting in constipation; diarrhea occurs with other antineoplastics and radiation therapy.
2 Alopecia is an expected side effect rather than a toxic response; it is not considered serious, and hair will regrow.
3 This is not a side effect of this drug but a toxic response to Cytoxan.

373. (2) Elevated uric acid levels from destroyed cells may lead to renal problems; increased fluid intake helps dilute urine. (3) (PT; AS)
1 This will have to be monitored, but it is not of primary importance at this time.
3 Same as answer 1.
4 Same as answer 1.

374. (2) Vincristine is neurotoxic; therefore, clients should be monitored for paresthesias, foot drop, bowel and bladder problems, and alterations of function of cranial nerves. (3) (PT; EV)

1 Emotional states may be altered by corticosteroid therapy; vincristine (Oncovin) does not affect the emotional status.

3 Respiratory problems are not associated with vincristine (Oncovin) therapy.

4 Problems such as anemia, thrombocytopenia, and leukopenia occur with drugs such as cyclophosphamide (Cytoxan) and doxorubicin hydrochloride (Adriamycin), not with vincristine (Oncovin).

375. (1) Constipation is a common side effect of vincristine because gastric motility is slowed. (3) (PT; EV)

2 This is not a toxic effect of prednisone.

3 An enlarged spleen would put pressure on the stomach and diaphragm, not on the large bowel.

4 This is not likely.

376. (3) The objective of the oral chemotherapy is to achieve a remission and prevent further relapses. (1) (PT; EV)

1 If the oral chemotherapy is effective, additional IV drugs will be unnecessary.

2 Oral chemotherapy is an adjunct to other therapies, not an alternative of other therapies.

4 The prime site of metastasis of ALL is to the central nervous system.

377. (2) Antineoplastic drugs exert their effect on rapidly dividing tissues such as hair follicles, resulting in alopecia. (2) (PT; IM)

1 This is not related to any of the side effects of the antineoplastics that are being used.

3 Same as answer 1.

4 Same as answer 1.

378. (4) AZT can cause life-threatening blood dyscrasias including thrombocytopenia. (3) (PT; EV)

1 With AZT toxicity, the child would demonstrate agitation, restlessness, and insomnia, not fatigue and lethargy.

2 This is usually a response to the disease rather than the therapy.

3 Urinary output is unrelated to AZT toxicity; a decreased urinary output can be related to a decreased fluid intake, vomiting, and diaphoresis associated with fever.

379. (3) Kidney function must be adequate to handle the lead being excreted; if kidney function is not adequate, nephrotoxicity or kidney damage may result. (2) (PT; PL)

1 There is no skin breakdown with chelation therapy.

2 A normal protein intake is adequate; excessive protein is not lost unless kidney damage is present.

4 This would not be a nursing function; liver damage does not occur with chelation therapy.

380. (3) EDTA removes calcium along with lead; serum calcium levels need to be monitored periodically. (3) (PT; EV)

1 This does not happen with chelation therapy.

2 Same as answer 1.

4 Same as answer 1.

381. (4) Chelating agents mobilize lead from blood and soft tissues into the urine. (2) (PT; AN)

1 Chelating agents act to deposit lead in the bone.

2 Free lead is excreted primarily through the urine, not feces; secretions are substances produced by glandular organs.

3 Chelating agents do not deposit iron in hemoglobin or displace appreciable quantities of iron from hemoglobin.

382. (4) Chelating agents are given parenterally (IM, IV, SC) for most efficient absorption; IM is the preferred route because it is safer than the IV route, and the large muscles can tolerate the drug better than the subcutaneous tissues. (3) (PT; PL)

1 This route is contraindicated; chelating agents result in unabsorbable iron complexes within the lumen of the GI tract.

2 Same as answer 1.

3 The subcutaneous route is superficial in comparison to the intramuscular route.

383. (3) This may reduce the risk of gingival hyperplasia, a common side effect of Dilantin. (2) (PT; EV)

1 The drug is strongly alkaline and should be administered with meals to avoid gastric irritation.

2 Avoidance of overeating and of overhydration may result in better seizure control.

4 A pink, not reddish-brown, color may occur during drug excretion; it is not a complication.

384. ③ This behavior is a form of denial that may occur once the seizures are controlled; status epilepticus may occur when the medication is not taken regularly. (2) (PT; EV)

1 Toxic reactions to Dilantin are not manifested by constant seizures.

2 This is desired; it should prevent status epilepticus at this level.

4 The dosage is not based on activity but on the type of seizure.

385. ① Pancrease must be taken with food and snacks because it acts on the nutrients and readies them for absorption. (3) (PT; PL)

2 The enzyme is useless when taken without food.

3 Same as answer 2.

4 The enzyme is needed only when food is eaten.

386. ④ Mannitol is an osmotic diuretic used to relieve cerebral edema. (2) (PT; AN)

1 The bladder is a storage basin and is not involved with filtration; mannitol acts in the kidneys.

2 Mannitol is an osmotic diuretic that does not reduce peripheral edema.

3 Mannitol is an osmotic diuretic that has no effect on the body's glucose.

387. ④ Phenylephrine, with frequent and continued use, can cause rebound congestion of mucous membranes. (2) (PT; IM)

1 Nasal polyps may be associated with allergies but are unrelated to phenylephrine.

2 Bleeding tendencies are related to inadequate clotting mechanisms; nasal bleeding may be associated with dry mucous membranes, but with rebound congestion the tissues are full of fluid.

3 These symptoms are unrelated to this drug; phenylephrine may cause hypotension, tachycardia, and tingling of the extremities.

388. ② Epinephrine relaxes the smooth muscles and promotes respiratory volume. (2) (PT; AN)

1 This drug has a beneficial anti-inflammatory effect, but it is not used in status asthmaticus.

3 Ceclor is an antibiotic and would be used for superimposed infections but not for status asthmaticus itself.

4 This is an antihistamine and is contraindicated in acute asthmatic attacks; it is used to prevent, not reverse, histamine-mediated reactions.

389. ② Vomiting is a classic sign of digoxin toxicity, and the physician must be notified. (3) (PT; EV)

1 This is rarely needed because Aldactone spares potassium.

3 This is rarely necessary except when the heart condition is severe; infants are usually not overactive.

4 There is no restriction on taking aspirin with Aldactone; however, infants and children are generally not given aspirin because of its association with Reye's syndrome.

390. ③ The chronotropic effect of Lanoxin, a cardiac glycoside, slows the heart rate; when the dose of Lanoxin is excessive, the child can develop bradycardia. (3) (PT; EV)

1 This is associated with congestive heart failure, not Lanoxin toxicity.

2 Same as answer 1.

4 This is associated with congestive heart failure, specifically right ventricular failure.

391. ② This would not affect the potency of the solution. (3) (PT; AN)

1 The drug should not be used if the solution is discolored.

3 The drug should not be used 24 hours or more after it is reconstituted from its powdered form.

4 The drug should be void of sediment; contamination is indicated if it is present.

392. ① Epinephrine is a sympathetic nervous system stimulant that causes tachycardia. (1) (PT; EV)

2 Hyperglycemia, not hypoglycemia, may result.

3 The pupils will be dilated, not constricted.

4 Epinephrine is more likely to cause hypertension.

393. ③ This would minimize oral discomfort; ulceration of the oral mucosa occurs as a result of the antineoplastic effect on the rapidly dividing GI epithelium. (2) (PT; PL)

1 Although pain may be present, aspirin would be avoided because doxorubicin is also being used, and a side effect of this medication is thrombocytopenia.

2 These would aggravate the stomatitis that is a common side effect of Cosmegen.

4 This is not related to the administration of Cosmegen.

394. ② Gradual weaning from prednisone is necessary to prevent adrenal insufficiency or adrenal crisis. (2) (PT; IM)

1   This may occur, but it is not life threatening.

3   Because of its depressing effect on the immune system, prednisone may increase susceptibility to infection.

4   Prednisone may cause a suppression of growth.

395. (2) Because steroids are irritating to the gastric mucosa, a peptic ulcer with bleeding may occur; stools should be checked for occult blood. (3) (PT; EV)

1   Steroids do not cause this to occur.

3   Same as answer 1.

4   Same as answer 1.

396. (4) This is an appropriate medication for hypertension because it acts to relax the smooth muscles of the arterioles. (2) (PT; PL)

1   This increases the contractility and output of the heart.

2   Diazepam is inappropriate; this relaxes the skeletal muscles, not the smooth muscles of the arterioles.

3   This is an anticonvulsant that might be prescribed later in the course of the disease if the antihypertensives were not effective and hypertensive encephalopathy caused convulsions.

397. (4) The purpose of administering lactobacilli, normally found in the GI tract, is to help recolonize the normal flora excreted with the diarrheal stools. (1) (PT; AN)

1   This is not the action of this medication.

2   Same as answer 1.

3   Same as answer 1.

398. (1) This is expected; clients should be advised so that they do not become alarmed and can protect clothing and bedding. (1) (PT; IM)

2   The drug will not affect rectal itching; it will eliminate the pinworms, and this takes some time to accomplish.

3   All family members should be treated.

4   Reinfestation is common; the drug may be needed again.

399. (4) Abdominal cramping and distention are associated with inadequate pancreatic enzyme replacement because foods are not being digested. (2) (PT; EV)

1   The opposite is true; they would have a voracious appetite.

2   Diarrhea, not constipation, would result.

3   There would be a weight loss because of decreased digestion and absorption, not a weight gain.

400. (3) True; this type of surgery results in considerable pain for several days afterward and requires a strong analgesic. (2) (PT; PL)

1   The first postoperative day is too early to begin weaning the client from meperidine.

2   Adolescents are no more prone to exaggerate discomfort than any other age group.

4   A more potent analgesic such as meperidine is necessary because the pain from this type of surgery is considerable.

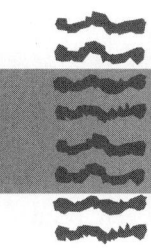

PEDIATRIC NURSING
QUIZ 1

1. A 3-year-old child is admitted to the hospital with the diagnosis of acute asthma. When formulating a nursing care plan the nurse should plan to:
   1. Maintain the child in the supine position
   2. Place the child in respiratory isolation for 24 hours
   3. Restrict the child's fluids to two-thirds of the usual intake
   4. Keep someone at the bedside when the child is short of breath

2. An infant who has exstrophy of the bladder had urinary diversion surgery prior to correction of the defect. Before discharge the nurse should plan to teach the parents to:
   1. Use rubber pants to contain urinary drainage
   2. Give tub baths after the umbilical cord has fallen off
   3. Keep the child in supine position to decrease urinary drainage
   4. Keep sterile petrolatum gauze over the bladder to prevent trauma

3. One month after abdominal surgery, a 6-month-old infant is brought to the pediatric clinic for a follow-up visit. The nurse observes the child performing all of the following tasks. The task that would be most unusual for a child of this age is:
   1. Putting clothespins in a plastic bottle
   2. Sitting alone for brief periods in a crib
   3. Playing with the toes during examination
   4. Showing stranger anxiety when the nurse approaches

4. A child with iron-deficiency anemia is placed on supplemental oral iron therapy. The parents of the child should be prepared to expect that the child may develop:
   1. Positive stool guaiac
   2. Staining of the teeth
   3. Orange-colored urine
   4. Greenish-black stools

5. Following a myringotomy for the treatment of otitis media, the nurse should expect the child to have:
   1. Symptoms of CNS irritation
   2. Irrigations to the lacrimal glands
   3. Difficulty voiding and slight hematuria
   4. Purulent drainage into the external auditory canal

6. When a toddler is required to have temporary dietary restrictions, the best way for the nurse to prevent any future problems is to have the restrictions:
   1. Handled in a matter-of-fact way
   2. Explained to the child by the dietitian
   3. Limited to foods that are not essential
   4. Administered by someone other than parents

7. When preparing a preschooler for a surgical procedure, the best time to tell the child what will happen is:
   1. 2 weeks before
   2. 3 to 4 days before
   3. 1 month in advance
   4. Just before it happens

8. Acute glomerulonephritis may occur secondary to impetigo. The treatment most beneficial in the prevention of this sequela is the:
   1. Use of an oil-based soap for bathing
   2. Administration of a systemic antibiotic
   3. Application of an antibiotic ointment to the lesions
   4. Removal of crusts with an antimicrobial liquid and water

9. When the nurse is teaching the parents about celiac disease, the mother sighs and states, "My neighbor told me that I would only need to watch the diet until my child is 8 years old. I'm so relieved; you know how kids are about eating!" The nurse's response should be based on the fact that:
   1. The basic defect of celiac disease is lifelong
   2. Susceptibility to celiac crisis lessens with age
   3. Though difficult at first, the diet is fairly easy to follow
   4. These children can tolerate small amounts of gluten by age 5

10. A mother of a child with diabetes mellitus had a sister who died of diabetic ketoacidosis. The mother asks the nurse for information about her daughter's disease. The nurse should recognize that the mother is:
    1. Exhibiting readiness for learning
    2. Too upset to learn new information
    3. Expressing her attitudes through her behavior
    4. Transferring attitudes about her sister to her daughter

11. After a prolonged period in the hospital, a 1½-year-old toddler becomes depressed, withdrawn, and apathetic toward the parents, especially the mother. Eventually, the toddler begins playing with toys and relating to others, even strangers. The nurse should recognize that the toddler has:
    1. Probably become somewhat detached because of this traumatic separation
    2. Accepted the hospitalization well and has matured because of the experience
    3. Finally recognized that the staff is not out to cause bodily harm to any of the children
    4. Grown out of the stage of separation and realizes that dependency on others is necessary

12. When asked by a 16-year-old what the best method of birth control is, it would be most therapeutic for the nurse to respond:
    1. "Oral contraceptives."
    2. "All are equally effective."
    3. "The one you will use consistently."
    4. "Do you really have a need for birth control?"

13. A child with juvenile rheumatoid arthritis complains of pains in the knees. To relieve the discomfort in the child's knees, the nurse should teach the parents to follow the physician's orders by:
    1. Applying moist heat
    2. Immobilizing the affected legs
    3. Supporting the knees with pillows
    4. Massaging the swollen areas gently

14. An infant is admitted to the hospital for surgery to correct pyloric stenosis. When performing a physical assessment of the abdomen of this child, the assessment should reveal:
    1. An impacted and distended colon
    2. Marked tenderness around the umbilicus
    3. An olive-sized mass in the right upper quadrant
    4. Rhythmic peristaltic waves in the lower abdomen

15. An adolescent with diabetes mellitus is brought to the emergency room in ketoacidosis. The client admits to not adhering to the diabetic regimen. As a first step in attempting to help the client develop some understanding of the importance of a diabetic regimen, the nurse should:
    1. Provide printed material about diabetes in teenagers
    2. Allow the client to express feelings about having diabetes
    3. Assume that the client has not been properly taught about diabetes
    4. Impress on the parents that it is their responsibility to demonstrate understanding

16. The nurse notes that on the second day after receiving severe, extensive burns, a young child has decreased urinary output with edema. The nurse, recognizing this as one sign of possible complications that commonly occur in the first 2 days, should watch for other signs, including:
    1. Rapid pulse and low BP
    2. Vomiting and bradycardia
    3. High fever and disorientation
    4. Subnormal temperature and slow pulse

17. A 6-year-old is hospitalized with a diagnosis of nephrotic syndrome, and the child's mother asks the nurse what she should bring her child to play with during the hospitalization. Considering the child's age, the nurse should suggest that the mother bring:
    1. Action toys such as a hula hoop
    2. Stuffed animals, large puzzles, and large blocks
    3. Table games, checkers, simple card games, and crayons
    4. A record player, transistor radio, and children's magazines

18. A 30-month-old child is brought to the emergency room in acute respiratory distress, and a diagnosis of laryngotracheobronchitis (viral croup) is made. Upon admitting this child to the pediatric unit, the nurse should anticipate the need for a:
    1. Cot so that a parent can stay
    2. Tracheotomy set at the bedside
    3. Pad for the side rails of the Croupette
    4. Quiet, cool room to facilitate breathing

19. A 6-month-old is admitted to the hospital with severe diarrhea. The nurse's assessment of the infant that is most indicative of dehydration would be a:
    1. Pulse rate of 120
    2. Generalized pallor
    3. Level anterior fontanel
    4. Decreased urine output

20. When monitoring the laboratory reports of a young infant with gastrointestinal distress, the nurse should be especially concerned about a decrease in sodium and:

1. Calcium
2. Chlorides
3. Potassium
4. Phosphates

21. When a young infant with gastrointestinal distress is permitted to receive a liquid diet in small amounts, it would be most beneficial for the nurse to give the infant:
    1. Skim milk
    2. Ginger ale
    3. Orange juice
    4. Liquefied gelatin

22. The physician has prescribed 30 mg of an antibiotic in 50 ml of D5W IVPB q6h for a school-age child. It is to run over 30 minutes. The drop factor of the set is 15 gtt/ml. The nurse should regulate the IV at:
    1. 13 gtt/min
    2. 25 gtt/min
    3. 50 gtt/min
    4. 60 gtt/min

23. An immunizing dose of tetanus immune globulin is administered to a child with an uncertain history of tetanus immunizations when a rusty nail pierces the sole of the foot. The nurse recognizes that tetanus immune globulin is prescribed because it:
    1. Produces lifelong passive immunity
    2. Induces longer-lasting active protection
    3. Confers immediate passive protection of short duration
    4. Immediately stimulates increased production of antibodies

24. The most important goal for a 1-year-old "boarder infant" who is in the pediatric unit for management of AIDS after being abandoned by the mother is the need for the:
    1. Assignment of a consistent nurse caregiver
    2. Development of plans for the infant's eventual adoption
    3. Use of play therapy to help the infant work through being abandoned
    4. Institutionalization of the infant with other AIDS infants in a state-operated facility

25. A 2½-year-old male child with cystic fibrosis is admitted to the pediatric unit with a severe upper respiratory infection. The nurse assesses that the child is very small for his age and attributes the slow growth to:
    1. The retention of CO_2
    2. Increased salt retention
    3. An atrial ventricular defect
    4. The absence of pancreatic enzymes

26. A child with a history of swollen lymph nodes, prolonged fever unresponsive to antibiotics, and erythema of the extremities is admitted to the hospital with a diagnosis of Kawasaki syndrome. The nurse is aware that this diagnosis was probably based on:
    1. An elevated ASO titer
    2. A combination of symptoms
    3. An elevated sedimentation rate
    4. A decreased serum protein level

27. A mother takes her child to the physician because the child is constantly scratching the anus. In caring for a child who is suspected of having a pinworm infestation, the initial nursing responsibility would be to:
    1. Prevent reinfection
    2. Identify the organism
    3. Collect a stool specimen
    4. Notify the health department

28. At 1 AM, a 28-month-old is admitted to the pediatric unit with meningitis. At 3 AM, after the child is settled down, the mother tells the nurse, "I have to leave now, but whenever I try to go my child gets upset and then I start to cry." The nurse should:
    1. Walk the mother to the elevator
    2. Encourage the mother to spend the night
    3. Stay with the child while the mother leaves
    4. Tell the mother to wait until the child falls asleep

29. One day in the clinic a mother asks when her newborn should be able to drink from a cup. The nurse responds:
    1. "Five months."
    2. "Seven months"
    3. "Twelve months"
    4. "Eighteen months."

30. Allopurinol (Zyloprim) is prescribed for a 6-year-old child on a chemotherapeutic regimen for cancer of the bone. When given the medication, the child asks, "Why do I have to take this pill?" The nurse's response should be:
    1. "To stop your sick white cells from going to other parts of your body."
    2. "Because this pill helps the other medicines get rid of the things that are making you sick."
    3. "To protect your body from developing other problems after your treatment has been stopped."
    4. "Because your doctor ordered it. The doctor would not order anything for you unless it was very important."

31. A child with a comminuted fracture is placed in skeletal traction. The nurse knows that the primary objective for performing pin site care with this kind of traction is to:
1. Decrease pain
2. Prevent infection
3. Decrease scarring
4. Prevent skin breakdown

32. After a tonsillectomy, a child is swallowing frequently. The nurse suspects that the child is:
1. Experiencing pain in the throat
2. Reacting from general anesthesia
3. Hemorrhaging from the operative site
4. In need of suctioning to keep the airway patent

33. A 10-month-old injured in an automobile accident is brought to the trauma center. The child, who is quiet but does not appear lethargic, has a large hematoma on the left temporal area. When assessing the child, the nurse should be particularly alert for signs and symptoms that may indicate neurologic involvement such as:
1. Persistent vomiting
2. A positive Babinski reflex
3. The development of a headache
4. A pulse rate of 110 beats per minute

34. The nurse recalls that some of the most common early signs of leukemia are:
1. Pallor, joint pain, anorexia, fever
2. Lethargy, petechiae, splenomegaly
3. Fatigue, pallor, alopecia, hemorrhage
4. Fatigue, mouth lesions, hepatomegaly

35. The side effect that is attributed to both the lead toxicity and the chelating agent, Ca EDTA, is:
1. Hypocalcemia
2. Nephrotoxicity
3. Bone marrow depression
4. Increased intracranial pressure

36. The orthopedic surgeon is anxious to have a child with cerebral palsy walk with crutches. When preparing this child for crutch walking, the nurse must determine:
1. The weight-bearing ability of the child's upper and lower extremities
2. Whether the child has the power in the trunk to drag the legs forward when erect
3. Whether the child's circulation can tolerate the body being placed in an erect position
4. The ability of the child's shoulder girdle to support the weight of the body when it leaves the floor

37. If a child with cerebral palsy has only slight sensory loss in the lower extremities, the child and the parents should be taught to:
1. Examine the skin for evidence of pressure points
2. Keep the braces in good repair and pad them well
3. Select shoes that have heels that are wide and low
4. Check that the brace joints are aligned with body joints

38. Because of a measles epidemic, a 6-month-old infant receives a measles vaccination. The nurse should help the mother understand that to ensure continuous protection against measles, the baby should be revaccinated at approximately:
1. 8 months of age
2. 10 months of age
3. 12 months of age
4. 18 months of age

39. Gamma benzene hexachloride 1% (Kwell) is ordered for a 5-year-old with pediculosis capitis (head lice). When teaching the mother about treatment for head lice, the nurse should include the fact that:
1. The shampoo must be repeated within 7 to 10 days
2. The medicated shampoo used has no known side effects
3. Clothing and personal belongings must be boiled or discarded
4. Other children should be kept away from the infected child for 1 week

40. When obtaining the health history of a 7-year-old admitted to the hospital with signs and symptoms suggestive of Reye's syndrome, the nurse should assess whether this child has:
1. Had a recent rash
2. A history of high fevers
3. An allergy to drugs or food
4. Had a recent viral infection

41. The question that the nurse will find to be most effective when eliciting information from a 5-year-old child regarding the reason for the child's hospitalization is:
1. "Do you know what place this is?"
2. "Why did you come to the hospital?"
3. "Do you know what is going to happen to you?"
4. "You do know why you are in the hospital, don't you?"

42. The nurse is aware that cleft lip defects are usually repaired early because they:
1. Tend to obstruct breathing
2. Usually create severe feeding problems
3. Have an emotional impact on the parents
4. Frequently cause respiratory tract infections

43. An infant's cleft lip is repaired 4 days after birth. During the postoperative period the nurse should avoid:
    1. Keeping the suture line clean and free of crusting
    2. Placing the infant on the abdomen to prevent aspiration
    3. Examining the tongue, lips, and mucous membranes for swelling
    4. Encouraging the parents to visit as much as possible to prevent the infant from crying

44. A nurse teaches a group of mothers in the community how HIV can be transmitted via blood transfusions. The nurse recognizes that further teaching is necessary when one of the mothers states:
    1. "Blood obtained from individuals in high-risk groups is discarded."
    2. "The process of donor screening has improved since HIV was discovered."
    3. "Once tested for antibodies, negatively tested blood cannot transmit the virus."
    4. "Blood that has tested negative for HIV can develop antibodies at a later time."

45. A 6-year-old is admitted to the hospital with nephrotic syndrome. The nurse should be aware that the changes in body fluid distribution that occur with this disorder are related to the fact that:
    1. Loss of sodium and water through an impaired basement membrane of the glomerulus results in hypovolemia
    2. The loss of body protein reduces oncotic pressure so that fluid moves from the intravascular to the interstitial space
    3. Hyperproteinemia results in increased oncotic pressure, which causes fluid to move from the intravascular to interstitial space
    4. The basement membrane of the glomerulus becomes selectively impermeable to water so that fluid is retained in the tissues

46. A 2½-year-old has a pediatric microdrip set up. A 500-ml bottle of D5 ⅓ NS is hung at 1 AM and is to infuse at 45 ml per hour. At 6 AM, the night nurse notes that the child's IV bottle has 125 ml left. The nurse correctly evaluates that the:
    1. IV has infused at the prescribed rate through the night
    2. Total amount of fluid infused is less than ordered for the child
    3. Total amount of fluid infused is more than the child should have received
    4. Amount of fluid infused should be calculated at the change of shift or 7 am

47. A 3-year-old with chronic lead poisoning must have 60 injections of Ca EDTA before the lead is chelated and removed from the body. The nurse should prepare the child for this painful treatment by:
    1. Rotating the injection sites and adding procaine to the chelating agents to lessen the discomfort
    2. Role playing with puppets dressed as physicians and nurses to minimize the child's fear of unfamiliar adults
    3. Allowing the child to play with a syringe and a doll before therapy is initiated and after receiving each injection
    4. Carefully explaining the rationale for the injections so that the child does not feel punished for eating paint chips

48. For a child with the diagnosis of lead poisoning, initial care should focus on the nursing diagnosis of:
    1. Constipation related to the excretion of lead
    2. Risk for injury related to the ingestion of lead
    3. Altered growth and development related to parental neglect
    4. Altered nutrition, less than body requirements, related to decreased iron intake

49. When the nurse is caring for a young toddler following a circumcision, a major nursing responsibility is to:
    1. Limit oral fluids
    2. Monitor IV fluids carefully
    3. Apply ice packs to the penis
    4. Observe for bleeding at the operative site

50. When teaching a young child with cerebral palsy to ambulate with crutches, the nurse should remember that:
    1. Learning is a result of adequate teaching
    2. Learning progresses on a line forward and upward
    3. The child must first understand normal walking patterns
    4. Because of the child's age, experiential background is limited

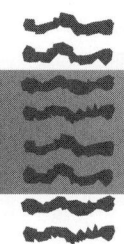

PEDIATRIC NURSING QUIZ 1
ANSWERS AND RATIONALES

1. (4) Keeping someone at the child's bedside may decrease anxiety, which would decrease oxygen requirements and respiratory distress. (2) (SI; PL; RE)
 1 The supine position will increase respiratory distress; some elevation of the torso will decrease pressure on the diaphragm and promote lung expansion.
 2 Respiratory isolation is not required for asthma; it is not an infectious disease.
 3 Fluids should be maintained to promote hydration and assist in liquefying secretions.

2. (4) This keeps the diaper from adhering and helps prevent infection. (2) (CC; IM; RG)
 1 This promotes infection and may cause trauma to the bladder.
 2 Sponge baths are given to prevent infection from bath water.
 3 The side-lying position is preferred to allow urinary drainage; the amount of urine produced is not affected by positioning.

3. (1) This is a behavior seen in 9- to 12-month-old children; this is the most advanced behavior listed and is the most unusual for this age. (2) (DF; AS; GD)
 2 By the age of 7 months an infant can sit alone, leaning forward on the hands for support. By 8 months the infant can sit unsupported.
 3 This is a coordination ability seen in 3- to 5-month-old children.
 4 This is a behavior seen in 6- to 8-month-old children.

4. (4) Iron is excreted in the feces, and the change in color results from the insoluble iron compound in the stool. (1) (PT; AS; DR)
 1 This is associated with GI bleeding, not iron administration.
 2 This should not occur with proper administration of the iron; iron elixir should be diluted in fluid and administered with a straw; the teeth should immediately be brushed after administration.

 3 This is not associated with iron; it occurs with Pyridium administration.

5. (4) A myringotomy relieves pressure and prevents spontaneous rupture of the eardrum by allowing pus and fluid to escape from the middle ear into the external auditory canal, from which the exudate drains. (2) (SI; EV; NM)
 1 A myringotomy involves drainage of the middle ear through an incision made in the eardrum; the CNS is not involved.
 2 The lacrimal glands are the tear-producing glands.
 3 These symptoms might be expected after surgery involving some part of the urinary tract.

6. (1) Toddlers are ritualistic and do not tolerate change well. Any change in diet should be done matter-of-factly. Because of their characteristic struggle for independence, toddlers should not be forced to eat. (2) (DF; IM; GI)
 2 The toddler does not have the cognitive capacity to understand reasons for behavior.
 3 This is not always possible in dietary restrictions.
 4 The toddler is still dependent on the parents and therefore responds better to them than to strangers.

7. (2) This will give the child adequate time to adjust and prepare for the surgery without presenting it too far in advance. (2) (DF; IM; GD)
 1 This is too far in advance; the child will forget or build up a great deal of fear.
 3 Same as answer 1.
 4 This is appropriate for a toddler.

8. (2) Systemic antibiotics are necessary to eradicate the streptococcal organism, which caused the primary infection, impetigo. (2) (SI; PL; RG)
 1 This would not prevent glomerulonephritis, a sequela of streptococcal infections.
 3 This would not prevent glomerulonephritis; this is part of the local therapy for impetigo.
 4 Same as answer 3.

9. (1) The diet must be followed forever because there will always be an absence of peptidase; some variations to this diet may be allowed, but this should not be promised. (2) (HI; IM; GI)

2 This is untrue; each phase of child development may have problems related to dietary management; follow-up care is needed to prevent crises.

3 This is untrue; a restricted diet is never easy to follow, especially for a growing child.

4 This is untrue; gluten must be avoided for a prolonged period of time and usually indefinitely.

10. (1) When a client or family member asks for information, this is a clue identifying readiness for learning. (3) (HI; AS; EN)

2 Asking this question indicates readiness to learn.

3 The question indicates a desire for more information.

4 Data are not evident to indicate this.

11. (1) The child has progressed to the third phase, detachment, in which there is a resignation to the loss of the mother and a superficial appearance of adjustment to the environment. (3) (SC; AS; EH)

2 This is often the mistaken interpretation of such behavior.

3 Staff members are usually viewed by toddlers as unfamiliar, frightening, and often threatening.

4 Eighteen-month-old children have not outgrown separation anxiety.

12. (3) Various birth control methods are effective, but the one used properly and consistently is the most effective for a given individual. (1) (DF; IM; RC)

1 This is premature advice; the subject should be explored further.

2 They are not all equally effective.

4 This response places the client on the defensive and cuts off communication.

13. (1) Moist heat increases circulation to the involved area, thereby reducing the swelling and relieving joint pain and stiffness. (2) (PC; IM; SK)

2 Immobilization will contribute to contracture deformity.

3 Pillows under the knees will promote flexion contractures, hence they are contraindicated in this disorder.

4 Massaging the swollen area will only aggravate the inflammation and cause more pain.

14. (3) The circular muscle of the pylorus is grossly enlarged as a result of hypertrophy and hyperplasia. (2) (HI; AS; GI)

1 The obstruction is above this area; there will be little or no fecal contents present.

2 These infants do not have any significant tenderness in the abdomen.

4 It is more likely that there will be an absence of peristalsis in the lower colon.

15. (2) Psychoemotional factors related to chronic illness often affect individual compliance with a medical regimen. These feelings must be explored and worked out for acceptance of the treatment plan. (2) (SC; IM; EH)

1 This may be helpful later, but the client needs an opportunity now to share feelings about the disease.

3 The client's knowledge and feelings should be explored.

4 Although this is important, adolescents need control and therefore the teaching must begin with the client; an emphasis should be placed on feelings first.

16. (3) High fever and disorientation may be initial indications of dehydration and/or early signs of hypoxia from respiratory complications. (3) (DA; AS; FE)

1 A decreased BP is rarely seen unless the client is in severe shock.

2 Tachycardia is normally associated with the early phases of burn injury; vomiting may be a sign of circulatory collapse or paralytic ileus or may be nonspecific.

4 High temperatures and an increased basal metabolic rate are seen with burns.

17. (3) School-age children enjoy competition, have manipulative skills, and are creative. (1) (DF; IM; GD)

1 This activity is inappropriate during the acute phase of nephrotic syndrome because it requires too much energy.

2 These are appropriate for the toddler who is developing fine motor skills.

4 Magazines would interest an older child who would be more proficient in reading.

18. (2) A patent airway is the priority, and the necessary equipment must be immediately available. (1) (DA; PL; RE)

1 Although helpful, this is not the priority.

3 This would be appropriate for convulsions, which are not associated with croup; respiratory promotion is the priority.

4 Same as answer 1.

19. (4) Dehydration leads to reduced blood volume, which in turn reduces kidney perfusion and results in decreased urine output. (2) (DA; AS; FE)
1 This is a normal rate for a 6-month-old child.
2 This can be indicative of many disorders; it is not specific for dehydration.
3 The anterior fontanel is normally level; it is depressed if the child is dehydrated.

20. (3) Sodium, potassium, and bicarbonate are the electrolytes most often lost because of diarrhea. They are excreted before they can be absorbed. (2) (CI; AS; FE)
1 Serum calcium levels are indicators of parathyroid function and calcium metabolism.
2 Chloride serves to maintain electrolyte neutrality and accompanies sodium losses or excesses.
4 Phosphorus levels are determined by calcium metabolism, parathormone, and to a lesser degree, intestinal absorption.

21. (4) Clear liquids with sugar added provide glucose without bulk. (3) (PC; PL; GI)
1 Clear liquids are best tolerated after a gastric upset.
2 Ginger ale adds unnecessary carbonation.
3 Fruit juices may cause fermentation, which can stimulate peristalsis.

22. (2) 50 ml multiplied by 15 (drop factor) equals 750, which is divided by 30 (the number of minutes the IV should run) and equals 25 gtt/min. (2) (PT; AN; FE)
1 This is an incorrect calculation.
3 Same as answer 1.
4 Same as answer 1.

23. (3) Tetanus immune globulin contains ready-made antibodies and only confers short-term passive immunity. (3) (SI; AN; BI)
1 Passive immunity is not lifelong.
2 Immune globulins confer passive artificial immunity.
4 Immune globulins are antibodies and therefore may block the formation of antibodies if given with the tetanus toxoid.

24. (1) The child's home and family are the pediatric unit and staff; a consistent caregiver gives the child a "mommy" or "daddy." (3) (DF; PL; GD)
2 The likelihood of the child's adoption may be slim; this may be unrealistic.
3 The child is too young for play therapy.
4 The child is institutionalized now.

25. (4) Nutrients such as protein, carbohydrates, and fats are not digested and are not properly absorbed by the intestinal mucosa. (2) (PC; AN; GI)
1 It is not the retention of carbon dioxide but the deprivation of nutrients and oxygen to all body cells that retards physical growth.
2 These children lose high concentrations of sodium and chloride in perspiration.
3 There is no evidence that the child has an atrial ventricular defect.

26. (2) The diagnosis is based on the presence of five out of six specific symptoms, including fever, trunk rash, enlarged cervical lymph nodes, bilateral congestion of the conjunctiva, edema, and redness of the extremities. (3) (HI; AS; NM)
1 An elevated ASO titer may be seen with a streptococcal infection.
3 This is nonspecific to Kawasaki syndrome; the sedimentation rate is elevated in the presence of inflammation with many disorders.
4 This is not present in Kawasaki syndrome.

27. (3) A specimen must be obtained and analyzed before the physician can prescribe treatment. (1) (SI; PL; GI)
1 The priority is collecting a specimen; then prevention techniques can be taught.
2 The organism can be identified only by laboratory analysis of a stool specimen.
4 This is not a reportable communicable disease.

28. (3) This action would encourage the mother to leave and reassure her that someone will be with and will comfort the child. (2) (SC; IM; EH)
1 The mother has indicated she is upset when the child is upset; this intervention meets no one's needs.
2 The mother has said she must leave; convincing her to stay will make her feel guilty about having to leave.
4 Same as answer 2.

29. (3) By 12 months of age a child can usually drink from a cup, although fluid may spill and a bottle may be preferred at times. (3) (DF; IM; GD)
1 The child is just beginning lip control at 5 months and cannot handle a cup.
2 At 7 months a child can handle a bottle but not a cup.
4 Since this skill is present at 12 months, by 18 months most children are quite proficient.

30. (2) This is the most accurate and age-appropriate response to the child's question. (3) (PT; IM; GD)
    1 This is inaccurate and may instill more fear.
    3 This is inaccurate; not being truthful interferes with the development of trust.
    4 This response is insensitive to the question and does not provide any explanation.

31. (2) Pin sites provide a direct avenue for organisms into the bone. (2) (SI; AN; SK)
    1 Analgesics, not pin care, will decrease pain.
    3 Some scarring will occur from the pin insertion.
    4 Skin has a tendency to grow around the pin rather than break down, as long as infection is prevented.

32. (3) If there is a trickle of blood from the operative site, the child will swallow frequently; this is usually the first sign of hemorrhage. (2) (CI; EV; RE)
    1 The child with a sore throat tries to minimize the frequency of swallowing.
    2 Swallowing frequently is not a specific reaction to general anesthesia.
    4 The child may need suctioning, but the presenting symptoms for this intervention would be restlessness or a color change, such as cyanosis.

33. (1) Vomiting frequently accompanies a head injury because of increased intracranial pressure and stimulation of the vomiting reflex. (1) (DA; AS; NM)
    2 A positive Babinski reflex is normal and expected in a 10-month-old child.
    3 A 10-month-old child would not complain of a headache; persistent vomiting is an objective sign.
    4 This would be within the normal range for a child this age.

34. (1) Vague symptoms commonly prolong the accurate diagnosis of the leukemic process; all of these are early signs related to the leukocyte alteration. (1) (HI; AS; BI)
    2 Splenomegaly is not a common sign of early leukemia; the symptoms are usually vague and nonspecific.
    3 Alopecia is not a common sign of early leukemia; the early symptoms are usually vague and nonspecific.
    4 Mouth lesions and an enlarged liver are not common symptoms of early leukemia; the early symptoms are usually vague and nonspecific.

35. (2) Lead toxicity and Ca EDTA both damage the proximal renal tubules, resulting in the abnormal excretion of protein and other substances. (3) (PT; EV; DR)
    1 This is attributable to some chelating agents only; it is not likely to occur with the Ca EDTA preparation, which replaces calcium.
    3 Bone marrow damage is caused by lead toxicity only.
    4 Lead encephalopathy causes serious elevation of intracranial pressure.

36. (1) The choice of gait is based on the weight-bearing capabilities of each of the four extremities. (2) (PC; AS; NM)
    2 The child with cerebral palsy uses upper extremity strength for crutch control and lower extremity strength to facilitate some movement.
    3 Under normal circumstances, orthostatic circulatory impairment is unlikely in the child with cerebral palsy.
    4 Because of the decreased muscle control in cerebral palsy, it is unlikely that the child would be able to utilize a gait involving complete support of body weight off the floor.

37. (1) When sensory perceptions are impaired, with resultant lack of effective specific motor responses, an individual will be more vulnerable to skin irritation and trauma. (1) (CC; IM; NM)
    2 A lack of sensation makes an individual more vulnerable to skin irritation from braces; observation of the skin takes priority in sensory loss.
    3 A wide, flat shoe will facilitate balance; observation of the skin takes priority in sensory loss.
    4 Alignment of brace joints to body joints is important to facilitate joint mobility; observation of the skin takes priority in sensory loss.

38. (3) The optimum age for measles vaccination is between 12 to 15 months; if a vaccination is given earlier than this because of exposure to a person with measles, it is not counted as one of the two required doses. (1) (SI; PL; BI)
    1 This is too early; the baby would still have antibodies from the previous vaccination.
    2 Same as answer 1.
    4 This is not the optimal time; the measles immunization should be given between 12 to 15 months of age.

39. (1) Reapplication destroys any surviving ova (nits), which eventually become lice. (1) (PT; IM; IT)
    2 This is untrue; excessive use can cause an eczematous rash and central nervous system disturbances.

3 Personal items can be soaked in pediculicidal solution; clothing and linen should be laundered in hot water and dried in a hot drier.

4 Once the hair has been shampooed, there is no reason to isolate the child.

40. (4) Although the cause is unknown, there is a strong relationship between Reye's syndrome and an antecedent viral infection such as varicella. (3) (HI; AS; NM)

1 A rash is not specific to Reye's syndrome; a rash can be caused by many things.

2 A fever may have been caused by many things.

3 This has no relationship to Reye's syndrome.

41. (2) This is an open-ended question that should elicit the desired information. (3) (SC; AS; GD)

1 This question can be answered with a yes or no and may not elicit the desired information.

3 Same as answer 1.

4 Same as answer 1.

42. (3) The tremendous visual impact of cleft lip on parents may significantly affect the parent-child attachment process and is often considered a reason for early surgical intervention. (3) (SC; AN; EH)

1 The baby uses the mouth to facilitate breathing; a cleft lip does not interfere.

2 Feeding can be accomplished by use of special equipment; this is not in itself an indication for early surgery.

4 Precautions other than surgery may be taken to prevent ear and upper respiratory tract infections.

43. (2) After a cleft lip repair, infants are always placed supine or slightly on their side to prevent damage to the suture line. (2) (CI; PL; GI)

1 For maximum healing and minimal scarring, a clean suture line is essential.

3 Swelling in or near the area traumatized by surgery is a complication that may occur postoperatively.

4 Crying may disturb the suture line; the parents should be encouraged to hold the infant to reduce crying.

44. (3) This statement is not true because the virus could still be transmitted if the result was a false negative. (3) (HI; EV; BI)

1 This is a true statement; members of high-risk groups are more apt to be HIV positive.

2 Blood screening has improved and has reduced the incidence of transmission, but it is still not perfect.

4 Blood that tests negative can seroconvert at a later time.

45. (2) The basement membrane of the glomerulus becomes permeable to protein that is lost in the urine; decreased serum protein reduces the oncotic pressure in the capillaries, which normally helps to hold fluid within the vascular system. (3) (DA; AN; FE)

1 Hypoproteinemia causes decreased oncotic pressure, which results in hypovolemia; sodium and water are retained to counter the hypovolemia.

3 Increased oncotic pressure pulls fluid from the interstitial space into the intravascular compartment; nephrosis is characterized by hypoproteinemia and therefore the opposite is true.

4 The basement membrane becomes permeable to protein, not impermeable to water.

46 (3) If the IV infused at the prescribed rate of 45 ml/hr for 5 hours, 225 ml would have infused, leaving 275 ml in the bottle. (2) (DA; EV; FE)

1 This is incorrect; an excessive amount of fluid has been infused.

2 Same as answer 1.

4 This is untrue; the IV should be delivered at 45 ml/hr and should be monitored continually.

47. (3) The child should be given an outlet for tension, and playing with a syringe and doll is the most appropriate in this situation. (1) (DF; PL; EH)

1 This may ease some of the discomfort, but an outlet for feelings should be provided.

2 The child's fear is not caused by unfamiliar adults but by the painful treatments.

4 This is part of the preparation, but not the most important; the child must be allowed to express feelings; the child is too young to comprehend too much.

48. (2) The most serious and irreversible effects of lead intoxication are on the nervous system; lead encephalopathy causes convulsions, mental retardation, paralysis, blindness, and ultimately coma and death. (2) (DA; AN; NM)

1 Although constipation can occur, it is not caused by the excretion of lead; lead is excreted via the urinary tract.

3 Although some studies have identified that some children with plumbism have received less than adequate child care, a cause-and-effect relationship between plumbism and parental neglect has not been established.

4 Altered nutrition is not caused by a decrease in the intake of iron; high serum blood levels of lead interfere with the biosynthesis of heme, preventing the formation of hemoglobin and resulting in anemia.

49. (4) It is difficult to apply adequate pressure to this vascular site; therefore, observation for early signs of bleeding is imperative. (2) (CI; EV; RG)

1 Fluids should be encouraged as soon as the child tolerates them.
2 This simple procedure usually requires minimal or no IV fluid therapy.
3 Ice packs are not necessary.

50. (4) Many behaviors require a background of knowledge, skills, and attitudes (experiential readiness). If this background is not available, then more learning must take place before the individual can function. (3) (DF; AN; GD)

1 Although teaching may facilitate learning, other factors related to readiness for learning and willingness to learn affect whether learning occurs.
2 Learning progress varies according to the individual; there are both progressive and regressive periods.
3 The child with a disability will be able to learn the four-point gait without understanding normal walking.

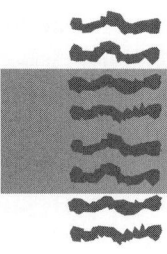

PEDIATRIC NURSING QUIZ 2

1. The nurse understands that a child with asthma is receiving albuterol (Proventil) to:
1. Relax smooth muscle
2. Thin pulmonary secretions
3. Enhance intercostal contractility
4. Stimulate respirations at the CNS level

2. The nurse knows that the treatment of choice in a 1-year-old child who has a risk of a hearing loss because severe otitis media has not responded to antibiotic therapy will probably be:
1. Steroids
2. Ear drops
3. A myringotomy
4. A mastoidectomy

3. When teaching a 12-year-old to use crutches, the nurse should teach the child to avoid:
1. Taking short steps of equal length
2. Looking forward to maintain balance
3. Maintaining an erect posture when walking
4. Looking down to watch where the crutches should be placed

4. A child is admitted to the hospital with acute lymphocytic leukemia (ALL). When performing the admission history and physical assessment, the nurse should assess the child for the common early signs of ALL including:
1. Nosebleeds and papilledema
2. Fever and areas of ecchymosis
3. Enlargement of the liver and spleen
4. Abdominal pain and reddened complexion

5. The physician orders a gavage feeding for an infant. To determine the length of tube needed to reach the stomach, the nurse should:
1. Advance the tube until resistance is met
2. Advance the tube as far as necessary to aspirate gastric contents
3. Measure from the mouth to the umbilicus and add half the distance
4. Measure the distance from the nose to the earlobe to the epigastric area of the abdomen

6. A Fredet-Ramstedt operation is performed on an infant who has pyloric stenosis. Postoperatively the nurse instructs the mother to:
1. Rock the child after feedings to reduce crying and ingestion of air
2. Feed the child in the supine position to reduce pressure on the sutures
3. Feed the child in the semi-Fowler's position and then afterward place the child prone
4. Feed the child in an upright position and then place the child on the right side with the head slightly elevated

7 A 5-year-old is admitted to the hospital with a diagnosis of acute glomerulonephritis. A nursing diagnosis of fluid volume excess would be correct if the assessment data included:
1. Dysuria, pruritus, weight loss
2. Diarrhea, polyuria, weight gain
3. Hypotension, tachycardia, hematuria
4. Periorbital edema, smoky urine, headaches

8 An 8-year-old who is experiencing sickle cell pain is admitted to the hospital. Appropriate nursing care during this period includes:
1. Titrating analgesics
2. Cold compresses to painful joints
3. Restricted fluids until the crisis is over
4. Active range-of-motion exercises to all joints

9. The mother of a school-age child asks the school nurse how her child could have gotten head lice. When replying the nurse should remember that:
1. Transmission occurs through household pets
2. Transmission just occurs through direct personal contact
3. Infestation is most common where crowded conditions exist
4. Infestation is more widespread among lower socioeconomic groups

10. A client brings her 3-week-old infant to the well-baby clinic for a checkup. The client is upset and states, "The baby cries all the time, and I don't know what to do." The nurse's initial response should be:
   1. "When did the baby eat last?"
   2. "What is the baby's daily schedule like?"
   3. "Did you sleep last night? You look exhausted."
   4. "Why don't you tell the pediatrician that the baby is colicky?"

11. Despite the physical distress and discomfort caused by the activity of weighing, a young child with burns must be accurately weighed because the weight provides a:
   1. Baseline for future growth
   2. Measure of the burned surface area
   3. Guideline for dietary and fluid management
   4. Basis for fluid replacement and medications

12. When teaching a child with the diagnosis of acute glomerulonephritis about the ordered diet, the nurse should explain that appropriate food choices would include:
   1. Corn on the cob, baked chicken, rice, apple, and milk
   2. Baked potato, ground beef, canned carrots, banana, and buttermilk
   3. Canned green beans, baked ham, bread and butter, peach, and milk
   4. Hot dog on a bun, potato chips, dill pickle slices, brownie, and buttermilk

13. The most appropriate therapeutic play activity for a 4-year-old whose arm is immobilized during IV therapy would be:
   1. Watching TV
   2. Comic books
   3. Jigsaw puzzles
   4. Cutting out paper dolls

14. A 1-year-old child with a very distended abdomen is admitted to the hospital with the diagnosis of Hirschsprung's disease. When planing nursing care, the most appropriate position for the child is:
   1. Prone
   2. Sitting
   3. Supine
   4. Side-lying

15. When teaching an adolescent client about diabetes and self-administration of insulin, the first step for the nurse should be to:
   1. Set specific and realistic short- and long-term goals

   2. Find out what the client knows about the health problem
   3. Begin the teaching program at the client's level of understanding
   4. Collect all the equipment needed to demonstrate giving an injection

16. When planning a menu for an adolescent with diabetes whose height and weight are average, the nurse considers a meal pattern that:
   1. Limits calories to prevent weight gain
   2. Avoids using potatoes, bread, and cereal
   3. Discourages substitutions on the menu pattern
   4. Allows for normal growth and developmental needs

17. An infant is born with talipes equinovarus (clubfoot) and the physician applies a cast to the lower extremity. The nurse knows that the mother will have to bring the baby back to the physician's office for a cast change:
   1. Each week
   2. Once a month
   3. Every other day
   4. When the cast is soiled

18. A 3-month-old with acute spasmodic bronchitis (spasmodic croup) is admitted to the hospital in severe dyspnea. The child has a temperature of 104° F. The child is placed in a high-humidity hood with cold mist. The nurse is aware that the main reason cold humidity is preferred to steam is that:
   1. Cold humidity dries up the mucosal secretions
   2. Cold humidity aids in reducing mucosal edema
   3. Cold mist produces a more comfortable environment
   4. Cool water vapor is more readily absorbed by the mucosa

19. A platelet transfusion is ordered for a child with acute lymphocytic leukemia, and an IV is started. The nurse should:
   1. Administer the platelets rapidly
   2. Administer the platelets over 2 hours
   3. Check vital signs 3 hours after the transfusion
   4. Flush the line with 5% dextrose and normal saline

20. An 18-month-old child is admitted to the hospital with an immunodeficiency syndrome. The toddler has never been hospitalized or separated from the mother prior to admission. Given this information, the nurse, during the initial admission, would expect the toddler to:
   1. Withdraw, sit quietly, and not be interested in playing

2. Cry relentlessly and be consoled by no one except the mother or father
3. Initially be unhappy and cry but become contented after meeting roommates
4. Cry when people enter the room but respond with a smile after a few minutes

21. To evaluate kidney function, the nurse must accurately measure the hourly urinary output for a 1½-year-old with severe, extensive burns. The minimum safe output per hour for this child would be:
    1. 10 to 20 ml
    2. 21 to 40 ml
    3. 41 to 60 ml
    4. 61 to 80 ml

22. A child with a fractured leg is to have no weight bearing on the affected leg. When measuring the client for crutches, the nurse knows that:
    1. The shoulders should be slightly stooped when the crutches are used
    2. The elbows should be in extension when crutches are held at the crossbar
    3. There should be a snug fit under the axillae when walking to provide support
    4. The tips of the crutches should be resting 6 inches from the feet when crutches are 2 inches below axillae

23. The parents of a newborn discuss with the nurse their child's need for immunizations. The nurse should explain the entire immunization schedule, including the fact that after 12 months of age, their child will receive the first immunization against:
    1. Polio
    2. Tetanus
    3. Measles
    4. Pertussis

24. The nurse can help a child with cystic fibrosis achieve normal growth and development by planning to teach the parents about the importance of a diet high in calories and:
    1. Low in protein
    2. Moderate in fat
    3. High in calcium
    4. High in potassium

25. Long-range management for children with asthma includes exercise. The nurse should teach the parents to encourage their son who has asthma to participate in the sport of:
    1. Soccer
    2. Baseball
    3. Wrestling
    4. Basketball

26. The nurse in a pediatric health clinic would be especially observant for signs of cerebral palsy in a 6-month-old infant who:
    1. Weighed less than 1500 g at birth
    2. Had a positive Moro reflex at birth
    3. Was born by elective cesarean delivery
    4. Was born to a mother older than 35 years

27. A preadolescent brings home a note from the school nurse informing the parents that the child should be evaluated for scoliosis. The mother calls the school nurse to ask what scoliosis is. To answer the question the nurse must understand that it is a:
    1. Pathologic process involving the vertebrae
    2. Concave lumbar curvature that is very exaggerated
    3. Lateral curvature of the spine with a rotary deformity
    4. Curvature of the thoracic spine that has an increased convex angulation

28. Following a successful craniotomy for the removal of a brain tumor in a 9-year-old, the nurse notes an area of serosanguineous drainage on the child's dressing about the size of a quarter. The nurse should:
    1. Notify the physician immediately
    2. Mark the area with nonabsorbable ink
    3. Reinforce the dressing with gauze pads
    4. Remove the dressing and check the sutures

29. An intravenous line is started in the scalp vein of an infant, and the mother asks why the IV is not placed in the hand as for an adult. The nurse responds:
    1. "Putting the IV in the scalp improves the absorption rate of the IV."
    2. "Inserting the IV in the scalp decreases the need to restrain the baby."
    3. "The IV solution is too irritating to be introduced through a vein in the hand."
    4. "Veins are closer to the surface of the scalp, making it easier to insert the IV."

30. A mother brings her 6-month-old infant to the emergency room with symptoms of severe gastrointestinal distress. The nurse should immediately prepare for:
    1. Insertion of an IV line
    2. Placement in an Isolette
    3. Blood type and crossmatch
    4. Intestinal intubation with continuous suction

31. The most important clinical indication of the degree of dehydration in a young infant would be:
    1. Dry skin
    2. Weight loss
    3. Sunken fontanel
    4. Decreased urine output

32. To facilitate optimum growth and development in a 10-week-old infant, the parents should be encouraged to provide age-appropriate activities such as:
    1. Giving the baby push-pull toys
    2. Playing pat-a-cake with the baby
    3. Placing the baby in an infant seat
    4. Keeping the baby in a carriage in the living room

33. To promote growth and development, the nurse instructs the mother of a 4-month-old to provide the infant with:
    1. Push-pull type toys
    2. Snap beads and strings
    3. Nesting blocks and cups
    4. Soft squeeze toys with squeakers

34. A 3-year-old, who was born prematurely and was a recipient of several blood transfusions while in the neonatal intensive care nursery, is admitted to the hospital with AIDS and is placed on bed rest with oxygen via a tent. The mother, having just been informed that her child has AIDS, is visibly upset at the child's bedside. The nurse's most therapeutic statement would be:
    1. "You must really feel like screaming."
    2. "Let me give you a referral for social service."
    3. "Your child will get the best care possible at this hospital."
    4. "It is a shame that your child needed so many blood transfusions."

35. A 9-year-old is admitted to the hospital for treatment of a badly infected wound of the toe. The aunt with whom the child now lives is uncertain whether the child has had a tetanus immunization. An immunizing dose of tetanus immune globulin is administered because it:
    1. Confers lifelong passive immunity
    2. Induces longer-lasting active protection
    3. Confers immediate passive protection of short duration
    4. Immediately stimulates increased production of antibodies

36. A male toddler is scheduled to receive methotrexate for treatment of leukemia. The mother asks the nurse whether the child can be started on vitamin supplements because he seems so weak. The nurse's best response would be:
    1. "Vitamin supplements won't help him feel any better right now."
    2. "That's a fine suggestion; I'll ask the physician to order some for him."
    3. "Yes, he will benefit from a vitamin supplement and will be receiving it soon."
    4. "Some vitamin preparations contain folic acid, which interferes with the drug."

37. A nurse should understand that when an adolescent with epilepsy is taking Dilantin and develops status epilepticus, the most common reason for its development is that the prescribed dosage of Dilantin:
    1. Was ineffective for the seizures
    2. Had reached a therapeutic level
    3. Was insufficient to cover activities
    4. Was probably not taken consistently

38. A 6-week-old infant has just had surgery for pyloric stenosis. An immediate postoperative nursing care priority for this infant would be:
    1. Giving the oral feedings very slowly
    2. Reporting any vomiting to the physician
    3. Checking the patency of the gastric tube
    4. Observing for signs of infection at the incisional site

39. An 8-year-old is admitted to orthopedics after falling out of a tree and suffering a supracondylar fracture of the right humerus. Traction is applied. With this type of fracture, the nurse should immediately report the presence of:
    1. Bruising near the right elbow
    2. Restlessness and apprehension
    3. A weak radial pulse in the right wrist
    4. Complaints of aching pain near the right elbow

40. The nurse teaches the mother of a 1-year-old who has frequent ear infections that the major cause of otitis media in young children is:
    1. Sinusitis
    2. Recurrent tonsillitis
    3. An inflamed mastoid process
    4. A malfunctioning eustachian tube

41. Phenobarbital is prescribed for an infant who experiences repeated seizures. One week later the child becomes lethargic and sleeps excessively. The nurse knows that:
    1. Another drug will be ordered to counteract this side effect
    2. Everyone has these responses, but the drug must be continued for life
    3. The child is getting too much medication and the dosage will be reduced
    4. This is probably a temporary response to the drug that will subside eventually

42. The nurse should be alert to identify the early signs of chronic lead poisoning (plumbism) in children such as:
    1. Anemia
    2. Oliguria
    3. Convulsions
    4. Mental retardation

43. The nurse can determine the success of chelation therapy in children by monitoring for:
    1. Elevated blood lead levels
    2. Increased fecal excretion of lead
    3. Increased urinary excretion of lead
    4. Decreased deposition of lead in the bones

44. A 4-year-old is admitted to the hospital for a diagnostic workup for pulmonic stenosis. The nurse understands that pulmonic stenosis is:
    1. Narrowing of the valve between the left atrium and left ventricle
    2. Hardening of the valve between the right atrium and right ventricle
    3. Hardening of the lining of the pulmonary artery at a point close to the lungs
    4. Narrowing of the valve between the right ventricle and the pulmonary artery

45. Three siblings are termed accident prone because of frequent incidents. The community health nurse involves the parents in a teaching program and instructs them to lower the risk of accidents by using discipline that emphasizes:
    1. Realistic rigidity
    2. Rational consistency
    3. Guarded indifference
    4. Serious overprotection

46. An infant is born with a cleft lip. The nursing care of this newborn, unlike that of other newborns, would include:
    1. Changing the infant's position frequently
    2. Using modified techniques to feed the infant
    3. Keeping the infant's head elevated at all times
    4. Maintaining the infant on strict intake and output

47. An 18-month-old with celiac disease is placed on a gluten-free diet. A teaching program about the diet is instituted with the child's parents. The nurse knows that the parents understand the teaching about food substances that must be avoided when they state that their child cannot have:
    1. Mashed corn
    2. Steamed rice
    3. Grilled frankfurter
    4. Fresh   applesauce

48. While informing the parents about the significance and the causes of cleft lip and palate, the nurse should:
    1. Emphasize that the two defects follow laws of mendelian genetics
    2. Assess the family history for presence of the defect in other siblings or relatives
    3. Stress that the defect is rare and will probably never happen twice in the same family
    4. Prepare the parents for the likelihood of mental and psychologic problems in the child

49. A 2-year-old is being prepared for surgery. As the nurse enters the room with a preoperative injection, the child begins to scream and flail about the bed. The father is sitting at the child's bedside and gets up to leave. The nurse could best handle this situation by:
    1. Allowing the child to say good-bye to the father before giving the injection
    2. Asking the father to stay to comfort and hold the child while the injection is given
    3. Telling the father that he can return to comfort the child after the injection is given
    4. Leaving the room and asking another nurse to come in to hold the child during the injection

50. The gluteal muscle is generally not used when giving IM injections to infants and young children because they:
    1. Fear intrusive procedures
    2. Associate this area with punishment
    3. Have an undeveloped muscle mass in this area
    4. Are able to wiggle and change position when placed on their abdomen

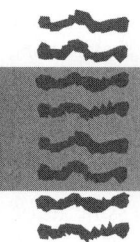

PEDIATRIC NURSING QUIZ 2
ANSWERS AND RATIONALES

1. (1) Albuterol is an adrenergic drug that stimulates beta receptors and leads to relaxation of the smooth muscles of the airway. (2) (PT; AN; DR)
   2 Albuterol does not affect the consistency of pulmonary secretions.
   3 Albuterol will not affect intercostal contractility.
   4 Albuterol provides no CNS stimulation.

2. (3) This is an opening into the eardrum to allow for drainage. (2) (CI; PL; NM)
   1 Antibiotics, not steroids, are used for an infectious process.
   2 These are not used because they obscure the view of the tympanic membrane.
   4 Removal of the mastoid will in no way relieve the pressure within inflamed ears.

3. (4) The client should maintain proper walking posture and not look down to ensure equilibrium and avoid losing balance. (2) (SI; PL; SK)
   1 This is a proper technique for safe ambulation while crutch walking.
   2 Same as answer 1.
   3 Same as answer 1.

4. (2) These symptoms are caused by unrestricted white blood cell proliferation and resultant decreased platelet production. (1) (HI; AS; BI)
   1 Papilledema is not a common presenting symptom because the blood-brain barrier is an initial deterrent.
   3 These are not the presenting symptoms; these occur through infiltration of the vascular organs of the reticuloendothelial system by immature WBCs.
   4 Pain is not an early symptom; the skin will be pale.

5. (4) Prior to inserting the gastric tube, the nurse measures the anatomic pathway the tube will follow—i.e., from the nose to the earlobe (corresponding to the nasopharynx) to the epigas-tric area of the abdomen (lower end of stomach). It is then marked and inserted to this point. (2) (PC; IM; GI)
   1 Resistance to the passage of a gastric tube may be felt, and rotation of the tube often changes placement enough to continue insertion to the point marked by measurement.
   2 This distance might not place the tube well into the stomach and would increase the risk of aspiration.
   3 This distance would be much too long.

6. (4) During and after feeding, the position most favoring gravity is employed to promote retention of fluid and prevent vomiting. (3) (CI; IM; GI)
   1 Vomiting may continue postoperatively so limited movement of the child after feedings is suggested.
   2 Feeding any child in the supine position greatly increases the risk of aspiration.
   3 Postoperative positioning with the head elevated helps to assist food passage; the prone position would promote vomiting.

7. (4) Periorbital edema is indicative of excess fluid; the kidneys are inflamed and the output of urine is lessened; hematuria can occur. (2) (HI; AS; RG)
   1 These symptoms do not indicate excess fluid.
   2 Diarrhea and polyuria would lead to fluid deficit and, perhaps, weight loss.
   3 Same as answer 1.

8. (1) The first priority is pain management; severe pain requires analgesics. (1) (PC; IM; BI)
   2 Cold will constrict blood vessels, further depleting oxygenation to affected parts; warmth is preferable.
   3 Increased hydration is necessary to promote hemodilution, improve circulation, and reverse sickling.
   4 There are too much swelling and pain in the joints during a crisis for this intervention.

9. (3) Transmission occurs through direct contact with infected individuals and indirect contact with contaminated articles; crowded conditions aid transmission. (1) (HI; IM; IT)
1 Lice are not carried or transmitted by household pets.
2 Pediculosis can also be transmitted by contact with infested clothing, personal articles, or bedding.
4 All socioeconomic groups are equally affected.

10. (2) This provides the client with the opportunity to express feelings and give the nurse more data. (3) (SC; AS; EH)
1 This could put blame for the crying on the mother by inferring that the baby is not being fed on time.
3 This is empathetic but gives the mother no opportunity to respond.
4 This is making a diagnosis without sufficient data collection.

11. (4) Body weight is used in the calculation of body surface area. It is the main criterion in determining drug dosage and fluid requirements. (3) (DA; AN; IT)
1 It is inappropriate to be concerned about growth in the face of an acute situation.
2 Measurement of the burned surface is determined by the parts of the body involved.
3 Dietary management is a later consideration.

12. (1) All these foods are permitted on low-sodium, low-potassium diets. (3) (PC; IM; RG)
2 Carrots are high in sodium, banana is high in potassium, and buttermilk is high in both.
3 Canned beans and ham are both very high in sodium.
4 All are fairly high in sodium and/or potassium.

13. (3) These are intellectually stimulating and can be done with an IV in place. (2) (DF; PL; GD)
1 This is a passive activity, which is not the most appropriate nor especially stimulating.
2 The child is not old enough to read.
4 The child needs both hands to do this activity.

14. (4) In this position the distended abdomen does not press against the diaphragm and lung expansion is facilitated. (3) (PC; PL; GI)
1 This position is not conducive to easier breathing, and it is difficult to assume because of abdominal distention.
2 The distended abdomen may press against the diaphragm and hinder full lung expansion.
3 The supine position will interfere with respiration because of abdominal distention.

15. (2) Before planning and instituting a teaching plan, the nurse must assess the client's attitudes, experience, knowledge, and understanding of the health problem. (2) (HI; IM; EN)
1 Before goals can be set, assessment must be carried out.
3 Before teaching can begin, assessment of the present level of knowledge is necessary.
4 Before teaching begins, assessment and planning must occur.

16. (4) As a result of hormones involved in growth and development, adolescents with type 1 diabetes have different and changing needs regarding nutrition and exogenous insulin. (1) (DF; PL; GD)
1 Adequate caloric intake is needed by the average-size adolescent because adolescence is a time of growth.
2 Potatoes, bread, and cereal contain necessary fiber and nutrients.
3 Some flexibility is needed to promote adherence to any dietary regimen.

17. (1) Casts are changed weekly to accommodate the rapid growth in early infancy. (2) (CI; PL; SK)
2 This is not frequent enough in early infancy; the cast could become too tight because of rapid growth of the infant.
3 This is too frequent; muscles and tendons would not be stretched and relaxed enough between cast changes to affect foot position.
4 Soiling is not usually a problem because casts for clubfoot do not extend to the perineal area.

18. (2) Cold causes vasoconstriction and reduces edema; it may also help to reduce fever. (2) (DA; AN; RE)
1 Heat dries secretions.
3 Cold mist is less comfortable because the environment is cold and damp.
4 Absorption via the mucosa is insignificant and cannot be considered fluid intake.

19. (1) Platelets are rapidly administered to prevent destruction after hanging the IV. (2) (CI; IM; BI)
2 Platelets should not hang for a long time because of their fragility.
3 This is too long an interval; during infusion of blood derivatives, vital signs are more closely monitored.
4 Dextrose solution is not appropriate for flushing a blood derivative line because it may become clogged.

20. ② The first phase of separation anxiety is protest, which is characterized by loud crying, rejection of all strangers, and inconsolable grief. (2) (DF; AS; EH)
 1 This would be indicative of despair, a more advanced stage of separation anxiety.
 3 Toddlers do not socialize well with peers.
 4 An 18-month-old is not so easily consoled when separated from parents.

21. ① The minimum safe urine volume is 10 to 20 ml per hour in children younger than 2 years. (2) (DA; EV; RG)
 2 This is the minimum safe output for children older than 2 years.
 3 This volume is more than the minimum for any age group.
 4 Same as answer 3.

22. ④ This provides a maximum base of support when the child ambulates and does not put pressure on the brachial plexus. (2) (CI; AN; SK)
 1 This would indicate that the crutches are too short.
 2 When holding the crossbar, the elbows should be flexed and the shoulders straight.
 3 This could cause a brachial plexus injury.

23. ③ This vaccine is not usually given before 12 months of age because of passive natural immunity from the mother. (2) (HI; IM; GD)
 1 This is given in the first 6 months of life.
 2 Same as answer 1.
 4 Same as answer 1.

24. ② Moderate fat is recommended because of caloric needs related to growth and development. (2) (PC; PL; GD)
 1 High protein is recommended to overcome protein maldigestion.
 3 High calcium is not recommended.
 4 Unless potassium levels are low, this would be contraindicated and dangerous.

25. ② Moderate exercise, including baseball and skiing, that involves starting and stopping does not overtax the respiratory mechanism; endurance exercise such as basketball, soccer, and wrestling are not well tolerated. (3) (SI; PL; RE)
 1 Endurance exercise overtaxes the respiratory mechanism.
 3 Same as answer 1.
 4 Same as answer 1.

26. ① Studies indicate that a large percentage of children with cerebral palsy had birth weights less than 1500 g. (2) (HI; AS; NM)
 2 The Moro reflex is normally present at birth.

 3 There is no greater incidence of cerebral palsy in children born by a cesarean delivery that is not done because of fetal distress.
 4 Studies do not indicate any greater incidence of cerebral palsy in children born to women older than 35 years.

27. ③ This is the correct definition. (1) (HI; AN; SK)
 1 There are no pathologic changes in the vertebrae.
 2 This is a description of lordosis.
 4 This is a description of kyphosis.

28. ② If the drainage progresses beyond the markings, it would enable the nurse to determine whether an abnormal amount of drainage was occurring. (1) (DA; EV; NM)
 1 This is not an emergency; some drainage is expected.
 3 This would not enable the nurse to monitor progression of the drainage.
 4 In the immediate postoperative period, the dressing is to be removed only by the neurosurgeon.

29. ④ This provides an accurate statement of why scalp veins are used in infants; other veins are not well developed for IV therapy. (2) (DF; IM; CV)
 1 The absorption rate through a peripheral vein is the same regardless of placement.
 2 The infant will still need to be restrained to prevent pulling out the IV or rolling over on it.
 3 Placement of the needle is not related to whether the solution is irritating.

30. ① GI distress causes a disturbance in intestinal motility and absorption, accelerating excretion. IV fluids are necessary. Severe dehydration and fluid and electrolyte imbalance occur rapidly and can lead to death in infancy because of the infant's large fluid content. (2) (DA; IM; FE)
 2 Isolettes serve to isolate, warm, and, if necessary, oxygenate small infants and newborns.
 3 This test is used to determine blood type when a transfusion is indicated.
 4 Intestinal intubation with suction is utilized to remove intestinal contents when there is a blockage or when it is necessary to have the GI tract clear of contents.

31. (2) Loss of fluid as a result of dehydration is most objectively assessed by measuring the infant's weight because body water accounts for approximately 60% of an infant's body weight. (3) (DA; AS; FE)
   1 Dry skin may be indicative of conditions other than dehydration.
   3 This is a clinical sign of dehydration but is not an accurate way to measure hydration.
   4 Decreased urine output is hard to measure in individuals who are not toilet trained.

32. (3) This is a suggested activity for a 2- to 3-month-old so that the infant can more readily observe the environment. (2) (DF; IM; GD)
   1 This is a suggested activity for a 12- to 24-month-old.
   2 This is a suggested activity for a 6- to 9-month-old.
   4 This is a suggested activity for a 4- to 6-month-old.

33. (4) This is appropriate for a 4-month-old; the child enjoys squeezing and hearing the sound of the squeaker. (1) (DF; IM; GD)
   1 This is appropriate for a child 12 to 24 months of age.
   2 This is appropriate for a child 10 to 12 months of age.
   3 This is appropriate for a child 16 months of age.

34. (1) This statement acknowledges the mother's feelings. (3) (SC; IM; EH)
   2 This statement does not address the mother's feelings.
   3 Although this statement may be true, it gives false hope for recovery.
   4 Same as answer 2.

35. (3) Tetanus immune globulin contains ready-made antibodies and only confers short-term passive immunity. (2) (SI; AN; BI)
   1 Passive immunity is temporary.
   2 Immune globulins confer passive artificial immunity.
   4 Immune globulins are antibodies and do not stimulate the formation of antibodies.

36. (4) Many vitamins contain folic acid and are contraindicated with methotrexate, a folic acid antagonist. (2) (PT; IM; DR)
   1 This is true, but it does not answer the question and leaves open vitamin use in the near future, which long-term chemotherapy contraindicates.
   2 Folic acid interferes with the action of methotrexate.
   3 This is incorrect; vitamin use is contraindicated.

37. (4) This behavior is a form of denial that may occur once the seizures are controlled; also, adolescents have a need to be like their peers, and they avoid anything that makes them seem different. (2) (DF; EV; GD)
   1 Drugs are prescribed according to the type of seizure.
   2 This is desired and indicates that the drug is effective.
   3 The dosage is not based on activity but on the type of seizure.

38. (3) A gastric tube is used postoperatively to decompress the stomach and limit tension on the suture line. (2) (CI; IM; GI)
   1 To limit pressure on the suture line, oral feedings should not be used in the immediate postoperative period when the gastric tube is in place.
   2 Vomiting indicates obstruction of the nasogastric tube; the initial action should be to check the patency of the nasogastric tube.
   4 This is too soon for signs of infection to occur.

39. (3) Edema at the fracture site may increase pressure on the radial artery, interrupting the blood supply to the hand of the affected extremity. (2) (CI; EV; SK)
   1 This is an expected occurrence caused by tissue trauma.
   2 By themselves these do not relate to the specific complication of radial artery pressure associated with a supracondylar fracture; with these findings further assessments regarding pain, bleeding, and respiratory distress should be made.
   4 Same as answer 1.

40. (4) A blocked eustachian tube impairs drainage and creates negative pressure; when the tube opens, bacteria are pulled into the middle ear. (2) (DF; AN; NM)
   1 This is usually not related to otitis media.
   2 This is caused by adenoiditis.
   3 This is a complication, not a cause, of otitis media.

41. (4) Drowsiness is a common side effect of barbiturate therapy because of its sedative properties. (2) (PT; EV; DR)
   1 Stimulants are not routinely administered because they would counteract the desired effect of seizure reduction.
   2 Although these are common responses, they are not expressed by everyone; the length of therapy depends on the client's status.
   3 This demonstrates little understanding of this drug therapy; this is a usual response to barbiturates.

42. (1) The bone marrow is most susceptible to lead toxicity; interference with hemoglobin biosynthesis leads to early signs of anemia. (3) (HI; AS; NM)

2 This is a late response indicating kidney shutdown; loss of protein and other substances occurs first.

3 This is a serious late response.

4 This is a late response indicating CNS involvement.

43. (3) The desired outcome is the increased excretion of lead in urine. (3) (PT; EV; DR)

1 This is expected when lead initially equilibrates to the blood; until lead is excreted in urine, the treatment is not considered a success.

2 Fecal elimination of lead is not as satisfactory as urinary elimination and is not as successful in ridding the soft tissues of lead.

4 This is a desirable effect, but it does not determine success of therapy; also, the amount is difficult to determine.

44. (4) The cusps of the valves may be fused or the infundibulum below may be hypertrophied, thus restricting blood flow to the lungs. (3) (HI; AN; CV)

1 The mitral valve is not involved in pulmonic stenosis.

2 The tricuspid valve is not involved in pulmonic stenosis.

3 This is untrue; pulmonic stenosis is a congenital condition that results from the failure of tissues to develop normally in utero.

45. (2) Unwavering adherence to the same principles and regulations promotes safe, firm limits. (1) (DF; PL; GD)

1 This stifles the child's natural development in learning to explore.

3 Injuries are promoted when there is lack of care.

4 This hinders the child's freedom to explore and enjoy the surroundings.

46. (2) With a cleft in the lip, the baby will be unable to suck like other newborns. (1) (PC; PL; GI)

1 This is common for all newborns, not just a baby with a cleft lip.

3 This is contraindicated in a newborn because the normal alignment of the spine will be interfered with if the head is elevated at all times.

4 This is not necessary because intake and output will be normal once a feeding method is established.

47. (3) Frankfurter has grain filler; parents should read labels and, unless they are sure of ingredients, should not feed the food to the child. (2) (PC; EV; GI)

1 This does not contain gluten; this is a substitute for grain foods.

2 Same as answer 1.

4 This does not contain gluten.

48. (2) Cleft lip and palate demonstrate a familial pattern of inheritance that is significantly increased when a close relative is similarly affected. (2) (DF; AS; GI)

1 Mendelian laws of inheritance do not apply to these defects.

3 The defects are familial; however, no exact pathogenesis has been found.

4 The way the young child responds to these defects depends on the parental response.

49. (2) The 2-year-old child is extremely fearful of separation as well as intrusive procedures. If parents are present, they should be encouraged to stay and give comfort. (2) (SC; IM; EH)

1 Two-year-olds still depend on their parents for comfort and control.

3 The father may provide comfort to the child during the procedure as well.

4 Two-year-olds are still dependent on their parents; the parent should be encouraged to participate in care.

50. (3) Infants and small children have small buttocks with a very proximal sciatic nerve. (2) (DF; AN; GD)

1 Children of this age fear the procedure no matter what site is chosen.

2 Preschoolers tend to associate many treatments, and illness itself, with punishment.

4 Children properly restrained can be held still in this position.

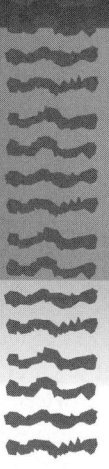

CHAPTER 3

Childbearing and Women's Health Nursing

REVIEW QUESTIONS

REPRODUCTIVE CHOICES (RC)

1. The nurse, as part of a teaching plan about contraception, tells the client that:
   1. The rim of the condom must be held in place while withdrawing the penis from the vagina
   2. Diaphragms are equally effective whether or not the partners choose to use a spermicidal cream
   3. No sperm can reach the ovum if the man uses coitus interruptus and withdraws before ejaculation
   4. Individuals using periodic abstinence should have intercourse on days when the woman has a rise in temperature

2. When obtaining the health history from a client who is seeking contraceptive information, the nurse should consider that oral contraceptives are contraindicated for a client who:
   1. Is older than 30 years
   2. Smokes a pack of cigarettes per day
   3. Has a history of borderline hypertension
   4. Has had at least one multiple pregnancy

3. A client is taking oral contraceptives. The nurse should inform the client to stop taking the contraceptive and report to the physician immediately if she experiences:
   1. Vertigo and nausea
   2. Weight loss and breast pain
   3. Hypotension and amenorrhea
   4. Headaches and visual disturbances

4. When counseling the client with diabetes mellitus who requests contraceptive information, it would be most therapeutic for the nurse to focus on:
   1. Rhythm
   2. The IUD
   3. A diaphragm
   4. Oral contraceptives

5. When teaching a client to use a diaphragm to prevent pregnancy, the nurse should tell the client that the diaphragm:
   1. May or may not be used with a spermicidal lubricant to be effective
   2. Must be inserted with the dome facing down to be maximally effective
   3. Often appears puckered, but this will not interfere with its effectiveness
   4. Should remain in place for at least 6 hours after intercourse to be effective

6. A couple tells the nurse that they wish to use the rhythm method of birth control. The wife tells the nurse that she menstruates every 32 days for 2 days. The nurse should teach the couple that, based on this cycle, ovulation probably occurs:
   1. On the 14th day of the cycle
   2. 10 days after the first day of bleeding
   3. 14 days before the start of the next menses
   4. 2 to 3 days after the last day of menstrual bleeding

7. A client states that she wishes to use the calendar method of birth control. The nurse is aware that the client understands how to calculate the beginning of the fertile period when she states, "I will:
   1. Subtract 11 days from the length of my longest cycle."
   2. Subtract 18 days from the length of my shortest cycle."
   3. Abstain from sexual intercourse after the 10th day of my cycle."
   4. Abstain from intercourse from the 10th day prior to the middle of my average cycle."

8. A client asks the nurse about the use of an intrauterine device (IUD) for contraception. When discussing this method with the client, the nurse includes that a common problem with IUDs is:
   1. Expulsion of the device
   2. Occasional dyspareunia
   3. Perforation of the uterus
   4. Frequent vaginal infections

9. The nurse informs a client contemplating implantable progestin (Norplant) as a means of contraception that the most common side effect is:
   1. Vertigo
   2. Dyspareunia
   3. An increase in breast size
   4. Irregular menstrual bleeding

10. A married client, 35 years old, who is to undergo a tubal ligation is assessed by the nurse to determine the client's possible emotional response to the procedure. A factor in the history that would contribute most to the healthy resolution of any emotional problem associated with sterilization would be that the client:
    1. Has a son and daughter and feels her family is complete
    2. Thinks the surgery will relieve her monthly dysmenorrhea
    3. Knows that her husband does not want her to have any more children
    4. Has just had a complicated delivery and never wants to undergo another birth again

11. A female client, undergoing infertility testing, is taught how to examine her cervical mucus. After listening to the instructions the client says, "That sounds gross. I don't think I can do it." The nurse is aware that:
    1. Having a baby is not that important to this client
    2. It is possible that the client is being unduly fastidious
    3. The client is afraid of finding out that the problem is her fault

4. Some women are uncomfortable touching their genitals and discharges

12. The birth control method that would be the safest and most reliable method for a postpartal client with type 1 diabetes would be:
    1. The rhythm method
    2. The vaginal sponge
    3. An oral contraceptive
    4. A diaphragm with spermicidal gel

13. A couple in their late thirties are expecting their third child and have been referred to the clinical genetics services. They have no family history of an inheritable problem and have some reservations about genetic counseling because they have heard that genetic clinics favor abortion when the studies reveal a defective fetus. The nurse can put them at ease by telling them that genetic counselors:
    1. Help families understand the diagnosis, the probable cause of the disorder, and how the condition can be managed
    2. Recommend abortion only if the fetus is diagnosed as having a condition that is not considered compatible with life
    3. Are able to predict which families are going to have defective children and then recommend contraception, not abortion
    4. Make the diagnosis of a defective fetus and then the family's own physician is responsible for taking the appropriate action on the diagnosis

14. A client has regular 30-day menstrual cycles. In teaching about the rhythm method, which the client and her husband have chosen to use for family planning, the nurse should emphasize that the client's most fertile days are:
    1. Days 9 to 11
    2. Days 12 to 14
    3. Days 15 to 17
    4. Days 18 to 20

15. When teaching about normal childbearing and contraceptive options, the nurse explains that fertilization of the ovum by the sperm occurs when:
    1. The male sperm count is high
    2. The ovum reaches the endometrium of the uterus
    3. One sperm successfully penetrates the wall of the ovum
    4. The sperm prevents the ovum from moving along the tube

16. A couple have been married for 5 years and would like to start a family. When talking with them regarding the timing and frequency of their sexual in-

tercourse, the husband says, "Well, I guess we are going to have to jump into bed three or four times a day, every day until it works." The nurse's best response would be to:
1. Tell them to continue relations as usual until the tests are completed
2. Instruct them on the frequency and timing of intercourse for conception
3. Discourage this behavior because sperm production decreases with frequent intercourse
4. Agree that the frequency of intercourse must increase, but three or four times daily is excessive

17. A young client who has become sexually active asks the nurse, "What is the most effective way to prevent a pregnancy?" The nurse's best response would be:
1. "Abstain from sex."
2. "Use birth control pills."
3. "Use a condom and foam."
4. "Have an intrauterine device inserted."

18. A client asks the nurse what she should do if she forgets to take the pill one day. The nurse's best response would be:
1. "Take your pills as instructed."
2. "Call the physician immediately."
3. "Continue as usual; missing one day is no problem."
4. "The next day take one pill in the morning and one before bedtime."

19. A client misses one period after being on the pill for 3 months. The nurse should assess her first for:
1. Illness
2. Anorexia
3. Pregnancy
4. Compliancy

REPRODUCTIVE PROBLEMS (RP)

20. A client, 16 weeks gestation, is being treated for *Trichomonas vaginalis*. The statement that best indicates to the nurse that the client has learned measures to prevent recurrence is, "I will:
1. Void immediately after intercourse."
2. Persuade my sexual partner to be treated."
3. Insert a vaginal suppository after intercourse."
4. Douche immediately after having sexual intercourse."

21. A client, admitted to the hospital for surgery for a ruptured ectopic pregnancy, asks the nurse why she has shoulder pain. The nurse would base a response on the fact that the pain is caused by:
1. Anxiety about the diagnosis
2. Cardiac changes from hypovolemia
3. Blood accumulation under the diaphragm
4. Rebound tenderness from the ruptured tube

22. The most appropriate nursing diagnosis for a client with a ruptured ectopic pregnancy would be:
1. Risk for infection
2. Fluid volume excess
3. Decreased cardiac output
4. Altered health maintenance

23. A 17-year-old client tells the nurse that her sister had a tubal pregnancy about 3 months ago and had to have her tube removed. The nurse knows that this young woman needs further explanation when she states:
1. "This kind of thing can happen to my sister again."
2. "I guess I'll have to wait awhile to become an aunt."
3. "This kind of thing can happen after a pelvic infection."
4. "My sister is lucky because she'll never have a period again."

24. After 5 years of unprotected intercourse, a childless couple comes to the infertility clinic. The husband tells the nurse that his parents have promised to make a down payment on a house for them if his wife gets pregnant this year. The nurse's best response to this comment would be:
1. "How do the two of you feel about having a baby?"
2. "You're lucky; I wish someone would give me a down payment for a house."
3. "Five years without a pregnancy is a long time. Do you think there is something wrong with both of you?"
4. "You know, you don't have to worry about satisfying your parents. Having a child should be a decision you make."

25. A 37-year-old female comes to the gynecologic clinic with a history of being unable to conceive after 4 years of unprotected intercourse. The client complains of recurring headaches with pain radiating down the neck. The physical examination reveals a blood pressure of 170/100 mm Hg. The client is advised to limit salt intake and take her basal temperature for 3 months. The health professionals are:
1. Not helping the client
2. Avoiding the major problem
3. Overlooking the hypertension
4. Encouraging self-responsibility

26. A client comes to the infertility clinic for a carbon dioxide insufflation test to determine whether her fallopian tubes are patent. As part of the teaching before the test the nurse tells the client:
    1. "You will receive a local anesthetic to lessen the pain of the test."
    2. "You will have to rest in bed for 8 hours after the test is completed."
    3. "You may have some persistent shoulder pain for 24 hours after the test."
    4. "You may become nauseated during the test, but the nausea will subside."

27. Because an infertility workup involves both partners, a male client is to have a semen analysis. As part of his instructions the nurse should tell him to:
    1. Use a condom to collect the semen specimen
    2. Make sure that the specimen is collected as soon as he awakens
    3. Ejaculate 2 to 3 days before collection to ensure a pure specimen
    4. Refrigerate the specimen until it can be delivered to the laboratory

28. On a return visit to the infertility clinic, a client whose temperature charts demonstrate an ovulatory pattern and a normal menstrual cycle despite an inability to become pregnant requests fertility drugs. The nurse should recognize that the client:
    1. Has a right to receive this drug
    2. Will require an endometrial biopsy
    3. Has to be scheduled for a culdoscopy
    4. Needs to bring her husband's semen in to be examined

29. A client who is to have a vacuum and curettage abortion at 10 weeks gestation should be told that:
    1. A general anesthetic will be used to insert the laminaria tent
    2. A fever just above 100° F is common for the first 24 to 48 hours
    3. The laminaria tent will have to be retained in the cervical canal for 4 to 24 hours
    4. A heavy amount of bleeding will be present for 3 to 5 hours after the abortion

30. After a first-trimester aspiration abortion, the nurse knows that the instructions are understood when the client states:
    1. "I will be able to resume intercourse in 4 to 5 days."
    2. "After 24 hours I can substitute tampons for sanitary pads."
    3. "I can expect my menstrual period to resume in 2 to 3 weeks."
    4. "I will call the physician if I must change my sanitary pad more than once in 4 hours."

31. A pregnant client who has one living child resulting from a full-term pregnancy has also had two spontaneous abortions. She is recorded as being:
    1. Gravida IV, para I
    2. Gravida I, para IV
    3. Gravida II, para III
    4. Gravida III, para II

EMOTIONAL NEEDS RELATED TO CHILDBEARING AND WOMEN'S HEALTH (EC)

32. The nurse is aware that during the taking-in phase of the postpartum period, the area of health teaching that the client will be most responsive to is:
    1. Perineal care
    2. Infant feeding
    3. Infant hygiene
    4. Family planning

33. A postpartum adolescent mother confides to the nurse that she hopes her baby will be good and sleep through the night. The nurse should plan to teach the mother to:
    1. Put a soft, cuddly toy next to the baby at bedtime
    2. Cuddle baby and talk softly when crying occurs
    3. Add cereal to the bedtime bottle to ensure deep sleep
    4. Keep the baby awake for longer periods during the day

34. The husband of a woman who had her fourth child 3 weeks ago states she has been irritable and crying since coming home from the hospital. The nurse tries to assist him in understanding the situation by stating that:
    1. His wife probably has postpartum blues that will pass soon
    2. Having four children is tiring and assistance may be needed
    3. This behavior is common after delivery and he should not be too concerned
    4. Often, women express themselves by crying and he should allow her to continue

35. A client comes to the infertility clinic for a workup and is told by the nurse to prepare for a Papanicolaou test. The client states, "I do not want this test. I want to speak to the doctor." The nurse should:
    1. Recognize that the client is uncooperative
    2. Inform the physician of the client's request
    3. Encourage the client to comply with clinic procedures
    4. Remind the client of the importance of cancer detection

36. A client in the infertility clinic is being treated for hypertension and obesity with a regimen of diet and exercise. During the past month she has lost 8 pounds and her blood pressure has gone down to 154/98 mm Hg. The client states that she is using self-control strategies to reduce her blood pressure and weight. The nurse should:
    1. Acknowledge the client's achievement and encourage continuation of the client's action program
    2. Encourage the client to take antihypertensive drugs until blood pressure is reduced to normal limits
    3. Emphasize the importance of the client's exercising in addition to the reduction of sodium and calories
    4. Point out to the client that her action program is inadequate because her blood pressure remains abnormal

37. A client with mild preeclampsia is told that she must remain on bed rest at home. The client starts to cry and tells the nurse that she has two small children at home who need her. The nurse's best response would be:
    1. "You'll need someone to care for the children."
    2. "You are worried about how you will be able to manage."
    3. "You can get a neighbor to help out, and your husband can do the housework in the evening."
    4. "You'll be able to fix light meals, and the children can go to nursery school a few hours each day."

38. The best nursing intervention to achieve cooperation of an extremely anxious client during her first pelvic examination to confirm pregnancy would be to:
    1. Assist the physician so that the examination can be finished quickly
    2. Distract the client by asking her preference as to the sex of the child
    3. Maintain eye contact, touch gently, and thoroughly explain the procedure
    4. Encourage the client to close her eyes and hold her breath during the pelvic examination

39. A pregnant client whose first child has Down syndrome is about to undergo an amniocentesis. The client tells the nurse that she does not know what she will do if this baby has the same diagnosis. The client asks the nurse, "Do you think abortion is the same as killing?" The nurse's best response would be:

1. "Some people think this is what an abortion is."
2. "No, I don't think so, but it's your decision to make."
3. "I really can't answer that question. You seem ambivalent about abortion."
4. "I don't want to answer that question at this time. How do you feel about it?"

40. A client who has just delivered an infant with Down syndrome tells the nurse that she could not possibly take a retarded child home and asks whether she should plan to place the child in an institution. An appropriate statement by the nurse at this time would be:
    1. "I understand how you feel, and I will notify the nursery personnel of your decision."
    2. "At this young age no one is able to predict your baby's ultimate level of functioning."
    3. "Give yourself time to get acquainted and you will see that your baby isn't retarded yet."
    4. "You should not make such a hasty decision, as your baby is like any other baby right now."

41. A client with severe pregnancy-induced hypertension (PIH) who has been admitted to the hospital anxiously asks the nurse, "Will my baby be all right?" The nurse's most appropriate response would be:
    1. "There is no way of telling at this time what the outcome will be."
    2. "The baby will probably be all right; it's protected by the amniotic fluid."
    3. "If you do what the doctor tells you to do, everything will progress normally."
    4. "We will be constantly monitoring your baby's condition. Would you like to listen to the baby's heartbeat?"

42. An infant born in a birthing center is experiencing respiratory distress and is being transferred to a regional neonatal intensive care unit. The nursing action that would best promote parent-infant attachment would be:
    1. Encouraging the parents to call their infant by name
    2. Allowing the parents to hold their infant before departure
    3. Giving the parents a picture of their infant in the intensive care unit
    4. Instructing the parents to phone the neonatal intensive care unit daily

43. The nurse should be aware of the stages of parental adjustment that follow delivery of an infant at risk who is in the neonatal intensive care unit. To better plan nursing care, nursing observations and assessments should be based on the recognition that the:
    1. Mother should not see the infant until she has completed the necessary grief work
    2. Parents should be encouraged to visit the newborn within the first 24 hours after birth
    3. Mother should be reunited with her infant as soon as possible to enhance adjustment
    4. Nurse should wait until the parents request to see the newborn before suggesting a visit

44. On her first visit to the neonatal intensive care unit to see her preterm newborn daughter, the mother stands 2 feet away and does not touch the infant. The mother's only comment to the nurse is, "She looks so fragile. Do you think she'll make it?" The most appropriate comment for the nurse to make would be:
    1. "The staff is confident because all preterm babies look like this at first."
    2. "The baby is small, but many infants born as small as she is have done just fine."
    3. "I can understand that she looks fragile to you. What have you learned about her condition?"
    4. "She's not as fragile as she might appear. You need to get used to her and then it won't be so frightening."

45. When first seeing her preterm infant in the neonatal intensive care unit (NICU), the mother immediately starts to cry and refuses to touch the baby. The nurse understands that this behavior represents:
    1. A normal detachment behavior
    2. An incomplete bonding behavior
    3. A normal reaction to the situation
    4. A reaction to the NICU environment

46. While watching her preterm infant son in the neonatal intensive care unit, a mother exclaims, "My baby is so little. How will I ever care for him?" The nurse should explain to the mother that she:
    1. Can watch his care to assist her in becoming familiar with the specific routines
    2. Should find someone with training in preterm care to help her at home the first week
    3. Will be able to care for him in a special nursery for a few days prior to his discharge
    4. Will be encouraged to participate in his care as much as possible from the beginning

47. The nurse observes a postpartum client's behavior but cannot decide whether the client is anxious about her baby or experiencing postpartum depression. The client's behaviors that would clarify this confusion for the nurse would be:
    1. Decreased appetite, crying, and insomnia
    2. Long periods of sleep, lethargy, and anorexia
    3. Ambivalence, lethargy, and increased appetite
    4. Increased appetite, insomnia, and ambivalence

48. A client, who has participated in caring for her infant in the neonatal intensive care unit for several days in preparation for the baby's discharge, comes to the unit on the last hospital day with alcohol on her breath and slurred speech. The nurse's most appropriate action would be to:
    1. Speak openly to the mother about her condition and have her see a social worker about discharge to a foster home
    2. Talk with the mother about her condition and assess her willingness to participate in an alternate plan for discharge
    3. Continue with the discharge procedure, alerting the home health nurse that immediate follow-up is needed for the mother
    4. Avoid confrontation by asking the mother to wait in the hospital lobby and calling the physician to cancel the discharge order

49. Before an amniocentesis, both parents express nervousness about the fetus' safety during the test. The nursing intervention that would best promote the parents' ability to cope is:
    1. Initiating a parent-physician conference
    2. Reassuring them that the procedure is safe
    3. Informing them about the procedure step by step
    4. Arranging for the father to be present during the test

50. During labor, a client tells the nurse that she and her husband are very concerned because the baby is coming a whole month early. The nurse's best response would be:
    1. "Your physician is very good; try not to worry about it now."
    2. "I don't blame you for feeling worried; tell me your concerns."
    3. "I can understand why you and your husband are so worried."
    4. "Don't worry; the care of preterm babies has greatly improved."

51. A neonate born at 32 weeks gestation and weighing 3 pounds is admitted to the neonatal intensive care nursery. The nurse plans to take the neonate's mother to visit the infant in the intensive care unit:
    1. As soon as the mother feels up to it
    2. When the infant's condition has stabilized

3. When the infant is out of immediate danger

4. After the physician writes an order permitting it

52. The mother of an infant in the neonatal intensive care nursery expresses concern about her infant. To best facilitate mother-infant bonding the nurse should:

1. Assure the mother that her baby will be fine

2. Have the mother stroke the baby whenever possible

3. Avoid discussing negative aspects of the baby's condition

4. Encourage the mother to let the nursing staff care for the baby

53. A common concern of the mother after an unexpected cesarean delivery that the nurse should anticipate is the:

1. Postoperative pain and scarring

2. Prolonged period of hospitalization

3. Inability to assume her mothering role

4. Sense of failure in the birthing process

54. After an unexpected emergency cesarean delivery, the client tells the nurse "I am a natural childbirth flunky." The postpartal phase of adjustment that this statement most closely typifies is:

1. Taking in

2. Letting go

3. Taking hold

4. Working through

55. A client who has just had her second child wishes to breast-feed. When the nurse brings the baby to be breast-fed, the mother asks whether she may drink a small glass of wine to help her relax. The nurse's best response would be:

1. "I'm sure that drinking one glass of wine would not cause any harm."

2. "Yes, it's relaxing, but I do think you should find another, better way to relax."

3. "You seem a little tense. Tell me about your past breast-feeding experiences."

4. "I'm sure a glass of wine would be OK, but you had better check with your physician."

56. On the first postpartum day, a client who is rooming-in asks the nurse to return her baby to the nursery and bring the baby to her only at feeding time. The best response by the nurse would be:

1. "I think you are having difficulties caring for the baby."

2. "All right, I will inform the other nurses of your decision."

3. "It seems you have changed your mind about rooming-in."

4. "Oh, you must be tired. I'll bring the baby back at feeding time."

57. The nurse is aware that babies born to very young mothers are at risk for neglect or abuse because adolescents characteristically:

1. Do not plan for their pregnancies

2. Cannot anticipate the baby's needs

3. Are involved in seeking their own identity

4. Resent having to give constant care to the baby

58. The parents of a male newborn ask the nurse whether they should have their son circumcised. The nurse's most appropriate response would be:

1. "It would probably be a good idea because circumcision is known to prevent penile cancer."

2. "That's something you both will have to decide after you discuss it thoroughly with your doctor."

3. "I'm sure you have discussed this with your doctor, but let's review the benefits and risks of circumcision."

4. "The Academy of Pediatrics recommends that circumcision not be done routinely because of the risks associated with the procedure."

59. The nurse is aware that a common nursing diagnosis for clients following an abdominal hysterectomy is:

1. Sexual dysfunction

2. Reflex incontinence

3. Risk for disuse syndrome

4. Altered growth and development

60. A client in preterm labor does not respond to therapy and must be delivered for maternal stabilization. The client begins to cry and says, "I'm so worried about my baby." The nurse should respond:

1. "You're receiving the best medical care available."

2. "All of this must leave you very confused and frightened."

3. "Think positively; your anxiety will increase your contractions."

4. "This hospital has a neonatal unit and can handle emergencies."

61. A client who suspects that she is 6 weeks pregnant appears mildly anxious as she is waiting for her first obstetrical appointment. A symptom the nurse would expect this client to exhibit is:

1. Dizziness

2. Breathlessness

3. Abdominal cramps

4. Increased alertness

62. When a client in the prenatal clinic is dressing at the completion of her pelvic examination, she states, "Why must I be pregnant now? It's the wrong time." It would be most therapeutic for the nurse to respond:
    1. "Why don't you want this pregnancy?"
    2. "This is a normal response to pregnancy."
    3. "No time is ever the right time to be pregnant."
    4. "You don't seem excited about this pregnancy."

63. A client is scheduled for a modified radical mastectomy. Initially, her preoperative nursing care should focus on:
    1. Allowing her to ventilate feelings about surgery
    2. Encourage range of motion of the upper extremities
    3. Arranging for a visit by a Reach for Recovery volunteer
    4. Increasing her knowledge about postoperative expectations

64. While waiting for his 38-year-old wife to change clothes following an amniocentesis, the husband says to the nurse, "I sure hope that they don't find anything wrong. I don't know how we would deal with a retarded child. We have two small children at home." The nurse's best response would be:
    1. "Your other children are healthy, chances are the third will be also."
    2. "It must be difficult worrying about whether or not the baby is healthy."
    3. "There are plenty of resources available to help families with retarded children."
    4. "The potential of Down syndrome children differs, making early decisions difficult."

65. The nurse is preparing to counsel a client whose two previous pregnancies were normal with term vaginal deliveries of healthy children. The nurse demonstrates an understanding of this client by recognizing that:
    1. Multiparas cope more successfully with pregnancy than do primigravidas
    2. Each pregnancy is a unique experience; therefore, this pregnancy is a crisis
    3. This pregnancy is a greater crisis because the client has two children to take care of at home
    4. Support persons are less important because the mother has gone through similar experiences before

66. A husband is sitting in the waiting room while his wife is getting her infertility prescription refilled by the clinic pharmacist. The nurse sits down beside him and he blurts out, "It's like there are three of us in bed—my wife, me, and the doctor." This is reflective of his feelings of:

1. Guilt
2. Anger
3. Depression
4. Unworthiness

DRUG-RELATED RESPONSES (DR)

67. During labor a client receives an epidural. The nurse should assess the client for a frequent side effect of this anesthesia, which is:
    1. Urinary frequency
    2. Respiratory distress
    3. Maternal tachycardia
    4. Hypotensive episodes

68. While receiving tocolytic therapy for preterm labor, the client begins to experience muscle tremors and nervousness and states, "My heart is racing." The nurse should:
    1. Get a medical order to discontinue the medication
    2. Recognize that these are the usual symptoms of preterm labor
    3. Review the client's laboratory results for electrolyte and glucose levels
    4. Reassure the client that these are expected side effects of the medication

69. The physician orders penicillin G benzathine suspension (Bicillin L-A) 2.45 million units for a client with a sexually transmitted disease. The drug is available in a multidose vial of 10 ml in which 1 ml = 300,000 units. The nurse should administer:
    1. 8.8 ml
    2. 8.2 ml
    3. 0.8 ml
    4. 0.008 ml

70. A client who is 33 weeks pregnant has contracted gonorrhea and is placed on probenecid (Benemid) and penicillin therapy. The nurse knows that the client understands the action of the drugs when she states that the probenecid will:
    1. "Minimize any allergy to penicillin."
    2. "Increase the penicillin in my blood."
    3. "Reduce the side effects of the disease."
    4. "Activate my immune defense mechanisms."

71. A pregnant client with an infection tells the nurse that she has taken tetracycline for infections on other occasions and would prefer to take it now. The nurse tells the client that tetracycline is avoided in the treatment of infection in pregnant women because it:
    1. Adversely affects breast-feeding
    2. Permanently stains the baby's teeth

3. Produces allergies to the drug in the baby

4. Increases the baby's tolerance to the drug

72. A client is placed on progesterone oral contraceptives (minipills) and is instructed by the nurse to take 1 pill daily:
    1. Throughout the menstrual cycle
    2. During the 5 days surrounding ovulation
    3. During the first 5 days of the menstrual cycle
    4. Throughout the first 21 days of the menstrual cycle

73. The nurse would know that a client taking oral contraceptives understood the teaching about the estrogen when the client indicates that the most common side effect of the estrogen would be:
    1. Amenorrhea
    2. Hypomenorrhea
    3. Nausea and vomiting
    4. Depression and lethargy

74. The nurse evaluates that a client understands the teaching about oral contraception when the client states that she will immediately cease taking the pill if she experiences:
    1. Chest pain
    2. Menorrhagia
    3. Mittelschmerz
    4. Increased leukorrhea

75. A pregnant client with iron-deficiency anemia is prescribed iron supplements daily. To increase iron absorption the nurse should suggest that the client eat foods high in:
    1. Vitamin C
    2. Fat content
    3. Water content
    4. Vitamin B complex

76. A 39-year-old who is Rh negative is seen by the physician during the first trimester of pregnancy. The nurse's teaching is effective if the client understands that she will first receive Rho(D) immune globulin (RhIg) at:
    1. 12 weeks gestation
    2. 28 weeks gestation
    3. 36 weeks gestation
    4. 40 weeks gestation

77. The nurse should be aware that the only anticoagulant drug that a pregnant client with thrombophlebitis can safely receive is:
    1. Dicumarol
    2. Anisindione
    3. Heparin sodium
    4. Warfarin sodium

78. A pregnant client on anticoagulant therapy has blood drawn daily for activated partial thromboplastin time (aPTT). One day her aPTT is 98 seconds. The nurse notifies the physician because the anticoagulant should be:
    1. Increased for better clotting results
    2. Discontinued because the aPTT is normal
    3. Changed to one of the other effective anticoagulants
    4. Omitted for today and the aPTT should be rechecked tomorrow

79. To halt preterm labor, a client is started on terbutaline (Brethine). The nurse is aware that a side effect of this drug is:
    1. Bradycardia
    2. Hyperkalemia
    3. Widening pulse pressure
    4. Hypertonic uterine contractions

80. When a client is receiving Pitocin, the nurse, aware of the adverse effects of this oxytocic drug, should carefully observe the client for:
    1. Intrauterine pressure of 60 mm Hg
    2. Contractions with a duration of 30 seconds
    3. A fetal heart rate of 120 to 150 beats per minute
    4. Contractions occurring more frequently than every 2 minutes

81. An adverse reaction from prolonged Pitocin administration for which a client in labor must be closely monitored is:
    1. Hyperventilation
    2. Water intoxication
    3. A change in affect
    4. An elevated temperature

82. A client with severe preeclampsia is receiving 2 grams/hour of IV magnesium sulfate. To evaluate the effectiveness of this therapy, the nurse should assess for:
    1. Excessive urinary output
    2. A decreased respiratory rate
    3. An increase in blood pressure
    4. Absence of clonus and hyperreflexia

83. The nurse understands that when magnesium sulfate is given to clients with pregnancy-induced hypertension, it can build to toxic levels. The nurse should withhold the drug and notify the physician if an assessment of the client reveals:
    1. Respirations of 10/min
    2. A BP of 140/100 mm Hg
    3. Urinary output of 30 ml/hr
    4. Deep tendon reflexes of +2

84. A client with severe preeclampsia who has a BP of 170/110 mm Hg, a pulse of 108 beats per minute, and respirations of 24/minute is placed on IV magnesium sulfate therapy. Eight hours later her BP is 150/110, the pulse is 98, and respirations are 10, and there is absence of the knee-jerk reflex. The nurse should:
   1. Stop the infusion of magnesium sulfate and notify the physician
   2. Administer calcium gluconate as an antidote for the magnesium sulfate
   3. Administer the next dose of magnesium sulfate because the blood pressure is still high
   4. Wait 1 hour, monitor the vital signs and reflexes again, and then administer the next dose

85. A client with preeclampsia who is receiving magnesium sulfate ($MgSO_4$) is showing signs of magnesium sulfate toxicity. The nurse is aware that these signs can be reversed by the administration of:
   1. Calcium gluconate (Kalcinate)
   2. Edetate disodium (Disodium EDTA)
   3. Hydralazine hydrochloride (Apresoline)
   4. Sodium polystyrene sulfonate (Kayexalate)

86. Before giving medications to a client who is 6 hours postpartum, the nurse assesses the client and notes the following findings: BP 178/110 mm Hg; TPR 98/60/18; fundus firm, one finger below umbilicus; episiotomy edematous, red, and approximated; and one Peri-pad saturated with lochia rubra in 6 hours. In light of these assessment findings, the nurse should contact the physician before administering:
   1. Cephradine (Velosef)
   2. Hydrocortisone acetate (Epifoam)
   3. Methylergonovine maleate (Methergine)
   4. Casanthranol and docusate sodium (Peri-Colace)

87. The nurse understands that a drug that is contraindicated for a woman who is breast-feeding would be:
   1. Heparin
   2. Propylthiouracil (PTU)
   3. Gentamicin (Garamycin)
   4. Diphenhydramine (Benadryl)

88. The nurse should withhold methylergonovine maleate (Methergine) from a postpartum client if the client is found to have a:
   1. Negative Homans' sign
   2. Negative Babinski reflex
   3. Blood pressure of 160/90 mm Hg
   4. Respiratory rate of 12 per minute

89. A preterm infant is started on digoxin (Lanoxin) and furosemide (Lasix) for persistent patent ductus arteriosus. The nursing assessment that would provide the best indication that the Lasix is effective is that the:
   1. Pedal edema is reduced
   2. Fontanels appear depressed
   3. Urine output exceeds fluid intake
   4. Drug has not precipitated digitalis toxicity

90. A client's preterm labor is to be treated with terbutaline (Brethine), a beta-adrenergic agent. The nurse understands that one of the maternal side effects of this drug is:
   1. Tachycardia
   2. Hyporeflexia
   3. Hyperkalemia
   4. Hypoglycemia

91. Based on knowledge of the side effects of terbutaline (Brethine), the nurse should expect that if antidote therapy is needed, the physician will order:
   1. Ritodrine (Yutopar)
   2. Levodopa (L-Dopa)
   3. Furosemide (Lasix)
   4. Propranolol (Inderal)

92. A client in her 30th week of gestation is in preterm labor and the physician orders betamethasone. The client asks the nurse why she is receiving the drug. The nurse explains that it is used to:
   1. Prevent chorioamnionitis
   2. Increase uteroplacental exchange
   3. Promote neonatal pulmonary maturity
   4. Treat fetal respiratory distress syndrome

93. A client with a large fetus is to have a pudendal block during the second stage of labor. The nurse plans to instruct the client that once the block is working she:
   1. Will not feel an episiotomy
   2. May lose the ability to push
   3. May lose bladder sensation
   4. Will no longer feel contractions

HEALTHY CHILDBEARING (HC)

94. A client comes to the antepartal clinic and states that her last menstrual period (LMP) was May 31. Using Nägele's rule the nurse would estimate the expected date of delivery (EDD) to be:
   1. March 7
   2. March 30
   3. February 24
   4. February 27

95. When discussing fetal development with a pregnant client, the nurse explains that:
1. If a baby is to be left-handed after birth, the structures on the left side of the embryo's body develop before those on the right side
2. Development proceeds from head to tail, so the embryo's brain, central nervous system, and heart are formed and functioning before limb buds appear
3. The brain and central nervous system are formed early in the embryonic period, and the right side of the brain starts functioning much earlier than the left side
4. Development of structures in the embryo proceeds from external to internal structures so the heart is formed and starts functioning after the limb buds appear

96. A positive early diagnosis of pregnancy is based on the presence of:
1. Quickening
2. Chadwick's sign
3. A fetal heart rate
4. Chorionic gonadotropin

97. In a client's 10th week of pregnancy, the presumptive signs of pregnancy that might be assessed by the nurse include:
1. Fatigue, abdominal enlargement, and HCG in her urine
2. Abdominal enlargement, urinary frequency, and nausea
3. Nausea and vomiting, urinary frequency, and amenorrhea
4. Breast changes, abdominal enlargement, and urinary frequency

98. A pregnant client is experiencing nausea and vomiting. The nurse is aware that this discomfort:
1. Is always present in early pregnancy
2. Will disappear when lightening occurs
3. Is a common response to an unwanted pregnancy
4. May be related to the HCG (human chorionic gonadotropin) level

99. The nurse is aware that absorption of drugs taken orally during pregnancy may be altered as the result of:
1. Delayed gastrointestinal emptying
2. Decreased glomerular filtration rates
3. Developing fetal-placental circulation
4. Increased secretion of hydrochloric acid

100. The nurse instructs a pregnant client about the sources of protein that assist in meeting the daily requirements of:

1. 15 grams
2. 30 grams
3. 45 grams
4. 60 grams

101. A client who is 8 weeks pregnant tells the nurse that she does not feel like making love to her husband and is concerned that her husband may not understand. The nurse could most appropriately respond:
1. "How long have you had this problem?"
2. "Why don't you feel like having intercourse?"
3. "I'm sure your husband eventually will understand your feelings are related to pregnancy."
4. "A decrease in libido is a normal occurrence in pregnant women during the first trimester."

102. A primigravida complains of morning sickness. The nurse should plan to teach her to:
1. Increase fluid intake
2. Increase calcium in her diet
3. Eat three small meals a day
4. Avoid long periods without food

103. The nurse discusses with a newly pregnant client, who is 5 feet, 3 inches tall and weighs 125 pounds, the recommended weight gain during pregnancy. The nurse is aware that, to fall within the recommended weight gain during pregnancy, at term, the client should weigh about:
1. 130 pounds
2. 135 pounds
3. 140 pounds
4. 150 pounds

104. A client who is in her 7th week of gestation asks the nurse when she can expect to feel her baby move. The nurse should reply that quickening usually occurs in the:
1. Twelfth week
2. Sixteenth week
3. Twentieth week
4. Twenty-fourth week

105. A prenatal client tells the nurse that she thinks she has developed an allergy, since her nose is often very congested and she is unable to breathe. The nurse states:
1. "Using a nasal decongestant once or twice a day will help."
2. "It is common for women to develop allergies during pregnancy."
3. "This is not normal; perhaps you have a chronic respiratory infection."
4. "It is a normal occurrence; the increased hormones are responsible for the congestion."

106. The nurse should explain to the newly pregnant primigravida that the fetal heartbeat will first be heard with:
1. A fetoscope around 8 weeks
2. A fetoscope at 12 to 14 weeks
3. An electronic Doppler after 17 weeks
4. An electronic Doppler at 10 to 12 weeks

107. A client who is pregnant returns for her second prenatal visit. She asks why she has to urinate so often. The nurse tells her that urinary frequency in the first trimester is:
1. Caused by the baby's head descending into the uterus
2. Influenced by the enlarging uterus, which is still contained in the pelvis
3. A result of the mother's kidneys filtering more waste products because of the growing fetus
4. Mostly a psychologic phenomenon that results from knowing that the pregnancy has occurred

108. A client at 10 weeks gestation tells the nurse that she voids often, without dysuria, and would like to know what to do. The nurse is aware that this client will have to:
1. Decrease her fluid intake during the day
2. Contact her physician as soon as possible
3. Maintain increased fluid intake during the day
4. Try to resist the urge to void as long as possible

109. Between 12 and 24 weeks gestation, applicable prenatal teaching for a pregnant client should include information about:
1. Preparation for the baby, travel to the hospital, signs of labor
2. Growth of the fetus, personal hygiene, and nutritional guidance
3. Growth of the fetus, interventions for nausea and vomiting, expectations for care
4. Danger signs of preeclampsia, relaxation breathing techniques, and signs of labor

110. A client is now in her second trimester. While listening to the fetal heart, the nurse hears a heartbeat at the rate of 136 in the right upper quadrant and also at the midline below the umbilicus. The sources of these sounds are:
1. Heart rates of two fetuses
2. Maternal and fetal heart rates
3. Fetal heart rate and funic souffle
4. Uterine souffle and fetal heart rate

111. A client, 28 weeks pregnant, has gained 13 pounds and tells the nurse in the antepartum clinic that she is glad she has not gained as much weight as her sis-ter during pregnancy. An appropriate response by the nurse to the client's comment would be:
1. "Do you think you are getting fat?"
2. "Are you trying to watch your figure?"
3. "You have to eat right during pregnancy."
4. "Tell me what you have been eating lately."

112. A client, 28 weeks gestation, experiences constant low back pain and asks the nurse to teach her exercises that will provide some relief. The nurse would consider the teaching effective if the client performs:
1. Leg lifts
2. Tailor sitting
3. Pelvic rocking
4. Kegel's exercises

113. The nurse tells a pregnant client that fetal weight gain during pregnancy is:
1. Begun in the second trimester
2. Greatest during the first trimester
3. Most marked in the third trimester
4. Equally distributed throughout pregnancy

114. The nurse explains to a pregnant couple that in childbirth classes the emphasis is on:
1. Birth as a family experience
2. Labor without using analgesics
3. Nutrition, relaxation, and breathing
4. Education, breathing, and exercise

115. A pregnant client, interested in childbirth education, asks how the Lamaze method differs from the Read method. The nurse explains that the Lamaze method:
1. Is a much easier method to teach and learn
2. Requires a good deal of prenatal preparation
3. Forbids the use of pain-relieving drugs during labor
4. Is a calm, relaxed approach based on "childbirth without pain"

116. During a childbirth class the nurse evaluates that the clients understand how to use effleurage correctly when the clients are observed:
1. Rocking gently back and forth on their knees
2. Practicing panting to avoid pushing during labor
3. Taking deep breaths before simulated contractions
4. Massaging their abdomens gently with their fingertips

117. The need to improve pubococcygeal muscle tone is explored during a childbirth class. The nurse recognizes that a client understands the instructions

about how to strengthen the muscle when she states she should do:
1. Tailor sitting
2. Pelvic rocking
3. Forward tilting
4. Kegel's exercises

118. The nurse would know that a pregnant client does not understand the teaching about fetal growth and development when the client states:
1. "The mother must observe proper nutrition."
2. "The fetus gets food from the amniotic fluid."
3. "There are two umbilical arteries and one vein."
4. "The baby's oxygen needs are provided for by the mother."

119. During a class for prepared childbirth, the nurse discusses the importance of the "spurt of energy" that occurs prior to labor. The nurse teaches the women to conserve this energy because:
1. Fatigue may influence the need for pain medication
2. Energy helps to increase the woman's progesterone level
3. This energy decreases the intensity of uterine contractions
4. Extra energy is needed to push during the first stage of labor

120. To ensure proper nutrition during pregnancy for a client who is a vegetarian, the nurse should:
1. Advise the client to include meat in her diet at least once daily
2. Help the client plan meals that use foods that she is willing to eat
3. Encourage the client to join a diet group to teach her good nutrition
4. Tell the client to discontinue the vegetarian practice entirely until she delivers

121. To evaluate whether a short-term goal for a low-income pregnant client with iron-deficiency anemia is being met, the nurse asks the client:
1. "Do you try to have beef liver twice a week?"
2. "Do you have red meat at least once a week?"
3. "Have you increased lean meat in your diet to four meals a week?"
4. "Did you like the three-bean casserole from the recipe I gave you?"

122. Folic acid supplements are prescribed for a prenatal client. The nurse is aware that these are necessary to prevent:
1. Pernicious anemia
2. Anaphylactic shock
3. Neural tube defects
4. Erythroblastosis fetalis

123. When discussing dietary needs during pregnancy, a client tells the nurse that milk constipates her at times. The nurse should explain that it is preferable to:
1. Substitute a variety of cheeses for the milk
2. Substitute skimmed or buttermilk for whole milk
3. Increase her prenatal capsules and omit the milk
4. Treat constipation in some way other than omitting milk

124. During a counseling discussion of nutrition, the nurse explains to a pregnant client that she will need additional calcium during pregnancy and that the best source is milk. The client states, "I never drink milk or eat milk products. They turn my stomach." The best reply by the nurse would be:
1. "How unfortunate; this may cause your teeth to loosen."
2. "Just make sure the rest of your diet is nutritionally sound."
3. "There are many mineral and vitamin supplements the physician can order for you."
4. "You will have to try and drink some milk so that your baby will have strong bones."

125. The nurse explains to a pregnant client who does not like milk that there are other foods that are good sources of calcium and advises the client to eat:
1. Corn
2. Liver
3. Broccoli
4. Lean meat

126. The nurse teaches a client that the increased need for vitamin A to meet rapid tissue growth during pregnancy may be met by using increased amounts of food such as:
1. Carrots
2. Nonfat milk
3. Citrus fruits
4. Extra egg whites

127. The nurse knows that a client in early pregnancy understands the need to increase her intake of complete proteins during her pregnancy when she reports she is eating more:
1. Nuts and seeds
2. Milk, eggs, and cheese
3. Beans, peas, and lentils
4. Whole grains and breads

128. A newly married client visits the clinic because she has not been feeling well. The nurse suspects that the client is probably pregnant when:
1. Her menses is already 7 days late
2. Her urine immunoassay test is positive
3. She relates that urinary frequency occurs every 2 to 3 hours
4. She complains of nausea and vomiting episodes every morning

129. A client who has type 1 diabetes had a nonstress test that was reactive. The nurse would recognize that the client understood what she was taught about the results when she is overheard telling her husband that the test was:
1. Normal, due to an increase in FHR with fetal movement
2. Abnormal, due to a decrease in FHR with fetal movement
3. Abnormal, due to an increase in FHR with maternal movement
4. Normal, due to the FHR remaining unchanged with maternal movement

130. At a routine monthly visit, while assessing a client who is in her 26th week of gestation, the nurse identifies the presence of striae gravidarum, which:
1. Are brownish blotches on the face
2. Is a bluish discoloration of the cervix
3. Are reddish streaks on the abdomen and breasts
4. Is a black line from the umbilicus to the mons veneris

131. The nurse understands that an ultrasound examination ordered for a pregnant client in the first trimester is used primarily to:
1. Estimate fetal age
2. Detect mental retardation
3. Rule out congenital defects
4. Approximate fetal linear growth

132. A pregnant client has a serum blood test for elevated alpha-fetoprotein (AFP). The nurse is aware that this test is done to detect the presence of:
1. Cystic fibrosis
2. Phenylketonuria
3. Down syndrome
4. Neural tube defects

133. A client in the 16th week of pregnancy is scheduled for ultrasonography. When preparing the client for the procedure, the nurse informs her that for this test it will be necessary to:
1. Have an enema the night before the examination
2. Monitor closely afterward for signs of precipitate labor

3. Fast for 12 hours to minimize the possibility of vomiting
4. Drink at least 1 liter of water 1 to 2 hours before the test and avoid urinating during that time

134. The nurse understands that when a contraction stress test is interpreted as negative it means:
1. The test should be repeated in 24 hours because examination results indicate hyperstimulation
2. The fetus at this time is likely to tolerate the stress of contractions but the test should be repeated weekly
3. Immediate delivery should be considered because there is no fetal heart acceleration with fetal movement
4. A trial induction should be started because fetal heart rate acceleration with movement is indicative of a false result

135. Nursing care for a pregnant client who is to have an oxytocin challenge test should include:
1. Having the client empty her bladder
2. Placing the client in a supine position
3. Keeping the client on nothing by mouth
4. Preparing the client for insertion of internal monitors

136. At 38 weeks gestation an amniocentesis is done to determine fetal maturity. The proper L/S ratio for lung maturity is:
1. 1:1
2. 1:2
3. 2:1
4. 1.5:1

137. The nurse is aware that a client at 40 weeks gestation is experiencing true labor if:
1. Cervical dilation has occurred
2. Her membranes have ruptured
3. The pains become more noticeable
4. The fetal heart rate baseline decreases

138. A primigravida, 36 weeks gestation, is admitted to the labor suite with ruptured membranes and a cervix that is 2 cm dilated and 75% effaced. A priority question the nurse should ask is:
1. "What is your expected date of delivery?"
2. "Are you planning to breast-feed or bottle-feed?"
3. "What time was your last meal and what did you eat?"
4. "How frequent are your contractions and how long do they last?"

139. The nurse uses nitrazine paper to test the pH of the leaking vaginal fluid from a laboring client. Amniotic fluid will turn nitrazine paper the color:
1. Red
2. Blue
3. Purple
4. Orange

140. For a client with a fetus in the left sacrum position, the nurse should place the fetal heart transducer on the client's abdomen in the:
1. Left lower quadrant
2. Left upper quadrant
3. Right upper quadrant
4. Midline of the lower quadrant

141. When doing Leopold's maneuvers on a client who has just been admitted to the labor room suite, the nurse notes the presence of a firm round prominence over the pubic symphysis, a smooth convex structure down her right side, irregular lumps down her left side, and a soft roundness in the fundus. The nurse should conclude that the fetal position is:
1. RSA
2. LOA
3. LOP
4. ROA

142. A laboring couple ask the nurse about the cause of low back pain during labor. The nurse replies, "This occurs most often when the position of the baby is:
1. Breech."
2. Transverse."
3. Mentum anterior."
4. Occiput posterior."

143. The nurse encourages the husband of a client who is experiencing back pain during labor to comfort her by:
1. Positioning her with her legs elevated
2. Having her perform a panting breathing pattern
3. Applying pressure to the sacrum during contractions
4. Encouraging her to do Kegel's exercises between contractions

144. In the 40th week of gestation, a client is admitted to the labor unit. On auscultation the fetal heart is heard in the lower right quadrant, and on palpation the occiput is identified as anterior. The nurse determines the fetal presentation to be:
1. LOP
2. LSA
3. ROA
4. RMA

145. The teaching plan for a father who is acting as a coach during labor would include the information that it would be best for him to:
1. Leave his wife alone periodically so that she can rest between contractions
2. Let his wife know the progress she is making and that she is doing a good job
3. See that his wife remains supine so that the monitoring equipment is not disturbed
4. Keep the conversation in the labor room to a minimum so that his wife can concentrate

146. For an actively laboring client whose cervix is 4 cm dilated and 100% effaced with the fetal head at 0 station, the best nursing intervention should be to:
1. Check the fetal heart rate every 5 minutes and record it on her chart
2. Call the anesthesia department and alert them to an imminent delivery
3. Ask the client how bad her pain is and whether she wants medication for it
4. Continue to assist the client's husband to coach her in the use of breathing techniques

147. The nurse would be aware that the husband of a client in the early phase of labor had successfully understood the teaching from childbirth classes when he is observed assisting his wife to:
1. Pant-blow breathe
2. Slow-chest breathe
3. Shallow-chest breathe
4. Accelerate-decelerate breathe

148. A husband who is coaching his wife during labor demonstrates an understanding of the transitional phase of labor when with each contraction he instructs his wife to:
1. Take cleansing breaths and push
2. Take quick, shallow breaths and blow
3. Use slow, rhythmic diaphragmatic breathing
4. Switch from accelerated to decelerated breathing

149. After being in labor for 6 hours a client is admitted to the labor unit. The client is 5 cm dilated and at -1 station. In the next hour her contractions gradually become irregular but are more uncomfortable. When caring for her, the nurse should first check for:
1. False labor
2. A full bladder
3. Uterine dysfunction
4. A breech presentation

150. A client in active labor is admitted to the birthing room. A vaginal examination reveals a 6 to 7 cm dilation. Based on this finding the nurse should expect that this client would:
    1. Have a profuse bloody show
    2. Appear unable to control her shaking legs
    3. Be uncomfortable because of nausea and vomiting
    4. Have contractions every 3 to 5 minutes of 60-second duration

151. As the nurse inspects the perineum of a client who has been admitted in early labor, the client suddenly turns pale and says she feels as if she is going to faint even though she is lying flat on her back. The nurse should:
    1. Elevate her feet
    2. Elevate her head
    3. Turn her on her left side
    4. Start oxygen and IV fluids

152. During labor, when the fetus is below the ischial spines, the nurse recognizes that the head is said to be at:
    1. +1
    2. −3
    3. Floating
    4. Station 0

153. During labor a client begins to experience dizziness and tingling of her hands. The nurse instructs the client to:
    1. Use a fast deep-breathing pattern
    2. Pant during the next three contractions
    3. Hold her breath with the next contraction
    4. Breathe into her cupped hands or a paper bag

154. A client in the active phase of labor states, "I feel all wet. I think I urinated." The nurse should first:
    1. Give her the bedpan
    2. Change the bed linens
    3. Inspect the perineal area
    4. Auscultate the fetal heart rate

155. The nurse is aware that when external fetal monitoring is compared to internal fetal monitoring, one advantage is that external fetal monitoring:
    1. Is simpler to read
    2. Allows the client freedom of movement
    3. More accurately monitors the fetal heart
    4. Does not introduce foreign materials into the uterus

156. When direct monitoring of the fetal heart rate during the first stage of labor shows an irregular baseline with variability, a nursing priority is to:
    1. Administer oxygen
    2. Notify the physician
    3. Change the client's position
    4. Continue to monitor the client

157. During labor a client has an internal fetal monitor applied. The nurse should take action in response to a fetal heart rate that:
    1. Does not drop during contractions
    2. Fluctuates from 130 to 140 beats per minute
    3. Uniformly drops to 120 bpm with each contraction
    4. Repeatedly drops abruptly to 90 bpm unrelated to contractions

158. A client in early labor is receiving oxytocin. When observing late decelerations in the fetal heart rate, the nurse should first:
    1. Administer oxygen
    2. Place her on her left side
    3. Check the blood pressure
    4. Discontinue the oxytocin infusion

159. The position a client should be taught to avoid when she experiences back pain during labor is the:
    1. Sitting position
    2. Supine position
    3. Side-lying position
    4. Knee-chest position

160. A client's membranes rupture. The nurse, observing an abrupt deceleration in the fetal heart rate, inspects the vaginal area and notes a prolapsed cord. The nurse should immediately:
    1. Administer oxygen by face mask at 7 L/minute
    2. Elevate the presenting part off the cord until delivery
    3. Notify the physician of the findings of the examination
    4. Instruct the client to assume a dorsal recumbent position

161. Priority nursing intervention for a laboring client with a sudden prolapse of the umbilical cord protruding from the vagina should focus on:
    1. Gently replacing the cord in the vaginal vault
    2. Checking the fetal heart rate every 15 minutes
    3. Covering the cord with a sterile moist dressing
    4. Starting oxygen at 10 L/minute via a tight face mask

162. After a client's membranes rupture spontaneously, the nurse observes the umbilical cord protruding from the vagina. The nursing action that should receive the highest priority in this situation is:
    1. Raising the foot of the bed
    2. Administering oxygen by mask

3. Auscultating the fetal heart tones
4. Preparing for a cesarean delivery

163. When assessing a laboring client for signs that the transitional phase is beginning, the nurse would expect the client to have:
1. Bulging of the perineum
2. Pinkish vaginal discharge
3. Crowning of the fetal head
4. Rectal pressure during contractions

164. The nurse is aware that the management of a client in transition is primarily directed toward:
1. Helping the client maintain control
2. Decreasing the client's fluid intake
3. Having the client breathe simple patterns
4. Reducing the client's discomfort with medication

165. While having contractions every 2 to 3 minutes that last from 60 to 90 seconds, a client complains of having rectal pressure. The nurse should:
1. Attach an external fetal heart monitor
2. Inspect the client's perineum for bulging
3. Determine when the client's labor began
4. Ask the client whether her membranes have ruptured

166. When a client experiences the urge to push at 9 cm dilation, the breathing pattern that the nurse should instruct the client to use is the:
1. Expulsion pattern
2. Slow-chest pattern
3. Slow-paced pattern
4. Panting or blowing pattern

167. When doing patterned, paced breathing during the transitional phase of labor, a client experiences tingling and numbness of the fingertips. The nurse should instruct her to breathe into:
1. A paper bag
2. An oxygen mask
3. The room's atmosphere
4. A trichloroethylene mask

168. The nursing action that has the highest priority for a client in the second stage of labor is to:
1. Check the fetal position
2. Help the client push effectively
3. Administer medication for the pain
4. Prepare the client to breast-feed on the delivery table

169. The nurse is aware that when a local anesthetic is used for delivery:
1. Labor is slowed after its administration

2. There is a danger of maternal aspiration
3. Maternal respirations may be depressed
4. Reactions such as vertigo and tinnitus may occur

170. A client in active labor starts screaming, "The baby is coming!" The first nursing action should be to:
1. Call the physician
2. Check the perineum
3. Tell her it is impossible
4. Keep the client's knees together

171. A client arrives in the labor room suite with the fetal head crowning. The nurse recognizes that delivery is imminent and tells the client to:
1. Push with all her power
2. Use the pant-breathing pattern
3. Assume the Trendelenburg position
4. Hold her breath and turn to the left side

172. A client's membranes rupture during the transition phase of labor and the amniotic fluid appears pale green. Because of this finding, at delivery the nurse should be prepared to:
1. Stimulate the baby to cry
2. Administer oxygen by face mask
3. Put a moist saline dressing on the cord stump
4. Provide for suctioning of the oropharynx when the head emerges

173. The nurse evaluates that a client's placenta has separated during the third stage of labor when:
1. There is a gush of blood
2. The uterus becomes boggy
3. The uterus decreases in size
4. There is an abrupt drop in BP

174. Five minutes after delivery, the nurse midwife assesses that the client's placenta is separating when:
1. The fundus becomes completely relaxed
2. There is a lengthening of the umbilical cord
3. The client complains of unbearable abdominal pain
4. Bright red blood continually seeps out of the vaginal opening

175. The primary outcome for client care in the third stage of labor would be:
1. An absence of discomfort
2. A firmly contracted fundus
3. An efficient FH beat to beat variability
4. A maternal respiratory rate between 16 and 20

176. After a cesarean delivery, the nurse performs fundal checks every 15 minutes. During one check the nurse notes that the fundus is soft and boggy. The priority nursing action at this time is to:
1. Elevate the client's legs
2. Massage the client's fundus
3. Increase the Pitocin drip rate
4. Examine the client's perineum for bleeding

177. The nurse should assess a client in the fourth stage of labor for:
1. Ability to relax
2. Level of maternal love
3. Distention of the bladder
4. Knowledge of newborn behavior

178. After delivery the mother's vital signs are T 99.4° F, P 80 regular and strong, R 16 slow and even, and BP 148/92 mm Hg. The assessment the nurse should monitor more frequently is the client's:
1. Pulse
2. Respiration
3. Temperature
4. Blood pressure

179. In the second hour after delivery a client's uterus is found to be firm, above the level of the umbilicus, and to the right of midline. The appropriate intervention would be to:
1. Observe for signs of retained secundines
2. Tell the client that this is a sign of uterine stabilization
3. Assist the client to the bathroom to empty her bladder
4. Massage her uterus vigorously to prevent hemorrhage

180. The nurse is aware that a client could be at increased risk for postpartum hemorrhage if the client:
1. Breast-fed in the delivery room
2. Received a pudendal block for delivery
3. Delivered a baby who weighed 9 lb, 8 oz
4. Had a third stage of labor that lasted 10 minutes

181. After delivery, a client tells the nurse, "I'm so cold, and I can't stop shaking." The nurse should tell the client:
1. "I am going to take your blood pressure and pulse."
2. "Let me check your fundus to see whether it is firm."
3. "Please turn on your side so I can check the amount of lochia."
4. "I will put some warm blankets on you; the chill will subside soon."

182. The best nursing intervention to minimize perineal edema following an episiotomy would be:
1. Hot sitz baths tid
2. Aspirin 10 grains po q4h
3. Ice packs to the perineum
4. Elevation of hips on a pillow

183. In the postpartum period, the nurse anticipates that a primipara with a second-degree laceration and repair is most likely to develop:
1. Posterior vaginal varicosities
2. Difficulty voiding spontaneously
3. Delayed onset of milk production
4. Maladaptive bonding and attachment

184. A primipara has a right mediolateral episiotomy following a vaginal delivery of an 8 lb baby. While assisting the client with a sitz bath, the nurse recalls that a mediolateral episiotomy is associated with:
1. Less swelling
2. More comfort
3. Less bleeding
4. More infections

185. A client required an extensive episiotomy because her baby was large. A priority nursing intervention that will minimize edema and lessen discomfort at the episiotomy site would be:
1. Spraying the area with a local anesthetic
2. Applying some form of cold to the perineum
3. Giving the client an oral analgesic immediately
4. Positioning the client on her side, off the incisional area

186. The nurse teaches a postpartum client how to care for her episiotomy at home. The nurse would know that the priority instruction was understood when the client states:
1. "I should discontinue the sitz baths once I am in my own home."
2. "I must not climb up or down stairs for at least 3 days after discharge."
3. "I must continue perineal care after I go to the bathroom until healing occurs."
4. "I should continue the sitz baths three times a day if it provides me with comfort."

187. Twelve hours after a normal spontaneous delivery, a client's temperature is 100.4° F. This elevation is most likely an indication of:
1. Mastitis
2. Dehydration
3. Puerperal infection
4. Urinary tract infection

188. Twenty-four hours after an uncomplicated labor and delivery, a client's CBC reveals a WBC count of 17,000/mm³. The nurse should interpret this WBC count as being indicative of:
1. A normal decrease in white blood cells
2. A normal response to the labor process
3. An acute sexually transmitted viral disease
4. A bacterial infection of the reproductive system

189. To help prevent postpartum infection, the most important discharge instruction given by the nurse would be:
1. "Don't take tub baths for at least 6 weeks after delivery."
2. "Wash your hands before and after changing your sanitary napkins."
3. "Douche gently with Betadine solution twice a day for a week or more after delivery."
4. "Tampons are better than sanitary napkins for inhibiting bacteria in the postpartum period."

190. After delivery, a client tells the nurse she was very uncomfortable during her pregnancy because of varicose veins. In light of this information, the nurse's assessment should include:
1. A daily clotting time
2. Tests for platelet fragility
3. Frequent Hgb and HCT values
4. Monitoring for signs of thrombophlebitis

191. The comment by a new mother of twins on the fourth postpartum day that indicates to the nurse the need for further assessment is:
1. "I hope I'll be a good mother to these sweet babies."
2. "I've been urinating large amounts ever since I delivered."
3. "My breasts feel full, heavy, and tingly before I feed the babies."
4. "My lochia is bright red with small brown clots the size of my thumb."

192. The type of lochia the nurse should expect to observe on the fifth postpartum day is:
1. Scant alba
2. Scant rubra
3. Moderate rubra
4. Moderate serosa

193. A client vaginally delivers a 7 pound, 2 ounce baby and has made the decision to breast-feed the infant. When instructing the client regarding breast-feeding, the nurse tells the client to expect that:
1. Weight loss will occur rapidly
2. Lochial flow will be increased
3. Uterine involution will be delayed
4. Use of heat will be contraindicated

194. When preparing a client to breast-feed, the nurse teaches the client that:
1. High levels of progesterone stimulate the secretion of oxytocin
2. High levels of estrogen stimulate secretions of lactogenic hormones
3. Milk secretion is under the control of hormones and starts immediately after delivery
4. Suckling stimulates the pituitary gland to release oxytocin, which initiates the let-down reflex

195. A breast-feeding mother asks the nurse how human milk compares with cow's milk. The information given to the client should include the following:
1. Lactose content is significantly higher in cow's milk
2. Protein content in human milk is higher than that in cow's milk
3. Immunologic and antiallergic factors are now found in cow's milk
4. Fat in human milk is easier to digest and absorb than that in cow's milk

196. A statement by a breast-feeding mother that indicates that the nurse's teaching about stimulating the breast-feeding let-down reflex has been successful is, "I will:
1. Drink at least 2 quarts of low-fat milk a day."
2. Take a cool shower before each breast-feeding."
3. Wear a snug-fitting breast binder 24 hours a day."
4. Apply warm moist packs and massage my breasts before each feeding."

197. As a result of her newborn's vigorous sucking, a client's nipples become sore and tender. The best nursing action would be to:
1. Apply continuous ice packs
2. Give analgesic medication as ordered
3. Expose the nipples to air several times a day
4. Remove the baby from the breasts for a few days

198. A breast-feeding mother asks the nurse what she can do to ease the discomfort caused by a cracked nipple. The nurse should instruct the client to:
1. Stop breast-feeding for 2 days to allow the nipple to heal
2. Manually express milk and feed it to the baby in a bottle
3. Use a breast shield to keep the baby from direct contact with the nipple
4. Feed the baby on the unaffected breast first until the affected breast heals

199. A new mother wishes to breast-feed her infant and asks the nurse whether she needs to alter her diet needs. The nurse can best respond:
1. "Just eat as you have been doing during your pregnancy."
2. "Just drink a lot of milk; you need the calcium to make your own milk."
3. "Don't worry about it, your body will produce the amount of milk your baby needs."
4. "You'll need greater amounts of the same foods you've been eating and more fluids."

200. A few weeks after discharge, a postpartal client develops mastitis and telephones for advice concerning breast-feeding. The nurse should tell the client to:
1. Start to wean the baby from the breast because it will reduce the pain
2. Get an antibiotic from the physician and start the baby on bottle feedings
3. Pump her breasts and wear a tight-fitting bra to suppress milk production
4. Breast-feed often because this will keep the breasts empty and reduce pain

201. The nurse discusses breast care with a mother who is bottle-feeding her infant. The nurse plans further teaching when the client states:
1. "May I have my medication for the discomfort in my breasts?"
2. "The discomfort I am feeling will go away in a couple of days."
3. "How should I apply heat to my breasts to help my milk dry up?"
4. "I must call my husband and ask him to bring my new brassiere."

202. Before discharge the nurse should teach the non-breast-feeding mother that if breast engorgement occurs, she should:
1. Wear a tightly fitted brassiere
2. Take 2 aspirins every 4 hours
3. Cease drinking milk for 2 weeks
4. Apply hot compresses to the breasts

203. A client who is not breast-feeding is taught how to care for her engorged breasts. The nurse realizes that the client does not understand the teaching when the client states:
1. "I'm wearing a well-fitting, tight brassiere."
2. "I am drinking 2 quarts of fluid every 24 hours."
3. "I'm expressing milk from my breasts every 3 or 4 hours."
4. "I do not let warm water run over my breasts when showering."

HIGH-RISK PREGNANCY (HP)

204. The nurse determines the fundal height of a client at 16 weeks gestation to be one finger above the umbilicus. The nurse should:
1. Assess for two distinct fetal heart rates
2. Ascertain birth weights of children of any siblings
3. Inform the client that she is mistaken about her dates
4. Instruct the client about appropriate weight gain during pregnancy

205. A client, gravida VI, with a history of four preterm births with fetal deaths and diabetes mellitus in her family, smokes 1½ packs of cigarettes a day. Both she and her husband work full time. Based on an analysis of this client's risk status, the nurse should:
1. Suggest to the client that she and her husband be seen by a geneticist
2. Explain to the client that she must stop smoking immediately for the health of the baby
3. Schedule this client for cerclage placement because of her history of multiple preterm births
4. Suggest to the client that she see a physician who specializes in the care of women with high-risk pregnancies

206. Shortly after delivery a grand multipara is in the postpartum unit. Because of the client's obstetric history, the nurse should monitor her for:
1. Uterine atony
2. Urinary distention
3. Profuse diaphoresis
4. Hypertensive episodes

207. Following delivery a client is transferred to the postpartum unit. Of the postpartum mothers on the unit, the one the nurse should observe most closely is:
1. A primipara who has delivered an 8 lb baby
2. A grand multipara who experienced a labor of only 1 hour
3. A primipara who received 100 mg of Demerol during her labor
4. A multipara whose placenta separated and who delivered in 10 minutes

208. During their first visit to the obstetrician, a young couple ask the nurse whether the wife should have an amniocentesis for genetic studies. The nurse responds that the indications for these studies include:
1. A recent history of drug use
2. Prior spontaneous abortions
3. First pregnancies in older women
4. A history of questionable genetic problems

209. A newly pregnant couple in their late thirties are having their first child and want to have an amniocentesis procedure. The nurse is aware that this procedure can be done:
    1. In the last trimester only
    2. Immediately after conception
    3. As soon as the mother feels life
    4. After the 14th week of pregnancy

210. The client who is scheduled for an amniocentesis states, "I'm glad this test will be able to tell whether my baby is well." The nurse's best response would be:
    1. "This is such a good test; work in this field is amazing."
    2. "A normal amniocentesis is a reliable indicator of a healthy baby."
    3. "This test is useful in detecting potential defects due to chromosomal errors."
    4. "Amniocentesis is a valuable tool for detecting congenital defects in the developing baby."

211. Before an amniocentesis, the nurse should:
    1. Initiate the intravenous therapy as ordered by the physician
    2. Assure that informed consent has been obtained from the client
    3. Inform the client that the procedure could precipitate an infection
    4. Perform a vaginal examination on the client to assess cervical dilation

212. When informed consent is obtained from a client who is to have an amniocentesis, it is unnecessary for the nurse to give the client:
    1. A copy of the Patient's Bill of Rights
    2. An explanation of any usable alternative procedures
    3. An offer to answer any questions about the procedure
    4. A complete description of the possible dangers and discomforts

213. During the third trimester a client has an amniocentesis to determine fetal lung maturity. To detect immediate complications, the nurse caring for the client should first:
    1. Assess fetal heart rate (FHR)
    2. Position the client on her left side
    3. Apply pressure for 5 minutes at the puncture site
    4. Check the client for vaginal bleeding or discharge

214. Following an amniocentesis, the nursing care should include:
    1. Giving perineal care
    2. Encouraging fluids every hour
    3. Changing the abdominal dressing
    4. Observing for signs of uterine contractions

215. A client at 16 weeks gestation is to have a sonogram followed by an amniocentesis. Nursing intervention would include directing this client to void:
    1. Just before each procedure is begun
    2. After the first sonogram images are obtained
    3. At least 1 hour before the procedures are scheduled to begin
    4. After the sonogram is completed and before the amniocentesis is begun

216. In her 32nd week of pregnancy, a client's ultrasonography reveals a low-lying placenta. The nurse is aware that when this client's pregnancy comes to term and labor begins, the client may experience:
    1. Vaginal bleeding
    2. Sharp abdominal pain
    3. Increased lower back pain
    4. Early rupture of membranes

217. The nurse explains to a pregnant client undergoing a nonstress test that the test is a way of evaluating the condition of the fetus by comparing the fetal heart rate with:
    1. Fetal lie
    2. Fetal movement
    3. Maternal blood pressure
    4. Maternal uterine contractions

218. The nurse understands that a positive oxytocin challenge test may be indicative of potential fetal compromise. The test demonstrates that during contractions the fetal heart rate shows:
    1. A normal baseline
    2. Late decelerations
    3. Early decelerations
    4. Variable decelerations

219. A client with a high-risk pregnancy is to undergo an oxytocin challenge test (OCT). The nurse understands that this test would not be done if the client had:
    1. Blurred vision
    2. Vaginal bleeding
    3. Sickle cell disease
    4. Increasing hypertension

220. Prior to the administration of RhIg, the nurse reviews the laboratory data of a pregnant client. RhIg is given to pregnant women who are:
    1. Rh positive and Coombs' positive
    2. Rh negative and Coombs' positive
    3. Rh positive and Coombs' negative
    4. Rh negative and Coombs' negative

221. During a prenatal visit, the nurse explains to a client who is Rh negative that RhoGAM will be administered:
    1. Weekly during the ninth month, because this is her third pregnancy
    2. Within 72 hours after delivery if infant is found to be Rh positive
    3. During the second trimester, if an amniocentesis indicates a problem
    4. To her infant immediately after delivery if the Coombs' test is positive

222. Laboratory studies reveal that a pregnant client's blood type is O and she is Rh positive. Problems related to incompatibility may develop in her infant if the infant is:
    1. Type A or B
    2. Rh negative
    3. Delivered preterm
    4. Type O, Rh positive

223. A client who is pregnant for the first time and is carrying twins is scheduled for a cesarean delivery. Preoperative teaching should include telling the client to expect to:
    1. Be discharged between 5 to 7 days postpartum
    2. Need an enema to have an effective bowel movement
    3. Be ambulating whenever desired the day after surgery
    4. Take sponge baths until the incision is completely healed

224. The nurse is aware that a critical outcome that would indicate an uncomplicated recovery after a multiple delivery is the woman's:
    1. Uterus being tightly contracted and in midline
    2. Capacity to breast-feed the babies immediately
    3. Request for sources of information on parenting twins
    4. Ability to rest quietly and discuss the birth of the babies

225. On the first postpartum day, a major concern of nursing intervention for a client who had a cesarean delivery would be:
    1. Promoting dietary intake
    2. Promoting bowel function
    3. Relieving postoperative pain
    4. Relieving gaseous distention

226. After the removal of a Foley catheter following a cesarean delivery, a client has difficulty voiding. The nurse can best evaluate whether the client has emptied her bladder by:
    1. Catheterizing the client for residual urine
    2. Gently palpating the client's bladder for distention
    3. Measuring the amount of urine the client has voided
    4. Asking the client whether she still feels the urge to void

227. A pregnant client who has a history of heart problems asks how she can relieve her occasional heartburn. As part of the teaching plan, the nurse should warn against taking antacids containing:
    1. Sodium
    2. Calcium
    3. Aluminum
    4. Magnesium

228. To facilitate delivery in a client with class III heart disease, the nurse would expect that the physician will probably:
    1. Use Pitocin induction
    2. Use forceps to assist delivery
    3. Schedule a cesarean delivery
    4. Do nothing and let nature proceed

229. The nurse is aware that placenta previa occurs when:
    1. There is premature separation of a normally implanted placenta
    2. The placenta is not implanted securely in place on the uterine wall
    3. There is premature aging of a placenta implanted in the uterine fundus
    4. The placenta is implanted in the lower uterine segment, covering part or all of the os

230. The nurse gently performs Leopold's maneuvers on a client with a suspected placenta previa and expects to find the:
    1. Fetal head firmly engaged
    2. Small fetal parts difficult to palpate
    3. Uterus hard and tetanically contracted
    4. Fetal presenting part high and floating

231. The best way for the nurse to assess the blood loss of a client with placenta previa is to:
    1. Count or weigh perineal pads
    2. Measure the height of the fundus
    3. Monitor pulse and blood pressure
    4. Check hemoglobin and hematocrit values

232. A client with heavy bleeding because of placenta previa is admitted to the hospital. The nurse places the client in a lateral Trendelenburg position to:
    1. Prevent shock
    2. Control bleeding
    3. Keep pressure off the cervix
    4. Move the placenta off the cervix

233. The attending physician prepares to do a speculum examination on a client with a confirmed diagnosis of placenta previa. When preparing for this examination the nurse should plan to:
   1. Have equipment ready for a fetal scalp pH after the examination
   2. Obtain a supply of IV magnesium sulfate in case early labor begins
   3. Arrange for a double setup that includes equipment for a cesarean delivery
   4. Attach the client to an internal monitor to closely watch the fetus' response

234. If the physician plans to do a speculum examination on a client with a marginal placenta previa, the nurse should have available:
   1. One unit of freeze-dried plasma
   2. Vitamin K for intramuscular injection
   3. Two units of typed and screened blood
   4. Heparin sodium for intravenous injection

235. A pregnant client is admitted to the hospital with abdominal pain and severe vaginal bleeding. After assessment, the nurse makes a nursing diagnosis of decreased cardiac output related to hemorrhage. The first nursing action should be to:
   1. Administer oxygen
   2. Elevate the head of the bed
   3. Draw blood for Hgb and HCT
   4. Give Demerol 50 mg IM for pain

236. The nurse is aware that the client most likely to be predisposed to placenta previa would be a:
   1. 19-year-old, gravida 1, para 0
   2. 25-year-old, gravida 2, para 1
   3. 40-year-old, gravida 2, para 1
   4. 30-year-old, gravida 6, para 5

237. The nurse instructs and encourages a client admitted to the hospital with placenta previa about the importance of:
   1. Breathing deeply to ensure that the fetus gets oxygen
   2. Keeping all movement to a minimum to diminish bleeding
   3. Remaining on her back to minimize pressure on the cervix
   4. Lying on her side to avoid putting pressure on the vena cava

238. When caring for a laboring client with a marginal placenta previa, it would be most important for the nurse to:
   1. Assess the fetal heart tones by fetoscope
   2. Frequently assess the height of the fundus
   3. Evaluate the external blood loss by pad count
   4. Perform frequent vaginal and rectal examinations

239. Dietary counseling for a pregnant client with sickle cell anemia should include supplemental folic acid. The nurse recognizes that this is important because it:
   1. Prevents sickle cell crises
   2. Decreases the sickling of RBCs
   3. Lessens the oxygen needs of cells
   4. Compensates for a rapid turnover of RBCs

240. When a pregnant client with sickle cell anemia comes to the clinic each month, in addition to the routine observations, the nurse should also assess her for:
   1. Signs of hypothyroidism
   2. Hyperemesis gravidarum
   3. Symptoms of pyelonephritis
   4. Complaints related to hypoglycemia

241. On admission to the hospital, a primigravida, 42 weeks gestation, complaining of back pain and fluid leaking from the vagina, is assessed. The findings are: contractions every 3 to 4 minutes, lasting 30 to 45 seconds; cervix 2 cm dilated and 70% effaced; presenting part floating; fetal heart rate 140 bpm in the RLQ; streaks of blood from the vagina; and a positive nitrazine test. The nurse is aware that the finding that indicates that a problem with delivery may occur is that the:
   1. Nitrazine test is positive
   2. Presenting part is floating
   3. Streaks of blood from the vagina
   4. Fetal heart tones of 140 beats/minute in the RLQ

242. When assessing a newly admitted laboring primigravida, the nurse notes that the fetal heartbeat is loudest in the upper left quadrant. The nurse suspects that the position of the fetus is probably left:
   1. Sacral anterior
   2. Occipital anterior
   3. Mentum anterior
   4. Occipital transverse

243. When a breech presentation is suspected, the nurse should diligently observe the client for signs of:
   1. Precipitate labor
   2. Prolapse of the cord
   3. Primary uterine inertia
   4. Progression of normal labor

244. The membranes of a laboring client whose fetus is in a breech position rupture spontaneously. The nurse then notes fresh meconium in the vaginal introitus and realizes that this:
   1. Indicates that the cord will prolapse
   2. Is evidence of fetal heart abnormalities
   3. Requires immediate notification of the physician
   4. Is a common occurrence in breech presentations

245. A client has a history of multiple preterm births followed by neonatal deaths. During the prenatal period, it is essential that the nurse teach this client that one of the most important danger signs to be aware of is:
1. Leg cramps
2. Pelvic pressure
3. Early morning headaches
4. Lack of fetal movement at 12 weeks

246. A client at 28 weeks gestation was admitted to the hospital in preterm labor yesterday and is receiving oral terbutaline (Brethine) 5 mg q6h. The contractions continue but a vaginal examination reveals no cervical dilation. The nurse anticipates that the physician will:
1. Discontinue the terbutaline for the present
2. Maintain the terbutaline as ordered previously
3. Increase the terbutaline to 10 mg every 6 hours
4. Increase the terbutaline to q4h for 24 to 48 hours

247. In her 30th week of gestation, a 16-year-old primigravida whose usual blood pressure is 120/70 mm Hg has a blood pressure of 130/88 mm Hg. She is admitted to the labor room and says, "I don't know why the doctor is so worried about my blood pressure. According to a book I have, it's normal." The nurse should respond:
1. "Your physician is just being very cautious."
2. "Your blood pressure is high for your age group."
3. "Your book is either for older women or outdated."
4. "Your blood pressure is slightly elevated by pregnancy guidelines."

248. A pregnant client who is experiencing contractions at 33 weeks gestation comes to the clinic. After a vaginal examination, the client is sent home and told to stay in bed. A teaching plan for this client should include the information that:
1. Blocks should be placed at the foot of the bed to raise it
2. She should lie on her side with her head raised on a small pillow
3. She should sit upright with several pillows behind her back
4. For 10 minutes every 2 hours she should assume the knee-chest position

249. The nurse explains to a client who is 33 weeks pregnant and is experiencing contractions that coitus:
1. Is permitted if penile penetration is not deep
2. Is safe as long as she is in the side-lying position

3. Need not be modified in any way by either partner
4. Should be restricted because it may stimulate uterine activity

250. A woman in preterm labor at 32 weeks gestation receives two IM injections of betamethasone (Celestone). The nurse is aware that this medication is given to:
1. Stop the labor process
2. Increase placental perfusion
3. Help mature the fetus' lungs
4. Reduce the intensity of contractions

251. A client who is at 34 weeks gestation has been receiving terbutaline (Brethine) subcutaneously. Her contractions increase to every 10 minutes, and her cervix dilates to 4 cm. The Brethine is discontinued. Priority nursing care during this time should be directed toward:
1. Promotion of maternal-fetal well-being during labor
2. Reduction of anxiety associated with preterm labor
3. Supportive communication with the client and her partner
4. Assisting the family to cope with the impending preterm birth

252. A client is admitted to the hospital in labor 24 hours after her membranes have ruptured. It is important for the nurse to assess the client's amniotic fluid for signs of potential:
1. Cord prolapse
2. Placenta previa
3. Chorioamnionitis
4. Abruptio placentae

253. A client, 35 weeks gestation, is admitted to the hospital with a small amount of bright red vaginal bleeding without contractions. After placing the client in bed, the nurse should:
1. Check fetal heart tones
2. Administer a Fleet enema
3. Obtain an amniotomy setup
4. Perform a vaginal examination

254. Aware of a client's history of narcotic abuse, the nurse's initial plans for providing pain relief measures during labor should include:
1. Scheduling pain medication at regular intervals
2. Administering the medication just when the pain is severe
3. Avoiding the administration of medication unless it is requested

4. Recognizing that she will not need as much pain medication as others

255. When planning care for a laboring client who is a drug user and has tested positive for HIV, the nurse should know that the:
1. Client will need invasive fetal monitoring
2. Client has acquired immunodeficiency syndrome
3. Incidence of HIV/AIDS in obstetrics is decreasing
4. HIV virus is transmitted primarily through body fluids

256. The nurse should be aware that a postpartum client with a history of drug abuse may be experiencing drug withdrawal if she develops:
1. Extreme hunger and thirst
2. Paranoia and evasiveness
3. Depression and tearfulness
4. Irritability and muscle tremors

257. A client is admitted to the hospital with uterine tenderness and minimal, dark red vaginal bleeding. She is diagnosed as having abruptio placentae. Upon admission, the priority assessment would include vital signs, skin color, urine output, and:
1. Her past obstetric history
2. Fundal height or abdominal girth
3. The time and amount of last meal
4. Family history of bleeding disorders

258. A client in the 38th week of gestation develops a slight increase in blood pressure. The physician advises her to remain in bed at home in a side-lying position. The client asks why this is important. The nurse's response would be based on the knowledge that this position:
1. Decreases intraabdominal pressure
2. Elevates the mean arterial pressure
3. Prevents the development of thromboses
4. Increases the circulation to the kidneys and uterus

259. A 16-year-old primigravida at 36 weeks gestation tells the nurse that her shoes and rings have been tight the past few mornings. Her blood pressure is found to be significantly elevated and there is 1+ proteinuria by dipstick. The client's blood pressure had been averaging 92/70 mm Hg during her previous prenatal visits. By the accepted definition of hypertension in pregnancy, a significant rise in blood pressure would be manifested by a reading of:
1. 100/76
2. 108/80
3. 110/84
4. 122/86

260. A nonstress test is scheduled for a client with pregnancy-induced hypertension. During the nonstress test the nurse should be aware that if nonperiodic accelerations of the fetal heart rate occur with fetal movement, it most likely indicates:
1. Fetal well-being
2. Head compression
3. Uteroplacental insufficiency
4. Umbilical cord compression

261. When providing health teaching for a client with pregnancy-induced hypertension, a therapeutic instruction that the nurse should give the client is:
1. Eat a sodium-free diet
2. Walk at least a mile a day
3. Limit fluid intake to 1000 ml a day
4. Rest often in the side-lying position

262. When performing an assessment of a client with worsening preeclampsia, the nurse should expect to find:
1. Diuresis
2. Vaginal spotting
3. Proteinuria of 3+
4. Blood pressure of 130/80 mm Hg

263. A nurse admits a client with severe preeclampsia to the hospital. After obtaining the vital signs, the nurse should:
1. Check the client's reflexes
2. Call the physician immediately
3. Determine the client's blood type
4. Administer intravenous normal saline

264. Before administering IV magnesium sulfate therapy to a client with preeclampsia, the nurse should assess the client's:
1. Urinary glucose, acetone, and specific gravity
2. Temperature, blood pressure, and respirations
3. Urinary output, respirations, and patellar reflexes
4. Level of consciousness, funduscopic appearance, and knee reflex

265. During her 6-month prenatal office visit, a pregnant client states that she is getting fat all over and that she even needs bigger shoes because her old ones are too tight. The nurse should:
1. Reassure the client that weight gain is expected
2. Encourage the client to use a comfortable walking shoe
3. Teach the client about the groupings in the food pyramid
4. Obtain the client's weight, blood pressure, and fundal height

266. When assessing the effectiveness of a teaching plan about self-care and conservative management of mild pregnancy-induced hypertension (PIH), the nurse would know that the client understood the teaching when the client recognized the importance of:
1. Eating a low-protein diet
2. Joining a weight-reduction program
3. Maintaining a normal sodium intake
4. Following the diuretic regimen as ordered

267. A client with worsening preeclampsia is hospitalized and placed in a private room. The nurse knows that this is important because a nonstimulating environment for a client with increased cerebral irritability:
1. Improves intracellular fluid reabsorption
2. Reduces the severity of frontal headaches
3. Reduces the probability of generalized convulsions
4. Prolongs the duration of hypotensive medications

268. A client with severe preeclampsia is suspected of having impending eclampsia. A characteristic symptom of eclampsia for which the nurse should assess is:
1. Anasarca
2. Convulsions
3. Excessive weight gain
4. Increased blood pressure

269. While a client is receiving IV magnesium sulfate for preeclampsia, a primary nursing intervention would be:
1. Limiting her fluid intake to 1000 ml/24 hours
2. Preparing for the possibility of a precipitate delivery
3. Restricting visitors and keeping the room darkened and quiet
4. Obtaining magnesium gluconate for use as an antagonist if necessary

270. A client is in the hospital undergoing therapy for severe pregnancy-induced hypertension. If eclampsia should occur, the nurse's first action should be to:
1. Assess fetal heart tones
2. Maintain an open airway
3. Protect the client from injury
4. Increase the infusion of magnesium sulfate immediately

271. A client with a history of phenylketonuria, who was maintained on a low-phenylalanine diet until 9 years of age, is now pregnant. The nurse teaches this client that:
1. Reinstitution of the low-phenylalanine diet will protect her baby from PKU
2. The baby will probably be mentally retarded because of her history of PKU

3. The fetus is at no risk prenatally but will require immediate care at birth to prevent PKU
4. Phenylalanine should be avoided even when not pregnant so that her body is able to support a pregnancy

272. Diet counseling for a breast-feeding client with a history of phenylketonuria should include providing a list displaying foods containing:
1. Lactose
2. Glucose
3. Fatty acids
4. Amino acids

273. The nurse knows that women with diabetes mellitus who become pregnant:
1. Have a 20% risk of fetal anomalies
2. Have decreased insulin requirements
3. Require intensive and thorough prenatal care
4. Should have their babies by cesarean delivery

274. The nurse explains to a newly pregnant client with diabetes mellitus that to minimize fetal neonatal complications, the most important action for her to take is to:
1. Check her blood glucose level as ordered
2. Keep all the physician's appointments made for her
3. Adhere strictly to the prescribed diet to limit weight gain
4. Comply with the management plan to maintain normal blood glucose levels

275. A client with diabetes mellitus has an amniocentesis in the 37th week of gestation; the L/S ratio of the amniotic fluid indicates adequate lung maturity. Based on this information the nurse assesses that:
1. Labor will probably be induced very shortly
2. The baby will have to be delivered immediately
3. There will be no need for further fetal monitoring
4. The baby should be free from major respiratory problems

276. Aware of the signs of an impending postpartal hemorrhage secondary to laceration of the cervix, the nurse assesses a postpartum client for a firm uterus and:
1. A decrease in pulse rate
2. Continuous trickling of blood
3. Persistent muscular twitching
4. An increase in blood pressure

277. A client who has undergone a cesarean delivery because of the presence of active genital herpes is transferred to the postpartum unit 2 hours after de-

livery. The nurse on the unit should plan to institute:
1. Strict isolation
2. Enteric isolation
3. Contact isolation
4. Protective isolation

278. The commonly used type of uterine incision that is now considered the preferred method of cesarean delivery is known as the:
1. Classical delivery
2. Pfannenstiel delivery
3. Low-segment delivery
4. Low-segment vertical delivery

279. In the 30th week of gestation a client delivers a still-born infant and is now 2 hours postpartum. The physical findings include: BP 80/40; TPR 98/100/22; fundus firm, four fingerbreadths above umbilicus; small spots of lochia rubra on Peri-pad; urinary bladder slightly distended. After catheterization the client's fundus remains firm and four fingerbreadths above the umbilicus. The nurse should:
1. Notify the client's physician immediately
2. Palpate the client's fundus every 2 hours
3. Recheck the client's vital signs again in 30 minutes
4. Catheterize the client again in 1 hour for residual urine

NORMAL NEONATE (NN)

280. Gloves are being used when handling newborns whether they are HIV positive or not. However, the nurse understands that it is usually not necessary to wear gloves with infants when:
1. Giving a feeding
2. Changing the diaper
3. Suctioning the infant
4. Doing an admission bath

281. At 1 minute after birth the nurse notes that an infant is crying, has a heart rate of 140, has acrocyanosis, resists the suction catheter, and keeps the arms extended. The nurse should assign this infant an Apgar score of:
1. 4
2. 6
3. 8
4. 10

282. The obstetrician hands the neonate to the nurse after delivery. The nurse's first action should be to:
1. Dry and place the infant in a warm environment
2. Administer oxygen by face mask until cyanosis clears

3. Cut the umbilical cord and attach a Hesseltine umbiliclip
4. Perform an abbreviated systematic physical assessment

283. During the physical assessment of a recently delivered newborn, the nurse palpates the infant's femoral pulses. This is done to detect the presence of:
1. Atrial septal defect
2. Coarctation of the aorta
3. Ventricular septal defect
4. Patent ductus arteriosus

284. An Apgar score recorded 5 minutes after birth helps to evaluate the:
1. Gestational age of the infant
2. Effectiveness of the labor and delivery
3. Adequacy of transition to extrauterine life
4. Possibility of respiratory distress syndrome

285. The nurse informs the parents of a newborn that the petechiae on their infant's face and neck are a result of:
1. A rash called erythema toxicum
2. Excessive superficial capillaries
3. Decreased vitamin K level in the newborn infant
4. Increased intravascular pressure during delivery

286. A newborn's total body response to noise or movement is often distressing to the parents. It is important for the nurse to teach the parents that this response is:
1. A reflex response that indicates normal development
2. An involuntary response that will remain for the first year of life
3. An automatic response that may indicate that the baby is hungry
4. A voluntary response that indicates insecurity in a new environment

287. A baby weighing 5 lb, 6 oz is born via cesarean delivery and is admitted to the newborn nursery. The nurse expects the newborn's respiratory rate to range between:
1. 20 to 40 per minute
2. 30 to 60 per minute
3. 60 to 80 per minute
4. 70 to 90 per minute

288. The nurse recognizes that in the healthy, full-term neonate heat production is accomplished by:
1. Oxidizing fatty acids
2. Shivering vigorously
3. Breaking down brown fat
4. Increasing muscular activity

289. Following the birth of her daughter, a mother states to the nurse, "I was told that my baby has to have an injection of vitamin K. She's so small to be getting a shot. Why does she have to have it?" The nurse's most appropriate response would be:
    1. "Your baby needs the injection to help her develop red blood cells."
    2. "Newborns are deficient in vitamin K. This treatment protects your baby from bleeding."
    3. "An injection of vitamin K will help to prevent your baby from becoming jaundiced."
    4. "A newborn's blood clots faster than it should. This injection helps decrease the clotting time."

290. A newborn is admitted to the nursery. When assessing for congenital hip dysplasia during the newborn assessment, the nurse should determine if the infant has extra skin folds in the:
    1. Thigh
    2. Abdomen
    3. Calf muscles
    4. Popliteal area

291. To best assist new parents to understand the unique characteristics of a newborn, the nurse should discuss with them the:
    1. Infant's response to routine feeding schedules
    2. Testing of the normal newborn's auditory acuity
    3. Newborn's behaviors and states of wakefulness
    4. Importance of reading about parent-infant bonding

292. When inspecting her newborn after delivery, a mother asks the nurse whether her newborn has flat feet. The nurse recalls that:
    1. Flat feet are common in children and infants
    2. This is difficult to assess because the feet are so small
    3. Flat feet are associated with major deformities of the bones of the feet such as clubfoot
    4. The arch of the newborn's foot is covered with a fat pad, giving the appearance of being flat

293. The nurse recognizes that survival in the neonatal period is largely related to:
    1. Gestational age and birth weight
    2. Reproductive history of the mother
    3. Timing and adequacy of prenatal care
    4. Parental health habits and social class

294. A newborn is delivered spontaneously. Immediately after delivery the neonate is active, cries lustily, and has a heart rate of 150 beats per minute, but the bottoms of the feet have a marked bluish tinge. The Apgar score for this infant at 1 minute would be:
    1. 3
    2. 5
    3. 7
    4. 9

295. The nurse, observing a sleeping newborn, notes periods of irregular breathing and occasional twitching movements of the arms and legs. The neonate's heart rate is 150 beats per minute; the respiratory rate is 50 breaths per minute; and the glucose strip reading is 60 mg/dl. The nurse's most appropriate assessment would be:
    1. The baby requires no intervention; all findings are normal
    2. The twitching movements suggest the baby may be having seizures
    3. The baby's blood sugar is low; twitching movements suggest hypoglycemia
    4. The rapid respiratory rate and irregular breathing suggest respiratory distress

296. When assessing a newborn, the nurse notes several areas of raised white spots on the chin and nose. The nurse is aware that these are known as:
    1. Milia
    2. Lanugo
    3. Vascular nevi
    4. Erythema toxicum

297. After pushing for an hour, a client delivers a full-term male infant with an 8/9 Apgar score. The immediate nursing care of this newborn should include:
    1. Assessing respirations, identifying the infant, and keeping him warm
    2. Rushing him to the nursery while aspirating the oropharynx, and stimulating him often
    3. Applying a prophylactic agent to the eyes, giving AquaMEPHYTON, and bathing him
    4. Weighing him, placing him in a crib, and leaving him near until the physician is finished with the mother

298. Phenylketonuria (PKU) testing is performed on a 4-day-old infant. The nurse plans to explain to the mother that the purpose of this genetic screening is to determine:
    1. If the infant is positive for PKU
    2. If the mother is a carrier for PKU
    3. The incidence of the disease in the population
    4. The risk for the infant's later development of PKU

299. A newborn infant has a PKU test after formula feedings have been initiated. This is done primarily to prevent:
    1. Failure to thrive
    2. Mental retardation

3. Growth retardation

4. Spread of the disease

300. A newborn male is circumcised. The nurse would recognize that the mother requires additional teaching regarding care of her son following the circumcision when she indicates that she plans to:
1. Change her son's diapers very frequently
2. Call the physician if there is excessive bleeding
3. Give the baby a tub bath the day after he is discharged
4. Place petrolatum gauze or A & D ointment on his penis with each diaper change

301. A client has her son circumcised. The statement that indicates to the nurse that the mother needs additional discharge instructions would be:
1. "I'll keep ice on the penis so it won't swell.
2. "I need to change the dressing four times a day."
3. "I'll call the pediatrician if I notice bleeding from the penis."
4. "I need to keep the diaper loose so it won't rub on the penis."

302. When changing her infant, a new mother notes a reddened area on the infant's buttock and reports it to the nurse. The nurse should:
1. Have the nursery staff change the infant's diaper
2. Use both lotion and powder to protect the involved area
3. Notify the pediatrician and request an order for a topical ointment
4. Encourage the new mother to cleanse and change the infant more often

303. A newborn develops jaundice 72 hours after birth. The nurse explains to the parents that the jaundice is probably a result of:
1. An allergic response to the feedings
2. Normal physiologic destruction of immature RBCs
3. An obstruction in the bile duct, which is common in newborns
4. Some Rh-negative blood that is still present in the baby's bloodstream

304. Twenty-four hours after delivery, a community health nurse visits an infant born at home to do a newborn assessment. When assessing the infant, the nurse notes slight jaundice of the face and trunk. The first nursing action should be to:
1. Plan for immediate admission to the hospital
2. Relate this finding to the baby's postdelivery age

3. Obtain a physician's order for bilirubin determination
4. Arrange for the baby to have phototherapy in the home

305. The nurse is aware that the nursing action that would best promote parent-infant attachment behaviors would be:
1. Restricting visitation on the postpartum unit
2. Supporting rooming-in with parental infant care
3. Encouraging the parents to choose breast-feeding
4. Keeping the new family together immediately postpartum

306. To help the parents proceed with bonding behaviors immediately after delivery, the nurse should:
1. Assess for proper parenting techniques
2. Demonstrate the desired behavior to the parents
3. Delay footprinting the baby until late evening
4. Delay administering the antibiotic to the baby's eyes

307. Following delivery a client wishes to begin breast-feeding her infant. The nurse assists the client by:
1. Positioning the infant to grasp the nipple to express milk
2. Giving the infant a bottle first to evaluate the baby's ability to suck
3. Leaving them alone and allowing the infant to nurse as long as desired
4. Touching the infant's cheek adjacent to the nipple to elicit the rooting reflex

308. The best indication that correct attachment to the breast has occurred is when the:
1. Baby's tongue is securely on top of the nipple
2. Baby's mouth covers most of the areolar surface
3. Baby sucks each breast vigorously for 5 minutes before falling asleep
4. Baby makes frequent loud clucking sounds while nursing at each breast

309. The nurse assures the breast-feeding mother that she will know that her infant is getting an adequate supply of breast milk if the infant gains weight and:
1. Rarely sucks on a pacifier
2. Has several hard stools daily
3. Voids six or more times a day
4. Awakens to feed every 4 hours

310. A mother is breast-feeding her newborn. She asks when she can switch the baby to a cup. The nurse would recognize that the mother understands the teaching about feeding when she says she will start to introduce a cup when the baby is:
    1. 4 months old
    2. 6 months old
    3. 12 months old
    4. 18 months old

311. The mother who is bottle-feeding her 1-month-old asks the nurse whether any vitamin or mineral supplements are required. The nurse bases the reply on the knowledge that babies who are bottle-fed with ready-to-use formula require:
    1. Iron
    2. Fluoride
    3. Vitamin K
    4. Vitamin B_{12}

312. A newborn has some defined ecchymotic-like areas across the buttocks. The nurse suspects that these discolorations are related to:
    1. An occult form of spina bifida
    2. Excessive handling at delivery
    3. The parent's ethnic background
    4. Pressure from a posterior breech presentation

313. The nurse teaches the new mother that neonatal weight loss in the first 3 days of life is most often the result of:
    1. Allergy to formula
    2. A hypoglycemic response
    3. Inadequate breast- or bottle-feeding
    4. Excretion of fluid via lungs, urinary bladder, and bowels

314. When teaching cord care, the nurse should explain to the parents about the appearance of the cord. The teaching should include:
    1. Taping a pressure dressing over the umbilicus to protect it until the cord dries and falls off
    2. Applying an antibiotic ointment to the base of the cord with each diaper change until the cord falls off
    3. Cleansing the base of the cord daily with an alcohol-soaked sponge until the cord dries and falls off
    4. Placing the diaper over the umbilical cord to prevent infection and to protect the umbilicus from injury until the cord falls off

HIGH-RISK NEONATE (HN)

315. A newborn of 30 weeks gestation has a heart rate of 86 beats per minute and has slow irregular respirations. The infant grimaces in response to suctioning, is cyanotic, and has flaccid muscle tone. The nurse should assign the infant an Apgar score of:
    1. 2
    2. 3
    3. 4
    4. 5

316. After delivery, a preterm neonate weighing 2300 g has Apgar scores of 3 and 8 at 1 and 5 minutes, respectively. These scores indicate that:
    1. Respiratory stimulation will be needed for the first 24 hours
    2. The newborn was responding according to the preterm gestational age
    3. Oxygen under pressure was not necessary because the 1-minute Apgar score was 3
    4. Initial resuscitation measures were effective and the infant responded well to the additional measures taken

317. The nurse would observe for symptoms of respiratory distress syndrome (RDS) in an infant whose mother:
    1. Is a type 1 diabetic
    2. Had previously used heroin
    3. Had been hypertensive during pregnancy
    4. Was preeclamptic during labor and delivery

318. Since preterm infants are prone to respiratory distress, a nurse caring for an infant of 33 weeks gestation should observe the infant for:
    1. Flaring nares
    2. Acrocyanosis
    3. Abdominal breathing
    4. Respirations of 40/minute

319. A client expresses a desire to breast-feed her preterm infant who is in the neonatal intensive care nursery. The nurse should:
    1. Tell the client this is not possible because the infant is being fed by gavage
    2. Discourage the client because of the time and effort it will take to pump her breasts
    3. Instruct the client that breast milk is inadequate for a preterm infant because it does not contain all the necessary nutrients
    4. Support the client's decision and explain that the infant may initially lose weight due to the energy expended when breast-feeding

320. A neonate, born at 33 weeks gestation, develops respiratory distress syndrome (RDS) at 6 hours of age. Assessment of the infant at this time would reveal:
    1. A high-pitched cry
    2. A heart rate of 140

3. Intercostal retractions
4. Respirations of 20 per minute

321. Respiratory acidosis is confirmed in an infant with respiratory distress syndrome when the nurse notes that the laboratory report of blood gases reveals:
1. A pH of 7.35
2. An elevated Pco_2
3. A potassium level of 4.6 mEq/L
4. An arterial O_2 level of 80 mm Hg

322. A preterm infant is receiving oxygen by an overhead hood. During the time the infant is under the hood, it would be appropriate for the nurse to:
1. Hydrate the infant q15 min
2. Put a hat on the infant's head
3. Keep the O_2 concentration consistent
4. Remove the infant q15 min for stimulation

323. Supplemental oxygen is ordered for a preterm infant with respiratory distress syndrome. To prevent retinopathy of prematurity, the nurse plans to:
1. Administer the oxygen by hood
2. Apply eye patches to both eyes
3. Warm and humidify all oxygen flow
4. Determine O_2 saturation frequently

324. A newborn with respiratory distress syndrome is placed on continuous positive-pressure ventilation therapy via an endotracheal tube. The nurse notes that the infant's breath sounds on the right side are diminished, and the point of maximum impulse (PMI) of the heartbeat is in the left axillary line. Interpretation of this assessment data should lead the nurse to understand that:
1. These are normal findings because infants with this disorder often have some degree of atelectasis
2. The inspiratory pressure on the ventilator is probably too low and needs to be increased for adequate ventilation
3. The infant may have a pneumothorax, and the physician should be contacted immediately so treatment can be instituted
4. The endotracheal tube has probably slipped into the left main stem bronchus and needs to be pulled back to ventilate both lungs

325. A nursing intervention for an infant with respiratory distress syndrome would be:
1. Position carefully to promote respiratory efforts
2. Set Isolette temperature at 85° F to prevent shivering
3. Observe carefully for possible congenital birth defects

4. Avoid handling to minimize overstimulation and conserve O_2

326. When estimating an infant's gestational age, the nurse should take into consideration:
1. Weight and length at birth
2. Rooting and sucking ability
3. Presence of a tonic neck reflex
4. Size of breast tissue and genitalia

327. When assessing a preterm infant, it is most important for the nurse to know the infant's gestational age and how it compares to the birth weight because:
1. This information will help identify potential newborn problems
2. This information must be documented on the admission record
3. The infant will lose 10% of birth weight during the first few days of life
4. Evaluation and classification records are necessary for health insurance

328. After an emergency cesarean delivery, a newborn born at 35 weeks gestation is admitted to the neonatal intensive care nursery. The neonate has a Silverman-Anderson score of 6, which reflects a need for:
1. Continuous cardiac monitoring
2. Increased caloric intake and fluids
3. Assessment of neurologic reflexes
4. Respiratory support and observation

329. A client delivers a baby girl by cesarean section, and the infant is transferred to the newborn nursery. The nurse monitors the baby's breathing, since infants born by cesarean delivery are more apt to have atelectasis because:
1. The rib cage is not compressed and then released during delivery
2. Oxygen deprivation of the infant occurs following a cesarean delivery
3. The sudden change in temperature at delivery causes the infant to aspirate
4. There is no chance during the delivery for gravity to drain the fluid from the lungs

330. The nurse is aware that in an infant of 32 weeks gestation the:
1. Areola and nipple are barely visible
2. Palms have clearly defined creases
3. Ear pinna springs back when folded
4. Square window sign shows a 0-degree angle

331. When assessing a newly delivered neonate, the nurse notes the following findings: arms and legs slightly flexed; skin smooth and transparent; abundant lanugo on the back; slow recoil of pinna; and few sole creases. In light of these findings, the care plan for this neonate should include nursing orders to monitor for:
    1. Polycythemia
    2. Hyperglycemia
    3. Postmaturity syndrome
    4. Respiratory distress syndrome

332. A newborn delivered at 39 weeks gestation, weighing 2450 g (5.5 lb), would be classified as being:
    1. Preterm and immature
    2. Small-for-gestational age
    3. Average-for-gestational age
    4. Average-for-gestational age but preterm

333. A small-for-gestational-age (SGA) newborn, who has just been admitted to the nursery, has a high-pitched cry, appears jittery, and has irregular respirations. The nurse is aware that these symptoms may be associated with:
    1. Hypovolemia
    2. Hypoglycemia
    3. Hypercalcemia
    4. Hypothyroidism

334. A finding in the physical assessment of a neonate that may indicate that the infant is preterm is:
    1. Flexion of extremities
    2. Many superficial veins
    3. Absent femoral pulses
    4. A positive Babinski reflex

335. The nurse suspects that a preterm infant receiving gastric feedings may have necrotizing enterocolitis (NEC) when:
    1. Circumoral pallor develops during gastric feeding
    2. An increased number of explosive stools are noted
    3. Several severe bouts of projectile vomiting are observed
    4. Large amounts of residual formula are withdrawn before gavage

336. Nursing care for an infant with necrotizing enterocolitis includes:
    1. Diluting the formula mixture as ordered
    2. Measuring abdominal girth every 2 hours
    3. Administering oxygen prior to gastric feeding
    4. Giving half-strength formula by gavage feeding

337. During the assessment of a preterm infant, the nurse determines that the infant is experiencing temperature instability. The nurse should:
    1. Rapidly rewarm the infant during the next hour until the temperature is stabilized
    2. Gradually rewarm the infant during the next several hours and monitor frequently
    3. Assess the infant for signs of hyperglycemia and begin temperature stabilization
    4. Assess and record the infant's skin temperature every hour until the temperature is stable

338. A mother delivers a male infant at 35 weeks gestation. When visiting her infant, whose condition has stabilized, in the neonatal intensive care nursery (NICU) for the first time, the mother asks, "When will I be able to breast-feed my son?" The nurse's most appropriate response would be:
    1. "Even though he is preterm, he is stable. You may try now if you would like."
    2. "Preterm infants should not breast-feed. It takes more calories than bottle-feeding."
    3. "Pump your breasts now and then feed him the milk by bottle to conserve his energy."
    4. "He is preterm and sucks weakly so it will be several weeks before you may breast-feed."

339. As a client watches, the nurse does a nasogastric feeding on the client's preterm infant son, who weighs 2350 g. The client asks, "Would it hurt my baby to suck on a pacifier during the feeding?" The nurse's most appropriate response would be:
    1. "It might tire him out because he's still so small. We don't want him to use up all his energy."
    2. "If he sucks on a pacifier a lot now, he may have problems learning how to suck from a bottle later."
    3. "There's no real benefit in using a pacifier and there is a relationship between the use of a pacifier and buck teeth."
    4. "Sucking on a pacifier during tube feedings may help him associate sucking with food so that he'll adjust better to bottle feedings."

340. The nurse caring for a 32-week appropriate-for-gestational-age (AGA) neonate establishes the following list of potential interventions for the infant. The intervention that should receive the highest priority is:
    1. Promoting bonding
    2. Preventing infection
    3. Maintaining respirations
    4. Supporting body temperature

341. When an apnea monitor sounds an alarm 10 seconds after cessation of respirations, the nurse should respond first by:
1. Assessing skin color and respirations
2. Using tactile stimuli on chest or extremities
3. Checking the device for signs of malfunction
4. Resuscitating with face mask and Ambu bag

342. The nurse is aware that a preterm infant may have a potential nutritional problem because of:
1. A poor sucking reflex
2. A decreased metabolic rate
3. Decreased caloric requirements
4. Increased absorption of nutrients

343. To prevent the development of retinopathy of prematurity, it would be most beneficial for the nurse to:
1. Place a shield over the neonate's eyes
2. Maintain a low level of lighting around the neonate
3. Assess the neonate every hour with a pulse oximeter
4. Position the neonate in an elevated side-lying position

344. The parents of a preterm infant are finally taking their child home. To evaluate the parents' competency level in the care of their infant, it would be most effective to:
1. Ask the parents what they plan to do at home
2. Determine the rationale behind the parents' actions
3. Observe the parents in a return demonstration of care
4. Give the parents a simple test on infant care practices

345. After delivery, a large-for-gestational-age (LGA) infant of a diabetic mother should be assessed for:
1. The presence of mongolian spots
2. A blood glucose level less than 45
3. A body temperature less than 98° F
4. Elevated bilirubin levels in the first 24 hours

346. The nurse transporting a newly delivered infant with Apgar scores of 9/10 to the transition nursery reports that this delivery was difficult because the infant has wide shoulders. Based on this information, the nursery nurse's priority assessment would be to check for a normal:
1. Moro reflex
2. Plantar grasp
3. Babinski reflex
4. Stepping reflex

347. An assessment that is typical for an infant of a diabetic mother would be:
1. Red-faced
2. Hyperreflexive
3. Long thin body
4. Excessive crying

348. A newborn whose mother has type 1 diabetes mellitus has been receiving a continuous infusion of fluids with glucose. When discontinuing the IV infusion, the nurse should:
1. Slowly decrease the rate
2. Observe for metabolic alkalosis
3. Withhold oral feedings for 4 to 6 hours
4. Check the urine for glucose every hour

349. A client developed a rubella infection during the fourth month of pregnancy. After delivery, the nursery nurse should:
1. Use enteric precautions
2. Institute blood precautions
3. Use standard precautions at all times
4. Place the baby in the isolation nursery

350. A preterm infant develops physiologic jaundice, and phototherapy is ordered to:
1. Activate the liver to dispose of the bilirubin
2. Break down the bilirubin into a conjugated form
3. Activate vitamin K to facilitate excretion of bilirubin
4. Dissolve the bilirubin and allow it to be excreted from the skin

351. When an infant is receiving phototherapy the nurse should plan to:
1. Use mineral oil on the skin to prevent excoriation
2. Cover the baby's head with a cap to minimize heat loss
3. Discontinue the therapy when feeding and hold the baby
4. Regulate radiant heat to keep the skin temperature at 99° F

352. When doing a newborn assessment of a male infant after a scheduled cesarean delivery, the nurse notes that the infant's head circumference is 4 centimeters smaller than his chest. The nurse is aware that this finding:
1. Is indicative of anencephaly
2. Could indicate microcephaly
3. Commonly occurs in babies born by cesarean delivery
4. Is normal in male newborns regardless of mode of birth

353. The nurse is aware that an ABO incompatibility is most common when the mother is:
1. Type A
2. Type B
3. Type O
4. Type AB

354. The nurse understands that the effect PKU has on development will depend on:
1. Diagnosis within the first 3 days after birth
2. The level of phenylalanine in the blood at birth
3. Compliance with a corrective diet and how early it is instituted
4. The presence of phenylpyruvic acid in the urine at 1 week of age

355. The nurse is aware that the child with PKU has a characteristic urinary odor best described as:
1. Fishy
2. Ammoniacal
3. Mousy or musty
4. Aromatic or pungent

356. The nurse assesses a newborn and observes central cyanosis. Central cyanosis is indicative of congenital heart defects that affect cardiac circulation by:
1. Shunting blood right to left
2. Shunting blood left to right
3. Obstructing flow of blood from the left side of the heart
4. Preventing shunting of blood between left and right sides of the heart

357. The nurse understands that congestive heart failure is the usual sequela to congenital cardiac defects that result from left-to-right shunting of blood in the heart. With this knowledge, the nurse would be aware that a sign that would be most indicative of early onset of congestive heart failure in the infant would be:
1. Cyanosis of skin
2. Decreased heart rate
3. Increased respiratory rate
4. Liver 2 cm below costal margin

358. A mother's laboratory results indicate the presence of cocaine and alcohol. The characteristic in the baby that would indicate to the nurse that the baby has been affected by fetal alcohol syndrome would be:
1. Cleft lip
2. Polydactyly
3. Umbilical hernia
4. Small upturned nose

359. A baby, born to an addicted mother who had her last opioid "fix" 28 hours ago, gradually becomes more jittery, cries a great deal, and has a shrill cry. The physician diagnoses acute opioid addiction of the neonate. The nursing priority would be to:
1. Reduce environmental stimuli
2. Administer narcotic antagonists
3. Reduce fluid intake to inhibit vomiting
4. Assess for appropriate developmental tasks

360. When assessing for signs of withdrawal in the newborn of a known drug user, the nurse expects to observe:
1. A weak, low-pitched cry
2. An increased Moro reflex
3. Hyperactivity and sneezing
4. Decreased deep tendon reflexes

361. A newborn has been exposed to HIV in utero. The finding that supports a diagnosis of HIV infection would be:
1. A delay in temperature regulation
2. Continued bleeding after circumcision
3. Hypoglycemia within 12 hours of delivery
4. Thrush that does not respond readily to treatment

362. A newborn is diagnosed as having a diaphragmatic hernia. An immediate nursing intervention after the neonate is admitted to the ICU nursery should be:
1. Hydrating the infant with isotonic enemas
2. Limiting formula feedings to small amounts
3. Placing the infant in the Trendelenburg position
4. Providing gastric decompression via nasogastric tube

363. During a vaginal delivery the nurse notes meconium-stained amniotic fluid. Upon birth of the baby the first nursing action should be to:
1. Thoroughly suction the airway
2. Vigorously stimulate the baby to cry
3. Use an Ambu bag with oxygen support
4. Position the baby in reverse Trendelenburg

364. The nurse in the newborn nursery notes that a newborn has not passed a meconium stool in the first 48 hours. The nurse should first assess the infant for:
1. Any episodes of vomiting
2. Presence of stool in rectum
3. Intake and output during the last 12 hours
4. Fussiness during feedings in the next 24 hours

365. A newborn male is admitted to the nursery. He weighs 10 pounds, 2 ounces, which is 2 pounds more than the birth weight of any of his siblings. Because of the baby's weight, the nurse:
1. Places him in a heated crib
2. Delays starting oral feedings

3. Has him seen by the pediatrician immediately
4. Performs a glucose test every hour for 8 hours

366. A preterm infant appears to have a strong sucking reflex. To prevent respiratory embarrassment, the nurse should plan to feed this infant:
1. Every 4 to 6 hours
2. Via a nasogastric feeding tube
3. Diluted formula more frequently
4. Small amounts of formula at each feeding

WOMEN'S HEALTH (WH)

367. When reviewing the role of hormones in the reproductive process, the nurse recalls that the corpus luteum secretes:
1. Cortisol
2. Prolactin
3. Oxytocin
4. Progesterone

368. A 13-year-old client whose menses began 2 years ago complains of having some mild upper abdominal pain between each period. The nurse explains that this:
1. Requires immediate medical attention
2. Usually occurs when menses first begin
3. Will disappear when ovulation is well established
4. Is a common occurrence known as mittelschmerz

369. A young client tells the nurse that her mother complains about having dysmenorrhea and asks the nurse what this means. The nurse can best describe dysmenorrhea as:
1. Cessation of menstruation
2. Abnormal vaginal bleeding
3. Uterine pain with menstruation
4. Spotting between menstrual periods

370. A 35-year-old woman presents to the Women's Health Care Center complaining of polyuria and pain and burning when urinating. Her medical diagnosis is a urinary tract infection. When assisting this client it is important to:
1. Have her void every 2 hours
2. Select the appropriate bedpan for voiding
3. Pour warm water over her vulva when she voids
4. Teach her to wash her hands before and after voiding

371. A pregnant client asks the nurse for information about toxoplasmosis during pregnancy. The nurse teaches the client that:
1. Pork and beef should be properly cooked before eating
2. Toxoplasmosis is a disease that is prevalent just in foreign countries

3. Eating salads with mayonnaise should be avoided during the summer
4. Raw shellfish are intermediary hosts and should be avoided during pregnancy

372. A client is receiving antibiotics and antifungal medication to treat a recurring vaginal infection. To compensate for the effect of these medications, the nurse should encourage the client to:
1. Eat extra fruit
2. Eat yogurt daily
3. Avoid spicy foods
4. Take a daily enema

373. A client, age 16, has a steady boyfriend and is having sexual relations with him. She is seeking advice as to how she can protect herself from contracting AIDS. The nurse advises her to:
1. Have her partner withdraw before ejaculating
2. Make certain their relationship is monogamous
3. Have her partner use a condom during sexual activity
4. Seek counseling about various contraceptive methods

374. A 16-year-old female comes to the clinic complaining of increased vaginal discharge and pain in her lower abdomen. She relates an active sexual history with multiple partners. The symptoms of illness have been present for several days. Based on this information, the nurse recognizes that the client most likely has:
1. Herpes
2. Syphilis
3. Gonorrhea
4. Human papilloma virus

375. The nurse understands that gonorrhea is difficult to control because:
1. The blood test is expensive and time consuming
2. The causative organism has become resistant to treatment
3. There is no specific diagnostic test for the causative organism
4. The symptoms are vague and the incubation period is relatively short

376. When assessing a female client who is suspected of having early syphilis, the nurse should expect the client to exhibit the early symptom of:
1. Flat wartlike plaques around the vagina and anus
2. An indurated painless nodule on the vulva that begins to drain
3. Glistening patches in the mouth covered with a yellow exudate
4. A maculopapular rash on the palms of the hands and soles of the feet

377. A client who has been diagnosed as having syphilis tells the nurse that it must have been contracted from a toilet seat. The nurse knows that this cannot be true because the causative agent of syphilis is:
1. Immobilized by body contact
2. Chelated by wood and plastic
3. Destroyed by warmth and moisture
4. Inactivated when exposed to dryness

378. The major body system affected in tertiary syphilis is the:
1. Reproductive
2. Integumentary
3. Cardiovascular
4. Lower respiratory

379. To perform breast self-examination correctly, a premenopausal female should be examining her breasts:
1. When she ovulates
2. The first of every month
3. The day her menses begins
4. Three days after her menses ends

380. When performing breast self-examination a client should be:
1. Squeezing the nipples to check for discharge
2. Using the right hand to examine the right breast
3. Placing a pillow under the shoulder opposite the side being examined
4. Pressing the palm of the hand against the breast to compress it to the chest wall

381. The factor in a female client's history that would have the nurse place the client in a potential high-risk category for breast cancer would be:
1. Early menopause
2. Low-income background
3. Delayed onset of menarche
4. Late beginning of childbearing

382. A 26-year-old female, whose sister recently had a mastectomy, calls the local women's health center for an appointment for a mammogram. To prepare for the test, the nurse should teach the client that:
1. The room will be darkened throughout the procedure
2. Each breast will be firmly compressed between two plates
3. Food and fluid must be avoided for 6 hours before the test
4. She does not need a mammogram until she is 50 years old

383. A client, age 56, is admitted for a biopsy of a lump in her left breast; the lump is suggestive of malignancy, based on mammography. The client reports that she found this lump 3 days ago and went right to the physician. The client states, "I know it can't be cancer, because it doesn't hurt." The most therapeutic response by the nurse would be:
1. "Well, let's hope it isn't malignant."
2. "Most lumps in the breast are not malignant."
3. "Tell me what you know about breast cancer."
4. "Has your physician told you that it wasn't cancer?"

384. A client is scheduled for a modified radical mastectomy for an adenocarcinoma of the right breast. It would be inappropriate for the nurse developing the client's preoperative teaching plan to include:
1. Allowing the client time to ventilate her feelings
2. Urging the client to have immediate reconstructive surgery
3. Explaining the dressings and drains that the client will have after surgery
4. Teaching the client the postoperative pulmonary routines that will be followed

385. The nurse teaches a client that in the immediate postoperative period after a modified radical mastectomy, the client can expect to have a:
1. Large pressure dressing
2. Sling to support her affected arm
3. Portable wound drainage system
4. Low-protein, low-carbohydrate diet

386. The nurse is aware that during the early postoperative period after a modified radical mastectomy, edema can be limited in the affected arm by having the client:
1. Turn to the unaffected side at least once every 2 hours
2. Maintain positive suction on the wound catheter and note the amount of drainage
3. Elevate the affected hand higher than the elbow and the elbow higher than the shoulder
4. Flex and extend the fingers of the affected hand and pronate and supinate the forearm

387. Before a client leaves the hospital after a mastectomy, it is most important that she learn to:
1. Apply the breast prosthesis
2. Curtail some of her usual activities
3. Avoid household tasks that require stretching
4. Examine her remaining breast for abnormalities

388. After a mastectomy or hysterectomy many clients feel incomplete as women. The statement that should alert the nurse to this feeling in a client following a total hysterectomy would be:
1. "I can't wait to see all my friends again."
2. "I feel washed out; there isn't much left."
3. "I can't wait to get home to see my grandchild."

4. "My husband plans for me to recuperate at our daughter's home."

389. Before a scheduled lumpectomy of the breast, the physician tells the client that any affected lymph nodes will be removed and radiation therapy will be given postoperatively. The client blurts out, "My cancer has spread! That's why the doctor is doing more surgery and giving me radiation." The nurse's most therapeutic initial response would be:
1. "Although the surgeon plans more extensive surgery and radiation therapy, you should not be upset at this time."
2. "Although more extensive surgery is planned, along with radiation therapy, it means you will be disease free 5 years from now."
3. "Biopsies of the axillary lymph nodes are needed to determine whether the cancer has spread. Radiation therapy will not be needed if the lymph nodes are negative."
4. "Biopsies of the axillary lymph nodes are necessary to determine whether they are cancer free, and the radiation therapy is frequently used as follow-up care after a lumpectomy."

390. A client has previously received estrogen replacement therapy as a treatment for osteoporosis. The nurse should recognize that the client has an increased risk of developing:
1. Endometrial cancer
2. Accelerated bone loss
3. Vaginal tissue atrophy
4. A myocardial infarction

391. A female client with Hodgkin's disease is to start total nodal irradiation. She and her husband have been trying to have a child and are quite concerned when they learn that radiation includes the pelvic nodal area. The nurse should refer them to the physician because the nurse should be aware that:
1. The ovaries can be surgically moved and placed in a shielded area
2. Intense radiation to the area always causes permanent sterilization
3. The radiation used is not radical enough to destroy ovarian function
4. Ovarian function will be temporarily destroyed but will return in time

392. A 45-year-old is admitted for an abdominal hysterectomy and bilateral salpingo-oophorectomy. The information from the admission interview that most likely reflects the reason the abdominal method was chosen over the vaginal method is that the:

1. Client has a prolapsed uterus
2. Client has large uterine fibroids
3. Client's cervical os has mild dysplasia
4. Client's bladder leaks urine when she coughs

393. When providing preoperative teaching to a 60-year-old who is admitted for a vaginal hysterectomy and an anterior and posterior repair of the vaginal wall, the nurse should prepare the client for the immediate postoperative use of a:
1. Douche
2. Pessary
3. Rectal tube
4. Urinary catheter

394. In planning care with a client during the postoperative recovery period following an abdominal hysterectomy and bilateral salpingo-oophorectomy, the nurse should include the explanation that:
1. Surgical menopause will occur
2. Urinary retention is a common problem
3. Depression is normal and should be expected
4. Weight gain is expected, and dietary plans are needed

395. After a vaginal hysterectomy and an anterior and posterior repair of the vaginal wall, a client is returned to her room. The nurse should plan to:
1. Check the client's vaginal packing
2. Observe the dressing for bleeding
3. Elevate the client's lower extremities
4. Initiate sitz baths the following morning

396. A 47-year-old school teacher is scheduled for a total abdominal hysterectomy because of uterine fibroids. The nurse, preparing the client for surgery, would recognize that the client understood the preoperative teaching when she says:
1. "Menstrual periods will no longer occur following surgery."
2. "I will not experience the symptoms of menopause at middle age."
3. "The incision will be made through the vagina into the pelvic cavity."
4. "The surgeon will just remove my ovaries and fallopian tubes during surgery."

397. Immediately following a rape, appropriate nursing care for a client should include:
1. Obtaining a gynecologic history from the client
2. Informing the police before the client is examined
3. Asking the client to collect a clean-catch urine specimen
4. Testing the client's urine for seminal alkaline phosphatase

398. When a client has surgery for cancer of the uterus, the nurse must be concerned with the prevention of postoperative thrombosis because the majority of pulmonary emboli begin as deep vein thromboses of the:
1. Calf of the leg
2. Thoracic cavity
3. Pelvis and thighs
4. Extremities and abdomen

399. A female client is tentatively diagnosed as having cystitis, pending laboratory results. The nurse recognizes that *Escherichia coli* is a common causative agent in cystitis because it is:

1. A particularly virulent bacteria
2. Commonly found in the kidneys
3. Usually found in the intestinal tract
4. A competitor with *Candida* for host sites

400. Postoperatively, the nursing care plan for a client who has had pelvic surgery should include:
1. Encouraging the client to ambulate in the hallway
2. Elevating the client's lower extremities by gatching the bed
3. Assisting the client to dangle the legs over the side of the bed
4. Maintaining the client on bed rest until the bandages are removed

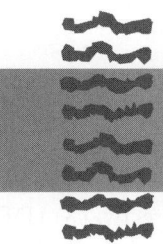

CHILDBEARING AND WOMEN'S HEALTH NURSING ANSWERS AND RATIONALES

REPRODUCTIVE CHOICES (RC)

1. (1) Unless the condom is held, it can be displaced, allowing the sperm to enter the vagina. (2) (DF; IM)

   2 Spermicidal cream is needed because the diaphragm may be displaced in some positions.

   3 This is not true; sperm can be deposited at the beginning of intercourse without the man's knowledge.

   4 When the woman has a rise in her basal temperature, she is most fertile and should avoid intercourse.

2. (3) Oral contraceptives may cause hypertension and place the client at risk for the development of a CVA. (2) (PT; AN)

   1 Oral contraceptives are contraindicated for women older than 40 years because of an increased risk of myocardial infarction.

   2 Clients should be strongly cautioned about smoking even 15 cigarettes a day.

   4 There is no relationship between oral contraceptives and multiple births.

3. (4) Headaches, either sudden or persistent, may indicate hypertension or a cardiovascular event; visual disorders, such as partial or complete loss of vision or double vision, may indicate neuro-ocular lesions, which are associated with the use of oral contraceptives. (2) (PT; EV)

   1 These are expected side effects, and the client may need an adjustment in the dose or have to change to another product.

   2 While there is controversy over the contribution of oral contraceptives to the development of breast cancer, the presence of breast pain, which may occur, is not a typical manifestation of a malignancy and is, therefore, not reportable; weight gain, not loss, occurs because of edema.

   3 Hypotension and amenorrhea do not occur; hypertension may occur with oral contraceptives and subsides when they are discontinued.

4. (3) This is the preferred method for clients with diabetes because it has no physiologic side effects. (1) (DF; PL)

   1 This requires a great deal of self-control and a strong desire to avoid pregnancy, and it is not as effective as a diaphragm.

   2 Because of the possibility of perforation, this method increases the risk of infection for diabetic women.

   4 Oral contraceptives have a diabetogenic effect; they alter carbohydrate metabolism, and insulin dosage must be adjusted.

5. (4) The diaphragm should remain in place for at least 6 hours after intercourse; if coitus occurs within those 6 hours, additional spermicide should be added and the 6-hour time frame begins again. (2) (DF; IM)

   1 The diaphragm must always be used with a spermicide to be effective.

   2 The diaphragm may be inserted with the dome facing either up or down and still be effective.

   3 Puckering, especially near the rim, could indicate thin spots that could rupture during intercourse; the diaphragm should not be used if puckering is noted.

6. (3) In a normal, regular cycle, ovulation occurs 2 weeks (14 days) before the onset of the next menses. (1) (DF; IM)

   1 This would occur in a woman who menstruates every 28 days.

   2 This is too early in the cycle.

   4 Same as answer 2.

7. ② The fertile period is determined by subtracting 18 days from the length of the shortest cycle to determine the first unsafe day and subtracting 11 days from the length of the longest cycle to determine the last unsafe day. (2) (DF; EV)

1 This is how the last day, not the first day, of the unsafe period is determined.

3 This is only true if the shortest cycle is 28 days; the date depends on a calculation based on the length of the woman's shortest and longest cycles.

4 This is incorrect; the longest and shortest cycles are used, not the average length of a cycle.

8. ① The presence of the IUD thread should be verified before coitus; the IUD causes an inflammatory response of the myometrium; the inflammatory cells may be spermicidal or damage the ova, interfering with fertilization; also, increased local prostaglandins inhibit implantation. (2) (DF; IM)

2 It is not common to have discomfort during coitus with an IUD in place; it is one of the warning signs that should be reported.

3 This is a rare, rather than a common, occurrence.

4 The incidence of vaginal infections is not increased with the use of an IUD; also, the risk of pelvic inflammatory disease is not increased if the relationship is mutually exclusive.

9. ④ This is the most common side effect of the implantation of nonbiodegradable Silastic capsules containing progestin. (2) (DF; IM)

1 Although this is a side effect, it is less common than irregular menstrual bleeding.

2 This may occur with some types of contraceptives that contain estrogen, not progestin.

3 This is a side effect of estrogen excess; this product contains progestin, not estrogen.

10. ① Many couples in their thirties who are happy with their family and feel their family is complete choose sterilization as their method of contraception. (1) (DF; AS)

2 Sterilization via tubal ligation should have no effect on dysmenorrhea.

3 The decision for sterilization should not be made by others, only by the woman herself.

4 Decisions regarding sterilization should be made when the client is not pregnant and is not under stress.

11. ④ This is true; some women find it emotionally unnerving to handle their genitals and discharges. (2) (DF; AS)

1 The data do not support this conclusion.

2 Same as answer 1.

3 Same as answer 1.

12. ④ This mechanical device offers the lowest risk with a high degree of reliability for a person with type 1 diabetes if used correctly. (3) (DF; AN)

1 This is not reliable because the menses during the postpartum and lactation periods are often irregular; partner cooperation is also required.

2 This can be used by people with diabetes mellitus but it is less reliable than the diaphragm with spermicidal gel.

3 Even a low-dose contraceptive increases the risk for vascular complications; people with type 1 diabetes mellitus are already at risk for vascular complications.

13. ① All of these, plus helping families to understand inheritance patterns and recurrence risks and then to select an appropriate course of action, are vital parts of genetic counseling. (2) (DF; IM)

2 Abortion is one management option given to families; it is not specifically stressed for any condition.

3 Laws of dominance, recessiveness, and probability, not predictability, govern which fetus will be defective; contraception may be recommended if risk for a defective fetus is calculated to be high.

4 Genetic counseling offers all aspects of care, not just making the diagnosis; many family physicians are not well versed in the ramifications and management of genetic illness.

14. ③ Ovulation occurs approximately 14 days before the next menses, about the 16th day in a 30-day cycle; the 15th to 17th days would be the best time to avoid sexual intercourse. (2) (DF; IM)

1 Too early; ovulation occurs 14 days before the next menstrual period.

2 Same as answer 1.

4 Too late; ovulation occurs 14 days before the next menstrual period.

15. ③ Conception occurs when one sperm penetrates one ovum and creates a viable zygote. (1) (DF; AN)

1 A high sperm count is optimum, but only one sperm is needed to penetrate the ovum.

2 Conception takes place in a fallopian tube, not the uterus.

4 The sperm penetrates the ovum in a fallopian tube, and then the impregnated ovum travels down the tube to the uterus.

16. (2) Instructing the couple to have intercourse four times a week with at least 12 to 24 hours between ejaculations will increase the chance of conception and will correct the client's misconceptions in a nonthreatening manner. (3) (DF; IM)
1 Their pattern of intercourse is not considered frequent enough for optimum sperm motility; continuing this pattern may promote their health problem and expose them to needless tests.
3 Infertility renders a couple very vulnerable; to openly discourage the mate could be harmful to a fragile relationship between the couple themselves or the couple and the nurse.
4 Too vague; specific instructions should be given in a nonthreatening manner.

17. (1) Absence of sexual intercourse is the most effective form of birth control (100% effective), because the egg and sperm do not come in contact with one another. (1) (DF; IM)
2 The birth control pill has a high, but not perfect (97% to 99%), effective rate when used correctly.
3 This is a fairly effective (82% to 98%) means of preventing pregnancy; effectiveness depends on correct, consistent use.
4 This is a fairly effective (94% to 99%) means of preventing pregnancy.

18. (4) The client should make up for the missed pill by taking two the next day; taking one in the morning and one in the evening lessens the chance of the client becoming nauseated. (3) (SI; IM)
1 This response does not tell the client what to do if a pill is missed; missing one pill can alter hormone levels and predispose the client to pregnancy.
2 It is unnecessary to call the physician unless other problems are noted.
3 This is wrong advice; missing one pill may alter hormone levels and predispose the client to pregnancy.

19. (4) Noncompliance with instructions can alter hormone levels; amenorrhea can occur. (2) (CC; AS)
1 Although some illnesses can cause amenorrhea, nothing in the data indicates that the client has been ill; compliance should be assessed first.
2 Nothing in the data indicates that the client is anorectic; women with anorexia become amenorrheic only after long periods of starvation.

3 This presumes the client has been sexually active; there are many reasons other than pregnancy that cause amenorrhea.

REPRODUCTIVE PROBLEMS (RP)

20. (2) The male should be treated to prevent the infection from passing back and forth between him and his sexual partner. (1) (DF; EV)
1 The organism is most likely present in the partner's urogenital tract; voiding will not prevent recurrence.
3 This is an ineffective remedy and will not prevent recurrence.
4 A douche is not recommended during pregnancy.

21. (3) Any blood from the rupture will accumulate, causing phrenic nerve irritation and pain. (3) (SC; AN)
1 Anxiety can cause many things, but shoulder pain is an atypical symptom.
2 The cardiac changes caused by hypovolemia do not cause shoulder pain.
4 A ruptured tube can cause rebound tenderness in the abdomen, not the shoulder.

22. (3) This is an appropriate nursing diagnosis; the bleeding is causing a decreased circulating blood volume and therefore a decreased cardiac output. (3) (CC; AN)
1 Infection could occur later but is not a problem at this time.
2 This would not be an appropriate nursing diagnosis for this client; there would be a fluid volume deficit, not excess.
4 This client is not incapable of making decisions for herself.

23. (4) Removing the tube does not bring a halt to menses; endometrial proliferation and shedding will occur as long as the ovaries and uterus are present. (1) (DF; EV)
1 This is a correct statement; there is evidence that clients who have one tubal pregnancy are highly susceptible to having another.
2 This is a correct statement; pregnancy should be delayed 6 to 12 months after a tubal pregnancy.
3 This is a correct statement; pelvic infections can lead to constriction of tubes, and a fertilized ovum may become trapped.

24. (1) This response encourages the clients to verbalize their feelings. (1) (SC; IM)
   2 The clients are not interested in the nurse's wishes; the focus should be on them.
   3 This is a very insensitive and incorrect statement; there may be nothing wrong with either client.
   4 The clients are not seeking advice about dealing with their parents.

25. (2) Only superficial attention has been directed to hypertension; ongoing health supervision is needed. (1) (HI; AN)
   1 Teaching and encouraging the client to engage in self-care is helping the client.
   3 This is untrue; however, limiting sodium intake is too superficial.
   4 Involving the client is important, but professional supervision is required.

26. (3) This is referred pain from the passage of carbon dioxide through the tubes; this is usually indicative of tubal patency. (2) (CI; IM)
   1 No anesthetic is given; the client's awareness of pain is necessary to evaluate whether carbon dioxide is able to pass through the tubes.
   2 The client can resume normal activities as soon as the test is over.
   4 The client does not usually experience nausea and/or vomiting.

27. (4) This is necessary to keep the sperm viable; if the specimen becomes warm, the sperm will die. (1) (CI; IM)
   1 Rubber solvents and preservatives may affect the semen specimen.
   2 The specimen can be collected at any time.
   3 This may lessen the amount of ejaculate needed for the specimen.

28. (4) Because the client is ovulating, the infertility may be a result of a seminal factor; the partner's semen should be examined before more extensive studies or treatments are begun with the woman. (2) (DF; AN)
   1 All other potential problems should be ruled out first; the client does not have a right to receive the drug unless it is appropriate for the problem.
   2 Other potential problems should be ruled out first.
   3 Same as answer 2.

29. (3) Since the laminaria tent is left in place for this length of time, it increases in size from absorption of moisture and dilates the cervix 2 to 3 times its original diameter before the suction procedure is done. (2) (DF; IM)
   1 A local anesthetic agent is usually injected into the cervix (paracervical block) and may cause mild cramping or light spotting.
   2 A temperature over 100° F is a danger sign and the client should be alerted to call her health care provider if this occurs.
   4 Cervical bleeding is reduced by the use of the laminaria tent and is usually equivalent to a heavy menstrual period; the client is usually observed for 1 to 3 hours following the procedure.

30. (4) This indicates that the bleeding is excessive and the physician should be notified. (1) (CI; EV)
   1 Although instructions vary among health care providers, sexual intercourse usually may be resumed in 1 to 3 weeks.
   2 Although instructions vary among health care providers, tampons usually are denied for 3 days to 3 weeks.
   3 The menstrual period will usually resume in 4 to 6 weeks.

31. (1) She is presently pregnant and has been pregnant three times previously, hence gravida IV; she has carried only one pregnancy past viability and delivered, hence para I. (1) (CI; AS)
   2 This answer reverses para and gravida.
   3 This fails to reflect the abortions and is incorrect for parity.
   4 This fails to reflect the present pregnancy and is incorrect for para.

EMOTIONAL NEEDS RELATED TO CHILDBEARING AND WOMEN'S HEALTH (EC)

32. (1) During the taking-in phase, a woman is primarily concerned with being cared for and being cared about. (2) (SC; PL)
   2 Infant feeding is best taught during the taking-hold phase of postpartum adjustment.
   3 Same as answer 2.
   4 This is not a primary concern during the immediate postpartum period.

33. (2) The mother needs to learn the realities of infant behavior and how to cope with them; holding and talking to the baby are consoling measures. (2) (DF; PL)
   1 At this age a toy would not be meaningful and would be an inadequate substitute for parental attention.
   3 The infant is too young to be given cereal at this time.

4 It is unhealthy to disrupt the baby's sleep pattern.

34. (2) This statement acknowledges the situation and suggests a possible solution to the problem. (3) (SC; IM)
    1 If it were true that the mother had postpartum depression, it would be inappropriate for the nurse to dismiss it so lightly
    3 This is not true, and the response does not address the problem that is evident in the situation.
    4 This is stereotyping and would not be therapeutic.

35. (2) The client has the right to refuse the Papanicolaou test; the nurse must accept the client's need to talk with the physician first. (2) (SC; IM)
    1 This is a subjective conclusion; the client has the right to refuse the test.
    3 This is inappropriate; this action is not client centered.
    4 The client's need must be recognized first.

36. (1) This recognizes achievement and reinforces the client's positive behavior. (2) (SC; IM)
    2 This is inappropriate; the client has been successful in reducing her blood pressure and weight with nonpharmacologic strategies.
    3 This implies that the client is not doing enough; the focus should be on the positive, and the gains should be reinforced.
    4 This focuses on the negative rather than the positive; small gains should be reinforced.

37. (1) The therapeutic regimen includes bed rest; peace of mind can best be achieved if the children are adequately cared for. (3) (SC; IM)
    2 This explores feelings without including a therapeutic regimen.
    3 This is giving solutions rather than exploring the situation with the client.
    4 Complete bed rest has been prescribed, and the suggested plan assumes that the client is able to afford it.

38. (3) Doing this will help the client relax and will lessen discomfort. (2) (SC; IM)
    1 The client may become more anxious if the procedure is hurried.
    2 This may distract the client but will not produce relaxation.
    4 This may make the client more anxious; holding the breath causes tightening of the perineum.

39. (3) This response is nonjudgmental; it permits the client to recognize her own feelings. (2) (SC; IM)

1 This is judgmental; it leaves no room for the client's feelings.
2 This is judgmental; it gives the nurse's opinion on a moral question for the client.
4 This response leaves the burden of the decision to the client without offering assistance.

40. (2) This is an accurate and nonjudgmental response. (2) (SC; IM)
    1 This response recognizes the client's feelings but cuts off communication because it ends the discussion.
    3 This is a judgmental response that questions the mother's decision-making and deals only with the present.
    4 Same as answer 3.

41. (4) This reassures the client that her baby is all right at the moment and that the nurses are aware of and monitoring the baby's status. (1) (SC; IM)
    1 This response does not provide the mother with any reassurance of the baby's status or that anything is being done to monitor the baby.
    2 This provides false reassurance; amniotic fluid will not protect the fetus if the mother has a seizure.
    3 This provides false reassurance; following instructions does not guarantee a healthy baby.

42. (2) Because seeing and touching the newborn infant are species-specific behaviors for human attachment, allowing them to hold the infant will promote bonding. (2) (SC; IM)
    1 Although this is a viable action, holding and touching will promote bonding more effectively.
    3 After touching and holding, this action can also contribute to bonding.
    4 Actual holding and touching promote bonding more than just hearing about the baby.

43. (3) The mother should be reunited with the baby at the first opportunity and when the mother is prepared and feeling well enough to do so. (2) (SC; PL)
    1 Grief work will go on for an extended period of time and has no relationship to when the baby is seen.
    2 There is no magic about the first 24 hours; some mothers are too ill or both parents may be too frightened to see the baby that soon.
    4 Some parents may be too frightened to think to ask to see the baby; the nurse can prepare the parents and then suggest a visit.

44. ③ This statement conveys acceptance by the nurse and encourages the mother to verbalize additional concerns; it also explores the mother's understanding of the physician's explanation. (2) (SC; IM)
1 This reply belittles the mother's concern and cuts off further communication.
2 Although this response does acknowledge part of the mother's concerns, it denies her the opportunity of further exploration of her fears that her infant may not respond to therapy.
4 Same as answer 1.

45. ③ To cry in this situation is a normal response; it is also normal to be frightened about touching a small preterm infant, but the nurse should provide support and encourage the mother to do so. (2) (SC; AN)
1 Bonding does not have a detachment behavior phase; the behavior indicates apprehension in a difficult situation.
2 This is not incomplete bonding but fear in a difficult situation.
4 The reaction to the baby is more complex than merely fear of the NICU.

46. ④ By participating in the infant's care, the client will gain confidence in her own ability to meet the infant's needs. (2) (SC; IM)
1 Watching the provision of care by others would only increase the client's sense of inadequacy.
2 There is no need for a specialist to care for the infant after discharge.
3 If she is not permitted to care for the infant sooner, the client will have to develop these skills under stress.

47. ① If the client demonstrates anorexia, crying, and insomnia, the nurse should recognize these as symptoms of depression, not anxiety. (2) (SC; AS)
2 Lethargy and anorexia are manifestations of depression, but they are usually accompanied by insomnia rather than prolonged sleep.
3 Ambivalence and lethargy can be associated with depression, but they are usually accompanied by anorexia, not increased appetite.
4 Insomnia and ambivalence may be seen with depression, but they are accompanied by anorexia, not increased appetite.

48. ② Confrontation about the active substance abuse and the mother's diminished ability to safely care for the infant at this time is necessary to help the mother get help and to also protect the baby. (2) (EP; IM)

1 This would be unsafe; the mother may not be capable of caring for the infant.
3 Same as answer 1.
4 Decisions should not be made without input from the mother.

49. ③ Giving the parents true information about what to expect during the procedure will help to allay their fears and encourage their cooperation. (2) (SC; IM)
1 The nurse should be able to provide information and interpretation of procedures for clients; delay in answering their questions may increase clients' concerns.
2 Reassurance is nontherapeutic; an amniocentesis is a low-risk procedure, but some complications may occur.
4 If the father is uninformed, viewing the procedure may increase his anxiety, even though his presence may be comforting to the mother.

50. ③ This response encourages the client to verbalize concerns; verbalization is an outlet for discharging tension. (1) (SC; IM)
1 This response denies the client's feelings and gives false reassurance.
2 This response reinforces the client's fears.
4 Same as answer 2.

51. ① The mother should see her infant as soon as possible so that she can acknowledge the reality of the delivery and begin bonding. (2) (SC; PL)
2 A delay retards maternal-infant bonding.
3 Same as answer 2.
4 This is an independent nursing action.

52. ② Touching and holding the infant are the most effective ways of promoting mother-infant bonding. (1) (EP; IM)
1 This provides false reassurance; the nurse does not know whether this is true.
3 This is an unreal approach; it does not prepare the mother to deal with her ill child.
4 This would prevent the mother from touching and holding the baby; the mother must do this to foster bonding.

53. ④ Unplanned cesarean delivery can result in guilt, disappointment, anger, and a sense of failure as a woman. (1) (SC; AN)
1 These are not usually common concerns.
2 The hospital stay is not exceptionally prolonged; the client usually is discharged within 2 to 5 days.
3 Mothers who deliver by cesarean delivery can assume the mothering role.

54. (1) By discussing the experience, the client is bringing it into reality; this is characteristic of the taking-in phase. (3) (SC; AN)

2 The client is not ready to assume the tasks of the letting-go phase until completing the tasks of the taking-in and taking-hold phases.

3 The taking-hold phase is marked by an increased desire to resume independence; this statement reveals the client is still in the taking-in phase.

4 The working-through phase is not a separate phase of adjustment to parenthood; this is not relevant.

55. (3) This recognizes the client's feelings, encourages ventilation, and does not encourage the use of alcohol for relaxation. (1) (SC; IM)

1 This gives false reassurance; the use of alcohol should not be encouraged for relaxation.

2 This does not recognize the client's underlying feelings and could put the client on the defensive.

4 The nurse cannot ensure the physician's response; the use of alcohol should not be encouraged for relaxation.

56. (3) This opens communication and allows the client to verbalize thoughts and feelings. (2) (SC; IM)

1 This is judgmental; there are not enough data to make this assumption.

2 This ignores the client's needs and cuts off communication.

4 This does not give the client the opportunity to verbalize feelings and needs.

57. (3) Adolescent parents are still involved in the developmental stage of resolving their own self-identity; they have not sequentially matured to intimacy and generativity. (1) (SC; AN)

1 Although this may be true, it is not the reason the baby is at risk for neglect or abuse.

2 Same as answer 1.

4 Same as answer 1.

58. (3) This statement allows parents an opportunity to obtain additional information and review their options to make an informed decision. (3) (SC; IM)

1 Recent studies do not support any connection between circumcision and penile cancer.

2 This information may have already been discussed with the physician; it may be more helpful for the nurse to review the information at this time.

4 This is a partially true statement; however, the academy primarily emphasizes that there are no medical indications for this procedure.

59. (1) Gynecologic surgery involves sexual/reproductive organs and may hold symbolic meaning relative to sexuality. (2) (DF; AN)

2 This is not caused by a hysterectomy or salpingo-oophorectomy.

3 This client is not at risk for deterioration of a body system; while there may be some sexual dysfunction, it does not mean sexual intercourse will be eliminated.

4 Would not be applicable because a hysterectomy is a physical alteration, not a physical disability.

60. (2) Focusing on the mother's feelings permits her to express fears and concerns. (3) (SC; IM)

1 This is subjective and cuts off further communication.

3 This answer will frighten the client and cut off further communication.

4 Same as answer 3.

61. (4) Increased alertness is a normal common behavior that occurs in new or different situations when a person is mildly anxious. (2)(SC; AS)

1 This is a common sign of moderate to severe anxiety.

2 Same as answer 1.

3 Same as answer 1.

62. (4) This is a reflective statement that opens the door for the client to express her feelings (3) (SC; IM)

1 Insufficient data for this conclusion to be drawn.

2 This false reassurance does not address the client's concern; it may close the door for further discussion.

3 Same as answer 2.

63. (1) The freedom to vent feelings about the loss of a body part and its meaning to the client influence the client's willingness to participate in the postoperative regimen, consequently affecting healing and rehabilitation. (1) (SC; PL)

2 This is not the initial preoperative focus.

3 Not an initial action; usually arranged after surgery.

4 Giving information without first identifying the client's level of anxiety is not using sound teaching principles.

64. ② This response denotes nonjudgmental interest in feelings and encourages open communication of thoughts. (1) (SC; IM)

1 This is erroneous; two healthy children do not influence outcome of a third, especially in regard to Down syndrome related to advanced maternal age.

3 This is giving unwarranted information because no diagnosis has been made; this response would close communication.

4 The diagnosis has not yet been made, so information is inappropriate; also this response could close further communication.

65. ② Each pregnancy creates a stress situation and is a developmental crisis. (3) (SC; AS)

1 It has not been determined that multigravidas are more successful in coping with pregnancy than primigravidas.

3 Each pregnancy is unique, and this pregnancy may or may not be more stressful than the last.

4 Support persons are important during any crisis or stressful situation.

66. ② Anger is a coping strategy that allows a person to gain a sense of control over life; the husband feels a loss of control over the spontaneity of his intimate relationship with his wife; intercourse is based on administration of the medications. (2) (SC; AN)

1 There is no evidence that client is feeling guilty; pregnancy is desired by both partners; anger is what is being expressed.

3 Client is not withdrawing or expressing sadness, dejection, or lethargy.

4 There is no evidence of the client feeling undeserving of an intimate relationship.

DRUG-RELATED RESPONSES (DR)

67. ④ This anesthesia creates a sympathetic blockade that causes loss of peripheral resistance and a decrease in venous return; this leads to a reduced cardiac output and results in lowered BP. (2) (PT; EV)

1 The urge to void is diminished with this kind of anesthesia; client should be encouraged to void to prevent distention.

2 This is not a frequent side effect of this anesthesia; it may occur if an accidental high placement of anesthetic occurs or excessive amount of anesthesia is used.

3 Bradycardia, not tachycardia, may result.

68. ④ Betamimetics have the unpleasant side effects of nervousness, tremors, and palpitations;

clients should be informed that these side effects are expected. (3) (PT; EV)

1 If contractions are lessened and the heart rate is less than 120 and regular, the medication is performing as expected and does not need to be discontinued.

2 These are not the usual symptoms of preterm labor.

3 There is no correlation between electrolyte levels and these symptoms.

69. ② Use ratio and proportion:
2,450,000 units : 300,000 units = X ml : 1 ml
300,000 X = 2,450,000
X = 8.2 ml
(2)(PT; AN)

1 This would deliver more than the ordered amount.

3 This would deliver less than the ordered amount.

4 Same as answer 3.

70. ② Probenecid reduces renal tubular excretion of penicillin. (2) (PT; EV)

1 This is unrelated to the concomitant administration of penicillin and probenecid.

3 Same as answer 1.

4 Same as answer 1.

71. ② Tetracycline has an affinity for calcium; if used during tooth bud development, it may cause discoloration of teeth. (2) (DF; IM)

1 This is untrue; it is associated only with the discoloration of teeth.

3 Same as answer 1.

4 Same as answer 1.

72. ① Maintenance of serum progesterone levels keeps cervical mucus thick and hostile to sperm at all times. (1) (DF; IM)

2 Insufficient information; the pill must be taken throughout the menstrual cycle.

3 Fertility drugs are often taken during the first part of the cycle to encourage ovulation, not contraception.

4 Combined estrogen and progesterone oral contraceptives are taken during the second, third, and fourth weeks of the cycle.

73. ③ Nausea and vomiting are related to excessive amounts of estrogen; these symptoms can usually be controlled by reducing the dose. (1) (PT; EV)

1 Amenorrhea is associated with pregnancy; breakthrough bleeding is more common than amenorrhea with estrogen.

2 Hypomenorrhea is caused by estrogen deficiency.

4 Depression and lethargy can be related to both excessive estrogen and excessive progesterone but are not common side effects.

74. (1) Oral contraceptives should be discontinued with any symptom that could be related to emboli. (1) (PT; EV)

2 Menorrhagia is a side effect related to excessive amounts of estrogen; immediate discontinuance of contraceptives is unnecessary.

3 Mittelschmerz is pain midway in the menstrual cycle, usually at ovulation.

4 This may be a sign of infection, not a side effect of oral contraceptives.

75. (1) Iron absorption is pH dependent; therefore, iron should be taken with a source of ascorbic acid to enhance duodenal absorption. (2) (PT; PL)

2 This is unrelated to the absorption of iron.

3 Same as answer 2.

4 Same as answer 2.

76. (2) RhIg administration during the 28th week of gestation reduces an active antibody response in an Rh-negative individual exposed to the positive blood; this drug is used during pregnancy. (3) (CI; EV)

1 An unsensitized pregnant woman receives RhIg at 28 weeks gestation as prophylaxis.

3 RhIg is given earlier in the pregnancy; it is a preventive measure.

4 Same as answer 1.

77. (3) Heparin can be used during pregnancy because it does not cross the placental barrier and will not cause hemorrhage in the fetus. (3) (SI; IM)

1 This drug can cross the placental barrier and cause hemorrhage in the fetus.

2 Same as answer 1.

4 Same as answer 1.

78. (4) Heparin should not be given because 98 seconds is almost 3 times the normal time it takes a fibrin clot to form (25 to 36 seconds) and prolonged bleeding may result; the therapeutic range with heparin is 1½ to 2 times the normal range. (3) (SI; EV)

1 Heparin must not be increased; the client already has received too much.

2 The aPTT is not normal but prolonged; it is almost 3 times the normal rate.

3 The medication does not need to be changed; it needs to be stopped.

79. (3) This is one commonly occurring side effect of this drug. (2) (PT; EV)

1 Tachycardia, not bradycardia, commonly occurs.

2 Hypokalemia, not hyperkalemia, is a potential side effect.

4 These do not occur with terbutaline.

80. (4) Frequent contractions with short relaxation periods may lead to fetal hypoxia. (2) (CI; AN)

1 This intensity is within the normal limits of 50 to 75 mm Hg.

2 An adverse response to Pitocin is a contraction lasting more than 90 seconds; contractions lasting 30 seconds usually occur in early labor.

3 This is within the normal fetal heart rate range of 120 to 160 during labor.

81. (2) Oxytocin (Pitocin) has an antidiuretic effect, acting to reabsorb water from the glomerular filtrate. (3) (PT; EV)

1 Hyperventilation is caused by inappropriate breathing patterns, not by prolonged use of Pitocin.

3 Affect is not altered by the use of Pitocin.

4 Fever occurs with infection or dehydration, not with prolonged use of Pitocin.

82. (4) Magnesium sulfate is used to depress CNS irritability; diminished reflexes would indicate the medication's effectiveness. (3) (PT; EV)

1 Magnesium sulfate is not a diuretic.

2 This is a sign of toxicity.

3 A transient lowering of blood pressure after a loading dose may occur, but within an hour the blood pressure returns to pretherapy levels; magnesium sulfate is an anticonvulsant drug.

83 (1) A side effect of this drug is respiratory depression; therefore, a reduction of respirations to this level indicates toxicity and the drug should be withheld. (3) (PT; EV)

2 The blood pressure is already known to be high, and this must be given to prevent a seizure from occurring.

3 This drug has a sodium retention effect, so urinary output is important, but 30 ml per hour is still within the acceptable range.

4 This is considered a therapeutic effect of this drug.

84. (1) Near-toxic levels of magnesium sulfate are suggested by the disappearance of the knee-jerk reflex and by depressed respirations (less than 12 per minute). (3) (SI; EV)
2 This is given as an antidote only when ordered by the physician.
3 This is unsafe; this would cause an overdose and exacerbate the toxic signs.
4 Waiting could put the client in jeopardy of respiratory arrest; toxic symptoms require medical intervention.

85. (1) Hypermagnesemia causes extreme muscle depression; calcium gluconate, the $MgSO_4$ antidote, promotes muscle function. (2) (PT; AN)
2 This is used in chelation therapy for lead poisoning.
3 This is an antihypertensive.
4 This is used for hyperkalemia.

86. (3) Methergine, an oxytocic, is used to promote uterine contractions; its vasoconstrictive action can also lead to hypertension, and it should not be used when hypertension is already present. (2) (SI; EV)
1 This medication generally is not given to a postpartum client unless an infection is present; there are no data to support the presence of an infection.
2 There is no contraindication to using this medication to relieve the discomfort of an episiotomy.
4 There is no contraindication for use of this drug; there are no data to support the fact that the client is constipated.

87. (2) The concentration of propylthiouracil excreted in breast milk is 3 to 12 times higher than its level in maternal serum; this may cause agranulocytosis or goiter in the infant. (3) (PT; IM)
1 Heparin is not excreted in breast milk.
3 The amount of breast milk excretion of gentamicin is unknown, but it can be given to infants directly without adverse effects.
4 Diphenhydramine is excreted in breast milk, but it does not adversely affect the infant when therapeutic doses are given to the mother.

88. (3) Methergine can cause hypertension and should not be given to a client with an elevated blood pressure. (2) (SI; EV)
1 This is a test for thrombophlebitis and is not related to Methergine use.
2 This is a normal adult finding.
4 Methergine does not affect respirations.

89. (3) This is the expected outcome; if output exceeds intake, it indicates that the infant is diuresing from the Lasix. (1) (PT; EV)
1 Although important to assess, this is subjective; intake and output would be an objective assessment.
2 This is not the desired outcome; this would indicate dehydration, which could occur with excessive administration of Lasix.
4 Although Lasix can cause hypokalemia, which can precipitate digitalis toxicity, this is not the desired effect of Lasix administration.

90. (1) Terbutaline stimulates beta receptors in the heart, which cause excitation resulting in tachycardia; pulse rates over 120 beats per minute should be reported to the physician, and the dosage will be decreased. (1) (PT; EV)
2 It is more likely that the client will experience tremors.
3 Hypokalemia is a maternal side effect.
4 Maternal hyperglycemia is a side effect; hypoglycemia can occur in the neonate.

91. (4) This is a beta-blocking agent that reverses the uterine inhibitory responses and cardiovascular effects of terbutaline. (3) (PT; PL)
1 This may cause symptoms similar to those of terbutaline; it is sometimes used to halt premature labor because it inhibits beta 2 receptors.
2 Not an antidote for terbutaline; used with Parkinson's disease.
3 This is a diuretic; it will not reverse cardiovascular effects.

92. (3) Betamethasone accelerates lung maturity and reduces intravascular hemorrhage and necrotizing enterocolitis in the preterm fetus if given 24 hours before delivery. (1) (PT; IM)
1 Chorioamnionitis will be treated with antibiotic therapy; this problem may occur if the membranes are ruptured prematurely and delivery does not occur within 24 hours.
2 The drug has no effect on uteroplacental exchange.
4 The neonate, not the fetus, develops respiratory distress syndrome (RDS); if given to the mother 24 hours before preterm delivery, the drug will decrease the severity and incidence of RDS in the neonate.

93. (1) A pudendal block provides anesthesia to the perineum. (1) (PT; IM)
2 This does not affect muscle control.
3 This affects only the perineum, not the bladder.
4 This anesthetizes only the perineum, not the cervix or body of uterus.

HEALTHY CHILDBEARING (HC)

94. ① Add 7 days to the beginning of the last menstrual period and subtract 3 months. (2) (DF; AN)

2 This is an incorrect calculation; this is too late.

3 This is an incorrect calculation; this is too early.

4 Same as answer 3.

95. ② Development proceeds in a cephalic to caudal progression. (1) (DF; IM)

1 Both sides of the brain develop at the same time; which side of the brain becomes dominant develops later.

3 Both sides of the brain develop and begin functioning at the same time.

4 Development proceeds in a cephalic to caudal progression, not caudal to cephalic.

96. ③ The presence of the fetal heart rate is a positive sign of pregnancy. (1) (DF; AS)

1 The feeling of movement is a presumptive sign of pregnancy.

2 The bluish color of the cervix caused by pelvic congestion and edema is a probable sign of pregnancy.

4 The presence of chorionic gonadotropin in the urine is a probable sign of pregnancy.

97. ③ These are all presumptive of pregnancy; they can be caused by other conditions and cannot be considered proof of pregnancy. (2) (DF; AS)

1 Abdominal enlargement and HCG in the urine are probable signs; their presence does not offer a definite diagnosis of pregnancy.

2 Abdominal enlargement is a probable sign; urinary frequency and nausea are presumptive signs.

4 Urinary frequency and breast changes, such as tingling and enlargement, are presumptive signs; abdominal changes are probable signs of pregnancy.

98. ④ Increased levels of HCG may cause nausea and vomiting, but the exact reason for this is unknown. (1) (PC; AN)

1 Some women do not experience nausea and vomiting.

2 Lightening occurs at the end of the third trimester; nausea and vomiting usually cease at the end of the first trimester.

3 Nausea and vomiting are unrelated to whether the pregnancy is desired or unwanted.

99. ① There is reduced GI motility during pregnancy because of the high level of placental progesterone and displacement of the stomach superiorly and the intestines laterally and posteriorly; absorption of some drugs, vitamins, and minerals may be increased. (3) (PT; AN)

2 The glomerular filtration rate increases during pregnancy.

3 This is unrelated to the absorption of drugs.

4 This is unrelated to the absorption of drugs; the amount of gastric secretion is somewhat lower in the first and second trimesters but increases dramatically in the third trimester.

100. ④ This amount is recommended by the Food and Nutrition Board of the National Academy of Sciences. (1) (DF; IM)

1 This is less than the recommended requirement.

2 Same as answer 1.

3 Same as answer 1

101. ④ Often there is a decrease in sexual desire in the first trimester, probably related to nausea and vomiting; if couples are informed about this, they are less likely to become distressed. (2) (DF; IM)

1 Calling the situation a problem may cause further anxiety in the client.

2 The client is asking the nurse for information; the client is unable to answer this question.

3 This does not tell the client why this feeling is occurring; this provides false reassurance.

102. ④ Fasting results in hypoglycemia, which can cause nausea; in addition, the developing fetus should not be deprived of nutrients for any length of time. (2) (PC; PL)

1 Fluids need not be increased but should be consumed between meals.

2 Calcium intake will not change the nausea.

3 This intake would not be sufficient to meet the normal nutritional needs of the mother and fetus.

103. ④ This is within the recommended weight gain for a woman who was of average weight for her height before pregnancy. (1) (DF; AN)

1 This is less than the recommended weight gain for a woman who was of average weight for her height before pregnancy.

2 Same as answer 1.

3 Same as answer 1.

104. ③ Most women feel movement by the 20th week of gestation. (2) (DF; IM)
    1 Twelve weeks is too early to feel movement.
    2 Multiparas may feel movement this early, but for most primigravidas movement is felt between 18 to 20 weeks.
    4 Very late to feel initial movement; lack of movement by the 24th week might indicate a problem.

105. ④ Estrogen and progesterone cause increased vascularization and resultant congestion of mucous membranes. (2) (PC; IM)
    1 Nasal decongestants are not advised during pregnancy; clients should consult a physician before using any medication.
    2 The contrary is true; it is not common for women to develop allergies during pregnancy.
    3 Untrue; this is very normal because higher estrogen and progesterone levels increase vascularization of mucous membranes.

106. ④ The fetal heartbeat can be heard with an electronic Doppler between 10 and 12 weeks gestation; a fetoscope cannot pick up the fetal heartbeat before the 17th week. (1) (DF; IM)
    1 This is too early for the heartbeat to be heard with a fetoscope.
    2 Same as answer 1.
    3 This is late; the fetal heart can be first heard with an electronic Doppler between 10 to 12 weeks.

107. ② The uterus remains in the pelvis until the second trimester, placing pressure on the bladder. (1) (DF; IM)
    1 The fetus is in the uterus; this is too early for the baby's head to descend, which occurs in the latter stages of pregnancy and can cause urinary frequency at that time.
    3 Fetal waste products are very slight at this time.
    4 Frequency is a physiologic, not a psychologic, symptom of pregnancy.

108. ③ During pregnancy the need for water is increased; it is related to the increased metabolic rate and expanded blood volume; there is no indication of urinary infection. (1) (SI; PL)
    1 Fluids must be increased, not decreased.
    2 This is unnecessary; there is no indication of urinary infection.
    4 The bladder needs to be emptied often to prevent urinary stasis and potential ascending infection into the kidneys.

109. ② The issue of pregnancy is resolved by this time; awareness of the fetus as a person and the body changes of pregnancy lead the client to desire to learn about fetal growth, body changes, and nutrition. (2) (DF; IM)
    1 This information would be appropriate for the last trimester.
    3 This information would be appropriate for the first trimester.
    4 Same as answer 1.

110. ③ The funic souffle is blood rushing through the fetal umbilical cord and is therefore the same rate as the fetal heart rate. (3) (DF; AN)
    1 Twins would have different heart rates.
    2 The maternal heart rate should be much slower than the fetal heart rate.
    4 The uterine souffle is blood moving through the maternal side of the placenta and is the same as the mother's heart rate; the maternal heart rate should be less than 100.

111. ④ This provides the opportunity for an evaluation of the client's food intake. (2) (PC; AS)
    1 This may prevent further exploration of the diet because the client may answer yes or no.
    2 Same as answer 1.
    3 This assumes the client is not eating properly.

112. ③ Pelvic rocking exercises in which the lower back can be pressed onto the floor by contracting the abdominal muscles can reduce back pain. (3) (PC; EV)
    1 Leg lifts help to strengthen the abdominal muscles and promote circulation in the lower extremities.
    2 Tailor sitting stretches the muscles of the inner thighs and tones the muscles of the outer thighs.
    4 Kegel's exercises are used to tone pelvic floor muscles.

113. ③ This is the time when the fetus is laying down fat deposits and gaining the most weight. (2) (DF; IM)
    1 There is weight gain throughout pregnancy, but it is most marked in the third trimester.
    2 There is little weight gain during this period of organ development.
    4 Same as answer 1.

114. ④ This is the content of childbirth classes to adequately prepare parents for childbirth. (2) (DF; IM)
    1 This is only part of the class content.

2 This is not an absolute; most childbirth methods inform parents that drugs are available if necessary.

3 Same as answer 1.

115. (2) There is much to be learned and practiced so that the client can vary the techniques through the stages of labor. (3) (DF; IM)

1 The Read method can be quickly taught to an "unprepared" woman in labor.

3 This is untrue, because small amounts of medication can be used if required.

4 The Read method focuses on naturalness and denial of pain.

116. (4) Effleurage is a gentle massage of the abdomen. (1) (PC; EV)

1 This is the pelvic rock; it is used during pregnancy to relieve backache.

2 This is a technique of breathing.

3 Same as answer 2.

117. (4) Kegel's exercises develop and strengthen the pubococcygeal muscle; they are done through repeated contractions of the vagina. (3) (CI; EV)

1 Tailor sitting aids in relaxing the muscles of the pelvic floor.

2 This is effective in relieving backaches.

3 Same as answer 2.

118. (2) The amniotic fluid is a protective environment; the fetus depends on the placenta, along with the umbilical blood vessels, for obtaining nutrients and oxygen. (2) (DF; EV)

1 This is a true statement and would not require further teaching.

3 Same as answer 1.

4 Same as answer 1.

119. (1) Fatigue will influence the successful use of other coping strategies such as distraction; this may lead to the client's requiring pain medication. (2) (DF; IM)

2 Progesterone is decreased at this time.

3 Energy will enhance the quality of contractions.

4 The client does not push during the first stage of labor; pushing is done during the second stage.

120. (2) Various foods can be substituted for meat or animal-related products in planning nutritious meals for the pregnant woman who is a vegetarian. (1) (PC; PL)

1 This would ignore the client's beliefs and lifestyle.

3 The client may know good nutrition; this client needs help to adapt the vegetarian diet to meet pregnancy needs.

4 Same as answer 1.

121. (4) Beans are economical and a high source of iron. (2) (PC; IM)

1 This is an iron-rich food but expensive and often unpalatable to many people.

2 This is a moderate source of iron, but the leaner cuts of meat are expensive and not as good a source of iron as the three-bean casserole.

3 This is expensive and a lower source of iron than a three-bean casserole.

122 (3) Folic acid supplements (0.4 mg/day) greatly reduce the incidence of neural tube defects. (2) (HI; AN)

1 This is not the action of folic acid.

2 Same as answer 1.

4 This is related to the Rh factor and is not prevented by folic acid.

123. (4) Unless a lactose intolerance is present, the client should drink milk; eating dried fruits and high-fiber foods and increasing fluids and activity will aid in lessening constipation. (2) (HI; IM)

1 These can cause constipation.

2 These are not as beneficial as whole milk and will cause constipation as well.

3 Megadoses of vitamins can be harmful; prenatal vitamins are not a substitute for milk.

124 (3) Calcium supplements are available for people who do not tolerate milk or milk products. (1) (HI; PL)

1 Good dental care and proper mouth hygiene will be more beneficial for maintaining healthy teeth.

2 Calcium is essential to the pregnant woman's diet for the development of the fetal skeleton; it must be supplemented if the client dislikes milk and milk products.

4 If milk makes the client ill, this would be poor advice and compliance would be suspect.

125. (3) Broccoli is a good source of calcium because it contains approximately 150 mg of calcium per 6 oz cup, compared with a 6 oz cup of milk, which contains 216 mg. (2) (PC; PL)

1 Corn contains about 5 mg calcium per 6 oz cup.

2 Liver contains about 18 mg calcium per 6 oz serving.

4 Lean meat contains about 20 mg calcium per 6 oz serving.

126. ① Carrots provide the precursor pigment, carotene, which the body converts to vitamin A. (1) (PC; AN)
    2 This contains only about half the needed vitamin A precursor.
    3 These contain a very small amount of vitamin A precursor.
    4 These do not contain any vitamin A precursor.

127. ② These animal proteins are complete proteins containing all eight essential amino acids. (2) (PC; EV)
    1 Plant proteins are incomplete proteins.
    3 Same as answer 1.
    4 These are incomplete proteins; also, comparatively small amounts of protein are contained in these foods.

128. ② A probable sign of pregnancy is a positive urine immunoassay pregnancy test because it is 95% accurate in detecting pregnancy; the basis for this test is the presence of HCG in the urine. (2) (CI; AS)
    1 This is a presumptive sign of pregnancy; there are many other causes of amenorrhea.
    3 This is a presumptive sign of pregnancy; there are other causes of frequency, such as urinary tract infection.
    4 This is a presumptive sign of pregnancy; nausea can occur during the first trimester because of the secretion of HCG; there are many causes of nausea other than pregnancy.

129. ① A reactive nonstress test is a normal finding; the nurse notes the occurrence of two or more increases in FHR greater than 15 beats/minute associated with fetal movement. (3) (CI; EV)
    2 A reactive nonstress test is a normal finding that suggests fetal well-being.
    3 Maternal movements have no bearing on nonstress test readings; fetal movements and fetal heart rate are monitored.
    4 Same as answer 3.

130. ③ This is a description of striae gravidarum; they occur from the stretching of the breast and abdominal skin. (1) (DF; AS)
    1 This is chloasma.
    2 This is Chadwick's sign.
    4 This is linea nigra.

131. ① Measurement of the fetal structures provides information that is useful in approximating fetal age. (1) (CI; AS)
    2 This test can detect only physical defects.
    3 Ultrasound can detect some, but not all, defects.

4 Ultrasound is done primarily to estimate fetal age, not to approximate linear growth.

132. ④ Elevated levels of alpha-fetoprotein in pregnant women have been found to reflect open neural tube defects such as spina bifida and anencephaly. (1) (DF; AS)
    1 This is not associated with alpha-fetoprotein levels; there are, as yet, no tests to determine if fetus has cystic fibrosis.
    2 Doing a Guthrie test on the infant soon after ingestion of formula is the only way to test for PKU; there are no tests available during pregnancy.
    3 This is not associated with elevated alpha-fetoprotein levels.

133. ④ A full bladder may improve resolution in women less than 20 weeks gestation; a full bladder serves as an anatomic landmark and elevates the uterus out of the pelvis for better visualization. (1) (CI; IM)
    1 This is unnecessary; the procedure does not involve the colon.
    2 This is a noninvasive procedure that does not irritate the uterus or initiate labor.
    3 This is a noninvasive procedure that does not affect the alimentary tract; fasting is contraindicated during pregnancy.

134. ② A negative test implies that placental support is adequate and the fetus is likely to tolerate the stress of labor contractions at that time. (2) (CI; AN)
    1 Interpretable data did not show signs of hyperstimulation if a negative result was reported.
    3 A positive test indicates that the fetus is at increased risk; the fetus has late decelerations with contractions.
    4 Fetal heart rate accelerations with movement do not require doing a trial induction.

135. ① Once the test is initiated the client will require continuous electronic monitoring and will be confined to bed; contractions are more uncomfortable with a full bladder. (1) (CI; IM)
    2 The client should be in the semi-Fowler's position to avoid supine hypotension.
    3 The client should eat so that the fetus does not become hyperactive.
    4 Only external monitoring is done.

136. ③ The lecithin concentration rises abruptly at 35 weeks, reaching a level that is twice the amount of sphingomyelin, which decreases concurrently. (1) (CI; AS)

1 At about 30 to 32 weeks gestation the amounts of lecithin and sphingomyelin are equal; this result indicates lung immaturity.

2 The ratio is 2:1 when lung maturity is adequate; it is only early in pregnancy that the sphingomyelin concentration is higher than the lecithin concentration.

4 It is not until the L/S ratio is 2:1 that fetal lung maturity is attained.

137. ① The first stage of true labor is cervical dilation. (1) (DF; AS)

2 It is not uncommon for membranes to rupture before true labor begins.

3 Client's perception is not an indication of true labor; because of admission to the hospital and loss of diversionary activities, the client perceives her contractions as becoming harder.

4 The fetal heart rate does not indicate true labor; the rate may be slowing because the fetus is resting or distress is occurring.

138. ④ The priority is to evaluate the progress of labor so that the nurse can plan care. (2) (SI; AS)

1 This question should be asked but is not the first priority.

2 Same as answer 1.

3 Same as answer 1.

139. ② Amniotic fluid is alkaline and turns nitrazine paper blue. (1) (CI; AN)

1 Amniotic fluid does not turn nitrazine paper this color.

3 Same as answer 1.

4 Same as answer 1.

140. ② The left sacrum anterior position indicates that the fetus is in a breech presentation and the head is in the fundus; fetal heart sounds are best heard in the left upper quadrant. (2) (SI; AS)

1 Fetal heart sounds would be in the left upper, not the lower, quadrant.

3 Fetal heart sounds would be in the left, not right, upper quadrant.

4 Same as answer 1.

141. ④ This is an ROA presentation because the prominence over the symphysis suggests a vertex and the fetal occiput and back are in the right anterior quadrant. (3) (SI; AN)

1 This is ruled out; this fetus is in a vertex, not a breech, presentation.

2 This is ruled out by the presence of irregular lumps on the left side, suggesting that the fetal back is in the mother's right quadrant.

3 The occiput is not located in the left posterior quadrant; the occiput and back are on the mother's right side.

142. ④ A persistent occiput-posterior position causes intense back pain because of fetal compression of the sacral nerves. (1) (DF; IM)

1 This would not cause back pain.

2 Same as answer 1.

3 Same as answer 1.

143. ③ Counterpressure alleviates some of the discomfort from the pressure of the fetal head on the sacrum. (1) (PC; IM)

1 Elevating the legs will increase tension and discomfort.

2 Panting may lead to hyperventilation, which will cause maternal respiratory alkalosis and fetal acidosis.

4 Kegel's exercises tone the pelvic musculature, not the back.

144. ③ This describes the right occiput-anterior presentation; the fetal heart is on the right and the head is anterior. (1) (SI; AS)

1 In the left occiput-posterior presentation, the back is on the left and the occiput is posterior.

2 In the left sacrum-anterior presentation, the breech is the presenting part, not the head.

4 In the right mentum-anterior presentation, the chin is the presenting part, not the occiput.

145. ② Identifying progress and providing encouragement motivate the client and promote positive feelings about the self. (1) (SC; PL)

1 A client in active labor should not be left alone.

3 Lying flat on her back may induce supine hypotension; side-lying should be encouraged to promote venous return.

4 During this early stage of labor, diversion is therapeutic because the client is not totally involved with herself.

146. ④ The client is in the early part of the first stage of labor, and it is important to help the partner with the role of coach. (2) (SC; IM)

1 It is not necessary to measure the fetal heart rate this often until the second stage of labor; the client may be on an external monitor, which constantly records FHR.

2 Delivery is not imminent at this time.

3 Suggesting that the pain is bad may increase anxiety and produce greater discomfort.

147. (2) This is used during the early phase of labor when mild contractions dilate the cervix to 3 cm. (2) (PC; EV)
1   This is used during the transition phase of labor.
3   This is used in combination with other breathing patterns; it is a part of the accelerated-decelerated pattern.
4   This is used during the active phase of the first stage of labor.

148. (2) This is done to prevent pushing, because full dilation has not yet occurred. (1) (SI; EV)
1   This is not done until full dilation; it may tire the mother and cause cervical edema.
3   This is done in the early part (preliminary phase) of the first stage of labor.
4   This is done in the middle part (accelerated phase) of the first stage of labor.

149. (2) A full bladder can impede the forces of labor and is uncomfortable for the woman. (2) (SI; AS)
1   The client has been dilating and is therefore in true, not false, labor.
3   Before this conclusion is considered, the client's bladder should be emptied to relieve the pressure of the presenting part on the uterus; the client can then be observed to see whether regular contractions resume.
4   This would have been established in the admission examination.

150. (4) This is a description of the contractions during the active portion of the first stage of labor. (2) (DF; AS)
1   This adaptation occurs in the transitional phase of the first stage of labor.
2   Same as answer 1.
3   Same as answer 1.

151. (3) The client is experiencing supine hypotension, which is caused by the gravid uterus compressing the large vessels; side-lying will relieve the pressure, increase venous return, improve cardiac output, and raise blood pressure. (1) (PC; IM)
1   This will not relieve uterine compression of large vessels; the client should be placed on her side.
2   Same as answer 1.
4   Same as answer 1.

152. (1) A plus station indicates that the head is below the ischial spines. (2) (DF; AS)
2   A minus station indicates that the head is above the level of the ischial spines.

3   This indicates that the head is still high and not engaged in the pelvis.
4   This indicates that the head is at the level of the ischial spines or engaged.

153. (4) The client is hyperventilating; these actions promote rebreathing of carbon dioxide, relieving respiratory alkalosis. (1) (CI; IM)
1   This may cause the client to hyperventilate further.
2   Same as answer 1.
3   This will not improve the client's respiratory alkalosis, the problem that is causing the client to hyperventilate.

154. (3) Inspection of the perineum is done to determine if rupture of the membranes has occurred; the fluid should be tested with nitrazine or ferning test. (2) (CI; AS)
1   The client probably does not have to void; rupture of the membranes should be documented.
2   This can be done after testing the fluid to rule out rupture of the membranes; color, amount, and odor of fluid are assessed before change of linen.
4   The fetal heart rate should be assessed after rupture of the membranes has been established.

155. (4) Internal monitoring requires the insertion of the probe into the fetal head inside the uterus, thus placing the mother and baby at greater risk for infection. (2) (CI; AN)
1   The monitor strips are the same.
2   Each time the client with external monitoring moves, the baseline must be adjusted.
3   Internal monitoring tends to be more accurate since the probe is not affected by the mother moving about.

156. (4) This is a normal occurrence caused by the interplay between the sympathetic and parasympathetic nervous systems. (3) (DF; IM)
1   There is no need for this intervention because this is a normal response.
2   Same as answer 1.
3   Same as answer 1.

157. (4) This fetal heart rate change is known as variable-type decelerations; this is indicative of umbilical cord compression that, if left uncorrected, may lead to fetal compromise; interventions are directed at improving umbilical circulation. (3) (CI; AN)
1   This is not an abnormal finding and, therefore, requires no action by the nurse.
2   This is a normal variation of the fetal heart rate

reflecting a well-oxygenated fetal nervous system.

3 These are recurrent early decelerations, a result of fetal head compression during a contraction; they are a nonhypoxic reflex response requiring no immediate intervention.

158. (4) The infusion should be stopped because it is placing the fetus in danger. (1) (SI; IM)
1 This can be done later; it is not the priority.
2 With late decelerations this action would not be sufficient.
3 Same as answer 2.

159. (2) Low back pain is aggravated when the client is in the supine position because of increased pressure from the fetal presenting part. (1) (PC; IM)
1 This position relieves back pain.
3 Same as answer 1.
4 Same as answer 1.

160. (2) If cord compression is allowed to occur, fetal hypoxia results in central nervous system damage or death; therefore, manual elevation of the presenting part is indicated. (2) (SI; IM)
1 This can be done after elevating the presenting part off the cord; this is necessary because with each contraction the umbilical cord becomes compressed between the maternal pelvis and the presenting part.
3 This would be the next step; first, intervention is necessary to relieve pressure on the umbilical cord and prevent fetal hypoxia.
4 This would be unsafe; compression of the cord would continue in this position; the knee-chest or Trendelenburg positions could be used to allow gravity to reduce pressure on the cord.

161. (3) This prevents the cord from drying and the umbilical vessel, which supplies oxygen to the fetus, from collapsing. (2) (PC; IM)
1 The cord is never handled because it can cause the cord to spasm, shutting down the fetal blood supply.
2 Priority interventions are directed toward relieving compression on the cord and delivering the fetus promptly, not on monitoring.
4 This is not the priority; the priority is to maintain cord integrity.

162. (1) This eases the pressure of the presenting part on the umbilical cord and receives the highest priority. (2) (SI; IM)
2 Oxygen can be given, but this is not the priority.

3 Time should not be wasted trying to locate the fetal heart when the fetus is at risk for hypoxia because of cord compression.
4 A cesarean delivery may be necessary; however, the priority is to ease the pressure on the cord to prevent fetal hypoxia.

163. (4) Pressure on the rectum during contractions would indicate that a laboring client is beginning transition. (3) (DF; AS)
1 This occurs when transition is complete, that is, when the client is fully dilated.
2 This occurs when labor begins, not in the beginning of the transitional phase.
3 Same as answer 1.

164. (1) This is the most difficult part of labor, and the client needs encouragement and support to cope. (2) (SI; AN)
2 Fluid management does not depend on the stage of labor.
3 Breathing patterns should be complex and should require a high level of concentration to distract the client.
4 Medication at this time will depress the infant's respirations and is contraindicated.

165. (2) All signs indicate impending delivery; the perineum should be inspected for the appearance of caput. (2) (SI; AS)
1 Assessment of fetal status is important; however, the nurse must first determine if delivery is imminent.
3 This is important to know, but it does not assess the client's complaint.
4 Same as answer 3.

166. (4) Clients should use a panting or blowing pattern to overcome the premature urge to push. (2) (SI; IM)
1 Expulsion breathing should not be used at this time because the cervix is not fully dilated and cervical edema and lacerations can occur.
2 Slow-chest breathing is used during the early phase of the first stage of labor.
3 Slow-paced breathing is used during the first stage of labor.

167. (1) A paper bag helps the client to rebreathe CO_2, which helps correct the respiratory alkalosis. (2) (CI; IM)
2 The client's O_2 level is already elevated; the client needs to elevate the CO_2 level.
3 CO_2 is too dilute in room atmosphere.
4 An anesthetic will further deplete CO_2 and is therefore contraindicated.

168. (2) Effective pushing will hasten the passage of the baby through the birth canal. (1) (DF; EV)
    1   The fetal position should have been established before the second stage.
    3   Delivery is imminent, and medication given at this time will depress the infant's respirations at birth.
    4   The mother will breast-feed in the fourth stage of labor.

169. (4) Mild toxic reactions occur because of vasodilation from direct action of these medications on maternal blood vessels; vertigo, dizziness, and hypotension may occur. (3) (PT; EV)
    1   Labor is not affected because there is no systemic effect.
    2   A local anesthetic will not lower the level of consciousness, thus the loss of the swallowing reflex is avoided.
    3   A local anesthetic does not affect the respiratory center in the central nervous system.

170. (2) The first action by the nurse should be to confirm whether delivery is imminent by seeing whether the presenting part is emerging. (1) (SI; IM)
    1   The priority is to confirm the client's sensation; the nurse should remain with the client and ask a colleague to call the physician.
    3   This response demeans the client; she may be correct.
    4   If delivery is imminent, this could cause injury to the baby.

171. (2) Panting will slow the process so the nurse can support the head as it is delivered. (1) (SI; IM)
    1   Pushing will speed up the delivery and could injure the mother and the baby.
    3   This will have no effect on the progress of the delivery.
    4   Usually holding the breath causes involuntary pushing, and it also depletes the mother and baby of oxygen.

172. (4) The color of the amniotic fluid is indicative of meconium staining; the practitioner must therefore prepare for the potential fetal aspiration of meconium. (1) (CI; IM)
    1   The newborn should not be stimulated to cry until the airway is cleared of meconium.
    2   The newborn is placed on the back to facilitate laryngoscopic visualization of the vocal cords and suctioning as necessary.
    3   There is no indication that an umbilical transfusion will be necessary.

173. (1) A gush of blood occurs when the placenta separates from the uterine wall before the normal physiologic clamping off of the vessels at this site. (2) (SI; AS)
    2   The uterus contracts and becomes firm when the placenta separates.
    3   The uterus appears to increase in size when the placenta separates.
    4   Untrue; blood pressure returns to prenatal status shortly after delivery; the decrease is gradual and unrelated to placental separation.

174. (2) As the placenta separates and drops down, the cord lengthens. (2) (DF; AS)
    1   The fundus contracts and becomes rounded and firmer.
    3   The client may feel a contraction, but it is not nearly as uncomfortable as the painful contractions at the end of the first stage of labor.
    4   Continual seepage occurs when there is hemorrhaging; a large sudden gush of blood heralds placental separation.

175. (2) The third stage of labor is from the delivery of the baby to the delivery of the placenta; a firmly contracted uterus is desired to minimize blood loss. (1) (DF; AN)
    1   Providing comfort is a desirable goal but is secondary to the life-threatening possibility of hemorrhage associated with a boggy uterus.
    3   This would be a concern in the first and second stages of labor; it is no longer applicable after the fetus is delivered.
    4   The maternal respiratory rate may vary above or below this range.

176. (2) Gentle massage stimulates muscle fibers and results in firming the tone of the fundus; it also helps expel any clots that may be interfering with the firming of the fundus. (1) (CI; IM)
    1   Elevating the client's legs would increase return of blood from the extremities, but it would not firm the tone of the client's fundus.
    3   This would be done only if massaging the uterus were ineffective; a physician's order is required.
    4   This would not be the first action at this time; gentle massage to firm the fundus is the priority.

177. (3) A distended bladder impedes uterine contractions, predisposing the client to hemorrhage. (1) (PC; AS)
    1   Relaxation is a priority before delivery; in the fourth stage the client is often euphoric.
    2   Love grows with care and responsibility.

4 The mother is egocentric at this point and is not yet totally involved with the infant.

178. (4) This blood pressure is elevated; intervention may be necessary. (1) (SI; AS)
1 This is within normal limits.
2 Same as answer 1.
3 This is a slight elevation, which is consistent with the physiology of labor.

179. (3) A full bladder commonly elevates the uterus and displaces it to the right; even though the uterus feels firm, it may relax enough to foster bleeding; therefore, the bladder needs to be emptied to increase uterine tone. (1) (PC; IM)
1 If part of the placenta, umbilical cord, or fetal membranes is not fully expelled during the third stage of labor, the retention limits uterine contraction and involution; a boggy uterus and bleeding would be evident.
2 This is not a sign of uterine stabilization; the uterus cannot remain contracted over a full bladder.
4 Vigorous massage tires the uterus, and even with massage the uterus cannot contract over a full bladder.

180. (3) Chances of postpartal hemorrhage are 5 times greater with large infants because uterine contractions may be impaired after delivery. (2) (PC; AN)
1 On the contrary, early breast-feeding will stimulate uterine contractions and lessen the chance of hemorrhage.
2 This does not contribute to postpartum hemorrhage because the anesthetic for a pudendal block does not affect uterine contractions.
4 This is a short third stage; a prolonged third stage of labor, 30 minutes or more, may lead to postpartum hemorrhage.

181. (4) A postpartum chill is a normal occurrence of unknown cause. (1) (PC; EV)
1 This measure is part of the routine postpartum assessment but does not need to be done in relation to the chill.
2 Same as answer 1.
3 Same as answer 1.

182. (3) Initially, cold therapy reduces edema and discomfort. (2) (PC; PL)
1 Heat therapy usually begins after 8 to 12 hours of cold therapy; warm, not hot, water would be used.

2 ASA is contraindicated in the puerperium; there is too great a risk for hemorrhage.
4 This provides little or minimal perineal relief.

183. (2) Voiding will be difficult because of periurethral edema and discomfort. (2) (CI; EV)
1 This rarely occurs with primiparas, even when a lot of pushing occurs.
3 A second-degree laceration is unrelated to lactation.
4 A second-degree tear is unrelated to maladaptive bonding and attachment.

184. (4) There is a greater chance for infection to occur with this type of episiotomy because there is more tissue damage and healing is slower. (2) (SI; AN)
1 There is more tissue damage so there would be more edema.
2 The mediolateral episiotomy cuts large muscles used in ambulation, so there is more discomfort with movement.
3 More blood vessels are injured, so there is more bleeding and bruising.

185. (2) Cold causes vasoconstriction and reduces edema by lessening the accumulation of blood and lymph at the episiotomy site; cold also deadens nerve endings and lessens the pain. (2) (PC; IM)
1 This is not used after an episiotomy.
3 This may diminish the pain but will not lessen the edema.
4 This position will not lessen pain or reduce edema.

186. (3) Prevention of infection is the priority. (2) (HI; EV)
1 It is not necessary to stop sitz baths as long as they provide comfort.
2 Stair climbing may cause some discomfort but is not detrimental to healing.
4 This provides comfort but is not the priority.

187. (2) A client's temperature may be elevated to 100.4° F during the first 24 hours postpartum as a result of dehydration from labor. (2) (PC; AN)
1 Mastitis usually develops after breast-feeding has been established and mature milk is present.
3 This usually begins with a fever of 100.4° F or more on 2 successive days, excluding the first 24 hours postpartum.
4 Urinary tract infections usually become evident later in the postpartum period.

188. (2) During the postpartum period a leukocytosis (WBC count of 15,000 to 20,000/mm³) is normal and related to the physical exertion experienced during labor and delivery, not infection. (2) (PC; EV)
   1 This is not a drop in the WBC count because the normal postpartal white blood cell count is between 15,000 and 20,000/mm³.
   3 The leukocytosis is normal and related to the physical exertion of labor and delivery, not infection.
   4 Same as answer 3.

189. (2) Infection is most commonly transmitted through contaminated hands. (2) (HI; IM)
   1 Tub baths are permitted.
   3 Douching is contraindicated until the cervix is closed, usually 6 weeks after delivery.
   4 This is contraindicated in the postpartum period until the cervix is completely closed.

190. (4) Varicose veins predispose the client to thrombophlebitis; warmth, redness, and pain in the calf are signs of thrombophlebitis. (1) (SI; AS)
   1 The clotting mechanism is not affected; clot formation results because of venous pooling and decreased venous return caused by the impaired vasculature.
   2 These tests, while concerned with clotting factors, are not related to the development of thrombophlebitis.
   3 These would be affected by the amount of bleeding incurred during delivery, which usually is not severe enough to impair circulatory competency.

191. (4) This indicates subinvolution and needs further assessment. (1) (SI; EV)
   1 This is a normal postpartal concern, especially with twins.
   2 This is normal postpartal diuresis.
   3 This is the normal milk let-down reflex.

192. (4) On the third to fourth day the uterine discharge becomes pink to brown; it lasts to approximately the tenth day. (1) (PC; AS)
   1 After about 10 days the uterine discharge becomes yellow to white; it may continue until 2 to 6 weeks after the birth of the baby.
   2 It is unusual to have scant lochia rubra.
   3 Lochia rubra lasts from the first to the third or fourth day; it is usually heavy but may be moderate after a few days.

193. (2) Breast-feeding stimulates oxytocin release and uterine contractions, resulting in increased lochial flow. (2) (PC; IM)

   1 Weight loss may occur more slowly in the breast-feeding mother because of increased nutritional and caloric needs.
   3 The increased levels of oxytocin and subsequent uterine contractions will enhance involution.
   4 Heat is not contraindicated, and the client may take warm showers; heat is also used if the mother experiences problems such as engorgement or sore nipples.

194. (4) If suckling or nipple stimulation is discontinued, acinic cells degenerate, regressive changes occur, and lactation ends. (2) (PC; IM)
   1 Other than suckling, the stimulant for oxytocin secretion is unknown.
   2 High levels of estrogen inhibit anterior pituitary gland secretion of lactogenic hormones.
   3 Milk secretion starts on the third or fourth day after delivery.

195. (4) This is a result of the arrangement of fatty acids on the glycerol molecule and is related to the natural lipase activity that is present in human milk that has not been heat treated. (2) (DF; AN)
   1 The converse is true; lactose content is higher in human milk.
   2 It is lower, but protein in human milk is easier for the infant to digest.
   3 This is untrue; these factors are found only in human milk.

196. (4) This dilates milk ducts, promotes emptying of the breasts, and stimulates further lactation. (1) (PC; EV)
   1 A large consumption of milk products is not required to stimulate the production of milk.
   2 This will contract the milk ducts and interfere with the milk let-down reflex.
   3 Breast binders may inhibit lactation; they fool the body into thinking that milk secretion is no longer needed.

197. (3) Exposure to air dries the nipples by evaporation; exposure also tends to harden the nipples, making them less tender. (2) (PC; IM)
   1 Continuous ice packs are used to relieve the discomfort caused by engorged breasts, not sore nipples.
   2 This may relieve discomfort but will do nothing to toughen the nipples.
   4 If kept from the breast for a prolonged period, the baby may become accustomed to the bottle and not wish to breast-feed again; in addition, absence of suckling will inhibit lactation.

198. (4) The most vigorous sucking will occur during the first few minutes of breast-feeding when the infant would be on the unaffected breast; later suckling is less traumatic. (3) (PC; IM)
 1 Stopping nursing for 2 days is unnecessary and would interfere with lactation.
 2 Manual expression may not completely empty the breast, interfering with lactation.
 3 A breast shield confuses an infant because it is necessary to use a different sucking pattern to obtain milk.

199. (4) Compared with the prenatal diet, the diet for lactation requires an increased intake of all food groups, vitamins, and minerals, plus increased fluid to replace that lost with milk secretion; calories should be increased by 500 daily and protein by 10 to 15 g daily. (2) (PC; IM)
 1 Breast-feeding mothers need to consume an additional 500 calories and 10 to 15 g of protein per day to maintain adequate milk production.
 2 The client needs a well-balanced diet, not just milk, and she must consume an additional 500 calories a day.
 3 This denies the client's concern; optimal nutrition is necessary to produce an adequate milk supply.

200. (4) This keeps the breasts as empty as possible, limiting pressure within the ducts, thereby reducing pain. (1) (PC; IM)
 1 Weaning will cause stasis of milk ducts and increase the fullness of the breasts at this time, thereby increasing pain.
 2 The causative organism is probably already in the baby's nose and mouth; weaning is not always necessary with mastitis.
 3 This is false; a tight-fitting bra will increase pain and suppress milk production; this will impede breast-feeding.

201. (3) Heat increases milk flow, and because the client is not breast-feeding, this is an undesired outcome; application of cold is recommended to restrict milk flow. (1) (CC; EV)
 1 This is a correct statement; analgesics will help lessen the discomfort of engorgement; no further teaching is needed.
 2 This is a correct statement; engorgement lasts 48 to 72 hours; no further teaching is needed.
 4 This is a correct statement; a tight, supportive bra will suppress milk production; no further teaching is needed.

202. (1) This is like binding the breasts; it reduces pain and prevents further engorgement. (1) (CC; IM)
 2 Medication would reduce pain but would not prevent further engorgement.
 3 Milk and fluids should not be restricted after delivery.
 4 Cold compresses would prevent further engorgement in the non-breast-feeding mother.

203. (3) This is an incorrect statement; if the client expresses milk from her breasts, she is stimulating milk production and this will not relieve engorgement; therefore additional teaching will be necessary. (2) (CC; EV)
 1 This is a correct statement; this measure is used to give the body the message that milk production is not needed.
 2 This is a correct statement; non-breast-feeding mothers do not need extra fluids.
 4 This is a correct statement; warm water will promote vasodilation, lead to emptying of breasts, and support further milk production.

HIGH-RISK PREGNANCY (HP)

204. (1) Twins should be suspected with a more rapid increase in fundal height than normal; the nurse should assess for two distinct heartbeats. (3) (CI; AS)
 2 Fundal height, not the size of the baby, should lead the nurse to suspect a multiple pregnancy.
 3 This cannot be determined until an ultrasound is done.
 4 Weight gain will not influence the height of the fundus.

205. (4) The many risk factors attributed to this client place her at high risk in this pregnancy, and a high-risk specialist is the provider of choice. (1) (CI; PL)
 1 A history of repeated early spontaneous abortions is not indicated, and it is not at all clear that a genetic evaluation is required.
 2 A suggestion to quit or cut down on smoking is in order, but one cannot insist that an action be taken by the client.
 3 There is no evidence that this client has an incompetent cervix, and a cerclage placement is a medical decision.

206. (1) Grand multiparas have decreased uterine muscle tone as a result of the repeated distention of pregnancy; consequently, the uterine muscles have difficulty in contracting after delivery. (1) (CI; PL)
2 This can occur in all clients after delivery; it is not specific to grand multiparas.
3 This occurs in all postpartal clients; it is the body's attempt to dispose of excess fluid now that the placenta and baby are no longer present.
4 These may be indicative of chronic hypertension, which is not specific to grand multiparas.

207. (2) Increased parity contributes to an increased incidence of uterine atony because the uterine muscle may not contract effectively, thus leading to postpartal hemorrhage; a 1-hour labor in a grand multipara is not uncommon. (3) (CI; AS)
1 A primipara should maintain a well-contracted uterus because with only one pregnancy the uterus usually maintains its tone.
3 100 mg of Demerol is not considered excessive for a primipara and would not contribute to uterine atony.
4 The delivery of the placenta 10 minutes after delivery of the fetus is normal and would not affect tone of the uterus; multiparity contributes to uterine atony, so the woman who is a grand multipara is at a higher risk for hemorrhage.

208. (4) An amniocentesis is commonly used to diagnose genetic problems, as well as fetal maturity and fetal hemolytic disease. (1) (CI; IM)
1 This is not a reason for doing this invasive procedure.
2 Same as answer 1.
3 An amniocentesis is no longer done routinely if the mother is an older primipara; a sonogram is usually done.

209. (4) In the 14th week, amniotic fluid is present, and small amounts can be withdrawn for testing. (1) (CI; IM)
1 This is untrue; an amniocentesis can be performed any time after the 14th week.
2 There is no amniotic fluid present at the time of conception.
3 It is more appropriate to do an amniocentesis before quickening is established.

210. (3) Amniocentesis has proved useful in detecting potential defects resulting from chromosomal errors, such as Down syndrome, Tay-Sachs disease, hemophilia, and thalassemia. (2) (CI; IM)

1 Not all fetal defects can be detected prior to birth; an amniocentesis can identify some.
2 This is false reassurance, and it may stop further communication by the mother.
4 An amniocentesis does not detect all congenital defects.

211. (2) An invasive procedure such as amniocentesis requires consent. (1) (CC; IM)
1 Intravenous therapy is unnecessary.
3 The infection rate is 1%; this is not an appropriate nursing intervention.
4 No vaginal examination is done before amniocentesis.

212. (1) A copy of the Patient's Bill of Rights is not necessary for informed consent for treatment. (2) (CC; IM)
2 Alternative treatment regimens should be discussed so that the client is able to make an informed choice about which course of treatment to pursue.
3 All questions should be answered honestly and in terms that the client can understand.
4 This information is required to give informed consent.

213. (1) This is done to determine whether any injury has occurred to the fetus or placenta during the procedure. (1) (CI; EV)
2 This position enhances placental perfusion, but it serves no purpose in detecting complications.
3 This is unnecessary; the puncture site seals by itself immediately after removal of the needle.
4 There is no entry into the vaginal canal with this procedure; bleeding or discharge is not expected.

214. (4) It is possible that stimulation of the uterus resulting from the amniocentesis may cause uterine contractions. (1) (CI; AS)
1 This is not necessary because an amniocentesis is not done via the vagina.
2 This is irrelevant because amniotic fluid is in no way influenced by the intake of fluid.
3 This should not be necessary because the pin-prick opening seals up immediately.

215. (4) A full bladder helps position the uterus so that the uterine contents can be visualized during the sonogram; the bladder should be emptied before an amniocentesis to prevent accidental puncture of the bladder during the procedure. (2) (CI; PL)
1 The client would not be asked to void before a sonogram.

2 The client voids only after all abdominal images are obtained.

3 The bladder may fill within 1 hour; a full bladder is undesirable before an amniocentesis.

216. (1) As the process of effacement occurs in the latter part of pregnancy, placental separation from the uterus may occur, causing bleeding. (2) (DA; AS)

2 This occurs in separation of a placenta from the uterine wall (abruptio placentae).

3 This is generally not associated with placenta previa.

4 This is not specific to a low-lying placenta.

217. (2) In a healthy well-oxygenated fetus the heart rate increases with fetal movement; the test looks for accelerations of 15 beats with fetal movements. (2) (CI; AS)

1 This is not a part of the evaluation of the fetus in the nonstress test.

3 Same as answer 1.

4 This is used in the contraction stress test (CST).

218. (2) The fetus with a borderline cardiac reserve will show hypoxia by a decreased heart rate when there is minimal stress, making the test positive. (2) (CI; AS)

1 A baseline fetal heart rate characteristically occurs between contractions and between accelerations and/or decelerations of the heart rate.

3 These decelerations are a response to head compression.

4 These are nonuniform drops in FHR before, during, or after a contraction; variable decelerations during an OCT do not make the test positive.

219. (2) Bleeding could indicate placenta previa or abruptio placentae, which would be aggravated by the contractions from the use of Pitocin. (1) (SI; AS)

1 An oxytocin challenge test is not contraindicated; blurred vision may indicate PIH; cardiac problems may diminish O_2 perfusion to the placenta and compromise the fetus during labor; fetal tolerance to the stress of contractions would be assessed.

3 An oxytocin challenge test is not contraindicated; sickling may reduce O_2 perfusion to the placenta and compromise the fetus during labor.

4 An oxytocin challenge test is not contraindicated; arteriolar spasms may diminish O_2 perfusion to the placenta and compromise the fetus during labor; fetal tolerance would be evaluated by this test.

220. (4) RhIg is given to prevent active formation of antibodies when an Rh-negative individual is at risk for sensitization; if given to an Rh-positive person, an injection of RhIg would cause hemolysis of RBCs. (3) (PT; AN)

1 RhIg is never given to an individual with Rh antibodies.

2 A positive Coombs' test indicates that the woman has Rh antibodies; RhIg is never given to an individual with Rh antibodies.

3 Administration of RhIg to an Rh-positive woman causes hemolysis of RBCs.

221. (2) RhoGAM will be given only if the infant is Rh positive and the Coombs' test is negative. (3) (PT; PL)

1 This is not done; however, a minimal dose of RhoGAM may be given prophylactically in the 28th week of gestation to decrease antibody response in the presence of transplacental bleeding.

3 RhoGAM might be given after the 28th week if the amniocentesis procedure resulted in the escape of some fetal blood into the maternal circulation.

4 RhoGAM is given only to Rh-negative mothers to prevent antibody formation and protect future pregnancies; it is never given to the baby.

222. (1) An ABO incompatibility may develop even in first-born infants since the mother has antibodies against the antigens of the A and B blood cells. These antibodies are transferred across the placenta and produce hemolysis of the fetal RBCs. The infant is AB and an incompatibility may also occur. (3) (CI; AN)

2 No problems will occur if the mother is Rh positive and the baby is Rh negative, only the other way around.

3 A preterm delivery will not produce an incompatibility; it may intensify problems if an incompatibility exists.

4 If the baby is the same type and has the same Rh factor as the mother, no incompatibility occurs.

223. (3) Early postoperative ambulation helps prevent many postpartal complications such as thrombophlebitis and constipation. (2) (SI; PL)

1 Clients are generally discharged by the third or fourth postpartal day.

2 A bowel movement can occur spontaneously if early ambulation and adequate fluids are encouraged.

4 Clients are permitted to shower after 48 hours.

224. ① A tightly contracted uterus in the midline reflects normal physiologic functioning following delivery of the fetuses and expulsion of the placenta; an atonic uterus is a common complication of a multiple delivery. (1) (PC; EV)
2 The woman may have complications but can still breast-feed her infants.
3 When considering recovery following a multiple delivery, physiologic stabilization takes precedence over psychologic concerns.
4 Resting comfortably does not indicate an uncomplicated delivery; a client can be resting quietly while hemorrhaging.

225. ③ Just as after any surgery, pain is a major postoperative problem during the first 24 hours after cesarean delivery. (3) (PC; AN)
1 Oral intake is usually limited for the first 24 to 48 hours postoperatively.
2 Bowels ordinarily do not move for 48 to 72 hours postoperatively.
4 Gaseous distention is more likely to occur on day 3.

226. ② Palpation will indicate whether bladder distention is present; the increased intra-abdominal space available after delivery can result in bladder distention without discomfort. (2) (CI; EV)
1 A physician's order is needed for catheterization.
3 Measurement alone is not sufficient for 24 to 48 hours postpartum.
4 Trauma to the area makes surrounding organs atonic; the client may have a full bladder and not feel the urge to void.

227. ① Excess fluid retention is an undesirable effect of sodium intake. (2) (PT; PL)
2 There is no concern about fluid retention when taking antacids that do not contain sodium.
3 Same as answer 2.
4 Same as answer 2.

228. ② This will decrease the workload of the heart during expulsion and permit a vaginal delivery. (3) (DA; PL)
1 This can increase the cardiac workload.
3 Many clients with cardiac problems can deliver vaginally when precautionary measures are instituted; it is preferable to avoid the secondary stresses that surgery may impose.
4 During the second stage, cardiac output can be increased and this might cause cardiac arrest; the client needs assistance to decrease the cardiac workload.

229. ④ This is the accepted definition of placenta previa. (2) (DA; AN)
1 This occurs in abruptio placentae.
2 Same as answer 1.
3 This may not lead to placenta previa but will place the fetus in jeopardy.

230. ④ With a low-implanted placenta (placenta previa) the presenting part may have difficulty entering the pelvis. (3) (CI; AS)
1 Engagement is difficult with a low-lying placenta.
2 Placenta previa does not make it difficult to palpate small fetal parts.
3 This occurs with abruptio placentae

231. ① An accurate measurement of the amount of blood loss may be obtained by counting or weighing pads, (1) (SI; AS)
2 The fundus may be higher than normal because the low-lying placenta prevents the descent of the fetus into the pelvis, but the height cannot be used to measure blood loss.
3 The vital signs will reflect the effects of the blood loss rather than the amount.
4 Laboratory results demonstrate the effects of the blood loss rather than the amount.

232. ① This position shunts blood to the upper body and vital organs. (3) (DA; AN)
2 The bleeding will continue regardless of this position.
3 Pressure on the cervix is thought to have no bearing on bleeding episodes.
4 The placenta is implanted and positioning will not move it off the cervix.

233. ③ A speculum examination may cause separation of the placenta resulting in profuse hemorrhage; an immediate cesarean delivery is done to prevent fetal demise. (2) (CI; PL)
1 This would do nothing to save the infant if total placental separation occurs.
2 Delaying labor would not be helpful; immediate cesarean delivery is necessary in this emergency.
4 Same as answer 1.

234. ③ A speculum examination may result in a sudden, severe hemorrhage because of the location of the placenta near the cervical os; blood should be ready for administration to prevent shock. (2) (CI; PL)
1 This is an incorrect response; fresh plasma may be used to restore coagulation factors when DIC occurs after severe blood loss.

2 Adults manufacture their own vitamin K, and an injection would not help to prevent bleeding from the placenta.

4 Giving heparin sodium is contraindicated in the presence of hemorrhage.

235. (1) The symptoms indicate loss of blood; to compensate for the decreased cardiac output, oxygen is needed to maintain the well-being of both the mother and the fetus. (1) (DA; IM)

2 This would decrease blood flow to the vital centers in the brain.

3 This would not be the first action; in view of the blood loss, providing oxygen is the priority.

4 This could mask abdominal pain and sedate an already compromised fetus.

236. (4) Multiple past pregnancies tend to make the endometrial lining more scarred and vulnerable to abnormal implantation. (1) (DF; AN)

1 Primigravidas are the least prone; the endometrium is receptive to normal implantation.

2 Two pregnancies have not compromised the endometrium too much; abnormal implantation is less likely to occur.

3 Age is not known to be a significant factor; also, two pregnancies have not compromised the endometrium too much.

237. (4) The side-lying position decreases pressure on the vena cava from the gravid uterus, ensuring more adequate oxygenation of the fetus. (1) (SI; IM)

1 Without proper positioning, breathing techniques will be less effective.

2 Although the client would probably have an order for bed rest, all movement would not be restricted.

3 Lying on the back will increase pressure on the vena cava, further compromising the fetus.

238. (3) As the cervix opens with labor, bleeding may ensue; blood loss is estimated and treatment is based on the maternal/fetal response. (2) (DA; AS)

1 A fetal monitor would be indicated because it more accurately records fetal well-being.

2 This is done in the postpartum period, not while the client is in labor.

4 This is contraindicated; these examinations may stimulate greater bleeding if the placenta is accidentally dislodged.

239. (4) Folic acid is needed to produce heme for hemoglobin. (3) (CI; AN)

1 Folic acid may reduce the risk of a sequestration crisis, but it will not prevent it.

2 There is no relationship between folic acid and the reduction of sickling.

3 There is no change in needs; sickling decreases the oxygen-carrying capacity of hemoglobin.

240. (3) These clients are particularly vulnerable to infections, especially of the genitourinary tract; urine cultures should be performed frequently. (2) (HI; AS)

1 Hypothyroidism affects 1 in 1500 women during pregnancy; women with sickle cell anemia are not at any higher risk for hypothyroidism than the general population.

2 Women with sickle cell anemia are not at an increased risk for this problem during pregnancy.

4 Same as answer 2.

241. (2) A floating fetal head in a primigravida of 42 weeks gestation who is in early labor is suggestive of disproportion because engagement usually occurs before labor in primigravidas. (3) (HI; AN)

1 This test confirms the presence of ruptured membranes, which should not cause any problems with delivery.

3 This occurs during early labor when the presenting part bears down on the capillary structure of the cervix.

4 This falls within the normal range of 120 to 160 beats per minute.

242. (1) If the heart is heard in the upper left quadrant, the baby must be lying in a breech position with the head upright and the heart uppermost. (2) (SI; AS)

2 Fetal heart tones are heard best in the lower quadrants of the abdomen in cephalic presentations.

3 Same as answer 2.

4 Same as answer 2.

243. (2) The feet or buttocks are not effective in blocking the cervical opening, and the cord may slip through and be compressed. (2) (CI; AS)

1 Rapid dilation and precipitate labor can occur with infants in cephalic positions as well.

3 Uterine inertia may result from fatigue or cephalopelvic disproportion and is not necessarily related to fetal position.

4 This is not specific to breech labors.

244. ④ This occurs because pressure on the fetal abdomen from the contractions forces meconium from the bowel. (2) (SI; AN)
  1 Cord prolapse is not an absolute, but it may occur if the presenting part does not fill the pelvic cavity.
  2 Fetal heart rate abnormalities are identified by auscultation or continuous electronic fetal monitoring, not by the presence of meconium.
  3 This is unnecessary; this is a normal occurrence caused by pressure on the fetal abdomen during contractions when the fetus is in the breech position.

245. ② Pelvic pressure or a feeling that the baby is pushing down is one symptom of preterm labor and should be taught to the client so that she can present herself early for care. (3) (DF; IM)
  1 This is not a danger sign of preterm labor.
  3 These are not danger signs of preterm labor.
  4 Fetal movement is not normally felt until approximately 16 weeks.

246. ④ Increasing the frequency may decrease or stop the contractions and prevent dilation. (3) (PT; PL)
  1 This would be unsafe because labor would continue.
  2 This will do nothing if labor is beginning.
  3 This could cause toxic side effects.

247. ④ This provides accurate information; an increase of 30 mm Hg in the systolic reading or an increase of 15 mm Hg in the diastolic reading indicates hypertension during pregnancy. (2) (DF; IM)
  1 This is false reassurance.
  2 This could be frightening and elevate the blood pressure even more.
  3 This response is demeaning.

248. ② Bed rest keeps the pressure of the fetal head off the cervix; the side-lying position keeps the gravid uterus from impeding major vessels, thus enhancing uterine perfusion. (2) (DF; PL)
  1 These are used only when the cord is prolapsed or the client is in shock.
  3 Sitting up in bed increases pressure on the cervix; this may lead to further dilation.
  4 This may aid in relieving pressure of the fetus on the cervix, but it will not enhance uterine perfusion.

249. ④ Prostaglandins in semen may stimulate labor, and penile contact with the cervix may increase myometrial contractility. (1) (DF; IM)

1 Sexual intercourse may cause labor to progress; delivery is not desired in the 33rd week of pregnancy.
2 Same as answer 1.
3 Same as answer 1.

250. ③ Corticosteroids stimulate surfactant production; they also have been shown to reduce the incidence of intraventricular hemorrhage. (2) (PT; AN)
  1 Betamethasone does not affect the labor process.
  2 Betamethasone does not increase placental perfusion.
  4 Betamethasone does not affect the intensity of contractions.

251. ① Labor is continuing, and the promotion of the well-being of the client and fetus is the most important priority for nursing care during this period. (2) (CI; PL)
  2 This response addresses only one aspect of this client's needs; this problem must be dealt with in the context of her other needs.
  3 Same as answer 2.
  4 Same as answer 2.

252. ③ The risk of developing chorioamnionitis (intra-amniotic infection) is increased with prolonged rupture of the membranes; foul-smelling fluid is a sign of infection. (3) (SI; AS)
  1 A prolapsed cord usually occurs shortly after the membranes rupture.
  2 This is an abnormally implanted placenta; it is totally unrelated to ruptured membranes.
  4 This is premature separation of a normally implanted placenta; it is totally unrelated to ruptured membranes.

253. ① In light of the vaginal bleeding, the priority nursing action is ascertaining whether a viable fetus is present. (2) (SI; AS)
  2 This is absolutely contraindicated; bright red bleeding is suggestive of placenta previa.
  3 Same as answer 2.
  4 Same as answer 2.

254. ① This client will have lower tolerance for pain and greater need for pain relief. (2) (PC; PL)
  2 Larger doses may be needed if this is done.
  3 Delays increase anxiety and discomfort, and larger doses are needed.
  4 Individuals who abuse drugs need more medication than do others because of tolerance.

255. ④ HIV is known to be transmitted through body fluids such as blood, semen, and vaginal secretions. (1) (SI; PL)

1 Invasive fetal monitoring is avoided to prevent vertical transmission.
2 HIV is the virus, not the disease. The diagnosis of AIDS is determined when the CD4T cell count drops below 200 per microliter.
3 The incidence is increasing.

256. (4) The earliest sign of drug withdrawal is CNS overstimulation. (2) (SI; AS)
1 These have no relation to drug abuse; most postpartum women are hungry and thirsty.
2 These are related to drug use, not withdrawal.
3 Same as answer 2.

257. (2) It is vital that a baseline measurement be obtained because increasing size is a sign of concealed hemorrhage; in abruptio placentae there is bleeding behind the placenta. (3) (DA; AS)
1 This would be an appropriate assessment, but it is not a priority at this time.
3 Same as answer 1.
4 Same as answer 1.

258. (4) This position moves the gravid uterus off the great vessels of the lower abdomen, which increases venous return, improves cardiac output, and promotes kidney and placental perfusion. (2) (DA; AN)
1 The side-lying position does not influence intraabdominal pressure.
2 While on bed rest the blood pressure decreases and the interstitial fluid is mobilized into the intravascular space.
3 The side-lying position does not prevent thromboses.

259. (4) The accepted definition of hypertension during pregnancy is an elevation of 30 mm Hg systolic and/or 15 mm Hg diastolic over previous normal levels; this reading is above that level and could be indicative of developing preeclampsia in a young primigravida. (2) (CI; AS)
1 These increases are within the acceptable elevations in blood pressure during pregnancy.
2 Same as answer 1.
3 Same as answer 1.

260. (1) Nonperiodic accelerations with fetal movement indicate fetal well-being. (3) (CI; AS)
2 Early decelerations are associated with fetal head compression.
3 Late decelerations are associated with uteroplacental insufficiency.
4 Variable decelerations are associated with cord compression.

261. (4) Rest is advised to reduce arteriolar spasm, and side-lying promotes more efficient venous return to the heart; this improves cardiac output and placental perfusion. (1) (HI; PL)
1 Sodium is necessary to maintain circulatory volume and is not removed from the diet.
2 This may increase general arteriolar spasm; also, venous return is inhibited in the upright position.
3 Because of the increased circulatory volume with pregnancy, the client needs 2000 ml of fluid per day.

262. (3) Blood pressure rises, edema increases, and degenerative changes of the kidney cause increasing proteinuria (3+) as preeclampsia worsens. (2) (CI; AS)
1 With worsening disease, oliguria, not diuresis, may occur.
2 Vaginal spotting is not a sign or symptom of worsening preeclampsia.
4 This is within normal limits; there is insufficient information to identify whether it is elevated in this client.

263 (1) The client is exhibiting symptoms of preeclampsia; the presence of hyperreflexia indicates central nervous system irritability, a sign of a worsening condition; this will help direct the physician to appropriate interventions while alerting the nurse to the possibility of seizures. (3) (SI; AS)
2 The physician will need to be called, but a complete assessment should be done first to provide the physician with the most information.
3 The client's blood type is not necessary at this time; assessment of the neurologic status is the priority.
4 An IV may need to be started but should not precede a proper assessment; normal saline would not be preferred.

264. (3) An adequate urinary output, an indicator of adequate renal function, is necessary to prevent toxicity because $MgSO_4$ is excreted by the kidneys; signs of $MgSO_4$ toxicity are reduced respirations and absent patellar reflexes; therefore baseline assessments should be done. (2) (SI; PL)
1 These are urine tests; they are not significant to $MgSO_4$ toxicity.
2 Deviations in temperature do not indicate $MgSO_4$ toxicity.
4 These are assessments that may indicate worsening preeclampsia, not $MgSO_4$ toxicity.

265. (4) From these assessments the nurse can determine unusual weight gain and an increase in blood pressure; both of these are early symptoms of pregnancy-induced hypertension. (3) (CI; IM)
   1 The data suggest greater than normal weight gain.
   2 This would be therapeutic for a backache; this answer ignores the possible edema and increased weight gain.
   3 The weight gain may not be caused by inappropriate dietary intake but rather by underlying pathology.

266. (3) Sodium is not restricted because restriction decreases blood volume, which in turn reduces placental perfusion. (3) (PC; EV)
   1 Women at risk for this condition are advised to eat a high-protein diet.
   2 Losing weight is contraindicated during pregnancy and does not reduce the incidence of pregnancy-induced hypertension.
   4 Diuretic therapy is dangerous because it decreases blood volume, which in turn reduces placental perfusion.

267. (3) Even minimal sensory stimuli can trigger exaggerated cerebral responses such as convulsions; therefore, a nonstimulating environment is therapeutic. (2) (SI; AN)
   1 This is an undesired action; intracellular volume should be increased during pregnancy.
   2 Nonstimulating environments do not reduce headaches resulting from hypertension.
   4 A nonstimulating environment has no relation to the length of time antihypertensive drugs must be given.

268. (2) This is the unique characteristic symptom of eclampsia that occurs because of CNS irritation. (2) (PC; AS)
   1 This is a symptom of preeclampsia.
   3 Same as answer 1.
   4 Same as answer 1.

269. (3) A quiet room helps to reduce stimuli, which is essential for limiting or preventing seizures. (2) (CI; PL)
   1 All infusions are closely monitored and are usually maintained at a volume of 125 ml/hr.
   2 Precipitous delivery is not a usual side effect of magnesium therapy.
   4 Calcium gluconate, not magnesium gluconate, is the antagonist for magnesium sulfate and should be on hand if symptoms of toxicity appear.

270. (3) When a client is eclamptic she will be experiencing seizures; protecting the client from injury is always the first priority with any seizure. (3) (CI; IM)
   1 Accurate assessment is improbable with a rigid abdomen and seizure activity occurring.
   2 This is done immediately following the seizure; injury can occur if force is applied to maintain the airway during the seizure.
   4 IV effects are immediate, and increasing the dose may cause immediate toxicity.

271. (1) The fetus is at risk for retardation prenatally from a buildup of metabolites in the PKU-affected mother if a prescribed diet is not followed by the mother. (3) (PC; IM)
   2 This will not occur if the proper diet is maintained by the mother.
   3 The fetus is at risk for mental retardation if the maternal diet contains phenylalanine; also, the infant can inherit phenylketonuria via an autosomal-recessive gene.
   4 The client should remain on a phenylalanine-restricted diet during pregnancy to prevent manifestations of the disease.

272. (4) PKU is an inborn error of metabolism involving an inability to properly metabolize phenylalanine, an essential amino acid. (2) (PC; AN)
   1 This is metabolized normally in those with PKU.
   2 Same as answer 1.
   3 Same as answer 1.

273. (3) There is a constant need for evaluation of diabetic status, fetal maturity, and placental functioning. (2) (CI; AN)
   1 The rate of anomalies for pregnant women with diabetes mellitus is 6% to 8%; this compares to an incidence of 2% to 3% in women without diabetes mellitus.
   2 Insulin requirements vary and are usually increased during the second and third trimesters of pregnancy.
   4 Many clients with diabetes deliver vaginally with no problems.

274. (4) The blood glucose level is important because hypoglycemia in early pregnancy can lead to congenital abnormalities; hyperglycemia in late pregnancy may lead to fetal hyperinsulinism and subsequent neonatal hypoglycemia. (2) (HI; PL)
   1 This is too limited a response; assessment without intervention is useless.

2 Appointments should be made by the client; an authoritative approach takes control away from the client and may increase anxiety.

3 Dietary regulation is usually minimal, with a restriction on excessive carbohydrate ingestion; a limited diet to control weight gain could jeopardize both the fetus and the mother's nutritional status.

275. (4) An L/S ratio indicates adequacy of pulmonary function, and the baby should be free from major respiratory problems. (1) (CI; AN)

1 There is no correlation between L/S ratio and the need for induced labor.

2 There is no indication of fetal distress; immediate delivery is unnecessary.

3 The L/S ratio only determines fetal lung maturity; further fetal monitoring will be necessary in the future as with any pregnancy.

276. (2) Blood pressure and pulse may not change significantly until large amounts of blood have been lost; the trickling of blood indicates continuous bleeding. (2) (SI; AS)

1 The pulse becomes very rapid, but not until a significant amount of blood is lost.

3 This is not a sign of impending hemorrhage.

4 Blood pressure is normotensive; it usually does not drop significantly until a large amount of blood is lost.

277. (2) Contact precautions include wearing gown and gloves; these protect the nurse from the virus. (1) (SI; PL)

1 A mask is not needed because the virus is not transmitted via the respiratory route.

3 Same as answer 1.

4 This is done for the client's protection, not the nurse's; when caring for a client with herpes, the nurse needs to be protected.

278. (3) A low-segment delivery heals easily and decreases the risk of rupture of the uterus in future pregnancies. (1) (CI; AN)

1 A classical cesarean delivery (a vertical midline incision into the body of the uterus) is rare and reserved for special situations including abnormal lie, extreme prematurity, significant abdominal adhesions, and other emergency situations.

2 This is the name of a horizontal skin incision.

4 This type of cesarean delivery is less common and is used when more space is needed for delivery.

279. (1) The physician should be notified because the increased height of the uterus may be due to accumulation of blood in the uterus from internal hemorrhaging; also, the blood pressure is low and the pulse is rapid, and this may be indicative of impending shock. (2) (CI; EV)

2 Further assessment to confirm hemorrhaging only delays immediate response to the problem; the physician should be notified immediately.

3 Same as answer 2.

4 The assessment points to possible hemorrhaging; with urinary distention uterus is relaxed and lochia is heavy.

NORMAL NEONATE (NN)

280. (1) Standard precautions do not include the use of gloves for feeding. (1) (SI; IM)

2 Wearing clean gloves for diaper changes in all newborns is a standard protocol.

3 Clean gloves should be worn when performing suctioning of an infant.

4 Clean gloves should be worn for all admission baths because the nurse will be exposed to blood and amniotic fluid.

281. (3) A perfect score is 10; 1 point is deducted for lessened muscle tone (the baby's arms do not flex) and 1 point for acrocyanosis, which is manifested by bluish hands and feet. (2) (HI; AS)

1 The infant must have a higher score based on the data.

2 Same as answer 1.

4 This infant would not have a perfect score of 10; the muscle tone is somewhat lessened, and there is acrocyanosis.

282. (1) Preventing heat loss conserves the infant's oxygen and glycogen reserves; this is a first priority. (2) (SI; IM)

2 Warming the infant will reduce cyanosis if no respiratory obstruction is present.

3 This can be done after provision has been made to prevent heat loss.

4 This is important but not a priority; assessment should be delayed until the infant is warm.

283. (2) Coarctation of the aorta results in diminished or absent femoral pulses. (3) (HI; AS)

1 This has no effect on the volume of peripheral circulation (minimal shunting occurs in the newborn period).

3 This has minimal effect on the volume of peripheral circulation (left-to-right shunt).

4 Same as answer 3.

284. (3) The score at 5 minutes evaluates the adequacy of the cardiac and respiratory systems' response to the environment. (1) (HI; AS)
1 The Dubowitz score relates to gestational age.
2 The score represents the neonate's response to the environment and has no relationship to the actual process of labor and delivery.
4 The Apgar score is not a diagnostic tool for this disorder of the lungs.

285. (4) Increased pressure during the birth process causes increased intravascular pressure, which may result in capillary rupture. (2) (DF; AN)
1 This is caused by the collection of eosinophils.
2 These are intact capillaries; they may be distinguished from petechiae if they disappear when the area is blanched.
3 Bloody stools or oozing from the umbilicus is the most common sign of vitamin K deficiency.

286. (1) This is a normal Moro reflex, which indicates an intact nervous system. (2) (DF; IM)
2 This total body reaction is the Moro reflex, which is normally not present after the third month of life; if it persists, there may be a neurologic disturbance.
3 The Moro reflex has no relationship to hunger.
4 The Moro reflex is an involuntary response to environmental stimuli.

287. (2) After the respirations are established, the rate ranges from 30 to 60 breaths per minute with short periods of apnea. (1) (DF; AS)
1 Twenty breaths per minute is too slow.
3 More than 60 breaths per minute is too rapid.
4 Same as answer 3.

288. (3) This metabolic process releases energy and increases heat production in the newborn. (3) (DF; AN)
1 Fatty acids are byproducts of the breakdown of brown fat.
2 Shivering is the mechanism of heat production for the adult, not for the newborn.
4 This will not be successful unless plentiful brown fat is present.

289. (2) The absence of normal intestinal flora in the newborn results in low levels of vitamin K, causing a transient blood coagulation deficiency; an injection of vitamin K is given prophylactically to all infants on the day of birth. (1) (HI; IM)
1 Vitamin K has no effect on erythropoiesis.
3 Vitamin K is important in the synthesis of clotting factor in the liver, but it will not prevent jaundice.

4 Newborns have a blood coagulation deficiency; the blood clots more slowly, not more quickly.

290. (1) With congenital hip dysplasia, there are extra folds found in the thigh as a result of the shortening of the leg. (2) (HI; AS)
2 There are no extra folds in this area in congenital hip dysplasia.
3 Same as answer 2.
4 Same as answer 2.

291. (3) This information assists parents to understand the unique features of their newborn and promotes interaction and care during periods of wakefulness. (3) (DF; PL)
1 Most infants are on a demand feeding schedule, not a routine schedule; demand feeding provides for individuality.
2 This is too limited; the parents need a broader discussion of infant behaviors.
4 Printed instructions are inadequate if unaccompanied by a discussion.

292. (4) The fat pad is present in newborns and infants; the arch develops when the child begins to walk. (2) (DF; AN)
1 Flat feet are no more common in children than in adults.
2 The size of the feet is not relevant; arch development is related to walking.
3 Flat feet are not associated with a deformity such as clubfoot.

293. (1) Adaptation to the extrauterine environment is largely dependent on the functional capacity of vital organ systems, which is established during intrauterine development; this is measurable in terms of gestational age and weight. (2) (DF; AN)
2 Although this factor may influence health, it is not critical to neonatal survival.
3 Although these factors may influence health, they are not critical to neonatal survival.
4 Same as answer 3.

294. (4) Using the Apgar score, a value of 1 will be assigned to the color category; the other four categories have values of 2, making the Apgar score 9, demonstrating a healthy baby. (2) (CI; AS)
1 This infant would be cyanotic and apneic and would have very diminished muscle tone and reflex responsiveness.
2 Diminished muscle tone and reflexes are characteristic of these infants; they would not be active or crying lustily.

3 This would apply to a healthy infant whose bluish color would be more generalized and who lost 1 point in one of the other categories.

295. ① During periods of active or irregular sleep it is normal for newborns to have some twitching movements and irregular respirations; the vital signs and blood glucose levels are normal. (1) (PC; AN)

2 Twitching is a common finding in normal neonates; it often occurs with crying or stimulation.

3 Hypoglycemia in normal newborns would be characterized by a blood glucose level less than 40 mg/dl.

4 The normal respiratory rate is 30 to 60; irregular breathing is normal.

296. ① These are raised sebaceous cysts commonly found on the chin and nose of a newborn; they disappear spontaneously in a few days or weeks. (1) (DF; AS)

2 This is the fine downy hair covering the back and arms of the newborn.

3 These are elevated lesions of immature capillaries and endothelial cells that regress over a period of years; they are commonly called birthmarks.

4 This is an innocuous pink papular neonatal rash; it appears within 24 to 48 hours after birth and resolves spontaneously within a few days.

297. ① Establishing a patent airway and diminishing cold stress are the priorities; identification is necessary before the infant leaves the delivery area. (1) (SI; PL)

2 These measures would be appropriate in a compromised infant; an 8/9 Apgar is indicative of a healthy uncompromised newborn.

3 Application of eye prophylaxis and administration of vitamin K are often delayed to allow the parents to bond with the infant; a bath at this time would increase the risk of cold stress.

4 The newborn needs constant monitoring and should be placed in a warmer rather than a crib; the infant can be weighed later.

298. ① The major purpose of genetic screening in this instance is to determine if the infant has phenylketonuria (PKU); PKU can be detected after the infant has started eating. (1) (DF; AN)

2 Determination of the carrier state of the mother is not the objective of the testing of the infant.

3 Epidemiologic information is a purpose of genetic screening, but in this instance the most

important determination is whether or not the infant has PKU.

4 Risk for later development is not the purpose of PKU testing; it is to determine if the infant in fact has the disease; development of PKU at a later time is not the general case.

299. ② Screening for the disease results in early diagnosis and treatment. which can prevent mental retardation. (1) (PT; PL)

1 These children have no problem with physical growth; their problem is mental retardation.

3 This is not the problem; the major manifestations are mental or neurologic in origin.

4 The disease is genetic and cannot be acquired other than by inheritance; testing is done for early identification and treatment.

300. ③ The newborn is not submerged in a tub; the penis should be gently cleaned with clear, warm water; in addition, sponge baths are given until the cord detaches. (1) (SI; EV)

1 The diaper should be changed frequently to prevent irritation from the urine.

2 There should be only minimal bleeding; excessive bleeding requires immediate attention.

4 Petrolatum gauze or A & D ointment prevents the diaper from adhering to the operative site.

301. ① Keeping ice on the infant is contraindicated because the infant's immature heat-regulating mechanisms can be easily disturbed; excessive swelling after circumcision does not usually occur. (2) (CC; EV)

2 This is a correct statement.

3 Same as answer 2.

4 Same as answer 2.

302. ④ Proper cleansing and frequent changing will limit the presence of irritating substances. (1) (SI; IM)

1 Having the nurses change the diaper may lower the mother's self-esteem.

2 Powder and lotion will cake and retain moisture in the area.

3 This is a nursing, not a medical, problem.

303. ② Once the infant breathes the oxygen in the air, the need for an increased amount of immature erythrocytes decreases. (2) (DF; AN)

1 Jaundice is not an allergic response.

3 This is not a common occurrence in newborns; also, symptoms would occur more quickly.

4 The infant and mother have independent blood supplies, and Rh-negative blood does not enter the baby's bloodstream.

304. (3) Jaundice that appears within 24 hours may be indicative of a pathologic process; if the bilirubin level is elevated, intervention is required. (2) (HI; IM)

1 Jaundice is not an indication for admission unless accompanied by a very high serum bilirubin level.

2 Since this is a 24-hour-after-delivery assessment, the nurse should be aware that this is pathologic, not physiologic, jaundice and get an order for a serum bilirubin determination.

4 The infant may require phototherapy after further assessment factors have been carried out; this is not the first action.

305. (4) Research strongly supports the theory that there is a sensitive period during the first few hours of life that is extremely important in the promotion of parent-infant attachment. (2) (EP; PL)

1 Contact with the entire family is important during the taking-in phase of postpartum adjustment.

2 Encouraging rooming-in is also helpful because it increases the amount of contact between the parents and the newborn; however, this contact is after the first few critical hours.

3 Contact with the baby can be achieved with breast-feeding or bottle-feeding; it is the contact, not the method, that promotes bonding.

306. (4) The parents need an opportunity for close, eye-to-eye contact in the first hour; any of the prophylactic eye medications may irritate the baby's eyes, preventing them from opening. (1) (DF; IM)

1 Assessment is appropriate but will not facilitate parent-child bonding; favorable conditions for bonding should be provided before assessment.

2 The nurse should assess, not demonstrate, behavior.

3 Footprinting should be done immediately to ensure proper identification of the baby.

307. (4) Stimulating the rooting reflex is effective in making the infant grasp the nipple. (1) (PC; IM)

1 For milk to be expressed the infant must grasp the entire areola, which contains the secretory ducts.

2 Bottle-feeding may interfere with the infant's learning to accept the breast.

3 The mother should be supervised for correct positioning of the infant's mouth on the nipple to avoid nipple soreness.

308. (2) This is the proper attachment and helps compress the milk glands. (1) (PC; EV)

1 The nipple must be on top of the tongue.

3 This is not a good indication; the infant may be sucking on the nipple only.

4 This indicates improper attachment.

309. (3) The presence of at least six to eight wet diapers each day indicates sufficient breast milk intake. (2) (PC; EV)

1 This is a poor indicator; not all babies need extra sucking stimulation.

2 This could indicate an inadequate amount of fluid ingestion.

4 This is not a reliable indicator; sleep patterns may vary.

310. (2) At about 6 months of age, infants are able to swallow independently of sucking and a cup can be introduced. (3) (DF; PL)

1 This would be inappropriate because the infant does not have the ability to swallow independently of sucking at this time.

3 Between 9 and 12 months of age, infants can swallow four to five times consecutively and hold and carry a cup to the mouth; introduction of a cup at 6 months of age makes the weaning easier at 9 to 12 months of age.

4 This is too late; by this time the child has teeth, and sucking on a bottle promotes the development of caries as well as a preference for milk over solid foods.

311. (2) Unless fluoridated water is used by the manufacturer, fluoride supplementation of 0.25 mg daily is required. (3) (PT; AN)

1 Commercial formulas are iron-fortified.

3 The supply of vitamin K is adequate after the first week of life.

4 This is unnecessary; vitamin B_{12} may be needed if the mother is a vegetarian and is breast-feeding.

312. (3) These areas of pigmentation tend to appear in children of Asian, southern European, or African descent, and though they may appear anywhere on the body, they most commonly are seen on the back and buttocks. (3) (DF; AS)

1 This is evidenced by a dimpling at the base of the spine, not by pigmentation.

2 This is highly unlikely; in addition, the pigmentation would not be confined only to the buttocks.

4 This is generally manifested by swollen genitalia, not defined pigmented areas.

313. (4) The immediate weight loss is because of fluid loss, not loss of body mass. (2) (DF; AN)
1 Weight loss is expected; there are no data to support an allergic response.
2 Weight loss is not related to hypoglycemia.
3 Neither breast- nor bottle-feeding will prevent the 10% weight loss that is expected in the first few days of life.

314. (3) The alcohol will dry the cord and facilitate its falling off. (2) (SI; PL)
1 Although this is practiced in some cultures, using a pressure dressing is ineffective and could promote infection because the air circulation needed to promote drying is decreased.
2 If no infection is present, antibiotic ointment is not used.
4 The warm, moist environment inside a diaper is a good medium for bacterial growth; the diaper should be turned down below the umbilicus.

HIGH-RISK NEONATE (HN)

315. (2) Heart rate less than 100 beats per minute = 1; slow and irregular respirations = 1; grimaces in response to suctioning = 1; flaccid muscle tone = 0; and cyanosis = 0; the Apgar score would total 3. (3) (DF; AN)
1 This score is too low; the infant should receive 1 point for heartbeat, 1 point for respirations, and 1 point for grimacing; thus the assigned score is 3.
3 This is too high; the infant should receive 1 point for heartbeat, 1 point for respirations, and 1 point for grimacing; thus the assigned score is 3.
4 Same as answer 3.

316. (4) The first Apgar score provides evidence of the need for the initial resuscitation efforts, which began immediately after birth, and the second Apgar score indicates that the intervention was successful. (2) (DA; AN)
1 Apgar scores do not determine needs this long after delivery.
2 Apgar measurements are for infants of all gestational ages; these scores indicate difficulty regardless of gestational age.
3 The Apgar score indicates that this neonate was severely depressed and in need of oxygen under pressure.

317. (1) Infants of diabetic mothers are at risk for respiratory distress syndrome due to delayed synthesis of surfactant caused by high serum levels of insulin; the presence of phosphatidylglycerol in the amniotic fluid is a better predictor of lung maturity in infants of diabetic mothers than the L/S ratio. (3) (DF; PL)

2 The use of heroin alone by the mother does not predispose the infant to RDS.
3 The baby of a mother with hypertension may be small for gestational age but not necessarily preterm and at risk for RDS.
4 Preeclampsia does not necessarily predispose the infant to the development of RDS.

318. (1) Flaring nares are a compensatory mechanism that attempts to lessen resistance of narrow nasal passages and increase oxygen intake. (1) (DF; AS)
2 Acrocyanosis is not related to respiratory distress but is caused by vasomotor instability; this is a normal occurrence in the newborn.
3 This is a normal finding in the newborn.
4 Same as answer 3.

319. (4) Weight loss results from the extra sucking effort required to obtain milk flow from the breast. (2) (PC; IM)
1 If the infant is being fed by gavage, the mother's breasts can be pumped and the breast milk can be used for gavage feedings.
2 Time consumption and effort are insufficient reasons to discourage breast-feeding.
3 Breast milk provides adequate nutrition, protects the infant from necrotizing enterocolitis, and provides antibodies.

320. (3) This is a classic sign of respiratory distress in the newborn. (1) (DA; AS)
1 This is associated with neurologic impairment, not respiratory distress.
2 This is within normal limits.
4 The respiratory rate increases, not decreases.

321. (2) In respiratory acidosis, the pH falls and the CO_2 level rises. (2) (CI; AN)
1 This is a normal pH.
3 This is normal but is unrelated to acidosis.
4 The arterial oxygen level may or may not change with acidosis.

322. (2) Oxygen has a cooling effect, and the baby should be kept warm so that metabolic activity and oxygen demands are not increased. (1) (PC; IM)
1 This could produce fluid overload, which would lead to increased cardiac output, an undesired outcome especially for an infant with respiratory distress.
3 Oxygen concentration is determined by blood gas levels and is changed accordingly.
4 This would tire the baby and increase the need for oxygen.

323. (4) This will help monitor the need for oxygen supplementation; prolonged use of oxygen concentrations exceeding those required to maintain adequate oxygenation have been found to contribute to the occurrence of retinopathy of prematurity. (1) (CI; PL)
1 Retinopathy of prematurity cannot be prevented by using any preferred route of oxygen administration.
2 Retinopathy of prematurity is caused by high blood concentrations of oxygen, not by eye exposure to oxygen.
3 Warming and humidifying oxygen will not affect its level in the environment.

324. (3) These are key signs of a pneumothorax, which can occur when an infant is receiving oxygen by positive pressure. (2) (CI; AN)
1 The findings are not normal and need immediate attention.
2 The findings do not indicate this occurrence.
4 Same as answer 2.

325. (1) Positioning with the head slightly hyperextended and changing the position every 1 to 2 hours helps to drain secretions and can increase O_2 available for use by promoting respiratory efforts in a premature infant with immature lung tissue. (2) (CI; AN)
2 This is too low; preterm infants do not shiver.
3 Congenital birth defects are observed for in all infants, not just those with RDS.
4 Extensive handling is not desired, but infants do need to be touched.

326. (4) The breast buds and genitalia develop at a specified rate and are good indicators of gestational age. (3) (DF; AS)
1 Weight and length may be influenced both by genetic factors and by prenatal stresses and are not accurate indicators of gestational age.
2 This is not a specific indicator of gestational age; rooting and sucking can be affected by neurologic deficits, injury, or CNS depression as well as maturity.
3 This provides information about the infant's neuromuscular status and is not a specific indicator of gestational age.

327. (1) A preterm, small-for-gestational-age infant is at risk for problems not seen in the term or average-for-gestational-age infant because of immaturity; this information will help the nurse to anticipate potential problems and aim interventions at prevention. (1) (HI; AN)

2 The information is documented on the infant's chart, but this is not the overriding reason for obtaining this data.
3 The infant will lose weight, but the comparison of weight and gestational age is important for the planning of appropriate nursing measures.
4 Same as answer 2.

328. (4) The Silverman-Anderson score is an index of neonatal respiratory distress. (3) (CI; EV)
1 The Silverman-Anderson score does not reflect cardiac function.
2 The Silverman-Anderson score does not reflect caloric needs.
3 The Silverman-Anderson score does not reflect neurologic status.

329. (1) The release following compression of the chest during vaginal delivery is the mechanism for expansion of the infant's lung; since this does not occur during a cesarean delivery, lung expansion is incomplete and atelectasis may result. (1) (CI; AN)
2 Not true; the infant is carefully monitored to prevent this.
3 Temperature change is not implicated in aspiration.
4 Gravity could be used following birth by cesarean delivery by holding the newborn upside down.

330. (1) Breast tissue is not palpable in an infant of less than 33 weeks gestation. (1) (DF; AS)
2 Creases in the palms and on the soles of the feet are not clearly defined until after the 37th week of gestation.
3 The ear pinna springs back in an infant of 36 weeks gestation.
4 A 0-degree square window sign is present in an infant of 40 to 42 weeks gestation.

331. (4) The assessment findings are indicative of a preterm infant; therefore the nurse should closely monitor the infant for signs of respiratory distress syndrome; this occurs commonly in preterm infants because their lungs are immature. (2) (DF; AS)
1 Preterm AGA infants do not develop polycythemia; preterm LGA infants may develop polycythemia, but there are no data to indicate the infant is LGA.
2 Preterm AGA infants may become hypoglycemic.
3 The neonate is preterm, not postterm.

332. (2) The infant would be classified as small-for-gestational age (SGA) because the weight is less than the 10th percentile on the growth curve for a term infant. (1) (DF; AN)

1 An infant is considered to be preterm if born before the end of the 37th week of gestation; the wording "small-for-gestational age" rather than "immature" is used.
3 This is untrue; the infant's weight is less than the 10th percentile for a term infant; the infant is SGA.
4 Same as answer 3.

333. (2) SGA infants may exhibit hypoglycemia, especially during the first 2 days of life, because of depleted glycogen stores and inhibited gluconeogenesis. (2) (DF; AN)

1 Decreased BP, pallor with cyanosis, tachycardia, retractions, lethargy, and weak cry are present in hypovolemia.
3 Hypercalcemia is uncommon in newborns.
4 These signs are unrelated to hypothyroidism; symptoms of hypothyroidism are difficult to identify in the newborn.

334. (2) Many superficial veins are common in the preterm infant because of the lack of subcutaneous fat deposits. (1) (DF; AS)

1 Flexion of extremities is the posturing of normal term infants; preterm infants usually posture with extremities extended and flaccid.
3 Absent femoral pulses are indicative of coarctation of the aorta, a congenital heart defect.
4 A positive Babinski reflex is a normal newborn reflex.

335. (4) Primary manifestations of NEC are feeding intolerance, increased gastric residual of undigested formula, and bile-stained emesis. (3) (DA; AS)

1 This may occur with a cardiac anomaly, not NEC.
2 This occurs with diarrhea; stools in those with NEC are generally reduced in number and contain glucose and blood.
3 This occurs with pyloric stenosis.

336. (2) Prolonged gastric emptying occurs when the baby has NEC; an increase in abdominal girth of greater than 1 cm in 4 hours is significant and needs immediate intervention. (2) (PC; PL)

1 Formula is stopped and the baby is placed on parenteral fluids.
3 This will have no therapeutic value for a child with NEC.
4 Same as answer 1.

337. (2) Gradually rewarming an infant experiencing cold stress is essential to avoid compromising the infant's cardiopulmonary status. (3) (SI; IM)

1 Rapid rewarming of an infant may result in apnea and neonatal stress.
3 An infant experiencing cold stress will become hypoglycemic; the infant uses up glycogen and glucose to maintain the core temperature.
4 Skin temperatures should be taken at least every 15 minutes until stable.

338. (1) A preterm infant may have a weak suck but usually can be breast-fed; the mother may at least attempt it, if the infant is stable. (2) (DF; IM)

2 It does not necessarily take more calories to breast-feed; also, there are immunologic benefits to the preterm infant who receives antibodies through breast milk.
3 Pumping the breasts may be necessary, but bottle-feeding is not needed because this only deters the mother and infant; at 35 weeks if the infant is stable and the mother so desires, breast-feeding should be attempted.
4 The suck may or may not be weak, but a supervised attempt to breast-feed may help the mother get to know the infant and feel competent in providing care.

339. (4) The pacifier may satisfy nonnutritive sucking needs and stimulate flow of saliva and digestive juices. (2) (PC; IM)

1 There is no evidence that nonnutritive sucking is harmful for a preterm infant this size.
2 On the contrary, sucking on a pacifier promotes adaptation to later bottle feedings.
3 Research has identified a benefit of nonnutritive sucking; buck teeth are associated with thumb sucking.

340 (3) If the airway is not patent and gas exchange is inadequate, life cannot be sustained, thus this must be the top priority. (3) (DA; PL)

1 Although bonding is important to the parent-child relationship, without oxygen, life could not be sustained.
2 Although preventing infection is important, without oxygen, life could not be sustained.
4 Although body temperature is important because the baby is lacking brown fat and other defense mechanisms needed to maintain temperature, without oxygen, life could not be sustained.

341. (2) The nurse applies tactile stimulation after validating that respirations are absent; this action may be sufficient to reestablish respirations in the high-risk neonate with frequent episodes of apnea. (2) (DA; IM)
1 Assessment will not interrupt the period of apnea; respirations must be immediately reestablished.
3 The monitor should be assessed for proper functioning before use.
4 These measures are too invasive and aggressive for initial intervention; gentle stimulation should be attempted first.

342. (1) The reflexes and muscles of sucking and swallowing are immature; this makes oral feeding ineffectual and exhausting. (3) (DF; AN)
2 The metabolic rate is increased because of fatigue and growth needs.
3 Caloric requirements are increased because of extra growth needs.
4 Absorption of nutrients is decreased because of immaturity of the intestines.

343. (3) Retinopathy of prematurity (ROP) is a complex disease of the premature infant; hyperoxemia is one of the numerous causes implicated, and it can be monitored for with pulse oximetry. (3) (CI; EV)
1 This will not prevent the development of ROP.
2 Light levels are not considered to be a factor in the cause of ROP.
4 Same as answer 1.

344. (3) Observing the care that the parents actually give the infant provides direct validation of their skill and comfort levels. (1) (SI; EV)
1 This action is helpful in anticipatory guidance but is only a small part of competency evaluation.
2 Although this is helpful in identifying empirical knowledge, it does not test their skill or comfort level.
4 Same as answer 2.

345. (2) At birth, circulating maternal glucose is removed; however, the infant still has a high level of insulin and may develop rebound hypoglycemia. (2) (CI; AS)
1 These are not related to diabetes mellitus in the mother; mongolian spots are pigmented nevi found primarily on the skin of the sacrum and buttocks of Asian and African American babies.
3 The temperature-regulating ability of a neonate born to a mother with diabetes mellitus is similar to that of a normal neonate unless the infant is preterm.

4 Pathologic jaundice is associated with hemolytic diseases such as Rh and ABO incompatibility and sepsis, not diabetes mellitus.

346. (1) A difficult delivery because of broad fetal shoulders may result in a fractured clavicle, which can be assessed by the findings of a knot or lump, limited arm movement, and a unilateral Moro reflex. (2) (HI; AS)
2 This is unrelated to a difficult delivery of a baby with broad shoulders.
3 This reflex involves the feet; it is in no way related to a difficult delivery because of broad shoulders.
4 Same as answer 3.

347. (1) Infants of diabetic mothers are polycythemic and therefore they appear flushed; the mechanism underlying this phenomenon is unknown. (3) (HI; AS)
2 These infants are limp, not hyperreflexive.
3 These infants are generally quite heavy due to macrosomia.
4 These infants tend to lie quietly and not cry.

348. (1) Decreasing IV glucose slowly is necessary to prevent a hypoglycemic response. (3) (CI; IM)
2 Metabolic alkalosis will not occur with discontinuation of the glucose; it occurs with excessive amounts of bicarbonate.
3 Withholding oral feedings while withdrawing IV glucose may result in hypoglycemia.
4 Glycosuria is unlikely to occur when decreasing the IV glucose because blood glucose levels will decrease.

349. (4) Because the virus is found in the respiratory tract and the urine, isolation is necessary; rubella is spread by droplets from the respiratory tract. (2) (SI; PL)
1 This is an outdated term that is no longer used; the techniques used with this precaution have been incorporated under contact precautions.
2 This is an outdated term that is no longer used; the techniques used with this precaution have been incorporated under standard precautions.
3 This would be unsafe; additional precautions must be taken to protect the nurse from droplet infection.

350. (2) Phototherapy changes unconjugated bilirubin in skin to conjugated bilirubin bound to protein, permitting excretion. (1) (PC; EV)
1 Phototherapy does not affect liver function; the liver does not dispose of bilirubin.
3 Vitamin K has no effect on bilirubin excretion; it is necessary for prothrombin formation.

4 The bilirubin is not excreted via the skin but in the urine and feces.

351. (3) This is necessary to provide for some normal psychosocial contact. (2) (PC; PL)

1 This may block light rays from acting on bilirubin deposits; frequent cleansing after voiding and defecation will prevent skin excoriation.

2 All parts of the body may contain bilirubin deposits and should be exposed to the light.

4 Radiant heaters are not used; only fluorescent bulbs are used.

352. (2) The head circumference is usually 2 centimeters larger than the chest; a head circumference 4 centimeters smaller than the chest could indicate microcephaly. (3) (DF; AS)

1 In anencephaly, the disparity between the head and chest circumference would be much larger than 4 cm.

3 No molding takes place in cesarean delivery; therefore the head should be about 2.5 cm larger than the chest at birth.

4 According to growth charts, the range of head circumference for boys is just slightly (1.25 cm) larger than the chest.

353. (3) Mothers with type O blood have anti-A and anti-B antibodies that are transferred across the placenta; this is the most common incompatibility because the mother is type O in 20% of all pregnancies. (2) (HI; IM)

1 This is usually not a problem.

2 Same as answer 1.

4 Same as answer 1.

354. (3) Adherence to the diet is necessary for optimal physical growth with no adverse effects on mental development; a diet that is instituted late will not reverse brain damage. (1) (PC; AN)

1 Detection cannot occur until the infant has taken milk or formula that contains phenylalanines for 24 hours and metabolites accumulate in the blood; behaviors indicating mental retardation and CNS involvement usually are evident by about 6 months of age in the untreated infant.

2 There is no phenylalanine at birth; it first becomes measurable after the infant ingests milk or formula.

4 This is untrue; it is related to compliance with the prescribed diet once the diagnosis is made.

355. (3) The term "phenylketonuria" is derived from phenylpyruvic acid, which gives urine a mousy, musty odor. (2) (HI; AS)

1 This odor is not present with phenylketonuria.

2 Same as answer 1.

4 Same as answer 1.

356. (1) Right-to-left shunts result in inadequate perfusion of blood; not enough blood flows to the lungs for oxygenation. (3) (DA; AN)

2 Left-to-right shunts result in too much blood flowing to the lungs; blood is adequately perfused.

3 Left-sided obstruction to the flow of blood results in decreased peripheral pulses, not cyanosis.

4 Normally there should be no shunting of blood between the right and left sides of the heart after the ductus arteriosus closes.

357. (3) Because the lungs are stressed by increased fluid, increased respirations are the first and best indicators of early congestive heart failure. (3) (HI; AS)

1 Cyanosis is not an early sign because there is adequate perfusion of blood.

2 The heart rate would not decrease; it would increase in an attempt to compensate.

4 This is a normal finding in the newborn.

358. (4) The abnormal facies associated with fetal alcohol syndrome includes a small, upturned nose, which is distinctive in these infants. (3) (HI; AS)

1 A cleft lip may occur without a precursor or with the trisomies.

2 Multiple fingers are associated with the trisomies.

3 An umbilical hernia can develop in early infancy and is not related to fetal alcohol syndrome.

359. (1) The addicted infant is very sensitive to lights, noise, and surrounding activities; the infant must be kept calm and comfortable to help reduce overreaction to stimuli. (2) (PC; PL)

2 These are only used to reduce narcotic levels in drug-addicted infants.

3 Fluid intake must be increased to prevent dehydration in the infant who vomits.

4 It is not necessary to assess for developmental status when physical needs take precedence.

360. (3) The signs of withdrawal in an infant of a narcotic-addicted mother are usually manifested by hyperactivity and sneezing. (2) (HI; AS)
1 During withdrawal the infant's cry is most closely described as being shrill and high pitched.
2 The Moro reflex is usually decreased as the signs of withdrawal become apparent.
4 The deep tendon reflexes are increased.

361. (4) Thrush, an oral infection caused by *Candida albicans*, is an opportunistic infection that may be indicative of HIV infection. (3) (HI; AS)
1 This symptom is more frequently associated with immaturity of the hypothalamus.
2 Bleeding after a circumcision is associated with a bleeding disorder such as hemophilia.
3 Hypoglycemia is usually associated with the infant born to a diabetic mother.

362. (4) When a diaphragmatic hernia is present, the intra-abdominal pressure must be minimized; this is accomplished by the use of gastric decompression. (3) (CI; IM)
1 This would not be a beneficial action; it might predispose the infant to diarrhea.
2 Decreasing the amount of formula will be injurious to the baby; the baby should receive the full amount of formula required; smaller feedings more frequently may be helpful.
3 Contraindicated; the abdominal organs would increase pressure on the diaphragm.

363. (1) This must be done to minimize the possibility of the neonate aspirating meconium into the lungs. (2) (SI; IM)
2 This action would increase the possibility of aspiration of meconium into the lungs.
3 Same as answer 2.
4 Same as answer 2.

364. (2) If the rectum is empty of feces, this may indicate intestinal obstruction present at birth. (2) (HI; AS)
1 Vomiting may occur, but vomiting may be related to any number of conditions.
3 Intake and output, while important, are not definitive information to making a diagnosis.
4 Fussiness during feeding is not definitive to making a diagnosis.

365. (4) Large babies may be the result of gestational diabetes; it is necessary to check the baby for hypoglycemia because maternal glucose is no longer available. (3) (HI; AS)

1 This would be indicated if the temperature were low and the newborn needed additional warmth.
2 To the contrary, the infant may be hypoglycemic and would require the glucose in an oral feeding immediately.
3 Unless there is a low blood glucose or some other indication of a problem, the infant can be seen whenever the pediatrician arrives.

366. (4) This prevents the neonate's stomach from becoming too distended and pressing upward against possibly compromised lungs. (2) (SI; PL)
1 This is too long a period between feedings; preterm infants should be fed every 2 to 3 hours because it takes this amount of time for the preterm infant's stomach to empty.
2 This would not prevent respiratory embarrassment; more important, however, is the fact that the infant with a strong sucking reflex should be fed with a nipple, otherwise the sucking reflex will disappear.
3 Preterm infants need the full caloric value of formula; giving diluted formula has no bearing on preventing respiratory embarrassment.

WOMEN'S HEALTH (WH)

367. (4) This is known as the hormone of pregnancy; together with estrogen it helps prepare the endometrium for the fertilized ovum, helps maintain pregnancy, and prepares the mammary glands for milk secretion. (1) (DF; AN)
1 This is secreted by the adrenal cortex, and it affects carbohydrate metabolism.
2 This is secreted by the anterior lobe of the pituitary gland; it starts and maintains milk secretion by the mammary glands.
3 This is secreted by the posterior pituitary gland; it stimulates labor contractions and contractile tissue around the nipple during nursing.

368. (4) Mittelschmerz is pain that sometimes occurs at ovulation when the ova erupts from the follicle. (1) (DF; AN)
1 The pain is mild, cyclic, and characteristic of mittelschmerz; it requires no medical intervention.
2 When menses first begin, the client is usually anovulatory and would not experience the pain known as mittelschmerz.
3 The pain will probably occur most often when ovulation is well established.

369. (3) This is the only correct definition of dysmenorrhea. (1) (DF; AS)

1 This occurs with menopause and during pregnancy, not dysmenorrhea.
2 This is any bleeding that occurs at any time other than during the menstrual period; there may or may not be any pain.
4 This is known as menometrorrhagia.

370. (4) This medical aseptic technique should limit the spread of microorganisms and help prevent future urinary tract infections if incorporated into her health practices. (1) (SI; IM)
1 This is unnecessary; also, the client is probably experiencing frequency.
2 The client does not have to use a bedpan; if intake and output are being measured, the use of a high hat to collect the urine is sufficient.
3 This is unnecessary while urinating; it may be employed as a part of perineal care after urinating.

371. (1) This avoids the possibility of ingesting infected cysts. (3) (SI; IM)
2 This disease, though more prevalent in foreign countries, is seen in the United States.
3 This is not related to toxoplasmosis.
4 Same as answer 3.

372. (2) Yogurt contains *Lactobacillus acidophilus* bacteria, which will replace those destroyed by antibiotics. (2) (PC; IM)
1 This is not relevant to antibiotics or intestinal flora.
3 Same as answer 1.
4 Same as answer 1.

373. (3) A condom covers the penis and contains the semen when it is ejaculated; semen contains a high percentage of HIV in infected individuals. (1) (SI; IM)
1 This is poor advice; preejaculatory fluid carries the HIV in an infected individual.
2 This is unsafe; although a monogamous relationship is less risky than having multiple sexual partners, if the one partner is HIV positive, the other person is at high risk for acquiring the HIV.
4 This is not what the client is asking; most contraceptives do not provide any protection from the HIV.

374. (3) The client has signs and symptoms indicative of pelvic inflammatory disease (PID), which is a complication of gonorrhea. (2) (HI; AN)
1 Herpes is noted for its painful lesions on the genitals; there are no data to indicate the presence of these lesions.

2 The client does not have the signs and symptoms associated with this disease.
4 Same as answer 2.

375. (4) Many clients are asymptomatic; the incubation period is 3 to 5 days. (3) (SI; AN)
1 There is no effective readily available blood test for gonorrhea.
2 Gonorrhea responds well to treatment; however, at times back-up secondary medications have to be utilized.
3 Urethral/vaginal smears or cultures are specific for the identification of the gonococcal organism

376. (2) This is a description of a chancre, which is the initial sign of syphilis. (3) (HI; AS)
1 These are condylomata lata, which are typical of the secondary stage.
3 This is typical of the secondary stage of systemic involvement, which occurs from 2 to 4 years after the disappearance of the chancre.
4 This is typical of the secondary stage.

377. (4) Dryness inactivates the *Treponema pallidum*, making it incapable of causing disease. (1) (SI; AN)
1 The organism is transferred by sexual contact; warm, moist body contact supports growth of the organism.
2 This is not true; nothing chelates the organism.
3 These support the growth of the organism.

378. (3) Syphilis is primarily a vascular disease; aortitis, valvular insufficiency, and aortic aneurysms are the most prevalent problems in tertiary syphilis. (2) (HI; AN)
1 Although lesions occur on the genitalia during primary and secondary syphilis, the reproductive system is not the major body system affected in tertiary syphilis.
2 A gumma skin lesion is the least commonly occurring lesion associated with tertiary syphilis; skin lesions such as macular and papular eruptions most commonly occur in secondary syphilis.
4 Although lesions can occur about the mouth (chancre in primary syphilis and mucous patches in secondary syphilis), the structures of the lower respiratory tract are not the major structures involved in tertiary syphilis.

379. ④ The least amount of breast engorgement occurs at this time, limiting lumps that may occur because of fluid accumulation. (1) (HI; AS)
   1 Breast engorgement begins before ovulation and does not subside until several days after menses ends; engorgement interferes with accurate palpation.
   2 Inaccurate assessment could result because examination would occur at different times of the menstrual cycle; accurate comparisons could not be made from month to month; this is appropriate for postmenopausal women.
   3 Same as answer 1.

380. ① Serous or bloody discharge from the nipple is abnormal. (2) (HI; AS)
   2 The right hand should examine the left breast because this allows the flattened fingers to palpate the entire breast including the tail (upper, outer quadrant toward the axilla) and axillary area.
   3 A small pillow or rolled towel should be place under the scapula of the side being examined.
   4 The flat part of the fingers, not the palm or fingertips, should be used for palpation.

381. ④ Advanced age at birth of first child is one of the high-risk factors for malignancy of the breast because of prolonged exposure to unopposed estrogen. (1) (HI; AN)
   1 This is not considered a high-risk factor.
   2 Same as answer 1.
   3 Same as answer 1.

382. ② Compression of the breast flattens mammary tissue and maximizes the penetration of the breast by x-rays; this is especially important for the dense breast tissue of adolescents, young nulliparous women, and women with large breasts. (1) (CI; IM)
   1 This is usually done with sonography.
   3 This is not necessary.
   4 The American Cancer Society recommends that women at high risk for breast cancer (the client's sister had breast cancer) should have routine mammographies regardless of age or relationship to menopause.

383. ③ This response allows the nurse to assess the client's knowledge and experience of breast cancer and to clarify any misconceptions. (3) (SC; AS)
   1 Avoids an opportunity to teach; a type of false reassurance; it may increase feelings of hopelessness if the lesion is determined to be malignant.

2 Although true, this statement provides a false sense of security and avoids an opportunity to teach.
   4 Too specific; focuses on what the physician has said rather than what the client knows and may limit further communication of feelings and beliefs.

384. ② Pressure to follow a course of therapy is never appropriate, especially at such a stressful time. (1) (PC; PL)
   1 This would be therapeutic.
   3 Knowledge of procedures decreases anxiety and would be therapeutic.
   4 This would help decrease postoperative complications and would be therapeutic.

385. ③ Suction (negative pressure) is used to remove blood and serum that may collect in the operative area and delay wound healing. (3) (CI; PL)
   1 Use of a portable drainage system has obviated the need for large pressure dressings; suction is more effective in preventing collection of fluid at the operative site.
   2 Should be avoided to prevent muscle stiffness; a regimen of gentle exercises helps restore muscle function and prevents a frozen shoulder.
   4 A high protein intake would be encouraged to promote tissue repair.

386. ③ Elevating the arm uses gravity to facilitate venous return and lymph drainage. (1) (CI; AN)
   1 Has no effect on drainage in the affected arm.
   2 Wound drainage uses negative pressure; positive pressure would not facilitate drainage.
   4 Acts primarily to decrease muscle tension and regain muscle function; secondary effect may be to prevent fluid accumulation.

387. ④ Clients who have cancer of one breast are at high risk for development of cancer in the other breast. (2) (HI; PL)
   1 A breast prosthesis is not used until healing has occurred.
   2 Most clients can resume full activity as strength returns.
   3 Stretching activities are considered helpful in regaining full movement.

388. ② The client's statement infers an emptiness with an associated loss. (1) (SC; EV)
   1 Resumption of social activities indicates an acceptance of her condition and a willingness to move on with life.
   3 This is a typical response of a grandmother anxious to resume her life.

4 The client is sharing planning and concern by her husband, not expressing a sense of loss.

389. (4) With infiltrating breast cancer, a surgical procedure that removes less than all of the breast tissue is combined with irradiation and removal of affected axillary lymph nodes. (1) (PC; IM)

1 This is an unfeeling response that denies the woman's feelings.

2 This gives false hope. It is not possible to state that the client will be disease free in 5 years.

3 This gives inaccurate information regarding the radiation therapy and false hope that radiation will not be needed.

390. (1) Statistics indicate a relationship between estrogen therapy and an increased incidence of endometrial cancer, although mortality is not increased. (3) (DA; AN)

2 Estrogen retards bone loss.

3 Estrogen maintains vaginal tissue turgor and lubrication.

4 Estrogen appears to play a protective role against myocardial infarction.

391. (1) Women in the childbearing years should be informed of all options available to preserve ovarian function. (3) (DA; AN)

2 This is an incorrect statement; "always" is too absolute.

3 This is an incorrect statement because radiation can influence or destroy ovarian functioning.

4 Once ova have been destroyed, they cannot regenerate.

392. (2) Attempting to remove a uterus with large uterine fibroids vaginally could cause trauma, resulting in hemorrhage. (2) (CI; AN)

1 Vaginal hysterectomy is indicated for prolapsed uterus because the uterus is usually collapsed into the vagina.

3 A hysterectomy is not the treatment of choice for mild cervical dysplasia; when a hysterectomy is necessary, the vaginal route is usually preferred.

4 When a cystocele also needs to be repaired, a vaginal hysterectomy, rather than an abdominal one, is indicated.

393. (4) Following surgery the urethral orifice may be distorted and edematous; a catheter keeps the bladder completely empty, limiting pressure on the operative site. (2) (PC; PL)

1 A cleansing douche would be ordered before surgery.

2 A pessary placed in the vagina is used for a displaced uterus; following an anterior/posterior repair (colporrhaphy), vaginal packing is used to support the surgical repair.

3 A rectal tube is used for abdominal distention caused by flatulence; it is rarely necessary and rarely used.

394 (1) When a bilateral oophorectomy is performed, both ovaries are excised, decreasing ovarian hormones and initiating menopause. (1) (DF; PL)

2 More often associated with vaginal hysterectomy.

3 Although depression may occur, it should not be expected; if it does occur, it requires intervention.

4 There is no physiologic reason for weight gain following a hysterectomy; overeating is sometimes associated with depression.

395. (1) Vaginal packing supports the repair and provides slight pressure to prevent bleeding; the packing needs to be checked for possible bleeding. (2) (CI; PL)

2 There is no dressing, only vaginal packing and a sanitary pad.

3 Elevation of the legs is unnecessary; leg exercises and a gradual increase in ambulation are encouraged to prevent pulmonary emboli.

4 Sitz baths are not instituted until the packing is removed; an ice pack and/or a heat lamp may be used to promote comfort.

396. (1) Menstrual flow is the shedding of the endometrial lining of the uterus, and in a hysterectomy the uterus is removed. (1) (DF; EV)

2 This would be associated with a total hysterectomy and bilateral salpingo-oophorectomy.

3 This occurs in a vaginal approach; the client will have an abdominal hysterectomy.

4 Only the uterus and cervix are removed in a total hysterectomy.

397. (1) This routine screening for information provides a basis for assessing trauma; in a younger client it also assesses risk for pregnancy. (3) (HI; AS)

2 Examination may precede reporting; the decision to report is mandated by law, not the client.

3 Inappropriate; using water or antiseptic solution would wash away spermatic or bloody evidence.

4 A test specifically for seminal acid phosphate, not alkaline phosphatase, is performed.

398. ③ More than 90% of pulmonary emboli originate in the deep veins of the pelvis and thighs because of the extensive vascular network. (2) (CI; AN)
   1 This is untrue; most pulmonary emboli originate in the pelvis or thighs.
   2 Same as answer 1.
   4 Same as answer 1.

399. ③ It is fact that *E. coli* is commonly found in the bowel and, because of close anatomic proximity and improper hygiene after bowel movements, may spread to the urethra. (1) (HI; AN)
   1 *E. coli* is no more virulent than other infective agents.

2 *E. coli* is not commonly found in the kidneys.
4 *E. coli* does not compete with *Candida* organisms for host sites.

400. ① Muscle contraction during ambulation improves venous return, which prevents venous stasis and thrombus formation. (2) (CI; PL)
   2 Gatching the bed places pressure on the popliteal spaces, limiting venous return and increasing the risk of thrombus formation.
   3 Dangling the legs places pressure on the popliteal spaces, limiting venous return and increasing the risk of thrombus formation.
   4 Bed rest is associated with venous stasis, which increases the risk of thrombus formation.

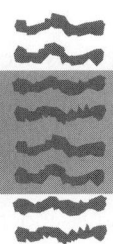

CHILDBEARING AND WOMEN'S HEALTH NURSING QUIZ 1

1. A client develops vaginitis for which the physician prescribes daily douches. The nurse provides the client with instructions for douching. The most important aspect of these instructions would be that the client should:
   1. Use a sterilized bulb syringe
   2. Insert the douche tip 5 to 6 inches
   3. Recline with head and shoulders elevated slightly
   4. Hold the labia together while instilling the douche solution

2. The physician tells the parents of a newborn that their child may have Down syndrome and that additional diagnostic studies will need to be done. In the plan of care, the nurse should prepare the infant for:
   1. Karyotyping
   2. A buccal smear
   3. An amniocentesis
   4. An enzyme assay

3. After doing a nursing assessment on a newly delivered male child, the nurse suspects that this neonate is postmature. It is most likely that this classification is based on the assessment finding of:
   1. Abundant lanugo and vernix caseosa
   2. Skin thick with desquamation over most of body
   3. Sole creases over anterior two thirds of each foot
   4. Testicles undescended with few rugae over scrotum

4. Five minutes after being born a newborn is pale, has irregular and slow respirations, has a heart rate of 120 beats per minute, and displays minimal flexion of extremities and minimal reflex responses. The nurse should expect this newborn's Apgar score and status to be reported as:
   1. 2, severely depressed
   2. 3, severely depressed
   3. 5, moderately depressed
   4. 8, moderately depressed

5. A client delivers an infant with a severe congenital heart defect. The most important nursing intervention to consider at this time is:

   1. Assisting the client and family with the grieving process
   2. Arranging for social services to provide help in the home
   3. Providing for physical assessments of the infant on every shift
   4. Obtaining an order for a sedative to help the client cope with the situation

6. A client who is 37 weeks pregnant is admitted to the hospital with sudden abdominal pain. Abruptio placentae is suspected. The nurse should assess the client for:
   1. Bright red bleeding and signs of shock
   2. Concealed hemorrhage and fetal accelerations
   3. Decrease in size of uterus and cessation of contractures
   4. Early incidence of increased fetal activity and uterine tenderness

7. The nurse teaches a breast-feeding mother about cleansing her nipples. The nurse emphasizes that the mother should:
   1. Wash the breasts and nipples daily with water
   2. Cleanse the nipples with sterile water before each feeding
   3. Thoroughly scrub nipples with soap and water before each feeding
   4. Cleanse the nipples with an alcohol sponge before and after each feeding

8. After taking clomiphene citrate (Clomid) for 3 months to treat anovulatory cycles, a client complains of difficulty with penetration during intercourse because of vaginal dryness. An appropriate response by the nurse would be:
   1. "Good, this means you are probably beginning to ovulate."
   2. "I know you are concerned about it, but this is only temporary."
   3. "Stop the Clomid immediately; the physician will have to prescribe another drug."
   4. "This is a common side effect; use a water-soluble lubricant to ease penetration."

9. As part of the physical examination of a newborn, the nurse palpates the baby's abdomen. The organ that the nurse would normally expect to palpate is the:
   1. Liver
   2. Stomach
   3. Pancreas
   4. Gallbladder

10. A preterm infant is placed in the neonatal intensive care unit. The nurse expects that a common concern the mother will express is the:
    1. Prolonged hospital stay
    2. Delayed ability to bond with the infant
    3. Fear of touching and handling the infant
    4. Inability to provide the infant with breast milk

11. The nurse recognizes that the laboring couple benefited from the childbirth preparation classes when, during the transition phase of labor, they use the breathing pattern known as:
    1. Pant-blow
    2. Slow-chest
    3. Shallow-chest
    4. Accelerated-decelerated

12. A pregnant woman expresses concern because she read that nutrition during pregnancy is important for proper growth and development of the baby. She asks the nurse about what foods she should be eating. The nurse should proceed by:
    1. Asking her what she usually eats at each meal
    2. Instructing her to continue eating a normal diet
    3. Giving her a list of foods to refer to in planning meals
    4. Emphasizing the importance of limiting salt and highly seasoned foods

13. A pregnant woman asks the nurse when she may expect her baby. She tells the nurse her last menstrual period began April 14. Her expected date of delivery would be:
    1. December 21
    2. January 7
    3. January 21
    4. February 1

14. A pregnant client complains of heartburn. When counseling the client, the nurse explains that during pregnancy:
    1. Gastric acidity and motility are increased
    2. The pyloric sphincter relaxes and acid is regurgitated
    3. The cardiac sphincter relaxes and acid is regurgitated
    4. There is increased gastric motility, which causes acid to be regurgitated

15. During labor a client states that she does not want eye drops or ointment placed in her baby's eyes immediately after delivery. The most appropriate response by the nurse would be:
    1. "You'll have to check with your doctor regarding this."
    2. "The medication protects the infant, and that's why it's used."
    3. "Is there a reason why you don't want the medication instilled?"
    4. "They are required and should be administered right after delivery."

16. A newborn is suspected of having toxoplasmosis. Toxoplasmosis, one of the TORCH diseases, may have been transmitted to the infant through:
    1. The placenta in utero
    2. Breast-feeding after delivery
    3. Contact with the maternal genitals
    4. A blood transfusion given to the mother

17. A primigravida, complaining of vaginal spotting and a sharp, shooting pain in the lower abdomen, is diagnosed as having a ruptured tubal pregnancy. When questioning the client about the initial appearance of symptoms, the nurse would expect the client to indicate that her symptoms started:
    1. About the sixth week of pregnancy
    2. At the beginning of the last trimester
    3. Midway through the second trimester
    4. Immediately after implantation occurred

18. A client admitted to the hospital with a threatened abortion anxiously asks the nurse, "Could this have happened because I had the flu?" The nurse's best response would be:
    1. "You feel that you did something to cause the bleeding? Tell me more about what you think."
    2. "We know that maternal infections sometimes result in miscarriages. Perhaps the flu did cause it."
    3. "The doctor will be here soon and at that time will tell you what is causing the bleeding. Wait until then."
    4. "I'm sure that there is absolutely nothing you could have done to cause this. You need not worry about that."

19. After a spontaneous vaginal delivery the obstetrician hands the neonate to the nurse. The nurse's first action should be to
    1. Stimulate the baby to cry
    2. Administer oxygen by mask
    3. Dry and place the infant in a warm environment
    4. Perform a physical assessment and instill eye drops

20. The nurse tells a primigravida who is attending the prenatal clinic for the first time that the most common emotional reaction to pregnancy in the first trimester is:
    1. Rejection
    2. Narcissism
    3. Depression
    4. Ambivalence

21. Clients undergoing panhysterectomy are often started on estrogen therapy to suppress:
    1. Dysmenorrhea
    2. Heat intolerance
    3. Premature aging
    4. Flushing of the face

22. During a contraction the nurse observes a 15-beat-per-minute acceleration of the FHR above the base-line rate. The nurse's most appropriate action would be to:
    1. Call the physician immediately and await orders
    2. Prepare for immediate delivery because of fetal distress
    3. Turn the mother on her left side to increase venous return
    4. Record this normal fetal response to contractions on the chart

23. When preparing a client to care for her episiotomy after discharge, the nurse should include, as a priority, instructions to:
    1. Discontinue the sitz baths once she is at home
    2. Continue perineal care after toileting until healing occurs
    3. Continue the sitz bath 3 times a day if it provides comfort
    4. Avoid stair climbing for at least a few days after discharge

24. A client who expected to deliver using Lamaze techniques has an emergency cesarean delivery. Three days later the client is found crying and tells the nurse she is extremely disappointed because a cesarean delivery was necessary. She asks the nurse why this happened to her. The nurse responds knowing that:
    1. The client's feelings will pass once she has bonded with the baby
    2. The client is probably suffering from postpartal depression and needs special care
    3. Cesarean delivery severely affects a woman's self-concept, and the client's statement reflects this
    4. Emergency cesarean delivery may be a traumatic psychologic experience in addition to an acute obstetric emergency

25. Five minutes after being born, a newborn receives an Apgar of 8. Twelve hours later the newborn becomes hyperactive and jittery, sneezes frequently, and has difficulty swallowing. The nurse should notify the physician because the infant may be exhibiting signs of:
    1. Cerebral palsy
    2. Neonatal syphilis
    3. Fetal alcohol syndrome
    4. Narcotic drug dependence

26. Following a modified radical mastectomy, a client has two portable wound drainage systems in place. When caring for these drains, the nurse should:
    1. Irrigate the drains with normal saline to ensure patency
    2. Leave them open to the air to ensure maximum drainage
    3. Attach the tubes to straight drainage to monitor the output
    4. Compress the drainage receptacles after emptying them to maintain suction

27. When teaching a client arm exercises following a right mastectomy, it is important for the nurse to:
    1. Exercise the right arm only
    2. Exercise both arms simultaneously
    3. Have the client wear a sling between exercise periods
    4. Wait until the incision has healed before actually doing the exercises

28. A sonogram demonstrates a low-lying placenta. Considering this condition, an adaptation that is usually exhibited by the client first would be:
    1. Sharp abdominal pain
    2. Painless vaginal bleeding
    3. Increased lower back pain
    4. Early rupture of membranes

29. A positive Babinski reflex in a newborn infant is a result of:
    1. Immaturity of the CNS
    2. Neurologic impairment
    3. Hypoxia during labor and delivery
    4. Hyperreflexia of the muscular system

30. A client attending a class in preparation for childbirth states, "I am sick and tired of wearing these same old clothes; how I wish all this would be done and over with." The nurse's best response would be:
    1. "Most women feel the same way you do at this time."
    2. "I understand how you feel; what do you know about labor?"
    3. "Yes, this is the most uncomfortable time during pregnancy."
    4. "Is there something bothering you? You sound discouraged."

31. The nurse uses the Leopold maneuvers when a client is admitted to the hospital in labor to palpate the abdomen and assess the:
    1. Station of the fetus
    2. Rate of uterine involution
    3. Position of the fetus in utero
    4. Strength and duration of contractions

32. A client is in active labor. Her contractions are now 2 to 3 minutes apart and last approximately 45 seconds. The fetal heart rate between contractions is about 100 beats per minute. The nurse should:
    1. Obtain the mother's vital signs
    2. Notify the physician immediately
    3. Record this normal fetal response
    4. Continue to monitor the fetal heart rate

33. A 15-year-old who is in true labor asks that her mother remain at her bedside but refuses her mother's assistance with comfort measures, stating, "I can do this myself." The client's conflicting behavior is representative of the adolescent's attempt to achieve the developmental task of:
    1. Identity
    2. Integrity
    3. Industry
    4. Intimacy

34. A 30-year-old client with a 35-day menstrual cycle is attempting to become pregnant. The client and her husband are counseled by the nurse about the optimal timing of intercourse during the cycle. The nurse would know that the counseling was effective when the couple state that they should have intercourse on the:
    1. Twelfth day of the cycle
    2. Fourteenth day of the cycle
    3. Twenty-first day of the cycle
    4. Twenty-fifth day of the cycle

35. During a contraction stress test the nurse should place the client in a:
    1. Sims' position to promote examination
    2. Lithotomy position to facilitate visualization
    3. Semi-Fowler's position to avoid hypotension
    4. Trendelenburg's position to prevent cervical pressure

36. The nurse knows that a client receiving oral contraceptives understands the related dietary teaching when the client states, "While on oral contraceptives I should increase my intake of:
    1. Calcium."
    2. Folic acid."
    3. Vitamin D."
    4. Vitamin B$_2$."

37. A primigravida has had a rupture of a fallopian tube. Following surgery, nursing care should include:
    1. Assuring the client that she is still capable of becoming pregnant
    2. Administering Rho(D) immune globulin (RhoGAM) to prevent isoimmunization
    3. Counseling about various methods of birth control to prevent another tubal pregnancy
    4. Telling the client to avoid douching after intercourse because this may dislodge the fertilized egg

38. A 37-year-old client with endometriosis complains of dysmenorrhea and dyspareunia. Dysmenorrhea is characterized by:
    1. Pain with menses
    2. Endometrial hyperplasia
    3. Bleeding between menses
    4. Heavy bleeding with menses

39. A sexually active woman with a history of a mucopurulent discharge and bleeding associated with cervical dysplasia, dysuria, and dyspareunia will receive tetracycline (Cyclopar) or erythromycin (Ilotycin) for the treatment of:
    1. Herpes simplex 2
    2. *Treponema pallidum*
    3. *Neisseria gonorrhoeae*
    4. *Chlamydia trachomatis*

40. During an emergency delivery, as the fetal head begins to crown, the nurse should:
    1. Press firmly on the fundus
    2. Apply gentle perineal pressure
    3. Suggest that the client push down vigorously
    4. Encourage the client to take prolonged deep breaths

41. After a client had a cesarean delivery, a nursing diagnosis of altered tissue perfusion is made related to the:
    1. Inability of the client to turn from side to side
    2. Unfavorable position of the fetus during labor
    3. Use of a regional anesthetic during the delivery
    4. Increased risk for hemorrhage after a cesarean delivery

42. Because preterm infants are at risk for respiratory distress syndrome, immediate nursing intervention is required when the infant develops:
    1. An expiratory grunt
    2. Substernal retractions
    3. Tachycardia of 160 beats per minute
    4. A respiratory rate of 50 breaths per minute

43. At 30 weeks gestation a client with Class 1 cardiac disease expresses concern about her delivery and asks the nurse what to expect. The nurse should be aware that the physician will probably:
    1. Induce her labor with Pitocin
    2. Prematurely rupture her membranes
    3. Perform an elective cesarean delivery
    4. Use prophylactic forceps after a pudendal block

44. The nurse caring for a woman in active labor monitors the fetal heart during a contraction and finds it is 115 beats per minute. The nurse should:
    1. Administer oxygen immediately
    2. Notify the physician immediately
    3. Take a second reading during decrement
    4. Take a second reading during the acme of the next contraction

45. The nurse knows that delivery may be imminent when the:
    1. Client's cervix is dilated to 6 cm
    2. Client feels the need to move her bowels
    3. Client becomes restless and thrashes about
    4. Client complains of sudden, intense back pain

46. During prenatal classes the nurse teaches the difference between true labor and false labor, including that:
    1. A bloody show is rare with false labor
    2. In true labor the cervix effaces and dilates
    3. Fetal movement is decreased in true labor
    4. The membranes rupture when true labor begins

47. A client who has been breast-feeding tells the nurse on the third postpartum day that she has a great deal of pain in her breasts and that she is afraid that the baby will hurt while grasping the nipple and sucking. It would be most appropriate for the nurse to:
    1. Administer a medication for pain
    2. Call the physician to obtain advice
    3. Help the client express some milk manually before putting the baby to breast
    4. Suggest that the client limit fluids and not try to nurse the baby for the next 2 days

48. While a client is receiving intravenous magnesium sulfate therapy, the nurse should have at the bedside:
    1. Adrenalin
    2. Solu-Cortef
    3. Calcium chloride
    4. Calcium gluconate

49. A client, gravida III, is admitted to the labor room. The nurse knows that gravida III means that the client:
    1. Had one preterm baby
    2. Is pregnant for the third time
    3. Has had an induced abortion
    4. Has three living children at home

50. In the 37th week of gestation, a client with diabetes mellitus undergoes an amniocentesis. The nurse is aware that this is being done primarily to determine the:
    1. Exact gestational age
    2. Lung maturity of the fetus
    3. Presence of genetic disorders
    4. Glucose level of the amniotic fluid

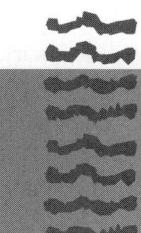

CHILDBEARING AND WOMEN'S HEALTH NURSING QUIZ 1 ANSWERS AND RATIONALES

1. ④ Because the vagina has no sphincter, holding the labia permits the solution to enter, fill, and irrigate the vaginal vault. (2) (SI; IM; WH)
   1 A bulb syringe should never be used because it may cause an air embolism.
   2 The douche tip should never be inserted more than 3 or 4 inches.
   3 This is a possible position to empty the vaginal vault after the fluid has been instilled; the client should be in the lithotomy position while the fluid is being instilled.

2. ① This is a pictorial analysis of chromosomes usually done on peripheral blood, which will show chromosomal abnormalities such as the translocation found in Down syndrome. (2) (CI; PL; HN)
   2 Karyotyping, not a buccal smear, is done to identify chromosomal aberrations.
   3 This is a test that is done before delivery.
   4 This test does not assess chromosomal aberrations.

3. ② The desquamation occurs from prolonged exposure to amniotic fluid, causing cracking, peeling, and drying of skin in the postterm baby. (2) (DF; AS; HN)
   1 This indicates a preterm infant.
   3 Creases would cover the entire sole of each foot.
   4 Same as answer 1.

4. ③ According to the Apgar scale, points would be lost in areas of muscle tone, reflex irritability, and color. (3) (CI; AS; HN)
   1 This is an inaccurate assessment; this number indicates only moderate heart activity and minimum breathing.
   2 Same as answer 1.
   4 This is an inaccurate assessment; this number indicates no difficulties.

5. ① Grieving is expected and necessary whenever an infant is born with severe problems. (2) (SC; IM; HN)

   2 This may be needed, but it is premature at this point.
   3 Depending on the condition and the status of the infant, physical assessments will probably be required more frequently.
   4 This may delay processing information and impede the grieving process.

6. ④ When the placenta initially separates, the fetus becomes hyperactive; the uterus is tender from the accumulation of blood at the abrupted placental site. (2) (DA; AS; HP)
   1 Bright red bleeding is associated with placenta previa.
   2 It is difficult to assess for concealed hemorrhage; the fetus may no longer be alive.
   3 The uterus generally enlarges because of an accumulation of blood at the placental site.

7. ① Daily washing of the breasts and nipples with water is sufficient for cleanliness; soap may be drying. (2) (PC; IM; HC)
   2 It is unnecessary to use sterile water; the inside of the baby's mouth is not sterile.
   3 Scrubbing as well as the use of soap may irritate and dry the nipples, which may already be tender.
   4 Alcohol is drying; also it may leave a taste that causes the infant to reject the nipple.

8. ④ This shows understanding and offers reassurance that this is a normal side effect; this response also offers a possible solution. (1) (PT; EV; RP)
   1 This side effect has nothing to do with ovulation.
   2 This side effect continues as long as the drug is continued.
   3 This indicates a problem where one probably does not exist.

9. ① In the newborn the liver is usually palpable 2 cm below the costal margin. (1) (HI; AS; NN)
2 The stomach is never palpable, and the borders are not detectable.
3 The pancreas is never palpable because it is located in the posterior portion of the abdominal cavity.
4 The gallbladder is not palpable because it is a posterior structure.

10. ③ Because these infants are so tiny and frail, mothers fear handling or touching them, but they should be encouraged to do so by the nursing staff. (2) (DF; AN; HN)
1 This is probably a concern but one not commonly expressed by mothers.
2 Bonding is possible; it does not have to be delayed.
4 Breasts can be pumped and milk given via gavage feedings.

11. ① Panting and blowing keep the glottis open so the mother cannot hold her breath and bear down. (1) (PC; EV; HC)
2 Even though this keeps the pressure of the diaphragm off the contracting uterus, it does not really help during transition.
3 This interferes with adequate oxygenation of the fetus because it limits the mother's oxygen intake.
4 Same as answer 2.

12. ① Successful dietary teaching usually incorporates, as much as possible, the client's food preferences and dietary patterns. (3) (PC; AS; HC)
2 This presupposes that the client has been eating a normal diet; it does not provide for the additional caloric or protein requirements of pregnancy.
3 This does not take into consideration the client's likes and dislikes or cultural preferences; it would not foster compliance.
4 Salt is not limited during pregnancy unless a disease process is present or develops; seasoned foods are permissible if the client does not experience discomfort from them.

13. ③ When calculating the expected date of delivery, subtract 3 months and add 7 days to the date of the last menstrual period. (2) (DF; AS; HC)
1 This is incorrect; count back 3 months and add 7 days.
2 Same as answer 1.
4 Same as answer 1.

14. ③ Tension may be the precipitating cause of gastroesophageal reflux early in pregnancy. Displacement of the stomach and delayed emptying of stomach contents because of uterine enlargement may be the cause of heartburn later in pregnancy. (3) (CI; IM; HC)
1 Gastric motility and acidity are generally decreased in pregnancy.
2 This would allow acid to pass into the small intestine.
4 Increased motility would utilize acid, leaving little or none to be regurgitated.

15. ③ This provides the mother with an opportunity to express her concerns regarding prophylactic eye medication. (2) (HI; IM; NN)
1 The nurse is capable of discussing this with the mother.
2 This does not respond to the mother's statement.
4 This blocks communication; instillation can be delayed for an hour.

16. ① This is the most common route of transmission. (2) (SI; AN; HN)
2 There is no evidence of toxoplasmosis being transmitted via this route.
3 The genital tract is not locally affected.
4 Same as answer 2.

17. ① At this time the tube is unable to expand to the size of the growing pregnancy. (1) (DA; AS; RP)
2 Tubal pregnancies cannot advance to this stage because of the tube's inability to expand with the growing pregnancy.
3 Same as answer 2.
4 The size of the fertilized egg at this time is minuscule and will cause no problem.

18. ① This response encourages the client to discuss her fears and anxieties. (1) (SC; IM; EC)
2 This gives inaccurate information; this conclusion has not been documented, and this response adds to the guilt felt by the client.
3 This response does not focus on the client's feelings; it cuts off communication between the nurse and the client.
4 This is false reassurance; it denies the client's feelings and cuts off communication.

19. ③ The first action should be to dry and place the infant under a warmer to prevent chilling, thereby decreasing the possibility of development of metabolic acidosis. (2) (PC; IM; NN)
1 This would not be done until after the baby is warmed.
2 This is useless without an adequate airway.

4   This is not the priority at this time; conserving body heat takes precedence.

20. (4) This is most common as the disbelief of actually being pregnant is experienced. (2) (SC; IM; EC)
1   Rejection is more commonly seen in the third trimester.
2   Narcissism is more commonly seen in the second and third trimesters.
3   Depression is more commonly seen in the third trimester and postpartum.

21. (4) The occurrence of hot flashes can be emotionally debilitating; low-dose estrogen therapy may reduce the frequency and degree of symptoms. (3) (PT; AN; WH)
1   There is no menstruation after this surgery.
2   Heat intolerance is not a symptom associated with estrogen deprivation.
3   The benefits derived must be weighed against the potential side effects; therapy for this reason is still controversial.

22. (4) Stimulation of the autonomic nervous system is a normal response to cord compression during uterine contraction. (1) (HI; IM; HC)
1   This requires no intervention at this time.
2   Same as answer 1.
3   This is done when deceleration occurs.

23. (2) The prevention of infection is the priority. (2) (SI; PL; HC)
1   It is not necessary to stop sitz baths as long as they provide comfort.
3   This provides comfort but is not the priority.
4   Stair climbing may cause some discomfort but is not detrimental to healing.

24. (4) The client's response is appropriate to the situation; the client is in the "why me" stage of the grieving process. (2) (SC; IM; EC)
1   The client's feelings are unrelated to bonding.
2   The client's statement is not indicative of depression.
3   This is untrue; self-concept is not severely affected.

25. (4) These adaptations indicate narcotic drug dependence; the infant should be watched for withdrawal signs during the first 24 hours. (3) (HI; AS; HN)
1   Symptoms of cerebral palsy are usually manifested later in infancy.
2   A low-grade fever and a copious serosanguineous discharge from the nose are symptoms of syphilis.

3   Growth deficiencies in length, weight, and head circumference are associated with fetal alcohol syndrome.

26. (4) Portable wound drainage systems are self-contained and can be emptied and compressed to reestablish negative pressure, which promotes drainage. (2) (CI; IM; WH)
1   Portable wound drainage systems are not irrigated; they just drain via negative pressure.
2   Portable wound drainage systems are self-contained, closed systems.
3   Portable wound drainage systems have collection chambers, so further drainage systems are not needed.

27. (2) Postmastectomy exercises should be bilateral, using both arms simultaneously to prevent shortening of muscles and contracture of joints. (2) (PC; IM; WH)
1   Both arms should be exercised to maintain symmetry of muscle tone and strength.
3   A sling immobilizes the arm, creating joint stiffness and loss of muscle tone.
4   Exercises of the affected arm are usually started within 24 hours to prevent contractures and muscle atony.

28. (2) As the process of effacement occurs in the latter part of pregnancy, placental separation from the uterus occurs, causing bleeding. (2) (DA; AN; HP)
1   This occurs in premature separation of a normally implanted placenta.
3   This is generally not associated with placenta previa at its onset; it may occur later.
4   This is not usually the first thing to occur; it may occur after the placenta separates.

29. (1) The newborn's immature neuromuscular development normally causes dorsiflexion of the big toe and fanning of the remaining toes, a positive Babinski sign. (2) (DF; AN; NN)
2   A positive Babinski sign is a normal response in an infant; its absence may demonstrate neural impairment.
3   CNS damage from hypoxia can result in a negative Babinski sign, an abnormal finding in an infant.
4   Hyperreflexia is an abnormal increase in reflexes; a positive Babinski sign is a normal response in the infant.

30. ① Near term, most mothers are tired of the pregnant state and anxious for labor to begin. It is helpful to know that this is a common reaction. (3) (SC; IM; EC)
2 This narrows the client's verbalization to what the nurse sees as the client's area of concern.
3 This does not encourage further verbalization, it merely closes off communication.
4 The client has just told the nurse what is bothering her; this response does not encourage the client to discuss her feelings further.

31. ③ The nurse palpates the abdomen to locate the head, back, and small parts of the fetus; the location of these parts reveals the position of the fetus. (1) (SI; AS; HC)
1 The station can be ascertained only on vaginal examination, which is not part of the Leopold maneuvers.
2 Uterine involution is measured during the postpartum period.
4 This can be done by lightly placing the hand on the fundus during a contraction; no other maneuver is necessary.

32. ② Bradycardia (baseline FHR less than 120 beats per minute) may indicate fetal distress and may require medical intervention. (2) (SI; IM; HC)
1 There is no indication of maternal distress; the fetus may be in distress.
3 The normal fetal heart rate is 120 to 160 beats per minute.
4 This is dangerous; the fetus may be in distress, and time should not be spent on monitoring.

33. ① According to Erikson, identity vs role confusion is the developmental conflict of the adolescent. (1) (DF; AN; EC)
2 Ego integrity vs despair is the developmental conflict of the older adult.
3 Industry vs inferiority is the developmental conflict of the school-age child.
4 Intimacy vs isolation is the developmental conflict of the young adult.

34. ③ Ovulation usually occurs 14 days before menses; in a 35-day cycle, ovulation may occur as late as the 21st day. (2) (DF; EV; RC)
1 This is the proliferative phase of the cycle; ovulation has not yet occurred.
2 If the woman has a 28-day cycle, ovulation could be expected at this time.
4 The ovum has already passed out of the fallopian tube and can no longer be fertilized.

35. ③ A semi-Fowler's position will avoid hypotension and is recommended for safety and comfort. (1) (CI; PL; HP)
1 The Sims' position would make monitoring very difficult.
2 Usually no vaginal examination is necessary, and the lithotomy position is very uncomfortable for long periods.
4 This position is used for shock or a prolapsed cord, not for a CST.

36. ② Oral contraceptives are thought to cause deficiencies of folic acid, vitamin C, vitamin B_6, and vitamin B_{12}. (3) (PT; EV; RC)
1 It is unnecessary to increase calcium intake when taking oral contraceptives.
3 No clinical evidence exists to link oral contraceptives to vitamin D deficiency.
4 It is unnecessary to increase the dietary intake of vitamin B_2 (riboflavin) while receiving oral contraceptives.

37. ① Removing a fallopian tube does not impair the ovarian ability to release an egg, which may be fertilized in the remaining tube. (1) (SC; IM; RP)
2 There is no information given that states the client is Rh negative.
3 There is no absolute way of knowing whether future pregnancies will be tubal pregnancies.
4 Douching in no way reaches the fertilized egg.

38. ① Dysmenorrhea is defined as pain with menses. (1) (CI; AS; WH)
2 Endometrial hyperplasia results from anovulation and persistent estrogen stimulation.
3 Bleeding between menses is called metrorrhagia.
4 Heavy bleeding with menses is termed menorrhagia.

39. ④ These are the signs and symptoms and the treatment for a chlamydial infection. (3) (PT; AN; WH)
1 Painful blisters on the genitalia, fever, malaise, dysuria, and dyspareunia are signs of herpes simplex virus 2 infection.
2 Chancre formation is a sign of primary syphilis; a symmetric rash with malaise, fever, anorexia, and headache are symptoms of secondary syphilis.
3 Dysuria, heavy greenish-yellow purulent discharge, and swollen Bartholin's glands are signs of gonorrhea.

40. (2) This prevents too rapid expulsion of the head, which can lead to increased intracranial pressure in the infant and a perineal laceration in the mother. (3) (CI; IM; HC)

1 This is unnecessary; precipitate delivery is caused by forceful uterine contractions that expel uterine contents.

3 Vigorous pushing will cause too rapid expulsion, leading to increased intracranial pressure in the infant and laceration in the mother.

4 At this time the urge to push is uncontrollable and she will be unable to take prolonged deep breaths.

41. (4) Abdominal surgery, especially pelvic surgery, predisposes the client to the risk of hemorrhage. (2) (DA; AN; HP)

1 An impaired gas exchange would be the appropriate diagnosis for a client who is immobilized.

2 The position of the fetus is unrelated to tissue perfusion.

3 A regional anesthetic is related to impaired physical mobility and sensory-perceptual alterations, not altered tissue perfusion.

42. (2) Retractions are a prominent feature of respiratory problems in preterm infants because of their compliant chest walls. (2) (HI; AS; HN)

1 This is more indicative of low body temperature, not respiratory distress.

3 This rate is still within the normal range of 120 to 160 per minute.

4 A rapid respiratory rate of 40 to 60 is normal in neonates.

43. (4) Forceps reduce the mother's need to push, thereby conserving energy; a regional anesthetic does not compromise cardiovascular function. (2) (SI; PL; HP)

1 Induced labor is often more stressful and painful than natural labor.

2 Same as answer 1.

3 Major abdominal surgery is performed on clients with cardiac problems only when absolutely necessary because surgery adds additional stress to a compromised heart.

44. (3) It is important to monitor the fetal heart rate during contractions; if the fetal heart rate slows during contractions and then resumes the normal rate within 30 seconds after acme, it is not serious. (3) (HI; EV; HC)

1 Unless the FHR remains decelerated, oxygen is not necessary.

2 Data are incomplete; during a contraction the fetal heart rate may be slowed; it needs to be monitored at the end of the contraction to see whether it remains decelerated.

4 The FHR should not be taken at the height of a contraction; wait until the contraction subsides to obtain the FHR to determine deceleration.

45. (2) The presenting part is low in the birth canal and may cause strong sensations of pressure on the rectum. (3) (SI; AS; HC)

1 This is the active phase of the first stage of labor.

3 Restlessness and thrashing are generally not present in the early or middle phases of labor; these behaviors usually begin during the transitional phase.

4 This may occur with persistent posterior pressure; it is usually not sudden and is not a sign of advanced labor.

46. (2) This is the only real difference between false and true labor. (1) (DF; AN; HC)

1 Some women have a bloody show without cervical dilation.

3 Untrue; fetal movement continues unchanged throughout labor.

4 Untrue; the membranes may rupture before labor begins or long after labor has started.

47. (3) The pressure and tenderness resulting from accumulated milk can be relieved by manually expressing some of the fluid. (2) (PC; IM; HC)

1 Pain medications would be offered if other measures were unsuccessful; medication can be transferred to the infant through breast milk.

2 The nurse is capable of independent action that involves client teaching.

4 This is unsound advice; the mother needs to drink fluid and keep nursing to ensure the flow of breast milk.

48. (4) Calcium gluconate is the antidote for magnesium sulfate toxicity. (1) (PT; PL; DR)

1 Adrenalin is a vasoconstrictor and is not used as an antidote for magnesium sulfate toxicity.

2 Solu-Cortef is a steroid and is not used to counteract magnesium sulfate toxicity.

3 Calcium gluconate, not calcium chloride, is the antidote.

49. (2) Gravid means pregnant, so gravida III indicates a third pregnancy. (1) (DF; AN; HC)

1 Para and gravida do not refer to the gestational age of babies at birth.

3 Gravida does not identify either induced or spontaneously aborted pregnancies.

4 Neither para nor gravida indicates whether there are living children.

50. (2) An amniocentesis at this stage of gestation is performed to determine L/S ratios, which indicate fetal lung maturity. (1) (CI; AS; HP)

1 This is determined by the less invasive procedure of ultrasonography.

3 Amniocentesis would be done between the 16th and 20th weeks to determine genetic disorders.

4 This is irrelevant; this is not the primary purpose for examining amniotic fluid at this time.

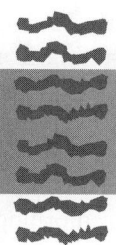

CHILDBEARING AND WOMEN'S HEALTH NURSING QUIZ 2

1. A client, 38 weeks gestation, who is having periods of bright red, painless bleeding, is in the high-risk unit because of a placenta previa. The nurse is aware that the client's labor has started when assessment demonstrates:
1. Decreased fetal heart rate
2. Increased vaginal bleeding
3. Decreased vaginal spotting
4. Rhythmic uterine contractions

2. After a cesarean delivery a client is transferred to the recovery area. In the first 2 hours following a cesarean delivery, priority nursing intervention is to:
1. Assess lochia to prevent hemorrhagic complications
2. Assess hydrational needs and provide fluid as ordered
3. Encourage bonding to promote parent-child interaction
4. Monitor the incision to prevent complications from infection

3. When planning nursing care for a client who delivered a preterm male infant, the nurse should understand that the mother will:
1. Suffer from feelings of guilt and withdrawal
2. Experience feelings of helplessness, failure, and loss of control
3. Be unable to form a healthy relationship with the baby until he is out of danger
4. Have increased feelings of attachment resulting from greater concern over the baby

4. The nurse understands that true labor is characterized by:
1. Contractions 10 minutes apart with no change in frequency over 2 hours; cervix closed
2. No contractions; cervix 3 cm dilated and 50% effaced; no change after 4 hours of walking and sitting
3. Contractions every 5 to 10 minutes; cervix dilated 2 cm and 75% effaced; dilation increased to 3 cm in 2 hours

4. Contractions irregular, 10 to 15 minutes; cervix dilated a fingertip and 50% effaced; no change with 4 hours of bed rest

5. If involution is progressing normally, immediately after birth the nurse should expect the fundus to be located:
1. At the level of the umbilicus
2. Two centimeters below the umbilicus
3. Three centimeters above the umbilicus
4. Two centimeters above the symphysis pubis

6. A newborn, delivered at 36 weeks gestation, must be carefully observed during the first 24 hours for:
1. Duration of cry
2. Respiratory distress
3. Frequency in voiding
4. Change in body temperature

7. The nurse is aware that a common adaptation during pregnancy is:
1. Increased ovarian activity
2. Increased pulmonary capacity
3. Decreased glomerular filtration rate
4. Decreased gastrointestinal motility

8. A client attending the prenatal clinic for the first time tells the nurse that her last menstrual cycle began on January 11. The client also states that she had 1 day of light spotting on February 7. The client's expected date of delivery is calculated to be:
1. October 14
2. October 18
3. November 14
4. November 18

9. A client with preeclampsia is placed on intravenous magnesium sulfate therapy. The nurse caring for this client should immediately notify the physician if the client's:
1. Blood level is 2.75 mEq/L
2. Respirations are 18 per minute
3. Patellar reflexes can be elicited
4. Output is less than 100 ml in 4 hours

10. A public health campaign is planned to immunize children and women in the childbearing age group. The nurse knows that the immunization that can be safely administered to a pregnant woman is:
    1. Polio
    2. Mumps
    3. Rubella
    4. Rubeola

11. A client who is at 4 months gestation is scheduled for a sonogram and an amniocentesis. The nurse instructs the client to drink 8 oz of fluid before the test. This is done to:
    1. Improve ultrasonic visualization of the fetus
    2. Hydrate the mother and increase circulation
    3. Hydrate the fetus and decrease fetal movement
    4. Replace fluid that may be lost during the procedure

12. When a new mother sees her newborn son assume a fencing position as she turns his head, she becomes concerned. The nurse should teach the mother that:
    1. This is a normal response
    2. This reflex disappears around 2 months of age
    3. The physician had been notified of this suspicious response
    4. The tonic neck reflex may indicate neurologic damage in the newborn

13. A woman in the family planning clinic asks the nurse why contraceptives are necessary since she is still breast-feeding her baby. The nurse's most appropriate response would be:
    1. "It is best to delay any sexual relations until you have your first menstrual period."
    2. "As long as you have no menstrual period, you won't have to worry about using contraceptives."
    3. "It is best to use contraceptive measures since ovulation may occur without a menstrual period."
    4. "Since lactation suppresses ovulation, you probably don't have to worry about becoming pregnant."

14. The nursing care for a drug-dependent infant in the nursery should include:
    1. Administering methadone
    2. Offering small, frequent feedings
    3. Increasing environmental stimuli
    4. Reducing elevated body temperature

15. While being prepared for a prenatal examination, a pregnant woman complains to the nurse of feeling very tired and being sick to her stomach, particularly in the morning. The best response by the nurse would be:

    1. "Perhaps you might discuss this with the doctor."
    2. "This is a common occurrence resulting from all the changes going on in your body."
    3. "This is a common occurrence during the early part of pregnancy, and you need not worry."
    4. "These are common occurrences. You say your sick feelings bother you most often in the morning?"

16. When assessing a client with an abruptio placentae, the nurse would expect to observe:
    1. A flaccid uterus
    2. Painless bleeding
    3. Bright red bleeding
    4. A boardlike abdomen

17. The nurse is aware that diabetes mellitus can affect pregnancy by:
    1. Promoting abnormal placental implantation
    2. Predisposing the client to hypertensive states
    3. Decreasing the amount of amniotic fluid present at term
    4. Increasing the appetite and causing excessive weight gain

18. Upon discovering the presence of a prolapsed cord, the nurse anticipates that the client's delivery will be:
    1. Induced with oxytocin
    2. Via a cesarean delivery
    3. A low forceps vaginal delivery
    4. Postponed as long as possible

19. In the eighth month of pregnancy, a client tells the nurse that she is experiencing dyspareunia. The nurse should plan to teach the client to:
    1. Avoid intercourse
    2. Try alternative positions
    3. Douche to lubricate the vaginal mucosa
    4. Consult a therapist for sexual counseling

20. Following a cesarean delivery, a client is receiving IV fluids and has a Foley catheter. The client's fluid intake should be increased when the nurse notes:
    1. Dark amber urine
    2. Urinary suppression
    3. Tinges of blood in the urine
    4. Blood pressure of 100/60 mm Hg

21. The physician prescribes and the nurse administers vitamin K intramuscularly to a newborn immediately after birth to:
    1. Substitute for normal bacterial flora
    2. Promote the synthesis of prothrombin
    3. Prevent increased levels of serum bilirubin
    4. Decrease calciferol until renal clearance can take over

22. The nurse should plan the postpartum care for a new mother with a history of rheumatic heart disease on the knowledge that:
    1. The client should increase her fluid intake, particularly if she is breast-feeding
    2. Clients with cardiac problems should be kept on bed rest for a minimum of 4 days
    3. The first 48 hours postpartum are the most stressful on the cardiopulmonary system
    4. The client is out of immediate danger because the stress associated with pregnancy is over

23. A client is receiving terbutaline (Brethine) tocolytic therapy. During the initial administration of Brethine, the nurse should:
    1. Check the client's reflexes every 2 hours
    2. Monitor the client's pulse every 15 minutes
    3. Insert a Foley catheter to monitor the urinary output
    4. Institute safety measures because of altered consciousness

24. The nurse prepares a client with a ruptured tubal pregnancy for immediate surgery. The nurse understands that an informed consent will have to include permission for a:
    1. Myomectomy
    2. Hysterectomy
    3. Salpingectomy
    4. Dilation and curettage

25. A strict vegetarian becomes pregnant and asks the nurse whether there is anything special she should do in relation to her diet during pregnancy. The nurse should teach the client the importance of:
    1. Eating at least 40 g of protein a day
    2. Drinking at least 1 quart of milk per day
    3. Taking a vitamin supplement and iron tablet daily
    4. Including a variety of vegetable proteins in her diet

26. A client with a possible ovarian tumor undergoes laparoscopic surgery. Following this surgery, the nurse should instruct the client that:
    1. Betadine douches should be used daily
    2. Vaginal bleeding may be present for 1 to 3 days
    3. Her usual activities may be resumed in 12 hours
    4. Shoulder pain may be present for 12 to 24 hours

27. One hour following delivery, the nurse finds that a client's uterus has become relaxed and boggy. The initial nursing response should be to:
    1. Check the BP
    2. Notify the physician
    3. Massage it until firm
    4. Observe the amount of bleeding

28. At 40 weeks gestation a client is scheduled for a contraction stress test, and the nurse explains the procedure. The nurse is satisfied that the mother understands the teaching when she says:
    1. "I hope this test does not cause my labor to begin early."
    2. "I hope the baby doesn't get too restless after this procedure."
    3. "I hate having needles in my arm, but now I understand why it's necessary."
    4. "If my baby's heart reacts normally during the test I will not need to have my labor induced today."

29. To support the body's natural defense mechanisms, a client who has had recurrent infections prior to and during pregnancy should be instructed to eat a nutrient-rich diet that emphasizes:
    1. The fat-soluble vitamins
    2. Dietary fiber and oat bran
    3. Vitamin A, C, E, and selenium
    4. Low fat with essential fatty acids

30. A baby has sustained an intracranial hemorrhage because of a tear in the tentorial membrane sustained during delivery. The nurse should expect this baby to display:
    1. Extreme lethargy
    2. Weak, timorous cry
    3. Generalized purpura
    4. Abnormal respirations

31. Before a client signs the informed consent for a modified radical mastectomy, the nurse should be certain that the client knows the surgery includes the removal of:
    1. Pectoral muscles
    2. Skin overlying the breast tissue
    3. Axillary lymph nodes on the affected side
    4. The involved half of the breast and nodes

32. An 18-year-old primigravida in the 36th week of gestation is admitted to the hospital with a diagnosis of pregnancy-induced hypertension. The nurse admitting this client is aware that nursing care measures will be directed chiefly toward reducing:
    1. Anxiety
    2. Bleeding
    3. Blood pressure
    4. Circulating blood volume

33. A client, wishing to postpone having children until she and her husband are financially sound, has been on oral contraceptive pills for several years. The nurse is aware that an assessment finding that would indicate a potential risk with continuing use of birth control pills is:
1. Dysmenorrhea
2. Midcycle bleeding
3. The lack of ovulation
4. A BP of 140/90 mm Hg

34. A client finds a lump in her breast and is hospitalized for a biopsy and possible mastectomy. As the nurse is preparing her for surgery, the client says, "I'm really scared. My mother and sister went through this. It was awful." The nurse's most appropriate response would be:
1. "You know, most breast lumps are benign."
2. "Breast cancer has an excellent cure rate."
3. "You are worried about the results tomorrow?"
4. "What happened with your mother and sister?"

35. Immediately after a client has completed the third stage of labor, the nurse administers 10 units of oxytocin (Pitocin) as ordered by the physician. The desired response from this medication will:
1. Lessen the discomfort of the episiotomy
2. Relax the uterus so that it can be emptied
3. Prevent bleeding following separation of the placenta
4. Stimulate the client's breasts so that breast-feeding can be started

36. A woman who is admitted to the hospital for uncontrolled gestational diabetes during her third pregnancy has two other children at home. An applicable nursing diagnosis is altered family processes related to hospitalization. Nursing care for this client should include:
1. Suggesting that she be allowed to leave the hospital every few days to visit with her other children
2. Providing family members the opportunity to discuss their feelings regarding the hospitalization
3. Suggesting that a social worker visit the family weekly to perform a home and child-care evaluation
4. Supporting her efforts to be discharged from the hospital as soon as she learns to administer insulin

37. A primigravida visits her obstetrician for the first time in the first trimester. The statement that illustrates a psychologic reaction to pregnancy that commonly occurs in the first trimester would be:
1. "I know I'm going to be a terrible mother. I'll forget the baby somewhere the first time I go shopping."
2. "I know I'm having a little girl. I dreamed she would be a doctor or a lawyer and be very successful."
3. "I want to be pregnant, and I'm excited about the baby, but I'm not sure I'm ready to become a mother."
4. "I'm so excited about this baby, but I'm so afraid of losing control during labor. I know I'll be a terrible patient."

38. In a preparation for childbirth class, the nurse would teach that labor:
1. Should be painless and uneventful
2. Will be painful, but clients will be taught how to tolerate it
3. Will be uncomfortable; however, medication will not be needed
4. May be uncomfortable, but medication is available when needed

39. The nurse realizes that the greatest influence on the perception of pain for a woman in labor is the:
1. Parity of the client
2. Length of the labor
3. Tension of the client
4. Difficulty of the labor

40. At 42 weeks gestation a client delivers an 8 lb, 5 oz baby. The nurse is aware that the characteristics that would be indicative of a postterm infant are:
1. Abundant lanugo, abundant vernix, and fine woolly hair
2. Smooth, edematous skin, with many blood vessels visible
3. Breast buds 5 mm, veins barely visible, and superficial peeling of skin
4. Thick, wrinkled and parchment-like skin with generalized peeling of epidermis

41. A client seeking family planning information asks the nurse during which phase of the menstrual cycle an IUD should be inserted. The nurse responds that the insertion is usually done on the:
1. First to fourth day
2. Fifth to eleventh day
3. Fourteenth to sixteenth day
4. Twenty-fifth to twenty-eighth day

42. When assessing a newborn in the nursery immediately after arrival from the delivery room, the nurse notes that the baby's skin is mottled. The nurse should first:

1. Administer oxygen
2. Notify the physician
3. Encourage an oral feeding
4. Place the baby under the radiant warmer

43. The nurse assesses the hips of a newborn for dislocation and is aware that dislocation would be indicated by:
1. Legs of equal length
2. Limitation in flexion of the hips
3. Limitation in abduction of either hip
4. Ability to abduct each hip 90 degrees

44. About 6 hours after delivery the nurse notes that a client's fundus is 2 fingerbreadths above the umbilicus and is deviated to the right of the midline. The nurse suspects that the client has:
1. Begun involution
2. Bladder distention
3. Second-degree uterine atony
4. Retained placental fragments

45. A client with endometriosis asks the nurse what side effects to expect from danazol (Danocrine). The nurse's response would be:
1. Weight gain
2. Frequent urination
3. Low blood pressure
4. Heavy menstrual bleeding

46. A client with a history of rheumatic heart disease starts labor. The nurse should encourage this client to assume the:
1. Supine position
2. Semi-Fowler's position
3. Trendelenburg's position
4. Left lateral recumbent position

47. A 60-year-old is admitted for a vaginal hysterectomy and an anterior and posterior repair of the vaginal wall. A current symptom the nurse would expect the client to divulge during the nursing history is:

1. Hematuria
2. Dysmenorrhea
3. Pain on urination
4. Stress incontinence

48. During preconceptual counseling, a couple with a history of congenital defects in the family have been advised to have an amniocentesis. They ask when the test will have to be performed. The nurse tells them that the test will be scheduled sometime during the:
1. Eighth to tenth week
2. Sixteenth to eighteenth week
3. Twentieth to twenty-fourth week
4. Thirty-second to thirty-sixth week

49. The day after a client's cesarean delivery, the Foley catheter is removed. The nurse can best evaluate that the client's urinary function has returned to normal when:
1. The client's daily urinary output is at least 1500 ml
2. The client's urinalysis indicates no bacteria present
3. The client has a residual urine of 90 ml after voiding
4. The client voids at least 300 ml 4 hours after catheter removal

50. A client who has just begun breast-feeding her infant complains that her nipples feel very sore. The mother should be encouraged to:
1. Apply continuous ice packs to her nipples to reduce the pain
2. Take the analgesic medication prescribed to limit the discomfort
3. Remove the baby from the breast for a few days to rest the nipples
4. Assume a different position when breast-feeding to adjust the infant's sucking

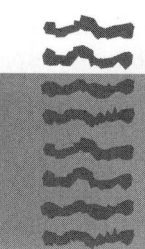

CHILDBEARING AND WOMEN'S HEALTH NURSING QUIZ 2 ANSWERS AND RATIONALES

1. ④ Rhythmic uterine contractions are positive signs of beginning labor. (2) (DF; AS; HP)
   1 This is not a sign of labor but a sign of fetal distress that demands medical attention.
   2 This has no relation to the onset of labor; it is more likely a sign of further placental separation.
   3 This is not a sign of labor; it may indicate that placental separation has lessened.

2. ① The amount and character of the lochia must be checked after cesarean delivery just as it is for vaginal delivery. (2) (CI; EV; HP)
   2 The mother is usually NPO in the early recovery period, although a small amount of ice chips may sometimes be given.
   3 Not a priority at this time; bonding is an important consideration after mother and infant are stable.
   4 The incisional area is observed for signs of hemorrhage; it is too early for evidence of infection.

3. ② Attachment theory states that the experience of the birth of a preterm infant carries with it feelings of loss of control for the mother. (3) (SC; AS; EC)
   1 Withdrawal from the situation is maladaptive and requires special help.
   3 A relationship can occur regardless of the baby's state of health.
   4 There is no basis to believe that increased attachment occurs.

4. ③ Progressive cervical dilation and regular contractions that get progressively closer and increase in intensity are indications of true labor. (2) (DF; AS; HP)
   1 These are not indications of true labor.
   2 Same as answer 1.
   4 Same as answer 1.

5. ① The first hour after delivery is called the fourth stage of labor. During this time the uterus is at the level of the umbilicus, and each day after de-livery it descends 1 fingerbreadth. (3) (HI; AS; HC)
   2 This will happen on the second day postpartum; the uterus descends 1 fingerbreadth each day.
   3 The uterus descends below the umbilicus, rising above it only if the bladder is full.
   4 This will happen on the fourth or fifth day postpartum; the uterus descends 1 fingerbreadth per day.

6. ② Respiratory distress is a common response indicative of possible immaturity of the infant's respiratory tract—such as a small lumen, weakness of the respiratory musculature, paucity of functional alveoli, or insufficient calcification of the bony thorax. (2) (DA; EV; HN)
   1 Tone (i.e., high, shrill) would be more pertinent than duration.
   3 This would not yet be important, since the baby's intake is limited in the first 24 hours.
   4 The temperature generally stabilizes within 24 hours unless the baby is in distress.

7. ④ The influence of progesterone and the pressure of the gravid uterus slows GI motility. (1) (DF; AN; HC)
   1 There is a decrease in ovarian activity.
   2 The pulmonary capacity stays relatively stable as the thoracic cage widens to accommodate lung volume.
   3 There is an increase in the glomerular filtration rate because of the increased fluid volume.

8. ② Using Nägele's rule to calculate, the estimated date of delivery (EDD) is the date of the last menstrual period (LMP) plus 7 days, minus 3 months; spotting is fairly common at the time of the expected menstrual period. (1) (DF; AN; HC)
   1 This is an incorrect calculation; this is too early.
   3 This is an incorrect calculation; this is too late.
   4 Same as answer 3.

9. ④ An output of 25 ml/hr would not permit proper excretion of the magnesium sulfate; excess magnesium levels can cause respiratory and cardiac depression. (2) (SI; EV; DR)
   1 This is within the therapeutic magnesium blood levels of 2.5 to 7.5 mEq/L.
   2 Respirations at this rate are normal; a rate of at least 16 breaths per minute should be present before each dose of magnesium sulfate.
   3 Loss of patellar reflex is suggestive of magnesium sulfate toxicity.

10. ① Salk vaccine (IPV) may be given because it is a killed virus vaccine and will not have a teratogenic effect on the fetus. (2) (HI; IM; HC)
   2 Mumps vaccine is not administered during pregnancy because it is an attenuated live virus vaccine and may be teratogenic to the fetus.
   3 Rubella vaccine is not administered during pregnancy because it is an attenuated live virus immunization and may be teratogenic to the fetus.
   4 Rubeola vaccine is contraindicated during pregnancy because it is an attenuated live virus vaccine and may be teratogenic to the fetus.

11. ① A full bladder helps support the uterus in proper position for the imaging during the scan. (2) (CI; AN; HP)
   2 Drinking a glass of water will not hydrate the mother or significantly affect the circulation.
   3 Drinking 8 ounces of water will have no significant effect on the fetus.
   4 The amniotic membranes will replenish the fluid that is removed during the procedure.

12. ① The tonic neck reflex is normal in the newborn and disappears within 3 to 4 months. (2) (DF; IM; NN)
   2 This response disappears between 3 and 4 months.
   3 This is a normal newborn response and does not need medical attention.
   4 Lack of this reflex may indicate neurologic impairment.

13. ③ Anovulation occurs in nursing mothers for varying periods. Lactation does affect the degree of infertility, but it is generally not a reliable method of birth control. (2) (DF; IM; RC)
   1 Menstrual periods may not occur for several months.
   2 Ovulation can occur without menstruation.
   4 This is untrue; lactation may delay menses but does not reliably suppress ovulation.

14. ② Drug-dependent newborns are poor feeders because of hyperactivity, nausea, vomiting, respiratory distress, excessive mucus, and pyrexia; small, frequent feedings are given to prevent dehydration. (3) (PC; PL; HN)
   1 These infants need supportive care during the time the drug is leaving their systems; methadone only modifies withdrawal, it does not prevent it.
   3 To minimize extraneous stimulation, environmental stimuli should be decreased.
   4 Infants of drug-addicted mothers are prone to hypothermia, not hyperthermia.

15. ④ Knowing that others share the same problems may be comforting. The second part of the nurse's statement is open-ended and allows the client to describe her physical and emotional feelings. (2) (DF; IM; HC)
   1 This discussion is within the purview of the nurse; the discussion should take place now.
   2 This is too factual; it closes off communication. The client needs to explore with the nurse the means for feeling better.
   3 Telling the client not to worry closes off communication; this does not allow exploration of avenues for further discussion.

16. ④ Extravasation of blood at the separation site into the myometrium causes a tetanic, boardlike uterus. (2) (HI; AS; HP)
   1 The uterus is rigid because of filling with blood and clots.
   2 This is associated with placenta previa; abdominal pain and uterine tenderness occur in abruptio placentae.
   3 Bleeding with abdominal pain can be concealed or apparent; the color is usually port wine or dark.

17. ② The likelihood of pregnancy-induced hypertension increases fourfold in clients with diabetes mellitus, probably because of a preexisting vascular condition. (1) (DA; AN; HP)
   1 Abnormal implantation may occur because of scarring or uterine abnormalities, not because of diabetes.
   3 Clients with diabetes have increased, rather than decreased, amniotic fluid.
   4 Most pregnant women have increased appetites; excessive weight gain may be caused by a macrosomic infant and hydramnios.

18. ② Immediate delivery is necessary to prevent fetal hypoxia and death. (1) (SI; PL; HP)
   1 This is unsafe; contractions would increase pressure on the cord, causing fetal hypoxia.

3　This is unsafe; this would increase pressure on the cord.

4　This is unsafe; the fetus is in distress, and immediate delivery is necessary.

19. ②　Pain caused by deep penetration by the male partner is common in late pregnancy and can be reduced by using alternative positions such as rear entry. (1) (DF; PL; HC)

1　This should not be suggested until other alternatives have been tried.

3　Douching is not recommended and does not lubricate the vagina; a water-soluble lubricant is more effective.

4　This is unnecessary because this is common during the third trimester.

20. ①　Dark amber or tea-colored urine indicates highly concentrated urine and requires additional hydration of the client. (1) (DA; IM; HP)

2　Increasing the IV rate in the presence of urinary suppression would be unsafe because it could cause hypervolemia.

3　Tinges of blood in the urine may indicate bladder injury and are not related to the client's fluid status.

4　This reading is meaningless unless a comparison with other readings indicates a decrease and the possibility of shock.

21. ②　Vitamin K stores are almost absent in the newborn because the intestinal flora that produce this vitamin are not present. Vitamin K is an essential precursor of prothrombin, which is part of the clotting mechanism. (2) (CI; IM; NN)

1　The normal flora develop as the newborn is exposed to extrauterine living conditions.

3　Increased bilirubin in the blood occurs in newborns because of the rapid breakdown of RBCs and the liver's difficulty in conjugating such large amounts; it is not influenced by vitamin K.

4　The young kidneys operate at a functional level appropriate to the needs of a healthy infant; this is not influenced by vitamin K.

22. ③　The rapid fluid shift after the placenta is delivered causes an increase in cardiac output and blood volume, making the first 48 hours postpartum crucial. (2) (DA; PL; HP)

1　This is not recommended because it will further increase the circulating blood volume and necessitate an increased cardiac output.

2　Progressive ambulation starting 48 hours after delivery is recommended.

4　This is false; the first 48 hours are crucial because of the rapid fluid shift and the increased cardiac output.

23. ②　Tachycardia is an expected side effect of terbutaline, a beta-mimetic agent; the pulse rate should be no greater than 120 beats per minute. (3) (PT; EV; DR)

1　The reflexes are not affected by this drug, which relaxes uterine musculature.

3　This drug does not have a diuretic effect.

4　This drug does not affect the client's level of consciousness.

24. ③　The ruptured fallopian tube will probably be removed rather than repaired; repair of the tube may result in scarring, predisposing the client to another tubal pregnancy. (1) (CC; PL; RP)

1　This is a procedure for removing myomas (fibroids) from the uterus.

2　The uterus is usually uninvolved in a tubal pregnancy and does not need to be removed.

4　A D&C would only be effective in cleaning out the uterine cavity; no pregnancy contents are in the uterus with a tubal pregnancy.

25. ④　A variety of incomplete proteins (vegetable proteins) can be combined to provide all the essential amino acids. (3) (PC; PL; HC)

1　The pregnant client should receive at least 60 g of protein in her diet.

2　Strict vegetarians do not drink milk.

3　These are not the most important factors in diet planning; other nutrients also must be provided.

26. ④　The postoperative teaching should include instructing the client to expect shoulder pain, secondary to the insufflated carbon dioxide, for 12 to 24 hours. (2) (PC; PL; WH)

1　There is no need to douche with Betadine postoperatively.

2　Vaginal bleeding is usually not present following laparoscopy.

3　Usual activities should not be resumed until 2 to 3 days postoperatively.

27. ③　Immediate action to prevent excessive bleeding is to massage the fundus until it is firm. This stimulates uterine muscle contraction. (1) (SI; IM; HC)

1　The immediate action is to promote uterine contraction; obtaining the BP would be indicated if a large amount of bleeding occurred or bleeding persisted.

2　This would not be necessary unless bleeding persists after massaging of the uterus.

4　If the uterus does not contract after massage, the nurse should notify the physician, not just observe the bleeding.

28. (4) This indicates that she understands that the well-being of the infant will be established by the testing. (2) (CI; EV; HP)
   1 This noninvasive procedure should not affect uterine musculature.
   2 The baby is not affected by the use of external monitoring.
   3 This test does not require any injections unless nipple stimulation fails.

29. (3) Vitamins A, C, and E and selenium are immune-stimulating nutrients. (2) (HI; IM; HC)
   1 Too much emphasis on the fat-soluble vitamins could result in an inadequate intake of important water-soluble vitamins and minerals.
   2 These have no known effect on natural defenses.
   4 Same as answer 2.

30. (4) Tears in the tentorial membrane cause bleeding into the cerebellum, pons, or medulla oblongata. The respiratory regulation centers are located in the medulla and pons. (3) (DA; AN; HN)
   1 Lethargy would be more indicative of cerebellar injury.
   2 A weak, timorous cry would be more indicative of cardiac or respiratory difficulty; a high-pitched, shrill cry is usually present with CNS difficulty.
   3 Purpura is unrelated to tentorial or other CNS injuries.

31. (3) Axillary lymph nodes are an early site of metastasis and thus are removed. (1) (CC; EV; WH)
   1 Pectoral muscles are not removed in a modified radical mastectomy.
   2 This is not removed in this type surgery; leaving the skin intact improves healing.
   4 The entire breast is removed in this type surgery.

32. (3) Treatment is directed primarily toward reducing the blood pressure and preventing seizures. (1) (DA; PL; HP)
   1 Anxiety may be present, but the blood pressure is the priority problem.
   2 Bleeding is not generally a problem with pregnancy-induced hypertension unless abruptio placentae occurs.
   4 In preeclampsia there is already a decrease in circulating blood volume, which causes hemoconcentration and decreased organ perfusion.

33. (4) The estrogen and progesterone in birth control pills increase the amount of renin, which in turn increases production of angiotensin, a potent pressor substance. (1) (SI; EV; RC)

   1 Dysmenorrhea does not occur.
   2 This is not usually serious; it often indicates a low hormone level; it is corrected by changes in the type of medication prescribed.
   3 Anovulation is the desired effect of oral contraceptives.

34. (3) Reflecting these feelings gives the client the opportunity to express fears and provides a chance to explore the family history. (2) (SC; IM; WH)
   1 Although true, this provides false reassurance.
   2 This supports the client's fears of cancer and blocks communication.
   4 This statement is probing and does not address the client's fears.

35. (3) Pitocin will cause the uterus to contract, preventing hemorrhage; if it is given before the placenta detaches, the placenta can be trapped. (2) (PT; EV; DR)
   1 Pitocin has no analgesic effect.
   2 Relaxation of the uterus is undesirable and would not result in separation of the placenta.
   4 Prolactin, not Pitocin, stimulates milk production.

36. (2) The appropriate nursing action is to assist the family in discussing their feelings about hospitalization and loss of the mother's presence. (3) (SC; IM; EC)
   1 Leaving the hospital while attempts are being made to stabilize her diabetes would be counterproductive.
   3 A social worker is necessary only if abnormal coping is assessed; it is not indicated in this situation.
   4 This is not related to the stated nursing diagnosis; the client also needs to understand the interaction between diet, medication, and exercise.

37. (3) This response reflects the ambivalence about the pregnancy that is typical during the first trimester. (1) (SC; EV; EC)
   1 This is a typical response during the third trimester when the client begins to doubt her ability to be a good parent.
   2 Fantasizing about the infant, its sex, and its future is common during the second trimester.
   4 Expressing fears about the labor and delivery experience and parenting is common during the third trimester.

38. (4) Preparation for parenthood classes should help couples develop realistic expectations of the laboring process, including associated discomfort and ways of dealing with it. (2) (CC; IM; HC)

1 This is untrue and unpredictable; contractions are uncomfortable, but childbirth preparation helps the client cope with discomfort.

2 Clients are taught what to expect and the proper exercises to expedite labor; the focus should not be on pain.

3 There is no way of predicting whether medication will be needed; however, the client should be assured of its availability.

39. ③ Tension in the woman prevents relaxation and has the greatest influence on how she will progress through labor and on her perception of pain. Tension is related to the expectation of pain, which is based on cultural norms and past experiences. (1) (SC; AN; EC)

1 Parity does not play a major role in the woman's perception of pain.

2 Although the woman often becomes more uncomfortable and tired near the end of labor, the length of labor does not play a major role in her perception of pain.

4 Although the difficulty of labor affects the amount of pain, it does not play a major role in the woman's perception of pain.

40. ④ This describes the skin of a postterm baby who has been exposed to amniotic fluid too long. (3) (DF; AS; HN)

1 These are characteristics of a preterm infant.

2 Same as answer 1.

3 These are characteristics of a term infant.

41. ① Intrauterine devices should be inserted during menses because the cervical os is slightly dilated and there is little chance of pregnancy. (3) (DF; IM; RC)

2 An IUD should not be inserted at this time; this is the proliferative phase of the menstrual cycle.

3 An IUD should not be inserted at this time because pregnancy may have occurred.

4 Same as answer 3.

42. ④ Mottling of the skin results from hypothermia in the newborn. (1) (PC; IM; NN)

1 This is not necessary; this is a normal phenomenon that usually indicates falling temperature; the baby requires warming.

2 This is not necessary; the baby requires warming.

3 Feeding will not increase temperature.

43. ③ Dislocation of the hip limits abduction to less than 90 degrees. (2) (CI; AS; HN)

1 It is normal for the legs to be of equal length.

2 Flexion of the hips is not affected by dislocation.

4 This is a normal finding in the newborn; maternal hormones cause loosening of ligaments, which allows abduction to 90 degrees.

44. ② Bladder distention causes uterine displacement, which interferes with uterine contractions and may lead to postpartum hemorrhage. (1) (CI; AS; HC)

1 During normal involution the uterus is not deviated to the right.

3 There is no such thing as second-degree uterine atony.

4 Retained placental fragments often cause bright red bleeding and a boggy uterus; there are no data to support this.

45. ① The nurse should teach the client that the side effects of danazol include weight gain, edema, and breakthrough bleeding. (3) (PT; EV; DR)

2 Frequent urination is not a side effect of danazol.

3 Hypotension is not a side effect of this medication.

4 Clients on danazol do not experience menstrual periods resulting from suppression of both FSH and LH.

46. ② This is correct because the head of the bed is elevated 45 degrees, and this allows for maximum chest expansion for ventilation. (1) (CI; IM; HP)

1 The laboring woman should not assume the supine position because she may develop hypotension because of impeded venous return.

3 Trendelenburg's position would interfere with optimal cardiac function during labor.

4 The head of the bed should be elevated, not flat.

47. ④ Increased intra-abdominal pressure associated with lifting, coughing, or laughing, in conjunction with a relaxed pelvic musculature and a bladder displaced into the vagina, results in an inability to control the urinary stream. (1) (HI; AS; WH)

1 Usually associated with urinary tract infection, bladder tumor, or renal calculi, not with a cystocele or rectocele.

2 Usually associated with pelvic inflammatory disease, endometriosis, or cervical stenosis, not with cystocele or rectocele; client is probably postmenopausal.

3 Usually associated with urinary infection, not with a cystocele or rectocele.

48. ② In pregnancies in which the fetus has a neural tube defect, alpha-fetoprotein leaks into the amniotic fluid, causing abnormally high levels; at 16 to 18 weeks there is a sufficient amount of amniotic fluid present. (3) (CI; AS; HP)

1 This is too early to safely obtain amniotic fluid.

3 This is not the optimal time for diagnostic testing for birth defects; the optimal time for an amniocentesis is when the uterus enters the abdominal cavity, there is sufficient amniotic fluid available, and the fetus is still small.

4 Same as answer 3.

49. ④ This would indicate that the urinary sphincter tone has not been affected by the catheter and urinary retention with overflow is not present. (2) (PC; EV; HP)

1 Although the total amount of urine indicates adequacy of kidney function, it does not reflect sphincter control or the possibility of retention.

2 The absence of bacteria indicates the absence of infection but does not portend the return of urinary function.

3 This indicates retention with overflow; the client urinates small amounts but does not completely empty the bladder.

50. ④ Altering the breast-feeding position may ensure that the entire nipple and as much of the areola as possible is in the infant's mouth; when the infant is latched on the nipple correctly and a finger is used to release suction at the end of a feeding, trauma to the nipple is reduced. (2) (PC; PL; HC)

1 This is contraindicated because it suppresses lactation; also, ice applications of any kind should never be continuous because they can result in tissue damage.

2 This is unnecessary; soreness is common; it usually occurs only at the beginning of a feeding and is temporary until the nipples become accustomed to the infant's sucking.

3 This is usually unnecessary; removing the infant from the breast will result in engorgement, which will increase the discomfort.

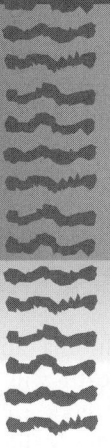

CHAPTER 4

Mental Health Nursing

REVIEW QUESTIONS

EMOTIONAL DISORDERS RELATED TO PHYSICAL HEALTH AND CHILDBEARING (ED)

1. A 3-year-old's parents have been unable to visit since the child was admitted to the hospital. The toddler has become quiet and withdrawn. To best help the child at this time, the nurse should:
1. Bring the child a doll or stuffed animal to cuddle
2. Encourage the child to play games with the other children
3. Assign the same nurse to care for the child whenever possible
4. Contact the child's parents and tell them to come immediately to visit

2. A 3½-year-old begins screaming and kicking when the laboratory technician comes to draw blood. The nurse recognizes that this reaction is primarily a result of the child's:
1. Fear of loss of control
2. Inability to localize pain
3. Fear of intrusive procedures
4. Past experience with this procedure

3. Just prior to a 6-year-old's physical examination by the physician, the nurse can best meet the child's developmental needs by:
1. Allowing the child to handle the examination equipment
2. Explaining exactly what will happen during the examination
3. Having the child talk to a child who has recently had an examination
4. Arranging to have one of the child's parents present during the examination

4. A 9-year-old boy, admitted to the hospital with a fractured femur, has just been told he must stay in the hospital in traction for at least 2 weeks. The nurse finds him crying and unwilling to talk. At this time, the nurse should give the highest priority to:
1. Giving him privacy and allowing him to cry
2. Telling him that his injury will not be permanent
3. Trying to distract him to prevent embarrassment
4. Arranging for him to have a tutor begin immediately

5. A client with a history of hypertension comes to the clinic where the intake physical reveals a blood pressure of 180/102 mm Hg. When the nurse asks whether the client has been taking any medications, the client replies, "I took the pills the doctor prescribed for a few weeks, but I didn't feel any different. So, I decided I'd just take them if I felt sick." The best initial response by the nurse would be:
1. "I'm glad to hear you felt well enough to stop the medication."
2. "You must be quite frightened about having high blood pressure."
3. "You really should try to take your medication; the physician felt it was needed."
4. "I think we should talk to the physician about a plan of treatment you can follow."

6. A male client, admitted to the hospital because of severe rectal bleeding with a history of ulcerative colitis, appears to be an angry, demanding person. One day the nursing assistant tells the nurse, "I've had it with that client and all his demands. I'm not going in there again." The nurse's best response to this statement would be:
   1. "You need to try to be patient with him. He's going through a lot right now."
   2. "I'll talk with him and see whether I can figure out the best way for us to handle this."
   3. "He's frightened and taking it out on the staff. Let's think how we can approach him."
   4. "Just ignore him and get on with the rest of your work. Let someone else take a turn."

7. A client calls out to every nursing staff member who passes by the door and asks them to do or get something. The nurse can best manage this behavior by:
   1. Closing the door to the room so the client cannot see the staff members as they pass by
   2. Assigning one staff member to approach the client regularly and spend time talking with the client
   3. Informing the client that one staff member will come in frequently to see whether the client has any requests
   4. Arranging for a variety of staff members to take turns going into the room to see whether the client has any requests

8. A client having presurgical testing prior to a possible colon resection and colostomy says to the nurse, "If I have to have this surgery, I know my husband will never come near me." The nurse's best initial response would be, "You're:
   1. Probably underestimating his love for you."
   2. Concerned that your husband will reject you."
   3. Wondering about the effect on your sexual relations."
   4. Worried that the surgery will change how others see you."

9. A client requiring surgery because of mitral valve incompetence is admitted to the hospital and states, "I need a new valve, and do an oil change, too!" The most therapeutic response by the nurse would be:
   1. "You really don't need to hide your anxieties."
   2. "You sure came to the right place for a valve job."
   3. "I'm glad to see you're handling the situation so well."
   4. "I'm sure you have a great deal to ask about your surgery."

10. A female client who has been told by her physician that she has untreatable metastatic carcinoma tells the nurse that she believes the physician has made an error and that she does not have cancer and is not going to die. The nurse evaluates that the client is experiencing the stage of death and dying known as:
    1. Anger
    2. Shock
    3. Bargaining
    4. Acceptance

11. A 68-year-old has metastatic carcinoma and has been told by the physician that death will occur within a month or two. The nurse enters the client's room after the physician leaves and finds the client crying. The nurse's action should take into consideration that:
    1. Crying relieves depression and helps the client face reality
    2. Crying releases tension, which frees psychic energy for coping
    3. Nurses should not interfere with a client's behavior and defenses
    4. Accepting a client's crying maintains and strengthens the nurse-client bond

12. A terminally ill 76-year-old client is very quiet and unwilling to have visitors. During the initial contact with the client, the nurse should:
    1. Attempt to understand what the death and dying process means to the client
    2. Avoid talking about the client's condition unless the client initiates the discussion
    3. Ascertain how much pain the client is experiencing and what medications have been ordered
    4. Explore the extent to which the client is aware of the prognosis and the client's feelings about the situation

13. A female client terminally ill with cancer says to the nurse, "My husband is avoiding me. He doesn't love me anymore because of this damn tumor!" The nurse's most appropriate response would be:
    1. "What makes you think he doesn't love you?"
    2. "Avoidance is a defense; he needs your help to cope."
    3. "He is probably having difficulty dealing with your illness."
    4. "You seem very upset. Tell me how your husband is avoiding you."

14. The major improvement in the body image of clients following early fitting with prostheses after amputation is usually related to:
    1. Their improved functional abilities

2. The feeling that they look more "whole"
3. The acceptance they receive from others
4. The fact that something is being done to help

15. A pregnant type 2 diabetic client who has a history of three spontaneous abortions is scheduled for an oxytocin challenge test. Before the test she begins to cry when answering the nurse's questions about her previous pregnancies. She states, "I know it's my diabetes. This baby will never live. It's all my fault." The nurse's best response would be:
1. "I understand that this must be very stressful for you."
2. "Diabetes and pregnancy are difficult to deal with together."
3. "This baby will live because it is being very closely monitored."
4. "I know you're worried, but getting upset can alter test findings."

16. The nurse identifies that a client who has had a myocardial infarction is struggling with an alteration in self-concept. The nurse intervenes to promote client autonomy. The behavior that would demonstrate an increase in client autonomy would be when the client:
1. Actively participates in planning self-care
2. Verbalizes realistic expectations of caregivers
3. Discusses necessary life-style changes with family members
4. States the conditions for recovery following a myocardial infarction

17. Before signing a consent for a total laryngectomy, a client asks, "Nurse, the doctor says he's going to take part of my throat out and put a hole in my neck to breathe. Will I be able to talk like before?" The nurse's best response would be:
1. "Would you like to talk to the doctor again to answer your questions?"
2. "We have lots of clients with this operation. You'll talk again. You'll see."
3. "Not like before but there is nothing to worry about. This procedure is done often."
4. "Why don't you tell me what you know about your surgery. You seem very concerned."

PERSONALITY DEVELOPMENT (PD)

18. One afternoon the nurse overhears a young female client having an argument with her boyfriend. A while later the client complains to the nurse that dinner is always late and the meals are terrible. The nurse recognizes that the defense mechanism the client is using is:

1. Projection
2. Dissociation
3. Displacement
4. Intellectualization

19 Although upset by a young client's continuous complaints about all aspects of care, the nurse ignores them and attempts to divert the conversation. Immediately following this exchange with the client, the nurse discusses with a friend the various stages of development of young adults. The defense mechanism the nurse is using is:
1. Substitution
2. Sublimation
3. Identification
4. Intellectualization

20. During an interview with the parents of an adolescent female, the nurse notices that her father continually defends and makes excuses for all of his daughter's actions whereas her mother seems to feel her daughter is just lazy and that there is nothing wrong with her that she couldn't fight with some effort. The nurse recognizes that the dynamics used by this family is known as:
1. Coalitions
2. Resignation
3. Scapegoating
4. Reaction formation

21. The nurse is aware that according to Erikson, a young child's increased vulnerability to anxiety in response to separations or pending separations from significant others results from failure to complete the developmental task called:
1. Trust
2. Identity
3. Initiative
4. Autonomy

22. The nurse knows that Erikson identified the developmental tasks of the preschool child from 3 to 5 years as:
1. Initiative vs guilt
2. Industry vs inferiority
3. Breaking away vs staying at home
4. Sexual impulses vs psychosexual development

23. According to Erikson, a young adult must accomplish the tasks associated with the stage known as:
1. Initiative vs guilt
2. Intimacy vs isolation
3. Industry vs inferiority
4. Generativity vs stagnation

24. A 65-year-old who immigrated from Cuba 25 years ago is admitted to the hospital with a history of depression. The client, who speaks little English and has few outside interests since retiring, states, "I feel useless and unneeded." According to Erikson, the client is in the developmental stage of:
    1. Initiative vs guilt
    2. Integrity vs despair
    3. Intimacy vs isolation
    4. Identity vs role confusion

25. A 7-year-old boy wakes up crying because he has wet his bed. It would be most appropriate for the nurse to:
    1. Allow him to change his bed and pajamas
    2. Change his bed while he changes his pajamas
    3. Take him to the bathroom and change his pajamas
    4. Remind him that he must call for the nurse the next time

26. The mother of an 18-year-old comes to the clinic at the local mental health center. She is extremely upset because her son has returned from his freshman year at college and is uncontrollable. He takes his brother's clothing, comes in at all hours, and refuses to get a job. Sometimes he is happy and outgoing, and other times he is withdrawn. The mother asks why her son is like this and speculates that college has done this to him. While contemplating this situation, the nurse understands that adolescents are usually:
    1. Anxious and unhappy
    2. Angry and irresponsible
    3. Impulsive and self-centered
    4. Hyperactive and self-destructive

27. According to Erikson, an individual who fails to master the maturational crisis of adolescence will most often:
    1. Experience role confusion
    2. Be interpersonally isolated
    3. Rebel at all parental orders
    4. Use drugs and alcohol to escape

28. A constructive and lengthy method of confronting the stress of adolescence and preventing a negative and unhealthy developmental outcome is:
    1. Role experimentation
    2. Adherence to all peer standards
    3. Sublimation through school work
    4. Development of dependency on parents

29. The parents of an overweight adolescent female tell the nurse that they are concerned that their daughter feels inferior to her sister who is an attractive, successful college senior. They ask the nurse what they can do about this problem. The nurse should:
    1. Suggest that they seem to be creating a problem where none exists
    2. Tell them to avoid talking about their older child's accomplishments
    3. Suggest they give this adolescent recognition for her own strong points
    4. Advise them to tell the adolescent to view her sister's success as a challenge

30. The nurse, along with an adolescent and the adolescent's parents, set bolstering the adolescent's self-esteem as a high-priority goal. The nursing action that would contribute to the achievement of this goal would be:
    1. Urging the adolescent to join a neighborhood social group
    2. Supporting the adolescent's interest in enrolling in a baby sitting course
    3. Suggesting that the mother give the adolescent lots of hugs and cuddling
    4. Encouraging the adolescent to talk about feelings of pride in successful siblings

31. The nurse would evaluate that the plan for bolstering an overweight adolescent's self-esteem was effective when, 3 months later, the adolescent's mother reports that the adolescent:
    1. Asks her to prepare a favorite dessert
    2. Seems to be doing average work in school
    3. Joined a dirt bicycle club that meets at the school
    4. Imitates an older sibling's manner of speech and dress

32. According to Erikson, a person's adjustment to the period of senescence will depend largely on the adjustment the individual made to the developmental stage of:
    1. Intimacy vs isolation
    2. Industry vs inferiority
    3. Generativity vs stagnation
    4. Identity vs identity diffusion

33. When helping the older adult (age 65 to 75 years) successfully complete Erikson's task of this stage, the nurse should assist the client to:
    1. Invest creative energies in promoting social welfare
    2. Redefine a role in society and offer something of value
    3. Look to recapture opportunities that were not started or completed
    4. Feel a sense of satisfaction when reflecting back on past achievements

34. The nurse's role in maintaining or promoting the health of the older adult should be based on the principle that:
1. There is a strong correlation between successful retirement and good health
2. Some of the physiologic changes that occur as a result of aging are reversible
3. Thoughts of impending death are frequent and depressing to most older adults
4. Older adults can better accept the dependent state chronic illness often causes

35. When planning care for an older client, the nurse is aware that normal aging has little effect on a client's:
1. Sense of taste or smell
2. Gastrointestinal motility
3. Muscle or motor strength
4. Ability to handle life's stresses

DISORDERS FIRST EVIDENT BEFORE ADULTHOOD (BA)

36. The nurse is aware that a child's emotional symptoms usually occur as a result of:
1. Rejection by the parents
2. Family pathologic factors
3. Authoritarian parenting style
4. Overbearing overprotectiveness

37. The nurse is aware that a child experiencing emotional problems would probably exhibit:
1. Impaired ability to reality-test
2. Passive and deliberate behavior
3. Overinvolvement with peer group
4. A mild to moderate level of anxiety

38. When assessing disturbed children, the clue that the nurse would find most indicative of severe emotional problems would be the child's:
1. Physical complaints
2. Behavioral outbursts
3. Poor school performance
4. Lack of response to the environment

39. A school-age child is brought to the clinic by the mother, who states, "Something is very wrong. My child never seems happy and refuses to play." When assessing this child for depressed behavior, the nurse should initially begin with the statement:
1. "Tell me about yourself."
2. "Let's talk about what you do after school."
3. "Can you tell me what is making you so unhappy?"
4. "Why does your mother think that you are unhappy?"

40. When implementing a tertiary preventive program for the mentally challenged, the nurse should:
1. Teach mentally challenged children how to feed themselves
2. Refer children for evaluation if they fail to meet developmental milestones
3. Encourage the use of birth control methods by women who are mentally challenged
4. Utilize the Denver Developmental Screening Test to evaluate children attending well-child clinics

41. A young child has a history of frequent temper tantrums. The mother asks how to limit this acting-out behavior. The nurse's most therapeutic response would focus on:
1. Restraining the child whenever a tantrum begins
2. Moving the child to a quiet area before the tantrum begins
3. Telling the mother to ignore the tantrum whenever possible
4. Asking the physician to order medication for behavioral control

42. With the diagnosis of a possible pervasive developmental autistic disorder, the nurse would find it most unusual for a 3-year-old to demonstrate:
1. Ritualistic behavior
2. An interest in music
3. An attachment to odd objects
4. A responsiveness to the parents

43. A 7-year-old is brought to the clinic by her mother, who tells the nurse that her child has been having trouble in school, has difficulty concentrating, and is falling behind in her school work since she and her husband separated 6 months ago. The mother reports that lately her daughter has not been eating her dinner and she often hears her crying in her room. The nurse realizes that the child:
1. Feels different from her classmates
2. Is working through her feelings of loss
3. Would probably be happier living with her father
4. Probably blames herself for her parents' breakup

44. When assessing the mental status of a 7- or 8-year-old child, it is most important for the nurse to:
1. Engage the child in a discussion about feelings
2. Listen to the parents' description of the child's behavior
3. Compare the child's functioning from one time to another
4. Use direct questions to determine the child's mental ability

45. An only child, who lives with the mother, begins demonstrating school and emotional problems after the parents' marital breakup. It is decided that the child would probably benefit most from family therapy. The nurse plans that the first group session will include:
    1. The parents
    2. The mother and the child
    3. The parents and the child
    4. The mother, the child, and the child's teacher

46. A 10-year-old child, who has a history of school failure and destructive acting out, is admitted to a child psychiatric unit with the diagnosis of attention deficit disorder with hyperactivity. The youngest of three children, the child is identified by both the parents and the siblings as the family problem. The parents tell the nurse that the child's behavior has resulted in severe marital problems. The nurse would be correct in identifying the family's pattern of relating to the child as:
    1. Controlling
    2. Patronizing
    3. Scapegoating
    4. Overburdening

47. To help a disturbed, acting-out child develop a trusting relationship, the nurse should:
    1. Inquire as to child's feelings about the parents
    2. Implement 30-minute one-to-one interactions daily
    3. Offer periodic support and emphasize safety in play activities
    4. Initiate limit-setting and explain the rules that must be followed

48. The nurse knows that, according to Erikson, the school-age child from 6 to 12 years old is in the latency period, which is characterized by:
    1. Initiative vs guilt
    2. Industry vs inferiority
    3. Breaking away vs staying at home
    4. Sexual urges vs psychosexual development

49. A 6-year-old recently started school but has been refusing to go for the past 3 weeks. The nurse is aware that an appropriate intervention for this child would be to:
    1. Delay the return to school for several months
    2. Enroll the child in a special education program
    3. Convince the child that school is a place to have fun
    4. Develop a behavior modification program with the child

50. The school nurse is requested to present an inservice program on attention deficit disorders to the teaching staff. The nurse should emphasize that:
    1. This problem is always evident before 4 years of age
    2. This disorder occurs more frequently in lower socioeconomic groups
    3. Children with this disorder sleep more than others because of their high activity level
    4. It is estimated that this problem affects around 5% to 10% of the school-age population

51. An acting-out, hyperactive 9-year-old boy is started on a behavior modification program. One day he is playing a game with his peers and becomes frustrated when he begins to lose. He begins to kick the other children under the table and to call them names. The nurse, using the most appropriate behavior modification technique, should:
    1. Require the child to have a time-out and regain control
    2. Negatively reinforce the child by taking two of his tokens
    3. Ignore the child's behavior with the intent of extinguishing it
    4. Engage the child in a conversation about good sportsmanship

52. After 1 month in a special school, a hyperactive 10-year-old is asked to leave the group therapy session because of disruptiveness. The child begins to cry when being led out. The nurse's best approach at this time would be to:
    1. Send the child for a time-out in a quiet room
    2. Engage the child in a talk about the school day
    3. Offer an interpretation of the child's self-defeating behavior
    4. Provide nurturance by sharing a snack and a glass of milk with the child

53. A hyperactive, self-destructive child is to be discharged from an inpatient setting in a couple of weeks. In preparation for the child's discharge, it is most important for the nurse to plan to:
    1. Establish, maintain, and/or enforce limits on behavior
    2. Meet with the child's teacher to review the child's needs
    3. Help the child begin to terminate relationships with the staff
    4. Schedule a home visit and a community trip with the child's family

ANXIETY, SOMATOFORM, FACTITIOUS, AND DISSOCIATIVE DISORDERS (AX)

54. The nurse can evaluate that a client has successfully achieved the long-term goal of mobilizing effective coping responses when the client states, "When I feel myself getting anxious I will:
1. Use meditation and relaxation exercises."
2. Get involved in some type of quiet activity."
3. Avoid the situation that precipitated the anxiety."
4. Carefully examine what precipitated my anxiety."

55. When the nurse considers a client's placement on the continuum of anxiety, a key in determining the degree of anxiety being experienced is the client's:
1. Memory state
2. Creativity level
3. Perceptual field
4. Delusional system

56. An elderly client who lives alone tells the nurse at the community health center, "I really don't need anyone to talk to, the TV is my best friend." The nurse recognizes that the client is using the defense mechanism known as:
1. Denial
2. Projection
3. Sublimation
4. Displacement

57. The symptoms that distinguish posttraumatic stress disorders from other anxiety disorders are:
1. Lack of interest in family and others
2. Reexperiencing the trauma in dreams or flashbacks
3. Avoidance of situations and certain activities that resemble the stress
4. Depression and a blunted affect when discussing the traumatic situation

58. A client with an anxiety disorder is hospitalized. The nurse realizes that an environment conducive to reducing emotional stress and providing psychologic safety is one in which:
1. All the client's needs are met
2. Needs are a primary concern
3. Realistic limits and controls are set
4. The physical environment is kept in order

59. A client comes to the hospital because of intense feelings of unrest, inability to sleep, and frequent episodes of panic. The client tells the nurse, "I admitted myself because I think I'm going crazy." The nurse should recognize the client's remark as a:
1. Plea for support
2. Symptom of depression
3. Reflection of insightfulness
4. Test of the nurse's trustworthiness

60. A nurse is accompanying a client with a diagnosis of anxiety disorder who is pacing the halls and crying. When the client's pacing and crying increase, the nurse suddenly feels uncomfortable and experiences a strong desire to leave. The probable reason for this feeling is:
1. An empathic communication of anxiety
2. A desire to go off duty after a busy day
3. A fear of the client becoming assaultive
4. An inability to tolerate any more bizarre behavior

61. The nurse is aware that nursing intervention for clients with anxiety disorders should include:
1. Promoting the verbalization of feelings by the clients
2. Promoting the suppression of anger/hostility by the clients
3. Insisting that the clients accept the role of psychologic factors
4. Limiting the involvement of the clients' families during the acute phase

62. A client is pacing the floor and appears extremely anxious. The nurse approaches in an attempt to alleviate the client's anxious feelings. The most therapeutic question by the nurse would be:
1. "Are you feeling upset right now?"
2. "Would you like me to walk with you?"
3. "Shall we sit and talk about your feelings?"
4. "Would you like to go to the gym and work out?"

63. Without knocking, the nurse enters the room of a young male client with the diagnosis of panic disorder and observes him masturbating. The nurse should:
1. Say, "Excuse me," and leave the room
2. Tactfully assess why he needs to masturbate
3. Pretend nothing was seen and carry out whatever task needs to be done
4. Explain in a calm, quiet manner that his behavior is inappropriate in the hospital

64. A 15-year-old client with the diagnosis of panic disorder jumps when spoken to, complains of feeling uneasy, and states, "It's as though something bad is going to happen." It would be most therapeutic for the nurse to:
1. Be physically present
2. Encourage the client to communicate with the staff
3. Allow the client to set the parameters for the interaction
4. Help the client to understand the cause of the feelings described

65. A client with a diagnosis of panic disorder, who had a panic attack on the previous day, says to the nurse, "That was a terrible feeling I had yesterday. I'm so afraid to talk about it." The nurse's most therapeutic response would be:
    1. "It's best that you try to talk about it."
    2. "OK, we don't have to talk about it."
    3. "What were you doing yesterday when you first noticed the feeling?"
    4. "I understand, but don't be concerned; that feeling probably won't come back."

66. A female client is admitted to an acute psychiatric unit with a diagnosis of panic disorder. During the initial assessment phase the nurse should focus on:
    1. Learning about the client's home life to facilitate planning care
    2. Reducing the client's level of anxiety so that further interviewing can occur
    3. Helping the client identify the source of her anxiety so the source can be avoided
    4. Suggesting to the client that she rest for awhile and informing her that they will talk later

67. When talking with a female client who displays many of the emotional and physiologic symptoms associated with a panic disorder, the nurse should:
    1. Describe for her the possible reasons for her anxiety
    2. Use short simple sentences and a firm authoritative voice
    3. Ask many questions, because she probably is not going to volunteer much information
    4. Suggest that she refrain from crying, because most of the time crying makes matters worse

68. The nurse plans to teach a client to use more healthy coping behaviors that can be consciously used to reduce anxiety, including:
    1. Eating, dissociation, fantasy
    2. Sublimation, fantasy, rationalization
    3. Repression, intellectualization, smoking
    4. Exercise, talking to friends, suppression

69. A male client asks one of the female staff members for a date. It is most likely that the:
    1. Staff member may have led the client on
    2. Client misinterpreted the staff member's friendliness
    3. Client may be trying to protect a threatened sexual identity
    4. Staff member may have been acting unprofessionally toward the client

70. The nurse could evaluate that the staff's approach to setting limits for a demanding, angry client was effective if the client:
    1. No longer calls the nursing staff for assistance
    2. Understands the reasons why frequent calls to the staff were made
    3. Apologizes for disrupting the unit's routine when something is needed
    4. Discusses concerns regarding the emotional condition that required hospitalization

71. The change in an elderly client's behavior that would indicate that the nurse should reassess the client's needs and current plan of care, which was attempting to maintain the client's independent living style, would be the development of:
    1. Confusion
    2. Hypochondriasis
    3. Additional complaints
    4. Increased socialization

72. During the first meeting of a therapy group, the members become quite uncomfortable. The nurse notes frequent periods of silence, tense laughter, and a good deal of nervous movement in the group. The nurse would assess that these responses:
    1. Require active leader intervention to relieve symptoms of obvious stress
    2. Indicate unhealthy group processes with an unwillingness to relate openly
    3. Are expected group behaviors because relationships are not yet established
    4. Should be pointed out and discussed so members will not become too uncomfortable

73. During the working phase of a therapy group, a female group member becomes tearful after being told by another member that she needs to change her behavior. The nurse evaluates that the client:
    1. Has had her feelings hurt by this response
    2. Feels too fragile to be challenged at this time
    3. Is angry about the confrontation with another member
    4. Has been depressed about this aspect of her behavior

CRISIS SITUATIONS (CS)

74. A crisis can best be defined as:
    1. An imbalance of life
    2 A threat to homeostasis
    3. The perception of the problem by the client
    4. A situation requiring help other than personal resources

75. When working with families encountering problems, it is most important that the nurse have a:
    1. Good memory for details
    2. Common social background

3. Warm nature and loving personality

4. Sense of self and empathy for others

76. An adolescent client seeks help at a crisis intervention clinic. The client relates, "I dropped out of college because the instructors were dumb and the kids acted like babies. I was a psychiatric aide for 6 weeks. I got into trouble because the staff's thinking was archaic. I tried waiting on tables but got fired. The boss said I was nasty to the customers. They were the nasty ones. Now I can't even pay my rent. If people were nicer, I wouldn't be in this mess." In relation to crisis theory, this client's stressful events can be seen as:

1. Experiential

2. Age-related and frequent

3. Usually noncrisis producing

4. Situational and maturational

77. A 60-year-old client complains of headaches, restlessness, and insomnia. During an interview, the nurse learns that the symptoms began 3 months ago after the client was forced into early retirement. The nurse recognizes that the client is probably experiencing:

1. A social crisis

2. A situational crisis

3. An economic crisis

4. A developmental crisis

78. Situational crises are usually resolved in a time period of:

1. 1 to 4 days

2. 2 to 3 weeks

3. 1 to 2 months

4. 2 to 6 months

79. According to crisis theory, the minimal long-term goal in crisis intervention is:

1. Relief of acute symptoms

2. Relief of panic-level anxiety

3. Restoration of the original functioning level

4. Reorganization and reordering of the personality

80. The most critical factor for the nurse to determine during crisis intervention is the client's:

1. Developmental history

2. Available situational supports

3. Underlying unconscious conflict

4. Willingness to restructure the personality

81. When intervening in a crisis situation, the nurse is initially concerned with:

1. What the precipitating factor was

2. How the individual is affecting others

3. How the client will deal with successive crises

4. Whether the individual can go back to daily activities

82. The nurse suggests a crisis intervention group to a client experiencing a developmental crisis. These groups are successful because the:

1. Client is encouraged to talk about personal problems

2. Crisis group supplies a workable solution to the client's problems

3. Crisis intervention worker is a psychologist and understands behavior patterns

4. Client is assisted to investigate alternative approaches to solving the identified problem

83. When talking with a client in crisis, the crisis intervention nurse should:

1. Restate the problem, putting it in the proper perspective

2. Respect the client and involve client in deciding what will be done and how it will be done

3. Explain to the client that the center has helped many other people with the same problem

4. Explore the client's religious and cultural beliefs so the instructions are within the client's value system

84. A young college student tells the nurse in the school's health service that his girlfriend's period is late and they both think she is pregnant. The client, with a broad smile on his face, states loudly and angrily, "If she is pregnant I will drop out of school, marry her, and get a full-time job." The nurse's best initial assessment of the client's verbal and nonverbal behavior would be that they are:

1. Uniform

2. Consistent

3. Appropriate

4. Incongruent

85. When talking to the nurse about his decision to drop out of school and marry his girlfriend who is pregnant, a young college student says, "It's really the best decision. It is important for a child to have two parents." The nurse recognizes that the client is using the defense mechanism known as:

1. Projection

2. Introspection

3. Displacement

4. Intellectualization

86. A 24-year-old secretary, pregnant for the first time, receives a letter from her boyfriend with a check for $500 and the news that he has left. The client is very upset, feels at the end of her rope, and calls the crisis intervention center for help. The nurse recognizes that the client is experiencing a crisis because:
    1. The client is under a great deal of stress
    2. The client is going to have to raise her child alone
    3. The client's boyfriend left her when she was pregnant
    4. The client's past methods of adapting are ineffective for this situation

87. A single, pregnant client who is attending a crisis intervention group has decided to go through with the pregnancy and keep the baby. Now the crisis intervention nurse's primary responsibility is to:
    1. Support the client for making a wise decision
    2. Explore other problems the client may be experiencing
    3. Make an appointment for the client to see a physician for prenatal care
    4. Provide information about other health resources where the client may receive additional assistance

88. A single, pregnant client, who has been attending a crisis intervention clinic, has decided to keep the baby and is looking forward to motherhood. The nurse recognizes that the decision to attend prenatal child-care classes is an example of:
    1. Intrinsic motivation
    2. Extrinsic motivation
    3. Operant conditioning
    4. Behavior modification

89. A couple in their late twenties are very happy to be expecting a child. The client has a cesarean delivery and is disappointed that the baby is a boy. The husband is pleased. On the second day after delivery, when the baby is brought to the client, she seems far away, as though she is daydreaming. The nurse calls the client's name and the client states, "I don't have a baby." She proceeds to rock her empty arms. The nurse recognizes that the precipitating factor for the client's emotional reaction is probably her:
    1. Alteration in role
    2. Husband's behavior
    3. Religious upbringing
    4. Desire for a baby girl

90. It is doubtful that a severely mentally challenged adult resident of a community home, who has had four pregnancies in 2 years, is able to exercise informed consent for sterilization. The nurse should be aware that the procedure cannot be performed without approval from the:
    1. Court
    2. Next of kin
    3. Legal guardian
    4. Court and legal guardian

91. When obtaining informed consent for sterilization from a mentally challenged adult client, the nurse must be sure that the:
    1. Parent or guardian signs the permit
    2. Client comprehends outcome of the procedure
    3. Client is fully able to explain what the procedure entails
    4. Parent or guardian has encouraged the client to make the decision

92. Child abuse is suspected in a 3-year-old girl admitted to the hospital with many poorly explained injuries. During a conversation, the statement by the mother that would further this suspicion would be:
    1. "When I get angry, I take her for a walk."
    2. "I send her to her room alone when she misbehaves."
    3. "I make her stand in the corner when she doesn't eat her dinner."
    4. "The other children were no problem; this one is stubborn and whiny."

93. A mother of four is remanded to the psychiatric unit by the court for observation. She was arrested and charged with abusing her 2-year-old son, who is in the pediatric intensive care unit in critical condition with a fractured skull and other injuries from a beating. When approaching the client, the nurse should expect the client to:
    1. Deny beating her son
    2. Express concern for her son
    3. Ask where her three other children are
    4. Avoid talking about the situation completely

94. When speaking with a mother accused of child abuse, the nurse should expect her to:
    1. Attempt to rationalize and explain her behavior
    2. Offer a detailed explanation of how her son was injured
    3. Reveal an overwhelming feeling that her children are worthless
    4. Ask how she may get permission to visit her son on the pediatric unit

95. One morning before work a nurse becomes very angry with her daughter, and both mother and daughter end the argument in tears. At the hospital the nurse is assigned to care for a mother accused of

severely beating her young child. The nurse has difficulty spending time with the client and tells another nurse about the argument and states, "This client makes me feel very uncomfortable today." The best explanation of what is happening with the nurse is that she:

1. Has identified with a client of the same sex
2. Would have difficulty caring for any client today
3. Is beginning to question her own potential for abuse
4. Is experiencing guilt that is causing her to be ineffective

96. The nurse may best assist an abusing parent to alter behavior toward an abused 2-year-old child by helping the client to:

1. Learn what behavior is appropriate for a 2-year-old child
2. Learn appropriate ways of punishing the child's inappropriate behavior
3. Identify the specific ways in which the child's behavior provokes frustration
4. Ignore the child's negative, nondestructive behavior and support acceptable behavior

97. A nurse on the pediatric unit is assigned to care for a 2-year-old child with a history of abuse. The nurse should expect the child to:

1. Smile readily at anyone who enters the room
2. Be wary of physical contact initiated by anyone
3. Pay little attention to the nurse standing at the bedside
4. Begin to cry and scream as the nurse nears the bedside

98. When the physician examines the genital area of a child suspected of being abused, the nurse can be most therapeutic by:

1. Explaining the procedure and remaining with the child during the examination
2. Telling the child that the doctor wants to see if there is "anything wrong down there."
3. Helping the mother explain the examination and the findings in terms the child will understand
4. Asking whether the child would prefer the nurse or the mother to be present during the examination

99. A young child suspected of being sexually abused says to the nurse, "Did I do something bad?" The nurse's most therapeutic reply would be:

1. "Who said you did something bad?"
2. "What do you mean, something bad?"
3. "Do you think that you did something bad?"
4. "I'm not sure I would say it was something bad?"

100. A recently married 22-year-old is brought to the trauma center by the police. She had been robbed, beaten, raped, and sodomized. The client, although very anxious and tearful, appears in control. The physician orders diazepam (Valium) 5 mg po prn for agitation. The nurse should administer this medication when the:

1. Client requests something to calm her
2. Client's crying and trembling seem to increase
3. Physician is ready to do a vaginal examination
4. Nurse determines the client's anxiety is increasing

101. The husband of a rape victim arrives at the hospital after being called by the police. After reassuring him about his wife's condition, the nurse should give priority to:

1. Discussing with him his own feelings about the situation
2. Calling the rape counselor in to immediately meet with the wife
3. Helping him to understand how his wife feels about the situation
4. Making him comfortable until the physician has completed examining his wife

102. A 13-year-old female is brought to the high school health office by two of her friends, who state, "We think she just took a handful of pills." The adolescent appears alert and refuses to speak. The school nurse's initial response should be to:

1. Ask the adolescent if she took any pills
2. Ask her friends where the adolescent got the pills
3. Call the rescue squad to stand by for an emergency
4. Call the adolescent's parents and tell them they must come immediately

103. The biggest problem for an elderly female client to deal with immediately after the death of her husband will probably be her:

1. Anger
2. Finances
3. Loneliness
4. Estrangement

104. A male client is brought to the psychiatric emergency room after attempting to jump off a bridge. The client's wife states that he lost his job several months ago and has been unable to get another job. The primary nursing intervention at this time would be to assess for:

1. A past history of depression
2. Feelings of excessive failure
3. Current plans to commit suicide
4. The presence of marital difficulties

105. In response to a parent's question about childhood suicide, the nurse's most appropriate response would be:
    1. "Children do not have readily available means to kill themselves."
    2. "Suicide threats and gestures in children should be taken seriously."
    3. "Children younger than age 6 may threaten but do not attempt suicide."
    4. "Suicide in young children is manipulative rather than self-destructive acting out."

106. A 7-year-old has been diagnosed as having acute myelogenous leukemia. The physician has discussed the diagnosis and prognosis with the parents. While the parents are sitting in the lounge after visiting their child, they have a severe argument over something trivial. The nurse should help them recognize that they are using the defense mechanism of:
    1. Denial
    2. Projection
    3. Displacement
    4. Compensation

107. An 8-year-old with a terminal illness is demanding of the staff. The child asks for many privileges that other children on the unit do not have, such as staying up later to watch TV and eating candy. The staff know the child does not have long to live. The nurse can best help the staff cope with the child's demands by encouraging them to:
    1. Give as many extra treats as possible since the child is dying
    2. Give the child some extra treats so they will feel less anxiety after the child dies
    3. Set reasonable limits to help the child and the family become more secure and content
    4. Recognize that the dying child has unique needs, and special privileges can provide the necessary security

108. When the parents visit their hospitalized child, the child continues to play and ignores their presence. The parents are extremely disturbed by the child's reaction to them. The nurse informs them that this behavior is common among hospitalized children and tells them this is called:
    1. Denial
    2. Undoing
    3. Repression
    4. Sublimation

109. A 7-year-old child dies after an explosion at school. The parents arrive at the hospital a few minutes later and are told what had happened. The parents ask the nurse whether they can see their child. The best response by the nurse would be:

    1. "It's best to wait a while."
    2. "I will take you in to see your child now."
    3. "Would you like to wait until the physician can be with you?"
    4. "It will be less traumatic if you wait to see your child at the funeral home."

110. A mother whose daughter is killed in a school bus accident tells the nurse that her daughter was just getting over the chickenpox and did not want to go to school, but she insisted that she go. The mother cries bitterly and says her child's death is her fault. The nurse should realize that perceiving a death as preventable will most often influence the grieving process in that:
    1. The loss may be easier to understand and to accept
    2. Bereavement may be of greater intensity and duration
    3. The grieving process may progress to a psychiatric illness
    4. It causes the mourner to experience a pathologic grief reaction

111. The initial nursing intervention for the significant others during the shock phase of a grief reaction should be focused on:
    1. Staying with the individuals involved
    2. Mobilizing the individuals' support systems
    3. Directing the individuals' activities at this time
    4. Presenting full reality of the loss to the individuals

112. Shortly after the death of her husband following a long illness, the wife visits the mental health clinic complaining of malaise, lethargy, and insomnia. The nurse, knowing that it is most important to help the wife cope with her husband's death, should attempt to determine the:
    1. Age of the wife
    2. Timing of the husband's death
    3. Socioeconomic status of the client
    4. Adequacy of the wife's support system

113. The nurse understands that an individual will probably have the greatest difficulty with the grieving process if the relationship with the significant other was:
    1 Loving
    2. Long-term
    3. Ambivalent
    4. Domineering

114. When an individual successfully completes the grieving process after the death of a significant other, the individual will be able to:
    1. Accept the inevitability of death
    2. Go on with life and forget the past

3. Remember the significant other realistically

4. Focus mainly on the good qualities of the person who died

115. The nurse discusses the plan of care with a depressed client whose husband has recently died. The nurse recognizes it would be most helpful to:

1. Involve the client in group outdoor games each morning

2. Encourage the client to talk about and plan for the future

3. Encourage the client to interact with male clients and the staff

4. Talk with the client about her husband and the details of his death

116. A depressed client, whose spouse recently died, attends an in-patient group therapy session in which a nurse is a co-leader. Toward the end of the session another client talks about being divorced and the resulting feelings of abandonment. As the members are leaving the session, the nurse notices that tears are running down the depressed client's face. Considering this client's depressed state, the nurse should:

1. Ask another client to stay and spend time talking with the client

2. Ask the group members to return and discuss this client's feelings

3. Observe the client's behavior carefully during the next several hours

4. Go to the client's room and encourage a discussion of thoughts and feelings

117. An elderly female comes to the mental health clinic. Assessment reveals that the client feels depressed and vaguely anxious and is unable to sleep at night. The client states, "I haven't felt right since my husband died 8 months ago." The nurse makes the nursing diagnosis of dysfunctional grieving associated with the loss of the husband. The nurse uses this diagnosis because of the client's:

1. Inability to talk about her loss

2. Difficulty in expressing her loss

3. Inability to sleep and the presence of symptoms of depression

4. Prolonged period of grief and mourning after her husband's death

118. A client with an inoperable occipital lobe tumor has been experiencing rather frightening visual hallucinations, especially when alone. The nurse can best help the client cope with these hallucinations by planning to:

1. Move the client to a four-bed room closer to the nurse's station

2. Suggest that the client turn on the radio or television when alone

3. Have family or friends remain with the client until hallucinations stop

4. Suggest that the client not be alone and work out a schedule for visitors

119. The nurse recognizes that a characteristic behavior often demonstrated in the initial stage of a client's coping with dying often includes:

1. Criticizing medical care

2. Sleeping for long periods

3. Asking for additional medical consultation

4. Ringing the call light as soon as the nurse leaves

120. The wife of a client who is dying tells the nurse that, although she wants to visit her husband daily, she can only visit twice a week because she works and has to take care of the house and their cat and dog. The nurse assesses that the wife's statement demonstrates the use of the defense mechanism known as:

1. Projection

2. Sublimation

3. Compensation

4. Rationalization

121. The nurse recognizes that at this time to assist a couple to deal with their feelings about the husband's terminal illness, it would be important to:

1. Assist the couple to express their feelings about his terminal illness to each other

2. Refer the husband to the psychotherapist for assistance in dealing with his anger about death

3. Place the couple in a couple's therapy group that deals with the terminal illness of one partner

4. Encourage the couple to verbalize their feelings to a therapist during their individual therapy sessions

122. The husband of a client who is dying tells the nurse that he knows that his wife is asking the nurses to leave her pain medication on her bedside table and fears she is saving it up for a suicide attempt. The nurse knows that many of the staff members have mixed feelings about the client's terminal status and prolonged pain. The nurse uses an approach that is ethically sound by:

1. Reporting the information to the supervisor and letting the supervisor handle it

2. Speaking to all of the nurses and telling them not to leave the medication at the bedside

3. Asking the head nurse to handle the problems of the client's medication and the staff's feelings

4. Suggesting a nursing conference be held to discuss staff feelings as well as the medication problem

123. An elderly female client, whose husband had been ill for a prolonged period of time, is visited by her husband's hospice nurse several weeks after the funeral for her husband. The nurse recognizes that the wife's biggest problem at this time will probably be her:
1. Loneliness and feelings of isolation
2. Anger at the husband for abandoning her
3. Financial worries about maintaining her life-style
4. Guilt over feelings of relief that her husband has died

124. The grieving wife of a client who has just died says to the nurse, "We should have spent more time together. I always felt the children's needs came first." The nurse recognizes that the wife is experiencing:
1. Displaced anger
2. Normal feelings of guilt
3. Shame for past behaviors
4. Ambivalent feelings about her husband

125. The nurse is aware that tranquilizers are rarely ordered for individuals undergoing acute grief because they:
1. Magnify depression and increase the risk of suicide
2. Suppress the brain activity needed to prevent depression
3. Extend the period of denial and suppress normal mourning
4. Cause lethargy and prevent the return to interpersonal activity

126. A female client is readmitted to the hospital in the terminal stage of cancer. When talking with the nurse the client states, "I don't understand why my husband won't tell me what's going on at home. His telling me not to worry, that everything is being taken care of, doesn't help." The most realistic interpretation of the husband's behavior is that he is:
1. Attempting to stop the client from worrying
2. Expressing his unacknowledged anger with her dying
3. Acting out his need for dominance in their relationship
4. Attempting to deal with his own needs and trying to cope without her input

SUBSTANCE ABUSE (SA)

127. The defense mechanism commonly used by clients who are alcoholics is:
1. Denial
2. Projection
3. Displacement
4. Compensation

128. A client who has been drinking heavily since the death of a child 3 years ago is brought to the mental health unit in a stupor by the spouse. Taking the client's history into consideration, the nurse makes a tentative nursing diagnosis of:
1. Dysfunctional grieving
2. Substance abuse, alcohol
3. Personal identity disturbance
4. Ineffective family coping: disabling

129. Within a few hours of alcohol withdrawal, the nurse should assess a client for the presence of:
1. Yawning, anxiety, convulsions
2. Tremors, fever, profuse diaphoresis
3. Disorientation, paranoia, tachycardia
4. Irritability, heightened alertness, jerky movements

130. A 42-year-old with a long history of alcohol abuse seeks help with the problem in one of the local hospitals. The nurse is aware that the major underlying factor for success in an alcohol treatment program will be the client's:
1. Family
2. Motivation
3. Psychiatrist
4. Self-esteem

131. The nurse understands that clients with a history of alcohol abuse ingest alcohol primarily because they:
1. Are dependent on it
2. Lack the motivation to stop
3. Have no other coping mechanism
4. Enjoy the associated socialization

132. On the third hospital day after being admitted for alcoholism, a client is confused, disoriented, and delusional. The nurse should be aware that the client may be developing alcoholic:
1. Amnesia
2. Dementia
3. Hallucinosis
4. Withdrawal delirium

133. A male client who has a long history of alcohol abuse is informed that he has extensive liver damage. The client, whose father was an alcoholic and died of cirrhosis, lives with his mother. The client, when told that he has approximately 1 year to live if alcohol abuse persists, becomes intensely depressed and one evening leaves the hospital and returns drunk. The nurse is aware that the client's behavior suggests that he:
1. Wants to punish his mother for his physical and emotional discomfort and cause her pain
2. Does not trust the judgment of the health professionals because he feels well physically

3. Believes he inherited the disease of alcoholism from his father and cannot stop drinking
4. Cannot associate his increasing depression and lack of fulfillment to his heavy use of alcohol

134. The nurse is aware that the reason some alcoholics are unable to stop drinking even though they begin to attend AA meetings is that they:
1. Enjoy the feeling caused by drinking alcohol
2. Physiologically require the substance in their body
3. Are trying to drastically alter a long-standing habit
4. Often have a character defect that defeats their willpower

135. When dealing with a client who is in an alcohol detoxification program, it would be most important for the nurse to:
1. Accept the client as a worthwhile person
2. Provide nurturing since the client needs it
3. Discuss the ill effects of alcohol with the client
4. Promote compliance by gently prodding the client

136. To give clients with long histories of alcohol abuse greater responsibility for self-control, the nurse should initially plan to:
1. Tell them about detoxification programs
2. Confront them with their substance abuse
3. Assist them to identify and adopt more healthful coping patterns
4. Administer their medications according to the prescribed schedule

137. Two days after admission to the detoxification program, a client with a long history of alcohol abuse tells the nurse, "I don't know why I came here." The nurse's best response would be:
1. "You feel you don't need this program?"
2. "You did admit yourself into the program."
3. "You realize you are trying to avoid your problem."
4. "Don't you remember why you decided to come here?"

138. A 40-year-old client has a long history of alcohol abuse. After an automobile accident the client is arrested for driving while intoxicated and is admitted to the hospital. When the client becomes angry and blames others, the nurse can be most therapeutic by stating:
1. "You know you are to blame for your alcohol abuse."
2. "You need help now or you are going to get even sicker."
3. "I can see that you are irritable and I want to help you feel better."
4. "I will talk to your family and friends about their behavior if you want me to."

139. On the third day of hospitalization a client with a history of heavy drinking develops delirium tremens. When the client is experiencing hallucinations, it would be most appropriate for the nurse to:
1. Do nothing because the client may just be having vivid dreams or nightmares
2. Pretend to see the imaginary things the client is talking about and do what the client asks
3. Tell the client that others do not sense or perceive the same things that seem to be so frightening
4. Ask the client to describe the sensation, then assure the client that the sensation is caused by the alcohol

140. A client has been in the alcohol detoxification unit for 5 days. One evening the client complains of numbness and tingling in the feet and legs. At this time it would be most appropriate for the nurse to:
1. Gently massage the client's lower extremities with lotion
2. Emphasize the need to rest and to keep the lower extremities elevated
3. Use mechanical aids to keep bed sheets off the client's lower extremities
4. Observe for the progression of symptoms and monitor the pedal pulses frequently

141. The family of a client who has completed alcohol detoxification relate that they are concerned about the client's behavior if drinking occurs. They state, "When the drinking starts, it really disrupts family life and we're not sure how to handle it." The nurse's best response would be:
1. "Try to maintain a normal home environment for your family."
2. "Include the client in the family's activities even when drinking has occurred."
3. "Search the house regularly for hidden alcohol and accompany the client outside."
4. "Help avoid embarrassment by making excuses for the client when functioning is impossible."

142. When caring for individuals with a history of abuse of multiple drugs, the nurse should be aware that the most serious life-threatening symptoms during withdrawal usually result from:
1. Heroin
2. Methadone
3. Barbiturates
4. Amphetamines

143. When planning care for a client who has abused multiple drugs and has completed the withdrawal period, the nurse should take into consideration that the client is probably:
1. Unable to give up drugs
2. Unconcerned with reality
3. Unable to delay gratification
4. Unaware of the dangers of drug addiction

144. With a tentative diagnosis of opiate addiction, the nurse should assess this recently hospitalized client for signs of opiate withdrawal. These signs would include:
1. Lacrimation, vomiting, drowsiness
2. Nausea, dilated pupils, constipation
3. Muscle aches, pupillary constriction, yawning
4. Rhinorrhea, convulsions, subnormal temperature

145. After a binge with cocaine, a young male is found unconscious and is admitted to the hospital with acute cocaine toxicity. Initial nursing action should be directed toward:
1. Giving support and understanding
2. Maintaining a drug-free environment
3. Establishing a therapeutic relationship
4. Providing the necessary physical care

146. After a visit from several friends the nurse finds a client with a known history of opiate addiction in a deep sleep and unresponsive to attempts at arousal. The nurse assesses the client's vital signs and would evaluate that an overdose of opiates had occurred if the findings showed a:
1. Blood pressure of 70/40 mm Hg, a pulse of 120, and respirations of 10
2. Blood pressure of 120/80 mm Hg, a pulse of 84, and respirations of 20
3. Blood pressure of 140/90 mm Hg, a pulse of 76, and respirations of 28
4. Blood pressure of 180/100 mm Hg, a pulse of 72, and respirations of 18

147. The nurse, when planning care for a client recovering from an opiate overdose, should take into consideration that the client's underlying problem is probably a feeling of:
1. Guilt with a rejection of reality
2 Hostility with a need for acceptance
3. Inferiority with strong dependency needs
4. Anger caused by an overwhelming need for independence

148. The nurse is aware that opiates are most commonly used because the individual:
1. Desires to become independent
2. Wants to fit in with the peer group
3. Attempts to blur reality and reduce stress
4. Enjoys the social interrelationships that occur

149. The nurse should know that the most common side effects of regular cocaine use include:
1. Nausea, fatigue, and extreme hunger
2. Anxiety, dysphoria, and suspiciousness
3. Seizures, hoarseness, and electrolyte imbalance
4. Lethargy, sexual arousal, and hormone imbalance

150. A client with a known history of opiate addiction is treated for multiple stab wounds to the abdomen. After surgical repair the nurse notes that the client's pain is not relieved by the prescribed IM meperidine hydrochloride (Demerol) injections. The nurse recognizes that the failure to achieve pain relief from the Demerol injections indicates that the client is probably experiencing the phenomenon of:
1. Tolerance
2. Habituation
3. Physical addiction
4. Psychologic addiction

151. At a staff meeting the question of a staff nurse returning to work after a drug rehabilitation program is discussed. The nursing supervisor helps the staff to decide that the most therapeutic way to handle the nurse's return would be to:
1. Offer the nurse support in a direct, straightforward manner
2. Avoid mentioning the problem unless the nurse brings up the topic
3. Assign another staff member to keep the nurse under close observation
4. Make certain the nurse is assigned to administer only non-narcotic medications

152. It is determined that a staff nurse has a drug abuse problem. As an initial intervention the staff nurse should be:
1. Counseled by the staff psychiatrist
2. Dismissed from the job immediately
3. Forced to promise to abstain from drugs
4. Referred to the employee assistance program

153. The nursing care coordinator in the surgical intensive care unit notes that a number of clients do not seem to be responding to meperidine (Demerol) that has been administered for pain. Later that evening the coordinator finds a staff nurse in the nurses' lounge dozing. On being awakened the staff nurse appears somewhat uncoordinated and drugged with slurred speech. The coordinator should:
1. Ask the other staff members whether they have noticed anything unusual

2. Tell the staff nurse that everyone now knows who has been stealing the Demerol

3. Call the nursing director and have the director present before confronting the staff nurse

4. Arrange to secretly observe the staff nurse the next time the staff nurse administers Demerol

154. A client with the dual diagnosis of major depression and polysubstance abuse has been attending group therapy. One day the client tells the nurse, "The things they talk about in group don't really pertain to me." At this time it would be most appropriate for the nurse to:

1. Confront the client with realistic feedback
2. Identify the client's stress-coping tolerance
3. Question what the client means by the statement
4. Communicate that the client needs to get involved

155. A client with a long history of alcohol abuse is placed on a diet high in vitamin B_1 (thiamine). The nurse would know that the diet is understood when the client states, "I will select something for each meal from among:

1. Fish, aged cheese, and breads."
2. Poultry, milk products, and eggs."
3. Lean pork, organ meat, and nuts."
4. Leafy and green vegetables and citrus fruits."

156. A client with a long history of alcohol abuse who has been in the hospital for 1 week tells the nurse, "I feel much better and will probably not require any further treatment." When evaluating the client's progress the nurse should recognize that:

1. The client has accepted the illness and now needs to use willpower to resist the alcohol
2. As long as the client's family remains supportive, the client will probably not use alcohol again
3. The client lacks insight about the emotional aspects of the illness and most likely needs continued supervision
4. The physician must be notified of the client's statement so that aversion therapy can be started before the client's discharge

EATING AND SLEEP DISORDERS (ES)

157. An adolescent who is extremely underweight talks constantly about being fat, refuses to eat hardly any food, and disappears into the bathroom after meals, angrily says to the nurse, "I don't need to be here. I don't have any problems. Stop watching me." To reduce the client's feeling of being threatened, the nurse would be most therapeutic by responding:

1. "I hear how frustrated you are to be here."
2. "If you do not follow the rules you will lose your privileges."

3. "Your feelings are part of your illness. You will feel different."
4. "I'll get you the medication your physician ordered for anxiety."

158. Many clients on the eating disorders unit have been admitted for anorexia nervosa. The signs and symptoms that would be most specific for this diagnosis are:

1. Slow pulse, 10% weight loss, and alopecia
2. Compulsive behaviors, excessive fears, and nausea
3. Excessive activity, memory lapses, and an increased pulse
4. Excessive weight loss, amenorrhea, and abdominal distention

159. A major recognizable difference between clients with anorexia nervosa and clients with bulimia nervosa is that clients with anorexia usually:

1. Tend to be more extroverted than clients with bulimia nervosa
2. Seek intimate relationships while clients with bulimia avoid them
3. Deny the problem while clients with bulimia generally recognize that their eating pattern is abnormal
4. Are at greater risk for physical problems such as fluid and electrolyte imbalance than are clients with bulimia

160. A young adolescent is diagnosed as having anorexia nervosa. The nurse is aware that anorexia nervosa is usually precipitated by:

1. The acting out of aggressive impulses, which results in feelings of hopelessness
2. An unconscious wish to punish a parent who tries to dominate the adolescent's life
3. The inability to deal with being the center of attention in the family and a desire for independence
4. An inaccurate perception of hunger stimuli and a struggle between dependence and independence

161. The nurse interviews a young female client with anorexia nervosa to obtain information for the nursing history. The client's history is likely to reveal:

1. A strong desire to improve her self-image
2. A close, supportive mother-daughter relationship
3. Low achievement in school, with little concern for grades
4. Satisfaction with and a desire to maintain her present weight

162. The nursing intervention that should receive the highest priority in the period immediately following an emaciated 13-year-old's admission to the hospital for starvation secondary to anorexia nervosa would be:
1. Providing adequate rest and nutrition
2. Monitoring client's fluid and electrolyte balance
3. Completing an assessment of the client's mental status
4. Obtaining more data about the client's diet and exercise program

163. The nurse is aware that the major health complication associated with severe anorexia nervosa is:
1. Protein depletion resulting in muscle wasting
2. Endocrine imbalance resulting in amenorrhea
3. Glucose intolerance resulting in hypoglycemia
4. Cardiac dysrhythmias resulting in cardiac arrest

164. While admitting a young client with severe anorexia nervosa to the unit, the nurse finds a bottle of assorted pills in the client's luggage. The client tells the nurse they are antacids for stomach pains. The best initial response by the nurse would be:
1. "Let's talk about your drug use."
2. "These pills don't look like antacids."
3. "Tell me more about these stomach pains."
4. "Some adolescents take pills to lose weight."

165. The parents of an adolescent female are very upset about their daughter's diagnosis of anorexia nervosa and the treatment plan proposed. The best intervention by the nurse when the client's parents ask to bring food in for the client is to state:
1. "Your concerns about food contribute to her problem."
2. "While in the hospital, she should eat the hospital food."
3. "For now, allow the hospital staff to handle her food needs."
4. "It is important that you bring in whatever you think she'll eat."

166. When interacting with an adolescent client with the diagnosis of anorexia nervosa, the nurse should:
1. Show empathy
2. Maintain control
3. Set and maintain limits
4. Focus on dietary nutrition

167. The multidisciplinary team decides to employ a behavior modification approach to a young female's problem with anorexia nervosa. A planned nursing intervention that would follow this approach would be to:

1. Have client role play interactions with her parents
2. Provide client with a high-calorie, high-protein diet
3. Restrict the client to her room until she gains 2 pounds
4. Force the client to talk about her favorite foods for 1 hour a day

168. When an adolescent female client with the diagnosis of anorexia nervosa starts to discuss food and eating, the nurse should plan to:
1. Tell her gently but firmly to direct all discussion of food to the dietitian
2. Use her current interest in food to encourage her to increase her intake
3. Listen closely to determine her favorite foods and secure these foods for her
4. Let her talk about food as long as she wants, but limit discussion about her eating

169. When talking with one of the day nurses, a client with the diagnosis of anorexia nervosa states that the day nurses give better care and are nicer than the night nurses. The client also asks a question that the day nurse is aware was answered by one of the night nurses. The nurse should recognize that the client:
1. Needs assistance in exploring and verbalizing feelings about the night nurses
2. Is trying to develop a bond of trust with a staff member that should be supported
3. Is attempting to divide the staff, and the behavior should be reported to the other staff members
4. Has negative feelings about the night nurses, and the nurses should be informed of these feelings

170. The nurse notes that a young female with anorexia nervosa telephones home just before each mealtime. She ignores reminders to eat and continues talking until the other clients are finished eating. She then refuses to eat "cold food." The nurse should:
1. Insist that the client eat the cold food
2. Remove the client's telephone privileges
3. Hang the telephone up when meals are served
4. Schedule a family meeting to discuss the problem

171. A young male client with anorexia nervosa telephones home just before mealtime. The client uses the phone calls to avoid eating. The nurse could evaluate that the nursing plan to set limits on this avoidance behavior was effective when the client:
1. Organizes an aerobic group for the clients
2. Arrives on time for meals without being called
3. Begins reading and clipping recipes from magazines

4. Contacts his family frequently by telephone between meals

172. The nurse is aware that the central issue regarding food and its consumption in anorexia nervosa is:
1. Love
2. Control
3. Security
4. Attention

173. The parents of a male adolescent diagnosed with anorexia nervosa ask the nurse how long their son's treatment will take. The most therapeutic response by the nurse would be:
1. "Treatment of anorexia nervosa takes a long time and setbacks are common."
2. "Most anorectic adolescents respond favorably to treatment within a few weeks."
3. "Your son's prognosis depends on your willingness to become involved in therapy."
4. "Your son's progress depends on determining what triggered his desire to lose weight."

SCHIZOPHRENIC DISORDERS (SD)

174. When working in a psychiatric setting, it is imperative that the nurse prevent clients from:
1. Breaking contracts
2. Using delusional thinking
3. Harming themselves or others
4. Increasing hallucinatory behaviors

175. When the rights of a client on a mental health unit are suspended, the nurse has the specific responsibility to:
1. Inform the client's family or guardian
2. Carefully monitor all pharmacologic intervention
3. Complete a rights denial form and forward it to the administrative officer
4. Document the client's behavior and the reason why specific rights were denied

176. A newly admitted client continues to be apathetic and exhibits an inappropriate affect. A diagnosis of acute schizophrenic reaction is made. Considering the diagnosis, a symptom the nurse would expect to observe in the client's communication or behavior is:
1. Suicidal preoccupation
2. Autistic magical thinking
3. Absence of self-criticism
4. Abstract and logical deductions

177. When a client's behavior becomes increasingly bizarre, the client is transferred to a psychiatric unit. When obtaining a history from the client's family, the nurse would expect to find that the premorbid behavior included:
1. Depression and anxiety
2. Irritability and circumstantiality
3. Suspiciousness and introversion
4. Extroversion and inflated self-esteem

178. On the afternoon of admission to a psychiatric unit, an adolescent client with the diagnosis of schizophrenia exposes his genitals to a female nurse. The most immediate therapeutic response by the nurse would be to:
1. Confront him with the behavior
2. Ignore the behavior at this time
3. Set limits by telling him that such behavior is abnormal
4. Tell him to come to the office later to discuss the behavior

179. When assessing a client with the diagnosis of an acute schizophrenic reaction, the nurse recognizes that the potential for recovery is better in a client whose history reveals the:
1. Occurrence of a precipitating event
2. Slow and insidious onset of the illness
3. Presence of a family history of schizophrenia
4. Presence of many poorly defined prepsychotic symptoms

180. When asking the family about the onset of problems in a young client with the diagnosis of acute schizophrenia, the nurse would expect that they would relate the client's difficulties began in:
1. Puberty
2. Adolescence
3. Late childhood
4. Early childhood

181. A female client, recently admitted to the hospital, is pacing the floor and acting aloof and suspicious. According to her husband she laughed in a silly manner when told her father was critically injured, and she has had difficulty with her colleagues at work, accusing them of sabotage. The client has stated that she is being controlled by others. The nurse, to be most helpful, should first:
1. Obtain a complete copy of the client's history
2. Review a textbook description of the schizophrenic client
3. Meet with the client's husband to learn why she was admitted
4. Observe and evaluate the behavior in terms of the client's needs

182. When making an assessment of a client's hallucinatory behavior, the nurse realizes that the most common type of hallucination is:
    1. Visual
    2. Tactile
    3. Auditory
    4. Olfactory

183. A client with the diagnosis of schizophrenia repeatedly says to the nurse, "No moley, jandu!" The statement "No moley, jandu!" is an example of:
    1. Echolalia
    2. Concretism
    3. A neologism
    4. Paleologic thinking

184. After 2 days on the unit, a female client with the diagnosis of acute schizophrenic reaction refuses to take a shower. It would be most appropriate for the nurse to:
    1. Have the staff give her a shower
    2. Simply state the client must shower now
    3. Tell her she can shower when she feels more comfortable
    4. Point out that her appearance is upsetting the other clients

185. A young client is admitted to the hospital with a diagnosis of acute schizophrenia. The family relates that one day the client looked at a linen sheet on a clothesline and thought it was a ghost. The nurse recognizes that this was:
    1 An illusion
    2. A delusion
    3. A hallucination
    4. A confabulation

186. A young male client with the diagnosis of schizophrenia states that he cannot eat because someone has taken his stomach. The nurse recognizes this as an example of:
    1. An illusion
    2. A hallucination
    3. Depersonalization
    4. A somatic delusion

187. Many clients with schizophrenia experience opposing emotions simultaneously. The nurse recognizes this phenomenon as:
    1. Double bind
    2. Ambivalence
    3. Loose association
    4. Inappropriate affect

188. Breaks with reality, such as those experienced by clients with schizophrenia, necessitate that the nurse first realize that:

    1. Extended institutional care is a necessary part of the treatment modality
    2. Clients believe what they feel they have undergone and are experiencing
    3. Electroconvulsive therapy produces remission in most clients with schizophrenia
    4. The clients' families must cooperate in the maintenance of the psychotherapeutic plan

189. While the nurse is assisting a client with the diagnosis of schizophrenia with morning care, the client suddenly throws off the covers and starts shouting, "My body is disintegrating; I am being pinched." The term that best describes the client's behavior is:
    1 Paranoid ideation
    2. Depersonalization
    3. Loose association
    4. Ideas of reference

190. The nurse is aware that a common nursing diagnosis for clients with a schizophrenic disorder is:
    1. Social isolation related to impaired ability to trust
    2. Sleep pattern disturbances related to impaired thinking ability
    3. Risk for violence directed at others related to hallucinations
    4. Impaired mobility related to fear of loss of control of hostile impulses

191. On admission a disturbed female client refuses to remove her clothing for psychiatric evaluation. To best meet the client's needs the nurse should:
    1. Provide her with two outfits to assist the client to make a simple decision
    2. Get assistance and remove her clothing to meet her basic hygiene needs
    3. Tell her she will look more attractive in clean clothes to increase her self-esteem
    4. Wait and allow her to undress when she is ready, to help the client maintain her identity

192. The central problem the nurse might face with a disturbed schizophrenic client is the client's:
    1. Continuous pacing
    2. Relationship with the family
    3. Concern about working with others
    4. High anxiety and suspicious feelings

193. A long-term client goal for a paranoid male client who has unjustifiably accused his wife of having many extramarital affairs would be to help the client develop:
    1. Faith in his wife
    2. Better self-control
    3. Feelings of self-worth
    4. Insight into his behavior

194. The most appropriate intervention for the nurse to take after finding an acting-out, disturbed client in the fetal position would be to:
1. Tap the client gently on the shoulder and say, "I'm here to spend time with you."
2. Sit down beside the client on the floor and say, "I'm here to spend time with you."
3. Go to the client and say, "I'll be waiting for you by the table and chairs so please get up and join me."
4. Leave the client alone because the behavior demonstrates the client is too regressed to benefit from talking with the nurse

195. The nurse believes an emotionally disturbed client is ready to begin participating in therapeutic activities. The nurse should initially suggest:
1. Participating on the softball team
2. Attending a class on medications
3. Drawing or painting with the nurse
4. Watching television in the dayroom

196. A client with a long history of disturbed behavior is unable to cope with the slightest change in the environment. To enhance the client's coping skills, the nurse should plan to:
1. Allow time for compulsive behavior
2. Maintain a low level of environmental stimuli
3. Provide ample opportunities for intellectual activities
4. Schedule short independent tasks that are achievable

197. To deal with a client's hallucinations therapeutically, the nurse plans to:
1. Reinforce the perceptual distortions until the client develops new defenses
2. Provide an unstructured environment and assign the client to a private room
3. Avoid helping the client make connections between anxiety-producing situations and hallucinations
4. Distract the client's attention by providing a competing stimulus that is stronger than the hallucinations

198. When planning activities for a withdrawn, hallucinatory client, the nurse should recognize that it would be most therapeutic for the client to:
1. Go for a walk with the nurse
2. Watch a movie with other clients
3. Play cards with a group of clients
4. Play solitaire alone in the dayroom

199. To plan care for a client with undifferentiated schizophrenia, the nurse should recognize that the client's delusions are a defense against underlying feelings of:

1. Guilt
2. Inferiority
3. Aggression
4. Persecution

200. A male client claims the voices he hears are clearly telling him what actions and decisions to make. It would be most therapeutic for the nurse to:
1. Play soft music when the client starts hearing voices
2. Begin talking to the client when he is hearing the voices
3. Demonstrate to the client that his perceptions are wrong
4. Recognize that the client is probably frightened by the voices

201. When a client with the diagnosis of schizophrenia talks about being controlled by others, the nurse should:
1. Express disbelief about the client's delusion
2. Arrange an interesting daily schedule for the client
3. Respond to the verbal content of the client's delusion
4. Respond to the feeling tone or theme of the client's delusion

202. A client on the psychiatric unit sits alone most of the day. No other clients ever seem to go near the client. The nurse, deciding to establish a relationship, approaches the client. As the nurse gets approximately 3 feet away, the client lets out a string of profanity and says, "Leave me alone; I don't want to talk to you." The most appropriate response for the nurse to make at this time would be:
1. "I'll leave for now, but I'll be back later."
2. "Why do you feel the need to greet me in this way?"
3. "Do not talk to me like that. I am here to spend time with you."
4. "I don't like it when you talk like that. Are you trying to push me away?"

203. One afternoon the nurse observes a male client rushing down the hall of the unit rapidly, hitting his fist against the wall as he goes. The best nursing action at this time would be to:
1. Forcefully use additional staff members to subdue the client and stop his acting-out behavior
2. Attempt to approach the client in a nonthreatening manner to determine the basis for his agitation
3. Observe the client to see whether this behavior escalates and may involve harm to other clients or staff
4. Immediately summon staff assistance to enable administration of medication prescribed for the client's agitation

204. When helping a client who has the diagnosis of an acute schizophrenic reaction, select foods for breakfast, the most therapeutic question by the nurse would be:
1. "What kind of foods do you like?"
2. "Which of these foods do you want?"
3. "Do you want boiled or scrambled eggs?"
4. "How do you want your eggs fixed today?"

205. A client with the diagnosis of schizophrenia watches the nurse pour juice for the morning medication from an almost empty pitcher and screams, "That juice is no good. It's poisoned." The nurse should:
1 Remark, "You sound frightened."
2. Assure the client that the juice is not poisoned
3. Pour the client a glass of juice from a full pitcher
4. Take a drink of the juice to show the client that it is OK

206. A disturbed client is scheduled to begin group therapy. The client refuses to attend. The nurse should:
1. Accept the client's decision without discussion
2. Have another client ask the client to reconsider
3. Tell the client that attendance at the meeting is required
4. Insist that the client join the group to help the socialization process

207. When a disturbed acting-out client's condition improves, the physician suggests giving a 1-day pass. The client's family is very nervous about the pass and is worried about what they will do if the client starts to act up. The nurse's best intervention at this time would be to:
1. Have the social worker talk with the family
2. Cancel the pass until the family is reassured
3. Have the client promise the family that acting out will not occur
4. Discuss this concern at a meeting with the client and the family present

208. One morning the nurse finds a disturbed client curled up in the fetal position in the corner of the dayroom. The most accurate initial evaluation of the behavior would be that the client is:
1. Feeling more anxious today
2. Attempting to hide from the staff
3. Tired and probably did not sleep well last night
4. Physically ill and experiencing abdominal discomfort

209. An extremely agitated male client hospitalized in a mental health unit begins to pace around the dayroom. The nurse should:
1. Lock the client in his room to limit external stimuli
2. Let the client pace in the hall away from other clients

3. Get the client involved in a card game to distract his thoughts
4. Encourage the client to work with another client on a unit task

210. The nurse should be aware that the defense mechanism a client with the diagnosis of schizophrenia, undifferentiated type, would most probably exhibit is:
1. Projection
2. Repression
3. Regression
4. Rationalization

211. The nurse notices a male client sitting alone in the corner smiling and talking to himself. Realizing that the client is hallucinating, the nurse should:
1. Ask the client why he is smiling
2. Leave the client alone until he stops talking
3. Invite the client to help to decorate the day room
4. Tell the client it is not good for him to talk to himself

212. To increase the self-esteem of a client with schizophrenia, the nurse should plan to:
1. Reward healthy behaviors
2. Identify various means of coping
3. Encourage good hygiene and grooming
4. Explain the diagnosis and treatment plan

213. A young client with schizophrenia states, "I am starting to hear voices." The nurse could best respond:
1. "What are the voices saying, and what do they mean to you?"
2. "You are the only one hearing the voices. Are you sure you hear them?"
3. "The other staff will observe your behavior and we will not leave you alone."
4. "I understand you are hearing voices talking to you, and that they are very real to you."

214. One morning a client tells the nurse, "My legs are turning to rubber because I have an incurable disease called schizophrenia." The nurse recognizes that this is an example of:
1. A hallucination
2. Paranoid thinking
3. Depersonalization
4. Autistic verbalization

215. A college student is admitted to the hospital with the diagnosis of schizophrenic disorder, paranoid type. The client is very guarded and suspicious and states, "My professors are trying to fail me because of my controversial ideas." A few hours after admission, another client sits down beside this client. The client jumps up and runs down the hall angrily shouting,

"Leave me alone! Don't you touch me!" The most accurate assessment of this behavior is that the client:
1. Fears close contact with people
2. Is responding to delusional thoughts
3. Is having a hallucinatory experience
4. Has confused the other client with a professor

216. When interacting with a confused, acting-out female client with a diagnosis of schizophrenia, the most therapeutic nursing intervention would be to:
1. Reassure the client that she will get better
2. Direct all of the client's activities on the unit
3. Help the client to clarify her experience and gain insight into her behavior
4. Provide the client with solutions to past and current problems she is experiencing

217. A client comes to emergency services and states, "The FBI is trying to kill me." The client is dressed in soiled clothes, wears no shoes, has on a pair of sunglasses, and has body odor. The client's symptoms are most typical of:
1. Shared paranoia
2. Paranoid disorder
3. Paranoid schizophrenia
4. Paranoid personality disorder

218. A male client's statement about a microcomputer implanted in his ear by a foreign agent would help the nurse recognize that the client is experiencing:
1 Illusions
2. Hallucinations
3. Neologistic thinking
4. Delusional thoughts

219. When establishing a nursing care plan, the nurse should understand that a male client's delusion that he is an important government adviser is most likely related to:
1. A psychotic loss of touch with his real identity
2. An attempt at wish fulfillment created to manipulate others
3. A need to feel a sense of importance and control over his environment
4. An attempt to compensate for feelings of depression about his problems

220. During the admission procedure, a client who has paranoid ideation refuses to answer the nurse's questions, stating, "You are in a conspiracy to kill me." The nurse understands these feelings are related to the client's:
1. Low self-esteem
2. Need to be alone
3. Need for attention
4. Lack of acceptance

221. The nurse recognizes that a paranoid client's accusations that the room is wired and the FBI is listening are an example of:
1. A delusion
2. A neologism
3. A hallucination
4. An idea of reference

222. When planning care for a client using paranoid ideation, the nurse should realize the importance of:
1. Not placing any demands on the client
2. Reducing all stress so the client can relax
3. Giving the client difficult tasks to provide stimulation
4. Providing the client with activities in which success can be achieved

223. A disturbed client is admitted to the hospital for psychiatric evaluation. When taking the nursing history, the nurse asks why the client came to the hospital. The client states, "They lied about me. They said I murdered my mother. You killed her. She died before I was born." The nurse recognizes that the client is experiencing:
1. Ideas of grandeur
2. Confusing illusions
3. Persecutory delusions
4. Auditory hallucinations

224. The nurse planning to establish a trusting relationship with a client who is using paranoid ideation should begin by:
1. Seeking the client out frequently to spend long blocks of time together
2. Sitting on the ward and observing the client's behavior throughout the day
3. Being available on the ward frequently but waiting for the client to approach
4. Calling the client into the office to establish a contract for regular therapy sessions

225. A client refuses to eat, stating, "The food is poisoned." The nurse should:
1. Ask the client what foods are desired so they can be ordered
2. Encourage the client's family to bring favorite foods from home
3. Suggest going to the cafeteria and selecting foods the client feels safe eating
4. Go with the client to the cafeteria and taste the food to show that it is not poisoned

226. During a team conference a family member suggests that a relative who has the diagnosis of schizophrenia, paranoid type, be assigned to group therapy. The nurse understands that:
1. Therapeutic group work tends to be threatening to individuals who are very suspicious
2. Individuals with schizophrenia, paranoid type, respond well to small therapeutic groups
3. Compliance with unit rules and medication regimens increases as the client's group involvement increases
4. Involvement in small therapeutic groups prevents the regression and dependency associated with institutionalization

DISORDERS OF PERSONALITY (PR)

227. The desensitization method that has been used successfully with clients experiencing phobias focuses on:
1. Imagery
2. Role playing
3. Modeling or imitation
4. Assertiveness training

228. A client comes to the psychiatric clinic for treatment of a phobia about cats. The nurse at the clinic should anticipate that this client will demonstrate:
1. Fear of discussing the phobia
2. Anger toward the feared object
3. Poor impulse control when threatened
4. Distortion of reality when completing daily routines

229. The mother of a 17-year-old female, hospitalized for extremely disturbed acting-out behavior, leaves a shopping bag at the desk saying, "This is for my daughter's birthday. I'm too busy to visit today." The gift is an unwrapped expensive pocketbook with the price tags attached. The daughter becomes extremely upset and tearful after being given the message and opening the package. The mother's behavior is an example of:
1. Maternal rejection
2. Projective behavior
3. A double-bind message
4. Passive-aggressive behavior

230. A nurse on a mental health unit has developed a therapeutic relationship with an acting-out, manipulative client. One day as the nurse is leaving, the client says, "Please stay. I'm afraid the evening staff doesn't like me. They often punish me." The nurse can best assist this client by saying:
1. "I'll ask the staff not to punish you."
2. "Tell me more about what you're feeling now."

3. "Don't worry, I told you everything would be all right."
4. "You know I leave at this time. We'll talk about this in the morning."

231. To maintain a therapeutic relationship with a client diagnosed as having a borderline personality disorder, the nurse on the psychiatric unit should:
1. Be firm, consistent, and understanding and focus on specific behaviors
2. Provide an unstructured environment for the client to promote self-expression
3. Use an authoritarian approach because this type of client needs to learn to conform to rules of society
4. Record but ignore marked shifts in mood, suicidal threats, and temper displays because they last only a few hours

232. After a conference with the psychiatrist, a client with a borderline personality disorder cries bitterly, pounds the bed in frustration, and threatens suicide. It would be most helpful for the nurse to:
1. Leave the client for a short period until the client regains control
2. Pat the client reassuringly on the back and say, "I know it is hard to bear."
3. Sit down and listen attentively if the client wishes to talk about the situation
4. Ask about the client's troubles and point out that other people also have problems

233. A client is exhibiting withdrawn patterns of behavior. The nurse is aware that this type of behavior eventually produces feelings of:
1. Anger
2. Paranoia
3. Loneliness
4. Repression

234. The nurse recognizes that a client's withdrawn behavior may temporarily provide a:
1 Defense against anxiety
2. Basis for emotional growth
3. Time for internal problem solving
4. Time to collect personal resources

235. For 2 weeks prior to admission to the hospital, a client spent hours each day performing a complicated handwashing ritual. The client's hands are raw and bloody. The nurse recognizes that the psychoanalytic interpretation is that the client's ritual represents a conflict with dirt, which is often associated with:
1. Aggression, with the ritual done to control rage
2. Freedom, with the ritual done to stay dependent

3. Initiative, with the ritual done to rebel against autonomy

4. Gender identity, with the ritual done to avoid homosexual panic

236. A client misses breakfast because of an elaborate handwashing ritual. During the early stage of the client's hospitalization, it would be most therapeutic for the nurse to:
1. Prevent the client from beginning the ritual until after breakfast is served
2. Encourage the client to interrupt the ritual for meals at the scheduled times
3. Allow the client to choose between eating breakfast or completing the ritual
4. Wake the client early so the ritual can be completed before breakfast is served

237. An executive secretary experiences an overwhelming impulse to count and arrange the rubber bands and paper clips in the desk. The client feels something dreadful will occur if the ritual is not carried out. In regard to the client's symptoms, the nurse is aware that:
1. Compulsive rituals are useful in our society as long as they can be controlled
2. Compulsive rituals serve to control anxiety resulting from unconscious impulses
3. The client is consciously controlling the symptoms, although this raises anxiety
4. The symptoms are a displacement of general anxiety onto an unrelated specific fear

238. A client who uses a time-consuming counting ritual tells the nurse, "I am spending 30 minutes counting each time and my boss is getting very upset. What should I do?" The nurse could best suggest that the client:
1. Limit the counting activity to only 20 minutes each time
2. Arrive at work 30 minutes early each morning for counting
3. Substitute another activity at home such as counting shoes or other objects
4. Talk with the boss and ask for tolerance until the psychiatric treatments help

239. A client who uses a complex ritual says to the nurse, "I feel so guilty. None of this makes any sense. Everyone must really think I'm crazy." The most therapeutic response by the nurse would be:
1. "Your behavior is bizarre, but it serves a useful purpose in your life."
2. "You are concerned about what other people are thinking about you."
3. "Guilt serves no useful purpose. It just helps you stay stuck where you are."

4. "I am sure people understand that you cannot help this behavior right now."

240. When attempts are made to prevent a client from carrying out ritualistic behavior, the nurse would expect that the client's response would be one of:
1. Relief
2. Anger
3. Gratitude
4. Embarrassment

241. An adolescent with a long history of drug abuse, stealing, refusal to comply with rules, and an inability to get along in any setting is admitted to an adolescent unit for evaluation. The most appropriate plan of care for the adolescent at this time would be for the nurse to:
1. Allow as much freedom as possible, setting few rules and minimum structure
2. Provide activities that ensure immediate gratification as well as social stimulation
3. Act as a role model for mature behavior while providing a very structured setting
4. Behave in a moralistic, punitive manner toward the adolescent when rules are not followed

242. A 22-year-old male is admitted to the psychiatric unit for observation for an antisocial personality disorder. He has been in repeated legal difficulty since he was a teenager and was recently arrested on charges of falsifying police records. As the client and the nurse are discussing his upcoming court date, the client states, "I know that everything has already been set. I don't have any witnesses so how can I prove I didn't do it. I know they'll convict me. I think I'll get out of town." Based on this statement the most appropriate nursing assessment would be that the client is using:
1. Escape ideation
2. Projective denial
3. Impaired judgment
4. Distorted reality testing

243. A male adolescent with the diagnosis of antisocial personality disorder spends a great deal of time with a female adolescent client on the unit. One day the nursing assistant enters the female client's room and finds them in her bed. Later the nursing assistant reports the incident to the nurse. The nurse should:
1. Arrange a discussion with both adolescents
2. Assign a staff member to observe both clients every 15 minutes
3. Lock the bedroom doors to keep the clients within view of the staff
4. Call a ward meeting to talk about sexual activity among all of the clients

244. An adolescent female with an antisocial personality disorder plans to live with her parents after discharge. The parents request advice on how to respond to their daughter's behavior. The nurse tells them it would be most therapeutic for them to:
1. Discuss her behavior with her and encourage her to develop self-control
2. Avoid setting expectations for her behavior and react to each situation as it arises
3. Help her find new friends and encourage her to get a job and assume responsibility for herself
4. Set clear limits, explain the consequences of disregarding them, and firmly and consistently apply them

245. A client with an antisocial personality disorder has been remanded to the inpatient psychiatric unit for approximately 1 week. The client refuses to discuss any problems with the nursing staff and the team has decided an appropriate intervention would be to use confrontation. One morning the nurse asks the client how things went the day before. The client states, "I didn't do much. I watched TV and read a little." The most appropriate confrontational response by the nurse would be:
1. "It seems that you're expecting us to wave a magic wand and cure you."
2. "That's not much for someone who wants to get out of the hospital soon."
3. "It seems that you're having difficulty facing up to your part in all your problems, and I wonder why that is."
4. "It doesn't sound to me like you've been doing much work on the problems that brought you into the hospital."

246. One afternoon a male client on the inpatient psychiatric service complains to the nurse that he has been waiting for over an hour for someone to accompany him to activities. The nurse replies, "We're doing the best we can. There are a lot of other people on the unit who need attention too." This response demonstrates the nurse's use of:
1. Impulse control
2. Defensive behavior
3. Limit-setting behavior
4. Reality reinforcement

247. Windows in the recreation room of the adolescent unit have been found broken on numerous occasions, and after group discussion, one of the adolescents indicates another male adolescent client has broken them. The nurse, using assertive intervention instead of aggressive confrontation, should:

1. Knock on the door of the adolescent's room and ask whether he would like to come out and talk about the situation
2. Confront the adolescent openly in the group, using a controlled voice and maintaining direct eye contact with him
3. Approach the adolescent when he is alone and, after making eye contact, inquire about his involvement in these incidents
4. Use a trusting approach to the adolescent, implying that the staff doubts his involvement but requests his denial for the record

248. An 18-year-old is admitted to the hospital after taking 20 tablets of diazepam (Valium). The client's diagnosis is antisocial personality disorder. When obtaining the history the nurse learns that the client had been arrested for drug use and is out on bail. During visiting hours, the nurse discovers the client and visitors smoking. By the odor, the nurse knows they are smoking marijuana. When confronted, the client responds, "I'm celebrating. Didn't you hear? I went to trial today and just got put on probation." The nurse's best response would be:
1. "You were lucky you just got probation, so don't get right back into trouble."
2. "I understand your relief about the trial, but pot smoking is against the rules."
3. "Why don't you and your friends come out and join the other clients and their visitors."
4. "If you can't follow the rules against pot smoking, your visiting privileges will be canceled."

249. An adolescent client with an antisocial personality disorder was admitted to the hospital because of drug abuse and repeated sexual acting-out behavior. The nurse could evaluate that nursing actions directed toward modifying the behavior of this client had been successful when the client:
1. Promises never to take drugs again
2. Discusses the need to seduce other adolescents
3. Recognizes the need to conform to society's norms
4. Identifies the feelings underlying the acting-out behavior

250. A nursing diagnosis for a client with a multiple personality disorder is self-esteem disturbance probably related to childhood abuse. The most appropriate short-term client outcome would be:
1 Engaging in object-oriented activities
2. Recognizing each existing personality
3. Eliminating defense mechanisms and fears
4. Verbalizing the need for antianxiety medications

251. A client with a multiple personality disorder is to be discharged after a 2-week hospitalization. The nurse evaluating the effectiveness of the short-term therapy would expect the client to verbalize:
1. That many of the personalities can be ignored
2. The ability to deal openly with feelings and fears
3. The need for long-term outpatient psychotherapy
4. That the personalities now serve no protective purpose

DISORDERS OF MOOD (MO)

252. A client's methods of coping are maladaptive. The nurse can best help the client develop healthier coping mechanisms by:
1 Promoting interpersonal relationships with peers
2. Providing a stress-free environment for the client
3. Allowing the client to assume responsibility for decisions
4. Setting realistic limits on the client's maladaptive behavior

253. The nurse can minimize agitation in a disturbed client by:
1. Ensuring constant client and staff contact
2. Increasing appropriate sensory stimulation
3. Discussing the reasons for suspicious beliefs
4. Limiting unnecessary interactions with the client

254. The nurse is aware that aside from feeling sad and having difficulty concentrating and sleeping, other common signs of depression include:
1. Rigidity and a narrowing of perception
2. Alternating episodes of fatigue and high energy levels
3. Diminished pleasure in activities and alteration in appetite
4. Fleeting participation and interest in activities of daily living

255. Extremely depressed clients seem to do best in settings where they have:
1. Many varied activities
2. A great deal of stimuli
3. A simple daily schedule
4. To make only simple decisions

256. When caring for a client with a major depression, the nurse usually has the most difficulty dealing with the:
1 Client's lack of energy
2. Negative nonverbal responses
3. Client's psychomotor retardation
4. Pervasive quality of the depression

257. When working with a client who is depressed, the nurse should initially:

1. Accept what the client says
2. Attempt to keep the client occupied
3. Try to keep the client from talking too much
4. Keep the client's surroundings bright and gay

258. When planning continuing care for the depressed client, the nurse should include:
1. Offering the client an opportunity to make some decisions
2. Encouraging the client to decide how to spend leisure time
3. Making all decisions to relieve the client of this responsibility
4. Allowing the client time to be alone to decide in which activities to engage

259. The nurse identifies establishing trust as a major nursing goal for a depressed client. This goal can best be accomplished by:
1. Spending the day with the client
2. Asking the client at least one question daily
3. Waiting for the client to initiate conversation
4. Spending short periods of time with the client every day

260. One day, the nurse sits by a depressed client's bed and states, "I will be spending some time with you today." The client responds angrily, "Go talk to someone else. They need you more." The most therapeutic response by the nurse would be:
1. "Why are you angry with me?"
2. "I'll go, but I will be back tomorrow."
3. "Don't say that. You are important, too."
4. "I will be spending the next 15 minutes with you."

261. One morning a client with the diagnosis of acute depression states to the nurse, "God is punishing me for my past sins." The nurse's best response would be:
1. "Why do you think that?"
2. "God is punishing you for your sins?"
3. "You really must feel upset about this."
4. "If you feel this way, you should talk to your clergyman."

262. As a client's depression begins to lift, the nurse encourages involvement with unit activities, primarily because this type of activity:
1. Provides for and encourages group interaction
2. Allows the client to verbalize repressed feelings of hostility
3. Supports the client's ego strengths and builds self-confidence
4. Keeps the client in view of the staff to limit suicide opportunities

263. On the fifth hospital day, the nurse observes that a depressed client remains lying on the bed when the clients are called to the dining room for lunch. To encourage the client to eat, the nurse should:
1. Simply state, "I will accompany you to the dining room."
2. Bring a tray to the client's room and leave it without comment
3. Provide information about the importance of eating to maintain health
4. Simply state, "All clients are expected to go to the dining room for meals."

264. When taking a history from a client with the diagnosis of a major depressive episode, with melancholic features, the nurse should expect to find that the client's premorbid behavioral characteristics would include:
1. No clear boundaries to the client's life-style
2. A history of jealousy and abrupt onset of symptoms
3. A narrow range of interests and over-meticulousness
4. The presence of nervous hypermanic and depressive symptoms

265. A 50-year-old homemaker is brought to the hospital with a history of weight loss, crying spells, restlessness, early morning insomnia, sitting in one place just staring into space, and picking at her skin. During the last 6 months, her husband of 30 years has been made president of his company and their children have both married. From the history, the nurse should realize the client is demonstrating the classic symptoms associated with:
1. Bipolar mood disorders
2. Major depressive episodes
3. Involutional induced reactions
4. Mood-incongruent manic disorders

266. The primary nursing diagnosis for a client with a medical diagnosis of major depression would be:
1. Spiritual distress related to depression
2. Self-esteem disturbance related to altered role
3. Powerlessness related to the loss of idealized self
4. Impaired verbal communication related to depression

267. A depressed female client appears to show sadness in her nonverbal behavior. The nurse should plan to help the client to:
1. Increase her structured physical activity
2. Be able to carry out activities of daily living
3. Improve her ability to communicate with significant others
4. Deal with painful feelings by sharing and expressing them

268. A depressed client is very resistive and complains about inabilities and worthlessness. The best approach by the nurse would be to:
1. Involve the client in activities in which success can be assured
2. Listen to the client and delay any planned activity for another time
3. Schedule activities for the client that can be implemented independently
4. Encourage the client to select an activity in which there is some interest

269. A woman who attempted suicide is brought to the hospital by her husband. During the admission interview the woman says, "I do not deserve to live. I am a bad mother and have mistreated my child." The husband later indicates that his wife has been a good mother and never hurt their child. He states that their marital relationship has deteriorated since the birth of their child 2 years ago and that he plans to get a divorce. The nurse should recognize that the most probable basis for the client's depression is her:
1. Use of withdrawal as a defense in an attempt to survive in a relationship with a man who wishes to desert her
2. Child-abusing behavior that has resulted in feelings of guilt and shame that are being relieved by self-punishment
3. Unmet dependency needs that have aroused unacceptable, hostile feelings that are turned inward against the self
4. Basic feeling of loneliness that has resulted in the use of self-abasement to evoke love, sympathy, and compliments from others

270. When finding a depressed client crying, the most therapeutic response by the nurse to help the client explore feelings would be:
1. "Does crying help?"
2. "I know you are upset."
3. "Tell me what has upset you."
4. "Do you want to tell me why you are crying?"

271. A client remains depressed even after an 8-week trial on several antidepressant medications. A decision to initiate electroconvulsive therapy is being considered by the treatment team. A contraindication for administering electroconvulsive therapy even with the use of a medication such as succinylcholine chloride (Anectine) would be:
1. The presence of a brain tumor
2. A current urinary tract infection
3. A history of hypothyroid disorder
4. The presence of diabetes mellitus

272. A client scheduled to begin electroconvulsive therapy to treat a severe depression that has not responded to any of the antidepressant medications tells the nurse, "I am frightened that there will be a permanent loss of memory after the treatment." The most therapeutic response by the nurse would be:
1. "Your memory loss may be permanent, but it is usually just temporary."
2. "You will not experience a permanent memory loss so there is no need to be frightened."
3. "You will experience only a temporary loss of memory and it is normal to feel frightened about this."
4. "Your memory loss will only be temporary, and it will help block out many of your painful past experiences."

273. A depressed client is admitted to the hospital after being found bleeding from a self-inflicted superficial gunshot wound. The client does not respond to any of the nurse's questions. To assess the client's current potential for suicide, the nurse should:
1. Ask the client why suicide was attempted
2. Determine whether there is a family history of suicide
3. Ask the family about any previous suicidal attempts or threats by the client
4. Observe the client for scars on the wrists or other signs of previous suicide attempts

274. To further assess a client's suicidal potential, the nurse should be especially alert to the client's expression of:
1. Anger and resentment
2. Anxiety and loneliness
3. Frustration and fear of death
4. Helplessness and hopelessness

275. A client whose wife recently died appears extremely depressed. The client states, "What's the use in talking? I'd rather be dead. I can't go on without my wife." The best response by the nurse would be:
1. "You'd rather be dead?"
2. "Are you thinking about killing yourself?"
3. "I can understand why you feel that way."
4. "Tell me, what does death mean to you?"

276. A female client has been hospitalized for 3 weeks while receiving a tricyclic medication for a severe depression. One day the client states to the nurse, "I'm really feeling better, my energy level is up. Did the nurse's aide tell you I gave her my designer purse?" The nurse recognizes that this statement may indicate:
1. An increased risk of suicide
2. An improved socialization level

3. A marked improvement in mood
4. A decreased need for continued observation

277 The nurse becomes aware of an elderly client's feeling of loneliness when the client states, "I only have a few friends. My daughter lives in another state and couldn't care whether I live or die. She doesn't even know I'm hospitalized." The nurse recognizes that the client's communication is probably a:
1. Clue to depression that is blocking motivation
2. Call for help to prevent acting on suicidal thoughts
3. Manipulative attempt to persuade the nurse to call the daughter
4. Request for information about community social support groups

278 A female client, who was admitted to the hospital because she attempted suicide, reveals that her desire for sex has diminished since her child's birth 3 years ago. The most accurate nursing diagnosis would probably be sexual dysfunction related to:
1. Decreased sexual desire associated with depression
2. Decreased sexual desire associated with dependency
3. Inadequate sexual desire associated with marital stress
4. Inadequate sexual desire associated with identity confusion

279. A depressed, suicidal client is placed on one-to-one observation. A short-term goal specific for this client's nursing care needs is that within:
1. Two days the client will go for a walk on the grounds with others
2. Two days the client will understand why there was a desire for suicide
3. Three days the client will verbally accept responsibility for own actions
4. Three days the client will understand the continued presence of a staff member

280. When talking with a depressed client who has recently lost a spouse, the client states, "I really see no purpose to life and sometimes feel like ending it all." The nurse's best response would be:
1. "How much consideration have you given to the method you would use to kill yourself?"
2. "Death is hard on everyone, but people make it through every day. Things will get better."
3. "Even though you feel that way now, you still have your whole life ahead of you. Make a new start!"
4. "It must be hard to lose someone you care about so much; it makes life seem not worth living right now."

281. A client who has been forced into early retirement is admitted to the hospital with severe depression. The client states, "I feel useless and have nothing to do." The best initial response by the nurse would be:
1. "Tell me what you would like to do."
2. "Your illness is adding to your feelings."
3. "Have you thought about volunteering?"
4. "You feel useless; tell me more about that."

282. When assessing a client with a diagnosis of mood disorder-manic episode, the nurse is aware that the manic episode is in reality an:
1. Exaggerated response to an elating situation
2. Uncontrolled acting out of uncensored id drives
3. Incorrect interpretation of environmental stimuli
4. Attempt to block unconscious feelings of depression

283. When caring for clients demonstrating manic behavior, the nurse must be aware of these clients' physical needs. This is particularly important because:
1. Left alone, these clients will withdraw to their rooms
2. These clients have difficulty making their needs known
3. These clients may gain too much weight from overeating
4. The danger of exhaustion is always present for these clients

284. When selecting a room for a client with the diagnosis of bipolar disorder, manic phase, who is hyperactive and talking nonstop in a loud demanding voice, the nurse recognizes that a most important factor would be that the:
1. Room have a pleasant view
2. Atmosphere be quiet and restful
3. Location be close to the nurses' station
4. Roommates have similar behavioral responses

285. When developing an initial nursing care plan for a female client with a bipolar disorder, manic phase, the nurse should plan to:
1. Increase her gym time
2. Isolate her from her peers
3. Provide food, fluids, and rest
4. Encourage her active participation in unit programs

286. A nursing care plan for a client with a bipolar disorder should include:
1 Providing a structured environment
2. Touching the client to provide reassurance
3. Engaging the client in conversation about current affairs

4. Designing activities that will require the client to maintain contact with reality

287. A client is admitted to the psychiatric unit wearing evening clothes and bright facial makeup. During the first 24 hours the client paces continually and laughs loudly. When approached by the nurse, the client refuses to cooperate with any requests, shouting, "I am in charge. I give the orders!" The nurse recognizes that in addition to neurotransmitter alterations, the client's manic symptoms can be viewed as:
1 The fulfillment of innate desires
2. An uncontrollable urge to relate
3. A response to an imagined loss
4. An attempt to ward off depression

288. The nurse realizes that the environment is very important when caring for a client with a diagnosis of mood disorder-manic episode. The nurse should therefore:
1. Put bright drapes in the client's room to cheer up the client
2. Place the client in a private room to provide a quiet atmosphere
3. Assign the client to a room with other clients to provide company
4. Assign the client to a room near the dayroom to provide access to activities

289. A client demonstrating manic behavior is elated and sarcastic. The client is constantly cursing and using foul language and has the other clients on the unit terrified. The nurse should:
1. Firmly tell the client that the behavior is unacceptable
2. Demand that the client stop the behavior immediately
3. Ask the client to identify what is precipitating the behavior
4. Increase the client's medication or get additional medication ordered

290. Encouragement and praise should be given to hyperactive clients to help them increase their feelings of self-esteem. When they have behaved well, the best way to let them know the staff is aware of their improvement is for the nurse to say:
1. "You behaved well today."
2. "I knew you could behave."
3. "Everyone likes you better when you behave like this."
4. "Your behavior today was much better than yesterday."

291. A male client with bipolar disorder, manic phase is admitted to the psychiatric unit. He has progres-

sively lost weight and does not take the time to eat his food. The nurse can best respond to this situation by:
1. Providing a tray for him in his room
2. Assuring him that he is deserving of food
3. Pointing out that the energy he is burning up must be replaced
4. Ordering food that he can hold in his hand to eat while moving around

292. A client with a diagnosis of mood disorder, manic type, is readmitted 4 months after discharge. The client has become increasingly hyperverbal, loud, and intrusive. The most therapeutic response by the nurse who had cared for the client during the previous hospitalization would be:
1. "When did you stop taking your Lithium?"
2. "You seem to have a need to interrupt me."
3. "How is your relationship with your spouse?"
4. "I have a feeling that you are under great stress."

DEMENTIA, DELIRIUM, AND OTHER COGNITIVE DISORDERS (DD)

293. When the nurse is communicating with a client with substance-induced persisting dementia, the client cannot remember facts and fills in the facts with imaginary information. The nurse is aware that this is typical of:
1. Concretism
2. Confabulation
3. Flight of ideas
4. Associative looseness

294. When taking a health history from a client with moderate dementia, the nurse would expect to note the presence of:
1. Hypervigilance
2. Increased inhibition
3. Enhanced intelligence
4. Accentuated premorbid traits

295. An elderly client is admitted to the hospital with the diagnoses of dementia of the Alzheimer's type and depression. The client has all of the following symptoms. The symptom that is unrelated to the depression would be:
1 Shallow or labile affect
2. Neglect of personal hygiene
3. "I don't know" answers to questions
4. Apathetic response to the environment

296. When planning activities for an elderly nursing home resident with a diagnosis of vascular dementia, the nurse should:

1. Plan varied activities that will keep the resident occupied
2. Provide familiar activities that the resident can successfully complete
3. Offer challenging activities to maintain the resident's contact with reality
4. Make sure that the resident actively participates in the unit's daily activities

297. An elderly client's family tells the nurse that the client has suffered some memory loss in the last few years. They say that the client is sensitive about not being able to remember and tries to cover up this loss to avoid embarrassment. When attempting to increase the client's self-esteem, the nurse should try to avoid discussing events that require memory of the client's:
1. Married life
2. Work years
3. Recent days
4. Young adulthood

298. During the first month in a nursing home, an elderly client demonstrates numerous behaviors related to disorientation and cognitive impairment. The nurse's plan for care should continue to take into consideration the:
1. Assessment of the client's orientation to time, place, and person
2. Realistic ability of the client to perform without becoming frustrated
3. Identification of stressors that appear to precipitate the client's disruptive behavior
4. Fact that the client's cognitive impairment will increase until adjustment to the home is accomplished

299. An elderly female client, who is quite confused and often does not recognize her children, is admitted to a nursing home. The client appears slovenly in attire, often soiling her clothing with feces and urine. The nurse can best manage this problem by:
1. Putting the client into orientation therapy
2. Toileting the client at least once every 2 hours
3. Supervising the client's bathroom activities closely
4. Explaining to the client how offensive her behavior is to others

300. To assess orientation to place in a client suspected of having Alzheimer's, the nurse should ask:
1. "Where are you?"
2. "Who brought you here?"
3. "Do you know where you are?"
4. "Do you know the day you arrived?"

301. An elderly nursing home resident with the diagnosis of Alzheimer's likes to talk about olden days and at times has a tendency to confabulate. The nurse should recognize that this behavior serves to:
1. Prevent regression
2. Increase self-esteem
3. Attract the attention of others
4. Reminisce about achievements

302. A priority of care for a client with a dementia resulting from AIDS would be:
1. Maintaining adequate nutrition
2. Planning for remotivational therapy
3. Arranging for long-term custodial care
4. Providing basic intellectual stimulation

303. When planning care for a 72-year-old client who has been admitted to the hospital because of bizarre behavior, forgetfulness, and confusion, the nurse should give priority to:
1. Preserving the dignity of the client
2. Promoting a structured environment
3. Limiting the acceleration of symptomatology
4. Determining or ruling out an organic etiology

304. An elderly nursing home resident with the diagnosis of dementia of the Alzheimer's type hoards leftover food from the meal tray and other seemingly valueless articles and stuffs them into pockets "so the others won't steal them." The nurse should plan to:
1. Remove unsafe and soiled articles from the resident's belongings during the night
2. Give the resident a small bag in which to place selected personal articles and food
3. Explain to the resident why the nursing home's policy for cleanliness and safety must be followed
4. Tell the resident that the staff is required to keep harmful objects out of reach in the resident's closet

305. A 54-year-old client has demonstrated increasing forgetfulness, irritability, and antisocial behavior. After being found disoriented and seminaked walking down a street, the client is admitted to the hospital, and a diagnosis of Alzheimer's disease is made. The client expresses fear and anxiety. Considering the client's diagnosis, the best approach would be for the nurse to:
1. Initiate a program of planned interaction and activity
2. Reassure the client by the frequent presence of staff
3. Explore in depth the reasons for the client's concerns
4. Explain the purpose of the unit and why admission was necessary

306. Nursing management of a forgetful, disoriented client with the diagnosis of Alzheimer's disease should be directed toward:
1. Rechanneling the client's excessive energies
2. Managing the client's somewhat bizarre behaviors
3. Restricting all gross motor activity to prevent injury
4. Preventing further deterioration in the client's condition

307. An elderly resident with the diagnosis of dementia of the Alzheimer's type frequently talks about the good old days at the ranch as a child. On the basis of an understanding of the resident's diagnosis, the nurse's most appropriate action at this time would be to:
1. Involve the resident in interesting diversional activities in a small group
2. Allow the resident to reminisce about the past and listen with interest to the stories
3. Gently remind the resident that those "good old days" are past and thinking should focus on the present
4. Introduce the resident to other residents of the same age so that they can mutually share their past experiences

308. An elderly female client is admitted to a nursing home from the general hospital with a diagnosis of dementia of the Alzheimer's type. One morning, after being in the nursing home for several days, the client is going to join a group of residents in recreational therapy. The nurse notes that the client has laid out several dresses on her bed but has not changed from her nightclothes. It would be most helpful for the nurse to:
1. Remind the client to dress more quickly to avoid delaying the other residents
2. Help the client select appropriate attire and offer her assistance in getting dressed
3. Help the client dress and tell her what time the residents are expected at the activity
4. Allow the client as much time as she needs but explain that she is too late to attend this activity

DRUG-RELATED RESPONSES (DR)

309. A client's family ask about the treatment of schizophrenia. The nurse, before responding, recalls that:
1. Family therapy has not proved to be effective in the treatment of clients with schizophrenia
2. Electroconvulsive therapy is more effective in treating schizophrenia than mood disorders
3. Insight therapy has proven to be highly successful in the treatment of clients with schizophrenia
4. Drug therapy, while not eliminating the underlying problem, reduces the symptoms of acute schizophrenia

310. Nurses should be aware that the use of opiates creates:
1. Psychologic addiction, tolerance, and physical addiction
2. Physical addiction, psychologic addiction, but no tolerance
3. Physical addiction, tolerance, but no psychologic addiction
4. Psychologic addiction, tolerance, but no physical addiction

311. To prevent life-threatening complications from the administration of chlorpromazine (Thorazine) to a disturbed, acting-out client, it is important that the nurse:
1. Provide adequate restraint
2. Monitor the client's vital signs
3. Protect against exposure to direct sunlight
4. Watch the client for extrapyramidal side effects

312. On the psychiatric unit, a client has been receiving high doses of Thorazine for 2 weeks. The client states, "I just can't sit still and I feel jittery." The nurse suspects that the client may be experiencing the side effect known as:
1. Akathisia
2. Torticollis
3. Tardive dyskinesia
4. Parkinsonian syndrome

313. In addition to hydration during delirium tremens, the physician prescribes parenteral administration of chlordiazepoxide (Librium) for the client. The nurse understands that chlordiazepoxide is given during detoxification primarily to:
1. Prevent physical injury to the client when convulsions occur
2. Enable the client to sleep and eat better during periods of agitation
3. Quiet the client and encourage cooperation and acceptance of treatment
4. Reduce the anxiety-tremor state and prevent more serious withdrawal symptoms

314. When lithium levels are scheduled to be done, the nurse should remember that a client's serum lithium concentration is more stable:
1. 2 to 4 hours after the last dose
2. 4 to 6 hours after the last dose
3. 6 to 8 hours after the last dose
4. 8 to 12 hours after the last dose

315. After a client has been receiving chlorpromazine (Thorazine), the nurse observes extrapyramidal symptoms and anticipates that the physician will limit these side effects by prescribing:

1. Hydroxyzine (Atarax)
2. Amobarbital (Amytal Sodium)
3. Benztropine mesylate (Cogentin)
4. Chlorzoxazone (Parafon Forte DSC)

316. The nurse is aware that haloperidol (Haldol) is most effective for clients who exhibit:
1. Manic-assaultive behavior
2. Excited-overactive behavior
3. Excited-depressed behavior
4. Withdrawn-secretive behavior

317. A client is receiving lithium for the treatment of a bipolar disorder, manic phase. When planning client teaching about this medication, the nurse understands that it is important for the client to know that:
1. A low-sodium diet must be followed every day
2. Lithium blood levels require periodic monitoring
3. It will be necessary to take a diuretic with the lithium
4. Lithium will need to be taken for the rest of the client's life

318. The immediate treatment for a client who has ingested a tricyclic antidepressant in an amount that is 20 to 30 times the daily recommended dose would include:
1. Dialysis or forced diuresis
2. Administration of physostigmine
3. IM or IV administration of an anticholinergic
4. Closer monitoring to prevent further suicidal attempts

319. A client is started on fluphenazine decanoate (Prolixin Decanoate). When teaching about this drug, the nurse should emphasize that:
1. Driving is forbidden while taking this drug
2. There will be a feeling of increased energy while on this drug
3. A sunscreen must be used for all outdoor activity on a year-round basis
4. The client's essential hypertension will indirectly be controlled by this drug

320. In a client suspected of and demonstrating the symptoms associated with opiate overdose, the nurse would expect the physician to prescribe:
1. Naloxone
2. Methadone
3. Epinephrine
4. Amphetamine

321. The nurse should teach a client receiving tranyl-cypromine (Parnate) that failure to adhere to the dietary restrictions can result in:
1. Syncope
2. Bradycardia
3. Hypertensive crisis
4. Hyperglycemic episodes

322. Therapy has helped a client with a long history of alcohol abuse make a fairly satisfactory adjustment while in the hospital. The client will continue to receive therapy on an outpatient basis and will be discharged with the drug disulfiram (Antabuse). The nurse should caution the client to avoid using:
1. Elixirs and liniments
2. Suntan lotions and oils
3. White sugars and all vinegars
4. Caffeine-containing coffee and strong tea

323. An antipsychotic, haloperidol (Haldol), is ordered three times a day for a male client who has been admitted to the psychiatric service because he has become more argumentative and physically and verbally abusive to his wife. The nurse would be aware that the client is responding favorably to the medication when his behavior changes and he becomes more:
1. Enthusiastic about eating the hospital diet
2. Aware of his behavior and its consequences
3. Involved with the activities of others on the unit
4. Preoccupied with his delusions, but less physically abusive

324. A client who is going home on a weekend pass has been receiving risperidone (Risperdal) 3 mg tid. The nurse should inform the client that:
1. The medication dosage can be reduced if the client feels better at home
2. The medication does not need to be taken during the time spent at home
3. No alcoholic beverages should be consumed while taking this medication
4. All the medication should be taken early in the day to be sure it is not forgotten

325. After talking with a client about the tricyclic antidepressant medication that has been prescribed, the nurse diagnoses the presence of a knowledge deficit when the client states:
1. "I notice I'm a little drowsy in the mornings."
2. "I'm expecting to feel somewhat better in 3 weeks."
3. "I've been on the medication for 8 days now and I don't feel any better."
4. "I know I will probably have to take this medication for at least a few months."

326. A client with an organic mental disorder becomes increasingly agitated and abusive. The physician orders haloperidol (Haldol). The nurse should assess the client for untoward effects including:
1. Jaundice and vomiting
2. Tardive dyskinesia and nausea
3. Hiccups and postural hypotension
4. Parkinsonism and agranulocytosis

327. A client with the diagnosis of schizophrenia is given one of the antipsychotic drugs. The nurse is aware that of all the extrapyramidal effects associated with these drugs, the one causing the most concern would be:
1. Akathisia
2. Tardive dyskinesia
3. Parkinsonian syndrome
4. An acute dystonic reaction

328. The nurse should continually assess a client receiving lithium for an early sign of lithium toxicity, which would be:
1. Tinnitus
2. Diarrhea
3. Akathisia
4. Torticollis

329. The nurse understands that after starting administration of diazepam (Valium), it is important to assess for potential side effects. Initially the nurse should:
1. Monitor the client's blood pressure
2. Measure the client's urinary output
3. Assess the client for abdominal distention
4. Check the client's pupil size every 4 hours

330. The physician orders haloperidol (Haldol) concentrate 10 mg po bid for a client who is also receiving phenytoin (Dilantin) for control of epilepsy. When planning the client's care, the nurse should be aware that anticonvulsants may interact with Haldol to:
1. Mask its therapeutic effect
2. Interfere with its absorption
3. Enhance its rate of metabolism
4. Potentiate its CNS depressant effect

331. A client is extremely depressed, and the physician orders a tricyclic antidepressant, imipramine hydrochloride (Tofranil). The client asks the nurse what the medication will do. The nurse's best response would be:
1. "This medication will help you forget why you are depressed."
2. "The medication will help increase your appetite and make you feel better."
3. "You will begin to feel much better after taking this medication for 2 to 3 days."
4. "When you take this along with phenelzine (Nardil), you'll feel less depressed."

332. A client with a mood disorder–manic episode is receiving lithium carbonate. The client complains of diarrhea, tremors, and drowsiness. The nurse should:
1. Withhold the medication
2. Get an order for a stimulant
3. Make certain the client stays on the special diet
4. Decrease the client's fluid intake to 2000 ml daily

333. A client who is receiving an MAO inhibitor is going home on a weekend pass. Considering this drug, the nurse plans to caution the client to avoid:
1. Pork, spinach, and fresh oysters
2. Milk, peanut butter, and meat tenderizers
3. Cheese, beer, and products with chocolate
4. Orange drinks, fresh apples, and ice cream

334. A client is receiving haloperidol (Haldol) for agitation. When observing the client for side effects, the nurse would recognize that the side effect that is unrelated to extrapyramidal tract symptoms would be:
1. Akathisia
2. Opisthotonos
3. Oculogyric crisis
4. Hypertensive crisis

335. A client is started on fluphenazine decanoate (Prolixin Decanoate). The nurse is aware that the primary advantage of this medication is that:
1. There are no side effects
2. It has a longer lasting effect
3. It is safe to use during pregnancy
4. The need for laboratory monitoring is reduced

336. A client receiving buspirone hydrochloride (BuSpar) is admitted to the hospital with the diagnosis of possible hepatitis. The nurse notices that the client's sclera looks yellow. The nurse's initial action regarding this medication should be to:
1. Withhold the BuSpar
2. Give the BuSpar with milk
3. Reduce the dosage of the BuSpar
4. Assure the client that the BuSpar can be given parenterally

337. A female client who is taking clozapine (Clozaril) calls the nurse to say she has suddenly developed a sore throat and has a high fever. The nurse, recognizing the drug's effects, evaluates the client's complaints and tells her to:
1. Stay in bed, force fluids, take aspirin, and skip the next two scheduled doses of Clozaril
2. Stop the medication immediately and see her physician when an appointment is available
3. Continue the medication, drink fluids, take aspirin, and see her physician if not improved in a few days
4. Skip the medication and, if her physician cannot see her today, go to the emergency room for evaluation

338. The physician plans to have a client continue on lithium after discharge. The nurse would recognize that the teaching about the medication plan was understood when the client states, "I know that this medication:
1. Should be stopped if illness is suspected."
2. May need to be taken for the rest of my life."
3. Must be increased at the first sign of a manic episode."
4. Rarely causes serious side effects when taken correctly."

339. An antianxiety medication is prescribed for an extremely anxious client. The client states, "I'm afraid to take these pills because I heard they're addicting." The nurse's response is based on the knowledge that antianxiety medications:
1. Rarely causes dependence when dosage is controlled
2. May result in psychologic but not physiologic dependence
3. May require increased dosage but rarely cause dependence
4. Have the potential for physiologic and psychologic dependence

340. A client is admitted to the hospital with a diagnosis of depression that has not responded to tricyclic antidepressants or outpatient ECT. The physician orders tranylcypromine (Parnate). The nurse would be aware that the teaching about the drug was understood when the client states, "While taking this medicine I should avoid eating:
1. Fish."
2. Red meat."
3. Chocolate."
4. Citrus fruit."

341. Forty-eight hours after starting on haloperidol (Haldol), a male client is observed standing by the nurse's station with his head arched sharply backward. The nurse should recognize that the client:
1. Is experiencing temporary side effects that usually disappear after several days
2. Is having pseudoparkinsonian side effects and needs to have the dosage adjusted
3. Needs to have the dosage increased since his psychotic behavior is not lessening
4. Needs immediate treatment because he is experiencing an acute dystonic reaction to the drug

THERAPEUTIC RELATIONSHIPS (TR)

342. An elderly female who has been a widow for 20 years comes to the community health center with a vague list of complaints. Her only child, a son, died at birth. She has lived alone since her husband's death and performs all of her own daily tasks of living. She has had a very active social life in the past but has outlived many of her friends and family members. When taking this client's health history, it is important for the nurse to ask:
1. "Do you feel all alone?"
2. "Do you still miss your husband?"
3. "What unfulfilled hopes do you have?"
4. "How did you feel when your son died?"

343. A client with the diagnosis of panic disorder is placed on alprazolam (Xanax), which the client refuses to take because of fear of addiction. Initially the nurse should:
1. Provide the client with information about Xanax
2. Further assess the client's knowledge and feelings about Xanax
3. Have the physician speak to the client about the safety of this drug
4. Speak with the physician regarding a change in the client's medication

344. In addition to hallucinating, a client yells and curses throughout the day. The nurse should:
1. Isolate the client until the behavior stops
2. Ignore the behavior exhibited by the client
3. Become aware of what the behavior means to the client
4. Be willing to explain the meaning of the behavior to the client

345. The nurse in charge gives permission to the hospital's inservice division to use a child's chart in a nursing seminar on the care of children with leukemia. The nurse believes releasing the information will help other nurses give better care to children with this illness. The legal right of the client that was violated is the right to:
1. Privacy
2. Freedom
3. Respectful care
4. Informed consent

346. At times a client's anxiety level is so high it blocks attempts at communication and the nurse is unsure of what is being said. To clarify understanding, the nurse states, "Let's see whether we both mean the same thing." This is an example of the technique of:
1. Reflecting feelings
2. Making observations

3. Seeking consensual validation
4. Attempting to place events in sequence

347. When a nurse revises a client's nursing care plan based on the client's responses that show evidence that goals were not attained, the phase of the nursing process being applied is:
1. Planning
2. Evaluation
3. Assessment
4. Implementation

348. The psychotherapeutic theory that uses hypnosis, dream interpretation, and free association as methods to release repressed feelings is the:
1. Behaviorist model
2. Psychoanalytic model
3. Psychobiologic model
4. Social-interpersonal model

349. The nurse is aware that the phase of the nurse-client relationship in which most of the problem solving occurs is called the:
1. Initial stage
2. Working stage
3. Planning stage
4. Evaluation/termination stage

350. A staff nurse on a medical-surgical unit has been assigned to have daily one-to-one interactions with a number of clients. Before making an initial contact with the clients, the nurse decides to review their individual medical records. This phase of the nurse-client relationship could best be referred to as the:
1. Working phase
2. Orientation phase
3. Termination phase
4. Preinteraction phase

351. The nurse is aware that in the working phase of the nurse-client relationship, clients:
1. Often focus the conversation on the nurse
2. Accept limits and initiate topics for discussion
3. Commonly exhibit testing behaviors such as flirtation and lateness
4. May repress emotionally charged material to avoid shocking the nurse

352. Three days after a stressful incident a client can no longer remember what there was to worry about. The nurse, in relating to this client, can be most therapeutic by recognizing that the inability to recall the situation is an example of the defense mechanism of:
1. Denial
2. Repression

3. Regression
4. Dissociation

353. A goal for a client with the nursing diagnosis, impaired verbal communication related to psychologic barriers, would be that the client will:
1. Be free of injury
2. Demonstrate decreased acting-out behavior
3. Interact with others in the external environment
4. Identify the consequences of acting-out behavior

354. A 15-year-old is admitted to an adolescent unit for evaluation. The adolescent has a long history of drug abuse, stealing, refusal to comply with rules, and an inability to get along in any setting. When collecting data related to the adolescent's life-style, the nurse may be prevented from accurately listening to what the client is saying by:
1. Personal cultural beliefs
2. The client's disease process
3. The pressure of time to complete care
4. A personal need to secure information

355. The condition of a child dying from leukemia deteriorates and the child becomes comatose. The parents state that a relative said they should not allow the child to be resuscitated but that they are unsure about this. The response by the nurse that would best demonstrate a recognition of the ethical issues involved is:
1. "Let me tell you about the implications of a DNR order, then you decide."
2. "Have you talked to your physician about this yet? I'll be happy to page him."
3. "You should discuss this thoroughly with your physician and with your religious adviser."
4. "The final decision must be made by you and your physician, but it is important to talk about it."

356. Before effectively responding to a rape victim on the phone, it is essential that the nurse in the rape intervention center:
1. Get the client's full name and address
2. Call for assistance from the psychiatrist
3. Know some myths and facts about rape
4. Be aware of any personal bias about rape

357. The nurse is aware that value clarification is a technique useful in therapeutic communication in that it helps:
1. Make clients aware of their personal values
2. Provide information related to clients' needs
3. Assist clients in making correct decisions related to their health
4. Alter clients' poor values to make them more socially acceptable

358. A reasonable short-term goal for clients who are functioning below the optimum level of mental health would be to help them become better able to:
1. Understand the dynamics behind the inadequate interpersonal relations
2. Discuss their feelings regarding significant others and their life experiences
3. Confront their inadequacies in interpersonal relations and be more sociable
4. Take actions that will increase their satisfaction with their relationships with others

359. A terminally ill client tells the nurse, "I would love to learn to speak German before I die." The nurse's response to the client's desire to learn a foreign language should be based on an awareness of the fact that:
1. Activities that support the client's denial should not be encouraged
2. Conversations and activities should focus on pleasant experiences
3. Clients should be encouraged to set meaningful goals for themselves
4. The energies expended on such an activity would not justify the outcome

360. It is most helpful to the nurse who is attempting to apply the principles of positive mental health to understand that:
1. Emotionally ill people can empathize easily with others
2. Emotionally healthy people function optimally in all settings
3. A sense of mastery of self and environment are crucial to emotional health
4. Mental illness is always characterized by observable signs or socially inappropriate behavior

361. An inexperienced nurse assigned to a mental health day-care setting elects to have a one-to-one therapeutic relationship with an elderly, depressed, withdrawn, female client. The nurse's selection was most likely based on the fear of being:
1. Hurt by a more active client
2. Rejected by a more alert client
3. Useless and then saying the wrong thing to a more alert client
4. Overly concerned for a younger client's well-being and mental status

362. The nurse should always take the time to keep a client's family informed about what is happening to the client. The main reason for this action is that informed families:
    1. Decrease the client's anxiety
    2. Commonly cause fewer nursing problems
    3. Are more relaxed and at ease with the client
    4. Are better equipped to undertake necessary family role changes

363. The nurse must recognize that when a client is a member of a different ethnic community it is important to:
    1. Ensure that the nurse's biases are understood by the family
    2. Offer a therapeutic regimen compatible to the life-style of the family
    3. Recognize that the client's responses will be different than other clients'
    4. Make plans to counteract both the client's and the family's misconceptions of family practice

364. After working with an elderly male client for a period of time, the client tells the nurse he always enjoyed working and playing with children. During discharge planning the nurse recommends that the client look into the volunteer foster grandparent program in his area. The nurse recognizes that this type of activity may help the client to:
    1. Be able to find new acquaintances with similar interests
    2. Forget his problems when he sees the problems of others
    3. Take better care of himself if he feels needed by someone
    4. Become motivated to become involved with younger people

365. The nurse is scheduled to be the co-leader of a therapy group to be formed in the mental health clinic. When planning for the first meeting, it is of primary importance that the nurse first consider the:
    1. Number of clients in the group
    2. Needs of the clients being included
    3. Diagnoses of the clients being included
    4. Socioeconomic status of the clients in the group

366. To further develop trust among members in a therapy group, the nurse plans to:
    1. Reveal some personal data as a role model for trusting behavior
    2. Remind group members about the need for confidentiality in the group
    3. Have group members reveal some personal information about themselves
    4. Bring up for discussion the need for and the importance of trusting each other

367. During the first session of a therapy group one of the clients asks, "What is supposed to happen in this group?" The most appropriate response by the nurse leader would be:
    1. "Before I answer that, I'd like for you to tell me what you want to happen."
    2. "This is your group and your participation will largely determine what happens."
    3. "The purpose of this group is to examine the way each of you interacts with the other."
    4. "You and the others are supposed to discuss any reality-based concerns you have about your illness."

368. Increased socialization and verbalization of reality-based concerns is the nurse's primary goal for a therapy group of clients with the diagnosis of schizophrenia. At the first session of the group, after introductions, the nurse could best begin by:
    1. Asking the clients what they hope to gain from the meetings
    2. Allowing the clients to discuss anything they wish to bring up
    3. Having each of the clients identify a specific concern to be discussed
    4. Sharing with the clients the purpose of the meetings and explaining the rules of behavior

369. The best initial approach to take with a self-accusatory, guilt-ridden client is to:
    1. Contradict the client's persecutory delusions
    2. Accept the client's statements as the client's beliefs
    3. Medicate the client when these thoughts are expressed
    4. Redirect the client whenever a negative topic is mentioned

370. The initial intervention strategy that is of primary importance in counseling an elderly female client who desires to remain independent and who is having increased difficulty in maintaining her independent living status is:
    1. Maintaining routines and supporting her usual habits
    2. Helping her secure assistance with cleaning and shopping
    3. Writing down and repeating important information for her use
    4. Setting clear goals and time limitations for her visits with the nurse

371. The nurse plans to use family therapy as a means of assisting a family to cope with their child's terminal illness. The nurse's basis for this choice is that:
1. It is more time-efficient to deal with the whole family together
2. The entire family is involved, since what happens to one member impacts all
3. The nurse can control manipulation and alliances better by using this mode of intervention
4. It will prevent the parents from deceiving each other about the true nature of their child's condition

372. A male client with advanced AIDS tells the nurse that all he wants is to pass his high school equivalency test before he dies. He asks the nurse whether this is possible. The nurse's best approach would be to:
1. Refocus the conversation on the things the client has already accomplished in his life
2. Attempt to get the client to understand that his wish is too taxing and somewhat unrealistic
3. Set up a study schedule with the client and offer to work with him in preparing for the test
4. Suggest to the client that he use this energy to work through his unexpressed anger at dying

373. The parents of an autistic child begin family therapy with a nurse therapist. The father states that the family wish to share their religion with the therapist. The nurse should:
1. Limit the father's discussion of religion
2. Invite the family's minister to a therapy session
3. Plan for a mutual discussion of religious beliefs
4. Keep the sessions focused on the family's concerns

374. A 17-year-old, admitted to the hospital because of weight loss and malnutrition, is diagnosed as having anorexia nervosa. After the client's physical condition is stabilized, the psychiatrist, in conjunction with the client and the parents, decides to institute a behavior modification program. The nurse is aware that a major component of behavior modification is that it:
1. Rewards positive behavior
2. Deconditions fear of weight gain
3. Decreases necessary restrictions
4. Reduces anxiety-producing situations

375. When caring for a client with a bipolar disorder-depressive episode, the nurse's first priority should be to help the client to:
1. Feel comfortable with the nurse
2. Investigate new leisure activities

3. Participate in small group activities
4. Initiate conversations about feelings

376. The nurse is planning a discharge conference with a psychiatric client and the client's family. The priority nursing action that should be included in the discharge plan is:
1. Obtaining a more complete family history
2. Teaching the client about the medication to be taken
3. Discussing new issues that could be worked on at home
4. Exploring what has been learned from this hospitalization

377. The most basic therapeutic tool used by the nurse to assist a client's psychologic coping is the:
1. Self
2. Milieu
3. Client's intellect
4. Helping process

378. In an attempt to remain objective and support a client during a crisis, the nurse uses imagination and determination to project the self into the client's emotions. The nurse accomplishes this by using the technique known as:
1. Empathy
2. Sympathy
3. Projection
4. Acceptance

379. Following a traumatic event a client is extremely upset and exhibits pressured and rambling speech. A therapeutic technique that the nurse can use when a client's communication rambles is:
1. Touch
2. Silence
3. Focusing
4. Summarizing

380. When a client with paranoid ideation tells the nurse about people coming through the doors to commit murder, the nurse should:
1. Listen to what the client is saying
2. Refuse to listen to the client's stories
3. Tell the client that no one can get through the door
4. Ask the client to explain where this information came from

381. When communicating with a client with a psychiatric diagnosis the nurse uses silence. When silence is used in therapeutic communication, clients should feel:
1. It is their turn to talk
2. Unhurried to answer
3. The nurse is thinking
4. There is nothing more to say

382. After speaking with the parents of a child dying from leukemia, the physician gives a verbal DNR order but refuses to put it in writing. The nurse should:
1. Follow the order as given by the physician
2. Refuse to follow the order, unless the nursing supervisor OKs it
3. Ask the physician to write the order in pencil on the client's chart before leaving
4. Determine whether the family is in accord with the physician and follow hospital policy

383. When speaking with a client diagnosed with schizophrenia, the nurse notices that the client keeps interjecting sentences that have nothing to do with the main thoughts being expressed. The client asks whether the nurse understands. The nurse should reply:
1. "You aren't making any sense; let's talk about something else."
2. "I'd like to understand what you are saying, but you are too confused now."
3. "Why don't you take a rest and then we can talk again later this afternoon."
4. "I'd like to understand what you are saying, but I'm having difficulty following you."

384. A mother visiting her hospitalized teenage daughter gets into an argument with her. Leaving her daughter's room in tears, the mother meets the nurse and relates the argument, stating, "I can't believe I got so angry I could have hit her." The most therapeutic response by the nurse would be:
1. "Sometimes we find it difficult to live up to our own expectations of ourselves."
2. "Why don't you bring a surprise for your daughter. It will make you both feel better."
3. "You can't compare yourself to an abusing parent. After all, you didn't beat your child."
4. "You're a wonderful mother. Everything will be OK. Teenagers can really drive you to distraction."

385. A husband is upset that his wife's delirium tremens have persisted for the second day. The initial response by the nurse that would be most appropriate is:
1. "I see that you are very worried. Medications are being used to lessen your wife's discomfort."

2. "This is totally normal. I suggest that you go home because there is nothing you can do to help at this time."
3. "Are you afraid that your wife may die? I assure you that very few alcoholics die during the detoxification process."
4. "The staff is making your wife comfortable while she is undergoing the withdrawal process. Your wife will not feel pain."

386. At a staff meeting discussing the return of one of the staff nurses from a drug rehabilitation program, one nurse states, "I don't know why we are wasting time on this. We all know that those people go back to using drugs as soon as the pressures increase." The nursing care coordinator's best response would be:
1. "It's important for us to share our feelings about staff members with problems."
2. "I know it's hard, but it's our professional obligation to work with these individuals."
3. "I guess you feel somewhat guilty that you failed to recognize that this nurse was addicted."
4. "Since you have such strong negative feelings, I don't think you should be assigned to work with this staff member."

387. A female nurse on the mental health unit has been assigned to work with a young male college student who was admitted the previous night. The client has never used any mental health services before. The nurse's most appropriate initial approach to this client would be to:
1. Address the client as Mr., saying, "Hello," and giving her first and last name
2. Address the client by his first name and state, "I've been assigned to care for you today."
3. Say good morning, addressing the client by his first name, stating, "I see you were admitted last night. Tell me what brought you to our unit."
4. Say good morning, addressing the client as Mr., and introducing herself by her first and last name, stating, "I am a nurse assigned to the mental health unit."

388. A newly admitted client looks at but does not respond to the nurse. The nurse's most appropriate action would be to state:
1. "I guess you would rather be alone for now; I will return later so we can talk."
2. "I am talking to you. Are you having trouble understanding what I am saying?"
3. "This is the mental health unit of the hospital. We have many services to offer. Let me tell you about them."

4. "I am here to tell you about the services available to you on the mental health unit and to offer you my help."

389. A newly admitted client quietly listens to a nurse's explanation of the services and activities available on the mental health unit. When the nurse is finished, the client looks around and states, "So this is where they keep the crazies." The nurse's most appropriate initial response would be:
1. "These people are sick. They are not crazy."
2. "Some people feel that way. Let's talk about mental health."
3. "No, that is not correct. Let me explain the purpose of a mental health unit."
4. "Are you feeling that a person has to be crazy to need mental health services?"

390. The nurse tells a client that talking with staff is part of the therapy program. The client responds, "I don't see how talking to you can possibly help." The nurse's most appropriate response would be:
1. "You will never know whether or not it is helpful unless you are willing to give it a try."
2. "I can see how you would feel that way now, but hopefully you'll change your mind."
3. "The one-to-one relationship has proven itself very helpful for others. Why don't you give it a try?"
4. "Hopefully, I can help you sort out your thoughts and feelings so you can better understand them."

391. The nurse can best handle the answering of personal questions asked by a client in any phase of the nurse-client relationship by:
1. Reviewing the positive and negative aspects of the subject
2. Providing brief, truthful answers and redirecting the focus of conversation
3. Offering an honest, brief expression of personal views on the subject raised
4. Gently reminding the client that the nurse's feelings are not the client's concern

392. The wife of a client, ultimately hospitalized with the diagnosis of Alzheimer's disease, appears tired and angry on her first visit with her husband. As she is leaving she says to the unit nurse in a sarcastic tone, "Let's see what you can do with him." The nurse's most therapeutic response would be:

1. "It must have been very difficult to care for him."
2. "I don't understand what you mean by that comment."
3. "It's too bad you didn't realize you needed help to care for him."
4. "We have experience in caring for clients such as your husband."

393. An overweight, 12-year-old male is brought to the clinic by his parents. The father states, "You've got to do something to help him. Just look at his size." The child tells the nurse that he dislikes school because his classmates tease him about his weight. He states rather sadly, "I'm always last when they choose up sides in gym." The nurse's most therapeutic response would be:
1. "That hurts a lot when you want to be liked."
2. "Not everybody's a great athlete. You have other strengths."
3. "Have you tried letting them know how that makes you feel?"
4. "Won't it be great when you lose weight and can do better in gym?"

394. During the meetings of a therapy group one member tends to monopolize the group discussions, and no one has confronted this behavior. The nurse could best handle this situation by:
1. Saying to the client, "You use too much of the time in our sessions."
2. Ignoring the behavior because the client may become upset if confronted
3. Encouraging other members of the group to do more talking by calling on various silent members
4. Saying to the group, "I'm wondering why the group is so willing to let this client do so much of the talking."

395. At a therapy group session a group member, using a teasing manner, makes several negative remarks about the nurse's appearance and behavior. The nurse could best respond by saying to:
1. The group, "What do you think this client is trying to tell me?"
2. The group, "Do you think this client's behavior is appropriate today?"
3. The client, "You seem very interested in my appearance and behavior. What's this all about?"
4. The client, "I cannot just sit here and let you talk about me this way. What have I done to make you angry?"

396. At a therapy group session a female client tearfully tells the other members that she lost her job as a receptionist during the past week. It would be most appropriate for the nurse to:
1. Ask her to look at the reasons this may have occurred
2. Quietly observe how the group responds to her statement
3. Suggest she check the help wanted advertisements in the local paper
4. Request that the group help her see how she may have precipitated the dismissal

397. At a therapy group session, after one of the members relates a traumatic incident that happened during the week, another client states with a smile, "Things haven't gone well in my life this week either." It would be most appropriate for the nurse to:
1. Ask the client to share what has been happening during this week
2. Make a note of the incongruity of the client's message but remain silent
3. Say to the client, "You say things have been bad this week, yet you are smiling."
4. Comment, "This seems to have been a bad week for a number of group members."

398. As a young male client is receiving a dialysis treatment, the nurse notes he is not talking with the other clients and his eyes are lowered and his jaw is clenched. The nurse states, "You look discouraged." The client replies, "I'm a bother. Not much good to anyone anymore. My wife would at least get some insurance money if I died." The nurse's most therapeutic response would be:
1. "I can understand how you feel."
2. "You feel so bad you wish you were dead."
3. "We all have days we feel like that. Let's talk about your diet."
4. "I know it's hard, but don't let it get you down or let your wife hear you."

399. When the behavior of a visiting family member agitates a client who is extremely disturbed, the nurse should:
1. Take the client to the coffee shop for a treat
2. Distract the client by providing another activity
3. Limit the client's contact with the family member
4. Discuss the family member's behavior with the client

400. The nurse would be aware that a therapy group had reached the working stage when the members:
1. Appear happy in their group interactions
2. Focus on a wide variety of needs and concerns
3. Show concern for the feelings of the group leaders
4. Say and do what is expected and wanted by the others

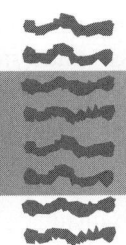

MENTAL HEALTH NURSING ANSWERS AND RATIONALES

EMOTIONAL DISORDERS RELATED TO PHYSICAL HEALTH AND CHILDBEARING (ED)

1. ③ This action would provide the child with a constant caregiver with whom the child could relate. (1) (SC; IM)
   1 Although this may provide some comfort, the child needs to receive love and attention from an adult.
   2 Same as answer 1.
   4 This would increase the parents' guilt and anxiety; data given assume parents have been unable, not unwilling, to visit the child.

2. ③ The preschooler is terrified by intrusive procedures and views them as a punishment for curiosity and fantasies. (2) (DF; AN)
   1 A child this age does not fear loss of control.
   2 A child this age can localize pain even if unable to express it.
   4 A child this age would be unlikely to recall the procedure; fear of intrusive procedures is primary.

3. ① This would permit the 6-year-old to investigate and become familiar with the equipment to be used. (2) (DF; IM)
   2 This would be beyond the comprehension of the average 6-year-old and would do little to reduce anxiety.
   3 This is beyond the ability of a 6-year-old and would do little to reduce anxiety.
   4 This would be supportive but is not always possible; even with the parent present, the child should be given an opportunity to handle the equipment.

4. ① The 9-year-old needs an opportunity to express emotions in private; talking about feelings after the child has regained control would be therapeutic. (2) (DF; AN)

   2 This is not of great concern to the child at this moment.
   3 This action would give the child a feeling that crying was wrong.
   4 Same as answer 2.

5. ④ This is a nonjudgmental response that does not pressure the client but does clearly indicate that treatment is necessary. (2) (CC; IM)
   1 This is an unrealistic response that gives approval to the client's behavior.
   2 This is an unrealistic response that is unsupported by the client's statement.
   3 This nonsupportive response tells the client that the physician knows best.

6. ③ This response interprets the client's behavior without belittling the nursing assistant's feelings; it encourages the assistant to get involved with plans for future care. (3) (CC; IM)
   1 Although this response recognizes the client's feelings, it does nothing to help the nursing assistant to deal with the client.
   2 This assumes the nursing assistant has nothing to contribute and only the nurse can deal with the problem.
   4 This statement does not help the nursing assistant with the situation or demonstrate any understanding of the client's feelings.

7. ② This action provides continuity and demonstrates to the client that the nursing staff is concerned; frequent contact reduces the client's need to call staff in. (2) (SC; PL)
   1 This would increase the client's anxiety and need for contact with staff.
   3 Telling the client is not the same as doing it; the client would not believe staff would come in frequently.
   4 This would not provide continuity of care.

8. ④ This is an open-ended response that encourages further discussion without focusing on an area that the nurse, not the client, feels is the problem. (3) (SC; IM)
   1 This response denies the client's feeling and can cause feelings of guilt for questioning the partner's love.
   2 This is too specific; the nurse does not have enough information to come to this conclusion.
   3 Same as answer 2.

9. ④ This response fosters open lines of communication with the client. (2) (SC; IM)
   1 This response would put the client on the defensive because it exposes the defensive behavior being used.
   2 This response does not recognize the client's concern and cuts off further communication.
   3 This could be interpreted as a sarcastic response that may cut off further communication.

10. ② The client has difficulty accepting the inevitability of death and attempts to deny the reality of it. (1) (EP; AN)
    1 In the anger stage the client strikes out with the "why me" and the "how could God do this" type of statements; the client is angry at life and still angrier to be removed from it by death.
    3 In this stage the client attempts to bargain for more time; the reality of death is no longer denied, but the client attempts to manipulate and extend the remaining time.
    4 In the acceptance stage the client accepts the inevitability of death and peaceably awaits it.

11. ② Crying is an expression of an emotion that, if not expressed, increases anxiety and tension; the increased anxiety and tension use additional psychic energy and hinder coping. (2) (SC; AN)
    1 Crying does not relieve depression, nor does it help a client face reality.
    3 This is not universally true; in most instances the client's defenses should not be taken away until they can be replaced by more healthful defenses; the nurse must always interfere with behavior and defenses that may place the client in danger; the client's current behavior creates no threat to the client.
    4 This is not always true; many clients are embarrassed by what they consider to be a "show of weakness" and have difficulty relating to the individual who witnessed it; the nurse must do more than just accept the crying to strengthen the nurse-client relationship.

12. ④ A starting point for working with all clients is ascertaining what is known, their understanding of their particular situation, and its meaning to them. (2) (SC; AS)
    1 It is not merely understanding what death and the dying process means, which is a philosophical discussion, but how the individual feels about the situation.
    2 Encouraging conversation about the situation tends to decrease anxiety.
    3 This may be part of the care plan, but it is not the most important piece of information during the initial contact.

13. ④ This response recognizes the client's feelings and encourages the client to look at the basis or reality of the expressed concern. (2) (SC; IM)
    1 This response goes in circles; the client has already told the nurse the basis for her feelings.
    2 This puts the responsibility for the husband's behavior on the client, who may not be able to handle it.
    3 This is a weak excuse for the behavior of the husband and may or may not be true.

14. ① Improved functioning relates most to improved body image, even if the prosthesis is not at all like the original body part. (3) (DF; AN)
    2 A slight improvement in body image occurs with a "normal" look, but the "normal" look usually occurs only when the prosthesis is covered by clothing.
    3 Acceptance by others does not necessarily guarantee acceptance by self.
    4 Although mood may be improved with an aggressive rehabilitation program, this in itself does not improve body image.

15. ① The nurse empathizes with the client and keeps the lines of communication open without being judgmental. (2) (SC; IM)
    2 This response does not address the client's feelings and may increase anxiety.
    3 This is false reassurance; close monitoring does not guarantee a live baby.
    4 This response denies the client's right to emotions and may evoke more feelings of guilt about her past obstetrical history.

16. ① Planning self-care demonstrates decision making by the client; participating in care enhances feelings of self-worth and autonomy. (1) (SC; EV)
    2 Expectations do not reflect autonomy.
    3 This does not reflect autonomy; it may also be intellectualization.
    4 Same as answer 3.

17. (4) The nurse should strive to clarify misconceptions and fears; this response promotes further communication and begins where the client is. (3) (SC; AS)

1 This avoids assuming the responsibility of answering the client's question; the client needs an immediate clarification.

2 The fact that others have had the surgery provides little solace; the remainder of the response is false reassurance and does not truthfully answer the client's question.

3 This denies the client's feelings and cuts off communication.

PERSONALITY DEVELOPMENT (PD)

18. (3) Displacement reduces anxiety by transferring the emotions associated with an object or person to another, emotionally safer object or person. (2) (SC; AS)

1 In projection, the individual attempts to deal with unacceptable feelings by attributing them to another.

2 Dissociation is an attempt by the person to detach emotional involvement or the self from an interaction or the environment.

4 Intellectualization is the use of facts or other logical reasoning rather than feelings to deal with the emotional impact of a problem; a form of denial.

19. (4) The nurse is using facts and knowledge to detach the self from the emotional impact of the client's problem and decrease the anxiety it is causing. (3) (SC; AS)

1 Substitution is similar to displacement; this reduces anxiety by transferring the emotions associated with an object or person to another, safer object or person.

2 Sublimation is the channeling of unacceptable thoughts or feelings into acceptable activity.

3 This is trying to unconsciously imitate the behavior of another who is considered important, in an attempt to incorporate this important other into the self.

20. (1) The father is siding with his daughter and supports her whereas the mother accuses her of negative behavior; this is an example of coalitions or alliances; in this instance the mother may also be in denial. (3) (SC; AS)

2 Resignation is evident when someone gives up.

3 Scapegoating is when an individual is labeled or blamed by other family members as the cause of the family's problems.

4 Reaction formation is a defensive mechanism that causes individuals to overtly behave in a manner that is exactly opposite to what they really feel in an attempt to conceal unacceptable feelings.

21. (1) Without the development of trust, the child has little confidence that the significant other will return; separation is considered abandonment by the child. (1) (SC; AN)

2 Without identity, the individual will have a problem in forming a social role and a sense of self; this results in identity diffusion and confusion.

3 Without initiative, the individual will experience the development of guilt when curiosity and fantasies about sexual roles occur.

4 Without autonomy, the individual has little self-confidence, develops a deep sense of shame and doubt, and learns to expect defeat.

22. (1) This is the developmental task of the preschool child; the child will feel guilty if initiative is stifled by others. (2) (SC; AN)

2 This is the task of the school-age child.

3 This is not a developmental task identified by Erikson.

4 Same as answer 3.

23. (2) The major tasks of young adulthood are centered around human closeness and sexual fulfillment; lack of love results in isolation. (1) (SC; AN)

1 This stage is associated with early childhood.

3 This stage is associated with middle childhood.

4 This stage is associated with middle adulthood.

24. (2) This is the task of the older adult; this client has not adapted to triumphs and disappointments, so there is no acceptance of what life is and was, and this results in feelings of despair and disgust. (2) (SC; AN)

1 This is the task of the preschool period.

3 This is the task of the young adult.

4 This is the task of the adolescent.

25. (2) This action would not call attention to the accident and would minimize the child's embarrassment. (2) (PC; IM)

1 The child would probably be unable to accomplish this task without assistance; failure to complete the task would add to embarrassment.

3 This would add to the child's embarrassment.

4 Same as answer 3.

26. ③ Adolescence is a time of great upheaval and maturation; before this maturational process is completed, adolescents act without thinking things through and are most concerned with their own needs, rather than the needs of others. (1) (DF; AN)
1 The rapid and complex biologic, social, and emotional changes during adolescence do not necessarily lead to these psychologic conflicts.
2 Same as answer 1.
4 Same as answer 1.

27. ① According to Erikson, adolescents are struggling with identity vs role confusion. (2) (EP; AN)
2 Adolescents tend to be group oriented, not isolated; they struggle to belong, not to escape.
3 This reflects part of the struggle for independence; it does not indicate failure to achieve the developmental task of adolescence; "all" is too inclusive.
4 This is untrue; most adolescents do not use drugs and alcohol to escape.

28. ① Adolescents learn about who they are by assuming and experiencing a variety of roles; experimentation results in the retention or rejection of behavior and roles. (2) (DF; EV)
2 This is not constructive; this would not allow for experimentation with a variety of roles.
3 Continuous sublimation would not be constructive and would delay and interfere with the successful completion of the struggle to formulate one's identity.
4 This is not constructive; it does not allow for the development of independence.

29. ③ This action would foster the development of an improved self-image. (2) (EP; IM)
1 A problem does exist; their child is unhappy.
2 Parents cannot avoid talking about the sibling but should avoid any comparisons.
4 The child already is doing this, and it has diminished her self-esteem.

30. ② This is an achievable goal that will bolster the child's self-esteem. (2) (HI; PL)
1 This would not improve the child's self-esteem.
3 Same as answer 1.
4 Same as answer 1.

31. ③ This would demonstrate a movement toward peer group activity and interests; exercise would also demonstrate an interest in an improved physical condition. (2) (HI; EV)
1 This would not demonstrate an increase in self-esteem.

2 There are no data to indicate that school was a problem.
4 Same as answer 1.

32. ③ Erikson theorized that how well people adapt to the present stage depends on how well they adapted to the stage immediately preceding it, in this instance adulthood. (2) (SC; AN)
1 This is the developmental task of an earlier stage of development; although Erikson believed that the strengths and weaknesses of each stage are present in some form in all succeeding stages, their influence decreases with time.
2 Same as answer 1.
4 Same as answer 1.

33. ④ This allows the client to accept what life is or was and helps avoid feelings of despair. (3) (DF; AN)
1 This could require a reversal in the client's past life-style; this is unlikely, if not impossible, for the client at this age.
2 This would be impossible to accomplish and denies the reality of what was or is in life.
3 Desire must come from the client.

34. ① The individual who can reflect back on life and accept it for what it was and is and who can adjust to and enjoy the changes retirement brings is less likely to develop health problems, especially stress-related health problems. (2) (DF; AN)
2 These changes are usually not reversible.
3 This is untrue; most emotionally healthy older individuals do not focus on these thoughts.
4 This is not true; dependency is often more threatening to this age group.

35. ④ An individual's ability to handle stress develops through experience with life; aging does not reduce this ability but often strengthens it. (1) (SC; PL)
1 The senses of taste and/or smell are often diminished in the aged individual.
2 Gastrointestinal motility is slowed in the aged individual.
3 Muscle or motor strength is diminished in the aged individual.

DISORDERS FIRST EVIDENT BEFORE ADULTHOOD (BA)

36. ② A child usually assumes a role in the family, and the child's symptoms reflect the pathologic factors that develop to fill that role. (3) (EP; AN)
1 This may create problems, but these problems usually develop later in life.

3 Same as answer 1.
4 Same as answer 1.

37. (1) Children with emotional problems usually have difficulty dealing with reality and tend to withdraw; they are afraid to use reality testing. (2) (EP; AS)
2 Behavior is more often disorganized rather than deliberate and aggressive rather than passive.
3 There is usually a withdrawal from the peer group.
4 The anxiety level is usually severe, often approaching the panic level.

38. (4) Unresponsiveness to the environment may be a serious indicator of childhood depression, autism, or possibly schizophrenia; all three are serious disorders. (3) (EP; AS)
1 This may be seen in children without emotional problems as well as in those with emotional problems; this behavior alone would not indicate severe emotional problems.
2 Same as answer 1.
3 Same as answer 1.

39. (2) This structured but nonthreatening statement avoids beginning with problems and may put the child at some ease, producing information that may be useful. (3) (EP; AS)
1 This statement is too open and global; the child would probably not know how to answer this question or know where to begin.
3 This question can produce a "yes" or "no" answer; also, the child may not know the answer to this question.
4 This question would probably produce an "I don't know" response; the focus should be on the child, not the mother.

40. (1) Tertiary prevention focuses on interventions that prevent complete disability or reduce the severity of a disorder or its associated disabilities. (3) (DF; PL)
2 This would be secondary prevention aimed at case-finding and early intervention.
3 This would be primary prevention.
4 Same as answer 2.

41. (2) This helps the child gain control by reducing stimuli while helping limit and prevent the use of tantrums by the child as an attention-getting behavior. (3) (SC; IM)
1 This would probably increase the behavior associated with the tantrum.

3 Although ignoring the temper tantrum may sometimes help, it often forces the child to act out further; using time-out is more successful because the child is removed and both the parent and child have a "cooling-off" period.
4 Medication is not the treatment of choice.

42. (4) One of the symptoms an autistic child displays is a lack of responsiveness to others; there is little or no extension to the external environment. (2) (EP; AS)
1 Repetitive behavior provides comfort.
2 Music is nonthreatening, comforting, and soothing.
3 Repetitive visual stimuli, such as a spinning top, are nonthreatening and soothing.

43. (4) Children usually blame themselves for their parents' marital problems, believing that they are the reason one parent leaves. (2) (SC; AN)
1 No data are presented to lead to this conclusion.
2 The child's response is not typical of grief work.
3 Same as answer 1.

44. (3) Comparison over time is the only way for the nurse to accurately assess the mental status of a child. (2) (DF; AS)
1 This would not be an accurate method because the child's ability to discuss feelings is limited.
2 This may be unrealistic and biased; the nurse should take the parents' description of behavior into consideration but should mainly rely on personal assessment and observation over time.
4 This would be threatening and may increase the child's anxiety.

45. (2) This is the family constellation as it is now constructed; without prior discussion and permission, an invitation to anyone else would be an intrusion of privacy. (3) (CC; AN)
1 In addition to needing the mother's permission to invite the father, the nurse must also include the child in family therapy.
3 The father cannot be invited without prior discussion with and permission of the mother.
4 The teacher is not part of the family constellation.

46. (3) When all the members of a family blame one member for all their problems, scapegoating is occurring. (2) (EP; AN)
1 There are no data to support identifying this pattern of relating.
2 Same as answer 1.
4 Same as answer 1.

47. ③ This action would set a foundation for trust because it allows the child to see that the nurse cares. (2) (EP; IM)
   1 This would be threatening at this stage of a relationship.
   2 This would be too infrequent to develop trust.
   4 Although this is necessary, limit-setting really does not support the development of a trusting relationship.

48. ② According to Erikson, this is the major conflict or task that must be dealt with by the school-aged child (6 to 12 years old). (3) (DF; AN)
   1 According to Erikson, this must be dealt with by the preschooler (3 to 5 years old).
   3 Not specific to any stage identified by Erikson.
   4 Part of Freud's theory, not Erikson's.

49. ④ A behavior modification program tailored for and developed with the individual child is the most appropriate approach at this time. (3) (SC; IM)
   1 Would serve no purpose at this time and could be viewed by the child as a reward for behavior.
   2 The child is not in need of special education at this time.
   3 May not be true; the child may not like school and may not think it is fun; having fun is not the purpose of school.

50. ④ Most statistical reports demonstrate an incidence of about 5% to 10% in school-age children, making this a rather serious problem. (3) (HI; IM)
   1 Untrue; this problem may not become evident until more environmental controls are imposed.
   2 Untrue; socioeconomic factors do not play a major role in the occurrence of this disorder.
   3 Untrue; these children have less need for sleep.

51. ① This response would be most successful because it provides a time period for the hyperactive child to regain control; it is neither a positive nor a negative reinforcement of acting-out behavior. (3) (EP; IM)
   2 The child would interpret removal of tokens as a punishment.
   3 Ignoring behavior can force the child to act out even more to gain attention.
   4 This action would reward acting-out behavior by providing special attention.

52. ③ This would help the child develop insight into reasons for acting-out behavior. (3) (EP; IM)
   1 This denies that the child's problem behavior is continuing and does not help the child develop insight.

   2 Same as answer 1.
   4 Same as answer 1.

53. ④ This would provide a trial opportunity for the child and the family to reunite outside the confines of the hospital. (2) (EP; PL)
   1 It is too late; this should have been done much earlier.
   2 This is not the responsibility of the nurse.
   3 Same as answer 1.

ANXIETY, SOMATOFORM, FACTITIOUS, AND DISSOCIATIVE DISORDERS (AX)

54. ① These are effective coping mechanisms to reduce stress. (2) (SC; EV)
   2 Not always possible; forced quiet activity may increase stress and anger rather than reduce it.
   3 Not always possible; stress can develop from a variety of feelings stimulated by many situations.
   4 Not always easy to identify; better to learn to deal with feelings once they develop.

55. ③ Perceptual fields are a key indicator of anxiety level because the perceptual fields decrease as anxiety increases. (2) (SC; AS)
   1 This is not related directly to anxiety levels.
   2 Same as answer 1.
   4 Same as answer 1.

56. ① The client's statement is an example of the use of denial, a defense that blocks problems by unconsciously refusing to admit they exist. (2) (SC; AS)
   2 The client is not using projection, a defense that is used to deny unacceptable feelings and emotions and attribute them to others.
   3 The client is not using sublimation, a defense that is used to substitute socially acceptable behavior for unacceptable instincts.
   4 The client is not using displacement, a defense mechanism that is used to allow the shifting of feeling from an emotionally charged person or object to a safe, substitute person or object.

57. ② Reexperiencing the actual trauma in dreams or flashbacks is the major symptom that distinguishes posttraumatic stress disorders from other anxiety disorders. (2) (EP; AN)
   1 This symptom is not usually associated with anxiety disorders.
   3 This symptom would be more common in phobic disorders.
   4 Although depression may be generated by discussion of the traumatic situation, the affect is usually exaggerated, not blunted.

58. ③ These actions make the environment as emotionally nonthreatening as realistically possible. (2) (EP; PL)

1 It is not possible or realistic to meet all of a person's needs.

2 All needs cannot be met; the person must learn how to deal with delaying gratification.

4 Order in the environment is of less importance; providing a nonthreatening environment is the priority action.

59. ① Anxiety is a threat to the identity of the individual; the client is seeking assurance that the fear and panic being experienced will not mean loss of control. (2) (EP; AN)

2 The client is not exhibiting depression but severe anxiety and panic.

3 This is not evidence of insightfulness but a plea for help in reducing the anxiety.

4 The client is not testing the nurse; the client is asking for help.

60. ① Because anxiety can be an interpersonal experience, it is contagious; the nurse then has a strong urge to get away. (2) (SC; AN)

2 The desire to go off duty would not suddenly make the nurse uncomfortable.

3 This is possible, but not probable; the client is exhibiting anxiety, not hostility, at this time.

4 There is no indication that this or any other behavior encountered has been bizarre.

61. ① Freedom to ventilate feelings acts as a safety valve to reduce the anxiety. (1) (EP; PL)

2 The suppression of anger or hostility would add to the clients' anxiety.

3 This would not be therapeutic; it might add to the anxiety the clients are feeling.

4 This may or may not be helpful; the clients' families may provide support to the clients.

62. ② The nurse's presence may provide the client with support and feelings of control. (2) (EP; IM)

1 It is evident that the client is upset; this question is not therapeutic and may lead to anger, which would interfere with the development of a therapeutic nurse-client relationship.

3 The client is too distraught to sit; to be therapeutic the nurse would have to be where the client is.

4 The client is in a panic; anger is not primary; there is no need to work off aggression.

63. ① The client has the right to privacy; his behavior is acceptable in the privacy of his room. (3) (DF; IM)

2 Masturbation is a normal sexual outlet; assessment is unnecessary unless the act is practiced to excess.

3 This can cause needless embarrassment to the client and may close off further communication.

4 His behavior is not inappropriate because he was in the privacy of his own room.

64. ① Fear can be overwhelming; the staff's presence provides protection from possible danger. (2) (EP; IM)

2 The client's anxiety level is interfering with the ability to communicate; anxiety must be reduced first.

3 The client's anxiety level is so high that sufficient emotional energy to set parameters would not be available.

4 This would only add to the client's anxiety at this time.

65. ③ This response helps the client focus on situations that precipitate frightening feelings. (2) (EP; AS)

1 This response would not help the client to focus on feelings.

2 The nurse cannot be certain what the client means about being afraid to talk about it; this response would not help the client to focus on feelings.

4 This is false reassurance; the nurse cannot guarantee that the feelings will not come back.

66. ② The client will be unable to concentrate or focus on the interview if anxiety is not reduced. (2) (EP; AS)

1 Not the priority at this time; anxiety must be reduced and the client's level of comfort increased.

3 Same as answer 1.

4 Not appropriate; the client could not rest until anxiety is reduced.

67. ② The attention span is shortened, making it difficult to follow long sentences; an authoritative voice lets the client know that the nurse is in control of the situation; the client is unable to set controls because of the anxiety level. (EP; IM)

1 This could increase the client's anxiety level even further.

3 Same as answer 1.

4 Crying is an outlet and should not be discouraged; telling someone not to cry usually increases the crying and adds to anxiety.

68. (4) These are positive coping behaviors that can consciously be used to promote mental health. (1) (SC; PL)
  1 These are not healthy coping behaviors, and their frequent use can lead to distortions of reality.
  2 Same as answer 1.
  3 Same as answer 1.

69. (3) Clients commonly respond in this manner when they feel their sexual identity is threatened; this behavior supports the client's ego integrity. (2) (DF; AN)
  1 The nurse's actions play only a minor role in these situations.
  2 This is not true; the client's feelings, not the situation, precipitated this response.
  4 Same as answer 1.

70. (4) This would document that the client feels comfortable enough to discuss the problems that have motivated the behavior. (3) (EP; EV)
  1 This does not demonstrate a resolution of problems underlying the behavior.
  2 Without discussion of the problems underlying the behavior, little would be accomplished.
  3 Same as answer 1.

71. (1) The development of confusion would indicate that the client's ability to maintain equilibrium has not been achieved and that further disequilibrium was occurring. (2) (DF; EV)
  2 This would not indicate the plan needed to be changed unless the client's history demonstrates no prior use of this defense.
  3 Same as answer 2.
  4 This would be a positive response to any plan of care but is not directly related to independence.

72. (3) Members have not established trust and are hesitant to discuss problems; the behaviors observed reflect anxiety and insecurity. (2) (SC; AS)
  1 This would add to the anxiety and insecurity of group members.
  2 These behaviors are expected in the early stage of the group.
  4 Same as answer 1.

73. (2) The client's response demonstrates an inability to deal with the other member's confrontational approach at this time. (2) (SC; EV)
  1 The group has reached the working stage, and if the client was able to deal with this area, the nurse would expect the client to state the feeling generated by the statement.

  3 Same as answer 1.
  4 There is not enough information to make this evaluation.

CRISIS SITUATIONS (CS)

74. (2) Caplan's theory states that a crisis is an internal disturbance caused by a stressful event that alters the usual way of coping with a threat to the self; this temporarily disturbs the homeostasis of the person involved. (3) (SC; AN)
  1 This is not the definition of a crisis.
  3 This is not the definition of a crisis; it is the assessment the nurse must make in the first phase of crisis intervention.
  4 This is not the definition of a crisis but is how a crisis is resolved.

75. (4) Awareness of limitations and the ability to place oneself in another's situation are essential to be able to intervene effectively. (1) (SC; AN)
  1 This is not a necessary characteristic to help families with problems and many times would be impossible to achieve; this is not a prerequisite for understanding.
  2 Although this may be helpful, it is not a priority.
  3 Same as answer 2.

76. (4) The data presented indicate developmentally related struggles and specific situations that are extremely stressful; multiple stresses can produce a crisis situation for the individual when past coping mechanisms are ineffective. (1) (DF; AS)
  1 It is not the experience but the individual's response to the experience that determines a crisis.
  2 A crisis is not an age-related problem; a crisis results when the individual's past coping mechanisms are no longer effective for dealing with a present stressful situation.
  3 The extent of the stress, although a factor, is not the most significant factor in identifying the presence of a crisis; the individual's inability to cope indicates a crisis.

77. (2) A situational crisis occurs when a specific external event upsets an individual's psychologic equilibrium. (2) (EP; AN)
  1 A social crisis occurs when a person feels discomfort with others or has a fear of being observed or scrutinized.
  3 There are insufficient data to come to this conclusion; the client may or may not be facing a financial crisis.
  4 Adjustment to retirement is a developmental task of later adulthood; the client was forced into early retirement, it was not a choice.

78. ③ A situational crisis is a sudden, unexpected event with which the individual is unable to cope using past coping behaviors; this time frame provides an opportunity for the individual to learn new coping behaviors. (2) (SC; AN)
1 This would be too short a period of time for the individual to develop new, successful coping mechanisms.
2 Same as answer 1.
4 This would be a longer than expected time period.

79. ③ The major goal of crisis intervention is to resolve the present crisis and return the client to the precrisis level of functioning. (1) (SC; AN)
1 This would be a short-term goal.
2 Same as answer 1.
4 This is a goal of psychotherapy, not crisis intervention.

80. ② Personal internal strengths and supportive individuals are critical factors that can be employed to assist the individual to cope with a crisis (2) (SC; AS)
1 Although this information may be helpful, it is not essential; factors concerning the present situation are paramount.
3 This is unrealistic; identifying unconscious conflicts takes a long time and is inappropriate for crisis intervention.
4 This is a goal of psychotherapy, not crisis intervention; it is usually not necessary to restructure the personality to resolve a crisis.

81. ④ The assessment of the client's present status and ability to perform ADL is the priority because it will influence the choice of an appropriate therapeutic regimen. (3) (SC; AS)
1 Although significant, it is not the priority.
2 Concern now is for the client, not how the client's behavior affects others.
3 The present crisis must be dealt with first.

82. ④ A crisis intervention group helps clients reestablish psychologic equilibrium by assisting them to explore new alternatives for coping; it considers realistic situations using rational and flexible problem-solving methods. (1) (SC; AN)
1 This is not an immediate goal of crisis intervention.
2 Clients are never given a solution; they are assisted in arriving at their own acceptable, workable solutions.
3 It is not necessary for crisis intervention workers to be psychologists.

83. ② Behavior that reflects the recognition of the intrinsic worth of each individual is essential in all

supportive relationships; problem-solving potential is increased when clients are involved in exploring alternatives that will affect the direction of their own lives. (1) (SC; PL)
1 The client, not the worker, is encouraged to identify the problem.
3 This is useless; the client is unable to empathize with others at this time.
4 This is impossible during the immediate intervention period; the client needs to find direction immediately.

84. ④ Although the client's facial expression suggests happiness, the client's tone of voice gives the message of anger; the behaviors do not go together. (1) (SC; AS)
1 The data given do not support this assessment.
2 Same as answer 1.
3 Same as answer 1.

85. ④ The client is using intellectual reasoning to block confronting the unconscious conflict and the stress of having to deal with his girlfriend's pregnancy. (2) (SC; AS)
1 No data demonstrate that the client is projecting blame on anyone else.
2 No data demonstrate that the client is concentrating thoughts and emotions on his inner self.
3 No data demonstrate the shifting of emotions from an emotionally charged object or person to a neutral one.

86. ④ A crisis is defined as a situation in which the client's previous methods of adaptation are inadequate to meet present needs. (2) (SC; AN)
1 A crisis is not necessarily related to the degree of stress; it occurs when past coping mechanisms are ineffective.
2 This is not the immediate stress for which the client has no coping mechanism.
3 This is not causing the crisis; the client's lack of coping mechanisms is.

87. ④ The crisis center nurse's main responsibility is to assist the client in using the problem-solving process; the client will be helped in exploring alternative solutions to a situation and will be given information regarding other agencies, facilities, and services. (2) (CC; PL)
1 Although the client's decision should be supported, this is a judgemental response.
2 This is not part of the immediate goal during the crisis; the client may be encouraged to seek help later for other problems.
3 This is one of many instructions for which the client must take primary responsibility.

88. (1) Intrinsic motivation is motivation that is stimulated from within the learner; it is most effective because the learner recognizes the need to know, is self-directed, and is ready to learn. (2) (DF; EV)
    2 This is stimulation from without and is very often ineffective; desire must come from within.
    3 There is no external reward for attending classes.
    4 This is a new behavior based on a new situation; there is no external reward for attending classes.

89. (1) Emotionally immature individuals are often unable to deal with the role changes associated with parenthood. (2) (DF; AN)
    2 This may have contributed to the crisis but did not precipitate it.
    3 Same as answer 2.
    4 Same as answer 2.

90. (4) According to the Guidelines for Sterilization of the American Association on Mental Deficiency, consent for sterilization in most states must be obtained from the court and legal guardian if the client is mentally incompetent and unable to give informed consent. (3) (CC; PL)
    1 The court alone cannot decide the issue; a legal guardian must be involved.
    2 The next of kin may place pressure on the client; therefore permission should also be obtained from the court and legal guardian.
    3 Legal guardians (who may be next of kin) may exert undue influence; therefore they alone cannot sign the consent.

91. (2) The client must be intellectually competent, that is, able to comprehend the outcome of the procedure, in order to give informed consent. (2) (CC; PL)
    1 Informed consent can only be obtained from a client who is intellectually competent to understand the outcome of the procedure.
    3 This is unrealistic; a mentally challenged client would be unable to fully explain what the procedure entails; it is more important for the client to understand the outcome of the procedure.
    4 The client should be free from the influence of a parent, legal guardian, or any others who might press to have the procedure performed.

92. (4) If one child in the family is identified as being different by the parents or siblings, coupled with other signs of abuse, abuse should be sus-

pected, and the situation warrants further investigation. (1) (EP; AS)
    1 Taking a walk would be helpful for both the mother and the child and would not indicate abuse.
    2 This is an acceptable punishment for misbehaviors.
    3 Although this is demeaning, it is not abuse.

93. (4) In most instances, the abusing parent avoids talking about the situation as a means of reducing guilt and repressing the action. (2) (EP; AS)
    1 Denying the beating requires the parent to fabricate a story about how the obvious physical injury occurred.
    2 Little concern is expressed for the child because this would require verbal expression and acceptance of the action.
    3 A parent's concern for the unabused children without the expression of concern about the abused child would document a different feeling about the abused child, which the parent would try to avoid.

94. (3) These underlying feelings commonly precipitate trying to improve the child's behavior by the beating. (2) (EP; AS)
    1 These parents usually do not admit their behavior, so they do not have a need to rationalize it.
    2 These parents offer many vague explanations of how the child was injured; rarely is the explanation detailed.
    4 This would be an unusual request because abusing parents do not usually ask to see their children.

95. (3) The nurse feels that an inability to deal with the daughter could have resulted in a loss of control, leading to possible abuse. (3) (SC; AN)
    1 The nurse is uncomfortable because the client's situation probably resulted from similar feelings.
    2 Same as answer 1.
    4 Nothing in the data presented leads to this conclusion; the argument could have been justified.

96. (3) By learning how the child's behavior provokes frustration, parents may develop more acceptable ways of responding. (3) (EP; IM)
    1 Although these parents need to learn what behavior is appropriate for a given age level, they must also learn how to respond correctly to less appropriate behavior.
    2 Punishment of a child for behavior is always wrong; it is an act of retribution, not an act of discipline.

4 The abusing parent responds to both negative and acceptable behavior with abuse.

97. (2) This child would distrust any approach because approaches commonly result in pain; abused children remain alert in an attempt to ward off an attack. (2) (EP; AS)
1 This child would not be open to an approach by a stranger; basic trust of others has not developed in abused children.
3 This child would be acutely aware of anyone coming near; abused children attempt to defend themselves by keeping alert to the possibility of attack.
4 This child would usually not cry out; abused children learn not to expect comforting or soothing of pain by others.

98. (1) This would provide reassurance and support for the child. (2) (EP; IM)
2 Using the phrase "anything wrong down there" could cause the child to have negative feelings about the self.
3 Depending on the mother's involvement in the situation, the mother's involvement might be threatening rather than supportive to the child.
4 Asking the child to make this decision at this time would be nontherapeutic and may be threatening.

99. (2) This response would elicit further clarification of what the child means by "bad." (3) (EP; AS)
1 This would not be helpful; it would do nothing to clarify the child's idea of what "bad" means or the child's feelings about what happened.
3 The nurse needs to determine what the child means by the word "bad" before reflecting the term back to the child.
4 This would be a nontherapeutic response because the uncertainty implied by the nurse would increase the child's feelings of guilt.

100. (1) Because rape is a threat to the sense of control over one's life, some control should be given back to the client as soon as possible. (3) (SC; PL)
2 This is a normal form of ventilating emotions; the client should be told that medication is available if desired.
3 This takes control away from the client; the client could view this as an additional assault on the body that increases feelings of vulnerability and anxiety and does not restore control.
4 Same as answer 3.

101. (1) Partners may themselves feel angry and abused; these feelings should be rapidly and openly discussed. (3) (SC; AN)

2 This should not be done yet; rape counselors deal with the victim and partner together.
3 The partner's feelings must be resolved before the partner can really help the client.
4 This may be reassuring, but it leaves the partner alone to deal with feelings.

102. (1) This is the most direct approach to ascertain if pills were ingested; client will usually respond to this type of direct question. (2) (EP; AS)
2 This would not provide any useful information.
3 This is not an initial response until a determination is made regarding the number of pills taken.
4 This would be appropriate later but is not an initial priority.

103. (1) Her anger at her husband for leaving her will make her feel guilty for having these feelings. (3) (SC; AS)
2 Money may or may not be a problem for this client.
3 Loneliness will be something she will have to deal with later, depending on her support system; it is not an immediate problem.
4 Estrangement may be something she will have to deal with later; it is not an immediate problem.

104. (3) Whether there is a suicide plan is a major criterion in assessing the client's determination to make another attempt. (1) (SI; AS)
1 Although this may be important for planning future therapeutic approaches, this does not explore the potential for suicide, the priority at this time.
2 Same as answer 1.
4 Same as answer 1.

105. (2) Suicide threats and gestures are a means of communicating anger, frustration, hopelessness, and despair to significant others and should always be taken seriously. (1) (EP; IM)
1 Children have many means readily available; many means of suicide are common objects around the home and playground.
3 Although suicide is the second leading cause of death in the 15- to 24-year-old age group, children younger than age 6 do attempt suicide and some succeed.
4 A suicide gesture is usually a cry for help, but a suicide attempt is usually self-destructive; a suicide attempt is usually carried out with the belief that death will result; neither a suicidal gesture nor a suicide attempt is manipulative; an impulsive act that is a rage response designed to punish others may be manipulative.

106. (3) The parents are focusing their feelings about their child's prognosis on someone or something else, in this case each other. (1) (SC; IM)
 1 Denial is ignoring, avoiding, or refusing to recognize painful realities.
 2 Projection is the attribution of one's own feelings to another person.
 4 Compensation is a defense in which one makes up for a perceived deficiency by emphasizing another feature perceived as an asset.

107. (3) Reasonable limits are necessary because they provide security and help to keep the child's behavior in acceptable bounds. (3) (SC; PL)
 1 This is an unrealistic approach that allows the child to manipulate the total situation.
 2 Care should be directed to help the child, not the staff.
 4 Relationships, not special privileges, should provide the necessary security.

108. (1) Children commonly use denial of parents as a stage of coping with separation; they avoid the fact that separation is real. (1) (SC; IM)
 2 Undoing is a behavior or communication technique calculated to neutralize earlier behavior or communication.
 3 Repression is the involuntary exclusion of painful thoughts, impulses, or memories; this child's behavior is voluntary.
 4 Sublimation is the act of substituting socially acceptable behavior for an unacceptable feeling or drive.

109. (2) Seeing their child as soon as possible will validate the death for them and initiate the grieving process. (2) (EP; IM)
 1 This will delay and prolong the grieving process; the response offers no explanation for waiting.
 3 This is unnecessary; the parents have asked to see their child now.
 4 This is untrue; it would be more traumatic to wait and delay the reality of the death.

110. (2) Deaths that are perceived as preventable cause more guilt for the mourners and therefore increase the intensity and length of the grieving process. (2) (EP; AN)
 1 This is untrue; it is usually more difficult.
 3 It may prolong and intensify the mourning process but will not necessarily result in a pathologic reaction.
 4 Same as answer 3.

111. (1) This provides support until the individuals' coping mechanisms and personal support systems can be mobilized. (3) (EP; IM)

 2 The individuals, not the nurse, must mobilize their support systems.
 3 This is not the role of the nurse.
 4 The individuals need time before the full reality of the death can be accepted.

112. (4) Support is most important when dealing with the crisis of death; the support system must be relied on for coping with the loss. (2) (EP; AS)
 1 The client's age may play a role in coping, but it is not the most important factor.
 2 The timing may be important if death is just one of a multiplicity of stresses, but it is not the most important factor in helping a client cope.
 3 The socioeconomic status may be important in long-term planning, but it is not the most important factor in the grieving process.

113. (3) When the relationship was ambivalent, there is both anger and guilt to work through. (1) (EP; AN)
 1 A loving relationship evokes fewer feelings of guilt and a less-complicated grieving process.
 2 The length of the relationship seems to have little to do with the ease or difficulty in completing the grieving process.
 4 Individuals in the subservient role usually have learned to accept directions and either find a new director or are relieved to have a chance to express their own feelings.

114. (3) Successful resolution means being able to remember the good as well as the bad qualities of the deceased and accepting them as part of being human. (1) (SC; EV)
 1 Resolution involves working through feelings, not just accepting what occurred.
 2 Resolution does not mean forgetting but rather realistically remembering what was.
 4 This is an unhealthy response that can become pathologic because of the unresolved feelings about the other person's qualities.

115. (4) Discussing the partner and the partner's death will help the client work through the grief process. (3) (EP; PL)
 1 This would refocus the client's attention away from dealing with feelings; the client would probably not have the physical or emotional energy to get involved with group activities.
 2 The client must deal with the past and present before dealing with the future.
 3 Same as answer 1.

116. (4) Helping a client deal with unresolved grief involves assisting the client to express thoughts

and feelings about the lost object or person as a necessary part of grief work. (1) (EP; IM)

1 This is the responsibility of the nurse; another client would not have the expertise to help the client.

2 This would be too threatening; at this point the client needs therapeutic one-to-one interaction.

3 The current nonverbal behavior indicates that the client is dealing with feelings; an opportunity should be provided for a verbal exploration.

117. ③ These are the defining characteristics of dysfunctional grieving. (3) (EP; AS)

1 The client's communication does not lead to this conclusion.

2 Same as answer 1.

4 Eight months does not constitute a prolonged period of mourning.

118. ② Such stimuli encourage the client to remain reality oriented; research has shown that competing stimuli are useful in controlling hallucinations. (2) (DA; PL)

1 This does not ensure that the client's needs will be met.

3 This is not very realistic and fosters greater dependency; it focuses on the client's inability to deal with the problem and increases the client's fear of being alone.

4 Same as answer 3.

119. ③ Denial may be handled by seeking other opinions in an attempt to prove an unacceptable one incorrect. (1) (EP; AS)

1 This occurs during the stage of anger, which is a later stage.

2 This occurs during the stage of depression, which is a later stage.

4 This is not associated with the initial stage; this behavior usually occurs after the client recognizes the inevitable outcome and is fearful of being alone.

120. ④ Rationalization is offering a socially acceptable or logical explanation to justify an unacceptable feeling or behavior. (1) (SC; AS)

1 Projection is the denial of emotionally unacceptable feelings and the attribution of the traits to another person.

2 Sublimation is the substitution of a socially acceptable behavior for an unacceptable feeling or drive.

3 Compensation is making up for a perceived deficiency by emphasizing another feature perceived as an asset.

121. ① It is important for the couple to discuss their feelings to maintain open communication and support each other. (2) (SC; PL)

2 This action would not meet the needs of this couple; it focuses only on the client's needs and ignores the partner's needs; in addition, most psychotherapy is a long-term process.

3 This may be useful in the future but is most likely premature; they need to deal with their own feelings first.

4 This may elicit feelings but will not improve communication; this is a rather long-term goal.

122. ④ This approach is positive because it attempts to deal with the staff's feelings and the problem without singling out people for guilt; the nurse therefore is taking ethically sound action without being moralistic or authoritarian. (1) (CC; PL)

1 This abdicates the nurse's responsibility and may create anger and guilt in the staff.

2 Same as answer 1.

3 Same as answer 1.

123. ① The client has lost a companion and a purpose in life; these feelings can be overwhelming until new activities are developed. (2) (SC; AS)

2 Anger should not be a major problem for this client; data do not address any problem in the husband-wife relationship.

3 Data do not address the financial status of this client.

4 If there is guilt over feelings of relief about the husband's death, they would be transitory and not a major problem.

124. ② The spouse is expressing the normal feelings of guilt associated with the death of a loved one; there is always initial guilt over what might have been. (1) (SC; AN)

1 No evidence supports this conclusion.

3 The spouse is expressing guilt, not shame.

4 Same as answer 1.

125. ③ With the medication, the individual does not face the reality of the loss and merely delays the onset of the pain associated with it; because most support is available at the time of the death and the funeral, tranquilizers at this time deny the individual the opportunity to use this assistance. (2) (SC; AN)

1 This is untrue; tranquilizers do not magnify the risk of suicide.

2 Brain activity does not cause a reactive depression.

4 Although tranquilizers may initially cause some lethargy, this is not the reason they are not ordered.

126. ④ The nurse should recognize that the husband's behavior represents anticipatory grieving. (3) (SC; AN)

1 Although this may be true, the husband is extremely involved with his own needs at this time.

2 The husband is beginning the grieving process; the husband's actions do not appear to be an expression of anger but rather an attempt to cope with the situation.

3 There are no data to substantiate this conclusion.

SUBSTANCE ABUSE (SA)

127. ① Denial is a method of resolving conflict or escaping unpleasant realities by ignoring their existence. (1) (SC; AN)

2 In projection a person faults another person for having unacceptable impulses, thoughts, or behaviors that are too uncomfortable to accept as one's own.

3 Displacement refers to the transfer of an emotion from one object or situation to another, usually safer, object.

4 In compensation the person makes up for personal inadequacies by emphasizing attributes to gain social approval.

128. ① The history of the loss of a child and the intensity of the client's drinking since the child's death should lead to this nursing diagnosis. (3) (EP; AN)

2 This is a symptom or a medical diagnosis, not a nursing diagnosis.

3 There is no documentation that the client is unable to distinguish between the self and nonself.

4 There is no documentation that the family has not attempted to provide support; it is the family who has brought the client for help.

129. ④ Alcohol is a central nervous system depressant; these symptoms are the body's neurologic adaptation to the withdrawal of alcohol. (2) (SI; AS)

1 Convulsions are not early signs of alcohol withdrawal; they would not occur before 48 to 72 hours of abstinence.

2 Fever and diaphoresis may be seen during prolonged periods of delirium and are a result of autonomic overactivity.

3 These are late signs of severe withdrawal that occur with delirium tremens; tachycardia results from autonomic overactivity.

130. ② Motivation is necessary to assist the client in withstanding the pain of giving up a defense; motivation is more influential in facilitating change than any external factor. (2) (SC; AN)

1 Although having family support is important, motivation to change is the most important factor.

3 This can be of assistance, but internal factors will have a greater impact on rehabilitation than will external factors.

4 Self-esteem will be useful if it precipitates abstinence behavior; however, people with alcoholism commonly have low self-esteem.

131. ① Alcohol causes both physical and psychologic dependence; the individual needs and depends on the alcohol to function. (2) (EP; AN)

2 This is untrue; alcoholism is a disease that entails physical and psychologic dependence.

3 This is a myth often associated with alcoholism; the individual needs to learn how to use other coping mechanisms more consistently and effectively.

4 People with alcoholism commonly drink alone or feel alone in a crowd.

132. ④ The data demonstrate that the classic symptoms of withdrawal delirium occur within a week after the cessation or reduction of alcohol intake. (2) (SI; AN)

1 The information does not demonstrate the presence of impaired short-term or long-term memory; this usually develops shortly after a period of prolonged heavy drinking.

2 There are insufficient data to identify dementia; impairment of thought processes, judgment, and intellectual abilities would have to continue for 3 weeks or longer.

3 The information does not demonstrate the presence of hallucinations; these usually develop within 48 hours of cessation or reduction of alcohol intake.

133. ④ This behavior indicates that the client lacks insight and is denying that alcohol is the problem. (1) (EP; AN)

1 There are no data to support this conclusion.

2 Same as answer 1.

3 Same as answer 1.

134. ③ To maintain sobriety, alcoholics must forever totally alter patterns of behavior that have been reinforced and used for prolonged periods of time. (2) (EP; AN)

1 Although drinking helps to reduce the pain of reality, it does not make the abuser feel good.

2 There is no known physiologic need for alcohol.

4 Alcoholics do not have character defects; drinking helps to blunt the pain of reality.

135. (1) Clients who abuse alcohol characteristically have lowered self-esteem; therefore, it is important for the nurse to accept the person as an individual with value. (1) (SC; PL)

2 Although nurturing is important, this client must learn self-reliance.

3 This would probably be an old story to this client and would have little positive effect.

4 This action would not provide an atmosphere that would help the client withstand the stress of the detoxification program.

136. (3) The client must learn to develop and use more healthful coping mechanisms if drinking is to be stopped; the responsibility is with the client because the client must do the changing. (2) (SC; PL)

1 This would tell the client what to expect but would not instill responsibility for change.

2 This will increase guilt and place the client on the defensive; it usually does not foster the development of a trusting relationship.

4 Medications do not provide the motivation for change; this must come from within the client.

137. (1) This statement recognizes the feeling of ambivalence associated with admitting that a problem with alcohol exists; this occurs early in treatment. (2) (SC; IM)

2 This places the client on the defensive and interferes with communication.

3 Same as answer 2.

4 Same as answer 2.

138. (3) This focuses on the client's feelings with a supportive, helpful approach. (2) (SC; IM)

1 This is a judgmental approach that alienates the client from the therapeutic process and prevents the establishment of rapport.

2 Same as answer 1.

4 This intervention reinforces the client's denial and avoidance of the problem.

139. (3) This strengthens the client's link with reality and reassures the client of safety because the hallucinations are usually frightening. (2) (EP; IM)

1 The nurse must respond to the client's behavior by attempting to point out reality and reduce anxiety.

2 Validation reinforces the client's distorted perceptions of reality, is not helpful, and may even be unsafe.

4 It is not helpful to argue or try to explain the hallucinations away because they are real to the client.

140. (3) Peripheral neuropathy is present, and this measure will limit tactile stimulation. (2) (PC; IM)

1 This would do little to relieve discomfort or reduce the occurrence of neurologic symptoms.

2 Same as answer 1.

4 This may be done periodically; however, these symptoms are not caused by impaired circulation but rather by peripheral neuropathy.

141. (1) This supports the family of the addicted individual and allows the family to continue on with life by reducing guilt. (2) (CC; IM)

2 The family has already stated that this is impossible when drinking occurs.

3 This places the burden for preventing drinking on the family and will create feelings of resentment and guilt.

4 This is enabling behavior, which does not help the abuser or the family.

142. (3) Symptoms begin with anxiety, shakiness, and insomnia; within 24 hours convulsions, delirium, tachycardia, and death can occur. (2) (PT; AS)

1 Heroin withdrawal is rarely life threatening but does cause severe discomfort, such as abdominal cramping and diarrhea.

2 Methadone withdrawal is rarely life threatening but does cause severe discomfort, such as abdominal cramping and diarrhea.

4 Amphetamine withdrawal does not cause life-threatening symptoms but does result in severe exhaustion and depression.

143. (3) The addict is unable to delay gratification because of inadequate ego strengths; addicts are concerned with the present and the self. (2) (EP; PL)

1 It is possible, although not easy, but it does require a change in attitude and a deconditioning process.

2 Drug users are concerned with reality; their drug use is an attempt to blur the pains of reality.

4 Education of the public has been extensive, but the new user of drugs does not believe addiction will occur.

144. (3) These symptoms are all associated with opiate withdrawal, which occurs after cessation or reduction of prolonged moderate or heavy use of opiates. (2) (PT; AS)
   1 Lacrimation and vomiting are present, but insomnia, not drowsiness, occurs with opiate withdrawal.
   2 Nausea is present, but diarrhea, not constipation, and constricted pupils, rather than dilated pupils, occur with opiate withdrawal.
   4 Rhinorrhea is present, but fever, rather than a subnormal temperature, and muscle aches, rather than convulsions, occur with opiate withdrawal.

145. (4) The client is unconscious and unable to meet physical needs. (2) (PC; PL)
   1 Support and understanding are important once the client's physical condition has stabilized.
   2 Maintaining a drug-free environment will be a priority later in the treatment program.
   3 Establishing a therapeutic relationship will increase in importance once the client's physical condition has stabilized.

146. (1) Opiates cause central nervous system depression, resulting in severe respiratory depression, hypotension, and unconsciousness. (2) (PT; AS)
   2 These findings, particularly the respirations, are not indicative of an overdose of an opiate.
   3 Same as answer 2.
   4 Same as answer 2.

147. (3) Addicted individuals often use a substance to increase their feelings of worth; the substance helps them appear bigger in their own eyes; they have strong unmet dependency needs. (2) (EP; AN)
   1 Although guilt about breaking society's code may be present, there is no rejection of reality, just an inability to deal with it.
   2 Although there is a need for acceptance, there is no underlying feeling of hostility.
   4 Although anger is present and internalized, there is no struggle for independence.

148. (3) Individuals often take drugs because they cannot deal with the pain of reality; the drug blurs the pain. (1) (EP; AN)
   1 Drugs increase dependency rather than fostering independence.
   2 Although this factor may encourage initial use by some adolescents, it is not the most common reason for use.
   4 The use of drugs fosters social isolation rather than social relationships.

149. (2) Stimulating the central nervous system with cocaine most commonly causes these responses, which can progress to fear, hallucinations, paranoid delusions, and violent behavior. (2) (EP; AS)
   1 Nausea is not a side effect; euphoria, rather than fatigue, and loss of appetite, rather than hunger, are side effects.
   3 These are not common side effects; nasal septal degeneration occurs with prolonged inhalation, and cardiac arrest is an ever-present danger.
   4 An increase in energy, rather than lethargy, occurs; some cocaine users believe it maximizes sexual experiences, but there is no documentation of this physiologic response.

150. (1) Tolerance is a phenomenon that occurs in addicted individuals and increases the amount of drug needed to satisfy their need; because this phenomenon is not permanent and may disappear without warning, overdose commonly occurs. (1) (PT; AN)
   2 The problem is not related to dependence and addiction; the failure to respond to an adult dose of an opiate is related to tolerance.
   3 Same as answer 2.
   4 Same as answer 2.

151. (1) This allows the individual to use the staff as a support system and removes an opportunity to deny the problem. (3) (CC; PL)
   2 This supports and permits denial; both the individual and the staff know a problem exists, and the individual must admit it.
   3 This is a nonprofessional approach that would be nontherapeutic for the individual.
   4 Although this may be part of a return-to-work contract, it is not necessarily therapeutic; it simply reduces legal risks.

152. (4) This is a nonpunitive approach that attempts to salvage the nurse as an individual and as a professional. (2) (CC; IM)
   1 This may be necessary for long-term therapy but would not be the initial approach.
   2 This is a punitive, nontherapeutic response that offers no chance for rehabilitation.
   3 The client is addicted; promises will not keep the client from abusing drugs.

153. (3) This is a serious charge, and confrontation should occur in the presence of a supervisor. (3) (EP; IM)
   1 This is unnecessary; as a professional, the nurse has enough information to confront the other nurse.

2 This is an assumption that may result in an altercation; a witness should be present.

4 This is not a professional approach; the nurse has a legal responsibility to intervene.

154. (1) The client is using denial to separate from the group members and needs realistic feedback to prevent withdrawal. (3) (EP; IM)

2 This would do nothing to help return the client to the group.

3 The client's meaning is clear; questioning the client at this point would be nontherapeutic.

4 This is inadequate; the client first needs to recognize that the problems being discussed are applicable.

155. (3) These provide high levels of thiamine; other sources include legumes, whole and enriched grains, and lean beef. (2) (PC; EV)

1 In this list, only fish is considered a source of thiamine.

2 In this list, only eggs are considered a source of thiamine; this list contains sources of protein.

4 Most vegetables contain only traces of thiamine; citrus fruits provide vitamin C.

156. (3) The client is still denying the illness and has not resolved the basic problem that led to the alcoholism. (2) (EP; EV)

1 This is incorrect because the client is still denying the illness; willpower alone will not keep the client away from alcohol.

2 This may be true, but it does not ensure compliance or successful rehabilitation.

4 This is not helpful unless the client understands the basis of the conflicts and his role in resolving them.

EATING AND SLEEP DISORDERS (ES)

157. (1) This is the best initial response; it encourages additional ventilation of feelings. (3) (EP; IM)

2 This is not necessarily true, and the response is somewhat threatening and non-therapeutic.

3 Same as answer 2.

4 This would not be therapeutic; the client is verbally expressing feelings, and the behavior does not require medication at this time.

158. (4) These are the major symptoms of anorexia nervosa; weight loss is excessive (15% of expected weight) and nutritional deficiencies result in the amenorrhea and a bloated abdomen. (3) (EP; AS)

1 Weight loss is greater than 10%.

2 Not associated with anorexia nervosa.

3 Memory lapses are not associated with anorexia nervosa; the other symptoms are more associated with anxiety.

159. (3) The client with anorexia nervosa simply denies the need for food or presence of hunger; the client with bulimia hides the behavior because there is a recognition that the behavior is a problem. (3) (EP; AN)

1 Untrue; clients with anorexia nervosa are more introverted and tend to avoid relationships.

2 Same as answer 1.

4 The clients with bulimia are at a greater risk for fluid and electrolyte problems because of the purging; clients with anorexia nervosa are more at risk for severe nutritional deficiencies.

160. (4) This is a theoretical explanation for the development of anorexia nervosa. (2) (EP; AN)

1 This does not play a role in the development of anorexia nervosa.

2 Same as answer 1.

3 The basis is the struggle between dependence and independence, not a desire for independence alone.

161. (1) Clients with anorexia nervosa have a disturbed self-image and always see themselves as fat and needing further reducing. (2) (EP; AS)

2 The relationship is usually not supportive; it is disturbed.

3 There is usually high achievement and great concern about grades.

4 There is usually dissatisfaction with weight and a desire to lose weight.

162 (2) These clients are usually severely malnourished and have severe fluid and electrolyte imbalances; unless these imbalances are corrected, cardiac irregularities and death can occur. (2) (DA; PL)

1 This is important, but it is not the highest priority at this time.

3 Same as answer 1.

4 Same as answer 1.

163. (4) These clients have severely depleted levels of potassium and sodium because of their starvation diet and energy expenditure; these electrolytes are necessary for proper cardiac functioning. (2) (DA; AN)

1 Although this may occur, it is not the major health problem.

2 Same as answer 1.

3 Same as answer 1.

164. ③ This is a nonthreatening, open-ended response that focuses discussion and leaves channels of communication open. (2) (EP; AS)
    1 Although this does not quite accuse the client of lying, it is a threatening response that questions the client's truthfulness.
    2 Same as answer 1.
    4 Same as answer 1.

165. ③ It is most therapeutic for the staff to control food needs, thus removing the parents from the struggle. (3) (CC; IM)
    1 This may be interpreted as accusatory and increase the parents' guilt.
    2 This is nontherapeutic; it cuts off the parents from future involvement.
    4 This is nontherapeutic; it only continues the struggle between the parents and the client.

166. ③ The client's security is increased by limit-setting; guidelines remove responsibility for behavior from the client and increase compliance with the regimen. (2) (EP; PL)
    1 The client needs control, not empathy.
    2 Simply maintaining control is not therapeutic and increases the power struggle.
    4 Emphasis on dietary intake increases the power struggle between the client and the staff.

167. ③ This action would be neither a positive nor a negative reinforcement of specific behavior; it would provide rewards for achievement of specific goals. (2) (EP; PL)
    1 This would not be included in a behavior modification program.
    2 Same as answer 1.
    4 Clients talk freely about food; the problem is with ingestion, not discussion.

168. ① All food issues should be discussed with the dietitian, thus removing a potential source of conflict between nurse and client. (2) (EP; PL)
    2 This would increase the conflict between nurse and client.
    3 This would accomplish little because the client's failure to eat is not based on likes or dislikes.
    4 This may be self-defeating because discussion of food would be the major focus of all nurse-client interactions.

169. ③ Clients with anorexia nervosa use manipulation to divide the nursing staff; sharing this knowledge alerts health team members. (3) (EP; AN)
    1 This would be counterproductive because it supports the client's manipulative behavior.

    2 The client is attempting to manipulate the staff; this is not how trust is established.
    4 Same as answer 1.

170. ④ By talking to the client on the telephone at mealtimes, the family is enabling the client to continue destructive behavior; the client and family must be included in discussion of and possible solutions to the problem. (2) (CC; PL)
    1 This would be a punitive approach that would not deal with the underlying problem.
    2 Same as answer 1.
    3 Same as answer 1.

171. ② This would demonstrate a change in behavior as well as a positive approach to meals. (2) (SC; EV)
    1 This would be typical behavior of a client with anorexia nervosa.
    3 The problem is not a lack of interest about food but a deliberate failure to ingest food.
    4 This behavior is unrelated to the behavior that needed to be changed.

172. ② Eating behavior may be the major area of life over which the client has power; the client also has a need to stay child-like. (1) (EP; AN)
    1 With anorexia nervosa, food is not equated with love.
    3 People with anorexia nervosa do not use food for security; overweight people tend to use food for security.
    4 People with anorexia nervosa do not stop eating to gain attention; it provides them with a sense of power or control over an area of their life.

173. ① Recovery necessitates major changes in self-esteem and body image, which require therapy over a long period of time. (2) (EP; IM)
    2 Untrue; long-term therapy is required.
    3 This is only one factor; the client must also be willing to work with the family and to accept the pain associated with change.
    4 This is too simplistic a response; anorexia nervosa is not just a desire to lose weight.

SCHIZOPHRENIC DISORDERS (SD)

174. ③ Physical safety of the client and others is the priority. (1) (SI; PL)
    1 Although it is important for the clients to live up to contracts, it is not imperative.
    2 The nurse cannot prevent or control the clients' use of defensive behavior.
    4 Same as answer 2.

175. (4) Seclusion or restraints are special procedures for dealing with acting-out, aggressive behavior for the protection of the client or others; clear documentation in the progress notes is essential when suspension of the client's rights is necessary. (1) (CC; IM)

1 This is not necessary because the use of restraints and/or seclusion would be included in the general consent form signed on admission.

2 This monitoring should be done for all clients.

3 There is no such form; however, documentation to justify the need for seclusion or the use of restraints is required.

176. (2) These clients are threatened by reality; withdrawal from reality and the use of magical thinking reduce anxiety. (2) (EP; AS)

1 Clients with schizophrenia are not preoccupied with suicidal thoughts.

3 Clients with schizophrenia have poor self-esteem and a low self-image and usually have feelings of guilt and self-blame.

4 The loosening of associative links that occurs in schizophrenia makes these impossible.

177. (3) These are the classic premorbid symptoms of schizophrenia. (2) (EP; AS)

1 These symptoms are usually not associated with schizophrenia.

2 These symptoms are not associated with any particular type of disorder.

4 Same as answer 1.

178. (2) Because he is newly admitted and unaware of the treatment plan, ignoring his behavior to avoid reinforcement is best at this time. (3) (EP; IM)

1 This may be seen as critical or condemning of the individual with an already low self-esteem.

3 This is judgmental; he may already be aware that the behavior is not normal.

4 It would serve little value to refocus on attention-getting behavior later; focus on appropriate behavior.

179. (1) The presence of ego strengths is demonstrated by some level of adjustment before the occurrence of the precipitating event; these ego strengths can be used to help the client reorganize the personality. (1) (EP; AS)

2 This would tend to contribute to a poor prognosis.

3 Same as answer 2.

4 Same as answer 2.

180 (2) The usual age of onset of schizophrenia is adolescence or early adulthood. (2) (EP; AS)

1 Symptoms usually do not appear this early.

3 Same as answer 1.

4 Same as answer 1.

181 (4) By observing the client, the nurse is able to adjust care and communications to reflect assessment of individual needs. (1) (PC; AS)

1 This is not vital to initially help the client; the nurse should meet the client where the client is now.

2 Specific clients differ; the nurse should meet the client where the client is now.

3 Same as answer 1.

182. (3) Most hallucinating clients hear voices without external stimuli. (1) (EP; AS)

1 Although hallucinating clients may see things without external stimuli, visual hallucinations are not as common as auditory hallucinations.

2 Tactile hallucinations are not very common.

4 Olfactory hallucinations are not very common.

183. (3) Neologisms are words that are invented and therefore understood only by the person using them. (1) (EP; AS)

1 Echolalia is the verbal repeating of exactly what is heard.

2 Concretism is a pattern of speech characterized by the absence of abstractions or generalizations.

4 Paleologic thinking is a disturbed system of logic in which subjects are made identical if two variables about them are the same.

184. (3) The client needs to feel comfortable in the environment before feeling enough trust to undress for showering; the nurse's statement allows the client to make the decision. (2) (EP; IM)

1 This action would add to the client's anxiety and feelings of loss of control; it could also add to any delusional thoughts the client may have.

2 Same as answer 1.

4 This statement would not help the client's self-image, and it really does not matter what other clients think.

185. (1) An illusion is a misinterpretation of an actual sensory stimulus. (2) (EP; AN)

2 A delusion is a fixed false belief.

3 A hallucination is a false sensory perception without a stimulus being present.

4 Confabulation is filling in blanks in memory.

186. (4) A somatic delusion is a fixed false belief pertaining to part of the body. (2) (EP; AN)
 1 An illusion is a misinterpretation of an actual sensory stimulus.
 2 Hallucinations are false sensory perceptions without stimuli being present.
 3 Depersonalization is a feeling of unreality concerning the self.

187. (2) Ambivalence describes the existence of two conflicting emotions, impulses, or desires. (2) (EP; AN)
 1 Double bind is two conflicting messages, not emotions, in a single communication.
 3 Loose associations are not two conflicting emotions but the loosening of connections between thoughts.
 4 Inappropriate affect is not two conflicting emotions but the inappropriate expression of emotions.

188. (2) Failure to accept the client and the client's fears establishes a barrier to effective communication. (1) (EP; AN)
 1 Today, mental health therapy is directed toward returning the client to the community as rapidly as possible.
 3 Electroconvulsive therapy (ECT) is not effective in clients with schizophrenia; in fact, it makes them more confused.
 4 Family cooperation is helpful but not an absolute necessity; the client can get well in spite of the family.

189. (2) Depersonalization is a feeling of change or unreality about the self or the environment caused by a loss of ego boundaries and a loss of reality testing. (3) (EP; AN)
 1 Paranoid ideations are beliefs that the individual is being singled out for unfair treatment.
 3 Loose associations are verbalizations that are difficult to understand because the links between thoughts are not apparent.
 4 Ideas of reference are false beliefs that the words and actions of others are concerned with or are directed toward the individual.

190. (1) The client cannot reach out to others because of lack of trust; withdrawal is used to defend against interpersonal threats and results in isolation. (2) (EP; AN)
 2 Sleep disturbances are not common because clients tend to use sleep to withdraw from reality.
 3 Most clients with schizophrenic disorders are not violent.
 4 This is usually not associated with this disorder.

191. (4) Any other approach would be threatening, increase anxiety, and probably result in a physical confrontation. (3) (EP; PL)
 1 This would increase anxiety, not foster decision making.
 2 This would increase the client's anxiety and probably result in a physical confrontation.
 3 This would increase anxiety, not increase self-esteem.

192. (4) The nurse must deal with these feelings and establish basic trust to promote a therapeutic milieu. (2) (EP; PL)
 1 Continuous pacing is not really a problem because the nurse can walk back and forth with the client.
 2 This may be of long-range importance but has little influence on the nurse's response to the client.
 3 Same as answer 2.

193. (3) Helping the client to develop feelings of self-worth would reduce the client's need to use pathologic defenses. (2) (EP; PL)
 1 Faith or the lack of faith is not the basic underlying problem but merely a symptom of it.
 2 Self-control or the lack of self-control is not the basic underlying problem but merely a symptom of it.
 4 Insight can only develop when the need to use the defense is reduced.

194. (2) This response accepts the client at the client's present level and, in addition, allows the client to set the pace of the relationship. (3) (EP; IM)
 1 This approach to any client can be misinterpreted and may precipitate an aggressive response.
 3 This response asks the client to reach out to the nurse; in the therapeutic relationship the nurse must reach out to the client.
 4 Even if the client is too withdrawn to respond, the nurse's physical presence can be reassuring.

195. (3) Participating with one trusted individual gradually diminishes the need for withdrawal. (2) (EP; PL)
 1 This activity fosters competition, which would not be helpful at this time.
 2 This would not increase socialization but rather would promote withdrawal.
 4 Same as answer 2.

196. (4) Providing opportunities to experience success in activities enhances coping abilities. (2) (SC; PL)
 1 This reinforces disturbed coping skills.
 2 A change in activities, not the level of stimuli, is what causes stress for this client.
 3 Success with tasks, not intellectual activities, enhances coping skills.

197. (4) This is very helpful in decreasing hallucinations because it provides another stimulus to compete for the client's attention. (2) (EP; PL)
 1 This would foster and support the hallucinations.
 2 Same as answer 1.
 3 Connections should be made to decrease the use of hallucinations.

198. (1) This would facilitate a one-to-one interaction and the development of a trusting relationship. (3) (EP; PL)
 2 This activity would allow the client to withdraw.
 3 This activity would foster competition and could increase anxiety.
 4 Same as answer 2.

199. (2) The delusional system contains grandiose ideation that allows the client to feel important rather than inferior. (2) (EP; AN)
 1 Although these individuals often feel guilty, feelings of inferiority, not guilt, precipitate delusions.
 3 These individuals are usually able to express aggressive feelings without difficulty.
 4 Delusions of persecution are not usually present in clients with undifferentiated schizophrenia.

200. (4) The client truly believes the voices are real because the voices reflect the client's thoughts; the voices are usually accusatory and derogatory and therefore very frightening. (2) (EP; IM)
 1 Soft music played after hallucinations have started will not be strong enough to compete for the client's attention.
 2 This would be too late; competing stimuli must be present to block the occurrence.
 3 This is incorrect; the client cannot be talked out of a hallucination.

201. (4) This helps the client explore underlying feelings and allows the client to see the message the verbalizations are communicating. (3) (EP; IM)
 1 This denies the client's feelings rather than accepting and working with them.
 2 Attempting to divert the client denies feelings rather than accepting and working with them.
 3 This focuses on the delusion itself rather than the feeling causing the delusion.

202. (1) This response accepts the client's behavior (desires) but lets the client know the nurse is not going to give up attempts to establish a relationship. (2) (EP; IM)
 2 This response requests insight that the client does not have at this point.
 3 This does not respect the client's wish for space at the present time.
 4 This statement on an initial encounter does not show respect for the client's space and is inappropriately interpretive.

203. (2) Attempting to approach the client in a nonthreatening manner and using a calm, consistent, nonviolent approach often helps the agitated client to gain more self-control. (2) (EP; IM)
 1 This action would increase, not decrease, agitation.
 3 Action should not be postponed; escalation must be prevented.
 4 This is premature; medication should not be used before trying to verbally calm the client.

204. (3) Because making decisions is often difficult, the nurse can best help the client by limiting the number and scope of choices. (2) (EP; IM)
 1 This is an unstructured question leaving too many choices; it would create great anxiety in the client.
 2 Same as answer 1.
 4 Same as answer 1.

205. (1) This response reflects the client's feelings and avoids focusing on the delusion. (2) (EP; IM)
 2 This will not change the client's feelings because the belief is real to the client.
 3 This will not change the client's feelings because the other pitcher could also be perceived as poisoned.
 4 This will not change the client's feelings; the client would believe that the nurse was not really drinking the juice.

206. (1) This is all the staff can do until trust is established and the client is able to give up some of these defenses; forcing the client to attend will disrupt the group. (2) (EP; IM)
2 This will serve only to create a confrontation between clients.
3 This will serve little purpose and will result in a confrontation; behavior cannot be altered by arguing.
4 Same as answer 3.

207. (4) This approach gives the client and family an opportunity to discuss their feelings together and clarifies their expectations. (2) (SC; IM)
1 This is the nurse's responsibility and should not be passed to someone else.
2 This is not the nurse's role; the family may never be reassured.
3 This would do little to reassure the family.

208. (1) The fetal position represents regressed behavior; regression is a way of responding to overwhelming anxiety. (1) (EP; EV)
2 Making this interpretation assumes that the nurse controls the client's behavior; the client is not responding to the nurse any differently than to anyone else who tries to establish reality contact.
3 There are no data to substantiate this; further assessment is necessary to make this interpretation.
4 Same as answer 3.

209. (2) This allows the client to work off energy without upsetting other clients. (1) (EP; IM)
1 This causes isolation and should only be used as a last resort if the client presents an actual danger to himself or others.
3 The client's present emotional state would limit concentration and prevent interaction with others.
4 Same as answer 3.

210. (3) Regression is the defense mechanism commonly used by clients with schizophrenia to reduce anxiety by returning to earlier behavior. (2) (EP; AN)
1 This organized defense is used by clients with schizophrenia, paranoid type, in which the delusional system is well systematized.
2 This unconscious forgetting is not a major defense used by clients with schizophrenia; if it were, they would not need to break with reality.
4 Rationalization, in which the individual blames others for problems and attempts to justify actions, is seldom used by clients with schizophrenia.

211. (3) This provides a stimulus that competes with and reduces hallucinations. (2) (EP; IM)
1 This is a direct question that the client probably could not answer; it would also increase anxiety.
2 If the nurse waits for the client to stop hallucinating, there may be no chance for contact with this client.
4 In addition to setting unrealistic standards, this response fails to recognize that the client believes the hallucinations are real.

212. (1) By realistically rewarding the healthy behaviors, the nurse provides secondary gains and encourages their continued use. (1) (EP; PL)
2 This would be important but would do little to increase the client's self-esteem.
3 Same as answer 2.
4 Same as answer 2.

213. (4) This response validates the presence of the client's hallucinations without agreeing with them, which communicates acceptance, and can form a foundation for trust; it may help the client return to reality. (3) (EP; IM)
1 Inappropriate; the client's contact with reality is too tenuous to push into this kind of analysis.
2 This response discounts the client, which blocks a trusting relationship and future communication.
3 This meets the staff's needs, not the client's needs; this response is condescending and impairs future communication.

214. (3) The state in which the client feels unreal or believes that parts of the body are distorted is known as depersonalization or loss of personal identity. (1) (EP; AN)
1 This is not an example of a hallucination; a hallucination is a sensory experience for which there is no external stimulus.
2 The client's statement does not indicate any feelings that others are out to do harm, are responsible for what is happening, or are in control of the situation.
4 The statement is not an example of autistic verbalization.

215. (1) Clients with paranoid-type schizophrenia usually become very anxious in social situations; the other client's closeness may have triggered latent homosexual feelings, which are thought to play a part in these clients' delusions. (2) (EP; AS)
2 This is an invalid interpretation given the data presented.

3 Same as answer 2.
4 Same as answer 2.

216. ③ Clients must be helped to develop insight into their behavior before the behavior can be altered; insight is an essential part of treatment. (1) (EP; PL)
1 Inappropriate; this would be false reassurance.
2 Directing all activities for the client is self-defeating and may precipitate aggression.
4 This is nontherapeutic and does not assist the client to develop insight.

217. ③ The client's physical appearance and presentation, as well as the feelings of paranoia, are indicative of this diagnosis. (1) (EP; AN)
1 There is no evidence in the history to demonstrate that the client's feelings are shared by anyone else.
2 The individual with a paranoid disorder usually remains organized in the other areas of functioning; the paranoia is usually isolated to only one area.
4 The individual usually has generalized feelings of suspiciousness and awareness of imagined wrongs, but delusions and disintegration are rarely present.

218. ④ The client's statement depicts the cognitive disturbance called a delusion, which is a fixed false belief that cannot be corrected by reason. (2) (EP; AS)
1 An illusion is a perceptual disturbance, a misperception of an actual environmental stimulus.
2 Hallucinations are sensory experiences without external stimuli.
3 Neologisms are made up words understood only by the maker.

219. ③ The client is fearful and suspicious; the feeling of being in a powerful position helps the client deal with anxiety. (2) (EP; AN)
1 The client is not out of touch with his real identity; he has given his real identity an important role.
2 This is incorrect; the client is compensating for feelings of inadequacy.
4 Same as answer 2.

220. ① Clients use a structured delusional system to justify and compensate for their feelings of worthlessness and low self-esteem. (2) (EP; AN)
2 Clients experiencing delusions of a paranoid nature are isolated and need contact with people to increase their contact with reality.

3 This is not the purpose of the delusional system.
4 There is nothing in the situation to indicate this client is not accepted by others.

221. ① The client has low self-esteem, which forms the basis for the delusion that others do not see the client as worthy; a delusion is a fixed false belief. (1) (EP; AN)
2 The client is not coining words with unusual meanings.
3 The client is not experiencing hallucinations, which are false sensory perceptions without external stimuli.
4 The client is not exhibiting ideas of reference, which are feelings that everything that is happening or is being said refers to the client.

222. ④ This will help the client develop self-esteem and reduce the use of paranoid ideation. (2) (EP; AN)
1 Because people must function in a social environment, it is almost impossible to avoid placing some demands on others.
2 It is impossible to remove all stress in any situation.
3 This will only succeed in supporting the client's ideas of persecution and will lower the client's self-esteem.

223. ③ The client's verbalization reflects feelings that others are blaming the client for negative actions. (1) (EP; AN)
1 No data demonstrate feelings of greatness or power.
2 No data demonstrate the client is experiencing confusing misinterpretations of stimuli.
4 No data demonstrate that the client is hearing voices at this time.

224. ③ The recommended approach for working with suspicious clients is to allow them to set the pace for the relationship. (3) (EP; PL)
1 This would be threatening and add to feelings of paranoia.
2 Same as answer 1.
4 Same as answer 1.

225. ③ Clients with paranoia often feel safer selecting foods from a cafeteria-type display that is prepared for the general population rather than eating from a tray specifically prepared for them. (3) (EP; IM)
1 This would not provide security because part of the food could still be poisoned.
2 Same as answer 1.
4 Same as answer 1.

226. (1) Suspicious individuals do not do well in groups because they are unable to tolerate the "give and take" that is necessary for successful group functioning. (3) (EP; AN)
2 Suspicious individuals do not trust others enough to do well in group therapy.
3 This is not always true for acutely ill psychiatric clients who may not be ready to accept reality.
4 This is not true; this is the purpose of remotivation, not a therapy group.

DISORDERS OF PERSONALITY (PR)

227. (1) Imagery is a therapeutic approach used to facilitate positive self-talk; mental pictures under the control of and initiated by the client may correct faulty cognitions. (3) (SC; AN)
2 This is a useful general behavioral approach but is not a specific desensitization technique.
3 These are useful general behavioral approaches but are not specific desensitization techniques.
4 Same as answer 2.

228. (1) A discussion of the feared object will trigger an emotional response to the object. (2) (EP; AN)
2 Extreme fear would be more of a problem than anger.
3 Clients with phobias generally have rigid impulse control.
4 Distortion of reality related to daily routines is usually not a problem for a client with a phobia.

229. (3) The mother's behavior sends two conflicting messages; one says "I care" and the other says "I don't care;" this behavior is often demonstrated by people with personality disorders. (3) (SC; AS)
1 If the mother were rejecting the daughter, she would not have brought a gift.
2 No evidence of a projection of feelings is given.
4 Passive-aggressive behavior is an indirect, rather than direct, expression of angry or hostile feelings.

230. (4) This response demonstrates acceptance of the client and sets limits on the client's manipulative behavior. (3) (SC; IM)
1 This is false reassurance and it assists in the attempt to manipulate the next shift.
2 The nurse would be allowing further manipulation by the client by not leaving when the shift was over.
3 This is false reassurance; the nurse cannot make everything all right.

231. (1) Consistency, limit-setting, and supportive confrontation are essential nursing interventions to provide a secure, therapeutic environment for this client. (1) (SC; PL)
2 To be therapeutic the environment needs structure and the staff must assist the client to set short-term goals for behavioral changes.
3 The use of an authoritarian approach will increase anxiety in this type of client, resulting in feelings of rejection and withdrawal.
4 Ignoring the client's behavior would be nontherapeutic and would reinforce the client's underlying fears of abandonment.

232. (3) Sitting with the client indicates acceptance and demonstrates that the nurse feels the client is worthy of the nurse's time. (2) (EP; IM)
1 It would be better to stay with the client quietly until control is regained; staying prevents a follow-through on the client's threat.
2 This would have the effect of closing off further communication.
4 This would provide little comfort for the client.

233. (3) The withdrawn pattern of behavior prevents the individual from reaching out to others for sharing; the isolation produces feelings of loneliness. (2) (EP; AN)
1 Feelings of anger may result in withdrawal, but withdrawal does not produce feelings of anger.
2 Feelings of paranoia may result in withdrawal, but withdrawal does not produce these feelings.
4 Repression is an unconscious defense whereby the individual excludes ideas, feelings, or situations from the conscious level of thought; this is not the result of withdrawal.

234. (1) Withdrawal provides a temporary defense against anxiety because it limits contact with reality and decreases the client's world. (2) (EP; AN)
2 Withdrawal does not accomplish this because feelings and anxieties are still present and little attempt is made to work through problems.
3 Same as answer 2.
4 Same as answer 2.

235. (1) This ritual is a process of undoing arising from unconscious conflicts from the anal stage in which dirt, or soiling oneself, is associated with an aggressive act against authority; the rage is controlled by rituals such as handwashing. (2) (EP; AN)
2 There is a desire for freedom and independence; the ritual is an act against authority.
3 There is initiative; the rebellion is against authority, not autonomy.

4 There is usually no problem with gender identity.

236. (4) In the early part of treatment, before new defenses are developed, enough time must be allowed for the client to complete the ritual to keep anxiety under control. (2) (EP; IM)
1 The ritual is a defense that cannot be interrupted or delayed; it is used until new defenses are developed.
2 Same as answer 1.
3 Same as answer 1.

237. (2) This is the psychoanalytic explanation for the development of obsessive-compulsive symptomatology. (2) (SC; AN)
1 Compulsive rituals commonly result in interference with activities of daily living and the individual becomes dysfunctional; rituals cannot be controlled.
3 The client is unable to consciously stop the behavior because anxiety would become overwhelming if the defense were not used.
4 This is not related to rituals but rather to phobias.

238. (1) This limits the time and still allows the ritual; until the underlying cause of anxiety can be dealt with, rituals should be allowed as much as possible. (2) (SC; PL)
2 This provides for only one time period and probably would result in increased anxiety.
3 One ritual cannot be substituted for another; this would interfere with the performance of the original ritual and could result in overwhelming anxiety.
4 This is the client's decision; the nurse should not recommend this action.

239. (2) Paraphrasing encourages further ventilation by the client. (2) (SC; IM)
1 This is a negative response that may increase the client's fears about being "crazy."
3 This response denies the client's feelings.
4 This provides false reassurance and implies that the client is out of control, which may increase fears.

240. (2) Clients use ritualistic behavior to control anxiety; when the defense is taken away, the client experiences the anxiety and becomes angry at the one who stopped the defense against it. (2) (EP; EV)
1 Because the anxiety increases discomfort, the client would not feel relief or gratitude when the behavior that controlled the anxiety was interrupted.

3 Same as answer 1.
4 Although these clients recognize that the ritualistic behavior is not necessary, they are unable to stop it and usually are not embarrassed by it.

241. (3) The client is unable to control impulses at this time, so controls must be provided for the client; the nurse's behavior provides a role model. (2) (EP; PL)
1 The client is not able to establish self-controls; freedom could prove frightening to a client who is not in control.
2 This is nontherapeutic; this would probably provoke even more acting-out behavior.
4 Same as answer 2.

242. (1) The client's comments indicate a desire to escape from the situation by the use of a loosely conceived, poorly developed plan. (2) (EP; AS)
2 There is nothing in the client's statement to indicate that this problem is occurring.
3 Same as answer 2.
4 Same as answer 2.

243. (1) Both clients must be included in a discussion about this behavior to make certain that limits on future behavior are understood by both of them; this action also places controls on the manipulative behavior often used by clients with an antisocial personality disorder. (2) (SC; PL)
2 This action would not set any limits on behavior but merely puts staff in the policing role.
3 This action would merely cause the clients to find another place to meet; the response sets no limits on behavior but only addresses location.
4 Although this may be necessary, the nurse must respond directly to the clients involved in this situation.

244. (4) This would be the most therapeutic thing the parents could do; the client must be made accountable for behavior and must know that manipulation and acting out will not be tolerated. (2) (SC; PL)
1 This would probably be a continuation of the parents' previous response to the client and would prove to be of little value.
2 This would probably cause the client to continue to act out to test the limit of the parents' endurance.
3 Same as answer 1.

245. (4) This response confronts the client with the fact that the client has not been working on personal problems. (2) (EP; IM)

1 This response is not confrontational but is sarcastic, judgmental, and attacking; this would tend to put the client on the defensive.

2 Same as answer 1.

3 This response requires insight from the client; this insight would be uncommon with this client's diagnosis.

246. (2) The nurse's response is not therapeutic because it does not recognize the client's needs but tries to make the client feel guilty for being demanding. (3) (SC; AN)

1 Untrue; impulse control would refer to a sudden driving force being constrained or held back.

3 Nothing in the nurse's statement would achieve this; the nurse is defensive, not therapeutic.

4 Same as answer 3.

247. (3) Private confrontation with reported facts provides for verification; a calm, direct manner is most assertive. (2) (EP; IM)

1 This action places control in the hands of the client rather than the nurse, which could lead to aggressive confrontation.

2 This is aggressive confrontation, not assertive intervention.

4 This is not assertive intervention; it is manipulation and is not truthful.

248. (4) This client needs firm, realistic limits set on behavior; this statement permits the client to make the choice and clearly states the consequences of behavior. (2) (EP; IM)

1 This is an unrealistic response; clients with this diagnosis do not learn from past errors.

2 This is an unrealistic response; this client would care very little about rules.

3 This is an unrealistic response; the client and visitors do not want to socialize with other clients and visitors.

249. (4) The expression of feelings by this individual would demonstrate the development of some insight and a willingness to at least begin to look at underlying causes of behavior. (2) (EP; EV)

1 These words would probably have little meaning to the client.

2 Same as answer 1.

3 Same as answer 1.

250. (2) The client must recognize the existence of the subpersonalities so that interpretation can occur. (3) (EP; AN)

1 This is not relevant to clients with multiple personality disorders; this outcome relates to clients with obsessive-compulsive behaviors.

3 This is not realistic; integration of the personalities generates fear and defensiveness in the client, and defense mechanisms will be used.

4 This is inappropriate; anxiety serves as a motivator for behavioral change, and antianxiety medications must be used judiciously.

251. (3) A multiple personality disorder is a complex, multifaceted problem that requires long-term therapy to achieve integration of the personalities. (3) (EP; EV)

1 None of the personalities can be ignored because their presence must be dealt with before integration can occur.

2 Each personality has the ability to deal openly with feelings and fears, but the personalities need to be integrated.

4 This is untrue; if they did not serve a protective purpose, they would be abandoned.

DISORDERS OF MOOD (MO)

252. (4) This provides structure and helps the client learn acceptable behavior. (2) (EP; PL)

1 The client may not be ready for this at the present time.

2 No environment will be stress free.

3 Same as answer 1.

253. (4) Limiting unnecessary interactions will decrease stimulation and thus agitation. (2) (SC; PL)

1 Constant client and staff contact increases stimulation and thus agitation.

2 This bombards the client's sensorium and increases agitation.

3 Not all disturbed clients are suspicious.

254. (3) Depression is characterized by feelings of hopelessness, helplessness, and despair, leaving little room for any pleasure; alteration in appetite (either decreased or increased) is common in depressed clients. (2) (EP; AS)

1 Although there is a narrowing of perception, rigidity would be uncommon.

2 Fatigue is continually present and does not alternate with high energy levels.

4 There is a loss of interest and little participation in activities of daily living.

255. (3) Depression is usually both emotional and physical, so a simple daily routine is the least stressful and least anxiety producing. (1) (CC; PL)

1 A depressed client has limited interest in simple activities; too many may increase the anxiety.

2 Too many stimuli increase the anxiety in a depressed client.
4 An extremely depressed client may be incapable of making even simple decisions.

256. (4) Depression is "contagious"; it affects the nurse as well as the client. (2) (EP; AN)
1 The client's lack of energy really does not make nursing care difficult.
2 These clients usually do not offer negative responses; they offer no responses.
3 Same as answer 1.

257. (1) Because clients cannot be argued out of their feelings, it is best to initially accept them; it also encourages communication. (2) (EP; PL)
2 This delays discussing the client's feelings and has little value.
3 The depressed client does very little talking and needs to be encouraged to communicate.
4 This has little effect on the depressed client; it can increase depression.

258. (1) Allowing the client to make those decisions that can be handled helps improve confidence. (2) (EP; PL)
2 The client is depressed, and this would probably result in total inactivity.
3 This action would demoralize the client; also, it is impossible for one individual to make all the decisions for another.
4 Same as answer 2.

259. (4) This action demonstrates to the client that the nurse feels the client is worth spending time with, and it helps restore and build trust. (2) (CC; PL)
1 This action would be impossible to carry out on a regular basis unless the client were potentially suicidal.
2 This action does little to establish communication between the nurse and the client and might be threatening.
3 The depressed client may never get around to speaking to the nurse and, left alone, will withdraw even further.

260. (4) The fact that the nurse spends time with the client conveys a feeling of importance and helps build the client's self-esteem. (3) (EP; IM)
1 This places the client on the defensive and does not respond to the feelings of worthlessness communicated by the client.

2 This infers agreement with the client's statement that the client is not worthy; the nurse should stay to convey a sense of self-worth to the client.
3 This response cuts off communication; the client responds better to actions than to words.

261. (3) This response focuses on the client's feelings rather than the statement, and it serves to open channels of communication. (2) (EP; IM)
1 This response asks the client to decide what is causing feelings; most people are unable to answer why they feel as they do.
2 Such a response simply echoes the client's statement and does not reflect feelings or stimulate further communication.
4 This response does nothing to stimulate further communication; in fact, it tells the client to talk about feelings with someone else.

262. (1) Group interaction provides a sense of belonging and fosters the assumption of responsibility. (2) (EP; AN)
2 The group is not the best arena for the expression of repressed hostility.
3 This is not assured by group interaction.
4 Same as answer 3.

263. (1) The client will be most likely to eat if accompanied and encouraged by an individual with whom a trusting relationship has been established. (1) (EP; IM)
2 This will not encourage the client to eat and will promote isolation.
3 This is inappropriate at this time; the client is not interested in maintaining health, nor is the client ready for any teaching.
4 This would be ineffective at this time; the client is too introspective to care.

264. (3) In a depressive episode with melancholic features, the range of interests becomes increasingly narrow, with eventual loss of pleasure in almost all activities; overmeticulousness occurs because the client cannot tolerate any change in the environment. (3) (EP; AS)
1 The boundaries would be increasingly narrowed because the client cannot tolerate change, which is viewed as a threat.
2 Jealousy is a symptom that is unrelated to a major depressive episode; the symptoms usually develop slowly over a period of time.
4 There is nothing to indicate that the client ever had a hypermanic episode, in which this behavior would be expected; hypermanic behavior does not occur in a major depressive episode.

265. ② This is a psychiatric problem in which there is no prior history of depression; it is related to age, changes in life-style, and feelings of not contributing and being worthless. (1) (EP; AS)
    1 In bipolar disorders, depression alternates with periods of extreme restlessness, hyperactivity, and flamboyance in dress and behavior; this is not evident here.
    3 Involutional melancholia, which was usually characterized by depression with agitation, is no longer considered a specific disorder.
    4 There is no inconsistency between the behavior and the mood.

266. ④ Depressed clients demonstrate decreased communication because of a lack of psychic or physical energy.(2) (EP; AN)
    1 There is insufficient evidence to identify this as the primary nursing diagnosis.
    2 Same as answer 1.
    3 Same as answer 1.

267. ④ Sharing painful feelings reduces the isolation and sense of uniqueness that feelings can cause; sharing of these feelings usually decreases depression. (2) (EP; PL)
    1 This would do little to decrease the client's sadness and does not consider the client's low level of energy.
    2 Same as answer 1.
    3 This may be important for the future, if a problem exists, but the expression and sharing of painful feelings are more important than improving communication.

268. ① Some success is important to increase the client's self-esteem. (3) (SC; IM)
    2 This would support the client's feelings of uselessness.
    3 The client who is in a major depression would not have the interest or energy to act independently.
    4 Same as answer 3.

269. ③ Caring for a child often refocuses a client's unmet dependency needs, resulting in resentment and anger; the feelings cause ambivalence and guilt, which are turned inward. (3) (EP; AS)
    1 There are no data to support this conclusion.
    2 There are no data to support the conclusion that child abuse occurred.
    4 Self-destructive behavior does not usually evoke love, sympathy, and compliments from others.

270. ③ A therapeutic response that encourages expression of the client's feelings. (2) (EP; IM)

    1 Nontherapeutic; does not explore feelings and the client may interpret it as a put-down.
    2 Nontherapeutic; nurse could not know client is upset unless client tells the nurse.
    4 Client may not know or may not wish to tell; in addition, response asks for facts and does not elicit feelings.

271. ① ECT would be contraindicated in the presence of a brain tumor because the treatment causes an increase in intracranial pressure. (2) (SI; AN)
    2 There would be no contraindication to ECT with this health problem.
    3 Same as answer 2.
    4 Same as answer 2.

272. ③ Giving the client simple facts and relating that it is normal to feel nervous help reduce the client's fears. (2) (CI; IM)
    1 This is untrue because memory loss is not permanent; this would unnecessarily worry the client.
    2 Although true about the memory loss, this response would serve to negate the client's feelings.
    4 This is untrue; ECT does not selectively block out painful experiences.

273. ③ Because the client refuses to talk, pertinent data must be obtained from the family. (2) (SI; AS)
    1 The client is not responding to questions; the client may not know the reason.
    2 This may or may not have influenced current behavior.
    4 The presence of scars is an inaccurate way to determine past behavior.

274. ④ The expression of these feelings may indicate that this client is unable to continue the struggle of life. (2) (SI; AS)
    1 These are not indications of potential suicide; the client is still responding to the world, not attempting to leave it.
    2 These are usually not sufficient to precipitate a suicide attempt.
    3 The client attempting suicide usually sees death as a release.

275. ② This is the most important assessment to make because suicide is a possibility with every depressed client. (2) (SI; AS)
    1 The client has already said this, and it responds to only part of the client's statement.
    3 The nurse does not have enough information to have this understanding; this response does not encourage communication.

4 This is a philosophic approach that would not encourage discussion of feelings.

276. (1) When energy levels improve in the depressed client, the risk of suicide increases; also, the client has given away a personal belonging, which may indicate a plan to commit suicide. (1) (SI; AS)
2 This may be true, but the gift of a personal belonging decreases the possibility that it is simply an improved socialization level.
3 This may be true, but the gift of a personal belonging decreases the possibility that it is simply an improvement in mood.
4 This may be true, but the situation should be explored further; the physician would ultimately make this decision.

277. (1) This statement provides clues that the client feels no one cares, so there is no reason the client should care. (2) (SI; AN)
2 The clues presented do not lead to this conclusion.
3 Same as answer 2.
4 Same as answer 2.

278. (1) Decreased sexual desire is a major symptom of clinical depression. (1) (DF; AN)
2 Although depression is often related to unmet dependency needs, decreased sexual desire is associated with the depression.
3 The sexual difficulties are associated with the depression, and the depression is the major cause of the marital stress.
4 Role confusion, not identity confusion, is associated with the depression.

279. (4) In 3 days the client should understand that the staff cares enough to prevent suicidal acting out. (3) (SI; AN)
1 This is unrealistic within the stated time period.
2 Same as answer 1.
3 Same as answer 1.

280. (1) Clients who have a plan are more apt to carry it out; safety is the priority. (2) (SI; IM)
2 This is a noncaring response and may be viewed as false reassurance.
3 This response is too upbeat for a depressed client and provides false reassurance.
4 Although this response is empathic, it does nothing to assist the client and may be viewed as a consent for suicide.

281. (4) This open-ended response encourages further discussion and allows for an exploration of feelings. (2) (EP; IM)

1 This response ignores the client's feelings expressed in the statement.
2 The depression is not adding to the feelings, the feelings are causing the depression.
3 Same as answer 1.

282. (4) The manic phase is the mirror image of the depressed phase; the behavior is an attempt to ward off depression by racing into reality. (2) (EP; AS)
1 It does not have to be an elating situation; the situation itself matters little, and the client responds to any stimulus.
2 It is an attempt to block feelings of depression, not an acting out of innate drives.
3 It is not an incorrect interpretation but an incorrect response to the stimulus.

283. (4) The elated client expends a great deal of energy; dehydration, oxygen deficits, cardiac problems, and death can occur. (2) (PC; AS)
1 The elated person does not withdraw from reality but continues to run headfirst into it.
2 The elated client has little difficulty verbalizing needs.
3 The elated client usually does not take time to eat while expending a great deal of energy, so weight loss is the problem.

284. (2) During the manic phase of the illness, the client responds to everything in the environment; therefore, it is important that the room be quiet and restful to decrease stimulation. (2) (CC; PL)
1 Not an important consideration at this time for this client.
3 This room would be too stimulating because of its location.
4 This would tend to increase both the client's and the roommate's behavioral acting out.

285. (3) The client in the manic phase of the illness often neglects basic needs; these needs are a priority to ensure adequate nutrition, fluid, and rest. (3) (EP; PL)
1 Although the client needs to expend excess energy, physical exhaustion and dehydration are real possibilities during the manic phase of the illness.
2 This would be counterproductive.
4 Client is unable to actively participate in structured activities at this time.

286. (1) Structure tends to decrease agitation and anxiety and to increase the client's feelings of security. (1) (CC; PL)
    2 Touching can be threatening for many clients and should not be used indiscriminately.
    3 Conversations should be kept simple; the client with a bipolar disorder, either depressed or manic phase, may have difficulty following involved conversations about current affairs.
    4 Clients with bipolar disorders are in contact with reality, so this activity would serve little purpose.

287. (4) The client expends a great amount of energy running headlong into reality in an attempt to ward off or avoid facing feelings of depression. (3) (EP; AN)
    1 The behavior is not an expression of innate desires but an attempt to avoid feelings of depression.
    2 The client has no difficulty relating to others; this behavior is an attempt to avoid feelings of depression.
    3 This client is not attempting to compensate for an imagined loss but is trying to avoid feelings of depression.

288. (2) The excited, overactive client needs a calm environment; external stimulation only serves to cause further excitation. (2) (CC; IM)
    1 The client needs reduced, not increased, external stimulation.
    3 Same as answer 1.
    4 Same as answer 1.

289. (1) A firm voice is most effective; the statement tells the client that it is the behavior, not the client, that is upsetting to others. (2) (EP; IM)
    2 Demanding that the client stop the current behavior is a useless action; the client is out of control and needs external control.
    3 The client does not know what is precipitating the behavior, and the question would be frustrating.
    4 This should only be done when there is real danger from exhaustion; external controls must be set.

290. (1) This response simply states a fact and delivers praise without making demands. (3) (SC; EV)
    2 This puts the total responsibility for control on a client who needs external controls set.
    3 This does not help the client separate the self from the behavior; it tells the client that acting-out behavior will result in rejection.

    4 The client may not recall what happened yesterday and would not be able to compare the differences.

291. (4) The hyperactive client cannot tolerate sitting long enough to eat an adequate meal. Hand foods will help to meet the client's nutritional needs and do not require the client to sit down. (2) (PC; PL)
    1 This client will most likely ignore the tray.
    2 Unworthy feelings are related to a depressive, not manic, episode.
    3 It is unlikely that this client would understand or care about this information.

292. (1) The symptoms exhibited are typical of the manic phase of a bipolar disorder, which would not have developed were the client taking the medication; this response states this fact in a nonchallenging, nonthreatening manner. (3) (PT; EV)
    2 This response is challenging and is not focused on assessing the problem.
    3 This response is not focused on the symptoms manifested.
    4 This response would do little to promote discussion with this client.

DEMENTIA, DELIRIUM, AND OTHER COGNITIVE DISORDERS (DD)

293. (2) Confabulation, or the filling in of memory gaps with imaginary facts, is a defense mechanism used by people experiencing memory deficits. (1) (EP; AS)
    1 Concretism is demonstrated by speech in which the major or salient point being made by the speaker is lost because it is buried in excessive verbal detail.
    3 Flight of ideas is demonstrated by speech that jumps from one topic to another with no obvious connection for either the speaker or the listener.
    4 Associative looseness is demonstrated by speech that is difficult to follow because the connections between the speaker's statements or train of thoughts are so loose that they are not obvious to the listener.

294. (4) Moderate dementia is characterized by increasing dependence on environmental and social structure and by increasing psychologic rigidity with accentuated previous traits and behaviors. (3) (EP; AS)
    1 Although paranoid attitudes may be exhibited, the decrease in cognitive functioning, disorien-

tation, and loss of memory usually do not lead to hypervigilance.

2 With the decrease in impulse control that is associated with dementia, decreased, not increased, inhibition would be present.

3 An enhancement of intelligence would not occur in dementia; initially, intellectual deterioration is subtle.

295. ① With a depression, there is little or no emotional involvement and therefore little alteration in affect. (3) (EP; AS)

2 This is associated with depression because of a low self-esteem.

3 People who are depressed do not have physical or emotional energy. "I don't know" answers require little thought and/or decision making.

4 This is associated with depression because a sense of futility leads to a lack of response to the environment.

296. ② Routines and familiarity with activities or the environment provide for a sense of security. (2) (DF; PL)

1 Change is poorly tolerated; frustration and the inability to accomplish tasks lead to lowered self-esteem.

3 Challenging activities can be frustrating and can lead to hostility or withdrawal.

4 Decreased physical capacity and attention span limit active participation; frustration can result.

297. ③ Clients with dementia have the greatest loss in the area of recent memory. (2) (DF; PL)

1 Memory of remote events usually remains fairly intact.

2 Same as answer 1.

4 Same as answer 1.

298. ② When the client is unable to perform a task, frustration occurs and results in more disorganized behavior. (3) (DF; PL)

1 The client's disorientation is documented and will not change, although some day-to-day variations may occur; most important is the assessment of the client's ability to function.

3 There is no documentation of disruptive behavior; frustration must be avoided.

4 The client will probably never adjust any further.

299. ② This client needs toileting every 2 hours to prevent soiling; physically seating the client on the toilet often prevents accidents and negates the use of diapers. (3) (DF; IM)

1 The client needs to be physically placed on the toilet; confusion limits effectiveness of other actions.

3 Same as answer 1.

4 Same as answer 1.

300. ① "Where are you?" is the best question to elicit information about the client's orientation to place because it encourages a response that can be assessed. (3) (DF; AS)

2 This question focuses on recent memory; it does not assess orientation to place.

3 This question would probably be answered by "yes" or "no"; this could not objectively determine the client's orientation.

4 This question focuses on orientation to time, not place.

301. ② Confabulation is used as a defense mechanism against embarrassment caused by lapse of memory; the client fills in the blanks in memory by making up details. (2) (DF; AN)

1 Regression is a defense mechanism in which the individual moves back to earlier developmental defenses.

3 Although the elderly fear being forgotten or losing others' affection, this is not the main reason for confabulation.

4 Confabulation is not used to reminisce about past achievement.

302. ④ This action maintains, for as long as possible, the client's remaining intellectual functions by providing an opportunity to use them. (3) (EP; PL)

1 All clients need adequate nutrition; this is not a particular priority of care for these clients.

2 The main priority should be directed toward maintaining intellectual functioning; remotivation is not always possible with extensive organic damage.

3 There are no data to indicate that the client needs custodial care at this time.

303. ② This client would require a structured environment, regardless of the cause of the behavior; this would help provide a safe environment. (2) (CC; PL)

1 This is important but is secondary to promotion of an environment conducive to safety and security.

3 Same as answer 1.

4 A battery of screening tests will probably be used in an attempt to determine the cause of the dementia; however, provision for safety is necessary first.

304. (2) This allows the client to exercise the right to decide which articles to keep and provides for safety and cleanliness. (2) (DF; PL)
  1 This deceives the client, limits judgment, and creates mistrust toward the staff.
  3 Explanations alone will not provide for safety or meet this client's needs because of a decreased attention span and memory.
  4 This does not help because all of the objects the client is hoarding are not harmful.

305. (2) The client needs constant reassurance because forgetfulness blocks previous explanations; frequent presence of staff serves as a continual reminder. (2) (DF; PL)
  1 The client needs continual reassurance and would not remember times for planned interactions or activities.
  3 This client would be unable to explain the reasons for concern.
  4 This client will not remember the explanation from one moment to the next.

306. (2) These clients require external controls to minimize danger of injury and to preserve human dignity. (2) (DF; PL)
  1 The client will not have excessive energy.
  3 The staff cannot prevent all gross motor activity; the client needs to use muscles or atrophy will occur.
  4 Further deterioration usually cannot be prevented in this disorder.

307. (2) This encourages verbalization, gives the client a feeling of security, and decreases the sense of isolation. (1) (DF; IM)
  1 This action discourages verbalization between client and nurse; the client may be unable to function in a small group because of increased anxiety.
  3 This discourages verbalization of feelings and will lead to feelings of being unwanted.
  4 Same as answer 1.

308. (2) This assists the client in decision making; new situations may be stressful and lead to ambivalent feelings. (2) (DF; IM)
  1 This would make the client feel guilty and add to anxiety.
  3 This is not sharing decision making; hurrying the client will lead to feelings of frustration and resentment.
  4 The client may perceive this action as punishment.

DRUG-RELATED RESPONSES (DR)

309. (4) Psychoactive drugs have been shown to be capable of interrupting the acute psychiatric process, making the client more amenable to other therapies. (1) (PT; AN)
  1 Family therapy is effective but is a long-term, costly therapy; symptoms must be reduced before the client can participate.
  2 This is untrue; ECT is more effective in treating depressed clients.
  3 Clients with schizophrenia usually have little insight into their problems; confronting the client through insight therapy will increase anxiety.

310. (1) The user has an emotional and physiologic dependence on the drug, and the phenomenon of tolerance (more drug needed to achieve the same effect) occurs. (2) (PT; AN)
  2 Tolerance, even to levels that would be lethal in the nonaddicted individual, occurs.
  3 There is an emotional dependence.
  4 There is a physical dependence.

311. (2) A hypotensive reaction is a common adverse effect of chlorpromazine. (2) (PT; AS)
  1 Restraints of any type may increase the client's anxiety and result in struggling and increased agitation.
  3 Photosensitivity occurs most commonly when clients are taking large doses and are spending time outdoors in the sun.
  4 Tardive dyskinesia usually results from prolonged large doses of phenothiazines in susceptible clients.

312. (1) Akathisia, a side effect of chlorpromazine, develops early in therapy and is characterized by restlessness and agitation. (2) (PT; EV)
  2 Torticollis is characterized by a stiff neck and is not a side effect of this drug.
  3 Tardive dyskinesia is characterized by gross involuntary movements of the extremities, tongue, and facial muscles that develop after prolonged chlorpromazine therapy.
  4 Pseudoparkinsonism, a side effect of chlorpromazine therapy, is characterized by motor retardation, rigidity, and tremors; the reaction resembles Parkinson's syndrome but usually responds to medications or stopping the chlorpromazine.

313. (4) Chlordiazepoxide suppresses central sympathetic activity, which reduces the discomfort of withdrawal and the risk of seizures. (3) (PT; AN)

1 This drug helps to reduce the risk of seizures but does not prevent physical injury during a convulsion.

2 Although these benefits may occur, they are not the primary objectives for using the drug.

3 The ability of the client to accept treatment depends on readiness to accept the reality of the problem.

314. (4) The lithium concentration is most stable at this time; absorption and excretion occur 8 to 12 hours after the last dose. (2) (PT; AN)

1 Absorption and excretion rates vary; concentrations may be falsely higher at this time, affecting the reliability of readings.

2 Same as answer 1.

3 Same as answer 1.

315. (3) Benztropine mesylate (Cogentin), an anticholinergic, helps balance neurotransmitter activity in the CNS and helps control extrapyramidal tract symptoms. (2) (PT; AN)

1 Hydroxyzine (Atarax) is a sedative that depresses activity in the subcortical areas in the CNS; it is used to reduce anxiety.

2 Amobarbital (Amytal Sodium), a barbiturate, interferes with transmission of impulses to the cerebral cortex; it is used for the treatment of insomnia.

4 Chlorzoxazone (Parafon Forte DSC), a skeletal muscle relaxant, depresses nerve transmission through polysynaptic pathways; it is used for the treatment of muscle spasms.

316. (2) Haloperidol (Haldol) reduces emotional tensions, excessive psychomotor activity, panic, and fear. (2) (PT; AN)

1 Clients exhibiting manic behaviors do not respond well to Haldol, and it can exacerbate their underlying feelings of depression.

3 Clients exhibiting excited-depressed behavior do not respond well to Haldol because it tends to increase the depression.

4 Haldol appears to have few stimulating effects and, in fact, increases feelings of lassitude and fatigue.

317. (2) Lithium's therapeutic window is very narrow and toxic levels could occur without routine monitoring of the blood drug level. (2) (SI; PL)

1 A low-sodium diet can lead to hyponatremia, which must be prevented because it limits the excretion of lithium and can result in toxicity.

3 Diuretics reduce serum sodium levels, and lithium is not excreted when sodium levels are decreased; the retention of lithium can result in toxic levels.

4 This may or may not be true.

318. (2) The drug physostigmine (Antilirium) is essential to manage an overdose of a tricyclic antidepressant; it increases acetylcholine at cholinergic nerve terminals and reverses central and peripheral anticholinergic effects. (3) (PT; PL)

1 Dialysis or forced diuresis is an ineffective treatment for an overdose of a tricyclic antidepressant; immediate administration of physostigmine to counteract the effects of the tricyclic antidepressant is necessary.

3 Acetylcholine is already depressed from the tricyclic antidepressant; anticholinergics are most effective in managing the side effects of antipsychotic/neuroleptic drugs.

4 Prevention of suicidal behavior is always advantageous; however, in this case, immediate emergency intervention is necessary.

319. (3) Extreme photosensitivity is a common side effect of Prolixin. (3) (PT; PL)

1 Once the client's medication is adjusted and CNS response is noted, driving may be permitted; drowsiness usually subsides after the first few weeks.

2 This is untrue; energy is usually decreased.

4 Although this drug can cause hypotension, it does not consistently lower blood pressure; a sudden drop can be dangerous to a client with hypertension.

320. (1) This drug is a narcotic antagonist that displaces narcotics from receptors in the brain, reversing respiratory depression. (2) (PT; PL)

2 This is a synthetic opiate that causes CNS depression; it would add to the problem of overdose.

3 This drug would have no effect on respiratory depression related to the presence of an overdose of a narcotic.

4 Same as answer 3.

321. (3) Monoamine oxidase uptake is inhibited by the medication, increasing concentrations of endogenous epinephrine, norepinephrine, serotonin, and dopamine in CNS storage sites; high levels of these transmitters in the presence of tyramine can cause hypertensive crisis. (2) (PT; PL)

1 This may be an adverse reaction to the drug but is not related to drug-food interaction.

2 Same as answer 1.

4 This is not related to drug-food interaction.

322. (1) These products often contain an alcohol base and can cause a severe reaction in the presence of Antabuse. (2) (PT; IM)
2 These do not contain alcohol and need not be avoided.
3 Only wine vinegar should be avoided.
4 Same as answer 2.

323. (2) As the therapeutic level is reached and maintained, the client's psychotic symptoms decrease and insight increases. (2) (PT; EV)
1 This in itself does not indicate that the client is responding therapeutically to the medication.
3 Same as answer 1.
4 This behavior may be an indication that the client is not responding to the medication because the symptoms are increasing.

324. (3) Risperdal potentiates the action of alcohol and can cause oversedation if the drug and alcohol are taken together. (1) (SI; IM)
1 This medication should be taken consistently to prevent recurrence of symptoms and maintain blood drug levels.
2 Same as answer 1.
4 Medication should be taken as ordered; taking it all at one time can interrupt the maintenance of a constant therapeutic blood level.

325. (3) This is too short a period of time; clients usually begin to feel a lightening of depression in approximately 14 to 20 days, with the full antidepressant effects being felt between 3 and 4 weeks. (1) (PT; EV)
1 Drowsiness usually occurs early in treatment but passes with time.
2 It usually takes this long before the full effects of the antidepressant are experienced.
4 Clients usually remain on these medications for a few months.

326. (4) The parkinsonian symptoms are related to extrapyramidal tract effects, and agranulocytosis is related to bone marrow depression. (2) (PT; EV)
1 Jaundice is an adverse reaction; vomiting is not.
2 Tardive dyskinesia is an adverse reaction; nausea is not.
3 Postural hypotension is an adverse reaction; hiccups are not.

327. (2) Tardive dyskinesia, an extrapyramidal response characterized by vermicular movements and protrusion of the tongue, chewing and puckering movements of the mouth, and a puffing of the cheeks, is often irreversible even when the

antipsychotic medication is withdrawn. (1) (PT; EV)
1 This can usually be treated with antiparkinsonian or anticholinergic drugs while the antipsychotic medication is continued.
3 Same as answer 1.
4 Same as answer 1.

328. (2) Diarrhea is an early sign of lithium toxicity, and the resulting loss of sodium tends to increase the lithium level even further. (2) (PT; EV)
1 Tinnitus can occur early in treatment, but unless other progressive neurologic symptoms develop, it is not considered a symptom of toxicity.
3 Akathisia is a symptom associated with the side effects of the antipsychotic drugs.
4 Torticollis is associated with the side effects of antipsychotic drugs.

329. (1) Hypotension is a major side effect of Valium that occurs early in therapy. (2) (PT; EV)
2 An alteration in urinary output is not a common side effect, but it may occur after prolonged use.
3 This is not a common side effect, but distention from constipation may occur after prolonged use.
4 CNS depression is not a common side effect, but it may occur after prolonged use.

330. (4) Anticonvulsants and Haldol exert a synergistic CNS depressant effect. (2) (PT; EV)
1 This is untrue; the effect is potentiated.
2 Anticonvulsants do not affect absorption or metabolism of Haldol.
3 Same as answer 2.

331. (2) This drug creates a general sense of well-being, increases appetite, and helps lift depression. (2) (PT; EV)
1 The client might not know the reason for depression, and the drug does not cause amnesia.
3 Symptomatic relief usually begins after 2 to 3 weeks of therapy.
4 Concomitant use of monoamine oxidase inhibitors and tricyclic antidepressants is usually contraindicated.

332. (1) Medication should be discontinued immediately if these symptoms of lithium toxicity occur. (1) (PT; EV)
2 These are symptoms of toxicity, and a stimulant will not alter them.
3 There is no special diet required; the client is encouraged to maintain a normal diet with normal salt and fluid intake.

4   The client receiving lithium is encouraged to maintain a normal salt and fluid intake.

333. ③ These foods are high in tyramine, which in the presence of an MAO inhibitor, such as Marplan, can cause an excessive epinephrine-type response that can result in a hypertensive crisis. (2) (PT; PL)
1   There is no relationship between these foods and this medication.
2   Same as answer 1.
4   Same as answer 1.

334. ④ A hypertensive crisis would not be associated with extrapyramidal tract symptoms. (2) (PT; EV)
1   Akathisia, characterized by restlessness and twitching or crawling sensations in muscles, is an extrapyramidal side effect.
2   Opisthotonos, characterized by hyperextension and arching of the back, is an extrapyramidal side effect.
3   Oculogyric crisis, characterized by the uncontrolled upward movement of the eyes, is an extrapyramidal side effect.

335. ② This medication can be taken every 2 weeks instead of every day. (1) (PT; AN)
1   Untrue; the side effects are the same as chlorpromazine (Thorazine).
3   The action of this drug during pregnancy is uncertain; animal studies have demonstrated an adverse effect on the fetus.
4   Same as answer 1.

336. ① The medication should be stopped immediately because the jaundice indicates possible liver damage, which would prolong elimination of the drug and may result in toxic accumulation. (1) (PT; EV)
2   Milk does not change the effect of the drug.
3   The drug must be stopped, not reduced.
4   The drug is only available in an oral form; in addition, the route of administration would not influence the occurrence of toxic accumulation.

337. ④ Infection can indicate agranulocytosis, a serious side effect that can occur with therapy and can cause death. (3) (PT; EV)
1   This would be unsafe because the client may be developing agranulocytosis, a potentially life-threatening side effect that needs immediate treatment.
2   Same as answer 1.
3   Same as answer 1.

338. ② For clients with bipolar disorders, it has been shown that long-term lithium therapy flattens the highs of the euphoric phase and the lows of the depressed phase. (1) (PT; EV)
1   This is untrue; the physician should be notified before medication is stopped.
3   This is untrue; clients should never adjust their own dosage of medication.
4   This is not true; the therapeutic level and the toxic level are very close, and serious side effects do occur.

339. ④ True; antianxiety medications have the potential for physiologic and/or psychologic dependence. (2) (PT; AN)
1   Untrue; physiologic and/or psychologic dependence can develop even when the dosage is controlled.
2   Untrue; both psychologic and physiologic dependence can develop.
3   Untrue; tolerance does not develop.

340. ③ Foods such as chocolate, aged cheese, and pickled herring and those with excessive caffeine have high levels of tyramine and cause dangerous hypertension in clients taking MAO inhibitors. (2) (SI; EV)
1   There is no need to limit intake of this food while taking MAO inhibitors.
2   Same as answer 1.
4   Same as answer 1.

341. ④ This acute dystonic reaction is a severe side effect of Haldol and requires intramuscular or intravenous administration of antiparkinsonian medication. (2) (PT; EV)
1   The dosage would not be increased but discontinued, and antiparkinsonian medication would be administered.
2   At this point symptoms are so severe that this medication must be discontinued.
3   These symptoms are more severe than pseudoparkinsonism and require more than an adjustment in this medication.

THERAPEUTIC RELATIONSHIPS (TR)

342. ③ The answer to this question will provide the nurse with an idea of the client's hopes and frustrations without being threatening or probing. (3) (DF; AS)
1   This question is probing, disregards the client's statement, and provides little information for the nurse to use in planning care.
2   Same as answer 1.
4   Same as answer 1.

343. ② Before deciding how to decrease the client's fears of addiction, the nurse must explore the full extent of the client's knowledge and feelings about taking the drug. (1) (SC; AS)
1 Information may or may not be helpful; the client's feelings are what must be addressed.
3 It is too early; exploration to find the basis of the fear is necessary first.
4 Same as answer 3.

344. ③ All behavior has meaning; the nurse must try to understand what the behavior means to the client. (3) (EP; AS)
1 The isolation of a client may increase anxiety and precipitate more acting-out behavior.
2 Ignoring behavior does little to alter it, and it may even cause further acting out.
4 The nurse cannot explain the meaning of the client's behavior; only the client can.

345. ① The client has the right to expect that all records of care will remain confidential; permission must be sought for use. (1) (CC; AN)
2 This right was not violated; this applies to the concept of holding clients against their will.
3 This right was not violated; this applies to the humane aspect of care.
4 This right was not violated; this applies to procedures and treatments to be performed on the client.

346. ③ This is a technique that prevents misunderstanding so that both the client and the nurse can work toward a common goal in the therapeutic relationship. (2) (SC; AN)
1 This would not provide for clarification or understanding.
2 Same as answer 1.
4 Same as answer 1.

347. ② Evaluation includes assessing the client's response to care, judging the effectiveness of the nursing care plan, and changing the plan as necessary. (1) (CC; AN)
1 Planning includes the development of a plan that focuses on specific goals and actions unique to the client's needs.
3 Assessing entails collecting and reviewing objective and subjective data about the client's health status.
4 Implementation includes performing specific actions designed to achieve the stated goals.

348. ② The psychoanalytic model studies the unconscious and uses the strategies of hypnosis, dream interpretation, and free association as a means of releasing repressed feelings. (1) (EP; AN)
1 The behaviorist model subscribes to the belief that the self and mental symptoms are viewed as learned behaviors that persist because they are consciously rewarding to the individual; this model deals with behaviors on a conscious level of awareness.
3 The psychobiologic model views emotional and behavioral disturbances as stemming from a physical disease; abnormal behavior is directly attributed to a disease process; this model deals with behaviors on a conscious level of awareness.
4 The social-interpersonal model affirms that crucial social processes are involved in the development and resolution of disturbed behavior; this model deals with behavior on a conscious level of awareness.

349. ② During the working stage goals are met, problems are resolved, and changes in behavior occur. (1) (CC; AN)
1 During the initial stage trust is the primary focus, goals and contracts are set, and problems are identified.
3 There is no such stage in the nurse-client relationship; this is a step in the nursing process.
4 The evaluation/termination stage focuses on accomplishments, reinforces new behaviors, and closes the relationship.

350. ④ The preinteraction phase is the period before or the preparatory phase of a planned therapeutic relationship. (2) (CC; AN)
1 The working phase is the period in a relationship when the individuals are occupied with achieving goals and sharing facts and feelings.
2 The orientation phase is the initial period of the actual interaction; it is the introductory or exploratory phase.
3 The termination phase is the period in a relationship when the individuals are beginning to separate and move toward independent paths.

351. ② This is a correct description of the working phase of a relationship; trust has been established and a relationship has been developed based on mutual respect. (2) (CC; AN)
1 This behavior would occur during the orientation phase before trust is established.
3 Same as answer 1.
4 Same as answer 1.

352. ② The client's inability to recall is an example of repression, which is the unconscious and involuntary forgetting of painful events, ideas, and conflicts. (2) (SC; AN)

1 There is nothing to demonstrate that denial, an unconscious refusal to admit an unacceptable behavior or idea, has occurred.

3 There is nothing to demonstrate that regression, a return to an earlier, more comfortable developmental level, has occurred.

4 There is nothing to demonstrate that dissociation, the separation and detachment of emotional affect and significance from a particular idea, situation, or incident, has occurred.

353. ③ This goal is related to the nursing diagnosis and is appropriate and measurable. (1) (EP; AN)

1 This is not related to the nursing diagnosis; this is true for everyone but the priority for this client is to facilitate interaction with others.

2 This is not related to the nursing diagnosis; acting-out behavior is not inherent in the situation.

4 This is not related to the nursing diagnosis and is not appropriate.

354. ① Without an awareness of personal beliefs, the nurse may unconsciously stop listening if the client expresses deviant beliefs. (1) (CI; AN)

2 Although this may create some anxiety, it usually does not interfere with accurate listening.

3 Same as answer 2.

4 Same as answer 2.

355. ④ This ethically sound response clearly defines who is involved in decision making and allows for parental expression of ideas and thoughts. (2) (CC; AN)

1 Discussion of the implication of a DNR order should not take place until after the family has spoken with the physician.

2 Although the answer promotes the physician-client relationship, it stops the nurse-client interaction.

3 This answer could be interpreted as negative; it also abdicates nursing responsibility.

356. ④ If nurses are unaware of their biases about rape, they will be unprepared to evaluate objectively and meet the client's needs. (1) (CI; AN)

1 This would interrupt communication; information can be solicited later.

2 The nurse should be able to deal with this client without assistance.

3 It is not necessary to know these to effectively respond to the client in a therapeutic manner.

357. ① Value clarification is a technique that uncovers individuals' values so that the individuals can be more aware of them and their effect on others. (2) (PC; AN)

2 This is untrue; it merely helps individuals become aware of values and their effect on others.

3 Same as answer 2.

4 Same as answer 2.

358. ② The ability to discuss feelings about others and life situations is necessary for positive mental health. (2) (EP; AN)

1 This is a long-term, not a short-term, goal.

3 Same as answer 1.

4 Same as answer 1.

359. ③ This is true; the client's goal is meaningful, and the nurse should do everything possible to help the client achieve it. (2) (CC; AN)

1 No evidence demonstrates the client is using denial.

2 There is no reason to attempt to move the client away from this meaningful goal.

4 If the client wants to work toward a goal, the energy expenditure is justified.

360. ③ An individual must feel a sense of control over self and the environment to feel secure, reduce anxiety, and function at an optimal level. (3) (HI; AN)

1 This is untrue; most emotionally ill people are too introspective to empathize with others.

2 No one functions optimally in all settings; the healthy individual can accept and handle temporary periods of confusion and loss of control.

4 This is not true; many individuals with mental illness do not demonstrate observable signs or socially inappropriate behavior.

361. ③ The greatest fear of an inexperienced nurse is of saying the wrong thing and doing harm to a client; it is important to recognize that it can actually become a therapeutic encounter whereby both the client and the nurse can learn from the situation. (3) (PC; AN)

1 Although this client presents a low risk of having the emotional or physical energy to hurt a caregiver, fear of injury is not as high as fear of saying the wrong thing.

2 Being rejected can occur with any client; fear of rejection usually is not an overwhelming concern of nurses.

4 This usually is not an overwhelming concern of nurses.

362. ④ Early notification provides an opportunity to prepare for change. (3) (CC; PL)
    1 This may be a secondary gain but not the primary purpose.
    2 Same as answer 1.
    3 Same as answer 1.

363. ② The client cannot be expected to accept or even respond to a plan that would be incompatible with the family's life-style. (1) (CC; PL)
    1 The family should not have to adjust to the nurse's biases; the nurse must deal with biases before they interfere with care.
    3 All individuals respond differently to situations.
    4 There is no documentation that misconceptions are present.

364. ③ Clients usually respond with better motivation for self-care if they feel someone is depending on them and they are needed. (2) (DF; PL)
    1 Clients need to feel needed, not just to establish new social contacts.
    2 This is untrue; emotionally healthy individuals do not feel better simply because others have more problems.
    4 Same as answer 1.

365. ② When planning a group, the nurse must ensure that clients have similar needs to foster relationships and interactions; diverse needs do not foster group process. (2) (EP; PL)
    1 Although important, this is not a primary consideration.
    3 Behavior and needs, rather than diagnoses, are of primary importance.
    4 This has little effect on group process.

366. ② Members must feel comfortable to discuss things in the group; there must be an awareness that what is discussed in the group will remain in the group. (2) (CC; PL)
    1 This will not establish trust and will increase anxiety because the members will feel their turn for exposure will come whether they want it or not.
    3 Same as answer 1.
    4 Talking about trust does little to foster it.

367. ① To achieve the greatest therapeutic value from a group session, the members must be involved in deciding what will be discussed. (3) (CC; PL)
    2 By this response the nurse leader abdicates the leadership role and places the entire responsibility for the success of the group on its members.

3 This response presents an extremely structured view of the purpose of a therapy group; the members must be involved in the selection of the topics to be discussed.
    4 Same as answer 3.

368. ④ This action by the leader would be most therapeutic because it sets both the parameters of discussion and limits on behavior. (2) (CC; IM)
    1 This would not be therapeutic for a group of clients with the diagnosis of schizophrenia; because of the disruption of cognitive processes, these clients would be unable to make this contribution.
    2 Same as answer 1.
    3 Same as answer 1.

369. ② The nurse must accept the client's statement and beliefs as real to the client to develop trust and move into a therapeutic relationship. (2) (EP; PL)
    1 Clients cannot be argued out of any delusions.
    3 These feelings and thoughts are constant; this would result in an overdose.
    4 Redirecting the client's conversation whenever negative topics are brought up adds to the client's feelings that negative thoughts are correct.

370. ① The client has been able to function well up to this time, and the client's usual behavior and routines should be supported. (2) (DF; PL)
    2 At this time the data presented do not identify this as a need.
    3 Same as answer 2.
    4 Same as answer 2.

371. ② Family therapy tries to view the whole (Gestalt) within the context in which the emotional problems are occurring. (3) (CC; PL)
    1 Time efficiency is not an adequate rationale for choosing a therapeutic approach.
    3 This may or may not be true; an astute nurse can control manipulation and alliance in any situation.
    4 Promotion of truthfulness is a secondary gain achieved through this mode.

372. ③ This is the client's desire, and the nurse should do everything possible to assist the client to achieve it. (2) (SC; PL)
    1 This would not be therapeutic; the client has an unmet need, and the nurse should not try to refocus the client away from the objective.
    2 The client should be encouraged, not discouraged; mental activity should not be too taxing

and it is not unrealistic if the client wishes to do it.
4 No data support the conclusion that the client needs to work through anger.

373. (4) If religion is a family concern, then the nurse should allow discussion of the family's thoughts and feelings on the subject. (2) (SC; PL)
1 If religion is a family concern, its discussion should be encouraged, not limited.
2 The minister is not part of the family unit; the minister would be invited only if requested by the family.
3 The role of the nurse is to facilitate and listen, not to have a mutual discussion.

374. (1) In behavior modification, positive behavior is reinforced and negative behavior is punished or not reinforced. (2) (CC; PL)
2 This may be a part of the program, but it is not a major component.
3 Same as answer 2.
4 Same as answer 2.

375. (1) Before therapy can begin, a trusting relationship must develop. (2) (SC; PL)
2 A client with major depression would not have the impetus or energy to investigate new leisure activities.
3 This is not appropriate initially; a trusting one-to-one relationship must be developed first.
4 This would not be successful unless the client had developed a trusting, comfortable relationship with the nurse.

376. (4) Evaluation and termination are the foci of a discharge planning conference; it is important for the nurse to assist the family in viewing the hospitalization as a learning experience. (3) (CC; PL)
1 This should have been discussed prior to the discharge conference, where evaluation and future planning are the foci.
2 Same as answer 1.
3 Same as answer 1.

377. (1) The use of self is often the only tool available to the nurse to help a client cope; the nurse must be present, actively listening, and attentive to be therapeutic. (1) (SC; IM)
2 The environment is important, but it is not the most basic tool.
3 The client's intellect is not necessarily a therapeutic tool used by the nurse.
4 The nurse must first use self before the helping process can begin.

378. (1) Empathy is the projection of self into another's emotions to share the emotions and the other's state of mind; this technique helps the nurse understand the meaning and significance of the experience to the client. (1) (SC; IM)
2 Sympathy is a shared expression of sorrow over a real or imagined loss.
3 Projection is an unconscious defense, not a therapeutic technique.
4 This approach does not require the nurse to project the self into the client's emotions but rather just to accept the client and the emotions.

379. (3) Focusing is indicated when communication is vague; the nurse attempts to concentrate or focus the client's communication on one specific aspect. (2) (SC; IM)
1 Touch would invade the client's space and would do nothing to help focus the client's communication.
2 Silence would only prolong the rambling communication; the client needs to be focused.
4 Until the concern is identified and explored, summarizing would be impossible.

380. (1) This demonstrates that the staff believes that what the client has to say is important; this also encourages verbalization. (2) (EP; IM)
2 This would only increase feelings of worthlessness and persecution and would cut off communication.
3 This would accomplish little if anything; individuals cannot be talked out of feelings.
4 Feelings cannot always be explained; this also forces the client to further develop the delusional system.

381. (2) Silence is a tool employed during therapeutic communication that indicates that the nurse is listening and receptive; it allows the client time to collect thoughts, gain control of emotions, or speak without hurrying. (2) (SC; IM)
1 Silence should be comfortable and should not create a feeling of pressure to break it by talking.
3 The nurse's facial expression should be projected outward, not inward.
4 This is incorrect; this would close communication.

382. (4) This verifies family and physician agreement and uses institutional policy developed by the ethics committee. (2) (CC; IM)
1 The nurse should not accept this inappropriate burden.

2   Same as answer 1.

3   The order must be part of the written record.

383. (4) This lets the client know the nurse is trying to understand; it increases the client's feeling of self-esteem and points out reality. (3) (SC; IM)

1   Clients with schizophrenia have problems with associative links, and these same problems will occur regardless of the topic.

2   This statement cuts off communication and tells the client that the nurse will only speak if the client's communication makes sense.

3   Same as answer 2.

384. (1) This is the best response that reflects back the feelings being expressed at this time. (2) (SC; IM)

2   This does not address the real concern; the mother's argument may have been justified and the daughter's behavior should not be rewarded.

3   This avoids the issue; the fear may be that next time control may be lost and abuse may occur.

4   False reassurance avoids the real issue.

385. (1) Recognizing the family's feelings and giving simple factual information help to allay anxiety. (2) (SC; IM)

2   This discourages further verbalization of concerns and promotes feelings of isolation and helplessness.

3   This is an inappropriate statement, especially during this time of stress; it also gives little assurance to the family.

4   This is false reassurance and does not allow the family to verbalize their anxieties or fears.

386. (1) Unless staff can share both positive and negative feelings, resentment, anger, and frustration will develop. (3) (CC; IM)

2   This response does little to foster communication and relationships among staff members.

3   This response attacks the speaker and cuts off communication in the group.

4   Same as answer 3.

387. (4) Establishment of a therapeutic relationship begins with introductions. (2) (SC; IM)

1   This provides no information except the speaker's name.

2   This does not provide the client with any information about who the speaker is; use of the client's first name demonstrates lack of respect.

3   In addition to not providing any information and using the client's first name, this greeting is probing and would place the client on the defensive.

388. (4) This response addresses the reality that the client is on the mental health unit and offers assistance. (2) (SC; IM)

1   On the basis of the information available, it would be too early to make this decision.

2   This is a rather hostile response that assumes the client is unable to follow conversation.

3   This response assumes the client is disoriented as to place; it sounds like the beginning of a lecture.

389. (4) This response addresses the client's misconceptions about mental health services and specific fear of being crazy. (2) (EP; IM)

1   This response ignores the feeling tones behind the client's statement and focuses on facts.

2   Same as answer 1.

3   Same as answer 1.

390. (4) This response is optimistic and supportive and clarifies the purpose of the relationship. (2) (SC; IM)

1   This statement diminishes the client's response and sets up a challenge; it does not foster a therapeutic relationship.

2   Same as answer 1.

3   Same as answer 1.

391. (2) Unless the nurse answers the question, the client will continue to focus on the nurse rather than on the self; the nurse can best redirect after a brief answer. (2) (SC; IM)

1   This moves the focus to the nurse's opinions rather than the client's feelings.

3   Same as answer 1.

4   This is not therapeutic; the client is being asked to share, and the nurse should also be willing.

392. (1) This response recognizes problems of the caregiver without a hint of blame for admission; it opens the channel of communication. (2) (SC; IM)

2   This is a somewhat hostile response that would place the caregiver on the defensive.

3   Same as answer 2.

4   Same as answer 2.

393. (1) This response identifies the child's feelings and lets the child know the nurse can understand them. (3) (SC; IM)

2   This denies the child's feelings and really does not offer support.

3 This is an unrealistic response; the child would probably be unable to express his feelings to peers.

4 This is an unrealistic response; the nurse cannot be sure that weight loss will improve the child's ability in gym.

394. (4) This response does not attack the client but places responsibility for allowing the behavior to continue on the group; the client will recognize the message without feeling increased anxiety. (3) (CC; IM)

1 This is not a confronting approach but an attack.

2 This is not therapeutic; it allows behavior to continue without limits being set.

3 This would increase anxiety by placing demands on other members; it does not deal with the identified problem.

395. (3) This response focuses the client on the behavior and what the client is trying to achieve by such behavior; it also helps the client to see how such behavior affects others. (3) (EP; IM)

1 The group would not know what the client was trying to tell the nurse; only the client would know and should be asked directly.

2 This response uses a nondirect approach to attack the client.

4 This is an attacking, defensive response made without really knowing what the client was attempting to accomplish.

396. (2) The leader should not intervene at this point; the client addressed the statement to the group, and the group response should be fostered. (2) (SC; IM)

1 This response would be viewed as an attack and would make other members fearful of contributing because they may be attacked.

3 This denies the client's feelings.

4 Same as answer 1.

397. (3) This is an open-ended, nonjudgmental response that points out incongruity between the client's verbal and nonverbal communication. (2) (SC; IM)

1 This would not help the client recognize the incongruity.

2 Same as answer 1.

4 Same as answer 1.

398. (2) This response uses paraphrasing to restate the content of the client's statement; it encourages further communication. (1) (SC; IM)

1 Feelings are personal and can really not be understood by others; this is an ineffective attempt to empathize and refocuses the attention on the nurse.

3 This response negates the client's feelings and changes the subject; the client needs to talk, and this response cuts off communication.

4 This response negates the client's feelings, makes the feelings impossible to share, may make the client feel guilty for the feelings, and tells the client how to behave and feel.

399. (4) Helping the client to understand the meaning of the family member's behavior reduces the family member's emotional control over the client. (2) (EP; IM)

1 This ignores the necessity of clarifying the family member's behavior.

2 Distraction is not a therapeutic way to deal with realistic feelings.

3 This is only a temporary measure and does not reduce the emotional conflict with the family member.

400. (2) This behavior is typical of the working stage of the group; trust has been established and a willingness to discuss any problems or needs is present. (2) (SC; EV)

1 This can occur at any stage; it occurs in social, as well as therapeutic, relationships.

3 This would occur in the early phase of the group before trust is established and when everyone is trying to fit in.

4 Same as answer 3.

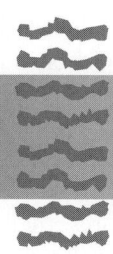

MENTAL HEALTH NURSING QUIZ 1

1. The nurse evaluates that a client has understood the teaching about the side effects and precautions associated with the neuroleptic drug haloperidol (Haldol) when the client states:
   1. "I will immediately report any diarrhea or vomiting to my doctor."
   2. "I will not eat any tyramine-containing foods while I'm taking Haldol."
   3. "I'll maintain an adequate fluid intake, since I may urinate more than usual."
   4. "I'll avoid direct sunlight and use sunburn preventatives when I go outdoors."

2. When caring for a middle-age female client who has lost about 20 pounds over the last 2 months, cries easily, sleeps poorly, and refuses to participate in any family or social activities that she previously enjoyed, it is very important for the nurse to:
   1. Provide the client with a high-calorie, high-protein diet
   2. Set firm consistent limits to reduce the client's crying episodes
   3. Assure the client that she will regain her usual function in a short time
   4. Allow the client to externalize her feelings, especially anger, in a safe manner

3. A client recently admitted to the hospital with the diagnosis of schizophrenia, paranoid type, says to the nurse, "I know they're spying on me in here, too. I'm not safe anywhere!" The most therapeutic response by the nurse would be:
   1. "Nobody's spying on you in here."
   2. "Why do you feel they'd want to follow you here?"
   3. "You don't feel safe anywhere, not even in the hospital."
   4. "You are safe in the hospital; nothing can happen to you here."

4. To foster a therapeutic relationship with a deeply depressed, unresponsive client who stares into space, remains curled up in bed, and refuses to talk, the nurse must first break through the client's withdrawal. Initially, this can best be achieved by:
   1. Sitting quietly next to the client for set periods of time each hour
   2. Urging the client to participate in simple games with other clients
   3. Gently touching the client on the arm when the opportunity arises
   4. Informing the client that dressing and going to the dayroom is required

5. When assessing an adolescent client with the diagnosis of schizophrenia, undifferentiated type, the nurse should expect to observe:
   1. Paranoid delusions, hallucinations, and hyperactivity
   2. Ritualistic behavior, inappropriate affect, and paranoia
   3. Loosened associations, bizarre behavior, and hallucinations
   4. Depression, disorders of thought, and loosened associations

6. The nurse counselor in the mental health clinic is working with a couple and their two sons, ages 14 and 16. One son has been in trouble at school because of truancy and poor grades. The other son appears quiet and withdrawn, and the parents report no problems. The father has been unemployed, and the mother works as a waitress. They have had severe marital problems for the past 10 years. The priority nursing diagnosis for this family at this time would be:
   1. Altered parenting R/T marital problems
   2. Impaired adjustment R/T children growing older
   3. Ineffective family coping R/T son's school problems
   4. Impaired social interaction R/T inability to form relationships

7. A teenage client is admitted to the hospital with a history of increasingly bizarre behavior. The client states, "I am wired to the television and it informed me that my family is out to kill me." The initial action by the admitting nurse that would be most therapeutic for this client would be:
   1. Taking the client to the dayroom and introducing the other clients on the unit
   2. Reassuring the client that the unit is safe and that the client will be protected from the family
   3. Telling the client that the door is locked and no one is permitted to enter the unit to harm any client
   4. Introducing the client to the primary nurse who will be assigned to work on a one-to-one basis with the client

8. When a disturbed client who has a history of using neologisms says to the nurse, "My lacket hss kelong mon," the nurse should respond by:
   1. Trying to learn the language of the client
   2. Telling the client that these words are not understood
   3. Communicating in simple terms directed toward the client
   4. Recognizing that the client needs a nurse who can understand the fantasies expressed

9. A disturbed client, unprovoked, attacks another client. A short-term goal for this client would be to:
   1. Get the client to apologize for the attack to the other client
   2. Have a staff member whom the client trusts stay with the client
   3. Protect others from the client's impulsive acts by secluding the client
   4. Keep the client actively participating in activities and in contact with reality

10. The nurse is aware that the medication used to prevent symptoms of withdrawal in clients with a long history of alcohol abuse is:
    1. Lorazepam (Ativan)
    2. Phenobarbital (Luminal)
    3. Chlorpromazine (Thorazine)
    4. Methadone hydrochloride (Methadone)

11. An adolescent is arrested for shoplifting and is brought to the psychiatric unit. Although describing her child as intelligent, witty, entertaining, and friendly, the mother states, "My child is somewhat unreliable, untruthful, and insincere." The client is diagnosed as having a personality disorder. The most accurate nursing diagnosis for the client would be:
    1. Ineffective individual coping
    2. Antisocial personality disorder
    3. Potential for introverted and mature behavior
    4. Impairment of common sense, feelings of guilt and remorse

12. An elderly client, accompanied by family members, is admitted to the hospital with the diagnosis of dementia. During the admission procedures, the initial approach by the nurse that would be most helpful to this client would be:
    1. "You are somewhat disoriented now, but do not worry. You will be all right in a few days."
    2. "Do not be frightened. I am the nurse, and everyone here in the hospital will help you get well."
    3. "I am the nurse on duty today. You are at the hospital. Your family can stay with you for a while."
    4. "Let me introduce you to the staff here before you get acquainted with our ward policy and routine."

13. The nurse is aware that a 6-year-old with normal psychosocial development would have achieved Erikson's developmental tasks of trust, autonomy, and:
    1. Identity
    2. Initiative
    3. Intimacy
    4. Belonging

14. The daycare treatment team decides it would be therapeutic for a client with an obsessive-compulsive personality disorder to get a part-time job. On the day of the job interview, the client comes to the center very anxious and displays an increase in compulsive behaviors. The nurse could best respond to these behavioral changes by stating:
    1. "I know you're anxious, but make yourself go and try to conquer your fear."
    2. "It must be that you really don't want that job after all. I think you should think more about it."
    3. "If going to an interview makes you this anxious, it seems to me that you're not ready to work."
    4. "Going for your interview triggered some feelings in you. Describe what you're feeling at this time."

15. A group of clients from a psychiatric unit are going to a professional ball game accompanied by staff members. The purpose of visits into the community under the supervision of staff members is to:
    1. Assist clients in adjusting to anxieties in the community
    2. Help the clients return to reality under controlled conditions
    3. Observe a client's ability to cope with a more complex society
    4. Broaden the client's experiences by providing exposure to cultural activities

16. To foster a healthy grieving response to the birth of a stillborn child, the nurse's best response to an expression of anger from the mother would be:
1. "You are young; wait and see, you'll have other children."
2. "You may be wondering if something you did caused this."
3. "It's God's will; we have to have faith that it was for the best."
4. "This often happens when something is wrong with the baby."

17. One day the nurse and a client sit together and draw. The client draws a face with horns on top of the head and says, "This is me. I'm a devil." The nurse should respond:
1. "I don't see a devil. Why do you see a devil?"
2. "Let's go to the mirror and see what you look like."
3. "You are not a devil. Don't talk about yourself like that."
4. "When I look at you, I see an attractive young person, not a devil."

18. The nurse is aware that most clients with phobias use the defense mechanisms of:
1. Dissociation and denial
2. Introjection and sublimation
3. Projection and displacement
4. Substitution and reaction formation

19. The nursing supervisor recognizes that one of the nurses in ICU may be experiencing burnout. The nursing supervisor should plan to help this nurse begin to confront the problem by:
1. Transferring to a primary nursing care unit
2. Choosing a nursing position on a low-stress unit
3. Attending inservice programs as often as possible
4. Identifying personal responses to daily work stresses

20. A client is admitted to a psychiatric unit with a history of sleeplessness, lack of interest in eating, and charging of excessive purchases to charge accounts. The symptoms that the nurse should expect the client to exhibit in the hospital would include:
1. Depressed mood and crying
2. Increased insight into behavior
3. Decreased psychomotor activity
4. Increased interest in the environment

21. A client leaves group therapy in the middle of the session. The nurse finds the client obviously upset and crying. The client tells the nurse that the group's discussion was too much to tolerate. The most therapeutic nursing action would be to:

1. Request kindly but firmly that the client return to the group to work out conflicts
2. Suggest that the client accompany the nurse to a quiet place so that they can talk about the situation
3. Ask the group leader what happened in the group session and base intervention on this additional information
4. Respect the client's right to decline therapy at this time and simply report the incident to the rest of the health team

22. A client sits huddled in a chair and leaves it only to let out a scream, run to the end of the hallway, and crouch in a corner. The nurse, observing this, realizes that this behavior is classified as:
1. Reactive
2. Regressive
3. Dissociative
4. Hallucinatory

23. When caring for a withdrawn, reclusive, autistic client, the priority goal would be for the client to develop:
1. Trust
2. Ego strengths
3. A sense of identity
4. An ability to socialize

24. When a client is receiving lithium, it is important for the nurse to monitor:
1. Weight
2. Visual acuity
3. Bowel sounds
4. Potassium level

25. A client has been experiencing delusions. The nurse understands that according to psychodynamic theory delusions are:
1. A defense against anxiety
2. Precipitated by external stimuli
3. The result of paleologic thinking
4. Subconscious expressions of anger

26. A nurse on the psychiatric unit is assigned to work with a client who appears reclusive and mistrustful of everyone. The nurse can help the client to develop trust by:
1. Attempting to be prompt for their scheduled meetings
2. Stating simply and sincerely that the nurse cares about the client's feelings
3. Handing the client medication and not watching to see whether it is swallowed
4. Listening attentively to the client's positive feelings and ignoring negative feelings

27. A female client in the terminal stage of cancer is admitted to the hospital in severe pain. The client refuses medication for the pain because it puts her to sleep and she wants to be awake. One day, despite the client's objection, a nurse administers the pain medication saying, "You know that this will make you more comfortable." The nurse in this situation could be charged with:
    1. Battery
    2. Assault
    3. Invasion of privacy
    4. Lack of informed consent

28. The nurse, recognizing the possible cause of alcohol-induced amnestic disorder, should take into consideration when planning a client's care that the client is probably experiencing:
    1. A deficiency in thiamine
    2. An iron intake reduction
    3. An increase in serotonin
    4. A riboflavin malabsorption

29. The day following the birth of their baby, the parents are very upset to learn that the baby has a heart defect. At this time it would be most helpful for the nurse to:
    1. Allow the parents to express their anger
    2. Explain the diagnosis in a variety of ways
    3. Encourage the parents to talk with other parents
    4. Assure the parents that surgery will correct the problem

30. An elderly client with primary degenerative dementia has difficulty following simple directions and selecting clothes to be worn for the day. The nurse recognizes that these problems are the result of:
    1. Impaired judgment
    2. Decreased attention span
    3. Clouding of consciousness
    4. Loss of abstract thinking ability

31. A 16-year-old female client is admitted to the adolescent psychiatric unit with a diagnosis of anorexia nervosa. The adolescent has lost 40 pounds during the last 6 months and her current weight is 75 pounds. When approaching this client, the nurse should initially:
    1. Point out how bad she looks
    2. Refrain from discussing her appearance
    3. Recognize that she is deliberately trying to kill herself
    4. State the rules about eating in a matter-of-fact manner

32. In addition to suicide, an awareness of serious health problems in adolescents requires that the school nurse teach the faculty that adolescents are at high risk for:
    1. Rubella and mononucleosis
    2. Heroin abuse and malnutrition
    3. Genital herpes and alcohol abuse
    4. Diabetes and the use of marijuana

33. During the admission interview of a client with a diagnosis of mood disorder—manic episode, the nurse would expect the client to demonstrate:
    1. Flight of ideas
    2. Ritualistic behaviors
    3. Associative looseness
    4. Delusions of persecution

34. To therapeutically relate to parents who are known to have abused their children, the nurse must first:
    1. Identify personal feelings about child abusers
    2. Recognize the emotional needs of the parents
    3. Call authorities to report the suspected incident
    4. Gather information about child's home environment

35. A young woman who has just lost her first job comes to the mental health clinic very upset and states, "Without warning, I just start crying without any reason." The nurse's initial response should be:
    1. "Do you know what makes you cry?"
    2. "Most of us need to cry from time to time."
    3. "Crying unexpectedly must be very upsetting."
    4. "Are you having any other problems at this time?"

36. A client who is receiving haloperidol (Haldol), 5 mg tid, complains of twitching of the fingers. The nurse should respond:
    1. "This is a temporary situation until your body adjusts to the medication."
    2. "You need the medication that we are giving you. You will soon get used to the side effects."
    3. "Let's wait a few days and see whether the side effects of the drug you are receiving go away."
    4. "I will get the doctor to order a medication that will help overcome this. It is a side effect of the drug you are taking."

37. The nurse, finding a client with schizophrenia lying under a bench in the hall, could best respond to the client's statement, "God told me to lie here," by stating:
    1. "I didn't hear anyone talking. Come with me to your room."
    2. "What you heard was in your head; it was your imagination."

3. "Come to the dayroom and watch television. You will feel better."

4. "God would not tell you to lie in the hall. God would want you to behave reasonably."

38. One day as the nurse sits next to a depressed client, the client states, "I want to tell you something but you must promise not to tell anyone." The best response by the nurse would be:
    1. "OK, I won't. It's good for you to talk about what's bothering you."
    2. "You can tell me if you want to, but I cannot give you that promise."
    3. "You have my promise not to tell. What's the secret you want to tell me?"
    4. "You seem to be more depressed today. Why don't you tell me what you are thinking?"

39. A male client with a history of marginally socially acceptable behavior is admitted to the hospital for evaluation. The client tells the nurse he has been married multiple times but never loved, nor would he support any of the women or the children he produced. The nurse could identify that the client's behavior, according to Freud, demonstrates a defect in:
    1. Id development
    2. Ego development
    3. Sexual development
    4. Superego development

40. When working with clients who use manipulative, socially acting-out behaviors, the nurse should be:
    1. Strict, punishing, and restrictive
    2. Sincere, cautious, and consistent
    3. Accepting, supportive, and friendly
    4. Sympathetic, motherly, and encouraging

41. A male client with the diagnosis of antisocial personality disorder takes a female nurse by the shoulder, suddenly kisses her, and shouts, "I like you." The most appropriate response by the nurse would be:
    1. "Thank you, I like you too."
    2. "I wish you wouldn't do that."
    3. "Don't ever touch me like that again."
    4. "I like you too, but please don't do that again."

42. A client with schizophrenia is started on an antipsychotic/neuroleptic medication. The nurse is aware that these drugs are used primarily to:
    1. Keep the client quiet and relaxed
    2. Reduce the need for physical restraints
    3. Control the client's behavior and reduce stress
    4. Make the client more receptive to psychotherapy

43. The nurse is aware that the major defense mechanism used by an individual with a phobic disorder is:

1. Projection
2. Avoidance
3. Regression
4. Repression

44. A young adult client is admitted to the hospital with a diagnosis of schizophrenia, paranoid type. The client's family is concerned about the client's safety and well-being because the client has been stating, "The voices are telling me to kill myself." The nursing diagnosis that should have first priority for this client is:
    1. Disturbed self-esteem
    2. Risk for self-directed violence
    3. Impaired verbal communication
    4. Sensory-perceptual alterations (auditory)

45. On a visit to the clinic, the husband of an alcoholic confides to the nurse that he and his wife are experiencing marital difficulties. He states, "After all the years of her drinking, I can't believe this is happening now that she's been sober for 6 months." An appropriate nursing diagnosis for the husband at this time would be:
    1. Altered role performance related to changes in wife's needs
    2. Impaired adjustment related to altered communication patterns
    3. Situational low self-esteem related to wife's increased independence
    4. Ineffective family coping: compromised, related to altered marital relationships

46. The nurse notices that each time the physician or head nurse visits a disturbed female client, she becomes extremely anxious. Today, after visiting with the physician, the client sits wringing her hands. The best initial response by the nurse would be:
    1. "How do you handle your anxiety?"
    2. "I notice that you are wringing your hands."
    3. "Tell me why you are afraid of authority figures."
    4. "Do you realize why you are wringing your hands?"

47. A female client is admitted to the psychiatric unit with the diagnosis of obsessive-compulsive disorder that is demonstrated by an increasing, consuming obsession with dirt. The client feels her hands are dirty and has a need to wash them about 70 to 80 times a day. The client's hands are red and raw with some bleeding. An immediate nursing goal for this client would be to get the client to:
    1. Understand that her hands are not dirty
    2. Develop insight into her emotional problems
    3. Stop washing her hands so the skin will heal
    4. Limit the number of times she washes her hands

48. A behavior that is most typical of the parents of the battered child is that they:
1. Become irritable about having the history taken
2. Show concern about the child's medical condition
3. Give contradictory explanations about what happened
4. Present many details related to how the trauma occurred

49. An inpatient therapy group on a psychiatric unit has as its goal helping clients participate in life more fully by gaining insight and changing behavior. The nurse leader can best help the group achieve this goal by using a leadership style that is:

1. Stimulating and guiding
2. Autocratic and directing
3. Democratic and controlling
4. Laissez-faire and observing

50. The nurse recognizes that behavior is usually viewed and accepted as normal if it:
1. Fits within standards accepted by one's society
2. Accurately expresses the individual's feelings and thoughts
3. Helps the person to reduce the use of defense mechanisms
4. Helps the individual to achieve short-term and long-term goals

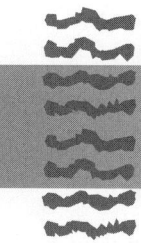

MENTAL HEALTH NURSING QUIZ 1
ANSWERS AND RATIONALES

1. (4) Photosensitivity is a side effect of many antipsychotic medications. (3) (PT; EV; DR)
   1 These symptoms are side effects of lithium, not of Haldol.
   2 Avoiding tyramine-containing foods is a precaution associated with MAO inhibitors, not with Haldol.
   3 This is a precaution associated with lithium, not with Haldol.

2. (4) The greatest danger associated with depression is self-inflicted injury when feelings, especially anger, are internalized. (2) (EP; PL; MO)
   1 A low-calorie diet would be more appropriate because of the client's decreased physical activity.
   2 The client is unable to control or regulate behavior at this time.
   3 This is false reassurance; this is not supportive of the client's feelings at this time.

3. (3) Rephrasing allows for further communication, expresses understanding, and does not belittle the client's feelings. (3) (EP; IM; SD)
   1 Presenting reality to the client at this time only raises anxiety and leads the client to defend the delusion.
   2 "Why" questions make a client defensive, and the wording implies the client's delusion could be true.
   4 This is false reassurance; in any event, a suspicious client would not believe the nurse.

4. (1) Sitting quietly with a severely withdrawn client can provide an opportunity for nonthreatening interaction. (2) (EP; PL; MO)
   2 The client is unable to deal with a one-to-one relationship at this time.
   3 Entering a withdrawn client's body space is intrusive and stressful; it often precipitates a need for further withdrawal.
   4 Placing demands on the withdrawn client causes a sense of threat, increased anxiety, and a need for additional withdrawal.

5. (3) These are the primary behaviors associated with a thought disorder, such as schizophrenia. (2) (EP; AS; SD)
   1 These symptoms are more common in paranoid-type schizophrenia than in the undifferentiated type.
   2 Ritualistic behavior is generally associated with obsessive-compulsive disorders, not schizophrenia.
   4 Depression is not characteristic of schizophrenia.

6. (1) The parents' ongoing marital problems appear to have interfered with their parental roles, with resulting behavior problems in their children. (3) (SC; AN; BA)
   2 There are no data to support this diagnosis.
   3 Same as answer 2.
   4 Same as answer 2.

7. (4) This is extremely important because the disturbed client can be assisted back to reality by a nurse who is interested in everything that happens and everything that the client feels. (2) (EP; IM; SD)
   1 This would be a later action.
   2 The nurse does not know the client will not be frightened; this would be false reassurance.
   3 This would have no effect since the client is involved with a strong delusional pattern.

8. (2) This is a simple statement that the client is not understood; it provides feedback and points out reality. (2) (EP; IM; SD)
   1 Neologisms have symbolic meaning only for the client.
   3 This will be of limited help and does not present reality.
   4 There is no one other than the client who can understand the fantasies.

9. (2) The client needs someone with whom there is a working and trusting relationship; this individual must observe, protect, anticipate, and prevent the client from acting out on destructive impulses. (3) (EP; PL; SD)
1 At this time the client cannot be held responsible for behavior.
3 It may still be necessary to do this, but staying with the client is a better immediate action.
4 The client may not be ready for participation, though there is a need to be kept in touch with reality.

10. (1) This drug is most effective in preventing delirium tremens. (2) (PT; PL; SA)
2 This drug is used to prevent withdrawal symptoms associated with barbiturate use.
3 This antipsychotic medication is often used to combat the physiologic effects of amphetamine use.
4 This drug is used to prevent withdrawal symptoms associated with narcotic use.

11. (1) History demonstrates that the client has had a difficult time controlling impulsive behavior and has consistently exhibited poor judgment; this indicates ineffective coping. (1) (EP; AN; PR)
2 This is a psychiatric, not a nursing, diagnosis.
3 This is not an acceptable nursing diagnosis.
4 Same as answer 3.

12. (3) Familiarity with the environment and orientation to the staff may help promote security and feelings of trust. (2) (DF; IM; DD)
1 This denies the client's feelings and provides false reassurance.
2 This statement denies feelings and is false reassurance because all personnel are not involved with the client.
4 A person under stress cannot assimilate much information; verbiage can only lead to more confusion and anger.

13. (2) A 6-year-old should have resolved the developmental task of initiative vs guilt. (1) (DF; AN; PD)
1 Resolution of identity vs role confusion occurs at adolescence.
3 Resolution of intimacy vs isolation occurs at adulthood.
4 This is part of Maslow's hierarchy of needs; the need for love and belonging arises once physical and safety needs are met.

14. (4) These symptoms are a defense against anxiety resulting from decision making, which triggers old fears; the client needs support. (2) (SC; EV; PR)
1 This denies the client's overwhelming anxiety and lacks realistic support.
2 This is judgmental; an increase in anxiety does not necessarily mean the client does not want to attain the goal.
3 This is judgmental; the client should be encouraged to work through the symptoms, not to avoid risk.

15. (3) The nurse's observations can help identify those clients who are ready to cope with outside stress and those who are not. (3) (EP; EV; SD)
1 A social evening in town would not accomplish this.
2 A social evening in town would not by itself constitute reality.
4 There is nothing to indicate that any of these clients needed broadening cultural experiences.

16. (2) The mother must be helped to identify feelings. (1) (SC; IM; CS)
1 False reassurance; this does not encourage the client to explore feelings.
3 This answer is based on the nurse's religious beliefs; there is no indication that the client has the same beliefs; this closes off communication.
4 Many stillborn children are apparently free of any defects.

17. (4) This response points out reality while attempting to let the client understand that the nurse sees the client as a person of worth. (3) (EP; IM; SD)
1 This asks the client to explain feelings, which may be an unrealistic goal.
2 This is nontherapeutic; the client may indeed view the self as a devil.
3 This is a somewhat punitive response; it cuts off communication.

18. (3) Clients using phobias deal with anxiety by placing it on specific persons, objects, or situations through the process of displacement and/or projection. (3) (EP; AS; AX)
1 The person with a phobia recognizes and admits the exaggerated fear as a real part of the self.
2 Neither introjection, whereby a person internalizes and incorporates the traits of another, nor sublimation, whereby socially acceptable behavior is substituted for unacceptable instincts, are related to phobic activity.
4 A less-valued object is not substituted for one more highly valued (substitution) nor are the expressed feelings opposite to the experienced feelings of fear (reaction formation).

19. ④ Identification of work stressors in the environment, coping strategies used, and evaluating the effectiveness of these strategies is the first step. (2) (SC; IM; CS)
    1 This may help, but prevention begins with knowing oneself and the effectiveness of one's coping strategies.
    2 Choosing a nursing position on a low-stress unit can help prevent burnout, but it is not the first step in confronting the problem once it occurs.
    3 Attending continuing education programs can help prevent burnout, but this is not the first step in confronting the problem once it occurs.

20. ④ In an attempt to ward off depression, the client in the hyperactive phase runs headlong into reality, becoming totally involved in everything that goes on in the environment. (2) (EP; AS; MO)
    1 This would be more indicative of the depressive episode.
    2 During this phase, there is no insight into behavior.
    3 Just the opposite is true; psychomotor activity is greatly increased.

21. ② This approach incorporates the principles of starting where the client is and helping the client verbalize feelings; it also provides for additional data collecting. (2) (EP; IM; TR)
    1 The client is obviously not ready to do this.
    3 This would be a later step, after the more appropriate nursing action was completed.
    4 This accepts the client's right not to be forced back into the group; however, another nursing intervention should be attempted at this time.

22. ② This behavior reflects the early fetal position. The individual curls up for both protection and security. (3) (EP; AN; SD)
    1 The client's behavior does not seem to be in response to an observable stimulus.
    3 The client's behavior does not indicate dissociation or depersonalization.
    4 The client gives no indication of a hallucinatory pattern.

23. ① Trust is basic to all therapies; without trust a therapeutic relationship cannot be established. (2) (EP; PL; BA)
    2 Treatment is to build on present ego strengths; development of new strengths is a long-term goal.
    3 There is nothing to indicate that the client does not have a sense of identity.
    4 Although helping the client relate to others is a part of the treatment, it is not a priority goal at this time.

24. ① Weight gain may be indicative of fluid retention; renal problems have been reported with lithium therapy. (3) (PT; EV; DR)
    2 This has no relationship to the administration of lithium.
    3 Same as answer 2.
    4 Same as answer 2.

25. ① Delusions are a way the unconscious defends the individual from real or imagined threats. (2) (EP; AN; SD)
    2 Illusions are false interpretations of actual external stimuli.
    3 This is logical thinking that is formulated from an illogical base
    4 Delusions are precipitated by feelings of anxiety, not anger.

26. ① This helps the client to feel important enough for the nurse to remember their meeting and be on time. (2) (EP; IM; TR)
    2 The client is mistrustful of others and will probably not believe the nurse; caring is best demonstrated through behavior.
    3 This would not only be unsafe but may make the client feel that the nurse does not care enough to stay.
    4 Feelings should never be ignored but should be accepted as important to the client.

27. ① This is the intentional touching of one person by another without permission of the person touched. (2) (CC; EV; CS)
    2 This is an intentional act without touching that makes a person fearful or produces reasonable apprehension of bodily harm.
    3 This refers to the right of clients to have their private affairs protected.
    4 This applies to written permission for procedures and treatments to be performed.

28. ① The deficiency of thiamine (vitamin B_1) is thought to be a primary cause of alcohol-induced amnestic disorder. (3) (PC; PL; SA)
    2 This is unrelated to alcohol-induced amnestic disorder.
    3 Same as answer 2.
    4 Same as answer 2.

29. ① Parents need to express and deal with their feelings; then perhaps they can move toward other coping strategies. (2) (SC; IM; TR)
    2 This does not focus on the need presented by the parents.
    3 This does not focus on their present concern but could be useful sometime in the future.
    4 This is premature, possibly false, reassurance.

30. (4) Impairment of abstract thinking interferes with interpreting and defining words in addition to following directions and selecting clothes. (3) (EP; AN; DD)
1 Following directions does not require skill in judgment or decision making.
2 The selection of clothes does not require an intact attention span.
3 Primary degenerative dementia does not cause a clouding of consciousness.

31. (2) Initially, the nurse should not discuss the client's appearance because this focuses on a symptom rather than on feelings. (2) (EP; IM; ES)
1 This focuses on a symptom rather than on feelings; in addition, the client does not believe she looks bad.
3 The client's objective is not suicide; the client has an unconscious desire to remain childlike.
4 Stating the rules would not accomplish anything and will not convince the client to eat.

32. (3) Adolescence is the period of development characterized by experimentation with adult-type behaviors; this experimentation often leads to unprotected sex and alcohol abuse. (2) (DF; AN; BA)
1 Mononucleosis is a problem for adolescents, but rubella is a health risk for young children and pregnant women, not adolescents.
2 Malnutrition is not a particular risk for adolescents unless other problems are present; heroin is not the drug of choice for adolescents.
4 Marijuana use continues to present some risks in adolescents, but the development of diabetes is not a particular risk for this age group.

33. (1) This is a fragmented, pressured, nonsequential pattern of speech typically used during a manic episode. (1) (EP; AS; MO)
2 These are repetitive, purposeful, and intentional behaviors that are carried out in a stereotyped fashion; they are typically found in clients with obsessive-compulsive disorders.
3 This is the pattern of speech found in clients with schizophrenia; usual connections between words and phrases are lost to the listener and meaningful only to the speaker.
4 These are fixed false beliefs that others are plotting to do harm; they are typically found in clients with paranoid-type schizophrenia.

34. (1) Self-awareness is an essential element in providing support, understanding, and empathy to others. (1) (EP; AS; CS)

2 Meeting emotional needs cannot be accomplished until an interpersonal relationship is established.
3 Although essential, this may in reality be a deterrent to the interpersonal relationship.
4 Although important, these data do not take priority at this time.

35. (3) This response identifies the client's feelings. (1) (SC; IM; AX)
1 This is an unrealistic question; the cause of anxiety is on an unconscious level.
2 This response moves the focus away from the client.
4 This disregards the client's comment and avoids feelings.

36. (4) This response reassures the client that the staff member is able to help and that the client's feelings are accurate. (2) (PT; EV; DR)
1 It is a reversible condition that can be treated with benztropine (Cogentin) or diphenhydramine (Benadryl).
2 It is not a symptom that requires adjusting to, rather one that must be treated.
3 Early treatment to reverse the symptom is important.

37. (1) The nurse is focusing on reality and trying to distract and refocus the client's attention on the original goal. (3) (EP; IM; SD)
2 This statement is too blunt and belittling; this approach will rarely work.
3 This is false reassurance; the nurse does not know that the client will feel better.
4 This may be interpreted as belittling or an attempt to convince the client that the behavior is irrational; it is usually ineffective.

38. (2) If a client tells the nurse something that should be shared, the nurse would have to break a promise or risk harm to the client or others. (2) (SI; IM; TR)
1 Encouraging discussion is helpful, but putting oneself in the position of possibly having to break a promise is not therapeutic.
3 This puts the nurse in the position of having to break a promise, which will destroy rapport and decrease trust.
4 This response ignores the client's statement and avoids the issue.

39. (4) Superego development reflects the internalized norms of the family and society; the antisocial personality has never achieved this internalization. (2) (EP; AN; PR)

1   A person with this behavior has an overdeveloped need to seek pleasure.
2   The ego is not the problem; the defect is in the superego.
3   There may be a defect in sexual attitudes but not in sexual development.

40. (2) A sincere, cautious, and consistent attitude limits this individual's ability to manipulate both situations and staff members. (3) (EP; PL; PR)
1   An attitude such as this would allow the client to rationalize the manipulative behavior to deal with the response of the nurse.
3   In accepting the person, the nurse should not support negative behavior; a friendly attitude may encourage further problem behavior.
4   This would only encourage clients to continue in their life-style rather than learn better ways to relate to their environment.

41. (4) This accepts the client while rejecting and setting limits on the behavior the client is using. (2) (EP; IM; PR)
1   This encourages this type of behavior instead of setting limits.
2   This sends a confusing message to the client because it is unclear what the nurse did not like.
3   This rejects the client instead of the behavior.

42. (4) Antipsychotic/neuroleptic medications help control anxiety and acting-out behavior, making the client more approachable. (3) (PT; AN; DR)
1   Although the medication may produce this effect, it is not the primary purpose of administration.
2   Same as answer 1.
3   Same as answer 1.

43. (2) The person transfers anxieties to objects, usually inanimate objects, which are then avoided to decrease anxiety. (1) (EP; AN; AX)
1   Projection, the attributing of undesirable traits or unacceptable feelings or motivations to others, is not the main defense mechanism used by someone with a phobia.
3   Regression, the return to an earlier, more comfortable level of development, is not the main defense mechanism used by someone with a phobia.
4   Repression, the pushing of unacceptable impulses or ideas into the unconscious, is not the main defense mechanism used by someone with a phobia.

44. (2) Client safety always has the highest priority over any other client needs. (1) (SI; PL; SD)

1   Although important, this diagnosis is not a priority at this time.
3   Same as answer 1.
4   Same as answer 1.

45. (1) Sobriety may alter how the alcoholic perceives the self and others, which can influence relationships; people who need to feel needed or who have assumed the role of the enabler may find this role altered as the alcoholic begins to change during recovery. (3) (EP; AN; SA)
2   This is not an acceptable nursing diagnosis.
3   There is no indication that there are negative feelings about the self or self-capabilities.
4   There are no data to support this nursing diagnosis.

46. (2) The nurse is making an observation; bringing it to the attention of the client is an initial step in understanding that behavior. (1) (EP; IM; TR)
1   This is premature because the client may not even be aware of anxiety.
3   This is premature and does not allow for self-recognition of feelings.
4   This is requesting an explanation from the client that the client probably is incapable of making.

47. (4) This action still permits the client to deal with feelings of anxiety while reducing the potential for skin damage. (2) (EP; AN; PR)
1   The anxiety is too great for the client to understand why handwashing is not necessary.
2   Recognition must precede the development of insight; neither can be done until the level of anxiety is reduced.
3   This will not allow the client any outlet for dealing with extreme anxiety, which is a priority need at this time.

48. (3) In an attempt to block the hospital staff from discovering what happened, abusing parents often provide inconsistent accounts. (3) (EP; AS; CS)
1   This may or may not occur.
2   Battering parents tend to show more concern about how the child's condition affects them than about how the child is affected.
4   Battering parents tend to be vague about the details of the accident but maintain the child was accidentally injured.

49. (1) This type of leader stimulates, guides, and assists the group to develop its maximum potential by facilitating and balancing group forces. (2) (EP; AN; TR)

2   This type of leader makes most of the decisions and controls the group, thus limiting group growth potential.

3   Democratic leadership allows for group growth, but controlling leadership limits it.

4   This type of leader allows group members to take over the group; if the members are not good leaders, little is gained from the group.

50. (1) An accepted practice in some parts of the world may well be considered abnormal behavior in others. (2) (SC; AN; PD)

2   Thoughts and feelings are under unconscious control and are seldom accurately expressed by behavior; they may be expressed in dreams.

3   All people need relief from tension from time to time and make use of defense mechanisms to accomplish this.

4   If the behavior were aggressive or destructive, even though it helped reach a goal, it could not be considered normal.

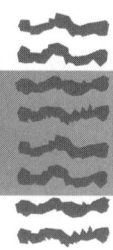

MENTAL HEALTH NURSING
QUIZ 2

1. A client, who is admitted to the hospital for an elective prostatectomy, is extremely anxious and has hand tremors. The client's wife informs the nurse that the client has been drinking heavily for the last 5 years. While the client is unpacking his suitcase, the nurse notices him hiding a bottle of whiskey in the rear of the drawer. The nurse's responsibility in relation to the alcohol includes:
   1. Trying to catch the client drinking the alcohol
   2. Asking the client how much alcohol he really drinks
   3. Confiscating the alcohol when the client is not looking
   4. Waiting for the client to bring up the subject of drinking

2. When a female client who has pressured speech mumbles incoherently, the most appropriate intervention would be for the nurse to:
   1. Set limits on the client's behavior by refusing to talk with her unless she speaks clearly
   2. Consistently ask the client to repeat what she said, so she will learn to recognize she is mumbling
   3. Ignore the client's mumbling since she is using this pathologic manner of speech to get attention
   4. Indicate to the client that she needs to slow down because what she says is important and cannot be understood

3. A 16-year-old, arrested for assault and robbery, has a history of truancy and prostitution. The client demonstrates little emotion and is unconcerned that this behavior has caused emotional distress to others. The diagnosis of antisocial personality disorder is made. According to Freud, the client's lack of remorse and repetitive behavior is probably related to an underdeveloped:
   1. Id
   2. Ego
   3. Superego
   4. Limbic system

4. The nurse can be most therapeutic in approaching an autistic toddler who is sitting in a corner rocking and spinning a top by:
   1. Gently stroking the toddler's arm to gain the child's attention
   2. Bending down and staring at the spinning top with the toddler
   3. Holding the toddler to provide a sense of support and security
   4. Waiting for the toddler to make the initial contact before moving close

5. When developing a plan of care for a client who is using ritualistic behavior, the nurse must recognize that the ritual:
   1. Is under conscious control
   2. Is used primarily for secondary gains
   3. Helps the client control the level of anxiety experienced
   4. Helps the client to focus on the inability to deal with reality

6. The level of achievement at 20 years of age, according to Erikson's developmental stages, should be:
   1. Having the capacity for love and a commitment to work
   2. Being creative and productive and having concern for others
   3. Having a coherent sense of self and plans for self-actualization
   4. Accepting the worth, integrity, and uniqueness of one's past and present life

7. A depressed client is brought to the emergency room after taking an overdose of a sedative. After a lavage, the client states, "Let me die. I'm no good." The nurse's most appropriate response would be:
   1. "Do you feel like telling me why you did this?"
   2. "Of course you're good; we'll take care of you."
   3. "You must have been upset to try to take your life."
   4. "You have been through a rough time; let me take care of you."

8. Clients receiving electroconvulsive therapy are usually given a muscle relaxant just before the treatment. The major disadvantage of this drug is that it inhibits the:
   1. Biceps and triceps muscles
   2. Facial and thoracic muscles
   3. Intercostal and diaphragmatic muscles
   4. Sternocleidomastoid and abdominal muscles

9. When planning care for a confused or delusional client, it is most important for the nurse to:
   1. Maintain quiet, dim surroundings to minimize stimuli
   2. Encourage realistic activity considering the client's ability
   3. Recognize that the client is completely unable to differentiate fantasy from reality
   4. Provide physical hygiene and comfort to demonstrate that the client is worthy of receiving care

10. A client is receiving an antipsychotic drug bid. Two thirds of the daily dose is given in the evening, one third in the morning. This is done to:
    1. Maintain diurnal rhythms
    2. Help the client sleep at night
    3. Reduce sedation during the daytime
    4. Reduce increased assaultiveness in the evening

11. In her eighth month of pregnancy, a 24-year-old client is brought to the hospital by the police, who were called when she barricaded herself in a ladies restroom of a restaurant. During the admission, the client shouts, "Don't come near me. My stomach is filled with bombs, and I'll blow up this place if anyone comes near me." This is an example of:
    1. Ideas of reference
    2. Delusional thinking
    3. Loose associations
    4. Tactile hallucinations

12. A client who appears dejected, barely responds to questions, and walks very slowly about the mental health unit tells the nurse in a barely audible voice that life is no longer worth living. The nurse's most therapeutic response to this statement would be:
    1. "Have you been thinking about suicide?"
    2. "What could be so bad to make you feel that way?"
    3. "We'll talk about your feelings after you have rested."
    4. "Let's talk about something pleasant, and you'll feel better."

13. A client with a diagnosis of mood disorder-manic episode is started on a regimen of chlorpromazine and lithium carbonate. The nurse is aware that the rationale behind this regimen is that the chlorpromazine:
    1. Potentiates the action of lithium for more effective results
    2. Acts with the lithium to prevent progression to the depressive phase
    3. Acts to quiet the client while allowing time for the lithium to take effect
    4. Helps decrease the incidence of lithium toxicity in the first week of therapy

14. After detoxification, a client with a long history of alcohol abuse has agreed to attend Alcoholics Anonymous meetings at the hospital. On the day of the second meeting, the client states, "I cannot attend the AA meeting today because I am expecting an important phone call." The nurse's most therapeutic response would be:
    1. "You are expected to go to the meeting."
    2. "Is your phone call really that important?"
    3. "You can go to the meeting after the call."
    4. "You can wait for the call and skip the meeting."

15. A client describes delusional material in minute detail to the nurse. The nurse should:
    1. Get the client involved in a repetitive project
    2. Accept this as the client's reality without argument
    3. Encourage the client to continue to discuss the delusional material
    4. Change the topic as soon as the client begins to discuss delusional material

16. A disturbed client who has been out of touch with reality has been hospitalized for a couple of weeks. One day the nurse notes the client's hair is dirty and asks whether the client would like to wash it. The client answers, "Yes, and I'd like to shower and change my clothes, too." The nurse uses this information to assess that the client:
    1. Is quite open to suggestions
    2. Has some feelings of self-worth
    3. May be entering a hyperactive phase
    4. Has a need for social reassurance and approval

17. A hyperactive client becomes loud and insulting and says to a staff member, "Get lost, you old buzzard!" The nurse could best handle this situation by:
    1. Telling the client that it is not necessary to be so rude
    2. Saying, "Here is something I feel you might be interested in."

3. Pointing out that the staff member is neither old nor a buzzard
4. Asking the client, "Could you please tell me why you are so angry?"

18. Thirty minutes after administering fluphenazine (Prolixin) to a client, the nurse notices the client's jaw is rigid, the tongue is thick, the client is drooling, and speech is slurred. There are a number of prn orders in the client's chart. The nurse should administer
    1. Diazepam (Valium), 10 mg PO
    2. Benztropine (Cogentin), 2 mg IM
    3. Trihexyphenidyl (Artane), 1 mg PO
    4. Chlorpromazine (Thorazine), 50 mg IM

19. The nurse understands that the unconscious basis of an obsessive-compulsive disorder is often:
    1. Feelings of guilt and inadequacy
    2. Problems with anger and remorse
    3. Problems with being too conscientious
    4. Feelings of unworthiness and hopelessness

20. A client with an obsessive-compulsive disorder continually walks up and down the hall, touching every other chair. If unable to do this, the client becomes quite upset. The nurse should:
    1. Keep talking to distract the client, which will help the client forget about touching the chairs
    2. Allow the client to continue to touch the chairs as long as desired and wait until fatigue sets in
    3. Remove all chairs from the hall, thereby relieving the client of the necessity to touch every other one
    4. Allow the behavior to continue for a specified time, letting the client help set the time limits to be imposed

21. To begin to establish a therapeutic relationship with a withdrawn, reclusive client, the nurse must:
    1. Obtain a complete history from the family
    2. Plan to keep the client's anxiety at a minimum
    3. Ascertain what topics are of interest to the client
    4. Protect the client from self-destructive tendencies or injury

22. The nurse understands that primary dementia results from:
    1. Cerebral atrophy specifically involving the frontal lobes
    2. A long history of poor nutrition and associated avitaminosis

3. A delayed response to severe emotional trauma in early adulthood
4. Anatomic changes in the brain that produce acute but transient symptoms

23. The parents of an infant born with a unilateral cleft lip and palate are very concerned about the defect and ask the nurse, "What caused our baby to be born deformed?" The nurse should reply:
    1. "Are you feeling guilty?"
    2. "I'm glad that you are able to ask these kinds of questions."
    3. "I don't know, but you don't need to worry because surgery can correct it."
    4. "It sounds as if you are wondering what you might have done to cause this situation."

24. When caring for a client with a somatoform disorder, conversion-type paralysis, the nurse should:
    1. Avoid discussing the paralysis
    2. Explain the reason for the paralysis
    3. Ask how the client feels about being paralyzed
    4. Encourage the client to get up, pointing out that walking is possible

25. A client with bulimia nervosa eats two sandwiches, two salads, and four desserts for lunch. After the meal the nurse would expect to observe the client:
    1. Performing excessive exercises
    2. Hoarding more food for a later binge
    3. Withdrawing from the group to the bathroom
    4. Actively socializing with small groups of clients

26. A client who has been raped, aware of the possible legal implications, decides to prosecute the rapist. The nurse carefully listens and documents all observations. This is done because with a charge of rape the burden of proof:
    1. Rests with the rape victim
    2. Rests with the health team
    3. Is on the defendant to prove innocence
    4. Must be established before the case will be heard

27. When the nurse enters the room to administer medication to an agitated and angry client with schizophrenia, paranoid type, the client shouts, "Get out of here!" The nurse should:
    1. Say, "You must take your medicine now."
    2. Say, "I'll be back in 15 minutes and we can talk."
    3. Explain why it is necessary to take the medication
    4. Get assistance and give the medication by injection

28. When talking with the nurse, a client with a mood disorder states, "I feel rotten and useless. I cannot think straight. I feel overwhelmed by everything. I don't know if I can go on." When recording this encounter in the client's record, the most objective description of the client's mood would be:
    1. Client appeared to be very depressed for most of the morning. Little interest in self or environment
    2. Client is not able to cope with problems and this hospitalization; states, "I cannot think straight."
    3. Client stated, "I feel rotten and useless. I feel overwhelmed by everything. I don't know if I can go on."
    4. Client expressed suicidal thoughts about not being able to go on and has decreased ability to think clearly

29. A client who is going to occupational therapy for the first time tells the nurse, "I do not want to go." The nurse, taking the client's diagnosis of schizophrenia into consideration, could best help the client by stating:
    1. "I will go with you to occupational therapy."
    2. "It is only for 1 hour, then you will be back."
    3. "Try it once; if you don't like it, you need not go back."
    4. "The doctor ordered it as part of your treatment. You should go."

30. A client is to begin lithium carbonate therapy. The nurse should ensure that prior to the drug's administration the client should have baseline:
    1. Renal studies
    2. Enzyme studies
    3. Neurologic studies
    4. Fluid and electrolyte studies

31. After lunch one afternoon, the nurse notes that a client with the diagnosis of dementia is alone in the dayroom away from other clients. When the nurse approaches, the client says, "I am all alone. No one has any use for me." The response by the nurse that would be most appropriate at this time would be:
    1. "You seem upset. Would you like to tell me what is bothering you?"
    2. "We need to be alone sometimes. It helps us get to know ourselves better."
    3. "You should focus on ways to change this. Let's play some games to improve your morale."
    4. "Have you done anything to avoid feeling lonely? I think you should socialize more with others."

32. A 5-year-old with an attention-deficit hyperactivity disorder exhibits a short attention span and demonstrates intermittent head banging, hair pulling, and excessive motor activity. The most important nursing diagnosis for this child at this time would be:
    1. Anxiety related to shortened attention span
    2. Sleep pattern disturbance related to hyperactivity
    3. Body image disturbance related to acting-out behavior
    4. Risk for violence: self-directed, related to self-destructive behavior

33. An elderly male is widowed suddenly when his wife is killed in an automobile accident. The first action by the nurse in the emergency room to best help the client at this time would be:
    1. Asking the clergyman of his faith to visit him
    2. Having the physician order a tranquilizer for him
    3. Referring him to a support group that meets near his home
    4. Assuring him that everything possible was done for his wife

34. The nurse should be aware that the statement by a client that would indicate an irreversible adverse response to long-term therapy with an antipsychotic medication would be:
    1. "My mouth is always dry."
    2. "I'm not eating like I should."
    3. "I can't seem to sleep at night."
    4. "My tongue and lips move themselves."

35. The staff of the psychiatric unit conduct a biweekly orientation meeting for newly admitted clients. When planning for this meeting the nurse recognizes that the beginning of the meeting should be directed toward defining the:
    1. Rules for client behavior
    2. Purpose of the group meeting
    3. Clients' role and the leader's expectations
    4. Development of trust between staff and clients

36. A disturbed client remains aloof and ridicules and is sarcastic to other clients. The client identifies with and only will respond to the staff. The care plan at this time should include:
    1. Encouraging group participation with withdrawn clients
    2. Assigning the client activities to be carried out with the staff
    3. Explaining that the client should be more accepting of others
    4. Accepting the client's negative behavior; praising positive responses

37. When a therapy group is achieving its objective, the nurse leader would expect to observe that all members:

1. Now attend every session of the group
2. Comment on each topic discussed by the group
3. Follow through on obeying rules governing behavior
4. Make an effort to include each other in the discussion

38. During crisis intervention therapy, it is most important for the nurse to assume the:
1. Passive listener role
2. Friendly adviser role
3. Active participant role
4. Participant observer role

39. The symptom that would cause the nurse to stop giving chlorpromazine (Thorazine) to a client until further laboratory work was done would be:
1. Grimacing
2. Shuffling gait
3. Yellow sclerae
4. Photosensitivity

40. A client who is to have a segmental mastectomy for cancer of the breast tells the nurse that she is worried about what she will look like after the surgery. The nurse's most appropriate initial response would be:
1. "Try not to think about the surgery now."
2. "I can understand that you'd be concerned."
3. "Why don't you discuss this with your husband?"
4. "Everyone having this surgery feels the same way."

41. A client comes to the mental health clinic complaining about feelings of extreme terror when attempting to ride in an elevator and feeling very uneasy in large crowds. The client feels uncomfortable about these fears and is beginning to experience difficulty concentrating at work. When assessing the situation, the nurse should understand that the client's symptoms are probably associated with:
1. The development of an obsession as a result of conflicts with society
2. Depression about life events, which frequently leads to unreasonable fears
3. A terrifying incident in an elevator in the past that has been repressed but unresolved
4. Generalized anxiety about conflicts that has been displaced into specific unreasonable fears

42. A husband and wife are admitted to the trauma unit with gunshot wounds sustained in a robbery. Shortly following admission, the husband dies from his wounds. A potential nursing diagnosis for the wife related to the death of the husband would be:

1. Defensive coping
2. Altered family processes
3. Ineffective individual coping
4. Personal identity disturbance

43. A 30-year-old high school dropout who is employed as a dishwasher and his 37-year-old wife of 9 years have five children between the ages of 2 months and 8 years. He has a long history of drinking heavily, especially on weekends, and it takes little or no provocation to send him into a rage, yelling obscenities, throwing and breaking furniture, and occasionally hitting his wife and the older children. The nurse recognizes that this abusive behavior is probably related to the client's:
1. Feeling trapped in a marriage
2. Living in the culture of poverty
3. Low socioeconomic background
4. Long-standing problem with alcohol

44. A client with a diagnosis of mood disorder-manic episode has pressured speech punctuated with profanity. It would be most therapeutic for the nurse to deal with this client's behavior by:
1. Thoroughly explaining the type of behavior allowed in the facility
2. Encouraging interaction with another client who is exhibiting similar behavior
3. Quietly stating that the use of profanity is unbecoming and will not be tolerated
4. Allowing for the expression of hostility in a safe manner without being judgmental

45. A 3-year-old child is admitted to a children's mental health unit with a diagnosis of infantile autism. The major goal of therapy for a child with this pervasive developmental disorder would be, the child will:
1. Develop language skills
2. Limit the use of regressive behavior
3. Be mainstreamed into a regular preschool group
4. Recognize the self as an independent person of worth

46. A client with the diagnosis of an antisocial personality disorder responds to a rebuff by a nurse by saying, "You sure do look messy today." The most appropriate response by the nurse would be:
1. "Don't you feel well today?"
2. "That's not a nice thing to say."
3. "I get the feeling you're angry with me."
4. "I really didn't think anyone would notice."

47. A nurse who has been working with a client with the diagnosis of borderline personality disorder is leaving for vacation in 2 weeks and tells the client. The nurse would recognize progress in the ability to maintain more mature relationships if the client:
1. States, "I must get well enough by then so I can also leave."
2. Wishes the nurse a safe trip and offers thanks for the help received
3. Responds, "I guess your leaving is just another loss I must adjust to."
4. Informs the nurse that there is no sense in waiting 2 weeks; the relationship can end today.

48. When a 4-year-old child is admitted to the hospital with traumatic injuries, a common clue that battering has occurred is that the child:
1. Does not cry during painful procedures
2. Shows no expectation of being comforted
3. Cries for long periods and has large hematomas
4. Ignores offers of toys and favors and cries when picked up

49. The nurse's first action when a schizophrenic client, who was admitted to a psychiatric facility involuntarily, runs away would be to notify the:
1. Client's family that the client has left the hospital
2. Law enforcement officers of the client's elopement
3. Client's psychiatrist after discovering the client has gone
4. Physicians who certified the client's need for hospitalization

50. A client receiving a phenothiazine drug is going on an all-day fishing outing with family members. It is important that the nurse:
1. Provide the client with sunscreen ointment
2. Caution the client to avoid excessive activity and overexertion
3. Give the client an additional dose of medication to take after lunch
4. Take the client's blood pressure before allowing the family to leave with the client

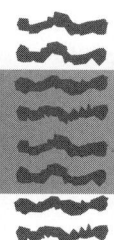

MENTAL HEALTH NURSING QUIZ 2
ANSWERS AND RATIONALES

1. (2) This assesses the client's level of alcohol abuse by direct questioning. (2) (CI; AS; SA)
1 This is a judgmental approach that uses manipulation and decreases the client's self-esteem.
3 This is not straightforward and will decrease trust.
4 The client probably would not bring up the subject because denial is often used to cope with alcohol abuse.

2. (4) This response provides feedback, which helps communication stay on track and demonstrates an interest in the client as a person. (2) (EP; IM; MO)
1 This response would not set limits but would create feelings of rejection.
2 This response would only increase the client's anxiety and anger; it may precipitate acting-out behavior.
3 Ignoring a client does not help the situation, nor does it demonstrate the nurse's interest or concern.

3. (3) Lack of remorse indicates weak superego, the aspect of personality concerned with prohibitions. (2) (EP; AN; PR)
1 This aspect of personality is not underdeveloped in this person; the id acts to achieve self-gratification.
2 The ego is not related to acting-out behavior.
4 This is not underdeveloped; it is related to achieving pleasure.

4. (2) Autistic children relate best with objects, which can be used as a bridge in interpersonal relationships; this begins at the child's level. (2) (EP; IM; BA)
1 Autistic children usually have difficulty tolerating being touched.
3 Autistic children often become agitated when movement is restricted and personal space is invaded.
4 Autistic children will not initiate contact or interactions.

5. (3) The rituals help control anxiety by maintaining a set pattern of action. (2) (EP; PL; PR)
1 The reason for the ritual is under unconscious control.
2 Rituals are used primarily to handle feelings of anxiety and are generally seen by the client as illogical; they provide few secondary gains.
4 Rituals are used primarily to handle feelings of anxiety and are a means of diverting attention from these feelings.

6. (1) Young adults ages 18 to 35 should be developing meaningful relationships and establishing themselves in a career. (1) (DF; AN; PD)
2 The stage of adulthood (generativity vs stagnation) is concerned with productivity, nurturance, and support of the next generation.
3 Having a coherent sense of self is a task of adolescence.
4 From age 65 to death, the individual should experience a feeling of the worth of one's life.

7. (3) Showing understanding of and identifying the client's feelings by giving feedback helps in establishing a therapeutic relationship. (2) (EP; IM; MO)
1 This is too direct; it does not allow the client time to reflect and explore feelings.
2 This negates the person's feelings and cuts off any further communication of feelings.
4 This encourages dependence; it does not allow for exploration of feelings.

8. (3) Succinylcholine (Anectine) causes paralysis of muscles, including the intercostals and diaphragm, so artificial support of respirations is required to sustain life. (1) (CI; AN; DR)
1 This is the purpose of the drug, not a disadvantage.
2 Same as answer 1.
4 Same as answer 1.

9. (2) These clients need sensory stimulation to maintain orientation and should be encouraged to do as much as possible for themselves, depending on their ability. (3) (EP; PL; SD)
   1 Surroundings should be bright to minimize confusion of stimuli.
   3 These clients are usually not completely out of contact with reality; it is important to differentiate fantasy from reality, but this would not take top priority in care.
   4 Although it is important to make certain that clients receive physical hygiene and comfort, they should be encouraged to help themselves as much as possible.

10. (3) Antipsychotic drugs tend to make the client listless or drowsy and can interfere with the ability to participate in the therapeutic regimen. (2) (PT; AN; SD)
    1 Antipsychotic drugs do not appreciably affect diurnal rhythms.
    2 Antipsychotic drugs do not really induce sleep, just listlessness.
    4 Assaultiveness is associated with increased anxiety and is unrelated to time.

11. (2) Delusions are false fixed beliefs that have a minimal reality base. (1) (EP; AN; SD)
    1 Ideas of reference are false beliefs that every statement or action of others relates to the individual.
    3 Loose associations are verbalizations that sound disjointed to the listener.
    4 Tactile hallucinations are false sensory perceptions of touch without external stimuli.

12. (1) It is important to determine whether the client is thinking about suicide; the direct approach is most appropriate. (1) (EP; AS; MO)
    2 This approach not only denies feelings but also tells the client it is not right to feel that way.
    3 This approach tells the client that her feelings do not have top priority.
    4 This approach denies the client's feelings and may be false reassurance.

13. (3) Antipsychotics are usually prescribed to calm agitated clients during the period it takes for the lithium to become effective. (3) (PT; AN; DR)
    1 Chlorpromazine is a drug that has a different, not a potentiating, mechanism of action.
    2 The drugs are used to control symptoms of mania, not to prevent depression.
    4 Chlorpromazine is a neuroleptic drug that has no effect on lithium toxicity.

14. (1) This helps the client recognize and adhere to established limits and goals. (2) (SC; IM; SA)
    2 This response can be degrading and reinforces the client's manipulative behavior.
    3 This reinforces the client's pattern of manipulation.
    4 Same as answer 3.

15. (2) The delusional client can never be argued out of a delusion because it serves as a defense against reality and is the client's reality. It is best to accept the delusion without discussion. (3) (EP; IM; SD)
    1 The client would have difficulty getting involved in a repetitive activity, and the activity would not stop the delusion.
    3 Encouraging discussion would give validity to the delusion rather than pointing out reality to the client.
    4 This action can lead to increased feelings of guilt about the delusion because the client will not understand why the nurse is changing the topic.

16. (2) When individuals express interest in physical appearance, it demonstrates a rebuilding of the self-image and the return of feelings of worth and concern for how others see them. (1) (EP; EV; SD)
    1 The client's response goes further than the nurse's implied suggestion; the client appears to be expressing own needs, not the nurse's.
    3 The client's response is well within the normal range; it does not indicate the beginning of a hyperactive phase.
    4 The information provided does not demonstrate a need for social reassurance or approval.

17. (2) Clients in the manic phase of a bipolar disorder are easily distracted. Rather than placing emphasis on their behavior, staff members should use the easy distractibility of these clients to redirect this behavior to more constructive channels. (1) (EP; IM; MO)
    1 This focuses on the behavior; it is a punitive response that does not foster communications.
    3 This encourages the client to defend the statement rather than what is behind the statement; it does not foster communications about feelings.
    4 The client would be unable to explain the basis for the expressed anger to the nurse.

18. ② Cogentin is an anticholinergic, antiparkinsonian drug used to treat drug-induced extrapyramidal symptoms associated with phenothiazine therapy; the IM route would reduce symptoms more rapidly. (2) (PT; IM; DR)

1 This medication is not effective in this situation.

3 While Artane is an appropriate medication, swallowing pills would be difficult for the client and the oral medication should not be administered.

4 This medication would produce parkinsonism, not relieve it.

19. ① Ritualistic behavior is aimed at controlling guilt and inadequacy by maintaining an absolute set pattern of action. (2) (EP; AS; PR)

2 These are not the major bases for this disorder.

3 The behavior and attitudes of clients with this disorder are contradictory; they are conscientious about some things and lax about others.

4 These are not the major bases for this disorder.

20. ④ It is important to set limits on the behavior, but it is also important to involve the client in the decision making. (3) (EP; IM; PR)

1 This would be nontherapeutic; rarely can a client be distracted from a ritual.

2 This is a nontherapeutic approach; some limits must be set by the client and nurse together.

3 This would increase anxiety because the client uses the ritual as a defense against anxiety.

21. ② When the client who is out of control feels that someone is assuming control, it promotes a feeling of security. As this continues, a sense of trust in this individual is established. (3) (SC; IM; SD)

1 This would be important in planning care but not in establishing a therapeutic relationship.

3 This is less important in the beginning phase of a relationship.

4 The client exhibits no self-destructive tendencies at this time.

22. ① The gross pathologic factor in primary degenerative dementia is brain tissue degeneration with loss of brain neurons. (3) (DF; AS; DD)

2 Inadequate nutrition may be one of the factors that brings about a general decline of health; however, there is no direct evidence that avitaminosis causes primary degenerative dementia.

3 Severe emotional trauma may contribute to but does not necessarily cause primary degenerative dementia.

4 This is false; nerve cells do not have the capacity to regenerate. Neural degeneration leads to permanent, not transient, changes.

23. ④ An almost universal reaction to birth of an imperfect child is guilt. Encouraging the parents to discuss such feelings, without actually asking whether they feel guilty, allows them an opportunity to express such thoughts. (2) (SC; IM; CS)

1 This statement lacks sensitivity in dealing with the parents' feelings.

2 This statement does not show recognition of the concern the parents are expressing.

3 This statement cuts off the parents' expression of feelings.

24. ① Discussion should not be initiated by the nurse; symptoms should be accepted but should not be the focus of discussion. (3) (EP; PL; AX)

2 This response would increase anxiety and take away the client's unconscious defense.

3 This response would increase anxiety because it focuses on unconscious feelings about the paralysis.

4 This response would increase anxiety and deny the client's symptoms; in reality this client cannot make the legs move to walk.

25. ③ Bulimia is characterized by the binge-purge cycle; most clients withdraw from others and vomit after an eating binge. (2) (EP; AS; ES)

1 Although some individuals with bulimia may perform excessive exercises, this is a more common finding with the diagnosis of anorexia nervosa.

2 Although individuals with bulimia may hoard food, this behavior commonly occurs later, when limits are put on their intake.

4 Most individuals with bulimia do not seek support or socialization after a binge, although they may socialize at other times.

26. ① When the rape victim chooses to prosecute the rapist, the victim must prove that rape occurred; the accused is innocent until proven guilty. (1) (CC; AN; CS)

2 The medical team may be asked to provide evidence at the trial, but the victim must prove that the rapist is guilty.

3 The perpetrator tries to establish innocence in a rape case; the victim must prove that the rapist is guilty.

4 Guilt or innocence will be established by a jury, with the burden of proof placed on the victim.

27. (2) This allows the angry client time to regain self-control; announcing a plan to return decreases fear of abandonment or retribution. (3) (EP; IM; SD)
   1 Staying and insisting that the client take the medication may provoke increased anger and further loss of control.
   3 Clients will not accept logical explanations when angry.
   4 This approach shows no respect for the client's feelings; it may decrease trust and increase anger.

28. (3) This statement directly quotes the client with no added value judgments. (2) (CC; AS; MO)
   1 This is a subjective judgment and an interpretation of what the client actually said.
   2 Same as answer 1.
   4 Same as answer 1.

29. (1) This statement indicates that the nurse sees the client as a person and is willing to help the client face a new experience. (2) (EP; IM; SD)
   2 This will do nothing to allay the client's anxiety about facing a new situation.
   3 This is untrue; even if the client does not like it, it is part of the therapy program and the client should be encouraged to go.
   4 Same as answer 2.

30. (1) Because of the severity of side effects and the stress it places on the renal as well as the cardiovascular system, its administration is contraindicated in clients with renal or cardiovascular disease. (2) (PT; AS; DR)
   2 This is necessary after the start of lithium administration.
   3 This is not necessary; lithium does not alter neurologic functions.
   4 Same as answer 2.

31. (1) This is a therapeutic approach that indicates awareness of the client's feelings and encourages verbalizations. (1) (EP; IM; TR)
   2 Moralizing is a roadblock to effective communication.
   3 This is diverting the client's attention to something else and ignoring the client's attempt to verbalize.
   4 This conveys a judgmental or critical attitude toward the client's actions.

32. (4) Excessive motor activity with intermittent head banging and hair pulling is self-destructive behavior that can result in injury; prevention of self-injury has the highest priority. (1) (SI; AN; BA)
   1 This is not the most important nursing diagnosis according to the data presented; prevention of self-injury is primary.
   2 Same as answer 1.
   3 Same as answer 1.

33. (4) This assurance helps allay guilt, reduces anxiety, and assists with coping. (3) (SC; IM; CS)
   1 The client should be consulted before a clergyman is called.
   2 This is a last resort because this delays the grieving process.
   3 It is too soon for this intervention.

34. (4) This is characteristic of tardive dyskinesia, an irreversible, antipsychotic, drug-induced, neurologic disorder. (2) (PT; EV; DR)
   1 This is an anticholinergic-like side effect that is not considered serious.
   2 This is unrelated to antipsychotic medications.
   3 This drug would cause sedation, not insomnia.

35. (2) Clients should know why they are there, what to expect, and what is to be accomplished. (1) (SC; PL; TR)
   1 This would come after the explanation of the purpose of the group.
   3 Same as answer 1.
   4 This is not necessary to define; this is a long-term goal.

36. (4) When the client realizes that staff are not responding to negative behaviors but are providing praise for positive accomplishments, the client's behavior should change, since recognition and self-esteem will have been gained for socially acceptable behavior. (2) (EP; PL; SD)
   1 The withdrawn clients must be protected from the client's ridicule and sarcasm just as this client must be protected.
   2 This is unrealistic because a goal is to function again in society.
   3 This is expecting too much of the client at the present time.

37. (4) This shows an increase in socialization and an awareness of the behavior of others. (2) (SC; EV; TR)
   1 Attendance alone is insufficient to evaluate the effectiveness of group therapy.
   2 The quantity and extent of comments are not what are significant.
   3 This may indicate a greater degree of impulse control on the part of the members, which is not the primary goal of group therapy.

38. ③ To intervene in a crisis the nurse must assume a direct, active role because the client's abilities to cope are lessened and help is needed to problem solve. (2) (EP; PL; CS)
1 This would be insufficient to help the client.
2 The role of the nurse should not include giving advice; this promotes dependence.
4 Same as answer 1.

39. ③ Yellow sclerae are a sign of jaundice, indicating liver damage, which can be irreversible if drug therapy is continued. (2) (PT; EV; DR)
1 Although this may be a sign of a serious side effect, it may just be a behavioral response of the disorder; the nurse should notify the physician rather than withhold the drug.
2 This is a Parkinson-type syndrome, which can be reversed with treatment; continuation of medication is permitted.
4 Photosensitivity is not a problem as long as the client is cautioned to stay out of the sun.

40. ② Women facing breast surgery often have many feelings relating to their sexuality, change in body image, etc. The nurse plays a vital role in helping the client verbalize feelings, and this response keeps channels of communication open. (2) (SC; IM; CS)
1 The client's concerns are real, and such a statement will only block further communication.
3 This can be interpreted as the nurse's reluctance to listen; the client may not be able to talk with the husband about this.
4 This does not focus on the importance of the client as an individual; each person feels differently.

41. ④ Phobias are specific fears that serve as a means of coping with generalized anxiety. (2) (EP; AN; AX)
1 Anxiety, not obsession, is related to phobias.
2 Anxiety, not depression, is related to phobias; finding a direct connection to life events is often difficult.
3 A direct connection to life events is often difficult to find.

42. ③ Because of the shock and trauma, it is expected that the client may have an alteration in normal coping mechanisms. (2) (SC; AN; CS)
1 The client will probably not project falsely positive self-opinions to enhance self-regard; it is more likely that the problems will be meeting life's demands and handling stress.
2 No family is mentioned in the situation.

4 No information is presented to indicate that the client cannot distinguish between the self and the nonself.

43. ④ Alcohol often reduces inhibitions and is one of the leading causes of violent behavior. (1) (EP; AS; SA)
1 This factor alone rarely precipitates abusive behavior.
2 Same as answer 1.
3 Same as answer 1.

44. ④ This shows acceptance and protects the client and others from possible aggressive behavior. (2) (SI; PL; MO)
1 Explanations are not helpful because the client's easy distractibility interferes with understanding.
2 Both clients would be very responsive to external stimuli; therefore this action could lead to loss of control for both.
3 This is a threatening approach that increases feelings of inadequacy.

45. ④ In a pervasive developmental disorder, the child does not have a self-concept or view the self as separate; until this goal is attained other therapies will produce little or no positive outcomes. (2) (DF; AN; BA)
1 The child with autism may have language skills but usually does not use them; nonverbal behavior is most generally associated with autism.
2 Regressive behavior should be anticipated as the child undergoes therapy, especially when working through earlier phases of development.
3 To be mainstreamed, the child must first have a developing independence.

46. ③ This helps the client focus on feelings rather than just pointing out that the current behavior is unacceptable. (2) (SC; IM; PR)
1 This gives the client an alibi for unacceptable behavior.
2 This points out the behavior in a negative way.
4 The nurse is becoming defensive rather than dealing with the problem directly.

47. ② This demonstrates the client's acceptance of the professional role of the nurse as well as the ability to end dependent relationships. (2) (SC; EV; PR)
1 This shows an inability to relate to other staff members involved with the client's care.
3 This still shows the existence of manipulation on the client's part.
4 This shows a childish need to punish the nurse for leaving.

48. (2) Children who have been abused quickly learn that attention-seeking behavior such as crying for comforting provokes more abuse. Therefore they learn to accept the battering silently. (2) (EP; AS; CS)

1   This may or may not be true.
3   Battered children seldom cry.
4   Battered children rarely cry; they tend to accept strangers.

49. (2) Legally it is the responsibility of the staff to notify law enforcement officers so that the client can be returned. (2) (CC; IM; SD)

1   The staff should notify the family, but it is not the first priority.
3   Although this would be done, it is not the first priority.

4   Although the physicians may appreciate being notified, it is not the first priority.

50. (1) Phenothiazines commonly cause photosensitivity, which can be controlled by the use of sunscreens. (1) (SI; IM; DR)

2   This is not a necessary precaution when taking phenothiazines; the client should be allowed to participate fully.
3   The medication should be administered as prescribed; additional doses should not be administered.
4   Participating in an outing should not affect the client's blood pressure negatively.

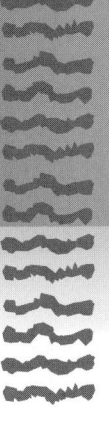

CHAPTER 5

Medical-Surgical Nursing

REVIEW QUESTIONS

GROWTH AND DEVELOPMENT (GD)

1. Developmentally, a 21-year-old male client who has sustained a spinal injury below the level of T6 will most likely have difficulty with:
1. Mastering his environment
2. Identifying with the male role
3. Developing meaningful relationships
4. Differentiating himself from the environment

2. When assessing a client with osteoporosis, the nurse should recognize that most observable changes will occur in:
1. Facial bones
2. The long bones
3. The vertebral column
4. Joints of the hands and feet

3. The nurse recognizes that a 70-year-old female can best limit further progression of osteoporosis by:
1. Taking supplemental calcium and vitamin D
2. Taking supplemental magnesium and vitamin E
3. Increasing the consumption of eggs and cheese
4. Increasing the consumption of milk and milk products

4. A 70-year-old male client asks the nurse whether he should be taking multivitamin and mineral supplements. The nurse's best response would be:
1. "Absolutely! They are a necessity as a person gets older."
2. "Older people usually do need more vitamins, especially vitamin A."
3. "There is no evidence that generally healthy older adults require added nutrient supplements."
4. "If you took the supplement you could cut down on your food and help keep your weight down."

5. A day after an explanation of the effects of surgery to create an ileostomy, a 68-year-old male client remarks to the nurse, "It will be difficult for my wife to care for a helpless old man." These comments by the client regarding himself are an example of Erikson's conflict of:
1. Initiative vs guilt
2. Integrity vs despair
3. Industry vs inferiority
4. Generativity vs stagnation

6. The nurse should recognize that a genitourinary factor that may contribute to urinary incontinence in the elderly is:
1. A sensory deprivation
2. A urinary tract infection
3. The frequent use of diuretics
4. The inaccessibility of a bathroom

7. When correcting myths about aging, the nurse should teach that older adults normally have:
1. An inflexible attitude
2. Periods of confusion
3. A slower reaction time
4. Some senile dementia

8. An 80-year-old female is admitted to the hospital because of complications associated with severe dehydration. The client's daughter asks the nurse how her mother could have become dehydrated because she is alert and able to care for herself. The nurse's best response would be:
   1. "Access to fluid may be limited and insufficient to meet the daily needs of the older adult."
   2. "The body's need for fluid decreases with age because of a change in the body composition."
   3. "Memory declines with age, and the older adult may forget to ingest adequate amounts of fluid."
   4. "The thirst reflex diminishes with age, and therefore the recognition of the need for fluid is decreased."

9. A 75-year-old female client tells the nurse that she read about a vitamin that may be related to aging because of its relationship to the structure of cell walls and wonders whether she should be taking it. The nurse should recognize that the client is probably referring to:
   1. Vitamin A
   2. Vitamin C
   3. Vitamin E
   4. Vitamin B_1

10. A female client's osteoporosis has progressed dramatically in the last 5 years, and she is especially prone to falling. The statement that best reflects the client's understanding of why there is a greater risk for falls would be:
    1. "I do not have the stamina that I used to have."
    2. "At my age, I'm more prone to dizziness and falling."
    3. "Because of the curvature of my spine, it is hard to keep my balance."
    4. "Because I am bent over, I look down instead of up while I'm walking."

11. Nursing actions for the older adult should include health education and promotion of self-care. When dealing with the older adult the nurse should:
    1. Encourage exercise and naps
    2. Strengthen the concept of ageism
    3. Reinforce the client's strengths and promote reminiscing
    4. Teach about a high-carbohydrate diet and focus on the present

12. The mental process most sensitive to deterioration with aging seems to be:
    1. Creativity
    2. Judgment
    3. Intelligence
    4. Short-term memory

13. The nurse is preparing a community health program for senior citizens. The nurse teaches the group that physical findings that are normal in older people include:
    1. A loss of skin elasticity and a decrease in libido
    2. Impaired fat digestion and increased salivary secretions
    3. An increase in body warmth and some swallowing difficulties
    4. Increased blood pressure and decreased hormone production in women

14. An 89-year-old client with osteoporosis is admitted to the hospital with a compression fracture of the spine. The nurse understands that a factor of special concern when caring for the older adult client is the client's:
    1. Inability to recall recent facts
    2. Irritability in response to deprivation
    3. Inability to maintain an optimal level of functioning
    4. Gradual memory loss resulting from change in environment

15. When considering Erikson's psychosocial developmental tasks, a nurse should focus care for middle-aged adults around their need to be:
    1. Productive
    2. Controlling
    3. Independent
    4. Autonomous

16. When nurses are conducting health assessment interviews with elderly clients, they should:
    1. Leave a written questionnaire for clients to complete at their leisure
    2. Ask family members rather than the clients to supply the necessary information
    3. Keep referring to previous questions to ascertain that the information given is correct
    4. Spend time in several short sessions to elicit more complete information from the clients

EMOTIONAL NEEDS RELATED TO HEALTH PROBLEMS (EH)

17. A female client has just spent 5 minutes complaining to the nurse about numerous aspects of her hospital stay. The best initial nursing response would be to:
    1. Attempt to explain the purpose of different hospital routines to the client
    2. Explain to the client that becoming so upset dangerously blocks her need for rest

3. Refocus the conversation on the client's fears, frustrations, and anger about her condition
4. Permit the client to release feelings and then promptly leave to allow her to regain composure

18. A male client who has had a myocardial infarction asks the nurse, "What's the chance of my having another heart attack if I watch my diet and stress levels carefully?" The most appropriate initial response would be for the nurse to:
1. Suggest he discuss his feelings of vulnerability with his physician
2. Avoid giving him direct information and help him explore his feelings
3. Tell him that he certainly needs to be especially careful in these areas
4. Recognize that he is frightened and suggest he talk with the psychiatric nurse

19. A client who has recently had an abdominoperineal resection and colostomy accuses the nurse of being uncomfortable during a dressing change because the "wound looks terrible." The nurse recognizes that the client is using the defense mechanism known as:
1. Projection
2. Sublimation
3. Intellectualization
4. Reaction formation

20. A male client who has had an abdominoperineal resection and colostomy refuses to allow his wife to see the incision or stoma and ignores most of his dietary instructions. The nurse, when assessing these data, can assume that the client is experiencing:
1. A reaction formation to his recent altered body image
2. Suicide thoughts and should be seen by a psychiatrist
3. A difficult time accepting reality and is in a state of denial
4. Impotency resulting from the surgery and needs sexual counseling

21. Three days after a below-the-knee amputation, a client is refusing to eat, talk, or perform any rehabilitative activities. The best initial nursing approach would be to:
1. Frequently explain why there is a need to quickly increase his activity
2. Emphasize repeatedly that with a prosthesis there will be a return to a normal life-style

3. Appear cheerful and noncritical regardless of the client's response to attempts at intervention
4. Accept and acknowledge that the client's withdrawal is a normal and necessary part of initial grieving

22. The key factor in accurately assessing how a client will cope with body image changes is the:
1. Suddenness of the change
2. Obviousness of the change
3. Extent of body change present
4. Client's perception of the change

23. A client who recently was diagnosed as having myelocytic leukemia discusses the diagnosis by referring to statistics, facts, and figures. The nurse recognizes that the client is using the defense mechanism known as:
1. Projection
2. Sublimation
3. Intellectualization
4. Reaction formation

24. A female client who is dying jokes about the situation even though she is becoming sicker and weaker. The nurse's most therapeutic response would be:
1. "Why are you always laughing?"
2. "Your laughter is a cover for your fear."
3. "Does it help to joke about your illness?"
4. "She who laughs on the outside, cries on the inside."

25. When caring for a dying client who is in the denial stage of grief, the best nursing approach is to:
1. Agree with and encourage the client's denial
2. Reassure the client that everything will be OK
3. Allow the denial but be available to discuss death
4. Leave the client alone to confront feelings of impending loss

26. The nurse notes that a female client seems to be depressed after a thymectomy for treatment of myasthenia gravis. The nursing action that would be most appropriate at this point would be:
1. Recognizing that depression often occurs after surgery
2. Asking her physician to arrange for a psychologic consultation
3. Reassuring the client that she will feel better when her discharge date is set
4. Talking with the client about her prognosis, emphasizing things that she can do

27. The home care nurse makes an initial visit to a 60-year-old widowed client with right ventricular failure who lives with her divorced, drug-addicted daughter and her seven grandchildren. The nurse finds the client feeding a 6-month-old granddaughter and preparing dinner for the rest of the family. A 14-year-old grandson, disabled and in a wheelchair, states his mother is sleeping. The nurse should proceed by:
    1. Sitting down with the client and exchanging identifying data and information
    2. Accepting coffee when offered by the client and socializing for a few minutes
    3. Asking the client whether it is all right to look around the apartment to evaluate environmental conditions
    4. Questioning the client to determine whether there is a private place to take a health history and perform an examination

28. A client is taught how to change the dressing and how to care for a recently inserted nephrostomy tube. On the day of discharge the client states, "I hope I can handle all this at home; it's a lot to remember." The best response by the nurse would be:
    1. "I'm sure you can do it!"
    2. "Oh, a family member can do it for you."
    3. "You seem to be nervous about going home."
    4. "Perhaps you can stay in the hospital another day."

29. An 83-year-old client with type 2 diabetes is admitted to the ambulatory surgery unit for elective cataract surgery. Before surgery the client asks the nurse, "How will my diabetes be managed while I am here?" The best response by the nurse would be:
    1. "The anesthesiologist will take care of it."
    2. "What did your surgeon or the anesthetist tell you?"
    3. "Your surgeon will write orders for fluids and insulin."
    4. "I'm not quite certain I understand what you are asking."

30. A female client with nodular poorly differentiated lymphocytic lymphoma (NLPD) who has been treated with multiple chemotherapeutic agents comes for her first radiation treatment. As the nurse prepares her for the therapy the client states, "I'm so discouraged" and starts to cry. The nurse should:
    1. Leave her alone so that she can regain her composure
    2. State, "It's difficult to deal with your diagnosis and treatment."

3. Complete the preparation and tell her, "We can talk about this later."
4. Explain the therapy and reiterate that it will cause only a little discomfort

31. A client's discouragement with the diagnosis of nodular poorly differentiated lymphocytic lymphoma (NLPD) continues during radiation therapy because of the long time required for treatment and its side effects. In assisting the client to plan for the future, the nurse should emphasize that:
    1. Antidepressant medication can be prescribed
    2. The client's feelings are normal and will lessen with time
    3. A positive outlook can influence the outcome of cancer therapy
    4. The prognosis for NLPD is more favorable than for other types of lymphoma

32. A client has just been diagnosed with multiple sclerosis. The client is obviously upset with the diagnosis and asks, "Am I going to die?" The nurse's best response would be:
    1. "Most individuals with your disease live a normal life span."
    2. "Is your family here? I would like to explain your disease to all of you."
    3. "The prognosis is variable; most individuals experience remissions and exacerbations."
    4. "Why don't you speak with your physician who can give you more details about your disease."

33. A client refuses to go to the twice-a-day prescribed sessions in physical therapy. The nurse might best approach this problem by:
    1. Having the client observe the progress of a more cooperative client with the same problem
    2. Being the client's advocate and asking the physician whether therapy can be decreased to once daily
    3. Assuring the client that pain medication will be administered before the scheduled physical therapy sessions
    4. Planning a conference with the client, the physical therapist, and the nurse present to discuss the client's feelings

34. A 22-year-old male client with AIDS signs a "do not resuscitate" (DNR) order when he is admitted to the hospital. When respiratory arrest occurs 3 weeks later, the client is not resuscitated. A true statement about the legal aspects of a DNR order would be:
    1. Age is an important factor in the decision not to resuscitate
    2. The decision not to resuscitate resides with the client's physician

3. The status of the DNR order is contingent on the policies of the institution

4. Once the order has been signed, it remains in force for the entire hospitalization

35. The physician places a client with an infected surgical incision on strict isolation. After being taught about isolation, the client is seen sneaking out of the room to make telephone calls on the public phone. The most effective nursing intervention would be to:
   1. Ensure regular visits by staff members
   2. Explore what isolation means to the client
   3. Report the situation to the infection control nurse
   4. Reteach the entire isolation procedure to the client

36. The community health nurse makes a home visit to the mother of a 13-year-old boy who is disabled and who has three siblings younger than 6 years old. The nurse observes that the 6-month-old sister lies quietly in her crib, rarely smiles or vocalizes, and barely has her basic needs attended. The nurse should:
   1. Place an aide in the home to assist the mother and care for the infant
   2. Advise the mother that the child will be retarded if she is not stimulated
   3. Ask the 13-year-old disabled brother to pay more attention to his sister
   4. Encourage the mother to purchase appropriate toys manufactured for the baby's age level

37. The nurse should suspect that a male client who has had a recent myocardial infarction is experiencing denial when he:
   1. Attempts to minimize his illness
   2. Lacks an emotional response to his illness
   3. Refuses to discuss his condition with his wife
   4. Expresses displeasure with his activity program

38. A client is scheduled for surgery. Legally, the client may not sign the operative consent if:
   1. Ambivalent feelings regarding the operation are present
   2. Any sedative type of medication has recently been given
   3. A discussion of alternatives with two physicians has not occurred
   4. A complete history and physical has not been performed and recorded

39. In the postanesthesia care unit after a below-the-knee amputation, a female client begins crying while feeling for her involved lower leg. The nurse should:
   1. Administer medication to induce sleep
   2. Allow the client to ventilate feelings of loss
   3. Ignore the behavior until the client is more alert
   4. Leave the client alone to provide time for privacy

40. The nurse raises a client's bedside rails at night. The use of side rails may affect a client's psychologic status, resulting in:
   1. A sense of security
   2. The prevention of falls
   3. Increased independence
   4. An alteration in proprioception

41. Teaching for clients who have sustained a sudden, traumatic, major loss is often most satisfactorily done during the acceptance or adaptation stage of coping. The nurse is aware that the rationale for this fact is that clients in this stage are:
   1. Ready for discharge and therefore in need of preparation
   2. At the peak of mental anguish and therefore open to change
   3. Less angry and therefore more compliant and easier to deal with
   4. Less anxious and more aware of reality and therefore ready to learn

42. On the fourth day after surgery for a fractured hip, a client appears angry and very restless and says, "I can't stand this another minute. There's a wrinkle in my sheet, and my water is warm." The client changes position frequently and does not maintain eye contact with the nurse. The best initial interpretation of the client's behavior is that it indicates:
   1. Severe discomfort in the hip
   2. An increased level of anxiety
   3. Anger at the poor nursing care
   4. Frustration with the need for leg abduction

43. A postsurgical client complains about a variety of minor environmental factors while frequently changing position and avoiding eye contact. The nurse responds to these observations by stating, "Let me get you some cold water and your pain pill, and you'll be much better." The nurse's approach demonstrates the:
   1. Empathetic recognition of anxiety
   2. Introduction of the needs approach
   3. Inappropriate use of the data presented
   4. Use of problem identification and clarification

44. A client in thyroid storm tells the nurse, "I know I'm going to die. I'm very sick." The best response by the nurse would be:
   1. "You must feel very sick and frightened."
   2. "Tell me why you feel you are going to die."
   3. "I can understand how you feel, but people do not die from this problem."
   4. "If you would like, I will call your family and tell them to come to the hospital."

45. The nurse should be aware that sensory restriction in a client who is blind can:
    1. Increase the use of daydreaming and fantasy
    2. Heighten the client's ability to make decisions
    3. Decrease the client's restlessness and lethargy
    4. Lead to the use of permanent neurotic behaviors

46. A young female client is diagnosed as having stage IIIA Hodgkin's disease with a grossly involved spleen and is scheduled for a splenectomy. After the nurse performs preoperative teaching the client appears very anxious. The best approach for the nurse to use at this time would be to:
    1. Allow the client to regress at this time and let her rest quietly
    2. State simply that she seems anxious and ask her whether she would like to talk for a while
    3. Consider her reaction an unconscious response and inquire about her relations with her mother
    4. Recognize that anxiety prevented the client from understanding, and repeat the information in simpler terms

47. A female client with chronic renal failure has been on hemodialysis for 2 years. She relates to the nurse in the dialysis unit in an angry, critical manner and is frequently noncompliant with medications and diet. The nurse can best intervene by first understanding that the client's behavior is most likely:
    1. An attempt to punish the nursing staff
    2. A constructive method of accepting reality
    3. A defense against underlying depression and fear
    4. An effort to maintain life to the fullest extent possible

48. The nurse observes another nurse changing a client's sterile dressing. The nurse changing the dressing uses the same pair of clean gloves to remove the soiled dressing and apply the new dressing. The observing nurse's initial action should be to:
    1. Discuss the incident with the nurse
    2. File an incident report about the action
    3. Offer to demonstrate the proper technique
    4. Report the individual to the nursing supervisor

49. Before major abdominal surgery for cancer the client says to the nurse, "I really don't think this is cancer at all. I'll bet they won't find anything." The nurse's most appropriate initial response would be:
    1. "I can understand why you'd like to believe that."
    2. "I hope you're right, but the tests do indicate cancer."
    3. "It must be difficult to be facing such serious surgery."
    4. "You think the physician may have made a wrong diagnosis?"

50. A female client is diagnosed as having cancer of the breast and is admitted to the hospital for a lumpectomy to be followed by radiation. While being admitted to ambulatory surgery by the nurse, the client has tears in her eyes and her chin is quivering. In a shaky voice the client says, "I can't believe this is happening." The nurse's best response would be:
    1. "You can't believe this is happening?"
    2. "This must be a very scary time for you."
    3. "Do you have any questions at this time?"
    4. "Cancer of the breast has a high cure rate."

51. A client who has been hospitalized for 2 days for renal colic is scheduled for an extracorporeal shock-wave lithotripsy. The night before surgery the client is extremely demanding, making frequent requests for about 2 hours. An empathetic nurse might best show understanding by saying:
    1. "You are being demanding tonight."
    2. "You are facing a rough day tomorrow."
    3. "Don't be so concerned, we'll take good care of you."
    4. "I know how scared you are, but this is routine surgery."

52. A nurse stops at the scene of an accident and finds a man with a deep laceration on his hand, a fractured arm and leg, and abdominal pain. The nurse wraps the man's hand in a soiled cloth and drives him to the nearest hospital. The nurse is:
    1. Negligent and can be sued for malpractice
    2. Practicing under guidelines of the Nurse Practice Act
    3. Protected for these actions, in most states, by Good Samaritan legislation
    4. Treating a health problem that can and should be handled by a physician

53. A male client is hospitalized following a major automobile accident. As the client is describing the accident to a friend, he becomes very restless, and his pulse and respirations increase sharply. It is most important for the nurse to recognize that these symptoms are probably related to:
    1. Bleeding from undiscovered injury
    2. The client's method of seeking sympathy
    3. Delayed psychologic response to trauma
    4. A parasympathetic nervous system response to anxiety

54. A male client with pancreatic cancer is aware of the terminal nature of the illness. Behaviors that would indicate that he is accepting the fact of his impending death are:
   1. Alternately crying and talking openly about death
   2. Getting second, third, and fourth medical opinions
   3. Making out his will and planning a visit to a good friend
   4. Refusing to follow treatments and stating, "I'm going to die anyway."

55. A female client with a terminal illness decides to donate her eyes for organ transplantation after she dies. Statutes that address organ transplantation attempt to prevent abuse by:
   1. Permitting active euthanasia when necessary
   2. Preventing children from giving organs to others
   3. Allowing physicians to control both donor and recipient
   4. Requiring participating institutions to have review boards

56. A male client with thrombophlebitis is apprehensive about the possibility of a clot reaching his heart and causing sudden death. The nurse's initial intervention should be to:
   1. Clarify his misconception
   2. Explain preventive measures
   3. Teach recognition of early symptoms
   4. Encourage discussion of his concern

57. The physician discusses the need for an abdominoperineal resection and a colostomy with a male client. After the physician leaves, the client tells the nurse that he is pleased only minor surgery is necessary. The nurse recognizes that the client's reaction is an example of:
   1. Reflection
   2. Regression
   3. Repudiation
   4. Reconciliation

58. To be most effective when teaching colostomy care to a client, the nurse must first:
   1. Wait until a family member is present
   2. Assess barriers to learning colostomy care
   3. Begin with simple written instructions concerning the care
   4. Wait until the client has accepted the change in body image

59. During the evening after a paracentesis, the nurse notices that the client, although denying any discomfort, seems very anxious. The best nursing approach would be to:
   1. Offer the client a back rub
   2. Administer the prescribed narcotic
   3. Reinforce the physician's explanation of the procedure
   4. Explore the client's concerns and administer the ordered hypnotic

60. Individual nurses are responsible for their own professional actions. In addition, a hospital can be held legally responsible for the actions of a nurse employed by the hospital under the doctrine of:
   1. *In loco parentis*
   2. *Respondeat superior*
   3. The state's department of health
   4. The American Hospital Association's Review Board

61. After receiving diabetic education, a client with recently diagnosed diabetes states, "I feel bad. I don't seem to get along with my husband. He does not care about my diabetes." The nurse's most appropriate response would be:
   1. "You don't get along with your husband."
   2. "I'm sorry, what can I do to make you feel better?"
   3. "It's probably just temporary; he needs more time to adjust."
   4. "You are unhappy. I wonder, have you tried to talk to your husband?"

62. A visitor from a room adjacent to a client asks the nurse what disease the client has. The nurse responds, "I will not discuss any client's illness with you. Are you concerned about it?" This response is based on the nurse's knowledge that to discuss a client's condition with someone not directly involved with that client is an example of:
   1. Libel
   2. Slander
   3. Negligence
   4. Breech of confidentiality

63. When approaching homosexual clients with AIDS, it is most important for nurses to:
   1. Have a strong sense of their own sexual identity
   2. Admit their feelings of uncomfortableness to the clients
   3. Pay particular attention to establishing a meaningful rapport
   4. Become aware of their own attitudes regarding homosexuality

64. A client with jaundiced skin and acute abdominal pain is refusing all visitors. The most therapeutic response would be to:
    1. Listen to the client's fears
    2. Encourage the client to socialize
    3. Grant the client's request about visitors
    4. Darken the client's room by pulling the drapes

65. A 26-year-old homosexual is diagnosed with AIDS. The primary nurse reports to the nursing team that the client wept when told of the diagnosis. One of the nursing assistants responds, "I don't feel sorry for him. He made his bed, and now he can lie in it." This comment is most likely a result of the nursing assistant's:
    1. Values and beliefs about sexual life-styles
    2. Anger and mistrust of homosexual males in general
    3. Discomfort with men who are unable to control their emotions
    4. Hostility over having to care for someone with a sexually related disease

66. A client with AIDS comments to the nurse, "There are so many rotten people around. Why couldn't one of them get AIDS instead of me?" The nurse could best respond:
    1. "It seems unfair that you should be so ill."
    2. "I can understand why you're afraid of death."
    3. "Have you thought of speaking with a minister?"
    4. "I'm sure you really don't wish this on someone else."

67. A client is to be transferred from the coronary care unit to a progressive care unit. The client asks the nurse, "Are you sure I'm ready for this move?" From this statement the nurse ascertains that the client is most likely experiencing:
    1. Fear
    2. Depression
    3. Dependency
    4. Ambivalence

68. A client who is suspected of having a brain tumor is scheduled for a CT scan. Before the test the nurse should:
    1. Withhold routine medication
    2. Describe the equipment involved
    3. Administer the prescribed sedative
    4. Explain that no radiation will be involved

69. During the assessment interview a client suspected of having an aldosteronoma states, "I don't know why the doctor doesn't just give me a prescription for high blood pressure pills. I'm missing work by

being here for these tests." The nurse's best response would be:
    1. "It might not be high blood pressure. We have to be sure."
    2. "I know it's frustrating, but you need to be here for these tests."
    3. "It's frustrating to miss work and not know for sure what's wrong."
    4. "Did you ask your physician whether the tests could be done on an outpatient basis?"

70. During a home visit the nurse discovers that a child in the household has a disability and has been experiencing convulsions. In addition the child's mother is unresponsive to the child's physical, emotional, or medical needs and seems to provoke convulsive episodes by harsh verbal exchanges with the child. The nurse believes that intervention by an appropriate community resource is indicated. A referral should be made to:
    1. The outpatient clinic
    2. The hospital pediatric unit
    3. The bureau of child welfare
    4. The bureau of the handicapped

FLUIDS AND ELECTROLYTES (FE)

71. A 74-year-old client comes to the emergency room complaining of weakness and dizziness. The blood pressure is 90/60, pulse is 92 and weak, and body weight reflects a 3-pound loss in 2 days. The weather has been very warm and the client's fluid intake has not increased. The most appropriate nursing diagnosis for this client would be:
    1. Fluid volume deficit related to insufficient fluid intake
    2. Self-care deficit, feeding related to altered thought processes
    3. Altered tissue perfusion related to decreased cardiac strength
    4. Altered nutrition: less than body requirements related to aging

72. During an 8-hour shift a client drinks two 6-ounce cups of tea and vomits 125 ml of fluid. Intravenous fluids absorbed equalled the urinary output. During this 8-hour period the client's fluid balance would be:
    1. +55 ml
    2. +137 ml
    3. +235 ml
    4. +485 ml

73. A client weighed 210 pounds on admission to the hospital. After 2 days of diuretic therapy the client

weighs 205.5 pounds. The nurse could estimate that the amount of fluid the client has lost is:
1. 0.5 L
2. 1.0 L
3. 2.0 L
4. 3.5 L

74. A male client with a history of heart failure and atrial fibrillation comes to the clinic for his regular 2-week visit. The client is 9 pounds heavier than his usual weight. The nurse interprets that the most likely cause of this sudden weight gain is:
1. Fluid retention
2. Urinary retention
3. Renal insufficiency
4. Abdominal distention

75. The dietary practice that will help a client reduce the dietary intake of sodium is:
1. Increasing the use of dairy products
2. Using an artificial sweetener in coffee
3. Avoiding the use of carbonated beverages
4. Using catsup for cooking and flavoring foods

76. An ECG is performed before a client is to have a cardiac catheterization, and hypokalemia is suspected. To confirm the presence of hypokalemia, the nurse would expect the physician to order:
1. Blood cultures × 3
2. A complete blood count
3. A serum electrolyte level
4. An x-ray film of long bones

77. A client is to receive an intravenous solution containing potassium chloride. When starting this IV infusion the nurse should select:
1. The antecubital space in the client's arm
2. The largest possible vein in the client's arm
3. A vein in the back of the client's dominant hand
4. A vein in the back of the client's nondominant hand

78. When observing a cardiac monitor, the nurse can ascertain data related to the client's serum potassium level by looking for:
1. Tall, peaked P waves
2. An extended PR interval
3. Tall or flattened T waves
4. Runs of trigeminy and bigeminy

79. When evaluating a client's response to fluid replacement therapy, the observation that indicates adequate tissue perfusion to vital organs is:
1. Urinary output of 30 ml per hour
2. Central venous pressure reading of 2 cm H_2O

3. Pulse rates of 120 and 110 in a 15-minute period
4. Blood pressure readings of 50/30 and 70/40 mm Hg within 30 minutes

80. When administering albumin intravenously, the nurse is aware that body water will shift from the:
1. Interstitial compartment to the extracellular compartment
2. Intravascular compartment to the interstitial compartment
3. Extracellular compartment to the intracellular compartment
4. Intracellular compartment to the intravascular compartment

81. A client with chronic renal failure cannot use salt substitutes in the diet because:
1. A person's body tends to retain fluid when a salt substitute is included in the diet
2. Limiting salt substitutes in the diet prevents a buildup of waste products in the blood
3. Salt substitutes contain potassium, which must be limited to prevent abnormal heartbeats
4. A substance in the salt substitute interferes with the transfer of fluid across capillary membranes, resulting in anasarca

82. The nurse is aware that total parenteral nutrition is a more desirable therapy than just intravenous fluids for clients with gastrointestinal problems. The nurse understands that clients receiving only IV fluids lose weight because of:
1. Lack of bulk in the diet
2. Deficient carbohydrate intake
3. Insufficient intake of water-soluble vitamins
4. Increased concentrations of electrolytes in cells

83. A client's clinical symptoms indicate a possible gastric ulcer. Considering the symptoms of epigastric pain, vomiting, dehydration, weakness, lethargy, and shallow respirations and the laboratory results that demonstrate metabolic alkalosis, the primary nursing diagnosis for this client would be:
1. Fluid volume deficit related to vomiting
2. Impaired gas exchange related to pain
3. Risk for injury related to increased weakness
4. Pain related to hypersecretion of gastric acids

84. A client with hypertension is being taught to restrict the intake of sodium. The nurse would know that the teaching about foods low in sodium was understood when the client states, "I can eat:
1. Broiled scallops."
2. A bologna sandwich."
3. Shredded wheat cereal."
4. Carrot and celery sticks."

85. An elderly client is diagnosed as having acute renal failure secondary to dehydration, and the physician orders an IV infusion of 50% glucose and regular insulin. The nurse understands that this is ordered for a client in renal failure to:
   1. Prevent cardiac arrest
   2. Increase urinary output
   3. Prevent respiratory acidosis
   4. Decrease serum calcium levels

86. When monitoring for hyponatremia, the nurse should assess the client for:
   1. Dry skin
   2. Confusion
   3. Tachycardia
   4. Pale coloring

87. A client with an acute episode of ulcerative colitis is admitted to the hospital. Blood studies reveal that the chloride level is low. This electrolyte deficiency can best be corrected by:
   1. A low-residue diet
   2. Intravenous therapy
   3. Total parenteral nutrition
   4. An oral electrolyte solution

88. Clients with type 1 diabetes may experience a fluid imbalance. The primary fluid shift that occurs in diabetes is:
   1. Intravascular to interstitial as a result of glycosuria
   2. Extracellular to interstitial as a result of hypoproteinemia
   3. Intracellular to intravascular as a result of hyperosmolarity
   4. Intercellular to intravascular as a result of increased hydrostatic pressure

89. A client with a history of cardiac dysrhythmias is admitted to the hospital with the diagnosis of dehydration. The nurse should anticipate that the physician will order:
   1. A glass of water every hour until hydrated
   2. Small frequent intake of juices, broth, or milk
   3. Short-term NG replacement of fluids and nutrients
   4. A rapid IV infusion of an electrolyte and glucose solution

90. A client is diagnosed as having renal failure. During the oliguric phase the nurse should assess the client for:
   1. Alkalosis
   2. Hyperkalemia
   3. Hypocalcemia
   4. Hypernatremia

91. A client with chronic renal failure who is receiving dialysis is prescribed a protein-restricted, sodium-restricted, and potassium-restricted diet. The nurse would know that the dietary teaching was effective when the client says:
   1. "I cannot add seasonings to my food."
   2. "I should avoid using salt substitutes."
   3. "I can eat canned, no-salt vegetables."
   4. "I should get the protein I eat from meat."

92. Before administering a prescribed intravenous solution that contains potassium chloride, the assessment by the nurse that should be brought to the physician's attention would be:
   1. Poor skin turgor with "tenting"
   2. Behaviors indicating irritability and confusion
   3. A urinary output of 200 ml during the previous shift
   4. An oral intake of 300 ml of fluid during the previous shift

93. A client with hyperpyrexia who has just been started on IV antibiotics has a diminished urine output. The nurse should recognize that this is probably the result of:
   1. A declining blood pressure
   2. Bacterial invasion of the kidneys
   3. Nephrotoxicity from antimicrobial agents
   4. A normal compensatory response to fever

94. A client is hospitalized with 50% of the body surface area burned. At the beginning of the 48-hour postburn period (diuretic phase) the client's urine specific gravity is 1.015, urine output is 50 ml, hematocrit is 32, albumin is 3.3 g/dL, and the pulmonary arterial wedge pressure is 10 mm Hg. These data indicate that:
   1. Albumin is critically low
   2. Hemoconcentration is occurring
   3. Fluid therapy has been successful
   4. Fluid replacement has been aggressive

95. The nurse, when assessing the adequacy of a client's fluid replacement during the first 2 to 3 days following full-thickness burns to the trunk and right thigh, would be aware that the most significant data would be obtained from recording:
   1. Weights every day
   2. Urinary output every hour
   3. Blood pressure every 15 minutes
   4. Extent of peripheral edema every 4 hours

96. A client's burns are being treated with silver nitrate 0.5% solution. A week after treatment is begun, the nurse notes that the client's sodium level is 135

mEq/L and the potassium level is 3.0 mEq/L. The nurse should notify the physician and expect to:
1. Add KCl to current IV of Ringer's lactate
2. Add NaCl to current IV of Ringer's lactate
3. Change the NaCl with 20 mEq KCl to 5% D/W
4. Change the 5% D/W with 40 mEq KCl to 5% D/W

97. A client with diabetes develops ketoacidosis. The arterial blood gas report that is representative of diabetic ketoacidosis is:
1. Pco_2 49, HCO_3 32, pH 7.50
2. Pco_2 26, HCO_3 20, pH 7.52
3. Pco_2 54, HCO_3 28, pH 7.30
4. Pco_2 28, HCO_3 18, pH 7.28

98. A client is placed on a low-sodium diet. The family asks whether they can bring some snacks from home. The nurse suggests that they bring foods low in sodium such as:
1. Ice cream
2. Celery sticks
3. Fresh oranges
4. Peanut butter cookies

99. A client with ascites has a paracentesis, and 1500 ml of fluid is removed. Immediately following the procedure it is most important for the nurse to observe for:
1. Decreased peristalsis
2. A rapid, thready pulse
3. Respiratory congestion
4. An increase in temperature

100. A client is receiving an IV infusion. The most serious complication of IV therapy would be:
1. Bleeding at the infusion site
2. Shortness of breath and wheezing
3. A feeling of warmth throughout the body
4. An infiltration at the catheter insertion site

101. A 79-year-old client is admitted for dehydration, and an IV infusion of normal saline at 125 ml/hr is started. One hour later the client begins screaming, "I can't breathe." The nurse should:
1. Call the physician to order a sedative
2. Discontinue the IV and call the physician
3. Elevate the head of the bed and obtain vital signs
4. Assess the client for allergies and change the IV to a heparin lock

102. A client has been receiving 2500 ml of IV fluid and 300 to 400 ml of oral intake daily for 2 days. The client's urine output has been decreasing and now has been less than 40 ml per hour for the past 3 hours. The nurse should initially:
1. Catheterize the client to empty the bladder
2. Assess breath sounds and obtain the client's vital signs
3. Check for dependent edema and continue to monitor I&O
4. Decrease the IV flow rate and increase oral fluids to compensate

103. The nurse is aware that the shift of body fluids associated with the intravenous administration of albumin occurs by the process of:
1. Filtration
2. Diffusion
3. Osmosis
4. Active transport

104. A hospitalized client on a 2-g sodium, 1600-calorie ADA diet complains about the bland food and refuses to eat dinner. The nurse should first:
1. Ask the client what foods are usually eaten at home
2. Explain to the client that the diet will eventually have to be accepted
3. Provide the client with several packets of lemon juice and one packet of salt
4. Urge the client to eat to become accustomed to the diet that must be eaten at home

105. After a gastrojejunostomy for cancer of the stomach, a client progresses to a regular diet. After eating lunch the client becomes diaphoretic and has palpitations. The symptoms are probably the result of:
1. An intolerance to fatty foods
2. The dehiscence of the surgical incision
3. An extracellular fluid shift into the bowel
4. Diminished peristalsis in the small intestine

106. After abdominal surgery a client should be encouraged to turn from side to side and to carry out deep-breathing exercises. These activities are essential to prevent:
1. Metabolic acidosis
2. Metabolic alkalosis
3. Respiratory acidosis
4. Respiratory alkalosis

107. A client's arterial blood gases (ABGs) show the following values: a Po_2 of 89 mm Hg; a Pco_2 of 35 mm Hg; and a pH of 7.37. These findings indicate that the client is experiencing:
1. Fluid balance
2. Oxygen depletion
3. Metabolic acidosis
4. Acid-base balance

108. A client is diagnosed as having metabolic acidosis. The nurse understands that in metabolic acidosis the:
   1. Blood pH level is increased
   2. Plasma bicarbonates are increased
   3. Respiratory center in the medulla is depressed
   4. Excess hydrogen ions are excreted in the urine

109. A client appears very anxious, with 40 shallow respirations per minute. The client complains of feeling dizzy and light-headed and of having tingling sensations of the fingertips and around the lips. The nurse should recognize that the client's complaints are probably related to:
   1. Eupnea
   2. Hyperventilation
   3. Kussmaul's respirations
   4. Carbon dioxide intoxication

110. Nursing intervention for a client who is hyperventilating should focus on providing reassurance and:
   1. Administering oxygen
   2. Using an incentive spirometer
   3. Having the client breathe into a paper bag
   4. Administering an IV containing bicarbonate ions

111. The physician orders serum electrolytes. To determine the effect of persistent vomiting the nurse should be most concerned with monitoring the:
   1. Sodium and chloride levels
   2. Bicarbonate and sulfate levels
   3. Magnesium and protein levels
   4. Calcium and phosphate levels

112. A client is hospitalized with epigastric pain, nausea, and vomiting of 4 days duration. The following laboratory values are noted: a plasma pH of 7.51; a Pco_2 of 50 mm Hg; a bicarbonate of 58 mEq/L; a chloride of 55 mEq/L; and a potassium of 3.8 mEq/L. The nurse recognizes that the collected data indicate:
   1. Hyperkalemia
   2. Hyperchloremia
   3. Metabolic alkalosis
   4. Respiratory acidosis

113. After a gastrectomy a client has a nasogastric tube to low continuous suction. The client begins to hyperventilate. The nurse should be aware that this pattern will alter the client's arterial blood gases by:
   1. Increasing the Po_2 level
   2. Decreasing the pH level
   3. Decreasing the Pco_2 level
   4. Increasing the HCO_3 level

114. When a client is in profound (late) hypovolemic shock, the nurse should assess the client's laboratory values, especially the arterial blood gases, because people in late shock will develop:
   1. Hypokalemia
   2. Metabolic acidosis
   3. Respiratory alkalosis
   4. A decreased Pco_2 level

115. The physician orders additional diagnostic studies to assess the client's acid-base status. The laboratory value that would indicate metabolic acidosis is:
   1. Urine pH of 8.4
   2. Gastric content pH of 6.0
   3. Venous serum pH of 7.28
   4. Arterial plasma pH of 7.40

116. The physician determines that a client has metabolic acidosis from severe dehydration. The characteristic respiration that the nurse would expect with metabolic acidosis is:
   1. Dyspnea
   2. Hyperpnea
   3. Kussmaul breathing
   4. Cheyne-Stokes breathing

117. When a client develops respiratory alkalosis, the nurse should expect the laboratory values to reflect:
   1. An elevated pH, elevated Pco_2
   2. A decreased pH, elevated Pco_2
   3. An elevated pH, decreased Pco_2
   4. A decreased pH, decreased Pco_2

118. When assessing a client's arterial blood gases the nurse would know that the client was in compensated respiratory acidosis when the pH value is 7.34 and the:
   1. Po_2 value is 80 mm Hg
   2. HCO_3 value is 50 mEq/L
   3. Pco_2 value is 60 mm Hg
   4. Serum potassium value is 4 mEq/L

119. Surgery is performed on a client with a parotid tumor. The postoperative arterial blood gas values are pH 7.32; Pco_2 53 mm Hg; HCO_3 25 mEq/L. The nurse should:
   1. Obtain a medical order for the administration of a diuretic
   2. Have the client breathe into a rebreather bag at a slow rate
   3. Encourage the client to cough productively and take deep breaths
   4. Obtain a medical order for the administration of sodium bicarbonate

120. After surgical clipping of a ruptured cerebral aneurysm, a client develops the syndrome of inappropriate secretion of antidiuretic hormone. Mani-

festations of excessive levels of antidiuretic hormone (ADH) are:
1. Increased BUN and hypotension
2. Hyperkalemia and poor skin turgor
3. Hyponatremia and decreased urine output
4. Polyuria and increased specific gravity of urine

121. A client has been diagnosed as having hepatitis B (HBV) and is admitted to a medical unit. Because of delayed treatment, the client has developed cirrhosis. One potential sequela of chronic liver disease is fluid and electrolyte imbalance. The nurse recognizes this may be attributed to a decrease in serum albumin level, which leads to:
1. Hemorrhage and subsequent anemia
2. Diminished resistance to bacterial insult
3. Malnutrition of all cells, especially hepatic cells
4. Reduction of colloidal osmotic pressure in the blood

122. An agitated, hyperventilating, semicomatose client is brought to the emergency service. The family indicates that they found empty bottles of 100 aspirin tablets and 50 cold capsules and assumes the client took them several hours ago. The client's vital signs are BP 160/100, pulse 140, respirations 40, and temperature 101.5° F. An oral airway is in place. The test considered the most vital in providing information that will guide the emergency treatment for this client would be:
1. A 24-hour urine
2. A blood glucose
3. Serum electrolytes
4. Liver function tests

123. After surgery a client is to receive an antibiotic by IV piggyback in 50 ml of D5W. The piggyback is to infuse in 20 minutes. The drop factor of the IV set is 10 gtt/ml. The nurse should set the piggyback to flow at:
1. 25 gtt/min
2. 30 gtt/min
3. 35 gtt/min
4. 45 gtt/min

124. The physician orders a client's IV fluids to be delivered at 80 ml/hour. To adjust the drip rate when administering the IV via gravity, the nurse must know the:
1. Total volume of fluid in the IV bag
2. Size of the needle or catheter in the vein
3. Drops per milliliter delivered by the infusion set
4. Diameter of the tubing being used to instill the fluid

125. The physician orders IV fluids and gentamicin sulfate (Garamycin) 100 mg IVPB q8h for a client.

The nurse, using gravity to instill the IV, hangs the piggyback gentamicin higher than the primary IV bag. When the piggyback bag is empty, the client observes air in the tubing of the IVPB and becomes frightened. The nurse should explain that:
1. Air in the tubing, even if it got into the vein, would not be fatal unless it was a large amount
2. The gentamicin and now the air are flowing into the large IV bag, not into the venous system directly
3. The solution from the large IV bag will begin to flow when the solution from the smaller bag ceases to flow
4. The clamps on the tubing leading from both bags can be closed for a few minutes to prevent air from entering the vein

BLOOD AND IMMUNITY (BI)

126. Nutritional support of a client's natural defense mechanisms would indicate the need for a diet high in:
1. The essential fatty acids
2. Dietary cellulose and fiber
3. The amino acid, tryptophan
4. Vitamins A, C, E, and selenium

127. Twelve hours after a female client is admitted to the critical care unit following a motorcycle injury, she begins to complain of increased abdominal pain in the left upper quadrant. A ruptured spleen is diagnosed, and she is scheduled for an emergency splenectomy. When preparing the client for surgery the nurse should emphasize the:
1. Complete safety of the procedure
2. Expectation of postoperative bleeding
3. Risk of the procedure with the other injuries
4. Presence of abdominal drainage for several days after the surgery

128. After abdominal surgery a client develops internal hemorrhaging. During further assessments, the nurse should expect the client to exhibit:
1. Polyuria
2. Bradypnea
3. Tachycardia
4. Hypertension

129. Immediately following abdominal surgery a client begins to hemorrhage. The nurse should observe the client for signs of progressive hypovolemia, which include:
1. Oliguria
2. Bradypnea
3. Pulse deficit
4. Hyperkalemia

130. A client is brought to the emergency service after an automobile accident. The client's blood pressure is 100/60 mm Hg, and the physical assessment suggests a ruptured spleen. Based on this information, the nurse should assess the client for an early sign of decreased arterial pressure, such as:
    1. Warm, flushed skin
    2. Confusion and lethargy
    3. Increased pulse pressure
    4. Reduced peripheral pulses

131. When a client is experiencing hypovolemic shock with decreased tissue perfusion, the body initially attempts to compensate by:
    1. Producing less ADH
    2. Producing more red blood cells
    3. Maintaining peripheral vasoconstriction
    4. Decreasing mineralocorticoid production

132. After sustaining multiple internal injuries from being hit by a car, the client's blood pressure suddenly drops to 80/60 mm Hg. The nurse should realize that this is probably caused by:
    1. A reduction in the circulating blood volume
    2. Diminished vasomotor stimulation to the arterial wall
    3. Vasodilation resulting from diminished vasoconstrictor tone
    4. Cardiac decompensation resulting from electrolyte imbalance

133. The nurse is aware that shock associated with an abdominal aneurysm is called:
    1. Vasogenic shock
    2. Neurogenic shock
    3. Cardiogenic shock
    4. Hypovolemic shock

134. A client has emergency surgery for a ruptured appendix. After assessing that the client is manifesting symptoms of shock, the nurse should:
    1. Prepare for a blood transfusion
    2. Notify the physician immediately
    3. Elevate the head of the bed 30 degrees
    4. Administer prescribed oxygen postoperatively

135. In the progressive stage of shock, anaerobic metabolism occurs. The nurse must be aware that this initially causes:
    1. Metabolic acidosis
    2. Metabolic alkalosis
    3. Respiratory acidosis
    4. Respiratory alkalosis

136. A client who is in hypovolemic shock has a hematocrit value of 20%. The nurse should anticipate that the physician will order:
    1. Serum albumin
    2. Ringer's lactate
    3. Blood replacement
    4. High molecular dextran

137. The physician orders 2 units of blood to be infused into a client who is bleeding. Before blood administration the nurse's highest priority should be:
    1. Obtaining the client's vital signs
    2. Allowing the blood to reach room temperature
    3. Monitoring the hemoglobin and hematocrit levels
    4. Determining proper typing and crossmatching of blood

138. A client who is scheduled for a modified radical mastectomy decides to have family members donate blood in the event it is needed. The client has type A negative blood. Blood could be used from relatives whose blood is:
    1. Type O positive
    2. Type AB negative
    3. Type A or O negative
    4. Type A or AB negative

139. When administering blood, it is important for the nurse to:
    1. Administer each unit within a 6-hour period
    2. Use a volume control infusion pump to administer the blood
    3. Run the blood at a slower rate during the first 5 to 10 minutes
    4. Draw blood samples from the client immediately after each unit is transfused

140. While receiving a blood transfusion a client develops flank pain, chills, fever, and hematuria. The nurse recognizes that the client is probably experiencing:
    1. An allergic transfusion reaction
    2. A hemolytic transfusion reaction
    3. A pyrogenic transfusion reaction
    4. An anaphylactic transfusion reaction

141. Halfway through the administration of a unit of blood, a client complains of lumbar pain. The nurse should:
    1. Obtain vital signs
    2. Stop the transfusion
    3. Assess the pain further
    4. Increase the flow of normal saline

142. A male client with chronic liver disease reports that his gums bleed spontaneously. In addition, the nurse

notes small hemorrhagic lesions on his face. The nurse recognizes that the client needs additional:
1. Bile salts
2. Folic acid
3. Vitamin A
4. Vitamin K

143. A client comes to the clinic complaining of weight loss, fatigue, and a low-grade fever. Physical examination reveals a slight enlargement of the cervical lymph nodes. When the nurse is assessing possible causes for the fever, it would be most appropriate for the nurse to initially ask:
1. "Have you been sexually active lately?"
2. "Do you have a sore throat at the present time?"
3. "When did you first notice that your temperature had gone up?"
4. "Have you been exposed recently to anyone with an infection?"

144. The nursing staff has a team conference on AIDS and discusses the routes of transmission of the human immunodeficiency virus (HIV). The discussion reveals that an individual has no risk of exposure to HIV when that individual:
1. Has intercourse with just the spouse
2. Makes a donation of a pint of whole blood
3. Limits sexual contact to those without HIV antibodies
4. Uses a condom each time there is sexual intercourse

145. The nurse knows that a positive diagnosis for HIV infection is made based on:
1. A history of high-risk sexual behaviors
2. Positive ELISA and Western blot tests
3. Evidence of extreme weight loss and high fever
4. Identification of an associated opportunistic infection

146. Blood screening tests of the immune system of a client with AIDS would indicate:
1. A decrease in CD4 T cells
2. An increase in thymic hormones
3. An increase in immunoglobulin E
4. A decrease in the serum level of glucose-6-phosphate dehydrogenase

147. When taking the blood pressure of a client who has AIDS, the nurse must:
1. Wear clean gloves
2. Use barrier techniques
3. Wear a mask and gown
4. Wash the hands thoroughly

148. A client with acquired immunodeficiency syndrome (AIDS) and *Cryptococcus* pneumonia is incontinent of feces and urine and is producing copious sputum. When providing care for this client the nurse's priority should be to:
1. Wear goggles when suctioning the client's airway
2. Use gown, mask, and gloves when bathing the client
3. Use gloves to administer oral medications to the client
4. Wear a gown when assisting the client with the bedpan

149. In addition to *Pneumocystis carinii,* a client with AIDS also has an ulcer 4 cm in diameter on the leg. Considering the client's total health status, the most critical nursing diagnosis would be:
1. Social isolation
2. Impaired skin integrity
3. Impaired gas exchange
4. Altered nutrition: less than body requirements

150. A client with AIDS is to receive palliative treatment. A palliative approach would involve planning measures to:
1. Restore the client's health
2. Promote the client's recovery
3. Relieve the client's discomfort
4. Support the client's significant others

151. A Schilling test is ordered for a client who is suspected of having pernicious anemia. The nurse recognizes that the primary purpose of the Schilling test is to determine the client's ability to:
1. Store vitamin B_{12}
2. Digest vitamin B_{12}
3. Absorb vitamin B_{12}
4. Produce vitamin B_{12}

152. When explaining the therapeutic regimen of vitamin B_{12} for pernicious anemia to a client, the nurse should explain that:
1. Weekly Z-track injections provide needed control
2. Daily intramuscular injections are required for control
3. Intramuscular injections once a month will maintain control
4. Oral tablets of vitamin B_{12} taken daily will control the symptoms

153. The nurse knows that the teaching regarding the use of vitamin B_{12} injections to treat pernicious anemia is understood when a client states, "I must take the drug:
1. When feeling fatigued."
2. Until my symptoms subside."
3. Monthly, for the rest of my life."
4. During exacerbations of anemia."

154. A client with Hodgkin's disease enters a remission period and remains symptom-free for 6 months, when a relapse occurs. The client is diagnosed at stage IV. The therapy option the nurse should expect to be implemented at this time is:
   1. Radiation therapy
   2. Combination chemotherapy
   3. Radiation with chemotherapy
   4. Surgical removal of the affected nodes

155. The physician has decided to use total nodal irradiation in conjunction with chemotherapy for a young female client with stage IIIA Hodgkin's disease. The client tells the nurse she wants to have children and is quite concerned that the radiation therapy includes the pelvic nodal areas. When questioned about this, the nurse should refer the client to the physician because the nurse should be aware that:
   1. The ovaries can be surgically moved and placed in a shielded area
   2. Intense radiation to the area always causes permanent sterilization
   3. The radiation used is not radical enough to destroy ovarian function
   4. Ovarian function will be temporarily destroyed but will return in about 6 months

156. The nurse should plan to teach a client with pancytopenia caused by chemotherapy to:
   1. Begin a program of aggressive, strict mouth care
   2. Avoid traumatic injuries and exposure to any infection
   3. Increase oral fluid intake to a minimum of 3000 ml daily
   4. Report any unusual muscle cramps or tingling sensations in the extremities

157. An elderly client develops severe bone marrow depression from chemotherapy for cancer of the prostate. The nurse should:
   1. Monitor for signs of alopecia
   2. Increase daily intake of fluids
   3. Monitor intake and output of fluids
   4. Use a soft toothbrush for oral hygiene

158. The laboratory results of a client following chemotherapy for cancer indicate bone marrow depression. The nurse should encourage the client to:
   1. Use an electric razor when shaving
   2. Drink citrus juices frequently for nourishment
   3. Sleep with the head of the bed slightly elevated
   4. Increase activity levels and ambulate frequently

159. A client has received three courses of chemotherapy and is admitted for tests before continuing with the fourth in the series. The physician decides to omit the treatment because the client demonstrates myelosuppression. When discussing this with the client, the nurse should explain that:
   1. Calcium must be increased in the diet because of the effects of myelosuppression
   2. The development of myelosuppression explains why the client has nausea, vomiting, anorexia, and alopecia
   3. Eating a balanced diet, resting, and care to avoid bleeding and infections are appropriate for the client at this time
   4. Frequent testing for restlessness, muscle control, and pupillary response will be necessary because the meninges may be irritable

160. A client who is suspected of having leukemia has a bone marrow aspiration. Immediately following the procedure, the nurse should:
   1. Apply brief pressure to the site
   2. Ask the client to lie on the affected side
   3. Swab the site with an antiseptic solution
   4. Monitor the vital signs every hour for 4 hours

161. When obtaining a health history from a young client with probable acute lymphocytic leukemia (ALL), the clinical manifestations the nurse should expect to be present are:
   1. Anorexia, insomnia
   2. Petechiae, alopecia
   3. Anorexia, petechiae
   4. Alopecia, bleeding gums

162. A 26-year-old client with a history of chronic myelogenous leukemia and splenomegaly is admitted to the hospital. The nurse should expect this client to have:
   1. An increased urinary output
   2. A tender mass in the left upper abdomen
   3. Increased erythrocytes, platelets, and granulocytes
   4. Polydipsia, increased appetite, and urinary frequency

163. The diagnostic finding associated with multiple myeloma is:
   1. Low serum calcium levels
   2. Bence-Jones protein in the urine
   3. Occult and frank blood in the stool
   4. Positive bacterial culture of sputum

164. The nurse understands that the most definitive test to confirm a diagnosis of multiple myeloma is:
   1. Bone marrow biopsy
   2. Serum test for hypercalcemia
   3. Urine test for Bence-Jones protein
   4. X-ray films of the ribs, spine, and skull

165. A client with multiple myeloma is scheduled to have a chest x-ray examination and a bone scan. For this client, the primary responsibility of the nursing and radiology staff is to:
1. Explain the procedure and its purpose
2. Observe the client for shortness of breath
3. Provide for rest periods during the procedure
4. Handle the client with supportive movements

166. A client who is newly diagnosed with multiple myeloma asks the physician what treatment will be necessary. The nurse should expect the physician to reply:
1. "Human leukocyte interferon therapy."
2. "Radiotherapy on an outpatient basis."
3. "Surgery to remove the lesion and lymph nodes."
4. "Chemotherapy employing a combination of drugs."

167. A client with multiple myeloma has a temperature that has risen 3° during a 6-hour period and is now 102.2° F. The nurse should:
1. Administer the prescribed antipyretic and notify the physician
2. Obtain the other vital signs and recheck the temperature in 1 hour
3. Assess the amount and color of urine and obtain a specimen for a urinalysis
4. Note the consistency of respiratory secretions and obtain a specimen for culture

168. A client with multiple myeloma asks how the disease may progress. When teaching this client, the nurse should discuss the possibility that:
1. Blood transfusions may be necessary
2. Frequent urinary tract infections may result
3. IV fluid therapy may be administered in the home
4. The disease is exacerbated by exposure to ultraviolet rays

169. The nurse is aware that a client is receiving azathioprine (Imuran), cyclosporine, and prednisone before kidney transplant surgery to:
1. Stimulate leukocytosis
2. Provide passive immunity
3. Prevent iatrogenic infection
4. Reduce antibody production

170. A farmer steps on a rusty nail and the puncture site becomes swollen and painful. Tetanus antitoxin is prescribed. The nurse explains that this is used because it:
1. Provides antibodies
2. Stimulates plasma cells

3. Produces active immunity
4. Facilitates long-lasting immunity

171. A tuberculin skin test with purified protein derivative (PPD) tuberculin is performed as part of a routine physical examination. The nurse should instruct the client to make an appointment so the test can be read in:
1. 3 days
2. 5 days
3. 7 days
4. 10 days

172. A client is admitted with cellulitis of the left leg and a temperature of 103° F. The physician orders IV antibiotics. Before instituting this therapy, the nurse should:
1. Determine whether the client has allergies
2. Apply a warm, moist dressing over the area
3. Measure the amount of swelling in the client's leg
4. Obtain the results of the culture and sensitivity tests

173. After multiple bee stings a client has an anaphylactic reaction. The nurse is aware that the symptoms the client is experiencing are caused by:
1. Respiratory depression and cardiac standstill
2. Bronchial constriction and decreased peripheral resistance
3. Decreased cardiac output and dilation of major blood vessels
4. Constriction of capillaries and decreased peripheral circulation

174. The plan of care for a postoperative client who has developed a pulmonary embolus includes monitoring and bed rest. The client asks why all activity is restricted. The nurse's response is based on the principle that bed rest:
1. Prevents further platelet aggregation
2. Enhances the peripheral circulation in the deep vessels
3. Decreases the potential for further development of emboli
4. Maximizes the amount of blood available to damaged tissues

175. A client comes to the clinic for a physical and asks to be tested for AIDS. The nurse explains that the initial screening for AIDS will be done via the:
1. Western blot test
2. CD4 T cell count
3. Polymerase chain reaction test
4. Enzyme-linked immunosorbent assay (ELISA)

176. A Schilling test is ordered for a client suspected of having cobalamin deficiency because of pernicious anemia. The nurse must plan to:
    1. Give all medications on time
    2. Order foods low in vitamin B_{12}
    3. Keep an accurate intake and output
    4. Collect a 24- to 48-hour urine specimen

INTEGUMENTARY (IT)

177. A client is experiencing stomatitis as a result of chemotherapy. An appropriate intervention related to this condition is to:
    1. Provide frequent saline mouthwashes
    2. Use karaya powder to decrease irritation
    3. Increase fluid intake to compensate for the diarrhea
    4. Provide meticulous skin care of abdomen with Betadine

178. When assessing the skin of an elderly client, a normal change that the nurse might identify is:
    1. Signs of ecchymosis
    2. Marked flaking of skin
    3. Hyperpigmented patches
    4. Scaling associated with dryness

179. A client has been in a coma for 2 months and is maintained on bed rest. The nurse understands that to prevent the effects of shearing force, the head of the bed should be at an angle of:
    1. 30 degrees
    2. 45 degrees
    3. 60 degrees
    4. 90 degrees

180. The physician orders bed rest for a client after surgery. The nurse is aware that the most beneficial method of preventing skin breakdown while the client is confined to bed is to:
    1. Massage the skin with cream
    2. Use a sheepskin pad on the bed
    3. Promote passive range of motion
    4. Encourage independent movement

181. The physician orders bed rest for a client with cellulitis of the leg. The nurse understands that the primary purpose of bed rest for this client is to:
    1. Decrease catabolism to promote healing at the site of injury
    2. Lower the metabolic rate in an attempt to help reduce the fever
    3. Reduce the energy demands on the body in the presence of infection
    4. Limit muscle contractions that would force causative organisms into the bloodstream

182. The nurse is preparing to change a client's dressing. The statement that best explains the basis of surgical asepsis that the nurse will perform in this procedure is:
    1. Keep the area free of microorganisms
    2. Protect self from microorganisms in the wound
    3. Confine the microorganisms to the surgical site
    4. Keep the number of opportunistic microorganisms to a minimum

183. The equipment that will be used by the nurse during central venous catheter site care for a client receiving total parenteral nutrition is:
    1. Double sterile gloves
    2. Mask and sterile gloves
    3. Gown and sterile gloves
    4. Mask, gown, and sterile gloves

184. An extremely obese client must self-administer insulin at home. The nurse should teach the client to:
    1. Pinch the tissue and inject at a 45-degree angle
    2. Pinch the tissue and inject at a 60-degree angle
    3. Spread the tissue and inject at a 45-degree angle
    4. Spread the tissue and inject at a 90-degree angle

185. A client develops an infection at a catheter insertion site. The nurse uses the term iatrogenic when describing this infection because it resulted from:
    1. Poor physical hygiene
    2. A therapeutic procedure
    3. Inadequate dietary patterns
    4. The client's developmental level

186. A client develops an infection of an abdominal incision and overhears the nurses say that it is a nosocomial infection. The client asks the nurse what this means. The nurse should reply:
    1. "The infection you had prior to hospitalization has flared up."
    2. "You acquired the infection after being admitted to the hospital."
    3. "This is a highly contagious infection requiring protective isolation."
    4. "As a result of medical treatment, you have developed a secondary infection."

187. A male client is admitted to the hospital for intravenous antibiotic therapy and an incision and drainage of an abscess that developed at the site of a puncture wound. The nurse should begin teaching wound care to the client:
    1. In the preoperative period
    2. A few days before discharge
    3. On the first postoperative day
    4. During the first dressing change

188. The nurse is aware that research has shown that malnutrition occurs in as many as 50% of hospitalized clients. In view of a postoperative client's poor appetite, the nurse should observe for:
1. Dependent edema
2. "Spoon-shaped" nails
3. Loose, decayed teeth
4. Delayed wound healing

189. The primary nurse tells a client with an infected wound that the nurse epidemiologist will visit daily. The client asks what the nurse epidemiologist does. The nurse could correctly explain the role by saying, "The nurse epidemiologist:
1. Helps providers of care to control infections."
2. Decides what antibiotics should be prescribed for infections."
3. Works in the laboratory to identify bacteria causing infection."
4. Is responsible for doing cultures of all infections and drainages."

190. After a choledocholithotomy, a client complains that the skin around the T-tube is raw and excoriated. After assessing the skin the nurse should plan to:
1. Reinforce the dressings when they are wet
2. Cleanse the area with an antiseptic solution
3. Use a skin barrier around the T-tube exit site
4. Change the type of adhesive tape used on the dressing

191. When teaching older adults how to limit the itching that results from dry skin, the nurse should instruct them to:
1. Use a moisturizer
2. Take hot tub baths
3. Wear warm clothes
4. Expose skin to the air

192. A client with a long history of bilateral varicose veins questions the nurse about the brownish discoloration of the skin of the lower extremities. The nurse should explain that this is probably the result of:
1. An inadequate arterial blood supply
2. Delayed healing of tissues after an injury
3. Leakage of RBCs through the vascular wall
4. Increased production of melanin in the area

193. A client arrives at the emergency room after being bitten by a stray dog. The bite involved tearing of skin and deep soft tissue injury. The client says the dog was foaming at the mouth and afterward ran away. The first nursing action is to:
1. Ask the client about horse serum allergy
2. Notify the police department to capture the dog

3. Assess the client's injury, vital signs, and past history
4. Inoculate the client with human rabies immune globulin

194. A client comes to the clinic after being bitten by a raccoon in the woods in an area where rabies is endemic. The nurse recalls that rabies is:
1. An acute bacterial infection characterized by encephalopathy and opisthotonos
2. An acute bacterial septicemia that results in convulsions and a morbid fear of water
3. A nonspecific immunoresponse to organisms deposited under the skin by an animal bite
4. An acute viral infection, characterized by convulsions and difficult swallowing, that affects the nervous system

195. A client who is to receive radiation therapy for cancer says to the nurse, "My family said I will get a radiation burn." The best response by the nurse would be:
1. "It will be no worse than a sunburn."
2. "A localized skin reaction usually occurs."
3. "Have they had experience with this type of radiation?"
4. "Daily application of an emollient will prevent the burn."

196. A client is to begin radiation therapy for cancer. When teaching about skin care to the irradiated area, the nurse should instruct the client to:
1. Apply warm compresses to the site
2. Apply no lotions or powders to the area
3. Cover the area with a sterile gauze bandage
4. Lie on the back and unaffected side when sleeping

197. Irradiation to the chest wall on an outpatient basis is prescribed for a client following removal of a tumor in the right lung. When teaching skin care to the client the nurse should emphasize:
1. Massaging 4 times a day to increase circulation
2. Frequent washing to remove desquamated cells
3. Keeping the skin dry and protected from abrasions
4. Using skin lotion twice daily to keep the skin supple

198. A client with scleroderma complains of having difficulty chewing and swallowing. When providing dietary counseling, the nurse should advise the client to:
1. Liquefy the food in a blender
2. Puree all foods before eating
3. Take frequent sips of water with meals
4. Use a local anesthetic mouthwash before eating

199. A female client with scleroderma tells the nurse that she often has numbness and tingling in her hands followed by blanching of her fingers. The nurse recognizes that the client has Raynaud's phenomenon, a condition commonly associated with scleroderma. The nurse should advise the client to:
    1. Bathe her hands frequently in hot water
    2. Keep her hands warm by wearing gloves
    3. Briskly rub her hands to increase circulation
    4. Take the anticoagulants that will be prescribed to prevent attacks

200. As part of the teaching plan for a client with scleroderma, the nurse should include the need for special skin care. The nurse should plan to advise the client to:
    1. Use calamine lotion for pruritus
    2. Keep the skin well lubricated with oil
    3. Apply warm soaks to the inflamed areas
    4. Take frequent baths to remove scaly lesions

201. The physician orders a regimen of daily exercises for a client with scleroderma. The nurse understands that with this client exercises are performed to:
    1. Preserve muscle strength
    2. Promote tissue regeneration
    3. Prevent spread of the disease
    4. Promote a sense of well-being

202. A young male client is admitted to the burn center after incurring electrical burns to both hands while playing golf during a lightning storm. When assessing the entrance and exit wounds, the nurse is aware that electrical injury:
    1. Causes severe nervous tissue destruction along a path of least resistance
    2. Results in severe tissue destruction when the burn is incurred by direct current
    3. Causes a line of destruction beginning at the grounding point to the point of contact
    4. Results in visible dermal wounds that denote the internal electrical current destruction

203. While an adult male is lighting a barbecue grill with lighter fluid, his shirt bursts into flames. To extinguish the flames most effectively, with as little further damage as possible, it is best to:
    1. Remove the burning clothes
    2. Slap the flames with the hands
    3. Log-roll the person in the grass
    4. Pour cold liquid over the flames

204. A woman's bathrobe ignites while she is cooking in the kitchen on a gas stove. Once the flames are extinguished, it is most important to:
    1. Give the person sips of water
    2. Assess the person's breathing
    3. Cover the person with a warm blanket
    4. Calculate the extent of the person's burns

205. A person sustains burns of the arms from a barbecue accident in the park. A bystander emerges from the crowd and suggests to a nurse who has come to the person's assistance that butter be applied to the burns. An appropriate response by the nurse would be, "Thanks, but:
    1. We'll just wait for the ambulance."
    2. It is better to use some first aid cream."
    3. I'll just use a tablecloth as a blanket for now."
    4. We should apply ice. Could you go get me some?"

206. A worker is involved in an explosion of a steam pipe and receives a scalding burn to the chest and arms. The burned areas are painful, mottled red, weeping, and edematous. These burns would be classified as:
    1. Eschar
    2. Full-thickness burns
    3. Deep partial-thickness burns
    4. Superficial partial-thickness burns

207. A client is seen in the emergency department after a barbecue accident. The physician diagnoses superficial partial-thickness burns. The family asks what is involved with this type of burn. The most accurate response by the nurse would be that the:
    1. Epidermis has been damaged
    2. Dermis has been partially damaged
    3. Epidermis and dermis have been destroyed
    4. Structures beneath the skin have been destroyed

208. A client is burned on the anterior part of both legs, from the knees to the feet. Using the rule of nines, the nurse estimates the burn surface to be:
    1. 4.5%
    2. 9%
    3. 18%
    4. 27%

209. A client is admitted for treatment of partial- and full-thickness burns of the entire right lower extremity and the anterior portion of the right upper extremity. Performing an immediate appraisal, using the rule of nines, the nurse estimates that the percent of body surface burned is:
    1. 4.5%
    2. 9%
    3. 18%
    4. 22.5%

210. When a female client who has partial-thickness burns on her chest, abdomen, and right leg from a fire at

her workplace arrives in the emergency department, the nurse's first responsibility should be to:
1. Carefully remove all of the client's clothing
2. Evaluate whether heat inhalation had occurred
3. Apply sterile saline dressings on all burned surfaces
4. Determine the extent of the burns, using the rule of nines

211. When a nurse is evaluating the condition of a client with burns of the upper body, an assessment that would indicate potential respiratory obstruction is:
1. Deep breathing
2. Pink-tinged, frothy sputum
3. Hoarse quality to the voice
4. Rapid abdominal breathing

212. During the first 48 hours after a thermal injury, the nurse should assess the client for:
1. Hypokalemia and hyponatremia
2. Hyperkalemia and hyponatremia
3. Hypokalemia and hypernatremia
4. Hyperkalemia and hypernatremia

213. A severely burned client has been hospitalized for 2 days. Until now recovery has been uneventful, but the client begins to exhibit extreme restlessness. The nurse recognizes that this most likely indicates that the client is developing:
1. Renal failure
2. Hypervolemia
3. Cerebral hypoxia
4. Metabolic acidosis

214. A client's partial-thickness burns differ from full-thickness burns in that with partial-thickness burns the burned area will:
1. Require grafting before it can heal
2. Be painful, reddened, and have blisters
3. Have total destruction of the epidermis and dermis
4. Take months of extensive treatment before healing

215. A client with burns develops a wound infection. The nurse knows that local wound infections are primarily treated with:
1. Oral antibiotics
2. Topical antibiotics
3. Intravenous antibiotics
4. Intramuscular antibiotics

216. The nurse identifies that a client in the acute phase of burns has eaten only a small portion of each meal.

Aware that malnutrition can have a variety of consequences, the nurse should assess the client for:
1. Dehydration
2. Dry, brittle hair
3. Prolonged wound healing
4. "Clubbing" of the finger tips

217. The method of treatment chosen for a client's burns is the exposure method with application of mafenide (Sulfamylon) bid. When applying this medication the nurse should plan to:
1. Use medical asepsis
2. Apply a dry sterile dressing
3. Monitor liver function studies
4. Give ordered pain medication

218. A client is admitted to the hospital with deep partial-thickness burns to both hands and forearms after an accident. The nurse applies the prescribed antimicrobial medication by placing it:
1. Directly on the dressing in a thick layer (2-4 cm) using clean gloves
2. Directly on the burn wound in a thin layer (2-4 mm) using sterile gloves
3. In the Hubbard tank and saturating sterile dressings with it before applying to the burn
4. In the Hubbard tank and allowing the client to soak in the tank for 20 minutes every day

219. When dressing the deep partial-thickness burns on a client's hands the nurse should use:
1. Cotton-backed gauze and fully extend the fingers with thumb in opposition
2. Non-cotton-backed gauze and place a hand roll with gauze between each finger
3. Non-cotton-backed gauze and extend fingers fully with gauze between each finger
4. Cotton-backed gauze and a hand roll, with fingers completely flexed and thumb in opposition

220. A client tells the nurse that the doctor mentioned doing skin grafts after a burn and asks when they will be done. The most appropriate response by the nurse would be:
1. "Within 7 days."
2. "What did your doctor tell you?"
3. "As soon as scar formation occurs."
4. "As soon as signs of infection disappear."

221. A client with a partial-thickness burn complains of chilling. To limit this adaptation, the nurse should:
1. Limit the occurrence of drafts
2. Keep room temperature more than 90° F
3. Maintain room humidity at less than 40%
4. Place a sterile top sheet over the client

222. The physician determines that a client's burn wounds need to be mechanically debrided and informs the client of this fact. Later the nurse should reinforce the physician's discussion by explaining that:
    1. The surgeon will surgically remove the dead tissue
    2. Incisions will be made along the length of the eschar
    3. Enzymatic agents will be applied daily to the wounds
    4. Mechanical devices will continually move the client's extremities

223. A temporary heterograft (pig skin) is used to treat burns because this graft will:
    1. Débride necrotic epithelium
    2. Be sutured in place for better adherence
    3. Relieve pain and promote rapid epithelialization
    4. Commonly be used concurrently with topical antimicrobials

224. When planning care to prevent deformities and contractures in a client with burns, the nurse should expect to begin range-of-motion exercises when the client's:
    1. Pain has lessened
    2. Vital signs are stable
    3. Skin grafts are healed
    4. Emotional status stabilizes

225. A client returns from surgery with an incisional dressing and a catheter that is attached to a portable wound drainage system exiting from the operative site. The nurse recognizes that the principle underlying the function of a portable drainage system is:
    1. Gravity
    2. Osmosis
    3. Active transport
    4. Negative pressure

226. After surgery for cancer, a client is to receive adjuvant chemotherapy. When teaching the client about the side effects of chemotherapy, the nurse should emphasize that the occurrence of alopecia is:
    1. Usually rare
    2. Never permanent
    3. Frequently prolonged
    4. Sometimes preventable

SKELETAL (SK)

227. The nurse recognizes that stimulation of calcium deposition in the bone after a distal femoral fracture is best achieved by:
    1. Resting the extremity
    2. Weight-bearing activity

3. The normal aging process
4. Ingesting foods high in calcium

228. A male college basketball player comes to the infirmary complaining of a "click" in his knee when walking. He states that it occasionally gives way when he is running and sometimes locks. He does not recall any specific injury. The nurse suspects that he may have:
    1. Cracked the patella
    2. A ruptured Achilles tendon
    3. Injured the cartilage in the knee
    4. A stress fracture of the tibial plateau

229. A client is scheduled for an arthroscopy of the knee in the morning and asks the nurse about the procedure. The statement by the nurse that best describes the procedure would be:
    1. "You will be anesthetized and not remember anything about the procedure."
    2. "It is a direct visualization of the joint to diagnose the extent of the knee injury."
    3. "It is a radiologic procedure that will help diagnose the extent of the knee injury."
    4. "The procedure will determine the type of treatments the physician will prescribe."

230. A client is scheduled for a bone scan to determine the presence of metastases. The nurse is aware that teaching prior to a scheduled bone scan was effective when the client states that:
    1. "X-rays will be taken to identify where I may have lost calcium from my bones."
    2. "A portion of my bone marrow will be removed and examined for cell composition."
    3. "A radioactive chemical will be injected into my vein that will destroy cancer cells present in my bones."
    4. "A substance of low radioactivity will be injected into my vein, and my body inspected by an instrument to detect where it is deposited."

231. Clients who have casts applied to an extremity must be monitored for complications. Therefore the nurse should assess the extremity for:
    1. Warmth
    2. Numbness
    3. Skin desquamation
    4. Generalized discomfort

232. A client's right tibia is fractured in an automobile accident, and a cast is applied. To assess for damage to major blood vessels from the fractured tibia, the nurse should monitor the client for:
    1. Swelling of the right thigh
    2. An increased blood pressure

3. A decreased dorsalis pedis pulse
4. Increased skin temperature of the foot

233. Three days after a cast is applied to a client's fractured tibia, the client states that there is a burning pain over the ankle. The cast over the ankle feels warm to the touch, and the pain is not relieved when the client changes position. The nurse's priority action should be to:
1. Obtain an order for an antibiotic
2. Explain that it is a typical response to a cast
3. Report the client's complaint to the physician
4. Administer the prescribed medication for pain

234. After a long leg cast is removed, the client should be instructed to:
1. Report any discomfort or stiffness of the ankle
2. Cleanse the leg by scrubbing with a brisk motion
3. Elevate the leg when sitting for long periods of time
4. Put the leg through a full range of motion once daily

235. The nurse performs full range of motion on a client's extremities. When putting an ankle through range of motion the nurse must perform:
1. Flexion, extension, and left and right rotation
2. Abduction, flexion, adduction, and extension
3. Pronation, supination, rotation, and extension
4. Dorsiflexion, plantar flexion, eversion, and inversion

236. The physician orders nonweight bearing with crutches for a client with a leg injury. The nurse understands that, before ambulation is begun, the most important activity to facilitate walking with crutches is:
1. Sitting up in a chair to help strengthen back muscles
2. Keeping the unaffected leg in extension and abduction
3. Exercising the triceps, finger flexors, and elbow extensors
4. Using the trapeze frequently to strengthen the biceps muscles

237. The nurse would recognize that the demonstration of crutch walking with a tripod gait was understood when the client places weight on the:
1. Axillary regions
2. Palms of the hands
3. Feet, which are set wide apart
4. Palms of the hands and axillary regions

238 An x-ray film of a client's arm reveals a comminuted fracture of the radial bone. The nurse understands that with a comminuted fracture:
1. There is a break in the skin and the bone is protruding
2. Splintering has occurred on one side and bending on the other
3. The bone has broken into several fragments, but the skin is intact
4. The bone is broken into two parts, and the skin may or may not be broken

239. A client with osteomyelitis of the leg is to have a débridement of the infected bone. When planning for postoperative care the nurse knows that:
1. Frequent range-of-motion exercises are needed
2. Septicemia is a common postoperative complication
3. The client's leg will be immobilized in a cast or splint
4. The client will be allowed out of bed after the first day

240. While performing a physical assessment of a client with gout of the great toe, the nurse should assess for additional tophi (urate deposits) on the:
1. Chin
2. Ears
3. Buttocks
4. Abdomen

241. When teaching about the dietary control of gout, the nurse is aware that the dietary teaching was understood when the client states; "I will avoid eating:
1. Eggs."
2. Shellfish."
3. Fried poultry."
4. Cottage cheese."

242. An elderly female client is experiencing frequency and uses the bathroom often during the night. One night while attempting to go to the bathroom without assistance, she develops severe pain and is found to have a vertebral compression fracture. The nurse recognizes that this is a:
1. Collapse of vertebral bodies
2. Demineralization of the spine
3. Wear and tear of the spinous processes
4. Bulging of the spinal cord from the vertebra

243. A client has painful swelling of multiple joints, and a tentative diagnosis of rheumatoid arthritis is made. During a subsequent visit the client tells the nurse, "I'm so confused. The doctor said I probably have arthritis, but my lab tests were negative. I don't see how that can be when I'm always so uncomfortable." The nurse's best response would be:
1. "It might help if you try not to think about your discomfort."
2. "Don't let that upset you; eventually the tests will be positive."
3. "Laboratory tests are often negative in the early stages of arthritis."
4. "Did the doctor say whether the laboratory tests were going to be repeated?"

244. A client with rheumatoid arthritis asks the nurse about ways to decrease morning stiffness. The nurse should suggest:
1. Wearing loose but warm clothing
2. Avoiding excessive physical stress and fatigue
3. Taking a hot tub bath or shower in the morning
4. Planning a rest break periodically for about 15 minutes

245. When assessing a client experiencing an acute episode of rheumatoid arthritis, the nurse observes that the client's finger joints are swollen. The nurse understands that this swelling is most likely related to:
1. Urate crystals in the synovial tissue
2. Inflammation in the joint's synovial lining
3. Formation of bony spurs on the joint surfaces
4. Escaped fluid from the capillaries, increasing interstitial fluids

246. As an acute episode of rheumatoid arthritis subsides, active and passive range-of-motion exercises are taught to the client's spouse. The nurse should encourage that direct pressure not be applied to the client's joints because this may precipitate:
1. Pain
2. Swelling
3. Nodule formation
4. Tophaceous deposits

247. The physician orders bed rest for a client with acute arthritis who has bilateral, painful, swollen knee and wrist joints. To prevent flexion deformities during the acute phase, the client's positioning schedule should include placement in the:
1. Sims' position
2. Prone position
3. Contour position
4. Trendelenburg's position

248. A client with acute rheumatoid arthritis improves, and the physician changes the order for bed rest to out of bed as tolerated. This client should be assisted out of bed to a:
1. Low soft lounge chair
2. Straight-back arm chair
3. Wheelchair with foot rests
4. Recliner chair with both legs elevated

249. After a painful exacerbation of rheumatoid arthritis, a client is to begin a walking and exercise program. An appropriate outcome would be that the client:
1. Avoids exercising when there is some discomfort
2. Is pain-free while engaging in activity program
3. Walks and exercises even when the pain is severe
4. Exercises unless the discomfort becomes too great

250. The nurse teaches a client with rheumatoid arthritis techniques to reduce stress to the joints. The statement by the nurse that best describes a technique to reduce joint stress would be:
1. "Respond to pain in your joints."
2. "Use your smaller muscles most frequently."
3. "Do your heavy tasks all at once to reduce muscle strain."
4. "If your joints are warm or swollen, increase exercise to reduce swelling."

251. A client with lower back pain is tentatively diagnosed as having a herniated intervertebral disc. When assessing this client's back pain, the nurse should ask:
1. "Is there any tenderness in the calf of your leg?"
2. "Have you had any burning sensation on urination?"
3. "Do you have any increase in pain during bowel movements?"
4. "Does the pain begin in your flank and move around to the groin?"

252. A client who is diagnosed as having a herniated intervertebral disc complains of pain. The nurse recognizes that the pain is caused by the:
1. The inflammation of the lamina of the involved vertebrae
2. Shifting of two adjacent vertebral bodies out of alignment
3. Compression of the spinal cord by the extruded nucleus pulposus
4. Increased pressure of cerebrospinal fluid within the vertebral column

253. A client is awaiting surgery for a ruptured lumbar nucleus pulposus. The nurse's teaching should in-

clude that the pain will most likely increase if the client:
1. Lies on the side
2. Flexes the knees
3. Sits for long periods
4. Coughs excessively

254. A male client who develops degenerative joint disease of the vertebral column is taught to turn himself from his back to his side, keeping his spine straight. The least effort will be exerted if he does this by crossing his arm over his chest and:
1. Pulling himself to one side by using his night table
2. Bending his top knee to the side to which he is turning
3. Crossing his ankles and turning with both his legs straight
4. Flexing his bottom knee to the side to which he wishes to turn

255. A client is scheduled for a laminectomy. Preoperatively the nurse should demonstrate the:
1. Use of a trapeze
2. Contour position
3. Traction apparatus
4. Logrolling technique

256. After a laminectomy, the nurse should monitor the client's vital signs and:
1. Logroll the client to the prone position
2. Check circulation and sensation of the feet
3. Observe bowel movements and voiding patterns
4. Encourage the client to drink a moderate amount of fluid

257. When two nurses are getting a client out of bed for the first time after a laminectomy, the client complains of feeling faint and light-headed. The nurses should have this client:
1. Slide slowly to the floor, to prevent a fall and injury
2. Sit on the edge of the bed and hold the client upright
3. Bend forward, because it will increase the blood flow to the brain
4. Lie down immediately, so they can take the client's blood pressure

258. When preparing a client for discharge after a laminectomy, the nurse would know that further health teaching is necessary when the client says, "I should:
1. Sleep on a firm mattress to support my back."
2. Spend most of the day sitting in a straight-back chair."
3. Put a pillow under my legs when sleeping on my back."
4. Avoid lifting heavy objects until the doctor tells me I can."

259. A back brace is prescribed for a client who has had a laminectomy. The nurse should include in the teaching plan instructions to:
1. Use the brace when the back feels tired
2. Apply the brace before getting out of bed
3. Put the brace on while in the sitting position
4. Wear the brace when performing twisting exercises

260. After an amputation of a limb, a client begins to experience extreme discomfort in the area where the limb once was. An appropriate nursing diagnosis at this time would be:
1. Ineffective coping related to surgery
2. Pain related to phantom phenomenon
3. Hopelessness related to altered life-style
4. Sensory-perceptual alteration related to bed rest

261. It has been 3 days since a client's surgical above-the-knee amputation. To prevent contractures it would be most appropriate for the nurse to place the client:
1. Lying on the abdomen
2. Sitting in the high Fowler's position
3. Supine with a pillow under the stump
4. Supine with pillows between the thighs

262. A 24-year-old college student had a right above-the-knee amputation secondary to trauma sustained in a motor vehicle accident. Six days after surgery the client falls while attempting to transfer to a chair unaided. A fall related to an amputation is most likely the result of:
1. A loss of balance
2. Phantom limb pain
3. Orthostatic hypotension
4. Decreased muscle strength

263. A client with an above-the-knee amputation asks why the stump needs to be wrapped with an elastic bandage. The nurse explains that it:
1. Decreases phantom limb pain
2. Limits the formation of blood clots
3. Prevents hemorrhage and covers the incision
4. Supports the soft tissue and minimizes edema

264. A client has an above-the-knee amputation because of a gangrenous leg ulcer. After the second postoperative day, to prevent deformities the nurse should:
1. Keep the client's stump elevated on a pillow
2. Place an abduction pillow between the client's legs
3. Encourage client to lie in the supine or prone position
4. Teach the client to press the stump against a hard surface

265. The nurse should teach a client with an above-the-knee amputation a variety of postoperative activities. The activity designed to aid in the use of crutches is:
 1. Stump care
 2. Weight lifting
 3. Changing bed position
 4. Phantom-limb exercise

266. The nurse is assisting a client with a full leg cast to use crutches. The nurse should interrupt the activity when the client demonstrates:
 1. A pulse of 100 and deep respirations
 2. Flushed skin and slowed respirations
 3. Profuse diaphoresis and rapid respirations
 4. A blood pressure of 130/90 and shallow respirations

267. A client requires a below-the-knee amputation. A major advantage of an immediate postoperative prosthesis is that it:
 1. Decreases phantom limb sensations
 2. Encourages a normal walking pattern
 3. Reduces the incidence of wound infection
 4. Allows for the fitting of the prosthesis before discharge

268. When assessing a client using a prosthesis following an above-the-knee amputation, the finding that indicates that the prosthesis fits correctly is:
 1. Shrinking of the stump
 2. Absence of phantom-limb pain
 3. Darkened skin areas on the stump
 4. Uneven wearing down of the heels

269. At the scene of an accident, the nurse can minimize the immediate life-threatening systemic complication of injury to the long bones of the injured person by:
 1. Elevating the affected limb
 2. Maintaining functional alignment
 3. Handling and transporting gently
 4. Encouraging deep breathing and coughing

270. An elderly client is admitted after falling and fracturing a hip. The physician applies Buck's extension until surgery can be performed to replace the head of the femur with a prosthesis. When checking the client's Buck's extension, the nurse should be aware that:
 1. The spreader bar should fit snugly around the foot
 2. Weights greater than 8 pounds will cause skin damage
 3. The Buck's boot should extend 3 inches above the ankle
 4. Tape must cover the malleoli to adequately secure the weights to the leg

271. After a client with multiple fractures of the left femur is admitted to the hospital for surgery, the client demonstrates cyanosis, tachycardia, dyspnea, and restlessness. Initially the nurse should:
 1. Administer oxygen by mask
 2. Immediately call the physician
 3. Place the client in the supine position
 4. Place the client in the high-Fowler's position

272. After surgery for a fractured hip a client complains of pain. The nurse should:
 1. Notify the physician
 2. Use distraction techniques
 3. Medicate the client as ordered
 4. Perform a complete pain assessment

273. A client has an open reduction and internal fixation of a fractured hip. To prevent the most common complication following this type of surgery, the nurse should expect the physician's order to state:
 1. "Turn from side to side q2h"
 2. "Apply sequential compression stockings"
 3. "Encourage isometric exercises to the extremities"
 4. "Perform passive range of motion to the affected extremity"

274. A client is admitted to the hospital for a total hip replacement. The nurse's preoperative teaching plan for the early postoperative period should include instructions related to:
 1. Abduction of the operative hip
 2. Adduction of the operative hip
 3. Turning 45 degrees onto the operative side
 4. Hip flexion of 90 degrees on the operative side

275. A client returns from surgery with a hip prosthesis. An abductor splint is in place. The nurse should remove the splint:
 1. When the client gets up in a chair
 2. When the client needs a change of position
 3. Once the client's operative pain has ceased
 4. To administer skin care and physical therapy

276. When assisting a client who has had a total hip replacement onto the bedpan on the first postoperative day, the nurse should instruct the client to:
 1. Turn toward the operative side
 2. Flex both knees and slowly lift the pelvis
 3. Extend both legs and pull on the trapeze to lift the pelvis
 4. Flex the unoperative knee and pull on the trapeze to lift the pelvis

277. A client with a distal femoral shaft fracture is at risk for developing a fat embolus. A distinguishing sign that is unique to a fat embolus is:
1. Oliguria
2. Dyspnea
3. Confusion
4. Petechiae

278. Discharge instructions by the nurse to a client who has had a left total hip replacement should include instructions that, when sitting, the client should use a:
1. Soft armchair with left leg straight out in front
2. Firm armchair with left leg elevated on a stool
3. Firm high chair with left foot flat on the floor's surface
4. Soft chair with enough pillows to keep the hip at a right angle

279. Following a cervical neck injury a client is placed in a halo brace with a body cast. A statement by the client that indicates that the nursing diagnosis, body image disturbance, has been successfully resolved is:
1. "I just hate having everyone else do things for me."
2. "I've been staying under 800 calories daily most of the time."
3. "I've gotten used to the brace. I may even miss it when it's gone."
4. "I can't get to sleep, but I make up for it in the morning by sleeping later."

280. A client with a distal femoral fracture is placed in skeletal traction. The nurse is aware that the weights would only be removed if:
1. There is a life-threatening situation
2. The client complains of intense pain
3. There is evidence of external rotation
4. The cords have become twisted during turning

NEUROMUSCULAR (NM)

281. When assessing a client's vagal nerve (cranial nerve X) function, the nurse will need:
1. A tuning fork
2. Tongue depressors
3. An ophthalmoscope
4. Cotton and a safety pin

282. When assessing trigeminal nerve function the nurse should evaluate:

1. Corneal sensation
2. Smiling and frowning
3. Ocular muscle movement
4. Shrugging of the shoulders

283. A client, recently diagnosed with Bell's palsy, has many questions about the course of the disease. The nurse should explain that:
1. Most clients recover within 3 to 5 weeks
2. Pain occurs with transient ischemic attacks
3. Body changes will occur with residual effects
4. Cool compresses decrease facial involvement

284. The nurse explains to a client with trigeminal neuralgia that a treatment that is effective on a temporary (6 to 18 months) basis is:
1. An intravenous injection of cobra venom
2. A lidocaine injection of the ventral root of the eleventh spinal nerve
3. Microvascular decompression of the blood vessels at the nerve root
4. An alcohol injection of the peripheral branch of the fifth cranial nerve

285. A client with pain and paresis of the left leg is scheduled for electromyography. Before the test, the nurse should explain that:
1. The involved area will be shaved immediately before testing
2. The client's heart rate and rhythm will be monitored frequently
3. Needles will be inserted into the affected muscles during the test
4. The client will be kept in a recumbent position after the procedure

286. A young female client goes to the physician because she has been experiencing fatigue and double vision. The physician suspects myasthenia gravis. When obtaining information from the client, the nurse would expect her to report that:
1. Her level of fatigue has been constant
2. The longer she rests, the weaker she feels
3. Her strength increases with progressive activity
4. The symptoms seem more severe in the evening

287. The basis of the nursing care plan for a client with myasthenia gravis is the fact that:
1. Muscle weakness decreases with hot baths
2. Muscle weakness decreases with muscle use
3. Muscle strength improves immediately after meals
4. Muscle strength decreases with repeated muscle use

288. When assisting a client who has myasthenia gravis with a bath, the nurse notices that the client's arms become weaker with sustained movement. The nurse should:
    1. Continue the bath while supporting the client's arms
    2. Encourage the client to rest for short periods of time
    3. Gradually increase the client's activity level each day
    4. Administer a dose of pyridostigmine bromide (Mestinon)

289. A client with myasthenia gravis comes to the neurology clinic at 4 pm for a routine visit. During an assessment the nurse should expect the client to report:
    1. Blurred vision and episodes of vertigo
    2. Tremors of the hands when attempting to lift objects
    3. Partial improvement of muscle strength with mild exercise
    4. Involvement of the distal muscles rather than the proximal muscles

290. During a routine clinic visit of a female client who has myasthenia gravis, the nurse reinforces previous teaching about the disease and self-care. The nurse would evaluate that the teaching was effective when the client recognizes that she should:
    1. Plan activities for later in the day
    2. Avoid people with respiratory infections
    3. Eat meals in a semi-recumbent position
    4. Take muscle relaxants when she is under stress

291. A client is scheduled to have a series of diagnostic studies for myasthenia gravis, including a Tensilon test. The nurse should explain to the client that the diagnosis of myasthenia gravis will be confirmed if the administration of Tensilon produces a:
    1. Brief exaggeration of symptoms
    2. Prolonged symptomatic improvement
    3. Rapid but brief symptomatic improvement
    4. Symptomatic improvement of just the ptosis

292. The physician has ordered a diagnostic workup for a client who may have myasthenia gravis. The initial nursing goal for the client during the diagnostic phase would be that, "The client will:
    1. Adhere to a teaching plan."
    2. Achieve psychologic adjustment."
    3. Maintain present muscle strength."
    4. Prepare for the appearance of myasthenic crisis."

293. The most significant initial nursing assessment associated with myasthenia gravis includes the client's:

1. Ability to chew and speak distinctly
2. Degree of anxiety about the diagnosis
3. Ability to smile and to open the eyelids
4. Respiratory exchange and ability to swallow

294. A client with myasthenia gravis begins to experience increased difficulty in swallowing. To prevent aspiration of food, the most effective nursing action would be to:
    1. Change client's diet order from soft to clear liquid
    2. Place an emergency tracheostomy set in client's room
    3. Assess the client's respiratory status before and after meals
    4. Coordinate meals with the peak effect of pyridostigmine bromide (Mestinon)

295. A client with myasthenia gravis is scheduled for a thymectomy. When preparing the client for surgery, the nurse should emphasize:
    1. A detailed explanation of the procedure
    2. The experimental nature of this procedure in myasthenia
    3. The difficulty of predicting the degree of improvement after the operation
    4. An explanation of the usual postoperative complications of a thymectomy

296. The nurse identifies that a client exhibits the characteristic gait associated with Parkinson's disease. When recording on the client's chart, the nurse should describe this gait as:
    1. Ataxic
    2. Spastic
    3. Shuffling
    4. Scissoring

297. While performing the history and physical examination of a client with Parkinson's disease, the nurse should assess the client for:
    1. Frequent bouts of diarrhea
    2. Hyperextension of the neck
    3. A low-pitched, monotonous voice
    4. A recent increase in appetite and weight gain

298. When assessing a client with Parkinson's disease, a common adaptation the nurse would expect to find is:
    1. Leaning toward the affected side
    2. Blank facies or lack of expression
    3. Tremors of the hand on movement
    4. Hyperextension of the affected extremity

299. While assessing a client with Parkinson's disease, the nurse identifies bradykinesia when the client exhibits:
    1. Muscle flaccidity

2. An intention tremor
3. Paralysis of the limbs
4. A lack of spontaneous movement

300. The nurse might expect a client with multiple sclerosis to complain about the most common initial symptom, which is:
1. Diarrhea
2. Headaches
3. Skin infections
4. Visual disturbances

301. A client with multiple sclerosis is informed that it is a chronic progressive neurologic condition. The client asks the nurse, "Will I experience pain?" The nurse's best response would be:
1. "Tell me about your fears regarding pain."
2. "Analgesics will be ordered to control the pain."
3. "Let's make a list of the things you need to ask your physician."
4. "Pain is not a characteristic symptom of this disease process."

302. A 28-year-old woman has known for the past 6 years that she has multiple sclerosis. She has two children, one of whom is an active 2½-year-old. The client is currently in remission. At the present time, it would be most important for the nurse to encourage the client to:
1. Schedule periodic quality time with her child
2. Provide support to other people with multiple sclerosis
3. Develop a flexible schedule for completion of routine daily activities
4. Meet with the psychotherapy group for people with multiple sclerosis

303. A client asks for an explanation about glaucoma. The nurse explains that with glaucoma there is:
1. An increase in the pressure within the eyeball
2. An opacity of the crystalline lens or its capsule
3. A curvature of the cornea that becomes unequal
4. A separation of the neural retina from the pigment retina

304. When obtaining the nursing history from a client who has open-angle (chronic) glaucoma, a complaint that the nurse should expect is:
1. Flashes of light
2. Intolerance to light
3. Seeing floating specks
4. Loss of peripheral vision

305. A client who has open-angle (chronic) glaucoma is scheduled for eye surgery to promote aqueous humor outflow. The nurse would know that the client understands the preoperative teaching about the first 24 hours after surgery when the client states, "I should:
1. Cough and deep breathe."
2. Lie on my unaffected side."
3. Move around freely in bed."
4. Elevate the head of my bed."

306. Preoperative teaching for a client who is to have an extracapsular cataract extraction with an intraocular lens implant should include the importance of:
1. Remaining in bed for 24 hours
2. Breathing and coughing deeply
3. Avoiding bending from the waist
4. Lying in the supine position for 12 hours

307. After cataract surgery a client complains of feeling nauseated. The nurse should:
1. Give the client some dry crackers to eat
2. Administer the antiemetic drug as ordered
3. Explain that this is expected following surgery
4. Instruct the client to deep breathe until the nausea subsides

308. A client is being prepared for discharge from an ambulatory surgical unit following a cataract removal with an intraocular lens implant. The statement by the client that suggests to the nurse that discharge teaching was effective is:
1. "I'm driving home since I feel so good."
2. "I can't wait until I get home to wash my hair."
3. "I can expect to see bright flashes of light for awhile."
4. "I'll call the surgeon if the analgesic doesn't relieve the pain."

309. After cataract surgery, a client is taught how to self-administer eye drops before discharge. The nurse approves the technique when the client:
1. Holds the dropper tip above the eye
2. Places the drops on the cornea of the eye
3. Raises the upper eyelid with gentle traction
4. Squeezes the eye shut after instilling the eye drops

310. After an automobile accident, a client complains of seeing frequent flashes of light. The nurse should suspect:
1. Scleroderma
2. Acute glaucoma
3. A detached retina
4. A cerebral concussion

311. A client with a detached retina is scheduled for surgery to reattach the retina. The nurse explains that the procedure employed involves the use of:
1. Radiation
2. Burr holes
3. Dermabrasion
4. Laser technique

312. A construction worker falls off the roof of a two-story building and is taken to the hospital unconscious. The nurse would be most concerned if the initial assessment reveals:
1. Reactive pupils
2. A depressed fontanel
3. Bleeding from the ears
4. An elevated temperature

313. During the immediate post-trauma period after injury to the frontal lobe of the brain, the nurse should place a client in the:
1. Supine position
2. Side-lying position
3. Low-Fowler's position
4. Trendelenburg position

314. A client has had spinal anesthesia for surgery. On the second day after surgery the client complains of a headache. The nurse should:
1. Begin an early ambulation program
2. Supply the client with several containers of juice
3. Assist the client to sit at the bedside and dangle the feet
4. Remove any elastic antiembolism stockings being worn

315. A 26-year-old, admitted with the diagnosis of subarachnoid hemorrhage, exhibits aphasia and hemiparesis. These neurologic deficits, which may be present immediately after a subarachnoid hemorrhage, are primarily caused by:
1. Blood loss
2. Tissue death
3. Vascular spasms
4. Electrolyte imbalance

316. The nurse should plan to position a client who has experienced a subarachnoid hemorrhage:
1. In the supine position
2. On the unaffected side
3. With the head of the bed elevated
4. With sandbags on either side of the head

317. Following an anterior fossa craniotomy, a client is placed on controlled mechanical ventilation. To ensure adequate cerebral blood flow the nurse should:
1. Clear the ear of draining fluid
2. Monitor the serum carbon dioxide

3. Discontinue anticonvulsant therapy
4. Elevate the head of the bed 30 degrees

318. A client has a supratentorial craniotomy for a tumor in the right frontal lobe of the cerebral cortex. Postoperatively, the position that would be most appropriate for this client would be:
1. High-Fowler's with knee gatch raised
2. Flat with a small pillow under the nape of the neck
3. Head of the bed elevated 20 degrees with the head turned to the operative side
4. Head of the bed elevated 45 degrees with a large pillow under the head and shoulders

319. A client undergoes a supratentorial craniotomy with burr holes after sustaining a head injury in an automobile accident. Because of the presence of burr holes, the client is at risk for developing an infection. An early clinical manifestation of meningeal irritation is:
1. Sunset eyes
2. Kernig's sign
3. Homans' sign
4. The plantar reflex

320. Following 3 months of rehabilitation after a craniotomy, a female client is still having some motor speech difficulty. To promote the client's use of speech the nurse should:
1. Correct her mistakes immediately
2. Respond to her crude efforts of speaking
3. Reexplain why she is having difficulty speaking
4. Speak to her in simple words and short sentences

321. A client with a history of hypertension is admitted to the hospital with aphasia. A bruit is heard over the left carotid artery, the pulse is irregular, and there is a pulse deficit. The nurse is aware that complete occlusion of the branches of the middle cerebral artery resulting in aphasia may occur because of:
1. A history of hypertensive disease
2. Emboli associated with atrial fibrillation
3. Developmental defects of the arterial wall
4. Inappropriate paroxysmal neural discharge

322. A client has a history of progressive carotid and cerebral atherosclerosis and transient ischemic attacks (TIAs). The nurse understands that TIAs are:
1. Temporary episodes of neurologic dysfunction
2. Transient attacks caused by multiple small emboli
3. Periods of alternating exacerbations and remissions
4. Ischemic attacks that result in progressive neurologic deterioration

323. A client has carotid atherosclerotic plaques, and a right carotid endarterectomy is performed. Two hours after surgery the client demonstrates progressive hypotension. The nurse should:
    1. Increase the IV flow rate
    2. Raise the head of the bed
    3. Notify the physician immediately
    4. Put the client in a slight Trendelenburg's position

324. After a carotid endarterectomy, the client should be monitored for the complication of cranial nerve dysfunction. To monitor for this complication, the nurse should assess the client for:
    1. Labored breathing
    2. Edema of the neck
    3. Difficulty in swallowing
    4. Alteration in blood pressure

325. A client is admitted to the hospital with weakness in the right extremities and a slight speech problem. Vital signs are normal. During the first 24 hours, the nurse should give priority to:
    1. Checking the client's temperature
    2. Evaluating the client's motor status
    3. Monitoring the client's blood pressure
    4. Obtaining a urine specimen from the client

326. On the evening before discharge from the hospital, a client has a hypertensive crisis and a cerebrovascular accident. Initially the nurse should place the client in a:
    1. Supine position
    2. Contour position
    3. Side-lying position
    4. Slight Trendelenburg's position

327. Initially after a cerebrovascular accident, a client's pupils are equal and reactive to light. Later the nurse assesses that the right pupil is reacting more slowly than the left and the systolic blood pressure is beginning to rise. The nurse recognizes that these adaptations are suggestive of:
    1. Spinal shock
    2. Hypovolemic shock
    3. Transtentorial herniation
    4. Increasing intracranial pressure

328. A client who has had a cerebrovascular accident is admitted to the hospital with right-sided hemiplegia. It is important for the nurse to consider any restrictions of mobility or neuromuscular abnormalities that are observed because:
    1. Shortening and eventual fibrosis of the muscles will occur
    2. Hypertrophy of the muscles will eventually result from disuse

    3. Rigid extension can occur, making therapy painful and difficult
    4. Decreased movement on the affected side predisposes the client to infection

329. A female client manifests right-sided hemianopia as a result of a cerebrovascular accident. The nurse should:
    1. Correct the client's misuse of equipment
    2. Instruct the client to scan her surroundings
    3. Provide tactile stimulation to the client's affected extremities
    4. Teach the client to look at the position of her left extremities

330. The wife of a client who has had a cerebrovascular accident tells the home health nurse that her husband cries easily and without provocation. She asks why he is so emotionally labile. The nurse should explain that:
    1. Her husband can remember only depressing events from the past
    2. This is a way of getting attention, and the behavior should be ignored
    3. Her husband feels guilty about the demands he is making on his family
    4. This behavior is a common response over which he has very little control

331. The husband of a client with aphasia as a result of a cerebrovascular accident asks whether his wife's speech will ever return. The nurse should respond:
    1. "You will have to ask your physician."
    2. "It should return to normal in 2 or 3 months."
    3. "It is hard to say how much improvement will occur."
    4. "This will probably be the extent of her speech from now on."

332. When assisting the family to help an aphasic member regain as much speech function as possible, the nurse should instruct them to:
    1. Speak louder than usual to the client during visits
    2. Tell the client to use the correct words when speaking
    3. Give positive reinforcement for correct communication
    4. Encourage the client to speak while being patient with all attempts

333. A client, employed as a carpenter, has trouble holding tools because of carpal tunnel syndrome. Because the client continues to work, the nursing diagnosis of most concern would be:
    1. Pain
    2. Anxiety
    3. Risk for injury
    4. Self-esteem disturbance

334. A client has a generalized seizure. The first intervention by the nurse during the tonic-clonic phase of the seizure should be to:
1. Call the client's physician
2. Protect the client's head from injury
3. Note the condition of the client's pupils
4. Check the client's pulse and respirations

335. A client sustains a vertebral fracture at the T1 level as a result of diving into shallow water. On admission to the emergency room a detailed neurologic assessment is performed. The nurse should expect to find:
1. Inability to move the lower arm
2. Normal biceps reflexes in the arm
3. Loss of pain sensation in the hands
4. Difficulty breathing and a flaccid diaphragm

336. A client who sustained a spinal cord injury at the T2 level should be assessed for signs of autonomic dysreflexia because:
1. The injury has resulted in loss of all reflexes
2. The injury is above the sixth thoracic vertebra
3. There has been a partial transection of the cord
4. There is a flaccid paralysis of the lower extremities

337. The nurse is aware that a client with a spinal cord injury is developing autonomic dysreflexia when the client has:
1. Flaccid paralysis and numbness
2. Absence of sweating and pyrexia
3. Escalating tachycardia and shock
4. Paroxysmal hypertension and bradycardia

338. During the first week after a spinal cord injury at the T3 level, a male client and the nurse identify a short-term goal. An appropriate short-term goal for this client would be, "The client will:
1. Understand his limitations."
2. Consider alternate life-styles."
3. Perform independent ambulation."
4. Carry out personal hygiene activities."

339. A client who is recuperating from a spinal cord injury at the T4 level wants to use a wheelchair. In preparation for this activity the client should be taught:
1. Leg lifts to prevent hip contractures
2. Push-ups to strengthen arm muscles
3. Balancing exercises to promote equilibrium
4. Quadriceps-setting exercises to maintain muscle tone

340. A client whose vertebral column at the level of T6 and T7 was completely crushed and whose left leg was traumatically amputated above the knee is admitted to the ICU. When performing an assessment the nurse would expect to find that the client was experiencing:
1. Difficulty with breathing
2. Pain in the stump of the left leg
3. Pain at the level of compression
4. Spastic paralysis of the arms and legs

ENDOCRINE (EN)

341. When obtaining a health history from a client recently diagnosed with type 2 diabetes, the nurse should expect the client to mention symptoms associated with the classic signs of diabetes, such as:
1. Polydipsia, polyuria, irritability
2. Polyphagia, confusion, polyuria
3. Polydipsia, polyphagia, polyuria
4. Polydipsia, nocturia, weight loss

342. A client has recently been diagnosed with type 1 diabetes. A glucose tolerance test is ordered. The order reads, "Administer glucose 1.0 g/kg." The client weighs 240 pounds. The nurse should administer:
1. 109 grams
2. 120 grams
3. 240 grams
4. 528 grams

343. A 50-year-old male with a history of diabetes has had progressive problems with venous stasis. He tells the nurse in the clinic that he bumped his leg a week ago and now has an open draining area just above the ankle. When exploring this client's health history, the most important assessment by the nurse would be to ascertain:
1. Whether he uses Humulin N insulin or Humulin R insulin
2. How many times a day the client voids and how often he moves his bowels
3. The type of treatment the client is receiving and how he has managed his care
4. How many children the client has and whether they are having similar problems

344. When assessing the laboratory values of a client with type 2 diabetes, the nurse should expect the results to reveal:
1. Ketones in the blood but not the urine
2. Glucose in the urine but not in the blood
3. Urine negative for ketones and glucose in the blood
4. Urine and blood positive for both glucose and ketones

345. The nurse should explain to a client with diabetes that self-monitoring of blood glucose is preferred to urine glucose testing because it is:
    1. More accurate
    2. Easier to perform
    3. Done by the client
    4. Not influenced by drugs

346. A client is diagnosed as having type 2 diabetes. A priority teaching goal would be, "The client will be able to:
    1. Perform foot care."
    2. Administer insulin."
    3. Test urine for sugar and acetone."
    4. Identify hypoglycemia/hyperglycemia."

347. The nurse teaches a client with type 2 diabetes how to provide self-care to prevent infections of the feet. The nurse recognizes that the teaching was effective when the client says, "I should:
    1. Massage my feet and legs with oil or lotion."
    2. Apply heat intermittently to my feet and legs."
    3. Eat foods high in protein and carbohydrate kilocalories."
    4. Control my diabetes through diet, exercise, and medication."

348. A client, newly diagnosed as having type 1 diabetes, is encouraged to exercise on a regular basis primarily because exercise has been shown to:
    1. Decrease insulin sensitivity
    2. Stimulate glucagon production
    3. Reduce cellular requirements for glucose
    4. Improve cell uptake of glucose without insulin

349. A client with type 2 diabetes is taking 1 glyburide (Micronase) tablet daily. The client asks whether an extra pill should be taken before exercise. The nurse should reply:
    1. "You will need to decrease your exercise."
    2. "An extra pill will help your body use glucose correctly."
    3. "Your diet and medicine will not be affected by exercise."
    4. "No, but observe for signs of hypoglycemia while exercising."

350. A client who is taking an oral hypoglycemic daily for type 2 diabetes develops the flu and is concerned about the need for special care. The nurse should advise the client to:
    1. Skip the oral hypoglycemic pill, drink plenty of fluids, and stay in bed

    2. Avoid food, drink clear liquids, take a daily temperature, and stay in bed
    3. Eat as much as possible, increase fluid intake, and call the office again the next day
    4. Take the oral hypoglycemic pill, drink warm fluids, and perform a serum glucose test ac and hs

351. An obese client with type 2 diabetes asks about the use of alcohol or special "dietetic" food in the diet. The client should be taught that:
    1. Alcohol can be used, with its calories accounted for in the diet
    2. Unlimited amounts of sugar substitutes can be used as desired
    3. Alcohol should not be used in cooking because it adds too many calories
    4. Special "dietetic" foods are needed because many regular foods cannot be used

352. A client with type 2 diabetes travels frequently and asks how to plan meals during trips. The nurse's most appropriate response would be:
    1. "You can order diabetic foods on most airlines and in restaurants."
    2. "You should plan your food ahead and carry it with you from home."
    3. "Make regular food choices, wherever you are, following your food plan."
    4. "You can monitor your blood glucose level frequently and can eat accordingly."

353. A client with newly diagnosed diabetes indicates a hatred for asparagus, broccoli, and mushrooms. When reviewing the exchange list with the client, the nurse would know that the teaching about the exchange list was understood when the client states, "Instead of these foods I can eat:
    1. String beans, beets, or carrots."
    2. Corn, lima beans, or dried peas."
    3. Baked beans, potatoes, or parsnips."
    4. Corn muffins, corn chips, or pretzels."

354. While hospitalized, a client with diabetes is observed picking at calluses on the feet. The nurse should immediately:
    1. Warn the client of the danger of infection
    2. Suggest that the client wear white cotton socks
    3. Check the client's shoes for proper fit in the area of the calluses
    4. Demonstrate and teach the importance of proper foot care to the client

355. Following a surgical procedure for cancer of the pancreas that included the removal of the stomach, the head of the pancreas, the distal end of the duodenum, and the spleen, the postoperative manifestation by the client that would require immediate attention by the nurse would be:
    1. Jaundice
    2. Indigestion
    3. Weight loss
    4. Hyperglycemia

356. Four hours after surgery, the blood glucose level of a client who has type 1 diabetes is elevated. The nurse should expect to:
    1. Administer an oral hypoglycemic
    2. Institute urine glucose monitoring
    3. Give supplemental doses of regular insulin
    4. Decrease the rate of the intravenous infusion

357. A client who has type 1 diabetes is admitted to the hospital for major surgery. Prior to surgery the client's insulin requirements are elevated but well controlled. Postoperatively the nurse would anticipate that the client's insulin requirements will:
    1. Fluctuate widely
    2. Increase sharply
    3. Remain elevated
    4. Decrease immediately

358. A client is admitted to the hospital with diabetic ketoacidosis. The nurse understands that the elevated ketone level present with this disorder is caused by the incomplete oxidation of:
    1. Fats
    2. Protein
    3. Potassium
    4. Carbohydrates

359. The serum potassium level of a client who has diabetic ketoacidosis is 5.4 mEq/L. When monitoring the ECG tracing, the nurse would expect to observe:
    1. Abnormal P waves and depressed T waves
    2. Peaked T waves and widened QRS complexes
    3. Abnormal Q waves and prolonged ST segments
    4. Peaked P waves and increased number of T waves

360. A client with type 1 diabetes is placed on an insulin pump. The most appropriate short-term goal in teaching this client to control the diabetes is: "The client will:
    1. Adhere to the medical regimen."
    2. Remain normoglycemic for 3 weeks."
    3. Demonstrate the correct use of the insulin pump."
    4. List three self-care activities necessary to control the diabetes."

361. When the nurse plans to teach a client with type 1 diabetes about the use of an insulin pump, it is of major importance that the client understand:
    1. That the needle needs to be changed every 24 hours
    2. That glucose monitoring will only be necessary once daily
    3. That the pump is an attempt to mimic the way a healthy pancreas works
    4. That the pump will be implanted in a subcutaneous pocket near the abdomen

362. For proper foot care, the nurse should provide a client with type 2 diabetes with instructions to:
    1. Remove all corns and stop smoking
    2. Always wear shoes and use natural fiber socks
    3. Wear nylon socks and wash feet in warm water
    4. Wear shoes that are slightly larger and avoid using corn removers

363. A client with type 1 diabetes of long duration takes Humulin N and Humulin R insulin every morning. At noon, before eating lunch, the client is admitted to the emergency department with an acute myocardial infarction. Two hours later, the client's serum glucose level drops to 30 mg/dl, and insulin coma is diagnosed. The nurse understands the reason for the development of acute hypoglycemia in this client is that:
    1. Glycogenolysis increased when lunch was not eaten after taking Humulin N insulin
    2. The stress brought on by the chest pain increases the use of serum glucose available to the client
    3. Glucose levels that are controlled by insulin drop more quickly than those controlled by oral antidiabetics
    4. The client's body became sensitive to the prescribed dose of insulin after long use, and the blood glucose level dropped erratically

364. A client with type 1 diabetes of several years' duration takes 40 U of Humulin N insulin and 20 U of Humulin R insulin each morning. The client's serum glucose level averages about 130 mg/dl. When the client complains of symptoms of hypoglycemia, the nurse should suspect that the client's serum glucose level is:
    1. About 100 mg/dl
    2. Between 50 and 70 mg/dl
    3. Between 100 and 120 mg/dl
    4. At least 20 mg/dl below the norm

365. When assisting a client with type 1 diabetes with care, the nurse notes a 5 cm nodule on the upper arm where the client states she has been injecting her insulin at home. The nurse is aware that the nodule, which is neither warm nor painful, is a result of:
    1. Keratosis
    2. An allergy
    3. An infection
    4. Lipodystrophy

366. When assessing a client who is experiencing hypoglycemia, the nurse should expect to find:
    1. Lethargy
    2. Tachycardia
    3. Warm, dry skin
    4. Increased respirations

367. A client with diabetic ketoacidosis who is receiving intravenous fluids and insulin complains of tingling and numbness of the fingers and toes and shortness of breath. The cardiac monitor shows the appearance of a U wave. The nurse should recognize that these symptoms indicate:
    1. Hypokalemia
    2. Hyponatremia
    3. Hypoglycemia
    4. Hypercalcemia

368. The nurse recognizes that a client with diabetes understands the teaching about the treatment of hypoglycemia when the client says, "If I become hypoglycemic I should initially eat:
    1. Hard candy and fruit juice."
    2. A slice of bread and sugar."
    3. Chocolate candy and a banana."
    4. Peanut butter crackers and a glass of milk."

369. A client with hypertension is scheduled for a medical workup including a scan for an aldosteronoma. The nurse recognizes that this scan is ordered to rule out disease of the:
    1. Thyroid
    2. Kidneys
    3. Pituitary
    4. Adrenals

370. A client who is scheduled to have surgery to remove an aldosterone-secreting adenoma wonders what will happen if he refuses to have the surgery. The nurse would base a response on the fact that:
    1. The tumor must be removed to prevent heart and kidney damage
    2. Surgery will prevent the tumor from metastasizing to other organs
    3. Radiation therapy can be just as effective as surgery if the tumor is small
    4. Chemotherapy is as reliable as surgery to treat adenomas of this type in some cases

371. Late in the postoperative period after the removal of an aldosteronoma the nurse would expect the client's blood pressure to:
    1. Gradually return to near normal levels
    2. Rise quickly above preoperative levels
    3. Fluctuate greatly during this entire period
    4. Drop very low, then rise rapidly to normal levels

372. After an adrenalectomy for an aldosteronoma, a client's wife states, "I hope this is the end of this problem, and my husband will be back to work soon." Based on an understanding of the health problem, the nurse should:
    1. Caution the wife against setting her expectations too high since the outcome for this problem is variable
    2. Explain that her husband's surgery is curative since the other adrenal gland is meeting the body's needs
    3. Advise the wife to investigate other occupational alternatives for her husband if he is planning to return to work
    4. Tell her that although her husband will require hormone supplements for the rest of his life, he should be able to work

373. A female college freshman visits the health center because she feels nervous, irritable, and extremely tired. She complains that, although she eats large amounts of food, she has frequent bouts of diarrhea and is losing weight. The nurse observes a fine hand tremor, an exaggerated reaction to external stimuli, and a wide-eyed expression. The laboratory tests that might be ordered to determine what may be causing this client's symptoms are the:
    1. PTT and PT
    2. T_3, T_4, and TSH
    3. VDRL and CBC
    4. Serum barbiturate levels

374. When assessing a client with Graves' disease, the nurse should expect to find:
    1. Constipation, dry skin, and weight gain
    2. Lethargy, weight gain, and forgetfulness
    3. Weight loss, exophthalmos, and restlessness
    4. Weight loss, protruding eyeballs, and lethargy

375. When assessing a client with Graves' disease (hyperthyroidism) the nurse would expect to find a history of:
    1. Diaphoresis
    2. Menorrhagia
    3. Dry, brittle hair
    4. Sensitivity to cold

376. The nurse teaches a client with exophthalmos how to reduce discomfort and prevent corneal ulceration. The nurse recognizes that the teaching is understood when the client states, "I should:
1. Eliminate excessive blinking."
2. Not move my extraocular muscles."
3. Elevate the head of my bed at night."
4. Avoid using a sleeping mask at night."

377. A client is scheduled to have a thyroidectomy for cancer of the thyroid. When providing preoperative teaching for the postoperative period, the nurse should teach the client to:
1. Cough and deep breathe every 2 hours
2. Perform range-of-motion exercises of the head and neck
3. Support the head with the hands when changing position
4. Apply gentle pressure against the incision when swallowing

378. When planning for a client's return from the operating room after a subtotal thyroidectomy, the nurse should understand that with this surgery:
1. The entire thyroid gland is removed
2. A small part of the gland is left intact
3. One parathyroid gland is also removed
4. A portion of the thyroid and four parathyroids are removed

379. The nurse plans to set up emergency equipment at the bedside of a client in the immediate postoperative period following a thyroidectomy. The nurse should include:
1. A crash cart with bed board
2. A tracheostomy set and oxygen
3. An airway and rebreathing mask
4. Two ampules of sodium bicarbonate

380. Immediately after a subtotal thyroidectomy the nurse plans to assess a female client for unilateral injury of the laryngeal nerve every 30 to 60 minutes by:
1. Checking her throat for swelling
2. Observing her for signs of tetany
3. Asking her to state her name out loud
4. Palpating the side of her neck for blood seepage

381. When planning care for a client who has had a thyroidectomy, the nursing action that should be given highest priority during the first 24 hours postoperatively is:
1. Humidifying the room air continuously
2. Performing range-of-motion neck exercises q4h

3. Checking vital signs every 2 hours after they have stabilized
4. Assessing for hoarseness and voice weakness every 2 hours

382. When planning care for a client in the first 24 hours after a thyroidectomy, the nurse should include:
1. Checking the back and sides of the operative dressing
2. Supporting the head during mild range-of-motion exercises
3. Encouraging the client to ventilate feelings about the surgery
4. Advising the client that normal activities can be resumed immediately

383. After a client has a thyroidectomy the nurse should observe for possible complications. The nurse should be aware that if tingling and numbness of the fingers and toes, muscle twitching, or muscle spasms occur, the client may be:
1. Hypokalemic
2. Hypocalcemic
3. In thyroid crisis
4. In hypovolemic shock

384. Following a thyroidectomy the client exhibits carpopedal spasm and some tremors. The client complains of tingling in the fingers and around the mouth. The nurse should notify the physician and expect to administer:
1. Potassium iodide
2. Calcium gluconate
3. Magnesium sulfate
4. Potassium chloride

385. After a thyroidectomy a client should be observed for the possible complication of thyroid crisis, which would be evidenced by:
1. An increased pulse deficit
2. A decreased blood pressure
3. A decreased pulse rate and respirations
4. An increased temperature and pulse rate

386. Following a thyroidectomy a client is treated with ^{131}I to eradicate residual thyroid tissue. Because of this treatment the nurse should:
1. Maintain standard precautions and limit visitors
2. Consider all discharges including urine and feces to be radioactive
3. Use strict isolation procedures to prevent contamination of the client
4. Provide frequent contact to prevent feelings of abandonment but maintain an appropriate distance

387. When preparing a client for discharge after a thyroidectomy, the nurse should teach the signs of hypothyroidism. The nurse would be aware that the client understands the teaching when the client says, "I should call my physician if I develop:
    1. Dry skin and an intolerance to cold."
    2. Muscle cramping and sluggishness."
    3. Fatigue and an increased pulse rate."
    4. Tachycardia and an increase in weight."

388. A client who has had a subtotal thyroidectomy does not understand how hypothyroidism could develop when the problem was hyperthyroidism. The nurse should base a response on the knowledge that:
    1. Hypothyroidism is a gradual slowing of the body's function
    2. There will be a decrease in pituitary thyroid-stimulating hormone
    3. There is less thyroid tissue to supply thyroid hormone after surgery
    4. Atrophy of tissue remaining after surgery reduces secretion of thyroid hormones

389. Prior to a client's discharge following a thyroidectomy, the nurse teaches the client to observe for signs of surgically induced hypothyroidism. The nurse would know that the teaching was understood when the client states that the physician should be notified if:
    1. Dry skin and fatigue occur
    2. Intolerance to heat is present
    3. Insomnia and excitability develop
    4. Progressive weight loss is evident

RESPIRATORY (RE)

390. The description that should be used for the soft swishing sounds of normal breathing heard when the nurse auscultates a client's chest would be:
    1. Fine crackles
    2. Adventitious sounds
    3. Vesicular breath sounds
    4. Diminished breath sounds

391. The nurse auscultates a client's lungs and notes a fine crackling sound in the left lower lung during respiration. If crackles and rhonchi in the left lower lung were charted on the nurse's notes, the notation would be:
    1. A nursing diagnosis
    2. A correct nursing notation
    3. An inaccurate interpretation
    4. Correct if palpation ruled out crepitus

392. The nurse's physical assessment of a client with congestive heart failure reveals tachypnea and bilateral crackles (rales). The nurse should:
    1. Initiate oxygen therapy
    2. Assess for a pleural friction rub
    3. Obtain a chest x-ray film immediately
    4. Position the client in the Fowler's position

393. The best method to assess a client for stridor in the immediate postoperative period after a radical neck dissection is to:
    1. Listen with a stethoscope over the trachea
    2. Determine the client's ability to do neck exercises
    3. Listen with a stethoscope over the base of the lungs
    4. Assess the client's ability to cough and deep breathe

394. A client arrives in the emergency room with multiple crushing wounds of the chest, abdomen, and legs. The assessments that assume the greatest priority are:
    1. Level of consciousness and pupil size
    2. Abdominal contusions and other wounds
    3. Pain, respiratory rate, and blood pressure
    4. Quality of respirations and presence of pulses

395. A client is admitted to the intensive care unit with a diagnosis of acute respiratory distress syndrome. When assessing this client the nurse should expect to find:
    1. Hypertension
    2. Tenacious sputum
    3. An altered mental status
    4. A slowed rate of breathing

396. A client's respiratory status necessitates endotracheal intubation and positive pressure ventilation. The most immediate nursing intervention for this client at this time would be to:
    1. Prepare the client for emergency surgery
    2. Facilitate the client's verbal communication
    3. Assess the client's response to the equipment
    4. Maintain sterility of the ventilation system the client is using

397. A client is placed on a ventilator. Because hyperventilation can occur when mechanical ventilation is used, the nurse should monitor the client for signs of:
    1. Hypoxia
    2. Hypercapnia
    3. Metabolic acidosis
    4. Respiratory alkalosis

398. A client is on a ventilator. One of the nurses asks what should be done when condensation resulting from humidity collects in the ventilator tubing. The best response to this question would be to:
1. "Notify the respiratory therapist."
2. "Empty the fluid from the tubing."
3. "Decrease the amount of humidity."
4. "Measure the fluid and record it on the I&O."

399. A tracheostomy is performed for a client with respiratory distress. After the procedure the client should be placed in the:
1. Supine position
2. Orthopneic position
3. High-Fowler's position
4. Semi-Fowler's position

400. A client with a pulmonary embolus is intubated and placed on mechanical ventilation. When suctioning the endotracheal tube, the nurse should:
1. Apply suction while inserting the catheter
2. Hyperoxygenate with 100% oxygen before and after suctioning
3. Use short, jabbing movements of the catheter to loosen secretions
4. Suction two to three times in quick succession to remove all secretions

401. The nurse knows that when a client has a tracheostomy tube with a high-volume, low-pressure cuff, it is used primarily to prevent:
1. Lung infection
2. Leakage of air
3. Mucosal necrosis
4. Tracheal secretion

402. A client with emphysema is short of breath and using accessory muscles of respiration. The nurse recognizes that the client's difficulty in breathing is caused by:
1. Spasm of the bronchi that traps the air
2. An increase in the vital capacity of the lungs
3. A too rapid expulsion of the air from the alveoli
4. Difficulty in expelling the air trapped in the alveoli

403. The nurse should observe a client with restrictive airway disease for early indications of respiratory acidosis, which include:
1. Bradypnea
2. Bradycardia
3. Restlessness
4. Light-headedness

404. The physician orders oxygen given in low concentration, rather than in high concentration and con-

tinuously, for a client with chronic obstructive pulmonary disease to prevent:
1. A decrease in red cell formation
2. Rupture of emphysematous bullae
3. Depression of the respiratory center
4. An excessive drying of the respiratory mucosa

405. The position that would provide for the greatest respiratory capacity for a client with dyspnea would be the:
1. Sims' position
2. Supine position
3. Orthopneic position
4. Semi-Fowler's position

406. The breathing exercises that the nurse teaches to a client with emphysema should include:
1. An inhalation that is longer than an exhalation
2. Abdominal exercises to limit the use of accessory muscles
3. Sit-ups to strengthen the abdominal and intercostal muscles
4. Diaphragmatic exercises to improve contraction of the diaphragm

407. The nurse is aware that a client understands the instructions about an appropriate breathing technique for chronic obstructive pulmonary disease when the client:
1. Inhales through the mouth
2. Increases the respiratory rate
3. Holds each breath for a second at the end of inspiration
4. Progressively increases the length of the inspiratory phase

408. A client who has had emphysema for many years develops an enlarged liver. The nurse understands that this results from:
1. Liver hypoxia
2. Hepatic acidosis
3. Esophageal varices
4. Portal hypertension

409. A client with chronic obstructive pulmonary disease complains of a weight gain of 5 pounds in 1 week. The complication that may have precipitated this weight gain is:
1. Polycythemia
2. Cor pulmonale
3. Left ventricular failure
4. Compensated acidosis

410. A client with a history of chronic obstructive pulmonary disease develops cor pulmonale. When teaching about nutrition, the nurse should encourage this client to:

1. Eat small meals 6 times a day to limit oxygen needs
2. Lie down after eating to permit energy to be used for digestion
3. Drink large amounts of fluids to help liquefy respiratory secretions
4. Increase protein intake to decrease intravascular hydrostatic pressure

411. A client with extensive cancer of the upper right lobe of the lung has been informed that a lobectomy will be performed. The statement by the client that indicates adequate understanding of the intended surgery is:
    1. "It shouldn't be that serious because the surgeon is going to use laser to destroy all the cells first."
    2. "I really won't need this portion that they're cutting out, because I still have three parts on the left to help me breathe."
    3. "I understand that several lung segments will be removed and that I'll have chest tubes in to help with drainage after surgery."
    4. "I understand that the remaining lung tissue will fill in the empty space and that I'll have chest tubes in to help with drainage after surgery."

412. When inspecting a dressing following a partial pneumonectomy for cancer of the lung, the nurse observes some puffiness of the tissue around the area. When the area is palpated, the tissue feels spongy and crackles. In charting, the nurse should describe this assessment as:
    1. Stridor
    2. Crepitus
    3. Pitting edema
    4. Chest distention

413. On the first day following a right pneumonectomy a male client suddenly sits straight up in bed. His respirations are labored, and he is making a crowing sound. His skin is pale, cool, and moist. Immediately the nurse should:
    1. Notify the physician
    2. Auscultate the left lung
    3. Inspect the incision for bleeding
    4. Check the chest tube for patency

414. When turning a client following a right pneumonectomy, the nurse should plan to place the client in either the:
    1. Right or left side-lying position
    2. High-Fowler's or supine position
    3. Supine or right side-lying position
    4. Left side-lying or low-Fowler's position

415. The nurse should be vigilant for the unique complications associated with a pneumonectomy by observing the client for:
    1. Signs of cardiac overload
    2. Increased pulse and respirations
    3. Cardiac irregularities with premature beats
    4. Increased BP, decreased temperature, and cold, moist skin

416. A female client develops increased respiratory secretions because of radiation therapy to the lung. When teaching postural drainage, the nurse should explain that the client will know that it is effective when she:
    1. Is free of crackles
    2. Can breathe deeply
    3. Has a productive cough
    4. Is able to expectorate saliva

417. A client with oat cell lung cancer is scheduled for a mediastinoscopy with biopsy. The nurse should:
    1. Tell the client that chest tubes will be present after the procedure
    2. Explain that procedure will visualize the lungs and the chest cavity
    3. Advise the client of the npo status after midnight the night before the test
    4. Inform the client that some pleural fluid will be removed during the procedure

418. A client is to have a thoracentesis for a pleural effusion. The nurse knows that:
    1. A thoracentesis generally is followed by instillation of a sclerosing agent
    2. A thoracentesis usually is contraindicated in clients with a history of emphysema
    3. The client usually has a temporary increase in dyspnea immediately after the procedure
    4. The rapid removal of large amounts of pleural fluid may precipitate cardiovascular collapse

419. When assessing a client with pleural effusion, the nurse should expect to find:
    1. Moist crackles at the posterior of the lungs
    2. Deviation of the trachea toward the involved side
    3. Reduced or absent breath sounds at the base of the lung
    4. Increased resonance with percussion of the involved area

420. A client with chronic obstructive pulmonary disease is admitted to the hospital with a tentative diagnosis of pleuritis. When caring for this client the nurse should plan to:
    1. Assess for signs of pneumonia
    2. Administer narcotics frequently
    3. Administer medication to suppress coughing
    4. Limit fluid intake to prevent pulmonary edema

421. After a thoracentesis for pleural effusion a client returns to the physician's office for a follow-up visit. The nurse would suspect a recurrence of pleural effusion when the client says:
1. "Lately I can only breathe well if I sit up."
2. "During the night I sometimes have a fever and chills."
3. "I get a sharp, stabbing pain when I take a deep breath."
4. "I'm coughing up larger amounts of thicker mucus for the last 2 days."

422. Before a scheduled bilateral herniorrhaphy, the nurse should teach the client that postoperatively the client will:
1. Have a nasogastric tube in place
2. Have a portable wound drainage system
3. Turn and change positions every 2 hours
4. Perform coughing and deep breathing exercises

423. The nurse performs preoperative teaching related to a subtotal thyroidectomy. The nurse would know that the client understands the teaching about the local effects of a general anesthetic when the client states, "Immediately after surgery I will experience:
1. Feelings of chilliness."
2. Transient headaches."
3. Discomfort swallowing."
4. Paroxysmal hiccoughs."

424. Following a gastroscopy the nurse should assess the client for the return of the gag reflex by:
1. Touching the pharynx with a tongue depressor
2. Giving a small amount of water using a syringe
3. Observing for when the client spits the airway out
4. Instructing the client to breathe deeply and cough gently

425. When caring for clients in the operating room, the nurse knows that the last physiologic function the client loses during the induction of an anesthetic is:
1. Gag reflex
2. Corneal reflex
3. Consciousness
4. Respiratory movement

426. A client has an aneurysm resected and replaced with a graft. On arrival in the postanesthesia care unit, the client is in shock. The nursing priority should be:
1. Assessing the client's respiratory status
2. Monitoring the client's hourly urine output
3. Putting several warm blankets on the client
4. Placing the client in the Trendelenburg's position

427. When a client returns from a bronchoscopy, the nurse should withhold food and fluid for several hours to prevent:
1. Aspiration
2. Projectile vomiting
3. Abdominal distention
4. Dysphasia and dyspepsia

428. A client has a bronchoscopy. The nurse should assess for return of the gag reflex by:
1. Having the client say a few words
2. Giving the client a small swallow of water
3. Touching the pharynx with a tongue depressor
4. Instructing the client to breathe deeply and cough gently

429. After a laryngectomy a client becomes concerned about frequent coughing episodes and copious production of secretions. The nurse is aware that this increase of coughing and secretions is due to:
1. An upper respiratory infection caused by allergies
2. Inadequate turning, coughing, and deep breathing
3. An irritation of the stoma by the tracheostomy tube
4. The mucous membranes' reaction to unwarmed, dry air

430. An important nursing intervention that ensures adequate ventilatory exchange postoperatively is:
1. Maintaining humidified oxygen via nasal cannula
2. Positioning the client laterally with the neck extended
3. Assessing for hypoventilation by auscultating the lungs
4. Removing the airway only when the client is fully conscious

431. After surgery, the physician orders an incentive spirometer for a client. The nurse would know that the client was using the spirometer correctly when observing that the client:
1. Coughs twice before inhaling deeply through the mouthpiece
2. Uses the incentive spirometer for 10 consecutive breaths per hour
3. Inhales deeply, seals the lips around the mouthpiece, and exhales
4. Inhales deeply through the mouthpiece, holds the breath for 2 seconds, and then exhales

432. The nurse expects that the initial treatment for a client who has a leak of the thoracic duct following radical neck surgery would include inserting a:
1. Gastrostomy tube to drain the fluid, a high-fat diet, and bed rest
2. Chest tube to drain the fluid, total parenteral nutrition, and bed rest

3. Rectal tube to prevent distention, a low-fat diet, and increased activity

4. Nasogastric tube to drain the fluid, a moderate-fat diet, and increased activity

433. A client is scheduled for coronary artery bypass surgery. The nurse explains to the client that chest tubes will be inserted during surgery to:
1. Drain fluid from the pericardial sac
2. Prevent atelectasis postoperatively
3. Reestablish negative intrapleural pressure
4. Monitor the amount of blood loss after surgery

434. When giving a client care on the second day postoperatively after coronary artery bypass surgery, the nurse notes that the fluid in the water-seal chamber of the drainage device (Pleur-Evac) stops fluctuating. The nurse should:
1. Increase the amount of suction
2. Look for obstructions of the tube
3. Add sterile water to the chamber
4. Consider this a normal occurrence

435. The nurse is instructed to measure and document the amount of drainage from a client's chest tube. The nurse should:
1. Aspirate fluid from the collection chamber of the closed chest drainage system (Pleur-Evac) with a needle and syringe and measure the drainage
2. Place a piece of tape on the outside of the collection chamber of the closed chest drainage system (Pleur-Evac) and mark the fluid level and the time of measurement
3. Clamp the chest tube, empty the collection chamber of the closed chest drainage system (Pleur-Evac) into a measuring cup, and reconnect the drainage system
4. Connect a new Pleur-Evac, measure the drainage in the collection chamber of the old closed chest drainage system (Pleur-Evac), and dispose of the old drainage system

436. After thoracic surgery a client has a chest tube connected to a closed chest drainage system. When excessive bubbling is noted in the water-seal chamber, the nurse should:
1. Check the system for air leaks
2. Decrease the amount of suction pressure
3. Recognize the system is functioning correctly
4. "Strip" the chest tube toward the collection chamber

437. A client is performing postthoracotomy exercises. The least productive exercise for this client would be:

1. Extending the arm up and back and out to the side and back
2. Climbing a wall with the fingers until the arm is fully extended
3. Tying a rope to a doorknob and swinging the arm in wide circles
4. Extending the arm out and bringing it up to touch the nose with the finger

438. A chest tube is inserted following a crushing chest injury. The observation that indicates a desired response to treatment of the client's chest injury would be:
1. Increased breath sounds
2. Increased respiratory rate
3. Crepitus detected on palpation of chest
4. Constant bubbling in collection chamber

439. A client sustains a stab wound to the chest, and a chest tube is inserted. Later the client's chest tube seems to be obstructed. The most appropriate nursing action would be to:
1. Gently squeeze the tube
2. Clamp the tube immediately
3. Prepare for chest tube removal
4. Arrange for a stat chest x-ray film

440. On the way to an x-ray examination a client with a chest tube becomes confused and pulls the chest tube out. The nurse's immediate action should be to:
1. Place the client in Trendelenburg's position
2. Hold the insertion site open with a Kelly clamp
3. Obtain sterile Vaseline gauze to cover the opening
4. Cover the opening with the cleanest material available

441. A client has a chest tube for a pneumothorax. The nurse finds the client in respiratory difficulty with the chest tube separated from the drainage system. The nurse should:
1. Obtain a new sterile drainage system
2. Clamp the drainage tubing with two clamps
3. Reconnect the client's tube to the drainage system
4. Place the client in the high-Fowler's position immediately

442. To promote continued improvement in a client's respiratory status after a chest tube is removed, the nurse should:
1. Continue observing for dyspnea and crepitus
2. Encourage frequent coughing and deep breathing
3. Turn the client from side to side at least every 2 hours
4. Encourage bed rest with active and passive range-of-motion exercises

443. During a routine physical examination, a client's chest x-ray film reveals a lesion in the right upper lobe. When the nurse obtains a history from the client, the information that supports the physician's tentative diagnosis of pulmonary tuberculosis is:
    1. Frothy sputum and fever
    2. Dry cough and pulmonary congestion
    3. Night sweats and blood-tinged sputum
    4. Productive cough and engorged neck veins

444. The nurse notes 12 mm of induration at the site of a Mantoux test when a client returns to the health office to have it read. The nurse should explain to the client that this:
    1. Test result is negative, and no follow-up is needed
    2. Test was used for screening and a tine test will now be given
    3. Skin test is inconclusive and will have to be repeated in 6 weeks
    4. Result indicates a need for further tests, including a chest x-ray film examination

445. As part of a yearly physical, a client has a Mantoux test. Within 48 hours the area of induration is 10 mm. This indicates that the client has:
    1. Contracted clinical tuberculosis
    2. Resistance to the tubercle bacillus
    3. A passive immunity to tuberculosis
    4. Been exposed to the tubercle bacillus

446. To make a definitive diagnosis of tuberculosis, the nurse understands that the physician must order a:
    1. Chest x-ray film
    2. Tuberculin skin test
    3. Pulmonary function test
    4. Sputum for acid-fast testing

447. A client with pulmonary tuberculosis is being treated on an outpatient basis. The nurse should expect that the physician will order a diet that:
    1. Includes liquid protein supplements
    2. Has frequent, small, high-calorie meals
    3. Is low in calories but high in carbohydrates
    4. Contains foods high in calories and low in protein

448. A client with pulmonary tuberculosis is being treated in the home. To help control the spread of the disease, the client should be instructed to:
    1. Have visitors sit at least 8 feet away
    2. Keep personal articles away from the rest of the family
    3. Open the windows slightly to allow a good airflow throughout the house
    4. Avoid putting used dishes in the dishwasher with the rest of the family's dishes

449. A client's tine test and chest x-ray film indicate pulmonary tuberculosis. The physician orders sputum specimens for acid-fast bacilli. The nurse knows that additional teaching is necessary when the client states that the sputum specimens must be:
    1. Coughed up from deep in the lungs
    2. Collected in the early morning hours
    3. Refrigerated until brought to the laboratory
    4. Brought to the clinic as soon as possible after collection

450. A client's sputum smears for acid-fast bacilli (AFB) are positive, and the client is placed on isolation. The nurse should instruct visitors to:
    1. Limit contact with nonexposed family members
    2. Avoid contact with any objects present in the client's room
    3. Wear an Ultra-Filter mask when they are in the client's room
    4. Put on a gown and gloves before going into the client's room

451. When teaching a client with tuberculosis about recovery after discharge from the hospital, the nurse should reinforce that the treatment measure with the highest priority is:
    1. Having sufficient rest
    2. Getting plenty of fresh air
    3. Changing the current life-style
    4. Consistently taking prescribed medication

452. A client with tuberculosis asks the nurse how long the chemotherapy must be continued. The nurse's most accurate reply would be:
    1. "1 to 2 weeks."
    2. "4 to 5 months."
    3. "A minimum of 9 months."
    4. "Probably 3 years or longer."

CARDIOVASCULAR (CV)

453. To determine the status of a client's carotid pulse, the nurse should palpate:
    1. In the lateral neck region
    2. Immediately below the mandible
    3. At the anterior neck, lateral to the trachea
    4. At the base of the neck, along the clavicle

454. During auscultation of the heart, the nurse would expect the first heart sound (S1) to be the loudest at the:
    1. Apex of the heart
    2. Base of the heart
    3. Left lateral border
    4. Right lateral border

455. When auscultating a client's heart, the nurse understands that the first heart sound is produced by the closure of the:
    1. Mitral and tricuspid valves
    2. Aortic and tricuspid valves
    3. Mitral and pulmonic valves
    4. Aortic and pulmonic valves

456. Thrombus formation is a danger for all postoperative clients. The nurse should act independently to prevent this complication by:
    1. Applying elastic stockings
    2. Performing in-bed exercises
    3. Massaging gently with lotion
    4. Encouraging adequate fluids

457. Oxygen by nasal cannula is prescribed for a client. The nurse plans to use safety precautions in the room because oxygen:
    1. Is flammable
    2. Supports combustion
    3. Has unstable properties
    4. Converts to an alternate form of matter

458. An electrocardiogram is ordered for a client complaining of chest pain. An early finding in the lead over an infarcted area would be:
    1. Flattened T waves
    2. Absence of P waves
    3. Elevated ST segments
    4. Disappearance of Q waves

459. A client has a Swan-Ganz catheter inserted for monitoring cardiovascular status. With the Swan-Ganz catheter the most accurate measurement of the client's left ventricular pressure would be the:
    1. Right atrial pressure
    2. Cardiac output by thermodilution
    3. Pulmonary artery diastolic pressure
    4. Pulmonary capillary wedge pressure

460. A thallium scan is performed on a client with a history of chest pain to:
    1. Monitor action of the heart valves
    2. Determine myocardial muscle viability
    3. Visualize ventricular systole and diastole
    4. Determine adequacy of electrical conductivity

461. A client's diet is modified to eliminate foods that act as cardiac stimulants. The nurse should teach the client to avoid:
    1. Yogurt
    2. Club soda
    3. Chocolate
    4. Red meats

462. A client with a family history of atherosclerosis is advised to follow a diet based on the U.S. Department of Agriculture's Food Guide Pyramid. The nurse should teach the client to eat:
    1. 4 to 6 servings of fruit daily
    2. 5 to 7 servings of vegetables daily
    3. 3 to 5 servings of meat, poultry, or fish daily
    4. 6 to 11 servings of bread, rice, or pasta daily

463. A female client tells the nurse that the physician just told her that her triglycerides and cholesterol are excessively elevated. The client appears discouraged and says, "Well, I guess I'd better cut out all the fat and cholesterol in my diet." The nurse's most appropriate response would be:
    1. "Well yes, that would certainly lower the amount of your blood fats."
    2. "That's good, but be sure to compensate by adding more proteins and carbohydrates."
    3. "You need some fat to supply a necessary fatty acid, so it's mainly just a need for cutting down the amount."
    4. "You need some cholesterol in your diet because your body cannot manufacture it, so just avoid excessive amounts."

464. To help reduce a client's risk factors for heart disease, the nurse, in discussing dietary guidelines, should teach the client to:
    1. Avoid eating between meals
    2. Decrease the amount of unsaturated fat
    3. Decrease the amount of fat-binding fiber
    4. Increase the ratio of complex carbohydrates

465. A client with chronic heart failure who is to receiving a 2-gram sodium diet complains about the lack of salt in the food. The nurse should explain that salt must be limited to:
    1. Prevent any rise in blood pressure from tissue edema
    2. Produce a diuretic effect and reduce the circulating blood volume
    3. Reduce the large amount of edema present, which interferes with heart action
    4. Prevent the possible accumulation of tissue fluid and the resulting increased cardiac workload

466. The nurse plans to teach a client receiving a 3-gram sodium-restricted diet that the foods lowest in sodium are:
    1. Meat and fish
    2. Milk and cheese
    3. Fruits and juices
    4. Dry cereals and grains

467. When preparing a client for a cardiac catheterization, the nurse should advise the client that:
1. The procedure will take 15 minutes
2. The client will be npo 6 to 8 hours before the procedure
3. Ambulation will be permitted within 1 hour after the procedure
4. Complete sedation will be maintained throughout the procedure

468. A client returns from a cardiac catheterization with a pressure dressing over the left groin. The client is to be flat in bed for 6 hours with the leg straight. These measures are important to prevent:
1. Orthostatic hypotension
2. Headache and disorientation
3. Bleeding at the arterial puncture site
4. Infiltration of radiopaque dye into tissue

469. For the first several hours after a cardiac catheterization, it would be most essential for the nurse to:
1. Keep the head of the client's bed elevated 45 degrees
2. Monitor the client's apical pulse and blood pressure frequently
3. Encourage the client to cough and deep breathe every 2 hours
4. Check the client's temperature every hour until it returns to normal

470. A male client is admitted for a cardiac catheterization via the femoral approach. The facts about this procedure that should be included as part of the nurse's teaching plan include:
1. His physician will immediately tell him about the results of the procedure
2. He will be permitted to get up and walk around as soon as he returns to his room
3. A general anesthetic will be given, and he will not be awake during the procedure
4. He will be kept on bed rest, in the supine position, with the affected leg extended for a number of hours

471. A client with a history of anginal pain is admitted to the surgical unit for cardiac catheterization. After the catheterization, the nurse notices that the client's urinary output is three times the intake. The client is stable otherwise. The nurse interprets this to be:
1. An increased cardiac output because of the procedure
2. A physiologic effect of the ordered IV rate of 50 ml/hour
3. A normal effect of the mercurial dye used during the procedure
4. An improvement in cardiac functioning following the catheterization

472. A client with continuous blood loss becomes increasingly diaphoretic, clammy, and pale. The client's blood pressure falls to 90/60. The nursing action that is of primary importance is to:
1. Place the client in a reclining position
2. Delay repositioning of the client for a few hours
3. Allow the client to determine the position of comfort
4. Assist the client to assume the semi-Fowler's position

473. The finding that would most significantly indicate that a client is hypertensive is:
1. An extended Korotkoff's sound
2. A regular pulse of 92 beats per minute
3. A systolic blood pressure ranging from 140 to 150 mm Hg
4. A diastolic blood pressure that remains greater than 90 mm Hg

474. After taking a client's blood pressure twice, 10 minutes apart, in one arm while the client is seated, the nurse in the blood pressure screening clinic records the two blood pressures of 172/104 and 164/98. The nurse's priority would be to:
1. Refer the client to a nutritionist after providing health teaching about a low-sodium diet
2. Place the client in a recumbent position and call the paramedics for transport to the hospital
3. Talk with the client to assess whether there is stress in the client's life and refer to a counseling service
4. Take the blood pressure in the other arm and then schedule a physician's appointment for the client as soon as possible

475. The nurse would expect a client diagnosed as having hypertension to report experiencing the most common symptom associated with this disorder, which is:
1. Fatigue
2. Headache
3. Nosebleeds
4. Flushed face

476. A businessman makes many long airplane trips. He confides to the nurse at his place of business that he is concerned because his legs swell on these long flights. The nurse should advise him to:
1. Relax in a reclining position
2. Sit upright with legs extended
3. Walk about the cabin at least once every hour
4. Sit in any position that relieves pressure on the legs

477. A client with a history of hypertension develops pedal edema and demonstrates dyspnea on exertion. The nurse recognizes that the client's dyspnea on exertion is probably:
1. Caused by cor pulmonale
2. A result of right atrial failure
3. A result of left ventricular failure
4. Associated with wheezing and coughing

478. A hospitalized client complains of chest pain that feels like a pressure or weight on the chest. The client also states, "I feel nauseated and very weak." The nurse should:
1. Perform a nutritional assessment
2. Summon medical help for a potential emergency
3. Explore for and discuss possible sources of stress
4. Provide reassurance while helping the client to focus on pleasant topics

479. A client who has been admitted to the cardiac care unit with a myocardial infarction complains of chest pain. The nursing intervention that would be most effective in relieving the client's pain would be to administer the ordered:
1. Morphine sulfate 2 mg IV
2. Oxygen per nasal cannula
3. Nitroglycerin sublingually
4. Lidocaine hydrochloride 50 mg IV bolus

480. The nurse admitting a client with a myocardial infarction to ICU understands that the pain the client is experiencing is a result of:
1. Compression of the heart muscle
2. Release of myocardial isoenzymes
3. Inadequate perfusion of the myocardium
4. Rapid vasodilation of the coronary arteries

481. A male client who is hospitalized following a myocardial infarction asks the nurse why he is receiving morphine. The nurse replies that morphine:
1. Dilates coronary blood vessels
2. Relieves pain and prevents shock
3. Decreases anxiety and restlessness
4. Helps prevent fibrillation of the heart

482. Isoenzyme laboratory studies are ordered for a client following a myocardial infarction. The isoenzyme test that is the most reliable early indicator of myocardial insult is:
1. AST
2. LDH
3. SGOT
4. CK-MB

483. When a client has a myocardial infarction, one of the major manifestations is a decrease in the conductive energy provided to the heart. When assessing this client the nurse understands that the existing action potential is in direct relationship to the:
1. Heart rate
2. Refractory period
3. Pulmonary pressure
4. Strength of contraction

484. Because a client with a myocardial infarction can develop left ventricular failure, the nurse should assess this client for:
1. Distended neck veins
2. Anorexia and weight loss
3. Paroxysmal nocturnal dyspnea
4. Right upper quadrant tenderness

485. A client who has had a myocardial infarction experiences a noticeably decreased pulse pressure. The nurse should immediately recognize this as a possible indication of:
1. Increased blood volume
2. Hyperactivity of the heart
3. Increased cardiac sufficiency
4. Decreased force of contraction

486. The nurse notes premature ventricular beats (PVBs) on a client's cardiac monitor and recognizes that these complexes are a sign of:
1. Atrial fibrillation
2. Cardiac irritability
3. Impending heart block
4. Ventricular tachycardia

487. The wife of a client who has had emergency coronary artery bypass surgery asks why her husband has a dressing on his left leg. The nurse explains that:
1. This is the access site for the heart lung machine
2. A filter is inserted in the leg to prevent embolization
3. The saphenous vein was used to bypass the coronary artery
4. The arteries in distal extremities are examined during surgery

488. It is determined that a client will require implantation of a permanent pacemaker to assist heart function. In response to the client's inquiries as to why this is necessary, the nurse's best response would be:
1. "It shocks the AV node to contract."
2. "It will cause a normal heartbeat to occur."
3. "It will work the valves of your heart better."
4. "It will slow down the heart to a more normal rate."

489. A client has had a ventricular demand pacemaker inserted. The priority nursing intervention immediately after the pacemaker insertion would be to:
1. Encourage fluids
2. Assess the implant site
3. Monitor the heart rate and rhythm
4. Encourage turning and deep breathing

490. The nurse recognizes that a pacemaker is indicated when a client is experiencing:
1. Angina
2. Chest pain
3. Heart block
4. Tachycardia

491. A client's wife arrives at the cardiac care unit and is informed that her husband needs a pacemaker. The wife expresses the concern that her husband could accidentally become electrocuted. The nurse's best response would be:
1. "No one has been electrocuted yet by a pacemaker."
2. "New technology prevents electrocution from occurring."
3. "Pacemakers are pretested for safety before they are inserted."
4. "The voltage emitted is not strong enough to electrocute him."

492. A client with a history of hypertension and left ventricular failure arrives for a scheduled clinic appointment and tells the nurse, "My feet are killing me. These shoes got so tight." The nurse's best initial action would be to:
1. Weigh the client
2. Notify the physician
3. Take the client's pulse rate
4. Listen to the client's breath sounds

493. A 76-year-old client is admitted with the diagnosis of mild chronic heart failure. The sounds indicative of chronic heart failure that the nurse will hear when listening to the client's lungs would be:
1. Stridor
2. Crackles
3. Wheezes
4. Friction rubs

494. When assessing a client with a diagnosis of left ventricular failure, the nurse should expect to find:
1. Crushing chest pain
2. Dyspnea on exertion
3. Jugular vein distention
4. Extensive peripheral edema

495. The teaching plan for a client receiving digoxin for left ventricular failure should include having the client:

1. Sleep flat in bed
2. Rest during the day
3. Follow a low-potassium diet
4. Take the pulse 3 times a day

496. A client is admitted to the intensive care unit with pulmonary edema. When performing the admission assessment, the nurse should expect:
1. A decreased blood pressure
2. Radiating anterior chest pain
3. A pulse that is weak and rapid
4. Crackles at the base of each lung

497. The physician orders "bathroom privileges only" for a client with pulmonary edema. The client becomes irritable and asks the nurse whether it is really necessary to stay in bed so much. The nurse's best reply would be:
1. "Why do you want to be out of bed?"
2. "Yes. Bed rest plays a role in most therapy."
3. "Rest helps your body direct energy to healing."
4. "Not always. Ask your physician to change the order."

498. When assessing a client for signs of right ventricular failure the nurse should expect to note:
1. A slowed pulse rate
2. A pleural friction rub
3. Neck vein distention
4. Increasing hypotension

499. The nurse suggests that a client with right ventricular failure should:
1. Take a hot bath before bedtime
2. Avoid sleeping in an air-conditioned room
3. Avoid emotionally stressful situations when possible
4. Exercise daily until the pulse rate exceeds 100 beats per minute

500. The home care nurse is assessing a client with cardiac insufficiency. The nurse notes that the client's pulse rate increases from 70 to 92 beats/minute while climbing the stairs. The nurse should immediately instruct the client to:
1. Continue climbing
2. Stand still and rest
3. Walk down the stairs
4. Climb at a slower rate

501. The client that would be considered at the highest risk for a dissecting aneurysm is:
1. A 50-year-old white male with moderate hypertension
2. A 42-year-old female with peripheral vascular disease

3. A 55-year-old black male with uncontrolled hypertension

4. A 40-year-old white female with uncontrolled hypertension

502. During a routine physical examination, an abdominal aortic aneurysm is diagnosed. The client is immediately admitted to the hospital, and surgery is scheduled for the next morning. When performing the admission assessment the nurse should expect:
1. Severe radiating abdominal pain
2. Cyanosis and symptoms of shock
3. A pattern of visible peristaltic waves
4. A palpable pulsating abdominal mass

503. A client is admitted for resection of an abdominal aortic aneurysm. During the evening before surgery, the client suddenly develops symptoms of shock. The nurse should:
1. Prepare for blood transfusions
2. Notify the physician immediately
3. Give the client nothing by mouth
4. Administer the prescribed sedative

504. To prevent thrombus formation after most surgeries, the nurse should plan to:
1. Keep the bed gatched to elevate client's knees
2. Have the client dangle the legs off the side of the bed
3. Have the client use an incentive spirometer every hour
4. Encourage the client to ambulate with assistance as needed

505. When performing a physical assessment the nurse identifies bilateral varicose veins. The symptom the nurse should expect the client to report is:
1. Increased sensitivity to cold
2. Pallor of the lower extremities
3. Calf pain when the foot is dorsiflexed
4. Increasing ankle edema over the day

506. The physician orders knee-length elastic support stockings for a client with varicose veins. When teaching the client about these stockings, the nurse should explain that:
1. It is best to apply them before getting out of bed
2. The stockings should come up to the middle of the knee

3. The stockings should be applied at the first sign of discomfort

4. Elastic bandages are more economical and may be substituted

507. When collecting data from a client with varicose veins who is to have sclerotherapy, the nurse should expect the client to report:
1. A feeling of heaviness in both legs
2. Calf pain on dorsiflexion of the foot
3. Intermittent claudication of the legs
4. Hematomas of the lower extremities

508. Before having sclerotherapy for varicose veins, a female client who states she is fearful of a chemical injection asks the nurse to explain what would be involved if she insisted on a ligation and stripping to correct the problem. The nurse should explain that this surgery involves:
1. Removing the dilated saphenous veins
2. Cleaning out plaque from within the vessels
3. Anastomosing superficial veins to deep veins
4. Placing an umbrella filter in the large affected veins

509. A client with a history of thrombophlebitis and varicosities is to have a herniorrhaphy for an incarcerated hernia. The client's past medical history and present diagnosis indicate to the nurse that a primary responsibility following surgery is to:
1. Raise the foot of the bed
2. Get the client out of bed twice daily
3. Maintain body alignment with firm support of the extremities
4. Encourage the client to turn often and to exercise the legs regularly

510. Before discharging a client who has had an inguinal herniorrhaphy, the nurse teaches the client about exercising to prevent venous stasis. For best results the nurse should:
1. Demonstrate specific exercises
2. Suggest frequent moving of the legs
3. Advise against sitting for prolonged periods
4. Suggest that the client change position frequently

511. During the initial assessment the nurse should specifically observe the client for:
1. Edema of the left leg
2. Mobility of the left leg
3. A positive left-sided Babinski's reflex
4. Presence of peripheral arterial pulses

512. A client reports a history of bilateral blanching and pain in the fingers on exposure to cold. When rewarmed the fingers become bright red and "tingly" with a slow return to normal color. The client smokes 1 to 2 packs of cigarettes per day. The nurse recognizes that the client has Raynaud's disease and not Raynaud's phenomenon because of the:
    1. Tingling sensation
    2. Skin color changes
    3. Bilateral involvement
    4. Changes in skin temperature

513. The nurse encourages a client with Raynaud's disease to stop smoking because it causes:
    1. Pain and tingling
    2. Cyanosis and necrosis
    3. Peripheral vasoconstriction
    4. Decreased blood oxygen content

514. A male client with intermittent claudication asks the nurse why he is not allowed to smoke. The nurse's best response would be:
    1. "The policy states that the hospital is a smoke-free environment."
    2. "Nicotine causes arteries to go into spasm, which decreases circulation."
    3. "Cigarette smoking is not allowed for clients who have venous problems."
    4. "The doctor may allow you to begin smoking again after you are feeling better."

515. An essential nursing function in the care of a client with arterial insufficiency in the left foot caused by generalized arteriosclerosis should be to:
    1. Maintain elevation of the legs
    2. Massage the legs when painful
    3. Check arterial pulses frequently
    4. Apply a hot water bottle to the feet

516. A client develops a nonhealing ulcer of the right lower extremity and complains of leg cramps after walking short distances. The client asks the nurse what causes these leg pains. The nurse's best response would be:
    1. "Muscle weakness occurs in the legs because of a lack of exercise."
    2. "Edema and cyanosis occur in the legs because they are dependent."
    3. "Pain occurs in the legs while walking because there is a lack of oxygen to the muscles."
    4. "Pressure occurs in the legs because of vasodilation and pooling of blood in the extremity."

517. The physician prescribes a progressive exercise program that includes walking for a client with a history of diminished arterial perfusion to the lower extremities. The nurse should explain to the client that if leg cramps occur while walking, the client should:
    1. Take 1 aspirin twice a day
    2. Stop and rest until the pain resolves
    3. Take 1 nitroglycerin tablet sublingually
    4. Walk more slowly while pain is present

518. A client comes to the outpatient clinic with a painful leg ulcer. The symptom that supports the diagnosis of arterial ulcer is:
    1. Pain at the ulcer site
    2. Bleeding around the ulcer area
    3. Dependent edema of the extremities
    4. Stasis dermatitis on affected extremity

519. A client is admitted to the hospital with a large leg ulcer, and a femoral angiogram is performed. After this procedure the nurse should:
    1. Provide passive ROM to all extremities
    2. Elevate the foot of the bed for 36 hours
    3. Assist the client to stand if unable to void
    4. Apply pressure to the catheter insertion site

520. A client with impaired peripheral pulses and signs of chronic hypoxia in a lower extremity has a femoral angiogram. After the angiogram the nurse should:
    1. Elevate the foot of the bed
    2. Have the client void within 2 hours
    3. Keep the client in the high-Fowler's position
    4. Perform a neurovascular assessment of the affected extremity

521. Six hours after a femoral-popliteal bypass graft, the client's blood pressure becomes severely elevated. The nurse should notify the physician primarily because the client's:
    1. Blood pressure could cause the graft to occlude
    2. Hypervolemia needs to be corrected immediately
    3. Intraabdominal pressure could compromise the viability of the graft
    4. Cardiovascular status could precipitate a cerebrovascular accident

522. A client with a history of severe intermittent claudication has a femoral-popliteal bypass graft. An appropriate postoperative intervention on the day after surgery would be to:
    1. Keep the client on bed rest
    2. Have the client sit in a chair
    3. Assist the client with ambulation
    4. Encourage the client to keep the legs elevated

523. The nurse provides discharge teaching for a client with a history of hypertension who has had a femoral-popliteal bypass graft. The nurse is aware that the teaching is effective when the client says, "I should:
    1. Massage my calves gently every day."
    2. Keep my foot elevated when I am in bed."

3. Sit in a hot bath for 25 minutes twice a day."
4. Assess the color and pulses of my legs every day."

524. The nurse is aware that during the early postoperative period after open heart surgery, adequate oxygenation is essential because:
1. Hypoxia can precipitate respiratory alkalosis
2. All clients have closed chest drainage in place
3. Hypoxia can stimulate dangerous dysrhythmias
4. An increased respiratory rate adds to postoperative pain

525. The nurse in the CCU is monitoring a client who has had an aortic valve replacement. The nurse is aware that a slow pulse rate during the early postoperative period following open heart surgery can indicate:
1. Shock
2. Hypoxia
3. Heart block
4. Heart failure

REPRODUCTIVE AND GENITOURINARY (RG)

526. A client is scheduled for an intravenous pyelogram (IVP). The nurse explains that on the day before the IVP the client must:
1. Eat a fat-free dinner
2. Drink a large amount of fluids
3. Omit dinner and limit beverages
4. Take a laxative before going to bed

527. During an exacerbation of multiple sclerosis a client complains of urinary urgency and frequency. The initial nursing measure should be to:
1. Palpate the suprapubic area
2. Begin teaching self-catheterization
3. Develop a plan to ensure high fluid intake
4. Initiate a regimen to monitor urinary output

528. To facilitate micturition in a male client, the nurse should instruct him to:
1. Use a urinal for voiding
2. Drink cranberry juice daily
3. Wash his hands after voiding
4. Assume the normal position for voiding

529. Urinary infection is a potential danger with an indwelling catheter. The nurse can best plan to avoid this complication by:
1. Assessing urine specific gravity
2. Maintaining the ordered hydration
3. Collecting a weekly urine specimen
4. Emptying the drainage bag frequently

530. As the nurse assesses a client's kidney function after surgery, the components of urine will be monitored. Essential to this process is the knowledge that the normal kidney filters a number of blood components and urine should not contain:
1. Sodium
2. Potassium
3. Urea nitrogen
4. Large proteins

531. A client with deep partial thickness burns appears to be recovering from the burn trauma but develops chills, fever, flank pain, and malaise. The diagnostic tests that the nurse expects the physician to order to identify the cause of these symptoms would be:
1. Urinalysis for C & S
2. Cystoscopy and bilirubin
3. Creatinine clearance and A/G ratio
4. Specific gravity and pH of the urine

532. A client complains of urinary problems. Cholinergic medications are prescribed. The nurse is aware that this type of medication is prescribed to prevent:
1. Kidney stones
2. A flaccid bladder
3. A spastic bladder
4. Urinary tract infections

533. Twenty-four hours after a penile implant the client's scrotum is edematous and painful. The nurse should:
1. Assist the client with a sitz bath
2. Apply warm soaks to the scrotum
3. Elevate the scrotum using a soft support
4. Prepare for a possible incision and drainage

534. A male client comes to the emergency room because he has a discharge from his penis. The physician suspects gonorrhea and asks the nurse to obtain a specimen and to send it for a culture. The nurse should:
1. Instruct the client to provide a semen specimen
2. Swab the discharge as it appears on the prepuce
3. Obtain a specimen of the drainage from the anterior urethra
4. Teach the client how to obtain a clean catch specimen of urine

535. A client comes to the infectious disease clinic because a sexual partner was recently diagnosed as having gonorrhea. The health history reveals that the client has engaged in receptive anal intercourse. The nurse should assess the client for:
1. Melena
2. Anal itching
3. Constipation
4. Ribbon-shaped stools

536. A young male client comes to the clinic complaining of a sore throat and a rash. Because of the client's active sexual history, serologic testing is performed to confirm the diagnosis of syphilis. The symptoms indicate that the client's syphilis can be classified as:
1. Latent
2. Tertiary
3. Primary
4. Secondary

537. The nurse should ask the client with secondary syphilis about sexual contacts during the past:
1. 21 days
2. 30 days
3. 3 months
4. 6 months

538. A client is suspected of having late-stage (tertiary) syphilis. When obtaining a health history, the nurse recognizes that the statement by the client that would most support this diagnosis would be:
1. "I noticed a wart on my penis."
2. "I have sores all over my mouth."
3. "I've been losing a lot of hair lately."
4. "I'm having trouble keeping my balance."

539. A male client, age 56, is being worked up for possible cancer of the urinary bladder. Of the client's symptoms, the one that is most significant for cancer of the urinary tract is:
1. Dysuria
2. Retention
3. Hesitancy
4. Hematuria

540. A client is diagnosed as having invasive cancer of the bladder, and radiation therapy is performed prior to surgery. When determining the success of radiation therapy, the nurse should expect the client to demonstrate:
1. A decrease in urine and feces
2. An increase in physical strength
3. A shrinkage of the tumor on scanning
4. An increase in the quantity of white blood cells

541. A 64-year-old client has been diagnosed as having cancer of the bladder and is scheduled for a total cystectomy and the formation of an ileal conduit. When assessing the client 8 hours after surgery, the nurse notes all of the following findings. The finding that should be promptly reported to the physician would be:
1. An edematous stoma
2. A dusky-colored stoma
3. Pink-tinged urinary drainage
4. The absence of bowel sounds

542. On the fourth postoperative day after a cystectomy and the formation of a continent diversion, the nurse notes mucus threads in a client's urine. The nurse should:
1. Send a specimen for culture and sensitivity
2. Report this to physician when making rounds
3. Recognize this assessment as a normal occurrence
4. Increase the client's oral intake for the next 12 hours

543. A client with an ileal conduit is being prepared for discharge. Before discharge the nurse should instruct the client to:
1. Abstain from beer and alcohol consumption
2. Maintain a daily fluid intake of at least 2000 ml
3. Notify the physician if the stoma size decreases
4. Avoid getting soap and water on the peristomal skin

544. A 45-year-old client develops acute glomerulonephritis following a streptococcal infection. When performing the health assessment the nurse would expect the client to report a history of:
1. Nocturia
2. Mild headache
3. A recent weight loss
4. An increased appetite

545. A client with acute glomerulonephritis complains of thirst. The nurse should offer:
1. Ginger ale
2. Hard candy
3. A milk shake
4. A cup of broth

546. To prevent future attacks of glomerulonephritis, the nurse should instruct a female client to:
1. Take showers instead of bubble baths
2. Avoid situations that involve physical activity
3. Continue the same restrictions on fluid intake
4. Seek early treatment for respiratory infections

547. The nurse is aware that the most serious complication for a client with acute renal failure is:
1. Anemia
2. Infection
3. Weight loss
4. Platelet dysfunction

548. A male client with chronic renal failure is admitted because of a severe infection and anemia. The client is depressed and lethargic. The client's wife approaches the nurse and asks about the therapy and nursing care. The nurse explains the rationale of the client's treatment by stating:

1. "The staff must do everything for your husband, because the infection makes him so lethargic and fatigued."
2. "Your husband is irritable and depressed, so the physician is giving him mood elevators to make him feel better."
3. "You may notice your husbands stools are dark because of all the iron he is receiving, but it is helping him overcome the anemia."
4. "The physician has restricted your husband's intake of meat, eggs, and cheese, so that his kidneys can clear the body of waste products."

549. The factor in a client's history that may have contributed to a present problem of renal calculi is:

1. A high-cholesterol diet
2. Excess ingestion of antacids
3. An excessive exercise program
4. Frequent consumption of alcohol

550. A client is admitted to the hospital with severe flank pain, nausea, and hematuria caused by a ureteral calculus. A lithotripsy is scheduled. The initial nursing action should be to:

1. Strain all urine output
2. Increase the oral fluid intake
3. Obtain a urine specimen for culture
4. Administer the prescribed analgesic

551. When taking the admitting history of a client with a left ureteral calculus who is scheduled for a transurethral ureterolithotomy, the nurse would expect the client to report:

1. A boring pain in the left flank
2. Pain that intensifies on urination
3. Pain that is dull and constant in the costovertebral angle
4. Spasmodic pain on the left side radiating to the suprapubis

552. The laboratory values of a client with renal calculi reveal a serum calcium within normal limits and an elevated serum purine. The nurse should recognize that the stone is probably composed of:

1. Cystine
2. Struvite
3. Oxalate
4. Uric acid

553. A female client is admitted to the hospital with severe renal colic caused by a ureteral calculus. Later that evening, the client's urinary output is much less than her intake. When it is noted that her bladder is not distended, the nurse should suspect the development of:

1. Oliguria
2. Hydroureter
3. Renal shutdown
4. Urethral obstruction

554. A client with a calculus in the right renal pelvis is admitted for its removal. The nurse prepares the client for the procedure by explaining that:

1. The right ureter will be removed
2. A suprapubic catheter will be in place
3. Surgery will be performed transurethrally
4. A small incision will be present in the flank area

555. A male client who has had recurring renal calculi has a ureterolithotomy. Before discharge the nurse discusses the need to avoid urinary tract infections. The nurse knows that the signs of infection are understood when the client says he will report:

1. Urgency or frequency of urination
2. The inability to maintain an erection
3. Pain radiating to the external genitalia
4. An increase in alkalinity or acidity of urine

556. The person at highest risk of developing prostate cancer is a:

1. 55-year-old Black male
2. 55-year-old Asian male
3. 45-year-old Hispanic male
4. 45-year-old Caucasian male

557. The nurse assesses a newly admitted client with kidney disease to determine subjective symptoms that may be present. A subjective symptom that the nurse should assess for is:

1. Uremia
2. Vomiting
3. Voiding at night
4. Flank discomfort

558. A client is scheduled for a transurethral resection of the prostate. As part of the preoperative teaching, the nurse should explain that after surgery:

1. Urinary control may be permanently lost to some degree
2. The client's ability to perform sexually will be permanently impaired
3. Urinary drainage will be dependent on an urethral catheter for 24 to 48 hours
4. Frequency and burning on urination will last while the cystotomy tube is in place

559. After a transurethral resection of the prostate, a client's nursing care should include:
1. Changing the abdominal dressing
2. Maintaining patency of the cystotomy tube
3. Maintaining patency of a three-way Foley catheter
4. Observing for hemorrhage and wound infection

560. After a transurethral resection of the prostate, the client's retention catheter is secured to his leg, causing slight traction of the inflatable balloon against the prostatic fossa. This is done to:
1. Limit discomfort
2. Provide hemostasis
3. Reduce bladder spasms
4. Promote urinary drainage

561. When planning care for a client with a continuous bladder irrigation the nurse should:
1. Measure the output hourly
2. Monitor the specific gravity of the urine
3. Irrigate the catheter with saline 3 times daily
4. Exclude the amount of irrigant instilled from the output

562. In the early postoperative period after a transurethral resection of the prostate, the most common complication the nurse should observe for is:
1. Sepsis
2. Hemorrhage
3. Leakage around the catheter
4. Urinary retention with overflow

563. Three days after prostate surgery a client's Foley catheter and continuous bladder irrigation (CBI) are to be removed, and the nurse discusses what to expect with the client. The nurse recognizes that the teaching has been understood when the client states, "After the catheter is removed I probably will:
1. Have dilute urine."
2. Exhibit dark red urine."
3. Be unable to pass my urine."
4. Experience some burning on urination."

564. The nurse would know that a client who has had a transurethral resection of the prostate understood his discharge teaching when he says, "I should:
1. Attempt to void every 3 hours when I'm awake."
2. Get out of bed into a chair for several hours daily."
3. Call the physician if my urinary stream decreases."
4. Avoid vigorous exercise for 6 months after surgery."

565. After a suprapubic prostatectomy, the nurse understands that a client's plan of care must include the prevention of postoperative deep vein thrombosis. This can best be achieved by increasing the:
1. Coagulability of the blood
2. Velocity of the venous return
3. Effectiveness of internal respiration
4. Oxygen-carrying capacity of the blood

566. An acute, life-threatening complication that the nurse should assess a client for in the early postoperative period after a partial nephrectomy would be:
1. Sepsis
2. Hemorrhage
3. Renal failure
4. Paralytic ileus

567. The nurse's postoperative plan of care for a client who has had a partial nephrectomy should include:
1. Giving the client a regular diet on the first postoperative day
2. Clamping the nephrostomy tube when the client is out of bed
3. Turning the client from the back to the operated side every 2 to 3 hours
4. Leaving the client's original dressing in place for at least the first 48 hours

568. After a partial nephrectomy the observation about the client's urinary output that the nurse should recognize as normal is that urine:
1. Output will be less than 30 ml per hour
2. Specific gravity will remain less than 1.010
3. Will remain dark red with clots for 3 to 4 days
4. Will drain from the wound for at least several days

569. After a partial nephrectomy a client is being discharged with the nephrostomy tube in place. The nurse should instruct the client to:
1. Limit the intake of fluids
2. Remain on bed rest at home
3. Irrigate the nephrostomy tube
4. Change the dressings frequently

570. A client with chronic renal failure is on a restricted protein diet and is taught about high-biologic-value protein foods. An understanding of the rationale for this diet is demonstrated when the client states that high-biologic-value protein foods are:
1. Needed to increase weight gain
2. Necessary to prevent muscle wasting
3. Used to increase urea blood products
4. Responsible for controlling hypertension

571. The nurse would be aware that a client with chronic renal failure recognizes an adequate source of high-

biologic-value protein when the food the client selected from the menu was:
1. Apple juice
2. Raw carrots
3. Cottage cheese
4. Whole wheat bread

572. A male client with a history of chronic renal failure is hospitalized. The nurse assesses the client for symptoms of related renal insufficiency, which include:
1. Facial flushing
2. Edema and pruritus
3. Dribbling after voiding
4. Diminished force and caliber of stream

573. A client with uremic syndrome has the potential to develop many complications. The complication that the nurse should anticipate in such a client is:
1. Hypotension
2. Hypokalemia
3. Flapping hand tremors
4. An elevated hematocrit value

574. The physician decides to treat a client with a history of chronic renal failure with continuous ambulatory peritoneal dialysis (CAPD). When assessing the client before the institution of CAPD, the nurse should be alert for the presence of:
1. Motivation
2. Dysrhythmias
3. Emotional lability
4. Pulmonary problems

575. A client who is to begin continuous ambulatory peritoneal dialysis (CAPD) asks the nurse what this treatment entails. The explanation should include information that:
1. Peritoneal dialysis is done in an ambulatory care clinic
2. There is continuous hemodialysis and peritoneal dialysis
3. There is continuous contact of dialysate with peritoneal membrane
4. About a quarter of a liter of dialysate is maintained intraperitoneally

576. Diet instruction for a client who is being treated with continuous ambulatory peritoneal dialysis (CAPD) for chronic glomerulonephritis should include the need for:
1. Low-calorie foods
2. High-quality protein
3. Increased fluid intake
4. Foods rich in potassium

577. A client with chronic glomerulonephritis will begin to perform continuous ambulatory peritoneal dialy-

sis (CAPD) at home. To decrease the discomfort of the peritoneal dialysis, the nurse should suggest that the client:
1. Eat a diet that is very low in fiber
2. Instill 2 liters of dialysate solution quickly
3. Keep the outflow bag level with the abdomen
4. Apply a heating pad to the abdomen during inflow

578. A client with renal insufficiency is to have peritoneal dialysis. Preparation for insertion of the peritoneal dialysis catheter includes:
1. IV pyelogram, vital signs, emptying the bladder
2. Catheterization, general anesthetic, IV pyelogram
3. Cleansing enema, consent form, general anesthetic
4. Cleansing enema, emptying the bladder, consent form

579. Nursing measures related to the inflow of dialysate fluid include:
1. Infusing the dialysate solution over 2 hours
2. Slightly warming the solution before instilling
3. Positioning the client in the side-lying position
4. Withholding medication until all solution is administered

580. If a client on peritoneal dialysis develops symptoms of severe respiratory difficulty during the infusion of the dialysate solution, the nurse should:
1. Slow the rate of infusion
2. Auscultate the lungs for breath sounds
3. Drain the fluid from the peritoneal cavity
4. Place the client in a low-Fowler's position

581. The nurse should monitor the client on peritoneal dialysis for complications such as:
1. Fever, oliguria, and hemorrhage
2. Pruritus, hemorrhage, and cloudy outflow
3. Abdominal pain, tachycardia, and pruritus
4. Tachycardia, cloudy outflow, and abdominal pain

582. A client is scheduled for the creation of an internal arteriovenous fistula and the placement of an external arteriovenous shunt to be used until the fistula matures. The nurse should keep in mind that:
1. Blood pressure readings will be higher in the arm with the fistula than in the arm with the shunt
2. The shunt is more subject to the complications of hemorrhage, clotting, and infection than the fistula
3. IV fluids should not be infused in the arm with the shunt, but they are permitted in the arm with the fistula
4. A light surgical dressing can be applied over the fistula incision, but the shunt should be thoroughly covered with a heavier dressing

583. A client has end-stage renal disease and is on hemodialysis. During dialysis, the client complains of nausea and a headache and appears confused. Operating on standing protocols, the nurse should:
1. Administer an antiemetic
2. Attempt to reorient the client
3. Decrease the rate of exchange
4. Monitor for changes in vital signs

584. A client with chronic renal failure is accepted for a kidney transplant and attends a group educational program for potential transplant candidates. The client asks the nurse which kidney will be removed. The nurse's best response would be:
1. "Neither of your kidneys will be removed unless they are infected."
2. "It is up to the surgeon as to which kidney is replaced with the new one."
3. "The kidney that is the most diseased is removed and replaced with the new one."
4. "Your right kidney will be removed because it has a longer renal vein, making transplant easier."

585. A client has end-stage renal disease and is admitted for a kidney transplant. The nurse is aware that the donor must:
1. Have the same blood type
2. Be a member of the same family
3. Be approximately the same body size
4. Have matching leukocyte antigen complexes

586. A client who is to have a kidney transplant should be taught to expect:
1. A colonoscopy
2. An appendectomy
3. A partial gastric resection
4. An intravenous cystogram

587. When a client returns from the recovery room after a kidney transplant, the nurse should plan to measure the client's urinary output every:
1. 15 minutes
2. 30 minutes
3. 1 hour
4. 2 hours

588. After a successful kidney transplant is completed, the nurse should anticipate that laboratory studies for a client with end-stage renal disease will demonstrate:
1. A correction of hypotension
2. An elevated serum albumin
3. An increased specific gravity
4. A decreasing serum creatinine

589. A client with a transplanted kidney is taught the signs of rejection. The nurse would know that the teaching was effective when the client says that a sign of rejection would be:
1. Weight loss
2. A subnormal temperature
3. An elevated blood pressure
4. An increased urinary output

590. A client who has had a kidney transplant develops leukopenia 3 weeks after surgery. The nurse should be aware that the leukopenia is probably caused by:
1. A bacterial infection
2. High creatinine levels
3. Rejection of the kidney
4. The antirejection medications

GASTROINTESTINAL (GI)

591. A client has decided to become a total vegetarian (vegan) and wishes to plan a diet to ensure adequate protein quality. To provide guidance, the nurse should instruct this client to:
1. Add milk to grains to provide complete proteins
2. Use eggs with plant foods to provide essential amino acids
3. Plan a careful mixture of plant proteins to provide a balance of amino acids
4. Add cheese to grains and beans to increase the quality of the protein consumed

592. An obese client asks the nurse how to lose weight. Before answering, the nurse should remember that long-term weight loss occurs best when:
1. Fats are limited in the diet
2. Eating patterns are altered
3. Carbohydrates are regulated
4. Exercise is part of the program

593. To motivate an obese client to eventually include aerobic exercises in a weight-reduction program, the nurse should discuss exercise and its relationship to weight loss. The nurse would know that this teaching was effective when the client states, "I know that exercise will:
1. Raise my heart rate."
2. Decrease my appetite."
3. Lower my metabolic rate."
4. Increase my lean body mass."

594. A client is admitted to the hospital with complaints of frequent loose, watery stools; anorexia; malaise; and a considerable weight loss during the past month. Laboratory findings indicate leukocytosis and an elevated sedimentation rate. The nurse recognizes that the presenting symptoms in conjunction with the laboratory findings could be indicative of:

1. The consistent, long-term use of an irritant-type laxative
2. An emotional response that has resulted in physical symptoms
3. Systemic responses of the body to a localized inflammatory process
4. Poor dietary practices that have resulted in an alteration of bowel function

595. When assessing a client's abdomen, the nurse palpates the area directly above the umbilicus. This area is known as the:
1. Iliac area
2. Epigastric area
3. Hypogastric area
4. Suprasternal area

596. An adaptation after a gastroscopy that indicates a major complication is:
1. Difficulty swallowing
2. Increased GI motility
3. Abdominal distention
4. Nausea and vomiting

597. A client is scheduled for a barium swallow; the nurse should:
1. Ask the client about allergies to iodine
2. Ensure a laxative is ordered after the test
3. Give only clear fluids on the day of the test
4. Administer cleansing enemas before the test

598. Routine postoperative intravenous fluids are designed to supply hydration and electrolytes and only limited energy. Since 1 L of a 5% dextrose solution contains 50 g of sugar, 3 L/day would supply approximately:
1. 400 kilocalories
2. 600 kilocalories
3. 800 kilocalories
4. 1000 kilocalories

599. After abdominal surgery a client is placed on a progressive postsurgical diet. This diet is characterized by progressive alterations in the:
1. Caloric content of food
2. Nutritional value of food
3. Texture and digestibility of food
4. Variety of food and fluids included

600. The diet ordered for a client permits 190 g of carbohydrates, 90 g of fat, and 100 g of protein. The nurse understands that this diet contains approximately:
1. 2200 calories
2. 2000 calories

3. 1300 calories
4. 1600 calories

601. A client's serum albumin value is 2.8 g/dl. The nurse should evaluate client teaching as successful when the client says, "For lunch I am going to have:
1. Fruit salad."
2. Sliced turkey."
3. Spinach salad."
4. Clear beef broth."

602. A client who is a heavy smoker is placed on a high-calorie, high-protein diet. In light of the history of smoking the client should also be encouraged to eat foods high in:
1. Niacin
2. Thiamin
3. Vitamin C
4. Vitamin B_{12}

603. A client is cautioned to avoid vitamin D toxicity while increasing protein intake. The nurse would know that the teaching was understood when the client states, "I must increase my intake of:
1. Tofu products."
2. Fruit and eggnog."
3. Powdered whole milk."
4. Cottage cheese custard."

604. When helping a client plan a therapeutic diet, the nurse is aware that an excellent food source of vitamin C is:
1. Apples
2. Lettuce
3. Broccoli
4. Apricots

605. The physician tells a client that an increase in vitamin E and beta-carotene is important for healthier skin. The nurse teaches the client that excellent food sources of both of these substances are:
1. Spinach and mangoes
2. Fish and peanut butter
3. Oranges and grapefruits
4. Carrots and sweet potatoes

606. Because of multiple physical injuries and emotional concerns, a hospitalized client is at high risk to develop a stress (Curling's) ulcer. The nurse should know that stress ulcers are usually evidenced by:
1. Unexplained shock
2. Melena for several days
3. Sudden massive hemorrhage
4. A gradual drop in the hematocrit value

607. A client should be instructed to avoid straining on defecation to prevent bleeding after pelvic surgery. The nurse is aware that the related teaching has been understood when the client states, "I must increase my intake of:
1. Milk products."
2. Ripe bananas."
3. Green vegetables."
4. Creamed potatoes."

608. A client with Parkinson's disease complains about a problem with elimination. The nurse should encourage the client to:
1. Eat a banana daily
2. Decrease fluid intake
3. Take cathartics regularly
4. Increase residue in the diet

609. The physician orders three stool specimens for occult blood from a client who complains of blood-streaked stools and a 10-pound weight loss in 1 month. To ensure valid test results the nurse should instruct the client to:
1. Avoid eating red meat before testing
2. Test the specimen while it is still warm
3. Discard the first stool of the day and use the next three stools
4. Take three specimens from different sections of the fecal sample

610. When a client develops steatorrhea, the nurse should describe this stool as:
1. Dry and rock-hard
2. Clay colored and pasty
3. Bulky and foul smelling
4. Black and blood-streaked

611. A client is admitted to the hospital with a diagnosis of acute pancreatitis. The physician's orders include nothing by mouth and total parenteral nutrition. When the client asks for an explanation for this therapy, the nurse's most accurate response would be:
1. "It is the easiest method for staff to administer needed nutrition."
2. "It is the safest method for meeting your nutritional requirements."
3. "It will satisfy your desire for food without the discomfort associated with eating."
4. "It will meet your nutritional needs without causing the discomfort associated with eating."

612. The physician orders total parenteral nutrition 1 L q12 hours. The primary nursing responsibility should be to monitor the client's:
1. Electrolytes
2. Urinary output
3. Administration rate
4. Serum glucose levels

613. When preparing a client to go home with total parenteral nutrition, the nurse should help the client plan:
1. Which days will be used for administration
2. For daily insertion of the circulatory access
3. For professional help to administer the TPN
4. A schedule of administration around normal activity

614. A client has symptoms associated with salmonellosis. Relevant data to gather from this client include a history of:
1. Rectal cancer in the family
2. All foods eaten in the past 24 hours
3. Any recent extreme emotional stress
4. An upper respiratory infection in the past 10 days

615. Contact precautions for a client with salmonellosis include:
1. Isolation in a private room
2. Wearing a gown if soiling is likely
3. Limiting visiting hours during the acute phase
4. Wearing a mask when emptying the client's bedpan

616. A client who is suspected of having salmonellosis asks how the diagnosis is confirmed. The nurse should respond that the diagnosis is established by a:
1. CBC
2. Urinalysis
3. Stool culture
4. Febrile agglutinin test

617. A client is admitted to the hospital with the diagnosis of acute salmonellosis. The nurse would expect that the client will be receiving:
1. Antacids
2. Electrolytes
3. Antidiarrheics
4. Antispasmodics

618. During a health symposium, a nurse teaches the group how to prevent food poisoning. The nurse evaluates that the teaching has been understood when one of the participants states:
1. "All meats and cream-based foods need to be refrigerated."
2. "Once most food is cooked it does not need to be refrigerated."
3. "Poultry should be stuffed and then refrigerated before cooking."
4. "Cooked food should be cooled before being put into the refrigerator."

619. The laboratory values of a client with cancer of the esophagus show a hemoglobin of 7 g/dl, hematocrit of 25%, and RBC count of 2.5 million/mm³. Considering these data, an appropriate nursing diagnosis for the client at this time is:
    1. Altered nutrition: less than body requirements related to dysphagia
    2. Ineffective airway clearance related to tumor growth and metastasis
    3. Pain related to pressure of the tumor on surrounding tissues and nerves
    4. Risk for injury related to possible metastasis and subsequent airway obstruction

620. Immediately after esophageal surgery the priority nursing assessment concerns the client's:
    1. Incision
    2. Respirations
    3. Level of pain
    4. Nasogastric tube

621. A client with achalasia is to have bougienage to dilate the lower esophagus and cardiac sphincter. After the procedure the nurse should assess the client for esophageal perforation, which is indicated by:
    1. Faintness and feelings of fullness
    2. Diaphoresis and cardiac palpitations
    3. Increased heart rate and abdominal pain
    4. Increased blood pressure and urinary output

622. A client with a hiatal hernia asks the nurse how to best prevent esophageal reflux. The nurse's best response would be:
    1. "Increase your intake of fat with each meal."
    2. "Lie down after eating to help your digestion."
    3. "Reduce your caloric intake to foster weight reduction."
    4. "Drink several glasses of fluid during each of your meals."

623. When performing the initial history and physical examination of a client with a tentative diagnosis of peptic ulcer, the nurse would expect the client to describe the pain as:
    1. Gnawing epigastric pain or boring pain in the back
    2. Sudden sharp abdominal pain, increasing in intensity
    3. Heartburn and substernal discomfort when lying down
    4. Located in the right shoulder and preceded by nausea

624. A client is diagnosed as having a peptic ulcer. The nurse would expect that the client's pain:
    1. Occurs 1 to 3 hours after meals
    2. Is intensified when vomiting occurs
    3. Increases when fatty foods are ingested
    4. Begins in the epigastrium and radiates to the abdomen

625. A traveling salesman develops gastric bleeding and is hospitalized. An important etiologic clue for the nurse to explore while taking this client's history would be:
    1. Any recent foreign travel
    2. The client's usual dietary pattern
    3. Any change in status of family relationships
    4. The medications that the client has been taking

626. Once a client's gastric bleeding is controlled, the physician orders individual dietary management. The meal pattern that would probably be most appropriate for this client would be:
    1. A flexible plan according to the client's appetite
    2. Limited food and fluid intake when pain is present
    3. Regular meals and snacks to limit gastric discomfort
    4. Three meals large enough to supply adequate energy

627. After an acute episode of upper GI bleeding, a client vomits undigested antacids and complains of severe epigastric pain. The nursing assessment reveals an absence of bowel sounds, pulse rate of 134, and shallow respirations of 32 per minute. In addition to calling the physician, the nurse should:
    1. Start O₂ per nasal cannula at 3 to 4 L per minute
    2. Keep the client npo in preparation for possible surgery
    3. Place the client in the supine position with the legs elevated
    4. Ask the client whether any red or black stools have been noted

628. Following a subtotal gastrectomy (Billroth I), a client begins to eat more food in varied forms. After meals the client experiences a cramping discomfort and a rapid pulse with waves of weakness, which are often followed by nausea and vomiting. The nurse recognizes that this response is known as the dumping syndrome and is caused by:
    1. A slowed passage of food dumping into the small intestine
    2. A rapid passage of dilute food mixture into the small intestine
    3. Rapid passage of hyperosmolar food solution into the small intestine
    4. Food that is less concentrated than surrounding extracellular fluid entering the small intestine

629. A client with gastric cancer asks whether this cancer will spread. The nurse recognizes that the client is looking for reassurance but knows that gastric cancers are most likely to metastasize to the:
    1. Liver and lung
    2. Bone and brain
    3. Pancreas and brain
    4. Lymph nodes and blood

630. Twelve hours after a subtotal gastrectomy, the nurse notes large amounts of bloody drainage from the client's nasogastric tube. The nurse should:
    1. Instill 30 ml of iced normal saline into the tube
    2. Clamp the tube and call the physician immediately
    3. Report the type and quantity of drainage to the physician
    4. Continue to monitor the drainage and record the observations

631. The nurse should assess for the development of pernicious anemia when a client has a history of:
    1. Hemorrhage
    2. Diabetes mellitus
    3. Poor dietary habits
    4. Having had a gastrectomy

632. After a client has a total gastrectomy, the nurse should plan to include in the discharge teaching the need for:
    1. Weekly injections of vitamin B_{12}
    2. Regular daily use of a stool softener
    3. Monthly injections of iron dextran (Imferon)
    4. Daily replacement therapy of pancreatic enzymes

633. The nurse discusses dietary needs with a client who has had a gastroduodenostomy (Billroth I). The nurse knows that the teaching was understood when the client states, "I will plan:
    1. Five or six small meals every day, limiting bulk."
    2. A diet of blenderized foods for an indefinite period."
    3. To gradually resume my normal eating routine and diet."
    4. A diet high in carbohydrates, proteins, and fats to replace lost nutrients."

634. After 2 months of self-management for symptoms of gastritis is unsuccessful, a client goes to the physician and extensive carcinoma of the stomach is diagnosed. The client asks the nurse how the disease got so advanced. The nurse's explanation should be based on the knowledge that carcinoma of the stomach is:
    1. Difficult to accurately diagnose until late in the disease process

2. Painful in early stages and often misdiagnosed as myocardial infarction
    3. Usually diagnosed following the discovery of enlarged lymph nodes in the epigastric area
    4. Rarely diagnosed early because the symptoms are usually nonspecific until late in the disease

635. A client with extensive gastric carcinoma is to be admitted to the hospital for an esophagojejunostomy. When preparing this client for surgery, the nurse should include information about the possibility that:
    1. A chest tube will be in place in addition to a nasogastric tube
    2. Liquids by mouth may be permitted the evening after surgery
    3. Complete bed rest may be necessary for 48 hours after surgery
    4. Trendelenburg's position will be used on the first day after surgery

636. When teaching a client how to avoid the dumping syndrome following a gastrectomy, the nurse should emphasize:
    1. Increasing activity after eating
    2. Avoiding excess fluids with meals
    3. Eating heavy meals to delay emptying
    4. Providing carbohydrates with each meal

637. Immediately after a subtotal gastrectomy a client returns to the surgical unit. The nurse identifies small blood clots in the gastric drainage. The nurse should:
    1. Clamp the tube
    2. Consider this a normal event
    3. Instill the tube with iced saline
    4. Notify the physician of this finding

638. On the third postoperative day after a subtotal gastrectomy, a client complains of severe abdominal pain. The nurse palpates the client's abdomen and notes rigidity. The nurse should first:
    1. Assist the client to ambulate
    2. Assess the client's vital signs
    3. Administer the prescribed analgesic
    4. Encourage the use of the spirometer

639. To determine when a client who has had a subtotal gastrectomy can begin oral feedings after surgery, the nurse must assess for the:
    1. Presence of flatulence
    2. Extent of incisional pain
    3. Stabilization of hematocrit levels
    4. Occurrence of dumping syndrome

640. A client who has had a gastric ulcer asks what to do if the epigastric pain occurs. The nurse would know

that the teaching was effective when the client states, "I will:
1. Increase my food intake."
2. Take the aspirin with milk."
3. Eliminate fluids with meals."
4. Take an antacid preparation."

641. A client is diagnosed as having a peptic ulcer. When teaching about peptic ulcers, the nurse should instruct the client to report any stools that appear:
1. Frothy
2. Ribbon shaped
3. Pale or clay colored
4. Dark brown or black

642. After abdominal surgery a client returns to the unit with a gastric tube to decompression. The physician has ordered an antiemetic q6h prn for nausea. When the client complains of nausea the first action by the nurse should be to:
1. Check for placement of the tube
2. Administer the ordered antiemetic
3. Irrigate the tube with normal saline
4. Notify the physician of the problem

643. A male client has a subtotal gastrectomy because of a perforated gastric ulcer. Postoperatively he has a nasogastric tube to low suction and IV fluids. Three hours after surgery the client complains of nausea and abdominal pain, his abdomen appears distended, and there are no bowel sounds. Considering the type of surgery and the orders for pain medication and instillation of the gastric tube, the nurse should first:
1. Instill saline into the tube
2. Give the prn pain medication
3. Check stools and gastric drainage for blood
4. Notify the physician of absent bowel sounds

644. The nurse notes some bright red blood in a client's gastric drainage 4 hours after a subtotal gastrectomy. The nurse should:
1. Clamp the tube and call the physician
2. Gently irrigate the tube with 30 ml normal saline
3. Continue to monitor the drainage from the tube and record the observations
4. Reduce the pressure of the suction and record observations of the drainage

645. The day before a client who has had a subtotal gastrectomy is to be discharged, the client complains of perspiring and having epigastric discomfort about a half hour after eating lunch. The symptoms disappear within a few minutes. The nurse recognizes that the teaching about prevention is under-

stood when the client states, "I can limit these symptoms by:
1. Increasing fluids with each meal."
2. Avoiding spicy, gas-forming meals."
3. Resting before and after each meal."
4. Eating small, low-carbohydrate meals."

646. After a subtotal gastrectomy a client develops the dumping syndrome. In addition, about 2 hours after the initial postmeal attack, the client experiences a second period of discomfort, feeling somewhat "shaky." The nurse recognizes that this later follow-up effect, which is precipitated by the dumping syndrome, is caused by:
1. The increased fat content and larger amount of seasoned food, creating digestive discomfort
2. The increased use of simple carbohydrates in meals, creating a more prolonged glucose rise
3. Hyperglycemia from a rapidly absorbed glucose load, which overwhelms the insulin-adjusting mechanism
4. Mild hypoglycemia from an overproduction of insulin that occurs in response to the postprandial blood glucose rise

647. The characteristics that would alert the nurse that a client is at increased risk of developing gallbladder disease would be:
1. Female, older than 40, obese
2. Male, younger than 40, past history of hepatitis
3. Male, older than 40, low serum cholesterol level
4. Female, younger than 40, family history of gallstones

648. A client with a tentative diagnosis of cholecystitis is discharged from the emergency room with instructions to make an appointment for a more definitive diagnostic workup. The recommendation that would produce the most valuable diagnostic information would be:
1. "Keep a journal related to your pain."
2. "Save all stool and urine for inspection."
3. "Follow the doctor's orders exactly without question."
4. "Keep a record of the amount of fluid you drink daily."

649. The nurse asks a client to make a list of the foods that cause dyspepsia. If the client has cholecystitis, the foods that are most likely to be included on this list would be:
1. Nuts and popcorn
2. Meatloaf and baked potato
3. Chocolate and boiled shrimp
4. Fried chicken and buttered corn

650. The physician orders a modified diet for a client with cholecystitis. The nurse understands that with cholecystitis:
1. Soft-textured foods are used to reduce the digestive burden
2. Low-cholesterol foods are used to avoid further formation of gallstones
3. Fat is decreased to avoid stimulation of the cholecystokinin mechanism for bile release
4. Increased protein and kilocalories are necessary to promote tissue healing and provide energy

651. A client develops a gallstone that becomes lodged in the common bile duct. The physician schedules an endoscopic sphincterotomy. Preoperative teaching should include information that for the procedure the client will:
1. Have a spinal anesthetic
2. Receive an epidural block
3. Have a general anesthetic
4. Receive an intravenous sedative

652. A client has cholelithiasis with possible obstruction of the common bile duct. Before the scheduled cholecystectomy, nutritional deficiencies and excesses should be corrected. A nutritional assessment should be conducted to determine whether the client:
1. Is deficient in vitamins A, D, and K
2. Eats adequate amounts of dietary fiber
3. Consumes excessive amounts of protein
4. Has excessive levels of potassium and folic acid

653. A client undergoes an abdominal cholecystectomy with common duct exploration. In the immediate postoperative period, the nursing action that should assume the highest priority for this client is:
1. Irrigating the T-tube frequently
2. Changing the dressing at least bid
3. Encouraging coughing and deep breathing
4. Promoting an adequate fluid intake by mouth

654. The nurse is aware that a client's T-tube has been inserted during a resection of the pancreas to:
1. Divert the bile flow to the cystic duct
2. Drain blood and pus from the operative site
3. Prevent postoperative infection at the site of the incision
4. Facilitate bile drainage while the common duct is edematous

655. A client with cholelithiasis has a laser laparoscopic cholecystectomy. Postoperatively it is most appropriate for the nurse to:
1. Wait about 24 hours to begin clear liquids
2. Monitor the abdominal incision for bleeding
3. Offer the client clear carbonated beverages
4. Instruct the client to resume moderate activity in 2 to 3 days

656. A client returns from surgery after a resection of the pancreas for cancer with a T-tube in place. The nurse recognizes that immediately after this surgery the position that would provide the most comfort for the client is the:
1. Sims' position
2. Supine position
3. Side-lying position
4. Low-Fowler's position

657. Because of prolonged bile drainage from a T-tube, a client may develop symptoms related to a lack of fat-soluble vitamins such as:
1. Easy bruising
2. Muscle twitching
3. Excessive jaundice
4. Tingling of the fingers

658. A client with obstruction of the common bile duct may show a prolonged bleeding and clotting time because:
1. Vitamin K is not absorbed
2. The ionized calcium level falls
3. The extrinsic factor is not absorbed
4. Bilirubin accumulates in the plasma

659. A client with cholelithiasis is scheduled for a lithotripsy. Preoperative teaching should include the information that:
1. Narcotics will be available for postoperative pain
2. A fever is a common response to this intervention
3. Heart palpitations commonly occur after the procedure
4. Analgesics and anesthetics are not necessary during the procedure

660. A client is to be discharged after a laser laparoscopic cholecystectomy. The nurse would recognize that the discharge instructions were understood when the client states:
1. "I can change the bandages every day."
2. "I should remain on a full liquid diet for 3 days."
3. "I should not bathe the surgical sites for a week."
4. "I may have mild shoulder pain for about a week."

661. After a cholecystectomy, a client asks whether there are any dietary restrictions that must be followed. The nurse would recognize that the dietary teaching was understood when the client tells a family member:

1. "I will need to avoid fatty foods for the rest of my life."
2. "I should not eat those foods that upset me before I had surgery."
3. "Most people can tolerate a regular diet after this type of surgery."
4. "Most people need to eat a high protein diet for several months after surgery."

662. A client has an extensive surgical revision of the head of the pancreas because of cancer. After surgery, to decrease the chance of hemorrhage at the operative site, the nurse should:
1. Keep the client in the supine position
2. Maintain patency of the nasogastric tube
3. Replace fat-soluble vitamins as necessary
4. Administer the ordered tube feeding slowly

663. When teaching a client about the diet following a Whipple procedure performed for cancer of the pancreas, the statement the nurse should include would be:
1. "There are no dietary restrictions; you may eat what you desire."
2. "Your diet should be low in calories to prevent taxing your diseased pancreas."
3. "Meals should be restricted in protein because of your compromised liver function."
4. "Low-fat meals should be eaten because of interference with your fat digestion mechanism."

664. A long-term complication that a client must be made aware of following a Whipple procedure for cancer of the pancreas is hypoinsulinism. The nurse would know that the teaching about hypoinsulinism is understood when the client states, "I should seek medical supervision if I experience:
1. Oliguria."
2. Anorexia."
3. Weight gain."
4. Increased thirst."

665. After revision of the pancreas because of cancer, total parenteral nutrition is instituted via a central venous infusion route. During the fourth hour of the first infusion the client complains of nausea, fatigue, and a headache. The hourly urine output is twice the amount of the previous hour. The nurse should call the physician and:
1. Stop the infusion and cover the insertion site
2. Slow the infusion and check the serum glucose level
3. Prepare the client for immediate surgery for possible bowel obstruction
4. Increase fluids via a peripheral IV route and give analgesics for the headache

666. A client is to be discharged with a percutaneous catheter for home administration of total parenteral nutrition (TPN). The nurse should help the client to:
1. Learn how to change the percutaneous catheter
2. Determine which days to self-administer the TPN
3. Schedule the TPN administration around meal times
4. Arrange professional help to monitor the administration of the TPN

667. After surgery for cancer of the pancreas, the client's nutritional and fluid regimen will be influenced by the remaining amount of functioning pancreatic tissue. Considering both the exocrine and the endocrine functions of the pancreas, the client's postoperative regimen would primarily include managing the intake of:
1. Alcohol and caffeine
2. Vitamins and minerals
3. Fluids and electrolytes
4. Fats and carbohydrates

668. A client with a 20-year history of excessive alcohol use is admitted to the hospital with jaundice and ascites. A priority nursing action during the first 48 hours after the client's admission will be to:
1. Monitor the client's vital signs
2. Increase the client's fluid intake
3. Improve the client's nutritional status
4. Identify the client's reasons for drinking

669. A male client with liver dysfunction reports that his gums bleed spontaneously. In addition, the nurse notes small hemorrhagic lesions on his face. The nurse recognizes that the client needs additional:
1. Bile salts
2. Folic acid
3. Vitamin A
4. Vitamin K

670. A client with ascites is scheduled for a paracentesis. To prepare the client for the abdominal paracentesis the nurse should:
1. Medicate the client for pain
2. Encourage the client to drink fluids
3. Shave and prep the client's abdomen
4. Instruct the client to empty the bladder

671. A client is diagnosed as having hepatitis A. The information from the health history that is most likely linked to hepatitis A is:
1. Working for a local plumber
2. Washing dishes at a local restaurant
3. Working in a hemodialysis unit of a hospital
4. Being exposed to arsenic compounds at work

672. The nurse should instruct a client with hepatitis A about untoward signs and symptoms that may develop and should be reported to the physician, especially:
1. Fatigue
2. Anorexia
3. Yellow urine
4. Clay-colored stools

673. A client with jaundice associated with hepatitis expresses concern over the change in skin color. The nurse should recognize that this color change is a result of:
1. Stimulation of the liver to produce an excess quantity of bile pigments
2. Inability of the liver to remove normal amounts of bilirubin from the blood
3. Increased destruction of red blood cells during the acute phase of the disease
4. Decreased prothrombin levels, leading to multiple sites of intradermal bleeding

674. After an acute episode of hepatitis, a client is to be discharged to continue recovery at home. The nurse should expect that the diet the physician will order will be:
1. Low calorie, high protein, low carbohydrate, low fat
2. High calorie, low protein, high carbohydrate, high fat
3. Low calorie, low protein, low carbohydrate, moderate fat
4. High calorie, high protein, high carbohydrate, moderate fat

675. A mother whose son has hepatitis A states that there is only one bathroom in their home and she is worried that other members of the family could get hepatitis. The nurse's best reply would be:
1. "I suggest that you buy a commode exclusively for your son's use."
2. "Your son may use the bathroom, but you need to use disposable toilet covers."
3. "There is no problem with your son sharing the same bathroom with everyone."
4. "It is important that your son and all family members wash their hands after using the bathroom."

676. The physician orders contact precautions for a client with hepatitis A. In addition to standard precautions, the isolation procedures that must be followed are:
1. A private room is required, and the door must be kept closed
2. Persons entering the room must wear a gown, a mask, and gloves

3. Gowns and gloves must be worn only when handling the client's soiled linen, dishes, or utensils
4. A gown and gloves must be worn when handling articles possibly contaminated by urine or feces

677. A client has a tentative diagnosis of primary biliary cirrhosis. Symptoms include jaundice, ascites, and peripheral edema. When performing the physical assessment, the skin change the nurse would expect to observe is:
1. Vitiligo
2. Hirsutism
3. Melanosis
4. Telangiectasis

678. Immediately following a liver biopsy, a client is placed on the right side. The nurse explains that this position should be maintained for 60 to 90 minutes because it will:
1. Help stop bleeding if any occurs
2. Restore circulating blood volume
3. Be the position of greatest comfort
4. Help reduce fluid trapped in the biliary ducts

679. Following a liver biopsy the nurse checks the client's dressing and notices a moderately large amount of bile-colored drainage. The client also complains of right upper quadrant pain. The nurse should recognize that:
1. Fluid is leaking into the intestine
2. The pancreas has been lacerated
3. A biliary vessel has been penetrated
4. This is the normal, expected response

680. The serum ammonia level of a client with cirrhosis is elevated. As a priority the nurse should:
1. Weigh the client daily
2. Observe for increasing confusion
3. Measure the urine specific gravity
4. Restrict the client's oral fluid intake

681. A client with Laënnec's cirrhosis and a Sengstaken-Blakemore tube in place becomes increasingly confused and tries to climb out of bed. The breath has become fetid. The nursing priority should be to:
1. Apply a safety jacket
2. Notify the physician immediately
3. Administer the prn sedative as ordered
4. Administer oxygen via a nasal catheter

682. A client with cirrhosis of the liver and malnutrition begins to develop slurred speech, confusion, drowsiness, and tremors. With these symptoms the diet would be limited to:
1. 20 g protein, 2000 calories
2. 80 g protein, 1000 calories

3. 100 g protein, 2500 calories
4. 150 g protein, 1200 calories

683. A client develops peritonitis and sepsis following the surgical repair of a ruptured diverticulum. The nurse should expect an assessment of the client to reveal:
1. Bradycardia
2. Hypertension
3. Abdominal rigidity
4. Increased bowel sounds

684. One month after abdominal surgery a client is readmitted to the hospital with recurrent abdominal pain and fever. The diagnosis is fistula formation with peritonitis. The nurse should place the client in the:
1. Supine position
2. Right Sims' position
3. Semi-Fowler's position
4. Position that is most comfortable

685. A client with colitis inquires as to whether surgery will ever be necessary. When teaching about the disease and its treatment, the nurse should emphasize that:
1. Medical treatment for colitis is curative, and surgery is not required
2. Surgery for colitis is considered only as a last resort for most clients
3. Surgery for colitis is done early in the course of the disease for most clients
4. Medical treatment is all that will be needed if the client can acquire some emotional stability

686. After surgery for creation of an ileostomy the client is to be discharged. Before discharge the primary nursing intervention should be to:
1. Coax the client into caring for the ileostomy alone
2. Evaluate the client's ability to care for the ileostomy
3. Ensure that the client understands the dietary limitations that must be followed
4. Have the client change the dry sterile dressing on the incision without assistance

687. When teaching a client about the signs of colorectal cancer, the nurse stresses that the most common complaint of persons with colorectal cancer is:
1. Abdominal pain
2. Rectal bleeding
3. Change in bowel habits
4. Change in caliber of stools

688. A client diagnosed with cancer of the sigmoid colon is to have an abdominoperineal resection with a permanent colostomy. Before surgery a low-residue diet is ordered. The nurse explains that this is necessary to:
1. Lower the bacteria count in the GI tract
2. Limit production of flatus in the intestine
3. Prevent irritation of the intestinal mucosa
4. Reduce the amount of stool in the large bowel

689. An abdominoperineal resection with a colostomy is scheduled for a client with cancer of the rectum. The nurse recognizes that the physician will need the client to sign a consent for a:
1. Permanent sigmoid colostomy
2. Permanent ascending colostomy
3. Temporary double-barrel colostomy
4. Temporary transverse loop colostomy

690. A client has a permanent sigmoid colostomy because of cancer. The physician orders daily colostomy irrigations. The nurse should explain to the client that the primary purpose of these irrigations is to:
1. Prevent straining at passage of stool
2. Establish a regular elimination schedule
3. Decrease the amount of flatus in the bowel
4. Limit the amount of fluid lost from the intestine

691. To promote perineal wound healing after an abdominoperineal resection, the nurse should encourage the client to assume the:
1. Knee-chest position
2. Left or right Sims' position
3. Dorsal recumbent position
4. Left or right side-lying position

692. On the second day following an abdominoperineal resection, the nurse anticipates that the colostomy stoma will appear:
1. Dry, pale pink, and flush with the skin
2. Moist, red, and raised above the skin surface
3. Dry, purple, and depressed below the skin surface
4. Moist, pink, flush with the skin, and painful when touched

693. The nurse plans to teach a client to irrigate a new colostomy when the:
1. Abdominal incision is closed and contamination is no longer a danger
2. Stool starts to become formed, around the seventh postoperative day
3. Client can lie on the side comfortably, about the third postoperative day
4. Perineal wound heals and the client can sit comfortably on the commode

694. A client returns from surgery with a permanent colostomy. During the first 24 hours the colostomy does not drain. The nurse should realize this is a result of:
1. Intestinal edema following surgery
2. A presurgical decrease in fluid intake
3. The absence of gastrointestinal motility
4. Proper functioning of nasogastric suction

695. When preparing to teach a client how to irrigate his colostomy, the nurse should plan to perform the procedure:
1. At least 2 hours before visitors
2. Prior to breakfast and morning care
3. After the client accepts the alteration in body image
4. When the client would have normally had a bowel movement

696. When observing a return demonstration of a colostomy irrigation, the nurse knows that more teaching is required if the client:
1. Clamps off the flow of fluid when feeling uncomfortable
2. Lubricates the tip of the catheter prior to inserting it into the stoma
3. Discontinues the insertion of fluid after only 500 ml of fluid has been instilled
4. Hangs the irrigation bag on the bathroom door clothes hook during fluid insertion

697. The nurse is aware that teaching about colostomy care is understood when the client states, "I will contact my physician and report:
1. If I notice a loss of sensation to touch in stoma tissue."
2. When mucus is passed from the stoma between irrigations."
3. The expulsion of flatus while the irrigating fluid is running out."
4. If I have any difficulty in inserting the irrigating tube in the stoma."

698. A client has a colostomy because of cancer of the colon. Postoperatively it would be most therapeutic for the nurse to:
1. Empty the colostomy bag when it is three-fourths full
2. Allow $\frac{1}{2}$ inch between the stoma and the colostomy bag
3. Help the client to remove the bag on the first day postoperatively
4. Apply stoma adhesive around the stoma before attaching the bag

699. The nurse would know that dietary teaching for a client with a colostomy had been effective when the client states, "It is important that I eat:
1. Food low in fiber so that there is less stool."
2. Bland foods so that my intestines do not become irritated."
3. Everything I ate before the operation, while avoiding foods that cause gas."
4. Soft foods that are more easily digested and absorbed by my large intestine."

700. A client is admitted for repair of bilateral inguinal hernias under a general anesthetic. Before surgery the nurse would assess the client for signs that strangulation may have occurred. An early sign of strangulation would be:
1. Increased flatus
2. Projectile vomiting
3. Sharp abdominal pain
4. Decreased bowel sounds

701. After a bilateral herniorrhaphy a male client should be observed for the development of:
1. Hydrocele
2. Paralytic ileus
3. Urinary retention
4. Thrombophlebitis

DRUG-RELATED RESPONSES (DR)

702. The physician orders 0.2 mg of cyanocobalamin (vitamin B_{12}) IM for a client with pernicious anemia. A vial of the drug labeled 1 ml = 100 mcg is available. The nurse should administer:
1. 0.5 ml
2. 1.0 ml
3. 1.5 ml
4. 2.0 ml

703. The physician orders lidocaine HCl 1.5 mg per minute for a client whose ECG tracing reveals multiple PVBs. The nurse adds 500 mg of lidocaine HCl to 100 ml of D5W. The drop factor of the IV set is 60 gtt/ml. To administer the correct amount of medication the nurse should set the IV at:
1. 10 gtt/minute
2. 12 gtt/minute
3. 14 gtt/minute
4. 18 gtt/minute

704. The physician orders cefazolin sodium (Kefzol) 375 mg IVPB every 8 hours. The vial of powder contains 500 mg of the drug. This must be reconstituted with 2 ml of 0.9% sodium chloride. In the resulting solution 1 ml equals 225 mg of Kefzol. The nurse should administer:

1. 1.2 ml
2. 1.4 ml
3. 1.7 ml
4. 2.2 ml

705. The physician orders aminocaproic acid (Amicar elixir) 4 grams po for a client with an intracerebral hemorrhage. The bottle is labeled 250 mg/ml. The nurse should administer:

1. 16 ml
2. 1.6 ml
3. 0.16 ml
4. 0.016 ml

706. The physician orders 250 mg of an antibiotic IVPB. A vial containing 1 g of the powdered form of the drug must be reconstituted with 2.8 ml of diluent to form a withdrawable volume of 3 ml. Using this solution, the nurse should administer:

1. 0.5 ml
2. 0.7 ml
3. 0.8 ml
4. 1.1 ml

707. The physician orders penicillin G benzathine suspension (Bicillin L-A) 2.45 million units for a client with a sexually transmitted disease. The drug is available in a multidose vial of 10 ml in which 1 ml = 300,000 units. The nurse should administer:

1. 8.8 ml
2. 8.2 ml
3. 0.8 ml
4. 0.008 ml

708. The physician orders 1000 mcg of procainamide (Pronestyl) IV per minute. The directions state 500 mg of the drug should be added to 500 ml of D5W. The IV set has a drop factor of 60 gtt/ml. To administer the medication correctly the nurse should set the flow rate at:

1. 30 gtt/min
2. 60 gtt/min
3. 90 gtt/min
4. 120 gtt/min

709. A client is hospitalized with joint pain, loss of hair, yellow pigmentation of the skin, and an enlarged liver. The nurse should suspect and direct further assessment toward an excess intake of:

1. Thiamin
2. Vitamin A
3. Vitamin C
4. Pyridoxine

710. Tuberculosis is confirmed and isoniazid (INH) and rifampin (Rifadin) therapy is prescribed for a client. The client says, "I've never had to take so much antibiotic for an infection before." The nurse should explain:

1. "Rifampin prevents side effects from INH."
2. "This type of organism is difficult to destroy."
3. "You'll need only one medication when you get better."
4. "Your infection is well advanced and needs aggressive therapy."

711. A client with tuberculosis is started on a chemotherapy protocol that includes rifampin (Rifadin). The nurse knows the teaching about rifampin was effective when the client states:

1. "I will have my hearing tested while I take this medicine."
2. "I will drink large amounts of fluid while I take this medicine."
3. "A skin rash is normal with rifampin and nothing to worry about."
4. "It's normal for my urine to be orange colored from this medication."

712. The physician prescribes vitamin B_6 and isoniazid as part of the chemotherapy protocol for a client with tuberculosis. The nurse understands that vitamin B_6 is used because it:

1. Improves the nutritional status of the client
2. Enhances the tuberculostatic effect of isoniazid
3. Accelerates destruction of dormant tubercular bacilli
4. Counteracts the peripheral neuritis that isoniazid may cause

713. A client is diagnosed as having pulmonary tuberculosis, and one of the drugs the physician orders is pyrazinamide (PZA). The nurse evaluates that the teaching concerning the drug was effective when the client says, "I will:

1. Drink at least 2 quarts of fluid a day."
2. Take the medication 2 hours after each meal."
3. Report any changes in vision to the physician."
4. Expect a discoloration of urine, sweat, and tears."

714. A client is diagnosed with tuberculosis associated with HIV infection. The test results that are crucial for the nurse to review before starting antitubercular pharmacotherapy are:

1. Liver function studies
2. Pulmonary function studies
3. Electrocardiogram and echocardiogram
4. White blood cell counts and sedimentation rate

715. A client with AIDS is receiving zidovudine (AZT). It is most important for the nurse to monitor the client's:
1. Cardiac enzymes
2. Serum electrolytes
3. HIV antibody levels
4. Complete blood count

716. A client with HIV-associated *Pneumocystis carinii* pneumonia is to receive pentamidine isethionate (Pentam 300) IV once daily. To ensure client safety the nurse should:
1. Mix the drug with sterile saline without a preservative
2. Administer the drug over a period of 20 to 30 minutes
3. Monitor the blood pressure for hypertension during therapy
4. Assess blood glucose levels daily during therapy and several times after therapy

717. A client who is receiving mechanical ventilation begins to "fight" the respirator, and the physician orders atracurium (Tracrium). The most important nursing action for a client receiving Tracrium is to:
1. Decrease anxiety
2. Monitor skin integrity
3. Promote urinary output
4. Maintain mechanical ventilation

718. The physician orders 50 U of insulin to be added to an IV of glucose and water for a client with diabetes mellitus. The nurse understands that the only insulin that can be used is:
1. Lente insulin
2. Ultralente insulin
3. Humulin N insulin
4. Humulin R insulin

719. A client, newly diagnosed with type 2 diabetes, is receiving glyburide (Micronase) and asks the nurse how this drug works. The nurse answers the question about Micronase by telling the client it acts by:
1. Stimulating the beta cells to produce insulin
2. Accelerating the liver's release of stored glycogen
3. Increasing glucose transport across the cell membrane
4. Lowering blood sugar in the absence of beta cell function

720. A client is diagnosed as having type 2 diabetes and the physician prescribes glipizide (Glucotrol). While taking this medication, the client should be taught to observe for:
1. Ketonuria
2. Weight loss
3. Ketoacidosis
4. Hypoglycemia

721. A client with type 1 diabetes self-administers Humulin N insulin every morning at 8:00. The nurse recognizes that the client understands the action of this insulin when the client says, "I should be alert for signs of hypoglycemia between:
1. 10 am and noon."
2. 1 pm and 3 pm."
3. 4 pm and 6 pm."
4. 8 pm and 10 pm."

722. A client with newly diagnosed type 1 diabetes is in a self-care teaching group. The nurse has confidence that the client is able to recognize hypoglycemia when the client states, "I will drink orange juice and eat a slice of bread when I feel:
1. Nervous and weak."
2. Flushed and short of breath."
3. Thirsty and have a headache."
4. Nauseated and have abdominal cramps."

723. A teaching program concerning diabetes and insulin therapy is begun for a client, newly diagnosed with type 2 diabetes. The client states, "I hate shots. Why can't I take the insulin in pill form?" The nurse's best response would be:
1. "Your diabetic condition is too serious for oral insulin."
2. "Insulin by mouth causes a high incidence of allergic reactions."
3. "Insulin is poorly absorbed and its action is erratic when given by mouth."
4. "Once your diabetes is controlled, your physician might consider a po drug."

724. A client with hyperthyroidism is treated first with propylthiouracil (PTU). When teaching the client about this medication, the nurse should include the information that:
1. This medication will have to be taken for the remainder of life
2. Milk should be taken with the medication so that gastric irritation does not occur
3. Symptoms may not subside for several days or weeks after the start of therapy
4. The medication should be taken between meals so that it is more readily absorbed

725. A client with hyperthyroidism is to receive methimazole (Tapazole). The nurse should instruct the client that:
1. The drug has few side effects
2. Initial improvement will take 2 to 4 weeks

3. The medication may be taken at any time during the day

4. A large dose is used to decrease the length of time drug therapy is necessary

726. On a clinic visit an elderly female client with rheumatoid arthritis tells the nurse that she has been taking aspirin to reduce the pain. The client asks whether the arthritis could have moved to the ears because both ears now are buzzing. The nurse recognizes that ringing in the ears is:
1. A symptom of otitis media
2. A normal part of the aging process
3. Evidence of eighth cranial nerve damage
4. Caused by an accumulation of cerumen in the ear

727. The physician orders ibuprofen (Motrin) 800 mg po tid for a client with rheumatoid arthritis. The nurse would know that the teaching about the side effects of Motrin was effective when the client:
1. Recognizes that exercise must be balanced with rest
2. Makes an appointment for blood work in 1 month
3. Realizes that position changes should be made slowly
4. Understands that the drug must be taken between meals

728. A client who has had a long leg cast applied is to be discharged. When discussing pain management, the nurse should advise the client to take the prescribed prn Tylenol with codeine:
1. Just as a last resort
2. Before going to sleep
3. When the discomfort begins
4. As the pain becomes intense

729. A client had a laminectomy and is receiving a skeletal muscle relaxant that will be continued after discharge. After teaching the client about the skeletal muscle relaxant, the nurse recognizes that no further health teaching is necessary when the client says:
1. "If the medication makes me sleepy, I'll stop taking it."
2. "I'm going to take the medication 3 hours after meals."
3. "If the medication upsets my stomach, I'll take it with milk."
4. "I'll take an extra dose of the medication before I do anything active."

730. The physician orders a low dosage of a narcotic to relieve the pain of a client with deep partial-thickness burns. The nurse recognizes that the preferred mode of administration is:
1. Oral
2. Rectal
3. Intravenous
4. Intramuscular

731. A female client is receiving ethylestrenol (Maxibolin), an anabolic steroid, for the treatment of the catabolic processes associated with a burn injury. The nurse should observe the client for signs of:
1. Lethargy
2. Virilization
3. Hyponatremia
4. Hyperglycemia

732. The nurse applies mafenide acetate cream (Sulfamylon) to a client's burns as ordered by the physician. This medication will:
1. Inhibit bacterial growth
2. Relieve pain from the burn
3. Prevent scar tissue formation
4. Provide chemical débridement

733. A client's burns are treated by the open exposure method and a topical antimicrobial is ordered. The nurse is aware that the statement that best explains the rationale for use and side effects of the antimicrobial ordered by the physician would be:
1. Silvadene is easy to apply and has good bacteriocidal action but it can depress granulocyte formation
2. Sulfamylon causes no side effects and is effective for both gram-positive and gram-negative organisms
3. Neomycin is a strong bacteriocidal agent effective against most organisms with the least number of toxic side effects for the client
4. Clients respond quickly to dry dressings with silver nitrate paste because it eliminates odors and has strong bacteriostatic properties

734. The nurse attempts to give a male client with chronic arterial insufficiency of the legs the prescribed dose of aspirin but the client refuses it stating, "My legs are not painful." The nurse should:
1. Explain the reason for the medication and encourage the client to take it
2. Withhold the medication at this time and return to check the client again in $1/2$ hour
3. Withhold the medication and tell the client to ask for it if his legs become uncomfortable
4. Request that the client take the medication to prevent him from being uncomfortable in the next few hours

735. The physician prescribes isosorbide dinitrate (Sorbitrate) 10 mg prn tid and a nitroglycerin transdermal disc once a day for a client with chronic angina pectoris. The client asks the nurse why the Sorbitrate is prescribed. The nurse's best response would be:
1. "It prevents the blood from clotting."
2. "It suppresses irritability in the ventricles."
3. "It allows more oxygen to get to heart tissue."
4. "It increases the force of contraction of the heart."

736. When teaching a client about isosorbide dinitrate (Sorbitrate), the nurse should plan to include instructions about how to prevent:
1. Bradycardia
2. Constipation
3. Respiratory distress
4. Postural hypotension

737. The nurse teaches a client about the side effects of furosemide (Lasix), which has just been prescribed. The nurse assesses that the teaching was understood when the client states, "I should:
1. Wear dark glasses."
2. Avoid lying flat in bed."
3. Avoid eating citrus fruits."
4. Change my position slowly."

738. At each clinic visit, the nurse should assess a client taking furosemide (Lasix) for:
1. Tinnitus
2. Xanthopsia
3. Hyporeflexia
4. Bronchospasm

739. A client with a history of hypertension comes to the emergency room with double vision and a blood pressure of 260/120 mm Hg. In addition to other drugs, the physician orders a sodium nitroprusside infusion. The nurse recognizes that this drug decreases blood pressure by:
1. Increasing cardiac output
2. Decreasing the heart rate
3. Increasing peripheral resistance
4. Relaxing venous and arterial muscles

740. A client is receiving clonidine (Catapres) for hypertension. The nurse would recognize that the discharge instructions were understood when the client states, "I will call the doctor if I develop:
1. Pruritus."
2. Diarrhea."
3. Euphoria."
4. Photosensitivity."

741. A client with hypertensive heart disease who had an acute episode of heart failure is to be discharged on a regimen of propranolol HCl (Inderal) and digoxin (Lanoxin). The nurse should be aware that Inderal, when administered with Lanoxin, may:
1. Produce headaches
2. Precipitate bradycardia
3. Increase blood pressure
4. Stimulate nodal conduction

742. When administering warfarin (Coumadin) the nurse should know that the antidote for Coumadin is:
1. Vitamin K
2. Fibrinogen
3. Prothrombin
4. Protamine sulfate

743. The physician orders furosemide (Lasix) for a client with hypervolemia. The nurse understands that Lasix exerts its effects in the:
1. Distal tubule
2. Collecting duct
3. Glomerulus of the nephron
4. Ascending limb of Henle's loop

744. A client has been receiving a cardiac glycoside, a diuretic, and a vasodilator and is on bed rest. The client's apical pulse rate is 44 beats per minute. The nurse concludes that this pulse rate is most likely a result of the:
1. Diuretic
2. Vasodilator
3. Bed rest regimen
4. Cardiac glycoside

745. A physician orders a tissue plasminogen activator to be administered intravenously over 1 hour for a client experiencing a myocardial infarction. The nursing priority that is specific to the use of this drug is the assessment of the client's:
1. Respiratory rate
2. Peripheral pulses
3. Level of consciousness
4. Intravenous insertion site

746. A 42-year-old is admitted to the cardiac care unit with an anterior lateral myocardial infarction. The physician orders 500 ml of D5W with 50 mg of nitroglycerin to be given intravenously to relieve pain. The most common side effect the nurse should observe for would be:
1. Nausea
2. Syncope
3. Bradycardia
4. Hypotension

747. Nitroglycerin is prescribed for a client with a history of a myocardial infarction and atrial tachycardia. The nurse instructs the client in the prophylactic use of these tablets. The statement by the client that indicates teaching was effective would be:
1. "I should take one tablet at 9 am, noon, and 4 pm."
2. "I should not undertake any activity I consider strenuous."
3. "I should have my wife take my pulse before I take a tablet."
4. "I should take one tablet before climbing two flights of stairs."

748. When administering the ordered digoxin (Lanoxin) 0.25 mg po qd to a client with a history of type 1 diabetes who is now in heart failure, the nurse should:
1. Give it with orange juice
2. Monitor for dysrhythmias
3. Administer it 1 hour after the am insulin
4. Withhold it if the apical pulse rate is 90 beats per minute

749. A client develops premature ventricular beats (PVBs). The nurse should anticipate that the client will receive:
1. Epinephrine
2. Atropine sulfate
3. Sodium bicarbonate
4. Lidocaine hydrochloride

750. The cardiac monitor indicates that a client's heart rate has increased to 150 beats per minute. Shortly after this increase, the nurse notices the client is in ventricular tachycardia (VT). After reporting this to the physician, the nurse anticipates the physician will order:
1. A bolus of lidocaine
2. Intracardiac epinephrine
3. Insertion of a pacemaker
4. Manual cardiopulmonary resuscitation (CPR)

751. A client is scheduled for an electrophysiology study (EPS) because of persistent ventricular tachycardia. Prior to the procedure the client is placed on a beta blocker. The client's response during the procedure that best indicates that the beta blocker is working effectively would be:
1. Increased anxiety
2. Sinus bradycardia
3. Ventricular fibrillation
4. Nausea and vomiting

752. When administering an intravenous titrated drip of lidocaine HCl to a client, a serious side effect that must be reported to the physician immediately is:

1. Tremors
2. Anorexia
3. Tachycardia
4. Hypertension

753. The physician prescribes furosemide (Lasix) 40 mg qd in conjunction with digoxin (Lanoxin) for a client. The nurse recognizes that potassium supplements are essential with this combination of drugs because:
1. Digitalis causes significant potassium depletion
2. Potassium is destroyed by the liver as digoxin is detoxified
3. Lasix requires adequate serum potassium to promote diuresis
4. Digitalis toxicity occurs rapidly in the presence of hypokalemia

754. The nurse would know that a client understands the side effects of hydrochlorothiazide (Esidrix) when the client states, "I should call the physician if I develop:
1. Insomnia."
2. A stuffy nose."
3. Increased thirst."
4. Generalized weakness."

755. A client experiences cardiac standstill. In addition to cardiac stimulants the nurse would expect to prepare:
1. Aminophylline
2. Furosemide (Lasix)
3. Sodium bicarbonate
4. Phenytoin (Dilantin)

756. A client with Hodgkin's disease is started on MOPP therapy. Mechlorethamine HCl (Mustargen) is given IV. With this drug the client is most likely to develop:
1. Hair loss
2. Transient nausea
3. Urinary incontinence
4. Neurologic dyskinesia

757. A client is to receive doxorubicin (Adriamycin) as part of a chemotherapy protocol. The major life-threatening side effect of Adriamycin that the nurse should assess the client for is:
1. Cardiotoxicity
2. Pancytopenia
3. Pulmonary fibrosis
4. Ulcerative stomatitis

758. Specific nursing intervention for a client who is receiving doxorubicin (Adriamycin) for acute myelogenous leukemia should include:
   1. Giving frequent oral hygiene and increasing oral fluids
   2. Serving hot liquids, such as broths or tea, with each meal
   3. Administering medications IM and increasing activity level
   4. Emphasizing that the disease will be cured with this treatment

759. A client with cancer develops pancytopenia during the course of chemotherapy. The client asks the nurse why this has occurred. The nurse should explain that:
   1. Normal cells are also susceptible to the effects of chemotherapeutic drugs
   2. Steroid hormones have a depressant effect on the spleen and bone marrow
   3. Lymph node activity is depressed by radiation therapy used prior to chemotherapy
   4. Dehydration caused by nausea, vomiting, and diarrhea results in hemoconcentration

760. A client is receiving busulfan (Myleran) for chronic myelogenous leukemia. When assessing for complications of alkylating agents, the nurse should be alert for:
   1. Stomatitis and nausea
   2. Feminization or masculinization
   3. Fluid retention and hyperglycemia
   4. Leukopenia and thrombocytopenia

761. When a client who is receiving methotrexate (Folex) for acute lymphocytic leukemia develops a temperature of 101° F, the physician is notified. Aspirin 650 mg q4h prn is ordered. The nurse should:
   1. Ask the physician for an antacid order
   2. Question the type of antipyretic ordered
   3. Question the dosage ordered by the physician
   4. Withhold the aspirin until the temperature reaches 102° F

762. A client with cancer is receiving a multiple chemotherapy protocol. Included in the protocol is leucovorin calcium (Wellcovorin). The nurse recognizes that this drug is administered to:
   1. Potentiate the effect of alkylating agents
   2. Diminish the toxicity of folic acid antagonists
   3. Limit the occurrence of nausea and vomiting associated with chemotherapy
   4. Interfere with cell division at a different stage of cell division than the other drugs

763. A client with metastatic melanoma is being treated with Interferon. The nurse is aware that the teaching about this drug is understood when the client states:
   1. "I will increase my fluid intake to 2 to 3 liters daily."
   2. "I need to discard any reconstituted solution at the end of the week."
   3. "I can continue driving my car as before, as long as I have the stamina."
   4. "I should be able to continue my usual activity while taking this medication."

764. The physician prescribes cimetidine (Tagamet) and Maalox for a client with a peptic ulcer. The nurse should teach the client to take these drugs:
   1. At the same time
   2. At least 1 hour apart
   3. One immediately before the other
   4. Together with milk or orange juice

765. A client is admitted to the hospital for abdominal surgery for cancer of the pancreas. Before surgery, meperidine hydrochloride (Demerol) is ordered for pain. The nurse recognizes that morphine sulfate would be contraindicated for this client because it:
   1. Causes respiratory excitement
   2. Stimulates pancreatic duct secretion
   3. Causes spasm of the pancreatic ducts
   4. Stimulates the sympathetic nervous system

766. The physician prescribes oral pancreatic enzymes for a client. The nurse would know that the teaching about the enzymes is understood when the client states, "I will take them:
   1. At bedtime."
   2. With meals."
   3. One hour before meals."
   4. On arising each morning."

767. Following a course of doxorubicin hydrochloride (Adriamycin), the physician decides to prescribe platinol (Cisplatin) for a client with metastatic cancer. To prevent toxic effects, the nurse should:
   1. Administer the ordered leucovorin
   2. Encourage regular vigorous oral care
   3. Increase hydration to promote diuresis
   4. Provide a high-protein, low-residue diet

768. The medication order that should be questioned by the nurse if it is prescribed for a client with acute pancreatitis is:
   1. Tagamet
   2. Phenergan
   3. Morphine sulfate
   4. Meperidine hydrochloride

769. Following an acute episode of GI bleeding, a peptic ulcer is confirmed by a gastroscopy and upper GI se-

ries. The physician orders ranitidine (Zantac) 150 mg bid with meals. The nurse should check this order with the physician because:
1. This is less than the recommended dose
2. Zantac is contraindicated for peptic ulcer
3. Zantac may be given by a variety of routes
4. This drug is usually given on an empty stomach

770. Prednisone, an adrenal steroid, is ordered for a client with an exacerbation of colitis. When administering the first dose, the nurse should stress that this drug:
1. Will protect the client from getting an infection
2. May decrease the client's appetite, causing weight loss
3. Is not curative but does cause a suppression of the inflammatory process
4. Is relatively slow in effecting a response but is effective in reducing symptoms

771. Immediately following a bilateral adrenalectomy, a client is placed on corticosteroids that are to be continued after discharge. The nurse should recognize a need for further teaching when the client states:
1. "I need to have periodic tests of my blood for sugar."
2. "I must take the pills between meals on an empty stomach."
3. "Hopefully, the dosage will be regulated so I can take them once or twice a day."
4. "I should tell the doctor if I am overly restless, depressed, or have trouble sleeping."

772. Neomycin is prescribed for a client with cirrhosis. The nurse recognizes that this drug is given to:
1. Reduce intestinal edema
2. Reduce abdominal distention
3. Provide a prophylactic antibiotic
4. Reduce the blood ammonia level

773. The physician orders docusate sodium (Colace) every day for a client. The nurse recognizes that this drug is ordered specifically to:
1. Lubricate the feces and GI tract
2. Create an osmotic effect in the GI tract
3. Stimulate the motor activity of the GI tract
4. Lower the surface tension of feces in the GI tract

774. The physician prescribes docusate calcium (Surfak) for a client with cardiac disease. The nurse is aware that this drug acts by:
1. Producing bulk
2. Softening feces
3. Lubricating feces
4. Stimulating peristalsis

775 The physician orders Maalox by mouth and cimetidine (Tagamet) by IVPB for a client with crushing injuries caused by a train accident. The client asks why these medications are being administered. The nurse's best response would be:
1. "They decrease irritability of the bowel."
2. "They limit acidity in the stomach and intestine."
3. "They're ordered for all clients with multiple trauma."
4. "They're what your doctor ordered to calm your stomach."

776. The physician orders bed rest, Maalox, and cimetidine (Tagamet) for a client who has just had major surgery. After several days of this regimen, the client complains of diarrhea. The nurse recognizes that the diarrhea is most likely caused by:
1. Bed rest
2. The Maalox
3. Diet alteration
4. The cimetidine

777. When caring for a client who has open-angle (chronic) glaucoma, the eye drops the nurse should expect to administer would be:
1. Tetracaine (Pontocaine)
2. Cyclopentolate (Cyclogyl)
3. Pilocarpine hydrochloride (Pilocar)
4. Atropine sulfate (Atropisol Ophthalmic)

778. A client has a tonic-clonic seizure that involves all extremities. The nurse anticipates that the physician will order the IV administration of:
1. Atropine
2. Epinephrine
3. Naloxone (Narcan)
4. Diazepam (Valium)

779. When a client's condition is assessed following a tonic-clonic seizure caused by an overdose of acetylsalicylic acid, the most appropriate nursing action would be to:
1. Insert a Foley catheter
2. Check reflexes every hour
3. Prepare a setup for a CVP line
4. Monitor vital signs every 15 minutes

780. A client taking levodopa should be taught about the signs of levodopa toxicity. The client and family should be instructed to contact the physician if they note the development of:
1. Nausea
2. Twitching
3. Dizziness
4. Constipation

781. Selegiline (Eldepryl) is prescribed for a client with Parkinson's disease who had had a poor response to levodopa therapy. When teaching the client and family about this drug, the nurse should explain that:
    1. If a severe headache occurs it should be reported to the physician immediately
    2. The side effects of levodopa will decrease when these drugs are taken concurrently
    3. Blood studies should be performed monthly to measure therapeutic blood levels of the drug
    4. The dosage of the drug can be adjusted daily depending on the client's response that day

782. Dexamethasone (Decadron) is ordered for the early management of a client's cerebral edema. This treatment is effective because it:
    1. Acts as a hyperosmotic diuretic
    2. Increases tissue resistance to infection
    3. Reduces the inflammatory response of tissues
    4. Decreases the formation of cerebrospinal fluid

783. When a client is receiving dexamethasone (Decadron), the nurse should observe for the development of a negative side effect by:
    1. Auscultating for bowel sounds
    2. Culturing respiratory secretions
    3. Measuring blood glucose levels
    4. Monitoring deep tendon reflexes

784. A client is receiving IV mannitol. An assessment specific to the safe administration of mannitol is:
    1. Vital signs q2h
    2. Body weight daily
    3. Urine output hourly
    4. Level of consciousness q8h

785. The wife of a client with an intracranial bleed asks why her husband is not receiving anticoagulant therapy. The nurse explains that in her husband's situation anticoagulant therapy:
    1. Will be started if necessary to enhance circulation
    2. May be necessary to prevent pulmonary thrombosis
    3. Is contraindicated because bleeding would be increased
    4. Is inadvisable because it would mask signs and symptoms

786. To prevent excessive bruising when administering heparin subcutaneously, the nurse should:
    1. Avoid rubbing injection site after the injection
    2. Dilute the heparin with 2 ml of sterile normal saline

    3. Inject the drug into the subcutaneous tissue quickly
    4. Use the Z-track technique for administering the injection

787. In the event of excessive bleeding in a client who is receiving heparin sodium, the nurse should be prepared to administer:
    1. Vitamin K
    2. Panheparin
    3. Warfarin sodium
    4. Protamine sulfate

788. The nurse knows discharge teaching in relation to warfarin sodium (Coumadin) therapy has been understood when the client states, "I will:
    1. Take Tylenol for my occasional headaches."
    2. Spend most of the day working at my desk."
    3. Make an appointment to have a CBC drawn."
    4. Ask the doctor for antibiotics before going to the dentist."

789. One month after an endarterectomy the physician instructs the client to take 5 grains of aspirin a day. The nurse evaluates that the reason for taking this drug is understood when the client says, "It will:
    1. Limit the inflammation around my incision."
    2. Help to prevent further clogging of my arteries."
    3. Lower the slight fever I have had since surgery."
    4. Reduce the discomfort I feel at the surgical site."

790. A client is diagnosed as having myasthenia gravis, and pyridostigmine bromide (Mestinon) therapy is started. During the first week of therapy, while the dosage is being adjusted, the nurse's priority intervention is to:
    1. Administer the medication exactly on time
    2. Administer the medication with food or milk
    3. Evaluate the client's psychologic responses between doses
    4. Evaluate the client's muscle strength hourly after administration

791. A client with myasthenia gravis is admitted to the emergency room in crisis. To distinguish between myasthenic crisis and cholinergic crisis, the nurse should expect the physician to administer:
    1. Atropine sulfate
    2. Protamine sulfate
    3. Naloxone (Narcan)
    4. Edrophonium chloride (Tensilon)

792. A client with myasthenia gravis improves and is discharged from the hospital. The discharge medications include pyridostigmine bromide (Mestinon) 10 mg every 6 hours. The nurse would know that

the drug regimen was understood when the client says, "I should:
1. Take milk with each dose of Mestinon."
2. Take the Mestinon on an empty stomach."
3. Set my alarm clock to take my medication."
4. Take my pulse and respirations before taking the drug."

793. A client with myasthenia gravis has been receiving pyridostigmine bromide (Mestinon). Because of inadequate symptomatic control, the physician begins long-term steroid therapy. When this type of therapy is being initiated, it is especially important to:
1. Increase the client's sodium intake
2. Place the client on protective isolation
3. Decrease the client's fluid intake to 1000 ml daily
4. Observe the client for an exacerbation of symptoms

794. The physician orders neostigmine (Prostigmin) for a client with myasthenia gravis. The nurse would know that the client understands the teaching about this drug when the client says:
1. "I should keep the drug in a tight container in the refrigerator."
2. "I should take the drug at the exact time specified by my doctor."
3. "The peak action of the drug occurs about 3 to 4 hours after ingestion."
4. "The drug should be taken between meals to promote absorption."

795. The physician orders oxacillin sodium (Bactocill) 500 mg every 6 hours for a client who is to be discharged. The nurse would know that the teaching about the Bactocill is understood when the client says:
1. "I will take the medication with meals."
2. "I should drink a glass of milk with each pill."
3. "I should drink at least 6 glasses of water every day."
4. "I will take the medication 1 hour before or 2 hours after meals."

796. Two days after surgery a client's temperature is normal and the IVPB of gentamicin q8h started earlier is discontinued. The physician orders a clear liquid diet and OOB ad lib. After lunch the client complains of dizziness when walking to go to the bathroom. The nurse should recognize that the dizziness is probably related to:
1. Bed rest
2. The liquid diet
3. The gentamicin
4. Postanesthesia hypotension

797. The physician orders peak and trough levels of gentamicin for a client who is receiving the medication IVPB. For peak levels the nurse should have the laboratory obtain a blood sample from the client:
1. Between 30 to 60 minutes after the IVPB
2. Halfway between 2 IVPB administrations
3. Immediately before administering the IVPB
4. Anytime it is convenient for the client and laboratory

798. When planning discharge teaching for a client who has had a kidney transplant, the nurse should explain the drug regimen that will be needed to prevent rejection of the kidney. The drugs of choice usually include:
1. Ancef, methotrexate, and citric acid
2. Lasix, neomycin, and cyclosporine (Sandimmune)
3. Methylprednisolone (Solu-Medrol), Dilantin, and insulin
4. Prednisone, cyclophosphamide (Cytoxan), and azathioprine (Imuran)

799. In relation to a client's treatment with the drug erythropoietin (Epogen), the finding that would be considered significant is an:
1. Elevated liver panel
2. Elevated hematocrit level
3. Increase in the WBC counts
4. Increase in Kaposi's sarcoma lesions

800. A client who is immunosuppressed is receiving filgrastim (Neupogen). When assessing the client's response to this medication the finding that would be considered significant is an increase in:
1. Platelets
2. Erythrocytes
3. Thrombocytes
4. White blood cells

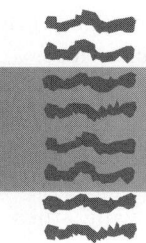

MEDICAL-SURGICAL NURSING ANSWERS AND RATIONALES

GROWTH AND DEVELOPMENT (GD)

1. (3) This is the young-adult task associated with intimacy vs isolation. (2) (DF; AN)
1 This is a toddler's task associated with autonomy vs shame and doubt.
2 This is a school-ager's task associated with initiative vs guilt.
4 Same as answer 1.

2. (3) Compression fractures of the vertebrae are the most common fractures in clients with osteoporosis; a gradual collapse of vertebrae may be asymptomatic and only observed as kyphosis. (1) (HI; AS)
1 This is untrue; it is not supported by statistics.
2 Same as answer 1.
4 Same as answer 1.

3. (1) Research demonstrates that women past menopause need 1500 mg of calcium a day, which is almost impossible to obtain through dietary sources because the average daily consumption of calcium is 300 to 500 mg; vitamin D promotes the deposition of calcium into the bone. (2) (PC; AN)
2 If large amounts of magnesium are present, calcium absorption is impeded because magnesium and calcium absorption are competitive; vitamin E is unrelated to osteoporosis.
3 These do not contain adequate calcium to meet requirements to prevent osteoporosis; these do not contain vitamin D unless fortified.
4 These do not contain adequate calcium and vitamin D to meet requirements to prevent osteoporosis.

4. (3) Data support supplementation in the elderly only in illness or debilitated states to help restore tissue integrity and function. (2) (HI; IM)
1 This is untrue; a well-balanced diet that contains a variety of foods, as recommended by the food pyramid, is adequate.
2 The elderly do not need supplemental vitamins and minerals as long as their diet is adequate.

4 This is untrue; the best source of all nutrients is from natural foods rather than supplements; loss of weight results from reduction of calories, not by the addition of vitamins and minerals.

5. (2) According to Erikson, poor self-concept and feelings of despair are conflicts manifested in the 65-and-older age group. (3) (DF; EV)
1 These are conflicts manifested in early childhood between 3 and 6 years of age.
3 These are conflicts manifested during the ages from 6 to 11 years.
4 These are conflicts manifested during middle adulthood, 45 to 65 years of age.

6. (2) Urinary tract infections affect the genitourinary tract and interfere with the voluntary control of micturition. (3) (PC; AS)
1 This is a neurologic, not a genitourinary, factor.
3 These are iatrogenic factors.
4 This is an environmental factor.

7. (3) A decrease in neuromuscular function slows reaction time. (1) (DF; IM)
1 The ability to be flexible has nothing to do with age but with character.
2 Confusion is not a normal process of aging, but it occurs for various reasons such as multiple stresses, perceptual changes, or medication side effects.
4 The majority of older adults do not have organic mental disease.

8. (4) For reasons that are still unclear, the thirst reflex diminishes with age and this leads to a concomitant decline in fluid intake. (1) (DF; IM)
1 This would not be true for an alert person who is able to meet the activities of daily living.
2 There are no data to support this statement.
3 Research does not support progressive memory loss in normal aging as a contributor to decreased fluid intake.

9. ③ Vitamin E hinders oxidative breakdown of structural lipid membranes in body tissues caused by free radicals in the cells. (2) (PC; AN)
   1 This assists in the formation of visual purple needed for night vision.
   2 This is used for formation of collagen, which is important for maintaining capillary strength, promoting wound healing, and resisting infection.
   4 This is necessary for protein and fat metabolism and normal function of the nervous system.

10. ③ Kyphosis alters the center of gravity, which contributes to alterations in balance and gait. (3) (SI; EV)
    1 Decreased endurance and fatigue should not change the center of gravity or alter the gait; a lack of stamina by itself should not cause falls.
    2 Age is incidental; one should not accept falls as an inescapable aspect of aging.
    4 Although kyphosis alters the line of vision downward, this by itself would not cause increased falls.

11. ③ Reinforcing strengths promotes self-esteem; reminiscing is a therapeutic tool that provides a life review that assists adaptation and helps achieve the task of integrity associated with older adults. (2) (SC; IM)
    1 Exercise should be encouraged, but naps tend to interfere with adequate sleep at night.
    2 Reinforcing ageism would enhance devaluation of the older adult.
    4 A well-balanced diet that also includes protein and fiber should be encouraged; the older adult needs to put the past in perspective and a positive self-assessment should be supported.

12. ④ In the aged, there is a progressive atrophy of the convolutions with a decrease in the blood flow to the brain, which may produce a tendency to become forgetful, a reduction in short-term memory, and susceptibility to personality changes. (1) (DF; AS)
    1 Creativity is not affected by aging; many people remain creative until very late in life.
    2 People with a normal aging process show little or no change in judgment.
    3 There is little or no intellectual deterioration; intelligence scores show no decline up to the ages of 75 to 80.

13. ④ With aging, there is a significant increase in the systolic blood pressure and a slight increase in the diastolic blood pressure; hormone production is decreased with menopause. (3) (CC; IM)
    1 There are no changes in libido with aging; there is a loss of skin elasticity.
    2 Salivary secretions decrease, not increase, causing more difficulty with swallowing; there is some impairment of fat digestion.
    3 There is a decrease in subcutaneous fat, decreasing body warmth; some swallowing difficulties occur because of decreased secretions.

14. ③ The onset of disabling illness will divert an aged person's energies, making it difficult to maintain an optimum level of functioning. (3) (SC; AN)
    1 This can result from the aging process and the change in environment; it is not as important as the loss of function.
    2 This would be an expected response.
    4 A gradual memory loss and some confusion can be expected; a sudden memory loss would be cause for alarm.

15. ① A psychosocial task for middle adulthood according to Erikson is generativity; this task is concerned with the sense of productivity and accomplishment. (3) (DF; PL)
    2 This is not involved in any task of middle adulthood identified by Erikson.
    3 Same as answer 2.
    4 Same as answer 2.

16. ④ Spending time in several short sessions reduces client fatigue and compensates for a shortened attention span, which is common in the elderly. (2) (DF; PL)
    1 Poor decision; questionnaire may never be completed.
    2 Degrading to client; certainly the client should be asked initially and, if necessary, family can be asked to fill in details later.
    3 May be overwhelming and create feelings of anger and resentment.

EMOTIONAL NEEDS RELATED TO HEALTH PROBLEMS (EH)

17. ③ This provides the opportunity for the client to verbalize the feelings underlying behavior. (2) (SC; IM)
    1 This has no effect on decreasing the client's anxiety or allowing ventilation.
    2 This explanation will not decrease anxiety so that the client can rest.
    4 Although allowing release of feelings is therapeutic, leaving immediately denies the

client the opportunity for verbalization and discussion.

18. (2) The nurse must analyze the feelings that are implied in the client's question and reflect these to help the client verbalize and explore them; the focus is on collecting more data. (1) (SC; IM)
    1 This is avoiding the responsibility of helping the client explore feelings; it cuts off communication.
    3 Although this may be true, it does not respond to the feelings implicit in the client's comment.
    4 No data presented at this time suggest that such a referral is warranted; this also cuts off communication when the client has expressed a need; the nurse is avoiding responsibility to assist the client.

19. (1) Projection is the attribution of unacceptable feelings and emotions to others. (2) (SC; AN)
    2 Sublimation is the substitution of socially acceptable feelings or instincts that, if expressed, would be threatening to the ego.
    3 Intellectualization is the use of mental reasoning processes to deny facing emotions and feelings involved in a situation.
    4 A reaction formation is the unconscious reversal of feelings or behavior unacceptable to the self-image and the assumption of opposite feelings or behavior.

20. (3) As long as no one else confirms the presence of the stoma and the client does not need to adhere to a prescribed regimen, the client's denial is supported. (2) (SC; AS)
    1 There is no evidence to document that reaction formation is being used.
    2 There is no evidence presented that suicidal thoughts are present or will be acted out.
    4 There are no data to support this conclusion; the client should be able to function sexually as before.

21. (4) The withdrawal provides time for the client to assimilate what has occurred and integrate the change in the body image. (2) (DF; IM)
    1 The client is not ready to hear these explanations until assimilation of the surgery has occurred.
    2 This does not acknowledge that the client must grieve; it also does not allow the client to express any feelings that may be present that life will never be normal again.
    3 The client might feel that the nurse had no comprehension of the situation or understanding of feelings.

22. (4) It is not reality, but the client's feeling about the change, that is the most important determinant in the ability to cope. (3) (DF; AN)
    1 This is not relevant to the client's ability to deal with a change in body image.
    2 Same as answer 1.
    3 The extent of change is not relevant; it is whether the client perceives the change as enormous or minuscule.

23. (3) Intellectualization is the use of reasoning and thought processes to avoid the emotional aspects of a situation; this is a defense against anxiety. (1) (SC; AN)
    1 Projection is denying unacceptable traits and regarding them as belonging to another person.
    2 Sublimation is a defense wherein the person redirects the energy of unacceptable impulses into socially acceptable behaviors or activities.
    4 Reaction formation is behavior exactly opposite to what the person is feeling.

24. (3) This nonjudgmentally points out the client's behavior. (2) (SC; IM)
    1 This is too confrontational; the client may not be able to answer the question.
    2 This is too confrontational and an assumption by the nurse.
    4 This is too judgmental, an assumption, and a stereotypic response.

25. (3) This does not take away the client's only way of coping, and it permits future movement through the grieving process when the client is ready. (2) (SC; PL)
    1 The client's denial should be neither supported nor taken away; encouraging denial is a form of false reassurance.
    2 This is false reassurance.
    4 The client must not be abandoned; the nurse's presence is a form of emotional support.

26. (4) Honest discussion with emphasis on functional and psychologic abilities helps promote adjustment. (2) (DF; IM)
    1 Postoperative depression is not a characteristic feature of thymectomy.
    2 This is too soon; it may eventually be necessary if the client has trouble adjusting to the chronicity of this condition.
    3 This provides false reassurance; there is no guarantee the client will feel better on discharge.

27. (4) The medical history could be obtained during assessment, and a relationship could be established if they were uninterrupted. (3) (CC; IM)
   1 Agency information and data could be obtained after the assessment data had been obtained and rapport established.
   2 Accepting coffee may be an imposition and is not the best way to develop trust.
   3 Assessment of the environment could be less obviously done while obtaining the history and physical data.

28. (3) Reflection conveys acceptance and encourages further communication. (2) (SC; IM)
   1 This is false reassurance that does not help to reduce anxiety.
   2 This provides false reassurance that also removes the focus from the client's needs.
   4 This is unrealistic, and it is a little late to think of this.

29. (4) The nurse needs to know specifically what the client is asking; this response permits clarification. (2) (DF; IM)
   1 This assumes that the client is referring to the diabetes in relation to surgery and sidesteps the question.
   2 Asking what the physician has said collects more information, but it will not clarify what the client wants to know.
   3 The nurse is making an assumption about medical management.

30. (2) This response focuses on the client's feelings of despair and provides the opportunity to talk about them. (1) (SC; IM)
   1 This abandons the client and leaves the client with no support.
   3 This response avoids a pressing problem and misses an opportunity for discussion of feelings.
   4 This focuses on the nurse's interpretation of the problem, not the client's.

31. (4) This is true; it could be the foundation for developing a positive mental outlook. (2) (SC; IM)
   1 There is no indication that the client needs drugs to combat depression.
   2 This is a patronizing response that does not recognize the despair.
   3 This is probably true, but this response belittles the client's actual concern and physical discomfort.

32. (3) This is a truthful answer that provides some realistic hope. (1) (SC; IM)

1 This provides false reassurance; remissions and exacerbations that may reduce the life span are common.
2 This response avoids the client's question; the family did not ask.
4 This avoids the client's question and transfers responsibility to the physician.

33. (4) This includes the client in the problem-solving process. (2) (CC; PL)
   1 This does not include the client in the problem-solving process; more data should be obtained from the client before deciding on an intervention, which may or may not be appropriate.
   2 Same as answer 1.
   3 Same as answer 1.

34. (3) Policies relative to DNR orders vary among hospitals and the nurse must adhere to the policies of the institution. (2) (CC; AN)
   1 This is untrue; the wish of the client is the deciding factor.
   2 The decision resides with the client.
   4 This may not be true for all hospitals or states; the information does not indicate this is so in this situation; these orders are reviewed periodically.

35. (2) Communication facilitates joint solution of the problem; the nurse must first determine the client's understanding and perceptions before solutions to the problem can be attempted. (2) (SI; AS)
   1 This will not collect data about why the client is leaving the room.
   3 This abdicates the responsibility of the primary nurse.
   4 This may be done but not until further assessment is done to determine the reason why the client is leaving the room.

36. (1) Placing an aide in the home will allow the mother to rest and provide the child with attention. (3) (CC; IM)
   2 Making the mother feel guilty will only increase anxiety and will not be constructive.
   3 The disabled sibling requires attention, and this may increase jealousy, rivalry, and resentment.
   4 Elaborate toys need not be employed for sensory stimulation; household objects can serve as well.

37. (1) This is a classic sign of denial; by reducing the importance or extent of the problem, the individual is able to cope; not acknowledging that it is really a problem is a form of denial. (2) (SC; AS)

2 This indicates repression of affect rather than denial.

3 Failure to communicate is insufficient evidence to diagnose denial; the husband/wife relationship may be strained, or the husband may be worried about upsetting the wife.

4 This usually indicates displacement of anger, not denial.

38. (2) Any client who has been sedated, or who is not fully conscious, may not sign the consent for a surgical procedure. (2) (CC; AN)

1 Many clients face contradictory feelings regarding their impending surgery, but their consent is legal unless they withdraw the consent.

3 A second opinion is not required for a consent to be legal.

4 A complete history and physical examination are needed before surgery, but they do not affect the legality of consent.

39. (2) Allowing the client to grieve for the lost limb often aids acceptance of the loss. (1) (SC; IM)

1 Sedation will prevent the client from facing the problem.

3 Behavior should never be ignored.

4 This behavior may be interpreted by the client as rejection.

40. (1) Because the hospital beds are narrower and higher than the beds clients sleep in at home, side rails often create a sense of security. (1) (SI; EV)

2 This relates to the physical status, not the psychologic status, of the client.

3 On the contrary, this will cause the client to feel more dependent.

4 Bed rails are unrelated to proprioception, which is knowing the location of a body part when it is out of the field of vision.

41. (4) This is a true statement; anxiety and/or anger associated with other stages interfere with learning. (3) (DF; AN)

1 This is too late to start preparation for discharge and teaching; many factors influence readiness for learning; planning for teaching must begin on the day of admission.

2 The anxiety associated with mental anguish would interfere with the ability to process new information; mental anguish is associated with an earlier stage.

3 Although clients in the acceptance or adaptation phase are less angry, the reason teaching is most effective is not because of their compliance but because new information can be processed more easily.

42. (2) When a client is anxious and has a decreased ability to cope, minor environmental irritants are magnified; eye contact is avoided to decrease additional stimuli. (3) (SC; EV)

1 This would be indicated by complaints of pain, splinting, refusal to move, and alteration in vital signs.

3 If the client were angry, eye contact would be maintained; prolonged eye contact may be used as a form of intimidation or aggression.

4 If this were so, the client would be verbalizing about the need to continue the abduction, not about a variety of other annoyances.

43. (3) The nurse never clarified whether the client was in pain; also, this offers false reassurance. (2) (CC; EV)

1 The nurse's response denies the client's anxiety; it identifies pain as the problem.

2 The nurse's response denies the client's needs; the client needs to discuss concerns and feelings.

4 The nurse inappropriately identifies the problem and never clarifies the need.

44. (1) This reflects the client's feelings and encourages a further exploration of concerns. (3) (SC; IM)

2 This response does not reflect the feeling tone of the client's statement; also, the client may not be able to answer this question.

3 This is false reassurance; thyroid storm is capable of causing death.

4 This could reinforce the client's anxiety and avoids discussing the client's concerns; it cuts off communication.

45. (1) Internal self-stimulation increases as external stimuli decrease. (3) (PC; AN)

2 Blindness is an added stress that could increase anxiety, which impairs decision making; lack of visual stimuli limits data for decision making.

3 Lack of visual stimuli would increase restlessness, lethargy, and apathy.

4 Blindness would not precipitate neurotic behavior unless other emotional factors were present.

46. (2) This provides an opportunity for the client to explore concerns with the nurse. (2) (SC; IM)

1 The data do not indicate regression; the client is anxious, not regressed.

3 The nurse is basing the response on an incorrect interpretation of the data.

4 The data do not indicate that the client does not understand; the nurse should attempt to provide for consensual validation before coming to this conclusion.

47. (3) Hostility and noncompliance are both forms of anger that are associated with grieving. (2) (SC; AN)

1 This is not a conscious attempt to hurt others but a way to relieve and reduce anxiety within the self.

2 This is a self-destructive method of coping, which could result in death.

4 This is an effort to maintain control over a situation that is really controlling the client; an unconscious method of coping and noncompliance may be a form of denial.

48. (1) This is the first action; the nurse should be aware that the technique is not safe, and it provides an opportunity for the offending nurse to correct the technique being used; the dressing should be immediately and correctly changed; the priority is to protect the client. (3) (SI; IM)

2 This depends on the policy of the institution and might be done later.

3 This may or may not be done by the observing nurse; it could be done by an inservice educator.

4 Same as answer 2.

49. (1) This response indicates recognition of the client's need to use denial and opens the way for discussion of feelings. (3) (SC; IM)

2 This forces reality on the client and blocks discussion of feelings.

3 This reply focuses on the surgery, which is not the concern expressed by the client.

4 This changes the subject and moves away from the client's feelings.

50. (2) This identifies the client's feelings and provides an opportunity for further discussion. (2) (SC; IM)

1 Although this echoes the client's statement, it does not identify a feeling.

3 This denies the client's feelings and focuses on information; the client may be too emotionally distraught to be able to construct or verbalize questions.

4 This provides false reassurance and cuts off communication; introduction of the word "cancer" could increase anxiety.

51. (2) This identifies the underlying client concern in a nonjudgmental way by not actually addressing the behavior; it provides the opportunity for an exploration of concerns and feelings. (2) (PC; IM)

1 Although it points out behavior, this is aggressive and may produce a defensive response instead of an exploration of concerns and feelings.

3 This response minimizes the client's feelings and provides false reassurance.

4 Same as answer 3.

52. (1) The nurse at the scene of an accident should function in a responsible and prudent manner; the use of a soiled cloth on an open wound is not prudent, nor is the independent transfer of an accident victim from the scene. (2) (CC; EV)

2 Although the Nurse Practice Act defines nursing, it does not provide detailed standards for practice; the nurse's action was not prudent.

3 The nurse's action was not what a reasonably prudent nurse would do and is therefore not protected.

4 The nurse's intervention was not prudent and placed the client in jeopardy; the nurse was not practicing medicine but attempting to provide first aid.

53. (3) Reliving the experience brings back the feelings, such as anxiety and fear, associated with it; the symptoms described reflect sympathetic nervous system activity. (2) (PC; AN)

1 The increased pulse and restlessness could indicate bleeding; however, the other data presented support anxiety; additional assessment would be necessary to identify bleeding.

2 Not enough data are present to recognize the client's usual method of seeking sympathy.

4 These symptoms are indicative of a sympathetic, not a parasympathetic, response.

54. (3) These are realistic, productive, and constructive ways of using this time. (1) (SC; AN)

1 These are signs of depression.

2 Going from physician to physician demonstrates disbelief, denial, or desperation.

4 This indicates anger and hopelessness, not acceptance.

55. (4) This is a legal requirement of participating institutions to protect the individuals involved. (1) (CC; AN)

1 Active euthanasia is a direct act to shorten a person's life and is illegal.

2 This is untrue; no age restrictions exist; a guardian's signature is required.

3 Legal statutes make certain the opposite is true.

56. (1) The client's misconception that the clot can cause sudden death when it reaches the heart must be corrected first; this may reduce the apprehension. (2) (DF; IM)

2 This was not the client's question and would not meet the need to explore feelings; this could be done later.

3 This is inappropriate; it disregards the client's expressed fears and would increase anxiety.

4 Once the client's understanding is corrected, this may or may not be necessary.

57. ③ A refusal to recognize anticipated loss in an attempt to protect oneself against the overpowering stress of illness is called repudiation. (2) (SC; AN)

1 The data do not suggest that the client has contemplated consequences related to the illness.

2 There are no data to support that the client is demonstrating behavior characteristic of an earlier stage of development.

4 The data do not suggest that the client has made a realistic adjustment to the illness.

58. ② Before a teaching plan can be developed, the factors that interfere with learning must be identified. (2) (CC; AS)

1 Although family members can be helpful, client involvement in care is important for promoting independence and self-esteem.

3 This is premature; assessment comes before intervention; written instructions may not be the most appropriate teaching modality.

4 This may be an unrealistic expectation; the client may never accept the change but must learn to manage care.

59. ④ Sharing concerns and talking about them often releases anxieties; giving the ordered hypnotic would produce relaxation. (1) (SC; IM)

1 This might relax the client but would do little to reduce the client's level of anxiety.

2 The client is not in pain at this time but needs to share concerns.

3 The procedure is over; this might be appropriate before the paracentesis; also, there are no data to support that this is the client's concern.

60. ② *Respondeat superior* is a doctrine that states employers may be held liable for torts committed by their employees. (3) (CC; AN)

1 *In loco parentis* refers to another person or agency assuming responsibility for a minor in the absence of the minor's parents.

3 The department of health sets standards and evaluates whether these are met; the hospital's responsibility for safe care under *respondeat superior* is a legal one.

4 This is unrelated to *respondeat superior* or to a hospital's legal responsibility to its clients.

61. ④ This response identifies the client's feelings and accepts them but also points out the responsibility of the client to take action. (1) (SC; IM)

1 Although this response identifies one of the client's concerns, the identification of the underlying feeling would be more therapeutic.

2 This response makes the nurse responsible for changing the situation, which is not appropriate or therapeutic.

3 This response denies the client's feelings and provides false reassurance.

62. ④ The release of information to an unauthorized person or gossiping about a client's activities constitutes a breech of confidentiality and an invasion of privacy. (1) (CC; AN)

1 Libel occurs when a person writes false statements about another that may injure the individual's reputation.

2 Slander occurs when a person verbally defames, detracts from, or maligns another's reputation.

3 Negligence is a careless act of omission or commission that results in injury to another.

63. ④ Before nurses can help others, they must understand themselves first, particularly on issues that may affect clients; this is the first step toward providing nonjudgmental care. (1) (DF; AS)

1 Although it is beneficial for nurses to understand themselves, this does not necessarily mean that the care will be nonjudgmental.

2 Although truthfulness is important in a therapeutic relationship, the nurse should attempt to be nonjudgmental; a nurse who feels uncomfortable should not be caring for the client in the first place.

3 Although this is important for all therapeutic relationships, it follows a thorough self-assessment of attitudes, values, and beliefs.

64. ① Voicing fears often reduces the associated anxiety. (1) (SC; IM)

2 Socialization when feelings need exploration is not therapeutic.

3 Although this should be done, simply accepting the client's wishes is not by itself therapeutic.

4 This avoids the problem and is not therapeutic.

65. ① This statement reflects values and beliefs regarding homosexuality as being bad and deserving of punishment. (2) (SC; EV)

2 There is not enough evidence presented to justify drawing this conclusion.

3 Same as answer 2.

4 Although this may be true, no information is given to suggest that the nursing assistant has been assigned to care for this client.

66. (1) The client is in the anger or "why me" stage; encouraging the expression of feelings will help the client resolve them and move toward acceptance. (2) (SC; IM)

2 This does not reflect on what the client said; introducing the topic of death may not be therapeutic.

3 This abdicates the responsibility of talking with the client; suggesting speaking with a minister ignores the client's present concerns.

4 This is judgmental, which may precipitate feelings of guilt and block the nurse-client relationship.

67. (1) Fear of recurrent myocardial infarct or sudden death is common when the client's environment is to be changed to one that appears less vigilant. (3) (PC; AN)

2 Depression is exhibited by withdrawal, crying, anorexia, and apathy and usually becomes more evident after discharge from the hospital.

3 Dependency would be exhibited by an unwillingness to increase exercise or perform tasks.

4 Ambivalence is exhibited by contrasting emotions; the client's statement does not demonstrate this.

68. (2) Knowing what to expect decreases anxiety. (2) (SC; IM)

1 Routine medications are not withheld unless ordered by the physician.

3 A sedative is not necessary for this test.

4 A small amount of radiation is emitted during the scan; the client should be assured the procedure is safe.

69. (3) This indicates the nurse has heard the client's verbal message and has empathy; it encourages further verbalization. (2) (PC; IM)

1 This may increase anxiety.

2 This minimizes the client's concerns.

4 This depersonalizes the client's concerns; it focuses on the tests and the setting rather than the client's concerns.

70. (3) The bureau of child welfare handles cases of child abuse and neglect, of which the child seems to be a victim. (2) (CC; IM)

1 The clinic would observe the client medically, but the bureau of child welfare would handle the child abuse and other social problems.

2 The hospital would probably not admit the child unless an immediate medical incident required it.

4 The bureau of the handicapped would be concerned with equipment and supplies required for the individual with a disability.

FLUIDS AND ELECTROLYTES (FE)

71. (1) The blood pressure indicates hypovolemia; the increased pulse is an attempt to maintain adequate oxygenation of tissues; the rapid weight loss reflects loss of body fluid. (1) (DA; AN)

2 The problem is an inadequate intake of fluid, not food; also, there are no data to support the presence of altered thought processes.

3 There are no data to support this diagnosis; the cardiac output is maintained by a compensatory increase in the heart beat.

4 The rapid weight loss reflects a loss of fluid, not a loss of body tissue.

72. (3) The client's intake was 360 (6 oz × 30 ml × 2 cups) and loss was 125 ml of fluid; loss is subtracted from intake. (1) (PC; EV)

1 This is an inaccurate calculation; the client has a more positive fluid balance.

2 Same as answer 1.

4 This is an inaccurate calculation; this answer added the output to the intake.

73. (3) One liter of fluid weighs approximately 2.2 pounds; therefore a 4.5-pound weight loss equals approximately 2 liters. (2) (PT; EV)

1 This is approximately a 1-pound weight loss.

2 This is approximately a 2.2-pound weight loss.

4 This is approximately a 7.5-pound weight loss.

74. (1) With the client's history and the large weight gain, this is the most likely cause of the increase in weight. (1) (DA; AN)

2 This occurs in the bladder, not the tissues, and would not account for the large weight gain.

3 This can occur with heart failure, but it is not the primary etiologic factor of the sudden weight gain.

4 Abdominal distention usually is caused by gas and should not contribute to this large a weight gain; if the abdomen is enlarged, assessment by ballottement should be done to determine whether enlargement is caused by fluid.

75. (3) Carbonated beverages are generally high in sodium and should be avoided. (1) (PC; PL)

1 Many of these products contain sodium.

2 Same as answer 1.

4 Same as answer 1.

76. (3) Hypokalemia is suspected when the T wave on an ECG tracing is depressed or flattened; a serum potassium level less than 3.5 mEq/L indicates hypokalemia. (1) (CI; AS)

1 This would have no significance in diagnosing a potassium deficit.

2 Same as answer 1.

4 Same as answer 1.

77. (2) Potassium is irritating to veins; larger veins can accommodate irritating substances for longer periods than small veins can. (2) (PT; IM)

1 Using the antecubital space would restrict client movement unnecessarily because the arm would have to be kept extended to prevent piercing the vein, which would result in an infiltration.

3 The back of the hand does not have veins large enough to accommodate a potassium infusion; using the dominant hand would restrict movement unnecessarily.

4 The back of the hand does not have veins large enough to accommodate a potassium infusion.

78. (3) T waves reflect serum potassium levels; flattened waves indicate hypokalemia, and tall or peaked waves indicate hyperkalemia. (3) (DA; AS)

1 Changes in P waves reflect atrial activity.

2 An extended PR interval indicates Lanoxin toxicity, not the potassium level.

4 Trigeminy and bigeminy reflect ventricular irritability, not the serum potassium level.

79. (1) A rate of 30 ml/hr is considered adequate for perfusion of the kidneys, heart, and brain. (1) (DA; EV)

2 A central venous pressure reading of 2 cm H_2O indicates hypovolemia.

3 This indicates improvement but not necessarily adequate tissue perfusion.

4 Same as answer 3.

80. (4) Albumin intravenously increases colloid osmotic pressure, resulting in a pull of fluid from the interstitial and intracellular compartments to the intravascular compartment. (3) (DA; AN)

1 The interstitial compartment is part of the extracellular compartment.

2 This is opposite to the actual shift of fluids.

3 Same as answer 2.

81. (3) Salt substitutes usually contain potassium, which would lead to symptoms of hyperkalemia, such as an abnormal heart beat. (3) (PC; AN)

1 Sodium in the diet causes retention of fluid.

2 Salt substitutes do not contain substances that influence BUN and creatinine levels; these are the result of protein metabolism.

4 There is no such substance in salt substitutes.

82. (2) Intravenous fluids supply minimal calories; a client on only IV therapy will lose weight and become malnourished. (1) (PT; EV)

1 This is not related to weight; lack of bulk in the diet results in constipation.

3 Vitamins are not related to weight loss.

4 Intracellular electrolytes are not related to weight loss.

83. (1) The stomach produces about 3 L of secretions per day; fluid lost through vomiting can produce a fluid volume deficit. The priority is fluid volume deficit, which can lead to dysrhythmias and death. (2) (DA; AN)

2 The shallow respirations are not related to a primary respiratory problem or pain; they are a compensatory mechanism to conserve CO_2 to combine with H^+ to form carbonic acid (H_2CO_3) and lower the plasma pH level.

3 Although this diagnosis is related, it is not the priority.

4 This would be true for duodenal ulcer; the gastric acid secretory rate is normal in persons with gastric ulcers; in gastric ulcers there is a decreased resistance of the gastric mucosa to acid-pepsin injury and a reflux of bile-containing duodenal contents back into the stomach.

84. (3) This has a low-sodium content. (3) (PC; EV)

1 Shellfish is high in sodium.

2 Processed meats are high in sodium.

4 These vegetables are high in sodium.

85. (1) This treats the hyperkalemia associated with renal failure; it moves potassium from the intravascular compartment into the intracellular compartment. (3) (DA; AN)

2 This would not increase urinary output.

3 This is not a treatment for respiratory acidosis.

4 Insulin and glucose do not decrease serum calcium levels.

86. (2) Cellular swelling and cerebral edema are associated with hyponatremia; as the extracellular sodium level decreases, the cellular fluid becomes relatively more concentrated and pulls water into cerebral cells. (2) (DA; AS)

1 This is not a symptom of hyponatremia; it may indicate dehydration.

3 This is associated with hypovolemia, not hyponatremia.

4 This is not a symptom of hyponatremia; it may indicate anemia.

87. (2) This ensures a well-controlled technique for electrolyte (chloride) replacement. (3) (PT; AN)
    1 There is no assurance that adequate chloride will be ingested and absorbed.
    3 Total parenteral nutrition is not necessary at this point, although it may eventually be used.
    4 This is not a well-controlled method to correct electrolyte deficiencies.

88. (3) The osmotic effect of hyperglycemia pulls fluid from cells, resulting in cellular dehydration. (3) (DA; AN)
    1 The opposite is true; hyperglycemia pulls fluid from the interstitial compartment to the intravascular compartment.
    2 Interstitial fluid is part of the extracellular compartment; the osmotic pull of glucose exceeds other osmotic forces.
    4 An increase in hydrostatic pressure results in an intravascular to interstitial shift.

89. (2) This would provide gradual replacement of both fluid and electrolytes without overloading the intravascular compartment. (2) (PC; PL)
    1 Water does not supply the necessary electrolytes, and hyponatremia could result.
    3 No data are presented to indicate that the client cannot take fluids orally; an NG tube is not necessary when the client can take fluids by mouth.
    4 This is unsafe; rapid correction of a fluid and electrolyte imbalance is dangerous; therapy should promote a gradual correction.

90. (2) During the oliguric phase of renal failure the kidneys retain potassium; an elevated potassium level is one of the main indicators of the need for dialysis. (3) (DA; AS)
    1 Metabolic acidosis would occur.
    3 Hypercalcemia would occur.
    4 Hyponatremia would occur.

91. (2) Commercially prepared salt substitutes are usually high in potassium. (2) (PC; EV)
    1 Nonsalt seasonings, such as lemon juice and pepper, can be used to make food more palatable.
    3 These contain high concentrations of potassium.
    4 Some complete protein foods must be included in the diet; however, on a restricted protein diet, proteins also will come from other sources.

92. (3) A decreased urinary output will result in the retention of potassium, causing hyperkalemia. (3) (DA; AS)
    1 Reporting this is unnecessary; this is a sign of dehydration, which can be corrected with appropriate hydration.
    2 Reporting this is unnecessary; these indicate dehydration, which is probably the rationale for the fluid ordered.
    4 Reporting this is unnecessary; this can precipitate dehydration or compound an existing dehydration and can be prevented by appropriate hydration.

93. (4) Pyrexia increases fluid loss through the skin; to maintain balance, the body compensates by reducing urinary output. (1) (DA; AN)
    1 Blood pressure is not directly affected by antimicrobials, and there are no data to suggest blood pressure is decreasing.
    2 This is unlikely; the client would have to become septic before this could occur.
    3 This represents a possible, but not probable, side effect of antimicrobial agents; also, nephrotoxicity would not be evident so soon.

94. (3) All the values provided are within normal limits. (3) (DA; EV)
    1 The albumin is normal.
    2 The hematocrit is normal; with hemoconcentration the hematocrit would be elevated.
    4 There is no evidence of overload; if the pulmonary arterial wedge pressure were higher, urine output increased, and hematocrit lower, it would indicate possible water intoxication.

95. (2) Urinary output reflects circulating blood volume; it is the most reliable, immediately available information to assess fluid needs. (1) (DA; EV)
    1 Daily weights reflect fluid retention or loss; however, other factors besides fluid affect weight; this is not as immediately accurate as hourly urines.
    3 This may indicate hypervolemia or hypovolemia but is not as accurate an indicator of insufficient fluid replacement as hourly urine output.
    4 Peripheral edema may have many causes; it is not an effective indicator of fluid balance.

96. (1) Silver nitrate can precipitate electrolyte imbalances; the client's potassium is below the normal level of 3.5 to 5 mEq/L and needs to be replaced. (3) (DA; PL)
    2 The client's sodium level is within the normal range of 135 to 145 mEq/L and there is no need to administer additional sodium chloride.
    3 Contraindicated; this would cause a further depletion of potassium.
    4 Same as answer 3.

97. (4) Low pH and bicarbonate values reflect metabolic acidosis; a low Pco$_2$ value indicates compensatory hyperventilation. (3) (CI; AS)
   1 Elevated pH and bicarbonate values reflect metabolic alkalosis; an elevated Pco$_2$ value indicates compensatory hypoventilation.
   2 Elevated pH and low Pco$_2$ values reflect hyperventilation and respiratory alkalosis.
   3 Low pH and elevated Pco$_2$ values reflect hypoventilation and respiratory acidosis.

98. (3) An orange only contains trace amounts of sodium. (3) (PC; IM)
   1 One cup of ice cream contains approximately 115 milligrams of sodium.
   2 One cup of celery contains approximately 106 milligrams of sodium.
   4 Four peanut butter cookies contain 142 milligrams of sodium.

99. (2) Fluid shifts from the intravascular compartment into the abdominal cavity, causing hypovolemia; a rapid, thready pulse is a compensatory response to this shift. (2) (DA; EV)
   1 Assessing for shock is the priority.
   3 After a paracentesis, intravascular fluid shifts into the abdominal cavity, not the lungs.
   4 Same as answer 1.

100. (2) Hypervolemia may precipitate pulmonary edema, which produces symptoms of shortness of breath, wheezing, cough, apprehension, and frothy sputum. (3) (DA; EV)
   1 Although this could occur, it is not the most serious complication; an altered respiratory status is the priority.
   3 This occurs with the IV administration of dye for diagnostic procedures; it does not occur with rehydration therapy.
   4 Same as answer 1.

101. (3) Verbalization indicates the client is breathing; elevating the head of the bed facilitates breathing by decreasing pressure against the diaphragm; checking the vital signs after this is the first step in assessing the cause of the distress. (2) (DA; PL)
   1 There is not enough information to support this option; further assessment is required.
   2 Discontinuing the IV access line may cause unnecessary discomfort and expense if it must be restarted; there are too little data to call the physician at this time.
   4 There is no information to support this option; assessment for allergies should be done on admission.

102. (2) The imbalance in intake and output, with a decreasing urinary output, may indicate renal failure with an increase of body fluid and the resulting development of congestive heart failure; assessing breath sounds and vital signs are the first steps to monitor for these complications. (3) (DA; EV)
   1 There are no data to support a problem with excretion of urine; the problem is with insufficient production.
   3 These are appropriate assessments after respirations and vital signs are evaluated.
   4 It is immaterial whether the fluid is given orally or intravenously; if there is hypervolemia, fluid intake would be decreased.

103. (3) Osmosis is the movement of fluid from an area of lesser solute concentration to an area of greater solute concentration. (3) (DA; AN)
   1 Filtration is the passage of fluid through a material that prevents the passage of certain constituents; hydrostatic pressure, the pressure exerted within a closed system, is known as filtration force; this force moves fluid by pressure and concentration gradients.
   2 Diffusion is the movement of particles across a semipermeable membrane from an area of greater concentration of particles to an area of lesser concentration of particles.
   4 In active transport, molecules move against a concentration gradient; this differs from diffusion and osmosis because metabolic energy is expended.

104. (1) This attempts to collect adequate data to plan the most appropriate intervention. (2) (PC; AS)
   2 This alone will not guarantee compliance once the client goes home; the client has the right to accept or reject therapy.
   3 Table salt is contraindicated on a 2-g sodium diet.
   4 Same as answer 2.

105. (3) Hypertonic food increases osmotic pressure and pulls fluid from the intravascular compartment into the intestine (dumping syndrome). (2) (DA; AN)
   1 Increased carbohydrates, not fats, are responsible for the increased osmotic pressure often associated with dumping syndrome.
   2 This is separation of the wound edges, usually accompanied by a gush of pink-tinged fluid.
   4 While peristalsis may be decreased because of surgery, it would not account for the symptoms.

106. (3) Shallow respirations, bronchial tree obstruction, and atelectasis compromise gas exchange in the lungs; an elevated carbon dioxide level leads to acidosis. (1) (DA; AN)
    1 Metabolic acidosis is seen in diarrhea with loss of base from the lower gastrointestinal tract.
    2 Metabolic alkalosis is caused by excessive loss of hydrogen ions from gastric decompression or excessive vomiting.
    4 Respiratory alkalosis is caused by increased expiration of carbon dioxide, a component of carbonic acid.

107. (4) All data are within normal limits; Po_2 is 80 to 100 mm Hg, Pco_2 is 35 to 45 mm Hg, and the pH is 7.35 to 7.45. (1) (DA; AN)
    1 None of the data is an indicator of fluid balance.
    2 Oxygen is within normal limits of 80 to 100 mm Hg.
    3 The pH would have to be less than 7.35

108. (4) The body fights acidosis by hydrogen exchange, which results in excretion of excess hydrogen in the urine, a compensatory mechanism to raise the serum pH level. (3) (DA; AN)
    1 In acidosis the pH level of the blood is decreased.
    2 Plasma bicarbonates would be decreased.
    3 In acidosis the respiratory center in the medulla is stimulated to increase respiration and blow off carbon dioxide, which is carried to the lung as carbonic acid; lowering carbonic acid raises the serum pH level.

109. (2) The client is hyperventilating and blowing off excessive carbon dioxide, which leads to these symptoms; if uninterrupted this could lead to respiratory alkalosis. (3) (DA; AS)
    1 Eupnea is normal, quiet breathing; the client has shallow, rapid breathing.
    3 Kussmaul's respirations are deep, gasping respirations associated with diabetic acidosis and coma, not hyperventilation associated with anxiety.
    4 These symptoms are related to a decreased carbon dioxide level in the body.

110. (3) Reassurance decreases anxiety and slows respirations; the bag is used so that exhaled carbon dioxide can be rebreathed to resolve respiratory alkalosis and return the client to acid-base balance. (3) (DA; IM)
    1 This is not necessary because there is no evidence of hypoxia.
    2 This is not necessary; this is used to prevent atelectasis.

    4 The client is already alkalotic; bicarbonate ions would increase the problem.

111. (1) Sodium, which is concerned with the regulation of extracellular fluid volume, is lost with vomiting; chloride, which balances cations in the extracellular compartment, is also lost with vomiting; because sodium and chloride are parallel electrolytes, hyponatremia will accompany hypochloremia. (2) (DA; PL)
    2 These values do not provide significant information in relation to the effects of vomiting.
    3 Same as answer 2.
    4 Same as answer 2.

112. (3) The normal plasma pH value is 7.35 to 7.45; the client is in alkalosis; the normal plasma bicarbonate value is 23 to 25 mEq/L; the client has an excess of base bicarbonate, indicating a metabolic cause for the alkalosis. (1) (DA; AS)
    1 The normal plasma potassium value is 3.5 to 5 mEq/L; the potassium value is within normal limits.
    2 The normal plasma chloride value is 95 to 105 mEq/L; the client has hypochloremia because of vomiting of gastric secretions.
    4 To be acidotic the plasma pH value would have to be less than 7.35.

113. (3) Hyperventilation results in the increased elimination of carbon dioxide from the blood. (2) (DA; AN)
    1 The Po_2 level would not be affected.
    2 The pH level will be increased.
    4 The carbonic acid level will be decreased.

114. (2) Decreased oxygen increases the conversion of pyruvic acid to lactic acid, resulting in metabolic acidosis. (3) (DA; AN)
    1 Hyperkalemia would occur because of renal shutdown; hypokalemia could occur in early shock.
    3 Respiratory alkalosis could occur in early shock because of rapid, shallow breathing, but in late shock metabolic or respiratory acidosis occurs.
    4 The Pco_2 level would be increased in profound shock.

115. (3) The normal range of arterial pH is 7.35 to 7.45; any condition that decreases bicarbonate anion concentration in extracellular fluid results in metabolic acidosis. (3) (DA; AN)
    1 This is not an accurate assessment for metabolic acidosis.
    2 Same as answer 1.
    4 This is within the normal range.

116. ③ Kussmaul breathing is an abnormally deep, very rapid, sighing type of respiratory pattern that develops as a compensatory response to metabolic acidosis and attempts to raise the pH of the blood by blowing off carbon dioxide. (2) (DA; AS)
1 Dyspnea is difficult breathing associated with subjective or objective distress in response to oxygen problems.
2 Hyperpnea is a deep, rapid rate of breathing without a subjective sense of extra effort, usually as a response to strenuous effort
4 Cheyne-Stokes respirations are characterized by a waxing and waning of breathing that is usually associated with pathology of the respiratory center in the brain.

117. ③ In respiratory alkalosis the pH level is elevated because of loss of hydrogen ions; the Pco_2 level is low because carbon dioxide is lost through hyperventilation. (3) (DA; AS)
1 This is partially compensated metabolic alkalosis.
2 This is respiratory acidosis.
4 This is metabolic acidosis with some compensation.

118. ② The urinary system compensates by retaining H+ ions, which become part of the bicarbonate ions; the bicarbonate level becomes elevated and raises the pH level to near normal; the normal HCO_3 value is 22 to 26 mEq/L, and the normal pH value is 7.35 to 7.45. (3) (DA; AS)
1 The normal Po_2 value is 80 to 100 mm Hg; this is within the normal range.
3 The normal Pco_2 value is 35 to 45 mm Hg; although in compensated respiratory acidosis the Pco_2 level may be elevated, it is the elevated HCO_3 level that indicates compensation.
4 This K+ level is within the normal range, which is 3.5 to 5 mEq/L; the serum potassium level is not significant in identifying compensated respiratory acidosis.

119. ③ The client is in respiratory acidosis, probably caused by the depressant effects of an anesthetic or a compromised airway; coughing clears the airway and deep breaths blow off CO_2. (3) (DA; IM)
1 This will not correct respiratory acidosis and may aggravate hypokalemia if present.
2 This is the treatment for respiratory alkalosis; the client is in respiratory acidosis.
4 This is not necessary if clearing of the airway rectifies the problem.

120. ③ Antidiuretic hormone (ADH) causes water retention, resulting in a decreased urine output and dilution of serum electrolytes. (3) (DA; AS)
1 Blood volume may increase, causing hypertension; diluting the nitrogenous wastes in the blood decreases rather than increases the BUN.
2 Water retention dilutes electrolytes; the client is overhydrated rather than underhydrated, so turgor is not poor.
4 ADH acts on the nephron to cause water to be reabsorbed from the glomerular filtrate, leading to reduced urine volume; the specific gravity is elevated as a result of increased concentration.

121. ④ Albumin is an essential component of the bloodstream that helps maintain both osmotic pressure and fluid and electrolyte balance. (3) (DA; AN)
1 This is not a cause of hemorrhage; blood components such as platelets, thrombin, and erythrocytes are involved in the prevention of hemorrhage or anemia.
2 Not directly involved with immunity and resistance; components such as T and B lymphocytes are the blood components involved in this process; the liver synthesizes specific proteins intrinsic to the function of antibodies.
3 Serum albumin is not related to nutrition of cells.

122. ③ This is important in determining whether or not the aspirin has affected the balance of electrolytes. (2) (DA; AS)
1 A test that requires 24 hours is not helpful in an emergency.
2 Glucose is unrelated to overdose of aspirin or cold tablets; a serum glucose would be done to confirm diabetic ketoacidosis; the history does not indicate that the client has diabetes mellitus.
4 Toxic hepatic effects of drug overdose are not apparent early after ingestion of the drugs.

123. ① This is the correct flow rate; multiply the amount to be infused (50 ml) by the drop factor (10) and divide the result by the amount of time in minutes (20). (2) (PT; AN)
2 This is an inaccurate calculation; this would deliver the solution too rapidly and would irritate the vein.
3 Same as answer 2.
4 Same as answer 2.

124. ③ Different infusion sets deliver different preset numbers of drops per milliliter; knowing this is a necessity for calculating the drip rate. (1) (PT; AN)
1 This does not determine the drip rate.
2 Same as answer 1.
4 This determines the size of the drop, not the drip rate.

125. (3) Air in the secondary line will not enter the vein; fluid from the primary bottle is under pressure and will flow before air from the secondary line could reach the port in the primary line. (1) (PT; IM)
   1 This is possibly true, but this answer would increase anxiety.
   2 Gentamicin bypasses the large IV bag because it is piggybacked into the primary line below the drip chamber and check valve.
   4 This is contraindicated; this stops the infusion, which can clog the lumen of the catheter that is inserted into the vein.

BLOOD AND IMMUNITY (BI)

126. (4) These nutrients stimulate the immune system. (2) (PC; AN)
   1 The role of fatty acids in natural defense mechanisms is uncertain.
   2 These have no known effect on natural defense mechanisms.
   3 This has no known immune-stimulating properties.

127. (4) Drains are usually inserted into the splenic bed to facilitate removal of fluid that could lead to abscess formation. (2) (CI; IM)
   1 Splenectomy has a low mortality rate (5%) except when multiple injuries are present (15% to 40%).
   2 Bleeding occurs more commonly with splenic repair than with removal.
   3 There is no need to frighten the client unnecessarily, but the operative risk increases with multiple injuries.

128. (3) With shock the heart rate accelerates to increase blood flow and oxygen to body tissues. (1) (CI; EV)
   1 With shock there would be a decreased urinary output because of the lowered glomerular filtration rate.
   2 With shock the respirations would be increased and shallow.
   4 With shock there would be hypotension.

129. (1) A decreased blood volume leads to decreased glomerular filtration; compensatory ADH and aldosterone secretion cause sodium and water retention. (2) (CI; EV)
   2 The respirations become rapid and shallow to compensate for decreased cellular oxygenation.
   3 The peripheral pulse rate may be rapid and thready, but it is the same rate as the apical rate.
   4 Hypokalemia occurs because as sodium is retained, potassium is excreted.

130. (4) Hypovolemia results in a decreased cardiac output and a decreased arterial pressure, which are reflected by a feeble, weak peripheral pulse. (3) (DA; AS)
   1 The skin would be cool and pale because of vasoconstriction.
   2 These are late signs of shock.
   3 The pulse pressure narrows with decreased cardiac pressure associated with hypovolemic shock.

131. (3) In shock, arteriolar vasoconstriction occurs, raising the total peripheral vascular resistance and shifting blood to the major organs. (2) (DA; AN)
   1 With shock more antidiuretic hormone (ADH) is produced to promote fluid retention, which will elevate the blood pressure.
   2 Although this is a response to hypoxia, peripheral vasoconstriction is a more effective compensatory mechanism.
   4 With shock the mineralocorticoids increase to promote fluid retention, which will elevate the blood pressure.

132. (1) A decreased intravascular volume results in hypovolemia and hypotension, which is evidenced by a decreased blood pressure and a decreased pulse pressure. (3) (DA; AN)
   2 Vasomotor stimulation to the arterial walls is increased with shock.
   3 This is a description of neurogenic shock, which is unlikely in this situation.
   4 Although electrolyte imbalances can precipitate cardiac decompensation, cardiogenic shock is unlikely in this situation.

133. (4) When an abdominal aneurysm ruptures, shock ensues because fluid volume depletion occurs as the heart continues to pump blood out of the ruptured vessel. (2) (DA; AN)
   1 This type of shock results from humoral or toxic substances acting directly on the blood vessels, causing vasodilation.
   2 This type of shock results from decreased neuromuscular tone, causing decreased vasoconstriction.
   3 This type of shock results from a decrease in cardiac output.

134. (2) Peritonitis and shock are potentially life-threatening complications following abdominal surgery; prompt, rigorous treatment is necessary. (1) (CI; IM)
   1 Fluids, not blood, would be needed to expand and maintain the circulating blood volume.

3 The head of the bed should be flat to increase tissue perfusion and oxygenation to the vital organs.

4 The physician should be notified; the oxygen was prescribed and the client should already be receiving it.

135. (1) This occurs during the progressive stage of shock as a result of accumulated lactic acid. (3) (DA; AN)

2 Metabolic alkalosis cannot occur with the buildup of lactic acid associated with the progressive stage of shock.

3 This can result from decreased respiratory function in late shock, further compounding metabolic acidosis.

4 This occurs as a result of hyperventilation during early shock.

136. (3) Blood replacement is needed to increase the oxygen-carrying capacity of the blood; the normal hematocrit value for women is 37% to 47% and for men is 45% to 52%. (1) (DA; PL)

1 Serum albumin helps maintain volume but does not affect the hematocrit level.

2 Ringer's lactate does not increase the oxygen-carrying capacity of the blood.

4 Dextran does expand blood volume, but it further decreases the hematocrit because it does not replace red blood cells.

137. (4) This is absolutely necessary to prevent an acute immunologic reaction if the donated blood is not compatible with the client's blood. (1) (PT; PL)

1 Although important, this is not the highest priority.

2 Blood must be kept cool until ready to use; if blood is at room temperature for 30 minutes prior to administration it should be returned to the blood bank; after it is started, blood must be administered within 4 hours.

3 This is not the highest priority; these laboratory results were part of the data used to determine the need for the blood.

138. (3) Both types are compatible; A negative is the same as her blood type and preferred; in an emergency, type O negative blood may also be given. (2) (PT; PL)

1 Although type O may be used, it would have to be Rh negative; Rh positive blood is incompatible with the client's and would cause hemolysis.

2 Type AB negative blood is incompatible with the client's blood and would cause hemolysis.

4 Type A negative is compatible with the client's blood but type AB negative is incompatible and may cause hemolysis.

139. (3) Transfusion reactions from mismatched blood will usually occur early during the transfusion (first 30 ml of blood); initially keeping the infusion at a slow rate decreases the amount of blood infused, so that the nurse has an opportunity to assess the recipient's response. (2) (PT; PL)

1 Blood must be administered within 4 hours to prevent bacterial contamination from occurring before infusion.

2 Blood should be administered via gravity; volume control and peristaltic infusion devices cause hemolysis.

4 For monitoring the client's response, blood would not be drawn until several hours after the infusion.

140. (2) This results from a recipient's antibodies that are incompatible with transfused red blood cells; it is also called a type II hypersensitivity; these signs result from RBC hemolysis, agglutination, and capillary plugging. (3) (PT; EV)

1 This results from an immune sensitivity to foreign serum protein; it is also called a type I hypersensitivity; signs include urticaria, wheezing, dyspnea, and shock.

3 Bacterial pyrogens are present in contaminated blood and can cause a febrile transfusion reaction; signs include fever and chills.

4 There is no transfusion reaction called by this name, although anaphylaxis occurs with an allergic transfusion reaction.

141. (2) This is a sign of an acute hemolytic transfusion reaction, indicating the recipient's blood is incompatible with the transfused blood; pain is caused by hemolysis, agglutination, and capillary plugging in the kidneys. (2) (PT; IM)

1 This is unsafe; this is a classic sign of a transfusion reaction, and the blood must be immediately stopped; while the assessment was being made, more incompatible blood would be infused, increasing the severity of the reaction.

3 Same as answer 1.

4 This is unsafe; blood must be stopped first, and then normal saline should be infused to keep the line patent and maintain blood volume.

142. ④ Fat-soluble vitamin K is essential for synthesis of prothrombin by the liver; a lack results in hypoprothrombinemia, poor coagulation, and hemorrhage. (2) (PT; EV)

1 Although cirrhosis may interfere with production of bile, which contains the bilirubin needed for optimum absorption of vitamin K, the best and quickest manner to counteract the bleeding is to provide vitamin K intramuscularly.

2 This is a coenzyme with vitamins B_{12} and C in the formation of nucleic acids and heme; thus a deficiency may lead to anemia, not bleeding.

3 A vitamin A deficiency contributes to development of polyneuritis and beriberi, not hemorrhage.

143. ③ The length of time a low-grade temperature is present, together with a history of night sweats and other physical findings, is valuable in making a diagnosis. (2) (HI; AS)

1 This is not immediately relevant to presenting signs and symptoms; more should be explored about the temperature itself before investigating causes of the temperature.

2 Same as answer 1.

4 Same as answer 1.

144. ② Equipment used is disposable; the donor does not come into contact with anyone else's blood. (1) (SI; AN)

1 The risk would depend on the spouse's prior behavior.

3 An individual may be infected for many weeks before testing positive for the antibodies; the individual could still transmit the virus.

4 Condoms offer some protection but are subject to failure because of condom rupture or improper use; risks of infection are present with any sexual contact.

145. ② These tests confirm the presence of HIV antibodies that occur in response to the presence of the human immunodeficiency virus. (1) (CI; AS)

1 This places someone at risk but does not constitute a positive diagnosis.

3 These do not confirm the presence of HIV; these adaptations are related to many disorders.

4 HIV infection is confirmed with the ELISA and Western blot tests; an opportunistic infection (included in the CDC surveillance case definition for AIDS) in the presence of HIV antibodies indicates that the individual has AIDS.

146. ① The HIV selectively infects helper T-cell lymphocytes; therefore 300 or fewer CD4 T cells per cubic millimeter of blood or CD4 cells accounting for less than 20% of lymphocytes is suggestive of AIDS. (1) (CI; AS)

2 The thymic hormones necessary for T-cell growth are decreased.

3 This finding is associated with allergies and parasitic infections.

4 This finding is associated with drug-induced hemolytic anemia and hemolytic disease of the newborn.

147. ④ Since this procedure normally does not involve contact with blood or secretions, additional protection is not indicated. (1) (SI; IM)

1 These are necessary only when there is risk of contact with blood or body fluid.

2 Same as answer 1.

3 A mask and gown would be indicated only if there were a danger of secretions or blood splattering on the nurse (for example, during suctioning).

148. ② These items prevent contact with feces, sputum, or other body fluids during intimate body care. (2) (SI; AN)

1 Goggles alone would be inadequate because the client is producing copious sputum.

3 Gloves are not necessary because touching body fluids when giving oral medication is not likely.

4 Only gloves are necessary when assisting the client with a bedpan.

149. ③ *Pneumocystis carinii* is a protozoan that causes pneumonia in immunosuppressed hosts, which can cause death in 60% of the clients; the client's respiratory status is the priority. (3) (CC; AN)

1 Although this is a concern, the client's respiratory status is the priority.

2 Same as answer 1.

4 There are no data to support this diagnosis.

150. ③ Palliative measures are aimed at relieving discomfort without curing the problem. (1) (PC; PL)

1 A cure or recovery is not part of palliative care; with a terminal disease this goal is unrealistic.

2 Same as answer 1.

4 Support of significant others is indicated, but it is not a palliative measure because it is not aimed at relieving the client's discomfort.

151. (3) Pernicious anemia is caused by the inability to absorb vitamin B_{12} resulting from a lack of intrinsic factor in gastric juices; for the Schilling test, radioactive vitamin B_{12} is administered and its absorption and excretion can be ascertained. (3) (CI; AS)
    1 Not measured by this test.
    2 Same as answer 1.
    4 Vitamin B_{12} is not produced in the body.

152. (3) IM injections bypass vitamin B_{12} absorption defect (lack of intrinsic factor, the transport carrier component of gastric juices); a monthly dose is usually sufficient since it is stored in active body tissues such as the liver, kidney, heart, muscles, blood, and bone marrow. (3) (CC; IM)
    1 The Z-track method need not be used as it is for iron dextran injections; injections once a month are usually sufficient.
    2 Since it is stored and only slowly depleted, it is not necessary to give injections this often.
    4 Vitamin B_{12} cannot be taken by mouth because of the lack of intrinsic factor.

153. (3) Since the intrinsic factor does not return to gastric secretions even with therapy, B_{12} injections will be required for the remainder of the client's life. (2) (CC; EV)
    1 It must be taken on a regular basis for the rest of the client's life.
    2 Same as answer 1.
    4 Same as answer 1.

154. (2) A protocol consisting of three or four chemotherapeutic agents that attack the dividing cells at various phases of development is the therapy of choice at this stage; alternating courses of different protocols may be used. (3) (DA; PL)
    1 Radiation, alone or in combination with chemotherapy, is used in stages IA, IB, IIA, IIB, and IIIA.
    3 This is recommended for use in stage IIIA.
    4 This is not a therapy for Hodgkin's disease at any stage; nodes may be removed for biopsy; nodes may be irradiated as part of therapy.

155. (1) Women in the childbearing years should be informed of all options available to preserve ovarian function. (3) (DA; PL)
    2 This is an incorrect statement; always is too absolute.
    3 This is an incorrect statement because radiation can influence or destroy ovarian functioning.

    4 Once the ovaries have been destroyed, they cannot regenerate.

156. (2) Reduced platelets increase the likelihood of uncontrolled bleeding; reduced lymphocytes increase susceptibility to infection. (2) (SI; PL)
    1 This would be helpful for stomatitis, not pancytopenia; aggressive oral hygiene could precipitate bleeding from the gums.
    3 Although fluids may be increased to flush out the toxic byproducts of chemotherapy, this would have no effect on pancytopenia.
    4 This is a sign of hypocalcemia that would not apply to pancytopenia.

157. (4) Thrombocytopenia occurs with most chemotherapy treatment programs; using a soft toothbrush helps prevent bleeding gums. (1) (CI; IM)
    1 While alopecia does occur, it is not related to bone marrow depression.
    2 Increasing fluids will neither reverse bone marrow depression nor stimulate hematopoiesis.
    3 This is not related to bone marrow depression.

158. (1) Suppression of bone marrow increases bleeding susceptibility associated with decreased platelets. (1) (CI; IM)
    2 This will not affect the bone marrow; citrus juices should be avoided by the client receiving chemotherapy because of the side effects of stomatitis.
    3 With bone marrow suppression the red blood cells are decreased in number and there is a decreased O_2-carrying capacity of the blood; this position will not increase the number of red blood cells.
    4 With bone marrow depression there would be a decrease in red blood cells; rest should be encouraged.

159. (3) Myelosuppression involves decreased red blood cells (anemia) resulting in less oxygen-carrying capacity of the blood and fatigue; decreased white blood cells (leukopenia) resulting in potential for infection; and decreased platelets (thrombocytopenia) resulting in potential for bleeding. (2) (CI; IM)
    1 Myelosuppression is not directly related to calcium deposition; myelosuppression, a reduction in bone marrow activity, results in decreased numbers of RBCs, WBCs, and platelets.
    2 Myelosuppression is not related to these responses.
    4 Myelosuppression is related to bone marrow activity, not the nervous system.

160. (1) Brief pressure is generally enough to prevent bleeding. (2) (CI; IM)
  2 Complications are rare; no special positions are required.
  3 The site is cleaned prior to aspiration.
  4 Complications are rare; frequent monitoring is unnecessary.

161. (3) Anemia with petechiae occurs because of bone marrow depression and rapidly proliferating leukocytes; all organs of the body are involved with the development of anorexia. (3) (HI; AS)
  1 While anorexia occurs, the client would more likely be sleeping excessively.
  2 There is no change in hair growth in the absence of chemotherapy.
  4 Same as answer 2.

162. (2) Splenomegaly usually accompanies chronic myelogenous leukemia; it is usually gross, palpable, and tender and necessitates removal; the spleen is located high in the abdomen on the left side and is not usually palpable unless it is enlarged. (3) (DA; AS)
  1 The urinary output is not affected.
  3 With leukemia and splenomegaly there is increased destruction of blood cells; the erythrocyte count will be low.
  4 These signs and symptoms are not associated with leukemia or splenomegaly but rather diabetes.

163. (2) This protein (monoclonal immunoglobulin) is thought to be the result of tumor cell metabolites; it is often present in clients with multiple myeloma. (2) (CI; AS)
  1 Hypercalcemia occurs with multiple myeloma because of bone erosion.
  3 Although this can be a late complication of multiple myeloma because of coagulation defects, it is not specific to multiple myeloma.
  4 Multiple myeloma is not caused by a bacterial infection.

164. (1) A definite confirmation of multiple myeloma can only be made through a bone marrow biopsy; this is a plasma cell malignancy with widespread bone destruction. (2) (CI; AN)
  2 While calcium is lost from bone tissue and hypercalcemia results, this is not a confirmation of the disease.
  3 While this protein is found in the urine, it does not confirm the disease.
  4 X-ray films will show the characteristic "punched-out" areas caused by the increase in the number of plasma cells, which contributes to the making of the diagnosis; the definitive diagnosis is made on biopsy.

165. (4) Because of bone erosion, pathologic fractures are a common complication of multiple myeloma. (2) (SI; PL)
  1 Although this would be done, the priority is to prevent injury.
  2 Although this is an adaptation to the associated anemia, it is not life threatening.
  3 Although this is important, preventing pathologic fractures is the priority.

166. (4) This is the treatment of choice; a variety of drugs affect rapidly dividing cells at different stages of cell division. (3) (DA; PL)
  1 Although this may be done, it is not the primary treatment.
  2 Although this may be used to alleviate pain and treat acute vertebral lesions, it is not the primary approach.
  3 Multiple myeloma is a diffuse disorder of the bone and no single lesion can be removed.

167. (1) Because an elevated temperature increases metabolic demands, the pyrexia must be treated immediately; the physician should be notified because this client is immunodeficient, from both the disease and the chemotherapy; a search for the cause of the pyrexia can then be initiated. (3) (HI; IM)
  2 More vigorous intervention is necessary; this client has a disease in which the immunoglobulins are ineffective and the therapy further suppresses the immune system.
  3 This is not the immediate priority, although it is important because the cause of the pyrexia must be determined; also, the increased amount of calcium and urates in the urine can cause renal complications if dehydration occurs.
  4 Not the priority, although important because respiratory tract infections are a common occurrence in clients with multiple myeloma.

168. (1) Blood products (packed RBCs or platelets) are administered when warranted. (1) (CC; IM)
  2 Renal insufficiency, not infections, may occur due to chronic hypercalcemia, proteinemia, and hyperuricemia.
  3 Dehydration may result in renal shutdown, and fluid replacement is provided in carefully supervised clinical settings.
  4 Untrue; ultraviolet rays are not related to exacerbations.

169. (4) These drugs suppress the immune system, decreasing the body's production of antibodies in response to the new organ which acts as an antigen; these drugs decrease the risk of rejection. (3) (PT; AN)
1 Leukocytosis is inhibited by these drugs.
2 These drugs do not provide immunity; they interfere with natural immune responses.
3 Because these drugs suppress the immune system, they increase the risk of infection.

170. (1) Tetanus antitoxin provides antibodies, which confer immediate passive immunity. (3) (HI; IM)
2 Antitoxin does not stimulate production of plasma cells, the precursors of antibodies.
3 Passive, not active, immunity occurs.
4 Passive immunity, by definition, is not long lasting.

171. (1) It takes this length of time for antibodies to respond to the antigen and form an indurated area. (1) (HI; AS)
2 This is longer than necessary; the site will reveal induration in 2 to 3 days.
3 Same as answer 2.
4 Same as answer 2.

172. (1) Drug hypersensitivity and anaphylaxis are most common with antimicrobial agents. (1) (SI; IM)
2 This is a dependent function; it is not crucial to starting antibiotic therapy.
3 This is an important assessment, but it is not crucial to starting antibiotic therapy.
4 Withholding treatment until culture results are available may extend the infection.

173. (2) Hypersensitivity to a foreign substance can cause an anaphylactic reaction; histamine is released, causing bronchial constriction, increased capillary permeability, and dilation of arterioles; this decreased peripheral resistance is associated with hypotension and inadequate circulation to major organs. (3) (DA; AN)
1 These are the problems that result from bronchial constriction and vascular collapse.
3 Dilation of arterioles occurs.
4 Arterioles dilate, capillary permeability increases, and eventually vascular collapse occurs.

174. (3) Activity may encourage the dislodgement of more microemboli. (2) (CI; AN)
1 Bed rest may enhance platelet aggregation and the formation of thrombi because of venous stasis.

2 Venous stasis, rather than enhanced circulation, is supported by bed rest.
4 Bed rest supports venous stasis rather than the circulation of blood to damaged tissues.

175. (4) This is the first screening test done to detect serum antibodies that bind to HIV antigens on test plates. (1) (CI; AS)
1 This screening test is not done first. The Western blot is done to validate repeatedly reactive ELISA results.
2 This is not a screening test; it is done to monitor the progression of HIV infection and response to treatment.
3 This is not an initial screening test; it is done when there are consistently inconclusive test results with previous screening tests.

176. (4) This test assesses parietal cell function. After radioactive cobalamin is administered, its excretion is measured. If cobalamin cannot be absorbed as in pernicious anemia, very little is excreted in the urine. (2) (CI; AS)
1 This test is not affected by medications.
2 The results of this test are not affected by food. With pernicious anemia there is a deficiency of intrinsic factor, which is necessary for vitamin B_{12} use.
3 Intake and output records are not necessary with a Shilling test.

INTEGUMENTARY (IT)

177. (1) This is soothing to the oral mucosa and helps prevent infection. (1) (PC; IM)
2 Stomatitis refers to the oral cavity; karaya is used to protect the skin around a stoma created on the abdomen.
3 There is no diarrhea or fluid loss caused by stomatitis.
4 The abdomen is not involved; stomatitis is an inflammation of the oral mucosa.

178. (3) Brown pigmentation spots (senile lentigines) increase in number, size, and distribution with aging. (2) (DF; AS)
1 Capillary fragility associated with aging contribute to senile purpura; however, ecchymosis indicates ineffective clotting or trauma.
2 Although sweat glands decrease in size, number, and function, contributing to skin dryness, there should not be marked flaking of skin.
4 Scaling of the skin is more often associated with psoriasis than aging.

179. ① Shearing force occurs when two surfaces move against each other; when the bed is at an angle greater than 30 degrees, the torso tends to slide and cause this phenomenon. (1) (SI; PL)
2 This would raise the head of the bed too high to prevent the client from sliding in bed.
3 Same as answer 2.
4 Same as answer 2.

180. ④ The client who is confined to bed should be encouraged to move in the bed to prevent prolonged pressure on any one skin surface. (2) (SI; PL)
1 This will help promote peripheral circulation, but prolonged pressure must be avoided.
2 This does not prevent prolonged pressure, although it can help decrease skin breakdown by allowing air to circulate beneath the client.
3 Range-of-motion exercises move joints to prevent contractures; they do not relieve prolonged pressure.

181. ④ Exercise would promote extension of the local infection from the leg into the circulation, causing septicemia. (3) (CI; AN)
1 This is not accomplished by bed rest.
2 Although bed rest does this, it is not the purpose for it in this situation.
3 Same as answer 2.

182. ① Surgical asepsis means that the defined area will contain no microorganisms. (2) (SI; AN)
2 This would be true of isolation procedures.
3 This would apply to isolation and medical asepsis.
4 Same as answer 3.

183. ② A mask and sterile gloves protect the infusion site against contamination with airborne and other pathogenic organisms. (2) (PT; PL)
1 Sterile gloves will not protect against airborne pathogens.
3 A gown and gloves will not protect against airborne pathogens.
4 A gown would serve no useful purpose during site care.

184. ④ In the obese individual this helps to inject the medication into subcutaneous tissue rather than adipose tissue, where its absorption would be poor. (2) (PT; IM)
1 This will result in the drug being injected into adipose tissue, where it will be poorly absorbed.
2 Same as answer 1.
3 Same as answer 1.

185. ② An iatrogenic infection is one caused by health care providers or therapy. (1) (SI; AN)
1 This is not the cause of an iatrogenic infection.
3 Same as answer 1.
4 Same as answer 1.

186. ② A nosocomial infection, by definition, is acquired during hospitalization. (1) (CI; IM)
1 Exposure to the pathogen must have occurred after admission to the hospital for classification as a nosocomial infection.
3 It may or may not be highly contagious; protective isolation protects the client from others; in this situation, others need to be protected from the client.
4 This nosocomial infection is a primary, not a secondary, infection.

187. ① Teaching for the postoperative period should begin as soon as the client is admitted; knowledge of what to expect decreases anxiety and may improve compliance with the treatment regimen. (1) (SI; PL)
2 This is too late; the client must have time to ask questions and demonstrate the ability to care for the wound; teaching begins preoperatively.
3 This is too late; at this time the client may be in too much discomfort to concentrate on learning.
4 Same as answer 3.

188. ④ Delayed wound healing is often caused by a lack of nutrients, such as protein and vitamin C, in the diet. (1) (PC; AS)
1 This usually occurs with severe protein deficiency and congestive heart failure.
2 This usually occurs with iron-deficiency anemia.
3 This usually indicates prolonged malnutrition.

189. ① The nurse epidemiologist acts as a consultant to help devise an infection control strategy. (1) (CC; IM)
2 This is the role of the physician.
3 This is the role of the laboratory technician or technologist.
4 This is usually done by the nurse.

190. ③ Barriers reduce the contact of bile with skin and limit excoriation. (2) (SI; PL)
1 Dressings should be changed when wet, not reinforced; usually the T-tube drainage empties into a collection chamber.
2 Antiseptics are drying and irritating to excoriated skin, and they sting when applied; this would require a physician's order.

4 This action may help, but the excoriation is probably caused by bile, not adhesive tape.

191. (1) Lubricating the skin with a moisturizer effectively relieves the dryness and thus the pruritus. (1) (PC; IM)

2 Warm or cool, not hot, tub baths would decrease the itching.

3 This would do nothing to lubricate the skin or relieve the pruritus.

4 Exposing the skin to air causes further drying and would not relieve the pruritus.

192. (3) Increased venous pressure alters the permeability of the veins, allowing extravasation of RBCs; lysis of RBCs causes brownish discoloration of the skin. (3) (DF; IM)

1 The arterial circulation is not affected by varicose veins.

2 Although tissue healing may be delayed, the brownish discoloration results from lysis of RBCs, not trauma.

4 There is no increase in melanocyte activity with varicose veins.

193. (3) To make effective decisions, baseline information on the client's condition, extent of injury, and significant past health history are needed. (3) (HI; PL)

1 This would be unnecessary; hyperimmune antirabies serum is not a preferred treatment because many people are sensitive to horse serum.

2 Notification of authorities is done after the injured person has received basic care.

4 Inoculation for establishment of short-term, passive immunity to rabies would be done after the initial assessment and treatment of the wound.

194. (4) This is a viral infection that enters the body through a break in the skin and is characterized by convulsions and choking. (3) (HI; PL)

1 Rabies is not bacterially caused; its outstanding symptoms are convulsions and choking.

2 Rabies is not associated with a bacterial septicemia; it is caused by a virus.

3 This is not true; the virus specifically attacks nervous tissue and is carried in the saliva of infected animals.

195. (2) Radiodermatitis occurs 3 to 6 weeks after the start of treatment. (2) (SC; IM)

1 The word "burn" should be avoided because it may increase anxiety.

3 This response does not address the client's concern.

4 Emollients are contraindicated; they may alter the calculated x-ray route and injure normal tissue.

196. (2) These contain alcohol and metals that can increase the skin reaction. (1) (DA; IM)

1 This is contraindicated; compresses can precipitate skin breakdown.

3 This is contraindicated; gauze and tape may irritate the skin.

4 This is not necessary, and controlling movement while sleeping is usually not possible.

197. (3) The skin is the first line of defense; keeping it dry and safe from injury promotes skin integrity. (1) (PC; PL)

1 This is unsafe; irradiated skin is fragile and subject to blistering and sloughing.

2 This is unsafe because irradiated skin is fragile; if soap is used, a film left after rinsing can change the angle and intensity of radiation.

4 This is unsafe; the skin should be free of emollients because they change the angle or degree of radiation.

198. (2) Scleroderma causes chronic hardening and shrinking of the connective tissues of any organ of the body including the esophagus and face; pureed foods limit the need to chew and this facilitates swallowing. (2) (PC; IM)

1 Liquefied foods are difficult to swallow; there is decreased esophageal peristalsis, and they are easily aspirated.

3 This will not help; it is equally difficult to swallow solids and liquids, and aspiration may result.

4 This is not necessary; no oral pain is associated with scleroderma.

199. (2) Raynaud's phenomenon is caused by vasospasm, precipitated by exposure to cold or emotional stress; Raynaud's is commonly associated with scleroderma, a connective tissue disorder. (1) (PC; PL)

1 Decreased blood flow would interfere with perception of temperature and increase the risk of burns.

3 Trauma to the hands must be avoided; nerve endings are affected by the diminished blood supply.

4 Vasodilators, not anticoagulants, are prescribed to counteract vasospasm and increase blood flow.

200. (2) With scleroderma the skin becomes dry because of interference with the underlying sweat glands. (3) (PC; PL)
1 There is no pruritus associated with scleroderma.
3 There are no inflamed areas associated with scleroderma.
4 There are no skin lesions associated with scleroderma.

201. (1) The changes in connective tissue associated with scleroderma lead to muscle weakness; optimal function must be preserved. (1) (PC; AN)
2 Tissue regeneration will not occur; this is a progressive degenerative disorder.
3 Prevention of extension is not possible in this systemic disease, which involves a number of organs.
4 Exercise of stiff and painful joints causes pain, not a sense of well-being.

202. (1) Nerves are the least resistant tissue, followed by blood vessels, skin, muscle, and bone. (3) (DA; AS)
2 Alternating electrical current, the most prevalent type in the United States, causes the most severe injuries.
3 Electrical current flows from the point of contact to the point of grounding.
4 It is difficult to track the path of electricity by external visualization; it often requires surgical exploration.

203. (3) This action effectively extinguishes the flames; avoiding additional intervention until the person receives medical attention prevents further injury. (2) (DA; IM)
1 This may protect the person from further burns but it could further injure the burned tissue.
2 This action does not deny the necessary oxygen and may in fact fan the flames.
4 This may extinguish the flames but it is not as effective as rolling in the grass.

204. (2) A patent airway is most vital; if the person is not breathing, CPR should be begun. (1) (DA; AS)
1 The person should be kept npo because large burns decrease intestinal peristalsis and the person may vomit and aspirate.
3 This is not done until the assessment for breathing has been completed.
4 This is not the priority; this assessment will be done after transfer to a medical facility.

205. (3) A tablecloth is nonfuzzy and nonadhering and will keep the burned person warm. (2) (DA; IM)

1 This is unsafe; body heat should be conserved with a nonadhering covering.
2 Cream is difficult to remove and may result in additional damage.
4 This is contraindicated because ice can result in additional tissue damage.

206. (3) In deep partial-thickness burns, destruction of the epidermis and part of the dermis occurs. (3) (DA; AS)
1 Eschar, a dry leathery covering of denatured protein, occurs with full-thickness burns.
2 In full-thickness burns, total destruction of the epidermis, dermis, and some underlying tissue occurs.
4 In superficial partial-thickness burns, the epidermis is destroyed or injured.

207. (1) This describes a superficial partial-thickness burn. (2) (DA; IM)
2 The entire epidermis and part of the dermis are affected with deep partial-thickness burns.
3 This describes full-thickness burns.
4 The statement is too vague a description of what is involved; in full-thickness burns the subcutaneous layer, muscle, and bone may or may not be damaged but are rarely destroyed.

208. (2) The anterior part of the lower legs are 4.5% each, which equals 9%. (2) (DA; AS)
1 The 4.5% would only represent the anterior part of one leg from the knee to the foot.
3 This would represent more than the anterior part of both legs from the knees to the feet.
4 Same as answer 3.

209. (4) The entire right lower extremity is 18%; the anterior portion of the right upper extremity is 4.5%. (2) (DA; AS)
1 This is less than the total percent of body surface burned.
2 Same as answer 1.
3 Same as answer 1.

210. (2) Heat inhalation can cause edema of the respiratory lumina, interfering with oxygenation; evaluation of respiratory status is a priority assessment. (3) (DA; AS)
1 This would be done after the client's respiratory status has been evaluated.
3 Burns should be evaluated by the physician and then the ordered medical therapy should be implemented; the airway has priority.
4 Same as answer 1.

211. (3) Hoarseness is a sign of potential respiratory insufficiency as a result of inhalation burns, which cause edema in the surrounding tissues, including the vocal cords. (3) (DA; AS)

1 This would indicate metabolic acidosis, not respiratory insufficiency.

2 Sputum would be sooty, not frothy; pink-tinged, frothy sputum is associated with pulmonary edema.

4 Same as answer 1.

212. (2) Massive amounts of potassium are released from the injured cells into the extracellular fluid; large amounts of sodium are lost in edema. (3) (DA; AS)

1 The serum potassium level will rise.

3 The serum potassium level will rise; the serum sodium level will fall.

4 The serum sodium level will fall.

213. (3) Extreme restlessness in a severely burned client usually indicates cerebral hypoxia. (3) (DA; AS)

1 With renal failure the client would become progressively confused and lethargic, not restless.

2 At this stage the client would be hypovolemic rather than hypervolemic.

4 With metabolic acidosis the client would be lethargic.

214. (2) Pain is from the loss of the protective covering of the nerve endings; blisters and redness occur because of the injury to the dermis and epidermis. (1) (DA; AS)

1 Because some epithelial cells remain, grafting is not needed with a partial-thickness burn unless it becomes infected and further tissue damage occurs.

3 Partial-thickness burns involve only the superficial layers of skin, unless they become infected.

4 Recovery from partial-thickness burns with no infection occurs in 2 to 3 weeks.

215. (2) Topical antibiotics are directly applied to the wound and are effective against many gram-positive and gram-negative organisms found on the skin. (2) (CI; PL)

1 Although these may be administered, they are most effective for systemic rather than local infections; the vasculature in and around a burn is impaired and the medication may not reach the organisms in the wound.

3 Same as answer 1.

4 Same as answer 1.

216. (3) Adequate intake of protein, carbohydrates, vitamin C, and minerals is necessary for tissue building and wound healing. (1) (CI; EV)

1 There are no data to come to this conclusion; although the client is not eating, the client may be drinking fluids.

2 These changes would take a prolonged period of time; they would not occur during a short period.

4 This is associated with prolonged hypoxia.

217. (4) Care of burns is a painful procedure; pain medication should be administered before care to limit discomfort. (1) (PT; PL)

1 Surgical asepsis should be used.

2 No dressings are applied when the exposure method is used.

3 This is unnecessary; Sulfamylon is not hepatotoxic.

218. (2) Sterile aseptic technique is necessary for an open wound, and a thin layer of ointment is applied directly to the affected area. (1) (SI; IM)

1 Surgically aseptic, not medically aseptic, technique is used.

3 Improper technique; while some medications may be placed directly in the tank, antimicrobial medications are placed directly on the affected area using surgically aseptic technique.

4 Same as answer 3.

219. (2) This gauze is less apt to adhere to the wound; the hand should be maintained in anatomical position of slight flexion with each finger separated. (3) (CI; IM)

1 Cotton filled or backed gauze should not be used because it may adhere to the wound; the hand should be in anatomical position with fingers slightly flexed.

3 The hand is not anatomically positioned when in full extension or full flexion; the hand should be slightly flexed.

4 Same as answer 1.

220 (2) The first step is to determine what the physician has told the client. (3) (CC; IM)

1 This statement would be inappropriate; grafting begins when the granulation bed is formed.

3 Same as answer 1.

4 Grafting will not be done until the wound is clean and granulation tissue has formed; there are no data to indicate the client has an infection.

221. (1) Clients with burns are sensitive to environmental changes; loss of the skin's microcirculation in the burned areas decreases the ability to retain body heat. (3) (PC; PL)
2 The room temperature should be kept at approximately 85° F or 24.4° C to limit loss of body heat.
3 This is too low; higher humidity (40% to 50% usually) is needed to maximize the warmth of the room.
4 A sterile sheet is not necessary; some clients may be treated by the open method and have burns exposed.

222. (1) Mechanical débridement means to physically remove dirt, damaged or dead tissue, and cellular debris from a wound or burn so that infection is prevented and healing is promoted. (2) (DA; IM)
2 This is not the definition of either mechanical or chemical débridement; this refers to grid escharotomy.
3 This is not the definition of mechanical débridement; enzymes, such as Travase, are used to débride chemically by dissolving and removing necrotic tissue.
4 This is continuous passive range of motion, not débridement.

223. (3) The graft covers nerve endings, which reduces pain and provides a framework for granulation. (3) (CI; AN)
1 The graft promotes epithelialization; enzymatic preparations or surgery débride wounds.
2 This is untrue; pig skin grafts are not sutured.
4 This is contraindicated; topical antimicrobials would soften the graft and impede healing.

224. (2) Range of motion should be instituted as soon as it will not compromise the individual's cardiopulmonary status. (2) (SI; PL)
1 Pain will continue for some time, and if ROM is delayed until it subsides, contractures will have already developed.
3 If ROM is delayed until skin grafts heal, contractures will have already developed.
4 Pain and inability to cope may be prolonged; if ROM is delayed, contractures will have already developed.

225. (4) The negative pressure of a portable wound drainage system exerts a sucking force that pulls fluid toward the collection chamber. (1) (DA; AN)
1 Gravity is the environmental force that pulls weight toward the center of the earth; an indwelling urinary catheter allows urine to flow by gravity from the bladder to a collection bag placed below the level of the bladder.
2 Osmosis occurs when solvent moves from a solution of lesser concentration to one of greater solute concentration when the two solutions are separated by a semipermeable membrane; fluid moving from the interstitial compartment into the intracellular compartment uses osmosis.
3 Active transport occurs when ions move across a cell membrane against a concentration gradient with the assistance of metabolic energy; the sodium-potassium pump uses active transport.

226. (2) Once the drugs that interfere with cell division are stopped, the hair will grow back; sometimes the hair will be a different color or texture. (1) (PT; IM)
1 Alopecia is a common side effect of chemotherapy.
3 Hair loss persists while the drugs are being received; once the drugs are withdrawn, the hair grows back.
4 Although ice caps on the head and rubber bands around the scalp have been used to try to limit alopecia, they have not been particularly effective.

SKELETAL (SK)

227. (2) Weight-bearing and the use of anti-gravity muscles stimulate bone formation or osteoblastic function. (1) (DF; AN)
1 This will result in bone demineralization, not calcium deposition in the bone.
3 The normal aging process contributes to a gradual and progressive demineralization of bone.
4 Calcium intake has a relationship to osteoclastic mechanisms in that it inhibits the withdrawal of calcium from bone.

228. (3) These symptoms are consistent with a torn cartilage, which is also a common injury with basketball. (1) (HI; AS)
1 A fractured patella would cause pain and usually manifests itself at the time of injury.
2 A ruptured Achilles tendon is painful and prevents plantar flexion of the foot; symptoms are usually manifested immediately upon injury.
4 A stress fracture is associated with pain, not with a clicking or locking of the knee.

229. (2) This is a truthful description of arthroscopy; the physician uses a scope to visualize the knee

structures to determine the extent of injury. (1) (CC; IM)
1 While this is true, it evades the client's concern and does not describe the procedure.
3 Arthroscopy is not a radiologic procedure.
4 Same as answer 1.

230. (4) A bone scan reflects the uptake of a bone-seeking radioactive isotope; an increased uptake is seen in metastatic bone disease, osteosarcoma, osteomyelitis, and certain fractures. (2) (CC; EV)
1 A bone scan measures the uptake of radioactive material, not the absence of calcium, which would be seen in an x-ray examination of bone.
2 This is a bone marrow aspiration, when a small amount of marrow is examined to determine the presence of abnormal cells in diseases such as leukemia.
3 A bone scan involves a small diagnostic dosage of a radioactive substance; it is not therapeutic.

231. (2) Numbness is a neurologic sign that should be reported immediately because it indicates pressure on the nerves and blood vessels. (2) (CI; EV)
1 Warmth is a normal reaction to a new cast.
3 This results from inadequate skin care but can be easily managed with lotion or oil.
4 Some degree of discomfort is expected following cast application.

232. (3) Damage to the blood vessels may decrease the circulatory perfusion of the foot; the absence of a pedis pulse would indicate a lack of blood supply to the lower extremity. (2) (CI; EV)
1 The break is between the knee and the ankle, not the thigh.
2 Damage to the major blood vessels would more likely cause a decrease in blood pressure from shock.
4 Decreased circulatory perfusion of the foot would cause the skin temperature to decrease, but the foot could also feel cool because of shock.

233. (3) This indicates tissue hypoxia or breakdown and should be reported to the physician. (1) (SI; IM)
1 Other data, such as elevated temperature or increased white blood cells, are not present to support the presence of an infection.
2 This is not a typical response to a cast and may indicate a complication.

4 The priority is to notify the physician; this could be done to provide relief to the client after the physician is notified.

234. (3) Elevation will help control the swelling that normally occurs. (3) (PC; IM)
1 Because the ankle has been at rest, discomfort and stiffness are expected after the cast is removed.
2 Because the skin has not been exposed, it needs gentle washing to prevent breaking the tissue.
4 The leg should be put through full range of motion more than once daily.

235. (4) These movements include all possible range of motion for the ankle joint. (1) (PC; IM)
1 Although the ankle can be moved in a circular motion, flexion and extension are more specifically called dorsiflexion and plantar flexion in relation to the ankle; also, eversion and inversion should be done when manipulating the ankle.
2 Flexion and extension are more specifically called dorsiflexion and plantar flexion in relation to the ankle; the ankle cannot be abducted or adducted; the ankle can be inverted and everted.
3 These motions refer to the upper extremities.

236. (3) These sets of muscles will be used in crutch walking and therefore need strengthening. (2) (SI; PL)
1 Although these muscles keep the person erect, the most important muscles for walking with crutches are the triceps, elbow extensors, finger flexors of the arms, and the muscles in the unaffected leg.
2 This will do nothing to strengthen the weight-bearing leg.
4 A pushing, not a pulling, motion is used with crutches; the triceps, not the biceps, are used.

237. (2) The palms should bear the client's weight to avoid damage to the nerves in the axilla (brachial plexus). (1) (SI; EV)
1 This would be unsafe; pressure on the axillary region could injure the nerves in the brachial plexus.
3 The physician ordered non–weight bearing on the affected leg.
4 Same as answer 1.

238. (3) In a comminuted fracture, the bone is splintered or crushed. (2) (HI; AN)
1 This is a compound fracture.
2 This is a greenstick fracture.
4 This is a complete fracture.

239. (3) The infected bone is placed at rest in a cast, splint, or traction to reduce pain and limit the spread of infection. (2) (SI; PL)
1 This is contraindicated; this would increase pain and spread the infection.
2 Osteomyelitis is usually caused by a microorganism traveling through the bloodstream to the bone, not the reverse; the client is already septic.
4 This is contraindicated; ambulation may facilitate the spread of infection.

240. (2) Uric acid has a low solubility; it tends to precipitate and form deposits at various sites where blood flow is least active, including cartilaginous tissue such as the ears. (3) (HI; AS)
1 Urate deposits will not form at this site because the blood flow is ample, and it is not cartilaginous tissue.
3 Same as answer 1.
4 Same as answer 1.

241. (2) Shellfish contains more than 100 mg of purine per 100 g. (2) (PC; EV)
1 This food is low in purine.
3 Same as answer 1.
4 Same as answer 1.

242. (1) Osteoporotic vertebrae collapse under the weight of the upper body or by improper or rapid turning, reaching, or lifting. (2) (CI; AN)
2 Bones, not the spine, demineralize in osteoporosis.
3 This occurs in osteoarthritis.
4 The spinal cord does not bulge; the nucleus pulposus bulges toward the spinal cord.

243. (3) The antibody called rheumatoid factor is present in 90% of clients with advanced arthritic changes; it is not found in the early stages of the disease process. (2) (CI; IM)
1 This denies the client's discomfort and does not deal with the stated confusion.
2 This denies the client's immediate feelings and blocks communication of them.
4 This response reinforces the client's felt discomfort and confusion over negative test results.

244. (3) Moist heat increases circulation and decreases muscle tension, which helps relieve chronic stiffness. (1) (PC; IM)
1 While this is advisable for someone with arthritis, it will not relieve morning stiffness.
2 This is related to muscle fatigue, not to stiffness of joints.
4 Inactivity promotes stiffness.

245. (2) This is caused by inflammation of the synovium, resulting in vascular congestion, fibrin exudate, and cellular infiltrate. (1) (PC; AN)
1 Urate crystals occur with gouty arthritis, not rheumatoid arthritis.
3 This is unrelated to rheumatoid arthritis.
4 Increased interstitial fluid (edema) is only one aspect of the inflammatory response.

246. (1) Palpation will elicit tenderness because pressure stimulates nerve endings and causes pain. (1) (PC; IM)
2 It is already swollen, and the pressure causes pain.
3 Nodules associated with rheumatoid arthritis are not caused by pressure; they occur spontaneously in about 25% of individuals with rheumatoid arthritis and are composed of collagen fibers, exudate, and cellular debris.
4 These are present in gout, not arthritis; they are composed of sodium urate.

247. (2) This position provides for extension of joints. (1) (CC; PL)
1 The side-lying position supports flexion of joints.
3 This creates continued flexion of joints.
4 This position does not influence position of joints.

248. (2) The hips and shoulders should be against the back of the chair; the thighs should be fully supported, with the knees and ankles at a 90-degree angle; this provides support and limits contractures. (2) (PC; IM)
1 In a low chair the hips will be flexed less than 90 degrees; the client needs a high-backed chair.
3 In this position the thighs and knees are not fully supported because they extend beyond the depth of the seat.
4 This would encourage flexion contractures.

249. (4) Some pain is to be expected, but the activity should not be continued when the pain becomes severe, because it can further traumatize the inflamed synovial membranes. (2) (PC; EV)
1 Some discomfort is expected; inactivity will promote the development of muscle atrophy and joint contracture.
2 This is unrealistic because some discomfort is expected.
3 This is unsafe; activity should be curtailed when pain becomes severe.

250. (1) Not neglecting joint pain protects the joints, especially if the pain lasts more than 1 or 2 hours after a particular activity. (3) (CI; IM)

2 The opposite would be true; the client should use large muscles, such as pushing doors open with arms rather than fingers.

3 This would increase joint stress; heavy and light tasks should be alternated.

4 When the inflammatory process is active, the joint should be used as little as possible to provide rest.

251. ③ The Valsalva maneuver raises cerebrospinal fluid pressure, thereby causing pain. (1) (PC; AS)

1 Calf tenderness is associated with thrombophlebitis, not disc problems.

2 Dysuria is associated with urinary problems, not disc problems.

4 This type of pain is not associated with intervertebral disc problems; it may occur with renal calculi.

252. ③ Pain results because herniation of a disc into the spinal column irritates the spinal cord or the roots of spinal nerves. (1) (PC; AN)

1 This is not involved; the lamina is that portion of the vertebrae removed during surgery to gain access to the disc.

2 The vertebral bodies themselves are not shifting.

4 Circulation of cerebrospinal fluid is not affected.

253. ④ Coughing places strain on the lumbar area, increasing the herniation of the disc. (2) (PC; IM)

1 This does not increase intervertebral pressure.

2 This will not increase pressure or cause pain; flexed knees are usually a more comfortable position.

3 This will not increase pressure or increase pain.

254. ② Putting the upper arm and leg toward the side to which the client is turning uses body weight to facilitate turning; the spine is kept straight. (3) (PC; AN)

1 This is unsafe; this would result in twisting the spinal column.

3 This could be done if another person were turning the client; when turning alone in this position, the client would have no leverage and turning would probably result in twisting the spinal column.

4 This would interfere with turning because the bent leg becomes an obstacle and provides a force opposite to the leverage needed to turn.

255. ④ This maintains vertebral alignment, decreasing trauma to the operative site. (2) (CI; IM)

1 This is contraindicated; it does not maintain vertebral alignment and may raise cerebrospinal fluid pressure.

2 This is contraindicated; the contour position flexes the vertebral column.

3 Traction is not used with this surgery.

256. ② Alteration in circulation and sensation indicates damage to the spinal cord and must be monitored. (2) (CI; EV)

1 Logrolling from side to side is preferred; the prone position may hyperextend the vertebral column.

3 Although observation is part of assessment, these assessments are not a priority.

4 Fluid intake is not a problem with these clients postoperatively.

257. ② Sitting maintains alignment of the back and allows the nurses to support the client until orthostatic hypotension subsides. (2) (SI; IM)

1 This would permit flexion of the vertebrae, which could traumatize the spinal cord.

3 Same as answer 1.

4 Rapid movement could flex the vertebrae, which would traumatize the spinal cord; safety is the priority; if the blood pressure were taken it would probably reveal orthostatic hypotension.

258. ② This is contraindicated; maintaining the sitting position for a prolonged period places excessive body weight and stress on the surgical area. (2) (CC; EV)

1 This maintains lordosis of the small of the back and provides proper support.

3 This relieves pressure on the back and promotes comfort in bed.

4 This prevents excessive pressure on the musculature and vertebral column.

259. ② This is done while in the supine position before the body is subjected to the force of gravity in a vertical position; anatomic landmarks are easier to locate for correct application of the brace, and intraabdominal organs have not shifted toward the pelvic floor by gravity. (2) (CI; PL)

1 This is unsafe; it should be applied while in the supine position before getting out of bed and should be worn the entire day for support.

3 The brace should be applied while in the supine position, not the sitting position.

4 Twisting exercises are contraindicated because they exert excessive pressure on the operative site.

260. (2) Phantom limb syndrome is a real experience with no known cause or cure. (1) (PC; AN)
    1 This may be an appropriate diagnosis for the client with an amputation, but it is not the diagnosis related to phantom limb syndrome.
    3 Same as answer 1.
    4 This is not an appropriate diagnosis for the individual with an amputation.

261. (1) Lying prone will prevent flexion contractures of the hip because it keeps the hip in extension. (1) (CI; IM)
    2 Sitting will promote the development of a flexion contracture, especially when prolonged, and should be avoided.
    3 Elevating the stump with pillows will cause a flexion contracture.
    4 Lying supine with pillows between the thighs will promote external rotation and hip abduction, contributing to a contracture.

262. (1) The loss of the limb has eliminated a wide base of support and altered the center of gravity. (1) (CI; AN)
    2 This is not related to falls.
    3 This could be a factor in any fall, not specifically one related to an amputation.
    4 While weakness may be a contributory factor in the elderly, it is unlikely in an otherwise healthy young adult.

263. (4) Pressure supports tissue, promotes venous return, and limits edema, thus promoting shrinkage. (2) (CI; IM)
    1 Activity, not bandaging, usually decreases the occurrence of phantom limb pain.
    2 Although it may limit clot formation, its primary purpose is to promote venous return, prevent edema, and shrink the stump.
    3 While pressure may prevent hemorrhage, its primary purpose is to prevent edema and shrink the stump.

264. (3) This stretches the flexor muscle and prevents a flexion contracture of the hip. (2) (CI; IM)
    1 This flexes the hip and may result in a hip flexion contracture; the stump is usually elevated for only 24 to 48 hours.
    2 This may result in an abduction deformity; the stump should be kept in functional alignment.
    4 This should be started on a soft surface with a physician's order approximately 5 days postoperatively.

265. (2) Preparation for crutch walking includes exercises to strengthen the arm and shoulder muscles. (1) (PC; IM)

1 This is important in healing and preparation for the prosthesis, not for crutch walking.
3 Position changes are to prevent hip flexion contractures, not to prepare for crutch walking.
4 Phantom limb phenomenon is a sensation that the absent limb is present; there are no such exercises.

266. (3) These indicate that the client has exceeded tolerance for the activity at this time. (1) (SI; EV)
    1 These are expected adaptations to the activity.
    2 Flushed skin is an expected response to activity; respirations would increase in depth rather than become slowed.
    4 An increase in blood pressure is an expected response to activity; respirations would increase in depth, not become shallow.

267. (2) Without the prosthesis, a walker or crutches would be necessary and require the readjustment of weight bearing on one leg. (3) (PC; AN)
    1 Early use of a prosthesis does not affect the development or presence of phantom limb pain, which occurs in about 10% of clients with an amputation.
    3 Early use of a prosthesis has no effect on wound infection.
    4 While this is true, it is not the major purpose; a prosthesis can easily be fitted after discharge when the stump is completely healed and no longer edematous.

268. (3) The even distribution of hemosiderin (iron-rich pigment) in the tissue in response to pressure of the prosthesis indicates proper fit. (2) (CI; EV)
    1 This would result in an improper fit.
    2 This has nothing to do with a proper fit.
    4 This indicates that the prosthesis is too long or too short.

269. (3) Gentle intervention reduces pain and shock and inhibits release of bone marrow into the system. (2) (CI; IM)
    1 This will not prevent fat emboli; it may limit edema and pain, a local effect.
    2 Maintaining alignment will not prevent fat emboli.
    4 This will not prevent fat emboli; it may prevent hypostatic pneumonia and atelectasis.

270. (2) Weight greater than 8 pounds causes excessive tension on the skin, leading to damage. (2) (CI; AN)

1 The spreader bar should be wide enough to keep materials away from the malleoli.

3 The Buck's boot should extend to the area just below the knee.

4 Tape is unnecessary, a Buck's boot is used.

271. (1) The client probably has a fat embolus; oxygen reduces surface tension of the fat globules and reduces hypoxia. (3) (DA; IM)

2 Oxygen should be administered and the client placed in semi-Fowler's position before the physician is called.

3 The client is not in shock resulting from hemorrhage but has a fat embolus; oxygen takes priority.

4 This will cause hip flexion, putting stress on the fractured femur; the semi-Fowler's position is preferred.

272. (4) A complete assessment must be performed to determine the location, characteristics, intensity, and duration of the pain; the pain could be incisional, result from a pulmonary embolus, or be caused by neurovascular trauma to the affected leg, and the intervention for each would be different. (1) (PC; EV)

1 This may be done after a complete assessment reveals that this would be the appropriate intervention; assessment is the priority.

2 Same as answer 1.

3 Same as answer 1.

273. (2) Compressed air inflates the padded plastic stockings systematically from ankle to calf to thigh and then deflates; this promotes venous return and prevents venous stasis and thromboembolism. (1) (CC; PL)

1 Turning on the operative side is contraindicated because it places tension on the hip joint and may traumatize the incision.

3 Although this may be ordered to promote muscle strength, the major complication, thromboembolism, is the priority.

4 This is contraindicated immediately after surgery.

274. (1) After surgery, abduction is maintained to reduce the chance of dislocation of the femoral head. (1) (CI; PL)

2 This can lead to dislocation of the femoral head.

3 This causes adduction, which can lead to dislocation of the femoral head and is contraindicated.

4 Same as answer 2.

275. (4) Until the order is written to discontinue the abduction splint, it is only removed for mobility such as physical therapy and hygiene; adduction to or beyond the midline is not permitted for 2 to 3 months. (1) (CI; IM)

1 It is needed at this time unless the client can be trusted to maintain abduction; flexing the hip with a prosthesis cannot be beyond 60 degrees for up to 10 days; from then on it cannot be beyond 90 degrees for up to 2 to 3 months.

2 It helps to maintain position and keep the hip prosthesis in the hip socket.

3 This is inappropriate; this is not the criterion for discontinuing abduction of the affected extremity.

276. (4) The pelvis is elevated by actions involving the unaffected upper extremities and unoperative leg. (2) (PC; IM)

1 This is not permitted because it causes adduction of the leg and can lead to dislocation of the femoral head.

2 No pressure is permitted on the operative hip because it can cause dislocation of the femoral head.

3 Lifting only with the arms requires strength; the use of both heels puts pressure on the operative hip.

277. (4) At the time of a fracture or orthopedic surgery, fat globules may move from the bone marrow into the bloodstream; also, elevated catecholamines cause mobilization of fatty acids and the development of fat globules; in addition to obstructing vessels in the lung, brain, and kidneys with systemic embolization from fat globules, petechiae are noted in the buccal membranes, conjunctival sacs, hard palate, chest, and anterior axillary folds; these adaptations only occur with a fat embolism. (3) (CI; AS)

1 This is a sign of an embolus, but it is not specific to a fat embolus.

2 Same as answer 1.

3 Same as answer 1.

278. (3) This puts the least strain on the prosthesis. (2) (CI; PL)

1 A soft chair would permit hip flexion greater than 90 degrees

2 Elevation of the leg places increased strain on the prosthesis.

4 A soft chair would permit hip flexion greater than 90 degrees.

279. (3) The client is demonstrating acceptance and is even looking toward the future. (1) (CI; EV)
   1 This relates to situational low self-esteem, not body image disturbance.
   2 This response may indicate that the client is trying to lose weight and may not accept present body weight.
   4 While this may indicate adaptability, it is not related to body image.

280. (1) Effective skeletal traction requires a continuous pull by the weights; the weights should be removed only in an emergency such as cardiac arrest or hospital evacuation. (2) (SI; IM)
   2 Contraindicated; removal of the weights would displace the fracture and increase pain.
   3 This problem can be corrected without removing the weights.
   4 Same as answer 3.

NEUROMUSCULAR (NM)

281. (2) These are used to observe the pharynx and larynx, assess the symmetry of the soft palate, and determine the presence of the gag reflex; they provide data about cranial nerve X (vagus nerve). (3) (HI; AS)
   1 This is used to assess cranial nerve VIII (auditory).
   3 This is used to assess cranial nerve II (optic).
   4 These are used to assess sensory function, that is, light touch and pain.

282. (1) The afferent sensory branch of the trigeminal nerve (cranial nerve V) innervates the cornea. (3) (HI; AS)
   2 This would test the function of cranial nerve VII.
   3 This would test the function of cranial nerves III, IV, and VI.
   4 This would test the function of cranial nerve XI.

283. (1) The client needs to be reassured that the symptoms are not caused by a stroke; the majority of clients recover in a few weeks. (2) (SC; IM)
   2 The symptoms are not caused by a TIA; paresis or paralysis of cranial nerve VII occurs; discomfort may or may not be present.
   3 The majority of clients recover without residual effects; occasionally some clients are left with evidence of Bell's palsy; exercises may help to maintain muscle tone; surgery may also be necessary.
   4 Moist heat to increase blood circulation to the nerve would be appropriate.

284. (4) A nerve block of the trigeminal (fifth cranial) nerve with alcohol is a conservative approach that lasts 6 to 18 months. (2) (DA; IM)
   1 This has been tried but provides little, if any, relief.
   2 Lidocaine is not a treatment used. Cranial nerve XI is the spinal accessory nerve, which innervates the sternocleidomastoid and trapezius muscles.
   3 This is not a conservative approach. This is the most commonly used surgical procedure for trigeminal neuralgia; neuralgia may recur in 30% of clients within 6 years.

285. (3) This is done to assess electrical activity and determine whether symptoms are primarily musculoskeletal or neurologic. (1) (HI; IM)
   1 No special preparation for an electromyography is required.
   2 No special care is required during the procedure.
   4 No special care is required after the procedure.

286. (4) Increased activity and stress precipitate exacerbation of symptoms because nerve impulses fail to pass to muscles at the myoneural junction; theories include inadequate acetylcholine, excessive cholinesterase, or a nonresponse of the muscle fibers to acetylcholine. (2) (HI; AS)
   1 Muscle weakness and fatigue come on quickly and disappear rapidly with rest in the initial stages of the disease.
   2 Rest promotes a decrease in symptoms because the demand for muscle contraction is reduced.
   3 Strength decreases with progressive activity.

287. (4) Because of myoneural junction defects, repeated use depletes acetylcholine, elevates cholinesterase, or exhausts acetylcholine receptor sites. (3) (PC; PL)
   1 Hot baths tend to increase muscle weakness.
   2 With myasthenia gravis, muscle weakness increases with muscle use.
   3 There is no evidence that eating meals will bring about improvement.

288. (2) Rest will decrease the demands at the synoptic membrane of the neuromuscular junction, reducing fatigue; activity should be paced to prevent fatigue before it begins. (1) (PC; IM)
   1 This will aggravate the fatigue; activity and rest should be delicately balanced to prevent fatigue.
   3 Same as answer 1.
   4 This cannot be done without a physician's order; rest will usually alleviate the fatigue.

289. (1) These are symptoms of myasthenia gravis and they are aggravated by physical activity. (3) (DA; AS)
2 Intention tremors are associated with multiple sclerosis.
3 Exercise decreases muscle strength.
4 The proximal muscles are more involved than the distal muscles.

290. (2) Respiratory infections place people with myasthenia gravis at high risk because they do not cough effectively and may develop pneumonia or airway obstruction. (1) (DA; EV)
1 Activity should be done earlier in the day before the energy reserve is depleted; periods of activity should be alternated with periods of rest.
3 This is unsafe; the client should eat sitting up to prevent aspiration.
4 This is contraindicated; these potentiate weakness because of their effect on the myoneural junction.

291. (3) Tensilon acts systemically to increase muscle strength with a peak effect in 30 seconds; it lasts several minutes. (3) (CI; IM)
1 Tensilon produces a brief increase in muscle strength; with a negative response the client would demonstrate no change in symptoms.
2 The duration of Tensilon's action is about 3 minutes.
4 Tensilon acts systemically on all muscles, rather than selectively on the eyelids.

292. (3) Until the diagnosis is confirmed, the primary goal should be to maintain adequate activity and prevent muscle atrophy. (3) (CC; AN)
1 It is too early to develop a teaching plan; the diagnosis is not yet established.
2 This is too early; the client cannot adjust if a diagnosis is not yet confirmed.
4 This is not a goal.

293. (4) Respiratory failure will require emergency intervention, and inability to swallow may lead to aspiration. (3) (DA; AS)
1 These are symptoms of myasthenia that may occur but are not life threatening.
2 This is a long-term problem that needs attention but is not life threatening.
3 Same as answer 1.

294. (4) Dysphagia should be minimized during peak effect of pyridostigmine bromide (Mestinon), thereby decreasing probability of aspiration. (2) (DA; IM)

1 There are insufficient data to know whether this is appropriate because liquids can also be aspirated.
2 This will not prevent aspiration.
3 This action will not prevent aspiration, although it is vital that the respiratory function be monitored.

295. (3) Results are unpredictable and symptoms may gradually return over several years. (3) (CI; IM)
1 This may increase the client's anxiety, and details of the actual surgery are not required with informed consent.
2 Thymectomy is a well-established treatment and is not considered experimental.
4 This may actually increase the client's anxiety; although complications should be mentioned, they should not be emphasized.

296. (3) Steps are short and dragging; this is seen with basal ganglia defects. (2) (HI; AS)
1 This is a staggering gait often associated with cerebellar disease.
2 This is associated with unilateral upper motor neuron disease.
4 This is associated with bilateral spastic paresis of the legs.

297. (3) Amplitude of the voice is reduced by neuromuscular involvement. (1) (DA; AS)
1 Constipation is a common problem because of a weakness of muscles used in defecation.
2 The tendency is for the head and neck to be drawn forward by loss of basal ganglia control.
4 Usually loss of weight occurs because of the embarrassment caused by slowness and untidiness in eating.

298. (2) There is a lack of neural control of individual muscle fibers, resulting in a characteristic masklike facies. (1) (DA; AS)
1 This is unrelated to Parkinson's disease; this is often associated with a CVA.
3 Movement usually abolishes tremors, which are known as nonintention tremors.
4 This does not occur; both arms fall rigidly to the sides and do not swing with a normal rhythm when walking.

299. (4) Bradykinesia is a slowing down in the initiation and execution of movement. (2) (DA; AS)
1 Cogwheel rigidity, not flaccidity, occurs because the disease causes sustained muscle contractions.
2 Tremors are more prominent at rest and are known as nonintention, not intention, tremors.
3 The limbs are rigid and move with a jerky quality; the limbs are not paralyzed.

300. (4) Visual disturbances such as diplopia and blurred vision are common initial symptoms of optic nerve lesions. (1) (DA; AS)
1 Constipation may occur late in the disease because of immobility.
2 Although this is a neuromuscular disorder, headaches are not a common symptom.
3 Pressure ulcers may become infected if they occur late in the disease because of immobility.

301. (4) This is a truthful answer that provides hope for the client. (3) (PC; IM)
1 This response avoids the client's question and could increase anxiety.
2 Analgesics are not commonly prescribed unless pain results from some other condition.
3 This avoids the client's question by suggesting a task to complete.

302. (3) The client must be flexible and adjust activities to provide for rest when necessary; activity should cease before the point of fatigue. (2) (PC; IM)
1 While quality time with children is important, it must be done on a flexible schedule to prevent fatigue in the client.
2 Laudable, but it cannot be done if the client is in need of support or if it overtaxes physical resources.
4 This may not be a need at this time, but prevention of fatigue is always important.

303. (1) An increase in intraocular pressure (IOP) results from a resistance of aqueous humor outflow; open-angle glaucoma, the most common type of glaucoma, results from increased resistance to aqueous humor outflow through the trabecular meshwork, Schlemm's canal, and the episcleral venous system. (1) (HI; IM)
2 This is the description of a cataract.
3 This is the description of astigmatism.
4 This is the description of a detached retina.

304. (4) Increased intraocular pressure damages the optic nerve, interfering with peripheral vision. (1) (DA; AS)
1 These may be associated with a detached retina.
2 There is difficulty in adjusting to darkness.
3 Same as answer 1.

305. (2) A major objective of early postoperative care is to prevent increased intraocular pressure; lying on the unaffected side or in the supine position will minimize intraocular pressure. (1) (CI; EV)
1 This is contraindicated; this increases intraocular pressure.

3 Same as answer 1.
4 Same as answer 1.

306. (3) Bending increases intraocular pressure and must be avoided. (2) (CI; IM)
1 Bed rest is not necessary.
2 This should be avoided because it raises intraocular pressure.
4 Same as answer 1.

307. (2) An antiemetic would prevent vomiting; vomiting increases intraocular pressure and should be avoided. (2) (CI; IM)
1 This is unsafe; vomiting increases intraocular pressure, and aggressive intervention is required.
3 Same as answer 1.
4 Same as answer 1.

308. (4) Postoperatively the client must check daily for signs of rejection, which include redness, irritation, discomfort, or vision loss; the surgeon should be notified if any of these appear; pain following a cataract extraction may indicate infection or hemorrhage and should be reported to the physician immediately. (3) (CI; EV)
1 Driving should be avoided until the client is given specific permission to do so by the physician.
2 Soap may irritate the eye, and showers or shampooing the hair should be avoided as instructed, usually from several days to 2 weeks.
3 This is a symptom of retinal detachment and would not be expected.

309. (1) To protect against physical injury and infection, the dropper tip should not touch the eye. (1) (SI; EV)
2 This is incorrect technique; drops should be placed within the lower lid.
3 This is incorrect technique; the lower lid should be retracted for placement of eye drops.
4 This is incorrect technique; it would squeeze medication out of the eye.

310. (3) This subjective symptom is caused by stimulation of retinal cells by ocular movement. (3) (DA; AS)
1 This is a disease of the connective tissue, not of the eye.
2 Glaucoma causes the individual to see halos around lights, not flashes of light.
4 Cerebral concussions do not result in this ocular symptom.

311. (4) The thermal inflammatory response, caused by a laser beam, results in a chorioretinal scar that holds the retina in place. (2) (CC; IM)
1 Radiation is not used because it destroys retinal tissue.
2 Burr holes are used in brain surgery.
3 Dermabrasion is used in acne vulgaris.

312. (3) Bleeding from the ears occurs only with basal skull fractures; bleeding from the ears is an assessment that assists in diagnosing the location of the injury. (2) (DA; AS)
1 This would be a positive response; pupils should react to light.
2 This would occur only in an infant in the presence of dehydration.
4 This occurs with increased intracranial pressure and pressure on the brain stem; it would be an expected response but not an immediate one.

313. (3) Elevating the head of the bed increases drainage of cerebrospinal fluid and decreases intracranial pressure. (1) (CI; IM)
1 This position would not promote cerebral drainage and may lead to increased intracranial pressure.
2 Same as answer 1.
4 This would promote retention of cerebrospinal fluid and increase intracranial pressure.

314. (2) Encouraging fluids will hydrate the client and contribute to the restoration of cerebrospinal fluid, which cushions the brain; this will ease the pain. (2) (PC; IM)
1 This will increase the pain. The client should be placed in the supine position and kept quiet.
3 Same as answer 1.
4 This is contraindicated for a postoperative client. The removal of antiembolism stockings will not influence a post–spinal anesthesia headache.

315. (3) In an attempt to stop the bleeding, adjacent arteries constrict; this in turn contributes to the ischemia responsible for the neurologic deficits. (3) (DA; AN)
1 The volume of blood loss is not great enough to significantly alter the oxygen-carrying capability of the remaining blood supply.
2 While prolonged ischemia may cause necrosis, many of the manifestations of cerebral ischemia are reversed as pressure diminishes, and there may be no permanent damage.
4 Severe electrolyte imbalance may cause generalized weakness; however, hemiparesis and aphasia are not the result of electrolyte loss.

316. (3) This utilizes the force of gravity to prevent additional intracranial pressure, which would intensify the ischemic manifestations of hemorrhage. (3) (CI; PL)
1 This position would not facilitate drainage of cerebral fluid; this position promotes accumulation of fluid, which increases intracranial pressure.
2 Same as answer 1.
4 Vomiting can occur with increased intracranial pressure, and placing sandbags to immobilize the head could result in aspiration.

317. (2) Controlled ventilation induces hypocapnia; subsequently it causes vasoconstriction and reduced cerebral blood flow. (3) (DA; EV)
1 The fluid may be cerebrospinal fluid; clearing the ear may cause further damage.
3 Because of manipulation during a craniotomy, anticonvulsants are given prophylactically to prevent seizures.
4 This would not increase cerebral blood flow and is inappropriate.

318. (4) This lessens the possibility of hemorrhage, provides for better circulation of cerebrospinal fluid, and promotes venous return. (3) (CI; IM)
1 Gatching the knees is contraindicated because it raises intracranial pressure.
2 This position is appropriate for infratentorial surgery.
3 A low-Fowler's position and turning the head impede venous return, which increases intracranial pressure.

319. (2) An inability to completely extend the legs is the classic sign of meningeal irritation. (3) (DA; AS)
1 With sunset eyes, the sclera is visible above the iris; this is often associated with hydrocephalus.
3 Homans' sign is pain caused by vascular irritability when the foot is flexed; this indicates thrombophlebitis.
4 The plantar reflex is a normal spinal cord reflex; it is not altered with meningeal irritation.

320. (2) Recognition of effort is motivating. (3) (SC; PL)
1 This may decrease both self-esteem and motivation.
3 Constantly focusing on the problem may decrease both self-esteem and motivation.
4 The problem is a motor, not a sensory (receptive), problem.

321. ② Emboli, occurring from atrial fibrillation, cause complete occlusion of vessels; usually middle cerebral arteries are involved; the infarct may cause hemiplegia, aphasia, or spatial perceptual deficits. (3) (DA; AN)
1 Hypertension is a disease that may cause spasm of the arteries, but it does not cause anatomic occlusion.
3 Developmental defects of the arterial wall are associated with saccular aneurysms.
4 Seizures are caused by inappropriate paroxysmal discharge.

322. ① Narrowing of arteries supplying the brain causes temporary neurologic deficits that last for a short period; between attacks neurologic functioning is normal. (2) (DA; AN)
2 Emboli result in a CVA; the damage is usually permanent.
3 This is not the description of a TIA; remissions and exacerbations occur with progressive degenerative neurologic disorders.
4 This occurs with multiple small cerebrovascular accidents; TIAs do not result in permanent damage.

323. ③ The cause of the hypotension must be evaluated by the physician. (3) (CC; IM)
1 This is a dependent function, and the physician must be notified first.
2 This is contraindicated; it would further decrease blood flow to the brain.
4 This is contraindicated; the physician must be notified first.

324. ③ Muscles used for swallowing are innervated by the ninth (glossopharyngeal) and tenth (vagus) cranial nerves. (3) (CI; EV)
1 This is unrelated to cranial nerves; this is associated with neck edema and potential compromise of the airway.
2 This is unrelated to cranial nerves; some edema is expected because of the inflammatory process at the site of surgery.
4 Alterations in blood pressure may occur but are not caused by cranial nerve dysfunction.

325. ② This assessment would indicate whether there is a progression of symptoms or improvement and assist the physician in determining the diagnosis. (2) (DA; AS)
1 An elevation in temperature is not an early sign of an extension of a cerebrovascular accident.

3 The data indicate the vital signs were normal and do not reflect hypertension; while vital signs would be monitored, the client's motor status in this instance is most significant.
4 This is not the priority assessment.

326. ③ This position will neither raise intracranial pressure nor interfere with respirations and will permit oral secretions to drain from the mouth by gravity. (2) (DA; IM)
1 This could compromise the airway by permitting the tongue to fall to the posterior pharynx and obstruct the airway.
2 Elevating the head of the bed could compromise vital functions by compressing the brain stem.
4 This is contraindicated because it may increase intracranial pressure.

327. ④ Increased intracranial pressure is manifested by sluggish pupils and elevation of the systolic blood pressure. (1) (CI; AN)
1 Spinal shock is manifested by a lowered systolic blood pressure with no pupillary changes.
2 Hypovolemic shock is indicated by a decrease in systolic pressure and tachycardia with no changes in pupillary reaction.
3 Transtentorial herniation is manifested by dilated pupils and severe posturing.

328. ① Shortening and eventual atrophy of muscles occur, resulting in contractures. (3) (CI; AN)
2 Muscles will atrophy, not hypertrophy, from disuse.
3 Flexion abnormalities, not extension rigidity, occur, resulting in contractures.
4 It does not predispose to infection but to atrophy and contractures.

329. ② The client has lost vision from the right visual field; scanning compensates for loss. (3) (SI; IM)
1 This is the approach to be used for apraxia.
3 This is the approach to be used for denial of the right side (unilateral neglect).
4 This aggravates neglect of the affected side.

330. ④ If the client exhibits emotional instability, it is usually caused by lesions affecting the thalamic area—the part of the neural system most responsible for emotions. (1) (SC; IM)
1 The client may have remote memory, but there is no selective process of what events are remembered.

2 This is associated with consistent behavior and cognitive thinking, of which the client is incapable at this time.

3 Same as answer 2.

331. ③ Recovery from aphasia is a continuous process; the amount of recovery cannot be predicted. (1) (SC; IM)

1 This response abdicates the nurse's responsibility; the physician cannot predict return of function.

2 This gives false reassurance; it may take a year or longer or may never return.

4 Speech return is a continuous process; it may take a year or longer or may never return.

332. ④ In addition to the extent of injury, a factor in relearning speech is the client's motivation and effort; the more the client attempts to talk, the more likely speech will progress to its optimal level; relearning is a slow process. (1) (SC; IM)

1 Clients with aphasia are not deaf.

2 This will cause frustration and anger in the client.

3 Although the nurse should instruct the family to approve and support every effort by the client to communicate, their action would provide external rather than internal motivation and is therefore not as effective.

333. ③ A weak grasp, pain, and uncoordinated movements could result in the dropping of tools, which could be dangerous. (3) (SI; AN)

1 Although this could be a problem, safety is the priority.

2 Same as answer 1.

4 Although in the future this could become an issue if the client can no longer work, at this time safety is the priority.

334. ② The tonic-clonic contractions that occur during a generalized seizure place the client at risk for developing a head injury; safety of the client takes precedence. (1) (SI; IM)

1 Safety must be established first.

3 During a seizure, changes in the pupils are not likely to occur.

4 This should be done but only after the client's safety is secured.

335. ② The nerves for arm innervation are above the injury level at C4. (3) (DA; AS)

1 Innervation of muscles used to move the lower arm is not affected by this injury; these muscles are innervated above C7.

3 Innervation for pain sensation of the hands is not affected by this injury; these muscles are innervated above C7.

4 Diaphragm innervation is not affected by this injury; the diaphragm is innervated above C4.

336. ② The T6 level is the sympathetic visceral outflow level, and any injury above this level results in autonomic dysreflexia. (3) (DA; AN)

1 The reflex arc remains after spinal cord injury.

3 The important point is not that the cord is totally transected but the level at which the injury occurs.

4 This is not related to autonomic dysreflexia; all cord injuries result in flaccid paralysis during the period of spinal shock; as the inflammation subsides, spasticity gradually increases.

337. ④ These symptoms occur as a result of exaggerated autonomic responses, and if autonomic dysreflexia is identified, immediate intervention is necessary to prevent serious complications. (3) (DA; AS)

1 Paralysis is related to transection, not dysreflexia; the client will have no sensation below the injury.

2 Profuse diaphoresis occurs.

3 Bradycardia occurs.

338. ④ The client has the capability; this maintains a positive identity and is necessary for progression to long-term goals. (2) (SC; AN)

1 This is a long-term goal.

2 Same as answer 1.

3 Same as answer 1.

339. ② Arm strength is necessary for transfers and activities of daily living and for the use of crutches or a wheelchair. (2) (PC; IM)

1 The client has no neurologic control of this activity.

3 Equilibrium is not a problem.

4 Same as answer 1.

340. ③ Injury and pressure at the nerve roots produce pain. (3) (DA; AS)

1 Injury at T6 and T7 is too low to cause paralysis of the respiratory muscles; control of respiration is in the medulla and the cervical plexus (phrenic nerve).

2 With complete crushing at the T6 and T7 level, there is no pain sensation in parts distal to the injury.

4 Initially, paralysis is flaccid; spasticity is a later manifestation.

ENDOCRINE (EN)

341. ③ Excessive thirst, excessive hunger, and frequent urination are caused by the body's inability to correctly metabolize glucose. (1) (HI; AS)
1 Lethargy, not irritability, results because of a lack of metabolized glucose for energy.
2 Confusion is related to both severe hypoglycemia and hyperglycemia.
4 Frequent urination occurs throughout a 24-hour period because glucose in the urine pulls fluid with it; weight loss occurs with type 1 diabetes, not with type 2 diabetes.

342. ① The equivalent of 2.2 pounds is 1 kilogram; therefore, the client weighs 109.1 kg; the nurse administers 1.0 g per kilogram. (3) (CI; AN)
2 This is more than the physician ordered.
3 Same as answer 2.
4 Same as answer 2.

343. ③ An open-ended question will elicit a variety of data such as medications, diet, and other aspects of care. (2) (CC; AS)
1 Although an important question, at this point it is too narrow and will elicit limited information.
2 Although this information should eventually be obtained, it is not significant at this point.
4 Same as answer 2.

344. ③ The reason for the lack of ketonuria in type 2 diabetes is unknown; one theory is that extremely high hyperglycemia and hyperosmolarity levels block the formation of ketones, stimulating lipogenesis, rather than lipolysis. (2) (CI; AS)
1 This does not occur with either type of diabetes.
2 This is impossible; if glycosuria is present, there must first be a level of glucose in the blood exceeding the renal threshold of 160 to 180 mg/dl.
4 This is expected in type 1 diabetes.

345. ① Blood glucose testing is a more direct and accurate measure; urine testing provides an indirect measure that can be influenced by kidney function and the amount of time the urine is retained in the bladder. (1) (CI; IM)
2 While both blood and urine testing are relatively simple, testing the blood involves additional knowledge of surgical aseptic technique.
3 Both procedures can be done by the client.
4 This would not be a factor; while some urine tests are influenced by drugs, there are methods to test urine to bypass this effect.

346. ④ Knowledge of the signs and treatment for hypoglycemia or hyperglycemia is critical to client health and well-being and essential for survival. (3) (CC; AN)
1 Although this is important, it is not the priority.
2 The client has type 2 diabetes; insulin injections are not necessary.
3 Self serum glucose monitoring is more accurate than urine S & A measurements to identify serum glucose levels.

347. ④ Controlling the diabetes decreases the risk of infection; this is the best prevention. (1) (SI; EV)
1 If not completely absorbed, these may provide a warm, moist environment for bacterial growth.
2 Coexisting neuropathy may result in injury from heat application.
3 Protein, carbohydrates, and fats must be in an appropriate balance; high carbohydrate intake could provide too many calories.

348. ④ Exercise increases the metabolic rate, and glucose is needed for cellular metabolism. (2) (CI; IM)
1 Regular vigorous exercise increases cell sensitivity to insulin.
2 Glucagon action raises blood glucose but does not affect cell uptake or utilization of glucose.
3 Cellular requirements for glucose increase with exercise.

349. ④ Exercise improves glucose metabolism; with exercise there is a risk of developing hypoglycemia, not hyperglycemia. (2) (PT; IM)
1 Exercise should not be decreased because it improves glucose metabolism.
2 An extra tablet would probably result in hypoglycemia because exercise alone improves glucose metabolism.
3 Control of glucose metabolism is achieved through a balance of diet, exercise, and pharmacologic therapy.

350. ④ Physiologic stress increases gluconeogenesis, requiring continued pharmacologic therapy despite an inability to eat; fluids prevent dehydration; monitoring serum glucose levels permits early intervention if necessary. (1) (CI; IM)
1 Skipping the oral hypoglycemic could precipitate hyperglycemia; serum glucose levels must be monitored.
2 Food intake should be attempted to prevent acidosis; oral hypoglycemics should be taken, and serum glucose levels should be monitored.

3 These are incomplete instructions; oral hypoglycemics should be taken, and serum glucose levels should be monitored; eating as much as possible could precipitate hyperglycemia.

351. (1) In the overweight individual with type 2 diabetes, occasional alcohol can be used with caloric substitution for equivalent fat exchanges in the diet because it is metabolized like fat. (2) (PC; IM)

2 Moderation is vital; these may not be used in unlimited quantities; they must be accounted for in the dietary calculations.

3 Alcohol can be used as long as it is accounted for in the diet.

4 This is untrue; regular foods can be used in the ADA diet.

352. (3) According to the individual's needs, consistency and regularity in the basic food plan should be maintained; this is a basic principle of dietary management of diabetes. (2) (PC; IM)

1 This is not necessary; the client can use the ADA food plan to make selections.

2 This is unrealistic; it cannot always be done; it is unnecessary because choices can be made within the ADA diet.

4 This is unnecessary because the client is not taking insulin.

353. (1) These vegetables are under the vegetable exchange, as are asparagus, broccoli, and mushrooms. (1) (CC; EV)

2 These are starchy vegetables and are listed as bread exchanges.

3 Same as answer 2.

4 These food items are from the bread exchange.

354. (4) Improper foot care can lead to skin breakdown, poor healing, and subsequent infection. (1) (SI; IM)

1 This potentially increases anxiety and reduces the client's ability to learn.

2 This is only one aspect of proper foot care; foot care must be more comprehensive.

3 Same as answer 2.

355. (4) When the head of the pancreas is removed, the client has a greatly reduced number of insulin-producing cells and hyperglycemia will occur; immediate treatment is necessary. (3) (CI; EV)

1 This is not immediately life threatening and would take time to develop.

2 Same as answer 1.

3 Same as answer 1.

356. (3) The blood glucose level needs to be reduced; regular insulin begins to act in 30 to 60 minutes. (1) (PT; PL)

1 Oral hypoglycemics are long acting and begin to act about 1 hour after administration; in addition, the client has type 1, not type 2, diabetes, and an oral hypoglycemic would be ineffective.

2 Blood glucose levels are far more accurate than urine glucose levels.

4 The rate may be increased because polyuria often accompanies hyperglycemia.

357. (3) Emotional and physical stress may cause insulin requirements to remain elevated in the postoperative period. (1) (CI; PL)

1 Fluctuating insulin requirements are usually associated with noncompliance, not surgery.

2 An increase in the client's insulin requirements could indicate sepsis, but this is not expected.

4 Insulin requirements would remain elevated rather than decrease.

358. (1) Incomplete oxidation of fat results in fatty acids that further break down to ketones. (3) (DA; AN)

2 Protein metabolism results in nitrogenous waste production, causing an elevated blood urea nitrogen (BUN).

3 Potassium is not oxidized; in hypokalemia or hyperkalemia no ketones are formed.

4 Carbohydrates do not contain fatty acids that are broken down into ketones.

359. (2) Potassium is the principal intracellular cation, and during ketoacidosis it moves out of cells into the extracellular compartment to replace potassium lost as a result of glucose-induced osmotic diuresis; overstimulation of the cardiac muscle results. (3) (DA; AS)

1 P waves are abnormal because the P-R interval may be prolonged and the P wave may be lost; however, the T wave is peaked, not depressed; the T wave is depressed in hypokalemia.

3 Initially, the QT segment is short, and as the potassium level rises, the QRS complex widens; the ST segment becomes depressed.

4 The P-R interval is prolonged, and the P wave may be lost; QRS complexes and thus T waves become irregular, and the rate does not necessarily change.

360. (3) This is a short-term goal, client-oriented, necessary for the client to control the diabetes, and measurable when the client performs a return demonstration for the nurse. (2) (CC; AN)
1 This is not a short-term goal.
2 This is measurable, but it is a long-term goal.
4 While this is measurable and is a short-term goal, it is not the one with the greatest priority when a client has an insulin pump that must be mastered prior to discharge.

361. (3) The basal infusion rate mimics the low rate of insulin secretion during fasting, and the bolus before meals mimics the high output after meals. (2) (CC; PL)
1 The subcutaneous needle may be left in place for as long as 3 days.
2 Blood glucose monitoring is done a minimum of 4 times a day.
4 Most insulin pumps are battery-driven syringes external to the body, which access the body via a subcutaneous needle.

362. (2) Wearing shoes protects the feet; they should fit well and be worn over socks that are wool or cotton to cushion the foot and absorb perspiration. (3) (SI; IM)
1 Smoking should be avoided, and self-removal of corns can result in injury to the feet.
3 Nylon is not a natural fiber, and it promotes perspiration.
4 Shoes that do not fit well will create friction and cause sores, blisters, and calluses; corn removers should be avoided.

363. (3) The dose of exogenous insulin causes a rapid drop in the blood glucose level, especially if food is not eaten; the oral hypoglycemic acts slowly, and there is time to take evasive measures if hypoglycemia begins to develop. (3) (DA; AN)
1 This would lead to hyperglycemia.
2 Stress usually contributes to hyperglycemia; the primary reason for this client's precipitous fall in the blood glucose level was because insulin had been taken and the client had not eaten at all.
4 The use of insulin over long periods does not build up tolerance or cause blood glucose levels to fluctuate dramatically.

364. (2) This is the point at which a client is generally hypoglycemic, resulting in increased sympathetic nervous system activity and deprivation of CNS glucose supply. (2) (CI; AS)
1 This is within the norm of 90 to 120 mg/dl.
3 Same as answer 1.

4 This is not a sufficient drop to cause hypoglycemia; hypoglycemia usually occurs at 50 to 70 mg/dl.

365. (4) Lipodystrophy is a noninflammatory reaction causing localized atrophy or hypertrophy and a localized increase in collagen deposits. (1) (HI; AN)
1 Injections of insulin would not cause a horny growth such as a wart or callus.
2 An allergic response would precipitate a localized or systemic inflammatory response.
3 Hyperthermia and localized heat, erythema, and pain would be associated with an infection.

366. (2) This occurs with low serum glucose levels because of sympathetic nervous system activity. (2) (DA; AS)
1 This is a sign of hyperglycemia and is related to metabolic acidosis and inadequate energy production.
3 This is a sign of hyperglycemia; it is caused by dehydration associated with osmotic diuresis related to glycosuria.
4 This is a sign of hyperglycemia and is related to metabolic acidosis; it is a compensatory response in an attempt to blow off CO_2 and raise the pH level.

367. (1) These are classic signs of hypokalemia that occur when potassium levels are reduced as potassium reenters cells with glucose. (3) (DA; EV)
2 Symptoms of hyponatremia are nausea, malaise, and changes in mental status.
3 Symptoms of hypoglycemia are weakness, nervousness, tachycardia, diaphoresis, irritability, and pallor.
4 Symptoms of hypercalcemia are lethargy, nausea, vomiting, paresthesias, and personality changes.

368. (2) The suggested treatment of hypoglycemia in a conscious client is a simple sugar (such as two packets of sugar), followed by a complex carbohydrates (such as a slice of bread), and last a protein (such as milk); the simple sugar elevates the blood glucose level rapidly; the complex carbohydrates and protein produce a more sustained response. (2) (CC; EV)
1 These are fast-acting sugars, and neither of them will provide a sustained response.
3 The fat content of chocolate candy decreases the rate of absorption of glucose.
4 Neither of these is a fast-acting sugar; peanut butter crackers and milk can be used to maintain the glucose level after it has been raised.

369. (4) An aldosteronoma is an aldosterone-secreting adenoma of the adrenal cortex. (2) (CI; AN)
   1 An aldosteronoma is a tumor of an adrenal gland, not the thyroid.
   2 An aldosteronoma is a tumor of an adrenal gland, not the kidneys.
   3 An aldosteronoma is a tumor of an adrenal gland, not the pituitary.

370. (1) Renal and cardiac complications will occur if the hypertension caused by the tumor is not arrested. (3) (CI; IM)
   2 An aldosteronoma is a benign tumor; metastasis is not possible.
   3 This is not true; surgery is required to remove the tumor.
   4 Drugs are not used; the tumor must be removed.

371. (1) Once the excessive secretion of aldosterone is stopped, the BP gradually drops to a near normal level. (3) (DA; EV)
   2 The BP drops gradually; it does not rise.
   3 The BP will only fluctuate if the hypervolemia is overcorrected, causing hypovolemia; this is not expected.
   4 The BP drops gradually in response to decreasing serum corticosteroid levels; a rapid drop immediately following surgery may indicate hemorrhage.

372. (2) This provides appropriate reassurance; the body has two adrenal glands, and an aldosteronoma is unilateral. (2) (SC; IM)
   1 The prognosis is usually excellent; this is unnecessarily alarming.
   3 This is unnecessary; the prognosis is usually excellent.
   4 No hormones are necessary; there is another adrenal gland.

373. (2) These tests provide a measure of thyroxine production, a disturbance that is associated with the client's symptoms. (1) (CI; AS)
   1 Prothrombin time (PT) and partial thromboplastin time (PTT) assess blood coagulation.
   3 The VDRL test is for syphilis; the CBC assesses the hematopoietic system.
   4 This measures the kind and amount of circulating barbiturates; the client's symptoms are not associated with barbiturate intake.

374. (3) These are the classic signs associated with hyperthyroidism; weight loss and restlessness occur because of an increased basal metabolic rate; exophthalmos occurs because of peribulbar edema. (2) (DA; AS)
   1 These are all associated with hypothyroidism because of the decreased metabolic rate.
   2 Lethargy and weight gain are associated with hypothyroidism as a result of a decreased metabolic rate; forgetfulness is not related.
   4 Although weight loss and exophthalmos occur with hyperthyroidism, the client would be hyperactive, not hypoactive.

375. (1) Increased basal metabolic rate, increased circulation, and vasodilation result in warm moist skin. (3) (DA; AS)
   2 This symptom is associated with hypothyroidism.
   3 Same as answer 2.
   4 Same as answer 2.

376. (4) The mask may irritate or scratch the eye if the client turns and lies on it during the night. (3) (SI; EV)
   1 Blinking of the eyes will bathe the eyes and prevent corneal ulceration.
   2 This will do nothing to relieve edema or prevent ulceration of the eye.
   3 Although this will help reduce periorbital edema, it will not prevent ulceration of the cornea.

377. (3) This relieves tension on the incision and limits the risk of dehiscence. (2) (CI; IM)
   1 Coughing should be avoided during the early postoperative period to prevent trauma to the operative site.
   2 This should be avoided until advised by the physician, usually after initial healing of the incision occurs.
   4 Pressure against the operative area is not necessary to promote the integrity of the incision, and it may act to inhibit swallowing.

378. (2) The remaining thyroid tissue may provide enough hormone for normal function. (1) (CI; AN)
   1 This would be a total, not a subtotal, thyroidectomy.
   3 No parathyroid glands should be removed in a thyroidectomy.
   4 Same as answer 3.

379. (2) Acute respiratory obstruction can result from edema, nerve damage, or tetany. (1) (CI; PL)

1 A cardiac arrest is not an expected response following thyroid surgery.

3 If the airway were obstructed by postoperative edema, the use of a mechanical airway would be ineffective because it would not reach the point of obstruction; a rebreathing mask would be used for clients with COPD, not a thyroidectomy.

4 Acidosis and cardiac arrest are not expected responses after a thyroidectomy.

380. (3) If the laryngeal nerve is damaged during surgery, the client will be hoarse and have difficulty speaking. (3) (CI; EV)

1 This would not indicate injury to the laryngeal nerve; this is part of the assessment for a compromised airway.

2 This would be an assessment for hypocalcemia resulting from inadvertent removal of the parathyroid glands.

4 This assesses for bleeding and possible hemorrhage, not laryngeal nerve injury.

381. (3) This detects complications such as thyroid storm, hemorrhage, and respiratory obstruction that occur early in the postoperative period. (3) (CI; EV)

1 This is contraindicated; humidifiers contribute to the spread of bacteria and infection.

2 This should not be begun until 2 to 4 days postoperatively because it can disrupt the suture line.

4 Hoarseness and voice weakness are usually temporary and not life threatening; the priority is to observe for thyroid storm, hemorrhage, and respiratory obstruction.

382. (1) Bleeding may occur, and blood will pool in the back of the neck because the blood will flow via gravity. (1) (CI; EV)

2 This is contraindicated; this would increase pain and would put tension on the suture line.

3 Talking should be avoided in the immediate postoperative period except to assess for a change in pitch or tone, which may indicate laryngeal nerve damage.

4 Activity should be gradually resumed and frequent rest periods encouraged.

383. (2) Injury to the parathyroid gland results in a deficiency of parathormone, which decreases calcium levels in the blood. (3) (CI; EV)

1 This is characterized by generalized weakness, a decrease in reflexes, shallow respirations, and cardiac dysrhythmias.

3 This is characterized by tachycardia, hyperpyrexia, and an exacerbation of thyroid symptoms.

4 This is characterized by a weak, thready pulse and hypotension.

384. (2) The client is exhibiting signs and symptoms of hypocalcemia, which occurs with accidental removal of the parathyroids; calcium gluconate is the treatment of choice. (3) (DA; PL)

1 This is prescribed for hyperthyroidism because it inhibits the release of thyroid hormones.

3 This is prescribed for hypomagnesemia or to prevent convulsions in eclampsia or preeclampsia.

4 This is prescribed for hypokalemia.

385. (4) Thyroid crisis is severe hyperthyroidism; excessive amounts of thyroxine increase the metabolic rate, thereby raising the pulse and temperature. (2) (CI; AS)

1 During thyroid crisis there is usually no increase in the difference between the apical and the peripheral pulse rates (pulse deficit).

2 The blood pressure will rise to meet the oxygen demand caused by the increased metabolic rate during thyroid crisis.

3 Because of the increased metabolic rate, the pulse and respiratory rates increase to meet the body's oxygen needs.

386. (2) The radioactive material will be excreted in urine, feces, perspiration, and other body discharges; disposal should be made according to hospital procedure to contain the radioactive discharges. (1) (SI; IM)

1 Standard precautions protect the staff from microorganisms, not radioactivity; these precautions would be used for this client as well as every other client; visitors would be limited when radioactive material is involved.

3 The purpose of isolation is not for protection of the client but for containment of the radioactivity and protection of staff, family, and friends.

4 Contact should be avoided, except to meet necessary needs; distance should be maintained.

387. (1) Dry, thickened skin and cold intolerance are characteristic adaptations to low serum thyroxine. (1) (HI; EV)

2 Muscle cramping is associated with hypocalcemia.

3 Low thyroxine levels reduce the metabolic rate, resulting in fatigue, and should not increase the pulse rate.

4 Low thyroxine levels reduce the metabolic rate, resulting in weight gain and bradycardia, not tachycardia.

388. (3) After a thyroidectomy the thyroxine output is usually inadequate to maintain an appropriate metabolic rate. (1) (HI; AN)

1 Hypothyroidism is a decrease in thyroid functioning, not a slowing of the entire body's functions.

2 In hypothyroidism the level of thyroid-stimulating hormone (TSH) from the pituitary is usually increased.

4 Atrophy of the thyroid tissue remaining after surgery does not occur.

389. (1) Dry skin is most likely caused by decreased glandular function, and fatigue is caused by a decreased metabolic rate. (2) (CC; EV)

2 This is associated with hyperthyroidism, not hypothyroidism.

3 Same as answer 2.

4 Same as answer 2.

RESPIRATORY (RE)

390. (3) These are normal respiratory sounds heard on auscultation as inspired air enters and leaves the alveoli. (1) (HI; AS)

1 These are fine crackling sounds heard at the end of an inspiration; they are associated with pulmonary edema.

2 Adventitious sounds is the general term for all abnormal breath sounds.

4 This is evidence of a reduction in the amount of air entering the alveoli, usually caused by obstruction or consolidation.

391. (3) Rhonchi are coarse sounds heard over the larger airways; including rhonchi in the notation makes it inaccurate. (2) (CC; AN)

1 Crackles and rhonchi are client adaptations, not a nursing diagnosis.

2 It would be incorrect to use the term *rhonchi* to refer to crackling sounds in the lower lung.

4 Crepitus, which indicates subcutaneous emphysema, is a condition unrelated to the breath sounds heard on auscultation.

392. (4) This position promotes lung expansion and gas exchange; it also decreases venous return and cardiac workload. (2) (CI; IM)

1 This may be done but positioning should be done first because it will have an immediate effect and time will be used in setting up for the delivery of the oxygen.

2 This will add nothing to the data already obtained; a friction rub is not related to congestive heart failure but to inflammation of the pleura.

3 Maintaining adequate oxygen exchange is the priority; an x-ray film will be obtained, but after breathing is supported.

393. (1) Stridor is a high-pitched, musical sound caused by an obstruction of the trachea or larynx. (2) (CI; AS)

2 Neck exercises are important for total rehabilitation, but they do not help identify stridor.

3 Auscultating the base of the lungs will determine the presence of vesicular breath sounds or crackles.

4 Although coughing and deep breathing are important, they do not help identify stridor.

394. (4) These are top priorities in trauma management; basic life functions must be maintained or reestablished. (2) (DA; PL)

1 This is an assessment for head injury that follows determination of respiratory and circulatory status.

2 This is an assessment for abdominal injury that follows determination of respiratory and circulatory status.

3 Pain assessment would follow the appraisal of airway, breathing, and circulation.

395. (3) This is secondary to cerebral hypoxia, which accompanies acute respiratory distress syndrome (ARDS); cognition and level of consciousness are reduced. (3) (DA; AS)

1 Hypotension occurs because of the hypoxia of the heart.

2 The sputum is not tenacious, but it may be frothy if pulmonary edema is present.

4 Breathing will be fast and shallow.

396. (3) Nothing is achieved if the equipment is working and the client is not responding. (3) (DA; EV)

1 This is presumptive; the database is incomplete for the assessment that surgery is necessary.

2 Endotracheal intubation does not permit verbal communication.

4 This is important but not the priority.

397. (4) Increased rate and depth of breathing result in excessive elimination of CO_2, and respiratory alkalosis results. (1) (DA; EV)

1 Hypoxia is associated with respiratory acidosis, not respiratory alkalosis, which is related to hyperventilation.

2 With hyperventilation, CO_2 levels will be decreased (hypocapnia), not elevated.

3 This results from excess hydrogen ions caused by a metabolic problem, not a respiratory problem.

398. (2) This is necessary to prevent flooding of the trachea with fluid; some systems have receptacles attached to the tubing to collect the fluid and others have to be temporarily disconnected while emptying the fluid. (3) (DA; IM)

1 This circumstance does not require assistance from a respiratory therapist.

3 This is unsafe; humidity is necessary to preserve moistness of the respiratory tract and liquefy secretions.

4 The amount of condensation is irrelevant in terms of recording the intake and output.

399. (4) This position promotes respirations by removing the pressure of abdominal organs on the diaphragm; it requires the least amount of energy to maintain, and it helps reduce edema at the tracheostomy site. (2) (DA; IM)

1 This position promotes collection of fluid around the operative site, which impedes healing; it also impedes respirations because of the pressure of the abdominal organs against the diaphragm.

2 It requires too much energy to maintain this position.

3 Same as answer 2.

400. (2) Suctioning also removes oxygen, which can cause cardiac dysrhythmias; the nurse should try to prevent this by hyperoxygenating the client prior to and after suctioning. (2) (DA; IM)

1 To prevent trauma to the trachea, suction should only be applied while removing the catheter.

3 This kind of movement could cause tracheal damage.

4 Suction only as needed; excessive suctioning irritates the mucosa, which increases secretion production.

401. (3) These cuffs do not compress the capillary beds and thus do not cause tracheal damage. (2) (CI; AN)

1 Surgical asepsis, not the use of these cuffs, prevents infection.

2 A minimal air leak is desirable to ensure the lowest possible pressure in the cuff while still maintaining placement of the tube.

4 Secretions will be increased because the cuff is a foreign body in the trachea.

402. (4) Emphysema involves destructive changes in the alveolar walls, leading to dilation of the air sacs; there is subsequent air trapping and difficulty with expiration. (1) (DA; AN)

1 Bronchospasm is characteristic of asthma, and it causes narrowing of the airways.

2 The vital capacity is increased to compensate for inefficient gaseous exchange; however, this is a secondary adaptation.

3 Although slow expiration tends to keep the airways open so there is less air trapping, rapid expulsion of air is not the primary problem.

403. (3) Restlessness is an early sign of cerebral hypoxia. (2) (DA; AS)

1 Tachypnea, not bradypnea, would occur.

2 Tachycardia, not bradycardia, would occur.

4 Light-headedness is a sign of respiratory alkalosis, not acidosis.

404. (3) Clients with chronic obstructive pulmonary disease (COPD) must be given only low concentrations of oxygen; a decreased oxygen blood level is the only stimulus for breathing for these clients. (2) (CI; AN)

1 Prolonged hypoxia will stimulate erythrocyte production; the goal of therapy is to relieve hypoxia.

2 The pressure, rather than the concentration, at which oxygen is administered increases this risk.

4 To prevent its drying effects on secretions and the mucosa, oxygen should be humidified.

405. (3) The orthopneic position lowers the diaphragm and provides for maximum thoracic expansion. (1) (DA; PL)

1 This would not facilitate thoracic expansion because it still permits abdominal organs to press against the diaphragm.

2 Same as answer 1.

4 Although this could help, it would not be as beneficial as the orthopneic position.

406. (4) With emphysema the diaphragm is flattened and weakened; strengthening the diaphragm is desirable. (2) (CC; PL)

1 The opposite is more desirable; clients with emphysema retain too much carbon dioxide, which eventually causes a barrel chest.

2 The abdominal muscles are accessory muscles of respiration, and their contraction and relaxation are involved in diaphragmatic breathing.

3 Sit-ups are too strenuous for clients with emphysema.

407. (3) This pause allows added time for gaseous exchange at the alveolar capillary beds. (3) (CC; EV)

1 Inhalation should be through the nose to moisten, filter, and warm the air.

2 This decreases the effectiveness of respirations.

4 The expiratory phase should be lengthened, and exhalation should be through pursed lips.

408. (4) The enlarged liver is caused by long-term respiratory acidosis with increased pulmonary pressure that eventually causes right ventricular enlargement and failure (cor pulmonale); the elevated pressure causes backup pressure in the hepatic circulation. (3) (DA; AN)

1 Liver hypoxia would cause atrophy and necrosis of cells, not enlargement.

2 Right ventricular failure with increased pressure in the ascending vena cava causes increased pressure in the hepatoportal system, resulting in an enlarged liver, not hepatic acidosis.

3 These are the result of hepatic portal hypertension, not the cause of an enlarged liver.

409. (2) A sudden weight gain is an initial symptom of right ventricular failure caused by chronic obstructive pulmonary disease (cor pulmonale). (3) (DA; AN)

1 This is associated with polycythemia vera, not chronic obstructive pulmonary disease.

3 Right, not left, ventricular failure occurs with chronic obstructive pulmonary disease.

4 A sudden weight gain is not associated with this condition.

410. (1) Eating small meals will decrease the amount of oxygen necessary for digestion at any one time. (2) (CC; PL)

2 Lying down increases intraabdominal pressure, pushing a full stomach against the diaphragm and limiting respiratory excursion.

3 While fluids do help liquefy secretions, they should not be encouraged in a client with right ventricular failure.

4 Protein maintains or increases hydrostatic pressure; it does not decrease it.

411. (4) This statement indicates understanding of what a lobectomy involves, including closed chest drainage; it shows that the client was given the information needed for an informed consent. (2) (CC; EV)

1 This client's lesion is extensive; laser endoscopic or laser bronchoscopic surgery can be used for some smaller lesions.

2 The right lung has three lobes and the left only two because of the position of the heart.

3 A segmental or wedge resection is used for small tumors near the surface of the lung.

412. (2) There is air in the tissues, and palpation results in a crackling sound referred to as crepitus. (2) (DA; EV)

1 This is a harsh, vibrating sound usually produced on inspiration because of airway obstruction.

3 This is excessive accumulation of fluid in tissue spaces.

4 The size of the chest is determined by the bony structure; a barrel chest with an increase in the AP diameter is associated with COPD, not cancer of the lung.

413. (2) A mediastinal shift with airway obstruction may occur because pressure builds up on the operative side, causing the trachea to deviate toward the unoperative side; assessment of the airway takes priority. (3) (DA; AS)

1 This is unsafe; the client needs immediate intervention; the airway is the priority.

3 Same as answer 1.

4 There is no need for a chest tube when a pneumonectomy is performed.

414. (3) These positions permit ventilation of the remaining lung and prevent fluid from draining into the sutured bronchial stump. (3) (CI; PL)

1 Lying on the unoperative side restricts left lung excursion and may allow fluid to drain into the right bronchial stump.

2 Although the high-Fowler's position promotes ventilation, it is extremely tiring.

4 Same as answer 1.

415. (1) Loss of the large vascular lung and/or the presence of a mediastinal shift can result in cardiac overload. (3) (CI; AS)

2 These signs are associated with hypoxia, which is a common complication of surgery and not unique to a pneumonectomy.

3 These are common complications of all thoracic surgeries and are not unique to a pneumonectomy.

4 An elevated BP may be associated with cardiac overload, but the other symptoms are not unique to a pneumonectomy.

416. (3) A productive cough indicates mucus is being raised from the lungs. (2) (CC; EV)

1 Crackles (rales) are unaffected by postural drainage or coughing.

2 The depth of respirations may not be altered by postural drainage.

4 Saliva comes from the mouth and does not indicate clearance of lungs.

417. ③ To prevent aspiration during the procedure, clients are required to be npo for at least 8 to 12 hours prior to the procedure. (2) (CI; IM)
  1 Chest tubes are not required unless the lungs are accidentally punctured; the client will have a small incision near the clavicle.
  2 A mediastinoscopy permits visualization of the anterior mediastinum or hilum extrapleurally; a bronchoscopy permits visualization of the main stem bronchus.
  4 Fluid is removed from the pleural space during a thoracentesis.

418. ④ The mechanism is unclear, but this is probably caused by fluid shifts. (2) (CI; AN)
  1 This is untrue; this is done only occasionally to slow development of new effusion in clients with recurrent effusions.
  2 This is untrue; it can provide dramatic relief and improvement.
  3 This is untrue; dyspnea should be immediately relieved; if dyspnea increases, pneumothorax should be suspected.

419. ③ Compression of the lung by fluid that accumulates at the base of the lungs reduces expansion and air exchange. (3) (DA; AS)
  1 There is no fluid in the alveoli, so no crackles are produced.
  2 If there is tracheal deviation, it is away from the involved side.
  4 Dullness is produced on percussion of the involved area.

420. ① Clients with pleuritic disease are prone to develop pneumonia because of impaired expansion, air exchange, and lung drainage. (2) (DA; AS)
  2 Contraindicated; narcotics depress respirations.
  3 Coughing should not be suppressed; it enhances expansion, air exchange, and lung drainage.
  4 Oral fluids are encouraged; pulmonary edema does not develop unless the client has severe cardiovascular disease.

421. ③ Tension is placed on the pleura at the height of inspiration and causes pain. (3) (HI; AS)
  1 This is typical of congestive heart failure.
  2 This may indicate pulmonary infection.
  4 Same as answer 2.

422. ④ After general anesthesia, these activities expand alveoli and prevent atelectasis. (1) (CI; IM)
  1 This is not necessary; the abdomen has not been entered, and there should be no interference with peristalsis.

  2 This is not necessary.
  3 This is not necessary; clients can ambulate after recovery from anesthesia.

423. ③ A general anesthetic is delivered via an endotracheal tube that irritates the posterior pharynx and larynx and causes discomfort when swallowing. (2) (CC; EV)
  1 Occasionally this may occur; however, it is a systemic, not a local, effect.
  2 This is not an effect of general anesthesia.
  4 Same as answer 2.

424. ① Both sides of the posterior pharynx should be touched to elicit the gag reflex; absence of the reflex indicates the client is at risk for aspiration of secretions or fluid. (1) (SI; EV)
  2 This is unsafe; if the gag reflex is absent, the client would aspirate.
  3 This is unsafe; this could happen even in the absence of a gag reflex.
  4 This is unsafe; the client might be able to breathe deeply and cough without an adequate gag reflex.

425. ④ There is no respiratory movement in stage 4 of anesthesia; before this stage, respirations are depressed but present. (3) (PT; AN)
  1 The gag reflex is lost in the first phase of stage 3 of anesthesia.
  2 The corneal reflex is lost in the second phase of stage 3 of anesthesia.
  3 Consciousness is lost in stage 2.

426. ① The respiratory process is essential to life and therefore is the first priority. (1) (DA; EV)
  2 This may eventually be done, but respiratory assessment is the priority.
  3 Same as answer 2.
  4 Trendelenburg's position decreases respiratory functioning and should be avoided.

427. ① To allow for the insertion of the bronchoscope, throat muscles are anesthetized, diminishing the protective gag reflex. (1) (CI; PL)
  2 This does not occur after a bronchoscopy.
  3 A general anesthetic is not usually used, therefore paralytic ileus is not a complication.
  4 Dysphasia is difficulty in talking and does not occur with a bronchoscopy; dyspepsia is disturbed digestion and is not the reason for withholding food or fluids.

428. ③ This is a safe and reliable method of testing the gag reflex. (1) (SI; EV)

1 Talking can occur without the gag reflex.

2 This could cause choking if the gag reflex has not returned.

4 Coughing can occur without the gag reflex.

429. (4) Normally, air is moisturized and warmed as it passes through the nasopharynx; with a laryngectomy this area is bypassed and the tracheobronchial tree compensates by producing copious amounts of secretions. (2) (DA; AN)

1 In this case it is not a response to bacteria but to the character of the air that is entering the tracheobronchial tract.

2 This occurs regardless of turning and deep breathing.

3 This would produce local irritation and a local response.

430. (2) In this position the tongue does not obstruct the airway, and this allows for drainage of secretions and oxygen and carbon dioxide exchange. (2) (CI; PL)

1 Oxygen is not always necessary; the client's airway is the priority.

3 This assesses ventilation but does nothing to correct it.

4 Once the pharyngeal reflex has returned, an in-place airway can cause the client to gag and vomit.

431. (4) These are correct techniques; deep inhalation promotes alveolar expansion, holding the breath promotes transfer of gases, and exhalation promotes lung recoil. (2) (CC; EV)

1 Coughing is done after deep breathing.

2 The breaths should not be in succession; they should be spaced by several normal breaths to avoid fatigue.

3 These are incorrect techniques; inhalation should be through the mouthpiece.

432. (2) A chest tube drains the leaking chyle from the thoracic area; TPN provides nutrition, boosts immune defenses, and decreases thoracic duct flow; bed rest is recommended because lymphatic flow increases with activity. (3) (CC; PL)

1 A gastrostomy tube will not drain fluid from the thoracic area; a high-fat diet is contraindicated, but bed rest is recommended.

3 A rectal tube has no relationship to the drainage of chyle from the thoracic area; a fat-poor diet and bed rest are recommended.

4 The nasogastric tube does not drain fluid from the thoracic area; a fat-poor diet and bed rest are recommended; a low-fat diet of medium

chain triglycerides will reduce the production and flow of chyle.

433. (3) During chest surgery, the negative pressure around the lung is disrupted and the lungs do not fill normally during inspiration; chest tubes are inserted to reestablish negative intrapleural pressure. (3) (CI; IM)

1 Chest tubes are inserted into the intrapleural space, not the pericardial sac.

2 Atelectasis refers to the collapse of alveoli or a lobule caused by a blockage of small airways; chest tubes do not cause or correct atelectasis.

4 Although the amount of drainage from chest tubes is measured, the reason for chest tubes is to reestablish negative intrapleural pressure.

434. (2) Fluid in the water-seal chamber should rise and fall as the client breathes in and out (tidaling) until the lungs have expanded completely; a lack of tidaling on the second postoperative day would indicate that the tube is obstructed. (1) (DA; EV)

1 This is contraindicated without a physician's order because it could traumatize pleural tissue.

3 The level of the fluid, as long as it covers the tube in the water-seal chamber, does not affect tidaling.

4 While full expansion of the lung will eliminate tidaling, this is unlikely on the second postoperative day; an obstruction of the tube should be ruled out.

435. (2) This allows for measuring the output without interrupting the closed drainage system. (1) (SI; IM)

1 This is only done to obtain a specimen for diagnostic procedures.

3 Clamping the chest tube is contraindicated because it can precipitate a pneumothorax; opening the system destroys the sterility of the closed drainage system.

4 This is only done when the drainage collection chamber is full and the closed chest drainage must continue.

436. (1) Excessive bubbling indicates an air leak, which must be eliminated to prevent a pneumothorax. (3) (SI; EV)

2 Excessive suction pressure results in excessive bubbling in the suction control bottle.

3 Excessive bubbling is not expected; it indicates a leak in the system.

4 Stasis or clots in the tubing will not result in excessive bubbling; "stripping" the tubing is contraindicated because it can cause a pneumothorax.

437. (4) This is ineffective; this exercises the elbow rather than the shoulder joint and muscles. (2) (CI; EV)
 1 This is effective; this exercises the trapezius muscle and shoulder joint.
 2 Same as answer 1.
 3 This is effective; this provides circular range of motion to the shoulder joint.

438. (1) The chest tube normalizes intrathoracic pressure, drains fluid and air from the pleural space, and improves pulmonary function. (2) (CI; EV)
 2 This may be a sign of pain, respiratory obstruction, or bleeding.
 3 This indicates that air has entered the subcutaneous tissue (subcutaneous emphysema).
 4 This indicates a probable leak in the drainage system.

439. (1) This is sufficient to move blood, fluid, or air, which may be obstructing drainage. (2) (CI; IM)
 2 This is contraindicated unless there is a break in the system.
 3 This is a medical decision and would not be done unless the tube could not be made patent.
 4 This is not indicated unless other symptoms, such as dyspnea, are present.

440. (4) This is an emergency situation and atmospheric air must be prevented from entering the thoracic cavity; the client's respiratory status takes priority over the potential for infection. (3) (DA; IM)
 1 This action is useless in this situation and would further impair the client's breathing.
 2 This is unsafe because it would allow atmospheric air to enter the thoracic cavity.
 3 Although an occlusive dressing such as Vaseline gauze is desirable, atmospheric air will enter the thoracic cavity while time is taken to obtain the occlusive dressing.

441. (3) To prevent further possibility of pneumothorax, the nurse should immediately reconnect the tube. (3) (DA; IM)
 1 This is unnecessary.
 2 Clamping is appropriate for changing a broken drainage system or to check for an air leak; it should not be done needlessly.
 4 The high-Fowler's position is appropriate for a client in respiratory distress, but this does not remedy this problem.

442. (2) This prevents atelectasis and collection of secretions and promotes respiratory exchange. (2) (CI; IM)

 1 Observing for dyspnea remains important, but crepitus is unlikely to occur with stabilization of respiratory status.
 3 This is important but not as conducive to improving respiratory status as are coughing and deep breathing.
 4 Activity should be promoted within limits of physical ability; bed rest is unnecessary.

443. (3) Blood-tinged sputum in the absence of pronounced coughing may be the presenting symptom; diaphoresis at night is a later symptom. (2) (DA; AS)
 1 Recurrent fever is present; however, frothy sputum is present with pulmonary edema, not tuberculosis.
 2 The cough would be productive, not dry.
 4 A productive cough may occur, but engorged neck veins are symptomatic of congestive heart failure.

444. (4) The Mantoux is the most accurate skin test because of the testing material used and the intradermal method; no other skin test would be appropriate as a follow-up; further tests are now warranted, including a chest x-ray film. (1) (HI; EV)
 1 The test result was positive, not negative; further testing is necessary.
 2 The tine test is less accurate than the Mantoux and would not be used as a follow-up test.
 3 More than 10 mm induration is a positive test result, not a doubtful test result.

445. (4) Induration measuring 10 mm or more in diameter is interpreted as significant; it does not indicate that active tuberculosis is present; about 90% of individuals who have significant induration do not develop the disease. (2) (HI; EV)
 1 It merely indicates exposure; infection can be past or present.
 2 It merely indicates exposure, not resistance.
 3 Passive immunity occurs when the body plays no part in the preparation of the antibodies; a positive Mantoux only indicates the presence of antibodies, not how they were formed.

446. (4) The tubercle bacilli can be stained with carbolfuchsin, an acid, when the stain is applied with heat; the bacilli resist discolorization when an acid-alcohol wash is applied. (1) (HI; AS)
 1 This reflects pulmonary status but does not identify the organism.
 2 This indicates the presence of antibodies but is not diagnostic for the presence of the disease.
 3 Same as answer 1.

447. (2) Clients with tuberculosis tend to lose weight and have anorexia; this will encourage food intake and provide calories for weight gain. (1) (PC; PL)
1 This is not necessary; protein and other nutrients can be obtained through natural foods.
3 This is not possible; carbohydrates contain calories.
4 Proteins are needed for tissue building.

448. (3) Fresh airflow into the house changes the air and lowers the concentration of microorganisms. (3) (SI; IM)
1 This is not necessary.
2 This is not necessary; only articles contaminated with infected sputum, such as used tissues, should be contained.
4 It is permissible to do this because the extreme heat used to process the dishes kills the mycobacteria.

449. (3) The directions need further clarification. Refrigeration is not necessary; however, the specimen needs to be taken to the laboratory at the earliest possible time after collection. (1) (CC; EV)
1 The specimen must represent phlegm containing the *Mycobacterium*, which is in the lung, not the oronasopharynx.
2 For the best results the sputum collection should be made when the client awakens in the morning when mucous secretions are more copious.
4 Delivery to the laboratory should be made on the same day as close as possible to the time of collection.

450. (3) Tubercle bacilli are transmitted through airborne droplets; therefore respiratory isolation with an Ultra-Filter mask is necessary. (1) (SI; IM)
1 This is unnecessary as long as appropriate isolation precautions are followed.
2 This would not be necessary unless objects are contaminated by respiratory secretions.
4 This is not necessary; tuberculosis is spread by airborne droplets; gloves are only necessary when touching articles contaminated by respiratory secretions.

451. (4) Tubercle bacilli are particularly resistant to treatment and can remain dormant for pro-longed periods; medication must be taken consistently as ordered for prolonged periods. (1) (CC; PL)
1 Although this is important, the microorganisms must be eliminated by the use of medication.
2 Same as answer 1.
3 Same as answer 1.

452. (3) The tubercle bacillus is a drug-resistant organism and takes a long time to be eradicated; a combination of two medications are usually used for a minimum of 9 months and at least 6 months beyond culture conversion. (2) (SI; IM)
1 This is too short a time for eradication of this organism.
2 Same as answer 1.
4 Usually, the organism can be eradicated in a shorter period of time, unless a resistant strain of the bacillus has developed.

CARDIOVASCULAR (CV)

453. (3) The carotid artery is located along the anterior edge of the sternocleidomastoid muscle at the level of the lower margin of the thyroid cartilage. (1) (HI; AS)
1 This is not the anatomic landmark for locating the carotid artery.
2 Same as answer 1.
4 Same as answer 1.

454. (1) The first heart sound is produced by closure of the mitral and tricuspid valves; it is best heard at the apex of the heart. (2) (HI; AS)
2 This is where the second heart sound (S2) is best heard; S2 is produced by closure of the aortic and pulmonic valves.
3 This border covers a large area; the auscultatory areas that lie near it are the pulmonic and mitral areas.
4 This border covers a large area; the only auscultatory area near it is the aortic area.

455. (1) Closure of the atrioventricular valves, the mitral and tricuspid, produces the first heart sound (S1). (3) (HI; AN)
2 These valves do not close simultaneously.
3 Same as answer 2.
4 These are the semilunar valves; closure of these valves produces the second heart sound (S2).

456. (2) Inactivity causes venous stasis, hypercoagulability, and external pressure against the veins, all of which lead to thrombus formation; early ambulation or exercise of the lower extremities reduces the occurrence of this phenomenon. (1) (CI; IM)

1 This will be helpful, but it is not an independent activity; elastic stockings require a physician's order.

3 Massaging would be contraindicated because any developing clot could be dislodged.

4 Although this may help, the primary intervention is to provide exercise of the extremities until ambulation is permitted.

457. (2) Oxygen is necessary for the production of fire. (1) (SI; PL)

1 Oxygen does not burn itself; it supports combustion.

3 This is irrelevant to the need for safety precautions.

4 Same as answer 3.

458. (3) This is an early typical finding after a myocardial infarct because of the altered contractility of the heart. (3) (CI; AS)

1 Flattened or depressed T waves indicate hypokalemia.

2 This occurs in atrial and ventricular fibrillation.

4 Q waves may become distorted with conduction or rhythm problems, but they do not disappear unless there is cardiac standstill.

459. (4) Pulmonary capillary wedge pressure is an indirect measure of left ventricular end diastolic pressure, an indication of ventricular contractility. (2) (CI; AN)

1 Right atrial pressure measures only the function of the right side of the heart and indirectly its ability to receive blood.

2 Cardiac output by thermodilution does not measure intracardiac pressures.

3 Pulmonary artery diastolic pressure may not be as accurate an indicator of left ventricular pressure if COPD or pulmonary hypertension exists.

460. (2) This is a radionuclear study that determines viability of myocardial tissue; necrotic or scar tissue does not extract the thallium isotope. (2) (CI; AN)

1 This information is available from a cardiac catheterization with an angiography.

3 These are determined by cardiac angiography.

4 This is determined by a 12-lead ECG.

461. (3) Chocolate has a high caffeine content, which may stimulate catecholamine release and act as a cardiac stimulant. (2) (CC; PL)

1 Yogurt is not a cardiac stimulant; it aids digestion if lactose intolerance is present.

2 Club soda contains sodium chloride but does not stimulate the myocardium.

4 Red meats do not stimulate the myocardium; red meats may be decreased or eliminated if serum cholesterol levels are elevated.

462. (4) This is the recommended allotment of servings from the bread, cereal, rice, and pasta group. (1) (PC; IM)

1 An adult should consume 2 to 4 servings of fruit daily.

2 An adult should consume 3 to 5 servings of vegetables daily.

3 An adult should consume 2 to 3 servings of meat, poultry, or fish daily.

463. (3) The essential fatty acid, linoleic acid, is necessary for muscle tissue integrity, especially of the myocardium. (2) (PC; IM)

1 All fats cannot and should not be eliminated from the diet.

2 Proteins and carbohydrates do not contain the essential fatty acid called linoleic acid.

4 The body does manufacture cholesterol.

464. (4) The fiber component of complex carbohydrates helps bind and eliminate dietary cholesterol and fosters growth of intestinal microorganisms to break down bile salts and release the cholesterol component for excretion. (3) (PC; IM)

1 It is what the client eats, not when the client eats, that is important.

2 Of the fats in the diet, saturated fats should be decreased.

3 Fat-binding fiber should be increased.

465. (4) An increase in total body water increases the intravascular volume and the cardiac workload; excess sodium intake contributes to fluid retention and edema. (1) (PC; IM)

1 Limiting sodium will not reduce the edema already present; it will prevent additional fluid retention.

2 Limiting sodium will not have a diuretic effect; it will prevent additional fluid retention.

3 Same as answer 1.

466. (3) Of all the basic food groups, fresh fruits and juices are the lowest in sodium. (2) (PC; PL)

1 Most fresh meat or fish contains approximately 50 to 120 mg of sodium per 100 g; processed

meats contain approximately 1700 to 2000 mg of sodium per 100 g.

2 Most cheeses contain approximately 300 to 1100 mg of sodium per 100 g of cheese.

4 Most dry cereals contain approximately 900 to 1100 mg of sodium per 100 g of cereal.

467. (2) Food and fluids are usually withheld for approximately 6 to 8 hours to prevent vomiting and aspiration. (1) (CI; PL)

1 The procedure takes approximately 2 hours.

3 Bed rest with legs extended and a weight applied to the femoral site is suggested for several hours after the femoral method of entry.

4 A mild sedative is used because the client must be alert enough during the procedure to follow directions.

468. (3) Immobilization of the left leg and pressure over the groin promote coagulation and healing at the puncture site of the femoral artery. (1) (CI; AN)

1 The catheterization does not decrease blood pressure in the presence of adequate fluid replacement.

2 These interventions cannot prevent or limit these symptoms; these symptoms are not usual.

4 A small amount of radiopaque dye is injected (via the catheter) directly into the heart, where it is diluted by the blood; it does not create a problem at the puncture site.

469. (2) An apical pulse is taken to detect dysrhythmias related to cardiac irritability; blood pressure is monitored to detect hypotension, which may indicate bleeding or shock. (1) (CI; EV)

1 This is contraindicated; flexion of the groin may compromise the clot at the femoral insertion site.

3 This is not necessary; the client did not undergo general anesthesia and will soon be ambulatory.

4 A temperature may indicate a bacterial invasion, but this will not be evident during the first few hours after the catheterization.

470. (4) Bed rest with the leg extended prevents trauma caused by hip flexion and provides time for the insertion site to heal. (2) (CI; PL)

1 The physician will thoroughly review test results and may consult other physicians before answering.

2 With the femoral approach, bed rest is maintained for several hours.

3 Mild sedation and local infiltration are used for adult clients. The client is conscious.

471. (3) The dye used is a mercurial derivative that has diuretic effects. (2) (CI; EV)

1 A cardiac catheterization is a diagnostic procedure, not a therapeutic one, that neither improves cardiac function nor increases cardiac output.

2 An IV rate of 50 ml/hr will not cause a urinary output three times the amount of intake.

4 Same as answer 1.

472. (1) A reclining position would be used to maintain blood supply to vital centers. (2) (DA; PL)

2 Unsafe; maintaining blood flow to vital centers is essential.

3 Maintaining blood flow to vital centers, not comfort, is the priority.

4 This is the opposite of what should be done; while semi-Fowler's position would facilitate breathing, it would not assist blood flow to vital centers.

473. (4) A sustained diastolic pressure exceeding 90 mm Hg reflects pathology and indicates hypertension. (1) (HI; AS)

1 This is unrelated to hypertension.

2 This reflects the heart rate, not the pressures within the artery.

3 This is not the most significant indicator; an elevated diastolic pressure is more important because it reflects the pressure while the heart is at rest.

474. (4) According to the US Department of Health and Human Services, both of these readings indicate hypertension and, thus, require further evaluation by a physician. Having a baseline for both arms can assist the physician with the medical diagnosis. (2) (HI; AS)

1 There are insufficient data to support this need at this time; the priority is control of the hypertension to help prevent a crisis.

2 Same as answer 1.

3 Although emotional stress can precipitate hypertension, physical causes should be ruled out first.

475. (2) Headache is the most common symptom because of the increased pressure within the arterial circulation. (1) (DA; AS)

1 Fatigue can be associated with hypertension in the elderly, but this is not common.

3 Nosebleeds are possible when the blood pressure does reach extremely elevated levels; however, this is not a common indication.

4 A flushed face may occur because of increased pressure, but it is not the most common symptom.

476. (3) Muscle contraction associated with walking prevents edema and pooling of blood in the extremities. (1) (PC; IM)
1 This is inactivity, and movement is required.
2 This is an inactive position, and no exercise is involved.
4 This does not include movement, which is essential to prevent thrombus formation.

477. (3) The failing left ventricle cannot accept blood returning from the lungs; this results in increased vascular pressure in the lungs. (3) (DA; AN)
1 This is associated with right ventricular failure.
2 This is no such thing as right atrial failure as a separate entity.
4 Wheezing and coughing are associated with paroxysmal nocturnal dyspnea and right ventricular failure.

478. (2) These are classic symptoms of a myocardial infarction; further medical evaluation is needed immediately. (1) (DA; IM)
1 This response presumes a dietary problem when a more serious situation may exist; the client needs medical intervention.
3 This response considers only an emotional source of the reported symptoms and ignores a potential medical emergency.
4 This provides false reassurance and ignores a potential medical emergency.

479. (1) Morphine is a narcotic analgesic that acts on the central nervous system by a sympathetic mechanism; it decreases systemic vascular resistance, which decreases left ventricular afterload, thus decreasing myocardial oxygen consumption. (2) (PC; IM)
2 Oxygen administration elevates arterial oxygen tension with the potential for improving tissue oxygenation; however, oxygen administration usually does not deliver enough oxygen to the myocardium to reverse the infarction and thus relieve the pain.
3 Nitroglycerin sublingually is effective in relieving anginal pain but not myocardial infarction pain.
4 Lidocaine is an antidysrhythmic, not an analgesic.

480. (3) Cessation of blood flow to the myocardium results in pain because of ischemia of the tissue, as in angina. (1) (DA; AN)
1 Neither myocardial infarction nor angina involves compression of the heart.
2 These are not related to pain or relief of pain; isoenzymes are indicators of myocardial damage.

4 Vasodilation would increase perfusion and contribute to pain relief.

481. (2) Morphine is a specific central nervous system depressant used to relieve the pain associated with myocardial infarction; it also decreases apprehension and prevents cardiogenic shock. (2) (PC; IM)
1 This is not the reason for the use of morphine.
3 This is not the primary reason for the use of morphine.
4 Lidocaine is given intravenously to accomplish this.

482. (4) Creatine kinase (CK) isoenzyme levels, especially the MB subunit, begin to rise in 3 to 6 hours, peak in 12 to 18 hours, and are elevated for 48 hours after the occurrence of the infarct; they are therefore most reliable in assisting with early diagnosis. (2) (CI; AS)
1 Serum aspartate aminotransferase (AST) levels begin to rise later and do not peak until 24 to 36 hours, so they are not as early an indicator.
2 Lactate dehydrogenase (LDH) isoenzyme levels, especially LDH1, do not begin to rise until 12 hours after the infarct and peak in 48 hours, so they are a much later indicator.
3 Serum glutamic-oxaloacetic transaminase (SGOT) isoenzyme is another name for AST.

483. (4) A direct relationship exists between the strength of cardiac contractions and the electrical conductions through the myocardium. (3) (DA; AN)
1 The heart rate is related to factors such as SA node function, partial pressures of oxygen and carbon dioxide, and emotions.
2 This is the period when the heart is at rest, not when it is contracting.
3 Pulmonary pressure does not influence action potential; it becomes elevated in the presence of left ventricular failure.

484. (3) Dyspnea at night, which usually requires the assumption of the orthopneic position, is a symptom of left ventricular failure; orthopnea, a compensatory mechanism, limits venous return, which decreases pulmonary congestion and promotes ventilation, easing the dyspnea. (2) (DA; AS)
1 This occurs with right ventricular failure because of hypervolemia.
2 Anorexia and nausea occur with right ventricular failure because of venous stasis and venous engorgement of abdominal viscera; weight gain would occur because of fluid retention.

4 This occurs with right ventricular failure because of portal hypertension and liver congestion.

485. (4) A direct relationship exists between the systolic blood pressure and the force of left ventricular contraction. (2) (DA; AN)
1 An increased blood volume would be indicated by hypertension, not a decreased pulse pressure.
2 Hyperactivity of the heart would be indicated by dysrhythmias and tachycardia.
3 A decreased pulse pressure would indicate decreased cardiac sufficiency.

486. (2) This is the cardinal reason for PVBs. (2) (DA; AS)
1 This is a type of dysrhythmia, not the cause of PVBs; the source of atrial fibrillation is the atrium, not the ventricles.
3 This type of dysrhythmia is associated with interference with the conduction system, not cardiac irritability.
4 This is a type of dysrhythmia, not the cause of PVBs.

487. (3) This response provides information and reduces anxiety. The greater saphenous vein from the leg is often removed and used to bypass the diseased coronary artery because one surgical team can obtain the vein while another team performs the chest surgery; this shortens the surgical time and lessens the risks of surgery; the internal mammary arteries are the grafts of choice but the surgery is usually longer because of the procedure of dissecting the arteries from the chest wall. (1) (SC; IM)
1 Cardiopulmonary bypass (extracorporeal circulation) is accomplished by placement of a cannula in the right atrium, vena cava, or femoral vein to withdraw blood from the body; blood is returned to the body via a cannula in the aorta or the femoral artery.
2 This is not done during a coronary artery bypass graft (CABG).
4 Same as answer 2.

488. (2) This type of pacemaker synchronizes impulses to the atria and ventricles to more closely simulate the normal action of the heart; it may be a fixed-rate or, most usually, a demand-mode pacemaker and may stimulate the atria, the ventricles, or both. (2) (CI; IM)
1 The physiologic pacemaker stimulates both the atria and ventricles to contract.
3 It affects the electrical conduction system of the heart, not the anatomic structures.

4 It will increase the heart beat to a more normal rate.

489. (3) Assessment of the heart's rate and rhythm monitors the functioning of the newly implanted pacemaker. (2) (DA; EV)
1 Unless the client is dehydrated, encouraging fluid will increase the workload of the heart.
2 Although this is an appropriate action, the priority is to assess the functioning of the pacemaker.
4 Although these are important actions, the priority is to assess the functioning of the pacemaker.

490. (3) This is the primary indication for a pacemaker because there is an interference with the electrical conduction system of the heart. (1) (DA; AN)
1 The primary treatment for this disorder is medication; this is not an indication for a pacemaker.
2 Same as answer 1.
4 Same as answer 1.

491. (4) This information will reduce anxiety. Milliamps are used, not volts of electricity; higher voltages are needed to electrocute. (1) (SC; IM)
1 This is a patronizing response and minimizes the stated concern.
2 The voltage used in pacemakers can never cause electrocution; technology is not related.
3 The voltage used can never cause electrocution; all pacemakers are pretested for accuracy.

492. (1) Shoes that become too tight indicate pedal edema, which is a sign of fluid retention; 2.2 pounds is equal to 1 liter of fluid. (3) (DA; AS)
2 Eventually the physician will be notified, but the nurse should have more data before calling.
3 With fluid retention the rate is not as significant as a bounding characteristic to the pulse.
4 Although left ventricular failure can proceed to right ventricular failure, the client has given no indication that pulmonary edema may be developing.

493. (2) Left-sided heart failure causes fluid accumulation in the capillary network of the lung; fluid eventually enters alveolar spaces and causes crackling sounds at the end of inspiration. (1) (DA; AS)
1 Not heard in chronic heart failure; usually associated with tracheal constriction or obstruction.
3 Not heard in chronic heart failure; usually associated with asthma.
4 Not heard in chronic heart failure; usually associated with pleurisy.

494. (2) Pulmonary congestion and edema occur because of fluid extravasation from the pulmonary capillary bed, resulting in difficult breathing. (1) (DA; AS)
   1 This is a hallmark of myocardial infarction; it is caused by inadequate oxygen supply to the myocardium.
   3 This results from increased pressure in the right atria associated with right ventricular failure.
   4 This is a sign of right, not left, ventricular failure; a weakened right ventricle causes venous congestion in the systemic circulation.

495. (2) Rest decreases demand on the heart and will also prevent fatigue. (1) (PC; PL)
   1 The client should sleep with the head slightly elevated to facilitate respiration.
   3 The client needs potassium; a low-potassium diet when the client is taking digoxin predisposes the client to toxicity and dangerous dysrhythmias.
   4 To avoid becoming obsessed with the pulse rate, the client should not be taught to take the pulse so often; once daily is adequate.

496. (4) Crackles are the sound of air passing through fluid in the alveolar spaces; in pulmonary edema, fluid moves from the intravascular compartment into the alveoli. (1) (DA; AS)
   1 The blood pressure is usually increased with hypervolemia.
   2 This would occur with angina or a myocardial infarction.
   3 The pulse would be bounding with hypervolemia.

497. (3) A client's knowledge about the treatment program enhances conformity and reduces stress. (1) (PC; IM)
   1 This response does not answer the client's question and might produce frustration.
   2 This is a general statement that does not focus on the specific client.
   4 This does not support the treatment regimen; it may cause more stress if the client interprets it as a conflict between the physician and the nurse.

498. (3) This is caused by hypervolemia and pulmonary hypertension. (1) (DA; AS)
   1 The pulse would most likely be rapid and bounding, not slowed.
   2 This is present in pleurisy, not heart failure.
   4 Hypertension, not hypotension, would occur because of hypervolemia.

499. (3) Stressful situations will increase the body's oxygen demands. (2) (SC; IM)
   1 Clients with low cardiac reserve cannot tolerate extremes of temperature; a hot bath will increase the body's oxygen demands.
   2 Hot, humid weather is not good for those with chronic heart disease; these individuals should use an air conditioner.
   4 The heart of a client with low cardiac reserve cannot tolerate a pulse rate this high.

500. (2) This pulse rate increase indicates activity tolerance is exceeded; rest brings the heart rate back to normal. (2) (SI; IM)
   1 Activity should be stopped, not continued.
   3 Though descending the stairs requires less energy than climbing, rest is essential to permit the heart rate to return to normal.
   4 This still constitutes activity, which aggravates the cardiac workload.

501. (3) The highest incidence is in people 40 to 70 years of age. It is seen four times more frequently in men than women, with a higher incidence in black, rather than white, men. It occurs most often in the older, hypertensive client. (2) (HI; AN)
   1 This client is not at as high a risk as a black male.
   2 Same as answer 1.
   4 Unless the 40-year-old female is pregnant or in labor, she is not at as great a risk as the black male.

502. (4) As the heart contracts, an expanding midline mass can be palpated to the left of the umbilicus. (2) (DA; AS)
   1 This is not definitive for abdominal aortic aneurysm.
   2 These are not definitive for abdominal aortic aneurysm; pallor is associated with shock.
   3 There is no disease in the intestinal tract; this finding is associated with intestinal obstruction.

503. (2) Immediate surgical intervention to clamp the aorta is necessary for survival; the aneurysm has ruptured. (2) (DA; IM)
   1 This may eventually be done, but notifying the physician is the priority.
   3 Same as answer 1.
   4 Sedatives mask important signs and symptoms.

504. (4) Ambulation is essential to promote venous return and prevent thrombus formation. (1) (CI; PL)

1 This causes increased popliteal pressure and impairs venous return.
2 Same as answer 1.
3 This helps prevent atelectasis, not thrombi.

505. (4) When the legs are dependent, gravity and incompetent valves promote increased hydrostatic pressure in leg veins, and as a result, fluid moves into the interstitial spaces. (2) (DA; AS)
1 This reflects inadequate arterial blood supply; arterial circulation is not affected by varicose veins.
2 Same as answer 1.
3 This pain is referred to as Homans' sign and is most often associated with thrombophlebitis.

506. (1) To prevent distention of the veins, the stockings should be applied before the legs are placed in a dependent position. (2) (SI; IM)
2 Knee-high stockings should end 2 inches below the knee to avoid popliteal pressure, which limits venous return.
3 The stockings should be used preventively before the discomfort associated with venous pressure and edema occurs.
4 The stockings apply uniform pressure; elastic bandages may slip, creating uneven pressure and constriction; edema may also result.

507. (1) Impaired venous return causes increased pressure, with subjective symptoms of fatigue and heaviness. (2) (DA; AS)
2 Homans' sign is indicative of thrombophlebitis.
3 Symptoms of hypoxia are related to impaired arterial, rather than venous, circulation.
4 Ecchymosis may occur in some individuals, but there is insufficient bleeding into tissue to cause hematomas.

508. (1) After ligation, the saphenous vein is removed. (1) (CI; IM)
2 Plaque is considered an arterial, rather than a venous, problem.
3 They are normally attached by communicating veins; surgery involves ligation to isolate the saphenous vein.
4 This prevents emboli from traveling to the lung; it is not a vein ligation and stripping.

509. (4) Because of the client's history and the site of the surgery, thrombi are likely to develop; activity is a preventive measure. (2) (CI; PL)
1 This alone will not prevent thrombi; activity is necessary.

2 Getting out of bed will provide little exercise if the client only sits in a chair; also, an order is needed.
3 Although body alignment is important for all clients, it will not discourage thrombus formation.

510. (1) Seeing the exercises demonstrated will reinforce the verbal explanations. (2) (CC; IM)
2 This statement is too vague; it does not explain how to move them.
3 This statement is too vague and thus may be ineffective; the time period should be stipulated.
4 This response is vague and open to interpretation; nonspecific instructions may be ineffective.

511. (1) Swelling of the extremity is indicative of thrombophlebitis. (2) (CI; AS)
2 Difficulty with mobility would occur with musculoskeletal or neuromuscular problems.
3 This is associated with neurologic deficits in the corticospinal tracts.
4 This assessment would be made to determine the status of the arterial, not venous, system.

512. (3) Raynaud's phenomenon has unilateral involvement, while Raynaud's disease has bilateral involvement. (3) (PC; AN)
1 Tingling of the fingers, indicating return of blood flow, is characteristic of both Raynaud's phenomenon and Raynaud's disease.
2 Blanching followed by redness, indicating blood return, is characteristic of both Raynaud's phenomenon and Raynaud's disease.
4 Coolness of the skin, indicating lack of blood supply, is characteristic of both Raynaud's phenomenon and Raynaud's disease.

513. (3) Nicotine causes spasms and constriction of the smooth muscles of the arterial vasculature, compromising blood flow to the distal extremities. (2) (CC; AN)
1 Nicotine does not directly cause pain and tingling, although these may occur as consequences of the nicotine-induced vasoconstriction.
2 Vasoconstriction from nicotine would not result in such severe adaptations.
4 Smoking raises the carboxyhemoglobin level in the blood; carbon monoxide combines with hemoglobin and occupies the sites on the hemoglobin molecule that normally binds with oxygen.

514. (2) This is a truthful answer that explains how nicotine is detrimental to physical status; nicotine also promotes platelet aggregation and clot formation. (3) (SI; IM)

1 This may be true, but it is not an appropriate explanation of why the client should not smoke.

3 Intermittent claudication is caused by impaired arterial, not venous, circulation.

4 The physician will probably advise against smoking because resuming smoking will continue to decrease oxygen to the lower extremities.

515. (3) An altered quality of a variety of pulses in the extremity is the earliest indication of limited circulation. (1) (CI; AS)

1 This would interfere with gravity that facilitates the flow of arterial blood to the legs and feet.

2 This could release an embolus into the circulation; it may also cause tissue trauma.

4 Altered sensation may limit sensitivity to heat, which could result in burns.

516. (3) Intermittent claudication is the pain that occurs during exercise because of a lack of oxygen to muscles in the involved extremities. (2) (CC; AN)

1 It is the exercise, not the lack of it, that precipitates the pain.

2 This is related to a venous problem, not an arterial one.

4 Same as answer 2.

517. (2) Decreasing the demand for oxygen by resting will relieve the pain. (1) (SI; IM)

1 Pain will not resolve as long as exercise and, thus muscle hypoxia, is continued, regardless of whether ASA is taken.

3 Sublingual nitroglycerin is not indicated in this situation.

4 This is appropriate for venous insufficiency, not arterial insufficiency.

518. (1) Arterial ulcers are painful because of the depth and interruption of the blood supply. (3) (PC; AN)

2 This is characteristic of venous ulcers.

3 Same as answer 2.

4 Same as answer 2.

519. (4) Pressure promotes coagulation and prevents the complication of bleeding. (2) (CI; IM)

1 Bending the operative leg may cause decreased perfusion to the leg or bleeding at the catheter insertion site.

2 Elevation would resist gravity flow of arterial blood, reducing oxygen to distal tissue.

3 Bed rest is required for 6 to 12 hours after the procedure.

520. (4) Because of the trauma associated with the insertion of the catheter during the procedure, the involved extremity should be assessed for sensation, motor ability, and arterial perfusion; hemorrhage or an arterial embolus could occur. (2) (CI; EV)

1 The client has an arterial problem, and perfusion is promoted by keeping the legs at the level of the heart.

2 A general anesthetic is not used; therefore voiding is really not a concern.

3 This is unsafe; this position increases pressure in the groin area, which could dislodge the clot at the catheter insertion site, resulting in bleeding; it also impedes arterial perfusion and venous return.

521. (3) The client is hypertensive, and the intraarterial pressure is elevated; this increased pressure could cause the arterial suture line to rupture. (2) (CC; AN)

1 This is unlikely because the blood pressure is elevated and the client is at risk for bleeding.

2 Hypervolemia would be an assumption; other causes, such as arterial constriction, can precipitate hypertension.

4 Although this could occur, the priority for this client is protecting the graft.

522. (3) Mobility will reduce venous stasis and edema as well as promote arterial perfusion and healing. (3) (CI; IM)

1 Bed rest is contraindicated because it promotes the development of thrombophlebitis and pulmonary emboli.

2 This is contraindicated; it constricts circulation at the hips and knees.

4 This would limit arterial perfusion.

523. (4) Presence of pulses and a normal skin color indicate adequate arterial perfusion and graft viability. (1) (CC; EV)

1 This is contraindicated in peripheral vascular disease because it may traumatize vessels; it could cause a thrombus to become an embolus.

2 This is appropriate for venous, not arterial, problems.

3 Clients with paresthesias cannot judge water temperatures; the peripheral dilation produced by a hot bath would increase the workload on

the heart, which is undesirable in a client with hypertension.

524. (3) Inadequate oxygenation can cause premature ventricular beats. (2) (PC; AN)
1 Hypoxia can precipitate respiratory acidosis; hyperventilation causes respiratory alkalosis.
2 While this may be true, it does not explain why adequate oxygenation is important.
4 The reverse is true; postoperative pain can increase the respiratory rate.

525. (3) During open heart surgery the conductive system of the heart can be damaged because of the trauma of surgery. (3) (DA; AN)
1 Shock results in a weak, rapid pulse.
2 Hypoxia causes tachycardia.
4 Heart failure causes a rapid pulse rate.

REPRODUCTIVE AND GENITOURINARY (RG)

526. (4) Laxatives remove feces and flatus, providing better visualization. (2) (CI; PL)
1 A light supper may be indicated; however, there is no restriction as to fat content.
2 Large amounts of fluids may dilute the dye, impairing visualization.
3 A light dinner and beverage are permitted.

527. (1) Assessment is the priority; the nurse should determine whether symptoms are caused by a full bladder. (2) (PC; AS)
2 This may eventually be necessary but is not the initial action.
3 This may be done to reduce urinary bacterial count and stone formation, but it is not the initial action.
4 This could be done, but it would not be the initial action.

528. (4) This uses gravity to allow urine to exert pressure on the area of the trigone, initiating relaxation of the urinary sphincter and facilitating micturition. (1) (PC; IM)
1 Although this may be important so that urine may be collected to be strained, it will not facilitate micturition.
2 An acid-ash diet may be used to prevent urinary infection and the formation of calcium stones; it will do nothing to facilitate micturition.
3 This is important after urination but will not help facilitate micturition.

529. (2) Promoting hydration maintains urine production at a higher rate, which flushes the bladder and prevents urinary stasis and possible infection. (2) (SI; PL)
1 Although this could help identify a urinary tract infection, it would not prevent it.
3 Same as answer 1.
4 The drainage bag is emptied once every shift unless it fills before then; changing the bag periodically, not emptying it, would help prevent infection.

530. (4) The glomerulus is not permeable to large proteins like albumin or to red blood cells. (1) (PC; EV)
1 The proximal tubules are responsible for regulating water, electrolytes (including sodium and potassium), urea nitrogen, and pH; the byproducts of this regulation will appear in normal urine.
2 Same as answer 1.
3 Same as answer 1.

531. (1) These responses may indicate a urinary tract infection; a culture and sensitivity of the urine would elicit the microorganisms. (3) (CI; PL)
2 A cystoscopy is too invasive as a screening procedure; altered bilirubin results would indicate liver or biliary problems.
3 Creatinine clearance reflects renal function; A/G ratio reflects liver function.
4 Although an increased urine specific gravity could indicate RBCs, WBCs, or casts in the urine, which are associated with urinary tract infection, it would not identify causative organisms.

532. (2) Cholinergics intensify and prolong the action of acetylcholine, which increases the tone in the genitourinary tract, preventing urinary retention. (3) (PT; AN)
1 Cholinergics will not prevent renal calculi.
3 Anticholinergics are prescribed for the frequency and urgency associated with a spastic bladder.
4 This would be a secondary gain because cholinergics help prevent urinary retention that can lead to a urinary tract infection, but this is not the primary purpose for administering these drugs.

533. (3) This increases lymphatic drainage, reducing edema and pain. (2) (PC; IM)
1 This increases circulation to the area, intensifying edema and pain.
2 Same as answer 1.
4 This is not indicated; scrotal swelling is caused by the trauma of surgery, not infection.

534. (3) This method obtains a specimen uncontaminated by environmental organisms. (3) (SI; IM)
1 This is not as accurate as obtaining the purulent discharge from the site of origin.
2 This would contaminate the specimen with organisms external to the body.
4 This would dilute and possibly contaminate the specimen.

535. (2) Anal itching and irritation result from erythema and edema of the anal crypts from the gonococci. (3) (HI; AS)
1 Frank rectal bleeding, not upper GI bleeding, occurs.
3 Diarrhea, not constipation, occurs.
4 Shape of formed stool does not change; however, diarrhea does occur.

536. (4) Secondary syphilis, occurring 1 to 3 months after healing of the primary lesion and lasting for several weeks to as long as a year, is the stage at which the individual is most infectious. (2) (DA; AN)
1 Latent syphilis occurs after the secondary stage and before the late stage of syphilis. In latent syphilis the immune system is able to suppress the infection and there are no clinical symptoms.
2 Tertiary syphilis, also known as late syphilis, is the final stage of syphilis.
3 Primary syphilis is the stage of initial infection; it precedes secondary syphilis.

537. (4) The client is in the secondary stage, which begins from 6 weeks to 6 months after primary contact; therefore, a 6-month history is needed to ensure that all possible contacts and original contacts are located. (3) (HI; IM)
1 Any time less than 6 months may miss contacts who could have become infected.
2 Same as answer 1.
3 Same as answer 1.

538. (4) Neurotoxicity, as manifested by ataxia, is evidence of tertiary syphilis, which may involve the CNS; other CNS signs include confusion, paralysis, delusions, impaired judgment, and slurred speech. (2) (HI; AS)
1 This occurs in the secondary stage.
2 Same as answer 1.
3 This is not a sign of late-stage syphilis.

539. (4) Research statistics indicate that hematuria is the most common early sign of cancer of the urinary system, probably because of its rich vascular network. (2) (HI; AN)
1 This is not specific for bladder cancer; it is usually associated with an enlarged prostate in the male.
2 Same as answer 1.
3 Same as answer 1.

540. (3) Radiation interferes with cell multiplication, which should control the growth and metastasis of cancerous tumors. (2) (CI; EV)
1 Radiation affects normal as well as abnormal cells; frequency and diarrhea could result.
2 Malaise is a side effect of radiation therapy.
4 Bone marrow sites may be affected by the radiation, resulting in a reduction of WBCs.

541. (2) This may denote a compromised blood supply to the stoma and impending necrosis. (2) (CI; EV)
1 This is expected in the early postoperative period following the surgery.
3 Splinting catheters may be in place in the stoma; pink-tinged urine may be present in the immediate postoperative period.
4 Same as answer 1.

542. (3) This is expected because mucus is normally secreted by the intestine. (3) (CI; EV)
1 Not necessary; at this point postsurgically the mucus is not an indication of infection; mucus in the urine following ureterostomy may indicate infection.
2 Not necessary; mucus is expected with an ileal conduit.
4 While fluids should be encouraged to maintain urine flow, this will not eliminate mucus, which is continually discharged from the intestinal segment.

543. (2) High fluid intake flushes the ileal conduit and prevents infection and obstruction caused by mucus or uric acid crystals. (3) (CI; IM)
1 These are not contraindicated with an ileal conduit.
3 This is expected; as edema decreases the stoma will become smaller.
4 Soap and water on the peristomal area help prevent irritation from waste products.

544. (2) The headaches are the result of the retention of fluid and hypertension. (3) (DA; AS)
1 The client would have oliguria, not nocturia.
3 The client would have a weight gain because of the retention of fluid.
4 The client would have anorexia related to elevated toxic substances in the blood.

545. (2) Sucking on a hard candy will relieve thirst and increase carbohydrates but does not supply extra fluid. (2) (PC; IM)

1 Carbonated beverages are high in sodium and provide additional fluid, which must be restricted.

3 A milk shake contains both fluid and protein, which must be restricted.

4 Broth contains sodium, which can compound the fluid retention problem.

546. (4) Streptococci, common in throat infections, initiate an antibody formation that damages the glomerulus. (2) (HI; IM)

1 The alkalinity of bubble baths has been linked to urethritis, not glomerulonephritis.

2 Moderate activity is helpful in preventing urinary stasis, which could precipitate an infection.

3 Any fluid restriction is moderated as the client improves; fluid is permitted to prevent urinary stasis.

547. (2) Infection is responsible for one third of the traumatic or surgically induced deaths of clients with renal failure, as well as for medically induced acute renal failure (ARF); resistance is reduced in clients with renal failure because of decreased phagocytosis, which makes them very susceptible to microorganisms. (3) (SI; AS)

1 Anemia occurs often with ARF, but it is not the most serious complication and should be treated in relation to the client's symptoms; erythropoietin and iron supplements are usually used.

3 Weight loss is not life threatening in and of itself.

4 Platelet dysfunction does occur because of decreased cell surface adhesiveness, but it is not as serious as infection.

548. (4) Restriction of protein intake limits nitrogenous wastes from protein metabolism. (3) (CI; IM)

1 The client is encouraged to be as active and independent as possible.

2 Medications are avoided because they may mask symptoms.

3 Iron supplements are not tolerated well by clients in renal failure and reduce the client's own stimulus to produce red blood cells; folate is usually given.

549. (2) An excessive use of antacids may result in hypercalciuria; most calculi contain calcium combined with phosphate or other substances. (3) (DA; AS)

1 Cholesterol is unrelated to the formation of renal calculi; cholesterol stones in the gallbladder are the result of increased cholesterol synthesis in the liver.

3 Immobility with the associated demineralization of bone, not exercise, may cause renal calculi.

4 Alcohol intake is unrelated to renal calculi formation.

550. (4) The pain of renal colic is excruciating; unless relief is obtained, the client is unable to cooperate with other therapy. (3) (PC; AN)

1 Any urine can be saved and strained after the client's priority needs are met.

2 Increasing fluid intake helps mobilize the stone, but a client who has severe pain may be nauseated and unable to drink.

3 Although a culture is generally ordered, it is not a priority when a client has severe pain.

551. (4) Pain with ureteral stones is caused by spasm and is excruciating and intermittent; it follows the path of the ureter to the bladder. (2) (DA; AS)

1 Pain is spasmodic and excruciating, not boring.

2 This is untrue; pain intensifies as the stone is caught in the ureter and spasms occur in an attempt to dislodge it.

3 This is typical of pain caused by a stone in the renal pelvis.

552. (4) Purines are precursors of uric acid. (3) (DA; AN)

1 Cystine stones are caused by a rare hereditary defect resulting in inadequate renal tubular reabsorption of cystine (inborn error of cystine metabolism).

2 A struvite stone is sometimes called a magnesium ammonium phosphate stone and is precipitated by recurrent urinary tract infections with coliform bacteria.

3 An oxalate stone would be composed of calcium oxalate.

553. (2) Calculi may obstruct the flow of urine to the bladder, allowing the urine to distend the ureter, causing hydroureter. (3) (DA; AN)

1 There is insufficient information to come to this conclusion even though output is less than intake; oliguria is present when the output is between 100 and 500 ml in a 24-hour period.

3 Calculi do not cause renal shutdown directly; they may obstruct the urinary tract and cause damage indirectly as a result of pressure from urine buildup.

4 If the urethra were obstructed, the bladder would be distended.

554. (4) If the calculus is in the renal pelvis, a percutaneous pyelolithotomy is performed; the stone is removed via a small flank incision. (2) (CI; AN)
1 This is not necessary.
2 This is usually unnecessary.
3 This route is used for stones in the distal portion of the ureters.

555. (1) These occur with a urinary tract infection because of bladder irritability; burning on urination and fever are additional signs of a UTI. (1) (CC; EV)
2 This is not related to a urinary tract infection.
3 This is a symptom of urinary calculus, not infection.
4 This is not a sign of a urinary tract infection; this may be caused by altering the diet to include foods that form acid ash or alkaline ash.

556. (1) Cancer of the prostate is rare before age 50 but increases with each decade; black men develop it twice as often as white men and at an earlier age. (3) (HI; AS)
2 This group of men have a lower incidence of prostatic cancer than white men, as well as a lower mortality rate
3 Same as answer 2.
4 White men develop prostatic cancer half as often as black men, but more commonly than Asian or Hispanic men.

557. (4) A subjective symptom can only be experienced and described by the client. Flank pain, pain on the side of the body between the ribs and the ileum, accompanies renal colic and is a subjective symptom. (2) (DA; AS)
1 This is an objective symptom that can be verified by observations or measurements.
2 Same as answer 1.
3 Same as answer 1.

558. (3) An indwelling urethral catheter is used because surgical trauma can cause urinary retention, leading to further complications such as bleeding. (2) (CI; IM)
1 Urinary control is not lost in most cases; loss of control is temporary if it does occur.
2 This is usually not affected; sexual ability is maintained if the client was able to perform before the surgery.
4 A cystotomy tube is not used if the client has a transurethral resection; however, it is used if a suprapubic resection is done.

559. (3) Patency promotes bladder decompression, which prevents distention and bleeding; continuous flow of fluid through the bladder limits

clot formation and promotes hemostasis. (2) (CI; PL)
1 There is no abdominal dressing with a TURP; surgery is performed via the urethra.
2 There is no cystotomy tube; a cystotomy tube is used when a suprapubic resection is performed.
4 There is no external wound because there is no abdominal incision.

560. (2) The pressure of the balloon against the small blood vessels of the prostate causes them to constrict, thereby preventing bleeding. (2) (CI; AN)
1 It may actually do this but is necessary to limit bleeding.
3 Same as answer 1.
4 It is not the balloon but the Foley catheter that promotes urinary drainage.

561. (4) The amount of irrigant instilled into the bladder must be deducted from the total output to determine the amount of urine produced. (1) (DA; IM)
1 Unless irrigant is subtracted from the output, the total would be inaccurate.
2 Specific gravity measures the concentration of urine; this measurement would be inaccurate because the urine is diluted with GU irrigant.
3 This is unnecessary; the bladder is constantly being irrigated with GU irrigant.

562. (2) After transurethral surgery, hemorrhage is common because of venous oozing and bleeding from many small arteries in the area. (1) (SI; EV)
1 Sepsis is unusual and occurs later in the postoperative course.
3 Leaking around the catheter is not a major complication.
4 Urinary retention is highly unlikely with an indwelling catheter in place.

563. (4) Because of the trauma to the mucous membranes of the urinary tract, burning on urination is an expected response that should gradually subside. (2) (CC; EV)
1 The urine may have a slight pink tinge because of the trauma from the surgery and the presence of the catheter, but it should no longer be dilute once the continuous bladder irrigation is removed.
2 This is a sign of hemorrhage that should not occur.
3 This should not occur unless the Foley catheter is removed too soon and there is still edema of the urethra.

564. (3) The urethral mucosa in the prostatic area is destroyed during surgery, and strictures may form with healing. (3) (CC; EV)
1 The client should void as the need arises; straining can cause pressure in the operative area, precipitating hemorrhage.
2 The client should be out of bed ambulating; sitting for several hours is contraindicated because it promotes venous stasis and thrombus formation.
4 Although vigorous exercise should be avoided, 6 months is too long for this restriction.

565. (2) Because venous stasis is the major predisposing factor of pulmonary emboli, venous flow velocity should be increased. (2) (DA; PL)
1 Increasing the coagulability of the blood would lead to the development of deep vein thrombosis.
3 This would not affect the development of deep vein thrombosis.
4 Same as answer 3.

566. (2) The kidney, an extremely vascular organ, receives a large percentage of the blood flow and hemorrhage from the incisional site can occur. (3) (DA; PL)
1 This may occur, but it would be later in the postoperative period.
3 This can occur but is not acute and develops later.
4 This can occur but is not life threatening.

567. (3) Turning facilitates drainage from the operative site. (3) (CI; PL)
1 Because clients are prone to develop paralytic ileus, they are kept npo or on clear fluid for at least 24 to 48 hours.
2 A nephrostomy tube should never be clamped unless specifically ordered by the physician.
4 The dressing should be changed often because the wound generally drains large amounts.

568. (4) The wound will drain urine until healing takes place, sometimes up to several weeks. (3) (CI; EV)
1 The hourly urine output must be 30 ml or more to be adequate.
2 A specific gravity this low indicates that the kidneys have lost their ability to concentrate urine.
3 Urine begins to clear within about 24 hours; urine that remains dark red and contains clots indicates abnormal bleeding.

569. (4) The dressing will need to be changed at home because drainage can persist for several weeks. (1) (SI; IM)
1 The client should be encouraged to take fluids.
2 The client should be up and about at home.
3 A nephrostomy tube is generally not irrigated unless specifically ordered by the physician.

570. (2) High-biologic-value (HBV) protein contains essential amino acids needed by the body for tissue building and repair. (1) (PC; EV)
1 A high-calorie diet would provide for weight gain.
3 Low-biologic-value (LBV) proteins avoid the accumulation of urea in the body.
4 This is not the purpose of high-biologic-value proteins; sodium restrictions would decrease blood pressure.

571. (3) One cup of cottage cheese contains approximately 225 calories, 27 grams of protein, 9 grams of fat, 30 milligrams of cholesterol, and 6 grams of carbohydrate; proteins of high biologic value (HBV) contain optimal levels of the amino acids essential for life. (1) (PC; EV)
1 Apple juice is a source of vitamins A and C, not protein.
2 Raw carrots are a carbohydrate source and contain beta-carotene.
4 Whole wheat bread is a source of carbohydrates and fiber.

572. (2) The accumulation of metabolic wastes in the blood (uremia) can cause pruritus; edema results from fluid overload caused by impaired renal excretion. (1) (DA; AS)
1 Pallor occurs with chronic renal failure as a result of the related anemia.
3 This is a urinary pattern that is not caused by chronic renal failure; this often occurs after prostate surgery.
4 This occurs with an enlarged prostate, not renal insufficiency.

573. (3) An elevation in uremic waste products causes irritation of the nerves, resulting in flapping hand tremors (asterixis, liver flap). (3) (DA; AS)
1 Hypertension results from kidney failure because of sodium and water retention.
2 The diseased kidney is unable to excrete potassium ions, resulting in hyperkalemia, not hypokalemia.
4 The hematocrit value will be low because of a decreased production of erythropoietin, a hormone synthesized in the kidney; erythropoietin regulates the production of erythrocytes.

574. (1) Lack of motivation is the most serious impediment to successful continuous ambulatory peritoneal dialysis (CAPD); CAPD is contraindicated for clients who are blind or have a colostomy, a psychosis, or PVD. (3) (CC; AS)
2 This is not a contraindication if the client is receiving medical supervision.
3 Same as answer 2.
4 Same as answer 2.

575. (3) Dialysate is introduced into the peritoneal cavity where fluids, electrolytes, and wastes are exchanged through the peritoneal membrane. (2) (CI; IM)
1 The client can dialyze alone in any location without the need for machinery and continuous technical supervision.
2 Hemodialysis is not necessary with this procedure.
4 About 2 L of dialysate is maintained intraperitoneally and exchanged by the client.

576. (2) While proteins may be restricted, those eaten should be high-quality proteins to replace proteins lost during the dialysis. (3) (PC; IM)
1 This client would be encouraged to eat a high-calorie diet.
3 This is inappropriate; there is usually a modest restriction of fluids when the client is on dialysis.
4 This is inappropriate; there is usually a restriction of high-potassium foods when the client is on dialysis.

577. (4) Warming of the dialysate before use or during infusion decreases discomfort. (2) (PC; IM)
1 A low-fiber diet may impede emptying of the bowel, which could then interfere with outflow of the dialysate.
2 It may take up to 2 weeks to tolerate a full 2 liters of exchange without leakage around the catheter; too quick an exchange may cause disequilibrium phenomenon.
3 The outflow bag should be lower than the abdomen to allow gravity flow of the return dialysate.

578. (4) A cleansing enema is given and the bladder emptied to lessen the possibility of bladder perforation; invasive procedures require consent. (2) (CC; PL)
1 An IV pyelogram is not required.
2 None of these is required.
3 A local anesthetic is used.

579. (2) The infusion should be warmed to lessen abdominal discomfort and allow for dilation of peritoneal vessels. (3) (CI; PL)

1 The infusion should be unrestricted, taking approximately 5 to 10 minutes.
3 The side-lying position may restrict fluid inflow and prevent maximum urea clearance; usually the client is in the semi-Fowler's position.
4 Medications are administered before the infusion.

580. (3) Pressure from the fluid may cause upward displacement of the diaphragm. (3) (DA; IM)
1 Additional fluid would aggravate the problem.
2 Auscultation is important, but it does not alleviate the problem.
4 The client is already in the semi-Fowler's position, which would normally relieve dyspnea; however, intraabdominal pressure must be reduced by draining the fluid.

581. (4) These symptoms may indicate peritonitis. (3) (DA; EV)
1 Oliguria is not a complication of this therapy; these are symptoms of illness, not complications of therapy.
2 Pruritus may be present as a result of the illness, not the therapy.
3 Same as answer 2.

582. (2) The external shunt may come apart; external temperatures make clotting a potential hazard; frequent handling increases risk of infection. (3) (CI; EV)
1 Neither the shunt nor the fistula will affect the blood pressure reading; obtaining a blood pressure reading in an arm with a fistula or shunt is contraindicated; blood pressure readings should be obtained using the other arm or the thighs.
3 An infusion should not be in the extremity with the shunt or fistula to avoid pressure from the tourniquet used during catheter insertion and lessen the chance of phlebitis.
4 The ends of the shunt cannula should be left exposed for rapid reconnection in the event of disruption.

583. (3) These are signs and symptoms of the disequilibrium phenomenon, which results from rapid changes in composition of the extracellular fluid and cerebral edema; the rate of exchange should be decreased. (3) (DA; EV)
1 This will not alleviate or reverse the cause of the symptoms, although it will provide relief from the nausea.
2 The cause of the confusion must be reversed, and this intervention will not accomplish this.
4 While these would be assessed, this action would do nothing to alleviate the symptoms.

584. (1) The recipient's own kidneys are not removed unless a chronic infection is present. (3) (CC; IM)

2 Both kidneys are left in place unless a chronic infection is present; the new kidney is placed in the right lower quadrant.

3 Same as answer 2.

4 Same as answer 2.

585. (4) Human leukocyte antigen compatibility provides the most specific predictions of the body's tendency to accept or reject foreign tissue. (2) (CI; AN)

1 Although ABO compatibility is necessary, the exact same blood type is not.

2 Unsafe unless the family member has matching leukocyte antigen complexes; this may increase the possibility of a match, but there is no guarantee that a family member will match.

3 Body size does not cause rejection.

586. (2) An appendectomy is performed to avoid appendicitis in the future, which might be confused with rejection of the renal transplant, which is usually placed in the right lower quadrant. (3) (CI; IM)

1 The intestines are not involved in a kidney transplant; therefore an examination of the colon is not necessary.

3 The stomach is not involved with renal function or with kidney transplant.

4 There is usually no problem with the bladder, but only with the kidneys.

587. (3) Hourly output is critical in assessing kidney function; decreasing urinary output is a sign of rejection. (2) (DA; EV)

1 It is not necessary to monitor this frequently.

2 Same as answer 1.

4 This is too infrequent to monitor output immediately after transplant; it is essential to monitor output more frequently to evaluate whether the new kidney is working or whether it is being rejected.

588. (4) As the transplanted organ functions, nitrogenous wastes are eliminated, lowering the serum creatinine. (2) (CI; EV)

1 With renal failure, fluid retention causes hypertension; there should be a correction of hypertension, not hypotension.

2 Serum albumin is not affected by kidney transplantation.

3 As more urine is produced by the transplanted kidney, the specific gravity or concentration of the urine will decrease.

589. (3) Hypertension is caused by a return of hypervolemia because of the failure of the new kidney. (2) (CC; EV)

1 There will be a weight gain because of fluid retention, which is indicative of failure of the transplanted kidney.

2 The client will have an elevated temperature exceeding 100° F.

4 Urine output will be decreased or absent, depending on the degree of failure.

590. (4) The WBC count can drop precipitously; if leukocytes are less than $3,000/mm^3$, the drug should be stopped to prevent irreversible bone marrow depression. (2) (PT; EV)

1 Leukocytosis, not leukopenia, would occur with an infection.

2 High creatinine levels are related to kidney failure but do not cause leukopenia.

3 The WBC count would be elevated, not decreased, if the kidney were being rejected.

GASTROINTESTINAL (GI)

591. (3) Complementary mixtures of essential amino acids in plant proteins provide complete dietary protein equivalents. (3) (PC; IM)

1 A total vegetarian does not consume flesh, milk, milk products, or eggs.

2 Same as answer 1.

4 Same as answer 1.

592. (2) A new dietary regimen, with a balance of foods from the food pyramid, must be established and continued for weight reduction to occur and be maintained. (3) (PC; AN)

1 Although this would be true in a weight reduction diet, this response does not address this nutrient's relationship to the other food groups.

3 Same as answer 1.

4 This is only one part of a weight reduction regimen; usually in obese individuals caloric intake exceeds energy expenditure.

593. (4) Increased exercise builds skeletal muscle mass and reduces excess fatty tissue. (3) (CC; EV)

1 This is unrelated to weight loss; during the aerobic exercise the heart rate will increase, but between periods of exercise the heart rate will decrease because of the development of collateral circulation.

2 Appetite is usually increased.

3 The metabolic rate will increase.

594. (3) With an inflammatory response the body increases its production of WBCs and fibrinogen, which increases the WBC count and blood sedimentation rate, respectively. (1) (DA; AN)
  1 This is untrue; this would not affect the white blood cell count or the sedimentation rate.
  2 Same as answer 1.
  4 Same as answer 1.

595. (2) The stomach is located within the sternal angle; thus the area is known as the epigastric area. (1) (HI; AS)
  1 This is in the area of iliac bones.
  3 This is the lowest middle abdominal area.
  4 This is the area above the sternum.

596. (3) Abdominal distention, which may be associated with pain, may indicate perforation, a complication that can lead to peritonitis. (3) (CI; EV)
  1 A local inflammatory response to insertion of the fiberoptic tube may result in a sore throat and dysphagia once the anesthetic wears off.
  2 This, together with cramping, is considered a normal response.
  4 These are not indicative of any particular problem in this situation.

597. (2) Barium will harden and may create an impaction; a laxative and increased fluids promote elimination of barium. (2) (CI; PL)
  1 Iodine is not used with barium.
  3 The client must be kept npo.
  4 This is not part of the preparation; feces in the lower GI tract would not interfere with visualization of the upper GI tract.

598. (2) Carbohydrates provide 4 kcal per gram; therefore 3 L × 50 g/L × 4 kcal/g = 600 kcal, only about a third of the basal energy need. (1) (PC; AN)
  1 This is less than the calories provided by the ordered IV fluid.
  3 This is more than the calories provided by the ordered IV fluid.
  4 Same as answer 3.

599. (3) This diet progresses from the one that makes the least metabolic demand on the client (clear liquid) to a regular diet, requiring the capability of unimpaired digestion. (1) (PC; AN)
  1 The caloric content is not the focus in a progressive postsurgical diet.
  2 Initially, a progressive diet has very little nutritional value; the focus is to rest the gastrointestinal tract immediately after surgery.
  4 Initially, a limited variety of fluids is presented to rest the gastrointestinal tract; food is not included until later.

600. (2) There are 9 calories in each gram of fat and 4 calories in each gram of carbohydrate and protein; this diet contains 1970 calories. (3) (PC; AN)
  1 This is too high for the diet prescribed.
  3 This is too low for the diet prescribed.
  4 Same as answer 3.

601. (2) This serum albumin value indicates severe depletion of visceral protein stores; the normal range for serum albumin is 3.5 to 5.5 g/dl; white meat turkey (two slices 4 × 2 × 1/4 inch) contains approximately 28 g of protein. (3) (CC; EV)
  1 A 6-oz serving of mixed fruit contains approximately 0.5 g of protein.
  3 A 3-oz serving of spinach salad contains approximately 9 g of protein.
  4 A 4-oz serving of beef broth contains approximately 2.4 g of protein.

602. (3) Smoking accelerates oxidation of tissue vitamin C, so smokers need 100 mg/day, whereas the adult RDA is 60 mg/day. (3) (PC; IM)
  1 This is not oxidized more rapidly in the smoker.
  2 Same as answer 1.
  4 Same as answer 1.

603. (1) Tofu products increase protein without increasing vitamin D because, unlike milk products, tofu does not contain vitamin D. (2) (PC; EV)
  2 Eggnog contains milk and should be avoided.
  3 This contains vitamin D and should be avoided.
  4 This contains milk, which has vitamin D, and should be avoided.

604. (3) Vitamin C (ascorbic acid), an antioxidant, is found in vegetables such as broccoli, tomatoes, and potatoes; 1 cup of broccoli has 140 mg of vitamin C. (2) (PC; AN)
  1 Apples, depending on their size, contain only 6 to 8 mg of vitamin C.
  2 An entire head of lettuce contains only 13 mg of vitamin C.
  4 Apricots contain only 11 mg of vitamin C; they are a source of beta-carotene.

605. (1) The antioxidants vitamin E and beta-carotene, which help inhibit oxidation and therefore tissue breakdown, are found in these foods. (3) (PC; AN)
  2 These are excellent sources of vitamin E, not beta-carotene.

3 These are excellent sources of vitamin C, not vitamin E and beta-carotene.

4 These are excellent sources of beta-carotene, not vitamin E.

606. ③ Stress ulcers are asymptomatic until they produce massive hematemesis and rectal bleeding. (2) (DA; AS)

1 Shock is the outcome of massive hemorrhage; it would not be unexplained because the sudden gastrointestinal bleeding would be seen.

2 Sudden massive bleeding occurs, not the slow oozing that causes melena.

4 A gradual drop in the hematocrit value indicates slow blood loss.

607. ③ Green vegetables contain fiber, which promotes defecation. (1) (CC; EV)

1 These have a binding effect and would cause constipation with resultant straining at stool.

2 Same as answer 1.

4 Same as answer 1.

608. ④ This produces bulk, which is a stimulant to defecation; the muscles used in defecation are weak in clients with Parkinson's disease, which often causes constipation. (2) (PC; IM)

1 Bananas are binding and will intensify the problem of constipation.

2 This will intensify the problem; fluids need to be increased.

3 Cathartics are irritating to the intestinal mucosa, and their regular administration promotes dependence.

609. ① Red meat can react with reagents used in the test to cause false-positive results. (2) (CI; AS)

2 This may apply for testing for ova and parasites, but not for occult blood.

3 If the correct procedure is followed, discarding the first specimen is unnecessary.

4 Random stool testing can be done but must be on three different bowel movements during the screening period.

610. ③ These characteristics describe steatorrhea, which results from impaired fat digestion. (3) (DA; AS)

1 This is descriptive of stools resulting from constipation.

2 This is descriptive of acholic stools occurring with biliary obstruction and resulting from an absence of urobilin.

4 This would be descriptive of upper and lower gastrointestinal bleeding.

611. ④ Nutrients by intravenous route eliminates pancreatic stimulation, therefore reducing the pain experienced in pancreatitis. (1) (PT; IM)

1 TPN meets the client's needs, not the nurses' needs.

2 TPN creates many safety risks for the client.

3 Hunger can be experienced with total parenteral nutrition therapy.

612. ③ The solution is hyperosmolar and a very concentrated source of glucose; too rapid infusion can cause a shift of fluid into the intravascular compartment, resulting in overload; an infusion device should be used as an added precaution. (3) (PT; PL)

1 Although important, it is not the priority.

2 Same as answer 1.

4 Same as answer 1.

613. ④ The less disruptive the procedure, the greater the acceptance by the client. (2) (PT; PL)

1 Most often, total parenteral nutrition is set up to run daily during sleeping hours.

2 Depending on the type of circulatory access used, it may not need to be changed for weeks.

3 The client or a significant other can be taught the principles of administration.

614. ② The *Salmonella* organism thrives in warm, moist environments; washing, cooking, and refrigeration of food limits the growth of or eliminates the organism. (1) (HI; AS)

1 Salmonellosis is unrelated to cancer.

3 Salmonellosis is caused by the *Salmonella* organism, not stress.

4 The *Salmonella* organism is ingested; it is not an airborne or blood-borne infection.

615. ② This medical aseptic technique reduces the possibility of transmitting the organisms to others. (2) (SI; PL)

1 Contact precautions for adults can be handled in a semiprivate room as long as fecally contaminated articles are disposed of appropriately.

3 The type of exposure, not the length of exposure, increases the risk of transmission; visitors are allowed, as long as those having close contact wear gowns and gloves.

4 The organism is not transmitted via the airborne route.

616. ③ The *Salmonella* bacilli can be visualized via microscopic examination. (1) (HI; AS)

1 Although this test might be done, it is not definitive for the diagnosis of salmonellosis.

2 Same as answer 1.

4 Same as answer 1.

617. (2) Fluids of dextrose and normal saline and electrolytes are administered to prevent profound dehydration caused by an excessive loss of water and electrolytes through diarrheal output. (3) (DA; PL)
 1 These are not necessary; salmonellosis is an infection, not a condition caused by hyperacidity.
 3 These are not used when there is a possibility of bacterial infection because slowed peristalsis decreases excretion of the *Salmonella* organism.
 4 Same as answer 3.

618. (1) A cold environment limits growth of microorganisms. (1) (CC; EV)
 2 All food should be refrigerated before and after it is cooked to limit the growth of microorganisms.
 3 This promotes the growth of microorganisms because the stuffing would still be warm for a period before the refrigerator's cold environment cooled the center of the bird; poultry should be stuffed immediately before being cooked.
 4 This promotes the growth of microorganisms because they thrive in warm, moist environments.

619. (1) The decreased hemoglobin and hematocrit levels and RBC count may be a result of malnutrition; also, cancer of the esophagus can cause dysphagia and anorexia. (3) (PC; AN)
 2 There are no data given related to airway obstruction or metastasis.
 3 There are no data given related to pressure on surrounding structures producing pain.
 4 There are no data given related to airway obstruction or injury.

620. (2) Because of the trauma of surgery and the proximity of the esophagus to the trachea, respiratory assessments become the priority. (1) (DA; EV)
 1 Although this is important, an adequate airway is the priority.
 3 Same as answer 1.
 4 Same as answer 1.

621. (3) An increased heart rate is related to an autonomic nervous system response; pain is related to the trauma of the perforation and possibly gastric reflux. (3) (CI; EV)
 1 These are signs of the dumping syndrome.
 2 Same as answer 1.
 4 An increased blood pressure may occur, but increased output has no relationship to esophageal perforation.

622. (3) Weight reduction decreases intraabdominal pressure, thereby decreasing the tendency to reflux into the esophagus. (3) (CC; IM)
 1 Fats decrease emptying of the stomach, extending the time period that reflux can occur; fats should be decreased.
 2 This increases the pressure against the diaphragmatic hernia, increasing symptoms.
 4 This would increase pressure; fluid should be discouraged with meals.

623. (1) Classic symptoms include gnawing, boring, or dull pain located in the midepigastrium or back; pain is caused by irritability and erosion of the mucosal lining. (2) (HI; AS)
 2 This type of pain is more characteristic of the complication of a perforated ulcer.
 3 This type of pain is more characteristic of a hiatal hernia.
 4 This type of pain is more characteristic of cholecystitis.

624. (1) Pain occurs after the stomach empties; eating stimulates gastric secretions, which act on the gastric mucosa of an empty stomach, causing gnawing pain. (2) (DA; AS)
 2 Vomiting temporarily alleviates pain because acid secretions are removed.
 3 There is no intolerance of fats; eating generally alleviates pain.
 4 Pain is sharply localized in the epigastrium; it only radiates to the abdomen if the ulcer has perforated.

625. (4) Some medications, such as aspirin and prednisone, irritate the stomach lining and may cause bleeding with prolonged use. (2) (HI; AS)
 1 This may be related to intestinal irritation, causing diarrhea and intestinal bleeding, not gastric bleeding.
 2 This is not the cause of gastric bleeding; it is important to ascertain dietary habits when teaching about dietary therapy.
 3 Although stress may play a part, the use of some medications has a more direct relationship.

626. (3) Presence of food in the stomach at regular intervals interacts with HCl, limiting acid mucosal irritation. (1) (PC; PL)
 1 The plan should be specific to try to keep food in the stomach at close intervals to limit mucosal irritation.
 2 Food will relieve the pain.
 4 Small, frequent meals or meals with planned snacks would be most appropriate; limiting intake to three large meals would leave the stomach empty for long periods.

627. (2) These are classic indicators of perforated ulcer, for which immediate surgery is indicated; this should be anticipated. (2) (SI; PL)
1 Tachycardia and tachypnea are related to pain and possible blood loss; keeping the client npo is the priority.
3 The symptoms are more indicative of perforation than shock.
4 Black, tarry stools or red stools indicate bleeding, not perforation.

628. (3) Without an adequate stomach reservoir, the hypertonic concentrated food mass "dumps" into the small intestine, drawing fluid from surrounding blood and tissue and causing hypovolemia and typical shock symptoms. (2) (CI; AN)
1 The opposite is true; the food passes too quickly into the small intestine.
2 The opposite is true; the food mass is more concentrated (hypertonic).
4 Same as answer 2.

629. (1) Statistics demonstrate that these are the most likely sites for metastasis of this tumor. (2) (HI; AN)
2 It is less likely that the tumor will spread to these areas.
3 Same as answer 2.
4 These are routes of metastasis, not sites.

630. (3) Large amounts of blood or excessive bloody drainage 12 hours postoperatively must be reported immediately because the client is hemorrhaging. (2) (CI; EV)
1 This must be ordered by the physician; this is not an independent function of the nurse; although ice might constrict vessels, 30 ml would not be effective; many physicians now order room temperature normal saline lavages to prevent lowering core body temperature.
2 This is contraindicated; accumulation of secretions would cause pressure on the suture line; this prevents further observation of drainage.
4 This is an unsafe intervention at this time; the physician should be notified.

631. (4) Removal of the fundus of the stomach destroys the parietal cells that secrete intrinsic factor (needed to complex with vitamin B_{12} as a preliminary to its absorption in the ileum). (3) (HI; AS)
1 Hemorrhaging may cause anemia; however, pernicious anemia occurs when intrinsic factor is not produced.
2 Diabetes has no effect on intrinsic factor.

3 Dietary intake has no effect on the production of intrinsic factor.

632. (1) Intrinsic factor is lost with removal of the stomach, and vitamin B_{12} is needed to maintain the hemoglobin level. (2) (CC; PL)
2 Adequate diet, fluid intake, and exercise should prevent constipation.
3 This would not be a routine expectation.
4 Pancreatic enzymes should be normal because surgery has not altered this function.

633. (1) The stomach's capacity is reduced because of surgery; smaller, frequent feedings with controlled bulk prevent the dumping syndrome. (1) (CC; EV)
2 This is not usually necessary; a diet of unblenderized foods is the goal.
3 This is not realistic.
4 The diet should be high in protein, low in carbohydrates, and normal in fats.

634. (4) This cancer is usually asymptomatic in the early stages; the stomach accommodates the mass. (1) (CC; AN)
1 This is untrue; it can be accurately diagnosed by gastric washings or biopsy.
2 This is untrue; this cancer is painless in its early stages.
3 This is untrue; this is typical of Hodgkin's disease, not gastric carcinoma.

635. (1) The thoracic cavity is usually entered for a complete resection, necessitating a chest tube. (2) (CC; PL)
2 Fluids are contraindicated until the suture line has healed and nasogastric suction is no longer being used.
3 The client would ambulate early to minimize the hazards of immobility.
4 There would be no physiologic necessity for this position.

636. (2) Taking fluids with meals causes rapid emptying of the food from the stomach into the jejunum before it has been adequately subjected to the digestive process; the hyperosmolar mixture causes a fluid shift to the jejunum. (1) (CI; IM)
1 Rest, not activity, after meals will assist in relieving dumping syndrome.
3 Small meals with low carbohydrate, moderate fat, and high protein are recommended; these are more readily digested and prevent rapid stomach emptying.
4 Low-carbohydrate meals are recommended to reduce the osmolarity of chyme as it empties into the jejunum.

637. (2) As a result of the trauma of surgery, some bleeding can be expected for 4 to 5 hours. (3) (CI; EV)
1 Clamping the tube would cause increased pressure on the gastric sutures from a buildup of gas and fluid.
3 Iced saline is rarely used because it causes vasoconstriction, local ischemia, and a reduction in body temperature.
4 This is not necessary; this is a normal occurrence.

638. (2) Rigidity and pain are hallmarks of bleeding from the suture line and/or of peritonitis; vital signs provide supporting data. (1) (CI; AS)
1 Ambulation would be indicated if pain were the result of flatulence; however, rigidity is clearly associated with bleeding or peritonitis and more data are needed.
3 An analgesic may mask the symptoms, delaying diagnosis.
4 This is unrelated to the symptoms presented.

639. (1) Bowel sounds and flatulence indicate the return of intestinal peristalsis; peristalsis is necessary for movement of nutrients through the GI tract. (1) (CI; EV)
2 Incisional pain is unrelated to intestinal peristalsis.
3 Hematocrit levels indicate blood loss; they are unaffected by GI functioning.
4 Dumping syndrome occurs after, not before, the ingestion of food and would not be an indication that the client was ready to ingest food.

640. (4) Over-the-counter antacid preparations are taken to neutralize gastric acid and relieve pain. (2) (CC; EV)
1 Although eating food initially prevents the gastric acids from irritating the gastric walls, it can precipitate acid production and weight gain.
2 This is contraindicated; aspirin irritates gastric mucosa and promotes bleeding by preventing platelet aggregation.
3 Reduction of fluids with meals does not affect pain; it does help control symptoms of dumping syndrome.

641. (4) Dark brown or black stools (melena) could indicate gastrointestinal bleeding. (1) (HI; AS)
1 Frothy stools are indicative of poor fat absorption and are associated with sprue.
2 Ribbon-shaped stools indicate a bowel mass or obstruction.
3 Clay-colored stools are usually related to problems causing a decrease in bile.

642. (1) Nausea with a gastric tube to decompression in place could mean tube displacement or obstruction; checking placement can determine whether it is in the stomach; once placement is verified, then fluid can be instilled to ensure patency. (3) (SI; IM)
2 This may relieve the discomfort but will not determine the cause.
3 The tube should never be irrigated because it would be too traumatic; if the tube is displaced, it could be in the trachea or bronchi and instillation of fluid would cause respiratory impairment.
4 The nurse should always assess a situation carefully before notifying the physician.

643. (1) Abdominal distention, nausea, and abdominal pain can be signs of nasogastric tube blockage. (2) (CI; IM)
2 Although narcotics are usually ordered postoperatively, they tend to decrease peristalsis and may increase abdominal distention and nausea.
3 There will be no stools for several days; gastric drainage may contain some blood from surgery.
4 No bowel sounds are expected for several days after stomach or intestinal surgery.

644. (3) Some bright blood at this point would be a normal finding that should be monitored; large amounts of blood or bleeding should be reported immediately. (2) (CI; EV)
1 This is contraindicated; secretions would accumulate and cause pressure on the suture line; this prevents observation of drainage.
2 If the tube is draining, there is no need to irrigate; also, irrigations are too traumatic.
4 Reducing suction would allow secretions to accumulate and cause pressure on the suture line.

645. (4) Symptoms characteristic of the dumping syndrome can be lessened with small, frequent meals that are low in carbohydrates and, therefore, not as hyperosmolar. (2) (CC; EV)
1 Fluids during meals should be limited; increased fluid intake speeds the emptying time of the stomach.
2 The client may need to avoid these foods to minimize irritating the gastric mucosa, but they are not the cause of the dumping syndrome.
3 Resting after meals might mitigate the symptoms of the dumping syndrome, but resting before meals will not prevent them; resting is nonspecific; the client should be lying down to slow the emptying of the stomach.

646. (4) The rapid absorption of sugars from the food mass causes elevation of blood sugar, and the insulin response often causes transient hypoglycemic symptoms. (2) (CI; AN)

1 This is unrelated; usually a bland diet is prescribed following a gastrectomy.

2 The response is rebound hypoglycemia, not hyperglycemia.

3 The insulin-adjusting mechanism is not overwhelmed but responds vigorously, causing rebound hypoglycemia.

647. (1) All these characteristics are well-established risk factors for gallbladder disease (female, overweight, and older than 40). (1) (HI; AS)

2 None of these factors is correctly associated with cholecystitis.

3 The age is correct, but clients are usually female and have an elevated serum cholesterol level.

4 Risk factors include being a female and having a family history of gallstones, but clients are usually older than 40.

648. (1) Pain is a cardinal symptom; it is helpful to have as much specific information about it as possible, particularly its description and its relationship to foods ingested. (2) (CC; AS)

2 It is not necessary to save all urine and stools, although changes in color should be reported.

3 The client should be free to question orders that are not understood or agreed with.

4 This would not add any valuable information.

649. (4) Cholecystitis is often accompanied by intolerance to fatty foods, including fried foods and butter. (2) (PC; AS)

1 These cause flatulence and pain for clients with lower intestinal problems such as diverticulosis.

2 These foods contain less fat than do fried foods or butter.

3 Neither chocolate nor boiled seafood has as much fat as fried chicken or butter.

650. (3) Local fat-stimulated duodenal glands dispatch the hormone, cholecystokinin, that signals the gallbladder to contract and release bile. (1) (PC; AN)

1 Soft-textured foods are unnecessary.

2 This would only be necessary if the cholecystitis were precipitated by cholelithiasis and the stones were composed of cholesterol.

4 This diet would be ordered after a cholecystectomy.

651. (4) During the procedure, a sedative is administered as required intravenously to help the client stay calm. (3) (CC; IM)

1 This is not used during this procedure.

2 Same as answer 1.

3 Same as answer 1.

652. (1) Bile promotes the absorption of the fat-soluble vitamins; an obstruction of the common bile duct limits the flow of bile to the duodenum. (3) (PC; AS)

2 This is not relevant to the situation.

3 This is unnecessary; protein would be desirable for wound healing.

4 These would be unexpected.

653. (3) Self-splinting results in shallow breathing, which does not aerate the lungs adequately, particularly the lower right lobe. (1) (CI; PL)

1 The T-tube is never irrigated; it drains by gravity until the edema in the operative area subsides; the tube is then removed by the physician.

2 The dressing is not changed by the nurse in the immediate postoperative period; the client's respiratory status takes first priority.

4 The client would be npo immediately after surgery.

654. (4) The inflammatory response occurs because of trauma of surgery; the T tube maintains patency of the common bile duct until edema subsides. (2) (CI; AN)

1 It diverts the bile out of the body into a collection bag.

2 This is the purpose of the portable wound drainage system, not of a T tube.

3 Surgical asepsis prevents infection at the site of the incision; the T tube is a portal of entry for microorganisms and places the client at risk for infection.

655. (4) Recovery will be quicker because there is no large abdominal incision. (2) (CC; IM)

1 Clear liquids may be started as soon as the client is awake and a gag reflex has returned.

2 With a laparoscopic cholecystectomy there will be several puncture wounds, not an incision, on the abdomen.

3 Carbonated beverages will create gas, which will distend the intestines and increase pain.

656. (4) Slight hip flexion reduces tension on the abdominal musculature and the operative site. (1) (PC; PL)

1 This position places pressure on the abdominal wall, causing pain.

2 This position causes tension on the abdominal musculature and operative site.

3 Same as answer 1.

657. (1) Vitamin K, a precursor for prothrombin, cannot be absorbed without bile. (2) (HI; AS)
2 This is commonly related to electrolyte imbalances, not fat-soluble vitamin deficiency.
3 Jaundice results from a backup of bile, not a deficiency of fat-soluble vitamins.
4 This may be related to electrolyte imbalances or deficiency of B vitamins, which are water soluble.

658. (1) Vitamin K, a fat-soluble vitamin, is not absorbed from the GI tract in the absence of bile. Bile enters the duodenum via the common bile duct. (1) (CI; AN)
2 Calcium is related to rhythmic muscle contraction, not coagulation.
3 The extrinsic factor (cyanocobalamin) is a water-soluble vitamin; bile is not necessary for its absorption.
4 Bilirubin is formed by the breakdown of hemoglobin and RBCs and is not related to coagulation.

659. (1) Biliary colic may occur in the postoperative period as a result of the passage of pulverized fragments of the calculi; this may occur 3 or more days after the lithotripsy. (3) (PT; IM)
2 Fever would indicate pancreatitis, which is a rare occurrence.
3 The delivery of shock waves during the procedure is synchronized with the heart beat to avoid initiation of dysrhythmias.
4 Light sedation may be used to keep the client comfortable and as still as possible.

660. (4) Mild shoulder pain is common up to 1 week after surgery because of diaphragmatic irritation secondary to abdominal stretching or residual carbon dioxide that was used to inflate the abdominal cavity during surgery. (3) (CC; EV)
1 This is not necessary; the bandages are removed the second day postoperatively.
2 This is not necessary; clients generally tolerate food after 24 to 48 hours.
3 This is not necessary; the client may bathe and shower as usual.

661. (3) It may take 4 to 6 months, but ultimately most people can eat anything they want. (2) (CC; EV)
1 Although fats may have to be gradually reintroduced, most people can tolerate them after a cholecystectomy.
2 Foods that caused gastric distress before surgery are usually tolerated after surgery.
4 Increased protein is only needed until healing has occurred.

662. (2) A patent nasogastric tube prevents distention and compression in the surgical area. (2) (CI; IM)

1 A low-Fowler's position is preferred to limit tension on the abdominal wall; movement should be encouraged.
3 Replacement of vitamins is a dependent function; vitamins must be ordered by the physician.
4 Tube feedings are contraindicated because peristalsis is absent and the feeding would place pressure on the suture line.

663. (4) The Whipple procedure leads to malabsorption because of impaired delivery of bile to the intestine; fat metabolism is interfered with, causing dyspepsia. (3) (PC; IM)
1 These clients are anorexic, require small frequent meals, and should eat high-calorie, high-protein, low-fat diets.
2 High-calorie meals are needed for energy and to promote use of protein for tissue repair.
3 High protein is required for tissue building; there is no problem with the liver in clients with cancer of the pancreas unless direct extension occurs.

664. (4) Polydipsia is characteristic of hypoinsulinism (diabetes mellitus) because of impaired carbohydrate metabolism. (2) (CC; EV)
1 Polyuria, not oliguria, is characteristic of diabetes mellitus because excess fluid is excreted with the glucose by the kidneys.
2 Increased appetite is characteristic of diabetes mellitus because of impaired metabolism.
3 Weight loss characterizes diabetes mellitus because of the use of body mass as a source of energy.

665. (2) Rapid administration can cause glucose overload, leading to osmotic diuresis and dehydration; slowing the infusion decreases glucose overload. (2) (PT; IM)
1 Stopping the flow would jeopardize the central line; slowing the infusion will give the client's body a chance to handle the excess glucose.
3 Signs of bowel obstruction are not present.
4 The client's headache should disappear with oral fluid replacement; analgesics are not indicated.

666. (4) Professional assistance would ensure correct administration, which may limit complications such as intravascular overload and sepsis. (3) (PT; IM)
1 This is usually done by a health care provider.
2 TPN is usually administered every day.
3 A TPN solution of 1000 to 2000 ml/day is usually administered at a prescribed rate of 50 to 125 ml/hour; the infusion usually runs continuously, not intermittently.

667. (4) Formation of lipase necessary for digestion of fats is an exocrine function; the endocrine function is to secrete insulin, which is a hormone essential in carbohydrate metabolism. (3) (PC; PL)

1 Although it is necessary to avoid alcohol, this is not related to the exocrine functioning of the pancreas.
2 Deficiencies of both may occur because of poor intake, but these deficiencies are not specifically related to exocrine or endocrine pancreatic functioning.
3 Fluid and electrolyte problems are not related specifically to exocrine or endocrine pancreatic functioning.

668. (1) A client's vital signs, especially the pulse and temperature, will rise before the client demonstrates any of the more severe symptoms of withdrawal from alcohol. (3) (CI; PL)
2 This is contraindicated initially because it may cause cerebral edema.
3 This becomes a priority after the problems of the withdrawal period have subsided.
4 This is not a priority until after the detoxification process.

669. (4) Petechiae are evidence of capillary bleeding; the diseased liver is no longer able to metabolize vitamin K, which is necessary to activate blood clotting factors. (1) (PC; AN)
1 Although bile is synthesized and secreted in the liver, bile salts are not involved in the clotting process.
2 Folic acid is stored in the liver but not involved in the clotting process.
3 Vitamin A is not involved in clotting, even though the transformation of carotene to vitamin A takes place in the liver.

670. (4) This keeps the bladder in the pelvic area and prevents puncture when the abdominal cavity is entered. (1) (CI; IM)
1 This is not necessary.
2 This is unsafe; the bladder will rise into the abdominal cavity and may be punctured.
3 Same as answer 1.

671. (1) Hepatitis A is primarily spread via a fecal-oral route; sewage-polluted water may harbor the virus. (3) (HI; AS)
2 This does not increase the risk of developing the disease but will increase the risk of an infected individual spreading the disease to others.
3 Hepatitis types B, C, and D are more often spread via the blood-borne route; using disposable equipment and proper handling of syringes decreases the risk of spreading the virus.
4 Exposure to arsenic or carbon tetrachloride can cause toxic hepatitis, which is not communicable.

672. (4) Clay-colored stools are indicative of hepatic obstruction and deterioration of the condition. (2) (CC; PL)

1 This is a characteristic of hepatitis from the onset of symptoms, and it need not be reported after discharge.
2 Same as answer 1.
3 This is normal; it need not be reported after discharge.

673. (2) Damage to liver cells affects the ability to remove bilirubin from the blood, with resulting deposition in the skin and sclera. (2) (DA; AN)
1 With hepatitis, the liver does not secrete excess bile.
3 There is no increased destruction of red blood cells in hepatitis.
4 This is unrelated; decreased prothrombin levels cause spontaneous bleeding, not jaundice.

674. (4) High carbohydrates provide calories for energy, high proteins provide for tissue repair, and fats are permitted as tolerated. (2) (PC; PL)
1 High carbohydrates are needed for body fuel, otherwise proteins will have to be used for metabolism rather than tissue repair; proteins will be high, and fats permitted as tolerated.
2 This diet does not offer adequate proteins, which are required for tissue repair in the client with hepatitis; carbohydrates will be high, and fats are permitted as tolerated.
3 Same as answer 2.

675. (4) Hepatitis A is spread via the fecal-oral route; transmission is interfered with by proper hand washing. (2) (SI; IM)
1 This may increase transmission because no provision is made for the cleaning of equipment or disposal of contaminated wastes.
2 This is inadequate; transmission is still possible via the roll of toilet tissue unless hand washing is done.
3 This is untrue; hepatitis can be transmitted via the fecal-oral route and precautions, such as hand washing, must be taken.

676. (4) Hepatitis A is transmitted via the fecal-oral route; contact precautions must be used when there are articles that have potential fecal and/or urine contamination. (2) (SI; PL)
1 Neither a private room nor a closed door is required; these are necessary only for respiratory (airborne) infections.
2 Hepatitis A is not transferred via the airborne route and therefore a mask is not necessary; a gown and gloves are required only when handling articles that may be contaminated.
3 This is too limited; a gown and gloves should also be worn when handling other fecally contaminated articles, such as a bedpan or rectal thermometer.

677. (4) This is a vascular lesion associated with cirrhosis; it is thought to be related to elevated estrogen levels. (3) (DA; AS)
    1 This refers to patches of depigmentation resulting from destruction of melanocytes.
    2 This is excessive growth of hair; in cirrhosis, endocrine disturbances result in loss of axillary and pubic hair.
    3 This word does not exist in medical or nursing terminology.

678. (1) Pressure applied to the puncture site compresses blood vessels, thereby preventing bleeding. (1) (CI; IM)
    2 The right side-lying position will not restore circulating blood volume.
    3 The semi- or high-Fowler's positions would be more comfortable because they decrease pressure on the diaphragm.
    4 This position will have little or no effect on biliary ducts.

679. (3) The flow of bile through the puncture site indicates a biliary vessel was punctured; this is a common complication following a liver biopsy. (2) (CI; AN)
    1 Fluid would leak through the puncture site or into the peritoneum, not into the intestine.
    2 The pancreas does not contain bile; also it is in the upper left quadrant, not the upper right quadrant.
    4 This is a complication, not an expected outcome.

680. (2) An increased serum ammonia level impairs the CNS, causing an altered level of consciousness. (2) (DA; PL)
    1 Rising ammonia levels are not related to weight; client safety is the priority.
    3 The priority is to protect the client; rising ammonia levels will precipitate hepatic encephalopathy.
    4 An alteration in the fluid intake will not affect the serum ammonia level.

681. (1) Measures must be taken immediately to ensure client safety. (2) (SI; PL)
    2 Although the physician should be notified, the nurse should first take measures to ensure client safety.
    3 This is contraindicated because sedatives mask the progressive signs of hepatic encephalopathy.
    4 Hepatic encephalopathy is caused by high serum ammonia levels, not hypoxia.

682. (1) The symptoms indicate hepatic coma. Protein is reduced according to tolerance, and calories are increased to prevent tissue catabolism. (1) (PC; PL)
    2 This represents a high-protein diet, which is contraindicated in hepatic coma.
    3 Same as answer 2.
    4 Same as answer 2.

683. (3) With increased intraabdominal pressure, the abdominal wall will become rigid and tender. (2) (CI; EV)
    1 The pulse rate will increase in an effort to compensate for impending septic shock.
    2 Hypovolemia and therefore hypotension result because of a loss of fluid, electrolytes, and protein into the peritoneal cavity.
    4 Peristalsis and associated bowel sounds will decrease or be absent in the presence of increased intraabdominal pressure.

684. (3) This position promotes localization of purulent material and inflammation and prevents an ascending infection. (1) (SI; IM)
    1 The risk of an ascending infection may be increased in this position because it allows fluid in the abdominal cavity to bathe the entire peritoneum.
    2 Same as answer 1.
    4 The client would probably prefer a Sims' position, which increases risk of an ascending infection.

685. (2) Medical treatment is directed toward reducing motility of the inflamed bowel, restoring nutrition, and treating and preventing infection; surgery is used selectively for those who are acutely ill or have excessive exacerbations. (2) (CC; IM)
    1 This is untrue; medical treatment is symptomatic, not curative.
    3 It is usually performed as a last resort.
    4 Although there is an emotional component, the physiologic adaptations determine whether surgery will be necessary.

686. (2) The client's feelings, knowledge, and skills concerning the ileostomy must be assessed before discharge. (1) (CC; EV)
    1 People should not be pressured into performing self-care before they are physically and emotionally ready.
    3 After an ileostomy the client is usually encouraged to eat a high-protein diet or a regular diet with supplemental protein; a high-fluid intake should be maintained.
    4 Often, the client no longer needs a dressing on the incision at the time of discharge; a collection pouch is used over the stoma.

687. (3) Constipation, diarrhea, and/or constipation alternating with diarrhea are the most common symptoms of colorectal cancer. (3) (HI; AS)

1 Pain is reported as a symptom in less than 25% of clients; also it is a late sign after other organs are invaded, intestinal obstruction occurs, or tissue necrosis develops.

2 This is the second most common complaint that results from destruction of the epithelial lining of the intestine.

4 This is a later sign that only becomes evident when the lumen of the intestine narrows as a result of the enlarging mass.

688. (4) This diet is low in fiber; after digestion and absorption there is only a small amount of residue to be eliminated. (1) (SI; AN)

1 This diet does not influence the flora of the intestine; antimicrobials, such as neomycin, are given to do this.

2 This diet does not promote peristalsis; the products of digestion remain in the intestine longer, and flatus is increased.

3 Although a low-residue diet is less irritating, this is not the primary reason for its use before surgery.

689. (1) When intestinal continuity cannot be restored after removal of the anus, rectum, and adjacent colon, a permanent colostomy is formed. (2) (CC; PL)

2 This segment of colon lies on the right side of the abdomen and has no anatomic proximity to the rectum.

3 This procedure is performed to allow a segment of colon to heal; intestinal continuity can be restored.

4 This procedure is commonly performed for inflammation of the colon when intestinal continuity can be restored.

690. (2) Irrigations regulate the bowel to function at a specific time for the convenience of the client. (2) (PC; AN)

1 Although irrigations will prevent straining, this is not the purpose of the irrigation.

3 Irrigations will facilitate expulsion of flatus but will not decrease the amount; avoidance of gas-forming foods will accomplish this.

4 This is not the function of the irrigation; most ingested fluid already has been absorbed in the large intestine by the time it reaches the sigmoid.

691. (4) The left or right side-lying position puts the least strain or pressure on the perineal suture line. (2) (SI; IM)

1 This position is difficult to maintain and would place stress on the suture line.

2 Flexion of one hip and knee would increase tension on the perineal suture line; depending on placement of the stoma, one of the Sims' positions would result in the client lying on the new colostomy, which would be traumatic.

3 This position places undue stress on the suture line and is the most uncomfortable position.

692. (2) The surface of a stoma is mucous membrane and should be dark pink to red, moist, and shiny; the stoma is usually raised beyond the skin surface. (1) (CI; EV)

1 The stoma should be moist, not dry. Pale pink indicates a low hemoglobin level. Although some stomas can be flush with the skin, a raised stoma is more common.

3 The stoma should be moist, not dry. A purple color indicates compromised circulation. A depressed stoma is prolapsed and abnormal.

4 Although the stoma should be moist and dark pink to red, it should not be flush with the skin or painful.

693. (2) There is no need to irrigate before the stool is formed. (2) (PC; PL)

1 If proper technique is used, fecal elimination should flow through the sleeve of the colostomy bag to the commode.

3 This is premature; the stool is not yet formed.

4 The perineal wound may take weeks to heal, and irrigations must be started when the stool is formed.

694. (3) This is caused by the trauma of intestinal manipulation and the depressive effects of anesthetics and analgesics. (1) (CI; AN)

1 Edema would not totally interfere with peristalsis, which may be less effective but would still result in some output.

2 Any ingested food or fluid initiates the gastrocolic reflex and would therefore result in some output.

4 A nasogastric tube decompresses the stomach; it does not influence intestinal motility at this time.

695. (4) Irrigation should be performed at the time the client normally defecated before the colostomy to maintain continuity in lifestyle. (2) (PC; PL)

1 This may be true for some people; however, irrigations usually can be completed in 1 hour.

2 Most people defecate after breakfast because the ingestion of food on an empty stomach initiates the gastrocolic reflex.

3 An irrigation cannot be postponed until the client accepts the altered body image because this may take weeks or months.

696. (4) The irrigation bag should be hung no higher than 12 to 18 inches above the level of the stoma, and a clothes hook is too high. (1) (CC; EV)
1 Fluid flowing into the intestines can cause distention and discomfort; clamping the tubing is an appropriate intervention.
2 The tip of the catheter should be lubricated to prevent trauma to mucosal tissue and to facilitate insertion.
3 There is not enough information provided to choose this response; the amount of fluid ordered is not included.

697. (4) This occurs with stenosis of the stoma; forcing insertion of the tube could cause injury. (2) (CC; EV)
1 This is an expected response.
2 Same as answer 1.
3 This is expected; feces and flatus accompany fluid expulsion.

698. (4) Stoma adhesive protects the skin and helps to keep the bag attached to the skin. (2) (PC; IM)
1 The bag should be emptied when it is one-third to one-half full.
2 This is too much space between the stoma and the bag; the enzymes in feces can erode the skin.
3 Initially the nurse should change the bag; self-care is usually more gradual.

699. (3) These clients can eat a regular diet; only gas-forming foods that cause distention and discomfort should be avoided. (1) (CC; EV)
1 The amount of stool does not have to be limited; therefore a low-residue diet is not necessary.
2 The affected tissue has been removed, and normal mucosal tissue lines the intestine and forms the stoma; therefore bland foods are not necessary.
4 Nutrients are absorbed by the small, not the large, intestine; a regular diet is usually easily digested and absorbed.

700. (3) Pain is wavelike, colicky, and sharp because of obstruction and localized bowel ischemia. (2) (DA; AS)
1 Flatus would be impeded by strangulation.
2 Vomiting is persistent, not projectile.
4 This is not an early sign of obstruction; decreased bowel sounds occur after gas and fluid accumulate.

701. (3) Because of pain and the proximity of the operative site to the lower urinary tract, voiding problems are common. (3) (CI; EV)

1 This is not a complication of herniorrhaphy.
2 The abdomen was not entered, and there should be no interference with peristalsis.
4 This should not be a complication of herniorrhaphy because early ambulation is permitted.

DRUG-RELATED RESPONSES (DR)

702. (4) First convert milligrams to micrograms and then use ratio and proportion (0.2 mg = 200 mcg)
200 mcg/100 mcg = X ml/1 ml
100 X = 200
X = 2 ml   (2) (PT; AN)
1 This is an inaccurate calculation; this is less than the desired dose.
2 Same as answer 1.
3 Same as answer 1.

703. (4) This is the correct flow rate; 500 mg of the drug added to 100 ml of IV fluid produces a solution in which 1 ml contains 5 mg of the drug; the required amount of the drug is contained in 0.3 ml of fluid; to calculate the dosage to be delivered, multiply the amount to be delivered (0.3 ml) by the drop factor (60) and divide the result by the amount of time in minutes (1 minute). (2) (PT; AN)
1 This is too slow; this would not deliver the ordered amount of lidocaine HCl.
2 Same as answer 1.
3 Same as answer 1.

704. (3) Use ratio and proportion to calculate:
375 mg /225 mg = X ml/1 ml
225 X = 375
X = 1.68 or 1.7 ml
(3) (PT; AN)
1 This is too little; it would not provide enough drug.
2 Same as answer 1.
3 This is too much; it would provide an excess of drug.

705. (1) 250 mg is equal to 0.25 g; therefore,
4 g/0.25 g = X ml/1 ml
0.25 X = 4
X = 16 ml   (2) (PT; AN)
2 This is an inaccurate calculation; it is too little to provide the ordered amount.
3 Same as answer 2.
4 Same as answer 2.

706. (3) Use ratio and proportion:
250 mg = 0.25 g
0.25 g/1 g = X ml/3 ml
X = 0.75 ml or 0.8 ml   (2) (PT; AN)

1 This is an incorrect calculation; too little medication will be administered.

2 Same as answer 1.

4 This is an incorrect calculation; too much medication will be administered.

707. (2) Use ratio and proportion:

$$2,450,000 \text{ U}/300,000 \text{ U} = X \text{ ml}/1 \text{ ml}$$
$$300,000 \text{ X} = 2,450,000$$
$$X = 8.2 \text{ ml}$$

(2) (PT; AN)

1 This would deliver more than the ordered amount.

3 This would deliver less than the ordered amount.

4 Same as answer 3.

708. (2) 1000 mcg = 1 mg; 500 mg of the drug added to 500 ml of IV fluid results in a solution in which 1 ml contains 1 mg of the drug; the physician's order is for 1 mg/minute, therefore, multiply total solution (1 ml) times the drop factor (60) and divide by total time in minutes (1 minute). (2) (PT; AN)

1 This is an incorrect calculation; this would deliver half the desired dose.

3 This is an incorrect calculation; this would deliver 50% more than the desired dose.

4 This is an incorrect calculation; this would deliver twice the desired dose.

709. (2) These signs as well as anemia, irritability, pruritus, and an enlarged spleen are some of the signs of vitamin A toxicity. (3) (CI; AS)

1 Excess thiamin is excreted in the urine and rarely, if ever, causes toxicity; excessive doses may elicit an allergic reaction in some individuals.

3 Excess vitamin C (ascorbic acid) does not cause these signs or toxicity; however, vitamin C may cause diarrhea or renal calculi.

4 Pyridoxine (vitamin B_6) is relatively nontoxic, and excess amounts are excreted in the urine.

710. (2) Organism mutation commonly results in drug resistance when treatment is inadequate. (2) (PT; IM)

1 Rifampin decreases tubercle bacillus' replication; pyridoxine (vitamin B_6) is used to prevent neuropathy associated with INH.

3 High concentrations of at least two antitubercular drugs are necessary for an extended period.

4 This may raise anxiety and may not be true; combination drug therapy is always used for tuberculosis.

711. (4) Rifampin causes the body fluids, such as sweat, tears, and urine, to turn orange. (2) (PT; EV)

1 Damage to the eighth cranial nerve is not a side effect of rifampin; it is a side effect of another drug, streptomycin sulfate, sometimes used to treat tuberculosis.

2 It is not necessary to drink large amounts of fluid with this drug; it is not nephrotoxic.

3 This is not a side effect of rifampin.

712. (4) One of the most common side effects of INH is peripheral neuritis, and vitamin B_6 will counteract this problem. (2) (PT; AN)

1 It does help nutrition, but that is not the specific reason it is given.

2 It counters the side effects of INH; it does not act to enhance its action.

3 It does not speed the destruction of the causative organism.

713. (1) This medication causes hyperuricemia, leading to joint swelling and pain; fluids dilute the urine and help remove the uric acid. (2) (CI; EV)

2 This medication causes GI irritation and should be taken with food.

3 This is not a side effect of this medication.

4 This is a side effect of rifampin (Rifadin), not pyrazinamide.

714. (1) Antitubercular drugs such as isoniazid (INH), rifampin (Rifadin), and aminosalicylate sodium are hepatotoxic. (3) (CI; AS)

2 These are not related to the administration of antitubercular drugs or to their side effects.

3 Same as answer 2.

4 The white blood cell count is expected to be higher in the presence of infection, but with AIDS the WBC count will be less than $2500/cm^3$ and helper T cells will be about $400/mm^3$; the T4/T8 ratio will be 1:2; these tests will not provide information relative to starting antitubercular therapy or to its side effects.

715. (4) AZT can cause anemia, leukopenia, and granulocytopenia; these blood dyscrasias can be life threatening, so the CBC is monitored. (2) (CI; EV)

1 These are not directly affected by the drug.

2 Same as answer 1.

3 Once infected, the client will continue to test positive for the antibody.

716. ④ Pentamidine can cause either hypoglycemia or hyperglycemia even after therapy is discontinued, and therefore blood glucose levels should be monitored. (3) (CI; PL)
    1 Pentamidine should be mixed with injectable sterile water or 5% dextrose injection and then further diluted before IV administration.
    2 This is too quick; the drug should be given over at least 60 minutes.
    3 Clients should be monitored closely for sudden, severe hypotension; clients should lie flat when receiving the drug.

717. ④ This drug relaxes the respiratory muscles; it inhibits transmission of nerve impulses by binding with cholinergic receptor sites and antagonizing the action of acetylcholine; the client will die without mechanical ventilation. (2) (PT; IM)
    1 This is not the priority.
    2 Same as answer 1.
    3 Same as answer 1.

718. ④ Humulin R insulin is the only insulin that acts rapidly and is compatible with intravenous solutions. (1) (PT; AN)
    1 This insulin is not compatible with intravenous solutions; it is an intermediate-acting insulin.
    2 Ultralente insulin is not compatible with intravenous solutions; it is a long-acting insulin.
    3 Same as answer 1.

719. ① Oral hypoglycemics stimulate endogenous insulin production by the beta cells of the pancreas. (1) (PT; IM)
    2 This occurs when serum glucose drops below normal levels.
    3 This occurs in the presence of insulin and potassium.
    4 Beta cells must have some function to enable this drug to be effective.

720. ④ Oral hypoglycemic agents decrease serum glucose levels. (1) (PT; IM)
    1 This occurs with type 1 diabetes.
    2 Weight gain is usually present in type 2 diabetes.
    3 Same as answer 1.

721. ③ Humulin N insulin's onset of action is 1 to 2 hours, peak action is 8 to 12 hours, and duration of action is 18 to 24 hours; if hypoglycemia were to occur, it would happen between 4 pm and 8 pm. (1) (PT; EV)
    1 Humulin R insulin peaks in 2 to 4 hours.
    2 Semilente insulin peaks in 4 to 7 hours.

    4 No insulin peaks in 12 to 14 hours; however, protamine zinc insulin and Ultralente insulin peak in 16 to 18 hours.

722. ① These are the most commonly reported signs of hypoglycemia and are related to increased sympathetic nervous system activity. (2) (PT; EV)
    2 These are signs of hyperglycemia.
    3 Same as answer 2.
    4 Same as answer 2.

723. ③ The chemical structure of insulin is altered by gastric secretions, rendering it ineffective. (2) (PT; IM)
    1 There is no such thing as oral insulin; this comment about the seriousness of the diabetic condition could raise anxiety.
    2 There are no data to support this statement, and insulin is given parenterally, not orally.
    4 Insulin is not absorbed but is destroyed by gastric secretions.

724. ③ This drug does not interfere with thyroxine already stored in the gland; symptoms remain until the hormone is depleted. (3) (PT; PL)
    1 Duration of therapy varies depending on the severity of the disease and the client's response to therapy.
    2 This drug is not irritating to mucosal tissue, and no special precautions are necessary.
    4 Absorption is not affected by the presence of food in the stomach.

725. ② Tapazole blocks thyroid hormone synthesis. The hormone stored in the thyroid gland must be reduced before improvement is seen. (2) (PT; IM)
    1 There are many common side effects that include nausea, vomiting, diarrhea, rash, urticaria, pruritus, alopecia, hyperpigmentation, drowsiness, headache, vertigo, and fever
    3 Tapazole should be spaced at regular intervals because blood levels are reduced in approximately 8 hours.
    4 Large doses cause toxic side effects that can be life threatening, including nephritis, hepatitis, agranulocytosis, leukopenia, thrombocytopenia, hypothrombinemia, and lymphadenopathy.

726. ③ ASA may damage the eighth cranial (acoustic) nerve, causing ringing in the ears and impairing hearing. (3) (PT; EV)
    1 Pain, not ringing in the ears, is a sign of otitis media; ASA toxicity affects the eighth cranial nerve, not the middle ear.
    2 Aging may cause decreasing acuity in the extremes of pitch, but it would not cause ringing in the ears.

4 Diminished hearing, not ringing, occurs because of mechanical obstruction of the outer ear.

727. (2) This is necessary because ibuprofen is nephrotoxic and hepatotoxic and prolongs the bleeding time. (2) (CI; EV)
  1 This is important for all clients with arthritis; it is not related to ibuprofen.
  3 Ibuprofen does not cause postural hypotension.
  4 Ibuprofen causes epigastric distress and occult bleeding; it should be taken with meals or milk to reduce these adverse reactions.

728. (3) Pain is most effectively relieved when the analgesic is administered at the onset of pain, before it becomes intense; this prevents a pain cycle from occurring. (1) (PT; PL)
  1 Analgesics are least effective when administered as pain is at its peak.
  2 This may or may not be necessary; the medication should be taken when the client begins to feel uncomfortable within the parameters specified by the physician's order.
  4 Same as answer 1.

729. (3) These drugs tend to irritate the gastric mucosa and should be taken with food or milk. (2) (CI; EV)
  1 This is an expected side effect; safety precautions are indicated, but the drug should not be discontinued.
  2 These drugs should be taken with food or milk to limit GI irritation.
  4 This could result in toxicity; the dosage should be prescribed by the physician.

730. (3) This provides for the quickest use of the narcotic, so that relief of pain can occur immediately. (3) (PT; PL)
  1 Nausea, vomiting, and paralytic ileus may occur postburn, making oral medication impractical.
  2 This route does not provide uniform absorption; also, relief of pain would be delayed.
  4 The medication may be sequestered in the tissues and, with fluid shifts, it is unknown when the medication will take effect.

731. (2) Anabolic agents are synthetic androgenic steroids, which may produce masculinizing effects in women. (3) (PT; EV)
  1 With an increase in muscle mass and stimulation of erythropoiesis, the client should have an increase in energy.
  3 The client may become hypernatremic, not hyponatremic; the client may become hypercalcemic as well.

4 This drug will not cause hyperglycemia; it may cause hypoglycemia in clients with diabetes mellitus.

732. (1) Sulfamylon is effective against a wide variety of gram-positive and gram-negative organisms including anaerobes. (1) (SI; AN)
  2 This is an antimicrobial, not an analgesic; topical application causes pain.
  3 It promotes healing and decreases the need for grafting.
  4 This medication is an antimicrobial; it does not provide chemical débridement.

733. (1) Silvadene is effective against gram-positive, gram-negative, and fungal organisms; however, it causes agranulocytosis, a serious side effect. (2) (PT; AN)
  2 Sulfamylon is effective against gram-negative and gram-positive organisms; however, it has serious side effects such as acidosis.
  3 Neomycin has extremely serious side effects; it is ototoxic, which can cause hearing loss, and it is nephrotoxic, which can cause kidney failure.
  4 Silver nitrate dressings must be kept moist; if they are allowed to dry, silver salts would precipitate into the wound, causing irritation.

734. (1) Aspirin is given here to prevent platelet aggregation and possible thrombosis and should be given on schedule. (2) (PT; IM)
  2 The ASA is not being given for comfort, and this should not be implied to the client; also, the nurse has not described the purpose of this medication.
  3 The ASA is not being given for comfort and should not be withheld.
  4 Same as answer 2.

735. (3) Sorbitrate dilates the coronary vasculature, improving the supply of oxygen to the hypoxic myocardium. (2) (PT; IM)
  1 This is the action of anticoagulants.
  2 This is the action of antidysrhythmics.
  4 This is the action of cardiac glycosides.

736. (4) Sorbitrate may produce vasodilation, resulting in postural hypotension from sudden changes in position. (2) (SI; PL)
  1 This drug may cause tachycardia, not bradycardia.
  2 The only gastrointestinal complications may be nausea and vomiting.
  3 This drug results in a more efficient cardiac output; therefore the lungs will not have to compensate for inadequate cardiac output by increasing the respiratory rate.

737. (4) Lasix may cause hypovolemia, which can result in orthostatic hypotension with sudden position changes. (1) (SI; EV)
1 Lasix does not cause photophobia.
2 This has no relationship to Lasix.
3 Citrus fruits, particularly oranges, are high in potassium and should be encouraged when the client is taking Lasix because this medication can cause hypokalemia.

738. (3) Lasix enhances the excretion of potassium, producing symptoms of hypokalemia, such as hyporeflexia. (3) (PT; EV)
1 This is not a side effect of Lasix.
2 Same as answer 1.
4 Same as answer 1.

739. (4) This drug decreases blood pressure by both of these mechanisms. (2) (PT; AN)
1 It decreases cardiac workload by decreasing preload and afterload.
2 Actually, this drug may increase the heart rate as a response to vasodilation.
3 It should decrease peripheral resistance by dilating peripheral vessels.

740. (1) Pruritus is a common side effect of clonidine HCl (Catapres). (3) (PT; EV)
2 This drug causes constipation, not diarrhea.
3 This drug may cause depression, anxiety, fatigue, and drowsiness, not euphoria.
4 This is not a side effect of this medication; photosensitivity occurs with chlorpromazine.

741. (2) Inderal and Lanoxin both exert a negative chronotropic effect. (2) (PT; AN)
1 Inderal reduces headaches associated with hypertension.
3 These drugs may cause hypotension.
4 These drugs may depress nodal conduction.

742. (1) Coumadin inhibits vitamin K; therefore vitamin K is the antidote for Coumadin. (2) (SI; PL)
2 This is a blood-clotting factor, not the antidote for Coumadin.
3 Same as answer 2.
4 This is the antidote for heparin.

743. (4) This is the site of action of Lasix. (3) (PT; AN)
1 Thiazides act here.
2 Potassium-sparing diuretics act here.
3 Plasma expanders and xanthines act here.

744. (4) A cardiac glycoside such as digitalis decreases the conduction speed within the myocardium and slows the heart rate. (2) (PT; EV)
1 The primary effect of a diuretic is on the kidneys, not the heart.
2 A vasodilator could cause tachycardia, not bradycardia, which is an adverse effect.
3 This does not drastically reduce the heart rate.

745. (4) The most common adverse effect of a tissue plasminogen activator is bleeding because of the thrombolytic action of the drug. (3) (PT; PL)
1 Although this is important for any client with a decreased cardiac output, it is not specific to the administration of a tissue plasminogen activator.
2 Same as answer 1.
3 Same as answer 1.

746. (4) The major action of intravenous nitroglycerin is venous and then arterial dilation, leading to a decrease in blood pressure. (2) (PT; EV)
1 Not a common side effect of intravenous nitroglycerin.
2 Infrequent when nitroglycerin is given intravenously.
3 Reflex tachycardia may occur with the decrease in blood pressure.

747. (4) This response indicates that the client understood the nurse's teaching because taking a nitroglycerin tablet before such an activity is prophylactic use. (2) (PT; EV)
1 This response does not indicate understanding; this is an example of scheduled administration, not prophylactic use.
2 The client does not understand the purpose of the medication; this indicates avoidance of activity rather than taking medication to prevent angina during the activity.
3 Incorrect; blood pressure, not pulse, is the parameter most affected by nitroglycerin.

748. (2) The speed of conduction is decreased when digoxin is given, and this can result in premature beats, atrial fibrillation, and first-degree heart block. (1) (PT; EV)
1 Digoxin does not deplete potassium and therefore orange juice does not need to be given; the orange juice is high in calories and would need to be calculated in the diet.
3 Insulin and digoxin can be given at the same time.
4 The purpose of the drug is to reduce a rapid heart rate.

749. (4) This drug suppresses ventricular activity; therefore it is used for the treatment of PVBs. (2) (PT; PL)

1 This drug increases myocardial contractility and heart rate; therefore it is contraindicated in the treatment of PVBs.
2 This drug blocks vagal stimulation; therefore it increases the heart rate; it is used for bradycardia.
3 This drug raises the serum pH level; therefore it combats metabolic acidosis.

750. (1) Lidocaine will interrupt the VT before it progresses to ventricular fibrillation. (3) (DA; PL)
2 Epinephrine is not used for VT; it is used for cardiac arrest and may even precipitate ventricular fibrillation.
3 A pacemaker is used for supraventricular tachycardia and bradycardia.
4 CPR is used for cardiac arrest, not VT.

751. (2) A decreased heart rate or sinus bradycardia is the normal response to a beta blocker. (3) (PT; EV)
1 Anxiety is not related to beta-blocker therapy.
3 Ventricular fibrillation during an electrophysiology study would indicate that the beta blocker was not working effectively.
4 Nausea and vomiting may occur as side effects but would not indicate the effectiveness of the medication.

752. (1) Tremors are a precursor to the major adverse effect of seizures or convulsions. (3) (PT; EV)
2 Although this can occur, it is not a serious side effect.
3 Bradycardia, which can lead to heart block, occurs, not tachycardia.
4 Hypotension, not hypertension, occurs.

753. (4) Lasix promotes potassium excretion; hypokalemia increases cardiac excitability; digitalis in the presence of low extracellular potassium produces ectopic pacemaker activity. (1) (DA; AN)
1 Digitalis does not affect potassium excretion; Lasix causes potassium excretion.
2 Potassium is excreted by the kidneys, not destroyed by the liver.
3 Lasix causes diuresis and consequent potassium loss regardless of the serum potassium level.

754. (4) Generalized weakness is a sign of significant hypokalemia, which may be a sequela to diuretic therapy. (1) (PT; EV)
1 Insomnia is not known to be related to hypokalemia or Esidrix therapy.
2 Although this is unrelated to hydrochlorothiazide therapy, it can occur with other antihypertensive drugs.

3 Increased thirst is associated with hypernatremia; because this drug increases excretion of water and sodium in addition to potassium and chloride, hyponatremia, not hypernatremia, may occur.

755. (3) This counteracts the acidosis that occurs secondary to tissue anoxia that accompanies cardiac standstill. (3) (DA; PL)
1 This medication is used after resuscitation to counteract convulsions if they occur.
2 This is a diuretic commonly used for congestive heart failure.
4 This medication relaxes smooth muscle of the bronchial airways and pulmonary blood vessels, which is not indicated in this instance.

756. (2) Nausea is the most common and immediate side effect of Mustargen therapy. (2) (PT; EV)
1 This is not a side effect of Mustargen therapy.
3 Mustargen is more likely to cause urinary retention than urinary incontinence.
4 Same as answer 1.

757. (1) Heart failure and dysrhythmias are the only life-threatening toxic effects unique to Adriamycin. (2) (PT; EV)
2 When bone marrow is depressed to precarious levels, the dose is altered and/or blood components administered.
3 This is not a side effect of Adriamycin nor of any of the other antineoplastic agents.
4 This is a very uncomfortable side effect, but it is not life threatening.

758. (1) Stomatitis and hyperuricemia are possible complications of therapy; therefore oral care and hydration are very important. (1) (PC; PL)
2 Hot substances are avoided because of the common occurrence of stomatitis.
3 Abnormal bleeding is a common problem and thus injections are contraindicated; rest is important for increased fatigability.
4 This is false reassurance; although complete remission occurs in 50% to 75% of treated clients, the median survival time is 2 to 3 years.

759. (1) Chemotherapy destroys normal erythrocytes, white blood cells, and platelets indiscriminately along with the neoplastic cells. (1) (PT; AN)
2 This is not a true description of the side effects of steroids.
3 This is not the cause for fewer erythrocytes, white blood cells, and platelets.
4 Although true, this does not explain pancytopenia.

760. (4) Alkylating drugs often cause severe bone marrow depression because they affect rapidly dividing cells. (3) (PT; EV)
1   These are complications that are more common with antimetabolites.
2   These are complications associated with hormonal therapy.
3   These are common side effects with the administration of steroids.

761. (2) Aspirin is contraindicated in the presence of bleeding tendencies, which often occur with ALL because of its inhibitory effect on platelet aggregation. (2) (SI; IM)
1   An antacid will reduce the gastric irritation common with aspirin but will not alter its effect on platelets.
3   The dosage is within acceptable limits.
4   Action needs to be taken before the temperature is 102° F.

762. (2) Leucovorin calcium limits toxicity of folic acid antagonists, such as methotrexate sodium, by competing for transport into cells. (3) (PT; AN)
1   This is not the action of leucovorin calcium.
3   This is the purpose of antiemetics such as prochlorperazine maleate (Compazine).
4   Leucovorin calcium does not interfere with cell division; this is the purpose of a multiple drug protocol.

763. (1) This helps flush the kidneys and prevent nephrotoxicity, especially during the early phase of treatment. (3) (CC; EV)
2   Reconstituted solution can be stored in the refrigerator for 1 month.
3   Confusion, dizziness, and hallucinations are side effects of this drug; the client should avoid hazardous tasks, such as driving or using machinery.
4   Activity may have to be altered because fatigue and other flulike symptoms are common with this drug.

764. (2) Antacids may interfere with complete absorption of Tagamet; therefore they should be administered at least 1 hour apart. (2) (CC; PL)
1   This would interfere with absorption of Tagamet.
3   Same as answer 1.
4   This would interfere with absorption of Tagamet, and orange juice may be irritating and slow the client's recovery; milk may be used because it may enhance effectiveness of the medication.

765. (3) Morphine sulfate increases spasm of smooth muscle and is contraindicated in all conditions in which there is obstruction of smooth muscle ducts. (3) (SI; AN)
1   Morphine sulfate and meperidine hydrochloride cause respiratory depression.
2   Ingestion of food stimulates pancreatic function; drugs do not have this effect.
4   Morphine sulfate and meperidine hydrochloride are central nervous system depressants.

766. (2) The pancreatic enzymes (amylase, trypsin, and lipase) must be present when food is ingested for digestion to take place. (2) (CC; EV)
1   At this time the food eaten for dinner has already passed beyond the duodenum; the enzyme would be given too late to aid in digestion.
3   The client would have no chyme in the duodenum for the enzyme to act on.
4   Same as answer 3.

767. (3) Cisplatin is nephrotoxic and can cause kidney damage unless the client is adequately hydrated to flush the kidneys. (3) (CI; PL)
1   This medication, a form of folic acid, is used to combat toxic effects of methotrexate; cisplatin does not interfere with folic acid metabolism.
2   Gentle, not vigorous, oral care is needed to cleanse the mouth without further aggravating the expected stomatitis.
4   Unnecessary; prolonged gastrointestinal irritation is not the major concern; nausea and vomiting last about 24 hours, and while diarrhea may occur and last longer, it is not the major concern.

768. (3) Morphine sulfate should be avoided because it causes spasms at Oddi's sphincter, thereby increasing pain. (3) (SI; IM)
1   Cimetidine (Tagamet) is useful in reducing gastric acid stimulation of pancreatic enzymes.
2   Promethazine HCl (Phenergan) is useful as an antiemetic for clients with pancreatitis.
4   Meperidine HCl (Demerol) would be useful for pain relief for clients with pancreatitis.

769. (3) It is necessary to clarify the route of administration because medication can be given po, IV, or IM. (2) (SI; IM)
1   This is the usual dose of ranitidine when given twice a day.
2   Ranitidine is used to decrease gastric acid and is helpful to clients with peptic ulcer.
4   Ranitidine is usually given with meals.

770. (3) Prednisone inhibits phagocytosis and suppresses other clinical phenomena of inflammation; this is a symptomatic treatment that is not curative. (1) (PT; IM)
1 The drug suppresses the immune response and increases the potential for infection.
2 The appetite is increased; weight gain may result from this or from fluid retention.
4 Generally, the response is rapid.

771. (2) Oral corticosteroids should be taken with food or antacids to prevent gastric irritation and gastric hemorrhage. (3) (CC; EV)
1 The client understands; this will have to be done because long-term administration of steroids leads to elevated blood glucose levels and possible steroid-caused diabetes.
3 The client understands; usually a larger dose is given at 8 am and the second dose is given prior to 4 pm to mimic the normal hormonal secretion and prevent insomnia.
4 The client understands; neurologic and emotional side effects include euphoria, mood swings, sleeplessness, and excitement.

772. (4) This reduces bacterial activity on blood and wastes in the GI tract, thereby reducing the level of blood ammonia, a by-product of protein metabolism. (2) (PT; AN)
1 Neomycin interferes with bacterial protein synthesis but has little or no effect on intestinal edema.
2 Neomycin reduces bacterial action in the GI tract but does not reduce abdominal distention.
3 Neomycin is an aminoglycoside antimicrobial used specifically against intestinal bacteria such as *Escherichia coli*.

773. (4) This detergent action promotes addition of fluid into stool to soften feces. (2) (PT; AN)
1 This is the action of lubricant laxatives such as mineral oil.
2 This is the action of saline laxatives such as milk of magnesia.
3 This is the action of peristaltic stimulants such as cascara and castor oil.

774. (2) A stool softener prevents constipation without irritating the colon, permitting the inflammation to subside. (1) (PC; AN)
1 Surfak is not a bulk cathartic; bulk-forming laxatives, such as Metamucil, form soft, pliant bulk that promotes physiologic peristalsis.
3 Surfak is not an emollient; emollient laxatives, such as mineral oil, lubricate the feces and decrease absorption of water from the intestinal tract.

4 Surfak is not a peristaltic stimulant; peristaltic stimulants, such as cascara, irritate the bowel, precipitating a bowel movement.

775. (2) Increased acidity caused by the stress occurring with burns and crushing injuries contributes to the formation of Curling's ulcer; Tagamet, a histamine H_2 antagonist, decreases the formation of gastric acid, and Maalox, an antacid, neutralizes it once it is formed. (1) (PT; IM)
1 These drugs do not decrease irritability of the bowel; their purpose is to decrease gastrointestinal acidity, not irritation.
3 This does not explain how these drugs work and why they are prescribed for this client.
4 Same as answer 3.

776. (2) Maalox can cause diarrhea and sometimes needs to be given alternately with another more constipating antacid. (2) (PT; EV)
1 Immobility causes constipation, not diarrhea.
3 Although diet can affect elimination, no data support this conclusion.
4 Cimetidine causes constipation, not diarrhea.

777. (3) This is a miotic that constricts the pupil, permitting fluid drainage, which reduces intraocular pressure. (2) (PT; PL)
1 This is a topical anesthetic; it will not reduce the increased intraocular pressure associated with glaucoma.
2 This is contraindicated; this dilates the pupil and paralyzes ciliary muscles.
4 This is contraindicated; this is a mydriatic that dilates the pupil, obstructing drainage, which increases intraocular pressure.

778. (4) Parenterally administered Valium is a benzodiazepine that has muscle relaxant and anticonvulsant effects that help limit massive muscular spasm. (2) (PT; PL)
1 Atropine is not used for convulsions; it is used for bradycardia resulting from vagal overstimulation but does not reverse bradycardia caused by metabolic changes such as acid-base imbalances or electrolyte imbalances.
2 Epinephrine does not limit convulsions; it increases contractility of the heart.
3 Narcan does not limit convulsions; it is a narcotic antagonist and is used for morphine, meperidine, and methadone overdose.

779. (4) Because of the lethal toxicity of aspirin overdose, hypotensive crisis and cardiac irregularities can occur. (2) (DA; EV)
   1 This is not the priority at this time.
   2 The central nervous system is not involved at the reflex level at this time.
   3 Central venous pressure readings are not indicated in this situation and are rarely used at the present time.

780. (2) Abnormal involuntary movements (dyskinesias), such as muscle twitching, rapid eye blinking, facial grimacing, head bobbing, and an exaggerated protrusion of the tongue, are signs of toxicity probably resulting from the body's failure to readjust properly to the disappearance of dopamine. (3) (CC; IM)
   1 This is a side effect of therapy, not toxicity.
   3 Same as answer 1.
   4 This is unrelated to levodopa toxicity.

781. (1) These are signs of an MAOI-induced hypertensive crisis and should be reported to the physician immediately. (3) (CC; IM)
   2 The opposite is true because increased amounts of dopamine react with supersensitive postsynaptic receptors.
   3 This is unnecessary; routine medical evaluations of the client should be done.
   4 This is unsafe; the recommended daily dose of 10 mg should not be exceeded.

782. (3) Corticosteroids act to decrease inflammation, which decreases edema. (2) (PT; AN)
   1 This is an antiinflammatory agent, not a diuretic; it does not cause diuresis by action on the kidney.
   2 Resistance to infection is decreased, but this is not pertinent in this situation.
   4 The problem is not with increased cerebrospinal fluid.

783. (3) Corticosteroids, such as Decadron, have a hyperglycemic effect. (2) (PT; EV)
   1 Corticosteroids are not known to precipitate cessation of gastrointestinal activity.
   2 Corticosteroids would not increase bacterial growth in the lungs.
   4 This is unnecessary; this is required when administering magnesium sulfate.

784. (3) The osmotic diuretic mannitol is contraindicated in the presence of inadequate renal function or heart failure because it increases the intravascular volume that must be filtered and excreted by the kidneys. (3) (DA; EV)
   1 This is not specific to mannitol, although vital signs would be monitored.

2 While loss of body fluid is expected with mannitol, a daily weight would be too infrequent to monitor for safe administration of this drug.
   4 While level of consciousness would be monitored with a head injury, this is not specific to mannitol and q8h is too infrequent an assessment.

785. (3) Administration of an anticoagulant to a client who is bleeding would interfere with clotting and increase hemorrhage. (2) (CC; IM)
   1 This is unsafe; it would not be used in this situation because it would increase bleeding; anticoagulants may be used with cerebral thrombosis.
   2 Same as answer 1.
   4 Although contraindicated, if given, it would increase signs and symptoms.

786. (1) This site should not be rubbed to avoid dispersion of the heparin around the site and subsequent bleeding into the area. (1) (SI; IM)
   2 This is not a routine practice; the extra volume would not be advantageous.
   3 The drug should be injected deep into the subcutaneous tissue slowly, not quickly.
   4 This technique and the intramuscular route are not used with heparin; subcutaneous injection or intravenous administration is used.

787. (4) This drug binds with heparin sodium to form a physiologically inert complex; it corrects clotting deficits. (3) (CI; PL)
   1 Vitamin K counteracts the effects of drugs like warfarin sodium (Coumadin).
   2 This is an alternate name for heparin sodium.
   3 This is an oral anticoagulant that interferes with the synthesis of prothrombin.

788. (1) Acetylsalicylic acid (aspirin) should be avoided because it interferes with platelet aggregation; acetaminophen (Tylenol) should be used instead. (2) (CC; EV)
   2 This causes venous pooling and could predispose the client to deep vein thrombosis.
   3 Prothrombin times, not complete blood counts, need to be done periodically.
   4 This is not necessary; this is done when clients have had cardiac problems such as rheumatic fever or surgery.

789. (2) Aspirin interferes with platelet aggregation and will impede the formation of thrombi. (2) (CC; EV)
   1 Although aspirin has antiinflammatory properties, 1 month after the surgery the edema has already subsided.
   3 This is not an expected response following surgery, and it would indicate infection; a prescription for antibiotics would be more appropriate.

4 At this point, there should no longer be discomfort at the surgical site.

790. (4) Peak response occurs 1 hour after administration and lasts up to 8 hours; the response will influence dosage levels. (3) (PT; PL)
1 This medication must be administered on time whether the dosage is already established or is being adjusted.
2 This reduces gastrointestinal upset whether the dosage is established or is being adjusted.
3 There are no psychologic side effects associated with Mestinon.

791. (4) A positive response to the administration of Tensilon indicates myasthenic crisis, whereas an increase in the severity of symptoms indicates cholinergic crisis. (3) (PT; PL)
1 This is the treatment for cholinergic crisis.
2 This is the antidote for heparin.
3 This is a narcotic antagonist.

792. (3) Mestinon is a vital drug that must be taken on time; missed or late doses can result in severe respiratory and neuromuscular consequences or even death. (3) (PT; EV)
1 This is unnecessary because it is not a gastric irritant.
2 This is unnecessary because food does not impair absorption of the drug.
4 This is unnecessary.

793. (4) Exacerbation of myasthenia may occur within 2 weeks of steroid therapy, causing respiratory impairment and dysphagia. (3) (PT; EV)
1 Steroids increase sodium retention, and this would be contraindicated.
2 Although clients should avoid contact with persons having upper respiratory infections, protective isolation is not required.
3 This is unnecessary; adequate fluid intake should be maintained.

794. (2) The drug should be taken as ordered, usually before meals, to limit dysphagia and possible aspiration. (1) (CC; EV)
1 This is not necessary; it may be kept at room temperature.
3 The action of the drug will begin within 30 minutes and start to peak in 1 hour.
4 This is unsafe; the drug should be taken with milk to prevent GI irritation; it is usually taken about 30 to 60 minutes before meals.

795. (4) Oxacillin is a form of penicillin and should be given on an empty stomach; food delays absorption. (2) (CC; EV)

1 This is incorrect; food or milk delays absorption of this drug.
2 Same as answer 1.
3 This is not necessary; however, it is appropriate with sulfonamides.

796. (3) Gentamicin can cause ototoxicity, resulting in vertigo, tinnitus, and hearing loss. (3) (PT; EV)
1 This is unlikely; the client has not been on prolonged bed rest.
2 A liquid diet and IV therapy may cause weakness because of a low-calorie intake, but dizziness does not occur.
4 This is unrelated; more than 24 hours have passed since an anesthetic was administered.

797. (1) Because the drug was just administered, the blood level of the drug would be at its highest level. (2) (CI; IM)
2 The result would reveal a drug blood level halfway between the peak and trough levels.
3 This would be done for a trough level when the drug level would be at its lowest.
4 This would produce inaccurate results; peak and trough levels are measured in relation to the time a drug is administered.

798. (4) These are the drugs of choice for suppression of normal immunologic response to prevent rejection of the donor kidney. (1) (DA; IM)
1 None of these drugs is used for immunosuppression.
2 Only cyclosporine (Sandimmune) is used for immunosuppression; this drug is commonly used in the elderly and those with diabetes mellitus.
3 Methylprednisolone (Solu-Medrol) in large doses is the only drug in this option that is used to prevent kidney rejection.

799. (2) Epogen stimulates red blood cell production, thereby elevating hematocrit levels. (3) (PT; EV)
1 An elevated liver panel may signify liver disease; it is not affected by Epogen.
3 WBC counts are not affected by Epogen.
4 Increased Kaposi's sarcoma lesions are significant of AIDS progression and are not affected by Epogen.

800. (4) Neupogen, a granulocyte colony-stimulating factor, increases the production of neutrophils with little effect on the production of hematopoietic cells. (3) (PT; EV)
1 The production of these cells is not influenced by Neupogen.
2. Same as answer 1.
3 Same as answer 1.

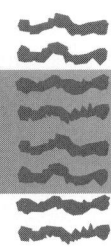

MEDICAL-SURGICAL NURSING
QUIZ 1

1. A client develops acute respiratory distress, and a tracheostomy is performed. An important intervention that should be included in the care plan is:
1. Encouraging a fluid intake of 3000 ml daily
2. Suctioning via the tracheostomy every hour
3. Applying an occlusive dressing over the site
4. Cleansing the stoma with peroxide and cotton balls

2. After a thyroidectomy the client should be placed in the:
1. Prone position
2. Supine position
3. Left Sims' position
4. Semi-Fowler's position

3. One week after beginning antithyroid medication for the treatment of hyperthyroidism, a client reports that diarrhea, abdominal pain, and fever have developed. The client is admitted for thyrotoxicosis. The most important goal of immediate treatment for this condition is:
1. Rapid reduction of body temperature and heart rate
2. Prevention of fluid overload by limiting client's intake
3. Observation for an exaggerated response to sedatives
4. Treatment of associated hyperglycemia and ketoacidosis

4. The nurse teaches a client how to use a nebulizer. The nurse would recognize that the nebulizer is not being used correctly and that additional instruction is needed when the client:
1. Holds the inspired breath for at least 3 seconds
2. Positions the tip of the nebulizer beyond the lips
3. Inhales with the lips tightly sealed around the mouthpiece
4. Exhales slowly through the mouth with lips pursed slightly

5. A client who has been admitted to the hospital with a possible small bowel obstruction is to have an intestinal tube inserted. Before assisting the physician with this procedure, the nurse should:
1. Place the client flat in bed, lying on the right side
2. Instruct the client in the techniques of mouth breathing
3. Spray the client's oropharynx with a local anesthetic solution
4. Reassure the client that the procedure will not cause discomfort

6. The nurse admits a client with deep partial-thickness burns of the face and neck to the burn unit. The nurse is aware that one of the first indications of heat inhalation is:
1. Changes in the chest x-ray examination
2. Sputum that contains particles of blood
3. Nasal discharge containing carbon particles
4. Changes in the arterial blood gases consistent with acidosis

7. A client with a severe respiratory infection who has a history of emphysema with chronic bronchitis and right-sided heart failure is receiving furosemide, digoxin, KCl, and aminophylline. One day as lunch is served, the client complains of nervousness and nausea. It would be most appropriate for the nurse to:
1. Remove the lunch tray immediately
2. Inform the physician about the client's complaints
3. Encourage the client to eat to help fight the infection
4. Stay with the client and encourage ventilation of anxiety

8. When preparing a plan of care for a client who is scheduled for a total laryngectomy, initially it is most important that the nurse include how to:
1. Communicate postoperatively
2. Do a tube feeding postoperatively
3. Care for a permanent tracheostomy
4. Cough and expectorate bronchial secretions

9. The urine output of a client in the oliguric phase of acute renal failure has been less than 1 cup for the last 12 hours. The nurse is aware that it will be necessary to limit oral fluids to 900 ml of water during the next 24 hours because this amount will:
   1. Equal the expected urinary output for the next 24 hours
   2. Prevent the development of complicating hypostatic pneumonia
   3. Prevent hyperkalemia, which could lead to serious cardiac dysrhythmias
   4. Compensate for both insensible and measured fluid losses over the next 24 hours

10. A client with a sucking chest wound has a large pressure dressing over the site. The nurse recognizes that the purpose of this dressing is to:
    1. Protect the pleura
    2. Seal off major vessels
    3. Prevent additional contamination
    4. Maintain negative intrathoracic pressure

11. A client with a history of emphysema with chronic bronchitis and right-sided heart failure develops a respiratory infection and is admitted in acute respiratory distress. The client's blood studies indicate a pH of 7.30, a Po_2 of 60 mm Hg, a Pco_2 of 55 mm Hg, and a HCO_3 of 23 mEq/L. The nurse is aware that the client is in:
    1. Hypocapnia
    2. Hyperkalemia
    3. Generalized anemia
    4. Respiratory acidosis

12. The nurse understands that edema caused by inadequate nutrition is a result of the:
    1. ADH mechanism
    2. Aldosterone mechanism
    3. Nitrogen balance mechanism
    4. Capillary fluid shift mechanism

13. The nurse can prevent a major reaction to total parenteral nutrition infusions by:
    1. Recording the intake and output
    2. Changing the site every 24 hours
    3. The slow administration of the fluid
    4. Checking the vital signs every 4 hours

14. Elderly individuals with type 2 diabetes mellitus:
    1. Seldom develop ketoacidosis
    2. Secrete no endogenous insulin
    3. Have a lower incidence of chronic complications
    4. Have a sudden and dramatic onset of symptoms

15. An elderly female admitted to the hospital with an acute episode of rheumatoid arthritis asks why her roommate, who also has arthritis, goes to physical therapy every day and she does not. The most appropriate response by the nurse would be:
    1. "It usually depends on who your physician is."
    2. "Her condition is much more advanced than yours."
    3. "Your joints are still inflamed and it would be harmful."
    4. "Physical therapy is an important aspect of rheumatologic care."

16. The client's spouse is present when the surgical consent is being obtained and asks how the coronary artery bypass surgery will help the client. The nurse bases a response on the knowledge that:
    1. This surgery significantly decreases symptoms in a large percentage of clients
    2. Studies have consistently shown that this surgery increases an individual's life span
    3. Surgery will permit the client to return to gainful employment after healing occurs
    4. Evidence substantiates that surgery can prevent progression of coronary artery disease

17. A female client with a history of rheumatic fever and a heart murmur has gained weight even though she has nausea and a loss of appetite. She wakes up short of breath several times nightly and notices that she gets tired and has trouble breathing while doing normal daily tasks. When hearing the client's symptoms the nurse should immediately seek additional critical data such as:
    1. A retrospective 24-hour calorie count
    2. Elimination pattern during the last month
    3. A complete gynecologic and sexual history
    4. The presence of a recent cough and pulmonary secretions

18. To assess an elderly client for a fracture of the hip, the nurse should:
    1. Observe for bruising over the affected hip
    2. Observe for shortening of the affected leg
    3. Move the affected leg to see if it causes pain
    4. Move the affected leg to feel and hear crepitus

19. While still in the postanesthesia unit after surgery to create a colostomy, a client requests that no one be allowed to visit. To support the client at this time, the nurse should first:
    1. Give assurance of respect for the client's wishes
    2. Determine the reason why visitors are not wanted

3. Explain that the surgery is over and everything is OK
4. Promote communication to find out how the client really feels

20. An air-conduction hearing aid increases hearing sensitivity in instances of:
    1. Destruction of the auditory nerve
    2. Diminished sensitivity of the cochlea
    3. Immobilization of the auditory ossicles
    4. Perforation of the tympanic membrane

21. The nurse should be aware that the most common complication of peptic ulcer is:
    1. Perforation
    2. Hemorrhage
    3. Pyloric obstruction
    4. Esophageal varices

22. The foods a client with a peptic ulcer would be permitted to eat include:
    1. Orange juice, fried eggs, sausage
    2. Applesauce, Cream of Wheat, milk
    3. Tomato juice, raisin bran cereal, tea
    4. Sliced oranges, pancakes with syrup, coffee

23. After a prostatectomy a client complains of painful bladder spasms. To limit these spasms the nurse should:
    1. Administer a narcotic every 4 hours
    2. Irrigate the Foley catheter with 60 ml of normal saline
    3. Encourage the client not to contract his muscles as if he were voiding
    4. Advance the catheter to relieve the pressure against the prostatic fossa

24. The physician schedules a paracentesis for a client with ascites. Immediately before the paracentesis, the nurse should:
    1. Instruct the client to void
    2. Position the client on the side
    3. Measure the client's abdominal girth
    4. Have the client drink a glass of water

25. The physician orders a stationary (nonrolling) walker for a client to aid in ambulation. The nurse plans to teach the client to:
    1. Place the back legs of the walker about 10 inches in front of the feet, shift the body weight to the walker, and step forward
    2. Move the walker about 8 inches forward while stepping forward to the walker with body weight on the walker and both legs

3. Place the walker flat on the floor with the front legs about 12 inches in front of the feet, shift the body weight to the walker, and step forward
4. Move the walker about 10 inches in front of the feet with only the front legs of the walker on the floor, then step forward and put the walker flat

26. A client with type 1 diabetes receives 30 units of Humulin N insulin at 7 AM. At 3:30 PM the client becomes diaphoretic, weak, and pale. The nurse should be aware that these are symptoms associated with:
    1. Diabetic coma
    2. Somogyi effect
    3. Diabetic ketoacidosis
    4. Hypoglycemic reaction

27. While setting up a male client's dinner tray, the nurse notices that he has a tremor of the right hand when it is lying in his lap but the tremor disappears when he reaches for his fork. The nurse recognizes that this is:
    1. A resting tremor
    2. A voluntary tremor
    3. An intention tremor
    4. An idiopathic tremor

28. An individual with serious burns requires frequent checks of potassium levels because:
    1. Damaged cells absorb potassium, causing hyperkalemia
    2. Damaged cells release potassium, causing hyperkalemia
    3. Potassium loss from damaged skin may cause hypokalemia
    4. Burned individuals excrete excessive potassium, resulting in hypokalemia

29. Following a total laryngectomy a client is receiving tube feedings. The nurse would know that the teaching was understood when the client states, "I will need tube feedings until:
    1. Healing is complete."
    2. The gag reflex returns."
    3. I can digest oral feedings."
    4. I develop the ability to belch."

30. A client who has inoperable cancer of the head of the pancreas involving the common bile duct has a T-tube inserted. During the first 48 hours after insertion, the nurse should:
    1. Maintain T-tube patency via gravity drainage
    2. Prevent the client from turning onto the right side
    3. Irrigate the T-tube with 30 ml of normal saline q2h
    4. Connect and maintain the T-tube to low intermittent suction

31. During peritoneal dialysis, the nurse observes that drainage of dialysate from the peritoneal cavity has ceased before the required amount has drained out. The nurse should assist the client to:
    1. Turn from side to side
    2. Drink 8 ounces of water
    3. Deep breathe and cough
    4. Periodically rotate the catheter

32. A client has a convulsive seizure at work and is brought to the emergency department. The question that will be most useful to the nurse in planning care related to the client's seizure pattern is:
    1. "Is your job demanding or stressful most of the time?"
    2. "Do you participate in any strenuous sports activities on a regular basis?"
    3. "Were you aware of anything different or unusual just before your seizure began?"
    4. "Does anyone in your family have a history of central nervous system health problems?"

33. A client has a basal cell epithelioma that is to be removed. The client tells the nurse about concerns that the cancer has spread. To best reduce the client's anxiety, the nurse should respond:
    1. "You are a good surgical risk."
    2. "I can understand how you must feel."
    3. "Basal cell tumors usually do not spread."
    4. "The physician probably caught it just in time."

34. A client with diabetes mellitus complains of difficulty seeing at night. The nurse knows this commonly occurs in clients with diabetes because of:
    1. Lack of glucose in the retina
    2. Neovascularization of the retina
    3. Poor glucose supply to rods and cones
    4. The effect of ketones on retinal metabolism

35. The paraplegic client often loses calcium from the skeletal system. A factor that contributes to this condition is:
    1. Decreased activity
    2. Inadequate fluid intake
    3. Decreased calcium intake
    4. Inadequate kidney function

36. Autonomic dysreflexia is a syndrome seen in clients with spinal cord injuries. The signs and symptoms include diaphoresis, pulsating headaches, and goose bumps. This syndrome may occur when:
    1. The client is upright on a tilt table
    2. The myelin sheath is deteriorating
    3. The bowel and/or bladder is distended
    4. The spinal cord is crushed rather than severed

37. The nurse plans rest periods for a client with rheumatoid arthritis and avoids scheduling multiple procedures on any one day. The nurse also monitors the client's vital signs and laboratory values. These nursing interventions are directed toward meeting the needs associated with the nursing diagnosis of:
    1. Anxiety related to an uncertain outcome of the disorder
    2. Activity intolerance related to fatigue, pain, and treatment regimen
    3. Impaired physical mobility related to joint pain, stiffness, and swelling
    4. Altered nutrition, less than body requirements, related to decreased intake

38. A male client with type 1 diabetes asks the nurse to explain why his shins have several brown spots on them. The nurse should reply:
    1. "These spots reflect the accumulation of blood fats in the skin; they should disappear."
    2. "These spots indicate a high glucose content in the skin and are likely to get infected if untreated."
    3. "These spots are the result of diseased small vessels in the shins and may spread if not treated soon."
    4. "These spots are due to small blood vessel damage; the blood contains iron, which leaves a brown spot."

39. A client with a history of angina asks the nurse, "When I get chest pain at home, how will I know whether I should call my doctor?" The nurse should teach the client to call the physician if the pain:
    1. Occurs after moderate exercise
    2. Radiates to the arms, neck, or jaw
    3. Is accompanied by mild diaphoresis
    4. Is not relieved by rest or by nitroglycerin

40. Primary preoperative teaching before open heart surgery should include:
    1. A thorough discussion of discharge plans
    2. A detailed description of the surgical procedure
    3. An explanation that just the chest and groin will be shaved
    4. Visits by postanesthesia care and/or intensive care nursing staff

41. The nurse is aware that a client understands the teaching about diaphragmatic breathing when the client states, "I will:
    1. Take deep breaths frequently."
    2. Breathe with my hands on my hips."
    3. Expand my abdomen on inhalation."
    4. Perform the exercises in the orthopneic position."

42. Measured intake and output is ordered for a postoperative client. The client requests permission to keep the intake and output record. The nurse should:
   1. Determine the client's willingness to really help
   2. Explain that this job is the responsibility of the nurse
   3. Identify the client's reason for wanting to do this task
   4. Assess the client's ability to measure the intake and output

43. A client with full-thickness burns on the chest has a skin graft. During the first 24 hours after a skin graft, care of the donor site includes immediately reporting:
   1. A small amount of yellowish green oozing
   2. A moderate area of serosanguineous oozing
   3. Epithelialization under the nonadherent dressing
   4. Separation of the edges of the nonadherent dressing

44. A client who has had a myocardial infarction receives 15 mg of morphine sulfate for chest pain. Fifteen minutes later, the client complains of feeling dizzy. The nurse should:
   1. Ask whether the client feels this is an allergic reaction
   2. Elevate the client's head and keep the extremities warm
   3. Place the client in the supine position and take the vital signs
   4. Tell the client this is a normal sensation after receiving morphine sulfate

45. After a diagnostic workup, the physician informs a client with hypertension that it is probably related to atherosclerosis. Later while talking with the nurse, the client wants to know more about atherosclerosis. The nurse bases the response on the fact that atherosclerosis is characterized by:
   1. Lipid plaque formation within the arterial vessels
   2. Mobilization of free fatty acid from adipose tissue
   3. Development of atheromas within the myocardium
   4. Gradual decrease in arterial pressure as a result of renin

46. A client who has had a colostomy is to be discharged in several days. A primary nursing goal should be to:
   1. Determine the client's ability to care for the colostomy
   2. Cajole the client into caring for the colostomy without assistance
   3. Teach the client about the special precautions concerning the diet
   4. Show the client how to change the dry sterile dressing on the incision

47. A client is to be discharged with prescriptions for digoxin (Lanoxin) and furosemide (Lasix) and with orders for a low-sodium diet. The nurse identifies that the discharge teaching was not understood when the client states:
   1. "I must check my pulse for its rate and rhythm every day."
   2. "I can gradually increase my exercise as long as I take rest periods."
   3. "I should call my physician if I have difficulty breathing when I am lying down."
   4. "I will use a little table salt on my food only when I do not use it when cooking."

48. A client with congestive heart failure is placed on digoxin and asks the nurse why this is necessary. The nurse bases the answer on the fact that digoxin:
   1. Increases ventricular contractions
   2. Reduces edema in extracellular spaces
   3. Slows and strengthens cardiac contractions
   4. Lengthens the refractory phase of the cardiac cycle

49. When caring for a client following a craniotomy, the nurse should:
   1. Take only axillary temperatures
   2. Encourage coughing but discourage deep breathing
   3. Administer narcotics and sedatives at the first sign of irritability
   4. Report yellow drainage on the dressing to the physician immediately

50. A client with a hiatal hernia frequently wakes up at night with heartburn. The suggestion by the nurse that would most likely reduce the symptoms of heartburn is:
   1. Eat a large meal at noontime
   2. Elevate the head of the bed on 6-inch blocks
   3. Have a light snack of orange juice and crackers at bedtime
   4. Take an intestinal sedative, such as hyoscyamine sulfate (Donnatal), at night

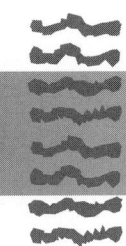

MEDICAL-SURGICAL NURSING QUIZ 1
ANSWERS AND RATIONALES

1. (1) Increasing fluids helps to liquefy secretions, enabling the client to clear the respiratory tract by coughing. (2) (DA; PL; RE)
   2 Excessive suctioning will irritate the mucosal lining of the respiratory tract and can actually result in more secretions.
   3 An occlusive dressing would totally block air exchange.
   4 The use of cotton balls around a tracheostomy introduces the risk of aspiration of one of the cotton fibers.

2. (4) This limits edema in the operative area and promotes respirations. (2) (DA; IM; EN)
   1 This would promote edema in the operative area, and the edema could compromise respirations.
   2 Same as answer 1.
   3 Same as answer 1.

3. (1) Immediate treatment in this emergency focuses on reduction of oxygen demands and thus cardiac workload to prevent cardiac decompensation. (2) (DA; PL; EN)
   2 The need is for an increase in fluid intake to compensate for that lost because of the very high metabolic rate.
   3 This is not likely because drugs are metabolized more rapidly in this condition; there is a danger of exaggerated effects with hypothyroidism.
   4 Clients with thyrotoxicosis are more apt to develop hypoglycemia from the high metabolic rate.

4. (3) This technique promotes nasal breathing, which negates the effects of aerosol medication; a loose seal around the mouthpiece allows for inhalation through the mouth. (3) (PT; EV; RE)
   1 This is a correct technique; it promotes contact of medication with the bronchial mucosa.
   2 This is a correct technique; the nebulizer tip must be past the lips to deliver medication.
   4 This is a correct technique; it prolongs and improves delivery of the medication to the respiratory mucosa.

5. (2) This helps to decrease the gag reflex, therefore easing the passage of the tube. (2) (CI; IM; GI)
   1 This does not take advantage of gravity; the side-lying position makes naso-oral cavities less accessible to the individual passing the tube.
   3 This will make it more comfortable but will interfere with swallowing, which is necessary during the procedure.
   4 This is false reassurance; the procedure is not painful, but it is uncomfortable.

6. (3) Singed nasal hair and sputum and nasal discharges that contain carbon are early warning signs of respiratory heat inhalation. (2) (DA; AS; RE)
   1 Changes in chest x-ray examination findings are a late sign of respiratory problems.
   2 This may be a sign of pneumonia or tuberculosis.
   4 Changes in arterial blood gases are late signs of respiratory problems.

7. (2) Aminophylline is a xanthine that can cause adverse reactions of the GI tract, such as nausea and vomiting, and the CNS, such as nervousness, restlessness, and convulsions; the physician should be notified. (2) (PT; EV; DR)
   1 Although this might be done, the priority is to notify the physician; the therapeutic blood level of the drug may have been exceeded.
   3 Unsafe; these are signs of aminophylline toxicity, not infection; eating when nauseated could result in vomiting.
   4 Same as answer 1.

8 (1) Communication is a priority; it facilitates interaction, limits anxiety, and promotes safety. (2) (SC; PL; RE)
   2 The need for tube feedings depends on the extent of surgery; if necessary, the client can be taught postoperatively.
   3 This is done postoperatively as the client begins to accept the laryngectomy.
   4 Clearing the airway will not be done the same way postoperatively as preoperatively.

9. (4) Insensible losses are 400 to 500 ml in 24 hours; the client's measured output is about 400 ml in 24 hours based on the available history. (3) (DA; PL; RG)

1 Based on the history, the expected urinary output should be about 400 ml in the next 24 hours.

2 This is insufficient fluid; at least 2500 ml daily is necessary to help prevent hypostatic pneumonia.

3 Hyperkalemia in acute renal failure is caused by inadequate glomerular filtration and is not related to fluid intake.

10. (4) A pressure dressing maintains negative intrathoracic pressure by limiting the amount of air rushing in from the outside. This wound can result in a total pneumothorax and mediastinal shift, severely embarrassing respiration. (2) (DA; AN; RE)

1 The dressing does not prevent injury to the pleura.

2 Major blood vessels are not occluded by a thoracic pressure dressing.

3 A sterile dressing without pressure would suffice in preventing contamination.

11. (4) The pH is below the norm of 7.35 to 7.45 indicating acidosis; the Po_2 is below the norm of 80 to 100 mm Hg; the Pco_2 is elevated above the norm of 35 to 45 mm Hg; and the HCO_3 is within the norm of 23 to 28 mEq/L, indicating a respiratory etiology. (3) (DA; AN; FE)

1 This client's carbon dioxide level is elevated, not decreased.

2 These values are unrelated to hyperkalemia; a serum potassium level above 5.0 mEq/L would indicate hyperkalemia.

3 These values are unrelated to anemia; decreased RBCs, Hgb, and Hct are related to anemia.

12. (4) When protein intake is inadequate, the liver is unable to manufacture albumin, the major plasma protein. There is a drop in intravascular colloidal osmotic pressure, with a resulting fluid shift to the interstitial spaces. (3) (DA; AN; FE)

1 ADH is secreted when the intravascular osmotic pressure increases, not decreases.

2 The renin/aldosterone mechanism is not directly linked to inadequate nutrition.

3 Nitrogen balance is the relationship between intake and output of nitrogen; it is linked only indirectly to fluid shifts.

13. (3) Total parenteral nutrition should be infused at a slow, constant rate. This will prevent both cellular dehydration from too rapid infusion of a hypertonic solution and hyperglycemia. (2) (PT; PL; GI)

1 Recording intake and output is essential because of the danger of fluid overload; however, monitoring will not prevent the complication.

2 Generally a major vein is selected for administration of total parenteral nutrition; the site is not changed every 24 hours.

4 Monitoring vital signs may indicate a complication such as infection; monitoring will not prevent a complication from occurring.

14. (1) Lipolysis is not a common response to meeting the metabolic needs of those with type 2 diabetes; therefore, ketones are not present in large enough amounts to cause ketoacidosis. (2) (HI; AN; EN)

2 Adults with type 2 diabetes do secrete endogenous insulin, but secretion is slow and subnormal.

3 The incidence of chronic complications depends on the level of glucose control, not the type of diabetes.

4 The onset of type 2 diabetes is usually slow, whereas in type 1 diabetes, it is sudden and dramatic.

15. (3) Rest is required during active inflammation of the joints to prevent injury; once active inflammation has receded, an activity and exercise regimen can begin. (1) (PC; AN; SK)

1 This is untrue; physical therapy would not be prescribed during a period of exacerbation.

2 The extent of the arthritis is not the determinant; whether the process is in exacerbation or remission is the deciding factor.

4 Although this is true, physical therapy is not performed during acute exacerbation of the arthritis.

16. (1) More than 80% of those who have this surgery have marked relief of their symptoms. (3) (DA; IM; CV)

2 So far, studies have failed to show that coronary artery bypass surgery affects life span.

3 This depends on the client's presurgical condition and occupation, not the surgery itself.

4 The surgery itself does not affect the disease process; clients must also reduce risk factors (obesity, smoking, and poor diet).

17. (4) These symptoms, in addition to a history of rheumatic fever, would require an assessment for other cardiopulmonary symptoms. (2) (HI; AS; CV)

1 Anorexia and weight gain do not indicate a nutritional problem but a fluid balance problem.

2 Loss of appetite in conjunction with shortness of breath and the history of rheumatic fever make gastrointestinal symptoms secondary in importance.

3 There is no reason to investigate the gynecologic and sexual history in relation to the current problem.

18. (2) Shortening of the affected leg occurs because of the overriding of bone fragments. (1) (CI; AS; SK)

1 Although bruising may be present with a fracture, it also may be present from soft tissue injury.

3 The affected leg should not be moved because it can cause further damage to nerves and blood vessels.

4 Same as answer 3.

19. (1) This meets the client's immediate personal needs; it also demonstrates respect and concern. (2) (SC; IM; EH)

2 This may be considered in future planning; it is not the priority at this time.

3 This provides false assurances that may block communication; it does not meet the client's immediate need.

4 This is inappropriate at this time; working with the client's feelings requires time.

20. (2) Since air-conduction hearing aids use the person's own middle ear, they increase hearing acuity in cases of diminished sensitivity of the cochlea. The amplified signal from the hearing aid gives the cochlea greater stimulation and promotes hearing. (2) (CI; AN; NM)

1 Destruction of the auditory nerve results in deafness because impulses cannot be transmitted to the brain's auditory center.

3 Immobilization of the ossicles prevents conduction of resonant vibrations from the tympanic membrane to the cochlea; air-conduction hearing aids will not correct this problem.

4 Perforation of the tympanic membrane prevents ossicular conduction, which involves transmission of resonant vibrations from the tympanic membrane to the ossicles to the cochlea; hearing aids will not correct this.

21. (2) Hemorrhage after erosion of blood vessel walls is often the first symptom that leads the client to seek medical assistance. (2) (CI; AS; GI)

1 Hemorrhage usually occurs before perforation.

3 This is not a common complication.

4 This occurs with portal hypertension, not peptic ulcers.

22. (2) Applesauce, Cream of Wheat, and milk are bland foods that do not irritate the gastric mucosa. (2) (PC; IM; GI)

1 This is not considered bland; it may be irritating to the mucosal lining.

3 Same as answer 1.

4 Same as answer 1.

23. (3) This action causes the bladder muscle to contract, initiating painful bladder spasms. (1) (PC; IM; RG)

1 Narcotics may dull the pain, but they will not necessarily limit muscle spasms.

2 Instillation of fluid will not decrease bladder spasms and may be irritating and precipitate additional spasms.

4 Manipulating the catheter may precipitate additional spasms.

24. (1) The bladder should be empty to avoid injury during insertion of the abdominal trocar. (2) (SI; IM; GI)

2 The upright position is assumed to allow accumulation of fluid in the lower abdomen by gravity.

3 Although regular monitoring of girth is important, it is not necessary immediately before this procedure.

4 This is unrelated to the procedure; however, it would be preferable to offer fluids after the procedure if allowed.

25. (3) Placing the walker flat on the floor provides stability; putting weight on the walker equalizes weight bearing on the upper and lower extremities. (2) (PC; IM; SK)

1 This is unsafe; this places the walker too far in front of the client for safe transfer of body weight, and all four legs should be flat on the ground.

2 It is not possible to move the walker and have it bear weight at the same time; the walker should be flat on the ground when the client is stepping forward.

4 This is unsafe; all four points of the walker should be flat on the ground when the client is stepping forward.

26. (4) These are sympathetic nervous system responses to hypoglycemia; the peak action of Humulin N insulin is 8 to 12 hours after administration, and 8½ hours have elapsed since it was given. (2) (PT; EV; EN)

1 The signs of diabetic coma are dry mucous membranes; hot, flushed skin; deep rapid respirations (Kussmaul breathing); acetone odor to the breath; nausea and vomiting; and, as in hypoglycemia, weakness.

2 The Somogyi effect includes wide swings in blood glucose levels between hyperglycemia and a profound hypoglycemia caused by insulin rebound.

3 Ketoacidosis results from excess use of fats for energy when carbohydrates cannot be used; lipids are incompletely metabolized and dehydration, acidosis (both ketotic and lactic), and electrolyte imbalance result; it is not the result of insulin administration.

27. (1) A resting tremor is typically present when the hand is not involved in a purposeful activity; also known as nonintention tremor; the tremor is caused by decreased neurotransmitter substance. (1) (HI; AS; NM)

2 This implies the tremor is under the client's control, which is not true.

3 An intention tremor is exhibited or intensified when purposeful movements are attempted.

4 The cause of the disease may be idiopathic, but the tremor is known as a resting or nonintention tremor.

28. (2) Cellular catabolism, as in burns, results in the loss of potassium from cells; this raises the K+ level in the extracellular fluid (normal range: 3.5 to 5.5 mEq/L). Such hyperkalemia may lead to dysrhythmias (peaked T wave at 6 mEq) and, as a terminal event, cardiac arrest. (3) (DA; AS; FE)

1 Hypokalemia may result on the fourth or fifth postburn day as potassium shifts from the extracellular fluid into the intracellular compartment.

3 Initially, potassium loss from damaged tissues results in hyperkalemia as the potassium shifts from the intracellular to the extracellular spaces.

4 The shift of potassium from the intracellular to the extracellular compartment results in hyperkalemia. In addition, to excrete means to eliminate waste matter by a normal discharge; the loss of fluids and electrolytes through a damaged integument is not a normal discharge.

29. (1) Food should be avoided until the area is totally healed. This will keep the area from becoming irritated and contaminated and will generally promote healing. (3) (SI; EV; GI)

2 Because of the alterations in structure, the gag reflex is no longer present.

3 The ability to tolerate oral feedings is not necessarily lost; such feedings are withheld to prevent irritation to the surgical site.

4 The ability to belch has no bearing on the decision to resume oral feedings.

30. (1) A T-tube drains by gravity into a small collection bag; gravity drainage is enhanced by the right side-lying or semi-Fowler's position. (2) (CI; PL; EN)

2 The right side-lying position facilitates drainage and should be encouraged.

3 A T-tube is never irrigated by the nurse; it drains by gravity.

4 A T-tube drains by gravity, not intermittent suction.

31. (1) Turning from side to side will change the position of the catheter, thereby freeing its drainage holes, which may be obstructed. (2) (CI; IM; RG)

2 Taking fluids into the gastrointestinal tract does not influence the drainage of dialysate from the peritoneal cavity; the client could also be on restricted fluids.

3 This improves pulmonary ventilation and helps in maintaining comfort but does not significantly improve the flow of dialysate from the catheter.

4 The position of the catheter should be changed only by the physician.

32. (3) Identification of a common experience that occurs before each seizure (aura) is helpful in planning care and in identifying ways to avoid a future attack. (2) (SI; AS; NM)

1 Although this may provide some information, it is not the most inclusive question the nurse can ask; also it limits the client's reply.

2 Same as answer 1.

4 Same as answer 1.

33. (3) This provides factual information and addresses the client's concern. (2) (SC; IM; EH)

1 This does not speak to the client's concern and may increase anxiety.

2 This may provide reassurance but does not permit further exploration of concern.

4 This reinforces the client's fears instead of pointing out reality.

34. (2) In diabetes mellitus, proliferation of fragile vessels, progressive thickening of the capillary basement membranes, and medial sclerosis of small arteries leading to the eyes gradually occlude the vessel lumina. Also the buildup of sorbitol in retinal tissue causes mural cell death. These changes significantly reduce retinal perfusion and lead to blindness. (2) (CI; AN; EN)

1 There is usually an increase in serum glucose with diabetes mellitus; thickening of the capillary basement membranes often occurs, even if the glucose level is maintained within normal limits.

3 Same as answer 1.

4 Ketones do not usually affect retinal metabolism; retinopathy is a result of vascular changes, retinal detachment, and hemorrhage within the eye.

35. (1) The bones respond to the stress of activity (walking, running, etc.) by laying down new bone substance along the lines of stress. Inactivity leads to reduced bone deposition and actual bone decalcification. (2) (PC; AN; SK)

2 Fluid intake has no effect on bone decalcification.

3 Calcium intake does not alter bone demineralization in the bedridden client.

4 Kidney function may be altered as bone decalcification occurs and stones are formed in the kidneys.

36. (3) Bowel or bladder distention causes autonomic nerve impulses to ascend via the cord to the point of injury. Here the reflex is completed, and autonomic outflow causes piloerection (goose bumps), sweating, and splanchnic vasoconstriction. The latter causes hypertension and a pounding headache. (3) (CI; AN; NM)

1 This is not involved in the autonomic dysreflexia phenomenon.

2 Same as answer 1.

4 Same as answer 1.

37. (2) Obtaining vital signs (particularly pulse, respirations, and blood pressure), monitoring laboratory values (particularly RBCs), and spacing activities all address a client's inability to tolerate activity; rest is a priority. (PC; PL; SK)

1 The intervention stated will not decrease anxiety related to the prognosis.

3 The intervention stated will decrease fatigue, not increase mobility; ROM and ambulation will improve mobility.

4 The stated intervention will not increase appetite and food intake.

38. (4) This is the only accurate explanation for the client's symptom, which is caused by deposit of hemosiderin in the tissue. (2) (HI; IM; IT)

1 This is the definition of a xanthoma.

2 This is not the cause of brown spots on the skin; increased glucose in the skin is not observable by inspection.

3 Brown spots result from the rupture of blood vessels and the deposit of iron, not disease of blood vessels; diseased vessels may be a sequel to diabetes but they do not cause brown spots.

39. (4) When neither rest nor nitroglycerin relieves the pain, there may be an acute myocardial infarction. (2) (CC; PL; CV)

1 This is expected; activity increases cardiac output, causing angina.

2 This is expected; anginal pain can, and often does, radiate.

3 This is expected; acute myocardial infarction causes profuse, not mild, diaphoresis, which should be reported.

40. (4) Clients should be familiar with these people and hear from them what will be experienced. (2) (SC; PL; EH)

1 Although discharge plans should be mentioned, they are not the primary focus at this time.

2 Most clients do not want or need a minutely detailed description.

3 The client's whole body may be prepped and shaved.

41. (3) This promotes descending of the diaphragm; therefore it is easier for air to enter and fill the lungs. (1) (CI; EV; RE)

1 Rapid breathing promotes respiratory acidosis in clients with COPD; diaphragmatic breathing includes slow, deep breathing.

2 This is not part of diaphragmatic breathing nor would it verify that it was being done correctly.

4 Diaphragmatic breathing may be performed in any position other than prone or Trendelenburg's.

42. (4) Clients should be allowed to maintain some control, depending on their ability to perform a given task. (2) (CC; AS; FE)

1 The client has indicated willingness by the request.

2 Able clients should be supported to perform self-care.

3 This is immaterial.

43. ① This indicates infection and should immediately be reported. (1) (SI; EV; IT)
2 Serosanguineous oozing is to be expected.
3 This indicates healing and is desirable.
4 This is not a problem.

44. ③ Vertigo may be a symptom of hypotension, a side effect of morphine sulfate. (2) (PT; EV; DR)
1 Hypotension is a side effect, not an allergic response to morphine sulfate.
2 Raising the client's head may aggravate symptoms.
4 Dizziness is not a normal sensation after morphine sulfate.

45. ① The term "atherosclerosis" means a thickening of the arterial lining by lipid plaques, which become atheromas. (1) (CI; AN; CV)
2 Mobilization of free fatty acids will produce an acid-base imbalance.
3 Atheromas develop within the lining of the artery, not within the muscle tissue.
4 Arterial pressure is increased as a result of renin.

46. ① The client's feelings, knowledge, and skills concerning the colostomy must be assessed before discharge. (2) (CC; AN; EH)
2 Individuals should not be coaxed into doing something they are not ready to do, particularly on a long-term basis.
3 After a colostomy the client is usually placed on a regular diet and told only to eliminate gas-producing foods.
4 Often the client no longer needs a dressing on the incision at this time.

47. ④ This is unsafe, which demonstrates that the client did not understand the discharge teaching; sodium helps retain fluid, which could cause a fluid volume excess that in turn could precipitate congestive heart failure. (1) (CC; EV; CV)
1 Digoxin should be held if the client's pulse is less than 60 or more than 120 beats per minute because these dysrhythmias are signs of digitalis

toxicity; the risk of digitalis toxicity is increased if the client develops hypokalemia as a result of receiving Lasix.
2 Slowly increasing activities and rest periods limits the stress on the heart and is desirable.
3 Orthopnea is a sign of pulmonary edema related to congestive heart failure, and the physician should be notified.

48. ③ Digoxin increases the strength of myocardial contractions (positive inotropic effect) and, by altering the electrophysiologic properties of the heart, slows the heart rate (negative chronotropic effect). (1) (PT; AN; DR)
1 This is too general; digoxin increases the strength of the contractions but decreases the heart rate.
2 Although this may result from the increased blood supply to the kidneys, it is not the primary reason for administering digoxin.
4 The PR interval is prolonged.

49. ④ Yellow drainage may be CSF and should be reported immediately. (1) (CI; EV; NM)
1 Temperature evaluation must be accurate; axillary temperatures are influenced by environmental conditions.
2 Deep breathing expands the lungs and mobilizes secretions to prevent respiratory complications; secretions may be removed by suctioning to avoid increased intracranial pressure associated with coughing.
3 Administration of narcotics makes accurate neurologic assessment impossible; narcotics depress the CNS.

50. ② Elevating the head of the bed will prevent reflux of gastric contents into the esophagus. (2) (PC; IM; GI)
1 Small, frequent meals would relieve symptoms of fullness and prevent gastric stimulation.
3 Eating at bedtime and drinking orange juice would stimulate gastric secretions.
4 This would delay emptying of the stomach, which would increase gastric secretions and the feeling of fullness.

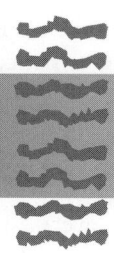

MEDICAL-SURGICAL NURSING QUIZ 2

1. The American Diabetes Association recommends that approximately 50% of the calories allowed come from carbohydrates. When discussing favorite foods with a client with type 1 diabetes the nurse indicates that the food lowest in carbohydrate is 1 cup of:
   1. Skim milk
   2. Apple juice
   3. Nonfat yogurt
   4. Fresh orange juice

2. An obese client has an abdominal cholecystectomy and returns from surgery with a nasogastric tube to low continuous suction, a T-tube, and a Foley catheter. The nurse should first:
   1. Irrigate each tube with normal saline
   2. Fasten each tubing to the bed sheets
   3. Empty the drainage from collection devices
   4. Check the drainage tubes and collection devices

3. After surgery to create an ileal conduit, a client is admitted to the postanesthesia care unit. During the first hour of the postoperative period, the nurse should notify the physician if:
   1. Vomiting occurs
   2. The stoma is swollen
   3. No urine output is noted
   4. Bowel sounds are diminished

4. A client is admitted to the trauma unit with multiple injuries including a crushed chest, abdominal trauma, probable head injury, and multiple fractures. In order of priority, the initial emergency care interventions for this client are to:
   1. Start an IV, get blood for typing and crossmatching, obtain a history
   2. Assess vital signs, obtain a history, arrange for emergency x-ray films
   3. Conduct a thorough physical assessment, assess vital signs, cover open wounds
   4. Assess vital signs, control accessible bleeding, determine the presence of critical injuries

5. When a client returns from surgery after a thyroidectomy, the nurse should position the client:
   1. Prone with the head turned to one side
   2. Supine with the knees flexed at 10 degrees
   3. Lateral with the head slightly flexed and elevated
   4. Supine with the head continuously hyperextended

6. A client with acute renal failure is to receive peritoneal dialysis and asks why this procedure is necessary. The nurse's best response would be, "It:
   1. Prevents the development of serious heart problems."
   2. Helps do some of the work usually done by the kidneys."
   3. Removes toxic chemicals from the body so you will not get worse."
   4. Speeds recovery because the kidneys are not responding to other therapy."

7. A female client who has had a colostomy asks, "How will my husband be able to care for me at home?" The nurse should recognize this statement as a:
   1. Readiness to discuss her colostomy
   2. Need for more time to think about the future
   3. Prediction of a change in a family relationship
   4. Beginning realization of implications for the future

8. A calcium supplement and an aluminum hydroxide gel are commonly ordered for clients with chronic renal failure because the ability to maintain calcium-phosphorus balance is lost with this disorder. The nurse should teach a client with these prescribed medications to take the aluminum hydroxide:
   1. With meals
   2. Whenever necessary
   3. One hour after meals
   4. One hour before meals

9. A client suffers deep partial- and full-thickness burns of the left leg, left arm, and face. The percentage of total skin surface that has been burned is:
   1. 18%
   2. 27%
   3. 36%
   4. 45%

10. The major objective during the emergent phase of a burn is to:
    1. Relieve pain
    2. Prevent infection
    3. Replace blood loss
    4. Restore fluid volume

11. The advantage of a gastrostomy tube feeding over a nasogastric tube feeding is that:
    1. There is less chance of aspiration
    2. The procedure does not require gravity
    3. The client can self-administer the feeding
    4. More tube feeding mixture can be given each time

12. A diagnostic test that would provide evidence that a client might have a peptic ulcer is a:
    1. Barium enema
    2. Gastric biopsy
    3. Gastric culture
    4. Stool examination

13. A client with chronic obstructive pulmonary disease has elevated hemoglobin and hematocrit levels. The nurse should recognize that this is a result of the fact that chronic:
    1. Hypoxia stimulates red blood cell production
    2. Infection stimulates white blood cell production
    3. Hypercapnia stimulates red blood cell production
    4. Infection promotes loss of extracellular fluid volume

14. A client with chronic obstructive pulmonary disease is admitted to the hospital with pneumonia. On the third day, the client complains of a sharp pain on the left side of the chest. The nurse suspects a left pneumothorax. When assessing breath sounds on the affected side, the nurse would expect to hear:
    1. Crackling sounds
    2. Wheezing sounds
    3. Adventitious sounds
    4. Decreased breath sounds

15. A client is receiving MOPP therapy for Hodgkin's disease. About halfway through the first 6-month course of treatment, the client complains of burning and tingling in the feet. The nurse should recognize that these complaints are symptoms of the:
    1. Side effects of prednisone therapy
    2. Neurotoxicity caused by vincristine
    3. Peripheral vasoconstriction caused by nitrogen mustard
    4. Electrolyte imbalances caused by anorexia and vomiting

16. A client with cardiac problems is to have a Holter monitor at home. The nurse should instruct the client to:
    1. Keep a diary of activities
    2. Stay away from microwave ovens
    3. Avoid taking any nitroglycerin that day
    4. Take both blood pressure and pulse every 2 hours

17. The nurse encourages a postoperative client who had an abdominal cholecystectomy and choledochostomy with a T-tube in place to perform deep breathing and coughing every hour. The client resists doing these exercises primarily because the:
    1. Nasogastric tube is irritating
    2. Pain at the incision site increases
    3. T-tube moves and causes cramps
    4. Bandage on the abdomen is constricting

18. When preparing a teaching plan for a client on hemodialysis, the nurse recalls that a substance that passes through the semipermeable membrane during hemodialysis is:
    1. RBCs
    2. Sodium
    3. Glucose
    4. Bacteria

19. After an accident in which there is a question of back injury, the individual involved:
    1. Can be transported in the sitting position, if necessary
    2. May be transported best when placed in the side-lying position
    3. Should be protected from flexion and hyperextension of the spine
    4. May be transported in any position because position is not important

20. When caring for a client with a spinal cord injury, the nurse should plan to provide a high intake of fluid to help:
    1. Prevent dehydration
    2. Maintain electrolyte balance
    3. Prevent urinary tract infection
    4. Prevent elevation of temperature

21. A 30-year-old steamfitter, who weighs 143 pounds and has 40% of the body surface area burned, is admitted to the burn unit. The physician orders fluid replacement of 7200 ml during the first 24 hours. The hourly IV fluid intake will be approximately:
    1. 125 ml/hr
    2. 200 ml/hr
    3. 300 ml/hr
    4. 425 ml/hr

22. A client who has been smoking for over 30 years has had several episodes of hemoptysis and is admitted to the hospital for a bronchoscopy and lung biopsy. The most important nursing measure following these procedures would be:
    1. Having the client rest in the supine position
    2. Assessing the client for signs of hemoptysis
    3. Keeping the client npo until the gag reflex returns
    4. Checking the client's level of consciousness every 2 hours

23. A couple, ages 82 and 80, live alone. During a home visit, the nurse finds that the husband has a prostate condition and at times is incontinent of urine. He is alert and cooperative but forgetful. The wife has diabetes that is controlled by oral medication and diet. She is severely arthritic and walks with difficulty. They both need some help with dressing, bathing, and meal preparation. The plan that appears most suitable for this couple would be to:
    1. Place them together in a health-related facility
    2. Place them together in a skilled nursing facility
    3. Keep them in their home with a long-term health care program
    4. Move them in with one of their children but allow them to choose which one

24. A client with cancer of the lung asks the nurse about immunotherapy treatment for cancer. In addition to referring the client to the physician for a specific answer, the nurse should base a response on the fact that immunotherapy is:
    1. Directed at altering the structure of the malignant cells
    2. Now a primary mode of therapy for a variety of cancers
    3. More likely effective if the tumor is first reduced by another mode of therapy
    4. Effective regardless of the type of tumor cells involved and that this is its major advantage

25. Outcome criteria on discharge from the clinic after a bacterial cystitis should include a statement that the client will:
    1. Understand the need for 7 to 8 liters of fluid per day
    2. Be able to plan menus to include any dietary restrictions
    3. Have relief of symptoms and show no loss of kidney function
    4. Be able to state activities to be avoided because of possible bleeding

26. When performing an assessment of a client with a hiatal hernia, the nurse should expect to find a history of:

    1. Obesity
    2. Bronchitis
    3. Alcoholism
    4. Esophageal varices

27. The nurse should assess for symptoms of hypovolemic shock, which is often associated with burns, because of:
    1. Decreased rate of glomerular filtration
    2. Excessive blood loss through the burned tissues
    3. Sodium retention as a result of the aldosterone mechanism
    4. A shift of proteins and water out of the intravascular compartment

28. A client with cancer of the lung is manifesting severe dyspnea. In lung cancer this is usually caused by:
    1. Obstruction or pleural effusion
    2. Abdominal distention or pressure
    3. Anxiety related to pain on inspiration
    4. Fluid retention caused by renal failure

29. For a client with a permanent tracheostomy, the nurse should include in the teaching plan:
    1. The importance of cleanliness around the site
    2. The necessity of covering the tube opening while swimming
    3. The establishment of a regular pattern for suctioning the tube
    4. The sterile technique necessary for care of the tracheostomy tube

30. A client with the diagnosis of multiple sclerosis develops increased visual problems, progressive muscular weakness, and frequent episodes of emotional lability, which are very distressing to the client. During a visit to the clinic, while having a discussion with the nurse, the client bursts into tears for no apparent reason. The nurse should:
    1. Ascertain why the client is upset
    2. Tell the client there is no reason to cry
    3. Let the client cry, then resume the discussion
    4. Tell the client it is perfectly normal to be upset

31. A client who was hospitalized with severe cirrhosis of the liver improves and is to be discharged. When planning the discharge with family members, the nurse tells them about:
    1. The need for a high-protein diet
    2. The use of chlorpromazine for relaxation
    3. The need for increased fluids to promote kidney output
    4. The importance of reporting personality changes to the physician

32. One of the medical orders the nurse would expect for a client who is to have a hemicolectomy would be:
    1. Give oil retention enemas daily for 2 days preoperatively
    2. Administer neomycin 1 gram q6h orally for 2 days preoperatively
    3. Have a Sengstaken-Blakemore tube at the bedside preoperatively
    4. Provide a high-protein, high-carbohydrate regular diet for 2 days preoperatively

33. A client visits the primary health care center because of persistent pyrexia, abdominal pain, and neuritis symptoms that the physician believes are suggestive of polyarteritis nodosa. When assessing the client the nurse should expect to find:
    1. An elevation in blood pressure
    2. Enlarged cervical lymph nodes
    3. An unexplained excessive gain in weight
    4. A hypersensitivity of the fifth cranial nerve

34. A client with a cardiac dysrhythmia is placed on digoxin (Lanoxin) and procainamide hydrochloride (Pronestyl). Because of the combined effect of Lanoxin and Pronestyl, the nurse should monitor the client for signs of increased:
    1. CNS depression
    2. Reflex stimulation
    3. Myocardial depression
    4. Respiratory stimulation

35. A client asks the nurse about the benefits that can be derived from surgery for a coronary artery bypass graft. The nurse bases a response on the knowledge that:
    1. Surgery will improve the chances of returning to gainful employment
    2. This surgery significantly decreases symptoms in a great number of individuals
    3. Studies have consistently shown that this surgery increases an individual's life span
    4. Evidence substantiates that surgery can prevent progression of coronary artery disease

36. In the immediate postoperative period following coronary artery bypass graft surgery, the nurse should be particularly alert for the common complication of:
    1. Graft closure with recurrence of angina-like chest pain
    2. Supraventricular dysrhythmias, especially atrial fibrillation
    3. Postpericardiotomy syndrome with fever and audible friction rub

    4. Elevation of hemoglobin and hematocrit levels with risk of embolization

37. The physician prescribes propranolol (Inderal) 20 mg po, qid for a client. In the teaching plan the nurse should include:
    1. Instructing the client not to abruptly discontinue the medication
    2. Advising the client to increase the medication if chest pain occurs
    3. Informing client to report a pulse rate less than 70 beats per minute
    4. Telling the client that alcoholic beverages may be taken in moderation

38. A client is admitted to the hospital for acute gastritis secondary to alcoholism and cirrhosis. When planning care, it is most important for the nurse to include an assessment for:
    1. Obstipation
    2. Blood in the stool
    3. Complaints of nausea
    4. Specific food intolerances

39. A client with cirrhosis develops ascites. The nurse understands that the ascites is most likely the result of:
    1. Impaired portal venous return
    2. Inadequate secretion of bile salts
    3. Excess production of serum albumin
    4. Decreased interstitial osmotic pressure

40. A client, admitted after vomiting bright red blood, is transferred to a medical/surgical unit after the vomiting ceases and the vital signs are stabilized. An assessment that best indicates that the client may still be bleeding would be:
    1. Lethargy
    2. Rapid pulse
    3. Deep breathing
    4. Abdominal pain

41. A client with bladder cancer is having diagnostic tests. The client asks the nurse, "How will I be able to urinate if my bladder is removed?" The best initial response by the nurse would be:
    1. "You can still function normally without a bladder."
    2. "I am sure this is very upsetting to you, but it will be over soon."
    3. "The tests will help to determine whether your bladder has to be removed."
    4. "I know you're upset, but there are alternatives to removing your bladder."

42. When removing an older client's meal tray, the nurse notices that the client did not eat any of the chicken. When asked why the chicken was not eaten, the client states, "I only eat meat once a week because old people don't need protein every day." Based on this statement the nurse knows that the client should be taught about the:
    1. Need for home-delivered meals
    2. Effect of aging on the need for some foods
    3. Foods included in the Food Guide Pyramid
    4. Need for meat at least once per day throughout life

43. A client with an intestinal obstruction has an intestinal tube (Miller-Abbott) inserted. The nurse is to maintain patency of the tube by instilling 30 ml of normal saline prn. When instilling this normal saline the nurse should:
    1. Subtract the 30 ml from the gastric output
    2. Record the 30 ml on the intake and output record
    3. Consider the amount too small to report on the I&O record
    4. Recognize that it will equal insensible losses and therefore is insignificant

44. When a client has a nephrostomy tube, the priority nursing care is to:
    1. Ensure drainage of urine
    2. "Milk" the tube every 2 hours
    3. Keep an accurate record of intake and output
    4. Instill 2 ml of normal saline solution every shift

45. A client is to receive 0.25 mg of digoxin IM. The ampule is labeled 0.5 mg = 2 ml. The nurse should administer:
    1. 0.8 ml
    2. 1 ml
    3. 1.2 ml
    4. 8 ml

46. The finding the nurse should consider unusual when performing an assessment of a client for increased intracranial pressure is:
    1. Rapid pulse
    2. Psychotic behavior
    3. Jacksonian seizures
    4. Nausea and vomiting

47. The nurse should recognize that the greatest danger of premature ventricular beats is that they can lead to ventricular fibrillation if they strike on the:
    1. T wave
    2. P wave
    3. PR interval
    4. QRS complex

48. A physician plans to administer chemotherapy 2 weeks after a client has had surgery for colon cancer. The delay in instituting the plan for drug therapy is because the drugs:
    1. Interfere with cell growth and delay wound healing
    2. Cause vomiting, which endangers the integrity of the incisional area
    3. Decrease red blood cell production, and the resultant anemia would add to postoperative fatigue
    4. Increase edema in areas distal to the incision by blocking lymph channels with destroyed lymphocytes

49. When planning care for a client with Cushing's syndrome, the nurse should take into consideration that the client would have:
    1. Hyperkalemia and edema
    2. Hypotension and sodium loss
    3. Muscle wasting and hypoglycemia
    4. Muscle weakness and frequent urination

50. Following an adrenalectomy, when teaching a client about the prescribed medications, the nurse should plan to emphasize the fact that:
    1. Steroid therapy will be given in conjunction with insulin
    2. Once regulated, the dosage will remain the same for life
    3. Taking the medications late in the evening may cause sleeplessness
    4. While taking the medications the client may have to restrict salt intake

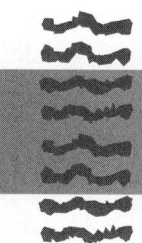

MEDICAL-SURGICAL NURSING QUIZ 2 ANSWERS AND RATIONALES

1. ① Skim milk contains about 12 grams of CHO per cup. (2) (PC; IM; EN)
   2 There are about 30 grams CHO in 1 cup of apple juice.
   3 There are about 16 grams CHO in 1 cup nonfat yogurt.
   4 There are about 25 grams CHO in 1 cup orange juice.

2. ④ All tubes should be immediately attached to appropriate collection devices to permit drainage; T-tube and nasogastric-tube drainage lessens tension on the operative site. (2) (CI; IM; GI)
   1 A T-tube drains by gravity and is never irrigated by the nurse.
   2 A T-tube should not be fastened to the bed sheets; a T-tube is surgically positioned in the common bile duct, and tension on the tube must be avoided to prevent accidental removal.
   3 This is not the priority at this time; this would be done later at the change of shift or when the collection devices get too full.

3. ③ Urine should drain continually from the conduit because there is no sphincter control unless a continent conduit is created. (2) (CI; EV; RG)
   1 Vomiting is a common occurrence after anesthesia.
   2 This is expected; the stoma may be swollen for several weeks after surgery.
   4 This is expected; bowel sounds should be diminished because of anesthesia and intestinal manipulation during surgery.

4. ④ Initial rapid assessment will determine priorities of care and subsequent actions. (3) (CC; PL; CV)
   1 Intravenous therapy and transfusions will be ordered, but baseline data are needed to assess the client's present condition and significance of future responses.
   2 Although important, obtaining a history and x-ray films can be postponed until bleeding is controlled and injuries are assessed.

3 A thorough physical assessment is too time-consuming initially; open wounds can be covered at a later time.

5. ③ Elevation decreases edema, promotes drainage, and facilitates respirations, while slight flexion prevents strain on the suture line. (2) (CI; IM; RE)
   1 This may increase edema, stasis of drainage, obstruction of the airway, and tension on the suture line.
   2 This may increase edema, stasis of drainage, and obstruction of the airway.
   4 Same as answer 1.

6. ② Dialysis removes chemicals, wastes, and fluids usually removed from the body by the kidneys. (3) (CC; IM; RG)
   1 The mention of heart problems is a threatening response and may cause increased fear or anxiety.
   3 This is a threatening response and can cause an increase in anxiety.
   4 Dialysis helps maintain fluid and electrolyte balance; there are no data to indicate the cause of the acute renal failure or previous therapy.

7. ④ Once survival needs are met and pain diminishes, there is a realization of numerous implications for the future. (2) (SC; AN; EH)
   1 The client is not talking about the colostomy.
   2 The client is expressing realistic concerns and needs to do more than think about the future.
   3 The information presented is not adequate to make this conclusion.

8. ① When taken with food, aluminum hydroxide binds phosphorus in the gut, preventing its absorption; this decreases serum phosphorus. (3) (PT; IM; DR)
   2 Aluminum hydroxide should be taken routinely with calcium supplements to maintain calcium-phosphorus balance.
   3 The desired action of aluminum hydroxide is altered when it is taken in this manner.
   4 Same as answer 3.

9. ③ By the rule of nines: each arm is 9%, each leg is 18%, and the head is 9%. Therefore the total percentage of burned area on this client is 36%. (1) (SI; AS; IT)
   1 This figure reflects a miscalculation of the rule of nines.
   2 Same as answer 1.
   4 Same as answer 1.

10. ④ In the first 48 hours after a severe burn, fluid moves into the tissues surrounding the injured area. Fluid is also lost in drainage and from evaporation. This results in a decreased blood volume and could lead to shock. (2) (DA; PL; FE)
   1 Although pain relief is an important aspect in the care of clients with burns, the immediate priority is to replace fluid losses to prevent death.
   2 The burn wound is generally considered sterile for the first 24 hours; if fluid losses are not replaced immediately, the client may die before infection can set in.
   3 Blood loss is usually minimal; the loss of fluid, colloids, and electrolytes is what causes the hypovolemia.

11. ① A gastrostomy is an opening made directly through the abdominal wall into the stomach. When tube feedings are given via this route, they bypass the upper GI tract and reduce the risk of tracheal aspiration. (2) (PC; AN; GI)
   2 Both methods use gravity.
   3 Clients can be taught to feed themselves with either method.
   4 The amount of feeding is not affected.

12. ③ This will enable the physician to identify the presence of *Helicobacter pylori*. Two thirds of individuals with gastric or duodenal ulcers are infected with this organism. (3) (CI; AS; GI)
   1 This will outline structural changes in the lower GI tract; it will not outline the stomach or duodenum.
   2 This will identify the presence of malignant cells.
   4 This may identify melena or parasites, but it is not definitive for peptic ulcers.

13. ① Hypoxia stimulates production of large quantities of erythrocytes in an attempt to compensate for the lack of oxygen. (2) (CI; AN; BI)
   2 White blood cell production increases with infection, but hemoglobin and hematocrit levels are not measures of white blood cell counts.
   3 Hypercapnia is an increase in PCO_2 in extracellular fluid; this has no direct effect on blood cell counts.

4 There is a loss of extracellular fluid in acute infections with a fever, but in a chronic condition this fluid is replenished.

14. ④ Because the affected lung will not expand, aeration of the lung will not be complete and breath sounds will be diminished. (1) (DA; AS; RE)
   1 This occurs with congestive heart failure, not with a pneumothorax; with a pneumothorax there is no air in the alveoli to produce crackles.
   2 This occurs with asthma, not with a pneumothorax.
   3 This is a broad term that includes all abnormal breath sounds; it is not specific to pneumothorax.

15. ② This is a common and expected side effect of vincristine. (3) (PT; EV; DR)
   1 Prednisone is not known to cause neurotoxicity.
   3 Burning and tingling are not related to vasoconstriction but rather to neurotoxicity.
   4 Tingling is associated with hypocalcemia, which is not induced by nausea and vomiting.

16. ① The purpose of monitoring is to correlate dysrhythmias with the client's reported activity. (2) (CI; IM; CV)
   2 A microwave oven would have no effect on the Holter monitor and would not affect the readings.
   3 The client should take nitroglycerin as needed and note it in the activities diary.
   4 Not required for correct interpretation of the test.

17. ② The incision is just below the diaphragm; deep breathing causes tension and pain. (2) (CI; EV; GI)
   1 Clients with nasogastric tubes mouth breathe, limiting nasal irritation.
   3 It would not move because it is sutured in place; it is unlikely to cause cramps because it is not in the intestine.
   4 Dressings do not encircle the abdomen; they should not be tight enough to restrict respirations.

18. ② Sodium is an electrolyte that can pass through the semipermeable membrane during hemodialysis. (2) (PC; PL; RG)
   1 Red blood cells do not pass through the semipermeable membrane during hemodialysis.
   3 Glucose does not pass through the semipermeable membrane during hemodialysis.
   4 Bacteria do not pass through the semipermeable membrane during hemodialysis.

19. (3) When transferring a suspected back-injured client, the client should be positioned to keep the vertebral column in perfect alignment (back straight) to prevent further spinal cord damage by vertebral (bone) movements. (1) (SI; PL; NM)
1 To prevent additional damage to the spinal cord, the vertebral column should be kept in alignment.
2 Same as answer 1.
4 Same as answer 1.

20. (3) Lack of or reduced movement predisposes the paraplegic or quadriplegic client to urinary tract infection and stone formation. (1) (PC; PL; NM)
1 All individuals require fluid to prevent dehydration; this is not why fluids are encouraged for this client.
2 Administration of fluids does not maintain electrolyte balance.
4 Fluids do not prevent temperature elevation, unless it is elevated because of dehydration.

21. (3) 7200 ml ÷ 24 hours = 300 ml/hour. (1) (DA; AN; FE)
1 This is an inadequate hourly rate; the rate should be 300 ml/hr.
2 Same as answer 1.
4 This is an excessive hourly rate; the rate should be 300 ml/hr.

22. (3) This will prevent aspiration. (2) (CI; PL; RE)
1 Unsafe; it could promote aspiration.
2 Although done because hemoptysis often occurs, it is most important to prevent aspiration and facilitate breathing.
4 This is unnecessary for the client who has returned to the unit following the procedure.

23. (3) A home-care program would be more efficient and cost effective; this couple can manage with proper assistance. (2) (CC; PL; GD)
1 Because the couple appear to be functioning, as long as some assistance at home is provided, it is not necessary to move them at this time.
2 There is nothing in the history to demonstrate that a skilled nursing facility is necessary.
4 Same as answer 1.

24. (3) True statement; currently immunotherapy is mainly useful as an adjuvant therapy. (2) (CC; AN; BI)
1 It does not directly affect tumor cells but stimulates the body's immune response.
2 Although used, surgery, chemotherapy, and radiation therapy are still considered primary modes of therapy for cancer.

4 Untrue; immunotherapy, to date, has shown promise with only a few types of malignancies, such as malignant melanoma, leukemia, and myeloma.

25. (3) These are measurable responses to therapy and are the desired outcomes. (2) (CC; EV; RG)
1 This is too much; it is approximately double the amount that is necessary.
2 No dietary restrictions are necessary.
4 There is no need to limit activities.

26. (1) Obesity causes stress on the diaphragmatic musculature, which weakens and allows the stomach to protrude into the thoracic cavity. (2) (HI; AS; GI)
2 Inflammation of the bronchi would not weaken the diaphragm.
3 This may cause an enlarged liver or pancreatitis, but not a hiatus hernia.
4 This results from increased portal pressure; it does not cause a hiatus hernia.

27. (4) The shift of plasma proteins into the burned area increases the shift of fluid from the intravascular to the interstitial compartment. The result is decreased blood volume and hypovolemic shock. (3) (DA; AN; FE)
1 Decreased glomerular filtration may occur because of hypovolemia; it is not involved in the cause of hypovolemia.
2 Extracellular fluid is lost through burned tissue.
3 Sodium passes to the burned area and helps cause blister formation.

28. (1) Proliferation of malignant cells may obstruct the bronchial tree or foster development of exudate in the pleural space, decreasing the availability of oxygen and increasing retention of carbon dioxide. (2) (DA; AN; RE)
2 A tumor of the lung does not cause abdominal distention or pressure.
3 While anxiety related to pain may increase the respiratory rate, it would not cause severe difficulty in breathing.
4 This is not related to cancer of the lung.

29. (1) The procedure should be explained so the client understands that the tracheostomy can serve as an entrance for bacteria and that cleanliness is imperative. (2) (SI; PL; RE)
2 Laryngectomized clients may no longer swim because water will flood the lungs.
3 Suctioning must be performed only as needed; a pattern is not necessary.
4 Sterile technique is not required; medical aseptic technique is adequate and realistic.

30. (3) Emotional outbursts are common and fleeting in these clients; it is best not to emphasize them. (2) (SC; IM; EH)
1 The client may be unaware of the reason; inappropriate emotional responses and outbursts are common.
2 The client is unable to control this emotion, and focusing on it may exaggerate the outburst.
4 The client is probably unaware of the cause of the crying, and saying it is normal is not reassuring.

31. (4) The damaged liver causes rising ammonia levels, resulting in CNS irritation, which produces behavioral changes. (3) (CC; PL; GI)
1 The liver cannot metabolize protein, and a low-protein diet is indicated.
2 Chlorpromazine is detoxified by the liver and is therefore contraindicated in severe hepatic disease.
3 Kidney function is usually not affected.

32. (2) This is an antibiotic given to decrease gram-negative bacteria in the colon, which should limit postoperative infection. (3) (SI; PL; GI)
1 These are used for constipation; oil-retention enemas would not be ordered before surgery; tap water enemas until clear might be ordered.
3 This would be used for a client with ruptured esophageal varices, not for one having a hemicolectomy.
4 This is contraindicated; a diet to decrease bulk and empty the colon would be ordered.

33. (1) This is a common manifestation caused by vascular disease, particularly renal vascular disease (renin/angiotensin mechanism). (3) (HI; AS; CV)
2 Lymphadenopathy does not generally occur.
3 Weight loss is common.
4 Peripheral nerves, not cranial nerves, can be involved.

34. (3) Both Lanoxin and Pronestyl decrease cardiac conduction, with resultant depression of the myocardium. (2) (PT; EV; DR)
1 These drugs do not influence the CNS.
2 These drugs do not influence the body's reflexes.
4 These drugs do not influence respirations.

35. (2) More than 80% of those who have this surgery initially have marked relief of their symptoms. (2) (CI; IM; CV)
1 This depends on the client's presurgical condition, not the surgery itself.
3 So far, studies have failed to show that bypass surgery affects life span.

4 The surgery itself does not affect the disease process; clients must also reduce risk factors (obesity, smoking, and poor diet).

36. (2) These dysrhythmias result from postoperative inflammation around the SA node area. (3) (CI; EV; CV)
1 This syndrome occurs later in the postoperative period.
3 This syndrome occurs later, not immediately.
4 Hgb and HCT levels usually fall; anemia can be a problem.

37. (1) Abrupt discontinuation of Inderal may cause an acute myocardial infarction. (2) (PT; PL; DR)
2 Clients should never increase medications without medical direction.
3 The pulse rate can go much lower as long as the client feels well and is not dizzy.
4 Alcohol is contraindicated for clients taking Inderal.

38. (2) Erosion of blood vessels may lead to hemorrhage, a life-threatening situation further complicated by decreased prothrombin production. (3) (CI; PL; GI)
1 Increased intraabdominal pressure may cause this; there is no immediate threat to life. Assessment for bleeding takes priority.
3 Same as answer 1.
4 While this may cause gastritis, there is no immediate threat to life; assessment for bleeding takes priority.

39. (1) An enlarged liver impairs venous return, leading to increased portal vein hydrostatic pressure and a fluid shift into the abdominal cavity. (2) (DA; AN; GI)
2 Bile plays an important role in digestion of fats but is not a major factor in fluid balance.
3 Increased serum albumin causes hypervolemia, not ascites.
4 Ascites is not related to the interstitial fluid compartment.

40. (2) Tachycardia is a cardiovascular compensatory mechanism as the effort to circulate the lowering blood volume is intensified. (1) (DA, AS; CV)
1 This is not an initial response to blood loss; client is more apt to be restless; lethargy may occur later.
3 Breathing may be rapid, but not deep.
4 This is not a response to blood loss

41. (4) This response offers the best combination of factual information and emotional support. (3) (SC; IM; EH)

1 This disregards the client's feelings; it is inaccurate information because if the bladder is removed, bladder function will not be normal.

2 Although this reflects the client's feelings, further communication is cut off by the second part of the response.

3 This response is factual but does not answer the question or offer emotional support; it may in fact increase anxiety.

42. ③ The need for the six basic food groups in the Food Guide Pyramid continues throughout life. (1) (DF; PL; GD)

1 The priority is to educate the client, although home-delivered meals may be one way to provide adequate nutrition.

2 Aging per se has no effect on the specific nutrients needed; however, it may influence digestion and/or absorption of food.

4 Protein is needed every day, but it need not be in the form of meat.

43. ② All fluid taken in by the client, regardless of the route, should be recorded on the intake and output record; documentation implies that the action was carried out. (2) (DA; IM; GI)

1 Fluid instilled must be added to the intake, not subtracted from the output; the tube is an intestinal tube, not a gastric tube.

3 The instillation is to be repeated as necessary and the total amount instilled may be significant.

4 No amount of fluid should be considered insignificant; insensible losses through the skin and lungs generally equal approximately 800 ml daily.

44. ① The tube must be kept patent to prevent urine backup, hydronephrosis, and kidney damage. (3) (CI; IM; RG)

2 This is unnecessary.

3 Although this is important, it is not the priority.

4 This is a dependent function of the nurse that requires a physician's order; this may be done by the physician rather than the nurse.

45. ② 0.25 mg/0.5 mg × X/2 ml
　　0.5 X = 0.5
　　　X = 1 ml　(1) (PT; IM; DR)

1 This would be less than the prescribed dose.

3 This would be more than the prescribed dose.

4 Same as answer 3.

46. ① When there is increased pressure within the cranial cavity, the body adapts by increasing the BP and decreasing the pulse. (3) (CI; EV; NM)

2 Signs and symptoms reflect the part of the brain upon which pressure is exerted, so that changes

in behavior, judgment, consciousness, and motor function may occur, as well as autonomic changes, such as pupil size and reactivity, vital signs, and vomiting.

3 Same as answer 2.

4 Same as answer 2.

47. ① The T wave is the period of repolarization of the ventricles. Stimulation of the ventricles during this vulnerable period often causes ventricular fibrillation. (3) (CI; AN; CV)

2 The P wave represents atrial contraction.

3 The PR interval represents the time it takes the impulse to travel from the SA node to the ventricular musculature.

4 The QRS complex represents ventricular contraction.

48. ① Chemotherapeutic agents are not specific for malignant cells. They generally interfere with protein synthesis and cell division in all rapidly dividing cells, including those regenerating traumatized tissue (as in wound healing), bone marrow, and cutaneous and alimentary tract epithelial tissue. (2) (SI; AN; DR)

2 Vomiting should not disturb the integrity of the area.

3 Decreased RBC levels caused by bone marrow depression can be corrected with transfusions.

4 Chemotherapy would not increase edema.

49. ④ Increased gluconeogenesis may lead to hyperglycemia and glycosuria, which can produce urinary frequency. Protein catabolism will cause muscle weakness. (3) (CI; PL; EN)

1 As sodium ions are retained, potassium is excreted; the result is hypokalemia. Edema occurs because of sodium retention.

2 These are symptoms of Addison's syndrome; in Cushing's syndrome retention of sodium and fluids leads to hypervolemia and hypertension.

3 Muscle wasting results from increased protein catabolism; however, hyperglycemia rather than hypoglycemia will result from increased gluconeogenesis.

50. ④ Administration of adrenocortical hormones causes sodium retention by the kidneys. Dietary intake of salt must therefore be limited. (2) (CI; PL; EN)

1 Since pancreatic function is unimpaired, insulin therapy is not normally indicated.

2 Since there is normally an increased secretion of glucocorticoids under stressful situations, the dosage must be adjusted accordingly.

3 Insomnia is not an adverse effect of adrenocortical hormones.

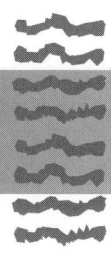

MEDICAL-SURGICAL NURSING
QUIZ 3

1. A client with cancer of the prostate has an extremely elevated serum alkaline phosphatase level. This finding should prompt the nurse to plan to:
   1. Institute seizure precautions
   2. Measure his intake and output
   3. Monitor his plasma pH for acidosis
   4. Handle him gently when turning him

2. When preparing a client for a liver biopsy, the nurse should instruct the client to:
   1. Turn on the left side after the procedure
   2. Breathe normally throughout the procedure
   3. Hold the breath at the moment of the actual biopsy
   4. Bear down (Valsalva maneuver) during the insertion of the biopsy needle

3. A 40-year-old visits a neurologist with complaints of blurred or double vision and muscular weakness. After multiple sclerosis is diagnosed the client, visibly upset, reports this diagnosis to a friend who is a nurse. The nurse could best respond:
   1. "Don't worry; early treatment often alleviates symptoms of the disease."
   2. "See another physician. I've heard of several treatments that aid recovery."
   3. "You should see a psychiatrist who will help you cope with this shocking news."
   4. "That must have really floored you. Tell me what the physician told you about it."

4. The major nursing problem in caring for a client with hyperthyroidism would be:
   1. Providing sufficient rest
   2. Modifying hospital routines
   3. Providing an adequate diet
   4. Keeping the bed linen neat

5. The physician orders propylthiouracil for a client with hyperthyroidism. The nurse explains that this drug:
   1. Increases the uptake of iodine
   2. Produces atrophy of the thyroid gland

   3. Interferes with the synthesis of thyroid hormone
   4. Decreases the secretion of thyroid-stimulating hormone

6. After a pneumonectomy, it would be most appropriate to have a client perform deep breathing and coughing exercises:
   1. Every hour for the first 24, then every 2 to 4 hours
   2. Every 2 hours for the first 24, then every 2 to 4 hours
   3. Every 2 hours for the first 24, then every 4 hours when awake
   4. Every 15 to 30 minutes for the first 24 hours, then every 2 to 4 hours

7. Stomatitis is a common side effect of chemotherapy. To limit this side effect, the teaching plan for a client who is receiving chemotherapy should include instructions to:
   1. Rinse the mouth 3 times a day with lemon juice and water
   2. Brush the teeth once daily and use dental floss after each meal
   3. Vigorously cleanse the mouth with toothpaste and a firm toothbrush
   4. Frequently cleanse the mouth with a soft toothbrush or a power spray

8. A client's appetite is returning approximately 1 week after a deep partial-thickness burn of the anterior chest. When helping the client plan a menu for the next day, the nurse should teach the client to include:
   1. Moderate caloric intake, liberal potassium intake, and 1 g protein/kg/day
   2. Restricted caloric intake, liberal potassium intake, and 3 g protein/kg/day
   3. 5000 or more calories per day, liberal potassium intake, and 3 g protein/kg/day
   4. 3000 or more calories per day, restricted potassium intake, and 1 g protein/kg/day

9. A client with a history of heavy alcohol use develops portal hypertension and an elevated serum aldosterone level. The nurse should carefully observe the client for:
   1. Chloride depletion and hypovolemia
   2. Potassium retention and dysrhythmias
   3. Sodium retention and fluid accumulation
   4. Calcium depletion and pathologic fractures

10. During the early postoperative period following open heart surgery, adequate oxygenation is essential because:
    1. All clients have a chest tube in place
    2. Hypoxia can precipitate respiratory alkalosis
    3. Hypoxia can stimulate dangerous dysrhythmias
    4. An increased respiratory rate adds to postoperative pain

11. The nurse notes that 2 weeks after severe burns, a client is losing 2 pounds of weight daily. The nurse's best action would be to adjust the client's diet by adding:
    1. Low-sodium milk
    2. High-protein drinks
    3. Fruit juices low in potassium
    4. 10% more calories in the form of fats

12. Prior to a total laryngectomy, an important aspect of preoperative nursing care includes:
    1. Having a speech therapist visit the client
    2. Adequate explanation of the nature of the surgery to be performed
    3. Basing instruction on those areas in which the client has questions
    4. Instruction in breathing exercises and/or equipment used postoperatively to prevent complications

13. A client has a tracheostomy tube attached to a tracheostomy collar for the delivery of humidified oxygen. The client will probably need suctioning primarily because the:
    1. Humidified oxygen is saturated with fluid
    2. Inner cannula of the tracheostomy tube irritates the mucosa
    3. Tracheostomy tube interferes with the ability to cough effectively
    4. Weaning process increases the amount of respiratory secretions

14. A client's serum albumin level is low, and the physician prescribes normal serum albumin 5% IV. Albumin replacement is expected to:
    1. Decrease capillary perfusion and BP
    2. Decrease ascites and the blood ammonia level

3. Decrease venous stasis and the blood urea nitrogen level
4. Decrease tissue fluid accumulation and the hematocrit level

15. When a client with chronic obstructive pulmonary disease becomes dyspneic and anxious, the nurse's first action to decrease dyspnea should be to:
    1. Increase oxygen to 6 L/min
    2. Check vital signs, including BP
    3. Encourage rhythmical breathing
    4. Have the client breathe into a bag

16. A client is admitted to the trauma unit of the hospital after a motor vehicle accident. Among other injuries the physician suspects that the client has a ruptured bladder. When interviewed, the statement by the client that would indicate an increased risk for rupture of the bladder is:
    1. "I have frequent bouts of cystitis."
    2. "I did not urinate before I got in the car."
    3. "I have a family history of bladder cancer."
    4. "I was drinking a can of soda while I was driving."

17. A client with the tentative diagnosis of Hodgkin's disease asks how the surgeon will know that it is definitely this disease. The nurse explains that the diagnosis is routinely confirmed by:
    1. Bone scan
    2. Lymph node biopsy
    3. Computerized tomography (CT) scan
    4. Radioactive iodine (^{131}I) uptake studies

18. The diet regimen that would be appropriate for a client with a gastric ulcer would be:
    1. A mechanical soft diet
    2. A low-fat, high-protein liquid diet
    3. Hourly feedings of cream and milk
    4. Regular meals that can be tolerated

19. A client with diabetes who has been diagnosed as having peripheral vascular disease tells the nurse that before hospitalization, exercise resulted in severe cramplike pain in both legs. The nurse should include in the teaching plan specific measures the client can use to increase arterial blood flow to the extremities. These measures should include:
    1. Exercises that promote muscular activity
    2. Meticulous care of minor skin breakdown
    3. Elevation of legs above the level of the heart
    4. Daily cleansing of feet by soaking in hot water

20. A client with liver disease is receiving neomycin sulfate (Mycifradin) orally. The purpose of administration of the drug is to:
    1. Increase urea digestive activity of enteric bacteria

2. Suppress ammonia-forming bacteria in the intestinal tract
3. Protect regenerative nodules in the liver from invading bacteria
4. Protect against infection while immune mechanisms are deficient

21. Immediately before an abdominal paracentesis, the nurse should ask the client to void because:
1. An empty bladder will decrease the intraabdominal pressure
2. A full bladder increases the danger of puncture of the bladder
3. A full bladder decreases the amount of fluid in the abdominal cavity
4. A urine specimen must be obtained at this time to check level of nonprotein nitrogen

22. After a myocardial infarction, a client's orders include strict bed rest and a clear liquid diet. The nurse should explain that the primary reason for this diet is to reduce the:
1. Client's weight
2. Acidity of gastric contents
3. Amount of fecal elimination
4. Metabolic workload of digestion

23. Two days following a myocardial infarction, a client's temperature is elevated. This indicates:
1. Pneumonia
2. Tissue necrosis
3. Possible infection
4. Pulmonary infarction

24. The primary responsibility of the nurse when caring for a client with a chest tube attached to a three-chamber underwater seal drainage system is to:
1. Maintain the closed system
2. Encourage deep breathing and coughing
3. Maintain mechanical suction to the system
4. Keep the client in the dorsal recumbent position

25. The nurse should teach clients with phosphatic calculi that their diets may include:
1. Apples
2. Chocolate
3. Rye bread
4. American cheese

26. When reviewing the laboratory results for a client in acute renal failure, the nurse notes that the serum potassium level is 6.2 mEq. The nurse should first:
1. Call the cardiac arrest team to alert them
2. Call the laboratory to schedule a repeat test
3. Obtain an ECG strip and have lidocaine available
4. Take the client's vital signs and notify the physician

27. One week after an above-the-knee amputation, a client refuses to go to physical therapy and tells the nurse, "I'll never be a whole person again!" The nurse's best response would be:
1. "Relax, you're still the same person you've always been."
2. "You've lost a part of yourself. That must be very difficult for you."
3. "You may feel that way, but I'm sure your family considers you a whole person."
4. "You must go to physical therapy every day or you will develop muscle contractures."

28. A 67-year-old client is diagnosed as having a right-sided cerebrovascular accident (CVA) and is admitted to the hospital. When preparing to care for this client, the nurse should plan to:
1. Use a bed cradle to prevent dorsiflexion of the feet
2. Apply elastic stockings to prevent flaccid leg muscles
3. Do passive range-of-motion exercises to prevent muscle atrophy
4. Use a hand roll and extend the left upper extremity on a pillow to prevent contractures

29. When teaching the client with arthritis about positioning, the nurse should encourage the client to:
1. Assume the Fowler's position
2. Maintain the limbs in extension
3. Place pillows beneath the knees
4. Assume a position that is most comfortable

30. A client is diagnosed as having type 2 diabetes, and the physician orders an oral hypoglycemic. The nurse should include in the teaching plan that people taking oral hypoglycemics:
1. Should not work where food is readily accessible
2. Need not be concerned about serious complications
3. Consciously or unconsciously may tend to relax dietary rules
4. Are not as threatened by their disease as those taking insulin

31. A client with Lyme disease is to self-administer ceftriaxone sodium (Rocephin) 2 g via IV at home. The drug is to be diluted in 100 ml of D5W and infused over a period of 35 minutes. The drop factor of the tubing the client will be using is 10 gtt/ml. The nurse should teach the client to set the flow at:
1. 17 gtt/min
2. 22 gtt/min
3. 29 gtt/min
4. 35 gtt/min

32. When discussing immunity with a prenatal client during her first visit to the prenatal clinic, the nurse recalls that active immunity occurs when:
    1. Protein antigens are formed in the blood to fight invading antibodies
    2. Protein substances are formed by the body to destroy or neutralize antigens
    3. Blood antigens are aided by phagocytes in defending the body against pathogens
    4. Sensitized lymphocytes from an immune donor act as antibodies against invading pathogens

33. After surgery a client develops pneumonia and atelectasis. To promote breathing the nurse should instruct the client to:
    1. Take short frequent breaths
    2. Exhale with the mouth open
    3. Plan to do the breathing exercises twice a day for 10 minutes
    4. Place the hands on the abdomen on inspiration and watch them rise

34. After surgery for a parotid tumor, the priority nursing intervention during the immediate postoperative period should be:
    1. Providing emotional support
    2. Observing for signs of infection
    3. Keeping the trachea free of secretions
    4. Promoting a means for communication

35. The nurse would expect a client with jaundice to also complain of:
    1. Pruritus
    2. Diarrhea
    3. Blurred vision
    4. Bleeding tendencies

36. During a teaching session about insulin injections, a client asks the nurse, "Why can't I take the insulin in pills instead of taking shots?" The nurse should respond:
    1. "Insulin cannot be manufactured in pill form."
    2. "Your doctor will order pills when you are ready."
    3. "The route of administration is decided on by the physician."
    4. "Insulin is destroyed by gastric juices, rendering it ineffective."

37. A 50-year-old client is admitted to the hospital with a suspected brain tumor. Based on the history of loss of equilibrium and coordination, the nurse suspects the tumor is located in the:
    1. Cerebellum
    2. Parietal lobe
    3. Basal ganglia
    4. Occipital lobe

38. The nurse checks for hypocalcemia by placing a blood pressure cuff on a client's arm and inflating it. After about 3 minutes the client develops carpopedal spasm. The nurse records this finding as a positive:
    1. Homans' sign
    2. Romberg's sign
    3. Chvostek's sign
    4. Trousseau's sign

39. A client is admitted for a coronary artery bypass graft. The client states that the physician said that pacemaker wires will be inserted during surgery as a precautionary measure. The nurse explains that the pacemaker is used to:
    1. Prevent a rapid heart rate
    2. Maintain a constant rhythm
    3. Provide access for defibrillation
    4. Manage an abnormally slow rate

40. A client with a cerebrovascular accident has regained control of bowel movements but is still incontinent of urine. To help reestablish bladder control, the nurse should encourage the client to:
    1. Assume a normal position for voiding
    2. Void every 4 hours and attempt to hold urine between set times
    3. Attempt to void more frequently in the afternoon than in the morning
    4. Drink a minimum of 4000 ml of fluid equally divided among the hours while awake

41. A client is taken to the emergency room with a sucking stab wound of the left thorax. The nurse should position the client:
    1. On the back with the feet elevated
    2. On the left side with the head elevated
    3. In a high-Fowler's position with the left side supported
    4. On the right side flat in bed with a pillow supporting the left arm

42. When assessing a client with a sucking stab wound of the chest, the nurse should be concerned primarily with the:
    1. Degree and level of pain
    2. Quality and depth of respirations
    3. Amount of serosanguineous drainage
    4. Blood pressure and pupillary response

43. A client is receiving total parenteral nutrition that contains glucose to which minerals, electrolytes, and vitamins are added. If the electrolyte, potassium, is not added, hypokalemia results. The nurse knows the signs and symptoms of hypokalemia include:
    1. Hyperventilation
    2. Metabolic alkalosis

3. Dysrhythmias such as heart block
4. A decreased serum potassium to 5.5 mEq/L

44. After emptying the contents of a portable wound drainage system, the nurse should squeeze the collection container and replace the stopper to:
1. Establish positive pressure
2. Decrease negative pressure
3. Maintain atmospheric pressure
4. Increase the difference in pressures

45. To promote independence in a debilitated, elderly male client with glaucoma who places great value on independence, the nurse should encourage the client to:
1. Prevent stressful events that can increase his symptoms
2. Be able to perform his own household chores and shopping
3. Conserve his eyesight by not reading or watching television
4. Learn to administer his prescribed eye medications correctly

46. A client undergoes removal of a pituitary tumor through a transsphenoidal approach. Postoperatively the nurse should plan to:
1. Keep the client npo until nasal packing is removed
2. Provide vigorous oral hygiene, including toothbrushing
3. Raise the head of the bed to a 30-degree angle at all times
4. Encourage the client to deep breathe and cough frequently

47. During preoperative teaching before a transurethral prostatectomy, the nurse tells the client that after the surgery:

1. His urine will be bright red for at least 24 to 48 hours
2. He can expect spasms of the bladder during the first 24 to 48 hours
3. Oral fluids should be avoided because of continuous urinary irrigation
4. He should use the Valsalva maneuver to decrease bladder contractions

48. A 52-year-old engineer is admitted to the hospital with Laënnec's cirrhosis and chronic pancreatitis. Bile salts are ordered, and the client asks why they are needed. The nurse replies that they:
1. Stimulate prothrombin production
2. Aid absorption of vitamins A, D, and K
3. Promote bilirubin secretion in the urine
4. Stimulate contraction of the common bile duct

49. Clients are encouraged to perform deep breathing exercises after surgery. The reason for this is that deep breathing exercises help to:
1. Increase blood volume
2. Expand the residual volume
3. Counteract respiratory acidosis
4. Decrease partial pressure of oxygen

50. An elderly client is hard of hearing and has severe painful rheumatoid arthritis. The client is admitted to a nursing home after becoming incontinent of urine; the family was no longer able to provide care. The primary consideration in the care of this client would be the need for:
1. Control of pain
2. Immobilization of joints
3. Motivation and teaching
4. Bladder control reeducation

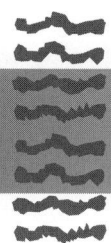

MEDICAL-SURGICAL NURSING QUIZ 3
ANSWERS AND RATIONALES

1. ④ This suggests metastasis to the bone, which results in pain and risk of pathologic fractures. (2) (CI; PL; SK)
 1 Seizure precautions are not necessary; an elevated serum alkaline phosphatase level indicates bone, not brain, involvement.
 2 Measuring intake and output is necessary for any client with prostatic cancer because of the high risk of obstruction, not just because of this test result.
 3 An elevated serum alkaline phosphatase level does not significantly affect the pH.

2. ④ This ensures that the liver does not move as it does with normal respiratory excursions. (2) (CI; IM; GI)
 1 Lying on the right side, not the left side, after the procedure applies pressure at the insertion site, preventing hemorrhage.
 2 Movement or breathing increases the danger of damage to the liver.
 3 There is no rationale for this, and it is difficult to carry out.

3. ④ This acknowledges the impact of this diagnosis on the client and should explore what is really known. (1) (SC; IM; EH)
 1 This statement is untrue and gives false reassurance.
 2 Same as answer 1.
 3 There is no evidence of ineffective coping.

4. ① Promotion of rest to reduce metabolic demands is a challenging but essential task with a client who has hyperthyroidism. (2) (PC; PL; EN)
 2 Hospital routines should not be considered a nursing problem; routines should always be flexible to meet client needs.
 3 The diet can be increased to meet metabolic demands; the client usually has an excellent appetite.
 4 The neatness of linen is not an important aspect of care; the client is usually hyperactive, and it is more important to promote rest.

5. ③ Propylthiouracil (Propyl-Thyracil), used in the treatment of hyperthyroidism, blocks the synthesis of thyroid hormones by preventing iodination of tyrosine. (2) (PT; IM; DR)
 1 Propylthiouracil does not increase the uptake of iodine.
 2 Iodine solutions reduce the size and vascularity of the thyroid gland.
 4 Thyroid-stimulating hormone (TSH), secreted by the anterior pituitary, is not affected by propylthiouracil.

6. ① Excessive endotracheal secretions after a pneumonectomy require coughing routines that are effective but not exhausting. (3) (CI; PL; RE)
 2 This is an ordinary routine for any client following anesthesia; it is not specific for pneumonectomy.
 3 The nurse should wake the client and maintain the routine.
 4 This is usually too exhausting.

7. ④ Chemotherapy destroys rapidly dividing cells of oral mucosa; frequent gentle oral hygiene limits additional trauma. (2) (CC; PL; GI)
 1 Lemon juice would be too caustic to the compromised mucosa.
 2 Flossing would disrupt and traumatize the gum surfaces; oral hygiene is needed more than once a day.
 3 Vigorous cleansing with hard materials would increase mucosal trauma.

8. ③ High calories are needed for increased BMR, potassium (after the first 48 hours) prevents hypokalemia, and protein promotes tissue repair. (3) (HI; PL; IT)
 1 This does not meet the nutritional needs of a client with burns.
 2 Same as answer 1.
 4 Same as answer 1.

9. (3) Aldosterone, a corticosteroid, causes sodium and water retention and potassium excretion by the kidneys. (2) (DA; AS; FE)
   1 Hypovolemia would not occur with increased aldosterone levels because sodium and water are retained.
   2 Potassium is excreted in the presence of aldosterone and therefore would not accumulate and cause dysrhythmias.
   4 Calcium is unaffected by aldosterone.

10. (3) Inadequate oxygenation can cause premature ventricular beats. (3) (CI; AN; CV)
   1 Although this may be true, it does not explain why adequate oxygenation is important.
   2 Hypoxia can precipitate respiratory acidosis; hyperventilation causes respiratory alkalosis.
   4 The reverse is true; postoperative pain can increase the respiratory rate.

11. (2) High-protein drinks have twice the calories per volume of other fluids and provide protein for wound healing. (2) (PC; PL; IT)
   1 Low-sodium milk does not contain adequate calories to help meet the high metabolic rate associated with burns.
   3 Fruit juices are comparably low in calories; potassium need not be restricted at this time; potassium is restricted during the hypovolemic stage (first 48 to 72 hours after burn injury).
   4 More fats are not indicated; increased calories in the form of protein and carbohydrates are needed.

12. (3) Before teaching, the nurse must determine the client's areas of concern. Learning increases when anxiety decreases, and it is hoped that knowledge will reduce anxiety. (3) (CC; IM; RE)
   1 This is not beneficial for all preoperative clients.
   2 Although this may be part of the information included in preoperative teaching, the teaching should begin at the client's level.
   4 The client will be dependent on a respirator postoperatively; breathing exercises are not required.

13. (3) Because the tracheostomy tube enters the trachea below the glottis, the client is unable to close the glottis to retain air in the lungs and raise the intrathoracic pressure and then open the glottis to expel an explosive cough. (2) (DA; AN; RE)
   1 This would decrease the need for suctioning because humidified oxygen liquefies secretions, which are then easier to expel.

2 The outer cannula of a tracheostomy tube irritates the mucosa; also, many disposable tracheostomy tubes do not have a removal cannula.
4 Weaning begins when the respiratory status improves and the amount of respiratory secretions subsides, not increases.

14. (4) Serum albumin is administered to maintain serum levels and normal oncotic pressures. It does this by pulling fluid from the interstitial spaces into the intravascular compartment, thus decreasing the hematocrit level. (3) (DA; EV; FE)
   1 The administration of albumin results in a shift of fluid from the interstitial to the intravascular compartment, which will probably increase the BP.
   2 Serum albumin will not affect blood ammonia levels; fluid accumulated in the abdominal cavity is best removed via paracentesis.
   3 Albumin administration does not affect venous stasis or blood urea nitrogen.

15. (3) This permits more complete exhalation and emptying of carbon dioxide from the lungs. (2) (CI; IM; RE)
   1 This is contraindicated; low amounts of oxygen are needed to prevent carbon dioxide narcosis.
   2 Although important, assessment will not decrease dyspnea.
   4 This is contraindicated in COPD because it increases carbon dioxide retention.

16. (2) The walls of a full bladder are stretched thinner and are more susceptible to rupture when traumatized. (3) (SI; AN; RG)
   1 This may predispose to developing infections, not to rupturing the bladder.
   3 This might increase the risk of cancer; however, it would not predispose to bladder rupture.
   4 This would not predispose the bladder to rupture when traumatized.

17. (2) The diagnosis depends on the identification of characteristic histologic features of an excised lymph node. (1) (HI; IM; BI)
   1 This is a diagnostic device to assess bony metastasis of cancers other than Hodgkin's disease.
   3 This is not used to diagnose Hodgkin's disease; it can identify the extent of the disease by locating abdominal or chest lesions.
   4 These are not indicated for Hodgkin's disease; they are usually used for radiotherapy or diagnosis of thyroid diseases.

18. (4) No specific diet is recommended; the client is encouraged to avoid meals that overdistend the stomach and foods that cause GI distress. (3) (PC; PL; GI)
    1 There is no need for a mechanical soft diet, which would be appropriate for those who have difficulty with chewing and swallowing.
    2 The client does not have to be restricted to a liquid diet.
    3 High-fat dairy products increase GI secretion and should be avoided.

19. (1) Exercise causes muscle contractions, which require an increase in arterial circulation to supply oxygen and nutrients for energy being expended. (1) (CC; PL; CV)
    2 This is important for the person with diabetes, but it does not improve arterial blood flow.
    3 This would reduce arterial blood flow; the legs should be kept dependent.
    4 Hot water is contraindicated because it can burn the skin and/or cause drying; also, individuals with diabetes may have neuropathies, which alter the perception of temperature.

20. (2) Neomycin aids in reducing the intestinal flora that act on protein substances, causing the production of ammonia. Ammonia is detoxified in the liver. When the liver is unable to perform this function adequately, blood ammonia levels rise, causing encephalopathy. (3) (PT; AN; DR)
    1 Bacteria in the intestines do not digest urea.
    3 This does not occur; neomycin controls protein metabolism, a severe problem in liver disease.
    4 Although immune mechanisms may be limited in disorders affecting the liver, the neomycin is administered to limit ammonia levels.

21. (2) When the bladder contains large amounts of urine, it becomes distended and may push upward into the abdominal cavity, where it can be punctured. (1) (SI; IM; GI)
    1 This is not the rationale for wanting the bladder to be empty.
    3 The amount of fluid in the bladder has no relationship to ascites.
    4 A urine specimen is not necessary; an empty bladder prevents accidental trauma to the bladder during a paracentesis.

22. (4) Acute care of the client with a myocardial infarction is aimed at reducing the cardiac workload. Foods that are easily digested help reduce this workload. Sympathetic nervous system involvement causes decreased peristalsis and gastric secretion, so limiting food intake will help prevent gastric distention. (2) (CI; AN; CV)

1 Weight control is not a concern immediately following coronary occlusion.
2 Gastric acidity is not reduced by a clear liquid diet.
3 Feces are in the intestine generally 4 days after the ingestion of foods. The client's desire to have a bowel movement will be the result of prehospitalization meals; the liquid diet will affect future elimination.

23. (2) The body's general inflammatory response to myocardial necrosis causes an elevation of temperature as well as leukocytosis within 24 to 48 hours. (2) (DA; EV; CV)
    1 This is not an expected complication after myocardial infarction.
    3 This is not an expected finding after tissue damage associated with an infarction.
    4 This is not a common complication after myocardial infarction; it is usually associated with phlebitis or multiple fractures.

24. (1) An airtight system is needed to reestablish negative pressure and reinflate the lung. (3) (DA; AN; RE)
    2 This is important, but not the priority.
    3 Drainage can be maintained without mechanical suction.
    4 Any position is acceptable as long as the tube is not compressed or pulled.

25. (1) These are low in phosphate. (3) (CI; PL; RG)
    2 This is not a preferred food.
    3 Same as answer 2.
    4 This is made with milk, which contains phosphate and should be avoided.

26. (4) Vital signs monitor the cardiorespiratory status; hyperkalemia can cause serious cardiac dysrhythmias that must be treated. (3) (CC; EV; FE)
    1 The cardiac arrest team is always on alert and will respond when called for a cardiac arrest.
    2 A repeat laboratory test would take time and probably reaffirm the original results.
    3 The priority is medical attention, and the physician should be notified immediately.

27. (2) This response acknowledges and reflects the client's feelings and encourages further communication. (2) (SC; IM; EH)
    1 This response negates the client's feelings.
    3 The nurse does not know how the client's family feels; this response takes the focus off the client.
    4 This is true, but telling the client this serves no therapeutic purpose at this time.

28. (4) These interventions maintain the affected left arm in functional alignment; the left side of the body will be affected with a right-sided CVA. (2) (CI; PL; NM)
1 Plantar flexion (footdrop), not dorsiflexion, may occur with a CVA; high-top sneakers or splints more appropriately prevent plantar flexion contractures.
2 Elastic stockings promote venous return rather than prevent flaccid muscles; also, this requires a physician's order.
3 Passive range-of-motion exercises prevent contractures rather than muscle atrophy; the institution of ROM should be discussed with the physician because activity during the acute phase can increase intracranial pressure; ROM may not be started until after the first 24 hours following the CVA.

29. (2) Maintenance of joints in extension offsets the deformity often seen in rheumatoid arthritis. It also frequently prevents potentially deforming contractures. (2) (HI; PL; SK)
1 The Fowler's position is not required; respiratory function is unimpaired.
3 Pressure on the popliteal space can impede venous return, increasing the risk of thrombus formation; it can also promote knee flexion contractures.
4 Clients often assume a fetal position, which results in flexion contractures of all extremities; this should be avoided.

30. (3) Taking a pill may give the client a false sense that the disease is under control, and this can lead to dietary indiscretions. (2) (HI; PL; EN)
1 A person's ability to follow dietary restrictions is highly individual; the employment setting is not a universal factor in such behaviors.
2 Diabetes is a chronic disease, and its complications can affect all individuals with the disease, particularly those who do not comply with the prescribed regimen.
4 Individuals receiving insulin may also believe that their diabetes is under control and may not adhere to their prescribed diet; having to take any medication every day can be threatening to some people.

31. (3) This is the correct flow rate; the drug is dissolved in 100 ml of D5W, so the amount to be infused (100) multiplied by the drop factor (10) equals 1000; this result divided by the amount of time in minutes (35) equals 28.57, which should be rounded off to 29 gtt/min. (2) (PT; AN; DR)

1 This rate is too slow; the drug would not be infused in 35 minutes.
2 Same as answer 1.
4 This rate is too fast; the drug would be infused in less than 35 minutes.

32. (2) Active immunity occurs when the individual's cells produce antibodies in response to an agent or its products; these antibodies will destroy the agent (antigen) should it enter the body again. (3) (HI; AN; BI)
1 Antigens do not fight antibodies; they trigger an antibody formation that in turn attacks the antigen.
3 Antigens are foreign substances that enter the body and trigger antibody formation.
4 Sensitized lymphocytes do not act as antibodies.

33. (4) Abdominal breathing improves lung expansion. (2) (SI; IM; RE)
1 Short breaths do not expand the lungs; deep slow breaths at a rate of 16 per minute should be encouraged.
2 This does not promote oxygen/carbon dioxide exchange in the alveoli; exhalation with pursed lips promotes expansion of alveoli.
3 Breathing exercises should be performed at least every 2 hours.

34. (3) A patent airway is always a priority concern; therefore removal of secretions is imperative. (2) (SI; PL; RE)
1 This is an important postoperative intervention but is not the priority immediately after surgery.
2 This is an important postoperative concern, but infection does not occur immediately.
4 Although this is important, it is not the priority.

35. (1) Itching associated with jaundice is believed to be caused by an accumulation of bile salts in the skin. (1) (PC; AS; IT)
2 This symptom is not related to jaundice.
3 Same as answer 2.
4 Same as answer 2.

36. (4) At this time, insulin in tablet form would be inactivated by gastric juices; insulin given by injection bypasses the destructive gastric juices. (2) (HI; IM; EN)
1 Insulin is not given orally at this time because it would be inactivated in the stomach; oral hypoglycemics contain substances such as sulfonylurea that stimulate beta cells to produce insulin.
2 This is incorrect information for a client who is currently insulin dependent; this provides false reassurance.

3 This does not answer the client's question; at this time insulin is administered IV or subcutaneously, and the route depends on the client's needs.

37. (1) The cerebellum is involved in synergistic control of the skeletal muscles and the coordination of voluntary movement. (1) (HI; AN; NM)
2 The parietal lobe is concerned with localization and two-point discrimination; tumors here cause motor seizures and sensory function loss.
3 Basal ganglia are concerned with large subconscious movements and muscle tone; damage here may cause paralysis, as in stroke or involuntary movements, and uncontrollable shaking, as in Parkinson's disease.
4 The occipital lobe is concerned with special sense perception; tumors here cause visual disturbances, visual agnosia, or hallucinations.

38. (4) Peripheral muscular hypoxia precipitates carpopedal spasm in the presence of hypocalcemia. (2) (DA; AS; FE)
1 This indicates thrombophlebitis when pain results from dorsiflexing the foot.
2 This indicates the loss of position sense; swaying results when the client stands still with the feet close together and the eyes closed.
3 Although this sign indicates hypocalcemia, it is elicited by tapping over the facial nerve.

39. (4) Vagal stimulation during surgery may cause a severe bradycardia. In anticipation, pacemaker wires are inserted into the right atrium to be used to initiate impulses if the natural rate falls below the preset rate of the pacemaker. (3) (CI; IM; CV)
1 A pacemaker initiates an impulse if the heart rate drops below a certain rate; the concept underlying a pacemaker is to speed up the heart, not to slow it down.
2 The rhythm can be irregular; however, if the pause between two beats is too long, the pacemaker will initiate an impulse.
3 The pacemaker wires are not used for defibrillation; defibrillator paddles are placed so that electricity affects the entire heart muscle.

40. (1) Assuming a normal position for voiding reduces tension (physical and psychologic), facilitates the movement of urine into the lower portion of the bladder, and relaxes the external sphincter (increasing pressure and initiating the micturition reflex). (3) (PC; IM; NM)
2 Bladder training should be instituted by encouraging voiding every 1 to 2 hours and progressively increasing the time between attempts.

3 Voiding should be encouraged at regular and frequent intervals during waking hours, not just in the afternoon.
4 This is an extremely large fluid intake and will result in a large volume of urine, probably increasing the frequency of incontinence.

41. (2) When the client lies on the affected side, the unaffected lung can expand to its fullest potential. Elevation of the head facilitates respirations by reducing the pressure of the abdominal organs on the diaphragm and allowing the diaphragm to descend with gravity. (3) (DA; IM; RE)
1 Maximum lung expansion is inhibited when the head is not elevated.
3 It is unclear as to what "left side supported" means. Although it facilitates diaphragmatic movement, it does not assist expansion of the right lung.
4 Pressure against the right thorax limits right intercostal expansion and gaseous exchange in the right lung; the supine position allows the abdominal viscera to restrict contraction of the diaphragm and does not permit the diaphragm to drop by gravity.

42. (2) The rate and characteristics of respiration should be assessed so that the amount of exertion required for breathing can be determined. The nurse should also evaluate signs such as unilateral chest movements that may indicate pneumothorax and tachypnea, which are associated with hypercapnia and acidosis. (2) (DA; AS; RE)
1 Although important, pain is not a life-threatening symptom.
3 Drainage may accumulate in the pleural space, which is inaccessible for direct assessment.
4 Excessive blood loss might affect the BP, but such bleeding would be indicated first by respiratory changes, because the blood would accumulate in the pleural space; pupillary response would be unaffected.

43. (3) Potassium is important for muscle contraction; the heart is a muscle and hypokalemia causes dysrhythmias. (3) (DA; EV; FE)
1 Decreased functioning of respiratory muscles may result in hypoventilation, not hyperventilation.
2 Decreased functioning of respiratory muscles may result in respiratory acidosis, not metabolic alkalosis.
4 A serum potassium level of 5.5 mEq/L is more than the normal range of 3.5 to 5 mEq/L.

44. (4) This creates negative pressure; the difference (pressure gradient) now promotes flow from the client to the container. (2) (SI; AN; IT)
1   This would prevent drainage.
2   This action will not decrease negative pressure.
3   Maintaining atmospheric pressure would not promote drainage.

45. (4) The responsibility for correctly doing this task will foster independence. (2) (DF; AN; GD)
1   This is a laudable goal but it does not necessarily lead to independence.
2   This goal is too ambitious for a debilitated, elderly client.
3   Moderate use of the eyes is not contraindicated in debilitated, elderly clients; also, these activities provide some diversion for these individuals.

46. (3) This decreases pressure on the sella turcica, as well as promoting venous return, thus limiting cerebral edema. (2) (SI; PL; NM)
1   There is no need to limit oral fluids because of the presence of nasal packing.
2   Gentle oral hygiene is performed, excluding brushing of teeth to prevent trauma to the surgical site.
4   Although deep breathing is encouraged, initially, coughing is discouraged to prevent increasing intracranial pressure.

47. (2) Spasms result from irritation of the bladder during surgery; they decrease in intensity and frequency as healing occurs. (3) (CC; IM; RG)
1   This is too long; it would indicate hemorrhage; after the first few hours, drainage should be dark red and then gradually turn pink.
3   The presence of a continuous bladder irrigation is unrelated to the amount of oral fluids that should be consumed; once the CBI is discontinued, oral fluids should be encouraged.
4   The Valsalva maneuver should be avoided because it may initiate prostatic bleeding.

48. (2) Bile salts are used to aid digestion of fats and absorption of the fat-soluble vitamins A, D, and K. (2) (PT; IM; DR)
1   Bile salts play no role in prothrombin production.
3   Bile salts are not necessary to stimulate the secretion of bilirubin.
4   Bile salts do not initiate the contraction of the common bile ducts.

49. (3) Retention of carbon dioxide in the blood lowers the pH, causing respiratory acidosis. Deep breathing maximizes gaseous exchange, ridding the body of excess carbon dioxide. (2) (DA; IM; RE)
1   Deep breathing improves oxygenation of the blood, but it does not affect volume.
2   Although regular deep breathing improves the vital capacity, residual volume is unaffected.
4   Deep breathing increases the partial pressure of oxygen.

50. (1) After the need to survive (air, food, water) the need for comfort and freedom from pain closely follow. Care should be given in order of the client's needs. (2) (PC; PL; SK)
2   Joints must be exercised to prevent stiffness, contractures, and muscle atony.
3   Motivation and learning will not occur unless basic needs, such as freedom from pain, are met.
4   Although bladder training should be included in care, it is not a priority when the client is in pain.

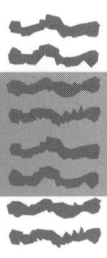

MEDICAL-SURGICAL NURSING
QUIZ 4

1. A client is admitted with 50% of the body surface area burned following an industrial explosion and fire. In evaluating the client's response to fluid replacement, according to Parkland's formula, the nurse should prepare to administer a colloid when the:
   1. Hematocrit is below 32
   2. Albumin is below 2 g/dl
   3. Urine specific gravity is 1.018
   4. Pulmonary wedge pressure is 10 mm Hg

2. A common early sign of laryngeal cancer for which the nurse should assess a client is:
   1. Aphasia
   2. Dyspnea
   3. Dysphagia
   4. Hoarseness

3. At the accident scene, emergency treatment for a client who has sustained partial- and full-thickness burns to the chest, right arm, and upper legs should include:
   1. Wrapping the client in a clean dry sheet
   2. Removing clothing from the burned areas
   3. Applying sterile dressings to burned areas
   4. Standing the client under a gentle-spray shower

4. The nurse's goal for a client with ureteral colic should be based on the knowledge that most factors contributing to the development of renal calculi can be limited by:
   1. Decreasing serum creatinine
   2. Excluding milk products from the diet
   3. Drinking 8 to 10 glasses of water daily
   4. Excreting 2000 ml of urine per 24 hours

5. The nurse suspects a pulmonary embolus when after surgery the client begins to cough violently, becomes short of breath, and complains of right-sided chest pain. In addition to notifying the physician, the nurse should:
   1. Elevate the head of the client's bed
   2. Auscultate the client's breath sounds

   3. Elevate both of the client's legs on pillows
   4. Place the client in alternating right and left side-lying positions

6. A client sustains a puncture wound of the chest wall as the result of an industrial accident. The nurse knows that a chest tube will be inserted to reestablish:
   1. Positive pressure in the pleural cavity
   2. Negative pressure in the pleural space
   3. Atmospheric pressure in the thoracic space
   4. Intrapulmonic pressure in the thoracic cavity

7. A client who is scheduled for a thyroidectomy is concerned that the surgery will interfere with her ability to become pregnant. When replying, the nurse bases the response on the following information:
   1. Hyperthyroidism can cause abortions or fetal anomalies
   2. Pregnancy is not advisable for the client with a thyroidectomy
   3. As long as medication is taken as ordered, ovulation will occur
   4. Pregnancy affects metabolism and will require greatly increased thyroid hormone

8. Following a thyroidectomy the nurse should carefully observe the client for signs of thyroid storm. These include:
   1. Hypothermia
   2. Elevated serum calcium
   3. Sudden drop in pulse rate
   4. Rapid heart beat and tremors

9. Burns classified as full thickness extend to involve destruction of the:
   1. Hair follicle
   2. Upper dermis
   3. Epidermal layer
   4. Subcutaneous tissue

10. A client with cancer of the larynx is to have a total laryngectomy with a radical neck dissection. When reinforcing the physician's explanation about what the surgery entails and what abilities will be lost, the nurse's discussion should also focus on what abilities the client will retain, such as the ability to:
    1. Blow the nose
    2. Sip through a straw
    3. Chew and swallow food
    4. Smell and differentiate odors

11. The nurse notes that a client, admitted to the hospital after experiencing an anterior septal myocardial infarction, does not have a prn order for a stool softener. The nurse decides to obtain such an order from the physician primarily because the client:
    1. Has poor peripheral perfusion
    2. Is likely to become constipated
    3. Will be kept npo for several days
    4. Is receiving digoxin, which decreases peristalsis

12. A client with chronic obstructive pulmonary disease (COPD) has been receiving low-flow oxygen by nasal cannula for 4 hours. The nurse notes that the client has increased restlessness and confusion followed by a decreased respiratory rate and lethargy. The nurse should:
    1. Percuss and vibrate the chest wall
    2. Increase the oxygen by 2% increments
    3. Decrease or discontinue the oxygen flow
    4. Quietly ask the client about the confusion

13. In a three-chamber underwater drainage system, the main purpose of the third chamber is to:
    1. Act as a drainage container
    2. Control the amount of suction
    3. Provide an air-tight water seal
    4. Allow for escape of air bubbles

14. When a client with kidney shutdown complains of thirst, the nurse should offer:
    1. Carbonated soda
    2. Sour ball candy
    3. A glass of milk
    4. A bowl of clear soup

15. When a client with severe rheumatoid arthritis reports that pain lasts for 2 to 3 hours after exercising, the nurse should teach the client to:
    1. Stop all exercises for 24 hours
    2. Increase the number of repetitions of each exercise
    3. Decrease progressive resistive exercises to 3 times a day
    4. Decrease the total time and number of repetitions of the exercise

16. After an above-the-knee amputation, a client complains of phantom limb pain. Initially the nurse should:
    1. Explain the psychologic component involved
    2. Reassure the client that these sensations will pass
    3. Encourage the client to get involved in diversional activities
    4. Describe the neurologic mechanism in language that the client understands

17. When assessing a client for respiratory involvement during the first 2 days following a burn injury, the nurse should observe for sputum that is:
    1. Sooty
    2. Frothy
    3. Yellow
    4. Tenacious

18. A client who is receiving multiple medications for a myocardial infarction complains of severe nausea, and the heartbeat is irregular and slow. The nurse should recognize these symptoms as toxic effects of:
    1. Lanoxin
    2. Lidocaine
    3. Morphine sulfate
    4. Meperidine hydrochloride

19. A female client with diabetes mellitus tells the nurse that she takes Robitussin (guaifenesin) cough syrup when she has a cold. The nurse should tell her that she:
    1. Can substitute an elixir for the syrup
    2. Must calculate the sugar in her daily carbohydrate allowance
    3. Can increase her fluid intake and humidify her bedroom to control a cough
    4. May take the cough syrup if her serum glucose level remains within an acceptable range

20. A client has surgery to remove a stone from the common bile duct. The nurse is aware that the bile flow into the duodenum has been reestablished after biliary surgery when:
    1. The liver is no longer tender
    2. Stools are normal brown in color
    3. Colic is absent after ingestion of fats
    4. The serum bilirubin level returns to normal

21. A 59-year-old diabetic client is admitted to the hospital in diabetic ketoacidosis. Under medical supervision, the client has been adjusting insulin dosage based on blood glucose levels at home. Of the statements made by the client during the health history, the nurse realizes that the best explanation of the etiologic factor of the ketoacidosis is that the client:
    1. Has a chronic postnasal drip
    2. Is going to turn 60 next week

3. Is planning to retire next year

4. Has been taking steroids for a rash

22. When receiving chemotherapy for non-Hodgkin's lymphoma, a client states, "I get so sick to my stomach. The medication is useless." The response by the nurse that employs the technique of paraphrasing is:
   1. "You get sick to your stomach?"
   2. "I'll get an order for an antiemetic."
   3. "Tell me more about how you feel."
   4. "You don't think the medication is helping you?"

23. After 1 week a client with acute renal failure moves into the diuretic phase. During this phase the client must be carefully assessed for signs of:
   1. Hypovolemia
   2. Hyperkalemia
   3. Metabolic acidosis
   4. Chronic renal failure

24. The physician makes the diagnosis of rheumatoid arthritis even though the client's laboratory tests are negative. The client appears confused and upset and says to the nurse, "How can the tests be negative when I am always in pain?" The nurse's best response would be:
   1. "It might help if you try not to think about your discomfort."
   2. "Don't let that upset you; eventually the tests will be positive."
   3. "Did the doctor say that the laboratory tests will be repeated?"
   4. "Laboratory tests are often negative in the early stages of arthritis."

25. When caring for a client who has had a total hip replacement, the nurse should position the client's affected limb in:
   1. Adduction and flexion
   2. Abduction and extension
   3. Adduction and internal rotation
   4. Abduction and external rotation

26. A male client who has had a cerebrovascular accident has varying moods. These moods have ranged from anger to depression and concern his aphasia, hemiparesis, and need for gavage feedings. The behavior observed by the nurse that would indicate the client's acceptance of his physical limitations would be:
   1. Performing his own tube feedings
   2. Smiling and becoming more extroverted

3. Allowing his wife to participate in his care

4. Ambulating in the hall and sitting in the lounge

27. A client with type 2 diabetes who is taking an oral hypoglycemic agent is to have a serum glucose test first thing in the morning. The nurse should instruct the client:
   1. "Eat your usual breakfast."
   2. "Have clear liquids for breakfast."
   3. "Take your medication before the test."
   4. "Do not take any food or fluids before the test."

28. After surgery a client's fever does not respond to antipyretics. The client is placed on a hypothermia blanket. One reaction to hypothermia that should be prevented is:
   1. Shivering
   2. Dehydration
   3. Hypotension
   4. Venous stasis

29. A client has an excision of a thrombosed external hemorrhoid. When cleaning the anus following a bowel movement, the client should be instructed to use:
   1. Betadine pads
   2. Soft facial tissue
   3. Moist cotton balls
   4. Sterile 4 × 4-inch gauze pads

30. A client with a rigid and painful abdomen is determined to have a perforated peptic ulcer. A nasogastric tube is inserted and surgery is scheduled. Before surgery the nurse should place the client in the:
   1. Sims' position
   2. Supine position
   3. Semi-Fowler's position
   4. Dorsal recumbent position

31. A client is scheduled for a cardiac catheterization, which is to be done via the femoral approach. When planning preoperative preparation for this client, the nurse should include information that:
   1. The physician will immediately tell the client about the results of the procedure
   2. The client will be permitted to get up and walk around after returning to the room
   3. The client will be on bed rest, with the affected leg extended for about 4 to 8 hours
   4. A general anesthetic will be given, and the client will be asleep during the procedure

32. A client with Crohn's disease is admitted to the hospital with a history of chronic, bloody diarrhea; weight loss; and signs of general malnutrition. The client also has edema, anemia, a low serum albumin level, and symptoms of negative nitrogen balance. The client's health status is related to a major deficiency of:
    1. Iron
    2. Protein
    3. Potassium
    4. Linoleic acid

33. The physician orders IV Ringer's lactate to infuse at the rate of 150 ml/hr. The nurse calculates that a liter bottle of this solution will infuse in:
    1. 6 hours and 40 minutes
    2. 6 hours and 50 minutes
    3. 7 hours and 10 minutes
    4. 7 hours and 30 minutes

34. A client is taught the signs and symptoms of a hypoglycemic reaction. The nurse would be aware that the teaching was effective when the client identifies that the signs of hypoglycemia include:
    1. Dehydration, abdominal pain, and nausea
    2. Increased urination, fatigue, and weight loss
    3. Tingling, paresthesia, and decreased respirations
    4. Nervousness, increased perspiration, and weakness

35. When assessing a client with a diagnosis of possible myocardial infarction for pain, the nurse would expect the client to describe it as:
    1. Severe, intense chest pain
    2. A burning sensation of short duration
    3. Mild chest pain, radiating down to the fingers
    4. A squeezing chest pain, relieved by nitroglycerin

36. A client who is on a 2-g sodium diet asks for a glass of juice. Since the only kinds of juice available are pear nectar, apple, and tomato, the nurse should:
    1. Suggest that the client drink the tomato juice
    2. Tell the client that juice is not permitted on a low-sodium diet
    3. Offer the apple juice or the pear nectar and ask for the client's preference
    4. Explain that the client cannot have juice between meals but may have a glass of water

37. The nurse is aware that the diagnosis of incarcerated hernia means that:
    1. The bowel has twisted on itself
    2. The protruding hernia cannot be reduced
    3. An erosion of the involved intestine has occurred
    4. The blood supply to the intestine has been cut off

38. To help limit a common complication following the repair of an inguinal hernia, the nurse should:
    1. Apply an abdominal binder
    2. Place a support under the scrotum
    3. Encourage a high-carbohydrate diet
    4. Have the client cough and deep breathe frequently

39. A client recovering from hepatitis A asks when a return to work will be permitted. The nurse's best response would be:
    1. "As soon as you're feeling less tired, you may go back to work."
    2. "Unfortunately, few people fully recover from hepatitis in less than 6 months."
    3. "You cannot return to work for 6 months because the virus will still be present in your stools for this period."
    4. "Gradually increase your activities. Relapses are very common in people who return to full activity too soon."

40. A client with cancer of the pancreas is scheduled for surgery. When arranging for the next appointment the client says to the nurse, "Wouldn't I be better off with some other treatment instead of surgery?" The nurse's best response would be:
    1. "Why don't you explore the other acceptable treatments for your cancer with the physician?"
    2. "Surgery is the recommended approach, but why don't you discuss this further with the physician?"
    3. "Maybe you would be more confident with a second opinion. Would you like a referral to another physician?"
    4. "With your disease your prognosis will improve if you follow the physician's suggestion and have the recommended surgery."

41. When teaching how to use a nebulizer, the nurse should instruct the client to:
    1. Hold the breath while spraying the medication carefully into each nostril
    2. Instill the medication from the nebulizer while exhaling through the nose
    3. Seal the lips around the mouthpiece taking rapid, shallow breaths through the mouth
    4. Loosely place the lips around the mouthpiece taking a slow, deep breath through the mouth

42. A 60-year-old schoolteacher is admitted to the hospital for a needle biopsy of the liver. The physician suspects cancer of the liver. When teaching about the needle biopsy, the nurse should inform the client that:
    1. A midline abdominal incision will be used
    2. Ambulation will be permitted after the procedure

3. Bed rest must be maintained after the procedure
4. General anesthesia will be used during the biopsy

43. During the immediate postoperative period following a total laryngectomy for cancer of the larynx, a priority nursing intervention would be to:
    1. Provide emotional support
    2. Observe for signs of infection
    3. Keep the trachea free of secretions
    4. Promote a means of communication

44. A client with Hodgkin's disease, stage III, is started on a MOPP regimen of nitrogen mustard, vincristine (Oncovin), procarbazine, and prednisone. The client wonders why so many drugs have to be used at once. The nurse should explain that:
    1. Using groups of drugs reduces the likelihood of serious side effects
    2. Each drug destroys the cancer cell at a different time in the cell cycle
    3. The drugs are used to destroy cells that are not susceptible to radiation therapy
    4. Since there are stages of Hodgkin's disease, if one drug is ineffective another will work

45. A client develops a thrombophlebitis in the right calf. Bed rest is prescribed, and heparin sodium IV is started. When describing the purpose of this drug to the client, the nurse should explain that it:
    1. Reduces the size of the thrombus
    2. Dissolves the blood clot in the vein
    3. Prevents extension of the blood clot
    4. Promotes absorption of red blood cells

46. The nurse is preparing a client, who has had a transurethral prostatectomy for cancer of the prostate, for discharge. The nurse is aware that the client understands the teaching when he states:
    1. "I will drink 8 cups of fluid daily, but none after 9 pm."
    2. "I will use stool softeners regularly for 1 to 2 months after I get home."

3. "I am so glad this is over. Now I don't have to keep going to the doctor."
4. "I was so worried and now I can hardly wait to get home and have sex with my wife."

47. A client has ear surgery. An early response that may be associated with possible damage to the motor branch of the facial nerve is:
    1. A bitter metallic taste
    2. Dryness of the mouth
    3. A sensation of pain behind the ear
    4. An inability to wrinkle the forehead

48. During the first 36 hours after surgery for cancer of the pancreas, a client complains of severe pain and is medicated every 4 hours with meperidine (Demerol) 75 mg IM. Because the client rests or sleeps between injections, the nurse can conclude that:
    1. Pain management is effective
    2. The dosage of the drug is excessive
    3. Another narcotic should be substituted
    4. The Demerol can probably be given orally

49. A client is admitted with a diagnosis of atrial fibrillation, and the physician suspects mitral stenosis. When obtaining a health history, it would be most significant if the client presented a history of:
    1. Rubella, age 6
    2. Cystitis, age 28
    3. Pleurisy, age 20
    4. Strep throat, age 12

50. A client undergoes a mitral valve replacement. Postoperatively the client's peripheral pulses must be checked frequently. The primary purpose of this is to detect:
    1. Atrial fibrillation
    2. Postsurgical bleeding
    3. Arteriovenous shunting
    4. The existence of emboli

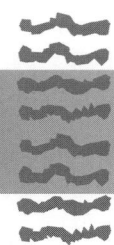

MEDICAL-SURGICAL NURSING QUIZ 4
ANSWERS AND RATIONALES

1. ② The Parkland formula calls for administration of a colloid only when the serum albumin falls below 2 g/dl; then, albumin must be administered to raise the level to the normal 3.5 to 5.5 g/dl; this raises the oncotic pressure and prevents the shift of intravascular fluid. (2) (DA; PL; FE)
 1 A hematocrit level of 32% is low and indicates overhydration; an administration of a colloid could increase this problem.
 3 This is within normal limits of 1.010 to 1.030 and would not indicate a need for serum albumin.
 4 This is within normal parameters and would indicate present therapy is achieving its goals.

2. ④ Hoarseness is caused by the inability of the vocal cords to come close during speech when a tumor exists. (2) (HI; AS; RE)
 1 Aphasia refers to an expressive or receptive communication deficit as a result of cerebral disease; it is not related to laryngeal cancer.
 2 Dyspnea is a late sign, occurring when the tumor is large enough to obstruct air flow.
 3 Dysphagia is a late sign, occurring when the tumor is large enough to compress the esophagus.

3. ① Covering exposed burned surfaces limits contamination by microorganisms and prevents exposure to air, which increases pain. (2) (SI; IM; IT)
 2 This is unsafe; this would traumatize the skin.
 3 Same as answer 2.
 4 Same as answer 2.

4. ③ This dilutes the urine, and crystals are less likely to coalesce. (3) (CI; PL; RG)
 1 This has no relationship to kidney stones.
 2 Calcium restriction is only necessary if calculi contain calcium phosphate or calcium oxalate.
 4 This is inadequate; urine output should be maintained above this amount to limit calculi formation.

5. ① This promotes breathing by reducing pressure of the abdominal organs on the diaphragm and increasing thoracic excursion. (1) (DA; IM; RE)

 2 This may confirm diminished breath sounds, but it does not help the situation.
 3 The client is not in shock.
 4 The client must be kept as quiet as possible to decrease oxygen needs.

6. ② The removal of air and fluid from the pleural space reestablishes negative pressure and lung expansion. (2) (DA; AN; RE)
 1 This causes collapse of the lung.
 3 Same as answer 1.
 4 This is the pressure within the lung itself, not the thoracic cavity.

7. ③ Medication is regulated to maintain the normal blood levels of thyroxine; therefore ovulation is not affected, and future pregnancy is possible. (3) (DF; IM; EN)
 1 The client will no longer be hyperthyroid after surgery because the overactive tissue is excised; therefore, this will not affect the fetus.
 2 Pregnancy is not contraindicated after a thyroidectomy.
 4 Thyroid hormones may have to be increased; however, the alterations in dosage are based on individual needs.

8. ④ Thyroid storm refers to a sudden and excessive release of thyroid hormones, which cause pyrexia, tachycardia, and exaggerated symptoms of thyrotoxicosis. Surgery, infection, and ablation therapy can precipitate this life-threatening condition. (3) (CI; EV; EN)
 1 The temperature would be elevated in thyroid storm because of the sudden excessive release of thyroid hormones, which elevate the basal metabolic rate.
 2 Hypercalcemia is not related to thyroid storm; hypocalcemia may result from accidental removal of the parathyroid glands.
 3 The pulse rate would be rapid in thyroid storm because of the sudden excessive release of thyroid hormones, which elevates the basal metabolic rate.

9. (4) Full-thickness burns are those extending into the subcutaneous tissue. They are not painful because of nerve destruction. (3) (HI; AS; IT)
   1 Burns extending through the epidermis and involving part of the dermis are deep partial-thickness burns.
   2 Same as answer 1.
   3 Burns affecting only the epidermis are superficial partial-thickness burns.

10. (3) Normal eating patterns are not lost when a laryngectomy is performed. (3) (DA; IM; RE)
   1 There is no passage of air from the lungs to the nose, and the ability to blow the nose is lost.
   2 There is no passage of air from the lungs to the mouth, and the ability to sip through a straw is lost.
   4 Air goes directly to the trachea, bypassing the nose and olfactory organs.

11. (2) Inactivity and narcotics decrease peristalsis, which may precipitate constipation; straining at stool should be avoided to prevent the Valsalva maneuver that increases demands on the heart. (2) (PC; PL; GI)
   1 This is unrelated to constipation; in addition there is no indication the client has poor peripheral circulation.
   3 Being npo is only one small factor that could contribute to constipation; narcotic analgesics and bed rest are the primary causes of constipation in this instance.
   4 Digoxin is unrelated to intestinal peristalsis; narcotic analgesics will decrease peristalsis.

12. (3) The decreased respiratory drive resulting from excessive O_2 administration indicates CO_2 narcosis. (3) (DA; IM; RE)
   1 There are no indications that secretions have increased.
   2 Increasing O_2 administration will further diminish respiratory drive until respiratory arrest occurs.
   4 This is inappropriate; a confused client cannot answer questions about the confusion.

13. (2) The first chamber collects drainage; the second chamber provides for the underwater seal; the third chamber controls the amount of suction. (3) (DA; AN; RE)
   1 The first chamber, not the third chamber, collects drainage.
   3 The second chamber, not the third chamber, provides an underwater seal.
   4 Although this occurs in a three-chamber system, the main purpose of the third chamber is to control suction.

14. (2) Sucking on candy will relieve thirst and provide calories without supplying extra fluid. (3) (PC; IM; RG)
   1 Carbonated beverages are high in sodium and provide additional fluid; both would be restricted.
   3 Milk contains both fluids and proteins, which must be restricted.
   4 Soup contains sodium, which can compound the problem of fluid retention and hypernatremia.

15. (4) Exercise should be to tolerance only. (2) (PC; IM; SK)
   1 Exercise should not be stopped.
   2 This will increase pain.
   3 Resistive exercises are not part of the usual exercise regimen for clients with rheumatoid arthritis.

16. (4) Explanation of the underlying mechanism usually helps reduce anxiety. (2) (SC; IM; SK)
   1 This reinforces the idea that there is something psychologically wrong with the client, which is untrue.
   2 This may be false reassurance.
   3 This may temporarily distract the client but does nothing to foster awareness of the cause.

17. (1) The mucous membranes of the respiratory tract may be charred after inhalation burns. This is evidenced by the production of sooty sputum. (2) (DA; EV; RE)
   2 Frothy sputum is usually indicative of pulmonary edema.
   3 This finding may indicate respiratory infection.
   4 Same as answer 3.

18. (1) Signs of digitalis toxicity include cardiac dysrhythmias, anorexia, nausea, vomiting, and visual disturbances. (2) (PT; EV; DR)
   2 Although nausea and heart block may occur with lidocaine, these symptoms are rarely seen; drowsiness and CNS disturbances are more common side effects of lidocaine.
   3 Toxic effects of morphine sulfate are slow, deep respirations; stupor; and constricted pupils. Nausea is a side effect, not a toxic effect.
   4 Nausea is a side effect of meperidine hydrochloride (Demerol); toxic effects include dilated pupils, tremor, confusion, and respiratory depression.

19. (2) Cough syrup contains a sugar base; the client should use a sugar-free product or account for the sugar. (2) (PT; IM; EN)
   1 Elixirs have an alcohol base and also contain sweeteners.

3 Although this will loosen secretions, it will not suppress a cough.

4 Additional glucose may increase serum glucose levels beyond the desired range; once control is achieved, it is unwise to alter dietary intake or medication without supervision.

20. (2) The return of brown color to the stool indicates that bile is entering the duodenum and being converted to urobilinogen by bacteria. (2) (CI; EV; GI)

1 Liver tenderness is unrelated to bile flow.

3 The absence of biliary colic is related to the removal of the stone, not the flow of bile.

4 The serum bilirubin level is not affected.

21. (4) Steroids cause gluconeogenesis and glycogenolysis, both of which raise blood glucose levels. (3) (PT; AN; EN)

1 This is a chronic condition that has probably been incorporated into the client's coping patterns; it would not cause sufficient stress to elevate the serum glucose level to this degree.

2 There are no data to indicate that this is a stressful event for the client.

3 This event is in the future and would not cause sufficient stress to elevate the serum glucose level to this degree at this time; retirement may be welcomed and may not cause stress.

22. (4) Rewording of the client's statement is paraphrasing that promotes further verbalization. (2) (SC; IM; EH)

1 This is not paraphrasing; this merely repeats the client's exact words.

2 This is not an interviewing technique; immediate movement to intervention cuts off communication.

3 This is clarifying, a therapeutic technique; it is not paraphrasing.

23. (1) In the diuretic phase, fluid retained during the oliguric phase is excreted and hypovolemia may occur. (3) (DA; AS; RG)

2 This develops in the oliguric phase when glomerular filtration is inadequate.

3 Same as answer 2.

4 Diuresing is an indication of improved renal functioning, not progression into chronic failure.

24. (4) The antibody called rheumatoid factor is present in 90% of clients with advanced arthritis; it is not found in the early stages of the disease. (2) (CI; IM; SK)

1 This diminishes the client's discomfort and does not deal with the stated confusion.

2 This denies immediate feelings and blocks further communication of feelings.

3 This reinforces confusion over negative test results and felt discomfort.

25. (2) This reduces stress on the joint capsule incision, preventing the prosthesis from becoming dislocated. (1) (SI; IM; SK)

1 Both strain the joint capsule, fostering dislocation.

3 Same as answer 1.

4 External rotation puts strain on the joint capsule, fostering dislocation.

26. (1) The best indicator of acceptance is when the client begins to participate in self-care. (2) (SC; EV; EH)

2 This may be an act and does not indicate acceptance.

3 Allowing others to provide care does not indicate acceptance.

4 This does not indicate acceptance and may be just an attempt to relieve boredom.

27. (4) Fasting prior to the test is indicated for accurate and reliable results. Food will elevate the serum glucose levels through metabolism of the nutrients. (2) (CI; IM; EN)

1 Food should not be ingested; it will raise the serum glucose level, negating the accuracy of the test.

2 Clear fluids contain carbohydrates, which will increase the serum glucose level.

3 Medications are withheld because of their influence on the serum glucose level.

28. (1) Shivering should be prevented. Peripheral vasoconstriction increases the temperature, the circulatory rate, and oxygen consumption. (2) (SI; IM; CV)

2 This is not a response to treatment with hypothermia.

3 Same as answer 2.

4 Same as answer 2.

29. (3) Moist cotton is not irritating and is the most soothing to the anal mucosa. (2) (PC; IM; GI)

1 Betadine may cause excessive drying and irritation; the rectum is normally contaminated; external cleansing with Betadine will not appreciably affect the bacteria present.

2 Moist cotton is softer and more effective in removing feces than is dry facial tissue, which is irritating and can cause trauma.

4 Sterile gauze pads are unnecessary; the rectal area is considered contaminated.

30. (3) The semi-Fowler's position will localize the spilled stomach contents in the lower part of the abdominal cavity. (2) (SI; IM; GI)
1   This position will exert pressure on the abdomen, which may be uncomfortable for the client.
2   This position exerts pressure against the diaphragm, inhibits breathing, and intensifies discomfort.
4   Same as answer 2.

31. (3) Bed rest with the leg extended prevents trauma caused by hip flexion and provides time for the insertion site to heal. (2) (SI; PL; CV)
1   The physician will thoroughly review test results and may consult other physicians before talking with the client.
2   With the femoral approach, bed rest is maintained for several hours.
4   Mild sedation and local infiltration are used for adults; the client is conscious.

32. (2) Protein deficiency causes a low serum albumin level, which permits fluid shifts from the intravascular to the interstitial compartment, causing edema; decreased protein also causes anemia; protein intake must be increased. (2) (DA; AN; GI)
1   Although a deficiency of iron would result in anemia, it would not cause the other symptoms.
3   Hypokalemia would not cause these symptoms; it causes cramps and weakness.
4   This is unrelated to these symptoms; it is an essential fatty acid.

33. (1) The amount of solution (1000) is divided by the hourly amount to be infused (150), which equals 6.66; the decimal portion of the result must be converted to minutes (0.66 × 60 min), which equals 40 minutes plus the 6 hours. (2) (PT; AN; FE)
2   The solution should be infused before this time.
3   Same as answer 2.
4   Same as answer 2.

34. (4) Nervousness and perspiration are caused by increased adrenergic activity and increased secretion of catecholamines, while weakness reflects CNS glucose deprivation. (2) (HI; EV; EN)
1   These are indications of diabetic ketoacidosis.
2   These are indications of diabetes.
3   These are indications of hypocalcemia.

35. (1) Blockage of myocardial blood supply causes accumulation of unoxidized metabolites that affect nerve endings and cause pain. (1) (PC; AS; CV)

2   This is not the type of pain associated with a myocardial infarction.
3   Same as answer 2.
4   Pain is unrelieved by nitroglycerin.

36. (3) Apple juice and pear nectar have low-sodium content and are therefore the better choices for this client. (2) (PC; IM; CV)
1   Tomato juice has a high-sodium content; it should be avoided to prevent fluid retention.
2   Low-sodium juices are not contraindicated.
4   The client is permitted juice between meals.

37. (2) When the intestine cannot be returned to the body cavity, the hernia is incarcerated. (2) (HI; AN; GI)
1   This is a volvulus.
3   Erosion of intestinal tissue may be caused by a variety of conditions; it usually results in perforation.
4   This is a strangulated hernia.

38. (2) After inguinal hernia repair, the scrotum commonly becomes edematous and painful. Drainage is facilitated by elevating the scrotum on rolled linen or using a scrotal support. (3) (CI; IM; GI)
1   An abdominal binder would not support the operative site; the incision is too low.
3   Obesity is a factor in the development of hernias; high-carbohydrate diets should not be encouraged.
4   Coughing increases intraabdominal pressure and may strain the operative site.

39. (4) Relapses are common. They occur after too early ambulation and too much physical activity. (2) (HI; IM; GI)
1   Fatigue is a cardinal symptom; if the client tires at rest, a return to work must be delayed.
2   The majority of clients recover in 3 to 16 weeks with no further problems.
3   Hepatitis is most communicable before the onset of symptoms and during the first few days of fever.

40. (2) This response provides needed information and establishes the opportunity for further discussion of surgery. (2) (SC; IM; EH)
1   This implies the other approaches are as effective as surgery; this places doubt in the client's mind that surgery is the most effective option.
3   This is an inappropriate response; the competence of the physician was not questioned, but there exists a need for a further discussion of the treatment; making this type of referral is not the nurse's role.

4 This is false reassurance; it cuts off communication and does not address the need for further discussion.

41. (4) This permits air to enter through the mouth along with the medication from the nebulizer; slow deep breaths deliver medication deep into the lung. (1) (SI; IM; RE)
1 This is a nasal spray that does not deliver medication to the lung.
2 This would not deliver the medication to the lungs but would deposit it in the oral cavity.
3 Shallow breaths are ineffective in delivering medication into the lung.

42. (3) There is a high risk of bleeding because of hepatic vascularity; bed rest supports healing and prevents injury. (2) (DA; IM; GI)
1 A needle biopsy requires a stab wound over the liver, not an abdominal incision.
2 The client is maintained on bed rest on the right side for a minimum of 6 to 8 hours following the biopsy.
4 The biopsy is done with local anesthesia.

43. (3) A patent airway is always a priority concern; therefore removal of secretions is imperative. (1) (CI; PL; RE)
1 This is an important postoperative intervention but is not the priority immediately after surgery.
2 This is an important postoperative concern, but infection does not occur immediately.
4 Same as answer 1.

44. (2) Cells are vulnerable to specific drugs through the stages of mitosis, and a combination bombards the malignant cells at various stages. (1) (PT; IM; DR)
1 This is not true; the side effects of a drug are not ameliorated by a combination with others.
3 This is true, but it is not the reason for using a combination of drugs.
4 There is more than one stage of Hodgkin's, but this is not the reason for using a combination of drugs.

45. (3) Heparin interferes with activation of prothrombin to thrombin and inhibits aggregation of platelets. (2) (PT; AN; DR)
1 This is not the action of heparin sodium.
2 Same as answer 1.
4 Same as answer 1.

46. (2) Straining at stool should be avoided for 4 to 6 weeks after surgery to permit healing and avoid initiating bleeding. (1) (CI; EV: RG)

1 This is insufficient fluid; at least 2500 ml/day should be consumed.
3 The client has carcinoma and needs continued medical supervision.
4 Sexual intercourse should be avoided for 3 to 4 weeks after surgery.

47. (4) The motor fibers of the facial nerve innervate the superficial muscles of the face and scalp. (2) (HI; AS; NM)
1 This response is usually not related to damage to the facial nerve but may indicate infection.
2 This is a sensory response that may be manifested when the injury is to the sensory branch of the facial nerve.
3 Same as answer 2.

48. (1) This behavior indicates that the client is comfortable; therefore the medication regimen is effective. (2) (PC; EV; NM)
2 This is the accepted dose; no data exist to indicate it is excessive for this client.
3 This is not necessary; the medication regimen is effective.
4 The efficacy of Demerol decreases when the drug is given orally; this is too soon after surgery to alter the route.

49. (4) Streptococcal infections occurring in childhood may result in damage to the heart valves. An autoimmune reaction occurs between antibodies made against the bacteria and the heart valves, particularly the mitral. (2) (HI; AS; CV)
1 The rubella virus would not affect the valves of the heart.
2 Cystitis is usually caused by *Escherichia coli*, which would not affect the heart valves.
3 Pleurisy usually follows pulmonary problems unrelated to streptococcal infection; it would not result in damage to the heart valve.

50. (4) Because blood pools in the extremities, there is an increased hazard of peripheral emboli in clients who have received a mitral valve replacement. (3) (CI; EV; CV)
1 A peripheral pulse alone will not reveal atrial fibrillation; to detect the presence of a pulse deficit, one must compare a peripheral pulse with the apical pulse.
2 Bleeding is detected by checking the wound dressing and observing for signs of shock (e.g., lowered BP, tachycardia, restlessness).
3 This is not a danger after mitral valve replacement.

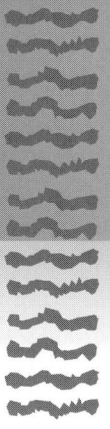

CHAPTER 6

Comprehensive Examination 1

PART A QUESTIONS

1. The hypertonicity of the muscles in an infant with cerebral palsy causes scissoring of the legs. The nurse should suggest to the infant's mother that the best way to carry the baby is in a sitting position:
   1. Astride one of her hips
   2. Strapped in an infant seat
   3. Wrapped tightly in a blanket
   4. Under the arm using a football hold

2. A client with portal hypertension and ascites is given 2 units of salt-poor albumin IV. The purpose of salt-poor albumin is to:
   1. Provide parenteral nutrients
   2. Increase the client's protein stores
   3. Increase the client's circulating blood volume
   4. Temporarily divert blood flow away from the liver

3. A client, 36 weeks gestation, with severe preeclampsia is admitted to the high-risk labor unit and placed on a fetal monitor as well as on an automated machine for blood pressure monitoring. Frequent monitoring of the blood pressure is essential to reduce the potential for:
   1. DIC
   2. Stroke
   3. Hemorrhage
   4. Precipitous labor

4. The most important aspect of a therapeutic contract is:
   1. Discussing and defining the client's goals for treatment
   2. Determining the time and place of meetings with the client
   3. Understanding the professional responsibilities of the nurse
   4. Planning the frequency and duration of meetings with the client

5. After a mastectomy, a client returns from surgery with a portable suction unit in place and a dry sterile dressing covering the site of the incision. When observing this client for signs of bleeding, the nurse should:
   1. Empty and measure the output in the portable suction unit hourly
   2. Inspect the bedclothes under the axillary area for signs of drainage
   3. Turn her on the affected side to inspect for blood that may flow backward
   4. Reinforce the operative site with a pressure dressing if any drainage appears on the dressing

6. The nurse empties a portable wound suction device when it is only half full because:
   1. It is easier and safer to empty the unit when it is only half full
   2. This facilitates a more accurate measurement of drainage output
   3. The negative pressure in the unit lessens as fluid accumulates in it, interfering with further drainage
   4. As fluid collects in the unit it exerts positive pressure, forcing drainage back up the tubing and into the wound

7. Before discharge after a myocardial infarction, a male client asks the nurse how long he should wait before having sexual relations with his wife. The nurse's best reply would be:
1. "It depends on how you are feeling."
2. "Two weeks is the usual waiting time."
3. "Have you discussed this with your physician?"
4. "You should wait until your heart feels stronger."

8. A female client and the nurse are standing next to each other in the mental health clinic when the client gets down on her hands and knees and says, "I am a table." It would be most effective for the nurse to:
1. State, "You were never a table before and you are not a table now."
2. State, "You are safe here in the clinic. You do not need to be a table."
3. Touch her arm while saying, "You must be very frightened to feel this way."
4. Offer a hand to help her up while saying, "You are not a table, you are a person."

9. To best help manage pain during burn dressing changes, the nurse can teach the client:
1. Deep breathing exercises
2. The importance of wound care
3. Active range-of-motion exercises
4. To alternately contract and relax muscles

10. A postpartum client who has type 1 diabetes states that she does not want another baby for at least 3 years. The nurse provides family planning instructions and is satisfied that the teaching has been understood when the client states, "Because of my diabetes, the best type of contraception is:
1. An oral contraceptive."
2. The intrauterine device."
3. A diaphragm with foam."
4. Tying the fallopian tubes."

11. After the administration of epinephrine to a child with asthma, the nurse would carefully monitor for the common side effect of:
1. Flushing
2. Dyspnea
3. Tachycardia
4. Hypotension

12. A client is admitted to the psychiatric unit with a history of abuse of multiple drugs. In planning care after the withdrawal period, the nurse should take into consideration that the client is probably:
1. Unable to give up drugs
2. Unconcerned with reality

3. Unable to delay gratification
4. Unaware of the dangers of drug addiction

13. A factor learned while obtaining the nursing history that probably predisposed a client to type 2 diabetes is:
1. Having diabetes insipidus
2. Eating low-cholesterol foods
3. Being 20 pounds overweight
4. Drinking a daily alcoholic beverage

14. After a pneumonectomy a client is receiving an intravenous infusion. In light of this particular surgery, the client is most vulnerable to the complication of:
1. Phlebitis
2. Infiltration
3. Air embolism
4. Pulmonary edema

15. The nurse tells a laboring client that she must avoid lying on her back during labor. The nurse bases this instruction on the knowledge that the supine position most significantly can:
1. Prolong labor
2. Cause decreased placental perfusion
3. Lead to transient episodes of hypertension
4. Interfere with free movement of the coccyx

16. When planning care for a child with leukemia, the nurse must keep in mind that the prognosis of a child with acute lymphocytic leukemia who is receiving therapy is:
1. Very poor, but the therapy keeps them pain free
2. Limited to a few months in 70% of the children affected
3. Extended to at least 5 years in 60% of the children treated
4. Very positive, with a probable cure in 90% of the children affected

17. Methotrexate is to be administered IV to a 14-year-old. The appropriate amount has been added to 500 ml of D5W. The physician has ordered the 500 ml to be absorbed in 4 hours. The nurse should set the infusion device at:
1. 50 ml/hr
2. 75 ml/hr
3. 100 ml/hr
4. 125 ml/hr

18. A subclavian catheter is inserted and the client is started on total parenteral nutrition (TPN). To prevent the most common complication of TPN, the nurse should plan to teach the client to:
1. Avoid touching the dressing as much as possible

2. Keep the head as still as possible whenever moving

3. Weigh daily at the same time, wearing the same clothing

4. Regulate the flow rate on the infusion pump as necessary

19. Care in a day treatment center is often indicated early in the treatment program for clients with incapacitating symptoms resulting from obsessive-compulsive personality disorders because it:

1. Limits the client's time to carry out the symptomatic rituals

2. Allows the therapeutic staff to exert control over the client's activities

3. Provides a neutral environment in which the client can work through conflicts

4. Resolves the client's anxiety since opportunities requiring decision making are reduced

20. To relieve the symptoms of parkinsonism, the nurse would expect the physician to order:

1. Levodopa

2. Dopamine

3. Vitamin B_6

4. Isocarboxazid

21. After a transurethral prostatectomy, a client returns to the recovery room with a three-way Foley catheter with continuous bladder irrigation. An initial nursing priority in the client's care plan would be to:

1. Observe for signs of confusion and agitation

2. Maintain the client in a semi-Fowler's position

3. Observe the suprapubic dressing for drainage

4. Force fluids by mouth as soon as the gag reflex returns

22. A client has had a cesarean delivery because of fetal heart rate tracing abnormalities. When doing postoperative coughing and deep breathing, she complains of localized pain in the incision, which subsides in a few moments. The nurse should:

1. Place her in the supine position and inspect the wound site

2. Instruct her to splint the wound with a pillow when coughing

3. Assess the intensity of the pain and give her the ordered analgesic

4. Call her physician immediately and then check for wound dehiscence

23. While doing a newborn assessment, the nurse suspects that the infant has a talipes equinovarus when the infant's toes are:

1. Lower than the heel with the foot pointing inward

2. Higher than the heel with the foot pointing inward

3. Lower than the heel with the foot pointing outward

4. Higher than the heel with the foot pointing outward

24. Dehiscence is a complication with gastrointestinal surgery. A factor that would further predispose an obese 55-year-old client who had an abdominal cholecystectomy to wound dehiscence would be:

1. The age of the client

2. The presence of a T-tube

3. The presence of excessive flatus

4. The weight is 25% more than the recommended level

25. A female client with chronic obstructive pulmonary disease (COPD) tells the nurse that although she smokes, she limits it to one or two cigarettes a day. She also tells the nurse that she has been lax in doing the prescribed pulmonary physiotherapy exercises because they are too tiring. The nurse should respond:

1. "Tell me more about your typical day before you changed your routine."

2. "Your being so sick is probably because of your smoking and not exercising."

3. "Smoking is probably a primary cause of the increased severity of your disease."

4. "I can't make you stop doing what you are doing. It's your choice to be sick or well."

26. A client with a diagnosis of bipolar disorder, manic episode is receiving lithium. During the third week of therapy with lithium, the nurse would be aware that the drug was effective when the client is:

1. More high-spirited

2. Able to laugh off criticism

3. More appropriately groomed

4. Able to spend the morning alone

27. The nurse is aware that an early sign of chronic lead poisoning (plumbism) in children is:

1. Anemia

2. Oliguria

3. Convulsions

4. Mental retardation

28. A neonate develops hyperbilirubinemia, and phototherapy is begun. The care plan for an infant during phototherapy should include:

1. Taking vital signs q1h

2. Giving additional fluids q2h

3. Drawing blood for a Guthrie test qd

4. Dressing the neonate in a light shirt and diaper

29. The nurse assesses a client for increasing intracranial pressure by monitoring the pulse pressure. The nurse understands that a client's pulse pressure is actually the:
   1. Force exerted against an arterial wall
   2. Difference between the apical and radial rates
   3. Difference between systolic and diastolic readings
   4. Degree of ventricular contraction in relation to output

30. At night an elderly client with dementia becomes more disoriented and sleeps very little. Confusion caused by sleep deprivation can often be decreased by:
   1. Applying restraints so the client will remain in bed
   2. Leaving a subdued light on in the client's room all night
   3. Giving the client a back rub and shutting the door tightly
   4. Administering the client's prescribed sedative medications

31. A pediatric client is receiving busulfan (Myleran). The child's blood tests should be monitored for side effects, which include:
   1. Polycythemia
   2. Hyperuricemia
   3. Aplastic anemia
   4. Hypoalbuminemia

32. The physician prescribes a low-fat, 2-gram sodium diet for a client with hypertension. The nurse understands that a low-sodium diet will:
   1. Chemically stimulate the loop of Henle
   2. Diminish the thirst response of the client
   3. Prevent reabsorption of water in the distal tubules
   4. Cause fluid to move toward the interstitial compartment

33. An open reduction and internal fixation is performed on a client with a fractured left hip sustained in a fall. The nurse plans postoperative care knowing that the client:
   1. Will not be permitted to lie on the affected side
   2. Should be positioned with the legs in adduction
   3. Cannot bear weight on the affected leg for 3 months
   4. Can walk for short distances 1 to 2 days following surgery

34. A woman with five children comes to the emergency room with multiple facial injuries. The client states her husband, who is an alcoholic, beat her up.

Because the client appears to be a victim of abuse the nurse should now:
   1. Discuss birth control with her
   2. Report her experiences to the police
   3. Inquire about her and the children's safety
   4. Discuss the possibility of her leaving him

35. On the fourth postpartum day, the client's vaginal discharge is pinkish and does not contain clots. This discharge is called:
   1. Lochia alba
   2. Lochia rubra
   3. Lochia serosa
   4. Lochia pinkosa

36. After an infant has the cast that was used to correct a talipes equinovarus (clubfoot) removed, the nurse teaches the mother how to exercise the baby's foot. The nurse would know that the mother understood the instructions when she says that she will exercise the foot:
   1. With each diaper change
   2. Every 4 hours without fail
   3. Twice a day, morning and evening
   4. Once a day, after the baby naps

37. A child is receiving chelation therapy for lead poisoning. The nurse should be aware that a condition that can be attributed both to lead toxicity and to the side effects of the chelating agent, Ca EDTA, is:
   1. Hypocalcemia
   2. Nephrotoxicity
   3. Bone marrow depression
   4. Increased intracranial pressure

38. Following a myocardial infarction a client begins a supervised, progressive jogging regimen and asks the nurse how to tell whether it is helping. The nurse should reply:
   1. The intermittent claudication will be reduced."
   2. "Your breathing will become regular and shallow."
   3. "You will be able to run progressively longer distances before tiring."
   4. "You will perspire less when you run, because you'll be using less energy."

39. A client returns from the operating room following a total laryngectomy with a laryngectomy tube in the permanent stoma. To facilitate respirations and promote comfort, the client should be placed in the:
   1. Side-lying position
   2. Orthopneic position
   3. High-Fowler's position
   4. Semi-Fowler's position

40. A neuromuscular blocking agent is administered to a client before ECT therapy. The nurse should observe the client for:
    1. Convulsions
    2. Loss of memory
    3. Nausea and vomiting
    4. Respiratory difficulties

41. An infant born with a cleft lip is to have a surgical repair of the lip at about 2½ months of age. In preparation for the postoperative period, the nurse should instruct the infant's mother to:
    1. Teach the infant to drink from a cup
    2. Keep the infant's arms in restraints at all times
    3. Burp the infant as little as possible after feeding
    4. Place the infant in the supine position for extended periods

42. A couple interested in delaying the start of a family discuss the various methods of family planning. Together, they decide to use the basal body temperature method. Before they begin using this method they should understand that the fertility period surrounding ovulation usually extends from:
    1. 12 hours before to 24 hours after ovulation
    2. 72 hours before to 24 hours after ovulation
    3. 24 to 48 hours before to 48 hours after ovulation
    4. 72 to 80 hours before to 72 hours after ovulation

43. A long-term goal for a male client experiencing dysfunctional grieving after the death of his wife would be that the client will be able to:
    1. Resume previously enjoyed activities
    2. Eat at least two meals a day with another person
    3. Decrease negativistic thinking about self and others
    4. Relocate to a state in which other family members reside

44. During a routine examination, an enlarged thyroid gland is discovered in a 44-year-old female, and hyperthyroidism is suspected. On assessment the nurse would expect this client to have:
    1. Thickened and coarse skin
    2. Tachycardia and palpitations
    3. Apathetic attitude and masklike facies
    4. Menstrual disturbances and loss of libido

45. A client with a history of chronic obstructive pulmonary disease develops a pneumothorax, and a chest tube is inserted. The primary purpose of the chest tube is to:
    1. Lessen the client's chest pain and discomfort
    2. Drain accumulated fluid from the pleural cavity
    3. Restore negative pressure in the pleural space
    4. Prevent subcutaneous emphysema in the chest wall

46. When assessing a young infant, the nurse uses inspection, palpation, percussion, and auscultation. The technique the nurse should perform last is:
    1. Percussion of the infant's lung fields
    2. Palpation of the infant's abdominal organs
    3. Inspection of the infant's general condition
    4. Auscultation of the infant's heart and breath sounds

47. Following gastrointestinal surgery, a client's condition improves and a regular diet is ordered. The food that will most likely be tolerated with little discomfort is:
    1. Fresh fruit
    2. Baked fish
    3. Whole milk
    4. Bran cereal

48. A pregnant client's last menstrual period was on February 11. By July 18, a physical assessment of the client should indicate that the top of the fundus is:
    1. Even with the umbilicus
    2. Just above the symphysis pubis
    3. Two fingerbreadths above the umbilicus
    4. Halfway between the symphysis and umbilicus

49. A 2-year-old requires close supervision to protect against potential accidents, because at this age the child's learning occurs primarily from:
    1. Playmates
    2. The parents
    3. Older siblings
    4. Trial and error

50. The most important nursing objective when planning care for a young, hyperactive child with an attention-deficit disorder who engages in self-destructive behavior would be to:
    1. Prevent child from inflicting any self-injury
    2. Help child to develop the ability to test reality
    3. Assist child to formulate realistic ego boundaries
    4. Provide child with opportunities to discharge energy

51. A client at 39 weeks gestation arrives in the Labor and Delivery Unit complaining of regular contractions. A vaginal examination identifies that the presentation is a double footling breech. A decision is made to proceed to cesarean delivery. An important nursing intervention to prevent postoperative complications would include:
    1. Providing scrupulous skin care
    2. Maintaining adequate hydration
    3. Notifying the neonatal intensive care unit
    4. Monitoring the maternal vital signs frequently

52. A client is to have a craniotomy for removal of a brain tumor. The client's preoperative preparation should include:
    1. Forcing nutrient fluids
    2. Gently shampooing the hair
    3. Administering morphine sulfate
    4. Administering cleansing enemas

53. Two days following a myocardial infarction, a client has a temperature of 100.2° F. The nurse should:
    1. Auscultate the chest for diminished breath sounds
    2. Notify the physician immediately about the temperature
    3. Encourage deep breathing and coughing every 2 hours
    4. Record the temperature and monitor vital signs at routine intervals

54. A client with newly diagnosed hyperthyroidism is treated with propylthiouracil, an antithyroid drug, along with potassium iodide. The nurse teaches the client about these medications with the knowledge that:
    1. Iodide solutions such as these must be taken on an empty stomach and diluted in water
    2. The client should carefully observe for signs of infection or bleeding while on this therapy
    3. The use of these drugs prior to thyroidectomy will increase the risk of postoperative hemorrhage
    4. These drugs will be discontinued as soon as the client's temperature and pulse rate return to normal

55. During a well-baby visit the nurse recognizes that an 18-month-old's growth and development is within the normal range when the child:
    1. Climbs up the stairs
    2. Pedals a tricycle easily
    3. Says 150 different words
    4. Builds a tower of 8 blocks

56. A client with phobias about elevators and large crowds comes to the clinic for help because of feelings of depression related to these fears. An appropriate short-term goal would be that the client will:
    1. Ride an elevator without anxiety when accompanied by the nurse
    2. Describe the thoughts and feelings experienced in terrifying situations
    3. Experience relief from feelings of depression and an elevation of mood
    4. Identify the early childhood conflicts leading to the development of the fears

57. A client is admitted to the hospital for a nonstress test in the 42nd week of gestation. The nurse is aware that this test is being done because a prolonged pregnancy can:
    1. Cause polyhydramnios
    2. Lead to placental insufficiency
    3. Indicate subclinical gestational diabetes
    4. Predispose the mother to postpartum infection

58. A client develops a small pressure ulcer on the sacral area. The nurse should plan to deal with this problem by:
    1. Keeping the area dry
    2. Applying moist dressings
    3. Providing a low-calorie diet
    4. Keeping the client on the right side

59. When implementing a screening program for scoliosis, the most appropriate site would be in a:
    1. Middle school
    2. Well-baby clinic
    3. Senior high school
    4. Preschool day-care center

60. The clinic nurse observes a 2-year-old girl sitting alone, rocking and staring at a small, shiny top she is spinning. Later, the mother relates to the nurse her concerns, stating, "She pushes me away. She does not speak and only shows feelings when I take her top away. Is it something I've done?" It would be most therapeutic for the nurse to respond by:
    1. Asking the mother about her relationship with her husband
    2. Asking the mother how she held the child when she was an infant
    3. Telling the mother that it's nothing she has done and sharing observations of the child
    4. Telling the mother not to be concerned, that the child will outgrow this phase of development

61. The nurse assesses a client with the diagnosis of bulimia nervosa for the psychologic characteristics commonly associated with this disorder, which include:
    1. A flattened affect
    2. Unmet dependency needs
    3. Rigidity of character and inflexibility
    4. An unwillingness to discuss problems

62. The physician prescribes propranolol (Inderal) 20 mg po qid to be taken after discharge for a client after double coronary artery bypass surgery. The teaching plan should include telling the client to:
    1. Avoid abruptly discontinuing the medication
    2. Increase the medication if chest pain occurs
    3. Only drink alcoholic beverages in moderation
    4. Report a pulse rate less than 70 beats per minute

63. A nurse is called upon to assist with an emergency home delivery. To facilitate uterine contractions after the baby is born, the nurse should:
    1. Push down vigorously on the fundus
    2. Have the mother breast-feed the baby
    3. Place gentle continuous tension on the cord
    4. Encourage the mother to vigorously bear down

64. As part of a high school sex education program, the school nurse discusses herpes genitalis. The nurse should tell the students that:
    1. Herpes genitalis is curable with penicillin
    2. The disease generally is painless in women
    3. The disease is transmitted via fomites such as toilet seats
    4. Herpes genitalis causes both local and systemic reactions

65. A client with a history of benign prostatic hypertrophy mentions that he has heard cranberry juice prevents bladder infection. The nurse replies that cranberry juice may be helpful because it:
    1. Increases the acidity of the urine
    2. Soothes the irritated bladder walls
    3. Improves the glomerular filtration rate
    4. Destroys microorganisms in the urinary tract

66. When a client is actively hallucinating, it would be most therapeutic for the nurse to:
    1. Ask to whom the client is speaking
    2. Request that the client speak softly
    3. Involve the client in a simple game of cards
    4. Allow the client to continue without interruption

67. When caring for a child with cystic fibrosis, the nurse plans to include times for postural drainage in the child's care plan. This therapy should be scheduled:
    1. Once a day, after breakfast
    2. Three times a day, before meals
    3. Three times a day, halfway between meals
    4. Two times a day, on awakening and at bedtime

68. When giving discharge instructions to the parents of a child with cystic fibrosis, the nurse realizes that further explanation about the problems caused by cystic fibrosis is needed when the parents state, "We will:
    1. Keep our child in an air-conditioned room."
    2. Give our child the pancreatic enzymes with meals."
    3. Provide our child skin care after each bowel movement."
    4. Move to Florida where the climate is better for our child."

69. The nurse is aware that one reason the dialysate used in peritoneal dialysis is warmed to body temperature before its instillation into the peritoneal cavity is to:
    1. Force potassium back into the cells, thereby decreasing serum levels
    2. Encourage removal of serum urea by dilating the peritoneal blood vessels
    3. Add extra warmth to the body because metabolic processes are disturbed
    4. Help prevent cardiac dysrhythmias by speeding removal of excess serum potassium

70. After exposure to a nephrotoxic substance, a client is hospitalized in the oliguric phase of acute renal failure. The client's estimated urine output for the last 24 hours is about 2 cups. The BUN level is 96 mg/dl. The physician orders 900 ml of water by mouth during the next 24 hours. In carrying out this order, the nurse realizes that the rationale for the order is that 900 ml of fluid will:
    1. Equal the expected urinary output for the next 24 hours
    2. Prevent the development of complicating hypostatic pneumonia
    3. Compensate for both insensible and measured fluid losses over the next 24 hours
    4. Prevent hyperkalemia, which could result in the client having serious cardiac dysrhythmias

71. A client, 38 weeks gestation, is experiencing painless bleeding and has been diagnosed as having placenta previa. The client is concerned that she may have done something to cause the bleeding. Recognizing that the client appears worried, it would be most therapeutic for the nurse to say:
    1. "You probably have a weak uterus."
    2. "It's not your fault; these things happen."
    3. "Don't worry; it's just a sign of beginning labor."
    4. "The placenta is lying low and separates when you dilate."

72. When formulating a plan of care for a client with a somatoform disorder, conversion type, the nurse must realize that:
    1. The illness is very real to the client and the client needs appropriate nursing care
    2. Although the client believes there is an illness, there is no cause to be concerned
    3. Good nursing care is needed even though the nurse recognizes the client is not ill
    4. There is no physiologic basis for the illness; therefore only emotional care is needed

73. A client with type 1 diabetes develops ketoacidosis. The laboratory value that would support a diagnosis of diabetic ketoacidosis would be:
1. A normal BUN
2. Elevated serum lipids
3. Low serum calcium levels
4. A decreased hematocrit level

74. A client has a below-the-knee amputation of the right leg. The nurse should understand that after this type of surgery:
1. Strict bed rest is usually maintained for at least several days
2. The stump dressing is usually changed daily by the physician
3. Hemorrhage rarely occurs during the early post-operative period
4. The client is usually positioned with the stump elevated for the first 24 hours

75. A client has a nasogastric tube after a gastric resection. The nurse should expect to observe:
1. Vomiting
2. Gastric distention
3. Intermittent periods of diarrhea
4. Bloody drainage for the first 12 hours

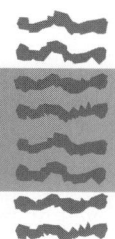

PART B QUESTIONS

76. A client who has had a continent urostomy created complains of postoperative pain. Initially, the nurse should:
    1. Interview the client for more data
    2. Tell the client to take deep breaths
    3. Measure the client's current vital signs
    4. Administer the prescribed analgesic to the client

77. A 35-year-old client, who has type 1 diabetes and has been maintaining glycemic control, is pregnant for the third time. Her first child is 4 years old, and her second pregnancy resulted in a stillbirth. She is seen in the Antepartum Testing Unit for a nonstress test (NST) at 33 weeks gestation. The nurse is aware that the client's history indicates that she is a candidate for a NST primarily because:
    1. The client is 35 years old
    2. The client is maintaining glycemic control
    3. A NST is indicated for high-risk clients with possible placental insufficiency
    4. A NST measures plasma levels of maternal estriols, which indicate fetal distress

78. A client is worried about what to expect after having a Whipple procedure for cancer of the pancreas. When assisting the client to plan, it would be most important for the nurse to know:
    1. Any history of alcohol or tobacco use
    2. The state and grade of the client's cancer
    3. Any previous exposure to known carcinogens
    4. The survival rate for individuals with pancreatic cancer

79. After a modified radical mastectomy, the nurse recognizes that a 36-year-old female client understands the schedule for self-examination of her remaining breast when she states she will carry out the procedure:
    1. Several days before an expected menstrual period
    2. Seven to ten days after the onset of each menstrual period
    3. Halfway between menstrual periods, preferably after taking a shower

    4. The same date every month, regardless of when menstruation occurs

80. The parents of a child who has recently been diagnosed with leukemia ask the nurse why the physician said that their child had too many white blood cells. The nurse's best response would be:
    1. "You seem to be focusing on your child's white blood cells."
    2. "Your doctor is the best one to answer that question for you."
    3. "It sounds like you really do not understand what occurs in leukemia."
    4. "The bone marrow isn't working properly and makes too many white blood cells."

81. An elderly couple, who live alone in their own home, both need help with dressing, bathing, and meal preparation. The condition that would rule out care in the home for this couple is:
    1. The home being located in a rural area
    2. The need for part-time multidisciplinary services
    3. The need for skilled nursing care on a 24-hour basis
    4. The need for part-time assistance for all personal care activities

82. Steroid therapy is ordered for a client with multiple sclerosis. The change that the nurse would expect to observe is decreased:
    1. Emotional lability
    2. Muscular contractions
    3. Pain in the extremities
    4. Episodes of vision loss

83. A client with the diagnosis of obsessive-compulsive disorder uses paper towels to open doors to avoid touching dirty doorknobs. The nurse's best response to this behavior would be to:
    1. Point out that the towels may be dirty
    2. Prevent the client from using the towels
    3. Ignore the behavior for the present time
    4. Quietly remove the paper towels from the area

84. After an automobile accident, a client who sustained multiple injuries is oriented as to person and place but is confused as to time. The client complains of a headache and drowsiness, but assessment reveals that the pupils are equal and reactive. A significant nursing intervention would be to:
    1. Keep the client alert and responsive
    2. Prevent unnecessary movement by the client
    3. Prepare the client for the administration of mannitol
    4. Monitor the client for symptoms of increased intracranial pressure

85. In discussions with the nurse, a child newly diagnosed with type 1 diabetes learns that insulin acts by:
    1. Preventing the glucose from being stored in the liver
    2. Helping to carry glucose into cells where it is burned for energy
    3. Helping to break down protein and fat to provide needed glucose
    4. Preventing the wasting of blood glucose by converting it to glycogen

86. After an uneventful pregnancy, a client delivers a baby with a meningomyelocele. The baby has an Apgar score of 9/10. An immediate priority of nursing care for this newborn would be:
    1. Administering oxygen by nasal catheter
    2. Protecting the sac with sterile, moist gauze
    3. Transferring the infant to the intensive care nursery
    4. Placing a name bracelet on the ankle and taking footprints

87. A client becomes increasingly agitated and screams at, curses at, fights with, and bites other clients. The physician orders a stat injection of haloperidol (Haldol). The nurse should carry out the order to administer the Haldol:
    1. Quickly with an attitude of concern
    2. After the client agrees to take the injection
    3. Before the client suspects what is happening
    4. Quietly, without any explanation about the reason for it

88. The laboratory finding that reflects a major problem commonly found in pregnant teenagers is:
    1. Urine glucose 2$^+$
    2. Hemoglobin 9.1 g
    3. Platelets 75,000/mm^3
    4. White blood cells 12,000/mm^3

89. A slightly overweight client is to be discharged from the hospital after a cholecystectomy. When teaching the client about nutrition, the priority intervention should be:

1. Listing those fatty foods that may be included in the diet
2. Explaining that fatty foods may not be tolerated for several weeks
3. Teaching the importance of a low-calorie diet to promote weight reduction
4. Encouraging the client to join a weight reduction program in the local community

90. The nurse is aware that 60% of brain tumors in children result in symptoms of increased intracranial pressure and are most commonly found in the:
    1. Cortex
    2. Cerebellum
    3. Temporal lobe
    4. Subarachnoid space

91. When caring for a client during the period of acute alcohol withdrawal, the nurse should:
    1. Apply restraints to provide security and keep the client calm
    2. Encourage the client to relate hallucinations and delusions in detail
    3. Continually reassure the client that the symptoms are part of the withdrawal syndrome
    4. Keep the room dimly lit to counter the visual distortions the client probably will experience

92. A class A diabetic primigravida attends the high-risk clinic. When assessing a client with class A diabetes, the nurse would probably find that the client has:
    1. Vascular changes in the eyes
    2. No demonstrable vascular disease
    3. A history of diabetes for at least 15 years
    4. Retinal changes seen on eye examination

93. In response to a client's question concerning the cause of polyarteritis nodosa, the nurse should respond that:
    1. The disease affects both males and females in equal numbers
    2. With current therapy, clients with this disease have an excellent prognosis
    3. The disease is considered one of hypersensitivity, but the exact cause is unknown
    4. Arteriolar pathology of the disease affects only the kidneys and the retina of the eyes

94. A client has an open reduction and internal fixation of the hip. The client is to be transferred to a chair on the second postoperative day for a half hour. Before the transfer, the nurse should:
    1. Assess the strength of the affected leg
    2. Explain how the transfer will be accomplished
    3. Instruct the client to bear weight evenly on both legs

4. Encourage the client to keep the affected leg elevated

95. A client is admitted to the hospital for acute gastritis and ascites secondary to alcoholism and cirrhosis. It is important for the nurse to routinely assess this client for:
   1. Obstipation
   2. Blood in the stool
   3. Any food intolerances
   4. Complaints of nausea

96. A client is at high risk for developing ascites. To assess for this condition the nurse should:
   1. Observe for signs of respiratory distress
   2. Percuss the abdomen, listening for dull sounds
   3. Palpate the lower extremities over the tibia and observe for edema
   4. Auscultate the abdomen, listening for decreased or absent bowel sounds

97. When the nurse approaches a delusional client, the client yells, "You're the one that made my lover leave me." The nurse should recognize that the client:
   1. Is confused and disoriented
   2. Is actively hallucinating at this time
   3. Feels a great sense of vulnerability
   4. Needs to have limits set after quieting down

98. A pregnant client with a history of delusions, hallucinations, and suspiciousness tells the nurse she is fearful about her upcoming delivery and the health of her baby. The most supportive approach by the nurse would be to:
   1. Reassure the client that she will have plenty of help with the delivery
   2. Commend the client on her ability to express her fears and concerns
   3. Share with the client the staff's concerns about how she will handle the infant
   4. Provide the client with a detailed explanation of what occurs during labor and delivery

99. A 6-year-old female is being discharged after diagnostic studies and treatment for frequent urinary tract infections. The statement by the girl's mother that would indicate that further teaching is necessary would be:
   1. "I guess I should not use the bubble bath I got my daughter for her birthday."
   2. "When it doesn't hurt my daughter to urinate, I can stop giving her these pills."
   3. "I hope my daughter can remember to always wipe from the front to the back."
   4. "I will tell my daughter's teacher to let her go to the toilet as soon as she needs to."

100. One day the mother of a 7-year-old, previously hospitalized with nephrotic syndrome several months before, calls the clinic nurse. She reports that for the past week her child has had a muddy pale appearance, a poor appetite, and been unusually tired after school. The nurse suspects that the child:
   1. Is not taking her medications
   2. Is developing a viral infection
   3. May be in impending renal failure
   4. May be overextending herself at school

101. A 5-month-old with tetralogy of Fallot is to be discharged with a prescription for digoxin (Lanoxin) and furosemide (Lasix). The nurse instructs the mother to notify the physician if:
   1. The child's pulse rate exceeds 100
   2. Feeding difficulties and vomiting occur
   3. Cyanosis occurs during periods of crying
   4. The child requires several naps each day

102. The nurse recognizes that a client understands how to appropriately take the antacids prescribed by the physician when the client states, "I will take my antacids:
   1. With the onset of pain."
   2. 30 minutes after meals."
   3. Every 4 hours around the clock."
   4. Every time I have anything to eat."

103. The mother of a preschooler with acute glomerulonephritis asks the nurse whether her child will have to stay in bed. The nurse should tell the mother that bed rest:
   1. Is not part of the usual treatment unless the child is seriously ill
   2. Will be necessary for 3 to 4 weeks, regardless of the response to therapy
   3. Is limited to 72 hours after the institution of antihypertensive drug therapy
   4. Will be necessary until the child's blood pressure is normal and the urine clear

104. A client who has been on IV magnesium sulfate therapy for preeclampsia delivers an infant weighing 4 lb in the 37th week of gestation. The nurse is aware that a finding in the newborn that may indicate magnesium sulfate toxicity is:
   1. Pallor
   2. Tremors
   4. Hypotonia
   3. A pulse of 200 beats/min

105. A teenager, hospitalized for pregnancy-induced hypertension, is anorectic and appears depressed. The nurse plans to further explore the client's emotional status when the client comments:
   1. "I'm tired of feeling so clumsy."
   2. "I'll be glad when I can sleep all night."
   3. "I dreamed my baby had only one arm."
   4. "I was really happy before I got pregnant."

106. A client who has been diagnosed as having atherosclerosis and hypertension has always been an active individual. The client is interested in measures that would help maintain health. The nurse explains that maintenance of vessel patency can be promoted by:
   1. Practicing relaxation techniques
   2. Decreasing the amount of exercise
   3. Increasing saturated fats in the diet
   4. Leading a more sedentary life-style

107. The nurse teaches a client with hypertension to reduce dietary sodium. The nurse should emphasize that:
   1. The taste for salt is inherent but it can be overcome with practice
   2. Salt-free natural seasonings can be used and taste the same as salt
   3. The taste for table salt is learned and increases over time, but it is not a biologic necessity
   4. Salt substitutes with potassium chloride bases can be used freely with foods to provide the same taste

108. The delivery room nurse does an Apgar on a newly delivered infant and knows that an Apgar value of 2 should be assigned to:
   1. A strong, lusty cry
   2. Body pink, extremities blue
   3. Arms and legs slightly flexed
   4. Heart rate of 90 beats per minute

109. When preparing a toddler with celiac disease for discharge, the nurse should caution the parents that the toddler characteristic that would make their child most susceptible to a celiac crisis is:
   1. Invention
   2. Autonomy
   3. Narcissism
   4. Negativism

110. When performing the initial assessment of a client in ketoacidosis the nurse notes:
   1. Nervousness and cool, pale skin
   2. Erythema toxicum rash and pruritus
   3. Diaphoresis and instability as with intoxication
   4. Deep respirations and fruity odor to the breath

111. A client develops bronchopneumonia. To help determine the effectiveness of therapy, the nurse should refer to the results of the client's:
   1. Lung scan
   2. Bronchoscopy
   3. Pulmonary function study
   4. Culture and sensitivity of sputum

112. Cor pulmonale is a common complication of COPD. The sign that would lead the nurse to suspect that the client is developing cor pulmonale would be:
   1. Peripheral edema
   2. A productive cough
   3. Twitching of the extremities
   4. Lethargy progressing to coma

113. A client has chest tubes inserted to treat a hemopneumothorax that resulted from a crushing chest injury. When connecting the two-chamber water-seal drainage system ordered, the nurse must understand that the chamber attached to the chest tube provides the:
   1. Suction
   2. Water seal
   3. Seal and suction
   4. Drainage receptacle

114. The nurse should plan to involve a client with anorexia nervosa in activities that will:
   1. Save the client's depleted energy
   2. Force the client into decision making
   3. Focus the client on sexual attractiveness
   4. Provide the client with peer group involvement

115. A client with the diagnosis of obsessive-compulsive disorder who has a need to wash the hands about 50 to 60 times a day tearfully tells the nurse, "I know my hands are not dirty, but I just can't stop washing them." The nurse's best response would be:
   1. "Let's talk about why you feel you must wash your hands."
   2. "I think you're getting better; you're beginning to understand your problem."
   3. "Don't worry about it; these actions are part of your illness, and these feelings will pass."
   4. "I understand that, but maybe we can work together to limit the number of times you wash them."

116. A child is admitted to the hospital for surgical incision and drainage of a puncture wound and intravenous antibiotic therapy. The nurse administers tetanus toxoid immunization. The rationale for this is that the toxoid confers:
   1. Lifelong passive immunity
   2. Longer-lasting active immunity

3. Lifelong active natural immunity
4. Temporary passive natural immunity

117. A client with the diagnosis of multiple sclerosis had a sudden loss of vision and asks the nurse what caused it. The nurse explains that the temporary blindness was probably caused by:
1. Virus-induced iritis
2. Intracranial pressure
3. Closed-angle glaucoma
4. Optic nerve inflammation

118. An elderly client is hospitalized with the diagnosis of dementia of the Alzheimer's type. Her daughter tearfully tells the nurse, "I should never have allowed my mother to live alone as she wanted to do. But she has not been this bad. I am to blame. She did not even recognize me immediately." The response by the nurse that would be the most helpful would be:
1. "I do not think that anybody could blame you. You did what she wanted. Your being here tells us that you care."
2. "I realize that you are upset now. You can visit again when she is more responsive. I am sure you'll see a change."
3. "Why do you think your mother's condition has deteriorated? Her forgetfulness is temporary. You'll help if you don't cry."
4. "This must be a difficult time for both of you. Would you like to share your other observations to help us plan for her care?"

119. A client in a birthing room is having contractions every 3 to 4 minutes that last about 35 seconds. A vaginal examination reveals that the cervix is 70% effaced and 6 cm dilated. From the results of the examination, the nurse knows that the client is:
1. Early in the first stage of labor
2. In the transitional phase of labor
3. Beginning the second stage of labor
4. Midway through the first stage of labor

120. A public health nurse visits the home of a female client in the terminal stage of cancer 3 days each week to provide physical care and emotional support. The nurse observes that the client's adolescent children are having difficulty talking with their mother. The nurse suggests family meetings with the hospice nurse, knowing that:
1. It is important to solve family problems before death occurs
2. They will be unable to deal with their feelings until after their mother dies
3. A deeper level of understanding will help the children comprehend what their mother is going through

4. Opening communication systems reduces the intensity of a family's emotional reaction to terminal illness

121. After a modified radical mastectomy a female client tells the nurse, "This diagnosis is as good as a death sentence, and I would rather go now than to suffer." At this time it would be most important for the nurse to:
1. Encourage her to admit herself to the psychiatric unit of the hospital
2. Determine whether she has experienced self-destructive or suicidal thoughts
3. Explore the possibility of a vacation after hospitalization to reduce her stress level
4. Encourage her to think positively and to try and focus on the good things in her life

122. A client with Crohn's disease is to have an upper gastrointestinal series. The nurse understands that an upper GI series with barium would be contraindicated if the client had:
1. Hemorrhoids
2. A perforation
3. Hyperkalemia
4. An inflamed colon

123. The teaching plan for a client with Crohn's disease should focus on:
1. Meeting nutritional needs
2. Anticipating sexual alteration
3. Controlling severe constipation
4. Preventing increased weakness

124. A client's contraction stress test (CST) is positive and indicates potential problems. The nurse understands that a positive test:
1. Indicates the need to perform a nonstress test
2. Showed late decelerations of the fetal heart rate with contractions
3. Showed a consistent fetal heart rate of 120 to 160 beats per minute
4. Indicates the need for immediate cesarean delivery to ensure viability of the fetus

125. A male client with the diagnosis of schizophrenia, paranoid type, often displays overt sexual behavior toward female clients and nurses. The nurses can best respond to the client's behavior by:
1. Refusing to speak with the client until he stops this behavior
2. Sending the client to his room when the behavior is observed
3. Matter-of-factly telling the client that his behavior is unacceptable
4. Ignoring this behavior until the client is more in control of his responses

126. Considering the anticholinergic-like side effects of many of the psychotropic drugs, the nurse should encourage clients taking these drugs to:
    1. Suck on hard candy
    2. Restrict their fluid intake
    3. Eat a diet high in carbohydrates
    4. Avoid products that contain aspirin

127. During the administration of an enema prior to gastrointestinal surgery, a client complains of cramps. The nurse should:
    1. Discontinue the enema and try again at a later time
    2. Close the lumen of the tubing until the client's cramps subside
    3. Lower the container to the floor to decrease intestinal pressure
    4. Administer the enema rapidly to permit a quicker evacuation of fluid

128. Nursing care for the child in the acute phase of nephrotic syndrome should include:
    1. Forcing fluids every hour
    2. Providing time for active play periods
    3. Encouraging frequent change of position
    4. Feeding low-protein, high-carbohydrate, and low-salt foods

129. The nurse teaches the mother of a child with a pinworm infestation how to collect a cellophane tape specimen. The nurse recommends that the specimen be collected:
    1. At night after the child has had a bath
    2. At night after the child has had a bowel movement
    3. In the late morning after the child has had a bowel movement
    4. In the morning before the child has had a bowel movement or a bath

130. A grand multigravida, 34 weeks gestation, is brought to the emergency room by her husband because she is experiencing vaginal bleeding. The nurse would suspect that the client had a placenta previa if the bleeding was:
    1. Painful vaginal bleeding in the first trimester
    2. Painful vaginal bleeding in the third trimester
    3. Painless vaginal bleeding in the first trimester
    4. Painless vaginal bleeding in the third trimester

131. A female client who has been delusional is found staring at the television set, which is on. The client suddenly rises and shouts, "Stop saying that. Who do you think you are?" It would be most therapeutic for the nurse to:
    1. Attempt to distract the client by taking her for a walk
    2. Point out the inappropriateness of her behavior to her
    3. Take the client to her room so she can have a quiet place to think
    4. Tell her that the voices she hears are coming from the television set

132. A female client is dying of cancer. Her family is concerned because she appears to be accepting less and less responsibility for her own care. To help the family plan for her care, the nurse should:
    1. Point out that denial is a normal response and will be temporary
    2. Assist them to identify methods to give her more control over the situation
    3. Encourage them to accept her regression until she can cope more effectively
    4. Explain that her anger is normal and identify ways for the family to deal with it

133. A client has a paracentesis, and the physician removes 1500 ml of fluid. It is essential that the nurse observe the client for:
    1. A hypertensive crisis
    2. Abdominal distention
    3. An increased pulse rate
    4. Dry mucous membranes

134. A client is scheduled for an amniocentesis. Before the procedure, the nurse should:
    1. Remind the client to empty her bladder
    2. Instruct the client to take nothing by mouth
    3. Give the client the ordered sedation to minimize fetal movement
    4. Administer a Fleet enema to the client to prevent contamination

135. An episiotomy is performed to facilitate delivery and prevent tearing of maternal tissues. The physician orders sitz baths three times a day. The nurse should include in the teaching plan that sitz baths promote healing by:
    1. Promoting vasodilation
    2. Softening the incision site
    3. Cleansing the perineal area
    4. Tightening the rectal sphincter

136. A client with osteoporosis is encouraged to drink milk. The client refuses the milk, explaining that it causes gassiness and bloating. A food rich in calcium that may be easily digested by clients who do not tolerate milk is:
    1. Eggs
    2. Yogurt
    3. Potatoes
    4. Bananas

137. A client receiving steroid therapy states, "I have had difficulty controlling my temper, which is so unlike me, and I don't know why this is happening." The nurse's best response would be to:
    1. Tell the client it is nothing to worry about
    2. Encourage the client to talk further about these findings
    3. Tell the client to attempt to avoid situations that cause irritation
    4. Try to determine whether the client has been experiencing other mood swings

138. A client who is scheduled for a muscle biopsy tells the nurse, "They better give me a general anesthetic. I don't want to feel anything." The most therapeutic response by the nurse would be:
    1. "You seem to be worried about the test."
    2. "Try not to think about it; you won't feel a thing."
    3. "This test is always done under local anesthesia."
    4. "Tell them when you have pain, and they'll take care of it."

139. A client who is experiencing acute alcohol withdrawal appears frightened and points toward the bed and says, "Bugs are crawling all over my bed and on me." The most therapeutic response by the nurse would be:
    1. "I do not see any bugs on your bed."
    2. "The bugs will go away when you feel better."
    3. "Do you want me to brush them away for you?"
    4. "Those are not bugs; it is just the design on the bedspread."

140. A client is to receive nitrogen mustard as part of a drug protocol for Hodgkin's disease. The nurse should explain that the nitrogen mustard is believed to act by:
    1. Interfering with cellular protein synthesis
    2. Inhibiting the synthesis of purine and pyrimidine
    3. Binding the DNA to interfere with RNA production
    4. Binding the DNA strands and interfering with cell replication

141. A biopsy of a brain tumor removed from a child identifies the tumor as a cerebellar astrocytoma. The nurse is aware that this tumor is:
    1. Fast growing and highly malignant
    2. Benign and associated with a high rate of cure
    3. A cause of pituitary malfunction as the child grows
    4. Close to vital centers and only partial excision is possible

142. A 9-year-old is admitted to the hospital with a diagnosis of possible infratentorial brain tumor. The nurse identifies the common presenting symptom of this type of tumor when the child demonstrates:
    1. Ataxia
    2. Seizures
    3. Papilledema
    4. Cranial enlargement

143. One morning a male client, whose thought processes are marked by ideas of reference and persecutory ideation, appears very upset. The client tells the nurse that the reporter on television told everyone that the client is "a queer." The most therapeutic response by the nurse would be:
    1. "It sounds to me like you're having some frightening feelings."
    2. "I will call the station and ask why the reporter said that about you."
    3. "You seem upset by this. Why do you think the reporter said that about you?"
    4. "Sometimes when we are unsure of ourselves we project our feelings on others."

144. When preparing a teaching plan for the parents of a child with celiac disease, the nurse recalls that the basic problem in celiac disease is the:
    1. Presence of meconium stool
    2. Clumping of the intestinal villi
    3. Absence of the enzyme peptidase
    4. Susceptibility to profound dehydration

145. The physician's order for a client in ketoacidosis states, add 500 units Humulin R insulin to 500 ml Ringer's lactate IV and run to administer 60 units of insulin in the next 30 minutes. To administer the correct amount of medication, the nurse should set the IV infusion device at:
    1. 30 ml/hr
    2. 60 ml/hr
    3. 90 ml/hr
    4. 120 ml/hr

146. The best nursing action to prevent heat loss in the neonate would be to:
    1. Administer oxygen to prevent shivering
    2. Dress the baby in a shirt and gown immediately
    3. Bathe the infant in warm water as soon as possible after birth
    4. Maintain skin-to-skin contact between mother and baby under a cover

147. A client thinks she is pregnant. The nurse is aware that a positive sign of pregnancy is:
    1. Hegar's sign
    2. Uterine enlargement
    3. A positive pregnancy test
    4. Fetal movements felt by the examiner

148. A young adult client with schizophrenia is started on haloperidol (Haldol). When the nurse gives the client the medication, the client asks, "What's this for?" The nurse could best respond, "This medication:
    1. Will help you to relax and think more clearly."
    2. Fights 'the blues' and keeps your thoughts together."
    3. Will raise your seizure threshold, letting you think more clearly."
    4. Maintains an even mood while keeping your temper under control."

149. A young pregnant client, who has been noncompliant with her prenatal visits, is seen in the triage area of the labor and delivery unit complaining of contractions. Assessment of the client reveals a nutritional deficit. A physical finding that can be related to inadequate protein intake is:
    1. Petechiae
    2. Bradycardia
    3. Bleeding gums
    4. Significant nondependent edema

150. An infant is admitted to the hospital nursery and is classified as being small-for-gestational age (SGA). A priority intervention for this infant would be to:
    1. Test the infant's stools for occult blood
    2. Monitor the infant's blood glucose levels
    3. Place the infant in the Trendelenburg's position
    4. Measure the infant's head circumference every shift

151. After several days on bed rest, a preschooler with the diagnosis of acute glomerulonephritis becomes demanding and will not listen to the nurses. The child was found in the playroom twice on the previous shift. To best meet the needs of this child the nurse should:
    1. Ask the child not to get up again and explain the reason for bed rest
    2. Place soft restraints on the child when family members cannot be present
    3. Have a color television set moved into the child's room as soon as possible
    4. Move the child into a room with another preschooler who has a fractured femur

152. A client with laryngeal cancer has a partial laryngectomy and tracheostomy. To facilitate communication postoperatively, the nurse should:
    1. Provide a pad and pencil for writing
    2. Allow more time for the client to articulate
    3. Face the client and speak slowly and distinctly
    4. Use visual clues such as gestures and objects

153. When planning the development of a nurse-client relationship, the nurse is aware that the most important aspect of this relationship during the early phase of its development is:
    1. Trust
    2. Empathy
    3. Personal rapport
    4. Open communication

154. A client with a peptic ulcer is scheduled for a subtotal gastrectomy. Nursing intervention directed toward minimizing the postoperative complication of dumping syndrome includes teaching the client to:
    1. Ambulate after every meal
    2. Remain on a diet low in fat
    3. Eat in a semirecumbent position
    4. Increase the fluid intake with meals

155. When teaching the parents of an 8-year-old with newly diagnosed type 1 diabetes, the nurse should tell them that with type 1 diabetes, it is common to develop:
    1. Obesity
    2. Ketoacidosis
    3. Resistance to treatment
    4. Hypersensitivity to other drugs

156. An 11-year-old has just been diagnosed as a type 1 diabetic. The child, who likes sweets, asks about sugar and sugar substitutes in the diet. The nurse and the dietitian tell the child that:
    1. Honey can be used as a natural sugar substitute
    2. Simple sugars such as sucrose or fructose should be avoided
    3. Sugar substitutes such as saccharin or aspartame can be used
    4. The sweet taste habit can be broken by eliminating sweets altogether

157. Before a client's discharge after having a generalized seizure, the nurse should reinforce previous teaching related to the anticonvulsant phenytoin (Dilantin). The nurse should instruct the client to:
    1. Report immediately any unsteadiness of gait
    2. Expect transient joint discomfort on occasion
    3. Avoid massaging the gums during oral hygiene
    4. Immediately discontinue the drug if a skin rash appears

158. A full-term neonate born with a meningomyelocele is scheduled for surgery. The priority preoperative nursing goal is to make certain that this baby:
    1. Remains sedated
    2. Gains 1 ounce per day
    3. Continues infection-free
    4. Develops a strong sucking reflex

159. The physician prescribes the "prudent diet" advocated by the American Heart Association for a male client with angina. The client's wife says to the nurse, "I guess I'm going to have to cook two meals, one for my husband and one for my daughter and myself." The most appropriate response by the nurse would be:
    1. "I wouldn't bother. This diet is really easy to follow; just cut down the salt you use and fry foods in peanut oil."
    2. "The diet that has been prescribed for your husband is a healthy diet that is recommended for all of us to follow."
    3. "You're right, and be careful not to cook his favorite meals because he will probably not want to adhere to the diet."
    4. "This is a very difficult diet to follow. I would recommend that you shop daily for food so there are no temptations in the kitchen."

160. When administering hydroxyzine hydrochloride (Vistaril), the nurse should monitor the clients for the common side effects of this drug, which include:
    1. Ataxia and confusion
    2. Drowsiness and dry mouth
    3. Vertigo and impaired vision
    4. Slurred speech and headache

161. After head and neck surgery, hyperextension of the neck for the first few postoperative days may cause:
    1. Cervical trauma
    2. Laryngeal spasm
    3. Laryngeal edema
    4. Wound dehiscence

162. When checking the cervical dilation of a client in labor, the nurse notices that the umbilical cord has prolapsed. The nurse's first action should be to:
    1. Take the fetal heart rate
    2. Turn the client on her side
    3. Cover the cord with sterile saline soaks
    4. Put the client in the Trendelenburg position

163. After abdominal surgery a goal is to have the client achieve alveolar expansion. This goal would be most effectively achieved by:
    1. Postural drainage
    2. Pursed-lip breathing
    3. Incentive spirometry
    4. Prolonged exhalation

164. A young woman comes to the mental health clinic because she has lost her job and does not know what to do. The client lives alone and her family lives in another state. The client states she feels like a failure. An appropriate nursing diagnosis for this client would probably be:
    1. Risk for violence self-directed related to panic state
    2. Altered thought processes related to impaired judgment
    3. Social isolation related to absence of satisfying personal relationships
    4. Personal identity disturbance related to inability to distinguish self from nonself

165. A client with advanced cancer of the bladder is scheduled for a cystectomy and ileal conduit. Physical preparation for the surgery should include:
    1. Insertion of a Foley catheter
    2. Well-balanced diet with vitamins
    3. Administration of neomycin sulfate
    4. Administration of urinary antiseptics

166. The nurse should be aware that the most therapeutic activity for a depressed client would probably be:
    1. Putting together a jigsaw puzzle
    2. Stuffing envelopes for a local charity
    3. Assembling and whip-stitching a wallet
    4. Participating in an aerobic exercise group

167. A few days after an automobile accident in which a child sustained a fractured femur, the laboratory reports of the child, who is now in Bryant's traction, indicate a slight decrease in hemoglobin and hematocrit levels. The most appropriate nursing action would be to:
    1. Order a type and crossmatch
    2. Notify the physician immediately
    3. Provide additional meat for dinner
    4. Assess the child's abdomen for internal bleeding

168. In her eighth month of gestation, a client spontaneously ruptures her membranes but still does not have contractions. Specific nursing care for clients with ruptured membranes should include:
    1. Observing for meconium staining
    2. Monitoring temperature frequently
    3. Watching for signs of preeclampsia
    4. Preparing for fetal scalp pH sampling

169. The physician prescribes the drug cholestyramine (Questran), an anion exchange resin used to treat a client's persistent diarrhea. This drug reduces the absorption of fat, which may produce a deficiency of:
    1. Thiamine
    2. Vitamin A
    3. Riboflavin
    4. Vitamin B$_6$

170. A client with colitis is started on steroid therapy. The nurse would know that teaching was effective when the client says, "When taking this medication, I should:
1. Take the medicine at bedtime with a snack."
2. Take the drug 1 hour before or 2 hours after eating."
3. Divide the dose into equal parts and take with each meal."
4. Take the drug in the early morning with food or an antacid."

171. When reviewing the breast self-examination procedure with a client, the comment by the client that the nurse should consider significant is:
1. "My bra feels tight when I am menstruating."
2. "My breasts feel lumpy just before menstruation."
3. "My left breast was always slightly larger than my right."
4. "My right breast feels and looks thicker than my left breast."

172. A terminally ill client is furious with one of the staff nurses. Over the next several days the client refuses the nurse's care and insists on doing self-care. A different nurse is assigned to care for the client. The nurse's initial step in revising the nursing care plan to meet the client's needs would be to:
1. Get a full report from the first nurse and adjust the plan accordingly
2. Ask the physician for a report on the client's condition and plan accordingly
3. Tell the client about the change in staff responsibilities and assess the client's reaction
4. Assess the client's present status and capabilities and include the client in a discussion of revisions of the care plan

173. The membranes of a laboring client rupture spontaneously. An observation by the nurse at this time that would necessitate additional newborn resuscitation measures would be:
1. Bloody show
2. Greenish fluid
3. Clear fluid with specks of mucus
4. Shortened intervals between contractions

174. The nursing care plan for a client who is being admitted to the hospital with a diagnosis of abruptio placentae should include careful assessment for signs and symptoms of:
1. Jaundice
2. Hypertension
3. Hypovolemic shock
4. Impending convulsions

175. When developing a plan of care for an elderly client with the diagnosis of dementia, the nurse should:
1. Be considerate of the client's various likes and dislikes
2. Be firm in dealing with the client's attitudes and behaviors
3. Explain to the client the details of the therapeutic regimen
4. Provide consistency in carrying out nursing activities for the client

176. After a suprapubic prostatectomy, a client returns from the operating room with a suprapubic tube and a three-way Foley catheter with continuous GU irrigant. The nurse realizes that the purpose of this therapy is to:
1. Promote continuous formation of urine
2. Prevent the formation of clots in the bladder
3. Facilitate the measurement of urinary output
4. Provide continuous pressure on the prostatic fossa

177. A child with a puncture wound, whose history of immunizations is uncertain, is given tetanus immune human globulin (Hyper-Tet). The nurse knows that the chief reason for using tetanus immune human globulin instead of tetanus antitoxin is that it:
1. Is as effective as the antitoxin
2. Is more convenient to administer
3. Is not likely to cause anaphylaxis
4. Can be safely given to everyone who needs it

178. A 2½-year-old is admitted to the hospital with a fever of 103° F, stiffness of the neck, and generalized malaise. The child is diagnosed as having acute bacterial meningitis. Priority nursing care for this child would include:
1. Hydrating the child
2. Administering oxygen
3. Giving the child a tepid sponge bath
4. Placing the child on droplet precautions

179. A client with a fractured tibia and fibula is to be discharged with a right leg cast and crutches. In addition to teaching the technical aspects of crutch walking, the nurse plans to advise the client to:
1. Avoid taking showers until the cast is removed
2. Increase intake of vitamin C to enhance healing
3. Gradually increase weight bearing on the injured leg
4. Remove loose rugs and rearrange furniture as necessary

180. X-ray films reveal that a client has closed fractures of the right femur and tibia. Multiple soft-tissue con-

tusions are also present. An important short-term intervention is to:
1. Prepare the client for application of skeletal traction
2. Reassure the client that these injuries are not that serious
3. Prepare the client for operative reduction of the injured extremity
4. Assess the circulatory, motor, and sensory status of the client's injured extremity

181. The primary goal of therapy for a client with chronic obstructive pulmonary disease is to:
1. Limit hydration
2. Improve ventilation
3. Increase oxygenation
4. Correct the bicarbonate deficit

182. The adaptation in a child with nephrotic syndrome that would necessitate that the nurse check vital signs, especially pulse quality and rate and blood pressure, is:
1. Hypovolemia
2. Hyperkalemia
3. Pulmonary emboli
4. Congestive heart failure

183. In a childbirth preparation class a client learned a number of exercises that could be performed on the first day after a cesarean delivery. The nurse on the postpartum floor recognizes that further teaching is necessary when the client states that one of the exercises learned was:
1. Leg bends
2. Foot circles
3. Pelvic rocking
4. Shoulder circles

184. A pregnant client comes to the emergency room because of vaginal bleeding. The nurse asks the client to approximate how much bleeding is occurring. The best gauge for the client to use is:
1. Whether or not clots are present
2. Any change in fetal activity with the bleeding
3. The amount in relation to her usual menstrual flow
4. How weak she has become since the bleeding began

185. After 3 weeks of mental health therapy a client tells the nurse, "I feel I am ready for discharge." The nurse can best evaluate the client's readiness for discharge by:
1. Testing the client's level of trust of self and staff
2. Exchanging views with the client about the prognosis

3. Asking the client to explain any changes in behavior since admission
4. Requiring the client to identify specific behaviors viewed as examples of wellness

186. The results of a biopsy indicate a client has a malignant sarcoma of the liver, and chemotherapy via regional perfusion is the treatment of choice. This method of drug administration probably was selected for this client because:
1. The drug therapy can be continued at home with little difficulty
2. Larger doses of drugs can be delivered to the actual site of the tumor
3. The toxic effects of the chemotherapeutic drugs will be confined to the area of the tumor
4. Combinations of drugs can be used to attack neoplastic cells at various stages of the cell cycle

187. When caring for a child in acute respiratory distress from laryngotracheobronchitis, who has a temperature of 103° F, the nurse should give priority to:
1. Delivering 40% humidified oxygen
2. Initiating measures to reduce fever
3. Constantly assessing the child's respiratory status
4. Providing support to reduce the child's apprehension

188. A client with diabetes develops persistent, alternating episodes of hyperglycemia and hypoglycemia despite increased amounts of insulin daily. The nurse recognizes that this effect is caused by:
1. Increased glycogen formation in the liver
2. Excessive insulin that causes glycogenolysis
3. Insufficient amounts of insulin to lower blood sugar
4. Excessive glucose intake that causes gluconeogenesis

189. After an open reduction and internal fixation of a hip fracture in an elderly individual, the assessment that would require immediate nursing intervention would be:
1. Complaints of pain in the chest 6 days postoperatively
2. An inability to cough productively 2 days postoperatively
3. A rectal temperature of 100.2° F 3 days postoperatively
4. Fatigue in the leg on the unaffected side 5 days postoperatively

190. When assessing the head of a 2-hour-old infant, the nurse would normally expect to find:
1. Closed suture lines
2. A sunken anterior fontanel
3. Open anterior and posterior fontanels
4. A soft fluctuant mass that outlines a bone

191. A 4-year-old admitted to the hospital with AIDS is placed on appropriate isolation precautions. These precautions include:
1. Gloves should be worn whenever approaching the bedside
2. Gloves should be worn when in contact with blood and body fluids
3. Limited physical contact should be made when care is administered
4. Gowns, masks, and gloves should be worn when providing direct care

192. The nurse is aware that the psychiatrist is concerned that one of the clients receiving haloperidol (Haldol) may be developing neuroleptic malignant syndrome. The nurse should carefully assess the client for symptoms that would include:
1. Jaundice, malaise, and pruritus
2. Sore throat, seizures, and tremors
3. Diaphoresis, muscle rigidity, and hyperpyrexia
4. Loss of visual acuity, dry skin, and hyperbilirubinemia

193. When auscultating the lungs of a client admitted with severe preeclampsia, the nurse notices the presence of crackles and is aware that this can be an indication that:
1. A convulsion is imminent
2. Pulmonary edema has developed
3. The stress of pregnancy precipitated bronchial constriction
4. The enlarged uterus has impaired diaphragmatic functioning

194. An infant's discharge is being delayed because of a rising reticulocyte count. The infant's mother, who is being discharged, asks the nurse why the baby must stay. The nurse's response is based on an understanding that the infant needs to be observed for:
1. Significant jaundice
2. A bacterial infection
3. Bleeding tendencies
4. Adequate oxygenation

195. A young female client admitted to the trauma center after being raped continues to talk about the rape. The primary nursing intervention should be directed toward:
1. Getting her involved with a rape therapy group

2. Remaining available and supportive to limit destructive anger
3. Exploring her feelings about men to promote future relationships
4. Helping her restore emotional control to expiate feelings of shame

196. A client develops subcutaneous emphysema after a laryngectomy. This is most readily detected by:
1. Palpating the neck or face
2. Auscultating the lung fields
3. Evaluating the blood gases
4. Reviewing the chest x-ray film

197. A client who complains of memory loss, nervousness, insomnia, and fear of going out of the house is admitted to the hospital after several days of increasing incapacitation. Initially, considering this client's history, the nurse should give priority to:
1. Evaluating the client's adjustment to the unit
2. Providing the client with a sense of security and safety
3. Exploring the client's memory loss and fear of going out
4. Assessing the precipitating factors for the client's hospitalization

198. After gastric surgery a client has a nasogastric tube in place. The nurse should plan to:
1. Change the tube at least once every 48 hours
2. Monitor the client for signs of electrolyte imbalance
3. Connect the nasogastric tube to high continuous suction
4. Assess placement by injecting 10 ml of water into the tube

199. A 9-year-old, recently diagnosed as having type 2 diabetes, attends the center for diabetic teaching with the parents. The nurse notes that the child's concerns about the illness center on:
1. How the parents will react to the diagnosis of the disease
2. How much school might have to be missed because of the disease
3. Whether or not the physician will be successful in controlling the diabetes
4. Whether having diabetes means it will be impossible to have children as an adult

200. A client comes to the clinic complaining of a productive cough with copious yellow sputum, fever, and chills for the past 2 days. The first thing the nurse should do when caring for this client is to:
1. Administer oxygen
2. Begin to push fluids

3. Collect a sputum specimen

4. Take the client's temperature

201. The nurse recognizes that the increased incidence of fractures with osteoporosis in the United States occurs because of the:
1. Dietary use of skim milk
2. Aging of the American population
3. Increased number of hysterectomies
4. Immobility associated with early retirement

202. A client's leg is placed in Buck's traction to reduce pain from muscle spasms until surgery can be performed. The nurse knows that Buck's traction is a type of:
1. Skin traction
2. Transfixation
3. Skeletal traction
4. Balanced suspension

203. When planning for a client's care during the detoxification phase of acute alcohol withdrawal, the nurse should anticipate the need to:
1. Supervise the client at all times
2. Keep the client's room lights dim
3. Speak to the client in a loud, clear voice
4. Restrain the client during periods of agitation

204. A client with Hodgkin's disease is to receive the cyclic antineoplastic vincristine (Oncovin) as part of a therapy protocol. The nurse explains that Oncovin helps destroy the malignant cells by:
1. Arresting mitosis in metaphase
2. Inhibiting the synthesis of pyrimidine
3. Alkylating nucleic acids needed for mitosis
4. Inactivating DNA and inhibiting RNA synthesis

205. During the postoperative period following a craniotomy for the removal of an astrocytoma, a child develops a sudden right pupillary dilation. The nurse recognizes this as a sign of:
1. Severe pain
2. Intense fear
3. Uncal herniation
4. Reduced intracranial pressure

206. A primigravida who is in the active phase of labor demonstrates irritability, an increased bloody show, and moderate to intense contractions every 2 minutes that last 60 to 90 seconds and complains of nausea. The nurse determines that the client is in the phase of labor known as:
1. Early first
2. Transition

3. Early third

4. Late second

207. A married, male client with three children has lost his job and feels useless. He is tearful, upset, and embarrassed. A goal of care appropriate for this client would be:
1. Limiting tearfulness
2. Increasing self-esteem
3. Controlling feelings of sadness
4. Promoting acceptance by others

208. Following major surgery, a 4-year-old girl is in the intensive care unit and is told that she can have medication to keep her comfortable. The nurse should tell her that when she hurts:
1. She will be given a pill to make the hurt better
2. The nurses will give her an injection to take the hurt away
3. The nurses will put medicine in her IV to make the hurt go away
4. She will be given medicines that will make all the hurt go right away

209. When lithium therapy is instituted, the nurse should teach the client to maintain a normal daily intake of:
1. Iron
2. Sodium
3. Potassium
4. Magnesium

210 To meet the needs of a client who has a seizure disorder and who has recently had a generalized seizure, the nurse should plan to:
1. Outline ways to prevent physical trauma from occurring during a seizure
2. Teach that anticonvulsant medications should be taken on an empty stomach
3. Explain that the client need not tell others of the illness because the medication will be increased
4. Teach the client that the symptoms and treatment of seizure disorders are similar, regardless of the cause

211. After surgery for a meningomyelocele, an infant is being fed by gavage feedings. When checking placement of the feeding tube, the nurse is unable to hear the air injected because of noisy breath sounds. The nurse should:
1. Notify the physician
2. Advance the tube 1 to 2 cm
3. Carefully insert 1 ml of formula
4. Attempt to aspirate the stomach contents

212. The physician orders 10 ml of a 10% solution of calcium gluconate for a client with a severely depressed serum calcium level. The client also is receiving an IV solution of D5W and digoxin (Lanoxin) 0.25 mg daily. The nurse should be aware that calcium gluconate:
    1. Is compatible with all IV solutions
    2. Is nonirritating to surrounding tissues
    3. Cannot be added to the IV of l000 ml D5W
    4. Potentiates the action of digitalis preparations

213. The nurse is aware that after a thyroidectomy the client should be observed carefully for symptoms indicating:
    1. Perforation of the lung
    2. Laceration of the esophagus
    3. Fracture of the laryngeal cartilage
    4. Accidental removal of the parathyroids

214. A 7-lb baby is delivered and admitted to the normal newborn nursery with an order for phytonadione (vitamin K) (AquaMEPHYTON) 1 mg IM. The nurse is aware that this treatment is administered to:
    1. Facilitate bilirubin excretion
    2. Stimulate normal bowel flora
    3. Increase liver glycogen stores
    4. Promote normal blood clotting

215. After surgery to create an ileal conduit, a client awakens and asks for a sip of water. The nurse informs the client that water by mouth cannot be given until the:
    1. Ileal loop begins to drain in 6 hours
    2. Intestinal anastomosis heals in 10 days
    3. Nasogastric suction is discontinued in 3 days
    4. Client leaves the recovery room in a few hours

216. A severely depressed male client responds to therapy and with the help of the staff begins to set some immediate goals. The goal that would most indicate improvement would be the client's plan to:
    1. Talk with at least one person on the unit daily
    2. Stay clear of people who seem to make him anxious
    3. Show the staff that he can do what they want him to do
    4. Take at least 3 to 4 hours daily to sit alone and think about things

217. A young child is placed in Bryant's traction for a fractured femur. The parents of the child ask the nurse about the traction. The nurse explains that with this system:
    1. A Thomas splint, a sling, and pulleys keep the femur aligned
    2. A system of pulleys keeps the affected leg elevated in a sling

    3. Traction is placed on both legs, even though the left femur is fractured
    4. Moleskin and weights are applied to the affected extremity to keep it extended

218. A client with active genital herpes has a cesarean delivery. The nurse teaches the mother how to limit transmission of the virus to her newborn. The nurse would evaluate that the instructions were understood when the mother states:
    1. "I should avoid kissing my baby on the lips."
    2. "I must wear gloves when I'm holding my baby."
    3. "I have to wash my clothes and my baby's clothes separately."
    4. "I should wash my hands thoroughly with soap and water before handling the baby."

219. A terminally ill client's oldest son is very concerned about his mother's condition. He asks the nurse, "Will she get better?" The nurse's most appropriate response would be:
    1. "Her vital signs are stable, so she is holding her own."
    2. "Of course she will. You can't give up. You must hope for the best."
    3. "I don't know. You'll have to ask the physician. I'll leave a note that you are here."
    4. "Her condition is very serious. Would you like to discuss your concerns with me?"

220. The nurse encourages a terminally ill client to make decisions about daily activities and care. The nurse would know that the client, who had been extremely angry with everything and everybody, had resolved some of the anger when the client states:
    1. "You've got a busy morning ahead of you! I'm really a mess."
    2. "What are you going to let me do this morning? You know I can help."
    3. "It's so hard to let someone do so much for me. It just doesn't seem right."
    4. "I can do my face, hands, arms, and chest today, but I think you'd better do the rest."

221. A male client, age 67, visits his physician complaining of urinary frequency, dysuria, nocturia, and difficulty starting his urinary stream. A cystoscopy and biopsy of the prostate gland are performed, and afterward the client is unable to void. The nurse should:
    1. Limit oral fluids until he voids
    2. Assure him that this is normal
    3. Insert a urinary retention catheter
    4. Palpate above the pubic symphysis

222. An IV of 800 ml/24 hours is ordered for a 2½-year-old. It is set up via a pediatric administration set (Buretrol). The solution should be set to infuse at:
1. 3 drops/minute
2. 6 drops/minute
3. 33 drops/minute
4. 60 drops/minute

223. A client is admitted to the emergency room with head and chest injuries received in an automobile accident. When evaluating the client's responses to the emergency room treatments, the assessments that indicate that the client can safely be transferred to a critical care unit are:
1. Alert but restless, stable vital signs, cyanosis
2. Stable vital signs, apprehension, complaints of pain
3. Drowsy but easily roused, improving tissue perfusion, fluctuating vital signs
4. Elevated temperature, slowing pulse and respirations, pain in the injured extremity

224. The nurse observes a client with acute bronchitis and chronic obstructive pulmonary disease sitting up in bed, appearing very anxious and dyspneic. The nurse should:
1. Provide oxygen at 2 L per minute
2. Administer the prescribed sedative
3. Have the client breathe into a paper bag
4. Encourage the client to cough and deep breathe

225. A primary nursing goal for a grand multipara in the event that a cesarean delivery is performed is:
1. Prevention of hemorrhage
2. Prevention of perineal infection
3. Avoidance of wound dehiscence
4. Assistance with mother-infant bonding

226. Several hours after admission to the hospital with laryngotracheobronchitis (viral croup), the nurse notes the child has developed tachypnea and tachycardia, accompanied by intercostal and substernal retractions and increased restlessness. The nurse should immediately:
1. Suction secretions from the child's trachea
2. Dislodge mucus by striking the child on the back
3. Report the child's respiratory status to the pediatrician
4. Increase the level of oxygen being delivered to the child

227. After surgery for a fractured hip, a client is restless and appears anxious. When the nurse and nursing assistant enter the room to provide evening care, the client becomes upset. The nurse should:
1. Explain that hip fractures are not life threatening
2. Reassure the client that they will be very careful
3. Describe the need for various personnel at this time
4. Suggest that the client can turn on the television for diversion

228. When a nurse suspects that a child has been abused, the nurse's primary responsibility must be to:
1. Treat the child's traumatic injuries
2. Confirm the suspected child abuse
3. Protect the child from any future abuse
4. Have the child examined by the physician

229. A 35-year-old male client is admitted to the hospital for confirmation of the diagnosis of Hodgkin's disease. The client and his wife are concerned that he may have cancer. The wife states, "Wouldn't it be unlikely for someone like my husband to have cancer?" Before responding to the emotional aspects of the question, the nurse should recall that:
1. It is impossible to predict who will develop cancer or when
2. Hodgkin's disease occurs most often in males ages 20 to 40
3. Hodgkin's disease is more common among women than men
4. Cancer is typically a disease of older, rather than younger, adults

230. The nurse avoids placing a client in the supine position during labor primarily because it can:
1. Lead to transient episodes of hypertension
2. Cause decreased perfusion of the placenta
3. Impede free movement of the symphysis pubis
4. Prolong labor because of the influence of gravity

231. A client who is bottle-feeding her infant complains of discomfort from her engorged breasts. The measure most appropriate to relieve the engorgement would be:
1. Expression of breast milk to relieve pressure
2. Application of ice packs and a binder to her breasts
3. Use of hot, moist towels as compresses on the breasts
4. Restriction of oral fluid intake to less than 1000 ml daily

232. The physician prescribes propylthiouracil for a client with hyperthyroidism. Two months after being started on the antithyroid medication, the client calls the nurse and complains of feeling tired and looking pale. The nurse should:
1. Advise the client to get more rest
2. Instruct the client to skip 1 dose daily
3. Schedule an appointment for the client
4. Tell the client to increase the medication

233. A client who is suspected of having had a silent myocardial infarction and is to have an electrocardiogram (ECG) asks the nurse, "Why are they doing this test?" The best reply by the nurse would be: "This test will:
1. Detect your heart sounds."
2. Reflect any heart damage."
3. Help us change your heart's rhythm."
4. Tell us how much stress you can tolerate."

234. Twelve hours after sustaining full-thickness burns to the chest and thighs, a client is complaining of severe thirst. The client's urinary output has been 60 ml/hr for the past 10 hours. No bowel sounds are heard. The nurse should:
1. Increase the client's IV flow rate
2. Give the client orange juice by mouth
3. Give the client 4 oz of water by mouth
4. Moisten the client's lips with wet gauze

235. The nursing diagnosis that takes priority in planning care for a client who has sustained partial- and full-thickness burns of the chest in a fire is:
1. Risk for infection related to burn trauma
2. Impaired physical mobility related to bed rest
3. Impaired gas exchange related to smoke inhalation
4. Risk for fluid volume deficit related to decreased intake

236. A depressed client frequently repeats doubts about going on and admits to thinking about suicide while denying a plan has been developed. During this period it is essential that the nurse:
1. Have a staff member stay with the client at all times
2. Plan to involve the client in activities that are interesting and absorbing
3. Explain in detail to the client how the staff will protect against self-harm
4. Use frequent, unobtrusive observations of the client's moods and activities

237. Eight days after a client starts antidepressant medication therapy, the nurse notices that the client who has been severely depressed is neatly dressed and well groomed. The client smiles at the nurse and states, "Things sure look better today." Based on an awareness of depressed behavior, the nurse should:
1. Begin preparing the client for discharge
2. Increase the client's privileges as a reward
3. Compliment both the client's appearance and smile
4. Assign a staff member to provide constant surveillance of the client

238. The nurse begins terminating the consistent one-to-one relationship with a client who is soon to be discharged. The nurse might expect the client to respond to termination by manifesting symptoms of:
1. Grief
2. Testing
3. Splitting
4. Manipulation

239. A young female client comes to the trauma center stating that she had been raped. She is disheveled, pale, and staring blankly. Taking what has occurred into consideration, the nurse asks the client to describe in detail what happened because:
1. It will help the nursing staff in giving legal advice and providing counseling
2. Talking about what led to the rape will help her see what may have precipitated it
3. It helps the victim put the event in better perspective and helps begin the resolution process
4. Discussing details will keep the victim from covering up the intimate happenings in the rape situation

240. A client's respiratory status may be affected after abdominal surgery. When planning care to provide for this problem, the behavioral objective for this client should be:
1. Respirations will improve with coughing and deep breathing
2. Demonstrates the technique of coughing and deep breathing
3. Coughing and deep breathing will facilitate output of secretions
4. Will cough and deep breathe five or six times every hour while awake

241. After abdominal surgery a client develops hyperpyrexia and a draining wound, and peritonitis is suspected. When caring for this client, the nurse should use standard precautions and:
1. Surgical asepsis
2. Protective isolation
3. Enteric precautions
4. Contact precautions

242. During the assessment of a pyrexic client who was admitted to the hospital because of a productive cough, fever, and chills, the nurse percusses an area of dullness over the right posterior lower lobe of the lung. Considering the presenting signs and symptoms, the nurse is aware that this finding may be indicative of:
1. Pleurisy
2. Bronchitis
3. Pneumonia
4. Emphysema

243. A male client with aortic stenosis is scheduled for a valve replacement in 2 days. He tells the nurse, "I told my wife all she needs to know if I don't make it." The response by the nurse that would be the most therapeutic would be:
1. "Men your age do very well."
2. "You are worried about dying."
3. "I'll get you a sleeping pill tonight, I know you will need it."
4. "I know you are concerned, but your physician is excellent."

244. To decrease or control the sensory and cognitive disturbances that can occur after a client has open heart surgery, the nurse should:
1. Restrict all visitors
2. Withhold analgesic medications
3. Plan for maximum periods of rest
4. Keep the room light on most of the time

245. A client with cancer of the sigmoid colon is to have an abdominoperineal resection with a permanent colostomy. In preparation for surgery, the client is placed on a low-residue diet to:
1. Limit the amount of flatus produced
2. Decrease the bacteria in the intestine
3. Reduce the amount of stool in the bowel
4. Prevent irritation of the intestinal mucosa

246. A 5-year-old girl is admitted to the hospital for a tonsillectomy and adenoidectomy. The nurse can best determine why the child thinks she is in the hospital by asking:
1. "Do you know what place this is?"
2. "Why did you come to the hospital?"
3. "Do you know what's going to happen to you?"
4. "You do know why you are in the hospital, don't you?"

247. A pregnant client at 37 weeks gestation is taught about signs that should be reported immediately to the obstetric care practitioner. The client indicates that she understands the information presented by stating that the sign that she would report immediately is:
1. Lower back pain
2. White vaginal discharge
3. Braxton Hicks contractions
4. Leakage of fluid from the vagina

248. A client with left ventricular failure is digitalized and placed on a maintenance dose of digoxin 0.25 mg qd. If a therapeutic effect is achieved, the nurse would expect to observe:
1. Decreased pulse rate, diuresis, decreased edema
2. Decreased pulse pressure, increased blood pressure, weight loss

3. Increased pulse rate, stable fluid balance, decreased blood pressure
4. Decreased pulse rate, reduced heart murmur, increased blood pressure

249. A child with acute spasmodic bronchitis who is receiving humidified air removes the face mask. During the bath, the nurse observes that the child has increased respiratory distress. The nurse should:
1. Do postural drainage and clap the child's chest
2. Discontinue the bath and replace the face mask
3. Suction the child's nasal passages to clear the airway
4. Put the child in the orthopneic position and call the physician

250. Oral feedings for an infant with acute spasmodic bronchitis are stopped, and intravenous fluids are ordered to:
1. Relieve laryngospasm
2. Lessen physical exertion
3. Decrease vagal stimulation
4. Meet the infant's caloric needs

251. A client has a permanent colostomy. During the first 24 hours there is no drainage from the colostomy. The nurse should realize that this is a result of the:
1. Edema following the surgery
2. Absence of intestinal peristalsis
3. Decrease in fluid intake prior to surgery
4. Proper functioning of the nasogastric tube

252. A client is admitted to the hospital for surgery for a possible intestinal carcinoma. The client complains of malaise, constipation, and stomach rumbling. When obtaining a health history, the nurse should expect the client to report changes in:
1. The shape of stools
2. The daily intake of fats
3. The amount of fluid intake
4. The use of spices in the diet

253. An infant who had been in a mist tent because of dyspnea caused by acute spasmodic laryngitis is being discharged. The parents ask the nurse about caring for their baby at home. The nurse's best response would be:
1. "Give 2 ounces of water after the feeding of formula."
2. "No one should be allowed to visit the baby for a while."
3. "All allergen producers such as animals should be avoided."
4. "No specific restrictions are necessary after your child goes home."

254. A child is receiving an infusion. The nurse notes some erythema at the infusion site. On palpation, the child cries and draws away. There is a blood return in the tubing when the solution container is held briefly below the insertion site. The nurse should:
1. Decrease the flow rate of the IV solution
2. Realize that additives may have this effect
3. Have the venipuncture site changed to a new area
4. Maintain the IV but continue to observe the insertion site

255. The nurse tells the parents of a child who is being discharged following a tonsillectomy and adenoidectomy that the child may have an objectionable mouth odor, slight ear pain, and a low-grade fever for the next few days. The nurse recommends that the parents:
1. Apply an ice collar for the pain
2. Give aspirin for pain as necessary
3. Let the child suck on peppermint candies
4. Encourage the child to gargle with warm salt solution

256. When caring for a client with an antisocial personality disorder, the staff should use a consistent approach that is:
1. Warm and firm without being punitive
2. Indifferent and detached but nonjudgmental
3. Conditionally acquiescent to client demands
4. Clearly communicative of personal resentment

257. The nurse is aware that group therapy with abusing parents should be designed to help these parents to:
1. Confront their own rage at being abused as children
2. Admit publicly and to themselves that they are child abusers
3. Recognize the long-range psychologic effects of child abuse
4. Share information about their parenting techniques and skills

258. The day after having a cesarean delivery, a client is encouraged to ambulate. The client angrily asks the nurse, "Why am I being made to walk so soon after surgery?" The nurse's best reply would be:
1. "Early walking lowers the incidence of urinary infection."
2. "You can hold your baby more quickly if you walk around."
3. "Walking about early will prevent your wound from opening."

4. "Walking keeps the blood from pooling in your legs and prevents clots."

259. A male client with a cerebrovascular accident (CVA) frequently cries when his family visits. The family members are obviously upset by the crying. The nurse should explain to the family that the client is:
1. Mourning his obvious loss of function
2. Having difficulty controlling his emotions
3. Demonstrating his usual premorbid personality
4. Conveying his unhappiness about the situation

260. A client is admitted to the hospital with a diagnosis of left ventricular failure secondary to rheumatic heart disease. The nurse should be aware that the symptoms of heart failure occur because the:
1. Excessive blood volume increases the workload of the heart muscles
2. Heart can no longer pump blood adequately in relation to venous return
3. Heart valves have become stenotic or regurgitant and impede blood flow
4. Arterial system is less flexible because of hypertension or arteriosclerosis

261. After a basal cell carcinoma is removed by fulguration, a client is given a topical steroid to apply to the surgical site. The nurse would recognize that the teaching regarding steroids and skin lesions was effective when the client states that the purpose of the medication is to:
1. Prevent infection of the wound
2. Increase fluid loss from the skin
3. Reduce inflammation at the surgical site
4. Limit itching around the area of the lesion

262. Intestinal infestation with *Enterobius vermicularis* (pinworm) is suspected in a 6-year-old. The nurse can assist in confirming the child's diagnosis by:
1. Asking the mother to collect stools for 3 consecutive days for culture
2. Having the mother bring in the child's stools for visual examination for 3 days
3. Instructing the mother to do an anal cellophane tape test early in the morning
4. Assisting the mother to schedule hypersensitivity tests of the child's blood serum

263. Considering the fact that a client with a right-sided cerebrovascular accident is right-handed, the task that will probably present the most difficulty is:
1. Eating meals
2. Writing letters
3. Combing the hair
4. Dressing every morning

264. The nurse would know that childbirth preparation classes were effective when during the transition phase of labor a couple who attended use the breathing pattern known as:
1. Pant-blow
2. Slow-chest
3. Shallow-chest
4. Accelerated-decelerated

265. The nipples of a client who is breast-feeding her baby become sore and cracked. The nurse should instruct the client to:
1. Apply continuous ice packs to her nipples
2. Take her analgesic medication as ordered
3. Remove the baby from the breast for a few days
4. Expose her nipples to the air several times a day

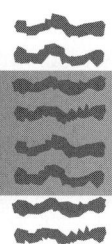

PART A ANSWERS AND RATIONALES

1. ① Straddling the hip would prevent scissoring by keeping the infant's legs abducted. (2) (PC; IM; PE; NM)
   2 An infant seat would not prevent scissoring.
   3 Tight wrapping would maintain the infant's legs in a scissored position.
   4 When the football hold is used, the infant is carried in a supine position with the legs adducted, which promotes scissoring.

2. ③ This increases oncotic pressure, which pulls interstitial fluid into the blood vessels, restoring blood volume and limiting ascites. (3) (DA; AN; MS; FE)
   1 Nutrients are provided by total parenteral nutrition, not salt-poor albumin.
   2 Salt-poor albumin is not given to increase protein stores.
   4 Salt-poor albumin has no effect on diverting blood flow away from the liver.

3. ② The likelihood of a cerebral hemorrhage increases with rising systemic blood pressure readings, particularly rising diastolic pressure readings. (1) (CI; EV; CW; HP)
   1 Fluctuations in blood pressure do not affect the status of the clotting factors.
   3 The degree of hypertension is not associated with hemorrhage.
   4 The course of labor is not affected by blood pressure changes except in the presence of abruptio placentae.

4. ① This gives direction to the relationship and provides a blueprint for future evaluation of progress. (2) (CC; AN; MH; TR)
   2 This is an aspect of the therapeutic relationship but is not the most important; what the client wants to achieve takes priority.
   3 Same as answer 2.
   4 Same as answer 2.

5. ② Drainage will flow with the force of gravity; therefore the dependent area should be checked for bleeding. (2) (SI; EV; CW; WH)
   1 The drainage is assessed for amount and characteristics; it is not emptied hourly, only once per shift or when necessary.
   3 Turning the client on the affected side is usually painful and should be avoided; the client should turn to the unaffected side.
   4 If the drainage on the dressing is excessive, the physician should be notified.

6. ③ As drainage collects and occupies space, the original level of negative pressure decreases; the lower the negative pressure, the less effective the drainage. (1) (SI; AN; MS; IT)
   1 It is easy and safe to empty, regardless of the amount of drainage in the unit.
   2 Drainage can be accurately measured by the calibrations on the unit or in a calibrated container after emptying.
   4 A one-way valve between the tubing and the collection chamber prevents drainage from entering the tubing and causing trauma to the wound.

7. ③ The physician must be consulted because the decision depends on the amount of damage and extent of healing. (3) (CC; IM; MS; EH)
   1 This is a medical decision that depends on the amount of damage, not how the client feels.
   2 It would be false reassurance to determine an exact time and date; this would be too early.
   4 Same as answer 1.

8. ④ This simply states reality without attempting to argue the client out of the delusion; actual physical contact should be initiated by the client. (3) (EP; IM; MH; SD)
   1 This denies the client's feelings and directly attacks the delusional system, forcing the client to defend it.
   2 This is false reassurance; the client does not feel safe, and the nurse's saying it does not make it so.
   3 Touching the client's arm can be frightening and overwhelming.

9. ① Deep breathing provides an active role in controlling pain; this is a positive coping skill. (3) (PC; PL; MS; NM)
  2 Understanding the importance of wound care will not reduce severe pain; health teaching should be initiated before, not during, a procedure.
  3 This could increase pain.
  4 Distraction techniques are usually ineffectual in the presence of severe pain; contraction may increase the pain.

10. ③ The properly fitted diaphragm, when used correctly with spermicidal gel, is 98% to 99% effective and has a low risk for complications in women with diabetes. (2) (CC; EV; CW; RC)
  1 Oral contraceptives are not recommended for women with diabetes because of their effect on carbohydrate metabolism and the risk of major cardiovascular complications.
  2 The intrauterine device carries a risk of infection, which is already increased in women with diabetes.
  4 The client is planning subsequent pregnancies.

11. ③ Epinephrine produces sympathetic nervous system side effects such as tachycardia and hypertension. (2) (PT; EV; PE; DR)
  1 Pallor, not flushing, is a common side effect.
  2 This is not a common side effect; this medication is given to decrease respiratory difficulty.
  4 Hypertension, not hypotension, is a common side effect.

12. ③ The person addicted to drugs is unable to adequately deal with reality; drugs help blur reality and reduce frustrations. (3) (EP; PL; MS; SA)
  1 It is possible, although not easy, but it does require a change in attitude and a deconditioning process.
  2 People who abuse drugs are concerned with reality; drug use is often an attempt to blur the pains of reality.
  4 Education of the public has been extensive, but the individual who experiments with drugs does not believe addiction will develop.

13. ③ Obesity is a known predisposing factor to type 2 diabetes; the exact relationship is unknown. (2) (HI; AS; MS; EN)
  1 Diabetes insipidus is a disease caused by too little ADH and has no relationship to type 2 diabetes.
  2 High-cholesterol diets and atherosclerotic heart disease are associated with type 2 diabetes.
  4 Alcohol intake is not known to predispose a person to type 2 diabetes, and alcohol lowers the blood glucose level.

14. ④ A lung is a highly vascular area that accommodates a large portion of intravascular volume; once a portion is removed, the remaining lung tissue is at risk for pulmonary congestion. (3) (CI; AN; MS; RE)
  1 When an intravenous solution is administered correctly, this should not occur.
  2 Same as answer 1.
  3 Same as answer 1.

15. ② In this position the gravid uterus impedes venous return; this causes decreased cardiac output and results in reduced placental circulation. (1) (SI; AN; CW; HC)
  1 Although this may be partially true, it is not the most significant reason for avoiding the supine position while in labor.
  3 This is false; it can lead to supine hypotension.
  4 Same as answer 1.

16. ③ Today, 5-year survival rates for children with acute lymphocytic leukemia exceed 60% in most treatment centers. (3) (PT; AN; PE; BI)
  1 This projected prognosis is too fatalistic; 5-year survival occurs in about 60% of children treated.
  2 Same as answer 1.
  4 This projected prognosis is too favorable; 5-year survival occurs in about 60% of children treated.

17. ④ 500 ml divided by 4 hours equals 125 ml per hour. (2) (PT; AN; PE; FE)
  1 This rate would deliver less than the prescribed rate in the prescribed time.
  2 Same as answer 1.
  3 Same as answer 1.

18. ① Infection is the most common complication; sterile technique at the catheter insertion site must be maintained. (2) (SI; PL; MS; GI)
  2 This is not necessary; the catheter is sutured in place, and reasonable movement is permitted.
  3 Excessive weight gain or loss is not a complication of total parenteral nutrition.
  4 The client should be taught to leave the infusion pump as set and to call the nurse if the alarm rings.

19. ③ These clients can better work through their underlying problems when the environment is controlled, demands are reduced, and the routine is simple. (2) (EP; AN; MH; PR)
  1 Preventing these clients from carrying out rituals can precipitate panic reactions.
  2 The intent of therapy should be to help the client gain control, not to enable others to do the controlling.

4 Because anxiety stems from unconscious conflicts, the environment alone is not enough to effect resolution.

20. ① Levodopa can cross the blood-brain barrier and be converted to dopamine, a substance depleted in Parkinson's disease. (2) (PT; PL; MS; DR)
   2 Dopamine is not given because it does not cross the blood-brain barrier.
   3 Vitamin B_6 can reverse the effects of some anti-Parkinson's disease medications and is contraindicated.
   4 Isocarboxazid is an MAO inhibitor used for the treatment of severe depression, not symptoms of parkinsonism.

21. ① Clients may develop cerebral edema caused by excessive absorption of irrigating solution by the venous sinusoids during surgery. (3) (CI; EV; MS; RG)
   2 The procedure is usually performed using spinal anesthesia; the supine position is maintained for 8 to 12 hours.
   3 The surgery is performed through the urinary meatus and urethra; there is no suprapubic incision.
   4 The client is initially npo and then advanced to a regular diet as tolerated; continuous irrigation supplies enough fluid to flush the bladder.

22. ② This relieves some of the pain because it provides support to the incised abdominal wall. (1) (PC; IM; CW; HP)
   1 The symptoms do not indicate a need for this action.
   3 Analgesics will not relieve the discomfort associated with coughing unless stress placed on the incision by coughing is relieved.
   4 Same as answer 1.

23. ① This is a correct description of talipes equinovarus. (2) (DF; AS; PE; SK)
   2 This describes talipes calcaneovarus.
   3 This describes talipes equinovalgus, also known as rocker-bottom foot.
   4 Same as answer 2.

24. ④ Being overweight is a predisposing factor to wound dehiscence because of decreased vascularity and fragility of adipose tissue. (2) (CI; AS; MS; GI)
   1 Age does not really contribute to dehiscence.
   2 This does not contribute to dehiscence; a T-tube helps remove bile from the common bile duct.
   3 This causes discomfort, not dehiscence.

25. ① More data are needed about activities of daily living to evaluate noncompliance before revising the care plan. (2) (CC; AS; MS; RE)
   2 This is nonproductive because it places blame for the illness on the client.
   3 Same as answer 2.
   4 Same as answer 2.

26. ③ The client will be able to attend to grooming because lithium effectively controls and reduces grandiosity, poor judgment, and psychomotor hyperactivity. (2) (EP; EV; MH; DR)
   1 This would not reflect an effective response to lithium.
   2 Same as answer 1.
   4 Same as answer 1.

27. ① Bone marrow is most susceptible to lead toxicity; interference with hemoglobin biosynthesis leads to early signs of anemia. (3) (HI; AS; PE; BI)
   2 This is a late response indicating kidney shutdown; loss of protein and other substances occurs first.
   3 This is a serious, life-threatening, late response.
   4 This is a late response indicating central nervous system involvement.

28. ② Insensible and intestinal fluid losses are increased during phototherapy; extra fluid prevents dehydration. (3) (CI; PL; CW; HN)
   1 This is unnecessary unless changes from the baseline occur.
   3 A Guthrie test is done for PKU screening.
   4 The total body needs to be exposed to light.

29. ③ The pulse pressure is obtained by subtracting the diastolic blood pressure reading from the systolic blood pressure reading; pulse pressure widens as intracranial pressure increases. (1) (CI; AN; MS; NM)
   1 This is reflected in the actual blood pressure readings and indicates cardiovascular function.
   2 This is the definition of a pulse deficit.
   4 This is determined by various diagnostic techniques used in cardiology; this is the role of the physician, not the nurse.

30. ② A small light in the room may prevent misinterpretation of shadows, which can heighten fear and alter the perception of the environment. (2) (EP; IM; MH; DD)
   1 Less restrictive intervention should be used first; restraints may not be necessary.
   3 This is unsafe; a disoriented and confused client should not be isolated but closely observed.
   4 This is unnecessary; sedatives should be used sparingly in the aged because they may further confuse and agitate the aged client.

31. ③ A serious side effect of busulfan is aplastic anemia, a common reaction to the drug; blood studies would monitor RBCs, as well as WBCs and thrombocytes. (3) (CI; EV; PE; DR)
  1 A decrease, not an increase, in RBCs occurs.
  2 This is unrelated to busulfan.
  4 Same as answer 2.

32. ③ Sodium absorbs water in the kidney's renal tubules; when dietary intake is decreased, water is not reabsorbed and edema is reduced. (2) (PC; AN; MS; FE)
  1 This is untrue; a decrease in sodium will prevent the reabsorption of water; furosemide stimulates the loop of Henle.
  2 Adequate hydration is the major factor that diminishes the thirst response.
  4 This will not be caused by a low-sodium diet; a low-sodium diet will cause fluid to enter the intravascular compartment.

33. ① Lying on the affected side causes adduction, which may result in dislocation of the femoral head. (3) (CI; PL; MS; SK)
  2 This is contraindicated; it would put too much pressure on the operative site; the leg on the operative side should be kept in abduction.
  3 This is too long a period; partial weight-bearing will be permitted sooner, depending on the extent of surgery and the progression of healing.
  4 Partial weight-bearing on the affected leg is generally contraindicated for at least 2 to 3 weeks.

34. ③ Safety for the victim and the children must be assessed as research shows children of an alcoholic parent are more frequently abused. (2) (EP; IM; MH; CS)
  1 This focuses on a topic that may not be the client's concern; the client should be permitted to direct the topic of conversation.
  2 This is not the legal responsibility of the nurse at this time.
  4 Same as answer 1.

35. ③ Lochia serosa is similar in appearance to serosanguineous drainage and generally appears on the third or fourth postpartum day. (1) (DF; AS; CW; HC)
  1 Lochia alba consists of leukocytes, decidua, epithelial cells, mucus, and serum; this appears 10 days postpartum and lasts 2 to 6 weeks.
  2 Lochia rubra is initially bright red, changing to dark red or reddish brown; it consists mainly of blood, decidua, and trophoblastic debris; it occurs immediately after delivery, lasting for 2 to 3 days.
  4 There is no such term.

36. ① Exercising should be done often; association with a specific activity makes it easier to incorporate it into the life-style, leading to increased compliance. (2) (CC; EV; PE; SK)
  2 Although this is frequent enough, such a rigid schedule is difficult to follow with an infant, and compliance becomes more difficult.
  3 This is not frequent enough.
  4 Same as answer 3.

37. ② Lead toxicity and Ca EDTA both damage the proximal renal tubules, resulting in the abnormal excretion of protein and other substances. (3) (PT; EV; PE; RG)
  1 This does not occur with lead toxicity, nor is it likely to occur with the Ca EDTA preparation, which replaces calcium.
  3 Bone marrow damage is caused by lead toxicity only.
  4 Lead encephalopathy causes serious elevation of intracranial pressure; this is not related to Ca EDTA.

38. ③ The ability to endure progressive activity indicates that collateral circulation has improved cardiopulmonary functioning. (3) (CI; IM; MS; CV)
  1 Intermittent claudication is related to peripheral arterial occlusive disease, not cardiopulmonary function.
  2 Breathing when jogging should be regular and deep to meet the oxygen demands of the body during exercise.
  4 Perspiration is an expected and desired adaptation to promote heat loss through evaporation.

39. ④ The semi-Fowler's position helps to maintain the head in proper body alignment and facilitates respiration. (2) (DA; PL; MS; RE)
  1 The side-lying position inhibits respiratory excursion.
  2 This position may cause flexion of the neck, which would inhibit respiration and place pressure on the suture line.
  3 Same as answer 2.

40. ④ Neuromuscular blockers, such as succinylcholine (Anectine) or pancuronium (Pavulon), produce respiratory depression because they inhibit contractions of respiratory muscles. (1) (DA; EV; MH; DR)
  1 As a muscle relaxant, the neuromuscular blocking agent prevents convulsions.
  2 The loss of memory results from the ECT treatment, not from the neuromuscular blocking agent.

3 Because the client is not permitted anything by mouth for 8 to 10 hours before the treatment, this is not a major problem.

41. (4) Because the child will be kept supine after surgery to prevent irritation to the lips, using this position early prepares the child for its use later. (2) (DF; IM; PE; GI)
1 The baby is too young and will miss out on some oral gratification with this method.
2 Constant restraint of the arms would be injurious to arm growth, as well as to the infant's need to move about.
3 Babies with a cleft lip need increased burping time because they tend to swallow large amounts of air during feeding.

42. (3) The ovum is fertilizable for 12 to 24 hours, and sperm remain motile for about 72 hours. (1) (DF; PL; CW; RC)
1 The fertility period is longer than this.
2 This time period is too long before ovulation and too short after ovulation.
4 The period of fertility is shorter than this.

43. (1) This is realistic, specific, and measurable; it relates to the client's stated nursing diagnosis. (1) (SC; PL; MH; CS)
2 This is an unrealistic goal.
3 There are no data to indicate that the client is thinking negatively about himself or others.
4 This may be unrealistic and cannot be made without involvement of the family.

44. (2) Hyperthyroidism raises the metabolic rate and need for oxygen; this results in increased heart rate and myocardial irritability. (2) (HI; AS; MS; EN)
1 These are signs associated with myxedema and hypothyroidism.
3 Same as answer 1.
4 Same as answer 1.

45. (3) Negative pressure is exerted by gravity drainage or by suction through the closed system. (1) (DA; AN; MS; RE)
1 Though the discomfort may be lessened, this is not the primary purpose.
2 In a pneumothorax associated with COPD, there is an accumulation of air, not fluid.
4 Subcutaneous emphysema in the chest wall is most commonly associated with clients receiving air under pressure, such as that received on a ventilator.

46. (2) Palpation is the most invasive technique of the assessment skills; infants do not tolerate invasive procedures as well as they do less-threatening maneuvers. (2) (SI; AS; PE; GD)
1 Percussion of the lung fields can be frightening to an infant but is generally tolerated better than palpation; it should be done toward the end of the assessment but before palpation.
3 Inspection is the least invasive procedure and should be done first.
4 Auscultation is less invasive than palpation; the accuracy of auscultation may be affected if it is done after a more invasive technique that produces crying, thus making auscultation more difficult.

47. (2) Baked fish is a low-residue, low-fat, high-protein, and non-gas-producing food that is usually well tolerated. (2) (PC; EV; MS; GI)
1 This has fiber and irritates the gastrointestinal tract.
3 This irritates the gastrointestinal tract and stimulates mucous production.
4 Same as answer 1.

48 (1) At about the 20th to 22nd week of gestation, the top of the fundus is at the level of the umbilicus. (3) (DF; AS; CW; HC)
2 In a normal pregnancy this would be too low for a pregnancy between the fifth and sixth months.
3 In a normal pregnancy, this would be too high for 20 to 22 weeks of gestation.
4 Same as answer 2.

49 (4) The child is developing autonomy, is curious, and learns from experience. (1) (DF; AN; PE; GD)
1 The toddler is still learning from experiences, not from others; this is the level of parallel, not interactive, play.
2 The toddler is still attempting to distinguish self as separate from the parents; the struggle for autonomy limits learning from parents.
3 The struggle for autonomy at this age limits learning from older siblings, even though the toddler attempts to copy their behaviors; older siblings often are not good role models because they tend to be careless.

50. (1) All nursing care should be directed toward preventing injury, particularly with a self-destructive client. (2) (EP; PL; MH; BA)
2 Although this is important, prevention of injury is the priority.
3 Same as answer 2.
4 Same as answer 2.

51. (2) Because of the administration of regional anesthesia and the potential for blood loss associated with cesarean delivery, the client should be well hydrated before surgery to maintain adequate blood volume. (3) (CI; PL; CW; HP)
   1 This is not relevant; just before surgery is done, the skin will be cleansed.
   3 Unless the fetus is extremely premature or in distress, there is no reason for admission to the neonatal intensive care unit.
   4 Only routine monitoring of vital signs is necessary unless there has been some indication of infection or pregnancy-induced hypertension

52. (2) Shampooing is done carefully to avoid scratching the scalp, which would provide a portal of entry for microorganisms. (2) (SI; PL; MS; IT)
   1 Fluids are withheld preoperatively; they may also be restricted because of cerebral edema.
   3 Narcotics are not used because of possible central nervous system depression.
   4 Enemas are contraindicated; the client should avoid straining because it increases intracranial pressure.

53. (4) Myocardial necrosis causes a rise in body temperature within the first 24 to 48 hours, which gradually returns to normal within a week. (2) (CI; PL; MS; CV)
   1 This temperature did not result from a respiratory complication.
   2 This is unnecessary; this is an expected response and not an emergency.
   3 Coughing necessitates the use of the Valsalva maneuver, which is contraindicated because it can precipitate dysrhythmias.

54. (2) Propylthiouracil can cause a depression of leukocytes and platelets. (3) (PT; EV; MS; DR)
   1 They are given with milk, juice, or food to prevent gastric irritation.
   3 Drug therapy decreases the risk of postoperative hemorrhage because it decreases the size and vascularity of the thyroid gland.
   4 Therapy will be continued for at least 6 to 8 weeks, even if the client's temperature and pulse return to normal.

55. (1) This is normal developmental behavior for 18-month-olds; however, they may have trouble coming down the stairs. (2) (DF; EV; PE; GD)
   2 This is above the level of an 18-month-old child.
   3 Same as answer 2.
   4 Same as answer 2.

56. (2) This is a realistic essential first step; the problem and related feelings must be thoroughly explored before solutions can be developed. (1) (SC; PL; MH; AX)
   1 This would be a long-term goal.
   3 Same as answer 1.
   4 This is an inappropriate goal; a direct connection to life events is often difficult to find.

57. (2) Placental function peaks at term (37 weeks) and declines slowly until 42 weeks; thus, continuation of the pregnancy past term places the fetus at risk secondary to placental insufficiency. (3) (DF; AN; CW; HP)
   1 Oligohydramnios (decreased amniotic fluid volume) can occur in postterm gestations; a normal decrease in fluid begins at 37 weeks and continues to decrease.
   3 This is unrelated.
   4 Same as answer 3.

58. (1) This encourages tissue regeneration and prevents creation of a moist area conducive to infection. (1) (SI; PL; MS; IT)
   2 This creates a warm, moist, protein-containing medium that is ideal for growing pathogens.
   3 A high-calorie diet is appropriate to provide energy for tissue repair.
   4 Placing the client in one position will promote development of additional pressure ulcers.

59. (1) Preadolescents and young adolescents are most at risk and can be most successfully treated. (1) (DF; PL; PE; SK)
   2 Scoliosis would not be identifiable in children of this age.
   3 Some students with scoliosis might be identified, but it may be too late for adequate treatment.
   4 About 80% of clients with idiopathic scoliosis are preadolescents and young adolescents, not preschoolers.

60. (3) The nurse provides support in a nonjudgmental way by sharing information and observations about the child. (3) (EP; IM; MH; BA)
   1 This indirectly indicates that the parent may be at fault; it negates the mother's need for support and increases her sense of guilt.
   2 Same as answer 1.
   4 This is false reassurance that does not provide support; the mother recognizes something is wrong.

61. (2) Individuals with bulimia have many unmet dependency needs that they attempt to meet by the use of food. (3) (SC; AS; MH; ES)

1 Individuals with bulimia tend to be extroverted rather than demonstrating a flattened affect.

3 Individuals with bulimia tend to be unable to make decisions; although they may appear obstinate at times, there is no rigidity of character.

4 Individuals with bulimia talk freely about their problems, although the discussion is usually on a superficial level.

62. (1) Abrupt discontinuation of Inderal may cause an acute myocardial infarction. (2) (CI; IM; MS; DR)

2 Clients should never increase medications without a physician's direction.

3 Alcohol is contraindicated for clients taking Inderal.

4 The pulse rate can go much lower as long as the client feels well and is not dizzy.

63. (2) Suckling will induce neural stimulation of the posterior pituitary gland, which in turn will release oxytocin and cause uterine contractions. (2) (CI; IM; CW; HC)

1 Vigorous fundal pressure should never be utilized; this could cause a uterine prolapse.

3 If the placenta is still attached to the uterine wall, this may disconnect the cord from the placenta or cause uterine prolapse

4 This could cause a uterine prolapse.

64. (4) Fever, malaise, and headache may accompany local reactions. (2) (SI; AS; CW; WH)

1 Herpes is of viral origin; there is no cure, and antibiotics are ineffective.

2 Vesicles on genitalia rupture, causing painful ulcerations.

3 Most transmissions occur through intimate sexual contact with acute or healing lesions.

65. (1) An acid-ash diet, including cranberries, lowers the pH of the urine and discourages pathogenic growth. (1) (SI; AN; MS; RG)

2 Acid urine does not soothe bladder walls.

3 The glomerular filtration rate is not affected.

4 An acid medium will discourage further growth but will not kill existing organisms.

66. (3) The nursing goal is to promote orientation by involving the client in activities with others; even though the client probably could not follow any game rules, the activity provides involvement with the nurse. (3) (EP; IM; MH; SD)

1 This is not therapeutic and implies that the client is actually talking to someone.

2 This would have no effect on the client's behavior.

4 This is not therapeutic; it allows for further withdrawal, rather than orienting the client to reality.

67. (3) Treatment is done several hours after meals to avoid regurgitation and several hours before meals so that unpleasant odor and taste do not affect eating. (2) (DA; PL; PE; RE)

1 Treatment is done more frequently than this.

2 Postural drainage should not be done before meals because the unpleasant odor and taste will interfere with eating.

4 Same as answer 1.

68. (4) Hot climates are contraindicated for children with cystic fibrosis because sweating brings about excessive loss of sodium chloride. (3) (CI; EV; PE; EN)

1 This is advisable because it will keep the child from sweating.

2 Pancreatic enzymes are essential to help in the digestion of nutrients so that they can be absorbed by the intestinal mucosa.

3 After passage of the feces associated with this disorder, the rectum may become inflamed if not properly cleaned.

69. (2) Heat promotes vasodilation, which aids the shift of urea, a large molecular substance, from blood vessels into the dialyzing solution. (3) (DA; AN; MS; RG)

1 Heat does not affect the shift of potassium into the cells.

3 The ability to remove metabolic wastes is affected in renal failure; the metabolic processes themselves are not disturbed.

4 Excess serum potassium is removed by dialyzing with a potassium-free solution, not by heat.

70. (3) The client's measured output is about 500 ml in 24 hours based on the available history; insensible losses are 400 to 500 ml in 24 hours. (3) (DA; AN; MS; FE)

1 Based on the history, the client's expected urinary output should be about 400 ml in the next 24 hours; this is far less than 900 ml.

2 This is insufficient fluid to help prevent hypostatic pneumonia; fluid intake is only one of many factors involved.

4 Hyperkalemia in acute renal failure is caused by inadequate glomerular filtration and is not related to fluid intake.

71. (4) This response presents facts that help to reduce guilt. (2) (SC; IM; CW; EC)
    1 This is not reassuring, as well as incorrect, because placenta previa can occur in a woman with a normal uterus.
    2 This is an inadequate explanation and gives the client no idea of what is happening.
    3 This is untrue, and labor may not be beginning at this time.

72. (1) Individuals who have somatoform disorders are really ill; they need care and a nonthreatening environment. (3) (EP; PL; MH; AX)
    2 The client requires physiologic and emotional care; without movement, venous stasis and atrophy of the "paralyzed" limbs can occur.
    3 The client is ill and requires good physical and emotional care.
    4 Same as answer 2.

73. (2) With diabetic ketoacidosis, serum lipid levels can go so high that the serum appears opalescent and creamy. (3) (CI; AS; MS; EN)
    1 With diabetic ketoacidosis, the BUN is generally elevated because of dehydration.

3 This is unrelated to diabetic ketoacidosis.
4 With diabetic ketoacidosis, the hematocrit level is generally elevated because of dehydration.

74. (4) Elevation in the first 24 hours helps prevent edema; continued elevation may lead to hip contractures. (1) (PC; PL; MS; SK)
    1 The client is usually out of bed on the second postoperative day.
    2 The stump dressing is usually a pressure dressing and it is not changed daily.
    3 Hemorrhage and infection are the two most common complications.

75. (4) Drainage is bright red initially and gradually becomes darker red during the first 24 hours. (2) (CI; EV; MS; GI)
    1 If the nasogastric tube is functioning correctly, secretions will be removed and vomiting will not occur.
    2 If the nasogastric tube is functioning correctly, gastric distention will not occur.
    3 Because the bowel was emptied before surgery and the client is now npo, there would be no expected intestinal activity.

How to Use Worksheet 1: Errors in Processing Information

Common errors in processing information are listed in the left-hand column of this worksheet. At the top of the worksheet is a row of blank spaces for inserting the number of the question missed. Directly below each number, check any errors you made in answering that question. You may have made more than one type of error in an answer.

Worksheet 1: Errors in Processing Information

Question number																								
Did not read situation/ question carefully																								
Missed important details																								
Confused major and minor points																								
Defined problem incorrectly																								
Could not remember terms/ facts/concepts/principles																								
Defined terms incorrectly																								
Focused on incomplete/incorrect data in assessing situation																								
Interpreted data incorrectly																								
Applied wrong concepts/ principles in situations																								
Drew incorrect conclusions																								
Identified wrong goals																								
Identified priorities incorrectly																								
Carried out plan incorrectly/ incompletely																								
Was unclear about criteria for evaluating success in achieving goals																								

How to Use Worksheet 2: Knowledge Gaps

Types of common knowledge gaps are listed along the top of this worksheet. Write a brief description of topics you want to review in the spaces provided. For example, if you missed a question on administration of a particular drug, write the drug name and problem (e.g., dosage) in the appropriate space under the column labeled *Pharmacology*.

Worksheet 2: Knowledge Gaps

Basic science	Skills/ procedures	Basic human needs	Growth and development	Normal nutrition	Psychosocial factors	Clinical area/ topic	Stressors/ coping mechanisms	Patho- physiology	Pharma- cology	Therapeutic nutrition	Legal implications	Other

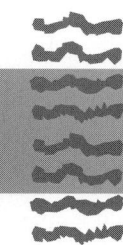

PART B ANSWERS AND RATIONALES

76. (1) The nurse should determine the location, intensity, and other characteristics of the pain before initiating intervention. (1) (PC; EV; MS; NM)
2 Assessment should occur before nursing intervention.
3 This is an incomplete assessment.
4 Same as answer 2.

77. (3) Pregnant women with diabetes have a tendency for developing placental insufficiency, which can threaten fetal well-being. (3) (CI; AN; CW; HP)
1 Although maternal age may be an indication for testing, the maternal medical condition places the client at the greatest risk.
2 The pregnant woman with diabetes requires fetal monitoring (NST) regardless of glycemic control.
4 The NST does not measure any plasma levels.

78. (2) This individualized information would be the best basis for predicting the outcome of therapy. (2) (CI; PL; MS; EH)
1 This knowledge would not be helpful in understanding the likelihood of additional problems associated with the current cancer.
3 Same as answer 1.
4 This would be useful, but it is not specific for this client.

79. (2) At this time breast swelling is minimal, and this provides a regular examination cycle. (1) (HI; EV; CW; WH)
1 Premenstrual breast engorgement may cause the breast to feel lumpy.
3 Ovulation is occurring, and hormones may influence breast consistency.
4 Breast consistency is altered by the menstrual cycle; the same date each month would fall in a different stage of the cycle. This method is ideal for women who do not menstruate.

80. (4) This accurately responds to the parents' question, reinforcing what the physician explained. (2) (SC; IM; PE; BI)
1 This is an insensitive response that places the parents on the defensive.
2 This would abdicate teaching responsibilities.
3 This statement reinforces parental insecurity about the information they have recently received from the physician; this may increase anxiety.

81. (3) It is most effective to provide skilled, professional care in a centralized facility rather than the home. (1) (DF; PL; MS; GD)
1 Clients prefer to stay in a familiar environment and will achieve their maximum potential regardless of the area.
2 Multidisciplinary services on a part-time basis in the home would be more therapeutic and cost-effective than placement in a nursing home.
4 Part-time personal care assistance in the home would be more therapeutic and cost-effective than placement in a nursing home.

82. (4) Steroids decrease the inflammatory process around the optic nerve, thus improving vision. (2) (PT; EV; MS; DR)
1 Steroids are usually associated with increased emotional lability.
2 Steroids are not effective in easing muscle contractions.
3 Pain in the extremities is not common unless spasms are present; steroids do not relieve spasms.

83. (3) A therapeutic relationship is easier to establish when anxiety is lowered; the use of paper towels may ultimately facilitate communication. (2) (EP; IM; MH; PR)
 1 This is untrue and only reinforces the use of dirt as a defense against real feelings and increases anxiety.
 2 This may increase anxiety further, thus hindering the development of a therapeutic relationship.
 4 Same as answer 2.

84. (4) Limitation of increased intracranial pressure and resultant brain damage depend on frequent, systematic observation. (3) (CI; AS; MS; NM)
 1 This is unrealistic; the state of consciousness should be observed, but otherwise rest is not contraindicated.
 2 There is no indication that hyperactivity is present.
 3 Mannitol is administered to reduce cerebral edema; there is no indication yet that this will be needed.

85. (2) Specialized insulin receptors on insulin-sensitive cells transport glucose through cell membranes, making it available for use. (1) (PT; AN; PE; DR)
 1 This is not the action of insulin.
 3 Same as answer 1.
 4 Same as answer 1.

86. (2) Preventing infection and trauma are the priorities; rupture of the sac may lead to meningitis. (1) (CI; IM; CW; HN)
 1 The Apgar is 9/10; there is no respiratory complication.
 3 The sac must be protected before this is done.
 4 This can be done before the infant leaves the delivery room; the priority is care of the sac.

87. (1) Quickness is used for safety; an attitude of concern may help to reduce client anxiety. (2) (SI; IM; MH; SD)
 2 A client this upset would never agree; the client may harm the self or others and must be sedated.
 3 The client must be told why sedation is being used.
 4 Same as answer 3.

88. (2) The iron needs of the adolescent are high because of growth requirements and often inadequate nutrition; the iron needs are increased by the demands of pregnancy. (2) (DF; AS; CW; HP)

 1 This is not common in pregnant adolescents.
 3 Same as answer 1.
 4 Same as answer 1.

89. (2) Bile, which aids in fat digestion, is not as concentrated as before surgery; once the body adapts to the absence of the gallbladder, the client should be able to tolerate a regular diet that contains fat. (2) (PC; IM; MS; GI)
 1 Initially, the client should still avoid fatty foods unless the physician indicates otherwise.
 3 This is an inappropriate priority at this time; a temporary avoidance of fatty foods with the gradual resumption of a regular diet is the priority.
 4 Same as answer 3.

90. (2) The cerebellum is the most common area; symptoms of increased intracranial pressure result from obstruction of cerebrospinal fluid flow. (2) (CI; AN; PE; NM)
 1 This is an uncommon site in children.
 3 Same as answer 1.
 4 Same as answer 1.

91. (3) This provides reality-based feedback for the client withdrawing from alcohol. (2) (SC; IM; MH; SA)
 1 Physical restraints will increase agitation and should be applied only as a last resort.
 2 This focuses on hallucinations and delusions rather than on reality.
 4 Shadows will increase the chance of distortions and illusions.

92. (2) Class A diabetes is gestational diabetes, and the woman demonstrates symptoms only when pregnant; this is the client's first pregnancy and her first sign of diabetes; she would not have symptoms associated with a chronic condition. (1) (DF; AS; CW; HP)
 1 Retinopathy is usually present in class D diabetes or higher.
 3 This is seen in class C diabetes. The onset of class C diabetes is between 10 and 20 years of age.
 4 Same as answer 1.

93. (3) Autoimmune response plays a role in the development of polyarteritis, although drugs and infections may precipitate it. (3) (HI; IM; MS; CV)
 1 Men are affected three times more often than women.
 2 The disease is often fatal, usually as a result of congestive heart failure or renal failure.
 4 Arteriolar pathology can affect any organ or system.

94. (2) The client needs to know details of transfer to assist appropriately and avoid injury. (1) (SI IM; MS; SK)

1 This is not advisable because this could disrupt the repair of the affected hip.

3 More commonly, no weight-bearing is permitted on the operative leg at first.

4 This is unnecessary; the client may touch the floor with the foot but may not bear weight on this extremity.

95. (2) Erosion of blood vessels may lead to hemorrhage, a life-threatening situation further complicated by decreased prothrombin production. (3) (CI; AS; MS; GI)

1 Although increased intraabdominal pressure may cause this, there is no immediate threat to life; assessment for bleeding takes priority.

3 Although this may cause gastritis, there is no immediate threat to life; assessment for bleeding takes priority.

4 Same as answer 1.

96. (2) Percussing over fluid produces a dull sound, not the normal tympanic sound. (3) (DA; AS; MS; GI)

1 Respiratory distress occurs with ascites but is not an early sign.

3 This would assess for dependent edema, not ascites.

4 Bowel sounds do not indicate much about early ascites; when ascites is extensive, bowel sounds may diminish.

97. (3) The client's low self-esteem causes doubt of lover's feelings; this statement reflects the low self-esteem. (3) (SC; AS; MH; SD)

1 The client's statements do not reflect confusion or disorientation but false beliefs.

2 The client's statements do not represent hallucinations because they are not false sensory perceptions.

4 Setting limits after the fact would not be effective in any situation; limits must be set when the situation occurs.

98. (2) Because suspicious clients lack trust and have difficulty sharing feelings, this healthy behavior should be recognized. (2) (SC; EV; MH; SD)

1 The client's feelings are rejected when blanket statements are given; this only responds to part of the client's concerns.

3 Focusing on the staff's concerns ignores the client's needs; the staff's attitude could decrease self-esteem.

4 A detailed description at this time may increase the client's fears.

99. (2) This would indicate a lack of understanding of the necessity of taking antibiotics for a full 10 to 14 days or until a follow-up urine culture is negative. (1) (PT; EV; PE; RG)

1 Bubble bath can be a source of irritation to the meatus, which can predispose females to a urinary tract infection and thus should be avoided.

3 Wiping from front to back prevents contamination of the urinary meatus with feces.

4 Stasis is the primary predisposing factor to a urinary tract infection, thus voiding when feeling the urge is important; the bladder should be emptied every 3 to 4 hours.

100. (3) Poor appetite and decreased energy are associated with an accumulation of toxic waste; associated anemia accounts for the pallor. (3) (DA; AN; PE; RG)

1 Discontinuing the corticosteroids and diuretics if prescribed might result in recurrence of edema in steroid-dependent children; pallor would not occur.

2 An elevated temperature would be present with an infection; an infection would not cause muddy pallor.

4 Once remission has occurred, usual activities can be resumed with discretion.

101. (2) Vomiting and feeding difficulties are early signs of digoxin toxicity in infants. (2) (PT; IM; PE; DR)

1 The pulse rate of an infant receiving digoxin should remain more than 100 beats per minute.

3 This is expected; tetralogy of Fallot causes cyanosis, which will be pronounced when crying occurs.

4 Infants routinely require several naps, and an infant with heart disease requires additional rest periods daily.

102. (2) Antacids are most effective when taken after digestion has started but before the stomach begins to empty. (2) (PT; EV; MS; DR)

1 Antacids should be taken before the onset of pain; pain indicates that gastric irritation has begun, and the aim of treatment is to protect the GI mucosa.

3 The antacids would interfere with the absorption of nutrients.

4 Same as answer 3.

103. (4) Bed rest promotes decreased cardiac output; it also decreases tissue catabolism, which lowers the workload of the kidneys, eventually increasing filtration by the renal glomeruli; limiting activity also helps lower the blood pressure. (2) (PC; PL; PE; RG)
    1 Bed rest is necessary during the acute phase of this illness.
    2 The activity level depends on the response to therapy.
    3 Same as answer 2.

104. (3) Hypotonia occurs in magnesium sulfate toxicity because of skeletal and smooth muscle relaxation. (3) (CI; EV; CW; DR)
    1 This is not a sign of magnesium sulfate toxicity.
    2 Same as answer 1.
    4 Same as answer 1.

105. (4) This indicates failure to resolve conflicting feelings about pregnancy that commonly occur in the first trimester. (3) (SC; PL; CW; EC)
    1 This is a normal feeling in the third trimester.
    2 Same as answer 1.
    3 Concerns about the expected baby having physical abnormalities are common in the third trimester.

106. (1) Research has shown that decreasing stress will slow the rate of atherosclerotic development. (1) (SC; PL; MS; CV)
    2 Exercise is thought to decrease atherosclerosis and the formation of lipid plaques.
    3 Saturated fats in the diet increase atherosclerosis.
    4 Same as answer 2.

107. (3) The taste for salt is learned from habitual use and can be unlearned or reduced with health improvement motivation and creative salt-free food preparation. (3) (PC; IM; MS; FE)
    1 The taste for salt is learned.
    2 This is untrue; substitutes often have a metallic taste.
    4 This is unsafe; abnormally high potassium levels can result.

108. (1) A strong cry indicates effective respiratory function and is assigned a value of 2. (1) (HI; AS; PA; NN)
    2 A value of 1 is assigned when the body is pink and the extremities are blue.
    3 If the flexion of the arms and legs is only slight and movement is diminished, the value assigned is 1.
    4 The heart rate should be more than 100 per minute and therefore is assigned a value of 1.

109. (2) Autonomy leads to exploring and self-feeding; the child may eat food not on the diet. (2) (DF; AS; PE; GD)
    1 Although this is a common characteristic of a toddler, it would not lead to the child eating restricted foods.
    3 Same as answer 1.
    4 Same as answer 1.

110. (4) These are classic symptoms of ketoacidosis because of the respiratory system's attempt to compensate by blowing off excess carbon dioxide, a component of carbonic acid. (2) (DA; AS; MS; EN)
    1 This is indicative of an insulin reaction (diabetic hypoglycemia).
    2 This is a hypersensitivity reaction; it is unrelated to diabetes.
    3 Diaphoresis is associated with an insulin reaction, not diabetic ketoacidosis.

111. (4) The aim of therapy is to eliminate the causative agent, which can be determined from culture and sensitivity tests of sputum. (1) (CI; EV; MS; RE)
    1 A lung scan permits visualization of lung vasculature but does not provide data on the condition of the lung tissue itself.
    2 Bronchoscopy shows the appearance of the bronchi but does not indicate the presence or absence of microorganisms.
    3 Pulmonary function studies indicate air volume that may fall within the expected range despite the presence of bronchopneumonia.

112. (1) Cor pulmonale is right ventricular failure caused by pulmonary congestion; edema results from increasing venous pressure. (3) (CI; AS; MS; CV)
    2 A productive cough is symptomatic of the original condition, COPD.
    3 This is caused by alterations in oxygen and hydrogen ion levels and their effects on the central nervous system.
    4 Same as answer 3.

113. (4) The chamber closest to the client collects drainage, and the second chamber acts as the water seal. (3) (DA; AN; MS; RE)
    1 Suction requires the addition of a third chamber or a modified system.
    2 The water seal in a two-chamber system is provided by the second chamber.
    3 In a one-chamber system the seal and drainage are combined and there is no suction chamber.

114. (4) These individuals need peer group relations to validate their feelings and to experience peer group pressures. (3) (EP; PL; MH; ES)

1 These individuals have a remarkable store of energy that does not reflect their malnourished state.

2 These individuals rarely have trouble making decisions, so they do not have to be forced into the role.

3 These individuals totally relate their attractiveness to thinness; they have diminished libido, and in addition, women have amenorrhea.

115. (4) The nurse shows an understanding of the client's needs by not totally restricting the handwashing. (3) (EP; IM; MH; PR)

1 At this time, the client is still too anxious and is incapable of dealing with the reasons for handwashing.

2 Continued handwashing does not reveal an understanding of the problem or a sign of progress.

3 This denies the client's feelings, is untrue, and will close off any communication.

116. (2) Toxoids are modified toxins that stimulate the body to form antibodies, lasting up to 10 years, against the specific disease. (3) (HI; AN; PE; BI)

1 Passive immunity is temporary; even the natural type derived from the mother does not last longer than the first year of life.

3 Only by having the disease can lifelong natural immunity become possible; toxoids confer active artificial immunity.

4 Toxoids give artificial active immunity.

117. (4) Optic nerve inflammation is an early effect of multiple sclerosis caused by lesions in the optic nerves or their connections. (1) (HI; IM; MS; NM)

1 At present there is no evidence of viral infection of the eyes in multiple sclerosis.

2 Tumors of the brain, not multiple sclerosis, cause increased intracranial pressure because the skull cannot expand.

3 This causes blindness as a result of increased intraocular pressure, not inflammation of the optic nerve.

118. (4) This lessens anxiety, promotes verbalization, reduces guilt, and helps the family feel useful. (2) (SC; IM; MH; TR)

1 This is a generalized personal opinion; the nurse at this time does not know about family relationships.

2 This is false reassurance.

3 The "why" creates defensiveness; the information is incorrect, and crying helps to relieve tension.

119. (4) The cervix is 70% effaced and 6 cm dilated; these findings are typical midway through the first stage of labor. (2) (DF; AS; CW; HC)

1 When the cervix is 6 cm dilated, the individual is beyond the early stage of labor.

2 The transitional phase of labor begins when the cervix is 8 cm dilated.

3 In the second stage of labor, the cervix is fully dilated and 100% effaced.

120. (4) Anxiety and stress tend to close communication; this in turn intensifies the reaction to illness and death. (3) (SC; PL; MH; CS)

1 The family system could not solve all its problems at this time because of the emotional turmoil of its members.

2 This is false; the family must begin to deal with feelings before death occurs.

3 This is true, but the focus should be on mutual understanding by all members.

121. (2) When clients in obvious crisis appear depressed, anxious, and desperate, the nurse should question them regarding the presence of suicidal thoughts. (2) (SI; IM; CW; WH)

1 Further assessment and exploration are needed before encouraging clients to admit themselves to a psychiatric facility.

3 Running away from problems does not help solve them, nor will escaping bring lasting relief.

4 When the client is overwhelmed with problems, it is difficult to think positively and to focus on the good things in life.

122. (2) When a client has perforated viscera, barium could leak out of the intestinal tract and cause inflammation and/or an abscess. (2) (SI; AN; MS; GI)

1 Although it may be irritating, this does not contraindicate barium studies.

3 Serum potassium would be unaffected; barium is insoluble and would not affect blood content.

4 Barium studies are not contraindicated and could be useful in diagnosing ulcerative colitis and Crohn's disease.

123. (1) To avoid symptoms, these clients often refuse to eat and become malnourished; a high-calorie, high-protein diet is advised. (1) (PC; PL; MS; GI)
    2 This is not a problem in Crohn's disease.
    3 Same as answer 2.
    4 This is a secondary problem that results from malnutrition; correcting the malnutrition will increase strength.

124. (2) This is the definition of a positive CST result. (3) (CI; AN; CW; HP)
    1 A CST is performed after, not before, a NST is nonreactive or equivocal.
    3 A normal fetal heart rate is 120 to 160 beats per minute.
    4 A positive CST result does not dictate a cesarean delivery; an expeditious vaginal delivery may be attempted.

125. (3) This rejects the behavior, not the client; it helps separate the client from the behavior. (3) (EP; IM; MH; SD)
    1 This does not help the client learn self-control; it rejects both the client and the behavior.
    2 Isolating this client limits learning more acceptable responses.
    4 Part of recovery is learning acceptable behavior; ignoring inappropriate behavior does not help.

126. (1) Hard candy may produce salivation, which helps alleviate the anticholinergic-like side effect of dry mouth that is experienced with phenothiazines. (2) (PC; PL; MH; DR)
    2 Fluids should be encouraged, not discouraged; fluids may alleviate the dry mouth.
    3 This is unnecessary.
    4 This is unnecessary; although these drugs can cause leukopenia and agranulocytosis, they do not cause thrombocytopenia.

127. (2) Stopping the flow reduces cramping caused by distention. (1) (PC; IM; MS; GI)
    1 There is no need to discontinue the enema; flow should be interrupted temporarily until discomfort subsides; an effective enema must be administered before surgery.
    3 This would result in the administration of a Harris flush; the purpose of the preoperative enema is to evacuate the bowel of feces, not flatus.
    4 This is unsafe; this would increase the discomfort.

128. (3) Severe edema is usually present, and change of position is necessary to prevent breakdown of tissue. (3) (SI; PL; PE; IT)
    1 Fluids are not forced and may even be restricted during periods of edema.
    2 Active play periods would be permitted during remission but not during the edema phase to limit energy expenditure.
    4 A high-protein diet minimizes negative nitrogen balance; a low-protein diet is used in renal failure with azotemia.

129. (4) The pinworm emerges when the client is asleep to lay its eggs; these can be collected from the perianal area by applying tape upon awakening. (1) (HI; PL; PE; GI)
    1 Any larvae on the perianal area would have been removed by the bath.
    2 The larvae would not have been deposited yet because the adult pinworm would still be in the bowel; it emerges when the client is asleep.
    3 Any larvae that were present in the morning would probably have been wiped away.

130. (4) As the lower uterine segment stretches and thins, tearing and bleeding occur at the lower implantation site. (2) (HI; AS; CW; HP)
    1 First trimester bleeding is associated with spontaneous abortion or inadequate implantation, not placenta previa.
    2 This is usually associated with abruptio placentae rather than placenta previa.
    3 Same as answer 1.

131. (4) This presents the reality of the situation and helps support the client during a threatening hallucination. (2) (EP; IM; MH; SD)
    1 Clients cannot be distracted from hallucinations without competing stimuli.
    2 This would have little effect on the client's behavior and would not stop the hallucinations.
    3 This would encourage withdrawal and isolation and would not stop the hallucinations.

132. (3) Regression to a more immature, helpless developmental level is normal and should be supported at this point. (2) (SC; PL; MS; EH)
    1 Denial is not the response described.
    2 The client's behavior is inconsistent with the need for more control.
    4 The client's behavior does not indicate anger.

133. (3) Fluid may shift from the intravascular space to the abdomen as fluid is removed, leading to hypovolemia and compensatory tachycardia. (3) (CI; EV; MS; GI)
    1 The fluid shift can cause hypovolemia with resulting hypotension, not hypertension.
    2 A paracentesis should decrease the degree of abdominal distention.

4 This sign of dehydration may occur, but it is not as vital or immediate as signs of shock.

134. (1) This reduces the possibility of bladder puncture. (1) (SI; IM; CW; HP)

2 The mother may eat and drink before the test.

3 Medications are not given because of their effect on the fetus.

4 This procedure is not done via the colon, so fecal material does not have to be expelled.

135. (1) Heat causes vasodilation and an increased blood supply to the area. (2) (PC; PL; CW; HC)

2 This is not the purpose of the sitz baths.

3 Cleansing is done immediately after voiding and defecating with a perineal bottle filled with cleansing solution.

4 Relaxation of the rectal sphincter is promoted by the sitz bath; this will provide comfort but will not increase healing.

136. (2) Yogurt, which contains calcium, is easily digested because it contains the enzyme lactase, which breaks down milk sugar. (1) (PC; AN; MS; GI)

1 These are deficient in calcium.

3 Same as answer 1.

4 Same as answer 1.

137. (4) Mood changes can occur as a side effect of steroid therapy. (3) (SC; EV; MS; DR)

1 This denies the value of the client's statement and offers false reassurance.

2 The client has already stated he does not know why this is happening.

3 This is difficult to do because the direction a situation will take cannot always be anticipated.

138. (1) This acknowledges the client's apprehension and encourages further communication. (1) (SC; IM; MS; EH)

2 This negates the client's feelings and promotes false reassurance.

3 This does not address the client's feelings and may cause more anxiety.

4 This is perhaps true, but it does not foster communication; the client may focus on the word "pain."

139. (1) This points out reality and does not support the client's hallucinations. (2) (EP; IM; MH; SA)

2 This feeds into the client's hallucination and provides false reassurance.

3 Same as answer 2.

4 The client has hallucinations, not illusions.

140. (4) This is responsible for the effectiveness of alkylating agents, of which nitrogen mustard is one. (3) (PT; AN; MS; DR)

1 Some drugs are believed to act in this way, but nitrogen mustard does not.

2 This is the mechanism of action of antimetabolites.

3 Antibiotics used in cancer chemotherapy are believed to act in this way.

141. (2) Cerebellar astrocytomas, unlike those in the cerebrum, are slow growing and benign, with a cure rate of 80% to 90% after surgery. (2) (HI; IM; PE; EH)

1 Other infratentorial tumors, such as medulloblastomas, are more rapid growing and highly malignant.

3 This is not true of a cerebellar astrocytoma.

4 Same as answer 3.

142. (1) This commonly begins with what parents describe as clumsiness that becomes progressively worse; the cerebellum controls coordination. (2) (HI; AS; PE; NM)

2 Seizures usually occur with cerebral or supracentorial tumors.

3 This is a very late sign of tumor involvement.

4 At this age, sutures are completely closed and the tumor would be contained in the cranium.

143. (4) Sexual anxiety and conflict occur with schizophrenia; uncertainty is projected onto others to defend the ego. (2) (EP; IM; MH; SD)

1 This avoids the issue and is a threatening response; this could increase anxiety.

2 This validates ideas of reference and is an inappropriate response.

3 The anxious client may not be able to handle being confronted with feelings; this may precipitate a panic reaction.

144. (3) Absence of peptidase results in an inability to metabolize the gliadin component of grains; this results in excessive glutamine that is toxic to the mucosal cells. (1) (HI; AN; PE; GI)

1 This is unrelated; meconium is present in the first bowel movements before the introduction of food.

2 This does not occur.

4 Fluid balance is not the basic problem with celiac disease; however, dehydration can occur in celiac crisis.

145. (4) This is the correct flow rate; 500 units of insulin added to 500 ml of IV fluid results in a solution in which 1 ml contains 1 unit of the drug; therefore 60 ml must be administered in 30 minutes; 120 ml per hour ÷ 2 = 60 units of insulin in 30 minutes. (2) (PT; AN; MS; DR)
    1 This would deliver an inadequate amount of insulin.
    2 Same as answer 1.
    3 Same as answer 1.

146. (4) Skin-to-skin contact between mother and infant is most effective in maintaining the infant's body temperature; heat is transferred by conduction. (1) (PC; IM; CW; NN)
    1 Oxygen is not administered unless the infant is experiencing respiratory and cardiac difficulties; oxygen has a cooling effect.
    2 This is not very effective and leaves much of the baby exposed; a blanket and warmer also would be necessary.
    3 Bathing the infant should be delayed until the body temperature is stabilized.

147. (4) An objective examiner confirms the fetal movements; this eliminates subjectivity by the client. (2) (DF; AS; CW; HC)
    1 This is a softening of the lower uterine segment, a presumptive sign of pregnancy.
    2 This could be caused by a variety of situations other than pregnancy; it is a probable sign of pregnancy.
    3 This is only a probable sign of pregnancy because of possible false-negative readings.

148. (1) This is an accurate and concise explanation of Haldol's effects; it blocks postsynaptic dopamine receptors in the brain. (1) (PT; IM; MH; DR)
    2 This is not true; it is a tranquilizer, and it does not alter mood.
    3 This drug lowers the seizure threshold.
    4 Same as answer 2.

149. (4) Inadequate protein decreases colloidal osmotic pressure within the vascular space, allowing fluid to shift into the interstitial spaces. (2) (PC; EV; CW; HP)
    1 Petechiae result from an abnormality in the clotting mechanism, such as thrombocytopenia.
    2 Protein does not affect heart rate.
    3 Bleeding gums result from a deficiency in vitamin C or platelets, not protein.

150. (2) SGA infants have little subcutaneous fat or glycogen stores. (2) (CI; AS; CW; HN)

1 Intestinal bleeding is not common in SGA infants.
3 This would provide no therapeutic value for this SGA infant.
4 A dynamic head circumference is not characteristic of SGA infants.

151. (4) Preschoolers are active, sociable individuals who enjoy the company of peers and become bored when isolated. (3) (DF; IM; PE; GD)
    1 Preschoolers have a limited ability to understand complex explanations of cause and effect; they employ concrete thinking.
    2 This will increase agitation and be punitive.
    3 Although this would provide some distraction, it is better to permit peer contacts.

152. (1) The client will be unable to speak because a tracheostomy tube is in place to prevent edema from blocking the airway; writing provides an alternate form of communication. (1) (CC; IM; MS; RE)
    2 The client cannot speak with a tracheostomy tube in place.
    3 This has no effect; the client's ability to hear or understand is not affected.
    4 Same as answer 3.

153. (1) Without trust, the nurse-client relationship will achieve nothing. (1) (SC; AN; PS; TR)
    2 Although this is important, trust must be developed first.
    3 Same as answer 2.
    4 Open communication will not occur until trust is developed.

154. (3) Eating in a semirecumbent position slows gastric emptying, thereby preventing premature gastric dumping of contents. (3) (CI; PL; MS; GI)
    1 This would speed gastric emptying and should be avoided.
    2 Same as answer 1.
    4 Same as answer 1.

155. (2) Children, because of the demands of growth and their dietary indiscretions, have a more fragile glucose balance. (1) (CI; IM; PE; EN)
    1 This is untrue; it is more often associated with type 2 diabetes.
    3 Resistance to treatment is not common; problems are related to the changing requirements associated with growth.
    4 This is untrue; hypersensitivity is unrelated to either type of diabetes.

156. (3) Saccharin is a non-nutritive substitute; aspartame is made of two amino acids, phenylalanine and aspartic acid, and is metabolized as such. (2) (PC; IM; PE; EN)

1 Honey, a fructose, provides 1.3 times as many kilocalories as does table sugar and must be calculated into the diet.

2 Simple sugars may be used in controlled amounts and must be calculated into the diet.

4 Foods do not have to taste sweet to contain sugar.

157. (1) Ataxia is a side effect of phenytoin, and drug continuation may cause cerebellar damage. (3) (PT; IM; MS; DR)

2 This sign of toxicity should be reported.

3 Massaging the gums should be done regularly to prevent gingival hyperplasia, which can result from phenytoin therapy.

4 The client should report the rash but keep taking the drug; withdrawal may precipitate a seizure.

158. (3) Prevention of infection continues as a priority both before and after the repair of the sac. (2) (SI; PL; PE; NM)

1 A neurologically impaired infant would not be given sedatives because they would interfere with accurate assessment.

2 This is unrealistic; newborns lose weight in the first few days of life.

4 This is not the priority at this time.

159. (2) The caloric distribution of the "prudent diet" proposed by the American Heart Association is 30% fat (less than 10% saturated fat), 50% CHO (35% complex carbohydrates), and 20% protein. (1) (PC; IM; MS; EH)

1 Fried foods are not advocated on the "prudent diet"; peanut oil is a monounsaturated fatty acid; these acids should not exceed 15% of the kilocalories of the diet.

3 This is untrue; this would be discouraging and encourage noncompliance.

4 Same as answer 3.

160. (2) This drug suppresses activity in key regions of the subcortical area of the CNS; it also has antihistaminic and anticholinergic effects. (1) (PT; EV; MH; DR)

1 These symptoms are not associated with Vistaril.

3 Same as answer 1.

4 Same as answer 1.

161. (4) Hyperextension of the neck places tension on the suture line. (2) (CI; AN; MS; RE)

1 The cervical vertebrae are designed to flex and hyperextend; there should be no ill effects.

2 Hyperextension would not cause this.

3 Same as answer 2.

162. (4) This position may prevent further prolapse and relieve pressure on the cord. (1) (SI; IM; CW; HP)

1 Although this may eventually be done, the priority is to relieve pressure on the cord.

2 This promotes placental perfusion but would not relieve pressure on the cord.

3 Same as answer 1.

163. (3) This expands collapsed alveoli and enhances surfactant activity, thereby preventing atelectasis. (3) (CI; PL; MS; RE)

1 This helps clear accumulated secretions from the pulmonary tree; it does not directly promote alveolar expansion.

2 This promotes sustained exhalation, not inhalation.

4 This promotes collapse of, not expansion of, alveoli.

164. (3) The lack of a social support system can precipitate feelings of isolation. (1) (SC; AN; MH; AX)

1 The data do not support this nursing diagnosis.

2 Same as answer 1.

4 Same as answer 1.

165. (3) Intestinal antibiotics and a complete cleansing of the bowel with enemas until returns are clear are necessary to reduce the possibility of fecal contamination when the bowel is resected. (1) (SI; PL; MS; GI)

1 The bladder will be removed, so there is no need for a Foley catheter.

2 A clear liquid diet is usually prescribed for several days with npo for at least 8 hours before surgery.

4 This is not necessary; there is no evidence of urinary infection.

166. (2) This provides for release of tension with no element of competition; success in a simple project increases the sense of accomplishment and worth. (2) (SC; PL; MH; MO)

1 This allows the client to withdraw further.

3 The client has psychomotor retardation and may not be able to cope with fine details at this time.

4 This is too cheerful; this could result in excessive stimuli and increase the client's irritation.

167. (3) This is an iron-rich food appropriate for slight anemia, which has probably occurred from blood loss into the tissues at the site of the injury. (2) (PC; IM; PE; BI)
    1 This is not necessary unless there is a marked decrease.
    2 Same as answer 1.
    4 If internal bleeding occurred, there would have been earlier signs other than a reduction in hemoglobin and hematocrit levels.

168. (2) The possibility of ascending infection increases when membranes are ruptured and delivery is not imminent; client must be monitored for signs of infection. (2) (SI; AS; CW; HP)
    1 This does not normally occur after membranes rupture unless fetal distress, accompanied by late fetal decelerations, is present or fetus is in breech position.
    3 This is unrelated to spontaneous rupture of the membranes.
    4 This is not indicated with spontaneous rupture of membranes; it is indicated if persistent late decelerations are observed on the fetal monitor.

169. (2) This fat-binding agent would also bind and eliminate other fat-soluble vitamins such as vitamins D, E, and K. (2) (PC; EV; MS; DR)
    1 This is not a fat-soluble vitamin and would be unaffected.
    3 Same as answer 1.
    4 Same as answer 1.

170. (4) Taking the drug in the early morning mimics normal adrenal secretions; food and/or antacid helps reduce gastric irritation. (1) (CC; EV; MS; DR)
    1 Diurnal rhythms may be altered, and steroids are ulcerogenic; they should be taken with more than just a snack.
    2 Steroids cause gastric irritation and should be taken with food.
    3 The food helps decrease gastric irritation; however, normal diurnal rhythms could be altered.

171. (4) Together, lack of symmetry and palpation of a thickening are abnormal findings. (1) (DA; EV; CW; WH)
    1 Engorgement is an expected response to menstrual hormones.
    2 Premenstrual engorgement may cause the breast to feel lumpy.
    3 This is a common deviation that is within normal limits.

172. (4) Because the client is feeling a loss of control, it would be important to include the client in revision of the plan. (2) (CC; PL; MH; CS)
    1 This does not consider changes in the client or the client's feelings.
    2 This is unnecessary; planning nursing care is within the nurse's function and judgment, not the physician's.
    3 This is very authoritarian and places total control with the nurse.

173. (2) Greenish fluid is indicative of meconium, which is released by the fetus; it is considered a possible indicator of fetal stress and potentially could be aspirated by the fetus. (1) (DA; PL; CW; HP)
    1 A bloody show is normally present and may increase at the end of the first stage of labor; this is a normal finding.
    3 This is normal; it is not indicative of any problem.
    4 This is a normal occurrence as labor progresses; it is not indicative of any problem.

174. (3) In abruptio placentae there is uterine bleeding that can result in massive internal hemorrhage, causing hypovolemic shock. (2) (DA; AS; CW; HP)
    1 Jaundice occurs only when there is a deposition of bilirubin into subcutaneous tissues and skin; this is not associated with abruptio placentae.
    2 It is more likely that with internal bleeding, blood pressure will fall rather than increase.
    4 Convulsions are associated with preeclampsia; there is no information indicating the presence of this condition.

175. (4) Familiarity with situations and continuity add to the client's sense of security and foster trust in the relationship. (2) (EP; PL; MH; DD)
    1 Although this helps individualize care, continuity is the priority.
    2 Some degree of flexibility by the nurse would help to individualize care.
    3 Detailed explanations will be forgotten; instructions should be simple and to the point and given when needed.

176. (2) A continuous flushing of the bladder dilutes the bloody urine and empties the bladder, preventing clots. (1) (CI; AN; MS; RG)
    1 Only the kidneys form urine; fluid instilled into the bladder does not affect kidney function.
    3 Urinary output can be easily measured regardless of the amount of fluid instilled.
    4 The continuous drip does not exert any additional pressure within the bladder.

177. (3) Tetanus immune human globulin should not cause anaphylaxis because it is derived from human serum. (2) (CI; AN; PE, BI)
   1 Tetanus immune human globulin is not as effective as the antitoxin, but it is not derived from horse serum and should not cause anaphylaxis.
   2 It is necessary to perform skin tests for both types of medications to determine the presence of sensitivity.
   4 These medications always carry the risk of hypersensitivity; it cannot be assumed they can be safely given to everyone.

178. (4) This prevents the spread of infection to others; isolation is a priority that should be immediately implemented. (1) (SI; IM; PE; NM)
   1 There is no indication that the child is dehydrated; fluid maintenance is a continuing goal.
   2 There is no indication that the child needs oxygen, and it would not be given routinely.
   3 This would not be given because these children are sensitive to stimuli, and movement causes increased discomfort.

179. (4) Furniture and loose rugs can interfere with crutch walking and should be removed to prevent further injury. (1) (SI; PL; MS; SK)
   1 The client may shower if the cast is protected from becoming wet.
   2 Calcium, rather than vitamin C, would be encouraged to enhance bone healing; vitamin C prevents capillary fragility.
   3 This is a medical, not a nursing, decision.

180. (4) Nerve and vascular injury and significant blood loss may be present. (2) (CI; AS; MS; SK)
   1 This is a medical decision that has not yet been made; closed fractures generally are reduced by manipulation.
   2 False reassurance is never appropriate.
   3 Closed fractures generally do not require operative reduction; they are usually reduced by manipulation.

181. (2) Improving ventilation provides comfort, maintains existing lung function, and prevents further lung damage. (1) (DA; PL; MS; RE)
   1 Maintaining hydration thins secretions so that there is less interference in achieving the goal of improved ventilation.
   3 Some decrease in hypoxia will promote comfort, but the primary problem is too much carbon dioxide rather than too little oxygen; oxygen should not exceed 2 liters per minute.
   4 Correcting the bicarbonate deficit will not help ventilation but will correct the accompanying respiratory acidosis.

182. (1) The shift of fluid predisposes the child to hypovolemia; an increased thready pulse and hypotension are signs of shock. (3) (DA; AS; PE; RG)
   2 Tubular reabsorption of sodium is increased to replenish vascular volume; therefore potassium would be excreted.
   3 This is not a complication of nephrotic syndrome, although pulmonary effusion may occur.
   4 This does not usually occur as an early complication of this disease; it is a major complication of glomerulonephritis.

183. (3) Pelvic rocking on the first postoperative day would be very painful and could traumatize the wound site. (1) (CC; EV; CW; HP)
   1 Leg bends promote circulation in the lower extremities and help to alleviate gas pains.
   2 Foot circles promote circulation in the lower extremities.
   4 Shoulder circles relieve neck stiffness and tension that may be present after delivery.

184. (3) This would give the client a familiar gauge in estimating the amount of bleeding she is experiencing. (1) (CC; AS; CW; HP)
   1 The presence of clots does not indicate the amount of bleeding.
   2 This may indicate a problem but does not relate to the amount of bleeding.
   4 Weakness is a subjective symptom and may not truly reflect blood loss.

185. (2) A sharing of views helps to support the client's trust in self and others; participation increases self-esteem. (3) (SC; EV; MH; SD)
   1 Clients are sensitive to others' feelings; this may be viewed as a lack of trust.
   3 Pressuring the client to explain behavioral changes increases anxiety and the need to use defenses.
   4 Asking clients to defend their point of view is threatening.

186. (2) This therapy permits relative isolation of the tumor area and saturation with the drug(s) selected. (2) (PT; AN; MS; GI)
   1 This is not true; this procedure requires medical and nursing supervision.
   3 These effects cannot be confined completely to the treated area; some migration still occurs.
   4 Combinations of drugs also can be administered via IV or oral routes.

187. ③ Laryngeal spasms can occur abruptly; patency of the airway is determined by constant assessment for symptoms of respiratory distress. (1) (DA; PL; PE; RE)
1 This is important, but maintenance of respiration has priority.
2 The fever should be treated, but it is not critical at 103° F; maintenance of respiration has priority.
4 Same as answer 1.

188. ② Hypoglycemia stimulates the production of ACTH, glucagon, glucocorticoids, growth hormone, and catecholamines; this produces glycogenolysis and gluconeogenesis and results in hyperglycemia. (2) (PT; AN; MS; EN)
1 This would result in hyperglycemia, not hypoglycemia.
3 Same as answer 1.
4 Excessive glucose intake causes hyperglycemia, not gluconeogenesis.

189. ① This is a prime time for symptoms of a pulmonary embolus to appear. (2) (CI; EV; MS; SK)
2 This would not require nursing intervention; a productive cough would indicate a respiratory infection.
3 This could result from the inflammatory process; the temperature-regulating mechanisms in the aged may be slightly compromised, and they may show a slight elevation in body temperature for a longer period after surgery.
4 Weight-bearing is being done by the unoperative leg at this time, and fatigue may be expected.

190. ③ The fontanels, anterior and posterior, are both open at birth. (2) (DF; AS; CW; NN)
1 Closed sutures are abnormal and may prevent normal brain growth during the first year.
2 This would indicate dehydration.
4 This is cephalhematoma, a deviation from normal indicating trauma.

191. ② The Centers for Disease Control and Prevention has determined that health professionals should use gloves when coming into direct contact with body fluids and blood; in HIV-infected individuals these fluids contain the virus. (1) (SI; PL; PE; BI)
1 Approaching the bedside does not expose the health care worker to the virus.
3 Contact should not be limited; this does not allow for optimal care of the client.

4 Gloves, mask, and an impervious gown are needed only when there is a potential for the health worker to be splashed with body fluids or blood.

192. ③ These are the classic symptoms of neuroleptic malignant syndrome, which is caused by neuroleptic-induced blockage of dopamine receptors. (3) (PT; EV; MH; DR)
1 These are side effects of Haldol, but they are not signs of neuroleptic malignant syndrome.
2 Same as answer 1.
4 Same as answer 1.

193. ② Pulmonary edema is associated with severe preeclampsia; as vasospasms worsen, capillary endothelial damage results in capillary leakage. (3) (DA; AN; CW; HP)
1 Crackles are not an indication of an impending seizure; signs of an impending seizure include hyperreflexia, developing or worsening clonus, severe headache, visual disturbances, and epigastric pain.
3 Pregnancy does not precipitate bronchial constriction; the hormones associated with pregnancy can cause nasal congestion.
4 This is a discomfort associated with pregnancy that may result in shortness of breath or dyspnea, not crackles.

194. ① A rising reticulocyte count indicates accelerated erythropoietic activity that may reflect increased RBC destruction; increased RBC destruction raises the bilirubin level, causing jaundice. (3) (CI; AN; CW; HN)
2 In this instance the sedimentation rate or WBCs, not the reticulocytes, would be elevated.
3 Although the reticulocyte count may be elevated with chronic blood loss, there are no data to indicate the baby is bleeding.
4 This test does not reflect respiratory functioning; however, ultimately with hemorrhage the respiratory rate will be elevated.

195. ④ The client needs to feel in control to prevent ego deterioration. (3) (SC; PL; MH; CS)
1 It is too soon after the rape to discuss this.
2 Although the nurse should always be available and supportive, feelings of anger are usually not the initial response.
3 Same as answer 1.

196. ① Subcutaneous emphysema refers to the presence of air in the tissue that surrounds an open-

ing in the normally closed respiratory tract; the tissue appears puffy, and a crackling sensation is detected under the fingertips because trapped air is compressed between the tissue. (2) (CI; AS; MS; RE)
2 The lungs are not affected.
3 Gas exchange, and thus blood gases, are not affected.
4 Same as answer 2.

197. (2) The client is nervous and afraid of leaving home; the priority is provision for safety and security needs. (3) (EP; PL; MH; AX)
1 Unless the client is provided with a sense of security, adjustment is likely to be unsatisfactory because the anxiety will most likely escalate.
3 This cannot be done until anxiety is reduced.
4 The client is experiencing memory loss and may not be able to remember what precipitated admission to the hospital; some memory loss may be a result of high anxiety and thought blocking.

198. (2) Gastric secretions, which are electrolyte rich, are lost through the NG tube; the imbalances that result could prove life threatening. (1) (DA; EV; MS; FE)
1 This is unnecessary and could damage the suture line.
3 This is unsafe; this could result in suture line disruption.
4 This is unsafe; if respiratory intubation has occurred, aspiration will result.

199. (2) School-age children are most concerned about school, if not for the academics, for the social aspects. (2) (DF; AS; PE; GD)
1 This may be of some concern, but not as much as school is.
3 School-age children generally look at physicians as authority figures and do not doubt their competence.
4 School-age children are generally not this future oriented.

200. (4) Baseline vital signs are extremely important; physical assessment precedes diagnostic measures and intervention. (2) (SI; AS; MS; RE)
1 This might be done after it is determined whether a specimen for blood gases is needed; this is not usually an independent function of the nurse; oxygen is only administered independently by the nurse in an emergency situation.
2 This would be done after the physician makes a medical diagnosis; this is not an independent function of the nurse.

3 A sputum specimen should be obtained after vital signs and before administration of antibiotics.

201. (2) Because more people are living longer, this problem of the elderly, especially elderly women, is increasing. (3) (DF; AN; MS; SK)
1 This is unrelated to osteoporosis; the fat that is removed in skim milk does not contain calcium.
3 This is unrelated to osteoporosis.
4 Early retirement does not necessarily imply inactivity or immobility.

202. (1) Buck's traction is an example of traction applied directly to the skin by traction tape or a foam boot. (2) (PC; AN; MS; SK)
2 There is no such traction.
3 Skeletal traction is applied directly to the bony skeleton.
4 Balanced suspension traction keeps the affected extremity elevated off the bed.

203. (1) During detoxification this provides safety and prevents suicide, which is a real threat. (2) (EP; PL; MH; SA)
2 Bright light is preferable to dim light, which creates shadows and increases illusions and misinterpretations.
3 This type of client does not usually lose the sense of hearing, so there is no need to shout.
4 Restraints tend to upset the client further; if at all possible, they should not be used.

204. (1) This is a plant alkaloid that is cell-cycle specific; it affects cell division during metaphase by interfering with spindle formation and causing cell death. (3) (PT; IM; MS; DR)
2 This is typical of antimetabolites such as fluorouracil.
3 Alkylating agents such as nitrogen mustard act in this way.
4 Antibiotics such as dactinomycin act in this way.

205. (3) This is an emergency situation that exerts sudden pressure on the third cranial nerve on the affected side from displacement of the tentorium or uncus. (3) (CI; AN; PE; NM)
1 Response to severe pain is generally equal in both eyes.
2 An autonomic response to fear would affect both pupils equally.
4 This is inaccurate; reduced pressure would not cause one pupil to dilate; more likely, increasing pressure would cause this response.

206. (2) The symptoms described, in conjunction with the longest and strongest contractions, occur in the transition phase of the first stage of labor. (1) (DF; AS; CW; HC)
   1 In this stage the contractions are well spaced, with time to relax in between.
   3 This is the period after the birth of the infant to the birth of the placenta.
   4 In this stage the fetal head begins to crown and would be seen at the perineum.

207. (2) The loss of a job can bruise the ego and decrease self-esteem. (1) (SC; PL; MH; AX)
   1 Feelings should be expressed, not limited; attempting to decrease crying often increases it.
   3 The crying is not necessarily an expression of sadness; there are other feelings involved.
   4 The focus should be on the client's self-acceptance.

208. (3) Intravenous analgesics will be used because after major surgery an infusion will be in place to prevent the discomfort of an injection. (3) (PC; PL; PE; NM)
   1 Oral analgesics are not as effective as IV analgesics.
   2 IM analgesics are not necessary because circulatory access will be available.
   4 This is false reassurance; it is rarely possible to relieve all pain immediately postoperatively.

209. (2) Decreased sodium intake can accelerate lithium retention, with subsequent toxicity. (2) (PC; IM; MH; DR)
   1 This is unrelated to the administration of lithium.
   3 Same as answer 1.
   4 Same as answer 1.

210. (1) The client may become injured in many ways during a seizure, and trauma prevention is a priority. (2) (SI; PL; MS; NM)
   2 Many anticonvulsants can cause GI disturbances, especially early in therapy; they should be taken with food.
   3 Others should be aware of the condition and taught how to help in case of a seizure.
   4 This is untrue; symptoms and treatment of seizure disorders vary greatly.

211. (4) Gastric returns would indicate correct placement. (1) (SI; IM; PE; GI)
   1 Further assessment is necessary.
   2 This could cause undue trauma regardless of where the tube is.

   3 This is unsafe until correct placement is verified; feeding could enter the lung if the tube is not in the stomach.

212. (4) Toxicity can result because the action of calcium ions is similar to that of digitalis. (3) (CI; AN; MS; DR)
   1 Calcium gluconate cannot be added to a solution containing carbonate or phosphate because a dangerous precipitation will occur.
   2 If calcium infiltrates, sloughing of tissue will result.
   3 Calcium can be added to this solution.

213. (4) Because of the anatomic position of the parathyroids, they may be accidentally removed during surgery. (3) (CI; EV; MS; EN)
   1 This is unlikely; the thoracic cavity is not entered.
   2 This is unlikely; the thyroid is nearer to the trachea.
   3 This is unlikely; this usually results from a blunt blow.

214. (4) The newborn's intestinal tract is sterile and therefore does not have intestinal flora to synthesize vitamin K, a precursor to prothrombin that is necessary for clotting. (2) (DF; AN; CW; NN)
   1 This is not affected by vitamin K.
   2 Same as answer 1.
   3 Same as answer 1.

215. (3) Nasogastric suction is maintained to prevent pressure on the intestinal anastomosis; no oral fluids are permitted until peristalsis resumes and nasogastric suction is stopped. (3) (CI; IM; MS; GI)
   1 This drains immediately unless it is a continent conduit; this has no bearing on when oral fluids can be started.
   2 This would be too long to wait; sips of water are permitted when peristalsis returns and the nasogastric tube is removed; this usually takes 2 to 3 days.
   4 This has no bearing; the return of peristalsis and removal of the nasogastric tube are determinants.

216. (1) Initiating interactions demonstrates that the depressed person is attempting to change behavior patterns. (2) (EP; EV; MH; MO)
   2 Avoiding people is a reinforcement of the depressed life-style.
   3 Clients who attempt to modify behavior to please others make only superficial changes.

4 Solitary activities are nonthreatening but do not deal with the problem of impaired relationships.

217. ③ This prevents the pelvis and hips from rotating and places equal stress on the growing extremities. (2) (CC; IM; PE; SK)
1 This is a balanced suspension traction that is used for a fractured femur in adults.
2 This is a description of Russell traction.
4 This is a description of Buck's traction.

218. ④ The virus disintegrates rapidly on contact with soap. (2) (SI; EV; CW; HP)
1 The lesion is in the genital area, not on the lips; kissing will not affect the infant.
2 This is unnecessary; only meticulous hand washing is required.
3 This is not necessary because soap effectively disintegrates the virus.

219. ④ This provides the family member with an opportunity to express feelings. (1) (SC; IM; MH; CS)
1 Although true, this statement really does not address the family members' concerns.
2 This is false reassurance and cuts off communication.
3 This shuts off communication and abdicates nursing responsibility toward the client.

220. ④ This demonstrates the client's diminished anger and is a realistic assessment and acceptance of present capabilities and limitations. (2) (SC; EV; MH; CS)
1 This shows dependency; the client either has given up or is being sarcastic.
2 Anger is still apparent at loss of decision making; there is no real evidence of sharing.
3 This shows dependency and suggests the client has given up.

221. ④ Urinary retention and distention are common problems after a cystoscopy because of urethral edema. (2) (CI; EV; MS; RG)
1 Fluids dilute the urine and reduce the chance of infection after a cystoscopy.
2 Although this can occur, an inability to void deserves further investigation because infection can follow retention of urine.
3 More conservative methods such as running water or a sitz bath should be attempted; catheterization carries a risk of infection.

222. ③ The formula is drop factor (60) ÷ time in minutes (60) × desired hourly volume (800 ÷ 24) =

drops per minute; the correct answer is 33 drops per minute. (2) (PT; AN; PE; FE)
1 This is too slow; a rate of 3 drops/min would deliver 72 ml in 24 hours.
2 This is too slow; a rate of 6 drops/min would deliver 144 ml in 24 hours.
4 This is too fast; a rate of 60 drops/min would deliver 1440 ml in 24 hours.

223. ② Stable vital signs are the major indicators that transfer will not jeopardize the client's condition; apprehension and complaints of pain do not place the client in jeopardy. (2) (SI; AS; MS; CV)
1 Restlessness may be a sign of shock; the client needs further assessment.
3 The vital signs are not stabilized, and transfer at this time is contraindicated.
4 These are signs of increased intracranial pressure, and the client should not be transferred at this time.

224. ① Low concentrations of oxygen do not reduce the stimulus to breathe and prevent carbon dioxide narcosis. (1) (DA; IM; MS; RE)
2 Sedatives would further depress respirations and increase the carbon dioxide level.
3 Chronic hypercapnia is present; additional carbon dioxide only adds to the problem and results in carbon dioxide narcosis.
4 Respiratory obstruction causes difficulty on expiration; deep breathing may aggravate this situation.

225. ① In grand multiparas, observation for failure of the uterus to contract after delivery is a priority. (3) (CI; PL; CW; HP)
2 This is not a priority goal specific to a grand multipara immediately after delivery; it also is not common after abdominal delivery.
3 This is an important goal, but it is not specific to a grand multipara.
4 This can be done only after the mother is in stable condition.

226. ③ These are signs of increasing hypoxia; a tracheotomy may be necessary to maintain an open airway. (2) (CI; IM; PE; RE)
1 The symptoms are not indicative of increased secretions; suctioning can precipitate sudden laryngospasm.
2 This is ineffective for laryngeal spasms.
4 Further assessment is warranted before increasing oxygen therapy.

227. (2) The client's major concern at this time is pain caused by inappropriate handling. (2) (SC; IM; MS; SK)
   1 This is not the major concern, and false reassurance should not be given.
   3 The number of personnel will not necessarily ensure careful handling; this does not address the client's primary concern.
   4 Diversion at this time would not be an appropriate response to the client's primary concern.

228. (3) Most injuries to abused children are not life threatening; protection takes priority over immediate treatment. (3) (EP; PL; MH; CS)
   1 Treatment of medical injuries is the physician's primary responsibility.
   2 An accurate diagnosis of child abuse may take time and must be fully investigated.
   4 The nurse is often the first individual to see the abused child and must establish protection even before the physician arrives.

229. (2) The disease is most common in this age group; it has a slight predilection for men. (2) (SC; AN; MS; BI)
   1 This is not completely accurate; sometimes it is possible.
   3 The opposite is true.
   4 Incidence is not limited to any age group.

230. (2) Pressure of the uterus against major blood vessels reduces circulation; decreased perfusion of the placenta can result. (1) (CI; AN; CW; HC)
   1 This is false; it can lead to supine hypotension.
   3 This does not interfere with the movement of the symphysis pubis.
   4 This could possibly prolong labor, but this is not the most essential reason.

231. (2) Application of cold relieves discomfort, and the binder provides support and aids in pressure atrophy of acini cells so that no more milk will be produced. (3) (PC; IM; CW; HC)
   1 This is suitable for the engorged breast-feeding mother because it promotes comfort and stimulates milk production.
   3 Same as answer 1.
   4 Severe restriction of fluid will not prevent engorgement and might cause dehydration.

232. (3) Anemia may result because of the possible bone marrow depressant effect of this drug. (1) (SI; IM; MS; DR)
   1 This would be unsafe; a physical examination and blood studies are necessary to determine the cause of the client's symptoms.

   2 This is unsafe; this is not the role of the nurse.
   4 An increase may result in toxic effects if the client is taking the maximum dose.

233. (2) Changes in an ECG will reflect the area of the heart that has been damaged because of hypoxia. (1) (CI; AN; MS; CV)
   1 A stethoscope is used to detect heart sounds.
   3 Medical interventions such as cardioversion or cardiac medications, not an ECG, can alter heart rhythm; an ECG will reflect heart rhythm, not change it.
   4 This is accomplished through a stress test; this uses an ECG in conjunction with physical exercise.

234. (4) No bowel sounds are present; therefore the client must remain npo. (2) (PC; IM; MS; IT)
   1 The urinary output is adequate; there is no need to increase IV fluids.
   2 This is unsafe; the client must be kept npo until bowel sounds are present.
   3 Same as answer 2.

235. (3) Maintaining a patent airway is the priority; inhalation burns may have occurred. (1) (DA; PL; MS; RE)
   1 This is extremely important but not the first priority.
   2 Same as answer 1.
   4 Same as answer 1.

236. (4) It is necessary to assess behavior changes that clue the nurse to impending suicidal acting out. (2) (EP; AS; MH; MO)
   1 Because there has been no overt acting out, continuous observation may threaten the client's ability to maintain self-control.
   2 The depressed client cannot follow involved interesting activities because of psychomotor retardation.
   3 Detailed explanations are inappropriate and overwhelming for a client with psychomotor retardation.

237. (4) A change in behavior may indicate the client has worked out a plan for suicide; the potential for acting out on suicide increases when physical energy returns. (2) (SI; IM; MH; MO)
   1 The client should not be considered for discharge simply because of a change in behavior.
   2 This is not indicated at this time.
   3 This may increase the client's feelings of inadequacy, because it implies the client did not look well before.

238. (1) Grief reactions to the impending loss and security present in the one-to-one relationship need to be anticipated. (3) (SC; EV; MH; TR)

2 This phase occurs early in the one-to-one relationship, not at termination.

3 There should not be a disintegration of the personality.

4 This behavior may occur in the early phase of a working relationship.

239. (3) Talking about what actually happened helps the client sort out the truth from confused thoughts and begins to help the client accept what happened as a part of the history. (2) (SC; IM; MH; CS)

1 The victim should be told of the legal services available; legal counsel should come from a legal authority.

2 Most rapes are planned in advance and are violent acts of the perpetrators who are responsible for their behavior; nevertheless, the victim often feels unjustified guilt related to the incident.

4 If the client does not want to discuss intimate details, this should be respected.

240. (4) This objective includes observable client behavior, which is specified by amount and time and therefore is measurable. (2) (CI; AN; MS; RE)

1 This objective is not stated in measurable terms.

2 Same as answer 1.

3 This is a statement, not an objective.

241. (4) The CDC Isolation Guidelines indicate contact precautions should be employed when a wound is infected to prevent the spread of infection to others. (2) (SI; IM; MS; IT)

1 This technique is used when changing dressings to protect the client from infection.

2 This may be used with clients who are immunosuppressed and at risk for infection.

3 Wound or peritoneal infections are transmitted through incisional drainage, not feces.

242. (3) The data presented indicate an infectious process within the lung. (1) (PC; AN; MS; RE)

1 The cardinal signs would be pain in the lower lobe at the height of inspiration and a pleural friction rub.

2 Although fever and chills can occur later in the disease, the cardinal signs are irritating cough, chest pain, and shortness of breath.

4 The cardinal signs would be barrel chest; resonance on percussion; and thick, tenacious sputum.

243. (2) This is a reflective statement that conveys acceptance and encourages further communication. (2) (SC; IM; MS; EH)

1 This is false reassurance that does not lessen anxiety.

3 The reliance on a pill to help the client in this instance evades the problem and cuts off further communication.

4 This is too direct; this statement does not encourage the client to discuss feelings.

244. (3) Sleep deprivation alone can cause these disturbances. (2) (CI; IM; MS; NM)

1 Lack of contact with significant others increases anxiety and feelings of isolation and can lead to disturbances in rest.

2 Pain limits or interrupts periods of sleep and rest.

4 Constant light limits sleep.

245. (3) This diet is low in fiber, and after digestion and absorption, there is very little substance to be eliminated. (2) (CI; AN; MS; GI)

1 This diet does not promote peristalsis; the products of digestion, although reduced, remain in the intestine longer, and flatus is increased.

2 This diet does not affect the bacterial milieu of the intestine.

4 Although a low-residue diet is less irritating, this is not the primary reason for its use before surgery.

246. (2) This is an open-ended question that should elicit the desired information. (2) (DF; AS; PE; GD)

1 This may establish that the child knows where she is but does not elicit whether she knows why she is there.

3 This question can be answered with a "yes" or "no" and does not elicit what the child thinks about the situation.

4 This is not an open-ended question because it can be answered with a "yes" or a "no" response.

247. (4) Leakage may indicate ruptured amniotic membranes; the client is at risk for an ascending infection from the vagina. (1) (CC; EV; CW; HP)

1 This is a common discomfort of pregnancy caused by the shift in the center of gravity because of the enlarged uterus.

2 Leukorrhea is common during pregnancy because of increased vascularity of the cervix and increased production of mucus.

3 Braxton Hicks contractions are contractions of the uterus occurring at irregular intervals throughout pregnancy.

248. (1) Digoxin slows the heart rate, which is reflected in a slowing of the pulse; it also increases kidney perfusion, which promotes urine formation, resulting in diuresis and decreased edema. (2) (PT; EV; MS; DR)
    2 Digoxin does not decrease the pulse pressure or directly influence the blood pressure; diuresis will lower the weight and blood pressure.
    3 Digoxin lowers the pulse rate and produces diuresis as a result of improved cardiac output.
    4 Digoxin will not increase the blood pressure or reduce a heart murmur.

249. (2) This reduces energy requirements, allows for rest, and lessens the demand for oxygen. (2) (PC; IM; PE; RE)
    1 Although this loosens secretions in the lungs, it should not be used when the child is in distress.
    3 This is not done unless respiratory distress is extremely severe; it increases restlessness and energy demands.
    4 This positional change will not reduce energy and oxygen demands.

250. (2) The function of sucking requires more oxygen and therefore tires the infant. (2) (PC; PL; PE; RE)
    1 Laryngospasm, or spasmodic closing of the glottis, which results from edema, is not related to oral feedings.
    3 IVs have no effect on vagal stimulation.
    4 Parenteral fluids do not provide complete caloric needs.

251. (2) This is caused by manipulation of abdominal contents and the depressant effects of anesthetics and analgesics. (1) (PC; AN; MS; GI)
    1 Edema would not totally interfere with peristalsis; edema may cause peristalsis to be less effective, but some output would still result.
    3 An absence of fiber has a greater effect on decreasing peristalsis than fluids do.
    4 A nasogastric tube decompresses the stomach; it does not cause cessation of peristalsis.

252. (1) A cancerous mass can grow into the lumen of the intestines, altering the shape of the stool; stools may be ribbon-like or pencil thin. (1) (HI; AS; MS; GI)
    2 This is not specific to intestinal cancer.
    3 Same as answer 2.
    4 Same as answer 2.

253. (4) Care for an infant after croup should be directed toward personal care, proper nutrition, and stimulation. (1) (HI; PL; PE; RE)

    1 The infant does not require additional fluids if all feedings are consumed.
    2 Young infants need environmental stimuli.
    3 Croup is not directly related to antigen-antibody responses.

254. (3) Redness and pain are signs of phlebitis; the site should be changed to prevent further inflammation and possible thrombus formation. (3) (CI; IM; PE; CV)
    1 Continuing administration can lead to further irritation and even permanent damage to the vein.
    2 Although this is generally true, there is no indication that this is what is occurring in this situation.
    4 Same as answer 1.

255. (1) This would numb the nerve endings and reduce pain; cold also produces vasoconstriction, limits edema, and prevents hemorrhage. (1) (PC; PL; PE; NM)
    2 Aspirin has anticoagulant properties that could increase the risk of bleeding and should therefore be avoided.
    3 Hard candies can scratch the operative site and dislodge the clot and should therefore be avoided.
    4 This could dislodge a clot and cause bleeding.

256. (1) The client needs positive relationships with other adults, but clear, consistent limits must be presented to minimize attempts at manipulation. (1) (EP; PL; MH; PR)
    2 This is not a therapeutic approach.
    3 This is not a therapeutic approach; clear, consistent limits are necessary to prevent manipulation.
    4 This is a judgmental attitude that should be avoided.

257. (1) Most child abusers have been abused themselves; they do not know how to deal with the rage they feel. (2) (EP; PL; MH; CS)
    2 Parents in abusing parent groups have already admitted and begun to seek help for their problem.
    3 This type of presentation usually alienates individuals who are struggling with their own present difficulties.
    4 Therapy should focus on and encourage the sharing of feelings, not personal parenting information.

258. (4) The muscular action during ambulation facilitates the return of venous blood to the heart. (1) (HI; IM; CW; HP)

1 Early ambulation would not prevent this complication.
2 This is unrelated; the baby is usually given to the mother in the delivery room to begin the bonding process.
3 Same as answer 1.

259. (2) A common complication of a cerebrovascular accident (CVA) is an inability to control emotional affect; clients may be depressed or apathetic and have a lability of mood. (2) (SC; IM; MS; NM)
1 There are no data to support this conclusion.
3 Same as answer 1.
4 Same as answer 1.

260. (2) The heart muscle begins to fail, and the muscle does not contract with enough strength to pump sufficient blood to meet the body's metabolic needs. (3) (CI; AN; MS; CV)
1 Hypervolemia can precipitate congestive heart failure in individuals with a diseased heart, but it is not specific to rheumatic heart disease.
3 Heart valves can become stenotic or regurgitant as a result of rheumatic fever; however, left ventricular failure will occur only when the heart can no longer pump an adequate amount of blood to maintain cardiac output.
4 Hypertension and arteriosclerosis can add stress to this situation, but they are not the reason for this client's congestive heart failure.

261. (3) Steroids are used for their antiinflammatory, vasoconstrictive, and antipruritic effects. (2) (CC; EV; MS; DR)
1 Steroids increase the incidence of infections by masking symptoms.
2 Steroids increase fluid retention.
4 Although steroids have an antipruritic effect, their major purpose after surgery is the antiinflammatory effect.

262 (3) Pinworms emerge nocturnally to lay eggs in the perianal area; eggs are caught on cellophane tape in the morning before toileting. (1) (HI; IM; PE; GI)
1 A stool culture will not reveal the presence of parasites.
2 The ova cannot be seen with the naked eye; the parasite is rarely observed in the stool.
4 There is no such test to diagnose pinworms.

263. (4) Many closures require the use of two hands, and some clothing requires movement of both sides of the body when dressing. (1) (CC; AP; MS; NM)
1 The client can continue to use the right hand to perform this activity.
2 Same as answer 1.
3 Same as answer 1.

264. (1) Panting and blowing keep the glottis open so the client cannot hold the breath and bear down. (2) (CC; EV; CW; HC)
2 While this pattern keeps the pressure of the diaphragm off the contracting uterus, it does not reduce the urge to push.
3 Shallow-chest breathing interferes with adequate oxygenation of the fetus because it limits the mother's oxygen intake.
4 Same as answer 2.

265. (4) Air drying the nipples several times a day hardens the nipples and reduces soreness; changing the position of the neonate while nursing will also relieve sore nipples. (1) (PC; PL; CW; HC)
1 This would cause vasoconstriction and lead to milk suppression.
2 This may reduce the discomfort, but it will not help the nipples to dry and harden.
3 This would inhibit lactation; the baby must suckle and empty the breasts regularly for milk production to continue.

How to Use Worksheet 1: Errors in Processing Information

Common errors in processing information are listed in the left-hand column of this worksheet. At the top of the worksheet is a row of blank spaces for inserting the number of the question missed. Directly below each number, check any errors you made in answering that question. You may have made more than one type of error in an answer.

Worksheet 1: Errors in Processing Information

Question number																									
Did not read situation/ question carefully																									
Missed important details																									
Confused major and minor points																									
Defined problem incorrectly																									
Could not remember terms/ facts/concepts/principles																									
Defined terms incorrectly																									
Focused on incomplete/incorrect data in assessing situation																									
Interpreted data incorrectly																									
Applied wrong concepts/ principles in situations																									
Drew incorrect conclusions																									
Identified wrong goals																									
Identified priorities incorrectly																									
Carried out plan incorrectly/ incompletely																									
Was unclear about criteria for evaluating success in achieving goals																									

How to Use Worksheet 2: Knowledge Gaps

Types of common knowledge gaps are listed along the top of this worksheet. Write a brief description of topics you want to review in the spaces provided. For example, if you missed a question on administration of a particular drug, write the drug name and problem (e.g., dosage) in the appropriate space under the column labeled *Pharmacology*.

Worksheet 2: Knowledge Gaps

Basic science	Skills/ procedures	Basic human needs	Growth and development	Normal nutrition	Psychosocial factors	Clinical area/ topic	Stressors/ coping mechanisms	Patho- physiology	Pharma- cology	Therapeutic nutrition	Legal implications	Other

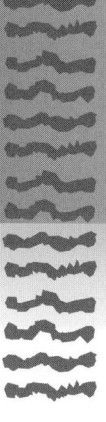

CHAPTER 7

Comprehensive Examination 2

PART A QUESTIONS

1. Two days after delivery, a neonate's head circumference is 16 inches (40.6 cm) and the chest circumference is 13 inches (33 cm). These measurements:
   1. Are both normal parameters at birth
   2. Suggest the presence of microcephaly
   3. Indicate that the baby's head is enlarged
   4. Demonstrate a smaller-than-normal chest

2. The nurse notes six premature ventricular beats (PVBs) in a row on the cardiac monitor of a client in the coronary care unit (CCU). Using the protocol generally established in most CCUs, the nurse should first:
   1. Encourage the client to cough and deep breathe
   2. Initiate cardiopulmonary resuscitation immediately
   3. Administer a 50- to 100-mg bolus of lidocaine (Xylocaine)
   4. Notify the client's physician of the findings immediately

3. A client paces back and forth across the floor, speaks incoherently, and spends a great deal of time talking and verbally fighting with persons who are not present. The initial therapeutic intervention by the nurse should be directed toward:
   1. Setting limits on the client's verbal aggression
   2. Isolating the client to decrease the aggressive behavior
   3. Engaging client in a structured, reality-oriented activity
   4. Establishing a relationship to reduce the client's loneliness

4. The laboratory calls to state that a client's lithium level is 1.9 mEq/L after 10 days of lithium therapy. The nurse should:
   1. Immediately report the finding to the physician since the level is dangerously high
   2. Monitor the client closely because the level of lithium in the blood is slightly elevated
   3. Continue to administer the medication as ordered because the level is within the therapeutic range
   4. Report the findings to the physician so the dosage can be increased because the level is below the therapeutic range

5. To check for wound hemorrhage after a client has a thyroidectomy, the nurse should:
   1. Loosen an edge of the dressing and lift it to visualize the wound
   2. Observe the dressing at the back of the neck for the presence of blood
   3. Outline the blood as it appears on the dressing to observe any progression
   4. Press gently around the incision to express accumulated blood from the wound

6. A 15-year-old primigravida comes to the labor suite. She is in her 38th week of gestation and states that she is in labor. To verify that the client is in true labor the nurse should:
   1. Obtain slides for a fern test
   2. Time any uterine contractions
   3. Prepare her for a pelvic examination
   4. Apply nitrazine paper to moist vaginal tissue

7. As part of a diagnostic workup for pulmonic stenosis, a child has a cardiac catheterization. The nurse is aware that children with pulmonic stenosis have increased pressure:
   1. In the pulmonary vein
   2. In the pulmonary artery
   3. On the left side of the heart
   4. On the right side of the heart

8. A client with cholecystitis is placed on a low-fat, high-protein diet. The nurse should teach the client that this diet can include:
   1. Boiled beef
   2. Skimmed milk
   3. Poached eggs
   4. Steamed broccoli

9. As a very anxious female client is talking to the nurse, she starts crying. She appears to be upset that she cannot control her crying. The most appropriate response by the nurse would be:
   1. "It is OK to cry; I'll just stay with you for now."
   2. "Is talking about your problem upsetting you?"
   3. "Sometimes it helps to get it out of your system."
   4. "You look upset; let's talk about why you are crying."

10. As a postpartum client is being prepared for discharge, she says to the nurse, "I don't think I'll be able to take care of the baby when I go home." The nurse's best response would be:
    1. "What is it that makes you think that?"
    2. "It will come naturally. Give yourself time."
    3. "Is there anything specific that concerns you?"
    4. "I know it can be frightening, but you'll do just fine."

11. The nurse initiates preparation of a 9-year-old girl for an infratentorial craniotomy. The nurse plans to:
    1. Encourage doll play with blunt tools and dressings
    2. Schedule role playing with others having similar surgery
    3. Have the child draw her concept of a brain and briefly clarify any misconceptions
    4. Provide a minutely detailed explanation of anatomy and the surgery to be performed

12. A client with Hodgkin's disease is placed on an ABVD combination chemotherapy regimen. Because doxorubicin (Adriamycin) is part of this therapy, the nurse should teach the client to:
    1. Cease taking any medication that contains vitamin D

    2. Keep the Adriamycin in a dark area, protected from light
    3. Expect urine to turn red for 1 to 2 days after taking this drug
    4. Take the Adriamycin on an empty stomach with plenty of fluids

13. After a muscle biopsy, the nurse should teach the client to:
    1. Change the dressing as needed
    2. Resume a usual diet as soon as desired
    3. Bathe or shower according to preference
    4. Expect a rise in body temperature for 48 hours

14. A client had an open reduction and internal fixation of the head of the femur. In the postanesthesia unit the client's vital signs remained stable for 1 hour, with a BP 130/78 mm Hg, P 68, and R 16. One hour after returning to the surgical unit, the client's vital signs are BP 100/60 mm Hg, P 74, R 22, and the client is restless. The nurse should:
    1. Increase the IV flow rate
    2. Raise the head of the bed
    3. Check the client's dressing
    4. Continue to monitor the vital signs

15. A fetal monitor is attached to a client in active labor. When the client has contractions, the nurse notes a 15-beat-per-minute uniform acceleration of the fetal heart rate above the baseline with each contraction. The nurse should:
    1. Change the maternal position
    2. Turn the client on her left side
    3. Call the physician immediately
    4. Prepare for immediate delivery

16. A male client receiving prolonged steroid therapy complains of always being thirsty and urinating frequently. The best initial action by the nurse would be to:
    1. Perform a finger stick to test the client's blood glucose level
    2. Have the physician assess the client for an enlarged prostate
    3. Obtain a urine specimen from the client for screening purposes
    4. Assess the client's lower extremities for the presence of pitting edema

17. A client is placed on the "prudent diet," which is advocated by the American Heart Association and designed to control the intake of saturated fats and cholesterol. To best explain the dietary nature of these two food substances, the nurse should teach the client that:

1. Cholesterol is a necessary body constituent and cannot be eliminated
2. Polyunsaturated fats come from animal foods such as meat and cheese
3. Plant sources of cholesterol must be controlled in the diet every day
4. The more-saturated fats come from plant foods such as seeds and grains

18. After 3 months of supplemental oral iron therapy, there is no significant rise in a child's hemoglobin level. The physician orders iron dextran (Imferon). When administering this medication, the nurse should:
   1. Use the Z-track method
   2. Massage the injection site
   3. Use a 25-gauge 5/8-inch needle
   4. Avoid aspiration before injection

19. When the neonate of a diabetic mother is admitted to the neonatal intensive care nursery after delivery, priority care for the neonate would include:
   1. Doing a Dextrostix test on heel blood
   2. Double clamping the cord immediately
   3. Starting an IV with 10% glucose in water
   4. Instilling prophylactic ophthalmic medication

20. An emergency tracheotomy is performed on a toddler in acute respiratory distress from laryngotracheobronchitis (viral croup). In addition to routine suctioning of the tracheotomy, the nurse should also suction if the toddler:
   1. Becomes diaphoretic and cyanotic
   2. Has severe substernal retractions and stridor
   3. Verbalizes an increased difficulty in breathing
   4. Becomes restless or pale or has an increased pulse rate

21. Based on a knowledge of possible complications after radical neck surgery, the nursing assessment with the highest priority would be:
   1. Pulmonary edema
   2. Cardiogenic shock
   3. Atrophy of chest muscles
   4. Rupture of the carotid artery

22. A client in the hospice home care program is experiencing severe pain. The physician orders morphine (MS Contin) for pain management. The nurse should explain to the client that:
   1. It is given automatically at regular intervals around the clock
   2. A heparin lock will be inserted for intermittent IV administration
   3. The medication must be requested before the pain becomes severe

4. The potential for dependency or addiction is decreased with this drug

23. After a lateral crushing chest injury, obvious right-sided paradoxic motion of a client's chest demonstrates multiple rib fractures, resulting in a flail chest. The complication the nurse should carefully observe for is:
   1. Mediastinal shift
   2. Tracheal laceration
   3. Open pneumothorax
   4. Pericardial tamponade

24. A client refuses to eat, saying, "They want to kill me." The most therapeutic response by the nurse would be:
   1. "No one is trying to harm you."
   2. "You feel someone is attempting to poison your food."
   3. "That's not true. It's the same food everyone else is eating."
   4. "If you want, I'll taste your food before you do so you'll know it's OK."

25. When planning care for a client at 30 weeks gestation admitted to the hospital after a vaginal bleed secondary to placenta previa, the nurse's primary objective is to:
   1. Provide a calm, quiet environment
   2. Prepare for an immediate cesarean delivery
   3. Prevent situations that may stimulate the cervix or uterus
   4. Ensure that the client has regular cervical examinations to assess for labor

26. A mother, exceedingly upset about her child being diagnosed as having a pinworm infestation, is taught by the public health nurse how the pinworm infestation is transmitted. The nurse evaluates that the teaching was effective when the mother states:
   1. "I'll need to be sure the cat stays off my child's bed at night."
   2. "I'll have to reinforce my child's handwashing, especially before eating or handling food."
   3. "I'll be sure to disinfect the toilet seat after my child's bowel movements for the next few days."
   4. "My child contracted this infestation from the dirty school toilets, and I'll report that to the school nurse."

27. When assessing the thorax of a client with chronic obstructive pulmonary disease, the nurse would expect to find:
   1. Decreased breath sounds
   2. Atrophic accessory muscles
   3. A shortened expiratory phase
   4. A decrease in the A-P diameter

28. A client has a chest tube inserted to treat a right hemopneumothorax. To facilitate chest drainage, the client should be encouraged to lie:
    1. In the supine position
    2. On the right (affected) side
    3. On the left (unaffected) side
    4. Immobilized as much as possible

29. An adolescent client with anorexia nervosa refuses to eat, stating, "I'll get too fat." The nurse can best respond to this behavior initially by:
    1. Not talking about the fact that the client is not eating
    2. Stopping all of the client's privileges until food is eaten
    3. Pointing out to the client that death can occur with malnutrition
    4. Telling the client that tube feedings will eventually be necessary

30. The nurse should plan to assist a client with an obsessive-compulsive disorder to control the use of ritualistic behavior by:
    1. Providing repetitive activities that require little thought
    2. Attempting to reduce or limit situations that increase anxiety
    3. Getting the client involved with activities that will provide distraction
    4. Allowing the client to perform menial tasks to expiate feelings of guilt

31. Immediately after a child is admitted to the hospital with acute bacterial meningitis, the nurse should plan to:
    1. Assess the child's vital signs every 4 hours
    2. Give oral antibiotic medications as ordered
    3. Check the child's level of consciousness every hour
    4. Restrict parental visiting until isolation is discontinued

32. A client with the diagnosis of otosclerosis undergoes a stapedectomy with insertion of a middle ear prosthesis. A few days after the operation the client is discouraged because there is no improvement in hearing. The nurse should bear in mind that this is most likely a result of:
    1. Swelling within the ear canal
    2. Damage to the organ of Corti
    3. Perforation of the tympanic membrane
    4. The graft having slipped out of position

33. After a transurethral resection of the prostate, a client has a three-way Foley catheter inserted with a continuous bladder irrigation. The client complains that he needs to void. The nurse should first:
    1. Obtain client's vital signs and notify the physician
    2. Assess the client's total intake and output for the day
    3. Explain to the client that the balloon inflated in the bladder causes this feeling
    4. Check the tubing connected to the client's collection bag to see whether it is draining

34. A client with the diagnosis of multiple sclerosis develops hand tremors. When performing a physical assessment, the nurse should take into consideration that the tremors associated with multiple sclerosis usually occur when the client:
    1. Is asleep
    2. Is inactive
    3. Gets nervous or upset
    4. Attempts to do something

35. An elderly client is hospitalized with the diagnosis of dementia. Considering the client's diagnosis, the behavior that the nurse would most likely observe is an:
    1. Increased attention span and perceptual disturbances
    2. Inability to learn new things and disturbance in thinking
    3. Acceptance of personality and social changes due to declining years
    4. Increased capacity for adaptation to the environment based on past life experiences

36. After delivery, the nurse assesses a client who had an abruptio placentae and suspects that disseminated intravascular coagulation (DIC) is occurring when assessments demonstrate:
    1. A boggy uterus
    2. Multiple vaginal clots
    3. Hypotension and tachycardia
    4. Bleeding from the venipuncture site

37. During a client's second stage of labor, the nurse should:
    1. Watch for bulging of the client's perineum
    2. Give the client pain medication as ordered
    3. Teach the client how to pant with each contraction
    4. Catheterize the client so that the head can be delivered

38. The nurse caring for a client with type 1 diabetes should be alert to the symptoms of an insulin reaction, which include:

1. Dry skin, drowsiness, and tachycardia
2. Excessive thirst, anorexia, and malaise
3. Headache, nervousness, and diaphoresis
4. Ataxia, dilated pupils, and Kussmaul respirations

39. The most beneficial between-meal snack for a client who is recovering from full-thickness burns would be a:
    1. Cheeseburger and a malted
    2. Piece of blueberry pie and milk
    3. Bacon and tomato sandwich and tea
    4. Chicken salad sandwich and soft drink

40. A client with a fractured right head of the femur and osteoporosis is placed in Buck's traction before surgical repair. Until surgery is performed, the nurse should plan to:
    1. Remove the weights from the traction every 2 hours to promote comfort
    2. Inspect the skin and circulation of the affected leg hourly to prevent trauma
    3. Turn the client from side to side every 4 hours to prevent pressure on the coccyx
    4. Raise and maintain the knee gatch on the bed to limit the shearing force of traction

41. The nurse recognizes that failure of a newborn to make the appropriate adaptation to extrauterine life would be indicated by:
    1. Cyanotic lips and face
    2. A respiratory rate of 40
    3. Cyanotic feet and hands
    4. A liver 2 cm below right costal margin

42. A 16-month-old has had large, frothy, foul-smelling stools since the introduction of table food and cow's milk into the diet. The child's mother also noted that the child changed from pleasant and outgoing to irritable and apathetic. The child is diagnosed as having celiac disease and is placed on a gluten-free diet. When evaluating the child's response to the diet after 2 days, the nurse anticipates the first change will be:
    1. A return of appetite
    2. An increase in weight
    3. A cessation of diarrhea
    4. An improved personality

43. A tricyclic antidepressant is prescribed for a depressed client. After 1 week, a member of the client's family comes to speak with the nurse and expresses concern that there does not seem to be much improvement after taking the medication. When responding to the family member the nurse explains that:

1. As the client's physical condition improves, the antidepressant medication will act more effectively
2. The client may require other drugs in addition to the antidepressants before behavioral changes are noted
3. In clients who have been depressed for a prolonged period, the drug takes additional time to be effective
4. The tricyclics are slow-acting drugs, and it may take 3 to 4 weeks until therapeutic effectiveness is achieved

44. A 2-year-old who is HIV-positive from multiple blood transfusions is examined in the emergency room. Presently, the child has been ill for a week with a temperature of 103° F. The child's mother states that the child is losing weight and has a whitish film on the gums. When doing the health history, the nurse determines that the child is at high risk for developing AIDS because of:
    1. A positive HIV antibody screening test
    2. An immature reticuloendothelial system
    3. The multiple blood transfusions received
    4. The presence of an opportunistic infection

45. A client stops taking birth control pills because she and her husband want to start a family. After 18 months of unsuccessful attempts at conception, the client is diagnosed as having primary infertility related to anovulatory cycles. Clomiphene citrate (Clomid) is prescribed for 6 months. The nurse knows that the teaching about the correct time to take the Clomid is understood when the client states, "I will begin to take the pills on the:
    1. Fifth day of my cycle."
    2. Last day of my period."
    3. First day I start my period."
    4. Fourteenth day of my cycle."

46. One day a young adult client's mother confides to the nurse that she is very troubled by her child's emotional illness. The nurse's most therapeutic initial response would be:
    1. "It is very important that you become involved in volunteer work at this time."
    2. "You must lessen your feelings of guilt and loneliness by seeing a psychiatrist."
    3. "I recognize it's hard to deal with this. Try to remember that this too shall pass."
    4. "Getting together with others who are coping with this problem can be quite helpful."

47. A pregnant client whose blood pressure has been averaging 92/64 mm Hg should be evaluated for pregnancy-induced hypertension when the blood pressure rises to:
1. 110/75 mm Hg
2. 112/74 mm Hg
3. 124/80 mm Hg
4. 140/90 mm Hg

48. While a client is on intravenous magnesium sulfate therapy for preeclampsia, it is essential for the nurse to monitor the client's deep tendon reflexes to:
1. Determine her level of consciousness
2. Evaluate the mobility of the extremities
3. Determine her response to painful stimuli
4. Prevent development of respiratory depression

49. The nurse makes notes on a female rape victim's record during the interview and physical examination because the record may go to court as evidence. It is most important when charting that the nurse include:
1. The client's verbatim statements about the rape and the rapist
2. Observations about the client's reaction to male staff members
3. General statements about the client's previous knowledge of the rapist
4. A summarized statement about the client's description of the rape and the rapist

50. A preschooler is admitted to the hospital with a diagnosis of acute glomerulonephritis. The child's history reveals a 5-pound weight gain in 1 week and periorbital edema. For the most accurate information on the status of the child's edema, nursing intervention should include:
1. Obtaining the child's daily weight
2. Doing a visual inspection of the child
3. Measuring the child's intake and output
4. Monitoring the child's electrolyte values

51. The physician orders the drug ranitidine (Zantac) to help treat a client's gastric ulcer. The nurse realizes this drug acts specifically by:
1. Lowering the gastric pH
2. Promoting the release of gastrin
3. Regenerating the gastric mucosa
4. Inhibiting the histamine H_2 receptors

52. The nurse teaches a 9-year-old and the child's parents about type 1 diabetes and the occurrence of hyperglycemia. The nurse would be aware that they understand the teaching when they state that ketoacidosis is most often precipitated by:
1. An infection

2. An insulin overdose
3. Decreased fluid intake
4. Excessive physical exercise

53. An adolescent is admitted to the hospital after having a generalized seizure at college. The client has a 2-year history of a seizure disorder, but the seizures have been well controlled by phenytoin (Dilantin) for the last 6 months. The client says to the nurse, "I am so upset. I didn't think I was going to have more seizures." The nurse's best response would be:
1. "Did you forget to take your medication?"
2. "You are worried about having more seizures?"
3. "You must be under a lot of stress at school right now."
4. "Don't be too concerned. Your medication needs to be increased."

54. Four hours after a liver biopsy, the client complains of pain, and the nurse notes that there is a leakage of a moderately large amount of bile on the dressing. Based on these findings, the nurse should:
1. Medicate the client for pain as ordered
2. Tell the client to remain flat on the back
3. Notify the client's physician immediately
4. Monitor the client's vital signs every 10 minutes

55. After surgery the physician orders gentamicin (Garamycin) 75 mg IVPB q6h and daily peak and trough levels. This is done primarily so that:
1. A drop in the client's fever may be correlated with the peak level
2. The blood culture can be obtained when gentamicin is at its lowest level
3. Any allergy that the client might have to the drug would be detected early
4. The presence of an adequate therapeutic level of the drug can be determined

56. On entering a depressed client's room one morning, the nurse finds the client still in bed. The client states, "I am unable to get dressed and go to breakfast." The nurse's best response would be:
1. "You cannot just lie in bed. You must get up now and go to breakfast."
2. "I'll get you dressed. I recognize that you have difficulty helping yourself."
3. "Take your time. It is not necessary to hurry. I'll help you if you need me to."
4. "You can lie there for awhile if you promise me you'll get dressed for lunch."

57. A 4-year-old boy with acute lymphocytic leukemia is to have a bone marrow aspiration. While involving the child in therapeutic play prior to the procedure, the nurse should help him understand that:

1. He needs to have a positive attitude
2. His parents are concerned about him
3. He did nothing to cause his present illness
4. His problem was caused by an environmental factor

58. The nurse recognizes that the most important factor in predicting a client's potential reaction to grief is the client's:
   1. Family interactions
   2. Emotional relationships
   3. Social support systems
   4. Earlier experiences with grief

59. The public health nurse presents a program on breast self-examination. The nurse realizes that certain aspects of the teaching program would have to be reviewed when, during the return demonstration, one of the women in the class:
   1. Palpates her breasts while in the sitting position
   2. Checks her nipples for alterations in size or shape
   3. Palpates her breast with the palmar surface of her extended fingers
   4. Observes her breasts for symmetry while holding her arms above her head

60. Based on the diagnosis of cancer of the pancreas, the nurse understands that a client's jaundice is caused by:
   1. Necrosis of the parenchyma caused by the neoplasm
   2. Excessive serum bilirubin caused by red blood cell destruction
   3. Obstruction of the common bile duct by the pancreatic neoplasm
   4. Impaired liver function resulting in incomplete bilirubin metabolism

61. Realizing that hypokalemia is a side effect of steroid therapy, the nurse should monitor a client taking steroid medication for:
   1. Hyperactive reflexes
   2. An increased pulse rate
   3. Nausea, vomiting, and diarrhea
   4. Leg weakness with muscle cramps

62. The nurse, when reviewing the history of a client admitted in possible preterm labor in her 30th week of gestation, suspects that the most likely cause of the preterm labor is that the client:
   1. Is having her first baby
   2. Is receiving medication for a seizure disorder
   3. Was diagnosed as having an android-shaped pelvis
   4. Has had three urinary tract infections during pregnancy

63. In addition to being nonconstipating, the diet for a child who is in traction should be:
   1. Low in calories and purine
   2. High in calories and phosphorus
   3. Moderate in calories and high in protein
   4. Adequate in calories and high in calcium

64. The behavior that would indicate to the nurse that the mental status of a client with the diagnosis of schizophrenia, paranoid type, was improving would be the client's:
   1. Absence of or freedom from anxiety
   2. Development of insight into the problem
   3. Decreased need to use defense mechanisms
   4. Ability to function effectively in activities of daily living

65. Because a severely depressed client has not responded to any of the antidepressant medications, the psychiatrist decides to try electroconvulsive therapy (ECT). Before the treatment the nurse should:
   1. Have the client speak with other clients receiving ECT
   2. Give the client a detailed explanation of the entire procedure
   3. Limit the client's intake to a light breakfast on the days of the treatment
   4. Provide a simple explanation of the procedure and continue to reassure the client

66. The nurse plans early postoperative ambulation for a client who has had a continent urostomy formed to prevent:
   1. Wound infection
   2. Urinary retention
   3. Abdominal distention
   4. Incisional evisceration

67. The day after a cesarean delivery, a client complains of abdominal pain and distention. The physician orders a Harris drip. After administering the Harris drip, the nurse can evaluate its effectiveness when:
   1. The client has a bowel movement
   2. The client's returns are finally clear
   3. The client's abdomen is less distended
   4. The client is able to retain 500 ml of fluid

68. A blond-haired, blue-eyed farmer goes to the physician because he has a large, crusty patch of skin on his cheek. The client states that it bleeds easily and has not gotten better, even after using different remedies. From the client's history, the nurse suspects skin cancer because the major precipitating factor associated with skin cancer is:
   1. Position of lesion
   2. Exposure to radiation
   3. Self-treatment of lesions
   4. Contact with soil contaminants

69. During a group therapy session, one of the clients asks a male client with the diagnosis of antisocial personality disorder why he is in the hospital. Considering this client's type of personality disorder, the nurse might expect him to respond:
    1. "I need a lot of help with my troubles."
    2. "Society makes people react in odd ways."
    3. "I decided that it's time I own up to my problems."
    4. "My life needs straightening out and this might help."

70. A child comes for a 6-week checkup following a tonsillectomy and adenoidectomy. In addition to assessing hearing, the nurse should include an assessment of the child's:
    1. Smell and taste
    2. Speech and taste
    3. Swallowing and smell
    4. Swallowing and speech

71. The nurse should plan to observe a client awaiting surgery for intestinal carcinoma for signs of:
    1. Diarrhea
    2. Dehydration
    3. Intestinal obstruction
    4. Abdominal peritonitis

72. A slow pulse rate during the early postoperative period following open heart surgery can be indicative of:
    1. Shock
    2. Hypoxia
    3. Heart block
    4. Congestive heart failure

73. When a 3-month-old baby is at the well-baby clinic for a checkup, the parents express concern that their baby still has a soft spot on the top of the head. The nurse explains that normal closure time for the anterior fontanel is between the ages of:
    1. 6 and 8 months
    2. 9 and 12 months
    3. 13 and 18 months
    4. 19 and 24 months

74. In a childbirth preparation class, the instructor teaches the clients to control their urge to push until the cervix is fully dilated by:
    1. Hyperventilating
    2. Doing pelvic rocking exercises
    3. Deep breathing between contractions
    4. Panting or blowing breathing patterns

75. A client who had a CVA is discharged with a hemiparesis but is able to ambulate with assistance. Whenever the client gets up from a lying down position, a feeling of being lightheaded and dizzy occurs. The nurse recognizes that this feeling is:
    1. Relieved by resting before performing activities
    2. Caused by blood pooling in the lower extremities
    3. A temporary response that will go away with time
    4. Precipitated by the medication, which may have to be changed

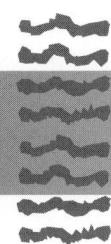

PART B QUESTIONS

76. Long-term care for parents of children with sickle cell anemia includes periodic conferences with groups of parents whose children have sickle cell disease to:
    1. Find special schooling facilities for the child
    2. Make plans for moving to a more therapeutic climate
    3. Choose a means of birth control to avoid future pregnancies
    4. Air their feelings regarding the transmission of the disease to the child

77. The nurse would expect a client with a somatoform disorder, conversion type, to display an affect that is:
    1. Happy and cheerful
    2. Sad and depressed
    3. Frightened and upset
    4. Calm and matter-of-fact

78. A client, 36 weeks gestation, is admitted to the labor and delivery unit because she is having painless periods of bright red bleeding. A tentative diagnosis of placenta previa is made. The nurse is aware that the diagnosis of placenta previa is usually confirmed by:
    1. Amnioscopy
    2. Laparoscopy
    3. Amniocentesis
    4. Ultrasonography

79. One morning a female client with the diagnosis of schizophrenia tells the nurse that she is Joan of Arc and is going to be burned at the stake. The most therapeutic response by the nurse would be:
    1. "Tell me more about being Joan of Arc."
    2. "You and I both know you are not Joan of Arc."
    3. "You are safe here, we won't let you be burned."
    4. "It seems like the world is a pretty scary place for you."

80. When a student in a sex education class in high school asks whether recurrent infection is possible with herpes genitalis, the nurse should reply that:

    1. Once herpes genitalis has been adequately treated, recurrence is rare
    2. The only sure way to prevent recurrent attacks is to abstain from sexual activity
    3. Unfortunately, recurrent attacks do occur but they are not as severe as the initial infection
    4. Good health practices such as getting adequate rest and maintaining good nutrition will prevent recurrences

81. The nurse prepares a client who has had coronary bypass surgery for discharge by teaching that there will be:
    1. No further drainage from the incisions after hospitalization
    2. Little incisional pain and tenderness 3 to 4 weeks after surgery
    3. A mild fever and extreme fatigue for several weeks after surgery
    4. Increased edema in the leg used for the donor graft with progressive activity

82. The nurse should observe a client with the diagnosis of bulimia nervosa for symptoms of:
    1. Weight gain
    2. Dehydration
    3. Hyperactivity
    4. Hyperglycemia

83. When discussing the frequency of tub baths with an elderly client, the prime factor that the nurse should consider is the:
    1. Condition of the client's skin
    2. Client's ability to provide self-care
    3. Client's history of allergic reactions
    4. Degree of the client's orientation to the environment

84. The nurse teaches a new mother how to clean and disinfect the base of her infant's umbilical cord. The nurse explains to the mother that the cord stump is a potential source of infection because:
   1. Wharton's jelly is no longer present
   2. It contains necrotic tissue and blood
   3. It is touched by blankets and clothing
   4. Newborns have no resistance to infection

85. The behavior that would indicate that nursing interventions have been effective for a 6-year-old male, hyperactive child with an attention deficit disorder would be that the child:
   1. Is not inhibited by rules or routines
   2. Enjoys playing with his cars by himself
   3. Is no longer enuretic during the nighttime
   4. Has an increased attention span in school

86. The nurse knows that the most important aspect of the preoperative care of a child with Wilms' tumor is:
   1. Checking the size of the child's liver
   2. Monitoring the child's blood pressure
   3. Maintaining the child in a prone position
   4. Collecting the child's urine for culture and sensitivity

87. The nurse's primary role in informed consent for surgery is one of client advocate. The nurse should recognize that in regard to informed consent:
   1. As a witness, the nurse must be present to see the client actually sign the form
   2. A consent is only valid for 7 days; if a longer time passes, a new form must be signed
   3. Clients need to know about the procedure; they need not be frightened with information about potential risks
   4. Either a nurse or a physician can give the client the necessary information to obtain informed consent for surgery

88. When caring for a client with a chest tube in place, the nurse should plan to:
   1. Administer cough suppressants at appropriate intervals as ordered
   2. Empty and measure the drainage in the collection chamber each shift
   3. Apply clamps below the insertion site whenever getting the client out of bed
   4. Encourage coughing, deep breathing, and ROM to the arm on the affected side

89. Twelve months following the traumatic death of a spouse, a client comes to the mental health clinic complaining of continuing depression. The client states, "I have not been seeing any of my friends or attending any of the activities I previously enjoyed. My married children live in another state, and I rarely have any contact with them." The most accurate nursing diagnosis for this client would be:
   1. Impaired verbal communication related to social isolation
   2. Ineffective family coping related to separation from children
   3. Ineffective individual coping related to low motivation to resume activities of daily living
   4. Dysfunctional grieving related to difficulty reestablishing life-style following the death of a spouse

90. The nurse in a family planning center knows that a client understands the discussion about the use of a diaphragm when the client states, "After intercourse, a diaphragm must be left in place for:
   1. 1 to 2 hours."
   2. 6 to 8 hours."
   3. 10 to 12 hours."
   4. 16 to 20 hours."

91. A client with a severe depression that has not responded to any of the antidepressant medications is to start electroconvulsive therapy (ECT) and asks how the treatment will help. The nurse should reply:
   1. "The important thing is it will help you."
   2. "It changes your usual behavior patterns."
   3. "Why don't you ask your physician to explain it better."
   4. "It works on chemical substances in the brain, and it does help."

92. When talking with a depressed client, the statement that would give the nurse a clue that the client may attempt suicide would be:
   1. "I don't feel too good today."
   2. "I feel much better; today is a lovely day."
   3. "I'm very tired today, and I'd like to be alone."
   4. "I feel a little better, but it probably will not last."

93. A client's angina is usually very intense. To hasten the absorption of the nitroglycerin tablet, the nurse should instruct the client to:
   1. Take the tablet with a glass of warm water
   2. Move the tablet around with the tip of the tongue
   3. Break up the tablet before placing it under the tongue
   4. Swallow saliva before placing the tablet under the tongue

94. Focusing attention on a group of parents who may be at high risk for child abuse is an example of:

1. Legal prevention
2. Tertiary prevention
3. Primary prevention
4. Secondary prevention

95. To answer client questions about fluid and electrolyte imbalances, the nurse must understand that in the human body, where a semipermeable membrane divides two solutions, the solution with the greatest number of particles:
1. Draws water in its direction
2. Draws particles in its direction
3. Does not contribute particles to the other side
4. Gives up water to the side with fewer particles

96. A client with severe hypertension and dependent edema is admitted to the hospital. The physician orders furosemide (Lasix) 20 mg IV push. When administering the Lasix, the nurse should explain that the client:
1. May have problems with hearing
2. Will need to void in 5 to 10 minutes
3. Will feel some burning at the IV site
4. Should drink 240 ml of orange juice

97. If all of the following beds are available on a pediatric unit, a nurse should plan to place a child admitted with meningitis in:
1. A semiprivate room in the middle of the unit
2. A corner of a four-bed room next to the nurses' station
3. A private room two doors away from the nurses' station
4. An isolation room away from activity at the far end of the unit

98. When a client is receiving dexamethasone (Decadron Phosphate), the nurse should plan nursing care considering the client's predisposition to:
1. Infection
2. Weight gain
3. Hypotension
4. Urinary stasis

99. A preterm newborn in the neonatal intensive care unit experiences occasional periods of apnea and is placed on an apnea monitor. If the apnea monitor sounds, the nurse's initial action should be to:
1. Provide oxygen by hood
2. Suction the nasopharynx
3. Provide light tactile stimulation
4. Institute resuscitative measures

100. The supervised activity that would be most therapeutic for a client during the early phase of hospitalization for a manic episode of a bipolar disorder would be:

1. Joining a brief swimming competition
2. Writing letters and doing a needlepoint
3. Playing a board game with another client
4. Walking around the hospital grounds with one of the nurses

101. A 3-year-old is admitted to the hospital for chelation therapy for lead poisoning. The child's mother asks the nurse how the child got lead poisoning. The nurse bases the response on the knowledge that this problem in children is:
1. Clearly related to the child's ingestion of nonfood substances
2. Attributed to an indigent and passive mother who fails to supervise them
3. Considered to be environmental, with lead available for oral exploration by children who are unsupervised
4. Associated with high-risk groups including those with pica and those exposed to hazards present in the environment

102. An intravenous solution of lactated Ringer's is ordered to replace the T-tube output of a client who has had a choledochostomy. To evaluate whether the solution was therapeutically effective, the nurse should observe the client for symptoms associated with:
1. Urinary stasis
2. A paralytic ileus
3. Metabolic acidosis
4. An increased potassium level

103. A mother whose child is born with a talipes equinovarus (clubfoot) tells the nurse that she would be afraid to have more children because they might have the same problem. The best intervention by the nurse would be to:
1. Discuss the certainty of the defect occurring in later children
2. Explore with her the environmental and hereditary factors involved
3. Reassure her that this is unlikely to happen again in the same family
4. Explain that future children will have a 1 in 4 chance of having the deformity

104. A 65-year-old retiree with prostatic carcinoma is scheduled for a transurethral prostatectomy. As he is being admitted to the surgical unit, he tells the nurse he is concerned that his operation will result in impotence. The nurse's best reply would be:
    1. "I can understand why you are worried; it is a very real possibility."
    2. "I can understand your concern, but this operation does not usually cause impotence."
    3. "Most men worry about their ability to function. You should speak with your physician."
    4. "You may be impotent for a while, but normal functioning will probably return within 5 months."

105. When developing a plan of care for a female client with an obsessive-compulsive personality disorder, the nurse should be aware that the client's anxiety level would be increased if the staff:
    1. Helped her to understand the nature of her anxiety
    2. Provided a nonjudgmental and accepting environment
    3. Involved her in establishing and implementing the therapeutic plan
    4. Permitted her ritualistic behaviors to be carried out at three set times during the day

106. After surgery for a brain tumor, the client is receiving IV mannitol (Osmitrol). During and after its administration, the nurse should assess the client for:
    1. Respiratory failure
    2. Cardiac dysrhythmias
    3. A decrease in the heart rate
    4. A rebound increase in intracranial pressure

107. On admission the nurse should assess a client with a possible spinal injury for signs of spinal shock, which include:
    1. Tachycardia
    2. Hypotension
    3. Spastic paralysis
    4. Increased pulse pressure

108. A client and her 2-week-old newborn are being seen by the visiting nurse at home. When the nurse arrives, the client appears exhausted and the baby is crying. The most appropriate comment by the nurse would be:
    1. "Oh, you're having a terrible day."
    2. "Tell me a little about your daily routine."
    3. "Is everything OK? You look exhausted."
    4. "When did the baby have the last feeding?"

109. A client with type 1 diabetes is found unconscious having Kussmaul respirations, an acetone odor to the breath, and dry, hot, flushed skin. The nurse suspects that the client is experiencing:
    1. Diabetic ketoacidosis
    2. A hypoglycemic reaction
    3. The Somogyi phenomenon
    4. Hyperosmolar, nonketotic coma

110. A client is diagnosed as having type 2 diabetes. The nurse should expect the client's fasting blood glucose level to be:
    1. 30 mg/100 ml
    2. 60 mg/100 ml
    3. 110 mg/100 ml
    4. 160 mg/100 ml

111. A young adult client who has a history of psychiatric problems and has been diagnosed as having an antisocial personality disorder is admitted to the hospital. The nurse should recognize that this client has probably had a long history of:
    1. Sexual aberrations
    2. Interpersonal difficulties
    3. Stringent parental discipline
    4. Diminished contact with reality

112. After delivery, a neonate is given an injection of vitamin K. The nurse knows that this is done to:
    1. Improve the absorption of biliary salts
    2. Promote liver formation of clotting factors
    3. Prolong the prothrombin time of the infant
    4. Replace necessary bacteria in the intestine

113. During the termination phase of a therapeutic relationship, the client misses a series of appointments without any explanation. The nurse should:
    1. Terminate the relationship immediately
    2. Explore personal feelings with the supervisor
    3. Contact the client to encourage another session
    4. Attend the remaining designated meetings and wait

114. A client with 36% of the body surface burned is receiving hydrotherapy. The best approach to wound care is to:
    1. Use a consistent approach to care and encourage the client's participation
    2. Prepare equipment while doing the procedure and explain interventions to the client
    3. Change staff every 4 to 5 days and have the client select the time for the procedure to be done
    4. Heat the water to 102° F to prevent loss of body temperature and prepare the equipment before starting

115. On admission to the mental health unit, a client with the diagnosis of schizophrenia, paranoid type, says, "They all are trying to kill me. They all are." The nurse's best response would be:
    1. "No one wants to hurt you."
    2. "We are here to protect you."
    3. "You are having very frightening thoughts."
    4. "Tell me more about their wanting to kill you."

116. A nurse performs Leopold's maneuvers on a client who has been admitted to the hospital in active labor. The examination reveals a soft, irregular, rounded mass at the fundus; a firm, smooth convex curvature on the right side of the abdomen; and bumpy projections on the left side. The fetal position is:
    1. LSP
    2. LOP
    3. ROA
    4. RScA

117. A major complication of hypertensive disease associated with pregnancy that the nurse should anticipate is:
    1. Placenta previa
    2. Polyhydramnios
    3. Isoimmunization
    4. Abruptio placentae

118. A 2-year-old child with cerebral palsy is admitted to the hospital for orthopedic surgery. During the admission interview the nurse should:
    1. Tell the mother that her child will be put in a private room
    2. Let the mother know at what hours she will be permitted to visit
    3. Explain to the mother that the hospital will provide a pureed diet
    4. Ask the mother about the therapy program used by the family at home

119. The nurse can determine that peristalsis has returned in a client who has had a gastrectomy when:
    1. Borborygmi are auscultated
    2. The feeling of nausea passes
    3. The client has a bowel movement
    4. The abdomen is no longer rigid and tender

120. An 8-year-old is being discharged after recovery from an episode of sickle cell pain. The nurse teaches the parents all the "do's and don'ts" concerning the child's care. The nurse is satisfied that the parents understand the principles of care when they state that they are:
    1. Not allowing the child to play with other children

    2. Keeping the child's fluid intake restricted at night
    3. Getting the child a private tutor to help with school work
    4. Not permitting the child to play soccer or go backpacking with the Scouts

121. Sterile warm saline soaks tid are ordered for a client with cellulitis from a puncture wound. A staff nurse has placed a clean basin, washcloth, and protective pad at the bedside in preparation for the soak but is unable to continue the procedure. The nurse assigned to complete the soak should:
    1. Continue the procedure as started
    2. Collect new supplies before starting
    3. Report the first nurse to the supervisor
    4. Discuss the type of soak with the physician

122. A client was hospitalized because of an inability to walk. After a physiologic basis for the problem was ruled out, a diagnosis of somatoform disorder, conversion type, was made. The nurse realizes that the client's paralysis is a:
    1. Nondisabling illness
    2. Way of getting attention
    3. Loss of contact with reality
    4. Result of intrapsychic conflict

123. The diagnosis of placenta previa indicates to the nurse that the placenta is:
    1. Infarcted
    2. Low-lying
    3. Immaturely developed
    4. Separating prematurely

124. A client is scheduled for peritoneal dialysis. When explaining this procedure to the client, the nurse should review that its purpose is to:
    1. Help do some of the work usually done by the kidneys
    2. Prevent the development of complicating heart problems
    3. Remove bad chemicals from the body so the condition will not worsen
    4. Speed recovery because the kidneys are not responding to other therapy

125. An early symptom that the nurse commonly observes in newborns who are later diagnosed as having cystic fibrosis is:
    1. Meconium ileus
    2. Imperforate anus
    3. Rapid respiratory rate
    4. Elevated bilirubin level

126. The verbalizations of a client with schizophrenia contain many words that sound like the client is speaking a different language. The nurse recognizes that this communication is an example of:
    1. Clanging
    2. Echolalia
    3. Echopraxia
    4. Neologisms

127. The physician prescribes sulfamethoxazole and trimethoprim (Septra) for a sexually active female who has been experiencing frequency and burning on urination for 24 hours. The nurse knows the teaching about this medication was understood when the client says she will:
    1. Limit her fluid intake
    2. Strain all urine for calculi
    3. Replace oral contraceptives with other methods
    4. Notify the physician if the urine becomes orange

128. During a precipitous labor, the nurse remembers that a complication associated with rapid descent of the fetus is:
    1. Microcephaly
    2. Fetal head trauma
    3. Fracture of the coccyx bone
    4. Prolonged retention of the placenta

129. In the immediate postoperative period after coronary bypass surgery, the nurse should be particularly alert for the common complication of:
    1. Graft closure with recurrence of angina-like chest pain
    2. Supraventricular dysrhythmias, especially atrial fibrillation
    3. Postpericardiotomy syndrome with fever and audible friction rub
    4. Elevation of hemoglobin and hematocrit levels with risk of embolization

130. A primary component of the nursing care plan for a client with bulimia nervosa would be:
    1. Intake and output
    2. Daily weighing before eating
    3. Careful observation after meals
    4. Daily room search for hoarded food

131. A client with a history of binge eating and purging is admitted to the eating disorder unit with a diagnosis of bulimia nervosa. The nurse is aware that bulimia nervosa can best be defined as the:
    1. Uncontrollable pilfering and hoarding of food for later consumption
    2. Uncontrollable ingestion of large quantities of food in a short period
    3. Refusal to eat in public and engaging in private, excessive overeating

    4. Mood swings, ranging from euphoria to depression, associated with food

132. A child, who is newly diagnosed with idiopathic scoliosis, has a mild structural curve. The child's mother asks whether the problem can be corrected with exercise and is told by the nurse that an exercise program will be:
    1. Used in conjunction with a Milwaukee brace
    2. Employed if the child appears highly motivated
    3. Avoided because it can exaggerate the curvature
    4. The only intervention needed to correct the curvature

133. A client with an acute exacerbation of rheumatoid arthritis is in severe pain and tells the nurse, "The only time I am pain free is when I lie perfectly still." The nurse should explain that the client needs to exercise every day to prevent:
    1. Paresthesias of the feet
    2. Osteoblastic development
    3. Shortening of the muscles
    4. Loss of muscular coordination

134. By the third day of life an infant weighs 5% less than at birth. The nurse is aware that the weight loss most likely results from:
    1. The development of sepsis
    2. An obstructive gastrointestinal anomaly
    3. Generalized muscle response to stimulation
    4. An imbalance between nutrient intake and fluid loss

135. An obese client becomes pregnant and continues to gain excessive weight during pregnancy. The nurse understands that treating a client's obesity during pregnancy is not advisable because:
    1. Weight loss can best be achieved after pregnancy
    2. Additional calories are needed for her increased activity
    3. Calcium utilization varies proportionately with calorie intake
    4. Protein is utilized for energy when carbohydrates are decreased

136. A primigravida is admitted to the hospital in active labor. The fetus is in a breech position. During this client's labor the nurse may most likely note:
    1. Heavy bleeding from the vagina
    2. Irregularities of the fetal heart rate
    3. Meconium staining of the amniotic fluid
    4. Decelerations at the end of contractions

137. The nurse begins planning for the discharge of a client who had a cerebrovascular accident with residual hemiparesis and hemianopia. Information that the nurse should include in the education program is the:
1. Importance of bed rest at home
2. Use of oxygen therapy at home
3. Importance of a safe environment
4. Need to decrease protein in the diet

138. The nurse is aware that a preschool child views death:
1. As permanent and irreversible
2. In a frightening and horrible way
3. As a departure from which the person returns
4. Without comprehending its meaning in any way

139. A 4-year-old is admitted to the hospital with the diagnosis of Wilms' tumor. The nurse should place a sign on the child's bed that states:
1. Strain all urine
2. No IV medications
3. Do not use diapers
4. Do not palpate abdomen

140. A malnourished client with a history of cirrhosis is admitted to the hospital with nausea, ascites, and gastrointestinal bleeding. The nurse recognizes that the ascites is primarily a result of the client's malnourished state, especially the lack of adequate:
1. Iron to prevent proper hemoglobin synthesis
2. Vitamins to maintain cell coenzyme functions
3. Sodium to maintain its proper concentration in tissue fluid
4. Plasma protein to maintain proper capillary-tissue circulation

141. One morning during the working phase of a therapeutic relationship, the client suddenly becomes very hostile. The most appropriate interpretation of this behavior is that:
1. The client is exercising assertiveness, which implies improvement
2. Hostility is being used as a defense because the nurse has come too close
3. Flare-ups often occur when the nurse and client have a good working relationship
4. This behavior is a form of regression and implies some deterioration in the client's condition

142. Nursing care for a child admitted to the hospital with acute glomerulonephritis should be directed toward:
1. Pushing fluids
2. Promoting diuresis
3. Enforcing strict bed rest
4. Eliminating sodium from the diet

143. A 5-month-old is brought to the pediatric clinic because of probable exposure to an adolescent sibling with measles. The infant's mother anxiously asks the clinic nurse whether the baby can be vaccinated against measles at this age. The nurse's reply should take into consideration the history of:
1. Immunization of this baby
2. Previous viral illnesses of this baby
3. Preexisting tuberculosis of the mother
4. Maternal diseases and immunizations

144. After attaching a cardiac monitor to a newly admitted client, the nurse notes six premature ventricular beats (PVBs) per minute. The nurse, following established protocols, should first:
1. Obtain an orthostatic blood pressure reading
2. Encourage the client to cough and deep breathe
3. Initiate cardiopulmonary resuscitation on the client
4. Administer an intravenous drip of lidocaine (Xylocaine)

145. A client who is a poorly controlled type 1 diabetic in her 34th week of pregnancy has been told by her physician that she will have an amniocentesis at 37 weeks to assess fetal lung maturity. Induction of labor will be initiated if the fetus' lungs are mature. The client asks the nurse why an early delivery may be necessary. The best initial reply by the nurse would be:
1. "You need to deliver early before the fetus gets too big to fit through the birth canal."
2. "After 36 weeks there is an increased risk of stillbirths despite reassuring fetal monitoring."
3. "You need to deliver early so that you do not develop pregnancy-induced hypertension."
4. "Early delivery will reduce the chance of your infant developing hypoglycemia."

146. A 23-year-old male is admitted to the surgical unit with superficial wounds of both wrists as the result of an abortive suicide attempt. When the nurse enters the client's room, he states, "I suppose you're going to ask me about my suicide attempt!" The nurse's best response would be:
1. "Do you want to talk about it?"
2. "Tell me how you feel about it."
3. "It's best not to dwell on it right now."
4. "Why do you think I'd ask you about it?"

147. A client is placed on a 1500-calorie diet. The client should be taught that 1 gram of carbohydrate contains:
    1. 2 calories
    2. 4 calories
    3. 9 calories
    4. 12 calories

148. A 22-year-old client with an antisocial personality disorder is being discharged after an abortive suicide attempt and is to continue psychotherapy on an outpatient basis. When evaluating chances for improvement, the nurse recognizes that:
    1. The client's prognosis for adjusting to a limited life-style is excellent
    2. The client will not change unless the client's parents are willing to set and keep firm limits
    3. The client requires intensive psychotherapy along with an anxiolytic drug to produce a remission
    4. The client will not change unless there is a readiness to accept the pain associated with change

149. After a normal vaginal delivery, the nurse should plan the postpartum care of a Class 1 cardiac client based on the knowledge that:
    1. The client should increase her fluid intake, particularly if she is breast-feeding
    2. The client is out of immediate danger because the stress of pregnancy is over
    3. Clients with cardiac problems should be kept on bed rest for a minimum of 7 days
    4. The first 24 hours postpartum are the most stressful on the cardiopulmonary system

150. A child with leukemia is receiving vincristine. Nursing care for this child includes checking and recording bowel sounds bid. This action is necessary because a side effect of this drug is:
    1. Decreased innervation of the GI tract
    2. Hyperactivity of the bowel from diarrhea
    3. Nausea that decreases fluid intake and produces constipation
    4. Increased antigen/antibody reactions, which cause edematous bowels

151. A client with lymphosarcoma is receiving allopurinol (Zyloprim) and methotrexate. The nurse can help the client prevent complications related to uric acid nephropathy by administering the:
    1. Allopurinol and promoting urine acidity
    2. Methotrexate after providing an antacid
    3. Allopurinol and encouraging the intake of fluid
    4. Methotrexate and restricting the intake of fluid

152. A client who had an above-the-knee amputation has an elastic bandage around the stump. The physician's orders include bathing the stump daily and rewrapping the elastic bandage prn. In regard to the bandage, the nurse should:
    1. Reapply it only if it slips off
    2. Apply it tightly so that it does not slip off
    3. Apply it smoothly without wrinkles or creases
    4. Reapply it while the stump is in the dependent position

153. A client receiving levodopa experiences orthostatic hypotension, a side effect of drug therapy. To help limit this adaptation, the nurse should:
    1. Initiate gait training
    2. Apply elastic stockings
    3. Withhold the next dose
    4. Increase the fluid intake

154. A pregnant client is scheduled for an ultrasound 8 weeks after her last menstrual period. When preparing the client for the sonography the nurse should tell her to:
    1. Avoid eating for 8 hours before the test
    2. Empty her bladder before the ultrasound
    3. Force fluids for 1 hour before the ultrasound
    4. Take a laxative the night before the sonogram

155. Before irrigating a client's nasogastric tube, the nurse must first:
    1. Assess breath sounds
    2. Instill 15 ml of normal saline
    3. Auscultate for bowel sounds
    4. Check the tube for placement

156. When helping to break the vicious cycle of fear, dyspnea, and avoidance of activities commonly found in clients with chronic obstructive pulmonary disease (COPD), the nurse should place primary emphasis in the teaching program on:
    1. Learning to control or prevent respiratory infections
    2. Education about the disease and breathing exercises
    3. Teaching about priorities in carrying out daily activities
    4. Judicious use of aerosol therapy, especially nebulizers

157. While caring for a client during a manic episode of a bipolar disorder, the priority in planning care would be:
    1. Arranging for the client to participate in a daily discussion group
    2. Encouraging frequent close contact with the staff and a few selected clients
    3. Reducing the number of people interacting with the client during any given day

    4. Varying the staff assigned to care for the client to reduce the opportunity for manipulation

158. When assessing the rate of involution of a client's uterus on the second postpartum day, the nurse palpating the fundus would expect it to be:
1. At the level of the umbilicus
2. One fingerbreadth above the umbilicus
3. One to two fingerbreadths below the umbilicus
4. Three to four fingerbreadths below the umbilicus

159. One evening, an elderly client with the diagnosis of dementia chokes on a piece of food and becomes panicky and cyanotic. The nurse performs the abdominal thrust maneuver and a wad of food pops out of the client's mouth. After several deep respirations, the client's cyanosis passes. It would be most appropriate at this time for the nurse to:
1. Inform the client that everything is fine
2. Stand the client up and check the pulse
3. Provide psychologic support to the client
4. Teach the client how to prevent future problems

160. A single teenage mother brings her 5-year-old daughter to the emergency department with a broken arm and contusions. It is decided that the child is a victim of abuse. The nurse suspects abuse because of the mother's:
1. Assertive and dominating personality
2. Effective use of defense mechanisms
3. Isolation and lack of emotional support
4. Lack of knowledge about child development

161. Several days after a client has had a total laryngectomy, the physician orders a progressive diet as tolerated. The nurse should:
1. Keep a suction apparatus readily available in case aspiration occurs
2. Administer the diet through a nasogastric tube until the suture line heals
3. Encourage intake of pureed foods because they promote the swallowing reflex
4. Administer pain medication as ordered 30 minutes before meals to limit discomfort

162. For a few days following the repair of a cleft lip, an infant should be fed by:
1. Nasogastric tube
2. Intravenous infusion
3. A rubber-tipped syringe
4. A bottle with a preemie nipple

163. The nurse suspects that a preterm baby, who is receiving gastric feedings, may be developing necrotizing enterocolitis (NEC) when the baby:
1. Develops persistent diarrhea
2. Vomits 5 ml after a feeding of 1 ounce or more
3. Exhibits a decreasing abdominal circumference
4. Has increasing nasogastric residual from prior feeding

164. A nurse assesses a client who is experiencing a crisis and formulates a nursing diagnosis. The nursing diagnosis that is made could be defined as a statement:
1. To explain the client's present needs
2. That describes the client's health status and related factors
3. Of the client's responses that are within the scope of nursing
4. That rewords the client's medical diagnosis in nursing terminology

165. Gamma globulin and aspirin are ordered for a 3-year-old with Kawasaki disease. When administering the aspirin to the child the nurse should:
1. Disguise the medication in the child's favorite food
2. Let the child decide how the medication will be taken
3. Offer the child a reward for taking the prescribed medication
4. Not respond to questions about the taste of the drug to avoid lying

166. A client is in labor. When her cervix is 3 to 4 cm dilated and 60% effaced and the vertex is at -1 station, the client has a sudden spurt of dark blood from the vagina. The uterus is irritable on palpation and does not relax well between contractions. The nurse should immediately:
1. Transport her to the delivery room without delay
2. Perform a vaginal examination to determine dilation
3. Monitor FHR, check her BP, and assess uterine activity
4. Change the client's underpad and position her on her back

167. A 50-year-old male client has difficulty communicating because of expressive aphasia following a cerebrovascular accident. When the nurse asks the client how he is feeling, his wife answers for him. The nurse should:
1. Ask the wife how she knows how the client feels
2. Instruct the wife to let the client answer for himself
3. Acknowledge the wife but look at the client for a response
4. Return later to speak to the client after the wife has gone home

168. The physician orders dexamethasone (Decadron Phosphate) for a client following a craniotomy for a brain tumor. The nurse recognizes that the expected response to this drug should be:
    1. Reduced cerebral edema
    2. Reduced cell proliferation
    3. Increased renal reabsorption
    4. Increased response to sedation

169. The physician orders aspirin therapy to be continued at home for a client with severe arthritis. When teaching regarding aspirin intake, the nurse should emphasize that the client should:
    1. Switch to Tylenol if tinnitus occurs
    2. See the dentist if bleeding gums develop
    3. Avoid spicy foods while taking the medication
    4. Take the medicine on a full stomach or with meals

170. A 10-year-old child who has head lice tells the school nurse, "My mother said I got lice because I don't keep myself clean." The nurse's best reply to this statement would be:
    1. "You feel that your mother is putting you down?"
    2. "You have problems getting along with your mother?"
    3. "There is no relationship between cleanliness and lice."
    4. "Lice are more common if you have poor personal hygiene."

171. Prior to obtaining informed consent for cardiac bypass surgery from a client, the nurse should:
    1. Explain to the client the risks involved in the surgery
    2. Explain to the client that obtaining the signature is routine for surgery
    3. Evaluate whether the client's knowledge level is sufficient to give consent
    4. Witness the client's signature, since this is what the nurse's signature documents

172. One of the adaptations associated with toxoplasmosis that the nurse might observe when assessing an infant affected with this disease is:
    1. Mongolian spots
    2. Caput succedaneum
    3. Irregular respirations when the infant is awake
    4. Chest circumference 1 to 2 cm larger than the head circumference

173. A client who has repeated episodes of cystitis is scheduled for a cystoscopy to determine the possibility of urinary abnormalities. In answer to the client's questions, the nurse describes the procedure as:

    1. A computerized scan that clearly outlines the bladder and surrounding tissue
    2. An x-ray film of the abdomen, kidneys, ureters, and bladder after administration of dye
    3. The visualization of the urinary tract through ureteral catheterization using a radiopaque material
    4. The visualization of the inside of the bladder with an instrument connected to a source of illumination

174. The nurse is aware that the most appropriate toy for a 2½-year-old would be a:
    1. Plastic mirror
    2. Set of nesting blocks
    3. Colorful hanging mobile
    4. Wooden puzzle with large pieces

175. A client with cirrhosis of the liver and ascites fails to respond to chlorothiazide, and spironolactone (Aldactone) is prescribed in addition to the chlorothiazide. The advantage of Aldactone is that it:
    1. Stimulates H_2O excretion
    2. Stimulates sodium excretion
    3. Helps prevent potassium loss
    4. Reduces arterial blood pressure

176. After her 6-week-old baby's surgery for pyloric stenosis, the mother is reluctant to resume feeding the baby but readily assists with all other aspects of infant care. The care plan for this infant specifically states that the mother should be encouraged to participate in the oral feedings. The rationale for the focus of this nursing care is that the mother is:
    1. Unaware that family involvement in care is permitted
    2. Uncertain how the thickened oral feedings are administered
    3. Reluctant to feed the baby because of the present need to use a special nipple
    4. Probably afraid to resume feedings because of the frequent vomiting before surgery

177. The nurse can best initially prepare a client for the termination of their relationship by:
    1. Periodically summarizing the client's progress during the working phase
    2. Stating that, if the client feels it is necessary, their meetings could be extended
    3. Telling the client how long they will be working together during their first meeting
    4. Waiting until the termination phase, then reminding the client periodically of the duration of their meetings

178. The nurse would know that a client who had a mastectomy understands the discharge teaching when she states she will report to the physician any:
1. Persistent itching around the incision
2. Swelling and erythema around the incision
3. Slightly irregular-appearing skin around the incision
4. Decreased sensations in the area around the incision

179. To assist a client with Bell's palsy, it would be most appropriate for the nurse to:
1. Prepare the client for surgery
2. Tape the affected eyelid open
3. Teach facial exercises to the client
4. Record the symmetrical progression of the paralysis

180. When obtaining a health history of a 7-year-old admitted to the hospital with acute glomerulonephritis, the nurse would expect the child's mother to report that:
1. The child had a sore throat 3 weeks ago
2. The child had just gotten over the measles
3. The child's father has a history of urinary infections
4. The child's immunizations for camp were completed last week

181. A 24-year-old client who has had type 1 diabetes for the past 6 years is concerned about how her pregnancy will affect her diabetes, especially her diet and insulin needs. The nurse should inform the client that pregnancy:
1. Decreases the need for insulin because the excess glucose will be used by the fetus for growth
2. Is a normal state, and that a revised diet and her usual insulin regimen will meet her and the fetus' needs
3. Increases the need for protein, and adjustments have to be made to meet the need for increasing doses of insulin
4. Will affect her diet and insulin needs, and she will have to monitor her blood glucose more often to make appropriate adjustments

182. A child with asthma is being discharged from the hospital on oral theophylline. The nurse recognizes that the parents understand the discharge teaching when they tell the nurse that they will monitor the child for the side effect of:
1. Apneic episodes
2. Frequent urination
3. Nausea and vomiting
4. Spasmodic hiccoughs

183. The primary difficulty in establishing long-range goals for clients with an antisocial personality disorder is related to the fact that such clients are usually:
1. Reluctant to change their life-style
2. Unwilling to accept their need for help
3. Unable to deal with their feelings of guilt
4. Resistant to any demonstration of feeling

184. A client with type 1 diabetes is taught the symptoms associated with hypoglycemia. The nurse would know that the teaching was understood when the client states, "I will know I am having an insulin reaction when I experience:
1. Thirst, excessive urination, and blurred vision."
2. Confusion, vomiting, and rapid deep breathing."
3. Generalized nervousness, headache, and perspiration."
4. Abdominal pain and nausea and have a fruity odor to my breath."

185. A 32-year-old client who had rheumatic fever as a child is now 34 weeks pregnant. She is classified as having Class 1 heart disease. This client will experience some symptoms during her pregnancy that are related to her cardiac disease. Of the symptoms that she may experience during her pregnancy, the one that is most likely due to her cardiac disease is:
1. Tachycardia
2. Dyspnea at rest
3. Progressive dependent edema
4. Shortness of breath on exertion

186. A 5-year-old boy with suspected leukemia is to have a bone marrow aspiration. Before the procedure the child should be told that:
1. He will be put to sleep and will not feel anything
2. After the test is over, he will have to stay in bed until supper
3. He will feel a little pressure, and when it's over a bandage will be applied
4. He will have a few stitches at the site, but he can get up and go to the playroom

187. A client is scheduled for an ileostomy. Preoperative teaching should include a statement that:
1. Skin irritation at the stoma site can occur easily
2. Fecal matter from the stoma can be controlled by daily irrigation
3. Regular bowel habits can be established within a few weeks postoperatively
4. Effluent discharge from the stoma will be formed fecal matter if the diet is regulated

188. A client with a seizure disorder is receiving phenytoin (Dilantin) and phenobarbital. The nurse is confident that the instructions regarding the medications are understood when the client states:
    1. "I will not have any seizures with these medications."
    2. "I must continue these medicines to prevent falls and injury."
    3. "Stopping the drugs can cause continuous seizures and I may die."
    4. "By my staying on the medicines I will prevent postseizure confusion."

189. Following a resection of the colon, a client returns from the operating room with a nasogastric tube in place. The nurse understands that the purpose of this tube is to:
    1. Monitor the acidity of the gastric secretions
    2. Provide a route for liquid tube feedings when possible
    3. Permit continuous decompression of the large intestine
    4. Remove fluids and gas from the upper gastrointestinal tract

190. While undergoing chemotherapy with doxorubicin (Adriamycin) for Hodgkin's disease, it is most essential for the client to immediately report to the nurse the presence of:
    1. Nausea
    2. Constipation
    3. A sore throat
    4. A loss of hair

191. A client with a chest tube is to be transported via a stretcher. When transporting the client, the nurse should keep the:
    1. Collection device attached to mechanical suction
    2. Collection device below the level of the client's chest
    3. End of the chest tube covered with sterile 4 × 4s securely taped
    4. Chest tube clamped between the water-seal chamber and the client

192. If antithyroid medication is not effective in decreasing the symptoms associated with hyperthyroidism, an additional health problem may develop as a result of continued tachycardia. The nurse should tell the client taking an antithyroid medication to immediately report the development of:
    1. Nervousness and irritability
    2. Weight gain or pedal edema
    3. Flushed skin and diaphoresis
    4. Changes in appetite or bowel habits

193. Studies have shown that the grieving process may last longer for people who have:
    1. Feelings of guilt
    2. Failed to remarry within 3 years
    3. Ambivalent feelings about the death
    4. Had a close relationship with their family

194. A client with a ruptured tubal pregnancy is experiencing sharp, shooting pains in the right lower quadrant of the abdomen. There is also vaginal spotting. In view of the symptoms, the nurse should expect to prepare the client for a:
    1. Myomectomy
    2. Hysterectomy
    3. Salpingectomy
    4. Dilation and curettage

195. A couple interested in family planning ask the nurse about the cervical mucus method of family planning. The nurse explains that with this method the couple who do not wish to become pregnant must avoid intercourse when the cervical mucus is:
    1. Clear and thick
    2. Yellow and thin
    3. Cloudy and viscid
    4. Clear and stretchable

196. Following an ECT treatment, a client complains of loss of memory. The nurse should reply:
    1. "I will help you try to remember when the treatments are over."
    2. "It is better if you forget what happened before you became ill."
    3. "The fact that you are getting well is the most important thing for you right now."
    4. "This is only temporary; your memory will return after the therapy is completed."

197. When caring for clients who are at risk for suicide, the nurse should be aware that:
    1. A client who fails in a suicide attempt will probably not try again
    2. It is best not to talk to clients about suicide because it may give them the idea
    3. The more formalized the plan, the greater the possibility that the client will attempt suicide
    4. Clients who talk about suicide never commit suicide; they just use the threat to gain attention

198. A wife, after visiting her husband who has been hospitalized with chest pain, says to the nurse, "He looks so pale." The nurse should reply:
    1. "Paleness is expected with heart problems."
    2. "You both must be terribly frightened by this."
    3. "I can understand why you are worried, but he will be all right."

4. "Other clients get pale and recover without any complications."

199. A client with Parkinson's disease begins carbidopa-levodopa (Sinemet). A nursing priority should be to:
1. Observe sleeping patterns weekly
2. Assess the vital signs every 4 hours
3. Perform a physical assessment daily
4. Monitor the intake and output every 8 hours

200. A client, pregnant for the first time, questions the nurse about all the changes in her body. The nurse, when explaining these changes, is aware that the most profound change of all during pregnancy occurs in the:
1. Urinary system
2. Endocrine system
3. Cardiovascular system
4. Gastrointestinal system

201. The total bile drainage from a T-tube 24 hours after gallbladder surgery is 150 ml. The nurse should:
1. Empty the drainage bag and record amount on the I&O record
2. Clamp the T-tube and drain small amounts of bile every 4 hours
3. Check the tube for kinks because the drainage is less than expected
4. Notify the physician immediately of the excessive amount of bile drainage

202. A client with asthma is receiving intravenous aminophylline. The adverse reaction for which the nurse should observe is:
1. Oliguria
2. Bradycardia
3. Hypotension
4. Hypoventilation

203. Chelation therapy injections cause local discomfort. To relieve some of this discomfort the nurse should:
1. Apply warm soaks to the affected area
2. Give the child a cool tub bath after each injection
3. Assist the child to ambulate immediately after each injection
4. Vigorously massage the affected injection site with an alcohol swab

204. A nurse observes a new mother timidly approach her preterm son for the first time in the neonatal intensive care nursery. To aid the mother in the bonding process, the nurse should say to the mother:
1. "He'll gain weight gradually and look better."
2. "I will give you instructions on how to care for him."

3. "It must be hard to let yourself love a baby when you're not sure he'll make it."
4. "Many mothers are shocked when they first see their babies; you'll see him grow."

205. When working with clients who are spontaneously aborting a pregnancy, it is important for nurses to first deal with their own feelings about abortion, death, and loss so that they can:
1. Share personal grief with the clients
2. Allow the clients to express grief fully
3. Maintain complete control of the situation
4. Teach the clients to cope more effectively

206. The nurse assists a client who has had a cerebrovascular accident with lunch. The nurse suspects left hemianopia when the client:
1. Asks to have all food moved to the left side of the tray
2. Drops the coffee cup when trying to use the right hand
3. Ignores the food on the left side of the tray when eating
4. Complains about not being able to use the left arm to help eat

207. To meet the nutritional needs of a 5-month-old with congenital heart disease, the nurse would instruct the mother to:
1. Use 1½ strength formula
2. Feed the infant slowly, allowing for rest
3. Administer iron supplements with meals
4. Avoid giving solid food until the infant is 1 year old

208. A client delivers an infant by cesarean birth. It is suspected that the infant has Down syndrome when the initial assessment reveals:
1. Low-set ears, micrognathia, and rocker-bottom feet
2. High-pitched catlike cry, microcephaly, and low-set ears
3. Hypotonia, low-set ears, simian crease, and epicanthal folds
4. Lymphedema of the dorsum of the hands and feet, webbed neck, and widely spaced nipples

209. A client is pregnant with her first child. About the eighth or ninth week of gestation, an assessment of the client's adaptations to pregnancy would reveal:
1. Lightening
2. Quickening
3. Goodell's sign
4. Braxton Hicks contractions

210. A client who has an adenocarcinoma of the descending colon with a partial obstruction is receiving doxorubicin (Adriamycin) IV to reduce the tumor mass. The nurse should assess for signs of toxicity, which include:
1. A minor skin rash
2. A blue tinge to the urine
3. An alteration in cardiac rhythm
4. An increased feeling of nervousness

211. The first child of adolescent parents is brought to the pediatric unit with a diagnosis of nonorganic failure to thrive (NFTT). In light of this diagnosis and the family structure, the priority nursing diagnosis for this family would be:
1. Risk for injury related to mothering failure
2. Altered parenting related to knowledge deficit
3. Altered nutrition: less than body requirements related to child abuse
4. Sensory-perceptual alterations (gustatory) related to infant deprivation

212. A client with an enlarged thyroid gland is referred to the laboratory for a radioactive iodine (RAI) uptake test. When teaching about this test, the nurse should explain that:
1. The urine will not contain any radioactive particles
2. Test results will not be influenced by any medications
3. This test is one of several needed for an accurate diagnosis
4. An accurate measure of thyroid hormone levels will be obtained

213. Formulating a nursing diagnosis for all clients, including clients experiencing a crisis, occurs during the stage of the nursing process known as:
1. Analysis
2. Planning
3. Evaluation
4. Assessment

214. A client with an abdominal aortic aneurysm being prepared for surgery complains of feeling lightheaded. The client is pale and has a very rapid pulse. The nurse assesses that the client may be:
1. Hyperventilating
2. Going into shock
3. Extremely anxious
4. Developing an infection

215. Following a cleft lip repair, the nurse places elbow restraints on a 10-week-old infant boy. The mother questions the nurse about the reason for the restraints. The nurse's best response would be:
1. "They are used routinely on all children with lip surgery."
2. "The surgeon insists that all postoperative infants have them on."
3. "By keeping the arms straight, it is difficult for the hands to touch the mouth."
4. "We can't be with your child all the time to watch that the hands do not touch the mouth."

216. The mother of a 10-week-old infant demonstrates good understanding of the infant's postoperative needs following a cleft lip repair when she:
1. Allows the infant to cry for short periods
2. Cleanses the suture line after each feeding
3. Offers a pacifier when the infant becomes restless
4. Gives the feeding while the infant is in a side-lying position

217. A client who has had a total laryngectomy is using a pad and pencil to communicate. The client becomes very frustrated and writes, "When can I learn how to speak again?" The nurse's best response would be:
1. "Every client is different. It's difficult to say just how long it will be."
2. "It must be difficult for you, but be patient. These things take time."
3. "You have to give the incision time to heal before going to speech therapy."
4. "Perhaps I can arrange for a member of the Laryngectomy Club to speak with you."

218. A client is placed on a low-sodium, weight-reducing diet following a myocardial infarction. The nurse knows that the dietary teaching was effective when the client chooses:
1. Lean steak and carrots
2. Tunafish salad with celery sticks
3. Baked chicken and mashed potatoes
4. Stir-fried Chinese vegetables and rice

219. On the day after delivery, a client mentions that her nipples are becoming sore from breast-feeding. The nurse should:
1. Instruct her to apply warm compresses before she begins to breast-feed
2. Provide a breast shield for her to keep the infant's mouth off the nipples
3. Assess her breast-feeding and hygiene techniques to identify possible causes
4. Instruct her to limit breast-feeding to 5 minutes per side until the soreness subsides

220. While caring for a client who had an open reduction and internal fixation of the hip, the nurse encourages active leg and foot exercises of the unaffected leg every 2 hours. This activity will specifically help to:
1. Reduce leg discomfort
2. Maintain muscle strength
3. Prevent formation of clots
4. Limit venous inflammation

221. When providing teaching about a 2-gram sodium diet, the nurse should instruct the client to:
1. Use lemon juice to season meat
2. Refrain from eating canned fruits
3. Restrict the intake of green vegetables
4. Drink carbonated beverages instead of decaffeinated coffee

222. A child is receiving an IV of 500 ml of D5 ½NS at 55 ml/hour via a microdrip set. The infusion rate should be:
1. 55 gtt/minute
2. 60 gtt/minute
3. 9 to10 gtt/minute
4. 13 to14 gtt/minute

223. When planning care for an 85-year-old newly admitted client with the diagnosis of dementia of the Alzheimer's type, the nurse should remember that confusion in the elderly with dementia:
1. Follows transfer to new surroundings
2. Is always progressive and will get worse
3. Results from brain changes and cannot be cured
4. Is a common finding and is to be expected with aging

224. The mother of a young man suspected of having Cushing's syndrome expresses anxiety about her son's condition. To help the mother better understand the illness, the nurse should inform her that:
1. He will need to take exogenous steroids for several months
2. His condition is improving as demonstrated by his weight gain
3. His physical changes are permanent but will improve with time
4. He may have mood swings or depression as a result of his disease

225. X-ray films reveal that a client has sustained an intracapsular fracture of the left hip during a fall. The client is temporarily placed in Buck's traction. When providing care the nurse should:
1. Turn the client from side to side every 2 hours
2. Monitor the client for tenderness in the left calf
3. Put each extremity through passive range of motion

4. Raise the head of the bed to a semi-Fowler's position

226. At 5 am, 2 hours after a long labor and vaginal delivery, a client is transferred to the postpartum unit. When planning morning care for this client, the nurse's highest priority would be:
1. Arranging an individual session in which the client can learn all that is necessary about successful breast-feeding
2. Preparing for the probability of hemorrhage and assessing the client's uterus every 15 minutes for firmness and position
3. Anticipating possible safety needs and instructing the client to remain in bed and call for assistance before attempting to ambulate
4. Planning activities so that the client has time to rest and providing mild analgesics as needed and other comfort measures to facilitate this rest

227. When making rounds, a nurse finds a client experiencing a seizure. The nurse should:
1. Hyperextend the client's neck
2. Move obstacles away from the client
3. Restrain the client's head and limb movements
4. Attempt to place an airway in the client's mouth

228. A child with meningitis suddenly assumes an opisthotonic position. The nurse should place the child in a:
1. Side-lying position
2. Knee-chest position
3. High-Fowler's position
4. Trendelenburg's position

229. When caring for a client after an ileostomy, the nurse should expect drainage from the ileostomy in the first 24 to 48 postoperative hours to be:
1. Fecal with flatus
2. Bloody with clots
3. Clear with mucoid shreds
4. Mucoid and serosanguineous

230. A client with sprue develops signs of tetany. The nurse is aware that this is caused by inadequate absorption of:
1. Sodium
2. Calcium
3. Potassium
4. Phosphorus

231. The physician has ordered alternate liters of D5W and D5RL at the rate of 175 ml/hour. The drop factor of the IV set is 15 gtt/ml. The nurse should adjust the flow to provide:
1. 40 gtt/min
2. 42 gtt/min
3. 44 gtt/min
4. 46 gtt/min

232. The ritual of a female client with an obsessive-compulsive personality disorder focuses on checking her pocketbook for her keys and other belongings every 4 minutes. If prevented from doing this, she gets quite upset. The nurse should:
1. Allow her to continue to check her pocketbook for as long as she wants and wait for her to get tired
2. Keep her actively involved in projects to distract her so she will forget about checking her pocketbook
3. Lock her pocketbook in a safe place in the morning, thereby relieving the client of the necessity to check it
4. Allow the behavior, but with the client's agreement, set limits on the amount of time and the frequency of the checks

233. A client returns from surgery after an abdominal cholecystectomy for a gangrenous gallbladder. The location of the surgical site high in the abdominal cavity places the client at high risk for the postoperative complication of:
1. Atelectasis
2. Hemorrhage
3. Paralytic ileus
4. Wound infection

234. A client with chronic obstructive pulmonary disease (COPD) continues to smoke and does not perform the prescribed chest physiotherapy exercises because they cause fatigue. The nursing diagnosis that is most indicated for this client is:
1. Self-care deficit related to fatigue
2. Altered thought processes related to cerebral hypoxia
3. Noncompliance with therapeutic regimen related to nonacceptance
4. Knowledge deficit related to causal relationship of smoking to lung disease

235. Chelation therapy with Ca EDTA is started on a child with chronic lead poisoning. The nurse can evaluate the success of this therapy by monitoring for:
1. Elevated blood lead levels
2. Increased fecal excretion of lead
3. Increased urinary excretion of lead
4. Decreased deposition of lead in the bones

236. While observing a mother visiting her preterm infant son in the neonatal intensive care nursery, the nurse recognizes that the mother has still not begun the normal bonding process when the mother says:
1. "It looks like such a tiny baby."
2. "Do you think he will make it?"
3. "He looks so much like my husband."
4. "Why does he need to be in an incubator?"

237. Clinical manifestations of Cushing's syndrome that the nurse might observe in a client who is suspected of having a pituitary tumor include:
1. Retention of sodium and water
2. Hypotension and a rapid, thready pulse
3. Increased fatty deposition in the extremities
4. Hypoglycemic episodes in the early morning

238. An elderly client, whose conformance to social norms, hygiene, and dress requirements have deteriorated over the last 3 months, is admitted to the psychiatric unit. The nurse notes the client seems quite anxious, frequently paces about, has a short attention span, and is very forgetful. The nurse can best cope with the behaviors associated with dementia by:
1. Providing a restrictive environment, including restraints, to prevent self-injury
2. Ignoring instances when denial is used as well as the client's memory lapses
3. Consistently having all staff members reinforce reality with each client contact
4. Focusing on the client's coping skills to prevent precipitating feelings of inadequacy

239. A 4-year-old is admitted to the hospital to rule out a leukemic process. On admission the most important nursing assessment is to determine:
1. The parents' ability to cope in stressful situations
2. What the child has been told about the diagnosis
3. The child's growth percentile and developmental abilities
4. The child's previous experience with illness and hospitalization

240. A client who has had a tremendous weight loss is told by the physician that more protein foods must be eaten to provide the needed essential amino acids. The client asks the nurse why these substances in protein foods are "essential." The nurse should respond:
1. "They will give you the added energy you need."
2. "They are essential for rebuilding your body tissue protein."

3. "They contain the necessary nitrogen you need for healing."
4. "Your body can't make them so you must get them in your food."

241. The physician prescribes regular insulin each morning for a client with type 1 diabetes. After administering the insulin at 8 am, the nurse should observe the client for a potential insulin reaction:
1. At breakfast
2. Before lunch
3. Before dinner
4. In the early afternoon

242. A male client with a history of an antisocial personality disorder is admitted to a medical unit. The client threatens the staff and other clients. He refuses to turn off his television set at bedtime and is often found in other clients' rooms. One client on the unit states that a watch is missing, and another has lost a wallet. The nurse's best response to this situation would be to:
1. Call the hospital security and tell them the client is stealing
2. Have the client transferred to the psychiatric unit or discharged
3. Tell the client that the staff knows he took the articles in question
4. Search the client's belongings for the missing articles without telling him

243. An 8-year-old with a history of severe asthma is admitted to the hospital after an asthma attack at home. The child is extremely short of breath. To facilitate breathing and to promote respiratory drainage, the nurse should place the child in a:
1. Supine position
2. Left lateral position
3. High-Fowler's position
4. Trendelenburg's position

244. A client who has been severely burned receives an autograft. One week after the graft the client, who has been taught to do dressing changes, notices the edges of the graft curling up and asks the nurse about it. The nurse's best response would be:
1. "May I take a look at it?"
2. "It's time for another graft."
3. "Is there any sign of redness?"
4. "Let me see whether it is infected."

245. A male client with the diagnosis of schizophrenia, paranoid type, appears very suspicious of the nurse. The most effective approach to this problem would be for the nurse to:

1. Assign various caregivers to the client
2. Make brief, frequent contacts with the client
3. Engage the client in a discussion about his thoughts
4. Allow the client to stay in his room without interruption

246. The coach of a primigravida in active labor for about 6 hours asks the nurse, "How much longer will this take? She is having a lot of back pain and is so uncomfortable." The nurse could best respond:
1. "It shouldn't be much longer now."
2. "I think you should take a break for awhile."
3. "Everything is progressing nicely as expected."
4. "Let me show you how to provide back pressure."

247. On the first postoperative day after a mastectomy, the nurse encourages the client to perform exercises such as flexion and extension of fingers and pronation and supination of the forearm. These interventions are done primarily to:
1. Preserve muscle tone
2. Prevent joint contractures
3. Assess extent of lymphedema
4. Stimulate peripheral circulation

248. A client with a history of a ruptured nucleus pulposus is scheduled for a total hip replacement. To prevent the most common complication associated with total hip replacement surgery, the nurse should teach the client to perform:
1. Straight leg raises
2. Buerger-Allen exercises
3. Deep-breathing and coughing
4. Plantar flexion and dorsiflexion exercises

249. A client with a chronic progressive disease repeatedly expresses the fear of becoming a burden and states, "I want to die." The client asks the nurse for help. Before responding, the nurse recognizes that a nurse actively or passively aiding a client with euthanasia is:
1. Liable to be tried for a crime
2. Acting within the law in most states
3. Practicing medicine without a license
4. Negligent in performing nursing duties

250. On the third postoperative day after a cesarean birth, a client tells the nurse that her breasts feel warm, firm, and tender. The skin appears shiny and taut. The nurse suspects that the cause of the client's breast discomfort is:
    1. Overdistention of the acini with milk
    2. Stasis of milk in the mammary ducts
    3. Inadequate emptying of the breasts during each feeding
    4. Increased lymphatic and venous circulation in the breasts

251. An elderly client with a chronic degenerative disease progresses to the stage where self-care is no longer possible, and admission to a nursing home becomes necessary. According to Erikson the major developmental conflict for this client would be:
    1. Intimacy vs isolation
    2. Ego integrity vs despair
    3. Identity vs role diffusion
    4. Generativity vs stagnation

252. A 13-year-old with newly diagnosed idiopathic scoliosis is very upset about the treatment regimen and is worried about being different from friends. To help the child develop a positive self-image during treatment, the nurse should:
    1. Refer the child for psychologic counseling until the treatment program is over
    2. Remind the child how crooked the back would be if treatment were not started
    3. Exaggerate the child's positive attributes and avoid focusing on negative attributes
    4. Assist the mother to help the child select appropriate clothing to minimize the condition

253. While walking to the examination room with the nurse, a toddler with autism suddenly runs over and begins head-banging on the wall. The nurse's initial action should be to:
    1. Allow the toddler to act out feelings
    2. Ask the toddler to stop this behavior
    3. Restrain the toddler to prevent head injury
    4. Tell the toddler that the behavior is unacceptable

254. After a cardiac catheterization, a client is discharged and is scheduled to return for coronary bypass surgery. The client asks, "When I get chest pain at home, how will I know whether I should call the doctor?" The nurse should teach the client to call the physician if the pain:
    1. Radiates to the arms, neck, or jaw
    2. Is accompanied by mild diaphoresis
    3. Repeatedly occurs after mild exercise
    4. Is not relieved by rest or by nitroglycerin

255. Before a client's scheduled open heart surgery, the nurse should plan to include in the teaching plan a:
    1. Thorough discussion of discharge plans
    2. Detailed description of the surgical procedure
    3. Visit by a postanesthesia unit nursing staff member
    4. Discussion of the specific areas of the body that will be shaved

256. Immediately after assisting with an unattended precipitous birth in the triage room of the Labor and Delivery Unit, the nurse should be especially concerned with:
    1. Expelling the placenta
    2. Keeping the infant warm
    3. Cutting the umbilical cord
    4. Controlling maternal bleeding

257. The nurse should expect the physician's orders for a client admitted to the hospital with mild pregnancy-induced hypertension to include:
    1. Diuretics, low-salt diet, intake and output
    2. Daily weights, glucose monitoring, bed rest
    3. Complete bed rest, intake and output, diuretics
    4. A 24-hour urine collection, daily weights, bed rest

258. A mother, whose infant was diagnosed with cerebral palsy at 6 months of age, asks why she was not told that her baby had cerebral palsy when the infant was born. The nurse should reply:
    1. "The joint deformities of cerebral palsy appear only after 6 months of age."
    2. "The health care personnel in the clinic did not want to alarm you until it was necessary."
    3. "The neurologic lesions responsible for the child's condition may have changed as the child matured."
    4. "Early diagnosis of cerebral palsy is difficult in infants until they develop control of voluntary movements."

259. The nurse is aware that the genotypic makeup of the parents of a child with sickle cell disease is:
    1. Mother-heterozygous (sickle trait); father-heterozygous (sickle trait)
    2. Mother-homozygous (no sickle trait); father-heterozygous (sickle trait)
    3. Mother-heterozygous (sickle trait); father-homozygous (no sickle trait)
    4. Mother-homozygous (has sickle cell disease); father-homozygous (no sickle trait)

260. A client with type 2 diabetes is admitted to the hospital for elective surgery. The client is now being given regular insulin, even though oral hypoglycemics were adequate prior to hospitalization. The nurse recognizes that regular insulin is now needed because:
1. The client will need a higher serum glucose level while on bed rest
2. The possibility of acidosis is greater when a client is on oral hypoglycemics
3. The dosage can be adjusted to changing needs during recovery from surgery
4. Regular insulin is readily available in the event of any complications after surgery

261. A client with an antisocial personality disorder continually uses manipulative behavior. The nurse is aware that this behavior can best be controlled by having the staff:
1. Avoid focusing on this aspect of the client's behavior
2. Develop and use a unified approach in response to this behavior
3. Assign members who are not easily manipulated to care for the client
4. Designate one staff member to approach the client about this behavior

262. A client in acute renal failure says to the nurse, "My doctor said I am going to be given some insulin. Do I also have diabetes?" The response by the nurse that best demonstrates an understanding of the use of insulin in acute renal failure would be:
1. "No, the insulin will help your body handle a chemical called potassium."
2. "Why don't you ask that question when your physician comes to see you today?"
3. "You probably had an elevated blood glucose level and your physician is cautious."
4. "No, but insulin will reduce the toxins in your blood by lowering your metabolic rate."

263. In view of the adaptations of cystic fibrosis, the diet the nurse should provide for a child affected with this disease should be:
1. Low in fat and low in salt
2. High in fat and high in protein
3. Low in fat and high in carbohydrates
4. Low in salt and high in carbohydrates

264. A 19-year-old college sophomore is admitted to the hospital with schizophrenia, undifferentiated type. The client sits in a corner for long periods rocking and responds to voices using words that the staff cannot understand. When developing the plan of care for this client, the nurse should:
1. Plan to spend short periods with the client
2. Include the client in a discussion group on the unit
3. Encourage the client to talk to other clients during the day
4. Allow the client to be alone, but maintain observation from a distance

265. After treatment for a bladder infection, a female client asks whether there is anything she can do to prevent cystitis in the future. The nurse tells her to plan to:
1. Avoid the regular use of tampons
2. Decrease her intake of orange juice
3. Increase her daily fluid consumption
4. Cleanse from vaginal orifice to urethra

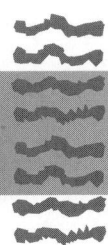

PART A ANSWERS AND RATIONALES

1. ③ Normal head circumference is 13 to 14 inches (33 to 35.5 cm) and about 1 inch greater than the chest circumference. (2) (DF; AS; PE; NM)
   1 The head is too large.
   2 Same as answer 1.
   4 The chest circumference is within normal limits.

2. ③ Multiple premature ventricular beats (PVBs) can eventually fall on a T wave, causing ventricular fibrillation; PVBs must be immediately interrupted with a drug that reduces cardiac irritability. (3) (DA; IM; MS; CV)
   1 This will not interrupt the dysrhythmia.
   2 This is unnecessary; a cardiac arrest has not occurred.
   4 This is unsafe; the dysrhythmia must be interrupted immediately; the physician can be notified later.

3. ③ Clients who have lost contact with reality can be assisted to reestablish contact by being provided structured activities. (2) (EP; PL; MH; SD)
   1 This client is responding to voices, not reality; limit-setting is reality oriented and is usually ineffective unless it involves directing the client to dismiss the voices.
   2 The client represents no immediate threat to the self or others; isolating the client would decrease contact with reality and would most likely increase the hallucinations.
   4 Although this may decrease the hallucinations, it takes a long time to establish such a relationship and the client needs help now.

4. ① Levels close to 2 mEq/L are dangerously close to the toxic level; immediate action must be taken. (3) (CC; EV; MH; DR)
   2 The level is dangerously high, and the priority is the immediate notification of the physician rather than continued monitoring.
   3 Same as answer 2.

4 The level is dangerously high; the therapeutic range for lithium is 0.6 to 1.2 mEq/L.

5. ② Drainage flows by gravity. (3) (SI; EV; MS; EN)
   1 This is unsafe; lifting and replacing a dressing would contaminate the surgical site.
   3 Although this might be done, it does not take into consideration that drainage flows by gravity.
   4 This is unsafe; this would interfere with the healing process.

6. ③ Pelvic examination would reveal dilation and effacement. (3) (HI; AS; CW; HC)
   1 Ferning is the characteristic frondlike pattern of crystallization in amniotic fluid when it dries. This test assesses for the rupture of membranes.
   2 Contractions are also present with false labor and are called Braxton Hicks contractions.
   4 This helps to differentiate between amniotic fluid and urine; rupture of the membranes can occur with or without the onset of true labor.

7. ④ Pulmonic stenosis increases resistance to blood flow, causing right ventricular hypertrophy; with right ventricular failure there is an increase in pressure on the right side of the heart. (3) (CI; AN; PE; CV)
   1 The pressure would be decreased in pulmonic stenosis.
   2 Same as answer 1.
   3 Same as answer 1.

8. ② During acute cholecystitis, low-fat liquids are permitted; skim milk is low in fat and contains protein, which will eventually promote healing. (3) (PC; PL; MS; GI)
   1 Beef, even if it is lean, contains fat and should be avoided.
   3 Egg yolks contain fat and should be avoided.
   4 Gas-forming vegetables should be avoided.

9. (1) This portrays a nonjudgmental attitude that recognizes the client's needs. (2) (SC; IM; MH; AX)
 2 The client is upset that she cannot control her crying.
 3 This is untrue; it implies that crying will make it all better.
 4 This is unrealistic; the cause of the anxiety, and hence the crying, is on an unconscious level.

10. (3) This offers the mother the opportunity to identify what is bothering her about caring for her baby. (3) (SC; IM; CW; HC)
 1 This response could put the client on the defensive.
 2 This minimizes the client's feelings and offers false reassurance.
 4 Same as answer 2.

11. (3) This indicates the child's level of understanding to the nurse, and an explanation can then proceed at this level. (3) (CC; PL; PE; GD)
 1 Doll play is more appropriate for younger children; it is inappropriate in this instance.
 2 Role playing is inappropriate and nontherapeutic at this time.
 4 Although the school-age child appreciates some detail, extensive detail is inappropriate.

12. (3) This occurs with the administration of Adriamycin and may alarm the client. (3) (CC; EV; MS; DR)
 1 This is true for mithramycin, not the drugs in this protocol.
 2 This is unnecessary.
 4 Adriamycin is not given orally, only via the IV route.

13. (2) As long as the client has no nausea or vomiting, there are no diet restrictions. (2) (PC; IM; MS; IT)
 1 The client should not disturb the dressing; it should be changed by the physician on a follow-up visit.
 3 The biopsy site will be sutured and should not get wet.
 4 Temperature elevation is not a usual response and may indicate infection.

14. (3) The data indicate impending shock; the dressing should be assessed for signs of hemorrhage. (3) (DA; EV; MS; CV)
 1 Although this may eventually be done, it is not the priority.
 2 There are no signs of respiratory distress; if hemorrhage is confirmed, the supine position is preferable.

4 This is unsafe; the client may be hemorrhaging and needs immediate assessment.

15. (1) Stimulation of the sympathetic nervous system is an initial response to mild hypoxia that accompanies partial cord compression (umbilical vein) during contractions. Changing the maternal position can alleviate the compression (3) (HI; EV; CW; HC)
 2 The client can be changed to any position that may facilitate movement of the fetus to alleviate cord compression.
 3 This is a compensatory physiologic response by a healthy fetus. The nurse should first intervene to assist with alleviating cord compression.
 4 Uniform accelerations are not indicative of the need for emergency delivery.

16. (1) The client has signs of diabetes, which may result from steroid therapy; testing the blood glucose level is a method of screening for diabetes, thus gathering more data. (2) (CI; AS; MS; EN)
 2 The symptoms are not those of benign prostatic hypertrophy.
 3 Assessing the urine for glucose and ketones is not as accurate as a blood glucose level.
 4 The symptoms presented are not those of fluid retention but of diabetes.

17. (1) Cholesterol is an essential precursor of body substances such as vitamin D and steroid hormones; although some cholesterol is synthesized by the body, a small amount is needed in the diet. (3) (PC; IM; MS; CV)
 2 Polyunsaturated fats come from vegetable sources; saturated fats come from animal sources, coconut oil, and palm oil.
 3 The diet must contain some cholesterol, and this should come from plant sources.
 4 Same as answer 2.

18. (1) This prevents seepage of Imferon up through the needle track, thereby limiting staining of the skin. (1) (PT; IM; PE; DR)
 2 This is unsafe; massage would force Imferon into the subcutaneous tissue, causing irritation and staining.
 3 This length of needle is too short to get into a muscle; a 1½- to 2-inch needle is required; the gauge of the needle is too small for the viscosity of Imferon; a 19- to 22-gauge needle is required.
 4 Aspiration would still be performed with the Z-track method.

19. (1) Hypoglycemia may be present because of sudden withdrawal of maternal glucose and increased insulin production by the infant. (3) (HI; PL; CW; HN)

2 The umbilical vein may be used for starting an IV; it should not be obliterated.

3 This is not the priority until the blood glucose level is determined.

4 This can be delayed for 2 hours; determination of the blood glucose level is the priority.

20. (4) These are some of the first signs of hypoxia; the airway must be kept patent to promote oxygenation. (2) (DA; AS; PE; RE)

1 These are late signs of hypoxia; suctioning should have been done well before this time.

2 These are late signs of respiratory difficulty; suctioning and other measures should have been done well before this time.

3 The client will not be able to communicate verbally after a tracheotomy.

21. (4) Because of the proximity of the carotid artery to the surgical area and the possibility that age or the disease process has weakened the carotid artery, the client should routinely be assessed for signs of hemorrhage because of carotid rupture. (3) (DA; EV; MS; RE)

1 This is related to heart decompensation, not rupture of a carotid artery.

2 Same as answer 1.

3 With a radical neck dissection the trapezius muscle, not chest muscles, may atrophy.

22. (1) A continual level of this drug is maintained to keep the terminally ill client comfortable. (1) (PC; PL; MS; NM)

2 This medication is not administered intermittently; in addition, for clients' management at home it is usually prescribed in liquid form and is taken orally.

3 The client should not have to request this medication; it should be given routinely.

4 Addiction is not a major concern for the terminally ill client.

23. (1) Mediastinal structures move toward the uninjured lung, reducing oxygenation and venous return. (3) (DA; AS; MS; RE)

2 This is unusual with a crushing chest injury.

3 Flail chest is a closed chest injury; open pneumothorax results from a penetrating injury to the chest wall.

4 This is a result of cardiac contusion and usually occurs from sternal, not lateral, compression.

24. (2) This reflects the client's feelings and opens channels of communication. (3) (EP; IM; MH; SD)

1 This response denies the client's feelings; this will not foster development of trust.

3 This provides false reassurance; it is really not the same food because it is on the client's plate.

4 This is false reassurance; the delusional client would believe the nurse knows what part to taste.

25. (2) If the bleeding continues, both mother and fetus are at risk and a cesarean delivery would be indicated. (1) (DA; PL; CW; HP)

1 This is desired in all situations, not just with placenta previa.

3 If bleeding is under control and the client is stable, delivery can be delayed.

4 Stimulation of the cervix or uterus may cause bleeding or hemorrhage and should be avoided.

26. (2) This infestation is transferred by the anal-oral route, and handwashing is the most effective method to prevent transmission. (1) (CC; EV; PE; GI)

1 Cats do not transmit this disease.

3 This is unnecessary; pinworms are found in the rectum or colon and travel to the perianal area only when the person sleeps.

4 This is not the usual mode of transmission.

27. (1) Decreased normal breath sounds result from reduced air flow, pleural effusion, and destruction of lung tissue. (3) (DA; AS; MS; RE)

2 There is enlargement of accessory muscles, which are used during the expiratory phase to help force air out.

3 There is an increased expiratory phase because of entrapment of air and collapse of airways.

4 There is an increased A-P diameter (barrel chest) because of air trapping and enlargement of the lungs with a loss of recoil ability.

23. (2) Lying on the affected side increases drainage and allows the unaffected lung to expand to the fullest extent. (3) (DA; IM; MS; RE)

1 This will not facilitate drainage and leads to pooling of drainage in the operative site.

3 Same as answer 1.

4 This is undesirable because this may not allow the unaffected lung to fully expand and provide maximum oxygenation.

29. (1) The client expects the nurse to focus on eating, but the emphasis should be placed on feelings rather than actions. (2) (EP; IM; MH; ES)
   2 This is a threat that will not convince the client to eat; privileges are associated with a contracted weight gain.
   3 This is a threat that will not convince the client to eat; the client is not concerned about the effects of malnutrition.
   4 Threats will not convince the client to eat; this response sets up a challenge that the nurse will not win.

30. (2) Reducing anxiety limits the need for these obsessive-compulsive actions. (3) (EP; PL; MH; PR)
   1 Simple repetitive activities are not therapeutic for this client and could increase anxiety.
   3 This is a temporary action that does not deal with the feelings that cause anxiety.
   4 Menial tasks may decrease feelings of self-worth; these individuals do not have a need to expiate guilt.

31. (3) This measure assesses for increasing intracranial pressure, which may occur if drainage channels are obstructed by the bacterial infection. (2) (DA; PL; PE; NM)
   1 This is insufficient monitoring; many changes could occur in this time span.
   2 Antibiotics are administered intravenously throughout the course of treatment.
   4 Parents can visit if they are taught how to carry out the isolation procedures.

32. (1) Edema associated with the inflammatory reaction after surgery limits the conduction of sound. (3) (CI; AN; MS; NM)
   2 This should not happen because the inner ear is not involved in this surgery.
   3 This is extremely unlikely.
   4 If the graft slips out, the hearing loss would be worse than before surgery.

33. (4) This action monitors the tube for patency; retained fluid raises intravesicular pressure, causing discomfort similar to the urge to void. (2) (SC; IM; MS; RG)
   1 The need to take vital signs is not indicated by the client's complaint; the physician is not called unless a blocked tube cannot be corrected.
   2 Total intake and output have no relationship to the client's present feeling that there is a need to void.
   3 This is true; however, the integrity of the gravity system should be ascertained before this reason can be assumed.

34. (4) Muscle contractions cause neuronal stimulation; multiple foci of demyelination cause interruption or distortion of the impulse, resulting in intention tremors. (2) (DA; AN; MS; NM)
   1 There are no tremors when the person is asleep.
   2 There are no tremors when the client is inactive; this is a resting tremor that is usually associated with parkinsonism.
   3 Intention tremors are associated with muscle contraction, not feelings; stress can exacerbate the symptoms of MS.

35. (2) Destruction of brain cells decreases cognitive abilities, and learning and thinking are compromised. (2) (EP; AS; MH; DD)
   1 The attention span would be decreased.
   3 This is not specific to dementia or aging; it depends on the individual.
   4 Psychologic stress and changes caused by aging result in a decreased capability to adapt to the environment; familiar routines provide security.

36. (4) This indicates afibrinogenemia; massive clotting in the area of the separation has resulted in a lowered circulating fibrinogen level. (2) (DA; AS; CW; HP)
   1 A boggy uterus indicates uterine atony.
   2 Blood clots indicate normal fibrinogen levels; vaginal clots may indicate a failure of the uterus to contract and should be explored further.
   3 These findings may be indicative of hypovolemic shock, which may occur after DIC; these findings are not unique to only DIC.

37. (1) Bulging of the perineum is caused by the presence of the fetal head and usually signifies imminent delivery. (2) (DF; IM; CW; HC)
   2 Pain medication at this time is harmful to the fetus; it crosses the placental barrier and causes respiratory distress.
   3 During the second stage of labor, the client is encouraged to push, not pant, with each contraction.
   4 Catheterization is indicated earlier in labor so that uterine contractions are not impeded; voiding will occur spontaneously with pushing.

38. (3) Hypoglycemia affects the central nervous system, causing headache and nervousness, and also affects the sympathetic nervous system, causing diaphoresis. (2) (CI; EV; MS; EN)
   1 Dry skin and drowsiness are associated with ketoacidosis.
   2 These are not signs of an insulin reaction; excessive thirst and malaise are clinical symptoms of diabetes; hunger, not anorexia, is related to an insulin reaction.

4 All of these signs are associated with ketoacidosis.

39. (1) Of the selections offered, this is the highest in calories and protein, which are needed for the increased basal metabolic rate and for tissue repair. (3) (PC; PL; MS; IT)
2 These foods do not provide as high an amount of calories and protein as the correct choice.
3 Same as answer 2.
4 Same as answer 2.

40. (2) Arterial perfusion and the presence of hemorrhage must be assessed hourly to prevent complications or to identify problems early. (1) (CI; PL; MS; SK)
1 This is unsafe; this will interfere with the pull of traction.
3 Same as answer 1.
4 Same as answer 1.

41. (1) Central cyanosis (blue lips and face) indicates lowered oxygen of the blood, caused by either decreased lung expansion or right-to-left shunting of blood. (2) (DA; AS; CW; HN)
2 This is normal in the newborn.
3 This is not an abnormal finding because peripheral circulation of the infant is unstable; blueness disappears when the baby is warm.
4 Same as answer 2.

42. (4) Favorable personality change within 1 to 2 days attests to the effectiveness of the diet; other improvements take longer. (3) (PT; EV; PE; GI)
1 Usually anorexia is not a problem; if it does occur, it does so during bouts of diarrhea.
2 This occurs after the personality change.
3 Same as answer 2.

43. (4) The effects of the tricyclic antidepressants are cumulative; it may be some time before improvement is noted. (2) (PT; IM; MH; DR)
1 This is false; antidepressants help relieve the physical and mental discomforts of the depressed client.
2 It is too early to arrive at this conclusion.
3 Antidepressant drugs are effective in the treatment, regardless of the length of the depression.

44. (1) A large percentage of HIV-positive individuals eventually develop AIDS. (2) (HI; AS; PE; BI)
2 An immature reticuloendothelial system does not indicate the potential development of AIDS.

3 Although blood transfusions can place someone at risk, a positive serologic status for HIV places a person at highest risk for AIDS.
4 The presence of an opportunistic infection alone is not an indication of AIDS.

45. (1) The objective is to stimulate ovulation near the fourteenth day of the menstrual cycle, which is achieved by taking the drug on the fifth through the ninth days; there is an increase in the pituitary gonadotrophins LH and FSH, with subsequent ovarian stimulation. (2) (DF; EV; CW; RP)
2 There are insufficient hormones for Clomid to be effective.
3 There is insufficient estrogen at this time for Clomid to be effective this early in the cycle.
4 This is too late in the cycle.

46. (4) Talking with others in similar circumstances provides support and allows for sharing of experiences. (1) (SC; IM; MH; TR)
1 This avoids the client's concerns and cuts off communication.
2 The feeling of guilt has not been expressed.
3 This is not therapeutic; it offers false reassurance.

47. (3) The client's systolic reading is 32 mm Hg above the baseline reading, and the diastolic reading is 16 mm Hg above the baseline; an increase of 30 mm Hg for the systolic reading or 15 mm Hg for the diastolic reading is suggestive of hypertension. (3) (HI; AS; CW; HP)
1 This is an insufficient rise to indicate a hypertensive disorder of pregnancy.
2 Same as answer 1.
4 Although this would indicate hypertension, the threshold to diagnose hypertension in a client who normally exhibits a low blood pressure is less.

48. (4) Respiratory distress or arrest may occur when the serum level of magnesium sulfate reaches 12 to 15 mg/dl; deep tendon reflexes disappear when the serum level is 10 to 12 mg/dl; the drug is withheld in the absence of deep tendon reflexes. (3) (PT; EV; CW; DR)
1 This is an inappropriate assessment to determine client response to magnesium sulfate; deep tendon reflexes need to be assessed.
2 Same as answer 1.
3 Same as answer 1.

49. (1) This eliminates the nurse's subjectivity from the report. (1) (CC; IM; MH; CS)
    2 This is unrelated to the rape itself; this would allow for subjectivity.
    3 This is not part of the responsibility of the nurse.
    4 This would allow for subjectivity.

50. (1) Weight monitoring is the most useful means of assessing fluid balance and changes in the edematous state; 1 liter of fluid weighs about 2.2 pounds. (2) (DA; PL; PE; FE)
    2 This would be subjective and inaccurate.
    3 This is not as accurate as daily weights; fluid can be trapped in the third compartment with no alteration in intake and output.
    4 This is unreliable; these may or may not be altered with fluid shifts.

51. (4) Ranitidine inhibits histamine at H2 receptor sites in parietal cells, which limits gastric secretion. (3) (PT; AN; MS; DR)
    1 This is not the direct action of this drug but it will eventually raise the pH.
    2 This would be undesirable; gastric hormones increase gastric acid secretion.
    3 This drug does not regenerate the gastric mucosa; the drug prevents its erosion by gastric secretions.

52. (1) Failure to meet a need for increased insulin caused by the elevated metabolic rate associated with infection eventually results in hyperglycemia. (3) (CC; EV; PE; EN)
    2 This would cause hypoglycemia.
    3 Same as answer 2.
    4 Same as answer 2.

53. (2) This recognizes the client's feelings and encourages communication. (2) (SC; IM; MS; EH)
    1 This question sounds accusatory; it ignores the client's feelings and discourages communication.
    3 Although this may be true, it does not encourage further communication.
    4 This statement negates the client's feelings and discourages communication.

54. (3) A small amount of bile-colored spotting is expected, but a moderately large amount is abnormal; the physician should be notified. (2) (CI; IM; MS; GI)
    1 The findings need further assessment; this intervention only treats one symptom and disregards the need for medical evaluation of the complication.
    2 The client should be on the right side to compress the liver capsule against the chest wall.

    4 Although this is important, the priority is to notify the physician.

55. (4) Drug dose and frequency are adjusted according to these results to enhance efficacy; therapeutic levels are maintained. (1) (CI; EV; MS; DR)
    1 A sustained drop in fever is the desired outcome, not reduction just at peak serum levels of the drug.
    2 Blood cultures are obtained when the client spikes a temperature; they are not related to peak and trough levels for gentamicin.
    3 Peak and trough levels reveal nothing about allergic reactions.

56. (3) This response recognizes capability without adding stress or increasing dependency. (3) (EP; IM; MH; MO)
    1 This does not address the client's needs and is punitive.
    2 This increases dependency, which is not therapeutic.
    4 This attempts to manipulate compliance; the client cannot accept responsibility for the future.

57. (3) This will help to elicit any fantasy the child may have; it helps the child understand that treatment is not a punishment. (3) (SC; IM; PE; EH)
    1 This is inappropriate for a 4-year-old and does not elicit feelings.
    2 This is inappropriate; it may be frightening.
    4 This is not currently supported as a cause; this is an inappropriate discussion for a 4-year-old.

58. (4) How people handled grief in the past provides clues to their coping patterns when dealing with current grieving. (2) (SC; AN; MH; CS)
    1 Although these are important, past experiences with grief are paramount.
    2 Same as answer 1.
    3 Same as answer 1.

59. (1) Breast palpation should be done in the supine position with a small towel roll under the shoulder of the palpated side. (1) (HI; EV; CW; WH)
    2 This is a correct procedure for breast self-examination.
    3 Same as answer 2.
    4 Same as answer 2.

60. (3) The common bile duct passes through the head of the pancreas; it is often constricted or obstructed by the neoplasm. (3) (CI; AN; MS; GI)

1 This would not cause jaundice.
2 This is the prehepatic cause of jaundice; it is not applicable in this situation.
4 This is a hepatic cause of jaundice; it is not applicable in this situation.

61. (4) Impulse conduction of skeletal muscles is impaired with decreased potassium levels; muscular weakness and cramps may occur with hypokalemia. (2) (PT; EV; MS; FE)
   1 Hyperactive reflexes indicate hyperkalemia, not hypokalemia.
   2 The pulse would be weak and irregular with hypokalemia because of an impaired conduction system in the cardiac muscle.
   3 Diarrhea is caused by hyperkalemia, not hypokalemia.

62. (4) Infections, especially urinary tract infections, are risk factors for preterm labor. (2) (HI; AN; CW; HP)
   1 Gravidity is not a risk factor for preterm labor.
   2 Clients with seizure disorders are not at an increased risk for preterm labor.
   3 An android-shaped pelvis is more likely to cause dystocia than preterm labor.

63. (4) Calcium promotes osteoblastic activity, and calories support the growth and energy needs of the child. (1) (PC; PL; PE; SK)
   1 The level of purine does not affect bone repair; a decrease in calories would not support growth and development.
   2 Extra calories are converted to adipose tissue; if calcium needs are met, sufficient phosphorus will be ingested.
   3 Bone tissue responds to sufficient calcium in the diet; if injury had occurred to the soft tissues, a high-protein diet would be necessary.

64. (4) A person who can handle the activities of daily living and function in society is considered mentally healthy. (2) (EP; EV; MH; SD)
   1 Some anxiety is necessary; anxiety causes problems when it is overwhelming for an extended period.
   2 Insight into one's problems is of no use if one is unable to function in society.
   3 Everyone uses defense mechanisms; the degree to which they are used determines mental health.

65. (4) The nurse should offer support and use clear, simple terms to allay the client's anxiety. (1) (EP; IM; MH; MO)
   1 This may be too frightening or confusing to the client.

2 When anxiety is high, the client cannot retain details, and details may lead to added fears.
3 The client generally is kept npo before ECT to prevent aspiration during the treatment.

66. (3) Bed rest weakens the perineal and abdominal muscles used in defecating; ambulation promotes peristalsis and improves muscle tone, thereby facilitating expulsion of flatus and promoting defecation. (1) (CI; PL; MS; GI)
   1 Early ambulation will not prevent this complication.
   2 There will be no retention because the surgery involved removal of the bladder and the creation of a permanent urinary diversion.
   4 Same as answer 1.

67. (3) The Harris drip, or flush, removes accumulated gas in the intestine, which reduces distention of the abdomen. (2) (PC; EV; CW; HP)
   1 Stimulating evacuation is not the purpose of a Harris drip; a bowel movement would indicate the procedure was done improperly.
   2 The returns of a Harris drip usually contain small amounts of fecal material; it is not for cleansing the bowel.
   4 The fluid is not retained; small amounts are instilled slowly and then permitted to return, taking gas with them.

68. (2) The major cause of skin cancer is exposure to the sun's ultraviolet light, a form of radiation. (1) (HI; AN; MS; IT)
   1 This is not a causative factor.
   3 Same as answer 1.
   4 Although environmental pollutants may have some bearing, they are not considered the major cause of skin cancer.

69. (2) The client is incapable of accepting responsibility for self-created problems and blames society for the behavior. (2) (EP; AS; MH; PR)
   1 This response demonstrates insight, and these individuals rarely develop insight into their problems.
   3 Same as answer 1.
   4 Same as answer 1.

70. (1) Swelling can obstruct nasal breathing, interfering with the senses of taste and smell. (2) (SI; EV; PE; RE)
   2 Speech should not be affected because the vocal cords are not within the operative area.
   3 The operative site should be healed and not cause discomfort when swallowing.
   4 The vocal cords are outside the operative area; the site should be healed and not cause discomfort when swallowing.

71. (3) The intestinal lumen is narrowed by the large mass, and contraction of the proliferate fibrous tissue causes an obstruction. (2) (CI; AS; MS; GI)
   1 Diarrhea may occur but usually alternates with constipation.
   2 Dehydration usually does not occur unless there is severe vomiting and/or diarrhea.
   4 This usually results from a perforation of the bowel that is caused by a buildup of pressure behind the obstruction.

72. (3) During open heart surgery the conductive system of the heart can be damaged because of the trauma of surgery. (3) (CI; EV; MS; CV)
   1 Shock results in a weak, rapid pulse, not bradycardia.
   2 Hypoxia causes tachycardia, not bradycardia.
   4 Congestive heart failure causes a rapid pulse rate, not bradycardia.

73. (3) According to the standards of normal growth and development, the anterior fontanel closes between 13 and 18 months of age. (3) (DF; IM; PE; GD)
   1 This is too early; early closure may impede the growth of the infant's brain, impairing mental development.
   2 Same as answer 1.

4 The closure should have occurred by 18 months; delayed closure may indicate neurologic difficulties.

74. (4) Panting or blowing breathing patterns may counteract the desire to push. (1) (HI; IM; CW; HC)
   1 Prolonged hyperventilation can cause respiratory alkalosis and a resulting decrease in uterine blood flow, which can put the fetus at risk.
   2 Pelvic-rocking exercises are used to alleviate backache.
   3 Deep-breathing exercises between contractions increase oxygenation and help maintain relaxation; this is unrelated to pushing.

75. (2) Dilation of blood vessels causes dependent pooling when the client moves to an upright position, resulting in cerebral hypoxia. (1) (SI; AN; MS; NM)
   1 Resting will not help; the client can limit this response by moving gradually when changing positions.
   3 This is false; the client needs teaching regarding changing positions gradually.
   4 This is an expected response when moving from a horizontal to a vertical position; this single symptom is not enough reason for changing any medications the client may be taking.

How to Use Worksheet 1: Errors in Processing Information

Common errors in processing information are listed in the left-hand column of this worksheet. At the top of the worksheet is a row of blank spaces for inserting the number of the question missed. Directly below each number, check any errors you made in answering that question. You may have made more than one type of error in an answer.

Worksheet 1: Errors in Processing Information

Question number																									
Did not read situation/ question carefully																									
Missed important details																									
Confused major and minor points																									
Defined problem incorrectly																									
Could not remember terms/ facts/concepts/principles																									
Defined terms incorrectly																									
Focused on incomplete/incorrect data in assessing situation																									
Interpreted data incorrectly																									
Applied wrong concepts/ principles in situations																									
Drew incorrect conclusions																									
Identified wrong goals																									
Identified priorities incorrectly																									
Carried out plan incorrectly/ incompletely																									
Was unclear about criteria for evaluating success in achieving goals																									

How to Use Worksheet 2: Knowledge Gaps

Types of common knowledge gaps are listed along the top of this worksheet. Write a brief description of topics you want to review in the spaces provided. For example, if you missed a question on administration of a particular drug, write the drug name and problem (e.g., dosage) in the appropriate space under the column labeled *Pharmacology*.

Worksheet 2: Knowledge Gaps

Basic science	Skills/ procedures	Basic human needs	Growth and development	Normal nutrition	Psychosocial factors	Clinical area/ topic	Stressors/ coping mechanisms	Patho- physiology	Pharma- cology	Therapeutic nutrition	Legal implications	Other

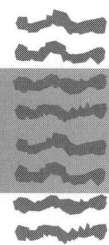

PART B ANSWERS AND RATIONALES

76. (4) Discussion with parents who have children with similar problems helps to reduce some of their discomfort and guilt. (2) (SC; PL; PE; EH)
   1 When not in crisis, the child should be allowed to attend school and should not be developmentally damaged by social isolation.
   2 There is no recommended therapeutic climate for these children, and moving may not be beneficial to the child or the family.
   3 Some parents do not choose to avoid future pregnancies and should not be forced to do so.

77. (4) The symptoms prevent the individual from being forced to move in either direction on the conflict; symptoms thus reduce anxiety and remove the conflict. (2) (EP; AS; MH; AX)
   1 The individual is not happy and cheerful but is relieved by the reduction in anxiety.
   2 The individual is not sad and depressed but is relieved by the reduction in anxiety.
   3 The individual is not frightened or upset because the symptoms relieve the anxiety.

78. (4) This is a noninvasive, harmless way to visualize the location of the placenta. (2) (CI; AN; CW; HP)
   1 This is a visualization of the amniotic fluid via vaginal examination; it is contraindicated because it may detach the placenta.
   2 This is an invasive surgical procedure used for diagnostic purposes other than the location of the placenta.
   3 This is used for removing amniotic fluid for diagnosis and fetal evaluation, not for diagnosing placenta previa.

79. (4) This statement attempts to understand the symbolism, reflects and acknowledges the client's feelings, and helps to preserve ego integrity. (3) (EP; IM; MH; SD)
   1 This validates the client's delusion and does not test reality.
   2 This rejects the client's feelings and does not address the client's fears of being harmed; clients cannot be argued out of delusions.

3 This is false reassurance; the nurse cannot fully understand the symbolism and therefore cannot make this promise.

80. (3) The initial attack is both local and systemic; recurrent attacks are milder and localized. (1) (SI; IM; CW; WH)
   1 There is recurrence in 50% of clients.
   2 Recurrent attacks are precipitated by physical and emotional stress, not by sexual activity.
   4 Although optimal health habits may limit recurrence, they will not totally prevent it.

81. (4) Because the client is up more at home, edema usually increases. (3) (CC; IM; MS; CV)
   1 Serosanguineous drainage will persist after discharge.
   2 These symptoms will persist longer because it takes 6 to 12 weeks for the sternum to heal.
   3 These should not be expected and are in fact signs of postpericardiotomy syndrome.

82. (2) The nurse should be alert for dehydration caused by fluid loss through vomiting in the binge-purge cycle. (2) (DA; AS; MH; ES)
   1 Weight gain would not be expected because of the purging that usually follows a binge.
   3 Hyperactivity would not be expected because most individuals with bulimia withdraw and vomit after a binge.
   4 Hyperglycemia would not be expected because of the vomiting that follows a binge.

83. (1) Aging causes reduction in skin lubrication, which results in dry skin. (1) (DF; PL; MS; IT)
   2 This influences how much assistance is necessary, not the frequency of bathing.
   3 This influences what bath products may be used, not the frequency of bathing.
   4 This influences safety factors applicable during the bath, not the frequency of bathing.

84. (2) Necrotic tissue and blood in an area that is moist, warm, and dark make excellent culture media. (2) (SI; AN; CW; NN)
    1 This is untrue; Wharton's jelly is present and provides a protective barrier.
    3 The diaper is kept below the level of the umbilicus; although the site may be touched by clothing, the clothing should be 100% cotton to allow for drying of the cord.
    4 This is untrue; newborns carry antibodies from the mother.

85. (4) An increased attention span in school indicates that the child has improved. (2) (SC; EV; MH; BA)
    1 This would indicate that the child has not made sufficient progress to be ready for discharge.
    2 This would indicate that the child has not made progress because children should enjoy playing with peers at this age.
    3 A child of age 6 is not usually enuretic at night, even without hyperactivity; there are no data to indicate enuresis in this situation.

86. (2) Because the tumor is of renal origin, the renin-angiotensin mechanism can be involved, and blood pressure monitoring is very important. (3) (CI; PL; PE; RG)
    1 This could put pressure on the involved area, causing rupture of the tumor and seeding of cancer cells.
    3 Same as answer 1.
    4 This is unnecessary; no infection is present.

87. (1) At the very minimum, the role of a witness is to attest to the validity of the signature. (1) (CC; AN; MS; EH)
    2 A consent is valid for the length of the hospital stay.
    3 Clients also need to know about the outcomes and risks of surgery.
    4 This is the responsibility of the physician.

88. (4) All these interventions promote aeration of the reexpanding lung and maintenance of function in the arm and shoulder on the affected side. (3) (CI; PL; MS; RE)
    1 Cough suppressants would not be indicated because coughing and deep breathing are to be encouraged.
    2 Drainage is marked with time tapes on the side of the device; the closed system is not entered for emptying; when full, the entire device is replaced.
    3 Clamps are not necessary and should be avoided in almost all instances because of the danger of tension pneumothorax.

89. (4) The client's grieving process is severe and extended, indicating dysfunction. (2) (SC; AN; MH; CS)
    1 The data do not support this; the client is communicating effectively with the nurse.
    2 There are not enough data to support this conclusion.
    3 This is not as specific as identifying grieving; also, low motivation is not the reason for the client's inability to cope.

90. (2) The diaphragm is used in conjunction with spermicidal jelly or cream, which remains active for only 6 to 8 hours. (1) (DF; EV; CW; RC)
    1 Removal this soon would allow motile sperm to pass through the cervical os.
    3 The diaphragm may be left in place as a mechanical barrier, but the spermicidal jelly or cream becomes inactive after 6 to 8 hours.
    4 The diaphragm may be left in place for this period but may cause an unpleasant odor.

91. (4) This is a true statement that assures the client that it is effective. (2) (EP; IM; MH; MO)
    1 This response does not answer the client's question.
    2 ECT does not change usual patterns of behavior, although it does interrupt disturbed or disturbing acting-out behavior.
    3 This response puts off answering the client's question by referring it to the physician.

92. (2) A rapid mood upswing and psychomotor change commonly signals that the client has made a decision and has developed a plan for suicide. (2) (EP; EV; MH; MO)
    1 This statement is typical in the depressed client and does not really signal a change in mood.
    3 Same as answer 1.
    4 Same as answer 1.

93. (3) Breaking up the tablet increases the surface area of the tablet, which permits it to dissolve faster. (2) (CC; IM; MS; DR)
    1 If taken with water, the tablet is held away from the site of absorption or swallowed; this tablet is taken sublingually.
    2 This does not hasten absorption and may even impair the tablet's ability to dissolve.
    4 This would slow absorption because saliva helps to dissolve the tablet.

94. (3) Identifying potential health problems and planning and implementing educational programs aimed at avoidance are primary prevention. (2) (EP; AN; MH; CS)

1 There is no such term; legal intervention would be a form of secondary prevention.
2 These are interventions such as long-term or permanent removal of the child from the home.
4 These are interventions such as screening or investigation of suspected child abuse.

95. (1) The solution with the greatest number of particles is hyperosmolar and exerts an osmotic force, pulling fluid in its direction. (2) (CC; AN; MS; FE)
2 This is untrue; the opposite occurs.
3 Same as answer 2.
4 On the contrary, water moves to the side with the greater number of particles.

96. (2) Furosemide (Lasix) is a diuretic with a rapid onset of action, particularly when given intravenously. (1) (CC; IM; MS; DR)
1 Hearing loss is associated only with high doses; 20 mg is a small dose.
3 Burning is not associated with IV administration of furosemide.
4 This is contraindicated; at this time, the client has a fluid volume excess; potassium needs will be addressed later.

97. (3) This private room will provide isolation; also, being close to the nurses' station will facilitate frequent neurologic monitoring. (2) (SI; PL; PE; NM)
1 This is unsafe; it would permit cross-infection; a private room is necessary to prevent transmission of airborne droplets.
2 Same as answer 1.
4 This is unsafe; it interferes with frequent monitoring of the client.

98. (1) Corticosteroids inhibit the inflammatory response and increase an individual's susceptibility to infection. (3) (SI; EV; MS; DR)
2 With therapy, the client may gain or lose weight, but weight itself is not a major factor in planning nursing care.
3 The client will be hypertensive, not hypotensive, until stabilized.
4 Urinary stasis is unrelated to Decadron; Decadron will promote diuresis.

99. (3) Immature neurologic and chemical respiratory mechanisms cause apnea in preterm infants; tactile stimulation should restart the newborn's breathing. (2) (DA; IM; CW; HN)
1 This is not done unless the infant is cyanotic or the blood Po₂ is low.

2 This is not done unless the infant shows signs of an obstructed airway.
4 This is done only if the infant does not respond to tactile stimulation.

100. (4) This involves no element of competition and still allows for channeling of excessive energy. (2) (EP; PL; MH; MO)
1 The sense of competition and increased stimulation may raise the client's anxiety.
2 This requires fine motor skill from a client who is hyperactive and whose attention span is limited.
3 The client is too hyperactive to complete this task and may respond with distractibility or aggressiveness toward others.

101. (4) Three factors appear to be related to lead poisoning: lead in the environment; toxins in the environment; and characteristics of the child and the parents. (3) (CC; IM; PE; NM)
1 This is only one of the three etiologic factors.
2 Same as answer 1.
3 Same as answer 1.

102. (3) Lactated Ringer's is an alkaline solution that replaces bicarbonate ions lost from T-tube bile drainage, thus preventing or treating acidosis. (3) (DA; EV; MS; FE)
1 This is unrelated to the effectiveness of this IV solution.
2 Same as answer 1.
4 Same as answer 1.

103. (2) Exploration allows the nurse to assess the client's knowledge and fears. (2) (CC; IM; PE; EH)
1 This is not true; this would imply a dominant mode of transmission.
3 This is false reassurance; it may occur again.
4 This is not true; this would imply an autosomal recessive mode of transmission.

104. (2) This response recognizes the concern and provides accurate information that may reduce anxiety. (2) (SC; IM; MS; EH)
1 This is inaccurate information; impotence usually does not result; it is possible after perineal prostatectomy.
3 This reply closes off communication and transfers responsibility to the physician.
4 This does not recognize feelings and provides inaccurate information; impotence rarely, if ever, occurs with this operation.

105. ④ This sets an unrealistic limit that would increase anxiety by removing a defense the client needs at this time; rituals cannot be this controlled until other defenses are developed to replace them. (2) (EP; EV; MH; PR)
1 This is done in therapy as the client's condition improves; insight must be developed slowly to minimize anxiety.
2 This would reduce, not increase, anxiety because the client would feel free to express feelings.
3 This would increase self-esteem and self-control, not increase anxiety.

106. ④ This is an adverse reaction that may occur about 8 to 12 hours after diuresis occurs. (3) (CI; EV; MS; DR)
1 Although mannitol may cause respiratory congestion, it does not cause respiratory failure.
2 This is not an expected response; mannitol does not directly affect the heart, but injudicious use can cause electrolyte imbalances, which can eventually cause dysrhythmias.
3 Tachycardia, not bradycardia, is an adverse reaction related to mannitol.

107. ② Initially there is a loss of vascular tone below the injury, resulting in hypotension. (2) (DA; AS; MS; NM)
1 Bradycardia, not tachycardia, is associated with spinal shock.
3 Initially flaccid paralysis is associated with spinal shock; as spinal shock subsides, the spastic paralysis develops.
4 There would be a decreased pulse pressure with the associated hypotension.

108. ② This provides for collection of more data. (1) (SC; AS; CW; HC)
1 This is a negative comment that closes communication.
3 This implies that things are not well and the mother may be to blame.
4 This could make the mother feel guilty about not meeting her baby's needs.

109. ① Ketoacidosis occurs when insulin is lacking and carbohydrates cannot be used for energy; this increases the breakdown of protein and fat, causing Kussmaul respirations, decreased alertness, decreased circulatory volume, metabolic acidosis, and an acetone breath. (1) (DA; EV; MS; EN)
2 Hypoglycemia is manifested by cool, moist skin, not hot, dry skin or Kussmaul respirations.

3 The Somogyi phenomenon is a rebound hyperglycemia induced by severe hypoglycemia; there are not enough data to determine whether this occurred.
4 Hyperosmolar nonketotic coma usually occurs in type 2 diabetes because available insulin prevents the breakdown of fat.

110. ④ The 160 mg/100 ml level exceeds the normal fasting blood glucose level and is highly indicative of diabetes. (2) (HI; AS; MS; EN)
1 This would indicate hypoglycemia, not hyperglycemia.
2 Same as answer 1.
3 This is an expected blood glucose level in the nondiabetic individual.

111. ② There is usually a history of interpersonal difficulties; clients are unable to engage in the give-and-take a relationship requires. (2) (EP; AN; MH; PR)
1 There is no direct relationship between antisocial personality disorders and sexual aberrations.
3 The parents of these individuals rarely impose any discipline or limits at all.
4 There is no diminished contact with reality; these clients are in contact with reality; they just do not care about it.

112. ② Because the newborn's GI tract is sterile, the infant does not have the bacteria necessary to synthesize vitamin K; vitamin K functions to stimulate the liver's production of clotting factors. (1) (DF; AN; CW; DR)
1 Vitamin K has no relation to the absorption of biliary salts.
3 A prolonged prothrombin time indicates potential clotting problems for the newborn and would not be deliberately produced.
4 The vitamin K does not replace the necessary bacteria; the bacteria will develop once oral feedings are established.

113. ③ An additional meeting is important to deal with the problem of termination with the client. (2) (SC; IM; MH; TR)
1 This would not be a therapeutic termination because the issues would not be resolved for the client.
2 The nurse may want to do this; however, the focus should be on the needs of the client.
4 The client is avoiding the nurse, and the nurse must reach out to help the client with the termination process.

114. (1) Client participation provides for a sense of control, and a consistent approach provides a routine with no surprises; these approaches may limit pain and promote compliance with the regimen. (2) (SC; PL; MS; IT)
2 Preparation of the equipment and explanation of the procedure should be performed before the procedure; when performed during the procedure, it wastes time, which can prolong pain.
3 Changing staff disrupts the client's routine and sense of trust.
4 This is too hot; the water should be approximately 100° F.

115. (3) This statement leaves communication lines open. (3) (EP; IM; MH; SD)
1 This response discounts the client's thoughts and may increase agitation.
2 This provides false reassurance; it does not provide security because the client may believe the nurse is one of those involved in the plot to kill.
4 This supports the client's delusional system.

116. (3) Right occiput anterior is a vertex presentation; the fundus has a soft, rounded mass, which is the buttock, and therefore this is a cephalic presentation; the irregular curvature on the right indicates the fetal spine, and the bumpy projections on left indicate the extremities. (2) (DF; AS; PA; HC)
1 Left sacrum posterior is a breech presentation; the fetal part in the fundus would be firm.
2 The fetal spine would be on the left side in left occiput posterior position.
4 This is a transverse lie with right scapula (shoulder) presenting anteriorly.

117. (4) Vasospasms of placental vessels occur because of elevated blood pressure, and the placenta may separate prematurely (abruptio placentae). (2) (SI; PL; CW; HP)
1 Placenta previa is an abnormal placental implantation and is not related to hypertension.
2 This is an excessive amount of amniotic fluid and is not associated with hypertensive disorders of pregnancy.
3 Isoimmunization in pregnancy is associated with Rh problems, not hypertension.

118. (4) The therapy program in use in the home should be incorporated into the care plan to maintain continuity. (1) (SI; PL; PE; SK)
1 The child has social needs, and interaction should be promoted; there is no need for a private room.
2 The parents should be encouraged to stay with the child and actively participate in care.

3 The child should have a regular diet appropriate for the developmental age.

119. (1) The nurse should auscultate the abdomen and listen for bowel sounds, which signify the passage of flatus. (1) (CI; IM; MS; GI)
2 Nausea may be present even though peristalsis has begun.
3 The first bowel movement occurs after peristalsis returns and usually after food is eaten.
4 Peristalsis should return before the tenderness of the abdomen subsides.

120. (4) Strenuous exercise leads to increased cellular metabolism, causing tissue hypoxia, which can precipitate sickling. (2) (CC; EV; PE; BI)
1 This is unnecessary unless the other children have an infectious disease; peer relationships should be encouraged.
2 Fluid should never be restricted; keeping the child well hydrated helps to prevent sickling.
3 This is detrimental to the child's developmental needs and may result in social isolation.

121. (2) An open wound needs sterile technique; the supplies at the bedside are not sterile and the physician ordered sterile soaks. (2) (SI; IM; MS; IT)
1 This is unsafe; a clean basin and washcloth are not sterile.
3 Client safety is the priority.
4 This is unnecessary; the physician has already indicated the type of soak desired.

122. (4) When action on either side of a conflict creates anxiety, a physical reason for not acting at all may unconsciously be used. (3) (SC; AN; MH; AX)
1 These disorders are disabling; the client truly believes the symptoms are real.
2 These individuals do not enjoy their illness; their anxiety is relieved by it.
3 These individuals are in contact with reality.

123. (2) Because normal implantation occurs in the upper third of the uterus, a low-lying placenta would be an abnormal implantation. (2) (SI; EV; CW; HP)
1 Infarctions may appear on a placenta because of some interference with the blood supply; this is not related to position.
3 Placenta previa indicates where the placenta has implanted and has no relationship to placental aging.
4 Abruptio placentae is the premature separation of a normally implanted placenta.

124. (1) Dialysis removes chemicals, wastes, and fluids normally removed from the body by the kidneys. (1) (CC; IM; MS; RG)
    2 The mention of heart problems is a threatening response and may cause increased fear or anxiety.
    3 This is a threatening response and can cause an increase in the client's level of anxiety.
    4 Dialysis does not speed recovery; it helps maintain fluid and electrolyte balance.

125. (1) The intestine is obstructed by thick, tenacious, pasty meconium. (2) (HI; AS; PE; EN)
    2 Imperforate anus is a congenital malformation in which the anal opening is obliterated; it is not associated with cystic fibrosis
    3 A rapid respiratory rate is normal in infants; respirations accelerate with movement and crying.
    4 This is common in most newborns because of destruction of immature erythrocytes; it may be physiologic or pathologic.

126. (4) Neologisms are newly coined words with personal meanings to the client with schizophrenia. (2) (EP; AN; MH; SD)
    1 Clanging is the association of words by sound rather than meaning.
    2 Echolalia is parrot-like echoing of spoken words or sounds.
    3 Echopraxia is mimicking the movements of others.

127. (3) This medication decreases the effectiveness of oral contraceptives and alternate contraceptive measures will have to be used. (2) (CC; EV; CW; WH)
    1 This is a sulfa-based medication; fluid intake should be increased to prevent crystalluria.
    2 Although the urine should be observed for crystals, straining is not necessary.
    4 The urine will not turn orange with this medication; if bloody or smoky urine were present prior to treatment, it would become a normal color once treatment was initiated.

128. (2) If there is any bony or soft tissue resistance to descent and delivery, trauma to the fetal head may occur. (3) (DF; AN; CW; HP)
    1 This is not associated with a birth event.
    3 This is not associated with a precipitous labor.
    4 Although this could occur, the placenta usually is expelled shortly after the fetus.

129. (2) These dysrhythmias may result from postoperative inflammation around the SA node. (3) (CI; EV; MS; CV)

1 This syndrome occurs later, not immediately.
3 Same as answer 1.
4 Hemoglobin and hematocrit levels usually fall; anemia can be a problem.

130. (3) This is the only way the nurse can be certain that purging does not occur. (3) (SC; PL; MH; ES)
    1 An accurate intake and output is difficult to maintain unless the individual is closely observed throughout the day.
    2 Weighing daily would not help to assess the individual's electrolyte or nutritional status.
    4 Searching for hoarded food establishes a negative relationship and documents lack of trust.

131. (2) This is the most complete definition of bulimia. (2) (SC; AN; MH; ES)
    1 This may occur with bulimia, but it is not the definition.
    3 Although clients with bulimia do consume large amounts of food in private, they do eat in public.
    4 Same as answer 1.

132. (1) An exercise program and the Milwaukee brace are the treatments of choice for mild structural scoliosis. (2) (DF; IM; PE; SK)
    2 Although compliance will affect the ultimate outcome of treatment, exercises alone are not helpful in this type of scoliosis.
    3 Exercises are to be encouraged, regardless of the type or extent of scoliosis.
    4 Exercises alone are used only with scoliosis that is related to posture, not structure.

133. (3) Flexion and extension prevent tightening of muscles and tendons. (2) (PC; PL; MS; SK)
    1 These are abnormal sensations and are related to neurologic, not musculoskeletal, alterations.
    2 Weight-bearing, not exercise, would promote the development of osteoblasts.
    4 This is the result of cerebellar changes; it is not related to immobility.

134. (4) The newborn's intake of milk is gradual and small, and, at the same time, there is loss of extracellular fluid primarily in the form of stool and urine. (2) (PC; AN; CW; NN)
    1 This is untrue; slight weight loss after delivery is a normal physiologic response.
    2 Same as answer 1.
    3 Same as answer 1.

135. (4) If carbohydrate intake is reduced, protein is utilized for energy, thereby lowering the recom-

mended elevated protein requirements for pregnancy. (2) (PC; AN; CW; HP)

1 This is not the reason for avoiding dieting during pregnancy.

2 Additional calories are needed to spare protein.

3 This is untrue.

136. ③ This is common in breech presentation because the contracting uterus exerts pressure on the lower colon, forcing out the meconium. (3) (DF; AS; CW; HC)

1 Mild bloody show is expected; a heavier flow is a deviation from normal and not a common finding with breech presentations

2 This is not a finding specific to breech presentations.

4 Late decelerations are not more common in the laboring client with a breech presentation.

137. ③ Safety becomes a priority when the client has hemiparesis and hemianopia. (2) (SI; PL; MS; NM)

1 Although a balance between activity and rest is important, the client does not have to maintain bed rest.

2 Oxygen is generally not necessary.

4 All the basic nutrients should be included in the diet; there is no need to reduce protein intake.

138. ③ Between the ages of 3 and 5 years, death is viewed as a departure or sleep, which is reversible. (3) (DF; AN; PE; GD)

1 This is the concept of death held by children 9 or 10 years of age; death is viewed as reversible by the preschooler.

2 The early school-age child of 6 or 7 years personifies death and sees it as horrible and frightening; this is consistent with the concrete thinking present at this age.

4 Children of all ages have some concept of death.

139. ④ Palpation would create a risk of rupturing the tumor mass. (2) (SI; IM; PE; RG)

1 This is unnecessary; no calculi are present.

2 There is no related contraindication for IV medication.

3 Diapers can be worn as long as there is no palpation.

140. ④ Malnutrition and liver damage lead to a reduced serum albumin level and failure of the capillary fluid shift mechanism, resulting in ascites. (2) (DA; AN; MS; FE)

1 Iron promotes hemoglobin synthesis; this is unrelated to cirrhosis.

2 Vitamins are unrelated to ascites.

3 The sodium level is usually excessive with cirrhosis.

141. ② Closeness increases anxiety, which cannot be tolerated; hostility is used to keep people away. (2) (SC; EV; MH; TR)

1 Hostility is more extreme than assertiveness and is not an indication of improvement.

3 This is true, but the expression of hostility is not really a flare-up in this situation.

4 Regressive behavior is the resumption of behavior characteristic of an earlier stage of development; hostility does not fit this definition.

142. ② With the reduction of edema the child's health improves, the appetite increases, and the blood pressure normalizes. (2) (DA; PL; PE; RG)

1 Fluids should not be pushed since the kidneys are inflamed and cannot tolerate large amounts.

3 Ambulation does not have an adverse effect on the disease; most children voluntarily restrict their activities during the acute phase.

4 Sodium is lowered, not eliminated; sodium restriction is not tolerated well by children and may further restrict their appetite.

143. ④ This is important in determining whether the baby has maternally transmitted antibodies against measles. (2) (DF; AS; PE; BI)

1 This baby has no vaccination against measles because this is not done until the baby is about 1 year of age.

2 This has no relationship to the present exposure to measles.

3 Same as answer 2.

144 ④ Xylocaine usually reduces the irritability of the heart. (3) (CI; IM; MS; CV)

1 Treating the PVBs is the first priority.

2 Stimulating the cough reflex will not affect an irritable heart muscle.

3 At present, manually stimulating the heart is not needed.

145 ② This response provides accurate information and answers the client's direct question. (3) (CC; IM; CW; HP)

1 Early induction does not improve the chances for vaginal birth among clients with large babies.

3 Labor is never induced for the sole purpose of preventing pregnancy-induced hypertension.

4 Neonates generally develop hypoglycemia shortly after birth; however, early delivery has no effect on this development.

146. (2) This moves the emphasis from facts to feelings; it focuses on the client without setting the direction for communication. (3) (SC; IM; MH; MO)
    1 This is a rhetorical question because the client has already brought up the topic.
    3 This denies the client's feelings and cuts off further communication.
    4 This asks a direct question that the client will probably be unable to answer.

147. (2) This is an accurate statement; each gram of carbohydrate contains 4 calories. (1) (PC; PL; MS; GI)
    1 This provides too few calories; carbohydrates contain 4 calories per gram.
    3 This provides too many calories; fat contains 9 calories per gram.
    4 This provides too many calories; no nutrient contains this many calories per gram.

148. (4) Change is always accompanied by some anxiety and pain; without motivation, change will not occur. (2) (SC; EV; MH; PR)
    1 The life-style of these individuals is rarely limited because they tend to be rather gregarious and outgoing; in reality, they attempt to live by their guile.
    2 The reactions of the client's parents would be of little influence at this age.
    3 These usually do not work unless the individual is motivated to change; there is no need for an anxiolytic to reduce anxiety.

149. (4) During the first 24 hours there is a rapid fluid shift, which causes an increase in cardiac output and blood volume; this taxes an already compromised heart. (2) (CI; PL; CW; HP)
    1 This is not recommended because this will further increase circulating blood volume and necessitate an increase in cardiac workload.
    2 This is false; the first 24 hours are crucial because of the rapid fluid shift and the stress of increased cardiac output on a compromised heart.
    3 Progressive ambulation starting during the first 24 hours after delivery is recommended because it decreases the risk of thromboembolic events.

150. (1) Constipation and paralytic ileus are common problems because of decreased nerve innervation of the GI tract; they are symptomatic of neurotoxicity. (2) (PT; EV; PE; DR)
    2 This is not a toxic effect; it is not necessary to check bowel sounds if diarrhea is present.

3 This is not a factor in the development of constipation; fluid can be given intravenously if nausea is present.
    4 Vincristine causes leukopenia, which increases susceptibility to infection; it does not cause antigen/antibody reactions.

151. (3) Allopurinol decreases serum uric acid levels before and during chemotherapy; increased fluid intake aids in the increased excretion of uric acid; allopurinol and increased fluids help prevent renal tubular impairment and kidney failure because of hyperuricemia. (3) (CI; PL; MS; DR)
    1 The client should be encouraged to follow a diet that promotes urine alkalinity.
    2 If the oral route is used, this will limit gastric irritation, not uric acid nephropathy.
    4 Fluid intake should be increased to 2 to 3 liters per day to prevent urate deposits and calculus formation.

152. (3) The elastic bandage must be applied smoothly without wrinkles, folds, or creases because these can cause excessive pressure or irritation. (2) (SI; PL; MS; SK)
    1 The bandage should be reapplied whenever necessary; this may be necessary if it slips off, if it is too tight or too loose, or if it has wrinkles or creases.
    2 This would be unsafe because it could impede circulation; the bandage should be snug, not tight.
    4 This would be unsafe because the dependent position allows the blood vessels to become engorged; the bandage should be applied with the leg level with the heart.

153. (2) Elastic stockings help decrease venous pooling of blood and help maintain systemic blood pressure when the client stands up. (2) (PT; IM; MS; CV)
    1 Orthostatic hypotension occurs on rising to an upright position; gait training will not affect this.
    3 An alteration in dosage may be ordered by the physician, but sudden withdrawal is dangerous and unwarranted.
    4 This may increase the intravascular fluid volume temporarily but will not affect reflexes involved in orthostatic hypotension.

154. (3) In the first trimester fluid fills the bladder, which helps to push the uterus up for optimal ultrasound viewing. (2) (CI; IM; CW; HC)
    1 This has no relation to ultrasound preparation.

2 The bladder must be full, not empty, for better visualization of the uterus.
4 Same as answer 1.

155. (4) This reduces the risk of introducing the irrigant into the lungs. (1) (SI; AS; MS; GI)
1 This is irrelevant to nasogastric tube irrigation.
2 This increases the risk of introducing irrigant into the lungs if tube placement is not checked first.
3 Same as answer 1.

156. (2) Clients need to understand the cycle as a physiologic event that can be dealt with, not as a life-threatening crisis. (2) (CC; PL; MS; RE)
1 This is insufficient to break the cycle.
3 Same as answer 1.
4 Though helpful, this is not primary in helping to break the cycle of fear, dyspnea, and inactivity.

157. (3) This will help decrease irritability and impulsiveness caused by the client's increased responsiveness to changing stimuli. (2) (EP; PL; MH; MO)
1 This is not helpful because of the client's easy distractibility and minimal attention span.
2 At this time, the client may misidentify this approach as threatening and may retaliate with aggression.
4 A strange environment and new staff tend to increase the anxiety level, not reduce manipulation.

158. (3) The fundus tends to stay at or slightly above the umbilicus for about 24 hours, then decreases in size about 1 fingerbreadth per day. (1) (DF; AS; CW; HC)
1 This would be the position in the first 24 hours postpartum.
2 Same as answer 1.
4 This would be the position on the fourth or fifth day.

159. (3) The client will need reassurance and support after this frightening experience. (1) (SC; IM; MH; CS)
1 This provides reassurance but gives no support.
2 This is inappropriate; the priority is to allay anxiety; also, there is no need to stand the client up to take the pulse.
4 This is inappropriate; the client has dementia and will have limited recall of recent teaching.

160. (3) Parental role support and contact with other adults is very important in parenting. (2) (EP; AS; MH; CS)

1 No personality type is specifically associated with abusive behavior.
2 This is untrue; present defenses are ineffective when an adult engages in child abuse.
4 Although lack of knowledge may lead to unrealistic expectations of the child, this factor alone does not significantly contribute to abusive behavior.

161. (1) Initial attempts at oral feeding may cause a choking feeling that may produce severe coughing and raise secretions. (3) (PC; IM; MS; RE)
2 Swallowing does not have an adverse effect on the suture line; a nasogastric tube would not be used because it could traumatize the suture line.
3 A progressive diet is started with liquids, not pureed foods.
4 The pain medication may cause a decrease in the client's respiratory effort and may also depress the cough reflex.

162. (3) This would minimize the sucking and yet not be irritating to the suture line. (2) (DF; PL; PE; GI)
1 This is not used because it would be irritating to the nostrils.
2 Intravenous infusions do not supply the necessary caloric intake.
4 No nipple should be used because the baby should not suck.

163. (4) An increasing residual without increasing intake indicates that absorption is decreasing, a symptom of NEC. (3) (CI; EV; CW; HN)
1 Diarrhea may or may not be related to NEC.
2 Small amounts of vomitus (spitting up) are common in the neonate because of laxity of the cardiac sphincter.
3 The abdominal circumference increases in NEC.

164. (2) The nursing diagnosis consists of two parts—the statement of the client's health status (health problem) and the related factors (probable causes). (3) (CC; AN; MH; CS)
1 The nursing diagnosis includes a statement of the problem; the client's needs are addressed in the plan of care.
3 Although the client's responses may reflect health status, this is only one part of the diagnosis.
4 A medical diagnosis describes a disease process; a nursing diagnosis describes a person's response to a disease process, condition, or situation.

165. (2) When children are allowed to have some control in their care, their cooperation and willingness to tolerate procedures and medications are enhanced. (2) (PT; PL; PE; EH)
    1 A child's favorite food should never be used to disguise medication because it will likely cause an aversion to that food and can affect the child nutritionally.
    3 Bribing sets up a nontherapeutic relationship between the child and the nurse and should not be used.
    4 The nurse should be truthful about the taste of medication so that the child can have an opportunity to suggest ways to deal with the taste.

166. (3) The client should be assessed further for signs of abruptio placentae by looking for cessation of uterine activity, fetal heart deceleration, and falling blood pressure. (2) (CI; IM; CW; HP)
    1 This is unsafe; the status of the fetus and mother must be assessed immediately.
    2 This is not the priority; the status of the fetus is paramount.
    4 This is unsafe; the status of the fetus is primary; in a partial abruptio placentae, placing the client in a supine position would further compromise blood flow to the fetus.

167. (3) The opportunity must be provided for the client to practice language skills; family participation must be accepted and recognized. (1) (SC; IM; MS; NM)
    1 This demeans the spouse and cuts off communication.
    2 Same as answer 1.
    4 The spouse should be included and involved in the client's care.

168. (1) Decadron is a corticosteroid with antiinflammatory effects. (1) (PT; EV; MS; DR)
    2 Decadron will not keep the tumor from growing; it will only reduce fluid content and therefore cell size, not the number of cells.
    3 Decadron does not promote fluid reabsorption, which is undesirable because it increases fluid retention and therefore cerebral edema.
    4 Decadron does not promote sedation; sedation is not desired because it could mask symptoms.

169. (4) ASA is irritating to the stomach lining and can cause ulceration; the presence of food, fluid, or antacids decreases this response. (1) (PT; IM; MS; DR)
    1 Tylenol does not contain the antiinflammatory properties present in aspirin; tinnitus should be reported to the physician.

2 This should be reported to the physician, not the dentist.
3 This is unnecessary as long as aspirin is taken with food.

170. (1) This statement focuses on the client's perceptions and promotes further communication. (3) (SC; IM; PE; EH)
    2 This reads too much into the client's statement and may be too emotionally charged.
    3 This is untrue; there is a higher incidence of lice in people with inadequate personal hygiene.
    4 This statement is valid but accusatory; it discourages further communication.

171. (3) Informed consent means the client must comprehend the surgery, the alternatives, and the consequences. (2) (CC; IM; MS; CV)
    1 This explanation is not within nursing's domain.
    2 This is true, but it does not determine the client's ability to give informed consent.
    4 The nurse's signature documents that the client has given informed consent; however, this follows after the nurse determines the client's comprehension.

172. (4) Microcephaly is one sign of congenital toxoplasmosis. If the head circumference is 1 to 2 cm less than the chest circumference, then microcephaly is considered. (3) (DF; AS; CW; HN)
    1 This is a normal variation characterized by bluish pigmentation over the lower back and buttocks seen in dark-skinned infants.
    2 Localized edematous swelling of the scalp that crosses the suture lines of the skull can occur after vertex vaginal births; this has no pathologic significance and is not specific to toxoplasmosis.
    3 This is a normal assessment in a healthy neonate.

173. (4) This answers the client's question and provides an accurate description of a cystoscopy. (1) (CC; AN; MS; RG)
    1 This is not a computerized examination.
    2 This procedure does not involve x-ray films or dye.
    3 Radiopaque material is not used and the catheter is inserted into the bladder via the urethra, not the ureters.

174. (4) A 2½-year-old is capable of fitting large wooden pieces into the puzzle; this activity challenges the child's ability to recognize shapes. (2) (DF; PL; PE; GD)

1 This is a toy suitable for the young infant.

2 This is more appropriate for the child of 12 months who is becoming adept at motor skills.

3 Same as answer 1.

175. ③ Aldactone is a potassium-sparing diuretic often used in conjunction with thiazide diuretics. (1) (PT; AN; MS; DR)

1 Both diuretics do this, so it is not a particular advantage of Aldactone.

2 Same as answer 1.

4 Same as answer 1.

176. ④ Previous experiences with projectile vomiting are frightening; an explanation that surgery has eliminated this as well as support and encouragement of the parents as they resume care of their infant are necessary. (1) (HI; EV; PE; EH)

1 This is untrue; the data indicate the mother is eager to assist with other aspects of care.

2 These are not used initially; oral feedings are reinstituted with clear liquids and electrolytes and progress to formula as tolerated.

3 No special nipple is required; these are used for infants with cleft lip and/or palate.

177. ③ A first step in any therapeutic relationship should include setting parameters of meetings, such as time and frequency. (3) (SC; PL; MH; TR)

1 This is part of the working phase of a therapeutic relationship and therefore not an initial intervention.

2 The nurse should not deviate from the original contract.

4 There is no need to carry this out during the termination phase because it is part of the initial ground rules.

178. ② These would be signs of infection and should be reported to the physician. (1) (CC; EV; CW; WH)

1 This is a sign of healing that is expected and normal.

3 There is little subcutaneous fat in the thoracic area, and the skin may be taut at the operative site, appearing irregular; this commonly occurs.

4 This results from severance of nerves and formation of scar tissue, which are expected and normal.

179. ③ Facial exercises help preserve muscle tone and prevent atrophy while inflammation or pressure on cranial nerve VII is present. (2) (CI; IM; MS; NM)

1 With conservative treatment, recovery occurs in 80% to 90% of clients; if a tumor were the cause, surgery would be necessary.

2 To the contrary, the eye may be closed or an eye patch worn to protect the cornea; taping the eyelid open would be done if the client had ptosis associated with myasthenia gravis.

4 The paralysis is usually unilateral, not bilateral.

180. ① Glomerulonephritis is associated with a history of prior streptococcal infection of the throat. (1) (CC; AS; PA; RG)

2 A streptococcal infection, not the measles virus, is associated with glomerulonephritis.

3 Glomerulonephritis is not an inherited disease; it usually follows a streptococcal infection.

4 There are no immunizations that would cause glomerulonephritis.

181. ④ Insulin requirements may decrease in early pregnancy because of increased fetal needs for nutrients and the possibility of maternal nausea and vomiting; insulin requirements increase in the second and third trimesters as a resistance to insulin develops; the blood glucose level is monitored to prevent ketoacidosis and ensuing harm to the fetus. (1) (CI; IM; CW; HP)

1 This is true only during early pregnancy; during the second and third trimesters of pregnancy, there is a resistance to insulin and more insulin is required.

2 This is untrue; even the nondiabetic woman makes a dietary adjustment to keep pace with the increased demands of pregnancy; in addition, insulin requirements increase in the second and third trimesters.

3 Most nutrient requirements, not just protein, increase in pregnancy.

182. ③ This drug is a local irritant to the GI tract. (3) (CC; EV; PE; DR)

1 This is not a side effect; theophylline is given to ease respirations.

2 This is not a side effect; frequent urination is expected because this drug often produces diuresis.

4 This is not a side effect.

183. ② Anxiety about the behavior is absent, as is motivation for change; these persons are unwilling to accept help. (2) (SC; AN; MH; PR)

1 More than life-style needs to change; the client's entire view of life and interpersonal response are involved.

3 These individuals do not experience feelings of guilt about their behavior.

4 It is not a resistance to demonstrating feelings but a total lack of feeling for others.

184. (3) These are early signs of hypoglycemia, or too much insulin; the client should be taught to take additional food or an oral glucose solution. (3) (CC; EV; MS; EN)
   1 These are symptoms of hyperglycemia.
   2 Same as answer 1.
   4 Same as answer 1.

185. (2) Dyspnea at rest is associated with cardiopulmonary conditions and may be a sign of impending decompensation. (2) (CI; AS; CW; HP)
   1 This is a normal physiologic change.
   3 Dependent edema is seen in 90% of uncomplicated pregnancies.
   4 In the third trimester, clients with uncomplicated pregnancies complain of shortness of breath on exertion; this occurs because of compression of the diaphragm by the enlarging uterus.

186. (3) The child needs to know that he will remain awake; he should be prepared to experience the pressure of the aspiration or the biopsy needle entry, and he should know that he will not be incapacitated following the test. (2) (CI; IM; PE; BI)
   1 This is a false statement; false statements must be avoided or the child will not trust what is said in the future.
   2 The child will be permitted to ambulate freely.
   4 A bone marrow specimen is obtained through a puncture wound; sutures are not necessary.

187. (1) Drainage from the small intestine contains residual digestive enzymes that cause the skin to break down. (1) (CI; PL; MS; GI)
   2 An ileostomy is not irrigated; the stool is liquid and drains unassisted.
   3 An ileostomy will continually drain liquid stool; control of fecal elimination is impossible.
   4 An ileostomy will continually drain liquid stool; this is unrelated to diet; the stool is excreted before fluid can be reabsorbed in the large intestine.

188. (3) Sudden withdrawal of antiepileptic medication can cause status epilepticus. (2) (PT; EV; MS; NM)
   1 It is important to take medication as prescribed to lessen the frequency of seizures; medication may or may not eliminate the seizures; stress may cause another seizure to occur.
   2 This is not why antiepileptics are prescribed.
   4 Same as answer 2.

189. (4) This procedure removes fluids and gas from the GI tract, which permits better healing of the surgical area and minimizes nausea. (1) (CI; AN; MS; GI)
   1 This is not the purpose in this situation; the tube is used to decompress the stomach.
   2 Tube feedings would be contraindicated after gastrointestinal surgery.
   3 The tube decompresses the stomach, not the large bowel.

190. (3) Respiratory tract infection may be the first clinical sign of bone marrow suppression. (2) (CI; EV; MS; DR)
   1 This is an expected side effect that is not life threatening.
   2 Same as answer 1.
   4 This is not a side effect of doxorubicin.

191. (2) The collection device must be kept below the level of the chest to prevent backflow of fluid into the pleural space. (1) (DA; IM; MS; RE)
   1 For transport, suction should be turned off and the suction tubing disconnected distal to the water-seal connection.
   3 There is no reason to disconnect the chest tube from the water-seal system; this would allow atmospheric air to enter the pleural space.
   4 The chest tube should almost never be clamped; this may precipitate a tension pneumothorax.

192. (2) These are signs of congestive heart failure, which may develop with the persistent tachycardia that is present with hyperthyroidism. (3) (CI; EV; MS; EN)
   1 These are expected to occur with hyperthyroidism and need not be reported immediately.
   3 Same as answer 1.
   4 Same as answer 1.

193. (1) Guilt feelings can prolong the grieving process because the individual is overwhelmed by both guilt and grief, and consequently the energy needed to cope with both is excessive. (3) (SC; AN; MH; CS)
   2 There are no research data to support this.
   3 Ambivalent feelings about the deceased, not the death itself, can prolong grief.
   4 Usually the opposite is true; the support provided would hasten resolution of grief.

194. (3) The ruptured tube usually will be removed; if the tube is repaired it may result in scarring, predisposing the client to another tubal pregnancy. (1) (SI; PL; CW; RP)
   1 This is a procedure for removing myomas (fibroids) from the uterus.

2 The uterus is usually uninvolved in a tubal pregnancy, and this would make the woman incapable of future pregnancy.

4 The D&C would be effective only in cleaning out the uterine cavity; no pregnancy contents are in the uterus with a tubal pregnancy.

195. (4) The cervical mucus is clear and stretchable (spinnbarkeit) at ovulation because of maximum estrogen stimulation. (3) (DF; IM; CW; RC)

1 These characteristics do not normally occur at any specific point during the cycle.

2 Same as answer 1.

3 Same as answer 1.

196. (4) This is a true statement that addresses the client's concern. (2) (SC; IM; MH; MO)

1 This approach denies the client's fears and feelings and really does not address the current concern.

2 This approach denies the client's fears and feelings and could be frightening and upsetting.

3 This approach denies the client's fears and feelings and may not be the most important thing to the client at this time.

197. (3) A formal plan demonstrates determination, concentration, and effort, with conclusions already thought out. (2) (EP; AS; MH; MO)

1 Failure to successfully complete the suicidal act can add to feelings of worthlessness and stimulate further acts.

2 Talking about suicide does not give clients the idea; verbalizing feelings may help reduce clients' need to act out.

4 Many clients verbalize their suicidal thoughts as they are working on their decision and plan of action; suicide is not always attempted for the attention it achieves.

198. (2) This response recognizes the client's and family's concerns and encourages further verbalization of feelings. (2) (SC; IM; MS; EH)

1 This response does not focus on the client's and family's underlying concerns and keeps the discussion on a physiologic level.

3 This provides false reassurance and cuts off further verbalization of feelings.

4 Same as answer 3.

199. (3) Because of the serious side effects of Sinemet, a thorough daily nursing assessment is a priority. (3) (PT; PL; MS; DR)

1 This is incomplete; it is only one part of a thorough nursing assessment.

2 This is an incomplete assessment; vital signs would be a priority if there were an additional abnormality or indication beyond the drug therapy.

4 This is an incomplete assessment; isolating this as a priority would be indicated if there were a fluid imbalance.

200. (3) Total blood volume increases 50%, which necessitates the heart pumping harder and working more to accommodate this increase. (2) (DF; AN; CW; HC)

1 Although the renal threshold is lowered, the major changes occur in the cardiovascular system.

2 Changes in hormonal levels occur but are not as profound as changes in the cardiovascular system.

4 There are few changes in this system, but pressure from the growing uterus can result in altered patterns of elimination.

201. (3) Bile drainage for the first 24 hours is usually 300 to 500 ml; kinks in the tubing hinder the flow. (3) (SI; IM; MS; GI)

1 Further intervention is necessary because this amount of bile is less than normal.

2 Clamping the tube is contraindicated in the first 24 hours.

4 Drainage of 150 ml is less than expected in the first 24 hours.

202. (3) Aminophylline acts as a vasodilator, and hypotension results when vessels are dilated. (3) (PT; EV; MS; DR)

1 Increased diuresis, not oliguria, is a common side effect.

2 Tachycardia, not bradycardia, is a common side effect.

4 Tachypnea, not hypoventilation, is a common side effect.

203. (1) Applying moist or dry heat relieves muscle pain through vasodilation, increases circulation to the area, and facilitates drug absorption. (2) (PC; PL; PE; NM)

2 This will prolong the discomfort by slowing the rate of absorption of the drug because of vasoconstriction.

3 Movement most likely will be difficult and cause more discomfort.

4 This will cause more discomfort because the injection site is tender.

204. (3) Focusing on the client's feelings permits her to work through her fears. (2) (SC; IM; CW; EC)
1 This statement does not encourage the client to focus on her feelings.
2 This closes off communication and does not allow the client to verbalize her feelings.
4 Same as answer 1.

205. (2) The nurse can be more sensitive to the needs of the client by dealing with personal emotions first. (3) (SC; AS; CW; EC)
1 The focus should be on the client's feelings, not the nurse's.
3 Complete control is not, and should not be, the goal of the nurse.
4 A time of crisis is not the time to teach; the client is not ready to learn.

206. (3) Clients with left hemianopia ignore whatever is in the left field of vision. (3) (SI; EV; MS; NM)
1 This would occur if the client had right hemianopia and wished to see better when eating.
2 This would occur with right hemiparesis, not with hemianopia.
4 This indicates hemiplegia, not hemianopia.

207. (2) Because of poor exercise tolerance and fatigue, these children become too tired to eat; allowing for rest and feeding slowly limit the fatigue associated with eating. (2) (PC; IM; PE; CV)
1 The child should be fed the same type of diet and foods as other infants; it is most important is to feed the child slowly and allow for periods of rest.
3 This will impede absorption; this should be given with orange juice between meals.
4 Same as answer 1

208. (3) These are characteristics associated with children who have Down syndrome; a slant of the eyes is also present. (1) (DF; AS; CW; HN)
1 Only low-set ears occur with Down syndrome; all of these symptoms occur with trisomy 18 syndrome.
2 Although low-set ears occur with Down syndrome, microcephaly and a high-pitched cry may indicate a variety of neurologic problems; the latter two symptoms are part of a syndrome known as cri du chat.
4 A webbed neck and widely spaced nipples are associated with Turner's syndrome.

209. (3) Goodell's sign, or softening of the cervix, occurs at 8 to 9 weeks gestation. (3) (DF; AS; CW; HC)

1 Lightening or settling of the fetal presenting part into the pelvis usually occurs 2 weeks before the onset of labor in nulliparas.
2 This refers to the fetal movements usually perceived by the mother between the 16th and 20th weeks of gestation.
4 Braxton Hicks are intermittent, cramplike contractions that usually occur toward the end of pregnancy and may be mistaken by the mother for the onset of labor.

210. (3) Doxorubicin is cardiotoxic and causes dysrhythmias. (2) (PT; EV; MS; DR)
1 Toxicity causes severe, not minor, dermatitis.
2 This is a side effect of doxorubicin, not a toxic effect.
4 Same as answer 2.

211. (2) This is the most appropriate family nursing diagnosis because it is the alteration in parenting that forms the basis for the other problems experienced by the child and family. (3) (DF; AN; PE; GD)
1 Risk for injury is not an actual problem at this time because there is no history or evidence of physical abuse.
3 Altered nutrition is a problem for the child but only indirectly for the family; the history does not support child abuse.
4 Sensory-perceptual alteration is a diagnosis that is related most specifically to the infant; the problem can be resolved by addressing the parenting problems.

212. (3) Additional studies such as T_3 and T_4 will be necessary to confirm hyperthyroidism; it is not reliable to base the diagnosis on an RAI uptake test alone. (3) (CC; IM; MS; EN)
1 This test uses ^{131}I, which emits radioactive particles for at least 24 hours; these are excreted in the urine.
2 Test results are affected by many medications, especially those containing iodine.
4 This test measures uptake of ^{131}I by the thyroid gland; it does not measure levels of thyroid hormones.

213. (1) During this stage the nurse comes to a conclusion about the collected data and makes a diagnosis. (3) (CC; AN; MH; CS)
2 During this stage, the nurse sets priorities, establishes goals, identifies outcome criteria, and develops a nursing care plan.
3 The client's response to nursing care is assessed in relation to the stated outcome criteria to determine whether goals have been met.

4 The nurse gathers and clusters data during this stage of the nursing process.

214. (2) These are early signs of shock; shock ensues rapidly in a ruptured aortic aneurysm because of profound hemorrhage; this is a surgical emergency. (3) (DA; EV; MS; CV)

1 The nurse can observe hyperventilation by watching the client's breathing patterns; rapid respirations, rather than a rapid pulse, would be expected.

3 Extreme anxiety is not usually associated with light-headedness unless there is hyperventilation.

4 The symptoms are not inclusive enough to indicate infection; there is no indication of fever or a rising white blood cell count.

215. (3) An explanation of how the restraints work and why may reassure the mother. (2) (DF; AN; PE; GI)

1 Using things routinely does not explain why they are being used now; this is an unsatisfactory response.

2 This implies strict adherence to the physician's wishes without any thinking or understanding by the nurse.

4 This is most unreassuring because it gives the mother the feeling that the baby is not being watched at all times.

216. (2) Cleansing after feeding keeps the suture line from becoming infected. (2) (CC; EV; PE; IT)

1 This exerts pressure on the suture line and may cause wound separation.

3 Same as answer 1.

4 The baby should be held and cuddled during feedings.

217. (4) This frustration reveals readiness to deal with the problem of speech that may be best demonstrated by a person with a laryngectomy. (3) (SC; PL; MS; EH)

1 This type of answer leaves the client in limbo and offers no plans for goal setting.

2 This closes off communication, and the client's frustration reveals a need for positive action.

3 The healing of the incision is not a factor in the initial activities of learning a new way of speaking; initially, discussions, demonstrations, and breathing exercises are performed.

218. (3) These foods are low in sodium and calories. (1) (CC; EV; MS; CV)

1 Beef is high in calories, and carrots are high in sodium.

2 Canned tunafish and celery are high in sodium.

4 Stir-fried Chinese vegetables are made with soy sauce, which is very high in sodium, and cornstarch, which is high in calories.

219. (3) Nipple soreness often occurs when there is incorrect positioning of the newborn's mouth on the breast; also, nipples still need to toughen in response to sucking. (2) (CI; AS; CW; HC)

1 This is premature; the cause of soreness must be determined first and will dictate what type of intervention is necessary.

2 Same as answer 1.

4 Same as answer 1.

220. (3) Active ROM increases venous return from the unaffected leg, preventing complications of immobility, including thrombophlebitis. (2) (CI; AN; MS; CV)

1 These isotonic exercises are being performed on the unaffected extremity; there should be no discomfort.

2 Although isotonic exercises do promote muscle strength, that is not the purpose at this time; the priority is to prevent thrombi from developing.

4 These activities will prevent, not limit, venous inflammation.

221. (1) Lemon juice adds flavor and is low in sodium. (1) (PC; IM; MS; FE)

2 This is unnecessary.

3 This is unnecessary; however, canned vegetables generally contain a large amount of sodium.

4 Carbonated beverages generally contain sodium; coffee, whether it is decaffeinated or not, does not contain sodium.

222. (1) With a microdrip set, the nurse should know that the number of ml/hr equals the number of microdrops/minute; thus 55 ml/hr = 55 microgtt/min. (2) (PT; AN; PE; FE)

2 This rate is too high; it would deliver more than the required amount of solution.

3 This is too low; this rate would not deliver the required amount of solution.

4 Same as answer 3.

223. (1) A change in environment and introduction of unfamiliar stimuli precipitate confusion in the elderly client with dementia-type disorders; with appropriate intervention, including frequent reorientation, confusion can be reduced. (2) (SC; AN; MH; DD)

2 Reality orientation can reduce confusion.

3 Same as answer 2.

4 This is a stereotype and is untrue.

224. (4) High levels of steroids result in emotional changes; the actual cause is unknown, but knowing the response may help the mother to better cope with her son's behavior. (2) (CC; IM; MS; EN)
  1 This is unnecessary; the problem has been excessive production of steroids.
  2 Weight loss, not weight gain, would indicate an improving condition.
  3 The changes are not permanent with adequate therapy.

225. (2) Thrombophlebitis is a common complication of immobility in situations related to the application of traction. (2) (CI; AS; MS; SK)
  1 This is contraindicated during use of Buck's traction; positioning with a pillow to relieve back pressure is permitted for short periods.
  3 This is contraindicated in the affected extremity; this could cause further soft tissue injury and pain for the client.
  4 This interferes with the pull of traction; it would cause further soft tissue injury and pain for the client.

226. (4) After laboring all night the client is tired. (3) (PC; PL; PA; HC)
  1 This would be premature; the client is not ready to learn.
  2 This assessment would be too frequent and would interfere with the client's rest.
  3 This is necessary only the first time the client ambulates; otherwise the client can ambulate ad lib.

227. (2) This helps the client avoid hitting obstacles and thus prevents trauma during the tonic-clonic phase of the seizure. (1) (SI; IM; MS; NM)
  1 This is contraindicated; it could injure the client.
  3 Same as answer 1.
  4 If done during the tonic-clonic phase of the seizure, it could cause injury; if necessary, this is done immediately after the seizure to establish an airway.

228. (1) The position of maximum safety and comfort is side-lying because the client's neck and back are hyperextended. (2) (SI; IM; PE; NM)
  2 This would be impossible because the child is in a rigid opisthotonic position; this could be injurious to the child.
  3 Same as answer 2.
  4 This is contraindicated; this would increase intracranial pressure.

229. (4) The stoma secretes mucus immediately following surgery and continues to secrete mucus mixed with serum and blood because of surgical trauma; fecal drainage begins in about 72 hours. (2) (PC; EV; MS; GI)
  1 This would not occur until about 72 hours after surgery.
  2 Drainage that is bloody with clots would indicate hemorrhage; the expected drainage is mucoid and serosanguineous.
  3 Drainage will not be clear; it will be serosanguineous because of the trauma of surgery.

230. (2) The normal muscle contraction-relaxation cycle requires a normal serum calcium-phosphorus ratio; the reduction of the ionized serum calcium level associated with sprue causes tetany (spastic muscle spasms). (2) (SI; AN; MS; GI)
  1 Sodium is the major extracellular cation; the major route of excretion is the kidneys, under the control of aldosterone; although it plays a part in neuromuscular transmission, it is not related to the development of tetany.
  3 Potassium is the major intracellular cation; it is part of the sodium-potassium pump and helps to balance the response of nerves to stimulation; it is not related to the development of tetany.
  4 Although phosphorus is closely related to calcium because they exist in a definite ratio, phosphorus is not related to the development of tetany.

231. (3) This is the correct flow rate; multiply the amount of fluid to be infused (175 ml) by the drop factor (15) and divide this result by the amount of time in minutes (1 hr × 60 min). (2) (PT; AN; MS; FE)
  1 This rate would deliver less than the prescribed amount of fluid.
  2 Same as answer 1.
  4 This rate would deliver more than the prescribed amount of fluid.

232. (4) It is important to set limits on behavior, but it is also important to involve the client in decision making. (2) (EP; IM; MH; PR)
  1 This is a nontherapeutic approach; some limits must be set by the client and the nurse together.
  2 This is a nontherapeutic approach; rarely can a client be distracted from a ritual.
  3 This would increase anxiety because the client uses the ritual as a defense against anxiety.

233. (1) Subcostal incisional pain causes the client to splint and avoid deep breathing, which impedes air exchange in the alveoli. (3) (CI; AN; MS; GI)
2 The location of the incision does not increase the risk of hemorrhage.
3 This can be a postoperative problem, but it is unrelated to the site of the incision.
4 The site is not specifically vulnerable to infection.

234. (3) The client's behaviors are contrary to the medical regimen and are not conducive to positive self-interests. (2) (SC; AN; MS; RE)
1 The behavior does not indicate a self-care deficit.
2 The client's behavior does not indicate altered thought processes.
4 There are no data to support this conclusion.

235. (3) The desired outcome is the increased excretion of lead in the urine. (2) (PT; EV; PE; DR)
1 This is expected when lead initially equilibrates to the blood; until the lead is excreted in the urine, the treatment is not considered a success.
2 The elimination of lead via the GI tract is less than via the urinary tract and would be an unsatisfactory measure of the success of the chelation therapy.
4 This is a desirable effect, but it does not determine the success of therapy; also, the amount is difficult to determine.

236. (1) By failing to acknowledge the baby as a person, the client indicates that she has not released her fantasy baby and accepted the real one. (1) (DF; AN; CW; EC)
2 The mother has acknowledged the infant by using "he," and her question denotes a relationship.
3 The mother incorporated the infant into the family by this statement.
4 Same as answer 2.

237. (1) Increased levels of steroids and aldosterone cause sodium and water retention. (3) (DA; AS; MS; EN)
2 Hypertension, not hypotension, would be expected because of sodium and water retention.
3 The extremities would be thin; subcutaneous fat deposits occur in the upper trunk, especially the back between the scapulae.
4 Hyperglycemia, not hypoglycemia, occurs because of increased secretion of glucocorticoids; hyperglycemia is sustained and not restricted to the morning hours.

238. (3) This will help to decrease the client's anxiety, provide a safe environment, and compensate for impaired cognition. (2) (EP; PL; MH; DD)
1 Restraints may increase confusion and agitation; they should be used only when absolutely unavoidable.
2 Reality orientation should be employed when necessary.
4 Focusing on coping skills would increase the client's feelings of inadequacy; coping skills are on the unconscious level.

239. (4) Positive and negative experiences connected with previous illness or hospitalization will influence the child's response and adaptation to this and subsequent hospitalization. (3) (DF; AS; PE; GD)
1 The priority care at this time should be directed to the child's coping abilities.
2 This will not be too meaningful because a 4-year-old may not have too great an understanding of the illness.
3 This is not a priority; the child's present acute illness must be attended to first.

240. (4) All amino acids are needed for the synthesis of various proteins, but the term "essential" refers to those amino acids the body cannot make, which are thus essential in the diet. (2) (PC; IM; MS; GI)
1 All amino acids in a protein contribute the same number of calories for energy.
2 All amino acids, not just essential amino acids, are necessary for rebuilding body tissue.
3 All amino acids, not just essential amino acids, contain nitrogen.

241. (2) Regular insulin is short acting, and it peaks in 2 to 4 hours, which in this case would be at or before lunch. (1) (CI; PL; MS; DR)
1 This is too soon; regular insulin peaks in 2 to 4 hours.
3 This is too late; regular insulin peaks in 2 to 4 hours.
4 Same as answer 3.

242. (2) This client does not need to be on the medical unit; the client's problem is emotional; it is unfair to upset the other clients; the client should be removed from this unit. (3) (EP; EV; MH; PR)
1 An individual cannot be accused of stealing unless actual proof is obtained.
3 This would accomplish little because the client would deny taking them.
4 This action supports the client's feelings that any means can be used to justify a desired goal.

243. (3) This position allows the lungs more room to expand, thus affording more comfort; this enables the child to breathe better. (1) (DA; IM; PE; RE)
  1 This position would increase difficulty in breathing; it does not allow for chest expansion.
  2 Same as answer 1.
  4 Same as answer 1.

244. (1) An autograft is a permanent graft that should not be rejected; the nurse needs to assess the site immediately. (1) (CI; EV; MS; IT)
  2 An autograft is a permanent graft that should not need to be replaced.
  3 The nurse needs to assess the site; the responsibility of assessment should not be left up to the client.
  4 This could raise the client's anxiety and draws a conclusion before assessment of the site; infection is usually associated with purulent drainage.

245. (2) Brief, frequent contacts are less threatening and help to build trust. (2) (EP; PL; MH; SD)
  1 This would increase suspiciousness; the client needs consistent caregivers to help increase the level of trust.
  3 This supports the client's delusional system, thus increasing suspiciousness.
  4 The client needs to be observed to prevent self-harm as a result of delusional thinking.

246. (4) Counter pressure against the sacrum during contractions affords some relief from the discomfort of back pain. (1) (PC; IM; CW; HC)
  1 It is difficult to predict the length of labor for any client.
  2 This does not respond to the situation; the coach should be included in providing comfort to the client.
  3 Same as answer 2.

247. (4) These movements require muscle contraction, putting pressure on blood vessels, increasing tissue oxygen, and thus promoting circulation. (3) (CI; AN; CW; WH)
  1 Muscle atrophy is not a common complication following a mastectomy.
  2 Contractures are a rare complication following a mastectomy.
  3 Lymphedema is assessed by measuring the circumference of the extremity, not by doing exercises.

248. (4) These exercises promote venous return, which helps prevent venous thrombi formation. (2) (SI; IM; MS; SK)

  1 This exercise would be contraindicated in a client who had a history of ruptured nucleus pulposus.
  2 These exercises stimulate collateral circulation for clients with peripheral vascular disease; they are seldom used because walking is considered a more effective exercise.
  3 Although these should be encouraged to prevent respiratory complications, thrombus formation is a more common complication than respiratory complications following a total hip replacement.

249. (1) Euthanasia is a crime and is against the law in most states. (1) (CC; AN; MS; EH)
  2 Euthanasia is against the law in most states.
  3 Nurses do not have the legal authority to perform acts of euthanasia.
  4 This is not negligence; it is a crime.

250. (4) This is breast engorgement, which immediately precedes milk production on the second to fourth day postpartum. (2) (DF; AN; CW; HC)
  1 Acini cells do not become overdistended because of the supply-and-demand nature of milk production; in addition, milk production is not yet established; the client is engorged.
  2 Milk production has not yet begun; this is engorgement, which precedes milk production.
  3 This is impossible because the breasts have not filled with milk yet; engorgement is occurring.

251. (2) The need for acceptance of life as fulfilling and meaningful is the major task of the elderly. (2) (DF; AN; MS; GD)
  1 This is the task of young adulthood (18 to 25 years); it involves establishment of an intimate relationship and occupation.
  3 The task of the adolescent (12 to 20 years) is establishing identity through work and development of relationships and an occupation.
  4 This is the task of adulthood (21 to 45 years); it involves establishment of a family and guiding of the next generation.

252. (4) Properly chosen clothes can minimize the appearance of the brace, especially if an effort is made to keep up with the current styles. (1) (SC; IM; PE; EH)
  1 There are no data to indicate that the child will not adjust to the treatment regimen.
  2 This has a negative connotation that emphasizes the client's problem.
  3 This may be misinterpreted as false praise, and a trusting nurse-client relationship may not develop.

253. ③ The autistic child needs protection from self-injury. (2) (EP; IM; MH; BA)
 1 This can only be permitted if it does not place the child in jeopardy.
 2 The autistic child has difficulty following directions, especially when out of control.
 4 The autistic child cannot separate self from behavior; a punitive approach will decrease the child's self-esteem further.

254. ④ When neither rest nor nitroglycerin relieves the pain, there may be an acute myocardial infarction. (1) (CI; IM: MS; CV)
 1 This is expected; anginal pain can, and often does, radiate.
 2 This is expected; acute myocardial infarction causes profuse, not mild, diaphoresis, which should be reported.
 3 This is expected; activity increases cardiac output, causing angina.

255. ③ Clients should be familiar with these staff members and hear from them what will be experienced. (2) (CC; PL; MS; CV)
 1 Although discharge plans should be mentioned, they are not the primary focus at this time.
 2 Most clients do not want or need a minutely detailed description.
 4 This is not an important part of the teaching plan at this time.

256. ② Immature thermoregulation necessitates keeping the infant warm to prevent neonatal hypothermia. (3) (DF; IM; CW; NN)
 1 There is no hurry; the placenta may not separate for 30 minutes without danger.
 3 The cord can be left intact once it is clamped; there is no urgency to cut the cord.
 4 It is too soon to evaluate the hemorrhagic condition of the mother; the placenta has not yet been delivered.

257. ④ Rapid weight gain may indicate an edematous state. While the client is on bed rest, the blood pressure decreases and interstitial fluid is mobilized into the intravascular space, enhancing flow to the uterus and kidneys. A 24-hour collection of urine for protein and creatinine clearance is more reflective of true renal status. (2) (CI; PL; CW; HP)
 1 Diuretics and low-salt diets will gradually decrease circulating fluid, creating hypovolemia, which decreases placental circulation. Lateral bed rest would be ordered.
 2 Elevated serum glucose levels are unrelated to pregnancy-induced hypertension. Daily weights and lateral bed rest would be ordered.

 3 Diuretics are contraindicated in pregnancy because they decrease circulating fluids and may impair fluid supply to the placenta. Lateral bed rest and intake and output would be ordered

258. ④ Cortical control of voluntary muscles occurs between 2 and 4 months. (3) (DF; IM; PE; NM)
 1 Cerebral palsy is not diagnosed by the presence of joint deformities; these may develop later because of spastic muscle imbalance.
 2 Parents have a right to be informed of their child's diagnosis as soon as possible.
 3 The neurologic lesions are fixed and will neither progress nor regress.

259. ① This is an autosomal recessive disorder; each parent contributes one affected gene. (2) (DF; AN; PE; BI)
 2 There is only a 50% chance that a child will have the sickle cell trait, not sickle cell disease.
 3 Same as answer 2.
 4 All of the children from these parents will have the sickle cell trait but not sickle cell disease.

260. ③ There is better control with short-acting (regular) insulin, and emergencies can be handled more quickly. (3) (CI; AN; MS; EN)
 1 This is untrue; the level of glucose must be maintained as close to normal as possible.
 2 This is untrue; the occurrence is greater when the client is receiving exogenous insulin.
 4 This is not the reason for using regular insulin; both oral hypoglycemics and insulin are available.

261. ② Limit setting must be consistent with a client who is using manipulative behavior; a unified approach is vital because the client will play the staff against each other. (1) (EP; IM; MH; PR)
 1 Limit setting is required to control inappropriate behavior.
 3 The most important concept is unity of approach, not the ability of a few to resist manipulation.
 4 This must be a group decision, not the responsibility of one staff member alone; the client is unable to set self-limits.

262. ① Insulin causes an increased flow of potassium into the cells, which will then reduce the circulating blood levels of potassium. (3) (PT; AN; MS; FE)
 2 This response halts communication and is nonsupportive.
 3 Blood glucose levels are usually not elevated in clients with acute renal failure.
 4 Insulin will not lower the metabolic rate.

263. (3) Impaired fat absorption necessitates lowering dietary fat; more calories are needed because of poor absorption of nutrients. (3) (CI; PL; PE; EN)

1 A low-fat diet is recommended, but these children need high-salt diets to replace the large amount of salt lost by sweating.

2 A high-protein diet is recommended, but fat must be avoided because fat absorption is impaired.

4 A high-carbohydrate diet is correct, but because salt depletion via sweating is a hazard, children are encouraged to use salt generously.

264. (1) Withdrawn clients can tolerate personal contact only for short periods. (2) (EP; PL; MH; SD)

2 The client could not function in this type of group.

3 The client has a problem with interpersonal relations, therefore this would not work.

4 Allowing the client to be alone would not relieve anxiety; it would foster withdrawal.

265. (3) Increasing fluids flushes the urinary tract of microorganisms. (3) (SI; PL; CW; WH)

1 Tampons do not increase the risk of cystitis.

2 Fluids should be increased, not decreased; the type of fluid is not contributory.

4 This promotes the transfer of microorganisms to the urethra, where they could ascend to the bladder.

How to Use Worksheet 1: Errors in Processing Information

Common errors in processing information are listed in the left-hand column of this worksheet. At the top of the worksheet is a row of blank spaces for inserting the number of the question missed. Directly below each number, check any errors you made in answering that question. You may have made more than one type of error in an answer.

Worksheet 1: Errors in Processing Information

Question number																								
Did not read situation/ question carefully																								
Missed important details																								
Confused major and minor points																								
Defined problem incorrectly																								
Could not remember terms/ facts/concepts/principles																								
Defined terms incorrectly																								
Focused on incomplete/incorrect data in assessing situation																								
Interpreted data incorrectly																								
Applied wrong concepts/ principles in situations																								
Drew incorrect conclusions																								
Identified wrong goals																								
Identified priorities incorrectly																								
Carried out plan incorrectly/ incompletely																								
Was unclear about criteria for evaluating success in achieving goals																								

How to Use Worksheet 2: Knowledge Gaps

Types of common knowledge gaps are listed along the top of this worksheet. Write a brief description of topics you want to review in the spaces provided. For example, if you missed a question on administration of a particular drug, write the drug name and problem (e.g., dosage) in the appropriate space under the column labeled *Pharmacology*.

Worksheet 2: Knowledge Gaps

Basic science	Skills/ procedures	Basic human needs	Growth and development	Normal nutrition	Psychosocial factors	Clinical area/ topic	Stressors/ coping mechanisms	Patho- physiology	Pharma- cology	Therapeutic nutrition	Legal implications	Other